ADX500 CAS:107-13-1 *HR: 3*

ACRYLONITRILE

DOT: UN 1093

mf: C_3H_3N mw: 53.07

PROP: Colorless, mobile liquid; mild odor. Sol in water. Mp: −82°, bp: 77.3°, fp: −83°, flash p: 30°F (TCC), lel: 3.1%, uel: 17%, d: 0.806 @ 20°/4°, autoign temp: 898°F, vap press: 100 mm @ 22.8°, vap d: 1.83, flash p: (of 5% aq sol): <50°F.

SYNS: ACRYLNITRIL (GERMAN, DUTCH) ◇ ACRYLONITRILE MONOMER ◇ AKRYLONITRYL (POLISH) ◇ CARBACRYL ◇ CIANURO di VINILE (ITALIAN) ◇ CYANOETHYLENE ◇ CYANURE de VINYLE (FRENCH) ◇ ENT 54 ◇ FUMIGRAIN ◇ NITRILE ACRILICO (ITALIAN) ◇ NITRILE ACRYLIQUE (FRENCH) ◇ PROPENENITRILE ◇ 2-PROPENENITRIL
◇ VENTOX ◇ VINY

TOXICITY DA
bfa-rat/sat 30
dns-rat:lvr 1 m
slt-dmg-orl 152
skn-hmn 500 m
skn-rbt 10 mg/
skn-rbt 500 mg
eye-rbt 20 mg
ipr-ham TDLo
TJADAB 23,325
orl-rat TDLo:
DOWCC* 03NO
orl-rat TDLo:
24,129,86
ihl-rat TCLo:
orl-rat LD:364
ihl-hmn TCLo
17,199,48
ihl-man LCLo
skn-chd LDLo
75,1087,50
orl-rat LD50
ihl-rat LCLo
skn-rat LD50
ihl-mus LCL
ipr-mus LD5
orl-mus LD5
scu-mus LD5
ihl-dog LCLo

CONSENSU
Carcinogens
EMDT 7,79,
19,73,79; An
Community
ardous Subst
tory.

OSHA PEL: TWA 2 ppm; CL 10 ppm/15M; Cancer Hazard.
ACGIH TLV: Suspected Human Carcinogen, TWA 2 ppm (skin).
DFG TRK: 3 ppm (7 mg/m^3), Animal Carcinogen, Suspected Human Carcinogen.
NIOSH REL: TWA 1 ppm; CL 10 ppm/15M
DOT Classification: Flammable Liquid and Poison.

Y PROFILE: Confirmed human carcinogen perimental carcinogenic, neoplastigenic, and tu-ic data. Poison by inhalation, ingestion, skin contact, and other routes. Human systemic effects by inhalation and skin contact: conjunctive irritation, somnolence, general anesthesia, cyanosis and diarrhea. An experimental teratogen. Other experimental reproductive effects. Human mutation data reported. Dangerous fire hazard when exposed to heat, flame, or oxidizers. Moderate explosion hazard when exposed to flame. Can react vigorously with oxidizing materials (see also CYANIDES).

HR: Hazard Rating indicating relativ... fire, and reacti... hazar...
See Introduction: paragra...

CAS: – The American Chemical Society's Chemical Abstracts Service number. A complete CAS number cross-index is located in Section 1.
See Introduction: paragraph 4, *p.* xiii.

Cited Reference: – A code which represents the cited reference for the toxicity data. Complete bibliographic citations in order are listed in Section 3.
See Introduction: paragraph 16, *p.* xxiv.

Safety Profiles: – These are text summaries of the toxicity, flammability, reactivity, incompatibility and other dangerous properties of the entry.
See Introduction: paragraph 19, *p.* xxvii.

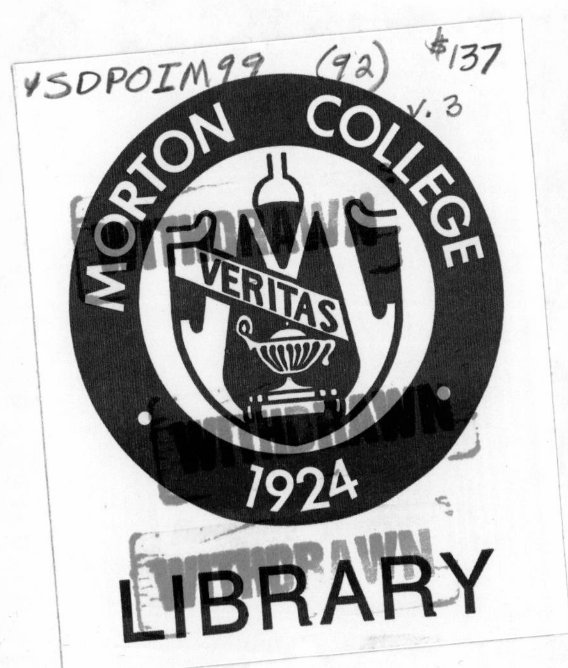

YSDPOIM99 (92) $137
v. 3
MORTON COLLEGE
VERITAS
1924
WITHDRAWN
LIBRARY

Sax's Dangerous Properties of Industrial Materials

Eighth Edition

Volume III

RICHARD J. LEWIS, SR.

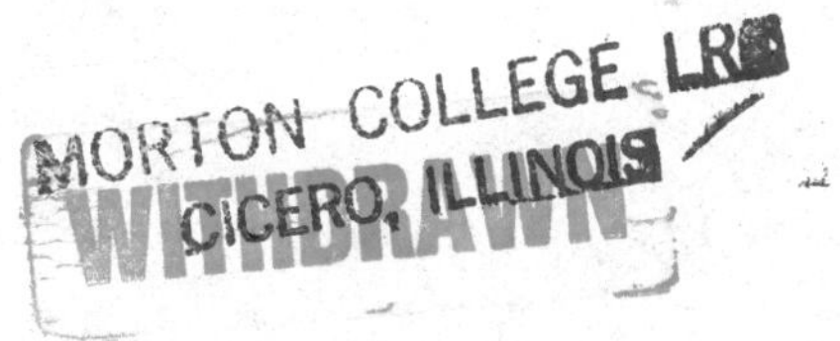

VNR VAN NOSTRAND REINHOLD
New York

DISCLAIMER

Extreme care has been taken in preparation of this work. However, neither the publisher nor the authors shall be held responsible or liable for any damages resulting in connection with or arising from the use of any of the information in this book.

Library of Congress Catalog Card Number 92-3896
ISBN VOLUME I 0-442-01276-4
ISBN VOLUME II 0-442-01277-2
ISBN VOLUME III 0-442-01278-0
ISBN SET 0-442-01132-6

Manufactured in the United States of America

Published by Van Nostrand Reinhold
115 Fifth Avenue
New York, NY 10003

Chapman and Hall
2-6 Boundary Row
London, SE 1 8HN

Thomas Nelson Australia
102 Dodds Street
South Melbourne 3205
Victoria, Australia

Nelson Canada
1120 Birchmont Road
Scarborough, Ontario M1K 5G4, Canada

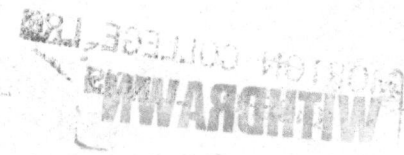

16 15 14 13 12 11 10 9 8 7 6 5 4 3 2 1

Library of Congress Cataloging-in-Publication Data
Lewis, Richard J., Sr.
Sax's dangerous properties of industrial materials / Richard J. Lewis, Sr.—8th ed.
p. cm.
Rev. ed. of: Dangerous properties of industrial materials /N. Irving Sax and Richard J. Lewis, Sr. 7th ed. ©1989.
Includes bibliographical references and index.
ISBN 0-442-01132-6
1. Hazardous substance—Handbooks, manuals, etc. I. Sax, N. Irving (Newton Irving). Dangerous properties of industrial materials. II. Title. III. Title: Dangerous properties of industrial materials.
T55.3.H3S3 1992
604.7—dc20 92-3896
CIP

Contents

Preface

This eighth edition of *Dangerous Properties of Industrial Materials*, a three volume set, represents a major revision and updating of the seventh edition. The objective of the book, however, remains the same: to promote safety by providing the most up-to-date hazard information available. The growth in the availability of toxicological and hazard control reports continues unabated. This book cannot contain all the published data and continue to provide the accessibility for which it is known. To continue to provide complete hazard assessments for the maximum number of entries, data for each entry has been selectively reduced. In particular, carcinogenic and reproductive data lines above those required to establish the hazard of the entry have been excluded. Complete data for these entries are available in the books *Carcinogenically Active Chemicals* and *Reproductively Active Chemicals*, both available from the publisher.

Over 14,000 entries have been revised for this edition, and 1,500 new entries have been added. Some less useful entries have been eliminated or combined with related entries. Emphasis has been placed on including new carcinogenic and reproductive entries. A special effort has resulted in a significant increase in entries containing skin and eye irritation data.

All carcinogenic entries were reviewed and have been categorized as either confirmed, suspected, or questionable. This assessment was based on a detailed classification scheme discussed under Safety Profiles in the Introduction.

All reproductive entries have been updated and new data added.

Numerous synonyms have been added to assist in locating the many materials which are known under a variety of systematic and common names. The synonym cross-index contains the entry name as well as each synonym. This index should be consulted first to locate a material by name. Synonyms are given in English, as well as other major languages such as French, German, Dutch, Polish, Japanese, and Italian.

Additional physical and chemical properties have been added. Whenever available, physical descriptions, formulas, molecular weights, melting points, boiling points, explosion limits, flash points, densities, autoignition temperatures, and the like, have been supplied.

The following classes of data have been updated for all entries for which they apply as follows:

1. IARC Group 1-4 classes and recent assessments.
2. OSHA revised standards which were published in January 1989 and take effect on December 31, 1992.
3. ACGIH TLVs and BEIs reflect the latest recommendations and now include "intended Changes."
4. NTP Fifth Annual Report on Carcinogens entries are identified.
5. DOT classifications now have the corresponding guide number for entries having multiple guide numbers.
6. CAS numbers are provided for additional entries.

Each entry concludes with a safety profile, a textual summary of the hazards presented by the entry. The discussion of human exposures includes target organs and specific effects reported. Carcinogenic and reproductive assessments have been completely revised for this edition.

Fire and explosion hazards are briefly summarized in terms of conditions of flammable or reactive hazard. Where feasible, fire-fighting materials and methods are discussed. Materials which are known to be incompatible with an entry are listed here.

Also included in the safety profile are comments on disaster hazards which serve to alert users of materials to the dangers that may be encountered on entering storage premises during a fire or other emergency. Although the presence of water, steam, acid fumes, or powerful vibrations can cause the decomposition of many materials into dangerous compounds, of particular concerned are high temperatures (such as those resulting from a fire) since these can cause many otherwise mild chemicals to emit highly toxic gases or vapors such as NO_x, SO_x, acids, and so forth, or evolve vapors of antimony, arsenic, mercury, and the like.

The book, which consists of three volumes, is divided as follows:

The first volume contains a CAS number cross-index, a synonym cross-index, and the complete citations for bibliographic references given in the data section.

Section 1 contains the CAS Number cross-index for CAS numbers for the listed materials.

Section 2 contains the prime name and synonym cross-index for the listed materials

Section 3 contains the complete bibliographic references.

The main section of the book is contained in the Volumes II and III. It lists and describes approximately 20,000 materials in alphabetical order by entry name.

Please refer to the Introduction in Volume I for an explanation of the sources of data and codes used.

Every effort was made to include the most current and complete information. The author welcomes comments or corrections to the data presented.

Richard J. Lewis, Sr.

Key to Abbreviations

abs - absolute
ACGIH - American Conference of Governmental Industrial Hygienists
alc - alcohol
alk - alkaline
amorph - amorphous
anhyd - anhydrous
approx - approximately
aq - aqueous
atm - atmosphere
autoign - autoignition
aw - atomic weight
af - atomic formula
BEI - ACGIH Biological Exposure Indexes
bp - boiling point
b range - boiling range
CAS - Chemical Abstracts Service
cc - cubic centimeter
CC - closed cup
CL - ceiling concentration
COC - Cleveland open cup
conc - concentration, concentrated
compd(s) - compound(s)
contg - containing
cryst, crys - crystal(s), crystalline
d - density
D - day(s)
decomp, dec - decomposition
deliq - deliquescent
dil - dilute
DOT - U.S. Department of Transportation
EPA - U.S. Environmental Protection Agency
eth - ether
(F) - Fahrenheit
FCC - Food Chemical Codex
FDA - U.S. Food and Drug Administration
flam - flammable
flash p - flash point
fp - freezing point
g, gm - gram
glac - glacial
gran - granular, granules
hygr - hygroscopic
H, hr - hour(s)
HR: - hazard rating
htd - heated
htg - heating
IARC - International Agency for Research on Cancer
immisc - immiscible
incomp - incompatible
insol - insoluble
IU - International Unit
kg - kilogram (one thousand grams)
L,l - liter
lel - lower explosive limit
liq - liquid
M - minute(s)
m^3 - cubic meter
mf - molecular formula
mg - milligram
misc - miscible
μ, u - micron
mL, ml - milliliter
mm - millimeter
mod - moderately
mp - melting point
mppcf - million particles per cubic foot
mw - molecular weight
ng - nanogram
NIOSH - National Institute for Occupational Safety and Health
nonflam - nonflammable
NTP - National Toxicology Program
OBS - obsolete
OC - open cup
org - organic
OSHA - Occupational Safety and Health Administration
Pa - Pascals
PEL - permissible exposure level
petr - petroleum
pg - picogram (one trillionth of a gram)
Pk - peak concentration
pmole - picomole
powd - powder
ppb - parts per billion (v/v)
pph - parts per hundred (v/v)(percent)
ppm - parts per million (v/v)
ppt - parts per trillion (v/v)
prep - preparation
PROP - properties
refr - refractive
rhomb - rhombic

S,sec – second(s)
sl, slt, sltly – slightly
sol – soluble
soln – solution
solv(s) – solvent(s)
spont – spontaneous(ly)
STEL – short term exposure limit
subl – sublimes
TCC – Tag closed cup
tech – technical
temp – temperature
TLV – Threshold Limit Value
TOC – Tag open cup
TWA – time weighted average
U, unk – unknown, unreported
μ, u – micron
uel – upper explosive limit
μg, ug – microgram
ULC, ulc – Underwriters Laboratory Classification
USDA – U.S. Department of Agriculture
vac – vacuum
vap – vapor
vap d – vapor density
vap press – vapor pressure
vol – volume
visc – viscosity
vsol – very soluble
W – week(s)
Y – year(s)
% – percent(age)
> – greater than
< – less than
<= – equal to or less than
=> – equal to or greater than
° – degrees of temperature in Celsius (Centigrade)
(F),°F – temperature in Fahrenheit

General Chemicals
G–Z

G

GAA100 CAS:51909-61-6 ***HR: 3***
G-52
mf: $C_{20}H_{39}N_5O_7$ mw: 461.64

SYNS: o-2-AMINO-2,3,4,6-TETRADEOXY-6-(METHYLAMINO)-α-d-glycero-HEX-4-ENOPYRANOSYL-(1-4)-o-(3-DEOXY-4-C-METHYL-3-(METHYLAMINO)-β-l-ARABINOPYRANOSYL-(1-6))-2-DEOXY-d-STREPTAMINE ◇ ANTIBIOTIC G-52 ◇ SCH-17726

TOXICITY DATA with REFERENCE
ipr-mus LD50:200 mg/kg 85GDA2 1,180,80
scu-mus LD50:400 mg/kg 85GDA2 1,180,80
ivn-mus LD50:50 mg/kg 85GDA2 1,180,80

SAFETY PROFILE: Poison by subcutaneous, intravenous, and intraperitoneal routes. When heated to decomposition it emits toxic fumes of NO_x.

GAA120 CAS:51922-16-8 ***HR: 3***
G 52 SULFATE
mf: $C_{20}H_{39}N_5O_7 \cdot (H_2O_4S)_7$ mw: 1148.07

SYNS: 4-o-(3-AMINO-3,4-DIHYDRO-6-((METHYLAMINO)METHYL)-2H-PYRAN-2-YL)-2-DEOXY-6-o-(3-DEOXY-4-C-METHYL-3-(METHYLAMINO)-β-l-ARABINOPYRANOSYL)-d-STREPTAMINE (2S-cis)-, SULFATE (salt) ◇ ANTIBIOTIC G-52 SULFATE

TOXICITY DATA with REFERENCE
ipr-mus LD50:200 mg/kg JANTAJ 29,483,76
scu-mus LD50:400 mg/kg JANTAJ 29,483,76
ivn-mus LD50:50 mg/kg JANTAJ 29,483,76

SAFETY PROFILE: Poison by subcutaneous, intravenous, and intraperitoneal routes. When heated to decomposition it emits toxic fumes of NO_x and SO_x.

GAC000 CAS:1952-11-0 ***HR: 3***
G 3063 HYDROCHLORIDE
mf: $C_{18}H_{25}NO_2 \cdot ClH$ mw: 323.90

SYNS: G 3063 ◇ 1-PHENYLCYCLOPENTANECARBOXYLIC ACID-1-METHYL-4-PIPERIDINYL ESTER HYDROCHLORIDE ◇ 1-PHENYLCYCLOPENTANECARBOXYLICACID-1-METHYL-4-PIPERIDYL ESTER HYDROCHLORIDE

TOXICITY DATA with REFERENCE
ims-rat LD50:2483 μg/kg BJPCBM 39,822,70
ims-mus LD50:730 μg/kg BJPCBM 39,822,70
ims-gpg LD50:157 μg/kg BJPCBM 39,822,70

SAFETY PROFILE: A deadly poison by intramuscular route. When heated to decomposition it emits toxic fumes of NO_x and HCl. See also ESTERS.

GAD000 CAS:3060-41-1 ***HR: 2***
p-GABA HYDROCHLORIDE
mf: $C_{10}H_{13}NO_2 \cdot ClH$ mw: 215.70

SYNS: β-(AMINOMETHYL)-BENZENEPROPANOIC ACID HYDROCHLORIDE ◇ β-(AMINOMETHYL)-HYDROCINNAMIC ACID HYDROCHLORIDE ◇ FENIBUT HYDROCHLORIDE ◇ FENIGAM HYDROCHLORIDE ◇ PHENIBUT HYDROCHLORIDE ◇ PHENIGAM HYDROCHLORIDE ◇ PHENIGAMA HYDROCHLORIDE ◇ PHENYBUT HYDROCHLORIDE ◇ PHENYGAM HYDROCHLORIDE ◇ PHENYL-GAMMA HYDROCHLORIDE ◇ PHGABA HYDROCHLORIDE

TOXICITY DATA with REFERENCE
ipr-rat LD50:900 mg/kg PCJOAU 10,1703,76
ipr-mus LD50:1000 mg/kg PCJOAU 10,1703,76

SAFETY PROFILE: Moderately toxic by intraperitoneal route. When heated to decomposition it emits very toxic fumes of Cl^- and NO_x.

GAD400 CAS:56974-61-9 ***HR: 3***
GABEXATE MESYLATE
mf: $C_{16}H_{23}N_3O_4 \cdot CH_4O_3S$ mw: 417.53

SYNS: ETHYL-p-(6-GUANIDINOHEXANOYLOXY)BENZOATE METHANESULFONATE ◇ FOY ◇ GABEXATE MESILATE

TOXICITY DATA with REFERENCE
ivn-mus TDLo:600 mg/kg (7-12D preg):TER OYYAA2 9,743,75
ivn-rat TDLo:560 mg/kg (14D pre):REP OYYAA2 9,743,75
orl-rat LD50:6480 mg/kg OYYAA2 9,743,75
scu-rat LD50:4020 mg/kg OYYAA2 9,743,75
ivn-rat LD50:79 mg/kg OYYAA2 9,743,75
orl-mus LD50:8 g/kg IYKEDH 8,680,77
scu-mus LD50:4550 mg/kg OYYAA2 9,743,75
ivn-mus LD50:248 mg/kg OYYAA2 9,743,75

SAFETY PROFILE: Poison by intravenous route. Mildly toxic by ingestion. Experimental teratogenic and reproductive effects. When heated to decomposition it emits toxic fumes of SO_x and NO_x.

GAF000 CAS:7440-54-2 ***HR: 3***
GADOLINIUM
af: Gd aw: 157.25

PROP: A yellow-white, malleable, and ductile metallic element. A rare earth, stable in dry air; reacts slowly with H_2O. Mp: 1312°, bp: 3233°, d: 7.898 @ 25°.

TOXICITY DATA with REFERENCE
imp-mus TDLo:25 g/kg:ETA PSEBAA 135,426,70

CONSENSUS REPORTS: Reported in EPA TSCA Inventory.

SAFETY PROFILE: Questionable carcinogen with experimental tumorigenic data. It may act as an anticoagulant. It can react violently with air and halogens. See also RARE EARTHS.

GAH000 CAS:10138-52-0 ***HR: 3***
GADOLINIUM CHLORIDE
mf: Cl_3Gd mw: 263.60

PROP: White, monoclinic crystals. D: 4.52 @ 0°, mp: approx 609°. Sol in water.

SYN: GADOLINIUM TRICHLORIDE

TOXICITY DATA with REFERENCE
skn-rbt 500 mg MOD BJPCAL 17,526,61
eye-rbt 1 mg/1H MLD BJPCAL 17,526,61
ipr-mus LD50:378 mg/kg AEHLAU 5,437,62

CONSENSUS REPORTS: Reported in EPA TSCA Inventory.

SAFETY PROFILE: Poison by intraperitoneal route. A skin and eye irritant. When heated to decomposition it emits very toxic fumes of Cl^-. See also CHLORIDES, GADOLINIUM, and RARE EARTHS.

GAJ000 CAS:3088-53-7 ***HR: 3***
GADOLINIUM CITRATE

TOXICITY DATA with REFERENCE
ipr-mus LD50:153 mg/kg AEHLAU 5,437,62
ipr-gpg LD50:60 mg/kg AEHLAU 5,437,62

SAFETY PROFILE: Poison by intraperitoneal route. When heated to decomposition it emits acrid smoke and irritating fumes. See also GADOLINIUM and RARE EARTHS.

GAL000 CAS:10168-81-7 ***HR: 3***
GADOLINIUM(III) NITRATE (1:3)
mf: N_3O_9•Gd mw: 343.28

SYN: NITRIC ACID, GADOLINIUM(3+) SALT

TOXICITY DATA with REFERENCE
orl-rat LD50:3805 mg/kg EQSSDX 1,1,75
ipr-rat LD50:175 mg/kg EQSSDX 1,1,75

CONSENSUS REPORTS: Reported in EPA TSCA Inventory.

SAFETY PROFILE: Poison by intraperitoneal route. Moderately toxic by ingestion. When heated to decomposition it emits toxic fumes of NO_x. See also GADOLINIUM, RARE EARTHS, and NITRATES.

GAN000 CAS:19598-90-4 ***HR: 3***
GADOLINIUM(III) NITRATE, HEXAHYDRATE (1:3:6)
mf: N_3O_9•Gd•$6H_2O$ mw: 451.40

PROP: Deliqescent, triclinic crystals. Mp: 91-92°, d: 2.406 @ 15°, d: 2.332. Sol in water and alc.

TOXICITY DATA with REFERENCE
ipr-rat LD50:230 mg/kg TXAPA9 5,750,63
ipr-mus LD50:300 mg/kg TXAPA9 5,750,63

SAFETY PROFILE: Poison by intraperitoneal route. When heated to decomposition it emits very toxic fumes of NO_x. See also RARE EARTHS, NITRATES, and GADOLINIUM.

GAP000 CAS:12064-62-9 ***HR: 2***
GADOLINIUM OXIDE
mf: Gd_2O_3 mw: 362.50

PROP: White-to cream-colored, hygroscopic powder. D: 7.407 @ 15°, mp: 2330°. Insol in water; sol in acids.

SYN: GADOLINIA

TOXICITY DATA with REFERENCE
ipr-rat LD50:1000 mg/kg CURL** 34,117,60

CONSENSUS REPORTS: Reported in EPA TSCA Inventory.

SAFETY PROFILE: Moderately toxic by intraperitoneal route. See also GAOLINIUM and RARE EARTHS.

GAP500 ***HR: D***
GAJAR, seed extract

SYNS: CARROT, SEED EXTRACT ◇ DAUCUS CAROTA LINN., SEED EXTRACT

TOXICITY DATA with REFERENCE
orl-rat TDLo:350 mg/kg (female 1-7D post):REP JRPFA4 31,143,72

SAFETY PROFILE: Experimental reproductive effects.

GAR000 CAS:526-99-8 ***HR: 1***
GALACTARIC ACID
mf: $C_6H_{10}O_8$ mw: 210.16

PROP: Crystalline powder, decomp @ approx 225° when rapidly heated. Practically insol in alc and ether.

SYNS: GALACTOSACCHARIC ACID ◇ MUCIC ACID ◇ SACCHAROLACTIC ACID ◇ SCHLEIMSAURE ◇ TETRAHYDROXYADIPIC ACID

TOXICITY DATA with REFERENCE
orl-mus LD50:8000 mg/kg BIJOAK 34,1196,40

CONSENSUS REPORTS: Reported in EPA TSCA Inventory.

SAFETY PROFILE: Mildly toxic by ingestion. When heated to decomposition it emits acrid smoke and irritating fumes.

GAT000 CAS:1772-03-8 ***HR: 3***
d-GALACTOSAMINE HYDROCHLORIDE
mf: $C_6H_{13}NO_5$•ClH mw: 215.66

SYN: 2-AMINO-2-DEOXY-d-GALACTOSEHYDROCHLORIDE

TOXICITY DATA with REFERENCE
ipr-rat TDLo:135 g/kg/77W-I:ETA VAAZA2 12,285,73
ipr-mus LD50:2660 mg/kg CTYAD9 11,262,80

CONSENSUS REPORTS: Reported in EPA TSCA Inventory.

SAFETY PROFILE: Moderately toxic by intraperitoneal route. Questionable carcinogen with experimental tumorigenic data. When heated to decomposition it emits very toxic fumes of NO_x and Cl^-.

GAV000 CAS:59-23-4 ***HR: D***
GALACTOSE
mf: $C_6H_{12}O_6$ mw: 180.18

PROP: (α form): Prisms from water or ethanol. Mp: 167°. Freely sol in hot water; sol in pyridine; sltly sol in alc. (β form): Crystals. Mp: 167°.

SYN: d-GALACTOSE

TOXICITY DATA with REFERENCE
orl-rat TDLo:1000 g/kg (female 3-22D post):TER SCIEAS 214,1145,81
orl-rat TDLo:440 g/kg (female 1-22D post):REP JOPDAB 67,438,65

CONSENSUS REPORTS: Reported in EPA TSCA Inventory.

SAFETY PROFILE: Experimental teratogenic and reproductive effects. When heated to decomposition it emits acrid smoke and irritating fumes.

GAV100 CAS:9031-11-2 ***HR: 2***
β-GALACTOSIDASE

SYNS: E.C. 3.2.1.23 ◇ LACTASE

TOXICITY DATA with REFERENCE
ipr-rat LD50:660 mg/kg KSRNAM 4,725,70
scu-rat LD50:4090 mg/kg KSRNAM 4,725,70
ipr-mus LD50:630 mg/kg KSRNAM 4,729,70
scu-mus LD50:1470 mg/kg KSRNAM 4,729,70

CONSENSUS REPORTS: Reported in EPA TSCA Inventory.

SAFETY PROFILE: Moderately toxic by subcutaneous and intraperitoneal routes.

GAX000 CAS:7681-28-9 ***HR: 2***
GALACTURONIC ACID with α-(6-METHOXY-4-QUINOLYL)-5-VINYL-2-QUINUCLIDINE-METHANOL
mf: $C_{20}H_{24}N_2O_2$•$xC_6H_{10}O_7$ mw: 1683.58

SYNS: CARDIOQUIN ◇ GALACTOQUIN ◇ α-(6-METHOXY-4-QUINOLYL)-5-VINYL-2-QUINUCLIDINEMETHANOLGALACTURONATE (salt) ◇ NATICARDINA ◇ QUINIDINE POLYGALACTURONATE ◇ SINEFLUTTER

TOXICITY DATA with REFERENCE
orl-rat LD50:3200 mg/kg ANTCAO 9,97,59
orl-mus LD50:2680 mg/kg ANTCAO 9,97,59

SAFETY PROFILE: Moderately toxic by ingestion. Used as a cardiac depressant. When heated to decomposition it emits toxic fumes of NO_x.

GAZ000 CAS:548-83-4 ***HR: D***
GALANGIN
mf: $C_{15}H_{10}O_5$ mw: 270.25

SYNS: NORIZALPININ ◇ 3,5,7-TRIHYDROXYFLAVONE ◇ 3,5,7-TRIHYDROXY-2-PHENYL-4H-BENZOPYRAN-4-ONE

TOXICITY DATA with REFERENCE
mnt-mus-ipr 1 g/kg MUREAV 124,255,83
msc-ham:ovr 20 mg/L MUREAV 113,45,83

SAFETY PROFILE: Mutation data reported. When heated to decomposition it emits acrid smoke and irritating fumes.

GBA000 CAS:69353-21-5 ***HR: 3***
GALANTHAMINE HYDROBROMIDE
mf: $C_{17}H_{21}NO_3$•BrH mw: 368.31

PROP: Crystals from water. Decomp @ 246-247°.

SYNS: JILKON HYDROBROMIDE ◇ LYCOREMINE HYDROBROMIDE ◇ NIVALIN ◇ 1,2,3,4,6,7,7a,11c-OCTAHYDRO-9-METHOXY-2-METHYL-BENZOFURO(4,3,2-efg)(2)BENZAZOCIN-6-OLHBr

TOXICITY DATA with REFERENCE
orl-mus LD50:18700 μg/kg MEIEDD 10,620,83
ipr-mus LD50:14900 μg/kg AAREAV 22,285,65
ivn-mus LD50:5200 μg/kg AAREAV 22,285,65

SAFETY PROFILE: Poison by ingestion, intravenous, and intraperitoneal routes. Used as a cholinesterase inhibitor. When heated to decomposition it emits very toxic fumes of HBr and NO_x.

GBB500 CAS:3691-74-5 **HR: 3**
GALATONE
mf: $C_{12}H_{13}N_3O_6$ mw: 295.28

PROP: Plates and rods from methanol, needles from abs ethanol. Decomp 150-160°. Freely sol in water, practically insol in cold alc.

SYNS: GATALONE ◇ GLUCAZIDE ◇ GLURONAZID ◇ GLURONAZIDE ◇ GLYCONIAZIDE ◇ GUIDAZIDE ◇ INH-G ◇ HYDRONSAN ◇ N-ISONICOTINOYL-N'-GLUCURONSAEURE-Γ-LACTON-HYDRAZON (GERMAN) ◇ MYCOBACTYL

TOXICITY DATA with REFERENCE
orl-rat LD50:2990 mg/kg ARZNAD 26,409,76
ivn-rat LD50:820 mg/kg ARZNAD 26,409,76
orl-mus LD50:348 mg/kg ARZNAD 26,409,76
ivn-mus LD50:298 mg/kg ARZNAD 26,409,76
orl-gpg LD50:530 mg/kg ARZNAD 26,409,76

SAFETY PROFILE: Poison by ingestion and intravenous routes. When heated to decomposition it emits toxic fumes of NO_x.

GBC000 CAS:8023-91-4 **HR: 1**
GALBANUM OIL

PROP: Found in the dried resinous exudate of *Ferula galbaniflua boiss & buhse* and other *Ferula* species (FCTXAV 16,637,78).

TOXICITY DATA with REFERENCE
skn-rbt 500 mg/24H MLD FCTXAV 16,765,78

CONSENSUS REPORTS: Reported in EPA TSCA Inventory.

SAFETY PROFILE: A skin irritant. When heated to decomposition it emits acrid smoke and irritating fumes.

GBE000 CAS:149-91-7 **HR: 3**
GALLIC ACID
mf: $C_7H_6O_5$ mw: 170.13

PROP: White- to pale fawn-colored, odorless crystals; somewhat water-sol. D: 1.694, mp: 225-250° (decomp), $-H_2O$ @ 100-120°.

SYN: 3,4,5-TRIHYDROXYBENZOIC ACID

TOXICITY DATA with REFERENCE
mmo-sat 100 μg/plate ABCHA6 45,327,81
mrc-smc 100 mg/L MUREAV 135,109,84
scu-rat TDLo:5 mg/kg (1D pre):REP ENDOAO 57,466,55
scu-rat LDLo:5 g/kg JAFCAU 17,497,69
ipr-mus LDLo:800 mg/kg RBPMAZ 22,1,52
ivn-mus LD50:320 mg/kg CSLNX* NX#02597
orl-rbt LD50:5 g/kg AJVRAH 23,1264,62

CONSENSUS REPORTS: Reported in EPA TSCA Inventory. EPA Genetic Toxicology Program.

SAFETY PROFILE: Poison by intravenous route. Moderately toxic by intraperitoneal route. Mildly toxic by ingestion. Experimental reproductive effects. Mutation data reported. When heated to decomposition it emits acrid smoke and irritating fumes.

GBG000 CAS:7440-55-3 **HR: 3**
GALLIUM
DOT: UN 2803
af: Ga aw: 69.72

PROP: A beautiful, lustrous, silvery liquid or metal or a gray solid. Mp: 29.78°, bp: 2403°, d (solid): 5.904 @ 29.6°, d (liquid): 6.905 @ 29.8°.

SYNS: GALLIUM METAL, solid (DOT) ◇ GALLIUM METAL, liquid (DOT)

CONSENSUS REPORTS: Reported in EPA TSCA Inventory.

DOT Classification: ORM-B; Label: None; liquid: ORM-B; Label: None; solid: Corrosive Material; Label: Corrosive.

SAFETY PROFILE: Poison by subcutaneous and intravenous routes. Corrosive; probably an eye, skin, and mucous membrane irritant. It has a metallic taste, causes dermatitis and depression of bone marrow function. Potentially explosive reaction with hydrogen peroxide + hydrochloric acid. Violent or vigorous reaction with halogens. Forms an amalgam with aluminum alloys. See also GALLIUM COMPOUNDS.

GBK000 CAS:1303-00-0 **HR: 3**
GALLIUM ARSENIDE
mf: AsGa mw: 144.64

PROP: Cubic crystals with dark gray metallic sheen. Mp: 1238°, d: 5.31.

SYN: GALLIUM MONOARSENIDE

TOXICITY DATA with REFERENCE
ipr-rat LD30:10 g/kg GTPZAB 24(3),45,80
ipr-mus LD50:4700 mg/kg GISAAA 45(10),13,80

CONSENSUS REPORTS: Reported in EPA TSCA Inventory. Arsenic and its compounds are on the Community Right-To-Know List.

OSHA PEL: Cancer Hazard
NIOSH REL: (Gallium Arsenide) CL 0.002 mg(As)/m^3/15M

SAFETY PROFILE: Confirmed carcinogen. Mildly toxic by intraperitoneal route. Most arsenic compounds are poisons. Can react with steam, acids and acid fumes to evolve the deadly poisonous arsine. Molten gallium arsenide attacks quartz. When heated to decomposition it emits very toxic fumes of As. See also ARSENIC COMPOUNDS and GALLIUM COMPOUNDS.

GBM000 CAS:13450-90-3 ***HR: 3***
GALLIUM (3+) CHLORIDE
mf: Cl_3Ga mw: 176.07

PROP: Colorless needles. Mp: 78°.

SYN: GALLIUM CHLORIDE

TOXICITY DATA with REFERENCE
ihl-rat LCLo:316 mg/m³/3H JPETAB 95,487,49
scu-rat LD50:306 mg/kg EQSSDX 1,1,75
ivn-rat LD50:47 mg/kg EQSSDX 1,1,75
ipr-mus LD50:93400 μg/kg COREAF 256,1043,63
ivn-dog LD50:41 mg/kg EQSSDX 1,1,75
scu-rbt LD50:245 mg/kg EQSSDX 1,1,75
ivn-rbt LD50:43 mg/kg EQSSDX 1,1,75

CONSENSUS REPORTS: Reported in EPA TSCA Inventory. EPA Extremely Hazardous Substances List.

SAFETY PROFILE: Poison by inhalation, subcutaneous, intravenous, and intraperitoneal routes. When heated to decomposition it emits very toxic fumes of Cl^-. See also GALLIUM COMPOUNDS.

GBO000 CAS:27905-02-8 ***HR: 3***
GALLIUM CITRATE
mf: $C_6H_5O_7$•Ga mw: 258.83

SYN: CITRIC ACID, GALLIUM SALT (1:1)

TOXICITY DATA with REFERENCE
scu-rat LD50:220 mg/kg 14CYAT 2,1040,63
scu-mus LD50:2250 mg/kg EQSSDX 1,1,75
scu-dog LD50:37500 μg/kg EQSSDX 1,1,75
scu-dom LD50:37500 μg/kg EQSSDX 1,1,75

SAFETY PROFILE: Poison by subcutaneous route. When heated to decomposition it emits acrid smoke and irritating fumes. See also GALLIUM COMPOUNDS.

GBO500 ***HR: 3***
GALLIUM COMPOUNDS

SAFETY PROFILE: Preliminary investigations were done with the oxide, tartrate, benzoate, and anthranilate, which were used by some investigators in the experimental treatment of syphilis. Amounts up to 15 mg/kg of body weight were injected intravenously and were tolerated without harm by laboratory animals. Larger doses produced hemorrhagic nephritis. In the case of gallium lactate, work done at the Naval Medical Research Institute showed that intravenous injections of about 40 mg of gallium per kg of body weight in rats or rabbits was lethal. Metallic gallium as well as the nitrate produced no skin injury, and subcutaneous injections of relatively large amounts could be tolerated both by rabbits and rats without evidence of injury. It has, however, been demonstrated that gallium remains in the tissues for long periods of time following intramuscular injections of soluble gallium salts. Tissue distribution experiments indicate that it behaves like bismuth and mercury in that one respect.

GBQ000 ***HR: 3***
GALLIUM LACTATE mixed with SODIUM LACTATE (1.6:1 moles)

TOXICITY DATA with REFERENCE
scu-rat LD50:720 mg/kg JPETAB 95,487,49
ivn-rat LD50:274 mg/kg JPETAB 95,487,49
scu-rbt LD50:583 mg/kg JPETAB 95,487,49
ivn-rbt LD50:256 mg/kg JPETAB 95,487,49

SAFETY PROFILE: Poison by intravenous route. Moderately toxic by subcutaneous route. When heated to decomposition it emits toxic fumes of Na_2O. See also GALLIUM COMPOUNDS.

GBS000 CAS:13494-90-1 ***HR: 3***
GALLIUM(III) NITRATE (1:3)
mf: N_3O_9•Ga mw: 255.75

PROP: White, deliquescent crystals. Mp: decomp @ 110°, bp: releases Ga_2O_3 @ 200°.

SYNS: GALLIUM NITRATE ◇ NITRIC ACID, GALLIUM(3+) SALT

TOXICITY DATA with REFERENCE
skn-mam 500 mg SEV GISAAA 45(10),13,80
dnd-mam:lym 40 μmol/L JCHODP 7,411,76
ivn-hmn TDLo:7 mg/kg:GIT,KID CTRRDO 62,1449,78
ivn-hmn TDLo:144 mg/kg:BLD CTRRDO 62,1449,78
ipr-rat LD50:67500 μg/kg EQSSDX 1,1,75
scu-rat LDLo:72 mg/kg INMEAF 12,7,43
orl-mus LD50:4360 mg/kg GISAAA 45(10),13,80
ipr-mus LD50:80 mg/kg EQSSDX 1,1,75
scu-mus LD50:600 mg/kg GISAAA 45(10),13,80

CONSENSUS REPORTS: Reported in EPA TSCA Inventory.

SAFETY PROFILE: Poison by intraperitoneal and subcutaneous routes. Moderately toxic by ingestion. Human systemic effects by intravenous route: nausea or vomiting, renal function changes, proteinuria, normocytic anemia and thrombocytopenia. A severe skin irritant. Mutation data reported. When heated to decomposition it emits toxic fumes of NO_x. See also GALLIUM COMPOUNDS and NITRATES.

GBS100 CAS:13494-91-2 ***HR: 2***
GALLIUM SULFATE
mf: $O_{12}S_3$•2Ga mw: 427.62

SYNS: DIGALLIUM TRISULFATE ◇ SULFURIC ACID, GALLIUM SALT (3:2)

TOXICITY DATA with REFERENCE
ivn-ham TDLo:40 mg/kg (female 8D post):REP TXAPA9 16,166,70
ipr-rat LD50:2 g/kg JICEEM 7,411,87
ipr-mus LD50:2330 mg/kg JICEEM 7,411,87

CONSENSUS REPORTS: Reported in EPA TSCA Inventory.

SAFETY PROFILE: Moderately toxic by intraperitoneal route. Experimental reproductive effects. When heated to decomposition it emits acrid smoke and irritating fumes.

GBU000 CAS:1222-05-5 ***HR: 1***
GALOXOLIDE
mf: $C_{18}H_{26}O$ mw: 258.44

SYN: 1,3,4,6,7,8-HEXAHYDRO-4,6,6,7,8,8-HEXAMETHYL-CYCLOPENTA-Γ-2-BENZOPYRAN

TOXICITY DATA with REFERENCE
skn-rbt 500 mg/24H MOD FCTXAV 14,793,76

CONSENSUS REPORTS: Reported in EPA TSCA Inventory.

SAFETY PROFILE: A skin irritant. When heated to decomposition it emits acrid smoke and irritating fumes.

GBU600 ***HR: 2***
GANCIDIN (unpurified)

PROP: Produced by the strain *Streptomyces sp.* AAK-84 (85ERAY 2,1354,78)

TOXICITY DATA with REFERENCE
ipr-mus LD50:1500 mg/kg 85ERAY 2,1354,78
scu-mus LD50:2800 mg/kg 85ERAY 2,1354,78
ivn-mus LD50:700 mg/kg 85ERAY 2,1354,78

SAFETY PROFILE: Moderately toxic by intravenous, intraperitoneal, and subcutaneous routes.

GBU800 ***HR: 1***
GARLIC OIL

PROP: From steam distillation of *Allium sativum* L. (Fam. *Liliaceae*). Clear to yellow liquid; strong odor and taste of garlic. Sol in fixed oils, mineral oil; insol in glycerin, alc, propylene glycol.

SAFETY PROFILE: An eye irritant. When heated to decomposition it emits acrid smoke and irritating fumes.

GBU850 ***HR: 2***
GARLIC POWDER

SYN: ALLIUM SATIVUM Linn., powder

TOXICITY DATA with REFERENCE
orl-rat TDLo:9 g/kg (male 45D pre):REP IJEBA6 20,534,82
ivn-mus LD50:1650 mg/kg PJPHEO 5(1),21,88

SAFETY PROFILE: Moderately toxic by intravenous route. Experimental reproductive effects. When heated to decomposition it emits acrid smoke and irritating fumes.

GBW000 CAS:64741-44-2 ***HR: 2***
GAS OIL

PROP: Yellow liquid. Flash p: 150°F, d: 1, lel: 6.0%, uel: 13.5%, autoign temp: 640°F, boiling range: 230-250°.

TOXICITY DATA with REFERENCE
skn-mus TDLo:114 g/kg/38W-I:CAR FAATDF3 7,228,86

CONSENSUS REPORTS: Reported in EPA TSCA Inventory.

SAFETY PROFILE: Questionable carcinogen with experimental carcinogenic data. Pulmonary aspiration can cause severe pneumonitis. Combustible when exposed to heat or flame; can react vigorously with oxidizing materials. A moderate explosion hazard when exposed to heat or flame. To fight fire, use foam, CO_2, dry chemical. See also KEROSENE.

GBW010 CAS:64742-86-5 ***HR: 2***
GAS OILS (petroleum), hydrodesulfurized heavy vacuum

SYN: HYDRODESULFURIZED HEAVY VACUUM GAS OIL

TOXICITY DATA with REFERENCE
skn-mus TDLo:104 g/kg/26W-I:NEO EPASR* 8EHQ-0887-0687

SAFETY PROFILE: Questionable carcinogen with experimental neoplastigenic data data. When heated to decomposition it emits acrid smoke and irritating fumes.

GBW025 CAS:64741-58-8 ***HR: 3***
GAS OILS (petroleum), light vacuum

SYN: LIGHT GAS OIL

TOXICITY DATA with REFERENCE
skn-mus TDLo:306 g/kg/2Y-I:CAR FAATDF 7,228,86

CONSENSUS REPORTS: Reported in EPA TSCA Inventory.

SAFETY PROFILE: Questionable carcinogen with experimental carcinogenic data. Pulmonary aspiration can cause severe pneumonitis. Flammable liquid. When heated to decomposition it emits acrid smoke and irritating fumes.

GBY000 CAS:8006-61-9 ***HR: 3***
GASOLINE
DOT: UN 1203/UN 1257

PROP: Clear, aromatic, volatile liquid; a mixture of aliphatic HC. Flash p: −50°F, d: <1.0, vap d: 3.0-4.0, ULC: 95-100, lel: 1.3%, uel: 6.0%, autoign temp: 536-853°F, bp: Initially 39°, after 10% distilled = 60°, after 50% = 110°, after 90% = 170°, final bp: 204°. Insol in water; freely sol in abs alc, ether, chloroform, and benzene.

SYNS: CASING HEAD GASOLINE (DOT) ◇ MOTOR FUEL (DOT) ◇ MOTOR SPIRIT (DOT) ◇ NATURAL GASOLINE (DOT) ◇ PETROL (DOT)

TOXICITY DATA with REFERENCE
eye-man 500 ppm/1H MOD AEHLAU 1,548,60
eye-hmn 140 ppm/8H MLD JIHTAB 25,225,43
ihl-man TCLo:900 ppm/1H:EYE,CNS,PUL JIHTAB 25,225,43
ihl-rat LC50:300 g/m^3/5M NTIS** PB158-508
ihl-mus LC50:300 g/m^3/5M NTIS** PB158-508
ihl-gpg LC50:300 g/m^3/5M NTIS** PB158-508
ihl-mam LCLo:30000 ppm/5M AEPPAE 138,65,28

CONSENSUS REPORTS: Reported in EPA TSCA Inventory.

OSHA PEL: TWA 300 ppm; STEL 500 ppm
ACGIH TLV: TWA 300 ppm; STEL 500 ppm
DOT Classification: Flammable Liquid; Label: Flammable Liquid.

SAFETY PROFILE: Mildly toxic by inhalation. Human systemic effects by inhalation: cough, conjunctiva irritation, hallucinations or distorted perceptions. Repeated or prolonged dermal exposure causes dermatitis. Can cause blistering of skin. Questionable carcinogen. Inhalation or ingestion can cause central nervous system depression. Pulmonary aspiration can cause severe pneumonitis. Some addiction has been reported from inhalation of fumes. Even brief inhalations of high concentrations can cause a fatal pulmonary edema. The vapors are considered to be moderately poisonous. If its concentration in air is sufficiently high to reduce the oxygen content below that needed to maintain life, it acts as a simple asphyxiant. A human eye irritant. Gasoline is a common air contaminant. A very dangerous fire and explosion hazard when exposed to heat or flame; can react vigorously with oxidizing materials. To fight fire, use foam, CO_2, dry chemical.

GCA000 ***HR: 3***
GASOLINE (100-130 octane)

PROP: Flash p: −50°F, autoign temp: 824°F, lel: 1.3%, uel: 7.1%.

SAFETY PROFILE: Moderately toxic by inhalation. Pulmonary aspiration can cause severe pneumonitis. A very dangerous fire and explosion hazard when exposed to heat, flame or oxidizers. To fight fire, use water spray or mist, CO_2, dry chemical. See also GASOLINE.

GCC000 ***HR: 3***
GASOLINE (115-145 octane)

PROP: Flash p: −50°F, autoign temp: 880°F, lel: 1.2%, uel: 7.1%.

SAFETY PROFILE: Moderately toxic by inhalation. Pulmonary aspiration can cause severe pneumonitis. A very dangerous fire and explosion hazard when exposed to heat or flame. To fight fire, use water spray or mist, CO_2, dry chemical. See also GASOLINE.

GCE000 ***HR: 3***
GASOLINE ENGINE EXHAUST "TAR"

SYNS: AUTOMOBILE EXHAUST CONDENSATE ◇ GASOLINE ENGINE EXHAUST CONDENSATE

TOXICITY DATA with REFERENCE
skn-mus TDLo:110 g/kg/69W-I:CAR CANCAR 15,103,62
itr-ham TDLo:469 mg/kg/60W-I:NEO CANCAR 40,203,77
itr-ham TD:420 mg/kg/78W-I:ETA JCROD7 105,24,83

SAFETY PROFILE: Questionable carcinogen with experimental carcinogenic, neoplastigenic, and tumorigenic data.

GCE100 ***HR: 3***
GASOLINE, UNLEADED

SYNS: UNLEADED GASOLINE ◇ UNLEADED MOTOR GASOLINE

TOXICITY DATA with REFERENCE
skn-rbt 500 mg/24H MLD 52MLA2 1,1,83
ihl-rat TCLo:1501 ppm/78W-C:CAR AETODY 7,65,84
ihl-mus TCLo:2056 ppm/6H/78W-I:CAR NTIS** PB86-209152

SAFETY PROFILE: Suspected carcinogen with experimental carcinogenic data. Moderately toxic by inhalation. Pulmonary aspiration can cause severe pneumonitis. Skin irritant. Flammable liquid. When heated to decomposition it emits acrid smoke and irritating fumes.

GCE500 CAS:29868-97-1 ***HR: 3***
GASTROZEPIN
mf: $C_{19}H_{21}N_5O_2 \cdot 2ClH$ mw: 424.37

SYNS: L-S 519 ◇ LS 519 DIHYDROCHLORIDE ◇ 5,11-DIHYDRO-11-((4-METHYL-1-PIPERAZINYL)ACETYL)-6H-PYRIDO(2,3-b)(1,4)BENZODIAZEPIN-6-ONE DIHYDROCHLORIDE ◇ PIRENZEPINE HYDROCHLORIDE

TOXICITY DATA with REFERENCE
ivn-rat TDLo:810 mg/kg (female 17-22D post):REP IYKEDH 17,859,86
orl-rbt TDLo:1125 mg/kg (female 8-16D post):TER OYYAA2 9,377,75
orl-rat LD50:5 g/kg IYKEDH 11,328,80
ipr-rat LD50:440 mg/kg YAKUD5 23,1999,81
scu-rat LD50:3 g/kg IYKEDH 19,735,88
ivn-rat LD50:92 mg/kg IYKEDH 19,735,88
orl-mus LD50:2600 mg/kg IYKEDH 11,328,80
ipr-mus LD50:407 mg/kg DRUGAY 6,645,82
scu-mus LD50:2100 mg/kg IYKEDH 19,735,88
ivn-mus LD50:96 mg/kg IYKEDH 19,735,88
orl-rbt LD50:3000 mg/kg OYYAA2 9,377,75

SAFETY PROFILE: Poison by intravenous route. Moderately toxic by ingestion and intraperitoneal route. An experimental teratogen. Experimental reproductive effects. When heated to decomposition it emits toxic fumes of NO_x and HCl. See also DIAZEPAM.

GCG300 CAS:427-01-0 ***HR: 3***
GEISSOSPERMINE
mf: $C_{40}H_{48}N_4O_3$ mw: 632.92

PROP: Anhydr form, crystals from abs acetone. Mp: 213-214° (decomp). Sltly sol in water, ether; sol in alc.

TOXICITY DATA with REFERENCE
orl-mus LD50:450 mg/kg APFRAD 19,104,61
ipr-mus LD50:125 mg/kg APFRAD 19,104,61
scu-mus LD50:150 mg/kg APFRAD 19,104,61
ivn-mus LD50:20 mg/kg APFRAD 19,104,61

SAFETY PROFILE: Poison by subcutaneous, intravenous, and intraperitoneal routes. Moderately toxic by ingestion. When heated to decomposition it emits toxic fumes of NO_x.

GCI000 CAS:30562-34-6 ***HR: 3***
GELDANAMYCIN
mf: $C_{29}H_{40}N_2O_9$ mw: 560.71

SYN: U-29135

TOXICITY DATA with REFERENCE
dni-mus:lym 1 mg/L JANTAJ 35,886,82
orl-rat LD50:2500 mg/kg 85ERAY 1,820,78
ipr-mus LD50:1 mg/kg UPJOH* 2(6),-,71

CONSENSUS REPORTS: EPA Genetic Toxicology Program.

SAFETY PROFILE: Poison by intraperitoneal route. Moderately toxic by ingestion. Mutation data reported. When heated to decomposition it emits toxic fumes of NO_x.

GCK000 CAS:509-15-9 ***HR: 3***
GELSEMINE
mf: $C_{20}H_{22}N_2O_2$ mw: 322.44

PROP: An alkaloid. Mp: 178°. Sltly sol in water; sol in alc, benzene, chloroform, ether, acetone, and dilute acids.

SYN: GELSEMIN

TOXICITY DATA with REFERENCE
ipr-mus LD50:49 mg/kg ARZNAD 7,349,57
scu-rbt LDLo:100 μg/kg MEIEDD 10,625,83

SAFETY PROFILE: A deadly poison by subcutaneous and intraperitoneal routes. A poisonous alkaloid. Can cause muscular weakness and respiratory arrest. Used as a central nervous system stimulant. When heated to decomposition it emits toxic fumes of NO_x.

GCK300 CAS:25812-30-0 ***HR: 3***
GEMFIBROZIL
mf: $C_{15}H_{22}O_3$ mw: 250.37

PROP: Crystals from hexane. Mp: 61-63°, bp: 158-159°.

SYNS: CI-719 ◇ 5-(2,5-DIMETHYLPHENOXY)-2,2-DIMETHYLPENTANOIC ACID (9CI) ◇ 2,2-DIMETHYL-5-(2,5-XYLYLOXY)VALERIC ACID ◇ GEVILON ◇ LIPUR ◇ LOPID

TOXICITY DATA with REFERENCE
orl-rat TDLo:2648 mg/kg (female 15-22D post):REP FAATDF 8,454,87
orl-rat TDLo:218 g/kg/2Y-C:CAR JJIND8 67,1105,81
orl-mus TDLo:16 g/kg/78W-C:ETA JJIND8 67,1105,81
orl-rat TD:22 g/kg/2Y-C:NEO JJIND8 67,1105,81
orl-rat LD50:4786 mg/kg PRSMA4 69(Suppl 2),15,76
orl-mus LD50:3162 mg/kg PRSMA4 69(Suppl 2),15,76

SAFETY PROFILE: Moderately toxic by ingestion. Experimental reproductive effects. Questionable carcinogen with experimental carcinogenic, neoplastigenic, and tumorigenic data.

GCM000 CAS:8023-80-1 ***HR: 1***
GENET ABSOLUTE

PROP: Extracted from the flowers of *Spartium junceum* (FCTXAV 14,659,76).

SYNS: BROOM ABSOLUTE ◇ SPANISH BROOM

TOXICITY DATA with REFERENCE
skn-rbt 500 mg/24H MOD FCTXAV 14,779,76

CONSENSUS REPORTS: Reported in EPA TSCA Inventory.

SAFETY PROFILE: A skin irritant. When heated to decomposition it emits acrid smoke and irritating fumes.

GCM300 CAS:6902-77-8 ***HR: 3***
GENIPIN
mf: $C_{11}H_{14}O_5$ mw: 226.25

TOXICITY DATA with REFERENCE
orl-rat TDLo:1750 mg/kg (35D male):REP OYYAA2 19,259,80
orl-mus LD50:237 mg/kg YKKZAJ 94,157,74
ipr-mus LD50:190 mg/kg YKKZAJ 94,157,74
ivn-mus LD50:153 mg/kg YKKZAJ 94,157,74

SAFETY PROFILE: Poison by ingestion, intravenous, and intraperitoneal routes. Experimental reproductive effects. When heated to decomposition it emits acrid smoke and irritating fumes. See also ESTERS.

GCM350 CAS:446-72-0 ***HR: D***
GENISTEIN
mf: $C_{15}H_{10}O_5$ mw: 270.23

PROP: Rectangular or six-sided rods from 60% alc. Dendritic needles from ether. Mp: 297-298° (slt decomp). Sol in the usual organic solvents; practically insol in water; sol in dil alkalies with yellow color.

SYNS: GENISTEOL ◇ GENISTERIN ◇ PRUNETOL ◇ SOPHORICOL ◇ 4′,5,7-TRIHYDROXYISOFLAVONE ◇ 5,7,4′-TRIHYDROXYISOFLAVONE

TOXICITY DATA with REFERENCE
orl-mus TDLo:18600 mg/kg (female 31D pre):REP JOENAK 13,94,55

CONSENSUS REPORTS: Reported in EPA TSCA Inventory.

SAFETY PROFILE: Experimental reproductive effects. When heated to decomposition it emits acrid smoke and irritating fumes.

GCO000 CAS:1403-66-3 ***HR: 3***
GENTAMICIN

SYNS: GARAMYCIN ◇ GENTAMYCIN ◇ GENTAMYCIN-CREME (GERMAN) ◇ UROMYCINE

TOXICITY DATA with REFERENCE
mmo-esc 250 μg/L MUREAV 140,13,84
dnd-rat-ipr 70 mg/kg/7D AMACCQ 24,586,83
scu-rat TDLo:660 mg/kg (female 10-15D post):REP TJADAB 33,13A,86
unr-rat TDLo:90 mg/kg (female 9-14D post):TER ANTBAL 27,926,82
ivn-wmn TDLo:8200 μg/kg/12D:EAR,CNS APMIBM 81(Suppl 241),73,73
ivn-hmn TDLo:2 mg/kg:KID NEJMAG 303,1002,80
ivn-hmn TDLo:23 mg/kg/1Y-I:PNS AIMDAP 138,1621,78
ims-man TDLo:8 mg/kg/2W-I:EYE,SKN JAMAAP 211,123,70
ims-inf TDLo:20 mg/kg:KID PEDIAU 77,848,86
orl-rat LD50:6600 mg/kg AMPMAR 39,259,78
ivn-rat LD50:70 mg/kg JIMRBV 2,100,74
ims-rat LD50:463 mg/kg JJANAX 35,461,82
orl-mus LD50:10 g/kg JJANAX 30,386,77
ipr-mus LD50:235 mg/kg JIMRBV 2,100,74
scu-mus LD50:274 mg/kg TXAPA9 25,398,73
ivn-mus LD50:51 mg/kg JJANAX 30,386,77
ims-mus LD50:167 mg/kg JJANAX 35,461,82

CONSENSUS REPORTS: EPA Genetic Toxicology Program.

SAFETY PROFILE: Poison by intravenous, intraperitoneal, intramuscular, and subcutaneous routes. Mildly toxic by ingestion. Experimental teratogenic and reproductive effects. Mutation data reported. Human systemic effects by intravenous route: vestibular function changes, hallucinations, distorted perceptions, motor activity changes, trigeminal nerve sensory changes, kidney changes, visual field changes, eye hemorrhage. Affects the peripheral nervous system by intravenous route. An antibiotic. When heated to decomposition it emits acrid smoke and irritating fumes. See also other gentamycin entries.

GCO200 CAS:11097-82-8 ***HR: 3***
GENTAMICIN C COMPLEX

TOXICITY DATA with REFERENCE
ipr-mus LD50:430 mg/kg AMACCQ 2,464,72
scu-mus LD50:485 mg/kg AMACCQ 2,464,72
ivn-mus LD50:55900 μg/kg JANTAJ 35,94,82

SAFETY PROFILE: Poison by intravenous route. Moderately toxic by intraperitoneal and subcutaneous routes. See also other gentamycin entries.

GCS000 CAS:1405-41-0 ***HR: 3***
GENTAMYCIN SULFATE

SYNS: GARAMYCIN ◇ GENOPTIC ◇ GENOPTIC S.O.P. ◇ GM SULFATE ◇ NSC-82261 ◇ SCH 9724

TOXICITY DATA with REFERENCE
dnd-esc 5 mg/L MUREAV 89,95,81
spm-rat-unr 9600 μg/kg/8D JOURAA 112,348,74
scu-rat TDLo:660 mg/kg (female 15-20D post):REP ARTODN 62,274,88
scu-rat TDLo:660 mg/kg (female 10-15D post):TER ARTODN 62,274,88
ivn-wmn TDLo:45 mg/kg/1W-I:SYS DICPBB 18,596,84
ivn-man TDLo:21 mg/kg/6D-I:SYS DICPBB 18,596,84
ipr-rat LD50:630 mg/kg JJANAX 26,221,73
ivn-rat LD50:96 mg/kg JZKEDZ 8,219,82
ims-rat LD50:384 mg/kg TXAPA9 18,185,71
ipr-mus LDLo:245 mg/kg JZKEDZ 8,219,82
scu-mus LD50:478 mg/kg JANTAJ 23,551,70

ivn-mus LD50:47 mg/kg JJANAX 39,3164,86
ims-mus LD50:250 mg/kg JZKEDZ 8,219,82

SAFETY PROFILE: Poison by intravenous, intraperitoneal, and intramuscular routes. Moderately toxic by subcutaneous route. Human systemic effects: level changes for metals other than Na/K/Fe/Ca/P/Cl. An experimental teratogen. Other experimental reproductive effects. Mutation data reported. When heated to decomposition it emits very toxic fumes of SO_x. See also other gentamycin entries.

GCU000 CAS:490-79-9 ***HR: 3***
GENTISIC ACID
mf: $C_7H_6O_4$ mw: 154.13

PROP: Needles, monoclinic prisms from water. Mp: 199-200°; sol in water (more so in hot water), alc, ether; practically insol in carbon disulfide, chloroform, ether.

SYNS: 2,5-DHBA ◇ 2,5-DIHYDROXYBENZOIC ACID ◇ GENTISATE ◇ HYDROQUINONECARBOXYLIC ACID ◇ 5-HYDROXYSALICYLIC ACID

TOXICITY DATA with REFERENCE
scu-rat TDLo:380 mg/kg (9D preg):REP BCPCA6 22,407,73
scu-rat TDLo:642 mg/kg (11D preg):TER RCOCB8 38,209,82
orl-rat LD50:800 mg/kg 14CYAT 2,1838,63
ipr-rat LD50:3000 mg/kg BCFAAI 112,53,73
orl-mus LD50:4500 mg/kg BJPCAL 8,30,53
ivn-mus LD50:374 mg/kg JAPMA8 42,254,53

CONSENSUS REPORTS: Reported in EPA TSCA Inventory.

SAFETY PROFILE: Poison by intravenous route. Moderately toxic by ingestion and intraperitoneal routes. Experimental teratogenic and reproductive effects. When heated to decomposition it emits acrid smoke and irritating fumes.

GCW000 CAS:459-80-3 ***HR: 2***
GERANIC ACID
mf: $C_{10}H_{16}O_2$ mw: 168.26

SYNS: 3,7-DIMETHYL-2,6-OCTADIENOIC ACID ◇ 3,7-DIMETHYL-2,7-OCTADIENOIC ACID

TOXICITY DATA with REFERENCE
skn-rbt 500 mg/24H MOD FCTXAV 17,785,79
orl-mus LD50:4000 mg/kg BIJOAK 34,1196,40
skn-rbt LD50:1750 mg/kg FCTXAV 17,785,79

CONSENSUS REPORTS: Reported in EPA TSCA Inventory.

SAFETY PROFILE: Moderately toxic by ingestion and skin contact. A skin irritant. When heated to decomposition it emits acrid smoke and irritating fumes.

GCY000 CAS:105-86-2 ***HR: 1***
GERANIOL FORMATE
mf: $C_{11}H_{18}O_2$ mw: 182.29

PROP: Colorless to pale yellow liquid; rose odor. D: 0.906-0.920, refr index: 1.457-1.466, flash p: 205°F. Sol in alc, fixed oils; insol in glycerin, propylene glycol, water @ 216°.

SYNS: trans-3,7-DIMETHYL-2,6-OCTADIEN-1-OL FORMATE ◇ 3,7-DIMETHYL-2,6-OCTADIENYL ESTER FORMIC ACID (E) ◇ trans-3,7-DIMETHYL-2,6-OCTADIEN-1-YL FORMATE ◇ FEMA No. 2514 ◇ FORMIC ACID, GERANIOL ESTER ◇ GERANYL FORMATE (FCC)

TOXICITY DATA with REFERENCE
skn-hmn 10 mg/48H MLD FCTXAV 12,893,74
eye-rbt 100 mg/24H MLD FCTXAV 12,893,74

CONSENSUS REPORTS: Reported in EPA TSCA Inventory.

SAFETY PROFILE: A human skin irritant and an experimental eye irritant. Combustible liquid. When heated to decomposition it emits acrid smoke and irritating fumes. See also ESTERS.

GDA000 CAS:8000-46-2 ***HR: 1***
GERANIUM OIL ALGERIAN TYPE

PROP: From steam distillation of leaves from *Pelargonium graveolens* l'Her (Fam. *Geraniaceae*). Contains geraniol and geranyl tiglate (FCTXAV 14,659,76). Yellow liquid; odor of rose and geraniol. D: 0.886-0.898, refr index: 1.454-1.472 @ 20°. Sol in fixed oils, mineral oil; insol in glycerin.

SYNS: GERANIUM OIL ◇ OIL of GERANIUM ◇ OIL of PELARGONIUM ◇ OIL of ROSE GERANIUM ◇ OIL ROSE GERANIUM ALGERIAN ◇ PELARGONIUM OIL ◇ ROSE GERANIUM OIL ALGERIAN

TOXICITY DATA with REFERENCE
skn-rbt 500 mg/24H MLD FCTXAV 14,781,76
skn-gpg 100% MLD FCTXAV 14,781,76

CONSENSUS REPORTS: Reported in EPA TSCA Inventory.

SAFETY PROFILE: A skin irritant. When heated to decomposition it emits acrid smoke and irritating fumes.

GDC000 ***HR: 2***
GERANIUM OIL BOURBON

PROP: Chief constituents are geraniol, citronellol (FCTXAV 12,807,74).

SYNS: OIL GERANIUM REUNION ◇ OIL ROSE GERANIUM

TOXICITY DATA with REFERENCE
skn-rbt 500 mg/24H MOD FCTXAV 12,883,74
skn-rbt LD50:2500 mg/kg FCTXAV 12,883,74

SAFETY PROFILE: Moderately toxic by skin contact. A skin irritant. When heated to decomposition it emits acrid smoke and irritating fumes.

GDE000 ***HR: 1***
GERANIUM OIL MOROCCAN

PROP: Found in the leaves and stems of *Pelargonium roseum* (FCTXAV 13,449,75).

TOXICITY DATA with REFERENCE
skn-rbt 500 mg/24H MLD FCTXAV 13,451,75
skn-gpg 100% MLD FCTXAV 13,451,75

SAFETY PROFILE: A skin irritant. When heated to decomposition it emits acrid smoke and irritating fumes.

GDE300 ***HR: 2***
GERANIUM THUNBERGII Sieb. et Zucc., extract

TOXICITY DATA with REFERENCE
scu-rat LD50:3700 mg/kg KSRNAM 11,458,77
ivn-rat LD50:461 mg/kg KSRNAM 11,458,77
scu-mus LD50:524 mg/kg KSRNAM 11,458,77
ivn-mus LD50:841 mg/kg KSRNAM 11,458,77

SAFETY PROFILE: Moderately toxic by intravenous and subcutaneous routes.

GDE400 CAS:689-67-8 ***HR: 1***
GERANYL ACETONE
mf: $C_{13}H_{22}O$ mw: 194.35

SYNS: DIHYDROPSEUDOIONONE ◇ α-β-DIHYDROPSEUDOIONONE ◇ 6,10-DIMETHYL-UNDECA-5,9-DIEN-2-ONE ◇ 5,9-UNDECADIEN-2-ONE, 6,10-DIMETHYL-

TOXICITY DATA with REFERENCE
skn-rbt 500 mg/24H MOD FCTXAV 17,787,79

CONSENSUS REPORTS: Reported in EPA TSCA Inventory.

SAFETY PROFILE: A skin irritant. When heated to decomposition it emits acrid smoke and irritating fumes.

GDE800 ***HR: 2***
GERANYL BENZOATE
mf: $C_{17}H_{22}O_2$ mw: 258.36

PROP: Sltly yellow liquid; floral odor resembling ylang ylang oil. D: 0.978-0.984, refr index: 1.513-1.518, flash p: +212°F. Misc in alc, chloroform; insol in water @ 305°.

SYNS: 3,7-DIMETHYL-2,6-OCTADIEN-1-YL BENZOATE ◇ FEMA No. 2511

SAFETY PROFILE: Combustible liquid. When heated to decomposition it emits acrid smoke and irritating fumes.

GDE825 ***HR: 1***
GERANYL BUTYRATE
mf: $C_{14}H_{24}O_2$ mw: 224.34

PROP: Colorless to pale yellow liquid; fruity, roselike odor. D: 0.889-0.904, refr index: 1.455-1.462, flash p: +199°F. Sol in alc, fixed oils; insol in glycerin, propylene glycol, water @ 253°.

SYNS: 3,7-DIMETHYL-2,6-OCTADIENE-1-YL BUTYRATE ◇ FEMA No. 2512

SAFETY PROFILE: Combustible liquid. When heated to decomposition it emits acrid smoke and irritating fumes.

GDG000 CAS:10032-02-7 ***HR: 1***
GERANYL CAPROATE
mf: $C_{16}H_{28}O_2$ mw: 252.44

SYNS: (E)-3,7-DIMETHYLOCTA-2,6-DIEN-1-YL ESTER, HEXANOIC ACID ◇ (E)-3,7-DIMETHYLOCTA-2,6-DIEN-1-YL-n-HEXANOATE ◇ GERANYL HEXANOATE

TOXICITY DATA with REFERENCE
skn-rbt 500 mg/24H MLD FCTXAV 14,783,76

CONSENSUS REPORTS: Reported in EPA TSCA Inventory.

SAFETY PROFILE: A skin irritant. When heated to decomposition it emits acrid smoke and irritating fumes.

GDG100 CAS:40267-72-9 ***HR: 1***
GERANYL ETHYL ETHER
mf: $C_{12}H_{22}O$ mw: 182.34

SYNS: 1-ETHOXY-3,7-DIMETHYL-2,6-OCTADIENE ◇ ETHYL GERANYL ETHER ◇ 2,6-OCTADIENE, 1-ETHOXY-3,7-DIMETHYL-

TOXICITY DATA with REFERENCE
skn-rbt 500 mg/24H MLD FCTOD7 20,693,82

CONSENSUS REPORTS: Reported in EPA TSCA Inventory.

SAFETY PROFILE: A skin irritant. When heated to decomposition it emits acrid smoke and irritating fumes.

GDI000 CAS:2345-26-8 ***HR: 1***
GERANYL ISOBUTYRATE
mf: $C_{14}H_{24}O_2$ mw: 224.38

SYNS: trans-3,7-DIMETHYL-2,6-OCTADIEN-1-OLISOBUTYRATE ◇ trans-3,7-DIMETHYL-2,6-OCTADIENYL ESTER ISOBUTYRIC ACID ◇ trans-3-7-DIMETHYL-2,6-OCTADIENYL ISOBUTYRATE ◇ (E)2-METHYL-3,7-DIMETHYL-2,6-OCTADIENYL ESTER PROPANOIC ACID

TOXICITY DATA with REFERENCE
skn-rbt 500 mg/24H MLD FCTXAV 13,45175

CONSENSUS REPORTS: Reported in EPA TSCA Inventory.

SAFETY PROFILE: A skin irritant. When heated to decomposition it emits acrid smoke and irritating fumes.

GDK000 CAS:109-20-6 ***HR: 1***
GERANYL ISOVALERATE
mf: $C_{15}H_{26}O_2$ mw: 238.41

SYNS: trans-3,7-DIMETHYL-2,6-OCTADIENYLISOPENTANOATE ◇ (E)-ISOVALERIC ACID-3,7-DIMETHYL-2,6-OCTADIENYL ESTER ◇ (E)-3-METHYLBUTYRIC ACID-3,7-DIMETHYL-2,6-OCTADIENYL ESTER

TOXICITY DATA with REFERENCE
skn-rbt 500 mg/24H MLD FCTXAV 14,785,76

CONSENSUS REPORTS: Reported in EPA TSCA Inventory.

SAFETY PROFILE: A skin irritant. When heated to decomposition it emits acrid smoke and irritating fumes. See also ESTERS.

GDM000 CAS:5585-39-7 ***HR: 2***
GERANYL NITRILE
mf: $C_{10}H_{15}N$ mw: 149.26

SYNS: (E)-3,7-DIMETHYL-2,6-OCTADIENENITRILE ◇ GERANONITRILE

TOXICITY DATA with REFERENCE
orl-rat LD50:3100 mg/kg FCTXAV 14,787,76
skn-rbt LD50:4300 mg/kg FCTXAV 14,787,76

CONSENSUS REPORTS: Reported in EPA TSCA Inventory. Cyanide and its compounds are on the Community Right-To-Know List.

SAFETY PROFILE: Moderately toxic by ingestion. Mildly toxic by skin contact. When heated to decomposition it emits toxic fumes of NO_x and CN^-. See also NITRILES.

GDM100 CAS:65405-73-4 ***HR: 1***
GERANYL OXYACETALDEHYDE
mf: $C_{12}H_{20}O_2$ mw: 196.32

SYNS: ACETALDEHYDE, ((3,7-DIMETHYL-2,6-OCTADIENYL)OXY)-, (E)- ◇ GERANOXY ACETALDEHYDE

TOXICITY DATA with REFERENCE
skn-rbt 500 mg/24H MOD FCTXAV 17,789,79

CONSENSUS REPORTS: Reported in EPA TSCA Inventory.

SAFETY PROFILE: A skin irritant. When heated to decomposition it emits acrid smoke and irritating fumes.

GDM400 ***HR: 1***
GERANYL PHENYLACETATE
mf: $C_{18}H_{24}O_2$ mw: 272.39

PROP: Yellow liquid; honey-rose odor. D: 0.971-0.978, refr index: 1.507-1.511, flash p: +212°F. Misc in alc, chloroform, ether; insol in water.

SYNS: 3,7-DIMETHYL-2,6-OCTADIEN-1-YLPHENYLACETATE ◇ FEMA No. 2516

SAFETY PROFILE: Combustible liquid. When heated to decomposition it emits acrid smoke and irritating fumes.

GDM450 ***HR: 1***
GERANYL PROPIONATE
mf: $C_{13}H_{22}O_2$ mw: 210.32

PROP: Colorless liquid; rosy, fruity odor. D: 0.896-0.913, refr index: 1.456-1.464, flash p: +212°F. Sol in alc, fixed oils; insol in glycerin, propylene glycol, water @ 253°.

SYNS: 3,7-DIMETHYL-2,6-OCTADADIEN-1-YLPROPIONATE ◇ FEMA No. 2517

SAFETY PROFILE: Combustible liquid. When heated to decomposition it emits acrid smoke and irritating fumes.

GDO000 CAS:7785-33-3 ***HR: 1***
GERANYL TIGLATE
mf: $C_{15}H_{24}O_2$ mw: 236.39

SYNS: cis-α,β-DIMETHYL ACRYLIC ACID, GERANIOL ESTER ◇ trans-3,7-DIMETHYL-2,6-OCTADIEN-1-YL cis-α,β-DIMETHYL ACRYLATE ◇ (E,E)-3,7-DIMETHYL-2,6-OCTADIENYL ESTER-2-METHYL-2-BUTENOIC ACID ◇ TIGLIC ACID, GERANIOL ESTER

TOXICITY DATA with REFERENCE
skn-rbt 500 mg/24H MLD FCTXAV 12,899,74

CONSENSUS REPORTS: Reported in EPA TSCA Inventory.

SAFETY PROFILE: A skin irritant. When heated to decomposition it emits acrid smoke and irritating fumes. See also ESTERS.

GDO800 CAS:39236-46-9 ***HR: 2***
GERMALL 115
mf: $C_{11}H_{16}N_8O_8$ mw: 388.35

SYNS: METHANEBIS(N,N'-(5-UREIDO-2,4-DIKETOTETRAHYDROIMIDAZOLE)-N,N-DIMETHYLOL) ◇ N,N''-METHYLENEBIS(N'-(1-(HYDROXYMETHYL)-2,5-DIOXO-4-IMIDAZOLIDINYL)UREA)

TOXICITY DATA with REFERENCE
orl-rat LD50:11300 mg/kg JEPTDQ 4(4),133,80
ipr-rat LDLo:4000 mg/kg JEPTDQ 4(4),133,80
orl-mus LD50:7200 mg/kg JEPTDQ 4(4),133,80

CONSENSUS REPORTS: Reported in EPA TSCA Inventory.

SAFETY PROFILE: Moderately toxic by intraperitoneal route. Mildly toxic by ingestion. When heated to decomposition it emits toxic fumes of NO_x.

GDS000 CAS:1310-53-8 ***HR: 3***
GERMANIC OXIDE (crystalline)

TOXICITY DATA with REFERENCE
ipr-gpg LDLo:300 mg/kg JPETAB 42,277,31

CONSENSUS REPORTS: Reported in EPA TSCA Inventory.

SAFETY PROFILE: Poison by intraperitoneal route. When heated to decomposition it emits acrid smoke and irritating fumes.

GDU000 CAS:7440-56-4 ***HR: 2***
GERMANIUM
af: Ge aw: 72.59

PROP: A gray-white metalloid, crystalline and brittle, stable @ room temp. Mp: 925-975° (937.2° best value), bp: 2830°, d: 5.323 @ 25°. Insol in water, hydrochloric acid, dilute alc, hydroxides. Relatively stable.

SAFETY PROFILE: Moderately toxic by subcutaneous route. Inhaled dust is rapidly absorbed. Explosive reaction when heated with potassium chlorate or potassium nitrate. Violent reaction with nitric acid. Ignites on contact with chlorine or bromine + heat. Incandescent reaction when heated with oxygen or potassium hydroxide. Incompatible with aqua regia; concentrated sulfuric acid; fused alkalies; nitrates or carbonates; halogens; oxidants. See also GERMANIUM COMPOUNDS.

GDW000 CAS:13450-92-5 ***HR: 3***
GERMANIUM BROMIDE
mf: Br_4Ge mw: 392.23

PROP: Gray-white crystals. Mp: 26.1°, bp: 186.5°, d: 3.232 @ 29°/29°.

SYN: GERMANIUM TETRABROMIDE

TOXICITY DATA with REFERENCE
ivn-mus LD50:56 mg/kg CSLNX* NX#02624

CONSENSUS REPORTS: Reported in EPA TSCA Inventory.

SAFETY PROFILE: Poison by intravenous route. When heated to decomposition it emits very toxic fumes of Br^-. See also GERMANIUM COMPOUNDS.

GDY000 CAS:10038-98-9 ***HR: 3***
GERMANIUM CHLORIDE
mf: Cl_4Ge mw: 214.39

PROP: Colorless, mobile liquid. Fumes in air. Peculiar acidic odor but can be distinguished from that of concentrated HCl. Volatile @ room temp. Sol in benzene, ether, other organic solvents. Mp: −49.5°, bp: 83.1°, d: 1.879 @ 20°/20°.

SYNS: EXTREMA ◇ GERMANIUM TETRACHLORIDE

TOXICITY DATA with REFERENCE
ihl-mus LC50:44 g/m³/2H GTPZAB 12(5),51,68
ivn-mus LD50:56 mg/kg CSLNX* NX#02625

CONSENSUS REPORTS: Reported in EPA TSCA Inventory. EPA Genetic Toxicology Program.

SAFETY PROFILE: Poison by intravenous route. Mildly toxic by inhalation. A skin, eye, and mucous membrane irritant. Will react violently with water or steam to produce toxic and corrosive fumes. When heated to decomposition it emits toxic fumes of Cl^-. See also GERMANIUM COMPOUNDS.

GEA000 ***HR: 2***
GERMANIUM COMPOUNDS

SAFETY PROFILE: Germanium compounds are considered to be of a low order of toxicity, but rare instances of poisoning have been reported in the literature. Experimental LD50 values are typically about 100-1000 mg/kg for parenteral route and 500-5000 mg/kg for ingestion. The animals suffer from hypothermia, diarrhea, and respiratory, and cardiac failure. Inhalation of large amounts of $GeCl_4$ by experimental animals causes necrosis of the tracheal epithelium, bronchitis, and interstitial pneumonia. These effects were not apparent with chronic inhalation of 7 mg/m³. The tetrachloride and tetrafluoride are eye, skin, and mucous membrane irritants. Alkyl germanium compounds are much less toxic than the corresponding tin or lead compounds. Tributyl germanium and germanium tetrachloride are mutagens. Dimethyl germanium is a teratogen. Chronic ingestion of 1000 ppm or 100 ppm of germanium dioxide in water has been shown to inhibit growth in chickens. No effect was seen at 5 ppm. It has been found that the dioxide stimulates the generation of red blood cells, but it is believed to be relatively nontoxic. Buffered germanium dioxides in solution have been found to be nonirritating to the skin. Germanium hydride is a hemolytic gas and has been shown to have toxic properties at a concentration of 100 ppm. It can cause death at a concentration of 150 ppm. Otherwise, little is known about the toxicity of organic germanium compounds except that they may resemble other organometals in having higher toxicity than inorganic forms. When germanium is given in sublethal amounts, it causes a pronounced tolerance to be exhibited. Interest is high in this material because of its close chemical relationship to arsenic.

GEC000 CAS:1310-53-8 ***HR: 3***
GERMANIUM DIOXIDE
mf: GeO_2 mw: 104.59

PROP: Soluble form: Hexagonal, colorless crystals. Mp: 1115.0°, d: 4.703 @ 18°. Insoluble form: Tetragonal crystals, mp: 1086 ± 5°; d: 6.239.

SYNS: GERMANIA ◇ GERMANIC ACID ◇ GERMANIUM OXIDE ◇ GERMANIUM OXIDE (GeO_2)

TOXICITY DATA with REFERENCE
unr-rat TDLo:133 mg/kg (7-13D preg):REP OYYAA2 21,797,81
unr-rat TDLo:266 mg/kg (7-13D preg):TER OYYAA2 21,773,81
orl-rat LD50:1250 mg/kg 31ZUAV -,95,64
ipr-rat LD50:750 mg/kg AMIHBC 8,466,53
scu-rat LD50:1910 mg/kg OYYAA2 21,773,81
orl-mus LD50:1250 mg/kg 85GMAT-,71,82
ipr-mus LD50:1550 mg/kg OYYAA2 16,671,78
scu-mus LD50:2550 mg/kg OYYAA2 21,773,81
scu-rbt LD50:845 mg/kg EQSSDX 1,1,75
ipr-gpg LDLo:400 mg/kg EQSSDX 1,1,75

CONSENSUS REPORTS: Reported in EPA TSCA Inventory.

SAFETY PROFILE: Poison by intraperitoneal route. Moderately toxic by ingestion and subcutaneous routes. Experimental teratogenic and reproductive effects. When heated to decomposition it emits acrid smoke and irritating fumes. See also GERMANIUM COMPOUNDS.

GEG000 ***HR: 3***
GERMANIUM MONOHYDRIDE
mf: $(GeH)_n$ mw: $(73.60)_n$
SYN: POLY(GERMANIUM MONOHYDRIDE)

SAFETY PROFILE: A very dangerous explosion hazard, solid polymeric hydride can decompose explosively when exposed to air. Self-ignites in air. See also HYDRIDES, GERMANIUM COMPOUNDS, and GERMANIUM.

GEI000 CAS:12025-32-0 ***HR: 3***
GERMANIUM(II) SULFIDE
mf: GeS mw: 104.64

SAFETY PROFILE: Explosive reaction when heated with potassium nitrate. When heated to decomposition it emits toxic fumes of SO_x. See also GERMANIUM COMPOUNDS.

GEI100 CAS:7782-65-2 ***HR: 3***
GERMANIUM TETRAHYDRIDE
DOT: UN 2192
mf: GeH_4 mw: 76.63

PROP: Colorless gas. Mp: −165°, bp: −90°, d: 1.523 @ −142°/4°. Sltly sol in hot HCl; decomp in nitric acid.

SYNS: GERMANE (DOT) ◇ GERMANIUM HYDRIDE ◇ MONOGERMANE

CONSENSUS REPORTS: Reported in EPA TSCA Inventory.

ACGIH TLV: TWA 0.2 ppm
DOT Classification: Poison A; Label: Poison Gas and Flammable Gas.

SAFETY PROFILE: Poison by inhalation. A hemolytic gas. Ignites spontaneously in air. Incompatible with Br_2. See also HYDRIDES, GERMANIUM COMPOUNDS, and GERMANIUM.

GEK000 CAS:306-67-2 ***HR: D***
GERONTINE TETRAHYDROCHLORIDE
mf: $C_{10}H_{26}N_4•4ClH$ mw: 348.24

SYNS: 1,4-BIS(AMINOPROPYL)BUTANEDIAMINETETRAHYDROCHLORIDE ◇ MUSCULAMINE TETRAHYDROCHLORIDE ◇ NEURIDINE TETRAHYDROCHLORIDE ◇ SPERMINE TETRAHYDROCHLORIDE

TOXICITY DATA with REFERENCE
dnd-esc 8900 nmol/L BIPMAA 5,227,67
dnd-srm 8900 nmol/L BIPMAA 5,227,67
dnd-mam:lym 16700 nmol/L BIPMAA 5,227,67

CONSENSUS REPORTS: Reported in EPA TSCA Inventory.

SAFETY PROFILE: Mutation data reported. When heated to decomposition it emits very toxic fumes of HCl and NO_x.

GEK200 CAS:137-86-0 ***HR: 3***
GEROSTOP
mf: $C_{23}H_{36}N_4O_5S_3$ mw: 544.81

PROP: Crystals. Mp: 106-109°.

SYNS: NEUVITAN ◇ OCTOTIAMINE ◇ TATD

TOXICITY DATA with REFERENCE
orl-mus LD50:2590 mg/kg NIIRDN 6,159,82
ipr-mus LD50:522 mg/kg NIIRDN 6,159,82
scu-mus LD50:1410 mg/kg NIIRDN 6,159,82
ivn-mus LD50:399 mg/kg NIIRDN 6,159,82

SAFETY PROFILE: Poison by intravenous route. Moderately toxic by ingestion, subcutaneous, and intraperitoneal routes. When heated to decomposition it emits toxic fumes of SO_x and NO_x. See also ESTERS.

GEK500 CAS:434-03-7 ***HR: 3***
GESTORAL
mf: $C_{21}H_{28}O_2$ mw: 312.49

PROP: Crystals from ethyl acetate. Mp: 269-275°. Practically insol in water; sltly sol in alc, acetone, ether, chloroform, vegetable oils.

SYNS: AETHISTERON ◇ ANHYDROHYDROXYPROGESTERONE ◇ ANHYDROXYPROGESTERONE ◇ COLUTOID ◇ ETHINONE ◇ ETHINYLTESTOSTERONE ◇ 17-ETHINYLTESTOSTERONE ◇ ETHISTERONE ◇ ETHYNYLTESTOSTERONE ◇ 17-α-ETHYNYLTESTOSTERONE ◇ 17-β-HYDROXY-17-α-ETHYNYL-4-ANDROSTER-3-ONE ◇ 17-HYDROXY-17-α-PREGN-4-EN-20-YN-3-ONE ◇ LUCORTEUM ORAL ◇ LUTIDON ORAL ◇ LUTOCYLOL ◇ NALUTORAL ◇ NUGESTORAL ◇ ORA-LUTIN ◇ PRANONE ◇ PREGNENINOLONE ◇ PREGNIN ◇ PRIMOLUT C ◇ PRODOXAN ◇ PRODROXAN ◇ PRODUXAN ◇ PROGESTAB ◇ PROGESTIN P ◇ PROGESTOLETS ◇ PROGESTORAL ◇ PROLUTOL ◇ PROLUTON C ◇ PRONE ◇ SYNGESTROTABS ◇ TROSINONE

TOXICITY DATA with REFERENCE
orl-wmn TDLo:168 mg/kg (3W pre):REP AJOGAH 76,626,58
orl-wmn TDLo:386 mg/kg (6-28W preg):TER AJOGAH 84,962,62

SAFETY PROFILE: Human female reproductive and teratogenic effects: menstrual cycle changes or disorders and developmental abnormalities of the fetal urogenital system. Other experimental reproductive effects. A steroid. When heated to decomposition it emits acrid smoke and irritating fumes.

GEK510 CAS:1253-28-7 ***HR: 3***
GESTRONOL CAPROATE
mf: $C_{26}H_{38}O_4$ mw: 414.64

PROP: Crystals. Mp: 123-124°.

SYNS: 17-β-ACETYL-17-HYDROXYESTR-4-ENE-3-ONEHEXANOATE ◇ DEPOSTAT ◇ GESTONORONE CAPROATE ◇ GESTONORONE CAPRONATE ◇ GESTRONOL HEXANOATE ◇ 17-HYDROXY-19-NOR-PREGN-4-ENE-3,20-DIONE HEXANOATE ◇ 17-HYDROXY-19-NOR-4-PREGNENE-3,20-DIONE HEXANOATE ◇ 17-α-HEXANOYLOXY-19-NOR-4-PREGNENE-3,20-DIONE ◇ 17-α-HYDROXY-19-NORPROGESTERONE CAPROATE ◇ PRIMOSTAT ◇ SH 582 ◇ SH 80582

TOXICITY DATA with REFERENCE
scu-rat TDLo:700 mg/kg (28W male):REP IYKEDH 9,307,78
ipr-rat LD50:6800 mg/kg IYKEDH 13,349,82
ims-rat LD50:400 mg/kg IYKEDH 8,296,77
scu-mus LD50:10 g/kg IYKEDH 13,349,82

SAFETY PROFILE: Poison by intramuscular route. Experimental reproductive effects. A steroid. When heated to decomposition it emits acrid smoke and irritating fumes.

GEK600 ***HR: 3***
GEUM ELATUM (Royle) Hook. f., extract

PROP: Indian plant belonging to the family *Rosaceae* (IJEBA6 18,594,80).

TOXICITY DATA with REFERENCE
orl-rat TDLo:150 mg/kg (12-14D preg):REP IJEBA6 18,594,80
ipr-mus LD50:375 mg/kg IJEBA6 18,594,80

SAFETY PROFILE: Poison by intraperitoneal route. Experimental reproductive effects.

GEK875 ***HR: 1***
GFX-E

PROP: Consists of glucose, fructose and xylitol (4:2:1) in a 23.3% solution with electrolytes (IYKEDH 16,328,85).

TOXICITY DATA with REFERENCE
ipr-rat LD50:42600 mg/kg IYKEDH 16,328,85
ivn-rat LD50:40600 mg/kg IYKEDH 16,328,85
ivn-rbt LD50:40 g/kg IYKEDH 16,328,85

SAFETY PROFILE: Mildly toxic by intravenous and intraperitoneal routes. When heated to decomposition it emits acrid smoke and irritating fumes. See also individual components.

GEK880 ***HR: 1***
GFX-ES

PROP: Consists of glucose, fructose and xylitol (4:2:1) in a 23.3% solution with electrolytes (IYKEDH 16,320,85).

TOXICITY DATA with REFERENCE
ipr-rat LD50:57600 mg/kg IYKEDH 16,320,85
ivn-rat LD50:61 g/kg IYKEDH 16,320,85
ivn-rbt LD50:59300 mg/kg IYKEDH 16,320,85

SAFETY PROFILE: Mildly toxic by intravenous and intraperitoneal routes. When heated to decomposition it emits acrid smoke and irritating fumes. See also individual components.

GEM000 CAS:77-06-5 ***HR: 3***
GIBBERELLIC ACID
mf: $C_{19}H_{22}O_6$ mw: 346.41

PROP: A plant growth-promoting hormone. White crystals or crystalline powder. Mp: 233-235°. Sltly sol in water, ether; sol in methanol, ethanol, acetone, aq solns of sodium bicarbonate and sodium acetate; moderately sol in ethyl acetate.

SYNS: BERELEX ◇ BRELLIN ◇ CEKUGIB ◇ FLORALTONE ◇ GA ◇ GIBBERELLIN ◇ GIBBREL ◇ GIB-SOL ◇ GIB-TABS ◇ GROCEL

◇ NCI-C55823 ◇ PRO-GIBB ◇ 2,4a,7-TRIHYDROXY-1-METHYL-8-METHYLENEGIBB-3-ENE-1,10-CARBOXYLIC ACID 1-4-LACTONE

TOXICITY DATA with REFERENCE
dnd-sal:spr 1 mmol/L PYTCAS 11,3135,72
dnd-mam:lym 1 mmol/L PYTCAS 11,3135,72
orl-mus TDLo:142 g/kg/78W-I:ETA NTIS** PB223-159
orl-rat LD50:6300 mg/kg 85ARAE 3,43,76/77

CONSENSUS REPORTS: EPA Genetic Toxicology Program. Reported in EPA TSCA Inventory.

SAFETY PROFILE: Mildly toxic by ingestion. Questionable carcinogen with experimental tumorigenic data. Mutation data reported. When heated to decomposition it emits acrid smoke and irritating fumes.

GEO000 CAS:12002-43-6 ***HR: 1***
GILSONITE

PROP: A black solid hydrocarbon mineral formed from petroleum millions of years ago by geologic processes.

SYN: NCI-C55185

SAFETY PROFILE: A skin, eye, and mucous membrane irritant. An allergen. Has been known to cause photosensitization of skin. Flammable when exposed to heat or open flame. To fight fire, use water, foam, dry chemical and CO_2. When heated to decomposition it emits acrid smoke and irritating fumes.

GEO200 CAS:77879-90-4 ***HR: 3***
GILVOCARCIN V
mf: $C_{27}H_{26}O_9$ mw: 494.53

SYNS: DC-38-A ◇ DC-38-V

TOXICITY DATA with REFERENCE
dni-bcs 500 μg/L JANTAJ 35,545,82
oms-bcs 500 μg/L JANTAJ 35,545,82
ivn-mus LD50:300 mg/kg JANTAJ 34,266,81

SAFETY PROFILE: Poison by intravenous route. Mutation data reported. When heated to decomposition it emits acrid smoke and fumes.

GEO600 CAS:4880-82-4 ***HR: 3***
GINDARINE HYDROCHLORIDE
mf: $C_{21}H_{25}NO_4$•ClH mw: 391.93

SYNS: GINDARIN HYDROCHLORIDE ◇ (S)-5,8,13,13a-TETRAHYDRO-2,3,9,10-TETRAMETHOXY-6H-DIZENZO(a,g)QUINOLIZINE HYDROCHLORIDE ◇ 2,3,9,10-TETRAMETHOXY-13a,α-BERBINE HYDROCHLORIDE

TOXICITY DATA with REFERENCE
orl-rat TDLo:20 mg/kg (female 1-20D post):REP FATOAO 46(4),107,83
orl-rat TDLo:1 g/kg (female 1-20D post):TER FATOAO 46(4),107,83
orl-rat LD50:580 mg/kg FATOAO 46(4),107,83
ipr-rat LD50:330 mg/kg FATOAO 46(4),107,83
orl-mus LD50:1190 mg/kg FATOAO 46(4),107,83
ipr-mus LD50:550 mg/kg IJMRAQ 60,472,72
ipr-gpg LD50:460 mg/kg FATOAO 46(4),107,83

SAFETY PROFILE: Poison by intraperitoneal route. Moderately toxic by ingestion. An experimental teratogen. Experimental reproductive effects. When heated to decomposition it emits toxic fumes of NO_x and HCl.

GEQ000 CAS:8007-08-7 ***HR: 1***
GINGER OIL

PROP: From steam distillation of ground rhizomes of *Zingiber officinale* Roscoe (Fam. *Zingiberaceae*) (FCTXAV 12,807,74). Yellow liquid; odor of ginger. D: 0.870-0.882, refr index: 1.488 @ 20°. Sol in fixed oils, mineral oil, alc; insol in glycerin, propylene glycol.

TOXICITY DATA with REFERENCE
skn-rbt 500 mg/24H MOD FCTXAV 12,901,74
dnr-bcs 5 μL/disc TOFOD5 8,91,85

CONSENSUS REPORTS: Reported in EPA TSCA Inventory.

SAFETY PROFILE: A skin irritant. Mutation data reported. When heated to decomposition it emits acrid smoke and irritating fumes.

GEQ400 ***HR: 3***
GINSENG

SYN: PANAX

TOXICITY DATA with REFERENCE
orl-rat LD50:750 mg/kg FRMTAL 31(6),45,82
orl-mus LD50:200 mg/kg FRMTAL 31(6),45,82
ipr-mus LD50:54 mg/kg FRMTAL 31(6),45,82

SAFETY PROFILE: Poison by ingestion and intraperitoneal routes.

GEQ425 ***HR: 3***
GINSENG ROOT-NEUTRAL SAPONINS

PROP: A mixture of neutral saponins composed of ginsenoside-rb$_1$, -rb$_2$, and -rc (JJPAAZ 23,29,73).

SYNS: GINSENG, ROOT EXTRACT ◇ GINSENGWURZEL, EXTRACT (GERMAN) ◇ GNS ◇ NEUTRAL SAPONINS of PANAX GINSENG ROOT ◇ PANAX GINSENG, ROOT EXTRACT ◇ SONG-SAM, ROOT EXTRACT

TOXICITY DATA with REFERENCE
oms-rat-ipr 50 mg/kg CPBTAL 25,1665,77
ipr-rat LDLo:200 mg/kg JJPAAZ 23,29,73
ipr-mus LD50:545 mg/kg JJPAAZ 23,29,73
ivn-mus LD50:367 mg/kg JJPAAZ 23,29,73
par-mus LDLo:32500 mg/kg AEPPAE 170,443,33
par-frg LDLo:8500 mg/kg AEPPAE 170,443,33

SAFETY PROFILE: Poison by intravenous and intraperitoneal routes. Mutation data reported. See also SAPONIN.

GES000 CAS:1391-75-9 ***HR: 3***
GITALIN
mf: $C_{35}H_{56}O_{12}$ mw: 668.91

PROP: Amorph powder. Readily sol in alc; slt to slowly sol in water.

SYN: VERODIGEN

TOXICITY DATA with REFERENCE
scu-mus LDLo:29 mg/kg 27ZWAY E.1,78,-
ivn-mus LDLo:2400 μg/kg ARZNAD 6,182,56
orl-cat LD50:1230 μg/kg JAPMA8 44,607,55
ivn-cat LDLo:1040 μg/kg JAPMA8 44,607,55
ivn-pgn LDLo:1 mg/kg JPHAA3 25,611,36

SAFETY PROFILE: Poison by ingestion, subcutaneous, and intravenous routes. When heated to decomposition it emits acrid smoke and irritating fumes.

GES100 CAS:3261-53-8 ***HR: 3***
GITALOXIN
mf: $C_{42}H_{64}O_{15}$ mw: 809.06

SYNS: 16-FORMYL-GITOXIN ◇ GITALOXIGENIN-TRIDIGITOXOSID (GERMAN) ◇ GITALOXIN-16-FORMATE ◇ GITOXIGENIN TRIDIGITOXOXIDE-16-FORMATE

TOXICITY DATA with REFERENCE
orl-mus LD50:2870 μg/kg AIPTAK 153,436,65
scu-mus LD50:2830 μg/kg AIPTAK 153,436,65
ivn-cat LDLo:900 μg/kg JMPCAS 5,988,62
orl-gpg LDLo:2455 μg/kg AIPTAK 153,436,65
ivn-gpg LDLo:982 μg/kg AIPTAK 153,436,65

SAFETY PROFILE: A deadly poison by ingestion, subcutaneous, and intravenous routes. When heated to decomposition it emits acrid smoke and irritating fumes.

GEU000 CAS:4562-36-1 ***HR: 3***
GITOXIN
mf: $C_{41}H_{64}O_{14}$ mw: 781.05

PROP: Stout prisms from chloroform and methanol. Decomp @ 285° (rapid heating). Almost insol in chloroform, ethyl acetate, and acetone. Dissolves in a mixture of chloroform and alc or pyridine or dil alc.

SYNS: ANHYDROGITALIN ◇ BIGITALIN ◇ GITOXIGENIN-TRIDIGITOXOSID (GERMAN) ◇ PSEUDODIGITOXIN

TOXICITY DATA with REFERENCE
orl-cat LDLo:880 μg/kg 27ZWAY E.1,78,-
par-cat LDLo:465 μg/kg ARZNAD 6,182,56
ivn-gpg LD50:68 mg/kg ARZNAD 6,182,56
par-pgn LDLo:1210 μg/kg CPBTAL 11,613,63

SAFETY PROFILE: Poison by ingestion, intravenous, and parenteral routes. When heated to decomposition it emits acrid smoke and irritating fumes.

GEW000 CAS:7242-04-8 ***HR: 3***
GITOXIN PENTAACETATE
mf: $C_{51}H_{74}O_{19}$ mw: 991.25

PROP: Rhombic crystals. Mp: 151-155°.

SYNS: CARNACID-COR ◇ CORDOVAL ◇ PENGITOXIN ◇ PENTAACETYLGITOXIN ◇ PENTA-o-ACETYLGITOXIN ◇ PENTAGIT

TOXICITY DATA with REFERENCE
ivn-rat LD50:21 mg/kg AIPTAK 155,165,65
ipr-mus LD50:6400 μg/kg AIPTAK 155,165,65
orl-cat LD50:200 μg/kg AIPTAK 159,1,66
ipr-cat LD50:230 μg/kg AIPTAK 159,1,66
orl-gpg LD50:1 mg/kg AIPTAK 159,1,66
ivn-gpg LD50:450 μg/kg ARZNAD 6,182,56

SAFETY PROFILE: A deadly poison by ingestion, intraperitoneal, and intravenous routes. Used as a cardiotonic agent. When heated to decomposition it emits acrid smoke and irritating fumes.

GEW700 CAS:1448-23-3 ***HR: 3***
GLARUBIN
mf: $C_{25}H_{36}O_{10}$ mw: 496.61

SYNS: GLAUCARUBIN ◇ GLAUMEBA ◇ α-KIRONDRIN ◇ β-KIRONDRIN ◇ MK-33 ◇ SIMARUBACEAE

TOXICITY DATA with REFERENCE
orl-rat LD50:800 mg/kg 85GDA2 8(2),172,82
orl-mus LD50:1200 mg/kg AIPTAK 114,307,58
scu-mus LD50:28 mg/kg 85GDA2 8(2),172,82

SAFETY PROFILE: Poison by subcutaneous route. Moderately toxic by ingestion. When heated to decomposition it emits acrid smoke and fumes. See also ESTERS.

GEW750 CAS:535-65-9 ***HR: D***
GLIPASOL
mf: $C_{12}H_{16}N_4O_2S_2$ mw: 312.44

SYNS: BENZENESULFONAMIDE,4-AMINO-N-(5-(1,1-DIMETHYLETHYL)-1,3,4-THIADIAZOL-2-YL)- (9CI) ◇ GLYBUTHIAZOL ◇ GLYBUTHIAZOLE ◇ GLYPASOL ◇ 2259 R.P. ◇ RP 2259 ◇ SULFANILAMIDE, N^1-(5-tert-BUTYL-1,3,4-THIADIAZOL-2-YL)-

TOXICITY DATA with REFERENCE
orl-rat TDLo:12 g/kg (female 1-12D post):REP COREAF 247,1134,58

SAFETY PROFILE: Experimental reproductive effects. When heated to decomposition it emits toxic fumes of NO_x and SO_x.

GEW800 ***HR: 2***
GLORY LILY

PROP: Climbing lilies with tuberous roots. The leaves have tendrils on the tips. The flowers are bright red and yellow with well separated petals. They are cultivated in Hawaii, the southern-most parts of the continental United States and the West Indies.

SYNS: CLIMBING LILY ◇ GLORIOSA LILY ◇ GLORIOSA ROTHSCHILDIANA ◇ GLORIOSA SUPERBA ◇ PIPA de TURCO (CUBA)

SAFETY PROFILE: The whole plant and especially the tubers contain the poison colchicine. Ingestion of any part of the plant causes a burning pain in the mouth, intense thirst, nausea, vomiting, abdominal cramps, severe diarrhea, and sometimes kidney damage. There may be extensive fluid and electrolyte loss. Colchicine is excreted slowly so the effects may persist for some time. See also COLCHICINE.

GEW875 CAS:9007-92-5 ***HR: 1***
GLUCAGON
mf: $C_{151}H_{224}N_{42}O_{50}S$ mw: 3460.23

PROP: A polypeptide hormone produced in the alpha cells of the islets of Langerhans in the pancreas. Rhombic dodecahedra. Stable. Practically insol in water; sol in acidic, basic media, in the range below pH 3 and above pH 9.5.

TOXICITY DATA with REFERENCE
dns-rat:lvr 75 mg/L EJBCAI 34,474,73
scu-rat TDLo:3600 μg/kg (16-21D preg):TER AJOGAH 106,656,70
ims-man TDLo:28 μg/kg AIPTAK 218,312,75

SAFETY PROFILE: Human systemic effects by intramuscular route: leukopenia (reduced white blood cell count). An experimental teratogen. Mutation data reported. When heated to decomposition it emits toxic fumes of SO_x and NO_x.

GFA000 CAS:15879-93-3 ***HR: 3***
α-d-GLUCOCHLORALOSE
mf: $C_8H_{11}Cl_3O_6$ mw: 309.54

SYNS: AGC ◇ ALFAMAT ◇ ANHYDROGLUCOCHLORAL ◇ APHOSAL ◇ CHLORALOSANE ◇ α-CHLORALOSE ◇ CHLOROALOSANE ◇ DULCIDOR ◇ GLUCOCHLORAL ◇ GLUCOCHLORALOSE ◇ KALMETTUMSOMNIFERUM ◇ MONOTRICHLOR-AETHYLIDEN-α-GLUCOSE (GERMAN) ◇ MUREX ◇ SOMIO ◇ 1,2-o-(2,2,2-TRICHLOROETHYLIDENE)-α-d-GLUCOFURANOSE

TOXICITY DATA with REFERENCE
scu-mus TDLo:215 mg/kg:ETA NTIS** PB223-159
orl-rat LDLo:400 mg/kg 85ESA3 9,260,76
scu-rat LDLo:200 mg/kg 27ZIAQ -,70,73
orl-mus LD50:32 mg/kg FMCHA2 -,C50,83
ipr-mus LD50:175 mg/kg ARZAND 21,1727,71
orl-dog LD50:250 mg/kg RMVEAG 154,137,78
ipr-dog LDLo:400 mg/kg AIPTAK 3,191,1897
orl-cat LD50:250 mg/kg RMVEAG 154,137,78
ipr-cat LDLo:150 mg/kg AIPTAK 3,191,1897

CONSENSUS REPORTS: Reported in EPA TSCA Inventory.

SAFETY PROFILE: Poison by ingestion, subcutaneous, and intraperitoneal routes. Questionable carcinogen with experimental tumorigenic data. When heated to decomposition it emits toxic fumes of Cl^-.

GFC000 CAS:124-99-2 ***HR: 3***
GLUCOPROSCILLARIDIN A
mf: $C_{36}H_{52}O_{13}$ mw: 692.88

SYNS: 14-β-HYDROXY-3-β-SCILLOBIOSIDOBUFA-4,20,22-TRIENOLIDE ◇ SCILLAGLYKOSID A (GERMAN) ◇ SCILLAREN A ◇ SCILLARENIN-3,6-DEOXY-4-o-β-d-GLUCOPYRANOSYL-α-l-MANNOPYRANOSIDE ◇ SCILLAREN & RHAMNOSE & GLUCOSE (GERMAN) ◇ 3-β-SCILLOBIOSIDO-14-β-HYDROXY-Δ-4,20,22-BUFATRIENOLID (GERMAN) ◇ TRANSVAALIN

TOXICITY DATA with REFERENCE
ivn-rat LD50:15 mg/kg ARZNAD 11,848,61
ivn-cat LDLo:143 mg/kg MEIEDD 10,1208,83
ivn-gpg LDLo:353 μg/kg AEPPAE 252,314,66

SAFETY PROFILE: Poison by intravenous route. When heated to decomposition it emits acrid smoke and irritating fumes.

GFC100 CAS:554-35-8 ***HR: 2***
2-(β-d-GLUCOPYRANOSYLOXY)ISOBUTYRONITRILE
mf: $C_{10}H_{17}NO_6$ mw: 247.28

SYNS: 2-(β-d-GLUCOPYRANOSYLOXY)-2-METHYLPROPANENITRILE ◇ LINAMARIN ◇ PHASEOLUNATIN ◇ PROPANENITRILE, 2-(β-d-GLUCOPYRANOSYLOXY)-2-METHYL-

TOXICITY DATA with REFERENCE
orl-ham TDLo:120 mg/kg (female 8D post):TER TJADAB 31,241,85
orl-rat LDLo:500 mg/kg TXAPA9 42,539,77

SAFETY PROFILE: Moderately toxic by ingestion. An experimental teratogen. When heated to decomposition it emits toxic fumes of NO_x.

GFG000 CAS:50-99-7 ***HR: 3***
d-GLUCOSE
mf: $C_6H_{12}O_6$ mw: 180.18

$$O(CHOH)_4CHCH_2OH$$

PROP: Colorless crystals or white crystalline or granular powder; odorless with sweet taste. D: 1.544, mp: 146°. Sol in water; sltly sol in alc. α Form: (monohydrate) crystals from water. Mp: 83°. α Form: (anhydrous) crystals from hot ethanol or water. Mp: 146°. Very sparingly sol in abs alc, ether, acetone; sol in hot glacial acetic acid, pyridine, aniline. β Form: crystals from hot H_2O + ethanol, from dil acetic acid or from pyridine; mp: 148-155°.

SYNS: CARTOSE ◇ CERELOSE ◇ CORN SUGAR ◇ DEXTROPUR ◇ DEXTROSE (FCC) ◇ DEXTROSE, anhydrous ◇ DEXTROSOL ◇ GLUCOLIN ◇ GLUCOSE ◇ d-GLUCOSE, anhydrous ◇ GLUCOSE LIQUID ◇ GRAPE SUGAR ◇ SIRUP

TOXICITY DATA with REFERENCE
mmo-sat 25 mg/plate NARHAD 12,2127,84
oms-omi 1 mol/L ARMKA7 91,305,73
ivn-wmn TDLo:2 g/kg (female 39W post):TER BJOGAS 89,27,82
ivn-wmn TDLo:2 g/kg (female 39W post):REP BJOGAS 89,27,82
scu-rat TDLo:15400 g/kg/22W-C:ETA GANNA2 30,419,36
orl-rat LD50:25800 mg/kg 85AIAL -,39,73
ipr-mus LDLo:18 g/kg PSEBAA 35,98,36
ivn-mus LD50:9 g/kg ARZNAD 18,666,68
orl-dog LDLo:8000 mg/kg HBTXAC 1,150,55
orl-rbt LDLo:20000 mg/kg HBTXAC 1,150,55
ivn-rbt LDLo:12000 mg/kg HBTXAC 1,150,55

CONSENSUS REPORTS: Reported in EPA TSCA Inventory. EPA Genetic Toxicology Program.

SAFETY PROFILE: Mildly toxic by ingestion. An experimental teratogen. Experimental reproductive effects. Questionable carcinogen with experimental tumorigenic data. Mutation data reported. Potentially explosive reaction with potassium nitrate + sodium peroxide when heated in a sealed container. Mixtures with alkali release carbon monoxide when heated. When heated to decomposition it emits acrid smoke and irritating fumes.

GFG100 CAS:9001-37-0 ***HR: 3***
GLUCOSE OXIDASE

PROP: Amorph powder or crystals. Freely sol in water giving yellowish-green solns.

SYNS: CORYLOPHYLINE ◇ DEOXIN-1 ◇ E.C. 1.1.3.4 ◇ GLUCOSE AERODEHYDROGENASE ◇ β-d-GLUCOSE OXIDASE ◇ MICROCID ◇ NOTATIN ◇ OXIDASE GLUCOSE ◇ PENATIN

TOXICITY DATA with REFERENCE
ipr-mus LD50:3 mg/kg 85GDA2 4(2),302,80
scu-mus LD50:4500 μg/kg 85GDA2 4(2),302,80
ivn-mus LD50:13 mg/kg BJPCAL 1,225,46
scu-rbt LD50:7500 μg/kg BJPCAL 1,225,46

CONSENSUS REPORTS: Reported in EPA TSCA Inventory.

SAFETY PROFILE: Poison by subcutaneous, intravenous, and intraperitoneal routes.

GFG200 ***HR: 1***
GLUCOSE-RINGER'S SOLUTION (23.3%)

PROP: Consists of Ringer's solution containing 93.2 g/400 mL glucose (IYKEDH 16,320,85).

SYNS: GR-23 ◇ RINGER'S GLUCOSE SOLUTION

TOXICITY DATA with REFERENCE
ipr-rat LD50:53300 mg/kg IYKEDH 16,320,85
ivn-rat LD50:57200 mg/kg IYKEDH 16,320,85
ivn-rbt LDLo:50 g/kg IYKEDH 16,320,85

SAFETY PROFILE: Very mildly toxic by intravenous and intraperitoneal routes.

GFG205 ***HR: 1***
GLUCOSE-RINGER'S SOLUTION (29.2%)

PROP: Consists of Ringer's solution containing 116.8 g/400 mL glucose (IYKEDH 16,328,85).

SYN: GR-29

TOXICITY DATA with REFERENCE
ipr-rat LD50:41500 mg/kg IYKEDH 16,328,85
ivn-rat LD50:38700 mg/kg IYKEDH 16,328,85
ivn-rbt LDLo:45 g/kg IYKEDH 16,328,85

SAFETY PROFILE: Very mildly toxic by intravenous and intraperitoneal routes.

GFK000 CAS:20408-97-3 ***HR: 1***
α-d-GLUCOTHIOPYRANOSE
mf: $C_6H_{12}O_5S$ mw: 196.24

SYNS: 5-THIO-α-d-GLUCOPYRANOSE ◇ 5-THIO-d-GLUCOSE

TOXICITY DATA with REFERENCE
spm-mus-orl 1400 mg/kg/35D JOHEA8 70,325,79
spm-mus-ipr 840 mg/kg/21D-C JOHEA8 72,347,81
orl-mus TDLo:2450 mg/kg (male 49D pre):REP CCPTAY 17,123,78
ipr-mus LDLo:5500 mg/kg IOBPD3 8,589,82

CONSENSUS REPORTS: EPA Genetic Toxicology Program.

SAFETY PROFILE: Mildly toxic by intraperitoneal route. Mutation data reported. Experimental reproductive effects. When heated to decomposition it emits very toxic fumes of SO_x.

GFM000 CAS:32449-92-6 ***HR: 2***
GLUCURONIC ACID LACTONE
mf: $C_6H_8O_6$ mw: 176.14

PROP: Crystals from ethanol. Mp: 176-178° (commercial grades, mp: 172°), d: 1.76. Sol in water; sltly sol in ethanol; very sltly sol in abs ethanol in glacial acetic acid.

SYNS: DICURONE ◇ GLUCOXY ◇ GLUCURON ◇ GLUCURONE ◇ GLUCURONIC ACID-Γ-LACTONE ◇ d-GLUCURONIC ACID LACTONE ◇ d-GLUCURONIC ACID-Γ-LACTONE ◇ GLUCURONOLACTONE ◇ d-GLUCURONOLACTONE ◇ GLUCURONOSAN ◇ GLYCURONE ◇ GURONSAN ◇ REULATT S.S.

TOXICITY DATA with REFERENCE
orl-rat LD50:10700 mg/kg NIIRDN 6,225,82
scu-rat LD50:4700 mg/kg NIIRDN 6,225,82
ivn-rat LD50:3200 mg/kg NIIRDN 6,225,82
ipr-mus LD50:5797 mg/kg NIIRDN 6,225,82
scu-dog LD50:4700 mg/kg NIIRDN 6,225,82
ivn-dog LD50:940 mg/kg NIIRDN 6,225,82
scu-rbt LD50:4700 mg/kg NIIRDN 6,225,82
ivn-rbt LD50:940 mg/kg NIIRDN 6,225,82

CONSENSUS REPORTS: Reported in EPA TSCA Inventory.

SAFETY PROFILE: Moderately toxic by intravenous route. Mildly toxic by ingestion and subcutaneous routes. When heated to decomposition it emits acrid smoke and irritating fumes.

GFM200 CAS:1492-02-0 ***HR: 3***
GLUDIASE
mf: $C_{12}H_{15}N_3O_2S_2$ mw: 297.42

PROP: Needles. Mp: 163°.

SYNS: AN 1324 ◇ 2-BENZENESULFONAMIDO-5-tert-BUTYL-1,3,4-THIADIAZOLE ◇ 2-BENZENESULFONAMIDO-5-TERTIOBUTYL-1-THIA-3,4-DIAZOLE ◇ N-(5-tert-BUTYL-1,3,4-THIADIAZOL-2-YL) BENZENESULFONAMIDE ◇ DESAGLYBUZOLE ◇ GLYBUZOLE ◇ RP 7891 ◇ TH-1395

TOXICITY DATA with REFERENCE
orl-mus TDLo:2100 mg/kg (female 7-13D post):TER YIKUAO 18,21,69
orl-rat LD50:500 mg/kg YIKUAO 18,21,69
ipr-rat LD50:219 mg/kg YIKUAO 18,21,69
scu-rat LD50:310 mg/kg YIKUAO 18,21,69
orl-mus LD50:730 mg/kg YIKUAO 18,21,69
ipr-mus LD50:235 mg/kg YIKUAO 18,21,69
scu-mus LD50:248 mg/kg YIKUAO 18,21,69
ivn-mus LD50:193 mg/kg YIKUAO 18,21,69
orl-rbt LD50:967 mg/kg YIKUAO 18,21,69
ipr-rbt LD50:300 mg/kg YIKUAO 18,21,69
ivn-rbt LD50:118 mg/kg YIKUAO 18,21,69

SAFETY PROFILE: Poison by subcutaneous, intravenous, and intraperitoneal routes. Moderately toxic by ingestion. An experimental teratogen. When heated to decomposition it emits toxic fumes of SO_x and NO_x.

GFO000 CAS:56-86-0 ***HR: 1***
l-GLUTAMIC ACID
mf: $C_5H_9NO_4$ mw: 147.15

PROP: A nonessential amino acid present in all complete proteins. White crystals or crystalline powder. Mp (dl form): 194°, d (dl form): 1.4601 @ 20°/4°, mp (l form): 224-225°, d (l form): 1.538 @ 20°/4°. Sltly sol in water.

SYNS: α-AMINOGLUTARIC ACID ◇ l-2-AMINOGLUTARIC ACID ◇ 2-AMINOPENTANEDIOIC ACID ◇ 1-AMINOPROPANE-1,3-DICARBOXYLIC ACID ◇ GLUSATE ◇ GLUTACID ◇ GLUTAMIC ACID ô α-GLUTAMIC ACID ◇ d-GLUTAMIENSUUR ◇ GLUTAMINIC ACID ◇ l-GLUTAMINIC ACID ◇ GLUTAMINOL ◇ GLUTATON

TOXICITY DATA with REFERENCE
orl-hmn TDLo:71 mg/kg:CNS SCIEAS 163,826,69
ivn-hmn TDLo:117 mg/kg:GIT AJMSA9 214,281,47

CONSENSUS REPORTS: Reported in EPA TSCA Inventory.

SAFETY PROFILE: Human systemic effects by ingestion and intravenous route: headache and nausea or vomiting. When heated to decomposition it emits toxic fumes of NO_x.

GFO025 CAS:138-15-8 ***HR: D***
l-GLUTAMIC ACID HYDROCHLORIDE
mf: $C_5H_9NO_4HCL$ mw: 183.59

PROP: White crystals or crystalline powder. Sol in water; insol in alc, ether.

SYNS: α-AMINOGLUTARIC ACID HYDROCHLORIDE ◇ l-2-AMINOGLUTARIC ACID HYDROCHLORIDE ◇ 2-AMINOPENTANEDIOIC ACID HYDROCHLORIDE ◇ 1-AMINOPROPANE-1,3-DICARBOXYLIC ACID HYDROCHLORIDE ◇ GLUTAMIC ACID HYDROCHLORIDE ◇ α-GLUTAMIC ACID HYDROCHLORIDE ◇ GLUTAMINIC ACID HYDROCHLORIDE ◇ l-GLUTAMINIC ACID HYDROCHLORIDE

SAFETY PROFILE: When heated to decomposition it emits toxic fumes of NO_x and Cl^-.

GFO050 CAS:56-85-9 ***HR: 1***
GLUTAMINE
mf: $C_5H_{10}N_2O_3$ mw: 146.17

PROP: l-Form (natural): Fine opaque needles from water or dil ethanol. Decomp 185-186°. Sol in water; practically insol in methanol, ethanol, ether, benzene, acetone, ethyl acetate, chloroform. dl-Form: prisms from dil acetone. Mp: 185-186°.

SYNS: 2-AMINOGLUTARAMIC ACID ◇ l-2-AMINOGLUTARAMIDIC ACID ◇ CEBROGEN ◇ GLUMIN ◇ GLUTAMIC ACID AMIDE ◇ GLUTAMIC ACID-5-AMIDE ◇ Γ-GLUTAMINE ◇ l-GLUTAMINE (9CI, FCC) ◇ LEVOGLUTAMID ◇ LEVOGLUTAMIDE ◇ STIMULINA

TOXICITY DATA with REFERENCE
orl-rat TDLo:300 g/kg (30D male):REP KSRNAM 8,902,74

orl-man TDLo:27 mg/kg/1W-I AJPSAO 141,1302,84
orl-rat LD50:7500 mg/kg NIIRDN 6,228,82
orl-mus LD50:21700 mg/kg NIIRDN 6,228,82

CONSENSUS REPORTS: Reported in EPA TSCA Inventory.

SAFETY PROFILE: Mildly toxic by ingestion. Experimental reproductive effects. When heated to decomposition it emits toxic fumes of NO_x.

GFO100 ***HR: 2***
N^2-(Γ-l-(+)-GLUTAMYL)-4-CARBOXYPHENYLHYDRAZINE
mf: $C_{12}H_{15}N_3O_5$ mw: 281.30

SYNS: ANTHGLUTIN ◇ l-GLUTAMIC ACID, 5-(2-(4-CARBOXYPHENYL)HYDRAZIDE)

TOXICITY DATA with REFERENCE
orl-mus TDLo:72800 mg/kg/52W-I:CAR ANTRD4 6,917,86

SAFETY PROFILE: Questionable carcinogen with experimental carcinogenic data. When heated to decomposition it emits toxic fumes of NO_x.

GFO200 CAS:60762-50-7 ***HR: 2***
1-(l-α-GLUTAMYL)-2-ISOPROPYLHYDRAZINE
mf: $C_8H_{17}N_3O_3$ mw: 203.28

SYN: RO 4-1385

TOXICITY DATA with REFERENCE
orl-mus LD50:500 mg/kg 27ZQAG -,429,72
scu-mus LD50:1400 mg/kg 27ZQAG -,429,72
ivn-mus LD50:1000 mg/kg 27ZQAG -,429,72

SAFETY PROFILE: Moderately toxic by ingestion, subcutaneous, and intravenous routes. When heated to decomposition it emits toxic fumes of NO_x. See also HYDRAZINE.

GFQ000 CAS:111-30-8 ***HR: 3***
GLUTARALDEHYDE
mf: $C_5H_8O_2$ mw: 100.13

SYNS: CIDEX ◇ GLUTARAL ◇ GLUTARALDEHYD (CZECH) ◇ GLUTARDIALDEHYDE ◇ GLUTARIC DIALDEHYDE ◇ NCI-C55425 ◇ 1,5-PENTANEDIAL ◇ 1,5-PENTANEDIONE ◇ POTENTIATED ACID GLUTARALDEHYDE ◇ SONACIDE

TOXICITY DATA with REFERENCE
skn-hmn 6 mg/3D-I SEV 85DKA8 -,127,77
skn-rbt 13 mg open MLD UCDS** 1/30/70
skn-rbt 500 mg/24H SEV 28ZPAK -,42,72
eye-rbt 1 mg SEV UCDS** 1/30/70
eye-rbt 250 μg/24H SEV 28ZPAK -,42,72
oms-nml:oth 50 mmol/L MUREAV 148,25,85
sce-ham:ovr 110 μg/L ENMUDM 7,1,85
orl-mus TDLo:8 g/kg (female 6-15D post):TER TJADAB 22,51,80
orl-rat TDLo:4370 mg/kg (female 35D pre):REP OYYAA2 12,11,76
orl-rat LD50:134 mg/kg OYYAA2 19,503,80
ipr-rat LD50:17900 μg/kg IYKEDH 10,232,79
scu-rat LD50:2390 mg/kg OYYAA2 19,503,80
ivn-rat LD50:15300 μg/kg OYYAA2 19,503,80
orl-mus LD50:100 mg/kg OYYAA2 19,503,80
ipr-mus LD50:13900 μg/kg IYKEDH 10,232,79
scu-mus LD50:1430 mg/kg OYYAA2 19,503,80
ivn-mus LD50:15400 μg/kg OYYAA2 19,503,80
skn-rbt LD50:2560 mg/kg AIHAAP 23,95,62

CONSENSUS REPORTS: Reported in EPA TSCA Inventory.

OSHA PEL: CL 0.2 ppm
ACGIH TLV: CL 0.2 ppm
DFG MAK: 0.2 ppm (0.8 mg/m^3)

SAFETY PROFILE: Poison by ingestion, intravenous, and intraperitoneal routes. Moderately toxic by inhalation, skin contact, and subcutaneous routes. Experimental teratogenic and reproductive effects. A severe eye and human skin irritant. Mutation data reported. When heated to decomposition it emits acrid smoke and irritating fumes. See also ALDEHYDES.

GFS000 CAS:110-94-1 ***HR: 1***
GLUTARIC ACID
mf: $C_5H_8O_4$ mw: 132.13

PROP: Colorless crystals. D: 1.429 @ 15°/4°, mp: 97.5°, bp: 200°. Very sol in abs alc and in ether; sol in benzene chloroform, alc, and ether. Large monoclinic prisms.

SYNS: PENTANDIOIC ACID ◇ PENTANEDIOIC ACID ◇ 1,5-PENTANEDIOIC ACID ◇ 1,3-PROPANEDICARBOXYLIC ACID

TOXICITY DATA with REFERENCE
orl-mus LD50:6000 mg/kg BIJOAK 34,1196,40

CONSENSUS REPORTS: Reported in EPA TSCA Inventory.

SAFETY PROFILE: Mildly toxic by ingestion. When heated to decomposition it emits acrid smoke and irritating fumes.

GFU000 CAS:108-55-4 ***HR: 2***
GLUTARIC ANHYDRIDE
mf: $C_5H_6O_3$ mw: 114.11

PROP: Sol in benzene and toluene; highly sol in water on complete hydrolysis. Bp: 144-146° @ 13 mm, d: 0.989.

TOXICITY DATA with REFERENCE
orl-rat LDLo:4460 mg/kg AIHAAP 23,95,62
skn-rbt LDLo:1780 mg/kg AIHAAP 23,95,62

CONSENSUS REPORTS: Reported in EPA TSCA Inventory.

SAFETY PROFILE: Moderately toxic by skin contact. Mildly toxic by ingestion. When heated to decomposition it emits acrid smoke and irritating fumes. See also ANHYDRIDES.

GFU200 CAS:64624-44-8 ***HR: 3***
GLUTARYL DIAZIDE
mf: $C_5H_6N_6O_2$ mw: 182.14

$N_3CO•(CH_2)_3CO•N_3$

SAFETY PROFILE: Explodes when heated. Upon decomposition it emits toxic fumes of NO_x. See also AZIDES.

GFW000 CAS:70-18-8 ***HR: 2***
GLUTATHIONE
mf: $C_{10}H_{17}N_3O_6S$ mw: 307.36

PROP: Colorless prisms out of alc. Mp: 195° decomp in hot water; insol in abs alc, ether, and acid. Freely sol in H_2O, dil alc, liquid ammonia, and dimethylformamide.

SYNS: COPREN ◇ DELTATHIONE ◇ GLUTATHIONE (reduced) ◇ GLUTATIOL ◇ GLUTATIONE ◇ GLUTIDE ◇ GLUTINAL ◇ GSH ◇ ISETHION ◇ NEUTHION ◇ TATHIONE ◇ TRIPTIDE

TOXICITY DATA with REFERENCE
mma-sat 6 mmol/L SCIEAS 220,961,83
dns-hmn:fbr 1 mmol/L CALEDQ 5,199,78
cyt-ham:ovr 1 mmol/L CALEDQ 5,199,78
sce-ham:ovr 100 μmol/L MUREAV 68,351,79
dni-mam:lym 10 mmol/L CBINA8 31,265,80
orl-rat TDLo:1250 mg/kg (1-22D preg):REP AJANA2 110,29,62
orl-mus LD50:5 g/kg 85IPAE -,93,72
ipr-mus LD50:4020 mg/kg 85IPAE -,93,72
scu-mus LD50:5 g/kg 85IPAE -,93,72
ivn-mus LD50:2238 mg/kg JJANAX 38,137,85

CONSENSUS REPORTS: EPA Genetic Toxicology Program. Reported in EPA TSCA Inventory.

SAFETY PROFILE: Moderately toxic by intravenous route. Experimental reproductive effects. Human mutation data reported. When heated to decomposition it emits very toxic fumes of SO_x and NO_x.

GFY100 CAS:26944-48-9 ***HR: 2***
GLUTRIL
mf: $C_{18}H_{26}N_2O_4S$ mw: 366.52

PROP: Crystals. Mp: 192-195° (ethanol loses water).

SYNS: GLIBORNURIDE ◇ GLUBORID ◇ 1-((1R)-2-endo-HYDROXY-3-endo-BORNYL)-3-(p-TOLYLSULFONYL)UREA ◇ RO 6-4563 ◇ RO-6-4563/8

TOXICITY DATA with REFERENCE
orl-rat TDLo:70 mg/kg (7-13D preg):REP KSRNAM 6,1968,72
orl-rat TDLo:70 mg/kg (7-13D preg):TER KSRNAM 6,1968,72
orl-rat LD50:18 g/kg KSRNAM 6,1925,72
ipr-rat LD50:1360 mg/kg KSRNAM 6,1925,72
scu-rat LD50:10800 mg/kg KSRNAM 6,1925,72
ipr-mus LD50:1530 mg/kg KSRNAM 6,1925,72
scu-mus LD50:20 g/kg KSRNAM 6,1925,72

SAFETY PROFILE: Moderately toxic by intraperitoneal route. Mildly toxic by ingestion. Experimental teratogenic and reproductive effects. When heated to decomposition it emits toxic fumes of SO_x and NO_x.

GGA000 CAS:56-81-5 ***HR: 2***
GLYCERIN
mf: $C_3H_8O_3$ mw: 92.11

$HOCH_2CHOHCH_2OH$

PROP: Colorless or pale yellow liquid; odorless, syrupy, sweet and warm taste. Mp: 17.9 (solidifies at a much lower temp), bp: 290°, ULC: 10-20, flash p: 320°F, d: 1.260 @ 20°/4°, autoign temp: 698°F, vap press: 0.0025 mm @ 50°, vap d: 3.17. Misc with water, alc; insol in chloroform, ether, oils.

SYNS: GLYCERIN, anhydrous ◇ GLYCERINE ◇ GLYCERIN, synthetic ◇ GLYCERITOL ◇ GLYCEROL ◇ GLYCYL ALCOHOL ◇ GROCOLENE ◇ MOON ◇ 1,2,3-PROPANETRIOL ◇ STAR ◇ SUPEROL ◇ SYNTHETIC GLYCERIN ◇ 90 TECHNICAL GLYCERINE ◇ TRIHYDROXYPROPANE ◇ 1,2,3-TRIHYDROXYPROPANE

TOXICITY DATA with REFERENCE
skn-rbt 500 mg/24H MLD 28ZPAK -,37,72
eye-rbt 126 mg MLD BIOFX* 9-4/70
eye-rbt 500 mg/24H MLD 28ZPAK -,37,72
dni-hmn:lym 200 mmol/L PNASA6 79,1171,82
itt-rat TDLo:1600 mg/kg (male 1D pre):REP CCPTAY 29,291,84
orl-hmn TDLo:1428 mg/kg:CNS,GIT 34ZIAG -,288,69
ipr-rat LD50:4420 mg/kg RCOCB8 56,125,87
scu-rat LD50:100 mg/kg NIIRDN 6,215,82
orl-mus LD50:4090 mg/kg FRZKAP (6),56,77
ipr-mus LD50:8982 mg/kg ARZNAD 26,1581,76
scu-mus LD50:91 mg/kg NIIRDN 6,215,82
ivn-mus LD50:4250 mg/kg JAPMA8 39,583,50
ivn-rbt LD50:53 g/kg NIIRDN 6,215,82
orl-gpg LD50:7750 mg/kg JIHTAB 23,259,41

CONSENSUS REPORTS: Reported in EPA TSCA Inventory.

OSHA PEL: (Transitional: TWA Total Mist: 15 mg/m^3;

Respirable Fraction: 5 mg/m^3) TWA Total Mist: 10 mg/m^3; Respirable Fraction: 5 mg/m^3
ACGIH TLV: TWA 10 mg/m^3 (vapor)

SAFETY PROFILE: Poison by subcutaneous route. Mildly toxic by ingestion. Human systemic effects by ingestion: headache and nausea or vomiting. Experimental reproductive effects. Human mutation data reported. A skin and eye irritant. In the form of mist it is a nuisance particulate and inhalation irritant.

Combustible liquid when exposed to heat, flame, or powerful oxidizers. Mixtures with hydrogen peroxide are highly explosive. Ignites on contact with potassium permanganate, calcium hypochlorite. Mixture with nitric acid + sulfuric acid forms the explosive glyceryl nitrate. Mixture with perchloric acid + lead oxide forms explosive perchlorate esters. Confined mixture with chlorine explodes if heated to 70-80°. Can react violently with acetic anhydride; aniline + nitrobenzene; $Ca(OCl)_2$; CrO_3; Cr_2O_3; F_2 + PbO; phosphorus triiodide; ethylene oxide + heat; $KMnO_4$; K_2O_2; $AgClO_4$; Na_2O_2; NaH. Energetic reaction with sodium hydride. Mixture with nitric acid + hydrofluoric acid is a storage hazard due to gas evolution. To fight fire, use alcohol foam, CO_2, dry chemical. When heated to decomposition it emits acrid smoke and fumes.

GGA800 CAS:9001-62-1 ***HR: 3***
GLYCEROL ESTER HYDROLASE

SYNS: AMANO N-AP ◇ BUTYRINASE ◇ E.C.3.1.1.3. ◇ GA 56 (ENZYME) ◇ LIPASE ◇ LIPAZIN ◇ MEITO MY 30 ◇ REMZYME PL 600 ◇ STEAPSIN ◇ TAKEDO 1969-4-9 ◇ TRIACETINASE ◇ TRIACYLGLYCEROL HYDROLASE ◇ TRIACYLGLYCEROL LIPASE ◇ TRIBUTYRASE ◇ TRIBUTYRINASE ◇ TRIBUTYRIN ESTERASE ◇ TRIGLYCERIDE HYDROLASE ◇ TRIGLYCERIDE LIPASE ◇ TRIOLEIN HYDROLASE ◇ TWEENASE ◇ TWEEN ESTERASE ◇ TWEEN HYDROLASE

TOXICITY DATA with REFERENCE
orl-mus TDLo:56 g/kg (female 7-13D post):REP NYKZAU 70,107,74
ipr-rat LD50:634 mg/kg NYKZAU 69,191,73
ipr-mus LD50:395 mg/kg NYKZAU 69,191,73
scu-mus LD50:2050 mg/kg NYKZAU 69,191,73

SAFETY PROFILE: Poison by intraperitoneal route. Moderately toxic by subcutaneous route. Experimental reproductive effects.

GGG000 CAS:96-11-7 ***HR: 2***
GLYCEROL TRIBROMOHYDRIN
mf: $C_3H_5Br_3$ mw: 280.81

SYNS: GLYCERYL TRIBROMOHYDRIN ◇ sym-TRIBROMOPROPANE ◇ 1,2,3-TRIBROMOPROPANE

TOXICITY DATA with REFERENCE
mmo-sat 1 μmol/plate ENMUDM 2,59,80
mma-sat 500 ng/plate ENMUDM 7(Suppl 3),15,85
orl-rat TDLo:250 mg/kg (5D male):REP MUREAV 101,321,82
orl-rat LDLo:500 mg/kg MUREAV 101,321,82

CONSENSUS REPORTS: EPA Genetic Toxicology Program. Reported in EPA TSCA Inventory.

SAFETY PROFILE: Moderately toxic by ingestion. Experimental reproductive effects. Mutation data reported. When heated to decomposition it emits toxic fumes of Br^-. See also BROMIDES.

GGI000 CAS:38571-73-2 ***HR: 3***
GLYCEROL (TRI(CHLOROMETHYL))ETHER
mf: $C_6H_{11}Cl_3O_3$ mw: 237.52

SYN: TRIS-1,2,3-(CHLOROMETHOXY)PROPANE

TOXICITY DATA with REFERENCE
skn-mus TDLo:8640 mg/kg/72W-I:ETA CNREA8 35,2553,75
ipr-mus TDLo:910 mg/kg/76W-I:NEO CNREA8 35,2553,75

CONSENSUS REPORTS: IARC Cancer Review: Group 2A IMEMDT 7,56,87; Animal Sufficient Evidence IMEMDT 15,301,77.

SAFETY PROFILE: Suspected carcinogen with experimental neoplastigenic and tumorigenic data. When heated to decomposition it emits toxic fumes of Cl^-. See also ETHERS.

GGK000 CAS:621-70-5 ***HR: 3***
GLYCEROL TRIHEXANOATE
mf: $C_{21}H_{38}O_6$ mw: 386.59

SYN: TRICAPRONIN

TOXICITY DATA with REFERENCE
ivn-mus LD50:122 mg/kg APSCAX 40,338,57

CONSENSUS REPORTS: Reported in EPA TSCA Inventory.

SAFETY PROFILE: Poison by intravenous route. When heated to decomposition it emits acrid smoke and irritating fumes.

GGM000 CAS:620-63-3 ***HR: 3***
GLYCEROL TRIISOPENTANOATE
mf: $C_{18}H_{32}O_6$ mw: 344.50

SYN: TRIISOVALERIN

TOXICITY DATA with REFERENCE
ivn-mus LD50:82 mg/kg APSCAX 40,338,57

CONSENSUS REPORTS: Reported in EPA TSCA Inventory.

SAFETY PROFILE: Poison by intravenous route. When heated to decomposition it emits acrid smoke and irritating fumes.

GGO000 CAS:106-61-6 ***HR: 2***
GLYCERYL ACETATE
mf: $C_5H_{10}O_4$ mw: 134.15

PROP: Colorless, very hygroscopic liquid; characteristic odor. D: 1.206 @ 20°/4°, bp: 158° @ 17 mm. Very sol in water and alc; sltly sol in ether; insol in benzene.

SYNS: 1-ACETATE-1,2,3-PROPANETRIOL ◇ ACETIC ACID, MONOGLYCERIDE ◇ ACETIN ◇ 2,3-DIHYDROXYPROPYL ACETATE ◇ GLYCEROL-1-ACETATE ◇ GLYCEROL MONOACETATE ◇ GLYCEROL-α-MONOACETATE ◇ GLYCEROL-1-MONOACETATE ◇ GLYCERYL MONOACETATE ◇ MONOACETIN ◇ α-MONOACETIN ◇ 1-MONOACETIN ◇ MONOACETYL GLYCERINE ◇ 1,2,3-PROPANETRIOL MONOACETATE

TOXICITY DATA with REFERENCE
mmo-sat 3333 μg/plate NTPTB* JAN 82
mma-sat 100 μg/plate NTPTB* JAN 82
scu-rat LD50:5500 mg/kg PSEBAA 46,26,41
scu-mus LD50:3500 mg/kg PSEBAA 46,26,41

SAFETY PROFILE: Moderately toxic by subcutaneous route. Mutation data reported. When heated to decomposition it emits acrid smoke and irritating fumes.

GGQ000 CAS:136-44-7 ***HR: 1***
GLYCERYL-p-AMINOBENZOATE
mf: $C_{10}H_{13}NO_4$ mw: 211.24

PROP: Semi-solid, waxy mass or syrup. Faint aromatic odor. Liquifies and congeals very slowly. Sol in methanol, ethanol, isopropanol, glycerol, propylene glycol; insol in water, oils, fats.

SYNS: p-AMINOBENZOIC ACID MONOGLYCERYL ESTER ◇ GLYCEROL-1-p-AMINOBENZOATE ◇ MONOGLYCEROL-p-AMINOBENZOATE

TOXICITY DATA with REFERENCE
skn-hmn 15 mg/3D-I MLD 85DKA8 -,127,77

CONSENSUS REPORTS: Reported in EPA TSCA Inventory.

SAFETY PROFILE: A human skin irritant. When heated to decomposition it emits toxic fumes of NO_x.

GGR100 CAS:50264-96-5 ***HR: D***
β-GLYCERYL 1-p-CHLOROBENZYL-1H-INDAZOLE-3-CARBOXYLATE
mf: $C_{18}H_{17}ClN_2O_4$ mw: 360.82

SYNS: 1-(p-CHLOROBENZYL)-1H-INDAZOLE-3-CARBOXYLICACID 1,3-DIHYDROXY-2-PROPYL ESTER ◇ 1H-INDAZOLE-3-CARBOXYLIC ACID, 1-(p-CHLOROBENZYL)-, 1,3-DIHYDROXY-2-PROPYL ESTER

TOXICITY DATA with REFERENCE
orl-rat TDLo:503 mg/kg (male 1D pre):REP JMCMAR 19,778,76

SAFETY PROFILE: Experimental reproductive effects. When heated to decomposition it emits toxic fumes of NO_x and Cl^-.

GGS000 CAS:59-47-2 ***HR: 3***
GLYCERYL-o-TOLYL ETHER
mf: $C_{10}H_{14}O_3$ mw: 182.24

SYNS: A 1141 ◇ AGEFLEX CGE ◇ ANATENSIN ◇ ANXINE ◇ ATENSIN ◇ AVESYL ◇ AVOXYL ◇ BDH 312 ◇ CRESODIOL ◇ o-CRESOL GLYCERYL ETHER ◇ CRESOSSIDIOLO ◇ CRESOSSIPROPANDIOLO ◇ CRESOXYDIOL ◇ CRESOXYPROPANEDIOL ◇ o-CRESYL-α-GLYCERYL ETHER ◇ CURARIL ◇ CURARYTHAN ◇ DASERD ◇ DASEROL ◇ DECONTRACTIL ◇ α,β-DIHYDROXY-Γ-(2-METHYLPHENOXY)PROPANE ◇ 1,2-DIHYDROXY-3-(2-METHYLPHENOXY)PROPANE ◇ DILOXOL ◇ FINDOLAR ◇ GLUKRESIN ◇ GLYOTOL ◇ KINAVOSYL ◇ o-KRESOL-GLYCERINAETHER (GERMAN) ◇ KRESOXYPROPANDIOL ◇ LISSENPHAN ◇ MC 2303 ◇ MEFENSINA ◇ MEPHATE ◇ MEPHEDAN ◇ MEPHELOR ◇ MEPHENSIN ◇ MEPHOSAL ◇ MEPHSON ◇ 3-(2-METHYLPHENOXY)-1,2-PROPANEDIOL ◇ MIANESINA ◇ MOCTYNOL ◇ MYANIL ◇ MYCOCURAN ◇ MYODETENSINE ◇ MYOLAX ◇ MYOPAN ◇ MYOSEROL ◇ MYOXANE ◇ NEMBUSEN ◇ NEPHELOR ◇ ORANIXON ◇ ORTOL ◇ PROLAX ◇ PROLOXIN ◇ RELAXANT ◇ RELAXAR ◇ RENARCOL ◇ REX REGULANS ◇ RP 3602 ◇ SANSDOLOR ◇ SECONESINZ ◇ SINAN ◇ SPARTOLOXYN ◇ SQ 1156 ◇ STILALGIN ◇ THIOXIDIL ◇ TOLANSIN ◇ TOLCIL ◇ TOLOFREN ◇ 3-o-TOLOXY-1,2-PROPANEDIOL ◇ TOLSEROL ◇ TOLULOX ◇ TOLYDRIN ◇ 1-o-TOLYLGLYCEROL ETHER ◇ α-(o-TOLYL)GLYCERYL ETHER ◇ 3-(o-TOLYLOXY)PROPANE-1,2-DIOL ◇ TOLYNOL ◇ TOLYSPAZ ◇ TORULOX ◇ WALKO-NESIN ◇ XERAL

TOXICITY DATA with REFERENCE
orl-rat LD50:625 mg/kg AIPTAK 130,280,61
ipr-rat LD50:283 mg/kg JPETAB 129,75,60
ivn-rat LD50:195 mg/kg PAREAQ 1,243,49
orl-mus LD50:720 mg/kg ARZNAD 17,242,67
ipr-mus LD50:470 mg/kg AIPTAK 155,69,65
scu-mus LD50:285 mg/kg APTOA6 19,247,62
ivn-mus LD50:175 mg/kg COREAF 248,3642,59
orl-rbt LDLo:2300 mg/kg AIPTAK 89,145,52
ivn-rbt LD50:125 mg/kg IJNEAQ 5,305,66
orl-ham LD50:821 mg/kg JPETAB 129,75,60
ipr-ham LD50:322 mg/kg JPETAB 129,75,60

SAFETY PROFILE: Poison by intraperitoneal, intravenous, and subcutaneous routes. Moderately toxic by ingestion. When heated to decomposition it emits acrid smoke and irritating fumes. See also ETHERS.

GGU000 CAS:614-33-5 ***HR: 1***
GLYCERYL TRIBENZOATE
mf: $C_{24}H_{20}O_6$ mw: 404.44

PROP: Colorless liquid. D: 1.032, mp: $< -75°$, bp: 305-309°. Insol in water; sol in alc and ether.

SYNS: BENZOFLEX S-404 ◇ BENZOIC ACID TRIESTER with GLYCERIN ◇ GTB ◇ TRIBENZOIN

TOXICITY DATA with REFERENCE
orl-rat LD50:11700 mg/kg NPIRI* 2,59,75

CONSENSUS REPORTS: Reported in EPA TSCA Inventory.

SAFETY PROFILE: Mildly toxic by ingestion. When heated to decomposition it emits acrid smoke and irritating fumes. See also ESTERS.

GGW000 CAS:765-34-4 ***HR: 3***
GLYCIDALDEHYDE
DOT: UN 2622
mf: $C_3H_4O_2$ mw: 72.07

PROP: Colorless liquid. Bp: 113°, d: 1.1403 @ 20°/4°.

SYNS: EPIHYDRINALDEHYDE ◇ EPIHYDRINE ALDEHYDE ◇ 2,3-EPOXYPROPANAL ◇ 2,3-EPOXY-1-PROPANAL ◇ 2,3-EPOXYPROPIONALDEHYDE ◇ GLYCIDAL ◇ GLYCIDYLALDEHDYE ◇ OXIRANE-CARBOXALDEHYDE ◇ RCRA WASTE NUMBER U126

TOXICITY DATA with REFERENCE
eye-hmn 1 ppm/5M MOD AEHLAU 2,23,61
mmo-esc 10 μg/plate ENMUDM 6(Suppl 2),1,84
mma-esc 33300 ng/plate ENMUDM 6(Suppl 2),1,84
scu-rat TDLo:13 g/kg/77W-I:CAR JNCIAM 39,1213,67
skn-mus TDLo:17 g/kg/48W-I:CAR JNCIAM 35,707,65
scu-mus TDLo:8844 mg/kg/67W-I:NEO JNCIAM 37,825,66
skn-mus TD:26 g/kg/22W-I:ETA JNCIAM 39,1217,67
ihl-hmn TCLo:5 ppm:BRN,EYE,CNS 34ZIAG -,289,69
orl-rat LDLo:50 mg/kg AJHYA2 76,209,62
ihl-rat LCLo:251 ppm/4H 14CYAT 2,1636,63
ipr-mus LD50:200 mg/kg JJIND8 62,911,79
skn-rbt LD50:249 mg/kg AEHLAU 2,23,61
ivn-rbt LDLo:20 mg/kg AEHLAU 2,23,61

CONSENSUS REPORTS: IARC Cancer Review: Group 2B IMEMDT 7,56,87; Animal Sufficient Evidence IMEMDT 11,175,76. EPA Genetic Toxicology Program.

DOT Classification: Flammable or Combustible Liquid; Label: Flammable and Poison.

SAFETY PROFILE: Suspected carcinogen with experimental carcinogenic, neoplastigenic, and tumorigenic data. Poison by ingestion, skin contact, intraperitoneal, and intravenous routes. Moderately toxic by inhalation. Human systemic effects by inhalation: changes in central nervous system electrical activity, olfactory changes, and excitement. Mutation data reported. A human eye irritant. Powerful skin sensitizer and mucous membrane irritant. Flammable when exposed to heat, flame, or oxidizing materials. When heated to decomposition it emits acrid smoke and irritating fumes. See also ALDEHYDES.

GGW500 CAS:556-52-5 ***HR: 3***
GLYCIDOL
mf: $C_3H_6O_2$ mw: 74.09

OCH_2CHCH_2OH

PROP: Colorless liquid. D: 1.165 @ 0°/4°, bp: 167° (decomp). Entirely sol in water, alc, and ether.

SYNS: EPIHYDRIN ALCOHOL ◇ 2,3-EPOXYPROPANOL ◇ 2,3-EPOXY-1-PROPANOL ◇ GLYCIDE ◇ GLYCIDYL ALCOHOL ◇ 3-HYDROXY-1,2-EPOXYPROPANE ◇ NCI-C55549 ◇ OXIRANEMETHANOL ◇ OXIRANYLMETHANOL

TOXICITY DATA with REFERENCE
skn-rbt 558 mg/3D MOD AMIHAB 14,250,56
mmo-klp 200 μmol/L MUREAV 89,269,81
cyt-hmn:lym 400 μmol/L MUREAV 91,243,81
orl-mus TDLo:2 g/kg (6-15D preg):TER JTEHD6 9,87,82
orl-rat TDLo:180 mg/kg (12D male):REP NATUAS 226,87,70
orl-rat LD50:420 mg/kg FCTXAV 19,347,81
ihl-rat LC50:580 ppm/8H AMIHAB 14,250,56
ipr-rat LD50:200 mg/kg FCTXAV 19,347,81
orl-mus LD50:431 mg/kg GTPZAB 24(3),42,80
ihl-mus LC50:450 ppm/4H AMIHAB 14,250,56
ipr-mus LDLo:500 mg/kg PSEBAA 35,98,36
skn-rbt LD50:1980 mg/kg AMIHAB 14,250,56

CONSENSUS REPORTS: Reported in EPA TSCA Inventory. EPA Genetic Toxicology Program.

OSHA PEL: (Transitional: TWA 50 ppm) TWA 25 ppm
ACGIH TLV: TWA 25 ppm
DFG MAK: 50 ppm (150 mg/m^3)

SAFETY PROFILE: Poison by intraperitoneal route. Moderately toxic by ingestion, inhalation, and skin contact. Experimental teratogenic and reproductive effects. A skin irritant. Human mutation data reported. Animal experiments suggest somewhat lower toxicity than related epoxy compounds. Readily absorbed through the skin. Causes nervous excitation followed by depression. Explodes when heated or in the presence of strong acids, bases, metals (e.g., copper, zinc) and metal salts (e.g., aluminum chloride, iron(III) chloride, tin(IV) chloride). When heated to decomposition it emits acrid smoke and fumes. See also DIGLYCIDYL ETHER.

GGW800 CAS:2917-91-1 ***HR: 2***
N-GLYCIDYL DIETHYL AMINE
mf: $C_7H_{15}NO$ mw: 129.23

SYNS: 3-DIETHYLAMINO-1,2-EPOXYPROPANE ◇ EPIHYDRINAMINE, N,N-DIETHYL- ◇ N-(2,3-EPOXYPROPYL)DIETHYLAMINE

◇ GLYCIDYLDIETHYLAMINE ◇ PROPYLAMINE, 2,3-EPOXY-N,N-DIETHYL-

TOXICITY DATA with REFERENCE
skn-rbt 100 μg/24H open AIHAAP 23,95,62
skn-rbt 5 mg/24H SEV 85JCAE-,774,86
eye-rbt 1 mg MLD UCDS** 5/13/59
eye-rbt 250 μg/24H SEV 85JCAE-,774,86
orl-rat LD50:420 mg/kg UCDS** 5/13/59
ihl-rat LCLo:2000 ppm/4H AIHAAP 23,95,62
skn-rbt LD50:790 mg/kg AIHAAP 23,95,62

SAFETY PROFILE: Moderately toxic by ingestion and skin contact. A severe skin and eye irritant. When heated to decomposition it emits toxic fumes of NO_x.

GGY000 CAS:17526-74-8 ***HR: 3***
GLYCIDYL ESTER of HEXANOIC ACID
mf: $C_9H_{16}O_3$ mw: 172.25

TOXICITY DATA with REFERENCE
scu-rat TDLo:2000 mg/kg/7W-I:ETA ANYAA9 68,750,58

SAFETY PROFILE: Questionable carcinogen with experimental tumorigenic data. When heated to decomposition it emits acrid smoke and irritating fumes. See also ESTERS.

GGY200 CAS:3033-77-0 ***HR: 3***
GLYCIDYLTRIMETHYLAMMONIUM CHLORIDE
mf: $C_6H_{14}NO{\bullet}Cl$ mw: 151.66

SYN: (2,3-EPOXYPROPYL)TRIMETHYLAMMONIUMCHLORIDE

TOXICITY DATA with REFERENCE
mmo-klp 2 mmol/L MUREAV 89,269,81
cyt-rat:lvr 10 mg/L MUREAV 153,57,85
scu-mus LD50:90 mg/kg JCSOA9 -,176,47

CONSENSUS REPORTS: Reported in EPA TSCA Inventory.

SAFETY PROFILE: Poison by subcutaneous route. Mutation data reported. When heated to decomposition it emits toxic fumes of Cl^-, NH_3, and NO_x. See also AMMONIUM CHLORIDE.

GHA000 CAS:56-40-6 ***HR: 2***
GLYCINE
mf: $C_2H_5NO_2$ mw: 75.08

PROP: The simplest amino acid and the principal amino acid in sugar cane. White crystals; odorless, sweet taste. Mp: 232-236° (decomp), d: 1.1607. Sol in water; insol in alc and ether.

SYNS: AMINOACETIC ACID ◇ GLYCOLIXIR ◇ HAMPSHIRE GLYCINE

TOXICITY DATA with REFERENCE
orl-rat LD50:7930 mg/kg YACHDS 5,1502,77
scu-rat LD50:5200 mg/kg YACHDS 5,1502,77
ivn-rat LD50:2600 mg/kg YACHDS 5,1502,77
orl-mus LD50:4920 mg/kg YACHDS 5,1502,77
ipr-mus LD50:4450 mg/kg YACHDS 5,1502,77
scu-mus LD50:5060 mg/kg YACHDS 5,1502,77
ivn-mus LD50:2370 mg/kg YACHDS 5,1502,77
ivn-cat LDLo:3000 mg/kg JAPMA8 31,306,42

CONSENSUS REPORTS: Reported in EPA TSCA Inventory.

SAFETY PROFILE: Moderately toxic by intravenous route. Mildly toxic by ingestion. When heated to decomposition it emits toxic fumes of NO_x.

GHE000 CAS:2619-97-8 ***HR: 3***
GLYCINE NITROGEN MUSTARD
mf: $C_6H_{11}Cl_2NO_2{\bullet}ClH$ mw: 236.54

SYNS: N,N-BIS(β-CHLOROETHYL)GLYCINE HYDROCHLORIDE ◇ N,N-BIS(2-CHLOROETHYL)GLYCINE HYDROCHLORIDE ◇ GLYCINE MUSTARD ◇ NSC 17661

TOXICITY DATA with REFERENCE
ice-rat LD50:113 μg/kg JPPMAB 18,760,66
ipr-rat LD50:15 mg/kg PHBUA9 2,275,54
ipr-mus LD50:9700 μg/kg NCISA* PH-43-63-1132
ivn-dog LDLo:1 mg/kg CCSUBJ 2,201,65
ivn-mky LDLo:1 mg/kg CCSUBJ 2,201,65

SAFETY PROFILE: A deadly poison by intraperitoneal, intravenous, and intracerebral routes. When heated to decomposition it emits very toxic fumes of Cl^- and NO_x.

GHG000 CAS:6000-44-8 ***HR: 2***
GLYCINE, SODIUM SALT
mf: $C_2H_4NO_2{\bullet}Na$ mw: 97.06

TOXICITY DATA with REFERENCE
ivn-mus LD50:564 mg/kg RPOBAR 2,292,70

CONSENSUS REPORTS: Reported in EPA TSCA Inventory.

SAFETY PROFILE: Moderately toxic by intravenous route. When heated to decomposition it emits toxic fumes of NO_x and Na_2O. See also GLYCINE.

GHI000 CAS:540-61-4 ***HR: D***
GLYCINONITRILE
mf: $C_2H_4N_2$ mw: 56.08

SYNS: AMINOACETONITRILE ◇ CYANOMETHYLAMINE ◇ GLYCINE NITRILE

TOXICITY DATA with REFERENCE
ipr-rat TDLo:300 mg/kg (female 15D post):TER VHAGAS 76,229,82

CONSENSUS REPORTS: Cyanide and its compounds are on the Community Right-To-Know List.

SAFETY PROFILE: Experimental teratogenic effects. When heated to decomposition it emits very toxic fumes of NO_x and CN^-. See also NITRILES.

GHI100 CAS:6011-14-9 ***HR: D***
GLYCINONITRILE HYDROCHLORIDE
mf: $C_2H_4N_2$•ClH mw: 92.54

SYNS: ACETONITRILE, AMINO-, MONOHYDROCHLORIDE (9CI) ◇ AMINOACETONITRILE HYDROCHLORIDE ◇ GLYCINONITRILE, MONOHYDROCHLORIDE

TOXICITY DATA with REFERENCE
orl-rat TDLo:600 mg/kg (female 3-7D post):REP TJADAB 24(2),41A,81
orl-rat TDLo:600 mg/kg (female 3-7D post):TER TJADAB 24(2),41A,81

CONSENSUS REPORTS: Reported in EPA TSCA Inventory.

SAFETY PROFILE: An experimental teratogen. Experimental reproductive effects. When heated to decomposition it emits toxic fumes of NO_x and HCl.

GHK000 CAS:6000-43-7 ***HR: 2***
GLYCOCOIL HYDROCHLORIDE
mf: $C_2H_5NO_2$•ClH mw: 111.54

SYN: GLYCOHYDROCHLORIDE

TOXICITY DATA with REFERENCE
orl-rat LD50:3340 mg/kg JPMSAE 62,49,73

CONSENSUS REPORTS: Reported in EPA TSCA Inventory.

SAFETY PROFILE: Moderately toxic by ingestion. When heated to decomposition it emits very toxic fumes of Cl^- and NO_x.

GHK200 CAS:3459-20-9 ***HR: 2***
GLYCODIAZINE SODIUM SALT
mf: $C_{13}H_{14}N_3O_4S$•Na mw: 331.35

SYNS: 2-BENZENESULFONAMIDO-5-(β-METHOXYETHOXY)PYRIMIDINE SODIUM SALT ◇ GLYCONORMAL ◇ GLYMIDINE SODIUM SALT ◇ GONDAFON ◇ LYCANOL ◇ N-(5-(2-METHOXYETHOXY)-2-PYRIMIDINYL)BENZENESULFONAMIDE SODIUM SALT ◇ REDUL ◇ SH 717

TOXICITY DATA with REFERENCE
orl-rat LD50:2850 mg/kg ARZNAD 14,377,64
ipr-rat LD50:3120 mg/kg NIIRDN 6,221,82
scu-rat LD50:2800 mg/kg NIIRDN 6,221,82
ivn-rat LD50:2000 mg/kg ARZNAD 14,377,64
orl-mus LD50:5300 mg/kg ARZNAD 14,377,64
ipr-mus LD50:3210 mg/kg NIIRDN 6,221,82
scu-mus LD50:3340 mg/kg NIIRDN 6,221,82
ivn-mus LD50:1480 mg/kg ARZNAD 14,377,64

SAFETY PROFILE: Moderately toxic by ingestion, intraperitoneal, subcutaneous, and intravenous routes. When heated to decomposition it emits toxic fumes of SO_x, NO_x, and Na_2O.

GHK500 CAS:4746-61-6 ***HR: 2***
GLYCOLANILIDE
mf: $C_8H_9NO_2$ mw: 151.18

SYN: 2-HYDROXY-N-PHENYLACETAMIDE

TOXICITY DATA with REFERENCE
orl-hmn TDLo:14286 μg/kg:PUL JAPMA8 35,50,46
orl-rat LD50:1700 mg/kg JAPMA8 35,50,46
orl-mus LD50:2300 mg/kg JAPMA8 35,50,46

SAFETY PROFILE: Moderately toxic by ingestion. Human systemic effects by ingestion: cyanosis. When heated to decomposition it emits toxic fumes of NO_x.

GHM000 CAS:96-49-1 ***HR: 2***
GLYCOL CARBONATE
mf: $C_3H_4O_3$ mw: 88.07

PROP: Colorless liquid or crystalline solid. Bp: 244° @ 740 mm, fp: 35.7°, flash p: 290°F (OC), d: 1.322 @ 40°/20°, vap press: 0.01 mm @ 20°, vap d: 3.04.

SYNS: CARBONIC ACID, CYCLIC ETHYLENE ESTER ◇ CYCLIC ETHYLENE CARBONATE ◇ 1,3-DIOXOLAN-2-ONE ◇ DIOXOLONE-2 ◇ ETHYLENE CARBONATE ◇ ETHYLENE CARBONIC ACID ◇ ETHYLENE GLYCOL CARBONATE ◇ ETHYLENE GLYCOL, CYCLIC CARBONATE

TOXICITY DATA with REFERENCE
skn-rbt 660 mg open MLD UCDS** 7/21/71
eye-rbt 20 mg open AMIHBC 10,61,54
eye-rbt 100 mg MOD 34ZIAG -,255,69
orl-rat LD50:10 g/kg UCDS** 7/21/71
ipr-mus LDLo:500 mg/kg CBCCT* 5,338,53

CONSENSUS REPORTS: Reported in EPA TSCA Inventory.

SAFETY PROFILE: Moderately toxic by intraperitoneal route. Mildly toxic by ingestion. A skin and eye irritant. Combustible when exposed to heat or flame; can react with oxidizing materials. To fight fire, use alcohol foam, CO_2, dry chemical. When heated to decomposition it emits acrid smoke and irritating fumes.

GHN000 ***HR: 3***
GLYCOL ETHERS

CONSENSUS REPORTS: Glycol ether compounds are on the Community Right-To-Know List.

SAFETY PROFILE: The acute toxic effects of ethylene

glycol monomethyl ether (2-methoxyethanol or 2ME) in humans are irritation of the eyes, nose and throat; drowsiness; weakness; and shaking. Ingestion of 2ME may be fatal. Prolonged or repeated exposures may cause headache, drowsiness, weakness, fatigue, staggering, personality change, and decreased mental ability. Exposed workers have suffered encephalopathy (degenerative brain disease), bone marrow depression and pancytopenia (reduced levels of all blood cells). 2ME and cellosolve (2-ethoxyethanol or 2EE) have the potential to cause adverse reproductive effects in male and female workers. They have been shown to cause embryotoxicity and other reproductive effects in several species of animals exposed by different routes of administration. The exposure of pregnant animals to concentrations of 2ME or 2EE at or below their OSHA permissable exposure limits lead to increased incidences of embryonic death, teratogenesis, or growth retardation. Exposure of male animals resulted in testicular atrophy and sterility. They can be absorbed through the skin. Structurally related glycol ethers are 2-methoxyethyl acetate; 2-ethoxyethyl acetate; 2-butoxyethanol; 2-phenoxyethanol; ethylene glycol dimethyl ether; bis(2-methoxyethyl)ether; 2-(2-ethoxyethoxy)ethanol; 1-methoxy-2-propanol; propylene glycol monomethyl ether. Although there is limited experimental information on the reproductive effects of these individual compounds, much of the information that is available is consistent with the reproductive effects caused by 2ME and 2EE. The acetate esters of 2ME and 2EE (2MEA and 2EEA) have caused male reproductive toxicity equivalent to that of 2ME and 2EEK in male mice. 2EEA appears to have fetotoxicity and teratogenicity equivalent to that of 2EE. Flammable or combustible when exposed to heat or flame; can react vigorously with oxidizing materials. When heated to decomposition they emit acrid smoke and fumes.

GHO000 CAS:79-14-1 ***HR: 2***
GLYCOLIC ACID
mf: $C_2H_4O_3$ mw: 76.06

PROP: Rhombic leaflets from ether. Odorless. Bp: decomp, mp (α): 63°, mp (β): 79°. Sol in H_2O, methanol, alc, acetone, acetic acid, ether.

SYNS: HYDROXYACETIC ACID ◇ HYDROXYETHANOIC ACID

TOXICITY DATA with REFERENCE
eye-rbt 2 mg SEV AJOPAA 29,1363,46
orl-rat LD50:1950 mg/kg JIHTAB 23,259,41
orl-gpg LD50:1920 mg/kg JIHTAB 23,259,41

CONSENSUS REPORTS: Reported in EPA TSCA Inventory.

SAFETY PROFILE: Moderately toxic by ingestion. A severe eye irritant. A skin and mucous membrane irritant. When heated to decomposition it emits acrid smoke and irritating fumes.

GHQ000 CAS:5847-48-3 ***HR: 3***
(GLYCOLOYLOXY)TRIBUTYLSTANNANE
mf: $C_{14}H_{30}O_3Sn$ mw: 365.13

SYNS: TRIBUTYL(GLYCOLOYLOXY)STANNANE ◇ TRIBUTYL (GLYCOLOYLOXY)TIN

TOXICITY DATA with REFERENCE
orl-mus LDLo:470 mg/kg AECTCV 14,111,85
ivn-mus LD50:18 mg/kg CSLNX* NX#03601

OSHA PEL: TWA 0.1 mg(Sn)/m^3 (skin)
ACGIH TLV: TWA 0.1 mg(Sn)/m^3 (skin) (Proposed: TWA 0.1 mg(Sn)/m^3; STEL 0.2 mg(Sn)/m^3 (skin))
NIOSH REL: (Organotin Compounds) TWA 0.1 mg(Sn)/m^3

SAFETY PROFILE: Poison by intravenous route. Moderately toxic by ingestion. Tributyl tin compounds are extremely toxic to marine life. When heated to decomposition it emits acrid smoke and irritating fumes. See also TIN COMPOUNDS.

GHR609 CAS:631-27-6 ***HR: 2***
GLYCOLPYRAMIDE
mf: $C_{11}H_{14}ClN_3O_3S$ mw: 303.79

SYNS: 4-CHLORO-N-((1-PYRROLIDINYLAMINO)CARBONYL) BENZENESULFONAMIDE (9CI) ◇ CPBU 7 ◇ DEAMELIN S

TOXICITY DATA with REFERENCE
orl-mus TDLo:560 mg/kg (female 7-13D post):REP KSRNAM 11,1620,77
orl-rat LD50:4100 mg/kg KSRNAM 11,1605,77
ipr-rat LD50:1620 mg/kg KSRNAM 11,1605,77
orl-mus LD50:8910 mg/kg KSRNAM 11,1605,77
ipr-mus LD50:1600 mg/kg KSRNAM 11,1605,77
scu-mus LD50:5500 mg/kg KSRNAM 11,1605,77

SAFETY PROFILE: Moderately toxic by intraperitoneal route. Mildly toxic by ingestion. Experimental reproductive effects. When heated to decomposition it emits toxic fumes of Cl^-, NO_x, and SO_x.

GHS000 CAS:9036-19-5 ***HR: 3***
GLYCOLS, POLYETHYLENE, MONO(1,1,3,3-TETRAMETHYLBUTYL)PHENYL) ETHER
mf: $(C_2H_4O)_n\ C_{14}H_{22}O$

SYNS: CHARGER E ◇ ETHOXYLATED OCTYL PHENOL ◇ ETHYLAN CP ◇ IGEPAL CA ◇ NEUTRONYX 622 ◇ NONIDET P40 ◇ NONION HS 206 ◇ OCTYLPHENOXYPOLY(ETHOXYETHANOL) ◇ tert-OCTYLPHENOXYPOLY(ETHOXYETHANOL) ◇ OCTYLPHENOXYPOLY(ETHYLENEOXY)ETHANOL ◇ tert-OCTYLPHENOXY POLY(OXYETHYLENE)ETHANOL ◇ OP 1062 ◇ POLYETHYLENE GLYCOL MONO(OCTYLPHENYL) ETHER ◇ POLYETHYLENE GLYCOL OCTYLPHENYL ETHER ◇ POLY(ETHYLENE OXIDE)OCTYLPHENYL ETHER ◇ POLYOXYETHYLENE MONOOCTYLPHENYL ETHER

◇ POLY(OXYETHYLENE)OCTYLPHENOL ETHER ◇ SECOPAL OP 20 ◇ SYNPERONIC OP ◇ T 45 (POLYGLYCOL) ◇ α-((1,1,3,3-TETRAMETHYLBUTYL)PHENYL)-omega-HYDROXY-POLY(OXY-1,2-ETHANEDIYL) ◇ TRITON X 15

TOXICITY DATA with REFERENCE
eye-rbt 1% SEV JAPMA8 38,428,49
dni-hmn:lym 5 ppm ENPBBC 5,84,75
dni-mus:oth 10 ppm ENPBBC 5,84,75
orl-rat LD50:4190 mg/kg FCTOD7 22,665,84
ipr-rat LD50:770 mg/kg FCTOD7 22,665,84
orl-mus LD50:3500 mg/kg JAPMA8 38,428,49
ivn-mus LD50:70 mg/kg JAPMA8 38,428,49

CONSENSUS REPORTS: Reported in EPA TSCA Inventory. Glycol ether compounds are on the Community Right-To-Know List.

SAFETY PROFILE: Poison by intravenous route. Moderately toxic by ingestion and intraperitoneal routes. Human mutation data reported. A severe eye irritant. When heated to decomposition it emits acrid smoke and irritating fumes. See also GLYCOL ETHERS.

GHU000 CAS:9008-57-5 ***HR: 2***
GLYCOLS, POLYETHYLENE MONO(TRIMETHYLNONYL)

SYN: TERGITOL TMN-6

TOXICITY DATA with REFERENCE
eye-rbt 100 mg SEV UCDS** 9/12/72
orl-rat LD50:7460 mg/kg UCDS** 9/12/72
skn-rbt LD50:8480 mg/kg UCDS** 9/12/72

SAFETY PROFILE: Mildly toxic by skin contact and ingestion. A severe eye irritant. When heated to decomposition it emits acrid smoke and irritating fumes.

GHY000 CAS:9038-95-3 ***HR: 1***
GLYCOLS, POLYETHYLENE POLYPROPYLENE, MONOBUTYL ETHER (nonionic)

SYNS: TERGITOL NONIONIC XD ◇ TERGITOL XD (nonionic)

TOXICITY DATA with REFERENCE
skn-rbt 500 mg open MLD UCDS** 9/22/64
orl-rat LD50:12 g/kg UCDS** 9/22/64

CONSENSUS REPORTS: Reported in EPA TSCA Inventory. Glycol ether compounds are on the Community Right-To-Know List.

SAFETY PROFILE: Mildly toxic by ingestion. A skin irritant. When heated to decomposition it emits acrid smoke and irritating fumes. See also GLYCOL ETHERS.

GIA000 CAS:36734-19-7 ***HR: 2***
GLYCOPHEN
mf: $C_{13}H_{13}Cl_2N_3O_3$ mw: 330.19

SYNS: CHIPCO 26019 ◇ 3-(3,5-DICHLOROPHENYL)-N-(1-METHYLETHYL)-2,4-DIOXO-1-IMIDAZOLIDINECARBOXAMIDE ◇ GLYCOPHENE ◇ IPRODIONE ◇ 1-ISOPROPYL CARBAMOYL-3-(3,5-DICHLOROPHENYL)-HYDANTOIN ◇ LFA 2043 ◇ MRC 910 ◇ PROMIDIONE ◇ ROP 500 F ◇ ROVRAL ◇ RP 26019

TOXICITY DATA with REFERENCE
orl-rat LD50:4400 mg/kg FMCHA2 -,C132,83
orl-mus LD50:4000 mg/kg FMCHA2 -,C132,83

SAFETY PROFILE: Moderately toxic by ingestion. When heated to decomposition it emits very toxic fumes of NO_x and Cl^-.

GIC000 CAS:596-51-0 ***HR: 3***
GLYCOPYRRONIUM BROMIDE
mf: $C_{19}H_{28}NO_3 \cdot Br$ mw: 398.39

PROP: Crystals from butanone. Mp: 193.2-194.5°.

SYNS: ASECRYL ◇ 1,1-DIMETHYL-3-HYDROXYPYRROLIDINIUM BROMIDE-α-CYCLOPENTYLMANDELATE ◇ GASTRODYN ◇ GLYCOPYRROLATE ◇ GLYCOPYRROLATE BROMIDE ◇ NODAPTON ◇ ROBANUL ◇ ROBINUL ◇ TARODYL ◇ TARODYN

TOXICITY DATA with REFERENCE
orl-rat TDLo:4004 mg/kg (26W pre):REP OYYAA2 7,627,73
orl-rat LD50:1150 mg/kg TXAPA9 17,361,70
ipr-rat LD50:196 mg/kg TXAPA9 17,361,70
scu-rat LD50:833 mg/kg NIIRDN 6,349,82
orl-mus LD50:570 mg/kg JMPCAS 2,523,60
ipr-mus LD50:90 mg/kg JMPCAS 2,523,60
scu-mus LD50:122 mg/kg YKYUA6 26,741,75
ivn-mus LD50:15 mg/kg 29ZVAB -,55,69
orl-rbt LD50:2360 mg/kg OYYAA2 7,627,73
ivn-rbt LD50:29100 μg/kg OYYAA2 7,627,73

SAFETY PROFILE: Poison by intravenous and intraperitoneal routes. Moderately toxic by ingestion and subcutaneous routes. Experimental reproductive effects. When heated to decomposition it emits very toxic fumes of NO_x and Br^-. See also BROMIDES.

GIE000 CAS:471-53-4 ***HR: 3***
α-GLYCYRRHETINIC ACID
mf: $C_{30}H_{46}O_4$ mw: 470.76

SYN: 3-β-HYDROXY-11-OXOOLEAN-12-EN-30-OICACID

TOXICITY DATA with REFERENCE
ivn-mus LD50:56 mg/kg CSLNX* NX#02067

CONSENSUS REPORTS: Reported in EPA TSCA Inventory.

SAFETY PROFILE: Poison by intravenous route.

When heated to decomposition it emits acrid smoke and irritating fumes.

GIE050 CAS:1449-05-4 ***HR: 3***
β-GLYCYRRHETINIC ACID
mf: $C_{30}H_{46}O_4$ mw: 470.76

SYN: 3-β-HYDROXY-11-OXO-18-α-OLEAN-12-EN-30-OIC ACID

TOXICITY DATA with REFERENCE
orl-mus LD50:560 mg/kg CPBTAL 28,3449,80
ipr-mus LD50:455 mg/kg CPBTAL 28,3449,80
ivn-mus LD50:100 mg/kg CSLNX* NX#02068

SAFETY PROFILE: Poison by intravenous route. Moderately toxic by ingestion and intraperitoneal routes. When heated to decomposition it emits acrid smoke and irritating fumes.

GIE100 CAS:53956-04-0 ***HR: 2***
GLYCYRRHIZIC ACID, AMMONIUM SALT
mf: $C_{42}H_{63}O_{16}•xH_3N$ mw: 943.33

SYNS: AMMONIATED GLYCYRRHIZIN ◇ AMMONIUM GLYCYRRHIZINATE ◇ α-D-GLUCOPYRANOSIDURONIC ACID, (3-β,20-β)-20-CARBOXY-11-OXO-30-NOROLEAN-12-EN-3-YL 2-O-β-D-GLUCOPYRANURONOSYL-, AMMONIATE ◇ MONOAMMONIUM GLYCYRRHIZINATE

TOXICITY DATA with REFERENCE
dlt-rat-orl 54600 mg/kg/10W-C ENMUDM 8,357,86
dlt-mus-orl 350 g/kg/10W-C NTIS** PB279-650
orl-rat TDLo:350 g/kg (male 10W pre):REP ENMUDM 8,357,86
orl-rat TDLo:256 mg/kg (female 7-17D post):TER FCTOD7 26,435,88
ipr-mus LDLo:1 g/kg YHTPAD 22,449,87
ivn-mus LD50:540 mg/kg YHTPAD 22,449,87

CONSENSUS REPORTS: Reported in EPA TSCA Inventory.

SAFETY PROFILE: Moderately toxic by intravenous route. An experimental teratogen. Experimental reproductive effects. Mutation data reported. When heated to decomposition it emits toxic fumes of NH_3.

GIG000 CAS:1405-86-3 ***HR: 3***
GLYCYRRHIZINIC ACID
mf: $C_{42}H_{62}O_{16}$ mw: 823.04

PROP: Crystals from glacial acetic acid. Intensely sweet taste. Freely sol in hot water, alc; practically insol in ether. The active component of liquorice (BMJOAE 1,488,77).

SYN: LIQUORICE

TOXICITY DATA with REFERENCE
orl-rat TDLo:4500 mg/kg (90D pre):REP YACHDS 5,2041,77
orl-hmn TDLo:280 mg/kg/4W:CNS,MET BMJOAE 1,488,77
orl-rat LDLo:3 g/kg YACHDS 5,2041,77
ipr-rat LDLo:2 g/kg YACHDS 5,2041,77
orl-mus LDLo:4 g/kg YACHDS 5,2041,77
ipr-mus LDLo:1 g/kg YACHDS 5,2041,77
ivn-mus LDLo:300 mg/kg YACHDS 5,2041,77

CONSENSUS REPORTS: Reported in EPA TSCA Inventory.

SAFETY PROFILE: Poison by intravenous route. Moderately toxic by ingestion and intraperitoneal route. Human systemic effects by ingestion: somnolence and changes in the metabolism of phosphorus. Experimental reproductive effects. When heated to decomposition it emits acrid smoke and irritating fumes.

GII000 CAS:556-22-9 ***HR: 2***
GLYODIN
mf: $C_{20}H_{40}N_2•C_2H_4O_2$ mw: 368.68

PROP: Light orange crystals. Mp: 62-68°, d: 1.035 @ 20°. The base is a soft greasy wax (mp: 94°). Insol in water, acetone, toluene; sol in isopropanol.

SYNS: CRAG 341 ◇ CRAG FRUIT FUNGICIDE 341 ◇ EXPERIMENTAL FUNGICIDE 341 ◇ GLYODIN ACETATE ◇ GLYOXIDE ◇ GLYOXIDE DRY ◇ 2-HEPTADECYL-4,5-DIHYDRO-1H-IMIDAZOLYL MONOACETATE ◇ 2-HEPTADECYL GLYOXALIDINE ACETATE ◇ 2-HEPTADECYL-2-IMIDAZOLINE ACETATE

TOXICITY DATA with REFERENCE
orl-rat LD50:4600 mg/kg FMCHA2 -,C116,83
unr-mam LD50:1000 mg/kg 30ZDA9 -,416,71

CONSENSUS REPORTS: Reported in EPA TSCA Inventory.

SAFETY PROFILE: Moderately toxic by unspecified route. Mildly toxic by ingestion. A skin, eye, and mucous membrane irritant. A fungicide which can damage the cornea. When heated to decomposition it emits toxic fumes of NO_x.

GIK000 CAS:107-22-2 ***HR: 3***
GLYOXAL
mf: $C_2H_2O_2$ mw: 58.04

PROP: Yellow prisms or irregular pieces turning white on cooling. D: 1.14 @ 20°/4°. Opaque @ 10°, mp: 15°, bp: (776) 51°. The vapors are green and burn with a purple flame, n (20.5/D) 1.3826. Sol in anhyd solvents, pH of a 40% aq soln: 2.1-2.7; d: (20/4) 1.27.

SYNS: AEROTEX GLYOXAL 40 ◇ BIFORMAL ◇ BIFORMYL ◇ DIFORMAL ◇ ETHANDIAL ◇ ETHANEDIAL ◇ 1,2-ETHANEDIONE ◇ GLYOXYLALDEHYDE ◇ OXAL ◇ OXALALDEHYDE

TOXICITY DATA with REFERENCE
skn-rbt 545 mg open MLD UCDS** 3/16/61

eye-rbt 1870 μg SEV AJOPAA 29,1363,46
mmo-sat 1 nmol/plate MUREAV 78,113,80
mma-sat 1 nmol/plate MUREAV 78,113,80
dns-rat-orl 300 mg/kg JJCREP 76,809,85
orl-rat LD50:1100 mg/kg 34ZIAG -,291,69
ipr-rat LDLo:100 mg/kg NCNSA6 5,9,53
ipr-mus LD50:200 mg/kg NTIS** AD438-895
orl-gpg LD50:760 mg/kg JIHTAB 23,259,41
skn-gpg LD50:6600 mg/kg 34ZIAG -,291,69

CONSENSUS REPORTS: Reported in EPA TSCA Inventory.

SAFETY PROFILE: Poison by intraperitoneal route. Moderately toxic by ingestion and skin contact. Mutation data reported. A skin and severe eye irritant. A powerful reducing agent. May explode on contact with air. Polymerizes violently on contact with water. During storage it may spontaneously polymerize and ignite. Reacts violently with chlorosulfonic acid; ethylene imine; HNO_3; oleum; NaOH; can cause violent reactions. Can explode during manufacture. When heated to decomposition it emits acrid smoke and irritating fumes. See also ALDEHYDES.

GIO000 CAS:105-28-2 ***HR: 2***
GLYOXIDE
mf: $C_{20}H_{40}N_2$ mw: 308.62

SYNS: CRAG FRUIT FUNGICIDE 34 ◇ 2-HEPTADECYL GLYOXALIDINE ◇ 2-HEPTADECYL-2-IMIDAZOLINE

TOXICITY DATA with REFERENCE
eye-rbt 750 μg SEV AJOPAA 29,1363,46
orl-rat LD50:3170 mg/kg WRPCA2 9,119,70

CONSENSUS REPORTS: Reported in EPA TSCA Inventory.

SAFETY PROFILE: Moderately toxic by ingestion. A severe eye irritant. A fruit fungicide. When heated to decomposition it emits toxic fumes of NO_x.

GIQ000 CAS:298-12-4 ***HR: 3***
GLYOXYLIC ACID
mf: $C_2H_2O_3$ mw: 74.04

PROP: In anhydrous form as monoclinic crystals from water. Deliquesces quickly and forms a syrup on short exposure to air. Freely sol in water; insol in ether and hydrocarbons. D: 1.42 @ 20°/4°, mp: 73°.

TOXICITY DATA with REFERENCE
mmo-sat 200 μg/plate ABCHA6 47,2461,83
ims-rat LDLo:25 mg/kg BSIBAC 36,1937,60

CONSENSUS REPORTS: Reported in EPA TSCA Inventory.

SAFETY PROFILE: Poison by intramuscular route. Mutation data reported. A skin, eye, and mucous membrane irritant. When heated to decomposition it emits acrid smoke and fumes.

GIS000 CAS:7440-57-5 ***HR: 1***
GOLD
af: Au aw: 196.97

PROP: Cubic, yellow, ductile, metallic crystals. Mp: 1064.76°, bp: 2700°, d: 19.3 (liquid) 17.0 @ 1063°, vap press: 1 mm @ 1869°, Hardness (Mohs') 2.5-3.0, (Brinell's) 18.5.

SYNS: BURNISH GOLD ◇ COLLOIDAL GOLD ◇ GOLD FLAKE ◇ GOLD LEAF ◇ GOLD POWDER ◇ MAGNESIUM GOLD PURPLE ◇ SHELL GOLD

TOXICITY DATA with REFERENCE
imp-rat TDLo:200 mg/kg:ETA NATWAY 42,75,55
ivn-rat LDLo:58 mg/kg ZEKBAI 63,586,60

CONSENSUS REPORTS: Reported in EPA TSCA Inventory.

SAFETY PROFILE: Poison by intravenous route. Questionable carcinogen with experimental tumorigenic data by implantation. Can form explosive compounds with NH_3, NH_4OH + aqua regia, H_2O_2. Incompatible with mixtures containing chlorides, bromides, or iodides (if they can generate nascent halogens), some oxidizing materials (especially those containing halogens), alkali cyanides, thiocyanate solutions, and double cyanides. See also GOLD COMPOUNDS.

GIT000 CAS:70950-00-4 ***HR: 3***
GOLD(I) ACETYLIDE
mf: C_2Au_2 mw: 417.96

AuC≡CAu

SAFETY PROFILE: An unstable explosive with high shattering power. It is easily detonated by light, impact, friction or rapid heating to 83°C. See also GOLD COMPOUNDS and EXPLOSIVES.

GIW176 CAS:13453-07-1 ***HR: 2***
GOLD CHLORIDE
mf: $AuCl_3$ mw: 303.33

PROP: Claret red crystals; mp: 254° (decomp); bp: subl 265°; d: 3.9.

SYNS: AURIC CHLORIDE ◇ GOLD(III) CHLORIDE ◇ GOLD TRICHLORIDE

TOXICITY DATA with REFERENCE
dns-hmn:lym 6 mg/L IAAAAM 77,459,85
scu-mus TDLo:22106 μg/kg (30D male):REP JRPFA4 7,21,64

CONSENSUS REPORTS: Reported in EPA TSCA Inventory.

SAFETY PROFILE: Experimental reproductive effects. Human mutation data reported. Reaction with ammonia or ammonium salts yields fulminating gold, a heat-, friction- and impact-sensitive explosive similar to mercury and silver fulminates. See also GOLD COMPOUNDS and CHLORIDES. When heated to decomposition it emits toxic fumes of Cl^-.

GIW179 ***HR: 1***
GOLD COMPOUNDS

SAFETY PROFILE: Gold poisoning is rare. The few recorded cases of fatalities are the result of therapeutic overdose. Human systemic effects are similar to those of arsenic exposure and include: violent diarrhea, gastritis, colitis, dermatitis, blood dyscrasias, leukopenia, aganulocytosis, and aplastic anemia. The therapeutic use of gold compounds has been associated with serious effects of the kidney, liver, and other vital organs. Gold sodium thiomalate, aurothioglucose, gold thioglycoanilid, and gold sodium thiosulfate are used to treat rheumatoid arthritis and lupus erythematosus. Generally, gold compounds are poorly absorbed when ingested. Effects are usually greater by intramuscular and intravenous routes.

GIW189 CAS:506-65-0 ***HR: 3***
GOLD(I) CYANIDE
mf: CAuN mw: 222.98

CONSENSUS REPORTS: Cyanide and its compounds are on the Community Right-To-Know List.

SAFETY PROFILE: Explosive reaction when heated with magnesium. When heated to decomposition it emits toxic fumes of CN^- and cyanogen. See also CYANIDE and GOLD COMPOUNDS.

GIW195 ***HR: 1***
GOLDEN CHAIN

PROP: A tree which may grow to 30 feet. It blooms in masses of golden flowers, and produces bean-type seedpods.

SYNS: BEAN TREE ◇ LABURNUM ◇ LABURNUM ANAGYROIDES

SAFETY PROFILE: All parts of the plant and particularly the seeds contain the poison cytisine (related to nicotine). Ingestion of any part of the plant rapidly causes vomiting, sleepiness, poor muscular control, headache, sweating, dilated pupils, and increased heart rate. See also CYTISINE, various cytisine compounds, and NICOTINE.

GIW200 ***HR: 3***
GOLDEN DEWDROP

PROP: A large shrub commonly cultivated as a hedge. It produces small light blue or white flowers and masses of orange berries. It grows wild in southern Florida and southern Texas of the United States, and the West Indies.

SYNS: AZOTA CABALLO ◇ BOIS JAMBETTE (HAITI) ◇ CUENTAS de ORO (PUERTO RICO) ◇ DURANTA REPENS ◇ GARBANCILLO (CUBA) ◇ MAIS BOUILLI (HAITI) ◇ PIGEON BERRY ◇ SKY FLOWER ◇ VELO de NOVIA (MEXICO)

SAFETY PROFILE: The berries contain poisonous saponins. Ingestion of the berries can cause sleepiness, fever, increased heart rate, and convulsions. See also SAPONIN.

GIW300 ***HR: 2***
GOLDEN SHOWER

PROP: A tropical tree (up to 30 feet tall) with long leaves composed of 4 to 8 pairs of 2- to 6-inch long leaflets. It produces large numbers of gold flowers on drooping racemes. The fruit is in the form of a long, thin pod holding up to 100 flat seeds embedded in a sticky matrix. The tree is cultivated in the United States in south Florida, the southern coast of California, and Hawaii, and the West Indies. Related species are found in the West Indies, Hawaii, and Guam.

SYNS: CANAFISTOLA (CUBA) ◇ CASSE (HAITI) ◇ CASSIA FISTULA ◇ GOLDEN RAIN ◇ INDIAN LABURNUM ◇ PUDDING-PIPE TREE ◇ PURGING FISTULA

SAFETY PROFILE: The sticky pulp of the seeds, and to a lesser extent the leaves and bark, contains toxic emodin glycosides. Human systemic effects by ingestion include nausea, vomiting, abdominal cramps, dizziness, and diarrhea. Emodin also causes a harmless discoloration of the urine.

GIX300 ***HR: 3***
GOLD(III) HYDROXIDE-AMMONIA
mf: $2AuH_3O_3{\cdot}3H_3N$ mw: 547.07

SAFETY PROFILE: A sensitive explosive. Heating in water forms the more explosive $Au_2O_3{\cdot}2NH_3$. Dry heating first forms the explosives $Au_2O_3{\cdot}3NH_3$ and $Au_2O{\cdot}4NH_3$. When heated to decomposition it emits toxic fumes of NH_3. See also GOLD COMPOUNDS and EXPLOSIVES.

GIY000 ***HR: 3***
GOLD NITRIDE AMMONIA
mf: $Au_3N{\cdot}NH_3$ mw: 621.94

SAFETY PROFILE: An unstable explosive. When

heated to decomposition it emits toxic fumes of NO_x and NH_3. See also GOLD COMPOUNDS and NITRIDES.

GIY300 ***HR: 3***
GOLD(I) NITRIDE-AMMONIA
mf: $Au_2N_3 \cdot H_3N$ mw: 452.98

SAFETY PROFILE: An explosive. Upon decomposition it emits toxic fumes of NO_x and NH_3. See also NITRIDES, EXPLOSIVES, and other gold compounds.

GIZ000 ***HR: 3***
GOLD(III) NITRIDE TRIHYDRATE
mf: $Au_3N_2 \cdot 3H_2O$ mw: 672.96

SAFETY PROFILE: Very explosive when dry. When heated to decomposition it emits toxic fumes of NO_x. See also NITRIDES, EXPLOSIVES, and GOLD COMPOUNDS.

GJC000 CAS:12244-57-4 ***HR: 3***
GOLD SODIUM THIOMALATE
mf: $C_4H_3AuO_4S \cdot 2Na$ mw: 390.08

SYNS: AuTM ◇ ((1,2-DICARBOXYETHYL)THIO)GOLD DISODIUM SALT ◇ (DIHYDROGEN MERCAPTOSUCCINATO)GOLD DISODIUM SALT ◇ DISODIUM AUROTHIOMALATE ◇ (MERCAPTOBUTANEDIOATO(1-))GOLD DISODIUM SALT ◇ MERCAPTOSUCCINIC ACID, GOLD SODIUM SALT ◇ MYOCHRYSINE ◇ MYOCRISIN ◇ SODIUM AUROTHIOMALATE ◇ TAURE(o)DON

TOXICITY DATA with REFERENCE
ipr-rat TDLo:75 mg/kg (4-6D preg):REP JPETAB 214,250,80
ipr-rat TDLo:75 mg/kg (4-6D preg):TER JPETAB 214,250,80
orl-man TDLo:5500 μg/kg:PNS,KID,BLD ARHEAW 19,936,76
ims-wmn TDLo:182 μg/kg:SKN BMJOAE 2,1294,76
ivn-wmn TDLo:2700 μg/kg/4W-I:CNS JRHUA9 11,235,84
ims-wmn TDLo:20 mg/kg/1Y:CNS,KID ARPAAQ 96,133,73
ims-man TDLo:5 mg/kg/13W:BLD,MSK JAMAAP 133,754,47
ims-wmn LDLo:600 μg/kg/1W-I:PUL,SYS JRSMD9 77,960,84
par-wmn TDLo:3366 μg/kg/2W:CNS,PNS NEURAI 20,455,70
par-man TDLo:12500 μg/kg/7W-I:CNS,PNS JRHUA9 11,233,84
par-man TDLo:7857 μg/kg/11W-I:SKN BMJOAE 286,1547,83
unr-cld TDLo:11900 μg/kg/18W-I: JRHUA9 13,224,86
unr-hmn TDLo:9750 μg/kg/23W-I:BLD BMJOAE 1,1266,76
unr-man TDLo:12 mg/kg:BLD JRHUA9 13,225,86
unr-wmn TDLo:1600 μg/kg/3W-I:LVR,BLD ANZJB8 16,72,86
unr-wmn TDLo:7 mg/kg/7W-I:BLD JRHUA9 12,180,85
unr-man TDLo:7857 μg/kg/11W-I:ALR BJRHDF 24,367,85
scu-rat LD50:303 mg/kg NIIRDN 6,208,82
ivn-rat LD50:440 mg/kg NIIRDN 6,208,82
ims-rat LDLo:185 mg/kg PSEBAA 49,121,42
scu-mus LD50:930 mg/kg NIIRDN 6,208,82
ivn-mus LD50:855 mg/kg NIIRDN 6,208,82
ims-mus LD50:800 mg/kg VTPHAK 15(Suppl.5),1,78

SAFETY PROFILE: Poison by subcutaneous and intramuscular routes. Moderately toxic by intravenous route. Human systemic effects: aggression, agranulocytosis, aplastic anemia, cell count changes, changes in circulation, cholestatic jaundice, dermatitis, encephalitis, fasciculations, flaccid paralysis without anesthesia, hemorrhage, hepatitis (hepatocellular necrosis), increased body temperature, interstitial fibrosis, muscle weakness, proteinuria, recording from peripheral motor nerve, renal function tests depressed, somnolence, structural changes in nerve sheath, thrombocytopenia, uncharacterized allergic reaction, changes in blood, teeth, and supporting structures. Experimental teratogenic and reproductive effects. When heated to decomposition it emits very toxic Na_2O and SO_x.

GJE000 CAS:10233-88-2 ***HR: 3***
GOLD SODIUM THIOSULFATE
mf: $O_6S_4 \cdot Au \cdot 3Na$ mw: 490.18

PROP: White crystals, odorless.

SYN: THIOSULFURIC ACID, GOLD(1+) SODIUM SALT (2:1:3)

TOXICITY DATA with REFERENCE
ipr-rat LD50:78 mg/kg JAPMA8 41,105,52
orl-mus LD50:35 mg/kg JINCAO 40,2081,78
ipr-mus LD50:245 mg/kg TXAPA9 49,41,79
ivn-mus LDLo:140 mg/kg JAPMA8 41,105,52

CONSENSUS REPORTS: Reported in EPA TSCA Inventory.

SAFETY PROFILE: Poison by ingestion, intraperitoneal, and intravenous routes. When heated to decomposition it emits very toxic fumes of SO_x and Na_2O. See also GOLD SODIUM THIOSULFATE DIHYDRATE.

GJG000 CAS:10210-36-3 ***HR: 3***
GOLD SODIUM THIOSULFATE DIHYDRATE
mf: $O_6S_4 \cdot Au \cdot 3Na \cdot 2H_2O$ mw: 526.22

SYNS: AURICIDINE ◇ AUROCIDIN ◇ AUROLIN ◇ AUROPEX ◇ AUROPIN ◇ AUROSAN ◇ AUROTHION ◇ CRISALBINE ◇ NOVACRYSIN ◇ SANOCHRYSINE ◇ SODIUM AUROTHIOSULPHATE DIHYDRATE ◇ SOLFOCRISOL ◇ THIOCHRYSINE ◇ THIOSULFURIC ACID, GOLD(1+) SODIUM SALT(2:1:3), DIHYDRATE

TOXICITY DATA with REFERENCE
ims-hmn TDLo:400 μg/kg/1W:SKN,BLD AIMEAS 37,323,52
scu-rat LDLo:30 mg/kg 27ZWAY 3.3,2134,-
ivn-rat LDLo:80 mg/kg 27ZWAY 3.3,2134,-
ims-rat LDLo:35 mg/kg PSEBAA 49,121,42
ipr-mus LD50:110 mg/kg JRHUA9 12,274,85
ivn-mus LDLo:100 mg/kg EMSUA8 3,146,45
ivn-rbt LDLo: 59 mg/kg 27ZWAY 3.3,2134,-
scu-gpg LDLo:60 mg/kg ZGEMAZ 57,77,27

SAFETY PROFILE: Poison by subcutaneous, intraperitoneal, intravenous, and intramuscular routes. Human systemic effects by intramuscular route: dermatitis, granulocytopenia, and thrombocytopenia. Used as an antirheumatic agent. When heated to decomposition it emits very toxic fumes of SO_x and Na_2O.

GJI075 ***HR: D***
GONADOGRAPHON LUTEINIZING HORMONE

SYN: LUTEINIZING HORMONE, GONADOGRAPHON RELEASING HORMONE

TOXICITY DATA with REFERENCE
par-dom TDLo:45 iu/kg (male 5D pre):REP IJEBA6 17,128,79

SAFETY PROFILE: Experimental reproductive effects. When heated to decomposition it emits acrid smoke and irritating fumes.

GJI100 ***HR: D***
GONADOTROPIN RELEASING HORMONE AGONIST

SYNS: D-TRP-LH-RH ◇ GnRH-A

TOXICITY DATA with REFERENCE
scu-rat TDLo:800 ng/kg (female 7-10D post):TER JSTBBK 26,1,87
scu-rat TDLo:800 ng/kg (female 7-10D post):REP JSTBBK 26,1,87

SAFETY PROFILE: An experimental teratogen. Experimental reproductive effects. When heated to decomposition it emits acrid smoke and irritating fumes.

GJI250 CAS:500-64-1 ***HR: 3***
GONOSAN
mf: $C_{14}H_{14}O_3$ mw: 230.28

PROP: (+)- Form: Rods from methanol + ether. Mp: 105-106°, bp: 195-197°. Practically insol in water; sol in acetone, ether, methanol; sltly sol in hexane. (+/-)-Form: Needles from methanol. Mp: 146-147°.

SYNS: (R)-5,6-DIHYDRO-4-METHOXY-6-STYRYL-2H-PYRAN-2-ONE ◇ 5-HYDROXY-3-METHOXY-7-PHENYL-2,6-HEPTADIENOIC ACID Γ-LACTONE ◇ KAVAIN ◇ (+)-KAVAIN ◇ KAWAIN ◇ 4-METHOXY-6-(β-PHENYLVINYL)-5,6-DIHYDRO-α-PYRONE

TOXICITY DATA with REFERENCE
orl-mus LD50:1130 AIPTAK 177,261,69
ipr-mus LD50:420 mg/kg AIPTAK 177,261,69
ivn-mus LD50:69 mg/kg AIPTAK 177,261,69

SAFETY PROFILE: Poison by intravenous route. Moderately toxic by ingestion and intraperitoneal routes. When heated to decomposition it emits acrid smoke and irritating fumes.

GJK000 CAS:50933-33-0 ***HR: 1***
GOSSYPLURE
mf: $C_{18}H_{32}O_2$ mw: 280.50

SYNS: GOSSYPLURE H.F. ◇ NOMATE PBW

TOXICITY DATA with REFERENCE
orl-rat LD50:15 g/kg FMCHA2 -,D219,80

CONSENSUS REPORTS: Reported in EPA TSCA Inventory.

SAFETY PROFILE: Mildly toxic by ingestion. When heated to decomposition it emits acrid smoke and irritating fumes.

GJM000 CAS:303-45-7 ***HR: 2***
GOSSYPOL
mf: $C_{30}H_{30}O_8$ mw: 518.60

PROP: A polyphenolic yellow pigment isolated from cottonseed pigment glands (JAOCA7 40,571,63). Mp: 180° from ether; mp: 199° from chloroform, mp: 214° from ligroin. Very sltly sol in methanol, ethanol, ether, chloroform, DMF; freely sol (with slow decomp) in dilute solns of ammonia, sodium carbonate; insol in water.

SYNS: 2,2′-BIS(1,6,7-TRIHYDROXY-3-METHYL-5-ISOPROPYL-8-ALDEHYDONAPHTHALENE ◇ 8-FORMYL-1,6,7-TRIHYDROXY-5-ISOPROPYL-3-METHYL-2,2′-BISNAPHTHALENE

TOXICITY DATA with REFERENCE
dni-hmn:hla 10 mg/L CNREA8 44,35,84
cyt-man:lym 9 μg/plate CPHPA5 12,293,81
sce-man:lym 1 μg/plate CPHPA5 12,293,81
orl-man TDLo:17 mg/kg (60D male):REP CMJODS 4,417,78
orl-rat LD50:2315 mg/kg JAOCA7 37,40,60
orl-pig LD50:550 mg/kg JAOCA7 40,571,63

CONSENSUS REPORTS: EPA Genetic Toxicology Program.

SAFETY PROFILE: Moderately toxic by ingestion. Human reproductive effects by ingestion: spermatogenesis and male fertility index changes. Experimental reproductive effects. Human mutation data reported. Can be irritating to the gastrointestinal tract. In experimental animals, large doses cause edema of lungs, shortness of breath, paralysis. When heated to decomposition it emits acrid smoke and irritating fumes.

GJM025 CAS:20300-26-9 ***HR: 3***
(+)-GOSSYPOL
mf: $C_{30}H_{30}O_8$ mw: 518.60

TOXICITY DATA with REFERENCE
orl-rat TDLo:50 mg/kg (1-5D preg):REP CUSCAM 50,64,81
ipr-mus LD50:35 mg/kg CUSCAM 50,64,81

SAFETY PROFILE: Poison by intraperitoneal route. Experimental reproductive effects. When heated to decomposition it emits acrid smoke and irritating fumes. See also GOSSYPOL.

GJM030 CAS:40112-23-0 ***HR: D***
racemic-GOSSYPOL
mf: $C_{30}H_{30}O_8$ mw: 518.60

SYNS: (2,2-BINAPHTHALENE)-8,8′-DICARBOXALDEHYDE, 1,1′,6,6′,7,7′-HEXAHYDROXY-3,3′-DIMETHYL-5,5′-BIS (1-METHYLETHYL)-, (±)- ◇ (±)-GOSSYPOL

TOXICITY DATA with REFERENCE
orl-rat TDLo:420 mg/kg (male 2W pre):REP JETHDA 20,21,87

SAFETY PROFILE: Experimental reproductive effects. When heated to decomposition it emits acrid smoke and irritating fumes.

GJM035 CAS:30719-67-6 ***HR: D***
GOSSYPOL ACETATE
mf: $C_{42}H_{42}O_{14}$ mw: 770.84

TOXICITY DATA with REFERENCE
scu-rat TDLo:100 mg/kg (female 20D pre):REP CCPTAY 32,491,85

SAFETY PROFILE: Experimental reproductive effects. When heated to decomposition it emits acrid smoke and irritating fumes. See also GOSSYPOL.

GJM259 CAS:12542-36-8 ***HR: 3***
GOSSYPOL ACETIC ACID

SYN: GOSSYPOL ACETATE

TOXICITY DATA with REFERENCE
dnd-hmn:leu 15 mg/L/1H MUREAV 164,71,86
sce-hmn:lym 1 mg/L ENMUDM 7(Suppl 3),66,85
sce-mus-ipr 20 mg/kg TCMUD8 6,83,86
orl-man TDLo:12800 μg/kg (male 12W pre):REP CCPTAY 37,153,88
orl-rat TDLo:10 mg/kg (female 10D post):TER TJADAB 32,251,85
orl-man TDLo:282 mg/kg/63W-I FESTAS 48,459,87
orl-mus LDLo:3 g/kg BIMADU 12,1,84
ivn-cat LDLo:75 mg/kg JAFCAU 17,497,69

SAFETY PROFILE: Poison by intravenous route. Human reproductive effects by ingestion: spermatogenesis. An experimental teratogen. Experimental reproductive effects. Human mutation data reported. See also GOSSYPOL.

GJO000 CAS:1405-97-6 ***HR: 3***
GRAMICIDIN
mf: $C_{148}H_{210}N_{30}O_{26}$ mw: 2825.88

PROP: Spear-shaped or lenticular platelets. Mp: 229-230°. Almost insol in water; sol in lower alc, acetic acid, and pyridine; practically insol in ether or hydrocarbons.

TOXICITY DATA with REFERENCE
ipr-mus LDLo:20 mg/kg JPETAB 74,75,42
ivn-mus LD50:1500 μg/kg SCIEAS 103,419,46
par-mus LD50:17 μg/kg 85ERAY 3,1542,78

SAFETY PROFILE: Poison by intravenous, intraperitoneal, and parenteral routes. An antibiotic. When heated to decomposition it emits very toxic fumes of NO_x.

GJO025 CAS:11029-61-1 ***HR: 3***
GRAMICIDIN A
mf: $C_{148}H_{210}N_{30}O_{26}$ mw: 2825.88

SYN: VALYL GRAMICIDIN A

TOXICITY DATA with REFERENCE
orl-mus LD50:1000 mg/kg 85GDA2 4(1),240,80
ipr-mus LD50:60 mg/kg 85GDA2 4(1),240,80
ivn-mus LD50:5 mg/kg 85GDA2 4(1),240,80

SAFETY PROFILE: Poison by intravenous and intraperitoneal routes. Moderately toxic by ingestion. An antibiotic. When heated to decomposition it emits toxic fumes of NO_x.

GJQ100 CAS:1481-70-5 ***HR: 3***
GRAMINIC ACID
mf: $C_{66}H_{87}N_{13}O_{13}\cdot ClH$ mw: 1307.12

SYN: TYROCIDINE A, HYDROCHLORIDE

TOXICITY DATA with REFERENCE
orl-mus LD50:1000 mg/kg 85GDA2 4(1),269,80
ipr-mus LD50:40 mg/kg 85GDA2 4(1),269,80
ivn-mus LD50:15 mg/kg 85GDA2 4(1),269,80

SAFETY PROFILE: Poison by intravenous and intraperitoneal routes. Moderately toxic by ingestion. When heated to decomposition it emits toxic fumes of NO_x and HCl.

GJS000 CAS:19879-06-2 ***HR: 3***
GRANATICIN
mf: $C_{22}H_{20}O_{10}$ mw: 444.42

PROP: Deep-red, garnet-like crystals from acetone. Decomp 204-206°.

SYNS: ANTIBIOTIC WR 141 ◇ GRANATICIN A ◇ LITMOMYCIN ◇ WR 141

TOXICITY DATA with REFERENCE
dni-bcs 600 nmol/L ZBPHA6 212,259,69
oms-bcs 600 nmol/L ZBPHA6 212,259,69
ipr-mus LD50:25 mg/kg 85ERAY 1,135,78
scu-mus LD50:250 mg/kg MEIEDD 10,652,83

SAFETY PROFILE: Poison by subcutaneous and intraperitoneal routes. Mutation data reported. An antibiotic. When heated to decomposition it emits acrid smoke and irritating fumes.

GJS200 CAS:22345-47-7 ***HR: 3***
GRANDAXIN
mf: $C_{22}H_{26}N_2O_4$ mw: 382.50

PROP: Colorless to light cream crystalline powder from isopropyl alc. Mp: 156-157°.

SYNS: 1-(3,4-DIMETHOXYPHENYL)-5-ETHYL-7,8-DIMETHOXY-4-METHYL-5H-2,3-BENZODIAZEPINE ◇ EGYT 341 ◇ SERIEL ◇ TF ◇ TOFISOPAM

TOXICITY DATA with REFERENCE
orl-rat TDLo:270 mg/kg (female 17-22D post):REP IYKEDH 12,565,81
orl-rat LD50:825 mg/kg IYKEDH 12,547,81
ipr-rat LD50:1270 mg/kg IYKEDH 12 547,81
ivn-rat LD50:103 mg/kg IYKEDH 12,547,81
orl-mus LD50:3800 mg/kg IYKEDH 12,547,81
ipr-mus LD50:1950 mg/kg IYKEDH 12,547,81
ivn-mus LD50:415 mg/kg IYKEDH 12,547,81

SAFETY PROFILE: Poison by intravenous route. Moderately toxic by ingestion and intraperitoneal route. Experimental reproductive effects. When heated to decomposition it emits toxic fumes of NO_x. See also DIAZEPAM.

GJU000 CAS:8016-20-4 ***HR: 3***
GRAPEFRUIT OIL

PROP: From the fresh peel of *Citrus paradisi* Macfayden (*Citrus decumana* L.). Yellow liquid. Sol in fixed oils, mineral oil; sltly sol in propylene glycol; insol in glycerin.

SYNS: GRAPEFRUIT OIL, coldpressed ◇ GRAPEFRUIT OIL, expressed ◇ OIL of GRAPEFRUIT ◇ OIL of SHADDOCK

TOXICITY DATA with REFERENCE
skn-rbt 500 mg/24H MLD FCTXAV 12,743,74
dnr-bcs 20 mg/disc TOFOD5 8,91,85
skn-mus TDLo:280 g/kg/33W-I:ETA JNCIAM 24,1389,60

CONSENSUS REPORTS: Reported in EPA TSCA Inventory.

SAFETY PROFILE: A skin irritant. Questionable carcinogen with experimental tumorigenic data. Mutation data reported. When heated to decomposition it emits acrid smoke and irritating fumes.

GJU460 ***HR: 3***
GREEN LILY

PROP: Perennial bulb-producing herbs. The grass-like leaves grow directly from the bulb. Pale green to yellow-white flowers grow in heavy clusters from a leafless stalk. The seed pod has 3 lobes and holds 4 or more seeds. The various species grow in the area from southern New Mexico to Peru.

SYNS: CEBOLLEJA (MEXICO) ◇ SCHOENOCAULON DRUMMONDII ◇ SCHOENOCAULON OFFICIANLIS ◇ SCHOENOCAULON TEXANUM

SAFETY PROFILE: The whole plant and especially the seeds are thought to contain poisonous veratridine alkaloids. Ingestion of the seeds may cause severe vomiting, catharsis, slowed heartbeat, low blood pressure and, in some cases, death. See also VERATRIDINE.

GJU475 ***HR: 2***
GREEN LOCUST

PROP: A large tree which may grow to 80 feet. The leaves are compound with many 1-inch leaflets. There are 2 thorns on the branch where each leaf stem is attached. It produces clusters of white, fragrant flowers and a flat, red-brown, 4-inch long seed pod which stays on the tree through the winter. It is native to the Smoky Mountains and the Ozarks of the United States. It is commonly cultivated in the temperate regions of the United States and southern Canada.

SYNS: BASTARD ACACIA ◇ BLACK ACACIA ◇ BLACK LOCUST ◇ FALSE ACACIA ◇ PEA FLOWER LOCUST ◇ POST LOCUST ◇ ROBINIA PSEUDOACACIA ◇ SILVER CHAIN ◇ TREESAIL ◇ WHITE HONEY FLOWER ◇ WHITE LOCUST ◇ WHYO TREE ◇ YELLOW LOCUST

SAFETY PROFILE: The bark, seeds, and leaves contain the poison robin, a plant lectin (toxalbumin) which inhibits protein synthesis in the intestinal wall. Ingestion of these plant parts may cause after a delay period of several hours: nausea, vomiting, diarrhea, and intestinal dysfunction. There is a potential for massive fluid and electrolyte loss. See also ABRIN as an example of toxalbumin.

GJU500 ***HR: 3***
GREEN MAMBA VENOM

SYNS: DENDROASPIS VIRIDIS VENOM ◇ VENOM, SNAKE, DENDROASPIS VIRIDIS

TOXICITY DATA with REFERENCE
scu-mus LD50:790 μg/kg 19DDA6 1,223,67
ivn-mus LD50:667 μg/kg 23EIAT 1,437,68
ivn-rbt LDLo:700 μg/kg TOXIA6 2,5,64

SAFETY PROFILE: A deadly poison by subcutaneous and intravenous routes.

GJU800 CAS:1391-82-8 ***HR: 3***
GRISEIN
mf: $C_{40}H_{61}FeN_{10}O_{20}S$ mw: 1090.02

PROP: Antibiotic substance produced by strains of *Streptomyces griseus*. Amorph, red powder. Sol in water; sltly sol in 95% alc; insol in abs alc, ether, acetone, chloroform, benzene.

SYNS: ALBOMYCIN A1 ◇ CORMOGRIZIN ◇ GRISIN ◇ KORMOGRIZEIN

TOXICITY DATA with REFERENCE
orl-mus LD50:600 mg/kg MEIEDD 10,653,83
ipr-mus LD50:600 mg/kg 85GDA2 4(1),442,80
scu-mus LD50:34 mg/kg MEIEDD 10,653,83

SAFETY PROFILE: Poison by subcutaneous route. Moderately toxic by ingestion and intraperitoneal routes. An antibiotic. When heated to decomposition it emits toxic fumes of SO_x and NO_x.

GJW000 CAS:2072-68-6 ***HR: 3***
GRISEOLUTEIN B
mf: $C_{17}H_{16}N_2O_6$ mw: 344.35

TOXICITY DATA with REFERENCE
orl-mus LD50:800 mg/kg 85ERAY 1,763,78
scu-mus LD50:400 mg/kg 85GDA2 5,144,81
ivn-mus LD50:200 mg/kg 85GDA2 5,144,81

SAFETY PROFILE: Poison by intravenous and subcutaneous routes. Moderately toxic by ingestion. When heated to decomposition it emits toxic fumes of NO_x.

GJY000 CAS:1393-89-1 ***HR: 3***
GRISEOMYCIN
mf: $C_{25}H_{46}ClNO_8$ mw: 524.17

PROP: Produced by *Streptomyces sp.* (ANTCAO 3,1243,53). Crystalline base, bitter taste, platelets, decomp @ 76-80°, alkaline reaction; freely sol in chloroform, ethanol, butanol, benzene, acetone, ethyl acetate.

SYN: LOMYCIN

TOXICITY DATA with REFERENCE
ipr-mus LD50:210 mg/kg 85ERAY 1,98,78
scu-mus LD50:1330 mg/kg 85ERAY 1,98,78

SAFETY PROFILE: Poison by intraperitoneal route. Moderately toxic by subcutaneous route. When heated to decomposition it emits very toxic fumes of NO_x and Cl^-.

GJY100 CAS:73297-70-8 ***HR: 3***
GRISEORUBIN COMPLEX

TOXICITY DATA with REFERENCE
ipr-mus LD50:50 mg/kg 85GDA2 3,202,80
scu-mus LD50:100 mg/kg 85GDA2 3,202,80
ivn-mus LD50:10 mg/kg 85GDA2 3,202,80

SAFETY PROFILE: Poison by subcutaneous, intravenous, and intraperitoneal routes.

GKA000 CAS:101563-93-3 ***HR: 3***
GRISEORUBIN I HYDROCHLORIDE

PROP: Produced by *Streptomyces griseus* (JANTAJ 31,78-111,78).

TOXICITY DATA with REFERENCE
ipr-mus LD50:50 mg/kg JANTAJ 31,78,78
scu-mus LD50:100 mg/kg JANTAJ 31,78-111,78
ivn-mus LD50:10 mg/kg JANTAJ 31,78,78

SAFETY PROFILE: Poison by intraperitoneal, subcutaneous and intravenous routes. When heated to decomposition it emits very toxic fumes of NO_x and HCl.

GKC000 CAS:53216-90-3 ***HR: 3***
GRISEOVIRIDIN
mf: $C_{22}H_{27}N_3O_7S$ mw: 477.58

PROP: Polymorphic crystals. Decomp @ 158-166°, 194-200° or 230-240°, depending on the crystal modification. Sol in pyridine; mod sol in lower alcs; very sltly sol in water and nonpolar solvents.

TOXICITY DATA with REFERENCE
ipr-mus LD50:100 mg/kg 85ERAY 1,322,78
scu-mus LD50:100 mg/kg 85ERAY 1,322,78
ivn-mus LD50:75 mg/kg ABANAE 2,790,54/55

SAFETY PROFILE: Poison by intraperitoneal, subcutaneous, and intravenous routes. When heated to decomposition it emits very toxic fumes of SO_x and NO_x.

GKE000 CAS:126-07-8 ***HR: 3***
GRISOFULVIN
mf: $C_{17}H_{17}ClO_6$ mw: 352.79

SYNS: AMUDANE ◇ BIOGRISIN-FP ◇ 7-CHLORO-4,6,2'-TRIMETHOXY-6'-METHYLGRIS-2'-EN-3,4'-DIONE ◇ DELMOFULVINA ◇ FULCIN ◇ FULCINE ◇ FULVICAN GRISACTIN ◇ FULVICIN

◇ FULVINA ◇ FULVISTATIN ◇ FUNGIVIN ◇ GREOSIN ◇ GRESFEED ◇ GRICIN ◇ GRIFULVIN ◇ GRISACTIN ◇ GRISCOFULVIN ◇ GRISEFULINE ◇ GRISEO ◇ (+)-GRISEOFULVIN ◇ GRISEOFULVIN-FORTE ◇ GRISEOFULVINUM ◇ GRISETIN ◇ GRISOVIN ◇ GRIS-PEG ◇ GRYSIO ◇ GUSERVIN ◇ LAMORYL ◇ LIKUDEN ◇ MURFULVIN ◇ NEO-FULCIN ◇ NSC 34533 ◇ PONCYL ◇ SPIROFULVIN ◇ SPOROSTATIN ◇ USAF SC-2

TOXICITY DATA with REFERENCE
dnr-bcs 100 μL/plate MUREAV 97,1,82
dni-hmn:fbr 20 mg/L/3D-C KAMJDW 2,127,76
dni-hmn:lym 20 mg/L/3D-C KAMJDW 2,127,76
cyt-hmn:lym 40 mg/L/3D MUREAV 25,123,74
cyt-ham:fbr 10 mg/L CRNGDP 3,499,82
orl-rat TDLo:1250 mg/kg (6-15D preg):REP SCIEAS 175,1483,72
orl-rat TDLo:500 mg/kg (female 9D post):TER ANTBAL 14,44,69
orl-rat TDLo:462 g/kg/2Y-I:NEO BJCAAI 38,237,78
ivn-rat LD50:400 mg/kg NATUAS 182,1320,58
ipr-mus LD50:200 mg/kg NTIS** AD277-689
scu-mus LD50:1200 mg/kg 85GDA2 6,290,81
ivn-mus LD50:280 mg/kg 85ERAY 3,1766,78

CONSENSUS REPORTS: IARC Cancer Review: Group 2B IMEMDT 7,56,87; Animal Sufficient Evidence IMEMDT 10,153,76. EPA Genetic Toxicology Program.

SAFETY PROFILE: Suspected carcinogen with experimental neoplastigenic and teratogenic data. Poison by intravenous and intraperitoneal routes. Moderately toxic by subcutaneous route. Human mutation data reported. Experimental reproductive effects. Used as a antibiotic, pharmaceutical and veterinary drug. When heated to decomposition it emits toxic fumes of Cl^-.

GKE900 CAS:14567-61-4 ***HR: 2***
GRUNERITE

TOXICITY DATA with REFERENCE
ipl-rat TDLo:80 mg/kg:ETA CNREA8 2,157,86

SAFETY PROFILE: Questionable carcinogen with experimental tumorigenic data.

GKG300 CAS:17471-82-8 ***HR: 3***
GUABENXANE
mf: $C_{10}H_{13}N_3O_2 \cdot 1/2H_2O_4S$ mw: 256.26

SYNS: GUANIDINO-6-METHYL-1,4-BENZODIOXANE◇ 6-GUANIDINOMETHYL 1,4-BENZODIOXANE SULFATE ◇ GUANIDINO METHYL-6-BENZODIOXANNE-1,4 (FRENCH) ◇ L'HEMISULFATE de GUANIDINO METHYL-6-BENZODIOXANNE-1,4 (FRENCH) ◇ LM 433

TOXICITY DATA with REFERENCE
ipr-rat LD50:243 mg/kg EJMCA5 12,241,77
orl-mus LD50:535 mg/kg EJMCA5 12,241,77
ipr-mus LD50:179 mg/kg EJMCA5 12,241,77
ivn-mus LD50:46 mg/kg EJMCA5 12,241,77

SAFETY PROFILE: Poison by intravenous and intraperitoneal routes. Moderately toxic by ingestion. When heated to decomposition it emits toxic fumes of NO_x and SO_x. See also SULFATES.

GKI000 CAS:90-05-1 ***HR: 3***
GUAIACOL
mf: $C_7H_8O_2$ mw: 124.15

PROP: Clear, pale yellow liquid or solid. Characteristic odor, darkens on exposure to air and light. D (crystals): 1.129, d (liquid): about 1.112. Misc with alc, chloroform, ether, oils, glacial acetic acid; sltly sol in petr ether; sol in NaOH soln, mp: 28°, bp: 202-209°, flash p: 180°F (OC), d: 1.097 @ 25°/25°.

SYNS: GUAICOL ◇ o-HYDROXYANISOLE ◇ 2-HYDROXYANISOLE ◇ 1-HYDROXY-2-METHOXYBENZENE ◇ o-METHOXYPHENOL ◇ 2-METHOXYPHENOL ◇ METHYLCATECHOL ◇ PYROGUAIAC ACID

TOXICITY DATA with REFERENCE
skn-rbt 500 mg/24H SEV FCTOD7 20(Suppl),697,82
eye-rbt 5 mg MLD FCTOD7 20(Suppl),697,82
sce-hmn:lym 500 μmol/L MUREAV 169,129,86
orl-hmn LDLo:43 mg/kg:CNS,GIT 34ZIAG -,295,69
orl-rat LD50:725 mg/kg TXAPA9 6,378,64
scu-rat LDLo:900 mg/kg FCTOD7 20(Suppl),697,82
orl-mus LD50:890 mg/kg KSGZA3 36,932,82
ihl-mus LC50:7570 mg/m³ FCTOD7 20(Suppl),697,82
ivn-mus LD50:170 mg/kg FCTOD7 20(Suppl),697,82
skn-rbt LD50:4600 mg/kg FCTOD7 20(Suppl),697,82
scu-gpg LDLo:900 mg/kg FCTOD7 20(Suppl),697,82

CONSENSUS REPORTS: Reported in EPA TSCA Inventory. EPA Genetic Toxicology Program.

SAFETY PROFILE: Human poison by ingestion. Experimental poison by intravenous route. Mildly toxic by skin contact and inhalation. Human systemic effects by ingestion: tremors and gastrointestinal changes. Human mutation data reported. An eye and severe skin irritant. Ingestion produces burning in the mouth and throat. Flammable when exposed to heat or flame; can react with oxidizing materials. To fight fire, use foam, CO_2, dry chemical. Protect from light. Used as an expectorant. When heated to decomposition it emits acrid smoke and irritating fumes. See also PHENOL.

GKK000 CAS:532-03-6 ***HR: 2***
GUAIACOL GLYCERYL ETHER CARBAMATE
mf: $C_{11}H_{15}NO_5$ mw: 241.27

SYNS: GLYCERYLGUAIACOLATE CARBAMATE ◇ GLYCERYLGUAJACOL-CARBAMAT ◇ GUIACOL-GLICERILETERE MONOCARBAMMATO ◇ 2-HYDROXY-3-(o-METHOXYPHENOXY)PROPYL-1-CARBAMATE ◇ 3-(2-METHOXYPHENOXY)-1-GLYCERYL CARBAMATE ◇ 3-(o-METHOXYPHENOXY-2-HYDROXYPROPYL CAR-

BAMATE ◇ 3-(o-METHOXYPHENOXY)-1,2-PROPANEDIOL-1- CARBAMATE

TOXICITY DATA with REFERENCE
orl-rat LD50:1320 mg/kg JPETAB 129,75,60
ipr-rat LD50:820 mg/kg JPETAB 129,75,60
orl-mus LD50:1530 mg/kg JPETAB 129,75,60
ipr-mus LD50:955 mg/kg ARZNAD 17,242,67
scu-mus LD50:780 mg/kg APTOA6 19,247,62
orl-dog LD50:2000 mg/kg 27ZQAG -,398,72
ivn-rbt LD50:680 mg/kg IJNEAQ 5,305,66
orl-ham LD50:1410 mg/kg JPETAB 129,75,60
ipr-ham LD50:1050 mg/kg JPETAB 129,75,60

CONSENSUS REPORTS: Reported in EPA TSCA Inventory.

SAFETY PROFILE: Moderately toxic by ingestion, intraperitoneal, subcutaneous, and intravenous routes. When heated to decomposition it emits toxic fumes of NO_x. See also ETHERS and CARBAMATES.

GKM000 CAS:8016-23-7 ***HR: 1***
GUAIAC WOOD OIL

PROP: From steam distillation of *Bulnesia sarmienti lor.* wood chips or sawdust (FCTXAV 12,807,74).

TOXICITY DATA with REFERENCE
skn-rbt 500 mg/24H MOD FCTXAV 12,905,74

SAFETY PROFILE: A skin irritant. When heated to decomposition it emits acrid smoke and irritating fumes.

GKO000 CAS:88-84-6 ***HR: 2***
GUAIA-1(5),7(11)-DIENE
mf: $C_{15}H_{24}$ mw: 204.39

SYNS: GUAIENE ◇ β-GUAIENE ◇ (1S,cis)1,2,3,4,5,6,7,8-OCTAHYDRO-1,4-DIMETHYL-7-(1-METHYLETHYLIDENE)-AZULENE

TOXICITY DATA with REFERENCE
skn-rbt 500 mg/24H SEV FCTXAV 17,371,79

CONSENSUS REPORTS: Reported in EPA TSCA Inventory.

SAFETY PROFILE: A severe skin irritant. When heated to decomposition it emits acrid smoke and irritating fumes.

GKO500 ***HR: 3***
GUAIMERCOL
mf: $C_{11}H_{11}Hg_2NO_8$ mw: 1102.18

SYN: 6-ACETOXYMERCURI-5-NITROGUAIACOL and 4,6-DIACETOXYMERCURI-5-NITROGUAIACOL

TOXICITY DATA with REFERENCE
orl-rat LD50:125 mg/kg JAPMA8 38,270,49
ipr-rat LD50:6 mg/kg JAPMA8 38,270,49
ivn-rbt LD50:2600 μg/kg JAPMA8 38,270,49

CONSENSUS REPORTS: Mercury and its compounds are on the Community Right-To-Know List.

OSHA PEL: (Transitional: CL 1 mg/10m^3) CL 0.1 mg(Hg)/m^3 (skin)
ACGIH TLV: TWA 0.1 mg(Hg)/m^3 (skin)
NIOSH REL: TWA 0.05 mg(Hg)/m^3

SAFETY PROFILE: Poison by ingestion, intravenous, and intraperitoneal routes. When heated to decomposition it emits toxic fumes of NO_x and Hg. See also MERCURY COMPOUNDS, ORGANIC.

GKO750 CAS:23256-50-0 ***HR: 3***
GUANABENZ ACETATE
mf: $C_8H_8Cl_2N_4 \cdot C_2H_4O_2$ mw: 291.16

SYNS: BR 750 ◇ ((2,6-DICHLOROBENZYLIDENE)AMINO)GUANIDINE ACETATE ◇ 2-((2,6-DICHLOROPHENYL)METHYLENE) HYDRAZINECARBOXIMIDAMIDE MONOACETATE (9CI) ◇ GUANABENZ ◇ WY-8678 ◇ WY 8678 ACETATE

TOXICITY DATA with REFERENCE
orl-rat TDLo:170 mg/kg (female 6-22D post):REP JTSCDR 7(Suppl 2),123,82
orl-rat TDLo:165 mg/kg (7-17D preg):TER JTSCDR 7(Suppl 2),107,82
orl-chd TDLo:1 mg/kg:CNS,CVS AIMEAS 102,787,85
orl-wmn TDLo:1 mg/kg:CNS,CVS AIMEAS 102,787,85
orl-rat LD50:238 mg/kg YACHDS 10,4571,82
ipr-rat LD50:62 mg/kg YACHDS 10,4571,82
scu-rat LD50:84 mg/kg YACHDS 10,4571,82
orl-mus LD50:260 mg/kg YACHDS 10,4571,82
ipr-mus LD50:75 mg/kg YACHDS 10,4571,82
scu-mus LD50:89 mg/kg YACHDS 10,4571,82

SAFETY PROFILE: Poison by ingestion, subcutaneous, and intraperitoneal routes. Human systemic effects by ingestion: sleep, pulse rate and blood pressure changes. An experimental teratogen. Experimental reproductive effects. When heated to decomposition it emits toxic fumes of Cl^- and NO_x. A hypotensive drug.

GKO800 CAS:32059-15-7 ***HR: 3***
GUANAZODINE
mf: $C_9H_{20}N_4$ mw: 184.33

SYNS: 1-AZACYCLOOCT-2-YL METHYL GUANIDINE ◇ ((OCTAHYDRO-2-AZOCINYL)METHYL)GUANIDINE

TOXICITY DATA with REFERENCE
orl-mus TDLo:150 mg/kg (7-12D preg):REP OYYAA2 15,333,78
orl-mus TDLo:2400 mg/kg (female 7-12D post):TER OYYAA2 15,333,78
ipr-mus LD50:136 mg/kg DRFUD4 2,592,77

SAFETY PROFILE: Poison by intraperitoneal route. An experimental teratogen. Experimental reproductive effects. When heated to decomposition it emits toxic fumes of NO_x.

GKQ000 CAS:55-65-2 ***HR: 3***
GUANETHIDINE
mf: $C_{10}H_{22}N_4$ mw: 198.36

SYNS: 1-(2-GUANIDINOETHYL)HEPTAMETHYLENIMINE ◇ 1-(2-GUANIDINOETHYL)OCTAHYDROAZOCINE ◇ ISMELIN ◇ OCTATENSIN ◇ OCTATENZINE ◇ SU 5864

TOXICITY DATA with REFERENCE
orl-mus LD50:845 mg/kg USXXAM #3856778
ipr-mus LD50:100 mg/kg DRFUD4 4,185,79
scu-mus LD50:224 mg/kg USXXAM #3856778
ivn-mus LD50:28 mg/kg PCJOAU 15,349,81
ivn-cat LDLo:50 mg/kg 27ZIAQ -,112,73
ivn-rbt LDLo:50 mg/kg 27ZIAQ -,112,73

SAFETY PROFILE: Poison by intraperitoneal, intravenous, and subcutaneous routes. Moderately toxic by ingestion. When heated to decomposition it emits toxic fumes of NO_x.

GKS000 CAS:60-02-6 ***HR: 3***
GUANETHIDINE BISULFATE
mf: $C_{20}H_{44}N_8 \cdot H_2O_4S$ mw: 494.80

SYNS: β-1-AZACYCLOOCTYLETHYLGUANIDINE SULFATE ◇ GUANETHIDINE SULFATE ◇ 1-(2-GUANIDINOETHYL)OCTAHYDROAZOCINE SULFATE (2:1) ◇ (2-(HEXAHYDRO-1(2H)-AZOCINYL)ETHYL)GUANIDINE SULFATE ◇ ISMELIN ◇ ISMELIN SULFATE ◇ ISOMELIN ◇ NSC-29863 ◇ (2-(OCTAHYDRO-1-AZOCINYL)ETHYL)GUANIDINE SULFATE ◇ SU-5864

TOXICITY DATA with REFERENCE
unr-man TDLo:391 mg/kg (78W male):REP JSXRAJ 3,69,67
orl-rat LD50:1000 mg/kg JPETAB 128,22,60
ivn-rat LD50:23 mg/kg JPETAB 128,22,60
ipr-mus LD50:137 mg/kg RPTOAN 32,11,69
scu-mus LD50:750 mg/kg YKKZAJ 83,629,63
ivn-mus LD50:18 mg/kg CSLNX* NX#08547

SAFETY PROFILE: Poison by intravenous and intraperitoneal route. Moderately toxic by ingestion and subcutaneous routes. Human reproductive effects by unspecified route. Experimental reproductive effects. Used to treat hypertension. When heated to decomposition it emits very toxic fumes of SO_x and NO_x.

GKU000 CAS:645-43-2 ***HR: 3***
GUANETHIDINE MONOSULFATE
mf: $C_{10}H_{22}N_4 \cdot H_2O_4S$ mw: 296.44

SYNS: N-(2-GUANIDINO ETHYL)HEPTAMETHYLENIMINE SULFATE ◇ (2-(HEXAHYDRO-1(2H)-AZOCINYL)ETHYL) GUANIDINE HYDROGEN SULFATE ◇ 2-(OCTAHYDRO-1-AZOCINYL)ETHYL GUANIDINE SULPHATE

TOXICITY DATA with REFERENCE
eye-hmn 420 mg/6W-I MLD BMJOAE 4,592,67
ipr-rat LD50:290 mg/kg TXAPA9 24,37,73
orl-mus LD50:1450 mg/kg BCFAAI 111,353,72
ipr-mus LD50:180 mg/kg BCFAAI 111,353,72
ivn-mus LD50:18500 μg/kg BCFAAI 111,353,72

SAFETY PROFILE: Poison by intraperitoneal and intravenous routes. Moderately toxic by ingestion. A human eye irritant. When heated to decomposition it emits very toxic fumes of SO_x and NO_x.

GKU300 CAS:29110-48-3 ***HR: 3***
GUANFACINE HYDROCHLORIDE
mf: $C_9H_9Cl_2N_3O \cdot ClH$ mw: 282.57

SYNS: N-AMIDINO-2-(2,6-DICHLOROPHENYL)ACETAMIDEHYDROCHLORIDE ◇ N-(AMINOIMINOMETHYL)-2,6-DICHLOROBENZENEACETAMIDE HYDROCHLORIDE ◇ BS100-141 ◇ ESTULIC ◇ LON 798

TOXICITY DATA with REFERENCE
orl-mus TDLo:10 mg/kg (female 6-15D post):REP JZKEDZ 6,105,80
orl-mus TDLo:20 mg/kg (female 6-15D post):TER JZKEDZ 6,105,80
orl-rat LD50:145 mg/kg IYKEDH 16,357,85
scu-rat LD50:114 mg/kg KSRNAM 14,4511,80
ivn-rat LD50:5800 μg/kg KSRNAM 14,4511,80
orl-mus LD50:15300 μg/kg IYKEDH 16,357,85
scu-mus LD50:46 mg/kg KSRNAM 14,4511,80
ivn-mus LD50:25 mg/kg KSRNAM 14,4511,80

SAFETY PROFILE: Poison by ingestion, subcutaneous and intravenous routes. Experimental teratogenic and reproductive effects. When heated to decomposition it emits toxic fumes of NO_x and HCl.

GKW000 CAS:113-00-8 ***HR: 3***
GUANIDINE
mf: CH_5N_3 mw: 59.09

PROP: Hygroscopic, colorless crystals. Mp: 50°. Very sol in water; sol in alc.

SYNS: AMINOFORMAMIDINE ◇ AMINOMETHANAMIDINE ◇ CARBAMAMIDINE ◇ CARBAMIDINE ◇ IMINOUREA

TOXICITY DATA with REFERENCE
ipr-rat LDLo:22500 μg/kg AEXPBL 90,129,21
scu-rat LDLo:150 mg/kg HBAMAK 4,1352,35
scu-mus LDLo:266 mg/kg HBAMAK 4,1352,35
ipr-cat LDLo:10 mg/kg AEXPBL 90,129,21
orl-rbt LDLo:500 mg/kg MEIEDD 10,657,83
scu-frg LDLo:3000 mg/kg HBAMAK 4,1352,35

SAFETY PROFILE: Poison by intraperitoneal and subcutaneous routes. Moderately toxic by ingestion. On heating to 160° it converts to melamine and NH_3. Keep well closed. When heated to decomposition it emits toxic fumes of NO_x.

GKY000 CAS:50-01-1 ***HR: 3***
GUANIDINE MONOHYDROCHLORIDE
mf: $CH_5N_3 \cdot HCl$ mw: 72.11

PROP: White powder. Mp: 183°. Freely sol in water, alc.

SYNS: GUANIDINIUM CHLORIDE ◇ USAF EK-749

TOXICITY DATA with REFERENCE
mmo-smc 4 mmol/L BBRCA9 53,531,73
scu-rat LDLo:404 mg/kg JPETAB 28,251,26
ipr-mus LD50:500 mg/kg NTIS** AD277-689
scu-dog LDLo:200 mg/kg HBAMAK 4,1352,35
scu-gpg LDLo:100 mg/kg HBAMAK 4,1352,35

CONSENSUS REPORTS: Reported in EPA TSCA Inventory.

SAFETY PROFILE: Poison by ingestion and subcutaneous routes. Moderately toxic by intraperitoneal route. Mutation data reported. Can cause nausea, diarrhea, and neurological disturbances. When heated to decomposition it emits highly toxic fumes of HCl and NO_x.

GLA000 CAS:506-93-4 ***HR: 3***
GUANIDINE MONONITRATE
DOT: UN 1467
mf: $CH_5N_3 \cdot HNO_3$ mw: 122.11

PROP: White granules. Mp: 214°.

SYN: GUANIDINE NITRATE (DOT)

CONSENSUS REPORTS: Reported in EPA TSCA Inventory.

DOT Classification: Oxidizer; Label: Oxidizer.

SAFETY PROFILE: A powerful oxidizer. Flammable when shocked or exposed to heat or flame. A stable, flashless, non-hygroscopic high explosive used as a blasting explosive in combination with charcoal and inorganic nitrates. Keep away from heat and open flame. When heated to decomposition it emits very toxic fumes of HNO_3 and NO_x. See also NITRATES, GUANIDINE MONOHYDROCHLORIDE, and EXPLOSIVES, HIGH.

GLB100 CAS:27698-99-3 ***HR: 3***
GUANIDINIUM DICHROMATE
mf: $C_2H_{12}Cr_2N_6O_7$ mw: 336.14

CONSENSUS REPORTS: Chromium and its compounds are on the Community Right-To-Know List.

SAFETY PROFILE: Explodes violently when heated in a sealed container. When heated to decomposition it emits toxic fumes of NO_x. See also CHROMIUM COMPOUNDS.

GLB300 CAS:52470-25-4 ***HR: 3***
GUANIDINIUM NITRATE
mf: $CH_6N_4O_3$ mw: 122.08

SAFETY PROFILE: Decomposes explosively when heated. When heated to decomposition it emits toxic fumes of NO_x. See also NITRATES.

GLC000 CAS:10308-84-6 ***HR: 3***
GUANIDINIUM PERCHLORATE
mf: $CH_6ClN_3O_4$ mw: 159.53

SAFETY PROFILE: A very sensitive, powerful and unstable explosive. Decomposes violently at 350°C. Mixtures with 10% iron are more thermally sensitive. Upon decomposition it emits toxic fumes of Cl^- and NO_x. See also PERCHLORATES.

GLC100 ***HR: 2***
p-GUANIDINOBENZOIC ACID 4-METHYL-2-OXO-2H-1-BENZOPYRAN-7-YL ESTER
mf: $C_{18}H_{15}N_3O_4$ mw: 337.36

SYNS: BENZOIC ACID, p-GUANIDINO-, 4-METHYL-2-OXO-2H-1-BENZOPYRAN-7-YL ESTER ◇ COUMARIN, 4-METHYL-7-HYDROXY-, p-GUANIDINOBENZOATE ◇ 4-METHYLUMBELLIFERONE 4-GUANIDINOBENZOATE ◇ MUGB

TOXICITY DATA with REFERENCE
imp-mus TDLo:2 mg/kg (female 2D pre):REP CCPTAY 26,137,82
ipr-mus LD50:600 mg/kg CCPTAY 26,137,82

SAFETY PROFILE: Moderately toxic by intraperitoneal route. Experimental reproductive effects. When heated to decomposition it emits toxic fumes of NO_x.

GLI000 CAS:73-40-5 ***HR: 3***
GUANINE
mf: $C_5H_5N_5O$ mw: 151.15

PROP: Usually amorph. Decomp: >360° with partial subl. Very sol in ammonia water; aq KOH solns, dil acids; very sltly sol in alc, ether; insol in water.

SYNS: 2-AMINOHYPOXANTHINE ◇ MEARLMAID

TOXICITY DATA with REFERENCE
sln-hmn:lym 30 μmol/L MUTAEX 1,99,86
cyt-mus-ipr 15 ng/kg NULSAK 19,40,76
scu-rat TDLo:1300 mg/kg/26W-I:ETA,REP CNREA8 27,925,67

CONSENSUS REPORTS: Reported in EPA TSCA Inventory. EPA Genetic Toxicology Program.

SAFETY PROFILE: Experimental reproductive effects.

Questionable carcinogen with experimental tumorigenic data. Human mutation data reported. When heated to decomposition it emits toxic fumes of NO_x.

GLK000 CAS:635-39-2 ***HR: 3***
GUANINE HYDROCHLORIDE
mf: $C_5H_5O \cdot ClH$ mw: 187.61

PROP: Crystalline powder. Practically insol in water, alc, ether; sol in acidulated H_2O.

SYN: USAF S-1

TOXICITY DATA with REFERENCE
ipr-mus LD50:200 mg/kg NTIS** AD277-689

CONSENSUS REPORTS: Reported in EPA TSCA Inventory.

SAFETY PROFILE: Poison by intraperitoneal route. When heated to decomposition it emits very toxic fumes of HCl.

GLK100 CAS:18905-29-8 ***HR: 3***
GUANINE-3-N-OXIDE
mf: $C_5H_5N_5O_2$ mw: 167.15

TOXICITY DATA with REFERENCE
ipr-rat TDLo:75 mg/kg (female 11D post):TER ARPAAQ 86,395,68
ipr-rat LD50:75 mg/kg ARPAAQ 86,395,68

SAFETY PROFILE: Poison by intraperitoneal route. An experimental teratogen. When heated to decomposition it emits toxic fumes of NO_x.

GLM000 CAS:5227-68-9 ***HR: 3***
GUANINE-7-N-OXIDE
mf: $C_5H_5N_5O_2$ mw: 167.15
SYN: 7-HYDROXYGUANINE

TOXICITY DATA with REFERENCE
scu-rat TDLo:390 mg/kg/26W-I:NEO CNREA8 27,925,67
ipr-rat LD50:95 mg/kg ADTEAS 3,181,68
ipr-mus LD50:40 mg/kg JANTAJ 38,972,85

SAFETY PROFILE: Poison by intraperitoneal route. Questionable carcinogen with experimental neoplastigenic data. When heated to decomposition it emits very toxic fumes of NO_x.

GLO000 CAS:19039-44-2 ***HR: 3***
GUANINE-3-N-OXIDE HEMIHYDROCHLORIDE
mf: $C_5H_5O_2 \cdot 1/2ClH$ mw: 115.33

TOXICITY DATA with REFERENCE
scu-rat TDLo:1040 mg/kg/24W-I:NEO CNREA8 30,184,70

SAFETY PROFILE: Questionable carcinogen with experimental neoplastigenic data. When heated to decomposition it emits toxic fumes of Cl^-.

GLQ000 CAS:39202-39-6 ***HR: 3***
GUANOCTINE
mf: $C_{18}H_{41}N_7 \cdot H_2O_4S$ mw: 453.74

SYNS: N,N′-DIAMIDINO-9-AZA-1,17-HEPTADECANEDIAMINEHYDROGEN SULFATE ◇ DI-(8-GUANIDINO-OCTYL)AMINE SULFATE ◇ (IMINOBIS(OCTAMETHYLENE))-DIGUANIDINE SULFATE

TOXICITY DATA with REFERENCE
orl-rat LD50:227 mg/kg FMCHA2 -,D230,80
skn-rbt LD50:1176 μg/kg FMCHA2 -,D230,80

SAFETY PROFILE: Poison by ingestion and skin contact. When heated to decomposition it emits very toxic fumes of SO_x and NO_x. See also SULFATES.

GLS000 CAS:118-00-3 ***HR: 3***
GUANOSINE
mf: $C_{10}H_{13}N_5O_5$ mw: 283.28

SYNS: 2-AMINOINOSINE ◇ GR ◇ GUANINE RIBOSIDE ◇ 2(3H)-IMINO-9-β-d-RIBOFURANOSYL-9H-PURIN-6(1H)-ONE◇ 9-β-d-RIBOFURANOSYLGUANINE ◇ USAF CB-11 ◇ VERNINE

TOXICITY DATA with REFERENCE
pic-esc 1 g/L ZAPOAK 12,583,72
oms-hmn:oth 1 mmol/L JIDEAE 65,52,75
ipr-mus TDLo:750 mg/kg (13D preg):TER JJPAAZ 22,201,72
ipr-mus LD50:500 mg/kg NTIS** AD277-689
ivn-mus LD50:180 mg/kg CSLNX* NX#03206

CONSENSUS REPORTS: Reported in EPA TSCA Inventory.

SAFETY PROFILE: Poison by intravenous route. Moderately toxic by intraperitoneal route. Experimental teratogenic effects. Human mutation data reported. When heated to decomposition it emits toxic fumes of NO_x.

GLS700 CAS:1021-11-0 ***HR: 3***
GUANOXYFEN SULFATE
mf: $C_{20}H_{30}N_6O_2 \cdot H_2O_4S$ mw: 484.61

SYNS: C.I. 515 ◇ EA 166 ◇ (3-PHENOXYPROPYL)GUANIDINE SULFATE

TOXICITY DATA with REFERENCE
orl-mus LD50:402 mg/kg JPETAB 143,374,64
ipr-mus LD50:123 mg/kg JPETAB 143,374,64
ivn-mus LD50:17 mg/kg JPETAB 143,374,64
ims-mus LD50:122 mg/kg JPETAB 143,374,64

SAFETY PROFILE: Poison by intravenous, intramuscular, and intraperitoneal routes. Moderately toxic by ingestion. See also SULFATES.

GLS800 CAS:5550-12-9 ***HR: 2***
GUANYLIC ACID SODIUM SALT
mf: $C_{10}H_{14}N_5O_8P \cdot 2Na$ mw: 409.24

PROP: Colorless to white crystals; characteristic taste. Sol in water; sltly sol in alc; insol in ether.

SYNS: DISODIUM GMP ◇ DISODIUM-5′-GMP ◇ DISODIUM GUANYLATE (FCC) ◇ DISODIUM-5′-GUANYLATE ◇ GMP DISODIUM SALT ◇ 5′-GMP DISODIUM SALT ◇ GMP SODIUM SALT ◇ SODIUM GMP ◇ SODIUM GUANOSINE-5′-MONOPHOSPHATE ◇ SODIUM GUANYLATE ◇ SODIUM-5′-GUANYLATE

TOXICITY DATA with REFERENCE
cyt-ham:fbr 1 g/L FCTOD7 22,623,84
ipr-rat LD50:3880 mg/kg AJINO* -,-,73
scu-rat LD50:3400 mg/kg AJINO* -,-,73
ivn-rat LD50:2720 mg/kg AJINO* -,-,73
orl-mus LD50:15 g/kg AJINO* -,-,73
ipr-mus LD50:5010 mg/kg AJINO* -,-,73
scu-mus LD50:5050 mg/kg AJINO* -,-,73
ivn-mus LD50:3580 mg/kg AJINO* -,-,73

CONSENSUS REPORTS: Reported in EPA TSCA Inventory.

SAFETY PROFILE: Moderately toxic by intraperitoneal, subcutaneous, and intravenous routes. Mildly toxic by ingestion. Mutation data reported. When heated to decomposition it emits toxic fumes of PO_x, NO_x, and Na_2O.

GLU000 CAS:9000-30-0 ***HR: 1***
GUAR GUM

PROP: Yellowish-white powder, dispersible in hot or cold water, obtained from the ground endosperms of *Cyanopsis tetragonoloan* L. Taub (Fam. *Leguminosae*). White powder; odorless. Sol in water; insol in oils, grease, hydrocarbons, ketones, esters.

SYNS: A-20D ◇ BURTONITE V-7-E ◇ CYAMOPSIS GUM ◇ DEALCA TP1 ◇ DECORPA ◇ GALACTASOL ◇ GENDRIV 162 ◇ GUAR ◇ GUAR FLOUR ◇ GUM CYAMOPSIS ◇ GUM GUAR ◇ INDALCA AG ◇ JAGUAR No. 124 ◇ JAGUAR GUM A-20-D ◇ JAGUAR PLUS ◇ LYCOID DR ◇ NCI-C50395 ◇ REGONOL ◇ REIN GUARIN ◇ SUPERCOL U POWDER ◇ SYNGUM D 46D ◇ UNI-GUAR

TOXICITY DATA with REFERENCE
orl-rat LD50:7060 mg/kg FCTXAV 19,287,81
orl-mus LD50:8100 mg/kg FDRLI* 124,-,76
orl-rbt LD50:7000 mg/kg FDRLI* 124,-,76
orl-ham LD50:6000 mg/kg FDRLI* 124,-,76

CONSENSUS REPORTS: NTP Carcinogenesis Bioassay (feed); No Evidence: mouse, rat NTPTR* NTP-TR-229,82. Reported in EPA TSCA Inventory. EPA Genetic Toxicology Program.

SAFETY PROFILE: Mildly toxic by ingestion. When heated to decomposition it emits acrid smoke and irritating fumes.

GLW000 ***HR: 3***
GUAVA

PROP: Material extracted with hot water from the unripe fruits of *P. guajava* (JNCIAM 60,683,78).

SYN: PSIDIUM GUAJAVA

TOXICITY DATA with REFERENCE
scu-rat TDLo:10 g/kg/72W-I:ETA JNCIAM 60,683,78

SAFETY PROFILE: Questionable carcinogen with experimental tumorigenic data. When heated to decomposition it emits acrid smoke and irritating fumes.

GLY000 CAS:9000-28-6 ***HR: 1***
GUM GHATTI

PROP: The gummy exudation from the stem of *Anogeissus latifolia*. Colorless to pale yellow tears; almost odorless. Sltly sol in water.

SYN: INDIAN GUM

TOXICITY DATA with REFERENCE
orl-rat LD50:17 g/kg FDRLI* 124,-,76
orl-rbt LD50:7000 mg/kg FDRLI* 124,-,76

CONSENSUS REPORTS: Reported in EPA TSCA Inventory.

SAFETY PROFILE: Mildly toxic by ingestion. When heated to decomposition it emits acrid smoke and irritating fumes.

GLY100 CAS:9000-29-7 ***HR: 2***
GUM GUAIAC

PROP: From wood of *guajacum officinale* L. or *Guajacum sanctum* L. (Fam. *Zygophyllaceae*). Brown solid; balsamic odor, sltly acrid taste. Sol in alc, ether, chloroform, solns of alkalies; sltly sol in carbon disulfide, benzene.

SYN: GUAIAC GUM

TOXICITY DATA with REFERENCE
orl-gpg LD50:1120 mg/kg AFREAW 3,197,51

SAFETY PROFILE: Moderately toxic by ingestion. When heated to decomposition it emits acrid smoke and irritating fumes.

GMA000 CAS:39300-88-4 ***HR: D***
GUM TARA

SYNS: NCI-C54364 ◇ TARA GUM

CONSENSUS REPORTS: NTP Carcinogenesis Bioassay (feed); No Evidence: mouse, rat NTPTR* NTP-TR-224,82.

SAFETY PROFILE: When heated to decomposition it emits acrid smoke and irritating fumes.

GMC000 CAS:73341-70-5 ***HR: 3***
GUNACIN
mf: $C_{17}H_{16}O_8$ mw: 348.33

TOXICITY DATA with REFERENCE
dni-omi 200 μg/L JANTAJ 32,1104,79
ipr-mus LD50:16 mg/kg JANTAJ 32,1104,79
ivn-mus LD50:12 mg/kg JANTAJ 32,1104,79

SAFETY PROFILE: Poison by intraperitoneal and intravenous routes. Mutation data reported. When heated to decomposition it emits acrid smoke and irritating fumes.

GME000 CAS:8030-55-5 ***HR: 1***
GURJUN BALSAM

PROP: Oleoresin from various species of *Dipterocarpus* (FCTXAV 14,789,76).

SYNS: BALSAM GURJUN ◇ EAST INDIAN COPAIBA ◇ WOOD OIL

TOXICITY DATA with REFERENCE
skn-rbt 500 mg/24H MOD FCTXAV 14,789,76

CONSENSUS REPORTS: Reported in EPA TSCA Inventory.

SAFETY PROFILE: A skin irritant. When heated to decomposition it emits acrid smoke and irritating fumes.

GME300 CAS:11048-92-3 ***HR: 3***
α-2-GUTTIFERIN
mf: $C_{33}H_{38}O_8$ mw: 562.71

SYNS: α-GUTTIFERIN (9CI) ◇ A''2-GUTTIFERIN ◇ Y-GUTTIFERIN

TOXICITY DATA with REFERENCE
ipr-rat LD50:91 mg/kg IJEBA6 5,96,67
scu-rat LD50:279 mg/kg IJEBA6 5,96,67
ivn-rat LD50:105 mg/kg IJEBA6 5,96,67
ipr-mus LD50:83 mg/kg IJEBA6 5,96,67
scu-mus LD50:400 mg/kg 85GDA2 8(1),331,82
ivn-mus LD50:97 mg/kg IJEBA6 5,96,67

SAFETY PROFILE: Poison by subcutaneous, intravenous, and intraperitoneal routes. When heated to decomposition it emits acrid smoke and fumes.

GMG000 CAS:639-14-5 ***HR: 3***
GYPSOGENIN
mf: $C_{30}H_{46}O_4$ mw: 470.76

PROP: Needles or leaflets from methanol. Mp: 274-276°.

SYNS: ALBSAPOGENIN ◇ ASTRANTIAGENIN D ◇ GITHAGENIN ◇ GYPSOPHILASAPOGENIN ◇ GYPSOPHILASAPONIN ◇ 3-β-HYDROXY-23-OXO-OLEAN-12-EN-28-OIC ACID ◇ SAPONIN-GYPSOPHILA

TOXICITY DATA with REFERENCE
orl-mus LDLo:2 g/kg HBAMAK 4,1289,35
scu-mus LDLo:100 mg/kg HBAMAK 4,1289,35
ivn-mus LDLo:15 mg/kg HBAMAK 4,1289,35

SAFETY PROFILE: Poison by subcutaneous and intravenous routes. Moderately toxic by ingestion. When heated to decomposition it emits acrid smoke and irritating fumes.

H

HAA300 CAS:35278-53-6 ***HR: D***
H-1075
mf: $C_{30}H_{35}NO_3 \cdot ClH$ mw: 494.12

SYNS: trans-1-(2-(p-(1,2-BIS(p-METHOXYPHENYL)-1-BUTENYL) PHENOXY)ETHYL)PYRROLIDINE HYDROCHLORIDE ◇ trans-1,2-BIS(p-METHOXYPHENYL)-1-(p-(2-N-PYRROLIDINOETHOXY)PHENYL)BUT-1-ENE HYDROCHLORIDE

TOXICITY DATA with REFERENCE
orl-rat TDLo:15 μg/kg (1-3D preg):REP JRPFA4 34,23,73

SAFETY PROFILE: Experimental reproductive effects. When heated to decomposition it emits toxic fumes of NO_x and HCl.

HAA310 CAS:42824-34-0 ***HR: D***
H-1286
mf: $C_{29}H_{35}NO_2 \cdot C_6H_8O_7$ mw: 621.79

SYN: 2-(p-(2-(p-METHOXYPHENYL)-1-PHENYL-1-BUTENYL) PHENOXY)TRIETHYLAMINE CITRATE

TOXICITY DATA with REFERENCE
orl-mus TDLo:480 μg/kg (female 4-6D post):REP JRPFA4 34,23,73

SAFETY PROFILE: Experimental reproductive effects. When heated to decomposition it emits toxic fumes of NO_x.

HAA320 CAS:104931-87-5 ***HR: 3***
HBK
mf: $C_{22}H_{44}N_6O_{10} \cdot xH_2O_4S$ mw: 1239.28

SYN: d-STREPTAMINE, O-3-AMINO-3-DEOXY-α-d-GLUCOPYRANOSYL-(1-6)-O-(2,6-DIAMINO-2,3,4,6-TETRADEOXY α-d-erythro-HEXOPYRANOSYL-(1-4))-N'-(4-AMINO-2-HYDROXY-1-OXOBUTYL)-2-DEOXY-, (S)-, SULFATE (salt)

TOXICITY DATA with REFERENCE
ivn-rat TDLo:1300 mg/kg (female 17-22D post):REP KSRNAM 20,7963,86
ims-rbt TDLo:780 mg/kg (female 6-18D post):TER KSRNAM 20,7875,86
ivn-rat LD50:150 mg/kg KSRNAM 20,7473,86
ipr-mus LD50:540 mg/kg KSRNAM 20,7473,86
scu-mus LD50:476 mg/kg KSRNAM 20,7473,86
ivn-mus LD50:82300 μg/kg KSRNAM 20,7473,86
ims-mus LD50:372 mg/kg KSRNAM 20,7473,86
ivn-dog LDLo:150 mg/kg KSRNAM 20,7587,86
ims-dog LDLo:400 mg/kg KSRNAM 20,7587,86

SAFETY PROFILE: Poison by intravenous and intramuscular routes. An experimental teratogen. Experimental reproductive effects. When heated to decomposition it emits toxic fumes of NO_x and SO_x

HAA325 ***HR: 3***
HE-HK-52
mf: $C_{18}H_{23}N_2OS \cdot Br$ mw: 395.40

SYNS: 1-((DIPHENYLSULFOXIMIDO)METHYL)-1-METHYLPYRROLIDINIUM BROMIDE ◇ N-(N'-METHYLPYRROLIDINIUMMETHYL)-2,2-DIPHENYL SULFOXIMIDE BROMIDE

TOXICITY DATA with REFERENCE
orl-rat LD50:381 mg/kg DRFUD4 7,478,82
ivn-rat LD50:9 mg/kg DRFUD4 7,478,82
orl-mus LD50:331 mg/kg DRFUD4 7,478,82
ivn-mus LD50:8 mg/kg DRFUD4 7,478,82

SAFETY PROFILE: Poison by ingestion and intravenous routes. When heated to decomposition it emits toxic fumes of Br^-, NO_x, and SO_x.

HAA340 ***HR: 3***
HGG-12
mf: $C_{20}H_{19}N_3O_3 \cdot 2I \cdot H_20$ mw: 621.24

SYN: 3'-BENZOYL-2-FORMYL-1,1'-(OXYDIMETHYLENE)DIPYRIDINIUM DIIODIDE DIHYDRATE

TOXICITY DATA with REFERENCE
ims-rat LD50:1179 mg/kg FAATDF 4(2, Pt 2),S106,84
ipr-mus LD50:136 mg/kg FAATDF 1,193,81
ims-dog LD50:60 mg/kg FAATDF 4(2, Pt 2),S106,84
ims-gpg LD50:281 mg/kg FAATDF 4(2, Pt 2),S106,84

SAFETY PROFILE: Poison by intramuscular and intraperitoneal routes. When heated to decomposition it emits toxic fumes of NO_x and I^-. See also IODIDES.

HAA345 CAS:34433-31-3 ***HR: 3***
HJ 6
mf: $C_{14}H_{16}N_4O_3 \cdot 2Cl$ mw: 359.24

SYNS: (((4-(AMINOCARBONYL)PYRIDINO)METHOXY)-2-((HYDROXYIMINO)METHYL)PYRIDINIUM DICHLORIDE ◇ 1-(((4-(AMINOCARBONYL)PYRIDINIO)METHOXY)METHYL)-2-((HYDROXYIMINO)METHYL) PYRIDINIUM 2Cl ◇ 4'-CARBAMOYL-2-FORMYL-1,1'-(OXYDIMETHYLENE)DI-PYRIDINIUM-DICHLORIDE-2-OXIME ◇ HI-6

TOXICITY DATA with REFERENCE
ims-rat LD50:819 mg/kg FAATDF 4(2, Pt 2),S106,84
ipr-mus LD50:514 mg/kg FAATDF 1,193,81

ivn-mus LD50:168 mg/kg FAATDF 2,88,82
ims-mus LD50:451 mg/kg ATXKA8 26,293,70
ims-dog LD50:333 mg/kg FAATDF 4(2, Pt 2),S106,84
ims-gpg LD50:476 mg/kg FAATDF 4(2, Pt 2),S106,84

SAFETY PROFILE: Poison by intramuscular and intravenous routes. Moderately toxic by intraperitoneal routes. When heated to decomposition it emits toxic fumes of Cl^- and NO_x.

HAA355 CAS:13104-70-6 ***HR: 2***
HP 1325
mf: $C_{10}H_{18}N_4O_2 \cdot 2ClH$ mw: 299.24

SYN: 1,1'-(p-PHENYLENEBIS(OXYETHYLENE))DIHYDRAZINEDI-HYDROCHLORIDE

TOXICITY DATA with REFERENCE
scu-mus TDLo:6 mg/kg (1-3D preg):REP JOENAK 27,147,63
orl-mus LD50:1000 mg/kg JOENAK 27,147,63
ipr-mus LD50:983 mg/kg 27ZQAG -,394,72

SAFETY PROFILE: Moderately toxic by ingestion and intraperitoneal routes. Experimental reproductive effects. When heated to decomposition it emits toxic fumes of NO_x and HCl. See also HYDRAZINE.

HAA360 CAS:20866-13-1 ***HR: 3***
HQ-275
mf: $C_{21}H_{23}N_3O_3 \cdot ClH$ mw: 401.93

SYN: 2,3-DIHYDRO-2-METHYL-1-(MORPHOLINOACETYL)-3-PHENYL-4(1H)-QUINAZOLINONE HYDROCHLORIDE

TOXICITY DATA with REFERENCE
orl-rat LD50:1708 mg/kg JJPAAZ 22,235,72
ipr-rat LD50:534 mg/kg JJPAAZ 22,235,72
scu-rat LD50:587 mg/kg JJPAAZ 22,235,72
orl-mus LD50:700 mg/kg JJPAAZ 22,235,72
scu-mus LD50:409 mg/kg JJPAAZ 22,235,72
ivn-mus LD50:387 mg/kg JJPAAZ 22,235,72

SAFETY PROFILE: Poison by intravenous route. Moderately toxic by ingestion, intraperitoneal, and subcutaneous routes. When heated to decomposition it emits toxic fumes of NO_x and HCl.

HAA370 CAS:25487-36-9 ***HR: 3***
HS 3
mf: $C_{14}H_{16}N_4O_3 \cdot 2Cl$ mw: 359.24

SYN: 2,4'-DIFORMYL-1,1'-(OXYDIMETHYLENE) DIPYRIDINIUM DICHLORIDE, DIOXIME

TOXICITY DATA with REFERENCE
ipr-rat LD50:149 mg/kg ARTODN 41,301,79
ivn-rat LD50:168 mg/kg ARTODN 41,301,79
ims-mus LD50:100 mg/kg ATXKA8 26,293,70

SAFETY PROFILE: Poison by intramuscular, intravenous, and intraperitoneal routes. When heated to decomposition it emits toxic fumes of Cl^- and NO_x.

HAB600 ***HR: 3***
2,4-HADIYNYLENE CHLOROFORMATE
mf: $C_8H_4Cl_2O_4$ mw: 235.02

$$(ClCO \cdot OCH_2C \equiv C-)_2$$

SAFETY PROFILE: Potentially explosive at 15°C/0.2 mbar. Upon decomposition it emits toxic fumes of Cl^-. See also ACETYLENE COMPOUNDS and CHLORIDES.

HAC000 CAS:7440-58-6 ***HR: 3***
HAFNIUM
af: Hf aw: 178.49
DOT: UN 1326/UN 2545

PROP: A silvery, ductile, lustrous metal. Mp: 2227°, bp: 4602°, d: 13.31 @ 20°.

SYNS: HAFNIUM METAL, dry (DOT) ◇ HAFNIUM METAL, wet (DOT) ◇ HAFNIUM, wet with not less than 25% water (DOT)

TOXICITY DATA with REFERENCE
unr-mus LD50:76 mg/kg DTLVS* 4,207,80

CONSENSUS REPORTS: Reported in EPA TSCA Inventory.

OSHA PEL: TWA 0.5 mg/m^3
ACGIH TLV: TWA 0.5 mg/m^3
DFG MAK: 0.5 mg/m^3
DOT Classification: Flammable Solid; Label: Flammable Solid (UN1326, UN2545); Flammable Solid; Label: Spontaneously Combustible (UN2545)

SAFETY PROFILE: A poison by an unspecified route. It is poorly soluble in water and thus is not absorbed efficiently by ingestion. Many hafnium compounds are poisons. Dangerous fire hazard. The powder ignites with friction, heat, sparks, or exposure to air. The damp powder burns explosively. The powder may self-explode. The powder can explode when heated with nitrogen, phosphorus, oxygen, sulfur, non-metals, or halogens. May explode on contact with hot nitric acid and other oxidants.

HAC500 CAS:7440-58-6 ***HR: 3***
HAFNIUM (wet)
af: Hf aw: 178.49

SYN: HAFNIUM METAL, WET (DOT)

CONSENSUS REPORTS: Reported in EPA TSCA Inventory.

DOT Classification: Label: Flammable Solid.

SAFETY PROFILE: Flammable in form of dust. See also HAFNIUM.

HAC800 CAS:37230-84-5 ***HR: 3***
HAFNIUM CHLORIDE
mf: Cl_4Hf mw: 320.3

SYN: HAFNIUM TETRACHLORIDE

TOXICITY DATA with REFERENCE
skn-rbt 500 mg MLD 34ZIAG -,296,69
orl-rat LD50:2362 mg/kg 85GMAT -,71,82
ipr-mus LD50:135 mg/kg EQSSDX 1,1,75

SAFETY PROFILE: Poison by intraperitoneal route. Moderately toxic by ingestion. A skin irritant. When heated to decomposition it emits toxic fumes of Cl^-. See also HAFNIUM and CHLORIDES.

HAD000 CAS:14456-34-9 ***HR: 3***
HAFNIUM CHLORIDE OXIDE OCTAHYDRATE
mf: $Cl_2HfO \cdot 8H_2O$ mw: 409.55

PROP: Colorless crystals.

SYN: HAFNIUM OXYCHLORIDE OCTAHYDRATE

TOXICITY DATA with REFERENCE
idr-mus TDLo:800 μg/kg:ETA CNREA8 33,287,73

SAFETY PROFILE: Questionable carcinogen with experimental tumorigenic data. See also HAFNIUM and HAFNIUM OXYCHLORIDE. When heated to decomposition it emits very toxic fumes of Cl^-.

HAD500 CAS:13759-17-6 ***HR: 3***
HAFNIUM OXYCHLORIDE
mf: Cl_2HfO mw: 265.39

SYN: HAFNIUM CHLORIDE OXIDE

TOXICITY DATA with REFERENCE
ipr-mus LD50:112 mg/kg TXAPA9 4,238,62
ipr-rbt LD50:112 mg/kg FEPRA7 19,389,60

CONSENSUS REPORTS: Reported in EPA TSCA Inventory.

SAFETY PROFILE: A poison by intraperitoneal route. See also HAFNIUM. When heated to decomposition it emits very toxic fumes of Cl^-.

HAE000 CAS:25869-93-6 ***HR: 1***
HAFNIUM(IV) TETRAHYDROBORATE
mf: $B_4H_{16}Hf$ mw: 237.85

SAFETY PROFILE: Violent reaction upon exposure to air. See also HAFNIUM and BORON COMPOUNDS.

HAE500 CAS:12116-66-4 ***HR: 3***
HAFNOCENE DICHLORIDE
mf: $C_{10}H_{10}Cl_2Hf$ mw: 379.59

SYNS: DICHLOROBIS(eta-CYCLOPENTADIENYL)HAFNIUM ◇ DICHLORODICYCLOPENTADIENYLHAFNIUM ◇ DICHLORODI-pi-CYCLOPENTADIENYLHAFNIUM ◇ DICYCLOPENTADIENYLHAFNIUM DICHLORIDE ◇ HAFNIUM DICYCLOPENTADIENE DICHLORIDE

TOXICITY DATA with REFERENCE
ivn-mus LD50:100 mg/kg CSLNX* NX#04787

CONSENSUS REPORTS: Reported in EPA TSCA Inventory.

SAFETY PROFILE: A poison by intravenous route. See also HAFNIUM. When heated to decomposition it emits very toxic fumes such as Cl^-.

HAF000 CAS:80-13-7 ***HR: 3***
HALAZONE
mf: $C_7H_5Cl_2NO_4S$ mw: 270.09

PROP: Crystals or white powder; decomp about 195°. Odor of chloroform; sltly sol in water, chloroform; sol in glacial acetic acid and in some solns of alkali, hydroxides, and of alkali carbonates with the formation of a salt.

SYNS: p-CARBOXYBENZENESULFONDICHLOROAMIDE ◇ p-DICHLOROSULFAMOYLBENZOIC ACID ◇ p-(N,N-DICHLOROSULFAMYL)BENZOIC ACID ◇ PANTOCID ◇ PARASULFONDICHLORAMIDO BENZOIC ACID ◇ p-SULFONDICHLORAMIDOBENZOIC ACID

TOXICITY DATA with REFERENCE
mma-sat 6666 μg/plate ENMUDM 8(Suppl 7),1,86
orl-rat LDLo:3500 mg/kg MEIEDD 10,661,83
ivn-rat LDLo:300 mg/kg PHRPA6 59,541,44

CONSENSUS REPORTS: Reported in EPA TSCA Inventory.

SAFETY PROFILE: Poison by intravenous route. Moderately toxic by ingestion. When heated to decomposition it emits very toxic fumes of Cl^-, NO_x, and SO_x.

HAF300 CAS:3093-35-4 ***HR: 3***
HALCIDERM
mf: $C_{24}H_{32}ClFO_5$ mw: 455.01

PROP: Crystals from acetone-petr ether. Mp: 264-265° (decomp). Sol in acetone, chloroform, DMSO; sltly sol in benzene, ethanol, ethyl ether, and methanol; insol in water, HCl, NaOH, and hexanes.

SYNS: 21-CHLORO-9-FLUORO-11-β,16-α-17-TRIHYDROXYPREGN-4-ENE-3,20-DIONE cyclic 16,17-ACETAL with ACETONE ◇ HALCIMAT ◇ HALCINONIDE ◇ HALCORT ◇ HALOG ◇ SQ-18,566

TOXICITY DATA with REFERENCE
dni-mus:oth 1 nmol/L ARZNAD 36,1782,86

scu-rat TDLo:4375 mg/kg (female 35D pre):REP YACHDS 7,1765,79
ipr-rat LD50:39800 μg/kg IYKEDH 13,349,82
scu-rat LD50:95500 μg/kg IYKEDH 13,349,82
ipr-mus LD50:150 mg/kg PBPSDY 1,215,77
scu-mus LD50:409 mg/kg IYKEDH 13,349,82

SAFETY PROFILE: Poison by subcutaneous and intraperitoneal routes. Experimental reproductive effects. Mutation data reported. When heated to decomposition it emits toxic fumes of F^- and Cl^-.

HAF375 CAS:12068-50-7 ***HR: 2***
HALLOYSITE

TOXICITY DATA with REFERENCE
imp-rat TDLo:200 mg/kg:ETA JJIND8 67,965,81

SAFETY PROFILE: Questionable carcinogen with experimental tumorigenic data.

HAF400 CAS:1480-19-9 ***HR: 3***
HALOANISONE
mf: $C_{21}H_{25}FN_2O_2$ mw: 356.48

PROP: Crystals. Mp: 67.5-68.5°. Sol in chloroform; sparingly sol in methanol; sltly sol in ether. Practically insol in water.

SYNS: ANTI-PICA ◇ FLUANISON ◇ FLUANISONE ◇ 4'-FLUORO-4-(4-(o-METHOXYPHENYL)-1-PIPERAZINYL)BUTYROPHENONE ◇ HALONISON ◇ HYPNORM ◇ MD 2028 ◇ 2028 MD ◇ 4-(4-(o-METHOXYPHENYL)-1-PIPERAZINYL)-p-FLUOROBUTYROPHENONE ◇ R 2028 ◇ R 2167 ◇ SEDALANDE ◇ SEDAVIC ◇ SOLUSEDIV 2%

TOXICITY DATA with REFERENCE
orl-rat TDLo:132 mg/kg (15D preg):TER PJPPAA 32,199,80
scu-rat LD50:420 mg/kg MDCHAG 4(2),199,67
orl-mus LD50:550 mg/kg APPHAX 40,159,83
ipr-mus LD50:200 mg/kg FRPSAX 35,605,80

SAFETY PROFILE: Poison by intraperitoneal route. Moderately toxic by ingestion and subcutaneous routes. An experimental teratogen. When heated to decomposition it emits toxic fumes of F^- and NO_x.

HAF500 CAS:36167-63-2 ***HR: 2***
HALOFANTRINE HYDROCHLORIDE
mf: $C_{26}H_{30}Cl_2F_3NO \cdot ClH$ mw: 536.93

SYNS: 1,3-DICHLORO-6-TRIFLUOROMETHYL-9-(3-(DIBUTYLAMINO)-1-HYDROXYPROPYL)PHENANTHRENE HCl ◇ 1-(1,3-DICHLORO-6-TRIFLUOROMETHYL-9-PHENANTHRYL)-3-(DI-N-BUTYLAMINO)PROPANOL HYDROCHLORIDE ◇ HALOFANTRINO (SPANISH) ◇ WR-171,669

TOXICITY DATA with REFERENCE
orl-rat TDLo:240 mg/kg (6D preg):TER TOXID9 4,85,84
orl-rat TDLo:240 mg/kg (6D preg):REP TOXID9 4,85,84
orl-man TDLo:18 mg/kg/D:GIT DRFUD4 5,547,80
orl-rat LD50:3400 mg/kg ACTRAQ 37,232,80
ipr-rat LD50:2050 mg/kg ACTRAQ 37,232,80

SAFETY PROFILE: Moderately toxic by ingestion and intraperitoneal routes. Human systemic effects by ingestion: nausea or vomiting and other gastrointestinal effects. An experimental teratogen. Experimental reproductive effects. When heated to decomposition it emits toxic fumes of F^-, NO_x, and HCl.

HAF825 ***HR: 2***
HALOMICIN

PROP: Produced by *Micromonospora halophytica sp. nov.* NRRL 2998 and *Micromonospora halophytica var. nov.* NIGRA NRRL 3097 (85ERAY 1,274,78).

TOXICITY DATA with REFERENCE
ipr-mus LD50:1259 mg/kg AACHAX -,435,67
scu-mus LD50:5650 mg/kg 85ERAY 1,274,78

SAFETY PROFILE: Moderately toxic by intraperitoneal route. Mildly toxic by subcutaneous route. When heated to decomposition it emits acrid smoke and fumes.

HAG300 CAS:74050-97-8 ***HR: 3***
HALOPERIDOL DECANOATE
mf: $C_{31}H_{41}ClFNO_3$ mw: 530.11

SYNS: DECANOICACID-4-(4-CHLOROPHENYL)-1-(4-(4-FLUOROPHENYL)-4-OXYBUTYL)-4-PIPERIDINYL ESTER ◇ KD-136

TOXICITY DATA with REFERENCE
ims-rat TDLo:30 mg/kg (female 17D post):REP KSRNAM 19,6827,85
ims-rat TDLo:90 mg/kg (7D preg):TER KSRNAM 19,6827,85
orl-rat LD50:1717 mg/kg KSRNAM 19,6731,85
ipr-rat LD50:328 mg/kg KSRNAM 19,6731,85
scu-rat LD50:780 mg/kg KSRNAM 19,6731,85
orl-mus LD50:739 mg/kg KSRNAM 19,6731,85
ipr-mus LD50:288 mg/kg KSRNAM 19,6731,85
scu-mus LD50:1990 mg/kg KSRNAM 19,6731,85
ims-mus LDLo:2550 mg/kg KSRNAM 19,6731,85
ims-dog LDLo:1 g/kg KSRNAM 19,6731,85

SAFETY PROFILE: Poison by intraperitoneal route. Moderately toxic by ingestion, subcutaneous, and intramuscular routes. An experimental teratogen. Other experimental reproductive effects. When heated to decomposition it emits toxic fumes of F^-, Cl^-, and NO_x.

HAG325 CAS:57781-14-3 ***HR: D***
HALOPREDONE ACETATE
mf: $C_{25}H_{29}BrF_2O_7$ mw: 559.45

SYNS: 17,21-BIS(ACETYLOXY)-2-BROMO-6-β,9-DIFLUORO-11-β-HYROXYPREGNA-1,4-DIEN-3,20-DIONE ◇ THS-201 ◇ TOPICON

TOXICITY DATA with REFERENCE
par-rbt TDLo:5200 μg/kg (female 6-18D post):TER YACHDS 14,6853,86
par-rbt TDLo:1300 μg/kg (female 6-18D post):REP YACHDS 14,6853,86

SAFETY PROFILE: An experimental teratogen. Experimental reproductive effects. A steroid. When heated to decomposition it emits toxic fumes of F^- and Br^-.

HAG500 CAS:151-67-7 ***HR: 3***
HALOTHANE
mf: $C_2HBrClF_3$ mw: 197.39

PROP: Nonflammable, highly volatile liquid; characteristic sweetish, not unpleasant odor. D: 1.871 @ 20°/4°, bp: 50.2°, 20° @ 243 mm. Sensitive to light. Misc with petr ether, other fat solvents; sltly sol in water.

SYNS: BROMOCHLOROTRIFLUOROETHANE ◇ 2-BROMO-2-CHLORO-1,1,1-TRIFLUOROETHANE ◇ CHALOTHANE ◇ FLUOROTANE ◇ FLUOTHANE ◇ FTOROTAN (RUSSIAN) ◇ HALOTAN ◇ HALSAN ◇ NARCOTANE ◇ NARCOTANN NE-SPOFA (RUSSIAN) ◇ 1,1,1-TRIFLUORO-2-BROMO-2-CHLOROETHANE ◇ 1,1,1-TRIFLUORO-2-CHLORO-2-BROMOETHANE ◇ 2,2,2-TRIFLUORO-1-CHLORO-1-BROMOETHANE

TOXICITY DATA with REFERENCE
eye-rbt 100 mg SEV FEPRA7 35,729,76
sln-dmg-ihl 1 pph/1H ANESAV 62,305,85
cyt-hmn:lym 1 pph/48H-C NSMZDZ 7,19,79
dns-rat-orl 10 g/kg JTEHD6 10,327,82
ihl-rat TCLo:12500 ppm/2H (3D preg):REP EVHPAZ 21,189,77
ihl-rat TCLo:10 ppm/8H (female 1-22D post):TER ENVRAL 11,40,76
ihl-hmn LCLo:7000 ppm/3H:LIV,GIT,MET ANESAV 24,29,63
ivn-man LDLo:129 mg/kg:CNS,CVS,PUL LANCAO 1,340,82
orl-rat LD50:5680 mg/kg GTPZAB 24(3),36,80
ihl-rat LC50:29000 ppm FATOAO 25,143,62
ihl-mus LC50:22000 ppm/10M JPETAB 86,197,46
orl-gpg LD50:6000 mg/kg GTPZAB 24(3),36,80

CONSENSUS REPORTS: IARC Cancer Review: Animal Inadequate Evidence IMEMDT 7,93,87. EPA Genetic Toxicology Program.

ACGIH TLV: TWA 50 ppm
DFG MAK: 5 ppm (40 mg/m^3); BAT: 250 μg/dL of trifluoroacetic acid in blood at end of shift.
NIOSH REL: (Waste Anesthetic Gases and Vapors) CL 2 ppm/1H

SAFETY PROFILE: A human poison by intravenous route. Human systemic effects by intravenous route: general anesthetic, heart rate change, cyanosis; by inhalation: hepatitis, nausea, fever. An experimental teratogen. Other experimental reproductive effects. A severe eye irritant. Questionable carcinogen. Human mutation data reported. Used as a clinical anesthetic. When heated to decomposition it emits very toxic fumes of F^-, Cl^-, and Br^-. See also CHLORINATED HYDROCARBONS, ALIPHATIC BROMIDES, and FLUORIDES.

HAG800 CAS:59128-97-1 ***HR: 2***
HALOXAZOLAM
mf: $C_{17}H_{14}BrFN_2O_2$ mw: 377.24

PROP: Colorless crystals. Mp: 185°. Sparingly sol in water.

SYNS: 10-BROMO-11b-(2-FLUOROPHENYL)2,3,7,11b-TETRAHYDRO-OXAZOLO(3,2-d)(1,4)BENZODIAZEPIN-6(5H)-ONE ◇ CS-430 ◇ SOMELIN

TOXICITY DATA with REFERENCE
orl-mus TDLo:3600 mg/kg (1-18D preg):TER SKKNAJ 27,64,75
orl-rat LD50:2858 mg/kg SKKNAJ 27,64,75
orl-mus LD50:1413 mg/kg SKKNAJ 27,64,75

SAFETY PROFILE: Moderately toxic by ingestion. When heated to decomposition it emits toxic fumes of F^-, Br^-, and NO_x. An experimental teratogen. A sedative and hypnotic agent. Note: This is a controlled substance (depressant) listed in the U.S. Code of Federal Regulations, Title 21 Part 1308.14 (1985). See also DIAZEPAM.

HAH000 CAS:13382-33-7 ***HR: 3***
HALVISOL
mf: $C_{21}H_{27}FN_2O_2$ mw: 358.50

SYNS: ANISOPIROL ◇ (±)-α-(p-FLUOROPHENYL)-4-(o-METHOXY-PHENYL)-1-PIPERAZINEBUTANOL ◇ dl-1-(4-FLUOROPHENYL)-4-(1-(4-(2-METHOXY-PHENYL))-PIPERAZINYL)BUTANOL

TOXICITY DATA with REFERENCE
scu-rat LD50:250 mg/kg 27ZQAG -,204,72
ivn-rat LD50:17 mg/kg 27ZQAG -,204,72
ivn-mus LD50:17 mg/kg 27ZQAG -,204,72

SAFETY PROFILE: A poison by subcutaneous and intravenous routes. When heated to decomposition it emits very toxic fumes of F^- and NO_x.

HAH800 CAS:1403-71-0 ***HR: 3***
HAMYCIN

PROP: Polyene antibiotic complex produced by *Streptomyces pimprina*. Yellow, amorph powder; decomp @ 160°. Almost insol in water, benzene, chloroform, dry lower aliphatic alcs, ether; sol in basic solvents such as pyridine, collidine, and in aq lower alcs.

SYN: PRIMAMYCIN

TOXICITY DATA with REFERENCE
orl-mus LD50:1200 mg/kg 85GDA2 2,258,80
ipr-mus LD50:8200 μg/kg AACHAX -,737,64
ivn-mus LD50:1120 μg/kg 85GDA2 2,258,80
ivn-dog LDLo:2780 μg/kg IJMRAQ 51,453,63

SAFETY PROFILE: Poison by intravenous and intraperitoneal routes. Moderately toxic by ingestion.

HAI300 CAS:6028-07-5 ***HR: 3***
HARMALOL HYDROCHLORIDE
mf: $C_{12}H_{12}N_2O{\cdot}ClH$ mw: 236.72

SYN: 4,9-DIHYDRO-1-METHYL-3H-PYRIDO(3,4-b)INDOL-7-OL MONOHYDROCHLORIDE

TOXICITY DATA with REFERENCE
ipr-mus LDLo:230 mg/kg QJPPAL 3,218,30
scu-mus LDLo:380 mg/kg QJPPAL 3,218,30
ivn-mus LD50:100 mg/kg CSLNX* NX#03207
scu-rbt LDLo:300 mg/kg QJPPAL 3,218,30
scu-gpg LDLo:300 mg/kg QJPPAL 3,218,30
scu-frg LDLo:250 mg/kg QJPPAL 3,218,30

SAFETY PROFILE: Poison by subcutaneous, intravenous, and intraperitoneal routes. When heated to decomposition it emits toxic fumes of NO_x and HCl.

HAI500 CAS:442-51-3 ***HR: 3***
HARMINE
mf: $C_{13}H_{12}N_2O$ mw: 212.27

PROP: An alkaloid isolated from *Banisteria caapi sp.*, a South American narcotic agent (AEPPAE 129,133,28).

SYNS: BANISTERINE ◇ LEUCOHARMINE ◇ 6-METHOXYHARMAN ◇ 7-METHOXY-1-METHYL-9H-PYRIDO(3,4-b)INDOLE ◇ 1-METHYL-7-METHOXY-β-CARBOLINE ◇ TELEPATHINE ◇ YAGEINE ◇ YAJEINE

TOXICITY DATA with REFERENCE
ims-man TDLo:3 mg/kg:CNS,GIT AEPPAE 129,133,28
scu-mus LD50:243 mg/kg PCJOAU 10,1171,76
ivn-mus LDLo:50 mg/kg AEPPAE 129,133,28
ivn-rat LDLo:10 mg/kg AEPPAE 129,133,28
scu-rbt LDLo:200 mg/kg AEPPAE 129,133,28
ivn-rbt LDLo:60 mg/kg NEPHBW 10,15,71
scu-gpg LDLo:100 mg/kg AEPPAE 129,133,28
scu-frg LDLo:300 mg/kg QJPPAL 9,37,36

SAFETY PROFILE: Poison by intravenous and subcutaneous routes. Human systemic effects by intramuscular route: sleep disturbance, tremors, nausea. When heated to decomposition it emits toxic fumes of NO_x.

HAJ500 CAS:34465-46-8 ***HR: 3***
HCDD
mf: $C_{12}H_2Cl_6O_2$ mw: 390.84

PROP: Colorless solid. Mp: 239°.

SYNS: HEXACHLORODIBENZO-p-DIOXIN ◇ 1,2,3,6,7,8-HEXACHLORODIBENZO-p-DIOXIN

TOXICITY DATA with REFERENCE
eye-rbt 2 mg MOD EVHPAZ 5,87,73
orl-rat TDLo:100 μg/kg (female 6-15D post):TER ADCSAJ 120,55,73
orl-rat TDLo:100 μg/kg (female 6-15D post):REP ADCSAJ 120,55,73
orl-rat LDLo:100 mg/kg ADCSAJ 120,55,73

CONSENSUS REPORTS: IARC Cancer Review: Animal Inadequate Evidence IMEMDT 15,41,77.

SAFETY PROFILE: A deadly poison by ingestion. An experimental teratogen. An eye irritant. When heated to decomposition it emits toxic fumes of Cl^-.

HAJ700 CAS:30007-39-7 ***HR: 3***
HEAT
mf: $C_{19}H_{21}NO_2{\cdot}ClH$ mw: 331.87

SYNS: 3,4-DIHYDRO-2-(((p-HYDROXYPHENETHYL)AMINO)METHYL)-1(2H)-NAPHTHALENONE HYDROCHLORIDE ◇ 3,4-DIHYDRO-2-(((2-(4-HYDROXYPHENYL)ETHYL)AMINO)METHYL)-1(2H)NAPHTHALENONE HYDROCHLORIDE ◇ 2-(β-(HYDROXYPHENYL)ETHYLAMINOMETHYL)TETRALONEHYDROCHLORIDE

TOXICITY DATA with REFERENCE
orl-rat LD50:650 mg/kg DRFUD4 7,231,82
ivn-rat LD50:22500 μg/kg DRFUD4 7,231,82
orl-mus LD50:730 mg/kg DRFUD4 7,231,82
ivn-mus LD50:33 mg/kg DRFUD4 7,231,82

SAFETY PROFILE: Poison by intravenous route. Moderately toxic by ingestion. When heated to decomposition it emits toxic fumes of NO_x and HCl.

HAJ750 ***HR: 1***
#6 HEAVY FUEL OILS

PROP: #6 Fuel oil, API gravity 5.2/1.2% S (52MLA2 1,1,83).

TOXICITY DATA with REFERENCE
skn-rbt 500 mg/24H MLD 52MLA2 1,1,83
eye-rbt 100 mg/30S MLD 52MLA2 1,1,83
orl-rat LD50:4700 mg/kg 52MLA2 1,1,83
skn-rbt LDLo:5 g/kg 52MLA2 1,1,83

SAFETY PROFILE: Mildly toxic by ingestion and skin contact. A mild eye and skin irritant. Flammable when exposed to heat or flame. When heated to decomposition it emits toxic fumes.

HAK000 CAS:7789-20-0 ***HR: D***
HEAVY WATER
mf: D_2O mw: 20.02

PROP: Mp: 3.81°, triple point temp 3.82°, bp: 101.42°.

Critical temp 371.5°, d: 1.1044. Heat is evolved on mixing with normal water.

SYNS: DEUTERIUM OXIDE ◇ DIDEUTERIUM OXIDE ◇ HEAVY WATER-d2 ◇ WATER-d2 (9CI) ◇ WATER²-H2

TOXICITY DATA with REFERENCE
orl-mus TDLo:3360 g/kg (female 56D pre):REP SCIEAS 127,1445,58

CONSENSUS REPORTS: EPA Genetic Toxicology Program. Reported in EPA TSCA Inventory.

SAFETY PROFILE: Experimental reproductive effects. See also WATER.

HAK300 CAS:6754-13-8 ***HR: 3***
HELENALIN
mf: $C_{15}H_{18}O_4$ mw: 262.33

PROP: Bitter, sternutative crystals from benzene. Mp: 167-168°. Sltly sol in water; sol in alc, chloroform, hot benzene.

TOXICITY DATA with REFERENCE
dnr-bcs 1 mg/plate RCOCB8 34,161,81
dni-mus-unr 15 mg/kg/3D JMCMAR 20,333,77
dni-mus:ast 2143 µmol/L JPMSAE 67,1235,78
cyt-ham:ovr 780 ppb BSECBU 13,365,85
orl-mus LD50:150 mg/kg MEIEDD 10,668,83
ipr-mus LD50:10 mg/kg RCOCB8 28,189,80

CONSENSUS REPORTS: EPA Genetic Toxicology Program.

SAFETY PROFILE: Poison by ingestion and intraperitoneal routes. Mutation data reported. Human Toxicity (Merck): Intensely poisonous, capable of causing paralysis of voluntary and cardiac musculature and fatal gastroenteritis. When heated to decomposition it emits toxic fumes.

HAK500 CAS:8023-95-8 ***HR: 1***
HELICHRYSUM OIL

PROP: The chief constituents are several diketones which possess powerful odors (FCTXAV 16,637,78).

SYNS: EVERLASTING FLOWER OIL ◇ IMMORTELLE

TOXICITY DATA with REFERENCE
skn-rbt 500 mg/24H MLD FCTXAV 16,637,78

CONSENSUS REPORTS: Reported in EPA TSCA Inventory.

SAFETY PROFILE: Mild skin irritant. When heated to decomposition it emits acrid smoke. See also KETONES.

HAL000 CAS:20004-62-0 ***HR: 3***
HELIOMYCIN
mf: $C_{22}H_{16}O_6$ mw: 376.38

PROP: Yellow needles from dioxane. Decomp 315°, sublimes at 200-205° (0.001 mm). Stable to hot conc H_2SO_4 or hot KOH. Weakly acid. Slt solubility in water; fair in ether, benzene, alc, acetone, acetic acid.

SYNS: A 3733A ◇ ANTIBIOTIC A 3733A ◇ CROCEOMYCIN ◇ GELIOMYCIN ◇ ITAMYCIN ◇ RESISTOMYCIN ◇ X-340

TOXICITY DATA with REFERENCE
imp-rat TDLo:20750 µg/kg:CAR JJIND8 71,539,83
orl-mus LD50:2000 mg/kg 85GDA2 3,354,80
ipr-mus LD50:20 mg/kg 85GDA2 3,351,80

SAFETY PROFILE: Poison by intraperitoneal route. Moderately toxic by ingestion. Questionable carcinogen with experimental carcinogenic data. When heated to decomposition it emits toxic fumes of NO_x.

HAL500 CAS:303-33-3 ***HR: 3***
HELIOTRINE
mf: $C_{16}H_{27}NO_5$ mw: 313.44

SYN: HELIOTRON

TOXICITY DATA with REFERENCE
mma-sat 1 mg/plate MUREAV 68,211,79
dni-hmn:lvr 25 mg/L IJEVAW 1,107,71
oms-hmn:lvr 10 mg/L IJEVAW 1,107,71
sce-ham:lng 1200 µg/L MUREAV 142,209,85
cyt-mam:leu 50 µmol/L AJBSAM 21,469,68
ipr-rat TDLo:100 mg/kg (9D preg):TER BJEPA5 42,369,61
ipr-rat TDLo:50 mg/kg (10D preg):REP BJEPA5 42,369,61
orl-rat TDLo:460 mg/kg/6D-I:ETA CNREA8 35,2020,75
orl-rat LDLo:50 mg/kg NATUAS 179,361,57
ipr-rat LD50:296 mg/kg JPBAA7 75,17,58
ivn-rat LD50:274 mg/kg AMPLAO 64,152,57
ivn-mus LD50:251 mg/kg AMPLAO 64,152,57
ipr-mus LD50:296 mg/kg JPBAA7 75,17,58
ipr-mam LDLo:250 mg/kg PAREAQ 22,429,70

CONSENSUS REPORTS: EPA Genetic Toxicology Program.

SAFETY PROFILE: A poison by ingestion, intravenous, and intraperitoneal routes. Experimental teratogenic and reproductive effects. Questionable carcinogen with experimental tumorigenic data. Human mutation data reported. When heated to decomposition it emits toxic fumes of NO_x.

HAM000 ***HR: 3***
HELIOTROPIUM SUPINUM L.

PROP: Crude alkaloidal fraction (CNREA8 30,2127, 70).

TOXICITY DATA with REFERENCE
orl-rat TDLo:300 mg/kg:ETA CNREA8 30,2127,70

SAFETY PROFILE: Questionable carcinogen with experimental tumorigenic data. When heated to decomposition it emits toxic fumes.

HAM500 CAS:7440-59-7 ***HR: 1***
HELIUM
DOT: UN 1046/UN 1963
af: He aw: 4.00

PROP: Colorless, odorless, tasteless, inert gas. Mp: −272.2° @ 26 atm, bp: −268.9°, d: (gas): 0.1785 g/L @ 0°, d: (liquid): 0.147 @ −270.8°.

SYNS: HELIUM, compressed (DOT) ◇ HELIUM, refrigerated liquid (DOT)

CONSENSUS REPORTS: Reported in EPA TSCA Inventory.

DOT Classification: Nonflammable Gas; Label: Nonflammable Gas.

SAFETY PROFILE: A simple asphyxiant. Nonflammable Gas. See ARGON for a description of simple asphyxiants.

HAN000 CAS:58933-55-4 ***HR: 1***
HELIUM-OXYGEN (mixture)
DOT: NA 1980
mf: HeO_2 mw: 36.00

SYN: HELIUM-OXYGEN mixture (DOT)

DOT Classification: Nonflammable Gas; Label: Nonflammable Gas.

SAFETY PROFILE: A simple asphyxiant. See also HELIUM and OXYGEN.

HAN500 CAS:1399-70-8 ***HR: 3***
HELLEBOREIN

PROP: Glucoside crystallizable in yellow prisms. Mp: 270°.

TOXICITY DATA with REFERENCE
ivn-cat LDLo:1900 μg/kg 27ZWAY E.1,78,-
scu-rbt LDLo:9 mg/kg HBAMAK 4,1289,35

SAFETY PROFILE: Poison by subcutaneous and intravenous routes. Irritating to skin, eyes, and mucous membranes. Combustible when exposed to heat or flame. When heated to decomposition it emits toxic fumes.

HAN600 CAS:13289-18-4 ***HR: 3***
HELLEGRIGENIN GLULCORHAMNOSIDE
mf: $C_{36}H_{52}O_{15}$ mw: 724.88

PROP: Crystals from hot methanol. Mp: 283-284°. Sol in dil alc; less sol in methanol, ethanol; sltly sol in water. Practically insol in ether.

SYNS: HELLEBRIGENIN-GLUCO-RHAMNOSID(GERMAN) ◇ HELLEBRIN

TOXICITY DATA with REFERENCE
ivn-rat LD50:21 mg/kg AIPTAK 155,165,65
ipr-mus LD50:8400 μg/kg AIPTAK 155,165,65
ivn-cat LD50:104 μg/kg JPETAB 99,395,50
ivn-gpg LDLo:616 μg/kg AEPPAE 252,314,66

SAFETY PROFILE: A deadly poison by intravenous and intraperitoneal routes. When heated to decomposition it emits acrid smoke and fumes.

HAN625 ***HR: 3***
HELODERMA SUSPECTUM VENOM

SYN: VENOM, LIZARD, HELODERMA SUSPECTUM

TOXICITY DATA with REFERENCE
par-rat LD50:1340 μg/kg TOXIA6 5,5,67
scu-mus LD50:1500 μg/kg 29QKAZ 2,499,72
ivn-mus LDLo:400 μg/kg TOXIA6 5,139,67

SAFETY PROFILE: Poison by subcutaneous, intraperitoneal, intravenous, and parenteral routes.

HAN800 CAS:630-64-8 ***HR: 3***
HELVETICOSIDE
mf: $C_{29}H_{42}O_9$ mw: 534.71

PROP: A β-glycoside consisting of one mole strophanthidin and one mole d-digitoxose.

SYNS: ERIZIMIN ◇ ERYSIMIN ◇ ERYSIMOTOXIN ◇ HELVETICOSID (GERMAN) ◇ STROPHANTHIDIN-β-d-DIGITOXOSID (GERMAN)

TOXICITY DATA with REFERENCE
ivn-rat LD50:54 mg/kg AIPTAK 155,165,65
ipr-mus LD50:7800 μg/kg AIPTAK 155,165,65
ivn-cat LDLo:50 μg/kg AIPTAK 155,165,65

SAFETY PROFILE: Poison by intravenous and intraperitoneal routes. When heated to decomposition it emits acrid smoke and fumes.

HAO000 CAS:13495-01-7 ***HR: 3***
HELVETICOSIDE DIHYDRATE
mf: $C_{29}H_{42}O_9 \cdot 2H_2O$ mw: 570.75

PROP: Needles from oil, methanol. Mp: 153-157°.

SYNS: ALLEOSIDE A DIHYDRATE ◇ ERISIMIN DIHYDRATE ◇ ERYSIMIN DIHYDRATE

TOXICITY DATA with REFERENCE
ivn-mky LDLo:103 μg/kg ARZNAD 13,412,63
ivn-cat LDLo:104 μg/kg ARZNAD 22,1854,72
ivn-gpg LDLo:867 μg/kg ARZNAD 13,412,63
ivn-pgn LDLo:285 μg/kg ARZNAD 13,412,63

SAFETY PROFILE: A deadly poison by intravenous route. When heated to decomposition it emits acrid smoke and fumes.

HAO500 ***HR: 3***
HEMACHATUS HAEMACHATUS VENOM

SYNS: HEMACHATUS HAEMACHATES VENOM ◇ H. HAEMACHATES VENOM ◇ VENOM, SNAKE, HEMACHATUS HAEMCHATES

TOXICITY DATA with REFERENCE
par-rat LD50:75 μg/kg TOXIA6 19,61,81
ipr-mus LD50:1600 μg/kg 19DDA6 1,283,67
scu-mus LD50:2625 μg/kg TOXIA6 5,47,67
ivn-mus LD50:1222 μg/kg 23EIAT 1,437,68

SAFETY PROFILE: Poison by subcutaneous, parenteral, intravenous, and intraperitoneal routes.

HAO875 CAS:1317-60-8 ***HR: 1***
HEMATITE

PROP: Consists mainly of Fe_2O_3 (IARC** 1,29,71).

SYNS: BLOOD STONE ◇ HAEMATITE ◇ IRON ORE ◇ RED IRON ORE

CONSENSUS REPORTS: Reported in EPA TSCA Inventory.

CONSENSUS REPORTS: IARC Cancer Review: Group 3, Indefinite IMSUPP 4,254,82. Reported in EPA TSCA Inventory.

SAFETY PROFILE: Questionable carcinogen.

HAO900 CAS:635-65-4 ***HR: D***
HEMATOIDIN
mf: $C_{33}H_{36}N_4O_6$ mw: 584.73

SYNS: BILINE-8,12-DIPROPIONIC ACID, 1,10,19,22,23,24-HEXAHYDRO-2,7,13,17-TETRAMETHYL-1,19-DIOXO-3,18-DIVINYL- ◇ BILIRUBIN ◇ BILIRUBIN IX-α ◇ HEMETOIDIN ◇ PRINCIPAL BILE PIGMENT

TOXICITY DATA with REFERENCE
ipr-rat TDLo:175 mg/kg (female 9-15D post):TER AJOGAH 127,497,77
ipr-rat TDLo:175 mg/kg (female 9-15D post):REP AJOGAH 127,497,77

CONSENSUS REPORTS: Reported in EPA TSCA Inventory.

SAFETY PROFILE: An experimental teratogen. Experimental reproductive effects. When heated to decomposition it emits toxic fumes of NO_x.

HAP000 CAS:15375-94-7 ***HR: 3***
HEMATOPORPHYRIN MERCURY DISODIUM SALT
mf: $C_{34}H_{34}HgN_4O_6•2Na$ mw: 841.29

SYNS: Hg-HEMATOPORPHYRIN-Na ◇ MERCURI-HEMATOPORPHYRIN DISODIUM SALT ◇ MERPHYLLIN ◇ MERPHYRIN ◇ M.H.

TOXICITY DATA with REFERENCE
ipr-rat TDLo:8700 μg/kg (1D male):REP HIKYAJ 12,1339,66
ipr-mus LD50:23 mg/kg NYKZAU 57,219,61
scu-mus LD50:30 mg/kg NYKZAU 57,219,61
ivn-mus LD50:11500 μg/kg NYKZAU 57,219,61
ice-mus LD50:380 μg/kg NYKZAU 57,219,61

CONSENSUS REPORTS: Mercury and its compounds are on the Community Right-To-Know List.

OSHA PEL: (Transitional: CL 1 mg/10m^3) CL 0.1 mg(Hg)/m^3 (skin)
ACGIH TLV: TWA 0.1 mg(Hg)/m^3 (skin)
NIOSH REL: (Inorganic Mercury) TWA 0.05 mg(Hg)/m^3

SAFETY PROFILE: Poison by intraperitoneal, subcutaneous, intravenous, and intracerebral routes. Experimental reproductive effects. When heated to decomposition it emits very toxic fumes of Hg and NO_x. See also MERCURY COMPOUNDS.

HAP500 CAS:517-28-2 ***HR: 3***
HEMATOXYLIN
mf: $C_{16}H_{14}O_6$ mw: 302.30

SYN: NCI-C55889

TOXICITY DATA with REFERENCE
orl-rat TDLo:400 mg/kg/26W-C:ETA GASTAB 23,1,53

CONSENSUS REPORTS: Reported in EPA TSCA Inventory.

SAFETY PROFILE: Questionable carcinogen with experimental tumorigenic data. When heated to decomposition it emits acrid smoke and fumes.

HAQ000 CAS:312-45-8 ***HR: 3***
HEMICHOLINIUM-3-DIBROMIDE
mf: $C_{24}H_{34}N_2O_4•2Br$ mw: 574.42

SYNS: 2,2'-(1,1'-BIPHENYL-4,4'-DIYLBIS(2-HYDROXY-4,4-DIMETHYL-MORPHOLINIUM DIBROMIDE ◇ HC-3 ◇ HEMICHOLINE

◇ HEMICHOLINIUM-3 ◇ HEMICHOLINIUM BROMIDE ◇ HEMICHOLINIUM-3-BROMIDE ◇ HEMICHOLINIUM DIBROMIDE

TOXICITY DATA with REFERENCE
ipr-rat LD50:160 μg/kg EJPHAZ 33,145,75
ipr-mus LD50:46 μg/kg EJPHAZ 33,145,75
scu-mus LDLo:150 μg/kg AIPTAK 152,253,64
ivn-mus LD50:80 μg/kg TXAPA9 27,666,74
ivn-dog LD50:750 μg/kg JMPCAS 4,505,61
ivn-cat LD50:300 μg/kg JMPCAS 4,505,61
ivn-rbt LD50:250 μg/kg AIPTAK 119,20,59
ipr-gpg LDLo:300 μg/kg IRNEAE 2,77,60

CONSENSUS REPORTS: Reported in EPA TSCA Inventory.

SAFETY PROFILE: A poison by subcutaneous, intraperitoneal, and intravenous routes. See also BROMIDES. When heated to decomposition it emits very toxic fumes of NO_x and Br^-.

HAQ100 ***HR: 3***
HENBANE

PROP: Biennial or annual weeds which grow to about 2 feet and are covered with fine hairs. The leaves are about 8 inches long with toothed edges. The flowers range in color from green-yellow to yellow-white with a purple throat and veins. They are native to Europe but can be found in scrub lands in the northeastern United States and southern Canada, and in sandy prairie areas across the United States.

SYNS: BEIENO (MEXICO) ◇ FETID NIGHTSHADE ◇ HYOSCYAMUS NIGER ◇ INSANE ROOT ◇ JUSQUIAME (CANADA) ◇ POISON TOBACCO ◇ STINKING NIGHTSHADE

SAFETY PROFILE: The seed contains poisonous belladonna alkaloids. Ingestion of the seeds can result in increased heart rate, fever, vision impairment, delirium, and hallucinations. See also BELLADONNA.

HAQ500 CAS:9005-49-6 ***HR: 2***
HEPARIN

SYNS: α-HEPARIN ◇ HEPARINATE ◇ HEPARINIC ACID ◇ HEPARIN SULFATE ◇ LIPO-HEPIN ◇ LIQUAEMIN ◇ LIQUEMIN ◇ NOVOHEPARIN ◇ SUBLINGULA ◇ THROMBOLIQUINE ◇ VETREN ◇ VITAMIN AB ◇ VITRUM AB

TOXICITY DATA with REFERENCE
dns-mus:lvr 33 mg/L AMOKAG 33,149,79
dns-mus:oth 10 mg/L JJIND8 71,615,83
dni-mus:ast 42 mg/L AMOKAG 33,149,79
dni-mus:lvr 167 mg/L AMOKAG 33,149,79
dnd-ham:ovr 200 mg/L BBACAQ 517,486,78
orl-rat LD50:1950 mg/kg GWXXBX #2636091
ipr-rat LDLo:420 mg/kg TXAPA9 1,156,59
ivn-mus LD50:1500 mg/kg AJMSA9 216,234,48

SAFETY PROFILE: Moderately toxic by ingestion, intraperitoneal, and intravenous routes. Mutation data reported. When heated to decomposition it emits toxic fumes of NO_x. See also HEPARIN SODIUM

HAQ550 CAS:9041-08-1 ***HR: 3***
HEPARIN SODIUM

SYNS: DEPO-HEPARIN ◇ HED-HEPARIN ◇ HEPATHROM ◇ LIQUAEMIN SODIUM ◇ LIQUEMIN ◇ PK 10169 ◇ PULARIN ◇ SODIUM ACID HEPARIN ◇ SODIUM HEPARIN ◇ SODIUM HEPARINATE

TOXICITY DATA with REFERENCE
scu-man TDLo:7 mg/kg/4D:CNS,SKN,BLD JAMAAP 244,1831,80
ivn-rat LD50:354 mg/kg DEBIDR 12,535,81
ivn-mus LD50:2800 mg/kg JPETAB 102,156,51
ivn-dog LD50:1 g/kg JPETAB 102,156,51

SAFETY PROFILE: Poison by intravenous route. Human systemic effects by subcutaneous route: hallucinations and distorted perceptions, hemorrhage and dermatitis. When heated to decomposition it emits toxic fumes of NO_x and Na_2O. See also HEPARIN.

HAR000 CAS:76-44-8 ***HR: 3***
HEPTACHLOR
mf: $C_{10}H_5Cl_7$ mw: 373.30

PROP: Crystals. Mp: 96°. Nearly insol in water; sol in organic solvents.

SYNS: AGROCERES ◇ 3-CHLOROCHLORDENE ◇ DRINOX ◇ E 3314 ◇ ENT 15,152 ◇ EPTACLORO (ITALIAN) ◇ 1,4,5,6,7,8,8-EPTACLORO-3a,4,7,7a-TETRAIDRO-4,7-endo-METANO-INDENE (ITALIAN) ◇ GPKh ◇ H-34 ◇ HEPTACHLOOR (DUTCH) ◇ 1,4,5,6,7,8,8-HEPTACHLOOR-3a,4,7,7a-TETRAHYDRO-4,7-endo-METHANO-INDEEN (DUTCH) ◇ HEPTACHLORE (FRENCH) ◇ 3,4,5,6,7,8,8-HEPTACHLORODICYCLOPENTADIENE ◇ 3,4,5,6,7,8,8a-HEPTACHLORODICYCLOPENTADIENE ◇ 1,4,5,6,7,10,10-HEPTACHLORO-4,7,8,9,-TETRAHYDRO-4,7-ENDOMETHYLENEINDENE ◇ 1,4,5,6,7,8,8-HEPTACHLORO-3a,4,7,7a-TETRAHYDRO-4,7-ENDOMETHANOINDENE ◇ 1,4,5,6,7,8,8a-HEPTACHLORO-3a,4,7,7a-TETRAHYDRO-4,7-METHANOINDANE ◇ 1,4,5,6,7,8,8-HEPTACHLORO-3a,4,7,7a-TETRAHYDRO-4,7-METHANOINDENE ◇ 1(3a),4,5,6,7,8,8-HEPTACHLORO-3a(1),4,7,7a-TETRAHYDRO-4,7-METHANOINDENE ◇ 1,4,5,6,7,8,8-HEPTACHLORO-3a,4,7,7a-TETRAHYDRO-4,7-METHANOL-1H-INDENE ◇ 1,4,5,6,7,8,8-HEPTACHLORO-3a,4,7,7,7a-TETRAHYDRO-4,7-METHYLENE INDENE ◇ 1,4,5,6,7,8,8-HEPTACHLOR-3a,4,7,7,7a-TETRAHYDRO-4,7-endo-METHANO-INDEN (GERMAN) ◇ HEPTAGRAN ◇ HEPTAMUL ◇ NCI-C00180 ◇ RCRA WASTE NUMBER P059 ◇ RHODIACHLOR ◇ VELSICOL 104

TOXICITY DATA with REFERENCE
mma-hmn:fbr 100 μmol/L MUREAV 42,161,77
cyt-rat-orl 60 μg/kg 34LXAP -,555,76
dlt-rat-orl 60 μg/kg 34LXAP -,555,76
cyt-mus-ipr 5200 μg/kg SOGEBZ 2,80,66
orl-mus TDLo:403 mg/kg/80W-C:CAR NCITR* NCI-CG-TR-9,77
orl-mus TD:930 mg/kg/80W-C:CAR NCITR* NCI-CG-TR-9,77

orl-rat LD50:40 mg/kg PHJOAV 185,361,60
skn-rat LD50:119 mg/kg SPEADM 78-1,12,78
ipr-rat LD50:27 mg/kg FCTXAV 11,63,73
orl-mus LD50:68 mg/kg SPEADM 78-1,12,78
ipr-mus LD50:130 mg/kg SOGEBZ 2,80,66
ivn-mus LDLo:20 mg/kg JPETAB 107,266,53
skn-rbt LD50:2000 mg/kg AFDOAQ 16,3,52
orl-gpg LD50:116 mg/kg PCOC** -,576,66
orl-ham LD50:100 mg/kg EJTXAZ 7,159,74

CONSENSUS REPORTS: IARC Cancer Review: Group 3 IMEMDT 7,146,87; Human Inadequate Evidence IMEMDT 20,129,79; Animal Inadequate Evidence IMEMDT 5,173,74; Animal Sufficient Evidence IMEMDT 20,129,79. NCI Carcinogenesis Bioassay (feed) Clear Evidence: Mouse NCITR* NCI-CG-TR-9,77; Results negative: rat NCITR* NCI-CG-TR-9,77. EPA Genetic Toxicology Program. Community Right-To-Know List.

OSHA PEL: TWA 0.5 mg/m^3 (skin)
ACGIH TLV: TWA 0.5 mg/m^3 (skin); Proposed: TWA 0.05 mg/m^3 (skin); Suspected Human Carcinogen
DFG MAK: 0.5 mg/m^3, Suspected Carcinogen.

SAFETY PROFILE: Suspected carcinogen with experimental carcinogenic data. A poison by ingestion, skin contact, intraperitoneal, and intravenous routes. Human mutation data reported. Acute exposure and chronic doses have caused liver damage. See also closely related chlordane. In man, a dose of 1-3 grams can cause serious symptoms, especially where liver impairment is the case. Acute symptoms include tremors, convulsions, kidney damage, respiratory collapse, and death. When heated to decomposition it emits toxic fumes of Cl^-.

NOTE: The EPA has canceled registration of pesticides containing heptachlor with the exception of its use for termite control by subsurface ground insertion external to the dwelling.

HAR500 ***HR: 3***
HEPTACHLOR (technical grade)

PROP: Mixture of 73% heptachlor, 22% trans-chlordane, and 5% nonachlor NCITR* NCI-CG-TR-9.

SYN: 1,4,5,6,7,8,8-HEPTACHLORO-3a,4,7,7a-TETRAHYDRO4,7-METHANOINDENE (technical grade)

TOXICITY DATA with REFERENCE
orl-mus TDLo:410 mg/kg/80W-C:CAR NCITR* NCI-CG-TR-9,77
orl-cat LD50:67 mg/kg 85GMAT -,71,82
skn-gpg LD50:627 mg/kg 85GMAT -,71,82
skn-rbt LD50:500 mg/kg 85GMAT -,71,82
orl-rat LD50:40 mg/kg KSKZAN 16(2),59,78

SAFETY PROFILE: Poison by ingestion. Moderately toxic by skin contact. Questionable carcinogen with experimental carcinogenic data. When heated to decomposition it emits toxic fumes of Cl^-. See also HEPTACHLOR.

HAS000 CAS:311-89-7 ***HR: 2***
HEPTACOSAFLUOROTRIBUTYLAMINE
mf: $C_{12}F_{27}N$ mw: 671.13

SYN: 1,1,2,2,3,3,4,4,4-NONAFLUORO-N,N-BIS(NONAFLUOROBUTYL)-1-BUTANAMINE

TOXICITY DATA with REFERENCE
ipr-mus LDLo:512 mg/kg CBCCT* 3,362,51

CONSENSUS REPORTS: Reported in EPA TSCA Inventory.

SAFETY PROFILE: Moderately toxic by intraperitoneal route. When heated to decomposition it emits very toxic fumes of F^- and NO_x.

HAS500 CAS:506-12-7 ***HR: 3***
HEPTADECANOIC ACID
mf: $C_{17}H_{34}O_2$ mw: 270.51

SYNS: n-HEPTADECOIC ACID ◇ n-HEPTADECYLIC ACID ◇ MARGARIC ACID

TOXICITY DATA with REFERENCE
ivn-mus LD50:36 mg/kg APTOA6 18,141,61

CONSENSUS REPORTS: Reported in EPA TSCA Inventory.

SAFETY PROFILE: Poison by intravenous route. When heated to decomposition it emits acrid smoke and fumes.

HAT000 CAS:52783-44-5 ***HR: 1***
HEPTADECANOL (mixed primary isomers)
mf: $C_{17}H_{36}O$ mw: 256.53

PROP: Mp: 54°, bp: 309°, flash p: 310°F (COC), d: 0.8475 @ 20°/20°, vap press: 0.01 mm @ 20°, vap d: 8.84.

TOXICITY DATA with REFERENCE
skn-rbt 8475 μg/24H open MLD AIHAAP 23,95,62
orl-rat LD50:51600 mg/kg AIHAAP 23,95,62
skn-rbt LD50:16800 mg/kg AIHAAP 23,95,62

SAFETY PROFILE: Mildly toxic by ingestion and skin contact. A skin irritant. Combustible when exposed to heat or flame; can react with oxidizing materials. To fight fire, use CO_2, dry chemical. When heated to decomposition it emits acrid smoke and fumes.

HAT500 CAS:95-19-2 ***HR: 2***
2-HEPTADECYL-2-IMIDAZOLINE-1-ETHANOL
mf: $C_{22}H_{46}N_2O$ mw: 354.70

SYN: 2-HEPTADECYL-1-HYDROXYETHYLIMIDAZOLINE

TOXICITY DATA with REFERENCE
orl-rat LD50:3800 mg/kg AMIHBC 4,494,51

CONSENSUS REPORTS: Reported in EPA TSCA Inventory.

SAFETY PROFILE: Moderately toxic by ingestion. When heated to decomposition it emits toxic fumes of NO_x.

HAU500 CAS:63982-03-6 ***HR: 2***
HEPTADECYLTRIMETHYLAMMONIUM METHYLSULFATE
mf: $C_{20}H_{44}N \cdot CH_3O_4S$ mw: 409.75

SYN: HEPTADECYL-TRIMETHYLAMMONIUMMETHYLSULFAT (CZECH)

TOXICITY DATA with REFERENCE
eye-rbt 250 μg/24H SEV 28ZPAK -,74,72
orl-rat LD50:3650 mg/kg 28ZPAK -,74,72

CONSENSUS REPORTS: Reported in EPA TSCA Inventory.

SAFETY PROFILE: Moderately toxic by ingestion. A severe eye irritant. When heated to decomposition it emits very toxic fumes of NH_3, NO_x, and SO_x. See also SULFATES.

HAV450 CAS:5910-85-0 ***HR: 3***
2,4-HEPTADIENAL
mf: $C_7H_{10}O$ mw: 110.17

PROP: Sltly yellow liquid; green odor. Refr index: 1.478-1.480, flash p: 140°F. Sol in alc, fixed oils, water.

SYNS: FEMA No. 3164 ◇ HEPTADIENAL-2,4 ◇ trans,trans-2,4-HEPTDIENAL ◇ 2,4-HEPTDIENAL

TOXICITY DATA with REFERENCE
skn-rbt 500 mg SEV FCTOD7 21,855,83
skn-gpg 500 mg SEV FCTOD7 21,855,83
orl-rat LD50:1150 mg/kg FCTOD7 21,855,83
skn-rbt LD50:313 mg/kg FCTOD7 21,855,83

SAFETY PROFILE: Poison by skin contact. Moderately toxic by ingestion. A severe skin irritant. Combustible liquid. When heated to decomposition it emits acrid smoke and fumes.

HAV500 CAS:2396-63-6 ***HR: 2***
1,6-HEPTADIYNE
mf: C_7H_8 mw: 92.15

TOXICITY DATA with REFERENCE
orl-rat LD50:2300 mg/kg FEPRA7 19,389,60
orl-dog LD50:3830 mg/kg FEPRA7 19,389,60
orl-rbt LD50:2620 mg/kg FEPRA7 19,389,60

CONSENSUS REPORTS: Reported in EPA TSCA Inventory.

SAFETY PROFILE: Moderately toxic by ingestion. When heated to decomposition it emits acrid smoke and fumes. See also ACETYLENE COMPOUNDS.

HAW000 CAS:3794-64-7 ***HR: 2***
HEPTAFLUOROBUTANOIC ACID, SILVER SALT
mf: $C_4F_7O_2 \cdot Ag$ mw: 320.91

SYN: HEPTAFLUORMASELNAN STRIBRNY (CZECH)

TOXICITY DATA with REFERENCE
skn-rbt 500 mg/24H MLD 28ZPAK -,8,72
eye-rbt 5 mg/24H SEV 28ZPAK -,8,72
orl-rat LD50:2140 mg/kg 28ZPAK -,8,72

CONSENSUS REPORTS: Silver and its compounds are on the Community Right-To-Know List. Reported in EPA TSCA Inventory.

OSHA PEL: TWA 0.01 mg(Ag)/m^3
ACGIH TLV: TWA 0.01 mg(Ag)/m^3

SAFETY PROFILE: Moderately toxic by ingestion. A skin and severe eye irritant. When heated to decomposition it emits toxic fumes of F^-. See also SILVER COMPOUNDS.

HAX000 CAS:662-50-0 ***HR: 3***
2,2,3,3,4,4,4-HEPTAFLUOROBUTYRAMIDE
mf: $C_4H_2F_7NO$ mw: 213.07

$F_7C_3CO \cdot NH_2$

TOXICITY DATA with REFERENCE
ipr-rbt LD50:126 mg/kg CBCCT* 2,299,50

CONSENSUS REPORTS: Reported in EPA TSCA Inventory.

SAFETY PROFILE: Poison by intraperitoneal route. Forms an unstable explosive complex with lithium tetrahydroaluminate. When heated to decomposition it emits very toxic fumes of F^- and NO_x.

HAX500 CAS:375-22-4 ***HR: 3***
HEPTAFLUOROBUTYRIC ACID
mf: $C_4HF_7O_2$ mw: 214.05

PROP: Colorless liquid; sharp, butyric acid odor. Bp: 210° @ 735 mm.

TOXICITY DATA with REFERENCE
ipr-mus LD50:153 mg/kg CBCCT* 2,56,50

CONSENSUS REPORTS: Reported in EPA TSCA Inventory.

SAFETY PROFILE: A poison by intraperitoneal route. Probably an eye, skin, and mucous membrane irritant.

Will react with water or steam to produce corrosive fumes. When heated to decomposition it emits toxic fumes of F^-.

HAY000 CAS:356-27-4 ***HR: 3***
HEPTAFLUOROBUTYRIC ACID, ETHYL ESTER
mf: $C_6H_5F_7O_2$ mw: 242.11

TOXICITY DATA with REFERENCE
ipr-rbt LD50:250 mg/kg CBCCT* 2,299,50

CONSENSUS REPORTS: Reported in EPA TSCA Inventory.

SAFETY PROFILE: Poison by intraperitoneal route. See also ESTERS. When heated to decomposition it emits toxic fumes of F^-.

HAY059 CAS:71359-62-1 ***HR: 3***
HEPTAFLUOROBUTYRYL HYPOCHLORITE
mf: $C_4ClF_7O_2$ mw: 248.48

$C_3F_7CO{\cdot}OCl$

SAFETY PROFILE: As a gas it explodes above 27-62 mbar. Thermally unstable above 22°C. Upon decomposition it emits toxic fumes of Cl^- and F^-. See also HYPOCHLORITES and FLUORIDES.

HAY100 ***HR: 3***
HEPTAFLUOROBUTYRYL HYPOFLUORITE
mf: $C_4F_8O_2$ mw: 232.03

$C_3F_7CO{\cdot}OF$

SAFETY PROFILE: Decomposes explosively upon exposure to spark or flame. When heated to decomposition it emits toxic fumes of F^-. See also FLUORIDES and HYPOCHLORITES.

HAY200 CAS:663-25-2 ***HR: 3***
HEPTAFLUOROBUTYRYL NITRATE
mf: $C_4F_7NO_3$ mw: 207.00

$C_3F_7CO{\cdot}ON{:}O$

SAFETY PROFILE: Potentially explosive. When heated to decomposition it emits toxic fumes of F^- and NO_x. See also NITRATES and FLUORIDES.

HAY300 CAS:27636-85-7 ***HR: 1***
HEPTAFLUOROIODOPROPANE
mf: C_3F_7I mw: 295.93

SYNS: HEPTAFLUORJODPROPAN ◇ IODOHEPTAFLUOROPROPANE ◇ PROPANE, HEPTAFLUOROIODO-

TOXICITY DATA with REFERENCE
eye-rbt 100 mg/24H MOD 85JCAE-,143,86

CONSENSUS REPORTS: Reported in EPA TSCA Inventory.

SAFETY PROFILE: An eye irritant. When heated to decomposition it emits toxic fumes of Cl^- and I^-.

HAY500 CAS:360-53-2 ***HR: 3***
HEPTAFLUOROISOBUTYLENE METHYL ETHER
mf: $C_5H_3F_7O$ mw: 212.08

TOXICITY DATA with REFERENCE
orl-mus LD50:1070 mg/kg TXAPA9 14,114,69
orl-rat LD50:1070 mg/kg TXAPA9 14,114,69
ipr-mus LD50:66 mg/kg TXAPA9 12,486,68
ivn-mus LD50:58 mg/kg TXAPA9 12,486,68

SAFETY PROFILE: Poison by intraperitoneal and intravenous routes. Moderately toxic by ingestion. When heated to decomposition it emits toxic fumes of F^-. See also ETHERS.

HAY600 CAS:87050-95-1 ***HR: 3***
2-HEPTAFLUOROPROPYL-1,3,4-DIOXAZOLONE
mf: $C_5F_7NO_3$ mw: 255.05

$F_7C_3C{=}NOCO{\cdot}O$ (ring: C to O)

SAFETY PROFILE: May explode when heated to 102°C. Upon decomposition it emits toxic fumes of F^- and NO_x. See also FLUORIDES.

HAY650 CAS:2203-57-8 ***HR: 3***
HEPTAFLUOROPROPYL HYPOFLUORITE
mf: C_3F_8O mw: 204.02

SAFETY PROFILE: An explosive. When heated to decomposition it emits toxic fumes of F^-. See also FLUORIDES and HYPOCHLORITES.

HBA259 CAS:64296-43-1 ***HR: 2***
2,3,3',4,4'5,7-HEPTAHYDROXYFLAVAN
mf: $C_{15}H_{12}O_8$ mw: 320.27

SYNS: 2-(3,4-DIHYDROXYPHENYL)-2,3,4,5,7-PENTAHYDROXY-1-BENZOPYRAN ◇ 2,3,3',4,4',5,7-HETAHYDROXYFLAVAN

TOXICITY DATA with REFERENCE
orl-mus LD50:2500 mg/kg EJMCA5 13,241,78
ipr-mus LD50:450 mg/kg EJMCA5 13,241,78

SAFETY PROFILE: Moderately toxic by ingestion and intraperitoneal routes. When heated to decomposition it emits acrid smoke and fumes.

HBA500 CAS:28016-59-3 ***HR: 3***
HEPTAKIS (DIMETHYLAMINO)TRIALUMINUM TRIBORON PENTAHYDRIDE
mf: $C_{14}H_{47}Al_3B_3N_7$ mw: 426.96

SAFETY PROFILE: Crystalline solid is spontaneously flammable in air. When heated to decomposition it emits toxic fumes of NO_x. See also BORON COMPOUNDS, ALUMINUM COMPOUNDS, and HYDRIDES.

HBA550 CAS:105-21-5 ***HR: 1***
γ-HEPTALACTONE
mf: $C_7H_{12}O_2$ mw: 128.19

PROP: Colorless, sltly oily liquid; coconut, sweet, malty, caramel odor. D: 0.997-1.004 @ 20°, refr index: 1.439-1.445. Misc in alc, fixed oils; very sltly sol in water.

SYNS: FEMA No. 2539 ◇ HEPTANOLIDE-4,1 ◇ HEPTANOLIDE-1,4 ◇ 4-HYDROXYHEPTANOIC ACID LACTONE ◇ 4-HYDROXY-HEPTANOIC ACID, Γ-LACTONE ◇ Γ-PROPIOBUTYROLACTONE

TOXICITY DATA with REFERENCE
skn-rbt 500 mg/24H MOD FCTOD7 20,703,82

CONSENSUS REPORTS: Reported in EPA TSCA Inventory.

SAFETY PROFILE: A skin irritant. When heated to decomposition it emits acrid smoke and irritating fumes.

HBA600 CAS:10448-09-6 ***HR: D***
HEPTAMETHYLPHENYLCYCLOTETRASILOXANE
mf: $C_{13}H_{26}O_4Si_4$ mw: 358.75

SYNS: MONOPHENYLHEPTAMETHYLCYCLOTETRASILOXANE ◇ $PM^{(1)}MM^{(3)}$

TOXICITY DATA with REFERENCE
orl-rat TDLo:132 mg/kg (female 16-21D post):REP TXAPA9 21,29,72
orl-rat TDLo:500 mg/kg (female 8-12D post):TER TXAPA9 21,29,72

SAFETY PROFILE: An experimental teratogen. Experimental reproductive effects. When heated to decomposition it emits acrid smoke and fumes.

HBB000 CAS:543-15-7 ***HR: 2***
HEPTAMINOL HYDROCHLORIDE
mf: $C_8H_{19}NO{\cdot}ClH$ mw: 181.74

PROP: Crystals. Mp: 178-180° (also reported as 150°). Freely sol in water; sol in alc; insol in acetone, benzene, ether.

SYNS: 6-AMINO-2-METHYL-2-HEPTANOL HYDROCHLORIDE ◇ HEPTAMYL HYDROCHLORIDE ◇ 2-METHYL-6-AMINO-2-HEPTANOL HYDROCHLORIDE ◇ 6-METHYL-2-AMINO-6-HEPTANOL HYDROCHLORIDE

TOXICITY DATA with REFERENCE
orl-hmn TDLo:28 μg/kg:CVS,KID,SKN JPETAB 103,178,51
ipr-mus LD50:900 mg/kg JPETAB 103,178,51

SAFETY PROFILE: Moderately toxic by intraperitoneal route. Human systemic effects by ingestion: pulse rate increase without fall in blood pressure, sweating, urine volume increase. When heated to decomposition it emits very toxic fumes of Cl^- and NO_x.

HBB500 CAS:111-71-7 ***HR: 1***
HEPTANAL
mf: $C_7H_{14}O$ mw: 114.18

PROP: Colorless liquid; penetrating, fruity odor. D: 0.814-0.819, refr index: 1.412-1.420, mp: −43.3°, bp: 152.8°, flash p: 93°F. Sol in alc, ether, fixed oils; sltly sol in water @ 153°; misc in alc, ether.

SYNS: ENANTHAL ◇ ENANTHALDEHYDE ◇ ENANTHOLE ◇ FEMA No. 2540 ◇ HEPTALDEHYDE ◇ OENANTHALDEHYDE ◇ OENANTHOL

TOXICITY DATA with REFERENCE
orl-rat LD50:14 g/kg FDRLI* 123,-,76
orl-mus LD50:20 g/kg BIJOAK 34,1196,40

CONSENSUS REPORTS: Reported in EPA TSCA Inventory.

SAFETY PROFILE: Mildly toxic by ingestion. Flammable liquid. When heated to decomposition it emits acrid smoke.

HBC000 CAS:1708-35-6 ***HR: 2***
HEPTANAL-1,2-GLYCERYL ACETAL
mf: $C_{10}H_{20}O_3$ mw: 188.30

SYN: HEPTANAL, CYCLIC (HYDROXYMETHYL)ETHYLENE ACETAL

TOXICITY DATA with REFERENCE
ipr-mus LD50:439 mg/kg AIPTAK 85,474,51

CONSENSUS REPORTS: Reported in EPA TSCA Inventory.

SAFETY PROFILE: Moderately toxic by intraperitoneal route. When heated to decomposition it emits acrid smoke and fumes.

HBC500 CAS:142-82-5 ***HR: 3***
HEPTANE
DOT: UN 1206
mf: C_7H_{16} mw: 100.23

PROP: Colorless liquid. Bp: 98.52, lel: 1.05%, uel: 6.7%, fp: −90.5°, flash p: 25°F (CC), d: 0.684 @ 20°/4°, autoign temp: 433.4° F, vap press: 40 mm @ 22.3°, vap d: 3.45. Sltly sol in alc; misc in ether and chloroform; insol in water.

SYNS: DIPROPYL METHANE ◇ EPTANI (ITALIAN) ◇ GETTYSOLVE-C ◇ HEPTAN (POLISH) ◇ n-HEPTANE ◇ HEPTANEN (DUTCH) ◇ HEPTYL HYDRIDE

TOXICITY DATA with REFERENCE
ihl-hmn TCLo:1000 ppm/6M:CNS BMRII* 2979,-,29
ivn-mus LD50:222 mg/kg JPMSAE 67,566,78

CONSENSUS REPORTS: Reported in EPA TSCA Inventory.

OSHA PEL: (Transitional: TWA 500 ppm) TWA 400 ppm; STEL 500 ppm
ACGIH TLV: TWA 400 ppm; STEL 500 ppm
DFG MAK: 500 ppm (2000 mg/m^3)
NIOSH REL: TWA (Alkanes) 350 mg/m^3
DOT Classification: Flammable Liquid; Label: Flammable Liquid.

SAFETY PROFILE: Poison by intravenous route. Human systemic effects by inhalation: hallucinations. Narcotic in high concentrations. A volatile, flammable liquid when exposed to heat or flame. Can react vigorously with oxidizing materials. Moderately explosive when exposed to heat or flame. Violent reaction with phosphorus + chlorine. To fight fire, use foam, CO_2, dry chemical. When heated to decomposition it emits acrid smoke and fumes.

HBD000 CAS:646-20-8 ***HR: 3***
HEPTANEDINITRILE
mf: $C_7H_{10}N_2$ mw: 122.19

SYNS: 1,5-DICYANOPENTANE ◇ PIMELIC ACID DINITRILE ◇ PIMELONITRILE

TOXICITY DATA with REFERENCE
orl-rat LDLo:500 mg/kg JPETAB 90,260,47
orl-mus LD50:126 mg/kg ARTODN 57,88,85

CONSENSUS REPORTS: Reported in EPA TSCA Inventory. Cyanide and its compounds are on the Community Right-To-Know List.

SAFETY PROFILE: Poison by ingestion. Many nitriles are poisons by ingestion and inhalation. When heated to decomposition it emits very toxic fumes of NO_x and CN^-. See also NITRILES.

HBD500 CAS:1639-09-4 ***HR: 3***
1-HEPTANETHIOL
mf: $C_7H_{16}S$ mw: 132.29

SYNS: HEPTYL MERCAPTAN ◇ n-HEPTYLMERCAPTAN ◇ USAF EK-2122

TOXICITY DATA with REFERENCE
ipr-mus LD50:200 mg/kg NTIS** AD277-689

CONSENSUS REPORTS: Reported in EPA TSCA Inventory.

NIOSH REL: (n-Alkane Mono Thiols) CL 0.5 ppm/15M

SAFETY PROFILE: A poison by intraperitoneal route. When heated to decomposition it emits very toxic fumes of SO_x. See also MERCAPTANS.

HBD650 CAS:20919-99-7 ***HR: 3***
1,1,1,3,5,5,5-HEPTANITROPENTANE
mf: $C_5H_5N_7O_{14}$ mw: 387.13

$$(O_2N)_3CCH_2CH(NO_2)CH_2C(NO_2)_3$$

SAFETY PROFILE: An explosive. A strong oxidant. When heated to decomposition it emits toxic fumes of NO_x. See also NITRO COMPOUNDS.

HBE000 CAS:111-14-8 ***HR: 2***
HEPTANOIC ACID
mf: $C_7H_{14}O_2$ mw: 130.21

PROP: Oily liquid; disagreeable, rancid odor; less odor when very pure. D: 0.9345 @ 0°/4°, mp: −7.5°, bp: 223.0°. Sol in alc, ether, and ethanol.

SYNS: ENANTHIC ACID ◇ ENANTHYLIC ACID ◇ HEPTHLIC ACID ◇ n-HEPTOIC ACID ◇ n-HEPTYLIC ACID ◇ HEXACID C-7 ◇ 1-HEXANECARBOXYLIC ACID ◇ OENANTHIC ACID ◇ OENANTHYLIC ACID

TOXICITY DATA with REFERENCE
orl-rat LD50:7000 mg/kg FDRLI* 123,-,76
orl-mus LD50:6400 mg/kg BIJOAK 34,1196,40
ivn-mus LD50:1200 mg/kg APTOA6 18,141,61

CONSENSUS REPORTS: Reported in EPA TSCA Inventory.

SAFETY PROFILE: Moderately toxic by intravenous route. Mildly toxic by ingestion. When heated to decomposition it emits acrid smoke and fumes.

HBE500 CAS:543-49-7 ***HR: 2***
2-HEPTANOL
mf: $C_7H_{16}O$ mw: 116.23

PROP: Liquid. Bp: 160.4°, flash p: 160°F (OC), d: 0.8344 @ 0°, vap press: 1 mm @ 14.6°, vap d: 4.01. Insol in water; sol in alc, ether, and benzene.

SYNS: AMYL METHYL CARBINOL ◇ HEPTANOL-2 ◇ 2-HYDROXYHEPTANE ◇ METHYL AMYL CARBINOL

TOXICITY DATA with REFERENCE
skn-rbt 10 mg/24H open MLD AMIHBC 10,61,54
eye-rbt 250 μg open SEV AMIHBC 10,61,54
orl-rat LD50:2580 mg/kg AMIHBC 10,61,54
skn-rbt LD50:1780 mg/kg AMIHBC 10,61,54

CONSENSUS REPORTS: Reported in EPA TSCA Inventory.

SAFETY PROFILE: Moderately toxic by ingestion and skin contact. A skin and severe eye irritant. Combustible when exposed to heat and flame; can react vigorously with oxidizers. To fight fire, use foam, CO_2, dry chemical. See also ALCOHOLS.

HBF000 CAS:589-82-2 ***HR: 2***
3-HEPTANOL
mf: $C_7H_{16}O$ mw: 116.23

PROP: Liquid. Bp: 156.2°, flash p: 140°F (COC), fp: −70°, d: 0.8224 @ 20°/20°, vap press: 0.5 mm @ 20°, vap d: 4.01.

SYN: 3-HYDROXYHEPTANE

TOXICITY DATA with REFERENCE
skn-rbt 10 mg/24H open MLD AMIHBC 4,119,51
eye-rbt 20 mg open SEV AMIHBC 4,119,51
orl-rat LD50:1870 mg/kg AMIHBC 4,119,51
skn-rbt LD50:4360 mg/kg AMIHBC 4,119,51

CONSENSUS REPORTS: Reported in EPA TSCA Inventory.

SAFETY PROFILE: Moderately toxic by ingestion. Moderately toxic skin contact. A moderate skin and severe eye irritant. Combustible when exposed to heat or flame; can react with oxidizing materials. To fight fire, use foam, CO_2, dry chemical. When heated to decomposition it emits acrid smoke and fumes. See also ALCOHOLS.

HBF500 CAS:63039-95-2 ***HR: 3***
3-HEPTANOL-6-METHYL-3-PHENYL-1-(N-PIPERIDYL) HYDROCHLORIDE
mf: $C_{19}H_{31}NO{\cdot}ClH$ mw: 325.97

SYNS: α-(3-METHYL-BUTANE)-α-(2-PIPERIDYLETHYL) BENZYL ALCOHOL HYDROCHLORIDE ◇ 1-(N-PIPERIDYL)-6-METHYL-3-PHENYL HEPTANOL HYDROCHLORIDE

TOXICITY DATA with REFERENCE
ipr-rat LD50:84 mg/kg JPETAB 96,151,49
ivn-rat LD50:25 mg/kg JPETAB 96,151,49
ipr-mus LD50:110 mg/kg JPETAB 96,151,49
ivn-mus LD50:31 mg/kg JPETAB 96,151,49

SAFETY PROFILE: Poison by intraperitoneal and intravenous routes. When heated to decomposition it emits very toxic fumes of HCl and NO_x.

HBI500 CAS:12258-22-9 ***HR: 3***
HEPTA SILVER NITRATE OCTAOXIDE
mf: Ag_7NO_{11} mw: 945.10

CONSENSUS REPORTS: Silver and its compounds are on the Community Right-To-Know List.

SAFETY PROFILE: Unstable. It explodes weakly at 110°C. Forms impact-sensitive explosive mixtures with phosphorus and sulfur. Ignites on contact with hydrogen. Ignites when ground with antimony trisulfide. See also SILVER COMPOUNDS.

HBI725 CAS:66486-68-8 ***HR: 3***
HEPTA-1,3,5-TRIYNE
mf: C_7H_4 mw: 88.11

$$H(C{\equiv}C)_3CH_3$$

SAFETY PROFILE: Explodes above 0°C in the absence of air. The residue from distillation explodes on contact with air. When heated to decomposition it emits acrid smoke and fumes. See also ACETYLENE COMPOUNDS.

HBI800 ***HR: 3***
cis-4-HEPTEN-1-AL
mf: $C_7H_{12}O$ mw: 112.17

PROP: Sltly yellow liquid; fatty odor. Refr index: 1.432-1.436, flash p: 68°F. Sol in alc, fixed oils; insol in water.

SYNS: FEMA No. 3289 ◇ 4-HEPTENAL ◇ n-PROPYLIDENE BUTYRALDEHYDE

SAFETY PROFILE: Flammable liquid. When heated to decomposition it emits acrid smoke and irritating fumes.

HBJ000 ***HR: 3***
n-HEPTENE
mf: C_7H_{14} mw: 98.19
DOT: UN 2278

PROP: Colorless liquid, insol in water, sol in ether. D: 0.6969 @ 20°, mp: −10°, bp: 93.6, flash p: <30.2°F, autoign temp: 707°F.

SYN: 1-HEPTYLENE

DOT Classification: Flammable Liquid; Label: Flammable Liquid.

SAFETY PROFILE: A simple asphyxiant. See ARGON for a description of simple asphyxiants. Dangerous fire hazard when exposed to heat, flame or oxidizers. Unknown explosion hazard. To fight fire, use foam, dry chemical, CO_2. When heated to decomposition it emits acrid smoke and fumes.

HBJ500 ***HR: 2***
2-HEPTENE
mf: C_7H_{14} mw: 98.19

PROP: Clear liquid. Bp: 98.2°, flash p: <30.2°F, d: 0.709 @ 20°/4°, vap d: 3.4.

SAFETY PROFILE: Probably mildly toxic by ingestion and inhalation. A simple asphyxiant. See ARGON for a description of simple asphyxiants. Dangerous fire hazard when exposed to heat or flame; can react vigorously with oxidizing materials. To fight fire, use dry chemical, CO_2, foam. When heated to decomposition it emits acrid smoke and fumes.

HBK350 CAS:25339-56-4 ***HR: 2***
3-HEPTENE (mixed isomers)
mf: C_7H_{14} mw: 98.21

PROP: Liquid (mixture of cis and trans isomers). Bp: 96°, flash p: <19.4°F, d: 0.705 @ 15.5°/25.5°, vap d: 3.38.

SYN: HEPTYLENE

SAFETY PROFILE: Probably irritating and narcotic in high concentration. Dangerous fire hazard when exposed to heat or flame; can react vigorously with oxidizing materials. To fight fire, use foam, CO_2, dry chemical. See also 2-HEPTENE.

HBK450 ***HR: 3***
1-HEPTENE-4,6-DIYNE
mf: C_7H_6 mw: 90.12

$$H_2C{=}CHCH_2(C{\equiv}C)_2H$$

SAFETY PROFILE: An unstable explosive. When heated to decomposition it emits acrid smoke and fumes. See also ACETYLENE COMPOUNDS.

HBK500 CAS:18999-28-5 ***HR: 2***
2-HEPTENOIC ACID
mf: $C_7H_{12}O_2$ mw: 128.19

TOXICITY DATA with REFERENCE
ipr-mus LD50:1600 mg/kg JPPMAB 21,85,69
scu-mus LD50:1600 mg/kg JPPMAB 21,85,69

SAFETY PROFILE: Moderately toxic by intraperitoneal and subcutaneous routes. When heated to decomposition it emits acrid smoke and fumes.

HBK700 CAS:23560-59-0 ***HR: 3***
HEPTENOPHOS
mf: $C_9H_{12}ClO_4P$ mw: 250.63

PROP: Pale amber liquid, bp: 94-95°, d: 1.294. Vap. press @ 20°: 0.00075 mm Hg. Sol in xylene, acetone, methanol.

SYNS: 7-CHLOROBICYCLO(3.2.0)HEPTA-2,6-DIEN-6-YLDIMETHYL PHOSPHATE ◇ O,O-DIMETHYL-O-(6-CHLOROBICYCLO(3.2.0) HEPTADIEN-1,5-YL)PHOSPHATE ◇ 5-(O,O-DIMETHYLPHOSPHORYL)-6-CHLOROBICYCLO(3.2.0)HEPTA-1,5-DIEN ◇ HOE 2982 ◇ HOE 2982 OJ ◇ HOSTAQUICK ◇ HOSTAVIK (RUSSIAN) ◇ RAGADAN ◇ XOE 2982 (RUSSIAN)

TOXICITY DATA with REFERENCE
orl-rat LD50:96 mg/kg 85ARAE 1,11,77
skn-rat LD50:2 g/kg GISAAA 49(4),18,84
skn-mus LD50:2 g/kg GISAAA 49(4),18,84

SAFETY PROFILE: Poison by ingestion route. Moderately toxic by skin contact. When heated to decomposition it emits toxic fumes of Cl^- and PO_x. See also PHOSPHATES.

HBL000 CAS:112-06-1 ***HR: 1***
n-HEPTYL ACETATE
mf: $C_9H_{18}O_2$ mw: 158.27

PROP: Colorless liquid. D: 0.875, mp: −50.2°, bp: 192.5°. Insol in water; sol in alc and ether.

SYNS: ACETATE C-7 ◇ HEPTANYL ACETATE ◇ HEPTYL ACETATE ◇ 1-HEPTYL ACETATE

TOXICITY DATA with REFERENCE
skn-rbt 500 mg/24H MLD FCTXAV 12,813,74

CONSENSUS REPORTS: Reported in EPA TSCA Inventory.

SAFETY PROFILE: Irritating to skin, eyes and mucous membranes. When heated to decomposition it emits acrid smoke and fumes. See also ESTERS.

HBL500 CAS:111-70-6 ***HR: 2***
HEPTYL ALCOHOL
mf: $C_7H_{16}O$ mw: 116.23

PROP: Colorless liquid; citrus odor. Mp: −34.6°, bp: 175.8°, d: 0.824 @ 20°/4°, refr index: 1.423-1.427, flash p: 160°F. Misc in alc, fixed oils, ether; sltly sol in water @ 175°.

SYNS: l'ALCOOL n-HEPTYLIQUE PRIMAIRE (FRENCH) ◇ ENANTHIC ALCOHOL ◇ FEMA No. 2548 ◇ n-HEPTANOL ◇ 1-HEPTANOL ◇ n-HEPTANOL-1 (FRENCH) ◇ 1-HYDROXYHEPTANE

TOXICITY DATA with REFERENCE
orl-rat LD50:500 mg/kg AMPMAR 35,501,74
orl-mus LD50:1500 mg/kg GISAAA 31,16,66
ihl-mus LC50:6600 mg/kg 85GMAT -,72,82
orl-rbt LD50:750 mg/kg HYSAAV 31,310,66
skn-rbt LD50:2 g/kg AMPMAR 35,501,74

CONSENSUS REPORTS: Reported in EPA TSCA Inventory.

SAFETY PROFILE: Moderately toxic by ingestion and skin contact. Mildly toxic by inhalation. Combustible liquid. Can react with oxidizing materials. When heated to decomposition it emits acrid smoke and fumes. See also ALCOHOLS.

HBM500 CAS:28292-42-4 ***HR: 3***
3-HEPTYLAMINE
mf: $C_7H_{17}N$ mw: 115.25

TOXICITY DATA with REFERENCE
ipr-mus LD50:70 mg/kg JAPMA8 30,623,41

CONSENSUS REPORTS: Reported in EPA TSCA Inventory.

SAFETY PROFILE: Poison by intraperitoneal route. Probably flammable. When heated to decomposition, it emits acrid smoke and fumes. See also AMINES.

HBN000 CAS:63019-32-9 ***HR: 3***
8-HEPTYLBENZ(a)ANTHRACENE
mf: $C_{25}H_{26}$ mw: 326.51

SYN: 5-n-HEPTYLBENZ(1:2)BENZANTHRACENE

TOXICITY DATA with REFERENCE
skn-mus TDLo:1970 mg/kg/82W-I:ETA PRLBA4 131,170,42

SAFETY PROFILE: Questionable carcinogen with experimental tumorigenic data. When heated to decomposition it emits acrid smoke and fumes.

HBN250 CAS:7409-44-1 ***HR: 2***
HEPTYL CELLOSOLVE
mf: $C_9H_{20}O_2$ mw: 160.29

SYNS: ETHANOL, 2-(HEPTYLOXY)- ◇ ETHYLENE GLYCOL MONOHEPTYL ETHER ◇ 2-(HEPTYLOXY)ETHANOL

TOXICITY DATA with REFERENCE
skn-rbt 500 mg MOD VHTODE 29,361,87
eye-rbt 100 mg SEV VHTODE 29,361,87
orl-rat LD50:2280 mg/kg VHTODE 29,361,87

SAFETY PROFILE: Moderately toxic by ingestion. A moderate skin and severe eye irritant. When heated to decomposition it emits acrid smoke and irritating fumes.

HBN500 CAS:137-03-1 ***HR: 1***
α-HEPTYL CYCLOPENTANONE
mf: $C_{12}H_{22}O$ mw: 182.34

SYN: 2-n-HEPTYLCYCLOPENTANONE

TOXICITY DATA with REFERENCE
skn-rbt 500 mg/24H MLD FCTXAV 13,449,75
skn-rbt LD50:5000 mg/kg FCTXAV 13,452,75

CONSENSUS REPORTS: Reported in EPA TSCA Inventory.

SAFETY PROFILE: Mildly toxic by skin contact. A skin irritant. When heated to decomposition it emits acrid smoke and fumes. See also KETONES.

HBN600 CAS:64049-21-4 ***HR: 3***
HEPTYLDICHLORARSINE
mf: $C_7H_{15}AsCl_2$ mw: 245.04

SYNS: ARSINE, DICHLOROHEPTYL- ◇ TL 229

TOXICITY DATA with REFERENCE
ihl-mus LC50:13100 mg/m^3/10M NTIS** PB158-508
skn-mus LD50:40 mg/kg NTIS** PB158-508

OSHA PEL: TWA 0.5 mg(As)/m^3

SAFETY PROFILE: Poison by skin contact. Slightly toxic by inhalation. When heated to decomposition it emits toxic fumes of As and Cl^-.

HBO000 CAS:629-64-1 ***HR: 2***
HEPTYL ETHER
mf: $C_{14}H_{30}O$ mw: 214.44

SYN: DIHEPTYL ETHER

TOXICITY DATA with REFERENCE
ivn-mus LD50:470 mg/kg JPMSAE 67,566,78

CONSENSUS REPORTS: Reported in EPA TSCA Inventory.

SAFETY PROFILE: Moderately toxic by intravenous route. When heated to decomposition it emits acrid smoke and fumes. See also ETHERS.

HBO500 CAS:112-23-2 ***HR: 1***
HEPTYL FORMATE
mf: $C_8H_{16}O_2$ mw: 144.24

SYNS: FORMIC ACID, HEPTYL ESTER ◇ HEPTANOL, FORMATE ◇ n-HEPTYL METHANOATE

TOXICITY DATA with REFERENCE
skn-rbt 500 mg/24H MOD FCTXAV 16,771,78

CONSENSUS REPORTS: Reported in EPA TSCA Inventory.

SAFETY PROFILE: A skin irritant. When heated to decomposition it emits acrid smoke and fumes. See also ESTERS.

HBO600 CAS:2656-72-6 ***HR: 3***
HEPTYL HYDRAZINE
mf: $C_7H_{18}N_2$ mw: 130.27

SYN: HYDRAZINE, HEPTYL-

TOXICITY DATA with REFERENCE
scu-mus TDLo:3 mg/kg (female 1-6D post):REP JOENAK 27,147,63
ipr-mus LD50:150 mg/kg JOENAK 27,147,63

SAFETY PROFILE: Poison by intraperitoneal route. Experimental reproductive effects. When heated to decomposition it emits toxic fumes of NO_x.

HBO700 ***HR: 2***
HEPTYLIDENE METHYL ANTHRANILATE
mf: $C_{15}H_{21}NO_2$ mw: 247.37

SYN: HEPTALDEHYDE METHYLANTHRANILATE, Schiff's base

TOXICITY DATA with REFERENCE
skn-rbt 500 mg/24H MOD FCTOD7 20(Suppl),705,82
orl-rat LD50:2600 mg/kg FCTOD7 20(Suppl),705,82
skn-rbt LD50:2500 mg/kg FCTOD7 20(Suppl),705,82

SAFETY PROFILE: Moderately toxic by skin contact and ingestion. A skin irritant. When heated to decomposition it emits toxic fumes of NO_x.

HBP000 CAS:16338-99-1 ***HR: 3***
HEPTYLMETHYLINITROSAMINE
mf: $C_8H_{18}N_2O$ mw: 158.28

SYNS: METHYLHEPTYLNITROSAMIN (GERMAN) ◇ N-METHYL-N-NITROSOHEPTYLAMINE ◇ N-NITROSO-N-METHYLHEPTYLAMINE

TOXICITY DATA with REFERENCE
mma-sat 50 μg/plate TCMUD8 1,295,80
orl-rat TDLo:1960 mg/kg/30W-I:ETA JCROD7 106,171,83
scu-rat LD50:420 mg/kg ZEKBAI 69,103,67

SAFETY PROFILE: Moderately toxic by subcutaneous route. Mutation data reported. Questionable carcinogen with experimental tumorigenic data. Many N-nitroso compounds are carcinogens. When heated to decomposition it emits toxic fumes of NO_x. See also N-NITROSO COMPOUNDS.

HBP250 CAS:24346-78-9 ***HR: 2***
1-HEPTYL-1-NITROSOUREA
mf: $C_8H_{17}N_3O_2$. mw: 187.28

SYN: n-HEPTYL NITROSUREA

TOXICITY DATA with REFERENCE
orl-rat TDLo:600 mg/kg:ETA ANYAA9 381,250,82

SAFETY PROFILE: Questionable carcinogen with experimental tumorigenic data. When heated to decomposition it emits toxic fumes of NO_x.

HBP275 CAS:25961-87-9 ***HR: 2***
2-(2-(HEPTYLOXY)ETHOXY)ETHANOL
mf: $C_{11}H_{24}O_3$ mw: 204.35

SYNS: DIETHYLENE GLYCOL MONOHEPTYL ETHER ◇ ETHANOL, 2-(2-(HEPTYLOXY)ETHOXY)-

TOXICITY DATA with REFERENCE
skn-rbt 500 mg MOD VHTODE 29,361,87
eye-rbt 100 mg SEV VHTODE 29,361,87
orl-rat LD50:2940 mg/kg VHTODE 29,361,87

SAFETY PROFILE: Moderately toxic by ingestion. A moderate skin and severe eye irritant. When heated to decomposition it emits acrid smoke and irritating fumes.

HBP400 CAS:3648-21-3 ***HR: D***
HEPTYL PHTHALATE
mf: $C_{22}H_{34}O_4$ mw: 362.56

SYNS: 1,2-BENZENEDICARBOXYLIC ACID, DIHEPTYL ESTER (9CI) ◇ DIHEPTYL PHTHALATE ◇ DI-n-HEPTYL PHTHALATE ◇ PHTHALIC ACID, DIHEPTYL ESTER

TOXICITY DATA with REFERENCE
orl-mus TDLo:2500 mg/kg (female 9D post):TER SEIJBO 17,380,77

CONSENSUS REPORTS: Reported in EPA TSCA Inventory.

SAFETY PROFILE: An experimental teratogen. When heated to decomposition it emits acrid smoke and irritating fumes.

HBP450 CAS:713-95-1 ***HR: 1***
n-HEPTYL-Δ-VALEROLACTONE
mf: $C_{12}H_{22}O_2$ mw: 198.34

SYNS: Δ-DODECALACTONE ◇ 5-HYDROXYDODECANOIC ACID LACTONE ◇ 5-HYDROXYDODECANOIC ACID Δ-LACTONE ◇ 2H-PYRAN-2-ONE, 6-HEPTYLTETRAHYDRO-

TOXICITY DATA with REFERENCE
skn-rbt 500 mg/24H MOD FCTXAV 17,773,79

CONSENSUS REPORTS: Reported in EPA TSCA Inventory.

SAFETY PROFILE: A skin irritant. When heated to decomposition it emits acrid smoke and irritating fumes.

HBS500 CAS:1002-36-4 ***HR: 3***
2-HEPTYN-1-OL
mf: $C_7H_{12}O$ mw: 112.17

SAFETY PROFILE: The residue from distillation is explosive. When heated to decomposition it emits acrid smoke and fumes. See also ACETYLENE COMPOUNDS.

HBT000 CAS:1057-81-4 ***HR: 3***
HEPZIDINE MALEATE
mf: $C_{21}H_{25}NO \cdot C_4H_4O_4$ mw: 423.55

PROP: Crystals from ethanol. Mp: 153-156°.

SYNS: BS 7051 ◇ 4-((10,11-DIHYDRO-5H-DIBENZO(a,d)CYCLOHEPTEN-5-YL)OXY)-1-METHYLPIPERIDINE HYDROGEN MALEATE ◇ 4-((10,11-DIHYDRO-5H-DIBENZO(a,d)CYCLOHEPTEN-5-YL)OXY)-1-METHYLPIPERIDINE, MALEATE (1:1)

TOXICITY DATA with REFERENCE
orl-rat LD50:1500 mg/kg AIPTAK 167,334,67
ivn-rat LDLo:37 mg/kg AIPTAK 162,497,66
orl-mus LD50:306 mg/kg AIPTAK 167,334,67
ipr-mus LD50:102 mg/kg AIPTAK 167,334,67
scu-mus LD50:156 mg/kg AIPTAK 167,334,67

ivn-mus LD50:30 mg/kg ARZNAD 16,1342,66
orl-dog LD50:588 mg/kg AIPTAK 167,334,67

SAFETY PROFILE: Poison by ingestion, intravenous, intraperitoneal, and subcutaneous routes. When heated to decomposition it emits toxic fumes of NO_x.

HBT500 CAS:561-27-3 ***HR: 3***
HEROIN
mf: $C_{21}H_{23}NO_5$ mw: 369.45

PROP: White, odorless, bitter crystals or crystalline powder. Mp: 173°, bp: 273° @ 12 mm.

SYNS: ACETOMORFINE ◇ ACETOMORPHINE ◇ ASPRON ◇ BOY ◇ DIACEPHIN ◇ DIACETYLMORFIN ◇ DIACETYLMORPHINE ◇ DIAMORFINA ◇ DIAMORPHINE ◇ DIAPHORM ◇ DIASETIEL-MORFIEN ◇ DIASETILMORFIN ◇ DIASETYLMORFIIMI ◇ DIAZETYL-MORPHINE ◇ 7,8-DIHYDRO-4,5-α-EPOXY-17-METHYLMORPHINAN-3,6-α-DIOL DIACETATE ◇ DOOJE ◇ ECLORION ◇ EROINA ◇ "H" ◇ HAIRY ◇ HARRY ◇ HEROIEN ◇ HEROIIN ◇ HEROLAN ◇ HORSE ◇ IEROIN ◇ IROINI ◇ JOY POWDER ◇ MORPHACETIN ◇ MORPHINE DIACETATE ◇ PREZA ◇ SCOT ◇ WHITE STUFF

TOXICITY DATA with REFERENCE
cyt-mky-ivn 141 mg/kg/26W-I MUREAV 118,77,83
sce-mky-ivn 141 mg/kg/26W-I MUREAV 118,77,83
ivn-wmn TDLo:32 mg/kg (1-39W preg):REP PEDIAU 24,288,59
ivn-rbt TDLo:9 mg/kg (female 24-26D post):TER JOPDAB 82,869,73
scu-mus LDLo:262 mg/kg JPETAB 53,430,35
ice-mus LD50:137 μg/kg EJPHAZ 85,317,82
scu-dog LDLo:25 mg/kg HBAMAK 4,1289,35
orl-cat LDLo:20 mg/kg HBAMAK 4,1289,35
scu-gpg LDLo:400 mg/kg HBAMAK 4,1289,35

SAFETY PROFILE: A poison by ingestion, intracerebral, and subcutaneous routes. The fatal dose is between 1/6 and 2 grains. Human reproductive effects by subcutaneous and intravenous routes: newborn drug dependence. An experimental teratogen. Experimental reproductive effects. Mutation data reported.

Resembles morphine in its general results, but acts more strongly on the respiration and is therefore more poisonous. Its depressant effects on the cerebrum appear to be greater than that of codeine. Large doses cause excitement and convulsions in animals and humans. The more common symptoms are headache; disturbance of vision; slow, small, regular pulse; restlessness; cramps in the extremities; slight cyanosis; slow and deep respiration, and death from respiratory paralysis. A poisonous, habit-forming drug. When heated to decomposition it emits toxic fumes of NO_x. See also MORPHINE.

HBU000 CAS:520-26-3 ***HR: 2***
HESPERIDIN
mf: $C_{28}H_{34}O_{15}$ mw: 610.62

PROP: Hygroscopic needles. Mp: 251° decomp; sltly sol in alc; insol in ether and benzene.

SYNS: CIRANTIN ◇ 7-(6-o-(6-DEOXY-α-l-MANNOPYRANOSYL)-β-d-GLUCOPYRANOSIDE)HESPERETIN ◇ HESPERIDOSIDE ◇ HESPERITIN-7-RHAMNOGLUCOSIDE ◇ USAF CF-3

TOXICITY DATA with REFERENCE
ipr-mus LD50:1000 mg/kg NTIS** AD277-689

CONSENSUS REPORTS: Reported in EPA TSCA Inventory.

SAFETY PROFILE: Moderately toxic by intraperitoneal route. When heated to decomposition it emits acrid smoke and fumes.

HBU400 CAS:24292-52-2 ***HR: D***
HESPERIDIN METHYLCHALCONE
mf: $C_{29}H_{36}O_{15}$ mw: 624.65

SYN: CHALCONE, 2′,3,4′-TRIHYDROXY-4,6′-DIMETHOXY-, 4′-(6-O-(6-DEOXY-α-l-MANNOPYRANOSYL)-β-d-GLUCOPYRANOSIDE)

TOXICITY DATA with REFERENCE
orl-mus TDLo:3 g/kg (male 10D pre):REP SCIEAS 118,657,53

CONSENSUS REPORTS: Reported in EPA TSCA Inventory.

SAFETY PROFILE: Experimental reproductive effects. When heated to decomposition it emits acrid smoke and irritating fumes.

HBU500 ***HR: 3***
HEXAAMMINECHROMIUM(III) NITRATE
mf: $CrH_{18}N_9O_9$ mw: 340.203

$$[(H_3N)_6Cr][NO_3]_3$$

CONSENSUS REPORTS: Chromium and its compounds are on the Community Right-To-Know List.

SAFETY PROFILE: A moderately impact-sensitive explosive which explodes when heated to 263°C. Upon decomposition it emits toxic fumes of NO_x. See also NITRATES and CHROMIUM COMPOUNDS.

HBV000 CAS:26156-56-9 ***HR: 3***
HEXAAMMINECOBALT(III) CHLORATE
mf: $Cl_3CoH_{18}N_6O_9$ mw: 411.48

$$[(H_3N)_6Co][ClO_3]_3$$

CONSENSUS REPORTS: Cobalt and its compounds are on the Community Right-To-Know List.

SAFETY PROFILE: Unstable and explosive. When heated to decomposition it emits very toxic fumes of Cl^- and NO_x. See also COBALT COMPOUNDS and CHLORATES.

HBV500 ***HR: 3***
HEXAAMMINECOBALT(III) CHLORITE
mf: $Cl_3CoH_{18}N_6O_6$ mw: 363.48

$[(H_3N)_6Co][ClO_2]_2$

CONSENSUS REPORTS: Cobalt and its compounds are on the Community Right-To-Know List.

SAFETY PROFILE: An impact-sensitive explosive. When heated to decomposition it emits very toxic fumes of NO_x and Cl^-. See also COBALT COMPOUNDS and CHLORITES.

HBW000 CAS:15742-33-3 ***HR: 3***
HEXAAMMINECOBALT(III) HEXANITRO-COBALTATE (3⁻)
mf: $Co_2H_{18}N_{12}O_{12}$ mw: 496.19

$[(H_3N)_6Co][Co(NO_2)_6]$

CONSENSUS REPORTS: Cobalt and its compounds are on the Community Right-To-Know List.

SAFETY PROFILE: An unstable, impact-sensitive explosive. Upon ignition it burns very rapidly. When heated to decomposition it emits toxic fumes of NO_x and NH_3. See also COBALT COMPOUNDS and AMINES.

HBW500 CAS:14589-65-2 ***HR: 3***
HEXAAMMINECOBALT(III) IODATE
mf: $CoH_{18}I_3N_6O_9$ mw: 685.81

$[(H_3N)_6Co][IO_3]_3$

CONSENSUS REPORTS: Cobalt and its compounds are on the Community Right-To-Know List.

SAFETY PROFILE: Explodes when heated to 335°C. It has low impact-sensitivity. Upon decomposition it emits very toxic fumes of NO_x and I^-. See also COBALT COMPOUNDS and IODATES.

HBX000 CAS:10534-86-8 ***HR: 3***
HEXAAMMINECOBALT(III) NITRATE
mf: $CoH_{18}N_9O_9$ mw: 347.14

$[(H_3N)_6Co][NO_3]_3$

CONSENSUS REPORTS: Cobalt and its compounds are on the Community Right-To-Know List.

SAFETY PROFILE: Explodes when heated to 295°C or upon impact. When heated to decomposition it emits toxic fumes of NO_x. See also COBALT COMPOUNDS and NITRATES.

HBX500 CAS:13820-83-2 ***HR: 3***
HEXAAMMINECOBALT(III) PERCHLORATE
mf: $Cl_3CoH_{18}N_6O_{12}$ mw: 459.45

$[(H_3N)_6Co][ClO_4]_3$

CONSENSUS REPORTS: Cobalt and its compounds are on the Community Right-To-Know List.

SAFETY PROFILE: Explodes when heated to 360°C. It is very impact-sensitive. When heated to decomposition it emits very toxic fumes of Cl^- and NO_x. See also COBALT COMPOUNDS and PERCHLORATES.

HBY000 CAS:22388-72-3 ***HR: 3***
HEXAAMMINECOBALT(III) PERMANGANATE
mf: $CoH_{18}Mn_3N_6O_{12}$ mw: 517.92

CONSENSUS REPORTS: Manganese and its compounds as well as cobalt and its compounds are on the Community Right-To-Know List.

SAFETY PROFILE: Explodes on impact or heating. It is extremely sensitive to impact. Upon decomposition it emits toxic fumes of NO_x. See also MANGANESE COMPOUNDS and COBALT COMPOUNDS.

HBY500 ***HR: 3***
HEXAAMMINETITANIUM(III) CHLORIDE
mf: $Cl_3H_{18}N_6Ti$ mw: 256.42

$[(H_3N)_6Ti]Cl_3$

SAFETY PROFILE: Violent reaction with water. When heated to decomposition it emits very toxic fumes of Cl^- and NO_x. See also TITANIUM COMPOUNDS and CHLORIDES.

HBZ000 ***HR: 3***
HEXAAQUACOBALT(II) PERCHLORATE
mf: $Cl_2CoH_{12}O_{14}$ mw: 365.98

CONSENSUS REPORTS: Cobalt and its compounds are on the Community Right-To-Know List.

SAFETY PROFILE: Explodes on impact. When heated to decomposition it emits toxic fumes of Cl^-. See also COBALT COMPOUNDS and PERCHLORATES.

HCA000 CAS:22295-99-4 ***HR: 3***
1,1,3,3,5,5-HEXAAZIDO-2,4,6-TRIAZA-1,3,5-TRIPHOSPHORINE
mf: $N_{21}P_3$ mw: 387.07

$(N_3)_2P{=}NP(N_3)_2{=}NP(N_3)_2{=}N$ (ring: terminal N bonded back to first P)

SAFETY PROFILE: A violent explosive sensitive to shock or friction. Very unstable. Upon decomposition it emits very toxic fumes of NO_x and PO_x. See also AZIDES.

HCA275 CAS:23777-80-2 *HR: 3*
HEXABORANE(10)
mf: B_6H_{10} mw: 74.94

SAFETY PROFILE: Ignites spontaneously in air. See also BORANES and BORON COMPOUNDS.

HCA285 CAS:28375-94-2 *HR: 3*
HEXABORANE(12)
mf: B_6H_{12} mw: 76.95

SAFETY PROFILE: An unstable material which ignites spontaneously in air. See also BORANES and BORON COMPOUNDS.

HCA500 CAS:36355-01-8 *HR: 3*
HEXABROMOBIPHENYL
mf: $C_{12}H_4Br_6$ mw: 627.62

SYNS: HBB ◇ NCI-C53634 ◇ POLYBROMINATED BIPHENYL

TOXICITY DATA with REFERENCE
orl-rat TDLo:530 mg/kg (30W pre):REP FCTOD7 22,743,84
orl-rat LD50:21500 mg/kg MFLRA3 38,709,73
skn-rbt LDLo:5 g/kg AIHAAP 38,307,77

CONSENSUS REPORTS: NTP Fifth Annual Report on Carcinogens. Polybrominated biphenyl compounds are on the Community Right-To-Know List. Reported in EPA TSCA Inventory.

SAFETY PROFILE: Experimental reproductive effects. Mildly toxic by skin contact. When heated to decomposition it emits toxic fumes of Br^-. See also POLYBROMINATED BIPHENYLS.

HCA600 CAS:56480-06-9 *HR: D*
HEXABROMONAPHTHALENE
mf: $C_{10}H_2Br_6$ mw: 601.58

SYN: NAPHTHALENE, HEXABROMO-

TOXICITY DATA with REFERENCE
orl-mus TDLo:50 mg/kg (female 6-15D post):TER FAATDF 7,398,86
orl-mus TDLo:50 mg/kg (female 6-15D post):REP FAATDF 7,398,86

SAFETY PROFILE: An experimental teratogen. Experimental reproductive effects. When heated to decomposition it emits toxic fumes of Br^-.

HCA700 CAS:4808-30-4 *HR: 3*
1,1,1,3,3,3-HEXABUTYLDISTANNTHIANE
mf: $C_{24}H_{54}SSn_2$ mw: 612.22

SYNS: BIS(TRIBUTYLTIN)SULFIDE ◇ DISTANNATHIANE, HEXABUTYL-(9CI) ◇ DISTANNTHIANE, HEXABUTYL- ◇ HEXABUTYLDISTANNTHIANE ◇ TRIBUTYLTIN SULFIDE

TOXICITY DATA with REFERENCE
orl-mus LDLo:470 mg/kg AECTCV 14,111,85
ipr-mus LD50:144 mg/kg RPTOAN 42,73,79

OSHA PEL: TWA 0.1 mg(Sn)/m^3
ACGIH TLV: TWA 0.1 mg(Sn)/m^3; STEL 0.2 mg/m^3 (skin)
NIOSH REL: (Organotin Compounds): 10H TWA 0.1 mg(Sn)/m^3

CONSENSUS REPORTS: Reported in EPA TSCA Inventory.

SAFETY PROFILE: Poison by intraperitoneal route. Moderately toxic by ingestion. When heated to decomposition it emits toxic fumes of SO_x and Sn.

HCB000 CAS:13007-92-6 *HR: 3*
HEXACARBONYLCHROMIUM
mf: C_6CrO_6 mw: 220.06

$(OC)_6Cr$

SYNS: CHROMIUM CARBONYL (MAK) ◇ CHROMIUM CARBONYL (OC-6-11) (9CI) ◇ CHROMIUM HEXACARBONYL ◇ HEXACARBONYL CHROMIUM

TOXICITY DATA with REFERENCE
imp-rat TDLo:75 mg/kg:ETA CNREA8 37,1476,77
ivn-mus LD50:30 mg/kg AQMOAC #70-15,70

CONSENSUS REPORTS: IARC Cancer Review: Animal Inadequate Evidence IMEMDT 23,205,80; Chromium and its compounds are on the Community Right-To-Know List. Reported in EPA TSCA Inventory.

OSHA PEL: CL 0.1 mg(CrO_3)/m^3
ACGIH TLV: TWA 0.05 mg(Cr)/m^3; Confirmed Human Carcinogen
DFG TRK: 0.1 mg/m^3 calculated as CrO_3 in that portion of dust that can possibly be inhaled; 0.2 mg/m^3 arc-welding by hand; others 0.1 mg/m^3. Animal Carcinogen, Suspected Human Carcinogen.
NIOSH REL: (Chromium(VI)) TWA 0.001 mg(Cr (VI))/m^3

SAFETY PROFILE: Confirmed carcinogen with experimental tumorigenic data. Poison by intravenous route. Explodes at 210°C. See also CHROMIUM COMPOUNDS and CARBONYLS.

HCB500 CAS:13939-06-5 *HR: 3*
HEXACARBONYLMOLYBDENUM
mf: C_6MoO_6 mw: 264.00

$(CO)_6Mo$

SAFETY PROFILE: Solutions in diethyl ether may explode in storage. When heated to decomposition it emits acrid smoke and fumes. See also MOLYBDENUM COMPOUNDS and CARBONYLS.

HCC000 CAS:14040-11-0 **HR: 2**
HEXACARBONYLTUNGSTEN
mf: C_6O_6W mw: 351.92

$(CO)_6W$

SAFETY PROFILE: Dangerous during preparation procedures. When heated to decomposition it emits toxic fumes of CO. See also TUNGSTEN COMPOUNDS and CARBONYLS.

HCC475 CAS:14024-00-1 **HR: 3**
HEXACARBONYL VANADIUM
mf: C_6O_6V mw: 219.00

$(CO)_6V$

SAFETY PROFILE: Ignites spontaneously in air. When heated to decomposition it emits acrid toxic fumes of VO_x. See also VANADIUM COMPOUNDS and CARBONYLS.

HCC500 CAS:118-74-1 **HR: 3**
HEXACHLOROBENZENE
DOT: UN 2729
mf: C_6Cl_6 mw: 284.76

PROP: Monoclinic prisms. Mp: 231°, bp: 323-326°, flash p: 468°F, vap press: 1 mm @ 114.4°, vap d: 9.8. D: 2.44. Insol in water; sol in benzene; very sltly sol in hot alc; sol in hot ether and chloroform.

SYNS: AMATIN ◇ ANTICARIE ◇ BUNT-CURE ◇ BUNT-NO-MORE ◇ CO-OP HEXA ◇ ESACHLOROBENZENE (ITALIAN) ◇ GRANOX NM ◇ HCB ◇ HEXA C.B. ◇ HEXACHLORBENZOL (GERMAN) ◇ JULIN'S CARBON CHLORIDE ◇ NO BUNT LIQUID ◇ PENTACHLOROPHENYL CHLORIDE ◇ PERCHLOROBENZENE ◇ PHENYL PERCHLORYL ◇ RCRA WASTE NUMBER U127 ◇ SAATBEIZFUNGIZID (GERMAN) ◇ SANOCIDE ◇ SMUT-GO ◇ SNIECIOTOX

TOXICITY DATA with REFERENCE
dnd-esc 20 μmol/L MUREAV 89,95,81
mmo-smc 100 ppm RSTUDV 6,161,76
orl-rat TDLo:212 mg/kg (female 14D pre-17D post):REP ARTODN 38,191,77
orl-rat TDLo:6450 mg/kg (female 1-22D post):TER DCTODJ 2,61,79
orl-rat TDLo:2738 mg/kg/2Y-C:CAR PAACA3 24,59,83
orl-mus TDLo:6972 mg/kg/83W-C:NEO IJCNAW 23,47,79
orl-ham TDLo:1000 mg/kg/18W-C:CAR NATUAS 269,510,77
unr-man LDLo:220 mg/kg 85DCAI 2,73,70
orl-rat LD50:10000 mg/kg 85DPAN -,-,71/76
ihl-rat LC50:3600 mg/m^3 85GMAT -,72,82
orl-mus LD50:4 g/kg 85GMAT -,72,82
ihl-mus LC50:4 g/m^3 85GMAT -,72,82
orl-cat LD50:1700 mg/kg 85GMAT -,72,82
ihl-cat LC50:1600 mg/m^3 85GMAT -,72,82
orl-rbt LD50:2600 mg/kg 85GMAT -,72,82
ihl-rbt LC50:1800 mg/m^3 85GMAT -,72,82
orl-mam LD50:1047 mg/kg NTIS** PB288-416

CONSENSUS REPORTS: NTP Fifth Annual Report on Carcinogens. IARC Cancer Review: Group 2B IMEMDT 7,219,87; Animal Sufficient Evidence IMEMDT 20,155,79; Human Limited Evidence IMEMDT 20,155,79. Community Right-To-Know List. Reported in EPA TSCA Inventory. EPA Genetic Toxicology Program.

ACGIH TLV: (Proposed TWA 0.025 mg/m^3; Suspected Human Carcinogen
DFG MAK: BAT: 15 μg/dL in plasma/serum.
DOT Classification: IMO: Poison B; Label: St. Andrews Cross.

SAFETY PROFILE: Confirmed carcinogen with experimental carcinogenic, neoplastigenic, and teratogenic data. A human poison by an unspecified route. A suspected human carcinogenic. Experimental reproductive effects. Mildly toxic experimentally by inhalation. Mutation data reported. A fungicide. Combustible when exposed to heat or flame. Violent reaction with dimethylformamide. To fight fire, use CO_2, dry chemical. When heated to decomposition it emits highly toxic fumes of Cl^-. See also CHLORINATED HYDROCARBONS, AROMATIC.

HCD000 CAS:35065-27-1 **HR: 3**
2,2',4,4',5'5'-HEXACHLORO-1,1'-BIPHENYL
mf: $C_{12}H_4Cl_6$ mw: 360.86

SYNS: 2,2',4,4'5,5'-HEXACHLOROBIPHENYL ◇ 2,4,5,2',4',5'-HEXACHLOROBIPHENYL

TOXICITY DATA with REFERENCE
dnd-mus-orl 36400 μg/kg/5D CBINA8 27,99,79
oms-mus-orl 36400 μg/kg/5D CBINA8 27,99,79
orl-rat TDLo:50 mg/kg (8-18D preg):REP TXAPA9 51,233,79
orl-mus TDLo:200 mg/kg (female 13-21D post):TER PHTXA6 61,220,87
orl-rat TDLo:3650 mg/kg/104W-C:ETA TXAPA9 48,A181,79

SAFETY PROFILE: An experimental teratogen. Experimental reproductive effects. Questionable carcinogen with experimental tumorigenic data. Mutation data reported. When heated to decomposition it emits toxic fumes of Cl^-. See also CHLORINATED HYDROCARBONS, AROMATIC.

HCD075 CAS:38411-22-2 ***HR: D***
2,2',3,3',6,6'-HEXACHLORO-1,1'-BIPHENYL
mf: $C_{12}H_4Cl_6$ mw: 360.86

SYN: 2,3,6,2',3',6'-HEXACHLOROBIPHENYL

TOXICITY DATA with REFERENCE
dnd-mus-orl 36400 μg/kg/5D CBINA8 27,99,79
oms-mus-orl 36400 μg/kg/5D CBINA8 27,99,79
orl-mam TDLo:563 μg/kg (30D pre):REP TOXID9 4,83,84

SAFETY PROFILE: Experimental reproductive effects. Mutation data reported. When heated to decomposition it emits toxic fumes of Cl^-. See also 3,3',4,4',5,5'-HEXACHLOROBIPHENYL; and CHLORINATED HYDROCARBONS, AROMATIC.

HCD100 CAS:32774-16-6 ***HR: 3***
3,3',4,4',5,5'-HEXACHLOROBIPHENYL
mf: $C_{12}H_4Cl_6$ mw: 360.86

SYN: 3,4,5,3',4',5'-HEXACHLOROBIPHENYL

TOXICITY DATA with REFERENCE
orl-mus TDLo:40 mg/kg (female 6-15D post):TER APTOD9 19,A22,80
orl-mus TDLo:40 mg/kg (female 6-15D post):REP APTOD9 19,A22,80
orl-gpg LD50:223 μg/kg PSSID2 5,367,82

SAFETY PROFILE: Deadly poison by ingestion. Experimental teratogenic and reproductive effects. When heated to decomposition it emits toxic fumes of Cl^-. See also CHLORINATED HYDROCARBONS, AROMATIC.

HCD250 CAS:87-68-3 ***HR: 3***
HEXACHLOROBUTADIENE
DOT: UN 2279
mf: C_4Cl_6 mw: 260.74

PROP: Autoign temp: 1130°F, vap d: 8.99.

SYNS: DOLEN-PUR ◇ GP-40-66:120 ◇ HCBD ◇ HEXACHLOR-1,3-BUTADIEN (CZECH) ◇ HEXACHLORO-1,3-BUTADIENE (MAK) ◇ 1,1,2,3,4,4-HEXACHLORO-1,3-BUTADIENE ◇ PERCHLOROBUTADIENE ◇ RCRA WASTE NUMBER U128

TOXICITY DATA with REFERENCE
skn-rbt 810 mg/24H MOD JETOAS 9,171,76
eye-rbt 162 mg MLD JETOAS 9,171,76
mma-sat 320 μg/plate CRNGDP 7,431,86
dns-rat-orl 77 g/kg/11W TXAPA9 60,287,81
dns-ham:emb 2 mg/L CALEDQ 23,297,84
otr-ham:emb 10 mg/L CALEDQ 23,297,84
orl-rat TDLo:45 mg/kg (male 13W pre):REP TXAPA9 42,387,77
ipr-rat TDLo:150 mg/kg (female 1-15D post):TER SWEHDO 7(Suppl 4),66,81
orl-rat TDLo:15 g/kg/2Y-C:CAR AIHAAP 38(1),589,77
orl-rat LD50:90 mg/kg HYSAAV 31,18,66
ipr-rat LD50:175 mg/kg JETOAS 8(3),180,75
orl-mus LD50:110 mg/kg SCCUR* -,5,61
ihl-mus LCLo:235 ppm/4H SCCUR* -,5,61
ipr-mus LD50:76 mg/kg JETOAS 7(4),247,74
skn-rbt LD50:1211 mg/kg APTOA6 43,346,78
orl-gpg LD50:90 mg/kg GISAAA 28,9,63
orl-ham LD50:960 mg/kg TXAPA9 48,A192,79

CONSENSUS REPORTS: IARC Cancer Review: Group 3 IMEMDT 7,56,87; Animal Suspected IMEMDT 20,179,79. Community Right-To-Know List. Reported in EPA TSCA Inventory.

OSHA PEL: TWA 0.02 ppm
ACGIH TLV: TWA 0.02 ppm (skin); Suspected Human Carcinogen.
DFG MAK: Suspected Carcinogen.
DOT Classification: Poison B; Label: St. Andrews Cross.

SAFETY PROFILE: Suspected carcinogen with experimental carcinogenic data. Poison by ingestion and intraperitoneal routes. Moderately toxic by inhalation and skin contact. A skin and eye irritant. An experimental teratogen. Experimental reproductive effects. Mutation data reported. Combustible when exposed to heat or flame; can react vigorously with oxidizing materials. To fight fire, use dry chemical, CO_2, alcohol foam, water spray, fog, mist. Reacts with bromine perchlorate to form an explosive product. When heated to decomposition it emits very toxic fumes of Cl^-. A solvent, heat transfer fluid, transformer, hydraulic fluid, and wash liquor.

HCD500 CAS:26523-63-7 ***HR: 2***
HEXACHLOROBUTANE
mf: $C_4H_4Cl_6$ mw: 264.78

SYN: PCB2

TOXICITY DATA with REFERENCE
orl-mus LD50:2000 mg/kg GISAAA 28,9,63
orl-gpg LD50:940 mg/kg GISAAA 28,9,63

SAFETY PROFILE: Moderately toxic by ingestion. When heated to decomposition it emits toxic fumes of Cl^-. See also CHLORINATED HYDROCARBONS, ALIPHATIC.

HCE000 CAS:599-52-0 ***HR: 3***
HEXACHLORO-2,5-CYCLOHEXADIEN-1-ONE
mf: C_6Cl_6O mw: 300.76

SYNS: 2,3,4,4,5,6-HEXACHLORCYKLOHEXA-2,5-DIEN-1-ON

(CZECH) ◇ HEXACHLORFENOL (CZECH) ◇ HEXACHLORO-2,5-CYCLOHEXADIENONE ◇ USAF DO-65

TOXICITY DATA with REFERENCE
skn-rbt 500 mg/24H SEV 28ZPAK -,86,72
eye-rbt 100 mg/24H MOD 28ZPAK -,86,72
orl-rat LD50:218 mg/kg 28ZPAK -,86,72
ipr-mus LD50:50 mg/kg NTIS** AD277-689

SAFETY PROFILE: A poison by ingestion and intraperitoneal routes. A eye and skin and eye irritant. When heated to decomposition it emits toxic fumes of Cl^-.

HCE400 ***HR: 2***
HEXACHLOROCYCLOHEXANE, delta and epsilon mixture
mf: $C_6H_6Cl_6$ mw: 290.82

SYN: Δ-HEXACHLOROCYCLOHEXANE mixed with epsilon-HEXACHLOROCYCLOHEXANE

TOXICITY DATA with REFERENCE
orl-mus TDLo:12960 mg/kg/26W-C:CAR CMSHAF 1,279,72

SAFETY PROFILE: Questionable carcinogen with experimental carcinogenic data. When heated to decomposition it emits toxic fumes of Cl^-.

HCE500 CAS:77-47-4 ***HR: 3***
HEXACHLOROCYCLOPENTADIENE
DOT: UN 2646
mf: C_5Cl_6 mw: 272.75

$$\overline{ClC{=}CClCCl_2CCl{=}CCl}$$

PROP: Yellow- to amber-colored liquid, pungent odor. Mp: 9.9°, bp: 239°, fp: −2°, flash p: none (OC), d: 1.715 @ 15.5°/15.5°, vap d: 9.42.

SYNS: C-56 ◇ HCCPD ◇ HEXACHLORCYKLOPENTADIEN (CZECH) ◇ NCI-C55607 ◇ PCL ◇ RCRA WASTE NUMBER U130

TOXICITY DATA with REFERENCE
skn-mky 10 mg SEV AMIHAB 11,459,55
skn-rbt 500 mg/4H SEV VELPB* 50101-2,76
eye-rbt 20 mg/24H MOD 28ZPAK -,30,72
eye-rbt 100 mg/5M SEV VELPB* 50101-2,76
skn-gpg 20 mg MLD AMIHAB 11,459,55
orl-rbt TDLo:975 mg/kg (6-18D preg):TER TXAPA9 53,497,80
orl-rat LD50:1300 mg/kg TSCAT* OTS0513386
ihl-rat LC50:1600 ppb/4H JTEHD6 9,743,82
orl-mus LD50:505 mg/kg JAFCAU 23,967,75
ihl-mus LCLo:1500 ppb/7H TXAPA9 53,497,80
orl-rbt LDLo:420 mg/kg PCOC** -,586,66
ihl-rbt LCLo:1500 ppb/7H TXAPA9 53,497,80
skn-rbt LD50:430 mg/kg 34ZIAG -,308,69

CONSENSUS REPORTS: EPA Extremely Hazardous Substances List. Community Right-To-Know List. Reported in EPA TSCA Inventory.

OSHA PEL: TWA 0.01 ppm
ACGIH TLV: TWA 0.01 ppm
DOT Classification: Corrosive Material; Label: Corrosive; IMO: Poison B; Label: Poison.

SAFETY PROFILE: A deadly poison by inhalation. Moderately toxic by ingestion and skin contact. Experimental teratogenic effects. Corrosive. A severe skin and eye irritant. May explode on contact with sodium. When heated to decomposition it emits toxic fumes of Cl^-. See also CHLORINATED HYDROCARBONS, ALIPHATIC.

HCF000 CAS:57653-85-7 ***HR: 3***
1,2,3,4,7,8-HEXACHLORODIBENZO-p-DIOXIN
mf: $C_{12}H_2Cl_6O_2$ mw: 390.84

TOXICITY DATA with REFERENCE
orl-mus LD50:825 μg/kg TXAPA9 44,335,78
orl-gpg LD50:72500 ng/kg TXAPA9 44,335,78

CONSENSUS REPORTS: IARC Cancer Review: Animal Inadequate Evidence IMEMDT 15,41,77.

SAFETY PROFILE: A deadly poison by ingestion. Questionable carcinogen. When heated to decomposition it emits toxic fumes of Cl^-.

HCF500 ***HR: 3***
1,2,3,6,7,8-HEXACHLORODIBENZO-p-DIOXIN mixed with 1,2,3,7,8,9-HEXACHLORODIBENZO-p-DIOXIN

PROP: Composed of 67% of 1,2,3,7,8,9-hexachlorodibenzo-p-dioxin and 31% of 1,2,3,6,7,8-hexachlorodibenzo-p-dioxin NCITR* NCI-CG-TR-198,80

SYNS: 1,2,3,7,8,9-HEXACHLORODIBENZO-p-DIOXIN mixed with 1,2,3,6,7,8-HEXACHLORODIBENZO-p-DIOXIN ◇ NCI-C03703

TOXICITY DATA with REFERENCE
orl-mus TDLo:520 μg/kg/2Y-I:CAR NCITR* NCI-CG-TR-198,80
orl-rat LD50:800 μg/kg NCITR* NCI-CG-TR-198,80
orl-mus LD50:500 μg/kg NCITR* NCI-CG-TR-198,80

CONSENSUS REPORTS: NCI Carcinogenesis Bioassay (gavage); Clear Evidence: mouse, rat NCITR* NCI-CG-TR-198,80. NCI Carcinogenesis Bioassay (dermal); No Evidence: mouse NCITR* NCI-CG-TR-202,80.

SAFETY PROFILE: Suspected carcinogen with experimental carcinogenic data. A deadly poison by ingestion. When heated to decomposition it emits very toxic fumes of Cl^- and dioxin.

HCG000 ***HR: 3***
1,2,3,6,7,8-HEXACHLORODIBENZODIOXIN mixed with PENTACHLORO ISOMERS and HEPTACHLORO ISOMERS 96.8% : 1.97% : 1.23%

TOXICITY DATA with REFERENCE
orl-rat LD50:750 μg/kg NCIIR* NO1-CP-12338
orl-mus LD50:500 μg/kg NCIIR* NO1-CP-12338

SAFETY PROFILE: A deadly poison by ingestion. When heated to decomposition it emits very toxic fumes of Cl^-.

HCH400 CAS:70648-26-9 ***HR: D***
1,2,3,4,7,8-HEXACHLORODIBENZOFURAN
mf: $C_{12}H_2Cl_6O$ mw: 374.84

SYN: DIBENZOFURAN,1,2,3,4,7,8-HEXACHLORO-

TOXICITY DATA with REFERENCE
orl-mus TDLo:400 μg/kg (female 10-13D post):TER TXAPA9 90,206,87

SAFETY PROFILE: Experimental reproductive effects. When heated to decomposition it emits toxic fumes of Cl^-.

HCH500 CAS:13465-77-5 ***HR: 3***
HEXACHLORODISILANE
mf: Cl_6Si_2 mw: 268.83

SAFETY PROFILE: Potentially explosive reaction with chlorine above 300°C. Corrosive irritant to the eyes, skin and mucous membranes. When heated to decomposition or on contact with water it emits toxic fumes of HCl and Cl^-. See also CHLOROSILANES.

HCI000 CAS:67-72-1 ***HR: 3***
HEXACHLOROETHANE
DOT: NA 9037
mf: C_2Cl_6 mw: 236.72

PROP: Rhombic, triclinic, or cubic crystals, colorless, camphor-like odor. Mp: 186.6° (subl), d: 2.091, vap press: 1 mm @ 32.7°, bp: 186.8° (triple point). Sol in alc, benzene, chloroform, ether, oils; insol in water.

SYNS: AVLOTANE ◇ CARBON HEXACHLORIDE ◇ DISTOKAL ◇ DISTOPAN ◇ DISTOPIN ◇ EGITOL ◇ ETHANE HEXACHLORIDE ◇ ETHYLENE HEXACHLORIDE ◇ FALKITOL ◇ FASCIOLIN ◇ HEXACHLOR-AETHAN (GERMAN) ◇ 1,1,1,2,2,2-HEXACHLOROETHANE ◇ HEXACHLOROETHYLENE ◇ MOTTENHEXE ◇ NCI-C04604 ◇ PERCHLOROETHANE ◇ PHENOHEP ◇ RCRA WASTE NUMBER U131

TOXICITY DATA with REFERENCE
orl-rat TDLo:5500 mg/kg (6-16D preg):REP AIHAAP 40,187,79
orl-mus TDLo:230 g/kg/78W-I:CAR NCITR* NCI-CG-TR-68,78
orl-mus TD:460 g/kg/78W-I:CAR NCITR* NCI-CG-TR-68,78
orl-rat LD50:4460 mg/kg AIHAAP 40,187,79
ipr-rat LDLo:2900 mg/kg AIHAAP 40,187,79
ipr-mus LD50:4500 mg/kg ARZNAD 11,902,61
ivn-dog LDLo:325 mg/kg QJPPAL 7,205,34
skn-rbt LD50:32 g/kg AIHAAP 40,187,79
scu-rbt LDLo:4000 mg/kg QJPPAL 7,205,34
orl-gpg LD50:4970 mg/kg AIHAAP 40,187,79

CONSENSUS REPORTS: IARC Cancer Review: Group 3 IMEMDT 7,56,87; Animal Limited Evidence IMEMDT 20,467,79. NCI Carcinogenesis Bioassay (gavage); Clear Evidence: mouse NCITR* NCI-CG-TR-68,78. NCI Carcinogenesis Bioassay (gavage); No Evidence: rat NCITR* NCI-CG-TR-68,78. Community Right-To-Know List. Reported in EPA TSCA Inventory. EPA Genetic Toxicology Program.

OSHA PEL: TWA 1 ppm (skin)
ACGIH TLV: TWA 1 ppm; (Proposed: TWA 1 ppm; Suspected Human Carcinogen
DFG MAK: 1 ppm (10 mg/m^3)
NIOSH REL: (Hexachloroethane) Reduce to lowest level
DOT Classification: ORM-A; Label: None.

SAFETY PROFILE: Suspected carcinogen with experimental carcinogenic data. A poison by intravenous route. Moderately toxic by intraperitoneal route. Mildly toxic by ingestion. Experimental reproductive effects. Liver injury has resulted from exposure to this material. An insecticide. Slightly explosive by spontaneous chemical reaction. Dehalogenation of this material by reaction with alkalies, metals, etc., will produce spontaneous explosive chloroacetylenes. When heated to decomposition it emits highly toxic fumes of Cl^- and phosgene. See also CHLORINATED HYDROCARBONS, ALIPHATIC.

HCI475 CAS:14458-95-8 ***HR: 3***
5,6,7,8,9,9-HEXACHLORO-1,4,4a,5,8,8a-HEXAHYDRO-1,4:5,8-DIMETHANOPHTHALAZINE-2-OXIDE
mf: $C_{10}H_6Cl_6N_{2O}$ mw: 382.88

SYNS: ENT 25,582 ◇ SD 3450 ◇ SHELL SD-3450

TOXICITY DATA with REFERENCE
orl-rat LD50:2800 μg/kg TXAPA9 21,315,72
orl-mus LDLo:5500 μg/kg AECTCV 14,111,85
orl-bwd LD50:25 mg/kg TXAPA9 21,315,72

SAFETY PROFILE: Poison by ingestion. When heated to decomposition it emits toxic fumes of Cl^- and NO_x.

HCK000 CAS:2592-62-3 ***HR: 3***
6,7,8,9,10,10-HEXACHLORO-1,5,5a,6,9,9a-HEXAHYDRO-3-METHYL-6,9-METHANO-2,4-BENZDIOXEPIN
mf: $C_{11}H_{10}Cl_6O_2$ mw: 386.91

SYNS: BAYER 38920 ◇ ENT 25,700-X ◇ 6,7,8,9,10,10-HEXACHLORO-1,5,5a,6,9,9a-HEXAHYDRO-6,9-METHANO-3-METHYL-2,4-BENZODIOXEPIN

TOXICITY DATA with REFERENCE
orl-rat LD50:120 mg/kg ARSIM* 20,4,66
orl-bwd LD50:50 mg/kg TXAPA9 21,315,72

SAFETY PROFILE: Poison by ingestion. When heated to decomposition it emits toxic fumes of Cl^-.

HCK500 CAS:1335-87-1 ***HR: 3***
HEXACHLORONAPHTHALENE
mf: $C_{10}H_2Cl_6$ mw: 334.82

PROP: White solid.

CONSENSUS REPORTS: Community Right-To-Know List. Reported in EPA TSCA Inventory.

OSHA PEL: TWA 0.2 mg/m^3 (skin)
ACGIH TLV: TWA 0.2 mg/m^3 (skin)

SAFETY PROFILE: A poison by ingestion, skin contact, and inhalation. Causes severe acne-form eruptions and toxic narcosis of liver. Absorbed by skin. When heated to decomposition it emits toxic fumes of Cl^-. See also CHLORINATED HYDROCARBONS, AROMATIC.

HCL000 CAS:70-30-4 ***HR: 3***
HEXACHLOROPHENE
DOT: UN 2875
mf: $C_{13}H_6Cl_6O_2$ mw: 406.89

PROP: Crystals, water insol. Mp: 165°. Sol in alc, acetone, ether, chloroform, propylene glycol, polyethylene glycols, olive oil, cottonseed oil, dil solns of alkalies.

SYNS: ACIGENA ◇ ALMEDERM ◇ AT 7 ◇ B32 ◇ BILEVON ◇ BIS(2-HYDROXY-3,5,6-TRICHLOROPHENYL)METHANE ◇ BIS-2,3,5-TRI-CHLOR-6-HYDROXYFENYLMETHAN (CZECH) ◇ BIS(3,5,6-TRI-CHLORO-2-HYDROXYPHENYL)METHANE ◇ COMPOUND G-11 ◇ COTOFILM ◇ DERMADEX ◇ 2,2'-DIHYDROXY-3,3',5,5',6,6'-HEXACHLORODIPHENYLMETHANE ◇ 2,2'-DIHYDROXY-3,5,6,3',5',6'-HEXACHLORODIPHENYLMETHANE ◇ EXOFENE ◇ FOMAC ◇ FOSTRIL ◇ G-11 ◇ GAMOPHENE ◇ G-ELEVEN ◇ GERMA-MEDICA ◇ HCP ◇ HEXABALM ◇ 2,2',3,3',5,5'-HEXACHLORO-6,6' -DIHYDROXYDIPHENYLMETHANE ◇ HEXACHLOROFEN (CZECH) ◇ HEXACHLOROPHANE ◇ HEXACHLOROPHEN ◇ HEXACHLOROPHENE (DOT) ◇ HEXAFEN ◇ HEXIDE ◇ HEXOPHENE ◇ HEXOSAN ◇ ISOBAC 20 ◇ 2,2'-METHYLENEBIS(3,4,6-TRICHLOROPHENOL) ◇ NABAC ◇ NCI-C02653 ◇ NEOSEPT ◇ PHISODANV ◇ PHISOHEX ◇ RCRA WASTE NUMBER U132 ◇ RITOSEPT ◇ SEPTISOL ◇ SEPTOFEN ◇ STERAL ◇ STERASKIN ◇ SURGI-CEN ◇ SUROFENE ◇ TERSASEPTIC ◇ TRICHLOROPHENE ◇ TURGEX

TOXICITY DATA with REFERENCE
skn-hmn 50 μg/24H MLD JSCCA5 25,113,74
eye-dog 3 mg/24H AAOPAF 53,817,55
skn-rbt 30 μg open MLD JSCCA5 25,113,74
eye-rbt 1500 mg/24H AAOPAF 53,817,55
skn-gpg 30 μg open MLD JSCCA5 25,113,74
skn-gpg 1250 μg/24H MLD JSCCA5 25,113,74
orl-rat TDLo:137 mg/kg (male 14W pre):REP JAFCAU 23,866,75
ivg-rat TDLo:300 mg/kg (female 8-11D post):TER AEHLAU 28,43,74
ihl-rat TCLo:33500 μg/m^3/52W:NEO VOONAW 33(1),62,87
skn-mus TDLo:8400 mg/kg/21W-I:ETA CNREA8 19,413,59
orl-inf TDLo:257 mg/kg/7D-I:CNS,GIT AJDCAI 111,333,66
orl-cld LDLo:250 mg/kg MJAUAJ 1,737,63
orl-wmn TDLo:600 mg/kg:CVS,GIT JAMAAP 181,587,62
orl-rat LD50:56 mg/kg TXAPA9 25,332,73
ihl-rat LC50:340 mg/m^3 GISAAA 49(8),25,84
skn-rat LD50:1840 mg/kg GISAAA 47(10),26,82
ipr-rat LD50:22 mg/kg TXAPA9 24,239,73
scu-rat LD50:14700 mg/kg 26UZAB 6,245,68/70
ivn-rat LD50:7500 μg/kg TXAPA9 25,332,73
orl-mus LD50:67 mg/kg CMJOAP 82,691,63
ihl-mus LC50:290 mg/m^3 GISAAA 49(8),25,84
skn-mus LD50:270 mg/kg GISAAA 47(10),26,82
ipr-mus LD50:20 mg/kg ZBPHA6 234,110,76
orl-dog LDLo:40 mg/kg VEARA6 35,35,65
ivn-dog LDLo:5 mg/kg SURGAZ 24,542,48

CONSENSUS REPORTS: IARC Cancer Review: Group 3 IMEMDT 7,56,87; Human Inadequate Evidence IMEMDT 20,241,79. NCI Carcinogenesis Bioassay (feed); No Evidence: rat NCITR* NCI-CG-TR-40,78. Reported in EPA TSCA Inventory. Chlorophenols are on the Community Right-To-Know List.

DOT Classification: IMO: Poison B; Label: St. Andrews Cross.

SAFETY PROFILE: A human poison by ingestion. An experimental poison by ingestion, intraperitoneal, and intravenous routes. Moderately toxic by skin contact. Human systemic effects by ingestion: cardiomyopathy (damage to the heart muscle), nausea or vomiting, diarrhea, shock. Unspecified human reproductive effects. Experimental teratogenic and reproductive effects. An eye and human skin irritant. Questionable carcinogen with experimental neoplastigenic and tumorigenic data. Strong concentrations may be irritating, but ordinary use of 1-2% solutions is not.

For many years, the toxicologic hazard of hexachlorophene was unrecognized and the compound had a wide and virtually unrestricted use. However, studies by FDA scientists demonstrated that brain lesions occur from exposure in both rats and monkeys treated at levels only slightly higher than those of persons using soaps, toothpaste, shampoos, and a variety of other household products and cosmetics containing it. The FDA has now restricted sale of hexachlorophene, and most preparations containing higher levels of the compound are available only through prescription. In the recent FDA studies, it was found that 2 weeks after onset of exposure, rats fed 500 ppm (25 mg/kg/day) of hexachlorophene in their diet showed weakness in their hindquarters which progressed to paralysis. Microscopic examination of the brain and spinal cord of these rats revealed a particular edema of the white matter resembling spongy degeneration noted in infants. When the animals were removed from the diet, they recovered gradually over a period of weeks. Similar symptoms were noted in the monkey. Following ingestion of hexachlorophene, early symptoms are primarily gastrointestinal in nature and include anorexia, nausea, vomiting, abdominal cramps, and diarrhea. Dehydration is sometimes severe and may be associated with shock.

Used as a germicidal agent. An additive permitted in the feed and drinking water of animals and/or for the treatment of food-producing animals. Also permitted in food for human consumption. When heated to decomposition it emits toxic fumes of Cl^-. See also CHLOROPHENOLS.

HCL300 ***HR: D***
HEXACHLOROPLATINATE(2-) DISODIUM HEXAHYDRATE
mf: $Cl_6Pt{\bullet}2Na{\bullet}6H_2O$ mw: 653.85

SYN: SODIUM HEXACHLOROPLATINATE HEXAHYDRATE

TOXICITY DATA with REFERENCE
scu-mus TDLo:56987 μg/kg (female 12D post):REP JTEHD6 13,879,84

OSHA PEL: TWA 0.002 mg(Pt)/m^3
ACGIH TLV: TWA 0.002 mg(Pt)/m^3

SAFETY PROFILE: Experimental reproductive effects. When heated to decomposition it emits toxic fumes of Cl^-.

HCL500 CAS:116-16-5 ***HR: 3***
HEXACHLORO-2-PROPANONE
DOT: UN 2661
mf: C_3Cl_6O mw: 264.73

PROP: Liquid. Bp: 203°, fp: −2°, vap d: 9.2.

SYNS: GC-1106 ◇ HCA ◇ HEXACHLOROACETONE (DOT) ◇ 1,1,1,3,3,3-HEXACHLORO-2-PROPANONE

TOXICITY DATA with REFERENCE
mmo-sat 100 mg/plate BECTA6 24,590,80
dnr-esc 700 μg/plate MUREAV 89,137,81
mrc-smc 10 uL/L MUREAV 155,53,85
orl-rat LD50:1290 mg/kg RREVAH 10,97,65
ihl-rat LC50:360 ppm/6H TXAPA9 7,592,65
skn-rat LD50:2980 mg/kg TXAPA9 7,592,65
ihl-mus LC50:920 mg/m^3/2M 85GMAT -,72,82
orl-dog LD50:700 mg/kg PCOC** -,580,66
skn-rbt LD50:2980 mg/kg 85DPAN -,-,71/76

CONSENSUS REPORTS: Reported in EPA TSCA Inventory.

DOT Classification: IMO: Poison B; Label: Poison.

SAFETY PROFILE: A poison. Moderately toxic by ingestion, inhalation, and skin contact. Mutation data reported. When heated to decomposition it emits toxic fumes of Cl^-.

HCM000 CAS:1888-71-7 ***HR: 3***
HEXACHLOROPROPENE
mf: C_3Cl_6 mw: 248.73

SYN: HEXACHLOROPROPYLENE

TOXICITY DATA with REFERENCE
ihl-rat LC50:425 ppm/30M XEURAQ MDDC-1715
ipr-rat LD50:400 mg/kg XEURAQ MDDC-1715
ihl-mus LCLo:300 ppm/30M XEURAQ MDDC-1715
ipr-mus LDLo:64 mg/kg CBCCT* 1,46,49
ihl-rbt LCLo:85 ppm/30M XEURAQ MDDC-1715

CONSENSUS REPORTS: Reported in EPA TSCA Inventory.

SAFETY PROFILE: A poison by inhalation and intraperitoneal routes. A powerful irritant. When heated to decomposition it emits toxic fumes of Cl^-. See also CHLORINATED HYDROCARBONS, ALIPHATIC.

HCM500 CAS:68-36-0 ***HR: 1***
α,α′-HEXACHLOROXYLENE
mf: $C_8H_4Cl_6$ mw: 312.82

SYNS: 1:4-BIS-TRICHLOROMETHYL BENZENE ◇ α,α,α,α′,α′,α′-HEXACHLORO-p-XYLENE

TOXICITY DATA with REFERENCE
orl-rat TDLo:2330 mg/kg (26W male):REP GNAMAP 21,34,82
orl-rat LD50:3200 mg/kg GNAMAP 21,34,82

CONSENSUS REPORTS: Reported in EPA TSCA Inventory.

SAFETY PROFILE: Moderately toxic by ingestion. Experimental reproductive effects. When heated to decomposition it emits toxic fumes of Cl^-. See also CHLORINATED HYDROCARBONS, AROMATIC.

HCN000 CAS:3734-48-3 ***HR: 2***
4,5,6,7,8,8-HEXACHLOR-$\Delta^{1,5}$-TETRAHYDRO-4,7-METHANOINDEN
mf: $C_{10}H_6Cl_6$ mw: 338.86

SYN: ADDUKT HEXACHLORCYKLOPENTADIENU S CYKLOPENTADIENEM (CZECH) ◇ CHLORDENE ◇ 4,5,6,7,8,8-HEXACHLORO-3a,4,7,7a-TETRAHYDRO-4,7-METHANOINDENE ◇ 4,7-METHANOINDENE,4,5,6,7,8,8-HEXACHLORO-3a,4,7,7a-TETRAHYDRO-

TOXICITY DATA with REFERENCE
eye-rbt 500 mg/24H MLD 85JCAE -,177,86
orl-rat TDLo:6 g/kg (female 6-15D post):TER EPASR* 8EHQ-1085-0570
orl-hmn LDLo:583 mg/kg YKYUA6 30,505,79
ihl-hmn LCLo:2 g/m^3 YKYUA6 30,505,79
skn-hmn LDLo:69 mg/kg YKYUA6 30,505,79
ihl-rat LC50:2 g/m^3 YKYUA6 30,505,79
skn-rat LD50:690 mg/kg YKYUA6 30,505,79

CONSENSUS REPORTS: Reported in EPA TSCA Inventory.

SAFETY PROFILE: Moderately toxic by ingestion, inhalation, and skin contact. An experimental teratogen. A severe eye irritant. When heated to decomposition it emits toxic fumes of Cl^-. See also CHLORINATED HYDROCARBONS, ALIPHATIC.

HCN100 CAS:56187-09-8 ***HR: 3***
HEXACYCLEN TRISULFATE
mf: $C_{12}H_{30}N_6 \cdot 3H_2O_4S$ mw: 360.51

SYN: 1,4,7,10,13,16-HEXAAZACYCLOOCTADECANE SULFATE (1:3)

TOXICITY DATA with REFERENCE
skn-rbt 100 mg/24H MLD DCTODJ 8,451,85
eye-rbt 50 mg MOD DCTODJ 8,451,85
ipr-rat LD50:852 mg/kg DCTODJ 8,451,85
ipr-mus LD50:327 mg/kg DCTODJ 8,451,85

SAFETY PROFILE: Poison by intraperitoneal route. A skin and eye irritant. When heated to decomposition it emits toxic fumes of SO_x and NO_x. See also SULFATES.

HCO000 CAS:376-18-1 ***HR: 3***
2,2,3,3,4,4,5,5,6,6,7,7,8,8,9,9-HEXADECAFLUORONONANOL
mf: $C_9H_4F_{16}O$ mw: 432.13

SYNS: HEXADECAFLUORO-1-NONANOL ◇ omega-h-HEXADEKAFLUORNONANOL-1 (GERMAN)

TOXICITY DATA with REFERENCE
orl-rat LDLo:4400 mg/kg ZHYGAM 26,9,80
ipr-mus LD50:251 mg/kg ZHYGAM 26,9,80

CONSENSUS REPORTS: Reported in EPA TSCA Inventory.

SAFETY PROFILE: A poison by intraperitoneal route. Mildly toxic by ingestion. When heated to decomposition it emits toxic fumes of F^-.

HCO500 CAS:143-27-1 ***HR: 3***
1-HEXADECANAMINE
mf: $C_{16}H_{35}N$ mw: 241.52

SYNS: ALAMINE 6 ◇ ARMEEN 16D ◇ CETYLAMIN (GERMAN) ◇ CETYLAMINE ◇ N-HEXADECYLAMINE ◇ PALMITYLAMINE

TOXICITY DATA with REFERENCE
orl-mus TDLo:40 mg/kg (9D preg):REP DZZEA7 32,861,77
orl-mus TDLo:40 mg/kg (9D preg):TER DZZEA7 32,861,77
ipr-mus LD50:200 mg/kg NTIS** AD691-490

CONSENSUS REPORTS: Reported in EPA TSCA Inventory.

SAFETY PROFILE: A poison by intraperitoneal route. An experimental teratogen. Other experimental reproductive effects. When heated to decomposition it emits toxic fumes of NO_x. See also AMINES.

HCP000 CAS:36653-82-4 ***HR: 3***
1-HEXADECANOL
mf: $C_{16}H_{34}O$ mw: 242.50

PROP: Solid or leaf-like crystals. Mp: 49.3°, bp: 190° @ 15 mm, d: 0.8176 @ 50°/4°. Insol in water; sol in alc, chloroform, ether.

SYNS: ADOL ◇ ALCOHOL C-16 ◇ ATALCO C ◇ CACHALOT C-50 ◇ CETAFFINE ◇ CETAL ◇ CETALOL CA ◇ CETYL ALCOHOL ◇ CETYLIC ALCOHOL ◇ CETYLOL ◇ CO-1670 ◇ CRODACOL-CAS ◇ CYCLAL CETYL ALCOHOL ◇ DYTOL F-11 ◇ EPAL 16NF ◇ ETHAL ◇ ETHOL ◇ HEXADECANOL ◇ n-HEXADECANOL ◇ HEXADECAN-1-OL ◇ HEXADECYL ALCOHOL ◇ n-HEXADECYL ALCOHOL ◇ LOROL 24 ◇ LOXANOL K ◇ PALMITYL ALCOHOL ◇ PRODUCT 308

TOXICITY DATA with REFERENCE
skn-hmn 75 mg/3D-I MLD 85DKA8 -,127,77
skn-rbt 2600 mg/kg/24H MLD AIHAAP 34,493,73
eye-rbt 82 mg MLD AIHAAP 34,493,73
skn-gpg 100 % MLD FCTXAV 16,683,78
orl-rat LD50:6400 mg/kg FCTXAV 16,683,78
ipr-rat LD50:1600 mg/kg FCTXAV 16,683,78
orl-mus LD50:3200 mg/kg FCTXAV 16,683,78
ipr-mus LD50:1600 mg/kg FCTXAV 16,683,78

CONSENSUS REPORTS: Reported in EPA TSCA Inventory.

SAFETY PROFILE: Moderately toxic by ingestion and intraperitoneal routes. An eye and human skin irritant. Flammable when exposed to heat or flame; can react with oxidizing materials. To fight fire, use foam, CO_2, dry chemical. When heated to decomposition it emits acrid smoke and fumes. See also ALCOHOLS.

HCP500 CAS:54460-46-7 **HR: 2**
HEXADECYL CYCLOPROPANECARBOXYLATE
mf: $C_{20}H_{38}O_2$ mw: 310.58

SYNS: CYCLOPRATE ◇ CYCLOPROPANECARBOXYLIC ACID, HEXADECYL ESTER ◇ ZARDEX ◇ ZR-856

TOXICITY DATA with REFERENCE
orl-rat LD50:12200 mg/kg 85ARAE 1,99,77
orl-dog LD50:2500 mg/kg SPEADM 78-1,19,78
skn-rbt LD50:2670 mg/kg SPEADM 78-1,19,78

SAFETY PROFILE: Moderately toxic by ingestion and skin contact. When heated to decomposition it emits acrid smoke and fumes. See also ESTERS.

HCP550 CAS:59130-69-7 **HR: 2**
HEXADECYL 2-ETHYLHEXANOATE
mf: $C_{24}H_{48}O_2$ mw: 368.72

SYNS: CETYL 2-ETHYLHEXANOATE ◇ HEXANOIC ACID, 2-ETHYL-, HEXADECYL ESTER ◇ PERCELINE OIL

TOXICITY DATA with REFERENCE
skn-man 50 mg/48H MLD CTOIDG 94(8),41,79
skn-rat 100 mg/24H MLD CTOIDG 94(8),41,79
skn-rbt 100 mg/24H SEV CTOIDG 94(8),41,79
skn-gpg 100 mg/24H MOD CTOIDG 94(8),41,79

CONSENSUS REPORTS: Reported in EPA TSCA Inventory.

SAFETY PROFILE: A severe skin irritant. When heated to decomposition it emits acrid smoke and irritating fumes.

HCP600 CAS:67749-11-5 **HR: 1**
HEXADECYL NEODECANOATE
mf: $C_{26}H_{52}O_2$ mw: 396.78

SYNS: HEXANATE D ◇ NEODECANOIC ACID, HEXADECYL ESTER ◇ OCTANOIC ACID, 2,2-DIMETHYL-, HEXADECYL ESTER

TOXICITY DATA with REFERENCE
skn-rat 100 mg/24H MLD CTOIDG 94(8),41,79
skn-rbt 100 mg/24H MOD CTOIDG 94(8),41,79
skn-gpg 100 mg/24H MOD CTOIDG 94(8),41,79

SAFETY PROFILE: A skin irritant. When heated to decomposition it emits acrid smoke and irritating fumes.

HCQ000 CAS:5894-60-0 **HR: 2**
HEXADECYLTRICHLOROSILANE
DOT: UN 1781
mf: $C_{16}H_{33}Cl_3Si$ mw: 359.93

PROP: Colorless to yellow liquid. D: 0.996 @ 25°/25°, bp: 269°, flash p: 295°F (COC).

SYN: TRICHLOROHEXADECYLSILANE

CONSENSUS REPORTS: Reported in EPA TSCA Inventory.

DOT Classification: Corrosive Material; Label: Corrosive.

SAFETY PROFILE: A corrosive irritant to skin, eyes, and mucous membranes. Combustible when exposed to heat or flame. When heated to decomposition or on contact with water it emits toxic fumes of Cl^- and HCl. See also CHLOROSILANES.

HCQ500 CAS:57-09-0 **HR: 3**
HEXADECYLTRIMETHYLAMMONIUM BROMIDE
mf: $C_{19}H_{42}N \cdot Br$ mw: 364.53

SYNS: ACETOQUAT CTAB ◇ BROMAT ◇ CEE DEE ◇ CENTIMIDE ◇ CETAB ◇ CETAROL ◇ CETAVLON ◇ CETRIMIDE ◇ CETRIMONIUM BROMIDE ◇ CETYLAMINE ◇ CETYLTRIMETHYLAMMONIUM BROMIDE ◇ N-CETYLTRIMETHYLAMMONIUM BROMIDE ◇ CIRRASOL-OD ◇ CTAB ◇ CYCLOTON V ◇ N-HEXADECYLTRIMETHYLAMMONIUM BROMIDE ◇ N-HEXADECYL-N,N,N-TRIMETHYLAMMONIUM BROMIDE ◇ (1-HEXADECYL)TRIMETHYLAMMONIUM BROMIDE ◇ LISSOLAMINE ◇ MICOL ◇ POLLACID ◇ QUAMONIUM ◇ SUTICIDE ◇ TRIMETHYLCETYLAMMONIUM BROMIDE ◇ N,N,N-TRIMETHYL-1-HEXADECANAMINIUM BROMIDE ◇ TRIMETHYLHEXADECYLAMMONIUM BROMIDE

TOXICITY DATA with REFERENCE
skn-mus 50 mg/1H open BJDEAZ 86,361,72
eye-rbt 450 mg SEV AROPAW 40,668,48
ipr-mus TDLo:35 mg/kg (female 12D post):TER FCTXAV 13,331,75
ipr-mus TDLo:35 mg/kg (8D preg):REP FCTXAV 13,331,75
orl-rat LD50:410 mg/kg YKYUA6 31,471,80
ivn-rat LD50:44 mg/kg APTOA6 47,17,80
ipr-mus LD50:106 mg/kg FCTXAV 13,331,75
ivn-mus LD50:32 mg/kg APTOA6 47,17,80
ipr-rbt LD50:125 mg/kg PCOC** -,210,66
scu-rbt LD50:125 mg/kg PCOC** -,210,66
scu-gpg LD50:100 mg/kg 28ZEAL 5,40,76

CONSENSUS REPORTS: Reported in EPA TSCA Inventory.

SAFETY PROFILE: A poison by ingestion, intravenous, intraperitoneal, and subcutaneous routes. Experimental teratogenic and reproductive effects. A skin and severe eye irritant. When heated to decomposition it emits very toxic fumes of NH_3, NO_x, and Br^-. See also BROMIDES.

HCR000 CAS:592-45-0 **HR: 3**
1,4-HEXADIENE
mf: C_6H_{10} mw: 82.15

$H_2C{=}CHCH_2CH{=}CHCH_3$

PROP: Flash p: −5.8°F; lel: 2.0%; uel: 6.1%.

SAFETY PROFILE: A very dangerous fire and explosion hazard when exposed to heat, flame or oxidizers. When heated to decomposition it emits acrid smoke and fumes.

HCR500 CAS:592-42-7 ***HR: 3***
1,5-HEXADIENE
mf: C_6H_{10} mw: 82.16

PROP: Liquid. D: 0.691, mp: −141°, bp: 59.6°. Flash p: −50.80°F. Insol in water.

SYNS: BIALLYL ◇ DIALLYL ◇ HEXA-1,5-DIENE

CONSENSUS REPORTS: Reported in EPA TSCA Inventory.

DOT Classification: Label: Flammable Liquid.

SAFETY PROFILE: A very dangerous fire and explosion hazard when exposed to heat, flame, or oxidizers. When heated to decomposition it emits acrid smoke and fumes.

HCS100 CAS:10420-90-3 ***HR: 3***
1,3-HEXADIENE-5-YNE
mf: C_6H_6 mw: 78.11

$H_2C{=}CHCH{=}CHC{\equiv}CH$

SAFETY PROFILE: An unstable material which may explode during distillation. When heated to decomposition it emits acrid smoke and fumes. See also ACETYLENE COMPOUNDS.

HCS500 CAS:111-28-4 ***HR: 2***
2,4-HEXADIEN-1-OL
mf: $C_6H_{10}O$ mw: 98.16

SYNS: HEXACOSE ◇ HEXADENOL ◇ 2,4-HEXADIENOL ◇ HEXAKOSE ◇ HEXENE-OL ◇ HEXENOL ◇ 1-HYDROXY-2,4-HEXADIENE ◇ SORBIC ALCOHOL ◇ SORBINIC ALCOHOL ◇ SORBYL ALCOHOL

TOXICITY DATA with REFERENCE
skn-hmn TDLo:14 mg/kg:SKN JAPMA8 34,221,45
orl-rat LD50:2140 mg/kg TXAPA9 28,313,74
skn-rat LD50:1010 mg/kg TXAPA9 28,313,74
scu-rbt LDLo:5 g/kg JAPMA8 34,221,45

CONSENSUS REPORTS: Reported in EPA TSCA Inventory.

SAFETY PROFILE: Moderately toxic by ingestion and skin contact. Human systemic effects by skin contact: sweating. When heated to decomposition it emits acrid smoke and fumes. See also ALCOHOLS.

HCT500 CAS:27310-21-0 ***HR: 2***
2-(2,4-HEXADIENYLOXY)ETHANOL
mf: $C_8H_{14}O_2$ mw: 142.22

SYN: ETHYLENE GLYCOL, MONO-2,4-HEXADIENE ETHER

TOXICITY DATA with REFERENCE
orl-rat LD50:3360 mg/kg TXAPA9 28,313,74
skn-rbt LD50:1010 mg/kg TXAPA9 28,313,74

CONSENSUS REPORTS: Glycol ether compounds are on the Community Right-To-Know List.

SAFETY PROFILE: Moderately toxic by ingestion and skin contact. When heated to decomposition it emits acrid smoke and fumes. See also GLYCOL ETHERS.

HCU500 CAS:821-08-9 ***HR: 3***
1,5-HEXADIEN-3-YNE
mf: C_6H_6 mw: 78.11

$H_2C{=}CHC{\equiv}CCH{=}CH_2$

PROP: Flash p: < −4°F, lel: 1.5%.

SYN: DIVINYL ACETYLENE

SAFETY PROFILE: Upon exposure to air it readily forms explosively unstable polymeric peroxides. A very dangerous fire and explosion hazard when exposed to heat, flame or oxidizing materials. When heated to decomposition it emits acrid smoke and fumes. See also ACETYLENE COMPOUNDS and PEROXIDES.

HCV000 CAS:2749-79-3 ***HR: 3***
4,5-HEXADIEN-2-YN-1-OL
mf: C_6H_6O mw: 94.114

$H_2C{=}C{=}CHC{\equiv}CCH_2OH$

SAFETY PROFILE: The concentrated residue from distillation may explode. When heated to decomposition it emits acrid smoke and fumes. See also ACETYLENE COMPOUNDS.

HCV500 CAS:9011-04-5 ***HR: 3***
HEXADIMETHRINE BROMIDE
mf: $C_{10}H_{24}N_2{\cdot}xC_3H_6Br_2$ mw: 1585.73

PROP: White hygros, amorph polymer. Sol in water up to 10%.

SYNS: 1,3-DIBROMOPROPANE polymer with N,N,N′,N′-TETRAMETHYL-1,6-HEXANEDIAMINE ◇ 1,5-DIMETHYL-1,5-d-DIAZAUNDECAMETHYLENE POLYMETHOBROMIDE ◇ POLYBREME ◇ POLY (N,N,N′,N′-TETRAMETHYL-N-TRIMETHYLENEHEXAMETHYLENEDIAMMONIUM DIBROMIDE) ◇ N,N,N′,N′-TETRAMETHYLHEXAMETHYLENEDIAMINE-1,3-DIBROMOPROPANEcopolymer

TOXICITY DATA with REFERENCE
ivn-rat LD50:20 mg/kg TXAPA9 1,185,59
ipr-mus LD50:62 mg/kg TXAPA9 1,185,59

ivn-mus LD50:28 mg/kg TXAPA9 1,185,59
ivn-dog LD50:15 mg/kg TXAPA9 1,185,59
ivn-rbt LD50:10 mg/kg TXAPA9 1,185,59

SAFETY PROFILE: A poison by intravenous and intraperitoneal routes. When heated to decomposition it emits very toxic fumes of NO_x and Br^-. See also BROMIDES.

HCV850 CAS:628-16-0 ***HR: 3***
1,5-HEXADIYNE
mf: C_6H_6 mw: 78.11

$(HC\equiv CCH_2\text{-})_2$

SAFETY PROFILE: Explodes when heated to 100°C. Upon decomposition it emits acrid smoke and fumes. See also ACETYLENE COMPOUNDS.

HCV875 ***HR: 3***
2,4-HEXADIYNE-1,6-DIOIC ACID
mf: $C_6H_2O_4$ mw: 138.08

$HOCO\cdot(C\equiv C)_2CO\cdot OH$

SAFETY PROFILE: Explodes when heated. Upon decomposition it emits acrid smoke and fumes. See also ACETYLENE COMPOUNDS.

HCV880 ***HR: 3***
1,5-HEXADIYNE-3-ONE
mf: C_6H_4O mw: 92.10

$HC\equiv CCO\cdot CH_2C\equiv CH$

SAFETY PROFILE: Explodes when heated. Upon decomposition it emits acrid smoke and fumes. See also ACETYLENE COMPOUNDS and KETONES.

HCW000 ***HR: 3***
2,4-HEXADIYNYLENE BISCHLOROFORMATE
mf: $C_8H_4Cl_2O_4$ mw: 235.03

$(ClCO\cdot OCH_2C\equiv C\text{-})_2$

SAFETY PROFILE: May explode at 15°C/0.2 mbar. When heated to decomposition it emits toxic fumes of Cl^-. See also ACETYLENE COMPOUNDS.

HCX000 CAS:604-88-6 ***HR: 3***
HEXAETHYLBENZENE
mf: $C_{18}H_{30}$ mw: 246.48

PROP: Colorless crystals. Mp: 130°, bp: 298°, d: 0.831 @ 130°, vap press: 10 mm @ 150.3°. Insol in water; very sol in benzene.

TOXICITY DATA with REFERENCE
imp-mus TDLo:1000 mg/kg:ETA CNREA8 26,105,66

CONSENSUS REPORTS: Reported in EPA TSCA Inventory.

SAFETY PROFILE: Questionable carcinogen with experimental tumorigenic data. Flammable when exposed to heat or flame; can react with oxidizing materials. When heated to decomposition it emits acrid smoke and fumes.

HCX050 CAS:1112-63-6 ***HR: 3***
HEXAETHYLDISTANNOXANE
mf: $C_{12}H_{30}OSn_2$ mw: 427.80

SYNS: DISTANNOXANE, HEXAETHYL- ◇ 1,1,1,3,3,3-HEXAETHYLDISTANNOXANE

TOXICITY DATA with REFERENCE
ipr-mus LD50:1 mg/kg RPTOAN 42,73,79

OSHA PEL: TWA 0.1 mg(Sn)/m^3
ACGIH TLV: TWA 0.1 mg(Sn)/m^3; STEL 0.2 mg/m^3 (skin)
NIOSH REL: (Organotin Compound): 10H TWA 0.1 mg(Sn)/m^3

SAFETY PROFILE: Poison by intraperitoneal route. When heated to decomposition it emits toxic fumes of Sn.

HCX100 CAS:994-50-3 ***HR: 3***
HEXAETHYLDISTANNTHIANE
mf: $C_{12}H_{30}SSn_2$ mw: 443.86

SYNS: DISTANNTHIANE, HEXAETHYL- ◇ 1,1,1,3,3,3-HEXAETHYLDISTANNTHIANE

TOXICITY DATA with REFERENCE
ipr-mus LD50:7 mg/kg RPTOAN 42,73,79

OSHA PEL: TWA 0.1 mg(Sn)/m^3
ACGIH TLV: TWA 0.1 mg(Sn)/m^3; STEL 0.2 mg/m^3 (skin)
NIOSH REL: (Organotin Compounds): 10H TWA 0.1 mg(Sn)/m^3

SAFETY PROFILE: Poison by intraperitoneal route. When heated to decomposition it emits toxic fumes of SO_x and Sn.

HCY000 CAS:757-58-4 ***HR: 3***
HEXAETHYL TETRAPHOSPHATE
DOT: UN 1611/NA 2783
mf: $C_{12}H_{30}O_{13}P_4$ mw: 506.30

PROP: Liquid. Mp: −40°; bp: decomp above 150°.

SYNS: BLADAN ◇ BLADAN BASE ◇ ETHYL TETRAPHOSPHATE ◇ HET ◇ HETP ◇ HEXAETHYL TETRAPHOSPHATE, liquid (DOT) ◇ HEXAETHYL TETRAPHOSPHATE, liquid, containing more than 25% hexaethyl tetraphosphate (DOT) ◇ HTP ◇ RCRA WASTE NUMBER P062 ◇ TETRAPHOSPHATE HEXAETHYLIQUE (FRENCH)

TOXICITY DATA with REFERENCE
orl-rat LD50:7 mg/kg FEPRA7 6,335,47
skn-rat LDLo:15 mg/kg APTOA6 4,143,48
ipr-rat LD50:2500 μg/kg JAMAAP 144,104,50
scu-rat LD50:640 μg/kg APTOA6 4,143,48
orl-mus LD50:56 mg/kg FEPRA7 6,335,47
ipr-mus LD50:6100 μg/kg JPETAB 92,173,48
scu-mus LDLo:1 mg/kg APTOA6 4,143,48
ivn-dog LDLo:1300 μg/kg JPETAB 92,173,48
ims-dog LDLo:1500 μg/kg JPETAB 92,173,48
scu-cat LDLo:3 mg/kg APTOA6 4,143,48
orl-rbt LD50:21 mg/kg FEPRA7 6,335,47
scu-rbt LDLo:2 mg/kg APTOA6 4,143,48
ivn-rbt LD50:690 μg/kg FEPRA7 6,335,47
orl-gpg LD50:16 mg/kg FEPRA7 6,335,47
scu-gpg LD50:1500 μg/kg APTOA6 4,143,48

DOT Classification: Poison B; Label: Poison.

SAFETY PROFILE: A poison by ingestion, skin contact, intraperitoneal, subcutaneous, intravenous, and intramuscular routes. When heated to decomposition it emits toxic fumes of PO_x. See also TETRAETHYL PYROPHOSPHATE.

HCY500 CAS:17548-36-6 ***HR: 3***
HEXAETHYLTRIALUMINUM TRITHIOCYANATE
mf: $C_{15}H_{30}Al_3N_3S_3$ mw: 429.56

SAFETY PROFILE: Explodes when heated to 210°C in a vacuum. When heated to decomposition it emits very toxic fumes of SO_x, NO_x, and CN^-. See also ALUMINUM COMPOUNDS and THIOCYANATES.

HCZ000 CAS:684-16-2 ***HR: 3***
HEXAFLUOROACETONE
DOT: UN 2420
mf: C_3F_6O mw: 166.03

PROP: A colorless, nonflammable solvent liquid. D: 1.65 @ 25°.

SYNS: 6FK ◇ NCI-C56440

TOXICITY DATA with REFERENCE
skn-rat TDLo:55 mg/kg (6-16D preg):TER TXAPA9 41,195,77
skn-rat TDLo:100 mg/kg (6-15D preg):REP FAATDF 2,73,82
orl-rat LDLo:191 mg/kg TXAPA9 6,341,64
ihl-rat LC50:275 ppm/3H TXAPA9 6,341,64

CONSENSUS REPORTS: Reported in EPA TSCA Inventory.

OSHA PEL: TWA 0.1 ppm (skin)
ACGIH TLV: TWA 0.1 ppm (skin)

DOT Classification: IMO: Poison A; Label: Poison Gas.

SAFETY PROFILE: A poison by ingestion. Moderately toxic by inhalation. A poisonous irritant to the skin, eyes, and mucous membranes. An experimental teratogen. Other experimental reproductive effects. When heated to decomposition it emits toxic fumes of F^-. See also FLUORINE and FLUORIDES.

HDA000 CAS:10543-95-0 ***HR: 3***
HEXAFLUOROACETONE HYDRATE
DOT: UN 2552
mf: $C_3F_6O \cdot H_2O$ mw: 184.05

SYN: HEXAFLUORO-2-PROPANONE HYDRATE

DOT Classification: IMO: Poison B; Label: Poison.

SAFETY PROFILE: Poison by ingestion and skin contact. When heated to decomposition it emits toxic fumes of F^-. See also HEXAFLUOROACETONE.

HDA500 CAS:34202-69-2 ***HR: 3***
HEXAFLUORO ACETONE TRIHYDRATE
mf: $C_3F_6O \cdot 3H_2O$ mw: 220.09

SYN: GC 7787

TOXICITY DATA with REFERENCE
skn-rat TDLo:275 mg/kg (6-16D preg):TER TXAPA9 47,35,79
skn-rat TDLo:275 mg/kg (6-16D preg):REP TXAPA9 47,35,79
orl-rat LD50:190 mg/kg 28ZEAL 5,128,76
skn-rbt LD50:113 mg/kg 28ZEAL 5,128,76

SAFETY PROFILE: A poison by ingestion and skin contact. An experimental teratogen. Experimental reproductive effects. When heated to decomposition it emits toxic fumes of F^-. See also HEXAFLUOROACETONE.

HDB000 CAS:392-56-3 ***HR: 3***
HEXAFLUOROBENZENE
mf: C_6F_6 mw: 186.06

PROP: Flash p: 50°F.

TOXICITY DATA with REFERENCE
ihl-mus LC50:95 g/m³/2H IZSBAI 3,91,65

CONSENSUS REPORTS: Reported in EPA TSCA Inventory.

SAFETY PROFILE: Mildly toxic by inhalation. Dangerous fire hazard when exposed to heat, flame, or oxidizers. Forms extremely heat-sensitive explosive complexes with metals (e.g., chromium, vanadium; and other transition metals). When heated to decomposition it emits toxic fumes of F^-. See also FLUORIDES.

HDB500 CAS:11111-49-2 ***HR: 3***
HEXAFLUORODICHLOROBUTENE
SYN: HFCB

TOXICITY DATA with REFERENCE
ihl-rat LC50:16 ppm/4H FLABAZ 5,4,72
ihl-mus LC50:26 ppm/4H FLABAZ 5,4,72
ihl-dog LC50:182 ppm/4H FLABAZ 5,4,72
ihl-mky LC50:54 ppm/3H FLABAZ 5,4,72
ihl-rbt LCLo:500 ppm/1H FLABAZ 5,4,72

SAFETY PROFILE: A poison by inhalation. When heated to decomposition it emits very toxic fumes of Cl^- and F^-.

HDC000 CAS:333-36-8 ***HR: 3***
HEXAFLUORODIETHYL ETHER
mf: $C_4H_4F_6O$ mw: 182.08

SYNS: BIS(TRIFLUOROETHYL)ETHER ◇ BIS(2,2,2-TRIFLUOROETHYL)ETHER ◇ FLUOROETHYL ◇ FLUOROTHYL ◇ FLUROTHYL ◇ HFE ◇ INDOKLON ◇ SF6539 ◇ SKF 6539

TOXICITY DATA with REFERENCE
ipr-rat LD50:1260 mg/kg AIPTAK 129,223,60
ivn-mus LD50:46 mg/kg AIPTAK 129,223,60

SAFETY PROFILE: A poison by intravenous route. Moderately toxic by intraperitoneal route. When heated to decomposition it emits toxic fumes of F^-. See also ETHERS and FLUORIDES.

HDC300 CAS:376-89-6 ***HR: 3***
HEXAFLUOROGLUTARONITRILE
mf: $C_5F_6N_2$ mw: 202.07

SYNS: PERFLUOROGLUTARIC ACID DINITRILE ◇ PERFLUOROGLUTARONITRILE

TOXICITY DATA with REFERENCE
orl-rat LD50:2600 mg/kg 85GMAT -,97,82
orl-mus LD50:997 mg/kg 85GMAT -,97,82
ihl-mus LC50:58 mg/m³/4H 85GMAT -,97,82

CONSENSUS REPORTS: Cyanide and its compounds are on the Community Right-To-Know List.

SAFETY PROFILE: Poison by inhalation. Moderately toxic by ingestion. When heated to decomposition it emits toxic fumes of F^-, CN^-, and NO_x. See also NITRILES.

HDC425 CAS:71359-64-3 ***HR: 3***
HEXAFLUOROGLUTARYL DIHYPOCHLORITE
mf: $C_5Cl_2F_6O_4$ mw: 308.95

SAFETY PROFILE: Explodes above −10°C. Upon decomposition it emits toxic fumes of F^- and Cl^-. See also FLUORIDES and HYPOCHLORITES.

HDC500 CAS:920-66-1 ***HR: 3***
HEXAFLUOROISOPROPANOL
mf: $C_3H_2F_6O$ mw: 168.05

SYNS: 1,1,1,3,3,3-HEXAFLUORO-2-PROPANOL ◇ HFIP

TOXICITY DATA with REFERENCE
eye-rbt 100 mg rns SEV 34ZIAG -,310,69
ihl-rat LCLo:3200 ppm/4H 34ZIAG -,310,69
orl-mus LD50:600 mg/kg JMCMAR 13,1215,70
ipr-mus LD50:300 mg/kg JMCMAR 13,1215,70
ivn-mus LD50:180 mg/kg CSLNX* NX#03623

CONSENSUS REPORTS: Reported in EPA TSCA Inventory.

SAFETY PROFILE: A poison by intravenous and intraperitoneal routes. Moderately toxic by ingestion. Mildly toxic by inhalation. A severe eye irritant. When heated to decomposition it emits toxic fumes of F^-. See also FLUORIDES.

HDD000 CAS:1645-75-6 ***HR: 3***
HEXAFLUOROISOPROPYLIDENEAMINE
mf: C_3HF_6N mw: 165.04

SYN: 2-IMINOHEXAFLUOROPROPANE

SAFETY PROFILE: Explosive reaction with concentrated solutions of butyllithium in hexane at 0°C. When heated to decomposition it emits toxic fumes of F^- and NO_x. See also FLUORIDES and AMINES.

HDD500 CAS:31340-36-0 ***HR: 3***
HEXAFLUOROISOPROPYLIDENEAMINOLITHIUM
mf: C_3F_6LiN mw: 170.97

$(F_3C)_2C{=}NLi$

SAFETY PROFILE: Explosive reaction with thionyl chloride. Violent reaction at 25°C with chloro- and fluoro-derivatives of arsenic, boron, phosphorus, silicon and sulfur. Incompatible with nonmetal halides. When heated to decomposition it emits very toxic fumes of NO_x and F^-. See also FLUORIDES and LITHIUM COMPOUNDS.

HDE000 CAS:16940-81-1 ***HR: 3***
HEXAFLUOROPHOSPHORIC ACID
DOT: UN 1782
mf: F_6HP mw: 145.98

PROP: Corrosive, colorless, clear liquid. Mp: 31°, d: 1.65.

SYN: HYDROGEN HEXAFLUOROPHOSPHATE

CONSENSUS REPORTS: Reported in EPA TSCA Inventory.

OSHA PEL: TWA 2.5 mg(F)/m^3
NIOSH REL: (Inorganic Fluorides) TWA 2.5 mg(F)/m^3
DOT Classification: Corrosive Material; Label: Corrosive.

SAFETY PROFILE: A poison by all routes. A corrosive irritant to skin, eyes, and mucous membranes. When heated to decomposition it emits highly toxic F^- and PO_x. See also HYDROFLUORIC ACID and PHOSPHORIC ACID.

HDE500 CAS:13098-39-0 ***HR: 3***
HEXAFLUORO-2-PROPANONE SESQUIHYDRATE
mf: $C_3F_6O \cdot 3/2H_2O$ mw: 193.06

SYN: HEXAFLUOROACETONE SESQUIHYDRATE

TOXICITY DATA with REFERENCE
skn-rat TDLo:1820 mg/kg (male 14D pre):REP TOXID9 4,135,84
orl-mus LD50:300 mg/kg JMCMAR 13,1215,70
ipr-mus LD50:250 mg/kg JMCMAR 13,1215,70
ivn-mus LD50:180 mg/kg CSLNX* NX#03172

SAFETY PROFILE: A poison by ingestion, intraperitoneal, and intravenous routes. Experimental reproductive effects. When heated to decomposition it emits toxic fumes of F^-. See also HEXAFLUOROACETONE.

HDF000 CAS:116-15-4 ***HR: 3***
HEXAFLUOROPROPENE
DOT: UN 1858
mf: C_3F_6 mw: 150.03

PROP: Gas. Mp: −156°, bp: −29°, d: 1.583 @ −40°/4°.

SYNS: HEXAFLUOROPROPYLENE (DOT) ◇ PERFLUOROPROPENE ◇ PERFLUOROPROPYLENE

TOXICITY DATA with REFERENCE
ihl-rat LC50:11200 mg/m^3/4H GTPZAB 15(2),38,71
ihl-mus LC50:750 ppm/4H AMPMAR 27,509,66

CONSENSUS REPORTS: Reported in EPA TSCA Inventory.

DOT Classification: Nonflammable Gas; Label: Nonflammable Gas.

SAFETY PROFILE: Mildly toxic by inhalation. Explosive reaction with Grignard reagents (e.g., phenylmagnesium bromide). Reacts with tetrafluorethylene + air to form explosive peroxides. When heated to decomposition it emits toxic fumes of F^-.

HDF100 CAS:61840-09-3 ***HR: 3***
HEXAFUNGIN

TOXICITY DATA with REFERENCE
orl-mus LD50:100 mg/kg 37XLA2 1,226,78
ipr-mus LD50:20 mg/kg 37XLA2 1,226,78
ivn-mus LD50:5 mg/kg 37XLA2 1,226,78

SAFETY PROFILE: Poison by ingestion, intravenous, and intraperitoneal routes.

HDF500 CAS:35281-34-6 ***HR: D***
1,2,3,7,8,9-HEXAHYDROANTHANTHRENE
mf: $C_{22}H_{20}$ mw: 284.42

SYN: 1,2,3,7,8,9-HEXAHYDRODIBENZO(def,mno)CHRYSENE

TOXICITY DATA with REFERENCE
mma-sat 100 μg/plate MUREAV 51,311,78

CONSENSUS REPORTS: EPA Genetic Toxicology Program.

SAFETY PROFILE: Mutation data reported. Many chrysene compounds are carcinogens. When heated to decomposition it emits acrid smoke and fumes.

HDG000 CAS:111-49-9 ***HR: 3***
HEXAHYDRO-1H-AZEPINE
DOT: UN 2493

PROP: Flash p: 71.6°F.
mf: $C_6H_{13}N$ mw: 99.20

SYNS: AZACYCLOHEPTANE ◇ 1-AZACYCLOHEPTANE ◇ CYCLOHEXAMETHYLENIMINE ◇ G 0 ◇ HEXAHYDROAZEPINE ◇ HEXAMETHYLENE IMINE (DOT) ◇ HEXAMETHYLENIMINE ◇ HOMOPIPERIDINE ◇ PERHYDROAZEPINE

TOXICITY DATA with REFERENCE
orl-rat LD50:410 mg/kg GTPZAB 18(2),29,74
ihl-rat LCLo:4800 ppm/4H 34ZIAG -,311,69
ihl-mus LC50:10800 mg/m^3/2H 85GMAT -,74,82
scu-mus LDLo:550 mg/kg AEXPBL 50,199,1903

CONSENSUS REPORTS: Reported in EPA TSCA Inventory.

DOT Classification: Corrosive Material; Label: Corrosive; IMO: Flammable Liquid; Label: Flammable Liquid, Corrosive.

SAFETY PROFILE: Moderately toxic by ingestion and subcutaneous routes. Mildly toxic by inhalation. A corrosive irritant to the eyes, skin, and mucous membranes. A dangerous fire hazard when exposed to heat or flame; can react vigorously with oxidizers. When heated to decomposition it emits toxic fumes of NO_x.

HDG600 ***HR: 3***
(2-HEXAHYDRO-1-AZEPINYL)ETHYL)GUANIDINE SULFATE (2:1)
mf: $C_{18}H_{40}N_8•H_2O_4S$ mw: 466.74

TOXICITY DATA with REFERENCE
orl-mus LD50:2 g/kg YKKZAJ 83,629,63
scu-mus LD50:887 mg/kg YKKZAJ 83,629,63
ivn-mus LD50:97400 μg/kg YKKZAJ 83,629,63

SAFETY PROFILE: Poison by intravenous route. Moderately toxic by ingestion and subcutaneous routes. When heated to decomposition it emits toxic fumes of NO_x and SO_x. See also SULFATES.

HDH000 CAS:63041-92-9 ***HR: 3***
1,2,4,5,6,7-HEXAHYDROBENZ(e)ACEANTHRYLENE
mf: $C_{20}H_{18}$ mw: 258.38

SYN: 1′,2′,3′,4′-TETRAHYDRO-4,10-ACE-1,2-BENZANTHRACENE

TOXICITY DATA with REFERENCE
scu-mus TDLo:200 mg/kg:ETA AJCAA7 33,499,38

SAFETY PROFILE: Questionable carcinogen with experimental tumorigenic data. When heated to decomposition it emits acrid smoke and fumes.

HDK000 CAS:153-32-2 ***HR: 3***
1,2,3,4,12,13-HEXAHYDRODIBENZ(a,h)ANTHRACENE
mf: $C_{22}H_{20}$ mw: 284.42

TOXICITY DATA with REFERENCE
mma-sat 1 μg/plate MUREAV 51,311,78
skn-mus TDLo:152 mg/kg/50W-I:NEO JNCIAM 34,1,65

SAFETY PROFILE: Questionable carcinogen with experimental neoplastigenic data. Mutation data reported. When heated to decomposition it emits acrid smoke and fumes.

HDO500 CAS:100447-46-9 ***HR: 3***
1,2,3a,4,5,9b-HEXAHYDRO-8-HYDROXY-3-METHYL-9b-PROPYL-3H-BENZ(e)INDOLE, HYDROCHLORIDE
mf: $C_{16}H_{23}NO•ClH$ mw: 281.86

TOXICITY DATA with REFERENCE
orl-mus LDLo:300 mg/kg CHTPBA 7,450,72
ipr-mus LDLo:100 mg/kg CHTPBA 7,450,72

SAFETY PROFILE: Poison by ingestion and intraperitoneal routes. When heated to decomposition it emits very toxic fumes of NO_x and HCl.

HDP500 CAS:18530-56-8 ***HR: 2***
3-(HEXAHYDRO-4,7-METHANOINDAN-5-YL)-1,1-DIMETHYLUREA
mf: $C_{13}H_{22}N_2O$ mw: 222.37

PROP: Crystals. Mp: 178°.

SYNS: ASEPTA HERBAN ◇ 1-(3a,4,5,6,7,7a-HEXAHYDRO-4,7-METHANO-5-INDANYL)-3,3-DIMETHYLUREA ◇ 1-(5-(3a,4,5,6,7,7a-HEXAHYDRO-4,7-METHANOINDANYL))-3,3-DIMETHYLUREA ◇ 3-(5-(3a,4,5,6,7,7a-HEXAHYDRO-4,6-METHANOINDANYL))-1,1-DIMETHYLUREA ◇ HERBAN ◇ HERCULES 7531 ◇ NAREA ◇ NOREA ◇ NORES ◇ NORURON ◇ 1-(TETRAHYDRODICYCLOPENTADIENYL)-3,3-DIMETHYLUREA

TOXICITY DATA with REFERENCE
orl-rat LD50:2000 mg/kg FMCHA2 -,C122,83
skn-rat LD50:23 g/kg PCOC** -,817,66
orl-mus LD50:4600 mg/kg PCOC** -,817,66
orl-dog LD50:3700 mg/kg PCOC** -,817,66
skn-rbt LD50:723 mg/kg 28ZEAL 5,167,76

SAFETY PROFILE: Moderately toxic by skin contact and ingestion. An herbicide. When heated to decomposition it emits toxic fumes of NO_x.

HDQ500 CAS:15923-42-9 ***HR: 3***
1,2,3,4,5,6-HEXAHYDRO-6-METHYLAZEPINO(4,5-b)INDOLE HYDROCHLORIDE
mf: $C_{13}H_{16}N_2•ClH$ mw: 236.77

SYN: U-22394A

TOXICITY DATA with REFERENCE
orl-rat LD50:96 mg/kg 27ZQAG -,140,72
ipr-rat LD50:58 mg/kg 27ZQAG -,140,72
ipr-mus LD50:112 mg/kg 27ZQAG -,140,72

SAFETY PROFILE: A poison by ingestion and intraperitoneal routes. When heated to decomposition it emits very toxic fumes of NO_x and HCl.

HDR500 CAS:35281-27-7 ***HR: 3***
6,7,8,9,10,12b-HEXAHYDRO-3-METHYLCHOLANTHRENE
mf: $C_{21}H_{22}$ mw: 274.43

SYN: 1,2,6,7,8,9,10,12b-OCTAHYDRO-3-METHYLBENZ(j)ACEANTHRYLENE

TOXICITY DATA with REFERENCE
mma-sat 1 μg/plate MUREAV 51,311,78
skn-mus TDLo:140 mg/kg/50W-I:CAR ZKKOBW 77,226,72

SAFETY PROFILE: Questionable carcinogen with experimental carcinogenic data. Mutation data reported. When heated to decomposition it emits acrid smoke and fumes.

HDS200 CAS:7458-65-3 ***HR: 3***
1,2,3,4,5,6-HEXAHYDRO-1-(2'-METHYL-3'-(N-METHYLAMINO)PROPYL)-1-BENZAZOCINE HYDROCHLORIDE
mf: $C_{16}H_{26}N_2 \cdot ClH$ mw: 282.90

SYN: UCB 4268

TOXICITY DATA with REFERENCE
orl-rat LD50:485 mg/kg 27ZQAG -,314,72
ivn-rat LD50:47 mg/kg 27ZQAG -,314,72
orl-mus LD50:280 mg/kg 27ZQAG -,314,72
ivn-mus LD50:37 mg/kg 27ZQAG -,314,72

SAFETY PROFILE: Poison by ingestion and intravenous routes. When heated to decomposition it emits toxic fumes of NO_x and HCl.

HDV500 CAS:13980-04-6 ***HR: 3***
HEXAHYDRO-1,3,5-s-TRIAZINE
mf: $C_3H_6N_6O_3$ mw: 174.15

$CH_2N(N:O)CH_2N(N:O)CH_2NN:O$ (ring)

SYNS: HEXAHYDRO-1,3,5-TRINITROSO-s-TRIAZINE ◇ HEXAHYDRO-1,3,5-TRINITROSO-1,3,5-TRIAZINE ◇ 1,3,5-TRINITROSO-1,3,5-TRIAZACYCLOHEXANE ◇ TRINITROSOTRIMETHYLENETRIAMINE ◇ TRINITROSOTRIMETHYLENTRIAMIN (GERMAN) ◇ TTT

TOXICITY DATA with REFERENCE
orl-rat TDLo:765 mg/kg/52W-I:ETA ARGEAR 46,657,76
orl-rat LD50:160 mg/kg ZEKBAI 69,103,67

SAFETY PROFILE: A poison by ingestion. Questionable carcinogen with experimental tumorigenic data. Explodes on contact with sulfuric acid. When heated to decomposition it emits toxic fumes of NO_x.

HDW000 CAS:7779-27-3 ***HR: 3***
HEXAHYDRO-1,3,5-TRIETHYL-s-TRIAZINE
mf: $C_9H_{21}N_3$ mw: 171.33

PROP: A light yellow liquid, sol in water. D: 0.89 @ 25°.

SYN: VANCIDE TH

TOXICITY DATA with REFERENCE
eye-rbt 100 mg SEV DCTODJ 1(1),1,78
eye-rbt 100 mg/2S rns SEV DCTODJ 1(1),1,78
orl-rat LD50:316 mg/kg IMSUAI 39,56,70
orl-mus LD50:370 mg/kg TXAPA9 10,404,67
skn-rbt LDLo:2000 mg/kg TXAPA9 10,404,67

CONSENSUS REPORTS: Reported in EPA TSCA Inventory.

SAFETY PROFILE: Poison by ingestion. Moderately toxic by skin contact. A severe eye irritant. When heated to decomposition it emits toxic fumes of NO_x.

HDX500 CAS:18501-44-5 ***HR: 3***
HEXAHYDROXYLAMINECOBALT(III) NITRATE
mf: $CoH_{18}N_9O_{15}$ mw: 443.14

$[(HONH_2)_6Co][NO_3]_3$

CONSENSUS REPORTS: Cobalt and its compounds are on the Community Right-To-Know List.

SAFETY PROFILE: Explodes during preparation or handling. When heated to decomposition it emits toxic fumes of NO_x. See also COBALT COMPOUNDS and NITRATES.

HDY000 CAS:531-18-0 ***HR: 3***
HEXA(HYDROXYMETHYL)MELAMINE
mf: $C_9H_{18}N_6O_6$ mw: 306.33

SYNS: CILAG 61 ◇ HEXAKIS(HYDROXYMETHYL)MELAMINE ◇ HEXAKIS(HYDROXYMETHYL)-1,3,5-TRIAZINE-2,4,6-TRIAMINE ◇ HEXAMETHYLOLMELAMIN (CZECH) ◇ HEXAMETHYLOLMELAMINE ◇ RESLOOM M 75 ◇ (1,3,5-TRIAZINE-2,4,6-TRIYLTRINITRILO) HEXAKIS METHANOL ◇ (s-TRIAZINE-2,4,6-TRIYLTRINITRILO) HEXAMETHANOL ◇ 2,4,6-TRIS(BIS(HYDROXYMETHYL)AMINO)-s-TRIAZINE ◇ 2,4,6-TRIS(DI(HYDROXYMETHYL)AMINO)-1,3,5-TRIAZINE

TOXICITY DATA with REFERENCE
skn-rbt 500 mg/24H MOD 28ZPAK -,154,72
eye-rbt 100 mg/24H MOD 28ZPAK -,154,72
ivn-mus LD50:180 mg/kg CSLNX* NX#04006

CONSENSUS REPORTS: Reported in EPA TSCA Inventory.

SAFETY PROFILE: A poison by intravenous route. A skin and eye irritant. When heated to decomposition it emits toxic fumes of NO_x.

HDY100 CAS:3750-18-3 ***HR: 3***
HEXAISOBUTYLDITIN
mf: $C_{24}H_{54}Sn_2$ mw: 580.16

SYNS: BIS(TRIISOBUTYLSTANNANE) ◇ DISTANNANE, HEXAISOPROPYL-

TOXICITY DATA with REFERENCE
ivn-mus LD50:10 mg/kg CSLNX* NX#05529

OSHA PEL: TWA 0.1 mg(Sn)/m^3
ACGIH TLV: TWA 0.1 mg(Sn)/m^3; STEL 0.2 mg/m^3 (skin)
NIOSH REL: (Organotin Compounds): 10H TWA 0.1 mg(Sn)/m^3

SAFETY PROFILE: Poison by intravenous route. When heated to decomposition it emits toxic fumes of Sn.

HDY500 CAS:3089-11-0 ***HR: 2***
***N,N,N',N',N'',N''*-HEXAKIS(METHOXYMETHYL)-1,3,5-TRIAZINE-2,4,6-TRIAMINE**
mf: $C_{15}H_{30}N_6O_6$ mw: 390.51

SYN: HEXAKIS-METHOXYMETHYLMELAMIN(CZECH)

TOXICITY DATA with REFERENCE
skn-rbt 500 mg/24H MLD 28ZPAK -,156,72
eye-rbt 100 mg/24H SEV 28ZPAK -,156,72
orl-rat LD50:3080 mg/kg 28ZPAK -,156,72
ipr-rat LD50:560 mg/kg ARZNAD 16,734,66
ipr-mus LD50:1150 mg/kg ARZNAD 16,734,66

CONSENSUS REPORTS: Reported in EPA TSCA Inventory.

SAFETY PROFILE: Moderately toxic by ingestion and intraperitoneal routes. A skin and severe eye irritant. When heated to decomposition it emits toxic fumes of NO_x. See also AMINES.

HDZ000 ***HR: 3***
HEXALITHIUM DISILICIDE
mf: Li_6Si_2 mw: 97.81

SAFETY PROFILE: Explodes on contact with nitric acid. Violent reaction with water evolves silanes which ignite. Similar reaction with diiute acids. Ignites on contact with fluorine or when heated with chlorine, bromine, and iodine. Vigorous reaction with sulfur. Incandescent reaction with non-metals (e.g., phosphorus; selenium; or tellurium); concentrated hydrochloric acid; sulfuric acid. See also LITHIUM COMPOUNDS.

HEA000 CAS:55-97-0 ***HR: 3***
HEXAMETHONIUM BROMIDE
mf: $C_{12}H_{30}N_2 \cdot 2Br$ mw: 362.26

PROP: Crystals. Mp: 274-276°. Sol in water, alc, acid to litmus. Aq solns are stable.

SYNS: α,omega-BIS(TRIMETHYL AMMONIUM)HEXANE DIBROMIDE ◇ C 6 ◇ ESAMETINA ◇ GANGLIOSTAT ◇ HB ◇ HEXAMETHIONIUM BROMIDE ◇ HEXAMETHONIUM DIBROMIDE ◇ HEXAMETHYLENEBIS(TRIMETHYLAMMONIUM)BROMIDE ◇ N,N,N,N',N',N'-HEXAMETHYL-1,6-HEXANEDIAMINIUM DIBROMIDE ◇ HEXAMETON ◇ HEXONIUM DIBROMIDE ◇ SIMPATOBLOCK ◇ VEGOLYSEN ◇ VEGOLYSIN

TOXICITY DATA with REFERENCE
scu-wmn TDLo:46 mg/kg (31-34W preg):REP ADCHAK 29,354,54
scu-wmn TDLo:46 mg/kg (31-34W preg):TER ADCHAK 29,354,54
orl-rat LD50:2891 mg/kg SKNEA7 10,15,60
ivn-rat LD50:64130 μg/kg SKNEA7 10,15,60
orl-mus LD50:838 mg/kg NIIRDN 6,356,82
ipr-mus LD50:70 mg/kg YKKZAJ 91,1307,71
scu-mus LD50:78 mg/kg YKKZAJ 91,1307,71
ivn-mus LD50:26300 μg/kg CPBTAL 23,1639,75
ivn-rbt LD50:50300 μg/kg SKNEA7 10,15,60

SAFETY PROFILE: A poison by intraperitoneal, subcutaneous, and intravenous routes. Moderately toxic by ingestion. Human reproductive and teratogenic effects by subcutaneous route: abnormal neonatal measurements and developmental abnormalities of the gastrointestinal system. Used to treat hypertension. When heated to decomposition it emits very toxic NH_3, NO_x, and Br^-. See also BROMIDES.

HEA500 CAS:60-25-3 ***HR: 3***
HEXAMETHONIUM DICHLORIDE
mf: $C_{12}H_{30}N_2 \cdot 2Cl$ mw: 273.34

PROP: Hygroscopic crystals. Decomp @ 289-292°. Very sol in water; insol in chloroform, ether.

SYNS: α,omega-BIS(TRIMETHYLAMMONIUM)HEXANEDICHLORIDE ◇ BISTRIUM CHLORIDE ◇ CHLOOR-HEXAVIET ◇ DEPRESSIN ◇ ESOMID CHLORIDE ◇ HESTRIUM CHLORIDE ◇ HEXAMETHIONIUM CHLORIDE ◇ HEXAMETHONIUM CHLORIDE ◇ HEXAMETHYLENE(BISTRIMETHYLAMMONIUM)CHLORIDE ◇ HEXAMETON CHLORIDE ◇ N,N,N,N',N',N'-HEXAMETHYL-1,6-HEXANEDIAMINIUM DICHLORIDE ◇ HEXON CHLORIDE ◇ HEXONE CHLORIDE ◇ HIOHEX CHLORIDE ◇ METHIUM CHLORIDE ◇ METON

TOXICITY DATA with REFERENCE
ipr-rat LD50:100 mg/kg JAPMA8 46,346,57
ipr-mus LD50:62 mg/kg AIPTAK 155,69,65
ivn-mus LD50:26700 μg/kg AIPTAK 127,1,60
ivn-dog LD50:35 mg/kg JAPMA8 46,346,57

CONSENSUS REPORTS: Reported in EPA TSCA Inventory.

SAFETY PROFILE: Poison by intravenous and intraperitoneal routes. When heated to decomposition it emits very toxic fumes of NH_3, NO_x, and Cl^-. See also CHLORIDES.

HEB000 CAS:870-62-2 ***HR: 3***
HEXAMETHONIUM DIIODIDE
mf: $C_{12}H_{30}N_2 \cdot 2I$ mw: 456.24

SYNS: ESAMETONIO IODURO (ITALIAN) ◇ HEXAMETHONIUM IODIDE ◇ HEXAMETHYLENEBIS(TRIMETHYLAMMONIUM IODIDE) ◇ N,N,N,N',N',N'-HEXAMETHYL-1,6-HEXANEDIAMINIUM DIIODIDE ◇ HEXATHIDE ◇ HEXONIUM DIIODIDE

TOXICITY DATA with REFERENCE
ivn-rat LD50:20 mg/kg JPETAB 115,172,55
orl-mus LD50:850 mg/kg FRPSAX 20,482,65
scu-mus LD50:210 mg/kg FRPSAX 20,482,65
ivn-mus LD50:47 mg/kg FATOAO 25,428,62

SAFETY PROFILE: A poison by subcutaneous and intravenous routes. Moderately toxic by ingestion. When heated to decomposition it emits very toxic fumes of NO_x, NH_3, and I^-. See also IODIDES.

HEC000 CAS:87-85-4 ***HR: 3***
HEXAMETHYLBENZENE
mf: $C_{12}H_{18}$ mw: 162.30

PROP: Plates from ethanol. Mp: 165.5°, bp: 265°. Insol in water; very sol in ether.

TOXICITY DATA with REFERENCE
imp-mus TDLo:1000 mg/kg:NEO CNREA8 26,105,66
orl-rat LDLo:5000 mg/kg AMIHAB 19,403,59

CONSENSUS REPORTS: Reported in EPA TSCA Inventory.

SAFETY PROFILE: Mildly toxic by ingestion. Questionable carcinogen with experimental neoplastigenic data. Potentially explosive reaction with nitromethane. When heated to decomposition it emits acrid smoke and fumes.

HEC500 CAS:7641-77-2 ***HR: 3***
HEXAMETHYLBICYCLO(2.2.0)HEXA-2,5-DIENE
mf: $C_{12}H_{18}$ mw: 162.30

SYNS: 2-BUTIN HEXAMETHYL-DEWAR-BENZOL (GERMAN) ◇ HEXAMETHYL-BICYCLO(2.2.0)HEXA-2,5-DIEN (GERMAN)

TOXICITY DATA with REFERENCE
scu-mus TDLo:800 mg/kg/15W-I:ETA ZEKBAI 74,100,70

SAFETY PROFILE: Questionable carcinogen with experimental tumorigenic data. When heated to decomposition it emits acrid smoke and fumes.

HED000 CAS:4711-74-4 ***HR: 3***
HEXAMETHYLDIPLATINUM
mf: $C_6H_{18}Pt_2$ mw: 480.37

$(CH_3)_3PtPt(CH_3)_3$

SAFETY PROFILE: Explodes when heated. Upon decomposition it emits acrid smoke and fumes. See also PLATINUM COMPOUNDS.

HED425 CAS:1450-14-2 ***HR: 3***
HEXAMETHYLDISILANE
mf: $C_6H_{18}Si_2$ mw: 146.38

$(CH_3)_3SiSi(CH_3)_3$

SAFETY PROFILE: Potentially explosive reaction with pyridine-N-oxide + tetrabutylammonium fluoride. When heated to decomposition it emits acrid smoke and fumes. See also SILANE.

HED500 CAS:999-97-3 ***HR: 3***
HEXAMETHYLDISILAZANE
mf: $C_6H_{19}NSi_2$ mw: 161.44

PROP: Flash p: 57.2°F.

SYNS: BIS(TRIMETHYLSILYL)AMINE ◇ HEXAMETHYLSILAZANE ◇ HMDS ◇ OAP ◇ 1,1,1-TRIMETHYL-N-(TRIMETHYLSILYL)SILANAMINE

TOXICITY DATA with REFERENCE
ipr-mus TDLo:1 g/kg/I:ETA JNCIAM 54,495,75
ipr-mus LDLo:650 mg/kg StoGD# 27May75

CONSENSUS REPORTS: Reported in EPA TSCA Inventory.

SAFETY PROFILE: Moderately toxic by intraperitoneal route. Questionable carcinogen with experimental tumorigenic data. A dangerous fire hazard when exposed to heat or flame; can react vigorously with oxidizing materials. When heated to decomposition it emits toxic fumes of NO_x.

HEE000 CAS:107-46-0 ***HR: 1***
HEXAMETHYLDISILOXANE
mf: $C_6H_{18}OSi_2$ mw: 162.42

PROP: Viscous liquid. Vap d: 5.5.

SYNS: DOW CORNING 200 ◇ OXYBIS(TRIMETHYLSILANE)

TOXICITY DATA with REFERENCE
skn-rbt 500 mg/24H MLD 28ZPAK -,219,72
ipr-mus LD50:4500 mg/kg RCRVAB 38,975,69
orl-gpg LDLo:50 g/kg JIHTAB 30,332,48

CONSENSUS REPORTS: Reported in EPA TSCA Inventory.

SAFETY PROFILE: Mildly toxic by ingestion and intraperitoneal routes. A skin irritant. When heated to decomposition it emits acrid smoke and fumes. See also SILANE.

HEE500 CAS:661-69-8 ***HR: 3***
HEXAMETHYLDITIN
mf: $C_6H_{18}Sn_2$ mw: 327.62

SYNS: HEXAMETHYLDISTANNANE ◇ PENNSALT TD 5032 ◇ TD-5032

TOXICITY DATA with REFERENCE
orl-rat LD50:25 mg/kg 28ZEAL 4,245,69

OSHA PEL: TWA 0.1 mg(Sn)/m^3 (skin)
ACGIH TLV: TWA 0.1 mg(Sn)/m^3 (skin) (Proposed: TWA 0.1 mg(Sn)/m^3; STEL 0.2 mg(Sn)/m^3 (skin))
NIOSH REL: (Organotin Compounds) TWA 0.1 mg(Sn)/m^3

SAFETY PROFILE: A poison by ingestion. When heated to decomposition it emits acrid smoke and fumes. See also TIN COMPOUNDS.

HEF200 CAS:3818-69-7 ***HR: 3***
HEXAMETHYLEN-1,6-(N-DIMETHYLCARBO-DESOXYMETHYL)AMMONIUM DICHLORIDE
mf: $C_{38}H_{78}N_2O_4 \cdot 2Cl$ mw: 698.08

SYNS: AMMONIUM,HEXAMETHYLENEBIS((CARBOXYMETHYL) DIMETHYL-, DICHLORIDE, DIDODECYL ESTER ◇ DODECONIUM ◇ HEXAMETHYLENEBIS((CARBOXYMETHYL)DIMETHYLAMMONIUM), DICHLORIDE, DIDODECYL ESTER ◇ 1,6-HEXANEDIAMINE, N,N'-DICARBOXYMETHYL-N,N'-DIMETHYL-, DIMETHOCHLORIDE, DIDODECYL ESTER ◇ 1,6-HEXANEDIAMINIUM, N,N'-BIS(2-(DODECYLOXY)-2-OXOETHYL)-N,N,N',N'-TETRAMETHYL-, DICHLORIDE ◇ PREPARATION C (the Russian Drug) ◇ PREPARATION S

TOXICITY DATA with REFERENCE
ipr-rat TDLo:20 mg/kg (female 4D post):REP FATOAO 40,219,77
par-uns LD50:104 mg/kg RPTOAN 33,239,70

SAFETY PROFILE: Poison by parenteral route. Experimental reproductive effects. When heated to decomposition it emits toxic fumes of NO_x and Cl^-.

HEF300 CAS:3613-89-6 ***HR: 2***
N,N'-HEXAMETHYLENEBIS(2,2-DICHLORO-N-ETHYLACETAMIDE)
mf: $C_{14}H_{24}Cl_4N_2O_2$ mw: 394.20

SYNS: N,N'-BIS(DICHLOROACETYL)-N,N'-DIETHYL-1,6-HEXANEDIAMINE ◇ WIN 17,416

TOXICITY DATA with REFERENCE
spm-hmn-orl 3650 mg/kg/28W TXAPA9 3,1,61
orl-man TDLo:3650 mg/kg (28W male):REP TXAPA9 3,1,61
ipr-mus LD50:4400 mg/kg ANTCAO 11,245,61

CONSENSUS REPORTS: EPA Genetic Toxicology Program.

SAFETY PROFILE: Moderately toxic by intraperitoneal route. Human reproductive effects by ingestion: impaired spermatogenesis. Experimental reproductive effects. Mutation data reported. When heated to decomposition it emits toxic fumes of Cl^- and NO_x.

HEG000 CAS:317-52-2 ***HR: 3***
HEXAMETHYLENE BIS(9-FLUORENYL DIMETHYLAMMONIUM)DIBROMIDE
mf: $C_{36}H_{42}N_2 \cdot Br$ mw: 582.71

SYNS: 1,6-BIS(9FLUORENYLDIMETHYL-AMMONIUM)HEXANE BROMIDE ◇ HEXAFLUORENIUM DIBROMIDE ◇ HEXAFLURONIUM BROMIDE ◇ HEXAMETHYLENEBIS(DIMETHYL-9-FLUORENYLAMMONIUM BROMIDE) ◇ HEXAMETHYLENEBIS(FLUOREN-9-YLDIMETHYLAMMONIUM BROMIDE) ◇ IN-117 ◇ MILAXEN ◇ MYLAXEN ◇ NSC-19477

TOXICITY DATA with REFERENCE
ivn-hmn TDLo:400 μg/kg:GIT,SYS 85IVAW 1,E1,82
ipr-rat LD50:20 mg/kg CLDND* 85,603,54
scu-mus LD50:240 mg/kg PSEBAA 85,603,54
ivn-mus LD50:1760 μg/kg BCPCA6 2,233,59
ivn-rbt LD50:80 μg/kg PHTXA6 22,200,59

SAFETY PROFILE: A poison by intraperitoneal, subcutaneous, and intravenous routes. Human systemic effects by intravenous route: nausea or vomiting, cholinesterase changes. When heated to decomposition it emits very toxic fumes of NH_3, NO_x, and Br^-. See also BROMIDES.

HEG100 CAS:971-60-8 ***HR: 3***
HEXAMETHYLENEBIS(TRIMETHYLAMMONIUM) DIBENZENESULFONATE
mf: $C_{12}H_{30}N_2 \cdot C_{12}H_{10}O_3S$ mw: 436.72

SYNS: AMMONIUM, HEXAMETHYLENEBIS(TRIMETHYL-, DIBENZENESULFONATE ◇ BENZOHEXONIUM ◇ 1,6-HEXANEDIAMINIUM, N,N,N,N',N',N'-HEXAMETHYL-, DIBENZENESULFONATE (9CI)

TOXICITY DATA with REFERENCE
unr-rbt TDLo:150 mg/kg (female 1-15D post):REP AKGIAO 43(12),10,67
scu-mus LD50:166 mg/kg RPTOAN 31,53,68

SAFETY PROFILE: Poison by subcutaneous route. Experimental reproductive effects. When heated to decomposition it emits toxic fumes of NO_x and SO_x.

HEI000 CAS:1169-26-2 ***HR: 3***
1-(2-HEXAMETHYLENEIMINOETHYL)-2-OXOCYCLOHEXANECARBOXYLIC ACID BENZYL ESTER HYDROCHLORIDE
mf: $C_{22}H_{31}NO_3 \cdot ClH$ mw: 394.00

SYN: 2-(β-HEXAMETHYLENIMINOAETHYL)CYCLOHEXANON-2-CARBONSAUREBENZYLESTER-HYDROCHLORIDE(GERMAN)

TOXICITY DATA with REFERENCE
orl-rat LD50:450 mg/kg ARZNAD 14,986,64
ipr-rat LD50:56 mg/kg ARZNAD 14,986,64
scu-rat LD50:72 mg/kg ARZNAD 14,986,64
orl-mus LD50:110 mg/kg ARZNAD 14,986,64
ipr-mus LD50:75 mg/kg ARZNAD 14,986,64
scu-mus LD50:10 g/kg ZGEMAZ 113,536,44
ivn-mus LD50:12 mg/kg ARZNAD 14,986,64

SAFETY PROFILE: A poison by ingestion, intraperitoneal, subcutaneous, and intravenous routes. When heated to decomposition it emits very toxic fumes of HCl and NO_x. See also ESTERS.

HEI500 CAS:100-97-0 ***HR: 3***
HEXAMETHYLENETETRAMINE
DOT: UN 1328
mf: $C_6H_{12}N_4$ mw: 140.22

PROP: Odorless, rhombic crystals from alc. Mp: 280°

(subl), flash p: 482°F, d: 1.33 @ −5°. Very sltly sol in hot ether.

SYNS: ACETO HMT ◇ AMINOFORM ◇ AMMOFORM ◇ AMMONIOFORMALDEHYDE ◇ CYSTAMIN ◇ CYSTOGEN ◇ ESAMETILENTETRAMINA (ITALIAN) ◇ FORMAMINE ◇ FORMIN ◇ HEXAFORM ◇ HEXAMETHYLENAMINE ◇ HEXAMETHYLENEAMINE ◇ HEXAMETHYLENETETRAAMINE ◇ HEXAMETHYLENTETRAMIN (GERMAN) ◇ HEXAMINE (DOT) ◇ HEXILMETHYLENAMINE ◇ HMT ◇ METHAMIN ◇ METHENAMINE ◇ PREPARATION AF ◇ RESOTROPIN ◇ 1,3,5,7-TETRAAZAADAMANTANE ◇ URITONE ◇ UROTROPIN ◇ UROTROPINE

TOXICITY DATA with REFERENCE
dnr-bcs 1 mg/disc SAIGBL 26,147,84
cyt-hmn:hla 1 mmol/L HUMAA7 4,112,67
otr-ham:kdy 10 mg/L CRNGDP 4,457,83
scu-rat TDLo:140 g/kg/78W-I:ETA FAONAU 50A,77,72
scu-rat LDLo:200 mg/kg HBTXAC 1,84,56
ivn-rat LD50:9200 mg/kg AEPPAE 221,166,54
orl-mus LDLo:512 mg/kg NTIS** AD-A066-307
ipr-mus LDLo:512 mg/kg CBCCT* 3,126,51
scu-mus LD50:215 mg/kg AEPPAE 225,428,55
scu-cat LDLo:200 mg/kg HBTXAC 1,84,55

CONSENSUS REPORTS: EPA Genetic Toxicology Program. Reported in EPA TSCA Inventory.

DOT Classification: IMO: Flammable Solid; Label: Flammable Solid.

SAFETY PROFILE: A poison by subcutaneous route. Moderately toxic by ingestion and intraperitoneal routes. Questionable carcinogen with experimental tumorigenic data. An irritant to skin, eyes, and mucous membranes. Some persons suffer a skin rash if they come in contact with this material or the fumes evolved when it is heated. Human mutation data reported. Pure hexamethylenetetramine may be taken internally in small amounts and has been used in medicine as a urinary antiseptic. Its major industrial use is in the manufacture of phenolic resins.

Combustible when exposed to heat or flame. Can react with oxidizing materials. Explosive reaction with acetic acid + acetic anhydride + ammonium nitrate + nitric acid; 1-bromopenta borane(9) (above 90°C); iodoform (at 178°C); iodine (at 138°C). Reaction with nitric acid + acetic anhydride forms the military explosives RDX and HMX. Reacts violently with Na_2O_2. When heated to decomposition it emits toxic fumes of formaldehyde and NO_x. See also AMINES.

HEI650 CAS:12001-65-9 ***HR: 3***
HEXAMETHYLENE TETRAMINE TETRAIODIDE
mf: $C_6H_{12}I_4N_4$ mw: 647.81

SAFETY PROFILE: The complex ignites or explodes weakly at 138°C. Upon decomposition it emits toxic fumes of I^- and NO_x. See also IODIDES and AMINES.

HEJ000 ***HR: 3***
HEXAMETHYLENETETRAMMONIUM TETRAPEROXOCHROMATE(V)
mf: $C_{18}H_{48}Cr_4N_{12}O_{32}$ mw: 1152.65

CONSENSUS REPORTS: Chromium and its compounds are on the Community Right-To-Know List.

SAFETY PROFILE: May explode spontaneously. When heated to decomposition it emits toxic fumes of NH_3 and NO_x. See also CHROMIUM COMPOUNDS and PEROXIDES.

HEJ350 CAS:66862-11-1 ***HR: 3***
HEXAMETHYLERBIUM-HEXAMETHYLETHYLENEDIAMINE LITHIUM COMPLEX
mf: $C_6H_{18}Er{\cdot}3C_2H_{16}N_2Li$ mw: 482.78

$$Li_3[((CH_3)_2NC_2H_4N(CH_3)_2)_3Er(CH_3)_6]$$

SAFETY PROFILE: Ignites spontaneously in air. Upon decomposition it emits toxic fumes of NO_x. See also LITHIUM COMPOUNDS, ERBIUM, and AMINES.

HEJ375 CAS:10369-17-2 ***HR: 3***
2,4,6,8,9,10-HEXAMETHYLHEXAAZA-1,3,5,7-TETRAPHOSPHAADAMANTANE
mf: $C_6H_{18}N_6P_4$ mw: 258.14

SAFETY PROFILE: Ignites on contact with strong oxidants. When heated to decomposition it emits toxic fumes of PO_x and NO_x.

HEJ500 CAS:645-05-6 ***HR: 3***
HEXAMETHYLMELAMINE
mf: $C_9H_{18}N_6$ mw: 210.33

PROP: A solid material. Insol in water; sol in acetone.

SYNS: ALTRETAMINE ◇ ENT 50,852 ◇ HEMEL ◇ N,N,N′,N′,N′,N″-HEXAMETHYL-1,3,5-TRIAZINE-2,4,6-TRIAMINE ◇ HEXASTAT ◇ HMM ◇ NCI-C50259 ◇ NSC 13875 ◇ 2,4,6-TRIS(DIMETHYLAMINO)-s-TRIAZINE ◇ 2,4,6-TRIS(DIMETHYLAMINO)-1,3,5-TRIAZINE

TOXICITY DATA with REFERENCE
mma-sat 200 μg/plate MUREAV 142,121,85
cyt-hmn:leu 250 μmol/L CHROAU 24,314,68
orl-rat TDLo:80 mg/kg (female 15-22D post):REP TXAPA9 72,245,84
orl-rat TDLo:320 mg/kg (female 9-12D post):TER TXAPA9 72,245,84
orl-wmn TDLo:3108 mg/kg/3Y-I:ETA AJMSA9 292,393,86
orl-rat TDLo:7 g/kg/46W-C:NEO JNCIAM 51,403,73

orl-hmn TDLo:8 mg/kg:GIT,BLD CCROBU 56,505,72
orl-rat LD50:350 mg/kg 85DZAJ -,315,68
ipr-rat LD50:265 mg/kg JPETAB 100,398,50
orl-mus LD50:437 mg/kg AIPTAK 160,83,66
ipr-mus LD10:200 mg/kg CNREA8 40,2762,80
ivn-mus LD50:171 mg/kg NTIS** PB293-046

CONSENSUS REPORTS: EPA Genetic Toxicology Program.

SAFETY PROFILE: A poison by ingestion, intraperitoneal, and intravenous routes. Experimental teratogenic and reproductive effects. Questionable carcinogen with experimental neoplastigenic data. Human mutation data reported. Human systemic effects by ingestion: nausea or vomiting and leukopenia (reduced white blood cell count). When heated to decomposition it emits toxic fumes of NO_x. See also AMINES.

HEK000 CAS:680-31-9 ***HR: 3***
HEXAMETHYL PHOSPHORAMIDE
mf: $C_6H_{18}N_3OP$ mw: 179.24

PROP: Clear, colorless, mobile liquid, spicy odor. Bp: 233°, fp: 6°, d: 1.024 @ 25°/25°, vap d: 6.18.

SYNS: ENT 50,882 ◇ HEXAMETAPOL ◇ HEXAMETHYLPHOSPHORIC ACID TRIAMIDE (MAK) ◇ HEXAMETHYLPHOSPHORIC TRIAMIDE ◇ N,N,N,N,N,N-HEXAMETHYLPHOSPHORIC TRIAMIDE ◇ HEXAMETHYLPHOSPHOROTRIAMIDE ◇ HEXAMETHYLPHOSPHOTRIAMIDE ◇ HEXMETHYLPHOSPHORAMIDE ◇ HMPA ◇ HMPT ◇ HPT ◇ MEMPA ◇ PHOSPHORIC TRIS(DIMETHYLAMIDE) ◇ PHOSPHORYL HEXAMETHYLTRIAMIDE ◇ TRI(DIMETHYLAMINO)PHOSPHINEOXIDE ◇ TRIS(DIMETHYLAMINO)PHOSPHINE OXIDE ◇ TRIS(DIMETHYLAMINO)PHOSPHORUS OXIDE

TOXICITY DATA with REFERENCE
dns-hmn:hla 125 mg/L PMRSDJ 5,375,85
cyt-hmn:lym 500 mg/L MUREAV 156,19,85
sce-rat:lvr 2 g/L PMRSDJ 5,287,85
orl-rat TDLo:2430 mg/kg (MGN):REP TXAPA9 18,499,71
ihl-rat TCLo:50 ppb/52W-C:CAR AJPAA4 106,8,82
ihl-rat TC:400 ppb/35W-I:CAR SCIEAS 190,422,75
ihl-rat TD:50 ppb/6H/52W-I:ETA TXAPA9 62,90,82
orl-rat LD50:2525 mg/kg 85DZAJ -,315,68
skn-rat LDLo:3500 mg/kg NATUAS 211,146,66
ipr-mus LD50:1600 mg/kg PMRSDJ 1,682,81
ivn-mus LD50:800 mg/kg CHINAG (36),1529,66
orl-rbt LDLo:1500 mg/kg JEENAI 48,139,55
skn-rbt LD50:2600 mg/kg TXAPA9 18,499,71
orl-gpg LD50:1600 mg/kg 85DZAJ -,315,68
skn-gpg LD50:1175 mg/kg 85DZAJ -,315,68

CONSENSUS REPORTS: NTP Fifth Annual Report on Carcinogens. IARC Cancer Review: Group 2B IMEMDT 7,56,87; Animal Sufficient Evidence IMEMDT 15,211,77. Community Right-To-Know List. Reported in EPA TSCA Inventory. EPA Genetic Toxicology Program.

ACGIH TLV: Suspected Human Carcinogen.
DFG MAK: Animal Carcinogen, Suspected Human Carcinogen.

SAFETY PROFILE: Confirmed carcinogen with experimental carcinogenic and tumorigenic data. Moderately toxic by ingestion, skin contact, intraperitoneal, and intravenous routes. Experimental reproductive effects. Human mutation data reported. When heated to decomposition it emits very toxic fumes of phosphine, PO_x, and NO_x.

HEK550 CAS:56090-02-9 ***HR: 3***
HEXAMETHYLRHENIUM
mf: $C_6H_{18}Re$ mw: 276.42

SAFETY PROFILE: A dangerous explosive. It is unstable above −20°C and explodes on contact with oxygen or moisture. See also RHENIUM.

HEL000 CAS:828-26-2 ***HR: 3***
2,2,4,4,6,6-HEXAMETHYLTRITHIANE
mf: $C_9H_{18}S_3$ mw: 222.44

$SC(CH_3)_2SC(CH_3)_2SC(CH_3)_2$

SAFETY PROFILE: Explosive reaction with nitric acid. When heated to decomposition it emits toxic fumes of SO_x.

HEL500 CAS:1164-33-6 ***HR: 3***
HEXAMID
mf: $C_{18}H_{25}N_3O_3 \cdot ClH$ mw: 367.92

SYNS: 3-(2-(DIETHYLAMINO)ETHYL)-5-ETHYL-5-PHENYLBARBITURIC ACID HYDROCHLORIDE ◇ F 156 ◇ 5,5-PHENYL-AETHYL-3-(β-DIAETHYLAMINO-AETHYL)-2,4,6-TRIOXO-HEXAHYDROPYRIMIDIN-HCl (GERMAN)

TOXICITY DATA with REFERENCE
orl-mus LD50:190 mg/kg ARZNAD 6,482,56
scu-mus LD50:490 mg/kg ARZNAD 6,482,56
ivn-mus LD50:94 mg/kg ARZNAD 6,482,56

SAFETY PROFILE: A poison by ingestion and intravenous routes. Moderately toxic by subcutaneous route. When heated to decomposition it emits very toxic fumes of HCl and NO_x. See also BARBITURATES.

HEM000 CAS:66-25-1 ***HR: 3***
1-HEXANAL
DOT: UN 1207
mf: $C_6H_{12}O$ mw: 100.18

PROP: Colorless liquid; powerful fatty-green odor. Reported in about a dozen essential oils (FCTXAV 11,95,73). Mp: −56.3°, bp: 128.7°, flash p: 90°F (OC), d: 0.808-0.812, refr index: 1.402-1.407, vap press: 8.6 mm @ 20°, vap d: 3.45. Sol in alc, fixed oils, propylene glycol; very sltly sol in water.

SYNS: ALDEHYDE C-6 ◇ CAPROALDEHYDE ◇ CAPROIC ALDEHYDE ◇ CAPRONALDEHYDE ◇ n-CAPROYLALDEHYDE ◇ FEMA No. 2557 ◇ HEXALDEHYDE (DOT) ◇ HEXANAL

TOXICITY DATA with REFERENCE
skn-rbt 10 mg/24H open MLD AMIHBC 10,61,54
skn-rbt 14178 μg/24H open MLD AIHAAP 23,95,62
eye-rbt 100 mg/24H MLD FCTXAV 11,95,73
orl-rat LD50:4890 mg/kg AMIHBC 10,61,54
ihl-rat LCLo:2000 ppm/4H AMIHBC 10,61,54

CONSENSUS REPORTS: Reported in EPA TSCA Inventory.

DOT Classification: Flammable Liquid; Label: Flammable Liquid; Flammable or Combustible Liquid; Label: Flammable Liquid

SAFETY PROFILE: Mildly toxic by ingestion and inhalation. An irritant to skin and eyes. Flammable liquid. A dangerous fire hazard when exposed to heat or flame; can react vigorously with oxidizing materials. When heated to decomposition it emits acrid smoke and fumes. See also ALDEHYDES.

HEM500 CAS:628-02-4 ***HR: 3***
HEXANAMIDE
mf: $C_6H_{13}NO$ mw: 115.20

SYNS: CAPROAMIDE ◇ CAPRONAMIDE ◇ NCI-C02142

TOXICITY DATA with REFERENCE
orl-mus TDLo:438 g/kg/1Y-C:CAR JEPTDQ 3(5-6),149,80
orl-rat LD50:1700 mg/kg NCIMR* NIH-71-E-2144,73

CONSENSUS REPORTS: Reported in EPA TSCA Inventory.

SAFETY PROFILE: Moderately toxic by ingestion. Questionable carcinogen with experimental carcinogenic data. When heated to decomposition it emits toxic fumes such as NO_x.

HEN000 CAS:110-54-3 ***HR: 3***
n-HEXANE
DOT: UN 1208
mf: C_6H_{14} mw: 86.20

PROP: Colorless clear liquid; faint odor. Bp: 69°, ULC: 90-95, lel: 1.2%, uel: 7.5%, fp: −95.6°, flash p: −9.4°F, d: 0.6603 @ 20°/4°, autoign temp: 437°F, vap press: 100 mm @ 15.8°, vap d: 2.97. Insol in water; misc in chloroform, ether, alc. Very volatile liquid.

SYNS: ESANI (ITALIAN) ◇ GETTYSOLVE-B ◇ HEKSAN (POLISH) ◇ HEXANE (DOT) ◇ HEXANEN (DUTCH) ◇ HEXANES (FCC) ◇ NCI-C60571

TOXICITY DATA with REFERENCE
eye-rbt 10 mg MLD TXAPA9 55,501,80
cyt-ham:fbr 500 mg/L FCTOD7 22,623,84
ihl-rat TCLo:10000 ppm/7H (female 15D pre):REP TOXID9 1,152,81
ihl-rat TCLo:5000 ppm/20H (female 6-19D post):TER NTIS** DE88-006812
ihl-hmn TCLo:190 ppm/8W:PNS AJIMD8 10,111,86
orl-rat LD50:28710 mg/kg TXAPA9 19,699,71
ipr-rat LDLo:9100 mg/kg TXAPA9 1,156,59
ihl-mus LCLo:120 g/m³ AEPPAE 143,223,29

CONSENSUS REPORTS: Reported in EPA TSCA Inventory.

OSHA PEL: (Transitional: TWA 500 mg/m³) TWA 50 ppm
ACGIH TLV: TWA 50 ppm; BEI: 5 mg(2,5-hexanedione)/L in urine at end of shift; 40 ppm n-hexane in end-exhaled air during shift.
DFG MAK: 50 ppm (180 mg/m³)
NIOSH REL: TWA (Alkanes) 350 mg/m³
DOT Classification: Flammable Liquid; Label: Flammable Liquid.

SAFETY PROFILE: Slightly toxic by ingestion and inhalation. Human systemic effects: hallucinations, structural change in nerve or sheath. Experimental teratogenic and reproductive effects. Mutation data reported. An eye irritant. Can cause a motor neuropathy in exposed workers. May be irritating to respiratory tract and narcotic in high concentrations. Inhalation of 5000 ppm for 1/6-hour produces marked vertigo; 2500-1000 ppm for 12 hours produces drowsiness, fatigue, loss of appetite, paresthesia in distal extremities; 2500-500 ppm produces muscle weakness, cold pulsation in extremities, blurred vision, headache, anorexia, and onset of polyneuropathy. 2000 ppm for 1/6-hour produces no symptoms. 1000-500 ppm for 3-6 months produces fatigue, loss of appetite, distal paresthesia. Dangerous if abused.

Flammable liquid. A very dangerous fire and explosion hazard when exposed to heat or flame; can react vigorously with oxidizing materials. Mixtures with dinitrogen tetraoxide may explode at 28°. To fight fire, use CO_2, dry chemical. When heated to decomposition it emits acrid smoke and fumes.

HEO000 CAS:124-09-4 ***HR: 3***
1,6-HEXANEDIAMINE
DOT: UN 1783/UN 2280
mf: $C_6H_{16}N_2$ mw: 116.24

PROP: Colorless leaflets; odor of piperidine. Mp: 39-

42°, bp: 205°. Absorbs water and CO_2 from air; very sol in water; sltly sol in alc, benzene.

SYNS: 1,6-DIAMINOHEXANE ◇ 1,6-HEXAMETHYLENEDIAMINE ◇ HEXAMETHYLENE DIAMINE, solid (DOT) ◇ HMDA ◇ NCI-C61405

TOXICITY DATA with REFERENCE
orl-rat TDLo:3 g/kg (female 6-15D post):TER JJATDK 7,259,87
orl-rat LD50:750 mg/kg TXAPA9 42,417,77
ihl-mus LCLo:750 mg/m^3/10M NDRC** NDCrc-132,Sept,42
ipr-mus LD50:320 mg/kg GISAAA 43(11),110,78
scu-mus LD50:1300 mg/kg 85GMAT -,74,82
ivn-mus LD50:180 mg/kg CSLNX* NX#02313
skn-rbt LD50:1110 mg/kg TXAPA9 42,417,77

CONSENSUS REPORTS: Reported in EPA TSCA Inventory.

DOT Classification: Corrosive Material; Label: Corrosive (UN1783, UN2280); Corrosive Material; Label: Corrosive, Poison (UN1783)

SAFETY PROFILE: Poison by intravenous and intraperitoneal routes. Moderately toxic by ingestion, inhalation, and skin contact. An experimental teratogen. A corrosive irritant to skin, eyes and mucous membranes. Combustible when exposed to heat or flame; can react with oxidizing materials. See also AMINES.

HEO500 CAS:124-09-4 ***HR: 2***
1,6-HEXANEDIAMINE (solution)
SYN: HEXAMETHYLENE DIAMINE, solution (DOT)

CONSENSUS REPORTS: Reported in EPA TSCA Inventory.

DOT Classification: Label: Corrosive.

SAFETY PROFILE: A corrosive liquid. A powerful skin, eye and mucous membrane irritant. When heated to decomposition it emits toxic fumes of NO_x. See also 1,6-HEXANE DIAMINE.

HEP000 CAS:142-88-1 ***HR: 2***
HEXANEDIOIC ACID, compound with PIPERAZINE (1:1)
mf: $C_{10}H_{20}N_2O_4$ mw: 232.32

SYNS: ADIPRAZINE ◇ DIETELMIN ◇ ENTACYL ◇ OXURASIN ◇ OXYPAAT ◇ OXYZIN (TABL.) ◇ PIPADOX ◇ PIPERAZINE ADIPATE ◇ VERMICOMPREN (TABL.) ◇ VERMILASS

TOXICITY DATA with REFERENCE
orl-rat LD50:7900 mg/kg APFRAD 13,539,55
orl-mus LD50:11400 mg/kg JPPMAB 6,711,54
ipr-mus LD50:1640 mg/kg 85GMAT -,100,82

CONSENSUS REPORTS: Reported in EPA TSCA Inventory.

SAFETY PROFILE: Moderately toxic by intraperitoneal route. Mildly toxic by ingestion. Used as an anthelmintic (anti-worm agent). When heated to decomposition it emits toxic fumes such as NO_x.

HEP500 CAS:629-11-8 ***HR: 2***
1,6-HEXANEDIOL
mf: $C_6H_{14}O_2$ mw: 118.20

PROP: Mp: 42°, bp: 250°, flash p: 266°F, d: 0.967 @ 0°/4°, vap d: 4.07.

SYNS: HDO ◇ HEXAMETHYLENE GLYCOL

TOXICITY DATA with REFERENCE
eye-rbt 100 mg IHFCAY 6,1,67
orl-rat LD50:3730 mg/kg TXAPA9 28,313,74

CONSENSUS REPORTS: Reported in EPA TSCA Inventory.

SAFETY PROFILE: Moderately toxic by ingestion. An eye irritant. Combustible when exposed to heat or flame; can react with oxidizing materials. To fight fire, use foam, CO_2, dry chemical. When heated to decomposition it emits acrid smoke and fumes.

HEQ000 CAS:2935-44-6 ***HR: 1***
2,5-HEXANEDIOL
mf: $C_6H_{14}O_2$ mw: 118.20

PROP: Liquid. Bp: 220.8°, flash p: 230°F, d: 0.9617 @ 20°/20°, vap d: 4.07.

TOXICITY DATA with REFERENCE
skn-rbt 10 mg/24H open JIHTAB 30,63,48
eye-rbt 100 mg open JIHTAB 30,63,48
orl-rat LD50:5000 mg/kg JIHTAB 30,63,48
orl-mus LD50:4846 mg/kg JAPMA8 45,669,56

CONSENSUS REPORTS: Reported in EPA TSCA Inventory.

SAFETY PROFILE: Mildly toxic by ingestion. A skin and eye irritant. Combustible when exposed to heat or flame; can react with oxidizing materials. To fight fire, use alcohol foam, CO_2, dry chemical. When heated to decomposition it emits acrid smoke and fumes.

HEQ200 CAS:3848-24-6 ***HR: 3***
2,3-HEXANEDIONE
mf: $C_6H_{10}O_2$ mw: 114.16

PROP: Bp: 128°, d: 0.934, Flash p: 83° F.

SYNS: ACETYLBUTYRYL ◇ METHYL PROPYL DIKETONE

TOXICITY DATA with REFERENCE
skn-rbt 500 mg/24H MOD FCTXAV 17,697,79

CONSENSUS REPORTS: Reported in EPA TSCA Inventory.

SAFETY PROFILE: A skin irritant. Flammable liquid. When heated to decomposition it emits acrid smoke and irritating fumes.

HEQ500 CAS:110-13-4 ***HR: 2***
2,5-HEXANEDIONE
mf: $C_6H_{10}O_2$ mw: 114.16

PROP: Colorless liquid. Gradually turns yellow. Mp: −9°, bp: 188°, flash p: 174°F (CC), d: 0.970 @ 20°/4°, autoign temp: 920°, vap d: 3.94.

SYNS: ACETONYL ACETONE ◇ α,β-DIACETYLETHANE ◇ 1,2-DIACETYLETHANE ◇ 2,5-DIKETOHEXANE

TOXICITY DATA with REFERENCE
eye-rbt 19 mg AJOPAA 29,1363,46
orl-rat TDLo:4200 mg/kg (42D male):REP TXAPA9 62,262,82
orl-rat LD50:2076 mg/kg IJEBA6 24,371,86
ihl-rat LCLo:2000 ppm/4H JIDHAN 31,343,49
orl-mus LD50:2386 mg/kg JAMPA8 45,669,56
skn-gpg LD50:6422 mg/kg JIHTAB 26,269,44

CONSENSUS REPORTS: Reported in EPA TSCA Inventory.

SAFETY PROFILE: Moderately toxic by ingestion and inhalation. Mildly toxic by skin contact. An eye irritant. Experimental reproductive effects. Combustible when exposed to heat or flame; can react with oxidizing materials. To fight fire, use CO_2, dry chemical, (multi-purpose dry chemical), water spray or mist, alcohol foam. When heated to decomposition it emits acrid smoke and fumes.

HER000 CAS:69-65-8 ***HR: 1***
1,2,3,4,5,6-HEXANEHEXOL
mf: $C_6H_{14}O_6$ mw: 182.20

PROP: White, crystalline powder; odorless. D: 1.52, mp: 165°-167°, bp: 290-295° @ 3-5 mm. Sol in water; sltly sol in lower alc and amines; almost insol in organic solvents.

SYNS: MANNA SUGAR ◇ MANNITE ◇ MANNITOL ◇ d-MANNITOL ◇ NCI-C50362 ◇ OSMITROL

TOXICITY DATA with REFERENCE
oms-omi 1 mol/L ARMKA7 91,305,73
dni-hmn:lym 50 mmol/L PNSAS6 79,1171,82
ivn-man TDLo:17143 mg/kg/2D-C:CVS,GIT,KID AIMDAP 144,2053,84
orl-rat LD50:13500 mg/kg YKYUA6 32,1367,81
ivn-rat LD50:9690 mg/kg YKYUA6 32,1367,81
orl-mus LD50:22 g/kg FAONAU 40,161,67
ipr-mus LD50:14 g/kg PSEBAA 35,98,36
ivn-mus LD50:7470 mg/kg YKYUA6 32,1367,81

CONSENSUS REPORTS: NTP Carcinogenesis Bioassay (feed); No Evidence: mouse, rat NTPTR* NTP-TR-236,82. Reported in EPA TSCA Inventory.

SAFETY PROFILE: Mildly toxic by ingestion, intraperitoneal, and intravenous routes. Human systemic effects by intravenous route: blood pressure elevation, bladder tubule changes, nausea or vomiting. Human mutation data reported. Used as a nutrient and/or dietary supplement food additive. When heated to decomposition it emits acrid smoke and fumes.

HER500 CAS:628-73-9 ***HR: 3***
HEXANENITRILE
mf: $C_6H_{11}N$ mw: 97.18

SYNS: CAPRONITRILE ◇ NC5

TOXICITY DATA with REFERENCE
orl-mus LD50:463 mg/kg NEZAAQ 39,423,84
ivn-rbt LDLo:42 mg/kg COREAF 153,895,11
scu-gpg LDLo:310 mg/kg COREAF 153,895,11

CONSENSUS REPORTS: Reported in EPA TSCA Inventory. Cyanide and its compounds are on the Community Right-To-Know List.

SAFETY PROFILE: A poison by intravenous and subcutaneous routes. Moderately toxic by ingestion. When heated to decomposition it emits toxic fumes of NO_x and CN^-. See also NITRILES.

HES000 CAS:111-31-9 ***HR: 3***
1-HEXANETHIOL
mf: $C_6H_{14}S$ mw: 118.26

SYNS: HEXYL MERCAPTAN ◇ USAF EK-4628

TOXICITY DATA with REFERENCE
orl-rat LD50:1254 mg/kg AIHAAP 19,171,58
ihl-rat LC50:1080 ppm/4H AIHAAP 19,171,58
ihl-mus LD50:528 mg/kg AIHAAP 19,171,58
ipr-mus LD50:200 mg/kg NTIS** AD277-689

CONSENSUS REPORTS: Reported in EPA TSCA Inventory.

NIOSH REL: (n-Alkane Mono Thiols) CL 0.5 ppm/15M

SAFETY PROFILE: A poison by intraperitoneal route. Moderately toxic by inhalation and ingestion. When heated to decomposition it emits very toxic fumes of SO_x. See also MERCAPTANS.

HES500 CAS:106-69-4 ***HR: 1***
1,2,6-HEXANETRIOL
mf: $C_6H_{14}O_3$ mw: 134.20

PROP: Colorless liquid. Bp: 178° @ 5 mm, fp: −20°, flash p: 375°F (COC), d: 1.1063 @ 20°/20°, vap press: <0.01 mm @ 20°, vap d: 4.63.

SYNS: HEXANETRIOL-1,2,6 ◇ HEXANE-1,2,6-TRIOL

TOXICITY DATA with REFERENCE
skn-rbt 555 mg open MLD UCDS** 7/13/71
eye-rbt 500 mg open AMIHBC 10,61,54
orl-rat LD50:15500 mg/kg 34ZIAG -,731,69
ipr-rat LD50:10 g/kg TXAPA9 15,282,69
ivn-rat LD50:5600 mg/kg TXAPA9 15,282,69
orl-mus LD50:11400 mg/kg SCCUR* -,5,61

CONSENSUS REPORTS: Reported in EPA TSCA Inventory.

SAFETY PROFILE: Mildly toxic by ingestion. An eye and skin irritant. Combustible when exposed to heat or flame; can react with oxidizing materials. To fight fire, use alcohol foam, spray, mist, dry chemical. When heated to decomposition it emits acrid smoke and fumes.

HET000 CAS:6091-44-7 ***HR: 3***
HEXANHYDROPYRIDINE HYDROCHLORIDE
mf: $C_5H_{11}N{\cdot}ClH$ mw: 121.63

SYN: PIPERIDINE HYDROCHLORIDE

TOXICITY DATA with REFERENCE
orl-rat LD50:133 mg/kg 27ZQAG -,289,72
ipr-mus LD50:330 mg/kg JJPAAZ 17,475,67
scu-mus LD50:430 mg/kg AIPTAK 112,36,57
ivn-mus LD50:160 mg/kg AIPTAK 112,36,57

CONSENSUS REPORTS: Reported in EPA TSCA Inventory.

SAFETY PROFILE: Poison by ingestion, intravenous, and intraperitoneal routes. Moderately toxic by subcutaneous route. When heated to decomposition it emits very toxic fumes of NO_x and HCl.

HET350 CAS:13232-74-1 ***HR: 3***
HEXANITROBENZENE
mf: $C_6N_6O_{12}$ mw: 270.11

$C_6(NO_2)_6$

SAFETY PROFILE: A powerful explosive. When heated to decomposition it emits toxic fumes of NO_x. See also EXPLOSIVES, NITRO COMPOUNDS of AROMATIC HYDROCARBONS, and NITROBENZENE.

HET500 CAS:131-73-7 ***HR: 3***
2,4,6,2′,4′,6′-HEXANITRODIPHENYLAMINE
mf: $C_{12}H_5N_7O_{12}$ mw: 439.24

SYNS: BIS(2,4,6-TRINITRO-PHENYL)-AMIN (GERMAN) ◇ DPA ◇ ESANITRODIFENILAMINA (ITALIAN) ◇ HEXANITRODIFENYLAMINE (DUTCH) ◇ HEXANITRODIPHENYLAMINE ◇ HEXANITRODIPHENYLAMINE (FRENCH) ◇ 2,2′,4,4′,6,6′-HEXANITRODIPHENYLAMINE ◇ HEXYL (GERMAN, DUTCH)

TOXICITY DATA with REFERENCE
mmo-sat 228 nmol/plate MUREAV 136,209,84
mma-sat 456 nmol/plate MUREAV 136,209,84
orl-rat TDLo:14 g/kg/76W-C:NEO NATUAS 180,509,57

CONSENSUS REPORTS: Reported in EPA TSCA Inventory.

SAFETY PROFILE: Questionable carcinogen with experimental neoplastigenic data. Mutation data reported. A powerful and violent explosive used as a booster explosive; its use is superior to TNT. It is not as good for this purpose as tetryl, but is extremely stable and much safer to handle. See also NITRO COMPOUNDS of AROMATIC HYDROCARBONS.

HET675 CAS:918-37-6 ***HR: 3***
HEXANITROETHANE
mf: $C_2N_6O_{12}$ mw: 300.06

CONSENSUS REPORTS: Reported in EPA TSCA Inventory.

DOT Classification: Forbidden

SAFETY PROFILE: A powerful oxidant which explodes above 140°C. Explosive reaction with boron. Hypergolic reaction with dimethyl hydrazine or other strong organic bases. Forms powerfully explosive mixtures with nitrogen containing organic compounds (e.g., 2-nitroaniline). Upon decomposition it emits toxic fumes of NO_x. See also NITRO COMPOUNDS.

HEU000 CAS:142-62-1 ***HR: 2***
HEXANOIC ACID
DOT: NA 1706
mf: $C_6H_{12}O_2$ mw: 116.18

PROP: Oily, colorless liquid; odor of Limburger cheese. Bp: 205.0°, fp: −3.4°, flash p: 215°F (COC), d: 0.9295 @ 20°/20°, refr index: 1.415-1.418, vap press: 0.18 mm @ 20°, vap d: 4.0, autoign temp: 716°F. Very sol in ether, fixed oils; sltly sol in water.

SYNS: BUTYLACETIC ACID ◇ CAPROIC ACID ◇ n-CAPROIC ACID ◇ CAPRONIC ACID ◇ FEMA No. 2559 ◇ HEXACID 698 ◇ n-HEXANOIC ACID ◇ n-HEXOIC ACID ◇ PENTIFORMIC ACID ◇ PENTYLFORMIC ACID

TOXICITY DATA with REFERENCE
skn-rbt 10 mg/24H open MLD AMIHBC 10,61,54

skn-rbt 465 mg open MLD UCDS** 11/2/71
eye-rbt 695 μg SEV AJOPAA 29,1363,46
oms-nml:oth 10 mmol/L CHROAU 40,1,73
cyt-nml:oth 10 mmol/L CHROAU 40,1,73
orl-rat LD50:3000 mg/kg JIHTAB 26,269,44
orl-mus LD50:5 g/kg 85GMAT -,32,82
ihl-mus LC50:4100 mg/m^3/2H 85GMAT -,32,82
ipr-mus LD50:3180 mg/kg JPPMAB 21,85,69
scu-mus LD50:3180 mg/kg JPPMAB 21,85,69
skn-rbt LD50:630 mg/kg AMIHBC 10,61,54
skn-gpg LD50:4635 mg/kg JIHTAB 26,269,44

CONSENSUS REPORTS: Reported in EPA TSCA Inventory.

DOT Classification: Corrosive Material; Label: Corrosive.

SAFETY PROFILE: Moderately toxic by ingestion, skin contact, intraperitoneal, and subcutaneous routes. Mutation data reported. Corrosive. A skin and severe eye irritant. Combustible when exposed to heat or flame; can react with oxidizing materials. To fight fire, use CO_2, dry chemical, fog, mist. When heated to decomposition it emits acrid smoke and fumes.

HEU500 ***HR: 1***
HEXANOIC ACID, VINYL ESTER (mixed isomers)
mf: $C_8H_{14}O_2$ mw: 142.22

TOXICITY DATA with REFERENCE
skn-rbt 10 mg/24H open MLD AIHAAP 23,95,62
orl-rat LD50:20 g/kg AIHAAP 23,95,62
ihl-rat LCLo:4000 ppm/4H AIHAAP 23,95,62

SAFETY PROFILE: Mildly toxic by ingestion and inhalation. A skin irritant. When heated to decomposition it emits acrid smoke and fumes. See also ESTERS.

HEV000 CAS:591-78-6 ***HR: 3***
2-HEXANONE
mf: $C_6H_{12}O$ mw: 100.18

PROP: Clear liquid. Mp: −56.9°, bp: 127.2°, lel: 1.22%, uel: 8.0%, flash p: 95°F (OC), d: 0.830 @ 0°/4°, vap press: 10 mm @ 38.8°, vap d: 3.45, autoign temp: 991°F. Sltly sol in water; sol in alc and ether.

SYNS: BUTYL METHYL KETONE ◇ n-BUTYL METHYL KETONE ◇ HEXANONE-2 ◇ MBK ◇ METHYL n-BUTYL KETONE (ACGIH) ◇ MNBK

TOXICITY DATA with REFERENCE
skn-rbt 500 mg/24H MLD 85JCAE -,283,86
eye-rbt 100 mg open AMIHBC 10,61,54
eye-rbt 500 mg/24H MLD 85JCAE -,283,86
ihl-rat TCLo:2000 ppm/6H (1-21D preg):REP EESADV 5,291,81
ihl-rat TCLo:1000 ppm/6H (1-21D preg):TER EESADV 5,291,81
ihl-hmn TCLo:1000 ppm:EYE,CNS,GIT NPIRI* 1,78,74
orl-rat LD50:2590 mg/kg AMIHBC 10,61,54
ihl-rat LC50:8000 ppm/4H NPIRI* 1,78,74
ipr-rat LDLo:914 mg/kg RalRL# 01MAR74
orl-mus LD50:2430 mg/kg TOLED5 30,13,86
skn-rbt LD50:4800 mg/kg NPIRI* 1,78,74
orl-gpg LDLo:914 mg/kg RalRL# 01MAR74

CONSENSUS REPORTS: Reported in EPA TSCA Inventory.

OSHA PEL: (Transitional: TWA 100 ppm) TWA 5 ppm
ACGIH TLV: TWA 5 ppm
DFG MAK: 5 ppm (21 mg/m^3)
NIOSH REL: (Ketones) TWA 4 mg/m^3

SAFETY PROFILE: Moderately toxic by ingestion and intraperitoneal routes. Mildly toxic by inhalation and skin contact. Experimental teratogenic and reproductive effects. Human systemic effects by inhalation: unspecified eye effects, headache, nausea or vomiting. A skin and eye irritant. Dangerous fire and explosion hazard when exposed to heat or flame; can react with oxidizing materials. To fight fire, use alcohol foam, CO_2, dry chemical. See also KETONES.

HEV500 CAS:589-38-8 ***HR: 3***
3-HEXANONE
mf: $C_6H_{12}O$ mw: 100.18

PROP: Colorless liquid. Bp: 124°, d: 0.813 @ 21.8°/4°, flash p: 57.2°F (OC).

SYNS: AETHYLPROPYLKETON (GERMAN) ◇ ETHYL PROPYL KETONE

TOXICITY DATA with REFERENCE
orl-rat LD50:3360 mg/kg TXAPA9 28,313,74
ihl-rat LCLo:4000 ppm/4H TXAPA9 28,313,74
skn-rbt LD50:3170 mg/kg TXAPA9 28,313,74
scu-gpg LDLo:700 mg/kg BDKS** -,-,34

CONSENSUS REPORTS: Reported in EPA TSCA Inventory.

SAFETY PROFILE: Moderately toxic by ingestion, skin contact, and subcutaneous routes. Mildly toxic by inhalation. Flammable liquid. A very dangerous fire hazard when exposed to heat or flame; can react vigorously with oxidizing materials. To fight fire, use foam, CO_2, dry chemical. When heated to decomposition it emits acrid smoke and fumes. See also KETONES.

HEW000 CAS:45776-10-1 ***HR: 3***
1-HEXANOYLAZIRIDINE
mf: $C_8H_{15}NO$ mw: 141.24

SYNS: 1-CAPROYLAZIRIDINE ◇ CAPROYLETHYLENEIMINE ◇ HEXANOYLETHYLENEIMINE

TOXICITY DATA with REFERENCE
cyt-rat-ipr 50 mg/kg BJPCAL 9,306,54
scu-rat TDLo:495 mg/kg/19W-I:NEO BJPCAL 9,306,54
scu-mus TDLo:360 mg/kg/41W-I:ETA BJPCAL 9,306,54

SAFETY PROFILE: Questionable carcinogen with experimental neoplastigenic and tumorigenic data. Mutation data reported. When heated to decomposition it emits toxic fumes of NO_x.

HEW100 CAS:2787-93-1 ***HR: 2***
1,1,1,3,3,3-HEXAOCTYLDISTANNOXANE
mf: $C_{48}H_{102}OSn_2$ mw: 932.88

SYNS: DISTANNOXANE, HEXAOCTYL- ◇ HEXAOCTYLDISTANNOXANE

TOXICITY DATA with REFERENCE
ipr-mus LD50:2725 mg/kg RPTOAN 42,73,79

OSHA PEL: TWA 0.1 mg(Sn)/m^3
ACGIH TLV: TWA 0.1 mg(Sn)/m^3; STEL 0.2 mg/m^3 (skin)
NIOSH REL: (Organotin Compounds): 10H TWA 0.1 mg(Sn)/m^3

SAFETY PROFILE: Moderately toxic by intraperitoneal route. When heated to decomposition it emits toxic fumes of Sn.

HEW150 CAS:13413-18-8 ***HR: 2***
HEXAOCTYLDISTANNTHIANE
mf: $C_{48}H_{102}SSn_2$ mw: 948.94

SYNS: DISTANNTHIANE, HEXAOCTYL- ◇ 1,1,1,3,3,3-HEXAOCTYLDISTANNTHIANE

TOXICITY DATA with REFERENCE
ipr-mus LD50:3069 mg/kg RPTOAN 42,73,79

OSHA PEL: TWA 0.1 mg(Sn)/m^3
ACGIH TLV: TWA 0.1 mg(Sn)/m^3; STEL 0.2 mg/m^3 (skin)
NIOSH REL: (Organotin Compounds): 10H TWA 0.1 mg(Sn)/m^3

SAFETY PROFILE: Moderately toxic by intraperitoneal route. When heated to decomposition it emits toxic fumes of SO_x and Sn.

HEW200 CAS:7328-05-4 ***HR: 3***
HEXAPROPYLDISTANNTHIANE
mf: $C_{18}H_{42}SSn_2$ mw: 528.04

SYNS: DISTANNTHIANE, HEXAPROPYL- ◇ 1,1,1,3,3,3-HEXAPROPYLDISTANNTHIANE

TOXICITY DATA with REFERENCE
ipr-mus LD50:97 mg/kg RPTOAN 42,73,79

OSHA PEL: TWA 0.1 mg(Sn)/m^3
ACGIH TLV: TWA 0.1 mg(Sn)/m^3; STEL 0.2 mg/m^3 (skin)
NIOSH REL: (Organotin Compounds): 10H TWA 0.1 mg(Sn)/m^3

SAFETY PROFILE: Poison by intraperitoneal route. When heated to decomposition it emits toxic fumes of SO_x and Sn.

HEY000 CAS:23129-50-2 ***HR: 3***
HEXAPYRIDINEIRON(II) TRIDECACARBONYL TETRAFERRATE(2^-)
mf: $C_{43}H_{30}Fe_5N_6O_{13}$ mw: 1117.9

SYN: HEXAKIS(PYRIDINE)IRON(II)TRIDECACARBONYLTETRAFERRATE(2-)

SAFETY PROFILE: Ignites spontaneously in air. When heated to decomposition it emits toxic fumes of NO_x. See also CARBONYLS.

HEY500 CAS:14986-84-6 ***HR: 2***
HEXASODIUM TETRAPHOSPHATE
mf: $Na_6O_{13}P_4$ mw: 469.82

SYNS: HEXANATRIUMTETRAPOLYPHOSPHAT (GERMAN) ◇ HEXASODIUM TETRAPOLYPHOSPHATE

TOXICITY DATA with REFERENCE
orl-mus LD50:3920 mg/kg ARZNAD 7,445,57
scu-mus LD50:875 mg/kg ARZNAD 7,445,57

CONSENSUS REPORTS: Reported in EPA TSCA Inventory.

SAFETY PROFILE: Moderately toxic by ingestion and subcutaneous routes. When heated to decomposition it emits toxic fumes of Na_2O and PO_x. See also PHOSPHATES.

HEZ000 CAS:2235-12-3 ***HR: 3***
1,3,5-HEXATRIENE
mf: C_6H_8 mw: 80.14

SYN: DIVINYLETHYLENE

TOXICITY DATA with REFERENCE
eye-rbt 369 mg IHFCAY 6,1,67
orl-rat LD50:210 mg/kg IHFCAY 6,1,67
ihl-rat LCLo:100000 ppm/15M IHFCAY 6,1,67
skn-rbt LD50:6730 mg/kg IHFCAY 6,1,67

CONSENSUS REPORTS: Reported in EPA TSCA Inventory.

SAFETY PROFILE: Poison by ingestion. Mildly toxic by skin contact and inhalation. An eye irritant. When heated to decomposition it emits acrid smoke and fumes.

HEZ375 CAS:3161-99-7 ***HR: 3***
1,3,5-HEXATRIYNE
mf: C_6H_2 mw: 74.08

$$H(C \equiv C)_3H$$

SAFETY PROFILE: Polymerizes rapidly in air at room temperature to form a friction-sensitive explosive solid which is probably a peroxide. When heated to decomposition it emits acrid smoke and fumes. See also ACETYLENE COMPOUNDS and PEROXIDES.

HEZ800 CAS:14023-01-9 ***HR: 3***
HEXAUREA CHROMIC CHLORIDE
mf: $C_6H_{24}CrN_{12}O_6 \cdot 3Cl$ mw: 518.77

SYNS: CHROMIUM CHLORIDE, HEXAUREA ◇ CHROMIUM(3 +), HEXAKIS(UREA-O)-, TRICHLORIDE, (OC-6-11)-(9CI) ◇ CHROMIUM(3 +), HEXAKIS(UREA)-, TRICHLORIDE (8CI) ◇ CHROMIUM(III) HEXA-UREA CHLORIDE

TOXICITY DATA with REFERENCE
ivn-rat LD50:180 mg/kg EQSFAP 1,1,75

OSHA PEL: TWA 0.5 mg(Cr)/m^3
ACGIH TLV: TWA 0.5 mg(Cr)/m^3

SAFETY PROFILE: Poison by intravenous route. When heated to decomposition it emits toxic fumes of NO_x, Cr, and Cl^-.

HFA000 CAS:22471-42-7 ***HR: 3***
HEXAUREACHROMIUM(III) NITRATE
mf: $C_6H_{24}CrN_{15}O_{15}$ mw: 598.34

$$[(H_2NCO \cdot NH_2)_6Cr][NO_3]_3$$

CONSENSUS REPORTS: Chromium and its compounds are on the Community Right-To-Know List.
SAFETY PROFILE: An explosive sensitive to impact and heating to 265°C. Upon decomposition it emits toxic fumes of NO_x. See also EXPLOSIVES, NITRATES, and CHROMIUM COMPOUNDS.

HFA225 CAS:31332-72-6 ***HR: 3***
HEXAUREAGALLIUM(III) PERCHLORATE
mf: $C_6H_{24}Cl_3GaN_{12}O_{18}$ mw: 728.40

$$[(H_2NCO \cdot NH_2)_6Ga][ClO_4]_3$$

SAFETY PROFILE: Explodes violently when heated above its melting point of 179°C. Upon decomposition it emits toxic fumes of Cl^- and NO_x. See also GALLIUM and PERCHLORATES.

HFA300 CAS:51235-04-2 ***HR: 2***
HEXAZINONE
mf: $C_{12}H_{20}N_4O_2$ mw: 252.36

SYNS: 3-CYCLOHEXYL-6-(DIMETHYLAMINO)-1-METHYL-s-TRIAZINE-2,4(1H,3H)-DIONE ◇ 3-CYCLOHEXYL-6-(DIMETHYLAMINO)-1-METHYL-1,3,5-TRIAZINE-2,4(1H,3H)-DIONE ◇ DPX 3674 ◇ VELPAR ◇ VELPAR WEED KILLER

TOXICITY DATA with REFERENCE
eye-rbt 48 mg MOD FAATDF 4,603,84
orl-rat TDLo:51700 mg/kg (94D male/94D pre):REP FAATDF 4,960,84
orl-rat LD50:1690 mg/kg 85ARAE 2,135,77
skn-rat LD50:5278 mg/kg FMCHA2 -,C126,83
ipr-rat LD50:530 mg/kg FAATDF 4,603,84
orl-gpg LD50:860 mg/kg PESTC* 9,21,80
ipr-qal LD50:2258 mg/kg FAATDF 4,603,84

CONSENSUS REPORTS: Reported in EPA TSCA Inventory.

SAFETY PROFILE: Moderately toxic by ingestion and intraperitoneal routes. Mildly toxic by skin contact. Experimental reproductive effects. An eye irritant. When heated to decomposition it emits toxic fumes of NO_x.

HFA500 CAS:505-57-7 ***HR: 3***
2-HEXENAL
mf: $C_6H_{10}O$ mw: 98.16

SYNS: HEX-2-ENAL ◇ HEX-2-EN-1-AL ◇ HEXYLENIC ALDEHYDE ◇ LEAF ALDEHYDE

TOXICITY DATA with REFERENCE
mmo-sat 500 nmol/plate MUREAV 148,25,85
ipr-mus LD50:290 mg/kg ZolH## 23OCT75

CONSENSUS REPORTS: Reported in EPA TSCA Inventory.

SAFETY PROFILE: A poison by intraperitoneal route. Mutation data reported. When heated to decomposition it emits acrid smoke and fumes. See also ALDEHYDES.

HFA525 ***HR: 3***
trans-2-HEXEN-1-AL
mf: $C_6H_{10}O$ mw: 98.15

PROP: Pale yellow liquid; fruity, vegetable odor. D: 0.841-0.848, refr index: 1.445-1.449, flash p: 100°F. Sol in alc, propylene glycol, fixed oils; very sltly sol in water.

SYN: FEMA No. 2560

SAFETY PROFILE: Flammable liquid. When heated to decomposition it emits acrid smoke and irritating fumes.

HFB000 CAS:592-41-6 ***HR: 3***
1-HEXENE
DOT: UN 2370
mf: C_6H_{12} mw: 84.158

PROP: Colorless liquid. Bp: 64.5°, mp: −139.9°, flash p: −14.8°F, d: 0.6732 @ 20°/4°, vap press: 310 mm @ 38°, vap d: 3.0, lel: 1.2%, uel: 6.9%

SYNS: BUTYL ETHYLENE ◇ HEXENE ◇ HEXYLENE

CONSENSUS REPORTS: Reported in EPA TSCA Inventory.

DOT Classification: Flammable Liquid; Label: Flammable Liquid.

SAFETY PROFILE: Moderately toxic irritant to skin, eyes, and mucous membranes. A very dangerous fire and explosion hazard when exposed to heat, flame, or oxidizers. Can react vigorously with oxidizing materials. To fight fire, use dry chemical, CO_2, foam. When heated to decomposition it emits acrid smoke and fumes.

HFB500 ***HR: 3***
2-HEXENE
mf: C_6H_{12} mw: 84.158

PROP: Flash p: −5.8°F.

SAFETY PROFILE: A very dangerous fire and explosion hazard when exposed to heat, flame, or oxidizers. To fight fire, use dry chemical, CO_2, foam. Incompatible with oxidizers, heat, flame. When heated to decomposition it emits acrid smoke and fumes.

HFC000 CAS:1119-85-3 ***HR: 3***
3-HEXENE DINITRILE
mf: $C_6H_6N_2$ mw: 106.14

TOXICITY DATA with REFERENCE
ivn-mus LD50:56 mg/kg CSLNX* NX#05212

CONSENSUS REPORTS: Reported in EPA TSCA Inventory. Cyanide and its compounds are on the Community Right-To-Know List.

SAFETY PROFILE: Poison by intravenous route. When heated to decomposition it emits toxic fumes of NO_x and CN^-. See also NITRILES.

HFC500 ***HR: 3***
trans-2-HEXENE OZONIDE
mf: $C_6H_{12}O_3$ mw: 132.16

OCH(CH_3)OOCHCH$_2$CH$_2$CH$_3$ (ring: O to CH)

SYN: 3-METHYL-5-PROPYL-1,2,4-TRIOXOLANE

SAFETY PROFILE: A powerful explosive. When heated to decomposition it emits acrid smoke and fumes.

HFD000 CAS:4219-24-3 ***HR: 2***
3-HEXENOIC ACID
mf: $C_6H_{10}O_2$ mw: 114.16

SYN: HYDROSORBIC ACID

TOXICITY DATA with REFERENCE
ipr-mus LD50:1840 mg/kg JPPMAB 21,85,69
scu-mus LD50:1840 mg/kg JPPMAB 21,85,69

CONSENSUS REPORTS: Reported in EPA TSCA Inventory.

SAFETY PROFILE: Moderately toxic by intraperitoneal and subcutaneous routes. When heated to decomposition it emits acrid smoke and fumes.

HFD500 CAS:928-95-0 ***HR: 2***
2-HEXEN-1-OL, (E)-
mf: $C_6H_{12}O$ mw: 100.18

PROP: Colorless liquid; fruity-green odor. D: 0.836-0.841, refr index: 0.437-1.442, flash p: 129°F. Sol in alc, propylene glycol, fixed oils; very sltly sol in water.

SYNS: FEMA No. 2562 ◇ trans-2-HEXENOL ◇ 2-HEXENOL ◇ trans-2-HEXEN-1-OL (FCC)

TOXICITY DATA with REFERENCE
skn-rbt 500 mg/24H FCTXAV 12,911,74
orl-rat LD50:3500 mg/kg FCTXAV 12,911,74
skn-rbt LD50:4500 mg/kg FCTXAV 12,911,74

CONSENSUS REPORTS: Reported in EPA TSCA Inventory.

SAFETY PROFILE: Moderately toxic by ingestion. Mildly toxic by skin contact. A skin irritant. Combustible liquid. When heated to decomposition it emits acrid smoke and fumes. See also ALCOHOLS.

HFE000 CAS:928-96-1 ***HR: 3***
cis-3-HEXENOL
mf: $C_6H_{12}O$ mw: 100.18

PROP: Colorless liquid; powerful grassy-green odor. D: 0.846-0.850, refr index: 1.43-1.441, bp: 137°, flash p: 111°F. Sol in alc, propylene glycol, fixed oils; very sltly sol in water.

SYNS: BLATTERALKOHOL ◇ FEMA No. 2563 ◇ β-Γ-HEXENOL ◇ cis-3-HEXEN-1-OL (FCC) ◇ LEAF ALCOHOL

TOXICITY DATA with REFERENCE
orl-rat LD50:4700 mg/kg FCTXAV 12,909,74
ipr-rat LD50:600 mg/kg FCTXAV 7,451,69
orl-mus LD50:7000 mg/kg FCTXAV 7,451,69
ipr-mus LD50:400 mg/kg FCTXAV 7,451,69

CONSENSUS REPORTS: Reported in EPA TSCA Inventory.

SAFETY PROFILE: A poison by intraperitoneal route. Mildly toxic by ingestion. Combustible liquid. When heated to decomposition it emits acrid smoke and fumes. See also ALCOHOLS.

HFE100 CAS:2497-18-9 ***HR: 1***
2-HEXEN-1-OL ACETATE
mf: $C_8H_{14}O_2$ mw: 142.22

SYNS: HEX-2-ENYL ACETATE ◇ 2-HEXENYL ACETATE ◇ 2-HEXEN-1-YL-ACETATE ◇ (E)-2-HEXENYL ACETATE ◇ trans-2-HEXENYL ACETATE

TOXICITY DATA with REFERENCE
skn-rbt 500 mg/24H MOD FCTXAV 17,793,79

CONSENSUS REPORTS: Reported in EPA TSCA Inventory.

SAFETY PROFILE: A skin irritant. When heated to decomposition it emits acrid smoke and irritating fumes.

HFE500 CAS:25152-85-6 ***HR: 1***
cis-3-HEXENYL BENZOATE
mf: $C_{13}H_{16}O_2$ mw: 204.29

SYN: 3-HEXENYL ESTER, BENZOIC ACID (Z)-

TOXICITY DATA with REFERENCE
skn-rbt 500 mg/24H MOD FCTXAV 16,773,78

CONSENSUS REPORTS: Reported in EPA TSCA Inventory.

SAFETY PROFILE: A skin irritant. When heated to decomposition it emits acrid smoke and fumes.

HFE520 CAS:41519-23-7 ***HR: 1***
cis-3-HEXENYL ISOBUTYRATE
mf: $C_{10}H_{18}O_2$ mw: 170.28

SYNS: ENT 33348 ◇ β,Γ-HEXENYL ISOBUTANOATE ◇ PROPANOIC ACID, 2-METHYL-, 3-HEXENYL ESTER, (Z)-

TOXICITY DATA with REFERENCE
skn-rbt 500 mg/24H MOD FCTXAV 17,799,79

CONSENSUS REPORTS: Reported in EPA TSCA Inventory.

SAFETY PROFILE: A skin irritant. When heated to decomposition it emits acrid smoke and irritating fumes.

HFE550 ***HR: 2***
cis-3-HEXENYL 2-METHYLBUTYRATE
mf: $C_{11}H_{20}O_2$ mw: 184.28

PROP: Colorless liquid; powerful, fruity odor like unripe apples. D: 0.876-0.880, refr index: 1.430, flash p: 153°F. Sol in alc, fixed oils; insol in water.

SYN: FEMA No. 3497

SAFETY PROFILE: Combustible liquid. When heated to decomposition it emits acrid smoke and irritating fumes.

HFE600 CAS:68133-72-2 ***HR: 2***
cis-HEXENYL OCYACETALDEHYDE
mf: $C_8H_{14}O_2$ mw: 142.22

TOXICITY DATA with REFERENCE
skn-rbt 500 mg/24H MOD FCTXAV 17,801,79
orl-rat LD50:4100 mg/kg FCTXAV 17,801,79
skn-rbt LD50:2350 mg/kg FCTXAV 17,801,79

CONSENSUS REPORTS: Reported in EPA TSCA Inventory.

SAFETY PROFILE: Moderately toxic by skin contact and ingestion. A skin irritant. See also ALDEHYDES.

HFE625 CAS:42436-07-7 ***HR: 1***
cis-3-HEXENYL PHENYLACETATE
mf: $C_{14}H_{18}O_2$ mw: 218.32

SYNS: BENZENEACETIC ACID, 3-HEXENYL ESTER, (Z)- ◇ β,Γ-HEXENYL α-TOLUATE

TOXICITY DATA with REFERENCE
skn-rbt 500 mg/24H MLD FCTXAV 17,803,79

CONSENSUS REPORTS: Reported in EPA TSCA Inventory.

SAFETY PROFILE: A skin irritant. When heated to decomposition it emits acrid smoke and irritating fumes.

HFE650 CAS:33467-74-2 ***HR: 1***
cis-3-HEXENYL PROPIONATE
mf: $C_9H_{16}O_2$ mw: 156.25

SYNS: 3-HEXEN-1-OL, PROPANOATE (Z)- ◇ β,Γ-HEXENYL PROPANOATE

TOXICITY DATA with REFERENCE
skn-rbt 500 mg/24H MOD FCTXAV 17,805,79

CONSENSUS REPORTS: Reported in EPA TSCA Inventory.

SAFETY PROFILE: A skin irritant. When heated to decomposition it emits acrid smoke and irritating fumes.

HFF000 CAS:10138-60-0 ***HR: 3***
4-HEXEN-1-YN-3-OL
mf: C_6H_8O mw: 96.14

SYN: 4-HEXENE-1-YNE-3-OL

TOXICITY DATA with REFERENCE
skn-rbt 9090 μg/24H open MLD AIHAAP 23,95,62
orl-rat LD50:34 mg/kg AIHAAP 23,95,62
ihl-rat LCLo:62 ppm/4H AIHAAP 23,95,62
skn-rbt LDLo:71 mg/kg AIHAAP 23,95,62

SAFETY PROFILE: A poison by ingestion, inhalation, and skin contact. A skin irritant. When heated to decomposition it emits acrid smoke and fumes. See also ACETYLENE COMPOUNDS and ALCOHOLS.

HFF300 CAS:13061-80-8 ***HR: 3***
4-HEXEN-1-YN-3-ONE
mf: C_6H_6O mw: 94.12

SYN: 4-HEXENE-1-YNE-3-ONE

TOXICITY DATA with REFERENCE
skn-rbt 10 mg/24H open MLD AIHAAP 23,95,62
orl-rat LD50:71 mg/kg AIHAAP 23,95,62
ihl-rat LCLo:13 ppm/4H AIHAAP 23,95,62
skn-rbt LD50:100 mg/kg AIHAAP 23,95,62

SAFETY PROFILE: A poison by ingestion, inhalation, and skin contact. A skin irritant. When heated to decomposition it emits acrid smoke and fumes. See also ACETYLENE COMPOUNDS and KETONES.

HFF500 CAS:50-62-4 ***HR: 3***
HEXOBENDINE DIHYDROCHLORIDE
mf: $C_{30}H_{44}N_2O_{10}•2ClH$ mw: 665.68

SYNS: ANDIAMINE ◇ N,N'-DIMETHYL-N,N'-BIS(3-(3',4',5'-TRIMETHOXYBENZOXY)PROPYL)ETHYLENEDIAMINEDIHYDROCHLORIDE ◇ REOXYL

TOXICITY DATA with REFERENCE
orl-rat LD50:2550 mg/kg GNRIDX 3,77,69
scu-rat LD50:930 mg/kg GNRIDX 3,77,69
ivn-rat LD50:52 mg/kg GNRIDX 3,77,69
orl-mus LD50:682 mg/kg GNRIDX 3,77,69
scu-mus LD50:328 mg/kg GNRIDX 3,77,69
ivn-mus LD50:35200 μg/kg GNRIDX 3,77,69

SAFETY PROFILE: Poison by subcutaneous and intravenous routes. Moderately toxic by ingestion. When heated to decomposition it emits toxic fumes of NO_x and HCl. See also ESTERS.

HFG000 CAS:6004-98-4 ***HR: 3***
HEXOCYCLIUM
mf: $C_{21}H_{36}N_2O_5S$ mw: 428.61

PROP: Crystals. Mp: 200-210°, sltly sol in chloroform, insol in ether, sol in H_2O.

SYNS: 4-(β-CYCLOHEXYL-β-HYDROXYPHENETHYL)-1,1-DIMETHYL PIPERAZINIUM SULFATE ◇ TRAL

TOXICITY DATA with REFERENCE
mmo-sat 32 μg/plate JEPTDQ 4,345,80
orl-mus LD50:600 mg/kg 27ZIAQ -,-,65
ipr-mus LD50:55 mg/kg 27ZIAQ -,119,73
scu-mus LD50:360 mg/kg 27ZIAQ -,119,73
ivn-mus LD50:11 mg/kg 27ZIAQ -,119,73

SAFETY PROFILE: Poison by intraperitoneal, subcutaneous, and intravenous routes. Moderately toxic by ingestion. Mutation data reported. When heated to decomposition it emits very toxic fumes of NO_x and SO_x.

HFG400 CAS:115-63-9 ***HR: 3***
HEXOCYCLIUM METHYLSULFATE
mf: $C_{20}H_{33}N_2O•CH_3O_4S$ mw: 428.65

SYNS: 4-(β-CYCLOHEXYL-β-HYDROXYPHENETHYL)-1,1-DIMETHYLPIPERAZINIUM METHYLSULFATE ◇ N-(β-CYCLOHEXYL-β-HYDROXY-β-PHENYLETHYL)-N'-METHYLPIPERAZINE DIMETHYLSULFATE ◇ PIPERAZINIUM, 4-(β-CYCLOHEXYL-β-HYDROXYPHENETHYL)-1,1-DIMETHYL-, METHYL SULFATE

TOXICITY DATA with REFERENCE
orl-rat TDLo:120 mg/kg (female 9-14D post):REP KSRNAM 5,1475,71
orl-rat TDLo:120 mg/kg (female 9-14D post):TER KSRNAM 5,1475,71
ivn-mus LD50:8900 μg/kg CSLNX* NX#01104

SAFETY PROFILE: Poison by intravenous route. An experimental teratogen. Experimental reproductive effects. When heated to decomposition it emits toxic fumes of NO_x and SO_x.

HFG500 CAS:108-10-1 ***HR: 3***
HEXONE
DOT: UN 1245
mf: $C_6H_{12}O$ mw: 100.18

$$CH_3CO•CH_2CH(CH_3)_2$$

PROP: Colorless mobile liquid; fruity, ethereal odor. Bp: 118°, lel: 1.4%, uel: 7.5%, flash p: 62.6°F, d: 0.796-0.799, fp: −80.2°, autoign temp: 858°F, vap press: 16 mm @ 20°. Misc with alc, ether; sol in alc.

SYNS: FEMA No. 2731 ◇ HEXON (CZECH) ◇ ISOBUTYL-METHYLKETON (CZECH) ◇ ISOBUTYL METHYL KETONE ◇ ISOPROPYLACETONE ◇ METHYL-ISOBUTYL-CETONE (FRENCH) ◇ METHYLISOBUTYLKETON (DUTCH, GERMAN) ◇ METHYL ISOBUTYL KETONE (ACGIH, DOT) ◇ 4-METHYL-PENTAN-2-ON (DUTCH, GERMAN) ◇ 4-METHYL-2-PENTANON (CZECH) ◇ 2-METHYL-4-PENTANONE ◇ 4-METHYL-2-PENTANONE (FCC) ◇ METILISOBUTILCHETONE (ITALIAN) ◇ 4-METILPENTAN-2-ONE (ITALIAN) ◇ METYLOIZOBUTYLOKETON (POLISH) ◇ MIBK ◇ MIK ◇ RCRA WASTE NUMBER U161 ◇ SHELL MIBK

TOXICITY DATA with REFERENCE
eye-hmn 200 ppm/15M JIHTAB 28,262,46
skn-rbt 500 mg/24H MLD 28ZPAK -,42,72
eye-rbt 40 mg SEV UCDS** 4/25/58
eye-rbt 500 mg/24H MLD 28ZPAK -,42,72
ihl-mus TCLo:3000 ppm/6H (female 6-15D post):TER FAATDF 8,310,87
orl-rat LD50:2080 mg/kg UCDS** 4/25/58
ipr-rat LD50:400 mg/kg 38MKAJ 2C,4748,82
orl-mus LD50:2671 mg/kg TOLED5 30,13,86
ihl-mus LC50:23300 mg/m^3 GTPZAB 17(11),52,73
ipr-mus LD50:268 mg/kg SCCUR* -,7,61
orl-gpg LD50:1600 mg/kg 38MKAJ 2C,4748,82

CONSENSUS REPORTS: Reported in EPA TSCA Inventory. Community Right-To-Know List.

OSHA PEL: (Transitional: TWA 100 mg/m^3) TWA 50 ppm; STEL 75 ppm
ACGIH TLV: TWA 50 ppm; STEL 75 ppm
NIOSH REL: (Ketones) TWA 200 mg/m^3;)Proposed: BEI 2 mg/L MIBK in urine, end of shift.)
DOT Classification: Flammable Liquid; Label: Flammable Liquid.

SAFETY PROFILE: A poison by intraperitoneal route. Moderately toxic by ingestion. Mildly toxic by inhalation. Very irritating to the skin, eyes, and mucous membranes. An experimental teratogen. A human systemic irritant by inhalation. Narcotic in high concentration. Flammable liquid when exposed to heat, flame, or oxidizers. Ignites on contact with potassium-tert-butoxide. Moderately explosive in the form of vapor when exposed to heat or flame. May form explosive peroxides upon exposure to air. Can react vigorously with reducing materials. To fight fire, use alcohol foam, CO_2, dry chemical. Incompatible with air; potassium-tert-butoxide. See also KETONES.

HFG550 CAS:6556-11-2 ***HR: 3***
HEXOPAL
mf: $C_{42}H_{30}N_6O_{12}$ mw: 810.71

PROP: Crystals. Mp: 254.3-254.9°. Practically insol in water; sol in dil acids.

SYNS: DILCIT ◇ DILEXPAL ◇ ESANTENE ◇ HAMOVANNID ◇ HEXANICIT ◇ HEXANICOTINOYL INOSITOL ◇ HEXANICOTOL ◇ HEXA-3-PYRIDINECARBOXYLATE-myo-INOSITOL (9CI) ◇ INOSITOL HEXANICOTINATE ◇ m-INOSITOL HEXANICOTINATE ◇ meso-INOSITOL HEXANICOTINATE ◇ myo-INOSITOL HEXANICOTINATE ◇ INOSITOL NIACINATE ◇ INOSITOL NICOTINATE ◇ LINODIL ◇ MESONEX ◇ MESOTAL ◇ PALOHEX

TOXICITY DATA with REFERENCE
scu-rat LD50:1180 mg/kg NIIRDN 6,77,82
ivn-rat LD50:268 mg/kg NIIRDN 6,77,82
ipr-mus LD50:6400 mg/kg OYYAA2 7,149,73
ivn-mus LD50:345 mg/mg NIIRDN 6,77,82

SAFETY PROFILE: Poison by intravenous route. Moderately toxic by subcutaneous route. When heated to decomposition it emits toxic fumes of NO_x.

HFG600 CAS:4323-43-7 ***HR: 3***
HEXOPRENALINE DIHYDROCHLORIDE
mf: $C_{22}H_{32}N_2O_6 \cdot 2ClH$ mw: 493.48

SYNS: N,N'-BIS(2-(3',4'-DIHYDROXYPHENYL)-2-HYDROXYETHYL) HEXAMETHYLENEDIAMINE DIHYDROCHLORIDE ◇ ST-1512 DIHYDROCHLORIDE

TOXICITY DATA with REFERENCE
ivn-rat TDLo:12 mg/kg (9-14D preg):REP KSRNAM 6,983,72
ivn-mus TDLo:1200 μg/kg (7-12D preg):TER KSRNAM 6,983,72
orl-rat LD50:10 g/kg OYYAA2 26,811,83
ipr-rat LD50:139 mg/kg KSRNAM 6,1286,72
scu-rat LD50:143 mg/kg KSRNAM 6,1286,72
ivn-rat LD50:58 mg/kg OYYAA2 26,811,83
orl-mus LD50:2036 mg/kg KSRNAM 6,1286,72
ipr-mus LD50:133 mg/kg KSRNAM 6,1286,72
scu-mus LD50:110 mg/kg KSRNAM 6,1286,72
ivn-mus LD50:88 mg/kg KSRNAM 6,1286,72

SAFETY PROFILE: Poison by subcutaneous, intravenous and intraperitoneal routes. Moderately toxic by ingestion. Experimental teratogenic and reproductive effects. When heated to decomposition it emits toxic fumes of NO_x and HCl.

HFG650 CAS:32266-10-7 ***HR: 3***
HEXOPRENALINE SULFATE
mf: $C_{22}H_{32}N_2O_6 \cdot H_2O_4S$ mw: 518.64

SYNS: N,N'-BIS(2-(3',4'-DIHYDROXYPHENYL)-2-HYDROXYETHYL)HEXAMETHYLENEDIAMINE SULFATE ◇ ST-1512 SULFATE

TOXICITY DATA with REFERENCE
orl-rat TDLo:66 mg/kg (7-17D preg):TER OYYAA2 27,239,84
ipr-rat LD50:145 mg/kg NIIRDN 6,745,82
scu-rat LD50:150 mg/kg NIIRDN 6,745,82
ipr-mus LD50:158 mg/kg YKYUA6 28,1451,77
scu-mus LD50:274 mg/kg NIIRDN 6,745,82

SAFETY PROFILE: Poison by subcutaneous and intraperitoneal routes. Experimental teratogenic effects. When heated to decomposition it emits toxic fumes of SO_x and NO_x. See also SULFATES.

HFG700 CAS:17597-95-4 ***HR: 1***
HEXOXYACETALDEHYDE DIMETHYLACETAL
mf: $C_{10}H_{22}O_3$ mw: 190.32

SYNS: ACETALDEHYDE, (HEXYLOXY)-, DIMETHYL ACETAL ◇ β-HEXOXYACETALDEHYDE DIMETHYLACETAL ◇ 2-HEXOXYACETALDEHYDE DIMETHYLACETAL

TOXICITY DATA with REFERENCE
skn-rbt 500 mg/24H MOD FCTXAV 17,811,79

CONSENSUS REPORTS: Reported in EPA TSCA Inventory.

SAFETY PROFILE: A skin irritant. When heated to decomposition it emits acrid smoke and irritating fumes.

HFH500 CAS:63916-83-6 ***HR: 3***
p-HEXOXYBENZOIC ACID-3-(2'-METHYLPIPERIDINO)PROPYL ESTER
mf: $C_{22}H_{35}NO_3$ mw: 361.58

TOXICITY DATA with REFERENCE
scu-mus LD50:222 mg/kg RCPRAN 15,143,54
ivn-mus LD50:23 mg/kg RCPRAN 15,143,54

SAFETY PROFILE: Poison by subcutaneous and intravenous routes. When heated to decomposition it emits toxic fumes of NO_x. See also ESTERS.

HFI500 CAS:142-92-7 ***HR: 1***
HEXYL ACETATE
mf: $C_8H_{16}O_2$ mw: 144.24

PROP: Colorless liquid; fruity odor. D: 0.878, mp: −60.9°, bp: 171.5°, refr index: 1.407, flash p: 109°F. Insol in water; very sol in alc and ether.

SYNS: ACETIC ACID HEXYL ESTER ◇ FEMA No. 2565 ◇ n-HEXYL ACETATE (FCC) ◇ 1-HEXYL ACETATE ◇ HEXYL ALCOHOL, ACETATE ◇ HEXYL ETHANOATE

TOXICITY DATA with REFERENCE
orl-rat LD50:42 g/kg TXAPA9 28,313,74

CONSENSUS REPORTS: Reported in EPA TSCA Inventory.

SAFETY PROFILE: Mildly toxic by ingestion. Combustible liquid. When heated to decomposition it emits acrid smoke and fumes. See also ESTERS.

HFJ000 CAS:108-84-9 ***HR: 2***
sec-HEXYL ACETATE
DOT: UN 1233
mf: $C_8H_{16}O_2$ mw: 144.24

PROP: Clear liquid, pleasant odor. Bp: 146.3°, fp: −63.8°, flash p: 113°F (COC), d: 0.8598 @ 20°/20°, vap press: 3.8 mm @ 20°, vap d: 4.97.

SYNS: ACETIC ACID-1,3-DIMETHYLBUTYL ESTER ◇ 1,3-DIMETHYLBUTYL ACETATE ◇ MAAC ◇ METHYLAMYL ACETATE ◇ METHYL AMYL ACETATE (DOT) ◇ METHYLISOAMYL ACETATE ◇ METHYLISOBUTYLCARBINOL ACETATE ◇ METHYLISOBUTYLCARBINYL ACETATE ◇ 4-METHYL-2-PENTANOL, ACETATE ◇ 4-METHYL-2-PENTYL ACETATE

TOXICITY DATA with REFERENCE
eye-hmn 100 ppm/15M JIHTAB 28,262,46
skn-rbt 500 mg open MLD UCDS** 7/28/66
eye-rbt 500 mg open AMIHBC 10,61,54
ihl-hmn TCLo:100 ppm:NOSE,EYE,PUL JIHTAB 28,262,46
orl-rat LD50:6160 mg/kg UCDS** 7/28/66
ihl-rat LCLo:2000 ppm/4H UCDS** 7/28/66
skn-rbt LD50:20 g/kg UCDS** 7/28/66

CONSENSUS REPORTS: Reported in EPA TSCA Inventory.

OSHA PEL: TWA 50 ppm
ACGIH TLV: TWA 50 ppm
DFG MAK: 50 ppm (300 mg/m^3)
DOT Classification: Flammable or Combustible Liquid; Label: Flammable Liquid.

SAFETY PROFILE: Mildly toxic by ingestion, skin contact and inhalation. Human systemic effects by inhalation: conjunctiva irritation, unspecified changes in olfactory and respiratory systems. A skin and human eye irritant. Flammable when exposed to heat or flame; can react with oxidizing materials. To fight fire, use alcohol foam, CO_2, dry chemical. See also ESTERS.

HFJ500 CAS:111-27-3 ***HR: 3***
n-HEXYL ALCOHOL
DOT: UN 2282
mf: $C_6H_{14}O$ mw: 102.20

PROP: Colorless liquid. Bp: 157.2°, fp: −44.6°, flash p: 145°F, d: 0.816-0.821, vap press: 1 mm @ 24.4°, vap d: 3.52. Misc in alc, ether; sltly sol in water.

SYNS: AMYLCARBINOL ◇ CAPROYL ALCOHOL ◇ EPAL 6 ◇ FEMA No. 2567 ◇ HEXANOL ◇ n-HEXANOL (DOT) ◇ 1-HEXANOL ◇ HEXYL ALCOHOL ◇ 1-HYDROXYHEXANE ◇ PENTYLCARBINOL

TOXICITY DATA with REFERENCE
skn-rbt 410 mg open MLD UCDS** 4/21/67
skn-rbt 10 mg/24H open MLD AMIHBC 4,119,51
eye-rbt 250 μg open SEV AMIHBC 4,119,51
orl-rat LD50:720 mg/kg SAMJAF 43,795,69
orl-mus LD50:1950 mg/kg HYSAAV 31,310,66
ivn-mus LD50:103 mg/kg AIPTAK 135,330,62
skn-rbt LD50:3100 mg/kg AMIHBC 4,119,51

CONSENSUS REPORTS: Reported in EPA TSCA Inventory.

DOT Classification: Flammable or Combustible Liquid; Label: Flammable Liquid.

SAFETY PROFILE: Poison by intravenous route. Moderately toxic by ingestion and skin contact. A skin and severe eye irritant. Combustible liquid. Can react with oxidizing materials. To fight fire, use alcohol foam, CO_2, dry chemical. See also ALCOHOLS.

HFJ600 CAS:26401-20-7 ***HR: 3***
tert-HEXYL ALCOHOL
DOT: UN 2282
mf: $C_6H_{14}O$ mw: 102.20

SYN: tert-HEXANOL (9CI, DOT)

TOXICITY DATA with REFERENCE
orl-rat LD50:500 mg/kg 85GMAT -,74,82
orl-mus LD50:350 mg/kg 85GMAT -,74,82
ivn-mus LD50:243 mg/kg AIPTAK 135,330,62

DOT Classification: Flammable or Combustible Liquid; Label: Flammable Liquid.

SAFETY PROFILE: Poison by ingestion and intravenous routes. When heated to decomposition it emits acrid smoke and fumes. Flammable when exposed to heat or flame; can react vigorously with oxidizing materials. A fire hazard. See also ALCOHOLS.

HFK000 CAS:111-26-2 ***HR: 3***
HEXYLAMINE
mf: $C_6H_{15}N$ mw: 101.22

PROP: Liquid. Mp: −22.9°, bp: 131.4°, flash p: 85°F (OC), d: 0.7675 @ 20°/20°, vap d: 3.49.

SYNS: 1-AMINOHEXANE ◇ 1-HEXANAMINE ◇ N-HEXYLAMINE ◇ MONO-N-HEXYLAMINE

TOXICITY DATA with REFERENCE
skn-rbt 10 mg/24H open SEV AMIHBC 10,61,54
skn-rbt 500 mg open SEV UCDS** 11/3/71
eye-rbt 5 mg MOD UCDS** 11/3/71
orl-rat LD50:670 mg/kg AMIHBC 10,61,54
ihl-rat LCLo:500 ppm/4H AEHLAU 1,343,60
ipr-mus LDLo:16 mg/kg CBCCT* 2,188,50
skn-rbt LD50:420 mg/kg AEHLAU 1,343,60

CONSENSUS REPORTS: Reported in EPA TSCA Inventory.

SAFETY PROFILE: A poison by intraperitoneal routes. Moderately toxic by ingestion, inhalation, and skin contact. A severe skin and eye irritant. Dangerous fire hazard when exposed to heat or flame; can react with oxidizing materials. To fight fire, use alcohol foam, CO_2, dry chemical. Upon decomposition it emits toxic fumes of NO_x. See also AMINES.

HFL000 CAS:63019-34-1 ***HR: 3***
5-n-HEXYL-1,2-BENZANTHRACENE
mf: $C_{24}H_{24}$ mw: 312.48

SYN: 8-HEXYL-BENZ(a)ANTHRACENE

TOXICITY DATA with REFERENCE
skn-mus TDLo:1050 mg/kg/44W-I:ETA PRLBA4 129,439,40

SAFETY PROFILE: Questionable carcinogen with experimental tumorigenic data. When heated to decomposition it emits acrid smoke and fumes.

HFL500 CAS:6789-88-4 ***HR: 1***
n-HEXYLBENZOATE
mf: $C_{13}H_{18}O_2$ mw: 206.31

SYNS: BENZOIC ACID, HEXYL ESTER ◇ HEXYL BENZOATE

TOXICITY DATA with REFERENCE
skn-rbt 10 mg/24H open MLD AMIHBC 4,119,51
skn-rbt 500 mg/24H MOD FCTXAV 17,813,79
eye-rbt 500 mg open AMIHBC 4,119,51
orl-rat LD50:12300 mg/kg AMIHBC 4,119,51
skn-rbt LD50:21000 mg/kg AMIHBC 4,119,51

CONSENSUS REPORTS: Reported in EPA TSCA Inventory.

SAFETY PROFILE: Mildly toxic by ingestion and skin contact. A skin and eye irritant. When heated to decomposition it emits acrid smoke and fumes. See also ESTERS.

HFM500 CAS:111-25-1 ***HR: 2***
HEXYL BROMIDE
mf: $C_6H_{13}Br$ mw: 165.10

SYN: BROMOHEXANE

TOXICITY DATA with REFERENCE
ihl-mam LC50:13600 mg/m³ GTPZAB 18(4),55,74
ipr-mam LD50:1226 mg/kg GTPZAB 18(4),55,74
ihl-rat LC50:550000 mg/m³/30M FAVUAI 7,35,75

CONSENSUS REPORTS: Reported in EPA TSCA Inventory.

SAFETY PROFILE: Moderately toxic by intraperitoneal route. Mildly toxic by inhalation. When heated to decomposition it emits very toxic fumes of Br^-. See also BROMIDES.

HFM600 CAS:19089-92-0 ***HR: 1***
HEXYL-2-BUTENOATE
mf: $C_{10}H_{18}O_2$ mw: 170.28

PROP: Colorless liquid; fruity odor. D: 0.880, refr index: 1.428-1.449. Sol in alc, fixed oils; insol in water, propylene glycol.

SYNS: 2-BUTENOIC ACID, HEXYL ESTER ◇ FEMA No. 3354 ◇ n-HEXYL 2-BUTENOATE ◇ HEXYL CROTONATE

TOXICITY DATA with REFERENCE
skn-rbt 500 mg/24H MOD FCTOD7 20,715,82

CONSENSUS REPORTS: Reported in EPA TSCA Inventory.

SAFETY PROFILE: A skin irritant. When heated to decomposition it emits acrid smoke and irritating fumes.

HFM700 CAS:2639-63-6 ***HR: 1***
1-HEXYL BUTYRATE
mf: $C_{10}H_{20}O_2$ mw: 172.30

SYNS: BUTYRIC ACID, HEXYL ESTER ◇ HEXYL BUTANOATE ◇ n-HEXYL BUTANOATE ◇ n-HEXYL n-BUTANOATE ◇ HEXYL BUTYRATE ◇ n-HEXYL BUTYRATE

TOXICITY DATA with REFERENCE
skn-rbt 500 mg/24H MLD FCTXAV 17,815,79

CONSENSUS REPORTS: Reported in EPA TSCA Inventory.

SAFETY PROFILE: A skin irritant. When heated to decomposition it emits acrid smoke and irritating fumes.

HFN000 CAS:112-59-4 ***HR: 2***
n-HEXYL CARBITOL
mf: $C_{10}H_{22}O_3$ mw: 190.32

PROP: Liquid. Bp: 258.2°, fp: −33.3°, flash p: 285°F (OC), d: 0.9346 @ 20°/20°, vap press: 0.01 mm @ 20°.

SYNS: DIETHYLENE GLYCOL-n-HEXYL ETHER ◇ DIETHYLENE GLYCOL MONOHEXYL ETHER ◇ 3,6-DIOXADODECANOL-1 ◇ n-HEXOXYETHOXYETHANOL ◇ HEXYL CARBITOL ◇ 2-((2-HEXYLOXY)ETHOXY)ETHANOL

TOXICITY DATA with REFERENCE
skn-rbt 10 mg/24H open MLD AMIHBC 4,119,51
skn-rbt 500 mg/24H SEV JPETAB 82,377,44
skn-rbt 500 mg open MLD UCDS** 11/3/71
eye-rbt 5 mg MOD UCDS** 11/3/71
orl-rat LD50:4920 mg/kg AMIHBC 4,119,51
skn-rbt LD50:1500 mg/kg AMIHBC 4,119,51

CONSENSUS REPORTS: Reported in EPA ISCA Inventory. Glycol ether compounds are on the Community Right-To-Know List.

SAFETY PROFILE: Moderately toxic by skin contact route. Mildly toxic by ingestion. An eye and severe skin irritant. Combustible when exposed to heat or flame; can react with oxidizing materials. To fight fire, use foam, CO_2, dry chemical. When heated to decomposition it emits acrid smoke and fumes. See also GLYCOL ETHERS.

HFN500 CAS:20740-05-0 ***HR: 3***
n-HEXYL CARBORANE
mf: $C_8H_{24}B_{10}$ mw: 228.42

SYNS: HEXYLDICARBADODECABORANE(12) ◇ NHC

TOXICITY DATA with REFERENCE
skn-rbt 500 mg/24H MLD AEHA** 51-044-74/76
orl-rat TDLo:1100 mg/kg (6-16D preg):TER AEHA** 51-044-74/76
ipr-rat LDLo:1900 mg/kg AEHA** 51-044-74/76
ivn-rbt LDLo:150 mg/kg NTIS** AD-A041-973

CONSENSUS REPORTS: Reported in EPA TSCA Inventory.

SAFETY PROFILE: Poison by intravenous route. Moderately toxic by intraperitoneal route. An experimental teratogen. A skin irritant. When heated to decomposition it emits toxic fumes of boron. See also BORANES and BORON COMPOUNDS.

HFO500 CAS:101-86-0 ***HR: 2***
HEXYL CINNAMALDEHYDE
mf: $C_{15}H_{20}O$ mw: 216.35

PROP: Pale yellow liquid; jasmine odor. D 0.953-0.959, refr index: 1.548-1.552. Sol in fixed oils; insol in propylene glycol, glycerin.

SYNS: FEMA No. 2569 ◇ α-HEXYLCINNAMALDEHYDE (FCC) ◇ HEXYL CINNAMIC ALDEHYDE ◇ α-HEXYLCINNAMIC ALDEHYDE ◇ α-n-HEXYL-β-PHENYLACROLEIN ◇ 2-(PHENYLMETHYLENE)OCTANOL

TOXICITY DATA with REFERENCE
skn-rbt 500 mg/24H MOD FCTXAV 12,915,74
orl-rat LD50:3100 mg/kg FCTXAV 12,915,74

CONSENSUS REPORTS: Reported in EPA TSCA Inventory.

SAFETY PROFILE: Moderately toxic by ingestion. A skin irritant. When heated to decomposition it emits acrid smoke and fumes. See also ALDEHYDES.

HFO700 CAS:95-41-0 ***HR: 3***
2-n-HEXYL-2-CYCLOPENTEN-1-ONE
mf: $C_{11}H_{18}O$ mw: 166.29

SYNS: 2-CYCLOPENTEN-1-ONE, 2-HEXYL- ◇ DIHYDROISOJASMONE

TOXICITY DATA with REFERENCE
skn-rbt 500 mg/24H MOD FCTXAV 16,713,78
ivn-mus LD50:320 mg/kg CSLNX* NX#00920

CONSENSUS REPORTS: Reported in EPA TSCA Inventory.

SAFETY PROFILE: Poison by intravenous route. A skin irritant. When heated to decomposition it emits acrid smoke and irritating fumes.

HFP500 CAS:25354-97-6 ***HR: 3***
2-HEXYLDECANOIC ACID
mf: $C_{16}H_{32}O_2$ mw: 256.48

PROP: Viscous oil. Bp: 140-150° @ 0.02 mm; sol in water.

SYNS: 2-HEXYLDECANSAEURE (GERMAN) ◇ 7-PENTADECANECARBOXYLIC ACID

TOXICITY DATA with REFERENCE
ipr-rat LDLo:300 mg/kg ARZNAD 3,86,53

CONSENSUS REPORTS: Reported in EPA TSCA Inventory.

SAFETY PROFILE: Poison by intraperitoneal route. When heated to decomposition it emits acrid smoke and fumes.

HFP600 CAS:64049-22-5 ***HR: 3***
HEXYLDICHLORARSINE
mf: $C_6H_{13}AsCl_2$ mw: 231.01

SYNS: ARSINE, DICHLOROHEXYL- ◇ TL 231

TOXICITY DATA with REFERENCE
ihl-mus LC50:1500 mg/m^3/10M NTIS** PB158-508
skn-mus LD50:20 mg/kg NTIS** PB158-508

OSHA PEL: TWA 0.5 mg(As)/m^3

SAFETY PROFILE: Poison by skin contact. When heated to decomposition it emits toxic fumes of As and Cl^-.

HFP875 CAS:107-41-5 *HR: 2*
HEXYLENE GLYCOL
mf: $C_6H_{14}O_2$ mw: 118.20

PROP: Mild odor, colorless liquid, water-sol. Bp: 197.1°, fp: −50°, flash p: 205°F (OC), d: 0.9234 @ 20°/20°, vap press: 0.05 mm @ 20°, d: 4.

SYNS: 2,4-DIHYDROXY-2-METHYLPENTANE ◇ DIOLANE ◇ 1,2-HEXANEDIOL ◇ ISOL ◇ 2-METHYL PENTANE-2,4-DIOL ◇ 2-METHYL-2,4-PENTANEDIOL ◇ PINAKON ◇ α,α,α'-TRIMETHYLTRIMETHYLENE GLYCOL

TOXICITY DATA with REFERENCE
skn-rbt 465 mg open MLD UCDS** 11/3/71
skn-rbt 465 mg/24H MOD JPETAB 82,377,44
skn-rbt 500 mg/24H MOD FCTXAV 16,777,78
eye-rbt 93 mg SEV BIOFX* 12-4/70
ihl-hmn TCLo:50 ppm/15M:EYE,NOSE,PUL 34ZIAG -,312,69
ihl-hmn TCLo:50 ppm:EYE,NOSE,PUL JIHTAB 28,262,46
orl-rat LD50:3700 mg/kg NPIRI* 1,68,74
ipr-rat LDLo:1500 mg/kg JPPMAB 11,150,59
orl-mus LD50:3097 mg/kg JAPMA8 45,669,56
ipr-mus LD50:1299 mg/kg SCCUR* -,5,61
orl-rbt LD50:3200 mg/kg FEPRA7 4,142,45
skn-rbt LD50:8560 mg/kg 34ZIAG -,731,69
scu-rbt LD50:13 g/kg FCTXAV 16,777,78
orl-gpg LD50:2800 mg/kg FEPRA7 4,142,45

CONSENSUS REPORTS: Reported in EPA TSCA Inventory.

OSHA PEL: CL 25 ppm
ACGIH TLV: CL 25 ppm

SAFETY PROFILE: Moderately toxic by ingestion and intraperitoneal routes. Mildly toxic by skin contact. Human systemic effects by inhalation: conjunctiva and other eye, olfactory and pulmonary changes. Mutation data reported. Combustible when exposed to heat or flame; can react with oxidizing materials. To fight fire, use foam, CO_2, dry chemicals. When heated to decomposition it emits acrid smoke and fumes. See also GLYCOLS.

HFQ000 CAS:1637-24-7 *HR: 2*
HEXYLENE GLYCOL DIACETATE
mf: $C_{10}H_{18}O_4$ mw: 202.28

PROP: Liquid. Bp: 95° @ 12 mm.

SYNS: ACETIC ACID, HEXYLENE GLYCOL ◇ ACETIC ACID, 2-METHYL-2,4-PENTANEDIOL DIESTER

TOXICITY DATA with REFERENCE
skn-rbt 500 mg open MLD UCDS** 12/13/67
orl-rat LD50:3360 mg/kg AIHAAP 30,470,69
skn-rbt LD50:16 g/kg AIHAAP 30,470,69

CONSENSUS REPORTS: Reported in EPA TSCA Inventory.

SAFETY PROFILE: Moderately toxic by ingestion. Mildly toxic by skin contact. A skin irritant. When heated to decomposition it emits acrid smoke and fumes.

HFQ500 CAS:6378-65-0 *HR: 1*
HEXYL HEXOATE
mf: $C_{12}H_{24}O_2$ mw: 200.36

SYNS: HEXANOIC ACID, HEXYL ESTER ◇ HEXYL CAPROATE ◇ n-HEXYL HEXANOATE

TOXICITY DATA with REFERENCE
skn-rbt 500 mg/24H MLD FCTXAV 16,775,78

CONSENSUS REPORTS: Reported in EPA TSCA Inventory.

SAFETY PROFILE: A skin irritant. When heated to decomposition it emits acrid smoke and fumes. See also ESTERS.

HFQ550 CAS:2349-07-7 *HR: 1*
n-HEXYL ISOBUTYRATE
mf: $C_{10}H_{20}O_2$ mw: 172.30

SYNS: HEXYL ISOBUTANOATE ◇ n-HEXYL ISOBUTANOATE ◇ HEXYL ISOBUTYRATE ◇ 1-HEXYL ISOBUTYRATE ◇ ISOBUTYRIC ACID, HEXYL ESTER ◇ PROPANOIC ACID, 2-METHYL-, HEXYL ESTER (9CI)

TOXICITY DATA with REFERENCE
skn-rbt 500 mg/24H MLD FCTOD7 20,719,82

CONSENSUS REPORTS: Reported in EPA TSCA Inventory.

SAFETY PROFILE: A skin irritant. When heated to decomposition it emits acrid smoke and irritating fumes.

HFR000 CAS:5431-31-2 *HR: 1*
HEXYL MANDELATE
mf: $C_{14}H_{20}O_3$ mw: 236.34

TOXICITY DATA with REFERENCE
skn-rbt 10 mg/24H open MLD AIHAAP 23,95,62

orl-rat LD50:17 g/kg AIHAAP 23,95,62
skn-rbt LD50:15 g/kg AIHAAP 23,95,62

SAFETY PROFILE: Mildly toxic by ingestion and skin contact. A skin irritant. When heated to decomposition it emits acrid smoke and fumes. See also ESTERS.

HFR100 CAS:18431-36-2 ***HR: 3***
n-HEXYLMERCURIC BROMIDE
mf: $C_6H_{13}BrHg$ mw: 365.69

SYNS: BROMOHEXYLMERCURY ◇ HEXYLMERCURIC BROMIDE ◇ HEXYL MERCURY BROMIDE ◇ HMB ◇ MERCURY, BROMOHEXYL

TOXICITY DATA with REFERENCE
cyt-ham:oth 1400 nmol/L HEREAY 79,306,75
scu-mus LD50:56100 μg/kg OSDIAF 5,388,56

OSHA PEL: (Transitional: 0.01 mg/10m³) 0.01 mg(Hg)/m³; CL 0.03 mg(Hg)/m³ (skin)
ACGIH TLV: TWA 0.01 mg(Hg)/m³; STEL 0.03 mg(Hg)/m³
NIOSH REL: (Mercury, Inorganic): 8H TWA 0.05 mg(Hg)/m³

SAFETY PROFILE: Poison by subcutaneous route. Mutation data reported. When heated to decomposition it emits toxic fumes of Hg and Br^-.

HFR200 CAS:10032-15-2 ***HR: 2***
HEXYL 2-METHYLBUTYRATE
mf: $C_{11}H_{22}O_2$ mw: 186.33

PROP: Colorless liquid; strong, fresh-green, fruity odor. D: 0.854, refr index: 1.416-1.421, flash p: 122°F. Sol in alc, fixed oils; insol in water.

SYN: FEMA No. 3499 ◇ 2-METHYLBUTANOIC ACID, n-HEXYL ESTER

TOXICITY DATA with REFERENCE
skn-rbt 500 mg/24H MLD FCTOD7 20,721,82

CONSENSUS REPORTS: Reported in EPA TSCA Inventory.

SAFETY PROFILE: A skin irritant. Combustible liquid. When heated to decomposition it emits acrid smoke and irritating fumes.

HFR500 CAS:4351-10-4 ***HR: 2***
2-HEXYL-4-METHYL-1,3-DIOXOLANE
mf: $C_{10}H_{20}O_2$ mw: 172.30

TOXICITY DATA with REFERENCE
ipr-mus LDLo:500 mg/kg CBCCT* 8,742,56

CONSENSUS REPORTS: Reported in EPA TSCA Inventory.

SAFETY PROFILE: Moderately toxic by intraperitoneal route. When heated to decomposition it emits acrid smoke and fumes.

HFS500 CAS:18774-85-1 ***HR: 3***
1-HEXYL-1-NITROSOUREA
mf: $C_7H_{15}N_3O_2$ mw: 173.25

SYN: NITROSO-N-HEXYLUREA

TOXICITY DATA with REFERENCE
mmo-sat 1 μg/plate MUREAV 68,1,79
mma-sat 10 μg/plate TCMUE9 1,13,84
orl-rat TDLo:550 mg/kg:ETA ANYAA9 381,250,82
skn-mus TDLo:693 mg/kg/50W-I:CAR JCROD7 102,13,81

CONSENSUS REPORTS: EPA Genetic Toxicology Program.

SAFETY PROFILE: Questionable carcinogen with carcinogenic and tumorigenic data. Mutation data reported. When heated to decomposition it emits toxic fumes of NO_x. See also N-NITROSO COMPOUNDS.

HFS759 CAS:53370-90-4 ***HR: 2***
o-HEXYLOXYBENZAMIDE
mf: $C_{13}H_{19}NO_2$ mw: 221.33

PROP: Crystals from ethanol. Mp: 71°. Sol in methanol, acetone, chloroform, benzene. Sltly sol in ether. Practically insol in water.

SYNS: EXALAMIDE ◇ 2-(HEXYLOXY)BENZAMIDE ◇ 2-n-HEXYLOXYBENZAMIDE ◇ H.P. 216 ◇ HYPERAN

TOXICITY DATA with REFERENCE
ipr-rat LD50:530 mg/kg NIIRDN 6,100,82
orl-mus LD50:13210 mg/kg NIIRDN 6,100,82
ipr-mus LD50:650 mg/kg NIIRDN 6,100,82

SAFETY PROFILE: Moderately toxic by intraperitoneal route. Mildly toxic by ingestion. When heated to decomposition it emits toxic fumes of NO_x. See also AMIDES.

HFT500 CAS:112-25-4 ***HR: 2***
2-(HEXYLOXY)ETHANOL
mf: $C_8H_{18}O_2$ mw: 146.26

PROP: Liquid. Bp: 208.3°, fp: −45.1°, flash p: 195°F (OC), d: 0.8894 @ 20°/20°, vap press: 0.1 mm @ 20°, vap d: 5.04.

SYNS: ETHYLENE GLYCOL-N-HEXYL ETHER ◇ ETHYLENE GLYCOL MONOHEXYL ETHER ◇ N-HEXYL CELLOSOLVE

TOXICITY DATA with REFERENCE
skn-rbt 500 mg open MLD UCDS** 1/19/58
eye-rbt 1 mg MLD UCDS** 1/19/58
orl-rat LD50:1480 mg/kg AMIHBC 10,61,54
skn-rbt LD50:890 mg/kg AMIHBC 10,61,54

CONSENSUS REPORTS: Reported in EPA TSCA Inventory. Glycol ether compounds are on the Community Right-To-Know List.

SAFETY PROFILE: Moderately toxic by ingestion and skin contact. A skin and eye irritant. Combustible when exposed to heat or flame; can react with oxidizing materials. To fight fire, use foam, CO_2, dry chemical. See also GLYCOL ETHERS.

HFU500 CAS:5289-93-0 ***HR: 3***
4-HEXYLOXY-β-(1-PIPERIDYL)PROPIOPHENONE HYDROCHLORIDE
mf: $C_{20}H_{31}NO_2$•ClH mw: 353.98

SYN: 4'-(HEXYLOXY)-3-PIPERIDINOPROPIOPHENONEHYDROCHLORIDE

TOXICITY DATA with REFERENCE
ipr-mus LD50:52 mg/kg JPETAB 115,419,55
ivn-mus LD50:15 mg/kg COREAF 241,1523,55

SAFETY PROFILE: Poison by intraperitoneal and intravenous routes. When heated to decomposition it emits very toxic fumes of NO_x and HCl.

HFV000 CAS:67049-51-8 ***HR: 3***
2-(1-HEXYL-3-PIPERIDYL)ETHYL ESTER, BENZOIC ACID HYDROCHLORIDE
mf: $C_{20}H_{31}NO_2$•ClH mw: 353.98

TOXICITY DATA with REFERENCE
ivn-rat LDLo:32 mg/kg JACSAT 55,816,33
scu-mus LDLo:1000 mg/kg JACSAT 55,816,33

SAFETY PROFILE: Poison by intravenous route. Moderately toxic by subcutaneous route. When heated to decomposition it emits very toxic fumes of HCl and NO_x. See also ESTERS.

HFV500 CAS:136-77-6 ***HR: 3***
HEXYLRESORCINOL
mf: $C_{12}H_{18}O_2$ mw: 194.30

PROP: Colorless liquid to pale yellow, heavy liquid becoming solid on standing at room temp; needles from benzene or petr ether. Pungent odor, sharp astringent taste. Bp: 179°, mp: 67.5-69°. Very sol in water; sol in benzene, ether, acetone, chloroform, alc, vegetable oils; sltly sol in petr ether.

SYNS: ASCARYL ◇ CAPROKOL ◇ CRYSTOIDS ◇ CYSTOIDS ANTHELMINTIC ◇ 4-HEXYL-1,3-BENZENEDIOL ◇ 4-HEXYL-1,3-DIHYDROXYBENZENE ◇ HEXYLRESORCIN (GERMAN) ◇ 4-HEXYLRESORCINE ◇ p-HEXYLRESORCINOL ◇ 4-HEXYLRESORCINOL ◇ 4-n-HEXYLRESORCINOL ◇ NCI-C55787 ◇ S.T. 37 ◇ SUCRETS ◇ WORM-AGEN

TOXICITY DATA with REFERENCE
eye-rbt 2 mg AEPPAE 219,119,53
dnr-esc 3 mg/disc MUREAV 188,111,87
mma-mus:lyms 5 μg/L NTPTR* NTP-TR-330,88
sce-ham:ovr 18 μg/L NTPTR* NTP-TR-330,88
scu-rat TDLo:5 mg/kg (1D pre):REP ENDOAO 57,466,55
orl-mus TDLo:64 g/kg/2Y-C:ETA NTPTR* NTP-TR-330,88
orl-rat LD50:550 mg/kg JPETAB 53,198,35
orl-mus LD50:1040 mg/kg PHARAT 30,147,75
ipr-mus LDLo:50 mg/kg HBTXAC 5,148,59
scu-mus LDLo:750 mg/kg HBTXAC 5,148,59
orl-gpg LDLo:400 mg/kg PSEBAA 28,609,31

CONSENSUS REPORTS: Reported in EPA TSCA Inventory.

SAFETY PROFILE: A poison by ingestion, and intraperitoneal routes. Moderately toxic by subcutaneous route. Experimental reproductive effects. Questionable carcinogen with experimental tumorigenic data. Mutation data reported. An eye irritant. Concentrated solutions can cause burns on the skin and mucous membranes in humans. An anthelmintic and topical antiseptic. When heated to decomposition it emits acrid smoke and fumes.

HFW500 CAS:41956-90-5 ***HR: 3***
3-(5-(HEXYLTHIO)PENTYL)THIAZOLIDINE HYDROCHLORIDE
mf: $C_{14}H_{29}NS_2$•ClH mw: 312.02

TOXICITY DATA with REFERENCE
orl-mus LD50:500 mg/kg JMCMAR 16,328,73
ipr-mus LD50:75 mg/kg JMCMAR 16,328,73

SAFETY PROFILE: Poison by intraperitoneal route. Moderately toxic by ingestion. When heated to decomposition it emits very toxic fumes of NO_x, SO_x, and HCl.

HFX000 CAS:16930-96-4 ***HR: 1***
HEXYL TILGLATE
mf: $C_{11}H_{20}O_2$ mw: 184.31

SYNS: ENT 33,335 ◇ n-HEXYL trans-2-METHYL-2-BUTENOATE ◇ n-HEXYL TIGLINATE ◇ 2-METHYL-2-BUTENOIC ACID, HEXYL ESTER

TOXICITY DATA with REFERENCE
skn-rbt 500 mg/24H MOD FCTXAV 16,779,78

CONSENSUS REPORTS: Reported in EPA TSCA Inventory.

SAFETY PROFILE: A skin irritant. When heated to decomposition it emits acrid smoke and fumes.

HFX500 CAS:928-65-4 ***HR: 3***
HEXYLTRICHLOROSILANE
DOT: UN 1784
mf: $C_6H_{13}Cl_3Si$ mw: 219.63

CONSENSUS REPORTS: Reported in EPA TSCA Inventory.

DOT Classification: Label: Corrosive.

SAFETY PROFILE: A poison by ingestion and inhalation. Corrosive. A severe irritant to skin, eyes, and mucous membranes. When heated to decomposition or in reaction with water or steam it produces toxic and corrosive fumes of Cl^- and HCl. See also CHLOROSILANES.

HFY000 CAS:21961-08-0 ***HR: 2***
n-HEXYL VINYL SULFONE
mf: $C_8H_{16}O_2S$ mw: 176.30

TOXICITY DATA with REFERENCE
orl-rat LDLo:570 mg/kg AIHAAP 30,470,69
skn-rbt LD50:840 mg/kg AIHAAP 30,470,69

SAFETY PROFILE: Moderately toxic by ingestion and skin contact. When heated to decomposition it emits toxic fumes of SO_x. See also SULFONATES.

HFY500 CAS:3031-66-1 ***HR: 2***
3-HEXYNE-2,5-DIOL
mf: $C_6H_{10}O_2$ mw: 114.16

SYN: HEXYNE-3-DIOL-2,5

TOXICITY DATA with REFERENCE
ipr-mus LDLo:500 mg/kg CBCCT* 4,378,52

CONSENSUS REPORTS: Reported in EPA TSCA Inventory.

SAFETY PROFILE: Moderately toxic by intraperitoneal route. When heated to decomposition it emits acrid smoke and fumes. See also ACETYLENE COMPOUNDS.

HFZ000 CAS:105-31-7 ***HR: 3***
1-HEXYN-3-OL
mf: $C_6H_{10}O$ mw: 98.16

TOXICITY DATA with REFERENCE
orl-mus LD50:210 mg/kg THERAP 11,692,56
ivn-mus LD50:56 mg/kg CSLNX* NX#00219

CONSENSUS REPORTS: Reported in EPA TSCA Inventory.

SAFETY PROFILE: Poison by ingestion and intravenous routes. When heated to decomposition it emits acrid smoke and fumes. See also ACETYLENE COMPOUNDS and ALCOHOLS.

HGA000 CAS:25898-71-9 ***HR: 3***
HEXYNOL
mf: $C_9H_{16}Cl_2OSi$ mw: 239.24

SYN: 1-(DICHLOROMETHYLDIMETHYLSILYL)-1-HEXYN-3-OL

TOXICITY DATA with REFERENCE
orl-mus LD50:175 μg/kg 37ASAA 1,244,78
ivn-mus LD50:56 mg/kg CSLNX* NX#02569

SAFETY PROFILE: Poison by ingestion and intravenous routes. When heated to decomposition it emits very toxic fumes of Cl^-. See also ACETYLENE COMPOUNDS and ALCOHOLS.

HGA100 CAS:68917-43-1 ***HR: 1***
HIBAWOOD OIL

SYN: OIL, HIBAWOOD

TOXICITY DATA with REFERENCE
skn-rbt 500 mg/24H MOD FCTXAV 17,817,79

CONSENSUS REPORTS: Reported in EPA TSCA Inventory.

SAFETY PROFILE: A skin irritant. When heated to decomposition it emits acrid smoke and irritating fumes.

HGA500 CAS:14612-92-1 ***HR: 3***
HIBERNON HYDROCHLORIDE
mf: $C_{16}H_{20}BrN_3{\cdot}ClH$ mw: 370.76

SYNS: 2-((p-BROMOBENZYL)(2-(DIMETHYLAMINO)ETHYL) AMINO)PYRIDINE HYDROCHLORIDE ◇ N-p-BROMOBENZYL-N',N'-DIMETHYL-N-2-PYRIDYLETHYLENE-DIAMINEHYDROCHLORIDE ◇ N-p-BROMBENZYL-N-α-PYRIDYL-N',N'-DIMETHYL-AETHYLEN-DIAMIN-HYDROCHLORIDE (GERMAN)

TOXICITY DATA with REFERENCE
orl-mus LD50:600 mg/kg ARZNAD 14,940,64
scu-mus LD50:165 mg/kg ARZNAD 14,940,64
ivn-mus LD50:25800 μg/kg ARZNAD 14,940,64

SAFETY PROFILE: Poison by subcutaneous and intravenous routes. Moderately toxic by ingestion. When heated to decomposition it emits very toxic fumes of Br^-, HCl, and NO_x.

HGA550 ***HR: 2***
HIBISCUS MANIHOT Linn., extract

PROP: Indian plant belonging to the family *Malvaceae* IJEBA6 22,312,84

SYN: ABELMOSCHUS MANIHOT (Linn.) Medik., extract

TOXICITY DATA with REFERENCE
orl-ham TDLo:500 mg/kg (female 1-5D post):REP IJEBA6 22,312,84
ipr-mus LD50:562 mg/kg IJEBA6 24,48,86

SAFETY PROFILE: Moderately toxic by intraperitoneal route. Experimental reproductive effects. When heated to decomposition it emits acrid smoke and irritating fumes.

HGA600 ***HR: D***
HIBISCUS ROSA-SINENSIS, flower extract

TOXICITY DATA with REFERENCE
orl-rat TDLo:500 mg/kg (female 1-5D post):REP IJMRAQ 59,777,71

SAFETY PROFILE: Experimental reproductive effects.

HGB000 ***HR: 3***
HICAL-2

PROP: Contains 10% decaborane, and a mixture of monoethyldecaborane, diethyl decaborane and triethyldecaborane (NTIS** AD224006)

TOXICITY DATA with REFERENCE
ihl-rat LC50:103 mg/m^3/4H NTIS** AD224-006
skn-rat LD50:502 mg/kg XAWPA2 CWL 2-10,58
ipr-rat LD50:71 mg/kg XAWPA2 CWL 2-10,58
ihl-mus LC50:90 mg/m^3/4H NTIS** AD224-006
skn-cat LD50:126 mg/kg XAWPA2 CWL 2-10,58
skn-rbt LD50:104 mg/kg XAWPA2 CWL 2-10,58
ivn-rbt LD50:4 mg/kg XAWPA2 CWL 2-10,58
skn-gpg LD50:251 mg/kg XAWPA2 CWL 2-10,58
ipr-gpg LD50:40 mg/kg XAWPA2 CWL 2-10,58

SAFETY PROFILE: Poison by inhalation, skin contact, intraperitoneal, and intravenous routes. When heated to decomposition it emits toxic fumes of BO_x. See also individual components.

HGB200 CAS:133-17-5 ***HR: 2***
HIPPODIN
mf: $C_9H_8INO_3$•Na mw: 328.07

PROP: Dihydrate, crystals. Freely sol in water; sol in alc and in dil solns of alkalies.

SYNS: HIPPURAN ◇ IODAIRAL ◇ N-(2-IODOBENZOYL)-GLYCIN MONOSODIUM SALT (9CI) ◇ IODOHIPPURA ◇ IODOHIPPURATE SODIUM ◇ o-IODOHIPPURATE SODIUM ◇ JODAIROL ◇ o-JODHIPPURSAEURE NATRIUM (GERMAN) ◇ MEDOPAQUE ◇ ORTHOIODIN ◇ RENUMBRAL ◇ SODIUM IODOHIPPURATE ◇ SODIUM o-IODOHIPPURATE ◇ UROCONTRAST

TOXICITY DATA with REFERENCE
ivn-rat LD50:4000 mg/kg JPETAB 116,394,56
ivn-mus LD50:3800 mg/kg MECHAN 6,290,63

SAFETY PROFILE: Moderately toxic by intravenous route. When heated to decomposition it emits toxic fumes of I^-, Na_2O, NO_x, and Na_2O.

HGB500 CAS:1403-74-3 ***HR: 3***
HIRSUTIC ACID N

TOXICITY DATA with REFERENCE
orl-mus LD50:1000 mg/kg 85GDA2 6,109,81
ivn-mus LDLo:25 mg/kg 85GDA2 6,109,81

SAFETY PROFILE: Poison by intravenous route. Moderately toxic by ingestion.

HGC000 CAS:1936-15-8 ***HR: D***
HISPACID FAST ORANGE 2G
mf: $C_{16}H_{10}N_2O_7S_2$•2Na mw: 452.38

SYNS: ACIDAL FAST ORANGE ◇ ACID FAST ORANGE EGG ◇ ACID LEATHER ORANGE PGW ◇ ACID LIGHT ORANGE G ◇ ACID ORANGE 10 ◇ ACILAN ORANGE GX ◇ APOCID ORANGE 2G ◇ ATUL ACID CRYSTAL ORANGE G ◇ BRASILAN ORANGE 2G ◇ BUCACID FAST ORANGE G ◇ CALCOCID FAST LIGHT ORANGE 2G ◇ CERTICOL ORANGE GS ◇ CETIL LIGHT ORANGE GG ◇ C.I. 27 ◇ C.I. ACID ORANGE 10 ◇ C.I. FOOD ORANGE 4 ◇ CRYSTAL ORANGE 2G ◇ D&C ORANGE No. 3 ◇ ENIACID LIGHT ORANGE G ◇ ERIO FAST ORANGE AS ◇ FAST LIGHT ORANGE GA ◇ HEXACOL ORANGE GG CRYSTALS ◇ HIDACID FAST ORANGE G ◇ 7-HYDROXY-8-(PHENYLAZO)-1,3-NAPHTHALENEDISULFONIC ACID, DISODIUM SALT ◇ 7-HYDROXY-8-(PHENYLAZO)-1,3-NAPHTHALENEDISULPHONIC ACID, DISODIUM SALT ◇ INK ORANGE JSN ◇ INTRACID FAST ORANGE G ◇ JAVA ORANGE 2G ◇ KITON FAST ORANGE G ◇ NAPHTHALENE FAST ORANGE 2GS ◇ NCI-C53838 ◇ NEKLACID FAST ORANGE 2G ◇ ORANGE #10 ◇ ORANGE G (biological stain) ◇ ORANGE G DYE ◇ ORANGE G (indicator) ◇ ORANZ G (POLISH) ◇ 1-PHENYLAZO-2-NAPHTHOL-6,8-DISULFONIC ACID, DISODIUM SALT ◇ 1-PHENYLAZO-2-NAPHTHOL-6,8-DISULPHONIC ACID, DISODIUM SALT ◇ SCHULTZ No. 39 ◇ SOLAR LIGHT ORANGE GX ◇ STANDACOL ORANGE G ◇ SULFACID LIGHT ORANGE J ◇ TERTRACID LIGHT ORANGE G ◇ UNITERTRACID LIGHT ORANGE G ◇ VENDACID LIGHT ORANGE 2G ◇ WOOL ORANGE 2G ◇ XYLENE FAST ORANGE G

TOXICITY DATA with REFERENCE
pic-esc 100 mmol/L MDMIAZ 31,11,79
cyt-ham:ovr 20 µmol/L/5H-C ENMUDM 1,27,79
orl-pig TDLo:28 g/kg (16W male):REP FCTXAV 11,367,73

CONSENSUS REPORTS: IARC Cancer Review: Group 3 IMEMDT 7,56,87; Animal Inadequate Evidence IMEMDT 8,181,75. Reported in EPA TSCA Inventory. EPA Genetic Toxicology Program.

SAFETY PROFILE: Experimental reproductive effects. Questionable carcinogen. Mutation data reported. Used as a drug and cosmetic colorant. When heated to decomposition it emits very toxic SO_x, Na_2O, and NO_x.

HGC400 ***HR: 2***
HISTAGLOBIN

TOXICITY DATA with REFERENCE
scu-rat TDLo:12985 mg/kg (35D male):REP KSRNAM 13,89,79
ivn-rat LD50:6110 mg/kg KSRNAM 13,89,79
scu-mus LD50:11960 mg/kg KSRNAM 13,89,79
ivn-mus LD50:3320 mg/kg KSRNAM 13,89,79

SAFETY PROFILE: Moderately toxic by intravenous route. Experimental reproductive effects.

HGC500 CAS:569-65-3 ***HR: 3***
HISTAMETHIZINE
mf: $C_{25}H_{27}ClN_2$ mw: 390.99

SYNS: ANCOLAN ◇ BONADETTES ◇ BONADOXIN ◇ BONAMINE ◇ CALMONAL ◇ CHICLIDA ◇ 1-(p-CHLOROBENZHYDRYL)-4-(m-METHYLBENZYL)DIETHYLENEDIAMINE ◇ 1-p-CHLOROBENZHYDRYL-4-m-METHYLBENZYLPIPERAZINE ◇ 1-(p-CHLORO-α-PHENYLBENZYL)-4-(m-METHYLBENZYL)PIPERAZINE ◇ HISTAMETHINE ◇ HISTAMETIZINE ◇ HISTAMETIZYNE ◇ ITINEROL ◇ LONGIFENE ◇ MAREX ◇ MECLIZINE ◇ MECLOZINE ◇ NAVICALM ◇ NEO-ISTAFENE ◇ NEO-SUPRIMAL ◇ NEO-SUPRIMEL ◇ PARACHLORAMINE ◇ PEREMESIN ◇ POSTAFEN ◇ SABARI ◇ SEA-LEGS ◇ SIGURAN ◇ SUBARI ◇ SUPRIMAL ◇ TRAVELON ◇ UCB 170 ◇ VIBAZINE ◇ VOMISSELS

TOXICITY DATA with REFERENCE
unr-wmn TDLo:1500 μg/kg (48-69D preg):REP LANCAO 1,222,63
orl-rat TDLo:1400 mg/kg (female 8-15D post):TER PSDTAP 9,134,68
orl-rat LD50:1750 mg/kg PSDTAP 9,134,68
ivn-rat LD50:75 mg/kg MEXPAG 4,145,61
orl-mus LD50:1650 mg/kg CLDND*
ims-mus LD50:625 mg/kg CLDND*

CONSENSUS REPORTS: Reported in EPA TSCA Inventory.

SAFETY PROFILE: A poison by intravenous route. Moderate toxicity by ingestion and intramuscular routes. Human reproductive effects by an unspecified route: reduced viability of newborn. Experimental teratogenic and reproductive effects. An antihistamine. When heated to decomposition it emits very toxic fumes of Cl^- and NO_x.

HGD000 CAS:51-45-6 ***HR: 3***
HISTAMINE
mf: $C_5H_9N_3$ mw: 111.17

PROP: White crystals. Mp: 83-84°, bp: 210° @ 18 mm. Very sol in water; sol in hot chloroform; insol in ether.

SYNS: β-AMINOETHYLGLYOXALINE ◇ β-AMINOETHYLIMIDAZOLE ◇ 4-(2-AMINOETHYL)IMIDAZOLE ◇ ERAMIN ◇ ERGAMINE ◇ ERGOTIDINE ◇ FREE HISTAMINE ◇ 1H-IMIDAZOLE-4-ETHANAMINE ◇ IMIDAZOLE-4-ETHYLAMINE ◇ 4-IMIDAZOLEETHYLAMINE ◇ 5-IMIDAZOLEETHYLAMINE ◇ β-IMIDAZOLYL-4-ETHYLAMINE ◇ 2-(4-IMIDAZOLYL)ETHYLAMINE ◇ 2-IMIDAZOL-4-YL-ETHYLAMINE ◇ THERAMINE

TOXICITY DATA with REFERENCE
dni-hmn:oth 2 μmol/L JIDEAE 65,400,75
cyt-mus:emb 200 mg/L DNAKAS 282,173,85
scu-mus TDLo:200 mg/kg (1D male):REP PSEBAA 113,161,63
scu-rat LDLo:250 mg/kg AEPPAE 185,461,37
ivn-rat LD50:630 mg/kg KSRNAM 13,89,79
ipr-mus LD50:725 mg/kg BJPCAL 17,137,61
scu-mus LD50:2500 mg/kg KSRNAM 13,89,79
ivn-mus LD50:385 mg/kg KSRNAM 13,89,79
ivn-dog LD50:7 mg/kg IVEJAC 57,31,80
scu-cat LDLo:34 mg/kg AEPPAE 185,461,37
ivn-rbt LDLo:100 mg/kg HBAMAK 4,1289,35
ivn-gpg LD50:180 μg/kg JPETAB 95,45,49

SAFETY PROFILE: A poison by intravenous and subcutaneous routes. Moderately toxic by intraperitoneal route. Experimental reproductive effects. Human mutation data reported. A neurotransmitter. The most potent capillary dilator known. Ingestion or inhalation produces the following effects: flushing followed by pallor, dizziness, fainting, fall in blood pressure, headache, rapid, weak pulse. Allergic effects on skin (hives) may occur. When heated to decomposition it emits toxic fumes of NO_x.

HGD500 CAS:56-92-8 ***HR: 3***
HISTAMINE DICHLORIDE
mf: $C_5H_9N_3•2ClH$ mw: 184.09

PROP: Prisms of aqueous ethyl alc. Mp: 239-246° (decomp). Sol in water, methanol; sltly sol in alc; insol in ether.

SYN: HISTAMINE DIHYDROCHLORIDE

TOXICITY DATA with REFERENCE
ipr-mus TDLo:2400 mg/kg (1-6D preg):REP JRPFA4 6,179,63
orl-mus LD50:2534 mg/kg ARTODN 2,371,79
ipr-mus LD50:1289 mg/kg JPETAB 77,54,43
scu-mus LDLo:1300 mg/kg AEPPAE 166,437,32
ivn-mus LD50:370 mg/kg ATSUDG 2,371,79
ipr-rbt LDLo:200 mg/kg JIDIAQ 42,473,28
ipr-gpg LD50:4602 μg/kg JPETAB 77,54,43
scu-gpg LD50:1250 μg/kg TXAPA9 8,339,66
ivn-gpg LD50:294 μg/kg JAPMA8 33,80,44
par-gpg LDLo:570 μg/kg AEPPAE 211,328,50

CONSENSUS REPORTS: Reported in EPA TSCA Inventory.

SAFETY PROFILE: Poison by subcutaneous, intravenous, intraperitoneal and parenteral routes. Moderately toxic by ingestion. Experimental reproductive effects. When heated to decomposition it emits very toxic fumes of Cl^- and NO_x. See also HISTAMINE.

HGE000 CAS:51-74-1 ***HR: 3***
HISTAMINE DIPHOSPHATE
mf: $C_5H_9N_3•2H_3O_4P$ mw: 307.17

SYNS: 4-(2-AMINOETHYL)IMIDAZOLE BIS(DIHYDROGEN PHOSPHATE) ◇ 4-(2-AMINOETHYL)IMIDAZOLE DI-ACID PHOSPHATE ◇ HISTAMINE ACID PHOSPHATE ◇ HISTAMINE PHOSPHATE (1:2) ◇ 1H-IMIDAZOLE-4-ETHANAMINE PHOSPHATE (1:2)

TOXICITY DATA with REFERENCE
par-mus TDLo:62500 μg/kg (female 8D post):TER TJADAB 32,45B,85
ipr-rat LD50:1781 mg/kg JPETAB 119,444,57
orl-mus LD50:807 mg/kg PSEBAA 122,685,66
ipr-mus LD50:913 mg/kg JPMSAE 59,1659,70
ivn-mus LD50:333 mg/kg PSEBAA 90,726,55
ivn-rbt LD50:2763 μg/kg JPETAB 75,299,42
ivn-gpg LD50:608 μg/kg JPMSAE 57,1543,68

CONSENSUS REPORTS: Reported in EPA TSCA Inventory.

SAFETY PROFILE: A deadly poison by intravenous and parenteral routes. Moderately toxic by ingestion and intraperitoneal routes. When heated to decomposition it emits very toxic fumes of NO_x and PO_x.

HGE500 ***HR: 3***
HISTAMINE HYDROCHLORIDE
mf: $C_5H_9N_3$•xClH mw: 366.39

SYNS: 4-AMINOETHYLIMIDAZOLE HYDROCHLORIDE ◇ CHLORHYDRATE d'HISTAMINE (FRENCH)

TOXICITY DATA with REFERENCE
scu-mus TDLo:1140 mg/kg/28W-I:ETA BAFEAG 36,305,49

SAFETY PROFILE: Questionable carcinogen with experimental tumorigenic data. When heated to decomposition it emits very toxic fumes of NO_x and HCl. See also HISTAMINE.

HGE700 CAS:71-00-1 ***HR: D***
HISTIDINE
mf: $C_6H_9N_3O_2$ mw: 155.18

PROP: l-Histidine, the natural form. White needles, plates, or crystalline powder; sltly bitter taste. Decomp 287° (softens at 277°). Solubility in water at 25°: 41.9 g/L. Sol in water; very sltly sol in alc; insol in ether.

SYNS: l-α-AMINO-4(OR 5)-IMIDAZOLEPROPIONIC ACID ◇ GLYOXALINE-5-ALANINE ◇ l-HISTIDINE (FCC)

TOXICITY DATA with REFERENCE
dni-hmn:oth 1 mmol/L JIDEAE 65,400,75
cyt-ham:lng 2500 ppm TOLED5 28,117,85
ipr-rat TDLo:7 g/kg (35D pre):REP OYYAA2 17,807,79

CONSENSUS REPORTS: Reported in EPA TSCA Inventory.

SAFETY PROFILE: Experimental reproductive effects. Human mutation data reported. When heated to decomposition it emits toxic fumes of NO_x.

HGF000 CAS:8022-91-1 ***HR: 2***
HO LEAF OIL

PROP: Chief contituent is Linalool, found in tree *Cinnamomum camphora L.* (FCTXAV 12,807,74).

TOXICITY DATA with REFERENCE
skn-rbt 500 mg/24H MOD FCTXAV 12,917,74
orl-rat LD50:3270 mg/kg FCTXAV 12,807,74

CONSENSUS REPORTS: Reported in EPA TSCA Inventory.

SAFETY PROFILE: Moderately toxic by ingestion. A skin irritant. When heated to decomposition it emits acrid smoke and fumes.

HGF100 ***HR: 2***
HOLLY

PROP: Evergreen tree with stiff leaves. Most have bright red berries but some trees may have yellow berries. They are native to the eastern United States from the Atlantic to Texas. Some are cultivated in the Pacific coast states and British Columbia.

SYNS: AMERICAN HOLLY ◇ APPALACHIAN TEA ◇ CAROLINA TEA ◇ CASSENA ◇ DEER BERRY ◇ EMETIC HOLLY ◇ ENGLISH HOLLY ◇ EVERGREEN CASSENA ◇ EUROPEAN HOLLY ◇ ILEX AQUIFOLIUM ◇ ILEX OPACA ◇ ILEX VOMITORIA ◇ INDIAN BLACK DRINK ◇ OREGON HOLLY ◇ YAUPON

SAFETY PROFILE: The berries contain poisonous saponins. Ingestion may cause nausea, severe vomiting, and diarrhea. See also SAPONIN.

HGF500 ***HR: 2***
HOLMIUM
af: Ho aw: 164.93

PROP: Bright metallic luster; soft, malleable metal; stable in dry air; oxidizes rapidly in moist air. Bp: 2720°, d: 8.78 @ 25°, vap press 2 mm @ 1630°.

SAFETY PROFILE: It may be an anticoagulant like the lanthanides. The toxicity (intravenous administration) of the salts decreases as follows: nitrate > sulfate > 3-sulfoisonicotinate > acetate > propionate > chloride. Can react violently with air or halogens. See also RARE EARTHS.

HGG000 CAS:10138-62-2 ***HR: 3***
HOLMIUM CHLORIDE
mf: Cl_3Ho mw: 271.28

PROP: Bright yellow, crystalline solid. Mp: 718°.

TOXICITY DATA with REFERENCE
orl-mus LD50:5165 mg/kg EQSSDX 1,1,75
ipr-mus LD50:312 mg/kg AEHLAU 5,437,62

CONSENSUS REPORTS: Reported in EPA TSCA Inventory.

SAFETY PROFILE: A poison by intraperitoneal route. Mildly toxic by ingestion. When heated to decomposition it emits highly toxic fumes of Cl^-. See also HOLMIUM and RARE EARTHS.

HGG500 CAS:13455-50-0 ***HR: 3***
HOLMIUM CITRATE
mf: $C_6H_5O_7{\cdot}Ho$ mw: 354.04

TOXICITY DATA with REFERENCE
ipr-rat LD50:117 mg/kg AEHLAU 5,437,62
ipr-gpg LD50:63 mg/kg AEHLAU 5,437,62

SAFETY PROFILE: A poison by intraperitoneal route. When heated to decomposition it emits acrid smoke and fumes. See also HOLMIUM and RARE EARTHS.

HGH000 CAS:35725-31-6 ***HR: 3***
HOLMIUM(III) NITRATE, HEXAHYDRATE (1:3:6)
mf: $N_3O_9{\cdot}Ho{\cdot}6H_2O$ mw: 459.08

SYN: NITRIC ACID, HOLMIUM(3⁻) SALT, HEXAHYDRATE

TOXICITY DATA with REFERENCE
orl-rat LD50:3000 mg/kg TXAPA9 5,750,63
ipr-rat LD50:270 mg/kg TXAPA9 5,750,63
ipr-mus LD50:320 mg/kg TXAPA9 5,750,63

SAFETY PROFILE: A poison by intraperitoneal route. Moderately toxic by ingestion. When heated to decomposition it emits toxic fumes of NO_x. See also HOLMIUM and RARE EARTHS.

HGH100 CAS:10168-82-8 ***HR: 3***
HOLMIUM TRINITRATE
mf: HoN_3O_9 mw: 350.96

SYN: HOLMIUM NITRATE

TOXICITY DATA with REFERENCE
orl-rat LD50:2313 mg/kg EQSSDX 1,1,75
ipr-rat LD50:208 mg/kg EQSSDX 1,1,75
ipr-mus LD50:247 mg/kg EQSSDX 1,1,75

SAFETY PROFILE: Poison by intraperitoneal route. Moderately toxic by ingestion. When heated to decomposition it emits toxic fumes of NO_x. See also HOLMIUM, RARE EARTHS, and NITRATES.

HGH200 ***HR: 2***
#2 HOME HEATING OILS

TOXICITY DATA with REFERENCE
skn-rbt 500 mg/24H MOD 52MLA2 1,1,83
eye-rbt 100 mg/30S MLD 52MLA2 1,1,83
orl-rat LD50:14500 mg/kg 52MLA2 1,1,83

SAFETY PROFILE: Mildly toxic by ingestion. A skin and eye irritant. Flammable when exposed to heat and flame; can react vigorously with oxidizers.

HGI000 CAS:602-52-8 ***HR: 3***
HOMIDIUM CHLORIDE
mf: $C_{21}H_{20}N_3{\cdot}Cl$ mw: 349.89

PROP: Dark red cryst powder. Sol in 5 parts water at room temp

SYNS: BABIDIUM CHLORIDE ◇ 3,8-DIAMINO-5-ETHYL-6-PHENYLPHENANTHRIDINIUM CHLORIDE ◇ 2,7-DIAMINO-9-PHENYL-10-ETHYLPHENANTHRIDINIUM CHLORIDE ◇ ETHIDIUM CHLORIDE ◇ NOVIDIUM CHLORIDE ◇ NSC-84423

TOXICITY DATA with REFERENCE
ivn-rat LD50:21 mg/kg NCINS* -,107,65
scu-mus LD50:56730 μg/kg NCISP* JAN86
ivn-mus LD50:14 mg/kg NCINS* -,107,65

SAFETY PROFILE: Poison by subcutaneous and intravenous routes. When heated to decomposition it emits very toxic fumes of NO_x and Cl^-.

HGI200 CAS:24342-55-0 ***HR: 3***
HOMOCHLOROCYCLIZINE HYDROCHLORIDE
mf: $C_{19}H_{23}ClN_2{\cdot}xClH$ mw: 570.11

SYNS: 1-(p-CHLORO-α-PHENYLBENZYL)HEXAHYDRO-4-METHYL)-1H-1,4-DIAZEPINE HYDROCHLORIDE ◇ 1-((4-CHLOROPHENYL) PHENYLMETHYL)HEXAHYDRO-4-METHYL-1H-1,4-DIAZEPINEHYDROCHLORIDE (9CI) ◇ HOMOCHLORCYCLIZINE HYDROCHLORIDE

TOXICITY DATA with REFERENCE
orl-rat LD50:490 mg/kg NIIRDN 6,790,82
ivn-rat LD50:36 mg/kg NIIRDN 6,790,82
orl-mus LD50:390 mg/kg NIIRDN 6,790,82
scu-mus LD50:135 mg/kg NIIRDN 6,790,82
ivn-mus LD50:47 mg/kg NIIRDN 6,790,82
ivn-dog LD50:20 mg/kg NIIRDN 6,790,82
ivn-rbt LD50:20 mg/kg NIIRDN 6,790,82

SAFETY PROFILE: Poison by ingestion, subcutaneous, and intravenous routes. When heated to decomposition it emits toxic fumes of NO_x and HCl.

HGI525 CAS:3566-25-4 ***HR: D***
HOMOFOLATE
mf: $C_{20}H_{21}N_7O_6$ mw: 455.48

SYNS: HOMOFOLIC ACID ◇ 9-METHYLFOLIC ACID

TOXICITY DATA with REFERENCE
dns-mus-scu 300 mg/kg LIFSAK 14,1541,74
orl-rat TDLo:12800 μg/kg (8-9D preg):REP TXAPA9 10,413,67
orl-mus TDLo:400 mg/kg (8D preg):TER NCIMAV 2,41,60

SAFETY PROFILE: An experimental teratogen. Experimental reproductive effects. Mutation data reported. When heated to decomposition it emits toxic fumes of NO_x.

HGI575 CAS:26833-87-4 ***HR: 3***
HOMOHARRINGTONINE
mf: $C_{29}H_{39}NO_9$ mw: 545.69

SYNS: (3(R))-CEPHALOTAXINE-4-METHYL-2-HYDROXY-2-(4-HYDROXY-4-METHYLPENTYL)BUTANEDIOATE (ESTER) ◇ NSC 141633

TOXICITY DATA with REFERENCE
orl-mus LD50:7456 μg/kg NCISP* JAN86
ipr-mus LD50:1960 μg/kg CMJODS 92,175,79
scu-mus LD50:3948 μg/kg NCISP* JAN86
ivn-mus LD50:6879 μg/kg NCISP* JAN86

SAFETY PROFILE: Poison by ingestion, subcutaneous, intravenous, and intraperitoneal routes. When heated to decomposition it emits toxic fumes of NO_x.

HGI700 CAS:19882-03-2 ***HR: 1***
18-HOMO-OESTRIOL
mf: $C_{19}H_{26}O_3$ mw: 302.45

SYNS: GONA-1,3,5(10)-TRIENE-3,16-α-17-β-TRIOL, 13-ETHYL- ◇ GONA-1,3,5(10)-TRIENE-3,16,17-TRIOL, 13-ETHYL-, (16-α-17-β)-(9CI) ◇ 18-HOMO-ESTRIOL ◇ WY-5090

TOXICITY DATA with REFERENCE
orl-rat TDLo:20 mg/kg (female 1-4D post):REP JRPFA4 26,235,71
orl-rat LD50:5010 mg/kg TXAPA9 18,185,71
orl-mus LD50:5010 mg/kg TXAPA9 18,185,71

SAFETY PROFILE: Mildly toxic by ingestion. Experimental reproductive effects. When heated to decomposition it emits acrid smoke and irritating fumes.

SAFETY PROFILE: Experimental reproductive effects.

HGI900 CAS:505-66-8 ***HR: 2***
HOMOPIPERAZINE
mf: $C_5H_{12}N_2$ mw: 100.19

PROP: Hygroscopic solid. Mp: 38-40°, bp: 169°, flash p: 148° F.

SYNS: 1,4-DIAZACYCLOHEPTANE ◇ 1H-1,4-DIAZEPINE, HEXAHYDRO- ◇ HEXAHYDRO-1,4-DIAZEPINE ◇ TRIMETHYLENE-ETHYLENEDIAMINE

TOXICITY DATA with REFERENCE
skn-rbt 100 μg/24H open AIHAAP 23,95,62
skn-rbt 5 mg/24H SEV 85JCAE-,882,86
eye-rbt 250 μg/24H SEV 85JCAE-,882,86
orl-rat LD50:2830 mg/kg AIHAAP 23,95,62
skn-rbt LD50:1050 mg/kg AIHAAP 23,95,62

CONSENSUS REPORTS: Reported in EPA TSCA Inventory.

SAFETY PROFILE: Moderately toxic by ingestion and skin contact. A corrosive and severe skin and eye irritant. Combustible liquid. When heated to decomposition it emits toxic fumes of NO_x.

HGK500 CAS:117-51-1 ***HR: 3***
3-HOMOTETRA HYDRO CANNIBINOL
mf: $C_{22}H_{32}O_2$ mw: 328.54

SYNS: 3-HEXYL-7,8,9,10-TETRAHYDRO-6,6,9-TRIMETHYL-6H-DIBENZO(B,D)PYRAN-1-OL ◇ 1-HYDROXY-3-N-HEXYL-6,6,9-TRIMETHYL-7,8,9,10-TETRAHYDRO-6-DIBENZOPYRAN ◇ PARAHEXYL ◇ PYRAHEXYL ◇ SYNHEXYL

TOXICITY DATA with REFERENCE
ivn-mus LD50:170 mg/kg JPETAB 88,154,46
orl-dog LDLo:930 mg/kg JPETAB 88,154,46
ivn-dog LD50:223 mg/kg JPETAB 88,154,46
ivn-rbt LD50:143 mg/kg JPETAB 88,154,46
ipr-gpg LD50:850 mg/kg JPETAB 88,154,46

SAFETY PROFILE: A poison by intravenous route. Moderately toxic by ingestion and intraperitoneal routes. When heated to decomposition it emits acrid smoke and fumes.

HGK700 ***HR: D***
HONEYSUCKLE BUSH

PROP: Shrubs or climbing woody vines with yellow, pink, white or rose flowers which produce red berries. They are cultivated and grow wild in Europe, the northeastern United States, Ontario, and Quebec.

SYNS: CHEVREFEUILLE (CANADA) ◇ FLY TATARIA ◇ LONICERA PERICLYMENUM ◇ LONICERA TATARICA ◇ LONICERA XYLOSTEUM ◇ MADRESELVA ◇ MEDADDY BUSH ◇ WOODBINE HONEYSUCKLE

SAFETY PROFILE: In Europe, ingestion of the berries reputedly causes severe vomiting, colic, diarrhea, hypovolemic shock, irregular heartbeat, convulsions, and respiratory failure. In the United States, berries from the same species are eaten with no toxic effects.

HGL575 ***HR: 1***
HORSE CHESTNUT

PROP: A large flowering tree found in temperate North America in an area bounded by the Gulf coast, California, and Newfoundland. The seeds are a waxy brown with a white spot and form in pods.

SYNS: A. CALIFORNICA ◇ AESCULUS (VARIOUS SPECIES) ◇ A. FLAVA ◇ A. GLABRA ◇ A. HIPPOCASTANUM ◇ BONGAY ◇ BUCKEYE ◇ CONQUERORS ◇ FISH POISON ◇ MARRONNIER (CANADA)

SAFETY PROFILE: The nuts and twigs contain escin, a cytotoxic mixture of saponins. This toxin is poorly absorbed by the gastrointestinal tract so effects are limited to severe inflammation of the stomach and intestines, resulting in fluid loss and electrolyte imbalance. Generally, multiple doses are required to cause severe poisoning or death. See also SAPONIN, α-ESCIN, β-ESCIN, and ESCIN, SODIUM SALT.

HGL800 ***HR: D***
HUMAN IMMUNOGLOBULIN COG-78

TOXICITY DATA with REFERENCE
ivn-rat TDLo:1375 mg/kg (7-17D preg):REP KSRNAM 16,3107,82
ivn-rbt TDLo:3250 mg/kg (female 6-18D post):TER KSRNAM 16,3107,82

SAFETY PROFILE: An experimental teratogen. Other experimental reproductive effects.

HGM000 ***HR: 1***
HUMAN SPERM

TOXICITY DATA with REFERENCE
ipr-mus TDLo:56 g/kg/15D-I:CAR 13BYAH -,279,62
ipr-mus LDLo:16 g/kg 13BYAH -,279,62

SAFETY PROFILE: Slightly toxic by intraperitoneal route. Questionable carcinogen with experimental carcinogenic data.

HGM500 CAS:101670-43-3 ***HR: 3***
HUMIDIN
mf: $C_{12}H_{20}O_4$ mw: 228.32

TOXICITY DATA with REFERENCE
orl-mus LD50:54 mg/kg 85ERAY 2,1122,78
ipr-mus LD50:4500 μg/kg 85FZAT -,331,67
ivn-mus LD50:1 mg/kg 85GDA2 2,348,80

SAFETY PROFILE: Poison by ingestion, intravenous, and intraperitoneal routes. When heated to decomposition it emits acrid smoke and fumes.

HGM600 ***HR: 3***
HUNDRED PACE SNAKE VENOM

SYNS: AGKISTRODON ACUTUS VENOM ◇ VENOM, SNAKE, AGKISTRODON ACUTUS

TOXICITY DATA with REFERENCE
ipr-mus LD50:2800 μg/kg TIHHAH 61,239,62
scu-mus LD50:9200 μg/kg TIHHAH 61,239,62
ivn-mus LD50:380 μg/kg BCPCA6 20,1549,71

SAFETY PROFILE: A deadly poison by subcutaneous, intravenous, and intraperitoneal routes.

HGN000 CAS:7722-73-8 ***HR: 3***
HX-868
mf: $C_{21}H_{27}N_3O_3$ mw: 369.51

SYNS: 1,3,5-TRIS(CARBONYL-2-ETHYL-1-AZIDINE)BENZENE ◇ 1,3,5-TRIS((2-ETHYL-AZIRIDINYL)-CARBONYL)BENZENE

TOXICITY DATA with REFERENCE
ipr-mus LD50:100 mg/kg NTIS** AD441-640

CONSENSUS REPORTS: Reported in EPA TSCA Inventory.

SAFETY PROFILE: Poison by intraperitoneal route. When heated to decomposition it emits toxic fumes of NO_x.

HGN500 CAS:8023-94-7 ***HR: 1***
HYACINTH ABSOLUTE

PROP: Extracted from the flowers of *Hyacinthus orientalis* (FCTXAV 14,659,76).

TOXICITY DATA with REFERENCE
skn-rbt 500 mg/24H MOD FCTXAV 14,795,76
orl-rat LD50:4200 mg/kg FCTXAV 14,795,76

CONSENSUS REPORTS: Reported in EPA TSCA Inventory.

SAFETY PROFILE: Mildly toxic by ingestion. A skin irritant. When heated to decomposition it emits acrid smoke and fumes.

HGN600 CAS:9067-32-7 ***HR: 2***
HYALURONIC ACID, SODIUM SALT

SYNS: HEALON ◇ SODIUM HYALURONATE ◇ SPH

TOXICITY DATA with REFERENCE
scu-rat TDLo:77 mg/kg (female 7-17D post):REP OYYAA2 29,111,85
scu-rat TDLo:189 mg/kg (multi) :TER OYYAA2 29,139,85
ipr-rat LD50:1770 mg/kg YACHDS 12,5369,84
ipr-mus LD50:1500 mg/kg OYYAA2 28,1013,84
ipr-rbt LD50:1820 mg/kg YACHDS 12,5369,84

SAFETY PROFILE: Moderately toxic by intraperitoneal route. An experimental teratogen. Other experimental reproductive effects. When heated to decomposition it emits toxic fumes of Na_2O.

HGO500 CAS:23255-93-8 ***HR: 3***
HYCANTHONE METHANESULFONATE
mf: $C_{20}H_{24}N_2O_2S \cdot CH_4O_3S$; mw: 452.63

SYNS: 1-((2-(DIETHYLAMINO)ETHYL)AMINO)-4-(HYDROXYMETHYL)-9H-THIOXANTHEN-9-ONEMONOMETHANE-SULFONATE (SALT) ◇ ETRENOL ◇ HCT ◇ HYCANTHONE MESYLATE ◇ HYCANTHONE METHANESULPHONATE ◇ HYCANTHONE MONOMETHANESULPHONATE

TOXICITY DATA with REFERENCE
dnr-esc 16 μg/well ENMUDM 3,429,81
cyt-hmn:leu 1 μmol/L MUREAV 21,287,73
cyt-hmn:lym 2400 μg/L JTEHD6 1,211,76

dni-hmn:lym 50 mg/L BCPCA6 22,1253,73
sln-dmg-par 4400 μmol/L MUREAV 82,111,81
sln-dmg-orl 4400 μmol/L MUREAV 82,111,81
otr-rat-ipr 30 mg/kg CNREA8 40,1157,80
ipr-rat TDLo:200 mg/kg (male 5D pre):TER JPETAB 187,437,73
ims-mus TDLo:50 mg/kg (female 7D post):REP NATUAS 239,107,72
ipr-mus TDLo:815 mg/kg/33W-I:CAR IJCNAW 23,97,79
ims-mus TDLo:350 mg/kg/30W-I:ETA IJCNAW 23,97,79
ims-hmn TDLo:3 mg/kg:CNS,GIT PACHAS 42,209,75
ipr-rat LDLo:100 mg/kg JPETAB 187,437,73
orl-mus LD50:565 mg/kg NCISP* JAN86
ipr-mus LD50:252 mg/kg NCISP* JAN86
scu-mus LD50:204 mg/kg NCISP* JAN86
ivn-mus LD50:79 mg/kg JPETAB 186,402,73
ims-mus LD50:320 mg/kg JPETAB 186,402,73

CONSENSUS REPORTS: IARC Cancer Review: Group 3 IMEMDT 7,56,87; Animal Inadequate Evidence IMEMDT 13,91,77. EPA Genetic Toxicology Program.

SAFETY PROFILE: A poison by intraperitoneal, subcutaneous, intravenous, and intramuscular routes. Moderately toxic by ingestion. Questionable carcinogen with experimental carcinogenic, tumorigenic, and teratogenic data. Other experimental reproductive effects. Human systemic effects by intramuscular route: hallucinations, muscle weakness, nausea or vomiting. Human mutation data reported. When heated to decomposition it emits very toxic fumes of SO_x. See also SULFONATES.

HGO600 CAS:461-72-3 ***HR: D***
HYDANTOIN
mf: $C_3H_4N_2O_2$ mw: 100.09

PROP: Needles from methanol. Mp: 220°. Sltly sol in water or ether; sol in alc and in solns of fixed alkali hydroxides.

SYNS: GLYCOLYLUREA ◇ 2,4-IMIDAZOLIDINEDIONE (9CI)

TOXICITY DATA with REFERENCE
ipr-mus TDLo:12 g/kg (11-13D preg):TER TXAPA9 64,271,82

CONSENSUS REPORTS: Reported in EPA TSCA Inventory.

SAFETY PROFILE: An experimental teratogen. When heated to decomposition it emits toxic fumes of NO_x.

HGP000 CAS:109-78-4 ***HR: 3***
HYDRACRYLONITRILE
mf: C_3H_5NO mw: 71.09

$N{\equiv}CC_2H_4OH$

PROP: Colorless to straw-colored liquid. Bp: 228° decomp, fp: −46°, flash p: 265°F (OC), d: 1.0404 @ 25°, vap press: 0.08 mm @ 25°, vap d: 2.45. Misc with water, acetone, methyl ethyl ketone, and ethanol. Sltly sol in ether; insol in benzene, petr ether, carbon disulfide, and carbon tetrachloride.

SYNS: 2-CYANOETHANOL ◇ 2-CYANOETHYL ALCOHOL ◇ ETHYLENE CYANOHYDRIN ◇ GLYCOL CYANOHYDRIN ◇ β-HPN ◇ 3-HYDROXYPROPANENITRILE ◇ β-HYDROXYPROPIONITRILE ◇ 3-HYDROXYPROPIONITRILE ◇ METHANOLACETONITRILE ◇ USAF RH-7

TOXICITY DATA with REFERENCE
skn-rbt 10 mg/24H open JIHTAB 26,269,44
skn-rbt 520 mg open MLD UCDS** 8/18/67
eye-rbt 500 mg AJOPAA 29,1363,46
orl-rat LD50:10 g/kg JTEHD6 2,31,76
orl-mus LD50:1800 mg/kg AMIHBC 8,371,53
ihl-mus LC33:300 mg/m³/2H 85GMAT -,66,82
ipr-mus LD50:500 mg/kg NTIS** AD277-689
orl-rbt LDLo:900 mg/kg AMIHBC 8,371,53
skn-rbt LD50:5000 mg/kg UCDS** 8/18/67

CONSENSUS REPORTS: Reported in EPA TSCA Inventory. Cyanide compounds are on the Community Right-To-Know List.

SAFETY PROFILE: Poison by inhalation. Moderately toxic by ingestion and intraperitoneal routes. Mildly toxic by skin contact. A skin and eye irritant. Combustible when exposed to heat or flame. Reacts violently with mineral acids (e.g., chlorosulfonic acid, oleum, sulfuric acid), amines or inorganic bases (e.g., NaOH). Reacts with water or steam to produce toxic and flammable vapors. To fight fire, use CO_2, dry chemical, alcohol foam. When heated to decompositon or on contact with acid or acid fumes it emits highly toxic fumes of CN^-. See also NITRILES.

HGP495 CAS:86-54-4 ***HR: 3***
HYDRALAZINE
mf: $C_8H_8N_4$ mw: 160.20

SYNS: APRESOLIN ◇ APPRESSIN ◇ APREZOLIN ◇ BA5968 ◇ C-5068 ◇ C 5968 ◇ CIBA 5968 ◇ HIDRALAZIN ◇ HIPOFTALIN ◇ HYDRALLAZINE ◇ HYDRAZINOPHTHALAZINE ◇ 1-HYDRAZINOPHTHALAZINE ◇ HYPOPHTHALIN ◇ IDRALAZINA (ITALIAN) ◇ 1(2H)-PHTHALAZINONE HYDRAZONE

TOXICITY DATA with REFERENCE
mma-sat 500 μg/plate MUREAV 66,247,79
slt-dmg-unr 200 mmol/L/6H MUREAV 120,233,83
unr-wmn TDLo:54 mg/kg (26-39W preg):TER NEJMAG 303,1235,80
unr-rat TDLo:330 mg/kg (female 1-22D post):REP ANREAK 193,174,79
orl-man TDLo:2086 mg/kg/2Y-I:SKN BMJOAE 289,410,84

orl-wmn TDLo:730 mg/kg/2Y-I:SKN BMJOAE 289,410,84
ims-man TDLo:89 μg/kg:CVS EJPEDT 145,318,86
orl-rat LD50:90 mg/kg PLRCAT 8,295,76
ipr-rat LD50:25 mg/kg JPETAB 143,7,64
ivn-rat LD50:34 mg/kg JAPMA8 40,559,51
orl-mus LD50:122 mg/kg PLRCAT 8,295,76
ipr-mus LD50:100 mg/kg JMCMAR 22,671,79
scu-mus LD50:150 mg/kg RPTOAN 31,53,68
ivn-dog LD50:50 mg/kg ARZNAD 35,818,85

CONSENSUS REPORTS: IARC Cancer Review: Group 3 IMEMDT 7,222,87; Human Inadequate Evidence IMEMDT 24,85,80.

SAFETY PROFILE: Poison by ingestion, intravenous, intraperitoneal, and subcutaneous routes. Human systemic effects by ingestion: allergic dermatitis, cardiomyopathy, changes in coronary arteries. Human teratogenic effects by an unspecified route: developmental abnormalities of the blood and lymphatic system. Questionable carcinogen. Mutation data reported. When heated to decomposition it emits toxic fumes of NO_x.

HGP500 CAS:304-20-1 ***HR: 3***
HYDRALAZINE HYDROCHLORIDE
mf: $C_8H_8N_4 \cdot ClH$ mw: 196.66

PROP: Yellow crystals. Decomp @ 273°. Very sltly sol in ether.

SYNS: AISELAZINE ◇ APPRESINUM ◇ APRELAZINE ◇ APRESAZIDE ◇ APRESINE ◇ APRESOLIN ◇ APRESOLINE-ESIDRIX ◇ APRESOLINE HYDROCHLORIDE ◇ APREZOLIN ◇ BA 5968 ◇ CIBA 5968 ◇ DRALZINE ◇ HIDRALAZIN ◇ HIPOFTALIN ◇ HYDRALAZINE CHLORIDE ◇ HYDRALAZINE MONOHYDROCHLORIDE ◇ HYDRALLAZINE HYDROCHLORIDE ◇ HYDRAPRESS ◇ 1-HYDRAZINOPHTHALAZINE HYDROCHLORIDE ◇ 1-HYDRAZINOPHTHLAZINE MONOHYDROCHLORIDE ◇ HYPERAZIN ◇ HYPOPHTHALIN ◇ HYPOS ◇ IPOLINA ◇ LOPRESS ◇ NOR-PRESS 25 ◇ 1(2H)-PHTHALAZINONE HYDRAZONE HYDROCHLORIDE ◇ 1(2H)-PHTHLAZINONE, HYDRAZONE, MONOHYDROCHLORIDE ◇ PRAPARAT 5968 ◇ ROLAZINE ◇ SERPASIL APRESOLINE No. 2

TOXICITY DATA with REFERENCE
mmo-sat 500 μg/plate RCOCB8 49,415,85
dni-hmn:hla 150 μmol/L MUREAV 92,427,82
orl-rat TDLo:176 mg/kg (female 1-22D post):TER TJADAB 34,469,86
orl-mus TDLo:2950 mg/kg/78W-C:NEO JJIND8 61,1363,78
ivn-rat LD50:34 mg/kg NIIRDN 6,619,82
orl-mus LD50:188 mg/kg OYYAA2 3,97,69
ipr-mus LD50:83 mg/kg JPETAB 101,368,51
scu-mus LD50:73 mg/kg NIIRDN 6,619,82
ivn-mus LD50:84 mg/kg OYYAA2 3,97,69

CONSENSUS REPORTS: IARC Cancer Review: Animal Limited Evidence IMEMDT 24,85,80. Reported in EPA TSCA Inventory.

SAFETY PROFILE: Suspected carcinogen with experimental neoplastigenic data. A poison by ingestion, subcutaneous, intravenous, and intraperitoneal routes. Human mutation data reported. An experimental teratogen. An antihypertensive agent. When heated to decomposition it emits very toxic NO_x and HCl.

HGP550 CAS:10592-13-9 ***HR: 3***
HYDRAMYCIN
mf: $C_{22}H_{24}N_2O_8 \cdot ClH$ mw: 480.94

SYNS: BIOCAMYCIN ◇ DOXIGALUMICINA ◇ DOXYCYCLINE HYCLATE ◇ DOXYCYCLINE HYDROCHLORIDE ◇ DOXY-II ◇ DOXY-TABLINEN ◇ ECODOX ◇ LIOMYCIN ◇ MESPAFIN ◇ MIDOXIN ◇ NIVOCILIN ◇ NOVADOX ◇ RETENS ◇ ROXIMYCIN ◇ SAMECIN ◇ TANAMICIN ◇ TECACIN ◇ TETRADOX ◇ VIBRADOX ◇ VIBRAMYCIN HYCLATE ◇ VIBRA-TABS

TOXICITY DATA with REFERENCE
iut-rat TDLo:50 mg/kg (1D pre):REP CCPTAY 29,553,84
orl-rat LD50:1700 mg/kg YACHDS 8,1447,80
ipr-rat LD50:262 mg/kg TXAPA9 18,185,71
scu-rat LD50:700 mg/kg YACHDS 8,1447,80
ivn-rat LD50:137 mg/kg YACHDS 8,1447,80
orl-mus LD50:1890 mg/kg NIIRDN 6,505,82
scu-mus LD50:700 mg/kg YACHDS 8,1447,80
ivn-mus LD50:290 mg/kg YACHDS 8,1447,80

SAFETY PROFILE: Poison by intravenous and intraperitoneal routes. Moderately toxic by ingestion and subcutaneous routes. Experimental reproductive effects. An antibacterial agent. When heated to decomposition it emits toxic fumes of NO_x and HCl.

HGP600 ***HR: 2***
HYDRANGEA

PROP: A large bush which grows to 15 feet tall with red-brown stems. The 6-inch leaves have a dark green top and a fuzzy grey bottom. It produces small white, rose, blue or green-white flowers which stay on the plant until they dry. It is used as an ornamental throughout the United States and Canada.

SYNS: HILLS-OF-SNOW ◇ HORTENSIA (CUBA) ◇ HYDRANGEA MACROPHYLLA ◇ POPO-HAU (HAWAII) ◇ SEVEN BARK

SAFETY PROFILE: The flower bud contains the poison hydrangin, a cyanogenetic glycoside. Ingestion of the buds may produce, after a delay period, the symptoms of cyanide poisoning: abdominal pain, vomiting, lethargy, sweating, coma, convulsions, and lack of muscle control. See also CYANIDE.

HGQ500 CAS:5936-28-7 ***HR: 2***
HYDRASTINE HYDROCHLORIDE
mf: $C_{21}H_{21}NO_6 \cdot ClH$ mw: 419.89

PROP: Hygroscopic powder. Mp: 116°, sltly sol in $CHCl_3$; very sltly sol in ether. Keep well closed.

TOXICITY DATA with REFERENCE
orl-rat LD50:1000 mg/kg 29ZVAB -,57,69
scu-rat LD50:1270 mg/kg 29ZVAB -,57,69

SAFETY PROFILE: Moderately toxic by ingestion and subcutaneous routes. When heated to decomposition it emits very toxic fumes of NO_x and HCl.

HGR000 CAS:6592-85-4 ***HR: 3***
HYDRASTININE
mf: $C_{11}H_{13}NO_3$ mw: 207.25

PROP: White-yellowish needles. Mp: 116-117°. Very sol in alc, chloroform, ether, dil acids;. insol in cold water; mod sol in hot water.

SYN: 5,6,7,8-TETRAHYDRO-6-METHYL-1,3-DIOXOLO-(4,5-g)ISOQUINOLIN-5-OL

TOXICITY DATA with REFERENCE
scu-rat LDLo:1 g/kg HBAMAK 4,1289,35
scu-rbt LDLo:300 mg/kg HBAMAK 4,1289,35

SAFETY PROFILE: Poison by subcutaneous route. Can cause paralysis of vasomotor nerves and vagus endings. When heated to decomposition it emits highly toxic fumes of NO_x.

HGR500 ***HR: 2***
HYDRASTIS CANADENSIS L., ROOT EXTRACT

PROP: Plant containing berberine-type alkaloids (YKKZAJ 82,726,62).

TOXICITY DATA with REFERENCE
orl-mus LD50:1620 mg/kg YKKZAJ 82,726,62

CONSENSUS REPORTS: Reported in EPA TSCA Inventory.

SAFETY PROFILE: Moderately toxic by ingestion. See also BERBERINE.

HGS000 CAS:302-01-2 ***HR: 3***
HYDRAZINE
DOT: UN 2029/UN 2030
mf: H_4N_2 mw: 32.06

H_2NNH_2

PROP: Colorless, oily, fuming liquid or white crystals. Mp: 1.4°, bp: 113.5°, flash p: 100°F (OC), d: 1.1011 @ 15° (liquid), autoign temp: can vary from 74°F in contact with iron rust, 270°F in contact with black iron, 313°F in contact with stainless steel, 518°F in contact with glass. Vap d: 1.1; lel: 4.7%, uel: 100%.

SYNS: DIAMIDE ◇ DIAMINE ◇ HYDRAZINE, anhydrous (DOT) ◇ HYDRAZINE, aqueous solution (DOT) ◇ HYDRAZINE BASE ◇ HYDRAZYNA (POLISH) ◇ RCRA WASTE NUMBER U133

TOXICITY DATA with REFERENCE
mmo-omi 70 µg/L MUREAV 173,233,86
dni-hmn:hla 50 µmol/L CNREA8 44,59,84
scu-rat TDLo:80 mg/kg (female 11-20D post):REP AEHLAU 21,615,70
ihl-rat TCLo:1 mg/m³/24H (female 1-11D post):TER GISAAA 39(10),23,74
ihl-rat TCLo:5 ppm/6H/1Y-I:CAR FAATDF 5,1050,85
orl-mus TDLo:1951 mg/kg/2Y-C:NEO IJCNAW 9,109,72
ihl-mus TCLo:1 ppm/6H/1Y-I:ETA PAACA3 21,74,80
ipr-mus TDLo:400 mg/kg/5W-I:CAR UICMAI 7,180,67
orl-rat LD50:60 mg/kg MEPAAX 24,71,73
ihl-rat LC50:570 ppm/4H AMIHAB 12,609,55
ipr-rat LD50:59 mg/kg MEPAAX 24,71,73
ivn-rat LD50:55 mg/kg MEPAAX 24,71,73
orl-mus LD50:59 mg/kg MEPAAX 24,71,73
ihl-mus LC50:252 ppm/4H AMIHAB 12,609,55
ipr-mus LD50:62 mg/kg MEPAAX 24,71,73
ivn-mus LD50:57 mg/kg MEPAAX 24,71,73
skn-dog LDLo:96 mg/kg TXAPA9 21,186,72
ivn-dog LD50:25 mg/kg MEPAAX 24,71,73
skn-rbt LD50:91 mg/kg AMIHBC 9,199,54
ivn-rbt LD50:20 mg/kg AMIHBC 9,199,54
skn-gpg LD50:190 mg/kg XAWPA2 CWL 2-10,58

CONSENSUS REPORTS: NTP Fifth Annual Report on Carcinogens. IARC Cancer Review: Group 2B IMEMDT 7,223,87; Animal Sufficient Evidence IMEMDT 4,127,74. EPA Extremely Hazardous Substances List. Community Right-To-Know List. Genetic Toxicology Program. Reported in EPA TSCA Inventory.

OSHA PEL: (Transitional: TWA 1 ppm (skin)) TWA 0.1 ppm (skin)
ACGIH TLV: TWA 0.1 ppm (skin); Suspected Human Carcinogen; (Proposed: 0.01 ppm (skin); Suspected Human Carcinogen)
DFG TRK: 0.1 ppm; Animal Carcinogen, Suspected Human Carcinogen.
NIOSH REL: (Hydrazines) CL 0.04 mg/m³/2H
DOT Classification: Flammable Liquid; Label: Flammable Liquid and Poison (UN2029); Corrosive Material; Label: Corrosive (UN2030); Flammable or Combustible Liquid; Label: Flammable Poison, Combustible (UN2029)

SAFETY PROFILE: Confirmed carcinogen with experimental carcinogenic, neoplastigenic, and tumorigenic data. A poison by ingestion, skin contact, intraperitoneal, and intravenous routes. Moderately toxic by inhalation. An experimental teratogen. Other experimental reproductive effects. Human mutation data reported. A powerful reducing agent which is corrosive to the eyes, skin, and mucous membranes. May cause skin sensitization as well as systemic poisoning. Hydrazine and some of its derivatives may cause damage to the liver and destruction of red blood cells. See also PHENYL HYDRAZINE.

Flammable liquid. A very dangerous fire hazard when exposed to heat, flame or oxidizing agents. Severe explosion hazard when exposed to heat or flame, or by chemical reaction. Explodes on contact with barium oxide; calcium oxide; chromate salts; chromium dioxide; dicyanofurazan; mercury oxide; trioxygen difluroide; N-haloimides; potassium; silver compounds; sodium hydroxide; titanium compounds (at 130°). Potentially explosive reactions with alkali metals, NH_3; Cl_2; chromates; CuO; Cu^{++} salts; F_2; metallic oxides; Ni; $Ni(ClO_4)_2$; O_2; liquid O_2; $K_2Cr_2O_7$; $Na_2Cr_2O_7$; tetryl; zinc diamide; $Zn(C_2H_5)_2$. Forms sensitive, explosive mixtures with 2-chloro-5-methylnitrobenzene; metal salts [e.g., cadmium perchlorate; copper chlorate (heat-sensitive); manganese nitrate (heat-sensitive); mercury(I) chloride; mercury(II) chloride; mercury(I) nitrate; mercury(II) nitrate; tin(II) chloride]; methanol + nitromethane; air; lithium perchlorate; sodium perchlorate; sodium. Ignites on contact with cotton waste + heavy metals; dinitrogen oxide; rhenium + alumina; catalysts; nitric acid; hydrogen peroxide; N,2,4,6-tetranitroaniline; rust + heat. Ignites spontaneously in air when absorbed on earth, asbestos, cloth, wood. Violent reaction with 1-chloro-2,4-dinitrobenzene; oxidants (e.g., iron oxide; chlorates; peroxides); thiocarbonyl azide thiocyanate. Vigorous reaction with benzene-seleninic acid or anhydride; carbon dioxide + stainless steel; copper oxide; lead oxide; potassium peroxodisulfate; ruthenium(III) chloride. On contact with metal catalysts (e.g., platinum black; Raney nickel; copper-iron oxide; molybdenum; molybdenum oxides; iridium), it decomposes to ammonia, hydrogen and nitrogen gases which may ignite or explode. A hypergolic reaction with dinitrogen tetraoxide is the basis of a liquid rocket fuel mixture. The vapor will burn without air. It is a powerful explosive. It is very sensitive and must not be used without full and complete instructions from the manufacturer for handling, storage, and disposal. Dangerous; when heated to decomposition it emits highly toxic fumes of NO_x and NH_3.

HGT500 ***HR: 3***
HYDRAZINE BISBORANE
mf: $B_2H_{10}N_2$ mw: 59.71

SAFETY PROFILE: Explodes on impact or above 100°C. Highly flammable. When heated to decomposition it emits toxic fumes of NO_x and BO_x. See also HYDRAZINE and BORANE.

HGU000 CAS:57-56-7 ***HR: 3***
HYDRAZINE CARBOXAMIDE
mf: CH_5N_3O mw: 75.09

SYNS: AMINOUREA ◇ CARBAMIC ACID HYDRAZIDE ◇ CARBAMOYLHYDRAZINE ◇ CARBAMYLHYDRAZINE ◇ CARBAZAMIDE ◇ SEMICARBAZIDE

TOXICITY DATA with REFERENCE
mmo-sat 67 μmol/plate CNREA8 41,1469,81
spm-grh-par 100 mmol/L
orl-mus TDLo:25 g/kg/30W-C:ETA GANNA2 51,83,60
ivn-man TDLo:40 mg/kg:CNS JPETAB 122,110,58
ipr-rat LD50:140 mg/kg APEPA2 257,296,67
scu-rat LD50:140 mg/kg ARZNAD 18,645,68
orl-mus LD50:176 mg/kg JPETAB 122,110,58
ipr-mus LD50:123 mg/kg JPETAB 122,110,58
scu-mus LD50:105 mg/kg JPETAB 119,444,57
ivn-mus LD50:126 mg/kg JPETAB 122,110,58

CONSENSUS REPORTS: Reported in EPA TSCA Inventory.

SAFETY PROFILE: A poison by ingestion, intraperitoneal, subcutaneous, and intravenous routes. Human systemic effects by intravenous route: convulsions. Questionable carcinogen with experimental tumorigenic data. Mutation data reported. When heated to decomposition it emits toxic fumes of NO_x. See also HYDRAZINE.

HGU100 CAS:13537-45-6 ***HR: 3***
HYDRAZINE DIFLUORIDE
mf: $H_4N_2 \cdot 2FH$ mw: 72.08

SYN: HYDRAZINE, DIHYDROFLUORIDE

TOXICITY DATA with REFERENCE
ivn-mus LD50:56 mg/kg CSLNX* NX#04254

OSHA PEL: TWA 2.5 mg(F)/m^3
ACGIH TLV: TWA 2.5 mg(F)/m^3
NIOSH REL: (Fluorides, Inorganic): 10H TWA 2.5 mg(F)/m^3

SAFETY PROFILE: Poison by intravenous route. When heated to decomposition it emits toxic fumes of NO_x and HF.

HGU500 CAS:7803-57-8 ***HR: 3***
HYDRAZINE HYDRATE
mf: $H_4N_2 \cdot H_2O$ mw: 50.08

PROP: Colorless fuming, refractive liquid. Mp: −51.7°, bp: 118.5° @ 740 mm. D: 1.03 @ 21°. Faint characteristic odor. A strong base, very corrosive; attacks glass, rubber, and cork. Very powerful reducing agent. Misc with water and alc; insol in chloroform and ether.

SYNS: HYDRAZINE MONOHYDRATE ◇ IDRAZINA IDRATA (ITALIAN)

TOXICITY DATA with REFERENCE
mmo-sat 10 μmol/plate CNREA8 41,1469,81
cyt-rat-ihl 850 μg/m^3/5H/16W-I GISAAA 49(9),25,84

dnd-mus-ipr 1560 nmol/kg CNREA8 41,1469,81
ims-ham TDLo:150 mg/kg (female 12D post):REP APTOD9 19,A71,80
skn-mus TDLo:80 g/kg/43W-I:CAR LAPPA5 25,149,65
orl-rat LD50:129 mg/kg HYSAAV 30,191,65
orl-mus LD50:83 mg/kg HYSAAV 30,191,65
orl-rbt LD50:55 mg/kg HYSAAV 30,191,65
orl-gpg LD50:40 mg/kg HYSAAV 30,191,65

CONSENSUS REPORTS: EPA Genetic Toxicology Program.

NIOSH REL: (Hydrazines) CL 0.04 mg/m^3/2H

SAFETY PROFILE: A poison by ingestion. Experimental reproductive effects. A corrosive irritant to the eyes, skin, and mucous membranes. Mutation data reported. Questionable carcinogen with experimental carcinogenic data. Incompatible with HgO; Na; $SnCl_2$; 2,4-dinitrochlorobenzene. When heated to decomposition it emits toxic fumes of NO_x. See also HYDRAZINE.

HGV000 CAS:2644-70-4 ***HR: 3***
HYDRAZINE HYDROCHLORIDE
mf: H_4N_2•ClH mw: 68.52

TOXICITY DATA with REFERENCE
orl-rat LD50:128 mg/kg AMIHAB 13,34,56
ipr-rat LD50:126 mg/kg AMIHAB 13,34,56
ivn-rat LD50:118 mg/kg AMIHAB 13,34,56
orl-mus LD50:126 mg/kg AMIHAB 13,34,56
ipr-mus LD50:133 mg/kg AMIHAB 13,34,56
ivn-mus LD50:122 mg/kg AMIHAB 13,34,56
ivn-dog LD50:53 mg/kg AMIHAB 13,34,56

CONSENSUS REPORTS: Reported in EPA TSCA Inventory.

NIOSH REL: (Hydrazines) CL 0.04 mg/m^3/2H

SAFETY PROFILE: A poison by ingestion, intravenous, and intraperitoneal routes. When heated to decomposition it emits very toxic fumes of Cl^- and NO_x. See also HYDRAZINE.

HGV500 ***HR: 3***
HYDRAZINE MONOBORANE
mf: BH_7N_2 mw: 45.86

SAFETY PROFILE: A shock-sensitive explosive. Highly flammable. When heated to decomposition it emits toxic fumes of NO_x and BO_x. See also HYDRAZINE and BORANES.

HGW000 CAS:63884-40-2 ***HR: 3***
HYDRAZINE PROPANEMETHANE SULFONATE
mf: $C_4H_{11}N_2O_3S$•Na mw: 190.22

SYNS: N′-ISOPROPYL HYDRAZINOMETHANESULFONIC ACID, SODIUM SALT ◇ K 653

TOXICITY DATA with REFERENCE
orl-rat LD50:330 mg/kg 27ZQAG -,430,72
ivn-mus LD50:500 mg/kg 27ZQAG -,430,72

SAFETY PROFILE: Poison by ingestion. Moderately toxic by intravenous route. When heated to decomposition it emits very toxic fumes of Na_2O, SO_x, and NO_x. See also HYDRAZINE and SULFONATES.

HGW500 CAS:10034-93-2 ***HR: 3***
HYDRAZINE SULFATE (1:1)
mf: H_4N_2•H_2O_4S mw: 130.14

PROP: Colorless crystals. D: 1.378, mp: 85°. Sol in water; insol in alc; very sol in hot water.

SYNS: HYDRAZINE HYDROGEN SULFATE ◇ HYDRAZINE MONOSULFATE ◇ HYDRAZINE SULPHATE ◇ HYDRAZINIUM SULFATE ◇ HYDRAZONIUM SULFATE ◇ HS ◇ IDRAZINA SOLFATO (ITALIAN) ◇ NSC-150014 ◇ SIRAN HYDRAZINU (CZECH)

TOXICITY DATA with REFERENCE
eye-rbt 20 mg/24H MOD 28ZPAK -,15,72
dnr-bcs 20 μL/disc MUREAV 97,1,82
mmo-omi 1500 μg/L MUREAV 173,233,86
dns-hmn:fbr 1 mg/L PMRSDJ 1,528,81
orl-rat TDLo:43 g/kg/85W-C:CAR JNCIAM 41,331,68
orl-mus TDLo:1892 mg/kg (MGN):NEO,TER JCROD7 105,258,83
orl-mus TDLo:19 g/kg/61W-C:CAR JNCIAM 41,331,68
ipr-mus TDLo:832 mg/kg/8W-I:ETA JNCIAM 42,337,69
orl-rat TD:38 g/kg/68W-I:NEO JNCIAM 41,331,68
orl-hmn TDLo:201 mg/kg/8D:GIT,PNS,CNS CCROBU 59,1151,75
orl-rat LD50:601 mg/kg 28ZPAK -,15,72
ipr-rat LD50:230 mg/kg RPTOAN 41,74,78
orl-mus LD50:740 mg/kg RPTOAN 41,74,78
ipr-mus LD50:152 mg/kg AEHLAU 17,315,68
orl-dog LDLo:100 mg/kg HBAMAK 4,1289,35
orl-rbt LDLo:100 mg/kg HBAMAK 4,1289,35

CONSENSUS REPORTS: NTP Fifth Annual Report on Carcinogens. IARC Cancer Review: Animal Sufficient Evidence IMEMDT 4,127,74. Community Right-To-Know List. Reported in EPA TSCA Inventory. EPA Genetic Toxicology Program.

NIOSH REL: (Hydrazines) CL 0.04 mg/m^3/2H

SAFETY PROFILE: Confirmed carcinogen with experimental carcinogenic, neoplastigenic, and tumorigenic data. A poison by ingestion and intraperitoneal routes. Human systemic effects by ingestion: paresthesia (abnormal sensations), somnolence, nausea or vomiting. An experimental teratogen. Human mutation data reported. An eye irritant. A reducing agent. When heated to decomposition it emits very toxic fumes of SO_x and NO_x. See also HYDRAZINE and SULFATES.

HGX000 ***HR: 3***
HYDRAZINIUM CHLORATE
mf: $ClH_5N_2O_3$ mw: 116.51

SAFETY PROFILE: Explodes violently when heated to its melting point 80°C. When heated to decomposition it emits very toxic fumes of Cl^- and NO_x. See also CHLORATES.

HGX500 ***HR: 3***
HYDRAZINIUM CHLORITE
mf: $ClH_5N_2O_2$ mw: 100.50

SAFETY PROFILE: It is spontaneously flammable when dry. When heated to decomposition it emits very toxic fumes of Cl^- and NO_x. See also CHLORITES.

HGY000 CAS:13812-39-0 ***HR: 2***
HYDRAZINIUM DIPERCHLORATE
mf: $Cl_2H_6N_2O_8$ mw: 232.97

SAFETY PROFILE: An explosive salt. Mixtures with metal compounds (e.g., copper chromate; copper chloride; nickel oxide; iron(III) oxide; magnesium oxide) have enhanced sensitivity to heat, impact or friction. Used as a rocket propellant component. When heated to decomposition it emits very toxic fumes of Cl^- and NO_x. See also PERCHLORATES.

HGY500 ***HR: 3***
HYDRAZINIUM HYDROGENSELENATE
mf: $H_6N_2O_4Se$ mw: 177.02

CONSENSUS REPORTS: Selenium and its compounds are on the Community Right-To-Know List.

OSHA PEL: TWA 0.2 mg(Se)/m^3
ACGIH TLV: TWA 0.2 mg(Se)/m^3
DFG MAK: 0.1 mg(Se)/m^3

SAFETY PROFILE: A heat-sensitive explosive. When heated to decomposition, it emits very toxic fumes of NO_x and Se. See also SELENIUM COMPOUNDS.

HGZ000 CAS:13464-97-6 ***HR: 3***
HYDRAZINIUM NITRATE
mf: $H_5N_3O_3$ mw: 95.06

$H_2NN^+H_3NO_3^-$

SAFETY PROFILE: Explodes if heated rapidly to 300°C or if heated in a sealed container. Explodes above 70°C on contact with metals (e.g., cobalt; copper; zinc; and most other metals) and metal compounds (e.g., metal acetylides; -nitrides; -oxides; -sulfides). Explosive reaction with potassium dichromate above 100°C. Mixture with 2-hydroxyethylamine + water is an impact-sensitive explosive. Upon decomposition it emits toxic fumes of NO_x. See also EXPLOSIVES and NITRATES.

HHA000 CAS:13762-80-6 ***HR: 3***
HYDRAZINIUM PERCHLORATE
mf: $ClH_5N_2O_4$ mw: 132.50

$H_2NN^+H_3ClO_4^-$

SAFETY PROFILE: An impact-sensitive explosive. Sensitivity to heat, impact, or friction is increased by the presence of metal salts or metal oxides (e.g., copper(II) chloride; copper chromite; copper chloride; nickel oxide; iron oxide; magnesium oxide). A component of some solid rocket fuels. Upon decomposition it emits toxic fumes of Cl^- and NO_x. See also PERCHLORATES.

HHA100 CAS:73953-53-4 ***HR: 3***
HYDRAZINIUM TRIFLUOROSTANNITE
mf: $H_4N_2{\cdot}F_3HSn$ mw: 208.76

SYN: HYDRAZINE, TRIFLUOROSTANNITE

TOXICITY DATA with REFERENCE
ivn-mus LD50:18 mg/kg CSLNX* NX#04255

OSHA PEL: TWA 2 mg(Sn)/m^3; TWA 2.5 mg(F)/m^3
ACGIH TLV: TWA 2 mg(Sn)/m^3
NIOSH REL: (Fluorides, Inorganic): 10H TWA 2.5 mg(F)/m^3

SAFETY PROFILE: Poison by intravenous route. When heated to decomposition it emits toxic fumes of NO_x, Sn, and F^-.

HHB000 CAS:26049-71-8 ***HR: 3***
2-HYDRAZINO-4-(p-AMINOPHENYL)THIAZOLE
mf: $C_9H_{10}N_4S$ mw: 206.29

SYN: 2-HYDRAZINO-4-(4-AMINOPHENYL)THIAZOLE

TOXICITY DATA with REFERENCE
orl-rat TDLo:12 g/kg/24W-C:CAR CNREA8 30,897,70
orl-mus TDLo:8000 mg/kg/46W-C:ETA CNREA8 33,1593,73

SAFETY PROFILE: Questionable carcinogen with experimental carcinogenic and tumorigenic data. When heated to decomposition it emits very toxic fumes of NO_x and SO_x.

HHB500 CAS:615-21-4 ***HR: 3***
2-HYDRAZINOBENZOTHIAZOLE
mf: $C_7H_7N_3S$ mw: 165.23

SYN: USAF EK-3967

TOXICITY DATA with REFERENCE
orl-rat LDLo:100 mg/kg NCNSA6 5,23,53
ipr-mus LD50:100 mg/kg NTIS** AD277-689
par-mus LDLo:200 mg/kg CBCCT* 7,686,55

CONSENSUS REPORTS: Reported in EPA TSCA Inventory.

SAFETY PROFILE: Poison by ingestion, intraperitoneal, and parenteral routes. When heated to decomposition it emits very toxic fumes of NO_x and SO_x.

HHC000 CAS:109-84-2 ***HR: 3***
2-HYDRAZINOETHANOL
mf: $C_2H_8N_2O$ mw: 76.12

PROP: Colorless, sltly viscous liquid. Mp: −70°, bp: 145-153° @ 25 mm, flash p: 224°F, vap d: 2.63, d: 1.11. Misc with water; sol in lower alcs; sltly sol in ether.

SYNS: BOH ◇ HYDROXYETHYL HYDRAZINE ◇ β-HYDROXYETHYLHYDRAZINE ◇ N-(2-HYDROXYETHYL)HYDRAZINE

TOXICITY DATA with REFERENCE
mmo-esc 1 pph MUREAV 40,19,76
mma-esc 500 μg/plate MUREAV 116,185,83
orl-mus TDLo:572 mg/kg/78W-I-C:CAR JNCIAM 42,1101,69
orl-mus LD50:139 mg/kg OYYAA2 2,76,68

CONSENSUS REPORTS: Reported in EPA TSCA Inventory. EPA Genetic Toxicology Program.

SAFETY PROFILE: Poison by ingestion. Questionable carcinogen with experimental carcinogenic data. Mutation data reported. Combustible when exposed to heat or flame; can react with oxidizing materials. To fight fire, use foam, CO_2, dry chemical. When heated to decomposition it emits toxic fumes such as NO_x.

HHD000 CAS:56393-22-7 ***HR: 3***
3-HYDRAZINO-6-((2-HYDROXYPROPYL)METHYLAMINO)PYRIDAZINE DIHYDROCHLORIDE
mf: $C_8H_{15}N_5O•2ClH$ mw: 270.20

SYNS: 3-IDRAZINO-6-(N-(2-IDROSSIPROPIL)METILAMINO)PIRIDAZINA DICLORIDRATO (ITALIAN) ◇ PROPILDAZINA (ITALIAN)

TOXICITY DATA with REFERENCE
orl-rat LD50:1230 mg/kg PLRCAT 8,295,76
ipr-rat LD50:355 mg/kg PLRCAT 8,295,76
orl-mus LD50:1170 mg/kg PLRCAT 8,295,76
ipr-mus LD50:600 mg/kg FRPSAX 34,299,79

SAFETY PROFILE: Poison by intraperitoneal route. Moderately toxic by ingestion. When heated to decomposition it emits very toxic fumes of NO_x and HCl.

HHD500 CAS:26049-68-3 ***HR: 3***
2-HYDRAZINO-4-(5-NITRO-2-FURYL)THIAZOLE
mf: $C_7H_6N_4O_3S$ mw: 226.23

SYNS: HNT ◇ 2-HYDRAZINO-4-(5-NITRO-2-FURANYL)THIAZOLE

TOXICITY DATA with REFERENCE
mmo-esc 10 μg/plate MUREAV 26,3,74
pic-esc 100 μg/L MUREAV 26,3,74
orl-rat TDLo:14 g/kg/44W-C:CAR PAACA3 10,15,69
orl-mus TDLo:38 g/kg/46W-C:NEO CNREA8 33,1593,73
orl-rat TD:21 g/kg/46W-C:CAR CNREA8 30,897,70

CONSENSUS REPORTS: EPA Genetic Toxicology Program.

SAFETY PROFILE: Questionable carcinogen with experimental carcinogenic and neoplastigenic data. Mutation data reported. When heated to decomposition it emits very toxic fumes of NO_x and SO_x.

HHE000 CAS:26049-70-7 ***HR: 3***
2-HYDRAZINO-4-(4-NITROPHENYL)THIAZOLE
mf: $C_9H_8N_4O_2S$ mw: 236.27

TOXICITY DATA with REFERENCE
orl-rat TDLo:2700 mg/kg/46W-C:CAR JNCIAM 51,403,73
orl-mus TDLo:8000 mg/kg/46W-C:ETA CNREA8 33,1593,73
orl-rat TD:37 g/kg/46W-C:CAR CNREA8 30,897,70

SAFETY PROFILE: Questionable carcinogen with experimental carcinogenic and tumorigenic data. When heated to decomposition it emits very toxic fumes of NO_x and SO_x. See also NITRO COMPOUNDS of AROMATIC HYDROCARBONS.

HHE500 CAS:56173-18-3 ***HR: D***
1-HYDRAZINOPHTHALAZINE ACETONE HYDRAZONE
mf: $C_{11}H_{12}N_4$ mw: 200.27

SYN: 1-PHTHLAZINYLHYDRAZONEACETONE

TOXICITY DATA with REFERENCE
mmo-sat 500 μg/plate MUREAV 68,79,79
mma-sat 500 μg/plate MUREAV 68,79,79
dnr-esc 200 μg/disc MUREAV 68,79,79

SAFETY PROFILE: Mutation data reported. When heated to decomposition it emits toxic fumes of NO_x.

HHF500 CAS:63981-09-9 ***HR: 3***
4-HYDRAZINO-2-THIOURACIL
mf: $C_4H_6N_4S$ mw: 142.20

SYN: 2-THIO-4-HYDRAZINOURACIL

TOXICITY DATA with REFERENCE
ipr-rat LD50:360 mg/kg NEOLA4 22,255,75
ipr-mus LD50:340 mg/kg NEOLA4 22,255,75

SAFETY PROFILE: A poison by intraperitoneal route. When heated to decomposition it emits very toxic NO_x and SO_x.

HHG000 CAS:122-66-7 ***HR: 3***
HYDRAZOBENZENE
mf: $C_{12}H_{12}N_2$ mw: 184.26

PROP: Light or yellow crystals from ethanol. D: 1.58, mp: 131°, bp: decomp. Very sltly sol in water; insol in acetylene.

SYNS: N,N'-BIANILINE ◇ sym-DIPHENYLHYDRAZINE ◇ 1,2-DIPHENYLHYDRAZINE ◇ HYDRAZOBENZEN (CZECH) ◇ HYDRAZODIBENZENE ◇ NCI-C01854 ◇ RCRA WASTE NUMBER U109

TOXICITY DATA with REFERENCE
mma-sat 10 ng/plate ENMUDM 7(Suppl 5),1,85
dni-mus-ipr 100 mg/kg MUREAV 46,305,77
orl-rat TDLo:2620 mg/kg/78W-C:CAR NCITR* NCI-CG-TR-92,78
scu-rat TDLo:6 g/kg/27W-I:ETA,REP VOONAW 20(4),53,74
orl-mus TDLo:26 g/kg/78W-C:CAR NCITR* NCI-CG-TR-92,78
skn-mus TDLo:5280 mg/kg/26W-I:ETA VOONAW 20(4),53,74
orl-rat TD:36 g/kg/53W-I:ETA,TER VOONAW 20(4),53,74
orl-rat LD50:301 mg/kg NCIMR* NIH-71-E-2144,73

CONSENSUS REPORTS: NTP Fifth Annual Report on Carcinogens. NCI Carcinogenesis Bioassay (feed); Clear Evidence: mouse, rat NCITR* NCI-CG-TR-92,78. Community Right-To-Know List. Reported in EPA TSCA Inventory.

SAFETY PROFILE: Confirmed carcinogen with experimental carcinogenic and tumorigenic data. Poison by ingestion. Experimental reproductive effects. Mutation data reported. When heated to decomposition it emits toxic fumes of NO_x.

HHG500 CAS:7782-79-8 ***HR: 3***
HYDRAZOIC ACID
mf: HN_3 mw: 43.04

PROP: Colorless liquid, very sol in water, intolerable pungent odor. Mp: −80°, bp: 37°, d: 1.09 @ 25°/4°.

SYNS: AZOIMIDE ◇ DIAZOIMIDE ◇ HYDROGEN AZIDE ◇ HYDRONITRIC ACID ◇ STICKSTOFFWASSERSTOFFSAEURE (GERMAN) ◇ TRIAZOIC ACID

TOXICITY DATA with REFERENCE
ihl-hmn TCLo:300 ppb:CNS,BRN,CVS JIHTAB 30,98,48
ihl-rat LCLo:1100 ppm/1H PHRPA6 58,607,43
ipr-mus LD50:22 mg/kg JIHTAB 30,98,48

CONSENSUS REPORTS: Reported in EPA TSCA Inventory. EPA Genetic Toxicology Program.

ACGIH TLV: CL 0.1 ppm (vapor)
DFG MAK: 0.1 ppm (0.27 mg/m^3)

SAFETY PROFILE: Poison by intraperitoneal route. Mildly toxic by inhalation. A severe irritant to skin, eyes, and mucous membranes. Continued inhalation causes central nervous system problems in humans (changes in EEG, somnolence, cough, headache, change in heart rate, fall in blood pressure, collapse, chills, and fever). High concentrations can cause fatal convulsions. Chronic exposure has been reported to cause injury to kidneys and spleen, hypotension, palpitation, ataxia, weakness. A dangerously sensitive explosive hazard when shocked or exposed to heat. Reacts with heavy metals to form very unstable heavy metal azides. Reacts violently with Cd, Cu, Ni, HNO_3, F_2. When heated to decomposition it emits toxic fumes of NO_x. See also AZIDES.

HHH000 CAS:13529-51-6 ***HR: 3***
2,2'-HYDRAZONODIETHANOL
mf: $C_4H_{12}N_2O_2$ mw: 120.18

SYNS: 1,1-BIS(2-HYDROXYETHYL)HYDRAZINE ◇ DEH ◇ 1,1-DIETHANOLHYDRAZINE

TOXICITY DATA with REFERENCE
scu-ham TDLo:742 mg/kg:ETA CALEDQ 4,55,77
scu-ham LD50:70 mg/kg CALEDQ 4,55,77

SAFETY PROFILE: A poison by subcutaneous route. Questionable carcinogen with experimental tumorigenic data. When heated to decomposition it emits toxic fumes such as NO_x.

HHH100 CAS:67255-31-6 ***HR: 2***
HYDREL
mf: $C_4H_{14}ClN_4OP$ mw: 200.64

SYN: p-(2-CHLOROETHYL)-2,2'-DIMETHYL PHOSPHONIC DIHYDRAZIDE (9CI)

TOXICITY DATA with REFERENCE
orl-rat LD50:2313 mg/kg FATOAO 47(2),57,84

SAFETY PROFILE: Moderately toxic by ingestion. When heated to decomposition it emits toxic fumes of Cl^-, PO_x, and NO_x.

HHH500 ***HR: 3***
HYDRIDES

SAFETY PROFILE: Variable toxicity. The highly toxic hydrides of phosphorus, arsenic, sulfur, selenium, tellurium, and boron produce local irritations and destroy red blood cells. They are particularly dangerous because of their volatility and ease of entry into the body. The hydrides of the alkali metals, alkaline earths, aluminum, zirconium, and titanium react with moisture to evolve hydrogen and leave behind the hydroxide of the metallic element which is usually caustic. See also SODIUM HY-

DROXIDE. The primary metallic hydrides include those of calcium, lithium, magnesium, potassium, sodium, and strontium. In the presence of moisture, they are readily converted to hydroxides which are highly irritating to the skin by caustic and thermal action. Similar effects can occur on contact with the eyes and respiratory mucous membranes. The volatile hydrides are flammable, some spontaneously so in air. All hydrides react violently on contact with powerful oxidizing agents. When heated or on contact with moisture or acids, an exothermic reaction evolving hydrogen occurs. Often, enough heat is evolved to cause ignition. Hydrides require special handling instructions which should be obtained from the manufacturers. The volatile hydrides (such as hydrides of boron, arsenic, phosphorus, selenium, tellurium) form explosive mixtures with air. The nonvolatile hydrides (such as sodium, lithium, calcium) readily liberate hydrogen when heated or on contact with moisture or acids. Furthermore, hydrides form dust clouds which can explode upon contact with flames, sparks, heat, or oxidizers. Highly dangerous; when heated, they can ignite at once or liberate explosive hydrogen. They react with moisture or acids to evolve heat and hydrogen. Violent reaction on contact with powerful oxidizers.

HHI000 CAS:5950-69-6 ***HR: 3***
HYDRINDANTIN, anhydrous
mf: $C_{18}H_{14}O_4$ mw: 294.32

SYN: 2,2',3,3,3',3'-HEXAHYDROXY(2,2'-BI INDAN)-1,1'-DIONE

TOXICITY DATA with REFERENCE
ivn-mus LD50:320 mg/kg CSLNX* NX#00810

CONSENSUS REPORTS: Reported in EPA TSCA Inventory.

SAFETY PROFILE: Poison by intravenous route. When heated to decomposition it emits acrid smoke and fumes.

HHI500 CAS:10034-85-2 ***HR: 3***
HYDRIODIC ACID
DOT: UN 1787/UN 2197
mf: HI mw: 127.91

PROP: Colorless when freshly made, but rapidly turns yellowish or brown on exposure to light or air. Keep protected from light and air, preferably not above 3°. Misc with water and alc. Mp: −50.8°, bp: −35.38° @ 5 atm, d: 5.66 g/L @ 0°.

SYNS: HYDRIODIC ACID, solution (DOT) ◇ HYDROGEN IODIDE ◇ HYDROGEN IODIDE, anhydrous (DOT) ◇ HYDROGEN IODIDE solution (DOT)

CONSENSUS REPORTS: Reported in EPA TSCA Inventory.

DOT Classification: Corrosive Material; Label: Corrosive; IMO: Nonflammable Gas; Label: Nonflammable Gas, Corrosive.

SAFETY PROFILE: Poison by ingestion and inhalation. A corrosive and poisonous irritant to skin, eyes, and mucous membranes. Explodes on contact with ethyl hydroperoxide. Ignites on contact with magnesium; perchloric acid; potassium + heat; potassium chlorate + heat; oxidants (e.g., fluorine; dinitrogen trioxide; dinitrogen tetraoxide; fuming nitric acid). Violent reaction with $HClO_4$ + Mg; O_3; metals. Potentially violent reaction with phosphorus. Reacts with water or steam to produce toxic and corrosive fumes. When heated to decomposition it emits highly toxic fumes of I^-. See also IODIDES.

HHJ000 CAS:10035-10-6 ***HR: 3***
HYDROBROMIC ACID
DOT: UN 1048/UN 1788
mf: BrH mw: 80.92

PROP: Colorless gas or pale yellow liquid. Mp: −87°, bp: −66.5°, d: 3.50 g/L @ 0°. Misc with water, alc. Keep protected from light.

SYNS: ACIDE BROMHYDRIQUE (FRENCH) ◇ ACIDO BROMIDRICO (ITALIAN) ◇ BROMOWODOR (POLISH) ◇ BROMWASSERSTOFF (GERMAN) ◇ BROOMWATERSTOF (DUTCH) ◇ HYDROBROMIC ACID, anhydrous (DOT) ◇ HYDROGEN BROMIDE (OSHA ACGIH, MAK, DOT)

TOXICITY DATA with REFERENCE
ihl-rat LC50:2858 ppm/1H NTIS** PB214-270
ihl-mus LC50:814 ppm/1H NTIS** PB214-270

CONSENSUS REPORTS: Reported in EPA TSCA Inventory.

OSHA PEL: (Transitional: TWA 3 ppm) CL 3 ppm
ACGIH TLV: CL 3 ppm
DFG MAK: 5 ppm (17 mg/m^3)
DOT Classification: Corrosive Material; Label: Corrosive (UN1788); Nonflammable Gas; Label: Nonflammable Gas (UN1048); Poison A; Label: Poison Gas, Corrosive (UN1048)

SAFETY PROFILE: A poison gas. A corrosive irritant to the eyes, skin, and mucous membranes. Reacts violently with F_2, Fe_2O_3, NH_3, O_3. When heated to decomposition or in reaction with water or steam it emits toxic and corrosive fumes of Br^- and HBr. See also BROMIDES.

HHJ500 ***HR: 3***
HYDROCARBON GAS
DOT: UN 1023/UN 1964/UN 1965

PROP: Contains hydrogen, methane, carbon monoxide, lel: 5.3%, uel: 31%, autoign temp: 1200°F.

SYNS: COAL GAS ◇ HYDROCARBON GAS, compressed (DOT) ◇ HYDROCARBON GAS, liquefied (DOT) ◇ HYDROCARBON GAS, nonliquefied (DOT)

DOT Classification: Flammable Gas; Label: Flammable Gas; IMO: Poison A; Label: Poison Gas and Flammable Gas.

SAFETY PROFILE: A poison by inhalation. Very dangerous fire hazard when exposed to heat or flame; can react vigorously with oxidizing materials. Moderately explosive when exposed to heat or flame. To fight fire, stop flow of gas; CO_2, dry chemical, or water spray. See also CARBON MONOXIDE, HYDROGEN, and METHANE.

HHK000 CAS:9034-34-8 ***HR: 3***
HYDROCELLULOSE

SYN: REGENERATED CELLULOSE

TOXICITY DATA with REFERENCE
imp-rat TDLo:2 film disc/rat:NEO ZENBAX 7B,353,52

SAFETY PROFILE: Questionable carcinogen with experimental neoplastigenic data. When heated to decomposition it emits acrid smoke and fumes.

HHK100 ***HR: 2***
HYDROCHLORBENZETHYLAMINE DIMALEATE
mf: $C_{23}H_{31}ClN_2O_3{\cdot}2C_4H_4O_4$ mw: 651.17

SYNS: 2-(2-(2-(4-(p-CHLORO-α-PHENYLBENZYL)-1-PIPERAZINYL) ETHOXY)ETHOXY)ETHANOL DIMALEATE ◇ ETHANOL, 2-(2-(2-(4-(p-CHLORO-α-PHENYLBENZYL)-1-PIPERAZINYL)ETHOXY)ETHOXY)-, DIMALEATE ◇ ETODROXIZINE DIMALEATE ◇ ETODROXYZINE DIMALEATE

TOXICITY DATA with REFERENCE
orl-rat TDLo:1 g/kg (female 8-15D post):REP PSDTAP 9,134,68
orl-rat TDLo:1 g/kg (female 8-15D post):TER PSDTAP 9,134,68
orl-rat LD50:920 mg/kg PSDTAP 9,134,68

SAFETY PROFILE: Moderately toxic by ingestion. An experimental teratogen. Experimental reproductive effects. When heated to decomposition it emits toxic fumes of NO_x and Cl^-.

HHL000 CAS:7647-01-0 ***HR: 3***
HYDROCHLORIC ACID
DOT: UN 1050/UN 1789/UN 2186
mf: ClH mw: 36.46

PROP: Colorless, fuming gas or colorless, fuming liquid; strongly corrosive with pungent odor. Mp: −114.3°, bp: −84.8°, d: 1.639 g/L (gas) @ 0°, 1.194 @ −26° (liquid), vap press: 4.0 atm @ 17.8°. Misc with water, alc.

SYNS: ACIDE CHLORHYDRIQUE (FRENCH) ◇ ACIDO CLORIDRICO (ITALIAN) ◇ CHLOORWATERSTOF (DUTCH) ◇ CHLOROHYDRIC ACID ◇ CHLOROWODOR (POLISH) ◇ CHLORWASSERSTOFF (GERMAN) ◇ HYDROCHLORIC ACID, anhydrous (DOT) ◇ HYDROCHLORIC ACID, solution, inhibited (DOT) ◇ HYDROCHLORIDE ◇ HYDROGEN CHLORIDE (OSHA, ACGIH, MAK, DOT) ◇ HYDROGEN CHLORIDE, anhydrous (DOT) ◇ HYDROGEN CHLORIDE, refrigerated liquid (DOT) ◇ MURIATIC ACID (DOT) ◇ SPIRITS of SALT (DOT)

TOXICITY DATA with REFERENCE
eye-rbt 100 mg rns MLD TXCYAC 23,281,82
dnr-esc 25 μg/well ENMUDM 3,429,81
cyt-grh-par 20 mg NULSAK 9,119,66
ihl-rat TCLo:450 mg/m³/1H (1D pre):TER AKGIAO 53(6),69,77
ihl-hmn LCLo:1300 ppm/30M 29ZWAE -,207,68
ihl-hmn LCLo:3000 ppm/5M TABIA2 3,231,33
unr-man LDLo:81 mg/kg 85DCAI 2,73,70
ihl-rat LC50:3124 ppm/1H AMRL** TR-74-78,74
ihl-mus LC50:1108 ppm/1H JCTODH 3,61,76
ipr-mus LD50:1449 mg/kg COREAF 256,1043,63
orl-rbt LD50:900 mg/kg BIZEA2 134,437,23
ihl-rbt LCLo:4416 ppm/30M JIHTAB 24,222,42

CONSENSUS REPORTS: EPA Extremely Hazardous Substances List. Community Right-To-Know List. Reported in EPA TSCA Inventory. EPA Genetic Toxicology Program.

OSHA PEL: CL 5 ppm
ACGIH TLV: CL 5 ppm
DFG MAK: 5 ppm (7 mg/m³)
DOT Classification: Nonflammable Gas; Label: Nonflammable Gas (UN1050, UN2186); Corrosive Material; Label: Corrosive (NA1789, UN1789); Flammable Gas; Label: Nonflammable Gas, Corrosive (UN1050)

SAFETY PROFILE: A human poison by an unspecified route. Mildly toxic to humans by inhalation. Moderately toxic experimentally by ingestion. A corrosive irritant to the skin, eyes, and mucous membranes. Mutation data reported. An experimental teratogen. A concentration of 35 ppm causes irritation of the throat after short exposure. In general, hydrochloric acid causes little trouble in industry other than from accidental splashes and burns. It is a common air contaminant and is heavily used in industry.

Nonflammable Gas. Explosive reaction with alcohols + hydrogen cyanide; potassium permanganate; sodium; tetraselenium tetranitride. Ignition on contact with fluorine; hexalithium disilicide; metal acetylides or carbides (e.g., cesium acetylide; rubidium acetylide). Violent reactions with acetic anhydride; 2-amino ethanol; NH_4OH; Ca_3P_2; chlorosulfonic acid; 1,1-difluoroethylene; ethylene diamine; ethylene imine; oleum; $HClO_4$; β-propiolactone; propylene oxide; ($AgClO_4$ + CCl_4); NaOH; H_2SO_4; U_3P_4; vinyl acetate; CaC_2; CsC_2H; Cs_2C_2; Mg_3B_2; $HgSO_4$; RbC_2H; Rb_2C_2; Na. Vigorous reaction with aluminum; chlorine + dinitroanilines (evolves gas). Potentially dangerous reaction with sulfuric acid releases HCl gas. When heated to decomposition it emits toxic fumes of Cl^-. See also HYDROGEN CHLORIDE.

HHL500 ***HR: 3***
HYDROCHLORIC ACID (mixture)
DOT: UN 1789

SYN: HYDROCHLORIC ACID MIXTURE (DOT)

DOT Classification: Corrosive Material; Label: Corrosive.

SAFETY PROFILE: A highly corrosive irritant to skin, eyes and mucous membranes. When heated to decomposition it emits very toxic fumes of HCl and Cl^-. See also HYDROCHLORIC ACID.

HHM000 CAS:8007-56-5 ***HR: 3***
HYDROCHLORIC ACID, mixed with NITRIC ACID (3:1)
DOT: UN 1798
mf: $ClH \cdot HNO_3$ mw: 99.48

PROP: Yellow, fuming, corrosive, volatile liquid; suffocating odor. Misc with water.

SYNS: AQUA REGIA ◇ NITROHYDROCHLORIC ACID (DOT) ◇ NITROHYDROCHLORIC ACID, diluted (DOT) ◇ NITROMURIATIC ACID (DOT)

DOT Classification: Corrosive Material; Label: Corrosive

SAFETY PROFILE: A corrosive irritant to the eyes, skin, and mucous membranes. When heated to decomposition it emits very toxic HCl, HNO_3, Cl^-, and NO_x. See also HYDROCHLORIC ACID, NITRIC ACID, and NITROSYL CHLORIDE.

HHM500 CAS:455-80-1 ***HR: 3***
HYDROCHLORIDE of DI-n-BUTYLAMINOPROPYL-3-IODO-4-FLUOROBENZOATE
mf: $C_{18}H_{27}FINO_2 \cdot ClH$ mw: 471.82

SYN: 4-FLUORO-3-IODOBENZOICACID-3-(DIBUTYLAMINO)PROPYL ESTER, HYDROCHLORIDE

TOXICITY DATA with REFERENCE
scu-mus LD50:3000 mg/kg JAPMA8 39,4,50
ivn-mus LDLo:55 mg/kg JAPMA8 39,4,50

SAFETY PROFILE: Poison by intravenous route. Moderately toxic by subcutaneous route. When heated to decomposition it emits very toxic fumes of F^-, NO_x, I^-, and HCl.

HHP000 CAS:104-53-0 ***HR: 3***
HYDROCINNAMALDEHYDE
mf: $C_9H_{10}O$ mw: 134.19

PROP: Colorless to sltly yellow liquid; strong floral, hyacinth odor. Bp: 221-224°, d: 1.010-1.020, refr index: 1.520-1.532, flash p: 203°F. Misc with alc, ether; insol in water.

SYNS: BENZENEPROPANAL ◇ BENZYLACETALDEHYDE ◇ DIHYDROCINNAMALDEHYDE ◇ FEMA No. 2887 ◇ HYDROCINNAMIC ALDEHYDE ◇ 3-PHENYLPROPANAL ◇ 3-PHENYL-1-PROPANAL ◇ 3-PHENYLPROPIONALDEHYDE (FCC) ◇ β-PHENYLPROPIONALDEHYDE ◇ 3-PHENYLPROPYL ALDEHYDE

TOXICITY DATA with REFERENCE
skn-hmn 100 % FCTXAV 12,967,74
ivn-mus LD50:56 mg/kg CSLNX* NX#05219

CONSENSUS REPORTS: Reported in EPA TSCA Inventory.

SAFETY PROFILE: A poison by intravenous route. A human skin irritant. Combustible liquid. When heated to decomposition it emits acrid smoke and fumes. See also ALDEHYDES.

HHP050 CAS:122-97-4 ***HR: 2***
HYDROCINNAMIC ALCOHOL
mf: $C_9H_{12}O$ mw: 136.21

PROP: Colorless sltly viscous liquid; sweet, hyacinth-mignonette odor. D:.998-1.002, refr index: 1.524-1.528, flash p: 228°F. Sol in fixed oils, propylene glycol; insol in glycerin.

SYNS: 3-BENZENEPROPANOL ◇ FEMA No. 2885 ◇ HYDROCINNAMYL ALCOHOL ◇ (3-HYDROXYPROPYL)BENZENE ◇ Γ-PHENYLPROPANOL ◇ 3-PHENYLPROPANOL ◇ 3-PHENYL-1-PROPANOL (FCC) ◇ PHENYLPROPYL ALCOHOL ◇ Γ-PHENYLPROPYL ALCOHOL ◇ 3-PHENYLPROPYL ALCOHOL

TOXICITY DATA with REFERENCE
skn-rbt 500 mg/24H MOD FCTXAV 17,893,79
orl-rat LD50:2300 mg/kg FCTXAV 17,893,79
skn-rbt LD50:5000 mg/kg FCTXAV 17,893,79

CONSENSUS REPORTS: Reported in EPA TSCA Inventory.

SAFETY PROFILE: Moderately toxic by ingestion. Mildly toxic by skin contact. A skin irritant. Combustible liquid. When heated to decomposition it emits toxic fumes. See also ALCOHOLS.

HHP100 CAS:645-59-0 ***HR: 3***
HYDROCINNAMONITRILE
mf: C_9H_9N mw: 131.19

SYNS: BENZENEPROPANENITRILE (9CI) ◇ BENZENEPROPIONITRILE ◇ (2-CYANOETHYL)BENZENE ◇ HYDROCINNAMIQUE NITRILE (FRENCH) ◇ PHENETHYL CYANIDE ◇ 2-PHENYLETHYL CYANIDE ◇ 3-PHENYLPROPANENITRILE ◇ PHENYLPROPIONITRILE ◇ β-PHENYLPROPIONITRILE ◇ 3-PHENYLPROPIONITRILE

TOXICITY DATA with REFERENCE
orl-mus LD50:116 mg/kg ARTODN 55,47,84
ivn-rbt LDLo:39 mg/kg COREAF 153,895,11
scu-gpg LDLo:150 mg/kg COREAF 153,895,11

CONSENSUS REPORTS: Cyanide and its compounds are on the Community Right-To-Know List.

SAFETY PROFILE: Poison by ingestion, intravenous, and subcutaneous routes. When heated to decomposition it emits toxic fumes of CN^-. See also NITRILES.

HHP500 CAS:122-72-5 ***HR: 1***
HYDROCINNAMYL ACETATE
mf: $C_{11}H_{14}O_2$ mw: 178.25

PROP: Colorless liquid; spicy, floral odor. D: 1.012, refr index: 1.494, flash p: +212°F. Sol in alc; insol in water.

SYNS: FEMA No. 2890 ◇ 3-PHENYL-1-PROPANOL ACETATE ◇ PHENYLPROPYL ACETATE ◇ 3-PHENYLPROPYL ACETATE (FCC) ◇ 3-PHENYL-1-PROPYL ACETATE

TOXICITY DATA with REFERENCE
orl-rat LD50:4700 mg/kg FCTXAV 12,965,74

CONSENSUS REPORTS: Reported in EPA TSCA Inventory.

SAFETY PROFILE: Mildly toxic by ingestion. Combustible liquid. When heated to decomposition it emits acrid smoke and fumes.

HHQ000 CAS:104-64-3 ***HR: 1***
HYDROCINNAMYL FORMATE
mf: $C_{10}H_{12}O_2$ mw: 164.22

SYNS: PHENYLPROPYL FORMATE ◇ 3-PHENYL-1-PROPYL FORMATE

TOXICITY DATA with REFERENCE
skn-rbt 500 mg/24H MLD FCTXAV 14,659,76
orl-rat LD50:4090 mg/kg FCTXAV 14,659,76

CONSENSUS REPORTS: Reported in EPA TSCA Inventory.

SAFETY PROFILE: Mildly toxic by ingestion. A skin irritant. When heated to decomposition it emits acrid smoke and fumes. See also FORMIC ACID.

HHQ500 CAS:103-58-2 ***HR: 1***
HYDROCINNAMYL ISOBUTYRATE
mf: $C_{13}H_{18}O_2$ mw: 206.31

SYNS: ISOBUTYRIC ACID-3-PHENYLPROPYL ESTER ◇ 3-PHENYLPROPYL ISOBUTYRATE

TOXICITY DATA with REFERENCE
skn-rbt 500 mg/24H FCTXAV 16,851,78

CONSENSUS REPORTS: Reported in EPA TSCA Inventory.

SAFETY PROFILE: A skin skin irritant. When heated to decomposition it emits acrid smoke and fumes. See also ESTERS.

HHQ550 CAS:122-74-7 ***HR: 1***
HYDROCINNAMYL PROPIONATE
mf: $C_{12}H_{16}O_2$ mw: 192.28

SYNS: BENZENEPROPANOL, PROPANOATE (9CI) ◇ PHENYLPROPYL PROPIONATE ◇ β-PHENYLPROPYL PROPIONATE ◇ 3-PHENYLPROPYL PROPIONATE ◇ 1-PROPANOL, 3-PHENYL-, PROPIONATE

TOXICITY DATA with REFERENCE
skn-rbt 500 mg/24H MOD FCTOD7 20,809,82

CONSENSUS REPORTS: Reported in EPA TSCA Inventory.

SAFETY PROFILE: A skin irritant. When heated to decomposition it emits acrid smoke and irritating fumes.

HHQ800 CAS:50-03-3 ***HR: 3***
HYDROCORTISONE-21-ACETATE
mf: $C_{23}H_{32}O_6$ mw: 404.55

PROP: Monoclinic, sphenoidal, tabular crystals from dil acetone; tasteless, somewhat hygros. D: 1.289, decomp @ 223°. Solubility in water: 1 mg/100 mL; in ethanol: 0.45 g/100 mL; in methanol: 3.9 mg/mL; in acetone: 1.1 mg/g; in ether: 0.15 mg/mL. One gram dissolves in about 200 mL chloroform. Very sol in DMF. Also sol in dioxane.

SYNS: ABBOCORT ◇ ACETATE-AS ◇ ACETO-CORT ◇ 21-ACETOXY-11-β,17-α-DIHYDROXYPREGN-4-ENE-3,20-DIONE ◇ (11-β)-21-(ACETOXY)-11,17-DIHYDROXY-PREGN-4-ENE-3,20-DIONE ◇ (11-β)-21-(ACETYLOXY)-11,17-DIHYDROXY-PREGN-4-ENE-3,20-DIONE(9CI) ◇ BAMBICORT ◇ BERLISON F ◇ BIOCORTAR ◇ CARMOL HC ◇ CHEMYSONE ◇ COLLUSUL-HC ◇ COMPOUND F ACETATE ◇ CORTACREAM ◇ CORTAID ◇ CORTEF ACETATE ◇ CORTELL ◇ CORTES ◇ CORTIFOAM ◇ CORTISOL ACETATE ◇ CORTRIL ACETATE ◇ CORTRIL ACETATE-AS ◇ 11-β,17-α-DIHYDROXY-21-ACETOXYPREGESTERONE ◇ EPIFOAM ◇ EYE-CORT ◇ FERNISONE ◇ HA ◇ HCA ◇ HYCORTOLE ACETATE ◇ HYDRIN-2 ◇ 17-HYDROXYCORTICOSTERONE 21-ACETATE ◇ 17-α-HYDROXYCORTICOSTERONE ACETATE ◇ HYSONE-A ◇ ISOPTO-HYDROCORTISONE

◇ LANACORT ◇ MYSONE ◇ NSC 741 ◇ PABRACORT ◇ 11-β,17,21-TRI-HYDROXY-PREGN-4-ENE-3,20-DIONE21-ACETATE

TOXICITY DATA with REFERENCE
dnd-mus-ipr 60 mg/kg OFAJAE 51,1,74
oms-mus-ipr 150 mg/kg BEXBAN 77,437,74
ims-mus TDLo:1200 mg/kg (female 10-21D post):REP NATUAS 169,665,52
orl-rat TDLo:1 g/kg (6-15D preg):TER BCFAAI 119,391,80
scu-rat LDLo:250 mg/kg ARZNAD 27,2102,77
ipr-mus LD50:2300 mg/kg CLDND* JAN86

SAFETY PROFILE: Poison by subcutaneous route. Moderately toxic by intraperitoneal route. Experimental teratogenic and reproductive effects. Mutation data reported. A steroid. When heated to decomposition it emits acrid smoke and fumes. See also other hydrocortisone entries.

HHQ825 CAS:13609-67-1 ***HR: 2***
HYDROCORTISONE-17-BUTYRATE
mf: $C_{25}H_{36}O_6$ mw: 432.61

SYNS: (11-β)-11,21-DIHYDROXY-17-(1-OXOBUTOXY)-PREGN-4-ENE-3,20-DIONE ◇ HB[17] ◇ H.17B ◇ HYDROCORTISONE BUTYRATE ◇ HYDROCORTISONE 17-α-BUYTRATE ◇ LOCOID

TOXICITY DATA with REFERENCE
skn-rat TDLo:16500 μg/kg (7-17D preg):REP YACHDS 9,3045,81
scu-rat TDLo:54 mg/kg (female 9-14D post):TER OYYAA2 8,1035,74
ipr-mus LD50:1550 mg/kg NIIRDN 6,628,82
scu-mus LD50:2150 mg/kg JTSCDR 6(Suppl),1,81

SAFETY PROFILE: Moderately toxic by subcutaneous and intraperitoneal routes. Experimental teratogenic and reproductive effects. A steroid. When heated to decomposition it emits acrid smoke and fumes. See also other hydrocortisone entries.

HHQ850 CAS:72590-77-3 ***HR: 2***
HYDROCORTISONE-17-BUTYRATE-21-PROPIONATE
mf: $C_{24}H_{40}O_7$ mw: 488.68

SYNS: 17-BUTYRLOXY-11-β-HYDROXY-21-PROPIONYLOXY-4-PREGNENE-3,20-DIONE ◇ HBP ◇ HYDROCORTISONE BUYTRATE PROPIONATE ◇ (11-β)-11-HYDROXY-17-(1-OXOBUTOXY)-21-(1-OXOPROPOXY)-PREGN-4-ENE-3,20-DIONE ◇ 11-β,17,21-TRIHYDROXY-PREGN-4-ENE-3,30-DIONE 17-BUTYRATE, 21-PROPIONATE

TOXICITY DATA with REFERENCE
skn-rbt 1000 ppm MLD YACHDS 9,3023,81
skn-rat TDLo:82500 μg/kg (7-17D preg):REP YACHDS 9,3045,81
scu-rbt TDLo:6500 μg/kg (female 6-18D post):TER OYYAA2 21,483,81
orl-rat LD50:5120 mg/kg JTSDCR 6(Suppl),1,81
ipr-rat LD50:1420 mg/kg JTSCDR 6(Suppl),1,81
scu-rat LD50:3260 mg/kg JTSCDR 6(Suppl),1,81
orl-mus LD50:6720 mg/kg JTSCDR 6(Suppl),1,81
ipr-mus LD50:1660 mg/kg JTSCDR 6(Suppl),1,81
scu-mus LD50:1980 mg/kg JTSCDR 6(Suppl),1,81

SAFETY PROFILE: Moderately toxic by subcutaneous and intraperitoneal routes. Mildly toxic by ingestion. Experimental reproductive effects. An experimental teratogen. A skin irritant. A steroid. When heated to decomposition it emits toxic fumes. See also other hydrocortisone entries.

HHQ875 CAS:3863-59-0 ***HR: D***
HYDROCORTISONE-21-PHOSPHATE
mf: $C_{21}H_{31}O_8P$ mw: 442.49

SYNS: CORPHOS ◇ CORTIPHATE INJECTABLE ◇ CORTISOL PHOSPHATE ◇ CORTISOL-21-PHOSPHATE ◇ (11-β)-11,17-DIHYDROXY-21-(PHOSPHONOOXY)-PREGN-4-ENE-3,20-DIONE(9CI) ◇ HYDROCORTISONE PHOSPHATE ◇ 21-HYDROCORTISONEPHOSPHORIC ACID

TOXICITY DATA with REFERENCE
ivn-rbt TDLo:900 mg/kg (female 8-16D post):TER TJADAB 12,195,75
ivn-rbt TDLo:1800 mg/kg (8-16D preg):REP TJADAB 12,195,72

SAFETY PROFILE: Experimental teratogenic and reproductive effects. A steroid. When heated to decomposition it emits toxic fumes of PO_x. See also other hydrocortisone entries.

HHR000 CAS:125-04-2 ***HR: 2***
HYDROCORTISONE SODIUM SUCCINATE
mf: $C_{25}H_{35}O_9 \cdot Na$ mw: 502.59

PROP: White, odorless, hygroscopic, amorph solid. Mp: 169-171°. Very sol in water and alc; insol in chloroform; very sltly sol in acetone.

SYNS: A-HYDROCORT ◇ BUCCALSONE ◇ CORLAN ◇ CORTISOL HEMISUCCINATE SODIUM SALT ◇ CORTISOL SODIUM HEMISUCCINATE ◇ CORTISOL SODIUM SUCCINATE ◇ CORTISOL-21-SODIUM SUCCINATE ◇ CORTISOL SUCCINATE, SODIUM SALT ◇ ELCORTELAN SOLUBLE ◇ EMI-CORLIN ◇ FLEBOCORTID ◇ HYCORACE ◇ HYDROCORTISONE-21-SODIUM SUCCINATE ◇ 21-(HYDROGEN SUCCINATE)CORTISOL, MONOSODIUM SALT ◇ INTRACORT ◇ NORDICORT ◇ ORALSONE ◇ SODIUM HYDROCORTISONE SUCCINATE ◇ SODIUM HYDROCORTISONE-21-SUCCINATE ◇ SOLU-CORTEF ◇ SOLU-GLYC ◇ U 4905

TOXICITY DATA with REFERENCE
scu-mus TDLo:1200 mg/kg (female 7-12D post):REP KSRNAM 4,2957,70
ipr-rat TDLo:1800 mg/kg (9-14D preg):TER KSRNAM 4,2969,70

ipr-rat LD50:1320 mg/kg NIIRDN 6,625,82
ipr-mus LD50:1050 mg/kg NIIRDN 6,625,82

SAFETY PROFILE: Moderately toxic by intraperitoneal route. Experimental teratogenic and reproductive effects. When heated to decomposition it emits toxic fumes of Na_2O. See also other hydrocortisone entries.

HHR500 CAS:119-84-6 ***HR: 3***
HYDROCOUMARIN
mf: $C_9H_8O_2$ mw: 148.17

PROP: Colorless to pale yellow liquid; coconut odor. D: 1.186, refr index: 1.555, flash p: 266°F.

SYNS: 1,2-BENZODIHYDROPYRONE (FCC) ◇ 2-CHROMANONE ◇ DIHYDROCOUMARIN ◇ 3,4-DIHYDROCOUMARIN ◇ o-HYDROXYHYDROCINNAMIC ACID-Δ-LACTONE ◇ FEMA No. 2381 ◇ MELILOTIN ◇ MELILOTOL ◇ NCI-C55890 ◇ 2-OXOCHROMAN ◇ USAF DO-12

TOXICITY DATA with REFERENCE
skn-rbt 500 mg/24H MOD FCTXAV 12,521,74
skn-gpg 1%/48H MOD JSCCA5 28,357,77
orl-rat LD50:1460 mg/kg FCTXAV 2,327,64
ipr-mus LD50:200 mg/kg NTIS** AD277-689
orl-gpg LD50:1760 mg/kg FCTXAV 2,327,64

CONSENSUS REPORTS: Reported in EPA TSCA Inventory.

SAFETY PROFILE: A poison by intraperitoneal route. Moderately toxic by ingestion. A skin irritant. Combustible liquid. When heated to decomposition it emits acrid smoke and fumes.

HHR700 CAS:522-60-1 ***HR: 3***
HYDROCUPREINE ETHYL ETHER
mf: $C_{21}H_{28}N_2O_2$ mw: 340.51

PROP: White, bitter, crystalline powder. Mp: 123-128° when solvent-free. Practically insol in water; sol in alc, benzene, chloroform, ether, dil acids, oils, fats.

SYNS: ETHYLHYDROCUPREINE ◇ NUMOQUIN ◇ OPTOCHIN ◇ OPTOQUINE

TOXICITY DATA with REFERENCE
eye-rbt 1% OPHTAD 143,154,62
scu-mus LDLo:5000 mg/kg JPETAB 8,53,16
orl-mus LDLo:400 mg/kg HBAMAK 4,1289,35
scu-mus LDLo:24 mg/kg HBAMAK 4,1289,35

SAFETY PROFILE: Poison by ingestion and subcutaneous routes. An eye irritant. When heated to decomposition it emits toxic fumes of NO_x. See also ETHERS.

HHS000 CAS:74-90-8 ***HR: 3***
HYDROCYANIC ACID
DOT: NA 1051/UN 1613/UN 1614
mf: CHN mw: 27.03

PROP: Odor of bitter almonds. Mp: −13.2°, bp: 25.7°, lel: 5.6%, uel: 40%, flash p: 0°F (CC), d: 0.6876 @ 20°/4°, autoign temp: 1000°F, vap press: 400 mm @ 9.8°, vap d: 0.932. Misc in water, alc, and ether.

SYNS: ACIDE CYANHYDRIQUE (FRENCH) ◇ ACIDO CIANIDRICO (ITALIAN) ◇ AERO liquid HCN ◇ BLAUSAEURE (GERMAN) ◇ BLAUWZUUR (DUTCH) ◇ CYAANWATERSTOF (DUTCH) ◇ CYANWASSERSTOFF (GERMAN) ◇ CYCLON ◇ CYCLONE B ◇ CYJANOWODOR (POLISH) ◇ HCN ◇ HYDROCYANIC ACID, liquefied (DOT) ◇ HYDROCYANIC ACID (PRUSSIC), unstabilized (DOT) ◇ HYDROGEN CYANIDE (OSHA, ACGIH) ◇ HYDROGEN CYANIDE, anhydrous, stabilized (DOT) ◇ PRUSSIC ACID (DOT) ◇ PRUSSIC ACID, unstabilized ◇ RCRA WASTE NUMBER P063 ◇ ZACLON DISCOIDS

TOXICITY DATA with REFERENCE
orl-hmn LDLo:570 μg/kg PCOC** -,596,66
ihl-hmn LCLo:200 ppm/5M TABIA2 3,231,33
ihl-hmn LCLo:120 mg/m³/1H JIHTAB 24,255,42
ihl-hmn LCLo:200 mg/m³/10M WHOTAC -,30,70
ihl-man LCLo:400 mg/m³/2M 85GMAT -,75,82
scu-hmn LDLo:1 mg/kg SCJUAD 4,33,67
ivn-hmn LD50:1 mg/kg SCJUAD 4,33,67
ivn-man TDLo:55 μg/kg:PUL NTIS** PB158-508
unr-man LDLo:1471 μg/kg 85DCAI 2,73,70
ihl-rat LC50:484 ppm/5M TXAPA9 42,417,77
ivn-rat LD50:810 μg/kg NTIS** AD-A028-501
orl-mus LD50:3700 μg/kg APFRAD 19,740,61
ihl-mus LC50:323 ppm/5M TXAPA9 42,417,77
ipr-mus LD50:2990 μg/kg BJPCAL 23,455,64
scu-mus LDLo:3 mg/kg HBAMAK 4,1340,35
ivn-mus LD50:990 μg/kg NTIS** AD-A028-501
ims-mus LD50:2700 μg/kg BJPCAL 23,455,64
orl-dog LDLo:4 mg/kg HBAMAK 4,1340,35
ihl-dog LC50:616 mg/m³/1M NTIS** AD-A028-501
scu-dog LDLo:1700 μg/kg HBAMAK 4,1340,35
ivn-dog LD50:1340 μg/kg NTIS** AD-A028-501
ihl-mky LC50:1616 mg/m³/1M NTIS** AD-A028-501

CONSENSUS REPORTS: EPA Extremely Hazardous Substances List. Community Right-To-Know List. Reported in EPA TSCA Inventory.

OSHA PEL: (Transitional: TWA 10 ppm (skin)); STEL 4.7 ppm (skin)
ACGIH TLV: CL 10 ppm (skin)
DFG MAK: 10 ppm (11 mg/m³)
NIOSH REL: (Cyanide) CL 5 mg(CN)/m³/10M
DOT Classification: Poison A; Label: Poison Gas and Flammable Gas; IMO: Poison B; Label: Poison (UN 1614); IMO: Poison B; Label: Flammable Liquid and Poison; Forbidden, Unstabilized.

SAFETY PROFILE: A deadly human and experimental poison by all routes. Hydrocyanic acid and the cyanides are true protoplasmic poisons, combining in the tissues with the enzymes associated with cellular oxidation. They thereby render the oxygen unavailable to the tissues and cause death through asphyxia. The suspension of tissue oxidation lasts only while the cyanide is present; upon its removal, normal function is restored provided death has not already occurred. HCN does not combine easily with hemoglobin, but it does combine readily with methemoglobin to form cyanmethemoglobin. This property is utilized in the treatment of cyanide poisoning when an attempt is made to induce methemoglobin formation. The presence of cherry-red venous blood in cases of cyanide poisoning is due to the inability of the tissues to remove the oxygen from the blood. Exposure to concentrations of 100-200 ppm for periods of 30-60 minutes can cause death. In cases of acute cyanide poisoning death is extremely rapid, although sometimes breathing may continue for a few minutes. In less acute cases, there is cyanosis, headache, dizziness, unsteadiness of gait, a feeling of suffocation, and nausea. Where the patient recovers, there is rarely any disability.

Very dangerous fire hazard when exposed to heat, flame or oxidizers. Can polymerize explosively at 50-60C° or in the presence of traces of alkali. Severe explosion hazard when exposed to heat or flame or by chemical reaction with oxidizers. The anhydrous liquid is stabilized at or below room temperature by the addition of acid. The gas forms explosive mixtures with air. Reacts violently with acetaldehyde. To fight fire, use CO_2, non-alkaline dry chemical, foam. When heated to decomposition or in reaction with water, steam, acid or acid fumes it produces highly toxic fumes of CN^-. An insecticide. See also CYANIDE.

HHS500 ***HR: 3***
HYDROCYANIC ACID, mixed with CYANOGEN CHLORIDE (3:2 BY WT)

SYN: HCN-CNCl MIXTURE

TOXICITY DATA with REFERENCE
ihl-mus LC50:820 mg/m^3/2M NDRC** No.9-4-1-9,43

NIOSH REL: CL (Cyanide) 5 mg(CN)/m^3/10M

CONSENSUS REPORTS: Cyanide and its compounds are on the Community Right-To-Know List.

SAFETY PROFILE: A poison by all routes. When heated to decomposition it emits very toxic fumes of HCN, NO_x, Cl^-, and CN^-. See also HYDROCYANIC ACID and CYANOGEN CHLORIDE.

HHT000 ***HR: 3***
HYDROCYANIC ACID, SALTS
DOT: UN 1588/UN 1935

SYNS: CYANIDE or CYANIDE mixture, dry (DOT) ◇ CYANIDES ◇ CYANIDE, solution (DOT)

CONSENSUS REPORTS: Cyanide and its compounds are on the Community Right-To-Know List.

OSHA PEL: TWA 5 mg(CN)/m^3
ACGIH TLV: TWA 5 mg(CN)/m^3 (skin)
DFG MAK: 5 mg/m^3
NIOSH REL: (Cyanide) CL 5 mg/m^3/10M
DOT Classification: Poison B; Label: Poison, dry; Poison B; Label: Poison, solution.

SAFETY PROFILE: A deadly poison. When heated to decomposition it emits toxic fumes of CN^-. See also CYANIDE and SODIUM CYANIDE.

HHU000 CAS:74-90-8 ***HR: 3***
HYDROCYANIC ACID (unstabilized)

SYNS: HYDROCYANIC ACID, UNSTABILIZED (DOT) ◇ PRUSSIC ACID, UNSTABILIZED

CONSENSUS REPORTS: Reported in EPA TSCA Inventory.

NIOSH REL: CL 5 mg$(CN)_2$/m^3/10M
DOT Classification: Forbidden.

SAFETY PROFILE: A deadly poison to living tissue by all routes of exposure. A very dangerous storage hazard. See also HYDROCYANIC ACID.

HHU500 CAS:7664-39-3 ***HR: 3***
HYDROFLUORIC ACID
DOT: UN 1052/UN 1790
mf: FH mw: 20.01

PROP: Clear, colorless, fuming, corrosive liquid or gas. Mp: −83.1°, bp: 19.54°, d: 0.901 g/L (gas); 0.699 @ 22° (liquid), vap press: 400 mm @ 2.5°.

SYNS: ACIDE FLUORHYDRIQUE (FRENCH) ◇ ACIDO FLUORIDRICO (ITALIAN) ◇ FLUOROWODOR (POLISH) ◇ FLUORWASSERSTOFF (GERMAN) ◇ FLUORWATERSTOF (DUTCH) ◇ HYDROFLUORIC ACID, anhydrous (DOT) ◇ HYDROFLUORIC ACID, solution (DOT) ◇ HYDROFLUORIDE ◇ HYDROGEN FLUORIDE (OSHA, ACGIH, MAK, DOT) ◇ RCRA WASTE NUMBER U134

TOXICITY DATA with REFERENCE
dnd-dmg-ihl 1300 ppb/6W ATENBP 5,117,71
sln-dmg-ihl 2900 ppb FLUOA4 4,25,71
ihl-rat TCLo:4980 μg/m^3/4H (1-22D preg):TER GTPZAB 19(3),57,75
ihl-rat TCLo:470 μg/m^3/4H (1-22D preg):REP GTPZAB 19(3),57,75
ihl-man TCLo:100 mg/m^3/1M:NOSE,EYE,PUL JIDHAN 16,129,34
ihl-hmn LCLo:50 ppm/30M 34ZIAG -,318,69
ihl-rat LC50:966 ppm/1H TXAPA9 42,417,77
ipr-rat LDLo:25 mg/kg TXAPA9 13,76,68
ihl-mus LC50:342 ppm/1H JCTODH 3,61,76

skn-mus LDLo:500 mg/kg SAIGBL 17,281,75
ihl-mky LC50:1774 ppm/1H AMRL** TR-70-77/70
ihl-rbt LCLo:260 mg/m^3/7H AMIHBC 2,716,50
ihl-gpg LC50:4327 ppm/15M AIHAAP 24,253,63
scu-frg LDLo:112 mg/kg CRSBAW 124,133,37

CONSENSUS REPORTS: EPA Extremely Hazardous Substances List. Community Right-To-Know List. EPA Genetic Toxicology Program. Reported in EPA TSCA Inventory.

OSHA PEL: (Transitional: TWA 3 ppm (F)) TWA 3 ppm; STEL 6 ppm (F)
ACGIH TLV: CL 3 ppm (F)
DFG MAK: 3 ppm (2 mg/m^3); BAT 7.0 mg/g creatinine in urine at end of shift.
NIOSH REL: (HF) TWA 2.5 mg(F)/m^3; CL 5.0 mg(F)/m^3/15M
DOT Classification: Corrosive Material; Label: Corrosive (UN1052, UN1790); Poison A; Label: Poison Gas, Corrosive (UN1052); Corrosive Material; Label: Corrosive, Poison (UN1790)

SAFETY PROFILE: A human poison by inhalation. A poison experimentally by inhalation, subcutaneous, and intraperitoneal routes. A corrosive irritant to skin, eyes (@ 0.05 mg/L), and mucous membranes. Experimental teratogenic effects. Experimental reproductive effects. Mutation data reported. Inhalation of the vapor may cause ulcers of the upper respiratory tract. Concentrations of 50-250 ppm are dangerous, even for brief exposures. Hydrofluoric acid produces severe skin burns which are slow in healing. The subcutaneous tissues may be affected, becoming blanched and bloodless. Gangrene of the affected areas may follow. It is a common air contaminant.

Explosive reaction with cyanogen fluoride; glycerol + nitric acid; sodium (with aqueous acid); methanesulfonic acid (evolves oxygen difluoride which explodes). Violent reaction with As_2O_3; P_2O_5; acetic anhydride; 2-amino ethanol; NH_4OH; $HBiO_3$; bismuthic acid (evolves oxygen); CaO; chlorosulfonic acid; ethylene diamine; ethylene imine; F_2; mercury(II) oxide + organic materials (above 0°C); n-phenylazopiperidine; potassium permanganate; potassium tetrafluorosilicate(2-)(evolves silicon tetrafluoride gas); (HNO_3 + lactic acid); oleum; β-propiolactone; propylene oxide; Na; NaOH; H_2SO_4; vinyl acetate; HgO; sodium tetrafluoro silicate; n-phenyl azo piperidine. Incandescent reaction of liquid HF with oxides (e.g., arsenic trioxide, calcium oxide). Dangerous storage hazard with nitric acid + lactic acid; nitric acid + propylene glycol (mixtures evolve gas which may burst a sealed container). Reacts with water or steam to produce toxic and corrosive fumes. When heated to decomposition it emits highly corrosive fumes of F^-. See also FLUORIDES.

HHV000 ***HR: 3***
HYDROFLUORIC ACID mixed with SULFURIC ACID
DOT: UN 1786

SYNS: HYDROFLUORIC and SULFURIC ACIDS, MIXTURE (DOT) ◇ SULFURIC AND HYDROFLUORIC ACIDS, MIXTURE (DOT)

DOT Classification: Corrosive Material; Label: Corrosive; IMO: Corrosive Material; Label: Corrosive, Poison.

SAFETY PROFILE: Poison by ingestion, inhalation, and skin contact. A corrosive irritant to the eyes, skin and mucous membranes. When heated to decomposition it emits very toxic fumes of HF and SO_x. See also HYDROFLUORIC ACID and SULFURIC ACID.

HHW000 CAS:109-82-0 ***HR: 3***
α-HYDROFORMAMINE CYANIDE
mf: $C_3H_4N_2$ mw: 68.09

SYNS: METHYLENEAMINOACETONITRILE ◇ N-METHYLENE GLYCINONITRILE ◇ METHYLENIMINOACETONITRILE ◇ USAF DO-5

TOXICITY DATA with REFERENCE
ipr-mus LD50:200 mg/kg NTIS** AD277-689
par-mus LDLo:2000 mg/kg CBCCT* 7,689,55

CONSENSUS REPORTS: Reported in EPA TSCA Inventory. Cyanide and its compounds are on The Community Right-To-Know List.

SAFETY PROFILE: Poison by intraperitoneal route. Moderately toxic by parenteral route. When heated to decomposition it emits toxic fumes of NO_x and CN^-. See also NITRILES.

HHW500 CAS:1333-74-0 ***HR: 3***
HYDROGEN
DOT: UN 1049/UN 1966
mf: H_2 mw: 2.02

PROP: Colorless, odorless, tasteless gas. Mp: −259.18°, bp: −252.8°, lel: 4.1%, uel: 74.2%, d: 0.0899 g/L, autoign temp: 752°F, vap d: 0.069.

SYNS: HYDROGEN (DOT) ◇ HYDROGEN, compressed (DOT) ◇ HYDROGEN, refrigerated liquid (DOT)

CONSENSUS REPORTS: Reported in EPA TSCA Inventory.

DOT Classification: Flammable Gas; Label: Flammable Gas.

SAFETY PROFILE: Practically no toxicity except that it may asphyxiate. Highly dangerous fire and severe explosion hazard when exposed to heat, flame or oxidizers. Flammable or explosive when mixed with air; O_2; chlorine. To fight fire, stop flow of gas.

Explodes on contact with bromine trifluoride; chlorine trifluoride; fluorine; hydrogen peroxide + catalysts; acetylene + ethylene. Explodes when heated with calcium carbonate + magnesium; 3,4-dichloronitrobenzene + catalysts; vegetable oils + catalysts; ethylene + nickel catalysts; difluorodiazene (above 90°C); 2-nitroanisole (above 250°C/34 bar + 12% catalyst); copper(II) oxide; nitryl fluoride (above 200°C); polycarbon monofluoride (above 500°C).

Forms sensitive explosive mixtures with bromine; chlorine; iodine heptafluoride (heat- or spark-sensitive); chlorine dioxide; dichlorine oxide; iodine heptafluoride (heat- or spark-sensitive); dinitrogen oxide; dinitrogen tetraoxide; oxygen (gas); 1,1,1-trisazidomethylethane + palladium catalyst. Mixtures with liquid nitrogen react with heat to form an explosive product.

Violent reaction or ignition with air + catalysts (platinum and similar metals containing adsorbed oxygen or hydrogen); bromine; iodine; dioxane + nickel; lithium; nitrogen trifluoride; oxygen difluoride; palladium + isopropyl alcohol; 3-methyl-2-penten-4-yn-1-ol; lead trifluoride; bromine fluoride (ignition on contact); nickel + oxygen; fluorine perchlorate (ignition on contact); xenon hexafluoride (violent reaction); nitrogen oxide + oxygen (ignition above 360°C); palladium powder + 2-propanol + air (spontaneous ignition); platinum catalyst; polycarbon monofluoride (ignition above 400°C).

Vigorous exothermic reaction with benzene + Raney nickel catalyst; metals (e.g., lithium; calcium; barium; strontium; sodium; potassium; above 300°C); palladium(II)oxide; palladium trifluoride; 1,1,1-tris(hydroxymethyl)nitromethane + nickel catalyst.

HHW509 ***HR: 3***
HYDROGENATED COAL OIL FRACTION 1

PROP: Centrifugation residue obtained through the direct hydrogenation of coal by Bergius process; liquid phase, highly viscous, black material (IMSUAI 25, 51,56).

SYN: BERGIUS COAL HYDROGENATION PRODUCTS FRACTION 1

TOXICITY DATA with REFERENCE
ims-rat TDLo:600 mg/kg/32W-I:ETA,REP IMSUAI 25,51,56
skn-mus TDLo:443 mg/kg/60W-I:ETA IMSUAI 25,51,56

SAFETY PROFILE: Experimental reproductive effects. Questionable carcinogen with experimental tumorigenic data by skin contact and intramuscular routes. A fire hazard. When heated to decomposition it emits acrid smoke and fumes.

HHW519 ***HR: 3***
HYDROGENATED COAL OIL FRACTION 3

PROP: Light oil bottoms obtained through the direct hydrogenation of coal by the Bergius process, liquid phase; a viscous, brown oil containing a scaly admixture (IMSUAI 25,51,56).

SYN: BERGIUS COAL HYDROGENATION PRODUCTS FRACTION 3

TOXICITY DATA with REFERENCE
ims-rat TDLo:600 mg/kg/32W-I:ETA,REP IMSUAI 25,51,56
skn-mus TDLo:480 mg/kg/60W-I:ETA IMSUAI 25,51,56

SAFETY PROFILE: Experimental reproductive effects. Questionable carcinogen with experimental tumorigenic data by skin contact and intramuscular routes. A fire hazard. When heated to decomposition it emits acrid smoke and fumes.

HHW529 ***HR: 3***
HYDROGENATED COAL OIL FRACTION 4

PROP: Middle oil obtained through the direct hydrogenation of coal by the Bergius process; liquid phase, a thin, reddish-brown oil having an aromatic odor (IMSUAI 25,51,56).

SYN: BERGIUS COAL HYDROGENATION PRODUCTS FRACTION 4

TOXICITY DATA with REFERENCE
skn-mus TDLo:343 mg/kg/43W-I:ETA IMSUAI 25,51,56

SAFETY PROFILE: Questionable carcinogen with experimental tumorigenic data by skin contact and intramuscular routes data. A fire hazard. When heated to decomposition it emits acrid smoke and fumes.

HHW539 ***HR: 3***
HYDROGENATED COAL OIL FRACTION 7

PROP: Raw gasoline obtained through the direct hydrogenation of coal by the Bergius process; a thin dark brown liquid which quickly evaporates leaving a brownish-red, oily residue (IMSUAI 25,51,56).

SYN: BERGIUS COAL HYDROGENATION PRODUCTS FRACTION 7

TOXICITY DATA with REFERENCE
ims-rat TDLo:600 mg/kg/32W-I:ETA IMSUAI 25,51,56

SAFETY PROFILE: Questionable carcinogen with experimental tumorigenic data by skin contact and intramuscular routes data. A dangerous fire hazard. When heated to decomposition it emits acrid smoke and fumes.

HHW549 ***HR: 3***
HYDROGENATED COAL OIL FRACTION 9

PROP: Pitch flash distillation residue obtained through the direct hydrogenation of coal by the Bergius process; a solid, black, coke-like material (IMSUAI 25,51,56).

SYN: BERGIUS COAL HYDROGENATION PRODUCTS FRACTION 4

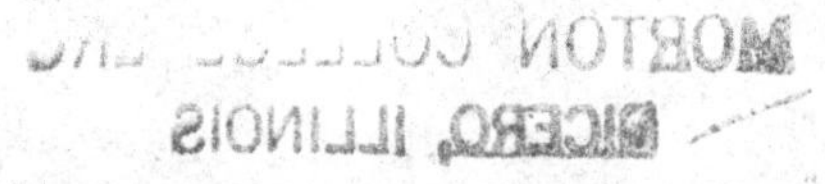

TOXICITY DATA with REFERENCE
ims-rat TDLo:600 mg/kg/32W-I:ETA,REP IMSUAI 25,51,56
skn-mus TDLo:206 mg/kg/26W-I:ETA IMSUAI 25,51,56

SAFETY PROFILE: Experimental reproductive effects. Questionable carcinogen with experimental tumorigenic data by skin contact and intramuscular routes. Flammable. When heated to decomposition it emits acrid smoke and fumes.

HHW800 CAS:61788-32-7 ***HR: 3***
HYDROGENATED TERPHENYLS

PROP: Complex mixtures of o-, m-, and p-terphenyls in various stages of hydrogenation. Five such stages exist for each of the three above isomers.

CONSENSUS REPORTS: Reported in EPA TSCA Inventory.

OSHA PEL: TWA 0.5 ppm
ACGIH TLV: TWA 0.5 ppm

SAFETY PROFILE: Contact with hot coolant can cause severe damage to lungs, skin, and eyes from burns. May cause chronic damage to liver, kidney, and blood-forming organs; metabolic disorders. Inhalation has caused bronchopneumonia. When heated to decomposition it emits acrid smoke and fumes.

HHX000 CAS:7647-01-0 ***HR: 3***
HYDROGEN CHLORIDE
mf: ClH mw: 36.46

PROP: Colorless, corrosive, nonflammable gas. Pungent odor, fumes in air. D: 1.639 @ −137.77°, bp: −154.37° @ 1.0 mm.

TOXICITY DATA with REFERENCE
ihl-rat LC50:4701 ppm/30M AIHAAP 35,623,74
ihl-mus LC50:2644 ppm/30M AIHAAP 35,623,74

CONSENSUS REPORTS: EPA Extremely Hazardous Substances List. EPA Genetic Toxicology Program. Reported in EPA TSCA Inventory.

OSHA PEL: CL 5 ppm
ACGIH TLV: CL 5 ppm
DFG MAK: 5 ppm (7 mg/m^3)

SAFETY PROFILE: A highly corrosive irritant to the eyes, skin, and mucous membranes. Mildly toxic by inhalation. Explosive reaction with alcohols + hydrogen cyanide; potassium permanganate; sodium (with aqueous HCl); tetraselenium tetranitride. Ignition on contact with aluminum-titanium alloys (with HCl vapor); fluorine; hexalithium disilicide; metal acetylides or carbides (e.g., cesium acetylide; rubidium acetylide). Violent reaction with 1,1-difluoroethylene. Vigorous reaction with aluminum; chlorine + dinitroanilines (evolves gas). Potentially dangerous reaction with sulfuric acid releases HCl gas. Adsorption of the acid onto silicon dioxide is exothermic. See also HYDROGEN CHLORIDE (AEROSOL), and HYDROCHLORIC ACID.

HHX500 CAS:7647-01-0 ***HR: 3***
HYDROGEN CHLORIDE (aerosol)
mf: ClH mw: 36.46

PROP: Saturated water aerosol mist (NTIS** AD744-829).

TOXICITY DATA with REFERENCE
ihl-rat LC50:5666 ppm/30M NTIS** AD744-829
ihl-mus LC50:2142 ppm/30M NTIS** AD744-829

CONSENSUS REPORTS: Reported in EPA TSCA Inventory. EPA Genetic Toxicology Program.

SAFETY PROFILE: Mildly toxic by inhalation. A very powerful human skin, eye, and mucous membrane irritant. See also HYDROGEN CHLORIDE.

HHY500 CAS:1333-74-0 ***HR: 3***
HYDROGEN, CRYOGENIC LIQUID (DOT)

CONSENSUS REPORTS: Reported in EPA TSCA Inventory.

DOT Classification: Label: Flammable Gas.

SAFETY PROFILE: Contact with the liquid can cause frostbite. A very flammable gas. Can explode in air. Forms an explosive mixture with ozone. See also HYDROGEN.

HHZ000 CAS:13465-07-1 ***HR: 2***
HYDROGEN DISULFIDE
mf: H_2S_2 mw: 66.14

PROP: Flash p: <71.6°F.

SAFETY PROFILE: A very dangerous fire hazard when exposed to heat or flame; can react vigorously with oxidizers. Decomposes violently in reaction with alkalies. When heated to decomposition it emits toxic fumes of SO_x. See also SULFIDES.

HIA000 CAS:17099-81-9 ***HR: 2***
(HYDROGEN(ETHYLENEDINITRILO)TETRA-ACETATO)IRON
mf: $C_{10}H_{13}FeN_2O_8$ mw: 345.10

SYN: ETHYLENEDIAMINE TETRAACETIC ACID, IRON(III) SALT

TOXICITY DATA with REFERENCE
ipr-mus LD50:600 mg/kg REPMBN 10,391,62

CONSENSUS REPORTS: Reported in EPA TSCA Inventory.

SAFETY PROFILE: Moderately toxic by intraperitoneal route. When heated to decomposition it emits toxic fumes of NO_x.

HIA500 CAS:16941-92-7 ***HR: 3***
HYDROGEN HEXACHLOROIRIDATE (4+)
mf: $Cl_6Ir•2H$ mw: 406.92

SYN: HEXACHLOROIRIDATE(2-) DIHYDROGEN, (OC-6-11)

TOXICITY DATA with REFERENCE
ipr-mus LD50:251 mg/kg COREAF 256,1043,63

CONSENSUS REPORTS: Reported in EPA TSCA Inventory. EPA Genetic Toxicology Program.

SAFETY PROFILE: Poison by intraperitoneal route. When heated to decomposition it emits toxic fumes of Cl^-. See also IRIDIUM.

HIB000 CAS:7722-84-1 ***HR: 3***
HYDROGEN PEROXIDE
DOT: UN 2014/UN 2015
mf: H_2O_2 mw: 34.02

PROP: Colorless, heavy liquid, or, at low temp, a crystalline solid; bitter taste. D: 1.71 @ −20°, 1.46 @ 0°, vap press: 1 mm @ 15.3°, unstable. Mp: −0.43°, bp: 152°. Misc with water, sol in ether; insol in petr ether. Decomposed by many organic solvents.

SYNS: ALBONE ◇ DIHYDROGEN DIOXIDE ◇ HIOXYL ◇ HYDROGEN DIOXIDE ◇ HYDROGEN PEROXIDE, solution (over 52% peroxide) (DOT) ◇ HYDROGEN PEROXIDE, stabilized (over 60% peroxide) (DOT) ◇ HYDROPEROXIDE ◇ INHIBINE ◇ OXYDOL ◇ PERHYDROL ◇ PERONE ◇ PEROSSIDO di IDROGENO (ITALIAN) ◇ PEROXAN ◇ PEROXIDE ◇ PEROXYDE d'HYDROGENE (FRENCH) ◇ SUPEROXOL ◇ T-STUFF ◇ WASSERSTOFFPEROXID (GERMAN) ◇ WATERSTOFPEROXYDE (DUTCH)

TOXICITY DATA with REFERENCE
dnd-hmn:oth 100 μmol/L CNREA8 45,2522,85
dni-hmn:oth 1200 μmol/L CNREA8 45,2522,85
orl-mus TDLo:144 g/kg/26W-C:ETA GANNA2 75,17,84
skn-rat LD50:4060 mg/kg GTPZAB 21(10),22,77
mul-rat LC50:2 g/m³/4H GTPZAB 21(10),22,77
orl-mus LD50:2 g/kg GISAAA 48(6),28,83
ihl-mus LCLo:227 ppm AEHLAU 4,327,62
skn-rbt LDLo:500 mg/kg MRLR** #75,51
ivn-rbt LD50:15 g/kg MRLR** #75,51
skn-pig LDLo:2 g/kg MRLR** #75,51

CONSENSUS REPORTS: IARC Cancer Review: Group 3 IMEMDT 7,56,87; Animal Limited Evidence IMEMDT 28,151,82. EPA Extremely Hazardous Substances List. Reported in EPA TSCA Inventory. EPA Genetic Toxicology Program.

OSHA PEL: TWA 1 ppm
ACGIH TLV: TWA 1 ppm
DFG MAK: 1 ppm (1.4 mg/m³)
DOT Classification: Oxidizer; Label: Oxidizer and Corrosive.

SAFETY PROFILE: Moderately toxic by inhalation, ingestion, and skin contact. A corrosive irritant to skin, eyes, and mucous membranes. Human mutation data reported. Questionable carcinogen with experimental tumorigenic data. A very powerful oxidizer.

Pure H_2O_2, its solutions, vapors, and mists are very irritating to body tissue. This irritation can vary from mild to severe depending upon the concentration of H_2O_2. For instance, solutions of H_2O_2 of 35 wt% and over can easily cause blistering of the skin. Irritation caused by H_2O_2 which does not subside upon flushing the affected part with water should be treated by a physician. The eyes are particularly sensitive to this material. It is a common air contaminant.

A dangerous fire hazard by chemical reaction with flammable materials. H_2O_2 is a powerful oxidizer, particularly in the concentrated state. It is important to keep containers covered because uncovered containers are much more prone to react with flammable vapors, gases, etc.; and if uncovered, the water from an H_2O_2 solution can evaporate, concentrating the material and thus increasing the fire hazard. For instance, solutions of H_2O_2 in concentration in excess of 65 wt% heat up spontaneously when decomposed to H_2O + 1/2 O_2. Thus, 90 wt% solutions, when caused to decompose rapidly due to the introduction of a catalytic decomposition agent, can get quite hot and perhaps start fires.

A severe explosion hazard when highly concentrated or when pure H_2O_2 is exposed to: heat, mechanical impact, detonation of a blasting cap, or caused to decompose catalytically by metals (in order of decreasing effectiveness: osmium; palladium; platinum; iridium; gold; silver; manganese; cobalt; copper; lead). Explodes on contact with alcohols + H_2SO_4; acetal + acetic acid + heat; acetic acid + n-heterocycles (above 50°); 2-amino-4-methyloxazole + iron(II) catalyst; aromatic hydrocarbons + trifluoroacetic acid; azeliac acid + sulfuric acid (above 45°); benzenesulfonic anhydride; tert-butanol + sulfuric acid; carboxylic acids; 3,5-dimethyl-3-hexanol + sulfuric acid; diphenyl diselenide (above 53°); 2-ethoxyethanol + polyacrylamide gel + toluene + heat; gadolinium hydroxide (above 80°); gallium + hydrochloric acid; hydrogen + palladium catalysts (has caused major industrial explosions); iron(II) sulfate + 2-methylpyridine + sulfuric acid; iron(II) sulfate + nitric acid + sodium carboxymethylcellulose (when evaporated); nitric acid + ketones (e.g., 2-butanone; 3-pentanone; cyclopentanone; cyclohexanone; 3-methylcyclohexanone); trioxane (sensitive to heat, shock, or on contact with lead); methanol + tert-amines + platinum catalysts; nitric acid + soils; nitrogenous bases (e.g., ammonia; hydrazine hydrate; 1,1-dimethylhydrazine); organic compounds (e.g., glycerol; acetic acid; ethanol; aniline; quinoline; 2-phenyl-1,1-dimethylethanol; cellulose; charcoal); organic materials + sulfuric acid (especially if confined); water + oxygenated compounds (e.g., acetaldehyde, acetic acid; acetone; ethanol; formaldehyde; formic acid; methanol; 2-propanol; propionaldehyde); sulfuric acid (during evaporation); tetrahydrothiophene; vinyl acetate; alcohols + tin chloride; P_2O_5; P; H_2O; HNO_3; Sb_2S_3; As_2S_3; Cl_2 + KOH; + chlorosulfonic acid; CuS; FeS; formic acid + organic matter; H_2Se; hydrazine; PbO_2; PbO; PbS; MnO_2; HgO; Hg_2O; MoS_2; organic matter; (2-methyl-1-phenyl-2-propanol + sulfuric acid); $KMnO_4$; $NaIO_3$; thiodiglycol; uns-dimethyl hydrazine; $FeSO_4$ + 2-methylpyridine + H_2SO_4; HgO + HNO_3.

Forms unstable explosive products in reaction with acetaldehyde + desiccants (forms polyethylidine peroxide); acetic acid (forms peracetic acid); acetic + 3-thietanol; acetic anhydride; acetone (forms explosive peroxides); alcohols (products are shock- and heat-sensitive); carboxylic acids (e.g., formic acid; acetic acid; tartaric acid); diethyl ether; ethyl acetate; formic acid + metaboric acid; ketene (forms diacetyl peroxide); mercury(II) oxide + nitric acid (forms mercury(II) peroxide); thiourea + nitric acid; polyacetoxyacrylic acid lactone + poly(2-hydroxyacrylic acid) + sodium hydroxide.

Ignition on contact with furfuryl alcohol; powdered metals (e.g., magnesium; iron); wood. Violent reaction with aluminum isopropoxide + heavy metal salts; charcoal; coal; dimethylphenylphosphine; hydrogen selenide; lithium tetrahydroaluminate; metals (e.g., potassium, sodium, lithium); metal oxides (e.g., cobalt oxide; iron oxide; lead oxide; lead hydroxide; manganese oxide; mercury oxide; nickel oxide); metal salts (e.g., calcium permanganate); methanol + phosphoric acid; 4-methyl-2,4,6-triazatricyclo $[5.2.2.0^{2,6}]$ undeca-8-ene-3,5-dione + potassium hydroxide; α-phenylselenoketones; phosphorus; phosphorus(V) oxide; tin(II) chloride; unsaturated organic compounds.

BEWARE: Although many mixtures of H_2O_2 and organic materials do not explode upon contact, the resultant combination is detonatable either upon catching fire or by impact. The detonation velocity of aqueous solutions of H_2O_2 has been found to be about 6500 m/second for solutions of between 96 wt% and 100 wt% H_2O_2. Another source of H_2O_2 explosions is from sealing the material in strong containers. Under such conditions, even gradual decomposition of H_2O_2 to H_2O + 1/2 O_2 can cause large pressures to build up in the containers which may then burst explosively. Highly dangerous; when heated, shocked, or contaminated, the concentrated material can explode or start fires.

HIB005 ***HR: 2***
HYDROGEN PEROXIDE, 8% to 20%
DOT: UN 2984
mf: H_2O_2 mw: 34.02

SYN: HYDROGEN PEROXIDE, solution, 8% to 20% (DOT)

TOXICITY DATA with REFERENCE
orl-rat TDLo:36432 mg/kg (male 12W pre):REP TOIZAG 23,531,76
orl-rat LD50:1518 mg/kg TOIZAG 23,531,76

DOT Classification: Oxidizer; Label: Oxidizer

SAFETY PROFILE: Moderately toxic by ingestion. Experimental reproductive effects. A moderate oxidizer.

HIB010 CAS:7722-84-1 ***HR: 3***
HYDROGEN PEROXIDE, 30%
DOT: UN 2014
mf: H_2O_2 mw: 34.02

SYNS: ALBONE 35 ◇ ALBONE 50 ◇ ALBONE 70 ◇ ALBONE 35CG ◇ ALBONE 50CG ◇ ALBONE 70CG ◇ HYDROGEN PEROXIDE, solution 30% ◇ HYDROGEN PEROXIDE, solution (8% to 40% PEROXIDE) (DOT) ◇ INTEROX ◇ KASTONE ◇ PERONE 30 ◇ PERONE 35 ◇ PERONE 50

TOXICITY DATA with REFERENCE
oth-hmn:emb 50 μmol/L MUREAV 172,245,86
cyt-hmn:emb 20 μmol/L MUREAV 172,245,86
msc-ham:lng 1 mmol/L MUREAV 192,65,87
orl-mus TDLo:622 g/kg/2Y-C:CAR HIUN** 17(1),53,74

CONSENSUS REPORTS: IARC Cancer Review: Group 3 IMEMDT 7,56,87, Animal Limited Evidence IMEMDT 36,285,85. Reported in EPA TSCA Inventory.

DOT Classification: Oxidizer; LABEL: Oxidizer

SAFETY PROFILE: Questionable carcinogen with experimental carcinogenic data. Mutation data reported.

HIB500 CAS:124-43-6 ***HR: 2***
HYDROGEN PEROXIDE with UREA (1:1)
DOT: UN 1511
mf: $CH_4N_2O \cdot H_2O_2$ mw: 94.09

PROP: White crystals. Mp: 75-85° (decomp).

SYNS: CARBAMIDE PEROXIDE ◇ GLY-OXIDE ◇ HYDROGEN PEROXIDE CARBAMIDE ◇ HYDROPERIT ◇ HYPEROL ◇ ORTIZON ◇ PERCARBAMIDE ◇ PERHYDRIT ◇ PERHYDROL-UREA ◇ THENARDOL ◇ UREA DIOXIDE ◇ UREA HYDROGEN PEROXIDE (DOT) ◇ UREA HYDROGEN PEROXIDE SALT ◇ UREA HYDROPEROXIDE ◇ UREA PEROXIDE (DOT)

CONSENSUS REPORTS: Reported in EPA TSCA Inventory.

DOT Classification: Organic Peroxide; Label: Organic Peroxide; IMO: Oxidizer; Label: Oxidizer.

SAFETY PROFILE: An irritant to skin, eyes, and mucous membranes. An FDA over-the-counter drug. When heated to decomposition it emits toxic fumes of NO_x. See also individual components and PEROXIDES, ORGANIC.

HIC000 CAS:7783-07-5 ***HR: 3***
HYDROGEN SELENIDE
DOT: UN 2202
mf: H_2Se mw: 80.98

PROP: Colorless gas. Mp: −64°, bp: −41.4°, d: 3.614 g/L (gas), 2.12 @ −42° (liquid), vap press: 10 atm @ 23.4°. Flammable. Disagreeable odor. Sol in carbonyl chloride and carbon disulfide.

SYNS: ELECTRONIC E-2 ◇ HYDROGEN SELENIDE, anhydrous (DOT) ◇ SELENIUM HYDRIDE

TOXICITY DATA with REFERENCE
ihl-rat LCLo:20 mg/m³/1H CTOXAO 17,171,80
ihl-gpg LC50:300 ppb/8H 34ZIAG -,320,69

CONSENSUS REPORTS: Selenium and its compounds are on the Community Right-To-Know List. EPA Extremely Hazardous Substances List. Reported in EPA TSCA Inventory.

OSHA PEL: TWA 0.05 ppm (Se)
ACGIH TLV: TWA 0.05 ppm (Se)
DFG MAK: 0.05 ppm (0.2 mg/m³)
DOT Classification: Flammable Gas; Label: Poison Gas and Flammable Gas; IMO: Poison A; Label: Poison Gas and Flammable Gas.

SAFETY PROFILE: A deadly poison by inhalation. Very poisonous irritant to skin, eyes, and mucous membranes. Causes central nervous system effects in humans. An allergen. Can cause damage to the lungs and liver as well as conjunctivitis. It has been found that repeated exposures to concentrations of 0.3 ppm prove fatal to experimental animals by causing a pneumonitis, as well as injury to the liver and spleen. Causes garlic odor of breath, dizziness, nausea. Concentrations of 0.3 ppm are readily detected by odor, but there is no noticeable irritant effect at that level. Concentrations of 1.5 ppm or higher are strongly irritating to the eyes and nasal passages.

As in the case of hydrogen sulfide, the odor of hydrogen selenide in concentrations below 1 ppm disappears rapidly because of olfactory fatigue. The odor and irritating effects do not offer a dependable warning to workmen who may be exposed to gradually increasing amounts and therefore become used to it. Due to its extreme toxicity and irritating effects, it seldom is allowed to reach a concentration in which it is flammable in air. Very little data are available on possible chronic effects of this material, but it is logical to assume that when the concentration of this gas is low enough to avoid the irritant effects, only the systemic effects will be noticeable.

Dangerous fire hazard when exposed to heat or flame; will react vigorously with powerful oxidizing agents, such as H_2O_2; HNO_3. Dangerous; forms explosive mixtures with air; keep away from heat and open flame. See also SELENIUM COMPOUNDS and HYDRIDES.

HIC500 CAS:7783-06-4 ***HR: 3***
HYDROGEN SULFIDE
DOT: UN 1053
mf: H_2S mw: 34.08

PROP: Colorless, flammable gas; offensive odor. Mp: −85.5°, bp: −60.4°, lel: 4%, uel: 46%, autoign temp: 500°F, d: 1.539 g/L @ 0°, vap press: 20 atm @ 25.5°, vap d: 1.189.

SYNS: ACIDE SULFHYDRIQUE (FRENCH) ◇ HYDROGENE SULFURE (FRENCH) ◇ HYDROGEN SULFURIC ACID ◇ IDROGENO SOLFORATO (ITALIAN) ◇ RCRA WASTE NUMBER U135 ◇ SCHWEFELWASSERSTOFF (GERMAN) ◇ SIARKOWODOR (POLISH) ◇ STINK DAMP ◇ SULFURETED HYDROGEN ◇ SULFUR HYDRIDE ◇ ZWAVELWATERSTOF (DUTCH)

TOXICITY DATA with REFERENCE
ihl-hmn LCLo:600 ppm/30M 29ZWAE -,207,68
ihl-man LDLo:5700 μg/kg:CNS,PUL AMPMAR 44,483,83
ihl-hmn LCLo:800 ppm/5M TABIA2 3,231,33
ihl-rat LC50:444 ppm LacHB# 09JUN78
ihl-mus LC50:673 ppm/1H NTIS** PB214-270
ihl-mam LCLo:800 ppm/5M AEPPAE 138,65,28

CONSENSUS REPORTS: EPA Extremely Hazardous Substances List. Reported in EPA TSCA Inventory.

OSHA PEL: (Transitional: CL 20 ppm; Pk 50/10M) TWA 10 ppm; STEL 15 ppm
ACGIH TLV: TWA 10 ppm; STEL 15 ppm
DFG MAK: 10 ppm (15 mg/m³)
NIOSH REL: (Hydrogen Sulfide) CL 15 mg/m³/10M
DOT Classification: Flammable Gas; Label: Poison Gas and Flammable Gas.

SAFETY PROFILE: A human poison by inhalation. A severe irritant to eyes and mucous membranes. An asphyxiant. Human systemic effects by inhalation: coma, chronic pulmonary edema. Low concentrations of 20-150 ppm cause irritation of the eyes; slightly higher concentrations may cause irritation of the upper respiratory tract, and if exposure is prolonged, pulmonary edema may result. The irritant action has been explained on the basis that H_2S combines with the alkali present in moist surface tissues to form sodium sulfide, a caustic. With higher concentration the action of the gas on the nervous system becomes more prominent. A 30 minute exposure to 500 ppm results in headache, dizziness, excitement, staggering gait, diarrhea and dysuria, followed sometimes by bronchitis or bronchopneumonia.

The action of small amounts on the nervous system is one of depression; in larger amounts, it stimulates, and with very high amounts the respiratory center is paralyzed. Exposures of 800-1000 ppm may be fatal in 30 minutes, and high concentrations are instantly fatal. Fatal hydrogen sulfide poisoning may occur even more rapidly than that following exposure to a similar concentration of HCN. H_2S does not combine with the hemoglobin of the blood; its asphyxiant action is due to paralysis of the respiratory center. With repeated exposures to low concentrations, conjunctivitis, photophobia, corneal bullae, tearing, pain, and blurred vision are the commonest findings. High concentrations may cause rhinitis, bronchitis, and occasionally pulmonary edema. Exposure to very high concentrations results in immediate death. Chronic poisoning results in headache, inflammation of the conjunctivae and eyelids, digestive disturbances, weight loss and general debility. It is a common air contaminant.

It is an insidious poison since sense of smell may be fatigued. The odor and irritating effects do not offer a dependable warning to workers who may be exposed to gradually increasing amounts and therefore become used to it.

Very dangerous fire hazard when exposed to heat, flame, or oxidizers. Moderately explosive when exposed to heat or flame. Explodes on contact with oxygen difluoride; nitrogen trichloride; bromine pentafluoride; chlorine trifluoride; dichlorine oxide; silver fulminate. Potentially explosive reaction with copper + oxygen. Explosive reaction when heated with perchloryl fluoride (above 100°C); oxygen (above 280°C). Reacts with 4-bromobenzenediazonium chloride to form an explosive product.

Ignites on contact with metal oxides (e.g., barium peroxide; chromium trioxide; copper oxide; lead dioxide; manganese dioxide; nickel oxide; silver(I) oxide; silver(II) oxide; sodium peroxide; thallium(III) oxide; mercury oxide; calcium oxide; nickel oxide); oxidants (e.g., silver bromate; heptasilver nitrate octaoxide; dibismuth dichromium nonaoxide; mercury(I) bromate; lead(II) hypochlorite; copper chromate; fluorine; nitric acid; sodium peroxide; lead(IV) oxide); rust; soda-lime + air. Reacts violently with NI_3; NF_3; p-bromobenzenediazonium chloride; OF_2; F_2; Cu; ClO; BrF_5; acetaldehyde; (BaO + Hg_2O + air); (BaO + NiO + air); hydrated iron oxide; phenyl diazonium chloride; (NaOH + CaO + air). Incandescent reaction with chromium trioxide. Vigorous reaction with metal powders (e.g., copper; tungsten). When heated to decomposition it emits highly toxic fumes of SO_x. To fight fire, stop flow of gas.

HIC600 CAS:15181-46-1 ***HR: D***
HYDROGEN SULFITE
mf: HO_3S mw: 81.07

SYNS: BISULFITE ◇ BISULPHITE ◇ HYDROSULFITE ANION ◇ SULFITE LYE

TOXICITY DATA with REFERENCE
mmo-esc 1 mol/L ENMUDM 2,239,80
otr-ham:emb 1 mmol/L CRNGDP 3,27,82

SAFETY PROFILE: Mutation data reported. When heated to decomposition it emits toxic fumes of SO_x. See also SULFITES.

HID000 CAS:13845-23-3 ***HR: 3***
HYDROGEN TRISULFIDE
mf: H_2S_3 mw: 98.22

SAFETY PROFILE: Explosive reaction with benzenediazonium chloride; silver oxide; nitrogen trichloride; pentanol. Ignites on contact with potassium permanganate; metal oxides (e.g., copper oxide; lead(II) oxide; lead(IV) oxide; mercury(II) oxide; tin(IV) oxide; iron(II,III) oxide). Incompatible with metal oxides; nitrogen trichloride; pentyl alcohol; potassium permanganate. When heated to decomposition it emits toxic fumes of SO_x. See also HYDROGEN SULFIDE and SULFIDES.

HID350 CAS:652-67-5 ***HR: 3***
HYDRONOL
mf: $C_6H_{10}O_4$ mw: 146.16

PROP: Crystals. Mp: 61-64°.

SYNS: AT 101 ◇ DEVICORAN ◇ d-1,4:3,6-DIANHYDROGLUCITOL ◇ 1,4:3,6-DIANHYDROSORBITOL ◇ ISMOTIC ◇ ISOBIDE ◇ ISOSORBIDE ◇ (+)-d-ISOSORBIDE

TOXICITY DATA with REFERENCE
orl-rat LD50:24150 mg/kg OYYAA2 3,15,69
ivn-rat LD50:11300 mg/kg NIIRDN 6,71,82
orl-mus LD50:289 mg/kg OYYAA2 3,187,69
ipr-mus LD50:13600 mg/kg NIIRDN 6,71,82
ivn-mus LD50:6870 mg/kg OYYAA2 3,15,69

SAFETY PROFILE: Poison by ingestion. When heated to decomposition it emits acrid smoke and fumes.

HID500 CAS:2207-76-3 ***HR: 3***
6-HYDROPEROXY-4-CHOLESTEN-3-ONE
mf: $C_{27}H_{44}O_3$ mw: 416.71

SYNS: 6-β-HYDROPEROXYCHOLEST-4-EN-3-ONE ◇ 6-β-HYDROPEROXY-Δ^4CHOLESTEN-3-ONE

TOXICITY DATA with REFERENCE
scu-mus TDLo:600 mg/kg/72W-I:CAR JNCIAM 19,977,57
scu-mus TD:600 mg/kg/W-I:ETA JCSOA9 77,3928,55

SAFETY PROFILE: Questionable carcinogen with experimental carcinogenic and tumorigenic data. When heated to decomposition it emits acrid smoke and fumes.

HIE000 CAS:4096-33-7 ***HR: 3***
1-HYDROPEROXYCYCLOHEX-3-ENE
mf: $C_6H_{10}O_2$ mw: 114.16

SYN: 1-HYDROPEROXY-3-CYCLOHEXENE

TOXICITY DATA with REFERENCE
skn-mus TDLo:6960 mg/kg/58W-I:NEO JNCIAM 35,707,65
skn-mus TD:8880 mg/kg/74W-I:ETA 14JTAF -,275,64

SAFETY PROFILE: Questionable carcinogen with experimental neoplastigenic and tumorigenic data. When heated to decomposition it emits acrid smoke and fumes.

HIE525 CAS:1238-54-6 ***HR: D***
10-β-HYDROPEROXY-17-α-ETHYNYL-4-ESTREN-17-β-OL-3-ONE
mf: $C_{20}H_{26}O_4$ mw: 330.46

SYN: SCH 10015

TOXICITY DATA with REFERENCE
orl-rat TDLo:10 mg/kg (female 2-3D post):REP JOENAK 33,241,65
scu-rbt TDLo:3 mg/kg (female 1-3D post):TER FESTAS 20,211,69

SAFETY PROFILE: An experimental teratogen. Experimental reproductive effects. A steroid. When heated to decomposition it emits acrid smoke and fumes.

HIE550 CAS:67292-63-1 ***HR: 3***
4-HYDROPEROXYIFOSFAMIDE
mf: $C_7H_{15}Cl_2N_2O_4P$ mw: 293.11

SYN: 3-(2-CHLOROETHYL)-2-(2-CHLOROETHYL)AMINO-4-HYDROPEROXYTETRAHYDRO-2H-1,3,2-OXAZAPHOSPHORINE

TOXICITY DATA with REFERENCE
ipr-rat LD50:180 mg/kg CTRRDO 60,361,76
ivn-rat LD50:220 mg/kg CTRRDO 60,361,76
ipr-mus LDLo:100 mg/kg CNREA8 36,2278,76
ivn-mus LD50:220 mg/kg CTRRDO 60,361,76

SAFETY PROFILE: Poison by intravenous and intraperitoneal routes. When heated to decomposition it emits toxic fumes of Cl^-, PO_x and NO_x.

HIE570 CAS:74955-23-0 ***HR: 2***
N-(HYDROPEROXYMETHYL)-N-NITROSOPROPYLAMINE
mf: $C_4H_{10}N_2O_3$ mw: 134.16

SYN: N-PROPYL-N-(HYDROPEROXYMETHYL)NITROSAMINE

TOXICITY DATA with REFERENCE
msc-ham:lng 100 μmol/L GANNA2 73,522,82
ivn-rat TDLo:50 mg/kg/10W-I:CAR CRNGDP 7,1313,86

SAFETY PROFILE: Questionable carcinogen with experimental carcinogenic data. Mutation data reported. When heated to decomposition it emits toxic fumes of NO_x.

HIE600 CAS:74940-23-1 ***HR: 3***
HYDROPEROXY-N-NITROSODIBUTYLAMINE
mf: $C_8H_{18}N_2O_3$ mw: 190.28

SYNS: BHPBN ◇ N-BUTYL-N-(1-HYDROPEROXYBUTYL)NITROSAMINE

TOXICITY DATA with REFERENCE
msc-ham:lng 50 μmol/L GANNA2 73,522,82
scu-rat TDLo:70500 μg/kg/10W-I:CAR GANNA2 73,687,82
scu-rat TD:72 mg/kg/10W-I:CAR IAPUDO 41,619,82

SAFETY PROFILE: Questionable carcinogen with experimental carcinogenic data. Mutation data reported. When heated to decomposition it emits toxic fumes of NO_x. See also NITROSAMINES.

HIE700 CAS:74940-26-4 ***HR: 2***
1-(HYDROPEROXY)-N-NITROSODIMETHYLAMINE
mf: $C_2H_6N_2O_3$ mw: 106.10

SYN: N-METHYL-N-(HYDROPEROXYMETHYL)NITROSAMINE

TOXICITY DATA with REFERENCE
msc-ham:lng 50 μmol/L GANNA2 73,522,82
ivn-rat TDLo:50 mg/kg/10W-I:CAR CRNGDP 7,1313,86

SAFETY PROFILE: Questionable carcinogen with experimental carcinogenic data. Mutation data reported. When heated to decomposition it emits toxic fumes of NO_x.

HIF000 CAS:67292-61-9 ***HR: 3***
4-HYDROPEROXYPHOSPHAMIDE
mf: $C_7H_{15}Cl_2N_2O_4P$ mw: 293.11

SYNS: 2-(BIS(2-CHLOROETHYL)AMINO)-4-HYDROPEROXYTETRAHYDRO-2H-1,3,2-OXAZAPHOSPHORINE ◇ 4-HYDROPEROXYCYCLOPHOSPHAMIDE ◇ NSC 181815 ◇ 4-PEROXY-CPA

TOXICITY DATA with REFERENCE
dnd-smc 100 μmol/L CBINA8 39,1,82
dni-rat:oth 6 μmol/L/1H CBINA8 39,191,82
ipr-mus LD50:76040 μg/kg NCISP* JAN86

SAFETY PROFILE: Poison by intraperitoneal route. Mutation data reported. When heated to decomposition it emits toxic fumes of Cl^-, PO_x, and NO_x.

HIF575 CAS:3736-26-3 ***HR: 3***
1-HYDROPEROXY-1-VINYLCYCLOHEX-3-ENE
mf: $C_8H_{12}O_2$ mw: 140.20

SYN: 4-HYDROPEROXY-4-VINYL-1-CYCLOHEXENE

TOXICITY DATA with REFERENCE
skn-mus TDLo:1440 mg/kg/24W-I:NEO 14JTAF -,275,64
skn-mus TD:1440 mg/kg/24W-I:ETA JNCIAM 31,41,63

SAFETY PROFILE: Questionable carcinogen with experimental neoplastigenic and tumorigenic data. When heated to decomposition it emits acrid smoke and fumes.

HIG000 ***HR: 3***
HYDROPHIS ELEGANS (AUSTRALIA) VENOM

SYN: VENOM, SEA SNAKE, HYDROPHIS ELEGANS

TOXICITY DATA with REFERENCE
scu-mus LD50:262 μg/kg TOXIA6 14,347,76
ivn-mus LD50:120 μg/kg 85EGD4 -,341,78
ims-mus LD50:120 μg/kg 85EGD4 -,341,78

SAFETY PROFILE: A deadly poison by intravenous, subcutaneous, and intramuscular routes.

HIG500 CAS:1435-55-8 ***HR: 3***
HYDROQUINIDINE
mf: $C_{20}H_{26}N_2O_2$ mw: 326.48

PROP: Plates from ether, needles from alc. Mp: 169°. Very sol in hot alc; sltly sol in water and ether.

SYNS: 10,11-DIHYDRO-6′-METHOXYCINCHONAN-9-OL ◇ DIHYDROQUINIDINE ◇ 10,11-DIHYDROQUINIDINE ◇ HYDROCONQUININE

TOXICITY DATA with REFERENCE
orl-rat LD50:369 mg/kg ARZNAD 27,589,77
ivn-rat LD50:32 mg/kg ARZNAD 27,589,77
ivn-mus LD50:56500 μg/kg JPETAB 84,184,45

SAFETY PROFILE: A poison by ingestion and intravenous routes. When heated to decomposition it emits toxic fumes of NO_x.

HIH000 CAS:123-31-9 ***HR: 3***
HYDROQUINONE
DOT: UN 2662
mf: $C_6H_6O_2$ mw: 110.12

PROP: Colorless, hexagonal prisms. Mp: 170.5°, bp: 286.2°, flash p: 329°F (CC), d: 1.358 @ 20°/4°, autoign temp: 960°F (CC), vap press: 1 mm @ 132.4°, vap d: 3.81. Very sol in alc and ether; sltly sol in benzene. Keep well closed and protected from light.

SYNS: ARCTUVIN ◇ p-BENZENEDIOL ◇ 1,4-BENZENEDIOL ◇ BENZOHYDROQUINONE ◇ BENZOQUINOL ◇ BLACK AND WHITE BLEACHING CREAM ◇ 1,4-DIHYDROXY-BENZEEN (DUTCH) ◇ 1,4-DIHYDROXYBENZEN (CZECH) ◇ DIHYDROXYBENZENE ◇ p-DIHYDROXYBENZENE ◇ 1,4-DIHYDROXYBENZENE ◇ 1,4-DIHYDROXY-BENZOL (GERMAN) ◇ 1,4-DIIDROBENZENE (ITALIAN) ◇ p-DIOXOBENZENE ◇ ELDOPAQUE ◇ ELDOQUIN ◇ HYDROCHINON (CZECH, POLISH) ◇ HYDROQUINOL ◇ α-HYDROQUINONE ◇ p-HYDROQUINONE ◇ p-HYDROXYPHENOL ◇ IDROCHINONE (ITALIAN) ◇ NCI-C55834 ◇ β-QUINOL ◇ TECQUINOL ◇ TENOX HQ ◇ USAF EK-356

TOXICITY DATA with REFERENCE
skn-hmn 2% MLD ARDEAC 93,589,66
skn-hmn 5% SEV ARDEAC 93,589,66
oms-hmn:lym 5 μmol/L CNREA8 45,2471,85
sce-hmn:lym 5 μmol/L CNREA8 45,2471,85
mnt-mus-orl 200 mg/kg AJIMD8 7,475,85
scu-rat TDLo:5100 mg/kg (male 51D pre):REP SSZBAC 12,491,65
orl-hmn LDLo:29 mg/kg 34ZIAG -,321,69
orl-hmn TDLo:170 mg/kg:CNS,CVS,PUL AMILAN 7,79,27
orl-rat LD50:320 mg/kg FEPRA7 8,348,49
ipr-rat LD50:170 mg/kg JIHTAB 31,79,49
scu-rat LDLo:300 mg/kg HBTXAC 1,162,56
ivn-rat LD50:115 mg/kg FEPRA7 8,348,49
orl-mus LD50:245 mg/kg GISAAA 38(8),6,73
ipr-mus LD50:100 mg/kg NTIS** AD414-344
scu-mus LD50:182 mg/kg ZGIMAL 2,333,47

CONSENSUS REPORTS: IARC Cancer Review: Group 3 IMEMDT 7,56,87; Animal Inadequate Evidence IMEMDT 15,155,77. Community Right-To-Know List. EPA Extremely Hazardous Substances List. EPA Genetic Toxicology Program. Reported in EPA TSCA Inventory.

OSHA PEL: TWA 2 mg/m^3
ACGIH TLV: TWA 2 mg/m^3
DFG MAK: 2 mg/m^3
NIOSH REL: (Hydroquinone) CL 2.0 mg/m^3/15M
DOT Classification: IMO: Poison B; Label: St. Andrews Cross.

SAFETY PROFILE: A human poison by ingestion. An experimental poison by ingestion, intraperitoneal, intravenous, and subcutaneous routes. Human systemic effects by ingestion: pulse rate increase without fall in blood pressure, cyanosis, coma. An active allergen and a strong skin irritant. Human mutation data reported. A severe human skin irritant. Experimental reproductive data. Questionable carcinogen.

Absorption of this material by tissues can cause symptoms of illness which resemble those induced by its o- or m- isomers. For instance, the ingestion of 1 gram by an adult or a smaller quantity by a child may induce tinnitis, nausea, dizziness, a sensation of suffocation, an increased rate of respiration, vomiting, pallor, muscular twitching, headache, dyspnea, cyanosis, delirium, and collapse. The literature contains reports of fatal cases which have been caused by the ingestion of 5-12 grams. Cases of dermatitis have resulted from skin contact, and have also followed the application of an antiseptic oil which apparently contained traces of hydroquinone added as an antioxidant. The report also contains cases of keratitis and discoloration of the conjunctiva among personnel exposed to this material in concentrations ranging from 10 to 30 mg of the vapor or dust per cubic meter of air. It is considered to be more toxic than phenol. The inhalation of vapors, particularly when liberated at high temperatures, must be avoided. If this material accidentally comes into contact with the skin, it should be removed at once and the affected area washed with plenty of soap and water.

Combustible when exposed to heat or flame; can react with oxidizing materials. Potentially explosive reaction with oxygen at 90°C/100 bar. Violent reaction with NaOH. Slight explosion hazard when exposed to heat. To fight fire, use water, CO_2, dry chemical.

HIH100 CAS:497-76-7 ***HR: D***
HYDROQUINONE-β-d-GLUCOPYRANOSIDE
mf: $C_{12}H_{16}O_7$ mw: 272.28

SYNS: ARBUTIN ◇ β-d-GLUCOPYRANOSIDE, 4-HYDROXYPHENYL-(9CI) ◇ URSIN ◇ UVASOL

TOXICITY DATA with REFERENCE
orl-rat TDLo:13600 mg/kg (female 14D pre):TER IYKEDH 19,282,88
orl-rat TDLo:13600 mg/kg (female 14D pre):REP IYKEDH 19,282,88

SAFETY PROFILE: An experimental teratogen. Experimental reproductive effects. When heated to decomposition it emits acrid smoke and irritating fumes.

HII000 CAS:1886-45-9 ***HR: 3***
HYDROTHIADENE
mf: $C_{19}H_{23}NS$ mw: 297.49

SYNS: dl-11-(3-DIMETHYLAMINOPROPYL)-6,11-DIHYDRODIBENZO(b,d)THIEPIN ◇ IDROTIADENE

TOXICITY DATA with REFERENCE
orl-mus LD50:290 mg/kg 27ZQAG -,76,72
ivn-mus LD50:36 mg/kg 27ZQAG -,76,72

SAFETY PROFILE: Poison by ingestion and intravenous routes. When heated to decomposition it emits very toxic fumes of NO_x and SO_x.

HII500 CAS:133-67-5 ***HR: 2***
HYDROTRICHLOROTHIAZIDE
mf: $C_8H_8Cl_3N_3O_4S_2$ mw: 380.66

SYNS: ACHLETIN ◇ ANATRAN ◇ ANISTADIN ◇ APONORIN ◇ CARVACRON ◇ 6-CHLORO-3-(DICHLOROMETHYL)-3,4-DIHYDRO-2H-1,2,4-BENZOTHIADIAZINE-7-SULFONAMIDE-1,1-DIOXIDE◇ 6-CHLORO-3-(DICHLOROMETHYL)3,4-DIHYDRO-7-SULFAMYL-1,2,4-BENZOTHIADIAZINE-1,1-DIOXIDE◇ 3-DICHLOROMETHYL-6-CHLORO-7-SULFAMOYL-3,4-DIHYDRO-1,2,4-BENZOTHIADIAZINE-1,1-DIOXIDE ◇ 3-DICHLOROMETHYL-6-CHLORO-7-SULFAMYL-3,4-DIHYDRO-1,2,4-BENZOTHIADIAZINE-1,1-DIOXIDE◇ DIURESE ◇ ESMARIN ◇ EURINOL ◇ FLUITRAN ◇ FLUTRA ◇ GANGESOL ◇ INTROMENE ◇ KUBACRON ◇ METAHYDRIN ◇ NAKVA ◇ NAQUA ◇ SALURIN ◇ TACHIONIN ◇ TOLCASONE ◇ TRICHLORMETAZID ◇ TRICHLORMETHIAZIDE ◇ TRICHLOROMETHIADIAZIDE ◇ TRICHLOROMETHIAZIDE ◇ TRICLORDIURIDE ◇ TRICLORMETIAZIDE (ITALIAN) ◇ TRIFLUMEN

TOXICITY DATA with REFERENCE
cyt-ham:fbr 750 mg/L ESKHA5 96,55,78
cyt-ham:lng 560 mg/L GMCRDC 27,95,81
orl-rat LD50:5600 mg/kg JPETAB 140,249,63
ivn-rat LD50:920 mg/kg 29ZVAB -,119,69
orl-mus LD50:2600 mg/kg FRPSAX 16,647,61
ipr-mus LD50:540 mg/kg FRPSAX 16,647,61
ivn-mus LD50:750 mg/kg 29ZVAB -,119,69

CONSENSUS REPORTS: Reported in EPA TSCA Inventory.

SAFETY PROFILE: Moderately toxic by ingestion, intraperitoneal, and intravenous routes. Mutation data reported. Used as a diuretic and antihypertensive drug. When heated to decomposition it emits very toxic fumes such as SO_x, Cl^- and NO_x.

HII600 CAS:90-87-9 ***HR: 2***
HYDROTROPIC ALDEHYDE DIMETHYL ACETAL
mf: $C_{11}H_{16}O_2$ mw: 180.27

SYNS: 1,1-DIMETHOXY-2-PHENYLPROPANE◇ HYDROTROPALDEHYDE DIMETHYL ACETAL ◇ 2-PHENYLPROPIONALDEHYDE DIMETHYL ACETAL

TOXICITY DATA with REFERENCE
skn-rbt 500 mg/24H MOD FCTXAV 17,819,79
orl-rat LD50:1850 mg/kg FCTXAV 17,819,79

CONSENSUS REPORTS: Reported in EPA TSCA Inventory.

SAFETY PROFILE: Moderately toxic by ingestion. A skin irritant. When heated to decomposition it emits acrid smoke and irritating fumes.

HIJ000 CAS:38234-12-7 ***HR: 3***
12-HYDROXYABIETIC ACID BIS(2-CHLOROETHYL)AMINE
mf: $C_{20}H_{30}O_3{\cdot}C_4H_9Cl_2N$ mw: 460.54

SYNS: 12-HYDROXYABIETIC ACID BIS(2-CHLOROETHYL)AMINE SALT ◇ 12-HYDROXY-13-ISOPROPYLPODOCARPA-7,13-DIEN-15-OIC ACID BIS(2-CHLOROETHYL)AMINE SALT

TOXICITY DATA with REFERENCE
ipr-rat LD50:200 mg/kg PCJOAU 6,647,72
ipr-mus LD50:400 mg/kg PCJOAU 6,647,72

SAFETY PROFILE: Poison by intraperitoneal route.

When heated to decomposition it emits toxic fumes of NO_x. See also AMINES.

HIJ400 CAS:2784-86-3 ***HR: 2***
1-HYDROXY-2-ACETAMIDOFLUORENE
mf: $C_{15}H_{13}NO_2$ mw: 239.29

SYN: N-(1-HYDROXY-2-FLUORENYL)ACETAMIDE

TOXICITY DATA with REFERENCE
imp-mus TDLo:96 mg/kg:CAR GANMAX 17,383,75
imp-mus TD:75 mg/kg:ETA CALEDQ 6,21,79

SAFETY PROFILE: Questionable carcinogen with experimental carcinogenic and tumorigenic data. When heated to decomposition it emits toxic fumes of NO_x.

HIJ500 CAS:5254-41-1 ***HR: 2***
HYDROXY-2-(β-(2′-ACETAMIDOPHENYL) ETHYLNAPHTHAMIDE
mf: $C_{21}H_{20}N_2O_3$ mw: 348.43

SYN: N-(o-(ACETYLAMINO)PHENETHYL)-1-HYDROXY-2-NAPHTHALENECARBOXAMIDE

TOXICITY DATA with REFERENCE
ipr-rat LDLo:800 mg/kg KODAK* -,-,71

CONSENSUS REPORTS: Reported in EPA TSCA Inventory.

SAFETY PROFILE: Moderately toxic by intraperitoneal route. When heated to decomposition it emits toxic fumes of NO_x.

HIK000 CAS:843-34-5 ***HR: 3***
trans-4′-HYDROXY-4-ACETAMIDOSTILBENE
mf: $C_{16}H_{15}NO_2$ mw: 253.32

SYNS: trans-4′-HYDROXY-AAS ◇ (E)-4′-(p-HYDROXYSTYRYL) ACETANILIDE

TOXICITY DATA with REFERENCE
orl-rat TDLo:713 mg/kg/36W-C:ETA BJCAAI 22,133,68

SAFETY PROFILE: Questionable carcinogen with experimental tumorigenic data. When heated to decomposition it emits toxic fumes of NO_x.

HIK500 CAS:363-49-5 ***HR: 3***
7-HYDROXY-2-ACETAMINOFLUORENE
mf: $C_{15}H_{13}NO_2$ mw: 239.290

SYNS: 7OH-2AAF ◇ 7-HYDROXY-N-2-FLUORENYLACETAMIDE

TOXICITY DATA with REFERENCE
mma-sat 250 μg/plate JJIND8 62,893,79
dnr-esc 10 mg/L JJIND8 62,873,79
orl-rat TDLo:14 g/kg/2Y-C:NEO BJCAAI 1,391,47

CONSENSUS REPORTS: EPA Genetic Toxicology Program.

SAFETY PROFILE: Questionable carcinogen with experimental neoplastigenic data. Mutation data reported. When heated to decomposition it emits toxic fumes of NO_x.

HIL000 CAS:614-80-2 ***HR: 1***
2′-HYDROXYACETANILIDE
mf: $C_8H_9NO_2$ mw: 151.18

SYNS: o-ACETAMIDOPHENOL ◇ 2-ACETAMIDOPHENOL ◇ ACET-o-AMINOFENOL (CZECH) ◇ 2-ACETAMINOPHENOL ◇ N-ACETYL-2-AMINOPHENOL ◇ o-(ACETYLAMINO)PHENOL ◇ 2-(ACETYLAMINO) PHENOL ◇ o-HYDROXYACETANILIDE

TOXICITY DATA with REFERENCE
eye-rbt 500 mg/24H MLD 28ZPAK -,106,72
orl-rat LD50:4960 mg/kg 28ZPAK -,106,72

CONSENSUS REPORTS: Reported in EPA TSCA Inventory.

SAFETY PROFILE: Mildly toxic by ingestion. An eye irritant. When heated to decomposition it emits toxic fumes of NO_x.

HIL500 CAS:621-42-1 ***HR: 2***
3-HYDROXYACETANILIDE
mf: $C_8H_9NO_2$ mw: 151.165

SYNS: m-ACETAMIDOPHENOL ◇ 3-ACETAMIDOPHENOL ◇ m-(ACETYLAMINO)PHENOL ◇ 3-(ACETYLAMINO)PHENOL ◇ m-HYDROXYACETANILIDE

TOXICITY DATA with REFERENCE
ipr-mus LD50:1025 mg/kg RCOCB8 28,447,80

CONSENSUS REPORTS: Reported in EPA TSCA Inventory.

SAFETY PROFILE: Moderately toxic by intraperitoneal route. When heated to decomposition it emits toxic fumes of NO_x.

HIM000 CAS:103-90-2 ***HR: 3***
4′-HYDROXYACETANILIDE
mf: $C_8H_9NO_2$ mw: 151.18

SYNS: ABENSANIL ◇ ACAMOL ◇ ACETAGESIC ◇ ACETALGIN ◇ p-ACETAMIDOPHENOL ◇ 4-ACETAMIDOPHENOL ◇ ACETAMINOPHEN ◇ p-ACETAMINOPHENOL ◇ N-ACETYL-p-AMINOPHENOL ◇ p-ACETYLAMINOPHENOL ◇ ALGOTROPYL ◇ ALPINYL ◇ ALVEDON ◇ AMADIL ◇ ANAFLON ◇ ANELIX ◇ ANHIBA ◇ APADON ◇ APAMIDE ◇ APAP ◇ BEN-U-RON ◇ BICKIE-MOL ◇ CALPOL ◇ CETADOL ◇ CLIXODYNE ◇ DATRIL ◇ DIAL-A-GESIC ◇ DIROX ◇ DOLIPRANE ◇ DYMADON ◇ ENELFA ◇ ENERIL ◇ EXDOL ◇ FEBRILIX ◇ FEBRO-GESIC ◇ FEBROLIN ◇ FENDON ◇ FINIMAL ◇ G 1 ◇ GELOCATIL ◇ HEDEX ◇ HOMOOLAN ◇ p-HYDROXYACETANILIDE ◇ 4-HYDROXYACETANILIDE ◇ N-(4-HYDROXYPHENYL)ACETAMIDE ◇ JANUPAP ◇ KORUM ◇ LESTEMP ◇ LIQUAGESIC ◇ LONARID ◇ LYTECA SYRUP ◇ MOMENTUM ◇ MULTIN ◇ NAPA ◇ NAPRINOL ◇ NCI-C55801 ◇ NOBEDON ◇ PACEMO ◇ PANADOL ◇ PANETS ◇ PANEX ◇ PANOFEN ◇ PARACETAMOLE ◇ PARACETAMOLO (ITALIAN) ◇ PARACETANOL ◇ PARAPAN ◇ PARASPEN ◇ PARMOL ◇ PEDRIC ◇ PYRINAZINE ◇ SK-Apap ◇ TABALGIN ◇ TAPAR ◇ TEMLO ◇ TEM-

PANAL ◇ TEMPRA ◇ TRALGON ◇ TUSSAPAP ◇ TYLENOL ◇ VALADOL ◇ VALGESIC

TOXICITY DATA with REFERENCE
oms-hmn:lym 200 mg/L NEZAAQ 37,673,82
cyt-hmn:lym 200 mg/L NEZAAQ 37,673,82
cyt-mus-orl 50 mg/kg CYTBAI 27,27,80
orl-wmn TDLo:650 mg/kg (female 29W post):REP ADCHAK 58,631,83
orl-rat TDLo:12500 mg/kg (female 14D pre):TER TJADAB 26(1),42A,82
orl-rat TDLo:164 g/kg/78W-C:CAR ACPADQ 93,367,85
orl-mus TDLo:135 g/kg/77W-C:CAR CRNGDP 4,363,83
orl-mus TD:270 g/kg/77W-C:ETA CRNGDP 4,363,83
orl-man LDLo:714 mg/kg:LVR HUTODJ 1,25,81
orl-inf TDLo:1440 mg/kg/6D:CNS,GIT,MET AJDCAI 137,386,83
orl-hmn LDLo:143 mg/kg:CNS BMJOAE 282,199,81
orl-chd LDLo:360 mg/kg/2D JOPDAB 92,832,78
orl-chd TDLo:801 mg/kg:CNS,GIT,LIV PEDIAU 61,68,78
orl-wmn TDLo:160 mg/kg:KID SAMJAF 67,791,85
orl-hmn LDLo:357 mg/kg LANCAO 1,66,73
orl-wmn LDLo:260 mg/kg JAMAAP 236,1874,76
orl-wmn TDLo:490 mg/kg:CNS,GIT,KID SMJOAV 71,906,78
orl-man TDLo:77 mg/kg:LIV CTOXAO 15,476,79
orl-rat LD50:2400 mg/kg TXAPA9 18,185,71
ipr-rat LD50:1205 mg/kg SSSEAK 57,561,79
orl-mus LD50:338 mg/kg TXAPA9 19,20,71
ipr-mus LD50:367 mg/kg ARZNAD 15,520,65
scu-mus LD50:310 mg/kg HUTODJ 3,13S,84
ivn-dog LDLo:826 mg/kg AIPTAK 149,571,64

CONSENSUS REPORTS: Reported in EPA TSCA Inventory. EPA Genetic Toxicology Program.

SAFETY PROFILE: Suspected carcinogen with experimental carcinogenic and tumorigenic data. A human poison by ingestion. An experimental poison by intraperitoneal and subcutaneous routes. Moderately toxic by intravenous routes. Human systemic effects by ingestion: changes in exocrine pancreas, diarrhea, nausea, irritability, somnolence, general anesthetic, fever, hepatitis, kidney tubule damage. Experimental teratogenic and reproductive effects. Human mutation data reported. Used as an analgesic and antipyretic. When heated to decomposition it emits toxic fumes of NO_x.

HIM500 CAS:107-16-4 ***HR: 3***
HYDROXYACETONITRILE
mf: C_2H_3NO mw: 57.06

SYNS: CYANOMETHANOL ◇ FORMALDEHYDE CYANOHYDRIN ◇ GLYCOLIC NITRILE ◇ GLYCOLONITRILE ◇ GLYCONITRILE ◇ 2-HYDROXYACETONITRILE ◇ HYDROXYMETHYLINITRILE ◇ USAF A-8565

TOXICITY DATA with REFERENCE
eye-rbt 26 mg MOD 34ZIAG -,290,69
orl-rat LD50:16 mg/kg AIHAAP 23,95,62
ihl-rat LCLo:27 ppm/8H 34ZIAG -,290,69
orl-mus LD50:10 mg/kg 34ZIAG -,290,69
ihl-mus LCLo:27 ppm/8H 34ZIAG -,290,69
ipr-mus LD50:10 mg/kg NTIS** AD277-689
scu-mus LDLo:15 mg/kg AIPTAK 12,447,04
skn-rbt LD50:5 mg/kg AIHAAP 23,95,62
ocu-rbt LDLo:13 mg/kg 34ZIAG -,290,69

CONSENSUS REPORTS: EPA Extremely Hazardous Substances List. Reported in EPA TSCA Inventory. Cyanide and its compounds are on The Community Right-To-Know List.

NIOSH REL: (Nitriles) CL 5 mg/m^3/15M

SAFETY PROFILE: A poison by ingestion, skin contact, inhalation, intraperitoneal, ocular, and subcutaneous routes. A eye irritant. May undergo spontaneous and violent decomposition. Traces of alkali promote violent polymerization. When heated to decomposition it emits toxic fumes of NO_x and CN^-. See also NITRILES.

HIN000 CAS:19315-64-1 ***HR: 3***
N-HYDROXY-p-ACETOPHENETIDIDE
mf: $C_{10}H_{13}NO_3$ mw: 195.24

SYNS: N-(4-ETHOXYPHENYL)ACETOHYDROXAMIC ACID ◇ N-(4-ETHOXYPHENYL)-N-HYDROXYACETAMIDE ◇ N-HYDROXYPHENACETIN

TOXICITY DATA with REFERENCE
mma-sat 500 μg/plate MUREAV 58,387,78
otr-mus:emb 100 μg/L NCIMAV 58,21,81
orl-rat TDLo:15 g/kg/73W-C:NEO PTLGAX 8,1,76
ipr-mus LD50:702 mg/kg ARTODN 56,96,84

CONSENSUS REPORTS: IARC Cancer Review: Animal Limited Evidence IMEMDT 13,141,77; IMEMDT 24,135,80.

SAFETY PROFILE: Moderately toxic by intraperitoneal route. Questionable carcinogen with experimental neoplastigenic data. Mutation data reported. When heated to decomposition it emits toxic fumes of NO_x.

HIN500 CAS:118-93-4 ***HR: 3***
o-HYDROXYACETOPHENONE
mf: $C_8H_8O_2$ mw: 136.16

PROP: Greenish-yellow liquid, highly refractive, minty odor. Mp: 95°, vap d: 4.69.

SYNS: o-ACETYLPHENOL ◇ 2-ACETYLPHENOL ◇ 2'-HYDROXYACETOPHENONE ◇ o-HYDROXYPHENYL METHYL KETONE ◇ USAF KE-20

TOXICITY DATA with REFERENCE
ipr-mus LD50:100 mg/kg NTIS** AD277-689

CONSENSUS REPORTS: Reported in EPA TSCA Inventory.

SAFETY PROFILE: Poison by intraperitoneal route. When heated to decomposition it emits acrid smoke and fumes. See also KETONES.

HIO000 CAS:99-93-4 ***HR: 3***
p-HYDROXYACETOPHENONE
mf: $C_8H_8O_2$ mw: 136.16

SYNS: p-ACETYLPHENOL ◇ 4-ACETYLPHENOL ◇ 4'-HYDROXYACETOPHENONE ◇ 1-(4-HYDROXYPHENYL)ETHANONE ◇ p-HYDROXYPHENYL METHYL KETONE ◇ METHYL-p-HYDROXYPHENYL KETONE ◇ p-OXYACETOPHENONE ◇ PICEOL ◇ USAF KF-15

TOXICITY DATA with REFERENCE
orl-mus LD50:1500 mg/kg MEXPAG 11,137,64
ipr-mus LD50:200 mg/kg NTIS** AD277-689

CONSENSUS REPORTS: Reported in EPA TSCA Inventory.

SAFETY PROFILE: Poison by intraperitoneal route. Moderately toxic by ingestion. Mutation data reported. When heated to decomposition it emits acrid smoke and fumes. See also KETONES.

HIO875 CAS:1838-56-8 ***HR: 3***
3-HYDROXY-N-ACETYL-2-AMINOFLUORENE
mf: $C_{15}H_{13}NO_2$ mw: 239.29

SYNS: 3-HO-AAF ◇ N-(3-HYDROXY-2-FLUORENYL)ACETAMIDE

TOXICITY DATA with REFERENCE
slt-dmg-par 10 mmol/L IJCNAW 9,284,72
imp-mus TDLo:96 mg/kg:CAR GMCRDC 17,383,75
imp-mus TD:75 mg/kg:ETA CALEDQ 6,21,79

CONSENSUS REPORTS: EPA Genetic Toxicology Program.

SAFETY PROFILE: Questionable carcinogen with experimental carcinogenic and tumorigenic data by implantation. Mutation data reported. When heated to decomposition it emits toxic fumes of NO_x.

HIP000 CAS:53-95-2 ***HR: 3***
N-HYDROXY-N-ACETYL-2-AMINOFLUORENE
mf: $C_{15}H_{13}NO_2$ mw: 239.29

SYNS: FLUORENYL-2-ACETHYDROXAMIC ACID ◇ N-FLUOREN-2-YL ACETOHYDROXAMIC ACID ◇ N-2-FLUORENYL ACETOHYDROXAMIC ACID ◇ N-HYDROXY-AAF ◇ N-HYDROXY-2-ACETAMIDOFLUORENE ◇ 2-(N-HYDROXYACETAMIDO)FLUORENE ◇ N-HYDROXY-2-ACETYLAMINOFLUORENE ◇ N-HYDROXY-2-FAA ◇ N-HYDROXY-N-(2-FLUORENYL)ACETAMIDE ◇ NOHFAA

TOXICITY DATA with REFERENCE
mmo-sat 1 μg/plate ENMUDM 6(Suppl 2),1,84
dnd-hmn:hla 25 μmol/L MUREAV 89,95,81
dns-hmn:oth 1 μmol/L JJIND8 72,847,84
dnd-mus:lvr 20 μmol/L CRNGDP 5,797,84
msc-ham:lng 25 μmol/L MUREAV 149,265,85
ipr-mus TDLo:150 mg/kg (1D male):REP TXAPA9 23,288,72
orl-rat TDLo:1000 mg/kg:CAR CNREA8 26,619,66
ipr-rat TDLo:120 mg/kg/4W-I:CAR CNREA8 32,1554,72
imp-rat TDLo:24 mg/kg:NEO CNREA8 37,111,77
orl-rbt TDLo:9800 mg/kg/56W-I:ETA CNREA8 27,838,67
ipr-gpg TDLo:4700 mg/kg/26W-I:CAR,TER CNREA8 24,2018,64
ipr-rat LD50:52 mg/kg CNREA8 35,2959,75
ipr-mus LD50:1500 mg/kg JJIND8 62,911,79

CONSENSUS REPORTS: EPA Genetic Toxicology Program.

SAFETY PROFILE: Suspected carcinogen with experimental carcinogenic, neoplastigenic, and tumorigenic data. A poison by intraperitoneal route. Experimental teratogenic and other reproductive effects. Human mutation data reported. When heated to decomposition it emits toxic fumes of NO_x.

HIP500 CAS:14751-90-7 ***HR: 3***
N-HYDROXY-2-ACETYLAMINOFLUORENE, CUPRIC CHELATE
mf: $C_{30}H_{24}N_2O_4 \cdot Cu$ mw: 540.10

SYNS: COPPER CHELATE of N-HYDROXY-2-ACETYLAMINOFLUORENE ◇ CUPRIC CHELATE of 2-N-HYDROXYFLUORENYL ACETAMIDE ◇ N-FLUOREN-2-YL ACETOHYDROXAMIC ACID, COPPER(2+) COMPLEX ◇ N-HYDROXY-N-2-FLUORENYL ACETAMIDE, CUPRIC CHELATE ◇ 2-N-HYDROXYFLUORENYL ACETAMIDE, CUPRIC CHELATE

TOXICITY DATA with REFERENCE
orl-rat TDLo:12 g/kg/17W-C:ETA CNREA8 25,527,65
scu-rat TDLo:160 mg/kg/4W-I:NEO CNREA8 25,527,65

CONSENSUS REPORTS: Copper and its compounds are on the Community Right-To-Know List.

SAFETY PROFILE: Questionable carcinogen with experimental tumorigenic and neoplastigenic data. When heated to decomposition it emits toxic fumes of NO_x. See also COPPER COMPOUNDS.

HIQ000 CAS:63904-81-4 ***HR: 3***
N-HYDROXY-2-ACETYLAMINOFLUORENE, FERRIC CHELATE
mf: $C_{45}H_{36}N_3O_6 \cdot Fe$ mw: 770.69

SYNS: N-FLUOREN-2-YL ACETOHYDROXAMIC ACID, IRON(3+) COMPLEX ◇ IRON complex with N-FLUOREN-2-YL ACETOHYDROXAMIC ACID

TOXICITY DATA with REFERENCE
scu-rat TDLo:160 mg/kg/4W-I:NEO CNREA8 25,527,65

SAFETY PROFILE: Questionable carcinogen with ex-

perimental neoplastigenic data. When heated to decomposition it emits toxic fumes of NO_x.

HIQ500 CAS:2495-54-7 ***HR: 3***
N-HYDROXY-2-ACETYLAMINOFLUORENE-o-GLUCURONIDE
mf: $C_{21}H_{21}NO_9$ mw: 431.43

SYN: ACETOHYDROXAMIC ACID, FLUOREN-2-YL-o-GLUCURONIDE

TOXICITY DATA with REFERENCE
scu-rat TDLo:850 mg/kg/9W-I:CAR CNREA8 31,1645,71

SAFETY PROFILE: Questionable carcinogen with experimental carcinogenic data. When heated to decomposition it emits toxic fumes of NO_x.

HIR000 CAS:14751-74-7 ***HR: 3***
N-HYDROXY-2-ACETYLAMINOFLUORENE, MANGANOUS CHELATE
mf: $C_{30}H_{24}N_2O_4$•Mn mw: 531.50

SYNS: N-FLUOREN-2-YL ACETOHYDROXAMIC ACID, MANGANESE(2+) COMPLEX ◇ MANGANESE complex with N-FLUOREN-2-YL ACETOHYDROXAMIC ACID

TOXICITY DATA with REFERENCE
scu-rat TDLo:160 mg/kg/4W-I:NEO CNREA8 25,527,65

CONSENSUS REPORTS: Manganese and its compounds are on the Community Right-To-Know List.

OSHA PEL: CL 5 mg(Mn)/m^3
ACGIH TLV: TWA 5 mg(Mn)/m^3

SAFETY PROFILE: Questionable carcinogen with experimental neoplastigenic data. When heated to decomposition it emits toxic fumes of NO_x. See also MANGANESE COMPOUNDS.

HIR500 CAS:14751-76-9 ***HR: 3***
N-HYDROXY-2-ACETYLAMINOFLUORENE, NICKELOUS CHELATE
mf: $C_{30}H_{24}N_2O_4$•Ni mw: 535.27

SYNS: N-FLUOREN-2-YL ACETOHYDROXAMIC ACID, NICKEL(2+) COMPLEX ◇ NICKEL COMPLEX with N-FLUOREN-2-YL ACETOHYDROXAMIC ACID

TOXICITY DATA with REFERENCE
scu-rat TDLo:160 mg/kg/4W-I:NEO CNREA8 25,527,65

CONSENSUS REPORTS: Nickel and its compounds are on the Community Right-To-Know List.

NIOSH REL: (Inorganic Nickel) TWA 0.015 mg(Ni)/m^3

SAFETY PROFILE: Questionable carcinogen with experimental neoplastigenic data. When heated to decomposition it emits toxic fumes of NO_x. See also NICKEL COMPOUNDS.

HIS000 CAS:6023-26-3 ***HR: 3***
N-HYDROXY-2-ACETYLAMINOFLUORENE, POTASSIUM SALT
mf: $C_{15}H_{13}NO_2$•K mw: 278.39

SYNS: N-FLUOREN-2-YL ACETOHYDROXAMIC ACID, POTASSIUM SALT ◇ POTASSIUM N-FLUOREN-2-YL ACETOHYDROXAMATE

TOXICITY DATA with REFERENCE
scu-rat TDLo:160 mg/kg/4W-I:ETA CNREA8 25,527,65

SAFETY PROFILE: Questionable carcinogen with experimental tumorigenic data. When heated to decomposition it emits toxic fumes of NO_x.

HIS300 CAS:79127-36-9 ***HR: 3***
2-HYDROXYACLACINOMYCIN A
mf: $C_{42}H_{53}NO_{16}$ mw: 827.96

TOXICITY DATA with REFERENCE
dni-mus:leu 1030 nmol/L JANTAJ 34,1596,81
oms-mus:leu 170 nmol/L JANTAJ 34,1596,81
ipr-mus LD50:50 mg/kg JANTAJ 35,82-107,82

SAFETY PROFILE: Poison by intraperitoneal route. Mutation data reported. When heated to decomposition it emits toxic fumes of NO_x.

HIT000 CAS:5667-20-9 ***HR: 2***
N-HYDROXYADENINE
mf: $C_5H_5N_5O$ mw: 151.15

SYNS: 6-HYDROXYAMINOPURINE ◇ 6-HYDROXYLAMINOPURINE ◇ 6-N-HYDROXYLAMINOPURINE ◇ N-HYDROXY-1H-PURIN-6-AMINE (9CI)

TOXICITY DATA with REFERENCE
mmo-sat 100 ng/plate MUREAV 144,231,85
mmo-nsc 600 mg/L MUREAV 124,61,83
dnd-ham:lng 5 mg/L MUREAV 82,355,81
msc-ham:emb 5 mg/L PNASA6 78,5685,81
ipr-rat TDLo:400 mg/kg (11D preg):TER ARPAAQ 86,395,68
ipr-rat LD50:800 mg/kg ARPAAQ 86,395,68

SAFETY PROFILE: Moderately toxic by intraperitoneal route. An experimental teratogen. Mutation data reported. When heated to decomposition it emits toxic fumes of NO_x.

HIT100 CAS:3414-62-8 ***HR: 2***
6-N-HYDROXYADENOSINE
mf: $C_{10}H_{13}N_5O_5$ mw: 283.28

SYNS: N-HYDROXYADENOSINE ◇ N^6-HYDROXYADENOSINE ◇ 6-HYDROXYADENOSINE ◇ 6-HYDROXYLAMINOPURINE RIBOSIDE ◇ N^6-HYDROXYLAMINOPURINE RIBOSIDE

TOXICITY DATA with REFERENCE
mmo-sat 100 ng/plate MUREAV 144,231,85

ipr-rat TDLo:700 mg/kg (11D preg):TER ARPAAQ 86,395,68
ipr-rat LD50:700 mg/kg ARPAAQ 86,395,68

SAFETY PROFILE: Moderately toxic by intraperitoneal route. An experimental teratogen. Mutation data reported. When heated to decomposition it emits toxic fumes of NO_x.

HIT500 CAS:141-31-1 ***HR: 1***
HYDROXYADIPALDEHYDE
mf: $C_6H_{10}O_3$ mw: 130.16

PROP: Liquid. Fp: −3.5°, d: 1.066 @ 20°/20°, vap press: 17 mm @ 20°.

SYNS: α-HYDROXYADIPALDEHYDE ◇ 2-HYDROXYADIPALDEHYDE ◇ 2-HYDROXYHEXANEDIAL

TOXICITY DATA with REFERENCE
skn-rbt 10 mg/24H open MLD AMIHBC 10,61,54
skn-rbt 500 mg MLD SCCUR* -,5,61
eye-rbt 500 mg open AMIHBC 10,61,54
orl-rat LD50:17 g/kg AMIHBC 10,61,54
ihl-rat LCLo:3000 ppm/4H SCCUR* -,5,61
orl-mus LD50:6 g/kg SCCUR* -,5,61
ihl-mus LCLo:1500 ppm/14H SCCUR* -,5,61

SAFETY PROFILE: Mildly toxic by inhalation and ingestion. A skin and eye irritant. When heated to decomposition it emits acrid smoke and fumes. See also ALDEHYDES.

HIU500 CAS:53-94-1 ***HR: 3***
N-HYDROXY-2-AMINOFLUORENE
mf: $C_{13}H_{11}NO$ mw: 197.25

SYNS: 2-FLUORENYL HYDROXYLAMINE ◇ N-FLUOREN-2-YLHYDROXYLAMINE

TOXICITY DATA with REFERENCE
mmo-sat 50 ng/plate CBINA8 54,71,85
dns-hmn:oth 100 nmol/L JJIND8 72,847,84
dnd-mus:lvr 5 μmol/L CRNGDP 5,797,84
scu-rat TDLo:420 mg/kg/9W-I:NEO CNREA8 31,1645,71
ipr-gpg TDLo:1600 mg/kg/17W-I:CAR CNREA8 24,2018,64

CONSENSUS REPORTS: EPA Genetic Toxicology Program.

SAFETY PROFILE: Questionable carcinogen with experimental carcinogenic and neoplastigenic data. Human mutation data reported. When heated to decomposition it emits toxic fumes of NO_x.

HIV000 CAS:13442-07-4 ***HR: 3***
4-(HYDROXYAMINO)-5-METHYLQUINOLINE-1-OXIDE
mf: $C_{10}H_{10}N_2O_2$ mw: 190.22

SYN: 5-METHYL-4-HYDROXYLAMINOQUINOLINE-1-OXIDE

TOXICITY DATA with REFERENCE
scu-mus TDLo:120 mg/kg/50D-I:ETA BCPCA6 16,631,67

SAFETY PROFILE: Questionable carcinogen with experimental tumorigenic data. When heated to decomposition it emits toxic fumes of NO_x.

HIV500 CAS:13442-08-5 ***HR: 3***
4-(HYDROXYAMINO)-6-METHYLQUINOLINE-1-OXIDE
mf: $C_{10}H_{10}N_2O_2$ mw: 190.22

SYN: 6-METHYL-4-HYDROXYLAMINOQUINOLINE-1-OXIDE

TOXICITY DATA with REFERENCE
pic-esc 5100 μg/L EXPEAM 24,1245,68
scu-mus TDLo:120 mg/kg/50D-I:ETA BCPCA6 16,631,67

SAFETY PROFILE: Questionable carcinogen with experimental tumorigenic data. Mutation data reported. When heated to decomposition it emits toxic fumes of NO_x.

HIW000 CAS:13442-09-6 ***HR: 3***
4-(HYDROXYAMINO)-7-METHYLQUINOLINE-1-OXIDE
mf: $C_{10}H_{10}N_2O_2$ mw: 190.22

SYN: 7-METHYL-4-HYDROXYLAMINOQUINOLINE-1-OXIDE

TOXICITY DATA with REFERENCE
scu-mus TDLo:120 mg/kg/50D-I:ETA BCPCA6 16,631,67

SAFETY PROFILE: Questionable carcinogen with experimental tumorigenic data. When heated to decomposition it emits toxic fumes of NO_x.

HIW500 CAS:13442-10-9 ***HR: 3***
4-(HYDROXYAMINO)-8-METHYLQUINOLINE-1-OXIDE
mf: $C_{10}H_{10}N_2O_2$ mw: 190.22

SYN: 8-METHYL-4-HYDROXYLAMINOQUINOLINE-1-OXIDE

TOXICITY DATA with REFERENCE
u-mus TDLo:120 mg/kg/50D-I:ETA BCPCA6 16,631,67

SAFETY PROFILE: Questionable carcinogen with experimental tumorigenic data. When heated to decomposition it emits toxic fumes of NO_x.

HIX000 CAS:607-30-7 ***HR: 3***
N-HYDROXY-1-AMINONAPHTHALENE
mf: $C_{10}H_9NO$ mw: 159.20

SYNS: N-HYDROXY-1-NAPHTHYLAMINE ◇ 1-NAPHTHYLHYDROXYLAMINE ◇ N-1-NAPHTHYLHYDROXYLAMINE

TOXICITY DATA with REFERENCE
mmo-sat 1 μg/plate MUREAV 122,243,83

mma-sat 1 μg/plate PNASA6 72,5135,75
otr-hmn:oth 2 mg/L ITCSAF 17,719,81
dnd-rat-scu 106 μmol/kg CNREA8 44,1172,84
oms-rat-scu 106 μmol/kg CNREA8 44,1172,84
ipr-rat TDLo:1300 mg/kg/13W-I:NEO CNREA8 28,535,68
ipr-rat TD:1200 mg/kg/12W-I:ETA CNREA8 31,1461,71

CONSENSUS REPORTS: EPA Genetic Toxicology Program.

SAFETY PROFILE: Questionable carcinogen with experimental neoplastigenic and tumorigenic data. Human mutation data reported. When heated to decomposition it emits toxic fumes of NO_x. See also AMINES.

HIX200 CAS:53130-67-9 ***HR: 3***
2-HYDROXYAMINO-1,4-NAPHTHOQUINONE
mf: $C_{10}H_7NO_3$ mw: 189.18

SYN: HANQ

TOXICITY DATA with REFERENCE
dnd-mus:ast 50 μmol/L CPBTAL 26,1031,78
dnd-mus-ipr 100 mg/kg CPBTAL 26,1031,78
ipr-mus LD50:318 mg/kg CPBTAL 23,1077,75

SAFETY PROFILE: Poison by intraperitoneal route. Mutation data reported. When heated to decomposition it emits toxic fumes of NO_x.

HIX500 CAS:13442-15-4 ***HR: 3***
4-(HYDROXYAMINO)-6-NITROQUINOLINE-1-OXIDE
mf: $C_9H_7N_3O_4$ mw: 221.19

SYN: 6-NITRO-4-HYDROXYLAMINOQUINOLINE-1-OXIDE

TOXICITY DATA with REFERENCE
scu-mus TDLo:60 mg/kg/I:ETA CPBTAL 17,544,69

SAFETY PROFILE: Questionable carcinogen with experimental tumorigenic data. When heated to decomposition it emits toxic fumes of NO_x.

HIY000 CAS:13442-16-5 ***HR: 3***
4-(HYDROXYAMINO)-7-NITROQUINOLINE-1-OXIDE
mf: $C_9H_7N_3O_4$ mw: 221.19

SYN: 7-NITRO-4-HYDROXYLAMINOQUINOLINE-1-OXIDE

TOXICITY DATA with REFERENCE
scu-mus TDLo:120 mg/kg/50D-I:ETA BCPCA6 16,631,67

SAFETY PROFILE: Questionable carcinogen with experimental tumorigenic data. When heated to decomposition it emits toxic fumes of NO_x.

HIY500 CAS:4637-56-3 ***HR: 3***
4-(HYDROXYAMINO)QUINOLINE-1-OXIDE
mf: $C_9H_8N_2O_2$ mw: 176.19

SYNS: 4HAQO ◇ N-(4-QUINOLYL)HYDROXYLAMINE-1′-OXIDE

TOXICITY DATA with REFERENCE
mmo-sat 300 ng/plate ENMUDM 6(Suppl 2),1,84
mma-esc 10 μg/plate ENMUDM 6(Suppl 2),1,84
dnd-hmn:fbr 10 μmol/L BBACAQ 781,273,84
dns-hmn:oth 100 nmol/L JJIND8 69,557,82
dns-rat-ivn 7 mg/kg CBINA8 56,125,85
dnd-ckn:emb 200 μmol/L CBINA8 47,123,83
ipr-rat TDLo:5 mg/kg:ETA JNCIAM 53,159,74
scu-rat TDLo:16 mg/kg/4W-I:CAR IGAYAY 120,1218,82
par-rat TDLo:10 mg/kg:CAR CANCAR 50,2057,82
skn-mus TDLo:1350 mg/kg/18W-C:NEO GANNA2 60,161,69
ivn-rat LDLo:5 mg/kg JNCIAM 53,159,74

CONSENSUS REPORTS: EPA Genetic Toxicology Program.

SAFETY PROFILE: A poison by intravenous route. Questionable carcinogen with experimental carcinogenic, neoplastigenic, and tumorigenic data. Human mutation data reported. When heated to decomposition it emits toxic fumes of NO_x.

HIZ000 CAS:1010-61-3 ***HR: 3***
4-(HYDROXYAMINO)QUINOLINE-1-OXIDE, HYDROCHLORIDE
mf: $C_9H_8N_2O_2$•ClH mw: 212.65

SYNS: 4-HYDROXYAMINOQUINOLINE-1-OXIDEHYDROCHLORIDE ◇ 4-HYDROXYAMINOQUINOLINE-1-OXIDE MONOHYDROCHLORIDE ◇ 4-(HYDROXYAMINO)QUINOLINE-1-OXIDE MONOHYDROCHLORIDE ◇ 4-HYDROXYQUINOLINAMINE-1-OXIDE MONOHYDROCHLORIDE ◇ N-HYDROXY-4-QUINOLINAMINE-1-OXIDE MONOHYDROCHLORIDE ◇ NSC 78572

TOXICITY DATA with REFERENCE
mmo-esc 25 mg/L CPBTAL 13,610,65
cyt-omi 495 μmol/L GANNA2 60,155,69
mmo-smc 26 mg/L TXAPA9 15,451,69
orl-rat TDLo:150 mg/kg/45W-I:ETA GANNA2 60,627,69
scu-rat TDLo:16 mg/kg/2W-I:NEO PSEBAA 136,1206,71
scu-mus TDLo:451 μg/kg:CAR JJEMAG 40,475,70
ipr-mus LD50:20440 μg/kg NCISP* JAN86

SAFETY PROFILE: Poison by intraperitoneal route. Questionable carcinogen with experimental carcinogenic, neoplastigenic, and tumorigenic data. Mutation data reported. When heated to decomposition it emits very toxic fumes of NO_x and HCl.

HJA000 CAS:60462-51-3 ***HR: 3***
trans-N-HYDROXY-4-AMINOSTILBENE
mf: $C_{14}H_{13}NO$ mw: 211.28

SYNS: (E)-N-(p-STYRYLPHENYL)HYDROXYLAMINE ◇ trans-N-(p-STYRYLPHENYL)HYDROXYLAMINE ◇ trans-N-(4-STYRYLPHENYL) HYDROXYLAMINE

TOXICITY DATA with REFERENCE
scu-rat TDLo:18 mg/kg/1W-I:NEO CNREA8 24,128,64

SAFETY PROFILE: Questionable carcinogen with experimental neoplastigenic data. When heated to decomposition it emits toxic fumes of NO_x. See also AMINES.

HJA500 CAS:623-10-9 ***HR: D***
4-(HYDROXYAMINO)TOLUENE
mf: C_7H_9NO mw: 123.17

SYNS: 4-METHYLPHENYLHYDROXYLAMINE ◇ N-(p-METHYLPHENYL)HYDROXYLAMINE ◇ N-(4-METHYLPHENYL)HYDROXYLAMINE ◇ p-TOLYL-HYDROXYLAMIN (GERMAN) ◇ N-(p-TOLYL)HYDROXYLAMINE ◇ N-p-TOLYHYDROXYLAMINE ◇ p-TOLYHYDROXYLAMINE

TOXICITY DATA with REFERENCE
mmo-sat 5 mg/L MUREAV 136,159,84
mmo-esc 3750 nmol/L MUREAV 151,201,85

SAFETY PROFILE: Mutation data reported. When heated to decomposition it emits toxic fumes of NO_x.

HJB050 CAS:53-41-8 ***HR: D***
3-α-HYDROXY-17-ANDROSTANONE
mf: $C_{19}H_{30}O_2$ mw: 290.49

SYNS: ANDROSTAN-17-ONE, 3-HYDROXY-, (3-α-5-α)- ◇ 5-α-ANDROSTAN-3-α-OL-17-ONE ◇ 5-α-ANDROSTAN-17-ONE, 3-α-HYDROXY- ◇ ANDROSTANON-3-α-OL-17-ONE ◇ ANDROSTERONE ◇ cis-ANDROSTERONE ◇ 3-EPIHYDROXYETIOALLOCHOLAN-17-ONE ◇ 3-α-HYDROXY-5-α-ANDROSTAN-17-ONE ◇ 3-α-HYDROXYETIOALLOCHOLAN-17-ONE

TOXICITY DATA with REFERENCE
scu-rat TDLo:8400 μg/kg (male 21D pre):REP ENDOAO 21,313,37
ims-rat TDLo:80 mg/kg (female 13-20D post):TER ENDOAO 99,1490,76

SAFETY PROFILE: An experimental teratogen. Experimental reproductive effects. When heated to decomposition it emits acrid smoke and irritating fumes.

HJB100 CAS:571-22-2 ***HR: 3***
17-β-HYDROXY-5-β-ANDROSTAN-3-ONE
mf: $C_{19}H_{30}O_2$ mw: 290.49

SYNS: 5-β-DHT ◇ 5-β-DIHYDROTESTOSTERONE ◇ ETIOCHOLANE-17-β-OL-3-ONE ◇ ETIOCHOLAN-17-β-OL-3-ONE

TOXICITY DATA with REFERENCE
dni-hmn:lym 50 μmol/L PSEBAA 146,401,74
scu-mus TDLo:160 mg/kg (12-15D preg):REP CNREA8 37,4456,77
scu-mus TDLo:160 mg/kg (12-15D preg):TER CNREA8 37,4456,77
scu-mus TDLo:240 mg/kg/5D-I:NEO CNREA8 37,4456,77

SAFETY PROFILE: Questionable carcinogen with experimental neoplastigenic and teratogenic data. Human mutation data reported. A steroid. When heated to decomposition it emits acrid smoke and irritating fumes.

HJB200 CAS:566-48-3 ***HR: D***
4-HYDROXY-4-ANDROSTENE-3,17-DIONE
mf: $C_{19}H_{26}O_3$ mw: 302.45

SYN: 4-HYDROXY-μ!MN$_4$-ANDROSTENEDIONE

TOXICITY DATA with REFERENCE
scu-rat TDLo:120 mg/kg (female 11-22D post):REP ECJPAE 36,29,89

SAFETY PROFILE: Experimental reproductive effects. When heated to decomposition it emits acrid smoke and irritating fumes.

HJB500 CAS:877-22-5 ***HR: 2***
2-HYDROXY-m-ANISIC ACID
mf: $C_8H_8O_4$ mw: 168.16

SYNS: ACIDE ORTHOVANILLIQUE ◇ 3-HYDROXY-m-ANISIC ACID ◇ 2-HYDROXY-3-METHOXYBENZOIC ACID ◇ 3-METHOXYSALICYLIC ACID ◇ o-VANILLIC ACID

TOXICITY DATA with REFERENCE
ipr-rat LD50:2056 mg/kg COREAF 243,609,56

SAFETY PROFILE: Moderately toxic by intraperitoneal route. When heated to decomposition it emits acrid smoke and fumes.

HJC000 CAS:645-08-9 ***HR: 2***
3-HYDROXYANISIC ACID
mf: $C_8H_8O_4$ mw: 168.16

SYNS: ACIDE ISOVANILLIQUIE (FRENCH) ◇ 3-HYDROXY-4-METHOXYBENZOIC ACID ◇ ISOVANILLIC ACID

TOXICITY DATA with REFERENCE
ipr-rat LD50:2974 mg/kg COREAF 243,609,56
ipr-mus LD50:3000 mg/kg COREAF 243,609,56

SAFETY PROFILE: Moderately toxic by intraperitoneal route. When heated to decomposition it emits acrid smoke and fumes.

HJC500 CAS:17672-21-8 ***HR: 3***
3-HYDROXYANTHRANILIC ACID METHYL ESTER
mf: $C_8H_9NO_3$ mw: 167.18

SYNS: 2-AMINO-3-HYDROXYBENZOIC ACID, METHYL ESTER ◇ METHYL-3-HYDROXYANTHRANILATE

TOXICITY DATA with REFERENCE
imp-mus TDLo:80 mg/kg:ETA BJCAAI 11,212,57

SAFETY PROFILE: Questionable carcinogen with experimental tumorigenic data. When heated to decomposition it emits toxic fumes of NO_x. See also ESTERS.

HJD000 CAS:484-78-6 ***HR: 3***
3-(3-HYDROXYANTHRANILOYL)ALANINE
mf: $C_{10}H_{12}N_2O_4$ mw: 224.24

SYNS: α,2-DIAMINO-3-HYDROXY-Γ-OXOBENZENEBUTANOIC ACID ◇ HYDROXYKYNURENINE ◇ 3-HYDROXYKYNURENINE

TOXICITY DATA with REFERENCE
cyt-hmn:emb 50 mg/L BEXBAN 67,200,69
cyt-mus-scu 10 mg/L NATUAS 222,484,69
imp-mus TDLo:80 mg/kg:NEO BJCAAI 11,212,57

SAFETY PROFILE: Questionable carcinogen with experimental neoplastigenic data. Human mutation data reported. When heated to decomposition it emits toxic fumes of NO_x.

HJD500 CAS:606-14-4 ***HR: 3***
3-(3-HYDROXYANTHRANILOYL)-l-ALANINE
mf: $C_{10}H_{12}N_2O_4$ mw: 224.24

SYNS: l-3-HYDROXYKYNURENINE ◇ 3-HYDROXY-l-KYNURENINE

TOXICITY DATA with REFERENCE
scu-mus TDLo:4408 mg/kg/21D-I:ETA JCROD7 96,163,80
imp-mus TDLo:160 mg/kg:CAR ANYAA9 108,924,63

SAFETY PROFILE: Questionable carcinogen with experimental carcinogenic and tumorigenic data. When heated to decomposition it emits toxic fumes of NO_x.

HJE000 CAS:129-43-1 ***HR: 1***
1-HYDROXYANTHRAQUINONE
mf: $C_{14}H_8O_3$ mw: 224.22

PROP: Orange-red in ethanol. Mp: 194-195°, bp: subl. Sol in alc, very sol in ether.

SYNS: 1-HYDROXY-9,10-ANTHRACENEDIONE ◇ 1-HYDROXY-ANTHRACHINON (CZECH) ◇ 1-HYDROXY-9,10-ANTHRAQUINONE

TOXICITY DATA with REFERENCE
eye-rbt 500 mg/24H MLD 28ZPAK -,101,72

CONSENSUS REPORTS: Reported in EPA TSCA Inventory.

SAFETY PROFILE: An eye irritant. When heated to decomposition it emits acrid smoke and fumes.

HJE400 CAS:6318-57-6 ***HR: 3***
2-HYDROXY-p-ARSANILIC ACID
mf: $C_6H_8AsNO_4$ mw: 233.07

SYNS: 4-AMINO-2-HYDROXYBENZENEARSONIC ACID ◇ BENZENEARSONIC ACID, 4-AMINO-2-HYDROXY-

TOXICITY DATA with REFERENCE
ivn-mus LD50:180 mg/kg CSLNX* NX#05106

OSHA PEL: TWA 0.5 mg(As)/m^3

SAFETY PROFILE: Poison by intravenous route. When heated to decomposition it emits toxic fumes of NO_x and As.

HJE500 CAS:2163-77-1 ***HR: 3***
4-HYDROXY-3-ARSANILIC ACID
mf: $C_6H_8AsNO_4$ mw: 233.07

SYNS: 3-AMINO-4-HYDROXYBENZENEARSONIC ACID ◇ 4-HYDROXY-m-ARSANILIC ACID

TOXICITY DATA with REFERENCE
orl-rat LDLo:5000 mg/kg JPETAB 63,122,38
ivn-rat LDLo:800 mg/kg JPETAB 63,122,38
ims-rat LD50:500 mg/kg JPETAB 63,122,38
ivn-mus LD50:180 mg/kg CSLNX* NX#05112

CONSENSUS REPORTS: Arsenic and its compounds are on the Community Right-To-Know List.

OSHA PEL: TWA 0.5 mg(As)/m^3

SAFETY PROFILE: Poison by intravenous route. Moderately toxic by intramuscular route. Mildly toxic by ingestion. When heated to decomposition it emits very toxic fumes of As and NO_x. See also ARSENIC COMPOUNDS.

HJE575 CAS:42028-33-1 ***HR: 2***
3-HYDROXY-8-AZAXANTHINE
mf: $C_4H_3N_5O_3$ mw: 169.12

SYN: 4-HYDROXY-3H-o-TRIAZOLO(4,5-d)PYRIMIDINE-5,7(4H-6H)-DIONE

TOXICITY DATA with REFERENCE
scu-rat TDLo:237 mg/kg/22W-I:ETA CNREA8 33,1113,73

SAFETY PROFILE: Questionable carcinogen with experimental tumorigenic data. When heated to decomposition it emits toxic fumes of NO_x.

HJE600 CAS:54301-19-8 ***HR: 3***
14-HYDROXYAZIDOMORPHINE
mf: $C_{17}H_{20}N_4O_3$ mw: 328.41

SYN: (5-α,6-β)-6-AZIDO-4,5-EPOXY-17-METHYL-MORPHINAN-3,14-DIOL

TOXICITY DATA with REFERENCE
orl-rat LD50:190 mg/kg JPPMAB 27,99,75
scu-rat LD50:85 mg/kg JPPMAB 27,99,75
ivn-rat LD50:45 mg/kg JPPMAB 27,99,75
orl-mus LD50:370 mg/kg JPPMAB 27,99,75
scu-mus LD50:185 mg/kg JPPMAB 27,99,75
ivn-mus LD50:45 mg/kg JPPMAB 27,99,75

SAFETY PROFILE: Poison by ingestion, subcutaneous, and intravenous routes. When heated to decomposition it emits toxic fumes of NO_x. See also MORPHINE and AZIDES.

HJF000 CAS:1689-82-3 ***HR: 3***
4-HYDROXYAZOBENZENE
mf: $C_{12}H_{10}N_2O$ mw: 198.24

PROP: Orange, rhombic crystals from ethanol. Mp: 155-156°, bp: 220-230°. Very sol in ether.

SYNS: p-BENZENEAZOPHENOL ◇ C.I. SOLVENT YELLOW 7 ◇ p-HYDROXYAZOBENZENE ◇ p-PHENYLAZOPHENOL ◇ 4-PHENYLAZOPHENOL

TOXICITY DATA with REFERENCE
ipr-mus LD50:75 mg/kg NTIS** AD691-490

CONSENSUS REPORTS: IARC Cancer Review: Group 3 IMEMDT 7,56,87; Animal Inadequate Evidence IMEMDT 8,157,75. Reported in EPA TSCA Inventory.

SAFETY PROFILE: A poison by intraperitoneal route. Questionable carcinogen. When heated to decomposition it emits toxic fumes of NO_x.

HJF500 CAS:3567-69-9 ***HR: 3***
4-HYDROXY-3,4'-AZODI-1-NAPHTHALENESULFONIC ACID, DISODIUM SALT
mf: $C_{20}H_{12}N_2O_7S_2 \cdot 2Na$ mw: 502.44

SYNS: ACETACID RED B ◇ ACID BRILLIANT RUBINE 2G ◇ ACID CHROME BLUE BA ◇ ACID FAST RED FB ◇ ACID RUBINE ◇ AIREDALE CARMOISINE ◇ AMACID CHROME BLUE R ◇ ATUL CRYSTAL RED F ◇ AZORUBIN ◇ BRASILAN AZO RUBINE 2NS ◇ BRILLIANT CRIMSON RED ◇ CARMOISIN (GERMAN) ◇ CARMOISINE ALUMINUM LAKE ◇ CARMOISINE SUPRA ◇ CERTICOL CARMOISINE S ◇ CHROME FAST BLUE 2R ◇ C.I. 14720 ◇ C.I. ACID RED 14, DISODIUM SALT ◇ C.I. FOOD RED 3 ◇ CRIMSON EMBL ◇ DIADEM CHROME BLUE R ◇ DISODIUM SALT of 2-(4-SULPHO-1-NAPHTHYLAZO)-1-NAPHTHOL-4-SULPHONIC ACID ◇ DISODIUM-2-(4-SULFO-1-NAPHTHYLAZO)-1-NAPHTHOL-4-SULFONATE◇ DISODIUM-2-(4-SULPHO-1-NAPHTHYLAZO)-1-NAPHTHOL-4-SULPHONATE ◇ EDICOL SUPRA CARMOISINE WS ◇ ENIACID BRILLIANT RUBINE 3B ◇ EUROCERT AZORUBINE ◇ EXTRACT D&C RED No. 10 ◇ FENAZO RED C ◇ FOOD RED 5 ◇ FRUIT RED A EXTRA YELLOWISH GEIGY ◇ HEXACOL CARMOISINE ◇ HIDACID AZO RUBINE ◇ 4-HYDROXY-3,4'-AZODI-1-NAPHTHALENESULPHONIC ACID, DISODIUM SALT ◇ 4-HYDROXY-3-((4-SULFO-1-NAPHTHALENYL) AZO)-1-NAPHTHALENESULFONIC ACID, DISODIUM SALT ◇ JAVA RUBINE N ◇ KARMESIN ◇ KENACHROME BLUE 2R ◇ KITON CRIMSON 2R ◇ LIGHTHOUSE CHROME BLUE 2R ◇ NACARAT A EXPORT ◇ NCI-C53849 ◇ NEKLACID RUBINE W ◇ NYLOMINE ACID RED P4B ◇ OMEGA CHROME BLUE FB ◇ POLOXAL RED 2B ◇ PONTACYL RUBINE R ◇ RED #14 ◇ 11959 RED ◇ SCHULTZ Nr. 208 (GERMAN) ◇ SOLAR RUBINE ◇ SOLOCHROME BLUE FB ◇ STANDACOL CARMOISINE ◇ 2-(4-SULFO-1-NAPHTHYLAZO)-1-NAPHTHOL-4-SULFONIC ACID, DISODIUM SALT ◇ TERTRACID RED CA ◇ TERTROCHROME BLUE FB

TOXICITY DATA with REFERENCE
mmo-sat 1 g/L MUREAV 53,289,78
mmo-esc 100 mg/L MUREAV 53,289,78
ipr-rat LD50:900 mg/kg FCTXAV 5,179,67
ivn-rat LD50:800 mg/kg APFRAD 15,402,57
ipr-mus LD50:900 mg/kg FCTXAV 5,179,67
ivn-mus LD50:800 mg/kg FCTXAV 19,413,81

CONSENSUS REPORTS: IARC Cancer Review: Group 3 IMEMDT 7,56,87; Animal Inadequate Evidence IMEMDT 8,83,75. NTP Carcinogenesis Bioassay (feed); No Evidence: mouse, rat NTPTR* NTP-TR-220,82. Reported in EPA TSCA Inventory. EPA Genetic Toxicology Program.

SAFETY PROFILE: Moderately toxic by intraperitoneal and intravenous routes. Questionable carcinogen. Mutation data reported. When heated to decomposition it emits very toxic fumes of SO_x, Na_2O, and NO_x.

HJG000 CAS:57598-00-2 ***HR: 3***
4'-HYDROXY-2,3'-AZOTOLUENE
mf: $C_{14}H_{14}N_2O$ mw: 226.30

SYN: p-(o-TOLYLAZO)-o-CRESOL

TOXICITY DATA with REFERENCE
mul-mus TDLo:400 mg/kg/I:ETA CNREA8 1,397,41

SAFETY PROFILE: Questionable carcinogen with experimental tumorigenic data. When heated to decomposition it emits toxic fumes of NO_x.

HJG100 CAS:66267-67-2 ***HR: 1***
α-HYDROXYBENZENEACETIC ACID 2-(2-ETHOXYETHOXY)ETHYL ESTER
mf: $C_{14}H_{20}O_5$ mw: 268.34

SYNS: AI3-36401 ◇ BENZENEACETIC ACID, α-HYDROXY-, 2-(2-ETHOXYETHOXY)ETHYL ESTER

TOXICITY DATA with REFERENCE
eye-rbt 100 mg MLD NTIS** AD-AO55-604

SAFETY PROFILE: An eye irritant. When heated to decomposition it emits acrid smoke and irritating fumes.

HJH000 CAS:68596-89-4 ***HR: 3***
4-HYDROXYBENZENEDIAZONIUM-3-CARBOXYLATE
mf: $C_7H_4N_2O_3$ mw: 164.12

SAFETY PROFILE: Explodes when heated above 155°C. Upon decomposition it emits toxic fumes of NO_x.

HJH500 CAS:7340-50-3 ***HR: 3***
N-HYDROXYBENZENESULFONANILIDE
mf: $C_{12}H_{11}NO_3S$ mw: 249.30

SYN: N-HYDROXYPHENYLBENZENESULFONAMIDE

TOXICITY DATA with REFERENCE
ipr-rat TDLo:623 mg/kg/4W-I:ETA CNREA8 30,1485,70

SAFETY PROFILE: Questionable carcinogen with experimental tumorigenic data. When heated to decomposition it emits very toxic fumes of NO_x and SO_x.

HJI000 CAS:6295-12-1 ***HR: 3***
2-HYDROXY-1,3,2-BENZODIOXASTIBOLE
mf: $C_6H_5O_3Sb$ mw: 246.86

TOXICITY DATA with REFERENCE
ipr-mus LDLo:63 mg/kg CBCCT* 4,227,52

CONSENSUS REPORTS: Antimony and its compounds are on the Community Right-To-Know List.

OSHA PEL: TWA 0.5 mg(Sb)/m^3
ACGIH TLV: TWA 0.5 mg(Sb)/m^3
NIOSH REL: (Antimony) TWA 0.5 mg/m^3

SAFETY PROFILE: Poison by intraperitoneal route. When heated to decomposition it emits toxic fumes of Sb. See also ANTIMONY COMPOUNDS.

HJI100 CAS:99-06-9 ***HR: 2***
3-HYDROXYBENZOIC ACID
mf: $C_7H_6O_3$ mw: 138.13

SYNS: ACIDO-m-IDROSSIBENZOICO (ITALIAN) ◇ 3-CARBOXYPHENOL ◇ m-HBA ◇ m-HYDROXYBENZOIC ACID ◇ m-SALICYLIC ACID

TOXICITY DATA with REFERENCE
scu-rat TDLo:400 mg/kg (11D preg):TER RCOCB8 38,209,82
ipr-rat LD50:3700 mg/kg BCFAAI 112,53,73
orl-mus LD50:2 g/kg QJPPAL 19,483,46

CONSENSUS REPORTS: Reported in EPA TSCA Inventory.

SAFETY PROFILE: Moderately toxic by ingestion and intraperitoneal routes. An experimental teratogen. When heated to decomposition it emits acrid smoke and fumes.

HJI500 CAS:36457-20-2 ***HR: 3***
p-HYDROXYBENZOIC ACID BUTYL ESTER, SODIUM SALT
mf: $C_{11}H_{13}O_3$•Na mw: 216.23

TOXICITY DATA with REFERENCE
orl-mus LD50:950 mg/kg JAPMA8 45,260,56
ipr-mus LD50:230 mg/kg JAPMA8 45,260,56

SAFETY PROFILE: Poison by intraperitoneal route. Moderately toxic by ingestion. When heated to decomposition it emits toxic fumes of Na_2O. See also ESTERS.

HJL000 CAS:120-47-8 ***HR: 2***
p-HYDROXYBENZOIC ACID ETHYL ESTER
mf: $C_9H_{10}O_3$ mw: 166.19

SYNS: ASEPTOFORM E ◇ BONOMOLD OE ◇ p-CARBETHOXYPHENOL ◇ EASEPTOL ◇ ETHYL-p-HYDROXYBENZOATE ◇ ETHYL PARABEN ◇ ETHYL PARASEPT ◇ p-HYDROXYBENZOIC ETHYL ESTER ◇ NIPAGIN A ◇ NIPAZIN A ◇ p-OXYBENZOESAEUREAETHYLESTER (GERMAN) ◇ SOLBROL A ◇ TEGOSEPT E

TOXICITY DATA with REFERENCE
mmo-esc 10 mmol/L ZBPIA9 112,226,59
cyt-ham:fbr 250 mg/L ESKHA5 96,55,78
cyt-ham:lng 440 mg/L GMCRDC 27,95,81
orl-rat TDLo:45600 mg/kg (8-15D preg):TER AOGLAR 22,94,75
orl-mus LD50:3 g/kg BCTKAG 14,301,84
ipr-mus LD50:520 mg/kg DRSTAT 20,89,52
orl-dog LDLo:5 g/kg AEPPAE 146,208,29
orl-rbt LDLo:5 g/kg AEPPAE 146,208,29
orl-gpg LDLo:2000 mg/kg FAONAU 40,20,67

CONSENSUS REPORTS: Reported in EPA TSCA Inventory.

SAFETY PROFILE: Moderately toxic by ingestion and intraperitoneal routes. Experimental teratogenic effects. Mutation data reported. When heated to decomposition it emits acrid smoke and fumes. See also ESTERS.

HJL500 CAS:99-76-3 ***HR: 2***
p-HYDROXYBENZOIC ACID METHYL ESTER
mf: $C_8H_8O_3$ mw: 152.16

PROP: Colorless crystals or white crystalline powder; faint odor and burning taste. Sol in alc, ether, and propylene glycol; sltly sol in water, glycerin, fixed oils, benzene, and carbon tetrachloride;

SYNS: ABIOL ◇ ASEPTOFORM ◇ MASEPTOL ◇ METHYLBEN ◇ METHYL CHEMOSEPT ◇ METHYL ESTER of p-HYDROXYBENZOIC ACID ◇ METHYL p-HYDROXYBENZOATE ◇ METHYL p-OXYBENZOATE ◇ METHYLPARABEN (FCC) ◇ METHYL PARAHYDROXYBENZOATE ◇ METHYL PARASEPT ◇ METOXYDE ◇ MOLDEX ◇ NIPAGIN ◇ p-OXYBENZOESAUREMETHYLESTER (GERMAN) ◇ PARABEN ◇ PARASEPT ◇ PARIDOL ◇ PRESERVAL M ◇ SEPTOS ◇ SOLBROL M ◇ TEGOSEPT M

TOXICITY DATA with REFERENCE
cyt-ham:lng 125 mg/L/27H MUREAV 66,277,79
cyt-ham:fbr 500 mg/L ESKHA5 96,55,78
ipr-mus LD50:960 mg/kg JAPMA8 45,260,56
scu-mus LD50:1200 mg/kg AIPTAK 128,135,60
orl-dog LD50:3000 mg/kg 14CYAT 2,1897,63
orl-rbt LDLo:3 g/kg AEPPAE 146,208,29
orl-gpg LDLo:3000 mg/kg FAONAU 40,23,67

CONSENSUS REPORTS: Reported in EPA TSCA Inventory.

SAFETY PROFILE: Moderately toxic by ingestion, subcutaneous and intraperitoneal routes. Mutation data reported. When heated to decomposition it emits acrid smoke and fumes. See also ESTERS.

HJM000 CAS:114-63-6 ***HR: 2***
p-HYDROXYBENZOIC ACID, SODIUM SALT
mf: $C_7H_5O_3 \cdot Na$ mw: 160.11

PROP: Colorless plates from ethanol.

TOXICITY DATA with REFERENCE
orl-mus LD50:2200 mg/kg JAPMA8 45,260,56
ivn-mus LD50:1200 mg/kg JAPMA8 45,260,56

CONSENSUS REPORTS: Reported in EPA TSCA Inventory.

SAFETY PROFILE: Moderately toxic by ingestion and intravenous routes. When heated to decomposition it emits toxic fumes of Na_2O.

HJN000 CAS:767-00-0 ***HR: 3***
4-HYDROXYBENZONITRILE
mf: C_7H_5NO mw: 119.13

PROP: Prisms of benzene. Mp: 97-98°, bp: 149°. Very sltly sol in water; very sol in alc and ether.

SYN: p-HYDROXYBENZONITRILE

TOXICITY DATA with REFERENCE
orl-mus LD50:450 mg/kg APFRAD 41,391,83
ipr-mus LD50:200 mg/kg NTIS** AD691-490

CONSENSUS REPORTS: Cyanide and its compounds are on the Community Right-To-Know List. Reported in EPA TSCA Inventory.

SAFETY PROFILE: Poison by intraperitoneal route. Moderately toxic by ingestion. When heated to decomposition it emits toxic fumes of NO_x and CN^-. See also NITRILES.

HJN500 CAS:37574-48-4 ***HR: D***
4-HYDROXYBENZO(a)PYRENE

SYNS: BENZO(a)PYREN-4-OL ◇ 4-HYDROXYBENZ(a)PYRENE

TOXICITY DATA with REFERENCE
mmo-sat 16 μg/plate MUREAV 36,379,76
dni-omi 200 μg/L PNASA6 74,1378,77

CONSENSUS REPORTS: EPA Genetic Toxicology Program.

SAFETY PROFILE: Mutation data reported. When heated to decomposition it emits toxic fumes of NO_x.

HJN650 CAS:2592-95-2 ***HR: 3***
1-HYDROXYBENZOTRIAZOLE
mf: $C_6H_5N_3O$ mw: 135.12

SAFETY PROFILE: Potentially explosive if heated above 160°C. Upon decomposition it emits toxic fumes of NO_x.

HJN875 CAS:60254-95-7 ***HR: D***
5-(α-HYDROXYBENZYL)-2-BENZIMIDAZOLE-CARBAMIC ACID METHYL ESTER
mf: $C_{16}H_{15}N_3O_3$ mw: 297.34

TOXICITY DATA with REFERENCE
oms-hmn:oth 2 mg/L THERAP 31,505,76
orl-rat TDLo:59200 μg/kg (8-15D preg):TER THERAP 31,505,76
ipr-mus LDLo:2 g/kg IJEBA6 25,871,87

SAFETY PROFILE: Experimental teratogenic effects. Human mutation data reported. When heated to decomposition it emits toxic fumes of NO_x. See also CARBAMATES and ESTERS.

HJO500 CAS:469-65-8 ***HR: 3***
(p-HYDROXYBENZYL)TARTARIC ACID
mf: $C_{11}H_{12}O_7$ mw: 256.23

PROP: From the bark of the Jamaica dogwood (AIPTAK 14,53,05).

SYNS: 2,3-DIHYDROXY-2-((4-HYDROXYPHENYL)METHYL)BUTANEDIOIC ACID ◇ PISCIDEIN ◇ PISCIDIC ACID

TOXICITY DATA with REFERENCE
ipr-cat LDLo:11 mg/kg AIPTAK 14,53,05
ipr-rbt LDLo:125 mg/kg AIPTAK 14,53,05
scu-gpg LDLo:200 mg/kg AIPTAK 14,53,05
par-frg LDLo:152 mg/kg AIPTAK 14,53,05

SAFETY PROFILE: A poison by intraperitoneal, subcutaneous, and parenteral routes. When heated to decomposition it emits acrid smoke and fumes.

HJP500 CAS:26690-77-7 ***HR: 3***
N-HYDROXY-4-BIPHENYLYLBENZAMIDE
mf: $C_{19}H_{15}NO_2$ mw: 289.35

SYNS: N-4-BIPHENYLYLBENZOHYDROXAMIC ACID ◇ N-HYDROXY-4-BIPHENYLBENZAMIDE

TOXICITY DATA with REFERENCE
ipr-rat TDLo:561 mg/kg/4W-I:CAR CNREA8 30,1485,70

SAFETY PROFILE: Questionable carcinogen with experimental carcinogenic data. When heated to decomposition it emits toxic fumes of NO_x.

HJP575 CAS:19899-80-0 ***HR: 3***
2-HYDROXY-4,6-BIS(NITROAMINO)-1,3,5-TRIAZINE
mf: $C_3H_3N_7O_5$ mw: 217.10

$$HOC{=}NC(NHNO_2){=}NC(NHNO_2){=}N$$

SAFETY PROFILE: An impact-sensitive explosive. Upon decomposition it emits toxic fumes of NO_x.

HJQ000 CAS:5809-59-6 ***HR: 3***
2-HYDROXY-3-BUTENENITRILE
mf: C_4H_5NO mw: 83.10

$N{\equiv}CCHOHCH{=}CH_2$

SYNS: ACROLEIN CYANOHYDRIN ◇ 1-CYANO-2-PROPEN-1-OL ◇ VINYLGLYCOLONITRILE

TOXICITY DATA with REFERENCE
skn-rbt 10 mg/24H open AMIHBC 10,61,54
eye-rbt 50 μg open SEV AMIHBC 10,61,54
orl-rat LD50:65 mg/kg AMIHBC 10,61,54
ihl-rat LC50:16 ppm/4H AMIHBC 10,61,54
skn-rbt LD50:7500 μg/kg AMIHBC 10,61,54

CONSENSUS REPORTS: Cyanide and its compounds are on the Community Right-To-Know List.

SAFETY PROFILE: A poison by ingestion, skin contact, and inhalation. A skin and severe eye irritant. May polymerize explosively when exposed to light and air above 25°C. A storage hazard. When heated to decomposition it emits toxic fumes of NO_x and CN^-. See also NITRILES.

HJQ350 CAS:3817-11-6 ***HR: 3***
4-HYDROXYBUTYLBUTYLNITROSAMINE
mf: $C_8H_{18}N_2O_2$ mw: 174.28

SYNS: BBN ◇ BBNOH ◇ BHBN ◇ BUTANOL (4)-BUTYL-NITROSAMINE ◇ BUTYL-BUTANOL(4)-NITROSAMIN ◇ BUTYL-BUTANOL-NITROSAMINE ◇ N-BUTYL-N-(4-HYDROXYBUTYL)NITROSAMINE ◇ n-BUTYL-(4-HYDROXYBUTYL)NITROSAMINE ◇ 4-(BUTYLNITROSAMINO)-1-BUTANOL ◇ 4-(n-BUTYLNITROSAMINO)-1-BUTANOL ◇ DIBUTYLAMINE, 4-HYDROXY-N-NITROSO- ◇ HBBN ◇ NBHA ◇ N-NITROSO-n-BUTYL-(4-HYDROXYBUTYL)AMINE ◇ OH-BBN

TOXICITY DATA with REFERENCE
mma-sat 10 μmol/plate MUREAV 140,147,84
sce-rat:lvr 1500 μmol/L MUREAV 93,409,82
cyt-ham:lng 900 mg/L/27H MUREAV 66,277,79
orl-rat TDLo:560 mg/kg/8W-C:CAR GANNA2 71,138,80
orl-mus TDLo:800 mg/kg/10W-I:CAR CRNGDP 2,251,81
orl-rat LD50:1800 mg/kg XENOBH 3,271,73
scu-ham LD50:3 g/kg CALEDQ 1,15,75

CONSENSUS REPORTS: IARC Cancer Review: Animal Sufficient Evidence IMEMDT 17,51,78

SAFETY PROFILE: Confirmed carcinogen with experimental carcinogenic and neoplastigenic data. Moderately toxic by ingestion. Mutation data reported. When heated to decomposition it emits toxic fumes of NO_x.

HJQ500 ***HR: 3***
1-HYDROXY-3-BUTYL HYDROPEROXIDE
mf: $C_4H_{10}O_3$ mw: 106.12

$HOC_2H_4CH(CH_3)OOH$

SAFETY PROFILE: An impact-sensitive explosive. When heated to decomposition it emits acrid smoke and fumes. See also PEROXIDES.

HJR000 CAS:63934-40-7 ***HR: D***
3-HYDROXYBUTYL-(2-HYDROXYPROPYL)-N-NITROSAMINE
mf: $C_7H_{16}N_2O_3$ mw: 176.25

SYN: 4-(N-(2-HYDROXYPROPYL)-N-NITROSOAMINO)-2-BUTANOL

TOXICITY DATA with REFERENCE
mmo-sat 50 μg/plate MUREAV 68,195,79

SAFETY PROFILE: Mutation data reported. Many N-nitroso compounds are carcinogens. When heated to decomposition it emits toxic fumes of NO_x. See also NITROSAMINES.

HJR500 CAS:63934-41-8 ***HR: D***
4-HYDROXYBUTYL-(2-HYDROXYPROPYL)-N-NITROSAMINE
mf: $C_7H_{16}N_2O_3$ mw: 176.25

SYN: 4-(N-(2-HYDROXYPROPYL)-N-NITROSOAMINO)-1-BUTANOL

TOXICITY DATA with REFERENCE
mmo-sat 100 μg/plate MUREAV 68,195,79

SAFETY PROFILE: Mutation data reported. Many N-nitroso compounds are carcinogens. When heated to decomposition it emits toxic fumes of NO_x. See also NITROSAMINES.

HJS000 CAS:62018-90-0 ***HR: D***
4-HYDROXYBUTYL-(3-HYDROXYPROPYL)-N-NITROSAMINE
mf: $C_7H_{16}N_2O_3$ mw: 176.25

SYN: 4-(N-(3-HYDROXYPROPYL)-N-NITROSOAMINO)-1-BUTANOL

TOXICITY DATA with REFERENCE
mmo-sat 1 mg/plate MUREAV 68,195,79

SAFETY PROFILE: Mutation data reported. Many N-nitroso compounds are carcinogens. When heated to decomposition it emits toxic fumes of NO_x. See also NITROSAMINES.

HJS400 CAS:61424-17-7 ***HR: 2***
4-HYDROXYBUTYL(2-PROPENYL)NITROSAMINE
mf: $C_7H_{14}N_2O_2$ mw: 158.23

SYNS: 4-(ALLYLNITROSAMINO)-1-BUTANOL ◇ 1-BUTANOL, 4-(NITROSO-2-PROPENYLAMINO)-

TOXICITY DATA with REFERENCE
scu-ham TDLo:22500 mg/kg/75W-I:CAR CDPRD4 4,79,81

SAFETY PROFILE: Questionable carcinogen with ex-

perimental carcinogenic data. When heated to decomposition it emits toxic fumes of NO_x.

HJS500 CAS:502-85-2 ***HR: 2***
4-HYDROXYBUTYRIC ACID SODIUM SALT
mf: $C_4H_7O_3$•Na mw: 126.10

SYNS: GAMMA OH ◇ Γ-HYDROXYBUTYRATE SODIUM SALT ◇ SODIUM-Γ-HYDROXYBUTYRATE ◇ SODIUM-4-HYDROXYBUTYRATE ◇ SODIUM OXYBATE ◇ SOMSANIT ◇ WY-3478

TOXICITY DATA with REFERENCE
ivn-gpg TDLo:40 mg/kg (68D preg):REP AJOGAH 100,72,68
orl-rat LD50:9690 mg/kg FATOAO 43,714,80
ipr-rat LD50:1650 mg/kg THERAP 32,375,77
ipr-mus LD50:3330 mg/kg JJPAAZ 17,30,67

CONSENSUS REPORTS: Reported in EPA TSCA Inventory.

SAFETY PROFILE: Moderately toxic by intraperitoneal route. Mildly toxic by ingestion. Experimental reproductive effects. When heated to decomposition it emits toxic fumes of Na_2O.

HJS850 CAS:1083-57-4 ***HR: 2***
3-HYDROXY-p-BUTYROPHENETIDIDE
mf: $C_{12}H_{17}NO_3$ mw: 223.30

SYNS: BETADID ◇ BUCETIN ◇ BUTANAMIDE, N-(4-ETHOXYPHENYL)-3-HYDROXY- ◇ BUTYRANILIDE, 4'-ETHOXY-3-HYDROXY- ◇ 4'-ETHOXY-3-HYDROXYBUTYRANILIDE ◇ β-HYDROXYBUTYRIC ACID-p-PHENETIDIDE ◇ β-OXYBUTTERSAEURE-p-PHENETIDID

TOXICITY DATA with REFERENCE
mma-sat 2 μmol/plate CPBTAL 33,2877,85
orl-mus TDLo:958 g/kg/76W-C:CAR JJIND8 79,1151,87
orl-mus TD:479 g/kg/76W-C:NEO JJIND8 79,1151,87
orl-mus LD50:2800 mg/kg NYKZAU 62,123,66
ipr-mus LD50:790 mg/kg NYKZAU 62,123,66

CONSENSUS REPORTS: Reported in EPA TSCA Inventory.

SAFETY PROFILE: Moderately toxic by ingestion and intraperitoneal routes. Questionable carcinogen with experimental carcinogenic and neoplastigenic data. Mutation data reported. When heated to decomposition it emits toxic fumes of NO_x.

HJU000 CAS:69103-96-4 ***HR: 3***
2-(3-(4-HYDROXY-4-p-CHLOROPHENYL-PIPERIDINO)-PROPYL)-3-METHYL-7-FLUOROCHROMONE
mf: $C_{24}H_{25}ClFNO_3$ mw: 429.95

SYN: 2-(3-(4-(p-CHLOROPHENYL)-4-HYDROXYPIPERIDINE)PROPYL)-7-FLUORO-3-METHYLCHROMONE

TOXICITY DATA with REFERENCE
ipr-mus LD50:175 mg/kg EJMCA5 13,387,78
scu-mus LD50:325 mg/kg EJMCA5 13,387,78

SAFETY PROFILE: Poison by intraperitoneal and subcutaneous routes. When heated to decomposition it emits toxic fumes of Cl^-, F^-, and NO_x.

HJV000 CAS:57651-82-8 ***HR: 3***
1-HYDROXYCHOLECALCIFEROL
mf: $C_{27}H_{44}O_2$ mw: 400.71

PROP: Mp: 134-136° or 138-139.5°.

SYNS: ALFACALCIDOL ◇ 1-α-DIHYDROXYVITAMIN D3 ◇ α-HCC ◇ HYDROXYCHOLECALCIFEROL ◇ 1-α-HYDROXYCHOLECALCIFEROL ◇ 1-α-HYDROXYVITAMIN D3 ◇ 1-α-OH-CC ◇ 1-α-OH-D^3 ◇ 1-α-OH VITAMIN D3 ◇ 9,10-SECOCHOLESTA-5,7,10(19)-TRIENE-1-α,3-β-DIOL ◇ VITAMIN D^3

TOXICITY DATA with REFERENCE
orl-rat TDLo:1100 ng/kg (7-17D preg):REP KSRNAM 12,32,78
orl-rat TDLo:1700 ng/kg (male 60D pre):TER KSRNAM 12,203,78
orl-rat LD50:200 μg/kg PSEBAA 178,668,81
scu-rat LD50:40 μg/kg IYKEDH 12,668,81
ivn-rat LD50:101 μg/kg IYKEDH 12,668,81
orl-mus LD50:440 μg/kg TXAPA9 36,323,76
scu-mus LD50:58 μg/kg IYKEDH 12,668,81
ivn-mus LD50:56 μg/kg TXAPA9 36,323,76
orl-dog LD50:500 μg/kg IYKEDH 9,103,78
ivn-dog LD50:200 μg/kg IYKEDH 9,103,78

SAFETY PROFILE: A deadly poison by ingestion, subcutaneous, and intravenous routes. Experimental teratogenic and reproductive effects. When heated to decomposition it emits acrid smoke and fumes.

HJV500 CAS:69853-71-0 ***HR: 3***
6-HYDROXYCHOLEST-4-EN-3-ONE
mf: $C_{27}H_{44}O_2$ mw: 400.71

TOXICITY DATA with REFERENCE
scu-mus TDLo:600 mg/kg/72W-I:ETA JNCIAM 19,977,57

SAFETY PROFILE: Questionable carcinogen with experimental tumorigenic data. When heated to decomposition it emits acrid smoke and fumes.

HJV700 ***HR: 1***
HYDROXYCITRONELLAL DIMETHYL ACETAL
mf: $C_{12}H_{26}O_3$ mw: 218.34

PROP: Colorless liquid; floral odor. D: 0.925, refr index: 1.441, flash p: +212°F. Sol in fixed oils, propylene glycol; insol in glycerin.

SYNS: FEMA No. 2585 ◇ 7-HYDROXY-3,7-DIMETHYL OCTANAL:-ACETAL

SAFETY PROFILE: Combustible liquid. When heated to decomposition it emits acrid smoke and irritating fumes.

HJX500 CAS:508-54-3 ***HR: 3***
14-HYDROXYCODEINONE
mf: $C_{18}H_{19}NO_4$ mw: 313.38

PROP: Decomp @ 275°; very sol in $CHCl_3$, methyl cellosolve, petr ether, ethyl acetate; sltly sol in alc; insol in H_2O, ether.

SYNS: 7,8-DIDEHYDRO-4,5-EPOXY-14-HYDROXY-3-METHOXY-17-METHYLMORPHINAN-5-α-6-ONE ◇ 7,8-DIDEHYDRO-4,5-α-EPOXY-14-HYDROXY-3-METHOXY-17-METHYLMORPHINAN-6-ONE ◇ HYDROXYCODEINONE ◇ 14-β-HYDROXYCODEINONE

TOXICITY DATA with REFERENCE
orl-mus LD50:1100 mg/kg JPPMAB 17,759,65
scu-mus LD50:23 mg/kg 28ZNAE 138,27,38
ivn-mus LD50:11800 μg/kg JPPMAB 16(Suppl),68T,64

SAFETY PROFILE: Poison by subcutaneous and intravenous routes. Moderately toxic by ingestion. When heated to decomposition it emits toxic fumes of NO_x.

HJX625 CAS:63643-78-7 ***HR: 3***
HYDROXYCOPPER(II) GLYOXIMATE
mf: $C_2H_4CuN_2O_3$ mw: 164.61

CONSENSUS REPORTS: Copper and its compounds are on the Community Right-To-Know List.

SAFETY PROFILE: Explosive decomposition occurs at 140°C. Upon decomposition it emits toxic fumes of NO_x. See also COPPER COMPOUNDS.

HJY000 CAS:1076-38-6 ***HR: 2***
4-HYDROXYCOUMARIN
mf: $C_9H_6O_3$ mw: 162.15

SYN: 4-HYDROXY-2H-1-BENZOPYRAN-2-ONE

TOXICITY DATA with REFERENCE
orl-mus LD50:2 g/kg MPHEAE 17,497,67
ipr-mus LD50:2000 mg/kg APTOA6 2,109,46
scu-mus LD50:750 mg/kg AIPTAK 128,126,60

CONSENSUS REPORTS: Reported in EPA TSCA Inventory.

SAFETY PROFILE: Moderately toxic by ingestion, subcutaneous, and intraperitoneal routes. When heated to decomposition it emits acrid smoke and fumes.

HJY500 CAS:63981-92-0 ***HR: 3***
(4-HYDROXY-o-CUMENYL)TRIMETHYLAMMONIUM CHLORIDE, METHYLCARBAMATE
mf: $C_{14}H_{23}N_2O_2{\bullet}Cl$ mw: 286.84

SYNS: N-METHYLCARBAMICACID-3-ISOPROPYL-4-DIMETHYLAMINOPHENYL ESTER, METHOCHLORIDE ◇ METHYLCARBAMIC ACID, (4-TRIMETHYLAMMONIO)-m-CUMENYL ESTER, CHLORIDE ◇ (4-METHYLCARBAMOYLOXY-o-CUMENYL)TRIMETHYLAMMONIUM CHLORIDE ◇ TL-1345

TOXICITY DATA with REFERENCE
scu-rat LD50:103 μg/kg NTIS** PB158-508
scu-mus LD50:47 μg/kg NTIS** PB158-508
scu-dog LD50:100 μg/kg NTIS** PB158-508
scu-mky LD50:150 μg/kg NTIS** PB158-508
scu-cat LD50:100 μg/kg NTIS** PB158-508
scu-rbt LD50:75 μg/kg NTIS** PB158-508
scu-gpg LD50:50 μg/kg NTIS** PB158-508

SAFETY PROFILE: A deadly poison by subcutaneous route. When heated to decomposition it emits toxic fumes of NO_x, NH_3 and Cl^-. See also CHLORIDES and CARBAMATES.

HJZ000 CAS:63981-53-3 ***HR: 3***
(4-HYDROXY-o-CUMENYL)TRIMETHYLAMMONIUM IODIDE, DIMETHYLCARBAMATE
mf: $C_{15}H_{25}N_2O_2{\bullet}I$ mw: 392.32

SYNS: ((4-(N,N-DIMETHYLCARBAMOYLOXY)-2-ISOPROPYL)PHENYL)TRIMETHYLAMMONIUM IODIDE ◇ N,N-DIMETHYL-4-DIMETHYLAMINO-3-ISOPROPYLPHENYL ESTER METHIODIDE, CARBAMIC ACID ◇ (N-(2-ISOPROPYL-4-DIMETHYLCARBAMOYLOXY)PHENYL)TRIMETHYLAMMONIUM IODIDE ◇ SB-8 ◇ ((4-TRIMETHYLAMMONIO)-3-ISOPROPYL)PHENYL ESTER, DIMETHYLCARBAMIC ACID IODIDE ◇ TL-599

TOXICITY DATA with REFERENCE
scu-rat LDLo:200 μg/kg NTIS** PB158-508
ipr-mus LD50:168 μg/kg NTIS** PB158-508
scu-mus LD50:75 μg/kg JACSAT 63,308,41
scu-dog LDLo:200 μg/kg NTIS** PB158-508
scu-cat LDLo:300 μg/kg NTIS** PB158-508
scu-rbt LDLo:200 μg/kg NTIS** PB158-508
scu-gpg LDLo:100 μg/kg NTIS** PB158-508

SAFETY PROFILE: A deadly poison by intraperitoneal and subcutaneous routes. When heated to decomposition it emits very toxic fumes of NO_x, NH_3, and I^-. See also IODIDES and CARBAMATES.

HKA000 CAS:931-97-5 ***HR: 3***
1-HYDROXYCYCLOHEPTANECARBONITRILE
mf: $C_8H_{13}NO$ mw: 139.22

SYN: 1-HYDROXY-CYCLOHEXANECARBONITRILE

TOXICITY DATA with REFERENCE
ivn-mus LD50:18 mg/kg CSLNX* NX#04232

CONSENSUS REPORTS: Cyanide and its compounds are on the Community Right-To-Know List. Reported in EPA TSCA Inventory.

SAFETY PROFILE: Poison by intravenous route.

When heated to decomposition it emits toxic fumes of CN^- and NO_x. See also NITRILES.

HKA109 CAS:2211-64-5 ***HR: D***
N-HYDROXYCYCLOHEXYLAMINE
mf: $C_6H_{13}NO$ mw: 115.20

SYN: N-CYCLOHEXYLHYDROXYLAMINE

TOXICITY DATA with REFERENCE
cyt-hmn:leu 10 μmol/L/5H MUREAV 39,1,76

SAFETY PROFILE: Human mutation data reported. When heated to decomposition it emits toxic fumes of NO_x.

HKA200 CAS:67292-62-0 ***HR: 3***
4-HYDROXYCYCLOPHOSPHAMIDE
mf: $C_7H_{15}Cl_2N_2O_3P$ mw: 277.11

SYNS: 2-(BIS(2-CHLOROETHYL)AMINO)-4-HYDROXYTETRAHYDRO-2H-1,3,2-OXAZAPHOSPHORINE ◇ 4-OH-CP

TOXICITY DATA with REFERENCE
mmo-sat 10 μg/plate CNREA8 42,3016,82
mma-sat 10 μg/plate CNREA8 42,3016,82
sce-hmn:lym 1 μmol/L MUREAV 129,47,84
dnd-mus:leu 50 μmol/L CNREA8 44,5156,84
ivn-rat LD50:139 mg/kg JMCMAR 18,376,75
ipr-mus LD50:141 mg/kg INNDDK 2,253,84

SAFETY PROFILE: Poison by intravenous and intraperitoneal routes. Human mutation data reported. When heated to decomposition it emits very toxic fumes of Cl^-, PO_x, and NO_x.

HKA300 CAS:25316-40-9 ***HR: 3***
HYDROXYDAUNORUBICIN HYDROCHLORIDE
mf: $C_{27}H_{29}NO_{11}$•ClH mw: 580.03

SYNS: ADM HYDROCHLORIDE ◇ ADR ◇ ADRIACIN ◇ ADRIAMYCIN ◇ ADRIAMYCIN, HYDROCHLORIDE ◇ ADRIBLASTIN ◇ ADRIBLASTINE ◇ DOX HYDROCHLORIDE ◇ DOXORUBICIN ◇ DOXORUBICIN HYDROCHLORIDE ◇ FI 106 ◇ FI 6804

TOXICITY DATA with REFERENCE
dni-mus:leu 1500 nmol/L JMCMAR 22,912,79
oms-mus:leu 580 nmol/L JMCMAR 22,912,79
dni-ckn:emb 900 nmol/L JMCMAR 26,638,83
ipr-rat TDLo:6 mg/kg (male 2W pre):REP REPTED 2,117,88
ivn-mky TDLo:27 mg/kg/2Y-I:ETA PHMGBN 20,9,80
ivn-man LDLo:2571 μg/kg/3W-I:KID AIMDAP 137,385,77
ivn-man TDLo:12 mg/kg/26W-I:CVS,PUL AIMEAS 106,814,87
ipr-rat LD50:16030 μg/kg YAKUD5 21,359,79
scu-rat LD50:21800 μg/kg NIIRDN 6,506,82
ivn-rat LD50:13100 μg/kg NIIRDN 6,506,82
ims-rat LD50:16 mg/kg NIIRDN 6,506,82
orl-mus LD50:698 mg/kg NIIRDN 6,506,82
ipr-mus LD50:11160 μg/kg NCISP* JAN86
scu-mus LD50:7678 μg/kg NCISP* JAN86
ivn-mus LD50:1245 μg/kg NCISP* JAN86
ims-mus LD50:13700 μg/kg NIIRDS 6,506,82
ivn-rbt LD50:6 mg/kg NIIRDN 6,506,82

CONSENSUS REPORTS: EPA Genetic Toxicology Program.

SAFETY PROFILE: Poison by subcutaneous, intramuscular, intravenous, and intraperitoneal routes. Moderately toxic by ingestion. Experimental reproductive effects. Human systemic effects: changes in kidney tubules, cardiomyopathy, acute pulmonary edema. Questionable carcinogen with experimental tumorigenic data. Mutation data reported. An antineoplastic and immunosuppressive agent. When heated to decomposition it emits toxic fumes of NO_x and HCl.

HKA500 CAS:706-14-9 ***HR: 3***
HYDROXYDECANOIC ACID-Γ-LACTONE
mf: $C_{10}H_{18}O_2$ mw: 170.28

SYNS: Γ-N-DECALACTONE ◇ DECANOLIDE-1,4 ◇ Γ-N-HEXYL-Γ-BUTYROLACTONE ◇ 5-HEXYLDIHYDRO-2(3H)-FURANONE

TOXICITY DATA with REFERENCE
skn-rbt 500 mg/24H MOD FCTXAV 14,741,76
ivn-mus LD50:56 mg/kg CSLNX* NX#02795

CONSENSUS REPORTS: Reported in EPA TSCA Inventory.

SAFETY PROFILE: Poison by intravenous route. A skin irritant. When heated to decomposition it emits acrid smoke and fumes.

HKA700 CAS:19115-30-1 ***HR: 2***
1'-HYDROXY-2',3'-DEHYDROESTRAGOLE
mf: $C_{10}H_{10}O_2$ mw: 162.20

SYNS: BENZENEMETHANOL, α-ETHYNYL-4-METHOXY- ◇ α-ETHYNYL-p-METHOXYBENZYL ALCOHOL

TOXICITY DATA with REFERENCE
dnd-mus-ipr 100 μmol/kg PAACA3 25,88,84
oth-mus-ipr 100 μmol/kg PAACA3 25,88,84
ipr-mus TDLo:1632 μg/kg:CAR CNREA8 47,2275,87

SAFETY PROFILE: Questionable carcinogen with experimental carcinogenic data. Mutation data reported. When heated to decomposition it emits acrid smoke and irritating fumes.

HKB000 CAS:71609-22-8 ***HR: 3***
N-HYDROXY-N,N'-DIACETYLBENZIDINE
mf: $C_{16}H_{16}N_2O_3$ mw: 284.34

SYNS: N-(4'-ACETAMIDOBIPHENYLYL)ACETOHYDROXAMIC ACID ◇ NOHDABZ

TOXICITY DATA with REFERENCE
mmo-sat 250 nmol/plate CNREA8 39,3107,79
mma-sat 250 nmol/plate CNREA8 39,3107,79
ipr-rat TDLo:68 mg/kg/4W-I:CAR CRNGDP 2,747,81

SAFETY PROFILE: Questionable carcinogen with experimental carcinogenic data. Mutation data reported. When heated to decomposition it emits toxic fumes of NO_x.

HKB200 CAS:63139-69-5 ***HR: 2***
HYDROXYDIHYDROCYCLOPENTADIENE
mf: $C_{10}H_{12}O$ mw: 148.22

SYN: TETRAHYDRO-4,7-METHANOINDENOL

TOXICITY DATA with REFERENCE
skn-rbt 500 mg/24H AIHAAP 30,470,69
orl-rat LD50:3250 mg/kg AIHAAP 30,470,69
skn-rbt LD50:3150 mg/kg AIHAAP 30,470,69

SAFETY PROFILE: Moderately toxic by ingestion and skin contact. A skin irritant. When heated to decomposition it emits acrid smoke and fumes.

HKB500 CAS:1689-83-4 ***HR: 3***
4-HYDROXY-3,5-DIIODOBENZONITRILE
mf: $C_7H_3I_2NO$ mw: 370.91

PROP: Colorless solid. Mp: 213°. Sltly sol in water.

SYNS: ACTRIL ◇ BANTROL ◇ BENTROL ◇ CERTROL ◇ 4-CYANO-2,6-DIIODOPHENOL ◇ 4-CYANO-2,6-DIJODPHENOL (GERMAN) ◇ 3,5-DIIODO-4-HYDROXYBENZONITRILE ◇ 3,5-DIJOD-4-HYDROXY-BENZONITRIL (GERMAN) ◇ IOTOX ◇ IOXYNIL ◇ LOXYNIL (GERMAN) ◇ M&B 8873 ◇ OXYTRIL ◇ TOTRIL

TOXICITY DATA with REFERENCE
orl-hmn LDLo:28 mg/kg ARTODN 37,241,77
orl-rat LD50:110 mg/kg WRPCA2 9,119,70
skn-rat LDLo:210 μg/kg GUCHAZ 6,304,73
orl-mus LD50:230 mg/kg 85GYAZ -,93,71
ivn-mus LD50:56 mg/kg CSLNX* NX#02818
orl-cat LD50:75 mg/kg 85GYAZ -,93,71
orl-rbt LD50:180 mg/kg 85GYAZ -,93,71
orl-gpg LD50:76 mg/kg 85GYAZ -,93,71
orl-ckn LD50:200 mg/kg 85GYAZ -,93,71

CONSENSUS REPORTS: Cyanide and its compounds are on the Community Right-To-Know List.

SAFETY PROFILE: A human poison by ingestion. Very poisonous experimentally by skin contact, ingestion, and intravenous routes. An herbicide. When heated to decomposition it emits toxic fumes of I^- and CN^-. See also NITRILES and IODIDES.

HKB600 CAS:21019-39-6 ***HR: 3***
18-α-HYDROXY-11, 17-α-DIMETHOXY-3-β, 20-α-YOHIMBAN-16-β-CARBOXYLIC ACID, METHYL ESTER, 4-HYDROXY-3,5-DIMETHOXYBENZOATE (ESTER), ETHYLCARBONATE (ESTER)
mf: $C_{35}H_{42}N_2O_{11}$ mw: 666.79

TOXICITY DATA with REFERENCE
ipr-rat LD50:286 mg/kg IYKEDH 6,386,75
orl-mus LD50:1293 mg/kg IYKEDH 6,386,75
ipr-mus LD50:101 mg/kg IYKEDH 6,386,75
scu-mus LD50:321 mg/kg IYKEDH 6,386,75

SAFETY PROFILE: Poison by subcutaneous and intraperitoneal routes. Moderately toxic by ingestion. When heated to decomposition it emits toxic fumes of NO_x.

HKB700 CAS:84371-65-3 ***HR: D***
17-β-HYDROXY-11-β-(4-DIMETHYLAMINOPHENYL-1)-17-α-(PROP-1-YNYL) OESTRA-4,9-DIEN-3-ONE
mf: $C_{29}H_{35}NO_2$ mw: 429.65

SYNS: MIFEPRISTONE ◇ R 38486 ◇ RU 486 ◇ RU 486-6 ◇ RU 38486 ◇ ESTRA-4,9-DIEN-3-ONE, 11-(4-(DIMETHYLAMINO)PHENYL)-17-HYDROXY-17-(1-PROPYNYL)-, (11-β, 17-β)-

TOXICITY DATA with REFERENCE
orl-wmn TDLo:12 mg/kg (female 6W post):REP EOGRAL 28,249,88

SAFETY PROFILE: Experimental reproductive effects. When heated to decomposition it emits toxic fumes of NO_x.

HKC000 CAS:75-60-5 ***HR: 3***
HYDROXYDIMETHYLARSINE OXIDE
DOT: UN 1572
mf: $C_2H_7AsO_2$ mw: 138.01

PROP: Colorless crystals; odorless. Mp: 192°. Sol in water.

SYNS: ACIDE CACODYLIQUE (FRENCH) ◇ ACIDE DIMETHYL-ARSINIQUE (FRENCH) ◇ AGENT BLUE ◇ ANSAR ◇ ARSAN ◇ BOLLS-EYE ◇ CACODYLIC ACID (DOT) ◇ CHEXMATE ◇ DILIC ◇ DIMETHYL-ARSENIC ACID ◇ DIMETHYLARSINIC ACID ◇ DMAA ◇ ERASE ◇ PHYTAR ◇ RAD-E-CATE 25 ◇ RCRA WASTE NUMBER U136 ◇ SALVO ◇ SILVISAR 510

TOXICITY DATA with REFERENCE
skn-rat 2600 mg/m^3/2H TXAPA9 37,165,76
eye-rat 2600 mg/m^3/2H TXAPA9 37,165,76
mma-smc 2 pph NTIS** PB84-138973
mnt-mus-ipr 7900 mg/kg/24H NTIS** PB84-138973
orl-mus TDLo:3 g/kg (female 8-12D post):REP TCMUD8 4,403,84
orl-mus TDLo:1600 mg/kg (female 8D post):TER TCMUD8 5,3,85

scu-mus TDLo:464 mg/kg:ETA NTIS** PB223-159
orl-rat LD50:644 mg/kg FAATDF 7,299,86
ipr-mus LDLo:500 mg/kg NTIS** AD295-864
unr-mus LD50:185 mg/kg 30ZDA9 -,393,71

CONSENSUS REPORTS: IARC Cancer Review: Animal Inadequate Evidence IMEMDT 23,39,80. Arsenic and its compounds are on the Community Right-To-Know List. Reported in EPA TSCA Inventory. EPA Genetic Toxicology Program.

OSHA PEL: TWA 0.5 mg(As)/m^3
ACGIH TLV: TWA 0.2 mg(As)/m^3
DOT Classification: IMO: Poison B; Label: Poison.

SAFETY PROFILE: Poison by an unspecified route. Moderately toxic by ingestion and intraperitoneal routes. Experimental teratogenic and reproductive effects. A skin and eye irritant. Questionable carcinogen with experimental tumorigenic data. Mutation data reported. Used as an herbicide, defoliant, and silvicide. Hazardous when water solution is in contact with active metals, i.e., Fe; Al; Zn. When heated to decomposition it emits toxic fumes of As. See also ARSINE and ARSENIC COMPOUNDS.

HKC500 CAS:124-65-2 ***HR: 3***
HYDROXYDIMETHYLARSINE OXIDE, SODIUM SALT
DOT: UN 1688
mf: $C_2H_6AsO_2{\cdot}Na$ mw: 159.99

SYNS: ALKARSODYL ◇ ANSAR 160 ◇ ARSECODILE ◇ ARSYCODILE ◇ BOLLS-EYE ◇ CACODYLATE de SODIUM (FRENCH) ◇ CACODYLIC ACID SODIUM SALT ◇ CHEMAID ◇ ((DIMETHYLARSINO)OXY)SODIUM-As-OXIDE ◇ DUTCH-TREAT ◇ PHYTAR 560 ◇ RAD-E CATE 16 ◇ SILVISAR ◇ SODIUM CACODYLATE (DOT) ◇ SODIUM DIMETHYLARSINATE ◇ SODIUM DIMETHYLARSONATE ◇ SODIUM SALT of CACODYLIC ACID

TOXICITY DATA with REFERENCE
ipr-ham TDLo:900 mg/kg (9D preg):TER BECTA6 26,679,82
ipr-ham TDLo:900 mg/kg (9D preg):REP BECTA6 26,679,82
orl-rat LD50:2600 mg/kg GUCHAZ 6,70,73
orl-mus LD50:4 mg/kg CLDND* 49,172.102,88

CONSENSUS REPORTS: Arsenic and its compounds are on the Community Right-To-Know List. EPA Extremely Hazardous Substances List. Reported in EPA TSCA Inventory.

OSHA PEL: TWA 0.5 mg(As)/m^3
ACGIH TLV: TWA 0.2 mg(As)/m^3
DOT Classification: IMO: Poison B; Label: Poison.

SAFETY PROFILE: Confirmed human carcinogen. Poison by ingestion. Experimental teratogenic and other reproductive effects. When heated to decomposition it emits toxic fumes of As and Na_2O. See also ARSINE and ARSENIC COMPOUNDS.

HKC550 CAS:6131-99-3 ***HR: 2***
HYDROXYDIMETHYLARSINE OXIDE, SODIUM SALT TRIHYDRATE
mf: $C_2H_6AsO_2{\cdot}Na{\cdot}3H_2O$ mw: 214.05

SYNS: ARSINE OXIDE, HYDROXYDIMETHYL-, SODIUM SALT, TRIHYDRATE ◇ DIMETHYLARSENIC ACID SODIUM SALT TRIHYDRATE ◇ DIMETHYLARSINIC ACID SODIUM SALT TRIHYDRATE ◇ SODIUM DIMETHYLARSINIC ACID TRIHYDRATE

TOXICITY DATA with REFERENCE
ivn-ham TDLo:20 mg/kg (female 8D post):REP EMPSAL 34,145,81

ACGIH TLV: TWA 0.2 mg(As)/m^3
OSHA PEL: 8H TWA 0.5 mg(As)/m^3

SAFETY PROFILE: Experimental reproductive effects. When heated to decomposition it emits toxic fumes of As.

HKD550 CAS:35869-74-0 ***HR: 3***
4-HYDROXY-3,5-DIMETHYL-1,2,4-TRIAZOLE
mf: $C_4H_7N_3O$ mw: 113.13

$HONC(CH_3){=}NN{=}CCH_3$ (ring)

SAFETY PROFILE: Explodes when heated to its melting point of 122°C. Upon decomposition it emits toxic fumes of NO_x.

HKE000 CAS:520-53-6 ***HR: 3***
4-HYDROXY-N,N-DIMETHYLTRYPTAMINE
mf: $C_{12}H_{16}N_2O$ mw: 204.30

SYNS: CX-59 ◇ 3-(2-(DIMETHYLAMINO)ETHYL)INDOL-4-OL ◇ PSILOCINE ◇ PSILOTSIN

TOXICITY DATA with REFERENCE
ivn-rat LD50:75 mg/kg PSCBAY 2,17,63
ipr-mus LD50:196 mg/kg JTEHD6 1,515,76
ivn-mus LD50:74 mg/kg 27ZQAG -,138,72
ivn-rbt LD50:7 mg/kg 27ZQAG -,138,72

SAFETY PROFILE: A poison by intravenous and intraperitoneal routes. When heated to decomposition it emits toxic fumes of NO_x.

HKE500 CAS:6269-50-7 ***HR: 3***
4-HYDROXY-3,5-DINITROBENZENEARSONIC ACID
mf: $C_6H_5AsN_2O_8$ mw: 308.04

$HO(O_2N)_2C_6H_2AsO(OH)_2$

CONSENSUS REPORTS: Arsenic and its compounds are on the Community Right-To-Know List.

SAFETY PROFILE: Arsenic compounds are generally poisons. Potentially explosive if heated. Upon decomposition it emits toxic fumes of As and NO_x. See also ARSENIC COMPOUNDS and NITRO COMPOUNDS of AROMATIC HYDROCARBONS.

HKE700 CAS:2980-33-8 ***HR: 3***
2-HYDROXY-3,5-DINITROPYRIDINE
mf: $C_5H_3N_3O_5$ mw: 185.10

$N{=}C(OH)C(NO_2){-}CHC(NO_2){=}CH$ (ring)

SAFETY PROFILE: An explosive. When heated to decomposition it emits toxic fumes of NO_x.

HKF000 CAS:101-73-5 ***HR: 3***
p-HYDROXYDIPHENYLAMINE ISOPROPYL ETHER
mf: $C_{15}H_{17}NO$ mw: 227.33

SYNS: AGERITE 150 ◇ AGERITE ISO ◇ p-ISOPROPOXYDIPHENYLAMINE ◇ 4-ISOPROPOXYDIPHENYLAMINE ◇ N-(4-ISOPROPOXYPHENYL)ANILINE

TOXICITY DATA with REFERENCE
orl-mus TDLo:332 g/kg/78W-I:NEO NTIS** PB223-159

CONSENSUS REPORTS: Reported in EPA TSCA Inventory.

SAFETY PROFILE: Questionable carcinogen with experimental neoplastigenic data. When heated to decomposition it emits toxic fumes of NO_x. See also AMINES and ETHERS.

HKF600 CAS:66427-01-8 ***HR: 3***
N-HYDROXYDITHIOCARBAMIC ACID
mf: CH_3NOS_2 mw: 109.16

SAFETY PROFILE: An unstable material which may explode at sub-zero temperatures. Upon decomposition it emits toxic fumes of SO_x and NO_x. See also CARBAMATES.

HKF875 CAS:1199-18-4 ***HR: D***
6-HYDROXYDOPAMINE
mf: $C_8H_{11}NO_3$ mw: 169.20

SYNS: OXIDOPAMINE ◇ 2,4,5-TRIHYDROXYPHENETHYLAMINE

TOXICITY DATA with REFERENCE
mmo-sat 100 μg/plate ABCHA6 45,327,81
mma-sat 100 μg/plate ABCHA6 45,327,81
ipr-rat TDLo:400 mg/kg (1D pre/1-22D preg):REP JPPMAB 26,518,74

SAFETY PROFILE: Experimental reproductive effects. Mutation data reported. When heated to decomposition it emits toxic fumes of NO_x.

HKG000 CAS:28094-15-7 ***HR: 3***
6-HYDROXYDOPAMINE HYDROCHLORIDE
mf: $C_8H_{11}NO_3{\cdot}ClH$ mw: 205.66

SYNS: 4-(2-AMINOETHYL)-1,2,3-BENZENETRIOLHYDROCHLORIDE ◇ 3,4,5-TRIHYDROXYPHENETHYLAMINE HYDROCHLORIDE

TOXICITY DATA with REFERENCE
scu-mus TDLo:100 mg/kg/52W-I:NEO RCOCB8 17,411,77

SAFETY PROFILE: Questionable carcinogen with experimental neoplastigenic data. When heated to decomposition it emits very toxic fumes of NO_x and HCl.

HKG500 CAS:5289-74-7 ***HR: 2***
20-HYDROXYECDYSONE
mf: $C_{27}H_{44}O_7$ mw: 480.71

SYNS: COMMISTERONE ◇ CRUSTECDYSON ◇ β-ECDYSONE ◇ ECDYSTERONE ◇ β-ECDYSTERONE ◇ 2-β,3-β,14,20,22,25-HEXAHYDROXY-5-β-CHOLET-7-EN-6-ONE ◇ ISOINOKOSTERONE ◇ POLYPODINE A ◇ THE-7 ◇ VITICOSTERONE

TOXICITY DATA with REFERENCE
cyt-slw-par 100 μmol/L ENZYAS 41,183,71
dnd-mus:lvr 100 nmol/L ENZYAS 41,183,71
scu-rat TDLo:350 μg/kg (7D male):REP NYKZAU 66,551,70
ipr-mus LD50:6400 mg/kg NYKZAU 66,551,70

SAFETY PROFILE: Moderately toxic by intraperitoneal route. Experimental reproductive effects. Mutation data reported. When heated to decomposition it emits acrid smoke and fumes.

HKH000 CAS:51131-85-2 ***HR: 3***
9-HYDROXYELLIPTICINE
mf: $C_{17}H_{14}N_2O$ mw: 262.33

SYNS: 5,11-DIMETHYL-6H-PYRIDO(4,3-b)CARBAZYL-9-OL ◇ 9-HYDROXYELLIPTICIN ◇ HYDROXY-9 ELLIPTICINE (FRENCH) ◇ IGIG 929

TOXICITY DATA with REFERENCE
mmo-sat 500 ng/plate CNREA8 43,3544,83
oms-ham:lng 4 μmol/L CNREA8 45,4229,85
ipr-mus TDLo:10 mg/kg (female 8D post):TER ABILAE 95,113,84
ipr-mus TDLo:20 mg/kg (8D preg):REP ABILAE 95,113,84
ipr-mus LD50:68 mg/kg TXAPA9 33,484,75
ivn-mus LD50:102 mg/kg TXAPA9 33,484,75

SAFETY PROFILE: Poison by intraperitoneal and intravenous routes. Experimental teratogenic and reproductive effects. Mutation data reported. When heated to decomposition it emits toxic fumes of NO_x.

HKH500 CAS:365-26-4 ***HR: 3***
p-HYDROXYEPHEDRINE
mf: $C_{10}H_{15}NO_2$ mw: 181.23

PROP: Crystalline powder. Mp: 152-154°. Very sltly sol in water, alc, and ether. Very sol in NaOH solns and dil acids.

SYNS: p-HYDROXYPHENYLMETHYLAMINOPROPANOL ◇ 1-(4-HYDROXYPHENYL)-2-METHYLAMINOPROPANOL ◇ α-(1-METHYLAMINOETHYL)-p-HYDROXYBENZYL ALCOHOL ◇ OXYEPHEDRINE ◇ SUPRIFEN ◇ SUPRIFENE

TOXICITY DATA with REFERENCE
ivn-rbt LDLo:150 mg/kg AEPPAE 160,127,31

SAFETY PROFILE: A poison by intravenous route. When heated to decomposition it emits toxic fumes of NO_x.

HKH850 CAS:5976-61-4 ***HR: 2***
4-HYDROXYESTRADIOL
mf: $C_{18}H_{24}O_3$ mw: 288.42

SYNS: 4-HYDROXY-17-β-ESTRADIOL ◇ 4-OH-E2 ◇ 4-OH-ESTRADIOL

TOXICITY DATA with REFERENCE
ivn-rat TDLo:10 μg/kg (1D pre):REP EXPEAM 42,165,86
imp-ham TDLo:900 mg/kg/90D-I:ETA JSTBBK 24,353,86

SAFETY PROFILE: Experimental reproductive data. Questionable carcinogen with experimental tumorigenic data. When heated to decomposition it emits acrid smoke and irritating fumes.

HKI000 CAS:51410-44-7 ***HR: 3***
1'-HYDROXYESTRAGOLE
mf: $C_{10}H_{12}O_2$ mw: 164.22

SYN: p-METHOXY-α-VINYLBENZYL ALCOHOL

TOXICITY DATA with REFERENCE
mmo-sat 1 μmol/plate MUREAV 60,143,79
mma-sat 1 μmol/plate MUREAV 60,143,79
orl-mus TDLo:106 g/kg/1Y-C:CAR CNREA8 43,1124,83
ipr-mus TDLo:16422 μg/kg:CAR CNREA8 47,2275,87

CONSENSUS REPORTS: EPA Genetic Toxicology Program.

SAFETY PROFILE: Questionable carcinogen with experimental carcinogenic data. Mutation data reported. When heated to decomposition it emits acrid smoke and fumes.

HKI075 CAS:730771-71-0 ***HR: 2***
1'-HYDROXY-ESTRAGOLE-2',3'-OXIDE
mf: $C_{10}H_{12}O_3$ mw: 180.22

SYN: α-(EPOXYETHYL)-p-METHOXYBENZYLALCOHOL

TOXICITY DATA with REFERENCE
mmo-sat 200 nmol/plate MUREAV 60,143,79
ipr-mus TDLo:43 mg/kg/12W-I:NEO CNREA8 43,1124,83

SAFETY PROFILE: Questionable carcinogen with experimental neoplastigenic data. Mutation data reported. When heated to decomposition it emits acrid smoke and irritating fumes.

HKI500 CAS:107-36-8 ***HR: 3***
2-HYDROXYETHANESULFONIC ACID
mf: $C_2H_6O_4S$ mw: 126.14

SYNS: ETHANOLSULFONIC ACID ◇ HYDROXYETHYLSULFONIC ACID ◇ ISETHIONIC ACID ◇ ISOTHIONIC ACID ◇ USAF DO-14

TOXICITY DATA with REFERENCE
ipr-mus LD50:50 mg/kg NTIS** AD277-689

CONSENSUS REPORTS: Reported in EPA TSCA Inventory.

SAFETY PROFILE: Poison by by intraperitoneal route. An irritant to skin, eyes and mucous membranes. See also SULFONATES.

HKJ000 CAS:106-11-6 ***HR: 3***
2-(2-HYDROXYETHOXY)ETHYL ESTER STEARIC ACID
mf: $C_{22}H_{44}O_4$ mw: 372.66

SYNS: AQUA CERA ◇ ATLAS G 2146 ◇ CERASYNT ◇ CLINDROL SDG ◇ DIETHYLENE GLYCOL, MONOESTER with STEARIC ACID ◇ DIETHYLENE GLYCOL MONOSTEARATE ◇ DIETHYLENE GLYCOL STEARATE ◇ DIGLYCOL MONOSTEARATE ◇ DIGLYCOL STEARATE ◇ EMCOL DS-50 CAD ◇ GLYCO STEARIN ◇ NONEX 411 ◇ PROMUL 5080 ◇ USAF KE-8

TOXICITY DATA with REFERENCE
ipr-mus LD50:200 mg/kg NTIS** AD277-689

CONSENSUS REPORTS: Reported in EPA TSCA Inventory.

SAFETY PROFILE: A poison by intraperitoneal route. When heated to decomposition it emits acrid smoke and fumes. See also ESTERS.

HKJ500 ***HR: 3***
2-(2-HYDROXYETHOXY)ETHYL PERCHLORATE
mf: $C_4H_9ClO_6$ mw: 188.56

$$HOC_2H_4OC_2H_4OClO_3$$

SAFETY PROFILE: Potentially explosive when heated.

Upon decomposition it emits toxic fumes of Cl^-. See also PERCHLORATES.

HKK000 CAS:30544-47-9 ***HR: 3***
2-(2-HYDROXYETHOXY)ETHYL-N-(α,α,α-TRIFLUORO-m-TOLYL)ANTHRANILATE
mf: $C_{18}H_{18}F_3NO_4$ mw: 369.37

SYNS: B 577 ◇ BAYROGEL ◇ ETOFENAMATE ◇ 2-(2-HYDROXY-AETHOXY)AETHYLESTER der FLUTENAMINSAEURE (GERMAN) ◇ RHEUMON ◇ RHEUMON GEL ◇ TV 485 ◇ TVX 485

TOXICITY DATA with REFERENCE
scu-rat TDLo:540 mg/kg (female 17-22D post):REP IYKEDH 13,910,82
scu-rat TDLo:3640 mg/kg (10W male/2W pre-7D preg):TER IYKEDH 13,886,82
orl-rat LD50:292 mg/kg ARZNAD 27,1333,77
ipr-rat LD50:373 mg/kg ARZNAD 27,1333,77
scu-rat LD50:568 mg/kg ARZNAD 27,1333,77
ivn-rat LD50:139 mg/kg YACHDS 10,5225,82
orl-mus LD50:743 mg/kg ARZNAD 27,1333,77
scu-mus LD50:1897 mg/kg YACHDS 10,5225,82
ivn-mus LD50:75 mg/kg YACHDS 10,5225,82
scu-rbt LD50:1532 mg/kg YACHDS 10,5225,82

SAFETY PROFILE: Poison by ingestion, intraperitoneal, and intravenous routes. Moderately toxic by subcutaneous routes. Experimental teratogenic and reproductive effects. Used as an anti-inflammatory agent. When heated to decomposition it emits very toxic fumes of F^- and NO_x. See also FLUORIDES.

HKM000 CAS:142-26-7 ***HR: 1***
N-β-HYDROXYETHYLACETAMIDE
mf: $C_4H_9NO_2$ mw: 103.14

PROP: Brown viscous liquid. Bp: 151°, flash p: 355°F (OC), fp: 15.8°, d: 1.122 @ 20°/20°, autoign temp: 860°F.

SYNS: 2-ACETAMIDOETHANOL ◇ 2-ACETYLAMINOETHANOL ◇ ACETYLCOLAMINE ◇ N-ACETYL ETHANOLAMINE ◇ N-ETHANOLACETAMIDE ◇ HYDROXYETHYL ACETAMIDE ◇ β-HYDROXYETHYLACETAMIDE ◇ N-(2-HYDROXYETHYL)ACETAMIDE

TOXICITY DATA with REFERENCE
skn-rbt 500 mg open MLD UCDS** 4/21/67
eye-rbt 100 mg AJOPAA 29,1363,46
orl-rat LD50:27660 mg/kg 34ZIAG -,706,69

CONSENSUS REPORTS: Reported in EPA TSCA Inventory.

SAFETY PROFILE: Mildly toxic by ingestion. A skin and eye irritant. Combustible when exposed to heat or flame; can react vigoroulsy with oxidizers. To fight fire, use alcohol foam, CO_2, dry chemical. When heated to decomposition it emits toxic fumes of NO_x.

HKM175 CAS:38092-76-1 ***HR: 3***
2-HYDROXYETHYLAMINIUM PERCHLORATE
mf: $C_2H_8ClNO_5$ mw: 161.54

SYN: ETHANOLAMINE PERCHLORATE

SAFETY PROFILE: A sensitive explosive even in aqueous solution. When heated to decomposition it emits toxic fumes of Cl^- and NO_x. See also PERCHLORATES.

HKM500 CAS:93-62-9 ***HR: 3***
2-HYDROXYETHYLAMINODIACETIC ACID
mf: $C_6H_{12}NO_5$ mw: 178.19

SYNS: ETHANOLAMINE-N,N-DIACETIC ACID ◇ (2-HYDROXYETHYL)IMINODIACETIC ACID ◇ USAF DO-37

TOXICITY DATA with REFERENCE
ipr-mus LD50:100 mg/kg NTIS** AD277-689

CONSENSUS REPORTS: Reported in EPA TSCA Inventory.

SAFETY PROFILE: Poison by intraperitoneal route. When heated to decomposition it emits toxic fumes of NO_x.

HKN500 ***HR: 3***
2-(β-HYDROXYETHYLAMINOMETHYL)-1,4-BENZODIOXANE HYDROCHLORIDE
mf: $C_{11}H_{15}NO_3•ClH$ mw: 245.73

SYN: 2-((1,4-BENZODIOXAN-2-YLMETHYL)AMINO)ETHANOLHYDROCHLORIDE

TOXICITY DATA with REFERENCE
ipr-rat LD50:355 mg/kg JAPMA8 44,302,55
ivn-rat LD50:102 mg/kg JAPMA8 44,302,55
ipr-mus LD50:385 mg/kg JAPMA8 44,302,55
ivn-mus LD50:74 mg/kg JAPMA8 44,302,55

SAFETY PROFILE: Poison by intravenous and intraperitoneal routes. When heated to decomposition it emits toxic fumes of NO_x and HCl.

HKN875 CAS:33229-34-4 ***HR: D***
2,2'-((4-((2-HYDROXYETHYL)AMINO)-3-NITROPHENYL)IMINO)DIETHANOL
mf: $C_{12}H_{19}N_3O_5$ mw: 285.34

SYNS: 2,2'-((4-((2-HYDROXYETHYL)AMINO-3-NITROPHENYL)IMINO)BISETHANOL ◇ HC BLUE No. 2 ◇ NCI-C54897 ◇ 3-NITRO-N^1,N^1,N^4-TRIS(2-HYDROXYETHYL)-p-PHENYLENEDIAMINE

TOXICITY DATA with REFERENCE
mmo-sat 3333 μg/plate NTPTR* NTP-TR-293,85
msc-mus:lym 150 mg/L NTPTR* NTP-TR-293,85

CONSENSUS REPORTS: NTP Carcinogenesis Studies

(feed); No Evidence: mouse, rat NTPTR* NTP-TR-293,85.

SAFETY PROFILE: Mutation data reported. When heated to decomposition it emits toxic fumes of NO_x. See also NITRO COMPOUNDS of AROMATIC HYDROCARBONS.

HKO000 CAS:33389-36-5 ***HR: 3***
4-(2-HYDROXYETHYLAMINO)-2-(5-NITRO-2-THIENYL)QUINAZOLINE
mf: $C_{14}H_{12}N_4O_3S$ mw: 316.36

TOXICITY DATA with REFERENCE
mma-sat 1250 μg/plate CNREA8 35,3611,75
orl-rat TDLo:13 g/kg/58W-C:CAR JNCIAM 57,277,76

SAFETY PROFILE: Questionable carcinogen with experimental carcinogenic data. Mutation data reported. When heated to decomposition it emits very toxic fumes of NO_x and SO_x.

HKQ000 CAS:5395-01-7 ***HR: D***
β-HYDROXYETHYLCARBAMATE
mf: $C_3H_7NO_3$ mw: 105.11

SYNS: CARBAMIC ACID-2-HYDROXYETHYL ESTER ◇ 2-HYDROXYETHYLCARBAMATE

TOXICITY DATA with REFERENCE
msc-ham:lng 1 g/L CRNGDP 3,1437,82
ipr-ham TDLo:505 mg/kg (8D preg):TER CNREA8 27,1696,67

SAFETY PROFILE: Experimental teratogenic effects. Mutation data reported. When heated to decomposition it emits toxic fumes of NO_x. See also CARBAMATES.

HKQ025 CAS:589-41-3 ***HR: 3***
N-HYDROXY ETHYL CARBAMATE
mf: $C_3H_7NO_3$ mw: 105.11

SYNS: ETHYL-N-HYDROXYCARBAMATE ◇ HYDROXYCARBAMIC ACID ETHYL ESTER ◇ N-HYDROXYURETHAN ◇ N-HYDROXYURETHANE ◇ NHU ◇ NSC-83629 ◇ NSC-71045 ◇ SQ 16819

TOXICITY DATA with REFERENCE
mmo-esc 10 mmol/L MUREAV 151,201,85
cyt-hmn:leu 333 μmol/L/48H CNREA8 25,980,65
ipr-ham TDLo:1009 mg/kg (female 8D post):TER CNREA8 27,1696,67
scu-mus TDLo:738 mg/kg (6-14D preg):REP NTIS** PB223-160
ipr-rat TDLo:500 mg/kg:NEO RRCRBU 52,29,75
scu-rat TDLo:8 g/kg/8W-I:ETA JNCIAM 43,749,75
unr-mus TDLo:1 mg/kg (21D preg):NEO,REP 40YJAX -,141,76
ipr-rat LD50:800 mg/kg ADTEAS 3,181,68

CONSENSUS REPORTS: EPA Genetic Toxicology Program.

SAFETY PROFILE: Moderately toxic by intraperitoneal route. Experimental teratogenic and reproductive effects. Questionable carcinogen with experimental neoplastigenic and tumorigenic data. Human mutation data reported. When heated to decomposition it emits toxic fumes of NO_x. See also CARBAMATES and ESTERS.

HKQ300 ***HR: 3***
HYDROXYETHYL CNU METHANESULFONATE
mf: $C_5H_{10}ClN_3O_3 \cdot CH_4O_3S$ mw: 291.74

SYNS: HECNU-MS ◇ HYDROXY CNU METHANESULPHONATE

TOXICITY DATA with REFERENCE
ivn-rat TDLo:16 mg/kg/60W-I:ETA DTESD7 8,273,80

SAFETY PROFILE: Questionable carcinogen with experimental tumorigenic data. When heated to decomposition it emits toxic fumes of Cl^-, SO_x, and NO_x. See also SULFONATES.

HKQ500 CAS:16179-44-5 ***HR: 3***
2-HYDROXYETHYL CYCLOHEXANECARBOXYLATE
mf: $C_9H_{16}O_3$ mw: 172.25

SYNS: A13-70087 ◇ CYCLOHEXANECARBOXYLIC ACID-(2-HYDROXYETHYL) ESTER

TOXICITY DATA with REFERENCE
skn-rbt 550 mg/15D MLD NTIS** AD-A022-908
eye-rbt 100 mg SEV NTIS** AD-A022-908
orl-rat LDLo:2874 mg/kg NTIS** AD-A022-908
ivn-rbt LDLo:75 mg/kg NTIS** AD-A022-908

SAFETY PROFILE: Poison by intravenous route. Moderately toxic by ingestion. A skin and severe eye irritant. When heated to decomposition it emits acrid smoke and fumes. See also ESTERS.

HKR000 CAS:1965-29-3 ***HR: 3***
N-(HYDROXYETHYL)DIETHYLENETRIAMINE
mf: $C_6H_{17}N_3O$ mw: 147.26

SYN: N-(2-HYDROXYETHYL)DIETHYLENETRIAMINE

TOXICITY DATA with REFERENCE
skn-rbt 10 mg/24H open MLD AMIHBC 4,119,51
eye-rbt 20 mg open SEV AMIHBC 4,119,51
orl-rat LD50:4630 mg/kg AMIHAB 17,129,58
ipr-rat LD50:146 mg/kg AMIHAB 17,129,58
ipr-mus LD50:320 mg/kg AMIHAB 17,129,58
skn-rbt LD50:1041 mg/kg JIHTAB 31,60,49

CONSENSUS REPORTS: Reported in EPA TSCA Inventory.

SAFETY PROFILE: Poison by intraperitoneal route. Moderately toxic by skin contact. Mildly toxic by ingestion. A skin and severe eye irritant. When heated to decomposition it emits toxic fumes of NO_x. See also AMINES.

HKR500 CAS:50-34-0 ***HR: 3***
(2-HYDROXYETHYL)DIISOPROPYLMETHYLAMMONIUMBROMIDE XANTHENE-9-CARBOXYLATE
mf: $C_{23}H_{30}NO_3 \cdot Br$ mw: 448.45

PROP: Crystals from isopropanol + ether. Mp: 159-161°. Very sol in water, alc, chloroform; practically insol in ether, benzene.

SYNS: CORRIGAST ◇ β-DIISOPROPYLAMINOETHYL-9-XANTHENECARBOXYLATE METHOBROMIDE ◇ DIISOPROPYL(2-HYDROXYETHYL)METHYLAMMONIUMBROMIDE with XANTHENE-9-CARBOXYLATE ◇ ERCORAX ◇ ERCOTINA ◇ GIQUEL ◇ KETAMAN ◇ KIVATIN ◇ NCI-C56257 ◇ NEOMETANTYL ◇ NEOPEPULSAN ◇ PANTAS ◇ PANTHELINE ◇ PERVAGAL ◇ PROBANTHINE ◇ PRODIXAMON ◇ PRO-GASTRON ◇ PROPANTEL ◇ PROPANTHELINE BROMIDE ◇ SC-3171 ◇ XANTHENE-9-CARBOXYLIC ACID, ESTER with (2-HYDROXYETHYL)DIISOPROPYLMETHYLAMMONIUM BROMIDE

TOXICITY DATA with REFERENCE
otr-ham:emb 1 g/L ENMUDM 9(Suppl 6),4,86
orl-rat LD50:370 mg/kg AIPTAK 180,155,69
ipr-rat LD50:25 mg/kg CLDND* 1,391,59
scu-rat LD50:298 mg/kg AIPTAK 180,155,69
idu-rat LD50:125 mg/kg AIPTAK 180,155,69
orl-mus LD50:445 mg/kg RPOBAR 2,319,70
ipr-mus LD50:78 mg/kg RPOBAR 2,319,70
ivn-mus LD50:6995 μg/kg AIPTAK 103,100,55
orl-rbt LD50:750 mg/kg GUCHAZ 6,36,73
ivn-gpg LDLo:51 mg/kg CRSBAW 151,614,57

CONSENSUS REPORTS: Reported in EPA TSCA Inventory.

SAFETY PROFILE: Poison by ingestion, intraperitoneal, subcutaneous, intraduodenal, and intravenous routes. Mutation data reported. When heated to decomposition it emits very toxic fumes of NO_x, NH_3, and Br^-. See also ESTERS.

HKR550 CAS:63886-56-6 ***HR: 2***
N-2-HYDROXYETHYL-3,4-DIMETHYLAZOLIDIN
mf: $C_8H_{17}NO$ mw: 143.26

SYNS: 3,4-DIMETHYLPYRROLIDINE ETHANOL ◇ PYRROLIDINEETHANOL, 3,4-DIMETHYL-

TOXICITY DATA with REFERENCE
skn-rbt 10 mg/24H open SEV AMIHBC 10,61,54
skn-rbt 5 mg/24H SEV 85JCAE-,820,86
eye-rbt 250 μg/24H SEV 85JCAE-,820,86
orl-rat LD50:1230 mg/kg AMIHBC 10,61,54
skn-rbt LD50:850 mg/kg AMIHBC 10,61,54

SAFETY PROFILE: Moderately toxic by ingestion and skin contact. A severe skin and eye irritant. When heated to decomposition it emits toxic fumes of NO_x.

HKS000 CAS:150-39-0 ***HR: 3***
N-HYDROXYETHYLENEDIAMINETRIACETIC ACID
mf: $C_{10}H_{18}N_2O_7$ mw: 278.30

SYNS: N-(CARBOXYMETHYL)-N'-(2-HYDROXYETHYL)-N,N'-ETHYLENEDIGLYCINE ◇ CHEM DM ACID ◇ HAMP-OL ACID ◇ HEDTA ◇ HEEDTA ◇ N-(β-HYDROXYETHYLETHYLENEDIAMINE)-N,N',N'-TRIACETIC ACID ◇ (N-HYDROXYETHYLETHYLENEDINITRILO)TRIACETIC ACID ◇ VERSENOL ◇ VERSENOL 120

TOXICITY DATA with REFERENCE
ipr-rat LD50:337 mg/kg AHRTAN 13,295,62
ipr-mus LD50:259 mg/kg ARTODN 57,212,85

CONSENSUS REPORTS: Reported in EPA TSCA Inventory.

SAFETY PROFILE: Poison by intraperitoneal route. When heated to decomposition it emits toxic fumes such as NO_x.

HKS300 CAS:63834-30-0 ***HR: 2***
2-HYDROXY-3-ETHYLHEPTANOIC ACID
mf: $C_9H_{18}O_3$ mw: 174.27

TOXICITY DATA with REFERENCE
skn-rbt 10 mg/24H open AIHAAP 23,95,62
orl-rat LD50:3400 mg/kg AIHAAP 23,95,62
skn-rbt LD50:1780 mg/kg AIHAAP 23,95,62

SAFETY PROFILE: Moderately toxic by ingestion and skin contact. A skin irritant. When heated to decomposition it emits acrid smoke and fumes.

HKS400 ***HR: 2***
N-HYDROXYETHYL-N-2-HYDROXYALKYLAMINE
mf: $C_{16}H_{35}NO_2 \cdot C_{14}H_{31}NO_2$ mw: 518.98

SYNS: DUSPAR 125B ◇ 2-TETRADECANOL, 1-(2-HYDROXYETHYLAMINO)-, and 1-(2-HYDROXYETHYL)-2-DODECANOL (1:1)

TOXICITY DATA with REFERENCE
orl-mus TDLo:27300 mg/kg (male 91D pre):REP KSRNAM 9,2903,75
orl-mus LD50:2130 mg/kg KSRNAM 9,2903,75

SAFETY PROFILE: Moderately toxic by ingestion. Experimental reproductive effects. When heated to decomposition it emits toxic fumes of NO_x.

HKS500 CAS:62018-89-7 ***HR: D***
N-(2-HYDROXYETHYL)-N-(4-HYDROXYBUTYL) NITROSAMINE
mf: $C_6H_{14}N_2O_3$ mw: 162.22

SYN: 4-((2-HYDROXYETHYL)NITROSOAMINO)BUTANOL

TOXICITY DATA with REFERENCE
mma-sat 12 μmol/plate CNREA8 37,399,77

SAFETY PROFILE: Mutation data reported. Many N-nitroso compounds are carcinogens. When heated to decomposition it emits toxic fumes of NO_x. See also NITROSAMINES.

HKS600 CAS:55636-92-5 ***HR: 3***
2-(1-HYDROXYETHYL)-7-(2-HYDROXY-3-ISOPROPYLAMINOPROPOXY)BENZOFURAN
mf: $C_{16}H_{23}NO_4$ mw: 293.40

SYN: 7-(2-HYDROXY-3-(ISOPROPYLAMINO)PROPOXY)α-METHYL-2-BENZOFURANMETHANOL

TOXICITY DATA with REFERENCE
orl-rat LD50:2130 mg/kg KSRNAM 13,4138,79
ipr-rat LD50:185 mg/kg KSRNAM 13,4138,79
scu-rat LD50:540 mg/kg KSRNAM 13,4138,79
orl-mus LD50:1275 mg/kg KSRNAM 13,4138,79
ipr-mus LD50:179 mg/kg KSRNAM 13,4138,79
scu-mus LD50:640 mg/kg KSRNAM 13,4138,79
ivn-mus LD50:68500 μg/kg KSRNAM 13,4138,79

SAFETY PROFILE: Poison by intravenous and intraperitoneal routes. Moderately toxic by ingestion and subcutaneous routes. When heated to decomposition it emits toxic fumes of NO_x.

HKS775 CAS:3279-95-6 ***HR: 3***
o-(2-HYDROXYETHYL)HYDROXYLAMINE
mf: $C_2H_7NO_2$ mw: 77.08

SAFETY PROFILE: Reacts explosively with concentrated sulfuric acid at 120°C. When heated to decomposition it emits toxic fumes of NO_x. See also AMINES.

HKS780 CAS:2809-21-4 ***HR: 3***
1-HYDROXYETHYLIDENE-1,1-DIPHOSPHONIC ACID
mf: $C_2H_8O_7P_2$ mw: 206.03

SYNS: DEQUEST 2010 ◇ DEQUEST 2015 ◇ DEQUEST Z 010 ◇ EHDP ◇ ETHANE-1-HYDROXY-1,1-DIPHOSPHONATE ◇ 1,1,1-ETHANETRIOL DIPHOSPHONATE ◇ ETIDRONIC ACID ◇ FERROFOS 510 ◇ HEDP ◇ 1-HYDROXY-1,1-DIPHOSPHONOETHANE ◇ HYDROXYETHANEDIPHOSPHONIC ACID ◇ 1-HYDROXYETHANEDIPHOSPHONIC ACID ◇ OXYETHYLIDENEDIPHOSPHONIC ACID ◇ PHOSPHONIC ACID, 1-HYDROXY-1,1-ETHANEDIYL ESTER ◇ PHOSPHONIC ACID, (1-HYDROXYETHYLIDENE)BIS- ◇ 1000SL ◇ TURPINAL SL

TOXICITY DATA with REFERENCE
ipr-mus TDLo:200 mg/kg (female 7D post):TER JODUA2 19,87,85
ipr-mus TDLo:40 mg/kg (female 7D post):REP SIGAAE 50,879,87
orl-mus LD50:1800 mg/kg ACIEAY 14,94,75

CONSENSUS REPORTS: Reported in EPA TSCA Inventory.

SAFETY PROFILE: Moderately toxic by ingestion. When heated above 200° it decomposes violently to produce toxic fumes of phosphine, phosphoric acid, and PO_x.

HKT200 CAS:51821-32-0 ***HR: 3***
2-HYDROXYETHYLMERCURY(II) NITRATE
mf: $C_2H_5HgNO_4$ mw: 307.66

CONSENSUS REPORTS: Mercury and its compounds are on the Community Right-To-Know List.

SAFETY PROFILE: Decomposes with a weak explosion when heated. Upon decomposition it emits toxic fumes of NO_x and Hg. See also MERCURY COMPOUNDS and NITRATES.

HKU000 CAS:13345-58-9 ***HR: 3***
7-(2-HYDROXYETHYL)-12-METHYLBENZ(a)ANTHRACENE
mf: $C_{21}H_{18}O$ mw: 286.39

SYN: 12-METHYLBENZ(A)ANTHRACENE-7-ETHANOL

TOXICITY DATA with REFERENCE
ims-rat TDLo:50 mg/kg:ETA BCPCA6 16,607,67

SAFETY PROFILE: Questionable carcinogen with experimental tumorigenic data. When heated to decomposition it emits acrid smoke and fumes.

HKU500 CAS:1331-41-5 ***HR: 2***
N-HYDROXYETHYL-α-METHYLBENZYLAMINE
mf: $C_{10}H_{15}NO$ mw: 165.26

SYN: ((α-METHYLBENZYL)AMINO)ETHANOL

TOXICITY DATA with REFERENCE
orl-rat LD50:2830 mg/kg AMIHBC 4,119,51
skn-rbt LD50:1540 mg/kg AMIHBC 4,119,51

SAFETY PROFILE: Moderately toxic by ingestion and skin contact. When heated to decomposition it emits toxic fumes such as NO_x.

HKV000 CAS:21600-45-3 ***HR: 3***
3-(2-HYDROXYETHYL)-3-METHYL-1-PHENYLTRIAZENE
mf: $C_9H_{13}N_3O$ mw: 179.25

SYNS: 1-PHENYL-3-METHYL-3-(2-HYDROXYAETHYL)-TRIAZEN (GERMAN) ◇ 1-PHENYL-3-METHYL-3-(2-HYDROXYETHYL)TRIAZENE

TOXICITY DATA with REFERENCE
scu-rat TDLo:1200 mg/kg/54W-I:CAR ZKKOBW 81,285,74
scu-rat LD50:360 mg/kg ZKKOBW 81,285,74

SAFETY PROFILE: A poison by subcutaneous route. Questionable carcinogen with experimental carcinogenic data. When heated to decomposition it emits toxic fumes of NO_x.

HKW300 CAS:41155-82-2 **HR: 2**
N-(β-HYDROXYETHYL)MORPHOLINE HYDROCHLORIDE
mf: $C_6H_{13}NO_2 \cdot ClH$ mw: 167.66

SYN: 4-MORPHOLINEETHANOL, HYDROCHLORIDE

TOXICITY DATA with REFERENCE
ipr-mus LD50:3957 mg/kg JPETAB 94,249,48
scu-mus LD50:3000 mg/kg AIPTAK 112,36,57
ivn-mus LD50:1420 mg/kg AIPTAK 112,36,57

SAFETY PROFILE: Moderately toxic by subcutaneous, intravenous, and intraperitoneal routes. When heated to decomposition it emits toxic fumes of NO_x and HCl.

HKW345 CAS:40343-32-6 **HR: 2**
N-(2-HYDROXYETHYL)-3-NITROBENZYLIDENIMINE N-OXIDE
mf: $C_9H_{10}N_2O_4$ mw: 210.21

SYN: 2-((m-NITROBENZYLIDENE)AMINO)ETHANOLN-OXIDE

TOXICITY DATA with REFERENCE
orl-rat TDLo:551 g/kg/46W-C:ETA FEPRA7 38,1403,79

SAFETY PROFILE: Questionable carcinogen with experimental tumorigenic data. When heated to decomposition it emits toxic fumes of NO_x.

HKW350 CAS:40343-30-4 **HR: 2**
N-(2-HYDROXYETHYL)-4-NITROBENZYLIDENIMINE N-OXIDE
mf: $C_9H_{10}N_2O_4$ mw: 210.21

SYN: 2-((p-NITROBENZYLIDENE)AMINO)ETHANOLN-OXIDE

TOXICITY DATA with REFERENCE
orl-rat TDLo:551 g/kg/46W-C:ETA FEPRA7 38,1403,79

SAFETY PROFILE: Questionable carcinogen with experimental tumorigenic data. When heated to decomposition it emits toxic fumes of NO_x.

HKW450 CAS:19561-70-7 **HR: 2**
N-(2-HYDROXYETHYL)-α-(5-NITRO-2-FURYL) NITRONE
mf: $C_7H_8N_2O_5$ mw: 200.17

SYNS: ETHANOL, 2-(((5-NITRO-2-FURANYL)METHYLENE)AMINO)-, N-OXIDE (9CI) ◇ ETHANOL, 2-((5-NITROFURFURYLIDENE)AMINO)-, N-OXIDE ◇ NIFURATRONE ◇ 2-(5-NITRO-2-FURFURYLIDENE) AMINOETHANOL N-OXIDE

TOXICITY DATA with REFERENCE
mmo-sat 100 ng/plate MUREAV 40,9,76
dnr-sat 500 nmol/well CNREA8 34,2266,74
mmo-esc 10 μg/plate MUREAV 26,3,74
dnr-esc 500 nmol/well CNREA8 34,2266,74
orl-rat TDLo:522 g/kg/46W-C:ETA FEPRA7 38,1403,79

SAFETY PROFILE: Questionable carcinogen with experimental tumorigenic data. Mutation data reported. When heated to decomposition it emits toxic fumes of NO_x.

HKW475 **HR: 2**
1-((2-HYDROXYETHYL)NITROSAMINO)-2-PROPANOL
mf: $C_5H_{12}N_2O_4$ mw: 164.19

SYNS: NIEA ◇ N-NITROSOETHANOLISOPROPANOLAMINE ◇ N-NITROSO-(2-HYDROXYPROPYL)-(2-HYDROXYETHYL)AMINE◇ NITROSOISOPROPANOL-ETHANOLAMINE

TOXICITY DATA with REFERENCE
mma-sat 500 μg/plate MUREAV 111,135,83
orl-rat TDLo:5 g/kg/50W-C:CAR CRNGDP 5,167,84

SAFETY PROFILE: Questionable carcinogen with experimental carcinogenic data. Mutation data reported. When heated to decomposition it emits toxic fumes of NO_x.

HKW500 CAS:13743-07-2 **HR: 3**
1-(2-HYDROXYETHYL)-1-NITROSOUREA
mf: $C_3H_7N_3O_3$ mw: 133.13

SYNS: HENU ◇ HNU ◇ NITROSO-2-HYDROXYETHYLUREA ◇ N-NITROSOHYDROXYETHYLUREA ◇ 1-NITROSO-1-(2-HYDROXYETHYL)UREA

TOXICITY DATA with REFERENCE
mmo-sat 41 μmol/L/48H MUREAV 48,131,77
mma-sat 2500 ng/plate TCMUE9 1,13,84
spm-mus-ipr 455 mg/kg MUREAV 108,337,83
ipr-mus TDLo:284 mg/kg (1D male):REP MUREAV 108,337,83
orl-rat TDLo:520 mg/kg/37W-I:CAR JJCREP 79,181,88
skn-mus TDLo:479 mg/kg/45W-I:CAR CNREA8 43,214,83
ipr-mus TDLo:8 mg/kg:ETA JJIND8 63,1469,79
ipr-rat LD50:120 mg/kg JNCIAM 56,445,76

CONSENSUS REPORTS: EPA Genetic Toxicology Program.

SAFETY PROFILE: Suspected carcinogen with experimental carcinogenic and tumorigenic data. A poison by intraperitoneal route. Experimental reproductive effects. Mutation data reported. When heated to decomposition it emits toxic fumes of NO_x. See also N-NITROSO COMPOUNDS.

HKX600 CAS:7416-48-0 ***HR: 3***
1-HYDROXYETHYL PEROXYACETATE
mf: $C_4H_8O_4$ mw: 120.10

$CH_3CH(OH)OOCO{\cdot}CH_3$

PROP: A low melting point solid.

SYN: 1-HYDROXYETHYL PERACETATE

SAFETY PROFILE: An explosive solid produced by the autooxidation of acetaldehyde. When heated to decomposition it emits acrid smoke and fumes. See also PEROXIDES and ACETALDEHYDE.

HKY000 CAS:58989-02-9 ***HR: 3***
N-HYDROXY-N-ETHYL-p-(PHENYLAZO) ANILINE
mf: $C_{14}H_{15}N_3O$ mw: 241.32

SYNS: N-ETHYL-N-(p-(PHENYLAZO)PHENYL)HYDROXYLAMINE ◇ N-HYDROXY-EAB ◇ N-HYDROXY-N-ETHYL-4-AMINOAZOBENZENE

TOXICITY DATA with REFERENCE
orl-rat TDLo:424 mg/kg/5W-I:ETA CNREA8 39,3411,79

SAFETY PROFILE: Questionable carcinogen with experimental tumorigenic data. When heated to decomposition it emits toxic fumes of NO_x.

HKY500 CAS:103-76-4 ***HR: 3***
1-(2-HYDROXYETHYL)PIPERAZINE
mf: $C_6H_{14}N_2O$ mw: 130.22

PROP: Light colored liquid. D: 1.10610, bp: 240°, flash p: 255°F (OC), vap d: 4.5.

SYNS: N-(β-HYDROXYETHYL)PIPERAZINE ◇ 1-PIPERAZINE-ETHANOL ◇ 2-(1-PIPERAZINYL)ETHANOL ◇ USAF DO-22

TOXICITY DATA with REFERENCE
skn-rbt 500 mg open MLD UCDS** 1/6/70
eye-rbt 100 mg MOD 34ZIAG -,689,69
orl-rat LD50:4920 mg/kg AIHAAP 23,95,62
ipr-mus LD50:100 mg/kg NTIS** AD277-689
orl-rbt LD50:3350 mg/kg GISAAA 45(5),67,80
orl-gpg LD50:3720 mg/kg GISAAA 45(5),67,80

CONSENSUS REPORTS: Reported in EPA TSCA Inventory.

SAFETY PROFILE: Poison by intraperitoneal route. Moderately toxic by ingestion. A skin and eye irritant. Combustible when exposed to heat or flame; can react vigorously with oxidizers. To fight fire, use foam, alcohol foam. When heated to decomposition it emits toxic fumes of NO_x.

HLB350 ***HR: 3***
4-(1-HYDROXYETHYL)PYRIDINE-N-OXIDE
mf: $C_7H_9NO_2$ mw: 139.15

$CH_3CH(OH)C_5H_4N{:}O$

SAFETY PROFILE: Can explode during vacuum distillation. When heated to decomposition it emits toxic fumes of NO_x.

HLB400 CAS:9005-27-0 ***HR: 1***
HYDROXYETHYL STARCH

SYNS: ESSEX 1360 ◇ ESSEX GUM 1360 ◇ ETHYLEX GUM 2020 ◇ HAS (GERMAN) ◇ HES ◇ HESPANDER ◇ HESPANDER INJECTION ◇ HYDROXYATHYLSTARKE (GERMAN) ◇ o-(HYDROXYETHYL) STARCH ◇ 2-HYDROXYETHYL STARCH ◇ o-(2-HYDROXYETHYL) STARCH ◇ 2-HYDROXYETHYL STARCH ETHER ◇ PENFORD 260 ◇ PENFORD 280 ◇ PENFORD 290 ◇ PENFORD P 208 ◇ PLASMA-STERIL ◇ STARCH HYDROXYETHYL ETHER ◇ TAPIOCA STARCH HYDROXYETHYL ETHER

TOXICITY DATA with REFERENCE
ivn-mus TDLo:420 g/kg (female 7-13D post):REP OYYAA2 6,1119,72
ivn-mus TDLo:675 g/kg (8-16D preg):TER OYYAA2 6,1119,72
ivn-rat LD50:11800 mg/kg OYYAA2 6,1023,72
ivn-mus LD50:20300 mg/kg OYYAA2 6,1023,72
ivn-rbt LD50:24100 mg/kg OYYAA2 6,1023,72

CONSENSUS REPORTS: Reported in EPA TSCA Inventory.

SAFETY PROFILE: Mildly toxic by intravenous route. Experimental teratogenic and reproductive effects. When heated to decomposition it emits acrid smoke and fumes.

HLB500 CAS:27375-52-6 ***HR: 1***
4'-(2-HYDROXYETHYLSULFONYL)ACETANILIDE
mf: $C_{10}H_{13}NO_4S$ mw: 243.30

SYNS: p-ACETAMINOFENYL-β-HYDROXYETHYLSULFON ◇ p-ACETAMINOFENYL-2-HYDROXYETHYLSULFON ◇ ACETANILIDE, 4'-(2-HYDROXYETHYLSULFONYL)-

TOXICITY DATA with REFERENCE
eye-rbt 500 mg/24H MLD 85JCAE-,1049,86

CONSENSUS REPORTS: Reported in EPA TSCA Inventory.

SAFETY PROFILE: An eye irritant. When heated to decomposition it emits toxic fumes of NO_x and SO_x

HLC000 CAS:519-37-9 ***HR: 3***
HYDROXYETHYLTHEOPHYLLINE
mf: $C_9H_{12}N_4O_3$ mw: 224.25

SYNS: AETHOPHYLLINUM ◇ ASCORPHYLLINE ◇ BIO-PHYLLINE ◇ CORDALIN ◇ COROPHYLLIN-N ◇ 3,7-DIHYDRO-7-(2-HYDROXYETHYL)-1,3-DIMETHYL-1H-PURINE-2,6-DIONE (9CI) ◇ 1,3-DIMETHYL-7-(2-HYDROXYETHYL)XANTHINE ◇ DILAPHYLLIN ◇ ETOFYLLINE ◇ FREKAPHYLLIN ◇ 7-(HYDROXYETHYL)THEOPHYLLINE ◇ 7-(β-HYDROXYETHYL)THEOPHYLLINE ◇ 7-(2-HYDROXYETHYL)THEOPHYLLINE ◇ 7-(2'-HYDROXYETHYL)THEOPHYLLINE ◇ OT ◇ OXYAETHYLTHEOPHYLLIN (GERMAN) ◇ OXYETHYLTHEOPHYLLINE ◇ OXYPHYLLINE ◇ OXYPHYLLINE (AMIDO) ◇ OXYTHEONYL ◇ PHYLLOCORMIN N ◇ SOLUPHYLLINE ◇ 7-THEOPHYLLINEETHANOL

TOXICITY DATA with REFERENCE
orl-rat TDLo:45 g/kg (90D male):REP YACHDS 8,383,80
orl-rat LD50:710 mg/kg ARZNAD 30,2023,80
scu-rat LD50:176 mg/kg AEPPAE 230,194,57
ivn-rat LD50:486 mg/kg AEPPAE 230,194,57
orl-mus LD50:400 mg/kg ARZNAD 27,1173,77
ipr-mus LD50:400 mg/kg ARZNAD 8,190,58
scu-mus LD50:400 mg/kg ARZNAD 4,649,54
ivn-mus LD50:344 mg/kg AEPPAE 230,194,57

SAFETY PROFILE: Poison by ingestion, subcutaneous, intravenous and intraperitoneal routes. Experimental reproductive effects. When heated to decomposition it emits toxic fumes of NO_x. See also THEOPHYLLINE.

HLC500 CAS:71-27-2 ***HR: 3***
(2-HYDROXYETHYL)TRIMETHYLAMMONIUM CHLORIDE SUCCINATE
mf: $C_{14}H_{30}N_2O_4 \cdot 2Cl$ mw: 361.36

SYNS: ANECTINE ◇ ANECTINE CHLORIDE ◇ BIS(2-DIMETHYLAMINOETHYL)SUCCINATE BIS(METHOCHLORIDE) ◇ BIS(SUCCINYLDICHLOROCHOLINE) ◇ CHLORSUCCINYLCHOLIN (GERMAN) ◇ CHLORURE de SUCCINILCOLINE (FRENCH) ◇ CHOLINE SUCCINATE DICHLORIDE ◇ CLORURO di SUCCINILCOLINA (ITALIAN) ◇ DIACETYLCHOLINE CHLORIDE ◇ DIACETYLCHOLINE DICHLORIDE ◇ 2-DIMETHYLAMINOETHYL SUCCINATE DIMETHOCHLORIDE ◇ DITILIN ◇ DITILINE ◇ 2,2'-((1,4-DIOXO-1,4-BUTANEDIYL)BIS(OXY)BIS(N,N,N-TRIMETHYLETHANAMINIUMDICHLORIDE ◇ LISTENON ◇ LYSTENON ◇ LYSTHENONE ◇ MIDARINE ◇ MYOPLEGINE ◇ PANTOLAX ◇ QUELICIN ◇ QUELICIN CHLORIDE ◇ SCH CHLORIDE ◇ SCOLINE ◇ SCOLINE CHLORIDE ◇ SKOLIN ◇ SUCCICURAN ◇ SUCCINIC ACID BIS(β-DIMETHYLAMINOETHYL) ESTER, DIHYDROCHLORIDE ◇ SUCCINIC ACID BIS(β-DIMETHYLAMINOETHYL)ESTER DIMETHOCHLORIDE ◇ SUCCINIC ACID DIESTER with CHOLINE CHLORIDE ◇ SUCCINOYLCHOLINE CHLORIDE ◇ SUCCINYL-ASTA ◇ SUCCINYL BISCHOLINE CHLORIDE ◇ SUCCINYLBISCHOLINE DICHLORIDE ◇ SUCCINYLCHOLINE CHLORIDE ◇ SUCCINYLCHOLINE DICHLORIDE ◇ SUCCINYLCHOLINE HYDROCHLORIDE ◇ SUCCINYLDICHOLINE CHLORIDE ◇ SUCCINYLFORTE ◇ SUCOSTRIN ◇ SUCOSTRIN CHLORIDE ◇ SURAMETHINIUM ◇ SUXAMETHIONIUM CHLORIDE ◇ SUXAMETHONIUM CHLORIDE ◇ SUXAMETHONIUM DICHLORIDE ◇ SUXCERT ◇ SUXETHONIUM CHLORIDE ◇ SUXINYL ◇ ULTRAPAL CHLORIDE

TOXICITY DATA with REFERENCE
ivn-man TDLo:7 mg/kg:CVS,MET AROTAA 96,464,72
ipr-mus LD50:1250 μg/kg NIIRDN 6,135,82
ivn-mus LD50:430 μg/kg JPETAB 109,83,53
ivn-rbt LD50:240 μg/kg IJNEAQ 5,305,66
ivn-ckn LD50:300 μg/kg AIPTAK 122,152,59

SAFETY PROFILE: A deadly poison by intravenous and intraperitoneal routes. Human systemic effects by intravenous route: metabolic changes in potassium level, cardiac arrythmias. Used as a skeletal muscle relaxant. When heated to decomposition it emits very toxic fumes of NO_x, NH_3, and Cl^-.

HLE000 CAS:1147-55-3 ***HR: D***
5-HYDROXY-N-2-FLUORENYLACETAMINE
mf: $C_{15}H_{13}NO_2$ mw: 239.29

TOXICITY DATA with REFERENCE
mma-sat 250 μg/plate JJIND8 62,839,79
dnr-esc 10 mg/L JJIND8 62,839,79

CONSENSUS REPORTS: EPA Genetic Toxicology Program.

SAFETY PROFILE: Mutation data reported. When heated to decomposition it emits very toxic fumes of NO_x.

HLE450 CAS:14461-87-1 ***HR: 2***
N-(7-HYDROXYFLUOREN-2-YL)ACETOHYDROXAMIC ACID
mf: $C_{15}H_{13}NO_3$ mw: 255.29

SYN: N-HYDROXY-N-(7-HYDROXY-2-FLUORENYL)ACETAMIDE

TOXICITY DATA with REFERENCE
ipr-rat TDLo:540 mg/kg/4W-I:CAR CNREA8 27,1443,67

SAFETY PROFILE: Questionable carcinogen with experimental carcinogenic data. When heated to decomposition it emits toxic fumes of NO_x.

HLE500 CAS:26630-60-4 ***HR: 3***
N-HYDROXY-2-FLUORENYLBENZENESULFONAMIDE
mf: $C_{19}H_{15}NO_3S$ mw: 337.41

SYN: N-HYDROXY-N-FLUORENYLBENZENESULFONAMIDE

TOXICITY DATA with REFERENCE
orl-rat TDLo:3660 mg/kg/29W:CAR CNREA8 31,778,71
ipr-rat TDLo:192 mg/kg/4W-I:CAR CNREA8 30,1485,70

SAFETY PROFILE: Questionable carcinogen with experimental carcinogenic data. When heated to decomposition it emits very toxic fumes of NO_x and SO_x.

HLE650 CAS:78281-06-8 ***HR: 2***
N-HYDROXY-4-FORMYLAMINOBIPHENYL
mf: $C_{13}H_{11}NO_2$ mw: 213.25

SYNS: N-(4-BIPHENYLYL)FORMOHYDROXAMIC ACID ◇ FORMAMIDE, N-(1,1′-BIPHENYL)-4-YL-N-HYDROXY- ◇ N-HYDROXY-FABP

TOXICITY DATA with REFERENCE
ipr-rat TDLo:85 mg/kg:CAR CNREA8 41,2450,81

SAFETY PROFILE: Questionable carcinogen with experimental carcinogenic data. When heated to decomposition it emits toxic fumes of NO_x.

HLE750 CAS:89947-76-2 ***HR: 2***
N-HYDROXY-N-GLUCURONOSYL-2-AMINOFLUORENE
mf: $C_{19}H_{19}NO_7$ mw: 373.39

SYNS: β-D-GLUCOPYRANURONIC ACID, 1-DEOXY-1-(9H-FLUOREN-2-YLHYDROXYAMINO)- ◇ GLUCURONIC ACID, 1-DEOXY-1-(2-FLUORENYLHYDROXYAMINO)- ◇ N-GLUCURONIDE of N-HYDROXY-2-AMINOFLUORENE ◇ N-HYDROXY-2-AMINOFLUORENE N-GLUCURONIDE

TOXICITY DATA with REFERENCE
dns-hmn:oth 100 nmol/L JJIND8 72,847,84
par-rat TDLo:5600 μg/kg/30W-C:CAR CNREA8 47,3406,87

SAFETY PROFILE: Questionable carcinogen with experimental carcinogenic data. Mutation data reported. When heated to decomposition it emits toxic fumes of NO_x.

HLF000 CAS:3393-34-8 ***HR: 3***
4-HYDROXYHEX-4-ENOIC ACID LACTONE
mf: $C_6H_8O_2$ mw: 112.14

SYN: 5-ETHYLIDENEDIHYDRO-2(3H)-FURANONE

TOXICITY DATA with REFERENCE
scu-rat TDLo:1160 mg/kg/58W-I:ETA BJCAAI 15,85,61

SAFETY PROFILE: Questionable carcinogen with experimental tumorigenic data. When heated to decomposition it emits acrid smoke and fumes.

HLF500 CAS:56395-66-5 ***HR: 3***
7-(5-HYDROXYHEXYL)-3-METHYL-1-PROPYL-XANTHINE
mf: $C_{15}H_{24}N_4O_3$ mw: 308.43

SYNS: A7301153 ◇ HWA 153

TOXICITY DATA with REFERENCE
orl-rat LD50:120 mg/kg ARZNAD 29,1013,79
ipr-rat LD50:241 mg/kg ARZNAD 29,1013,79
ivn-rat LD50:94 mg/kg ARZNAD 29,1013,79
ims-rat LD50:292 mg/kg ARZNAD 29,1013,79
orl-mus LD50:1345 mg/kg ARZNAD 29,1013,79
ipr-mus LD50:260 mg/kg ARZNAD 29,1013,79
scu-mus LD50:378 mg/kg ARZNAD 29,1013,79
orl-gpg LD50:695 mg/kg ARZNAD 29,1013,79
ivn-gpg LD50:203 mg/kg ARZNAD 29,1013,79

SAFETY PROFILE: Poison by ingestion, intraperitoneal, intravenous, intramuscular, and subcutaneous routes. When heated to decomposition it emits toxic fumes of NO_x.

HLH000 CAS:70145-54-9 ***HR: 2***
5-HYDROXY-2-(HYDROXYMETHYL)-4H-PYRAN-4-ONE SODIUM SALT
mf: $C_6H_5O_4$•Na mw: 164.10

TOXICITY DATA with REFERENCE
ivn-rat LDLo:1000 mg/kg SCIEAS 80,34,34
ivn-dog LDLo:1000 mg/kg SCIEAS 80,34,34
ivn-rbt LDLo:1000 mg/kg SCIEAS 80,34,34

SAFETY PROFILE: Moderately toxic by intravenous route. When heated to decomposition it emits toxic fumes of Na_2O.

HLH500 CAS:501-30-4 ***HR: 3***
5-HYDROXY-2-(HYDROXYMETHYL)-4-PYRONE
mf: $C_6H_6O_4$ mw: 142.12

SYNS: 5-HYDROXY-2-(HYDROXYMETHYL)-4H-PYRAN-4-ONE ◇ KOJIC ACID

TOXICITY DATA with REFERENCE
mmo-sat 1 mg/plate MUREAV 67,367,79
mma-sat 4 mg/plate JTSCDR 7,255,82
ivn-rat LDLo:1000 mg/kg 85ERAY 3,1755,78
orl-mus LDLo:4 g/kg NATUAS 155,302,45
ipr-mus LD50:250 mg/kg 85GDA2 5,393,81
scu-mus LDLo:1500 mg/kg NATUAS 155,302,45
ivn-mus LDLo:1500 mg/kg NATUAS 155,302,45
ivn-dog LDLo:1000 mg/kg 85ERAY 3,1755,78
ivn-rbt LDLo:1000 mg/kg 85ERAY 3,1755,78

CONSENSUS REPORTS: EPA Genetic Toxicology Program.

SAFETY PROFILE: Poison by intraperitoneal route. Moderately toxic by ingestion, intravenous, and subcutaneous routes. Mutation data reported. When heated to decomposition it emits acrid smoke and fumes.

HLI000 CAS:2092-55-9 ***HR: D***
4-HYDROXY-3-((2-HYDROXY-1-NAPHTHALENYL)AZO)BENZENESULFONIC ACID, MONOSODIUM SALT
mf: $C_{16}H_{12}N_2O_5S$•Na mw: 367.35

SYNS: ACID ALIZARINE VIOLET ◇ AIZEN CHROME VIOLET BH ◇ C.I. 15670

TOXICITY DATA with REFERENCE
mma-sat 500 μg/plate MUREAV 56,249,78

CONSENSUS REPORTS: Reported in EPA TSCA Inventory.

SAFETY PROFILE: Mutation data reported. When heated to decomposition it emits very toxic fumes of SO_x, Na_2O, and NO_x.

HLI300 ***HR: 3***
1-HYDROXYIMIDAZOLE-2-CARBOXALDOXIME-3-OXIDE
mf: $C_4H_5N_3O_3$ mw: 143.10

HONC(CH:NOH)=N(O)CH=CH

SAFETY PROFILE: The solid is an explosive. When heated to decomposition it emits toxic fumes of NO_x.

HLI325 CAS:35321-46-1 ***HR: 3***
1-HYDROXYIMIDAZOL-N-OXIDE
mf: $C_3H_4N_2O_2$ mw: 100.08

HONCH=CHN(O)=CH

SAFETY PROFILE: A sensitive explosive. When heated to decomposition it emits toxic fumes of NO_x.

HLI500 CAS:100-64-1 ***HR: 3***
(HYDROXYIMINO)CYCLOHEXANE
mf: $C_6H_{11}NO$ mw: 113.18

$CH_2(CH_2)_4C$:NOH

PROP: Prisms or liquid. Mp: 89-90°, bp: 204° (slt decomp), sol in water, sltly sol in ligroin, very sol in alc and ether.

SYNS: ANTIOXIDANT D ◇ CYCLOHEXANONE OXIME

TOXICITY DATA with REFERENCE
ipr-mus LD50:250 mg/kg NTIS** AD691-490

CONSENSUS REPORTS: Reported in EPA TSCA Inventory. EPA Genetic Toxicology Program.

SAFETY PROFILE: Poison by intraperitoneal route. Violent reaction with oleum (fuming sulfuric acid) above 150°C. When heated to decomposition it emits toxic fumes of NO_x.

HLJ000 CAS:54-16-0 ***HR: 3***
5-HYDROXYINDOLYLACETIC ACID
mf: $C_{10}H_9NO_3$ mw: 191.20

SYN: 5-HYDROXYINDOLEACETIC ACID

TOXICITY DATA with REFERENCE
scu-mus TDLo:2000 mg/kg/20W-I:ETA AICCA6 19,660,63
ipr-mus LD50:1125 mg/kg JTEHD6 1,515,76

SAFETY PROFILE: Moderately toxic by intraperitoneal route. Questionable carcinogen with experimental tumorigenic data. When heated to decomposition it emits toxic fumes of NO_x.

HLJ500 CAS:1689-89-0 ***HR: 3***
4-HYDROXY-3-IODO-5-NITROBENZONITRILE
mf: $C_7H_3IN_2O_3$ mw: 290.02

SYNS: DOVENIX ◇ NITROXYNIL ◇ TRODAX

TOXICITY DATA with REFERENCE
orl-mam LDLo:125 mg/kg FAZMAE 17,108,73
par-mam LDLo:50 mg/kg FAZMAE 17,108,73

CONSENSUS REPORTS: Cyanide and its compounds are on the Community Right-To-Know List.

SAFETY PROFILE: A poison by ingestion and parenteral routes. When heated to decomposition it emits very toxic I^-, NO_x, and CN^-. See also NITRILES.

HLK000 CAS:7376-66-1 ***HR: 3***
4-HYDROXY-α-ISOPROPYLAMINOMETHYLBENZYL ALCOHOL
mf: $C_{11}H_{17}NO_2$ mw: 195.29

SYNS: p-HYDROXY-α-ISOPROPYLAMINOMETHYLBENZYLALCOHOL ◇ WIN 833

TOXICITY DATA with REFERENCE
ipr-mus LD50:370 mg/kg JPETAB 106,341,52
ivn-mus LD50:144 mg/kg JPETAB 106,341,52

SAFETY PROFILE: Poison by intraperitoneal and intravenous routes. When heated to decomposition it emits toxic fumes such as NO_x.

HLK500 CAS:55688-38-5 ***HR: 3***
3-HYDROXY-α-ISOPROPYLAMINOMETHYLBENZYL ALCOHOL HYDROCHLORIDE
mf: $C_{11}H_{17}NO_2$•ClH mw: 231.75

SYN: 1-(3-HYDROXYPHENYL)-2-ISOPROPYLAMINOETHANOLHYDROCHLORIDE

TOXICITY DATA with REFERENCE
ipr-mus LD50:320 mg/kg JPETAB 106,440,52
ivn-mus LD50:130 mg/kg JPETAB 106,440,52
par-mus LD50:750 mg/kg PCJOAU 17,543,83

SAFETY PROFILE: Poison by intraperitoneal and intravenous routes. Moderately toxic by parenteral route. When heated to decomposition it emits very toxic fumes such as HCl and NO_x.

HLK800 ***HR: 3***
8-(2-HYDROXY-3-ISOPROPYLAMINO)PROPOXY-2H-1-BENZOPYRAN
mf: $C_{15}H_{21}NO_3$ mw: 263.34

SYN: 1-(2H-1-BENZOPYRAN-8-YLOXY)-3-ISOPROPYLAMINO-2-PROPANOL

TOXICITY DATA with REFERENCE
orl-rat TDLo:10500 mg/kg (35D male):REP OYYAA2 30,573,85
orl-rat LD50:473 mg/kg OYYAA2 30,573,85
ivn-rat LD50:39 mg/kg OYYAA2 30,573,85

SAFETY PROFILE: Poison by intravenous route. Moderately toxic by ingestion. Experimental reproductive effects. When heated to decomposition it emits toxic fumes of NO_x.

HLM000 CAS:10004-44-1 ***HR: 2***
HYDROXYISOXAZOLE
mf: $C_4H_5NO_2$ mw: 99.10

SYNS: 3-HYDROXY-5-METHYLISOXAZOLE ◇ HYMEXAZOL ◇ ITACHIGARDEN ◇ 5-METHYL-3-ISOXAZOLOL ◇ 5-METHYL-3(2H)-ISOXAZOLONE ◇ SF-6505 ◇ TACHIGAREN

TOXICITY DATA with REFERENCE
oms-omi 3 mg/L NNGADV 8,173,83
orl-rat LD50:3112 mg/kg GUCHAZ 6,302,73
scu-rat LD50:1884 mg/kg NYKGA7 2,165,75
orl-mus LD50:1968 mg/kg FMCHA2 -,C227,83
scu-mus LD50:1167 mg/kg NYKGA7 2,165,75
ivn-mus LD50:445 mg/kg NYKGA7 2,165,75

SAFETY PROFILE: Moderately toxic by ingestion, subcutaneous and intravenous routes. Mutation data reported. When heated to decomposition it emits toxic fumes of NO_x.

HLM500 CAS:7803-49-8 ***HR: 3***
HYDROXYLAMINE
mf: H_3NO mw: 33.04

PROP: Colorless liquid or white needles. Unstable, hygroscopic, decomp rapidly at room temp. Mp: 34.0°, bp: 110.0°, flash p: explodes at 265°F, d: 1.227, vap press: 10 mm @ 47.2°. Decomp in hot water; very sol in liquid ammonia and methanol; very sltly sol in ether, benzene, carbon disulfide, and chloroform.

SYN: OXAMMONIUM

TOXICITY DATA with REFERENCE
mmo-omi 1 mg/L MUREAV 74,113,80
slt-dmg-unr 45 mmol/L/48H MUREAV 120,233,83
dni-hmn:hla 10 mmol/L MUREAV 93,447,82
sce-ham:lng 5 mmol/L HUGEDQ 54,155,80
ipr-rat LD50:59 mg/kg CNREA8 26,1448,66
scu-rat LD50:29 mg/kg JPETAB 119,444,57
ipr-mus LD50:60 mg/kg JPETAB 165,30,69

CONSENSUS REPORTS: Reported in EPA TSCA Inventory. EPA Genetic Toxicology Program.

SAFETY PROFILE: A poison by intraperitoneal, and subcutaneous routes. A corrosive irritant to the eye, skin, and mucous membranes. Locally it is irritating, and systemically it can cause methemoglobinemia. Human mutation data reported. Dangerous fire hazard when exposed to heat, flame, and oxidizers. May ignite spontaneously in air if a large surface area is exposed (e.g., precipitate on paper). Explodes in air when heated above 70°C. Explosive reaction with potassium dichromate; chromium trioxide; powdered zinc + heat. Forms the heat-sensitive explosive bis(hydroxylamide) in reaction with zinc or calcium. Ignites on contact with copper(II) sulfate; metals (e.g., sodium); oxidants (e.g., barium peroxide; barium oxide; lead dioxide; potassium permanganate; chlorine); phosphorus chlorides (e.g., phosphorus trichloride; phosphorus pentachloride). Incompatible with carbonyls; pyridine. Vigorous reaction with hypochlorites. When heated to decomposition it emits toxic fumes of NO_x. See also AMINES.

HLN000 CAS:5470-11-1 ***HR: 3***
HYDROXYLAMINE HYDROCHLORIDE
mf: $H_3NO•ClH$ mw: 69.50

PROP: Monoclinic crystals. Decomp when moist. Mp: 151°, bp: decomp, d: 1.67 @ 17°.

SYN: OXAMMONIUM HYDROCHLORIDE

TOXICITY DATA with REFERENCE
dnr-esc 500 μg/plate JNCIAM 62,873,79
mmo-omi 1 mol/L MUREAV 73,1,80
mrc-smc 200 ppm JNCIAM 62,901,79
otr-rat:emb 19500 ng/plate JJATDK 1,190,81
cyt-mam:lym 500 mg/L MUREAV 81,63,81
orl-mus LD50:408 mg/kg APTOA6 6,285,50
ipr-mus LD50:10 mg/kg NTIS** AD277-689
scu-dog LDLo:70 mg/kg HBAMAK 4,1289,35

CONSENSUS REPORTS: Reported in EPA TSCA Inventory. EPA Genetic Toxicology Program.

SAFETY PROFILE: Poison by intraperitoneal and subcutaneous routes. Moderately toxic by ingestion. Mutation data reported. When heated to decomposition it emits very toxic fumes of HCl and NO_x. See also HYDROXYLAMINE.

HLN500 ***HR: 3***
HYDROXYLAMMONIUM PHOSPHINATE
mf: H_6NO_3P mw: 99.02

SAFETY PROFILE: Salt detonates above its melting point of 92°C. When heated to decomposition it emits very toxic fumes of PO_x, NH_3, and NO_x.

HLO300 CAS:67466-58-4 ***HR: 3***
o-(N-3-HYDROXYMERCURI-2-HYDROXYETHOXYPROPYLCARBAMYL)PHENOXY-ACETIC ACID, SODIUM SALT
mf: $C_{14}H_{18}HgNO_7$ mw: 535.91

SYN: (3-(α-CARBOXY-o-ANISAMIDO)-2-(2-HYDROXYETHOXY)PROPYL)HYDROXY-MERCURY MONOSODIUM SALT

TOXICITY DATA with REFERENCE
ivn-rat LD50:32 mg/kg JAPMA8 40,249,51
ims-rat LD50:35 mg/kg JAPMA8 40,249,51
ivn-mus LD50:113 mg/kg JAPMA8 40,249,51
ims-mus LD50:118 mg/kg JAPMA8 40,249,51
ims-rbt LD50:26 mg/kg JAPMA8 40,249,51

CONSENSUS REPORTS: Mercury and its compounds are on the Community Right-To-Know List.

OSHA PEL: (Transitional: CL 1 mg/10m^3) CL 0.1 mg(Hg)/m^3 (skin)
ACGIH TLV: TWA 0.1 mg(Hg)/m^3 (skin)
NIOSH REL: (Mercury, inorganic): TWA 0.05 mg(Hg)/m^3

SAFETY PROFILE: Poison by intramuscular and intravenous routes. When heated to decomposition it emits toxic fumes of NO_x, Na_2O and Hg. See also MERCURY COMPOUNDS, ORGANIC.

HLO400 CAS:61792-05-0 ***HR: 3***
HYDROXYMERCURI-o-NITROPHENOL
mf: $C_6H_5HgNO_4$ mw: 355.71

SYNS: HYDROXY(4-HYDROXY-3-NITROPHENYL)MERCURY ◇ MERCURY, HYDROXY(4-HYDROXY-3-NITROPHENYL)-

TOXICITY DATA with REFERENCE
ipr-mus LDLo:26 mg/kg JPETAB 31,87,27

ACGIH TLV: TWA 0.1 mg(Hg)/m^3 (skin)

SAFETY PROFILE: Poison by intraperitoneal route. When heated to decomposition it emits toxic fumes of NO_x and Hg.

HLO500 CAS:63869-04-5 ***HR: 3***
o-(HYDROXYMERCURI)PHENOL
mf: $C_6H_6HgO_2$ mw: 310.71

SYN: (2-HYDROXYPHENYL)HYDROXYMERCURY

TOXICITY DATA with REFERENCE
sln-dmg-orl 250 mg/L HEREAY 57,446,67
orl-rat LDLo:100 mg/kg NCNSA6 5,36,53
ipr-rat LDLo:25 mg/kg NCNSA6 5,36,53

CONSENSUS REPORTS: Mercury and its compounds are on the Community Right-To-Know List.

OSHA PEL: (Transitional: CL 1 mg/10m^3) CL 0.1 mg(Hg)/m^3 (skin)
ACGIH TLV: TWA 0.1 mg(Hg)/m^3 (skin)
NIOSH REL: (Inorganic Mercury) TWA 0.05 mg(Hg)/m^3

SAFETY PROFILE: Poison by ingestion and intraperitoneal routes. Mutation data reported. When heated to decomposition it emits toxic fumes of Hg. See also MERCURY COMPOUNDS.

HLP000 CAS:63868-96-2 ***HR: 3***
HYDROXYMERCURIPROPANOLAMIDE of m-CARBOXYPHENOXYACETIC ACID
mf: $C_{12}H_{15}HgNO_6$ mw: 469.87

SYN: (m-((2-HYDROXY-3-HYDROXYMERCURI)PROPYL)PHENOXYACETIC ACID

TOXICITY DATA with REFERENCE
ivn-rbt LDLo:12 mg/kg JPETAB 41,21,31

CONSENSUS REPORTS: Mercury and its compounds are on the Community Right-To-Know List.

OSHA PEL: (Transitional: CL 1 mg/10m^3) CL 0.1 mg(Hg)/m^3 (skin)
ACGIH TLV: TWA 0.1 mg(Hg)/m^3 (skin)
NIOSH REL: (Inorganic Mercury) TWA 0.05 mg(Hg)/m^3

SAFETY PROFILE: Poison by intravenous route. When heated to decomposition it emits very toxic fumes of Hg and NO_x. See also MERCURY COMPOUNDS.

HLP500 CAS:63868-98-4 ***HR: 3***
HYDROXYMERCURIPROPANOLAMIDE of p-CARBOXYPHENOXYACETIC ACID
mf: $C_{12}H_{15}HgNO_6$ mw: 469.87

SYNS: (3-(α-CARBOXY-p-ANISAMIDO)-2-HYDROXYPROPYL) HYDROXYMERCURY ◇ (p-((2-HYDROXY-3-HYDROXYMERCURI)PROPYL)CARBAMOYL)PHENOXYACETIC ACID

TOXICITY DATA with REFERENCE
ivn-rbt LDLo:9 mg/kg JPETAB 41,21,31

CONSENSUS REPORTS: Mercury and its compounds are on the Community Right-To-Know List.

OSHA PEL: (Transitional: CL 1 mg/10m^3) CL 0.1 mg(Hg)/m^3 (skin)
ACGIH TLV: TWA 0.1 mg(Hg)/m^3 (skin)
NIOSH REL: (Inorganic Mercury) TWA 0.05 mg(Hg)/m^3

SAFETY PROFILE: Poison by intravenous route. When heated to decomposition it emits very toxic fumes of Hg and NO_x. See also MERCURY COMPOUNDS.

HLQ000 CAS:64048-08-4 ***HR: 3***
(5-(HYDROXYMERCURI)-2-THIENYL)MERCURY ACETATE
mf: $C_6H_6Hg_2O_3S$ mw: 559.36

SYNS: ACETATO((5-HYDROXYMERCURI)-2-THIENYL)MERCURY ◇ (ACETATO-o)HYDROXY-mu-2,5-THIOPHENEDIYLDIMERCURY ◇ 2-(ACETOXYMERCURI)-5-(HYDROXYMERCURI)THIOPHENE ◇ (THIOPHENE-2,5-DIYL)BIS(HYDROXYMERCURY)MONOACETATE

TOXICITY DATA with REFERENCE
ipr-rat LDLo:50 mg/kg NCNSA6 5,31,53
ivn-mus LD50:18 mg/kg CSLNX* NX#05138

CONSENSUS REPORTS: Mercury and its compounds are on the Community Right-To-Know List.

OSHA PEL: (Transitional: CL 1 mg/10m^3) CL 0.1 mg(Hg)/m^3 (skin)
ACGIH TLV: TWA 0.1 mg(Hg)/m^3 (skin)
NIOSH REL: (Inorganic Mercury) TWA 0.05 mg(Hg)/m^3

SAFETY PROFILE: Poison by intraperitoneal and intravenous routes. When heated to decomposition it emits very toxic fumes of Hg and SO_x. See also MERCURY COMPOUNDS.

HLQ500 CAS:498-02-2 ***HR: 2***
4'-HYDROXY-3'-METHOXYACETOPHENONE
mf: $C_9H_{10}O_3$ mw: 166.19

SYNS: ACETOGUAIACONE ◇ ACETOVANILLONE ◇ ACETOVANILONE ◇ ACETOVANYLLON ◇ APOCYNINE ◇ 1-(4-HYDROXY-3-METHOXYPHENYL)ETHANONE

TOXICITY DATA with REFERENCE
mmo-smc 400 mg/L MUREAV 119,272,83
orl-mus LD50:9 g/kg BCTKAG 14,301,84
ipr-mus LD50:650 mg/kg JMCMAR 7,178,64

CONSENSUS REPORTS: Reported in EPA TSCA Inventory.

SAFETY PROFILE: Moderately toxic by intraperitoneal route. Mildly toxic by ingestion. Mutation data reported. When heated to decomposition it emits acrid smoke and fumes. See also KETONES.

HLR000 CAS:78265-95-9 ***HR: 3***
N-HYDROXY-3-METHOXY-4-AMINOAZOBENZENE
mf: $C_{13}H_{13}N_3O_2$ mw: 243.29

SYN: 4-HYDROXY-3-METHOXY-4-AMINOAZOBENZENE ◇ N-HYDROXY-2-METHOXY-4-(PHENYLAZO)BENZENAMINE ◇ N-(2-METHOXY-4-(PHENYLAZO)PHENYL)HYDROXYLAMINE

TOXICITY DATA with REFERENCE
mmo-sat 100 nmol/plate GANNA2 68,373,77
mma-sat 50 nmol/plate GANNA2 72,921,81
scu-mus TDLo:623 mg/kg/8W-I:CAR GANNA2 73,136,82
scu-mus TD:1246 mg/kg/8W-I:ETA GANNA2 73,136,82

SAFETY PROFILE: Questionable carcinogen with experimental carcinogenic and tumorigenic data. Mutation data reported. When heated to decomposition it emits toxic fumes of NO_x.

HLR500 CAS:63040-24-4 ***HR: 3***
3-HYDROXY-4'-METHOXY-4-AMINODIPHENYL
mf: $C_{13}H_{13}NO_2$ mw: 215.27

SYN: 4-AMINO-4'-METHOXY-3-BIPHENYLOL

TOXICITY DATA with REFERENCE
imp-mus TDLo:67 mg/kg:CAR AICCA6 18,538,62

SAFETY PROFILE: Questionable carcinogen with experimental carcinogenic data. When heated to decomposition it emits toxic fumes of NO_x.

HLS500 CAS:4756-45-0 ***HR: 3***
o-(2-HYDROXY-4-METHOXYBENZOYL)BENZOIC ACID
mf: $C_{15}H_{12}O_5$ mw: 272.27

SYN: 2'-CARBOXY-2-HYDROXY-4-METHOXYBENZOPHENONE(o-(2-HYDROXY-p-ANISOYL)BENZOIC ACID)

TOXICITY DATA with REFERENCE
ivn-mus LD50:320 mg/kg CSLNX* NX#01618

CONSENSUS REPORTS: Reported in EPA TSCA Inventory.

SAFETY PROFILE: A poison by intravenous route. When heated to decomposition it emits acrid smoke and fumes.

HLT000 CAS:298-45-3 ***HR: 3***
4-HYDROXY-3-METHOXY-5,6-METHYLENEDIOXY-APORPHIN
mf: $C_{19}H_{19}NO_4$ mw: 325.39

SYNS: BULBOCAPNINE ◇ d-BULBOCAPNINE ◇ 19-METHOXY-1,2-(METHYLENEDIOXY)-6a-α-APORPHIN-11-OL

TOXICITY DATA with REFERENCE
scu-mus LD50:195 mg/kg JPETAB 56,85,36
ipr-mus LD50:145 mg/kg AIPTAK 155,69,65
scu-mus LD50:195 mg/kg MEIEDD 10,205,83
ivn-mus LD50:79 mg/kg CSLNX* NX#12068

SAFETY PROFILE: Poison by subcutaneous, intraperitoneal, and intravenous routes. When heated to decomposition it emits toxic fumes of NO_x.

HLT300 CAS:25395-41-9 ***HR: 2***
2-HYDROXY-3-(o-METHOXYPHENOXY)PROPYL NICOTINATE
mf: $C_{16}H_{17}NO_5$ mw: 303.34

SYNS: 3-(o-METHOXYPHENOXY)-2-HYDROXYPROPYL-1-NICOTINATE ◇ NICOTINIC ACID-2-HYDROXY-3-(o-METHOXYPHENOXY)PROPYL ESTER ◇ 3-PYRIDINECARBOXYLIC ACID-2-HYDROXY-3-(2-METHOXYPHENOXY)PROPYL ESTER

TOXICITY DATA with REFERENCE
orl-rat LD50:7700 mg/kg OYYAA2 7,189,73
orl-mus LD50:9800 mg/kg OYYAA2 7,149,73
ipr-mus LD50:3100 mg/kg OYYAA2 7,189,73

SAFETY PROFILE: Moderately toxic by intraperitoneal route. Mildly toxic by ingestion. When heated to decomposition it emits toxic fumes of NO_x. See also ESTERS.

HLT500 CAS:2929-14-8 ***HR: 3***
4-HYDROXY-8-METHOXYQUINALDIC ACID
mf: $C_{11}H_9NO_4$ mw: 219.21

SYNS: 8-METHOXY-4-HYDROXYQUINOLINE-2-CARBOXYLICACID ◇ 8-METHYL ETHER of XANTHURENIC ACID ◇ XANTHURENIC ACID-8-METHYL ETHER

TOXICITY DATA with REFERENCE
scu-mus TDLo:5280 mg/kg/44W-I:CAR CNREA8 28,183,68
imp-mus TDLo:89 mg/kg:NEO BJCAAI 11,212,57
imp-mus TD:80 mg/kg:ETA BJCAAI 11,212,57

SAFETY PROFILE: Questionable carcinogen with experimental carcinogenic, neoplastigenic, and tumorigenic data. When heated to decomposition it emits toxic fumes of NO_x.

HLU000 CAS:70145-56-1 ***HR: 3***
HYDROXY(2-METHOXY-3-(2,4,6-TRIOXO-(1H,3H,5H)PYRIMID-5-YL)PROPYL)MERCURY SODIUM SALT
mf: $C_8H_{11}HgN_2O_5 \cdot Na$ mw: 438.79

SYN: 5-(3-HYDROXYMERCURI-2-METHOXYPROPYL)BARBITURIC ACID SODIUM SALT

TOXICITY DATA with REFERENCE
ipr-rat LD50:19500 μg/kg JAPMA8 39,297,50

CONSENSUS REPORTS: Mercury and its compounds are on the Community Right-To-Know List.

OSHA PEL: (Transitional: CL 1 mg/10m^3) CL 0.1 mg(Hg)/m^3 (skin)
ACGIH TLV: TWA 0.1 mg(Hg)/m^3 (skin)
NIOSH REL: (Inorganic Mercury) TWA 0.05 mg(Hg)/m^3

SAFETY PROFILE: Poison by intraperitoneal route. When heated to decomposition it emits very toxic fumes of Hg, Na_2O, and NO_x. See also MERCURY COMPOUNDS.

HLU500 CAS:924-42-5 ***HR: 2***
N-(HYDROXYMETHYL)ACRYLAMIDE
mf: $C_4H_7NO_2$ mw: 101.12

$HOCH_2NHCO \cdot CH=CH_2$

SYNS: N-(HYDROXYMETHYL)-2-PROPENAMIDE ◇ N-METHANOLACRYLAMIDE ◇ N-METHYLOLACRYLAMIDE ◇ MONOMETHYLOLACRYLAMIDE ◇ NCI-C60333 ◇ URAMINE T 80

TOXICITY DATA with REFERENCE
orl-mus TDLo:730 g/kg (male 6W pre):REP ARTODN 59,201,86
orl-rat LD50:474 mg/kg NEZAAQ 34,183,79
ipr-rat LD50:563 mg/kg BCPCA6 19,2591,70
orl-mus LD50:420 mg/kg 37ASAA 1,306,78

CONSENSUS REPORTS: Reported in EPA TSCA Inventory.

SAFETY PROFILE: Moderately toxic by ingestion and intraperitoneal routes. Experimental reproductive effects. May undergo spontaneous combustion in storage. When heated to decomposition it emits toxic fumes of NO_x.

HLV000 CAS:1910-36-7 ***HR: 3***
N-HYDROXY-N-METHYL-4-AMINOAZOBENZENE
mf: $C_{13}H_{13}N_3O$ mw: 227.29

SYNS: N-HYDROXY-MAB ◇ N-METHYL-N-(p-(PHENYLAZO)PHENYL)HYDROXYLAMINE

TOXICITY DATA with REFERENCE
mmo-sat 100 nmol/plate CALEDQ 1,91,75
mma-sat 100 nmol/plate CALEDQ 1,91,75
orl-rat TDLo:466 mg/kg/5W-I:CAR CNREA8 39,3411,79
ipr-rat TDLo:28 mg/kg:ETA CNREA8 39,3411,79

SAFETY PROFILE: Questionable carcinogen with experimental carcinogenic and tumorigenic data. Mutation data reported. When heated to decomposition it emits toxic fumes of NO_x.

HLV500 CAS:94-07-5 ***HR: 3***
p-HYDROXY-α-((METHYLAMINO)METHYL)BENZYL ALCOHOL
mf: $C_9H_{13}NO_2$ mw: 167.23

SYNS: ANALEPTIN ◇ ETHAPHENE ◇ 4-HYDROXY-α-((METHYLAMINO)METHYL)BENZENEMETHANOL ◇ p-HYDROXYPHENYLMETHYLAMINOETHANOL ◇ 1-(4-HYDROXYPHENYL)-2-METHYLAMINOETHANOL ◇ 1-(4-HYDROXYPHENYL)-N-METHYLETHANOLAMINE ◇ p-METHYLAMINOETHANOLPHENOL ◇ β-METHYLAMINO-α-(4-HYDROXYPHENYL)ETHYL ALCOHOL ◇ METHYLAMINOMETHYL(4-HYDROXYPHENYL)CARBINOL ◇ OXEDRINE ◇ p-OXEDRINE ◇ PARASYMPATOL ◇ SIMPALON ◇ SIMPATOL ◇ SYMPATHOL ◇ SYNEPHRIN ◇ p-SYNEPHRINE ◇ SYNTHENATE

TOXICITY DATA with REFERENCE
scu-rat LDLo:1500 mg/kg AIPTAK 101,81,55
ipr-mus LD50:1000 mg/kg JPETAB 89,297,47
ivn-mus LD50:270 mg/kg JPETAB 106,341,52
ivn-rbt LDLo:150 mg/kg AIPTAK 101,81,55

SAFETY PROFILE: Poison by intravenous route. Moderately toxic by subcutaneous and intraperitoneal routes. Used as an adrenergic agent and vasopressor. When heated to decomposition it emits toxic fumes such as NO_x.

HLX000 CAS:22346-43-6 ***HR: 1***
4-HYDROXY-7-(METHYLAMINO)-2-NAPHTHALENESULFONIC ACID
mf: $C_{11}H_{11}NO_4S$ mw: 253.29

SYNS: KYSELINA2-METHYLAMINO-5-NAFTOL-7-SULFONOVA (CZECH) ◇ KYSELINA-N-METHYL-I (CZECH)

TOXICITY DATA with REFERENCE
eye-rbt 500 mg/24H MLD 28ZPAK -,190,72

CONSENSUS REPORTS: Reported in EPA TSCA Inventory.

SAFETY PROFILE: An eye irritant. When heated to decomposition it emits very toxic fumes of NO_x and SO_x. See also SULFONATES.

HLX500 CAS:955-48-6 ***HR: 2***
4'-(1-HYDROXY-2-(METHYLAMINO)PROPYL)METHANESULFONANILIDE HYDROCHLORIDE
mf: $C_{11}H_{18}N_2O_3S{\cdot}ClH$ mw: 294.83

SYN: MJ 1998

TOXICITY DATA with REFERENCE
ipr-rat LD50:850 mg/kg JPETAB 149,161,65
ipr-mus LD50:785 mg/kg JPETAB 149,161,65
ipr-dog LD50:485 mg/kg JPETAB 149,161,65
orl-rbt LD50:1250 mg/kg JPETAB 149,161,65

SAFETY PROFILE: Moderately toxic by ingestion and intraperitoneal routes. When heated to decomposition it emits very toxic fumes of NO_x, SO_x, and HCl. See also SULFONATES.

HLX550 CAS:69321-16-0 ***HR: 2***
4-(N-HYDROXY-N-METHYLAMINO)QUINOLINE 1-OXIDE
mf: $C_{10}H_{10}N_2O_2$ mw: 190.22

SYNS: HYDROXYLAMINE, N-METHYL-N-(4-QUINOLINYL)-, 1-OXIDE ◇ 4-(N-METHYLHYDROXYAMINO)-QUINOLINE-1-OXIDE ◇ 4-QUINOLINAMINE, N-HYDROXY-N-METHYL-, 1-OXIDE

TOXICITY DATA with REFERENCE
mma-esc 200 μmol/L CPBTAL 34,1755,86
scu-mus TDLo:60 mg/kg/10W-I:ETA GANNA2 69,835,78

SAFETY PROFILE: Questionable carcinogen with experimental tumorigenic data. Mutation data reported. When heated to decomposition it emits toxic fumes of NO_x.

HLX600 CAS:50880-57-4 ***HR: D***
17-β-HYDROXY-7-α-METHYLANDROST-5-ENE-3-ONE
mf: $C_{20}H_{30}O_2$ mw: 302.50

SYN: RMI 12,936

TOXICITY DATA with REFERENCE
scu-rat TDLo:10 mg/kg (female 8D post):TER CCPTAY 19,481,79
scu-rat TDLo:2500 μg/kg (female 8D post):REP PRGLBA 4,121,80

SAFETY PROFILE: An experimental teratogen. Experimental reproductive effects. A steroid. When heated to decomposition it emits acrid smoke and irritating fumes.

HLX900 ***HR: 2***
4-(HYDROXYMETHYL)BENZENEDIAZONIUM SULFATE
mf: $C_7H_7N_2O{\cdot}1/2H_2O_4S$ mw: 184.15

SYNS: HMBD ◇ BENZENEDIAZONIUM, 4-(HYDROXYMETHYL)-, SULFATE (2:1)

TOXICITY DATA with REFERENCE
scu-mus TDLo:1300 mg/kg/26W-I:CAR PAACA3 25,115,84

SAFETY PROFILE: Questionable carcinogen with experimental carcinogenic data. When heated to decomposition it emits toxic fumes of NO_x and SO_x.

HLX925 CAS:78246-54-5 ***HR: 3***
4-(HYDROXYMETHYL)BENZENEDIAZONIUM TETRAFLUOROBORATE
mf: $C_7H_7N_2O{\cdot}BF_4$ mw: 221.97

SYNS: HMBD ◇ BENZENEDIAZONIUM, 4-(HYDROXYMETHYL)-, TETRAFLUOROBORATE(1-)

TOXICITY DATA with REFERENCE
mmo-sat 1 μmol/plate ZLUFAR 183,85,86
mma-sat 1 μmol/plate ZLUFAR 183,85,86
orl-mus TDLo:400 mg/kg:CAR BJCAAI 46,417,82
scu-mus TDLo:1300 mg/kg/26W-I:CAR CNREA8 41,2444,81

SAFETY PROFILE: Suspected carcinogen with experimental carcinogenic data. Mutation data reported.

When heated to decomposition it emits toxic fumes of NO_x, B, and F^-.

HLY000 CAS:68041-18-9 ***HR: 3***
6-HYDROXYMETHYLBENZO(a)PYRENESULFATE ESTER (SODIUM SALT)
mf: $C_{21}H_{13}O_4S \cdot Na$ mw: 384.39

TOXICITY DATA with REFERENCE
mmo-sat 1 nmol/plate CBINA8 58,253,86
skn-mus TDLo:56 mg/kg/40W-I:NEO CBINA8 22(1),53,78

SAFETY PROFILE: Questionable carcinogen with experimental neoplastigenic data. Mutagenic data reported. When heated to decomposition it emits toxic fumes of SO_x and Na_2O. See also ESTERS.

HLY400 CAS:3597-91-9 ***HR: D***
4-(HYDROXYMETHYL)BIPHENYL
mf: $C_{13}H_{12}O$ mw: 184.25

SYNS: (1,1'-BIPHENYL)-4-METHANOL ◇ 4-BIPHENYLMETHANOL ◇ 4HMB ◇ p-PHENYLBENZYL ALCOHOL

TOXICITY DATA with REFERENCE
ipr-mus TDLo:500 mg/kg (male 5D pre):REP MUREAV 100,345,82

SAFETY PROFILE: Experimental reproductive effects. When heated to decomposition it emits acrid smoke and irritating fumes.

HLY500 CAS:84-79-7 ***HR: 3***
2-HYDROXY-3-(3-METHYL-2-BUTENYL)-1,4-NAPHTHOQUINONE
mf: $C_{15}H_{14}O_3$ mw: 242.29

SYNS: BETHABARRA WOOD ◇ C.I. 75490 ◇ C.I. NATURAL YELLOW 16 ◇ GREENHARTEN ◇ 2-HYDROXY-3-(3-METHYL-2-BUTENYL)-1,4-NAPHTHALENEDIONE ◇ IPE-TOBACCO WOOD ◇ LAPACHIC ACID ◇ LAPACHOL ◇ LAPACHOL WOOD ◇ NSC-11905 ◇ SURINAM GREENHEART WOOD ◇ TAIGUIC ACID ◇ TAIGU WOOD ◇ TECOMIN

TOXICITY DATA with REFERENCE
orl-rat LDLo:1200 mg/kg TXAPA9 17,1,70
orl-mus LD50:487 mg/kg TXAPA9 17,1,70
ipr-mus LD50:400 mg/kg JMCMAR 26,570,83

SAFETY PROFILE: Poison by intraperitoneal route. Moderately toxic by ingestion. When heated to decomposition it emits acrid smoke and fumes.

HMA000 CAS:3342-98-1 ***HR: 3***
1-HYDROXY-3-METHYLCHOLANTHRENE
mf: $C_{21}H_{16}O$ mw: 284.37

SYNS: 15-HYDROXY-20-METHYLCHOLANTHRENE ◇ 3-METHYLCHOLANTHREN-1-OL ◇ 3-METHYL-1-CHOLANTHRENOL

TOXICITY DATA with REFERENCE
mma-sat 20 nmol/plate CNREA8 38,3398,78
skn-mus TDLo:91 mg/kg/20W-I:CAR,REP CBINA8 22(1),69,78
scu-mus TDLo:120 mg/kg/6W-I:NEO IJCNAW 2,505,67

CONSENSUS REPORTS: EPA Genetic Toxicology Program.

SAFETY PROFILE: Experimental reproductive effects. Questionable carcinogen with experimental carcinogenic and neoplastigenic data. Mutation data reported. When heated to decomposition it emits acrid smoke and fumes.

HMA500 CAS:3308-64-3 ***HR: 3***
2-HYDROXY-3-METHYLCHOLANTHRENE
mf: $C_{21}H_{16}O$ mw: 284.37

SYN: 3-METHYLCHOLANTHREN-2-OL

TOXICITY DATA with REFERENCE
mma-sat 20 nmol/plate CNREA8 38,3398,78
mma-ham:lng 15 nmol/plate CNREA8 38,3398,78
skn-mus TDLo:91 mg/kg/20W-I:CAR CBINA8 22,69,78
scu-mus TDLo:120 mg/kg/6W-I:NEO IJCNAW 2,505,67

CONSENSUS REPORTS: EPA Genetic Toxicology Program.

SAFETY PROFILE: Questionable carcinogen with experimental carcinogenic and neoplastigenic data. Mutation data reported. When heated to decomposition it emits acrid smoke and fumes.

HMB000 CAS:5980-33-6 ***HR: 3***
7-HYDROXY-4-METHYLCOUMARIN SODIUM
mf: $C_{10}H_7O_3 \cdot Na$ mw: 198.16

SYNS: CANTABILINE SODIUM ◇ HYMECROMONE SODIUM ◇ METHYL-4-OMBELLIFERONE SODEE (FRENCH) ◇ METHYL-4-UMBELLIFERONE SODIUM

TOXICITY DATA with REFERENCE
ipr-rat LD50:750 mg/kg THERAP 23,359,68
ivn-rat LDLo:292 mg/kg THERAP 23,359,68
ipr-mus LD50:325 mg/kg THERAP 23,359,68
ivn-mus LD50:250 mg/kg THERAP 23,359,68
ivn-dog LDLo:283 mg/kg THERAP 23,359,68
ivn-rbt LDLo:235 mg/kg THERAP 23,359,68

SAFETY PROFILE: A poison by intraperitoneal and intravenous routes. When heated to decomposition it emits toxic fumes of Na_2O.

HMB500 CAS:80-71-7 ***HR: 2***
2-HYDROXY-3-METHYL-2-CYCLOPENTEN-1-ONE
mf: $C_6H_8O_2$ mw: 112.14

PROP: White crystalline power; nutty odor, maple in dilute solutions. Flash p: +212°F. Sol in alc, propylene glycol; sltly sol in fixed oils, water.

SYNS: CORYLON ◇ CORYLONE ◇ CYCLOTEN ◇ FEMA No. 2700 ◇ MAPLE LACTONE ◇ 3-METHYLCYCLOPENTANE-1,2-DIONE ◇ METHYL CYCLOPENTENOLONE (FCC)

TOXICITY DATA with REFERENCE
sce-hmn:lym 1500 μmol/L MUREAV 169,129,86
ipr-rat LDLo:500 mg/kg FCTXAV 14,809,76
ipr-mus LDLo:500 mg/kg FCTXAV 14,809,76
orl-gpg LD50:1400 mg/kg FCTXAV 14,809,76

CONSENSUS REPORTS: Reported in EPA TSCA Inventory.

SAFETY PROFILE: Moderately toxic by ingestion and intraperitoneal routes. Human mutation data reported. Combustible liquid. When heated to decomposition it emits acrid smoke and fumes.

HMB550 CAS:5116-24-5 ***HR: D***
5-HYDROXYMETHYLDEOXYURIDINE
mf: $C_{10}H_{14}N_2O_6$ mw: 258.26

SYNS: 2'-DEOXY-5-(HYDROXYMETHYL)URIDINE ◇ 5-HYDROXYMETHYL-2'-DEOXYURIDINE ◇ α-HYDROXY-THYMIDINE (9CI)

TOXICITY DATA with REFERENCE
mmo-sat 50 μg/plate MUREAV 169,123,86
mma-sat 50 μg/plate MUREAV 169,123,86
sce-hmn:lng 5 mg/L MUREAV 117,317,83

SAFETY PROFILE: Human mutation data reported. When heated to decomposition it emits toxic fumes of NO_x.

HMB595 CAS:35282-68-9 ***HR: 2***
2'-HYDROXYMETHYL-N,N-DIMETHYL-4-AMINOAZOBENZENE
mf: $C_{15}H_{17}N_3O$ mw: 255.35

SYNS: BENZENEMETHANOL,2-((4-(DIMETHYLAMINO)PHENYL)AZO)-(9CI) ◇ o-((p-DIMETHYLAMINOPHENYL)AZO)BENZYL ALCOHOL

TOXICITY DATA with REFERENCE
mma-sat 500 nmol/plate MUREAV 121,95,83
dns-rat:lvr 1 μmol/L CNREA8 46,1654,86
orl-rat TDLo:3600 mg/kg/17W-C:NEO JJIND8 79,1159,87

SAFETY PROFILE: Questionable carcinogen with experimental neoplastigenic data. Mutation data reported. When heated to decomposition it emits toxic fumes of NO_x.

HMB600 CAS:35282-69-0 ***HR: 3***
3'-HYDROXYMETHYL-N,N-DIMETHYL-4-AMINOAZOBENZENE
mf: $C_{15}H_{17}N_3O$ mw: 255.35

SYNS: m-((p-DIMETHYLAMINOPHENYL)AZO)BENZYLALCOHOL ◇ 3-((p-DIMETHYLAMINO)PHENYLAZO)BENZYL ALCOHOL ◇ 3'-HYDROXYMETHYL-4-(DIMETHYLAMINO)AZOBENZENE

TOXICITY DATA with REFERENCE
mmo-sat 1 μmol/plate CRNGDP 4,1487,83
mma-sat 1 μmol/L CPBTAL 32,3641,84
dns-rat:lvr 1 μmol/L CNREA8 46,1654,86
orl-rat TDLo:2280 mg/kg/13W-C:NEO GANNA2 72,160,81
orl-rat TD:5240 mg/kg/90D-C:CAR CRNGDP 1,533,80
orl-rat TD:3600 mg/kg/17W-C:CAR JJIND8 79,1159,87

SAFETY PROFILE: Questionable carcinogen with experimental carcinogenic and neoplastigenic data. Mutation data reported. When heated to decomposition it emits toxic fumes of NO_x.

HMB650 CAS:7759-35-5 ***HR: D***
17-HYDROXY-16-METHYLENE-19-NORPREGN-4-ENE-3,20-DIONE ACETATE
mf: $C_{23}H_{30}O_4$ mw: 370.53

SYNS: 16-METHYLENE-17-α-ACETOXY-19-NOR-4-PREGNENE-3,20-DIONE ◇ ST-1435

TOXICITY DATA with REFERENCE
imp-wmn TDLo:2200 μg/kg (female 43W pre):REP CCPTAY 18,411,78

SAFETY PROFILE: Human female reproductive effects by implant: changes in menstrual cycle and fertility. When heated to decomposition it emits acrid smoke and fumes.

HMC000 CAS:503-49-1 ***HR: 2***
β-HYDROXY-β-METHYLGLUTARIC ACID
mf: $C_6H_{10}O_5$ mw: 162.16

SYNS: CB-337 ◇ DICROTALIC ACID ◇ HMG ◇ HMGA ◇ 3-HYDROXY-3-METHYLGLUTARIC ACID ◇ 3-HYDROXY-3-METHYLPENTANEDIOIC ACID ◇ LIPOGLUTAREN ◇ MEDROGLUTARIC ACID ◇ MEGLUTOL

TOXICITY DATA with REFERENCE
orl-mus LD50:7330 mg/kg DRFUD4 3,114,78
ipr-mus LD50:3230 mg/kg DRFUD4 3,114,78

SAFETY PROFILE: Moderately toxic by intraperitoneal route. Mildly toxic by ingestion. Used as an antihyperlipoproteinemic agent. When heated to decomposition it emits acrid smoke and fumes.

HMC500 CAS:63885-07-4 ***HR: 3***
3-HYDROXY-1-METHYLGUANINE
mf: $C_6H_7N_5O_2$ mw: 181.18

TOXICITY DATA with REFERENCE
scu-rat TDLo:96 mg/kg/8W-I:NEO CNREA8 33,1113,73

SAFETY PROFILE: Questionable carcinogen with experimental neoplastigenic data. When heated to decomposition it emits toxic fumes of NO_x.

HMD000 CAS:30345-27-8 ***HR: 3***
3-HYDROXY-7-METHYLGUANINE
mf: $C_6H_7N_5O_2$ mw: 181.18

SYN: 2-AMINO-3-HYDROXY-1,7-DIHYDRO-7-METHYL-6H-PURIN-6-ONE

TOXICITY DATA with REFERENCE
scu-rat TDLo:96 mg/kg/8W-I:ETA CNREA8 33,1113,73

SAFETY PROFILE: Questionable carcinogen with experimental tumorigenic data. When heated to decomposition it emits toxic fumes of NO_x.

HMD500 CAS:30345-28-9 ***HR: 3***
3-HYDROXY-9-METHYLGUANINE
mf: $C_6H_7N_5O_2$ mw: 181.18

SYN: 2-AMINO-3-HYDROXY-1,7-DIHYDRO-8-METHYL-6H-PURIN-6-ONE

TOXICITY DATA with REFERENCE
scu-rat TDLo:96 mg/kg/8W-I:ETA CNREA8 33,1113,73

SAFETY PROFILE: Questionable carcinogen with experimental tumorigenic data. When heated to decomposition it emits toxic fumes of NO_x.

HME000 CAS:15932-89-5 ***HR: 3***
HYDROXYMETHYL HYDROPEROXIDE
mf: CH_4O_3 mw: 64.04

PROP: Explodes when heated, but is relatively insensitive to friction. A powerful oxidizer. Upon decomposition it emits toxic fumes. See also PEROXIDES.

HMF000 CAS:568-75-2 ***HR: 3***
7-HYDROXYMETHYL-12-METHYLBENZ(a)ANTHRACENE
mf: $C_{20}H_{16}O$ mw: 272.36

SYNS: 7-HM-12-MBA ◇ 12-METHYBENZ(a)ANTHRACENE-7-METHANOL ◇ 7-OHM-MBA ◇ 7-OHM-12-MBA

TOXICITY DATA with REFERENCE
dnd-rat-ipr 100 μmol/kg CRNGDP 3,297,82
dnd-mus-skn 16 μmol/kg CNREA8 43,4221,83
ivn-rat TDLo:25 mg/kg (12D preg):TER IARCCD 4,149,73
ivn-rat TDLo:25 mg/kg (female 8D post):REP BJCAAI 24,548,70
orl-rat TDLo:50 mg/kg:ETA BJCAAI 22,122,68
scu-rat TDLo:150 mg/kg/39D-I:NEO CNREA8 31,1951,71
scu-mus TDLo:400 mg/kg/10W-I:CAR IJCNAW 2,500,67
ivn-rat LD50:73 mg/kg KIDZAK 23(Suppl 1),35,71

CONSENSUS REPORTS: EPA Genetic Toxicology Program.

SAFETY PROFILE: Poison by intravenous route. Experimental teratogenic and reproductive effects. Questionable carcinogen with experimental carcinogenic, neoplastigenic, and tumorigenic data. Mutation data reported. When heated to decomposition it emits acrid smoke and fumes.

HMF500 CAS:568-70-7 ***HR: 3***
12-HYDROXYMETHYL-7-METHYLBENZ(a)ANTHRACENE
mf: $C_{20}H_{16}O$ mw: 272.36

SYNS: 12-HM-7-MBA ◇ 7-METHYLBENZ(a)ANTHRACENE-12-METHANOL ◇ 7-METHYL-12-HYDROXYMETHYLBENZ(a)ANTHRACENE

TOXICITY DATA with REFERENCE
mma-sat 10 nmol/plate 46OJAN -,675,81
mma-ham:lng 400 nmol/L PNASA6 76,862,79
orl-rat TDLo:150 mg/kg:ETA BJCAAI 22,122,68
scu-mus TDLo:400 mg/kg/10W-I:CAR IJCNAW 2,500,67

CONSENSUS REPORTS: EPA Genetic Toxicology Program.

SAFETY PROFILE: Questionable carcinogen with experimental carcinogenic and tumorigenic data. Mutation data reported. When heated to decomposition it emits acrid smoke and fumes.

HMG000 CAS:590-96-5 ***HR: 3***
1-HYDROXYMETHYL-2-METHYLDITMIDE-2-OXIDE
mf: $C_2H_6N_2O_2$ mw: 90.10

SYNS: MAM ◇ METHYLAZOXYMETHANOL ◇ (METHYL-ONNAZOXY)METHANOL

TOXICITY DATA with REFERENCE
mma-sat 10 μmol/plate CNREA8 39,3780,79
hma-mus/sat 1200 mg/kg 22XWAN -,260,70
ipr-rat TDLo:20 mg/kg (female 14-16D post):REP SCIEAS 163,88,69
ivn-ham TDLo:23 mg/kg (female 8D post):TER PSEBAA 124,476,67
ipr-rat TDLo:20 mg/kg/2W-I:ETA JNCIAM 37,217,66

CONSENSUS REPORTS: IARC Cancer Review: Group 2B IMEMDT 7,56,87; Animal Sufficient Evidence IMEMDT 10,121,76. EPA Genetic Toxicology Program.

SAFETY PROFILE: Suspected carcinogen with experimental tumorigenic and teratogenic data. Other experimental reproductive effects. Mutation data reported.

When heated to decomposition it emits toxic fumes of NO_x.

HMG500 CAS:61413-61-4 ***HR: 3***
1-(HYDROXYMETHYL)-3-METHYLIMIDAZOLINIUM CHLORIDE, DODECANOATE
mf: $C_{17}H_{33}N_2O_2•Cl$ mw: 332.97

SYN: 1-((n-DODECANOYLOXY)METHYL)-3-METHYLIMIDAZOLIUM CHLORIDE

TOXICITY DATA with REFERENCE
orl-mus LD50:4110 mg/kg JMCMAR 23,469,80
ipr-mus LD50:155 mg/kg JMCMAR 23,469,80
ivn-mus LD50:75 mg/kg JMCMAR 23,469,80

SAFETY PROFILE: Poison by intravenous and intraperitoneal routes. Mildly toxic by ingestion. When heated to decomposition it emits very toxic fumes of NO_x and Cl^-.

HMG600 ***HR: 3***
3-(HYDROXYMETHYL)-1-METHYLIMIDAZOLINIUM CHLORIDE LAURATE (ESTER)
mf: $C_{17}H_{33}N_2O_2•Cl$ mw: 332.97

TOXICITY DATA with REFERENCE
orl-mus LD50:4110 mg/kg JMCMAR 23,469,80
ipr-mus LD50:155 mg/kg JMCMAR 23,469,80
ivn-mus LD50:75 mg/kg JMCMAR 23,469,80

SAFETY PROFILE: Poison by intravenous and intraperitoneal routes. Mildly toxic by ingestion. When heated to decomposition it emits toxic fumes of NO_x and Cl^-. See also ESTERS.

HMH000 ***HR: 3***
HYDROXYMETHYL METHYL PEROXIDE
mf: $C_2H_6O_3$ mw: 78.07

$HOCH_2OOCH_3$

SAFETY PROFILE: Violently explosive, impact-sensitive when heated. When heated to decomposition it emits acrid smoke and fumes. See also PEROXIDES.

HMH300 CAS:147-61-5 ***HR: 1***
5-HYDROXYMETHYL-4-METHYLURACIL
mf: $C_6H_8N_2O_3$ mw: 156.16

SYNS: 5-HYDROXYMETHYL-6-METHYL-2,4(1H,3H)-PYRIMIDINEDIONE ◇ 5-HYDROXYMETHYL-6-METHYLURACIL ◇ 4-METHYL-5-HYDROXYMETHYLURACIL ◇ 4-METHYL-5-OXYMETHYLURACIL ◇ PENTOKSIL ◇ PENTOXIL ◇ PENTOXYL ◇ 2,4(1H,3H)-PYRIMIDINEDIONE, 5-(HYDROXYMETHYL)-6-METHYL- (9CI) ◇ URACIL, 5-(HYDROXYMETHYL)-6-METHYL-

TOXICITY DATA with REFERENCE
par-rat TDLo:500 mg/kg (female 10D post):TER AAGEAA 65(11),19,73
orl-rat LD50:5 g/kg RPTOAN 51,100,88

SAFETY PROFILE: Slightly toxic by ingestion. An experimental teratogen. When heated to decomposition it emits toxic fumes of NO_x.

HMH500 CAS:5985-35-3 ***HR: 3***
dl-3-HYDROXY-N-METHYLMORPHINAN HYDROBROMIDE
mf: $C_{17}H_{23}NO•BrH$ mw: 338.33

PROP: Crystals. Mp: 193-195°. Sol in water; very sltly sol in alc; insol in ether.

SYNS: DROMORAN-HYDROBROMIDE ◇ (+ −)-17-METHYLMORPHINAN-3-OL HYDROBROMIDE

TOXICITY DATA with REFERENCE
orl-rat LD50:350 mg/kg JPETAB 109,189,53
scu-rat LD50:108 mg/kg AIPTAK 85,387,51
orl-mus LD50:375 mg/kg JPETAB 109,189,53
ipr-mus LD50:120 mg/kg JPETAB 99,163,50
scu-mus LD50:108 mg/kg ANESAV 12,225,51
ivn-mus LD50:33 mg/kg AIPTAK 85,387,51
ivn-rbt LD50:19 mg/kg AIPTAK 85,387,51

SAFETY PROFILE: Poison by ingestion, subcutaneous, intraperitoneal and intravenous routes. A narcotic analgesic. When heated to decomposition it emits very toxic fumes of NO_x and HBr. See also BROMIDES.

HMI000 CAS:483-55-6 ***HR: 3***
2-HYDROXY-3-METHYL-1,4-NAPHTHOQUINONE
mf: $C_{11}H_8O_3$ mw: 188.19

SYNS: 3-HYDROXY-2-METHYL-1,4-NAPHTHOQUINONE ◇ 2-METHYL-3-HYDROXY-1,4-NAPHTHOQUINONE ◇ PHTHIOCOL

TOXICITY DATA with REFERENCE
dni-hmn:hla 4 mmol/L MUREAV 93,447,82
orl-mus LDLo:200 mg/kg PSEBAA 43,125,40
ipr-mus LDLo:150 mg/kg PSEBAA 43,125,40
ipr-ckn LDLo:100 mg/kg PSEBAA 43,125,40

SAFETY PROFILE: Poison by ingestion and intraperitoneal routes. Human mutation data reported. When heated to decomposition it emits acrid smoke and fumes.

HMI500 CAS:18857-59-5 ***HR: 3***
3-HYDROXYMETHYL-1-((3-(5-NITRO-2-FURYL)ALLYLIDENE)AMINO)HYDANTOIN
mf: $C_{11}H_{10}N_4O_6$ mw: 294.25

TOXICITY DATA with REFERENCE
orl-rat TDLo:51 g/kg/49W-C:CAR CNREA8 33,2894,73

SAFETY PROFILE: Questionable carcinogen with experimental carcinogenic data. When heated to decomposition it emits toxic fumes of NO_x.

HMJ000 CAS:4812-40-2 ***HR: D***
2-HYDROXYMETHYL-5-NITROIMIDAZOLE-1-ETHANOL
mf: $C_6H_9N_3O_4$ mw: 187.18

SYN: 1-(2-HYDROXYETHYL)-2-HYDROXYMETHYL-5-NITRO-IMIDAZOLE

TOXICITY DATA with REFERENCE
mmo-sat 500 μg/L AMACCQ 10,476,76
mma-sat 100 μg/plate CMMUAO 4,171,76
mmo-klp 10 μmol/L/20H MUREAV 66,207,79

SAFETY PROFILE: Mutation data reported. When heated to decomposition it emits toxic fumes of NO_x.

HMJ500 CAS:126-11-4 ***HR: 3***
2-HYDROXYMETHYL-2-NITROPROPANE-1,3-DIOL
mf: $C_4H_9NO_5$ mw: 151.14

PROP: Crystalline. Mp: 165-170°, bp: decomp.

SYNS: 2-(HYDROXYMETHYL)-2-NITRO-1,3-PROPANEDIOL ◇ 2-NITRO-2-(HYDROXYMETHYL)-1,3-PROPANEDIOL ◇ TRIHYDROXYMETHYLNITROMETHANE ◇ TRIMETHYLOLNITROMETHANE ◇ TRIS(HYDROXYMETHYL)NITROMETHANE

TOXICITY DATA with REFERENCE
orl-rat LD50:1900 mg/kg IMSUAI 39,56,70
orl-mus LD50:1900 mg/kg PCOC** -,1198,66
ipr-mus LD50:4000 mg/kg KHFZAN 11(1),73,77
orl-rbt LDLo:250 mg/kg JIHTAB 22,315,40

CONSENSUS REPORTS: Reported in EPA TSCA Inventory.

SAFETY PROFILE: Poison by ingestion. Moderately toxic by intraperitoneal route. Probably an irritant. When heated to decomposition it emits toxic fumes of NO_x. See also NITRO COMPOUNDS.

HMK000 CAS:24356-60-3 ***HR: 2***
3-(HYDROXYMETHYL)-8-OXO-7-(2-(4-PYRIDYLTHIO)ACETAMIDO)-5-THIA-1-AZABICYCLO(4,2,0)OCT-2-ENE-2-CARBOXYLIC ACID ACETATE(ESTER), MONOSODIUM SALT
mf: $C_{17}H_{17}N_3O_6S_2 \cdot Na$ mw: 446.48

SYNS: BLP 1322 ◇ CEFADYL ◇ CEFALOJECT ◇ CEFAPIRIN SODIUM ◇ CEFAPRIN SODIUM ◇ CEFATREXYL ◇ CEPHALOTHIN SODIUM ◇ CEPHATREXYL ◇ SODIUM CEFAPIRIN ◇ SODIUM CEPHAPIRIN

TOXICITY DATA with REFERENCE
ims-dog TDLo:24 g/kg (30D pre):REP OYYAA2 9,11,75
ivn-man TDLo:7429 μg/kg/8W-I:KID DICPBB 21,380,87
ims-man TDLo:350 mg/kg/6D:BLD,MET AMACCQ 1,174,72
orl-rat LD50:16356 mg/kg NIIRDN 6,405,82
ipr-rat LD50:7850 mg/kg NIIRDN 6,405,82
ivn-rat LD50:4580 mg/kg ARZNAD 29,424,79
orl-mus LD50:26088 mg/kg NIIRDN 6,405,82
ipr-mus LD50:8899 mg/kg NIIRDN 6,405,82
scu-mus LD50:13556 mg/kg NIIRDN 6,405,82
ivn-mus LD50:4600 mg/kg ARZNAD 29,424,79
ivn-dog LD50:2500 mg/kg ARZNAD 29,424,79
ivn-rbt LD50:3000 mg/kg ARZNAD 29,424,79

SAFETY PROFILE: Moderately toxic by intravenous route. Human systemic effects by intramuscular route: fever, white blood cell effects (angranulocytosis), interstitial nephritis. Mildly toxic by ingestion. Experimental reproductive effects. When heated to decomposition it emits very toxic fumes of NO_x, Na_2O, and SO_x. See also ESTERS.

HMK100 CAS:90-01-7 ***HR: D***
2-HYDROXYMETHYLPHENOL
mf: $C_7H_8O_2$ mw: 124.15

SYNS: BENZENEMETHANOL, 2-HYDROXY- (9CI) ◇ BENZYL ALCOHOL, o-HYDROXY- ◇ DIATHESIN ◇ α-2-DIHYDROXYTOLUENE ◇ 2-HYDROXYBENZENEMETHANOL ◇ o-HYDROXYBENZYL ALCOHOL ◇ 2-HYDROXYBENZYL ALCOHOL ◇ o-(HYDROXYMETHYL)PHENOL ◇ o-METHYLOLPHENOL ◇ 2-METHYLOLPHENOL ◇ SAL ◇ SALICYL ALCOHOL ◇ SALIGENIN ◇ SALIGENOL

TOXICITY DATA with REFERENCE
scu-rat TDLo:400 mg/kg (female 11D post):REP RCOCB8 38,209,82
scu-rat TDLo:400 mg/kg (female 11D post):TER RCOCB8 38,209,82

CONSENSUS REPORTS: Reported in EPA TSCA Inventory.

SAFETY PROFILE: An experimental teratogen. Experimental reproductive effects. When heated to decomposition it emits acrid smoke and irritating fumes.

HMK200 CAS:13911-65-4 ***HR: 3***
HYDROXYMETHYLPHENYLARSINE OXIDE
mf: $C_7H_9AsO_2$ mw: 200.08

SYNS: ARSENIC ACID, METHYLPHENYL-(9CI) ◇ ARSINE OXIDE, HYDROXYMETHYLPHENYL- ◇ METHYLPHENYLARSINIC ACID ◇ METHYLPHENYLARSONIC ACID

TOXICITY DATA with REFERENCE
ivn-mus LD50:56 mg/kg CSLNX* NX#05108

OSHA PEL: TWA 0.5 mg(As)/m^3

SAFETY PROFILE: Poison by intravenous route. When heated to decomposition it emits toxic fumes of As.

HML500 CAS:2440-22-4 ***HR: 1***
2-(2-HYDROXY-5-METHYLPHENYL)BENZOTRIAZOLE
mf: $C_{13}H_{11}N_3O$ mw: 225.27

SYNS: BENAZOL P ◇ 2-(2H-BENZOTRIAZOL-2-YL)-4-METHYLPHENOL ◇ DROMETRIZOLE ◇ TIN P ◇ TINUVIN P ◇ UV ABSORBER-1

TOXICITY DATA with REFERENCE
eye-rbt 500 mg/24H MLD 28ZPAK -,146,72
orl-mus LD50:6500 mg/kg GTPZAB 10(3),49,66

CONSENSUS REPORTS: Reported in EPA TSCA Inventory.

SAFETY PROFILE: Mildly toxic by ingestion. An eye irritant. When heated to decomposition it emits toxic fumes of NO_x.

HMM500 CAS:32780-64-6 ***HR: 3***
5-(1-HYDROXY-2-((1-METHYL-3-PHENYLPROPYL)AMINO)ETHYL)SALICYLAMIDE HYDROCHLORIDE
mf: $C_{19}H_{24}N_2O_3\cdot ClH$ mw: 364.91

SYNS: AH 5158A ◇ LABETALOL HYDROCHLORIDE ◇ PRESDATE ◇ SCH 15719w ◇ TRANDATE

TOXICITY DATA with REFERENCE
orl-rat TDLo:1 g/kg (female 17-22D post):REP YACHDS 9,879,81
orl-rat TDLo:3850 mg/kg (female 7-17D post):TER YACHDS 9,851,81
orl-rat LD50:2114 mg/kg KSRNAM 21,6307,87
ipr-rat LD50:107 mg/kg HOIZAK 53,15,78
ivn-rat LD50:53 mg/kg HOIZAK 53,15,78
orl-mus LD50:1450 mg/kg HOIZAK 53,15,78
ipr-mus LD50:114 mg/kg HOIZAK 53,15,78
ivn-mus LD50:47 mg/kg HOIZAK 53,15,78
orl-rbt LD50:1250 mg/kg HOIZAK 53,15,78
ivn-rbt LD50:41 mg/kg HOIZAK 53,15,78

SAFETY PROFILE: Poison by intraperitoneal and intravenous routes. Moderately toxic by ingestion. Experimental teratogenic and reproductive effects. Used as an antihypertensive agent. When heated to decomposition it emits very toxic fumes of HCl and NO_x.

HMN000 CAS:32949-37-4 ***HR: 3***
2-HYDROXYMETHYL-6-PHENYL-3-PYRIDAZONE
mf: $C_{11}H_{10}N_2O_2$ mw: 202.23

SYNS: 2-(HYDROXYMETHYL)-6-PHENYL-3(2H)-PYRIDAZINONE ◇ PP-12

TOXICITY DATA with REFERENCE
orl-mus LD50:1260 mg/kg OYYAA2 4,195,70
ipr-mus LD50:405 mg/kg OYYAA2 4,195,70

SAFETY PROFILE: Moderately toxic by ingestion and intraperitoneal routes. A powerful explosive which becomes impact-sensitive when heated. When heated to decomposition it emits toxic fumes of NO_x.

HMP000 CAS:5756-69-4 ***HR: 3***
3-HYDROXY-3-METHYL-1-PHENYLTRIAZENE
mf: $C_7H_9N_3O$ mw: 151.19

SYNS: 1-PHENYL-3-METHYL-3-HYDROXY-TRIAZEN(GERMAN) ◇ 1-PHENYL-3-METHYL-3-HYDROXY-TRIAZENE

TOXICITY DATA with REFERENCE
scu-rat TDLo:420 mg/kg:CAR ZKKOBW 81,285,74
scu-rat TD:3300 mg/kg/I:CAR ZKKOBW 81,285,74
scu-rat LD50:550 mg/kg ZKKOBW 81,285,74

SAFETY PROFILE: Moderately toxic by subcutaneous route. Questionable carcinogen with experimental carcinogenic data. When heated to decomposition it emits toxic fumes of NO_x.

HMP100 CAS:118-29-6 ***HR: D***
N-(HYDROXYMETHYL)PHTHALIMIDE
mf: $C_9H_7NO_3$ mw: 177.17

SYNS: HYDROXYMETHYLPHTHALIMIDE ◇ 1H-ISOINDOLE-1,3(2H)-DIONE, 2-(HYDROXYMETHYL)- ◇ N-METHYLOLPHTHALIMIDE ◇ OXYMETHYLPHTHALIMIDE ◇ PHTHALIMIDE, N-(HYDROXYMETHYL)- ◇ PHTHALIMIDOMETHYL ALCOHOL

TOXICITY DATA with REFERENCE
unr-rat TDLo:300 mg/kg (female 13D post):TER GISAAA 48(10),64,83

CONSENSUS REPORTS: Reported in EPA TSCA Inventory.

SAFETY PROFILE: An experimental teratogen. When heated to decomposition it emits toxic fumes of NO_x.

HMQ000 CAS:1172-82-3 ***HR: D***
17-HYDROXY-6-METHYLPREGN-4-ENE-3,20-DIONE ACETATE
mf: $C_{23}H_{34}O_4$ mw: 385.57

TOXICITY DATA with REFERENCE
scu-rat TDLo:17500 µg/kg (35D pre):REP RPHRA6 20,395,64

SAFETY PROFILE: Experimental reproductive effects. A steroid. When heated to decomposition it emits acrid smoke and fumes.

HMQ500 CAS:65229-18-7 ***HR: 3***
d-N,N'-(1-HYDROXYMETHYLPROPYL)ETHYLENEDINITROSAMINE
mf: $C_{10}H_{22}N_4O_4$ mw: 262.36

SYNS: DDETA ◇ 2,2'-(ETHYLENEBIS(NITROSOIMINO))BISBUTANOL ◇ d-N,N'-(1-IDROSSIMETIL PROPIL)-ETILENDINITROSAMINA (ITALIAN)

TOXICITY DATA with REFERENCE
scu-mus TDLo:4320 mg/kg/36W-C:CAR LAPPA5 35,45,75

SAFETY PROFILE: Questionable carcinogen with experimental carcinogenic data. When heated to decomposition it emits toxic fumes of NO_x. See also NITROSAMINES.

HMR500 CAS:1677-46-9 ***HR: 3***
4-HYDROXY-1-METHYL-2-QUINOLONE
mf: $C_{10}H_9NO_2$ mw: 175.20

SYNS: 4-HYDROXY-1-METHYLCARBOSTYRIL ◇ 1-METHYL-4-HYDROXY-2-CHINOLON (CZECH) ◇ N-METHYL-4-HYDROXYKARBOSTYRIL (CZECH)

TOXICITY DATA with REFERENCE
eye-rbt 100 mg/24H MOD 28ZPAK -,149,72
orl-rat LDLo:7410 mg/kg 28ZPAK -,149,72
ivn-mus LD50:180 mg/kg CSLNX* NX#01354

CONSENSUS REPORTS: Reported in EPA TSCA Inventory.

SAFETY PROFILE: Poison by intravenous route. Mildly toxic by ingestion. An eye irritant. When heated to decomposition it emits toxic fumes of NO_x.

HMS500 CAS:23395-20-2 ***HR: 2***
1-(HYDROXYMETHYL)-2,8,9-TRIOXA-5-AZA-1-SILABICYCLO(3.3.3) UNDECANEMETHACRYLATE
mf: $C_{11}H_{19}NO_5Si$ mw: 273.40

SYN: METHAKRYLOXYMETHYLSILATRAN (CZECH)

TOXICITY DATA with REFERENCE
skn-rbt 500 mg/24H MOD 28ZPAK -,220,72
eye-rbt 100 mg/24H SEV 28ZPAK -,220,72
orl-rat LD50:7950 mg/kg 28ZPAK -,220,72

SAFETY PROFILE: Mildly toxic by ingestion. A skin and severe eye irritant. When heated to decomposition it emits toxic fumes of NO_x.

HMS875 ***HR: 3***
3-HYDROXY-α-METHYL-l-TYROSINE-1-(2,2-DIMETHYL-1-OXOPROPOXY)ETHYL ESTER PHOSPHATE HYDRATE
mf: $C_{17}H_{25}NO_6 \cdot H_3O_4P$ mw: 455.45

SYN: MK-872

TOXICITY DATA with REFERENCE
orl-rat LD50:1879 mg/kg DRFUD4 10,563,85
ipr-rat LD50:150 mg/kg DRFUD4 10,563,85
orl-mus LD50:2780 mg/kg DRFUD4 10,563,85
ivn-mus LD50:187 mg/kg DRFUD4 10,563,85

SAFETY PROFILE: Poison by intravenous and intraperitoneal routes. Moderately toxic by ingestion. When heated to decomposition it emits toxic fumes of PO_x and NO_x.

HMU000 CAS:32766-75-9 ***HR: 3***
N-HYDROXY-N-MYRISTOYL-2-AMINOFLUORENE
mf: $C_{27}H_{37}NO_2$ mw: 407.65

SYNS: N-FLUOREN-2-YL-N-TETRADECANOYLHYDROXAMIC ACID ◇ N-HYDROXY-N-TETRADECANOYL-2-AMINOFLUORENE

TOXICITY DATA with REFERENCE
mma-sat 20 nmol/plate CNREA8 37,1461,77
scu-rat TDLo:104 mg/kg/5W-I:NEO CNREA8 37,1461,77

CONSENSUS REPORTS: EPA Genetic Toxicology Program.

SAFETY PROFILE: Questionable carcinogen with experimental neoplastigenic data. Mutation data reported. When heated to decomposition it emits toxic fumes of NO_x.

HMU500 CAS:93-01-6 ***HR: 3***
6-HYDROXY-2-NAPHTHALENESULFONIC ACID
mf: $C_{10}H_8O_4S$ mw: 224.24

SYNS: 2-HYDROXY-6-NAPHTHALENESULFONIC ACID ◇ β-NAPHTHOLSULFONIC ACID S ◇ β-NAPHTHOL-6-SULFONIC ACID ◇ 2-NAPHTHOL-6-SULFONIC ACID ◇ 2-NAPHTOL-6-SULFOSAURE (GERMAN) ◇ SCHAEFFER'S-β-NAPHTHOLSULFONIC ACID

TOXICITY DATA with REFERENCE
orl-mus LDLo:250 mg/kg ZHINAV 64,113,1909
scu-mus LDLo:650 mg/kg ZHINAV 64,113,1909

CONSENSUS REPORTS: Reported in EPA TSCA Inventory.

SAFETY PROFILE: Poison by ingestion. Moderately toxic by subcutaneous route. When heated to decomposition it emits toxic fumes of SO_x. See also SULFONATES.

HMV000 CAS:1491-41-4 ***HR: 3***
N-HYDROXYNAPHTHALIMIDE, DIETHYL PHOSPHATE
mf: $C_{16}H_{16}NO_6P$ mw: 349.30

PROP: Tan, crystalline powder. Mp: 177.0°. Sol in

methylene chloride; difficultly sol in most organic solvents.

SYNS: B-9002 ◇ BAYER 25820 ◇ CHEMAGRO B-9002 ◇ 2-((DIETHOXYPHOSPHINYL)OXY)-1H-BENZ(de)ISOQUINOLINE-1,3(2H)-DIONE ◇ O,O-DIETHYL N-HYDROXYNAPHTHALIMIDE PHOSPHATE ◇ ENT 25,567 ◇ N-HYDROXYNAPHTHYLIMIDE DIETHYL PHOSPHATE ◇ MARETIN ◇ NAFTALOFOS ◇ NAPHTHALOPHOS ◇ PHOSPHORIC ACID, DIETHYL ESTER-N-NAPHTHALIMIDE deriv. ◇ PHOSPHORIC ACID, DIETHYL ESTER, NAPHTHALIMIDO deriv. ◇ PHTALOPHOS ◇ RAMETIN ◇ RAWETIN ◇ S 940

TOXICITY DATA with REFERENCE
orl-rat LD50:70 mg/kg WRPCA2 9,119,70
skn-rat LD50:140 mg/kg SPEADM 78-1,45,78
orl-mus LD50:50 mg/kg ARSIM* 20,3,66
orl-ckn LD50:43 mg/kg TXAPA9 11,49,67
orl-dom LDLo:200 mg/kg FAZMAE 17,108,73

SAFETY PROFILE: A poison by ingestion and skin contact. A pesticide used in veterinary medicine as a ruminant antihelminitic. A cholinesterase inhibitor. When heated to decomposition it emits toxic fumes of NO_x and PO_x. See also PARATHION.

HMW500 CAS:3665-51-8 ***HR: 2***
3-HYDROXY-2-NAPHTHAMIDE
mf: $C_{11}H_9NO_2$ mw: 187.21

TOXICITY DATA with REFERENCE
ipr-rat LD50:2000 mg/kg JPETAB 108,450,53

CONSENSUS REPORTS: Reported in EPA TSCA Inventory.

SAFETY PROFILE: Moderately toxic by intraperitoneal route. When heated to decomposition it emits toxic fumes of NO_x.

HMX000 CAS:567-47-5 ***HR: 2***
2-HYDROXYNAPHTHENESULFONIC ACID
mf: $C_{10}H_8O_4S$ mw: 224.24

SYN: KYSELINA-2-NAFTOL-1-SULFONOVA (CZECH)

TOXICITY DATA with REFERENCE
skn-rbt 500 mg/24H MLD 28ZPAK -,186,72
eye-rbt 100 mg/24H MOD 28ZPAK -,186,72
orl-rat LD50:3170 mg/kg 28ZPAK -,186,72

SAFETY PROFILE: Moderately toxic by ingestion. A skin and eye irritant. When heated to decomposition it emits toxic fumes of SO_x.

HMX500 CAS:2283-08-1 ***HR: 3***
2-HYDROXY-1-NAPHTHOIC ACID
mf: $C_{11}H_8O_3$ mw: 188.19

PROP: Yellow, needle-like crystals. Mp: 216°. Sltly sol in hot water; sol in alc, benzene, ether, and chloroform.

SYNS: 2-NAPHTHOL-3-CARBOXYLIC ACID ◇ β-OXYNAPHTHOIC ACID

TOXICITY DATA with REFERENCE
ipr-mus LD50:125 mg/kg 14CYAT 2,1840,63

CONSENSUS REPORTS: Reported in EPA TSCA Inventory.

SAFETY PROFILE: Poison by intraperitoneal route. When heated to decomposition it emits acrid smoke and fumes.

HMY000 CAS:121-19-7 ***HR: 3***
4-HYDROXY-3-NITROBENZENEARSONIC ACID
mf: $C_6H_6AsNO_6$ mw: 263.05

SYNS: AKLOMIX-3 ◇ 4-HYDROXY-3-NITROPHENYLARSONIC ACID ◇ NCI-C56508 ◇ 3N4HPA ◇ NITRO ACID 100 percent ◇ 2-NITRO-1-HYDROXYBENZENE-4-ARSONIC ACID ◇ 3-NITRO-4-HYDROXYBENZENEARSONIC ACID ◇ 3-NITRO-4-HYDROXYPHENYLARSONIC ACID ◇ NITROPHENOLARSONIC ACID ◇ NSC-2101 ◇ REN O-SAL ◇ RISTAT ◇ ROXARSONE (USDA)

TOXICITY DATA with REFERENCE
orl-rat LD50:155 mg/kg TXAPA9 5,507,63
ipr-rat LD50:66 mg/kg TXAPA9 5,507,63
orl-dog LDLo:50 mg/kg TXAPA9 5,507,63
orl-ckn LD50:110 mg/kg TXAPA9 5,507,63
ipr-ckn LD50:34 mg/kg TXAPA9 5,507,63
orl-trk LD50:61 mg/kg TXAPA9 5,507,63

CONSENSUS REPORTS: Arsenic and its compounds are on the Community Right-To-Know List. Reported in EPA TSCA Inventory.

OSHA PEL: TWA 0.5 mg(As)/m^3
ACGIH TLV: TWA 0.2 mg(As)/m^3

SAFETY PROFILE: Poison by ingestion and intraperitoneal routes. When heated to decomposition it emits very toxic fumes of NO_x and As. See also ARSENIC COMPOUNDS.

HMY050 ***HR: 3***
4-HYDROXY-3-NITROBENZENESULFONYL CHLORIDE
mf: $C_6H_4ClNO_5S$ mw: 237.61

$$HO(O_2N)C_6H_3SO_2Cl$$

SAFETY PROFILE: Decomposes very exothermically at 24°C. Upon decomposition it emits toxic fumes of Cl^-, SO_x, and NO_x.

HMY100 ***HR: D***
3-HYDROXY-2-(5-NITRO-2-FURYL)-2H-1,3-BENZOXAZIN-4(3H)-ONE
mf: $C_{12}H_8N_2O_6$ mw: 276.22

SYN: 4H-2,3-DIHYDRO-2-(5'-NITRO-2'-FURYL)-3-HYDROXY-1,3-BENZOXALINE-4-ONE

TOXICITY DATA with REFERENCE
mmo-sat 20 μg/plate BBIADT 44,485,85
mmo-esc 20 μg/plate BBIADT 44,485,85
dnr-bcs 5 μg/plate BBIADT 44,485,85

SAFETY PROFILE: Mutation data reported. When heated to decomposition it emits toxic fumes of NO_x.

HMY500 CAS:96-67-3 ***HR: 1***
2-HYDROXY-5-NITROMETANILIC ACID
mf: $C_6H_6N_2O_6S$ mw: 234.20

SYNS: 3-AMINO-2-HYDROXY-5-NITRO-BENZENESULFONICACID ◇ KYSELINA-4-NITRO-2-AMINOFENOL-6-SULFONOVA (CZECH)

TOXICITY DATA with REFERENCE
eye-rbt 100 mg/24H MOD 28ZPAK -,181,72
orl-rat LD50:6920 mg/kg 28ZPAK -,181,72

CONSENSUS REPORTS: Reported in EPA TSCA Inventory.

SAFETY PROFILE: Mildly toxic by ingestion. An eye irritant. When heated to decomposition it emits very toxic fumes of NO_x and SO_x. See also SULFONATES.

HMZ100 CAS:100836-60-0 ***HR: D***
3-HYDROXYNITROSOCARBOFURAN
mf: $C_{12}H_{14}N_2O_5$ mw: 266.28

SYN: METHYLNITROSO-CARBAMICACID-2,3-DIHYDRO-2,2-DIMETHYL-3-HYDROXY-7-BENZOFURANYLESTER

TOXICITY DATA with REFERENCE
mmo-sat 1 μg/plate JTEHD6 7,519,81
cyt-ham:ovr 50 nmol/L JTEHD6 7,519,81
sce-ham:ovr 500 nmol/L JTEHD6 7,519,81

SAFETY PROFILE: Mutation data reported. Many N-nitroso compounds are carcinogens. When heated to decomposition it emits toxic fumes of NO_x. See also N-NITROSO COMPOUNDS, CARBAMATES, and ESTERS.

HNA000 CAS:100700-12-8 ***HR: D***
3-HYDROXY-4-(NITROSOCYANAMIDO) BUTYRAMIDE
mf: $C_5H_8N_4O_3$ mw: 172.17

SYNS: Γ-GUANIDINO-β-HYDROXYBUTYRIC ACID AMIDE, NITROSATED ◇ 4-(NITROSOCYANAMIDO)-3-HYDROXYBUTYRAMIDE

TOXICITY DATA with REFERENCE
mmo-sat 400 nmol/L GANNA2 65,45,74

SAFETY PROFILE: Mutation data reported. Many N-nitroso compounds are carcinogens. When heated to decomposition it emits toxic fumes of NO_x. See also N-NITROSO COMPOUNDS.

HNA500 CAS:101913-96-6 ***HR: D***
2-HYDROXY-6-(NITROSOCYANAMIDO) HEXANAMIDE
mf: $C_7H_{12}N_4O_3$ mw: 200.23

SYN: HOMOARGININE, nitrosated

TOXICITY DATA with REFERENCE
mmo-sat 10 μmol/L GANNA2 65,45,74

SAFETY PROFILE: Mutation data reported. Many N-nitroso compounds are carcinogens. When heated to decomposition it emits toxic fumes of NO_x. See also N-NITROSO COMPOUNDS.

HNB000 CAS:525-05-3 ***HR: 3***
2-HYDROXY-3-NITROSO-2,7-NAPHTHALENE-DISULFONIC ACID DISODIUM SALT
mf: $C_{10}H_5NO_8S_2•2Na$ mw: 377.26

TOXICITY DATA with REFERENCE
ivn-mus LD50:320 mg/kg CSLNX* NX#05619

CONSENSUS REPORTS: Reported in EPA TSCA Inventory. EPA Genetic Toxicology Program.

SAFETY PROFILE: Poison by intravenous route. When heated to decomposition it emits very toxic fumes of NO_x, Na_2O, and SO_x. See also SULFONATES.

HNB500 CAS:30310-80-6 ***HR: 3***
trans-4-HYDROXY-1-NITROSO-l-PROLINE
mf: $C_5H_8N_2O_4$ mw: 160.15

SYN: N-NITROSOHYDROXYPROLINE

CONSENSUS REPORTS: IARC Cancer Review: Group 3 IMEMDT 7,56,87; Animal Inadequate Evidence IMEMDT 17,303,78.

SAFETY PROFILE: Questionable carcinogen. When heated to decomposition it emits toxic fumes of NO_x.

HNB600 CAS:29343-52-0 ***HR: 3***
4-HYDROXY-2-NONENAL
mf: $C_9H_{16}O_2$ mw: 156.25

SYN: 4-HYDROXYNONENAL

TOXICITY DATA with REFERENCE
dni-gpg:oth 100 μmol/L CBINA8 52,233,84
oms-gpg:oth 100 μmol/L CBINA8 52,233,84
ipr-rat LD50:35 mg/kg TOXID9 4,100,84
ipr-mus LD50:69 mg/kg ZolH## 23OCT75

SAFETY PROFILE: Poison by intraperitoneal route.

Mutation data reported. When heated to decomposition it emits acrid smoke and fumes.

HNB875 CAS:54-49-9 ***HR: 3***
HYDROXYNOREPHEDRINE
mf: $C_9H_{13}NO_2$ mw: 167.23

SYNS: 1-α-(1-AMINOETHYL)-m-HYDROXYBENZYLALCOHOL ◇ ARAMINE ◇ BENZENEMETHANOL, α-(1-AMINOETHYL)-3-HYDROXY-, (R-(R*,S*))-(9CI) ◇ m-HYDROXY NOREPHEDRINE ◇ m-HYDROXYPROPADRINE ◇ ICORAL B ◇ METARADRINE ◇ METARAMINOL ◇ (-)-METARAMINOL ◇ 1-METARAMINOL ◇ PRESSONEX

TOXICITY DATA with REFERENCE
orl-rat LD50:240 mg/kg 27ZIAQ -,155,73
ipr-rat LD50:41 mg/kg 27ZIAQ -,155,73
scu-rat LD50:117 mg/kg 27ZIAQ -,155,73
orl-mus LD50:99 mg/kg 27ZIAQ -,155,73
scu-mus LD50:92 mg/kg 27ZIAQ -,155,73
ivn-mus LD50:51 mg/kg 27ZIAQ -,155,73

SAFETY PROFILE: A poison by ingestion, intraperitoneal, subcutaneous, and intravenous routes. When heated to decomposition it emits toxic fumes of NO_x.

HNC000 CAS:33402-03-8 ***HR: 3***
1-m-HYDROXYNOREPHEDRINE
mf: $C_9H_{13}NO_2 \cdot C_4H_6O_6$ mw: 317.33

SYNS: 1-α-(1-AMINOETHYL)-m-HYDROXYBENZYL ALCOHOL BITARTRATE ◇ (−)-α-(1-AMINOETHYL)-m-HYDROXYBENZYL ALCOHOL BITARTRATE ◇ 1-α-(1-AMINOETHYL)-m-HYDROXYBENZYL ALCOHOL HYDROGEN-d-TARTRATE ◇ ARAMINE ◇ l-1-(m-HYDROXYPHENYL)-2-AMINO-1-PROPANOL-d-HYDROGENTARTRATE ◇ METARAMINOL BITARTRATE ◇ METRAMINOL BITARTRATE ◇ d-(−)-METRAMINOL BITARTRATE ◇ 1-METRAMINOL BITARTRATE ◇ (−)-METRAMINOL (+)-BITARTRATE ◇ METARAMINOL TARTRATE (1:1) ◇ PRESSONEX ◇ PRESSONEX BITARTRATE ◇ PRESSOROL

TOXICITY DATA with REFERENCE
orl-rat LD50:240 mg/kg NIIRDN 6,820,82
ipr-rat LD50:41 mg/kg NIIRDN 6,820,82
scu-rat LD50:117 mg/kg NIIRDN 6,820,82
orl-mus LD50:99 mg/kg NIIRDN 6,820,82
ipr-mus LD50:680 mg/kg TXAPA9 15,304,69
ivn-mus LD50:39 mg/kg NIIRDN 6,820,82

SAFETY PROFILE: Poison by ingestion, subcutaneous, intravenous, and intraperitoneal routes. When heated to decomposition it emits toxic fumes of NO_x.

HND000 CAS:78128-84-4 ***HR: 3***
4-HYDROXY-5-OCTYL-2(5H)FURANONE
mf: $C_{12}H_{20}O_3$ mw: 212.32

SYN: Γ-OCTYL-β-HYDROXY-Δ^{α},β-BUTENOLID (GERMAN)

TOXICITY DATA with REFERENCE
scu-mus LD50:550 mg/kg ARZNAD 11,277,61
ivn-mus LD50:160 mg/kg ARZNAD 11,277,61

SAFETY PROFILE: Poison by intravenous route. Moderately toxic by subcutaneous route. When heated to decomposition it emits acrid smoke and fumes.

HND375 ***HR: 3***
HYDROXYOXOPHENYL IODANIUM PERCHLORATE
mf: $C_6H_6ClIO_6$ mw: 336.47

$$C_6H_5I^+(O)OHClO_4^-$$

SAFETY PROFILE: An unstable explosive. Upon decomposition it emits toxic fumes of Cl^- and I^-. See also PERCHLORATES and IODIDES.

HND800 ***HR: D***
3-β-HYDROXY-20-OXO-17-α-PREGN-5-ENE-16-β-CARBOXAMIDE ACETATE
mf: $C_{24}H_{34}NO_4$ mw: 400.59

SYN: 16-β-CARBOXAMIDE-3-β-ACETOXY-Δ^5-(17-α)-ISOPREGNENE-20-ONE

TOXICITY DATA with REFERENCE
orl-rbt TDLo:1 mg/kg (female 1D pre):REP ACEDAB 73,17,63

SAFETY PROFILE: Experimental reproductive effects. When heated to decomposition it emits toxic fumes of NO_x.

HNE400 CAS:101565-05-3 ***HR: D***
1-HYDROXY-3-n-PENTYL-Δ^8-TETRAHYDROCANNABINOL
mf: $C_{21}H_{30}O_2$ mw: 314.51

SYNS: ABN-Δ^8-THC ◇ 1-PENTYL-6,6,9-TRIMETHYL-6a,7,10,10a-TETRAHYDRO-6H-DIBENZO(b,d)PYRAN-3-OL

TOXICITY DATA with REFERENCE
dni-mus:lng 1480 nmol/L CNREA8 36,95,76
dni-mus:leu 5 μmol/L CNREA8 36,95,76
dni-mus:bmr 3560 nmol/L CNREA8 36,95,76

SAFETY PROFILE: Mutation data reported. When heated to decomposition it emits acrid smoke and fumes. See also TETRAHYDROCANNABINOL.

HNF000 CAS:7568-93-6 ***HR: 3***
β-HYDROXYPHENETHYLAMINE
mf: $C_8H_{11}NO$ mw: 137.20

PROP: Leaves of benzene. Mp: 161°, bp: 175-181°; sol in water, alc, benzene; sltly sol in hot xylene.

SYN: 2-AMINO-1-PHENYL-1-ETHANOL

TOXICITY DATA with REFERENCE
ipr-mus LD50:250 mg/kg NTIS** AD691-490
scu-mus LDLo:686 mg/kg AIPTAK 47,96,34
ivn-rbt LDLo:23 mg/kg AIPTAK 47,96,34

CONSENSUS REPORTS: Reported in EPA TSCA Inventory.

SAFETY PROFILE: Poison by intraperitoneal and intravenous routes. Moderately toxic by subcutaneous route. When heated to decomposition it emits toxic fumes of NO_x. See also AMINES.

HNG000 CAS:66967-84-8 ***HR: 3***
(m-HYDROXYPHENETHYL)TRIMETHYLAMMONIUM PICRATE
mf: $C_{11}H_{17}NO{\cdot}C_6H_3N_3O_7$ mw: 408.41

SYNS: 3-HYDROXY-N,N,N-TRIMETHYLBENZENEETHANAMINIUM PICRATE ◇ LEPTODACTYLINE PICRATE

TOXICITY DATA with REFERENCE
orl-mus LD50:325 mg/kg MEIEDD 10,781,83
ivn-mus LD50:3300 μg/kg MEIEDD 10,781,83

SAFETY PROFILE: Poison by ingestion and intravenous routes. Used as a neuromuscular blocker. When heated to decomposition it emits toxic fumes of NH_3 and NO_x. See also PICRIC ACID.

HNG100 CAS:38246-95-6 ***HR: D***
N-HYDROXY-p-PHENETIDINE
mf: $C_8H_{11}NO_2$ mw: 153.20

SYNS: 4-ETHOXY-N-HYDROXY-BENZENAMINE ◇ N-(p-ETHOXYPHENYL)HYDROXYLAMINE ◇ (p-ETHOXYPHENYL)HYDROXYLAMINE ◇ N-HYDROXYPHENETIDINE

TOXICITY DATA with REFERENCE
mmo-sat 50 μg/plate CRNGDP 3,167,82
mma-sat 1 μmol/plate CPBTAL 33,2877,85
dnd-rat:lvr 50 μmol/L CNREA8 44,1098,84

SAFETY PROFILE: Mutation data reported. When heated to decomposition it emits toxic fumes of NO_x.

HNG500 CAS:550-82-3 ***HR: 2***
7-HYDROXY-3H-PHENOXAZIN-3-ONE-10-OXIDE
mf: $C_{12}H_7NO_4$ mw: 229.20

SYNS: AZORESORCIN ◇ DIAZORESORCINOL

TOXICITY DATA with REFERENCE
orl-rat LDLo:500 mg/kg NCNSA6 5,24,53

CONSENSUS REPORTS: Reported in EPA TSCA Inventory.

SAFETY PROFILE: Moderately toxic by ingestion. When heated to decomposition it emits toxic fumes of NO_x.

HNG700 CAS:306-23-0 ***HR: 3***
p-HYDROXYPHENYLACTIC ACID
mf: $C_9H_{10}O_4$ mw: 182.19

TOXICITY DATA with REFERENCE
scu-mus TDLo:800 mg/kg (15-21D post):CAR,TER BEXBAN 87,46,79
scu-mus TDLo:800 mg/kg (15-21D post):CAR,TER BEXBAN 87,46,79
scu-mus TDLo:1600 mg/kg/8W-I:CAR VOONAW 22(6),47,76
scu-mus TD:2800 mg/kg/10W-I:CAR GETRE8 28(2),50,83

SAFETY PROFILE: An experimental teratogen. Questionable carcinogen with experimental carcinogenic data. When heated to decomposition it emits acrid smoke and fumes.

HNG800 CAS:5453-66-7 ***HR: 1***
4-HYDROXYPHENYLARSENOUS ACID
mf: $C_6H_5AsO_2$ mw: 184.03

SYNS: p-ARSENOSOPHENOL ◇ PHENOL, p-ARSENOSO-

TOXICITY DATA with REFERENCE
ivn-mus LDLo:2 g/kg PHBUA9 2,19,54

OSHA PEL: TWA 0.5 mg(As)/m^3
ACGIH TLV: TWA 0.2 mg(As)/m^3

SAFETY PROFILE: Mildly toxic by intravenous route. When heated to decomposition it emits toxic fumes of As.

HNH475 CAS:29050-86-0 ***HR: D***
6-((p-HYDROXYPHENYL)AZO)URACIL
mf: $C_{10}H_8N_4O_3$ mw: 232.22

TOXICITY DATA with REFERENCE
oms-bcs 20 mg/L JOBAAY 126,108,76
dni-bcs 25 mg/L JOBAAY 126,108,76
pic-bcs 100 μmol/L JOVIAM 14,1470,74
dni-omi 6 μmol/L FOMIAZ 27,7,82

SAFETY PROFILE: Mutation data reported. When heated to decomposition it emits toxic fumes of NO_x.

HNH500 CAS:136-36-7 ***HR: 3***
3-HYDROXYPHENYL BENZOATE
mf: $C_{13}H_{10}O_3$ mw: 214.23

SYNS: BENZOIC ACID-m-HYDROXYPHENYL ESTER ◇ EASTMAN INHIBITOR RMB ◇ RESORCINOL, MONOBENZOATE

TOXICITY DATA with REFERENCE
eye-rbt 5% MILD JAPMA846,185,57
orl-rat LD50:800 mg/kg 14CYAT 2,1897,63
ipr-rat LD50:400 mg/kg 14CYAT 2,1897,63
ipr-mus LD50:710 mg/kg JAPMA8 46,185,57

CONSENSUS REPORTS: Reported in EPA TSCA Inventory.

SAFETY PROFILE: Poison by intraperitoneal route. Moderately toxic by ingestion. An eye irritant. When heated to decomposition it emits acrid smoke and fumes. See also ESTERS.

HNI000 CAS:835-64-3 ***HR: 2***
2-(o-HYDROXYPHENYL)BENZOXAZOLE
mf: $C_{13}H_9NO_2$ mw: 211.23

SYN: USAF EK-6754

TOXICITY DATA with REFERENCE
ipr-mus LD50:2000 mg/kg NTIS** AD277-689

CONSENSUS REPORTS: Reported in EPA TSCA Inventory.

SAFETY PROFILE: Moderately toxic by intraperitoneal route. When heated to decomposition it emits toxic fumes of NO_x.

HNI500 CAS:129-20-4 ***HR: 3***
p-HYDROXYPHENYLBUTAZONE
mf: $C_{19}H_{20}N_2O_3$ mw: 324.41

SYNS: ARTROFLOG ◇ BM 1 ◇ BUTAFLOGIN ◇ BUTANOVA ◇ BUTAPIRONE ◇ BUTILENE ◇ 4-BUTYL-2-(4-HYDROXYPHENYL)-1-PHENYL-3,5-DIOXOPYRAZOLIDINE ◇ 4-BUTYL-1-(p-HYDROXYPHENYL)-2-PHENYL-3,5-PYRAZOLIDINEDIONE ◇ 4-BUTYL-1-(4-HYDROXYPHENYL)-2-PHENYL-3,5-PYRAZOLIDINEDIONE ◇ 4-BUTYL-2-(p-HYDROXYPHENYL)-1-PHENYL-3,5-PYRAZOLIDINEDIONE ◇ CROVARIL ◇ DEFLOGIN ◇ 3,5-DIOXO-1-PHENYL-2-(p-HYDROXYPHENYL)-4-N-BUTYLPYRAZOLIDENE ◇ ETROZOLIDINA ◇ FLAMARIL ◇ FLANARIL ◇ GLOGAL ◇ FLOGHENE ◇ FLOGISTIN ◇ FLOGITOLO ◇ FLOGODIN ◇ FLOGORIL ◇ FLOGOSTOP ◇ FLOPIRINA ◇ FRABEL ◇ G 27202 ◇ 1-(p-HYDROXYPHENYL)-2-PHENYL-4-BUTYL-3,5-PYRAZOLIDINEDIONE ◇ 1-p-HYDROXYPHENYL-2-PHENYL-3,5-DIOXO-4-N-BUTYLPYRAZOLIDINE ◇ IDROBUTAZINA ◇ INFAMIL ◇ IPABUTONA ◇ IRIDIL ◇ ISOBUTAZINA ◇ ISOBUTIL ◇ METABOLITE I ◇ NEO-FARMADOL ◇ NEOFEN ◇ OFFITRIL ◇ OXALID ◇ OXAZOLIDIN ◇ OXAZOLIDIN-GEIGY ◇ OXIBUTOL ◇ OXIFENIBUTOL ◇ OXIFENYLBUTAZON ◇ OXYPHENBUTAZONE ◇ OXYPHENYLBUTAZONE ◇ 1-PHENYL-2-(p-HYDROXYPHENYL)-3,5-DIOXO-4-BUTYLPYRAZOLIDINE ◇ PIRABUTINA ◇ PIRAFLOGIN ◇ POLIFLOGIL ◇ REMAZIN ◇ REUMOX ◇ RUMAPAX ◇ TANDACOTE ◇ TANDALGESIC ◇ TANDEARIL ◇ TANDERAL ◇ TELIDAL ◇ TENDEARIL ◇ VALIOIL ◇ VISUBUTINA ◇ USAF GE-14

TOXICITY DATA with REFERENCE
sln-asn 1 g/L MUREAV 26,159,74
orl-rat TDLo:700 mg/kg (female 10-16D post):REP JJPAAZ 28,909,78
orl-chd LDLo:420 mg/kg/4W-I:LIV,BLD BMJOAE 2,1517,62
orl-man TDLo:27 mg/kg/14D:GIT,BLD LAKAA3 63,53,66
orl-wmn TDLo:50 mg/kg/7D:SKN,MET LAKAA3 63,53,66
orl-wmn TDLo:24 mg/kg/4D:GIT,END JAMAAP 238,1399,77
orl-rat LD50:350 mg/kg MPHEAE 16,536,67
ivn-rat LD50:68 mg/kg MEXPAG 6,88,62
orl-mus LD50:350 mg/kg BCFAAI 103,245,64
ipr-mus LD50:100 mg/kg NTIS** AD414-344
ivn-mus LD50:52 mg/kg ARZNAD 10,129,60
ivn-dog LD50:178 mg/kg AIPTAK 149,571,64
ivn-rbt LD50:104 mg/kg ANYAA9 86,263,60
orl-gpg LD50:2720 mg/kg ATSUDG 7,365,84
orl-ham LD50:1180 mg/kg ATSUDG 7,365,84

CONSENSUS REPORTS: IARC Cancer Review: Group 3 IMEMDT 7,56,87; Human Inadequate Evidence IMEMDT 13,183,77.

SAFETY PROFILE: A poison by ingestion, intraperitoneal, and intravenous routes. Moderately toxic to humans by ingestion. Human systemic effects by ingestion: salivary gland changes, diarrhea, nausea or vomiting, hepatitis, hemorrhage, agranulocytosis, thrombocytopenia, dermatitis, fever, and unspecified endocrine system effects. Experimental reproductive effects. Questionable carcinogen. Mutation data reported. Used as an anti-inflammatory agent. When heated to decomposition it emits toxic fumes of NO_x.

HNJ000 CAS:50-19-1 ***HR: 2***
2-HYDROXY-2-PHENYLBUTYL CARBAMATE
mf: $C_{11}H_{15}NO_3$ mw: 209.27

SYNS: β-ETHYL-β-HYDROXYPHENETHYL CARBAMATE ◇ β-ETHYL-β-HYDROXYPHENETHYL CARBAMIC ACID ESTER ◇ HIDROXIFENAMATO ◇ HYDROXYPHENAMATE ◇ LISTICA ◇ NSC-108034 ◇ OXYFENAMATE ◇ OXYPHENAMATE ◇ P 301 ◇ PHENYLBUTAMATE ◇ 2-PHENYL-1,2-BUTANEDIOL 1-CARBAMATE ◇ TENSIFEN

TOXICITY DATA with REFERENCE
orl-rat LD50:607 mg/kg 27ZQAG -,394,72
orl-mus LD50:830 mg/kg MEIEDD 10,705,83
ivn-mus LD50:425 mg/kg DNSSAW 22,9,61

SAFETY PROFILE: Moderately toxic by ingestion and intravenous routes. A minor tranquilizer. When heated to decomposition it emits toxic fumes of NO_x. See also CARBAMATES.

HNJ500 CAS:5418-32-6 ***HR: 3***
4-(p-HYDROXYPHENYL)-2,5-CYCLOHEXADIENE-1-ONE SODIUM SALT
mf: $C_{12}H_8NO_2 \cdot Na$ mw: 221.20

TOXICITY DATA with REFERENCE
ipr-mus LD50:50 mg/kg JMCMAR 21,11,78

CONSENSUS REPORTS: Reported in EPA TSCA Inventory.

SAFETY PROFILE: Poison by intraperitoneal route. When heated to decomposition it emits toxic fumes of NO_x and Na_2O.

HNK000 CAS:27068-06-0 ***HR: 3***
(3-HYDROXYPHENYL)DIETHYLMETHYL-AMMONIUM BROMIDE
mf: $C_{11}H_{18}NO \cdot Br$ mw: 260.21

SYNS: DIETHYL(m-HYDROXYPHENYL)METHYLAMMONIUMBROMIDE ◇ RO 2-2980

TOXICITY DATA with REFERENCE
orl-mus LD50:690 mg/kg JPETAB 100,83,50
ipr-mus LD50:26 mg/kg JPETAB 100,83,50
scu-mus LD50:61 mg/kg JPETAB 100,83,50
ivn-mus LD50:10 mg/kg JPETAB 100,83,50
ivn-dog LD50:20 mg/kg JPETAB 100,83,50
ivn-rbt LD50:27 mg/kg JPETAB 100,83,50

SAFETY PROFILE: Poison by intraperitoneal, subcutaneous, and intravenous routes. Moderately toxic by ingestion. When heated to decomposition it emits very toxic fumes of NH_3, NO_x, and Br^-. See also BROMIDES.

HNK500 CAS:6249-65-6 ***HR: 3***
(m-HYDROXYPHENYL)DIETHYLMETHYLAMMONIUM IODIDE, DIMETHYLCARBAMATE
mf: $C_{14}H_{23}N_2O_2 \cdot I$ mw: 378.29

SYNS: DIETHYL(m-HYDROXYPHENYL)METHYLAMMONIUMIODIDE DIMETHYLCARBAMATE ◇ N,N-DIMETHYLCARBAMIC ACID-3-DIETHYLAMINOPHENYL ESTER METHIODIDE ◇ DIMETHYLCARBAMIC ACID-m-(DIETHYLMETHYLAMINO)PHENYL ESTER IODIDE ◇ (3-(DIMETHYLCARBAMOYLOXY)PHENYL)DIETHYLMETHYLAMMONIUM IODIDE ◇ TL-1238

TOXICITY DATA with REFERENCE
scu-rat LDLo:400 μg/kg NTIS** PB158-508
scu-mus LD50:175 μg/kg NTIS** PB158-508
ivn-mus LD50:60 μg/kg NTIS** PB158-508
scu-dog LDLo:300 μg/kg NTIS** PB158-508
scu-cat LDLo:200 μg/kg NTIS** PB158-508
scu-rbt LDLo:150 μg/kg NTIS** PB158-508
scu-gpg LDLo:100 μg/kg NTIS** PB158-508

SAFETY PROFILE: A deadly poison by subcutaneous and intravenous routes. When heated to decomposition it emits very toxic fumes of NH_3, NO_x, and I^-. See also CARBAMATES and IODIDES.

HNK550 CAS:63957-59-5 ***HR: 3***
(m-HYDROXYPHENYL)DIETHYLMETHYLAMMONIUM METHOSULFATE METHYLCARBAMATE
mf: $C_{13}H_{21}N_2O_2 \cdot CH_3O_4S$ mw: 348.46

SYNS: (3-(N-CARBAMOYLOXY)PHENYL)DIETHYLMETHYL-AMMONIUM METHOSULFATE ◇ N-METHYL-CARBAMIC ACID-3-DIETHYLAMINOPHENYL ESTER, DIMETHYLSULFATE ◇ N-METHYL-CARBAMIC ACID-3-DIETHYLAMINOPHENYL ESTER, METHOSULFATE ◇ METHYL-CARBAMIC ACID-m-(DIETHYLMETHYLAMMONIO)PHENYL ESTER, METHOSULFATE ◇ TL-1317

TOXICITY DATA with REFERENCE
scu-mus LD50:100 μg/kg NTIS** PB 158-508
scu-cat LDLo:100 μg/kg NTIS** PB158-508
scu-rbt LDLo:50 μg/kg NTIS** PB158-508
scu-gpg LDLo:50 μg/kg NTIS** PB158-508

SAFETY PROFILE: A deadly poison by subcutaneous route. When heated to decomposition it emits toxic fumes of SO_x, NH_3, and NO_x. See also CARBAMATES.

HNK575 CAS:99071-30-4 ***HR: 3***
(4-HYDROXY-m-PHENYLENE)BIS-(ACETATOMERCURY)
mf: $C_{10}H_{10}Hg_2O_5$ mw: 611.38

SYNS: DIACETOXYMERCURIPHENOL ◇ MERCURY, (4-HYDROXY-m-PHENYLENE)BIS(ACETATO- ◇ PHENOL, 2,4-BIS(ACETOXYMERCURI)-

TOXICITY DATA with REFERENCE
ipr-mus LDLo:113 mg/kg JPETAB 31,87,27

ACGIH TLV: TWA 0.1 mg(Hg)/m^3 (skin)

SAFETY PROFILE: Poison by intraperitoneal route. When heated to decomposition it emits toxic fumes of Hg.

HNK600 ***HR: 2***
7-HYDROXY-4-PHENYL-3-(4-HYDROXY-PHENYL)COUMARIN
mf: $C_{21}H_{14}O_4$ mw: 330.35

SYNS: COUMARIN,7-HYDROXY-3-(p-HYDROXYPHENYL)-4-PHENYL- ◇ OV_1)

TOXICITY DATA with REFERENCE
orl-rat TDLo:100 mg/kg (female 1-2D post):REP IJEBA6 23,638,85
ipr-mus LD50:1410 mg/kg IJEBA6 25,450,87

SAFETY PROFILE: Moderately toxic by intraperitoneal route. Experimental reproductive effects. When heated to decomposition it emits acrid smoke and irritating fumes.

HNL100 CAS:34920-64-4 ***HR: 3***
1-(p-HYDROXYPHENYL)-2-(3'-PHENYLTHIO-PROPYLAMINO)-1-PROPANOL HYDROCHLORIDE
mf: $C_{18}H_{23}NO_2S \cdot ClH$ mw: 353.94

SYNS: 4-HYDROXY-α-(1-((3-(PHENYLTHIO)PROPYL)AMINO) ETHYL)-BENZENEMETHANOL HYDROCHLORIDE ◇ p-HYDROXY-α-

(1-((3-(PHENYLTHIO)PROPYL)AMINO)ETHYL)BENZYLALCOHOL HYDROCHLORIDE

TOXICITY DATA with REFERENCE
orl-rat LD50:650 mg/kg CHTPBA 6,474,71
orl-mus LD50:650 mg/kg CHTPBA 6,474,71
scu-mus LD50:125 mg/kg CHTPBA 6,474,71
scu-dog LDLo:300 mg/kg CHTPBA 6,474,71
ivn-dog LDLo:65 mg/kg CHTPBA 6,474,71
ivn-rbt LD50:27 mg/kg CHTPBA 6,474,71

SAFETY PROFILE: Poison by subcutaneous and intravenous routes. Moderately toxic by ingestion. When heated to decomposition it emits toxic fumes of SO_x, NO_x, and HCl.

HNL500 CAS:156-39-8 ***HR: 3***
p-HYDROXYPHENYLPYRUVIC ACID
mf: $C_9H_8O_4$ mw: 180.17

TOXICITY DATA with REFERENCE
scu-mus TDLo:1600 mg/kg/8W-I:ETA VOONAW 22(6),47,76

SAFETY PROFILE: Questionable carcinogen with experimental tumorigenic data. When heated to decomposition it emits acrid smoke and fumes.

HNM000 CAS:144-12-7 ***HR: 3***
4-(3-HYDROXY-3-PHENYL-3-(2-THIENYL)PROPYL)-4-METHYLMORPHOLINIUMIODIDE
mf: $C_{18}H_{24}NO_2S•I$ mw: 445.39

SYNS: 114 C.E. ◇ CERFA 114 ◇ l'IODURE de α-THIENYL-1 PHENYL-1 N-METHYL MORPHOLINIUM-3 PROPANOL-1 (FRENCH) ◇ TE 114 ◇ 1-α-THIENYL-1-PHENYL-3-N-METHYLMORPHOLINIUM-1-PROPANOL IODIDE ◇ TIEMONIUM IODIDE ◇ TIEMOZYL ◇ VISCERALGIN ◇ VISCERALGINE

TOXICITY DATA with REFERENCE
orl-rat LD50:2295 mg/kg OYYAA2 3,390,69
ipr-rat LD50:240 mg/kg OYYAA2 3,390,69
scu-rat LD50:1350 mg/kg OYYAA2 3,390,69
ivn-rat LD50:30 mg/kg AIPTAK 141,465,63
orl-mus LD50:1800 mg/kg OYYAA2 3,390,69
ipr-mus LD50:130 mg/kg OYYAA2 3,390,69
scu-mus LD50:450 mg/kg OYYAA2 3,390,69
ivn-mus LD50:30 mg/kg NIIRDN 6,865,82

SAFETY PROFILE: Poison by intravenous and intraperitoneal routes. Moderately toxic by ingestion and subcutaneous routes. When heated to decomposition it emits very toxic fumes of NO_x, SO_x, and I^-. See also IODIDES.

HNN000 CAS:64051-06-5 ***HR: 3***
(m-HYDROXYPHENYL)TRIMETHYLAMMONIUM CHLORIDE, METHYLCARBAMATE
mf: $C_{11}H_{17}N_2O_2•Cl$ mw: 244.75

SYNS: METHOCHLORIDE of N-METHYLURETHANE of 3-DIMETHYLAMINOPHENOL ◇ N-METHYLCARBAMIC ACID-3-DIMETHYLAMINOPHENYL ESTER METHOCHLORIDE ◇ METHYLCARBAMIC ACID, (m-(TRIMETHYLAMMONIO)PHENYL) ESTER CHLORIDE ◇ (3-(METHYLCARBAMOYLOXY)PHENYL)TRIMETHYLAMMONIUM CHLORIDE ◇ T-1690 ◇ TL-1226

TOXICITY DATA with REFERENCE
scu-mus LD50:140 μg/kg NTIS** PB158-508
ivn-mus LD50:70 μg/kg NTIS** PB158-508

SAFETY PROFILE: A deadly poison by subcutaneous and intravenous routes. When heated to decomposition it emits very toxic fumes of NH_3, NO_x, and Cl^-. See also CARBAMATES.

HNN500 CAS:2498-27-3 ***HR: 3***
(m-HYDROXYPHENYL)TRIMETHYLAMMONIUM IODIDE
mf: $C_9H_{14}NO•I$ mw: 279.14

SYNS: m-(DIMETHYLAMINO)PHENOL METHIODIDE ◇ 3-HYDROXY-N,N,N-TRIMETHYLBENZENAMINIUM IODIDE ◇ 3-OXYPHENYL TRIMETHYLAMMONIUM IODIDE ◇ TMPH ◇ 3-(TRIMETHYLAMMONIO)PHENOL IODIDE

TOXICITY DATA with REFERENCE
orl-mus LDLo:200 mg/kg JPETAB 43,413,31
ipr-mus LD50:65 mg/kg PCJOAU 10,327,76
ivn-mus LDLo:25 mg/kg JPETAB 43,413,31

SAFETY PROFILE: Poison by ingestion, intraperitoneal, and intravenous routes. When heated to decomposition it emits very toxic fumes of NH_3, NO_x, and I^-. See also IODIDES.

HNO000 CAS:64051-08-7 ***HR: 3***
(m-HYDROXYPHENYL)TRIMETHYLAMMONIUM IODIDE, BENZYLCARBAMATE
mf: $C_{17}H_{21}N_2O_2•I$ mw: 412.30

SYNS: N-BENZYLCARBAMICACID-3-DIMETHYLAMINOPHENYL ESTER METHIODIDE ◇ BENZYL CARBAMIC ACID-m-(TRIMETHYLAMMONIO)PHENYL ESTER IODIDE ◇ (3-(BENZYLCARBAMOYLOXY)PHENYL)TRIMETHYLAMMONIUM IODIDE ◇ METHIODIDE of N-BENZYLURETHANE of 3-DIMETHYLAMINOPHENOL ◇ T-1125

TOXICITY DATA with REFERENCE
scu-mus LD50:350 μg/kg NTIS** PB158-508
scu-rbt LD50:200 μg/kg NTIS** PB158-508

SAFETY PROFILE: A deadly poison by subcutaneous route. When heated to decomposition it emits very toxic fumes of NH_3, NO_x and I^-. See also IODIDES and CARBAMATES.

HNO500 CAS:3983-39-9 ***HR: 3***
(m-HYDROXYPHENYL)TRIMETHYLAMMONIUM IODIDE, METHYLCARBAMATE
mf: $C_{11}H_{17}N_2O_2•I$ mw: 336.20

SYNS: CARBAMICACID-N-METHYL-3-DIMETHYLAMINOPHENYL

ESTER METHIODIDE ◇ METHIODIDE of N-METHYLURETHANE of 3-DIMETHYLAMINOPHENOL ◇ METHYLCARBAMIC ACID, (m-(TRIMETHYLAMMONIO)PHENYL)ESTER, IODIDE ◇ (3-(METHYLCARBAMOYLOXY)PHENYL)TRIMETHYLAMMONIUM IODIDE ◇ T-1152 ◇ TL 1178

TOXICITY DATA with REFERENCE
scu-rat LDLo:500 μg/kg NTIS** PB158-508
scu-mus LD50:270 μg/kg NTIS** PB158-508
ivn-mus LD50:115 μg/kg NTIS** PB158-508
scu-dog LDLo:1 mg/kg NDRC** No.9-4-1-19,44
scu-cat LDLo:500 μg/kg NDRC** No.9-4-1-19,44
scu-rbt LD50:260 μg/kg NTIS** PB158-508
scu-gpg LDLo:250 μg/kg NDRC** No.9-4-1-19,44

SAFETY PROFILE: A deadly poison by subcutaneous and intravenous routes. When heated to decomposition it emits very toxic fumes of NH_3, NO_x, and I^-. See also IODIDES, CARBAMATES, and ESTERS.

HNP000 CAS:3983-40-2 ***HR: 3***
(p-HYDROXYPHENYL)TRIMETHYLAMMONIUM IODIDE, METHYLCARBAMATE
mf: $C_{11}H_{17}N_2O_2 \cdot I$ mw: 336.20

SYNS: AR-17 ◇ METHIODIDE of N-METHYLURETHANE of 4-DIMETHYLAMINOPHENOL ◇ N-METHYL CARBAMIC ACID, 4-DIMETHYLAMINOPHENYL ESTER METHIODIDE ◇ N-METHYLCARBAMIC ACID-p-(TRIMETHYLAMMONIO)PHENYL ESTER IODIDE ◇ METHYLCARBAMIC ESTER of p-OXYPHENYLTRIMETHYLAMMONIUM IODIDE ◇ (4-(N-METHYLCARBAMOYLOXY)PHENYL) TRIMETHYLAMMONIUM IODIDE ◇ ((p-METHYLCARBAMOYLOXY) PHENYL)TRIETHYLAMMONIUM IODIDE ◇ T-1088 ◇ TL-1097

TOXICITY DATA with REFERENCE
orl-mus LDLo:50 mg/kg JPETAB 43,413,31
scu-mus LD50:50 mg/kg NTIS** PB158-508
ivn-mus LD80:2 mg/kg NTIS** PB158-508

SAFETY PROFILE: Poison by ingestion, subcutaneous and intravenous routes. When heated to decomposition it emits very toxic fumes of NH_3, NO_x and I^-. See also IODIDES and CARBAMATES.

HNP500 CAS:64051-18-9 ***HR: 3***
(m-HYDROXYPHENYL)TRIMETHYLAMMONIUM METHYLSULFATE, CARBAMATE
mf: $C_{10}H_{15}N_2O_2 \cdot CH_3O_4S$ mw: 306.37

SYNS: AR-11 ◇ CARBAMIC ACID-3-DIMETHYLAMINOPHENYL ESTER, METHOSULFATE ◇ CARBAMIC ACID, (m-TRIMETHYLAMMONIO)PHENYL ESTER, METHYLSULFATE ◇ CARBAMIC ESTER of 3-OXYPHENYLTRIMETHYLAMMONIUM METHYLSULFATE ◇ ((m-CARBAMOYLOXY)PHENYL)TRIMETHYLAMMONIUMMETHYLSULFATE

TOXICITY DATA with REFERENCE
orl-mus LDLo:500 mg/kg JPETAB 43,413,31
ivn-mus LD80:700 μg/kg NTIS** PB158-508

SAFETY PROFILE: A deadly poison by intravenous route. Moderately toxic by ingestion. When heated to decomposition it emits very toxic fumes of NH_3, NO_x, and SO_x. See also SULFATES and CABAMATES.

HNQ000 CAS:6033-07-4 ***HR: 3***
(m-HYDROXYPHENYL)TRIMETHYLAMMONIUM METHYLSULFATE DIETHYLCARBAMATE
mf: $C_{14}H_{23}N_2O_2 \cdot CH_3O_4S$ mw: 362.49

SYNS: AR-33 ◇ DIETHYLCARBAMIC ACID ESTER with (m-HYDROXYPHENYL)TRIMETHYLAMMONIUMMETHYLSULFATE ◇ N,N-DIETHYLCARBAMIC ACID-3-(TRIMETHYLAMMONIO)PHENYL ESTER, METHYLSULFATE ◇ DIETHYLCARBAMIC ESTER of 3-OXYPHENYLTRIMETHYLAMMONIUM METHYLSULFATE ◇ (3-(N′,N′-DIETHYLCARBAMOYLOXY)PHENYL)TRIMETHYLAMMONIUM METHYLSULFATE

TOXICITY DATA with REFERENCE
orl-mus LDLo:71 mg/kg JPETAB 43,413,31
ivn-mus LD80:8 mg/kg NTIS** PB158-508

SAFETY PROFILE: Poison by ingestion and intravenous routes. When heated to decomposition it emits very toxic fumes of NH_3, NO_x, and SO_x. See also SULFATES, ESTERS, and CARBAMATES.

HNQ500 CAS:64051-20-3 ***HR: 3***
(m-HYDROXYPHENYL)TRIMETHYLAMMONIUM METHYLSULFATE, ETHYLCARBAMATE
mf: $C_{12}H_{19}N_2O_2 \cdot CH_3O_4S$ mw: 334.43

SYNS: AR-21 ◇ N-ETHYLCARBAMIC ACID-3-DIMETHYLAMINOPHENYL ESTER, METHOSULFATE ◇ N-ETHYLCARBAMIC ACID-3-(TRIMETHYLAMMONIO)PHENYL ESTER, METHYLSULFATE ◇ ETHYLCARBAMIC ESTER of m-OXYPHENYLTRIMETHYLAMMONIUM METHYLSULFATE ◇ (3-(N-ETHYLCARBAMOYLOXY)PHENYL)TRIMETHYLAMMONIUMMETHYLSULFATE

TOXICITY DATA with REFERENCE
orl-mus LDLo:100 mg/kg JPETAB 43,413,31
ivn-mus LD80:1 mg/kg NTIS** PB158-508

SAFETY PROFILE: Poison by ingestion and intravenous routes. When heated to decomposition it emits very toxic fumes of NH_3, NO_x, and SO_x. See also SULFATES, ESTERS, and CARBAMATES.

HNR000 CAS:64050-77-7 ***HR: 3***
(m-HYDROXYPHENYL)TRIMETHYLAMMONIUM METHYLSULFATE, METHYLCARBAMATE
mf: $C_{11}H_{17}N_2O_2 \cdot CH_3O_4S$ mw: 320.40

SYNS: AR-13 ◇ N-METHYLCARBAMIC ACID-3-DIMETHYLAMINOPHENYL ESTER, METHOSULFATE ◇ N-METHYLCARBAMIC ACID-(3-(TRIMETHYLAMMONIO)PHENYL) ESTER, METHYLSULFATE ◇ METHYLCARBAMIC ESTER of 3-OXYPHENYLTRIMETHYLAMMONIUM METHYLSULFATE ◇ (3-(N-METHYLCARBAMOYLOXY)PHENYL)TRIMETHYLAMMONIUMMETHYLSULFATE

TOXICITY DATA with REFERENCE
orl-mus LDLo:2500 μg/kg JPETAB 43,413,31
ivn-mus LD50:300 μg/kg THERAP 8,714,53

SAFETY PROFILE: Very poisonous by ingestion and intravenous routes. When heated to decomposition it emits very toxic fumes of NH_3, NO_x, and SO_x. See also SULFATES, ESTERS, and CARBAMATES.

HNR500 CAS:64050-79-9 ***HR: 3***
(m-HYDROXYPHENYL)TRIMETHYLAMMONIUM METHYLSULFATE METHYLPHENYLCARBAMATE
mf: $C_{17}H_{21}N_2O_2 \cdot CH_3O_4S$ mw: 396.50

SYN: METHYLPHENYLCARBAMIC ESTER OF 3-OXYPHENYLTRIMETHYLAMMONIUM METHYLSULFATE

TOXICITY DATA with REFERENCE
orl-mus LDLo:75 mg/kg JPETAB 43,413,31
scu-mus LDLo:3 mg/kg JPETAB 43,413,31
ivn-mus LDLo:3500 μg/kg JPETAB 43,413,31
scu-rbt LDLo:1 mg/kg JPETAB 43,413,31
ivn-rbt LDLo:500 μg/kg JPETAB 43,413,31

SAFETY PROFILE: A poison by ingestion, subcutaneous, and intravenous routes. When heated to decomposition it emits very toxic fumes of NH_3, NO_x, and SO_x. See also SULFATES, ESTERS, and CARBAMATES.

HNS000 CAS:64050-81-3 ***HR: 3***
(m-HYDROXYPHENYL)TRIMETHYLAMMONIUM METHYLSULFATE, PENTAMETHYLENECARBAMATE
mf: $C_{15}H_{23}N_2O_2 \cdot CH_3O_4S$ mw: 374.50

SYNS: AR-35 ◇ N,N-PENTAMETHYLENECARBAMIC ACID-3-DIMETHYLAMINOPHENYL ESTER, METHOSULFATE ◇ PENTAMETHYLENECARBAMICACID-m-(TRIMETHYLAMMONIO)PHENYL ESTER, METHYLSULFATE ◇ PENTAMETHYLENECARBAMIC ESTER of 3-OXYPHENYLTRIMETHYLAMMONIUM METHYLSULFATE ◇ (3-(PENTAMETHYLENECARBAMOYLOXY)PHENYL)TRIMETHYLAMMONIUM METHYLSULFATE

TOXICITY DATA with REFERENCE
orl-mus LDLo:500 mg/kg JPETAB 43,413,31
ivn-mus LD80:6 mg/kg NTIS** PB158-508

SAFETY PROFILE: Poison by intravenous route. Moderately toxic by ingestion. When heated to decomposition it emits very toxic fumes of NH_3, NO_x, and SO_x. See also CARBAMATES, ESTERS, and SULFATES.

HNS500 CAS:64050-83-5 ***HR: 3***
(m-HYDROXYPHENYL)TRIMETHYLAMMONIUM METHYLSULFATE, PHENYLCARBAMATE
mf: $C_{16}H_{19}N_2O_2 \cdot CH_3O_4S$ mw: 382.47

SYNS: AR-25 ◇ N-PHENYLCARBAMIC ACID-3-(TRIMETHYLAMMONIO)PHENYL ESTER, METHYLSULFATE ◇ PHENYLCARBAMIC ESTER of 3-OXYPHENYLTRIMETHYLAMMONIUM METHYLSULFATE ◇ (3-(N-PHENYLCARBAMOYLOXY)PHENYL)TRIMETHYLAMMONIUM METHYLSULFATE

TOXICITY DATA with REFERENCE
orl-mus LDLo:125 mg/kg JPETAB 43,413,31
ivn-mus LD80:2 mg/kg NTIS** PB158-508

SAFETY PROFILE: Poison by ingestion and intravenous routes. When heated to decomposition it emits very toxic fumes of NH_3, NO_x and SO_x. See also SULFATES; ESTERS and CARBAMATES.

HNT075 CAS:14838-45-0 ***HR: 2***
3-(1-HYDROXY-2-PIPERIDINOETHYL)-5-PHENYLISOXAZOLE CITRATE
mf: $C_{16}H_{20}N_2O_2 \cdot C_6H_8O_7$ mw: 464.52

SYNS: α-(PIPERIDINOMETHYL)-5-PHENYL-3-ISOXAZOLEMETHANOL CITRATE ◇ 31252-S

TOXICITY DATA with REFERENCE
orl-rat TDLo:900 mg/kg (female 9-14D post):REP SKNEA7 22,109,72
orl-mus TDLo:900 mg/kg (female 7-12D post):TER SKNEA7 22,109,72
scu-mus LD50:416 mg/kg JMCMAR 10,411,67

SAFETY PROFILE: Moderately toxic by subcutaneous route. An experimental teratogen. Experimental reproductive effects. When heated to decomposition it emits toxic fumes of NO_x.

HNT100 CAS:93793-83-0 ***HR: 3***
2-HYDROXY-N-(3-(m-(PIPERIDINOMETHYL) PHENOXY)PROPYL) ACETAMIDE ACETATE (ester) HYDROCHLORIDE
mf: $C_{19}H_{28}N_2O_4 \cdot ClH$ mw: 384.95

SYNS: 2-ACETOXY-N-(3-(m-(1-PIPERIDINYLMETHYL)PHENOXY) PROPYL)ACETAMIDE HYDROCHLORIDE ◇ TZU-0460

TOXICITY DATA with REFERENCE
orl-rat TDLo:1100 mg/kg (female 7-17D post):REP YACHDS 13,1325,85
orl-rat TDLo:1100 mg/kg (female 7-17D post):TER YACHDS 13,1325,85
orl-rat LD50:755 mg/kg YACHDS 13,1167,85
ipr-rat LD50:227 mg/kg YACHDS 13,1167,85
scu-rat LD50:595 mg/kg YACHDS 13,1167,85
ivn-rat LD50:110 mg/kg YACHDS 13,1167,85
orl-mus LD50:509 mg/kg YACHDS 13,1167,85
scu-mus LD50:384 mg/kg YACHDS 13,1167,85
ivn-mus LD50:83 mg/kg YACHDS 13,1167,85
orl-dog LD50:100 mg/kg YACHDS 13,1167,85
ivn-dog LD50:75 mg/kg YACHDS 13,1167,85

ivn-mky LD50:50 mg/kg YACHDS 13,1167,85
orl-rbt LD50:900 mg/kg YACHDS 13,1167,85

SAFETY PROFILE: Poison by ingestion, subcutaneous, intravenous, and intraperitoneal routes. An experimental teratogen. Experimental reproductive effects. When heated to decomposition it emits toxic fumes of NO_x and HCl. See also ESTERS.

HNT500 CAS:630-56-8 ***HR: 3***
HYDROXYPROGESTERONE CAPROATE
mf: $C_{27}H_{40}O_4$ mw: 428.67

PROP: Dense needles. Mp: 119-121°.

SYNS: CAPRON ◇ CORLUTIN L.A. ◇ DELALUTIN ◇ DEPO-PROLUTON ◇ DURALUTON ◇ ESTRALUTIN ◇ GESTEROL L.A. ◇ 17-α-HEXANOYLOXYPREGN-4-ENE-3,20-DIONE ◇ HORMOFORT ◇ HPC ◇ 17-HYDROXYPREGN-4-ENE-3,20-DIONE HEXANOATE ◇ 17-α-HYDROXYPROGESTERONE CAPROATE ◇ 17-α-HYDROXY PROGESTERONE-N-CAPROATE ◇ 17-α-HYDROXYPROGESTERONE HEXANOATE ◇ HYDROXON ◇ HYLUTIN ◇ HYPROVAL-PA ◇ IDROGESTENE ◇ LUETOCRIN DEPOT ◇ LUTATE ◇ LUTEOCRIN ◇ LUTOPRON ◇ NEOLUTIN ◇ NSC-17592 ◇ 17-((1-OXOHEXYL)OXY)PREGN-4-ENE-3,20-DIONE ◇ PRIMOLUT DEPOT ◇ PROGESTERONE CAPROATE ◇ PROGESTERONE RETARD PHARLON ◇ PROLUTON DEPOT ◇ RELUTIN ◇ SQUIBB ◇ SYNGYNON ◇ TERALUTIL

TOXICITY DATA with REFERENCE
unr-wmn TDLo:12 mg/kg (female 6-20W post):REP SCIEAS 211,1171,81
ims-mky TDLo:180 mg/kg (female 3-21W post):TER TJADAB 35,129,87

CONSENSUS REPORTS: IARC Cancer Review: Animal Inadequate Evidence IMEMDT 21,399,79.

SAFETY PROFILE: Human reproductive effects by an unknown route: behavioral effects on newborn. Experimental teratogenic and reproductive effects. Questionable carcinogen. A steroid. Used to treat menstrual disorders, threatened abortion, and sterility. When heated to decomposition it emits acrid smoke and fumes.

HNT600 CAS:999-61-1 ***HR: 3***
2-HYDROXYPROPYL ACRYLATE
mf: $C_6H_{10}O_3$ mw: 130.16

SYNS: ACRYLIC ACID-2-HYDROXYPROPYL ESTER ◇ β-HYDROXYPROPYL ACRYLATE ◇ 1,2-PROPANEDIOL-1-ACRYLATE ◇ 2-PROPENOIC ACID-2-HYDROXYPROPYL ESTER ◇ PROPYLENE GLYCOL MONOACRYLATE

TOXICITY DATA with REFERENCE
orl-rat LD50:250 mg/kg DTLVS* 4,227,80
orl-mus LD50:1056 mg/kg TOLED5 11,125,82
scu-rbt LD50:160 mg/kg AIHAAP 30,470,69

OSHA PEL: TWA 0.5 ppm (skin)
ACGIH TLV: TWA 0.5 ppm (skin)

SAFETY PROFILE: Poison by ingestion and subcutaneous routes. See also ESTERS. When heated to decomposition it emits acrid smoke and fumes.

HNU500 CAS:94-13-3 ***HR: 3***
p-HYDROXYPROPYL BENZOATE
mf: $C_{10}H_{12}O_3$ mw: 180.22

PROP: Colorless crystals or white powder. Sltly sol in water; sol in alc, ether.

SYNS: ASEPTOFORM P ◇ BETACIDE P ◇ BONOMOLD OP ◇ 4-HYDROXYBENZOIC ACID PROPYL ESTER ◇ p-HYDROXYBENZOIC ACID PROPYL ESTER ◇ NIPASOL ◇ p-OXYBENZOESAUREPROPYLESTER (GERMAN) ◇ PARABEN ◇ PARASEPT ◇ PASEPTOL ◇ PRESERVAL P ◇ PROPYL p-HYDROXYBENZOATE ◇ n-PROPYL p-HYDROXYBENZOATE ◇ PROPYLPARABEN (FCC) ◇ PROPYLPARASEPT ◇ PROTABEN P ◇ TEGOSEPT P

TOXICITY DATA with REFERENCE
ipr-mus LD50:200 mg/kg NTIS** AD691-490
scu-mus LD50:1650 mg/kg AIPTAK 128,135,60
orl-dog LD50:6000 mg/kg 14CYAT 2,1897,63
orl-rbt LDLo:6 g/kg AEPPAE 146,208,29

CONSENSUS REPORTS: Reported in EPA TSCA Inventory. EPA Genetic Toxicology Program.

SAFETY PROFILE: Poison by intraperitoneal routes. Moderately toxic by subcutaneous route. Mildly toxic by ingestion. An allergen. When heated to decomposition it emits acrid smoke and fumes.

HNV000 CAS:9004-64-2 ***HR: 1***
HYDROXYPROPYL CELLULOSE

PROP: White powder. Sol in water and organic solvents.

SYNS: HYDROXYPROPYL ETHER of CELLULOSE ◇ KLUCEL

TOXICITY DATA with REFERENCE
orl-rat LD50:10200 mg/kg FAONAU 46A,131,69

CONSENSUS REPORTS: Reported in EPA TSCA Inventory.

SAFETY PROFILE: Slightly toxic by ingestion. When heated to decomposition it emits acrid smoke and fumes.

HNX000 CAS:9004-65-3 ***HR: 1***
HYDROXYPROPYL METHYLCELLULOSE

PROP: White fibrous or granular powder. Sol in water, organic solvents; insol in anhyd alc, ether, and chloroform.

SYN: METHOCEL HG

TOXICITY DATA with REFERENCE
ipr-rat LD50:5200 mg/kg JPETAB 99,112,50
ipr-mus LD50:5000 mg/kg JPETAB 99,112,50

CONSENSUS REPORTS: Reported in EPA TSCA Inventory.

SAFETY PROFILE: Mildly toxic by intraperitoneal route. When heated to decomposition it emits acrid smoke and fumes.

HNX500 CAS:61499-28-3 ***HR: 3***
1-((2-HYDROXYPROPYL)NITROSO)AMINO)ACETONE
mf: $C_6H_{12}N_2O_3$ mw: 160.20

SYNS: HPOP ◇ 1-((2-HYDROXYPROPYL)NITROSOAMINO)-2-PROPANONE ◇ N-NITROSO(2-HYDROXYPROPYL)(2-OXOPROPYL)AMINE

TOXICITY DATA with REFERENCE
dnd-rat:oth 25 mg/L CBINA8 48,59,84
dns-rat:lvr 100 μmol/L MUREAV 144,197,85
orl-rat TDLo:881 mg/kg/41W-I:ETA IAPUDO 57,617,84
ipr-rat TDLo:160 g/kg:CAR JJIND8 74,209,85
skn-ham TDLo:1900 mg/kg/50W-I:CAR CALEDQ 10,163,80
scu-ham TD:500 mg/kg/53W-I:NEO CNREA8 39,3828,79
scu-ham LD50:354 mg/kg CNREA8 39,3828,79

SAFETY PROFILE: Suspected carcinogen with experimental carcinogenic, neoplastigenic, and tumorigenic data. A poison by subcutaneous route. Mutation data reported. When heated to decomposition it emits toxic fumes of NO_x. See also NITROSAMINES.

HNX600 CAS:21905-32-8 ***HR: 3***
2-HYDROXYPROPYL PHENYL ARSINIC ACID
mf: $C_9H_{13}AsO_3$ mw: 244.14

SYNS: ARSINE OXIDE, HYDROXY(2-HYDROXYPROPYL)PHENYL- ◇ HYDROXY(2-HYDROXYPROPYL)PHENYLARSINE OXIDE

TOXICITY DATA with REFERENCE
ivn-mus LD50:100 mg/kg CSLNX* NX#06914

OSHA PEL: TWA 0.5 mg(As)/m^3
OSHA PEL FINAL:8H TWA 0.5 mg(As)/m^3

SAFETY PROFILE: Poison by intravenous route. When heated to decomposition it emits toxic fumes of As.

HNX800 CAS:10171-78-5 ***HR: 2***
N-(3-HYDROXYPROPYL)-1,2-PROPANEDIAMINE
mf: $C_6H_{16}N_2O$ mw: 132.24

TOXICITY DATA with REFERENCE
skn-rbt 100 μg/24H open AIHAAP 23,95,62
orl-rat LD50:5660 mg/kg AIHAAP 23,95,62
skn-rbt LD50:2120 mg/kg AIHAAP 23,95,62

SAFETY PROFILE: Moderately toxic by skin contact. Mildly toxic by ingestion. A skin irritant. When heated to decomposition it emits toxic fumes of NO_x. See also AMINES.

HNY000 CAS:9049-76-7 ***HR: 3***
HYDROXYPROPYL STARCH

TOXICITY DATA with REFERENCE
orl-rat LD50:218 mg/kg FAONAU 50A,32,72
orl-dog LDLo:200 mg/kg FAONAU 50A,32,72

CONSENSUS REPORTS: Reported in EPA TSCA Inventory.

SAFETY PROFILE: Poison by ingestion. When heated to decomposition it emits acrid smoke and fumes.

HNY500 CAS:50-39-5 ***HR: 2***
1-(2-HYDROXYPROPYL)THEOBROMINE
mf: $C_{10}H_{14}N_4O_3$ mw: 238.28

PROP: Crystals from isopropanol. Mp: 140-142°. Very sol in water; sol in chloroform, hot ethanol, warm glycerol; insol in ether.

SYNS: BONICOR ◇ CORDABROMIN ◇ CORDALEROMIN ◇ CORODIL ◇ 3,7-DIHYDRO-1-(2-HYDROXYPROPYL)-3,7-DIMETHYL-1H-PURINE-2,6-DIONE ◇ 1-(2-HYDROXYPROPYL)-3,7-DIMETHYLXANTHINE ◇ 1-(β-HYDROXYPROPYL)THEOBROMINE ◇ MTB ◇ β-OXYPRO- PYLTHEOBROMINE ◇ PANTOBROMINO ◇ PRO-COR ◇ PROTHEOBROMINE ◇ TEBE ◇ THEOCOR ◇ VASCOPIL

TOXICITY DATA with REFERENCE
ipr-mus LDLo:780 mg/kg ARZNAD 6,457,56
scu-mus LD50:580 mg/kg ARZNAD 6,601,56

SAFETY PROFILE: Moderately toxic by intraperitoneal and subcutaneous routes. Used as a diuretic agent. When heated to decomposition it emits toxic fumes of NO_x. See also THEOBROMINE.

HNZ000 CAS:59413-14-8 ***HR: 2***
1-(3-HYDROXYPROPYL)THEOBROMINE
mf: $C_{10}H_{14}N_4O_3$ mw: 238.28

SYNS: 3,7-DIHYDRO-1-(3-HYDROXYPROPYL)-3,7-DIMETHYL-1H-PURINE-2,6-DIONE ◇ 1-(3-HYDROXYPROPYL)-3,7-DIMETHYLXANTHINE ◇ 1-(3-HYDROXYPROPYL)THEOBROMINE ◇ Γ-OXYPROPYL-THEOBROMIN (GERMAN) ◇ Γ-(Γ-OXYPROPYL)-THEOBROMIN (GERMAN)

TOXICITY DATA with REFERENCE
ipr-mus LDLo:1200 mg/kg ARZNAD 6,457,56
scu-mus LD50:1055 mg/kg ARZNAD 6,601,56

SAFETY PROFILE: Moderately toxic by intraperitoneal and subcutaneous routes. When heated to decomposition it emits toxic fumes of NO_x. See also THEOBROMINE.

HOA000 CAS:603-00-9 ***HR: 2***
β-HYDROXYPROPYLTHEOPHYLLINE
mf: $C_{10}H_{14}N_4O_3$ mw: 238.28

SYNS: BRONTYL ◇ 3,7-DIHYDRO-1,3-DIMETHYL-7-(2-HYDROXYPROPYL)-1H-PURINE-2,6-DIONE ◇ 3,7-DIHYDRO-7-(2-HYDROXYPROPYL)-1,3-DIMETHYL-1H-PURINE-2,6-DIONE ◇ HYDROXYPROPYLTHEOPHYLLINE ◇ 7-(β-HYDROXYPROPYL)THEOPHYLLINE ◇ 7-(2-HYDROXYPROPYL))THEOPHYLLINE ◇ MONOPHYLLINE ◇ MT ◇ OXYPROPYLTHEOPHYLLINE ◇ β-OXYPROPYLTHEOPHYLLIN ◇ PROXIPHYLLINE ◇ PROXYPHYLLINE ◇ PUROPHYLLIN ◇ SANWAPHYLLIN ◇ SIGOPHYL ◇ SPASMOLYSIN ◇ THEAN ◇ THEODEN ◇ THEON

TOXICITY DATA with REFERENCE
orl-rat LD50:460 mg/kg OYYAA2 19,845,80
ipr-rat LD50:445 mg/kg NIIRDN 6,720,82
ivn-rat LD50:430 mg/kg OYYAA2 19,845,80
orl-mus LD50:730 mg/kg NIIRDN 6,720,82
ipr-mus LD50:505 mg/kg ARZNAD 27,14,77
scu-mus LD50:410 mg/kg ARZNAD 4,649,54
ivn-mus LD50:475 mg/kg NIIRDN 6,720,82

SAFETY PROFILE: Moderately toxic by ingestion, intraperitoneal, subcutaneous and intravenous routes. When heated to decomposition it emits toxic fumes of NO_x. See also THEOPHYLLINE.

HOA500 CAS:590-63-6 ***HR: 3***
(2-HYDROXYPROPYL)TRIMETHYLAMMONIUM CHLORIDE CARBAMATE
mf: $C_7H_{17}N_2O_2 \cdot Cl$ mw: 196.71

SYNS: 2-((AMINOCARBONYL)OXY)-N,N,N-TRIMETHYL-1-PROPANAMINIUM CHLORIDE ◇ BETHAINE CHOLINE CHLORIDE ◇ BETHANECHOL CHLORIDE ◇ 2-CARBAMOYLOXYPROPYLTRIMETHYLAMMONIUM CHLORIDE ◇ CARBAMYLMETHYLCHOLINE CHLORIDE ◇ DUVOID ◇ MECHOTHANE ◇ β-METHYLCHOLINE CHLORIDE CARBAMINOYL ◇ β-METHYLCHOLINE CHLORIDE URETHAN ◇ MYOCHOLINE ◇ URECHOLINE ◇ URECHOLINE CHLORIDE

TOXICITY DATA with REFERENCE
scu-hmn TDLo:130 mg/kg:SKN 34ZIAG -,130,69
scu-man TDLo:1071 μg/kg/3D-I:SKN ARDEAC 95,499,67
orl-rat LD50:1500 mg/kg JPETAB 58,337,36
scu-rat LD50:175 mg/kg JPETAB 58,337,36
ivn-rat LD50:21 mg/kg JPETAB 58,337,36
ims-rat LD50:220 mg/kg NIIRDN 6,139,82
orl-mus LD50:250 mg/kg JPETAB 58,337,36
scu-mus LD50:120 mg/kg JPETAB 58,337,36
ivn-mus LD50:10 mg/kg JPETAB 58,337,36
ims-mus LD50:192 mg/kg NIIRDN 6,139,82
ivn-gpg LDLo:13 mg/kg AIPTAK 106,245,56

CONSENSUS REPORTS: Reported in EPA TSCA Inventory.

SAFETY PROFILE: Poison by ingestion, subcutaneous, intramuscular and intravenous routes. Human systemic effects by subcutaneous route: allergic dermatitis and sweating. Used as a synthetic cholinergic drug. When heated to decomposition it emits very toxic fumes of NH_3, NO_x, and Cl^-. See also CARBAMATES.

HOA575 CAS:114-03-4 ***HR: 2***
dl-HYDROXYTRYPTOPHAN
mf: $C_{11}H_{12}N_2O_3$ mw: 220.25

PROP: Rods or needles. Decomposes @ 298-300°. Sol in water; sol in 50% boiling alc.

SYNS: (±)-5-HYDROXYTRYPTOPHAN ◇ dl-5-HYDROXYTRYPTOPHAN ◇ 5-HYDROXYTRYPTOPHAN ◇ PRETONINE

TOXICITY DATA with REFERENCE
ipr-rat TDLo:100 mg/kg (female 21D post):TER JPBAA7 94,113,67
ipr-mus LD50:1080 mg/kg NYKZAU 69,523,73

CONSENSUS REPORTS: Reported in EPA TSCA Inventory.

SAFETY PROFILE: Moderately toxic by intraperitoneal route. An experimental teratogen. When heated to decomposition it emits toxic fumes of NO_x.

HOA600 CAS:4350-09-8 ***HR: 3***
5-HYDROXY-l-TRYPTOPHAN
mf: $C_{11}H_{12}N_2O_3$ mw: 220.25

SYNS: l-5-HTP ◇ l-5-HYDROXYTRYPTOPHAN ◇ l-TRYPTOPHAN, 5-HYDROXY-, (9CI)

TOXICITY DATA with REFERENCE
orl-rat TDLo:700 mg/kg (female 7-13D post):TER IYKEDH 6,356,75
orl-rat TDLo:700 mg/kg (female 7-13D post):REP IYKEDH 6,356,75
orl-rat LD50:243 mg/kg NYKZAU 69,523,73
ipr-rat LD50:91 mg/kg NYKZAU 69,523,73
scu-rat LD50:149 mg/kg IYKEDH 6,307,75
ivn-rat LD50:27 mg/kg IYKEDH 6,307,75
ipr-mus LD50:298 mg/kg IYKEDH 6,307,75
ivn-mus LD50:375 mg/kg NYKZAU 69,523,73

CONSENSUS REPORTS: Reported in EPA TSCA Inventory.

SAFETY PROFILE: Poison by ingestion, intraperitoneal, subcutaneous, and intravenous routes. An experimental teratogen. Other experimental reproductive effects. When heated to decomposition it emits toxic fumes of NO_x.

HOB000 CAS:54643-52-6 ***HR: 3***
3-HYDROXYPURIN-2(3H)-ONE
mf: $C_5H_4O_2$ mw: 96.09

SYN: 3-HYDROXY-2-OXOPURINE

TOXICITY DATA with REFERENCE
scu-rat TDLo:96 mg/kg/8W-I:ETA CBINA8 25,369,79

SAFETY PROFILE: Questionable carcinogen with experimental tumorigenic data. When heated to decomposition it emits acrid smoke and fumes.

HOB500 CAS:1121-30-8 ***HR: 3***
1-HYDROXY-2-PYRIDINETHIONE
mf: C_5H_5NOS mw: 127.17

SYNS: 1-HYDROXY-2-(1H)-PYRIDINETHIONE ◇ OMADINE ◇ PTO ◇ PYRITHIONE ◇ SQ 2113

TOXICITY DATA with REFERENCE
orl-mus LD50:535 mg/kg TXAPA9 2,156,60
ipr-mus LD50:165 mg/kg TXAPA9 2,156,60
scu-mus LD50:450 mg/kg TXAPA9 2,156,60
ivn-mus LD50:340 mg/kg TXAPA9 2,156,60
orl-cat LD50:108 mg/kg TOANDB 3,1,79
orl-bwd LD50:100 mg/kg AECTCV 12,355,83

SAFETY PROFILE: Poison by ingestion, intraperitoneal, and intravenous routes. Moderately toxic by subcutaneous route. When heated to decomposition it emits very toxic fumes of NO_x and SO_x.

HOC000 CAS:15922-78-8 ***HR: 3***
1-HYDROXY-2-(1H)-PYRIDINETHIONE SODIUM SALT
mf: $C_5H_5NOS•Na$ mw: 150.16

SYNS: OMACIDE 24 ◇ SODIUM OMADINE ◇ SODIUM PYRIDINETHIONE ◇ SODIUM PYRITHIONE SQ 3277 ◇ SQ 3277

TOXICITY DATA with REFERENCE
orl-rat LDLo:745 mg/kg TXAPA9 2,156,60
ipr-rat LDLo:385 mg/kg TXAPA9 2,156,60
ipr-mus LD50:265 mg/kg TXAPA9 2,156,60
scu-mus LD50:428 mg/kg OYYAA2 8,1067,84
ivn-mus LD50:335 mg/kg TXAPA9 2,156,60
ivn-rbt LDLo:200 mg/kg TXAPA9 9,269,66

SAFETY PROFILE: Poison by intraperitoneal and intravenous routes. Moderately toxic by ingestion and subcutaneous routes. When heated to decomposition it emits very toxic fumes of Na_2O, NO_x, and SO_x.

HOC500 CAS:10182-82-8 ***HR: D***
β-(N-(3-HYDROXY-4-PYRIDONE))-α-AMINOPROPIONIC ACID
mf: $C_8H_{10}N_2O_4$ mw: 198.20

PROP: Amino acid extracted from *Leucaena leucocephala* (TJADAB 3,21,70).

SYNS: LECENINE ◇ LEUCAENINE ◇ LEUCAENOL ◇ LEUCENOL ◇ MIMOSINE

TOXICITY DATA with REFERENCE
dni-hmn:oth 200 umol/L TOXIA6 9,241,71
oth-hmn:oth 200 umol/L TOXIA6 9,241,71
orl-rat TDLo:22750 mg/kg (female 42D pre):REP BCPCA6 14,1167,65
orl-rat TDLo:5558 mg/kg (female 1-19D post):TER TJADAB 3,21,70

SAFETY PROFILE: An experimental teratogen. Experimental reproductive effects. Human mutation data reported. When heated to decomposition it emits toxic fumes of NO_x.

HOE000 CAS:1571-30-8 ***HR: 3***
8-HYDROXYQUINALDIC ACID
mf: $C_{10}H_7NO_3$ mw: 189.18

TOXICITY DATA with REFERENCE
imp-mus TDLo:160 mg/kg:NEO ANYAA9 108,924,63

SAFETY PROFILE: Questionable carcinogen with experimental neoplastigenic data. When heated to decomposition it emits toxic fumes of NO_x.

HOE500 CAS:59901-91-6 ***HR: 3***
1'-HYDROXYSAFROLE-2',3'-OXIDE
mf: $C_{10}H_{10}O_4$ mw: 194.20

SYN: α-EPOXYETHYL-1,3-BENZODIOXOLE-5-METHANOL

TOXICITY DATA with REFERENCE
mmo-sat 200 nmol/plate MUREAV 60,143,79
mma-sat 20 nmol/plate CRSBAW 171,1041,77
scu-rat TDLo:1554 mg/kg/10W-I:CAR CNREA8 43,1124,83
skn-mus TDLo:700 mg/kg/6W-I:NEO CNREA8 37,1883,77

CONSENSUS REPORTS: EPA Genetic Toxicology Program.

SAFETY PROFILE: Questionable carcinogen with experimental carcinogenic and neoplastigenic data. Mutation data reported. When heated to decomposition it emits acrid smoke and fumes.

HOE600 CAS:89-86-1 ***HR: D***
4-HYDROXYSALICYLIC ACID
mf: $C_7H_6O_4$ mw: 154.13

SYNS: BENZOIC ACID, 2,4-DIHYDROXY- (9CI) ◇ 4-CARBOXYRESORCINOL ◇ 2,4-DHBA ◇ 2,4-DIHYDROXYBENZOIC ACID ◇ p-HYDROXYSALICYLIC ACID ◇ β-RESORCINOLIC ACID ◇ β-RESORCYLIC ACID

TOXICITY DATA with REFERENCE
scu-rat TDLo:642 mg/kg (female 11D post):TER RCOCB8 38,209,82

CONSENSUS REPORTS: Reported in EPA TSCA Inventory.

SAFETY PROFILE: An experimental teratogen. When

heated to decomposition it emits acrid smoke and irritating fumes.

HOF000 CAS:26782-43-4 ***HR: 3***
HYDROXYSENKIRKINE
mf: $C_{19}H_{27}NO_7$ mw: 381.47

PROP: Isolated from the plant *Crotalaria laburnifolia*.

SYN: 8,12,18-TRIHYDROXY-4-METHYL-11,16-DIOXOSENECIONANIUM

TOXICITY DATA with REFERENCE
ipr-rat TDLo:300 mg/kg:ETA JNCIAM 49,665,72
orl-rat LDLo:200 mg/kg NATUAS 227,401,70

CONSENSUS REPORTS: IARC Cancer Review: Group 3 IMEMDT 7,56,87; Animal Inadequate Evidence IMEMDT 10,265,76.

SAFETY PROFILE: Poison by ingestion. Questionable carcinogen with experimental tumorigenic data. When heated to decomposition it emits toxic fumes of NO_x.

HOF500 ***HR: 3***
5-HYDROXY-1(N-SODIO-5-TETRAZOLYLAZO) TETRAZOLE
mf: $C_2HN_{10}NaO$ mw: 204.09

SYN: SODIUM-5(5'-HYDROXYTETRAZOL-3'-YLAZO)TETRAZOLIDE

SAFETY PROFILE: Highly explosive when heated. When heated to decomposition it emits very toxic fumes of NO_x and Na_2O.

HOG000 CAS:106-14-9 ***HR: 3***
12-HYDROXYSTEARIC ACID
mf: $C_{18}H_{36}O_3$ mw: 300.54

TOXICITY DATA with REFERENCE
scu-mus TDLo:160 mg/kg/40W-I:NEO CNREA8 30,1037,70

CONSENSUS REPORTS: Reported in EPA TSCA Inventory.

SAFETY PROFILE: Questionable carcinogen with experimental neoplastigenic data. When heated to decomposition it emits acrid smoke and fumes.

HOG500 CAS:141-23-1 ***HR: 3***
12-HYDROXYSTEARIC ACID, METHYL ESTER
mf: $C_{19}H_{38}O_3$ mw: 314.57

SYN: METHYL-12-HYDROXYSTEARATE

TOXICITY DATA with REFERENCE
scu-mus TDLo:1600 mg/kg/40W-I:ETA CNREA8 30,1037,70

CONSENSUS REPORTS: Reported in EPA TSCA Inventory.

SAFETY PROFILE: Questionable carcinogen with experimental tumorigenic data. When heated to decomposition it emits acrid smoke and fumes. See also ESTERS.

HOH000 CAS:65520-53-8 ***HR: D***
1-HYDROXY-2-(3-SULFOPROPOXY)ANTHRAQUINONE SODIUM SALT
mf: $C_{17}H_{13}O_7S \cdot Na$ mw: 384.35

TOXICITY DATA with REFERENCE
mmo-sat 500 μg/plate BCSTB5 5,1489,77
mma-sat 500 μg/plate BCSTB5 5,1489,77

SAFETY PROFILE: Mutation data reported. When heated to decomposition it emits toxic fumes of SO_x and Na_2O.

HOH500 CAS:79-57-2 ***HR: 3***
5-HYDROXYTETRACYCLINE
mf: $C_{22}H_{24}N_2O_9$ mw: 460.48

SYNS: BIOSTAT ◇ NCI-C56473 ◇ OTC ◇ OXITETRACYCLIN ◇ OXYMYKOIN ◇ OXYTERRACINE ◇ OXYTETRACYCLINE ◇ OXYTETRACYCLINE AMPHOTERIC ◇ RIOMITSIN ◇ RYOMYCIN ◇ TAOMYCIN ◇ TAOMYXIN ◇ TERRAFUNGINE ◇ TERRAMITSIN ◇ TERRAMYCIN ◇ TETRAN

TOXICITY DATA with REFERENCE
spm-rat-unr 80 mg/kg/8D JOURAA 112,348,74
unr-wmn TDLo:420 mg/kg (1-39W preg):REP JAMAAP 188,178,64
ivn-rat TDLo:800 mg/kg (female 10-14D post):TER AEMBAP 27,291,72
orl-man TDLo:114 mg/kg/4D:SKN,BLD JAMAAP 231,734,75
par-inf TDLo:136 mg/kg:MSK LANCAO 1,827,62
orl-rat LD50:4800 mg/kg 85ERAY 1,501,78
ivn-rat LD50:260 mg/kg 85ERAY 1,501,78
orl-mus LD50:2240 mg/kg ARZNAD 5,1,55
ipr-mus LD50:5706 mg/kg ANTBAL 20,793,75
scu-mus LD50:700 mg/kg 85GDA2 3,44,80
ivn-mus LD50:140 mg/kg ARZNAD 5,1,55
ivn-dog LDLo:220 mg/kg ANTCAO 3,1015,53
ipr-gpg LDLo:2250 mg/kg ANTBAL 20,793,75

CONSENSUS REPORTS: Reported in EPA TSCA Inventory.

SAFETY PROFILE: A poison by intravenous route. Moderately toxic by ingestion, subcutaneous, and intraperitoneal routes. Human systemic effects by ingestion: hemorrhage, dermatitis, and unspecified effects on teeth and supporting structures. Human reproductive effects by an unspecified route: abnormal postnatal measures or effects. Experimental teratogenic and reproductive effects. Mutation data reported. When heated to decomposition it emits toxic fumes of NO_x. See also TETRACYCLINE and various tetracycline derivatives.

HOI000 CAS:2058-46-0 ***HR: 3***
5-HYDROXYTETRACYCLINE HYDROCHLORIDE
mf: $C_{22}H_{24}N_2O_9{\cdot}ClH$ mw: 496.94

SYNS: BISOLVOMYCIN ◇ HYDROCYCLIN ◇ LIQUAMYCIN INJECTABLE ◇ NSC 9169 ◇ OTETRYN ◇ OXLOPAR ◇ OXYJECT 100 ◇ OXYTETRACYCLINE HYDROCHLORIDE ◇ TERAMYCIN HYDROCHLORIDE ◇ TETRAMINE ◇ TETRAN HYDROCHLORIDE

TOXICITY DATA with REFERENCE
dnd-bcs 10 μmol/L BIORAK 39,587,74
mnt-mus-orl 100 mg/kg/24H-I MUREAV 117,193,83
par-wmn TDLo:20 mg/kg (female 20W post):REP GOBIDS 26,177,88
orl-rat TDLo:11250 mg/kg (female 6-15D post):TER NTIS** PB83-182469
orl-rat TDLo:1802 g/kg/2Y:ETA NTPTR* NTP-TR-315,87
scu-rat LD50:800 mg/kg ARZNAD 9,711,59
ivn-rat LD50:302 mg/kg JPETAB 99,234,50
orl-mus LD50:6696 mg/kg ANYAA9 53,238,50
scu-mus LD50:963 mg/kg JPETAB 99,234,50
ivn-mus LD50:100 mg/kg 85FZAT -,490,67
ivn-dog LDLo:100 mg/kg ANYAA9 53,238,50
ivn-rbt LDLo:80 mg/kg JPETAB 99,234,50

CONSENSUS REPORTS: NTP Carcinogenesis Studies (feed); Equivocal Evidence: rat NTPTR* NTP-TR-315,87. NTP Carcinogenesis Studies (feed); No Evidence: mouse NTPTR* NTP-TR-315,87. Reported in EPA TSCA Inventory.

SAFETY PROFILE: Poison by intravenous route. Moderately toxic by subcutaneous route. Mildly toxic by ingestion. Experimental teratogenic and reproductive effects. Questionable carcinogen with experimental tumorigenic data. Mutation data reported. When heated to decomposition it emits very toxic fumes of HCl and NO_x. See also TETRACYCLINE and various tetracycline derivatives.

HOJ000 CAS:1012-82-4 ***HR: 3***
7-HYDROXYTHEOPHYLLINE
mf: $C_7H_8N_4O_3$ mw: 196.19

SYNS: 3,7-DIHYDRO-7-HYDROXY-1,3-DIMETHYL-1H-PURINE-2,6-DIONE (9CI) ◇ 7-HYDROXYTHEOPHYLLIN (GERMAN)

TOXICITY DATA with REFERENCE
scu-mus TDLo:700 mg/kg/24W-I:ETA ARZFAN 21,356,71

SAFETY PROFILE: Questionable carcinogen with experimental tumorigenic data. When heated to decomposition it emits toxic fumes of NO_x. See also various theophylline entries.

HOK000 CAS:81-48-1 ***HR: 2***
1-HYDROXY-4-(p-TOLUIDINO)ANTHRAQUINONE
mf: $C_{21}H_{15}NO_3$ mw: 329.37

SYNS: AHCOQUINONE BLUE IR BASE ◇ ALIZARINE VIOLET 3B BASE ◇ C.I. SOLVENT VIOLET 13 ◇ D+C VIOLET No. 2 ◇ N-(4-HYDROXY-1-ANTHRAQUINONYL)-4-METHYLANILINE ◇ N-(4-HYDROXY-1-ANTHRAQUINONYL)-p-TOLUIDINE ◇ IRISOL BASE ◇ OIL VIOLET IRS ◇ N-(p-TOLYL)-4-HYDROXY-1-ANTHRAQUINONYLAMINE ◇ 11092 VIOLET ◇ WAXOLINE PURPLE A

TOXICITY DATA with REFERENCE
mmo-sat 25 μg/plate NTIS** AD-A142-106
ipr-mus LDLo:512 mg/kg CBCCT* 2,59,50

CONSENSUS REPORTS: Reported in EPA TSCA Inventory.

SAFETY PROFILE: Moderately toxic by intraperitoneal route. Mutation data reported. When heated to decomposition it emits toxic fumes of NO_x.

HOL000 CAS:64050-03-9 ***HR: 3***
(3-HYDROXY-p-TOLYL)TRIMETHYLAMMONIUM CHLORIDE,METHYLCARBAMATE
mf: $C_{12}H_{19}N_2O_2{\cdot}Cl$ mw: 258.78

SYN: METHYLCARBAMICACID-5-(TRIMETHYLAMMONIO)-o-TOLYL ESTER, CHLORIDE

TOXICITY DATA with REFERENCE
orl-rat LD50:2500 μg/kg NTIS** PB158-508
ipr-rat LD50:78 μg/kg NTIS** PB158-508
scu-rat LD50:100 μg/kg NTIS** PB158-508
ipr-mus LD50:88 μg/kg NTIS** PB158-508
scu-mus LD50:64 μg/kg NTIS** PB158-508
ivn-mus LD50:35 μg/kg NTIS** PB158-508
scu-dog LDLo:2 mg/kg NTIS** PB158-508
imp-mky LDLo:50 μg/kg NTIS** PB158-508

SAFETY PROFILE: A deadly poison by ingestion, intraperitoneal, subcutaneous, intravenous, and implantation routes. When heated to decomposition it emits very toxic fumes of NO_x and Cl^-. See also CARBAMATES and ESTERS.

HOM259 CAS:13074-00-5 ***HR: D***
17-β-HYDROXY-4,4,17-α-TRIMETHYL-ANDROST-5-ENE(2,3-d)ISOXAZOLE
mf: $C_{23}H_{33}NO_2$ mw: 355.57

SYNS: ANDROSTA-2,5-DIENO(2,3-d)ISOXAZOL-17-OL,4,4,17-TRIMETHYL-, (17-β)-(9CI) ◇ AZASTENE ◇ ISOXAZOL ◇ 4,4,17-α-TRIMETHYLANDROST-5-ENO(2,3-d)ISOXAZOL-17-OL ◇ WIN 17625

TOXICITY DATA with REFERENCE
orl-rat TDLo:9 mg/kg (female 10D post):TER AJOGAH 133,176,79
orl-mky TDLo:500 mg/kg (female 26-30D post):REP FESTAS 30,343,78

SAFETY PROFILE: Experimental teratogenic and reproductive effects. A steroid. When heated to decomposition it emits toxic fumes of NO_x.

HON000 CAS:76-87-9 ***HR: 3***
HYDROXYTRIPHENYLSTANNANE
mf: $C_{18}H_{16}OSn$ mw: 367.03

PROP: Mp: 122°.

SYNS: DOWCO 186 ◇ DU-TER ◇ ENT 28,009 ◇ FENOLOVO ◇ FINTINE HYDROXYDE (FRENCH) ◇ FINTIN HYDROXID (GERMAN) ◇ FENTIN HYDROXIDE ◇ FINTIN HYDROXYDE (DUTCH) ◇ FINTIN IDROSSIDO (ITALIAN) ◇ HAITIN ◇ HYDROXYDE de TRIPHENYLETAIN (FRENCH) ◇ HYDROXYTRIPHENYLTIN ◇ IDROSSIDO DI STAGNO TRIFENILE (ITALIAN) ◇ NCI-C00260 ◇ SUZU H ◇ TPTH ◇ TRIFENYL-TINHYDROXYDE (DUTCH) ◇ TRIPHENYLTIN HYDROXIDE (USDA) ◇ TRIPHENYLTIN OXIDE ◇ TRIPHENYL-ZINNHYDROXID (GERMAN) ◇ TUBOTIN ◇ VANCIDE KS

TOXICITY DATA with REFERENCE
eye-rbt 10 mg CTOXAO 13,281,78
eye-gpg 100%/2S rns SEV TXAPA9 14,628,69
otr-rat:emb 19 ng/plate JJATDK 1,190,81
orl-rat TDLo:105 mg/kg (8-14D preg):TER CTOXAO 13,281,78
orl-rat TDLo:140 mg/kg (1-7D preg):REP TOANDB 3,281,79
orl-rat LD50:46 mg/kg NCILB* NIH-NCI-E-C-72-3252
ipr-rat LDLo:100 mg/kg NCNSA6 5,46,53
orl-mus LD50:209 mg/kg YKYUA6 30,505,79
ipr-mus LDLo:8500 μg/kg TXAPA9 23,288,72

CONSENSUS REPORTS: NCI Carcinogenesis Bioassay (feed); No Evidence: mouse, rat NCITR* NCI-CG-TR-139,78. Reported in EPA TSCA Inventory.

OSHA PEL: TWA 0.1 mg(Sn)/m^3 (skin)
ACGIH TLV: TWA 0.1 mg(Sn)/m^3 (skin) (Proposed: TWA 0.1 mg(Sn)/m^3; STEL 0.2 mg(Sn)/m^3 (skin))
NIOSH REL: (Organotin Compounds) TWA 0.1 mg(Sn)/m^3

SAFETY PROFILE: A poison by ingestion and intraperitoneal routes. Moderately toxic by an unspecified route. A severe eye irritant. Experimental teratogenic and reproductive effects. Mutation data reported. When heated to decomposition it emits acrid smoke and fumes. See also TIN COMPOUNDS.

HON500 ***HR: 3***
3-HYDROXYTROPOLONE
mf: $C_7H_6O_3$ mw: 138.13

SYN: 2,3-DIHYDROXY-2,4,6-CYCLOHEPTATRIEN-1-ONE

TOXICITY DATA with REFERENCE
ipr-mus LD50:345 mg/kg YKKZAJ 91,550,71
scu-mus LD50:520 mg/kg YKKZAJ 91,550,71
ivn-mus LD50:177 mg/kg YKKZAJ 91,550,71

SAFETY PROFILE: Poison by intravenous and intraperitoneal routes. Moderately toxic by subcutaneous route. When heated to decomposition it emits acrid smoke and fumes.

HON800 CAS:114-03-4 ***HR: 3***
dl-HYDROXYTRYPTOPHAN
mf: $C_{11}H_{12}N_2O_3$ mw: 220.25

SYNS: (±)-5-HYDROXYTRYPTOPHAN ◇ 5-HYDROXYTRYPTOPHAN

TOXICITY DATA with REFERENCE
ipr-rat TDLo:100 mg/kg (female 21D post):TER JPBAA7 94,113,67
ipr-mus LD50:1080 mg/kg NYKZAU 69,523,73

CONSENSUS REPORTS: Reported in EPA TSCA Inventory.

SAFETY PROFILE: Poison by intraperitoneal route. An experimental teratogen. When heated to decomposition it emits toxic fumes of NO_x.

HOO000 CAS:4350-09-8 ***HR: 3***
5-HYDROXY-l-TRYPTOPHAN
mf: $C_{11}H_{12}N_2O_3$ mw: 220.25

SYNS: l-5-HTP ◇ l-5-HYDROXYTRYPTOPHAN

TOXICITY DATA with REFERENCE
orl-mus TDLo:2100 mg/kg (female 7-13D post):TER IYKEDH 6,356,75
orl-mus TDLo:2100 mg/kg (female 7-13D post):REP IYKEDH 6,356,75
orl-rat LD50:243 mg/kg NYKZAU 69,523,73
ipr-rat LD50:91 mg/kg NYKZAU 69,523,73
scu-rat LD50:149 mg/kg IYKEDH 6,307,75
ivn-rat LD50:27 mg/kg IYKEDH 6,307,75
orl-mus LD50:1708 mg/kg IYKEDH 6,307,75
ipr-mus LD50:298 mg/kg IYKEDH 6,307,75
scu-mus LD50:418 mg/kg IYKEDH 6,307,75
orl-rbt LD50:285 mg/kg IYKEDH 6,307,75

CONSENSUS REPORTS: Reported in EPA TSCA Inventory.

SAFETY PROFILE: Poison by ingestion, subcutaneous, intravenous, and intraperitoneal routes. Experimental teratogenic and reproductive effects. When heated to decomposition it emits toxic fumes of NO_x.

HOO100 CAS:56-69-9 ***HR: 3***
5-HYDROXYTRYPTOPHANE
mf: $C_{11}H_{12}N_2O_3$ mw: 220.25

PROP: dl Form: minute rods or needles, decomp @ 298-300°; l form: crystals; d form: crystals.

SYNS: 5-HTP ◇ HYDROXYTRYPTOPHAN ◇ 5-HYDROXYTRYPTOPHAN ◇ NCI-C56644 ◇ USAF CB-96

TOXICITY DATA with REFERENCE
ipr-rat TDLo:30 mg/kg (1D male):REP ANENAG 24(Suppl 3),1,63
ipr-mus LD50:200 mg/kg NTIS** AD277-689

CONSENSUS REPORTS: Reported in EPA TSCA Inventory.

SAFETY PROFILE: Poison by intraperitoneal route. Experimental reproductive effects. When heated to decomposition it emits toxic fumes of NO_x.

HOO500 CAS:127-07-1 ***HR: 3***
HYDROXYUREA
mf: $CH_4N_2O_2$ mw: 76.07

PROP: Needles from ethanol. Mp: 133-136°, bp: decomp. Very sol in water; sol in hot alc.

SYNS: N-(AMINOCARBONYL)HYDROXYLAMINE ◇ BIOSUPRESSIN ◇ CARBAMOHYDROXAMIC ACID ◇ CARBAMOHYDROXIMIC ACID ◇ CARBAMOHYDROXYAMIC ACID ◇ CARBAMOYL OXIME ◇ N-CARBAMOYLHYDROXYLAMINE ◇ CARBAMYL HYDROXAMATE ◇ HIDRIX ◇ HYDREA ◇ HYDROXYCARBAMINE ◇ HYDROXYLUREA ◇ N-HYDROXYUREA ◇ HYDURA ◇ LITALER ◇ NCI-C04831 ◇ NSC 32065 ◇ ONCO-CARBIDE ◇ OXYUREA ◇ SK 22591 ◇ SQ 1089

TOXICITY DATA with REFERENCE
skn-hmn 218 mg/3W CTRRDO 63,619,79
mmo-smc 7600 mg/L MUREAV 160,19,86
dni-hmn-orl 50 mg/kg/10H-C PAACA3 24,262,83
ipr-rat TDLo:200 mg/kg (female 9-12D post):REP SEIJBO 25,23,85
ipr-rat TDLo:1 g/kg (female 12D post):TER TJADAB 26,71,82
ipr-rat TDLo:2500 mg/kg/7W-I:ETA CANCAR 40(Suppl 4),1935,77
orl-hmn TDLo:80 mg/kg/D:BLD CNCRA6 29,103,63
ivn-hmn TDLo:86 mg/kg:BLD,GIT CCROBU 57,369,73
ipr-mus LD50:5800 mg/kg CNREA8 39,3575,79
scu-mus LD10:2400 mg/kg EJCAAH 10,667,74

CONSENSUS REPORTS: NCI Carcinogenesis Studies (ipr); No Evidence: mouse CANCAR 40,1935,77. NCI Carcinogenesis Studies (ipr); Equivocal Evidence: rat CANCAR 40,1935,77. EPA Genetic Toxicology Program.

SAFETY PROFILE: Moderately toxic by subcutaneous routes. Human systemic effects by ingestion and intravenous routes: nausea or vomiting, microcytosis (smaller than normal red blood cells), normocytic anemia (reduced red blood cell count), leukopenia (reduced white blood cell count), thrombocytopenia (decrease in the number of blood platelets), and other blood effects. Experimental teratogenic and reproductive effects. Human mutation data reported. Questionable carcinogen with experimental tumorigenic data. When heated to decomposition it emits toxic fumes of NO_x.

HOO875 CAS:22151-75-3 ***HR: 2***
3-HYDROXYURIC ACID
mf: $C_5H_4N_4O_4$ mw: 184.13

TOXICITY DATA with REFERENCE
scu-rat TDLo:440 mg/kg/22W-I:ETA BICHAW 10,4463,71

SAFETY PROFILE: Questionable carcinogen with experimental tumorigenic data. When heated to decomposition it emits toxic fumes of NO_x.

HOP000 CAS:13479-29-3 ***HR: 3***
3-HYDROXYXANTHINE
mf: $C_5H_4N_4O_3$ mw: 168.13

SYNS: 3,7-DIHYDRO-3-HYDROXY-1H-PURINE-2,6-DIONE ◇ XANTHINE-x-N-OXIDE ◇ XANTHINE-3-N-OXIDE

TOXICITY DATA with REFERENCE
dnd-rat-ipr 50 mg/kg CBINA8 10,19,75
ipr-rat TDLo:500 mg/kg (19-22D preg):CAR,TER JJIND8 61,1405,78
ipr-rat TDLo:96 mg/kg/8W-I:CAR CNREA8 38,2038,78
scu-rat TDLo:9600 μg/kg/8W-I:NEO CNREA8 38,2038,78
scu-rat TDLo:500 mg/kg (19-22D preg):NEO,TER JJIND8 61,1411,78
ipr-rat LD50:100 mg/kg ARPAAQ 86,395,68
scu-rat LDLo:1540 mg/kg CNREA8 30,179,70

SAFETY PROFILE: A poison by intraperitoneal route. Moderately toxic by subcutaneous route. Experimental teratogenic effects. Questionable carcinogen with experimental carcinogenic and neoplastigenic data. Mutation data reported. When heated to decomposition it emits toxic fumes of NO_x. See also other hydroxyxanthine entries.

HOP259 CAS:16870-90-9 ***HR: 3***
7-HYDROXYXANTHINE
mf: $C_5H_4N_4O_3$ mw: 168.13

SYN: XANTHINE-7-N-OXIDE

TOXICITY DATA with REFERENCE
ipr-rat TDLo:96 mg/kg/8W-I:CAR CNREA8 38,2038,78
scu-rat TDLo:96 mg/kg/8W-I:NEO CNREA8 38,2038,78
ipr-rat LD50:200 mg/kg ADTEAS 3,181,68

SAFETY PROFILE: Poison by intraperitoneal route. Questionable carcinogen with experimental carcinogenic and neoplastigenic data. When heated to decomposition it emits toxic fumes of NO_x. See also other hydroxyxanthine entries.

HOQ000 CAS:64038-48-8 ***HR: 3***
1-HYDROXYXANTHINE DIHYDRATE
mf: $C_5H_4N_4O_3$•$2H_2O$ mw: 204.17

TOXICITY DATA with REFERENCE
scu-rat TDLo:1040 mg/kg/26W-I:NEO CNREA8 30,184,70

SAFETY PROFILE: Questionable carcinogen with experimental neoplastigenic data. When heated to decomposition it emits toxic fumes of NO_x. See also other hydroxyxanthine entries.

HOQ500 CAS:64038-49-9 ***HR: 3***
3-HYDROXYXANTHINE HYDRATE
mf: $C_5H_4N_4O_3$•H_2O mw: 186.15

SYN: 3-HYDROXYXANTHINE

TOXICITY DATA with REFERENCE
scu-rat TDLo:60 mg/kg/24W-I:NEO CNREA8 30,184,70

SAFETY PROFILE: Questionable carcinogen with experimental neoplastigenic data. When heated to decomposition it emits toxic fumes of NO_x. See also other hydroxyxanthine entries.

HOR500 CAS:5978-92-7 ***HR: 3***
HYDROXYZINE PAMOATE
mf: $C_{44}H_{41}ClN_2O_7$ mw: 745.32

PROP: Crystals, insol in H_2O.

SYN: 4,4'-METHYLENEBIS(3-HYDROXY-2-NAPTHOIC ACID ESTER with 2-(2-(4-(p-CHLORO-α-PHENYLBENZYL)-1-PIPERAZINYL)-ETHOXY)ETHANOL

TOXICITY DATA with REFERENCE
orl-rat TDLo:600 mg/kg (10-15D preg):TER AJOGAH 95,109,66
orl-rat LD50:1740 mg/kg NIIRDN 6,621,82
orl-mus LD50:1840 mg/kg NIIRDN 6,621,82
ipr-mus LD50:360 mg/kg NIIRDN 6,621,82

SAFETY PROFILE: A poison by intraperitoneal route. Moderately toxic by ingestion. Experimental teratogenic effects. When heated to decomposition it emits very toxic fumes of Cl^- and NO_x.

HOS500 CAS:71767-91-4 ***HR: 3***
HYGROSTATIN

PROP: An antibiotic produced by *Streptomyces hygrostaticus* (85ERAY 2,1123,78).

TOXICITY DATA with REFERENCE
orl-mus LD50:530 mg/kg 85ERAY 2,1123,78
ipr-mus LD50:22 mg/kg 85ERAY 2,1123,78
scu-mus LD50:247 mg/kg 85ERAY 2,1123,78
ivn-mus LD50:9 mg/kg 85ERAY 2,1123,78

SAFETY PROFILE: Poison by intraperitoneal, subcutaneous, and intravenous routes. Moderately toxic by ingestion. When heated to decomposition it emits acrid smoke and fumes.

HOT000 CAS:57074-51-8 ***HR: D***
HYMENOVIN
mf: $C_{15}H_{22}O_5$ mw: 282.37

TOXICITY DATA with REFERENCE
mmo-sat 350 μmol/plate TXAPA9 45,629,78
mma-sat 3 μg/plate FCTXAV 15,225,77

SAFETY PROFILE: Mutation data reported. When heated to decomposition it emits acrid smoke and fumes.

HOT200 CAS:57377-32-9 ***HR: 3***
HYMENOXON
mf: $C_{15}H_{22}O_5$ mw: 282.37

SYN: HYMENOXONE

TOXICITY DATA with REFERENCE
mmo-sat 50 μg/plate TOLED5 9,395,81
dnr-bcs 1 mg/plate RCOCB8 34,161,81
orl-mus LD50:241 mg/kg TOLED5 9,395,81
ipr-mus LD50:16240 μg/kg RCOCB8 28,189,80
ivn-dog LD50:30 mg/kg RCOCB8 8,381,74
orl-dom LD50:75 mg/kg RCOCB8 31,181,81

SAFETY PROFILE: Poison by ingestion, intravenous, and intraperitoneal routes. Mutation data reported. When heated to decomposition it emits acrid smoke and fumes.

HOT500 CAS:114-49-8 ***HR: 3***
HYOSCINE HYDROBROMIDE
mf: $C_{17}H_{21}NO_4$•BrH mw: 384.31

SYNS: BELDAVRIN ◇ EUSCOPOL ◇ HYDROSCINE HYDROBROMIDE ◇ HYOSCINE BROMIDE ◇ HYOSCINE F HYDROBROMIDE ◇ (−)-HYOSCINE HYDROBROMIDE ◇ 1-HYOSCINE HYDROBROMIDE ◇ HYOSCYINE HYDROBROMIDE ◇ HYSCO ◇ ISOSCOPIL ◇ KWELLS ◇ SCOPAMIN ◇ SCOPOLAMINE BROMIDE ◇ (−)-SCOPOLAMINE BROMIDE ◇ SCOPOLAMINE HYDROBROMIDE ◇ (−)-SCOPOLAMINE HYDROBROMIDE ◇ SCOPOLAMINIUM BROMIDE ◇ SCOPOLAMMONIUM BROMIDE ◇ SCOPOS ◇ SEREEN ◇ TRIPTONE

TOXICITY DATA with REFERENCE
cyt-hmn:hla 1 pph/5H HUMAA7 4,371,67
cyt-hmn:hla 1 pph HUMAA7 4,371,67
scu-rat TDLo:500 μg/kg (1D male):REP PSYPAG 10,44,66
orl-rbt TDLo:2365 mg/kg (10-14D preg):TER AJBSAM 35,173,82
orl-rat LD50:1270 mg/kg AIPTAK 180,155,69
scu-rat LD50:3800 mg/kg AIPTAK 180,155,69
ims-rat LD50:486 μg/kg BJPCBM 39,822,70
idu-rat LD50:670 mg/kg AIPTAK 180,155,69
orl-mus LD50:1880 mg/kg ARZNAD 18,1132,68

ipr-mus LD50:650 mg/kg CLDND* 79,127,43
ims-mus LD50:974 μg/kg BJPCBM 39,822,70
ivn-cat LDLo:80 mg/kg AEPPAE 120,189,27
ivn-rbt LDLo:100 mg/kg AEPPAE 120,189,27

CONSENSUS REPORTS: Reported in EPA TSCA Inventory. EPA Genetic Toxicology Program.

SAFETY PROFILE: A poison by intravenous and intramuscular routes. Moderately toxic by ingestion, subcutaneous, intraduodenal, and intraperitoneal routes. Experimental teratogenic and reproductive effects. Human mutation data reported. When heated to decomposition it emits very toxic fumes of NO_x and HBr. See also SCOPOLAMINE.

HOU000 CAS:101-31-5 ***HR: 3***
(−)-HYOSCYAMINE
mf: $C_{17}H_{23}NO_3$ mw: 289.41

PROP: Mp: 106-108°. Very sol in alc, dil acids. White, crystalline alkaloid.

SYNS: (−)-ATROPINE ◇ DATURINE ◇ HYOSCYAMINE ◇ 1-HYOSCYAMINE ◇ (−)-TROPIC ACID ESTER with TROPINE

TOXICITY DATA with REFERENCE
unr-man LDLo:1471 μg/kg 85DCAI 2,73,70
ivn-mus LD50:95 mg/kg BJPCAL 24,138,65

CONSENSUS REPORTS: Reported in EPA TSCA Inventory.

SAFETY PROFILE: A deadly human poison by an unspecified route. An experimental poison by intravenous route. This is one of the atropine alkaloids and is very toxic, acting very much like atropine. It has the same effect on the central nervous system but twice the effect on the peripheral nerves. The symptoms of poisoning are dryness of the throat and mouth, marked difficulty in swallowing, and a sensation of burning and thirst. The vision becomes impaired through dilation and loss of accommodation, and the eyes present a rather prominent, brilliant, staring appearance. The voice is husky and the tongue is red. When heated to decomposition it emits highly toxic fumes of NO_x.

HOU059 CAS:1977-11-3 ***HR: 3***
HYPNODIN
mf: $C_{19}H_{21}N_3$ mw: 291.40

PROP: Yellow, prismatic crystals from acetone-petrol ether. Mp: 136-138°.

SYNS: AW-14'2333 ◇ HF-2333 ◇ 6-(4-METHYL-1-PIPERAZINYL)-11H-DIBENZ(b,e)AZEPINE ◇ 6-(4-METHYL-1-PIPERAZINYL)MORPHANTHRIDINE ◇ PERLAPINE

TOXICITY DATA with REFERENCE
orl-rat LD50:660 mg/kg IYKEDH 4,193,73
scu-rat LD50:420 mg/kg IYKEDH 4,193,73
ivn-rat LD50:60 mg/kg IYKEDH 4,193,73
orl-mus LD50:270 mg/kg NIIRDN 6,770,82
scu-mus LD50:250 mg/kg IYKEDH 4,193,73
ivn-mus LD50:61 mg/kg IYKEDH 4,193,73

SAFETY PROFILE: Poison by ingestion, subcutaneous, and intravenous routes. A hypnotic agent. When heated to decomposition it emits toxic fumes of NO_x.

HOU100 CAS:33125-97-2 ***HR: 3***
HYPNOMIDATE
mf: $C_{14}H_{16}N_2O_2{\cdot}H_2O_4S$ mw: 342.40

PROP: Crystals from diisopropyl ether. Mp: 67°. Solubility in water at 25°: 0.0045 mg/100 mL. Sol in chloroform, methanol, ethanol, propylene glycol, acetone.

SYNS: AMIDATE ◇ R-(+)-ETHYL-1-(1-PHENYLETHYL)-1H-IMIDAZOLE-5-CARBOXYLATE SULFATE ◇ ETOMIDATE ◇ (R)-(+)-1-(α-METHYLBENZYL)IMIDAZOLE-5-CARBOXYLIC ACID ETHYL ESTER ◇ 1-(α-METHYLBENZYL)-1H-IMIDAZOLE-5-CARBOXYLIC ACID ETHYL ESTER SULFATE ◇ 1-(1-PHENYLETHYL)-1H-IMIDAZOLE-5-CARBOXYLIC ACID ETHYL ESTER ◇ R 16659

TOXICITY DATA with REFERENCE
ivn-hmn TDLo:300 μg/kg:PUL,CNS AACRAT 55,730,76
ivn-rat LD50:14800 μg/kg AIPTAK 214,92,75
ivn-mus LD50:29500 μg/kg AIPTAK 214,92,75

SAFETY PROFILE: A deadly human poison by intravenous route. Human systemic effects by intravenous route: somnolence, muscle spasms and respiratory effects. A hypnotic agent. When heated to decomposition it emits toxic fumes of SO_x and NO_x. See also ESTERS and SULFATES.

HOU500 ***HR: 3***
HYPOCHLORITES

PROP: Salts of hypochlorous acid.

SAFETY PROFILE: Toxic by ingestion and inhalation. Powerful irritants to the skin, eyes, and mucous membranes. Flammable by chemical reaction with reducing agents. These are powerful oxidizers particularly at higher temperatures, when chlorine and then oxygen are evolved, or in the presence of moisture or carbon dioxide. With urea, it forms the highly explosive NCl_3. Dangerous; when heated or on contact with acid or acid fumes, they emit highly toxic fumes of Cl^-. React with water or steam to produce toxic and corrosive fumes of Cl^- and HCl. See also HYPOCHLOROUS ACID and CALCIUM SALT for more reactivity information.

HOV000 CAS:7790-92-3 ***HR: 3***
HYPOCHLOROUS ACID
mf: ClHO mw: 52.46

PROP: Greenish-yellow liquid (aqueous soln). Decomp to Cl_2, O_2 and $HClO_4$; very weak acid. Strong oxidizing agent. Protect from light.

TOXICITY DATA with REFERENCE
orl-rat TDLo:960 mg/kg (female 76D pre):TER
JJATDK 2,156,82

CONSENSUS REPORTS: Reported in EPA TSCA Inventory.

SAFETY PROFILE: An experimental teratogen. Explodes on contact with ammonia. Ignites on contact with arsenic. Mixture with acetic anhydride is a sensitive explosive. Incompatible with alcohols. When heated to decomposition it emits toxic fumes of Cl^-. See also HYPOCHLORITES.

HOV500 CAS:7778-54-3 ***HR: 3***
HYPOCHLOROUS ACID, CALCIUM SALT
DOT: UN 1748/UN 2208
mf: Cl_2O_2•Ca mw: 142.98

PROP: White powder. Compound contains 39% or less available chlorine (FEREAC 41,15972,76).

SYNS: B-K POWDER ◇ BLEACHING POWDER ◇ BLEACHING POWDER, containing 39% or less chlorine (DOT) ◇ CALCIUM CHLOROHYDROCHLORITE ◇ CALCIUM HYPOCHLORIDE ◇ CALCIUM HYPOCHLORITE ◇ CALCIUM OXYCHLORIDE ◇ CAPORIT ◇ CCH ◇ CHLORIDE of LIME (DOT) ◇ CHLORINATED LIME (DOT) ◇ HTH ◇ HY-CHLOR ◇ LIME CHLORIDE ◇ LO-BAX ◇ LOSANTIN ◇ PERCHLORON ◇ PITTCHLOR ◇ PITTCIDE ◇ PITTCLOR ◇ SENTRY

TOXICITY DATA with REFERENCE
mma-sat 1 mg/plate FCTOD7 22,623,84
cyt-ham:fbr 4 g/L FCTOD7 22,623,84
orl-rat LD50:850 mg/kg PESTC* 9,21,80

CONSENSUS REPORTS: Reported in EPA TSCA Inventory.

DOT Classification: ORM-C; Label: None (UN2208); Oxidizer; Label: Oxidizer (UN1748, UN2208).

SAFETY PROFILE: Moderately toxic by ingestion. Can cause severe irritation of skin and mucous membranes and emit fumes capable of causing pulmonary edema. Mutation data reported. A powerful oxidizer.

The bulk material may ignite or explode in storage. Traces of water may initiate the reaction. A rapid exothermic decomposition above 175°C releases oxygen and chlorine. Moderately explosive in its solid form when heated. Explosive reaction with acetic acid + potassium cyanide; amines; ammonium chloride; carbon or charcoal + heat; carbon tetrachloride + heat; N,N-dichloromethylamine + heat; ethanol; methanol; iron oxide; rust; 1-propanethiol; isobutanethiol; turpentine. Potentially explosive reaction with sodium hydrogen sulfate + starch + sodium carbonate. Reaction with acetylene or nitrogenous bases forms explosive products.

Ignites on contact with algacide; hydroxy compounds (e.g., glycerol; diethylene glycol monomethyl ether; phenol); organic sulfur compounds. Violent reaction with organic matter (above 100°C); sulfur. Vigorous reaction with nitromethane; reducing materials. Flammable by chemical reaction with combustible materials, i.e., anthracene; grease; oil; mercaptans; methyl carbitol; nitromethane; organic matter; propylmercaptan.

Deflagration occurs in contact with combustible substances. Dangerous; when heated to decomposition or on contact with acid or acid fumes, it emits highly toxic fumes of HCl and explodes. Reacts with water or steam to produce toxic and corrosive fumes or Cl^- and HCl.

HOW000 ***HR: 3***
HYPOCHLOROUS ACID, CALCIUM SALT (dry mixture)
mf: $Cl_2H_2O_2$•Ca mw: 145.00

PROP: Compound contains more than 39% available chlorine (FEREAC 41,15972,76).

SYN: CALCIUM HYPOCHLORITE MIXTURE, dry (DOT)

CONSENSUS REPORTS: Reported in EPA TSCA Inventory.

DOT Classification: Label: Oxidizer.

SAFETY PROFILE: A powerful irritant and oxidizer. Potentially explosive. When heated to decomposition or in reaction with water or acids it emits toxic fumes of Cl^-. See also HYPOCHLOROUS ACID, CALCIUM SALT.

HOW100 CAS:502-37-4 ***HR: 3***
HYPOGLYCINE B
mf: $C_{12}H_{18}N_2O_5$ mw: 270.32

SYNS: ALANINE, N-l-Γ-GLUTAMYL-3-(METHYLENECYCLOPROPYL)- ◇ GLUTAMINE, N-(1-CARBOXY-2-(METHYLENECYCLOPROPYL)ETHYL)- (7CI) ◇ HYPOGLYCIN B

TOXICITY DATA with REFERENCE
iut-rat TDLo:500 μg/kg (female 14D post):REP
TJADAB 4,246,71
iut-rat TDLo:500 μg/kg (female 14D post):TER
TJADAB 4,246,71
ivn-rat LDLo:200 mg/kg JPETAB 121,272,57
ivn-mus LDLo:160 mg/kg JPETAB 121,272,57
ivn-rbt LDLo:25 mg/kg JPETAB 121,272,57

SAFETY PROFILE: Poison by intravenous route. An experimental teratogen. Other experimental reproductive effects. When heated to decomposition it emits toxic fumes of NO_x.

HOW500 CAS:14448-38-5 ***HR: 3***
HYPONITROUS ACID

SYN: N-NITROSOHYDROXYLAMINE
mf: $H_2N_2O_2$ mw: 62.03

PROP: Solid.

DOT Classification: Forbidden.

SAFETY PROFILE: Many N-nitroso compounds are carcinogens. Incompatible with potassium hydroxide. When heated to decomposition it emits toxic fumes of NO_x. See also N-NITROSO COMPOUNDS.

HOX000 CAS:8006-83-5 ***HR: 2***
HYSSOP OIL

PROP: Consists of 50% pinene, small quantities of aromatic alcohol and some sesquiterpenes (FCTXAV 16,637,78).

TOXICITY DATA with REFERENCE
orl-mus LD50:1400 mg/kg FCTXAV 16,783,78
skn-rbt LD50:5000 mg/kg FCTXAV 16,783,78

CONSENSUS REPORTS: Reported in EPA TSCA Inventory.

SAFETY PROFILE: Moderately toxic by ingestion. Mildly toxic by skin contact. When heated to decomposition it emits acrid smoke and fumes. See also individual components.

I

IAB000 CAS:64622-45-3 ***HR: 2***
IBUPROFEN PICONOL
mf: $C_{19}H_{23}NO_2$ mw: 297.43

SYNS: 2-(p-ISOBUTYLPHENYL)PROPIONIC ACID-2-PYRIDYLMETHYL ESTER ◇ α-METHYL-4-(2-METHYLPROPYL) BENZENEACETIC ACID-2-PYRIDINYLMETHYL ESTER ◇ PIMEPROFEN ◇ 2-PYRIDYLMETHYL-2-(p-(2-METHYLPROPYL)PHENYL)PROPIONATE

TOXICITY DATA with REFERENCE
skn-rbt 25 mg MOD OYYAA2 24,231,82
eye-rbt 25 mg OYYAA2 24,221,82
scu-rat TDLo:2800 mg/kg (17-22D preg/21D post):REP OYYAA2 24,21,82
scu-rbt TDLo:2275 mg/kg (female 6-18D post):TER OYYAA2 24,37,82
orl-rat LD50:1440 mg/kg OYYAA2 23,691,82
ipr-rat LD50:1410 mg/kg OYYAA2 23,691,82
scu-rat LD50:1400 mg/kg OYYAA2 23,691,82
orl-mus LD50:1980 mg/kg OYYAA2 23,691,82
ipr-mus LD50:709 mg/kg OYYAA2 23,691,82
scu-mus LD50:1250 mg/kg OYYAA2 23,691,82
scu-dog LD50:986 mg/kg OYYAA2 23,691,82

SAFETY PROFILE: Moderately toxic by ingestion, intraperitoneal, and subcutaneous routes. Experimental teratogenic and reproductive effects. A skin and eye irritant. When heated to decomposition it emits toxic fumes of NO_x. See also p-ISOBUTYLHYDRATROPIC ACID (IBUPROFEN).

IAB100 CAS:31121-93-4 ***HR: D***
IBUPROFEN SODIUM
mf: $C_{13}H_{17}O_2$•Na mw: 228.29

SYNS: BENZENEACETIC ACID, α-METHYL-4-(2-METHYLPROPYL)-, SODIUM SALT (9CI) ◇ HYDRATROPIC ACID, p-ISOBUTYL-, SODIUM SALT ◇ p-ISOBUTYLHYDRATROPIC ACID SODIUM SALT ◇ SODIUM IBUPROFEN

TOXICITY DATA with REFERENCE
orl-rbt TDLo:1 g/kg (female 1D pre):REP FESTAS 38,238,82

SAFETY PROFILE: Experimental reproductive effects. When heated to decomposition it emits acrid smoke and irritating fumes.

IAC000 CAS:553-68-4 ***HR: 3***
IBYLCAINE HYDROCHLORIDE
mf: $C_{13}H_{20}N_2O_2$•ClH mw: 272.81

SYNS: BUTETHAMINE HYDROCHLORIDE ◇ 2-ISOBUTYLAMINOETHANOL HYDROCHLORIDE ACID SALT, p-AMINOBENZOIC ACID ESTER ◇ 2-(ISOBUTYLAMINO)ETHYL-p-AMINOBENZOATE HYDROCHLORIDE ◇ MONOCAINE HYDROCHLORIDE

TOXICITY DATA with REFERENCE
ipr-rat LD50:182 mg/kg ANESAV 3,398,42
ivn-rat LD50:28 mg/kg ANESAV 3,398,42
ipr-mus LD50:169 mg/kg AIPTAK 115,483,58
scu-mus LD50:449 mg/kg ANESAV 3,398,42
ivn-mus LD50:36 mg/kg AIPTAK 115,483,58
ivn-rbt LDLo:30 mg/kg JPETAB 62,69,38
scu-gpg LDLo:215 mg/kg JPETAB 62,69,38

SAFETY PROFILE: A poison by intraperitoneal, intravenous, and subcutaneous routes. Used in veterinary medicine as local anesthetic. When heated to decomposition it emits very toxic fumes of Cl^- and NO_x.

IAD000 CAS:8029-68-3 ***HR: 3***
ICHTHAMMOL

PROP: Pale yellow or (usually) brownish-black, thick, viscous liquid. Bituminous odor; misc with H_2O, glycerol, propylene glycol, fats, oils, etc.; sltly sol in alc, ether.

SYNS: AMMONIUM BITHIOLICUM ◇ AMMONIUM ICHTHOSULFONATE ◇ AMMONIUM SULFOICHTHYOLATE ◇ AMSUBIT ◇ FUNGICHTHOL ◇ HIRATHIOL ◇ ICHDEN ◇ ICHTAMMON ◇ ICHTHADONE ◇ ICHTHALUM ◇ ICHTHAMMONIUM ◇ ICHTHIUM ◇ ICHTHOSAN ◇ ICHTHOSAURAN ◇ ICHTHOSULFOL ◇ ICHTHYMALL ◇ ICHTHYNAT ◇ ICHTHYOL ◇ ICHTHYOPON ◇ ICHTHYSALLE ◇ LEUKICHTHOL ◇ LITHOL ◇ PERICHTHOL ◇ PETROSULPHO ◇ PISCAROL ◇ PISCIOL ◇ SAUROL ◇ SUBITOL ◇ SULFOGENOL ◇ THILAVEN ◇ THIOLIN ◇ THIOZIN ◇ TRASULPHANE ◇ TUMENOL

TOXICITY DATA with REFERENCE
orl-wmn LDLo:125 mg/kg:CNS,PUL,GIT AMLTAS 37,248,57

CONSENSUS REPORTS: Reported in EPA TSCA Inventory.

SAFETY PROFILE: A human poison which causes these systemic effects: nausea, acute pulminary edema, coma. An FDA over-the-counter drug. When heated to decomposition it emits very toxic fumes of NH_3, NO_x, and SO_x.

IAD100 CAS:40666-04-4 ***HR: 3***
ICI 79,939
mf: $C_{22}H_{29}FO_6$ mw: 408.51

SYNS: 7-(2-(4-(4-FLUOROPHENOXY)-3-HYDROXY-1-BUTENYL)-3,5-DIHYDROXYCYCLOPENTYL)-(1-α-(Z),2-β-(1E,3S*),3-α,5-α)-5-HEPTENOIC ACID ◇ racemic-ICI 79,939

TOXICITY DATA with REFERENCE
ims-dom TDLo:267 ng/kg (female 20W post):REP JRPFA4 43,403,75
unr-rat LDLo:100 μg/kg PRGLBA 10,5,75

SAFETY PROFILE: Deadly poison by an unspecified route. Experimental reproductive effects. When heated to decomposition it emits toxic fumes of F^-.

IAE000 CAS:38915-28-5 ***HR: 3***
ICR 340
mf: $C_{20}H_{24}Cl_2N_4O \cdot 2ClH$ mw: 480.30

SYN: 7-CHLORO-10-(3-(N-(2-CHLOROETHYL)-N-ETHYL)AMINOPROPYLAMINO)-2-METHOXY-BENZO(B)(1,5)NAPHTHYRIDINE DIHYDROCHLORIDE

TOXICITY DATA with REFERENCE
mmo-sat 500 ng/plate MUREAV 136,185,84
msc-ham:ovr 1 μmol/L CNREA8 39,4875,79
ivn-mus TDLo:4800 μg/kg:NEO CNREA8 36,2423,76
ivn-mus LDLo:7 mg/kg CNREA8 36,2423,76

SAFETY PROFILE: Poison by intravenous route. Questionable carcinogen with experimental neoplastigenic data. Mutation data reported. When heated to decomposition it emits very toxic fumes of HCl and NO_x.

IAG300 CAS:33691-06-4 ***HR: 2***
ID-622
mf: $C_{18}H_{16}ClFN_2O_3S$ mw: 394.87

TOXICITY DATA with REFERENCE
orl-mus LD50:2700 mg/kg ARZNAD 25,534,75
ipr-mus LD50:1350 mg/kg ARZNAD 25,534,75
scu-mus LD50:5 g/kg ARZNAD 25,534,75

SAFETY PROFILE: Moderately toxic by ingestion and intraperitoneal routes. When heated to decomposition it emits toxic fumes of F^-, Cl^-, SO_x, and NO_x.

IAG600 CAS:23210-56-2 ***HR: 3***
IFENPRODIL
mf: $C_{21}H_{27}NO_2$ mw: 325.49

PROP: Mp: 114°.

SYNS: 4-BENZYL-α-(p-HYDROXYPHENOL)-β-METHYL-1-PIPERIDINEETHANOL ◇ α-(4-HYDROXYPHENYL)-β-METHYL-4-(PHENYLMETHYL)-1-PIPERIDINEETHANOL (9CI) ◇ RC 61-91

TOXICITY DATA with REFERENCE
orl-mus LD50:320 mg/kg ARZNAD 21,1992,71
orl-dog LD50:461 mg/kg OYYAA2 14,415,77
ivn-dog LD50:41 mg/kg OYYAA2 14,415,77

SAFETY PROFILE: Poison by ingestion and intravenous routes. When heated to decomposition it emits toxic fumes of NO_x. See also IFENPRODIL TARTRATE.

IAG625 CAS:23210-58-4 ***HR: 3***
IFENPRODIL TARTRATE
mf: $C_{42}H_{54}N_2O_4 \cdot C_4H_6O_6$ mw: 801.08

SYNS: 4-BENZYL-α-(p-HYDROXYPHENYL)-β-METHYL-1-PIPERIDINEETHANOL TARTRATE ◇ 4-BENZYL-α-(p-HYDROXYPHENYL)-β-METHYL-1-PIPERIDINE-ETHANOL-(L)-(+)-TARTRATE ◇ 2-(4-BENZYL-PIPERIDINO)-1-(4-HYDROXYPHENYL)-1-PROPANOL TARTRATE (2:1) ◇ IFENPRODIL TARTRATE (2:1) ◇ IFENPRODIL l-(+)-TARTRATE ◇ VADILEX

TOXICITY DATA with REFERENCE
orl-rat TDLo:70 mg/kg (female 9-15D post):REP OYYAA2 10,819,75
orl-rat TDLo:2160 mg/kg (female 17-22D post):TER OYYAA2 18,1071,79
orl-rat LD50:538 mg/kg OYYAA2 10,785,75
ipr-rat LD50:138 mg/kg OYYAA2 10,785,75
ivn-rat LD50:50 mg/kg NIIRDN 6,79,82
ims-rat LD50:240 mg/kg OYYAA2 10,785,75
orl-mus LD50:17 mg/kg USXXAM #3509164
ipr-mus LD50:120 mg/kg USXXAM #3509164
ivn-mus LD50:44 mg/kg OYYAA2 10,785,75
ims-mus LD50:222 mg/kg OYYAA2 10,785,75
orl-dog LD50:461 mg/kg OYYAA2 14,415,77
ivn-dog LD50:34200 μg/kg OYYAA2 10,785,75
orl-rbt LD50:500 mg/kg OYYAA2 10,785,75
ivn-rbt LD50:18300 μg/kg OYYAA2 10,785,75
ims-rbt LD50:77600 μg/kg OYYAA2 10,785,75

SAFETY PROFILE: Poison by ingestion, intramuscular, intravenous, and intraperitoneal routes. An experimental teratogen. Experimental reproductive effects. When heated to decomposition it emits toxic fumes of NO_x.

IAG700 CAS:35607-13-7 ***HR: 3***
IF ROM 203
mf: $C_{20}H_{28}N_2P_4 \cdot 2ClH$ mw: 433.42

SYNS: N,N'-BIS(2-HYDROXY-3-PHENOXYPROPYL)ETHYLENEDIAMINE DIHYDROCHLORIDE ◇ N,N'-DI((3-PHENOXY-2-HYDROXYPROPYL)ETHYLENEDIAMINEDIHYDROCHLORIDE ◇ ROM 203

TOXICITY DATA with REFERENCE
orl-rat LD50:2920 mg/kg DPHFAK 24,103,72
ivn-rat LD50:71 mg/kg DPHFAK 24,103,72
orl-mus LD50:2080 mg/kg DPHFAK 24,103,72
ivn-mus LD50:61250 μg/kg DPHFAK 24,103,72

ivn-rbt LD50:55 mg/kg DPHFAK 24,103,72
ivn-gpg LD50:20100 μg/kg DPHFAK 24,103,72

SAFETY PROFILE: Poison by intravenous route. Moderately toxic by ingestion. When heated to decomposition it emits toxic fumes of NO_x and HCl.

IAH000 CAS:9036-19-5 ***HR: 3***
IGEPAL GAS

PROP: Polymerized ethylene oxide condensate (JAPMA 8 38,428,49).

TOXICITY DATA with REFERENCE
eye-rbt 1% SEV JAPMA8 38,428,49
orl-mus LD50:3500 mg/kg JAPMA8 38,428,49
ivn-mus LD50:70 mg/kg JAPMA8 38,428,49

CONSENSUS REPORTS: Reported in EPA TSCA Inventory.

SAFETY PROFILE: Poison by intravenous route. Moderately toxic by ingestion. A severe eye irritant. When heated to decomposition it emits acrid smoke and fumes.

IAK100 CAS:25843-64-5 ***HR: D***
IMET 3106
mf: $C_{21}H_{23}Cl_2N_3•Cl$ mw: 423.82

SYNS: 2-(((4-(BIS(2-CHLOROETHYL)AMINO)PHENYL)IMINO) METHYL)-1-METHYL-QUINOLINIUM CHLORIDE (9CI) ◇ QUINOLINE-2-ALDEHYDECHLOROMETHYLATE-(p-(BIS-β-CHLOROETHYL)-AMINO)ANIL) ◇ ZIMET 3106

TOXICITY DATA with REFERENCE
ipr-mus TDLo:125 mg/kg (7-11D preg):REP ZPPLBF 110,1067,71
ipr-mus TDLo:125 mg/kg (7-11D preg):TER ARZNAD 22,122,72

SAFETY PROFILE: Experimental teratogenic and reproductive effects. When heated to decomposition it emits toxic fumes of Cl^- and NO_x.

IAL000 CAS:288-32-4 ***HR: 3***
IMIDAZOLE
mf: $C_3H_4N_2$ mw: 68.09

PROP: Prisms. Mp: 90-91°, bp: 257°. Sol in water, ether, chloroform; very sol in alc, pyridine; sltly sol in benzene.

SYNS: 1,3-DIAZA-2,4-CYCLOPENTADIENE ◇ 1,3-DIAZOLE ◇ GLYOXALIN ◇ GLYOXALINE ◇ IMIDAZOL ◇ IMINAZOLE ◇ IMUTEX ◇ MIAZOLE ◇ PYRRO(b)MONAZOLE ◇ USAF EK-4733 ◇ N,N'-VINYLENEFORMAMIDINE

TOXICITY DATA with REFERENCE
dni-hmn:oth 1 mmol/L JIDEAE 65,400,75
scu-rat LD50:626 mg/kg JPETAB 119,444,57
orl-mus LD50:880 mg/kg GWXXBX #3046325
ipr-mus LD50:300 mg/kg NTIS** AD277-689
scu-mus LD50:817 mg/kg JPETAB 119,444,57
ivn-mus LD50:475 mg/kg ARZNAD 33,716,83
scu-cat LDLo:125 mg/kg AEXPBL 84,155,18

CONSENSUS REPORTS: Reported in EPA TSCA Inventory.

SAFETY PROFILE: A poison by subcutaneous and intraperitoneal routes. Moderately toxic by ingestion and intravenous routes. Human mutation data reported. When heated to decomposition it emits toxic fumes of NO_x.

IAM000 CAS:570-22-9 ***HR: 2***
IMIDAZOLE-4,5-DICARBOXYLIC ACID
mf: $C_5H_4N_2O_4$ mw: 156.11

SYNS: 4,5-DICARBOXYIMIDAZOLE ◇ GLYCOXALINEDICARBOXYLIC ACID ◇ α-β-IMIDAZOLECARBOXYLIC ACID ◇ 4,5-IMIDAZOLEDICARBOXYLATE

TOXICITY DATA with REFERENCE
ipr-mus LD50:1800 mg/kg RPTOAN 41,249,78

CONSENSUS REPORTS: Reported in EPA TSCA Inventory.

SAFETY PROFILE: Moderately toxic by intraperitoneal route. When heated to decomposition it emits toxic fumes of NO_x.

IAM100 ***HR: 2***
IMIDAZOLE-2-HYDROXYBENZOATE
mf: $C_7H_6O_3•C_3H_4N_2$ mw: 206.22

SYNS: IMIDAZOLE with SALICYLIC ACID ◇ IMIDAZOL-2-HYDROXYBENZOAT (GERMAN) ◇ ITF 182

TOXICITY DATA with REFERENCE
orl-rat LD50:1211 mg/kg ARZNAD 33,716,83
scu-rat LD50:724 mg/kg ARZNAD 33,716,83
inv-rat LD50:422 mg/kg ARZNAD 33,716,83
orl-mus LD50:1034 mg/kg ARZNAD 33,716,83
scu-mus LD50:595 mg/kg ARZNAD 33,716,83
ivn-mus LD50:435 mg/kg ARZNAD 33,716,83

SAFETY PROFILE: Moderately toxic by ingestion, subcutaneous and intravenous routes. When heated to decomposition it emits toxic fumes of NO_x. See also IMIDAZOLE and SALICYLIC ACID.

IAN000 CAS:5034-77-5 ***HR: 3***
IMIDAZOLE MUSTARD
mf: $C_8H_{12}Cl_2N_6O$ mw: 279.16

SYNS: BIC ◇ 5-(3,3-BIS(2-CHLOROETHYL)-1-TRIAZENO)IMIDAZOLE-4-CARBOXAMIDE ◇ NCI-C01616 ◇ NSC-82196 ◇ SRI 2489 ◇ TIC MUSTARD

TOXICITY DATA with REFERENCE
ipr-rat TDLo:900 mg/kg (12D preg):TER TJADAB 12,259,75
ipr-rat TDLo:600 mg/kg/7W-I:ETA CANCAR 40(Suppl 4),1935,77
ipr-mus TDLo:1500 mg/kg/8W-I:NEO CNREA8 33,3069,73
ivn-hmn TDLo:9 g/kg:GIT CCROBU 56,671,72
orl-cat LD50:267 mg/kg NCISP* JAN86
ipr-rat LD50:210 mg/kg NCISP* JAN86
ipr-mus LD50:60060 μg/kg NCISP* JAN86

CONSENSUS REPORTS: NCI Carcinogenesis Studies (ipr); Clear Evidence: mouse, rat CANCAR 40,1935,77.

SAFETY PROFILE: Suspected carcinogen with experimental neoplastigenic and tumorigenic data. Poison by ingestion and intraperitoneal routes. Experimental teratogenic effects. Human systemic effects by intravenous route: nausea. When heated to decomposition it emits very toxic fumes of Cl^- and NO_x.

IAN100 CAS:6714-29-0 ***HR: 3***
IMIDAZOLEPYRAZOLE
mf: $C_5H_7N_3$ mw: 109.15

SYNS: BA 21381 ◇ 2,3-DIHYDRO-1H-IMIDAZO(1,2-b)PYRAZOLE ◇ 2,3-DIHYDRO-1H-PYRAZOLO(2,3-a)IMIDAZOLE ◇ IMPY ◇ NSC 51143 ◇ PYRAZOLO(2,3-a)IMIDAZOLIDINE

TOXICITY DATA with REFERENCE
dni-hmn:leu 3 mmol/L CNREA8 43,5093,83
dni-mus:fbr 300 μmol/L BCPCA6 20,2639,71
ipr-mus LD50:410 mg/kg NCISP* JAN86
ivn-mus LD50:993 mg/kg CTRRDO 64,1031,80
ivn-dog LDLo:400 mg/kg CTRRDO 64,1031,80

SAFETY PROFILE: Poison by intravenous route. Moderately toxic by intraperitoneal route. Human mutation data reported. When heated to decomposition it emits toxic fumes of NO_x.

IAO000 CAS:872-35-5 ***HR: 3***
IMIDAZOLE-2-THIOL
mf: $C_3H_3N_2S$ mw: 99.14

SYNS: 2-MERCAPTOIMIDAZOLE ◇ USAF EL-57

TOXICITY DATA with REFERENCE
ipr-mus LD50:200 mg/kg NTIS** AD277-689

CONSENSUS REPORTS: Reported in EPA TSCA Inventory.

SAFETY PROFILE: Poison by intraperitoneal route. When heated to decomposition it emits very toxic fumes of NO_x and SO_x. See also MERCAPTANS.

IAP000 CAS:556-90-1 ***HR: 3***
4,5-IMIDAZOLIDINEDITHIONE
mf: $C_3H_4N_2S_2$ mw: 132.21

SYNS: PSEUDO-THIOHYDANTOIN ◇ USAF BE-4-5 ◇ USAF DM-1

TOXICITY DATA with REFERENCE
ipr-mus LD50:100 mg/kg NTIS** AD277-689
ivn-mus LD50:320 mg/kg CSLNX* NX#02090

CONSENSUS REPORTS: Reported in EPA TSCA Inventory.

SAFETY PROFILE: Poison by intraperitoneal and intravenous routes. When heated to decomposition it emits very toxic fumes of NO_x and SO_x.

IAQ000 CAS:96-45-7 ***HR: 3***
2-IMIDAZOLIDINETHIONE
mf: $C_3H_6N_2S$ mw: 102.17

PROP: White crystals. Water solubility: 9 g/100 mL @ 30°. Often occurs as a main degradation product of the metal salts of ethylene bis-dithiocarbamic acid.

SYNS: 4,5-DIHYDROIMIDAZOLE-2(3H)-THIONE ◇ ETHYLENE THIOUREA ◇ N,N'-ETHYLENETHIOUREA ◇ 1,3-ETHYLENE-2-THIOUREA ◇ l'ETHYLENE THIOUREE (FRENCH) ◇ ETU ◇ 2-MERCAPTOIMIDAZOLINE ◇ 2-MERKAPTOIMIDAZOLIN (CZECH) ◇ NA-22 ◇ NCI-C03372 ◇ PENNAC CRA ◇ RCRA WASTE NUMBER U116 ◇ RODANIN S-62 (CZECH) ◇ SODIUM-22 NEOPRENE ACCELERATOR ◇ 2-THIOL-DIHYDROGLYOXALINE ◇ USAF EL-62 ◇ VULKACIT NPV/C2 ◇ WARECURE C

TOXICITY DATA with REFERENCE
eye-rbt 500 mg/24H MLD 28ZPAK -,167,72
mma-sat 3333 μg/plate ENMUDM 8(Suppl 7),1,86
mma-esc 200 mg/L PMRSDJ 1,396,81
dnd-bcs 2 mg/disc PMRSDJ 1,175,81
otr-ham:kdy 80 μg/L BJCAAI 37,873,78
orl-mus TDLo:1600 mg/kg (female 12D post):REP JTEHD6 13,747,84
ihl-rat TCLo:120 mg/m^3/3H (female 7-14D post):TER NTIS** PB277-077
orl-rat TDLo:5306 mg/kg/77W-C:CAR JNCIAM 49,583,72
orl-mus TDLo:77 g/kg/82W-C:CAR JNCIAM 42,1101,69
orl-rat TD:44 g/kg/2Y-C:CAR,TER EJTXAZ 9,303,76
orl-rat LD:5470 mg/kg/26W-C:ETA GTPZAB 32(2),25,88
orl-mus LD50:3000 mg/kg TJADAB 21,71,80
ipr-mus LD50:200 mg/kg NTIS** AD277-689

CONSENSUS REPORTS: NTP Fifth Annual Report on Carcinogens. IARC Cancer Review: Group 2B IMEMDT 7,207,87; Animal Sufficient Evidence IMEMDT 7,45,74. Community Right-To-Know List. EPA Genetic Toxicology Program. Reported in EPA TSCA Inventory.

NIOSH REL: (ETU) Use encapsulated form; minimize exposure.

SAFETY PROFILE: Confirmed carcinogen with experimental carcinogenic data. Poison by ingestion and intraperitoneal routes. Experimental teratogenic and reproductive effects. Mutation data reported. An eye irritant. When heated to decomposition it emits very toxic fumes of NO_x and SO_x.

IAR000 ***HR: 3***
2-IMIDAZOLIDINETHIONE mixed with SODIUM NITRITE

SYNS: ETHYLENETHIOUREA mixed with SODIUM NITRITE ◇ SODIUM NITRITE mixed with ETHYLENETHIOUREA

TOXICITY DATA with REFERENCE
mmo-sat 1 μL/plate MUREAV 48,225,77
dlt-mus-orl 1000 mg/kg/5D-I MUREAV 56,335,78
hma-mus/sat 5 mg/kg MUREAV 106,27,82
orl-rat TDLo:140 mg/kg (15D preg):REP FCTOD7 20,273,82
orl-mus TDLo:600 mg/kg (female 8D post):TER TJADAB 21,71,80

SAFETY PROFILE: Suspected carcinogen. 2-Imidazolidinethione and sodium nitrite are experimental carcinogens. Experimental teratogenic and reproductive data. Sodium nitrite is a poison. Mutation data reported. When heated to decomposition it emits very toxic fumes of SO_x, Na_2O, and NO_x. See also SODIUM NITRITE and 2-IMIDAZOLIDINETHIONE.

IAS000 CAS:120-93-4 ***HR: 3***
2-IMIDAZOLIDINONE
mf: $C_3H_6N_2O$ mw: 86.11

PROP: Needles. Mp: 131°. Sol in water and in hot alc; very sltly sol in ether.

SYNS: 1,3-ETHYLENE UREA ◇ ETHYLENE UREA ◇ 2-IMIDAZOLIDONE

TOXICITY DATA with REFERENCE
ipr-rat TDLo:15 mg/kg (1D male):REP 85GUAJ -,37,66
scu-mus TDLo:1000 mg/kg:CAR NTIS** PB223-159
ipr-mus LD50:500 mg/kg EJMCA5 17,235,82

CONSENSUS REPORTS: Reported in EPA TSCA Inventory.

SAFETY PROFILE: Moderately toxic by intraperitoneal route. Experimental reproductive effects. Questionable carcinogen with experimental carcinogenic data. When heated to decomposition it emits toxic fumes of NO_x.

IAT000 CAS:504-75-6 ***HR: 3***
IMIDAZOLINE
mf: $C_3H_6N_2$ mw: 70.11

SYN: 2-IMIDAZOLINE

TOXICITY DATA with REFERENCE
ipr-mus LD50:50 mg/kg CURL** -,89,62

CONSENSUS REPORTS: Reported in EPA TSCA Inventory.

SAFETY PROFILE: Poison by intraperitoneal route. When heated to decomposition it emits toxic fumes of NO_x.

IAT100 CAS:5789-17-3 ***HR: 3***
IMIDAZOLINE-2,4-DITHIONE
mf: $C_3H_4N_2S_2$ mw: 132.20

HNC(:S)NHC(:S)CH_2

SYN: DITHIOHYDANTOIN

SAFETY PROFILE: A storage hazard; it may explode during long-term storage at room temperature. When heated to decomposition it emits toxic fumes of SO_x and NO_x.

IAY000 CAS:55843-86-2 ***HR: 3***
4-IMIDAZO(1,2-a)PYRIDIN-2-YL-α-METHYLBENZENEACETIC ACID
mf: $C_{16}H_{14}N_2O_2$ mw: 266.32

SYNS: 2-(4-(IMIDAZO(1,2-a)PYRIDIN-2-YL)PHENYL)PROPIONIC ACID ◇ 2-(p-(2-IMIDAZO(1,2-a)PYRIDYL)PHENYL)PROPIONIC ACID ◇ MIROPROFEN ◇ Y-9213

TOXICITY DATA with REFERENCE
orl-rat TDLo:275 mg/kg (female 7-17D post):REP IYKEDH 12,808,81
orl-rbt TDLo:1300 mg/kg (female 6-18D post):TER IYKEDH 12,808,81
orl-rat LD50:292 mg/kg IYKEDH 12,993,81
ipr-rat LD50:365 mg/kg IYKEDH 12,993,81
scu-rat LD50:372 mg/kg IYKEDH 12,993,81
orl-mus LD50:570 mg/kg DRFUD4 4,373,79
ipr-mus LD50:696 mg/kg IYKEDH 12,993,81
scu-mus LD50:687 mg/kg IYKEDH 12,993,81

SAFETY PROFILE: Poison by ingestion, intraperitoneal, and subcutaneous routes. Experimental teratogenic and reproductive effects. Used as an analgesic and anti-inflammatory agent. When heated to decomposition it emits toxic fumes of NO_x.

IBA000 CAS:2465-27-2 ***HR: 3***
4,4'-(IMIDOCARBONYL)BIS(N,N-DIMETHYLAMINE) MONOHYDROCHLORIDE
mf: $C_{17}H_{21}N_3 \cdot ClH \cdot H_2O$ mw: 321.89

SYNS: ADC AURAMINE O ◇ AIZEN AURAMINE ◇ AURAMINE (MAK) ◇ AURAMINE HYDROCHLORIDE ◇ AURAMINE O (BIOLOGICAL STAIN) ◇ AURAMINE YELLOW ◇ 4,4′-BIS(DIMETHYLAMINO) BENZHYDRYLIDENIMINE HYDROCHLORIDE ◇ 4,4′-BIS(DIMETHYL AMINO)BENZOPHENONE-IMINE HYDROCHLORIDE ◇ 1,1-BIS(p-DIMETHYLAMINOPHENYL)METHYLENIMINEHYDROCHLORIDE ◇ CALCOZINE YELLOW OX ◇ 4,4′-CARBONIMIDOYLBIS(N,N-DIMETHYLBENZENAMINE)MONOHYDROCHLORIDE ◇ C.I. 41000 ◇ C.I. BASIC YELLOW 2 ◇ C.I. BASIC YELLOW 2, MONOHYDROCHLORIDE ◇ MITSUI AURAMINE O

TOXICITY DATA with REFERENCE
mma-sat 2 mg/plate CRNGDP 2,1317,81
dnd-esc 30 ppm MUREAV 89,95,81
dnd-hmn:fbr 300 μmol/L JTEHD6 9,941,82
dnd-rat:lvr 3 μmol/L SinJF# 26OCT82
sce-mus-ipr 7500 μg/kg JTEHD6 9,941,82
orl-rat TDLo:40 g/kg/87W-C:NEO BJCAAI 16,87,62
scu-rat TDLo:440 mg/kg/21W-I:ETA BJCAAI 16,87,62
orl-rat LDLo:1500 mg/kg CNREA8 26,619,66
ipr-rat LD50:135 mg/kg NALSDJ 60,745,83
orl-mus LD50:480 mg/kg CRDLP* 4-59,64
skn-mus LD50:300 mg/kg CRDLP* 4-59,64
orl-car LD50:150 mg/kg CWLTM* 47-6,59
orl-dom LD50:150 mg/kg CWLTM* 47-6,59

CONSENSUS REPORTS: Reported in EPA TSCA Inventory. EPA Genetic Toxicology Program.

DFG MAK: Animal Carcinogen, Suspected Human Carcinogen.

SAFETY PROFILE: Confirmed carcinogen with experimental neoplastigenic and tumorigenic data. Poison by skin contact, ingestion, and intraperitoneal routes. Human mutation data reported. A chelating agent which might disturb trace element metabolism if taken into the body. Used as a biological stain. When heated to decomposition it emits very toxic fumes of NO_x and HCl.

IBB000 CAS:492-80-8 ***HR: 3***
4,4é-(IMIDOCARBONYL)BIS(N,N-DIMETHYLANILINE)
mf: $C_{17}H_{21}N_3$ mw: 267.41

PROP: Yellow needles. Mp: 136°. Insol in water.

SYNS: APYONINE AURAMINE BASE ◇ AURAMINE (MAK) ◇ AURAMINE BASE ◇ BIS(p-DIMETHYLAMINOPHENYL)METHYLENEIMINE ◇ BRILLIANT OIL YELLOW ◇ 4,4′-CARBONIMIDOYLBIS(N,N-DIMETHYLBENZENAMINE) ◇ C.I. 41000B ◇ C.I. BASIC YELLOW 2, FREE BASE ◇ C.I. SOLVENT YELLOW 34 ◇ 4,4′-DIMETHYLAMINOBENZOPHENONIMIDE ◇ GLAURAMINE ◇ RCRA WASTE NUMBER U014 ◇ TETRAMETHYLDIAMINODIPHENYLACETIMINE ◇ WAXOLINE YELLOW O ◇ YELLOW PYOCTANINE

TOXICITY DATA with REFERENCE
mma-sat 250 μg/plate PMRSDJ 1,333,81
dns-hmn:fbr 20 mg/L TXCYAC 21,151,81
otr-rat-orl 150 mg/kg CNREA8 40,1157,80
otr-ham:kdy 13100 μg/L PMRSDJ 1,626,81
otr-ham:emb 2 mg/L NCIMAV 58,243,81
orl-rat TDLo:37 g/kg/87W-C:ETA AICCA6 19,483,63
orl-mus TDLo:29 g/kg/52W:NEO BJCAAI 10,653,56
ipr-mus LD50:103 mg/kg PMRSDJ 1,682,81

CONSENSUS REPORTS: IARC Cancer Review: Group 2B IMEMDT 7,118,87; Human Sufficient Evidence IMEMDT 1,69,72; Animal Sufficient Evidence IMEMDT 1,69,72. Community Right-To-Know List. Reported in EPA TSCA Inventory.

DFG MAK: Animal Carcinogen, Suspected Human Carcinogen.

SAFETY PROFILE: Confirmed human carcinogen with experimental carcinogenic, neoplastigenic, and tumorigenic data. Poison by intraperitoneal route. Human mutation data reported. Used as an antiseptic. When heated to decomposition it emits toxic fumes of NO_x.

IBC000 CAS:4375-11-5 ***HR: 3***
IMIDODICARBOXYLIC ACID, DIHYDRAZIDE
mf: $C_2H_7N_5O_2$ mw: 133.14

SYNS: DIAMINOBIURET ◇ 1,5-DIAMINOBIURET ◇ 1,5-DIAMINOBIURET DIHYDRAZIDE ◇ DICARBAZAMIDE ◇ IMIDODICARBONIC DIHYDRAZIDE (9 CI) ◇ NSC 3095 ◇ X 34

TOXICITY DATA with REFERENCE
mmo-esc 5 g/L/1H CRSUBM 3,69,55
ipr-mus LD50:267 mg/kg NCISP* JAN86

CONSENSUS REPORTS: EPA Genetic Toxicology Program.

SAFETY PROFILE: A poison by intraperitoneal route. Mutation data reported. When heated to decomposition it emits toxic fumes of NO_x.

IBD000 CAS:128-87-0 ***HR: 1***
1,1′-IMINOBIS(4-AMINOANTHRAQUINONE)
mf: $C_{28}H_{17}N_3O_4$ mw: 459.48

SYNS: BIS(4-AMINO-1-ANTHRAQUINONYL)AMINE ◇ 4,4′-DIAMINO-1,1′-ANIHRIMIDE ◇ 4,4′-DIAMINO-1,1′-DIANTHRAQUINONYLAMINE ◇ 4,4′-DIAMINO-1,1′-DIANTHRAQUINONYLIMINE ◇ 4,4′-DIAMINO-1,1′-DIANTHRIMID (CZECH) ◇ 4,4′-DIAMINO-1,1′-DIANTHRIMIDE ◇ 4,4′-DIAMINO-IMINO-1,1′BIANTHRAQUINONE ◇ 4,4′-DIAMINO-1,1′-IMINOBISANTHRAQUINONE ◇ 1,1′-IMINOBIS(4-AMINO-9,10-ANTHRACENEDIONE)

TOXICITY DATA with REFERENCE
eye-rbt 500 mg/24H MLD 28ZPAK -,125,72

CONSENSUS REPORTS: Reported in EPA TSCA Inventory.

SAFETY PROFILE: An eye irritant. When heated to decomposition it emits toxic fumes of NO_x. See also AMINES.

IBE000 CAS:128-89-2 ***HR: 1***
4,5'-IMINOBIS(4-BENZAMIDOANTHRAQUINONE)
mf: $C_{42}H_{25}N_3O_6$ mw: 667.70

SYN: 4,5'-BIS-BENZOYLAMINO-1,1'-DIANTHRIMID (CZECH)

TOXICITY DATA with REFERENCE
eye-rbt 500 mg/24H MLD 28ZPAK -,127,72

CONSENSUS REPORTS: Reported in EPA TSCA Inventory.

SAFETY PROFILE: An eye irritant. When heated to decomposition it emits toxic fumes of NO_x.

IBF000 CAS:73816-77-0 ***HR: 2***
5,6'-IMINOBIS(1-HYDROXY-2-NAPHTHALENESULFONIC ACID)
mf: $C_{20}H_{15}NO_8S_2$ mw: 461.48

SYNS: BIS-(5-HYDROXY-7-SULFO-2NAFTYL)AMIN(CZECH) ◇ KYSELINA DI-I (CZECH)

TOXICITY DATA with REFERENCE
eye-rbt 500 mg/24H SEV 28ZPAK -,195,72
orl-rat LD50:11 g/kg 28ZPAK -,195,72

SAFETY PROFILE: A severe eye irritant. Mildly toxic by ingestion. When heated to decomposition it emits very toxic fumes of SO_x and NO_x. See also SULFONATES.

IBH000 CAS:142-73-4 ***HR: 3***
IMINODIACETIC ACID
mf: $C_4H_7NO_4$ mw: 133.12

PROP: Orthorhombic crystals. Decomp @ 247.5°. Insol in alc and ether. Mp: 220-250° (commercial grade). Insol in acetone, methanol, ether, benzene, carbon tetrachloride and heptanone.

SYNS: N-(CARBOXYMETHYL)GLYCINE ◇ DIGLYCIN ◇ DIGLYCINE ◇ DIGLYKOKOLL ◇ HAMPSHIRE ◇ IDA ◇ IMINOBIS(ACETIC ACID) ◇ 2,2'-IMINODIACETIC ACID ◇ IMINODIETHANOIC ACID ◇ USAF DO-55

TOXICITY DATA with REFERENCE
ipr-mus LD50:250 mg/kg NTIS** AD277-689

CONSENSUS REPORTS: Reported in EPA TSCA Inventory.

SAFETY PROFILE: Poison by intraperitoneal route. When heated to decomposition it emits toxic fumes of NO_x.

IBI000 CAS:82-22-4 ***HR: 1***
1,1'-IMINODIANTHRAQUINONE
mf: $C_{28}H_{15}NO_4$ mw: 429.44

SYNS: ANTHRAQUINONYLAMINOANTHRAQUINONE ◇ ANTHRIMIDE ◇ DI-1,1'-ANTHRACHINONYLAMIN (CZECH) ◇ DIANTHRAQUINONYLAMINE ◇ 1,1'-DIANTHRAQUINONYLAMINE ◇ 1,1-DIANTHRIMID (CZECH) ◇ DIANTHRIMIDE ◇ 1,1'-DIANTHRIMIDE ◇ IMINO-1,1'-BIANTHRAQUINONE ◇ 1,1'-IMINOBIS-9,10-ANTHRACENEDIONE

TOXICITY DATA with REFERENCE
skn-rbt 500 mg/24H MLD 28ZPAK -,125,72
eye-rbt 500 mg/24H MLD 28ZPAK -,125,72
orl-rat LD50:16200 mg/kg 28ZPAK -,125,72

CONSENSUS REPORTS: Reported in EPA TSCA Inventory.

SAFETY PROFILE: Mildly toxic by ingestion. A skin and eye irritant. When heated to decomposition it emits toxic fumes of NO_x.

IBJ000 CAS:128-79-0 ***HR: 2***
N,N'-(IMINODI-4,1-ANTHRAQUINONYLENE)BISBENZAMIDE
mf: $C_{42}H_{25}N_3O_6$ mw: 667.70

SYNS: 4,4'-BIS-BENZOYLAMINO-1,1'-DIANTHRIMID (CZECH) ◇ C.I. 65010 ◇ C.I. VAT BLACK 28 ◇ 4,4'-DIBENZAMIDO-1,1'-DIANTHRIMIDE ◇ 1,1'-IMINOBIS(4-BENZAMIDOANTHRAQUINONE) ◇ MIKETHRENE GREY K ◇ OLIVE AR, ANTHRIMIDE ◇ OLIVE R BASE ◇ ROMANTRENE GREY K

TOXICITY DATA with REFERENCE
eye-rbt 500 mg/24H SEV 28ZPAK -,126,72
orl-rat LD50:19300 mg/kg 28ZPAK -,126,72

CONSENSUS REPORTS: Reported in EPA TSCA Inventory.

SAFETY PROFILE: A severe eye irritant. Mildly toxic by ingestion. When heated to decomposition it emits toxic fumes of NO_x.

IBL000 CAS:6846-35-1 ***HR: 3***
5-IMINO-1,2,4-DITHIAZOLIDINE-3-THIONE
mf: $C_2H_2N_2S_3$ mw: 150.24

SYNS: 5-AMINO-1,2,4-DITHIAZOLE-3-THIONE ◇ ISOPERTHIOCYANIC ACID ◇ XANTHAHYDROGEN

TOXICITY DATA with REFERENCE
orl-rat LD50:3050 mg/kg GISAAA 49(4),90,84
orl-mus LD50:1200 mg/kg GISAAA 49(4),90,84
ivn-mus LD50:56 mg/kg CSLNX* NX#00229

SAFETY PROFILE: Poison by intravenous route. Moderately toxic by ingestion. When heated to decomposition it emits very toxic fumes of NO_x and SO_x.

IBM000 CAS:2152-34-3 ***HR: 3***
2-IMINO-5-PHENYL-4-OXAZOLIDINONE
mf: $C_9H_8N_2O_2$ mw: 176.19

SYNS: PHENOXAZOLE ◇ 5-PHENYL-2-IMINO-4-OXAZOLIDINONE ◇ 5-PHENYL-2-IMINO-4-OXOOXAZOLIDINE ◇ PHENYL ISOHYDANTOIN ◇ PHENYLPSEUDOHYDANTOIN

TOXICITY DATA with REFERENCE
ipr-mus TDLo:10 mg/kg (10D preg):TER CAJPBD 3,2,63
orl-man TDLo:536 μg/kg JCPYDR 3,331,83
unr-cld TDLo:66 μg/kg/4W-I:LVR SMJOAV 77,938,84
orl-rat LD50:436 mg/kg PBPSDY 1,33,77 TXAPA9 21,315,72
orl-mus LD50:375 mg/kg PBPSDY 1,33,77
ipr-mus LD50:365 mg/kg PBPSDY 1,33,77
orl-bwd LD50:100 mg/kg TXAPA9 21,315,72

SAFETY PROFILE: A poison by ingestion and intraperitoneal routes. Human systemic effects: hepatitis. An experimental teratogen. A central nervous system stimulant. When heated to decomposition it emits toxic fumes of NO_x.

IBP000 CAS:19864-71-2 ***HR: 3***
IMIPRAMINE-N-OXIDE HYDROCHLORIDE
mf: $C_{19}H_{24}N_2O\bullet ClH$ mw: 332.91

PROP: White crystals. Mp: 153.1-155° (decomp).

SYN: 5-(3-DIMETHYLAMINO) PROPYL-10,11-DIHYDRO 5H-DIBENZ(b,f)AZEPINE, 5-OXIDE MONOHYDROCHLORIDE

TOXICITY DATA with REFERENCE
scu-rbt TDLo:240 mg/kg (female 3-18D post):REP APTOA6 20,186,63
orl-rat LD50:770 mg/kg OYYAA2 11,601,76
ipr-rat LD50:90 mg/kg MEIEDD 10,716,83
scu-rat LD50:260 mg/kg OYYAA2 11,601,76
orl-mus LD50:640 mg/kg OYYAA2 11,601,76
ipr-mus LD50:150 mg/kg MEIEDD 10,716,83
scu-mus LD50:255 mg/kg OYYAA2 11,601,76
ivn-mus LD50:86 mg/kg OYYAA2 11,601,76

SAFETY PROFILE: A poison by intraperitoneal, subcutaneous, and intravenous routes. Moderately toxic by ingestion. Experimental reproductive effects. Used as an antidepressant. When heated to decomposition it emits very toxic fumes of HCl and NO_x.

IBP200 CAS:561-43-3 ***HR: 3***
IMMETROPAN
mf: $C_{21}H_{32}NO_3\bullet Br$ mw: 426.45

SYNS: BM 3055 ◇ BRL 556 ◇ IMETRO ◇ L.D. 3055 ◇ 1-METHYL-2-(PYRROLID-2-YL)METHYL PHENYLCYCLOHEXYL GLYCOLLATE METHOBROMIDE ◇ OXYPYRRONIUM ◇ OXYPYRRONIUM BROMIDE

TOXICITY DATA with REFERENCE
orl-mus LD50:1040 mg/kg AIPTAK 147,552,64
scu-mus LD50:303 mg/kg AIPTAK 147,552,64
ivn-mus LD50:18 mg/kg AIPTAK 147,552,64

SAFETY PROFILE: Poison by subcutaneous and intravenous routes. Moderately toxic by ingestion. When heated to decomposition it emits toxic fumes of Br^- and NO_x.

IBP309 CAS:2207-85-4 ***HR: 3***
IMPIRAMINE-N-OXIDE
mf: $C_{19}H_{24}N_2O$ mw: 296.45

PROP: White needle-like crystals. Mp: 120-123° (decomp); sol in methanol, ether, acetone, and benzene. Hygroscopic.

SYNS: 5-(3-(DIMETHYLAMINO)PROPYL)-10,11-DIHYDRO-5H-DIBENZ(b,f)AZEPINE-5-OXIDE ◇ GP 38383 ◇ IPNO

TOXICITY DATA with REFERENCE
orl-rat LD50:102 mg/kg MPHEAE 15,187,66
ipr-rat LD50:90 mg/kg 27ZQAG -,79,72
ipr-mus LD50:150 mg/kg 27ZQAG -,79,72

SAFETY PROFILE: A poison by ingestion and intraperitoneal routes. When heated to decomposition it emits toxic fumes of NO_x. See also IMIPRAMINE-N-OXIDE HYDROCHLORIDE.

IBQ075 CAS:65573-02-6 ***HR: 3***
IMPROMIDINE HYDROCHLORIDE
mf: $C_{14}H_{23}N_7S\bullet 3ClH$ mw: 430.88

SYNS: IMPROMIDINE TRIHYDROCHLORIDE ◇ SK&F 92676

TOXICITY DATA with REFERENCE
ivn-rat LD50:25 mg/kg JJATDK 2,265,82
ivn-mus LD50:9600 μg/kg JJATDK 2,265,82
ivn-dog LDLo:27700 μg/kg JJATDK 2,265,82

SAFETY PROFILE: Poison by intravenous route. A histamine H2-receptor agonist. When heated to decomposition it emits toxic fumes of SO_x, NO_x, and HCl.

IBQ100 CAS:32784-82-0 ***HR: 3***
IMPROSULFAN TOSYLATE
mf: $C_8H_{19}NO_6S_2\bullet C_7H_8O_3S$ mw: 461.61

SYNS: BIS(3-METHYLSULFONYLOXYPROPYL)AMINE p-TOLUENESULFONATE ◇ 3,3'-IMINOBIS-1-PROPANOL DIMETHANESULFONATE (ester), 4-METHYLBENZENESULFONATE (salt) ◇ IMINODIPROPYL DIMETHANESULFONATE 4-TOLUENESULPHONATE ◇ IMPROSULFAN-p-TOLUENESULFONATE ◇ IMPROSULFAN TOSILATE ◇ NSC-140117 ◇ PROTECTON ◇ 864T

TOXICITY DATA with REFERENCE
mma-esc 50 μg/plate TAKHAA 44,96,85
orl-rat TDLo:150 mg/kg (female 9-14D post):REP IYKEDH 5,473,74
orl-mus TDLo:150 mg/kg (female 7-12D post):TER IYKEDH 5,473,74
orl-rat LD50:108 mg/kg IYKEDH 10,232,79
ipr-rat LD50:104 mg/kg IYKEDH 10,232,79
scu-rat LD50:147 mg/kg IYKEDH 10,232,79
ivn-rat LD50:70 mg/kg GTKRDX 6,183,79
orl-mus LD50:211 mg/kg IYKEDH 10,232,79
ipr-mus LD50:210 mg/kg IYKEDH 10,232,79
scu-mus LD50:228 mg/kg IYKEDH 5,444,74
ivn-mus LD50:149 mg/kg IYKEDH 10,232,79

SAFETY PROFILE: Poison by ingestion, subcutaneous, intravenous and intraperitoneal routes. An experimental teratogen. Experimental reproductive effects. Mutation data reported. When heated to decomposition it emits toxic fumes of NO_x and SO_x.

IBQ300 CAS:53734-79-5 ***HR: 3***
INCASAN
mf: $C_{15}H_{17}N_3O \cdot ClH$ mw: 291.81

SYNS: INCAZANE ◇ INKASAN ◇ 3-METHYL-8-METHOXY-3H,1,2,5,6-TETRAHYDROPYRAZINO-(1,2,3-ab)-β-CARBOLINE HYDROCHLORIDE

TOXICITY DATA with REFERENCE
orl-mus LD50:445 mg/kg FATOAO 43,133,80
scu-mus LD50:250 mg/kg FATOAO 43,133,80
ivn-mus LD50:85 mg/kg FATOAO 43,133,80

SAFETY PROFILE: Poison by subcutaneous and intravenous routes. Moderately toxic by ingestion. When heated to decomposition it emits toxic fumes of NO_x and HCl.

IBQ400 CAS:57296-63-6 ***HR: 3***
INDACRINONE
mf: $C_{18}H_{14}Cl_2O_4$ mw: 365.22

SYNS: ACETIC ACID, ((2,3-DIHYDRO-6,7-DICHLORO-2-METHYL-1-OXO-2-PHENYL-1H-INDEN-5-YL)OXY)-,(±)- ◇ 6,7-DICHLORO-2-METHYL-1-OXO-2-PHENYL-5-INDANYLOXYACETIC ACID ◇ MK 196 ◇ (±)-MK 196

TOXICITY DATA with REFERENCE
orl-rat TDLo:720 mg/kg (female 6-17D post):TER TJADAB 34,468,86
orl-mus LD50:592 mg/kg DRFUD4 13,257,88
ipr-mus LD50:246 mg/kg DRFUD4 13,257,88

SAFETY PROFILE: Poison by intraperitoneal route. Moderately toxic by ingestion. An experimental teratogen. When heated to decomposition it emits toxic fumes of Cl^-.

IBR000 CAS:496-11-7 ***HR: 1***
INDAN
mf: C_9H_{10} mw: 118.19

PROP: Colorless liquid. D: 0.963, bp: 176.5°, mp: −51.4°. Insol in water, misc in alc and ether, sol in organic solvents.

SYNS: 2,3-DIHYDROINDENE ◇ 1,2-HYDRINDENE ◇ HYDRINDONAPHTHENE ◇ INDANE

TOXICITY DATA with REFERENCE
orl-rat LDLo:5000 mg/kg 28ZRAQ -,55,60

CONSENSUS REPORTS: Reported in EPA TSCA Inventory.

SAFETY PROFILE: Mildly toxic by ingestion. When heated to decomposition it emits acrid smoke and fumes.

IBR200 CAS:40507-78-6 ***HR: 3***
INDANAZOLINE HYDROCHLORIDE
mf: $C_{12}H_{15}N_3 \cdot ClH$ mw: 237.76

PROP: Crystals from petr ether. Mp: 109-113°.

SYNS: FARIAL ◇ N-(2-IMIDAZOLINE-2-YL)-N-(4-INDANYL)AMINMONOHYDROCHLORID (GERMAN) ◇ N-(2-IMIDAZOLINE-2-YL)-N-(4-INDANYL)AMINE MONOHYDROCHLORIDE ◇ INDANAZOLIN (GERMAN) ◇ INDANAZOLINE

TOXICITY DATA with REFERENCE
orl-rat TDLo:2520 mg/kg (16-22D preg/21D post):REP ARZNAD 30,1760,80
orl-rat TDLo:900 mg/kg (6-15D preg):TER ARZNAD 30,1760,80
orl-rat LD50:481 mg/kg ARZNAD 30,1760,80
ivn-rat LD50:16 mg/kg ARZNAD 30,1760,80
orl-mus LD50:179 mg/kg ARZNAD 30,1760,80
ivn-mus LD50:22 mg/kg ARZNAD 30,1760,80
orl-dog LDLo:316 mg/kg ARZNAD 30,1760,80

SAFETY PROFILE: Poison by ingestion and intravenous routes. Experimental teratogenic and reproductive effects. When heated to decomposition it emits toxic fumes of NO_x and HCl. A nasal decongestant and vasoconstrictor.

IBS000 CAS:606-23-5 ***HR: 3***
1,3-INDANDIONE
mf: $C_9H_6O_2$ mw: 146.15

PROP: Crystal or liquid. Mp: 129-131° decomp. Very sltly sol in cold water; sol in hot alc, benzene.

SYNS: 1,3-DIKETOHYDRINDENE ◇ 1H-INDENE-1,3(2H)-DIONE

TOXICITY DATA with REFERENCE
ipr-mus TDLo:25 mg/kg (9D preg):REP ARTODN 33,191,75
ipr-mus TDLo:25 mg/kg (9D preg):TER ARTODN 33,191,75
ipr-mus LDLo:100 mg/kg ARTODN 33,191,75

CONSENSUS REPORTS: Reported in EPA TSCA Inventory.

SAFETY PROFILE: A poison by intraperitoneal route. Experimental teratogenic and reproductive effects. When heated to decomposition it emits acrid smoke and fumes.

IBU000 CAS:1470-94-6 ***HR: 2***
5-INDANOL
mf: $C_9H_{10}O$ mw: 134.19

SYNS: 5-HYDROXYHYDRINDENE ◇ INDAN-5-OL

TOXICITY DATA with REFERENCE
skn-rbt 100 μg/24H open AIHAAP 23,95,62
orl-rat LD50:3250 mg/kg AIHAAP 23,95,62
skn-rbt LD50:450 mg/kg AIHAAP 23,95,62

CONSENSUS REPORTS: Reported in EPA TSCA Inventory.

SAFETY PROFILE: Moderately toxic by ingestion and skin contact. A skin irritant. When heated to decomposition it emits acrid smoke and fumes.

IBV000 CAS:615-13-4 ***HR: 3***
2-INDANONE
mf: C_9H_8O mw: 132.17

SYN: 1,3-DIHYDRO-2H-INDEN-2-ONE

TOXICITY DATA with REFERENCE
ivn-mus LD50:56 mg/kg CSLNX* NX#08390

CONSENSUS REPORTS: Reported in EPA TSCA Inventory.

SAFETY PROFILE: Poison by intravenous route. When heated to decomposition it emits acrid smoke and fumes.

IBV100 CAS:26807-65-8 ***HR: 3***
INDAPAMIDE
mf: $C_{16}H_{16}ClN_3O_3S$ mw: 365.86

PROP: Crystals from isopropanol/water. Mp: 160-162°.

SYNS: 3-(AMINOSULFONYL)-4-CHLORO-N-(2,3-DIHYDRO-2-METHYL-1H-INDOL-1-YL)-BENZAMIDE (9CI) ◇ BAJATEN ◇ 4-CHLORO-N-(2-METHYL-1-INDOLINYL)-3-SULFAMOYLBENZAMIDE ◇ FLUDEX ◇ INDAFLEX ◇ INDAMOL ◇ IPAMIX ◇ LOZOL ◇ NATRILIX ◇ NORANAT ◇ TANDIX ◇ S 1520 ◇ SE-1520

TOXICITY DATA with REFERENCE
orl-rat TDLo:270 mg/kg (17-22D preg/21D post):REP YACHDS 10,1363,82
orl-rat TDLo:110 mg/kg (female 7-17D post):TER YACHDS 10,1337,82
ipr-rat LD50:393 mg/kg ARZNAD 25,1491,75
ivn-rat LD50:394 mg/kg ARZNAD 25,1491,75
ipr-mus LD50:410 mg/kg ARZNAD 25,1491,75
ivn-mus LD50:577 mg/kg ARZNAD 25,1491,75
ivn-rbt LD50:337 mg/kg IYKEDH 12,1110,81
ipr-gpg LD50:347 mg/kg ARZNAD 25,1491,75
ivn-gpg LD50:272 mg/kg ARZNAD 25,1491,75

SAFETY PROFILE: Poison by intravenous and intraperitoneal routes. Experimental teratogenic and reproductive effects. When heated to decomposition it emits toxic fumes of Cl^-, SO_x, and NO_x.

IBW100 CAS:81265-54-5 ***HR: 2***
2-(p-(2H-INDAZOL-2-YL)PHENYL)PROPIONIC ACID
mf: $C_{16}H_{14}N_2O_2$ mw: 266.32

SYN: M.G.18755

TOXICITY DATA with REFERENCE
orl-rat LD50:1550 mg/kg FRPSAX 36,1037,81
orl-mus LD50:1150 mg/kg FRPSAX 36,1037,81
ipr-mus LD50:435 mg/kg FRPSAX 36,1037,81

SAFETY PROFILE: Moderately toxic by ingestion and intraperitoneal routes. When heated to decomposition it emits toxic fumes of NO_x.

IBW400 ***HR: 3***
INDECAINIDE HYDROCHLORIDE
mf: $C_{20}H_{24}N_2O•ClH$ mw: 344.92

SYN: 9-(3-((1-METHYLETHYL)AMINO)PROPYL)-9H-FLUORENE-9-CARBOXAMIDE HYDROCHLORIDE

TOXICITY DATA with REFERENCE
orl-rat LD50:82 mg/kg FAATDF 5,175,85
ivn-rat LD50:10 mg/kg TOLED5 26,107,85
orl-mus LD50:96 mg/kg FAATDF 5,175,85
orl-dog LD50:25 mg/kg FAATDF 5,175,85
ivn-dog LDLo:10 mg/kg TOLED5 26,107,85

SAFETY PROFILE: Poison by ingestion and intravenous routes. When heated to decomposition it emits toxic fumes of NO_x and HCl.

IBW500 CAS:65043-22-3 ***HR: 3***
INDELOXAZINE HYDROCHLORIDE
mf: $C_{14}H_{17}NO_2•ClH$ mw: 267.78

PROP: (±)-Form: Polymorphic; pale yellow needles from methanol. Mp: 169-170°. Colorless, acicular crystals from acetone. Mp: 155-156°. (+)-Form: Crystals from ethanol. Mp: 112-113°. (−)-Form: Crystals from isopropanol. Mp: 142-142.5°.

SYNS: ELEN ◇ 2-(7-INDENYLOXYMETHYL)MORPHOLINE HYDROCHLORIDE ◇ 2-((1H-INDEN-7-YLOXY)METHYL)MORPHOLINE HYDROCHLORIDE ◇ YM-08054 ◇ YM-08054-1

TOXICITY DATA with REFERENCE
orl-rat TDLo:810 mg/kg (female 17-22D post):REP KSRNAM 19,6067,85
orl-rat TDLo:1100 mg/kg (7-17D preg):TER OYYAA2 30,429,85
orl-rat LD50:502 mg/kg KSRNAM 19,5687,85
scu-rat LD50:206 mg/kg KSRNAM 19,5687,85
ivn-rat LD50:77300 μg/kg KSRNAM 19,5687,85
orl-mus LD50:444 mg/kg KSRNAM 19,5687,85
scu-mus LD50:245 mg/kg KSRNAM 19,5687,85
ivn-mus LD50:47 mg/kg AIPTAK 238,81,79

SAFETY PROFILE: Poison by subcutaneous and intravenous routes. Moderately toxic by ingestion. An experimental teratogen. Experimental reproductive effects. When heated to decomposition it emits toxic fumes of NO_x and HCl.

IBX000 CAS:95-13-6 **HR: 3**
INDENE
mf: C_9H_8 mw: 116.17

PROP: Liquid from coal tars. D: 0.9968 @ 20°/4°, mp: −1.8°, bp: 181.6°. Water-insol, but misc in organic solvents.

SYN: INDONAPHTHENE

CONSENSUS REPORTS: Reported in EPA TSCA Inventory.

OSHA PEL: TWA 10 ppm
ACGIH TLV: TWA 10 ppm

SAFETY PROFILE: Moderately toxic by ingestion, inhalation, and subcutaneous routes. Irritating to skin, eyes, and mucous membranes. It has exploded during nitration with (H_2SO_4 + HNO_3). When heated to decomposition it emits acrid smoke and fumes.

IBY000 **HR: 2**
INDENE TRIPROPYLAMINE

TOXICITY DATA with REFERENCE
eye-rbt 100 mg SEV IHFCAY 6,1,67
orl-rat LD50:540 mg/kg IHFCAY 6,1,67

SAFETY PROFILE: Moderately toxic by ingestion. A severe eye irritant. When heated to decomposition it emits toxic fumes of NO_x.

IBY600 CAS:81789-85-7 **HR: 3**
INDENOLOL HYDROCHLORIDE
mf: $C_{15}H_{21}NO_2$•ClH mw: 283.83

SYN: 1-(4(or7)-INDENYLOXY)-3-(ISOPROPYLAMINO)-2-PROPANOL HYDROCHLORIDE

TOXICITY DATA with REFERENCE
orl-rat LD50:549 mg/kg NIIRDN 6,89,82
ipr-rat LD50:62 mg/kg NIIRDN 6,89,82
scu-rat LD50:345 mg/kg NIIRDN 6,89,82
ivn-rat LD50:29 mg/kg NIIRDN 6,89,82
orl-mus LD50:388 mg/kg NIIRDN 6,89,82
ipr-mus LD50:107 mg/kg NIIRDN 6,89,82
scu-mus LD50:236 mg/kg NIIRDN 6,89,82
ivn-mus LD50:27 mg/kg NIIRDN 6,89,82

SAFETY PROFILE: Poison by ingestion, subcutaneous, intravenous, and intraperitoneal routes. When heated to decomposition it emits toxic fumes of NO_x and HCl.

IBZ000 CAS:193-39-5 **HR: 3**
INDENO(1,2,3-cd)PYRENE
mf: $C_{22}H_{12}$ mw: 276.34

SYNS: 2,3-PHENYLENEPYRENE ◇ 2,3-o-PHENYLENEPYRENE ◇ 1,10-(o-PHENYLENE)PYRENE ◇ 1,10-(1,2-PHENYLENE)PYRENE ◇ RCRA WASTE NUMBER U137

TOXICITY DATA with REFERENCE
mma-sat 3 μg/plate/48H FCTXAV 17,141,79
otr-ham:lng 100 μg/L TXCYAC 17,149,80
imp-rat TDLo:4150 μg/kg:CAR JJIND8 71,539,83
skn-mus TDLo:40 mg/kg/20D-I:ETA CRNGDP 7,1761,86
scu-mus TDLo:72 mg/kg/9W-I:CAR AICCA6 19,490,63

CONSENSUS REPORTS: NTP Fifth Annual Report on Carcinogens. IARC Cancer Review: Group 2B IMEMDT 7,56,87; Animal Sufficient Evidence IMEMDT 32,373,83; IMEMDT 3,229,73. Reported in EPA TSCA Inventory.

SAFETY PROFILE: Confirmed carcinogen with experimental carcinogenic and tumorigenic data. Mutation data reported. When heated to decomposition it emits acrid smoke and fumes.

IBZ100 CAS:99520-58-8 **HR: 2**
INDENO(1,2,3-cd)PYREN-8-OL
mf: $C_{22}H_{12}O$ mw: 292.34

SYN: 8-HYDROXYINDENO(1,2,3-cd)PYRENE

TOXICITY DATA with REFERENCE
mma-sat 10 μg/plate CNREA8 45,5421,85
skn-mus TDLo:40 mg/kg/20D-I:ETA CRNGDP 7,1761,86

SAFETY PROFILE: Questionable carcinogen with experimental tumorigenic data. Mutation data reported. When heated to decomposition it emits acrid smoke and irritating fumes.

ICA000 CAS:36576-14-4 **HR: 3**
(±)-1-(7-INDENYLOXY)-3-ISOPROPYLAMINO-PROPAN-2-OL HYDROCHLORIDE
mf: $C_{15}H_{21}NO_2$•ClH mw: 283.83

SYNS: dl-YB2 ◇ INDENOLOL HYDROCHLORIDE ◇ PULSAN

TOXICITY DATA with REFERENCE
orl-mus LD50:232 mg/kg AIPTAK 202,79,73
ivn-mus LD50:26 mg/kg ARZNAD 27,1022,77

SAFETY PROFILE: Poison by ingestion and intravenous routes. Used as a beta-adrenergic blocker. When heated to decomposition it emits very toxic fumes of NO_x and HCl.

ICB000 CAS:525-66-6 **HR: 3**
INDERAL
mf: $C_{16}H_{21}NO_2$ mw: 259.38

SYNS: AY 64043 ◇ DOCITON ◇ ICI 45520 ◇ 1-ISOPROPYLAMINE-3-(1-NAPHTHYLOXY)-2-PROPANOL ◇ 1-ISOPROPYLAMINO-3-(1-NAPHTHYLOXY)-2-PROPANOL ◇ NSC-91523 ◇ PROPANALOL ◇ PROPRANOLOL

TOXICITY DATA with REFERENCE
oms-rat-ipr 2 mg/kg IJEBA6 13,4,75
mnt-mus-ipr 74500 μg/kg MUREAV 173,207,86
orl-wmn TDLo:1096 mg/kg (1-40W preg):REP JOPDAB 86,962,75
orl-wmn TDLo:1096 mg/kg (1-40W preg):TER JOPDAB 86,962,75
orl-wmn TDLo:3200 μg/kg:GLN IJMDAI 18,725,82
orl-hmn TDLo:2300 μg/kg:CNS BMJOAE 1,1182,78
orl-wmn LDLo:120 mg/kg APTOA6 41,190,77
orl-cld LDLo:800 μg/kg/12H BMJOAE 2,254,78
orl-chd TDLo:400 mg/kg:CNS,CVS MJAUAJ 1,82,81
orl-man TDLo:8343 mg/kg/4Y-I:SYS AIMDAP 143,2193,83
ivn-hmn LDLo:71 μg/kg LANCAO 1,165,67
orl-rat LD50:660 mg/kg PHARAT 31,635,76
ivn-rat LD50:23 mg/kg PHARAT 31,635,76
orl-mus LD50:289 mg/kg ARZNAD 30,1831,80
ipr-mus LD50:42 mg/kg AIMJA9 30,23,79
scu-mus LD50:150 mg/kg FATOAD 44,342,81
ivn-mus LD50:1900 μg/kg CYLPDN 5,251,84
orl-dog LDLo:120 mg/kg NYKZAU 69(4),262P,73
ivn-gpg LD50:26 mg/kg DPHFAK 24,103,72

SAFETY PROFILE: A deadly human poison by ingestion and intravenous routes. Poison experimentally by ingestion, intraperitoneal, intravenous, and subcutaneous routes. Human systemic effects by ingestion: cardiac arrythmias, hallucinations, hypoglycemia, convulsions; thyroid malfunction. Human reproductive and teratogenic effects by ingestion: extra embryonic structures, abnormal apgar score in newborn, and abnormal growth statistics. Mutation data reported. When heated to decomposition it emits toxic fumes of NO_x. See also INDERAL HYDROCHLORIDE.

ICC000 CAS:318-98-9 ***HR: 3***
INDERAL HYDROCHLORIDE
mf: $C_{16}H_{21}NO_2{\cdot}ClH$ mw: 295.84

SYNS: ANAPRILIN ◇ AY 64043 ◇ BERKOLOL ◇ BETA-NEG ◇ CARIDOROL ◇ DERALIN ◇ DOCITON ◇ FREKVEN ◇ ICI 45520 ◇ INDEREX ◇ INDEROL ◇ INDOBLOC ◇ 1-(ISOPROPYLAMINO)-3-(α-NAPHTHOXY)-2-PROPANOL HYDROCHLORIDE ◇ 1-ISOPROPYLAMINO-3-(1-NAPHTHOXY)-PROPAN-2-OL HYDROCHLORIDE ◇ 1-(ISOPROPYLAMINO)-3-(1-NAPHTHYLOXY)PROPAN-2-OLHYDROCHLORIDE ◇ 1-(ISOPROPYLAMINO)-3-(1-NAPTHYLOXY)-2-PROPANOL HYDROCHLORIDE ◇ KEMI ◇ 1-((1-METHYLETHYL)AMINO)-3-(1-NAPHTHALENYLOXY)-2-PROPANOL HYDROCHLORIDE ◇ 1-(1-NAPHTHYLOXY)-2-HYDROXY-3-ISOPROPYLAMINOPROPANEHYDROCHLORIDE ◇ NSC-91523 ◇ OPOSIM ◇ PROPRANOLOL ◇ PROPRANOLOL HYDROCHLORIDE ◇ PYLAPRON ◇ RAPYNOGEN ◇ TESNOL

TOXICITY DATA with REFERENCE
orl-wmn TDLo:48 mg/kg (female 25-34W post):REP JOPDAB 91,808,77
orl-man TDLo:4 mg/kg/1W-I:PUL, DICPBB 16,776,82
orl-man TDLo:857 μg/kg/3D-I:SKN ARDEAC 121,1326,85
orl-man TDLo:43 mg/kg SAMJAF 67,1062,85
orl-man TDLo:417 mg/kg/1Y-I:PUL PRACAK 229,663,85
orl-wmn TDLo:77 mg/kg:EYE,PSY AEMED3 14,161,85
orl-wmn TDLo:22 mg/kg/4W-I:CNS JCGADC 8,74,86
ivn-wmn LDLo:40 μg/kg:CVS,PUL AIMDAP 144,173,84
ivn-man TDLo:29 μg/kg/5M-I:CVS,BPR AEMED3 14,1112,85
orl-rat LD50:466 mg/kg ARZNAD 35,1236,85
ipr-rat LD50:76 mg/kg ARZNAD 26,506,76
scu-rat LD50:115 mg/kg YACHDS 12(Suppl 6),969,84
ivn-rat LD50:21 mg/kg YACHDS 12(Suppl 6),969,84
orl-mus LD50:320 mg/kg OYYAA2 2,70,68
ipr-mus LD50:80 mg/kg JDGRAX 16,171,85
scu-mus LD50:208 mg/kg YACHDS 12(Suppl 6),969,84
ivn-mus LD50:18 mg/kg ARZNAD 27,1022,77
orl-rbt LD50:600 mg/kg NIIRDN 6,733,82

SAFETY PROFILE: A deadly human poison by intravenous route. Poison experimentally by ingestion, intraperitoneal, subcutaneous, and intravenous routes. Experimental reproductive effects. Human systemic effects: acute pulminary edema, altered sleep time, blood pressure elevation, bronchiolar constriction, cardiac arrythmias, convulsions, cough, dermatitis, dyspnea, encephalitis, mydriasis, pulse rate increase, somnolence. A beta-adrenergic blocker. When heated to decomposition it emits very toxic fumes of NO_x and HCl. See also INDERAL.

ICC700 ***HR: 3***
INDIAN COBRA VENOM

SYNS: NAJA NAJA NAJA VENOM ◇ N. NAJA NAJA VENOM ◇ VENOM, INDIAN COBRA, NAJA NAJA NAJA

TOXICITY DATA with REFERENCE
ipr-mus LD50:200 μg/kg TOXIA6 19,295,81
ivn-mus LD50:244 μg/kg TOXIA6 9,131,71
ivn-rbt LDLo:25 μg/kg TOXIA6 19,295,81

SAFETY PROFILE: Deadly poison by intravenous and intraperitoneal routes.

ICD000 CAS:2642-37-7 ***HR: 3***
INDICAN (POTASSIUM SALT)
mf: $C_8H_6NO_4S{\cdot}K$ mw: 251.31

PROP: Light brown plates from aq alc. Decomp @ 179-180° (subl), very sol in water; insol in cold alc.

SYNS: INDOL-3-OL, HYDROGEN SULFATE (ESTER), POTASSIUM SALT ◇ INDOL-3-OL, POTASSIUM SULFATE ◇ INDOL-3-YL POTASSIUM SULFATE ◇ INDOL-3-YL SULFATE, POTASSIUM SALT ◇ POTASSIUM INDOL-3-YL SULFATE ◇ URINARY INDICAN

TOXICITY DATA with REFERENCE
scu-mus TDLo:3500 mg/kg/26W-I:ETA KLWOAZ 36,1056,58

SAFETY PROFILE: Questionable carcinogen with experimental tumorigenic data. When heated to decomposition it emits very toxic fumes of NO_x, K_2O, and SO_x.

ICD100 CAS:41708-76-3 ***HR: 2***
INDICINE-N-OXIDE
mf: $C_{15}H_{25}NO_6$ mw: 315.41

SYNS: INDI ◇ NSC-132319

TOXICITY DATA with REFERENCE
ipr-mus LD50:1917 mg/kg NCISP* JAN86
ivn-mus LD50:3733 mg/kg NTIS** PB84-161694
ivn-dog LD50:1200 mg/kg DRFUD4 7,633,82

SAFETY PROFILE: Moderately toxic by intraperitoneal and intravenous routes. When heated to decomposition it emits toxic fumes of NO_x. See also ESTERS.

ICE000 CAS:520-18-3 ***HR: D***
INDIGO YELLOW
mf: $C_{15}H_{10}O_6$ mw: 286.25

SYNS: PELARGIDENOLON ◇ RHAMNOLUTEIN ◇ RHAMNOLUTIN ◇ 3,4',5,7-TETRAHYDROXYFLAVONE ◇ 5,7,4'-TRIHYDROXYFLAVONOL

TOXICITY DATA with REFERENCE
mma-sat 166 nmol/plate MUREAV 54,297,78
mnt-mus-ipr 200 mg/kg MUREAV 89,69,81

CONSENSUS REPORTS: IARC Cancer Review: Group 3 IMEMDT 7,56,87; Animal Inadequate Evidence IMEMDT 31,171,83.

SAFETY PROFILE: Questionable carcinogen. Mutation data reported. When heated to decomposition it emits acrid smoke and fumes.

ICF000 CAS:7440-74-6 ***HR: 3***
INDIUM
af: In aw: 114.82

PROP: Soft, silvery-white metal. Mp: 156.61°, bp: 2080°, d: 7.31 @ 20°.

TOXICITY DATA with REFERENCE
scu-mus LDLo:10 mg/kg 28ZLA8 -,144,61

CONSENSUS REPORTS: Reported in EPA TSCA Inventory.

OSHA PEL: TWA 0.1 mg(In)/m^3
ACGIH TLV: TWA 0.1 mg(In)/m^3

SAFETY PROFILE: A poison by subcutaneous route. It affects the liver, heart, kidneys, and the blood. Teratogenic effects. Inhalation of indium compounds may cause damage to the respiratory system. Hydrated indium oxide is a poison by intravenous route. Flammable in the form of dust when exposed to heat or flame. Incandesces. Explosive reaction with dinitrogen tetraoxide + acetonitrile. Violent reaction with mercury(II) bromide at 350°C. MIxtures with sulfur ignite when heated.

ICG000 CAS:14405-45-9 ***HR: 3***
INDIUM ACETYLACETONATE
mf: $C_{15}H_{21}InO_6$ mw: 412.18

SYN: TRIS(2,4-PENTANEDIONATO)INDIUM

TOXICITY DATA with REFERENCE
ivn-mus LD50:79 mg/kg CSLNX* NX#05186

CONSENSUS REPORTS: Reported in EPA TSCA Inventory.

SAFETY PROFILE: Poison by intravenous route. When heated to decomposition it emits acrid smoke and fumes. See also INDIUM.

ICH000 CAS:4194-69-8 ***HR: 2***
INDIUM CITRATE
mf: $C_{18}H_{15}O_{21}$•In mw: 682.15

TOXICITY DATA with REFERENCE
scu-mus LDLo:600 mg/kg 28ZLA8 -,144,61

SAFETY PROFILE: Moderately toxic by subcutaneous route. When heated to decomposition it emits acrid smoke and fumes. See also INDIUM.

ICI000 CAS:13770-61-1 ***HR: 2***
INDIUM NITRATE
mf: InN_3O_9 mw: 300.85

TOXICITY DATA with REFERENCE
skn-mam 500 mg SEV GISAAA 45(10),13,80
ivn-ham TDLo:2 mg/kg (8D preg):TER TXAPA9 16,166,70
ivn-ham TDLo:1 mg/kg (8D preg):REP TXAPA9 16,166,70

CONSENSUS REPORTS: Reported in EPA TSCA Inventory.

SAFETY PROFILE: Experimental teratogenic and reproductive effects. A severe skin irritant. When heated to decomposition it emits toxic fumes of NO_x. See also INDIUM and NITRATES.

ICJ000 CAS:13464-82-9 ***HR: 3***
INDIUM SULFATE
mf: $O_{12}S_3$•In_2 mw: 517.82

PROP: Grayish-white, hygroscopic powder. D: 3.44. Sol in water. Keep well-closed.

SYNS: INDISULFAT (GERMAN) ◇ SULFURIC ACID, INDIUM SALT

TOXICITY DATA with REFERENCE
orl-rat LDLo:1200 mg/kg AEPPAE 173,458,33
scu-rat LD50:22500 μg/kg EQSSDX 1,1,75
ivn-rat LD50:5630 μg/kg EQSSDX 1,1,75
orl-rbt LDLo:1300 mg/kg AEPPAE 173,458,33
scu-rbt LDLo:2600 μg/kg EQSSDX 1,1,75
ivn-rbt LDLo:670 μg/kg EQSSDX 1,1,75
scu-frg LDLo:600 mg/kg AEPPAE 173,458,33

CONSENSUS REPORTS: Reported in EPA TSCA Inventory.

SAFETY PROFILE: A poison by intravenous and subcutaneous routes. Moderately toxic by ingestion. When heated to decomposition it emits toxic fumes of SO_x. See also INDIUM and SULFATES.

ICK000 CAS:10025-82-8 ***HR: 3***
INDIUM TRICHLORIDE
mf: Cl_3In mw: 221.17

PROP: Yellowish, deliquescent crystals. D: 4.0, mp: 586°, sublimes @ 500°, bp: volatile @ 600°. Very sol in water. Keep tightly closed.

SYN: INDIUM CHLORIDE

TOXICITY DATA with REFERENCE
dnd-mam:lym 40 μmol/L JCHODP 7,411,76
scu-rat LDLo:10 mg/kg JIHTAB 24,243,42
ipr-mus LD50:9500 μg/kg COREAF 256,1043,63
scu-mus LDLo:60 mg/kg EQSSDX 1,1,75
scu-rbt LDLo:2350 μg/kg EQSSDX 1,1,75
ivn-rbt LDLo:640 μg/kg EQSSDX 1,1,75

CONSENSUS REPORTS: Reported in EPA TSCA Inventory.

SAFETY PROFILE: A poison by subcutaneous, intraperitoneal, and intravenous routes. Mutation data reported. When heated to decomposition it emits toxic fumes of Cl^-. See also INDIUM.

ICL000 ***HR: 3***
INDOCYANINE GREEN
mf: $C_{43}H_{47}N_2O_6S_2{\cdot}Na_2{\cdot}I$ mw: 924.92

SYN: 2-(7-(1,1-DIMETHYL-3-(4-SULFOBUTYL)BENZ(e)INDOLIN-2-YLIDENE)-1,3,5-HEPTATRIENYL)-1,1-DIMETHYL-3-(4-SULFOBUTYL)1H-BENZ(e)INDOLIUM IODIDE, INNER SALT, SODIUM SALT

TOXICITY DATA with REFERENCE
ivn-rat LD50:95 mg/kg OYYAA2 3,68,69
scu-mus LD50:1000 mg/kg OYYAA2 3,68,69
ivn-mus LD50:71 mg/kg OYYAA2 3,68,69
ivn-dog LD50:90 mg/kg OYYAA2 3,68,69

SAFETY PROFILE: Poison by intravenous route. Moderately toxic by subcutaneous route. When heated to decomposition it emits very toxic fumes of NO_x, Na_2O, SO_x, and I^-.

ICL500 ***HR: 3***
INDOLAPRIL HYDROCHLORIDE
mf: $C_{24}H_{34}N_2O_5{\cdot}ClH$ mw: 467.00

SYNS: (2S-(1(4*(4*)),2-α,3-α-β,7-α-β))-OCTAHYDRO-1-(2-((1-(ETHOXYCARBONYL)-3-PHENYLPROPYL)AMINO)-1-OXYPROPYL)-1H-INDOLE-2-CARBOXYLIC ACID MONOHYDROCHLORIDE ◇ SCH 31846 HYDROCHLORIDE

TOXICITY DATA with REFERENCE
orl-rat LD50:2500 mg/kg TXAPA9 82,104,86
ivn-rat LD50:150 mg/kg TXAPA9 82,104,86
orl-mus LD50:1800 mg/kg TXAPA9 82,104,86
ivn-mus LD50:450 mg/kg TXAPA9 82,104,86

SAFETY PROFILE: Poison by intravenous route. Moderately toxic by ingestion. When heated to decomposition it emits toxic fumes of NO_x and HCl.

ICM000 CAS:120-72-9 ***HR: 3***
INDOLE
mf: C_8H_7N mw: 117.16

PROP: Colorless to yellowish scales; intense fecal odor. Mp: 52°, bp: 253°; volatile with steam. Sol in hot water, alc, ether, petroleum ether; insol in mineral oil, glycerin.

SYNS: 1-AZAINDENE ◇ 1-BENZAZOLE ◇ BENZOPYRROLE ◇ 2,3-BENZOPYRROLE ◇ FEMA No. 2593 ◇ INDOL (GERMAN) ◇ KETOLE

TOXICITY DATA with REFERENCE
scu-mus TDLo:1000 mg/kg/25W-I:CAR KLWOAZ 35,504,57
scu-mus TD:2000 mg/kg/20W-I:ETA AICCA6 19,660,63
orl-rat LD50:1000 mg/kg AIHAAP 23,95,62
orl-mus LDLo:1070 mg/kg AECTCV 14,111,85
ipr-mus LD50:117 mg/kg YKKZAJ 94,1620,74
scu-mus LD50:225 mg/kg KLWOAZ 35,504,57
skn-rbt LD50:790 mg/kg AIHAAP 23,95,62

CONSENSUS REPORTS: Reported in EPA TSCA Inventory.

SAFETY PROFILE: A poison by intraperitoneal and subcutaneous routes. Moderately toxic by ingestion and skin contact. Questionable carcinogen with experimental carcinogenic and tumorigenic data. When heated to decomposition it emits toxic fumes of NO_x.

ICN000 CAS:87-51-4 ***HR: 3***
1H-INDOLE-3-ACETIC ACID
mf: $C_{10}H_9NO_2$ mw: 175.20

PROP: Colorless leaves from benzene. Mp: 165-168°. Very sltly sol in cold water; sol in alc, ether, and acetic acid; insol in chloroform.

SYNS: HETEROAUXIN ◇ IAA ◇ β-INDOLEACETIC ACID ◇ β-IN-

DOLE-3-ACETIC ACID ◇ 3-INDOLEACETIC ACID ◇ INDOLYACETIC ACID ◇ α-INDOL-3-YL-ACETIC ACID ◇ β-INDOLYLACETIC ACID ◇ INDOLYL-3-ACETIC ACID ◇ 3-INDOLYLACETIC ACID ◇ RHIZOPIN ◇ ω-SKATOLE CARBOXYLIC ACID

TOXICITY DATA with REFERENCE
mrc-asn 1150 μmol/L CRNGDP 4,1409,83
sln-asn 1150 μmol/L CRNGDP 4,1409,83
dnd-sal:spr 250 μmol/L PYTCAS 11,3135,72
dnd-mam:lym 250 μmol/L PYTCAS 11,3135,72
orl-mus TDLo:4500 mg/kg (female 7-15D post):TER TJADAB 19,321,79
scu-mus TDLo:2000 mg/kg/20W-I:ETA AICCA6 19,660,63
ipr-mus LD50:150 mg/kg NTIS** AD691-490

CONSENSUS REPORTS: Reported in EPA TSCA Inventory.

SAFETY PROFILE: A poison by intraperitoneal route. Questionable carcinogen with experimental tumorigenic and teratogenic data. Mutation data reported. When heated to decomposition it emits toxic fumes of NO_x.

ICO000 CAS:1204-06-4 ***HR: 3***
INDOLE-3-ACRYLIC ACID
mf: $C_{11}H_9NO_2$ mw: 187.21

SYNS: 3-INDOLYLACRYLIC ACID ◇ 3-(1-H-INDOL-3-YL)-2-PROPENOIC ACID

TOXICITY DATA with REFERENCE
cyt-hmn:leu 100 mg/L TSITAQ 15,1505,73
cyt-mky-scu 15 mg/kg/2D-I TSITAQ 15,1505,73
scu-mus TDLo:1600 mg/kg/8W-I:CAR VOONAW 22(6),47,76
scu-gpg TDLo:1100 mg/kg/13W-I:ETA BEXBAN 84,1156,77
scu-ham TDLo:1600 mg/kg/10W-I:NEO VOONAW 22(6),52,76

SAFETY PROFILE: Questionable carcinogen with experimental carcinogenic, neoplastigenic, and tumorigenic data. Human mutation data reported. When heated to decomposition it emits toxic fumes of NO_x.

ICP000 CAS:133-32-4 ***HR: 3***
1H-INDOLE-3-BUTANOIC ACID
mf: $C_{12}H_{13}NO_2$ mw: 203.26

PROP: White crystals or powder. Mp: 124°. Sol in acetone and ether; insol in water and chloroform.

SYNS: HORMEX ROOTING POWDER ◇ HORMODIN ◇ IBA ◇ INDOLE BUTYRIC ◇ INDOLE BUTYRIC ACID ◇ β-INDOLEBUTYRIC ACID ◇ Γ-(INDOLE-3)-BUTYRIC ACID ◇ 3-INDOLEBUTYRIC ACID ◇ 3-INDOLYL-Γ-BUTYRIC ACID ◇ Γ-(3-INDOLYL)BUTYRIC ACID ◇ Γ-(INDOL-3-YL)BUTYRIC ACID ◇ INDOLYL-3-BUTYRIC ACID ◇ 4-(INDOL-3-YL)BUTYRIC ACID ◇ 4-(INDOLYL)BUTYRIC ACID ◇ 4-(3-INDOLYL)BUTYRIC ACID ◇ JIFFY GROW ◇ ROOTONE

TOXICITY DATA with REFERENCE
mrc-asn 1 mmol/L CRNGDP 4,1409,83
sln-asn 1 mmol/L CRNGDP 4,1409,83
orl-mus LD50:100 mg/kg 85ARAE 3,76,76/77
ipr-mus LDLo:100 mg/kg PCOC** -,614,66

CONSENSUS REPORTS: Reported in EPA TSCA Inventory.

SAFETY PROFILE: A poison by ingestion and intraperitoneal routes. Mutation data reported. Used for promoting and accelerating root formation of plant clippings. When heated to decomposition it emits toxic fumes of NO_x.

ICP100 CAS:700-06-1 ***HR: 2***
INDOLE-3-CARBINOL
mf: C_9H_9NO mw: 147.19

SYNS: 3-HYDROXYMETHYLINDOLE ◇ INDOLE-3-METHANOL ◇ 1H-INDOLE-3-METHANOL (9CI) ◇ 3-INDOLYLCARBINOL

TOXICITY DATA with REFERENCE
scu-rat TDLo:400 mg/kg (female 8-9D post):TER FCTXAV 18,159,80
scu-rat LDLo:500 mg/kg FCTXAV 18,159,80

SAFETY PROFILE: Moderately toxic by subcutaneous route. An experimental teratogen. When heated to decomposition it emits toxic fumes of NO_x.

ICR000 CAS:91-56-5 ***HR: 3***
INDOLE-2,3-DIONE
mf: $C_8H_5NO_2$ mw: 147.14

SYNS: o-AMINOBENZOYLFORMIC ANHYDRIDE ◇ 2,3-DIKETOINDOLINE ◇ 2,3-DIOXOINDOLINE ◇ 2,3-INDOLINEDIONE ◇ ISATIC ACID LACTAM ◇ ISATIN ◇ ISATINIC ACID ANHYDRIDE ◇ 2,3-KETOINDOLINE

TOXICITY DATA with REFERENCE
orl-rat TDLo:200 mg/kg (1D male):REP IJMRAQ 67,73,78
orl-rat LDLo:5 g/kg IJPPAZ 6,145,62
orl-mus LD50:300 mg/kg NYKZAU 55,1514,59
ipr-mus LD50:563 mg/kg PCJOAU 15,858,81

CONSENSUS REPORTS: Reported in EPA TSCA Inventory.

SAFETY PROFILE: Poison by ingestion. Moderately toxic by intraperitoneal route. Experimental reproductive effects. When heated to decomposition it emits toxic fumes of NO_x. See also ANHYDRIDES.

ICS000 CAS:526-55-6 ***HR: 3***
INDOLE ETHANOL
mf: $C_{10}H_{11}NO$ mw: 161.22

SYN: 3-INDOLYLETHANOL

TOXICITY DATA with REFERENCE
ipr-mus LD50:351 mg/kg JTEHD6 1,515,76
ivn-mus LD50:180 mg/kg CSLNX* NX#00777

CONSENSUS REPORTS: Reported in EPA TSCA Inventory.

SAFETY PROFILE: Poison by intravenous and intraperitoneal routes. When heated to decomposition it emits toxic fumes such as NO_x.

ICS100 CAS:68527-79-7 ***HR: 1***
INDOLENE
mf: $C_{18}H_{25}NO$ mw: 271.44

SYNS: HYDROXYCITRONELLAL-INDOLE (SCHIFF BASE) ◇ HYDROXYCITRONELLYLIDENE-INDOLE ◇ 7-OCTEN-2-OL, 2,6-DIMETHYL-8-(1H-INDOL-1-YL)-

TOXICITY DATA with REFERENCE
skn-gpg 500 mg/24H MLD FCTOD7 21,857,83

CONSENSUS REPORTS: Reported in EPA TSCA Inventory.

SAFETY PROFILE: A skin irritant. When heated to decomposition it emits toxic fumes of NO_x.

ICU100 ***HR: 3***
INDOL-N-METHYLHARMINE HYDROCHLORIDE
mf: $C_{14}H_{14}N_2O \cdot ClH$ mw: 262.76

SYN: 1,9-DIMETHYL-7-METHOXY-9H-PYRIDO(3,4-b)INDOLEHYDROCHLORIDE

TOXICITY DATA with REFERENCE
ipr-mus LDLo:64 mg/kg QJPPAL 9,37,36
scu-gpg LDLo:38 mg/kg QJPPAL 9,37,36
scu-frg LDLo:113 mg/kg QJPPAL 9,37,36

SAFETY PROFILE: Poison by subcutaneous and intraperitoneal routes. When heated to decomposition it emits toxic fumes of NO_x and HCl.

ICW000 CAS:771-51-7 ***HR: 3***
3-INDOLYLACETONITRILE
mf: $C_{10}H_8N_2$ mw: 156.20

SYNS: 3-(CYANOMETHYL)INDOLE ◇ 3-INDOLACETONITRILE ◇ INDOLEACETONITRILE ◇ INDOLE-3-ACETONITRILE ◇ 1H-INDOLE-3-ACETONITRILE ◇ INDOLYLACETONITRILE ◇ USAF CB-29

TOXICITY DATA with REFERENCE
scu-rat LD50:255 mg/kg FCTXAV 18,159,80
ipr-mus LD50:200 mg/kg NTIS** AD277-689

CONSENSUS REPORTS: Reported in EPA TSCA Inventory. Cyanide and its compounds are on the Community Right-To-Know List.

SAFETY PROFILE: A poison by intraperitoneal and subcutaneous routes. When heated to decomposition it emits toxic NO_x and CN^-. See also NITRILES.

ICY000 CAS:35412-68-1 ***HR: 3***
1-(INDOLYL-3)-2-METHYLAMINOETHANOL-1 RACEMATE
mf: $C_{11}H_{14}N_2O$ mw: 190.27

SYN: (±)-1-(3-INDOLYL)-2-METHYLAMINOETHANOL

TOXICITY DATA with REFERENCE
orl-mus LD50:1700 mg/kg RPTOAN 35(3),109,72
ipr-mus LD50:144 mg/kg RPTOAN 35(3),109,72
scu-mus LD50:265 mg/kg RPTOAN 35(3),109,72
ivn-mus LD50:96 mg/kg RPTOAN 35(3),109,72

SAFETY PROFILE: Poison by intraperitoneal, subcutaneous, and intravenous routes. Moderately toxic by ingestion. When heated to decomposition it emits toxic fumes of NO_x.

ICZ100 CAS:75410-87-6 ***HR: 3***
1-(3'-INDOLYLMETHYL)-4-(2'-QUINOLYL) PIPERAZINE DIMALEATE
mf: $C_{22}H_{22}N_4 \cdot 2C_4H_4O_4$ mw: 574.64

SYN: 2-(4-(3-INCOLYLMETHYL)-1-PIPERAZINYL)-QUINOLINE DIMALEATE

TOXICITY DATA with REFERENCE
orl-mus LD50:620 mg/kg FATOAO 43(5),530,80
ipr-mus LD50:152 mg/kg FATOAO 43(5),530,80
ivn-mus LD50:93 mg/kg FATOAO 43(5),530,80

SAFETY PROFILE: Poison by intravenous and intraperitoneal routes. Moderately toxic by ingestion. When heated to decomposition it emits toxic fumes of NO_x.

IDA000 CAS:53-86-1 ***HR: 3***
INDOMETHACIN
mf: $C_{19}H_{16}ClNO_4$ mw: 357.81

PROP: Crystals. One form: mp: 155°; another form: mp: 162°. Sol in ethanol, ether, acetone, and castor oil; insol in water.

SYNS: AMUNO ◇ ARTRACIN ◇ ARTRINOVO ◇ ARTRIVIA ◇ N-p-CHLORBENZOYL-5-METHOXY-2-METHYLINDOLE-3-ACETICACID ◇ 1-(p-CHLOROBENZOYL)-5-METHOXY-2-METHYLINDOLE-3-ACETIC ACID ◇ 1-(p-CHLOROBENZOYL)-2-METHYL-5-METHOXYINDOLE-3-ACETIC ACID ◇ 1-(p-CHLOROBENZOYL)-2-METHYL-5-METHOXY-3-INDOLE-ACETIC ACID ◇ α-(1-(p-CHLOROBENZOYL)-2-METHYL-5-METHOXY-3-INDOLYL)ACETIC ACID ◇ 1-p-CLORO-BENZOIL-5-METOXI-2-METILINDOL-3-ACIDO ACETICO (SPANISH) ◇ CONFORTID ◇ DOLOVIN ◇ IDOMETHINE ◇ IMBRILON ◇ INACID ◇ INDOCID ◇ INDOMECOL ◇ INDOMED ◇ INDOMETHAZINE ◇ INDOMETICINA (SPANISH) ◇ INDOPTIC ◇ INDO-RECTOLMIN ◇ INDO-TABLINEN ◇ INFLAZON ◇ INTEBAN SP ◇ LAUSIT ◇ METACEN ◇ METARTRIL ◇ METHAZINE ◇ METINDOL ◇ MEZOLIN ◇ MIKAMETAN ◇ MOBILAN ◇ NCI-C56144 ◇ REUMACIDE ◇ SADOREUM ◇ TANNEX

TOXICITY DATA with REFERENCE
dni-mus-skn 44 μmol/kg RCOCB8 24,533,79
orl-wmn TDLo:1500 μg/kg (female 36W post):REP JOPDAB 92,478,78
rec-wmn TDLo:12 mg/kg (female 28W post):TER JOPDAB 113,738,88
orl-man LDLo:15 mg/kg/2W-I:BLD IJMDAI 17,433,81
orl-inf TDLo:400 μg/kg/2D-I:GIT JOPDAB 107,484,85
orl-wmn TDLo:2098 μg/kg/1D-I:KID SMJOAV 78,1390,85
orl-man TDLo:22500 μg/kg/3W-I:LVR BMJOAE 3,155,67
orl-hmn TDLo:113 mg/kg/8W-I:GIT ARZNAD 33,636,83
ivn-inf TDLo:200 μg/kg:BLD JOPDAB 107,312,85
unr-man TDLo:499 mg/kg/87W-I ARHEAW 20,917,77
rec-man TDLo:2586 mg/kg/3.5Y-I:EYE AJOPAA 73,846,72
mul-man TDLo:3557 mg/kg/5Y-I:EYE AJOPAA 73,846,72
orl-rat LD50:2420 μg/kg ARZNAD 25,1526,75
ipr-rat LD50:13 mg/kg TXAPA9 38,127,76
scu-rat LD50:12 mg/kg OYYAA2 2,70,68
ivn-rat LD50:35 mg/kg OYYAA2 7,333,73
orl-mus LD50:13 mg/kg ARZNAD 30,1398,80
ipr-mus LD50:15 mg/kg EJMCA5 24,91,89
scu-mus LD50:18300 μg/kg ARZNAD 30,1398,80
ivn-mus LD50:30 mg/kg ARZNAD 19,1198,69
orl-dog LD50:160 mg/kg OYYAA2 2,70,68
ivn-cat LDLo:20200 μg/kg ARZNAD 33,726,83

CONSENSUS REPORTS: Reported in EPA TSCA Inventory.

SAFETY PROFILE: A poison by ingestion, intravenous, intraperitoneal, and subcutaneous routes. Human systemic effects by ingestion: aplastic anemia, changes in kidney tubules, decreased urine volume, diarrhea, fibrous hepatitis, hemorrhage, hypermotility, liver changes, necrotic stomach changes, retinal changes. Human teratogenic effects by ingestion and intravenous routes: developmental abnormalities of the respiratory system and urogenital system; homeostasis, other neonatal effects. Experimental teratogenic and reproductive effects. Mutation data reported. When heated to decomposition it emits very toxic Cl^- and NO_x.

IDA400 CAS:31842-01-0 ***HR: 3***
INDOPROFEN
mf: $C_{17}H_{15}NO_3$ mw: 281.33

PROP: Colorless scales from ethanol. Mp: 213-214°.

SYNS: BOR-IND ◇ 2-(4-(1-CARBOXYETHYL)PHENYL)-1-ISOINDOLINONE ◇ 4-(1,3-DIHYDRO-1-OXO-2H-ISOINDOL-2-YL)-α-METHYLBENZENEACETIC ACID ◇ FLOSIN ◇ FLOSINT ◇ ISINDONE ◇ K 4277 ◇ p-(1-OXO-2-ISOINDOLINYL)-HYDRATROPIC ACID ◇ α-(4-(1-OXO-2-ISO-INDOLINYL)-PHENYL)-PROPIONIC ACID ◇ 2-(p-(1-OXO-2-ISOINDOLINYL)PHENYL)-PROPIONIC ACID ◇ 1-OXO-2-(p-((α-METHYL)CARBOXYMETHYL)PHENYL)ISOINDOLINE ◇ PRAXIS ◇ REUMOFENE

TOXICITY DATA with REFERENCE
orl-rat LD50:84 mg/kg ARZNAD 23,1100,73
orl-mus LD50:700 mg/kg ARZNAD 23,1100,73

SAFETY PROFILE: Poison by ingestion. An anti-inflammatory agent and analgesic. When heated to decomposition it emits toxic fumes of NO_x.

IDA500 CAS:3568-23-8 ***HR: 3***
INDORM
mf: $C_{20}H_{24}N_2OS \cdot C_4H_4O_4$ mw: 456.60

SYNS: 1678 CB ◇ 10-DIMETHYLAMINOISOPROPYL-2-PROPIONYLPHENOTHIAZINE MALEATE ◇ 1-(10-(2-DIMETHYLAMINOPROPYL)-PHENOTHIAZIN-2-YL)-1-PROPANONEMALEATE ◇ 10-(2-DIMETHYLAMINOPROPYL)-2-PROPIONYLPHENOTHIAZINE MALEATE ◇ DOREVANE ◇ PROPAVAN ◇ PROPIOMAZINE MALEATE ◇ 3-PROPIONYL-10-DIMETHYLAMINO-ISOPROPYLPHENOTHIAZINE MALEATE ◇ PROPIONYLPROMETHAZINE MALEATE

TOXICITY DATA with REFERENCE
orl-rat LD50:500 mg/kg AIPTAK 123,78,59
scu-mus LD50:288 mg/kg AIPTAK 119,367,59
ivn-mus LD50:67 mg/kg AIPTAK 123,78,59

SAFETY PROFILE: Poison by subcutaneous and intravenous routes. Moderately toxic by ingestion. When heated to decomposition it emits toxic fumes of NO_x and SO_x.

IDB000 CAS:37394-33-5 ***HR: 3***
INGENANE HEXADECANOATE
mf: $C_{36}H_{58}O_6$ mw: 586.94

SYN: 2,5,5a,6,9,10,10a,1a-OCTAHYDRO-4-HYDROXYMETHYL-1,1,7,9-TETRAMETHYL-5,5a-6-TRIHYDROXY-1H-2,8a-METHANOCYCLOPENTA(a)CYCLOPROPA(e)CYCLODECEN-11-ONE-5-HEXADECANOATE

TOXICITY DATA with REFERENCE
skin-mus 82 ng MLD 85CVA2 5,213,70
skin-mus TDLo:56 mg/kg/12W-I:ETA 85CVA2 5,213,70

SAFETY PROFILE: Questionable carcinogen with experimental tumorigenic data. A skin irritant. When heated to decomposition it emits acrid smoke and irritating fumes.

IDB100 ***HR: 2***
INHG-SODIUM
mf: $C_{12}H_{14}N_3O_7 \cdot Na \cdot 2H_2O$ mw: 371.32

SYNS: N-ISONICOTINOYL-N'-GLUCURONID-HYDRAZIN-NATRIUMSALZ DIHYDRAT (GERMAN) ◇ 2-(2-ISONICOTINOYL-HYDRAZINO)-d-GLUCOPYRANURONIC ACID SODIUM SALT DIHYDRATE

TOXICITY DATA with REFERENCE
orl-rat LD50:7500 mg/kg ARZNAD 26,409,76
ivn-rat LD50:3150 mg/kg ARZNAD 26,409,76
orl-mus LD50:483 mg/kg ARZNAD 26,409,76
ivn-mus LD50:2150 mg/kg ARZNAD 26,409,76
orl-gpg LD50:665 mg/kg ARZNAD 26,409,76

SAFETY PROFILE: Moderately toxic by ingestion and intravenous routes. When heated to decomposition it emits toxic fumes of NO_x and Na_2O.

IDD000 CAS:15130-85-5 ***HR: D***
INOKOSTERONE
mf: $C_{27}H_{44}O_7$ mw: 480.71

SYN: (22R,25RS)-2-β,3-β,14,20,22,26-HEXAHYDROXY-5-β-CHOLEST-7-EN-6-ONE

TOXICITY DATA with REFERENCE
cyt-dmg-par 300 μmol/L NNBYA7 230,222,71
ipr-mus LD50:7800 mg/kg NYKZAU 66,551,70

SAFETY PROFILE: Mutation data reported. When heated to decomposition it emits acrid smoke and fumes.

IDD100 CAS:18559-59-6 ***HR: 3***
INOLIN
mf: $C_{19}H_{23}NO_5•ClH$ mw: 381.89

PROP: dl-Form (hydrochloride): Pale yellow crystals from methanol + ether. Decomp 224.5-226°. l-Form (hydrochloride): Pale yellow crystals, freely sol in water; sol in alc.

SYNS: TRIMETHOQUINOL ◇ (−)-TRIMETHOQUINOL ◇ 1-1-(3,4,5-TRIMETHOXYBENZYL)-6,7-DIHYDROXY-1,2,3,4-TETRAHYDROISO-QUINOLINE HYDROCHLORIDE ◇ TRIMETOQUINOL ◇ 1-TRIMETO-QUINOL ◇ TRIMETOQUINOL HYDROCHLORIDE ◇ TRIQUINOL

TOXICITY DATA with REFERENCE
orl-rat TDLo:2 g/kg (7-14D preg):REP OYYAA2 2,383,68
orl-rat TDLo:2 g/kg (7-14D preg):TER OYYAA2 2,383,68
orl-rat LD50:2 g/kg YAKUD5 24,2331,82
ipr-rat LD50:298 mg/kg EJPHAZ 5,303,68
scu-rat LD50:1100 mg/kg EJPHAZ 5,303,68
ivn-rat LD50:164 mg/kg EJPHAZ 5,303,68
orl-mus LD50:2250 mg/kg NIIRDN 6,529,82
ipr-mus LD50:370 mg/kg EJPHAZ 5,303,68
scu-mus LD50:2000 mg/kg EJPHAZ 5,303,68
ivn-mus LD50:120 mg/kg NIIRDN 6,529,82
ivn-dog LD50:160 mg/kg EJPHAZ 5,303,68
ipr-gpg LD50:505 mg/kg EJPHAZ 5,303,68
scu-gpg LD50:1470 mg/kg EJPHAZ 5,303,68

SAFETY PROFILE: Poison by intravenous and intraperitoneal routes. Moderately toxic by ingestion and subcutaneous routes. Experimental teratogenic and reproductive effects. When heated to decomposition it emits toxic fumes of NO_x and HCl.

IDE000 CAS:58-63-9 ***HR: 2***
INOSINE
mf: $C_{10}H_{12}N_4O_5$ mw: 268.26

SYNS: ATOREL ◇ HXR ◇ HYPOXANTHINE NUCLEOSIDE ◇ HYPO-XANTHINE RIBONUCLEOSIDE ◇ HYPOXANTHINE RIBOSIDE ◇ HY-POXANTHINE-d-RIBOSIDE ◇ HYPOXANTHOSINE ◇ INO ◇ INOSIE ◇ β-INOSINE ◇ OXIAMIN ◇ PANTHOLIC-L ◇ RIBONOSINE ◇ SEL-FER ◇ TROPHICARDYL

TOXICITY DATA with REFERENCE
dnd-mam:lym 60 mmol/L PNASA6 48,686,62
orl-rat TDLo:558 g/kg (93D pre):REP OYYAA2 15,199,78
ipr-rat LD50:2900 mg/kg NIIRDN 6,77,82
ipr-mus LD50:3175 mg/kg PCJOAU 20,160,86
scu-mus LD50:5000 mg/kg NIIRDN 6,77,82

CONSENSUS REPORTS: Reported in EPA TSCA Inventory.

SAFETY PROFILE: Moderately toxic by intraperitoneal route. Experimental reproductive effects. Mutation data reported. When heated to decomposition it emits toxic fumes of NO_x.

IDE200 CAS:131-99-7 ***HR: 2***
INOSINIC ACID
mf: $C_{10}H_{13}N_4O_8P$ mw: 348.22

PROP: Syrup, solidifies to a glass when dried over H_2SO_4; agreeable sour taste. Freely sol in water, formic acid; very sparingly sol in alc, ether.

SYNS: IMP ◇ 5′-IMP ◇ INOSINE-5′-MONOPHOSPHATE ◇ INOSINE-5′-MONOPHOSPHORIC ACID ◇ INOSINE-5′-PHOSPHATE

TOXICITY DATA with REFERENCE
orl-rat LD50:16 g/kg ARTODN 47,77,81
ipr-rat LD50:4850 mg/kg ARTODN 47,77,81
scu-rat LD50:3900 mg/kg ARTODN 47,77,81
ivn-rat LD50:2730 mg/kg ARTODN 47,77,81
orl-mus LD50:12 g/kg ARTODN 47,77,81
ipr-mus LD50:5400 mg/kg ARTODN 47,77,81
scu-mus LD50:5480 mg/kg ARTODN 47,77,81
ivn-mus LD50:3300 mg/kg ARTODN 47,77,81

CONSENSUS REPORTS: Reported in EPA TSCA Inventory.

SAFETY PROFILE: Moderately toxic by subcutaneous and intravenous routes. When heated to decomposition it emits toxic fumes of PO_x and NO_x.

IDF000 CAS:909-39-7 ***HR: 3***
INSIDON DIHYDROCHLORIDE
mf: $C_{23}H_{29}N_3O•2ClH$ mw: 436.47

PROP: Long, rectangular plates from water. Mp: 90°.

SYNS: 4-(3-(5H-DIBENZ(b,f)AZEPIN-5-YL)PROPYL)-1-PIPERAZINE-ETHANOL DIHYDROCHLORIDE ◇ 5-(Γ-(β-HYDROXYETHYL-PIPERAZINO)PROPYL)-5H-DIBENZO(b,f)AZEPINE DIHYDRO-CHLORIDE ◇ OPIPRAMOL DIHYDROCHLORIDE

TOXICITY DATA with REFERENCE
orl-hmn TDLo:36 mg/kg/25D-C:CNS AJPSAO 119,465,62
orl-rat LD50:1110 mg/kg RSPSA2 52,204,63
ipr-rat LD50:95 mg/kg AIPTAK 148,560,64

scu-rat LD50:497 mg/kg AIPTAK 148,560,64
ivn-rat LD50:32 mg/kg AIPTAK 148,560,64
orl-mus LD50:443 mg/kg RSPSA2 52,204,63
ipr-mus LD50:120 mg/kg AIPTAK 148,560,64
scu-mus LD50:315 mg/kg AIPTAK 148,560,64
ivn-mus LD50:45 mg/kg AIPTAK 148,560,64
ivn-rbt LD50:11 mg/kg AIPTAK 148,560,64

SAFETY PROFILE: A poison by intraperitoneal, intravenous, and subcutaneous routes. Moderately toxic by ingestion. Human systemic effects by ingestion: somnolence. When heated to decomposition it emits very toxic NO_x and HCl.

IDF300 CAS:9004-10-8 ***HR: 3***
INSULIN

PROP: Crystals, hexagonal system, usually obtained as flat rhombohedra and containing 0.4% Zn. Readily sol in dil acids and alkalies.

SYNS: ACTRAPID ◇ DECURVON ◇ ENDOPANCRINE ◇ ILETIN ◇ INSULAR ◇ INSULIN INJECTION ◇ INSULYL ◇ ISZILIN ◇ OPTISULIN LONG

TOXICITY DATA with REFERENCE
dns-hmn:oth 10 mg/L CNREA8 46,2545,86
dns-mus:emb 50 μg/L ECREAL 158,311,85
ivn-rbt TDLo:278 μg/kg (female 9D post):REP BPAAAG 115,439,55
ivn-rbt TDLo:556 μg/kg (female 6-7D post):TER BPAAAG 115,439,55
ivn-cld TDLo:313 μg/kg PEDIAU 81,526,88
ivn-mus LD50:6300 units/kg PSEBAA 118,756,65

SAFETY PROFILE: Poison by intravenous route. An experimental teratogen. Other experimental reproductive effects. Human mutation data reported. A hormone which regulates sugar metabolism. When heated to decomposition it emits toxic fumes of NO_x and SO_x.

IDF325 CAS:9004-17-5 ***HR: 3***
INSULIN PROTAMINE ZINC

PROP: White suspension of rod-shaped crystals.

SYNS: DEPO-INSULIN ◇ DEPOSULIN ◇ HUMULIN I ◇ INSULATARD ◇ INSULIN RETARD RI ◇ INSULIN ZINC PROTAMINATE ◇ INSULIN ZINC PROTAMINE ◇ INSULYL-RETARD ◇ ISOPHANE INSULIN ◇ ISOPHANE INSULIN INJECTION ◇ ISOPHANE INSULIN SUSPENSION ◇ I.P.Z. ◇ NPH 50 INSULIN ◇ NPH INSULIN ◇ NPH ILETIN ◇ PROTAMINE ZINC INSULIN ◇ PROTAMINE ZINC INSULIN INJECTION ◇ PROTAMINE ZINC INSULIN SUSPENSION ◇ ZINC PROTAMINE INSULIN

TOXICITY DATA with REFERENCE
scu-rat TDLo:1458 μg/kg (15-21D preg):REP DIAEAZ 13,44,64
ipr-mus TDLo:167 μg/kg (8D preg):TER ANREAK 124,441,56
scu-mus LD50:195 iu/kg AIPTAK 153,379,65

CONSENSUS REPORTS: Zinc and its compounds are on the Community Right-To-Know List.

SAFETY PROFILE: Poison by subcutaneous route. Experimental teratogenic and reproductive effects. When heated to decomposition it emits toxic fumes of NO_x, SO_x. A hypoglycemic agent. See also various insulin compounds.

IDG000 CAS:480-79-5 ***HR: 3***
INTEGERRIMINE
mf: $C_{18}H_{25}NO_5$ mw: 335.44

PROP: An alkaloid isolated from *S. Senecio integerrimus* (RETOAE 5,55,49).

SYNS: 3-ETHYLIDENE-3,4,5,6,9,11,13,14,14A,14B-DECAHYDRO-6-HYDROXY-5,6-DIMETHYL(1,6)DIOXACYCLODODECINO(2,3,4-GH)-PYRROLIZINE-2,7-DIONE ◇ (15E)-12-HYDROXY-SENECIONAN-11,16-DIONE (9CI) ◇ SQUALIDIN ◇ SQUALIDINE

TOXICITY DATA with REFERENCE
ipr-rat LDLo:250 mg/kg NCNSA6 5,47,53
ipr-mus LD50:75 mg/kg IJEBA6 9,177,71
ivn-mus LD50:78 mg/kg JPETAB 75,69,42

SAFETY PROFILE: Poison by intraperitoneal and intravenous routes. When heated to decomposition it emits toxic fumes of NO_x.

IDH000 CAS:75432-59-6 ***HR: D***
INTRACRON BLUE 3G

TOXICITY DATA with REFERENCE
mmo-sat 50 μg/plate ENMUDM 2,405,80
mma-sat 500 μg/plate ENMUDM 2,405,80

SAFETY PROFILE: Mutation data reported.

IDJ500 CAS:54605-45-7 ***HR: 3***
IOCARMATE MEGLUMINE
mf: $C_{24}H_{20}I_6N_4O_8•2C_7H_{17}NO_5$ mw: 1644.38

SYNS: 5,5′-(ADIPOLYDIMINO)BIS(2,4,6-TRIIODO-N-METHYLISOPHTHALAMIC ACID) ◇ BIS-CONRAY ◇ DB 2041 ◇ DIMEGLUMINE IOCARMATE ◇ DIMERAY ◇ DIMER X ◇ DIRAX ◇ IOCARMIC ACID DI-N-METHYLGLUCAMINE SALT ◇ LM 280 ◇ MEGLUMINE IOCARMATE ◇ MYELOTRAST DI-N-METHYLGLUCAMINE SALT

TOXICITY DATA with REFERENCE
ipr-rat TDLo:3600 mg/kg (female 9-14D post):REP KSRNAM 8,612,74
ipr-rat TDLo:36 g/kg (female 9-14D post):TER KSRNAM 8,612,74
par-wmn TDLo:140 mg/kg:CNS,BPR,MSK BMJOAE 1,692,78
ivn-rat LD50:13300 mg/kg KSRNAM 8,595,74
ice-rat LD50:610 mg/kg KSRNAM 8,595,74
ivn-mus LD50:10900 mg/kg KSRNAM 8,595,74

ice-mus LD50:697 mg/kg KSRNAM 8,595,74
unr-rbt LD50:70 mg/kg FRPSAX 32,835,77

SAFETY PROFILE: Poison by an unspecified route. Moderately toxic by intracerebral route. Human systemic effects by parenteral route: muscle spasms, blood pressure depression and musculo-skeletal changes. An experimental teratogen. Experimental reproductive effects. When heated to decomposition it emits toxic fumes of I^- and NO_x.

IDJ600 CAS:18656-21-8 ***HR: 1***
IODAMIDE MEGLUMINE
mf: $C_{12}H_{11}I_3N_2O_4{\cdot}C_7H_{17}NO_5$ mw: 823.20

SYNS: CONRAXIN H ◇ IODAMIDE 380 ◇ IODAMIDE METHYLGLUCAMINE ◇ MEGLUMINE IODAMIDE

TOXICITY DATA with REFERENCE
ipr-rat LD50:17900 mg/kg NIIRDN 6,871,82
ivn-rat LD50:11400 mg/kg NIIRDN 6,871,82
ivn-mus LD50:9000 mg/kg NIIRDN 6,871,82
ivn-rbt LD50:13200 mg/kg NIIRDN 6,871,82
ipr-gpg LD50:15000 mg/kg NIIRDN 6,871,82

SAFETY PROFILE: Mildly toxic. When heated to decomposition it emits toxic fumes of I^- and NO_x.

IDJ700 ***HR: 1***
IODATES

SAFETY PROFILE: Salts of iodic acid. Variable toxicity. Generally eye, skin, and mucous membrane irritants. Powerful oxidizers. Similar to bromates and chlorates. Contamination of iodates with organic matter may produce explosive mixtures. Iodate is used in bread as an improving agent for the dough. When heated to decomposition they emit toxic fumes of I^-. See also specific compounds.

IDK000 CAS:7782-68-5 ***HR: 3***
IODIC ACID
mf: HIO_3 mw: 175.91

PROP: Orthorhombic crystals which darken upon exposure to light. D: 4.629 @ 0°/4°, mp: 110° (decomp), very sol in water; insol in abs alc, ether, chloroform. Keep well closed and away from light.

SAFETY PROFILE: A powerful oxidizer. Probably a severe eye, skin and mucous membrane irritant. Dangerous reactions with non-metals, e.g., boron (vigorous reaction); charcoal; phosphorus; sulfur (ignition on heating). Incompatible with non-metals; phosphonium iodide. When heated to decomposition it emits toxic fumes of I^-. See also IODINE.

IDL000 ***HR: 2***
IODIDES

SAFETY PROFILE: Similar in toxicity to bromides. Prolonged absorption of iodides may produce "iodism" which is manifested by skin rash, running nose, headache and irritation of mucous membranes. In severe cases, the skin may show pimples, boils, redness, black and blue spots, hives, and blisters. Weakness, anemia, loss of weight and general depression may occur. Generally very soluble in water and easily absorbed into the body. The iodides of copper(I); lead(II), silver(I) and mercury(II) are poorly soluble in water. When heated to decomposition they can emit highly toxic fumes of I^- and iodine compounds. See also IODINE.

IDM000 CAS:7553-56-2 ***HR: 3***
IODINE
mf: I_2 mw: 253.80

PROP: Rhombic, violet-black crystals, metallic luster. Mp: 113.5°, bp: 185.24°, d: 4.93 (Solid @ 25°), vap press: 1 mm @ 38.7°. Characteristic odor, sharp acrid taste, vap press (solid): 0.030 mm @ 0°, very sol in aq solns of HI and iodides.

SYNS: IODE (FRENCH) ◇ IODINE CRYSTALS ◇ IODINE SUBLIMED ◇ IODIO (ITALIAN) ◇ JOD (GERMAN, POLISH) ◇ JOOD (DUTCH)

TOXICITY DATA with REFERENCE
orl-rat TDLo:2750 mg/kg (female 1-22D post):REP JONUAI 84,107,64
orl-hmn LDLo:28 mg/kg:GIT 34ZIAG -,330,69
orl-wmn TDLo:26 mg/kg/1Y-I:SYS PGMJAO 62,661,86
unr-man LDLo:29 mg/kg 85DCAI 2,73,70
orl-rat LD50:14 g/kg DRFUD4 4,876,79
ihl-rat LCLo:800 mg/m^3/1H 85GMAT -,76,82
orl-mus LD50:22 g/kg DRFUD4 4,876,79
orl-dog LDLo:800 mg/kg HBAMAK 4,1289,35
ivn-dog LDLo:40 mg/kg HBTXAC 5,76,59
orl-rbt LD50:10 g/kg DRFUD4 4,876,79
scu-rbt LDLo:175 mg/kg HBTXAC 5,76,59

CONSENSUS REPORTS: Reported in EPA TSCA Inventory.

OSHA PEL: CL 0.1 ppm
ACGIH TLV: CL 0.1 ppm
DFG MAK: 0.1 ppm (1 mg/m^3)

SAFETY PROFILE: A human poison by ingestion and possibly other routes. An experimental poison by intravenous and subcutaneous routes. Moderately toxic by inhalation. Human systemic effects by ingestion: diarrhea, evidence of thyroid hyperfunction. Experimental reproductive effects. Mutation data reported. The effect of iodine vapor upon the body is similar to that of chlorine and bromine, but it is more irritating to the lungs. Serious exposures are seldom encountered in industry due to the low volatility of the solid at ordinary room temperatures. Signs and symptoms are irritation and burning of the eyes, lachrimation, coughing and irritation of the nose and throat. Ingestion of large quantities causes abdominal pain, nausea, vomiting, diarrhea. In severe cases, purging, excessive thirst, and circulatory failure may develop. Doses of 2-3 grams have been fatal. Chronic ingestion of large amounts (200 mg/day) results in thyroid disease.

Explosive reaction with acetylene; antimony powder; hafnium powder + heat; tetraamine copper(II) sulfate + ethanol; trioxygen difluoride (possibly ignition); polyacetylene (at 113°C). Forms sensitive, explosive mixtures with potassium (impact- and heat-sensitive); sodium (shock-sensitive); oxygen difluoride (heat-sensitive). Reacts to form explosive products with ammonia; ammonia + lithium 1-heptynide; ammonia + potassium; butadiene + ethanol + mercuric oxide; silver azide.

Ignition on contact with bromine pentafluoride (or violent reaction); chlorine trifluoride; fluorine; metals (powdered) + water; aluminum-titanium alloys + heat; metal acetylides (e.g., cesium acetylide; copper(I) acetylide; lithium acetylide; rubidium acetylide); non-metals (e.g., boron ignites at 700°C); phosphorus; sodium phosphinate. Violent reaction with acetaldehyde; aluminum + diethyl ether; dipropylmercury; titanium (above 113°C). Incandescent reaction with cesium oxide (above 150°C); bromine trifluoride; metal acetylides or carbides [e.g., barium acetylide (above 122°C); calcium acetylide (above 305°C); strontium acetylide (above 182°C); zirconium acetylide (above 400°C)].

Incompatible with ethanol; ethanol + butadiene; ethanol + phosphorus; ethanol + methanol + HgO; formamide + pyridine + sulfur trioxide; formamide; halogens or interhalogens (e.g., chlorine); mercuric oxide; metals (e.g., aluminum; lithium; magnesium); metal carbides (e.g., lithium carbide; zirconium carbide); oxygen; pyridine; sodium hydride; sulfides.

When heated to decomposition it emits toxic fumes of I^- and various iodine compounds. Reacts vigorously with reducing materials. See also IODIDES.

IDN000 CAS:14696-82-3 ***HR: 3***
IODINE AZIDE
mf: IN_3 mw: 168.93

SYNS: IODINE(I) AZIDE ◇ IODOAZIDE ◇ NITROGEN IODIDE

DOT Classification: Forbidden.

SAFETY PROFILE: A very shock- and friction-sensitive explosive. Incompatible with sulfur-containing alkenes. When heated to decomposition it emits very toxic fumes of I^- and NO_x. See also IODINE.

IDN200 CAS:7789-33-5 ***HR: 3***
IODINE BROMIDE
mf: BrI mw: 206.809

SAFETY PROFILE: Explosive or violent reaction with sodium (by impact); potassium (when heated); tin; phosphorus (when heated). When heated to decomposition it emits toxic fumes of Br^- and I^-. See also BROMIDES and IODIDES.

IDP000 CAS:25402-50-0 ***HR: 3***
IODINE DIOXYGEN TRIFLUORIDE
mf: F_3IO_2 mw: 215.9

SYN: IODINE DIOXIDE TRIFLUORIDE

SAFETY PROFILE: Ignites on contact with flammable organic materials. When heated to decomposition it emits very toxic fumes of F^- and I^-. See also IODINE and FLUORIDES.

IDQ000 CAS:16921-96-3 ***HR: 3***
IODINE HEPTAFLUORIDE
mf: F_7I mw: 259.89

PROP: Colorless gas, crystals when solid. Moldy acrid odor; attacks glass and quartz. D: (liquid): 2.8 @ 6°, sublimes @ 4.77°.

SAFETY PROFILE: Mixtures with hydrogen explode on exposure to sparks or heat. Ignition on contact with carbon; combustible gases (e.g., methane or carbon monoxide); other organic compounds (e.g., benzene; light petroleum; ethanol; ether; cellulose; grease; oils). Reacts vigorously with metals (e.g., barium; potassium; sodium; aluminum + heat; magnesium + heat; tin + heat); and organic materials (e.g., acetic acid; acetone; ethyl acetate). Incompatible with ammonium bromide; ammonium chloride; ammonium iodide; sulfuric acid; water. When heated to decomposition it emits very toxic fumes of F^- and I^-. See also IODINE and FLUORIDES.

IDR000 CAS:3607-48-5 ***HR: 3***
IODINE ISOCYANATE
mf: ClNO mw: 65.46

SAFETY PROFILE: A storage hazard. Solutions gradually precipitate a touch-sensitive explosive solid which is probably cyanogen peroxide. Mildly explosive. When heated to decomposition it emits very toxic fumes of NO_x, Cl^-, and CN^-. See also IODINE and CYANATES.

IDS000 CAS:7790-99-0 ***HR: 3***
IODINE MONOCHLORIDE
DOT: UN 1792
mf: ClI mw: 162.38

PROP: Black crystals or reddish-brown liquid. Exists in α, β forms; crystals α form (stable) black needles; sol in water, alc, ether, CS_2, acetic acid. Red-brown crystals or oily liquid. Mp: (α): 27°, (β): 14°, bp: 97.4 decomp @ 100°; d (α): 3.1822 @ 0°, (β): 3.24 @ 34°.

SYNS: IODINE CHLORIDE ◇ PROTOCHLORURE D'IODE (FRENCH) ◇ WIJS' CHLORIDE

TOXICITY DATA with REFERENCE
orl-rat LDLo:50 mg/kg KODAK* 21MAY71
skn-rat LDLo:500 mg/kg KODAK* 21MAY71

CONSENSUS REPORTS: Reported in EPA TSCA Inventory.

DOT Classification: Corrosive Material; Label: Corrosive.

SAFETY PROFILE: A poison by ingestion. Moderately toxic by skin contact. A corrosive irritant to skin, eyes, and mucous membranes. Moderately explosive when exposed to heat. Reacts with water or steam to produce toxic and corrosive fumes. Dangerous reactions with metals e.g., sodium (mixture explodes on impact); potassium (explodes on contact); aluminum (ignition after a delay period). Reacts violently with Al foil; CdS; PbS; organic matter; P; PCl_3; rubber; Ag_2S; ZnS. When heated to decomposition it emits highly toxic fumes of Cl^- and I^- and may explode. See also IODINE and CHLORIDES.

IDS300 CAS:12029-98-0 ***HR: 3***
IODINE(V) OXIDE
mf: I_2O_5 mw: 333.81

SAFETY PROFILE: Explosive reaction when warmed with non-metals (e.g., carbon; sulfur; rosin; sugar; and other easily combustible materials). Reacts violently with bromine pentafluoride; hydrazine (at high temperatures and pressures). Ignites on contact with aluminum powder. When heated to decomposition it emits toxic fumes of I^-. See also IODATES.

IDT000 CAS:7783-66-6 ***HR: 3***
IODINE PENTAFLUORIDE
DOT: UN 2495
mf: F_5I mw: 221.90

PROP: Liquid. Mp: 9.43°, bp: 100.5°, d: 3.19 @ 25°. Fumes in air attacks glass, especially when hot.

SYN: PENTAFLUOROIODINE

CONSENSUS REPORTS: Reported in EPA TSCA Inventory.

OSHA PEL: TWA 2.5 mg(F)/m^3
ACGIH TLV: TWA 2.5 mg(F)/m^3
DOT Classification: Oxidizer; Label: Oxidizer and Poison.

SAFETY PROFILE: A poison. Probably an irritant to the eyes, skin, and mucous membranes. A powerful oxidizer. Explosive reaction with benzene (above 50°C); diethylaminotrimethyl silane; dimethyl sulfoxide; limonene + tetrafluoroethylene (polymerization); potassium; molten sodium; tetraiodoethylene. Reaction with organic compounds results in charring and then ignition. Violent reaction with water; potassium hydroxide. Incandescent reaction with calcium carbide; potassium hydride; metals and non-metals (e.g., boron; silicon; red phosphorus; sulfur; arsenic; antimony; bismuth; molybdenum; tungsten). When heated to decomposition it emits very toxic fumes of F^- and I^-. See also IODINE and FLUORIDES.

IDU000 CAS:38005-31-1 ***HR: 3***
IODINE(III) PERCHLORATE
mf: Cl_3IO_{12} mw: 425.26

$I(ClO_4)_3$

SAFETY PROFILE: Exploded on laser irradiation. When heated to decomposition it emits very toxic fumes of Cl^- and I^-. See also IODINE and PERCHLORATES.

IDV000 CAS:6540-76-7 ***HR: 3***
IODINE TRIACETATE
mf: $C_6H_9IO_6$ mw: 304.0

$I(OCO{\bullet}CH_3)_3$

SAFETY PROFILE: Explodes when heated to 140°C. Upon decomposition it emits toxic fumes of I^-.

IDW000 CAS:144-48-9 ***HR: 3***
IODOACETAMIDE
mf: C_2H_4INO mw: 184.97

SYNS: α-IODOACETAMIDE ◇ 2-IODOACETAMIDE ◇ MONOIODOACETAMIDE ◇ SURAUTO ◇ USAF D-1

TOXICITY DATA with REFERENCE
dnd-esc 20 ppm MUREAV 24,365,74
dni-hmn:lym 10 μmol/L STBIBN 50,97,75
skn-mus TDLo:1480 mg/kg/20W-I:ETA BJCAAI 7,482,53
orl-mus LD50:74 mg/kg ARTODN 47,09,81
ipr-mus LDLo:50 mg/kg NTIS** AD277-689
ivn-mus LD50:56 mg/kg CSLNX* NX#02060

CONSENSUS REPORTS: EPA Genetic Toxicology Program. Reported in EPA TSCA Inventory.

SAFETY PROFILE: A poison by ingestion, intraperitoneal, and intravenous routes. Questionable carcinogen with experimental tumorigenic data. Human mutation data reported. When heated to decomposition it emits very toxic fumes of I^- and NO_x.

IDY000 CAS:622-50-4 ***HR: 3***
4'-IODOACETANILIDE
mf: C_8H_8INO mw: 261.07

SYNS: p-IODOACETANILIDE ◇ 4-IODOACETANILIDE

TOXICITY DATA with REFERENCE
vn-mus LD50:320 mg/kg CSLNX* NX#02848

CONSENSUS REPORTS: Reported in EPA TSCA Inventory.

SAFETY PROFILE: Poison by intravenous route. When heated to decomposition it emits very toxic fumes of I^- and NO_x.

IDZ000 CAS:64-69-7 ***HR: 3***
IODOACETIC ACID
mf: $C_2H_3IO_2$ mw: 185.95

PROP: Colorless or white crystals. Mp: 82-83°. Sol in water and alc; very sltly sol in ether.

SYNS: IA ◇ IODOACETATE ◇ MIA ◇ MONOIODOACETATE ◇ MONOIODOACETIC ACID

TOXICITY DATA with REFERENCE
cyt-smc 5 mg/L NATUAS 294,263,81
dni-hmn:hla 500 μmol/L RAREAE 37,334,69
dni-mus:ast 2143 μmol/L JPMSAE 67,1235,78
ims-mus TDLo:20 mg/kg (13D preg):TER TJADAB 7,177,73
skn-mus TDLo:5800 mg/kg/27W-I:NEO BJCAAI 7,482,53
scu-mus TDLo:480 mg/kg/14W-I:ETA GANNA2 45,601,54
ihl-rat LCLo:94 g/m^3/30M RPTOAN 41,113,78
ipr-rat LD50:75 mg/kg RPTOAN 41,113,78
scu-rat LD50:60 mg/kg TXAPA9 26,93,73
orl-mus LD50:83 mg/kg JPETAB 86,336,46
ivn-dog LD50:45 mg/kg JNCIAM 31,297,63
scu-rbt LDLo:60 mg/kg JPHYA7 80,360,34

CONSENSUS REPORTS: Reported in EPA TSCA Inventory.

SAFETY PROFILE: A poison by ingestion, subcutaneous, and intravenous routes. Experimental teratogenic effects. Questionable carcinogen with experimental neoplastigenic and tumorigenic data. Human mutation data reported. When heated to decomposition it emits toxic fumes of I^-. See also IODINE.

IDZ400 CAS:14545-08-5 ***HR: 3***
IODOACETYLENE
mf: C_2HI mw: 151.93

SAFETY PROFILE: Explodes when heated above 85°C. Upon decomposition it emits toxic fumes of I^-. See also IODIDES and ACETYLENE COMPOUNDS.

IEB000 CAS:626-01-7 ***HR: 3***
m-IODOANILINE
mf: C_6H_6IN mw: 219.03

PROP: Leaves or needles. Mp: 33°, insol in water, sol in alc.

SYNS: m-AMINOIODOBENZENE ◇ 3-AMINONITROBENZENE ◇ 3-IODOANILINE

TOXICITY DATA with REFERENCE
ivn-mus LD50:100 mg/kg CSLNX* NX#06766

CONSENSUS REPORTS: Reported in EPA TSCA Inventory.

SAFETY PROFILE: Poison by intravenous route. When heated to decomposition it emits very toxic fumes of NO_x and I^-.

IEC000 CAS:540-37-4 ***HR: 3***
p-IODOANILINE
mf: C_6H_6IN mw: 219.03

PROP: Needles from water. Mp: 67-68°; sltly sol in water; sol in alc, ether, and chloroform.

SYNS: p-AMINOPHENYL IODIDE ◇ 4-IODOANILINE ◇ 4-IODOBENZENAMINE

TOXICITY DATA with REFERENCE
orl-rat LD50:523 mg/kg CEHYAN 23,168,78
orl-bwd LD50:100 mg/kg TXAPA9 21,315,72

CONSENSUS REPORTS: Reported in EPA TSCA Inventory. EPA Genetic Toxicology Program.

SAFETY PROFILE: Poison by ingestion. When heated to decomposition it emits very toxic fumes of NO_x and I^-.

IEC500 CAS:591-50-4 ***HR: 3***
IODOBENZENE
mf: C_6H_5I mw: 204.01

SAFETY PROFILE: Explodes when heated above 200°C. Upon decomposition it emits toxic fumes of I^-. See also IODIDES.

IED000 ***HR: 3***
4-IODOBENZENEDIAZONIUM-2-CARBOXYLATE
mf: $C_7H_3IN_2O_2$ mw: 274.0

SAFETY PROFILE: A highly explosive solid. When

heated to decomposition it emits very toxic fumes of I^- and NO_x. See also IODIDES.

IEE000 CAS:88-67-5 ***HR: 3***
o-IODOBENZOIC ACID
mf: $C_7H_5IO_2$ mw: 248.02

PROP: White needles. D: 2.25, mp: 162°. Very sltly sol in water; sol in alc and ether.

SYN: USAF EK-572

TOXICITY DATA with REFERENCE
ipr-mus LD50:200 mg/kg NTIS** AD277-689

CONSENSUS REPORTS: Reported in EPA TSCA Inventory.

SAFETY PROFILE: Poison by intraperitoneal route. When heated to decomposition it emits toxic fumes of I^-.

IEE050 CAS:2532-17-4 ***HR: 2***
o-IODOBENZOIC ACID SODIUM SALT
mf: $C_7H_4IO_2$•Na mw: 270.00

SYNS: SODIUM-o-IODOBENZOATE ◇ SODIUM-2-IODOHIPPURATE

TOXICITY DATA with REFERENCE
orl-mus LD50:1500 mg/kg JAPMA8 43,495,54
ivn-mus LD50:1010 mg/kg MECHAN 6,343,63
ims-mus LD50:864 mg/kg JPETAB 117,307,56

SAFETY PROFILE: Moderately toxic by ingestion, intravenous, and intramuscular routes. When heated to decomposition it emits toxic fumes of I^- and Na_2O.

IEE100 CAS:1005-30-7 ***HR: 2***
p-IODOBENZOIC ACID SODIUM SALT
mf: $C_{15}H_{23}NO_5$ mw: 297.39

SYN: SODIUM-p-IODOBENZOATE

TOXICITY DATA with REFERENCE
ivn-rat LD50:786 mg/kg JAPMA8 32,44,43
orl-mus LD50:2500 mg/kg JAPMA8 43,495,54
ims-mus LD50:540 mg/kg JPETAB 117,307,56

SAFETY PROFILE: Moderately toxic by ingestion, intravenous, and intramuscular routes. When heated to decomposition it emits toxic fumes of I^- and Na_2O.

IEF000 CAS:73927-94-3 ***HR: 3***
(o-IODOBENZOYLOXY)TRIPROPYLSTANNANE
mf: $C_{16}H_{25}IO_2Sn$ mw: 495.00

SYN: TRIPROPYLTIN-o-IODOBENZOATE

TOXICITY DATA with REFERENCE
ivn-mus LD50:32 mg/kg CSLNX* NX#03672

OSHA PEL: TWA 0.1 mg(Sn)/m³ (skin)
ACGIH TLV: TWA 0.1 mg(Sn)/m³ (skin) (Proposed: TWA 0.1 mg(Sn)/m³; STEL 0.2 mg(Sn)/m³ (skin))
NIOSH REL: (Organotin Compounds) TWA 0.1 mg(Sn)/m³

SAFETY PROFILE: Poison by intravenous route. When heated to decomposition it emits toxic fumes of I^-. See also TIN COMPOUNDS.

IEG000 CAS:6088-91-1 ***HR: 3***
1-IODO-1,3-BUTADIYNE
mf: C_4HI mw: 175.94

IC≡CC≡CH

SAFETY PROFILE: Will explode when heated to 35°C or scratched. Do not handle above 30°C. When heated to decomposition it emits toxic fumes of I^-. See also IODINE and ACETYLENE COMPOUNDS.

IEH000 CAS:513-48-4 ***HR: 3***
2-IODOBUTANE
DOT: UN 2390
mf: C_4H_9I mw: 184.03

PROP: Flash p: 14°F.

SYN: sec-BUTYL IODIDE

TOXICITY DATA with REFERENCE
dnr-esc 25 μL/well/16H CBINA8 15,219,76
ipr-mus TDLo:6000 mg/kg/8W-I:NEO CNREA8 35,1411,75

CONSENSUS REPORTS: Reported in EPA TSCA Inventory. EPA Genetic Toxicology Program.

DOT Classification: Flammable Liquid; Label: Flammable Liquid.

SAFETY PROFILE: Questionable carcinogen with experimental neoplastigenic data. Mutation data reported. A very dangerous fire hazard when exposed to heat or flame; can react vigorously with oxidizing materials. When heated to decomposition it emits toxic fumes of I^-. See also IODIDES.

IEI000 CAS:676-75-5 ***HR: 3***
IODODIMETHYLARSINE
mf: C_2H_6AsI mw: 231.89

CONSENSUS REPORTS: Arsenic and its compounds are on the Community Right-To-Know List.

SAFETY PROFILE: Incompatible with air. Residue from distillation explodes at 72°C. When heated to decomposition it emits toxic fumes of As and I^-. See also ARSENIC COMPOUNDS, IODIDES, and ACETYLENE COMPOUNDS.

IEI600 CAS:10557-85-4 ***HR: 3***
4-IODO-3,5-DIMETHYLISOXAZOLE
mf: C_5H_6INO mw: 223.01

SAFETY PROFILE: Reaction with peroxytrifluoroacetic acid forms an explosive product. When heated to decomposition it emits toxic fumes of I^- and NO_x. See also IODIDES.

IEJ000 ***HR: 3***
2-IODO-3,5-DINITROBIPHENYL
mf: $C_{12}H_7IN_2O_4$ mw: 370.05

SAFETY PROFILE: Explosive reaction with the sodium salt of ethyl acetoacetate. When heated to decomposition it emits very toxic fumes of NO_x and I^-. See also IODIDES and NITRO COMPOUNDS of AROMATIC HYDROCARBONS.

IEL000 CAS:624-76-0 ***HR: D***
2-IODOETHANOL
mf: C_2H_5IO mw: 171.97

PROP: Colorless liquid. D: 2.197 @ 20°/4°, bp: 85° @ 25 mm, sol in water and alc.

SYNS: ETHYLENE IODOHYDRIN ◇ IODOETHANOL

TOXICITY DATA with REFERENCE
mmo-sat 2 μmol/plate MUREAV 26,367,74
dnr-esc 1 μmol/L EVHPAZ 21,79,77
mmo-klp 15 μmol/L EXPEAM 25,85,69

CONSENSUS REPORTS: EPA Genetic Toxicology Program.

SAFETY PROFILE: Mutation data reported. When heated to decomposition it emits toxic fumes of I^-. See also IODIDES and ALCOHOLS.

IEL700 CAS:83665-55-8 ***HR: 3***
2-(2-IODOETHYL)-1,3-DIOXOLANE
mf: $C_5H_9IO_2$ mw: 228.03

$IC_2H_4CHOC_2H_4O$ (ring: CH–O…O)

SAFETY PROFILE: Decomposes violently at 55°C/1.6 mbar. Upon decomposition it emits toxic fumes of I^-. See also IODIDES.

IEM300 CAS:5110-69-0 ***HR: 3***
(2-IODOETHYL)TRIMETHYLAMMONIUM IODIDE
mf: $C_5H_{13}IN \cdot I$ mw: 340.99

TOXICITY DATA with REFERENCE
orl-rat LD50:700 mg/kg QJPPAL 20,81,47
scu-rat LD50:200 mg/kg QJJPAL 20,81,47
orl-mus LD50:1500 mg/kg QJJPAL 20,81,47
ipr-mus LD50:150 mg/kg QJJPAL 20,81,47
scu-mus LD50:200 mg/kg QJJPAL 20,81,47
ims-mus LD50:130 mg/kg QJJPAL 20,81,47

SAFETY PROFILE: Poison by subcutaneous, intramuscular, and intraperitoneal routes. Moderately toxic by ingestion. When heated to decomposition it emits toxic fumes of I^-, NH_3, and NO_x. See also IODIDES.

IEN000 CAS:18181-70-9 ***HR: 2***
IODOFENOPHOS
mf: $C_8H_8Cl_2IO_3PS$ mw: 412.99

PROP: A crystalline powder, sol in kerosene.

SYNS: ALFACRON ◇ C-9491 ◇ CIBA C-9491 ◇ CIBA 9491 ◇ CIBA-GEIGY C-9491 ◇ COMPOUND C-9491 ◇ O-(2,5-DICHLORO-4-IODOPHENYL) O,O-DIMETHYL PHOSPHOROTHIOATE ◇ 3,4-DICHLOROPHENOL, O-ESTER with O-METHYL METHYLPHOSPHORAMIDOTHIOATE ◇ O,O-DIMETHYL-O-(2,5-DICHLOR-4-JODPHENYL)-THIONOPHOSPHAT (GERMAN) ◇ O,O-DIMETHYL-O-(2,5-DICHLOR-4-JODPHENYL)-MONOTHIOPHOSPHAT(GERMAN) ◇ O,O-DIMETHYL-O-2,5-DICHLORO-4-IODOPHENYL THIOPHOSPHATE ◇ ENT 27,408 ◇ IODOPHOS ◇ JODFENPHOS ◇ NSC 190998 ◇ NUVANOL N ◇ OMS-1211

TOXICITY DATA with REFERENCE
orl-rat LD50:2000 mg/kg GUCHAZ 5,197,68
skn-rat LD50:2150 mg/kg SPEADM 78-1,45,74
orl-mus LD50:3 g/kg 85DPAN -,-,71/76
orl-dog LD50:3 g/kg 85DPAN -,-,71/76
orl-rbt LD50:2 g/kg 85DPAN -,-,71/76
skn-rbt LD50:500 mg/kg 85DPAN -,-,71/76

CONSENSUS REPORTS: Chlorophenol compounds are on The Community Right-To-Know List.

SAFETY PROFILE: Moderately toxic by ingestion and skin contact. Used as a pesticide. When heated to decomposition it emits very toxic fumes of Cl^-, I^-, PO_x, and SO_x. See also CHLOROPHENOLS.

IEO000 CAS:14722-22-6 ***HR: 3***
N-(3-IODO-2-FLUORENYL)ACETAMIDE
mf: $C_{15}H_{12}INO$ mw: 349.18

SYN: 3-IODO-2-FAA

TOXICITY DATA with REFERENCE
orl-rat TDLo:3400 mg/kg/44W-C:CAR JNCIAM 24,149,60

SAFETY PROFILE: Questionable carcinogen with experimental carcinogenic data. When heated to decomposition it emits very toxic fumes of I^- and NO_x.

IEP000 CAS:75-47-8 ***HR: 3***
IODOFORM
mf: CHI_3 mw: 393.72

PROP: Yellow powder or crystals, disagreeable odor. D:

4.1, mp: 120° (approx), bp: subl. Decomp @ high temp evolving iodine, volatile with steam. Very sol in water, benzene, acetone; sltly sol in petr ether.

SYNS: NCI-C04568 ◇ TRIIODOMETHANE

TOXICITY DATA with REFERENCE
mmo-sat 67 μg/plate ENMUDM 5(Suppl 1),3,83
mma-sat 100 μg/plate ENMUDM 5(Suppl 1),3,83
orl-rat LD50:355 mg/kg ZDTUA6 27(5),9,83
ihl-rat LC50:165 ppm/7H JTEHD6 8,59,81
skn-rat LD50:1184 mg/kg ZDTUAB 27(5),9,83
orl-mus LD50:810 mg/kg ZDTUAB 27(5),9,83
scu-mus LD50:630 mg/kg TXAPA9 4,354,62
orl-dog LDLo:1000 mg/kg ZEPTAT 1,446,05
orl-rbt LD50:450 mg/kg ZDTUAB 27(5),9,83
scu-rbt LDLo:500 mg/kg ZEPTAT 1,446,05

CONSENSUS REPORTS: NCI Carcinogenesis Bioassay (gavage); No Evidence: mouse, rat NCITR* NCI-CG-TR-110,78. Reported in EPA TSCA Inventory.

OSHA PEL: TWA 0.6 ppm (skin)
ACGIH TLV: TWA 0.6 ppm

SAFETY PROFILE: A poison by ingestion. Moderately toxic by inhalation, skin contact, and subcutaneous routes. Mutation data reported. Used as an antiseptic, disinfectant on superficial wounds and in female reproductive tract. 1:1 mixtures with hexamethylenetetramine explode at 178°C. Incompatible with mercuric oxide; calomel; silver nitrate; tannin; Balsam Peru directly mixed; Li; acetone. When heated to decomposition it emits toxic fumes of I^-. See also IODIDES.

IEP200 CAS:547-91-1 ***HR: 2***
7-IODO-8-HYDROYQUINOLINE-5-SULFONIC ACID
mf: $C_9H_6INO_4S$ mw: 351.12

PROP: Sulfur yellow, almost odorless and tasteless, crystalline powder. Mp: 260-270° (decomp). One gram dissolves in 500 mL cold water, 170 mL boiling water; sltly sol in alc; practically insol in ether or oils.

SYNS: ANAYODIN ◇ CHINIOFON ◇ FERRON ◇ 8-HYDROXY-7-IODOQUINOLINE SULFONATE ◇ 8-HYDROXY-7-IODO-5-QUINOLINESULFONIC ACID ◇ 8-HYDROXY-7-IODOQUINOLINESULFONIC ACID ◇ LORETIN ◇ MEDITRENE ◇ QUINIPHEN ◇ QUINOXYL ◇ SEFONA ◇ YATREN ◇ YELLON

TOXICITY DATA with REFERENCE
ivn-rat LDLo:500 mg/kg JPETAB 63,122,38
ims-rat LDLo:1000 mg/kg JPETAB 63,122,38
orl-mus LD50:4000 mg/kg JAPMA8 43,495,54

CONSENSUS REPORTS: Reported in EPA TSCA Inventory.

SAFETY PROFILE: Moderately toxic by ingestion, intravenous, and intramuscular routes. When heated to decomposition it emits toxic fumes of NO_x and SO_x. Note: The name Anayodin is also used to designate sodium iodide.

IER000 CAS:27018-50-4 ***HR: 3***
7-IODOMETHYL-12-METHYLBENZ(a)ANTHRACENE
mf: $C_{20}H_{15}I$ mw: 382.25

TOXICITY DATA with REFERENCE
scu-rat TDLo:150 mg/kg/39D-I:NEO CNREA8 31,1951,71

SAFETY PROFILE: Questionable carcinogen with experimental neoplastigenic data. When heated to decomposition it emits toxic fumes of I^-.

IET000 CAS:64049-02-1 ***HR: 3***
IODOMETHYLTRIMETHYLARSONIUM IODIDE

TOXICITY DATA with REFERENCE
scu-mus LDLo:220 mg/kg JPETAB 25,315,25

CONSENSUS REPORTS: Arsenic and its compounds are on the Community Right-To-Know List.

OSHA PEL: TWA 0.5 mg(As)/m^3

SAFETY PROFILE: Poison by subcutaneous route. When heated to decomposition it emits very toxic fumes of As and I^-. See also ARSENIC COMPOUNDS.

IEU000 ***HR: 3***
1-IODO-3-PENTEN-1-YNE
mf: C_5H_5I mw: 192.00

$$IC{\equiv}CCH{=}CHCH_3$$

SAFETY PROFILE: Residue from distillation explodes at 72°C. When heated to decomposition it emits toxic fumes of I^-. See also IODIDES and ACETYLENE COMPOUNDS.

IEU100 ***HR: 3***
IODOPHENE
mf: $C_{20}H_{10}I_4O_4$ mw: 821.90

SYNS: TETRAIODOPHENOLPHTHALEIN ◇ 3′,3″,5′,5″-TETRAIODOPHENOLPHTHALEIN ◇ TETRAJODPHENOLPHTHALEIN (GERMAN)

TOXICITY DATA with REFERENCE
orl-rat LD50:2800 mg/kg KLWOAZ 20,125,41
ivn-rat LD50:310 mg/kg KLWOAZ 20,125,41
orl-dck LD50:79200 μg/kg VETNAL 54(5),64,78

SAFETY PROFILE: Poison by ingestion and intravenous routes. When heated to decomposition it emits toxic fumes of I^-.

IEV000 CAS:533-58-4 ***HR: 2***
2-IODOPHENOL
mf: C_6H_5IO mw: 220.01

PROP: Needles or plates. D: 1.876, mp: 43° bp: 186-187°. Sol in hot water; very sol in alc, ether, chloroform, and benzene.

SYN: o-IODOPHENOL

TOXICITY DATA with REFERENCE
scu-rat LDLo:4000 mg/kg RMSRA6 16,449,1896

CONSENSUS REPORTS: Reported in EPA TSCA Inventory.

SAFETY PROFILE: Moderately toxic by subcutaneous route. When heated to decomposition it emits toxic fumes of I^-. See also IODIDES and PHENOL.

IEW000 CAS:540-38-5 ***HR: 3***
4-IODOPHENOL
mf: C_6H_5IO mw: 220.01

PROP: Needles or water. Mp: 93-94°, d: 1.857, bp: decomp. Sltly sol in water; very sol in alc and ether.

SYN: p-IODOPHENOL

TOXICITY DATA with REFERENCE
skn-mus TDLo:7200 mg/kg/18W-I:NEO CNREA8 19,413,59
ipr-mus LDLo:700 mg/kg JPMSAE 67,1154,78

CONSENSUS REPORTS: Reported in EPA TSCA Inventory.

SAFETY PROFILE: Questionable carcinogen with experimental neoplastigenic data. Moderately toxic by intraperitoneal route. When heated to decomposition it emits toxic fumes of I^-. See also IODIDES and PHENOL.

IEX000 ***HR: 3***
1-IODO-3-PHENYL-2-PROPYNE
mf: C_9H_7I mw: 242.01

$$ICH_2C \equiv CC_6H_5$$

SAFETY PROFILE: Detonates on distillation. When heated to decomposition it emits toxic fumes of I^-. See also IODIDES and ACETYLENE COMPOUNDS.

IEY000 CAS:141-76-4 ***HR: 3***
3-IODOPROPIONIC ACID
mf: $C_3H_5IO_2$ mw: 199.98

PROP: (a) Needles from water. D: 1.857, mp: 93-94°, bp: decomp. Sltly sol in water; very sol in alc and ether. (b) Needles. Mp: 44.5-45.5°, bp: 105°. Very sltly sol in water; sol in alc, ether.

TOXICITY DATA with REFERENCE
mmo-sat 50 μg/plate DHEFDK FDA-78-1046,78
skn-mus TDLo:5700 mg/kg/3W-I:ETA CNREA8 28,653,68
skn-mus LDLo:1900 mg/kg CNREA8 28,653,68

CONSENSUS REPORTS: Reported in EPA TSCA Inventory.

SAFETY PROFILE: Moderately toxic by skin contact. Questionable carcinogen with experimental tumorigenic data. Mutation data reported. When heated to decomposition it emits toxic fumes of I^-.

IEZ800 CAS:659-86-9 ***HR: 3***
3-IODOPROPYNE
mf: C_3H_3I mw: 165.96

$$HC \equiv CCH_2I$$

SAFETY PROFILE: Explodes when heated to 180°C. Upon decomposition it emits toxic fumes of I^-. See also IODIDES and ACETYLENE COMPOUNDS.

IFA000 CAS:777-11-7 ***HR: 3***
3-IODO-2-PROPYNYL-2,4,5-TRICHLOROPHENYL ETHER
mf: $C_9H_4Cl_3IO$ mw: 361.38

SYNS: 2,4,5-TRICHLOROPHENYL-Γ-IODOPROPARGIL ETHER ◇ 2,4,5-TRICHLOROPHENYL IODOPROPARGYL ETHER

TOXICITY DATA with REFERENCE
skn-rbt 10 mg/72H MLD TXAPA9 22,375,72
eye-rbt 1 mg/24H MLD TXAPA9 22,375,72
skn-pig 5 mg/24H MLD TXAPA9 22,375,72
ipr-rat LD50:152 mg/kg TXAPA9 22,375,72
ipr-mus LD50:183 mg/kg TXAPA9 22,375,72
ipr-dog LD50:250 mg/kg TXAPA9 22,375,72
orl-rbt LD50:1625 mg/kg TXAPA9 22,375,72
ipr-rbt LD50:137 mg/kg TXAPA9 22,375,72

CONSENSUS REPORTS: EPA Genetic Toxicology Program.

SAFETY PROFILE: A poison by intraperitoneal route. Moderately toxic by ingestion. A skin and eye irritant. An FDA over-the-counter drug. An antibacterial agent. When heated to decomposition it emits very toxic Cl^- and I^-.

IFC000 CAS:536-80-1 ***HR: 3***
IODOSOBENZENE
mf: C_6H_5IO mw: 220.01

TOXICITY DATA with REFERENCE
ipr-mus LDLo:63 mg/kg CBCCT* 4,44,52
ivn-mus LD50:180 mg/kg CSLNX* NX#01479

CONSENSUS REPORTS: Reported in EPA TSCA Inventory.

SAFETY PROFILE: Poison by intraperitoneal and intravenous routes. Explodes when heated to 210°C. When heated to decomposition it emits toxic fumes of I^-.

IFD000 CAS:3240-34-4 ***HR: 3***
IODOSOBENZENE DIACETATE
mf: $C_8H_8IO{\bullet}C_2H_3O_2$ mw: 322.11

TOXICITY DATA with REFERENCE
ivn-mus LD50:56 mg/kg CSLNX* NX#02976

CONSENSUS REPORTS: Reported in EPA TSCA Inventory.

SAFETY PROFILE: Poison by intravenous route. When heated to decomposition it emits toxic fumes of I^-.

IFE000 CAS:696-33-3 ***HR: 3***
IODOSYLBENZENE
mf: C_6H_5IO mw: 219.98

$O{:}IC_6H_5$

SAFETY PROFILE: Explodes at 210°C. When heated to decomposition it emits toxic fumes of I^-.

IFE875 CAS:69180-59-2 ***HR: 3***
4-IODOSYLTOLUENE
mf: C_7H_7IO mw: 234.04

$CH_3C_6H_4I{:}O$

SAFETY PROFILE: Explodes when heated above 175°C. Upon decomposition it emits toxic fumes of I^-. See also IODIDES.

IFE879 ***HR: 3***
2-IODOSYLVINYL CHLORIDE
mf: C_2H_2ClIO mw: 204.5

$ClCH{=}CHI{:}O$

SAFETY PROFILE: Explodes when heated to 63°C. Reacts with water to form the more explosive 2-iodylvinyl chloride. Upon decomposition it emits toxic fumes of Cl^- and I^-. See also IODIDES and CHLORIDES.

IFG000 CAS:17236-22-5 ***HR: 3***
3-IODOTETRAHYDROTHIOPHENE-1,1-DIOXIDE
mf: $C_4H_7IO_2S$ mw: 246.07

SYN: TETRAHYDRO-3-IODOTHIOPHENE-1,1-DIOXIDE

TOXICITY DATA with REFERENCE
orl-mus LD50:100 mg/kg AIPTAK 119,423,59
ipr-mus LD50:33 mg/kg AIPTAK 119,423,59
ivn-mus LD50:9300 μg/kg AIPTAK 119,423,59

SAFETY PROFILE: A poison by ingestion, intraperitoneal, and intravenous routes. When heated to decomposition it emits very toxic I^- and SO_x.

IFK509 CAS:624-31-7 ***HR: 3***
4-IODOTOLUENE
mf: C_7H_7I mw: 218.04

SAFETY PROFILE: Explodes when heated above 200°C. Upon decomposition it emits toxic fumes of I^-. See also IODIDES.

IFL000 CAS:26037-72-9 ***HR: 3***
IODO(p-TOLYL)MERCURY
mf: C_7H_7HgI mw: 418.63

SYN: p-TOLYMERCURY IODIDE

TOXICITY DATA with REFERENCE
ivn-mus LD50:18 mg/kg CSLNX* NX#05140

CONSENSUS REPORTS: Mercury and its compounds are on the Community Right-To-Know List.
ACGIH TLV: TWA 0.1 mg(Hg)/m^3 (skin)
NIOSH REL: (Mercury, inorganic) TWA 0.05 mg(Hg)/m^3

SAFETY PROFILE: Poison by intravenous route. When heated to decomposition it emits very toxic fumes of Hg and I^-. See also MERCURY COMPOUNDS and IODIDES.

IFM000 CAS:7342-47-4 ***HR: 3***
IODOTRIBUTYLSTANNANE
mf: $C_{12}H_{27}ISn$ mw: 416.98

SYN: TRI-N-BUTYL TIN IODIDE

TOXICITY DATA with REFERENCE
ihl-mus LCLo:1340 mg/m^3 NDRC** NDCro-132,Feb,42
orl-rbt LDLo:100 mg/kg SAIGBL 15,3,73
skn-rbt LDLo:200 mg/kg SAIGBL 15,3,73

OSHA PEL: TWA 0.1 mg(Sn)/m^3 (skin)
ACGIH TLV: TWA 0.1 mg(Sn)/m^3 (skin) (Proposed: TWA 0.1 mg(Sn)/m^3; STEL 0.2 mg(Sn)/m^3 (skin))
NIOSH REL: (Organotin Compounds) TWA 0.1 mg(Sn)/m^3

SAFETY PROFILE: Poison by ingestion and skin contact. Moderately toxic by inhalation. Tributyl tin compounds are extremely toxic to marine life. When heated to decomposition it emits toxic fumes of I^-. See also IODIDES and TIN COMPOUNDS.

IFN000 CAS:811-73-4 ***HR: 3***
IODOTRIMETHYLTIN
mf: C_3H_9ISn mw: 290.71

PROP: White powder, insol in water and organic solvents.

SYNS: IODOTRIMETHYLSTANNANE ◇ TRIMETHYLSTANNYL IODINE ◇ TRIMETHYLTIN IODIDE

TOXICITY DATA with REFERENCE
ivn-mus LD50:4500 μg/kg CSLNX* NX#02982

OSHA PEL: TWA 0.1 mg(Sn)/m^3 (skin)
ACGIH TLV: TWA 0.1 mg(Sn)/m^3 (skin) (Proposed: TWA 0.1 mg(Sn)/m^3; STEL 0.2 mg(Sn)/m^3 (skin))
NIOSH REL: (Organotin Compounds) TWA 0.1 mg(Sn)/m^3

SAFETY PROFILE: A poison by intravenous route. When heated to decomposition it emits toxic fumes of I^-. See also IODIDES and TIN COMPOUNDS.

IFO000 CAS:894-09-7 ***HR: 3***
IODOTRIPHENYLSTANNANE
mf: $C_{18}H_{15}ISn$ mw: 476.92

SYN: TRIPHENYLTIN IODIDE

TOXICITY DATA with REFERENCE
ivn-mus LD50:56 mg/kg CSLNX* NX#02203

OSHA PEL: TWA 0.1 mg(Sn)/m^3 (skin)
ACGIH TLV: TWA 0.1 mg(Sn)/m^3 (skin) (Proposed: TWA 0.1 mg(Sn)/m^3; STEL 0.2 mg(Sn)/m^3 (skin))
NIOSH REL: (Organotin Compounds) TWA 0.1 mg(Sn)/m^3

SAFETY PROFILE: Poison by intravenous route. When heated to decomposition it emits toxic fumes of I^-. See also TIN COMPOUNDS and IODIDES.

IFO700 ***HR: 3***
IODO-UNDECINIC ACID
mf: $C_{11}H_{17}IO_2$ mw: 308.18

SYNS: 11-IODO-10-UNDECINIC ACID ◇ 11-IODO-10-UNDECYNOIC ACID

TOXICITY DATA with REFERENCE
scu-rat LD50:149 mg/kg OYYAA2 2,70,68
orl-mus LD50:225 mg/kg OYYAA2 2,70,68
scu-mus LD50:117 mg/kg OYYAA2 2,70,68

SAFETY PROFILE: Poison by ingestion and subcutaneous routes. When heated to decomposition it emits toxic fumes of I^-.

IFP000 CAS:696-07-1 ***HR: 3***
5-IODOURACIL
mf: $C_4H_3IN_2O_2$ mw: 237.99

TOXICITY DATA with REFERENCE
ipr-mus LDLo:300 mg/kg BCPCA6 13,1249,64

CONSENSUS REPORTS: Reported in EPA TSCA Inventory.

SAFETY PROFILE: Poison by intraperitoneal route. When heated to decomposition it emits very toxic fumes of I^- and NO_x.

IFP800 CAS:51764-33-1 ***HR: 2***
IODOXAMATE MEGLUMINE
mf: $C_{26}H_{26}I_6N_2O_{10}\cdot 2C_7H_{17}NO_5$ mw: 1678.44

SYNS: CHOLOVUE ◇ ENDOBIL ◇ IODOXAMIC ACID MEGLUMINE SALT ◇ MEGLUMINE IODOXAMATE

TOXICITY DATA with REFERENCE
ivn-rat LD50:4500 mg/kg NIIRDN 6,871,82
par-rat LD50:2928 mg/kg FRPSAX 28,1011,73
ivn-mus LD50:5300 mg/kg NIIRDN 6,871,82
ivn-dog LD50:4850 mg/kg FRPSAX 28,996,73
par-rbt LD50:5398 mg/kg FRPSAX 28,1011,73

SAFETY PROFILE: Moderately toxic by parenteral route. When heated to decomposition it emits toxic fumes of I^- and NO_x.

IFQ775 CAS:16825-74-4 ***HR: 3***
4-IODYLANISOLE
mf: $C_7H_7IO_3$ mw: 266.04

$CH_3OC_6H_4I(O){:}O$

SAFETY PROFILE: Explodes when heated to 225°C. Upon decomposition it emits toxic fumes of I^-. See also IODIDES.

IFR000 CAS:696-33-3 ***HR: 3***
IODYLBENZENE
mf: $C_6H_5IO_2$ mw: 236.0

SAFETY PROFILE: An explosive sensitive to impact and heating to 230°C. Upon decomposition it emits toxic fumes of I^-. See also IODIDES.

IFS000 ***HR: 3***
IODYLBENZENE PERCHLORATE
mf: $C_6H_6ClIO_6$ mw: 336.47

$C_6H_5I^+(O)OH\ ClO_4^-$

SYN: (HYDROXY)(OXO)(PHENYL)-LAMBDA^3-IODANIUM PERCHLORATE

SAFETY PROFILE: While damp it exploded violently. When heated to decomposition it emits toxic fumes of Cl^-. See also IODIDES and PERCHLORATES.

IFS350 CAS:16825-72-2 ***HR: 3***
4-IODYL TOLUENE
mf: $C_7H_7IO_2$ mw: 250.04

$CH_3C_6H_4I(O){:}O$

SAFETY PROFILE: Explodes when heated above 200°C. Upon decomposition it emits toxic fumes of I^-. See also IODIDES.

IFS385 ***HR: 3***
2-IODYLVINYL CHLORIDE
mf: $C_2H_2ClIO_2$ mw: 220.39

ClCH=CHI(:O)$_2$

SAFETY PROFILE: A powerful explosive sensitive to impact, friction, or heating to 135°C. Upon decomposition it emits toxic fumes of Cl^- and I^-. See also IODIDES and CHLORIDES.

IFS400 CAS:63941-74-2 ***HR: 3***
IOGLUCOMIDE
mf: $C_{20}H_{28}I_3N_3O_{13}$ mw: 899.21

SYNS: 3,5-BIS-d-GLUCONAMIDO-2,4,6-TRIIODO-N-METHYLBENZAMIDE ◇ N,N'-(2,4,6-TRIIODO-5-((METHYLAMINO) CARBONYL)-1,3-PHENYLENE)BIS-d-GLUCONAMIDE

TOXICITY DATA with REFERENCE
ivn-rat LD50:15600 mg/kg RADLAX 140,713,81
ice-rat LD50:365 mg/kg RADLAX 140,713,81
ivn-mus LD50:16200 mg/kg RADLAX 140,713,81
ivn-dog LD50:22700 mg/kg RADLAX 140,713,81
ivn-rbt LD50:23800 mg/kg RADLAX 140,713,81

SAFETY PROFILE: Poison by intracerebral route. When heated to decomposition it emits toxic fumes of I^- and NO_x.

IFT100 CAS:57285-10-6 ***HR: 3***
IOMEX

TOXICITY DATA with REFERENCE
orl-rat TDLo:60 g/kg (24D male):REP TXCYAC 7,57,77
orl-rat LDLo:10 g/kg BECTA6 14,241,75
ipr-rat LD50:5 mg/kg SRTCDF -,142,77
itr-rat LDLo:250 mg/kg BECTA6 14,241,75
orl-mus LDLo:10 g/kg BECTA6 14,241,75
ipr-mus LDLo:7500 mg/kg BECTA6 14,241,75
orl-rbt LDLo:10 g/kg BECTA6 14,241,75

SAFETY PROFILE: Poison by intraperitoneal and intratracheal routes. Mildly toxic by ingestion. Experimental reproductive effects. A weed killer.

IFT300 CAS:28728-55-4 ***HR: 3***
6,3-IONENE BROMIDE
mf: $(C_{13}H_{30}Br_2N_2)_n$

SYN:
POLY((DIMETHYLIMINIO)HEXAMETHYLENE(DIMETHYLIMINO)TRIMETHYLENE DIBROMIDE)

TOXICITY DATA with REFERENCE
ivn-rat LD50:20 mg/kg USXXAM #4013507
ipr-mus LD50:30 mg/kg RPTOAN 37,267,74
ivn-mus LD50:28 mg/kg USXXAM #4013507

SAFETY PROFILE: Poison by intravenous and intraperitoneal routes. When heated to decomposition it emits toxic fumes of NO_x and Br^-.

IFV000 CAS:8013-90-9 ***HR: 2***
IONONE
mf: $C_{13}H_{20}O$ mw: 192.33

PROP: Liquid. Odor of cedarwood. D: 0.933-0.937 @ 25°/25°, bp: 126-128° @ 12 mm. Misc with abs alc; very sltly sol in H_2O; moderate sol in alc, ether, chloroform, benzene.

SYNS: IRALDEINE ◇ IRISONE

TOXICITY DATA with REFERENCE
orl-rat LD50:4590 mg/kg FCTXAV 2,327,64
scu-mus LD50:2605 mg/kg JAPMA8 46,77,57

CONSENSUS REPORTS: Reported in EPA TSCA Inventory.

SAFETY PROFILE: Moderately toxic by subcutaneous route. Mildly toxic by ingestion. Caution: May cause allergic reactions. When heated to decomposition it emits acrid smoke and fumes.

IFW000 CAS:127-41-3 ***HR: 1***
α-IONONE
mf: $C_{13}H_{20}O$ mw: 192.33

PROP: Colorless oil; woody, violet odor. D: 0.930, refr index: 1.497-1.502, bp: 136.1. Sol in alc, fixed oils propylene glycol; sltly sol in water; misc in ether; insol in glycerin.

SYNS: α-CYCLOCITRYLIDENEACETONE ◇ FEMA No. 2594 ◇ 4-(2,6,6-TRIMETHYL-2-CYCLOHEXEN-1-YL)-3-BUTEN-2-ONE

TOXICITY DATA with REFERENCE
orl-rat LD50:4590 mg/kg FCTXAV 2,327,64

CONSENSUS REPORTS: Reported in EPA TSCA Inventory.

SAFETY PROFILE: Mildly toxic by ingestion. When heated to decomposition it emits acrid smoke and fumes. See also IONONE.

IFX000 CAS:14901-07-6 ***HR: 1***
β-IONONE
mf: $C_{13}H_{20}O$ mw: 192.33

PROP: Colorless oil; woody odor. D: 0.944, refr index: 1.517-1.522, bp: 140°, flash p: +234°F. Sol in alc, fixed oils, propylene glycol; sltly sol in water; misc in ether; insol in glycerin.

SYNS: β-CYCLOCITRYLIDENEACETONE ◇ FEMA No. 2595 ◇ 4-(2,6,6-TRIMETHYL-1-CYCLOHEXEN-1-YL)-3-BUTEN-2-ONE

TOXICITY DATA with REFERENCE
orl-rat LD50:4590 mg/kg FCTXAV 2,327,64

CONSENSUS REPORTS: Reported in EPA TSCA Inventory.

SAFETY PROFILE: Mildly toxic by ingestion. Combustible liquid. When heated to decomposition it emits acrid smoke and fumes. See also IONONE.

IFX200 CAS:88-26-6 ***HR: 1***
IONOX 100
mf: $C_{15}H_{24}O_2$ mw: 236.39

SYNS: ANTIOXIDANT 754 ◇ AO 754 ◇ 3,5-BIS(1,1-DIMETHYLETHYL)-4-HYDROXY-BENZENEMETHANOL◇ 3,5-DI-tert-BUTYL-4-HYDROXYBENZYL ALCOHOL ◇ 2,6-DI-tert-BUTYL-4-HYDROXYMETHYLPHENOL ◇ IONOX 100 ANTIOXIDANT

TOXICITY DATA with REFERENCE
dni-hmn:lym 25 μmol/L BBRCA9 80,963,78
orl-rat LDLo:7 g/kg TXAPA9 17,669,70
orl-mus LDLo:7 g/kg TXAPA9 17,669,70

CONSENSUS REPORTS: Reported in EPA TSCA Inventory.

SAFETY PROFILE: Mildly toxic by ingestion. Human mutation data reported. When heated to decomposition it emits acrid smoke and fumes.

IFY000 CAS:60166-93-0 ***HR: 2***
IOPAMIDOL
mf: $C_{17}H_{22}I_3N_3O_8$ mw: 777.12

SYNS: B-15000 ◇ l-(+)-N,N'BIS(2-HYDROXY)-1-HYDROXYMETHYL-ETHYL)-2,4,6-TRIIODO-5-LACTAMIDE ISOPHTHALAMIDE ◇ l-5α-HYDROXYPROPIONYLAMINO-2,4,6-TRIIODOISOPHTHALICACID DI(1,3-DIHYDROXY-2-PROPYLAMIDE)◇ l-5α-IDROSSIPROPIONIL-AMINO-2,4,6-TRIIODOISOFTAL-DI(1,3-DIIDROSSI-2-PROPILAMIDE) ◇ IOPAMIRON ◇ NIOPAM ◇ SOLUTRAST ◇ SQ 13396

TOXICITY DATA with REFERENCE
ipr-rat LD50:35 g/kg YACHDS 12(Suppl 1),11,84
ivn-rat LD50:25 g/kg YACHDS 12(Suppl 1),11,84
iat-rat LD50:13268 mg/kg USXXAM #4001323
ipr-mus LD50:40825 mg/kg USXXAM #4001323
ivn-mus LD50:33 g/kg YACHDS 12(Suppl 1),11,84
ice-mus LD50:3 g/kg FRPSAX 32,835,77
ivn-dog LD50:35 g/kg FRPSAX 32,835,77
ivn-rbt LD50:20 g/kg FRPSAX 32,835,77
par-rbt LD50:510 mg/kg USXXAM #4001323

SAFETY PROFILE: Moderately toxic by intracerebral and parenteral routes. Mildly toxic by other routes. When heated to decomposition it emits very toxic fumes of I^- and NO_x.

IFY100 CAS:96-83-3 ***HR: 3***
IOPANOIC ACID
mf: $C_{11}H_{12}I_3NO_2$ mw: 570.94

PROP: dl-Form: Cream-colored solid. Mp: 155.2-157°. Insol in water; sol in dil alkali, in 95% alc, and in other organic solvents. l-Form: Crystals. Mp: 162-163°. d-Form: Crystals. Mp: 162°.

SYNS: 3-AMINO-α-ETHYL-2,4,6-TRIIODOHYDROCINNAMIC ACID ◇ 2-(3-AMINO-2,4,6-TRIIODOBENZYL)BUTYRIC ACID ◇ 3-(3-AMINO-2,4,6-TRIIODOPHENYL)-2-ETHYLPROPANOIC ACID ◇ β-(3-AMINO-2,4,6-TRIIODOPHENYL)-α-ETHYLPROPIONIC ACID ◇ COLEPAX ◇ COPANOIC ◇ 2-ETHYL-3-(3-AMINO-2,4,6-TRIIODOPHENYL)PROPIONIC ACID ◇ IODOPANIC ACID ◇ IODOPANOIC ACID ◇ JOPAGNOST ◇ TELEPAQUE ◇ TELETRAST

TOXICITY DATA with REFERENCE
orl-rat TDLo:2200 mg/kg (1-22D preg):REP ATSUDG 7,425,84
orl-man TDLo:86 mg/kg/1W-I:SKN ARDEAC 123,387,87
orl-rat LD50:1540 mg/kg TXAPA9 14,232,69
ivn-rat LD50:280 mg/kg TXAPA9 14,232,69
orl-mus LD50:6600 mg/kg JMCMAR 13,997,70
ivn-mus LD50:320 mg/kg JMCMAR 13,997,70

SAFETY PROFILE: Poison by intravenous route. Moderately toxic by ingestion and intraperitoneal routes. Human systemic effects by ingestion: dermatitis. Experimental reproductive effects. When heated to decomposition it emits toxic fumes of I^- and NO_x.

IFZ800 CAS:96-84-4 ***HR: 3***
IOPHENOXIC ACID
mf: $C_{11}H_{11}I_3O_3$ mw: 571.92

PROP: Crystals from benzene + petr ether. Mp: 143-144°.

SYNS: α-ETHYL-3-HYDROXY-2,4,6-TRIIODOHYDROCINNAMIC ACID ◇ α-ETHYL-β-(3-HYDROXY-2,4,6-TRIIODOPHENYL)PROPIONIC ACID ◇ TERIDAX ◇ TRIIODOETHIONIC ACID ◇ α-(2,4,6-TRIIODO-3-HYDROXYBENZYL)BUTYRIC ACID

TOXICITY DATA with REFERENCE
ipr-rat LD50:648 mg/kg JAPMA8 42,476,53
orl-mus LD50:1850 mg/kg JAPMA8 42,476,53
ipr-mus LD50:440 mg/kg JAPMA8 42,476,53
ivn-mus LD50:374 mg/kg JMCMAR 13,997,70
ivn-dog LD50:203 mg/kg JAPMA8 42,476,53
ipr-gpg LD50:570 mg/kg JAPMA8 42,476,53

SAFETY PROFILE: Poison by intravenous route. Moderately toxic by ingestion and intraperitoneal routes. When heated to decomposition it emits toxic fumes of I^-.

IFZ900 CAS:26786-32-3 ***HR: 3***
IOPRAMINE HYDROCHLORIDE
mf: $C_{26}H_{27}ClN_2O \cdot ClH$ mw: 455.46

SYNS: 5-(3-((p-CHLOROBENZOYLMETHYL)-N-

METHYLAMINO)PROPYL)-5H-DIBENZ(b,f)AZEPINE ◇ CLOPEPRAMINE HYDROCHLORIDE ◇ GAMONIL ◇ LEO 640 HYDROCHLORIDE ◇ LOFEPRAMINE HYDROCHLORIDE ◇ LOPRAMINE HYDROCHLORIDE

TOXICITY DATA with REFERENCE
orl-rat TDLo:180 mg/kg (17-22D preg):REP KSRNAM 10,2186,76
orl-rat TDLo:150 mg/kg (female 9-14D post):TER KSRNAM 10,2186,76
ipr-mus LD50:4370 μg/kg YKYUA6 32,1279,81

SAFETY PROFILE: Poison by intraperitoneal route. Experimental reproductive effects. An experimental teratogen. When heated to decomposition it emits toxic fumes of NO_x and HCl. See also AMINES.

IGA000 CAS:37723-78-7 ***HR: 2***
IOPRONIC ACID
mf: $C_{15}H_{18}I_3NO_5$ mw: 673.04

SYNS: 3-(2-(3-ACETYLAMINO-2,4,6-TRIIODOPHENOXY)ETHOXY)-2-ETHYLPROPIONIC ACID ◇ 2-((2-(3-(ACETYLAMINO)-2,4,6-TRIIODOPHENOXY)ETHOXY)METHYL)BUTANOIC ACID ◇ B-11420 ◇ BILIMIRO ◇ BILIMIRON ◇ ORAVUE ◇ SQ-21983 ◇ VIDEOBIL

TOXICITY DATA with REFERENCE
orl-rat TDLo:3300 mg/kg (1-22D preg):REP ATSUDG 7,425,84
orl-rat LD50:5650 mg/kg FRPPAO 31,397,76
ivn-rat LD50:1 g/kg MEIEDD 10,732,83
orl-mus LD50:1950 mg/kg FRPPAO 31,397,76
ivn-mus LD50:1090 mg/kg 43FLAV 4(3),1153,80
ivn-dog LD50:835 mg/kg MEIEDD 10,732,83

SAFETY PROFILE: Moderately toxic by ingestion and intravenous routes. Experimental reproductive effects. When heated to decomposition it emits very toxic fumes of I^- and NO_x.

IGC000 CAS:13087-53-1 ***HR: 3***
IOTHALAMATE METHYLGLUCAMINE
mf: $C_{11}H_9I_3N_2O_4 \cdot C_7H_{17}NO_5$ mw: 809.17

SYNS: CONRAY ◇ CONRAY 30 ◇ CONRAY 60 ◇ CONRAY 280 ◇ CONRAY MEGLUMIN ◇ CONRAY MEGLUMINE 282 ◇ CONTRIX 28 ◇ CYSTO-CONRAY ◇ 1-DEOXY-1-(METHYLAMINO)-d-GLUCITOL 5-ACETAMIDO-2,4,6-TRIIODO-N-METHYLISOPHTHALAMATE(SALT) ◇ IOTALAMATE de METHYLGLUCAMINE (FRENCH) ◇ IOTHALAMATE MEGLUMINE ◇ IOTHALAMATE METHYLGLUCAMINE SALT ◇ MEGLUMINE CONRAY ◇ MEGLUMINE IOTHALAMATE ◇ MEGLUMINE ISOTHALAMATE ◇ METHYLGLUCAMINE IOTALAMATE ◇ METHYLGLUCAMINE IOTHALAMATE

TOXICITY DATA with REFERENCE
ivn-rat LD50:13600 mg/kg NIIRDN 6,64,82
ice-rat LD50:205 mg/kg NIIRDN 6,64,82
ivn-mus LD50:11500 mg/kg THERAP 26,595,71
ice-mus LD50:300 mg/kg THERAP 26,595,71

SAFETY PROFILE: Poison intracerebral routes. When heated to decomposition it emits very toxic fumes of NO_x and I^-.

IGD000 CAS:2276-90-6 ***HR: 3***
IOTHALAMIC ACID
mf: $C_{11}H_9I_3N_2O_4$ mw: 613.92

PROP: Crystals. Decomp @ about 285°.

SYNS: 5-ACETYLAMINO-N-METHYL-2,4,6-TRIIODOISOPHTHALAMIC ACID ◇ JOTA (GERMAN) ◇ JOTALAMSAEURE (GERMAN)

TOXICITY DATA with REFERENCE
ipr-rat LD50:19800 mg/kg ARZNAD 15,222,65
ivn-rat LD50:10500 mg/kg ARZNAD 15,222,65
ivn-mus LD50:19 mg/kg JMCMAR 6,24,63
ipr-gpg LD50:14 g/kg ARZNAD 15,222,65

SAFETY PROFILE: Poison by intravenous route. When heated to decomposition it emits very toxic fumes of NO_x and I^-.

IGD075 CAS:72704-51-9 ***HR: 1***
IOTROXATE MEGLUMINE
mf: $C_{22}H_{18}I_6N_2O_9 \cdot C_7H_{17}NO_5$ mw: 1411.07

SYNS: IOTROXATE METHYLGLUCMINE SALT ◇ MEGLUMINE IOTROXATE

TOXICITY DATA with REFERENCE
ivn-rat TDLo:1468 mg/kg (female 7-17D post):REP NIIHAO 26,119,81
ipr-rat TDLo:4670 mg/kg (female 1-7D post):TER NIIHAO 26,110,81
scu-rat LD50:78944 mg/kg NIIRDN 6,APP-1,82
ivn-rat LD50:46588 mg/kg NIIRDN 6,APP-1,82
ipr-mus LD50:58374 mg/kg NIIRDN 6,APP-1,82
scu-mus LD50:68381 mg/kg NIIRDN 6,APP-1,82
ivn-mus LD50:31355 mg/kg NIIRDN 6,APP-1,82
ivn-rbt LD50:45698 mg/kg NIIRDN 6,APP-1,82

SAFETY PROFILE: Mildly toxic by several routes. An experimental teratogen. Experimental reproductive effects. When heated to decomposition it emits toxic fumes of I^- and NO_x.

IGD100 CAS:51022-74-3 ***HR: 2***
IOTROXIC ACID
mf: $C_{22}H_{18}I_6N_2O_9$ mw: 1215.82

SYNS: BILISCOPIN ◇ IOTROXINSAEURE (GERMAN) ◇ 3,3'-(2,2'-OXYDIETHYLENEDIOXYBISACETAMIDO)BIS(2,4,6-TRIIODOBENZOIC ACID) ◇ SH 213AB ◇ 3,3'-(3,6,9-TRIOXAUNDECANEDIOYLDIAMINO)BIS(2,4,6-TRIIODOBENZOICACID)

TOXICITY DATA with REFERENCE
scu-rat LD50:7100 mg/kg IYKEDH 13,637,82
ivn-rat LD50:4190 mg/kg IYKEDH 13,637,82
ipr-mus LD50:5250 mg/kg IYKEDH 13,637,82

scu-mus LD50:6500 mg/kg IYKEDH 13,637,82
ivn-mus LD50:2820 mg/kg IYKEDH 13,637,82

SAFETY PROFILE: Moderately toxic by intravenous route. When heated to decomposition it emits toxic fumes of I^- and NO_x.

IGD200 CAS:59017-64-0 ***HR: 1***
IOXAGLIC ACID
mf: $C_{24}H_{21}I_6N_5O_8$ mw: 1268.90

PROP: Mixture of ioxaglate meglumine and ioxaglate sodium. Mp: 302°.

SYNS: HEXABRIX ◇ P 286 (contrast medium)

TOXICITY DATA with REFERENCE
ivn-rat TDLo:4320 mg/kg (17-22D preg/21D post):REP KSRNAM 19,2463,85
ivn-rat LD50:13300 mg/kg KSRNAM 19,2411,85

SAFETY PROFILE: Mildly toxic by intravenous route. Experimental reproductive effects. When heated to decomposition it emits toxic fumes of I^-, Na_2O, and NO_x. A herbicide.

IGE100 CAS:89367-92-0 ***HR: 3***
IP-10
mf: $C_{13}H_9N_4O_3S$ mw: 301.30

SYNS: 5-((p-HYDROXYBENZYLIDENE)AMINO)-3-METHYLISOTHIAZOLO(5,4-d)PYRIMIDINE-4,6(5H,7H)-DIONE◇ 5-((4'-HYDROXYBENZYLIDENOIMINO)-3-METHYLISOTHIAZOLO(5,4-d)PYRIMIDINE-(7H)-4,6)-DIONE

TOXICITY DATA with REFERENCE
ivn-rat LD50:535 mg/kg AITEAT 31,769,83
ipr-mus LD50:330 mg/kg AITEAT 31,769,83
ivn-mus LD50:275 mg/kg AITEAT 31,769,83

SAFETY PROFILE: Poison by intravenous and intraperitoneal routes. When heated to decomposition it emits toxic fumes of SO_x and NO_x.

IGF000 CAS:8012-96-2 ***HR: 1***
IPECAC SYRUP

PROP: Dried rhizome and roots of Rio or Brazilian ipecac. Contains emetine, cephaline, emetamine, ipecacuanic acid, psychotrine, methyl psychotaine, resin.

SYNS: DIHYDROTACHY STEROL ◇ IPECACUANHA ◇ SYRUP of IPECAC, U.S.P.

TOXICITY DATA with REFERENCE
orl-hmn TDLo:70 mg/kg:GIT JPMSAE 65,1398,76
orl-wmn LDLo:113 mg/kg/13W-C:CVS:PUL JAMAAP 243,1927,80
orl-rat LD50:7800 mg/kg 36THAV -,175,77
orl-dog LDLo:5 g/kg 36THAV -,175,77

SAFETY PROFILE: Mildly toxic by ingestion. A centrally acting emetic. Human systemic effects by ingestion: nausea, vomiting, blood pressure lowering, change in heart rate, dyspnea. Has caused fatalities after prolonged ingestion. An FDA over-the-counter drug. See also EMETINE.

IGF200 CAS:53011-73-7 ***HR: 3***
1,4-IPOMEADIOL
mf: $C_9H_{14}O_3$ mw: 170.23

SYN: 1-(3-FURANYL)-2,4-PENTANEDIOL

TOXICITY DATA with REFERENCE
orl-mus LD50:104 mg/kg BBACAQ 337,184,74
ipr-mus LD50:67 mg/kg BBACAQ 337,184,74
ivn-mus LD50:68 mg/kg BBACAQ 337,184,74

SAFETY PROFILE: Poison by ingestion, intravenous and intraperitoneal routes. When heated to decomposition it emits acrid smoke and fumes.

IGF300 CAS:496-06-0 ***HR: 3***
IPOMEANINE
mf: $C_9H_{10}O_3$ mw: 166.19

SYNS: 1-(3-FURANYL)-1,4-PENTANEDIONE◇ 1-(3-FURYL)-1,4-PENTANEDIONE ◇ IPOMEANIN ◇ β-(Γ-OXOVALEROYL)FURAN

TOXICITY DATA with REFERENCE
orl-mus LD50:26 mg/kg BBACAQ 337,184,74
ipr-mus LD50:25 mg/kg BBACAQ 337,184,74
ivn-mus LD50:14 mg/kg BBACAQ 337,184,74

SAFETY PROFILE: Poison by ingestion, intravenous, and intraperitoneal routes. When heated to decomposition it emits acrid smoke and fumes.

IGF325 CAS:34435-70-6 ***HR: 3***
IPOMEANOL
mf: $C_9H_{12}O_3$ mw: 168.21

SYNS: 5-(3-FURANYL)-5-HYDROXY-2-PENTANONE◇ 5-(3-FURYL)-5-HYDROXY-2-PENTANONE

TOXICITY DATA with REFERENCE
orl-mus LD50:79 mg/kg BBACAQ 337,184,74
ipr-mus LD50:49 mg/kg BBACAQ 337,184,74
ivn-mus LD50:35 mg/kg BBACAQ 337,184,74

SAFETY PROFILE: Poison by ingestion, intravenous and intraperitoneal routes. When heated to decomposition it emits acrid smoke and fumes. See also KETONES.

IGG000 CAS:22254-24-6 ***HR: 3***
IPRATROPIUM BROMIDE
mf: $C_{20}H_{30}NO_3•Br$ mw: 412.42

SYNS: ATEM ◇ ATROVENT ◇ 3-α-HYDROXY-8-ISOPROPYL-1-α-H,5-α-H-TROPANIUM BROMIDE (±)-TROPATE ◇ (8r)-3-α-HYDROXY-8-ISOPROPYL-1-α-H,5-α-H-TROPIUMBROMIDE-(±)-TROPATE

◇ IPRATROPIUMBROMID (GERMAN) ◇ 8-ISOPROPYLNORATROPINE METHOBROMIDE ◇ N-ISOPROPYLNORATROPINIUM BROMOMETHYLATE ◇ ITROP ◇ Sch 1000

TOXICITY DATA with REFERENCE
eye-rbt 100 mg/24H MOD KSRNAM 21,5692,87
orl-rat TDLo:5500 mg/kg (female 7-17D post):REP IYKEDH 9,971,78
orl-rat TDLo:55 mg/kg (female 7-17D post):TER IYKEDH 9,971,78
ihl-man TCLo:1 μg/kg:GIT BMJOAE 292,380,86
orl-rat LD50:1663 mg/kg ARZNAD 26,985,76
ipr-rat LD50:113 mg/kg NIIRDN 6,348,82
scu-rat LD50:634 mg/kg NIIRDN 6,348,82
ivn-rat LD50:15700 μg/kg PBPSDY 2,489,79
orl-mus LD50:1001 mg/kg ARZNAD 26,985,76
ipr-mus LD50:72 mg/kg IYKEDH 9,417,78
scu-mus LD50:300 mg/kg ARZNAD 26,985,76
ivn-mus LD50:12 mg/kg ARZNAD 26,985,76

SAFETY PROFILE: A poison by intravenous, intraperitoneal, and subcutaneous routes. Moderately toxic by ingestion. Human systemic effects by inhalation of very small amounts: gastrointestinal changes. An experimental teratogen. Experimental reproductive effects. An eye irritant. Used as a bronchodilator. When heated to decomposition it emits toxic fumes of NO_x and Br^-. See also BROMIDES.

IGG300 CAS:4013-92-7 ***HR: 3***
IPROHEPTINE HYDROCHLORIDE
mf: $C_{11}H_{25}N{\cdot}ClH$ mw: 207.83

SYNS: N-ISOPROPYL-6-METHYL-2-HEPTYLAMINEHYDROCHLORIDE ◇ 6-METHYL-N-(1-METHYLETHYL)-2-HEPTANAMINE HYDROCHLORIDE

TOXICITY DATA with REFERENCE
orl-mus TDLo:448 mg/kg NIIRDN 6,81,82
scu-mus LD50:223 mg/kg NIIRDN 6,81,82
ivn-mus LD50:31700 μg/kg NIIRDN 6,81,82

SAFETY PROFILE: Poison by subcutaneous and intravenous routes. Moderately toxic by ingestion. When heated to decomposition it emits toxic fumes of NO_x and HCl.

IGG600 CAS:6011-62-7 ***HR: 2***
IPRONIAZID HYDROCHLORIDE
mf: $C_9H_{13}N_3O{\cdot}2ClH$ mw: 252.17

SYNS: IPRONIAZID DIHYDROCHLORIDE ◇ 2-(1-METHYLETHYL) HYDRAZIDE 4-PYRIDINECARBOXYLIC ACID DIHYDROCHLORIDE (9CI)

TOXICITY DATA with REFERENCE
orl-rat TDLo:3233 mg/kg (61D pre):REP ENDOAO 67,511,60
orl-rat LD50:1470 mg/kg ENDOAO 67,511,60

SAFETY PROFILE: Moderately toxic by ingestion. Experimental reproductive effects. When heated to decomposition it emits toxic fumes of NO_x and HCl.

IGG700 CAS:305-33-9 ***HR: 2***
IPRONIAZID PHOSPHATE
mf: $C_9H_{13}N_3O{\cdot}H_3O_4P$ mw: 277.25

SYN: ISONICOTINIC ACID, 2-ISOPROPYLHYDRAZIDE, PHOSPHATE

TOXICITY DATA with REFERENCE
oth-bcs 10 mmol/L MUREAV 5,343,68
sce-mus-ipr 640 mg/kg JTEHD6 9,287,82
scu-rat TDLo:130 mg/kg (female 10-22D post):REP NETOD7 7,493,85
ipr-rat LD50:442 mg/kg ABMGAJ 18,617,67
scu-gpg LD50:730 mg/kg AIPTAK 137,375,62

SAFETY PROFILE: Moderately toxic by intraperitoneal and subcutaneous routes. Experimental reproductive effects. Mutation data reported. When heated to decomposition it emits toxic fumes of NO_x and PO_x.

IGG775 CAS:62928-11-4 ***HR: 3***
IPROPLATIN
mf: $C_6H_{20}Cl_2N_2O_2Pt$ mw: 418.27

SYNS: CHIP ◇ cis-DICHLORO-trans-DIHYDROXYBISISOPROPYLAMINE PLATINUM (IV) ◇ DIISOPROPYLAMMINE-trans-DIHYDROXYMALONATOPLATINUM(IV) ◇ JM-28 ◇ NSD 256927

TOXICITY DATA with REFERENCE
mmo-esc 10 μmol/L MUREAV 173,13,86
mnt-ham:lng 8250 nmol/L NEOLA4 31,655,84
ipr-rat LD50:60 mg/kg BJCAAI 42,668,80
scu-rat LD50:92 mg/kg BJCAAI 42,668,80
ivn-rat LD50:30 mg/kg EJCODS 20,1087,84
ipr-mus LD50:60 mg/kg EJCODS 20,1087,84
ivn-mus LD50:45 mg/kg EJCODS 20,1087,84

SAFETY PROFILE: Poison by subcutaneous, intravenous, and intraperitoneal routes. Mutation data reported. When heated to decomposition it emits toxic fumes of NO_x and Cl^-. See also PLATINUM COMPOUNDS.

IGH000 CAS:14885-29-1 ***HR: 2***
IPROPRAN
mf: $C_7H_{11}N_3O_2$ mw: 169.21

SYNS: IPRONIDAZOLE (USDA) ◇ 2-ISOPROPYL-1-METHYL-5-NITROIMIDAZOLE ◇ 1-METHYL-2-(1-METHYLETHYL)-5-NITRO-1H-IMIDAZOLE ◇ RO 7-1554

TOXICITY DATA with REFERENCE
mmo-sat 1 μmol/L TCMUD8 3,429,83
mmo-esc 50 μmol/L MUREAV 48,155,77
mmo-klp 20 μmol/L/20H MUREAV 66,207,79
mmo-omi 20 μmol/L MUREAV 48,155,77

mmo-smc 5 ppm MUREAV 86,243,81
orl-trk LD50:640 mg/kg POSCAL 49,92,70

SAFETY PROFILE: Moderately toxic by ingestion. Mutation data reported. When heated to decomposition it emits toxic fumes of NO_x.

IGH700 CAS:3614-57-1 ***HR: D***
IREHDIAMINE A
mf: $C_{21}H_{36}N_2$ mw: 316.59

SYN: PREGN-5-ENE-3-β,20-α-DIAMINE

TOXICITY DATA with REFERENCE
mmo-omi 200 μg/plate PNASA6 58,256,67
dni-omi 60 μmol/L PNASA6 58,256,67

SAFETY PROFILE: Mutation data reported. When heated to decomposition it emits toxic fumes of NO_x.

IGI000 CAS:8064-79-7 ***HR: 3***
IRGAPYRIN
mf: $C_{19}H_{20}N_2O_2{\bullet}C_{13}H_{17}N_3O$ mw: 539.74

SYNS: 4-BUTYL-1-1,2-DIPHENYL-3,5-PYRAZOLIDINEDIONE with 4-(DIMETHYLAMINO)-1,2-DIHYDRO-1,5-DIMETHYL-2-PHENYL-3H-PYRAZOL-3-ONE ◇ IRGAPYRINE ◇ PABIALGIN ◇ RHEOPYRINE

TOXICITY DATA with REFERENCE
orl-rat LD50:1375 mg/kg SMWOAS 79,577,49
ipr-rat LD50:290 mg/kg JPETAB 109,387,53
ivn-rat LD50:160 mg/kg SMWOAS 79,577,49
orl-mus LD50:700 mg/kg SMWOAS 79,577,49
ipr-mus LD50:412 mg/kg DPHFAK 23,363,71
ivn-mus LD50:155 mg/kg JPETAB 109,387,53
ims-mus LD50:560 mg/kg OYYAA2 13,79,77
ivn-rbt LD50:145 mg/kg SMWOAS 79,577,49

SAFETY PROFILE: Poison by intraperitoneal and intravenous routes. Moderately toxic by ingestion and intramuscular routes. When heated to decomposition it emits toxic fumes of NO_x.

IGJ000 CAS:7439-88-5 ***HR: 3***
IRIDIUM
af: Ir aw: 192.2

PROP: Silver-white very hard metallic element. Mp: 2450°, bp: approx 4500°, d: 22.65 @ 20°/4°. Highest specific gravity of all elements.

SAFETY PROFILE: The pure metal is clinically inert. Most of its compounds are poorly soluble in water and thus are not absorbed efficiently by the body. The chlorides are poison or moderately toxic by ingestion and are eye and skin irritants. There are no reports of acute or chronic health effects to workers handling iridium and its compounds. The ^{190}Ir and ^{192}Ir radioisotopes are used in clinical radiography and most references to the toxicity of iridium relate to these isotopes.

A catalytic metal. The powdered metal may ignite spontaneously in air. Violent reaction or ignition on contact with interhalogens (e.g., bromine pentafluoride; chlorine trifluoride). Alloys with zinc, after extraction with acids, leave heat-sensitive explosive residues. Is attacked by F_2, Cl_2 at red heat; by potassium sulfate or a mixture of potassium hydroxide and nitrate; on fusion; lead; zinc; tin.

IGJ300 CAS:12645-45-3 ***HR: 2***
IRIDIUM CHLORIDE

SYN: IRIDIUM MURIATE

TOXICITY DATA with REFERENCE
itt-rat TDLo:18214 μg/kg (male 1D pre):REP JRPFA4 7,21,64
ivn-dog LDLo:778 mg/kg SMSJAR 26,131,1826

SAFETY PROFILE: Moderately toxic by intravenous route. Experimental reproductive effects.

IGJ499 CAS:10025-97-5 ***HR: 3***
IRIDIUM TETRACHLORIDE
mf: Cl_4Ir mw: 334.00

SYN: IRIDIUM(IV) CHLORIDE

TOXICITY DATA with REFERENCE
orl-rat LD50:8115 μg/kg GTPZAB 21(7),55,77

CONSENSUS REPORTS: Reported in EPA TSCA Inventory.

SAFETY PROFILE: Poison by ingestion. When heated to decomposition it emits toxic fumes of Cl^-. See also IRIDIUM.

IGK800 CAS:7439-89-6 ***HR: 3***
IRON
af: Fe aw: 55.85

PROP: From decomposition of iron pentacarbonyl: dark grey powder. From electrodeposition: lusterless, gray black powder. From chemical reduction: gray-black powder.

SYNS: ANCOR EN 80/150 ◇ ARMCO IRON ◇ CARBONYL IRON ◇ IRON, CARBONYL (FCC) ◇ IRON, ELECTROLYTIC ◇ IRON, ELEMENTAL ◇ IRON, REDUCED (FCC)

TOXICITY DATA with REFERENCE
itr-rat TDLo:450 mg/kg/15W-I:ETA SAIGBL 16,380,74
orl-rat LD50:30 g/kg IJPAAO 13,240,51
ipr-rbt LDLo:20 mg/kg NTIS** PB158-508

CONSENSUS REPORTS: Reported in EPA TSCA Inventory.

SAFETY PROFILE: Poison by intraperitoneal route. Questionable carcinogen with experimental tumorigenic data. Iron is potentially toxic in all forms and by all routes of exposure. The inhalation of large amounts of iron dust results in iron pneumoconiosis (arc welder's lung). Chronic exposure to excess levels of iron (> 50-100 mg Fe/day) can result in pathological deposition of iron in the body tissues, the symptoms of which are fibrosis of the pancreas, diabetes mellitus, and liver cirrhosis.

As with other metals, it becomes more reactive as it is more finely divided. Ultrafine iron powder is pyrophoric and potentially explosive. Explosive or violent reaction with ammonium nitrate + heat; ammonium peroxodisulfate; chloric acid; chlorine trifluoride; chloroformamidinium nitrate; bromine pentafluoride + heat (with iron powder); air + oil (with iron dust); sodium acetylide. Ignites on contact with chlorine; dinitrogen tetraoxide; liquid fluorine; hydrogen peroxide (with iron powder); nitryl fluoride + heat; peroxyformic acid; potassium perchlorate; potassium dichromate; sodium peroxide (at 240°); polystyrene + friction or spark (iron powder). Mixtures of iron dust with air + water may ignite on drying. Reduced iron reacts with water to produce explosive hydrogen gas. Catalyzes the exothermic polymerization of acetaldehyde. See also IRON COMPOUNDS, IRON DUST, and FERROUS ION.

IGL000 CAS:14024-18-1 ***HR: 3***
IRON ACETYLACETONATE
mf: $C_{15}H_{21}FeO_6$ mw: 353.21

SYNS: FERRIC ACETYLACETONATE ◇ FERRIC TRIACETYLACETONATE ◇ TRIS(2,4-PENTANEDIONATO)IRON

TOXICITY DATA with REFERENCE
ivn-mus LD50:100 mg/kg CSLNX* NX#02372

CONSENSUS REPORTS: Reported in EPA TSCA Inventory.

SAFETY PROFILE: Poison by intravenous route. When heated to decomposition it emits acrid smoke and fumes. See also IRON COMPOUNDS.

IGM000 CAS:10102-50-8 ***HR: 3***
IRON(II) ARSENATE (3:2)
DOT: UN 1608
mf: $As_2O_8 \cdot 3Fe$ mw: 445.39

SYNS: ARSENATE of IRON, FERROUS ◇ FERROUS ARSENATE (DOT) ◇ FERROUS ARSENATE, solid (DOT) ◇ IRON ARSENATE (DOT)

CONSENSUS REPORTS: Arsenic and its compounds are on the Community Right-To-Know List.

OSHA PEL: Cancer Hazard
ACGIH TLV: TWA 0.2 mg(As)/m^3; TWA 1 mg/(Fe)/m^3
NIOSH REL: (Inorganic Arsenic) CL 0.002 mg(As)/m^3/15M
DOT Classification: Poison B; Label: Poison.

SAFETY PROFILE: Confirmed human carcinogen. A deadly poison by various routes. A pesticide. When heated to decomposition it emits toxic fumes of As. See also ARSENIC COMPOUNDS and IRON COMPOUNDS.

IGN000 CAS:10102-49-5 ***HR: 3***
IRON(III) ARSENATE (1:1)
DOT: UN 1606
mf: $AsO_4 \cdot Fe$ mw: 194.77

SYNS: ARSENATE of IRON, FERRIC ◇ FERRIC ARSENATE, solid (DOT)

CONSENSUS REPORTS: Arsenic and its compounds are on the Community Right-To-Know List.

OSHA PEL: Cancer Hazard
ACGIH TLV: TWA 0.2 mg(As)/m^3; TWA 1 mg/(Fe)/m^3
NIOSH REL: (Inorganic Arsenic) CL 0.002 mg(As)/m^3/15M
DOT Classification: Poison B; Label: Poison.

SAFETY PROFILE: Confirmed human carcinogen. A deadly poison. A pesticide. When heated to decomposition it emits toxic fumes of As. See also ARSENIC COMPOUNDS and IRON COMPOUNDS.

IGO000 CAS:63989-69-5 ***HR: 3***
IRON(III)-o-ARSENITE PENTAHYDRATE
DOT: UN 1607
mf: $As_2Fe_2O_6 \cdot Fe_2O_3 \cdot 5H_2O$ mw: 607.34

PROP: Brown-yellow powder.

SYNS: FERRIC ARSENITE, BASIC ◇ FERRIC ARSENITE, solid (DOT)

CONSENSUS REPORTS: Arsenic and its compounds are on the Community Right-To-Know List.

OSHA PEL: TWA 0.01 mg(As)/m^3; Cancer Hazard
ACGIH TLV: TWA 0.2 mg(As)/m^3; TWA 1 mg/(Fe)/m^3
NIOSH REL: (Inorganic Arsenic) CL 0.002 mg(As)/m^3/15M
DOT Classification: Poison B; Label: Poison.

SAFETY PROFILE: Confirmed human carcinogen. A deadly poison. When heated to decomposition it emits toxic fumes of As. See also ARSENIC COMPOUNDS and IRON COMPOUNDS.

IGP000 CAS:7789-46-0 ***HR: 3***
IRON(II) BROMIDE
mf: Br_2Fe mw: 215.67

SAFETY PROFILE: Mixtures with potassium or sodium are shock-sensitive explosives. When heated to decomposition it emits toxic fumes of Br^-. See also IRON COMPOUNDS and BROMIDES.

IGQ000 CAS:10031-26-2 ***HR: 3***
IRON(III) BROMIDE
mf: Br_3Fe mw: 295.58

SAFETY PROFILE: Mixtures with potassium or sodium are shock-sensitive explosives. When heated to decomposition it emits toxic fumes of Br^-. See also IRON COMPOUNDS and BROMIDES.

IGQ750 CAS:12011-67-5 ***HR: 2***
IRON CARBIDE
mf: CFe_3 mw: 179.55

SAFETY PROFILE: Reacts vigorously with halogens (e.g., chlorine below 100°C; bromine at 100°C). When heated to decomposition it emits acrid smoke and fumes. See also IRON COMPOUNDS.

IGR499 ***HR: D***
IRON COMPOUNDS

SAFETY PROFILE: Of varying toxicity. Exposure to iron oxides is potentially a serious risk in all industrial settings. Some iron compounds are suspected carcinogens. In general, ferrous compounds are more toxic than ferric compounds. Acute exposure to excessive levels of ferrous compounds can cause liver and kidney damage, altered respiratory rates, and convulsions. Accidental ingestion of medicinal iron preparations results in thousands of intoxications per year in the United States. Iron pentacarbonyl is a poison. Ferbam, the iron salt of dimethyldithiocarbamic acid is very toxic. Chelated iron compounds (e.g., sodium iron ethylenediaminetetraacetate) are less toxic than the ferrous salts. Intramuscular injections of iron dextran can cause severe anaphylactic reactions. See also IRON.

IGS000 CAS:9004-66-4 ***HR: 3***
IRON-DEXTRAN COMPLEX

PROP: For human use, it is a sterile dark brown colloidal solvent, water-soluble. Approximate molecular weight is 180,000 (IARC** 2,161,72).

SYNS: A 100 (pharmaceutical) ◇ CHINOFER ◇ DEXTRAN ION COMPLEX ◇ DEXTROFER 75 ◇ EISENDEXTRAN (GERMAN) ◇ Fe-DEXTRAN ◇ FENATE ◇ FERDEX 100 ◇ FERRIC DEXTRAN ◇ FERRIDEXTRAN ◇ FERRODEXTRAN ◇ FERROFLUKIN 75 ◇ FERROGLUCIN ◇ FERROGLUKIN 75 ◇ IMFERON ◇ IMPOSIL ◇ IRO-JEX ◇ IRON DEXTRAN ◇ IRON DEXTRAN INJECTION ◇ IRON HYDROGENATED DEXTRAN ◇ IRONORM INJECTION ◇ MYOFER 100 ◇ POLYFER ◇ PROLONGAL ◇ RCRA WASTE NUMBER U139 ◇ URSOFERRAN

TOXICITY DATA with REFERENCE
ims-rat TDLo:240 mg/kg (female 6W pre):REP BPNSBY 10,49,77
ims-rbt TDLo:650 mg(Fe)/kg (female 6-18D post):TER SJHSBD 32,69,77
ims-wmn TDLo:20 mg/kg/3Y-I:NEO BMJOAE 2,277,73
scu-rat TDLo:750 mg(Fe)/kg/4W-I:CAR IJCNAW 2,370,67
scu-mus TDLo:2300 mg/kg/I:ETA BECCAN 40,30,62
scu-rat TD:1500 mg(Fe)/kg/8W-I:CAR IJCNAW 2,370,67
ipr-rat LD50:3 gm(Fe)/kg TXAPA9 18,185,71
ims-rat LDLo:1617 mg(Fe)/kg ACVTA8 29,21,79
orl-mus LD50:1 g(Fe)/kg BJPCAL 24,352,65
ivn-mus LD50:460 mg(Fe)/kg APPHAX 18,149,61

CONSENSUS REPORTS: NTP Fifth Annual Report on Carcinogens. IARC Cancer Review: Group 2B IMEMDT 7,226,87; Human Inadequate Evidence IMEMDT 2,161,73; Animal Sufficient Evidence IMEMDT 2,161,73.

SAFETY PROFILE: Confirmed carcinogen producing tumors at site of application. Experimental carcinogenic, neoplastigenic, tumorigenic, and teratogenic data. Moderately toxic by ingestion and several other routes. Other experimental reproductive effects. See also IRON COMPOUNDS.

IGT000 ***HR: 3***
IRON DEXTRAN GLYCEROL GLYCOSIDE

TOXICITY DATA with REFERENCE
scu-rat TDLo:2500 mg(Fe)/kg/24W-I:NEO BJCAAI 22,521,68
ipr-mus LD50:2486 mg(Fe)/kg NNAPBA 270(Suppl),R50,71

SAFETY PROFILE: Moderately toxic by intraperitoneal route. Questionable carcinogen with experimental neoplastigenic data. See also IRON COMPOUNDS.

IGU000 CAS:9004-51-7 ***HR: 3***
IRON-DEXTRIN COMPLEX

PROP: For human use, it is a clear, brown, colloidal solvent. Approximate molecular weight is 230,000 (IARC** 2,161,72).

SYNS: ASTRAFER ◇ DEXTRIFERRON ◇ DEXTRIFERRON INJECTION ◇ FERRIGEN ◇ IRON CARBOHYDRATE COMPLEX ◇ IRON DEXTRIN INJECTION

TOXICITY DATA with REFERENCE
ims-rat TDLo:1150 mg/kg/17W-I:NEO BJCAAI 15,838,61
scu-mus TDLo:1200 mg(Fe)/kg/27W-I:ETA BMJOAE 1,1800,62
ipr-mus LD50:980 mg/kg AJMSA9 241,296,61
ivn-mus LD50:175 mg/kg AJMSA9 241,296,61
ivn-dog LD50:94 mg/kg AJMSA9 241,296,61

CONSENSUS REPORTS: IARC Cancer Review: Group 3 IMEMDT 7,227,87; Animal Sufficient Evidence IMEMDT 2,161,73.

SAFETY PROFILE: A poison by intravenous route. Moderately toxic by intraperitoneal route. Questionable carcinogen with experimental neoplastigenic and tumorigenic data. See also IRON COMPOUNDS.

IGV000 CAS:12068-85-8 ***HR: 3***
IRON DISULFIDE
mf: FeS_2 mw: 119.97

SYNS: IRON PYRITES ◇ IRON SULFIDE

CONSENSUS REPORTS: Reported in EPA TSCA Inventory.

SAFETY PROFILE: A poison by inhalation and ingestion. The powdered sulfide ignites spontaneously in air and some air-powder mixtures may be explosive. Trace carbon lowers the ignition temperature in air to 228°C and increases the sensitivity of air-dust mixtures. Heats up spontaneously and ignites with combustibles. Incompatible with water. When heated to decomposition or in reaction with acid or acid fumes it emits very toxic fumes of SO_x. See also IRON COMPOUNDS, SULFIDES, and HYDROGEN SULFIDE.

IGW000 ***HR: 3***
IRON DUST

PROP: Silvery-white, tenacious, lustrous, ductile metal. Mp: 1535°, bp: 3000°, d: 7.86, vap press: 1 mm @ 1787°. Iron dust from open hearth furnace contained 52% iron (85AGAF -,480,76).

TOXICITY DATA with REFERENCE
itr-rat TDLo:506 mg/kg/15W-I:NEO 85AGAF -,480,76

SAFETY PROFILE: Iron dust can cause conjunctivitis, choroiditis, retinitis, and siderosis of tissues if iron contacts and remains in these tissues. Iron ore dust can cause palpebral conjunctivitis, massive pulmonary fibrosis, and an increased incidence of lung cancer. Questionable carcinogen with experimental neoplastigenic data.

Flammable in the form of dust when exposed to heat or flame. Reacts violently with Cl_2; ClF_3; F_2; H_2O_2; NO_2; P; Na2C2; H_2SO_4; air; water; polystyrene. Moderately explosive in the form of dust when exposed to heat or flame. To fight fire, use special mixtures of dry chemical. See also IRON.

IGW500 CAS:79-69-6 ***HR: 2***
α-IRONE
mf: $C_{14}H_{22}O$ mw: 206.36

SYNS: 3-BUTEN-2-ONE,4-(2,5,6,6-TETRAMETHYL-2-CYCLOHEXEN-1-YL)-(9CI) ◇ 4-(2,5,6,6-TETRAMETHYL-2-CYCLO-HEXEN-1-YL)-3-BUTEN-2-ONE

TOXICITY DATA with REFERENCE
ipr-mus TDLo:1950 mg/kg/8W-I:ETA CNREA8 33,3069,73

CONSENSUS REPORTS: Reported in EPA TSCA Inventory.

SAFETY PROFILE: Questionable carcinogen with experimental tumorigenic data. When heated to decomposition it emits acrid smoke and irritating fumes.

IGX000 CAS:21393-59-9 ***HR: 3***
IRON(II) EDTA COMPLEX
mf: $C_{10}H_{12}FeN_2O_8{\cdot}2H$ mw: 346.11

TOXICITY DATA with REFERENCE
ipr-mus LD50:40 mg(Fe)/kg PABIAQ 11,853,63

CONSENSUS REPORTS: Reported in EPA TSCA Inventory.

SAFETY PROFILE: Poison by intraperitoneal route. When heated to decomposition it emits toxic fumes of NO_x. See also IRON COMPOUNDS.

IGX875 CAS:21626-24-4 ***HR: 3***
IRON(III)-EDTA SODIUM
mf: $C_{10}H_{12}FeN_2O_8{\cdot}2Na$ mw: 390.07

SYN: EDTA FERRIC SODIUM SALT

TOXICITY DATA with REFERENCE
cyt-ham:ovr 2 mmol/L CNREA8 41,1628,81
orl-mus LD50:3305 mg/kg ARZNAD 24,880,74
ipr-mus LD50:264 mg/kg ARZNAD 24,880,74
ivn-mus LD50:264 mg/kg ARZNAD 24,880,74
ims-mus LD50:1190 mg/kg ARZNAD 24,880,74

SAFETY PROFILE: Poison by intravenous and intraperitoneal routes. Moderately toxic by ingestion and intramuscular routes. Mutation data reported. When heated to decomposition it emits toxic fumes of NO_x and Na_2O. See also IRON COMPOUNDS.

IGY000 CAS:14038-43-8 ***HR: 3***
IRON(III) HEXACYANOFERRATE(4⁻)
mf: $C_{18}Fe_7N_{18}$ mw: 859.25

$$Fe_4[Fe(CN)_6]_3$$

SYN: IRON BLUE

CONSENSUS REPORTS: Cyanide and its compounds are on The Community Right-To-Know List.

OSHA PEL: TWA 5 mg(CN)/m^3
ACGIH TLV: TWA 5 mg(CN)/m^3 (skin)
DFG MAK: 5 mg/m^3
NIOSH REL: (Cyanide) CL 5 mg(CN)/m^3/10M

SAFETY PROFILE: Mixture with blown castor oil + tukey red oil (sulfonated castor oil) may ignite spontaneously in air. Reaction with ethylene oxide forms a product which ignites spontaneously in air. May ignite spontaneously in storage with lead chromate. When heated to decomposition it emits toxic fumes of NO_x and CN^-. See also IRON COMPOUNDS.

IHA000 ***HR: 3***
IRON(III)HYDROXIDE-POLYMALTOSE

SYN: EISEN-III-HYDROXID-POLYMALTOSE(GERMAN)

TOXICITY DATA with REFERENCE
imp-rat TDLo:1000 mg(Fe)/kg/10W-I:ETA SMWOAS 92,130,62

SAFETY PROFILE: Questionable carcinogen with experimental tumorigenic data by implant. See also IRON COMPOUNDS.

IHB675 CAS:7705-12-6 ***HR: 3***
IRON(II) MALEATE
mf: $C_4H_2FeO_4$ mw: 169.90

Fe(−OCO•CH:)$_2$

SAFETY PROFILE: When dispersed in the air the powder ignites above 150°C. It has been involved in industrial fires. When heated to decomposition it emits acrid smoke and fumes. See also IRON COMPOUNDS.

IHB680 CAS:33972-75-7 ***HR: 2***
IRON METHANEARSONATE
mf: CH_5AsO_3•xFe mw: 530.93

SYNS: ARSONIC ACID, METHYL-, IRON SALT (9CI) ◇ METHANEARSONIC ACID, IRON SALT

TOXICITY DATA with REFERENCE
mmo-sat 5 mg/plate MUREAV 116,185,83 85INA8 5,328,86

OSHA PEL: TWA 0.5 mg(As)/m^3
ACGIH TLV: TWA 0.2 mg(As)/m^3; TWA 1 mg/(Fe)/m^3

SAFETY PROFILE: Mutation data reported. When heated to decomposition it emits toxic fumes of As.

IHB800 CAS:12645-50-0 ***HR: D***
IRON NICKEL ZINC OXIDE

SYNS: NICKEL ZINC FERRATE ◇ NICKEL ZINC FERRITE ◇ 1000 NN FERRITE

TOXICITY DATA with REFERENCE
ihl-rat TCLo:560 µg/m^3/24H (female 1-22D post):TER GISAAA 49(5),79,84
ihl-rat TCLo:560 µg/m^3/24H (female 1-22D post):REP GISAAA 49(5),79,84

CONSENSUS REPORTS: Reported in EPA TSCA Inventory.

SAFETY PROFILE: An experimental teratogen. Experimental reproductive effects.

IHC000 CAS:7782-61-8 ***HR: 2***
IRON(III) NITRATE, NONAHYDRATE (1:3:9)
mf: N_3O_9•Fe•$9H_2O$ mw: 404.06

PROP: Crystals.

SYNS: FERRIC NITRATE, NONAHYDRATE ◇ NITRIC ACID, IRON (3−) SALT, NONAHYDRATE

TOXICITY DATA with REFERENCE
orl-rat LD50:3250 mg/kg AIHAAP 30,470,69

OSHA PEL: TWA 1 mg/(Fe)/m^3
ACGIH TLV: TWA 1 mg/(Fe)/m^3

SAFETY PROFILE: Moderately toxic by ingestion. When heated to decomposition it emits toxic fumes of NO_x. See also NITRATES and IRON COMPOUNDS.

IHC100 CAS:16448-54-7 ***HR: 2***
IRON NITRILOTRIACETATE
mf: $C_6H_6FeNO_6$ mw: 243.98

SYNS: ACETIC ACID, NITRILOTRI-, IRON(III) chelate ◇ FERRIC NITRILOTRIACETATE ◇ IRON, (N,N-BIS(CARBOXYMETHYL) GLYCINATO(3-)-N,O,O',O'')-, (T-4)-(9CI) ◇ IRON-NITRILOTRIACETATE CHELATE ◇ IRON(3+) NTA

TOXICITY DATA with REFERENCE
cyt-ham:ovr 2 mmol/L CNREA8 41,1628,81
ipr-rat TDLo:1988 mg/kg/65D-C:CAR JJIND8 76,107,86
ipr-mus TDLo:566 mg/kg/12W-I:CAR CNREA8 47,1867,87

SAFETY PROFILE: Questionable carcinogen with experimental carcinogenic data. Mutation data reported. When heated to decomposition it emits toxic fumes of NO_x.

IHC500 CAS:1345-25-1 ***HR: 3***
IRON(II) OXIDE
mf: FeO mw: 71.85

SAFETY PROFILE: Ignites when heated in air above 200°C. The powdered oxide may be pyrophoric. Incandescent or hazardous reaction with nitric acid (with powdered oxide); hydrogen peroxide; sulfur dioxide + heat. See also IRON and IRON COMPOUNDS.

IHC550 CAS:1317-61-9 ***HR: 3***
IRON(II,III) OXIDE
mf: Fe_3O_4 mw: 231.54

$FeO \cdot Fe_2O_3$

SYN: MAGNETITE

SAFETY PROFILE: Mixtures with aluminum + calcium silicide + sodium nitrate may explode if ignited. Mixtures with aluminum + sulfur react violently if heated. Ignites on contact with hydrogen trisulfide. See also IRON and IRON COMPOUNDS.

IHD000 CAS:1309-37-1 ***HR: 3***
IRON OXIDE
mf: Fe_2O_3 mw: 159.70

SYNS: ANCHRED STANDARD ◇ ANHYDROUS IRON OXIDE ◇ ANHYDROUS OXIDE of IRON ◇ ARMENIAN BOLE ◇ BAUXITE RESIDUE ◇ BLACK OXIDE of IRON ◇ BLENDED RED OXIDES of IRON ◇ BURNTISLAND RED ◇ BURNT SIENNA ◇ BURNT UMBER ◇ CALCOTONE RED ◇ CAPUT MORTUUM ◇ C.I. 77491 ◇ C.I. PIGMENT RED 101 ◇ COLCOTHAR ◇ COLLOIDAL FERRIC OXIDE ◇ CROCUS MARTIS ADSTRINGENS ◇ DEANOX ◇ EISENOXYD ◇ ENGLISH RED ◇ FERRIC OXIDE ◇ FERRUGO ◇ INDIAN RED ◇ IRON(III) OXIDE ◇ IRON OXIDE RED ◇ IRON SESQUIOXIDE ◇ JEWELER'S ROUGE ◇ LEVANOX RED 130A ◇ LIGHT RED ◇ MANUFACTURED IRON OXIDES ◇ MARS BROWN ◇ MARS RED ◇ NATURAL IRON OXIDES ◇ NATURAL RED OXIDE ◇ OCHRE ◇ PRUSSIAN BROWN ◇ RADDLE ◇ 11554 RED ◇ RED IRON OXIDE ◇ RED OCHRE ◇ ROUGE ◇ RUBIGO ◇ SIENNA ◇ SPECULAR IRON ◇ STONE RED ◇ SUPRA ◇ SYNTHETIC IRON OXIDE ◇ VENETIAN RED ◇ VITRIOL RED ◇ VOGEL'S IRON RED ◇ YELLOW FERRIC OXIDE ◇ YELLOW OXIDE of IRON

TOXICITY DATA with REFERENCE
scu-rat TDLo:135 mg/kg:ETA PBPHAW 14,47,78
ipr-rat LD50:5500 mg/kg GTPZAB 26(4),23,82
ipr-mus LD50:5400 mg/kg GTPZAB 26(4),23,82
scu-dog LDLo:30 mg/kg HBAMAK 4,1289,35

CONSENSUS REPORTS: IARC Cancer Review: Group 3 IMEMDT 7,216,87; Human Limited Evidence IMEMDT 1,29,72; Animal No Evidence IMEMDT 1,29,72. Reported in EPA TSCA Inventory.

OSHA PEL: Dust and Fume: TWA 10 mg(Fe)/m^3; Rouge (Transitional: Total Dust: 15 mg/m^3; Respirable Fraction: 5 mg/m^3) TWA Total Dust: 10 mg/m^3; Respirable Fraction: 5 mg/m^3
ACGIH TLV: TWA 5 mg(Fe)/m^3 (vapor, dust); Rouge: 10 mg/m^3
DFG MAK: 6 mg/m^3

SAFETY PROFILE: A poison by subcutaneous route. Questionable carcinogen with experimental tumorigenic data. Catalyzes the potentially explosive polymerization of ethylene oxide. Explosive reaction when heated with guanidinium perchlorate. Reaction with carbon monoxide may form an explosive product. Potentially violent reaction with hydrogen peroxide. The wet oxide reacts explosively with molten aluminum-magnesium alloys. Violent reaction when heated with powdered aluminum; calcium disilicide; magnesium; metal acetylides (e.g., calcium acetylide + iron(III) chloride (on ignition); cesium acetylide (incandescent reaction when warmed); rubidium acetylide). Reacts violently with Al; $Ca(OCl)_2$; N_2H_4; ethylene oxide. See also IRON and IRON COMPOUNDS.

IHE000 ***HR: 3***
IRON OXIDE, CHROMIUM OXIDE, and NICKEL OXIDE FUME

PROP: Fume composed of iron(+3) oxide:chromium(+3) oxide:nickel(+2)oxide, 6:1:1 (BJIMAG 29,169,72).

SYNS: CHROMIUM OXIDE, NICKEL OXIDE, and IRON OXIDE FUME ◇ NICKEL OXIDE, IRON OXIDE, and CHROMIUM OXIDE FUME

CONSENSUS REPORTS: Nickel and its compounds, as well as chromium and its compounds, are on The Community Right-To-Know List.

OSHA PEL: TWA 1 mg(Ni)/m^3; CL 0.1 mg (CrO_3)/m^3
ACGIH TLV: TWA 1 mg(Ni)/m^3; (Proposed: TWA 0.05 mg(Ni)/m^3; Human Carcinogen); 0.05 mg(Cr)/m^3
NIOSH REL: (Chromium (VI)) TWA 0.025 mg(Cr (VI))/m^3; CL 0.05/15M; (Inorganic Nickel) TWA 0.015 mg(Ni)/m^3

SAFETY PROFILE: Confirmed human carcinogen. See also individual components; NICKEL COMPOUNDS, IRON COMPOUNDS, and CHROMIUM COMPOUNDS.

IHF000 CAS:1309-37-1 ***HR: 3***
IRON OXIDE FUME
mf: Fe_2O_3 mw: 159.70

SYN: ZELAZA TLENKI (POLISH)

OSHA PEL: TWA 10 mg/m^3
ACGIH TLV: TWA 5 mg/m^3, welding fumes.

SAFETY PROFILE: Questionable carcinogen. See IRON(III) OXIDE.

IHG000 CAS:8047-67-4 ***HR: 3***
IRON OXIDE, SACCHARATED

PROP: Saccharated oxide of iron (JNCIAM 24,109,60).

SYNS: COLLIRON I.V. ◇ FEOJECTIN ◇ FERRIC OXIDE, SACCHARATED ◇ FERRIC SACCHARATE IRON OXIDE (MIX.) ◇ FERRIVENIN ◇ IRON SACCHARATE ◇ IRON SUGAR ◇ IVIRON ◇ NEO-FERRUM ◇ PROFERRIN ◇ SACCHARATED FERRIC OXIDE ◇ SACCHARATED IRON ◇ SUCROFER

TOXICITY DATA with REFERENCE
ims-rat TDLo:148 g/kg/74W:NEO BMJOAE 1,947,59
scu-mus TDLo:104 g/kg/13W-I:ETA JNCIAM 24,109,60
ivn-mus LD50:180 mg/kg NIIRDN 6,193,82

CONSENSUS REPORTS: IARC Cancer Review: Animal Sufficient Evidence IMEMDT 2,161,73.

SAFETY PROFILE: Confirmed carcinogen with experimental neoplastigenic and tumorigenic data. A poison by intravenous route. See also IRON COMPOUNDS.

IHG500 CAS:13463-40-6 ***HR: 3***
IRON PENTACARBONYL
DOT: UN 1994
mf: C_5FeO_5 mw: 195.90

$$Fe(OC)_5$$

PROP: Yellow to dark red viscous liquid. Mp: −25°, bp: 103.0°, flash p: 5°F, d: 1.453 @ 25°/4°, vap press: 40 mm @ 30.3°.

SYNS: FER PENTACARBONYLE (FRENCH) ◇ IRON CARBONYL ◇ PENTACARBONYLIRON

TOXICITY DATA with REFERENCE
ihl-rat LCLo:33 ppm/5.5H BJIMAG 27,1,70
ihl-mus LC50:7 g/m^3/10M NTIS** PB158-508
orl-rbt LD50:12 mg/kg JIHTAB 25,415,43
ihl-rbt LCLo:250 ppm/45M 34ZIAG -,335,69
skn-rbt LD50:240 mg/kg 34ZIAG -,335,69
scu-rbt LD50:240 mg/kg JIHTAB 25,415,43
ivn-rbt LD50:11 mg/kg JIHTAB 25,415,43
orl-gpg LD50:22 mg/kg JIHTAB 25,415,43

CONSENSUS REPORTS: EPA Extremely Hazardous Substances List. Reported in EPA TSCA Inventory.

OSHA PEL: TWA 0.1 ppm (Fe); STEL 0.2 ppm
ACGIH TLV: TWA 0.1 ppm (Fe); STEL 0.2 ppm
DFG MAK: 0.1 ppm (0.8 mg/m^3)
DOT Classification: Poison B; Label: Flammable Liquid and Poison

SAFETY PROFILE: A poison by inhalation, skin contact, ingestion, subcutaneous, and intravenous routes. Inhalation causes dizziness, nausea, and vomiting. If continued, unconsciousness follows. Often there is a delayed reaction of chest pain, cough, and difficult breathing. There may be cyanosis and circulatory collapse. In fatal cases, death occurs from the fourth to eleventh day with pneumonitis and injury to kidneys, liver, and brain. Iron carbonyl is less toxic than nickel carbonyl.

A very dangerous fire and moderate explosion hazard when exposed to heat or flame; can react vigorously with oxidizing materials. Pyrophoric in air!! Mixtures with nitrogen oxide explode above 50°C. Violent reaction with zinc + transition metal halides (e.g., cobalt halides; rhodium halides; ruthenium halides). Mixtures with acetic acid + water produce a pyrophoric powder. To fight fire, use water, foam, CO_2, dry chemical. See also CARBONYLS and IRON COMPOUNDS.

IHH000 ***HR: 3***
IRON-POLYSACCHARIDE COMPLEX

PROP: Solution of iron and synthetically prepared polysaccharide with a mean molecular weight of about 20,000 (BJCAAI 21,448,67).

SYN: MUSCULARON

TOXICITY DATA with REFERENCE
scu-mus TDLo:1000 mg(Fe)/kg/9W-I:ETA BJCAAI 21,448,67
ipr-mus LD50:318 mg(Fe)/kg AJMSA9 241,296,61
ivn-mus LD50:170 mg(Fe)/kg BJCAAI 241,296,61

SAFETY PROFILE: A poison by intravenous and intraperitoneal routes. Questionable carcinogen with experimental tumorigenic data. When heated to decomposition it emits acrid smoke and fumes. See also IRON COMPOUNDS.

IHH300 CAS:73361-47-4 ***HR: 2***
IRON-POLY(SORBITOL-GLUCONIC ACID) COMPLEX

TOXICITY DATA with REFERENCE
ims-rat TDLo:200 mg(Fe)/kg (12-15D preg):TER SJHSBD 32,69,77
ims-rat TDLo:200 mg(Fe)/kg (12-15D preg):REP SJHSBD 32,69,77
ivn-rat LD50:920 mg(Fe)/kg SJHSBD 32,58,77
scu-mus LD50:2098 mg(Fe)/kg SJHSBD 32,58,77
ivn-mus LD50:1160 mg(Fe)/kg SJHSBD 32,58,77

SAFETY PROFILE: Moderately toxic by intravenous and subcutaneous routes. Experimental teratogenic and reproductive effects. When heated to decomposition it emits acrid smoke and fumes. See also IRON COMPOUNDS.

IHJ000 CAS:50645-52-8 ***HR: 3***
IRON-SILICON
mf: Fe-Si mw: 83.93

SYN: FERROSILICON

SAFETY PROFILE: Dangerously flammable and explosive when ground. Incandescent reaction with solid sodium hydroxide when water is added. Incompatible with water, causing evolution of poisonous arsine; combustible acetylene; and spontaneously flammable phosphine, due to impurities. See also IRON COMPOUNDS.

IHK000 ***HR: 3***
IRON SODIUM GLUCONATE

SYNS: FERRIC SODIUM GLUCONATE COMPLEX ◇ OSMOFERRIN

TOXICITY DATA with REFERENCE
scu-mus TDLo:3000 mg(Fe)/kg/16W-I:NEO BJCAAI 22,521,68
scu-mus TD:40 g/kg/I:ETA BECCAN 40,30,62

SAFETY PROFILE: Questionable carcinogen with experimental neoplastigenic and tumorigenic data. When heated to decomposition it emits toxic fumes of Na_2O. See also IRON COMPOUNDS.

IHK100 CAS:62765-90-6 ***HR: 3***
IRON-SORBITOL

SYN: JECTOFER

TOXICITY DATA with REFERENCE
ims-rat TDLo:200 mg(Fe)/kg (12-15D preg):TER SJHSBD 32,69,77
ipr-mus LD50:25 mg(Fe)/kg NNAPBA 270(Suppl),R50,71
ivn-mus LD50:44 mg(Fe)/kg BJPCAL 27,114,66

SAFETY PROFILE: Poison by intravenous and intraperitoneal routes. An experimental teratogen. When heated to decomposition it emits acrid smoke and fumes. See also IRON COMPOUNDS.

IHL000 CAS:1338-16-5 ***HR: 3***
IRON SORBITOL CITRATE
mf: $C_6H_{14}O_6 \cdot C_6H_8O_7 \cdot xFe$ mw: 765.29

SYNS: ESZ ◇ IRON SORBITEX ◇ IRON-SORBITOL-CITRIC ACID

TOXICITY DATA with REFERENCE
orl-mus LD50:1370 mg/kg ARZNAD 24,880,74
scu-mus LD50:140 mg/kg ARZNAD 20,1795,70
ivn-mus LD50:174 mg/kg ARZNAD 20,1795,70
ims-mus LD50:918 mg/kg ARZNAD 24,880,74

CONSENSUS REPORTS: IARC Cancer Review: Group 3 IMEMDT 7,56,87; Animal No Evidence IMEMDT 2,161,73.

OSHA PEL: TWA 1 mg/(Fe)/m^3
ACGIH TLV: TWA 1 mg/(Fe)/m^3

SAFETY PROFILE: A poison by subcutaneous and intravenous routes. Moderately toxic by ingestion and intramuscular routes. Questionable carcinogen. When heated to decomposition it emits acrid smoke and fumes. See also IRON COMPOUNDS.

IHN000 CAS:1317-37-9 ***HR: 3***
IRON(II) SULFIDE
mf: FeS mw: 87.91

SAFETY PROFILE: Use of steel equipment in conjunction with materials containing hydrogen sulfide or volatile sulfur compounds will cause it to spontaneously explode in air. The moist sulfide may react incandescently with air. The impure sulfide ignites spontaneously in air. Violent reaction with lithium when heated above 260°C. When heated to decomposition it emits toxic fumes of SO_x. See also IRON COMPOUNDS and SULFIDES.

IHN050 CAS:12063-27-3 ***HR: 3***
IRON(III) SULFIDE
mf: Fe_2S_3 mw: 207.87

SAFETY PROFILE: Flammable when heated. When heated to decomposition it emits toxic fumes of SO_x. See also IRON COMPOUNDS and SULFIDES.

IHN200 CAS:118-48-9 ***HR: D***
ISATOIC ACID ANHYDRIDE
mf: $C_8H_5NO_3$ mw: 163.14

SYNS: 2H-3,1-BENZOXAZINE-2,4(1H)-DIONE ◇ IA ◇ ISATOIC ANHYDRIDE

TOXICITY DATA with REFERENCE
orl-mus TDLo:34800 mg/kg (male 8D pre):REP MPHEAE 15,7,66

CONSENSUS REPORTS: Reported in EPA TSCA Inventory.

SAFETY PROFILE: Experimental reproductive effects. When heated to decomposition it emits toxic fumes of NO_x.

IHO200 CAS:10075-36-2 ***HR: 3***
ISOAMINILE CYCLAMATE
mf: $C_{16}H_{24}N_2 \cdot C_6H_{13}NO_3S$ mw: 423.68

SYNS: α-(ISOPROPYL)-α-(β-DIMETHYLAMINOPROPYL)PHENYLACETONITRILECYCLAMATE ◇ MUCALAN

TOXICITY DATA with REFERENCE
orl-rat LD50:270 mg/kg KSRANM 5,2212,71
scu-rat LD50:138 mg/kg KSRNAM 5,2212,71
orl-mus LD50:298 mg/kg KSRNAM 5,2212,71
ivn-mus LD50:57 mg/kg KSRNAM 5,2212,71

CONSENSUS REPORTS: Cyanide and its compounds are on the Community Right-To-Know List.

SAFETY PROFILE: Poison by ingestion, subcutaneous, and intravenous routes. When heated to decomposition it emits toxic fumes of SO_x, CN^-, and NO_x.

IHO700 CAS:51371-34-7 ***HR: D***
ISOAMYGDALIN
mf: $C_{20}H_{27N}O_{11}$ mw: 457.48

SYNS: d,l-AMYGDALIN ◇ d,l-MANDELONITRILE-β-d-GLUCOSIDO-6-β-GLUCOSIDE ◇ NSC 251222

TOXICITY DATA with REFERENCE
orl-ham TDLo:225 mg/kg (female 8D post):TER SCIEAS 215,1513,82

ipr-rat LD50:19582 mg/kg NTIS** PB288-558

CONSENSUS REPORTS: Cyanide and its compounds are on the Community Right-To-Know List.

SAFETY PROFILE: An experimental teratogen. When heated to decomposition it emits toxic fumes of NO_x and CN^-.

IHO850 CAS:123-92-2 ***HR: 3***
ISOAMYL ACETATE
mf: $C_7H_{14}O_2$ mw: 130.21

PROP: Colorless liquid; banana-like odor. Bp: 142.0°, ULC: 55-60, lel: 1% @ 212°F, uel: 7.5%, flash p: 77°F, d: 0.876, refr index: 1.400, autoign temp: 680°F, vap d: 4.49. Misc in alc, ether, ethyl acetate, fixed oils; sltly sol in water; insol in glycerin, propylene glycol.

SYNS: ACETIC ACID, ISOPENTYL ESTER ◇ BANANA OIL ◇ FEMA No. 2055 ◇ ISOAMYL ETHANOATE ◇ ISOPENTYL ACETATE ◇ ISOPENTYL ALCOHOL ACETATE ◇ 3-METHYLBUTYL ACETATE ◇ 3-METHYL-1-BUTYL ACETATE ◇ 3-METHYLBUTYL ETHANOATE ◇ PEAR OIL

TOXICITY DATA with REFERENCE
orl-rat LD50:16600 mg/kg YKYUA6 32,1241,81
ihl-cat LCLo:35000 mg/m^3 AGGHAR 5,1,33
orl-rbt LD50:7422 mg/kg IMSUAI 41,31,72
scu-gpg LDLo:5000 mg/kg AGGHAR 5,1,33

CONSENSUS REPORTS: Reported in EPA TSCA Inventory.

OSHA PEL: TWA 100 ppm
ACGIH TLV: TWA 100 ppm

SAFETY PROFILE: Mildly toxic by ingestion, inhalation, and subcutaneous routes. Exposure to concentrations of about 1,000 ppm for 1 hour can cause headache, fatigue, pulmonary irritation and serious toxicity effects. Highly flammable liquid when exposed to heat or flame; can react vigorously with reducing materials. Moderately explosive in the form of vapor when exposed to heat or flame. To fight fire, use alcohol foam, CO_2, dry chemical. When heated to decomposition it emits acrid smoke and fumes.

IHP000 CAS:123-51-3 ***HR: 3***
ISOAMYL ALCOHOL
DOT: UN 1105
mf: $C_5H_{12}O$ mw: 88.17

$$(CH_3)_2CHC_2H_4OH$$

PROP: Clear liquid; pungent, repulsive taste. Bp: 132°, ULC: 35-40, lel: 1.2%, uel: 9.0% @ 212°F, flash p: 109°F (CC), d: 0.813, autoign temp: 662°F, vap d: 3.04, mp: −117.2°. Sol in water @ 14°; misc in alc and ether.

SYNS: ALCOOL AMILICO (ITALIAN) ◇ ALCOOL ISOAMYLIQUE (FRENCH) ◇ AMYLOWY ALKOHOL (POLISH) ◇ FERMENTATION AMYL ALCOHOL ◇ ISOAMYL ALKOHOL (CZECH) ◇ ISO-AMYLALKOHOL (GERMAN) ◇ ISOAMYLOL ◇ ISOBUTYLCARBINOL ◇ ISOPENTANOL ◇ ISOPENTYL ALCOHOL ◇ 2-METHYL-4-BUTANOL ◇ 3-METHYL BUTANOL ◇ 3-METHYLBUTAN-1-OL ◇ 3-METHYL-1-BUTANOL (CZECH) ◇ 3-METIL-BUTANOLO (ITALIAN)

TOXICITY DATA with REFERENCE
eye-hmn 150 ppm JIHTAB 25,282,43
skn-rbt 500 mg/24H MOD 28ZPAK -,36,72
eye-rbt 20 mg/24H MOD 28ZPAK -,36,72
cyt-smc 10 mmol/tube HEREAY 33,457,47
orl-rat TDLo:27 g/kg/75W-I:CAR ARGEAR 45,19,75
scu-rat TDLo:3800 mg/kg/85W-I:CAR ARGEAR 45,19,75
ihl-hmn TCLo:150 ppm:NOSE,EYE,PUL JIHTAB 25,282,43
orl-rat LD50:1300 mg/kg SAMJAF 43,795,69
ipr-rat LDLo:813 mg/kg AEPPAE 132,214,28
ipr-mus LDLo:233 mg/kg FCTXAV 16,785,78
scu-mus LDLo:7480 mg/kg FCTXAV 16,785,78
ivn-mus LD50:234 mg/kg AIPTAK 135,330,62
ivn-cat LDLo:210 mg/kg FCTXAV 16,785,78
orl-rbt LDLo:4250 mg/kg JLCMAK 10,985,25
skn-rbt LD50:3212 mg/kg AIHAAP 30,470,69
ivn-rbt LDLo:1570 mg/kg FCTXAV 16,785,78

CONSENSUS REPORTS: Reported in EPA TSCA Inventory.

OSHA PEL: TWA 100 ppm; STEL 125 ppm
ACGIH TLV: TWA 100 ppm; STEL 125 ppm
DFG MAK: 100 ppm (360 mg/m^3)
DOT Classification: Flammable or Combustible Liquid; Label: Flammable Liquid.

SAFETY PROFILE: A poison by intraperitoneal and intravenous routes. Moderately toxic by ingestion and skin contact. A skin and human eye irritant. Human systemic effects by inhalation: olfactory effects, conjunctiva irritation, respiratory changes. Questionable carcinogen with experimental carcinogenic data. Mutation data reported. Flammable when exposed to heat or flame; can react vigorously with reducing materials. Slight explosion hazard when exposed to flame. Explosive reaction with hydrogen trisulfide. To fight fire, use alcohol foam, CO_2, dry chemical. When heated to decomposition it emits acrid smoke and fumes. Used as a flotation agent, a solvent, and in organic synthesis.

IHP010 CAS:584-02-1 ***HR: 3***
ISOAMYL ALCOHOL
DOT: UN 2706
mf: $C_5H_{12}O$ mw: 88.17

PROP: Liquid; acetone-like odor. Bp: 115.6°, d: 0.815 @ 25°/4°, flash p: 66°F, lel: 1.2%, uel: 9%. Sol alc, ether; sltly sol in water.

SYNS: DIETHYL CARBINOL ◇ DIETHYLCARBINOL (DOT) ◇ 3-PENTANOL ◇ PENTANOL-3 ◇ PENTAN-3-OL

TOXICITY DATA with REFERENCE
skn-rbt 10 mg/24H open MLD AMIHBC 10,61,54
eye-rbt 5 mg open SEV AMIHBC 10,61,54
orl-rat LD50:1870 mg/kg AMIHBC 10,61,54
ipr-rat LDLo:1950 mg/kg JIHTAB 27,1,45
skn-rbt LD50:2520 mg/kg AMIHBC 10,61,54

CONSENSUS REPORTS: Reported in EPA TSCA Inventory.

OSHA PEL: TWA 100 ppm; STEL 125 ppm
ACGIH TLV: TWA 100 ppm; STEL 125 ppm
DFG MAK: 100 ppm (360 mg/m^3)
DOT Classification: Flammable or Combustible Liquid; Label: Flammable Liquid.

SAFETY PROFILE: Moderately toxic by ingestion, skin contact, and intraperitoneal routes. A severe eye and mild skin irritant. Dangerous fire and explosion hazard when exposed to heat, flame, or oxidizing materials. Used as a flotation agent, a solvent, and in organic synthesis. When heated to decomposition it emits acrid smoke and irritating fumes. See also ALCOHOLS.

IHP100 CAS:94-46-2 ***HR: 1***
ISOAMYL BENZOATE
mf: $C_{12}H_{16}O$ mw: 176.28

PROP: Colorless to pale yellow liquid; pungent, fruity odor. D: 0.986-0.992, refr index: 1.492, flash p: +212°F.

SYNS: AMYL BENZOATE ◇ BENZOIC ACID, 1-(3-METHYL)BUTYL ESTER ◇ FEMA No. 2058 ◇ ISOPENTYL BENZOATE ◇ 1-(3-METHYL) BUTYL BENZOATE

TOXICITY DATA with REFERENCE
skn-rbt 500 mg/24H MLD FCTXAV 11,1079,73
orl-rat LD50:6330 mg/kg FCTXAV 11,477,73

CONSENSUS REPORTS: Reported in EPA TSCA Inventory.

SAFETY PROFILE: Mildly toxic by ingestion. A skin irritant. Combustible liquid. When heated to decomposition it emits acrid smoke and irritating fumes. See also ESTERS.

IHP400 CAS:106-27-4 ***HR: 2***
ISOAMYL BUTYRATE
mf: $C_9H_{18}O_2$ mw: 158.24

PROP: Colorless liquid; fruity odor. D: 0.860, refr index: 1.409-1.414, flash p: 149°F. Sol in alc, fixed oils; insol in glycerin, propylene glycol, water @ 179°.

SYNS: AMYL BUTYRATE ◇ FEMA No. 2060

DOT Classification: Flammable or Combustible Liquid; Label: Flammable Liquid

SAFETY PROFILE: Combustible liquid. When heated to decomposition it emits acrid smoke and irritating fumes.

IHP500 CAS:2035-99-6 ***HR: 1***
ISOAMYL CAPRYLATE
mf: $C_{13}H_{26}O_2$ mw: 214.39

SYNS: ISOAMYL OCTANOATE ◇ ISOPENTYL OCTANOATE ◇ OCTANOIC ACID, ISOPENTYL ESTER

TOXICITY DATA with REFERENCE
skn-rbt 500 mg/24H MLD FCTXAV 17,827,79

CONSENSUS REPORTS: Reported in EPA TSCA Inventory.

SAFETY PROFILE: A skin irritant. When heated to decomposition it emits acrid smoke and irritating fumes.

IHQ000 CAS:543-86-2 ***HR: D***
ISOAMYL CARBAMATE
mf: $C_6H_{13}NO_2$ mw: 131.20

PROP: White or sltly yellowish crystal leaflets. Mp: 59°, bp: 220° +. Sltly sol in cold water; sol in boiling water, alc and ether.

SYNS: CARBAMATE (ISOAMYL) ◇ ISOAMYL AMINOFORMATE ◇ 3-METHYL-1-BUTANOL CARBAMATE

TOXICITY DATA with REFERENCE
mmo-esc 3000 ppm/3H AMNTA4 85,119,51

SAFETY PROFILE: Mutation data reported. When heated to decomposition it emits toxic fumes of NO_x. See also CARBAMATES.

IHQ100 CAS:64049-23-6 ***HR: 2***
ISOAMYLDICHLOROARSINE
mf: $C_5H_{11}AsCl_2$ mw: 216.98

SYNS: ARSINE, DICHLOROISOPENTYL- ◇ B-343 ◇ ISOAMYLDICHLORARSINE

TOXICITY DATA with REFERENCE
ihl-mus LC50:2 g/m^3/10M NTIS** PB158-508

OSHA PEL: TWA 0.5 mg(As)/m^3

SAFETY PROFILE: Moderately toxic by inhalation. When heated to decomposition it emits toxic fumes of As and Cl^-.

IHR200 CAS:73080-51-0 ***HR: 2***
ISOAMYL 5,6-DIHYDRO-7,8-DIMETHYL-4,5-DIOXO-4H-PYRANO(3,2-c)QUINOLINE-2-CARBOXYLATE
mf: $C_{20}H_{21}NO_5$ mw: 355.42

SYNS: MY-5116 ◇ 4H-PYRANO(3,2-c)QUINOLINE-2-CARBOXYLIC ACID, 5,6-DIHYDRO-7,8-DIMETHYL-4,5-DIOXO-, ISOPENTYL ESTER

TOXICITY DATA with REFERENCE
orl-rbt TDLo:5200 mg/kg (female 6-18D post):TER IYKEDH 17,51,86
orl-rbt TDLo:5200 mg/kg (female 6-18D post):REP IYKEDH 17,51,86
ipr-rat LD50:2783 mg/kg IYKEDH 19,164,88
ipr-mus LD50:1435 mg/kg IYKEDH 19,164,88

SAFETY PROFILE: Moderately toxic by intraperitoneal route. An experimental teratogen. Experimental reproductive effects. When heated to decomposition it emits toxic fumes of NO_x.

IHR300 CAS:482-44-0 ***HR: 2***
8-ISOAMYLENOXYPSORALEN
mf: $C_{16}H_{14}O_4$ mw: 270.30

PROP: From roots of *Imperatoria osthruthium L., Umbelliferae.* Prisms from ether, long fine needles from hot water. Mp: 102°. Practically insol in cold water; very sparingly sol in boiling water; freely sol in chloroform; sol in benzene, alc, ether, petrol ether, alkali hydroxides.

SYNS: AMMIDIN ◇ IMPERATORIN ◇ 8-ISOPENTENYLOXYPSORALENE ◇ MARMELOSIN ◇ 9-((3-METHYL-2-BUTENYL)OXY)-7H-FURO (3,2-g)(1)BENZOPYRAN-7-ONE

TOXICITY DATA with REFERENCE
cyt-hmn:lym 10 mg/L MUREAV 169,51,86
sce-hmn:lym 20 mg/L MUREAV 169,51,86
par-mus LDLo:600 mg/kg CBCCT* 7,689,55

SAFETY PROFILE: Moderately toxic by parenteral route. Human mutation data reported. When heated to decomposition it emits acrid smoke and fumes.

IHS000 CAS:110-45-2 ***HR: 2***
ISOAMYL FORMATE
DOT: UN 1109
mf: $C_6H_{12}O_2$ mw: 116.18

PROP: Clear liquid; fruity odor. Bp: 123.3°, d: 0.877 @ 20°, refr index: 1.396, vap press: 10 mm @ 17.1°, flash p: 127°F. Misc with alc, ether, propylene glycol; very sltly sol in water; insol in glycerin.

SYNS: FEMA No. 2069 ◇ FORMIC ACID, ISOPENTYL ESTER ◇ ISOAMYL METHANOATE ◇ ISOPENTYL ALCOHOL, FORMATE ◇ ISOPENTYL FORMATE ◇ 3-METHYLBUTYL FORMATE

TOXICITY DATA with REFERENCE
skn-rbt 500 mg/24H MOD FCTXAV 17,829,79
orl-rat LD50:9840 mg/kg FCTXAV 2,327,64
orl-rbt LD50:3020 mg/kg IMSUAI 41,31,72

CONSENSUS REPORTS: Reported in EPA TSCA Inventory.

DOT Classification: Flammable or Combustible Liquid; Label: Flammable Liquid.

SAFETY PROFILE: Moderately toxic by ingestion. A skin irritant. This material is very irritating and can cause narcosis. The symptoms are usually transient in nature, but it is possible upon severe or prolonged exposure to have serious consequences. Combustible liquid. Can react with oxidizing materials. When heated to decomposition it emits acrid smoke and fumes.

IHS100 CAS:68133-73-3 ***HR: 1***
ISOAMYL GERANATE
mf: $C_{15}H_{26}O_2$ mw: 238.41

SYN: 2,6-OCTADIENOIC ACID, 3,7-DIMETHYL-, ISOPENTYL ESTER, (E)-

TOXICITY DATA with REFERENCE
skn-rbt 500 mg/24H MOD FCTXAV 17,831,79

CONSENSUS REPORTS: Reported in EPA TSCA Inventory.

SAFETY PROFILE: A skin irritant. When heated to decomposition it emits acrid smoke and irritating fumes.

IHT000 CAS:627-92-9 ***HR: 2***
1-ISOAMYL GLYCEROL ETHER
mf: $C_8H_{18}O_3$ mw: 162.26

PROP: Colorless liquid. D: 0.987, bp: 260-262°, sol in water, misc in alc and ether.

TOXICITY DATA with REFERENCE
ipr-mus LDLo:1000 mg/kg CMDI** -,-,49
scu-mus LD50:2106 mg/kg JPETAB 93,470,48

SAFETY PROFILE: Moderately toxic by intraperitoneal and subcutaneous routes. When heated to decomposition it emits acrid smoke and fumes. See also ETHERS.

IHU100 CAS:2198-61-0 ***HR: 2***
ISOAMYL HEXANOATE
mf: $C_{11}H_{22}O_2$ mw: 186.33

PROP: Colorless liquid; fruity odor. D: 0.858-0.863, refr index: 1.418-1.422, flash p: 190°F. Sol in alc, fixed oils; insol in glycerin, propylene glycol, water @ 222°.

SYNS: AMYL HEXANOATE ◇ FEMA No. 2075 ◇ ISOAMYL CAPROATE

TOXICITY DATA with REFERENCE
skn-rbt 500 mg/24H MLD FCTXAV 17,825,79

CONSENSUS REPORTS: Reported in EPA TSCA Inventory.

SAFETY PROFILE: A mild skin irritant. Combustible liquid. When heated to decomposition it emits acrid smoke and irritating fumes.

IHV000 CAS:102-19-2 ***HR: 1***
ISOAMYL PHENYLACETATE
mf: $C_{13}H_{18}O_2$ mw: 206.31

SYNS: BENZENEACETIC ACID, 3-METHYLBUTYL ESTER (9CI) ◇ ISOPENTYLPHENYLACETATE

TOXICITY DATA with REFERENCE
skn-rbt 500 mg/24H MOD FCTXAV 16,637,78
mrc-bcs 20 μg/disc OEKSDJ 9,177,78

CONSENSUS REPORTS: Reported in EPA TSCA Inventory.

SAFETY PROFILE: A skin irritant. Mutation data reported. When heated to decomposition it emits acrid smoke and fumes. See also ESTERS.

IHV050 CAS:56011-02-0 ***HR: 1***
ISOAMYL PHENYLETHYL ETHER
mf: $C_{13}H_{20}O$ mw: 192.33

SYNS: ANTHER ◇ BENZENE, (2-(3-METHYLBUTOXY)ETHYL)- ◇ (2-(3-METHYLBUTOXY)ETHYL)BENZENE ◇ PHENYLETHYL ISOAMYL ETHER

TOXICITY DATA with REFERENCE
skn-rbt 500 mg/24H MLD FCTOD7 21,873,83

CONSENSUS REPORTS: Reported in EPA TSCA Inventory.

SAFETY PROFILE: A skin irritant. When heated to decomposition it emits acrid smoke and irritating fumes.

IHX200 CAS:1024-65-3 ***HR: 3***
1-ISOAMYL THEOBROMINE
mf: $C_{12}H_{18}N_4O_2$ mw: 250.34

SYN: 1-ISOPENTYL-THEOBROMINE

TOXICITY DATA with REFERENCE
orl-mus LD50:772 mg/kg JPETAB 116,343,56
ipr-mus LD50:222 mg/kg JPETAB 116,343,56
ivn-mus LD50:200 mg/kg JPETAB 86,113,46

SAFETY PROFILE: Poison by intravenous and intraperitoneal routes. Moderately toxic by ingestion. When heated to decomposition it emits toxic fumes of NO_x. See also THEOBROMINE.

IHX400 CAS:2883-98-9 ***HR: 2***
trans-ISOASARONE
mf: $C_{12}H_{16}O_3$ mw: 208.28

SYNS: ASARON ◇ ASARONE ◇ ASARONE, trans- ◇ α-ASARONE ◇ trans-ASARONE ◇ ASARUM CAMPHOR ◇ BENZENE, 1,2,4-TRIMETHOXY-5-PROPENYL-, (E)- ◇ BENZENE, 1,2,4-TRIMETHOXY-5-PROPENYL-, trans- ◇ ETHEROPHENOL

TOXICITY DATA with REFERENCE
ipr-mus TDLo:156 mg/kg:CAR CNREA8 47,2275,87
orl-mus LD50:418 mg/kg FATOAO 48(6),17,85
ipr-mus LD50:310 mg/kg FATOAO 48(6),17,85
ivn-mus LDLo:66 mg/kg YHTPAD 23,52,88

SAFETY PROFILE: Moderately toxic by ingestion, intravenous, and intraperitoneal routes. Questionable carcinogen with experimental carcinogenic data. When heated to decomposition it emits acrid smoke and irritating fumes.

IHX600 ***HR: 1***
ISOBORNYL ACETATE
mf: $C_{12}H_{20}O_2$ mw: 196.29

PROP: Colorless liquid; camphoraceous, piney, balsamic odor. D: 0.980, refr index: 1.462, flash p: +212°F. Sol in alc, fixed oils; sltly sol in propylene glycol; insol in water @ 227°.

SYN: FEMA No. 2160

SAFETY PROFILE: Combustible liquid. When heated to decomposition it emits acrid smoke and irritating fumes.

IHY000 CAS:124-76-5 ***HR: 3***
ISOBORNYL ALCOHOL
mf: $C_{10}H_{18}O$ mw: 154.28

PROP: A geometrical isomer of borneol. White solid, camphor odor, more sol in most solvents than borneol. Mp: 216° (subl).

SYNS: ISOBORNEOL ◇ dl-ISOBORNEOL ◇ ISOCAMPHOL

TOXICITY DATA with REFERENCE
skn-rbt 500 mg/24H MOD FCTXAV 17,509,79
orl-rat LD50:5200 mg/kg FCTXAV 17,509,79
ivn-mus LD50:56 mg/kg CSLNX* NX#03209

CONSENSUS REPORTS: Reported in EPA TSCA Inventory.

SAFETY PROFILE: Poison by intravenous route. Mildly toxic by ingestion. A skin irritant. When heated to decomposition it emits acrid smoke and irritating fumes.

IHZ000 CAS:115-31-1 ***HR: 2***
ISOBORNYL THIOCYANATOACETATE
mf: $C_{13}H_{19}NO_2S$ mw: 253.39

PROP: Yellow, oily liquid; terpene-like odor. D: 1.1465 @ 25°/4°, bp: 95° @ 0.06 mm, flash p: 82°C (180°F). Very sol in alc, benzene, chloroform, and ether; insol in water.

SYNS: BORNATE ◇ CIDALON ◇ ENT 92 ◇ ISOBORNEOL THIOCYANATOACETATE ◇ ISOBORNYL THIOCYANOACETATE ◇ TERPINYL THIOCYANOACETATE ◇ THANISOL ◇ THANITE ◇ THIOCYANATOACETIC ACID ISOBORNYL ESTER

TOXICITY DATA with REFERENCE
orl-rat LD50:1 g/kg YKYUA6 32,605,81
orl-rbt LD50:630 mg/kg JPETAB 82,377,44
skn-rbt LD50:6000 mg/kg WRPCA2 7,135,68
orl-gpg LD50:551 mg/kg FMCHA2 -,C233,83,80

CONSENSUS REPORTS: Reported in EPA TSCA Inventory.

SAFETY PROFILE: Moderately toxic by ingestion. Sltly toxic by skin contact. Very irritating to eyes, mucous membranes, and skin. Flammable when exposed to heat or flame; can react vigorously with oxidizing materials. When heated to decomp it emits very toxic fumes of NO_x and SO_x. See also THIOCYANATES and ESTERS. Used as an FDA over-the-counter drug; an insecticide and fly spray.

IIA000 CAS:124-68-5 ***HR: 2***
ISOBUTANOL-2-AMINE
mf: $C_4H_{11}NO$ mw: 89.16

PROP: Colorless liquid or crystalline mass. Mp: 30-31°, bp: 165°, flash p: 153°F (TOC) d: 0.934 @ 20°/20° vap d: 3.04. Misc with water, sol in alcs.

SYNS: 2-AMINODIMETHYLETHANOL ◇ β-AMINOISOBUTANOL ◇ 2-AMINO-2-METHYLPROPANOL ◇ 2-AMINO-2-METHYLPROPAN-1-OL ◇ 2-AMINO-2-METHYL-1-PROPANOL ◇ ISOBUTANOLAMINE

TOXICITY DATA with REFERENCE
orl-rbt LDLo:1000 mg/kg JIHTAB 22,315,40

CONSENSUS REPORTS: Reported in EPA TSCA Inventory.

SAFETY PROFILE: Moderately toxic by ingestion. Flammable when exposed to heat or flame, can react with oxidizing materials. To fight fire, use alcohol foam, dry chemical, mist or spray. When heated to decomposition it emits toxic fumes of NO_x. See also AMINES.

IIC000 CAS:115-11-7 ***HR: 3***
ISOBUTENE
DOT: UN 1055/UN 1075
mf: C_4H_8 mw: 56.12

PROP: Volatile liquid or easily liquefied gas. Bp: −6.9°, fp: −140.3°, flash p: <14°F, d: 0.600, autoign temp: 869°F, lel: 1.8%, uel: 9.6%. Insol in water, very sol in alc, ether, sulfuric acid.

SYNS: Γ-BUTYLENE ◇ ISOBUTYLENE (DOT) ◇ LIQUEFIED PETROLEUM GAS (DOT) ◇ 2-METHYLPROPENE

TOXICITY DATA with REFERENCE
ihl-rat LC50:620 g/m^3/4H FATOAO 30,102,67
ihl-mus LC50:415 g/m^3/2H FATOAO 30,102,67

CONSENSUS REPORTS: Reported in EPA TSCA Inventory.

DOT Classification: Flammable Gas; Label: Flammable Gas.

SAFETY PROFILE: A simple asphyxiant; may have narcotizing action. A very dangerous fire and explosion hazard when exposed to heat or flame. Can react vigorously with oxidizing materials. To fight fire, stop flow of gas. When heated to decomposition it emits acrid smoke and fumes.

IIE000 CAS:16006-09-0 ***HR: 2***
(2-ISOBUTOXYETHYL)CARBAMATE
mf: $C_7H_{15}NO_3$ mw: 161.23

SYN: CARBAMIC ACID (2-ISOBUTOXYETHYL) ESTER

TOXICITY DATA with REFERENCE
ipr-mus LDLo:500 mg/kg UCPHAQ 1,93,38

CONSENSUS REPORTS: Reported in EPA TSCA Inventory.

SAFETY PROFILE: Moderately toxic by intraperitoneal route. When heated to decomposition it emits toxic fumes of NO_x. See also CARBAMATES.

IIG000 CAS:23436-19-3 ***HR: 1***
1-ISOBUTOXY-2-PROPANOL
mf: $C_7H_{16}O_2$ mw: 132.23

SYNS: 1-(2-METHYLPROPOXY)-2-PROPANOL (9CI) ◇ PROPYLENE GLYCOL ISOBUTYL ETHER

TOXICITY DATA with REFERENCE
orl-rat LD50:4290 mg/kg NPIRI* 1,103,74
skn-rbt LD50:8000 mg/kg NPIRI* 1,103,74

CONSENSUS REPORTS: Glycol ether compounds are on the Community Right-To-Know List. Reported in EPA TSCA Inventory.

SAFETY PROFILE: Mildly toxic by ingestion and skin contact. When heated to decomposition it emits acrid smoke and fumes. See also GLYCOL ETHERS.

III000 CAS:32767-68-3 ***HR: 2***
2-ISOBUTOXY TETRAHYDROPYRAN
mf: $C_9H_{18}O_2$ mw: 158.27

TOXICITY DATA with REFERENCE
orl-mus LD50:2600 mg/kg SCCUR* -,5,61
ipr-mus LD50:780 mg/kg SCCUR* -,5,61

SAFETY PROFILE: Moderately toxic by ingestion and intraperitoneal routes. When heated to decomposition it emits acrid smoke and fumes.

IIJ000 CAS:110-19-0 ***HR: 3***
ISOBUTYL ACETATE
DOT: UN 1213
mf: $C_6H_{12}O_2$ mw: 116.18

PROP: Colorless, neutral liquid; fruit-like odor. Mp: −98.9°, bp: 118°, flash p: 64°F (CC) (18°), d: 0.8685 @ 15°, refr index: 1.389, vap press: 10 mm @ 12.8°, autoign temp: 793°F, vap d: 4.0, lel: 2.4%, uel: 10.5%. Very sol in alc, fixed oils, propylene glycol; sltly sol in water.

SYNS: ACETATE d'ISOBUTYLE (FRENCH) ◇ ACETIC ACID, ISOBUTYL ESTER ◇ ACETIC ACID-2-METHYLPROPYL ESTER ◇ FEMA No. 2175 ◇ 2-METHYLPROPYL ACETATE ◇ 2-METHYL-1-PROPYL ACETATE ◇ β-METHYLPROPYL ETHANOATE

TOXICITY DATA with REFERENCE
skn-rbt 500 mg open MLD UCDS** 11/3/71
skn-rbt 500 mg/24H MOD FCTXAV 16,637,78
eye-rbt 500 mg/24H MOD FCTXAV 16,637,78
orl-rat LD50:13400 mg/kg NPIRI* 1,8,74
ihl-rat LCLo:8000 ppm/4H AIHAAP 23,95,62
orl-rbt LD50:4763 mg/kg IMSUAI 41,31,72

CONSENSUS REPORTS: Reported in EPA TSCA Inventory.

OSHA PEL: TWA 150 ppm
ACGIH TLV: TWA 150 ppm
DFG MAK: 200 ppm (950 mg/m^3)
DOT Classification: Flammable Liquid; Label: Flammable Liquid.

SAFETY PROFILE: Mildly toxic by ingestion and inhalation. A skin and eye irritant. Upon absorption by the body it can hydrolyze to acetic acid and isobutanol. Highly flammable liquid. A very dangerous fire and moderate explosion hazard when exposed to heat, flame, or oxidizers. To fight fire, use alcohol foam, CO_2, dry chemical. When heated to decomposition it emits acrid smoke and fumes. See also ESTERS and N-BUTYL ACETATE.

IIK000 CAS:106-63-8 ***HR: 3***
ISOBUTYL ACRYLATE
DOT: UN 2527
mf: $C_7H_{12}O_2$ mw: 128.19

SYNS: ACRYLIC ACID ISOBUTYL ESTER ◇ ISOBUTYL ACRYLATE, INHIBITED (DOT) ◇ ISOBUTYL PROPENOATE ◇ ISOBUTYL-2-PROPENOATE ◇ Z-METHYLPROPYL ACRYLATE ◇ 2-PROPENOIC ACID-2-METHYLPROPYL ESTER

TOXICITY DATA with REFERENCE
skn-rbt 500 mg open MLD UCDS** 3/28/68
orl-rat LD50:7070 mg/kg TXAPA9 28,313,74
ihl-rat LCLo:2000 ppm/4H UCDS** 3/28/68
ipr-rat LD50:654 mg/kg AMPMAR 36,58,75
orl-mus LD50:6106 mg/kg TOLED5 11,125,82
ipr-mus LD50:760 mg/kg JDREAF 51,526,72
skn-rbt LD50:890 mg/kg UCDS** 3/28/68

CONSENSUS REPORTS: Reported in EPA TSCA Inventory.

DOT Classification: Flammable or Combustible Liquid; Label: Flammable Liquid.

SAFETY PROFILE: Moderately toxic by skin contact and intraperitoneal routes. Mildly toxic by inhalation and ingestion. A skin irritant. Flammable when exposed to heat or flame; can react vigorously with oxidizing materials. When heated to decomposition it emits acrid smoke and toxic fumes. See also ESTERS.

IIL000 CAS:78-83-1 ***HR: 3***
ISOBUTYL ALCOHOL
DOT: UN 1212
mf: $C_4H_{10}O$ mw: 74.14

$HOCH_2CH_2CH_2CH_3$

PROP: Clear mobile liquid; sweet odor. Bp: 107.90°, flash p: 82°F, ULC: 40-45, lel: 1.2%, uel: 10.9% @ 212°F, fp: −108°, d: 0.800, autoign temp: 800°F, vap press: 10 mm @ 21.7°, vap d: 2.55. Sltly sol in water; misc with alc and ether.

SYNS: ALCOOL ISOBUTYLIQUE (FRENCH) ◇ FEMA No. 2179 ◇ FERMENTATION BUTYL ALCOHOL ◇ 1-HYDROXYMETHYLPROPANE ◇ ISOBUTANOL (DOT) ◇ ISOBUTYLALKOHOL (CZECH) ◇ ISOPROPYLCARBINOL ◇ 2-METHYL PROPANOL ◇ 2-METHYL-1-PROPANOL ◇ 2-METHYLPROPAN-1-OL ◇ 2-METHYLPROPYL ALCOHOL ◇ RCRA WASTE NUMBER U140

TOXICITY DATA with REFERENCE
skn-rbt 500 mg/24H SEV 28ZPAK -,35,72
eye-rbt 2 mg open SEV AMIHBC 10,61,54
eye-rbt 20 mg/24H MOD 28ZPAK -,35,72
mmo-esc 25000 ppm ABMGAJ 23,843,69
cyt-smc 20 mmol/tube HEREAY 33,457,47
orl-rat TDLo:29 g/kg/I:ETA ARGEAR 45,19,75
scu-rat TDLo:9 g/kg/I:CAR ARGEAR 45,19,75
orl-rat LD50:2460 mg/kg AMIHBC 10,61,54
ihl-rat LCLo:8000 ppm/4H AMIHBC 10,61,54
ipr-rat LD50:720 mg/kg EVHPAZ 61,321,85
ivn-rat LD50:340 mg/kg EVHPAZ 61,321,85
ipr-mus LD50:1801 mg/kg EVHPAZ 61,321,85
ivn-mus LD50:417 mg/kg EVHPAZ 61,321,85
ivn-cat LDLo:725 mg/kg JPETAB 16,1,20
orl-rbt LDLo:3750 mg/kg JLCMAK 10,985,25
skn-rbt LD50:3400 mg/kg NPIRI* 1,11,74
ipr-rbt LD50:323 mg/kg EVHPAZ 61,321,85

CONSENSUS REPORTS: Reported in EPA TSCA Inventory.

OSHA PEL: (Transitional: TWA 100 mg/m^3) TWA 50 ppm
ACGIH TLV: TWA 50 ppm
DFG MAK: 100 ppm (300 mg/m^3)
DOT Classification: Flammable or Combustible Liquid; Label: Flammable Liquid.

SAFETY PROFILE: Poison by intravenous and in-

traperitoneal routes. Moderately toxic by ingestion and skin contact. Mildly toxic by inhalation. A severe skin and eye irritant. Questionable carcinogen with experimental carcinogenic and tumorigenic data. Mutation data reported. Flammable liquid. Dangerous fire hazard when exposed to heat or flame. Moderately explosive in the form of vapor when exposed to heat, flame, or oxidizers. Ignites on contact with chromium trioxide. Reacts with aluminum at 100° to form explosive hydrogen gas. Keep away from heat and open flame. To fight fire, use alcohol foam, CO_2, dry chemical. When heated to decomposition it emits acrid smoke and fumes. See also ALCOHOLS.

IIM000 CAS:78-81-9 ***HR: 3***
ISOBUTYLAMINE
DOT: UN 1214
mf: $C_4H_{11}N$ mw: 73.16

PROP: Colorless liquid. Mp: −85.5°, bp: 68.6°, flash p: 15°F, d: 0.731 @ 20°/20°, vap press: 100 mm @ 18.8°, autoign temp: 712°F, vap d: 2.5. Misc with water, alc, and ether.

SYNS: 1-AMINO-2-METHYLPROPANE ◇ MONOISOBUTYLAMINE ◇ VALAMINE

TOXICITY DATA with REFERENCE
orl-rat LD50:228 mg/kg TXAPA9 63,150,82

CONSENSUS REPORTS: Reported in EPA TSCA Inventory.

DFG MAK: 5 ppm (15 mg/m^3)
DOT Classification: Flammable Liquid; Label: Flammable Liquid.

SAFETY PROFILE: A poison by ingestion. A powerful irritant to skin, eyes, and mucous membranes. Skin contact can cause blistering. Inhalation can cause headache and dryness of nose and throat. A very dangerous fire hazard when exposed to heat or flame. Can react vigorously with oxidizing materials. To fight fire, use dry chemical, foam, CO_2, alcohol foam. When heated to decomposition it emits toxic fumes of NO_x. See also AMINES.

IIM100 CAS:3562-15-0 ***HR: 3***
2-(ISOBUTYLAMINO)-2-METHYL-1-PROPANOL BENZOATE HYDROCHLORIDE
mf: $C_{15}H_{23}NO_2$•ClH mw: 285.85

SYN: 2-(ISOBUTYLAMINO)-2-METHYL-1-PROPANOLBENZOATE (ESTER), HYDROCHLORIDE

TOXICITY DATA with REFERENCE
ipr-mus LD50:243 mg/kg AIPTAK 115,483,58
scu-mus LD50:298 mg/kg AIPTAK 115,483,58
ivn-mus LD50:19 mg/kg AIPTAK 115,483,58

SAFETY PROFILE: Poison by subcutaneous, intravenous, and intraperitoneal routes. When heated to decomposition it emits toxic fumes of NO_x and HCl.

IIN000 CAS:538-93-2 ***HR: 2***
ISOBUTYLBENZENE
mf: $C_{10}H_{14}$ mw: 134.24

PROP: Liquid. Insol in water; sol in alc and ether. Mp: −51.5°, bp: 170.5°, flash p: 131°F (CC), d: 0.867 @ 20°/4°, autoign temp: 806°F, vap press: 1 mm @ 14.1°, vap d: 4.62, lel: 0.8%, uel: 6.0%.

SYN: 2-METHYL-1-PHENYLPROPANE

TOXICITY DATA with REFERENCE
orl-rat LDLo:5000 mg/kg 28ZRAQ -,56,60

CONSENSUS REPORTS: Reported in EPA TSCA Inventory.

DOT Classification: Flammable or Combustible; Label: Flammable Liquid.

SAFETY PROFILE: Mildly toxic by ingestion. An irritant and possibly narcotic. Combustible when exposed to heat or flame; can react with oxidizing materials. Moderate explosion hazard when exposed to heat or flame. To fight fire, use foam, CO_2, dry chemical. When heated to decomposition it emits acrid smoke and fumes.

IIO750 ***HR: 3***
(2-(p-(5-(ISOBUTYLCARBOZMOYL)-2-OCTYLOXYBENZAMIDO)BENZAMIDO)ETHYL)-TRIETHYLAMMONIUM IODIDE
mf: $C_{35}H_{55}N_4O_4$•I mw: 722.84

TOXICITY DATA with REFERENCE
orl-mus LD50:3 g/kg FRPSAX 39,3,84
ipr-mus LD50:65 mg/kg FRPSAX 39,3,84
scu-mus LD50:750 mg/kg FRPSAX 39,3,84

SAFETY PROFILE: Poison by intraperitoneal route. Moderately toxic by ingestion and subcutaneous routes. When heated to decomposition it emits toxic fumes of I^-, NH_3 and NO_x. See also IODIDES.

IIP000 CAS:4439-24-1 ***HR: 3***
ISOBUTYL CELLOSOLVE
mf: $C_6H_{14}O_2$ mw: 118.20

PROP: Colorless liquid. D: 0.903 @ 20°/4°, bp: 171.2°; misc in water, alc, ether.

SYNS: EKTASOLVE EIB ◇ ETHYLENE GLYCOL MONOISOBUTYL ETHER ◇ 2-ISOBUTOXYETHANOL

TOXICITY DATA with REFERENCE
skn-rbt 500 mg open MLD UCDS** 3/4/69
orl-rat LDLo:400 mg/kg KODAK* 21MAY71

ihl-rat LCLo:1000 ppm/4H UCDS** 3/4/69
skn-rbt LD50:710 mg/kg UCDS** 3/4/69

CONSENSUS REPORTS: Glycol ether compounds are on the Community Right-To-Know List. Reported in EPA TSCA Inventory.

SAFETY PROFILE: Poison by ingestion. Moderately toxic by skin contact. Mildly toxic by inhalation. A skin irritant. When heated to decomposition it emits acrid smoke and fumes. See also GLYCOL ETHERS.

IIQ000 CAS:122-67-8 ***HR: 1***
ISOBUTYL CINNAMATE
mf: $C_{13}H_{16}O_2$ mw: 204.29

PROP: Colorless liquid; sweet, fruity odor. D: 1.001, refr index: 1.539-1.541, flash p: +212°F. Misc with alc, chloroform, ether, fixed oils; insol in water.

SYNS: CINNAMIC ACID, ISOBUTYL ESTER ◇ FEMA No. 2193 ◇ LABDANOL ◇ 3-PHENYL-2-PROPENOIC ACID, 2-METHYLPROPYL ESTER

TOXICITY DATA with REFERENCE
skn-rbt 500 mg/24H MLD FCTXAV 14,799,76

CONSENSUS REPORTS: Reported in EPA TSCA Inventory.

SAFETY PROFILE: A skin irritant. Combustible liquid. When heated to decomposition it emits acrid smoke and fumes. See also ESTERS.

IIQ200 ***HR: 3***
ISOBUTYLENE CHLORIDE
mf: $C_4H_6Cl_2$ mw: 125.00

SYN: 1,2-DICHLORO-2-METHYLPROPENE

TOXICITY DATA with REFERENCE
orl-rat LD50:1501 mg/kg 85GMAT -,77,82
ihl-rat LC50:400 mg/m³/4H 85GMAT -,77,82
orl-mus LD50:205 mg/kg 85GMAT -,77,82

SAFETY PROFILE: Poison by inhalation and ingestion. When heated to decomposition it emits toxic fumes of Cl^-.

IIR000 CAS:542-55-2 ***HR: 3***
ISOBUTYL FORMATE
DOT: UN 2393
mf: $C_5H_{10}O_2$ mw: 102.15

PROP: Liquid. D: 0.885 @ 20°/4°, mp: −95.3°, bp: 98.2°, flash p: <70°F, autoign temp: 608°F, lel: 2.0%, uel: 8%. Sol in water @ 22°; misc in alc and ether.

SYNS: FORMIC ACID, ISOBUTYL ESTER ◇ TETRYL FORMATE

TOXICITY DATA with REFERENCE
orl-rbt LD50:3064 mg/kg IMSUAI 41,31,72

CONSENSUS REPORTS: Reported in EPA TSCA Inventory.

DOT Classification: Flammable Liquid; Label: Flammable Liquid.

SAFETY PROFILE: Moderately toxic by ingestion. A very dangerous fire hazard when exposed to heat, open flame, or oxidizers. A moderate explosion hazard when exposed to heat or flame. To fight fire, use water spray, foam, CO_2, dry chemical. When heated to decomposition it emits acrid smoke and fumes. See also ESTERS.

IIR100 CAS:105-01-1 ***HR: 2***
ISOBUTYL FURYLPROPIONATE
mf: $C_{11}H_{16}O_3$ mw: 196.27

SYN: ISOBUTYL-2-FURANPROPIONATE

TOXICITY DATA with REFERENCE
skn-rbt 500 mg/24H MLD FCTXAV 17,835,79
orl-rat LD50:1950 mg/kg FCTXAV 17,835,79
skn-rbt LD50:2000 mg/kg FCTXAV 17,835,79

CONSENSUS REPORTS: Reported in EPA TSCA Inventory.

SAFETY PROFILE: Moderately toxic by ingestion and skin contact. A skin irritant. When heated to decomposition it emits acrid smoke and fumes. See also ESTERS.

IIS000 CAS:7779-80-8 ***HR: 1***
ISOBUTYL HEPTYLATE
mf: $C_{11}H_{22}O_2$ mw: 186.33

SYNS: HEPTANOIC ACID, ISOBUTYL ESTER ◇ HEPTANOIC ACID, 2-METHYLPROPYL ESTER ◇ ISOBUTYL HEPTANOATE

TOXICITY DATA with REFERENCE
skn-rbt 500 mg/24H MOD FCTXAV 16,799,78

CONSENSUS REPORTS: Reported in EPA TSCA Inventory.

SAFETY PROFILE: A skin irritant. When heated to decomposition it emits acrid smoke and fumes. See also ESTERS.

IIT000 CAS:105-79-3 ***HR: 1***
ISOBUTYL HEXANOATE
mf: $C_{10}H_{20}O_2$ mw: 172.30

SYNS: HEXANOIC ACID, ISOBUTYL ESTER ◇ HEXANOIC ACID, 2-METHYLPROPYL ESTER ◇ ISOBUTYL CAPROATE ◇ 2-METHYLPROPYL HEXANOATE

TOXICITY DATA with REFERENCE
skn-rbt 500 mg/24H MOD FCTXAV 16,797,78

CONSENSUS REPORTS: Reported in EPA TSCA Inventory.

SAFETY PROFILE: A skin irritant. When heated to decomposition it emits acrid smoke and fumes. See also ESTERS.

IIU000 CAS:15687-27-1 ***HR: 3***
p-ISOBUTYLHYDRATROPIC ACID
mf: $C_{13}H_{18}O_2$ mw: 206.31

SYNS: ACIDE (ISOBUTYL-4 PHENYL)-2 PROPIONIQUE (FRENCH) ◇ ADRAN ◇ ANFLAGEN ◇ ARTRIL 300 ◇ BLUTON ◇ BRUFANIC ◇ BRUFEN ◇ BUBURONE ◇ BUTYLENIN ◇ DOLGIN ◇ EMODIN ◇ EPOBRON ◇ IBUFEN ◇ IBUPROCIN ◇ IBUPROFEN ◇ IP-82 ◇ 4-ISOBUTYLHYDRATROPIC ACID ◇ 2-(4-ISOBUTYLPHENYL)PROPANOIC ACID ◇ α-p-ISOBUTYLPHENYLPROPIONIC ACID ◇ α-(4-ISOBUTYLPHENYL)PROPIONIC ACID ◇ 2-(p-ISOBUTYLPHENYL) PROPIONIC ACID ◇ LAMIDON ◇ LIPTAN ◇ α-METHYL-4-(2-METHYLPROPYL)BENZENEACETIC ACID ◇ MOTRIN ◇ MYNOSEDIN ◇ NAPACETIN ◇ NOBFELON ◇ NOBFEN ◇ NOBGEN ◇ R.D. 13621 ◇ REBUGEN ◇ ROIDENIN

TOXICITY DATA with REFERENCE
orl-wmn TDLo:8 mg/kg (1D pre):REP JRPMAP 28,592,83
orl-mus TDLo:1260 mg/kg (female 7-13D post):TER KSRNAM 4,1115,70
orl-man TDLo:120 mg/kg/W-I:EYE,SYS NYSJAM 78,1239,78
orl-man LDLo:171 mg/kg BMJOAE 281,1458,80
orl-wmn TDLo:8 mg/kg JAMAAP 239,1062,78
orl-cld TDLo:480 mg/kg/17D-I:LIV,ALR JOPDAB 90,657,77
unr-wmn TDLo:96 mg/kg/3D-I:EYE JAMAAP 248,649,82
orl-cld LDLo:469 mg/kg AEMED3 15,1308,86
orl-rat LD50:636 mg/kg ARZNAD 34,280,84
ipr-rat LD50:626 mg/kg ARZNAD 27,1006,77
scu-rat LD50:740 mg/kg OYYAA2 24,415,82
rec-rat LD50:530 mg/kg OYYAA2 24,415,82
orl-mus LD50:740 mg/kg PCJOAU 14,119,80
ipr-mus LD50:320 mg/kg TXAPA9 15,310,69
scu-mus LD50:395 mg/kg OYYAA2 24,415,82
rec-mus LD50:620 mg/kg OYYAA2 24,415,82

CONSENSUS REPORTS: Reported in EPA TSCA Inventory.

SAFETY PROFILE: A human poison by ingestion. Poison experimentally by subcutaneous and intraperitoneal routes. Moderately toxic by ingestion and rectal routes. Human systemic effects: eye effects, dermatitis, increased body temperature, hepatitis, allergic with multiple organ involvement, diplopia. Human reproductive effects by ingestion: menstrual cycle changes or disorders. Experimental teratogenic and reproductive effects. An FDA-over the counter drug used as an analgesic and anti-inflammatory agent. When heated to decomposition it emits acrid smoke and fumes.

IIV000 CAS:6104-30-9 ***HR: 3***
ISOBUTYLIDENEDIUREA
mf: $C_6H_{14}N_4O_2$ mw: 174.24

SYNS: 1,1-DIUREIDISOBUTANE ◇ DIUREIDOISOBUTANE ◇ IBDU ◇ ISOBUTYLDIUREA ◇ ISOBUTYLENEDIUREA ◇ 1,1′-ISOBUTYLIDENEBISUREA ◇ ISODUR ◇ N,N″-(2-METHYLPROPYLIDENE) BISUREA (9CI)

TOXICITY DATA with REFERENCE
orl-rat TDLo:64 g/kg/20W-I:ETA VPITAR 37(2),72,79

CONSENSUS REPORTS: Reported in EPA TSCA Inventory.

SAFETY PROFILE: Questionable carcinogen with experimental tumorigenic data. When heated to decomposition it emits toxic fumes of NO_x.

IIV509 CAS:513-38-2 ***HR: 2***
ISOBUTYL IODIDE
mf: C_4H_9I mw: 184.03

PROP: Liquid. D: 1.603, mp: −93.5°, bp: 121.0°, flash p: 32°F, insol in water, misc in alc and ether.

SYNS: 1-IODO-2-METHYLPROPANE ◇ PRIMARY ISOBUTYL IODIDE

TOXICITY DATA with REFERENCE
ihl-rat LC50:6700 mg/m³/4H 34ZIAG -,756,69
ipr-rat LD50:1241 mg/kg 34ZIAG -,756,69
ipr-mus LD50:594 mg/kg 34ZIAG - 756,69

CONSENSUS REPORTS: EPA Genetic Toxicology Program. Reported in EPA TSCA Inventory.

SAFETY PROFILE: Moderately toxic by intraperitoneal route. Mildly toxic by inhalation. A very dangerous fire hazard when exposed to heat or flame; can react vigorously with oxidizing materials. When heated to decomposition it emits toxic fumes of I^-. See also IODIDES.

IIW000 CAS:97-85-8 ***HR: 2***
ISOBUTYL ISOBUTYRATE
DOT: UN 2528
mf: $C_8H_{16}O_2$ mw: 144.24

PROP: Liquid with fruity odor. Mp: −81°, bp: 147.5°, d: 0.850-0.860 @ 20°/20°, vap press: 10 mm @ 39.9°. Insol in water; misc with alc.

SYNS: ISOBUTYLISOBUTYRATE (DOT) ◇ ISOBUTYRIC ACID, ISOBUTYL ESTER ◇ 2-METHYLPROPYL ISOBUTYRATE ◇ 2-METHYLPROPYLPROPANOIC ACID-2-METHYLPROPYL ESTER (9CI)

TOXICITY DATA with REFERENCE
orl-rat LD50:12800 mg/kg NPIRI* 1,13,74
ihl-rat LC50:5000 ppm/6H NPIRI* 1,13,74
orl-mus LDLo:12800 mg/kg FCTXAV 16,337,78

CONSENSUS REPORTS: Reported in EPA TSCA Inventory.

DOT Classification: Flammable or Combustible Liquid; Label: Flammable Liquid.

SAFETY PROFILE: Mildly toxic by ingestion and inhalation. An insect-repellent. Combustible when exposed to heat or flame. Can react with oxidizing materials. When heated to decomposition it emits acrid smoke and fumes. See also ISOBUTYL ALCOHOL.

IIX000 CAS:513-44-0 ***HR: 2***
ISOBUTYL MERCAPTAN
mf: $C_4H_{10}S$ mw: 90.20

$(CH_3)_2CHCH_2SH$

PROP: Liquid, heavy skunk-like odor. D: 0.836 @ 20°/4°, mp: −79°, bp: 88°. Very sltly sol in water; sol in alc, liquid hydrogen sulfide, and ether.

SYN: 2-METHYLPROPANETHIOL

TOXICITY DATA with REFERENCE
eye-rbt 84 mg AIHAAP 19,171,58
ipr-rat LD50:917 mg/kg AIHAAP 19,171,58

CONSENSUS REPORTS: Reported in EPA TSCA Inventory.

SAFETY PROFILE: Moderately toxic by intraperitoneal route. An eye irritant. Flammable when exposed to heat or flame. Explosive reaction with calcium hypochlorite. When heated to decomposition it emits toxic fumes of SO_x. See also MERCAPTANS.

IIY000 CAS:97-86-9 ***HR: 2***
ISOBUTYL METHACRYLATE
DOT: UN 2283
mf: $C_8H_{14}O_2$ mw: 142.22

SYNS: ISOBUTYL-α-METHACRYLATE ◇ METHACRYLIC ACID, ISOBUTYL ESTER ◇ 2-METHYL-2-PROPENOIC ACID-2-METHYLPROPYL ESTER ◇ 2-METHYLPROPYL METHACRYLATE

TOXICITY DATA with REFERENCE
ipr-rat TDLo:420 mg/kg (5-15D preg):TER JDREAF 51,1632,72
ipr-rat TDLo:1401 mg/kg (5-15D preg):REP JDREAF 51,1632,72
orl-rat LDLo:6400 mg/kg 14CYAT 2,1880,63
orl-mus LD50:11990 mg/kg TOLED5 11,125,82
ipr-mus LD50:1340 mg/kg JPMSAE 62,778,73

CONSENSUS REPORTS: Reported in EPA TSCA Inventory.

DOT Classification: Flammable or Combustible Liquid; Label: Flammable Liquid.

SAFETY PROFILE: Moderately toxic by intraperitoneal route. Mildly toxic by ingestion. Experimental teratogenic and reproductive effects. Flammable when exposed to heat or flame. When heated to decomposition it emits acrid smoke and fumes. See also ESTERS.

IJA000 CAS:10086-50-7 ***HR: 2***
2-(ISOBUTYL-3-METHYLBUTOXY)ETHANOL
mf: $C_{11}H_{24}O_2$ mw: 188.35

SYN: 2-(ISOPENTYLOXY)-4-METHYL-1-PENTANOL

TOXICITY DATA with REFERENCE
skn-rbt 10 mg/24H open MLD AMIHBC 10,61,54
eye-rbt 20 mg open SEV AMIHBC 10,61,54
orl-rat LD50:5410 mg/kg AMIHBC 10,61,54

SAFETY PROFILE: Mildly toxic by ingestion. A skin and severe eye irritant. When heated to decomposition it emits acrid smoke and fumes.

IJB000 CAS:63980-62-1 ***HR: 2***
2-(2-(1-ISOBUTYL-3-METHYLBUTOXY)ETHOXY)ETHANOL
mf: $C_{13}H_{28}O_3$ mw: 232.41

TOXICITY DATA with REFERENCE
skn-rbt 10 mg/24H open MLD AMIHBC 10,61,54
eye-rbt 20 mg open SEV AMIHBC 10,61,54
orl-rat LD50:8680 mg/kg AMIHBC 10,61,54
skn-rbt LD50:3000 mg/kg AMIHBC 10,61,54

SAFETY PROFILE: Moderately toxic by skin contact. Mildly toxic by ingestion. A skin and severe eye irritant. When heated to decomposition it emits acrid smoke and fumes.

IJD000 CAS:542-56-3 ***HR: 3***
ISOBUTYL NITRITE
mf: $C_4H_9NO_2$ mw: 103.14

PROP: Liquid. D: 0.870 @ 22°/4°, bp: 67-68°. Sltly sol in and decomp in water; misc in alc.

SYNS: IBN ◇ NCI-C61052 ◇ NITROUS ACID, ISOBUTYL ESTER ◇ NITROUS ACID, 2-METHYLPROPYL ESTER

TOXICITY DATA with REFERENCE
mmo-sat 1 mg/plate BSIBAC 56,816,80
mma-sat 333 μg/plate NTPTB* APR 82
orl-man TDLo:120 mg/kg:BLD,CVS AIMEAS 92,637,80
orl-rat LD50:410 mg/kg FEPRA7 41,1583,82
ihl-rat LC50:960 ppm/1H FEPRA7 41,1583,82
orl-mus LD50:205 mg/kg RCSADO 3,233,82
ihl-mus LC50:1033 ppm/1H FAATDF 1,448,81
ipr-mus LD50:169 mg/kg TXAPA9 48,A43,79

CONSENSUS REPORTS: Reported in EPA TSCA Inventory.

SAFETY PROFILE: A poison by ingestion and intraperitoneal routes. Mildly toxic by inhalation. Human systemic effects by ingestion: carboxhemoglobinemia, blood pressure lowering, change in heart rate. Mutation data reported. When heated to decomposition it emits toxic fumes of NO_x. See also ESTERS and NITRITES.

IJE000 CAS:5461-85-8 ***HR: D***
N-ISOBUTYL-N′-NITRO-N-NITROSOGUANIDINE
mf: $C_5H_{11}N_5O_3$ mw: 189.21

SYN: 1-ISOBUTYL-3-NITRO-1-NITROSOGUANIDINE

TOXICITY DATA with REFERENCE
mmo-sat 100 nmol/plate IDZAAW 50,403,75
dnr-smc 2 μmol/well IDZAAW 50,403,75
cyt-ham:fbr 30 mg/L/24H MUREAV 48,337,77
cyt-ham:lng 13 mg/L GMCRDC 27,95,81

CONSENSUS REPORTS: EPA Genetic Toxicology Program.

SAFETY PROFILE: Mutation data reported. Many N-nitroso compounds are carcinogens. When heated to decomposition it emits very toxic fumes of NO_x. See also N-NITROSO COMPOUNDS.

IJF000 CAS:760-60-1 ***HR: 3***
N-ISOBUTYL-N-NITROSOUREA
mf: $C_5H_{11}N_3O_2$ mw: 145.19

SYNS: ISO-BNU ◇ 1-ISO-BUTYL-1-NITROSOUREA ◇ N-(2-METHYLPROPYL)-N-NITROSOUREA ◇ N-NITROSO-ISO-BUTYLUREA

TOXICITY DATA with REFERENCE
mmo-sat 1 μg/plate MUREAV 68,1,79
mma-sat 10 μg/plate TCMUE9 1,13,84
cyt-ham:fbr 25 mg/L/48H MUREAV 48,337,77
orl-rat TDLo:447 mg/kg/64W-C:CAR GANNA2 74,342,83
orl-rat TD:675 mg/kg/49W-C:CAR GANNA2 74,342,83

CONSENSUS REPORTS: EPA Genetic Toxicology Program.

SAFETY PROFILE: Questionable carcinogen with experimental carcinogenic data. Mutation data reported. When heated to decomposition it emits toxic fumes of NO_x. See also N-NITROSO COMPOUNDS.

IJF400 ***HR: 1***
ISOBUTYL PHENYLACETATE
mf: $C_{12}H_{16}O_2$ mw: 192.23

PROP: Colorless liquid; rose, honey-like odor. D: 0.984-0.988, refr index: 1.486, flash p: 241°F. Sol in alc, fixed oils; insol in glycerin, propylene glycol, water.

SYN: FEMA No. 2210

SAFETY PROFILE: Combustible liquid. When heated to decomposition it emits acrid smoke and irritating fumes.

IJG000 CAS:1553-60-2 ***HR: 2***
4-ISOBUTYLPHENYLACETIC ACID
mf: $C_{12}H_{16}O_2$ mw: 192.28

SYNS: DYTRANSIN ◇ IBUFENAC ◇ IBUNAC ◇ (p-ISOBUTYLPHENYL)ACETIC ACID ◇ p-ISOBUTYL-α-TOLUIC ACID ◇ ISODILAN ◇ MEDIREX ◇ 4-(2-METHYLPROPYL)BENZENEACETIC ACID ◇ RD 11654

TOXICITY DATA with REFERENCE
orl-rat LD50:3 g/kg OYYAA2 8,481,74
ipr-rat LD50:860 mg/kg OYYAA2 2,22,68
orl-mus LD50:1800 mg/kg NATUAS 200,271,63
ipr-mus LD50:670 mg/kg OYYAA2 8,481,74
scu-mus LD50:1130 mg/kg OYYAA2 2,22,68

SAFETY PROFILE: Moderately toxic by ingestion, intraperitoneal, and subcutaneous routes. An analgesic anti-inflammatory agent. When heated to decomposition it emits acrid smoke and fumes.

IJH000 CAS:55837-18-8 ***HR: 2***
2-(4-ISOBUTYLPHENYL)BUTYRIC ACID
mf: $C_{10}H_{20}O_2$ mw: 172.30

SYNS: BUTIBUFEN ◇ BUTILOPAN ◇ α-ETHYL-4-(2-METHYLPROPYL)BENZENEACETIC ACID ◇ FF 106

TOXICITY DATA with REFERENCE
orl-rat LD50:1600 mg/kg DRFUD4 2,156,77
orl-mus LD50:810 mg/kg EJMCA5 13,77,78

SAFETY PROFILE: Moderately toxic by ingestion. Used as an anti-inflammatory agent. When heated to decomposition it emits acrid smoke and fumes.

IJJ000 CAS:66332-77-2 ***HR: 2***
2-(p-ISOBUTYLPHENYL)PROPIONIC ACID-o-METHOXYPHENYL ESTER
mf: $C_{20}H_{25}O_3$ mw: 313.45

SYNS: AF 2259 ◇ IBUPROFEN GUAIACOL ESTER

TOXICITY DATA with REFERENCE
orl-rat LD50:2062 mg/kg TXAPA9 54,332,80
orl-mus LD50:1624 mg/kg TXAPA9 54,332,80

SAFETY PROFILE: Moderately toxic by ingestion. When heated to decomposition it emits acrid smoke and fumes. See also ESTERS.

IJM000 CAS:93-19-6 ***HR: 2***
α-ISOBUTYLQUINOLINE
mf: $C_{13}H_{15}N$ mw: 185.29

SYN: 2-ISOBUTYLQUINOLINE

TOXICITY DATA with REFERENCE
orl-rat LD50:1020 mg/kg FCTXAV 14,307,76

CONSENSUS REPORTS: Reported in EPA TSCA Inventory.

SAFETY PROFILE: Moderately toxic by ingestion. When heated to decomposition it emits toxic fumes of NO_x.

IJN000 CAS:87-19-4 *HR: 2*
ISOBUTYL SALICYLATE
mf: $C_{11}H_{14}O_3$ mw: 194.25

PROP: Colorless liquid; orchid odor. D: 1.062-1.066, refr index: 1.507, flash p: 250°F. Sol in fixed oils; insol in glycerin, propylene glycol.

SYNS: FEMA No. 2213 ◇ ISOBUTYL-o-HYDROXYBENZOATE ◇ SALICYLIC ACID, ISOBUTYL ESTER

TOXICITY DATA with REFERENCE
orl-rat LD50:1560 mg/kg FCTXAV 13,681,75

CONSENSUS REPORTS: Reported in EPA TSCA Inventory.

SAFETY PROFILE: Moderately toxic by ingestion. Combustible liquid. When heated to decomposition it emits acrid smoke and fumes. See also ESTERS.

IJO000 CAS:592-65-4 *HR: 3*
ISOBUTYLSULFHYDRATE
mf: $C_8H_{18}S(H_2O)_x$

SYN: DIISOBUTYLSULFIDE HYDRATE

TOXICITY DATA with REFERENCE
scu-mus LDLo:250 mg/kg AIPTAK 12,447,04

CONSENSUS REPORTS: Reported in EPA TSCA Inventory.

SAFETY PROFILE: Poison by subcutaneous route. When heated to decomposition it emits toxic fumes of SO_x.

IJQ000 CAS:109-53-5 *HR: 3*
ISOBUTYL VINYL ETHER
DOT: UN 1304
mf: $C_6H_{12}O$ mw: 100.18

$(CH_3)_2CHCH_2OCH=CH_2$

PROP: Liquid. Mp: −112°, bp: 82.9-83.2°, flash p: 16°F, d: 0.76 @ 25°/4°, vap d: 3.45.

SYNS: IVE ◇ VINOFLEX MO 400* ◇ VINYL ISOBUTYL ETHER (DOT) ◇ VINYL ISOBUTYL ETHER, inhibited (DOT)

TOXICITY DATA with REFERENCE
orl-rat LD50:17 g/kg AIHAAP 23,95,62
ihl-rat LCLo:16000 ppm/4H AIHAAP 23,95,62
skn-rbt LD50:20 g/kg AIHAAP 23,95,62

CONSENSUS REPORTS: Reported in EPA TSCA Inventory.

DOT Classification: Flammable Liquid; Label: Flammable Liquid.

SAFETY PROFILE: Very mildly toxic by ingestion, inhalation, and skin contact. A very dangerous fire hazard when exposed to heat, flame, oxidizers. Severe explosion hazard when exposed to sparks or open flame. Can react vigorously with oxidizing materials. When heated to decomposition it emits acrid smoke and fumes. To fight fire, use alcohol foam, CO_2, dry chemical. See also ETHERS.

IJR000 CAS:63916-90-5 *HR: 3*
p-ISOBUTYOXYBENZOIC ACID-3-(2'-METHYLPIPERIDINO)PROPYL ESTER
mf: $C_{20}H_{31}NO_3$ mw: 333.52

TOXICITY DATA with REFERENCE
scu-mus LD50:161 mg/kg RCPRAN 15,143,54
ivn-mus LD50:32 mg/kg RCPRAN 15,143,54

SAFETY PROFILE: A poison by subcutaneous and intravenous routes. When heated to decomposition it emits toxic fumes of NO_x.

IJS000 CAS:78-84-2 *HR: 3*
ISOBUTYRALDEHYDE
DOT: UN 2045
mf: C_4H_8O mw: 72.12

PROP: Transparent, colorless, highly refractive liquid; pungent odor. Mp: −65°, bp: 64°, flash p: −40°F (CC), d: 0.783-0.788, autoign temp: 434°F, lel: 1.6%, uel: 10.6%, vap d: 2.5. Sol in water; misc in alc, ether, benzene, carbon disulfide, acetone, toluene, chloroform.

SYNS: FEMA No. 2220 ◇ ISOBUTANAL ◇ ISOBUTYLALDEHYDE ◇ ISOBUTYL ALDEHYDE (DOT) ◇ ISOBUTYRALDEHYD (CZECH) ◇ ISOBUTYRIC ALDEHYDE ◇ 2-METHYLPROPANAL ◇ 2-METHYL-1-PROPANAL ◇ 2-METHYLPROPIONALDEHYDE ◇ NCI-C60968 ◇ VALINE ALDEHDYE

TOXICITY DATA with REFERENCE
skn-rbt 397 mg open MLD UCDS** 11/3/71
skn-rbt 500 mg/24H SEV 28ZPAK -,41,72
eye-rbt 20 mg open SEV AMIHBC 10,61,54
eye-rbt 100 mg/24H MOD 28ZPAK -,41,72
orl-rat LD50:2810 mg/kg 28ZPAK -,41,72
ihl-rat LCLo:8000 ppm/4H AMIHBC 10,61,54
ihl-mus LC50:39500 mg/m³/2H 85GMAT -,77,82
skn-rbt LD50:7130 mg/kg AMIHBC 10,61,54

CONSENSUS REPORTS: Community Right-To-Know List. Reported in EPA TSCA Inventory.

DOT Classification: Flammable Liquid; Label: Flammable Liquid.

SAFETY PROFILE: Moderately toxic by ingestion. Mildly toxic by skin contact and inhalation. A severe skin and eye irritant. Highly flammable liquid. A very dangerous fire hazard when exposed to heat, flame, or oxidizers. Moderately explosive in the form of vapor when exposed to heat or flame. Can react vigorously with reducing materials. When heated to decomposition it emits acrid smoke and fumes. To fight fire, use dry chemical, CO_2, mist, foam. See also ALDEHYDES.

IJT000 CAS:151-00-8 ***HR: 3***
ISOBUTYRALDEHYDE, OXIME
mf: C_4H_9NO mw: 87.14

$CH_3CH_2CH_2CH{=}NOH$

SYNS: 1-HYDROXYIMINOBUTANE ◇ 2-METHYL-1-PROPANAL OXIME ◇ USAF AM-8

TOXICITY DATA with REFERENCE
ipr-mus LD50:100 mg/kg NTIS** AD277-689

CONSENSUS REPORTS: Reported in EPA TSCA Inventory.

SAFETY PROFILE: Poison by intraperitoneal route. May explode when heated above 90°C. When heated to decomposition it emits toxic fumes of NO_x. See also ALDLEHYDES.

IJU000 CAS:79-31-2 ***HR: 3***
ISOBUTYRIC ACID
DOT: UN 2529
mf: $C_4H_8O_2$ mw: 88.12

PROP: Colorless liquid; pungent odor of rancid butter. Mp: −47°, bp: 154.5°, flash p: 132°F (TOC), d: 0.949 @ 20°/4°, refr index: 1.392, vap press: 1 mm @ 14.7°, vap d: 3.04, autoign temp: 935°F. Misc with alc, chloroform and ether. Misc with alc, fixed oils, glycerin, propylene glycol; insol in water.

SYNS: DIMETHYLACETIC ACID ◇ FEMA No. 2222 ◇ ISOPROPYLFORMIC ACID ◇ α-METHYLPROPIONIC ACID ◇ 2-METHYLPROPANOIC ACID ◇ 2-METHYLPROPIONIC ACID

TOXICITY DATA with REFERENCE
skn-rbt 139 μg/24H open AIHAAP 23,95,62
orl-rat LD50:280 mg/kg AIHAAP 23,95,62
skn-rbt LD50:500 mg/kg AIHAAP 23,95,62

CONSENSUS REPORTS: Reported in EPA TSCA Inventory.

DOT Classification: Corrosive Material; Label: Corrosive; Flammable or Combustible Liquid; Label: Flammable Liquid.

SAFETY PROFILE: A poison by ingestion. Moderately toxic by skin contact. A corrosive irritant to the eyes, skin and mucous membranes. Flammable liquid when exposed to heat or flame; can react with oxidizing materials. To fight fire, use alcohol foam, CO_2, dry chemical. When heated to decomposition it emits acrid smoke and fumes.

IJV000 CAS:103-28-6 ***HR: 2***
ISOBUTYRIC ACID, BENZYL ESTER
mf: $C_{11}H_{14}O_2$ mw: 178.25

PROP: Colorless liquid; floral, jasmine odor. D: 1.001-1.005, refr index: 1.489, flash p: 212°F. Sol in alc, fixed oils; sltly sol in propylene glycol; insol in glycerin @ 229°.

SYNS: BENZYL ISOBUTYRATE (FCC) ◇ BENZYL-2-METHYL PROPIONATE ◇ FEMA No. 2141

TOXICITY DATA with REFERENCE
orl-rat LD50:2850 mg/kg FCTXAV 11,1023,73

CONSENSUS REPORTS: Reported in EPA TSCA Inventory.

SAFETY PROFILE: Moderately toxic by ingestion. Combustible liquid. When heated to decomposition it emits acrid smoke and fumes. See also ESTERS.

IJW000 CAS:97-72-3 ***HR: 2***
ISOBUTYRIC ANHYDRIDE
DOT: UN 2530
mf: $C_8H_{14}O_3$ mw: 158.22

PROP: Liquid, decomp in water. Bp: 360°F, d: 0.951-0.956 @ 20°/20°, vap d: 5.5, flash p: 139°F, autoign temp: 665°F.

CONSENSUS REPORTS: Reported in EPA TSCA Inventory.

DOT Classification: Corrosive Material; Label: Corrosive; Flammable or Combustible Liquid; Label: Flammable Liquid.

SAFETY PROFILE: A corrosive irritant to skin, eyes, and mucous membranes. Flammable when exposed to heat, flame, or oxidizers. To fight fire, use alcohol foam, fog, dry chemical, CO_2. When heated to decomposition it emits acrid smoke and fumes. See also ANHYDRIDES.

IJX000 CAS:78-82-0 ***HR: 3***
ISOBUTYRONITRILE
DOT: UN 2284
mf: C_4H_7N mw: 69.12

PROP: Colorless liquid, sltly sol in water, very sol in alc and ether. D: 0.773 @ 20°/20°, bp: 107°, mp: −75°, flash p: 46.4°F.

SYNS: ISOPROPYL CYANIDE ◇ 2-METHYLPROPIONITRILE

TOXICITY DATA with REFERENCE
skn-rbt 380 mg open MLD UCDS** 10/29/59
orl-rat LD50:102 mg/kg UCDS** 10/29/59
ihl-rat LCLo:1000 ppm/4H AIHAAP 23,95,62
orl-mus LD50:25 mg/kg ARTODN 55,47,84

skn-rbt LD50:310 mg/kg AIHAAP 23,95,62
scu-rbt LDLo:9 mg/kg AIPTAK 5,161,1899
scu-frg LDLo:4800 mg/kg AIPTAK 5,161,1899

CONSENSUS REPORTS: Cyanide and its compounds are on the Community Right-To-Know List. EPA Extremely Hazardous Substances List. Reported in EPA TSCA Inventory.

NIOSH REL: (Nitriles) TWA 22 mg/m^3
DOT Classification: Flammable Liquid; Label: Flammable Liquid and Poison.

SAFETY PROFILE: A poison by ingestion, skin contact, and subcutaneous routes. Mildly toxic by inhalation. A skin irritant. A very dangerous fire hazard when exposed to heat or flame. When heated to decomposition it emits toxic fumes of NO_x and CN^-. See also NITRILES.

IJZ000 CAS:533-28-8 ***HR: 3***
ISOCAINE
mf: $C_{16}H_{23}NO_2{\cdot}ClH$ mw: 297.86

SYNS: o-AMINOBENZOYL DI(ISOPROPYLAMINO)ETHANOL HYDROCHLORIDE ◇ 3-BENZOXY-1-(2-METHYLPIPERIDINO)PROPANE HYDROCHLORIDE ◇ dl-3-BENZOXY-1-(2-METHYLPIPERIDINO)PROPANE HYDROCHLORIDE ◇ BENZOYL-Γ-(2-METHYLPIPERIDINE)PROPANOL HYDROCHLORIDE ◇ METHCAINE HYDROCHLORIDE ◇ 2-METHYL-1-PIPERIDINEPROPANOL BENZOATE HYDROCHLORIDE ◇ 3-(2-METHYLPIPERIDINO)PROPYL BENZOATE HYDROCHLORIDE ◇ dl-(2-METHYLPIPERIDINO)PROPYL BENZOATE HYDROCHLORIDE ◇ Γ-(2-METHYLPIPERIDINO)PROPYL BENZOATE HYDROCHLORIDE ◇ (+ −)-Γ-(2-METHYLPIPERIDYL)PROPYL BENZOATE HYDROCHLORIDE ◇ NEOTHESIN HYDROCHLORIDE ◇ PIPEROCAINE HYDROCHLORIDE ◇ PIPEROCAINIUM CHLORIDE

TOXICITY DATA with REFERENCE
ipr-rat LD50:120 mg/kg JLCMAK 15,731,30
scu-rat LD50:1300 mg/kg MEIEDD 10,1078,83
ivn-rat LD50:20 mg/kg MEIEDD 10,1078,83
ipr-mus LD50:182 mg/kg JPETAB 94,299,48
scu-mus LD50:569 mg/kg ANESAV 5,605,44
ivn-mus LD50:18200 μg/kg TXAPA9 1,454,56
scu-cat LDLo:200 mg/kg JPETAB 24,167,25
scu-rbt LDLo:300 mg/kg JPETAB 24,167,25
ivn-rbt LD50:28 mg/kg JLCMAK 15,731,30
isp-rbt LDLo:10 mg/kg JPETAB 57,221,36

SAFETY PROFILE: A poison by intraperitoneal, intravenous, subcutaneous, and intraspinal routes. Used as a local anesthetic. An FDA over-the-counter drug. When heated to decomposition it emits very toxic fumes of NO_x and HCl.

IKA000 CAS:80748-58-9 ***HR: 1***
ISOCAMPHYL CYCLOHEXANOL (mixed isomers)

SYNS: INDISAN ◇ ISOCAMPHANYL CYCLOHEXANOL (mixed isomers)

TOXICITY DATA with REFERENCE
skn-hmn 20%/48H FCTXAV 14,801,76
skn-rbt 500 mg/24H MOD FCTXAV 14,801,76

SAFETY PROFILE: A human skin irritant. When heated to decomposition it emits acrid smoke and fumes.

IKB000 CAS:581-88-4 ***HR: 3***
ISOCARAMIDINE SULFATE
mf: $C_{10}H_{13}N_3{\cdot}1/2H_2O_4S$ mw: 224.30

SYNS: DEBRISOQUIN SULFATE ◇ DECLINAX ◇ 3,4-DIHYDRO-2(1H)-ISOQUINOLINECARBOXIMIDAMIDE SULFATE (2:1) ◇ RO 5-3307/1 ◇ TENDOR

TOXICITY DATA with REFERENCE
orl-rat LD50:610 mg/kg OYYAA2 17,129,79
orl-mus LD50:235 mg/kg CTCEA9 6,299,64
ipr-mus LD50:132 mg/kg CTCEA9 6,299,64
scu-mus LD50:136 mg/kg CTCEA9 6,299,64
ivn-mus LD50:31700 μg/kg CTCEA9 6,299,64

SAFETY PROFILE: Poison by ingestion, subcutaneous, intravenous, and intraperitoneal routes. An antihypertensive agent. When heated to decomposition it emits toxic fumes of NO_x and SO_x.

IKC000 CAS:59-63-2 ***HR: 3***
ISOCARBOXAZID
mf: $C_{12}H_{13}N_3O_2$ mw: 231.28

PROP: Crystals from methanol, practically tasteless. Mp: 106°, very sltly sol in hot water; sltly sol in alc, glycerol, and propylene glycol.

SYNS: BENAZIDE ◇ 1-BENZYL-2-(5-METHYL-3-ISOXAZOIYL-CARBONYL)HYDRAZINE ◇ 1-BENZYL-1-(5-METHYL-3-ISOXAZOIYLCARBONYL)HYDRAZINE ◇ N'-BENZYL N-METHYL-5-ISOXAZOLECARBOXYLHYDRAZIDE-3 ◇ BMIH ◇ ENERZER ◇ ISOCARBONAZID ◇ ISOCARBOSSAZIDE ◇ ISOCARBOXAZIDE ◇ ISOCARBOXYZID ◇ MARAPLAN ◇ MARPLAN ◇ MARPLON ◇ 5-METHYL-3-ISOXAZOLECARBOXYLIC ACID-2-BENZYLHYDRAZIDE ◇ RO 5-0831

TOXICITY DATA with REFERENCE
dnd-mus-ipr 350 μmol/kg CNREA8 41,1469,81
dnd-mus-orl 322 mg/kg/5D-C JTEHD6 9,287,82
ipr-mus TDLo:100 mg/kg (10D preg):REP CAJPBD 3,2,63
ipr-mus TDLo:100 mg/kg (female 10D post):TER CAJPBD 3,2,63
orl-wmn TDLo:4800 μg/kg/21W-I:SYS JCPYDR 3,42,83
orl-rat LD50:280 mg/kg ANYAA9 80,626,59
ipr-rat LD50:199 mg/kg 27ZQAG -,240,72
orl-mus LD50:193 mg/kg ANYAA9 80,626,59
ipr-mus LD50:138 mg/kg JMPCAS 2,133,60
scu-mus LD50:150 mg/kg TXAPA9 39,141,77
orl-dog LDLo:40 mg/kg ANYAA9 80,626,59

SAFETY PROFILE: A poison by ingestion, intraperitoneal, and subcutaneous routes. An experimental teratogen. Other experimental reproductive effects. Human systemic effects: monoamine oxidase. Mutation data reported. A pharmaceutical and veterinary drug. When heated to decomposition it emits toxic fumes of NO_x.

IKE000 CAS:513-37-1 ***HR: 3***
ISOCROTYL CHLORIDE
mf: C_4H_7Cl mw: 90.56

PROP: Liquid. D: 0.919 @ 20°/4°, bp: 68°.

SYNS: α-CHLOROISOBUTYLENE ◇ 1-CHLORO-2-METHYLPROPENE ◇ 1-CHLORO-2-METHYL-1-PROPENE ◇ β,β-DIMETHYLVINYL CHLORIDE ◇ NCI-C54819

TOXICITY DATA with REFERENCE
trn-oin-dmg 12750 ppm/3D C NTPTR* NTP-TR-316,86
msc-mus:lyms 400 μg/L NTPTR* NTP-TR-316,86
sce-ham:ovr 500 mg/L NTPTR* NTP-TR-316,86
orl-rat TDLo:51500 mg/kg/2Y-I:CAR NTPTR* NTP-TR-316,86
orl-mus TDLo:51 g/kg/2Y-I:CAR NTPTR* NTP-TR-316,86
orl-mus TD:102 g/kg/2Y-I:NEO NTPTR* NTP-TR-316,86
ihl-rat LC50:400 mg/m^3/4H 85JCAE-,112,86
ihl-mus LCLo:181 g/m^3/10M UCPHAQ 1,119,38

CONSENSUS REPORTS: NTP Fifth Annual Report on Carcinogens. NTP Carcinogenesis Studies (gavage); Clear Evidence: mouse, rat NTPTR* NTP-TR-316,86.

SAFETY PROFILE: Confirmed carcinogen with experimental carcinogenic and neoplastigenic data. Mildly toxic by inhalation. A local irritant and narcotic in high concentration. When heated to decomposition it emits toxic fumes of Cl^-.

IKF000 CAS:11071-47-9 ***HR: 3***
ISOCTENE
DOT: UN 1216
mf: C_8H_{16} mw: 112.24

SYN: 2,2,4-TRIMETHYL-1-PENTENE

DOT Classification: Label: Flammable Liquid.

SAFETY PROFILE: A dangerous fire hazard when exposed to heat or flame. When heated to decomposition it emits acrid smoke and fumes.

IKG000 CAS:103-65-1 ***HR: 3***
ISOCUMENE
DOT: UN 2364
mf: C_9H_{12} mw: 120.21

PROP: Clear liquid. Insol in water; misc in alc and ether. Mp: −99.5°, bp: 159.2°, flash p: 86°F (CC), d: 0.862, vap press: 10 mm @ 43.4°, vap d: 4.14, autoign temp: 842°F, lel: 0.8%, uel: 6%.

SYNS: 1-PHENYLPROPANE ◇ n-PROPYLBENZENE ◇ PROPYL BENZENE (DOT)

TOXICITY DATA with REFERENCE
orl-rat LD50:6040 mg/kg FCTXAV 2,327,64
ihl-mus LCLo:20 g/m^3 AEPPAE 143,223,29

CONSENSUS REPORTS: Reported in EPA TSCA Inventory.

DOT Classification: Flammable or Combustible Liquid; Label: Flammable Liquid.

SAFETY PROFILE: Mildly toxic by ingestion and inhalation. A very dangerous fire hazard when exposed to heat, flame, or oxidizers; can react with oxidizing materials. A moderate explosion hazard in the form of vapor when exposed to heat or flame. To fight fire, use foam, CO_2, dry chemical. Emitted from modern building materials. (CENEAR 6922,91) When heated to decomposition it emits acrid smoke and fumes.

IKG349 ***HR: D***
ISOCYANATES

SAFETY PROFILE: Compounds containing the isocyanate radical −NCO. Derivatives of isocyanic acid (cyanic acid). Usually the term refers to a diisocyanate. Inorganic isocyanates are only slightly toxic. Organic isocyanates (diisocyanates) can cause local irritation and allergic reactions. When heated to decomposition they emit toxic fumes of NO_x.

IKG700 CAS:30674-80-7 ***HR: 3***
2-ISOCYANATOETHYL METHACRYLATE
mf: $C_7H_9NO_3$ mw: 155.17

SYNS: β-ISOCYANATOETHYL METHACRYLATE ◇ METHACRYLOYLOXYETHYL ISOCYANATE

TOXICITY DATA with REFERENCE
ihl-rat TCLo:80 ppb/6H (49D male):REP DCTODJ 3,381,80
orl-rat LD50:670 mg/kg DCTODJ 3,381,80
ihl-rat LC50:4 ppm/6H DCTODJ 3,381,80

CONSENSUS REPORTS: EPA Extremely Hazardous Substances List. Reported in EPA TSCA Inventory.

SAFETY PROFILE: Poison by inhalation. Moderately toxic by ingestion. Experimental reproductive effects. When heated to decomposition it emits toxic fumes of NO_x. See also ESTERS and ISOCYANATES.

IKH000 CAS:1943-83-5 ***HR: 2***
ISOCYANIC ACID-2-CHLOROETHYL ESTER
mf: C_3H_4ClNO mw: 105.53

TOXICITY DATA with REFERENCE
dni-hmn:fbr 75 μmol/L CNREA8 38,1067,78

dnd-ham:lng 13 μmol/L CNREA8 38,3379,78
ihl-mus LCLo:1000 mg/m³/10M NDRC** -,11,42

CONSENSUS REPORTS: EPA Genetic Toxicology Program. Reported in EPA TSCA Inventory.

SAFETY PROFILE: Moderately toxic by inhalation. Human mutation data reported. When heated to decomposition it emits very toxic fumes of Cl^- and NO_x. See also ISOCYANATES and ESTERS.

IKH099 CAS:102-36-3 ***HR: 3***
ISOCYANIC ACID-3,4-DICHLOROPHENYL ESTER
mf: $C_7H_3Cl_2NO$ mw: 188.01

SYN: 3,4-DICHLOROPHENYL ISOCYANATE

TOXICITY DATA with REFERENCE
ihl-mus LCLo:140 mg/m³/2M GTPZAB 13(4),50,69
ihl-rat LCLo:140 mg/m³/4M GTPZAB 13(4),50,69

CONSENSUS REPORTS: EPA Extremely Hazardous Substances List.

SAFETY PROFILE: Poison by inhalation. When heated to decomposition it emits toxic fumes of Cl^- and NO_x. See also ISOCYANATES and ESTERS.

IKH339 ***HR: 3***
ISOCYANIDES

CONSENSUS REPORTS: Cyanide and its compounds are on the Community Right-To-Know List.

SAFETY PROFILE: Compounds of the form RN≡C:, also called carbylamines. The acid catalyzed hydrolysis of isocyanides to primary amines and formic acid is very rapid, sometimes explosive. When heated to decomposition they emit toxic fumes of CN^-.

IKH669 CAS:4702-38-9 ***HR: 3***
ISOCYANOAMIDE
mf: CH_2N_2 mw: 42.04

SYN: ISODIAZOMETHANE

SAFETY PROFILE: A thermally unstable liquid which explodes at 35°C. Upon decomposition it emits toxic fumes of NO_x. See also AMIDES.

IKI000 CAS:62967-27-5 ***HR: D***
4-ISOCYANO-4'-NITRODIPHENYLAMINE
mf: $C_{13}H_9N_3O_3$ mw: 255.25

SYNS: 4-ISOCYANATO-N-(4-NITROPHENYLBENZENAMINE(9CI) ◇ (p-(p-NITROANILINO)PHENYL)ISOCYANIC ACID

TOXICITY DATA with REFERENCE
mmo-sat 17 nmol/plate JMCMAR 20,981,77
mma-sat 17 nmol/plate JMCMAR 20,981,77

CONSENSUS REPORTS: EPA Genetic Toxicology Program.

SAFETY PROFILE: Mutation data reported. When heated to decomposition it emits toxic fumes of CN^- and NO_x. See also ISOCYANATES and NITRO COMPOUNDS of AROMATIC HYDROCARBONS.

IKJ000 CAS:1335-66-6 ***HR: 1***
ISOCYCLOCITRAL

SYN: 1-FORMYL-3,5,6-TRIMETHYL-3-CYCLOHEXENE and 1-FORMYL-2,4,6-TRIMETHYL-3-CYCLOHEXENE

TOXICITY DATA with REFERENCE
skn-rbt 500 mg/24H MLD FCTXAV 14,313,76
orl-rat LD50:4500 mg/kg FCTXAV 14,313,76

CONSENSUS REPORTS: Reported in EPA TSCA Inventory.

SAFETY PROFILE: Mildly toxic by ingestion. A skin irritant. When heated to decomposition it emits acrid smoke and fumes.

IKK000 CAS:25339-17-7 ***HR: 2***
ISODECANOL
mf: $C_{10}H_{22}O$ mw: 158.32

PROP: Insol in water. D: 0.8395, bp: 220°, flash p: 220°F.

SYN: ISODECYL ALCOHOL

TOXICITY DATA with REFERENCE
skn-rbt 415 mg open MLD UCDS** 4/1/68
orl-rat LD50:6400 mg/kg UCDS** 4/1/68
skn-rbt LD50:3150 mg/kg 31ZTAS -,72,68

CONSENSUS REPORTS: Reported in EPA TSCA Inventory.

SAFETY PROFILE: Moderately toxic by skin contact. Mildly toxic by ingestion. A skin irritant. Combustible when exposed to heat or flame; can react vigorously with oxidizing materials. To fight fire, use dry chemical, CO_2, foam, mist. When heated to decomposition it emits acrid smoke and fumes.

IKL000 CAS:1330-61-6 ***HR: 2***
ISODECYL ACRYLATE
mf: $C_{13}H_{24}O_2$ mw: 212.37

SYNS: ACRYLIC ACID, ISODECYL ESTER ◇ AGEFLEX FA-10 ◇ ISODECYL ALCOHOL ACRYLATE ◇ ISODECYL PROPENOATE ◇ 2-PROPENOIC ACID ISODECYL ESTER (9CI)

TOXICITY DATA with REFERENCE
skn-rbt 500 mg open MLD UCDS** 7/20/62
orl-rat LD50:12 g/kg UCDS** 7/20/62
skn-rbt LD50:3540 mg/kg UCDS** 7/20/62

CONSENSUS REPORTS: Reported in EPA TSCA Inventory.

SAFETY PROFILE: Moderately toxic by skin contact. Mildly toxic by ingestion. A skin irritant. When heated to decomposition it emits acrid smoke and fumes. See also ESTERS.

IKM000 CAS:29964-84-9 ***HR: 2***
ISODECYL METHACRYLATE
mf: $C_{14}H_{26}O_2$ mw: 226.40

SYNS: AGELFLEX FM-10 ◇ METHACRYLIC ACID, ISODECYL ESTER

TOXICITY DATA with REFERENCE
ipr-rat TDLo:741 mg/kg (5-15D preg):TER JDREAF 51,1632,72
ipr-rat TDLo:1479 mg/kg (5-15D preg):REP JDREAF 51,1632,72
ipr-rat LD50:2467 mg/kg JDREAF 51,1632,72
ipr-mus LD50:3688 mg/kg JPMSAE 62,778,73

CONSENSUS REPORTS: Reported in EPA TSCA Inventory.

SAFETY PROFILE: Moderately toxic by intraperitoneal route. Experimental teratogenic and reproductive effects. When heated to decomposition it emits acrid smoke and fumes. See also ESTERS.

IKM100 CAS:35158-25-9 ***HR: 1***
ISODIHYDROLAVANDULYL ALDEHYDE
mf: $C_{10}H_{18}O$ mw: 154.28

SYNS: 2-HEXEN-1-AL, 2-ISOPROPYL-5-METHYL- ◇ 2-HEXEN-1-AL, 5-METHYL-2-(1-METHYLETHYL)- ◇ 2-ISOPROPYL-5-METHYL-2-HEXEN-1-AL

TOXICITY DATA with REFERENCE
skn-rbt 500 mg/24H MOD FCTXAV 14,317,76

CONSENSUS REPORTS: Reported in EPA TSCA Inventory.

SAFETY PROFILE: A skin irritant. When heated to decomposition it emits acrid smoke and irritating fumes.

IKN200 CAS:24168-96-5 ***HR: 2***
ISODONAZOLE NITRATE
mf: $C_{18}H_{14}Cl_4N_2O•HNO_3$ mw: 479.16

SYNS: 1-(2-((2,6-DICHLOROBENZYL)OXY)-2-(2,4-DICHLOROPHENYL)ETHYL)IMIDAZOLE NITRATE ◇ 1-(2,4-DICHLORO-β-(2,6-DICHLOROBENZYLOXY)PHENETHYL)IMIDAZOLE NITRATE ◇ 1-(2-(2,4-DICHLOROPHENYL)-2-(2,6-DICHLOROPHENYL)METHOXY)ETHYL)-1H-IMIDAZOLEMONONITRATE

TOXICITY DATA with REFERENCE
scu-rat TDLo:330 mg/kg (female 7-17D post):REP IYKEDH 12,762,81
scu-rat TDLo:630 mg/kg (14D pre/1-7D preg):TER IYKEDH 12,762,81
orl-rat LD50:5600 mg/kg IYKEDH 13,1128,82
ipr-rat LD50:720 mg/kg IYKEDH 13,1128,82
ipr-mus LD50:2000 mg/kg IYKEDH 13,1128,82
ipr-mus LD50:560 mg/kg IYKEDH 13,1128,82

SAFETY PROFILE: Moderately toxic by intraperitoneal route. Mildly toxic by ingestion. An experimental teratogen. Other experimental reproductive effects. When heated to decomposition it emits toxic fumes of Cl^- and NO_x. A broad spectrum anti-fungal agent. See also NITRATES.

IKO000 CAS:465-73-6 ***HR: 3***
ISODRIN
mf: $C_{12}H_8Cl_6$ mw: 364.90

PROP: Crystals. Mp: 240-242°.

SYNS: COMPOUND 711 ◇ ENT 19,244 ◇ EXPERIMENTAL INSECTICIDE 711 ◇ 1,2,3,4,10,10-HEXACHLORO-1,4,4a,5,8,8a-HEXAHYDRO-1,4,5,8-endo,endo-DIMETHANONAPHTHALENE◇ 1,2,3,4,10,10-HEXACHLORO-1,4,4a,5,8,8a-HEXAHYDRO-1,4-endo,endo-5,8-DIMETHANONAPHTHALENE ◇ RCRA WASTE NUMBER P060

TOXICITY DATA with REFERENCE
orl-rat LD50:7 mg/kg WRPCA2 9,119,70
skn-rat LD50:23 mg/kg WRPCA2 9,119,70
orl-mus LD50:8800 μg/kg GIPZAB 8(4),30,64
ipr-mus LDLo:6400 mg/kg TXAPA9 23,288,72

CONSENSUS REPORTS: EPA Extremely Hazardous Substances List. Reported in EPA TSCA Inventory. Chlorophenol compounds are on the Community Right-To-Know List.

SAFETY PROFILE: A poison by ingestion and skin contact. Causes liver injury, acne, and skin rashes. When heated to decomposition it emits toxic fumes of Cl^-. See also CHLOROPHENOLS and ALDRIN.

IKQ000 CAS:97-54-1 ***HR: 2***
ISOEUGENOL
mf: $C_{10}H_{12}O_2$ mw: 164.22

PROP: Pale yellow oil; carnation odor. D: 1.079-1.085, refr index: 1.572-1.577, mp: −10°, bp: 266°. cis Form: liquid, bp: 133° @ 11 mm, d: 1.088 @ 20°/4°. trans Form: crystals, mp: 33°; bp: 140° @ 12 mm; d: 1.087 @ 20°/4°, flash p: +212°F. Sol in fixed oils, propylene glycol; very sltly sol in water; misc in alc and ether; insol in glycerin.

SYNS: FEMA No. 2468 ◇ 1-HYDROXY-2-METHOXY-4-PROPENYLBENZENE ◇ 4-HYDROXY-3-METHOXY-1-PROPENYLBENZENE ◇ 2-METHOXY-4-PROPENYLPHENOL ◇ NCI-C60979 ◇ 4-PROPENYLGUAIACOL

TOXICITY DATA with REFERENCE
sce-hmn:lym 250 μmol/L MUREAV 169,129,86
orl-rat LD50:1560 mg/kg TXAPA9 6,378,64
orl-gpg LD50:1410 mg/kg FCTXAV 2,327,64

CONSENSUS REPORTS: Reported in EPA TSCA Inventory.

SAFETY PROFILE: Moderately toxic by ingestion. Human mutation data reported. Combustible liquid. When heated to decomposition it emits acrid smoke and fumes. See also EUGENOL.

IKR000 CAS:93-16-3 ***HR: 3***
1,3,4-ISOEUGENOL METHYL ETHER
mf: $C_{11}H_{14}O_2$ mw: 178.25

PROP: Colorless to pale yellow liquid; clove-carnation odor. D: 1.047, refr index: 1.566, flash p: +212°F. Sol in fixed oils; insol in glycerin, propylene glycol.

SYNS: 1,2-DIMETHOXY-4-PROPENYLBENZENE ◇ FEMA No. 2476 ◇ ISOEUGENYL METHYL ETHER ◇ ISOHOMOGENOL ◇ METHYL ISOEUGENOL (FCC) ◇ 4-PROPENYL VERATROLE

TOXICITY DATA with REFERENCE
skn-rbt 500 mg/24H FCTXAV 13,865,75
ipr-mus LD50:570 mg/kg AIPTAK 199,226,72
ivn-mus LD50:181 mg/kg AIPTAK 199,226,72

CONSENSUS REPORTS: Reported in EPA TSCA Inventory.

SAFETY PROFILE: Poison by intravenous route. Moderately toxic by intraperitoneal route. A skin irritant. Combustible liquid. When heated to decomposition it emits acrid smoke and fumes. See also EUGENOL and ETHERS.

IKS400 CAS:26675-46-7 ***HR: 2***
ISOFLURANE
mf: $C_3H_2ClF_5O$ mw: 184.50

SYNS: ETHANE,2-CHLORO-2-(DIFLUOROMETHOXY)-1,1,1-TRIFLUORO- (9CI) ◇ ETHER, 1-CHLORO-2,2,2-TRIFLUOROETHYL DIFLUOROMETHYL ◇ FORANE

TOXICITY DATA with REFERENCE
ihl-rat TDLo:12000 ppm/3H (female 17-20D post):REP KSRNAM 21,3109,87
ihl-rat TCLo:10500 ppm/6H (female 14-16D post):TER ANESAV 64,339,86
orl-rat LD50:4770 mg/kg KSRNAM 21,3031,87
ihl-rat LC50:15300 ppm/3H KSRNAM 21,3031,87
ipr-rat LD50:4280 mg/kg KSRNAM 21,3031,87
orl-mus LD50:5080 mg/kg KSRNAM 21,3031,87
ihl-mus LC50:16800 ppm/3H KSRNAM 21,3031,87
ipr-mus LD50:3030 mg/kg KSRNAM 21,3031,87

SAFETY PROFILE: Moderately toxic by ingestion and intraperitoneal route. Slightly toxic by inhalation. An experimental teratogen. Experimental reproductive effects. When heated to decomposition it emits toxic fumes of F^- and Cl^-.

IKS500 CAS:26833-86-3 ***HR: 3***
ISOHARRINGTONINE
mf: $C_{28}H_{37}NO_9$ mw: 531.66

SYNS: NSC 141634 ◇ (3-(2R,3S))-4-METHYL-2,3-DIHYDROXY-2-(3-METHYLBUTYL)BUTANEDIOATE(ESTER)CEPHALOTAXINE

TOXICITY DATA with REFERENCE
orl-mus LD50:27950 μg/kg NCISP* JAN86
ipr-mus LD50:47930 μg/kg NCISP* JAN86
scu-mus LD50:26110 μg/kg NCISP* JAN86

SAFETY PROFILE: Poison by ingestion, subcutaneous, and intraperitoneal routes. When heated to decomposition it emits toxic fumes of NO_x.

IKS600 CAS:107-83-5 ***HR: 3***
ISOHEXANE
DOT: UN 1208/UN 2462
mf: C_6H_{14} mw: 86.20

PROP: Liquid. Bp: 54-60°, lel: 1.0%, uel: 7.0%, flash p: 20°F (CC), d: 0.669, vap d: 3.00, autoign temp: 583°F.

SYN: 2-METHYLPENTANE

CONSENSUS REPORTS: Reported in EPA TSCA Inventory.

OSHA PEL: TWA 500 ppm; STEL 1000 ppm
ACGIH TLV: TWA 500 ppm; STEL 1000 ppm (hexane isomer)
NIOSH REL: (Alkanes) TWA 350 mg/m^3
DOT Classification: Flammable Liquid; Label: Flammable Liquid .

SAFETY PROFILE: A human eye irritant. A very dangerous fire hazard when exposed to heat, flame, or oxidizers. Severe explosion hazard when exposed to heat or flame. Explosive in the form of vapor when exposed to heat or flame. Keep away from sparks, heat, or open flame; can react vigorously with oxidizing materials. To fight fire, use foam, CO_2, dry chemical. When heated to decomposition it emits acrid smoke and irritating fumes. See also HEXANE.

IKT000 CAS:646-07-1 ***HR: 2***
ISOHEXANOIC ACID (mixed isomers)
mf: $C_6H_{12}O_2$ mw: 116.18

TOXICITY DATA with REFERENCE
skn-rbt 465 mg open MOD UCDS** 2/4/59
eye-rbt 930 μg MLD UCDS** 2/4/59
orl-rat LD50:2050 mg/kg UCDS** 2/4/59
skn-rbt LD50:1050 mg/kg UCDS** 2/4/59

CONSENSUS REPORTS: Reported in EPA TSCA Inventory.

SAFETY PROFILE: Moderately toxic by ingestion and

skin contact. A skin and eye irritant. When heated to decomposition it emits acrid smoke and fumes.

IKV000 CAS:11050-62-7 ***HR: 1***
ISOJASMONE
mf: $C_{11}H_{16}O$ mw: 164.27

TOXICITY DATA with REFERENCE
skn-rbt 500 mg/24H MOD FCTXAV 16,801,78

CONSENSUS REPORTS: Reported in EPA TSCA Inventory.

SAFETY PROFILE: A skin irritant. When heated to decomposition it emits acrid smoke and fumes.

IKW000 CAS:54156-67-1 ***HR: 3***
ISOLASALOCID A
mf: $C_{34}H_{54}O_8$ mw: 590.88

TOXICITY DATA with REFERENCE
orl-mus LD50:1000 mg/kg 37ASAA 3,47,78
ipr-mus LD50:250 mg/kg 37ASAA 3,47,78

SAFETY PROFILE: Poison by intraperitoneal route. Moderately toxic by ingestion. When heated to decomposition it emits acrid smoke and fumes.

IKX000 CAS:73-32-5 ***HR: 1***
ISOLEUCINE
mf: $C_6H_{13}NO_2$ mw: 131.17

PROP: An essential amino acid; many isomeric forms. White crystalline powder; bitter taste. Mp: (dl): 292° (decomp), (l): 283-284° (decomp). Sltly sol in water; nearly insol in alc; insol in ether.

SYNS: 2-AMINO-3-METHYLPENTANOIC ACID ◇ α-AMINO-β-METHYLVALERIC ACID ◇ l-ISOLEUCINE (FCC)

TOXICITY DATA with REFERENCE
ipr-rat LD50:6822 mg/kg ABBIA4 58,253,55

CONSENSUS REPORTS: Reported in EPA TSCA Inventory.

SAFETY PROFILE: Mildly toxic by intraperitoneal route. When heated to decomposition it emits toxic fumes of NO_x.

IKY000 CAS:491-07-6 ***HR: 1***
ISOMENTHONE
mf: $C_{10}H_{18}O$ mw: 154.28

SYNS: 2-ISOPROPYL-5-METHYL-CYCLOHEXANONE ◇ (Z)-p-MENTHAN-3-ONE ◇ 5-METHYL-2-(1-METHYLETHYL)CYCLOHEXANONE ◇ (Z)-5-METHYL-2-(1-METHYLETHYL)CYCLO- HEXANONE

TOXICITY DATA with REFERENCE
skn-rbt 500 mg/24H MLD FCTXAV 14,315,76

CONSENSUS REPORTS: Reported in EPA TSCA Inventory.

SAFETY PROFILE: A skin irritant. When heated to decomposition it emits acrid smoke and fumes.

IKZ000 CAS:466-40-0 ***HR: 3***
ISOMETHADONE
mf: $C_{21}H_{27}NO$ mw: 309.49

SYNS: 6-DIMETHYLAMINO-5-METHYL-4,4-DIPHENYL-3-HEXANONE ◇ 1,1-DIPHENYL-1-(DIMETHYLAMINOISOPROPYL)BUTANONE-2 ◇ ISOAMIDONE II

TOXICITY DATA with REFERENCE
orl-mus LD50:400 mg/kg AIPTAK 120,450,59
ivn-mus LD50:17 mg/kg SCIEAS 104,587,46

SAFETY PROFILE: A poison by ingestion and intravenous routes. When heated to decomposition it emits toxic fumes of NO_x.

ILB000 CAS:1200-00-6 ***HR: 3***
ISONICOTINALDEHYDE THIOSEMICARBAZONE

SYNS: 4-FORMYLPYRIDINE THIOSEMICARBAZONE ◇ PYRIDINE-4-CARBOXALDEHYDETHIOSEMICARBAZONE

TOXICITY DATA with REFERENCE
orl-mus LD50:200 mg/kg JMCMAR 8,676,65
ipr-mus LD50:150 mg/kg NTIS** AD691-490
ivn-mus LD50:320 mg/kg CSLNX* NX#01772

SAFETY PROFILE: Poison by ingestion, intraperitoneal, and intravenous routes. When heated to decomposition it emits toxic fumes of NO_x. See also ALDEHYDES.

ILB150 CAS:31279-70-6 ***HR: 3***
ISONICOTINAMIDE PENTAAMMINE RUTHENIUM(II) PERCHLORATE
mf: $C_6H_{21}Cl_2N_7O_9Ru$ mw: 507.25

$$[C_6H_6N_2ORu(NH_3)_5][ClO_4]_2$$

SAFETY PROFILE: An explosive. When heated to decomposition it emits toxic fumes of Cl^- and NO_x. See also RUTHENIUM, PERCHLORATES, and AMINES.

ILC000 CAS:55-22-1 ***HR: 2***
ISONICOTINIC ACID
mf: $C_6H_5NO_2$ mw: 123.11

PROP: Platelets. Mp: 319°, sublimes @ 260° @ 15 mm pressure. Sltly sol in cold water; sol in hot water; insol in benzene, ether, boiling alc.

SYNS: ACIDE ISO-NICOTINIQUE (FRENCH) ◇ 4-CARBOXYPYRIDINE ◇ α-PICOLINIC ACID ◇ 4-PYRIDINECARBOXYLIC ACID

TOXICITY DATA with REFERENCE
orl-rat LD50:5000 mg/kg PCJOAU 11,481,77
orl-mus LD50:3123 mg/kg PCJOAU 11,481,77
ipr-mus LD50:436 mg/kg PCJOAU 11,481,77
ivn-mus LD50:5000 mg/kg THERAP 23,1343,68

CONSENSUS REPORTS: Reported in EPA TSCA Inventory.

SAFETY PROFILE: Moderately toxic by ingestion and intraperitoneal routes. Mildly toxic by intravenous route. When heated to decomposition it emits toxic fumes of NO_x.

ILD000 CAS:54-85-3 ***HR: 3***
ISONICOTINIC ACID HYDRAZIDE
mf: $C_6H_7N_3O$ mw: 137.16

PROP: Consists of 12% w/v each of dodecylamine hydrochloride, trimethyl alkyl ammonium chloride, and methyl alkyl dipolyoxypropylene ammonium methyl sulfate (TXAPA9 4,44,62).

SYNS: AMIDON ◇ ANTIMICINA ◇ ANTITUBERKULOSUM ◇ ATCOTIBINE ◇ AZUREN ◇ BACILLIN ◇ CEMIDON ◇ CHEMIAZID ◇ CORTINAZINE ◇ COTINAZIN ◇ DEFONIN ◇ DIBUTIN ◇ DINACRIN ◇ DITUBIN ◇ ERALON ◇ ERTUBAN ◇ EVALON ◇ FIMALENE ◇ HIDRANIZIL ◇ HIDRULTA ◇ HYCOZID ◇ HYDRAZID ◇ HYOZID ◇ HYZYD ◇ IDRAZIDE DELL'ACIDO ISONICOTINICO ◇ IDRAZIL ◇ ISIDRINA ◇ ISMAZIDE ◇ ISOCID ◇ ISOCOTIN ◇ ISOLYN ◇ ISONEX ◇ ISONIACID ◇ ISONIAZIDE ◇ ISONICAZIDE ◇ ISONICO ◇ ISONICOTAN ◇ ISONICOTIN-HYDRAZID ◇ ISONICOTINOYL HYDRAZIDE ◇ ISONICOTINSAEURE-HYDRAZID ◇ ISONICOTINYL HYDRAZIDE ◇ ISONIDE ◇ ISONIKAZID ◇ ISONIN ◇ ISONIRIT ◇ ISONIZIDE ◇ ISOTEBEZID ◇ ISOZIDE ◇ LANIAZID ◇ MYBASAN ◇ NEOXIN ◇ NEVIN ◇ NICAZIDE ◇ NICIZINA ◇ NICOTIBINA ◇ NICOZIDE ◇ NIDRAZID ◇ NIPLEN ◇ NITEBAN ◇ NSC 9659 ◇ NYDRAZID ◇ PELAZID ◇ PERCIN ◇ PYCAZIDE ◇ PYRICIDIN ◇ 4-PYRIDINECARBOXYLIC ACID, HYDRAZIDE ◇ RAUMANON ◇ RETOZIDE ◇ RIMICID ◇ RIMITSID ◇ ROBISELLIN ◇ SANOHIDRAZINA ◇ SAUTERZID ◇ TEBECID ◇ TEBEXIN ◇ TEEBACONIN ◇ TIBAZIDE ◇ TIBINIDE ◇ TIBIVIS ◇ TISIN ◇ TUBAZIDE ◇ TUBERCID ◇ TUBICON ◇ TYVID ◇ UNICOCYDE ◇ USAF CB-2 ◇ VAZADRINE ◇ VEDERON ◇ ZINADON ◇ ZONAZIDE

TOXICITY DATA with REFERENCE
skn-rbt 500 mg/24H MOD TXAPA9 4,44,62
mmo-sat 1 mg/plate CRNGDP 5,391,84
dns-hmn:lvr 100 μmol/L CALEDQ 30,103,86
dni-hmn:fbr 10 mmol/L MUREAV 89,9,81
orl-mus TDLo:1848 mg/kg (female 1-21D post):REP IJEBA6 18,1104,80
orl-mus TDLo:1848 mg/kg (female 1-21D post):TER IJEBA6 18,1104,80
orl-rat TDLo:55 g/kg/45W-C:NEO JNCIAM 41,331,68
orl-mus TDLo:1892 mg/k (multi) :CAR JCREA8 105,258,83
orl-mus TDLo:18524 mg/kg/84W-I:CAR JCREA8 105,258,83
orl-mus TD:8880 mg/kg/19W-C:ETA GANNA2 50,107,59
orl-man TDLo:430 mg/kg:CNS NEURAI 20,299,70
orl-man TDLo:100 mg/kg/3W-C:CNS,SKN,KID ARDSBL 106,849,72
orl-cld TDLo:1299 mg/kg/16W-I:CNS CPEDAM 22,518,83
orl-wmn TDLo:12 mg/kg/2D-I:PSY NEURAI 34,703,84
orl-man TDLo:39 mg/kg/9D-I:SKN SMJOAV 75,81,82
orl-hmn LDLo:100 mg/kg ARDSBL 105,206,72
orl-rat LD50:1250 mg/kg ARZNAD 26,409,76
ipr-rat LD50:335 mg/kg SZTPA5 9,226,52
scu-rat LD50:329 mg/kg JPETAB 119,444,57
ivn-rat LD50:365 mg/kg ARZNAD 26,409,76
ims-rat LD50:400 mg/kg THERAP 8,62,53
orl-mus LD50:133 mg/kg ARTUA4 65,376,52
ipr-mus LD50:100 mg/kg NTIS** AD277-689
scu-mus LD50:125 mg/kg YKKZAJ 81,1225,61
ivn-mus LD50:149 mg/kg JPETAB 122,110,58
ims-mus LD50:137 mg/kg ARTUA4 65,376,52
orl-dog LD50:50 mg/kg ARTUA4 65,392,52
ipr-cat LD50:325 mg/kg KLWOAZ 30,959,52
orl-rbt LD50:250 mg/kg ARZNAD 12,22,62
ipr-rbt LD50:147 mg/kg SZTPA5 9,226,52

CONSENSUS REPORTS: IARC Cancer Review: Group 3 IMEMDT 7,227,87, Animal Sufficient Evidence IMEMDT 4,159,74. EPA Genetic Toxicology Program.

SAFETY PROFILE: A human poison by ingestion. An experimental poison by ingestion, intravenous, subcutaneous, intraperitoneal, and intramuscular routes. Experimental teratogenic and reproductive effects. Human systemic effects by ingestion: peripheral nerve sensory changes, somnolence, respiratory depression, anorexia, sweating, urine changes, toxic psychosis, hepatitis, dermatitis. Human mutation data reported. A skin irritant. Questionable carcinogen with experimental carcinogenic, neoplastigenic, and tumorigenic data. Used as an antitubercular, antibacterial, and anti-actinomycotic agent. When heated to decomposition it emits toxic fumes of NO_x and NH_3. See also individual components.

ILE000 CAS:54-92-2 ***HR: 3***
ISONICOTINIC ACID-2-ISOPROPYLHYDRAZIDE
mf: $C_9H_{13}N_3O$ mw: 179.25

SYNS: EUPHOZID ◇ FOSFAZIDE ◇ IIH ◇ IPN ◇ IPRAZID ◇ IPRONIAZID ◇ IPRONID ◇ IPRONIN ◇ 1-ISONICOTINOYL-2-ISOPROPYLHYDRAZINE ◇ 1-ISONICOTINYL-2-ISOPROPYLHYDRAZINE ◇ N-ISOPROPYL ISONICOTINHYDRAZIDE ◇ LH ◇ MARSALID ◇ MARSILID ◇ P 887 ◇ RIVIVOL ◇ RO 2-4572 ◇ YATROZIDE

TOXICITY DATA with REFERENCE
oms-bcs 10 mmol/L MUREAV 5,343,68
orl-man TDLo:45 mg/kg (21D male):REP ANPBAZ 59,977,59

scu-mus TDLo:1200 mg/kg (female 6-11D post):TER SCIEAS 131,1101,60
orl-mus TDLo:10 g/kg/36W-I:ETA 34ZRA9 -,869,66
orl-hmn TDLo:2143 μg/kg/D:GIT,KID,MET ANPBAZ 59,977,59
orl-wmn LDLo:14 mg/kg/2W-I:LIV CMAJAX 78,131,58
orl-rat LD50:365 mg/kg NATUAS 185,532,60
ipr-rat LD50:375 mg/kg JJPAAZ 13,186,63
scu-rat LD50:538 mg/kg JPETAB 119,444,57
orl-mus LD50:681 mg/kg BJPCBM 34,236,68
ipr-mus LD50:475 mg/kg JJPAAZ 13,186,63
scu-mus LD50:750 mg/kg ARTUA4 65,376,52
ivn-mus LD50:725 mg/kg ARTUA4 65,376,52
ims-mus LD50:683 mg/kg ARTUA4 65,376,52
orl-dog LD50:95 mg/kg ANYAA9 80,626,59
orl-mky LD50:640 mg/kg ANYAA9 80,626,59

SAFETY PROFILE: A human poison by ingestion. An experimental poison by ingestion, intraperitoneal, and intravenous routes. Moderately toxic by skin contact, intramuscular, and subcutaneous routes. Human systemic effects by ingestion: constipation, anuria, metabolic changes, change in liver function. Human reproductive effects by ingestion: impotence. Experimental reproductive effects. Mutation data reported. Questionable carcinogen with experimental tumorigenic and teratogenic data. Used as an antidepressant. When heated to decomposition it emits toxic fumes of NO_x.

ILF000 CAS:16887-79-9 ***HR: 3***
ISONICOTINIC ACID, SODIUM SALT
mf: $C_6H_4NO_2 \cdot Na$ mw: 145.10

SYN: SODIUM SALT of ISONICOTINIC ACID

TOXICITY DATA with REFERENCE
orl-mus TDLo:13 g/kg/46W-I:ETA NATUAS 194,488,62

SAFETY PROFILE: Questionable carcinogen with experimental tumorigenic data. When heated to decomposition it emits toxic fumes of NO_x and Na_2O.

ILG000 CAS:63041-19-0 ***HR: 3***
4-(ISONICOTINOYLHYDRAZONE)PIMELIC ACID
mf: $C_{13}H_{15}N_3O_5$ mw: 293.31

SYNS: ACIDO-4-(ISONICOTINIL-IDRAZONE)PIMELICO(ITALIAN) ◇ 4-OXOHEPTANEDIOIC ACID, ISONICOTINOYL HYDRAZONE

TOXICITY DATA with REFERENCE
orl-mus TDLo:12 g/kg/30W-I:ETA LAPPA5 24,39,64

SAFETY PROFILE: Questionable carcinogen with experimental tumorigenic data. When heated to decomposition it emits toxic fumes of NO_x.

ILH000 CAS:532-54-7 ***HR: 3***
ISONITROSOACETOPHENONE
mf: $C_8H_7NO_2$ mw: 149.16

PROP: Plates or prisms. Mp: 126-128°. Sltly sol in cold water; sol in hot water, alkalies and alkali carbonates.

SYN: 2-HYDROXYIMINOACETOPHENONE

TOXICITY DATA with REFERENCE
orl-rat LDLo:100 mg/kg NCNSA6 5,27,53
ipr-mus LD50:75 mg/kg JPETAB 119,522,57

CONSENSUS REPORTS: Reported in EPA TSCA Inventory.

SAFETY PROFILE: Poison by ingestion and intraperitoneal routes. When heated to decomposition it emits toxic fumes of NO_x.

ILI000 CAS:119-51-7 ***HR: 3***
ISONITROSOPROPIOPHENONE
mf: $C_9H_9NO_2$ mw: 163.19

TOXICITY DATA with REFERENCE
orl-rat LD50:4240 mg/kg TXAPA9 28,313,74
ipr-mus LD50:250 mg/kg NTIS** AD691-490

CONSENSUS REPORTS: Reported in EPA TSCA Inventory.

SAFETY PROFILE: Poison by intraperitoneal route. Mildly toxic by ingestion. When heated to decomposition it emits toxic fumes of NO_x.

ILJ000 CAS:2430-22-0 ***HR: 2***
ISONONYL ALCOHOL
mf: $C_9H_{20}O$ mw: 144.29

SYN: 7-METHYL-1-OCTANOL

TOXICITY DATA with REFERENCE
skn-rbt 3200 mg/kg/24H SEV AIHAAP 34,493,73
eye-rbt 100 mg SEV AIHAAP 34,493,73
orl-rat LD50:2980 mg/kg AIHAAP 34,493,73
ihl-rat LCLo:21700 mg/m^3/6H AIHAAP 34,493,73

SAFETY PROFILE: Moderately toxic by ingestion. Mildly toxic by inhalation. A severe skin and eye irritant. When heated to decomposition it emits acrid smoke and fumes.

ILK000 CAS:503-01-5 ***HR: 3***
ISONYL
mf: $C_9H_{19}N$ mw: 141.29

PROP: Colorless, oily liquid; characteristic amine odor, water-insol.

SYNS: ISOMETHEPTENE ◇ 6-METHYLAMINO-2-METHYLHEPTENE ◇ METHYLISOOCTENYLAMINE ◇ 2-METHYL-6-METHYLAMINO-2-HEPTENE ◇ METHYLOCTENYLAMINE ◇ OCTANIL ◇ OCTIN ◇ OCTINUM ◇ OCTON ◇ N-1,5-TRIMETHYL-4-HEXENYLAMINE

TOXICITY DATA with REFERENCE
orl-mus LD50:134 mg/kg JPETAB 116,377,56
ipr-mus LD50:42 mg/kg JPETAB 116,377,56
scu-mus LD50:100 mg/kg FDMU** -,-,35
ivn-rbt LDLo:50 mg/kg FDMU** -,-,35

CONSENSUS REPORTS: Reported in EPA TSCA Inventory.

SAFETY PROFILE: A poison by ingestion, intravenous, intraperitoneal, and subcutaneous routes. Can cause headache, nausea, and dizziness in humans. When heated to decomposition it emits toxic fumes of NO_x.

ILL000 CAS:26952-21-6 ***HR: 2***
ISOOCTYL ALCOHOL
mf: $C_8H_{18}O$ mw: 130.26

SYN: ISOOCTANOL

TOXICITY DATA with REFERENCE
skn-rbt 2600 mg/kg/24H MOD AIHAAP 34,493,73
eye-rbt 100 mg SEV AIHAAP 34,493,73
orl-rat LD50:1480 mg/kg AIHAAP 34,493,73
orl-mus LD50:1670 mg/kg 85GMAT -,77,82
skn-rbt LD50:2520 mg/kg 31ZTAS -,72,68

CONSENSUS REPORTS: Reported in EPA TSCA Inventory.

OSHA PEL: TWA 50 ppm (skin)
ACGIH TLV: TWA 50 ppm (skin)

SAFETY PROFILE: Moderately toxic by ingestion and skin contact. A skin and severe eye irritant. When heated to decomposition it emits acrid smoke and fumes. See also ALCOHOLS.

ILM000 CAS:543-82-8 ***HR: 3***
2-ISOOCTYL AMINE
mf: $C_8H_{19}N$ mw: 129.28

PROP: dl-Form: Viscous liquid, fishy odor. Bp: 154-156°, n (24/D) 1.4200.

SYNS: AMIDRINE ◇ 2-AMINO-6-METHYLHEPTANE ◇ 6-AMINO-2-METHYLHEPTANE ◇ α,ε-DIMETHYLHEXYLAMINE ◇ 1,5-DIMETHYLHEXYLAMINE ◇ 2-METHYL-6-AMINOHEPTANE ◇ 2-METHYL-2-HEPTYLAMINE ◇ 6-METHYL-2-HEPTYLAMINE ◇ 2-METIL-6-AMINO-EPTANO (ITALIAN) ◇ OCTODRINE ◇ SKF 51 ◇ VAPORPAC

TOXICITY DATA with REFERENCE
orl-rat LD50:538 mg/kg BSIBAC 27,354,51
ipr-rat LD50:42 mg/kg JPETAB 90,351,47
ims-rat LD50:146 mg/kg BSIBAC 27,354,51
ipr-mus LD50:59 mg/kg JPETAB 90,351,47
scu-mus LDLo:2000 mg/kg KLWOAZ 15,1164,36
ipr-rbt LD50:44 mg/kg JPETAB 90,351,47
ipr-gpg LD50:39 mg/kg JPETAB 90,351,47

CONSENSUS REPORTS: Reported in EPA TSCA Inventory.

SAFETY PROFILE: A poison by intraperitoneal and intramuscular routes. Moderately toxic by ingestion and subcutaneous routes. When heated to decomposition it emits toxic fumes of NO_x. See also AMINES.

ILO000 CAS:25168-26-7 ***HR: 3***
ISOOCTYL-2,4-DICHLOROPHENOXYACETATE
mf: $C_{16}H_{22}Cl_2O_3$ mw: 333.28

SYNS: 2,4-DICHLOROPHENOXYACETIC ACID ISOOCTYL ESTER ◇ ISOOCTYL ALCOHOL (2,4-DICHLOROPHENOXY)ACETATE ◇ 2,4-D ISOOCTYL ESTER ◇ REED LV 2,4-D ◇ REED LV 400 2,4-D ◇ REED LV 600 2,4-D ◇ WEEDTRINE-II

TOXICITY DATA with REFERENCE
sce-hmn:lym 50 nL/L DBABEF 8,105,84
orl-rat TDLo:188 mg/kg (female 6-15D post):REP FCTXAV 9,801,71
scu-mus TDLo:1170 mg/kg (female 6-14D post):TER NTIS** PB223-160
orl-mus TDLo:14 g/kg/78W-I:ETA NTIS** PB223-159
scu-mus TDLo:21 mg/kg:CAR NTIS** PB223-159
orl-rat LD50:300 mg/kg FMCHA2 -,C309,89
orl-rbt LD50:300 mg/kg FMCHA2 -,C255,83
orl-gpg LD50:300 mg/kg FMCHA2 -,C255,83

CONSENSUS REPORTS: IARC Cancer Review: Animal Inadequate Evidence IMEMDT 15,111,77.

SAFETY PROFILE: A poison by ingestion. Other experimental reproductive effects. Questionable carcinogen with experimental carcinogenic, tumorigenic, and teratogenic data. Human mutation data reported. An herbicide. When heated to decomposition it emits toxic fumes of Cl^-. See also 2,4-DICHLOROPHENOXY ACETIC ACID.

ILR000 CAS:25103-09-7 ***HR: 3***
ISOOCTYL MERCAPTOACETATE
mf: $C_{10}H_{20}O_2S$ mw: 204.36

PROP: A clear, water-white liquid; fruity odor. Bp: 52° @ 17 mm, d: 0.9736 @ 25°.

SYNS: ISOOCTYL ESTER, MERCAPTOACETATE ACID ◇ ISOOCTYL THIOGLYCOLATE

TOXICITY DATA with REFERENCE
orl-rat LD50:348 mg/kg TRIPA7 -,1,73
orl-rbt LDLo:1200 mg/kg JEENAI 48,139,55

CONSENSUS REPORTS: Reported in EPA TSCA Inventory.

SAFETY PROFILE: Poison by ingestion. Can react vigorously with oxidizers. When heated to decomposition it emits toxic fumes of SO_x. See also 2-MERCAPTO-ACETIC ACID and MERPCAPTANS.

ILR100 CAS:27554-26-3 ***HR: 2***
ISOOCTYL PHTHALATE
mf: $C_{24}H_{38}O_4$ mw: 390.62

SYNS: 1,2-BENZENEDICARBOXYLIC ACID, DIISOOCTYL ESTER ◇ BIS(6-METHYLHEPTYL)ESTER of PHTHALIC ACID ◇ CORFLEX 880 ◇ DIISOOCTYL PHTHALATE ◇ FLEXOL PLASTICIZER DIP ◇ HEXAPLAS M/O

TOXICITY DATA with REFERENCE
skn-rbt 500 mg open MLD UCDS** 6/11/65
orl-rat LD50:22 g/kg EVHPAZ 3,61,73
orl-mus LD50:2769 mg/kg GTPZAB 17(11),51,73
skn-rbt LD50:13 g/kg UCDS** 6/11/65

CONSENSUS REPORTS: Reported in EPA TSCA Inventory.

SAFETY PROFILE: Moderately toxic by ingestion. Mildly toxic by skin contact. A skin irritant. When heated to decomposition it emits acrid smoke and irritating fumes.

ILT000 CAS:67051-25-6 ***HR: 3***
5-(1-ISOPENTENYL)-5-ISOPROPYLBARBITURIC ACID
mf: $C_{12}H_{18}N_2O_3$ mw: 238.32

TOXICITY DATA with REFERENCE
orl-mus LD50:280 mg/kg JACSAT 62,1199,40
ipr-mus LD50:200 mg/kg JACSAT 62,1199,40

SAFETY PROFILE: Poison by ingestion and intraperitoneal routes. When heated to decomposition it emits toxic fumes of NO_x. See also BARBITURATES.

ILU000 CAS:67051-27-8 ***HR: 3***
5-(2-ISOPENTENYL)-5-ISOPROPYL-1-METHYLBARBITURIC ACID
mf: $C_{13}H_{20}N_2O_3$ mw: 252.35

TOXICITY DATA with REFERENCE
scu-mus LD50:70 mg/kg JACSAT 72,4319,50
ivn-mus LD50:12 mg/kg JACSAT 72,4319,50
ivn-rbt LD50:5 mg/kg JACSAT 72,4319,50

SAFETY PROFILE: Poison by subcutaneous and intravenous routes. When heated to decomposition it emits toxic fumes of NO_x. See also BARBITURATES.

ILW000 CAS:105-68-0 ***HR: 1***
ISOPENTYL ALCOHOL, PROPIONATE
mf: $C_8H_{16}O_2$ mw: 144.24

PROP: Found in Cocoa bean and Bulgarian peppermint (FCTXAV 13,681,75).

SYNS: ISOAMYL PROPIONATE ◇ ISOPENTYL PROPIONATE ◇ PROPIONIC ACID, ISOPENTYL ESTER

TOXICITY DATA with REFERENCE
orl-rbt LD50:6924 mg/kg IMSUAI 41,31,72

CONSENSUS REPORTS: Reported in EPA TSCA Inventory.

SAFETY PROFILE: Mildly toxic by ingestion. When heated to decomposition it emits acrid smoke and fumes. See also ESTERS.

IMB000 CAS:110-46-3 ***HR: 3***
ISOPENTYL NITRITE
mf: $C_5H_{11}NO_2$ mw: 117.17

PROP: Transparent, flammable liquid; penetrating, fragrant odor. Unstable; decomp on exposure to air and light. D: 0.872 @ 20°/4°, bp: 97-99°, autoign temp: 408°F, vap d: 4.0, flash p: < 73.4°F.

SYNS: ISOAMYL NITRITE ◇ ISOPENTYL ALCOHOL NITRITE ◇ 3-METHYLBUTANOL NITRITE ◇ 3-METHYLBUTYL NITRITE ◇ NITROUS ACID-3-METHYL BUTYL ESTER ◇ VAPOROLE

TOXICITY DATA with REFERENCE
mmo-sat 333 μg/plate ENMUDM 8(Suppl 7),1,86
mma-sat 100 μg/plate ENMUDM 8(Suppl 7),1,86
orl-rat LD50:505 mg/kg FEPRA7 41,1583,82
ihl-rat LC50:1274 ppm/1H FEPRA7 41,1583,82
ipr-mus LD50:130 mg/kg RCOCB8 5,889,73
ivn-mus LD50:51 mg/kg RCOCB8 5,889,73
ivn-dog LDLo:167 mg/kg RCOCB8 5,889,73

CONSENSUS REPORTS: Cyanide and its compounds are on the Community Right-To-Know List. Reported in EPA TSCA Inventory.

SAFETY PROFILE: Poison by intravenous and intraperitoneal routes. Moderately toxic by ingestion. Mildly toxic by inhalation. Mutation data reported. A recreational drug said to enhance sexual enjoyment in humans by inhalation. Dangerous fire hazard when exposed to spark, heat, oxidizers or flame. Forms an explosive mixture in air or O_2. Vapors will explode when heated. When heated to decomposition it emits toxic fumes of NO_x. See also NITRITES.

IME000 CAS:87-20-7 ***HR: 2***
ISOPENTYL SALICYLATE
mf: $C_{12}H_{16}O_3$ mw: 208.28

PROP: Coorless liquid; pleasant odor. D: 1.047, refr index: 1.503-1.509, flash p: 271°F. Misc with alc, chloroform, ether, fixed oils; insol in glycerin, propylene glycol, water.

SYNS: FEMA No. 2084 ◇ ISOAMYL o-HYDROXYBENZOATE ◇ ISOAMYL SALICYLATE (FCC) ◇ ISOPENTYL-2-HYDROXYPHENYL METHANOATE ◇ 3-METHYLBUTYL 2-HYDROXYBENZOATE ◇ SALICYLIC ACID, ISOPENTYL ESTER

TOXICITY DATA with REFERENCE
orl-rat TDLo:12600 mg/kg (42D male):REP FCTXAV 13,185,75
ivn-dog LD50:500 mg/kg 14CYAT 2,1847,63

CONSENSUS REPORTS: Reported in EPA TSCA Inventory.

SAFETY PROFILE: Moderately toxic by intravenous route. Experimental reproductive effects. Combustible liquid. When heated to decomposition it emits acrid smoke and fumes.

IMF300 CAS:25311-71-1 ***HR: 3***
ISOPHENPHOS
mf: $C_{15}H_{24}NO_4PS$ mw: 345.43

PROP: Oil. Bp: 120°, d: (20°/4°) 1.13. Vap press at 20°: 0.000004 mm Hg. Solubility in water at 20°: 23.8 mg/kg. Sol in dichloromethane, cyclohexanone, acetone, alc, ether, benzene.

SYNS: 2-(O-AETHYL-N-ISOPROPYLAMINDOTHIOPHOSPHORYLOXY)-BENZOSAEURE-ISOPROPYLESTER (GERMAN) ◇ AMAZE ◇ BAY-92114 ◇ BAY-SRA-12869 ◇ 2-((ETHOXY((1-METHYLETHYL)AMINO)PHOSPHINOTHIOYL)OXY)BENZOIC ACID 1-METHYLETHYL ESTER ◇ O-ETHYL-O-(2-ISOPROPOXY-CARBONYL)-PHENYL ISOPROPYLPHOSPHORAMIDOTHIOATE ◇ ISOFENPHOS ◇ ISOPROPYL-PHOSPHORAMIDOTHIOIC ACID O-ETHYL O-(2-ISOPROPOXYCARBONYLPHENYL) ESTER ◇ ISOPROPYL SALICYLATE O-ESTER with O-ETHYLISOPROPYLPHOSPHORAMIDOTHIOATE ◇ 1-METHYLETHYL-2-((ETHOXY((1-METHYLETHYL)AMINO)PHOSPHINOTHIOYL)OXY)BENZOATE ◇ OFTANOL ◇ 40 SD ◇ SALICYLIC ACID ISOPROPYL ESTER O-ESTER with O-ETHYL ISOPROPYLPHOSPHORAMIDOTHIOATE ◇ SRA 12869

TOXICITY DATA with REFERENCE
orl-rat LD50:28 mg/kg 85ARAE 1,118,77
skn-rat LD50:188 mg/kg SPEADM 78-1,39,78
orl-mus LD50:91300 μg/kg 85DPAN -,-,71/76
skn-rbt LD50:162 mg/kg FMCHA2 -,C133,83
orl-ckn LD50:3 mg/kg BECTA6 33,386,84
orl-qal LD50:13 mg/kg EESADV 8,551,84

SAFETY PROFILE: Poison by ingestion and skin contact. When heated to decomposition it emits toxic fumes of PO_x, SO_x, and NO_x.

IMF400 CAS:78-59-1 ***HR: 3***
ISOPHORONE
mf: $C_9H_{14}O$ mw: 138.23

PROP: Practically water-white liquid. Bp: 215.2°, flash p: 184°F (OC), d: 0.9229, autoign temp: 864°F, vap press: 1 mm @ 38.0°, vap d: 4.77, lel: 0.8%, uel: 3.8%.

SYNS: ISOACETOPHORONE ◇ ISOFORON ◇ ISOFORONE (ITALIAN) ◇ IZOFORON (POLISH) ◇ NCI-C55618 ◇ 1,1,3-TRIMETHYL-3-CYCLOHEXENE-5-ONE ◇ 3,5,5-TRIMETHYL-2-CYCLOHEXENE-1-ONE ◇ 3,5,5-TRIMETHYL-2-CYCLOHEXEN-1-ON (GERMAN, DUTCH) ◇ 3,5,5-TRIMETIL-2-CICLOESEN-1-ONE (ITALIAN)

TOXICITY DATA with REFERENCE
eye-hmn 25 ppm/15M JIHIAB 28,262,46
skn-rbt 100 mg/24H MLD JETOAS 5,31,72
eye-rbt 920 μg SEV UCDS** 11/15/71
eye-gpg 840 ppm/4H SEV JIHTAB 22,477,40
msc-mus:lym 1 g/L NTPTR* NTP-TR-291,86
sce-ham:ovr 1 g/L NTPTR* NTP-TR-291,86
orl-mus TDLo:258 g/kg/2Y-I:CAR TXCYAC 39,207,86
ihl-hmn TCLo:25 ppm:NOSE,EYE,PUL JIHTAB 28,262,46
orl-rat LD50:2330 mg/kg TXAPA9 17,498,70
ihl-rat LCLo:1840 ppm/4H JIHTAB 22,477,40
skn-rbt LD50:1500 mg/kg UCDS** 11/15/71

CONSENSUS REPORTS: NTP Carcinogenesis Studies (gavage); Some Evidence: rat NTPTR* NTP-TR-291,86; (gavage); Equivocal Evidence: mouse NTPTR* NTP-TR-291,86. Reported in EPA TSCA Inventory.

OSHA PEL: (Transitional: TWA 25 ppm) TWA 4 ppm
ACGIH TLV: CL 5 ppm
DFG MAK: 5 ppm (28 mg/m^3)
NIOSH REL: TWA (Ketones) 23 mg/m^3

SAFETY PROFILE: Moderately toxic by ingestion and skin contact. Mildly toxic by inhalation. Human systemic effects by inhalation: olfactory changes, conjunctiva irritation, and respiratory changes. Human systemic irritant by inhalation. A skin and severe eye irritant. Questionable carcinogen with experimental carcinogenic data. Mutation data reported. Considered to be more toxic than mesityl oxide. However, due to its low volatility, it is not a dangerous industrial hazard. The response of guinea pigs and rats to repeated inhalation of the vapors indicates that it is one of the most toxic of the ketones. It is chiefly a kidney poison. It can cause irritation, lachrimation, possible opacity of the cornea, and necrosis of the cornea (experimental). It is irritating at the level of 25 ppm to humans. In animal experiments death during exposure was usually due to narcosis, but occasionally due to irritation of the lungs.

Flammable and explosive when exposed to heat or flame; can react with oxidizing materials. To fight fire, use foam, CO_2, dry chemical. See also KETONES.

IMG000 CAS:4098-71-9 ***HR: 3***
ISOPHORONE DIISOCYANATE
DOT: UN 2290
mf: $C_{12}H_{18}N_2O_2$ mw: 222.32

SYNS: 3-ISOCYANATOMETHYL-3,5,5-TRIMETHYLCYCLOHEXYLISOCYANATE ◇ ISOPHORONE DIAMINE DIISOCYANATE

TOXICITY DATA with REFERENCE
ihl-rat LC50:260 mg/m^3/4H DTLVS* 4,236,80
skn-rat LD50:1060 mg/kg DTLVS* 4,236,80

CONSENSUS REPORTS: EPA Extremely Hazardous Substances List. Reported in EPA TSCA Inventory.

OSHA PEL: TWA 0.005 ppm (skin)
ACGIH TLV: TWA 0.005 ppm (skin)
DFG MAK: 0.01 ppm (0.09 mg/m^3)
NIOSH REL: (Diisocyanates):10H TWA 0.005 ppm; CL 0.02 ppm/10M

SAFETY PROFILE: Poison by inhalation. Moderately toxic by skin contact. When heated to decomposition it emits toxic fumes of NO_x. See also ISOCYANATES.

IMH000 CAS:3778-73-2 ***HR: 3***
ISOPHOSPHAMIDE
mf: $C_7H_{15}Cl_2N_2O_2P$ mw: 261.11

SYNS: A 4942 ◇ ASTA Z 4942 ◇ N,N-BIS(β-CHLOROETHYL)-AMINO-N'-O-PROPYLENE-PHOSPHORIC ACID ESTER DIAMIDE ◇ 2,3-(N,N(1)-BIS(2-CHLOROETHYL)DIAMIDO-1,3,2-OXAZAPHOSPHORIDINOXY ◇ N,3-BIS(2-CHLOROETHYL)TETRAHYDRO-2H-1,3,2-OXAZAPHOSPHORIN-2-AMINE 2-OXIDE ◇ N-(2-CHLORAETHYL)-N'-(2 CHLOROETHYL)-N'-o-PROPYLEN-PHOSPHORSAUREESTER-DIAMID(GERMAN) ◇ 3-(2-CHLOROETHYL)-2-((2-CHLOROETHYL)AMINO)PERHYDRO-2H-1,3,2-OXAZAPHOSPHORINE OXIDE ◇ 3-(2-CHLOROETHYL)-2-((2-CHLOROETHYL)AMINO)TETRAHYDRO-2H-1,3,2-OXAZAPHOSPHORINE-2-OXIDE ◇ N-(2-CHLOROETHYL)-N'-(2-CHLOROETHYL)-N',O-PROPYLENEPHOSPHORIC ACID DIAMIDE ◇ N-(2-CHLOROETHYL)-N'-(2-CHLOROETHYL)-N',O-PROPYLENEPHOSPHORIC ACID ESTER DIAMIDE ◇ CYFOS ◇ HOLOXAN ◇ IFOSFAMID ◇ IFOSFAMIDE ◇ IPHOSPHAMIDE ◇ ISOENDOXAN ◇ ISOFOSFAMIDE ◇ MITOXANA ◇ MJF 9325 ◇ NAXAMIDE ◇ NCI-C01638 ◇ NSC-109724 ◇ Z 4942

TOXICITY DATA with REFERENCE
mma-sat 400 μg/plate TCMUD8 5,319,85
cyt-hmn:leu 130 mg/L HUMAA7 5,321,68
bfa-rat/sat 2 g/kg HIKYAJ 26,813,80
ivn-rat TDLo:260 mg/kg (female 17-21D post):REP KSRNAM 16,553,82
ivn-rat TDLo:13750 μg/kg (female 7-17D post):TER KSRNAM 16,517,82
ipr-rat TDLo:940 mg/kg/1Y-I:CAR NCITR* NCI-TR-32,78
ipr-mus TDLo:450 mg/kg/8W-I:NEO CNREA8 33,3069,73
scu-mus TDLo:2600 mg/kg/65W-I:CAR ARZNAD 29,483,79
orl-hmn TDLo:150 mg/kg:GIT,KID,SKN CNREA8 32,921,72
orl-hmn TDLo:100 mg/kg:BLD CCYPBY 3,33,72
ivn-hmn TDLo:2298 mg/kg/3D-I:KID,BLD EJCAAH 12,195,76
ivn-hmn TDLo:1915 mg/kg/2W-I:KID LANCAO 2,657,80
ivn-hmn TDLo:130 mg/kg/13D-I:KID,GIT CNREA8 32,921,72
orl-rat LD50:143 mg/kg KSRNAM 16,431,82
ipr-rat LD50:140 mg/kg USXXAM #3732340
scu-rat LD50:160 mg/kg KSRNAM 16,431,82
ivn-rat LD50:190 mg/kg KSRNAM 16,431,82
orl-mus LD50:1005 mg/kg KSRNAM 16,431,82
ipr-mus LD50:397 mg/kg ARZNAD 24,1149,74
scu-mus LD50:656 mg/kg KSRNAM 16,431,82
ivn-mus LD50:338 mg/kg KSRNAM 16,431,82
ipr-dog LDLo:50 mg/kg KSRNAM 16,431,82

CONSENSUS REPORTS: IARC Cancer Review: Group 3 IMEMDT 7,56,87; Animal Limited Evidence IMEMDT 26,237,81. NCI Carcinogenesis Bioassay (ipr); Clear Evidence: mouse, rat NCITR* NCI-CG-TR-32,77. EPA Genetic Toxicology Program.

SAFETY PROFILE: Suspected carcinogen with experimental carcinogenic and neoplastigenic data. A poison by ingestion, intraperitoneal, subcutaneous, and intravenous routes. Human systemic effects by ingestion and intravenous routes: nausea or vomiting; proteinuria, hematuria, inflammation, necrosis or scarring of the bladder, and other kidney, ureter, or bladder changes; changes in hair covering the skin; leukopenia (decreased white blood cell count), thrombocytopenia (decrease in the number of blood platelets); hallucinations, distorted perceptions; tumorigenic effects (active as an anti-cancer agent) data. Experimental teratogenic and reproductive effects. Human mutation data reported. When heated to decomposition it emits very toxic Cl^-, NO_x, and PO_x.

IMI000 CAS:626-19-7 ***HR: 3***
ISOPHTHALALDEHYDE
mf: $C_8H_6O_2$ mw: 134.14

SYN: ISOPHTALDEHYDES (FRENCH)

TOXICITY DATA with REFERENCE
ipr-mus LDLo:696 mg/kg COREAF 246,851,58
ivn-mus LD50:100 mg/kg CSLNX* NX#07922

SAFETY PROFILE: Poison by intravenous route. Moderately toxic by intraperitoneal route. When heated to decomposition it emits acrid smoke and fumes. See also ALDEHYDES.

IMJ000 CAS:121-91-5 ***HR: 1***
ISOPHTHALIC ACID
mf: $C_8H_6O_4$ mw: 166.14

PROP: Colorless crystals. Sltly sol in water, sol in alc and acetic acid, insol in benzene and petroleum ether. Mp: 345-348°. Subl without decomp.

SYNS: ACIDE ISOPHTALIQUE (FRENCH) ◇ ENZENE-1,3-DICARBOXYLIC ACID ◇ m-BENZENEDICARBOXYLIC ACID ◇ IPA ◇ KYSELINA ISOFTALOVA (CZECH)

TOXICITY DATA with REFERENCE
eye-rbt 100 mg/24H MLD 28ZPAK -,51,72
orl-rat LD50:10400 mg/kg 28ZPAK -,51,72
ipr-mus LD50:4200 mg/kg COREAF 246,851,58

CONSENSUS REPORTS: Reported in EPA TSCA Inventory.

SAFETY PROFILE: Mildly toxic by ingestion and intraperitoneal routes. An eye irritant. When heated to decomposition it emits acrid smoke and fumes.

IMK000 CAS:1087-21-4 ***HR: 3***
ISOPHTHALIC ACID, DIALLYL ESTER
mf: $C_{14}H_{14}O_4$ mw: 246.28

TOXICITY DATA with REFERENCE
ipr-mus LDLo:64 mg/kg CBCCT* 2,302,50

CONSENSUS REPORTS: Reported in EPA TSCA Inventory.

SAFETY PROFILE: Poison by intraperitoneal route. When heated to decomposition it emits acrid smoke and fumes. See also ALLYL COMPOUNDS and ESTERS.

IML000 CAS:1459-93-4 ***HR: 2***
ISOPHTHALIC ACID, DIMETHYL ESTER
mf: $C_{10}H_{10}O_4$ mw: 194.20

SYNS: 1,3-BENZENEDICARBOXYLIC ACID, DIMETHYL ESTER ◇ DIMETHYLESTER KYSELINY TEREFTALOVE (CZECH) ◇ DIMETHYL ISOPHTHALATE

TOXICITY DATA with REFERENCE
eye-rbt 500 mg/24H MLD 28ZPAK -,47,72
ipr-mus LDLo:971 mg/kg JPMSAE 56,1446,67

CONSENSUS REPORTS: Reported in EPA TSCA Inventory.

SAFETY PROFILE: Moderately toxic by intraperitoneal route. An eye irritant. When heated to decomposition it emits acrid smoke and fumes. See also ESTERS.

IMM000 CAS:20986-33-8 ***HR: 3***
3,3'-(ISOPHTHALOYLBIS(IMINO-p-PHENYLENECARBONYLIMINO))BIS(1-ETHYLPYRIDINIUM, DI-p-TOLUENESULFONATE
mf: $C_{36}H_{34}N_6O_4•2C_7H_7O_3S$ mw: 957.16

TOXICITY DATA with REFERENCE
dnd-mus:lym 900 nmol/L JMCMAR 22,134,79
ipr-mus LD10:12 mg/kg JMCMAR 22,134,79

SAFETY PROFILE: Poison by intraperitoneal route. Mutation data reported. When heated to decomposition it emits very toxic fumes of NO_x and SO_x.

IMO000 CAS:99-63-8 ***HR: 2***
ISOPHTHALOYL CHLORIDE
mf: $C_8H_4Cl_2O_2$ mw: 203.02

$(CO•Cl)C_6H_4(CO•Cl)$

PROP: White, crystalline solid. Sol in benzene and carbon tetrachloride; Bp: 276°, fp: 43.3°, d: 1.387 @ 46.9°, flash p: 356°F (COC), vap d: 6.9. Mp: 41°; decomp in water and alc.

SYNS: 1,3-BENZENEDICARBONYL CHLORIDE ◇ ISOPHTHALIC ACID CHLORIDE ◇ ISOPHTHALIC ACID DICHLORIDE ◇ ISOPHTHALOYL DICHLORIDE ◇ ISOPHTHALYL DICHLORIDE ◇ m-PHTHALIC DICHLORIDE ◇ m-PHTHALOYL CHLORIDE

TOXICITY DATA with REFERENCE
skn-rbt 200 mg open MOD UCDS** 11/3/71
eye-rbt 40 mg MLD UCDS** 11/3/71
eye-rbt 3 mg MOD 34ZIAG -,339,69
orl-rat LD50:2200 mg/kg 34ZIAG -,475,69
orl-mus LD50:2221 mg/kg GISAAA 47(7),75,82
orl-rbt LD50:1175 mg/kg GISAAA 47(7),75,82
skn-rbt LD50:1410 mg/kg TXAPA9 28,313,74

CONSENSUS REPORTS: Reported in EPA TSCA Inventory.

SAFETY PROFILE: Moderately toxic by ingestion and skin contact. A skin and eye irritant. Combustible when exposed to heat or flame. Reacts violently with methanol. To fight fire, use dry chemical, CO_2, spray, mist. When heated to decomposition it emits toxic fumes of Cl^-.

IMQ000 CAS:76-00-6 ***HR: 3***
ISOPRAL
mf: $C_3H_5Cl_3O$ mw: 163.43

PROP: Crystals, camphor-like odor, pungent taste, water-sol. Mp: 50°, bp: 162°.

SYNS: TRICHLOROISOPROPANOL ◇ 1,1,1-TRICHLOROISOPROPYL ALCOHOL ◇ 1,1,1-TRICHLORO-2-PROPANOL

TOXICITY DATA with REFERENCE
orl-rat LD50:1 g/kg JPETAB 63,183,38
scu-mus LDLo:500 mg/kg HDTU**-,-,33

SAFETY PROFILE: Moderately toxic by subscutaneous route. See also ALCOHOLS and CHLORINATED HYDROCARBONS, ALIPHATIC.

IMR000 CAS:51-30-9 ***HR: 3***
ISOPRENALINE HYDROCHLORIDE
mf: $C_{11}H_{17}NO_3ClH$ mw: 247.75

SYNS: 3,4-DIHYDROXY-α-((ISOPROPYLAMINO)METHYL)BENZYL ALCOHOL HYDROCHLORIDE ◇ EUSPIRAN ◇ 4-(1-HYDROXY-2-((1-METHYLETHYL)AMINO)ETHYL)-1,2-BENZENEDIOLHYDROCHLORIDE ◇ ISADRINE ◇ ISADRINE-HYDROCHLORIDE ◇ ISOPRENALINE CHLORIDE ◇ α-(ISOPROPYLAMINOMETHYL)-3,4-DIHYDROXYBENZYL ALCOHOL HYDROCHLORIDE ◇ ISOPROPYLARTERENOL HYDROCHLORIDE ◇ ISOPROPYLNOREPINEPHRINE-HYDROCHLORIDE ◇ ISOPROTERENOL HYDROCHLORIDE ◇ ISOPROTERNOL MONOHYDROCHLORIDE ◇ ISUPREL ◇ ISUPREL HYDROCHLORIDE ◇ IZADRIN ◇ NCI-C55630 ◇ NORISODRINE HYDROCHLORIDE ◇ SAVENTRINE ◇ VAPO-ISO

TOXICITY DATA with REFERENCE
oms-rat:oth 5 μmol/L INOPAO 13,210,74
scu-rat TDLo:10 mg/kg (6-15D preg):TER NTIS** PB83-211466

scu-rat TDLo:10 mg/kg (6-15D preg):REP NTIS** PB83-211466
orl-rat LD50:2221 mg/kg TXAPA9 18,185,71
ipr-rat LD50:203 mg/kg NIIRDN 6,74,82
scu-rat LD20:100 μg/kg FAATDF 1,443,81
ivn-rat LD50:26900 μg/kg YACHDS 7,627,79
orl-mus LD50:1645 mg/kg NIIRDN 6,74,82
ipr-mus LD50:450 mg/kg JPETAB 90,110,47
scu-mus LD50:60 mg/kg YKYUA6 24,431,73
ivn-mus LD50:96300 μg/kg TXAPA9 23,537,72

SAFETY PROFILE: Poison by intraperitoneal, subcutaneous, and intravenous routes. Moderately toxic by ingestion. Experimental teratogenic and reproductive effects. Mutation data reported. When heated to decomposition it emits very toxic fumes of HCl and NO_x.

IMS000 CAS:78-79-5 ***HR: 3***
ISOPRENE
DOT: UN 1218
mf: C_5H_8 mw: 68.13

$H_2C{=}C(CH_3)CH{=}CH_2$

PROP: Colorless, volatile liquid. Mp: −146.7°, bp: 34°, flash p: −65°F, d: 0.6806 @ 20°/4°, autoign temp: 428°F, vap press: 400 mm @ 15.4°, vap d: 2.35; fp: −145.95°. Insol in water; misc in alc and ether.

SYNS: ISOPRENE, INHIBITED (DOT) ◇ β-METHYLBIVINYL ◇ 2-METHYLBUTADIENE ◇ 2-METHYL-1,3-BUTADIENE (DOT)

TOXICITY DATA with REFERENCE
ihl-rat LC50:180 g/m³/4H RPTOAN 31,162,68
ihl-mus LC50:139 g/m³/2H GTPZAB 9(1),36,65

CONSENSUS REPORTS: Reported in EPA TSCA Inventory.

DOT Classification: Flammable Liquid; Label: Flammable Liquid.

SAFETY PROFILE: Mildly toxic by inhalation. Irritating to skin, eyes, and mucous membranes. A concentration of 2% in air is not narcotic to mice but produces bronchial irritation. Highly dangerous fire hazard when exposed to heat, flame, or oxidizers. Reacts with air to form dangerously unstable peroxides which can explode after concentration by evaporation. Ignites on contact with oxygen + ozone. Reacts with ozone to form explosive peroxides. Explosive reaction with vinylamine. Violent reaction with chlorosulfonic acid, HNO_3, oleum, H_2SO_4. Can react vigorously with reducing materials. To fight fire, use CO_2, dry chemical. When heated to decomposition it emits acrid smoke and fumes. See also PEROXIDES.

IMS300 CAS:64506-49-6 ***HR: D***
ISOPRENYL CHALCONE
mf: $C_{27}H_{30}O_6$ mw: 450.57

PROP: Light yellow needles from ethanol. Mp: 143-144°.

SYNS: 2'-CARBOXYMETHOXY-4,4'-BIS(3-METHYL-2-BUTENYLOXY)CHALCONE ◇ SOFALCONE ◇ SOLON ◇ SU-88

TOXICITY DATA with REFERENCE
orl-rat TDLo:1375 mg/kg (7-17D preg):REP OYYAA2 19,525,80
orl-rat TDLo:5500 mg/kg (7-17D preg):TER OYYAA2 19,525,80

SAFETY PROFILE: Experimental teratogenic and reproductive effects. When heated to decomposition it emits acrid smoke and fumes.

IMT000 CAS:64-39-1 ***HR: 3***
ISOPROMEDOL
mf: $C_{17}H_{25}NO_2$ mw: 275.43

SYNS: 4-PHENYL-1,2,5-TRIMETHYL-4-PIPERIDINOLPROPIONATE ◇ PROMEDOL ◇ TRIMEPERDINE ◇ 1,2,5-TRIMETHYL-4-PHENYL-4-PIPERIDINOL, PROPIONATE (ESTER) ◇ 1,2,5-TRIMETHYL-4-PHENYL-4-PROPIONOXYPIPERIDINE

TOXICITY DATA with REFERENCE
scu-rat LD50:38 mg/kg PCJOAU 14,850,80
ivn-rat LD50:22 mg/kg PCJOAU 8,189,74
ipr-mus LD50:137 mg/kg PCJOAU 10,1193,76
scu-mus LD50:200 mg/kg PCJOAU 10,1193,76
ivn-mus LD50:38 mg/kg RPTOAN 31,318,68

SAFETY PROFILE: Poison by intravenous, intraperitoneal, and subcutaneous routes. When heated to decomposition it emits toxic fumes of NO_x. See also ESTERS.

IMU000 CAS:75-33-2 ***HR: 3***
ISOPROPANETHIOL
DOT: UN 2402
mf: C_3H_8S mw: 76.17

$(CH_3)_2CHSH$

PROP: Liquid, extremely powerful unpleasant odor. Mp: −130.7°, bp: 58-60°, d: 0.814 @ 60°/60°F, boiling range: 51-55°, flash p: −30°F. Sltly sol in water; misc in alc and ether.

SYNS: ISOPROPYL MERCAPTAN (DOT) ◇ ISOPROPYLTHIOL ◇ 2-MERCAPTOPROPANE ◇ 1-METHYLETHANETHIOL ◇ 2-PROPANETHIOL ◇ 2-PROPYL MERCAPTAN

CONSENSUS REPORTS: Reported in EPA TSCA Inventory.

DOT Classification: Flammable Liquid; Label: Flammable Liquid.

SAFETY PROFILE: Probably moderately toxic by inhalation. A very dangerous fire hazard when exposed to heat, flame, or oxidizers. When heated to decomposition it emits highly toxic fumes of SO_x. See also MERCAPTANS.

IMV000 **HR: 2**
ISOPROPANOLAMINES

PROP: Clear liquid. Mp: 29.5°, bp: 159°, flash p: 160°F (OC), d: 0.962, vap d: 2.58. Very sol in water; sol in alc.; insol in ether. 12% mono-, 44% di-, 44% triisopropanolamine (JIHTAB 23,259,41).

SYN: ISOPROPANOLAMINES, mixed

TOXICITY DATA with REFERENCE
eye-rbt 2 mg SEV AJOPAA 29,1363,46
orl-rat LD50:5240 mg/kg UCDS** 4/25/58
skn-rbt LD50:8900 mg/kg UCDS** 4/25/58
orl-gpg LD50:1520 mg/kg JIHTAB 23,259,41

SAFETY PROFILE: Moderately toxic by ingestion. Mildly toxic by skin contact. A severe eye irritant. Combustible when exposed to heat or flame; can react with oxidizing materials. To fight fire, use foam, CO_2, dry chemical. When heated to decomposition it emits toxic fumes of NO_x. See also AMINES.

IMW000 CAS:513-42-8 **HR: 2**
ISOPROPENYL CARBINOL
DOT: UN 2614
mf: C_4H_8O mw: 72.12

PROP: Liquid. D: 0.852 @ 20°/4°C, bp: 114.5°. Sol in water @ 25°.

SYNS: METHALLYL ALCOHOL (DOT) ◇ 2-METHYL-2-PROPEN-1-OL

TOXICITY DATA with REFERENCE
skn-rbt 500 mg MOD SCCUR* -,6,61
orl-mus LDLo:500 mg/kg SCCUR* -,6,61
ihl-mus LCLo:2924 ppm/2H SCCUR* -,6,61
skn-rbt LDLo:2000 mg/kg SCCUR* -,6,61

CONSENSUS REPORTS: Reported in EPA TSCA Inventory.

DOT Classification: Flammable or Combustible Liquid; Label: Flammable Liquid.

SAFETY PROFILE: Moderately toxic by ingestion and skin contact. Mildly toxic by inhalation. A skin irritant. Combustible when exposed to heat or flame. When heated to decomposition it emits acrid smoke and fumes. See also ALCOHOLS and ALLYL COMPOUNDS.

IMX000 CAS:67262-74-2 **HR: 3**
2'-ISOPROPENYL-2-(2-METHOXYETHYLAMINO)PROPRIONANILIDE OXALATE
mf: $C_{15}H_{22}N_2O_2 \cdot C_2H_2O_4$ mw: 352.43

TOXICITY DATA with REFERENCE
ipr-mus LD50:350 mg/kg JPMSAE 67,595,78
ivn-mus LD50:55 mg/kg JPMSAE 67,595,78

SAFETY PROFILE: Poison by intravenous and intraperitoneal routes. When heated to decomposition it emits toxic fumes of NO_x. See also OXALATES.

INA200 CAS:29026-74-2 **HR: 2**
o-ISOPROPOXYANILINE
mf: $C_9H_{13}NO$ mw: 151.23

SYNS: o-IPA ◇ o-IZOPROPOKSYANILINA (POLISH) ◇ 2-(1-METHYLETHOXY)-BENZENAMINE

TOXICITY DATA with REFERENCE
orl-rat LD50:840 mg/kg MEPAAX 28,157,77
skn-rat LDLo:2200 mg/kg MEPAAX 28,157,77
ipr-rat LDLo:200 mg/kg MEPAAX 28,157,77

SAFETY PROFILE: Moderately toxic by skin contact, ingestion, and intraperitoneal routes. When heated to decomposition it emits toxic fumes of NO_x. See also ANILINE DYES.

INA400 CAS:83053-59-2 **HR: 2**
11-ISOPROPOXY-15,16-DIHYDRO-17-CYCLOPENTA(a)PHENANTHREN-17-ONE
mf: $C_{20}H_{18}O_2$ mw: 290.38

TOXICITY DATA with REFERENCE
mma-sat 20 μg/plate CRNGDP 3,677,82
skn-mus TDLo:1600 μg/kg:ETA CRNGDP 3,677,82

SAFETY PROFILE: Questionable carcinogen with experimental tumorigenic data. Mutation data reported. When heated to decomposition it emits acrid smoke and irritating fumes.

INA500 CAS:109-59-1 **HR: 2**
2-ISOPROPOXYETHANOL
mf: $C_5H_{12}O_2$ mw: 104.17

SYNS: DOWANOL EIPAT ◇ ETHYLENE GLYCOL ISOPROPYL ETHER ◇ ETHYLENE GLYCOL, MONOISOPROPYL ETHER ◇ β-HYDROXYETHYL ISOPROPYL ETHER ◇ ISOPROPYL CELLOSOLVE ◇ ISOPROPYL GLYCOL ◇ MONOISOPROPYL ETHER of ETHYLENE GLYCOL

TOXICITY DATA with REFERENCE
orl-rat LD50:5660 mg/kg AIHAAP 30,470,69
ihl-rat LCLo:4000 ppm/4H AIHAAP 30,470,69

ipr-rat LD50:800 mg/kg 85GMAT -,67,82
orl-mus LD50:4900 mg/kg 85GMAT -,67,82
ihl-mus LC50:1930 ppm/7H JIHTAB 25,157,43
ipr-mus LD50:1860 mg/kg 85GMAT -,67,82
skn-rbt LD50:1600 mg/kg AIHAAP 30,470,69

CONSENSUS REPORTS: Glycol ether compounds are on the Community Right-To-Know List. Reported in EPA TSCA Inventory.

OSHA PEL: TWA 25 ppm
ACGIH TLV: TWA 25 ppm

SAFETY PROFILE: Moderately toxic by skin contact and intraperitoneal routes. Mildly toxic by inhalation and ingestion. When heated to decomposition it emits acrid smoke and fumes. See also GLYCOL ETHERS.

IND000 CAS:67952-46-9 ***HR: 2***
(2-ISOPROPOXYETHYL)CARBAMATE
mf: $C_6H_{13}NO_3$ mw: 147.20

TOXICITY DATA with REFERENCE
ipr-mus LDLo:4000 mg/kg UCPHAQ 1,93,38

CONSENSUS REPORTS: Reported in EPA TSCA Inventory.

SAFETY PROFILE: Moderately toxic by intraperitoneal route. When heated to decomposition it emits toxic fumes of NO_x. See also CARBAMATES.

INE000 CAS:67465-43-4 ***HR: 3***
N-(2-ISOPROPOXY-3-HYDROXYMERCURIPROPYL)BARBITAL
mf: $C_{14}H_{24}HgN_2O_5$ mw: 500.99

SYN: HYDROXY(3-(5,5-DIETHYL-2,4,6-TRIOXO-(1H,3H,5H) PYRIMIDINO)-2-ISOPROPOXYPROPYL)MERCURY

TOXICITY DATA with REFERENCE
ivn-rat TDLo:23100 μg/kg JAPMA8 37,333,48

CONSENSUS REPORTS: Mercury and its compounds are on the Community Right-To-Know List.

OSHA PEL: (Transitional: CL 1 mg/10m^3) CL 0.1 mg(Hg)/m^3 (skin)
ACGIH TLV: TWA 0.1 mg(Hg)/m^3 (skin)
NIOSH REL: (Inorganic Mercury) TWA 0.05 mg(Hg)/m^3

SAFETY PROFILE: Poison by intravenous route. When heated to decomposition it emits very toxic fumes of Hg and NO_x. See also MERCURY COMPOUNDS.

INE100 CAS:108-21-4 ***HR: 3***
ISOPROPYL ACETATE
DOT: UN 1220
mf: $C_5H_{10}O_2$ mw: 102.15

PROP: Colorless, aromatic liquid. Mp: −73°, bp: 88.4°, lel: 1.8%, uel: 7.8%, fp: −69.3°, flash p: 40°F, d: 0.874 @ 20°/20°, autoign temp: 860°F, vap press: 40 mm @ 17.0°, d: 3.52. Sltly sol in water; misc in alc, ether, fixed oils.

SYNS: ACETATE d'ISOPROPYLE (FRENCH) ◇ ACETIC ACID ISOPROPYL ESTER ◇ ACETIC ACID-1-METHYLETHYL ESTER (9CI) ◇ 2-ACETOXYPROPANE ◇ FEMA No. 2926 ◇ ISOPROPILE (ACETATO di) (ITALIAN) ◇ ISOPROPYLACETAAT (DUTCH) ◇ ISOPROPYLACETAT (GERMAN) ◇ ISOPROPYL (ACETATE d') (FRENCH) ◇ 2-PROPYL ACETATE

TOXICITY DATA with REFERENCE
eye-hmn 200 ppm/15M JIHTAB 28,262,46
eye-rbt 500 mg AMIHBC 10,61,54
ihl-hmn TCLo:200 ppm:IRR AMIHAB 21,28,60
unk-hmn TCLo:200 ppm:EYE JIHTAB 28,262,46
orl-rat LD50:3000 mg/kg 14CYAT 2,1879,63
ihl-rat LCLo:32000 ppm/4H AMIHBC 10,61,54
orl-rbt LD50:6946 mg/kg IMSUAI 41,31,72

CONSENSUS REPORTS: Reported in EPA TSCA Inventory.

OSHA PEL: (Transitional: TWA 250 ppm) TWA 250 ppm; STEL 310 ppm
ACGIH TLV: TWA 250 ppm; STEL 310 ppm
DFG MAK: 200 ppm (840 mg/m^3)
DOT Classification: Flammable Liquid; Label: Flammable Liquid.

SAFETY PROFILE: Moderately toxic by ingestion. Mildly toxic by inhalation. See also ESTERS. Human systemic irritant effects by inhalation and systemic eye effects by an unspecified route. Narcotic in high concentration. Chronic exposure can cause liver damage. Highly flammable liquid. Dangerous fire hazard when exposed to heat, flame, or oxidizers. Moderately explosive when exposed to heat or flame. Dangerous; keep away from heat and open flame; can react vigorously with oxidizing materials. To fight fire, use foam, CO_2, dry chemical. See also ESTERS.

ING300 CAS:70715-91-2 ***HR: D***
N-ISOPROPYL-N-(ACETOXYMETHYL)NITROSAMINE
mf: $C_6H_{12}N_2O_3$ mw: 160.20

SYN: ((1-METHYLETHYL)NITROSOAMINO)-METHANOLACETATE (ESTER) (9CI)

TOXICITY DATA with REFERENCE
mmo-sat 5 μmol/plate GANNA2 70,663,79
mmo-esc 25 μmol/plate GANNA2 70,663,79

SAFETY PROFILE: Mutation data reported. Many nitrosamines are carcinogens. When heated to decomposition it emits toxic fumes of NO_x. See also NITROSAMINES.

ING400 CAS:4212-94-6 ***HR: 2***
ISOPROPYL-N-ACETOXY-N-PHENYLCARBAMATE
mf: $C_{12}H_{15}NO_4$ mw: 237.28

SYNS: o-ACETYL-N-CARBOXY-N-PHENYL-HYDROXYLAMINE ISOPROPYL ESTER ◇ (ACETYLOXY)PHENYL-CARBAMIC ACID ◇ ACYLATE ◇ ACYLATE-1 ◇ N-CARBOISOPROPOXY-o-ACETYL-N-PHENYL CARBAMATE

TOXICITY DATA with REFERENCE
orl-rat LD50:3400 mg/kg 85GMAT -,32,82
ihl-rat LC50:1170 mg/m^3/6H 85GMAT -,32,82
orl-mus LD50:2075 mg/kg 85GMAT -,32,82

SAFETY PROFILE: Moderately toxic by inhalation and ingestion routes. When heated to decomposition it emits toxic fumes of NO_x. See also CARBAMATES and ESTERS.

ING509 ***HR: 3***
2-ISOPROPYL ACRYLALDEHYDE OXIME
mf: $C_6H_{11}NO$ mw: 113.16

$H_2C{=}C[CH(CH_3)_2]CH{=}NOH$

SAFETY PROFILE: Reacts in air to form heat-sensitive explosive peroxides. When heated to decomposition it emits toxic fumes of NO_x. See also ALDEHYDES.

INH000 CAS:2210-25-5 ***HR: 3***
N-ISOPROPYLACRYLAMIDE
mf: $C_6H_{11}NO$ mw: 113.18

SYNS: ISOPROPYL ACRYLAMIDE ◇ NIPAM

TOXICITY DATA with REFERENCE
orl-mus TDLo:238 g/kg (male 5W pre):REP ARTODN 59,201,86
orl-rat LD50:350 mg/kg 14CYAT 2,1834,63
orl-mus LD50:419 mg/kg ARTODN 47,179,81
orl-rat LD50:350 mg/kg 14CYAT 2,1834,63
ipr-mus LDLo:500 mg/kg CBCCT* 6,51,54

CONSENSUS REPORTS: Reported in EPA TSCA Inventory.

SAFETY PROFILE: Poison by ingestion. Moderately toxic by intraperitoneal route. Experimental reproductive effects. When heated to decomposition it emits toxic fumes of NO_x. See also AMIDES.

INJ000 CAS:67-63-0 ***HR: 3***
ISOPROPYL ALCOHOL
DOT: UN 1219
mf: C_3H_8O mw: 60.11

$(CH_3)_2CHOH$

PROP: Clear, colorless liquid; slt odor, sltly bitter taste. Mp: −88.5 to −89.5°, bp: 82.5°, lel: 2.5%, uel: 12%, flash p: 53°F (CC), d: 0.7854 @ 20°/4°, refr index: 1.377 @ 20°, vap d: 2.07, ULC: 70. fp: −89.5°; autoign temp: 852°F. Misc with water, alc, ether, chloroform; insol in salt solns.

SYNS: ALCOOL ISOPROPILICO (ITALIAN) ◇ ALCOOL ISOPROPYLIQUE (FRENCH) ◇ DIMETHYLCARBINOL ◇ ISOHOL ◇ ISOPROPANOL (DOT) ◇ ISO-PROPYLALKOHOL (GERMAN) ◇ LUTOSOL ◇ PETROHOL ◇ PROPAN-2-OL ◇ 2-PROPANOL ◇ i-PROPANOL (GERMAN) ◇ sec-PROPYL ALCOHOL (DOT) ◇ i-PROPYLALKOHOL (GERMAN) ◇ SPECTRAR

TOXICITY DATA with REFERENCE
skn-rbt 500 mg MLD NTIS** AD-A106-944
eye-rbt 16 mg AJOPAA 29,1363,46
eye-rbt 10 mg MOD TXAPA9 55,501,80
cyt-smc 200 mmol/tube HEREAY 33,457,47
cyt-rat-ihl 1030 μg/m^3/16W-I GTPZAB 25(7),33,81
orl-rat TDLo:6480 mg/kg (male 26W pre):REP GISAAA 43(1),8,78
ihl-rat TCLo:10000 ppm/7H (female 1-19D post):TER FCTOD7 26,247,88
orl-man TDLo:14432 mg/kg:CNS,CVS,PUL NEJMAG 277,699,67
orl-hmn TDLo:223 mg/kg:CNS,CVS JLCMAK 12,326,27
orl-man LDLo:5272 mg/kg AJCPAI 38,144,62
orl-hmn LDLo:3570 mg/kg:CNS,PUL,GIT 34ZIAG -,339,69
unr-man LDLo:2770 mg/kg 85DCAI 2,73,70
orl-rat LD50:5045 mg/kg GISAAA 43(1),8,78
ihl-rat LCLo:16000 ppm/4H JIDHAN 31,343,49
ipr-rat LD50:2735 mg/kg EVHPAZ 61,321,85
ivn-rat LD50:1099 mg/kg EVHPAZ 61,321,85
orl-mus LD50:3600 mg/kg GISAAA 43(1),8,78
ihl-mus LCLo:12800 ppm/3H IAEC** 17JUN74
ipr-mus LD50:4477 mg/kg EVHPAZ 61,321,85
scu-mus LDLo:6000 mg/kg HBTXAC 1,172,56
ivn-mus LD50:1509 mg/kg EVHPAZ 61,321,85
orl-dog LD50:4797 mg/kg JLCMAK 29,561,44
ivn-dog LDLo:5120 mg/kg JLCMAK 29,561,44
ivn-cat LDLo:1963 mg/kg HBTXAC 1,172,55
orl-rbt LD50:6410 mg/kg FAONAU 48A,114,70
skn-rbt LD50:12800 mg/kg NPIRI* 1,100,74

CONSENSUS REPORTS: IARC Cancer Review: Group 3 IMEMDT 7,229,87. The isopropyl alcohol strong acid manufacturing process is on the Community Right-To-Know List. EPA Genetic Toxicology Program. Reported in EPA TSCA Inventory.

OSHA PEL: (Transitional: TWA 400 ppm) TWA 400 ppm; STEL 500 ppm
ACGIH TLV: TWA 400 ppm; STEL 500 ppm
DFG MAK: 400 ppm (980 mg/m^3)
NIOSH REL: (Isopropyl Alcohol) TWA 400 ppm; CL 800 ppm/15M
DOT Classification: Flammable Liquid; Label: Flammable Liquid.

SAFETY PROFILE: Moderately toxic to humans by an unspecified route. Moderately toxic experimentally by intravenous and intraperitoneal routes. Mildly toxic by skin contact. Human systemic effects by ingestion or inhalation: flushing, pulse rate decrease, blood pressure lowering, anesthesia, narcosis, headache, dizziness, mental depression, hallucinations, distorted perceptions, dyspnea, respiratory depression, nausea or vomiting, coma. Experimental teratogenic and reproductive effects. Mutation data reported. An eye and skin irritant. Questionable carcinogen.

The single lethal dose for a human adult is about 250 mL although as little as 100 mL can be fatal. It can cause corneal burns and eye damage. Acts as a local respiratory irritant and in high concentration as a narcotic. It has good warning properties because it causes a mild irritation of the eyes, nose, and throat at a concentration level of 400 ppm. It may induce a mild narcosis, the effects of which are usually transient, and it is somewhat less toxic than the normal isomer, but twice as volatile.

There is some evidence that humans can acquire a slight tolerance to this material. It is absorbed by the skin, but single or repeated applications on the skin of rats, rabbits, dogs, or human beings induced no untoward effects. It acts very much like ethanol in regard to absorption, metabolism, and elimination but with a stronger narcotic action. Chronic injuries have been detected in animals. Workers producing isopropanol show an excess of sinus and laryngeal cancers. This may all or in part be due to the by-product, isopropyl oil. Humans have ingested up to 20 mL diluted with water and noticed only a sensation of heat and slight lowering of the blood pressure. There are, however, reports of serious illness from as little as 10 mL taken internally. A common air contaminant.

Flammable liquid. A very dangerous fire hazard when exposed to heat, flame, or oxidizers. Moderately explosive when exposed to heat or flame. Reacts with air to form dangerous peroxides. The presence of 2-butanone increases the reaction rate for peroxide formation. Hydrogen peroxide sharply reduces the autoignition temperature. Violent explosive reaction when heated with aluminum isopropoxide + crotonaldehyde + heat. Forms explosive mixtures with trinitromethane; hydrogen peroxide (similar in power and sensitivity to glyceryl nitrate). Reacts with barium perchlorate to form the highly explosive propyl perchlorate. Ignites on contact with dioxgenyl tetrafluoroborate; chromium trioxide; potassium tert-butoxide (after a delay). Reacts with oxygen to form dangerously unstable peroxides. Vigorous reaction with sodium dichromate + sulfuric acid; aluminum (after a delay period). Reacts violently with H_2 + Pd; nitroform; oleum; $COCl_2$; Al triisopropoxide; oxidants. Can react vigorously with oxidizing materials. To fight fire, use CO_2, dry chemical, alcohol foam. When heated to decomposition it emits acrid smoke and fumes. See also ALCOHOLS.

INK000 CAS:75-31-0 ***HR: 3***
ISOPROPYLAMINE
DOT: UN 1221
mf: C_3H_9N mw: 59.13

$$(CH_3)_2CHNH_2$$

PROP: Colorless liquid; amino odor. Mp: −101.2°, flash p: −35°F (OC), d: 0.694 @ 15°/4°, autoign temp: 756°F, d: 2.03, bp: 33-34°, lel: 2.3%, uel: 10.4%. Misc with water, alc, and ether.

SYNS: 2-AMINO-PROPAAN (DUTCH) ◇ 2-AMINOPROPAN (GERMAN) ◇ 2-AMINOPROPANE ◇ 2-AMINO-PROPANO (ITALIAN) ◇ ISOPROPILAMINA (ITALIAN) ◇ 1-METHYLETHYLAMINE ◇ MONOISOPROPYLAMINE ◇ 2-PROPANAMINE ◇ sec-PROPYLAMINE ◇ 2-PROPYLAMINE

TOXICITY DATA with REFERENCE
skn-rbt 500 mg/24H SEV 28ZPAK -,62,72
skn-rbt 345 mg open MOD UCDS** 11/15/71
skn-rbt 10 mg/24H open SEV AMIHBC 4,119,51
eye-rbt 50 μg open SEV AMIHBC 4,119,51
eye-rbt 50 μg/24H SEV 28ZPAK -,62,72
orl-rat LD50:820 mg/kg UCDS** 11/15/71
ihl-rat LC50:4000 ppm/4H IAEC** 17JUN74
orl-mus LD50:2200 mg/kg GISAAA 45(3),79,80
ihl-mus LCLo:7000 ppm/40M SCCUR* -,7,61
orl-rbt LD50:3200 mg/kg GISAAA 45(3),79,80
skn-rbt LD50:380 mg/kg IAEC** 17JUN74
skn-rbt LD50:550 mg/kg IAEC** 17JUN74
orl-gpg LD50:2700 mg/kg GISAAA 45(3),79,80
ihl-mam LC50:1800 mg/m^3 TPKVAL 14,80,75

CONSENSUS REPORTS: Reported in EPA TSCA Inventory.

OSHA PEL: (Transitional: TWA 5 ppm) TWA 5 ppm; STEL 10 ppm
ACGIH TLV: TWA 5 ppm; STEL 10 ppm
DFG MAK: 5 ppm (12 mg/m^3)
DOT Classification: Flammable Liquid; Label: Flammable Liquid.

SAFETY PROFILE: Poison by skin contact. Moderately toxic by ingestion. Mildly toxic by inhalation. A severe skin and eye irritant. Occasionally contact causes sensitization. Narcotic in high concentration. Very dangerous fire hazard and moderate explosion hazard when exposed to sparks, heat, flame or oxidizers. Can react vigorously with oxidizing materials. Reacts with perchloryl fluoride to form an explosive liquid. Incompatible with 1-chloro-1,3-epoxypropane. To fight fire, use alcohol foam, foam, CO_2, dry chemical. When heated to decomposition it emits toxic fumes of NO_x. See also AMINES.

INL000 CAS:15572-56-2 ***HR: 2***
ISOPROPYLAMINE HYDROCHLORIDE
mf: $C_3H_9N{\cdot}ClH$ mw: 95.59

SYN: MONOISOPROPYLAMINE HYDROCHLORIDE

TOXICITY DATA with REFERENCE
ipr-mus LDLo:800 mg/kg SCCUR* -,7,61

CONSENSUS REPORTS: Reported in EPA TSCA Inventory.

SAFETY PROFILE: Moderately toxic by intraperitoneal route. When heated to decomposition it emits very toxic fumes of NO_x and HCl. See also AMINES.

INM000 CAS:3615-24-5 ***HR: 3***
4-(ISOPROPYLAMINO)ANTIPYRINE
mf: $C_{14}H_{19}N_3O$ mw: 245.36

SYNS: 1,2-DIHYDRO-1,5-DIMETHYL-4-((1-METHYLETHYL)AMINO)-2-PHENYL-3H-PYRAZOL-3-ONE ◇ ISOPIRINA ◇ ISOPROPYLAMINOANTIPYRINE ◇ 4-ISOPROPYLAMINO-2,3-DIMETHYL-1-PHENYL-3-PYRAZOLIN-5-ONE ◇ ISOPROPYLAMINOPHENAZON ◇ ISOPROPYLAMINOPHENAZONE ◇ 4-ISOPROPYLAMINO-1-PHENYL-2,3-DIMETHYL-3-PYRAZOLIN-5-ONE ◇ ISOPYRIN ◇ ISOPYRINE ◇ 4-MONOISOPROPYLAMINO-1-PHENYL-2,3-DIMETHYL-5-PYRAZOLONE ◇ 1-PHENYL-2,3-DIMETHYL-4-(ISOPROPYLAMINO)-2-PYRAZOLIN-5-ONE ◇ 1-PHENYL-2,3-DIMETHYL-4-ISOPROPYLAMINOPYRAZOLONE ◇ TOMANOL-WIRKSTOFF

TOXICITY DATA with REFERENCE
ipr-rat LD50:715 mg/kg ARZNAD 20,1024,70
ivn-rat LD50:450 mg/kg ARZNAD 10,665,60
ims-rat LD50:820 mg/kg ARZNAD 10,665,60
orl-mus LD50:1070 mg/kg ARZNAD 20,1024,70
ipr-mus LD50:690 mg/kg ARZNAD 10,665,60
ivn-mus LD50:370 mg/kg AEPPAE 233,365,58
ipr-ham LD50:567 mg/kg ARZNAD 20,1024,70

SAFETY PROFILE: Poison by intravenous route. Moderately toxic by ingestion, intraperitoneal, and intramuscular routes. An analgesic, antipyretic, and anti-inflammatory agent. When heated to decomposition it emits toxic fumes of NO_x.

INN400 CAS:109-56-8 ***HR: 2***
N-ISOPROPYLAMINOETHANOL
mf: $C_5H_{13}NO$ mw: 103.19

SYNS: ETHANOLISOPROPYLAMINE ◇ ETHANOL, 2-(ISOPROPYLAMINO)- ◇ ETHANOL, 2-((1-METHYLETHYL)AMINO)-(9CI) ◇ (N-HYDROXYETHYL)ISOPROPYLAMINE ◇ ISOPROPYLAMINOETHANOL ◇ 2-ISOPROPYLAMINOETHANOL ◇ N-ISOPROPYLETHANOLAMINE ◇ MONOISOPROPYLAMINOETHANOL

TOXICITY DATA with REFERENCE
orl-rat TDLo:50 mg/kg (female 5D post):REP JRPFA4 35,625,73
orl-mus LD50:1250 mg/kg CPBTAL 31,4116,83

CONSENSUS REPORTS: Reported in EPA TSCA Inventory.

SAFETY PROFILE: Moderately toxic by ingestion. Experimental reproductive effects. When heated to decomposition it emits toxic fumes of NO_x.

INN500 CAS:54472-62-7 ***HR: 2***
ISOPROPYLAMINO ETHANOL HYDROCHLORIDE
mf: $C_5H_{13}NO{\cdot}ClH$ mw: 139.65

SYNS: 2-(ISOPROPYLAMINO)ETHANOL HYDROCHLORIDE ◇ PAE

TOXICITY DATA with REFERENCE
orl-mus TDLo:200 mg/kg (female 3D post):REP TIYADG 9,553,81
orl-mus LD50:434 mg/kg TIYADG 9,553,81
ipr-mus LD50:442 mg/kg TIYADG 9,553,81

SAFETY PROFILE: Moderately toxic by ingestion and intraperitoneal routes. Experimental reproductive effects. When heated to decomposition it emits toxic fumes of NO_x and HCl.

INP100 CAS:27524-97-6 ***HR: 3***
7-(ISOPROPYLAMINOISOPROPYL)THEOPHYLLINE HYDROCHLORIDE
mf: $C_{13}H_{21}N_5O_2{\cdot}ClH$ mw: 315.85

SYNS: IPT (GERMAN) ◇ 7-(2-ISOPROPYLAMINO-2-METHYLETHYL)THEOPHYLLINE HYDROCHLORIDE ◇ 7-(2-(ISOPROPYLAMINO)PROPYL)-THEOPHYLLINE MONOHYDRO- CHLORIDE

TOXICITY DATA with REFERENCE
orl-rat LD50:4200 mg/kg AEPPAE 230,194,57
ipr-rat LD50:573 mg/kg AEPPAE 230,194,57
scu-rat LD50:879 mg/kg AEPPAE 230,194,57
ivn-rat LD50:347 mg/kg AEPPAE 230,194,57
orl-mus LD50:973 mg/kg AEPPAE 230,194,57
ipr-mus LD50:1090 μg/kg ARZNAD 9,198,59

ivn-mus LD50:361 mg/kg AEPPAE 230,194,57
ivn-rbt LD50:215 mg/kg AEPPAE 230,194,57
scu-gpg LD50:376 mg/kg AEPPAE 230,194,57

SAFETY PROFILE: Poison by subcutaneous, intravenous, and intraperitoneal routes. Moderately toxic by ingestion. When heated to decomposition it emits toxic fumes of NO_x and HCl.

INQ000 CAS:841-06-5 ***HR: 2***
2-ISOPROPYLAMINO-4-(3-METHOXYPROPYLAMINO)-6-METHYLTHIOs-TRIAZINE
mf: $C_{11}H_{21}N_5OS$ mw: 271.43

SYNS: GESARAN ◇ 2-ISOPROPYLAMINO-4-(3-METHOXYPROPYLAMINO)-6-METHYLTHIO-1,3,5-TRIAZIN (GERMAN) ◇ 4-ISOPROPYLAMINO-6-(3'-METHOXYPROPYLAMINO)-2-METHYTHIO-1,3,5-TRIAZIN (GERMAN) ◇ METHOPROPTRYNE

TOXICITY DATA with REFERENCE
orl-rat LD50:5000 mg/kg 85ARAE 2,125,77
orl-mus LD50:2400 mg/kg 28ZEAL 4,272,69

SAFETY PROFILE: Moderately toxic by ingestion. Used as a pesticide. When heated to decomposition it emits very toxic fumes of NO_x and SO_x.

INR000 CAS:1014-69-3 ***HR: 2***
2-ISOPROPYLAMINO-4-METHYLAMINO-6-METHYLMERCAPTO-s-TRIAZINE
mf: $C_8H_{15}N_5S$ mw: 213.34

SYNS: DESMETRYN (GERMAN, DUTCH) ◇ DESMETRYNE ◇ G 34360 ◇ GS 34360 ◇ 2-ISOPROPYLAMINO-4-METHYLAMINO-6-METHYLTHIO-1,3,5-TRIAZINE ◇ 2-ISOPROPILAMINO-4-METILAMINO-6-METILTIO-1,3,5-TRIAZINA (ITALIAN) ◇ 2-METHYLAMINO-4-METHYLTHIO-6-ISOPROPYLAMINO-1,3,5-TRIAZINE ◇ METHYLMERCAPTO-4-ISOPROPYLAMINO-6-METHYLAMINO-s-TRIAZINE ◇ 2-METHYLTHIO-4-ISOPROPYLAMINO-6-METHYLAMINO-s- TRIAZINE ◇ 2-(METHYLTHIO)-4-(METHYLAMINO)-6-(ISOPROPYLAMINO)-s-TRIAZINE ◇ SAMURON ◇ SEMERON ◇ TOPUSYN

TOXICITY DATA with REFERENCE
orl-rat LD50:1390 mg/kg 28ZEAL 5,71,76
orl-mus LD50:1750 mg/kg 85GYAZ -,110,71

SAFETY PROFILE: Moderately toxic by ingestion. An herbicide used for postemergence control of annual broadleaf and grassy weeds. When heated to decomposition it emits very toxic fumes of NO_x and SO_x. See also MERCAPTANS.

INS000 CAS:54-80-8 ***HR: 3***
α-((ISOPROPYLAMINO)METHYL)-2-NAPHTHALENEMETHANOL
mf: $C_{15}H_{19}NO$ mw: 229.35

SYNS: ALDERLIN ◇ COMPOUND 38,174 ◇ INETOL ◇ 2-ISOPROPYLAMINO-1-(NAPHTH-2-YL)ETHANOL ◇ 2-ISOPROPYL- AMINO-1-(2-NAPHTHYL)ETHANOL ◇ (2-NAPHTHYL)-1-ISOPROPYLAMINOETHANOL ◇ NAPHTHYLISOPROTERENOL ◇ NEATHALIDE ◇ NETALID ◇ NETH ◇ NETHALIDE ◇ PRONETALOL ◇ PRONETHALOL

TOXICITY DATA with REFERENCE
dns-hmn:hla 100 μmol/L CNREA8 38,2621,78
orl-rat TDLo:15 g/kg/30W-C:CAR PSDTAP 10,175,69
orl-mus TDLo:28080 mg/kg/33W-C:ETA PSDTAP 10,175,69
orl-mus TD:72 g/kg/43W-C:ETA,REP PSDTAP 4,30,64
orl-rat LD50:900 mg/kg LANCAO 2,311,62
ivn-rat LD50:50 mg/kg LANCAO 2,311,62
orl-mus LD50:337 mg/kg JPETAB 149,161,65
ipr-mus LD50:124 mg/kg JPETAB 149,161,65
ivn-mus LD50:28800 μg/kg ARZNAD 28,794,78

CONSENSUS REPORTS: EPA Genetic Toxicology Program.

SAFETY PROFILE: Poison by ingestion, intravenous, and intraperitoneal routes. Questionable carcinogen with experimental carcinogenic and tumorigenic data. Experimental reproductive effects.Human mutation data reported. When heated to decomposition it emits toxic fumes of NO_x.

INT000 CAS:51-02-5 ***HR: 3***
α-((ISOPROPYLAMINO)METHYL)NAPHTHALENEMETHANOL, HYDROCHLORIDE
mf: $C_{15}H_{19}NO_2$•ClH mw: 265.81

SYNS: ALDERLIN HYDROCHLORIDE ◇ ICI 38174 ◇ I.C.I. HYDROCHLORIDE ◇ INETOL ◇ 2-ISOPROPYLAMINO-1-(2-NAPHTHYL)ETHANOL HYDROCHLORIDE ◇ α-(((1-METHYLETHYL)AMINO)METHYL)-2-NAPHTHALENEMETHANOL, HYDROCHLORIDE ◇ NAPHTHYLISOPROTERENOL HYDROCHLORIDE ◇ NETHALIDE HYDROCHLORIDE ◇ PRONETHALOL ◇ PRONETHALOL HYDROCHLORIDE

TOXICITY DATA with REFERENCE
orl-mus TDLo:10 g/kg/25W-C:ETA NATUAS 207,594,65
orl-mus LD50:512 mg/kg BJPCAL 25,577,65
ipr-mus LD50:145 mg/kg AIPTAK 195,57,72
ivn-mus LD50:45 mg/kg BJPCAL 25,577,65

CONSENSUS REPORTS: IARC Cancer Review: Group 3 IMEMDT 7,56,87; Animal Limited Evidence IMEMDT 13,227,77.

SAFETY PROFILE: A poison by intravenous and intraperitoneal routes. Moderately toxic by ingestion. Questionable carcinogen with experimental tumorigenic data. When heated to decomposition it emits very toxic fumes of NO_x and HCl.

INU000 CAS:5054-57-9 ***HR: 3***
α-(ISOPROPYLAMINOMETHYL)-4-NITROBENZYL ALCOHOL
mf: $C_{11}H_{16}N_2O_3$ mw: 224.29

SYNS: α-(((1-METHYLETHYL)AMINO)METHYL)-4-NITROBENZENEMETHANOL ◇ NIFENALOL

TOXICITY DATA with REFERENCE
ipr-mus LD50:207 mg/kg FATOAO 35,29,72
ivn-mus LD50:70 mg/kg ARZNAD 27,1022,77

SAFETY PROFILE: Poison by intraperitoneal and intravenous routes. When heated to decomposition it emits toxic fumes of NO_x.

INU200 CAS:21299-86-5 ***HR: 3***
(±)-1-(ISOPROPYLAMINO)-3-(o-PHENOXY-PHENOXY)-2-PROPANOL HYDROCHLORIDE
mf: $C_{18}H_{23}NO_3 \cdot ClH$ mw: 337.88

SYN: Ph-QA 33

TOXICITY DATA with REFERENCE
orl-mus LD50:810 mg/kg APTOA6 26,343,68
ipr-mus LD50:110 mg/kg APTOA6 27,453,69
ivn-mus LD50:36 mg/kg APTOA6 26,343,68

SAFETY PROFILE: Poison by intravenous and intraperitoneal routes. Moderately toxic by ingestion. When heated to decomposition it emits toxic fumes of NO_x and HCl.

INW000 CAS:63710-43-0 ***HR: 3***
9-((3-(ISOPROPYLAMINO)PROPYL)AMINO)-1-NITROACRIDINE DIHYDROCHLORIDE
mf: $C_{19}H_{22}N_4O_2 \cdot 2ClH$ mw: 411.37

SYNS: N-(1-METHYLETHYL)-N'-(1-NITRO-9-ACRIDINYL)-1,3-PROPANEDIAMINE DIHYDROCHLORIDE ◇ 1-NITRO-9-(3-ISOPROPYLAMINOPROPYLAMINE)-ACRIDINE DIHYDROCHLORIDE ◇ 1-NITRO-9-(3-ISOPROPYLAMINOPROPYLAMINO)-ACRIDINE DIHYDROCHLORIDE

TOXICITY DATA with REFERENCE
dnd-hmn:hla 2500 nmol/L CBINA8 49,311,84
dns-hmn:hla 1 μmol/L CBINA8 49,311,84
dni-hmn:hla 30 nmol/L BBACAQ 825,244,85
sce-hmn:lym 80 nmol/L/69H MUREAV 67,93,79
orl-rat LD50:127 mg/kg AITEAT 28,735,80
ivn-rat LD50:900 μg/kg MMDPA6 8,252,76
orl-mus LD50:126 mg/kg AITEAT 28,735,80
ivn-mus LD50:2 mg/kg AITEAT 28,735,80
ivn-pgn LD50:2100 μg/kg AITEAT 28,777,80

CONSENSUS REPORTS: EPA Genetic Toxicology Program.

SAFETY PROFILE: A poison by ingestion and intravenous routes. Human mutation data reported. When heated to decomposition it emits very toxic fumes of NO_x and HCl.

INX000 CAS:768-52-5 ***HR: 2***
N-ISOPROPYLANILINE
mf: $C_9H_{13}N$ mw: 135.23

SYNS: o-AMINOISOPROPYLBENZENE ◇ 2-AMINOISOPROPYLBENZENE ◇ o-CUMIDINE ◇ o-ISOPROPYLANILINE ◇ 2-ISOPROPYL ANILINE ◇ 2-(1-METHYLETHYL)BENZENAMINE

TOXICITY DATA with REFERENCE
orl-rat LD50:1180 mg/kg TXAPA9 22,153,72

CONSENSUS REPORTS: Reported in EPA TSCA Inventory.

OSHA PEL: TWA 2 ppm (skin)
ACGIH TLV: TWA 2 ppm (skin)

SAFETY PROFILE: Moderately toxic by ingestion. When heated to decomposition it emits toxic fumes of NO_x.

INY000 CAS:479-92-5 ***HR: 3***
4-ISOPROPYLANTIPYRINE
mf: $C_{14}H_{18}N_2O$ mw: 230.34

SYNS: 1,2-DIHYDRO-1,5-DIMETHYL-4-((1-METHYLETHYL)AMINO)-2-PHENYL-3H-PYRAZOL-3-ONE ◇ ISOPROPYLANTIPYRIN ◇ ISOPROPYLANTIPYRINE ◇ 4-ISOPROPYL-2,3-DIMETHYL-1-PHENYL-3-PYRAZOLIN-5-ONE ◇ ISOPROPYLPHENAZONE ◇ ISOPYRINE ◇ LARODON ◇ 1-PHENYL-2,3-DIMETHYL-4-ISOPROPYL-3-PYRAZOLIN-5-ONE ◇ 1-PHENYL-2,3-DIMETHYL-4-ISOPROPYLPYRAZOL-5-ONE ◇ PROPYPHENAZONE

TOXICITY DATA with REFERENCE
orl-rat LD50:860 mg/kg YKKZAJ 97,601,77
orl-mus LD50:960 mg/kg ARZNAD 9,401,59
ipr-mus LD50:295 mg/kg AIPTAK 122,434,59
orl-cat LDLo:150 mg/kg JPETAB 61,205,37
orl-rbt LDLo:500 mg/kg JPETAB 61,205,37
orl-gpg LD50:1050 mg/kg SMWOAS 84,351,54

CONSENSUS REPORTS: Reported in EPA TSCA Inventory.

SAFETY PROFILE: A poison by ingestion and intraperitoneal routes. When heated to decomposition it emits toxic fumes of NO_x.

INY100 CAS:82464-70-8 ***HR: 2***
ISOPROPYLANTIPYRINE with ETHENZAMIDE and CAFFEINE MONOHYDRATE
mf: $C_{14}H_{18}N_2O \cdot C_9H_{11}NO_2 \cdot C_8H_{10}N_4O_2 \cdot H_2O$ mw: 607.79

SYN: Ro 04-7683

TOXICITY DATA with REFERENCE
orl-mus LD50:1250 mg/kg YACHDS 10,1407,82
ipr-mus LD50:405 mg/kg YACHDS 10,1407,82
scu-mus LD50:1400 mg/kg YACHDS 10,1407,82

SAFETY PROFILE: Moderately toxic by ingestion, intraperitoneal and subcutaneous routes. When heated to decomposition it emits toxic fumes of NO_x. See also 4-ISOPROPYLANTIPYRINE and CAFFEINE.

INZ000 CAS:63020-47-3 ***HR: 3***
5-ISOPROPYL-1:2-BENZANTHRACENE
mf: $C_{21}H_{18}$ mw: 270.39

SYN: 8-ISOPROPYLBENZ(a)ANTHRACENE

TOXICITY DATA with REFERENCE
skn-mus TDLo:840 mg/kg/35W-I:ETA PRLBA4 129,439,40

SAFETY PROFILE: Questionable carcinogen with experimental tumorigenic data. When heated to decomposition it emits acrid smoke and fumes.

IOA000 CAS:63020-48-4 **HR: 3**
6-ISOPROPYL-1:2-BENZANTHRACENE
mf: $C_{21}H_{18}$ mw: 270.39

SYN: 9-ISOPROPYLBENZ(a)ANTHRACENE

TOXICITY DATA with REFERENCE
skn-mus TDLo:700 mg/kg/29W-I:ETA PRLBA4 111,485,32

SAFETY PROFILE: Questionable carcinogen with experimental tumorigenic data. When heated to decomposition it emits acrid smoke and fumes.

IOB000 CAS:80-15-9 **HR: 3**
ISOPROPYLBENZENE HYDROPEROXIDE
DOT: UN 2116
mf: $C_9H_{12}O_2$ mw: 152.21

PROP: Bp: 153°, flash p: 175°F, d: 1.05. The hydroperoxide of cumene.

SYNS: CUMEENHYDROPEROXYDE (DUTCH) ◇ CUMENE HYDROPEROXIDE (DOT) ◇ CUMENE HYDROPEROXIDE, TECHNICALLY PURE (DOT) ◇ CUMENT HYDROPEROXIDE ◇ CUMENYL HYDROPEROXIDE ◇ CUMOLHYDROPEROXID (GERMAN) ◇ CUMYL HYDROPEROXIDE ◇ α-CUMYL HYDROPEROXIDE ◇ CUMYL HYDROPEROXIDE, TECHNICAL PURE (DOT) ◇ α,α-DIMETHYLBENZYL HYDROPEROXIDE (MAK) ◇ HYDROPEROXYDE de CUMENE (FRENCH) ◇ HYDROPEROXYDE de CUMYLE (FRENCH) ◇ IDROPEROSSIDO di CUMENE (ITALIAN) ◇ IDROPEROSSIDO di CUMOLO (ITALIAN) ◇ RCRA WASTE NUMBER U096

TOXICITY DATA with REFERENCE
skn-rbt 500 mg AIHAAP 19,205,58
skn-rbt 500 mg MLD SCCUR* -,3,61
eye-rbt 1 mg AIHAAP 19,205,58
mmo-sat 100 μg/plate PNASA6 79,7445,82
mma-sat 100 μg/plate ABCHA6 44,1989,00
scu-mus TDLo:8844 mg/kg/67W-I:ETA JNCIAM 37,825,66
orl-rat LD50:382 mg/kg AIHAAP 19,205,58
ihl-rat LC50:220 ppm/4H AIHAAP 19,205,58
skn-rat LD50:500 mg/kg AEHLAU 30,1,75
ipr-rat LD50:95 mg/kg AIHAAP 19,205,58
orl-mus LDLo:5000 mg/kg SCCUR* -,3,61
ihl-mus LC50:200 ppm/4H AIHAAP 19,205,58
ipr-mus LDLo:90 mg/kg TXAPA9 23,288,72
scu-mus LD50:490 mg/kg GISAAA 26(12),22,61

CONSENSUS REPORTS: Community Right-To-Know List. Reported in EPA TSCA Inventory. EPA Genetic Toxicology Program.

DFG MAK: Moderate Skin Effects.
DOT Classification: Organic Peroxide; Label: Organic Peroxide.

SAFETY PROFILE: A poison by ingestion and intraperitoneal routes. Moderately toxic by skin contact, inhalation and subcutaneous routes. Mutation data reported. A skin and eye irritant. Questionable carcinogen with experimental tumorigenic data. A strong oxidizing agent. Flammable when exposed to heat or flame; can react with reducing materials. Its use in industry has resulted in many explosions. Storage above 109°C may cause explosive decomposition. Potentially explosive reactions with acids or reductants. Violent or explosive reaction when heated with solutions of 1,2-dibromo-1,2-diisocyanatoethane polymers in benzene. Violent decomposition on contact with cobalt; copper; copper alloys; lead alloys; mineral acids. Vigorous exothermic reaction on contact with charcoal. When heated to decomposition it emits acrid smoke and fumes. To fight fire, use foam, CO_2, dry chemical. See also PEROXIDES.

IOD000 CAS:939-48-0 **HR: 2**
ISOPROPYL BENZOATE
mf: $C_{10}H_{12}O_2$ mw: 164.22

PROP: Liquid. Mp: −26.4°, bp: 219°, flash p: 210°F (OC), d: 1.0112 at 25°/25°, vap press: 0.12 mm @ 20°, vap d: 5.67. Insol in water; sol in alc and ether.

SYN: BENZOIC ACID, ISOPROPYL ESTER

TOXICITY DATA with REFERENCE
skn-rbt 10 mg/24H open MLD AMIHBC 4,119,51
eye-rbt 500 mg open AMIHBC 4,119,51
orl-rat LD50:3730 mg/kg AMIHBC 4,119,51
skn-rbt LD50:20 g/kg AMIHBC 4,119,51

CONSENSUS REPORTS: Reported in EPA TSCA Inventory.

SAFETY PROFILE: Moderately toxic by ingestion. Mildly toxic by skin contact. A skin and eye irritant. Combustible when exposed to heat or flame; can react with oxidizing materials. To fight fire, use water, spray, CO_2, dry chemical. See also ESTERS.

IOD050 CAS:13816-33-6 **HR: 2**
p-ISOPROPYLBENZONITRILE
mf: $C_{10}H_{11}N$ mw: 145.22

SYNS: BENZONITRILE, p-ISOPROPYL- ◇ BENZONITRILE, 4-(1-METHYLETHYL)- ◇ CUMINYL NITRILE ◇ p-CYANOCUMENE ◇ 4-(1-METHYLETHYL)BENZONITRILE

TOXICITY DATA with REFERENCE
skn-rbt 500 mg/24H MOD FCTOD7 21,837,83
orl-rat LD50:3900 mg/kg FCTOD7 21,837,83

CONSENSUS REPORTS: Reported in EPA TSCA Inventory.

SAFETY PROFILE: Moderately toxic by ingestion. A skin irritant. When heated to decomposition it emits toxic fumes of NO_x.

IOE000 CAS:63904-87-0 ***HR: 3***
N-ISOPROPYL BENZOTHIAZOLE SULFONAMIDE

SYN: ISOCYCLEX

TOXICITY DATA with REFERENCE
skn-hmn 250 mg/48H MOD AMIHBC 5,311,52
skn-rbt 500 mg MOD AMIHBC 5,311,52
ipr-mam LD50:250 mg/kg AMIHBC 5,311,52

SAFETY PROFILE: Poison by intraperitoneal route. A human skin irritant. When heated to decomposition it emits toxic fumes of SO_x and NO_x.

IOF000 CAS:63020-53-1 ***HR: 3***
2-ISOPROPYL-3:4-BENZPHENANTHRENE
mf: $C_{21}H_{18}$ mw: 270.39

TOXICITY DATA with REFERENCE
skn-mus TDLo:720 mg/kg/30W-I:ETA PRLBA4 129,439,40

SAFETY PROFILE: Questionable carcinogen with experimental tumorigenic data. When heated to decomposition it emits acrid smoke and fumes.

IOF100 CAS:4762-14-5 ***HR: 3***
1-ISOPROPYLBIGUANIDE HYDROCHLORIDE
mf: $C_5H_{13}N_5{\cdot}ClH$ mw: 179.69

SYNS: 1-ISOPROPILBIGUANIDE CLORIDRATO (ITALIAN) ◇ N-ISOPROPYLBIGUANIDE HYDROCHLORIDE

TOXICITY DATA with REFERENCE
orl-rat LD50:2472 mg/kg FRPSAX 15,521,60
ipr-rat LD50:296 mg/kg FRPSAX 15,521,60
ipr-mus LD50:475 mg/kg JAJAAA 18,196,65

SAFETY PROFILE: Poison by intraperitoneal route. Moderately toxic by ingestion. When heated to decomposition it emits toxic fumes of NO_x and HCl.

IOF200 CAS:25640-78-2 ***HR: 2***
ISOPROPYLBIPHENYL
mf: $C_{15}H_{16}$ mw: 196.29

SYNS: ISOPROPYLDIPHENYL ◇ MONOISOPROPYLBIPHENYL ◇ WEMCOL

TOXICITY DATA with REFERENCE
cyt-rat-orl 2 g/kg CHYCDW 19,132,85
orl-rat LD50:4570 mg/kg CHYCDW 19,132,85
orl-mus LD50:2610 mg/kg CHYCDW 19,132,85
orl-rbt LDLo:4 g/kg CHYCDW 19,132,85

CONSENSUS REPORTS: Reported in EPA TSCA Inventory.

SAFETY PROFILE: Moderately toxic by ingestion. Mutation data reported. When heated to decomposition it emits acrid smoke and fumes.

IOF300 ***HR: 3***
ISOPROPYL-BIS(β-CHLOROETHYL)AMINE
mf: $C_7H_{15}Cl_2N$ mw: 184.13

SYNS: 2,2'-DICHLORO-1''-METHYLTRIETHYLAMINE ◇ ISOPROPYL-S ◇ TL-301

TOXICITY DATA with REFERENCE
scu-rat LD50:1 mg/kg NTIS** PB158-507
ivn-rat LD50:500 μg/kg NTIS** PB158-507
orl-mus LD50:22 mg/kg NTIS** PB158-507
scu-mus LD50:500 μg/kg NTIS** PB158-507
ivn-rbt LD50:2 mg/kg NTIS** PB158-507

SAFETY PROFILE: Poison by ingestion, subcutaneous, and intravenous routes. When heated to decomposition it emits toxic fumes of Cl^- and NO_x. See also AMINES.

IOI000 CAS:5419-55-6 ***HR: 3***
ISOPROPYL BORATE
DOT: UN 2616
mf: $C_9H_{21}BO_3$ mw: 188.11

PROP: Colorless liquid. Mp: −59°, bp: 141.0-142.4°, flash p: 82°F (TCC), d: 0.8138 @ 25°.

SYN: TRIISOPROPYL BORATE

TOXICITY DATA with REFERENCE
eye-rbt 100 mg MLD 14KTAK -,706,64
orl-mus LD50:2500 mg/kg 14KTAK -,706,64
ivn-mus LD50:100 mg/kg CSLNX* NX#00382

CONSENSUS REPORTS: Reported in EPA TSCA Inventory.

DOT Classification: Flammable or Combustible Liquid; Label: Flammable Liquid

SAFETY PROFILE: A poison by intravenous route. Moderately toxic by ingestion. An eye irritant. Dangerous fire hazard when exposed to heat, flame, or oxidizers. Can react vigorously with oxidizing materials. To fight fire, use foam, CO_2, dry chemical. See also ESTERS and BORON COMPOUNDS.

IOI600 ***HR: 3***
ISOPROPYLCAINE HYDROCHLORIDE
mf: $C_{16}H_{25}NO_2{\cdot}ClH$ mw: 299.88

SYNS: β-DIETHYLAMINOETHYL CUMATE HYDROCHLORIDE ◇ p-

ISOPROPYLBENZOIC ACID-2-(DIETHYLAMINO)ETHYL ESTER HYDROCHLORIDE

TOXICITY DATA with REFERENCE
ipr-mus LD50:370 mg/kg JAPMA8 40,449,51
scu-mus LD50:950 mg/kg APFRAD 40,133,82
ivn-mus LD50:78 mg/kg THERAP 8,934,53

SAFETY PROFILE: Poison by intravenous and intraperitoneal routes. Moderately toxic by subcutaneous route. When heated to decomposition it emits toxic fumes of NO_x and HCl. See also ESTERS.

IOJ000 CAS:1746-77-6 ***HR: 3***
ISOPROPYL CARBAMATE
mf: $C_4H_9NO_2$ mw: 103.14

PROP: Prisms. Mp: 60-61°, bp: 200°C. Very sol in water, alc, and ether.

SYNS: CARBAMIC ACID, ISOPROPYL ESTER ◇ CARBAMIC ACID-1-METHYLETHYL ESTER

TOXICITY DATA with REFERENCE
mmo-esc 25000 ppm CRSBAW 143,776,49
dni-mus-ipr 1 g/kg CNREA8 29,994,69
sce-mus-ipr 4400 μmol/kg CNREA8 41,4489,81
ipr-mus TDLo:2400 mg/kg/4W-I:NEO CNREA8 29,2184,69
scu-mus LD50:1280 mg/kg AJEBAK 45,507,67

CONSENSUS REPORTS: Reported in EPA TSCA Inventory.

SAFETY PROFILE: Moderately toxic by subcutaneous route. Mutation data reported. Questionable carcinogen with experimental neoplastigenic data. When heated to decomposition it emits toxic fumes of NO_x. See also CARBAMATES.

IOK000 CAS:2138-43-4 ***HR: 3***
4-ISOPROPYLCATECHOL
mf: $C_9H_{12}O_2$ mw: 152.21

SYN: 4-(1-METHYLETHYL)-1,2-BENZENEDIOL

TOXICITY DATA with REFERENCE
skn-man 1%/48H BJDEAZ 94,687,76
ivn-mus LD50:56 mg/kg CSLNX* NX#07862

SAFETY PROFILE: Poison by intravenous route. A human skin irritant. When heated to decomposition it emits acrid smoke and fumes.

IOL000 CAS:108-23-6 ***HR: 3***
ISOPROPYL CHLOROCARBONATE
DOT: UN 2407
mf: $C_4H_7ClO_2$ mw: 122.56

$(CH_3)_2CHOCO \cdot Cl$

PROP: A clear, colorless, volatile liquid with a pungent, irritating odor. D: 1.078 @ 20°/4°, bp: 105°, flash p: 28°C (TOC), 20°C (TCC), fp: −80°, fire p: 40°C (TOC), autoign temp: >500°, vap d: 4.2 @ 20°, refr index: 1.3974 @ 20°, lel: 4%, uel: 15%, vap press: 72 mm @ 70°F, bulk d: 9.0 lbs/gal, % volatile: 100%. Sol in aromatic or aliphatic hydrocarbon solvents, ethyl ether, acetone, and chloroform. Insol in water and alc. Decomp slowly in cold water, faster in hot water. Misc in ether and benzene. A phosgene derivative.

SYNS: CARBONOCHLORIDE ACID-1-METHYL ESTER ◇ CHLOROFORMIC ACID ISOPROPYL ESTER ◇ ISOPROPYL CHLOROFORMATE ◇ ISOPROPYL CHLOROMETHANOATE

TOXICITY DATA with REFERENCE
skn-rbt 500 mg IHFCAY 6,1,67
eye-rbt 500 mg SEV IHFCAY 6,1,67
orl-rat LD50:1070 mg/kg IHFCAY 6,1,67
ihl-rat LCLo:200 ppm/5H BJIMAG 27,1,70
orl-mus LD50:178 mg/kg 37ASAA 4,758,78
ihl-mus LD50:299 ppm/1H 37ASAA 4,758,78
skn-mus LD50:12 mg/kg 37ASAA 4,758,78
skn-rbt LD50:11300 mg/kg IHFCAY 6,1,67

CONSENSUS REPORTS: EPA Extremely Hazardous Substances List. Reported in EPA TSCA Inventory.

DOT Classification: Flammable Liquid; Label: Flammable Liquid.

SAFETY PROFILE: A poison by skin contact and ingestion. Moderately toxic by inhalation. Ingestion of even small amounts can be fatal. A skin and severe eye irritant. Inhalation of a small amount can cause immediate lachrimation, coughing, choking, and respiratory distress. Death may result from pulmonary edema which may not appear for several hours after exposure. A skin and severe eye irritant. A dangerous fire and moderate explosion hazard when exposed to heat, spark, or flame. Self-reactive. Iron salts may catalyze a potentially explosive thermal decomposition. Incompatible with water; iron; metal salts; acids; alkalies; amines; alcohols. Stable under refrigeration below 20°, but one reference (1973) reports that it has exploded while stored in a refrigerator. Present day formulations appear to be more stable. Temperatures above 20° can cause decomposition. When heated to decomposition it emits acrid smoke and fumes.

ION000 CAS:63041-70-3 ***HR: 3***
20-ISOPROPYLCHOLANTHRENE
mf: $C_{23}H_{20}$ mw: 296.43

SYN: 3-ISOPROPYLCHOLANTHRENE

TOXICITY DATA with REFERENCE
scu-mus TDLo:80 mg/kg:ETA JNCIAM 2,99,41

SAFETY PROFILE: Questionable carcinogen with experimental tumorigenic data. When heated to decomposition it emits acrid smoke and fumes.

IOO000 CAS:7780-06-5 ***HR: 2***
ISOPROPYL CINNAMATE
mf: $C_{12}H_{14}O_2$ mw: 190.26

SYN: CINNAMIC ACID, ISOPROPYL ESTER

TOXICITY DATA with REFERENCE
orl-gpg LD50:2700 mg/kg JPETAB 93,26,48

CONSENSUS REPORTS: Reported in EPA TSCA Inventory.

SAFETY PROFILE: Moderately toxic by ingestion. When heated to decomposition it emits acrid smoke and fumes. See also ESTERS.

IOO300 CAS:4621-04-9 ***HR: 2***
4-ISOPROPYLCYCLOHEXANOL
mf: $C_9H_{18}O$ mw: 142.27

SYNS: CYCLOHEXANOL, p-ISOPROPYL- ◇ p-ISOPROPYLCYCLOHEXANOL

TOXICITY DATA with REFERENCE
skn-rbt 500 mg/24H MOD FCTXAV 16,803,78
orl-rat LD50:2750 mg/kg FCTXAV 16,803,78

CONSENSUS REPORTS: Reported in EPA TSCA Inventory.

SAFETY PROFILE: Moderately toxic by ingestion. A skin irritant. When heated to decomposition it emits acrid smoke and irritating fumes.

IOR000 CAS:10457-59-7 ***HR: 3***
14-ISOPROPYLDIBENZ(a,j)ACRIDINE
mf: $C_{24}H_{19}N$ mw: 321.44

SYN: 10-ISOPROPYL-3,4,5,6-DIBENZACRIDINE(FRENCH)

TOXICITY DATA with REFERENCE
skn-mus TDLo:636 mg/kg/53W-I:ETA ACRSAJ 4,315,56

SAFETY PROFILE: Questionable carcinogen with experimental tumorigenic data. When heated to decomposition it emits toxic fumes of NO_x.

IOS000 CAS:18181-80-1 ***HR: 1***
ISOPROPYL-4,4'-DIBROMOBENZILATE
mf: $C_{17}H_{16}Br_2O_3$ mw: 428.15

SYNS: ACAROL ◇ 4-BROMO-α-(4-BROMOPHENYL)-α-HYDROXY-BENZENEACETIC ACID-1-METHYLETHYL ESTER ◇ BROMOPROPYLATE ◇ CIBA-GEIGY GS 19851 ◇ 4,4'-DIBROMO- BENZILIC ACID ISOPROPYL ESTER ◇ ENT 27,552 ◇ GEIGY GS-19851 ◇ 1-METHYLETHYL 4-BROMO-α-(4-BROMOPHENYL)-α-HYDROXY- BENZENEACETATE ◇ NEORON ◇ NSC 195087 ◇ PHENISOBROMOLATE

TOXICITY DATA with REFERENCE
skn-rbt 121 mg open MOD CIGET* -,-,77
eye-rbt 600 μg MLD CIGET* -,-,77
orl-rat LD50:5000 mg/kg BESAAT 15,97,69
orl-mus LD50:8000 mg/kg 28ZEAL 5,29,76
skn-rbt LD50:10200 mg/kg CIGET* -,-,77

SAFETY PROFILE: Mildly toxic by skin contact and ingestion. A skin and eye irritant. A miticide for citrus, cotton, fruits and ornamentals. When heated to decomposition it emits toxic fumes of Br^-.

IOT000 CAS:2275-18-5 ***HR: 3***
ISOPROPYL DIETHYLDITHIOPHOSPHORYLACETAMIDE
mf: $C_9H_{20}NO_3PS_2$ mw: 285.39

SYNS: AC 18682 ◇ AMERICAN CYANAMID 18682 ◇ O,O-DIETHYLDITHIOPHOSPHORYLACETICACID-N-MONOISOPROPYLAMIDE ◇ O,O-DIETHYL-S-(N-ISOPROPYLCARBAMOYLMETHYL) DITHIOPHOSPHATE ◇ O,O-DIETHYL-S-ISOPROPYLCARBAMOYLMETHYL PHOSPHORODITHIOATE ◇ O,O-DIETHYL-S-(N-ISOPROPYLCARBAMOYLMETHYL) PHOSPHORODITHIOATE ◇ ENT 24,652 ◇ FAC ◇ FAC 20 ◇ FOSTION ◇ N-ISOPROPYL-2-MERCAPTO-ACETAMIDE-S-ESTER with O,O-DIETHYL PHOSPHORODITHIOATE ◇ L343 ◇ N-MONOISOPROPYLAMIDE of O,O-DIETHYLDITHIOPHOSPHORYLACETIC ACID ◇ OLEOFAC ◇ PHOSPHORODITHIOIC ACID-O,O-DIETHYL ESTER-S-ESTER with N-ISOPROPYL-2-MERCAPTO-ACETAMIDE ◇ PROTHOATE ◇ PROTOAT (HUNGARIAN) ◇ TELEFOS ◇ TRIMETHOATE

TOXICITY DATA with REFERENCE
orl-rat LD50:8 mg/kg GUCHAZ 6,439,73
ihl-rat LD50:165 mg/m³/4H EGESAQ 24,173,80
skn-rat LD50:100 mg/kg WRPCA2 9,119,70
orl-mus LD50:8 mg/kg ARSIM* 20,1,66
orl-dog LD50:15 mg/kg 28ZEAL 4,333,69
orl-rbt LD50:8500 μg/kg 28ZEAL 4,333,69
skn-rbt LD50:14 mg/kg SPEADM 78-1,32,78

CONSENSUS REPORTS: EPA Extremely Hazardous Substances List.

SAFETY PROFILE: A poison by ingestion, inhalation, and skin contact. An insecticide. When heated to decomposition it emits very toxic NO_x, PO_x, and SO_x.

IOT875 CAS:24596-38-1 ***HR: 2***
4'-ISOPROPYL-4-DIMETHYLAMINOAZOBENZENE
mf: $C_{17}H_{21}N_3$ mw: 267.41

SYN: p-(p-CUMENYLAZO)-N,N-DIMETHYLANILINE

TOXICITY DATA with REFERENCE
orl-rat TDLo:2142 mg/kg/17W-C:ETA JNCIAM 27,663,61

SAFETY PROFILE: Questionable carcinogen with experimental tumorigenic data. When heated to decomposition it emits toxic fumes of NO_x.

IOU000 CAS:1505-95-9 ***HR: 3***
α-ISOPROPYL-α-(2-DIMETHYLAMINOETHYL)-1-NAPHTHACETAMIDE
mf: $C_{19}H_{26}N_2O$ mw: 298.47

SYNS: α-ISOPROPYL-α-(2-(DIMETHYLAMINO)ETHYL)-1-NAPHTHALENEACETAMIDE ◇ α-ISOPROPYL-α-(2-DIMETHYLAMINOETHYL)-1-NAPHTHYLACETAMIDE ◇ NAFTIPRAMIDE ◇ NAFTYPRAMIDE ◇ NAPHTHIPRAMIDE ◇ MAPHTHYPRAMIDE

TOXICITY DATA with REFERENCE
orl-rat LD50:1030 mg/kg JMCMAR 16,720,73
ipr-rat LD50:269 mg/kg AIPTAK 162,378,66
orl-mus LD50:1086 mg/kg AIPTAK 162,378,66
ipr-mus LDLo:250 mg/kg JMCMAR 8,594,65
ivn-mus LD50:72 mg/kg EXPEAM 20,457,64

SAFETY PROFILE: Poison by intraperitoneal and intravenous routes. Moderately toxic by ingestion. When heated to decomposition it emits toxic fumes of NO_x.

IOU200 ***HR: 3***
m-ISOPROPYL-p-DIMETHYL-AMINO-PHENOL-DIMETHYL-URETHANE METHIODIDE
mf: $C_{15}H_{25}N_2O_2$•I mw: 392.32

SYN: (CARBOXYMETHYL)TRIMETHYLAMMONIUMIODIDE-4-(DIMETHYLAMINO)-3-ISOPROPYLPHENYLESTER

TOXICITY DATA with REFERENCE
orl-mus LDLo:39800 μg/kg FEPRA7 5,184,46
scu-mus LDLo:1500 ng/kg FEPRA7 5,184,46
scu-dog LDLo:153 μg/kg FEPRA7 5,184,46

SAFETY PROFILE: A deadly poison by ingestion and subcutaneous routes. When heated to decomposition it emits toxic fumes of I^-, NH_3, and NO_x.

IOV000 CAS:14211-01-9 ***HR: 3***
ISOPROPYL-O,O-DIMETHYLDITHIOPHOSPHORYL-1-PHENYLACETATE
mf: $C_{13}H_{19}O_4PS_2$ mw: 334.41

SYNS: ISOPROPYL ESTER MERCAPTOPHENYLACETIC ACID S-ESTER with O-O-DIMETHYL PHOSPHORODITHIOATE ◇ ISOPROPYL MERCAPTOPHENYLACETATE-O,O-DIMETHYLPHOSPHORODITHIOATE ◇ M 1703 ◇ OMS-1092

TOXICITY DATA with REFERENCE
orl-rat LD50:205 mg/kg JAFCAU 20,944,72
orl-mus LD50:700 mg/kg BESAAT 15,119,69

SAFETY PROFILE: Poison by ingestion. When heated to decomposition it emits very toxic fumes of PO_x and SO_x.

IOW000 CAS:63905-13-5 ***HR: 2***
N'-ISOPROPYL-N,N'-DIMETHYL-1,3-PROPANEDIAMINE
mf: $C_8H_{20}N_2$ mw: 144.30

TOXICITY DATA with REFERENCE
orl-rat LD50:1300 mg/kg TXAPA9 28,313,74
skn-rbt LD50:450 mg/kg TXAPA9 28,313,74

SAFETY PROFILE: Moderately toxic by ingestion and skin contact. When heated to decomposition it emits toxic fumes of NO_x.

IOW500 CAS:36170-25-9 ***HR: 3***
3-ISOPROPYL-5,7-DIMORPHOLINOMETHYLTROPOLONE DIHYDROCHLORIDE
mf: $C_{20}H_{30}N_2O_4$•2ClH mw: 435.44

SYNS: 5,7-BIS(MORPHOLINOMETHYL)-2-HYDROXY-3-ISOPROPYL-2,4,6-CYCLOHEPTATRIEN-1-ONE DIHYDROCHLORIDE ◇ IDMT

TOXICITY DATA with REFERENCE
ipr-mus LD50:75 mg/kg YKKZAJ 92,570,72
scu-mus LD50:120 mg/kg YKKZAJ 92,570,72
ivn-mus LD50:48 mg/kg YKKZAJ 92,570,72

SAFETY PROFILE: Poison by subcutaneous, intravenous, and intraperitoneal routes. When heated to decomposition it emits toxic fumes of NO_x and HCl.

IOX000 CAS:4097-47-6 ***HR: 3***
4-ISOPROPYL-2,6-DINITROPHENOL
mf: $C_9H_{10}N_2O_5$ mw: 226.21

TOXICITY DATA with REFERENCE
ipr-mus LDLo:31 mg/kg CBCCT* 6,60,54

CONSENSUS REPORTS: Reported in EPA TSCA Inventory.

SAFETY PROFILE: Poison by intraperitoneal route. When heated to decomposition it emits toxic fumes of NO_x. See also NITRO COMPOUNDS of AROMATIC HYDROCARBONS.

IOY000 CAS:94-11-1 ***HR: 3***
ISOPROPYL-2,4-D ESTER
mf: $C_{11}H_{12}Cl_2O_3$ mw: 263.13

SYNS: (2,4-DICHLOROPHENOXY)ACETIC ACID, ISOPROPYL ESTER ◇ (2-4-DICHLOROPHENOXY)ACETIC ACID-1-METHYLETHYL ESTER (9CI) ◇ 2,4-D ISOPROPYL ESTER ◇ ESTERON 44 ◇ WEEDONE 128

TOXICITY DATA with REFERENCE
scu-mus TDLo:846 mg/kg (6-14D preg):TER NTIS** PB223-160
scu-mus TDLo:414 mg/kg (6-14D preg):REP NTIS** PB223-160
orl-mus TDLo:12 g/kg/78W-I:ETA NTIS** PB223-159
orl-rat LD50:375 mg/kg FMCHA2 -,C166,89
orl-mus LD50:541 mg/kg AJVRAH 15,622,54

CONSENSUS REPORTS: IARC Cancer Review: Animal Inadequate Evidence IMEMDT 15,111,77.

SAFETY PROFILE: Poison by ingestion. Experimental teratogenic and reproductive effects. Questionable carcinogen with experimental tumorigenic data. Used as a pesticide. When heated to decomposition it emits toxic fumes of Cl^-. See also ESTERS.

IOZ750 CAS:108-20-3 ***HR: 3***
ISOPROPYL ETHER
DOT: UN 1159
mf: $C_6H_{14}O$ mw: 102.20

$(CH_3)_2CHOCH(CH_3)_2$

PROP: Colorless liquid; ethereal odor. Mp: −60°, bp: 68.5°, lel: 1.4%, uel: 7.9%, flash p: −18°F (CC), d: 0.719 @ 25°, autoign temp: 830°F, vap press: 150 mm @ 25°, vap d: 3.52. Misc in water.

SYNS: DIISOPROPYL ETHER ◇ DIISOPROPYL OXIDE ◇ ETHER ISOPROPYLIQUE (FRENCH) ◇ 2-ISOPROPOXYPROPANE ◇ IZOPROPYLOWY ETER (POLISH)

TOXICITY DATA with REFERENCE
skn-rbt 363 mg open MLD UCDS** 4/10/68
orl-rat LD50:8470 mg/kg UCDS** 4/10/68
ihl-rat LC50:162 g/m^3 GTPZAB 19(10),55,75
ihl-mus LC50:131 g/m^3 GTPZAB 19(10),55,75
ipr-mus LD50:812 mg/kg SCCUR* -,5,61
ihl-rbt LC50:121 g/m^3 GTPZAB 19(10),55,75
skn-rbt LD50:20 g/kg UCDS** 4/10/68

OSHA PEL: TWA 500 ppm
ACGIH TLV: TWA 250 ppm; STEL 310 ppm
DFG MAK: 500 ppm (2100 mg/m^3)
DOT Classification: Flammable Liquid; Label: Flammable Liquid.

SAFETY PROFILE: Moderately toxic by intraperitoneal route. Mildly toxic by ingestion, inhalation, and skin contact. A skin irritant. A very dangerous fire hazard and severe explosion hazard when exposed to heat, flame, sparks, or oxidizers. Under some conditions shock will explode it. Dangerous; on exposure to air it rapidly forms very sensitive, explosive peroxides which precipitate as crystals. Violent reaction with chlorosulfonic acid; HNO_3. Potentially dangerous reaction with propionyl chloride can burst a sealed container. Reacts vigorously with oxidizing materials. To fight fire, use alcohol foam, CO_2, foam, dry chemical. When heated to decomposition it emits acrid smoke and fumes. See also ETHERS.

IOZ800 ***HR: 3***
ISOPROPYL 3-(((ETHYLAMINO)METHOXYPHOSPHINOTHIOYL)OXY)CROTONIC ACID
mf: $C_{10}H_{20}NO_4PS$ mw: 281.34

TOXICITY DATA with REFERENCE
orl-rat LD50:94200 μg/kg IYKEDH 13,1128,82
scu-rat LD50:140 mg/kg IYKEDH 13,1128,82
orl-mus LD50:62400 μg/kg IYKEDH 13,1128,82
scu-mus LD50:117 mg/kg IYKEDH 13,1128,82

SAFETY PROFILE: Poison by ingestion and subcutaneous routes. When heated to decomposition it emits toxic fumes of PO_x, SO_x, and NO_x. See also ESTERS.

IPA000 CAS:2594-20-9 ***HR: 3***
ISOPROPYL ETHYL URETHAN
mf: $C_6H_{13}NO_2$ mw: 131.20

PROP: Liquid. Bp: 192-193°.

SYN: ISOPROPYLCARBAMIC ACID, ETHYL ESTER

TOXICITY DATA with REFERENCE
ipr-mus TDLo:6500 mg/kg/13W-I:ETA JNCIAM 9,35,48

SAFETY PROFILE: Questionable carcinogen with experimental tumorigenic data. When heated to decomposition it emits toxic fumes of NO_x. See also ESTERS and CARBAMATES.

IPC000 CAS:625-55-8 ***HR: 3***
ISOPROPYL FORMATE
DOT: UN 1281
mf: $C_4H_8O_2$ mw: 88.12

PROP: Clear liquid. Bp: 68.3°, flash p: 22°F (CC), d: 0.873, autoign temp: 905°F, vap press: 100 mm @ 17.8°, vap d: 3.03.

SYN: FORMIC ACID, ISOPROPYL ESTER

TOXICITY DATA with REFERENCE
orl-gpg LD50:1400 μg/kg 28ZEAL 4,256,69

CONSENSUS REPORTS: EPA Extremely Hazardous Substances List. Reported in EPA TSCA Inventory.

DOT Classification: Flammable Liquid; Label: Flammable Liquid.

SAFETY PROFILE: A poison by ingestion. A toxic fumigant. A very dangerous fire hazard when exposed to heat, spark or flame. Can react vigorously with oxidizing materials. When heated to decomposition it emits acrid smoke and fumes. To fight fire, use alcohol foam, foam, CO_2, dry chemical. See also ESTERS.

IPD000 CAS:4016-14-2 ***HR: 2***
ISOPROPYL GLYCIDYL ETHER
mf: $C_6H_{12}O_2$ mw: 116.18

PROP: A liquid.

SYNS: IGE ◇ NCI-C56439

TOXICITY DATA with REFERENCE
skn-rbt 459 mg/3D MOD AMIHAB 14,250,56
eye-rbt 92 mg MOD AMIHAB 14,250,56

mmo-esc 20 μmol/L ARTODN 46,277,80
dnd-esc 1 μmol/L ARTODN 46,277,80
orl-rat LD50:4200 mg/kg AMIHAB 14,250,56
ihl-rat LC50:1100 ppm/8H AMIHAB 14,250,56
orl-mus LD50:1300 mg/kg AMIHAB 14,250,56
ihl-mus LC50:1500 ppm/4H AMIHAB 14,250,56
skn-rbt LD50:9650 mg/kg AMIHAB 14,250,56

CONSENSUS REPORTS: Glycol ether compounds are on the Community Right-To-Know List. Reported in EPA TSCA Inventory.

OSHA PEL: (Transitional: TWA 50 ppm) TWA 50 ppm; STEL 75 ppm
ACGIH TLV: TWA 50 ppm; STEL 75 ppm
DFG MAK: 50 ppm (240 mg/m^3)
NIOSH REL: (Glycidyl Ethers) CL 240 mg/m^3/15M

SAFETY PROFILE: Moderately toxic by ingestion. Mildly toxic by inhalation and skin contact. A skin and eye irritant. Mutation data reported. When heated to decomposition it emits acrid smoke and fumes. See also GLYCOL ETHERS.

IPG000 CAS:24426-36-6 ***HR: 3***
ISOPROPYL-S HYDROCHLORIDE
mf: $C_7H_{15}Cl_2N{\cdot}ClH$ mw: 220.59

SYNS: N,N-BIS(2-CHLOROETHYL)ISOPROPYLAMINEHYDROCHLORIDE ◇ 2,2'-DICHLORO-N-ISOPROPYLDIETHYLAMINE HYDROCHLORIDE ◇ 2,2'-DICHLORO-1''-METHYLTRIETHYLAMINE HYDROCHLORIDE ◇ ISOPROPYL BIS(β-CHLOROETHYL)AMINE HYDROCHLORIDE ◇ TL 301 HYDROCHLORIDE

TOXICITY DATA with REFERENCE
orl-rat LDLo:25 mg/kg NCNSA6 5,11,53
ivn-rat LD50:500 μg/kg JPETAB 91,224,47
orl-mus LD50:22 mg/kg JPETAB 91,224,47
ipr-mus LD50:1330 μg/kg CANCAR 2,1075,49
scu-mus LD50:1100 μg/kg JPETAB 91,224,47
ivn-rbt LD50:2 mg/kg JPETAB 91,224,47

SAFETY PROFILE: A poison by ingestion, subcutaneous, intraperitoneal, and intravenous routes. When heated to decomposition it emits very toxic fumes of HCl and NO_x. See also AMINES.

IPI000 CAS:3031-75-2 ***HR: 3***
ISOPROPYL HYDROPEROXIDE
mf: $C_3H_8O_2$ mw: 76.09

$(CH_3)_2CHOOH$

SYN: 2-HYDROPEROXYPROPANE

SAFETY PROFILE: Explodes when heated above its boiling point of 107-109°C. Upon decomposition it emits acrid smoke and fumes. See also PEROXIDES.

IPI350 CAS:53578-07-7 ***HR: 3***
ISOPROPYL HYPOCHLORITE
mf: C_3H_7ClO mw: 94.54

$(CH_3)_2CHOCl$

SAFETY PROFILE: An extremely unstable explosive sensitive to heat and light. Upon decomposition it emits toxic fumes of Cl^-. See also HYPOCHLORITES and EXPLOSIVES.

IPJ000 CAS:79-96-9 ***HR: 3***
4,4'-ISOPROPYLIDENE-BIS(2-tert-BUTYLPHENOL)
mf: $C_{23}H_{32}O_2$ mw: 340.55

TOXICITY DATA with REFERENCE
ipr-mus LD50:40 mg/kg NTIS** AD691-490

CONSENSUS REPORTS: Reported in EPA TSCA Inventory.

SAFETY PROFILE: Poison by intraperitoneal route. When heated to decomposition it emits acrid smoke and fumes.

IPK000 CAS:79-97-0 ***HR: 3***
4,4'-ISOPROPYLIDENEDI-o-CRESOL
mf: $C_{17}H_{20}O_2$ mw: 256.37

TOXICITY DATA with REFERENCE
ivn-mus LD50:25 mg/kg CBCCT* 6,139,54

CONSENSUS REPORTS: Reported in EPA TSCA Inventory.

SAFETY PROFILE: Poison by intravenous route. When heated to decomposition it emits acrid smoke and fumes.

IPL000 CAS:3173-79-3 ***HR: 3***
4-ISOPROPYLIDENE-3,3-DIMETHYL-2-OXETANONE
mf: $C_8H_{12}O_2$ mw: 140.20

SYN: 3-HYDROXY-2,2,4-TRIMETHYL-3-PENTENOIC ACID, β-LACTONE

TOXICITY DATA with REFERENCE
scu-rat TDLo:39 g/kg/78W-I:ETA JNCIAM 39,1213,67

SAFETY PROFILE: Questionable carcinogen with experimental tumorigenic data. When heated to decomposition it emits acrid smoke and fumes.

IPM000 CAS:25068-38-6 ***HR: D***
4,4'-ISOPROPYLIDENEDIPHENOL DIMER with 1-CHLORO-2,3-EPOXYPROPANE
mf: $C_{30}H_{32}O_4{\cdot}C_6H_{10}Cl_2O_2$ mw: 641.68

SYNS: E1001 ◇ EPIKOTE 1001

TOXICITY DATA with REFERENCE
mmo-sat 1 μmol/plate NATUAS 276,391,78
mmo-esc 20 umol ARTODN 46,277,80

CONSENSUS REPORTS: EPA Genetic Toxicology Program. Reported in EPA TSCA Inventory.

SAFETY PROFILE: Mutation data reported. When heated to decomposition it emits toxic fumes of Cl^-. See also other 4,4'-isopropylidenediphenol entries.

IPN000 CAS:25068-38-6 ***HR: D***
4,4'-ISOPROPYLIDENEDIPHENOL, MONOMER with 1-CHLORO-2,3-EPOXYPROPANE
mf: $C_{15}H_{16}O_2 \cdot C_3H_5ClO$ mw: 320.84

SYN: EPIKOTE 828

TOXICITY DATA with REFERENCE
mmo-sat 500 nmol/plate NATUAS 276,391,78
mma-sat 1350 nmol/plate NATUAS 276,391,78

CONSENSUS REPORTS: Reported in EPA TSCA Inventory.

SAFETY PROFILE: Mutation data reported. When heated to decomposition it emits toxic fumes of Cl^-. See also other 4,4'-isopropylidenediphenol entries.

IPO000 CAS:25068-38-6 ***HR: 2***
4,4'-ISOPROPYLIDENEDIPHENOL, POLYMER with 1-CHLORO-2,3-EPOXYPROPANE
mf: $(C_{15}H_{16}O_2 \cdot C_3H_5ClO)x$

SYNS: EPIDIAN 5 ◇ EPON 828

TOXICITY DATA with REFERENCE
eye-rbt 100 MLD AMIHAB 17,129,58
skn-gpg 2750 mg/55D-I AITEAT 23,155,75
skn-gpg TDLo:6111 mg/kg (10-66D preg):TER AITEAT 23,155,75
skn-gpg TDLo:6111 mg/kg (10-66D preg):REP AITEAT 23,155,75
orl-rat LD50:11400 mg/kg AMIHAB 17,129,58
ipr-rat LD50:2400 mg/kg AMIHAB 17,129,58
orl-mus LD50:15600 mg/kg AMIHAB 17,129,58
ipr-mus LD50:4000 mg/kg AMIHAB 17,129,58

CONSENSUS REPORTS: EPA Genetic Toxicology Program. Reported in EPA TSCA Inventory.

SAFETY PROFILE: Moderately toxic by intraperitoneal route. Very slightly toxic by ingestion. Experimental teratogenic effects. Other experimental reproductive effects. A skin and eye irritant. When heated to decomposition it emits toxic fumes of Cl^-. See also other 4,4'-isopropylidenediphenol entries.

IPP000 CAS:25068-38-6 ***HR: D***
4,4'-ISOPROPYLIDENEDIPHENOL, TETRAMER with 1-CHLORO-2,3-EPOXYPROPANE
mf: $C_{60}H_{64}O_8 \cdot C_{12}H_{20}Cl_4O_4$ mw: 1283.36

SYNS: E 1004 ◇ EPIKOTE 1004

TOXICITY DATA with REFERENCE
mmo-sat 1 μmol/plate NATUAS 276,391,78
dnd-esc 1 μmol/L ARTODN 46,277,80

CONSENSUS REPORTS: EPA Genetic Toxicology Program. Reported in EPA TSCA Inventory.

SAFETY PROFILE: Mutation data reported. When heated to decomposition it emits toxic fumes of Cl^-. See also other 4,4'-isopropylidenediphenol entries.

IPR000 CAS:15964-31-5 ***HR: D***
ISOPROPYLIDINE AZASTREPTONIGRIN
mf: $C_{28}H_{27}N_5O_7$ mw: 545.60

SYN: NSC-62709

TOXICITY DATA with REFERENCE
ipr-rat TDLo:30 mg/kg (9D preg):TER CCROBU 53,23,69

SAFETY PROFILE: Experimental teratogenic effects. When heated to decomposition it emits toxic fumes of NO_x.

IPS000 CAS:75-30-9 ***HR: 3***
ISOPROPYL IODIDE
mf: C_3H_7I mw: 170.00

PROP: Colorless liquid. Readily discolors in air and light. D: 1.703 @ 20°/4°, mp: −90°, bp: 89-90°. Sltly sol in water; misc with alc, benzene, chloroform, and ether.

SYNS: 2-IODOPROPANE ◇ i-PROPYL IODIDE

TOXICITY DATA with REFERENCE
ipr-mus TDLo:1190 mg/kg/8W-I:NEO CNREA8 35,1411,75
ihl-rat LC50:320000 mg/m³/30M FAVUAI 7,35,75
ipr-rat LD50:1850 mg/kg 85GMAT -,78,82
ipr-mus LD50:1300 mg/kg 34ZIAG -,756,69

CONSENSUS REPORTS: Reported in EPA TSCA Inventory.

SAFETY PROFILE: Moderately toxic by intraperitoneal route. Mildly toxic by inhalation. Questionable carcinogen with experimental neoplastigenic data. When heated to decomposition it emits toxic fumes of I^-. See also IODIDES.

IPS100 CAS:73791-43-2 ***HR: 3***
ISOPROPYL ISOBUTYL ARSINIC ACID
mf: $C_7H_{17}AsO_2$ mw: 208.16

SYNS: ARSINE OXIDE, HYDROXYISOBUTYLISOPROPYL- ◇ HYDROXYISOBUTYLISOPROPYLARSINE OXIDE

TOXICITY DATA with REFERENCE
ivn-mus LD50:56 mg/kg CSLNX* NX#06282

OSHA PEL: TWA 0.5 mg(As)/m^3

SAFETY PROFILE: Poison by intravenous route. When heated to decomposition it emits toxic fumes of As.

IPS400 CAS:29119-58-2 ***HR: 3***
ISOPROPYL ISOCYANIDE DICHLORIDE
mf: $C_4H_7Cl_2N$ mw: 140.01

$(CH_3)_2CHN{=}CCl_2$

CONSENSUS REPORTS: Cyanide and its compounds are on the Community Right-To-Know List.

SAFETY PROFILE: Reacts violently with water. Violent reaction with iron(III) chloride + metal oxides (e.g., calcium oxide; mercury oxide; or silver oxide). When heated to decomposition it emits toxic fumes of CN^-, Cl^-, and NO_x. See also CHLORIDES and ISOCYANIDES.

IPS500 ***HR: 1***
ISOPROPYL LANOLATE

PROP: A mixture of isopropyl esters of lanolin acids (JEPTDQ 4(4),121,80).

SYNS: AMERLATE P ◇ AMERLATE W ◇ ETHYLAN ◇ LANALENE L ◇ LANALENE P ◇ LANALENE S ◇ LANESTA L ◇ LANESTA P.S.

TOXICITY DATA with REFERENCE
skn-rbt 500 mg MLD JEPTDQ 4(4),121,80
eye-rbt 2 pph MLD JEPTDQ 4(4),121,80
skn-gpg 500 mg MLD JEPTDQ 4(4),121,80

SAFETY PROFILE: An eye and skin irritant. When heated to decomposition it emits acrid smoke and fumes.

IPT000 CAS:4118-51-8 ***HR: 2***
ISOPROPYL MANDELATE
mf: $C_{11}H_{13}O_2$ mw: 177.24

TOXICITY DATA with REFERENCE
eye-rbt 5 mg SEV AJOPAA 29,1363,46

SAFETY PROFILE: A severe eye irritant. When heated to decomposition it emits acrid smoke and fumes. See also ESTERS.

IPU000 CAS:78-44-4 ***HR: 3***
ISOPROPYL MEPROBAMATE
mf: $C_{12}H_{24}N_2O_4$ mw: 260.38

SYNS: APESAN ◇ ARUSAL ◇ BRIANIL ◇ CAPRODAT ◇ CARBAMIC ACID, ESTER with 2-(HDYROXYMETHYL)-1-METHYLPENTYL-ISOPROPYLCARBAMATE ◇ CARBAMIC ACID, ESTER with 2-METHYL-2-PROPYL-1,3-PROPANEDIOL ISOPROPYL-CARBAMATE ◇ CARISOL ◇ CARISOMA ◇ CARISOPRODATE ◇ CARISOPRODATUM ◇ CARISOPRODOL ◇ CARLSODAL ◇ CARLSOMA ◇ CARSODOL ◇ CB 8019 ◇ DIOLENE ◇ DOMARAX ◇ FLEXAL ◇ FLEXARTAL ◇ FLEXARTEL ◇ ISOBAMATE ◇ ISOMEPROBAMATE ◇ ISOPROPYLCARBAMIC ACID, ESTER with 2-(HYDROXYMETHYL)-2-METHYLPENTYL CARBAMATE ◇ N-ISOPROPYL-2-METHYL-2-PROPYL-1,3-PROPANEDIOL DICARBAMATE ◇ MEDIQUIL ◇ (1-METHYLETHYL)CARBAMIC ACID 2-(((AMINOCARBONYL)OXY) METHYL)-2-METHYLPENTYL ESTER ◇ 2-METHYL-2-PROPYL-1,3-PROPANEDIOL CARBAMATE ISOPROPYLCARBAMATE ◇ MIOARTRINA ◇ MIOLISODAL ◇ MIOLISODOL ◇ MIORATRINA ◇ MIORIL ◇ MIORIODOL ◇ NCI-C56235 ◇ NOSPASM ◇ RELA ◇ RELASOM ◇ RELAX ◇ SANOMA ◇ SCH 7307 ◇ SOMA ◇ SOMADRIL ◇ SOMALGIT ◇ SOMANIL ◇ TONOLYT ISOPROPYL MEPROBAMATE

TOXICITY DATA with REFERENCE
orl-rat LD50:1320 mg/kg JPETAB 127,66,59
ipr-rat LD50:450 mg/kg JPETAB 127,66,59
ivn-rat LD50:450 mg/kg PSCBAY 2,17,63
orl-mus LD50:1900 mg/kg AEPPAE 238,92,60
ipr-mus LD50:800 mg/kg AEPPAE 238,92,60
ivn-mus LD50:165 mg/kg JPETAB 127,66,59
ivn-rbt LD50:124 mg/kg IJNEAQ 5,305,66

SAFETY PROFILE: A poison by intravenous route. Moderately toxic by ingestion and intraperitoneal routes. A skeletal muscle relaxant. When heated to decomposition it emits toxic fumes of NO_x. See also MILTOWN.

IPW000 CAS:33020-34-7 ***HR: 3***
ISOPROPYLMERCURY HYDROXIDE
mf: C_3H_8HgO mw: 260.70

SYN: HYDROXYISOPROPYLMERCURY

TOXICITY DATA with REFERENCE
ipr-mus LD50:16 mg/kg OCHRAI 15,5,63

CONSENSUS REPORTS: Mercury and its compounds are on the Community Right-To-Know List.

OSHA PEL: (Transitional: CL 1 mg/10m^3) TWA 0.01 mg(Hg)/m^3; STEL 0.03 mg/m^3 (skin)
ACGIH TLV: TWA 0.01 mg(Hg)/m^3; STEL 0.03(Hg)/m^3
NIOSH REL: (Inorganic Mercury) TWA 0.05 mg(Hg)/m^3

SAFETY PROFILE: Poison by intraperitoneal route. When heated to decomposition it emits toxic fumes of Hg. See also MERCURY COMPOUNDS.

IPX000 CAS:107-44-8 ***HR: 3***
ISOPROPYL METHANEFLUOROPHOSPHONATE
mf: $C_4H_{10}FO_2P$ mw: 140.11

PROP: Bp: 147°, fp: −58°, d: 1.100 @ 20° vap press: 1.57 mm @ 20°, vap d: 4.86.

SYNS: FLUOROISOPROPOXYMETHYLPHOSPINE OXIDE ◇ GB

◇ IMPF ◇ ISOPROPHYL METHYLPHOSPHONOFLUORIDATE ◇ ISOPROPOXYMETHYLPHORYL, FLUORIDE ◇ ISOPROPYL METHYLFLUOROPHOSPHATE ◇ ISOPROPYL METHYLPHOSPHONOFLUORIDATE ◇ O-ISOPROPYL METHYLPHOSPHONOFLUORIDATE ◇ ISOPROPYL-METHYL-PHOSPHORYL FLUORIDE ◇ METHYLFLUOROPHOSPHORIC ACID, ISOPROPYL ESTER ◇ METHYLFLUORPHOSPHORSAEUREISOPROPYLESTER (GERMAN) ◇ METHYLPHOSPHONOFLUORIDIC ACID ISOPROPYL ESTER ◇ METHYLPHOSPHONOFLUORIDIC ACID-1-METHYLETHYL ESTER ◇ MFI ◇ SARIN ◇ SARIN II ◇ T-144 ◇ T-2106 ◇ TL 1618 ◇ TRILONE 46

TOXICITY DATA with REFERENCE
orl-hmn TDLo:2 μg/kg:CNS,PUL,GIT JCINAO 37,350,58
skn-hmn TDLo:103 μg/kg:PNS,EYE,BIO 27ZXA3 -,106,80
ihl-man TCLo:90 μg/m³:EYE,BIO ARTODN 56,201,85
skn-hmn LD50:28 mg/kg SCJUAD 4,33,67
ihl-hmn LC50:70 mg/m³ SCJUAD 4,33,67
orl-rat LD50:550 μg/kg NTIS** PB158-508
ihl-rat LD50:150 mg/m³/10M NTIS** PB158-508
ipr-rat LD50:303 μg/kg FAATDF 5,S84,85
scu-rat LD50:113 μg/kg TXAPA9 16,40,70
ivn-rat LD50:45 μg/kg NTIS** PB158-508
ims-rat LD50:108 μg/kg FAATDF 5,S84,85
ihl-mus LD50:5 mg/m³/30M DEGEA3 15,2179,60
skn-mus LD50:1080 μg/kg NTIS** PB158-508
ipr-mus LD50:420 μg/kg AIPTAK 113,101,57
scu-mus LD50:60 μg/kg DEGEA3 15,2179,60
ivn-mus LD50:113 μg/kg AIPTAK 204,110,73
ims-mus LD50:222 μg/kg AIPTAK 204,110,73
ihl-dog LD50:100 mg/m³/10M NTIS** PB158-508
ivn-dog LD50:19 μg/kg JPETAB 132,50,61
ihl-mky LD50:100 mg/m³/10M NTIS** PB158-508
ims-mky LD50:22300 ng/kg FAATDF 5,S169,85

CONSENSUS REPORTS: EPA Extremely Hazardous Substances List. Reported in EPA TSCA Inventory.

SAFETY PROFILE: A deadly human poison by skin contact and inhalation. (A small drop on the skin can kill a man.) A deadly experimental poison by ingestion, inhalation, skin contact, subcutaneous, intravenous, intramuscular, and intraperitoneal routes. Human systemic effects by inhalation and ingestion: muscle weakness, bronchiolar constriction (including asthma), nausea or vomiting, flaccid paralysis without anesthesia, miosis (pupillary constriction), cholinesterase inhibition. A "nerve gas" used as a chemical warfare agent. To fight fire, use foam, CO_2, dry chemical. When heated to decomposition or reacted with steam, it emits very toxic fumes of F^- and PO_x. See also TABUN and PARATHION.

IPY000 CAS:926-06-7 ***HR: 3***
ISOPROPYLMETHANESULFONATE
mf: $C_4H_{10}O_3S$ mw: 138.20

SYNS: IMS ◇ ISOPROPYL MESYLATE ◇ ISOPROPYL METHANE SULPHONATE ◇ METHANESULFONIC ACID-1-METHYLETHYL ESTER

TOXICITY DATA with REFERENCE
dnd-hmn:fbr 8 mmol/L CRNGDP 3,7,82
msc-ham:lng 1 g/L CNREA8 44,3720,84
ipr-rat LDLo:50 mg/kg (16D preg):REP JRPFA4 18,15,69
ipr-rat LDLo:50 mg/kg (16D preg):TER JRPFA4 18,15,69
skn-mus TDLo:20200 mg/kg/60W-I:CAR CNREA8 47,3402,87
scu-mus TDLo:1760 mg/kg/29W-I:ETA CNREA8 47,3402,87
orl-bwd LD50:287 mg/kg AECTCV 12,355,83

CONSENSUS REPORTS: EPA Genetic Toxicology Program. Reported in EPA TSCA Inventory.

SAFETY PROFILE: Poison by ingestion. An experimental teratogen. Experimental reproductive effects. Human mutation data reported. When heated to decomposition it emits toxic fumes of SO_x. See also SULFONATES.

IPZ000 CAS:67262-75-3 ***HR: 3***
2'-ISOPROPYL-2-(2-METHOXYETHYLAMINO)-PROPIONANILIDE HYDROCHLORIDE
mf: $C_{15}H_{24}N_2O_2$•ClH mw: 300.87

TOXICITY DATA with REFERENCE
ipr-mus LD50:450 mg/kg JPMSAE 67,595,78
ivn-mus LD50:65 mg/kg JPMSAE 67,595,78

SAFETY PROFILE: Poison by intravenous route. Moderately toxic by intraperitoneal route. When heated to decomposition it emits very toxic fumes of NO_x and HCl.

IPZ500 ***HR: 3***
8-((ISOPROPYLMETHYLAMINO)METHYL)QUINOLINE
mf: $C_{14}H_{18}N_2$ mw: 214.34

SYN: N-(8-QUINOLYLMETHYL)-N-METHYL-2-PROPYLAMINE

TOXICITY DATA with REFERENCE
scu-mus LD50:330 mg/kg PCJOAU 11,318,77
ivn-mus LD50:135 mg/kg PCJOAU 11,318,77

SAFETY PROFILE: Poison by subcutaneous and intravenous routes. When heated to decomposition it emits toxic fumes of NO_x.

IQB000 CAS:2235-59-8 ***HR: 3***
N-ISOPROPYL-α-(2-METHYLAZO)-p-TOLUAMIDE
mf: $C_{12}H_{17}N_3O$ mw: 219.32

TOXICITY DATA with REFERENCE
ipr-mus TDLo:255 mg/kg (male 1D pre):REP CNREA8 47,1547,87
orl-mus TDLo:2800 mg/kg/8W-I:NEO JNCIAM 42,337,69

SAFETY PROFILE: Questionable carcinogen with experimental neoplastigenic data. Experimental reproductive effects. When heated to decomposition it emits toxic fumes of NO_x.

IQD000 CAS:74926-98-0 ***HR: 3***
2-ISOPROPYL-6-(1-METHYLBUTYL)PHENOL
mf: $C_{14}H_{22}O$ mw: 206.2

TOXICITY DATA with REFERENCE
ivn-mus LD50:80 mg/kg JMCMAR 23,1350,80
ivn-rbt LD50:20 mg/kg JMCMAR 23,1350,80

SAFETY PROFILE: Poison by intravenous route. When heated to decomposition it emits acrid smoke and fumes.

IQE000 CAS:52061-60-6 ***HR: 2***
1-ISOPROPYL-4-METHYLCYCLOHANE HYDROPEROXIDE
DOT: UN 2125
mf: $C_{10}H_{20}O_2$ mw: 172.30

SYNS: HEXAHYDRO-p-CYMENE HYDROPEROXIDE ◇ p-MENTHANE HYDROPEROXIDE, TECHNICALLY PURE (DOT) ◇ PARAMENTHANE HYDROPEROXIDE (DOT)

DOT Classification: Organic Peroxide; Label: Organic Peroxide.

SAFETY PROFILE: A severe irritant to skin, eyes, and mucous membranes. A very powerful oxidizer. When heated to decomposition it emits acrid smoke and fumes. See also PEROXIDES, ORGANIC.

IQF000 CAS:490-91-5 ***HR: 3***
5-ISOPROPYL-2-METHYL-2,5-CYCLOHEXADIENE-1,4-DIONE
mf: $C_{10}H_{12}O_2$ mw: 164.22

PROP: Yellow triclinic. Mp: 46-47°, bp: 232°. Very sltly sol in water; sol in alc and ether.

SYN: THYMOQUINONE

TOXICITY DATA with REFERENCE
ipr-rat LD50:10 mg/kg ARZNAD 15,1227,65

CONSENSUS REPORTS: Reported in EPA TSCA Inventory.

SAFETY PROFILE: Poison by intraperitoneal route. When heated to decomposition it emits acrid smoke and fumes.

IQH000 CAS:40853-53-0 ***HR: 1***
2-ISOPROPYL-5-METHYL-2-HEXEN-1-OL
mf: $C_{10}H_{20}O$ mw: 156.30

SYNS: ISODIHYDROLAVANDULOL ◇ 5-METHYL-2-(1-METHYLETHYL)-2-HEXEN-1-OL

TOXICITY DATA with REFERENCE
skn-rbt 500 mg/24H MOD FCTXAV 14,319,76

CONSENSUS REPORTS: Reported in EPA TSCA Inventory.

SAFETY PROFILE: A skin irritant. When heated to decomposition it emits acrid smoke and fumes.

IQJ000 CAS:4427-56-9 ***HR: 3***
2-ISOPROPYL-4-METHYLPHENOL
mf: $C_{10}H_{14}O$ mw: 150.24

TOXICITY DATA with REFERENCE
skn-mus TDLo:3840 mg/kg/12W-I:ETA CNREA8 19,413,59

CONSENSUS REPORTS: Reported in EPA TSCA Inventory.

SAFETY PROFILE: Questionable carcinogen with experimental tumorigenic data. When heated to decomposition it emits acrid smoke and fumes.

IQL000 CAS:2814-20-2 ***HR: 2***
2-ISOPROPYL-4-METHYL-6-PYRIMIDOL
mf: $C_8H_{12}N_2O$ mw: 152.22

SYN: 2-ISOPROPYL-4-METHYL-6-HYDROXYPYRIMIDINE

TOXICITY DATA with REFERENCE
orl-rat LD50:2700 mg/kg JAFCAU 18,208,70

CONSENSUS REPORTS: Reported in EPA TSCA Inventory.

SAFETY PROFILE: Moderately toxic by ingestion. When heated to decomposition it emits toxic fumes of NO_x.

IQM000 CAS:1331-24-4 ***HR: 3***
ISOPROPYLMORPHOLINE
mf: $C_7H_{15}NO$ mw: 129.23

TOXICITY DATA with REFERENCE
skn-rbt 500 mg open MOD UCDS** 4/23/63
eye-rbt 15 mg SEV UCDS** 4/23/63
orl-rat LD50:710 mg/kg AIHAAP 30,470,69
ihl-rat LCLo:500 ppm/4H AIHAAP 30,470,69
skn-rbt LD50:100 mg/kg AIHAAP 30,470,69

SAFETY PROFILE: Poison by skin contact. Moderately toxic by ingestion. Mildly toxic by inhalation. A skin and severe eye irritant. When heated to decomposition it emits toxic fumes of NO_x.

IQN000 CAS:110-27-0 ***HR: 3***
ISOPROPYL MYRISTATE
mf: $C_{17}H_{34}O_2$ mw: 270.44

PROP: Liquid of low viscosity, odorless. Bp: 192.6° @

20 mm, decomp @ 208°, d: 0.8532 @ 20°. Sol in castor oil, cottonseed oil, acetone, chloroform, ethyl acetate, ethanol, toluene, and mineral oil. Insol in water, glycerol, and propylene glycol. Dissolves many waxes.

SYNS: ISOMYST ◇ KESSCOMIR ◇ TETRADECANOIC ACID, ISOPROPYL

TOXICITY DATA with REFERENCE
skn-hmn 85 mg/3D-I MLD 85DKA8 -,127,77
skn-rbt 426 mg/24H MLD TXAPA9 19,276,71

CONSENSUS REPORTS: Reported in EPA TSCA Inventory.

SAFETY PROFILE: A human skin irritant. When heated to decomposition it emits toxic smoke and fumes.

IQP000 CAS:1712-64-7 ***HR: 3***
ISOPROPYL NITRATE
DOT: UN 1222
mf: $C_3H_7NO_3$ mw: 105.11

$(CH_3)_2CHONO_2$

PROP: Liquid. Bp: 102°, flash p: 51.8°F, uel: 100%.

SYN: NITRIC ACID, ISOPROPYL ESTER

TOXICITY DATA with REFERENCE
ihl-mus LC50:65 g/m^3/2H 85GMAT -,78,82

CONSENSUS REPORTS: Reported in EPA TSCA Inventory.

DOT Classification: Flammable Liquid; Label: Flammable Liquid.

SAFETY PROFILE: Mildly toxic by inhalation. A dangerous fire hazard when exposed to heat, spark, or flames. Ignites spontaneously when compressed. The pure vapor ignites spontaneously at very low temperatures and pressures. An explosive of low sensitivity. It can be used as a rocket monopropellant. When heated to decomposition it emits toxic fumes of NO_x. Incompatible with Lewis acids. See also NITRATES.

IQQ000 CAS:541-42-4 ***HR: 3***
ISOPROPYL NITRITE
mf: $C_3H_7NO_2$ mw: 89.10

PROP: Flash p: < 50°F.

SYNS: 1-METHYL ETHYL ESTER NITROUS ACID (9CI) ◇ NITROUS ACID, ISOPROPYL ESTER ◇ PROPANOL NITRITE

TOXICITY DATA with REFERENCE
ihl-rat LC50:1250 mg/m^3/4H 85GMAT -,79-82
ihl-mus LC50:2800 mg/m^3/2H 85GMAT -,79-82

CONSENSUS REPORTS: Reported in EPA TSCA Inventory.

SAFETY PROFILE: Moderately toxic by inhalation. Can cause vasodilation with fall in blood pressure, tachycardia, headache. Large doses can cause methemoglobinuria and cyanosis. Severe poisoning results in shock which can be fatal. A very dangerous fire hazard when exposed to heat, spark, or flame. When heated to decomposition it emits toxic fumes of NO_x. See also NITRITES.

IQS000 CAS:72505-65-8 ***HR: D***
2-ISOPROPYL-3-NITROSOTHIAZOLIDINE
mf: $C_6H_{12}N_2OS$ mw: 160.2

SYN: N-NITROSOPROPYLTHIAZOLIDINE

TOXICITY DATA with REFERENCE
mmo-sat 1 mg/L JAFCAU 28,62,80
mma-sat 1 mg/L JAFCAU 28,62,80

SAFETY PROFILE: Mutation data reported. Many N-nitroso compounds are carcinogens. When heated to decomposition it emits very toxic fumes of NO_x and SO_x. See also N-NITROSO COMPOUNDS.

IQS500 CAS:949-36-0 ***HR: 3***
dl-N-ISOPROPYLNORADRENALINE HYDROCHLORIDE
mf: $C_{11}H_{17}NO_3$•ClH mw: 247.75

SYNS: (±)-3,4-DIHYDROXY-α-((ISOPROPYLAMINO)METHYL)BENZYL ALCOHOL HYDROCHLORIDE ◇ (±)1-(3,4-DIHYDROXYPHENYL)-2-ISOPROPYLAMINOETHANOL HYDROCHLORIDE ◇ (±)-4-(1-HYDROXY-2-((1-METHYLETHYL)AMINO)ETHYL)-1,2-BENZENEDIOL HYDROCHLORIDE ◇ dl-ISADRINE HYDROCHLORIDE ◇ (±)-ISOPRENALINE HYDROCHLORIDE ◇ dl-ISOPRENALINE HYDROCHLORIDE ◇ dl-ISOPROPYLNORADRENALINE HYDROCHLORIDE ◇ dl-ISOPROPYLNOREPINEPHRINE HYDROCHLORIDE ◇ (±)-ISOPROTERENOL HYDROCHLORIDE ◇ dl-ISOPROTERENOL HYDROCHLORIDE ◇ dl(±)-ISOPROTERENOL HYDROCHLORIDE ◇ racemic ISOPRENALINE HYDROCHLORIDE ◇ racemic ISOPROTERENOL HDYROCHLORIDE

TOXICITY DATA with REFERENCE
dns-mus-ipr 10 umol/kg CNREA8 39,2751,79
ivg-rbt TDLo:5 mg/kg (1D pre):REP CCPTAY 17,309,78
orl-hmn TDLo:250 μg/kg:BPR JPETAB 92,108,48
ipr-rat LD50:303 mg/kg NIIRDN 6,74,82
scu-rat LD50:435 mg/kg NIIRDN 6,74,82
ivn-rat LD50:24040 μg/kg NIIRDN 6,74,82
orl-mus LD50:2420 mg/kg ARZNAD 13,51,63
ipr-mus LD50:250 mg/kg NIIRDN 6,74,82
scu-mus LD50:298 mg/kg NIIRDN 6,74,82
ivn-mus LD50:23500 μg/kg NIIRDN 6,74,82

SAFETY PROFILE: Poison by subcutaneous, intravenous, and intraperitoneal routes. Experimental reproductive effects. Human systemic effects: blood pressure elevation. Mutation data reported. When heated to decomposition it emits toxic fumes of NO_x and HCl.

IQT000 CAS:13329-71-0 ***HR: 2***
ISOPROPYLOCTADECYLAMINE
mf: $C_{21}H_{46}N$ mw: 312.68

SYN: N-ISOPROPYLOCTADECYLAMINE

TOXICITY DATA with REFERENCE
orl-rat LD50:1270 mg/kg HYSAAV 34(1-3),129,69
orl-mus LD50:2100 mg/kg HYSAAV 34(1-3),129,69
orl-gpg LD50:950 mg/kg HYSAAV 34(1-3),129,69

CONSENSUS REPORTS: Reported in EPA TSCA Inventory.

SAFETY PROFILE: Moderately toxic by ingestion. When heated to decomposition it emits toxic fumes of NO_x. See also AMINES.

IQU000 ***HR: 3***
ISOPROPYL OILS

PROP: A by-product of isopropyl alcohol manufacture composed of trimeric and tetrameric polypropylene + small amounts of benzene, toluene, alkyl benzenes, polyaromatic ring compounds, hexane, heptane, acetone, ethanol, isopropyl ether, and isopropyl alcohol (IARC** 15,225,77).

TOXICITY DATA with REFERENCE
scu-mus TDLo:20 g/kg/20W-I:NEO AMIHBC 5,535,52

CONSENSUS REPORTS: IARC Cancer Review: Animal Inadequate Evidence IMEMDT 15,223,77; Human Limited Evidence IMEMDT 15,223,77.
DFG MAK: Suspected Carcinogen.

SAFETY PROFILE: Suspected carcinogen with experimental neoplastigenic data. When heated to decomposition they emit acrid smoke and fumes.

IQW000 CAS:142-91-6 ***HR: 3***
ISOPROPYL PALMITATE
mf: $C_{19}H_{38}O_2$ mw: 298.57

SYNS: CRODAMOL IPP ◇ DELTYL ◇ DELTYL PRIME ◇ EMCOL-IP ◇ EMEREST 2316 ◇ ESTOL 103 ◇ HEXADECANOIC ACID, ISOPROPYL ESTER ◇ ISOPAL ◇ ISOPROPYL HEXADECANOATE ◇ ISOPROPYL-n-HEXADECANOATE ◇ JA-FA IPP ◇ KESSCO ISOPROPYL ◇ PLYMOUTH IPP ◇ PROPAL ◇ STARFOL IPP ◇ STEPAN D-70 ◇ TEGESTER ISOPALM ◇ UNIMATE IPP ◇ USAF KE-5 ◇ WICKENOL 111

TOXICITY DATA with REFERENCE
skn-hmn 84 mg/3D-I MLD 85DKA8 -,127,77
skn-rbt 500 mg/24H MOD FCTOD7 20 (Suppl),727,82
ipr-mus LD50:100 mg/kg NTIS** AD277-689

CONSENSUS REPORTS: Reported in EPA TSCA Inventory.

SAFETY PROFILE: Poison by intraperitoneal route. A human skin irritant. Used in cosmetics as a physical stabilizer in deodorants, an emollient, emulsifier. When heated to decomposition it emits acrid smoke and fumes. See also ESTERS.

IQX000 CAS:528-92-7 ***HR: 3***
(2-ISOPROPYL-4-PENTENOYL)UREA
mf: $C_9H_{16}N_2O_2$ mw: 184.27

PROP: Needles from ethyl alcohol. Mp: 194°; sol in water @ 0.03 cold, 0.5 hot, sol in alc @ 10, sol in ether @ 1.3.

SYNS: ALLYLISOPROPYLACETYLCARBAMIDE ◇ ALLYLISOPROPYLACETYLUREA ◇ N-(AMINOCARBONYL)-2-(1-METHYLETHYL)-4-PENTENAMIDE ◇ APRONAL ◇ APRONALIDE ◇ ISODORMID ◇ ISOPROPYLALLYLAZETYLKARBAMID (GERMAN) ◇ SEDORMID

TOXICITY DATA with REFERENCE
orl-rat LD50:1050 mg/kg OYYAA2 19,323,80
orl-mus LD50:1220 mg/kg OYYAA2 19,323,80
orl-dog LDLo:300 mg/kg DMWOAX 54,1166,28

SAFETY PROFILE: Poison by ingestion. A sedative. When heated to decomposition it emits toxic fumes of NO_x.

IQZ000 CAS:99-89-8 ***HR: 3***
p-ISOPROPYLPHENOL
mf: $C_9H_{12}O$ mw: 136.21

PROP: Needles. D: 0.990 @ 20°, mp: 61°, bp: 228.2-229.2°. Very sltly sol in water; sol in alc @ 25°, sol in ether @ 25°.

SYNS: AUSTRALOL ◇ p-CUMENOL ◇ 4-ISOPROPYLPHENOL ◇ 4-(1-METHYLETHYL)PHENOL ◇ PRODOX 133

TOXICITY DATA with REFERENCE
ipr-mus LDLo:250 mg/kg CBCCI* 5,339,53
orl-mus LD50:875 mg/kg GISAAA 46(1),94,80
ivn-mus LD50:40 mg/kg JMCMAR 23,1350,80

CONSENSUS REPORTS: Reported in EPA TSCA Inventory.

SAFETY PROFILE: A poison by intraperitoneal and intravenous routes. Moderately toxic by ingestion. When heated to decomposition it emits acrid smoke and fumes. See also PHENOL.

IRA000 CAS:4395-92-0 ***HR: 2***
4-ISOPROPYL PHENYLACETALDEHYDE
mf: $C_{11}H_{14}O$ mw: 162.25

SYNS: CUMINIC ACETALDEHYDE ◇ CUMYL ACETALDEHYDE ◇ p-CYMENE-7-CARBOXALDEHYDE

TOXICITY DATA with REFERENCE
skn-rbt 500 mg/24H SEV FCTXAV 17,509,79
orl-rat LD50:4100 mg/kg FCTXAV 17,509,79

CONSENSUS REPORTS: Reported in EPA TSCA Inventory.

SAFETY PROFILE: Mildly toxic by ingestion. A severe skin irritant. When heated to decomposition it emits acrid smoke and fumes. See also ALDEHYDES.

IRC000 CAS:10099-57-7 ***HR: 2***
p-ISOPROPYLPHENYLETHYL ALCOHOL
mf: $C_{11}H_{16}O$ mw: 164.27

TOXICITY DATA with REFERENCE
orl-rat LD50:1800 mg/kg JPETAB 93,26,48
orl-mus LD50:3900 mg/kg JPETAB 93,26,48

CONSENSUS REPORTS: Reported in EPA TSCA Inventory.

SAFETY PROFILE: Moderately toxic by ingestion. When heated to decomposition it emits acrid smoke and fumes. See also ALCOHOLS.

IRF000 CAS:55-91-4 ***HR: 3***
ISOPROPYL PHOSPHOROFLUORIDATE
mf: $C_6H_{14}FO_3P$ mw: 184.17

PROP: Oily liquid. Mp: −82°, bp: 46° @ 5 mm, d: 1.07 (approx), vap d: 5.24.

SYNS: DFP ◇ DIFLUPYL ◇ DIFLUROPHATE ◇ DIISOPROPOXY-PHOSPHORYL FLUORIDE ◇ DIISOPROPYL FLUOROPHOSPHATE ◇ O,O-DIISOPROPYL FLUOROPHOSPHATE ◇ DIISOPROPYL FLUOROPHOSPHONATE ◇ DIISOPROPYL- FLUOROPHOSPHORIC ACID ESTER ◇ DIISOPROPYLFLUORPHOS- PHORSAEUREESTER (GERMAN) ◇ DIISOPROPYL PHOSPHOFLUOR- IDATE ◇ DIISOPROPYL PHOSPHOROFLUORIDATE ◇ O,O'-DIISOPROPYL PHOSPHORYL FLUORIDE ◇ DYFLOS ◇ FLOROPRYL ◇ FLUOPHOS-PHORIC ACID, DIISOPROPYL ESTER ◇ FLUORODI- ISOPROPYL PHOSPHATE ◇ FLUOROPRYL ◇ FLUOSTIGMINE ◇ ISOFLUOROPH-ATE ◇ ISOFLUROPHATE ◇ ISOPROPYL FLUOPHOSPHATE ◇ NEOGLAUCIT ◇ PF-3 ◇ PHOSPHOROFLUORIDIC ACID, DIISOPROPYL ESTER ◇ RCRA WASTE NUMBER P043 ◇ T-1703 ◇ TL 466

TOXICITY DATA with REFERENCE
ipr-mus TDLo:500 μg/kg (13D preg):REP FEPRA7 31,596,72
ihl-hmn TCLo:8200 μg/m³/10M:EYE,CNS NTIS** PB158-508
orl-rat LD50:5 mg/kg NTIS** PB158-508
ihl-rat LC50:360 mg/m³/10M JCSOA9 -,695,48
ipr-rat LD50:1280 μg/kg ARZNAD 14,85,64
scu-rat LD50:1750 μg/kg ARZNAD 27,1983,77
ims-rat LD50:1800 μg/kg JCINAO 37,350,58
orl-mus LD50:2 mg/kg NTIS** PB158-508
ihl-mus LC50:440 mg/m³/10M:EYE NATUAS 157,287,46
skn-mus LD50:72 mg/kg JPETAB 87,414,46
ipr-mus LD50:2450 μg/kg AITEAT 23,769,75
scu-mus LD50:3 mg/kg JPPMAB 34,603,82
ivn-mus LD50:3200 μg/kg BCPCA6 15,169,66
ihl-dog LD50:5 g/m³/10M NTIS** PB158-508
ihl-mky LD50:500 mg/m³/2M NTIS** PB158-508

CONSENSUS REPORTS: EPA Extremely Hazardous Substances List. Reported in EPA TSCA Inventory.

SAFETY PROFILE: A poison by ingestion, inhalation, skin contact, intraperitoneal, subcutaneous, intramuscular, ocular, and intravenous routes. Moderately toxic by parenteral route. Human systemic effects by inhalation: miosis (pupillary constriction), and headache. Experimental reproductive effects. Used as a basis for "nerve gases." An insecticide. Ingestion can cause damage to eyes, nausea, vomiting, diarrhea, and central nervous system disturbances. An FDA proprietary drug. Used as a miotic agent. When heated to decomposition it emits toxic fumes of F^- and PO_x. See also PARATHION.

IRG000 CAS:304-17-6 ***HR: 2***
N-ISOPROPYLPHTHALIMIDE
mf: $C_{11}H_{11}NO_2$ mw: 189.23

TOXICITY DATA with REFERENCE
orl-rat LDLo:500 mg/kg NCNSA6 5,27,53
ipr-mus LDLo:512 mg/kg CBCCT* 2,135,50

CONSENSUS REPORTS: Reported in EPA TSCA Inventory.

SAFETY PROFILE: Moderately toxic by ingestion and intraperitoneal routes. When heated to decomposition it emits toxic fumes of NO_x.

IRG100 CAS:3772-26-7 ***HR: 2***
1-ISOPROPYL-2-PYRROLIDINONE
mf: $C_7H_{13}NO$ mw: 127.21

PROP: Liquid. Mp: 18°, bp: 216°, d: 0.971, flash p: 212° F.

SYNS: N-ISOPROPYLPYRROLIDINONE ◇ 2-PYRROLIDINONE, 1-ISOPROPYL- ◇ 2-PYRROLIDINONE, 1-(1-METHYLETHYL)-(9CI)

TOXICITY DATA with REFERENCE
skn-rbt 500 mg/24H MLD FCTOD7 26,475,88
eye-rbt 100 mg MOD FCTOD7 26,475,88
orl-rat LD50:2900 mg/kg FCTOD7 26,475,88

CONSENSUS REPORTS: Reported in EPA TSCA Inventory.

SAFETY PROFILE: Moderately toxic by ingestion. A skin and eye irritant. Combustible liquid. When heated to decomposition it emits toxic fumes of NO_x.

IRL000 CAS:1333-53-5 ***HR: 3***
ISOPROPYL QUINOLINE
mf: $C_{12}H_{13}N$ mw: 171.26

SYNS: p-ISOPROPYL QUINOLINE ◇ 6-ISOPROPYL QUINOLINE ◇ LICHENOL

TOXICITY DATA with REFERENCE
orl-rat LD50:940 mg/kg FCTXAV 13,821,75
skn-rbt LD50:160 mg/kg FCTXAV 13,821,75

CONSENSUS REPORTS: Reported in EPA TSCA Inventory.

SAFETY PROFILE: Poison by skin contact. Moderately toxic by ingestion. When heated to decomposition it emits toxic fumes of NO_x.

IRN000 CAS:779-47-5 ***HR: 3***
N-ISOPROPYL TEREPHTHALAMIC ACID
mf: $C_{11}H_{13}NO_3$ mw: 207.25

SYNS: 4-(((1-METHYLETHYL)AMINO)CARBONYL)-BENZOICACID (9CI) ◇ TEREPHTHALIC ACID ISOPROPYLAMIDE

TOXICITY DATA with REFERENCE
orl-mus TDLo:1936 mg/kg/8W-I:ETA JNCIAM 42,337,69

SAFETY PROFILE: Questionable carcinogen with experimental tumorigenic data. When heated to decomposition it emits toxic fumes of NO_x.

IRN100 CAS:1733-25-1 ***HR: 1***
ISOPROPYL TIGLATE
mf: $C_8H_{14}O_2$ mw: 142.22

SYNS: 2-BUTENOIC ACID, 2-METHYL, 1-ISOPROPYL ESTER (E)- ◇ ISOPROPYL 2-METHYL-2-BUTENOATE ◇ ISOPROPYL α-METHYL CROTONATE

TOXICITY DATA with REFERENCE
skn-rbt 500 mg/24H MOD FCTXAV 17,839,79

CONSENSUS REPORTS: Reported in EPA TSCA Inventory.

SAFETY PROFILE: A skin irritant. When heated to decomposition it emits acrid smoke and irritating fumes.

IRP000 CAS:26629-87-8 ***HR: 3***
4-ISOPROPYL-2-(α,α,α-TRIFLUORO-m-TOLYL) MORPHOLINE
mf: $C_{14}H_{18}F_3NO$ mw: 273.33

SYNS: CERM-1766 ◇ 4-(1-METHYLETHYL)-2-(3-(TRIFLUOROMETHYL)PHENYL)MORPHOLINE ◇ OXAFLOZANE ◇ OXAFLOZANO (SPANISH) ◇ 2-((3-TRIFLUOROMETHYL)PHENYL)-4-ISOPROPYL-TETRAHYDRO-1,4-OXAZINE

TOXICITY DATA with REFERENCE
orl-mus LD50:420 mg/kg DRFUD4 3,667,78
ivn-mus LD50:80 mg/kg DRFUD4 3,667,78

SAFETY PROFILE: Poison by intravenous route. Moderately toxic by ingestion. When heated to decomposition it emits very toxic fumes of F^- and NO_x.

IRQ000 CAS:26328-00-7 ***HR: 2***
N-ISOPROPYL-4-(3,4,5-TRIMETHOXY-CINNAMOYL)-1-PIPERAZINEACETAMIDE MALEATE
mf: $C_{21}H_{31}N_3O_5 \cdot C_4H_4O_4$ mw: 521.63

SYNS: MALEATE de CINPROPAZIDE ◇ 68111 M.D. ◇ (3′,4′,5′-TRIMETHOXYCINNAMOYL)-1-(N-ISOPROPYL AMINO CARBONYL METHYL)-4 PIPERAZINE MALEATE

TOXICITY DATA with REFERENCE
orl-rat LD50:1295 mg/kg THERAP 29,233,74
orl-mus LD50:589 mg/kg EJMCA5 10,373,75
ivn-mus LD50:475 mg/kg THERAP 29,233,74

SAFETY PROFILE: Moderately toxic by ingestion and intravenous routes. When heated to decomposition it emits toxic fumes of NO_x.

IRQ100 CAS:67590-57-2 ***HR: 3***
4-ISOPROPYL-2,6,7-TRIOXA-1-ARSABICYCLO (2.2.2)OCTANE
mf: $C_7H_{13}AsO_3$ mw: 220.12

SYN: 2,6,7-TRIOXA-1-ARSABICYCLO(2.2.2)OCTANE,4-ISOPROPYL-

TOXICITY DATA with REFERENCE
ivn-mus LD50:35 mg/kg EJMCA5 13,207,78

OSHA PEL: TWA 0.5 mg(As)/m^3

SAFETY PROFILE: Poison by intravenous route. When heated to decomposition it emits toxic fumes of As.

IRR000 CAS:499-44-5 ***HR: 3***
4-ISOPROPYLTROPOLONE
mf: $C_{10}H_{12}O_2$ mw: 164.22

PROP: An isomeric isopropyltropolone which occurs in the heartwood of western red cedar, or *Thaja plicata D. Don* (JAPMA8 48,722,59).

SYNS: HINOKITIOL ◇ HINOKITOL ◇ 2-HYDROXY-4-ISOPROPYL-2,4,6-CYCLOHEPTATRIEN-1-ONE ◇ 2-HYDROXY-4-(1-METHYL-ETHYL)-2,4,6-CYCLOHEPTATRIEN-1-ONE ◇ β-ISOPROPYL-TROPOLON ◇ β-THUJAPLICIN ◇ β-THUJAPLICINE

TOXICITY DATA with REFERENCE
ipr-mus LD50:85 mg/kg JMCMAR 6,755,63
scu-mus LD50:541 mg/kg YKKZAJ 91,550,71
ivn-mus LD50:128 mg/kg YKKZAJ 91,550,71
scu-gpg LDLo:500 mg/kg JAPMA8 48,722,59

CONSENSUS REPORTS: Reported in EPA TSCA Inventory.

SAFETY PROFILE: Poison by intravenous and intraperitoneal routes. Moderately toxic by subcutaneous route. When heated to decomposition it emits acrid smoke and fumes.

IRS000 CAS:926-65-8 ***HR: 2***
ISOPROPYL VINYL ETHER
mf: $C_5H_{10}O$ mw: 86.14

$(CH_3)_2CHOCH=CH_2$

PROP: Flash p: −25.6°F.

SAFETY PROFILE: A very dangerous fire hazard when exposed to heat, spark, or flame. When heated to decomposition it emits acrid smoke and fumes. See also ETHERS.

IRU000 CAS:114-45-4 ***HR: 3***
(±)-ISOPROTERENOL SULFATE
mf: $C_{22}H_{34}N_2O_6 \cdot H_2O_4S$ mw: 520.66

PROP: (dl form): Crystals from (acetone + methanol). Mp: 128° (some decomp). Sltly sol in alc; insol in chloroform, ether, benzene. (l Form): Crystals. Mp: 164-165°.

SYNS: dl-α-3,4-DIHYDROXYPHENYL-β-ISOPROPYLAMINO-ETHANOL SULFATE ◇ dl-ISOPRENALINE SULFATE ◇ (±)-ISOPRENALINE SULFATE ◇ dl-ISOPROTERENOL SULFATE

TOXICITY DATA with REFERENCE
oms-rat-ipr 20 g/L JNCSAI 16,309,74
orl-rat LD50:3602 mg/kg JPETAB 189,167,74
ivn-rat LD50:335 mg/kg JPETAB 189,167,74
orl-mus LD50:320 mg/kg APTOA6 31,49,72
ipr-mus LD50:365 mg/kg APTOA6 31,43,72
scu-mus LD50:72 mg/kg NIIRDN 6,75,82
ivn-mus LD50:230 mg/kg JPETAB 189,167,74
orl-dog LD50:600 mg/kg NIIRDN 6,75,82
ivn-dog LD50:50 mg/kg NIIRDN 6,75,82
orl-rbt LD50:3070 mg/kg NIIRDN 6,75,82
ivn-rbt LD50:27 mg/kg NIIRDN 6,75,82
orl-gpg LD50:282 mg/kg JPETAB 189,167,74
scu-gpg LD50:610 μg/kg JPETAB 189,167,74
ihl-gpg LC50:4100 ppm/40M JPETAB 189,167,74

SAFETY PROFILE: A poison by ingestion, intravenous, intraperitoneal, and subcutaneous routes. Mildly toxic by inhalation. Mutation data reported. When heated to decomposition it emits very toxic fumes of NO_x and SO_x.

IRV000 CAS:52-53-9 ***HR: 3***
ISOPTIN
mf: $C_{27}H_{38}N_2O_4$ mw: 454.67

SYNS: CP-16533-1 ◇ D-365 ◇ DILACORAN ◇ 5-((3,4-DIMETHOXY-PHENETHYL)METHYLAMINO)-2-(3,4-DIMETHOXYPHENYL)-2-ISOPROPYLVALERONITRILE ◇ IPROVERATRIL ◇ α-(N-METHYL-N-HOMOVERATRYL)-Γ-AMINOPROPYL)-3,4-DIMETHOXYPHENYL-ACETONITRILE ◇ VASOLAN ◇ VERAPAMIL

TOXICITY DATA with REFERENCE
ivn-man TDLo:1429 μg/kg/5M-C:CVS,PUL,SKN NEJMAG 306,238,82
ivn-man TDLo:71 μg/kg:CVS,PUL AHJOA2 111,622,86
orl-man TDLo:48 mg/kg/2W-I:LIV NEJMAG 306,612,82
orl-wmn TDLo:64 mg/kg:CVS BMJOAE 2,1127,78
orl-man LDLo:2 g/kg CHETBF 75,200,79
orl-rat LD50:114 mg/kg ARZNAD 31,1401,81
ipr-rat LD50:67 mg/kg ARZNAD 31,1401,81
scu-rat LD50:107 mg/kg ARZNAD 31,1401,81
ivn-rat LD50:16 mg/kg ARZNAD 31,1401,81
orl-mus LD50:163 mg/kg ARZNAD 31,1401,81
ipr-mus LD50:68 mg/kg ARZNAD 31,1401,81
scu-mus LD50:30770 μg/kg JPPMAB 34,329,82
ivn-mus LD50:1520 μg/kg EJTXAZ 8,188,75

CONSENSUS REPORTS: Cyanide and its compounds are on the Community Right-To-Know List.

SAFETY PROFILE: A poison by ingestion, subcutaneous, intraperitoneal, and intravenous routes. Human systemic effects by ingestion: hepatitis, hepatocellular necrosis (diffuse), pulse rate increased, blood pressure lowering, change in cardiac rate, sweating, dyspnea, cyanosis. When heated to decomposition it emits toxic fumes of NO_x and CN^-. See also NITRILES.

IRV300 CAS:2365-26-6 ***HR: 2***
ISOQUINALDEHYDE THIOSEMICARBAZONE
mf: $C_{11}H_{10}N_4S$ mw: 230.31

SYNS: 1-FORMYLISOQUINOLINE THIOSEMICARBAZONE ◇ IQ 1 ◇ 2-(1-ISOQUINOLINYMETHYLENE)-HYDRAZINECARBOTHIOAMIDE (9CI)

TOXICITY DATA with REFERENCE
oms-esc 2500 μmol/L BCPCA6 21,3213,72
dnd-hmn:hla 40 μmol/L ANYAA9 284,525,77
dnd-mus/ast 30 mg/kg BCPCA6 24,1631,75
ipr-mus LD50:800 mg/kg JMCMAR 9,585,66

SAFETY PROFILE: Moderately toxic by intraperitoneal route. Human mutation data reported. When heated to decomposition it emits toxic fumes of NO_x and SO_x.

IRW000 CAS:7492-29-7 ***HR: 3***
ISOQUINAZEPON
mf: $C_{18}H_{17}ClN_2O$ mw: 312.82

SYNS: 2-CHLORO-5-METHYL-6,7,9,10-TETRAHYDRO-5H-ISOQUINO(2,1-D)(1,4)BENZODIAZEPIN-6ONE ◇ 5,9,10,14B-TETRAHYDRO-2-CHLORO-5-METHYLISOQUINO(2,1-D)(1,4)BENZODIAZEPIN-6(7H)-ONE

TOXICITY DATA with REFERENCE
orl-rat LD50:1730 mg/kg 27ZQAG -,163,72
ipr-rat LD50:465 mg/kg 27ZQAG -,163,72
orl-mus LD50:1160 mg/kg 27ZQAG -,163,72
ipr-mus LD50:233 mg/kg JMCMAR 11,777,68

SAFETY PROFILE: Poison by intraperitoneal route. Moderately toxic by ingestion. When heated to decomposition it emits very toxic fumes of Cl^- and NO_x.

IRX000 CAS:119-65-3 ***HR: 3***
ISOQUINOLINE
mf: C_9H_7N mw: 129.17

PROP: Liquid, pungent odor, almost insol in water, misc with many organic solvents, sol in dilute acids. Hygroscopic platelets when solid. D: 1.09101 @ 30°/4°, mp: 26.48°, bp: 243°.

SYNS: 2-AZANAPHTHALENE ◇ 2-BENZAZINE ◇ BENZO(c)PYRIDINE ◇ LEUCOLINE

TOXICITY DATA with REFERENCE
skn-rbt 10 mg/24H open SEV AMIHBC 4,119,51
eye-rbt 250 μg open SEV AMIHBC 4,119,51
orl-rat LD50:360 mg/kg AMIHBC 4,119,51
ipr-mus LDLo:128 mg/kg CBCCT* 2,189,50
skn-rbt LD50:590 mg/kg AMIHBC 4,119,51

CONSENSUS REPORTS: Reported in EPA TSCA Inventory. EPA Genetic Toxicology Program.

SAFETY PROFILE: A poison by ingestion and intraperitoneal routes. Moderately toxic by skin contact. A severe skin and eye irritant. When heated to decomposition it emits toxic fumes of NO_x.

IRY000 CAS:94-86-0 ***HR: 2***
ISOSAFROEUGENOL
mf: $C_{11}H_{14}O_2$ mw: 178.25

PROP: White crystalline powder; vanilla odor. Flash p: +212°F. Sol fixed oils; insol in water.

SYNS: 6-ETHOXY-m-ANOL ◇ 1-ETHOXY-2-HYDROXY-4-PROPENYLBENZENE ◇ FEMA No. 2922 ◇ HYDROXY METHYL ANETHOL ◇ PROPENYLGUAETHOL (FCC)

TOXICITY DATA with REFERENCE
orl-rat LD50:2400 mg/kg AFDOAQ 15,82,51

CONSENSUS REPORTS: Reported in EPA TSCA Inventory.
SAFETY PROFILE: Moderately toxic by ingestion. Combustible liquid. When heated to decomposition it emits acrid smoke and fumes.

IRZ000 CAS:120-58-1 ***HR: 3***
ISOSAFROLE
mf: $C_{10}H_{10}O_2$ mw: 162.20

PROP: Liquid, odor of anise. Bp: 253°, mp: 8.2°.

SYNS: 1,2-METHYLENEDIOXY-4-PROPENYLBENZENE ◇ 3,4-METHYLENEDIOXY-1-PROPENYL BENZENE ◇ 5-(1-PROPENYL)-1,3-BENZODIOXOLE ◇ 4-PROPENYLCATECHOL METHYLENE ETHER ◇ 4-PROPENYL-1,2-METHYLENEDIOXYBENZENE ◇ RCRA WASTE NUMBER U141

TOXICITY DATA with REFERENCE
skn-rbt 500 mg/24H closed MOD FCTXAV 14,307,76
orl-mus TDLo:61 g/kg/81W-C:CAR JNCIAM 42,1101,69
orl-mus TD:101 g/kg/81W-C:ETA FCTXAV 19,130,81
orl-rat LD50:1340 mg/kg TXAPA9 7,18,65
orl-mus LD50:2470 mg/kg FCTXAV 2,327,64
ipr-mus LDLo:256 mg/kg CBCCT* 1,45,49
scu-cat LDLo:2 g/kg AEXPBL 35,342,1895
ivn-rbt LDLo:300 mg/kg AEXPBL 35,342,1895

CONSENSUS REPORTS: IARC Cancer Review: Group 3 IMEMDT 7,56,87; Animal Sufficient Evidence IMEMDT 1,169,72. Reported in EPA TSCA Inventory. EPA Genetic Toxicology Program.

SAFETY PROFILE: Poison by intraperitoneal and intravenous routes. Moderately toxic by ingestion and subcutaneous routes. A skin irritant. Questionable carcinogen with experimental carcinogenic and tumorigenic data. Used as a pesticide. When heated to decomposition it emits acrid smoke and fumes.

ISA000 CAS:120-62-7 ***HR: 3***
ISOSAFROLE-n-OCTYLSULFOXIDE
mf: $C_{18}H_{28}O_3S$ mw: 324.52

PROP: Water-insol; sltly sol in petr oils; sol in most organic solvents.

SYNS: ENT 16,634 ◇ ISOSAFROLE, OCTYL SULFOXIDE ◇ 1,2-(METHYLENEDIOXY)-4-(2-(OCTYLSULFINYL)PROPYL)BENZENE◇ 1-METHYL-2-(3,4-METHYLENEDIOXYPHENYL)ETHYL OCTYL SULFOXIDE ◇ NCI-C02824 ◇ n-OCTYLISOSAFROLE SULFOXIDE ◇ PIPERONYL SULFOXIDE ◇ SULFOX-CIDE ◇ SULFOXIDE ◇ SULFOXYL ◇ SULPHOXIDE

TOXICITY DATA with REFERENCE
mma-mus:lyms 2500 μg/L MUREAV 196,61,88
scu-mus TDLo:90 mg/kg(6-14D preg):TER NTIS** PB223-160
orl-mus TDLo:31 g/kg/2Y-C:CAR NCITR* NCI-CG-TR-124,79
orl-mus TD:62 g/kg/2Y-C:ETA NCITR* NCI-CG-TR-124,79
orl-rat LD50:2000 mg/kg ARSIM* 20,24,66
skn-rbt LD50:9000 mg/kg 85DPAN -,-,71/76

CONSENSUS REPORTS: NCI Carcinogenesis Bioassay (feed); No Evidence: rat NCITR* NCI-CG-TR-124,79. NCI Carcinogenesis Bioassay (feed); Clear Evidence: mouse NCITR* NCI-CG-TR-124,79.

SAFETY PROFILE: Moderately toxic by ingestion. Slightly toxic by skin contact. Questionable carcinogen with experimental carcinogenic, tumorigenic, and teratogenic data. Reacts violently with $HClO_4$. An insecticide. Mutation data reported. When heated to decomposition it emits highly toxic fumes of SO_x.

ISC500 CAS:16051-77-7 ***HR: 2***
ISOSORBIDE 5-NITRATE
mf: $C_6H_9NO_6$ mw: 191.16

SYNS: 5-ISMN ◇ ISOSORBIDE 5-MONONITRATE

TOXICITY DATA with REFERENCE
orl-rat TDLo:27 g/kg (female 17-22D post):REP KSRNAM 20,6911,86
orl-rat LD50:2010 mg/kg OYYAA2 29,327,85
ipr-rat LD50:1760 mg/kg OYYAA2 29,327,85
ivn-rat LD50:1750 mg/kg OYYAA2 29,327,85
orl-mus LD50:2910 mg/kg OYYAA2 29,327,85
ipr-mus LD50:1810 mg/kg OYYAA2 29,327,85
ivn-mus LD50:1820 mg/kg OYYAA2 29,327,85

SAFETY PROFILE: Moderately toxic by ingestion, intraperitoneal, and intravenous routes. Experimental reproductive effects. When heated to decomposition it emits toxic fumes of NO_x. See also NITRATES.

ISC550 CAS:58958-60-4 ***HR: 1***
ISOSTEARYL NEOPENTANOATE
mf: $C_{23}H_{46}O_2$ mw: 354.69

SYNS: CERAPHYL 375 ◇ CYCLOCHEM INEO ◇ 2,2-DIMETHYLPROPANOIC ACID ISOOCTADECYL ESTER ◇ PROPANOIC ACID, 2,2-DIMETHYL-, ISOOCTADECYL ESTER ◇ SCHERCEMOL 85

TOXICITY DATA with REFERENCE
eye-rbt 100 mg MLD JACTDZ 4(3),1,85

CONSENSUS REPORTS: Reported in EPA TSCA Inventory.

SAFETY PROFILE: An eye irritant. When heated to decomposition it emits acrid smoke and irritating fumes.

ISD000 CAS:2496-92-6 ***HR: 3***
ISO SYSTOX SULFOXIDE
mf: $C_8H_{19}O_4PS_2$ mw: 274.36

SYNS: O,O-DIETHYL-S-(2-ETHTHIONYLETHYL)PHOSPHOROTHIOATE ◇ DIETHYL-S-(2-ETHTHIONYLETHYL) THIOPHOSPHATE ◇ O,O-DIETHYL-S-ETHYL-2-ETHYLMERCAPTO PHOSPHOROTHIOLATE SULFOXIDE

TOXICITY DATA with REFERENCE
orl-rat LD50:2000 μg/kg AEPPAE 234,352,58
ipr-rat LD50:1500 μg/kg AMIHAB 13,606,56
ipr-mus LD50:5600 μg/kg PAREAQ 11,636,59
ipr-gpg LD50:5000 μg/kg AMIHAB 13,606,56

SAFETY PROFILE: A poison by ingestion and intraperitoneal routes. When heated to decomposition it emits very toxic fumes of PO_x and SO_x.

ISE000 CAS:556-61-6 ***HR: 3***
ISOTHIOCYANATOMETHANE
DOT: UN 2477
mf: C_2H_3NS mw: 73.12

PROP: Crystalline. Bp: 119°, d: 1.069. Very sltly sol in water; misc in alc and ether.

SYNS: EP-161E ◇ ISOTHIOCYANATE de METHYLE (FRENCH) ◇ ISOTHIOCYANIC ACID, METHYL ESTER ◇ ISOTIOCIANATO di METILE (ITALIAN) ◇ METHYLISOTHIOCYANAAT (DUTCH) ◇ METHYLISOTHIOCYANAT (GERMAN) ◇ METHYL ISOTHIOCYANATE (DOT) ◇ METHYL MUSTARD OIL ◇ METHYLSENFOEL (GERMAN) ◇ MIC ◇ MIT ◇ MITC ◇ MORTON WP-161E ◇ TRAPEX ◇ TRAPEXIDE ◇ VORLEX ◇ VORTEX ◇ WN 12

TOXICITY DATA with REFERENCE
orl-wmn LDLo:1 g/kg:CNS BMJOAE 283,18,81
orl-rat LD50:97 mg/kg ASCHAN 32,183,59
skn-rat LD50:2780 mg/kg 85DPAN -,-,71/76
orl-mus LD50:97 mg/kg PCOC** -,729,66
skn-mus LD50:1820 mg/kg 85DPAN -,-,71/76
scu-mus LD50:50 mg/kg ARZNAD 5,505,55
skn-rbt LD50:33 mg/kg TXAPA9 42,417,77

CONSENSUS REPORTS: EPA Extremely Hazardous Substances List. Reported in EPA TSCA Inventory.

DOT Classification: Flammable Liquid; Label: Flammable Liquid and Poison.

SAFETY PROFILE: A poison by ingestion, skin contact, and subcutaneous routes. Very irritating to skin, eyes, and mucous membranes. Human systemic effects by ingestion: convulsions or effects on seizure threshold, change in motor activity, coma. An agricultural chemical and pesticide. Flammable when exposed to heat or flame; can react vigorously with oxidizing materials. When heated to decomposition it emits very toxic fumes of NO_x and SO_x. See also THIOCYANATES.

ISF000 CAS:15597-43-0 ***HR: 3***
ISOTHIOCYANATOTRIMETHYLTIN
mf: C_4H_9NSSn mw: 221.89

SYNS: (ISOTHIOCYANATO)TRIMETHYLSTANNE ◇ TRIMETHYLTIN ISOTHIOCYANATE

TOXICITY DATA with REFERENCE
ivn-mus LD50:3600 μg/kg CSLNX* NX#03472

OSHA PEL: TWA 0.1 mg(Sn)/m^3 (skin)
ACGIH TLV: TWA 0.1 mg(Sn)/m^3 (skin) (Proposed: TWA 0.1 mg(Sn)/m^3; STEL 0.2 mg(Sn)/m^3 (skin))
NIOSH REL: (Organotin Compounds) TWA 0.1 mg(Sn)/m^3

SAFETY PROFILE: Poison by intravenous route. When heated to decomposition it emits very toxic fumes of NO_x and SO_x. See also TIN COMPOUNDS and THIOCYANATES.

ISG000 CAS:3137-83-5 ***HR: 3***
ISOTHIOCYANIC ACID-m-ACETAMIDOPHENYL ESTER
mf: $C_9H_8N_2OS$ mw: 192.25

SYN: 3-ACETAMIDOPHENYL ISOTHIOCYANATE

TOXICITY DATA with REFERENCE
ivn-mus LD50:56 mg/kg CSLNX* NX#07851

CONSENSUS REPORTS: Reported in EPA TSCA Inventory.

SAFETY PROFILE: Poison by intravenous route. When heated to decomposition it emits toxic fumes of SO_x. See also THIOCYANATES.

ISH000 CAS:2131-55-7 ***HR: 3***
ISOTHIOCYANIC ACID, p-CHLOROPHENYL ESTER
mf: C_7H_4ClNS mw: 169.63

SYN: 4-CHLOR-PHENYL-ISOTHIOCYANAT(GERMAN)

TOXICITY DATA with REFERENCE
ipr-rat LDLo:100 mg/kg ARZNAD 16,870,66
ipr-mus LDLo:100 mg/kg ARZNAD 21,121,71

SAFETY PROFILE: Poison by intraperitoneal route. When heated to decomposition it emits very toxic fumes of Cl^-, NO_x, and SO_x. See also THIOCYANATES.

ISI000 CAS:2719-32-6 ***HR: 3***
ISOTHIOCYANIC ACID-p-CYANOPHENYL ESTER
mf: $C_8H_4N_2S$ mw: 160.20

SYNS: p-CYANOPHENYL ISOTHIOCYANATE ◇ 4-CYANOPHENYL ISOTHIOCYANATE

TOXICITY DATA with REFERENCE
ivn-mus LD50:24 mg/kg CSLNX* NX#07920

CONSENSUS REPORTS: Cyanide and its compounds are on the Community Right-To-Know List. Reported in EPA TSCA Inventory.

SAFETY PROFILE: Poison by intravenous route. When heated to decomposition it emits very toxic fumes of CN^-, NO_x, and SO_x. See also THIOCYANATES and CYANIDES.

ISJ000 CAS:1122-82-3 ***HR: 3***
ISOTHIOCYANIC ACID, CYCLOHEXYL ESTER
mf: $C_7H_{11}NS$ mw: 141.25

SYN: CYCLOHEXYL-ISOTHIOCYANAT(GERMAN)

TOXICITY DATA with REFERENCE
ipr-rat LDLo:300 mg/kg ARZNAD 16,870,66

CONSENSUS REPORTS: Reported in EPA TSCA Inventory.

SAFETY PROFILE: Poison by intraperitoneal route. When heated to decomposition it emits very toxic fumes of NO_x and SO_x. See also THIOCYANATES.

ISK000 CAS:3688-08-2 ***HR: 3***
ISOTHIOCYANIC ACID, ETHYLENE ESTER
mf: $C_4H_4N_2S_2$ mw: 144.22

SYNS: AETHYLEN-BIS-THIURAMMONOSULFID(GERMAN) ◇ AETHYLSENFOEL (GERMAN) ◇ AETM (GERMAN) ◇ 1,2-DIISOTHIOCYANATOETHANE ◇ DIMETHYLENE DIISOTHIOCYANATE ◇ EBI ◇ EBIS ◇ ETHYLENEBISISOTHIOCYANATE ◇ ETHYLENE-BIS-THIURAMMONO-SULFIDE ◇ ETHYLENE DIISOTHIOCYANATE ◇ SENFOL (GERMAN)

TOXICITY DATA with REFERENCE
orl-rat TDLo:465 mg/kg (female 7-22D post):REP JTEHD6 5,821,79
orl-rat TDLo:500 mg/kg (female 6-15D post):TER BECTA6 17,159,77
orl-rat LD50:112 mg/kg BECTA6 17,159,77
scu-mus LD50:110 mg/kg ARZNAD 5,505,55

SAFETY PROFILE: A poison by ingestion and subcutaneous routes. Experimental teratogenic and reproductive effects. When heated to decomposition it emits very toxic fumes of NO_x, NH_3, and SO_x. See also THIOCYANATES.

ISL000 CAS:404-72-8 ***HR: 3***
ISOTHIOCYANIC ACID-m-FLUOROPHENYL ESTER
mf: C_7H_4FNS mw: 153.18

SYNS: 3-FLUOROPHENYL ISOTHIOCYANATE ◇ m-FLUOROPHENYL ISOTHIOCYANATE

TOXICITY DATA with REFERENCE
ivn-mus LD50:56 mg/kg CSLNX* NX#00949

CONSENSUS REPORTS: Reported in EPA TSCA Inventory.

SAFETY PROFILE: Poison by intravenous route. When heated to decomposition it emits very toxic fumes of F^-, NO_x, and SO_x. See also THIOCYANATES and FLUORIDES.

ISM000 CAS:1544-68-9 ***HR: 3***
ISOTHIOCYANIC ACID-p-FLUOROPHENYL ESTER
mf: C_7H_4FNS mw: 153.18

SYNS: p-FLUOROPHENYL ISOTHIOCYANATE ◇ 4-FLUOROPHENYL ISOTHIOCYANATE

TOXICITY DATA with REFERENCE
ivn-mus LD50:56 mg/kg CSLNX* NX#00950

CONSENSUS REPORTS: Reported in EPA TSCA Inventory.

SAFETY PROFILE: Poison by intravenous route. When heated to decomposition it emits very toxic fumes of F^-, NO_x, and SO_x. See also THIOCYANATES.

ISN000 CAS:551-06-4 ***HR: 3***
ISOTHIOCYANIC ACID-1-NAPHTHYL ESTER
mf: $C_{11}H_7NS$ mw: 185.25

PROP: White, odorless, tasteless needles. Mp: 58°, d: 1.81.

SYNS: ANI ◇ ANIT ◇ 1-ISOTHIOCYANATE-NAPHTHALENE ◇ 1-ISOTHIOCYANATONAPHTHALENE ◇ KESSCOCIDE ◇ α-NAPHTHYL ISOTHIOCYANATE ◇ 1-NAPHTHYL ISOTHIOCYANATE

TOXICITY DATA with REFERENCE
mmo-sat 100 μg/plate ABCHA6 44,3017,80
dns-rat:lvr 10 μmol/L ENMUDM 3,11,81
orl-rat LD50:200 mg/kg JPBAA7 76,175,58
orl-mus LD50:245 mg/kg TXAPA9 7,804,65
ipr-mus LD50:152 mg/kg JJIND8 62,911,79

CONSENSUS REPORTS: EPA Genetic Toxicology Program. Reported in EPA TSCA Inventory.

SAFETY PROFILE: Poison by ingestion and intraperitoneal routes. May cause dermatitis, chills, fever, and kidney injury. Can be absorbed by the intact skin when in solution. Mutation data reported. When heated to decomposition it emits very toxic fumes of NO_x and SO_x. See also THIOCYANATES.

ISO000 CAS:3529-82-6 ***HR: 3***
ISOTHIOCYANIC ACID-m-NITROPHENYL ESTER
mf: $C_7H_4N_2O_2S$ mw: 180.19

SYNS: m-NITROPHENYL ISOTHIOCYANATE ◇ 3-NITROPHENYL ISOTHIOCYANATE

TOXICITY DATA with REFERENCE
ivn-mus LD50:75 mg/kg CSLNX* NX#07672

CONSENSUS REPORTS: Reported in EPA TSCA Inventory.

SAFETY PROFILE: Poison by intravenous route. When heated to decomposition it emits very toxic fumes of NO_x and SO_x. See also THIOCYANATES.

ISP000 CAS:2257-09-2 ***HR: 3***
ISOTHIOCYANIC ACID, PHENETHYL ESTER
mf: C_9H_9NS mw: 163.25

SYNS: (2-ISOTHIOCYANATOETHYL)BENZENE ◇ PHENYLAETHYLSENFOEL (GERMAN) ◇ β-PHENETHYL ISOTHIOCYANATE ◇ PHENYLETHYL ISOTHIOCYANATE ◇ β-PHENYLETHYL ISOTHIOCYANATE ◇ 2-PHENYLETHYL ISOTHIOCYANATE ◇ PHENYLETHYL MUSTARD OIL

TOXICITY DATA with REFERENCE
ipr-rat LDLo:100 mg/kg ARZNAD 19,558,69
ipr-mus LDLo:100 mg/kg ARNZAD 21,121,71
scu-mus LD50:250 mg/kg ARZNAD 5,505,55

CONSENSUS REPORTS: Reported in EPA TSCA Inventory.

SAFETY PROFILE: Poison by intraperitoneal and subcutaneous routes. When heated to decomposition it emits very toxic fumes of NO_x and SO_x. See also THIOCYANATES.

ISQ000 CAS:103-72-0 ***HR: 3***
ISOTHIOCYANIC ACID, PHENYL ESTER
mf: C_7H_5NS mw: 135.19

PROP: Pale yellow liquid. Mp: −21°, bp: 221°, d: 1.1282. Insol in water; sol in alc and ether.

SYNS: BENZENE-1-ISOTHIOCYANATE ◇ ISOTHIOCYANATOBENZENE ◇ PHENYL ISOTHIOCYANATE ◇ PHENYL MUSTARD OIL ◇ PHENYLSENFOEL (GERMAN) ◇ PITC ◇ THIOCARBANIL ◇ USAF M-4

TOXICITY DATA with REFERENCE
scu-mus TDLo:225 mg/kg (6-14D preg):TER NTIS** PB223-160
ipr-rat LDLo:150 mg/kg ARZNAD 16,870,66
orl-mus LD50:87 mg/kg AGACBH 8,610,78
ipr-mus LD50:100 mg/kg NTIS** AD277-689
scu-mus LD50:250 mg/kg ARZNAD 5,505,55

CONSENSUS REPORTS: Reported in EPA TSCA Inventory.

SAFETY PROFILE: A poison by ingestion, intraperitoneal, and subcutaneous routes. An experimental teratogen. When heated to decomposition, or on contact with acid or acid fumes, it emits highly toxic fumes of cyanides and SO_x. See also THIOCYANATES.

ISR000 CAS:62-56-6 ***HR: 3***
ISOTHIOUREA
DOT: UN 2877
mf: CH_4N_2S mw: 76.13

PROP: White powder or crystals. Mp: 177°, bp: decomp, d: 1.405.

SYNS: PSEUDOTHIOUREA ◇ RCRA WASTE NUMBER U219 ◇ SULOUREA ◇ THIOCARBAMATE ◇ THIOCARBAMIDE ◇ β-THIOPSEUDOUREA ◇ THIOUREA (DOT) ◇ 2-THIOUREA ◇ THU ◇ TSIZP 34 ◇ USAF EK-497

TOXICITY DATA with REFERENCE
mmo-sat 150 μg/plate ABCHA6 44,3017,80
dni-hmn:lym 20 mmol/L PNASA6 79,1171,82
orl-rat TDLo:4800 mg/kg (female 17-22D post):TER TJADAB 31,57A,85
orl-rat TDLo:1 g/kg (female 12D post):REP TJADAB 23,335,81
orl-rat TDLo:78 g/kg/56W-C:CAR CNREA8 17,302,57
mul-rat TDLo:151 g/kg/52W-I:CAR CNREA8 17,302,57
orl-rat TD:18 g/kg/2Y-C:NEO SCIEAS 108,626,48
orl-rat TD:40 g/kg/2Y-C:ETA SCIEAS 108,626,48
orl-wmn TDLo:1660 mg/kg/5W:BLD LANCAO 246,179,44
unr-man LDLo:147 mg/kg 85DCAI 2,73,70
orl-rat LD50:125 mg/kg HBTXAC 5,177,59

CONSENSUS REPORTS: NTP Fifth Annual Report on Carcinogens. IARC Cancer Review: Group 2B IMEMDT 7,56,87; Animal Sufficient Evidence IMEMDT 7,95,74. EPA Genetic Toxicology Program. Reported in EPA TSCA Inventory.

DOT Classification: Poison B; Label: St. Andrews Cross.

SAFETY PROFILE: Confirmed carcinogen with experimental carcinogenic, neoplastigenic, and tumorigenic data. A human poison by an unspecified route. An experimental poison by ingestion. Human mutation data reported. Human systemic effects by ingestion: hemorrhage, granulocytopenia (reduction in number of granulocytes), and changes in cell count (unspecified). May cause depression of bone marrow with anemia, leukopenia, and thrombocytopenia. May also cause allergic skin eruptions. Causes hepatic tumors upon chronic administration. Experimental teratogenic and reproductive effects. May react violently with acrolein. Incompatible with acrylaldehyde; H_2O_2; HNO_3. When heated to decomposition it emits very toxic fumes of NO_x and SO_x.

ISU000 CAS:503-74-2 ***HR: 3***
ISOVALERIC ACID
DOT: NA 1760
mf: $C_5H_{10}O_2$ mw: 102.15

PROP: Colorless liquid; acid taste, disagreeable rancid-cheese odor. Solidifies @ −37°, d: 0.931 @ 20°/4°, refr index: 1.403, mp: −34.5° (−50°), bp: 175-177°. Sol in water @ 16°; misc in alc, chloroform, ether.

SYNS: DELPHINIC ACID ◇ FEMA No. 3102 ◇ ISOPENTANOIC ACID (DOT) ◇ ISOPROPYLACETIC ACID ◇ ISOVALERIANIC AICD ◇ 3-METHYLBUTANOIC ACID ◇ β-METHYLBUTYRIC ACID ◇ 3-METHYLBUTYRIC ACID

TOXICITY DATA with REFERENCE
skn-rbt 470 mg open MOD UCDS** 1/31/72
skn-rbt 500 mg/24H MOD FCTXAV 17,841,79
eye-rbt 940 μg MLD UCDS** 1/31/72
orl-rat LD50:2000 mg/kg UCDS** 1/31/72
ivn-mus LD50:1120 mg/kg APTOA6 18,141,61
skn-rbt LD50:310 mg/kg UCDS** 1/31/72

CONSENSUS REPORTS: Reported in EPA TSCA Inventory.

DOT Classification: Corrosive Material; Label: Corrosive.

SAFETY PROFILE: A poison by skin contact. Moderately toxic by ingestion and intravenous routes. A corrosive skin and eye irritant. When heated to decomposition it emits acrid smoke and fumes.

ISV000 CAS:2835-39-4 ***HR: 3***
ISOVALERIC ACID, ALLYL ESTER
mf: $C_8H_{14}O_2$ mw: 142.22

SYNS: ALLYL ISOVALERATE ◇ ALLYL ISOVALERIANATE ◇ ALLYL 3-METHYLBUTYRATE ◇ FEMA No. 2045 ◇ 3-METHYLBUTANOIC ACID, 2-PROPENYL ESTER ◇ 3-METHYLBUTYRIC ACID, ALLYL ESTER ◇ NCI-C54717 ◇ 2-PROPENYL ISOVALERATE ◇ 2-PROPENYL 3-METHYLBUTANOATE

TOXICITY DATA with REFERENCE
skn-rbt 500 mg/24H MOD FCTXAV 17,703,79
orl-rat TDLo:31930 mg/kg/2Y-I:CAR NTPTR* NTP-TR-253,83
orl-mus TDLo:31930 mg/kg/2Y-I:CAR NTPTR* NTP-TR-253,83
orl-rat TD:15065 mg/kg/2Y-I:ETA NTPTR* NTP-TR-253,83
orl-rat LD50:230 mg/kg FCTXAV 17,703,79
skn-rbt LD50:560 mg/kg FCTXAV 17,703,79

CONSENSUS REPORTS: IARC Cancer Review: Group 3 IMEMDT 7,56,87; Animal Limited Evidence IMEMDT 36,69,85. NTP Carcinogenesis Studies (gavage); Clear Evidence: mouse, rat NTPTR* NTP-TR-253,83. Reported in EPA TSCA Inventory.

SAFETY PROFILE: Suspected carcinogen with experimental carcinogenic and tumorigenic data. A poison by ingestion. Moderately toxic by skin contact. A skin irritant. When heated to decomposition it emits acrid smoke and fumes. See also ALLYL COMPOUNDS and ESTERS.

ISW000 CAS:103-38-8 ***HR: 1***
ISOVALERIC ACID, BENZYL ESTER
mf: $C_{12}H_{16}O_2$ mw: 192.28

PROP: Colorless liquid; fruity apple odor. D: 0.985-0.9911, refr index: 1.486, flash p: +212°F. Sol in alc, fixed oils; sltly sol in propylene glycol; insol in glycerin, water @ 246°.

SYNS: BENZYL ISOVALERATE (FCC) ◇ BENZYL-3-METHYLBUTANOATE ◇ BENZYL-3-METHYL BUTYRATE ◇ FEMA No. 2152 ◇ ISOPENTANOIC ACID, PHENYLMETHYL ESTER ◇ ISOPROPYL ACETIC ACID, BENZYL ESTER ◇ 3-METHYLBUTANOIC ACID, PHENYLETHYL ESTER

TOXICITY DATA with REFERENCE
skn-rbt 500 mg/24H MLD FCTXAV 12,829,74

CONSENSUS REPORTS: Reported in EPA TSCA Inventory.

SAFETY PROFILE: A skin irritant. Combustible liquid. When heated to decomposition it emits acrid smoke and fumes. See also ESTERS.

ISX000 CAS:109-19-3 ***HR: 1***
ISOVALERIC ACID, BUTYL ESTER
mf: $C_9H_{18}O_2$ mw: 158.27

PROP: Colorless to pale yellow liquid; fruity odor. Vap d: 5.45, bp: 150°, d: 0.851-0.857, refr index: 1.407. Misc with alc, fixed oils; sltly sol in propylene glycol; insol in water.

SYNS: n-BUTYL ISOPENTANOATE ◇ n-BUTYL ISOVALERATE ◇ 1-BUTYL ISOVALERATE ◇ BUTYL ISOVALERIANATE ◇ BUTYL 3-METHYLBUTYRATE ◇ FEMA No. 2218 ◇ 3-METHYLBUTANOIC ACID, BUTYL ESTER

TOXICITY DATA with REFERENCE
skn-rbt 500 mg/24H MLD FCTXAV 18,659,80
orl-rbt LD50:8200 mg/kg FCTXAV 18,659,80

CONSENSUS REPORTS: Reported in EPA TSCA Inventory.

SAFETY PROFILE: Mildly toxic by ingestion. A skin irritant. Flammable when exposed to heat, flame, sparks, and oxidizers. To fight fire, use alcohol foam, dry chemical, spray, mist, fog. When heated to decomposition it emits acrid smoke and fumes. See also ESTERS.

ISY000 CAS:108-64-5 ***HR: 3***
ISOVALERIC ACID, ETHYL ESTER
mf: $C_7H_{14}O_2$ mw: 130.21

PROP: Colorless, oily liquid; apple odor. Flash p: 77°F, d: 0.868 @ 20°/20°, refr index: 1.395-1.399, bp: 135°, mp: −99°. Misc with alc, fixed oils, benzene, ether; sol in propylene glycol; sltly sol in water @ 135°.

SYNS: ETHYL ISOVALERATE (FCC) ◇ FEMA No. 2463 ◇ 3-METHYL-BUTANOIC ACID, ETHYL ESTER ◇ 3-METHYLBUTYRIC ACID, ETHYL ESTER

TOXICITY DATA with REFERENCE
skn-rbt 500 mg/24H MLD FCTXAV 16,743,78
ipr-rat LD50:1200 mg/kg FCTXAV 16,743,78
orl-rbt LD50:7031 mg/kg IMSUAI 41,31,72

CONSENSUS REPORTS: Reported in EPA TSCA Inventory.

SAFETY PROFILE: Moderately toxic by intraperitoneal route. Mildly toxic by ingestion. A skin irritant. Flammable liquid when exposed to heat, flame, or sparks. When heated to decomposition it emits acrid smoke and fumes. See also ESTERS.

ISZ000 CAS:35154-45-1 ***HR: 1***
(Z)-ISOVALERIC ACID-3-HEXENYL
mf: $C_{11}H_{20}O_2$ mw: 184.31

PROP: Colorless liquid; sweet, apple odor. D: 0.869-0.874, refr index: 1439-1.435. Sol in alc, propylene glycol, fixed oils; insol in water.

SYNS: AI3-35966 ◇ FEMA No. 3498 ◇ cis-3-HEXENYL ISOVALERATE (FCC)

TOXICITY DATA with REFERENCE
skn-rbt 500 mg/24H MLD NTIS** AD-A053-884

CONSENSUS REPORTS: Reported in EPA TSCA Inventory.

SAFETY PROFILE: A skin irritant. When heated to decomposition it emits acrid smoke and fumes.

ITA000 CAS:589-59-3 ***HR: 1***
ISOVALERIC ACID, ISOBUTYL ESTER
mf: $C_9H_{18}O_2$ mw: 158.27

SYNS: ISOBUTYL ISOPENTANOATE ◇ ISOBUTYL ISOVALERATE ◇ 3-METHYLBUTANOIC ACID-2-METHYLPROPYL ESTER ◇ 2-METHYLPROPYL ISOVALERATE ◇ 2-METHYLPROPYL-3-METHYLBUTYRATE

TOXICITY DATA with REFERENCE
skn-rbt 500 mg/24H MOD FCTOD7 20(Suppl),725,82
orl-rbt LD50:7755 mg/kg IMSUAI 41,31,72

CONSENSUS REPORTS: Reported in EPA TSCA Inventory.

SAFETY PROFILE: Mildly toxic by ingestion. A skin irritant. When heated to decomposition it emits acrid smoke and fumes. See also ESTERS.

ITB000 CAS:659-70-1 ***HR: 1***
ISOVALERIC ACID, ISOPENTYL ESTER
mf: $C_{10}H_{20}O_2$ mw: 172.30

PROP: Colorless liquid; apple odor. D: 0.851-0.857, refr index: 1.411, flash p: 162°F. Misc in alc, fixed oils; sltly sol in propylene glycol; insol in water.

SYNS: FEMA No. 2085 ◇ ISOAMYL ISOVALERATE (FCC) ◇ ISOPENTYL ISOVALERATE

TOXICITY DATA with REFERENCE
skn-rbt 500 mg/24H MOD FCTXAV 16,789,78
orl-rbt LD50:13956 mg/kg IMSUAI 41,31,72

CONSENSUS REPORTS: Reported in EPA TSCA Inventory.

SAFETY PROFILE: Mildly toxic by ingestion. A skin irritant. Combustible liquid. When heated to decomposition it emits acrid smoke and fumes. See also ESTERS.

ITC000 CAS:556-24-1 ***HR: 3***
ISOVALERIC ACID, METHYL ESTER
DOT: UN 2400
mf: $C_6H_{12}O_2$ mw: 116.18

SYNS: 3-METHYLBUTANOIC ACID, METHYL ESTER ◇ METHYL

ISOPENTANOATE ◇ METHYL ISOVALERATE ◇ METHYLISOVALERATE (DOT) ◇ METHYL-3-METHYLBUTANOATE ◇ METHYL-3-METHYLBUTYRATE

TOXICITY DATA with REFERENCE
ihl-mus LC50:20250 mg/m^3/2H 85GMAT -,84,82
orl-rbt LD50:5693 mg/kg IMSUAI 41,31,72

CONSENSUS REPORTS: Reported in EPA TSCA Inventory.

DOT Classification: Flammable Liquid; Label: Flammable Liquid.

SAFETY PROFILE: Mildly toxic by ingestion and very slightly toxic by inhalation. Flammable when exposed to heat or flame; can react vigorously with oxidizing materials. When heated to decomposition it emits acrid smoke and fumes. See also ESTERS.

ITD000 CAS:625-28-5 ***HR: 3***
ISOVALERONITRILE
mf: C_5H_9N mw: 83.15

PROP: Colorless liquid. D: 0.795 @ 15°/4°, mp: −100.9°, bp: 130.3°.

SYNS: ISOAMYLNITRILE ◇ 2-METHYLBUTANE SECONDARY MONONITRILE ◇ 2-METHYLBUTYRONITRILE ◇ 3-METHYLBUTYRONITRILE

TOXICITY DATA with REFERENCE
par-mus LDLo:400 mg/kg CBCCT* 7,690,55
orl-mus LD50:233 mg/kg NEZAAQ 39,423,84

CONSENSUS REPORTS: Cyanide and its compounds are on the Community Right-To-Know List. Reported in EPA TSCA Inventory.

SAFETY PROFILE: A poison by ingestion and parenteral routes. When heated to decomposition it emits toxic fumes of NO_x and CN^-. See also NITRILES.

ITD050 ***HR: 2***
ITALIAN ARUM

PROP: Stemless plants with large oval leaves and tuberous roots. A dull purple flower contains a spike which bears brilliant red berries. They are native to Europe and the near East, and are common house plants that grow outdoors in the southern United States.

SYNS: ADAM AND EVE ◇ A. ITALICUM ◇ A. MACULATUM ◇ A. PALAESTINUM ◇ ARUM (VARIOUS SPECIES) ◇ BLACK CALLA ◇ CUCKOOPINT ◇ LORDS-AND-LADIES ◇ SOLOMON'S LILY

SAFETY PROFILE: The whole plant contains toxic calcium oxalate raphides. Chewing any part of the plant results in burning pain in the lips, mouth and throat, possibly followed by inflammation and blistering. Systemic effects are usually not seen because of the insolubility of calcium oxalate, however ingestion may cause inflammation of the stomach and intestines. See also OXALATES.

ITD100 CAS:38357-93-6 ***HR: 3***
ITF 1016
mf: $C_{22}H_{25}F_3N_4O•ClH$ mw: 454.97

SYN: 3-(3-(HEXAHYDRO-1H-AZEPIN-1-YL)PROPOXY)-1(α,α,α-TRIFLUORO-m-TOLYL)-1H-PYRAZOLO(3,4-b)PYRIDINE MONOHYDROCHLORIDE

TOXICITY DATA with REFERENCE
orl-rat LD50:780 mg/kg BCFAAI 111,167,72
ipr-rat LD50:200 mg/kg BCFAAI 111,167,72
orl-mus LD50:750 mg/kg BCFAAI 111,167,72
ipr-mus LD50:380 mg/kg BCFAAI 111,167,72

SAFETY PROFILE: Poison by intraperitoneal route. Moderately toxic by ingestion. When heated to decomposition it emits toxic fumes of F^-, NO_x, and HCl.

ITD875 CAS:70288-86-7 ***HR: 3***
IVERMECTIN

SYNS: 22,23-DIHYDROAVERMECTIN B1 ◇ HYVERMECTIN ◇ MK 933

TOXICITY DATA with REFERENCE
scu-ctl LDLo:8 mg/kg SCIEAS 221,823,83

SAFETY PROFILE: Poison by subcutaneous route. When heated to decomposition emits toxic fumes of NO_x.

ITE000 ***HR: 3***
IVORY

TOXICITY DATA with REFERENCE
imp-rat TDLo:2330 mg/kg:ETA NATWAY 42,75,55

SAFETY PROFILE: Questionable carcinogen with experimental tumorigenic data by implant.

ITF000 CAS:101809-55-6 ***HR: 3***
IYLOMYCIN

PROP: Iylomycin is obtained from *Str. phaeoverticillatus* and belongs to a group of peptide antibiotics (ARZNAD 17,693,67)

TOXICITY DATA with REFERENCE
ipr-mus LDLo:15 mg/kg ARZNAD 17,693,67
ivn-mus LDLo:4400 μg/kg ARZNAD 17,693,67

SAFETY PROFILE: Poison by intraperitoneal and intravenous routes. When heated to decomposition it emits toxic fumes of NO_x.

ITG000 CAS:11006-64-7 ***HR: 3***
IYOMYCIN

PROP: An anti-tumor antibiotic produced by the strain *Streptomyces phaeoverticillatus* (85ERAY 2,1304,78).

TOXICITY DATA with REFERENCE
ipr-mus LD50:180 mg/kg 85ERAY 2,1304,78
ivn-mus LD50:150 mg/kg 85ERAY 2,1304,78

SAFETY PROFILE: Poison by intraperitoneal and intravenous routes. When heated to decomposition it emits toxic fumes of NO_x.

ITH000 CAS:11030-13-0 ***HR: 3***
IYOMYCIN B1

PROP: An anti-tumor antibiotic produced by the strain *Streptomyces phaeoverticillatus* (85ERAY 2,1304,78).

TOXICITY DATA with REFERENCE
ipr-mus LD50:12500 μg/kg 85ERAY 2,1304,78
ivn-mus LD50:4400 μg/kg 85ERAY 2,1304,78
ivn-dog LD50:1500 μg/kg 85ERAY 2,1304,78

SAFETY PROFILE: Poison by intraperitoneal and intravenous routes. When heated to decomposition it emits toxic fumes of NO_x.

J

JAJ000 ***HR: 1***
JACK-IN-THE-PULPIT

PROP: An erect plant with a 3-foot stem and just a few leaves near the top. The flower is pulpit-shaped, covered with a hood and has a spike-like spadix inside. The plant blooms from late spring to early autumn, and bears red or orange-red berries. They are found in moist, shady areas in the region roughly bounded by Ontario, Florida, Texas, and Minnesota.

SYNS: A. DRACONTIUM ◇ ARISAEMA (VARIOUS SPECIES) ◇ A. TRIPHYLLUM ◇ BOG ONION ◇ BROWN DRAGON ◇ CUCKOO PLANT ◇ DRAGON ARUM ◇ DRAGON ROOT ◇ DRAGONS HEAD ◇ DRAGON TAIL ◇ GREEN DRAGON ◇ INDIAN JACK-IN-THE-PULPIT ◇ MEMORY ROOT ◇ PEPPER TURNIP ◇ PETIT PRECHEUR (CANADA) ◇ PRIESTS PENTLE ◇ SMALL JACK-IN-THE-PULPIT ◇ STARCHWORT ◇ THREE-LEAVED INDIAN TURNIP ◇ WAKE ROBIN

SAFETY PROFILE: The whole plant contains toxic calcium oxalate raphides. Chewing any part of the plant results in burning pain in the lips, mouth and throat, possibly followed by inflammation and blistering. Systemic effects are usually not seen because of the insolubility of calcium oxalate. See also OXALATES.

JAK000 CAS:6870-67-3 ***HR: 3***
JACOBINE
mf: $C_{18}H_{25}NO_6$ mw: 351.44

PROP: An alkaloid isolated from *S. Jacobaea* (RETOAE 5,55,49).

SYNS: 15,20-EPOXY-15,30-DIHYDRO-12-HYDROXYSENECIONAN-11,16-DIONE ◇ NSC 89936

TOXICITY DATA with REFERENCE
sln-dmg-par 20 μmol/L ZEVBA5 91,74,60
dns-rat:lvr 1 μmol/L CNREA8 45,3125,85
ivn-mus LD50:77 mg/kg JPETAB 75,69,42

CONSENSUS REPORTS: IARC Cancer Review: Group 3 IMEMDT 7,56,87; Animal Inadequate Evidence IMEMDT 10,275,76. EPA Genetic Toxicology Program.

SAFETY PROFILE: Poison by intravenous route. Questionable carcinogen. Mutation data reported. When heated to decomposition it emits toxic fumes of NO_x.

JAT000 CAS:128-58-5 ***HR: 1***
JADE GREEN BASE
mf: $C_{36}H_{20}O_4$ mw: 516.56

SYNS: C.I. 59825 ◇ DIMETHOXYVIOLANTHRONE ◇ 16,17-DIMETHOXYVIOLANTHRONE ◇ ZELEN OSTANTHRENOVA BRILANTNI FFB (CZECH)

TOXICITY DATA with REFERENCE
eye-rbt 500 mg/24H MLD 28ZPAK -,248,72

CONSENSUS REPORTS: Reported in EPA TSCA Inventory.

SAFETY PROFILE: An eye irritant. When heated to decomposition it emits acrid smoke and fumes.

JBA000 ***HR: 2***
JAMAICA GINGER EXTRACT

PROP: Adulterated with tri-o-cresyl phosphate (JHHBAI 52,39,33).

SYNS: EXTRACT of JAMAICA GINGER ◇ FLUID-EXTRACT of JAMAICA GINGER U.S.P.

TOXICITY DATA with REFERENCE
orl-man TDLo:12 g/kg/7D:CNS ANPSAI 25,29,31
orl-wmn LDLo:284 g/kg/15W JHHBAI 52,39,33

SAFETY PROFILE: Mildly toxic to humans by ingestion. Human systemic effects by ingestion: demyelination of spinal cord. When heated to decomposition it emits toxic fumes of PO_x. See also TRI-2-TOLYL PHOSPHATE (tri-o-cresyl phosphate).

JBS050 ***HR: 1***
JAPANESE AUCUBA

PROP: A large, bushy evergreen with purple flowers at the ends of the branches. The red fruit ripens in winter. It is cultivated as an ornamental in the Pacific, Gulf and Atlantic Coastal states of the United States.

SYNS: AUCUBA JAPONICA ◇ JAPANESE LAUREL

SAFETY PROFILE: The whole plant contains the toxic aucubin glycoside, however, only the fruit is known to have caused poisoning. Ingestion of the fruit may cause vomiting and fever.

JBS100 ***HR: 3***
JAPANESE LANTERN PLANT

PROP: Many species of this plant are cultivated for the

beauty of their fruit pod which has a paper-like shell containing a berry. The ripe berries of some species are edible either raw or cooked. They are native to and cultivated in most of the United States, central and eastern Canada, Hawaii, Guam, and the West Indies.

SYNS: ALQUEQUENJE (PUERTO RICO) ◇ BARBADOS GOOSEBERRY ◇ BATTRE AUTOUR (HAITI) ◇ CAPE GOOSEBERRY ◇ CHINESE LANTERN PLANT ◇ COQUE MOLLE (HAITI) ◇ COQUERET (CANADA) ◇ FAROLITO (CUBA) ◇ GOOSEBERRY TOMATO ◇ HUEVO de GATO (CUBA) ◇ JAMBERRY ◇ MAMAN LAMAN (HAITI) ◇ MEXICAN HUSK TOMATO ◇ PA'INA (HAWAII) ◇ PHYSALIS (VARIOUS SPECIES) ◇ POHA (HAWAII) ◇ SACABUCHE (PUERTO RICO) ◇ STRAWBERRY TOMATO ◇ TOMATES (MEXICO) ◇ TOPE-TOPES (DOMINICAN REPUBLIC) ◇ VEJIGA de PERRO (CUBA) ◇ WINTER CHERRY ◇ YELLOW HENBANE

SAFETY PROFILE: The immature berries contain poisonous solanine glycoalkaloids. Ingestion of the berries has only minor effects on adults, but has caused fatalities in children. Human systemic effects by ingestion: gastroenteric irritation, diarrhea, and fever. See also SOLANINE.

JCA000 ***HR: 2***
JAPAN LACQUER

SAFETY PROFILE: A moderately toxic irritant to skin, eyes, and mucous membranes. An allergen. Dermatitis is frequently caused by natural Japan lacquer due to a highly irritating chemical, urushiol. Synthetic Japan lacquer contains linseed oil, lead oxide and pigments, and solvents such as kerosene or turpentine. Flammable when exposed to heat or flame. Incompatible with oxidizing materials. When heated to decomposition it emits acrid smoke and fumes.

JCS000 CAS:469-59-0 ***HR: 3***
JERVINE
mf: $C_{27}H_{39}NO_3$ mw: 425.67

PROP: Needles from (methanol + water). Mp: 243.5-244.5°. An alkamine isolated from *Veratrum album*.

TOXICITY DATA with REFERENCE
orl-rbt TDLo:66700 μg/kg (female 7D post):TER PSEBAA 136,1174,71
orl-ham TDLo:127 mg/kg (female 7D post):REP 41CIAR -,409,78
orl-rat LDLo:240 mg/kg PSEBAA 149,302,75
orl-mus LDLo:180 mg/kg PSEBAA 149,302,75
scu-mus LD50:29 mg/kg JPETAB 113,89,55
ivn-mus LD50:9300 μg/kg JPETAB 82,167,44
orl-ham LDLo:80 mg/kg TJADAB 17,327,78

SAFETY PROFILE: Poison by ingestion, intravenous, and subcutaneous routes. An experimental teratogen. Experimental reproductive effects. A natural toxin found in some plants. When heated to decomposition it emits toxic fumes of NO_x.

JDA000 CAS:14788-78-4 ***HR: D***
JERVINE-3-ACETATE
mf: $C_{29}H_{40}NO_4$ mw: 466.70

SYN: 3-o-ACETYLJERVINE

TOXICITY DATA with REFERENCE
orl-ham TDLo:150 mg/kg (female 7D post):REP 41CIAR -,409,78
orl-ham TDLo:150 mg/kg (7D preg):TER JAFCAU 26,561,78

SAFETY PROFILE: An experimental teratogen. Other experimental reproductive effects. When heated to decomposition it emits toxic fumes of NO_x. See also JERVINE.

JDA075 ***HR: 2***
JET BEAD

PROP: A shrubby rose which grows to 6 feet. It produces white, 2-inch flowers and a small black berry. It is grown as an ornamental in the northern United States.

SYNS: RHODOTYPOS SCANDENS ◇ WHITE KERRIA

SAFETY PROFILE: The berries contain an unknown toxin. Ingestion of the berries may result in severe low blood sugar levels, ketosis, fever, and convulsions.

JDA100 ***HR: 3***
JET FUEL HEF-2

PROP: Mixture of alkylpentaborane derivatives (CRDLR* 3035,60).

SYN: HEF-2

TOXICITY DATA with REFERENCE
orl-rat LD50:240 mg/kg 14ktak -,693,64
ihl-rat LC50:12 ppm/4H CRDLR* 3035,60
skn-rat LD50:2 g/kg XAWPA2 CWL 2-10,58
ipr-rat LD50:63 mg/kg XAWPA2 CWL 2-10,58
scu-rat LD50:89 mg/kg XAWPA2 CWL 2-10,58
ivn-rat LD50:5 mg/kg XAWPA2 CWL 2-10,58
ihl-mus LC50:11 ppm/4H 14KTAK -,693,64
ihl-dog LCLo:10 ppm/4H CRDLR* 3035,60
skn-rbt LD50:500 mg/kg XAWPA2 CWL 2-10,58
scu-rbt LD50:32 mg/kg XAWPA2 CWL 2-10,58
ivn-rbt LD50:7 mg/kg 14KTAK -,693,64
orl-gpg LD50:316 mg/kg XAWPA2 CWL 2-10,58
skn-gpg LD50:1 g/kg XAWPA2 CWL 2-10,58
ipr-gpg LD50:79 mg/kg XAWPA2 CWL 2-10,58
scu-gpg LD50:100 mg/kg XAWPA2 CWL 2-10,58

SAFETY PROFILE: Poison by ingestion, inhalation, intravenous, subcutaneous, and intraperitoneal routes.

Moderately toxic by skin contact. Flammable when exposed to heat or flame. When heated to decomposition it emits toxic fumes of boron and acrid smoke and fumes. See also BORON COMPOUNDS, BORON HYDRIDE, and PENTABORANE(9).

JDA125 ***HR: 3***
JET FUEL HEF-3

PROP: Boron hydride fuel (14KTAK -,693,64).

SYNS: HEF-3 ◇ HI-CAL 3

TOXICITY DATA with REFERENCE
orl-rat LD50:40 mg/kg 14KTAK -,693,64
ihl-rat LC50:23 ppm/4H 14KTAK -,693,64
skn-rat LD50:317 mg/kg 14KTAK -,693,64
ipr-rat LD50:20 mg/kg 14KTAK -,693,64
ivn-rat LD50:13 mg/kg 14KTAK -,693,64
ihl-mus LD50:6 ppm/4H 14KTAK -,693,64
ivn-mus LD50:16 mg/kg XAWPA2 CWL 2-10,58
skn-cat LD50:126 mg/kg 14KTAK -,693,64
skn-rbt LD50:79 mg/kg XAWPA2 CWL 2-10,58
ipr-rbt LD50:10 mg/kg XAWPA2 CWL 2-10,58
ivn-rbt LD50:3 mg/kg XAWPA2 CWL 2-10,58
skn-gpg LD50:158 mg/kg XAWPA2 CWL 2-10,58
ipr-gpg LD50:18 mg/kg 14KTAK -,693,64

SAFETY PROFILE: Poison by ingestion, inhalation, skin contact, intravenous, and intraperitoneal routes. Flammable when exposed to heat or flame. When heated to decomposition it emits toxic fumes of boron and acrid smoke and fumes. See also BORON COMPOUNDS, BORON HYDRIDE, and PENTABORANE(9).

JDA135 ***HR: 2***
JET FUEL JP-4

PROP: A mixture of aliphatic and aromatic hydrocarbon compounds which meet the requirement of military specification MIL-J-5624E (AMRL** TR-74-78,74)

TOXICITY DATA with REFERENCE
ihl-rat TCLo:500 mg/m^3/2Y:CAR JETPEZ 108,387,86
orl-mus LD50:500 mg/kg AMRL** TR-74-78,74

SAFETY PROFILE: Moderately toxic by ingestion. Questionable carcinogen with experimental carcinogenic data. When heated to decomposition it emits acrid smoke and irritating fumes.

JDJ000 ***HR: 3***
JET FUELS

PROP: Petroleum products similar to kerosene; a number of different types are used. *Jet A and Jet A-1*: flash p: 110-150°F; bp: 400-550°F. *Jet B*: Flash p: -16 to −30°F. *JP-1*: Flash p: 95-145°F, autoign temp: 442°F. *JP-4*: 65% gasoline, 35% light petroleum distillate. Flash p: −10 to 30°F, autoign temp: 468°F, lel: 1.3%, uel: 8%. *JP-5*: Specially refined kerosene. flash p: 95-145°F, autoign temp: approx 475°F. *JP-6*: A higher kerosene cut than JP-4, with fewer impurities. Flash p: 100°F (OC), autoign temp: 446°F, lel: 0.6%, uel: 3.7%, d: 0.8, vap d: < 1, bp: 250°F.

SAFETY PROFILE: *Jet A and Jet A-1*: Combustible. *Jet B*: Dangerous fire hazard. *JP-1*: Flammable to combustible. *JP-4*: Dangerous fire hazard and moderate explosion hazard in the form of vapor. *JP-5*. Flammable to combustible. *JP-6*: Dangerous fire hazard and a moderate explosion hazard in the form of vapor. Violent reaction with F_2. When heated to decomposition they emits acrid smoke and fumes. See also KEROSENE.

JDJ100 ***HR: 2***
JOGEN

PROP: A solution of freeze-dried powder extracted from a culture medium of *Lentinus edodes* (OYYAA2 23,661,82).

SYN: CORTINELLUS SHIITAKE EXTRACT (JAPANESE)

TOXICITY DATA with REFERENCE
orl-rat LD50:15300 mg/kg OYYAA2 23,661,82
orl-mus LD50:16700 mg/kg OYYAA2 23,661,82
ipr-mus LD50:3400 mg/kg OYYAA2 23,661,82
scu-mus LD50:4300 mg/kg OYYAA2 23,661,82
ivn-mus LD50:1600 mg/kg OYYAA2 23,661,82

SAFETY PROFILE: Moderately toxic by intravenous and intraperitoneal routes.

JDS000 CAS:12688-25-4 ***HR: 3***
JOLIPEPTIN

TOXICITY DATA with REFERENCE
ipr-mus LD50:63 mg/kg 85ERAY 3,1642,78
ivn-mus LD50:5210 μg/kg 85ERAY 3,1642,78

SAFETY PROFILE: Poison by intraperitoneal and intravenous routes.

JDS200 CAS:16846-24-5 ***HR: 3***
JOSAMYCIN
mf: $C_{42}H_{69}NO_{15}$ mw: 828.02

PROP: Colorless needles from benzene. Mp: 130-133° (after drying under reduced pressure at 100° for 5 hours). Very sol in methanol, ethanol, acetone, chloroform, ethyl acetate, dioxane, and acidic water; sol in butanol, ether, CCl_4, benzene, and toluene; practically insol in water, petr ether, ligroin, and n-hexane.

SYNS: ANTIBIOTIC YL-704 A3 ◇ IOSALIDE ◇ JOMYBEL ◇ JOSAMINA ◇ KITASAMYCIN A3 ◇ LEUCOMYCIN A3 ◇ TURIMYCIN A5 ◇ YL-704 A3

TOXICITY DATA with REFERENCE
orl-mus TDLo:21 g/kg (7-13D preg):TER JJANAX 22,219,69
orl-mus LD50:6400 mg/kg JJANAX 37,1565,84
ipr-mus LD50:780 mg/kg 85GDA2 2,85,80
ivn-mus LD50:385 mg/kg NIPDAD (35),41,78

SAFETY PROFILE: Poison by intravenous route. Moderately toxic by intraperitoneal route. Mildly toxic by ingestion. An experimental teratogen. When heated to decomposition it emits toxic fumes of NO_x.

JEA000 CAS:8012-91-7 ***HR: 2***
JUNIPER BERRY OIL

PROP: A volatile oil. Principal constituents include d-pinene, camphene, 1-terpineol-4, and other oxygenated constituents. From steam distillation of the fruit of *Juniperus communis* L. (Fam. *Cupressaceae*). (FCTXAV 14,307,76). Colorless to faint green-yellow liquid; aromatic bitter taste. Sol in fixed oils, mineral oil; insol in glycerin, propylene glycol.

SYNS: OIL of JUNIPER BERRY ◇ WACHOLDERBEER OEL (GERMAN)

TOXICITY DATA with REFERENCE
skn-hmn 100% FCTXAV 14,333,76
skn-rbt 500 mg/24H MOD FCTXAV 14,333,76
orl-rat LD50:6280 mg/kg PHARAT 14,435,59

CONSENSUS REPORTS: Reported in EPA TSCA Inventory.

SAFETY PROFILE: Mildly toxic by ingestion. A human skin irritant. An allergen. A systemic irritant. If taken internally, a severe kidney irritation similar to that caused by turpentine may result. When heated to decomposition it emits acrid smoke and fumes. See also individual components.

JEJ000 CAS:8013-10-3 ***HR: 1***
JUNIPER TAR

PROP: Dark brown, viscous, volatile oil. D: 0.950-1.055 @ 25°/25°. Smoky odor; acrid, sltly aromatic taste; very sltly sol in water; sol in 3 vols ether, in chloroform, amyl alc, glacial acetic acid, oil terpentine; sltly sol in alc and petr ether. Main constituents are d-cadinene, 1-cadinol and prepared by destructive distillation of chopped wood of *Juniperus oxycedrus* (FCTXAV 13,681,75).

SYN: CADE OIL RECTIFIED

TOXICITY DATA with REFERENCE
dnr-bcs 8 mg/disc SKEZAP 25,378,84
orl-rat LD50:8014 mg/kg FCTXAV 2,327,64

CONSENSUS REPORTS: Reported in EPA TSCA Inventory.

SAFETY PROFILE: Mildly toxic by ingestion. Mutation data reported. An allergen. A combustible material; can react with oxidizing materials. An FDA-over the counter drug. When heated to decomposition it emits acrid smoke and fumes. See also individual components.

JEJ100 ***HR: 3***
JUNIPERUS COMMUNIS Linn. var. SAXATILIS Pallas, extract excluding roots

PROP: Indian plant belonging to the family *Cupressaceae* IJEBA6 22,487,84

SYN: JUNIPERUS COMMUNIS auct. non. Linn., extract excluding roots

TOXICITY DATA with REFERENCE
orl-rat TDLo:150 mg/kg (female 12-14D post):REP IJEBA6 22,487,84
ipr-mus LD50:100 mg/kg IJEBA6 22,487,84

SAFETY PROFILE: Poison by intraperitoneal route. Experimental reproductive effects. When heated to decomposition it emits acrid smoke and irritating fumes.

K

KAH000 ***HR: 3***
K315
mf: $C_{23}H_{24}N_2O_3 \cdot CH_3SO_4$ mw: 487.59

SYN: 15,16,17,18,19,20-HEXAHYDRO-18-HYDROXY-17-METHOXY-YOHIMBAN PROPIONATE METHANESULFONATE

TOXICITY DATA with REFERENCE
orl-rat LD50:600 mg/kg AIPTAK 110,20,57
ipr-rat LD50:210 mg/kg AIPTAK 110,20,57
ipr-mus LD50:290 mg/kg AIPTAK 110,20,57

SAFETY PROFILE: Poison by intraperitoneal route. Moderately toxic by ingestion. When heated to decomposition it emits toxic fumes of SO_x and NO_x. See also SULFONATES.

KAJ000 CAS:40596-69-8 ***HR: 2***
KABAT
mf: $C_{19}H_{34}O_3$ mw: 310.53

PROP: Amber liquid. Bp: 100°. Solubility in water: 1.39 ppm. Sol in most organic solvents.

SYNS: ALTOSID ◇ ALTOSID IGR ◇ ALTOSID SR 10 ◇ ENT 70,460 ◇ ISOPROPYL(2E,4E)-11-METHOXY-3,7,11-TRIMETHYL-2,4-DODECADIENOATE ◇ MANTA ◇ METHOPRENE ◇ (E,E)-11-METHOXY-3,7,11-TRIMETHYL-2,4-DODECANDIENOATE ◇ ZR 515

TOXICITY DATA with REFERENCE
dni-oin:ovr 100 μmol/L ABCHA6 43,1285,79
oms-oin:ovr 100 μmol/L ABCHA6 43,1285,79
orl-dog LD50:5000 mg/kg EVHPAZ 14,119,76
skn-rbt LD50:3000 mg/kg EVHPAZ 14,119,76

SAFETY PROFILE: Moderately toxic by skin contact. Mildly toxic by ingestion. Mutation data reported. When heated to decomposition it emits acrid smoke and fumes.

KAJ100 ***HR: 1***
KAFFIR LILLY

PROP: These plants have long thin leaves, many of which may grow in the shape of a fan. The flowers are orange or red and form clusters on a leafless stem. They produce red berries. They are native to Africa, but are grown as houseplants and cultivated outdoors in subtropical climates.

SYN: CLIVIA (VARIOUS SPECIES)

SAFETY PROFILE: The whole plant contains the poison lycorine. Ingestion of large amounts can cause nausea, vomiting, and diarrhea.

KAJ500 CAS:82-08-6 ***HR: 2***
KAMALIN
mf: $C_{30}H_{28}O_8$ mw: 516.58

PROP: Light yellow prisms from toluene. Mp: 206-207°. Sol in ether, chloroform, alc, benzene, ethyl acetate; sltly sol in glacial acetic acid; practically insol in water.

SYNS: MALLOTOXIN ◇ ROTTLERIN

TOXICITY DATA with REFERENCE
orl-rat TDLo:150 mg/kg (6D pre):REP IJPPAZ 3,168,59
orl-rat LDLo:750 mg/kg IJPPAZ 3,168,59

SAFETY PROFILE: Moderately toxic by ingestion. Experimental reproductive effects. When heated to decomposition it emits acrid smoke and fumes.

KAL000 CAS:59-01-8 ***HR: 3***
KANAMYCIN
mf: $C_{18}H_{36}N_4O_{11}$ mw: 484.58

SYNS: CANTREX ◇ 4,6-DIAMINO-2-HYDROXY-1,3-CYCLOHEXANE-3,6'-DIAMINO-3,6'-DIDEOXYDI-α-d-GLUCOSIDE ◇ 4,6-DIAMINO-2-HYDROXY-1,3-CYCLOHEXYLENE 3,6'-DIAMINO-3,6'-DIDEOXYDI-d-GLUCOPYRANOSIDE ◇ KANAMICINA (ITALIAN) ◇ KANAMYCIN A ◇ KANAMYTREX ◇ KANTREX ◇ KM ◇ KM (the antibiotic)

TOXICITY DATA with REFERENCE
dnr-bcs 27 μg/L WATRAG 14,1613,80
scu-rat TDLo:3600 mg/kg (1-12D preg):TER ANTBAL 13,344,68
scu-rat TDLo:3600 mg/kg (1-12D preg):REP ANTBAL 13,344,68
ipr-rat LD50:2283 mg/kg NKRZAZ 29(Suppl 2),137,81
ivn-rat LD50:437 mg/kg NKRZAZ 29(Suppl 2),137,81
orl-mus LD50:20500 mg/kg NKRZAZ 29(Suppl 2),137,81
ipr-mus LD50:794 mg/kg CHTHBK 16,371,71
scu-mus LD50:1350 mg/kg AACHAX -,341,67
ivn-mus LD50:115 mg/kg MIFAAB 11,108,62
ims-mus LD50:54 mg/kg BCFAAI 98,224,59
ivn-rbt LD50:150 mg/kg 85ERAY 1,690,78

CONSENSUS REPORTS: Reported in EPA TSCA Inventory. EPA Genetic Toxicology Program.

SAFETY PROFILE: Poison by intravenous and intramuscular routes. Moderately toxic by ingestion, intraperitoneal, and subcutaneous routes. An experimental teratogen. Experimental reproductive effects. Mutation data reported. When heated to decomposition it emits toxic fumes of NO_x.

KAM000 CAS:25389-94-0 ***HR: 3***
KANAMYCIN SULFATE
mf: $C_{18}H_{36}N_4O_{11}•O_4S$ mw: 580.64

PROP: Irregular prisms. Decomp over wide range above 250°. Very sol in water; insol in common alc and nonpolar solvents.

SYNS: KANNASYN ◇ KANTREX SULFATE

TOXICITY DATA with REFERENCE
ipr-rat LD50:3200 mg/kg JJANAX 28,415,75
ivn-rat LD50:225 mg/kg JJANAX 28,415,75
ipr-mus LDLo:1914 mg/kg AIMDAP 119,493,67
scu-mus LD50:2150 mg/kg JJANAX 28,415,75
ivn-mus LD50:180 mg/kg ANTBAL 24,60,79
ims-mus LDLo:1914 mg/kg AIMDAP 119,493,67
ivn-rbt LD50:550 mg/kg JJANAX 28,415,75

SAFETY PROFILE: A poison by intravenous route. Moderately toxic by intraperitoneal, subcutaneous, and intramuscular routes. When heated to decomposition it emits very toxic NO_x and SO_x. See also KANAMYCIN SULFATE (1 : 1) SALT.

KAV000 CAS:25389-94-0 ***HR: 3***
KANAMYCIN SULFATE (1 : 1) SALT
mf: $C_{18}H_{36}N_4O_{11}•H_2O_4S$ mw: 582.66

PROP: Irregular prisms. Decomp above 250°, very sol in water, insol in common alcs and nonpolar solvents.

SYNS: CANTREX ◇ CRISTALOMICINA ◇ KAMYCIN ◇ KAMYNEX ◇ KANABRISTOL ◇ KANACEDIN ◇ KANAMYCIN A SULFATE ◇ KANAMYCIN MONOSULFATE ◇ KANAMYCIN SULFATE ◇ KANAMYTREX ◇ KANAQUA ◇ KANASIG ◇ KANATROL ◇ KANESCIN ◇ KANNASYN ◇ KANO ◇ KANTREX ◇ KANTREXIL ◇ KANTROX ◇ KLEBCIL ◇ OPHTALMOKALIKAN ◇ OTOKALIXIN ◇ RESISTOMYCIN (BAYER)

TOXICITY DATA with REFERENCE
scu-rat TDLo:980 mg/kg (8-14D preg):TER JJANAX 28,372,75
ims-chd TDLo:390 mg/kg:EAR JOPDAB 60,230,62
ipr-rat LD50:3200 mg/kg JJANAX 28,415,75
ivn-rat LD50:225 mg/kg JJANAX 28,415,75
orl-mus LD50:17500 mg/kg NIIRDN 6,173,82
ipr-mus LD50:1648 mg/kg 85ERAY 1,690,78
scu-mus LD50:1648 mg/kg 85ERAY 1,690,78
ivn-mus LD50:240 mg/kg NIIRDN 6,173,82
ims-mus LD50:1190 mg/kg NIIRDN 6,173,82
ivn-rbt LD50:550 mg/kg JJANAX 28,415,75

SAFETY PROFILE: Poison by intravenous route. Moderately toxic by subcutaneous, intraperitoneal, and intramuscular routes. Mildly toxic by ingestion. Human systemic effects by intramuscular route: acuity changes. An experimental teratogen. An FDA proprietary drug. Used as an antibacterial agent. When heated to decomposition it emits very toxic fumes of NO_x and SO_x.

KBA100 CAS:29701-07-3 ***HR: 3***
KANENDOMYCIN SULFATE
mf: $C_{18}H_{37}N_5O_{10}•H_2O_4S$ mw: 581.68

SYNS: o-3-AMINO-3-DEOXY-α-d-GLUCOPYRANOSYL-(1-6)-o-(2,6-DIAMINO-2,6-DIDEOXY-α-d-GLUCOPYRANOSYL-(1-4))-2-DEOXY-d-STREPTAMINE SULFATE (1:1) ◇ AMINODEOXYKANAMYCIN SULFATE ◇ BEKANAMYCIN SULFATE ◇ KANAMYCIN B SULFATE ◇ KANAMYCIN B, SULFATE (1:1) (SALT)

TOXICITY DATA with REFERENCE
ipr-rat LD50:1400 mg/kg NIIRDN 6,742,82
scu-rat LD50:1900 mg/kg NIIRDN 6,742,82
ivn-rat LD50:141 mg/kg NIIRDN 6,742,82
ims-rat LD50:1420 mg/kg NIIRDN 6,742,82
ipr-mus LD50:760 mg/kg NIIRDN 6,742,82
scu-mus LD50:740 mg/kg NIIRDN 6,742,82
ivn-mus LD50:112 mg/kg NIIRDN 6,742,82
ims-mus LD50:628 mg/kg NIIRDN 6,742,82

SAFETY PROFILE: Poison by intravenous route. Moderately toxic by subcutaneous, intramuscular, and intraperitoneal routes. When heated to decomposition it emits toxic fumes of SO_x and NO_x.

KBB000 CAS:32891-29-5 ***HR: 3***
KAO 264
mf: $C_{21}H_{27}NO_3S_2•ClH•H_2O$ mw: 460.09

SYN: PG 501

TOXICITY DATA with REFERENCE
orl-rat TDLo:1200 mg/kg (9-14D preg):TER OYYAA2 8,1213,74
orl-rat LD50:1182 mg/kg KSRNAM 6,2448,72
ipr-rat LD50:104 mg/kg KSRNAM 6,2448,72
orl-mus LD50:858 mg/kg KSRNAM 6,2448,72
ipr-mus LD50:126 mg/kg KSRNAM 6,2448,72
scu-mus LD50:341 mg/kg KSRNAM 6,2448,72
ivn-mus LD50:20200 μg/kg KSRNAM 6,2448,72

SAFETY PROFILE: Poison by subcutaneous, intravenous, and intraperitoneal routes. Moderately toxic by ingestion. An experimental teratogen. When heated to decomposition it emits toxic fumes of SO_x, NO_x, and HCl. See also ESTERS.

KBB600 CAS:1332-58-7 ***HR: 2***
KAOLIN

PROP: Fine white to light yellow powder; earth taste. Insol in ether, alc, dil acids, and alkali solutions.

SYN: ALTOWHITES ◇ BENTONE ◇ CONTINENTAL ◇ DIXIE ◇ EMATHLITE ◇ FITROL ◇ FITROL DESICCITE 25 ◇ GLOMAX

◇ HYDRITE ◇ KAOPAOUS ◇ KAOPHILLS-2 ◇ LANGFORD ◇ MCNAMEE ◇ PARCLAY ◇ PEERLESS ◇ SNOW TEX

TOXICITY DATA with REFERENCE
orl-rat TDLo:590 g/kg (female 37D pre):REP JONUAI 107,2020,77

OSHA PEL: (Transitional: TWA Respirable Fraction: 15 mg/m^3; Respirable Fraction: 5 mg/m^3) TWA Total Dust: 10 mg/m^3; Respirable Fraction: 5 mg/m^3
ACGIH TLV: TWA (nuisance particulate) 10 mg/m^3 of total dust (when toxic impurities are not present, e.g., quartz < 1%); (Proposed: 2 mg/m^3; Respirable Fraction).

SAFETY PROFILE: A nuisance dust.

KBK000 CAS:9000-36-6 ***HR: 1***
KARAYA GUM

PROP: Dried exudate of the tree, *Sterculia ureus* Roxburgh (Fam. *Sterculiaceae*). Fine, white powder; slt odor of acetic acid. Insol in alc; swells in water to a gel.

TOXICITY DATA with REFERENCE
orl-rat LDLo:30 g/kg FOREAE 13,29,48

CONSENSUS REPORTS: Reported in EPA TSCA Inventory.

SAFETY PROFILE: Very mildly toxic by ingestion. A mild allergen.

KBU000 CAS:39472-31-6 ***HR: 3***
KARMINOMYCIN

SYNS: CARMINOMYCIN ◇ o-DEMETHYLDAUNOMYCIN

TOXICITY DATA with REFERENCE
mmo-sat 1 μg/plate ENMUDM 7,129,85
oms-rat:lvr 4870 pmol/L CNREA8 40,387,80
ipr-mus TDLo:400 μg/kg (female 9D post):TER RCOCB8 28,497,80
ipr-mus TDLo:300 μg/kg (7D preg):REP RCOCB8 28,497,80
ivn-hmn TDLo:12 mg/kg/5D:BLD,CNS,GIT CTRRDO 61,1705,77
ivn-man TDLo:1071 μg/kg/17D:CNS,GIT,CVS VOONAW 25(2),63,79
ipr-rat LD50:1550 μg/kg ANTBAL 24,218,79
ivn-rat LD50:1600 μg/kg ANTBAL 23,1005,78
orl-mus LD50:7300 μg/kg CTRRDO 61,1705,77
ipr-mus LD50:1100 μg/kg ANTBAL 20(10),897,75
scu-mus LD50:2500 μg/kg ANTBAL 21,1030,76
ivn-mus LD50:3200 μg/kg ANTBAL 23,1005,78
orl-dog LD50:2700 μg/kg CTRRDO 63,899,79
ivn-dog LDLo:600 μg/kg CTRRDO 63,899,79
orl-mky LD50:288 mg/kg DCTODJ 4,383,81

SAFETY PROFILE: Poison by ingestion, intravenous, intraperitoneal, and subcutaneous routes. Human systemic effects by intravenous route: anorexia, hallucinations and distorted perceptions, thromobosis, nausea or vomiting, fatty liver degeneration, impaired liver function, endocrine changes, and leukopenia (reduced white blood cell count). An experimental teratogen. Experimental reproductive effects. Mutation data reported. When heated to decomposition it emits acrid smoke and fumes.

KCA000 CAS:52794-97-5 ***HR: 3***
KARMINOMYCIN HYDROCHLORIDE

SYN: CARMINOMYCIN HYDROCHLORIDE

TOXICITY DATA with REFERENCE
orl-rat LD50:23000 μg/kg ANTBAL 23,128,78
ivn-rat LD50:860 μg/kg ANTBAL 23,128,78
orl-mus LD50:7300 μg/kg ANTBAL 23,128,78
ivn-mus LD50:3700 μg/kg ANTBAL 23,128,78
orl-dog LD50:3000 μg/kg ANTBAL 23,128,78
ivn-dog LD50:1000 μg/kg ANTBAL 23,128,78
orl-rbt LD50:12000 μg/kg ANTBAL 23,128,78
ivn-rbt LD50:1500 μg/kg ANTBAL 23,128,78
orl-gpg LD50:3000 μg/kg ANTBAL 23,128,78
ivn-gpg LD50:1400 μg/kg ANTBAL 23,128,78

SAFETY PROFILE: Poison by ingestion and intravenous routes. When heated to decomposition it emits toxic fumes of HCl. See also KARMINOMYCIN.

KCA100 ***HR: 2***
KASH, LEAF EXTRACT

PROP: Indian plant belonging to the family *Taxaceae* (IJMRAQ 57,237,69).

SYNS: BIRMI, LEAF EXTRACT ◇ DINGSABLCH, LEAF EXTRACT ◇ TAXUS BACCATA LINN., LEAF EXTRACT

TOXICITY DATA with REFERENCE
orl-rbt TDLo:200 mg/kg (1D pre):REP JRPFA4 22,151,70
ipr-rat LDLo:700 mg/kg THERAP 19,1021,64

SAFETY PROFILE: Moderately toxic by intraperitoneal route. Experimental reproductive effects.

KCK000 CAS:19408-46-9 ***HR: 3***
KASUGAMYCIN HYDROCHLORIDE
mf: $C_{14}H_{25}N_3O_9 \cdot ClH$ mw: 415.88

SYNS: KASUGAMYCIN MONOHYDROCHLORIDE ◇ KASUMIN

TOXICITY DATA with REFERENCE
orl-mus LD50:20500 μg/kg JJANAX 21,206,68
ipr-mus LD50:7600 mg/kg AACHAX -,225,69
scu-mus LD50:12 g/kg JJANAX 21,206,68
ivn-mus LD50:3850 mg/kg AACHAX -,225,69

SAFETY PROFILE: Poison by ingestion. Moderately

toxic by intravenous route. When heated to decomposition it emits very toxic fumes of HCl and NO_x. See also other kasugamycin entries.

KCU000 CAS:101651-86-9 ***HR: 1***
KASUGAMYCIN PHOSPHATE
mf: $C_{14}H_{25}N_3O_9 \cdot xH_3O_4P$ mw: 1065.42

TOXICITY DATA with REFERENCE
orl-mus LD50:21 g/kg JJANAX 21,206,68
ipr-mus LD50:12500 mg/kg JJANAX 21,206,68
scu-mus LD50:13500 mg/kg JJANAX 21,206,68
ivn-mus LD50:5400 mg/kg JJANAX 21,206,68

SAFETY PROFILE: Mildly toxic by ingestion and intravenous routes. When heated to decomposition it emits very toxic fumes of PO_x and NO_x. See also other kasugamycin entries.

KDA000 CAS:78822-08-9 ***HR: 1***
KASUGAMYCIN SULFATE
mf: $C_{14}H_{25}N_3O_9 \cdot H_2O_4S$ mw: 477.50

TOXICITY DATA with REFERENCE
orl-rat LD50:22 g/kg JJANAX 21,206,68
ipr-rat LD50:12 g/kg JJANAX 21,206,68
scu-rat LD50:15500 mg/kg JJANAX 21,206,68
ivn-rat LD50:4900 mg/kg JJANAX 21,206,68
orl-mus LD50:20700 mg/kg JJANAX 21,206,68
ipr-mus LD50:10 g/kg JJANAX 21,206,68
scu-mus LD50:12 g/kg JJANAX 21,206,68
ivn-mus LD50:5500 mg/kg JJANAX 21,206,68
scu-dog LD50:10500 mg/kg JJANAX 21,206,68
ivn-dog LD50:4500 mg/kg JJANAX 21,206,68
scu-rbt LDLo:9200 mg/kg JJANAX 21,206,68
ivn-rbt LD50:4500 mg/kg JJANAX 21,206,68

SAFETY PROFILE: Mildly toxic by ingestion and several other routes. When heated to decomposition it emits very toxic fumes of SO_x and NO_x. See also other kasugamycin entries.

KDA025 CAS:11121-08-7 ***HR: 2***
KATEXOL 300

TOXICITY DATA with REFERENCE
skn-rbt 500 mg/24H SEV 28ZPAK -,281,72
eye-rbt 50 µg/24H SEV 28ZPAK -,281,72
orl-rat LD50:2390 mg/kg 28ZPAK -,281,72

SAFETY PROFILE: Moderately toxic by ingestion. A severe eye and skin irritant.

KDA035 ***HR: D***
KATHON BIOCIDE

PROP: Containing a mixture of 5-chloro-2-methyl-4-isothiazolin-3-one and 2-methyl-4-isothiazolin-3-one (MUREAV 118,129,83).

SYN: KATHON MW 886 BIOCIDE

TOXICITY DATA with REFERENCE
mma-sat 268 ng/plate MUREAV 118,129,83
msc-mus:lym 198 µg/L MUREAV 118,129,83

SAFETY PROFILE: Mutation data reported. An antimicrobial agent. When heated to decomposition it emits toxic fumes of Cl^-, SO_x, and NO_x.

KDA100 ***HR: 3***
KC-404
mf: $C_{14}H_{18}N_2O$ mw: 230.34

SYN: 3-ISOBUTYRYL-2-ISOPROPYLPYRAZOLO(1,5-a)PYRIDINE

TOXICITY DATA with REFERENCE
orl-rat TDLo:405 mg/kg (female 17-22D post):REP KSRNAM 20,101,86
orl-rat LD50:1340 mg/kg KSRNAM 19,5503,85
ipr-rat LD50:419 mg/kg KSRNAM 19,5503,85
scu-rat LD50:1300 mg/kg KSRNAM 19,5503,85
ivn-rat LD50:42500 µg/kg KSRNAM 19,5503,85
orl-mus LD50:1860 mg/kg KSRNAM 19,5503,85
ipr-mus LD50:460 mg/kg KSRNAM 19,5503,85
scu-mus LD50:3100 mg/kg KSRNAM 19,5503,85
ivn-mus LD50:146 mg/kg KSRNAM 19,5503,85

SAFETY PROFILE: Poison by intravenous route. Moderately toxic by ingestion, intraperitoneal, and subcutaneous routes. Experimental reproductive effects. When heated to decomposition it emits toxic fumes of NO_x.

KDK000 CAS:9002-83-9 ***HR: 3***
KEL-F
mf: $(C_2ClF_3)_n$

SYN: VOLTALEF 10

TOXICITY DATA with REFERENCE
imp-rat TDLo:36 mg/kg:ETA CNREA8 15,333,55

SAFETY PROFILE: Questionable carcinogen with experimental tumorigenic data. A relatively inert chlorofluorocarbon polymer. When heated to decomposition it emits very toxic fumes of Cl^- and F^-.

KDU100 ***HR: 3***
KEMAMINE S 190
mf: $C_{19}H_{29}NO_2 \cdot ClH$ mw: 339.95

SYNS: β-DIETHYLAMINOETHYL-2-PHENYLHEXAHYDROBENZOATE HYDROCHLORIDE ◇ 2-PHENYLCYCLOHEXANECARBOXYLIC ACID 2-DIETHYLAMINOETHYL ESTER HYDROCHLORIDE ◇ S 190

TOXICITY DATA with REFERENCE
orl-rat LDLo:1600 mg/kg APPNAH 1,4,50

scu-rat LDLo:1600 mg/kg APPNAH 1,4,50
ivn-rat LDLo:60 mg/kg APPNAH 1,4,50
ivn-rbt LDLo:20 mg/kg APPNAH 1,4,50
scu-gpg LDLo:267 mg/kg APPNAH 1,4,50
ivn-gpg LDLo:27 mg/kg APPNAH 1,4,50

SAFETY PROFILE: Poison by subcutaneous and intravenous routes. Moderately toxic by ingestion. When heated to decomposition it emits toxic fumes of NO_x and HCl.

KEA000 CAS:143-50-0 ***HR: 3***
KEPONE
mf: $C_{10}Cl_{10}O$ mw: 490.60

PROP: A chlorinated polycyclic ketone. A crystalline material. Mp: decomp @ 350°. Sltly water-sol; sol in alc, ketones, and acetic acid. Readily hydrates on exposure to room temperature and humidity; normally used as a mono- to trihydrate (NCIBR*).

SYNS: CHLORDECONE ◇ CIBA 8514 ◇ COMPOUND 1189 ◇ 1,2,3,5,6,7,8,9,10,10-DECACHLORO(5.2.1.0$^{(2,6)}$.0$^{(3,9)}$.0$^{(5,8)}$)DECANO-4-ONE ◇ DECACHLOROKETONE ◇ DECACHLORO-1,3,4-METHENO-2H-CYCLOBUTA(cd)PENTALEN-2-ONE ◇ DECACHLOROOCTAHYDROKEPONE-2-ONE ◇ DECACHLOROOCTAHYDRO-1,3,4-METHENO-2H-CYCLOBUTA(cd)PENTALEN-2-ONE ◇ 1,1a,3,3a,4,5,5,5a,5b,6-DECACHLOROOCTAHYDRO-1,3,4-METHENO-2H-CYCLOBUTA(cd)PENTALEN-2-ONE ◇ DECACHLOROPENTACYCLO (5.2.1.0$^{(2,6)}$.0$^{(3,9)}$.0$^{(5,8)}$)DECAN-4-ONE ◇ DECACHLOROPENTACYCLO (5.3.0.0$^{(2,6)}$.0$^{(4,10)}$.0$^{(5,9)}$)DECAN-3-ONE ◇ DECACHLOROTETRACYCLODECANONE ◇ DECACHLOROTETRAHYDRO-4,7-METHANOINDENEONE ◇ ENT 16,391 ◇ GENERAL CHEMICALS 1189 ◇ MEREX ◇ NCI-C00191 ◇ RCRA WASTE NUMBER U142

TOXICITY DATA with REFERENCE
spm-qal-orl 200 ppm/42D-C TXAPA9 43,535,78
orl-rat TDLo:17 mg/kg (female 1-22D post):REP FEPRA7 40,667,81
orl-mus TDLo:120 mg/kg (female 7-16D post):TER TXAPA9 38,189,76
orl-rat TDLo:200 mg/kg/2Y-C:CAR NEOLA4 26,231,79
orl-mus TDLo:1200 mg/kg/80W-I:CAR NCITR* NCI-CG-TR-b,76
orl-rat LD50:95 mg/kg GUCHAZ 6,96,73
orl-dog LD50:250 mg/kg TXAPA9 45,331,78
orl-rbt LD50:65 mg/kg PCOC** -,642,66
skn-rbt LD50:345 mg/kg GUCHAZ 6,96,73

CONSENSUS REPORTS: NTP Fifth Annual Report on Carcinogens. IARC Cancer Review: Group 2B IMEMDT 7,56,87; Human Limited Evidence IMEMDT 20,67,79; Animal Sufficient Evidence IMEMDT 20,67,79. EPA Genetic Toxicology Program.
DFG MAK: Suspected Carcinogen.
NIOSH REL: (Kepone) CL 0.001 mg/m^3/15M

SAFETY PROFILE: Confirmed carcinogen with experimental carcinogenic data. Poison by ingestion, skin contact. Experimental teratogenic and reproductive effects. Mutation data reported. Inhalation, absorption or ingestion by humans can lead to central nervous system, liver and kidney damage, including bizarre symptoms caused by damage to the nervous system. Usually, the symptoms are tremors, ataxia, skin changes, hyperexcitability, hyperactivity, muscle spasms, testicular atrophy, low sperm count, estrogenic effects, sterility, breast enlargement, liver lesions and cancer. An insecticide and fungicide. Registration suspended by the USEPA.

KEA300 CAS:10405-02-4 ***HR: 3***
KEPTAN
mf: $C_{25}H_{30}NO_3$•Cl mw: 428.01

SYNS: AS XVII ◇ AZONIASPIRO(3-α-BENZILOYLOXY-NORTROPAN-8,1'-PYRROLIDINE)-CHLORIDE ◇ AZONIASPIRO COMPOUND XVII ◇ 8-α-BENZILOYLOXY-6,10-ETHANO-5-AZONIASPIRO (4.5)DECANE CHLORIDE ◇ 3-α-HYDROXY-SPIRO (1-α-H,5-α-H-NORTROPANE-8,1'-PYRROLIDINIUM) CHLORIDE BENZILATE ◇ SPASMEX ◇ SPASMO 3 ◇ TROSPIUM CHLORIDE

TOXICITY DATA with REFERENCE
orl-rat LD50:1510 mg/kg NIIRDN 6,137,82
ipr-rat LD50:97700 μg/kg NIIRDN 6,137,82
scu-rat LD50:707 mg/kg NIIRDN 6,137,82
ivn-rat LD50:15500 μg/kg NIIRDN 6,137,82
idu-rat LDLo:500 mg/kg ARZNAD 25,1037,75
orl-mus LD50:812 mg/kg NIIRDN 6,137,82
ipr-mus LD50:50 mg/kg NIIRDN 6,137,82
scu-mus LD50:203 mg/kg NIIRDN 6,137,82
ivn-mus LD50:11200 μg/kg NIIRDN 6,137,82
scu-dog LDLo:160 mg/kg OYYAA2 8,199,74
orl-rbt LDLo:3200 mg/kg OYYAA2 8,199,74
ivn-rbt LDLo:20 mg/kg OYYAA2 8,199,74

SAFETY PROFILE: Poison by subcutaneous, intravenous and intraperitoneal, routes. Moderately toxic by ingestion and intraduodenal routes. When heated to decomposition it emits toxic fumes of Cl^- and NO_x.

KEA350 CAS:63659-19-8 ***HR: 3***
KERLONE
mf: $C_{18}H_{29}NO_3$•ClH mw: 343.94

SYNS: BETAXOLOL HYDROCHLORIDE ◇ 1-(4-(2-(CYCLOPROPYL-METHOXY)ETHYL)PHENOXY)-3-ISOPROPYL-AMINOPROPAN-2-OL HYDROCHLORIDE ◇ 2-PROPANOL, 1-(4-(2-(CYCLOPROPYL-METHOXY)ETHYL)PHENOXY)-3-((1-METHYL-ETHYL)AMINO)-, HYDROCHLORIDE ◇ SL-75212 ◇ SL-D.212

TOXICITY DATA with REFERENCE
orl-rat TDLo:2048 mg/kg (female 15-22D post):REP LMSED6 1,43,83
orl-rat TDLo:4 g/kg (female 6-15D post):TER LMSED6 1,43,83
ivn-rat LD50:30 mg/kg LMSED6 1,43,83
orl-mus LD50:48 mg/kg LMSED6 1,43,83
ivn-mus LD50:37 mg/kg USXXAM #4252984

SAFETY PROFILE: Poison by ingestion and intravenous routes. An experimental teratogen. Experimental reproductive effects. When heated to decomposition it emits toxic fumes of NO_x and HCl.

KEK000 CAS:8008-20-6 ***HR: 3***
KEROSENE
DOT: UN 1223

PROP: A pale yellow to water-white, oily liquid. Bp: 175-325°, ULC: 40, flash p: 150-185°F, d: 0.80 to <1.0, lel: 0.7%, uel: 5.0%, autoign temp: 410°F, vap d: 4.5. Insol in water; misc with other petr solvents. A mixture of petroleum hydrocarbons, chiefly of the methane series having from 10-16 carbon atoms per molecule.

SYNS: COAL OIL ◇ COAL OIL (export shipment only) (DOT) ◇ DEOBASE ◇ STRAIGHT-RUN KEROSENE

TOXICITY DATA with REFERENCE
skn-rbt 500 mg SEV JACTDZ 1,30,90
mmo-sat 25 uL/plate CBTOE2 2,63,86
orl-man TDLo:3570 mg/kg:PUL,GIT,MET TORAAK 15,263,66
orl-man LDLo:500 mg/kg YAKUD5 22,883,80
ivn-man TDLo:403 mg/kg:CNS CTOXAO 10,283,77
unr-man LDLo:1176 mg/kg 85DCAI 2,73,70
ipr-rat LDLo:10700 mg/kg TXAPA9 1,156,59
itr-rat LDLo:800 mg/kg TXAPA9 1,462,59
orl-dog LDLo:4 g/kg AJMSA9 221,531,51
ivn-dog LDLo:200 mg/kg AJMSA9 221,531,51
itr-dog LDLo:800 mg/kg AJMSA9 221,531,51
orl-rbt LD50:28 g/kg TXAPA9 3,689,61
ipr-rbt LD50:6600 mg/kg AIMEAS 21,803,44
ivn-rbt LD50:180 mg/kg AIMEAS 21,803,44
itr-rbt LD50:200 mg/kg TXAPA9 3,689,61
orl-gpg LD50:20 g/kg AIMEAS 21,803,44

CONSENSUS REPORTS: IARC Cancer Review: Group 2A IMEMDT 45,39,89; Animal Limited Evidence IMEMDT 45,39,89. Reported in EPA TSCA Inventory.

NIOSH REL: (Kerosene) TWA 100 mg/m^3
DOT Classification: Combustible Liquid; Label: None; Flammable or Combustible Liquid; Label: Flammable Liquid.

SAFETY PROFILE: Suspected carcinogen. Poison by intravenous and intratracheal routes. Moderately toxic to animals by ingestion. A severe skin irritant. Mutation data reported. Human systemic effects by ingestion and intravenous routes: somnolence, hallucinations and distorted perceptions, coughing, nausea or vomiting, and fever. Aspiration of vomitus can cause serious pneumonitis, particularly in young children. Combustible when exposed to heat or flame; can react with oxidizing materials. Moderately explosive in the form of vapor when exposed to heat or flame. When heated to decomposition it emits acrid smoke and fumes. To fight fire, use foam, CO_2, dry chemical.

KEK200 CAS:6740-88-1 ***HR: 3***
KETAMINE
mf: $C_{13}H_{16}ClNO$ mw: 237.75

PROP: Crystals from pentane-ether. Mp: 92-93°.

SYNS: 2-(o-CHLOROPHENYL)-2-(METHYLAMINO)-CYCLOHEXANONE ◇ 2-(METHYLAMINO)-2-(2-CHLOROPHENYL)CYCLOHEXANONE

TOXICITY DATA with REFERENCE
par-mky TDLo:1 mg/kg (23W preg):TER BJANAD 47,917,75
ipr-mus LD50:400 mg/kg BJANAD 55,457,83
ivn-mus LD50:77 mg/kg JZKEDZ 1,119,75

SAFETY PROFILE: Poison by intravenous and intraperitoneal routes. An experimental teratogen. A general anesthetic. When heated to decomposition it emits toxic fumes of Cl^- and NO_x.

KEU000 CAS:463-51-4 ***HR: 3***
KETENE
mf: C_2H_2O mw: 42.04

$H_2C{=}C{=}O$

PROP: Colorless gas with disagreeable taste. Decomp in water. Mp: $-150°$, bp: $-56°$, vap d: 1.45. Decomp in alc; sol in ether and acetone.

SYNS: CARBOMETHENE ◇ ETHENONE ◇ KETO-ETHYLENE

TOXICITY DATA with REFERENCE
orl-rat LD50:1300 mg/kg UCDS**
ihl-mus LCLo:23 ppm/30M JIHTAB 31,209,49
ihl-rat LCLo:53 ppm/100M JIHTAB 31,209,49
ihl-mky LCLo:200 ppm/10M JIHTAB 31,209,49
ihl-cat LCLo:750 ppm/10M JIHTAB 31,209,49
ihl-rbt LCLo:53 ppm/2H JIHTAB 31,209,49
ihl-gpg LCLo:53 ppm/100M JIHTAB 31,209,49

CONSENSUS REPORTS: Reported in EPA TSCA Inventory. EPA Genetic Toxicology Program.

OSHA PEL: (Transitional: TWA 0.5 ppm) TWA 0.5 ppm; STEL 1.5 ppm
ACGIH TLV: TWA 0.5 ppm; STEL 1.5 ppm
DFG MAK: 0.5 ppm (0.9 mg/m^3)

SAFETY PROFILE: Poison by inhalation. Moderately toxic by ingestion. Can cause pulmonary edema. Reacts with hydrogen peroxide to form the explosive diacetyl peroxide. When heated to decomposition it emits acrid smoke and fumes. See also ALDLEHYDES.

KFA000 CAS:674-82-8 ***HR: 3***
KETENE DIMER
DOT: UN 2521
mf: $C_4H_4O_2$ mw: 84.08

H_2!J1C=CCH_2CO•!J2O

PROP: Colorless, nonhygroscopic liquid; pungent odor. Mp: −6.5°, bp: 127.4°, d: 1.0897, vap d: 2.9, flash p: 93°F(TOC). Decomp in water.

SYNS: 3-BUTENO-β-LACTONE ◇ DIKETENE ◇ DIKETENE, inhibited (DOT) ◇ 4-METHYLENE-2-OXETANONE

TOXICITY DATA with REFERENCE
skn-rbt 500 mg open MLD UCDS** 2/20/63
skn-rbt 537 mg IHFCAY 6,1,67
eye-rbt 25 mg SEV UCDS** 2/20/63
eye-rbt 537 mg IHFCAY 6,1,67
orl-rat LD50:560 mg/kg TXAPA9 28,313,74
ihl-rat LCLo:20000 ppm/1H IHFCAY 6,1,67
orl-mus LDLo:800 mg/kg KODAK* 21MAY71
skn-rbt LD50:2830 mg/kg TXAPA9 28,313,74

CONSENSUS REPORTS: Reported in EPA TSCA Inventory.

DOT Classification: Flammable or Combustible Liquid; Label: Flammable Liquid.

SAFETY PROFILE: Moderately toxic by ingestion and skin contact. A skin and severe eye irritant. Flammable when exposed to heat or flame; can react vigorously with oxidizing materials. A violent polymerization reaction is catalyzed by acids, bases, or sodium acetate. A storage hazard. Self-initiated exothermic dimerization is explosive. To fight fire, use alcohol foam. When heated to decomposition it emits acrid smoke and fumes.

KFA100 CAS:2507-91-7 ***HR: 3***
KETHOXAL-BIS-THIOSEMICARBAZIDE
mf: $C_8H_{16}N_6OS_2$ mw: 276.42

SYNS: B.W. 356-C-61 ◇ 356C61 ◇ CONTRAPAR ◇ 1,1′-((1-ETHOXYETHYL)ETHANEDIYLIDENE)BIS(3-THIOSEMICARBAZIDE) ◇ (1-ETHOXYETHYL)GLYOXAL BIS(THIOSEMICARBAZONE) ◇ 3-ETHOXY-2-OXOBUTYRALDEHYDEBIS(THIOSEMICARBAZONE) ◇ GLOXAZON ◇ GLOXAZONE ◇ HYDRAZINECARBOTHIOAMIDE, 2,2′-(1-(1-ETHOXYETHYL)-1,2-ETHANEDIYLIDENE)BIS◇ 2-KETO-3-ETHOXY-BUTYRALDEHYDE-BIS(THIOSEMICARBAZONE)◇ KTS (PHARMACEUTICAL)◇ NSC-82116 ◇ U-7726 ◇ WR 9838

TOXICITY DATA with REFERENCE
dni-hmn:hla 59 μmol/l MUREAV 92,427,82
ivn-rat LD50:182 mg/kg CHTHBK 13,339,68
ipr-mus LDLo:1000 mg/kg CNCRA6 30,9,63

SAFETY PROFILE: Poison by intravenous route. Moderately toxic by intraperitoneal route. Human mutation data reported. When heated to decomposition it emits toxic fumes of SO_x and NO_x.

KFK000 CAS:5965-49-1 ***HR: 3***
KETOBEMIDONE HYDROCHLORIDE
mf: $C_{15}H_{21}NO_2$ mw: 247.37

SYNS: CLIRADON HYDROCHLORIDE ◇ 1-METHYL-4-(m-HYDROXYPHENYL)PIPERIDINE-4-ETHYLKETONE HYDROCHLORIDE

ivn-rat LD50:40 mg/kg AIPTAK 115,213,58
orl-rat LD50:215 mg/kg AIPTAK 115,213,58

SAFETY PROFILE: Poison by ingestion and intravenous routes. When heated to decomposition it emits toxic fumes of NO_x. See also KETONES.

KFK100 CAS:65277-42-1 ***HR: 3***
KETOCONAZOLE
mf: $C_{26}H_{28}Cl_2H_4O_4$ mw: 531.48

PROP: Crystals from 4-methyl-2-pentanone. Mp: 146°.

SYNS: cis-1-ACETYL-4-(4-((2-(2,4-DICHLOROPHENYL)-2-(1H-IMIDAZOL-1-YLMETHYL)-1,3-DIOXOLAN-4-YL)METHOXY)PHENYL)-PIPERAZINE ◇ NIZORAL

TOXICITY DATA with REFERENCE
dni-hmn:lym 10 mg/L AMACCQ 24,478,83
orl-man TDLo:1029 mg/kg (male 60D pre):REP ACENA7 110,276,85
orl-man LDLo:45 mg/kg/17D-I GUTTAK 26,636,86
orl-wmn LDLo:412 mg/kg/15W-I GUTTAK 26,636,86
orl-wmn LDLo:264 mg/kg/66D-I:LIV GASTAB 86,503,84
orl-cld TDLo:450 mg/kg/90D-I:SYS BMJOAE 293,993,86
orl-man TDLo:49 mg/kg/17D-I:LIV GASTAB 86,503,84
orl-wmn TDLo:60 mg/kg:LIV GASTAB 86,503,84
orl-rat LD50:166 mg/kg MDACAP 17,373,81
ivn-rat LD50:86 mg/kg DRUGAY 23,1,82
orl-mus LD50:618 mg/kg MDACAP 17,373,81
ivn-mus LD50:41500 μg/kg MDACAP 17,373,81
orl-dog LD50:178 mg/kg MDACAP 17,373,81

SAFETY PROFILE: Poison by ingestion and intravenous routes. Experimental reproductive effects. Human systemic effects: liver changes, evidence of thyroid hypofunction. Human mutation data reported. Note: May be associated with hepatic toxicity. When heated to decomposition it emits toxic fumes of Cl^- and NO_x.

KGA000 ***HR: D***
KETONES

PROP: Liquid organic compounds containing the carbonyl group C=O attached to two alkyl groups. Derived from secondary alcohols by oxidation. Acetone, which is dimethyl ketone, is the most familiar of this group of compounds.

SAFETY PROFILE: No general statement can be made as to the toxicity of ketones. Some are highly volatile and hence may have narcotic or anesthetic effects. Skin absorption, as well as inhalation, may be an important route of entry into the body. None of the ketones has been shown to have a high degree of chronic toxicity. Some are dangerous fire hazards. They react violently with aldehydes; HNO_3; $HNO_3 + H_2O_2$; $HClO_4$. A variety of peroxides can be formed from the reactions of ketones and hydrogen peroxide. Many of these peroxides are explosives sensitive to heat and shock. Common air contaminants. See also ACETONE, DIETHYL KETONE, and METHYL ETHYL KETONE.

KGA100 CAS:7039-09-0 ***HR: 3***
4-KETONIRIDAZOLE
mf: $C_6H_4N_4O_4S$ mw: 228.20

SYN: 1-(5-NITRO-2-THIAZOLYL)HYDANTOIN

TOXICITY DATA with REFERENCE
mmo-sat 1 nmol/plate ENMUDM 4,320,82
mma-sat 1 nmol/plate ENMUDM 4,320,82
ipr-mus LD50:55 mg/kg JPETAB 228,662,84

SAFETY PROFILE: Poison by intraperitoneal route. Mutation data reported. When heated to decomposition it emits toxic fumes of SO_x and NO_x.

KGK000 CAS:853-34-9 ***HR: 2***
KETOPHENYLBUTAZONE
mf: $C_{19}H_{18}N_2O_3$ mw: 322.39

SYNS: CHEBUTAN ◇ CHEPIROL ◇ CHETAZOLIDIN ◇ CHETIL ◇ COPIRENE ◇ 1,2-DIPHENYL-4-(Γ-KETOBUTYL)-3,5-PYRAZOLIDINEDIONE ◇ 1,2-DIPHENYL-4-(3′-OXOBUTYL)-3,5-DIOXOPYRAZOLIDINE ◇ HICHILLOS ◇ KEBUZONE ◇ KEOBUTANE-JADE ◇ KETASON ◇ KETAZONE ◇ KPB ◇ 4-(3-OXOBUTYL)-1,2-DIPHENYL-3,5-PYRAZOLIDINEDIONE ◇ PECNON ◇ RECHETON

TOXICITY DATA with REFERENCE
orl-rat LD50:1551 mg/kg OYYAA2 3,390,69
ipr-rat LD50:683 mg/kg OYYAA2 3,390,69
ivn-rat LD50:616 mg/kg NIIRDN 6,265,82
ipr-mus LD50:662 mg/kg OYYAA2 3,390,69
scu-mus LD50:620 mg/kg OYYAA2 3,390,69
ivn-mus LD50:580 mg/kg OYYAA2 3,390,69

SAFETY PROFILE: Moderately toxic by ingestion, intraperitoneal, intravenous, and subcutaneous routes. When heated to decomposition it emits toxic fumes of NO_x.

KGK100 CAS:57495-14-4 ***HR: 3***
KETOPROFEN SODIUM
mf: $C_{16}H_{14}O_3$•Na mw: 277.28

SYNS: 3-BENZOYL-α-METHYL-BENZENEACETIC ACID SODIUM SALT ◇ 2-(3-BENZOYLPHENYL)PROPIONATE SODIUM ◇ 19583RP-Na ◇ 19583 RP SODIUM ◇ SODIUM KETOPROFEN

TOXICITY DATA with REFERENCE
scu-mus TDLo:600 mg/kg (7-13D preg):TER JZKEDZ 1,101,75
orl-rat LD50:215 mg/kg OYYAA2 13,709,77
ipr-rat LD50:66 mg/kg OYYAA2 13,709,77
scu-rat LD50:154 mg/kg JZKEDZ 1,101,75
ivn-rat LD50:343 mg/kg OYYAA2 13,709,77
ims-rat LD50:125 mg/kg OYYAA2 13,709,77
orl-mus LD50:735 mg/kg OYYAA2 13,709,77
ipr-mus LD50:520 mg/kg OYYAA2 13,709,77

SAFETY PROFILE: Poison by ingestion, subcutaneous, intramuscular, intravenous, and intraperitoneal routes. An experimental teratogen. When heated to decomposition it emits toxic fumes of Na_2O.

KGK150 CAS:16694-30-7 ***HR: 2***
4-KETOSTEARIC ACID
mf: $C_{18}H_{34}O_3$ mw: 298.52

TOXICITY DATA with REFERENCE
scu-mus TDLo:2000 mg/kg/25W-I:ETA CNREA8 30,1037,70

SAFETY PROFILE: Questionable carcinogen with experimental tumorigenic data. When heated to decomposition it emits acrid smoke and irritating fumes.

KGK200 CAS:34580-14-8 ***HR: 3***
KETOTIFEN FUMARATE
mf: $C_{19}H_{19}NOS$•$C_4H_4O_4$ mw: 425.53

SYNS: HC 20-511 FUMARATE ◇ ZADITEN

TOXICITY DATA with REFERENCE
orl-rat TDLo:9800 μg/kg (9W male/2W pre-21D preg):TER KSRNAM 13,4096,79
orl-rat TDLo:140 mg/kg (35D pre):REP KSRNAM 13,4069,79
orl-rat LD50:360 mg/kg KSRNAM 13,4069,79
scu-rat LD50:370 mg/kg KSRNAM 13,4069,79
ivn-rat LD50:5100 μg/kg KSRNAM 13,4069,79
orl-mus LD50:585 mg/kg KSRNAM 13,4069,79
scu-mus LD50:820 mg/kg KSRNAM 13,4069,79
ivn-mus LD50:18800 μg/kg KSRNAM 13,4069,79

SAFETY PROFILE: Poison by ingestion, subcutaneous, and intravenous routes. An experimental teratogen. Experimental reproductive effects. When heated to decomposition it emits toxic fumes of SO_x and NO_x.

KGK300 CAS:54240-36-7 ***HR: 3***
KF-868
mf: $C_{13}H_{18}ClF_3N_2O$•ClH mw: 347.24

SYNS: 4-AMINO-α-((tert-BUTYLAMINO)METHYL)-3-CHLORO-5-(TRIFLUOROMETHYL)BENZYL ALCOHOL HCl ◇ 4-AMINO-3-CHLOROTRIFLUOROMETHYL-α-((tert-BUTYLAMINO)METHYL) BENZYLALCOHOL HYDROCHLORIDE

TOXICITY DATA with REFERENCE
orl-rat TDLo:440 mg/kg (female 7-17D post):REP JACTDZ 4(1),91,85
orl-rat TDLo:440 mg/kg (7-17D preg):TER JACTDZ 4,91,85
ivn-mus LD50:36500 μg/kg ARZNAD 34,1625,84

SAFETY PROFILE: Poison by intravenous route. An experimental teratogen. Experimental reproductive effects. When heated to decomposition it emits toxic fumes of F^-, NO_x, and HCl.

KGK350 ***HR: D***
KHAT LEAF EXTRACT

PROP: A common chewing material in East Africa containing (-)-cathinone as the active ingredient RCSADO 6,179,85

SYN: CATHA EDULIS Forsk, leaf extract

TOXICITY DATA with REFERENCE
mmo-sat 25 uL/plate RCSADO 6,179,85
dni-hmn:fbr 45 μg/L MUREAV 204,317,88
orl-gpg TDLo:106 g/kg (female 45-68D post):REP JOETD7 23,11,88
orl-gpg TDLo:106 g/kg (female 45-68D post):TER JOETD7 23,11,88

SAFETY PROFILE: An experimental teratogen. Experimental reproductive effects. Mutation data reported. When heated to decomposition it emits acrid smoke and irritating fumes.

KGK400 CAS:56583-56-3 ***HR: 3***
KHINOTILIN
mf: $C_{26}H_{26}N_2O_4 \cdot 2I$ mw: 684.34

SYNS: CHINOTILIN ◇ 1,1'-((1,4-DIOXO-1,4-BUTANEDIYL)BIS(OXY-2,1-ETHANEDIYL))BIS-QUINOLINIUM DIIODIDE ◇ QUINOTILINE

TOXICITY DATA with REFERENCE
orl-mus LD50:320 mg/kg EKFMA7 11,111,82
ipr-mus LD50:5600 μg/kg EKFMA7 11,111,82
ivn-mus LD50:2200 μg/kg PCJOAU 16,164,82

SAFETY PROFILE: Poison by ingestion, intravenous, and intraperitoneal routes. When heated to decomposition it emits toxic fumes of NO_x and I^-.

KGU100 ***HR: 2***
KM-1146
mf: $C_{12}H_{15}NO_3S$ mw: 253.34

SYN: 2-(3,4-DIMETHOXYPHENYL)-5-ETHYLTHIAZOLIDIN-4-ONE

TOXICITY DATA with REFERENCE
orl-rat TDLo:825 mg/kg (7-17D preg):REP OYYAA2 28,261,84
orl-rat TDLo:825 mg/kg (7-17D preg):TER OYYAA2 28,261,84
ipr-rat LD50:2700 mg/kg OYYAA2 26,701,83
ipr-mus LD50:4300 mg/kg OYYAA2 26,701,83

SAFETY PROFILE: Moderately toxic by intraperitoneal routes. An experimental teratogen. Experimental reproductive effects. When heated to decomposition it emits toxic fumes of SO_x and NO_x.

KHK000 ***HR: 2***
KOTORAN

TOXICITY DATA with REFERENCE
orl-rat LD50:1450 mg/kg FATOAO 35,352,72
orl-mus LD50:1465 mg/kg 17QLAD 12,85,77

CONSENSUS REPORTS: Reported in EPA TSCA Inventory.

SAFETY PROFILE: Moderately toxic by ingestion.

KHK100 CAS:73379-85-8 ***HR: 3***
KPE
mf: $C_{20}H_{32}O_5 \cdot C_{29}H_{36}O_{15}$ mw: 977.17

SYN: PROSTOGLANDIN E_2-METHYLHESPERIDIN COMPLEX

TOXICITY DATA with REFERENCE
orl-rat TDLo:54 g/kg (17-22D preg/21D post):REP YACHDS 9,1395,81
orl-mus TDLo:20 g/kg (female 6-15D post):TER YACHDS 9,1369,81
orl-rat LD50:12 g/kg YACHDS 9,1409,81
scu-rat LD50:818 mg/kg YACHDS 9,1409,81
ivn-rat LD50:308 mg/kg YACHDS 9,1409,81
orl-mus LD50:15 g/kg YACHDS 9,1409,81
scu-mus LD50:999 mg/kg YACHDS 9,1409,81
ivn-mus LD50:417 mg/kg YACHDS 9,1409,81

SAFETY PROFILE: Poison by intravenous route. Moderately toxic by subcutaneous route. Mildly toxic by ingestion. An experimental teratogen. Other experimental reproductive effects. When heated to decomposition it emits acrid smoke and fumes. See also various prostaglandins.

KHU000 CAS:74278-22-1 ***HR: 3***
KROMAD

PROP: Contains 5% cadmium sebacate, 5% potassium chromate, 1% malachite green, and 16% thiram (FMCHA2-,D176,80).

TOXICITY DATA with REFERENCE
orl-rat LD50:400 mg/kg FMCHA2 -,D176,80
skn-rbt LD50:1 g/kg FMCHA2 -,D176,80

CONSENSUS REPORTS: Cadmium and its compounds, as well as chromium and its compounds, are on the Community Right-To-Know List.

OSHA PEL: TWA 0.1 mg(Cd)/m^3; CL 0.6 mg(Cd)/m^3 (fume)
ACGIH TLV: TWA 0.05 mg(Cd)/m^3 (Proposed: TWA 0.01 mg(Cd)/m^3 (dust), Suspected Human Carcinogen; 0.002 mg(Cd)/m^3 (respirable dust), Suspected Human Carcinogen); BEI: 10 μg/g creatinine in urine; 10 μg/L in blood.
DFG BAT: Blood 1.5 μg/dL; Urine 15 μg/dL, Suspected Carcinogen.
NIOSH REL: (Cadmium) Reduce to lowest feasible level

SAFETY PROFILE: Confirmed human carcinogen. Poison by ingestion. Moderately toxic by skin contact. When heated to decomposition it emits toxic fumes of K_2O, Cd, and Cr. See also CADMIUM COMPOUNDS, POTASSIUM CHROMATE, and BIS(DIMETHYLTHIOCARBAMYL)DISULFIDE (thiram).

KHU100 CAS:28557-25-7 ***HR: 1***
KT 136
mf: $C_{12}H_{15}N_5O_2$ mw: 261.32

SYNS: 7,8-DIHYDRO-1,3-DIMETHYL-8-(2-PROPENYL)-1H-IMIDAZO(2,1-f)PURINE-2,4(3H,6H)-DIONE ◇ 1H-IMIDAZO(2,1-f)PURINE-2,4(3H,6H)-DIONE,7,8-DIHYDRO-8-ALLYL-1,3-DIMETHYL-(8CI)

TOXICITY DATA with REFERENCE
skn-rbt 500 mg MLD YACHDS 17(Suppl 1),75,89

SAFETY PROFILE: A skin irritant. When heated to decomposition it emits toxic fumes of NO_x.

KHU136 CAS:1400-17-5 ***HR: 3***
KURCHICINE
mf: $C_{20}H_{36}N_2O$ mw: 320.52

TOXICITY DATA with REFERENCE
ipr-mus LDLo:195 mg/kg JPETAB 58,361,36
scu-gpg LDLo:88 mg/kg JPETAB 58,361,36
scu-frg LDLo:110 mg/kg JPETAB 58,361,36

SAFETY PROFILE: Poison by intraperitoneal and subcutaneous routes. When heated to decomposition it emits toxic fumes of NO_x.

L

LAC000 CAS:8016-26-0 ***HR: 1***
LABDANUM OIL

PROP: Main constituents are acetophenone, 1,5,5-trimethyl-6-cyclohexanone and ladaniol found in the gum of the shrub *Cistus ladaniferus* L. (Fam. *Cistaceae)*. Prepared by steam distillation of the crude gum. Yellow, viscous liquid; powerful balsamic odor. D: 0.905-0.993, refr index: 1.492-1.507 @ 20°, flash p: 187°F. Sol in fixed oils, mineral oil; insol in glycerin, propylene glycol.

SYN: OIL of LABDANUM

TOXICITY DATA with REFERENCE
skn-rbt 500 mg/24H MOD FCTXAV 14,307,76
orl-rat LD50:8980 mg/kg FCTXAV 14,307,76

SAFETY PROFILE: Mildly toxic by ingestion. A skin irritant. Combustible liquid. When heated to decomposition it emits acrid smoke and fumes. See also ACETOPHENONE.

LAD000 ***HR: 3***
LACQUERS

PROP: Solutions of resins, gums or plastics in an organic solvent. Flash p: 0-80°F.

SAFETY PROFILE: Variable toxicity. They may have allergic effects. A very dangerous fire hazard when exposed to heat or flame. A large part of the dangerous fire hazard is due to the highly flammable solvents commonly used. A severe explosion hazard in the form of vapor when exposed to flame. Keep away from heat and open flame. Incompatible with oxidizing materials. To fight fire, use CO_2, dry chemical, water spray.

LAE000 ***HR: 3***
LACQUERS, NITROCELLULOSE

PROP: Flash p: 40°F.

SAFETY PROFILE: Variable toxicity. They may have allergic effects. A very dangerous fire hazard when exposed to heat or flame, even when solvent-free. Moderately explosive when exposed to heat or flame. To fight fire, use CO_2, dry chemical. When heated to decomposition they emit highly toxic fumes.

LAG000 CAS:50-21-5 ***HR: 2***
LACTIC ACID
mf: $C_3H_6O_3$ mw: 90.09

$CH_3CHOHCO{\bullet}OH$

PROP: Yellow to colorless crystals or syrupy 50% liquid. Mp: 16.8°, bp: 122° @ 15 mm, d: 1.249 @ 15°. Volatile with superheated steam. Sol in eater, alc, and furfurol; sltly sol in ether; insol in chloroform, petr ether, carbon disulfide. Misc in water, (alc + ether).

SYNS: ACETONIC ACID ◇ ETHYLIDENELACTIC ACID ◇ 1-HYDROXYETHANECARBOXYLIC ACID ◇ 2-HYDROXYPROPANOIC ACID ◇ 2-HYDROXYPROPIONIC ACID ◇ α-HYDROXYPROPIONIC ACID ◇ KYSELINA MLECNA (CZECH) ◇ dl-LACTIC ACID ◇ MILCHSAURE (GERMAN) ◇ MILK ACID ◇ ORDINARY LACTIC ACID ◇ racemic LACTIC ACID

TOXICITY DATA with REFERENCE
skn-rbt 500 mg/24H SEV 28ZPAK -,105,72
eye-rbt 750 μg SEV AJOPAA 29,1363,46
eye-rbt 750 μg/24H SEV 28ZPAK -,105,72
mmo-esc 210 ppm/3H AMNTA4 85,119,51
orl-rat LD50:3730 mg/kg JIHTAB 23,259,41
orl-mus LD50:4875 mg/kg FAONAU 40,146,67
scu-mus LD50:4500 mg/kg ZGEMAZ 113,536,44
orl-rbt LDLo:500 mg/kg IECHAD 15,628,23
rec-rbt LDLo:1200 mg/kg CRSBAW 83,136,20
orl-gpg LD50:1810 mg/kg JIHTAB 23,259,41

CONSENSUS REPORTS: Reported in EPA TSCA Inventory.

SAFETY PROFILE: Moderately toxic by ingestion and rectal routes. Mutation data reported. A severe skin and eye irritant. Mixtures with nitric acid + hydrofluoric acid may react vigorously and are storage hazards. When heated to decomposition it emits acrid smoke and irritating fumes.

LAH000 CAS:64059-26-3 ***HR: 3***
LACTIC ACID, BERYLLIUM SALT

SYN: BERYLLIUM LACTATE

TOXICITY DATA with REFERENCE
ivn-rat LD50:11 mg/kg BJEPA5 30,375,49
ivn-mus LD50:7600 μg/kg BJEPA5 30,375,49
ivn-rbt LDLo:10 mg/kg BJEPA5 30,375,49

CONSENSUS REPORTS: Beryllium and its compounds are on The Community Right-To-Know List.

OSHA PEL: (Transitional: TWA 0.002 mg(Be)/m^3; CL 0.005; Pk 0.025/30M/8H) TWA 0.002 mg(Be)/m^3; STEL 0.005 mg(Be)/m^3/30M; CL 0.025 mg(Be)/m^3
ACGIH TLV: TWA 0.002 mg(Be)m^3, Suspected Carcinogen
NIOSH REL: CL (Beryllium) not to exceed 0.0005 mg(Be)/m^3

SAFETY PROFILE: Confirmed carcinogen. Poison by intravenous route. When heated to decomposition it emits very toxic fumes of Be. See also BERYLLIUM COMPOUNDS.

LAJ000 CAS:97-64-3 ***HR: 2***
LACTIC ACID, ETHYL ESTER
DOT: UN 1192
mf: $C_5H_{10}O_3$ mw: 118.15

PROP: Colorless liquid; mild odor. Bp: 154°, ULC: 30-35%, lel: 1.55% @ 212°F, flash p: 115°F (CC), flash p (technical): 131°F, d: 1.029-1.032, refr index: 1.410-1.420, autoign temp: 752°F, vap d: 4.07. Very sol in alc, ether, chloroform, water.

SYNS: ACTYLOL ◇ ACYTOL ◇ ETHYL α-HYDROXYPROPIONATE ◇ ETHYL 2-HYDROXYPROPIONATE ◇ ETHYL LACTATE (DOT,FCC) ◇ FEMA No. 2440 ◇ LACTATE d'ETHYLE (FRENCH) ◇ SOLACTOL

TOXICITY DATA with REFERENCE
ipr-rat LDLo:1000 mg/kg JPPMAB 11,150,59
orl-mus LD50:2500 mg/kg JPETAB 65,89,39
scu-mus LD50:2500 mg/kg JPETAB 65,89,39
ivn-mus LD50:600 mg/kg JPETAB 65,89,39

CONSENSUS REPORTS: Reported in EPA TSCA Inventory.

DOT Classification: Flammable or Combustible Liquid; Label: Flammable Liquid; Combustible Liquid; Label: None.

SAFETY PROFILE: Moderately toxic by ingestion, intraperitoneal, subcutaneous, and intravenous routes. Combustible or combustible liquid when exposed to heat or flame; can react with oxidizing materials. Slight explosion hazard in the form of vapor when exposed to flame. To fight fire, use foam, CO_2, dry chemical. When heated to decomposition it emits acrid smoke and irritating fumes. See also IRON COMPOUNDS.

LAL000 CAS:5905-52-2 ***HR: 3***
LACTIC ACID, IRON(2+) SALT (2:1)
mf: $C_6H_{10}O_6$•Fe mw: 234.01

PROP: Greenish-white crystals; slight peculiar odor. Moderately sol in water; sltly sol in alc.

SYNS: FERROUS LACTATE ◇ IRON(2+) LACTATE

TOXICITY DATA with REFERENCE
scu-mus TDLo:4200 mg/kg/21W-I:ETA JNCIAM 24,109,60
orl-mus LD50:147 mg/kg JPMSAE 54-1211,65

OSHA PEL: TWA 1 mg(Fe)/m^3
ACGIH TLV: TWA 1 mg(Fe)/m^3

SAFETY PROFILE: Poison by ingestion. Questionable carcinogen with experimental tumorigenic data. When heated to decomposition it emits acrid smoke and irritating fumes. See also IRON COMPOUNDS.

LAM000 CAS:72-17-3 ***HR: 2***
LACTIC ACID, MONOSODIUM SALT
mf: $C_3H_5O_3$•Na mw: 112.07

PROP: Hygroscopic solid; slt salt taste.

SYNS: 2-HYDROXYPROPANOIC ACID MONOSODIUM SALT ◇ LACOLIN ◇ LACTIC ACID SODIUM SALT ◇ PER-GLYCERIN ◇ SODIUM LACTATE

TOXICITY DATA with REFERENCE
eye-rbt 100 mg MLD FCTOD7 20,573,82
ipr-rat LD50:2000 mg/kg FAONAU 40,146,67

CONSENSUS REPORTS: Reported in EPA TSCA Inventory.

SAFETY PROFILE: Moderately toxic by intraperitoneal route. An eye irritant. When heated to decomposition it emits toxic fumes of Na_2O.

LAN000 CAS:19042-19-4 ***HR: 3***
LACTIC ACID, NEODYMIUM SALT
mf: $C_9H_{15}NdO_9$ mw: 411.48

SYNS: NEODYMIUM LACTATE ◇ NEODYMLACTAT (GERMAN)

TOXICITY DATA with REFERENCE
scu-mus LD50:10 g/kg ZGEMAZ 113,536,44
ivn-cat LDLo:64 mg/kg ZGEMAZ 113,536,44

SAFETY PROFILE: Poison by intravenous route. When heated to decomposition it emits acrid smoke and fumes. See also NEODYMIUM.

LAO000 CAS:10377-98-7 ***HR: 2***
LACTIC ACID, SODIUM ZIRCONIUM SALT (4:4:1)
mf: $C_{12}H_{20}O_{12}$•4Na•Zr mw: 539.50

SYNS: SODIUM ZIRCONIUM LACTATE ◇ ZIRCONIUM SODIUM LACTATE

TOXICITY DATA with REFERENCE
idr-man TDLo:170 μg/kg/I:SKN JIDEAE 38,223,62

OSHA PEL: (Transitional: TWA 5 mg(Zr)/m^3) TWA 5 mg(Zr)/m^3; STEL 10 mg(Zr)/m^3
ACGIH TLV: TWA 5 mg(Zr)/m^3; STEL 10 mg(Zr)/m^3
DFG MAK: 5 mg(Zr)/m^3

SAFETY PROFILE: Human systemic effects by intradermal route: primary skin irritation, skin corrosion and dermatitis. When heated to decomposition it emits toxic fumes of Na_2O. See also ZIRCONIUM COMPOUNDS.

LAO300 ***HR: D***
LACTIC DEHYDROGENASE X

SYNS: ANTISERUM AGAINST THE ISOZYME of LACTATE DEHYDROGENASE ◇ ANTISERUM TO SPERM SPECIFIC LACTATE DEHYDROGENASE ◇ LDH-X ◇ SPERM SPECIFIC ISOZYME of LACTATE DEHYDROGENASE ◇ LACTATE DEHYDROGENASE X

TOXICITY DATA with REFERENCE
scu-mus TDLo:16 g/kg (1-4D preg):REP SCIEAS 176,686,72

SAFETY PROFILE: Experimental reproductive effects.

LAP000 CAS:502-44-3 ***HR: 2***
ε-LACTONE HEXANOIC ACID
mf: $C_6H_{10}O_2$ mw:114.16

SYNS: CAPROLACTONE ◇ epsilon-CAPROLACTONE ◇ 6-HEXANOLACTONE ◇ 1,6-HEXANOLIDE ◇ 6-HYDROXYHEXANOIC ACID LACTONE ◇ 2-OXEPANONE (8CI, 9CI)

TOXICITY DATA with REFERENCE
eye-rbt 750 μg open SEV AMIHBC 10,61,54
orl-rat LD50:4290 mg/kg AMIHBC 10,61,54
ipr-mus LD50:1300 mg/kg JJIND8 62,911,79
skn-rbt LD50:5990 mg/kg AMIHBC 10,61,54

CONSENSUS REPORTS: Reported in EPA TSCA Inventory. EPA Genetic Toxicology Program.

SAFETY PROFILE: Moderately toxic by intraperitoneal route. Mildly toxic by ingestion. A severe eye irritant. When heated to decomposition it emits acrid smoke and irritating fumes.

LAQ000 CAS:78-97-7 ***HR: 3***
LACTONITRILE
mf: C_3H_5NO mw: 71.09

PROP: Straw-colored liquid. Mp: −40°, bp: 103° @ 50 mm, fp: −34°, flash p: 170°F (TCC), d: 0.9834 @ 25°, vap d: 2.45.

SYN: 2-HYDROXYPROPIONITRILE

TOXICITY DATA with REFERENCE
orl-rat LD50:87 mg/kg AIHAAP 30,470,69
ihl-rat LCLo:125 ppm/4H AIHAAP 30,470,69
skn-rbt LD50:20 mg/kg AIHAAP 30,470,69
scu-rbt LDLo:5 mg/kg AIPTAK 5,161,1899
scu-frg LDLo:200 mg/kg AIPTAK 5,161,1899

CONSENSUS REPORTS: EPA Extremely Hazardous Substances List. Reported in EPA TSCA Inventory. Cyanide and its compounds are on the Community Right-To-Know List.

SAFETY PROFILE: Poison by ingestion, skin contact, and subcutaneous routes. Moderately toxic by inhalation. In the presence of alkali, it evolves HCN. Combustible when exposed to heat or flame; can react vigorously with oxidizing materials. To fight fire, use foam, CO_2, dry chemical. When heated to decomposition it emits toxic fumes of CN^- and NO_x. See also NITRILES.

LAR000 CAS:63-42-3 ***HR: 2***
LACTOSE
mf: $C_{12}H_{22}O_{11}$ mw: 342.34

$C_6H_7O(OH)_4OC_6H_7O(OH)_4$

PROP: Colorless, rhombic crystals; faintly sweet taste. D: 1.525 @ 20°, mp: 202° (anhyd), bp: decomp. Sol in water; insol in alc and ether.

SYNS: 4-(β-d-GALACTOSIDO)-d-GLUCOSE ◇ LACTIN ◇ LACTOBIOSE ◇ d-LACTOSE ◇ MILK SUGAR ◇ SACCHARUM LACTIN

TOXICITY DATA with REFERENCE
orl-rat TDLo:375 mg/kg (4-18D preg):TER TJADAB 4,497,71
scu-mus TDLo:1000 g/kg/29w-C:ETA application GANNA2 46,363,55
ivn-dog LDLo:1500 mg/kg HBAMAK 4,1289,35

CONSENSUS REPORTS: Reported in EPA TSCA Inventory.

SAFETY PROFILE: Moderately toxic by intravenous route. Questionable carcinogen with experimental tumorigenic and teratogenic data. Mixtures with oxidants (e.g., potassium chlorate, potassium nitrate, or potassium perchlorate) may be explosion hazards. When heated to decomposition it emits acrid smoke and irritating fumes.

LAR100 CAS:576-08-9 ***HR: 1***
LACTULOSE
mf: $C_{12}H_{22}O_{11}$ mw: 342.34

SYNS: 4-o-β-d-GALACTOPYRANOSYL-d-FRUCTOFURANOSE ◇ 4-o-β-d-GALATTOPIRANOSIL-d-FRUTTOFURANOSIO(ITALIAN) ◇ LATTULOSIO (ITALIAN)

TOXICITY DATA with REFERENCE
orl-rat LD50:25 g/kg NIIRDN 6,875,82
ipr-rat LD50:16 g/kg BCFAAI 115,596,76
scu-rat LD50:33 g/kg NIIRDN 6,875,82
ivn-rat LD50:14 g/kg NIIRDN 6,875,82

orl-mus LD50:31 g/kg NIIRDN 6,875,82
ipr-mus LD50:16 g/kg NIIRDN 6,875,82
scu-mus LD50:30 g/kg NIIRDN 6,875,82
ivn-mus LD50:10 g/kg NIIRDN 6,875,82

SAFETY PROFILE: Very mildly toxic by ingestion and other routes. When heated to decomposition it emits acrid smoke and irritating fumes.

LAR500 ***HR: 3***
LADY LAUREL

PROP: Deciduous shrubs which grow to a height of 4 or 5 feet with leaves about 3.5 inches long and 0.75 inches wide. They produce clusters of lilac or white flowers and red or yellow fruit with a pit. They are native to Eurasia and grow wild in the northeastern United States and eastern Canada.

SYNS: BOIS GENTIL (CANADA) ◇ BOIS JOLI (CANADA) ◇ DAPHNE MEZEREUM ◇ DWARF BAY ◇ FEBRUARY DAPHNE ◇ FLAX OLIVE ◇ MEZEREUM ◇ SPURGE LAUREL ◇ SPURGE OLIVE

SAFETY PROFILE: The whole plant and especially the fruit and seeds contains the poisons daphnetoxin, mezerein, and diterpene alcohols. Ingestion causes blistering of the lips, mouth and throat, followed by abdominal pain, vomiting, bloody diarrhea, and possibly kidney damage, and death.

LAS000 CAS:1332-94-1 ***HR: 3***
LAETRILE
mf: $C_{14}H_{15}NO_7$ mw: 309.30

PROP: Solid. Mp: 214-216°.

SYNS: CYANOPHENYLMETHYL-β-d-GLUCOPYRANOSIDURONIC ACID ◇ 1-MANDELONITRILE-β-GLUCURONIC ACID

TOXICITY DATA with REFERENCE
mma-sat 9600 pmol/plate SCIEAS 198,625,77
orl-man TDLo:1286 mg/kg/24W:CNS JAMAAP 238,1361,77
mul-wmn TDLo:4170 mg/kg/8W:GIT CTRRDO 62,169,78
orl-wmn LDLo:198 mg/kg JAMAAP 239,1532,78
ipr-rat LDLo:250 mg/kg JAMAAP 242,169,79

CONSENSUS REPORTS: Cyanide and its compounds are on The Community Right-To-Know List.

SAFETY PROFILE: Human poison by ingestion. Experimental poison by intraperitoneal route. Human systemic effects by ingestion and multiple routes: central nervous system and gastrointestinal changes. Mutation data reported. A controversial treatment for cancer. When heated to decomposition it emits toxic fumes of NO_x and CN^-. See also NITRILES.

LAS500 ***HR: 2***
LAMIUM ALBUM LINN., EXTRACT

PROP: Indian plant belonging to the family *Labiatae* (IJEBA6 18,594,80).

TOXICITY DATA with REFERENCE
orl-rat TDLo:150 mg/kg (12-14D preg):REP IJEBA6 18,594,80
ipr-mus LD50:750 mg/kg IJEBA6 18,594,80

SAFETY PROFILE: Moderately toxic by intraperitoneal route. Experimental reproductive effects.

LAT000 CAS:17575-20-1 ***HR: 3***
LANATOSIDE A
mf: $C_{49}H_{76}O_{19}$ mw: 969.25

PROP: Large, flat prisms from methanol, decomp @ 245-248°. Sol in 20 parts methanol, 40 parts alc, 225 parts chloroform, 16,000 parts of water.

SYNS: DIGILANID A ◇ DIGITOXIGENIN + 2 DIGITOXOSE + ACETYL-DIGILANIDOBOSE (GERMAN) ◇ LANATOSID A (GERMAN)

TOXICITY DATA with REFERENCE
ivn-rat LD50:16 mg/kg AIPTAK 155,165,65
ipr-mus LD50:20 mg/kg AIPTAK 155,165,65
ivn-cat LDLo:220 μg/kg AEPPAE 184,181,37
orl-gpg LD50:100 mg/kg ARZNAD 15,481,65
ivn-gpg LDLo:1100 μg/kg ARZNAD 15,481,65

SAFETY PROFILE: A deadly poison by ingestion, intraperitoneal, and intravenous routes. When heated to decomposition it emits acrid smoke and irritating fumes.

LAT500 CAS:17575-21-2 ***HR: 3***
LANATOSIDE B
mf: $C_{49}H_{76}O_{20}$ mw: 985.10

PROP: Long, flat prisms from alc. Decomp 245-248° after drying in high vacuum at 150°. Sol in 20 parts methanol, 40 parts alc, 550 parts chloroform. Nearly insol in water. Desacetyldigilanide B (Purpurea glycoside B) is a glycoside from *Digitalis purpurea.* It contains no acetyl group.

SYNS: DIGILANID B ◇ DIGILANIDE B ◇ LANATOSID B (GERMAN)

TOXICITY DATA with REFERENCE
ivn-rat LDLo:12 mg/kg AEPPAE 177,60,34
ivn-cat LD50:388 μg/kg JAPMA8 31,236,42
ivn-gpg LDLo:3695 μg/kg AEPPAE 252,314,66

SAFETY PROFILE: Poison by intravenous route. When heated to decomposition it emits acrid smoke and irritating fumes.

LAU000 CAS:17575-22-3 ***HR: 3***
LANATOSIDE C
mf: $C_{49}H_{76}O_{20}$ mw: 985.25

PROP: Long, flat prisms from alc. Decomp @ 248-250° after drying in high vacuum @ 150°. Very sol in pyridine and dioxane, insol in ether, petr ether.

SYNS: CEDILANID ◇ DIGILANID C ◇ ISOLANID ◇ ISOLANIDE ◇ LANATOSID C (GERMAN)

TOXICITY DATA with REFERENCE
ipr-rat LDLo:10 mg/kg TXAPA9 1,156,59
ivn-rat LD50:30 mg/kg AIPTAK 155,165,65
ipr-mus LD50:7 mg/kg AIPTAK 155,165,65
ivn-mus LD50:8100 μg/kg NIIRDN 6,877,82
ivn-dog LD50:340 μg/kg IVEJAC 57,31,80
ivn-cat LDLo:230 μg/kg ARZNAD 19,657,69
orl-gpg LD50:100 mg/kg ARZNAD 15,481,65
ivn-gpg LDLo:502 μg/kg ARZNAD 17,1237,67

SAFETY PROFILE: Poison by ingestion, intraperitoneal, and intravenous routes. When heated to decomposition it emits acrid smoke and irritating fumes.

LAU400 CAS:11014-59-8 ***HR: 3***
LANATOSIDES

SYNS: ABC LANATOSIDE COMPLEX ◇ CARDIOLANATA ◇ CORDILAN ◇ DIGILANIDES ◇ DIGIMED ◇ LANOSTABIL ◇ LANTOSIDE ◇ PANDIGAL ◇ PANLANAT

TOXICITY DATA with REFERENCE
orl-man TDLo:240 μg/kg:GIT SAVEAB 10,121A,39
ivn-rat LDLo:12 mg/kg AEPPAE 177,60,34
ivn-cat LDLo:343 mg/kg JPHAA3 27,761,38

SAFETY PROFILE: Poison by intravenous route. Human systemic effects by ingestion: nausea and vomiting.

LAU600 ***HR: 1***
LANTANA

PROP: Low shrubs with prickly stems and coarse leaves that have a strong odor when crushed. The flowers grow in flat clusters and change color from yellow to orange to red within 24 hours after opening. The outer flowers in a cluster open first so that the overall effect is one of concentric rings of colored flowers, yellow in the center and red on the edge. They grow as weeds in southern Florida, Texas, California, Hawaii, Guam, and the West Indies. They are also grown as ornamentals in more northerly areas.

SYNS: BONBONNIER (HAITI) ◇ CARIAQUILLO (PUERTO RICO) ◇ CINCO NEGRITOS (MEXICO) ◇ FILIGRANA (CUBA) ◇ HERBE A PLOMB (HAITI) ◇ LAKANA, MIKINOLIA-HIHIU (HAWAII) ◇ LANTANA CAMARA ◇ SHRUB VERBENA ◇ YELLOW SAGE

SAFETY PROFILE: The unripened berries contain an unknown poison. Ingestion of this fruit can cause within 6 hours vomiting, diarrhea, dilation of the pupil, weakness and slowed breathing.

LAV000 CAS:7439-91-0 ***HR: 3***
LANTHANUM
af: La aw: 138.91

PROP: Silvery-white, malleable and ductile metal element soft enough to cut with a knife. Very reactive rare earth metal. Mp: 920°, bp: 3454°, d: 6.166 @ 25°.

SAFETY PROFILE: Poison by intravenous route. Lanthanum and other lanthanons can cause delayed blood clotting leading to hemorrhages. Has caused liver injury in experimental animals. The dust is a dangerous fire hazard when exposed to flame; can react vigorously with oxidizing materials. Violent reaction with nitric acid; phosphorus (above 400°C); air; halogens. Moderately explosive in the form of dust when exposed to flame or by chemical reaction. Incompatible with H_2O; C; N; B; Se; Si; S. See also RARE EARTHS and POWDERED METALS.

LAW000 CAS:917-70-4 ***HR: 2***
LANTHANUM ACETATE
mf: $C_2H_4O_2 \cdot xLa$ mw: 1032.43

SYNS: LANTHANACETAT (GERMAN) ◇ LANTHANUM TRIACETATE

TOXICITY DATA with REFERENCE
dnd-esc 100 μmol/L MUREAV 89,95,81
orl-rat LD50:32700 mg/kg EQSSDX 1,1,75
ipr-rat LD50:1553 mg/kg EQSSDX 1,1,75
scu-mus LD50:3500 mg/kg ZGEMAZ 113,536,44

CONSENSUS REPORTS: Reported in EPA TSCA Inventory.

SAFETY PROFILE: Moderately toxic by intraperitoneal and subcutaneous routes. Mildly toxic by ingestion. Mutation data reported. When heated to decomposition it emits acrid smoke and irritating fumes. See also LANTHANUM and RARE EARTHS.

LAX000 CAS:10099-58-8 ***HR: 3***
LANTHANUM CHLORIDE
mf: Cl_3La mw: 245.26

PROP: Heptahydrate: triclinic crystals. Sol in water and alc.

TOXICITY DATA with REFERENCE
otr-mus:oth 30 umol/L CRNGDP 7,1949,86
spm-dom-itr 5 mg/kg IJEBA6 11,143,73
ipr-mus TDLo:112 mg/kg (12D preg):REP TXCYAC 34,315,85
orl-rat LD50:4184 mg/kg EQSSDX 1,1,75
ipr-rat LD50:106 mg/kg AMIHAB 16,475,57
ivn-rat LDLo:4 mg/kg AMIHAB 16,475,57
ipr-mus LD50:213 mg/kg COREAF 256,1043,63
scu-mus LD50:2424 mg/kg EQSSDX 1,1,75
ivn-mus LD50:18 mg/kg JNCIAM 13,559,52
ivn-rbt LD50:148 mg/kg EQSSDX 1,1,75

CONSENSUS REPORTS: Reported in EPA TSCA Inventory. EPA Genetic Toxicology Program.

SAFETY PROFILE: Poison by intraperitoneal and intravenous routes. Moderately toxic by subcutaneous route. Mildly toxic by ingestion. Experimental reproductive effects. Mutation data reported. When heated to decomposition it emits toxic fumes of Cl^-. See also RARE EARTHS, LANTHANUM, and CHLORIDES.

LAY499 CAS:13823-36-4 ***HR: 3***
LANTHANUM DIHYDRIDE
mf: H_2La mw: 140.92

SAFETY PROFILE: Ignites spontaneously in air. See also HYDRIDES, LANTHANUM, and RARE EARTHS.

LAZ000 CAS:11138-87-7 ***HR: 3***
LANTHANUM EDETATE

TOXICITY DATA with REFERENCE
ipr-mus LD50:37 mg/kg AEHLAU 5,437,62
ipr-gpg LD50:83 mg/kg AEHLAU 5,437,62

SAFETY PROFILE: Poison by intraperitoneal route. See also LANTHANUM and RARE EARTHS.

LBA000 CAS:10099-59-9 ***HR: 3***
LANTHANUM NITRATE
mf: $N_3O_9 \cdot La$ mw: 324.94

PROP: Hexahydrate; white, deliquescent crystals. Mp: approx 40°, bp: 126°. Very sol in water, alc. Keep well stoppered.

TOXICITY DATA with REFERENCE
sln-smc 33300 ppb ANYAA9 407,186,83
scu-rat TDLo:25995 μg/kg (1D male):REP JRPFA4 7,21,64
orl-rat LD50:4500 mg/kg AIHOAX 1,637,50
ipr-rat LD50:450 mg/kg AIHOAX 1,637,50
ipr-mus LD50:309 mg/kg EQSSDX 1,1,75

CONSENSUS REPORTS: Reported in EPA TSCA Inventory.

SAFETY PROFILE: Poison by intraperitoneal route. Mildly toxic by ingestion. Experimental reproductive effects. Mutation data reported. When heated to decomposition it emits toxic fumes of NO_x. See also NITRATES, LANTHANUM, and RARE EARTHS.

LBB000 CAS:10099-60-2 ***HR: 3***
LANTHANUM(III) SULFATE (2:3)
mf: $O_{12}S_3 \cdot 2La$ mw: 566.00

PROP: White, hygroscopic powder. Mp: 1150° (decomp), d: 3.60 @ 15°.

SYNS: LANTHANUM SULFATE ◇ SULFURIC ACID LANTHANUM(3+) SALT (3:2)

TOXICITY DATA with REFERENCE
ipr-rat LD50:275 mg/kg AIHOAX 1,637,50

CONSENSUS REPORTS: Reported in EPA TSCA Inventory.

SAFETY PROFILE: Poison by intraperitoneal route. Sulfuric acid is formed upon hydrolysis of this material. When heated to decomposition it emits toxic fumes of SO_x. See also LANTHANUM and RARE EARTHS.

LBC000 CAS:13864-01-2 ***HR: 3***
LANTHANUM TRIHYDRIDE
mf: H_3La mw: 141.93

SYN: LANTHANUM HYDRIDE

SAFETY PROFILE: Ignites spontaneously in air. See also HYDRIDES, LANTHANUM, and RARE EARTHS.

LBD000 CAS:32854-75-4 ***HR: 3***
LAPPACONITINE
mf: $C_{32}H_{44}N_2O_8$ mw: 584.78

PROP: Bitter crystals. Mp: 217-218°; sol in benzene; sltly sol in alc, ether; insol in water. Chief alkaloid in aconitum septentrionale.

TOXICITY DATA with REFERENCE
ipr-rat LD50:9900 μg/kg CYLPDN 8,301,87
orl-mus LD50:20 mg/kg APTOA6 7,337,51
ipr-mus LD50:10500 μg/kg CYLPDN 8,301,87
ivn-mus LD50:6900 μg/kg APTOA6 7,337,51

SAFETY PROFILE: Poison by ingestion, intraperitoneal, and intravenous routes. When heated to decomposition it emits toxic fumes of NO_x.

LBE000 ***HR: 3***
LARAHA

PROP: Aqueous extract from the dried leaves of the plant.

SYN: CITRUS AURANTIUM

TOXICITY DATA with REFERENCE
ims-rat TDLo:45 g/kg/1Y-I:ETA JNCIAM 46,1131,71

SAFETY PROFILE: Questionable carcinogen with expermental tumorigenic data. When heated to decomposition it emits acrid smoke and irritating fumes.

LBF000 ***HR: 3***
LARKSPUR

PROP: Dried, ripe seeds.

SYNS: DELPHINIUM ◇ STAGGER WEED ◇ KNIGHT'S SPUR

SAFETY PROFILE: Poison by ingestion and inhalation. An allergen. Poisoning from percutaneous absorption may occur. Combustible when exposed to heat or flame.

LBF500 CAS:11054-70-9 ***HR: 3***
LASALOCID
mf: $C_{35}H_{54}O_8$ mw: 602.89

SYN: ANTIBIOTIC X 537

TOXICITY DATA with REFERENCE
skn-rbt 100 mg/24H MLD DCTODJ 8,451,85
eye-rbt 50 mg MOD DCTODJ 8,451,85
ipr-rat LD50:8 mg/kg DCTODJ 8,451,85
orl-hor LD50:22 mg/kg AJVRAH 42,456,81

SAFETY PROFILE: Poison by ingestion and intraperitoneal routes. An eye and skin irritant. When heated to decomposition it emits acrid smoke and irritating fumes.

LBG000 CAS:303-34-4 ***HR: 3***
LASIOCARPINE
mf: $C_{21}H_{33}NO_7$ mw: 411.55

PROP: An alkaloid isolated from *H. Lasiocarpum.*

SYNS: HELIOTRIDINE ESTER with LASIOCARPUM and ANGELIC ACID ◇ NCI-C01478 ◇ RCRA WASTE NUMBER U143

TOXICITY DATA with REFERENCE
sln-dmg-orl 750 ppm ENMUDM 7,349,85
trn-dmg-orl 750 ppm ENMUDM 7,349,85
orl-rat TDLo:255 mg/kg/2Y-C:CAR NCITR* NCI-CG-TR-39,78
ipr-rat TDLo:470 mg/kg/56W-I:CAR CNREA8 32,908,72
orl-rat LD50:150 mg/kg TXAPA9 17,290,70
ipr-rat LD50:78 mg/kg CNREA8 32,908,72
ivn-rat LD50:88 mg/kg JPETAB 126,179,59
par-rat LD50:80 mg/kg NATUAS 223,1269,69
ivn-mus LDLo:85 mg/kg JPETAB 126,179,59
ivn-mky LDLo:20 mg/kg JPETAB 126,179,59
ivn-gpg LDLo:50 mg/kg JPETAB 68,123,40

CONSENSUS REPORTS: IARC Cancer Review: Group 2B IMEMDT 7,56,87; Animal Limited Evidence IMEMDT 10,281,76. NCI Carcinogenesis Bioassay (feed); No Evidence: mouse, rat NCITR* NCI-CG-TR-39,78. EPA Genetic Toxicology Program.

SAFETY PROFILE: Suspected carcinogen with experimental carcinogenic data. Poison by ingestion, intravenous, intraperitoneal, and parenteral routes. Human mutation data reported. When heated to decomposition it emits toxic fumes of NO_x. See also ESTERS.

LBH200 CAS:64953-12-4 ***HR: 1***
LATAMOXEF SODIUM
mf: $C_{20}H_{18}N_6O_9S•2Na$ mw: 564.48

SYNS: ANTIBIOTIC 6059-S ◇ DISODIUM LATAMOXEF ◇ LY 127935 ◇ MOXALACTAM DISODIUM ◇ MOXAM ◇ 6059S ◇ SHIONOGI 6059S

TOXICITY DATA with REFERENCE
ivn-rat TDLo:11 g/kg (female 7-17D post):REP NKRZAZ 28(Suppl 7),1119,80
ivn-rat TDLo:5500 mg/kg (female 7-17D post):TER NKRZAZ 28(Suppl 7),1119,80
orl-man TDLo:571 mg/kg/20D-I:GIT JAMAAP 250,730,83
ivn-man LDLo:371 mg/kg/17D-I:BLD DICPBB 18,140,84
ivn-man TDLo:1143 mg/kg/10D-I:BLD DICPBB 18,721,84
ipr-rat LD50:8100 mg/kg NIIRDN 6,APP-21,82
scu-rat LD50:9000 mg/kg NIIRDN 6,APP-21,82
ivn-rat LD50:5500 mg/kg NIIRDN 6,APP-21,82
ipr-mus LD50:8100 mg/kg NIIRDN 6,APP-21,82
scu-mus LD50:9000 mg/kg NIIRDN 6,APP-21,82
ivn-mus LD50:5500 mg/kg NIIRDN 6,APP-21,82

SAFETY PROFILE: Mildly toxic by several routes. An experimental teratogen. Other experimental reproductive effects. Human systemic effects: ulceration or bleeding from small and large intestine, hemorrhage, thrombocytopenia. When heated to decomposition it emits toxic fumes of SO_x, NO_x, and Na_2O.

LBI000 ***HR: 3***
LATICAUDA SEMIFASCIATA VENOM

SYNS: LATICATOXIN ◇ VENOM, SEA SNAKE, LATICAUDA SEMIFASCIATA

TOXICITY DATA with REFERENCE
scu-mus LD50:200 μg/kg 85EGD4 5,426,78
ivn-mus LD50:211 mg/kg TIHHAH 58,182,59
ims-mus LD50:500 μg/kg BIJOAK 99,624,66
scu-rbt LD50:49500 ng/kg TIHHAH 58,182,59
ivn-rbt LD50:48600 ng/kg TIHHAH 58,182,59
scu-gpg LD50:89700 ng/kg TIHHAH 58,182,59
ivn-gpg LD50:63100 ng/kg TIHHAH 58,182,59
par-frg LDLo:10 mg/kg TIHHAH 58,182,59

SAFETY PROFILE: Poison by subcutaneous, intravenous, intramuscular, and parenteral routes.

LBK000 CAS:8006-78-8 ***HR: 2***
LAUREL LEAF OIL

PROP: Main constituent is cineole. From steam distillation of the leaves of *Laurus nobilis* L. (Fam. *Lauraceae).* Yellow liquid; aromatic and spicy odor. D: 0.905-0.929, refr index: 1.465 at 20°. Sol in fixed oils, mineral oil, propylene glycol; insol in glycerin.

SYNS: BAY LEAF OIL ◇ OIL of LAUREL LEAF

TOXICITY DATA with REFERENCE
skn-rbt 500 mg/24H MOD FCTXAV 14,337,76
orl-rat LD50:3950 mg/kg FCTXAV 14,337,76

CONSENSUS REPORTS: Reported in EPA TSCA Inventory.

SAFETY PROFILE: Moderately toxic by ingestion. A skin irritant. When heated to decomposition it emits acrid smoke and irritating fumes. See also CAJEPUTOL.

LBL000 CAS:143-07-7 ***HR: 3***
LAURIC ACID
mf: $C_{12}H_{24}O_2$ mw: 200.36

PROP: Colorless, needle-like crystals; slt odor of bay oil. Mp: 48°, bp: 225° @ 100 mm, d: 0.883, vap press: 1 mm @ 121.0°. Insol in water; sol in chloroform, benzene, alc, ether, and petroleum ether.

SYNS: DODECANOIC ACID ◇ DODECOIC ACID ◇ DUODECYLIC ACID ◇ HYDROFOL ACID 1255 ◇ HYSTRENE 9512 ◇ LAUROSTEARIC ACID ◇ NEO-FAT 12 ◇ NINOL AA-62 EXTRA ◇ 1-UNDECANECARBOXYLIC ACID ◇ WECOLINE 1295

TOXICITY DATA with REFERENCE
cyt-smc 10 mg/L NATUAS 294,263,81
skn-mus TDLo:108 g/kg/15W-I:NEO APMIAL 46,51,59
ivn-mus LD50:131 mg/kg APTOA6 18,141,61
orl-rat LD50:12 g/kg FDRLI* 123,-,76

CONSENSUS REPORTS: Reported in EPA TSCA Inventory.

SAFETY PROFILE: Poison by intravenous route. Mildly toxic by ingestion. Questionable carcinogen with experimental neoplastigenic data. Mutation data reported. Combustible when exposed to heat or flame; can react with oxidizing materials. When heated to decomposition it emits acrid smoke and irritating fumes.

LBM000 CAS:1984-77-6 ***HR: 3***
LAURIC ACID-2,3-EPOXYPROPYL ESTER
mf: $C_{15}H_{28}O_3$ mw: 256.43

SYN: GLYCIDYL LAURATE

TOXICITY DATA with REFERENCE
scu-mus TDLo:16 mg/kg/40W-I:NEO,REP CNREA8 30,1037,70

SAFETY PROFILE: Experimental reproductive effects. Questionable carcinogen with experimental neoplastigenic data. When heated to decomposition it emits acrid smoke and irritating fumes. See also ESTERS.

LBN000 CAS:629-25-4 ***HR: 3***
LAURIC ACID, SODIUM SALT
mf: $C_{12}H_{24}O_2 \cdot Na$ mw: 223.35

SYNS: SODIUM DODECANOATE ◇ SODIUM LAURATE

TOXICITY DATA with REFERENCE
dni-gpg:kdy 100 µmol/L FCTXAV 14,431,76
skn-rat 28 mg/24H SEV JSCCA5 26,29,75
unr-mus LDLo:400 mg/kg ATMPA2 32,177,38

CONSENSUS REPORTS: Reported in EPA TSCA Inventory.

SAFETY PROFILE: Poison by unspecified routes. A severe skin irritant. Mutation data reported. When heated to decomposition it emits toxic fumes of Na_2O.

LBO000 CAS:301-11-1 ***HR: 2***
LAURIC ACID, 2-THIOCYANATOETHYL ESTER
mf: $C_{15}H_{27}NO_2S$ mw: 285.49

SYNS: DODECANOIC ACID, 2-THIOCYANATOETHYL ESTER ◇ ENT 5 ◇ LAURIC ACID ESTER with 2-HYDROXYETHYL THIOCYANATE ◇ LETHANE 60 ◇ THIOCYANIC ACID, 2-HYDROXYETHYL ESTER, LAURATE ◇ 2-THIOCYANOETHYL COCONATE ◇ 2-THIOCYANOETHYL DODECANOATE ◇ β-THIOCYANOETHYL LAURATE ◇ 2-THIOCYANOETHYL LAURATE

TOXICITY DATA with REFERENCE
orl-rat LD50:500 mg/kg ARSIM* 20,13,66
scu-rat LD50:4300 mg/kg INMEAF 11,-,42
ipr-gpg LD50:1480 mg/kg INMEAF 11,-,42

SAFETY PROFILE: Moderately toxic by ingestion and intraperitoneal routes. When heated to decomposition it emits very toxic fumes of SO_x and NO_x. See also ESTERS and THIOCYANATES.

LBO100 CAS:5890-18-6 ***HR: 3***
LAUROLITSINE
mf: $C_{18}H_{19}NO_4$ mw: 313.38

SYNS: DIMETHOXY-1,10 DIHYDROXY-2,9 NOR-APORPHINE (FRENCH) ◇ 1,10-DIMETHOXY-6a-α-NORAPORPHINE-2,9-DIOL ◇ NORBOLDINE ◇ (+)-NORBOLDINE

TOXICITY DATA with REFERENCE
orl-mus LD50:450 mg/kg APFRAD 38,537,80
ipr-mus LD50:170 mg/kg APFRAD 38,537,80
ivn-mus LD50:90 mg/kg APFRAD 38,537,80

SAFETY PROFILE: Poison by intravenous and intraperitoneal routes. Moderately toxic by ingestion. Used in insecticides. When heated to decomposition it emits toxic fumes of NO_x.

LBO200 CAS:128-76-7 ***HR: 3***
LAUROTETANIN
mf: $C_{19}H_{21}NO_4$ mw: 327.41

SYNS: LAUROTETANINE ◇ (+)-LAUROTETANINE ◇ LITSOEINE ◇ 1,2,10-TRIMETHOXY-6a-α-NORAPORPHIN-6-OL

TOXICITY DATA with REFERENCE
orl-mus LD50:450 mg/kg APFRAD 38,537,80
ipr-mus LD50:170 mg/kg APFRAD 38,537,80
ivn-mus LD50:90 mg/kg APFRAD 38,537,80

SAFETY PROFILE: Poison by intravenous and intraperitoneal routes. Moderately toxic by ingestion. When heated to decomposition it emits toxic fumes of NO_x.

LBQ000 CAS:48163-10-6 ***HR: 3***
LAUROYLETHYLENEIMINE
mf: $C_{14}H_{27}NO$ mw: 225.42

SYN: 1-LAUROYLAZIRIDINE

TOXICITY DATA with REFERENCE
cyt-rat-ipr 100 mg/kg BJPCAL 9,306,54
scu-rat TDLo:1000 mg/kg/32D-I:ETA BJPCAL 9,306,54

SAFETY PROFILE: Questionable carcinogen with experimental tumorigenic data. Mutation data reported. When heated to decomposition it emits toxic fumes of NO_x.

LBR000 CAS:105-74-8 ***HR: 3***
LAUROYL PEROXIDE
DOT: UN 2124/UN 2893
mf: $C_{24}H_{46}O_4$ mw: 398.70

PROP: White, tasteless, coarse powder; faint odor. Mp: 53-55°.

SYNS: ALPEROX C ◇ BIS(1-OXODODECYL)PEROXIDE ◇ DILAUROYL PEROXIDE ◇ DILAUROYL PEROXIDE, TECHNICAL PURE (DOT) ◇ DODECANOYL PEROXIDE ◇ DYP-97 F ◇ LAUROX ◇ LAUROYL PEROXIDE, TECHNICALLY PURE (DOT) ◇ LAURYDOL ◇ LYP 97 ◇ PEROXYDE de LAUROYLE (FRENCH)

TOXICITY DATA with REFERENCE
eye-rbt 500 mg/24H MLD 28ZPAK -,53,72
scu-mus TDLo:184 mg/kg/46W-I:ETA JNCIAM 37,825,66

CONSENSUS REPORTS: IARC Cancer Review: Group 3 IMEMDT 7,56,87; Animal Inadequate Evidence IMEMDT 36,315,85. Reported in EPA TSCA Inventory.
DFG MAK: Mild skin effects.
DOT Classification: Organic Peroxide; Label: Organic Peroxide.

SAFETY PROFILE: Questionable carcinogen with experimental tumorigenic data. A powerful oxidizing agent. It is a corrosive irritant to the eyes and mucous membranes and can cause burns. A dangerous fire hazard. When heated to decomposition it emits acrid smoke and fumes. See also PEROXIDES, ORGANIC.

LBS000 CAS:9002-92-0 ***HR: 1***
LAURYL ALCOHOL EO (4)
mf: $C_2H_4O_n{\cdot}C_{12}H_{26}O$

SYNS: DODECYL ALCOHOL CONDENSED with 4 MOLES ETHYLENE OXIDE ◇ LAURYL ALCOHOL CONDENSED with 4 MOLES ETHYLENE OXIDE

TOXICITY DATA with REFERENCE
orl-rat LD50:8600 mg/kg SPCOAH 38,47,65
orl-mus LD50:4940 mg/kg APRCAS 77,35,62

CONSENSUS REPORTS: Reported in EPA TSCA Inventory.

SAFETY PROFILE: Mildly toxic by ingestion. When heated to decomposition it emits acrid smoke and irritating fumes.

LBT000 CAS:9002-92-0 ***HR: 3***
LAURYL ALCOHOL EO (7)
mf: $C_2H_4O_n{\cdot}C_{12}H_{26}O$

SYNS: DODECYL ALCOHOL CONDENSED with 7 MOLES ETHYLENE OXIDE ◇ PED ◇ POLYOXYETHYLENE DODECANOL

TOXICITY DATA with REFERENCE
skn-rbt 100 mg/kg TXAPA9 2,133,60
eye-rbt 10 mg TXAPA9 2,133,60
orl-rat LD50:4150 mg/kg TXAPA9 2,133,60
orl-mus LD50:1170 mg/kg TXAPA9 2,133,60
ivn-rat LD50:390 mg/kg TXAPA9 2,133,60

CONSENSUS REPORTS: Reported in EPA TSCA Inventory.

SAFETY PROFILE: Poison by intravenous route. Moderately toxic by ingestion. A skin and eye irritant. When heated to decomposition it emits acrid smoke and irritating fumes.

LBU000 CAS:9002-92-0 ***HR: 2***
LAURYL ALCOHOL EO (23)
mf: $C_2H_4O_n{\cdot}C_{12}H_{26}O$

SYNS: DODECYL ALCOHOL CONDENSED with 23 MOLES ETHYLENE OXIDE ◇ LAURYL ALCOHOL CONDENSED with 23 MOLES ETHYLENE OXIDE

TOXICITY DATA with REFERENCE
orl-rat LD50:8600 mg/kg SPCOAH 38,47,65
orl-mus LD50:3500 mg/kg APRCAS 77,35,62

CONSENSUS REPORTS: Reported in EPA TSCA Inventory.

SAFETY PROFILE: Moderately toxic by ingestion. When heated to decomposition it emits acrid smoke and irritating fumes.

LBV000 CAS:5760-73-6 ***HR: 3***
LAURYLDIETHYLENETRIAMINE

TOXICITY DATA with REFERENCE
skn-rat LDLo:1500 mg/kg JIHTAB 22,488,40
scu-rat LDLo:1500 mg/kg JIHTAB 22,488,40
orl-rbt LDLo:400 mg/kg JIHTAB 22,488,40
ivn-rbt LDLo:10 mg/kg JIHTAB 22,488,40
skn-gpg LDLo:1800 mg/kg JIHTAB 22,488,40

CONSENSUS REPORTS: Reported in EPA TSCA Inventory.

SAFETY PROFILE: Poison by ingestion and intravenous routes. Moderately toxic by skin contact and subcutaneous routes. When heated to decomposition it emits toxic fumes of NO_x.

LBW000 CAS:93-23-2 ***HR: 3***
LAURYLISOQUINOLINIUM BROMIDE
mf: $C_{21}H_{32}N{\bullet}Br$ mw: 378.45

PROP: Deep amber, water-sol liquid; pleasant, characteristic odor.

SYNS: 2-DODECYLISOQUINOLINIUM BROMIDE ◇ INTEXSAN LQ75 ◇ ISOTHAN

TOXICITY DATA with REFERENCE
eye-mus 2 mg SEV FCTXAV 15,131,77
eye-mky 2 mg FCTXAV 15,131,77
eye-rbt 2 mg MOD FCTXAV 15,131,77
eye-gpg 2 mg FCTXAV 15,131,77
eye-ham 2 mg SEV FCTXAV 15,131,77
orl-rat LD50:230 mg/kg SSCHAH 25,125,49
orl-gpg LD50:200 mg/kg SSCHAH 25,125,49

CONSENSUS REPORTS: Reported in EPA TSCA Inventory.

SAFETY PROFILE: Poison by ingestion. A severe eye irritant. Combustible when exposed to heat or flame. Incompatible with oxidizing materials. An FDA over-the-counter drug. When heated to decomposition emits toxic fumes of Br^- and NO_x. See also BROMIDES.

LBX000 CAS:112-55-0 ***HR: 3***
LAURYL MERCAPTAN
mf: $C_{12}H_{26}S$ mw: 202.44

PROP: Water-white to pale yellow liquid. Mp: −7°, bp: 115-177°, flash p: 262°F (OC), d: 0.849 @ 15.5°/15.5°.

SYNS: 1-DODECANETHIOL ◇ DODECYL MERCAPTAN ◇ m-DODECYL MERCAPTAN ◇ 1-DODECYL MERCAPTAN ◇ m-LAURYL MERCAPTAN ◇ 1-MERCAPTODODECANE ◇ NCI-C60935 ◇ PENNFLOAT M ◇ PENNFLOAT S

TOXICITY DATA with REFERENCE
cyt-rat-ihl 5020 $\mu g/m^3$/16W BZARAZ 27,102,74

CONSENSUS REPORTS: Reported in EPA TSCA Inventory.

NIOSH REL: (n-Alkane Mono Thiols) CL 0.5 ppm/15M

SAFETY PROFILE: Mutation data reported. Combustible when exposed to heat or flame. To fight fire, use alcohol foam. When heated to decomposition it emits toxic fumes of SO_x. See also MERCAPTANS.

LCA000 CAS:8022-15-9 ***HR: 1***
LAVANDIN OIL

PROP: Main constituent is Linalool. Prepared by steam distillation of the flowering stalks of the plants *Lavanoula hybrida reverchon*, Lavandula abrialis (Fam. *Labiatae*), *Lavandula officinalis*, or *Lavandula latifolia*. Yellow liquid; camphoraceous odor of lavender. D: 0.885, refr index: 1.460 @ 20°. Sol in fixed oils, propylene glycol, mineral oil; insol in glycerin.

SYN: OIL of LAVANDIN, ABRIAL TYPE

TOXICITY DATA with REFERENCE
skn-rbt 500 mg/24H MLD FCTXAV 14,443,76

CONSENSUS REPORTS: Reported in EPA TSCA Inventory.

SAFETY PROFILE: A skin irritant. When heated to decomposition it emits acrid smoke and irritating fumes.

LCA100 CAS:20777-39-3 ***HR: 1***
LAVANDULYL ACETATE
mf: $C_{12}H_{20}O_2$ mw: 196.32

SYNS: 4-HEXEN-1-OL, 5-METHYL-2-(1-METHYLETHENYL)-, ACETATE ◇ 5-METHYL-2-(1-METHYLETHENYL)-4-HEXEN-1-OL ACETATE

TOXICITY DATA with REFERENCE
skn-rbt 500 mg/24H MOD FCTXAV 16,805,78

CONSENSUS REPORTS: Reported in EPA TSCA Inventory.

SAFETY PROFILE: A skin irritant. When heated to decomposition it emits acrid smoke and irritating fumes.

LCC000 CAS:8000-28-0 ***HR: 1***
LAVENDER ABSOLUTE

PROP: Found in the flowers of *Lavandula officinalis chaix*. The main constituent is linalyl acetate. A dark green liquid prepared from alcoholic extract of a residue which is extracted from plant material using an organic solvent.

TOXICITY DATA with REFERENCE
skn-rbt 500 mg/24H MLD FCTXAV 14,449,76
orl-rat LD50:4250 mg/kg FCTXAV 14,449,76

SAFETY PROFILE: Mildly toxic by ingestion. A skin irritant. When heated to decomposition it emits acrid smoke and irritating fumes. See also 3,7-DIMETHYL-1,6-OCTADIEN-3-OL ACETATE.

LCD000 CAS:8000-28-0 ***HR: 1***
LAVENDER OIL

PROP: Found in the flowers of *Lavandula officinalis* Chaix et Villars (*Lavabdula vera* De Candolle (Fam. *Labiatae*). The main constituent is linalyl acetate. A colorless to yellow liquid; characteristic odor and taste of lavender flowers. D: 0.875, refr index: 1.459-1.470 @ 20°.

SYNS: LAVENDEL OEL (GERMAN) ◇ OIL of LAVENDER

TOXICITY DATA with REFERENCE
skn-rbt 500 mg/24H MLD FCTXAV 14,451,76
orl-rat LD50:9040 mg/kg PHARAT 14,435,59

CONSENSUS REPORTS: Reported in EPA TSCA Inventory.

SAFETY PROFILE: Mildly toxic by ingestion. A skin irritant. When heated to decomposition it emits acrid smoke and irritating fumes. See also 3,7-DIMETHYL-1,6-OCTADIEN-3-OL ACETATE.

LCE000 CAS:64083-05-2 ***HR: 3***
LD-813

PROP: Commercial mixture of aromatic amines containing approx 40% MOCA.

TOXICITY DATA with REFERENCE
orl-rat TDLo:37 g/kg/2Y-C:CAR TXAPA9 31,159,75

SAFETY PROFILE: Questionable carcinogen with experimental carcinogenic data. When heated to decomposition it emits toxic fumes of NO_x. See also AROMATIC AMINES.

LCF000 CAS:7439-92-1 ***HR: 3***
LEAD
af: Pb aw: 207.19

PROP: Bluish-gray, soft metal. Mp: 327.43°, bp: 1740°, d: 11.34 @ 20°/4°. vap press: 1 mm @ 973°.

SYNS: C.I. 77575 ◇ C.I. PIGMENT METAL 4 ◇ GLOVER ◇ LEAD FLAKE ◇ LEAD S2 ◇ OLOW (POLISH) ◇ OMAHA ◇ OMAHA & GRANT ◇ SI ◇ SO

TOXICITY DATA with REFERENCE
cyt-hmn-unr 50 $\mu g/m^3$ MUREAV 147,301,85
cyt-rat-ihl 23 $\mu g/m^3$/16W GTPZAB 26(10),38,82
cyt-mky-orl 42 mg/kg/30W TOLED5 8,165,81
orl-dom TDLo:662 mg/kg (female 1-21W post):REP TXAPA9 25,466,73
orl-mus TDLo:4800 mg/kg (female 1-16D post):TER BECTA6 18,271,77
orl-wmn TDLo:450 mg/kg/6Y:PNS:CNS JAMAAP 237,2627,77
ihl-hmn TCLo:10 $\mu g/m^3$:GIT:LIV VRDEA5 (5),107,81
ipr-rat LDLo:1000 mg/kg EQSSDX 1,1,75
orl-pgn LDLo:160 mg/kg HBAMAK 4,1289,35

CONSENSUS REPORTS: IARC Cancer Review: Group 2B IMEMDT 7,230,87; Animal Inadequate Evidence IMEMDT 23,325,80. Lead and its compounds are on the Community Right-To-Know List. Reported in EPA TSCA Inventory. EPA Genetic Toxicology Program.

OSHA PEL: TWA 0.05 mg(Pb)/m^3
ACGIH TLV: TWA 0.15 mg(Pb)/m^3; BEI: 50 μg(lead)/L in blood; 150 μg(lead)/g creatinine in urine.
DFG MAK: 0.1 mg/m^3; BAT: 70 μg(lead)/L in blood, 30 μg(lead)/L in blood of women less than 45 years old.
NIOSH REL: TWA (Inorganic Lead) 0.10 mg(Pb)/m^3

SAFETY PROFILE: Suspected carcinogen. Poison by ingestion. Moderately toxic by intraperitoneal route. Human systemic effects by ingestion and inhalation: loss of appetite, anemia, malaise, insomnia, headache, irritability, muscle and joint pains, tremors, flaccid paralysis without anesthesia, hallucinations and distorted perceptions, muscle weakness, gastritis and liver changes. The major organ systems affected are the nervous system, blood system, and kidneys. Lead encephalopathy is accompanied by severe cerebral edema, increase in cerebral spinal fluid pressure, proliferation and swelling of endothelial cells in capillaries and arterioles, proliferation of glial cells, neuronal degeneration and areas of focal cortical necrosis in fatal cases. Experimental evidence now suggests that blood levels of lead below 10 μg/dl can have the effect of diminishing the IQ scores of children. Low levels of lead impair neurotransmission and immune system function and may increase systolic blood pressure. Reversible kidney damage can occur from acute exposure. Chronic exposure can lead to irreversible vascular schlerosis, tubular cell atrophy, interstitial fibrosis, and glomerular sclerosis. Severe toxicity can cause sterility, abortion and neonatal mortality and morbidity. An experimental teratogen. Experimental reproductive effects. Human mutation data reported. Very heavy intoxication can sometimes be detected by formation of a dark line on the gum margins, the so-called "lead line."

When lead is ingested, much of it passes through the body unabsorbed, and is eliminated in the feces. The greater portion of the lead that is absorbed is caught by the liver and excreted, in part, in the bile. For this reason, larger amounts of lead are necessary to cause toxic effects by this route, and a longer period of exposure is usually necessary to produce symptoms. On the other hand, upon inhalation, absorption takes place easily from the respiratory tract and symptoms tend to develop more quickly. For industry, inhalation is much more important than is ingestion. For the general population, exposure to lead occurs from inhaled air, dust of various types, and food and water with an approximate 50/50 division between inhalation and ingestion routes. Lead occurs in water in either dissolved or particulate form. At low pH, lead is more easily dissolved. Chemical treatment to soften water increases the solubility of lead. Adults absorb about 5-15% of ingested lead and retain less than 5%. Children absorb about 50% and retain about 30%.

Lead produces a brittleness of the red blood cells so that they hemolyze with but slight trauma; the hemoglobin is not affected. Due to their increased fragility, the red cells are destroyed more rapidly in the body than is normal, producing an anemia which is rarely severe. The loss of circulating red cells stimulates the production of new young cells which, on entering the blood stream, are acted upon by the circulating lead, with resultant coagulation of their basophilic material. These cells after suitable staining, are recognized as "stippled cells." There is no uniformity of opinion regarding the effect of lead on the white blood cells.

In addition to its effect on the red blood cells, lead produces a damaging effect on the organs or tissues with which it comes in contact. No specific or characteristic lesion is produced. Autopsies in deaths attributed to lead poisoning and experimental work on animals have shown pathological lesions of the kidneys, liver, male gonads, nervous system, blood vessels and other tissues. None of these changes, however, has been found consistently. In cases of severe lead poisoning, the amount of lead found in the blood is frequently in excess of 0.07 mg per 100 cc of whole blood. The urinary lead excretion generally exceeds 0.1 mg per liter of urine.

Flammable in the form of dust when exposed to heat or flame. Moderately explosive in the form of dust when exposed to heat or flame. Mixtures of hydrogen peroxide + trioxane explode on contact with lead. Rubber gloves containing lead may ignite in nitric acid. Violent reaction on ignition with chlorine trifluoride; concentrated hydrogen peroxide; ammonium nitrate (below 200° with powdered lead); sodium acetylide (with powdered lead). Incompatible with NaN_3; Zr; disodium acetylide; oxidants. Can react vigorously with oxidizing materials. A common air contaminant. When heated to decomposition it emits highly toxic fumes of Pb. See also LEAD COMPOUNDS.

LCG000 CAS:15347-57-6 ***HR: D***
LEAD ACETATE
mf: $C_2H_4O_2 \cdot xPb$ mw: 1510.39

SYN: ACETIC ACID, LEAD SALT

TOXICITY DATA with REFERENCE
cyt-hmn:lyms 1 umol/L TECSDY 8,39,84
cyt-hmn:leu 10 umol/L EXPEAM 30,1006,74
orl-rat TDLo:18900 μg/kg (male 9W pre):REP ANDRDQ 21,161,89

SAFETY PROFILE: Experimental reproductive effects. Mutation data reported. When heated to decomposition it emits toxic fumes of Pb.

LCG500 ***HR: 3***
LEAD(IV) ACETATE AZIDE
mf: $C_6H_9N_3O_6Pb$ mw: 426.35

$Pb(OOCCH_3)_3N_3$

CONSENSUS REPORTS: Lead and its compounds are on the Community Right-To-Know List.

SAFETY PROFILE: Above 0°C it decomposes to nitrogen and the explosive lead(II) azide. When heated to decomposition it emits toxic fumes of NO_x. See also LEAD COMPOUNDS and AZIDES.

LCH000 CAS:1335-32-6 ***HR: 3***
LEAD ACETATE, BASIC
mf: $C_4H_{10}O_8Pb_3$ mw: 807.71

PROP: White powder.

SYNS: BASIC LEAD ACETATE ◇ BIS(ACETO)DIHYDROXYTRILEAD ◇ BIS(ACETATO)TETRAHYDROXYTRILEAD ◇ BLA ◇ LEAD MONOSUBACETATE ◇ LEAD SUBACETATE ◇ MONOBASIC LEAD ACETATE ◇ RCRA WASTE NUMBER U146 ◇ SUBACETATE LEAD

TOXICITY DATA with REFERENCE
mmo-sat 250 mg/L ENMUDM 2,234,80
orl-mus TDLo:258 g/kg (28D male):REP EXPEAM 30,486,74
orl-rat TDLo:350 g/kg/90W-C:CAR BJCAAI 16,289,62
orl-mus TDLo:90 g/kg/2Y-C:ETA BJCAAI 23,765,69
ipr-mus TDLo:38 mg/kg/6W-I:NEO TXAPA9 82,19,86

CONSENSUS REPORTS: IARC Cancer Review: Group 3 IMEMDT 7,230,87; Animal Sufficient Evidence IMEMDT 23,325,80; IMEMDT 1,40,72; Human Limited Evidence IMEMDT 23,325,80. Lead and its compounds are on the Community Right-To-Know List. Reported in EPA TSCA Inventory. EPA Genetic Toxicology Program.

SAFETY PROFILE: Experimental reproductive effects. Questionable carcinogen with experimental carcinogenic, neoplastigenic, and tumorigenic data. Mutation data reported. When heated to decomposition it emits toxic fumes of Pb. See also LEAD and LEAD COMPOUNDS.

LCI000 ***HR: 3***
LEAD ACETATE BROMATE
mf: $C_2H_3BrO_5Pb$ mw: 394.15

CONSENSUS REPORTS: Lead and its compounds are on the Community Right-To-Know List.

SAFETY PROFILE: A poison. A very friction-sensitive explosive. Upon decomposition it emits toxic fumes of Pb and Br^-. See also LEAD COMPOUNDS and BROMATES.

LCI600 ***HR: 3***
LEAD ACETATE-LEAD BROMITE
mf: $C_4H_6O_4Pb \cdot Br_2O_6Pb$ mw: 788.29

$Pb(OOCCH_3)_2 \cdot Pb(BrO_3)_2$

CONSENSUS REPORTS: Lead and its compounds are on the Community Right-To-Know List.

SAFETY PROFILE: A friction sensitive explosive. When heated to decomposition it emits toxic fumes of Br^-. See also LEAD COMPOUNDS.

LCJ000 CAS:6080-56-4 ***HR: 3***
LEAD ACETATE(II), TRIHYDRATE
mf: $C_4H_6O_4 \cdot Pb \cdot 3H_2O$ mw: 379.35

SYNS: ACETIC ACID, LEAD(+2) SALT TRIHYDRATE ◇ BIS(ACETATO)TRIHYDROXYTRILEAD ◇ BLEIAZETAT (GERMAN) ◇ LEAD ACETATE TRIHYDRATE ◇ LEAD DIACETATE TRIHYDRATE ◇ PLUMBOUS ACETATE

TOXICITY DATA with REFERENCE
dni-mus-ipr 20 g/kg ARGEAR 51,605,81
orl-rat TDLo:2219 mg/kg (18D post):REP NTOTDY 4,105,82
orl-rat TDLo:8524 mg/kg/78W-C:CAR,TER ZAPPAN 111(1),1,68
ipr-rat LD50:200 mg/kg INMEAF 10,15,41
scu-gpg LDLo:2100 mg/kg BMJOAE 2,217,13

CONSENSUS REPORTS: NTP Fifth Annual Report on Carcinogens. IARC Cancer Review: Animal Sufficient Evidence IMEMDT 1,40,72. EPA Genetic Toxicology Program. Lead and its compounds are on the Community Right-To-Know List.

OSHA PEL: TWA 0.05 mg(Pb)/m^3
NIOSH REL: (Inorganic Lead) TWA 0.10 mg(Pb)/m^3

SAFETY PROFILE: Confirmed carcinogen with experimental carcinogenic and teratogenic data. Poison by intraperitoneal route. Moderately toxic by subcutaneous route. Experimental reproductive effects. Mutation data reported. When heated to decomposition it emits toxic fumes of Pb. See also LEAD COMPOUNDS.

LCK000 CAS:7784-40-9 ***HR: 3***
LEAD ACID ARSENATE
DOT: UN 1617
mf: $AsHO_4 \cdot Pb$ mw: 347.12

PROP: White crystals.

SYNS: ACID LEAD ARSENATE ◇ ACID LEAD ORTHOARSENATE ◇ ARSENATE of LEAD ◇ ARSINETTE ◇ DIBASIC LEAD ARSENATE ◇ GYPSINE ◇ LEAD ARSENATE ◇ LEAD ARSENATE, solid (DOT) ◇ LEAD ARSENATE (standard) ◇ ORTHO L10 DUST ◇ ORTHO L40 DUST ◇ SCHULTENITE ◇ SECURITY ◇ SOPRABEL ◇ STANDARD LEAD ARSENATE ◇ TALBOT

TOXICITY DATA with REFERENCE
unr-man LDLo:1050 mg/kg FMCHA2 -,C139,83
orl-rat LD50:100 mg/kg PCOC** -,653,66
orl-rbt LDLo:75 mg/kg JPETAB 39,246,30
orl-ckn LD50:450 mg/kg PCOC** -,653,66

CONSENSUS REPORTS: NTP Fifth Annual Report on Carcinogens. IARC Cancer Review: Human Sufficient Evidence IMEMDT 23,39,80; Animal Inadequate Evidence IMEMDT 1,40,72; IMEMDT 1,40,72. Arsenic and its compounds, as well as lead and its compounds, are on the Community Right-To-Know List. Reported in EPA TSCA Inventory.

OSHA PEL: TWA 0.05 mg(Pb)/m^3; 0.01 mg(As)/m^3; Cancer Hazard
ACGIH TLV: TWA 0.15 mg(Pb)/m^3; 0.2 mg(As)/m^3
NIOSH REL: (Inorganic Lead) TWA 0.10 mg(Pb)/m^3; (Inorganic Arsenic) CL 0.002 mg(As)/m^3/15M
DOT Classification: Poison B; Label: Poison.

SAFETY PROFILE: Confirmed human carcinogen. A poison by ingestion. Moderately toxic to humans by an unspecified route. Used as an insecticide and herbicide. When heated to decomposition it emits very toxic fumes of As and Pb. See also ARSENIC COMPOUNDS and LEAD COMPOUNDS.

LCL000 CAS:10031-13-7 ***HR: 3***
LEAD(II) ARSENITE
DOT: UN 1618
mf: $As_2O_4 \cdot Pb$ mw: 421.03

PROP: White powder. D: 5.85. Insol in water; sol in dil HNO_3.

SYN: LEAD ARSENITE, solid (DOT)

CONSENSUS REPORTS: Arsenic and its compounds, as well as lead and its compounds, are on the Community Right-To-Know List.

OSHA PEL: TWA 0.05 mg(Pb)/m^3; 0.01 mg(As)/m^3; Cancer Hazard
ACGIH TLV: TWA 0.15 mg(Pb)/m^3; 0.2 mg(As)/m^3
NIOSH REL: (Inorganic Lead) TWA 0.10 mg(Pb)/m^3; (Inorganic Arsenic) CL 0.002 mg(As)/m^3/15M
DOT Classification: Poison B; Label: Poison.

SAFETY PROFILE: Confirmed human carcinogen. A poison. When heated to decomposition it emits very toxic fumes of Pb and As. See also LEAD COMPOUNDS and ARSENIC COMPOUNDS.

LCM000 CAS:13424-46-9 ***HR: 3***
LEAD(II) AZIDE
DOT: UN 0129
mf: N_6Pb mw: 291.25

PROP: Colorless needles or white powder. Explodes @ 350° or when shocked; very sol in acetic acid; insol in NH_4OH.

SYNS: INITIATING EXPLOSIVE LEAD AZIDE, DEXTRINATED TYPE ONLY (DOT) ◇ LEAD AZIDE, DRY (DOT)

CONSENSUS REPORTS: Lead and its compounds are on the Community Right-To-Know List. Reported in EPA TSCA Inventory.

OSHA PEL: TWA 0.05 mg(Pb)/m^3
ACGIH TLV: TWA 0.15 mg(Pb)/m^3
NIOSH REL: (Inorganic Lead) TWA 0.10 mg(Pb)/m^3
DOT Classification: Class A Explosive; Label: Explosive A: Forbidden, Dry.

SAFETY PROFILE: A deadly poison. An explosive sensitive to shock or heating to 250°C. Will explode spontaneously during crystallization. Mixtures with calcium stearate may explode spontaneously. May explode spontaneously after prolonged contact with copper; zinc; or their alloys (e.g., brass). Incompatible with CS_2. Used in commercial blasting caps and military ammunition. When heated it emits highly toxic fumes of Pb and NO_x. See also LEAD COMPOUNDS, AZIDES, and EXPLOSIVES, HIGH.

LCN000 ***HR: 3***
LEAD(IV) AZIDE
mf: $N_{12}Pb$ mw: 275.27

CONSENSUS REPORTS: Lead and its compounds are on the Community Right-To-Know List.

SAFETY PROFILE: Crystalline material may explode spontaneously. Upon decomposition it emits very toxic fumes of NO_x and Pb. See also AZIDES and LEAD COMPOUNDS.

LCO000 CAS:34018-28-5 ***HR: 3***
LEAD BROMATE
mf: Br_2O_6Pb mw: 463.01

PROP: Colorless, monoclinic crystals. Mp: 180° (decomp), d: 5.53. Sltly sol in cold water; sol in hot water.

CONSENSUS REPORTS: Lead and its compounds are on the Community Right-To-Know List.

SAFETY PROFILE: A posion. Explosive. Pure lead bromate is stable to 180°C. Upon decomposition it emits very toxic fume of Br^- and Pb. See also LEAD COMPOUNDS and BROMATES.

LCP000 CAS:598-63-0 ***HR: 2***
LEAD CARBONATE
mf: CO_3•Pb mw: 267.20

PROP: White, heavy powder. D: 6.61, decomp @ 400° leaving residue of PbO. Insol in water, alc; sol in acetic acid, dil HNO_3 (effervescence).

SYNS: CARBONIC ACID, LEAD(2+) SALT (1:1) ◇ CERUSSETE ◇ DIBASIC LEAD CARBONATE ◇ LEAD(2+) CARBONATE ◇ WHITE LEAD

TOXICITY DATA with REFERENCE
orl-rat TDLo:40 g/kg (16D post):REP TXAPA9 37,160,76
orl-man TDLo:214 mg/kg/4W:GIT,LIV NEJMAG 303,459,80
orl-hmn LDLo:571 mg/kg:CNS,PSY,GIT IPSTB3 3,93,76
orl-gpg LDLo:1000 mg/kg EQSSDX 1,1,75

CONSENSUS REPORTS: IARC Cancer Review: Animal Inadequate Evidence IMEMDT 23,325,80; IMEMDT 1,40,72. Lead and its compounds are on the Community Right-To-Know List. Reported in EPA TSCA Inventory.

OSHA PEL: TWA 0.05 mg(Pb)/m^3
ACGIH TLV: TWA 0.15 mg(Pb)/m^3
NIOSH REL: (Inorganic Lead) TWA 0.10 mg(Pb)/m^3

SAFETY PROFILE: Moderately toxic by ingestion. Human systemic effects by ingestion: gastrointestinal contractions, jaundice, brain degenerative changes, convulsions, nausea or vomiting. Experimental reproductive effects. Questionable carcinogen. Ignites spontaneously and burns fiercely in fluorine. When heated to decomposition it emits toxic fumes of Pb. See also LEAD COMPOUNDS.

LCQ000 CAS:7758-95-4 ***HR: 3***
LEAD CHLORIDE
DOT: NA 2291
mf: Cl_2Pb mw: 278.09

PROP: White crystals. Mp: 501°, bp: 950°, d: 5.85, vap press: 1 mm @ 547°. Somewhat sol in cold water, more sol in hot water; very sol in ammonium chloride, NH_4NO_3, alkali hydroxides.

SYNS: LEAD (2+) CHLORIDE ◇ LEAD (II) CHLORIDE ◇ LEAD DICHLORIDE ◇ PLUMBOUS CHLORIDE

TOXICITY DATA with REFERENCE
mmo-smc 1 mmol/L CPBTAL 33,1571,85
oms-hmn:hla 250 μmol/L TXCYAC 5,167,75
dni-mus:fbr 20 μmol/L ZHPMAT 161,26,75
orl-rat TDLo:570 mg/kg (14D pre-21D post):REP PBBHAU 11,95,79
ivn-mus TDLo:20 mg/kg (female 8D post):TER TJADAB 34,207,86
orl-gpg LDLo:1500 mg/kg MEIEDD 10,777,83

CONSENSUS REPORTS: IARC Cancer Review: Animal Inadequate Evidence IMEMDT 23,325,80. Lead and its compounds are on the Community Right-To-Know List. Reported in EPA TSCA Inventory. EPA Genetic Toxicology Program.

OSHA PEL: TWA 0.05 mg(Pb)/m^3
ACGIH TLV: TWA 0.15 mg(Pb)/m^3
NIOSH REL: (Inorganic Lead) TWA 0.10 mg(Pb)/m^3
DOT Classification: ORM-B; Label: None.

SAFETY PROFILE: Moderately toxic by ingestion. An experimental teratogen. Experimental reproductive effects. Questionable carcinogen. Human mutation data reported. Explosive reaction with calcium when heated slightly. When heated to decomposition it emits very toxic fumes of Pb and Cl^-. See also LEAD and LEAD COMPOUNDS.

LCQ300 CAS:13453-57-1 ***HR: 3***
LEAD(II) CHLORITE
mf: Cl_2O_4Pb mw: 342.10

$Pb(ClO_2)_2$

CONSENSUS REPORTS: Lead and its compounds are on the Community Right-To-Know List.

SAFETY PROFILE: Explodes when heated above 100°C. Mixtures with antimony sulfide or sulfur are friction-sensitive explosives. Reacts violently with non-metals (e.g., carbon, red phosphorus, sulfur). When heated to decomposition it emits toxic fumes of Pb and Cl^-. See also LEAD COMPOUNDS and CHLORITES.

LCR000 CAS:7758-97-6 ***HR: 3***
LEAD CHROMATE
mf: CrO_4•Pb mw: 323.19

PROP: Yellow or orange-yellow powder. One of the most insol salts. Insol in acetic acid; sol in solns of fixed alkali hydroxides, dil HNO_3. Mp: 844°. Bp: decomp, d: 6.3.

SYNS: CANARY CHROME YELLOW 40-2250 ◇ CHROMATE de PLOMB (FRENCH) ◇ CHROME GREEN ◇ CHROME LEMON ◇ CHROME YELLOW ◇ CHROMIC ACID, LEAD(2+) SALT (1:1) ◇ CHROMIUM YELLOW ◇ C.I. 77600 ◇ C.I. PIGMENT YELLOW 34 ◇ COLOGNE YELLOW ◇ C.P. CHROME YELLOW LIGHT ◇ CROCOITE ◇ DAINICHI CHROME YELLOW G ◇ GIALLO CROMO (ITALIAN) ◇ KING'S YELLOW ◇ LEAD CHROMATE(VI) ◇ LEIPZIG YELLOW ◇ LEMON YELLOW ◇ PARIS YELLOW ◇ PIGMENT GREEN 15 ◇ PLUMBOUS CHROMATE ◇ PURE LEMON CHROME L3GS

TOXICITY DATA with REFERENCE
cyt-hmn:lym 13 μmol/L MUREAV 77,157,80
mnt-mus-ipr 500 mg/kg TJEMAO 146,373,85 TUMOAB 57,213,71
ims-rat TDLo:324 mg/kg/39W-I:NEO CNREA8 36,1779,76
scu-rat TD:135 mg/kg:ETA PBPHAW 14,47,78
orl-mus LD50:12 g/kg OYYAA2 2,76,68
ipr-gpg LD75:156 mg/kg MEIEDD 10,777,83

CONSENSUS REPORTS: NTP Fifth Annual Report on Carcinogens. IARC Cancer Review: Group 1 IMEMDT 7,165,87; Animal Inadequate Evidence IMEMDT 2,100,73; Animal Sufficient Evidence IMEMDT 23,205,80; Human Sufficient Evidence IMEMDT 23,205,80. Lead and its compounds, as well as chromium and its compounds, are on the Community Right-To-Know List. Reported in EPA TSCA Inventory. EPA Genetic Toxicology Program.

OSHA PEL: TWA 0.05 mg(Pb)/m^3; CL 0.1 mg(CrO_3)/m^3
ACGIH TLV: 0.05 mg(Cr)/m^3; Human Carcinogen
DFG MAK: Suspected Carcinogen.
NIOSH REL: (Chromium(VI)) TWA 0.001 mg(Cr(VI))/m^3; (Inorganic Lead) TWA 0.10 mg(Pb)/m^3

SAFETY PROFILE: Confirmed carcinogen with experimental neoplastigenic and tumorigenic data. Poison by intraperitoneal route. Mildly toxic by ingestion. Human mutation data reported. Potentially explosive reactions with azo-dye stuffs (e.g., dinitroaniline orange; chlorinated para red). Violent reaction with aluminum + dinitronaphthalene + heat. Forms pyrophoric mixtures with sulfur; tantalum; and iron(III) hexacyanoferrate(4^-) (e.g., brunswick green pigment; prussian blue pigment). When heated to decomposition it emits toxic fumes of Pb. See also LEAD COMPOUNDS and CHROMIUM COMPOUNDS.

LCS000 CAS:18454-12-1 ***HR: 3***
LEAD CHROMATE, BASIC
mf: CrO_4Pb•OPb mw: 546.38

PROP: Red, amorphous or crystalline solid. Mp: 920°.

SYNS: ARANCIO CROMO (ITALIAN) ◇ AUSTRIAN CINNABAR ◇ BASIC LEAD CHROMATE ◇ CHINESE RED ◇ CHROME ORANGE ◇ CHROMIUM LEAD OXIDE ◇ C.I. 77601 ◇ C.I. PIGMENT ORANGE 21 ◇ C.I. PIGMENT RED ◇ C.P. CHROME LIGHT 2010 ◇ C.P. CHROME ORANGE DARK 2030 ◇ C.P. CHROME ORANGE MEDIUM 2020 ◇ DAINICHI CHROME ORANGE R ◇ GENUINE ACETATE CHROME ORANGE ◇ GENUINE ORANGE CHROME ◇ INDIAN RED ◇ INTERNATIONAL ORANGE 2221 ◇ IRGACHROME ORANGE OS ◇ LEAD CHROMATE OXIDE (MAK) ◇ LEAD CHROMATE, RED ◇ LIGHT ORANGE CHROME ◇ No. 156 ORANGE CHROME ◇ ORANGE CHROME ◇ ORANGE NITRATE CHROME ◇ PALE ORANGE CHROME ◇ PERSIAN RED ◇ PURE ORANGE CHROME M ◇ RED LEAD CHROMATE ◇ VYNAMON ORANGE CR

TOXICITY DATA with REFERENCE
oms-hmn:oth 500 mg/L BJCAAI 44,219,81
dni-ham:kdy 150 mg/L BJCAAI 44,219,81
scu-rat TDLo:135 mg/kg:CAR ANYAA9 271,431,76
scu-rat TD:135 mg/kg:ETA PBPHAW 14,47,78
scu-rat TD:135 mg/kg:NEO TUMOAB 57,213,71

CONSENSUS REPORTS: NTP Fifth Annual Report on Carcinogens. IARC Cancer Review: Human Sufficient Evidence IMEMDT 23,205,80; Animal Limited Evidence IMEMDT 23,205,80. Lead and its compounds, as well as chromium and its compounds are on the Community Right-To-Know List. Reported in EPA TSCA Inventory.

OSHA PEL: TWA 0.05 mg(Pb)/m^3; CL 0.1 mg(CrO_3)/m^3
ACGIH TLV: TWA 0.05 mg(Cr)/m^3; TWA 0.15 mg (Pb)/m^3
DFG MAK: Suspected Carcinogen.
NIOSH REL: (Chromium(VI)) TWA 0.001 mg(Cr(VI))/m^3; (Inorganic Lead) TWA 0.10 mg(Pb)/m^3

SAFETY PROFILE: Confirmed human carcinogen with experimental carcinogenic, neoplastigenic, and tumorigenic data. Human mutation data reported. When heated to decomposition it emits very toxic fumes of Pb. See also LEAD COMPOUNDS and CHROMIUM COMPOUNDS.

LCT000 ***HR: 3***
LEAD COMPOUNDS

CONSENSUS REPORTS: Lead and its compounds are on the Community Right-To-Know List.

SAFETY PROFILE: Some are experimental neoplastigens and tumorigens. Lead poisoning is one of the commonest of occupational diseases. The presence of lead-bearing materials or lead compounds in an industrial plant does not necessarily result in exposure on the part of the worker. The lead must be in such form, and so distributed, as to gain entrance into the body or tissues of the worker in measurable quantity, otherwise no exposure can be said to exist. Some lead compounds are carcinogens of the lungs and kidneys.

Mode of entry into body: 1. By inhalation of the dust, fumes, mists or vapors. (Common air contaminants). 2. By ingestion of lead compounds trapped in the upper respiratory tract or introduced into the mouth on food, tobacco, fingers, or other objects. 3. Through the skin; this route is of special importance in the case of organic compounds of lead, as lead tetraethyl. In the case of the inorganic forms of lead, this route is of no practical importance. Significant quantities of lead can be ingested from water that has been sitting in pipes with lead solder. Some water coolers may also have this type of solder.

Lead is a cumulative poison. Increasing amounts build up in the body and eventually reach a point where symptoms and disability occur. See LEAD for symptoms of overexposure.

The toxicity of the various lead compounds appears to depend upon several factors: (1) the solubility of the compound in the body fluids; (2) the fineness of the particles of the compound; solubility is greater in proportion to the fineness of the particles; (3) conditions under which the compound is being used. Where a lead compound is used as a powder, contamination of the atmosphere will be much less if the powder is kept damp. Of the various lead compounds, the carbonate, the monoxide, and the sulfate are considered to be more toxic than metallic lead or other lead compounds. Lead arsenate is very toxic due to the presence of the arsenic radical. Organolead compounds are rapidly absorbed by the respiratory and gastrointestinal systems and through the skin. Tetraethyl lead is converted in the body to triethyl lead which is a more severe neurotoxin than inorganic lead. Diagnostic mobilization of lead with calcium EDTA may be useful in questionable cases. When heated to decomposition they emit toxic fumes of Pb. See also LEAD and specific compounds.

LCU000 CAS:592-05-2 ***HR: 3***
LEAD(II) CYANIDE
DOT: UN 1620
mf: C_2N_2Pb mw: 259.23

PROP: White powder.

SYNS: C.I. 77610 ◇ C.I. PIGMENT YELLOW 48 ◇ CYANURE de PLOMB (FRENCH) ◇ LEAD CYANIDE (DOT)

TOXICITY DATA with REFERENCE
ipr-rat LDLo:100 mg/kg NCNSA6 5,27,53

CONSENSUS REPORTS: Lead and its compounds, as well as cyanide and its compounds, are on the Community Right-To-Know List.

OSHA PEL: TWA 0.05 mg(Pb)/m^3
ACGIH TLV: TWA 0.15 mg(Pb)/m^3
NIOSH REL: (Inorganic Lead) TWA 0.10 mg(Pb)/m^3
DOT Classification: Poison B; Label: Poison.

SAFETY PROFILE: Poison by intraperitoneal route. Violent reaction with Mg. A fire hazard and a powerful oxidizer. When heated to decomposition it emits very toxic fumes of Pb, CN^-, and NO_x. See also LEAD COMPOUNDS and CYANIDES.

LCV000 CAS:301-04-2 ***HR: 3***
LEAD DIACETATE
DOT: UN 1616
mf: $C_4H_6O_4 \cdot Pb$ mw: 325.29

PROP: Trihydrate: colorless crystals or white granules or powder. Sltly acetic odor, slowly effloresces. D: 2.55, mp: 75° (when rapidly heated), decomp above 200°. Very sol in glycerol.

SYNS: ACETATE de PLOMB (FRENCH) ◇ ACETIC ACID LEAD (2+) SALT ◇ BLEIACETAT (GERMAN) ◇ DIBASIC LEAD ACETATE ◇ LEAD ACETATE ◇ LEAD (2+) ACETATE ◇ LEAD(II) ACETATE ◇ LEAD DIBASIC ACETATE ◇ NORMAL LEAD ACETATE ◇ PLUMBOUS ACETATE ◇ RCRA WASTE NUMBER U144 ◇ SALT of SATURN ◇ SUGAR of LEAD

TOXICITY DATA with REFERENCE
sln-smc 250 μmol/L MUTAEX 1,21,86
cyt-hmn:lym 1 mmol/L/24H TXCYAC 10,67,78

orl-mky TDLo:765 mg/kg (female 90D pre):REP NETOD7 5,391,83
ipr-mus TDLo:35 mg/kg (female 8D post):TER BIMDB3 30,223,79
orl-rat TDLo:900 mg/kg/60D-C:NEO ENVRAL 24,391,81
orl-rat TD:250 g/kg/47W-C:ETA BJCAAI 16,283,62
ipr-rat LD50:150 mg/kg EQSSDX 1,1,75
ipr-mus LD50:189 mg/kg COREAF 256,1043,63
ivn-mus LD50:104 mg/kg IGSBAL 93,461,77
orl-dog LDLo:300 mg/kg HBAMAK 4,1289,35
scu-dog LDLo:80 mg/kg HBAMAK 4,1289,35
ivn-dog LDLo:300 mg/kg EQSSDX 1,1,75

CONSENSUS REPORTS: NTP Fifth Annual Report on Carcinogens. IARC Cancer Review: Group 3 IMEMDT 7,230,87; Animal Sufficient Evidence IMEMDT 23,325,80; IMEMDT 1,40,72; Human Limited Evidence IMEMDT 23,325,80. Lead and its compounds are on the Community Right-To-Know List. Reported in EPA TSCA Inventory. EPA Genetic Toxicology Program.

OSHA PEL: TWA 0.05 mg(Pb)/m^3
ACGIH TLV: TWA 0.15 mg(Pb)/m^3
NIOSH REL: (Inorganic Lead) TWA 0.10 mg(Pb)/m^3
DOT Classification: ORM-E; Label: None; Poison B; Label: St. Andrews Cross.

SAFETY PROFILE: Confirmed carcinogen with experimental neoplastigenic, tumorigenic, and teratogenic data. Poison by ingestion, intraperitoneal, subcutaneous, and intravenous routes. Experimental reproductive effects. Human mutation data reported. Used as color additive in hair dyes, an insecticide, an astringent, and sedative. Incompatible with $KBrO_3$, acids, soluble sulfates, citrates, tartrates, chlorides, carbonates, alkalies, tannin phosphates, resorcinol, salicylic acid, phenol, chloral hydrate, sulfites, vegetable infusions, tinctures. When heated to decomposition it emits toxic fumes of Pb. See also LEAD COMPOUNDS.

LCW000 CAS:19010-66-3 ***HR: 2***
LEAD DIMETHYLDITHOCARBAMATE
mf: $C_6H_{12}N_2S_4$•Pb mw: 447.63

PROP: Solid. Mp: 258°, d: 2.5.

SYNS: BIS(DIMETHYLCARBAMODITHIOATO-S,S')LEAD ◇ BIS(DIMETHYLDITHIOCARBAMIATO)LEAD ◇ DIMETHYLDITHIOCARBAMIC ACID, LEAD SALT ◇ METHYL LEDATE ◇ NCI-C02891

TOXICITY DATA with REFERENCE
mmo-sat 100 μg/plate ENMUDM 5(Suppl 1),3,83
mma-sat 33 μg/plate NTPTB* JAN 82
scu-mus TDLo:1000 mg/kg:ETA NTIS** PB223-159

CONSENSUS REPORTS: IARC Cancer Review: Group 3 IMEMDT 7,230,87; Animal Inadequate Evidence IMEMDT 12,131,76. NCI Carcinogenesis Bioassay (feed); No Evidence: mouse, rat NCITR* NCI-CG-TR-151,79. Lead and its compounds are on the Community Right-To-Know List. Reported in EPA TSCA Inventory.
NIOSH REL: (Inorganic Lead) TWA 0.10 mg(Pb)/m^3

SAFETY PROFILE: Questionable carcinogen with experimental tumorigenic data. Mutation data reported. Combustible when exposed to heat or flame. When heated to decomposition it emits very toxic fumes of Pb, NO_x, and SO_x. See also LEAD COMPOUNDS and CARBAMATES.

LCX000 CAS:1309-60-0 ***HR: 3***
LEAD DIOXIDE
DOT: UN 1872
mf: O_2Pb mw: 239.19

PROP: Brown, hexagonal crystals or dark brown powder. Mp: decomp @ 290°, d: 9.375. Liberates O_2 when heated. Insol in water; sol in HCl evolving chlorine, sol in alkali iodide solns liberating iodine, sol in hot caustic alkali soln.

SYNS: BIOXYDE de PLOMB (FRENCH) ◇ C.I. 77580 ◇ LEAD BROWN ◇ LEAD(IV) OXIDE ◇ LEAD OXIDE BROWN ◇ LEAD PEROXIDE (DOT) ◇ LEAD SUPEROXIDE ◇ PEROXYDE de PLOMB (FRENCH)

TOXICITY DATA with REFERENCE
ipr-gpg LD50:220 mg/kg EQSSDX 1,1,75

CONSENSUS REPORTS: Reported in EPA TSCA Inventory. Lead and its compounds are on the Community Right-To-Know List.

OSHA PEL: TWA 0.05 mg(Pb)/m^3
ACGIH TLV: TWA 0.15 mg(Pb)/m^3
NIOSH REL: (Inorganic Lead) TWA 0.10 mg(Pb)/m^3
DOT Classification: Oxidizer; Label: Oxidizer.

SAFETY PROFILE: Poison by intraperitoneal route. A powerful oxidizer. Probably a severe eye, skin, and mucous membrane irritant. Explosive reaction with warm potassium or sodium; cesium acetylide at 350°C; boron (when ground); yellow phosphorus (when ground); sulfinyl dichloride. Mixtures with silicon (2:1 silicon/lead dioxide) are used as initiators and heat to 1100°C when exposed to flame. Mixtures with zirconium can deflagrate (burn explosively) and are sensitive to friction, ignition, and static electricity. Violent reaction or ignition with chlorine trifluoride; hydrogen sulfide; nitrogen compounds (e.g., hydroxylamine); red phosphorus; sulfur (when ground); sulfur + sulfuric acid; peroxyformic acid. Violent reactions with powdered aluminum; Al_4C_3; metal acetylides or carbides H_2O_2; magnesium; nonmetal halides; performic acid; phenyl hydrazine; $S(OCl)_2$. Vigorous reaction with seleninyl chloride; metal sulfides + heat (e.g., calcium sulfide; strontium sulfide; or barium sulfide). Incandescent reaction with powdered molybdenum or tungsten when heated; warm phosphorus trichloride; sulfur dioxide. Metal oxides increase the explosive sensitivity of nitroalkanes (e.g., nitromethane, nitroethane). Can react vigorously with reducing materials. When heated to decomposition it emits toxic fumes of Pb. See also LEAD COMPOUNDS and PEROXIDES.

LCZ000 CAS:56764-40-0 ***HR: 3***
LEAD DIPHENYL ACID PROPIONATE

TOXICITY DATA with REFERENCE
ipr-rat LDLo:55 mg(Pb)/kg JPETAB 38,161,30

NIOSH REL: TWA 0.10 mg(Pb)/m^3

CONSENSUS REPORTS: Lead and its compounds are on the Community Right-To-Know List.

SAFETY PROFILE: Poison by intraperitoneal route. When heated to decomposition it emits toxic fumes of Pb. See also LEAD COMPOUNDS.

LDA000 CAS:41825-28-9 ***HR: 3***
LEAD DIPHENYL NITRATE

TOXICITY DATA with REFERENCE
ipr-rat LDLo:20 mg(Pb)/kg JPETAB 38,161,30

CONSENSUS REPORTS: Lead and its compounds are on the Community Right-To-Know List.

NIOSH REL: TWA 0.10 mg(Pb)/m^3

SAFETY PROFILE: Poison by intraperitoneal route. When heated to decomposition it emits very toxic fumes of Pb and NO_x. See also LEAD COMPOUNDS and NITRATES.

LDA500 CAS:16824-81-0 ***HR: 3***
LEAD DIPICRATE
mf: $C_{12}H_4N_6O_{14}Pb$ mw: 696.40

CONSENSUS REPORTS: Lead and its compounds are on the Community Right-To-Know List.

SAFETY PROFILE: An explosive very sensitive to heat, friction, or sparks. Upon decomposition it emits very toxic fumes of Pb and NO_x. See also LEAD COMPOUNDS and PICRATES.

LDB000 CAS:22904-40-1 ***HR: 2***
LEAD DISODIUM ETHYLENEDINITRILOTETRACETATE
mf: $C_{10}H_{12}N_2O_8\cdot 2Na\cdot Pb$ mw: 541.41

SYN: LEAD DISODIUM EDTA

TOXICITY DATA with REFERENCE
ipr-rbt LD50:915 mg/kg FEPRA7 11,321,52
ivn-rbt LD50:2613 mg/kg FEPRA7 11,321,52

CONSENSUS REPORTS: Reported in EPA TSCA Inventory. Lead and its compounds are on the Community Right-To-Know List.

NIOSH REL: (Lead, Inorganic): 10H TWA 0.10 mg (Pb)/m^3

SAFETY PROFILE: Moderately toxic by intraperitoneal and intravenous routes. When heated to decomposition it emits very toxic fumes of NO_x, Na_2O, and Pb. See also LEAD COMPOUNDS.

LDC000 CAS:69029-52-3 ***HR: 3***
LEAD DROSS (DOT)
DOT: UN 1794

SYNS: LEAD DROSS (containing 3% or more free acid) (DOT) ◇ LEAD SCRAP (DOT)

CONSENSUS REPORTS: Reported in EPA TSCA Inventory. Lead and its compounds are on the Community Right-To-Know List.

OSHA PEL: TWA 0.05 mg(Pb)/m^3
ACGIH TLV: TWA 0.15 mg(Pb)/m^3
NIOSH REL: (Inorganic Lead) TWA 0.10 mg(Pb)/m^3
DOT Classification: ORM-C; Label: None: Corrosive Material; Label: Corrosive.

SAFETY PROFILE: A corrosive irritant to the eyes, skin and mucous membranes. When heated to decomposition it emits toxic fumes of lead. See also LEAD.

LDD000 CAS:15954-94-6 ***HR: 3***
LEAD(II) EDTA COMPLEX

SYN: (ETHYLENEDINITRILO)TETRA ACETIC ACID, LEAD(II) COMPLEX

TOXICITY DATA with REFERENCE
ipr-mus LD50:642 mg(Pb)/kg PABIAQ 11,853,63

CONSENSUS REPORTS: Lead and its compounds are on the Community Right-To-Know List.

NIOSH REL: (Lead, Inorganic): 10H TWA 0.10 mg (Pb)/m^3

SAFETY PROFILE: Moderately toxic by intraperitoneal route. When heated to decomposition it emits very toxic fumes of Pb and NO_x. See also LEAD COMPOUNDS.

LDE000 CAS:13814-96-5 ***HR: 3***
LEAD FLUOBORATE
DOT: NA 2291
mf: $B_2F_8\cdot Pb$ mw: 380.81

SYN: TETRAFLUORO BORATE(1-) LEAD (2+)

TOXICITY DATA with REFERENCE
orl-rat LDLo:50 mg/kg KODAK* -,-,71

CONSENSUS REPORTS: Reported in EPA TSCA Inventory. Lead and its compounds are on the Community Right-To-Know List.

OSHA PEL: TWA 0.05 mg(Pb)/m^3; TWA 2.5 mg(Pb)/m^3
ACGIH TLV: TWA 0.15 mg(Pb)/m^3
NIOSH REL: (Lead, Inorganic):10H TWA 0.10 mg (Pb)/m^3
DOT Classification: ORM-B; Label: None.

SAFETY PROFILE: Poison by ingestion. When heated to decomposition it emits very toxic fumes of Pb, F^-, and BO_x. See also LEAD COMPOUNDS and FLUORIDES.

LDF000 CAS:7783-46-2 ***HR: 3***
LEAD(II) FLUORIDE
DOT: NA 2811
mf: F_2Pb mw: 245.19

PROP: Colorless solid. D: (orthorhombic) 8.445, d: (cubic) 7.750, mp: 824°, bp: 1293°, vap press: 10 mm @ 904°. Low solubility in water. Solubility increases in presence of HNO_3 or nitrates.

SYNS: LEAD DIFLUORIDE ◇ LEAD FLUORIDE (DOT) ◇ PLOMB FLUORURE (FRENCH) ◇ PLUMBOUS FLUORIDE

TOXICITY DATA with REFERENCE
scu-gpg LDLo:2800 mg/kg CRSBAW 124,133,37

CONSENSUS REPORTS: Lead and its compounds are on the Community Right-To-Know List. Reported in EPA TSCA Inventory.

OSHA PEL: TWA 0.05 mg(Pb)/m^3; TWA 2.5 mg(F)/m^3
ACGIH TLV: TWA 0.15 mg(Pb)/m^3; TWA 2.5 mg (F)/m^3
NIOSH REL: (Inorganic Lead) TWA 0.10 mg(Pb)/m^3
DOT Classification: OMB-B; Label: None.

SAFETY PROFILE: Moderately toxic by subcutaneous route. Vigorous reaction with fluorine. Incompatible with CaC_2. When heated to decomposition it emits very toxic fumes of Pb and F^-. See also LEAD COMPOUNDS and FLUORIDES.

LDG000 CAS:25808-74-6 ***HR: 3***
LEAD(II) FLUOROSILICATE
mf: $F_6Si{\cdot}Pb{\cdot}2H_2O$ mw: 385.32

PROP: Monoclinic, colorless powder. Mp: decomp.

SYN: HEXAFLUOROSILICATE (2-1) LEAD(II) SALT DIHYDRATE

TOXICITY DATA with REFERENCE
orl-rat LDLo:250 mg/kg NCNSA6 5,27,53

CONSENSUS REPORTS: Lead and its compounds are on the Community Right-To-Know List. Reported in EPA TSCA Inventory.

OSHA PEL: TWA 0.05 mg(Pb)/m^3
ACGIH TLV: TWA 0.15 mg(Pb)/m^3
NIOSH REL: (Lead, Inorganic):10H TWA 0.10 mg(Pb)/m^3

SAFETY PROFILE: Poison by ingestion. When heated to decomposition it emits very toxic fumes of F^- and Pb. See also LEAD COMPOUNDS.

LDH000 ***HR: 3***
LEAD GLYCERONITRATE

SYN: GLYCEROPLUMBONITRATE

TOXICITY DATA with REFERENCE
ipr-rat LDLo:50 mg(Pb)/kg JPETAB 38,161,30

CONSENSUS REPORTS: Lead and its compounds are on the Community Right-To-Know List.

SAFETY PROFILE: Poison by intraperitoneal route. When heated to decomposition it emits very toxic fumes of Pb and NO_x. See also LEAD COMPOUNDS and NITRATES.

LDI000 CAS:19423-89-3 ***HR: 3***
LEAD HYPONITRITE
mf: N_2O_2Pb mw: 267.20

CONSENSUS REPORTS: Lead and its compounds are on the Community Right-To-Know List.

SAFETY PROFILE: Explodes when heated to 150-160°C. Incompatible with phosphine; phosphorus. When heated to decomposition it emits very toxic fumes of Pb and NO_x. See also LEAD COMPOUNDS and NITRATES.

LDJ000 ***HR: 3***
LEAD HYPOPHOSPHITE
mf: $H_4O_4P_2Pb$ mw: 337.20

$Pb(OP(O)H_2)_2$

PROP: Hygroscopic, crystalline powder. Decomposes at high temps. Sltly sol in cold water, sol in hot water, insol in alc.

SYN: LEAD(II) PHOSPHINATE

CONSENSUS REPORTS: Lead and its compounds are on the Community Right-To-Know List.

SAFETY PROFILE: Poisonous. An impact-sensitive explosive used as a primer. Incompatible with $Pb(NO_3)_2$. See also LEAD COMPOUNDS.

LDK000 ***HR: 3***
LEAD IMIDE
mf: HNPb mw: 222.21

CONSENSUS REPORTS: Lead and its compounds are on the Community Right-To-Know List.

SAFETY PROFILE: Explodes when heated or on contact with H_2O or dilute acid. Upon decomposition it emits very toxic fumes of NO_x and Pb. See also LEAD COMPOUNDS.

LDL000 CAS:18917-82-3 ***HR: 2***
LEAD LACTATE
mf: $C_3H_4O_6 \cdot Pb$ mw: 343.26

PROP: White, heavy, crystalline powder. Sol in water, hot alc. Keep well closed.

SYN: LACTIC ACID, LEAD(2+) SALT (2:1)

TOXICITY DATA with REFERENCE
orl-gpg LDLo:1000 mg/kg AHBAAM 125,273,41

CONSENSUS REPORTS: Lead and its compounds are on the Community Right-To-Know List.
NIOSH REL: (Inorganic Lead) TWA 0.10 mg(Pb)/m^3

SAFETY PROFILE: Moderately toxic by ingestion. When heated to decomposition it emits toxic fumes of Pb. See also LEAD COMPOUNDS.

LDM000 CAS:12709-98-7 ***HR: 3***
LEAD-MOLYBDENUM CHROMATE
SYNS: CHROMIC ACID, LEAD and MOLYBDENUM SALT ◇ CHROMIC ACID LEAD SALT with LEAD MOLYBDATE ◇ C.I. PIGMENT RED 104 ◇ LEAD CHROMATE, SULPHATE and MOLYBDATE ◇ MOLYBDENUM-LEAD CHROMATE ◇ MOLYBDENUM ORANGE

TOXICITY DATA with REFERENCE
mmo-sat 2 mg/plate CRNGDP 2,283,81
oms-hmn:oth 500 mg/L BJCAAI 44,219,81
cyt-hmn:oth 500 mg/L BJCAAI 44,219,81
dni-ham:kdy 150 mg/L BJCAAI 44,219,81
oms-ham:kdy 150 mg/L BJCAAI 44,219,81
cyt-ham:ovr 5 mg/L BJCAAI 44,219,81
sce-ham:ovr 100 μg/L MUREAV 156,219,85
scu-rat TDLo:135 mg/kg:NEO ANYAA9 271,431,81
scu-rat TD:135 mg/kg:ETA PBPHAW 14,47,78

CONSENSUS REPORTS: Lead and its compounds, as well as chromium and its compounds, are on the Community Right-To-Know List.

OSHA PEL: TWA CL 0.1 mg(CrO_3)/m^3; TWA 0.05 mg (Pb)/m^3; TWA 5 mg(Mo)/m^3
ACGIH TLV: TWA 0.05 mg(Cr)/m^3; TWA 5 mg(Mo)/m^3; TWA 0.15 mg(Pb)/m^3
NIOSH REL: (Chromium(VI)) TWA 0.001 mg(Cr (VI))/m^3; (Inorganic Lead) TWA 0.10 mg(Pb)/m^3

SAFETY PROFILE: Questionable carcinogen with experimental neoplastigenic and tumorigenic data. Human mutation data reported. A powerful oxidizer. Probably a severe eye, skin, and mucous membrane irritant. When heated to decomposition it emits toxic fumes of Pb, chromium trioxide, and Mo. See also LEAD COMPOUNDS, MOLYBDENUM COMPOUNDS, and CHROMIUM COMPOUNDS.

LDN000 CAS:1317-36-8 ***HR: 2***
LEAD MONOXIDE
mf: OPb mw: 223.19

PROP: Exists in 2 forms: (1) red to reddish-yellow, tetragonal crystals; stable at ordinary temps. (2) Yellow, orthorhombic crystals; stable > 489°. D: 9.53, mp: 888°. Insol in water, alc; sol in acetic acid, dil HNO_3, warm solns of fixed alkali hydroxides.

SYNS: C.I. 77577 ◇ C.I. PIGMENT YELLOW 46 ◇ LEAD OXIDE ◇ LEAD(II) OXIDE ◇ LEAD OXIDE YELLOW ◇ LEAD PROTOXIDE ◇ LITHARGE ◇ LITHARGE YELLOW L-28 ◇ MASSICOT ◇ MASSICOTITE ◇ PLUMBOUS OXIDE ◇ YELLOW LEAD OCHER

TOXICITY DATA with REFERENCE
skn-rbt 100 mg/24H MLD AEHLAU 30,168,75
otr-ham:emb 50 μmol/L CNREA8 39,193,79
dnd-ham:emb 50 μmol/L CNREA8 39,193,79
ipr-rat LDLo:430 mg/kg INMEAF 10(2),15,41
orl-dog LDLo:1400 mg/kg HBAMAK 4,1289,35

CONSENSUS REPORTS: IARC Cancer Review: Animal Inadequate Evidence IMEMDT 23,325,80. Reported in EPA TSCA Inventory. EPA Genetic Toxicology Program. Lead and its compounds are on the Community Right-To-Know List.

OSHA PEL: TWA 0.05 mg(Pb)/m^3
ACGIH TLV: TWA 0.15 mg(Pb)/m^3
NIOSH REL: (Inorganic Lead) TWA 0.10 mg(Pb)/m^3

SAFETY PROFILE: Moderately toxic by ingestion and intraperitoneal routes. Mutation data reported. A skin irritant. Questionable carcinogen. Avoid breathing dust. Wash thoroughly after contact with the material and before eating or smoking. Explosive reaction with rubidium acetylide at 200°C; zirconium + heat; silicon + aluminum + heat; chlorine + ethylene (at 100°C); perchloric acid + glycerol. Violent or explosive thermite reaction when heated with aluminum powder. Violent or explosive reaction with chlorinated rubber (above 200°C); fluoroelastomers (at 200°C); peroxyformic acid. Violent reaction or ignition with hydrogen trisulfide. May ignite spontaneously with linseed oil; dichloromethylsilane; fluorine + glycerol. Vigorous reaction with silicon + heat. Incandescent reaction with warm aluminum carbide; lithium acetylide; boron; seleninyl chloride. Incompatible with chlorine; perchloric acid; metal acetylides; metals, non-metals. Mixtures of lead oxide with glycerol have been used as a jointing compound and may explode when exposed to powerful oxidizers. When heated to decomposition it emits toxic fumes of Pb. Used in manufacturing of storage batteries, ceramic products, paints, and rubber. See also LEAD COMPOUNDS.

LDO000 CAS:10099-74-8 ***HR: 3***
LEAD(II) NITRATE (1:2)
DOT: UN 1469
mf: $N_2O_6 \cdot Pb$ mw: 331.21

PROP: White crystals. Mp: decomp @ 470°, d: 4.53 @ 20°.

SYNS: LEAD DINITRATE ◇ LEAD NITRATE ◇ LEAD (2+) NITRATE ◇ LEAD(II) NITRATE ◇ NITRATE de PLOMB (FRENCH) ◇ NITRIC ACID, LEAD (2+) SALT

TOXICITY DATA with REFERENCE
pic-esc 320 μmol/L ENMUDM 6,59,84
cyt-mus-par 200 μg/kg MILEDM 17,29,81
ivn-rat TDLo:39964 μg/kg (female 17D post):REP BNEOBV 41,193,82
ivn-ham TDLo:50 mg/kg (female 8D post):TER LAINAW 37,369,77
ipr-rat LDLo:270 mg/kg EQSSDX 1,1,75
ivn-rat LD50:93 mg/kg PSEBAA 92,331,56
ipr-mus LD50:74 mg/kg BECTA6 9,80,73
orl-gpg LDLo:500 mg/kg AHBAAM 125,273,41

CONSENSUS REPORTS: IARC Cancer Review: Animal Inadequate Evidence IMEMDT 23,325,80. Reported in EPA TSCA Inventory. Lead and its compounds are on the Community Right-To-Know List.

OSHA PEL: TWA 0.05 $mg(Pb)/m^3$
ACGIH TLV: TWA 0.15 $mg(Pb)/m^3$
NIOSH REL: (Inorganic Lead) TWA 0.10 $mg(Pb)/m^3$
DOT Classification: Oxidizer; Label: Oxidizer, Poison.

SAFETY PROFILE: Poison by intravenous and intraperitoneal routes. Moderately toxic by ingestion. Experimental teratogenic and reproductive effects. Questionable carcinogen. Probably a severe eye, skin, and mucous membrane irritant. Mutation data reported. A powerful oxidizer. Explodes on contact with red hot carbon; cyclopentadienylsodium (at 100-130°C); potassium acetate + heat. Reacts violently with ammonium thiocyanate; carbon; lead hypophosphite. When heated to decomposition it emits very toxic fumes of Pb and NO_x. Used as a mordant, a chemical reagent, and in production of matches and pyrotechnics. See also LEAD COMPOUNDS and NITRATES.

LDP000 CAS:51317-24-9 ***HR: 3***
LEAD NITRORESORCINATE
mf: $C_6H_5NO_4 \cdot xPb$ mw: 1605.45

SYNS: INITIATING EXPLOSIVE LEAD MONONITRORESORCINATE (DOT) ◇ LEAD MONONITRORESORCINATE (DRY) (DOT)

CONSENSUS REPORTS: Lead and its compounds are on the Community Right-To-Know List.

DOT Classification: Forbidden, Dry; Class A Explosive; Label: Explosive A.

SAFETY PROFILE: Poison by ingestion and inhalation. An explosive. When heated to decomposition it emits very toxic fumes of NO_x and Pb. See also LEAD TRINITRORESORCINATE, LEAD COMPOUNDS, NITRO COMPOUNDS of AROMATIC HYDROCARBONS, and EXPLOSIVES, HIGH.

LDQ000 CAS:1120-46-3 ***HR: 2***
LEAD(II) OLEATE (1:2)
mf: $C_{36}H_{66}O_4 \cdot Pb$ mw: 770.21

PROP: White, ointment-like granules or mass. Insol in water; sol in alc, benzene, ether, oil, turpentine.

SYNS: OLEIC ACID LEAD SALT ◇ OLEIC ACID, LEAD(2+) SALT (2:1)

TOXICITY DATA with REFERENCE
orl-gpg LDLo:4000 mg/kg AHBAAM 125,273,41

CONSENSUS REPORTS: Lead and its compounds are on the Community Right-To-Know List. Reported in EPA TSCA Inventory.

OSHA PEL: TWA 0.05 $mg(Pb)/m^3$
ACGIH TLV: TWA 0.15 $mg(Pb)/m^3$
NIOSH REL: (Inorganic Lead) TWA 0.10 $mg(Pb)/m^3$

SAFETY PROFILE: Moderately toxic by ingestion. Used as a grease it may explode in hot-running bearings. When heated to decomposition it emits toxic fumes of Pb. Used in varnishes and high pressure lubricants. See also LEAD COMPOUNDS.

LDS000 CAS:1314-41-6 ***HR: 3***
LEAD OXIDE RED
mf: O_4Pb_3 mw: 685.57

PROP: Bright red powder. Mp: 890° (decomp), bp: 1472°, d: 8.32-9.16, vap press: 1 mm @ 943°.

SYNS: C.I. 77578 ◇ C.I. PIGMENT RED 105 ◇ DILEAD(II) LEAD(IV) OXIDE ◇ GOLD SATINOBRE ◇ LEAD ORTHOPLUMBATE ◇ LEAD TETRAOXIDE ◇ MINERAL ORANGE ◇ MINERAL RED ◇ MINIUM ◇ MINIUM NON-SETTING RL-95 ◇ ORANGE LEAD ◇ PARIS RED ◇ PLUMBOPLUMBIC OXIDE ◇ RED LEAD ◇ RED LEAD OXIDE ◇ SANDIX ◇ SATURN RED ◇ TRILEAD TETROXIDE

TOXICITY DATA with REFERENCE
ipr-rat LD50:630 mg/kg GTPZAB 19(3),30,75
orl-gpg LDLo:1000 mg/kg AHBAAM 125,273,41
ipr-gpg LD50:220 mg/kg MEIEDD 11,854,89

CONSENSUS REPORTS: Lead and its compounds are on the Community Right-To-Know List. Reported in EPA TSCA Inventory.

OSHA PEL: TWA 0.05 mg(Pb)/m^3
ACGIH TLV: TWA 0.15 mg(Pb)/m^3
NIOSH REL: (Inorganic Lead) TWA 0.10 mg(Pb)/m^3

SAFETY PROFILE: Poison by intraperitoneal route. Moderately toxic by ingestion. Combustible by chemical reaction with reducing agents. An oxidizing agent. Explodes on contact with peroxyformic acid. Ignites on contact with dichloromethylsilane. Incandescent reaction with seleninyl chloride. One percent fresh red lead decreases the explosion temperature of 2,4,6-trinitrotoluene to 192°C. Incompatible with Al; $CsHC_2$; (F_2 + glycerol); H_2S_3; (glycerin + $HClO_4$); $RbHC_2$; (Si + Al); Na; SO_3; Ti; Zr. Mixtures of lead oxide with glycerol have been used as a jointing compound and may explode when exposed to powerful oxidizers. When heated to decomposition it emits toxic fumes of Pb. See also LEAD COMPOUNDS.

LDS499 CAS:13637-76-8 ***HR: 3***
LEAD(II) PERCHLORATE
DOT: UN 1470
mf: Cl_2O_8Pb mw: 406.10

$Pb(ClO_4)_2$

CONSENSUS REPORTS: Lead and its compounds are on the Community Right-To-Know List. Reported in EPA TSCA Inventory.

DOT Classification: Oxidizer; Label:Oxidizer and Poison

SAFETY PROFILE: Solutions in methanol are sensitive explosives when no moisture is present. When heated to decomposition it emits toxic fumes of Cl^- and Pb. Used as a corrosion inhibiting pigment in primers and paints, and in making storage batteries. See also LEAD COMPOUNDS and PERCHLORATES.

LDT000 CAS:63916-96-1 ***HR: 3***
LEAD(II) PERCHLORATE, HEXAHYDRATE (1:2:6)
mf: $Cl_2O_8{\bullet}Pb{\bullet}6H_2O$ mw: 514.21

SYN: PERCHLORATE ACID, LEAD SALT, HEXAHYDRATE

TOXICITY DATA with REFERENCE
ipr-mus LDLo:275 mg/kg JAFCAU 14,512,66

OSHA PEL: TWA 0.05 mg(Pb)/m^3
ACGIH TLV: TWA 0.15 mg(Pb)/m^3
NIOSH REL: (Inorganic Lead) TWA 0.10 mg(Pb)/m^3

CONSENSUS REPORTS: Lead and its compounds are on the Community Right-To-Know List.

SAFETY PROFILE: Poison by intraperitoneal route. When heated to decomposition it emits very toxic fumes of Pb and Cl^-. See also LEAD COMPOUNDS and PERCHLORATES.

LDU000 CAS:7446-27-7 ***HR: 3***
LEAD(II) PHOSPHATE (3:2)
mf: $O_8P_2{\bullet}3Pb$ mw: 811.51

PROP: Hexagonal, colorless crystals or white powder. Mp: 1014, d: 6.9-7.3. Insol in water, alc; sol in HNO_3, fixed alkali hydroxides.

SYNS: BLEIPHOSPHAT (GERMAN) ◇ C.I. 77622 ◇ LEAD ORTHOPHOSPHATE ◇ LEAD PHOSPHATE ◇ LEAD PHOSPHATE (3:2) ◇ LEAD (2+) PHOSPHATE ◇ NORMAL LEAD ORTHOPHOSPHATE ◇ PHOSPHORIC ACID, LEAD (2+) SALT (2:3) ◇ PLUMBOUS PHOSPHATE ◇ TRILEAD PHOSPHATE

TOXICITY DATA with REFERENCE
scu-rat TDLo:540 mg/kg/6W-I:ETA APAVAY 323,694,53
par-rat TDLo:580 mg/kg/34W-I:CAR BJCAAI 19,860,65

CONSENSUS REPORTS: NTP Fifth Annual Report on Carcinogens. IARC Cancer Review: Group 2B IMEMDT 7,230,87; Animal Sufficient Evidence IMEMDT 23,325,80; IMEMDT 1,40,72; Human Limited Evidence IMEMDT 23,325,80. Lead and its compounds are on the Community Right-To-Know List. Reported in EPA TSCA Inventory.

OSHA PEL: TWA 0.05 mg(Pb)/m^3
ACGIH TLV: TWA 0.15 mg(Pb)/m^3
NIOSH REL: TWA 0.10 mg(Pb)/m^3

SAFETY PROFILE: Confirmed carcinogen with experimental carcinogenic and tumorigenic data. A suspected human carcinogenic. When heated to decomposition it emits very toxic fumes of Pb and PO_x. See also LEAD COMPOUNDS.

LDV000 CAS:63916-97-2 ***HR: 3***
LEAD POTASSIUM THIOCYANATE

TOXICITY DATA with REFERENCE
ipr-rat LDLo:42 mg(Pb)/kg JPETAB 38,161,30

CONSENSUS REPORTS: Lead and its compounds are on the Community Right-To-Know List.

OSHA PEL: TWA 0.05 mg(Pb)/m^3
ACGIH TLV: TWA 0.15 mg(Pb)/m^3
NIOSH REL: (Inorganic Lead) TWA 0.10 mg(Pb)/m^3

SAFETY PROFILE: Poison by intraperitoneal route. When heated to decomposition it emits very toxic fumes of Pb, SO_x, NO_x, K_2O and CN^-. See also LEAD COMPOUNDS and THIOCYANATES.

LDW000 CAS:10099-76-0 ***HR: 2***
LEAD SILICATE
mf: $O_3Si{\bullet}Pb$ mw: 283.28

PROP: White crystals. Mp: 766°; d: 6.49.

CONSENSUS REPORTS: Lead and its compounds are

on the Community Right-To-Know List. Reported in EPA TSCA Inventory.

OSHA PEL: TWA 0.05 mg(Pb)/m^3
ACGIH TLV: TWA 0.15 mg(Pb)/m^3
NIOSH REL: (Inorganic Lead) TWA 0.10 mg(Pb)/m^3

SAFETY PROFILE: When heated to decomposition it emits toxic fumes of Pb. Used in paints, electrode position process in the automotive industry, as a heating stabilizer. See also LEAD COMPOUNDS.

LDX000 CAS:7428-48-0 ***HR: 1***
LEAD STEARATE
mf: $C_{18}H_{36}O_2 \cdot xPb$ mw: 1734.87

PROP: White powder. Insol in water, sol in hot alc. Mp: 115.7°.

SYNS: BLEISTEARAT (GERMAN) ◇ STEARIC ACID, LEAD SALT

TOXICITY DATA with REFERENCE
orl-gpg LDLo:6000 mg/kg AHBAAM 125,273,41

CONSENSUS REPORTS: Lead and its compounds are on the Community Right-To-Know List. Reported in EPA TSCA Inventory.
ACGIH TLV: TWA 0.15 mg(Pb)/m^3
NIOSH REL: (Lead, inorganic) TWA 0.10 mg(Pb)/m^3

SAFETY PROFILE: Mildly toxic by ingestion. When heated to decomposition it emits toxic fumes of Pb. See also LEAD COMPOUNDS.

LDY000 CAS:7446-14-2 ***HR: 3***
LEAD(II) SULFATE (1:1)
DOT: UN 1794
mf: $O_4S \cdot Pb$ mw: 303.25

PROP: White crystals. Mp: decomp @ 1000°, d: 6.2. Insol in alc; sol in NaOH, ammonium acetate, or tartrate soln, concentrated HI. Practically insol in water; somewhat more sol in dil HCl or HNO_3.

SYNS: ANGLISLITE ◇ BLEISULFAT (GERMAN) ◇ C.I. 77630 ◇ C.I. PIGMENT WHITE 3 ◇ FAST WHITE ◇ FREEMANS WHITE LEAD ◇ LEAD BOTTOMS ◇ LEAD DROSS (DOT) ◇ LEAD SULFATE, solid, containing more than 3% free acid (DOT) ◇ MILK WHITE ◇ MULHOUSE WHITE ◇ SULFATE de PLOMB (FRENCH) ◇ SULFURIC ACID, LEAD (2+) SALT (1:1)

TOXICITY DATA with REFERENCE
sce-hmn:leu 23 μmol/L DMBUAE 27,40,80
sce-ham:ovr 5 μmol/L ENMUDM 7,381,85
orl-dog LDLo:2 g/kg HBAMAK 4,1289,35
orl-gpg LDLo:30 g/kg AHBAAM 125,273,41
ipr-gpg LDLo:290 mg/kg MEIEDD 10,779,83

CONSENSUS REPORTS: Lead and its compounds are on the Community Right-To-Know List. Reported in EPA TSCA Inventory.

OSHA PEL: TWA 0.05 mg(Pb)/m^3
ACGIH TLV: TWA 0.15 mg(Pb)/m^3
NIOSH REL: (Inorganic Lead) TWA 0.10 mg(Pb)/m^3
DOT Classification: ORM-E; Label: None; Corrosive Material; Label: Corrosive, Solid.

SAFETY PROFILE: Poison by intraperitoneal route. Moderately toxic by ingestion. Human mutation data reported. A corrosive irritant to skin, eyes and mucous membranes. Violent or explosive reaction with potassium. When heated to decomposition it emits very toxic fumes of Pb and SO_x. Used in batteries, lithography, rapid drying oil varnishes, weighting fabrics. See also LEAD COMPOUNDS and SULFLATES.

LDZ000 CAS:1314-87-0 ***HR: 2***
LEAD SULFIDE
mf: PbS mw: 239.25

PROP: Silvery, metallic crystals or black powder. Mp: 1114°, bp: 1281° (subl), d: 7.5, vap press: 1 mm @ 852°. Insol in water; sol in HNO_3, hot dil HCl.

SYNS: C.I. 77640 ◇ GALENA ◇ NATURAL LEAD SULFIDE ◇ PLUMBOUS SULFIDE

TOXICITY DATA with REFERENCE
ipr-rat LDLo:1847 mg/kg INMEAF 10(2),15,41
orl-gpg LDLo:10 g/kg AHBAAM 125,273,41

CONSENSUS REPORTS: Lead and its compounds are on the Community Right-To-Know List. Reported in EPA TSCA Inventory.

OSHA PEL: TWA 0.05 mg(Pb)/m^3
ACGIH TLV: TWA 0.15 mg(Pb)/m^3
NIOSH REL: TWA 0.10 mg(Pb)/m^3

SAFETY PROFILE: Moderately toxic by intraperitoneal route. Mildly toxic by ingestion. Violent reaction with ICl; H_2O_2. When heated to decomposition it emits very toxic fumes of Pb and SO_x. Used in glazing earthenware, as a friction additive in clutch facings and disc brakes. See also SULFIDES and LEAD COMPOUNDS.

LEA000 CAS:815-84-9 ***HR: 2***
LEAD(II) TARTRATE (1:1)
mf: $C_4H_4O_6 \cdot Pb$ mw: 355.27

PROP: White, crystalline powder. D: 2.54 @ 19°.

TOXICITY DATA with REFERENCE
ipr-rat LDLo:1200 mg/kg INMEAF 10(2),15,41

CONSENSUS REPORTS: Lead and its compounds are on the Community Right-To-Know List. Reported in EPA TSCA Inventory.

NIOSH REL: (Inorganic Lead) TWA 0.10 mg(Pb)/m^3

SAFETY PROFILE: Moderately toxic by intraperi-

toneal route. When heated to decomposition it emits toxic fumes of Pb. See also LEAD COMPOUNDS.

LEB000 ***HR: 1***
LEAD TETRACETATE
mf: $C_8H_{12}O_8Pb$ mw: 443.39

PROP: Colorless, monoclinic prisms from glacial acetic acid. Mp: 175-180°; d: 2.228 @ 17°/4°. Sol in hot glacial acetic acid, benzene, chloroform, tetrachloroethane, nitrobenzene.

CONSENSUS REPORTS: Lead and its compounds are on the Community Right-To-Know List.

SAFETY PROFILE: Unstable in air. Hydrolysis liberates brown lead dioxide and acetic acid. A skin irritant. See also LEAD COMPOUNDS.

LEC000 CAS:13463-30-4 ***HR: 3***
LEAD TETRACHLORIDE
mf: Cl_4Pb mw: 349.00

PROP: Yellow, oily liquid. Mp: −15°, bp: explodes @ 105°, d: 3.18 @ 0°.

CONSENSUS REPORTS: Lead and its compounds are on the Community Right-To-Know List.

SAFETY PROFILE: May explode when heated to 100°C. Explodes on contact with potassium; dilute sulfuric acid + heat; concentrated sulfuric acid + chlorine + heat. When heated to decomposition it emits very toxic fumes of Pb and Cl^-. See also LEAD COMPOUNDS and HYDROCHLORIC ACID.

LEC500 CAS:592-87-0 ***HR: 3***
LEAD(II) THIOCYANATE
mf: $C_2N_2PbS_2$ mw: 323.36

CONSENSUS REPORTS: Lead and its compounds are on the Community Right-To-Know List.

SAFETY PROFILE: An explosive. When heated to decomposition it emits toxic fumes of Pb, SO_x, and NO_x. See also LEAD COMPOUNDS and THIOCYANATES.

LED000 CAS:12060-00-3 ***HR: 2***
LEAD TITANATE
mf: $O_3Ti•Pb$ mw: 303.09

PROP: Pale yellow solid. D: 7.52.

SYN: TITANIC ACID, LEAD SALT

TOXICITY DATA with REFERENCE
ipr-rat LD50:2000 mg/kg IMSUAI 31,302,62

CONSENSUS REPORTS: Lead and its compounds are on the Community Right-To-Know List. Reported in EPA TSCA Inventory.

OSHA PEL: TWA 0.05 mg(Pb)/m^3
ACGIH TLV: TWA 0.15 mg(Pb)/m^3
NIOSH REL: TWA 0.10 mg(Pb)/m^3

SAFETY PROFILE: Moderately toxic by intraperitoneal route. When heated to decomposition it emits toxic fumes of Pb. See also LEAD COMPOUNDS and TITANIUM COMPOUNDS.

LED500 ***HR: 2***
LEAD TREE

PROP: Shrubs or small trees with feathery leaves and clusters of off-white flowers. The flat seed pods turn red when mature and hold about 20 brown seeds. They grow wild in Florida, Texas, Hawaii, Guam, the Bahamas, and the West Indies. The seeds are used in jewelry.

SYNS: ACACIA PALIDA (PUERTO RICO) ◇ AROMA BLANCA (CUBA, HAWAII) ◇ CAMPECHE (PUERTO RICO) ◇ COWBUSH (BAHAMAS) ◇ EKOA (HAWAII) ◇ FALSE KOA (HAWAII) ◇ GRAINS de LIN PAYS (HAITI) ◇ GRANALINO (DOMINICAN REPUBLIC) ◇ GUACIS (MEXICO) ◇ HEDIONDILLA (PUERTO RICO) ◇ JIMBAY BEAN (BAHAMAS) ◇ JUMP-AND-GO (BAHAMAS) ◇ KOA-HAOLE (HAWAII) ◇ LEUCAENA LEUCOCEPHAIA ◇ TANTAN (PUERTO RICO) ◇ WHITE POPINAC ◇ WILD TAMARIND (HAWAII, PUERTO RICO) ◇ ZARCILLA (PUERTO RICO)

SAFETY PROFILE: The whole plant and especially the mature seed pod contains the poisonous amino acid mimosine which inhibits DNA synthesis. The toxin is destroyed by cooking. Ingestion causes loss of hair within 48 hours. May cause cataract formation and may inhibit growth. See also β-(N-(3-HYDROXY-4-PYRIDONE))-α-AMINOPROPIONIC ACID.

LEE000 CAS:63918-97-8 ***HR: 3***
LEAD TRINITRORESORCINATE
DOT: UN 0130
mf: $C_6HN_3O_8Pb$ mw: 450.29

PROP: Orange-yellow, monoclinic crystals. Mp: explodes @ 311°, d: 3.1-2.9.

SYNS: INITIATING EXPLOSIVE LEAD STYPHNATE (DOT) ◇ INITIATING EXPLOSIVE LEAD TRINITRORESORCINATE (DOT) ◇ LEAD STYPHNATE (DRY) (DOT) ◇ LEAD TRINITRORESORCINATE (DOT) ◇ LEAD 2,4,6-TRINITRORESORCINOXIDE ◇ STYPHNATE of LEAD (DOT)

CONSENSUS REPORTS: Lead and its compounds are on the Community Right-To-Know List. Reported in EPA TSCA Inventory.

DOT Classification: Class A Explosive; Label: Explosive A; Forbidden, Dry.

SAFETY PROFILE: A poisonous material. A very shock-, heat- and friction-sensitive priming explosive. It has detonated spontaneously when dry. Explodes when heated to 311°. Upon decomposition it emits very toxic fumes of NO_x and Pb. See also LEAD COMPOUNDS, NITRATES, and EXPLOSIVES, HIGH.

LEF000 ***HR: 3***
LEAD(II) TRINITROSOPHLOROGLUCINOLATE
mf: $C_{12}N_6O_{12}Pb_3$ mw: 1041.74

SYN: LEAD(II)TRINITROSOBENZENE-1,3,5-TRIOXIDE

CONSENSUS REPORTS: Lead and its compounds are on the Community Right-To-Know List.

SAFETY PROFILE: Has exploded when disturbed (air-dried). When heated to decomposition it emits very toxic fumes of NO_x and Pb.

LEF100 ***HR: 1***
LEATHERWOOD

PROP: A deciduous shrub which grows to 6 feet tall with elliptical leaves 2 to 3 inches long. Light yellow flowers grow before the leaves appear. The small berries range in color from green to red. It grows wild in wooded areas in the region bounded by New Brunswick, Florida, Louisiana, Oklahoma, and Ontario.

SYNS: AMERICAN MEZEREON ◇ BOIS de PLOMB (CANADA) ◇ DIRCA PALUSTRIS ◇ LEATHER BUSH ◇ LEAVER WOOD ◇ MOOSEWOOD ◇ ROPE BARK ◇ SWAMP WOOD ◇ WICKERBY BUSH ◇ WICKUP ◇ WICOPY

SAFETY PROFILE: The whole plant and especially the bark contain an unknown poison. The bark can produce severe dermatitis. Chewing any part of the plant can cause blistering of the lips, mouth, and throat.

LEF200 ***HR: 3***
LECITHIN IODIDE

SYNS: LBI ◇ LECITHIN-BOUND IODINE

TOXICITY DATA with REFERENCE
orl-rat TDLo:60 mg/kg (9-14D preg):REP TOIZAG 23,525,76
orl-rat TDLo:60 mg/kg (9-14D preg):TER TOIZAG 23,525,76
ipr-rat LD50:122 mg/kg TOIZAG 23,582,76
orl-mus LD50:1070 mg/kg TOIZAG 23,582,76
ipr-mus LD50:81 mg/kg TOIZAG 23,582,76
scu-mus LD50:205 mg/kg TOIZAG 23,582,76

SAFETY PROFILE: Poison by subcutaneous and intraperitoneal routes. Moderately toxic by ingestion. An experimental teratogen. Experimental reproductive effects. When heated to decomposition it emits toxic fumes of I^-. See also IODIDES.

LEF300 CAS:6514-85-8 ***HR: 3***
LEDAKRIN
mf: $C_{18}H_{20}N_4O_2{\cdot}2ClH$ mw: 397.34

SYNS: C 283 ◇ N,N-DIMETHYL-N'-(1-NITRO-9-ACRIDINYL)-1,3-PROPANEDIAMINE DIHYDROCHLORIDE (9CI) ◇ 1-NITRO-9-(3-DIMETHYLAMINOPROPYLAMINO)-ACRIDINEDIHYDROCHLORIDE

TOXICITY DATA with REFERENCE
mmo-sat 2500 pg/plate PJPPAA 31,661,79
pic-esc 600 ng/plate PJPPAA 31,661,79
hma-mus/sat 500 μg/kg PJPPAA 31,661,79
scu-mus LD50:1 mg/kg PJPPAA 31,661,79
ivn-pgn LD50:1010 μg/kg AITEAT 28,777,80

SAFETY PROFILE: Poison by subcutaneous and intravenous routes. Mutation data reported. When heated to decomposition it emits toxic fumes of NO_x and HCl.

LEF400 CAS:5633-16-9 ***HR: 3***
LEIOPYRROLE
mf: $C_{23}H_{28}N_2O$ mw: 348.47

SYNS: N,N-DIETHYL-2-(2-(2-METHYL-5-PHENYL-1H-PYRROL-1-YL)PHENOXY)-ETHANAMINE ◇ DV 714 ◇ LEIOPLEGIL

TOXICITY DATA with REFERENCE
orl-rat LD50:700 mg/kg AAREAV 20,371,63
ipr-rat LD50:180 mg/kg AAREAV 20,371,63
orl-mus LD50:475 mg/kg AAREAV 20,371,63

SAFETY PROFILE: Poison by intraperitoneal route. Moderately toxic by ingestion. When heated to decomposition it emits toxic fumes of NO_x.

LEF800 CAS:53043-29-1 ***HR: D***
LEMMATOXIN
mf: $C_{48}H_{78}O_{18}$ mw: 943.26

SYN: OLEANOGLYCOTOXIN B

TOXICITY DATA with REFERENCE
iut-rat TDLo:250 μg/kg (female 1D post):REP CCPTAY 14,39,76

SAFETY PROFILE: Experimental reproductive effects. When heated to decomposition it emits acrid smoke and irritating fumes.

LEG000 ***HR: 1***
LEMONGRASS OIL EAST INDIAN

PROP: From steam distillation of the freshly cut and partially dried grasses of *Cymbopogon flexuosus* and *Andropogon nardus var. flexuosus.* The main constituent is citral. Dark yellow to brown-red liquid; heavy lemon odor. D: 0.894-0.902, refr index: 1.483. Sol in mineral oil, propylene glycol, alc; insol in water, glycerin.

SYNS: BRITISH EAST INDIAN LEMONGRASS OIL ◇ COCHIN ◇ EAST INDIAN LEMONGRASS OIL ◇ LEMONGRAS OEL (GERMAN) ◇ OIL of LEMONGRASS, EAST INDIAN

TOXICITY DATA with REFERENCE
skn-mus 100% MLD FCTXAV 14,455,76
skn-rbt 500 mg/24H MOD FCTXAV 14,443,76

skn-pig 100% MLD FCTXAV 14,455,76
orl-rat LD50:5600 mg/kg FCTXAV 14(5),443,76

SAFETY PROFILE: Mildly toxic by ingestion. A skin irritant. When heated to decomposition it emits acrid smoke and irritating fumes. See also 3,7-DIMETHYL-2,6-OCTADIENAL.

LEH000 CAS:8007-02-1 ***HR: 1***
LEMONGRASS OIL WEST INDIAN

PROP: Main constituent is citral. From steam distillation of freshly cut and partially dried grasses of *Cymbopogon citratus* (STAPF) and *Andropogon nardus var. ceriferus* (Hack). Light yellow to brown liquid; light lemon odor. D: 0.869-0.894, refr index: 1.483. Sol in mineral oil, propylene glycol; insol in water,

SYNS: GUATEMALA LEMONGRASS OIL ◇ MADAGASCAR LEMONGRASS OIL ◇ OIL of LEMONGRASS, WEST INDIAN ◇ WEST INDIAN LEMONGRASS OIL

TOXICITY DATA with REFERENCE
skn-mus 100% MLD FCTXAV 14,443,76
skn-rbt 500 mg/24H MOD FCTXAV 14,443,76
skn-pig 100% MLD FCTXAV 14,443,76

CONSENSUS REPORTS: Reported in EPA TSCA Inventory.

SAFETY PROFILE: A skin irritant. When heated to decomposition it emits acrid smoke and irritating fumes. See also 3,7-DIMETHYL-2,6-OCTADIENAL.

LEI000 CAS:8008-56-8 ***HR: 2***
LEMON OIL

PROP: Expressed from the peel of the fruit of *Citrus limon* L. Burmann filius (Fam. *Rutaceae*). Pale yellow liquid; taste and odor of lemon peel. D: 0.849, refr index: 1.473 @ 20°. Misc with dehydrated alc, glacial acetic acid.

SYNS: CEDRO OIL ◇ LEMON OIL, COLDPRESSED (FCC) ◇ LEMON OIL, EXPRESSED ◇ OIL of LEMON ◇ ZITRONEN OEL (GERMAN)

TOXICITY DATA with REFERENCE
skn-mus 100% MLD FCTXAV 12,703,74
skn-rbt 500 mg/24H MOD FCTXAV 12,703,74
skn-mus TDLo:280 g/kg/33W-I:ETA JNCIAM 24,1389,60
orl-rat LD50:2840 mg/kg PHARAT 14,435,59

CONSENSUS REPORTS: Reported in EPA TSCA Inventory.

SAFETY PROFILE: Moderately toxic by ingestion. A skin irritant. Questionable carcinogen with experimental tumorigenic data. When heated to decomposition it emits acrid smoke and irritating fumes.

LEI025 ***HR: 1***
LEMON OIL, desert type, coldpressed

PROP: Expressed without heat from the peel of the fruit of *Citrus limon* L. Burmann filius (Fam. *Rutaceae*). Pale yellow liquid; taste and odor of lemon peel. D: 0.846, refr index: 1.473 @ 20°. Misc with dehydrated alc, glacial acetic acid.

SYN: OIL of LEMON, desert type, coldpressed

SAFETY PROFILE: A skin irritant. When heated to decomposition it emits acrid smoke and irritating fumes.

LEI030 ***HR: 1***
LEMON OIL, distilled

PROP: From distillation of fresh peel from *Citrus limon* L. Burmann filius (Fam. *Rutaceae*). Pale yellow liquid; taste and odor of fresh lemon peel. D: 0.842, refr index: 1.470 @ 20°. Misc with dehydrated alc, glacial acetic acid.

SYN: OIL of LEMON, distilled

SAFETY PROFILE: A skin irritant. When heated to decomposition it emits acrid smoke and irritating fumes.

LEJ000 CAS:8008-56-8 ***HR: 1***
LEMON PETITGRAIN OIL

PROP: The main constituents include d-α-pinene, camphene, d-limonene, dipentene, l-linalool, geraniol and nerol and the corresponding acetate, esterified cineol and citral.

TOXICITY DATA with REFERENCE
skn-mus 100% MLD FCTXAV 16,637,78

CONSENSUS REPORTS: Reported in EPA TSCA Inventory.

SAFETY PROFILE: A skin irritant. When heated to decomposition it emits acrid smoke and irritating fumes. See also individual components.

LEJ500 CAS:80734-02-7 ***HR: 2***
LENAMPICILLIN HYDROCHLORIDE
mf: $C_{21}H_{23}N_3O_7S \cdot ClH$ mw: 497.99

SYNS: KBT-1585 ◇ (5-METHYL-2-OXO-1,3-DIOXOLEN-4-YL)METHYL-d-α-AMINOBENZYLPENICILLINATE HYDROCHLORIDE

TOXICITY DATA with REFERENCE
orl-rat TDLo:13 g/kg (female 17-22D post):REP KSRNAM 19,863,85
orl-rat TDLo:15 g/kg (17-22D preg):TER KSRNAM 19,863,85
orl-rat LD50:10 g/kg NKRZAZ 32(Suppl 8),31,84
scu-rat LD50:4362 mg/kg NKRZAZ 32(Suppl 8),31,84
ivn-rat LD50:838 mg/kg NKRZAZ 32(Suppl 8),31,84

orl-mus LD50:8294 mg/kg NKRZAZ 32(Suppl 8),31,84
scu-mus LD50:3576 mg/kg NKRZAZ 32(Suppl 8),31,84
ivn-mus LD50:711 mg/kg NKRZAZ 32(Suppl 8),31,84

SAFETY PROFILE: Moderately toxic by intravenous and subcutaneous routes. Mildly toxic by ingestion. An experimental teratogen. Experimental reproductive effects. When heated to decomposition it emits toxic fumes of SO_x, NO_x, and HCl. See also ESTERS.

LEJ600 CAS:57801-81-7 ***HR: 1***
LENDORMIN
mf: $C_{15}H_{10}BrClN_4S$ mw: 393.71

PROP: Colorless crystals from ethanol. Mp: 212-214°.

SYNS: 2-BROMO-4-(2-CHLOROPHENYL)-9-METHYL-6H-THIENO(3,2-f)(1,2,4)TRIAZOLO(4,3-a)(1,4)DIAZEPINE ◇ BROTIZOLAM ◇ LENDORM ◇ WE 941 ◇ WE 941-BS

TOXICITY DATA with REFERENCE
orl-rat TDLo:675 mg/kg (17-22D preg/21D post):REP IYKEDH 16,818,85
orl-rat TDLo:2750 mg/kg (female 7-17D post):TER IYKEDH 16,818,85
orl-rat LDLo:3456 mg/kg ARZNAD 36,540,86
orl-mus LDLo:2800 mg/kg ARZNAD 36,587,86
ipr-mus LD50:920 mg/kg ARZNAD 36,592,86

SAFETY PROFILE: Mildly toxic by ingestion. An experimental teratogen. Experimental reproductive effects. When heated to decomposition it emits toxic fumes of Cl^-, Br^-, SO_x, and NO_x.

LEJ700 CAS:51257-84-2 ***HR: 3***
LENOREMYCIN
mf: $C_{47}H_{78}O_{13}$ mw: 851.25

SYNS: ANTIBIOTIC A 130A ◇ ANTIBIOTIC Ro 21 6150 ◇ (11R(2R,5S,6R),12R)-10-DEMETHYL-19-DE((TETRAHYDRO-5-METHOXY-6-METHYL-2H-PYRAN-2-YL)OXY)-12-METHYL-11-o-(TETRAHYDRO-5-METHOXY-6-METHYL-2H-PYRAN-2-YL)-DIANEMYCIN

TOXICITY DATA with REFERENCE
orl-mus LD50:55 mg/kg 37ASAA 3,47,78
ipr-mus LD50:2520 μg/kg 85GDA2 5,500,81
scu-mus LD50:34300 μg/kg 85GDA2 5,500,81

SAFETY PROFILE: Poison by ingestion, subcutaneous, and intraperitoneal routes. When heated to decomposition it emits acrid smoke and irritating fumes.

LEK000 CAS:8049-62-5 ***HR: 3***
LENTE INSULIN

SYNS: EXTENDED ZINC INSULIN SUSPENSION ◇ ILETIN U 40 ◇ INSULIN LENTE ◇ INSULIN NOVO LENTE ◇ INSULIN ZINC COMPLEX ◇ INSULIN ZINC SUSPENSION ◇ IZSAB ◇ LENTE ◇ MONOTARD ◇ PROMPT INSULIN ZINC SUSPENSION ◇ ULTRALENTE INSULIN ◇ ULTRA LENTE ISZILIN ◇ ZINC INSULIN

TOXICITY DATA with REFERENCE
scu-rat TDLo:31250 μg/kg (6-8D post):TER SEIJBO 19,291,79
ims-wmn TDLo:2100 mg/kg/9.5Y I JAMAAP 233,985,75
scu-mus LDLo:37 mg/kg DRUGAY 6,84,82
scu-rbt LDLo:75 mg/kg DRUGAY 6,84,82

CONSENSUS REPORTS: Zinc and its compounds are on the Community Right-To-Know List.

SAFETY PROFILE: An experimental teratogen. Questionable carcinogen with experimental carcinogenic data. When heated to decomposition it emits toxic fumes of ZnO. See also INSULIN.

LEK100 CAS:37339-90-5 ***HR: 3***
LENTINAN

TOXICITY DATA with REFERENCE
ivn-rat TDLo:10 mg/kg (male 9W pre):REP TOLED5 9,55,81
ivn-rat TDLo:960 μg/kg (male 9W pre):TER TOLED5 9,55,81
ivn-rat LD50:250 mg/kg IYKEDH 17,365,86
ivn-mus LD50:250 mg/kg IYKEDH 17,365,86

SAFETY PROFILE: Poison by intravenous route. An experimental teratogen. Experimental reproductive effects.

LEL000 CAS:491-35-0 ***HR: D***
4-LEPIDINE
mf: $C_{10}H_9N$ mw: 143.20

PROP: Colorless, oily liquid; quinoline odor. Turns reddish-brown in light. Bp: 261-263°. Sltly sol in water; misc with alc, benzene, ether. Protect from light.

SYNS: CINCHOLEPIDINE ◇ Γ-METHYLQUINOLINE ◇ 4-METHYLQUINOLINE

TOXICITY DATA with REFERENCE
mma-sat 90 μmol/L/2H CNREA8 39,4152,79

CONSENSUS REPORTS: Reported in EPA TSCA Inventory.

SAFETY PROFILE: Mutation data reported. When heated to decomposition it emits toxic fumes of NO_x.

LEM000 CAS:4053-40-1 ***HR: D***
LEPIDINE-1-OXIDE
mf: $C_{10}H_9NO$ mw: 175.1

SYNS: LEPIDINE-N-OXIDE ◇ 4-METHYLQUINOLINE OXIDE

TOXICITY DATA with REFERENCE
mma-sat 7500 μg/plate CPBTAL 27,1954,79

SAFETY PROFILE: Mutation data reported. When heated to decomposition it emits toxic fumes of NO_x.

LEN000 CAS:21609-90-5 ***HR: 3***
LEPTOPHOS
mf: $C_{13}H_{10}BrCl_2O_2PS$ mw: 412.07

SYNS: ABAR ◇ O-(4-BROMO-2,5-DICHLOROPHENYL)-O-METHYL PHENYLPHOSPHONOTHIOATE ◇ O-(2,5-DICHLORO-4-BROMO-PHENYL)-O-METHYL PHENYLTHIOPHOSPHONATE ◇ FOSVEL ◇ K62-105 ◇ MBCP ◇ O-METHYL-O-(4-BROMO-2,5-DICHLOROPHENYL) PHENYL THIOPHOSPHONATE ◇ O-METHYL-O-2,5-DICHLORO-4-BROMOPHENYL PHENYLTHIOPHOSPHONATE ◇ NK 711 ◇ PHENYLPHOSPHONOTHIOIC ACID O-(4-BROMO-2,5-BROMO-2,5-DICHLOROPHENYL) O-METHYL ESTER ◇ PHOSVEL ◇ PSL ◇ VELSICOL 506 ◇ VELSICOL VCS 506

TOXICITY DATA with REFERENCE
sce-ham:ovr 300 μmol/L JTEHD6 8,939,81
orl-rat TDLo:8125 μg/kg (8-20D preg):TER OYYAA2 22,373,81
orl-rat LD50:19 mg/kg FAATDF 7,299,86
skn-rat LD50:44 mg/kg FAATDF 7,299,86
ipr-rat LD50:135 mg/kg OYYAA2 22,373,81
orl-mus LD50:65 mg/kg JAFCAU 27,1197,79
scu-mus LD50:120 mg/kg OYYAA2 3,74,69
orl-rbt LD50:124 mg/kg JETOAS 6,70,73
skn-rbt LD50:800 mg/kg JETOAS 6,70,73

CONSENSUS REPORTS: EPA Extremely Hazardous Substances List.

SAFETY PROFILE: Poison by ingestion, skin contact, intraperitoneal, and subcutaneous routes. An experimental teratogen. Mutation data reported. Used in insecticides. When heated to decomposition it emits very toxic fumes of SO_x, PO_x, Br^-, and Cl^-.

LEO000 CAS:13093-88-4 ***HR: 3***
LEPTRYL
mf: $C_{22}H_{28}N_2O_2S$ mw: 384.58

SYNS: 2-METHOXY-10-(3-(4-HYDROXYPIPERIDINO)-2-METHYL-PROPYL) PHENOTHIAZINE ◇ 3-METHOXY-10-(3-(4-HYDROXYPIPER-IDYL)-2-METHYLPROPYL)PHENOTHIAZINE ◇ 2-METHOXY-10-(2-METHYL-3-(4-HYDROXYPIPERIDINO)PROPYL)PHENOTHIAZINE ◇ 1-(3-(2-METHOXYPHENOTHIAZIN-10-YL)-2-METHYLPROPYL)-4-PIPERIDINOL ◇ PERIMETAZINE ◇ PERIMETHAZINE ◇ RP 9159 ◇ 9159 RP

TOXICITY DATA with REFERENCE
orl-mus LD50:310 mg/kg 27ZQAG -,36,72
ipr-mus LD50:140 mg/kg 27ZQAG -,36,72
scu-mus LD50:330 mg/kg 27ZQAG -,36,72
ivn-mus LD50:115 mg/kg 27ZQAG -,36,72

SAFETY PROFILE: Poison by ingestion, intraperitoneal, subcutaneous, and intravenous routes. When heated to decomposition it emits very toxic fumes of NO_x and SO_x.

LEP000 CAS:51473-23-5 ***HR: 3***
LERGOTRILE MESYLATE
mf: $C_{17}H_{20}ClN_3 \cdot CH_4O_3S$ mw: 397.96

SYN: d-2-CHLORO-6-METHYLERGOLINE-8-β-ACETONITRILE METHANESULFONIC ACID SALT

TOXICITY DATA with REFERENCE
orl-rat LD50:290 mg/kg TXAPA9 33,197,75
ipr-rat LD50:96 mg/kg TXAPA9 33,197,75
orl-mus LD50:275 mg/kg TXAPA9 33,197,75

CONSENSUS REPORTS: Cyanide and its compounds are on the Community Right To Know List.

SAFETY PROFILE: Poison by ingestion and intraperitoneal routes. When heated to decomposition it emits toxic fumes of SO_x, Cl^-, CN^-, and NO_x.

LEQ000 CAS:63917-01-1 ***HR: 3***
LETHANE (special)

PROP: A liquid. Bp: 160-190° @ 0.1 mm, bp: 120-125° @ 0.28 mm. Insol in water; misc with hydrocarbons and most organic solvents. A mixture of Lethane 60 (3 parts) and Lethane 384 (1 part).

TOXICITY DATA with REFERENCE
orl-rat LD50:400 mg/kg AFDOAQ 15,122,51
skn-rat LD50:2500 mg/kg INMEAF 11,-,42
ipr-rat LD50:320 mg/kg INMEAF 11,-,42
orl-dog LD50:500 mg/kg INMEAF 11,-,42
ihl-dog LCLo:9800 mg/m³/H INMEAF 11,-,42
scu-dog LD50:1250 mg/kg INMEAF 11,-,42
orl-rbt LD50:120 mg/kg INMEAF 11,-,42
skn-rbt LD50:400 mg/kg INMEAF 11,-,42
scu-rbt LD50:500 mg/kg INMEAF 11,-,42
ihl-gpg LCLo:9800 mg/m³/H INMEAF 11,-,42

SAFETY PROFILE: Poison by ingestion, skin contact, and intraperitoneal routes. Moderately toxic by subcutaneous route. Mildly toxic by inhalation. Insecticides with n-butyl carbitolthiocyanate, etc., in a light petroleum base. Accidental and suicidal poisonings have occurred. Symptoms include drowsiness followed by coma, the limbs becoming flaccid and the appearance of twitching and convulsions. The pupils may dilate and respiration may become labored. Cyanosis and vomiting occur. The lethanes are mild irritants and, in higher doses, narcotic. Can be absorbed by intact skin. When heated to decomposition it emits very toxic fumes of SO_x and NO_x. See also THIOCYANATES.

LEQ300 CAS:55-03-8 ***HR: 3***
LEVOTHYROXINE SODIUM
mf: $C_{15}H_{11}I_4NO_4 \cdot Na$ mw: 799.86

PROP: Pentahydrate, triclinic crystals or cream-colored powder; odorless and tasteless. Somewhat hygroscopic. D: 2.381. Solubility @ 25° in water: about 15 mg/100 mL. Sol in mineral acids and in solns of alkali hydroxides and carbonates. More sol in alc; very sltly sol in chloroform and ether.

SYNS: DATHROID ◇ EFEROX ◇ ELTROXIN ◇ EUTHYROX ◇ LAEVOXIN ◇ LETTER ◇ LEVAXIN ◇ LEVOROXINE ◇ LEVOTHROID ◇ LEVOTHYROX ◇ LEVOTHYROXINE SODIUM ◇ MONOSODIUM THYROXINE ◇ OROXINE ◇ SODIUM LEVOTHYROXINE ◇ SODIUM THYROXIN ◇ SODIUM THYROXINATE ◇ SODIUM THYROXINE ◇ SODIUM l-THYROXINE ◇ SYNTHROID ◇ SYNTHROID SODIUM ◇ 3,3',5,5'-TETRAIODO-l-THYRONINE, SODIUM SALT ◇ THYROXEVAN ◇ l-THYROXINE MONOSODIUM SALT ◇ THYROXINE SODIUM SALT ◇ l-THYROXINE SODIUM SALT

TOXICITY DATA with REFERENCE
scu-rat TDLo:6750 μg/kg (10D pre):REP BNEOBV 31,71,77
scu-mus TDLo:2 mg/kg (9-10D preg):TER DPTHDL 2,17,81
orl-cld TDLo:449 μg/kg:CVS,BPR AEMED3 14,1114,85
orl-wmn TDLo:117 μg/kg/60D-I AJDCAI 138,927,84
orl-cld TDLo:20 μg/kg AJDCAI 141,1025,87
ipr-rat LD50:20 mg/kg NIIRDN 6,905,82
scu-rat LD50:50 mg/kg NIIRDN 6,905,82

CONSENSUS REPORTS: Reported in EPA TSCA Inventory.

SAFETY PROFILE: Poison by ingestion, subcutaneous, and intraperitoneal routes. An experimental teratogen. Experimental reproductive effects. Human systemic effects by ingestion: pulse rate increase, blood pressure elevation. When heated to decomposition it emits toxic fumes of I^-, NO_x, and Na_2O.

LER000 CAS:328-38-1 ***HR: 1***
dl-LEUCINE
mf: $C_6H_{13}NO_2$ mw: 131.17

PROP: dl Form (synthetic form): Leaflets from water; odorless with sweet taste. Mp: 290 (decomp). Sol in water; sltly sol in alc; insol in ether.

SYN: dl-2-AMINO-4-METHYLVALERIC ACID

TOXICITY DATA with REFERENCE
ipr-rat LD50:6429 mg/kg ABBIA4 64,319,56

CONSENSUS REPORTS: Reported in EPA TSCA Inventory.

SAFETY PROFILE: Mildly toxic by intraperitoneal route. When heated to decomposition it emits toxic fumes of NO_x.

LES000 CAS:61-90-5 ***HR: 2***
l-LEUCINE
mf: $C_6H_{13}NO_2$ mw: 131.20

PROP: An essential amino acid; occurs in isomeric forms. White crystals. Mp (dl): 332° with decomp, mp (l): 295°, d: 1.239 @ 18°/4°. l Form (natural): glistening, hexagonal plates from aq alc. D: 1.291 @ 18°, subl @ 145-148°, decomp @ 293-295°. Sol in water; sltly sol in alc; insol in ether.

SYNS: α-AMINOISOCAPROIC ACID ◇ 2-AMINO-4-METHYLPENTANOIC ACID ◇ 2-AMINO-4-METHYLVALERIC ACID ◇ l-2-AMINO-4-METHYLVALERIC ACID ◇ α-AMINO-Γ-METHYLVALERIC ACID ◇ LEUCIN (GERMAN) ◇ LEUCINE ◇ 4-METHYLNORVALINE

TOXICITY DATA with REFERENCE
orl-rat TDLo:138 g/kg (5-15D preg):TER JONUAI 74,93,61
orl-rat TDLo:138 g/kg (5-15D preg):REP JONUAI 74,93,61
ipr-rat LD50:5379 mg/kg ABBIA4 58,253,55
scu-rbt LDLo:2620 mg/kg AEXPBL 40,313,1898

CONSENSUS REPORTS: Reported in EPA TSCA Inventory.

SAFETY PROFILE: Moderately toxic by subcutaneous route. An experimental teratogen. Experimental reproductive effects. When heated to decomposition it emits toxic fumes of NO_x.

LET000 CAS:92-23-9 ***HR: 3***
LEUCINOCAINE
mf: $C_{17}H_{28}N_2O_2$ mw: 292.47

SYN: 2-(DIETHYLAMINO)-4-METHYL-1-PENTANOL-p-AMINOBENZOATE (ESTER)

TOXICITY DATA with REFERENCE
scu-gpg LDLo:250 mg/kg AEPPAE 144,197,29
scu-rbt LDLo:160 mg/kg AEPPAE 144,197,29
ivn-rbt LDLo:20 mg/kg AEPPAE 144,197,29
ivn-gpg LDLo:140 mg/kg AEPPAE 144,197,29
ivn-gpg LDLo:20 mg/kg AEPPAE 144,197,29

SAFETY PROFILE: Poison by subcutaneous and intravenous routes. When heated to decomposition it emits toxic fumes of NO_x.

LEU000 CAS:135-44-4 ***HR: 3***
LEUCINOCAINE METHANESULFONATE
mf: $C_{17}H_{28}N_2O_2 \cdot CH_4O_3S$ mw: 388.58

SYNS: p-AMINOBENZOICACID-N-1-DIETHYLAMINO-1-ISOBUTYLETHANOL METHANESULFONATE ◇ p-AMINOBENZOIC ACID-β-DIETHYLAMINOISOHEXYL ESTER METHANESULFONATE ◇ p-AMINOBENZOIC ACID-N,N-DIETHYLLEUCINOL ESTER METHANE- SULFONATE ◇ 2-(DIETHYLAMINO)-4-METHYL-1-PENTANOL, p-AMINOBENZOATE ESTER, METHANESULFONATE ◇ N,N-DIETHYLLEUCINON-p-AMINOBENZOIC ACID METHANESULFONATE ◇ LEUCINOCAINE MESYLATE ◇ METHANESULFONIC ACID, COMPOUND with 2-(DIETHYLAMINO)-4-METHYLPENTYL PAMINO- BENZOATE (1:1) ◇ 2-METHYL-4-DIETHYLAMINOPENTAN-5-OL p-AMINOBENZOATE ◇ PANTHESIN ◇ PANTHESINE ◇ S.P. 147

TOXICITY DATA with REFERENCE
scu-mus LDLo:300 mg/kg PHREA7 12,190,32

ivn-mus LDLo:49 mg/kg WDMU** -,-,36
scu-rbt LDLo:240 mg/kg PHREA7 12,190,32
ivn-rbt LDLo:20 mg/kg JPETAB 57,221,36
isp-rbt LDLo:11 mg/kg JPETAB 57,221,36
scu-gpg LDLo:112 mg/kg PHREA7 12,190,32
ivn-gpg LDLo:20 mg/kg PHREA7 12,190,32

SAFETY PROFILE: Poison by subcutaneous, intravenous, and intraspinal routes. Used as a topical and infiltration anesthetic. When heated to decomposition it emits very toxic fumes of SO_x and NO_x.

LEV025 CAS:18361-48-3 ***HR: 2***
LEUCOMYCIN A6
mf: $C_{40}H_{65}NO_{15}$ mw: 800.06

SYNS: 3-ACETAET 4B-PROPANOATE LEUCOMYCIN V ◇ ANTIBIOTIC YL-704 B3 ◇ LEUKOMYCIN A6 ◇ MIDECAMYCIN ◇ PLATENOMYCIN-B3 ◇ TURIMYCIN A3 ◇ YL 704 B3

TOXICITY DATA with REFERENCE
orl-rat LD50:9600 mg/kg NIIRDN 6,810,82
orl-mus LD50:4900 mg/kg JJANAX 37,1565,84
ipr-mus LD50:800 mg/kg 85GDA2 2,27,80

SAFETY PROFILE: Moderately toxic by intraperitoneal route. Mildly toxic by ingestion. When heated to decomposition it emits toxic fumes of NO_x.

LEW000 CAS:102648-39-5 ***HR: 3***
LEUCOMYCIN B

PROP: Produced by *Streptomyces kitasatoensis*.

TOXICITY DATA with REFERENCE
ipr-mus LD50:616 mg/kg 85ERAY 1,106,78
scu-mus LD50:641 mg/kg 85ERAY 1,106,78
ivn-mus LD50:208 mg/kg 85ERAY 1,106,78

SAFETY PROFILE: Poison by intravenous route. Moderately toxic by intraperitoneal and subcutaneous routes. When heated to decomposition it emits acrid smoke and irritating fumes.

LEX000 CAS:37280-56-1 ***HR: 2***
LEUCOMYCIN TARTRATE

SYN: KITASAMYCIN TARTRATE

TOXICITY DATA with REFERENCE
orl-mus LD50:2000 mg/kg 85ERAY 1,106,78
scu-mus LD50:650 mg/kg 85ERAY 1,106,78
ivn-mus LD50:650 mg/kg NIIRDN 6,199,82

SAFETY PROFILE: Moderately toxic by ingestion, subcutaneous and intravenous routes.

LEX400 CAS:24365-47-7 ***HR: 3***
LEUPEPTIN Ac-LL
mf: $C_{20}H_{38}N_6O_4$ mw: 426.64

SYNS: 2-(2-ACETAMIDO-4-METHYLVALERAMIDO)-N-(1-FORMYL-4-GUANIDINOBUTYL)-4-METHYLVALERAMIDE ◇ N-ACETYL-l-LEUCYL-N-(4-((AMINOIMINOMETHYL)AMINO)-1-FORMYLBUTYL)-l-LEUCINAMIDE (9CI) ◇ N-ACETYL-l-LEUCYL-l-LEUCYL-l-ARGINAL ◇ NK-381

TOXICITY DATA with REFERENCE
orl-rat LD50:720 mg/kg JZKEDZ 3,9,77
scu-rat LD50:1100 mg/kg JZKEDZ 3,9,77
ivn-rat LD50:80 mg/kg JZKEDZ 3,9,77
orl-mus LD50:740 mg/kg JZKEDZ 3,9,77
scu-mus LD50:540 mg/kg JZKEDZ 3,9,77
ivn-mus LD50:74 mg/kg JZKEDZ 3,9,77

SAFETY PROFILE: Poison by intravenous route. Moderately toxic by ingestion and subcutaneous routes. When heated to decomposition it emits toxic fumes of NO_x.

LEY000 CAS:57-22-7 ***HR: 3***
LEUROCRISTINE
mf: $C_{46}H_{56}N_4O_{10}$ mw: 825.06

SYNS: LCR ◇ NCI-C04864 ◇ NSC-67574 ◇ ONCOVIN ◇ 22-OXOVINCALEUKOBLASTINE ◇ VCR ◇ VINCRISTINE ◇ VINCRYSTINE ◇ VINKRISTIN

TOXICITY DATA with REFERENCE
sce-hmn:leu 20 μg/L MUREAV 138,55,84
spm-mus-ipr 160 μg/kg MUREAV 138,55,84
ipr-rat TDLo:125 μg/kg (9D preg):TER ARINAU 15,61,67
ipr-rat TDLo:125 μg/kg (9D preg):REP ARINAU 15,61,67
ipr-rat TDLo:1200 μg/kg/7W-I:ETA CANCAR 40(Suppl 4),1935,77
par-chd LDLo:8290 μg/kg:CNS,GIT,BLD ADCHAK 51,289,76
ivn-chd TDLo:500 μg/kg:CNS,GIT JOPDAB 81,90,72
ivn-hmn TDLo:120 μg/kg/8W:PNS,PUL BMJOAE 1,1251,77
ivn-wmn TDLo:40 μg/kg MJAUAJ 143,305,85
ipr-rat LD50:1250 μg/kg ADTEAS 3,181,68
ivn-rat LD50:1 mg/kg JMCMAR 28,1079,85
ipr-mus LD50:1300 μg/kg CTRRDO 65,1049,81
ivn-mus LD50:1700 μg/kg CNREA8 39,3575,79
par-ham LD10:350 μg/kg JSONAU 15,355,80

CONSENSUS REPORTS: NCI Carcinogenesis Studies (ipr); No Evidence: mouse, rat CANCAR 40,1935,77. EPA Genetic Toxicology Program.

SAFETY PROFILE: Poison by parenteral, intraperitoneal, and intravenous routes. Human systemic effects: sensory change involving peripheral nerves, flaccid paralysis without anesthesia, somnolence, anorexia, convulsions or effect on seizure threshold, nausea or vomiting, changes in blood cell count and bone marrow, pulmonary and gastrointestinal changes. Experimental reproductive effects. Questionable carcinogen with experimental tumorigenic and teratogenic data. Human mutation data reported. A skin irritant. When heated to decomposition it emits toxic fumes of NO_x.

LEZ000 CAS:2068-78-2 ***HR: 3***
LEUROCRISTINE SULFATE (1:1)
mf: $C_{46}H_{56}N_4O_{10}$•H_2O_4S mw: 923.14

SYNS: KYOCRISTINE ◇ LILLY 37231 ◇ NSC 67574 ◇ ONCOVIN ◇ VCR SULFATE ◇ VINCRISTINE SULFATE ONCORIN ◇ VINCRISTIN-SULFAT (GERMAN) ◇ VINCRISUL

TOXICITY DATA with REFERENCE
dni-hmn:otr 69120 pmol/L CNREA8 38,560,78
mnt-ham-ipr 200 μg/kg HEREAY 93,329,80
otr-ham:emb 1 μg/L CRNGDP 7,131,86
sce-ham:ovr 50 μg/L ENMUDM 4,65,82
sln-ham:emb 3 μg/L CRNGDP 7,131,86
ipr-mus TDLo:250 μg/kg (female 9D post):TER AJANA2 126,291,69
ipr-mus TDLo:250 μg/kg (female 9D post):REP TJADAB 2,235,69
ipr-rat LD50:1900 μg/kg NIIRDN 6,648,82
ivn-rat LD50:1010 μg/kg KSRNAM 17,1549,83
ipr-mus LD50:3 mg/kg NIIRDN 6,648,82
ivn-mus LD50:2100 μg/kg NIIRDN 6,648,82

CONSENSUS REPORTS: IARC Cancer Review: Group 3 IMEMDT 7,372,87; Human Inadequate Evidence IMEMDT 26,365,81; Animal Inadequate Evidence IMEMDT 26,365,81.

SAFETY PROFILE: Poison by intraperitoneal and intravenous routes. An experimental teratogen. Experimental reproductive effects. Questionable carcinogen. Human mutation data reported. When heated to decomposition it emits very toxic fumes of NO_x and SO_x.

LFA000 CAS:6649-23-6 ***HR: 3***
LEVAMISOLE
mf: $C_{11}H_{12}N_2S$ mw: 204.31

SYNS: 6-PHENYL-2,3,5,6-TETRAHYDROIMIDAZO(2,1-b)THIAZOLE ◇ 2,3,5,6-TETRAHYDRO-6-PHENYLIMIDAZO(2,1-b)THIAZOLE

TOXICITY DATA with REFERENCE
orl-chd TDLo:40 mg/kg/8D:CNS,SKN,MET JOPDAB 93,304,78
orl-rat LD50:345 mg/kg DRUGAY 20,89,80
scu-rat LD50:89 mg/kg DRUGAY 20,89,80
ivn-rat LD50:28 mg/kg DRUGAY 20,89,80
orl-mus LD50:285 mg/kg DRUGAY 20,89,80
ipr-mus LD50:35 mg/kg JAVMA4 176,1166,80
scu-mus LD50:121 mg/kg DRUGAY 20,89,80
ivn-mus LD50:28 mg/kg DRUGAY 20,89,80
scu-pig LD50:39800 μg/kg AJVRAH 42,1912,81

SAFETY PROFILE: Poison by ingestion, intravenous, intraperitoneal, and subcutaneous routes. Human systemic effects by ingestion: coma, skin dermatitis and irritation, and fever. When heated to decomposition it emits very toxic fumes of NO_x and SO_x.

LFA020 CAS:16595-80-5 ***HR: 3***
LEVAMISOLE HYDROCHLORIDE
mf: $C_{11}H_{12}N_2S$•ClH mw: 240.77

SYNS: CITARIN L ◇ DECARIS ◇ IMIDAZO(2,1-β)THIAZOLE MONOHYDROCHLORIDE ◇ KW-2-LE-T ◇ LEVAMISOLE ◇ LEV HYDROCHLORIDE ◇ LEVOMYSOL HYDROCHLORIDE ◇ NEMICIDE ◇ NIRATIC HYDROCHLORIDE ◇ NIRATIC-PURON HYDROCHLORIDE ◇ NSC-177023 ◇ R-12,564 ◇ RIPERCOL-L ◇ SOLASKIL ◇ STIMAMIZOL HYDROCHLORIDE ◇ (−)-2,3,5,6-TETRAHYDRO-6-PHENYLIMIDAZO(2,1-b)THIAZOLE HYDROCHLORIDE ◇ l-(−)-2,3,5,6-TETRAHYDRO-6-PHENYL-IMIDAZO(2,1-B)THIAZOLEHYDROCHLORIDE ◇ 1-TETRAMISOLE HYDROCHLORIDE ◇ TRAMISOL ◇ TRAMISOLE ◇ WORM-CHEK

TOXICITY DATA with REFERENCE
dns-mus-unr 10 mg/kg CCROBU 59,531,75
orl-rat TDLo:750 mg/kg (female 19-22D post):REP YACHDS 10,3155,82
orl-rat TDLo:330 mg/kg (7-17D preg):TER YACHDS 10,3155,82
orl-wmn TDLo:180 mg/kg/36D:BLD BMJOAE 2(6086),555,77
orl-rat LD50:180 mg/kg ARTODN 54,275,83
ipr-rat LD50:42 mg/kg YACHDS 10,3141,82
scu-rat LD50:80 mg/kg YACHDS 10,3141,82
ivn-rat LD50:26 mg/kg YACHDS 10,3141,82
orl-mus LD50:223 mg/kg YACHDS 10,3141,82
ipr-mus LD50:34 mg/kg YACHDS 10,3141,82
scu-mus LD50:52 mg/kg YACHDS 10,3141,82
ivn-mus LD50:18 mg/kg YACHDS 10,3141,82
ims-mus LD50:121 mg/kg FMDZAR 94,793,76

SAFETY PROFILE: Poison by ingestion, intraperitoneal, subcutaneous, intravenous, and intramuscular routes. An experimental teratogen. Human systemic effects by ingestion: thrombocytopenia. Experimental reproductive effects. Mutation data reported. When heated to decomposition it emits very toxic fumes of NO_x, SO_x, and HCl.

LFC000 CAS:14641-96-4 ***HR: 3***
LEVOMEPATE HYDROCHLORIDE
mf: $C_{18}H_{25}NO_3$•ClH mw: 339.90

PROP: Crystals from ethyl acetate. Mp: 210-212°.

SYNS: (−)-α-METHYLHYOSCYAMINE HYDROCHLORIDE ◇ TROPINE (−)-α-METHYLTROPATE HYDROCHLORIDE ◇ 3-α-TROPANYL (−)-2-METHYL-2-PHENYLHYDRACRYLATEHYDROCHLORIDE

TOXICITY DATA with REFERENCE
orl-rat TDLo:4 g/kg (2-21D preg):REP TXAPA9 10,424,67
orl-rat LD50:1100 mg/kg RPOBAR 2,297,70
ipr-rat LD50:226 mg/kg RPOBAR 2,297,70
orl-mus LD50:425 mg/kg RPOBAR 2,296,70
ipr-mus LD50:182 mg/kg RPOBAR 2,296,70

ivn-mus LD50:108 mg/kg TXAPA9 10,424,67
ivn-dog LD50:55 mg/kg TXAPA9 10,424,67
ipr-rbt LD50:200 mg/kg TXAPA9 10,424,67
scu-rbt LD50:400 mg/kg TXAPA9 10,424,67
ivn-rbt LD50:70 mg/kg TXAPA9 10,424,67

SAFETY PROFILE: Poison by intraperitoneal, intravenous, and subcutaneous routes. Moderately toxic by ingestion. Experimental reproductive effects. When heated to decomposition it emits very toxic fumes of Cl^- and NO_x.

LFD200 CAS:125-68-8 ***HR: 3***
LEVOMETHORPHAN HYDROBROMIDE
mf: $C_{18}H_{25}NO \cdot BrH$ mw: 352.61

SYNS: 3-METHOXY-17-METHYLMORPHINANHYDROBROMIDE ◇ RO 1-7788 ◇ RO 1-5470/6

TOXICITY DATA with REFERENCE
orl-rat LD50:242 mg/kg JPETAB 109,189,53
scu-rat LD50:363 mg/kg JPETAB 109,189,53
orl-mus LD50:145 mg/kg JPETAB 109,189,53
scu-mus LD50:103 mg/kg JPETAB 109,189,53
ivn-mus LD50:31 mg/kg JPETAB 109,189,53
ivn-rbt LD50:11500 μg/kg JPETAB 109,189,53

SAFETY PROFILE: Poison by ingestion, subcutaneous, and intravenous routes. When heated to decomposition it emits toxic fumes of NO_x and HBr.

LFF000 CAS:1403-17-4 ***HR: 3***
LEVORIN
mf: $C_{63}H_{85}N_{21}O_{19}$ mw: 1440.69

SYNS: CANDEPTIN ◇ CANDIMON ◇ VANOBID

TOXICITY DATA with REFERENCE
orl-rbt TDLo:650 mg/kg (female 6-18D post):TER ANTBAL 17,742,72
orl-rbt TDLo:130 mg/kg (female 6-18D post):REP ANTBAL 17,742,72
orl-rat LD50:2900 mg/kg ANTBAL 14,932,69
ipr-mus LD50:14 mg/kg MEIEDD 10,240,83
scu-mus LD50:160 mg/kg PMDCAY 14,105,77

SAFETY PROFILE: Poison by intraperitoneal route. Moderately toxic by ingestion and subcutaneous routes. Experimental teratogenic data reported. Other experimental reproductive effects. When heated to decomposition it emits toxic fumes of NO_x.

LFG000 CAS:77-07-6 ***HR: 3***
LEVORPHANOL
mf: $C_{17}H_{23}NO$ mw: 257.41

SYNS: levo-DROMORAN ◇ (−)-3-HYDROXY-N-METHYLMORPHINAN ◇ LEVORPHAN

TOXICITY DATA with REFERENCE
scu-rat TDLo:25750 μg/kg (female 5-12D post):REP DABBBA 36,6107,75
orl-rat LD50:150 mg/kg 27ZIAQ -,126,73
scu-rat LD50:110 mg/kg 27ZIAQ -,-,65
orl-mus LD50:285 mg/kg 27ZIAQ -,126,73
scu-mus LD50:187 mg/kg 27ZIAQ -,-,65
ivn-mus LD50:41 mg/kg 27ZIAQ -,-,65

SAFETY PROFILE: Poison by ingestion, subcutaneous, and intravenous routes. Experimental reproductive effects reported. When heated to decomposition it emits toxic fumes of NO_x.

LFG100 CAS:23257-58-1 ***HR: 3***
LEVOXADROL HYDROCHLORIDE
mf: $C_{20}H_{23}NO_2 \cdot ClH$ mw: 345.90

SYNS: CL 912C ◇ 1-DIOXADROL HYDROCHLORIDE ◇ 1-2-(2,2-DIPHENYL-1,3-DIOXOLAN-4-YL)PIPERIDINE HYDROCHLORIDE ◇ 1-2,2-DIPHENYL-4-(2-PIPERIDYL)-1,3-DIOXOLANEHYDROCHLORIDE ◇ LEVOSAN ◇ U-22,304A

TOXICITY DATA with REFERENCE
orl-rat LD50:310 mg/kg AIPTAK 153,105,65
orl-mus LD60:230 mg/kg AIPTAK 153,105,65
ivn-rbt LD50:25 mg/kg AIPTAK 153,105,65

SAFETY PROFILE: Poison by ingestion and intravenous routes. When heated to decomposition it emits toxic fumes of NO_x and HCl.

LFH000 CAS:123-76-2 ***HR: 2***
LEVULINIC ACID
mf: $C_5H_8O_3$ mw: 116.13

PROP: Deliq, amorph plates or leaflets. D: 1.447, mp: 33-35°, bp: 245-246°. Very sol in water; very sltly sol in alc; insol in ether.

SYNS: ACETOPROPIONIC ACID ◇ β-ACETYLPROPIONIC ACID ◇ Γ-KETOVALERIC ACID ◇ 4-KETOVALERIC ACID ◇ LAEVULIC ACID ◇ LAEVULINIC ACID ◇ LEVULIC ACID ◇ 4-OXOPENTANOIC ACID ◇ 4-OXOVALERIC ACID ◇ USAF CZ-1

TOXICITY DATA with REFERENCE
skn-rbt 500 mg/24H MOD FCTXAV 17(Suppl.),695,79
orl-rat LD50:1850 mg/kg FCTXAV 17,695,79
ipr-mus LD50:450 mg/kg NTIS** AD607-952

CONSENSUS REPORTS: Reported in EPA TSCA Inventory.

SAFETY PROFILE: Moderately toxic by ingestion and intraperitoneal routes. A skin irritant. When heated to decomposition it emits acrid smoke and irritating fumes.

LFI000 CAS:7660-25-5 ***HR: 3***
LEVULOSE
mf: $C_6H_{12}O_6$ mw: 180.18

PROP: White, hygroscopic crystals or crystalline powder; odorless with sweet taste. D: 1.6. Sol in methanol, ethanol, water.

SYNS: FRUCTOSE (FCC) ◇ FRUIT SUGAR ◇ FRUTABS ◇ LAEVORAL ◇ LAEVOSAN ◇ LEVUGEN

TOXICITY DATA with REFERENCE
scu-mus TDLo:5000 mg/kg:ETA GANNA2 46,371,55

SAFETY PROFILE: Questionable carcinogen with experimental tumorigenic data. When heated to decomposition it emits acrid smoke and fumes.

LFJ000 CAS:982-43-4 ***HR: 3***
LIBEXIN
mf: $C_{23}H_{27}N_3O•ClH$ mw: 397.99

PROP: Crystals from ethanol. Mp: 192-193°.

SYNS: 3-(2,2-DIPHENYLAETHYL)-5-(2-PIPERIDINOAETHYL)-1,2,4-OXADIAZOL (GERMAN) ◇ 1-(2-(3-(2,2-DIPHENYLETHYL)-1,2,4-OXADIAZOL-5-YL)ETHYL)PIPERIDINE MONOHYDROCHLORIDE ◇ HK 256 ◇ LOMAPECT ◇ PRENOXDIAZINE HYDROCHLORIDE ◇ TIBEXIN

TOXICITY DATA with REFERENCE
ivn-rat LD50:32 mg/kg ARZNAD 16,617,66
orl-mus LD50:920 mg/kg ARZNAD 16,617,66
scu-mus LD50:540 mg/kg BCFAAI 112,691,73
ivn-mus LD50:34 mg/kg ARZNAD 16,617,66

SAFETY PROFILE: Poison by intravenous route. Moderately toxic by ingestion and subcutaneous routes. Used as an antitussive agent. When heated to decomposition it emits very toxic fumes of NO_x and HCl.

LFK000 CAS:58-25-3 ***HR: 3***
LIBRIUM
mf: $C_{16}H_{14}ClN_3O$ mw: 299.78

SYNS: CD 2 ◇ CDP ◇ CHLORDIAZEPOXIDE ◇ CHLORIDIAZEPIDE ◇ CHLORIDIAZEPOXIDE ◇ CHLORODIAZEPOXIDE ◇ 7-CHLORO-2-METHYLAMINO-5-PHENYL-3H-1,4-BENZODIAZEPINE 4-OXIDE ◇ 7-CHLORO-2-METHYLAMINO-5-PHENYL-3H-1,4-BENZODIAZEPIN4-OXIDE ◇ 7-CHLORO-N-METHYL-5-PHENYL-3H-1,4-BENZODIAZEPIN-2-AMINE-4-OXIDE ◇ CLOPOXIDE ◇ CLORDIAZEPOSSIDO (ITALIAN) ◇ 7-CLORO-2-METILAMINO-5-FENIL-3H-1,4-BENZOIDIAZEPINA 4-OSSIDO (ITALIAN) ◇ DECACIL ◇ EDEN ◇ ELENIUM ◇ IFIBRIUM ◇ KALMOCAPS ◇ LIBRAX ◇ LIBRININ ◇ LIBRITABS ◇ MESURAL ◇ METHAMINODIAZEPOXIDE ◇ MILDMEN ◇ NAPOTON ◇ PSICOSAN ◇ RADEPUR ◇ VIOPSICOL

TOXICITY DATA with REFERENCE
cyt-mus-orl 20 mg/kg CYTBAI 36,73,83
spm-mus-orl 300 mg/kg/15D-C CYTBAI 36,45,83
orl-man TDLo:286 μg/kg (1D male):REP AJPSAO 121,610,64
scu-mus TDLo:150 mg/kg (female 9D post):TER DGDFA5 22,61,80
orl-wmn TDLo:4 mg/kg:CNS TXAPA9 3,619,61
orl-hmn TDLo:857 μg/kg:CNS DMBUAE 13,170,66
orl-hmn TDLo:2 mg/kg/2D:CNS JLSMAW 112,142,60
orl-rat LD50:548 mg/kg TXAPA9 18,185,71
ipr-rat LD50:143 mg/kg ARZNAD 17,242,67
ivn-rat LD50:165 mg/kg CICEA9 7,590,65
orl-mus LD50:260 mg/kg BCFAAI 111,293,72
ipr-mus LD50:207 mg/kg OYYAA2 7,381,73
scu-mus LD50:392 mg/kg APTOA6 19,247,62
ivn-mus LD50:95 mg/kg AIPTAK 178,216,69
ims-mus LD50:366 mg/kg NIIRDN 6,248,82

CONSENSUS REPORTS: EPA Genetic Toxicology Program.

SAFETY PROFILE: Poison by ingestion, intraperitoneal, intravenous, subcutaneous, and intramuscular routes. Human male reproductive effects by ingestion: impotence. Human systemic effects by ingestion: sleep, euphoria, somnolence, ataxia, and antianxiety. An experimental teratogen. Experimental reproductive effects. Mutation data reported. Has been implicated in development of asplastic anemia. Used as a pharmaceutical and veterinary drug. When heated to decomposition it emits very toxic fumes of NO_x.

LFK200 ***HR: 3***
LICABILE HYDROCHLORIDE
mf: $C_{24}H_{42}N_2•ClH$ mw: 395.14

SYNS: 1,3-BIS(DIETHYLAMINO)-2-(α-PHENYL-α-CYCLOHEXYLMETHYL)PROPANE HYDROCHLORIDE ◇ 2-(α-CYCLOHEXYLBENZYL)-N,N,N',N'-TETRAETHYL-1,3-PROPANEDIAMINE HYDROCHLORIDE ◇ GIACOSIL HYDROCHLORIDE ◇ LICARAN HYDROCHLORIDE ◇ PHENETAMINE HYDROCHLORIDE ◇ PHENETHAMINE HYDROCHLORIDE ◇ 1-PHENYL-1-CYCLOHEXYL 2-2 BIS (DIETHYLAMINOMETHYL)ETHANE CHLORHYDRATE (FRENCH) ◇ SPASMEXAN HYDROCHLORIDE ◇ UCB 1545 HYDROCHLORIDE

TOXICITY DATA with REFERENCE
orl-rat LD50:900 mg/kg AIPTAK 123,264,60
scu-rat LD50:400 mg/kg AIPTAK 123,264,60
ivn-rat LD50:23500 μg/kg AIPTAK 123,264,60
orl-mus LD50:240 mg/kg AIPTAK 123,264,60
scu-mus LD50:175 mg/kg AIPTAK 123,264,60
ivn-mus LD50:32 mg/kg AIPTAK 123,264,60

SAFETY PROFILE: Poison by ingestion, subcutaneous, and intravenous routes. When heated to decomposition it emits toxic fumes of HCl and NO_x.

LFL500 CAS:1403-89-0 ***HR: 3***
LICHENIFORMIN A

TOXICITY DATA with REFERENCE
ipr-mus LD50:375 mg/kg 85GDA2 4(2),225,80
scu-mus LD50:1000 mg/kg 85GDA2 4(2),225,80
ivn-mus LD50:250 mg/kg 85GDA2 4(2),225,80

SAFETY PROFILE: Poison by intravenous and intraperitoneal routes. Moderately toxic by subcutaneous route.

LFN000 CAS:11114-18-4 ***HR: 3***
LICORICE COMPONENT FM 100

PROP: One of the crude drugs which is used in combination with paeony.

SYN: FM 100

TOXICITY DATA with REFERENCE
ipr-mus LD50:641 mg/kg YKKZAJ 89,879,69
ivn-mus LD50:251 mg/kg YKKZAJ 89,879,69

SAFETY PROFILE: Poison by intravenous route. Moderately toxic by intraperitoneal route.

LFN300 CAS:8008-94-4 ***HR: 2***
LICORICE ROOT EXTRACT

SYNS: GLYCYRRHIZA ◇ GLYCYRRHIZAE (LATIN) ◇ GLYCYRRHIZA EXTRACT ◇ GLYCYRRHIZINA ◇ KANZO (JAPANESE) ◇ LICORICE ◇ LICORICE EXTRACT ◇ LICORICE ROOT

TOXICITY DATA with REFERENCE
dnr-bcs 100 g/L MUREAV 97,81,82
orl-rat LD50:14200 mg/kg OYYAA2 14,535,77
ipr-rat LD50:1420 mg/kg OYYAA2 14,535,77
scu-rat LD50:4200 mg/kg OYYAA2 14,535,77
ipr-mus LD50:1500 mg/kg OYYAA2 14,535,77
scu-mus LD50:4000 mg/kg OYYAA2 14,535,77

SAFETY PROFILE: Moderately toxic by intraperitoneal and subcutaneous routes. Mildly toxic by ingestion. Mutation data reported. When heated to decomposition it emits acrid smoke and irritating fumes.

LFO000 CAS:23257-56-9 ***HR: 3***
LIDEPRAN HYDROCHLORIDE
mf: $C_{14}H_{19}NO_2 \cdot ClH$ mw: 269.80

SYNS: LEVOPHACETOPERANE HYDROCHLORIDE ◇ LEVOPHACETOPERAN HYDROCHLORIDE ◇ α-PHENYL-2-PIPERIDINE- METHANOL ACETATE HYDROCHLORIDE ◇ 1-PHENYL-1-(2-PIPERIDYL)-1-ACETOXYMETHANE HYDROCHLORIDE ◇ PHENYL-(2-PIPERIDYL) METHYL ACETATE HYDROCHLORIDE ◇ RP 8228

TOXICITY DATA with REFERENCE
orl-rat LD50:400 mg/kg 27ZQAG -,254,72
ipr-rat LD50:180 mg/kg 27ZQAG -,254,72
scu-rat LD50:400 mg/kg 27ZQAG -,254,72
ivn-rat LD50:80 mg/kg 27ZQAG -,254,72
orl-mus LD50:390 mg/kg 27ZQAG -,254,72
ipr-mus LD50:140 mg/kg 27ZQAG -,254,72
scu-mus LD50:220 mg/kg AEPPAE 241,182,61
ivn-mus LD50:77 mg/kg 27ZQAG -,254,72

SAFETY PROFILE: Poison by ingestion, intraperitoneal, subcutaneous, and intravenous routes. When heated to decomposition it emits very toxic fumes of NO_x and HCl.

LFO300 CAS:59160-29-1 ***HR: 3***
LIDOFENIN
mf: $C_{14}H_{18}N_2O_5$ mw: 294.34

SYNS: N-(CARBOXYMETHYL)-N-(2-(2,6-DIMETHYLPHENYL) AMINO)-2-OXOETHYL)-GLYCINE (9CI) ◇ N-(2,6-DIMETHYLPHENYLCARBAMOYLMETHYL)-IMINODIACETIC ACID ◇ N-(N'-(2,6-DIMETHYLPHENYL)CARBAMOYLMETHYL)IMINODIACETIC ACID ◇ HIDA ◇ (((2,6-XYLYLCARBAMOYL)METHYL)IMINO)DI-ACETIC ACID

TOXICITY DATA with REFERENCE
ivn-rat LD50:88 mg/kg YACHDS 6,1331,78
ipr-mus LD50:1100 mg/kg EJNMD9 3,41,78
ivn-mus LD50:168 mg/kg DRFUD4 4,342,79

CONSENSUS REPORTS: Reported in EPA TSCA Inventory.

SAFETY PROFILE: Poison by intravenous route. Moderately toxic by intraperitoneal route. When heated to decomposition it emits toxic fumes of NO_x.

LFP000 CAS:12710-02-0 ***HR: 3***
LIENOMYCIN

PROP: Produced by the strain *Actinomyces diastatochromogenes var. lienomycini.*

SYN: CRYSTALLINE LIENOMYCIN

TOXICITY DATA with REFERENCE
orl-mus LD50:129 mg/kg 85ERAY 2,1462,78
ipr-mus LD50:3 mg/kg 85ERAY 2,1462,78
scu-mus LD50:12 mg/kg 85ERAY 2,1462,78
ivn-mus LD50:2 mg/kg 85ERAY 2,1462,78

SAFETY PROFILE: Poison by ingestion, intraperitoneal, subcutaneous, and intravenous routes.

LFQ000 CAS:23978-09-8 ***HR: 3***
LIGAND 222
mf: $C_{18}H_{36}N_2O_6$ mw: 376.56

SYN: 13,16,21,24-HEXAOXA-1,10-DIAZABICYCLO-(8,8,8)-HEXACOSANE

TOXICITY DATA with REFERENCE
ipr-rat LD50:110 mg/kg TXAPA9 41,113,77
ivn-rat LD50:35 mg/kg TXAPA9 41,113,77
ipr-mus LD50:153 mg/kg TXAPA9 41,113,77
ivn-mus LD50:32 mg/kg TXAPA9 41,113,77

SAFETY PROFILE: Poison by intraperitoneal and intravenous routes. When heated to decomposition it emits toxic fumes of NO_x.

LFT000 CAS:80-54-6 ***HR: 2***
LILIAL
mf: $C_{14}H_{20}O$ mw: 204.34

SYNS: p-tert-BUTYL-α-METHYLHYDROCINNAMALDEHYDE ◇ p-tert-BUTYL-α-METHYLHYDROCINNAMIC ALDEHYDE ◇ β-(4-tert-BUTYLPHENYL)-α-METHYLPROPIONALDEHYDE ◇ 4-(1,1-DIMETHYLETHYL)-α-METHYLBENZENEPROPANAL ◇ LILYAL ◇ α-METHYL-p-(tert-BUTYL)HYDROCINNAMALDEHYDE ◇ α-METHYL-β-(p-tert-BUTYLPHENYL)PROPIONALDEHYDE

TOXICITY DATA with REFERENCE
skn-rbt 500 mg/24H MOD FCTXAV 16,637,78
orl-rat LD50:3700 mg/kg FCTXAV 16,637,78

CONSENSUS REPORTS: Reported in EPA TSCA Inventory.

SAFETY PROFILE: Moderately toxic by ingestion. A skin irritant. When heated to decomposition it emits acrid smoke and irritating fumes. See also ALDEHYDES.

LFT100 CAS:91-51-0 ***HR: 1***
LILIAL-METHYLANTHRANILATE, Schiff's base
mf: $C_{22}H_{27}NO_2$ mw: 337.50

SYNS: ANTHRANILIC ACID, N-(3-(p-tert-BUTYLPHENYL)-2-METHYLPROPYLIDENE)-, METHYL ESTER ◇ BENZOIC ACID,2-((3-(4-(1,1-DIMETHYLETHYL)PHENYL)-2-METHYLPROPYLIDENE)AMINO)-, METHYLESTER ◇ METHYL-N-(p-tert-BUTYL-α-METHYL-HYDROCINNAMYLIDENE) ANTHRANILATE ◇ VERDANTIOL

TOXICITY DATA with REFERENCE
skn-rbt 500 mg/24H MOD FCTOD7 20,729,82
orl-rat LD50:3 g/kg FCTOD7 20,729,82

CONSENSUS REPORTS: Reported in EPA TSCA Inventory.

SAFETY PROFILE: Mildly toxic by ingestion. A skin irritant. When heated to decomposition it emits toxic fumes of NO_x.

LFT700 ***HR: 3***
LILLY-OF-THE-VALLEY

PROP: Small perennials which grow 2 oblong leaves, a single flower stalk with bell-shaped white flowers, and sometimes orange-red berries, found in dense clusters. They are native to Eurasia, but now grow wild in the northern United States and eastern Canada.

SYNS: CONVALLARIA MAJALIS ◇ CONVAL LILY ◇ LILIA-O-KE-AWAWA (HAWAII) ◇ MAYFLOWER ◇ MUGUET (CANADA)

SAFETY PROFILE: The whole plant contains poisonous digitalis-like glycosides and irritant saponins. Human systemic effects by ingestion include: mouth pain, nausea, vomiting, abdominal pain and cramps, and diarrhea. Cardiac glycosides may cause death by their effect on heart function. See also DIGITALIS and SAPONIN.

LFU000 CAS:5989-27-5 ***HR: 3***
d-LIMONENE
mf: $C_{10}H_{16}$ mw: 136.26

PROP: Colorless liquid; citrus odor. Bp: 175.5-176°, d: 0.8402 @ 25°/4°, refr index: 1.471. Misc with alc, fixed oils; sltly sol in glycerin; insol in propylene glycol, water.

SYNS: FEMA No. 2633 ◇ (+)-4-ISOPROPENYL-1-METHYLCYCLOHEXENE ◇ d-(+)-LIMONENE ◇ (+)-R-LIMONENE ◇ d-p-MENTHA-1,8-DIENE ◇ p-MENTHA-1,8-DIENE ◇ (R)-1-METHYL-4-(1-METHYLETHENYL)-CYCLOHEXENE ◇ NCI-C55572

TOXICITY DATA with REFERENCE
orl-rat TDLo:20083 mg/kg (9-15D preg):REP OYYAA2 10,179,75
orl-rat TDLo:20083 mg/kg (9-15D preg):TER OYYAA2 10,179,75
orl-mus TDLo:67 g/kg/39W-I:ETA JNCIAM 35,771,65
orl-rat LD50:4400 mg/kg NIIRDN 6,887,82
ipr-rat LD50:3600 mg/kg NIIRDN 6,887,82
ivn-rat LD50:110 mg/kg NIIRDN 6,887,82
orl-mus LD50:5600 mg/kg NIIRDN 6,887,82
ipr-mus LD50:600 mg/kg OYYAA2 8,1439,74
idu-mus LDLo:1 g/kg OYYAA2 8,1439,74

CONSENSUS REPORTS: Reported in EPA TSCA Inventory.

SAFETY PROFILE: Poison by intravenous route. Moderately toxic by intraperitoneal and intraduodenal routes. Mildly toxic by ingestion. Experimental reproductive effects. Questionable carcinogen with experimental tumorigenic and teratogenic data. Reacts explosively with iodine pentafluoride + tetrafluoroethylene (the pentafluoride reacts exothermically with the inhibitor and initiates explosive polymerization of the TFE). When heated to decomposition it emits acrid smoke and irritating fumes. See also DIPENTENE. Used as a food additive, flavor agent, packaging material, as an inhibitor of tetrafluoroethylene polymerization, and as a gallstone solubilizer.

LFV000 CAS:96-08-2 ***HR: 3***
LIMONENE DIOXIDE
mf: $C_{10}H_{16}O_2$ mw: 168.26

SYNS: 1,2,8,9-DIEPOXYLIMONENE ◇ 1,2:8,9-DIEPOXYMENTHANE ◇ 1,2:8,9-DIEPOXY-p-MENTHANE ◇ DIPENTENE DIOXIDE ◇ EPOXIDE 269 ◇ 4-(1,2-EPOXY-1-METHYLETHYL)-1-METHYL-7-OXABICYCLO(4.1.0)HEPTANE ◇ UNOXAT EPOXIDE 269

TOXICITY DATA with REFERENCE
skn-rbt 10 mg/24H open MLD AIHAAP 23,95,62
unr-mus TDLo:6700 mg/kg:ETA RARSAM 3,193,63
orl-rat LD50:5630 mg/kg UCDS** 8/22/61
ims-mus LD50:600 mg/kg JSICAZ 21,342,62
skn-rbt LDLo:1770 mg/kg AIHAAP 23,95,62

CONSENSUS REPORTS: Reported in EPA TSCA Inventory.

SAFETY PROFILE: Moderately toxic by skin contact and intramuscular routes. Mildly toxic by ingestion. A skin irritant. Questionable carcinogen with experimental tumorigenic data. When heated to decomposition it emits acrid smoke and irritating fumes.

LFW000 CAS:1317-63-1 ***HR: D***
LIMONITE

PROP: Consists mainly of hydrated sesquioxide of iron (IARC** 1,29,71).

SYNS: BROWN HEMATITE ◇ BROWN IRON ORE ◇ BROWN IRONSTONE CLAY ◇ IRON SESQUIOXIDE HYDRATED

CONSENSUS REPORTS: IARC Cancer Review: Animal Inadequate Evidence IMEMDT 1,29,72

SAFETY PROFILE: Questionable carcinogen. See also IRON COMPOUNDS.

LFW300 CAS:5928-69-8 ***HR: 3***
LINADRYL HYDROCHLORIDE
mf: $C_{19}H_{23}NO_2 \cdot ClH$ mw: 333.89

SYNS: A 446 ◇ A 446 HYDROCHLORIDE ◇ β-MORPHOLINOETHYL BENZHYDRYL ETHER HYDROCHLORIDE

TOXICITY DATA with REFERENCE
orl-rat LD50:916 mg/kg JPETAB 89,227,47
ipr-rat LD50:185 mg/kg JPETAB 102,250,51
ivn-rat LD50:35 mg/kg JPETAB 89,227,47
orl-mus LD50:327 mg/kg JPETAB 89,227,47
ipr-mus LD50:185 mg/kg JPETAB 89,227,47
scu-mus LD50:440 mg/kg JPETAB 89,227,47
ivn-dog LD50:70 mg/kg JPETAB 89,227,47
ivn-rbt LD50:21 mg/kg JPETAB 89,227,47
ipr-gpg LD50:160 mg/kg JPETAB 83,120,45

SAFETY PROFILE: Poison by ingestion, intravenous, and intraperitoneal routes. Moderately toxic by subcutaneous route. When heated to decomposition it emits toxic fumes of NO_x and HCl.

LFX000 CAS:78-70-6 ***HR: 2***
LINALOOL
mf: $C_{10}H_{18}O$ mw: 154.28

PROP: Colorless liquid; odor similar to that of bergamot oil and French lavender. D: 0.858-0.868 @ 25°, refr index: 1.461, bp: 195-199°, flash p: 172°F. Sol in alc, ether, fixed oils, propylene glycol; insol in glycerin.

SYNS: ALLO-OCIMENOL ◇ 2,6-DIMETHYL-2,7-OCTADIENE-6-OL ◇ 2,6-DIMETHYLOCTA-2,7-DIEN-6-OL ◇ 3,7-DIMETHYLOCTA-1,6-DIEN-3-OL ◇ 3,7-DIMETHYL-1,6-OCTADIEN-3-OL ◇ FEMA No. 2635 ◇ LINALOL ◇ LINALYL ALCOHOL

TOXICITY DATA with REFERENCE
skn-rbt 500 mg/24H MLD FCTXAV 14,673,76
orl-rat LD50:2790 mg/kg FCTXAV 2,327,64
ims-mus LD50:8000 mg/kg JSICAZ 21,342,62
skn-rbt LD50:5610 mg/kg FCTXAV 13,827,75

CONSENSUS REPORTS: Reported in EPA TSCA Inventory.

SAFETY PROFILE: Moderately toxic by ingestion. Mildly toxic by skin contact. A skin irritant. A synthetic flavoring substance and adjuvant. When heated to decomposition it emits acrid smoke and irritating fumes.

LFY000 CAS:126-91-0 ***HR: 3***
p-LINALOOL
mf: $C_{10}H_{18}O$ mw: 154.28

SYN: 3,7-DIMETHYL-(−)-1,6-OCTADIEN-3-OL

TOXICITY DATA with REFERENCE
ivn-mus LD50:180 mg/kg CSLNX* NX#01477

CONSENSUS REPORTS: Reported in EPA TSCA Inventory.

SAFETY PROFILE: Poison by intravenous route. When heated to decomposition it emits acrid smoke and irritating fumes. See also LINALOOL.

LFY100 CAS:115-95-7 ***HR: 1***
LINALYL ACETATE
mf: $C_{12}H_{20}O_2$ mw: 196.32

PROP: Clear, colorless, oily liquid; odor of bergamot. Bp: 108-110°, d: 0.898-0.914, flash p: 185°F. Sol in alc, ether, diethyl phthalate, benzyl benzoate, mineral oil, fixed oils; sltly sol in propylene glycol; insol in water, glycerin.

SYNS: ACETIC ACID LINALOOL ESTER ◇ BERGAMIOL ◇ 3,7-DIMETHYL-1,6-OCTADIEN-3-OL ACETATE ◇ 3,7-DIMETHYL-1,6-OCTADIEN-3-YL ACETATE ◇ FEMA No. 2636 ◇ LICAREOL ACETATE ◇ LINALOL ACETATE ◇ LINALOOL ACETATE

TOXICITY DATA with REFERENCE
orl-rat LD50:14550 mg/kg FCTXAV 2,327,64
orl-mus LD50:13360 mg/kg FCTXAV 2,327,64

CONSENSUS REPORTS: Reported in EPA TSCA Inventory.

SAFETY PROFILE: Mildly toxic by ingestion. Combustible liquid. When heated to decomposition it emits acrid smoke and irritating fumes. See also ESTERS.

LFY500 CAS:60047-17-8 ***HR: 2***
LINALOOL OXIDE
mf: $C_{10}H_{18}O_2$ mw: 170.28

SYN: EPOXYDIHYDROLINALOOL, mixed isomers

TOXICITY DATA with REFERENCE
skn-rbt 500 mg/24H SEV FCTOD7 21,863,83
orl-rat LD50:1150 mg/kg FCTOD7 21,863,83
skn-rbt LD50:2500 mg/kg FCTOD7 21,863,83

CONSENSUS REPORTS: Reported in EPA TSCA Inventory.

SAFETY PROFILE: Moderately toxic by skin contact and ingestion. A severe skin irritant. When heated to decomposition it emits acrid smoke and irritating fumes.

LFZ000 CAS:126-64-7 ***HR: 1***
LINALYL BENZOATE
mf: $C_{17}H_{22}O_2$ mw: 258.39

PROP: Found in the essential oils of Ylang-Ylang and Tuberose (FCTXAV 14,443,76). Yellow to brown-yellow liquid; tuberose odor. D: 0.980-0.999, refr index: 1.505-1.520, flash p: 208°F. Sol in chloroform, alc, ether; insol in water.

SYNS: 3,7-DIMETHYL-1,6-OCTADIEN-3-OL BENZOATE ◇ 3,7-DIMETHYL-1,6-OCTADIEN-3-YL BENZOATE ◇ 1,5-DIMETHYL-1-VINYL-4-HEXEN-1-OL BENZOATE ◇ 1,5-DIMETHYL-1-VINYL-4-HEXEN-1-YL BENZOATE ◇ FEMA No. 2638

TOXICITY DATA with REFERENCE
skn-rbt 500 mg/24H MLD FCTXAV 14,461,76

CONSENSUS REPORTS: Reported in EPA TSCA Inventory.

SAFETY PROFILE: A skin irritant. Combustible liquid. When heated to decomposition it emits acrid smoke and irritating fumes.

LGA000 CAS:78-37-5 ***HR: 1***
LINALYL CINNAMATE
mf: $C_{19}H_{24}O_2$ mw: 284.43

SYNS: CINNAMICACID-1,5-DIMETHYL-1-VINYL-4-HEXENYL ESTER ◇ CINNAMIC ACID-1,5-DIMETHYL-1-VINYL-4-HEXEN-1-YL ESTER ◇ CINNAMIC ACID, LINALYL ESTER ◇ 3,7-DIMETHYL-1,6-OCTADIEN-3-OL CINNAMATE ◇ 1,5-DIMETHYL-1-VINYL-4-HEXEN-1-OL CINNAMATE ◇ 1,5-DIMETHYL-1-VINYL-4-HEXEN-1-YL CINNAMATE ◇ 3-PHENYL-2-PROPENOIC ACID-1,5,DIMETHYL-1-VINYL-4-HEXEN-1-YL ESTER ◇ 3-PHENYL-2-PROPENOIC ACID-1-ETHENYL-1,5-DIMETHYL-4-HEXENYL ESTER

TOXICITY DATA with REFERENCE
skn-rbt 500 mg/24H MLD FCTXAV 14,463,76
orl-rat LD50:9960 mg/kg FCTXAV 14,463,76

CONSENSUS REPORTS: Reported in EPA TSCA Inventory.

SAFETY PROFILE: Mildly toxic by ingestion. A skin irritant. When heated to decomposition it emits acrid smoke and irritating fumes.

LGA050 ***HR: 2***
LINALYL FORMATE
mf: $C_{11}H_{18}O_2$ mw: 182.26

PROP: Colorless liquid; citrus, herbaceous odor. D: 0.910-0.918, refr index: 1.453-1.458, flash p: 189°F. Sol in alc, fixed oils; sltly sol in propylene glycol, water; insol in glycerin @ 202°.

SYNS: 3,7-DIMETHYL-1,6-OCTADIEN-3-YL FORMATE ◇ FEMA No. 2642

SAFETY PROFILE: Combustible liquid. When heated to decomposition it emits acrid smoke and irritating fumes.

LGB000 CAS:78-35-3 ***HR: 1***
LINALYL ISOBUTYRATE
mf: $C_{14}H_{24}O_2$ mw: 224.38

PROP: Colorless liquid; fresh, rosy odor. D: 0.882-0.888, refr index: 1.446-1.451, flash p: +212°F. Misc with alc, chloroform, ether; insol in water @ 20°.

SYNS: 3,7-DIMETHYL-1,6-OCTADIEN-3-OL ISOBUTYRATE ◇ 3,7-DIMETHYL-1,6-OCTADIEN-3-YL ISOBUTYRATE ◇ 1,5-DIMETHYL-1-VINYL-4-HEXENYL ESTER, ISOBUTYRIC ACID ◇ FEMA No. 2640 ◇ LINALOOL ISOBUTYRATE

TOXICITY DATA with REFERENCE
orl-mus LD50:15100 mg/kg FCTXAV 2,327,64

CONSENSUS REPORTS: Reported in EPA TSCA Inventory.

SAFETY PROFILE: Combustible liquid. Mildly toxic by ingestion. When heated to decomposition it emits acrid smoke and irritating fumes.

LGC000 CAS:1118-27-0 ***HR: 1***
LINALYL ISOVALERATE
mf: $C_{15}H_{26}O_2$ mw: 238.41

SYNS: 4,7-DIMETHYL-1,6-OCTADIEN-3-OL ISOVALERATE ◇ 3,7-DIMETHYL-1,6-OCTADIEN-3-YL ISOVALERATE ◇ ISOVALERIC ACID, (4,7-DIMETHYL-1,6-OCTADIEN-3-YL)ESTER

TOXICITY DATA with REFERENCE
skn-rbt 500 mg/24H MOD FCTXAV 16,811,78

CONSENSUS REPORTS: Reported in EPA TSCA Inventory.

SAFETY PROFILE: A skin irritant. When heated to decomposition it emits acrid smoke and irritating fumes. See also ESTERS.

LGC100 ***HR: 2***
LINALYL PROPIONATE
mf: $C_{13}H_{22}O_2$ mw: 210.32

PROP: Colorless liquid; fresh, pear odor. D: 0.893-

0.902, refr index: 1.449-1.454, flash p: 189°F. Sol in alc, fixed oils; insol in glycerin @ 226°.

SYN: FEMA No. 2645

SAFETY PROFILE: Combustible liquid. When heated to decomposition it emits acrid smoke and irritating fumes.

LGC200 ***HR: 3***
LINCOCIN HYDROCHLORIDE HYDRATE
mf: $C_{18}H_{34}N_2O_6S{\cdot}ClH{\cdot}1/2H_2O$ mw: 452.00

SYNS: FRADEMICINA ◇ LINCOCIN ◇ LINCOMYCIN, HYDROCHLORIDE, HEMIHYDRATE ◇ LINCOMYCIN, HYDROCHLORIDE HYDRATE ◇ LINCOMYCIN HYDROCHLORIDE MONOHYDRATE ◇ MYCIVIN ◇ WAYNECOMYCIN

TOXICITY DATA with REFERENCE
scu-rat LD50:9778 mg/kg 29ZVAB -,65,69
ivn-rat LD50:342 mg/kg NIIRDN 6,891,82
ipr-mus LD50:1000 mg/kg UPHOH* 2(6),-,71
ivn-mus LD50:214 mg/kg NIIRDN 6,891,82

SAFETY PROFILE: Poison by intravenous route. Moderately toxic by intraperitoneal route. Used as an antibacterial agent. When heated to decomposition it emits toxic fumes of SO_x, NO_x, and HCl.

LGD000 CAS:154-21-2 ***HR: 3***
LINCOMYCIN
mf: $C_{18}H_{34}N_2O_6S$ mw: 406.60

PROP: Sol in methanol, ethanol, butanol, isopropanol, ethyl acetate, n-butyl acetate, amylacetate, etc. Moderately sol in water.

SYNS: ALBIOTIC ◇ LINCOCIN ◇ LINCOLCINA ◇ LINCOLNENSIN ◇ LINCOMYCINE (FRENCH) ◇ NSC-70731 ◇ U-10149

TOXICITY DATA with REFERENCE
orl-rat LD50:1000 mg/kg 85ERAY 1,186,78
scu-rat LD50:9780 mg/kg TXAPA9 18,185,71
ipr-mus LD50:1000 mg/kg 85ERAY 1,186,78
ims-rbt LDLo:200 µg/kg RMVEAG 156,915,80

CONSENSUS REPORTS: EPA Genetic Toxicology Program.

SAFETY PROFILE: Poison by intramuscular route. Moderately toxic by ingestion and intraperitoneal routes. When heated to decomposition it emits very toxic fumes of SO_x and NO_x.

LGE000 CAS:859-18-7 ***HR: 3***
LINCOMYCIN HYDROCHLORIDE
mf: $C_{18}H_{34}N_2O_6S{\cdot}ClH$ mw: 443.06

SYN: LINCOCIN

TOXICITY DATA with REFERENCE
scu-rat LD50:9778 mg/kg TXAPA9 9,445,66
ivn-rat LD50:342 mg/kg TXAPA9 6,476,64
ipr-mus LD50:1000 mg/kg TXAPA9 6,476,64
ivn-mus LD50:214 mg/kg TXAPA9 6,476,64

SAFETY PROFILE: Poison by intravenous route. Moderately toxic by intraperitoneal route. When heated to decomposition it emits very toxic fumes of Cl^-, SO_x, and NO_x.

LGF800 ***HR: 3***
LINEAR ALKYLBENZENESULFONATE, MAGNESIUM SALT

PROP: Linear alkyl derivative containing from C-10 to C-14 (TOIZAG 25,850,78).

SYNS: LAS-Mg ◇ LAS, MAGNESIUM SALT ◇ MAGNESIUM LINEAR ALKYLBENZENESULFONATE

TOXICITY DATA with REFERENCE
orl-rat LD50:1840 mg/kg TOIZAG 25,850,78
scu-rat LD50:710 mg/kg TOIZAG 25,850,78
ivn-rat LD50:27200 µg/kg TOIZAG 25,850,78
orl-mus LD50:2108 mg/kg GISAAA 48(6),81,83
scu-mus LD50:1520 mg/kg TOIZAG 25,850,78
ivn-mus LD50:98 mg/kg TOIZAG 25,850,78
orl-gpg LD50:1900 mg/kg GISAAA 48(6),81,83

SAFETY PROFILE: Poison by intravenous route. Moderately toxic by ingestion and subcutaneous routes. When heated to decomposition it emits toxic fumes of SO_x. See also SULFONATES and MAGNESIUM COMPOUNDS.

LGF825 CAS:68411-30-3 ***HR: 3***
LINEAR ALKYLBENZENESULFONATE, SODIUM SALT

SYNS: LAS-Na ◇ LAS, SODIUM SALT ◇ STRAIGHT-CHAIN ALKYL BENZENE SULFONATE

TOXICITY DATA with REFERENCE
orl-rat TDLo:53 g/kg (MGN):TER TXAPA9 18,83,71
orl-rat LD50:404 mg/kg:EYE,GIT TRENAF 24,397,72
scu-rat LD50:810 mg/kg TOIZAG 25,850,78
ivn-rat LD50:119 mg/kg TOIZAG 25,850,78
orl-mus LD50:1575 mg/kg TRENAF 24,397,72
scu-mus LD50:1250 mg/kg TOIZAG 25,850,78
ivn-mus LD50:207 mg/kg TOIZAG 25,850,78

CONSENSUS REPORTS: Reported in EPA TSCA Inventory.

SAFETY PROFILE: Poison by intravenous route. Moderately toxic by ingestion and subcutaneous routes. An experimental teratogen. Human systemic effects: lacrimation, somnolence, hypermotility, diarrhea. When heated to decomposition it emits toxic fumes of SO_x and Na_2O. See also SULFONATES.

LGF875 CAS:83968-18-7 ***HR: 1***
LINEVOL 7-9

SYNS: ALCOHOLS, C7-9 ◇ DA79P ◇ DIALKYLPHTHALATE C7-C9 ◇ DIALKYL 79 PHTHALATE ◇ LINEVOL 79

TOXICITY DATA with REFERENCE
orl-rat TDLo:52500 mg/kg (21D male):REP TXAPA9 61,205,81
orl-rat LD50:11100 mg/kg ATXKA8 26,84,70
orl-mus LD50:5900 mg/kg ATXKA8 26,84,70

SAFETY PROFILE: Mildly toxic by ingestion. Experimental reproductive effects. When heated to decomposition it emits acrid smoke and irritating fumes. See also ALCOHOLS.

LGG000 CAS:60-33-3 ***HR: 1***
LINOLEIC ACID
mf: $C_{18}H_{32}O_2$ mw: 280.50

PROP: Colorless oil, easily oxidized by air. D: 0.9038 @ 18°/4°, mp: −12°, bp: 230° @ 16 mm. Sol in ether and ethanol; misc with dimethyl formamide, fat solvents, oils.

SYNS: LEINOLEIC ACID ◇ 9,12-LINOLEIC ACID ◇ cis,cis-9,12-OCTADECADIENOIC ACID ◇ cis-9,cis-12-OCTADECADIENOIC ACID ◇ 9,12-OCTADECADIENOIC ACID

TOXICITY DATA with REFERENCE
skn-hmn 75 mg/3D-I MOD 85DKA8 -,127,77

CONSENSUS REPORTS: Reported in EPA TSCA Inventory.

SAFETY PROFILE: A human skin irritant. Ingestion can cause nausea and vomiting. When heated to decomposition it emits acrid smoke and irritating fumes.

LGH000 ***HR: D***
LINOLEIC ACID (oxidized)

PROP: Linoleic acid was oxidized until about 30% conjugated diene was presented.

SYNS: OXIDIZED LINOLEATE ◇ OXIDIZED LINOLEIC ACID

TOXICITY DATA with REFERENCE
orl-rat TDLo:166 g/kg (39D pre/1-22D preg):TER FCTXAV 11,935,73

SAFETY PROFILE: An experimental teratogen. When heated to decomposition it emits acrid smoke and irritating fumes. See also LINOLEIC ACID.

LGI000 ***HR: 3***
LINOLEIC ACID mixed with OLEIC ACID

SYNS: (Z)-9-OCTADECENOIC ACID mixed with (Z,Z)-9,12-OCTADECADIENOIC ACID ◇ OLEIC ACID mixed with LINOLEIC ACID

TOXICITY DATA with REFERENCE
orl-mus TDLo:33 g/kg/52W-C:ETA PAACA3 14,35,73

SAFETY PROFILE: Questionable carcinogen with experimental tumorigenic data. When heated to decomposition it emits acrid smoke and irritating fumes. See also individual components.

LGJ000 CAS:24124-25-2 ***HR: 2***
(LINOLEOYLOXY)TRIBUTYLSTANNANE
mf: $C_{30}H_{58}O_2Sn$ mw: 569.57

SYN: TRIBUTYL TIN LINOLEATE

TOXICITY DATA with REFERENCE
unr-rat LD50:1370 mg/kg TIUSAD 107,1,76

OSHA PEL: TWA 0.1 mg(Sn)/m^3 (skin)
ACGIH TLV: TWA 0.1 mg(Sn)/m^3 (skin) (Proposed: TWA 0.1 mg(Sn)/m^3; STEL 0.2 mg(Sn)/m^3 (skin))
NIOSH REL: (Organotin Compounds) TWA 0.1 mg(Sn)/m^3

SAFETY PROFILE: Moderately toxic by unspecified route. Tributyl tin compounds are extremely toxic to marine life. When heated to decomposition it emits acrid smoke and irritting fumes. See also TIN COMPOUNDS.

LGK000 CAS:8001-26-1 ***HR: 1***
LINSEED OIL

PROP: Yellowish liquid, peculiar odor, bland taste. Sltly sol in alc; misc with chloroform, ether, petr ether, carbon disulfide, oil, turpentine. Bp: 343°, mp: −19°, d: 0.93, flash p: (raw oil) 432°F (CC), flash p: (boiled) 403°F (CC), autoign temp: 650°F. From seed of *Linum usitatissimum*.

SYNS: GROCO ◇ L-310

TOXICITY DATA with REFERENCE
skn-hmn 300 mg/3D-I MOD 85DKA8 -,127,77

CONSENSUS REPORTS: Reported in EPA TSCA Inventory.

SAFETY PROFILE: An allergen and skin irritant to humans. Combustible liquid when exposed to heat or flame; can react with oxidizing materials. Subject to spontaneous heating. Violent reaction with Cl_2. To fight fire, use CO_2, dry chemical.

LGK050 CAS:6893-02-3 ***HR: 1***
LIOTHYRONINE
mf: $C_{15}H_{12}I_3NO_4$ mw: 650.98

SYNS: ALANINE,3-(4-(4-HYDROXY-3-IODOPHENOXY)-3,5-DIIODOPHENYL)-, l- ◇ l-3-(4-(4-HYDROXY-3-IODOPHENOXY)-3,5-DIIODOPHENYL)ALANINE ◇ O-(4-HYDROXY-3-IODOPHENYL)-3,5-DIIODO-l-TYROSINE ◇ 4-(3-IODO-4-HYDROXYPHENOXY)-3,

5-DIIODOPHENYLALANINE ◇ LIOTHYRONIN ◇ l-LIOTHYRONINE ◇ T3 ◇ T(sub 3) ◇ L-T3 ◇ TRESITOPE ◇ TRIIODOTHYRONINE ◇ TRIIODO-L-THYRONINE ◇ l-TRIIODOTHYRONINE ◇ 3,3',5-TRIIODOTHYRONINE ◇ 3,5,3'-TRIIODOTHYRONINE ◇ TRIOTHYRONE ◇ l-TYROSINE, O-(4-HYDROXY-3-IODOPHENYL)-3,5-DIIODO- (9CI)

TOXICITY DATA with REFERENCE
scu-rat TDLo:1400 μg/kg (female 16-22D post):TER HMMRA2 10,425,78
unr-rat TDLo:75 μg/kg (female 3D pre):REP BIREBV 11,529,74
orl-rat LDLo:7500 mg/kg RPZHAW 32,197,81

SAFETY PROFILE: Mildly toxic by ingestion. An experimental teratogen. Experimental reproductive effects. When heated to decomposition it emits toxic fumes of NO_x, I^-, and Cl^-.

LGK100 CAS:59547-52-3 ***HR: 2***
4-(dl-α-LIPAMIDO)BUTYRIC ACID

SYNS: 4-((5-(1,2-DITHIOLAN-3-YL)-1-OXOPENTYL)AMINO)BUTANOIC ACID ◇ LABA

TOXICITY DATA with REFERENCE
orl-mus LD50:3860 mg/kg YKKZAJ 85,463,65
ipr-mus LD50:891 mg/kg YKKZAJ 85,463,65
scu-mus LD50:832 mg/kg YKKZAJ 85,463,65
ivn-mus LD50:889 mg/kg YKKZAJ 85,463,65

SAFETY PROFILE: Moderately toxic by ingestion, intravenous, subcutaneous, and intraperitoneal routes. When heated to decomposition it emits toxic fumes of SO_x and NO_x.

LGK150 ***HR: 3***
LIPASE AP6

TOXICITY DATA with REFERENCE
orl-rat LD50:18200 mg/kg KSRNAM 8,3387,74
ipr-rat LD50:771 mg/kg KSRNAM 8,3387,74
scu-rat LD50:1640 mg/kg KSRNAM 8,3387,74
orl-mus LD50:17330 mg/kg KSRNAM 8,3387,74
ipr-mus LD50:345 mg/kg KSRNAM 8,3387,74
scu-mus LD50:1640 mg/kg KSRNAM 8,3387,74

SAFETY PROFILE: Poison by intraperitoneal route. Moderately toxic by subcutaneous route. Mildly toxic by ingestion.

LGK200 CAS:26717-47-5 ***HR: 2***
LIPENAN
mf: $C_{16}H_{22}ClNO_4$ mw: 327.84

SYNS: 2-(4-CHLOROPHENOXY)-2-METHYL-PROPIONIC ACID 4-(DIMETHYLAMINO)-4-OXOBUTYL ESTER (9CI) ◇ CHLORO-4 PHENOXYISOBUTYRATE D'HYDROXY-4N-DIMETHYLBUTYRAMIDE (FRENCH) ◇ N-DIMETHYL-4-(p-CHLOROPHENOXY-1,1'-DIMETHYLACETATE)BUTYRAMIDE ◇ N-DIMETHYL-4-(1,4'-CHLOROPHENOXY-1,1'-DIMETHYLACETATE)BUTYRAMIDE ◇ N-DIMETHYL-4-(p-CHLOROPHENOXYISOBUTYRATE)BUTYRAMIDE ◇ N-DIMETHYL-4-(1,4'-CHLOROPHENOXYISOBUTYRATE)BUTYRAMIDE ◇ 4-HYDROXY-N-DIMETHYLBUTYRAMIDE-4-CHLOROPHENOXYISOBUTYRATE ◇ MG 46

TOXICITY DATA with REFERENCE
orl-rat TDLo:14 g/kg (7D male/7D pre/1-22D preg):REP JETOAS 5,239,72
orl-rat LD50:2485 mg/kg JETOAS 5,239,72
ipr-rat LD50:1175 mg/kg JETOAS 5,239,72
orl-mus LD50:1080 mg/kg JETOAS 5,239,72
ipr-mus LD50:960 mg/kg JETOAS 5,239,72

SAFETY PROFILE: Moderately toxic by ingestion and intraperitoneal routes. Experimental reproductive effects. When heated to decomposition it emits toxic fumes of Cl^- and NO_x.

LGK300 ***HR: 3***
LIPOPOLYSACCHARIDE, ESCHERICHIA COLI

TOXICITY DATA with REFERENCE
ipr-rat TDLo:500 μg/kg (15D preg):TER PSEBAA 109,429,62
ivn-rat TDLo:1 μg/kg (female 12D post):REP ESKHA5 (99),68,81
ipr-rat LD50:10 mg/kg PSEBAA 109,429,62
ivn-mus LD50:7670 μg/kg MIIMDV 26,455,82

SAFETY PROFILE: Poison by intravenous and intraperitoneal routes. An experimental teratogen. Experimental reproductive effects.

LGK350 ***HR: 3***
LIPOPOLYSACCHARIDE, from B. ABORTUS Bang.

SYN: B. ABORTUS Bang. LIPOPOLYSACCHARIDE

TOXICITY DATA with REFERENCE
ipr-rat TDLo:20 mg/kg (female 19D post):TER PSEBAA 109,429,62
ipr-rat TDLo:20 mg/kg (female 15D post):REP PSEBAA 109,429,62
ipr-rat LD50:60 mg/kg PSEBAA 109,429,62

SAFETY PROFILE: Poison by intraperitoneal route. An experimental teratogen. Experimental reproductive effects. When heated to decomposition it emits acrid smoke and irritating fumes.

LGL000 ***HR: 2***
LIQUEFIED CARBON DIOXIDE
mf: CO_2 mw: 44.0

PROP: Heavy gas or liquid under pressure. Mp: −56.6° @ 3952 mm; bp: −78.5° (subl); d: 1.977 g/L @ 0°; (liquid) 1.101 @ −37°.

SYN: LIQUID CARBONIC GAS

SAFETY PROFILE: Contact with skin or living tissue can cause frostbite-like burns. This material is stable when very cold. Solid CO_2 goes directly (sublimes) to gaseous CO_2 which is mainly an asphyxiant. See also CARBON DIOXIDE.

LGM000 CAS:68476-85-7 ***HR: 3***
LIQUEFIED PETROLEUM GAS
DOT: UN 1075

SYNS: LPG ◇ L.P.G. (OSHA, ACGIH) ◇ PETROLEUM GAS, LIQUEFIED

CONSENSUS REPORTS: Reported in EPA TSCA Inventory.

OSHA PEL: TWA 1000 ppm
NIOSH REL: TWA 350 mg/m^3; CL 1800 mg/m^3/15M
ACGIH TLV: TWA 1000 ppm
DOT Classification: Flammable Gas; Label: Flammable Gas.

SAFETY PROFILE: Olefinic impurities may lend a narcotic effect or it may act as a simple asphyxiant. A very dangerous fire hazard when exposed to heat or flame. Can react with oxidizing materials. To fight fire, use CO_2, dry chemical, water spray. Used as a fuel refrigerant, propellant, and raw material in chemical synthesis.

LGM200 ***HR: 2***
LIQUIPRON

PROP: A yeast (CANDIDA MALTOSA) protein concentrate TOERD9 3,305,81

TOXICITY DATA with REFERENCE
orl-rat TDLo:15525 g/kg/3Y-C:CAR TOERD9 3,305,81
orl-rat TD:2628 g/kg/2Y-C:ETA,REP TOERD9 3,305,81

SAFETY PROFILE: Experimental reproductive effects. Questionable carcinogen with experimental carcinogenic and tumorigenic data. When heated to decomposition it emits acrid smoke and irritating fumes.

LGO000 CAS:7439-93-2 ***HR: 3***
LITHIUM
DOT: UN 1415
af: Li aw: 6.94

PROP: Silver-colored, light metal; mixture of isotopes Li^6 and Li^7. Mp: 179°, bp: 1317°, d: 0.534 @ 25°, vap press: 1 mm @ 723°. Sol in liquid ammonia. Keep under mineral oil or other liquid free from O_2 or water.

SYNS: LITHIUM METAL (DOT) ◇ LITHIUM METAL, IN CARTRIDGES (DOT)

CONSENSUS REPORTS: Reported in EPA TSCA Inventory.

DOT Classification: Flammable Solid; Label: Flammable Solid and Dangerous When Wet.

SAFETY PROFILE: See LITHIUM COMPOUNDS for a discussion of the toxicity of the lithium ion. See SODIUM for a discussion of the toxicity of metallic lithium.

A very dangerous fire hazard when exposed to heat or flame. The powder may ignite spontaneously in air. The solid metal ignites above 180°C. It will burn in oxygen, nitrogen, or carbon dioxide, and will continue to burn in sand or sodium carbonate. The use of most types of fire extinguishers (e.g., water, foam, carbon dioxide, halocarbons, sodium carbonate, sodium chloride, and other dry powders) may cause an explosion. Molten lithium is extremely reactive and attacks such inert materials as sand, concrete, and ceramics.

Explosive reaction with bromobenzene; carbon + lithium tetrachloroaluminate + sulfinyl chloride; diazomethane. Forms very friction- and impact-sensitive explosive mixtures with halogens [e.g., bromine; iodine (above 200°C)]; halocarbons (e.g., bromoform; carbon tetrabromide; carbon tetrachloride; carbon tetraiodide; chloroform; dichloromethane; diiodomethane; fluorotrichloromethane; tetrachloroethylene; trichloroethylene; 1,1,2-trichloro-trifluoroethane).

Violent reaction with acetonitrile; sulfur; mercury (potentially explosive); metal oxides [e.g., chromium (III) oxide (at 185°C); molybdenum trioxide (at 180°C); niobium pentoxide (at 320°C); titanium dioxide (at 200-400°C); tungsten trioxide (at 200°C); vanadium pentoxide (at 394°C)]; iron(II) sulfide (at 260°C); manganese telluride (at 230°C); hot water; bromine pentafluoride (may ignite with lithium powder); platinum (at about 540°C); trifluoromethyl hypofluorite (at about 170°C); arsenic; beryllium; maleic anhydride; carbides; carbon dioxide; carbon monoxide + water; chlorine; chromium; chromium trichloride; cobalt alloys; iron sulfide; diborane; manganese alloys; nickel alloys; nitric acid; nitrogen; organic matter; oxygen; phosphorus; rubber; silicates; $NaNO_2$; Ta_2O_5; Fe alloys; V; $ZrCl_4$; CHI_3; trifluoromethylhypofluorite.

Ignition on contact with carbon + sulfinyl chloride (when ground); nitric acid (becomes violent); viton (poly(1,1-difluorethylene-hexafluoropropylene); chlorine tri- and penta-fluorides (hypergolic reaction); diborane (forms a complex which is pyrophyoric); hydrogen (above 300°C).

Incandescent reaction with ethylene + heat; nitrogen + metal chlorides [e.g., chromium trichloride; zirconium tetrachloride; nitryl fluoride (at 200°C)]. Incompatible with atmospheric gases; bromine pentafluoride; diazomethane; metal chlorides; metal oxides; non-metal oxides.

When burned it emits toxic fumes of LiO_2 and hydroxide. Reacts vigorously with water or steam to produce heat and hydrogen. Can react vigorously with oxidizing materials. To fight fire, use special mixtures of dry chemical, soda ash, graphite. Note: water, sand, carbon tetrachloride and carbon dioxide are ineffective.

LGP875 CAS:1070-75-3 ***HR: 3***
LITHIUM ACETYLIDE
mf: C_2Li_2 mw: 37.90

SAFETY PROFILE: Ignites and burns vigorously in fluorine; chlorine; phosphorus; selenium; or sulfur vapors. Ignites when heated in bromine or iodine vapors. When heated to decomposition it emits acrid smoke and fumes. See also LITHIUM COMPOUNDS and ACETYLIDES.

LGQ000 CAS:50475-76-8 ***HR: 3***
LITHIUM ACETYLIDE COMPLEXED with ETHYLENEDIAMINE
DOT: NA 2813

SYN: LITHIUM ACETYLIDE-ETHYLENEDIAMINE COMPLEX

DOT Classification: Flammable Solid; Label: Flammable Solid and Dangerous When Wet.

SAFETY PROFILE: A very flammable, unstable mixture. When heated to decomposition it emits toxic fumes of NO_x. See also LITHIUM ACETYLIDE and 1,2-ETHANEDIAMINE.

LGS000 CAS:17476-04-9 ***HR: 3***
LITHIUM ALUMINUM TRI-tert-BUTOXYHYDRIDE
mf: $C_{12}H_{18}AlLiO_3$ mw: 244.22

TOXICITY DATA with REFERENCE
ivn-mus LD50:32 mg/kg CSLNX* NX#00620

CONSENSUS REPORTS: Reported in EPA TSCA Inventory.

ACGIH TLV: TWA 2 mg(Al)/m^3

SAFETY PROFILE: Poison by intravenous route. When heated to decomposition it emits acrid smoke and irritating fumes. See also ALUMINUM and LITHIUM COMPOUNDS.

LGT000 CAS:7782-89-0 ***HR: 3***
LITHIUM AMIDE
DOT: UN 1412
mf: H_2LiN mw: 22.97

PROP: White, crystalline solid or powder. Subl in NH_3 current. Insol in anhydrous ether, benzene, toluene. Mp: 380-400°. D: 1.178 @ 17.50.

SYNS: LITHAMIDE ◇ LITHIUM AMIDE, POWDERED (DOT)

CONSENSUS REPORTS: Reported in EPA TSCA Inventory.

DOT Classification: Flammable Solid; Label: Dangerous When Wet; Flammable Solid; Label: Flammable Solid.

SAFETY PROFILE: A powerful irritant to skin, eyes, and mucous membranes. Flammable when exposed to heat or flame. Ammonia is liberated and lithium hydroxide is formed when this compound is exposed to moisture. Reacts violently with water or steam to produce toxic and flammable vapors. Vigorous reaction with oxidizing materials. Exothermic reaction with acid or acid fumes. When heated to decomposition it emits very toxic fumes of LiO, NH_3, and NO_x. Used in synthesis of drugs, vitamins, steroids, and other organics. See also LITHIUM COMPOUNDS, AMIDES, AMMONIA, and LITHIUM HYDROXIDE.

LGU000 CAS:305-97-5 ***HR: 3***
LITHIUM ANTIMONY THIOMALATE
mf: $C_{12}H_9O_{12}S_3Sb \cdot 6Li$ mw: 604.78

SYNS: ANTHIOLIMINE ◇ ANTHIOMALINE ◇ LITHIUM ANTIMONIOTHIOMALATE ◇ MERCAPTOSUCCINIC ACID ANTIMONATE(III) HEXALITHIUM SALT ◇ MERCAPTOSUCCINIC ACID-S-ANTIMONY DERIVATIVE LITHIUM SALT ◇ MERCAPTOSUCCINIC ACID, THIOANTHIMONATE(III), DILITHIUM SALT ◇ 2,2',2''-(STIBILIDYNETRIS(THIO)TRIS-BUTANEDIOIC ACID HEXALITHIUM SALT

TOXICITY DATA with REFERENCE
orl-hmn TDLo:11 mg/kg:GIT,MET JAMAAP 125,952,44
ivn-man TDLo:262 mg/kg/5W-I:CNS,SKN METRA2 19,103,59
ipr-mus LD50:82 mg/kg AJTMAQ 25,263,45
ivn-mus LD50:181 mg/kg JPETAB 81,224,44

CONSENSUS REPORTS: Antimony and its compounds are on the Community Right-To-Know List.

OSHA PEL: TWA 0.5 mg(Sb)/m^3
ACGIH TLV: TWA 0.5 mg(Sb)/m^3
NIOSH REL: (Antimony) TWA 0.5 mg(Sb)/m^3

SAFETY PROFILE: Poison by intraperitoneal and intravenous routes. Human systemic effects by ingestion and intravenous routes: hallucinations, distorted perceptions, nausea or vomiting, skin dermatitis and fever. An anthelmintic agent. When heated to decomposition it emits very toxic fumes of SO_x, Sb, and Li_2O. See also ANTIMONY COMPOUNDS and LITHIUM COMPOUNDS.

LGV000 CAS:19597-69-4 ***HR: 3***
LITHIUM AZIDE
mf: LiN_3 mw: 48.96

SAFETY PROFILE: The moist or dry salt explodes

when heated to 115-298°C. Forms very shock-sensitive explosive mixtures with alkyl nitrates or dimethylformamaide above 200°C. Incompatible with CS_2. When heated to decomposition it emits very toxic fumes of Li_2O and NO_x. See also AZIDES and LITHIUM COMPOUNDS.

LGV700 ***HR: 3***
LITHIUM BENZENEHEXOXIDE
mf: $C_6Li_6O_6$ mw: 209.71

SAFETY PROFILE: Explodes on contact with water. When heated to decomposition it emits acrid smoke and irritating fumes. See also LITHIUM COMPOUNDS.

LGW000 CAS:553-54-8 ***HR: 2***
LITHIUM BENZOATE
mf: $C_7H_5O_2$•Li mw: 128.06

PROP: White, crystalline powder. Fairly sol in water. Somewhat sol in alc.

SYN: BENZOIC ACID, LITHIUM SALT

TOXICITY DATA with REFERENCE
orl-mus LD50:1198 mg/kg RPTOAN 33,266,70
scu-mus LD50:964 mg/kg RPTOAN 33,266,70

CONSENSUS REPORTS: Reported in EPA TSCA Inventory.

SAFETY PROFILE: Moderately toxic by ingestion and subcutaneous routes. When heated to decomposition it emits acrid smoke and irritating fumes. See also LITHIUM COMPOUNDS.

LGX000 CAS:4039-32-1 ***HR: 3***
LITHIUM BIS(TRIMETHYLSILYL)AMIDE
mf: $C_6H_{18}LiNSi_2$ mw: 167.33

$LiN[Si(CH_3)_3]_2$

SAFETY PROFILE: Unstable in air. Ignites when compressed. Upon decomposition it emits toxic fumes of LI_2O and NO_x. See also LITHIUM COMPOUNDS, SILANE, and AMIDES.

LGY000 CAS:7550-35-8 ***HR: 2***
LITHIUM BROMIDE
mf: BrLi mw: 86.85

PROP: White, hygroscopic, granular powder; sltly bitter taste. Mp: 549°, bp: 1265°, d: 3.46 @ 25°, vap press: 1 mm @ 748°. Very sol in alc, glycol; sol in ether, amyl alc. Keep well closed.

TOXICITY DATA with REFERENCE
scu-mus LD50:1680 mg/kg RPTOAN 33,266,70

CONSENSUS REPORTS: Reported in EPA TSCA Inventory.

SAFETY PROFILE: Moderately toxic by subcutaneous route. Large doses may cause central nervous system depression in humans. Chronic absorption may cause skin eruptions and central nervous system disturbances due to bromide. May also cause disturbed blood electrolyte balance. See also BROMIDES and LITHIUM COMPOUNDS.

LGZ000 CAS:554-13-2 ***HR: 3***
LITHIUM CARBONATE (2:1)
mf: CO_3•2Li mw: 73.89

PROP: White, light alkaline, crystalline powder. D: 2.11 @ 17.5°; mp: 618°. Insol in alc. @ 17.5°.

SYNS: CAMCOLIT ◇ CANDAMIDE ◇ CARBOLITH ◇ CARBONIC ACID, DILITHIUM SALT ◇ CARBONIC ACID LITHIUM SALT ◇ CEGLUTION ◇ CP-15467-61 ◇ DILITHIUM CARBONATE ◇ ESKALITH ◇ HYPNOREX ◇ LIMAS ◇ LISKONUM ◇ LITHANE ◇ LITHICARB ◇ LITHINATE ◇ LITHIUM CARBONATE ◇ LITHOBID ◇ LITHONATE ◇ LITHOTABS ◇ NSC-16895 ◇ PLENUR ◇ PRIADEL ◇ QUILONUM RETARD

TOXICITY DATA with REFERENCE
dnd-hmn:fbr 500 mg/L MUREAV 169,171,86
msc-ham:lng 2 g/L MUREAV 169,171,86
orl-wmn TDLo:4256 mg/kg (1-38W preg):REP BMJOAE 3,233,73
orl-wmn TDLo:4900 mg/kg (1-35W preg):TER LANCAO 2,595,74
orl-wmn TDLo:3600 mg/kg/21W-C:CAR,BLD NEJMAG 302,808,80
orl-wmn TD:21 g/kg/3.5Y-C:CAR,END ANZJB8 10,62,80
orl-wmn TD:5940 mg/kg/47W-C:CAR,BLD AIMEAS 92,262,80
orl-man TD:6132 mg/kg/2Y-C:CAR,BLD HAEMAX 67,944,82
orl-man TDLo:8 mg/kg:GIT,SKN AJPSAO 141,909,84
orl-hmn TDLo:4111 mg/kg:CNS,GIT NEJMAG 287,867,72
orl-man TDLo:54 mg/kg NZMJAX 97,23,84
orl-wmn TDLo:120 mg/kg/10D-I JCLPDE 48,81,87
orl-man TDLo:1080 mg/kg/13W-I:SKN JCLPDE 47,330,86
unr-wmn TDLo:556 mg/kg/32D JAMAAP 213,865,70
orl-rat LD50:525 mg/kg KSRNAM 7,1273,73
ipr-rat LD50:156 mg/kg KSRNAM 7,1273,73
scu-rat LD50:434 mg/kg KSRNAM 7,1273.73
ivn-rat LD50:241 mg/kg KSRNAM 7,1273,73
orl-mus LD50:531 mg/kg RPTOAN 33,266,70
ipr-mus LD50:236 mg/kg KSRNAM 7,1273,73
scu-mus LD50:413 mg/kg RPTOAN 33,266,70
ivn-mus LD50:497 mg/kg KSRNAM 7,1273,73
orl-dog LD50:500 mg/kg 27ZQAG -,436,72

CONSENSUS REPORTS: Reported in EPA TSCA Inventory.

SAFETY PROFILE: Human carcinogenic data. Poison by intraperitoneal and intravenous routes. Moderately toxic by ingestion, and subcutaneous routes. Human systemic effects by ingestion: toxic psychosis, tremors, changes in fluid intake, muscle weakness, increased urine volume, nausea or vomiting, allergic dermatitis. Human reproductive effects by ingestion: effects on newborn including apgar score changes and other neonatal measures or effects. Human teratogenic effects by ingestion: developmental abnormalities of the cardiovascular system, central nervous system, musculoskeletal and gastrointestinal systems. An experimental teratogen. Experimental reproductive effects. Questionable carcinogen producing leukemia and thyroid tumors. Human mutation data reported. Used in the treatment of manic-depressive psychoses. Incompatible with fluorine. See also LITHIUM COMPOUNDS.

LHA000 CAS:12772-56-4 ***HR: D***
LITHIUM CARMINE
mf: $C_{22}H_{20}O_{13}\cdot Li$ mw: 499.36

TOXICITY DATA with REFERENCE
ipr-mus TDLo:100 mg/kg (female 8D post):TER NNAPBA 270,56,71
ipr-mus TDLo:100 mg/kg (female 8D post):REP NNAPBA 270,56,71

SAFETY PROFILE: An experimental teratogen. Experimental reproductive effects. When heated to decomposition it emits acrid smoke and irritating fumes. See also LITHIUM COMPOUNDS.

LHB000 CAS:7447-41-8 ***HR: 3***
LITHIUM CHLORIDE
mf: ClLi mw: 42.39

PROP: Cubic, white, deliquescent crystals. Mp: 605°, bp: 1350°, d: 2.068 @ 25°, vap press: 1 mm @ 547°.

SYNS: CHLORKU LITU (POLISH) ◇ CHLORURE de LITHIUM (FRENCH)

TOXICITY DATA with REFERENCE
skn-rbt 500 mg/24H SEV 28ZPAK -,7,72
eye-rbt 100 mg/24H MOD 28ZPAK -,7,72
mrc-smc 9 mmol/L MUTAEX 1,21,86
dni-hmn:hla 70 mmol/L MUREAV 92,427,82
ipr-mus TDLo:320 mg/kg (female 6-7D post):REP JPETAB 101,362,51
ipr-rat TDLo:1 g/kg (female 7-11D post):TER CRSBAW 167,183,73
ipr-mus TDLo:882 mg/kg/7D-I:NEO PWPSA8 22,343,79
orl-hmn LDLo:200 mg/kg/3D JAMAAP 139,688,49
orl-hmn TDLo:243 mg/kg/13D:CNS,GIT JAMAAP 139,688,49
orl-rat LD50:526 mg/kg APTOA6 47,351,80
ipr-rat LD50:514 mg/kg PetKP# 22DEC77
scu-rat LD50:499 mg/kg PetKP# 22DEC77
ice-rat LD50:4800 μg/kg PJPPAA 26,399,74
orl-mus LD50:1165 mg/kg RPTOAN 33,266,70
ipr-mus LD50:600 mg/kg JTBIDS 6,87,81
scu-mus LD50:828 mg/kg OYYAA2 7,413,73
ivn-mus LD50:363 mg/kg OYYAA2 7,413,73
ipr-cat LD50:492 mg/kg RPTOAN 42,9,79
scu-cat LDLo:450 mg/kg EQSSDX 1,1,75
orl-rbt LD90:850 mg/kg BEXBAN 74,914,73
scu-rbt LDLo:531 mg/kg EQSSDX 1,1,75
scu-gpg LDLo:620 mg/kg EQSSDX 1,1,75
scu-pgn LDLo:513 mg/kg HBAMAK 4,1289,35
orl-qal LD50:422 mg/kg AECTCV 12,355,83

CONSENSUS REPORTS: Reported in EPA TSCA Inventory. EPA Genetic Toxicology Program.

SAFETY PROFILE: Human poison by ingestion. Experimental poison by intravenous and intracerebral routes. Moderately toxic by subcutaneous and intraperitoneal routes. Experimental teratogenic and reproductive effects. Human systemic effects by ingestion: somnolence, tremors, nausea or vomiting. An eye and severe skin irritant. Human mutation data reported. Questionable carcinogen with experimental neoplastigenic data. This material has been recommended and used as a substitute for sodium chloride in "salt-free" diets, but cases have been reported in which the ingestion of lithium chloride has produced dizziness, ringing in the ears, visual disturbances, tremors and mental confusion. In most cases, the symptoms disappeared when use was discontinued. Prolonged absorption may cause disturbed electrolyte balance, impaired renal function. Reaction is violent with BrF_3. When heated to decomposition it emits toxic fumes of Cl^-. Used for dehumidification in the air conditioning industry. Also used to obtain lithium metal. See also LITHIUM COMPOUNDS.

LHC000 CAS:6180-21-8 ***HR: 3***
LITHIUM CHLOROACETYLIDE
mf: C_2ClLi mw: 66.41

$$LiC{\equiv}CCl$$

SYN: LITHIUMCHLOROETHYNIDE

SAFETY PROFILE: Violently explosive when dry. When heated to decomposition it emits very toxic fumes of Li_2O and Cl^-. See also LITHIUM COMPOUNDS and ACETYLIDES.

LHD000 ***HR: 3***
LITHIUM CHROMATE
mf: CrH_2O_4•2Li mw: 131.90

PROP: Yellow, crystalline, deliquescent powder. Mp: $-2H_2O$ @ 150°.

SYNS: CHROMIC ACID, DILITHIUM SALT ◇ CHROMIUM LITHIUM OXIDE ◇ DILITHIUM CHROMATE

CONSENSUS REPORTS: Chromium and its compounds are on the Community Right-To-Know List.

ACGIH TLV: TWA 0.05 mg(Cr)/m^3, Confirmed Human Carcinogen.

SAFETY PROFILE: A toxic material. Combustible when exposed to heat or flame. An oxidizer. It can react with reducing materials. Potentially explosive reaction with zirconium at 450-600°C. When heated to decomposition it emits toxic fumes of Li_2O. See also LITHIUM COMPOUNDS and CHROMIUM COMPOUNDS.

LHD099 ***HR: 3***
LITHIUM CHROMATE(VI)
mf: $CrLi_2O_4$ mw: 129.87

CONSENSUS REPORTS: Chromium and its compounds are on The Community Right-To-Know List.

SAFETY PROFILE: Potentially explosive reaction with zirconium when heated above 400°C. See also LITHIUM COMPOUNDS and CHROMIUM COMPOUNDS.

LHE000 ***HR: D***
LITHIUM COMPOUNDS

SAFETY PROFILE: Lithium oxide, hydroxide, carbonate, etc., are strong bases and their solutions in water are very caustic. Otherwise, toxicity of lithium compounds is a function of their solubility in water. The halide salts, except the fluoride, are highly soluble in water. The carbonate, phosphate, oxalate, and fluoride are relatively insoluble in water. Lithium ion has central nervous system toxicity. In industry, the most hazardous lithium compound is the hydride. It produces large amounts of hydrogen gas when exposed to water; this reaction can cause severe damage to exposed tissue. Some lithium compounds, particularly the carbonate, are used in psychiatry. The difference between therapeutic levels of lithium and toxic levels is small. Plasma lithium concentrations of 2 mmol/L are associated with toxic symptoms. Concentrations of 4 mmol/L can be fatal.

The initial effects of lithium exposure are tremors of the hands, nausea, micturition, slurred speech, sluggishness, sleepiness, vertigo, thirst, and increased urine volume. Effects from continued exposure are apathy, anorexia, fatigue, lethargy, muscular weakness, and changes in ECG. Long-term exposure leads to hypothyroidism, leukocytosis, edema, weight gain, polydipsia/polyuria (increased water intake leading to increased urinary output), memory impairment, seizures, kidney damage, shock, hypotension, cardiac arrhythmias, coma, death. Have been implicated in development of aplastic anemia. See also specific compounds, LITHIUM, and POTASSIUM COMPOUNDS.

LHE450 CAS:67880-27-7 ***HR: 3***
LITHIUM DIAZOMETHANIDE
mf: $CHLiN_2$ mw: 47.97

SAFETY PROFILE: The dry material is very explosive when exposed to air. When heated to decomposition it emits toxic fumes of NO_x. See also LITHIUM COMPOUNDS.

LHE475 CAS:816-43-3 ***HR: 3***
LITHIUM DIETHYL AMIDE
mf: $C_4H_{10}LiN$ mw: 79.07

$LiN(CH_2CH_3)_2$

SAFETY PROFILE: Ignites spontaneously in air. When heated to decomposition it emits toxic fumes of NO_x. See also LITHIUM COMPOUNDS and AMIDES.

LHE525 CAS:13529-75-4 ***HR: 3***
LITHIUM-2,2-DIMETHYLTRIMETHYLSILYL HYDRAZIDE
mf: $C_5H_{15}LiN_2Si$ mw: 138.21

$LiN[Si(CH_3)_3]N[CH_3]_2$

SAFETY PROFILE: Explosive reaction with 1:1 mixture of nitric and sulfuric acids; liquid ozone + oxygen. Hypergolic reaction with fuming nitric acid. Ignites on contact with fluorine. When heated to decomposition it emits toxic fumes of NO_x. See also LITHIUM COMPOUNDS and HYDRAZINE.

LHF000 CAS:7789-24-4 ***HR: 3***
LITHIUM FLUORIDE
mf: FLi mw: 25.94

PROP: Fine, white powder. Mp: 845°, bp: 1681°, d: 2.635 @ 20°, vap press: 1 mm @ 1047°. Sol in acids.

SYNS: LITHIUM FLUORURE (FRENCH) ◇ TLD 100

TOXICITY DATA with REFERENCE
orl-gpg LDLo:200 mg/kg MEIEDD 10,793,83
scu-frg LDLo:280 mg/kg CRSBAW 124,133,37

CONSENSUS REPORTS: Reported in EPA TSCA Inventory.

OSHA PEL: TWA 2.5 mg(F)/m^3
ACGIH TLV: TWA 2.5 mg(F)/m^3
NIOSH REL: (Inorganic Fluorides) TWA 2.5 mg(F)/m^3

SAFETY PROFILE: Poison by ingestion and subcutaneous routes. When heated to decomposition it emits toxic fumes of F^-. Used as a flux in enamils, glasses, glazes, and welding. See also FLUORIDES and LITHIUM COMPOUNDS.

LHF625 CAS:42017-07-2 ***HR: 3***
LITHIUM-1-HEPTYNIDE
mf: $C_7H_{11}Li$ mw: 102.10

$LiC{\equiv}CC_5H_{11}$

SAFETY PROFILE: Reaction with ammonia + iodine forms an explosive product. When heated to decomposition it emits acrid smoke and irritating fumes. See also LITHIUM COMPOUNDS and ACETYLENE COMPOUNDS.

LHG000 ***HR: 3***
LITHIUM HEXAAZIDOCUPRATE(4⁻)
mf: $CuLi_4N_{18}$ mw: 287.39

CONSENSUS REPORTS: Copper and its compounds are on The Community Right-To-Know List.

SAFETY PROFILE: A powerful explosive used as an initiating detonator. Upon decomposition it emits very toxic fumes of Li_2O and NO_x. See also COPPER COMPOUNDS and LITHIUM COMPOUNDS.

LHH000 CAS:7580-67-8 ***HR: 3***
LITHIUM HYDRIDE
DOT: UN 1414/UN 2805
mf: HLi mw: 7.95

PROP: White, translucent, crystals. Mp: 680°, d: 0.76-0.77. Darkens rapidly on exposure to light. Decomp in water liberating LiOH and H_2.

SYN: HYDRURE de LITHIUM (FRENCH)

TOXICITY DATA with REFERENCE
eye-rbt 5 mg/m^3 AMIHAB 14,468,56
eye-gpg 5 mg/m^3 AMIHAB 14,468,56
ihl-rat LCLo:10 mg/m^3/4H AMIHAB 14,468,56

CONSENSUS REPORTS: Reported in EPA TSCA Inventory. EPA Extremely Hazardous Substances List.

OSHA PEL: TWA 0.025 mg/m^3
ACGIH TLV: TWA 0.025 mg/m^3
DFG MAK: 0.025 mg/m^3
DOT Classification: Flammable Solid; Label: Flammable Solid and Dangerous When Wet.

SAFETY PROFILE: Poison by inhalation. A severe eye, skin, and mucous membrane irritant. Upon contact with moisture, lithium hydroxide is formed. The LiOH formed is very caustic and therefore highly toxic, particularly to lungs and respiratory tract skin and mucous membranes. The powder ignites spontaneously in air. The solid can ignite spontaneously in moist air. Mixtures of the powder with liquid oxygen are explosive. Ignites on contact with dinitrogen oxide; oxygen + moisture. To fight fire, use special mixtures of dry chemical. See also LITHIUM COMPOUNDS and HYDRIDES.

LHI000 CAS:7580-67-8 ***HR: 3***
LITHIUM HYDRIDE (fused solid form)

SYN: LITHIUM HYDRIDE in fused solid form (DOT)

CONSENSUS REPORTS: Reported in EPA TSCA Inventory.

DOT Classification: Flammable Solid; Label: Flammable Solid and Dangerous When Wet.

SAFETY PROFILE: Highly flammable. Used as a dessicant, source of hydrogen, and a condensing agent in organic synthesis. See also LITHIUM HYDRIDE and LITHIUM COMPOUNDS.

LHJ000 CAS:13840-33-0 ***HR: 1***
LITHIUM HYPOCHLORITE
DOT: UN 1471
mf: ClO•Li mw: 58.39

SYN: LITHIUM HYPOCHLORITE COMPOUND, dry, containing more than 39% available chlorine (DOT)

CONSENSUS REPORTS: Reported in EPA TSCA Inventory.

DOT Classification: Oxidizer; Label: Oxidizer.

SAFETY PROFILE: A powerful oxidizer. An eye, skin, and mucous membrane irritant. When heated to decomposition it emits very toxic fumes of Li_2O and Cl^-. Used for swimming pool chlorination, and as a laundry bleach. See also LITHIUM COMPOUNDS and HYPOCHLORITES.

LHK000 CAS:64082-35-5 ***HR: 3***
LITHIUM IRON SILICON
DOT: UN 2830
mf: FeLiSi mw: 90.88

PROP: Dark, crystalline, brittle, metallic lumps or powder; evolves a flammable gas when in contact with moisture.

SYN: LITHIUM FERRO SILICON

DOT Classification: Flammable Solid; Label: Flammable Solid and Dangerous When Wet.

SAFETY PROFILE: Flammable solid which evolves a flammable gas when exposed to moisture, steam, or acid fumes. Flammable when exposed to heat or flame. See also LITHIUM COMPOUNDS.

LHL000 CAS:867-55-0 ***HR: 2***
LITHIUM LACTATE
mf: $C_3H_5O_3$•Li mw: 96.02

SYN: LACTIC ACID LITHIUM SALT

TOXICITY DATA with REFERENCE
orl-mus LD50:2100 mg/kg RPTOAN 33,266,70
scu-mus LD50:1530 mg/kg RPTOAN 33,266,70

CONSENSUS REPORTS: Reported in EPA TSCA Inventory.

SAFETY PROFILE: Moderately toxic by ingestion and subcutaneous routes. When heated to decomposition it emits acrid smoke and irritating fumes. See also LITHIUM COMPOUNDS.

LHM000 CAS:26134-62-3 ***HR: 3***
LITHIUM NITRIDE
DOT: UN 2806
mf: Li_3N mw: 34.82

PROP: Brownish-red, hexagonal crystals; slowly decomp on contact with moisture. D: 1.3, mp: 845°.

CONSENSUS REPORTS: Reported in EPA TSCA Inventory.

DOT Classification: Flammable Solid; Label: Flammable Solid and Dangerous When Wet.

SAFETY PROFILE: A powerful reducing agent. Upon contact with moisture, it decomposes into lithium hydroxide, lithium compounds and ammonia. The powder may ignite spontaneously in moist air. Flammable at elevated temperatures; ignites and burns intensely in air. Violent reaction with silicon tetrafluoride; copper(I) chloride + heat. To fight fire, use dry chemical, sand, graphite; avoid use of water or carbon tetrachloride. When heated to decomposition it emits very toxic fumes of Li_2O and NO_x. Used as a strong reducing agent in organic synthesis and a solid electrolyte in lithium batteries. See also LITHIUM COMPOUNDS and NITRIDES.

LHM750 CAS:78350-94-4 ***HR: 3***
LITHIUM-4-NITROTHIOPHENOXIDE
mf: $C_6H_4LiNO_2S$ mw: 161.10

SAFETY PROFILE: The dry salt explodes on contact with air. When heated to decomposition it emits toxic fumes of SO_x and NO_x. See also LITHIUM COMPOUNDS.

LHM850 CAS:50662-24-3 ***HR: 3***
LITHIUM PENTAMETHYLTITANATE-BIS(2,2'-BIPYRIDINE)
mf: $C_5H_{15}LiTi$•$2C_{10}H_8N_2$ mw: 442.37

$Li[Ti(CH_3)_5]$•$2C_{10}H_8N_2$

SAFETY PROFILE: A friction-sensitive explosive. When heated to decomposition it emits toxic fumes of NO_x. See also LITHIUM COMPOUNDS and TITANIUM COMPOUNDS.

LHM875 ***HR: 3***
LITHIUM PERCHLORATE
mf: $ClLiO_4$ mw: 106.39

SAFETY PROFILE: Reacts with hydrazine to form a friction-sensitive explosive product. Battery electrolyte systems with 1,3-dioxolane or nitromethane are potentially explosive. When heated to decomposition it emits toxic fumes of Cl^-. See also PERCHLORATES and LITHIUM COMPOUNDS.

LHN000 CAS:13453-78-6 ***HR: 2***
LITHIUM PERCHLORATE TRIHYDRATE
mf: $ClLiO_4$•$3H_2O$ mw: 160.46

PROP: Colorless, deliquescent crystals. Mp: 236°, bp: decomp @ 430°, d: 2.429.

TOXICITY DATA with REFERENCE
ipr-mus LD50:1160 mg/kg JAFCAU 14,512,66

SAFETY PROFILE: Moderately toxic by intraperitoneal route. A skin, eye, and mucous membrane irritant. Incompatible with nitromethane, acetone; platinium; hydrogen; oxygen. When heated to decomposition it emits very toxic fumes of Cl^- and Li_2O. See also PERCHLORATES and LITHIUM COMPOUNDS.

LHO000 CAS:12031-80-0 ***HR: 2***
LITHIUM PEROXIDE
DOT: UN 1472
mf: Li_2O_2 mw: 45.88

PROP: Fine, white powder or sandy yellow, granular material. Mp: decomp, d: 2.14 @ 20°.

CONSENSUS REPORTS: Reported in EPA TSCA Inventory.

DOT Classification: Oxidizer; Label: Oxidizer.

SAFETY PROFILE: A powerful oxidizer and irritant to skin, eyes, and mucous membranes. A very dangerous fire hazard because it is an extremely powerful oxidizing agent. Will react with water or steam to produce heat; on contact with reducing materials, can react vigorously. See also LITHIUM COMPOUNDS, PEROXIDES, and PEROXIDES, INORGANIC.

LHP000 CAS:68848-64-6 ***HR: 3***
LITHIUM SILICON
DOT: UN 1417

PROP: Solid. Composition: Li + Si.

DOT Classification: Flammable Solid; Label: Flammable Solid and Dangerous When Wet.

SAFETY PROFILE: A very dangerous fire hazard in the form of dust when exposed to heat or flame or by chemical reaction with moisture or acids. In contact with water, silane and hydrogen are evolved. Slightly explosive in the form of dust when exposed to flame. Will react with water or steam to produce flammable vapors; on contact with oxidizing materials, can react vigorously; on contact with acid or acid fumes, can emit toxic and flammable fumes. To fight fire, use CO_2, dry chemical. See also LITHIUM, SILICON, and POWDERED METALS.

LHQ000 ***HR: 3***
LITHIUM SODIUM NITROXYLATE
mf: $LiNNaO_2$ mw: 75.94

SAFETY PROFILE: Decomposes violently. When heated to decomposition it emits very toxic fumes of Li_2O, NO_x, and Na_2O. See also LITHIUM COMPOUNDS.

LHR000 CAS:10377-48-7 ***HR: 2***
LITHIUM SULFATE (2:1)
mf: $O_4S{\bullet}2Li$ mw: 109.94

SYNS: LITHIUM SULPHATE ◇ SULFURIC ACID, DILITHIUM SALT ◇ SULFURIC ACID, LITHIUM SALT (1:2)

TOXICITY DATA with REFERENCE
mmo-smc 100 mmol/L MUREAV 117,149,83
mrc-smc 100 mmol/L MUREAV 117,149,83
orl-mus LD50:1190 mg/kg RPTOAN 33,266,70
scu-mus LD50:953 mg/kg RPTOAN 33,266,70

CONSENSUS REPORTS: Reported in EPA TSCA Inventory.

SAFETY PROFILE: Moderately toxic by ingestion and subcutaneous routes. Mutation data reported. When heated to decomposition it emits toxic fumes of SO_x. Used in photographic developer compositions and special high strength glass. See also LITHIUM SALTS and SULFATES.

LHR650 CAS:67849-02-9 ***HR: 3***
LITHIUM TETRAAZIDOALUMINATE
mf: $AlLiN_{12}$ mw: 202.00

SAFETY PROFILE: A shock-sensitive explosive. When heated to decomposition it emits toxic fumes of NO_x. See also AZIDES, LITHIUM COMPOUNDS, and ALUMINUM COMPOUNDS.

LHR675 ***HR: 3***
LITHIUM TETRAAZIDOBORATE
mf: $BLiN_{12}$ mw: 185.83

$Li[B(N_3)_4]$

SAFETY PROFILE: A powerful explosive sensitive to heat, impact, and friction. When heated to decomposition it emits toxic fumes of NO_x. See also AZIDES, LITHIUM COMPOUNDS, and BORON COMPOUNDS.

LHR700 CAS:14128-54-2 ***HR: 3***
LITHIUM TETRADEUTEROALUMINATE
mf: AlD_4Li mw: 41.99

SAFETY PROFILE: Ignites spontaneously in moist air. See also LITHIUM COMPOUNDS and ALUMINUM COMPOUNDS.

LHS000 CAS:16853-85-3 ***HR: 3***
LITHIUM TETRAHYDROALUMINATE
DOT: UN 1410/UN 1411
mf: $AlH_4{\bullet}Li$ mw: 37.96

PROP: White, microcrystalline lumps.

SYNS: ALUMINUM LITHIUM HYDRIDE ◇ LITHIUM ALANATE ◇ LITHIUM ALUMINOHYDRIDE ◇ LITHIUM ALUMINUM HYDRIDE (DOT) ◇ LITHIUM ALUMINUM HYDRIDE, ETHEREAL (DOT) ◇ LITHIUM ALUMINUM TETRAHYDRIDE

CONSENSUS REPORTS: Reported in EPA TSCA Inventory.
ACGIH TLV: TWA 2 mg(Al)/m^3

DOT Classification: Flammable Solid; Label: Flammable Solid & Danger When Wet (UN1410); DOT Class: Flammable Liquid; Label: Flammable Liquid (UN1411); DOT Class: Flammable Solid; Label: Danger When Wet, Flammable Liquid (UN1411)

SAFETY PROFILE: Stable in dry air at room temperature. It decomposes above 125° forming Al, H_2 and lithium hydride. Very powerful reducer. Can ignite if pulverized even in a dry box. Reacts violently with air; acids; alcohols; benzoyl peroxide; boron trifluoride etherate; (2-chloromethyl furan + ethyl acetate); diethylene glycol dimethyl ether; diethyl ether; 1,2-dimethoxyethane, dimethyl ether; methyl ethyl ether; (nitriles + H_2O); perfluoro-succinamide; (perfluorosuccinamide + H_2O); tetrahydrofuran; water. To fight fire, use dry chemical, including special formulations of dry chemicals as recommended by the supplier of the lithium aluminum hydride. Do not use water, fog, spray or mist. Incompatible with bis(2-methoxy-ethyl)ether;

CO_2; BF_3; diethyl etherate; dibenzoyl peroxide; 3,5-dibromocyclopentene; 1,2-dimethoxy ethane; ethyl acetate; fluoro amides; pyridine; tetrahydrofuran. Used as a reducing agent in the preparation of pharmaceuticals. See also ALUMINUM, LITHIUM COMPOUNDS, and HYDRIDES.

LHT000 CAS:16949-15-8 ***HR: 3***
LITHIUM TETRAHYDROBORATE
DOT: UN 1413
mf: BH_4•Li mw: 21.79

SYN: LITHIUM BOROHYDRIDE (DOT)

CONSENSUS REPORTS: Reported in EPA TSCA Inventory.

DOT Classification: Flammable Solid; Label: Flammable Solid and Dangerous When Wet.

SAFETY PROFILE: Poison by ingestion, inhalation, and skin contact. Flammable; can liberate H_2. Incompatible H_2O as moisture on fibers of cellulose or as liquid. See also LITHIUM, BORON COMPOUNDS, and HYDRIDES.

LHT400 CAS:2169-38-2 ***HR: 3***
LITHIUM TETRAMETHYLBORATE
mf: $C_4H_{12}BLi$ mw: 77.89

$Li[(CH_3)_4B]$

SAFETY PROFILE: Ignites spontaneously in moist air. When heated to decomposition it emits acrid smoke and irritating fumes. See also LITHIUM COMPOUNDS and BORON COMPOUNDS.

LHT425 ***HR: 3***
LITHIUM TETRAMETHYL CHROMATE(II)
mf: $C_4H_{12}CrLiO_4$ mw: 183.07

CONSENSUS REPORTS: Chromium and its compounds are on the Community Right-To-Know List.

SAFETY PROFILE: Ignites spontaneously in air. When heated to decomposition it emits acrid smoke and irritating fumes. See also CHROMIUM COMPOUNDS and LITHIUM COMPOUNDS.

LHU000 ***HR: 3***
LITHIUM TETRAZIDO ALUMINATE
mf: $AlLiN_{12}$ mw: 202

SAFETY PROFILE: A shock-sensitive explosive. When heated to decomposition it emits very toxic fumes of Li_2O and NO_x. See also LITHIUM and ALUMINUM COMPOUNDS.

LHV000 ***HR: 3***
LITHIUM-TIN ALLOY
mf: LiSn mw: 125.63

SAFETY PROFILE: Ignites spontaneously in air. When heated to decomposition it emits toxic fumes of Li_2O. See also LITHIUM and TIN COMPOUNDS.

LHV500 ***HR: 3***
LITHIUM TRIETHYLSILYL AMIDE
mf: $C_6H_{16}LiNSi$ mw: 137.23

$LiNHSi(CH_2CH_3)_3$

SAFETY PROFILE: Explosive or violent reaction with strong oxidants (e.g., fluorine (hypergolic); fuming nitric acid (hypergolic); ozone (explodes). When heated to decomposition it emits toxic fumes of NO_x. See also AMIDES and LITHIUM COMPOUNDS.

LHW000 CAS:434-13-9 ***HR: 2***
LITHOCHOLIC ACID
mf: $C_{24}H_{40}O_3$ mw: 376.64

PROP: Hexagonal leaflets from alc; prisms from acetic acid. Mp: 184-186°. Very sol in hot alc; sltly sol in glacial acetic acid; insol in petr ether, gasoline, ligroin, water.

SYNS: 3-α-HYDROXYCHOLANIC ACID ◇ 3-α-HYDROXY-5-β-CHOLANIC ACID ◇ (3-α,5-β)-3-HYDROXY-CHOLAN-24-OIC ACID ◇ 17-β-(1-METHYL-3-CARBOXYPROPYL)ETHIOCHOLAN-3α-OL ◇ NCI-C03861

TOXICITY DATA with REFERENCE
sln-smc 100 mg/L CRNGDP 5,447,84
dnd-mus:oth 2500 μmol/L CBINA8 52,311,85
orl-mus LD50:3900 mg/kg NCILB* NIH-NCI-E-C-72-3252

CONSENSUS REPORTS: NCI Carcinogenesis Bioassay (gavage); No Evidence: mouse, rat NCITR* NCI-CG-TR-175,79. EPA Genetic Toxicology Program.

SAFETY PROFILE: Moderately toxic by ingestion. Mutation data reported. When heated to decomposition it emits acrid smoke and irritating fumes.

LHX000 CAS:1345-05-7 ***HR: 3***
LITHOPONE

PROP: White powder. Mixture of zinc sulfide, barium sulfate and zinc oxide.

SYN: GRIFFITH'S ZINC WHITE

CONSENSUS REPORTS: Zinc and its compounds, as well as barium and its compounds, are on the Community Right-To-Know List. Reported in EPA TSCA Inventory.

SAFETY PROFILE: Poison because it can liberate hydrogen sulfide upon decomposition by heat, moisture, and acids. When heated to decomposition it emits highly toxic fumes of SO_x, ZnO, and H_2S. See also ZINC, BARIUM COMPOUNDS, SULFIDES, and HYDROGEN SULFIDE.

LHX300 ***HR: D***
LITHOSPERMUM RUDERALE, root extract

TOXICITY DATA with REFERENCE
scu-rat TDLo:250 μg/kg (1D preg):REP JOENAK 10,212,54

SAFETY PROFILE: Experimental reproductive effects.

LHX325 ***HR: 2***
LITSEA CUBEBA OIL

PROP: Found in fruits of the tree *Litsea cubeba* (FCTOD7 20(Suppl),731,82).

TOXICITY DATA with REFERENCE
skn-mus 100% FCTOD7 20(Suppl),731,82
skn-rbt 500 mg/24H SEV FCTOD7 20(Suppl),731,82
skn-rbt LD50:4800 mg/kg FCTOD7 20(Suppl),731,82

SAFETY PROFILE: Mildly toxic by skin contact. A severe skin irritant.

LHX350 CAS:11111-23-2 ***HR: 3***
LIVIDOMYCIN

SYNS: LIVODYMYCIN ◇ LVM

TOXICITY DATA with REFERENCE
ims-rbt TDLo:1 g/kg (8-17D preg):REP OYYAA2 7,1241,73
ims-rbt TDLo:1 g/kg (8-17D preg):TER OYYAA2 7,1241,73
scu-rat LD50:1819 mg/kg OYYAA2 6,787,72
ivn-rat LD50:365 mg/kg OYYAA2 6,787,72
ims-rat LD50:1750 mg/kg OYYAA2 6,787,72
scu-mus LD50:1249 mg/kg OYYAA2 6,787,72
ivn-mus LD50:225 mg/kg OYYAA2 6,787,72
ims-mus LD50:1348 mg/kg OYYAA2 6,787,72

SAFETY PROFILE: Poison by intravenous route. Moderately toxic by subcutaneous and intramuscular routes. An experimental teratogen. Experimental reproductive effects.

LHX360 CAS:37229-14-4 ***HR: 3***
LIVIDOMYCIN SULFATE

TOXICITY DATA with REFERENCE
orl-rat LD50:10 g/kg OYYAA2 9,601,75
scu-rat LD50:1819 mg/kg OYYAA2 9,601,75
ivn-rat LD50:365 mg/kg OYYAA2 9,601,75
ims-rat LD50:1750 mg/kg OYYAA2 9,601,75
orl-mus LD50:10 g/kg OYYAA2 9,601,75
scu-mus LD50:1249 mg/kg OYYAA2 9,601,75
ivn-mus LD50:225 mg/kg OYYAA2 9,601,75
ims-mus LD50:1348 mg/kg OYYAA2 9,601,75

SAFETY PROFILE: Poison by intravenous route. Moderately toxic by subcutaneous and intramuscular routes. See also LIVIDOMYCIN.

LHX495 CAS:24570-10-3 ***HR: 3***
LM 2911
mf: $C_{20}H_{23}N_2O{\bullet}I$ mw: 434.35

SYN: (2-((5H-DIBENZO(a,d)CYCLOHEPTEN-5-YLIDENEAMINO)OXY)ETHYL)TRIMETHYLAMMONIUMIODIDE

TOXICITY DATA with REFERENCE
orl-mus LD50:750 mg/kg FRPSAX 24,685,69
scu-mus LD50:150 mg/kg FRPSAX 24,685,69
ivn-mus LD50:1500 μg/kg FRPSAX 24,685,69

SAFETY PROFILE: Poison by subcutaneous and intravenous routes. Moderately toxic by ingestion. When heated to decomposition it emits toxic fumes of NH_3, I^-, and NO_x.

LHX498 CAS:25410-69-9 ***HR: 3***
LM 2910
mf: $C_{20}H_{22}N_{2O}{\bullet}ClH$ mw: 342.90

SYN: o-(2-DIMETHYLAMINO)PROPYL)OXIME-5H-DIBENZO(a,d)CYCLOHEPTEN-5-ONEMONOHYDROCHLORIDE

TOXICITY DATA with REFERENCE
orl-mus LD50:75 mg/kg FRPSAX 24,685,69
ipr-mus LD50:45 mg/kg FRPSAX 24,685,69
scu-mus LD50:60 mg/kg FRPSAX 24,685,69
ivn-mus LD50:17 mg/kg FRPSAX 24,685,69

SAFETY PROFILE: Poison by ingestion, subcutaneous, intravenous, and intraperitoneal routes. When heated to decomposition it emits toxic fumes of NO_x and HCl.

LHX500 CAS:25450-02-6 ***HR: 3***
LM 2916
mf: $C_{20}H_{22}ClN_2O{\bullet}I$ mw: 468.79

TOXICITY DATA with REFERENCE
orl-mus LD50:450 mg/kg FRPSAX 24,685,69
scu-mus LD50:300 mg/kg FRPSAX 24,685,69
ivn-mus LD50:900 μg/kg FRPSAX 24,685,69

SAFETY PROFILE: Poison by subcutaneous and intravenous routes. Moderately toxic by ingestion. When heated to decomposition it emits toxic fumes of Cl^-, I^-, and NO_x.

LHX510 CAS:24570-12-5 ***HR: 3***
LM 2917
mf: $C_{21}H_{25}N_2O{\bullet}I$ mw: 448.38

SYN: (2-((5H-DIBENZO(a,d)CYCLOHEPTEN-5-YLIDENEAMINO)OXY)-1-METHYLETHYL)TRIMETHYLAMMONIUM

TOXICITY DATA with REFERENCE
orl-mus LD50:450 mg/kg FRPSAX 24,685,69
ipr-mus LD50:60 mg/kg FRPSAX 24,685,69
ivn-mus LD50:1200 μg/kg FRPSAX 24,685,69

SAFETY PROFILE: Poison by intravenous and intraperitoneal routes. Moderately toxic by ingestion. When heated to decomposition it emits toxic fumes of NH_3, I^-, and NO_x.

LHX515 CAS:25410-64-4 ***HR: 3***
LM 2918
mf: $C_{21}H_{24}N_{20}$•ClH mw: 356.93

SYN: o-(2-(DIETHYLAMINO)ETHYL)OXIME-5H-DIBENZO(a,d) CYCLOHEPTEN-5-ONEMONOHYDROCHLORIDE

TOXICITY DATA with REFERENCE
orl-mus LD50:200 mg/kg FRPSAX 24,685,69
scu-mus LD50:120 mg/kg FRPSAX 24,685,69
ivn-rbt LD50:12 mg/kg FRPSAX 24,685,69

SAFETY PROFILE: Poison by ingestion, subcutaneous, and intravenous routes. When heated to decomposition it emits toxic fumes of NO_x and HCl.

LHX600 CAS:24570-11-4 ***HR: 3***
LM 2930
mf: $C_{22}H_{27}N_2O$•I mw: 462.41

SYN: (2-((5H-DIBENZO(a,d)CYCLOHEPTEN-5-YLIDENEAMINO) OXY)-1-METHYLETHYL)ETHYLDIMETHYLAMMONIUM

TOXICITY DATA with REFERENCE
orl-mus LD50:500 mg/kg FRPSAX 24,685,69
ipr-mus LD50:40 mg/kg FRPSAX 24,685,69
scu-mus LD50:425 mg/kg FRPSAX 24,685,69
ivn-mus LD50:1500 μg/kg FRPSAX 24,685,69

SAFETY PROFILE: Poison by intravenous and intraperitoneal routes. Moderately toxic by ingestion and subcutaneous routes. When heated to decomposition it emits toxic fumes of NH_3, I^- and NO_x.

LHX700 ***HR: 2***
LOBELIA NICOTIANIFOLIA Roth ex R. & S., extract

PROP: Indian plant belonging to the family *Campanulaceae* IJEBA6 22,487,84.

TOXICITY DATA with REFERENCE
orl-rat TDLo:150 mg/kg (female 12-14D post):REP IJEBA6 22,487,84
ipr-mus LD50:562 mg/kg IJEBA6 22,487,84

SAFETY PROFILE: Moderately toxic by intraperitoneal route. Experimental reproductive effects. When heated to decomposition it emits acrid smoke and irritating fumes.

LHY000 CAS:90-69-7 ***HR: 3***
LOBELINE
mf: $C_{22}H_{27}NO_2$ mw: 337.46

PROP: Mp: 131°.

SYNS: 8,10-DIPHENYL LOBELIONOL ◇ INFLATINE ◇ LOBELIA, INDIAN TOBACCO ◇ LOBNICO

TOXICITY DATA with REFERENCE
ipr-mus LD50:107 mg/kg JPETAB 67,153,39
ivn-mus LD50:6300 μg/kg AIPTAK 103,146,55

SAFETY PROFILE: Poison by intraperitoneal and intravenous routes. Causes stimulation which leads to convulsions in severe cases. Nausea and vomiting are frequent. When heated to decomposition it emits toxic fumes of NO_x.

LHZ000 CAS:134-63-4 ***HR: 3***
LOBELINE HYDROCHLORIDE
mf: $C_{22}H_{27}NO_2$•ClH mw: 373.96

SYNS: (−)-2-(6-(β-HYDROXYPHENETHYL)-1-METHYL-2-PIPERIDYL)-ACETOPHENONE HYDROCHLORIDE ◇ (−)-LOBELINE HYDROCHLORIDE ◇ LOBELIN HYDROCHLORIDE

TOXICITY DATA with REFERENCE
ipr-mus LD50:39900 μg/kg NIIRDN 6,913,82
scu-mus LD50:87500 μg/kg NIIRDN 6,913,82
ivn-mus LD50:7800 μg/kg NIIRDN 6,913,82

SAFETY PROFILE: Poison by subcutaneous, intravenous, and intraperitoneal routes. When heated to decomposition it emits toxic fumes of NO_x and HCl. See also LOBELINE and CARDINAL FLOWER.

LHZ600 CAS:64808-48-6 ***HR: 2***
LOBENZARIT DISODIUM
mf: $C_{14}H_8ClNO_4$•2Na mw: 335.66

SYNS: BENZOIC ACID, 2-((2-CARBOXYPHENYL)AMINO)-4-CHLORO-, DISODIUM SALT ◇ 2-((2-CARBOXYPHENYL)AMINO)-4-CHLOROBENZOIC ACID DISODIUM SALT ◇ CCA ◇ LOBENZARIT SODIUM

TOXICITY DATA with REFERENCE
orl-rat TDLo:2700 mg/kg (female 17-22D post):REP YACHDS 15,5193,87
orl-rat TDLo:8100 mg/kg (male 8W pre):TER YACHDS 15,5193,87
orl-rat LD50:1150 mg/kg YACHDS 15,4579,87
ipr-rat LD50:263 mg/kg YACHDS 15,4579,87
scu-rat LD50:314 mg/kg YACHDS 15,4579,87
orl-mus LD50:740 mg/kg YACHDS 15,4579,87
ivn-mus LD50:400 mg/kg YACHDS 15,4579,87

SAFETY PROFILE: Poison by intravenous, subcutaneous, and intraperitoneal routes. Moderately toxic by ingestion. An experimental teratogen. Experimental reproductive effects. When heated to decomposition it emits toxic fumes of NO_x and Cl^-.

LHZ700 ***HR: D***
LOCOWEED

PROP: Found in mountain regions of western United States from *Astragalus lentiginosus* and *Astragalus wootini* (TXAPA9 41,139,77).

TOXICITY DATA with REFERENCE
orl-rat TDLo:55 g/kg (7-17D preg):REP TXAPA9 41,139,77
orl-dom TDLo:218 g/kg (female 10-14W post):TER AJVRAH 36,825,75

SAFETY PROFILE: An experimental teratogen. Experimental reproductive effects.

LIA000 CAS:9000-40-2 ***HR: 1***
LOCUST BEAN GUM

PROP: From the ground endosperms of *Ceratonia ailiqua* (L.) Taub. (Fam. *Leguminosae*). White powder; odorless and tasteless but acquires a leguminous taste when boiled in water. A galactomannan polysaccharide. Mw: 310,000 (approx). Insol in most organic solvents.

SYNS: ALGAROBA ◇ CAROB BEAN GUM ◇ CAROB FLOUR ◇ NCI-C50419 ◇ ST. JOHN'S BREAD ◇ SUPERCOL

TOXICITY DATA with REFERENCE
orl-rat LD50:13 g/kg FDRLI* 124,-,76
orl-mus LD50:13 g/kg FDRLI* 124,-,76
orl-rbt LD50:9100 mg/kg FDRLI* 124,-,76
orl-ham LD50:10 g/kg FDRLI* 124,-,76

CONSENSUS REPORTS: NTP Carcinogenesis Bioassay (feed); No Evidence: mouse, rat NTPTR* NTP-TR-221,82. Reported in EPA TSCA Inventory.

SAFETY PROFILE: Mildly toxic by ingestion. When heated to decomposition it emits acrid smoke and irritating fumes.

LIA400 CAS:21498-08-8 ***HR: 3***
LOFETENSIN HYDROCHLORIDE
mf: $C_{11}H_{12}Cl_2N_2O{\bullet}ClH$ mw: 295.61

SYNS: BA 168 ◇ 2-(1-(2,6-DICHLORPHENOXY)ATHYL)-2-IMIDAZOLIN-HYDROCHLORID (GERMAN) ◇ 2-(1-(2,6-DICHLOROPHENOXY)ETHYL)-4,5-DIHYDRO-1H-IMIDAZOLEMONOHYDROCHLORIDE ◇ 2-(1-(2,6-DICHLOROPHENOXY)ETHYL)-2-IMIDAZOLINE HYDROCHLORIDE ◇ LOFETENSIN ◇ LOFEXIDINE HYDROCHLORIDE ◇ LOXACOR ◇ LOXACOR HYDROCHLORIDE ◇ RMI-14042A

TOXICITY DATA with REFERENCE
orl-rat TDLo:30 mg/kg (7-16D preg):TER ARZNAD 32,962,82
orl-rbt TDLo:218 mg/kg (female 14D pre):REP ARZNAD 32,962,82
orl-hmn TDLo:270 μg/kg/6W-I ARZNAD 32,976,82
orl-rat LD50:70 mg/kg ARZNAD 32,955,82
ivn-rat LD50:8 mg/kg ARZNAD 32,955,82
orl-mus LD50:54 mg/kg ARZNAD 32,966,82
ivn-mus LD50:8 mg/kg ARZNAD 32,955,82

SAFETY PROFILE: Poison by ingestion and intravenous routes. An experimental teratogen. Experimental reproductive effects. An antihypertensive agent. When heated to decomposition it emits toxic fumes of NO_x and HCl.

LIB000 CAS:3810-80-8 ***HR: 3***
LOMOTIL
mf: $C_{30}H_{32}N_2O_2{\bullet}ClH$ mw: 489.10

SYNS: 1-(3-CYANO-3,3-DIPHENYLPROPYL)-4-PHENYLISONIPECOTIC ACID ETHYL ESTER HYDROCHLORIDE ◇ DIPHENOXYLATE HYDROCHLORIDE

TOXICITY DATA with REFERENCE
orl-chd TDLo:2300 μg/kg:PUL ADCHAK 54,222,79
orl-chd TDLo:2300 μg/kg:CNS ADCHAK 54,222,79
orl-chd LDLo:1515 μg/kg:CVS,PUL,MET ADCHAK 54,222,79
orl-rat LD50:221 mg/kg ARZNAD 24,1633,74

SAFETY PROFILE: A human and experimental poison by ingestion. Human systemic effects by ingestion: cardiac effects, dyspnea, body temperature decrease, general anesthetic, somnolence, respiratory depression. When heated to decomposition it emits very toxic fumes of HCl and NO_x.

LIC000 CAS:8012-74-6 ***HR: 3***
LONDON PURPLE
DOT: UN 1621

SYN: LONDON PURPLE, solid (DOT)

DOT Classification: Poison B; Label: Poison.

SAFETY PROFILE: A poison. When heated to decomposition it emits very toxic fumes of As and NO_x. See also ARSENIC and ANILINE.

LID000 CAS:1897-96-7 ***HR: 2***
LONETHYL
mf: $C_{17}H_{16}N_2O_2$ mw: 280.35

SYNS: 3-(4-ETHOXYPHENYL)-2-METHYL-4(3H)-QUINAZOLINONE ◇ LONETIL

TOXICITY DATA with REFERENCE
orl-rat LD50:4800 mg/kg ARTODN 1,379,78
ipr-rat LD50:1900 mg/kg ARTODN 1,379,78
orl-mus LD50:3420 mg/kg ARTODN 1,379,78

ipr-mus LD50:1 g/kg ATSUDG 1,379,78
par-mus LD50:3000 mg/kg PCJOAU 7,626,73

SAFETY PROFILE: Moderately toxic by ingestion, intraperitoneal and parenteral routes. When heated to decomposition it emits toxic fumes of NO_x.

LIF000 CAS:58785-63-0 ***HR: 3***
LONOMYCIN
mf: $C_{44}H_{76}O_{14}$ mw: 829.20

SYNS: ANTIBIOTIC DE 3936 ◇ ANTIBIOTIC TM 481 ◇ EMERICID

TOXICITY DATA with REFERENCE
orl-mus LD50:45800 μg/kg 85GDA2 5,482,81
ipr-mus LD50:8280 μg/kg 85ERAY 1,801,78
scu-mus LD50:37500 μg/kg 85GDA2 5,482,81
ivn-mus LD50:4860 μg/kg 85ERAY 1,801,78
orl-ckn LD50:150 mg/kg JANTAJ 29,76-99,76

SAFETY PROFILE: Poison by ingestion, intravenous, intraperitoneal, and subcutaneous routes. When heated to decomposition it emits acrid smoke and irritating fumes.

LIG000 CAS:58845-80-0 ***HR: 3***
LONOMYCIN, SODIUM SALT
mf: $C_{44}H_{76}O_{14}\cdot Na$ mw: 852.19

SYN: LONOMYCIN, MONOSODIUM SALT

TOXICITY DATA with REFERENCE
orl-mus LD50:46 mg/kg 85ERAY 1,801,78
ipr-mus LD50:13 mg/kg 37ASAA 3,47,78
scu-mus LD50:38 mg/kg 85ERAY 1,801,78

SAFETY PROFILE: Poison by ingestion, subcutaneous, and intraperitoneal routes. When heated to decomposition it emits toxic fumes of Na_2O.

LIH000 CAS:34552-83-5 ***HR: 3***
LOPERAMIDE HYDROCHLORIDE
mf: $C_{29}H_{33}ClN_2O_2\cdot ClH$ mw: 513.55

PROP: Crystals from isopropanol. Mp: 222-223°. Insol in water, stable.

SYNS: BREK ◇ 4-(4-CHLOROPHENYL)-4-HYDROXY-N,N-DIMETHYL-α,α-DIPHENYL-1-PIPERIDINEBUTANAMIDE HYDROCHLORIDE ◇ 4-(p-CHLOROPHENYL)-4-HYDROXY-N,N-DIMETHYL-α,α-DIPHENYL-1-PIPERIDINE BUTYRAMIDE HCl ◇ 4-(4-(p-CHLOROPHENYL)-4-HYDROXY-1-PIPERIDYL)-N,N-DIMETHYL-2,2-DIPHENYLBUTYRAMIDE HCl ◇ DISSENTEN ◇ IMODIUM ◇ LOPEMID ◇ LOPEMIN ◇ LOPERYL ◇ PJ185 ◇ R 18553 HYDROCHLORIDE ◇ SUPRASEC

TOXICITY DATA with REFERENCE
orl-rat TDLo:580 mg/kg (15-22D preg/21D post):REP ARZNAD 24,1645,75
orl-cld TDLo:125 μg/kg BMJOAE 294,1383,87
orl-rat LD50:185 mg/kg JMCMAR 16,782,73
scu-rat LD50:78700 μg/kg NIIRDN 6,913,82
ivn-rat LD50:7490 μg/kg NIIRDN 6,913,82
orl-mus LD50:105 mg/kg ARZNAD 24,1633,74
ipr-mus LD50:28 mg/kg ARZNAD 24,1633,74
scu-mus LD50:75 mg/kg ARZNAD 24,1636,74
ivn-mus LD50:12600 μg/kg NIIRDN 6,913,82
orl-gpg LD50:41500 μg/kg IYKEDH 12,1204,81

SAFETY PROFILE: Poison by ingestion, intravenous, intraperitoneal, and subcutaneous routes. Experimental reproductive effects. When heated to decomposition it emits very toxic fumes of Cl^-, NO_x, and HCl.

LIH200 ***HR: 3***
LOQUAT

PROP: An evergreen which grows to about 20 feet with large, rough leaves 8 to 12 inches long. It produces clusters of fragrant, off-white flowers and pear-shaped, yellow fruit about 3 inches long. It is cultivated in California, Florida, Hawaii, the Gulf Coastal states, and the West Indies.

SYNS: ERIOBOTRYA JAPONICA ◇ JAPANESE MEDLAR ◇ JAPANESE PLUM ◇ NISPERO DEL JAPON (CUBA, PUERTO RICO)

SAFETY PROFILE: The insides of the seeds contain a poisonous cyanogenetic glycoside. Ingestion of chewed or otherwise broken seeds can cause after a delay period abdominal pain, vomiting, coma, convulsions, and other symptoms of cyanosis. See also CYANIDE.

LIH300 CAS:71-82-9 ***HR: 3***
LORFAN TARTRATE
mf: $C_{19}H_{25}NO\cdot C_4H_6O_6$ mw: 433.55

SYNS: LEVALLORPHAN TARTRATE ◇ l-LEVALLORPHAN TARTRATE ◇ LEVALLORPHINE TARTRATE ◇ LORFAN

TOXICITY DATA with REFERENCE
orl-rat LD50:850 mg/kg NIIRDN 6,904,82
scu-rat LD50:870 mg/kg NIIRDN 6,904,82
ivn-rat LD50:40 mg/kg NIIRDN 6,904,82
orl-mus LD50:350 mg/kg NIIRDN 6,904,82
ipr-mus LD50:168 mg/kg NIIRDN 6,904,82
scu-mus LD50:240 mg/kg NIIRDN 6,904,82
ivn-mus LD50:42 mg/kg NIIRDN 6,904,82

SAFETY PROFILE: Poison by ingestion, subcutaneous, intravenous, and intraperitoneal routes. When heated to decomposition it emits toxic fumes of NO_x.

LII000 CAS:8016-31-7 ***HR: 2***
LOVAGE OIL

PROP: The constituents include d-α-terpineol, butyl dihydrophthalides, butyl tetrahydrophthalides, coumarin, aldehydes, and acetic and isovaleric acid. From steam distillation of fresh root of *Levisticum officinale*

L. Koch syn. *Angelica levisticum*, Baillon (Fam. *Umbelliferae*). Yellow to green to brown liquid; strong odor and taste. D: 1.034-1.057, refr index: 1.536-1.554 @ 20°. Sol in fixed oils; sltly sol in mineral oil; insol in glycerin, propylene glycol.

TOXICITY DATA with REFERENCE
skn-rbt 500 mg/24H MOD FCTXAV 16,813,78
skn-gpg 100% MLD FCTXAV 16,813,78
orl-mus LD50:3400 mg/kg FCTXAV 16,637,78

CONSENSUS REPORTS: Reported in EPA TSCA Inventory.

SAFETY PROFILE: Moderately toxic by ingestion. A skin irritant. When heated to decomposition it emits acrid smoke and irritating fumes. See also constituents as listed.

LII300 CAS:80382-23-6 ***HR: 3***
LOXOPROFEN SODIUM DIHYDRATE
mf: $C_{15}H_{17}O_3 \cdot Na \cdot 2H_2O$ mw: 304.35

SYN: CS-600

TOXICITY DATA with REFERENCE
orl-rat TDLo:8 mg/kg (21D preg):REP SKKNAJ 36,1,84
orl-rat LD50:145 mg/kg SKKNAJ 36,1,84
ipr-rat LD50:245 mg/kg SKKNAJ 36,1,84
scu-rat LD50:285 mg/kg SKKNAJ 36,1,84
ivn-rat LD50:155 mg/kg SKKNAJ 36,1,84
orl-mus LD50:3030 mg/kg SKKNAJ 36,1,84
ipr-mus LD50:1020 mg/kg SKKNAJ 36,1,84
scu-mus LD50:1070 mg/kg SKKNAJ 36,1,84
ivn-mus LD50:740 mg/kg SKKNAJ 36,1,84

SAFETY PROFILE: Poison by ingestion, subcutaneous, intravenous, and intraperitoneal routes. Experimental reproductive effects. When heated to decomposition it emits toxic fumes of Na_2O.

LII400 ***HR: 3***
LOZILUREA
mf: $C_{10}H_{13}ClN_2O$ mw: 212.70

SYNS: N-3'-CHLOROBENZYL-N'-ETHYLUREA ◇ ITA 312

TOXICITY DATA with REFERENCE
ipr-rat LD50:418 mg/kg ARZNAD 33,1655,83
orl-mus LD50:3 g/kg ARZNAD 33,1655,83
ipr-mus LD50:328 mg/kg ARZNAD 33,1655,83

SAFETY PROFILE: Poison by intraperitoneal route. Moderately toxic by ingestion. When heated to decomposition it emits toxic fumes of Cl^- and NO_x.

LIJ000 CAS:39456-76-3 ***HR: 3***
LSP 1

SYNS: LAC LSP-1 ◇ OIL-SHALE PYROLYSE LAC LSP-1

TOXICITY DATA with REFERENCE
skn-mus TDLo:80 g/kg/25W-I:ETA GTPPAF 8,175,72

SAFETY PROFILE: Questionable carcinogen with experimental tumorigenic data.

LIK000 ***HR: 1***
LUBRICATING OIL

PROP: Flash p: 315-366°F, d: < 1.00, autoign temp: 783°F.

SYNS: STRAW OIL ◇ LUBRICATING OIL, CYLINDER ◇ LUBRICATING OIL (mainly mineral) ◇ LUBRICATING OIL, MOTORS ◇ LUBRICATING OIL, SPINDLE ◇ LUBRICATING OIL, TURBINE

SAFETY PROFILE: Can cause dermatitis. Combustible when exposed to heat or flame. Incompatible with oxidizing materials. To fight fire, use spray, foam, CO_2, dry chemical. See also PETROLEUM.

LIM000 CAS:3105-97-3 ***HR: 3***
LUCANTHONE METABOLITE
mf: $C_{20}H_{24}N_2O_2S$ mw: 356.52

SYNS: 1-((2-(DIETHYLAMINO)ETHYL)AMINO)-4-(HYDROXYMETHYL)THIOXANTHEN-9-ONE ◇ 1-((2-(DIETHYLAMINO)ETHYL)AMINO)-4-(HYDROXYMETHYL)9H-THIOXANTHEN-9-ONE ◇ HYCANTHON ◇ HYCANTHONE ◇ NSC-134434 ◇ WIN 24933

TOXICITY DATA with REFERENCE
sln-smc 20 μmol/L ENMUDM 7,121,85
oms-hmn:lym 5 mg/L BCPCA6 22,1253,73
par-mus TDLo:50 mg/kg (10D preg):TER EVHPAZ 24,113,78
ims-mus TDLo:180 mg/kg/60D-I:CAR JPETAB 197,703,76
orl-rat LD50:980 mg/kg EJBLAB 1(2),181,74
scu-rat LD50:286 mg/kg EJBLAB 1(2),181,74
ivn-rat LD50:75 mg/kg EJBLAB 1(2),181,74
orl-mus LD50:1120 mg/kg EJBLAB 1(2),181,74
scu-mus LD50:270 mg/kg EJBLAB 1(2),181,74
ims-mus LD50:253 mg/kg JPETAB 200,1,77

CONSENSUS REPORTS: EPA Genetic Toxicology Program.

SAFETY PROFILE: Poison by subcutaneous, intravenous, and intramuscular routes. Moderately toxic by ingestion. Experimental teratogenic effects. Human mutation data reported. Questionable carcinogen with experimental carcinogenic data. When heated to decomposition it emits very toxic fumes of NO_x and SO_x.

LIN000 CAS:13058-67-8 ***HR: 3***
LUCENSOMYCIN
mf: $C_{36}H_{53}NO_{13}$ mw: 707.90

PROP: Crystalline powder. Insol in water, anhydrous

alc, nonpolar solvents; sol in pyridine, dimethyl formamide, Unstable beyond pH 6-8 and to heat, light, or air.

SYNS: ANTIBIOTIC 1163 F.I. ◇ ETRUSCOMICINA ◇ ETRUSCOMYCIN ◇ FI 1163

TOXICITY DATA with REFERENCE
dnd-esc 20 μmol/L MUREAV 89,95,81
orl-mus LD50:1263 mg/kg MEIEDD 11,879,89
ipr-mus LD50:37 mg/kg 85ERAY 2,967,78
ivn-mus LD50:45 mg/kg 85ERAY 2,967,78

SAFETY PROFILE: Poison by intraperitoneal and intravenous routes. Moderately toxic by ingestion. Mutation data reported. Used as an antibiotic. When heated to decomposition it emits toxic fumes of NO_x.

LIN400 CAS:1716-09-2 ***HR: 3***
LUCIJET
mf: $C_{12}H_{19}O_3PS_2$ mw: 306.40

SYNS: BAY 29492 ◇ BAYER 29492 ◇ ENT 25,636 ◇ OM-1455 ◇ O,O-DIETHYL-O-(3-METHYL-4-(METHYLTHIO)PHENYL)PHOSPHOROTHIOATE

TOXICITY DATA with REFERENCE
orl-rat LD50:14 mg/kg PSEBAA 107,908,61
ipr-rat LD50:22 mg/kg PSEBAA 107,908,61
ipr-mus LD50:25 mg/kg PSEBAA 107,908,61
ipr-gpg LD50:30 mg/kg PSEBAA 107,908,61

SAFETY PROFILE: Poison by ingestion and intraperitoneal routes. When heated to decomposition it emits toxic fumes of PO_x and SO_x.

LIN600 CAS:61912-76-3 ***HR: 3***
LUCKNOMYCIN
mf: $C_{61}H_{96}N_2O_{24}$ mw: 1241.59

TOXICITY DATA with REFERENCE
ipr-mus LD50:5 mg/kg JANTAJ 32,79-4,79
scu-mus LD50:200 mg/kg 85GDA2 2,282,80
ivn-mus LD50:10 mg/kg JANTAJ 32,79-4,79

SAFETY PROFILE: Poison by subcutaneous, intravenous and intraperitoneal routes. When heated to decomposition it emits toxic fumes of NO_x.

LIN800 CAS:10262-69-8 ***HR: 3***
LUDIOMIL
mf: $C_{20}H_{23}N$ mw: 277.44

PROP: Mp: 92-94°.

SYNS: 276-Ba ◇ 3-(9,10-DIHYDRO-9,10-ETHANOANTHRACEN-9-YL)PROPYLMETHYLAMINE ◇ MAPROTILINE

TOXICITY DATA with REFERENCE
orl-chd TDLo:26 mg/kg:CNS,CVS BMJOAE 2(6081),260,77
orl-rat LD50:760 mg/kg HEPHD2 55,527,80
ivn-rat LD50:38 mg/kg HEPHD2 55,527,80
orl-mus LD50:660 mg/kg HEPHD2 55,527,80
ivn-mus LD50:31 mg/kg HEPHD2 55,527,80

SAFETY PROFILE: Poison by intravenous route. Moderately toxic by ingestion. Human systemic effects by ingestion: somnolence, coma, and blood pressure elevation. When heated to decomposition it emits toxic fumes of NO_x.

LIN850 ***HR: 2***
LUMNITZERA RACEMOSA Willd., extract excluding roots

PROP: Indian plant belonging to the family *Combretaceae* IJEBA6 22,487,84

TOXICITY DATA with REFERENCE
orl-rat TDLo:150 mg/kg (female 12-14D post):REP IJEBA6 22,487,84
ipr-mus LD50:681 mg/kg IJEBA6 22,487,84

SAFETY PROFILE: Moderately toxic by intraperitoneal route. Experimental reproductive effects. When heated to decomposition it emits acrid smoke and irritating fumes.

LIO600 CAS:1149-99-1 ***HR: 3***
LUNAMYCIN
mf: $C_{15}H_{20}O_4$ mw: 264.35

SYNS: (2′S,3′R,6′R)-DIHYDROXY-2′-(HYDROXYMETHYL)-2′,4′,6′-TRIMETHYL-SPIRO(CYCLOPROPANE-1,5′-(5H)INDEN)-7(6′H)-ONE, 2′,3′-DIHYDRO-3′,6′- ◇ ILLUDIN S ◇ ILLUDINE S ◇ LAMPTEROL

TOXICITY DATA with REFERENCE
pic-esc 200 ng/plate CNREA8 43,2819,83
ivn-mus LD50:30 mg/kg 85GDA2 6,113,81

SAFETY PROFILE: Poison by intravenous route. Mutation data reported. When heated to decomposition it emits acrid smoke and irritating fumes.

LIP000 CAS:64036-86-8 ***HR: 3***
LUNARINE HYDROCHLORIDE
mf: $C_{25}H_{33}N_3O_5{\cdot}ClH$ mw: 492.07

PROP: Alkaloid isolated from *Lunaria biennis* (JAPMA8 39,516,50).

SYN: 22H-BENZOFURO(3A,3-H)(1,5,10)TRIAZACYCLOEICOSINE-3,14,22-TRIONE,4,5,6,7,8,9,10,11,12,13,20A,21,23,24-TETRADECAHYDRO-17,19-ETHENO-, HYDROCHLORIDE

TOXICITY DATA with REFERENCE
ivn-mus LD50:62 mg/kg JAPMA8 39,516,50
ivn-rbt LDLo:70 mg/kg JAPMA8 39,516,50

SAFETY PROFILE: Poison by intravenous route. When heated to decomposition it emits very toxic fumes of HCl and NO_x.

LIQ800 CAS:550-90-3 ***HR: 3***
LUPANINE
mf: $C_{15}H_{24}N_2O$ mw: 248.36

PROP: Racemic lupanine is found in white lupins, d-lupanine is found in blue lupins, l-lupanine has been prepared from the natural racemic form. dl-Form: Orthorhombic prisms from acetone. Mp: 98-99°, bp: 185-195°. Sol in water, alc, ether, chloroform; insol in petr ether. d-Form: Syrupy liquid crystallizing with difficulty in hygroscopic needles. Mp: 40-44°, bp: 190-193°, n (24/D) 1.5444. Freely sol in water, alc, chloroform, ether, sol in petr ether. l-Form: Viscous oil. Bp: 186-188°.

SYNS: LUPANIN ◇ (+)-LUPANINE ◇ d-LUPANINE ◇ 2-OXOSPARTEINE

TOXICITY DATA with REFERENCE
orl-rat LD50:1440 mg/kg AIPTAK 210,27,74
orl-mus LD50:410 mg/kg PLMEAA 50,420,84
ipr-mus LD50:175 mg/kg PLMEAA 50,420,84
ipr-gpg LDLo:22 mg/kg JAGRAC 32,51,26
ivn-gpg LDLo:78 mg/kg PLMEAA 50,420,84

SAFETY PROFILE: Poison by intravenous and intraperitoneal routes. Moderately toxic by ingestion. When heated to decomposition it emits toxic fumes of NO_x.

LIT000 ***HR: 2***
LUPINUS

PROP: Dried plant (JTEHD6 1,887,76).

SYN: LUPIN ◇ LUPINUS ANGUSTIFOLIUS, seed alkaloid mixture

TOXICITY DATA with REFERENCE
orl-rat LD50:2279 mg/kg JJATDK 7,51,87

SAFETY PROFILE: Moderately toxic by ingestiion. When heated to decomposition it emits acrid smoke and irritating fumes.

LIU000 CAS:468-28-0 ***HR: 3***
LUPULONE
mf: $C_{26}H_{38}O_4$ mw: 414.64

PROP: Prisms from 90% methanol. Mp: 92-94°. Bitter taste. Stable in vacuum. Sol in methanol, ethanol, petr ether, hexane, isooctane; sltly sol in neutral or acidic aq solns.

SYNS: B''-ACID ◇ β-BITTER ACID ◇ 3,5-DIHYDROXY-4-ISOVALERYL-2,6,6-TRIS(3-METHYL-2-BUTENYL)-2,4-CYCLO-HEXADIEN-1-ONE ◇ 3,5-DIHYDROXY-2,6,6-TRIS(3-METHYL-2-BUTENYL)-4-(3-METHYL-1-OXOBUTYL)-2,4-CYCLOHEXADIEN-1-ONE ◇ β-LUPULIC ACID ◇ LUPULON

TOXICITY DATA with REFERENCE
orl-rat LD50:1800 mg/kg FEPRA7 8,281,49
ims-rat LD50:330 mg/kg FEPRA7 8,281,49
orl-mus LD50:525 mg/kg ARZNAD 17,79,67
scu-mus LD50:600 mg/kg 85GDA2 8(2),39,82
ims-mus LD50:600 mg/kg FEPRA7 8,281,49
orl-rbt LDLo:1000 mg/kg AIPTAK 82,1,50
orl-gpg LD50:130 mg/kg FEPRA7 8,281,49

SAFETY PROFILE: Poison by ingestion and intramuscular routes. Moderately toxic by subcutaneous route. When heated to decomposition it emits acrid smoke and irritating fumes.

LIU100 ***HR: 2***
LUTAMIN
mf: $C_{10}H_{12}N_2{\cdot}ClH$ mw: 196.70

SYNS: 3-(1-AMINOETHYL)INDOLE HYDROCHLORIDE ◇ α-INDOLAETHYLAMIN SALZSAEURE (GERMAN) ◇ α-INDOLEETHYLAMINE HYDROCHLORIDE

TOXICITY DATA with REFERENCE
scu-rat LDLo:800 mg/kg TKIZAM 36,117,22
scu-mus LDLo:700 mg/kg TKIZAM 36,117,22
scu-frg LDLo:800 mg/kg TKIZAM 36,117,22

SAFETY PROFILE: Moderately toxic by subcutaneous route. When heated to decomposition it emits toxic fumes of NO_x and HCl.

LIU300 CAS:9002-67-9 ***HR: D***
LUTEINIZING HORMONE
mw: 30,000

PROP: A glycoprotein gonadotrophic hormone found in the anterior lobe of the pituitary gland. White powder. Sol in water.

SYNS: ICCSH ◇ ICSH ◇ INTERSTITIAL CELL STIMULATING HORMONE ◇ LH ◇ LUTEINIZING GONADOTROPIC HORMONE ◇ LUTEOZIMAN ◇ LUTROPIN ◇ NIH-LH-B 9 ◇ OVINE PITUITARY INTERSTITIAL CELL STIMULATING HORMONE ◇ PITUITARY LEUTEINIZING HORMONE ◇ PLH

TOXICITY DATA with REFERENCE
scu-rat TDLo:1375 μg/kg (female 12-22D post):REP ENDOAO 86,874,70

SAFETY PROFILE: Experimental reproductive effects.

LIU350 ***HR: D***
LUTEINIZING HORMONE ANTISERUM

SYNS: ANTISERUM TO LUTEINIZING HORMONE ◇ AS-LH ◇ LH ANTISERUM ◇ LHAS ◇ RABBIT ANTISERUM TO OVINE LUTEINIZING HORMONE

TOXICITY DATA with REFERENCE
scu-ham TDLo:1600 mg/kg (female 6D post):TER JRPFA4 29,239,72
scu-mus TDLo:4 g/kg (female 1D pre):REP BIREBV 15,311,76

SAFETY PROFILE: An experimental teratogen. Experimental reproductive effects.

LIU360 CAS:9034-40-6 ***HR: 3***
LUTEINIZING HORMONE-RELEASING HORMONE
mf: $C_{55}H_{75}N_{17}O_{13}$ mw: 1182.33

PROP: Neurohumoral hormone produced in the hypothalamus which stimulates the secretion of the pituitary hormones, LH (luteinizing hormone) and FSH (follicle-stimulating hormone), which in turn produce changes resulting in the induction of ovulation.

SYNS: AY 24034 ◇ CYSTORELIN ◇ FERTIRAL ◇ Gn-RH ◇ GONADORELIN ◇ GONADOTROPIN-RELEASING FACTOR ◇ GONADOTROPIN RELEASING HORMONE ◇ KRYPTOCUR ◇ LH RELEASING FACTOR ◇ LH-RELEASING HORMONE ◇ LH-RF ◇ LHRH ◇ LH-RH ◇ LH-RH/FSH-RH ◇ LRF ◇ LRH ◇ LULIBERIN ◇ LUTEINIZING HORMONE-RELEASING FACTOR ◇ LUTEOSTIMULIN ◇ LUTRELEF ◇ OVARELIN ◇ RELEFACT LH-RH ◇ SYNTHETIC LH-RH

TOXICITY DATA with REFERENCE
scu-wmn TDLo:25 μg/kg (2D pre):REP INJFA3 25,203,80
scu-rat TDLo:5 mg/kg (female 11D post):TER BIREBV 16,333,77

CONSENSUS REPORTS: Reported in EPA TSCA Inventory. EPA Genetic Toxicology Program.

SAFETY PROFILE: An experimental teratogen. Human reproductive effects in women by subcutaneous route: menstrual cycle changes and other unspecified effects. Experimental reproductive effects. Used in the treatment of oligospermia and male infertility. See also LUTEINIZING HORMONE and other luteinizing hormone releasing hormone entries.

LIU380 CAS:71447-49-9 ***HR: 3***
LUTEINIZING HORMONE-RELEASING HORMONE, DIACETATE (SALT)
mf: $C_{55}H_{75}N_{17}O_{13}•2C_2H_4O_2$ mw: 1302.59

SYNS: GONADORELIN DIACETATE ◇ LH-RH ◇ LHRH DIACETATE

TOXICITY DATA with REFERENCE
ivn-rat TDLo:5400 μg/kg (female 17-22D post):REP KSRNAM 20,8871,86
ivn-rbt TDLo:26 mg/kg (female 6-18D post):TER KSRNAM 20,8865,86
ivn-rat LD50:154 mg/kg DRUGAY 6,273,82
ivn-mus LD50:303 mg/kg DRUGAY 6,273,82

SAFETY PROFILE: Poison by intravenous route. An experimental teratogen. Experimental reproductive effects. When heated to decomposition it emits toxic fumes of NO_x.

LIU400 ***HR: 3***
LUTEINIZING HORMONE-RELEASING HORMONE, DIACETATE, TETRAHYDRATE
mf: $C_{55}H_{75}N_{17}O_{13}•2C_2H_4O_2•4H_2O$ mw: 1342.67

SYN: LHRH DIACETATE TETRAHYDRATE

TOXICITY DATA with REFERENCE
ipr-rat TDLo:168 μg/kg (male 9W pre):TER OYYAA2 20,149,80
ipr-rat TDLo:154 μg/kg (male 63D pre):REP OYYAA2 20,149,80
ivn-rat LD50:203 mg/kg OYYAA2 8,605,74
ivn-mus LD50:416 mg/kg OYYAA2 8,605,74

SAFETY PROFILE: Poison by intravenous route. An experimental teratogen. Experimental reproductive effects. When heated to decomposition it emits toxic fumes of NO_x.

LIU420 CAS:57982-77-1 ***HR: 1***
LUTEINIZING HORMONE-RELEASING HORMONE (PIG), 6-(O-(1,1-DIMETHYLETHYL)-d-SERINE)-9-(N-ETHYL-l-PROLINAMIDE)-10-DEGLYCINAMIDE-
mf: $C_{60}H_{86}N_{16}O_{13}$ mw: 1239.62

SYNS: BUSERELIN ◇ HOE 766 ◇ HOE 766A ◇ ICI 123215 ◇ LUTEINIZING HORMONE-RELEASING HORMONE, (d-SER(TBU)^6-EA^{10})- ◇ (d-SER(BU^{t6}))-LH-RH(1-9)NONAPEPTIDE-ETHYLAMIDE ◇ (d-SER(TBU)^6-EA^{10})-LHRH ◇ (d-SER(TBU)^6-EA^{10})-LUTEINIZING HORMONE-RELEASING HORMONE ◇ d-SER(TBU^6)-LH-RH-(1-9)-NONAPEPTIDE ETHYLAMIDE ◇ SUPREFACT

TOXICITY DATA with REFERENCE
ihl-wmn TDLo:1080 μg/kg (female 90D pre):REP LANCAO 2,215,79
ihl-wmn TCLo:1620 μg/kg/13W-I:BPR BMJOAE 294,1101,87

SAFETY PROFILE: Human systemic effects by inhalation: blood pressure elevation. Experimental reproductive effects. When heated to decomposition it emits toxic fumes of NO_x.

LIV000 CAS:21884-44-6 ***HR: 3***
LUTEOSKYRIN
mf: $C_{30}H_{22}O_{12}$ mw: 574.52

PROP: Yellow rectangular crystals. Mp: 278° (decomp). Anthraquinoid hepatotoxin of *Penicillium islandicum sopp* (JJEMAG 41,177,71).

SYNS: 5H,6H-6,5A,13A,14-(1,2,3,4)BUTANETETRAYCYCLOOCTA(1,2-B:5,6-B')DINAPHTHALENE ◇ 8,8'-DIHYDROXY-RUGULOSIN ◇ FLAVOMYCELIN ◇ (−)-LUTEOSKYRIN

TOXICITY DATA with REFERENCE
pic-esc 500 ng/plate CNREA8 43,2819,83
dni-mus:ast 1 mg/L ECREAL 57,19,69
orl-mus TDLo:1356 mg/kg/32W-C:NEO FCTXAV 10,193,72
orl-mus TD:1200 mg/kg/27W-I:ETA NGGKED 32,187,73
orl-mus LD50:220 mg/kg ALLVAR 50,77,62
ipr-mus LD50:41 mg/kg JJEMAG 41,177,71
scu-mus LD50:146 mg/kg JJEMAG 42,91,72
ivn-mus LD50:6650 μg/kg FCTXAV 10,193,72

CONSENSUS REPORTS: IARC Cancer Review: Group 3 IMEMDT 7,56,87; Animal Limited Evidence IMEMDT 10,163,76.

SAFETY PROFILE: Poison by ingestion, intraperitoneal, subcutaneous, and intravenous routes. Questionable carcinogen with experimental carcinogenic and tumorigenic data. Human mutation data reported. When heated to decomposition it emits acrid smoke and irritating fumes.

LIW000 CAS:10099-66-8 ***HR: 3***
LUTETIUM CHLORIDE
mf: Cl_3Lu mw: 281.32

PROP: Colorless crystals. Subl above 750°, mp: 892° ± 2°, sol in water.

TOXICITY DATA with REFERENCE
orl-mus LD50:7074 mg/kg EQSSDX 1,1,75
ipr-mus LD50:315 mg/kg JPMSAE 53,1186,64
ipr-gpg LD50:161 mg/kg AEHLAU 5,437,62

CONSENSUS REPORTS: Reported in EPA TSCA Inventory.

SAFETY PROFILE: Poison by intraperitoneal route. Mildly toxic by ingestion. When heated to decomposition it emits toxic fumes of Cl^-. See also RARE EARTHS.

LIX000 CAS:63917-04-4 ***HR: 3***
LUTETIUM CITRATE

TOXICITY DATA with REFERENCE
ipr-mus LD50:135 mg/kg AEHLAU 5,437,62
ipr-gpg LD50:81 mg/kg AEHLAU 5,437,62

SAFETY PROFILE: Poison by intraperitoneal route. When heated to decomposition it emits acrid smoke and irritating fumes. See also RARE EARTHS.

LIY000 CAS:10099-67-9 ***HR: 3***
LUTETIUM(III) NITRATE (1:3)
mf: $N_3O_9 \cdot Lu$ mw: 361.00

SYN: NITRIC ACID, LUTETIUM(3+) SALT

TOXICITY DATA with REFERENCE
ipr-rat LD50:258 mg/kg EQSSDX 1,1,75
ipr-mus LD50:223 mg/kg EQSSDX 1,1,75

CONSENSUS REPORTS: Reported in EPA TSCA Inventory.

SAFETY PROFILE: Poison by intraperitoneal route. When heated to decomposition emits toxic fumes of NO_x. See also NITRATES and RARE EARTHS.

LJA000 CAS:589-93-5 ***HR: 2***
2,5-LUTIDINE
mf: C_7H_9N mw: 107.17

PROP: Liquid. D: 0.938, bp: 156.2°. Sol in water @ 25 parts cold water; sltly sol in hot water; misc in alc and ether.

SYN: 2,5-DIMETHYLPYRIDINE

TOXICITY DATA with REFERENCE
orl-rat LD50:800 mg/kg HYSAAV 33,341,68
orl-mus LD50:670 mg/kg HYSAAV 33,341,68
orl-gpg LD50:827 mg/kg HYSAAV 33,341,68

CONSENSUS REPORTS: Reported in EPA TSCA Inventory.

SAFETY PROFILE: Moderately toxic by ingestion. When heated to decomposition it emits toxic fumes of NO_x.

LJB000 CAS:583-58-4 ***HR: 3***
3,4-LUTIDINE
mf: C_7H_9N mw: 107.17

PROP: Liquid. Bp: 163.5-164.5°.

SYN: 3,4-DIMETHYLPYRIDINE

TOXICITY DATA with REFERENCE
orl-rat LD50:677 mg/kg AIHAAP 30,470,69
ihl-rat LCLo:500 ppm/4H AIHAAP 30,470,69
skn-rbt LD50:134 mg/kg AIHAAP 30,470,69

CONSENSUS REPORTS: Reported in EPA TSCA Inventory.

SAFETY PROFILE: Poison by skin contact. Moderately toxic by ingestion and inhalation. When heated to decomposition it emits toxic fumes of NO_x.

LJB700 CAS:33390-21-5 ***HR: D***
LYCOPERSIN
mf: $C_{20}H_{14}O_8$ mw: 382.34

SYNS: BIKAVERIN ◇ 6,11-DIHYDROXY-3,8-DIMETHOXY-1-METHYL-10H-BENZO(b)XANTHENE-7,10,12-TRIONE

TOXICITY DATA with REFERENCE
dni-mus:ast 500 μg/L NEOLA4 22,335,75
dni-mus:leu 1400 μg/L NEOLA4 22,335,75

SAFETY PROFILE: Mutation data reported. When heated to decomposition it emits acrid smoke and irritating fumes.

LJB800 CAS:2121-12-2 ***HR: 3***
LYCORANIUM, 1,2,3,3a,6,7,12b,12c-OCTADEHYDRO-2-HYDROXY-
mf: $C_{16}H_{12}NO_3$ mw: 266.29

SYNS: 4,5-DIHYDRO-2-HYDROXY-(1,3)DIOXOLO(4,5-j)PYRROLO (3.2.1-de)PHENANTHRIDINIUM ◇ (1,3)DIOXOLO(4,5-j)PYRROLO(3,2,1-de)PHENANTHRIDINIUM,4,5-DIHYDRO-2-HYDROXY-(9CI) ◇ LYCOBETAINE ◇ 1,2,3,3a,6,7,12b,12c-OCTADEHYDRO-2-HYDROXYLYCORANIUM ◇ UNGEREMINE

TOXICITY DATA with REFERENCE
ipr-mus TDLo:75 mg/kg (male 3D pre):REP PLMEAA 54,114,88
orl-rat LD50:90 mg/kg YHTPAD 23,316,88
ipr-mus LD50:72 mg/kg YHTPAD 23,316,88

SAFETY PROFILE: Poison by ingestion and intraperitoneal route. Experimental reproductive effects. When heated to decomposition it emits toxic fumes of NO_x.

LJC000 CAS:29477-83-6 ***HR: 3***
LYCORCIDINOL
mf: $C_{14}H_{13}NO_7$ mw: 307.28

SYNS: LYCORICIDIN-A ◇ LYCORICIDINOL ◇ NACRICLASINE ◇ NARCICLASINE

TOXICITY DATA with REFERENCE
dnd-mam:lym 30 mg/L FRPSAX 32,67,77
scu-mus LD50:5 mg/kg 85GDA2 8(1),126,82

SAFETY PROFILE: Poison by subcutaneous route. Mutation data reported. When heated to decomposition it emits toxic fumes of NO_x.

LJD500 CAS:4148-16-7 ***HR: 3***
LYCURIM
mf: $C_{10}H_{24}N_2O_8S_2$ mw: 364.48

SYNS: 1,4-BIS(2'-MESYLOXYETHYLAMINO)-1,4-DIDEOXYMESO-ERYTHRITOL ◇ 1,4-DIDEOXY-1,4-BIS((2-HYDROXYETHYL)AMINO) ERYTHRITOL 1,4-DIMETHANESULFONATE (ESTER) ◇ 1,4-DI(MESYLOXYETHYLAMINO)ERYTHRITOL ◇ (R*,S*)-3,14-DIOXA-2,15-DITHIA-6,11-DIAZEHEXADECANE-8,9-DIOL,2,2,15,15-TETRAOXIDE (9CI) ◇ LYKURIM ◇ 1,4-(METHYLSULFONYLOXYETHYLAMINO)-1,4-DIDEOXY-ERYTHRIOLDIMETHYLSULFONATE ◇ NSC-122402 ◇ R 74 BASE ◇ RITROSULFAN

TOXICITY DATA with REFERENCE
mmo-sat 10 mg/plate CNREA8 43,4530,83
mma-sat 10 mg/plate CNREA8 43,4530,83
sce-ham:oth 1300 ng/L CNREA8 43,4530,83
orl-mus LD50:113 mg/kg NCISP* JAN86
ipr-mus LD50:54210 μg/kg NCISP* JAN86

SAFETY PROFILE: Poison by ingestion and intraperitoneal routes. Mutation data reported. When heated to decomposition it emits toxic fumes of SO_x and NO_x.

LJD600 ***HR: 2***
LYGODIUM FLEXUOSUM (LINN.) SWARTZ., EXTRACT

PROP: Indian plant belonging to the family *Schizaeacae* (IJEBA6 22,312,84).

TOXICITY DATA with REFERENCE
orl-rat TDLo:1 g/kg (4-5D preg):REP IJMRAQ 72,597,80
ipr-mus LD50:464 mg/kg IJEBA6 22,312,84

SAFETY PROFILE: Moderately toxic by intraperitoneal route. Experimental reproductive effects.

LJE000 CAS:8015-14-3 ***HR: 3***
LYNDIOL
mf: $C_{21}H_{26}O_2•C_{20}H_{28}O$ mw:594.95

SYNS: LYNESTRENOL mixed with MESTRANOL ◇ LYNESTROL mixed with MESTRANOL ◇ LYNOESTRENOL mixed with MESTRANOL ◇ MESTRANOL mixed with LYNESTRENOL ◇ MESTRANOL mixed with LYNESTROL ◇ NORACYCLINE ◇ OVANON ◇ OVARIOSTAT (FRENCH) ◇ RESTOVAR ◇ SISTOMETRENOL

TOXICITY DATA with REFERENCE
orl-mus TDLo:649 μg/kg (3D pre):TER IRLCDZ 6,307,78
orl-rat TDLo:75 mg/kg (1-5D preg):REP INDRBA 21,141,84
orl-wmn TDLo:32 mg/kg/130W-I:CAR,LIV MJAUAJ 2,223,78
orl-wmn TD:104 mg/kg/10Y-I:NEO,LIV MJAUAJ 2,223,78
orl-wmn TD:91 mg/kg/7Y-I:NEO,LIV NPMDAD 5,3014,76
orl-wmn TD:34 mg/kg/2Y-I:CAR,LIV HEGAD4 29,187,82
orl-wmn TDLo:22 mg/kg/2Y-I:PUL,GIT,MET LANCAO 1,1479,73

CONSENSUS REPORTS: EPA Genetic Toxicology Program.

SAFETY PROFILE: Suspected human carcinogen producing liver tumors. An experimental teratogen. Human systemic effects by ingestion: dyspnea, nausea or vomiting, and fever. Experimental reproductive effects. Used as an oral contraceptive. When heated to decomposition it emits acrid smoke and irritating fumes.

LJE100 CAS:31136-61-5 ***HR: 3***
LYONIATOXIN
mf: $C_{22}H_{34}O_7$ mw: 410.56

SYNS: 2-β,3-β,6-β,7-α)-2,3-EPOXY-GRAYANOTOXANE-5,6,7,10,16-PENTOL-6-ACETATE ◇ LYONIOL A

TOXICITY DATA with REFERENCE
ipr-mus LD50:3010 μg/kg TXAPA9 35,303,76
ivn-gpg LD50:400 μg/kg ARTODN 44,259,80

SAFETY PROFILE: Poison by intravenous and intraperitoneal routes. When heated to decomposition it emits acrid smoke and irritating fumes.

LJE500 CAS:19875-60-6 ***HR: 3***
LYSENYL HYDROGEN MALEATE
mf: $C_{20}H_{26}N_4O•C_4H_4O_4$ mw: 454.58

SYNS: CUVALIT ◇ N'-((8-α)-9,10-DIDEHYDRO-6-METHYLERGOLIN-8-YL)-N,N-DIETHYL-UREA (Z)-2-BETENEDIOATE ◇ 3-(9,10-DIDEHYDRO-6-METHYLERGOLIN-8-YL)-1,1-DIETHYLUREAHYDROGEN MALEATE ◇ 3-(9,10-DIDEHYDRO-6-METHYLERGOLIN-8-α-YL)-1,1-DIETHYLUREA MALEATE (1:1) ◇ LISURIDE HYDROGEN MALEATE ◇ LYSENYL ◇ LYSENYL BIMALEATE

TOXICITY DATA with REFERENCE
orl-rat TDLo:242 μg/kg (lactating female 4D post):REP CCCCAK 49,2828,84
orl-mus TDLo:75 mg/kg (multi) :TER KSRNAM 15,2346,81
orl-hmn TDLo:714 mg/kg:CNS,GIT ARZNAD 16,220,66
ivn-man TDLo:1429 ng/kg/3M-C:CVS,GIT CHETBF 91,792,87
orl-rat LD50:138 mg/kg KSRNAM 15,1165,81
scu-rat LD50:12200 μg/kg KSRNAM 15,1165,81
ivn-rat LD50:2900 μg/kg KSRNAM 15,1165,81
orl-mus LD50:405 mg/kg KSRNAM 15,1165,81
scu-mus LD50:530 mg/kg KSRNAM 15,1165,81
ivn-mus LD50:14800 μg/kg KSRNAM 15,1165,81
orl-rbt LD50:123 mg/kg KSRNAM 15,1165,81

SAFETY PROFILE: Poison by ingestion, subcutaneous, and intravenous routes. Human systemic effects by ingestion: headache, nausea or vomiting, cardiac changes, sweating. Experimental reproductive effects. An experimental teratogen. When heated to decomposition it emits toxic fumes of NO_x.

LJF000 ***HR: 3***
d-LYSERGIC ACID DIETHYLAMIDE
mf: $C_{20}H_{25}N_3O$ mw: 323.48

SYN: LYSERGIC ACID DIETHYLAMIDE, 1-ISOMER

TOXICITY DATA with REFERENCE
ivn-rbt LD50:17 mg/kg 27ZQAG -,98,72

SAFETY PROFILE: Poison by intravenous route. When heated to decomposition it emits toxic fumes of NO_x.

LJG000 ***HR: 3***
d-LYSERGIC ACID DIETHYLAMIDE TARTRATE
mf: $C_{40}H_{50}N_6O_2 \cdot C_4H_6O_6 \cdot 2CH_4O$ mw: 861.16

SYNS: 9,10-DIDEHYDRO-N,N-DIETHYL-6-METHYL-ERGOLINE-8-β-CARBOXAMIDE-d- TARTRATE with METHANOL (1:2) ◇ LSD TARTRATE ◇ LYSERGIC ACID DIETHYLAMIDE TARTRATE

TOXICITY DATA with REFERENCE
cyt-hmn:lym 10 mg/L/72H MUREAV 51,403,78
dlt-mus-ipr 10 μg/kg MUREAV 26,517,74
scu-rat TDLo:5 μg/kg (4D preg):REP JPETAB 173,48,70
ipr-mus TDLo:2 μg/kg (female 7D post):TER SCIEAS 157,1325,67
ipr-rat LDLo:5 mg/kg JPMSAE 60,304,71

SAFETY PROFILE: Poison by intraperitoneal route. An experimental teratogen. Experimental reproductive effects. Human mutation data reported. When heated to decomposition it emits toxic fumes of NO_x.

LJH000 CAS:4238-84-0 ***HR: 3***
d-LYSERGIC ACID DIMETHYLAMIDE
mf: $C_{18}H_{21}N_3O$ mw: 295.42

SYNS: DAM-57 ◇ 9,10-DIDEHYDRO-N,N,6-TRIMETHYLERGOLINE-8-β-CARBOXAMIDE

TOXICITY DATA with REFERENCE
orl-hmn TDLo:10 μg/kg:CNS,CVS PSYPAG 1,20,59
ivn-rbt LD50:400 μg/kg 27ZQAG -,95,72

SAFETY PROFILE: Poison by intravenous route. Human systemic effects by ingestion: hallucinations, distorted perceptions, toxic psychosis, arteriolar or venous dilation. When heated to decomposition it emits toxic fumes of NO_x.

LJI000 CAS:478-99-9 ***HR: 3***
LYSERGIC ACID ETHYLAMIDE
mf: $C_{18}H_{21}N_3O$ mw: 295.42

SYNS: 9,10-DIDEHYDRO-N-ETHYL-6-METHYLERGOLINE-8-β-CARBOXAMIDE, N-ETHYLLYSERGAMIDE ◇ LAE-32 ◇ d-LYSERGIC ACID MONOETHYLAMIDE

TOXICITY DATA with REFERENCE
orl-hmn TDLo:20 μg/kg:CNS PSYPAG 1,20,59
ivn-mus LD50:44 mg/kg 27ZQAG -,98,72
ivn-rbt LD50:900 μg/kg 27ZQAG -,98,72

SAFETY PROFILE: Poison by intravenous route. Human systemic effects by ingestion: hallucinations, distorted perceptions, toxic psychosis. When heated to decomposition it emits toxic fumes of NO_x.

LJJ000 CAS:4314-63-0 ***HR: 3***
LYSERGIC ACID MORPHOLIDE
mf: $C_{20}H_{23}N_3O_2$ mw: 337.46

SYNS: LSM-775 ◇ d-LYSERGIC ACID MORPHOLIDE

TOXICITY DATA with REFERENCE
orl-hmn TDLo:9 μg/kg:CNS,CVS PSYPAG 1,20,59
ivn-rbt LD50:700 μg/kg 27ZQAG -,97,72

SAFETY PROFILE: Poison by intravenous route. Human systemic effects by ingestion: hallucinations, distorted perceptions, toxic psychosis, arteriolar or venous dilation. When heated to decomposition it emits toxic fumes of NO_x.

LJK000 CAS:63938-26-1 ***HR: 3***
d-LYSERGIC ACID MORPHOLIDE, TARTARIC ACID SALT

SYN: 9,10-DIDEHYDRO-6-METHYLERGOLINE-8-β-CARBOXYLIC ACID MORPHOLIDE, TARTARIC ACID SALT

TOXICITY DATA with REFERENCE
orl-hmn TDLo:1 μg/kg:CNS JPETAB 120,340,57
ivn-mus LD50:55.5 mg/kg JPETAB 120,340,57
ivn-cat LDLo:6400 μg/kg JPETAB 120,340,57
ivn-rbt LDLo:400 μg/kg JPETAB 120,340,57

SAFETY PROFILE: Poison by intravenous route. Human systemic effects by ingestion: central nervous system effects.

LJL000 CAS:60-79-7 ***HR: 3***
d-LYSERGIC ACID-l,2-PROPANOLAMIDE
mf: $C_{19}H_{23}N_3O_2$ mw: 325.45

SYNS: BASERGIN ◇ CORNOCENTIN ◇ 9,10-DIDEHYDRO-N-(α-(HYDROXYMETHYL)ETHYL)-6-METHYLERGOLINE -8-β-CARBOXAMIDE ◇ ERGOATETRINE ◇ ERGOBASINE ◇ ERGOKLININE ◇ ERGOMETRINE ◇ ERGONOVINE ◇ ERGOTOCINE ◇ ERGOTRATE ◇ ERMETRINE ◇ N-(α-(HYDROXYMETHYL)ETHYL)-d-LYSERGOMIDE ◇ N-(1-(HYDROXYMETHYL)ETHYL)-d-LYSERGOMIDE ◇ d-LYSERGIC ACID-1-HYDROXYMETHYLETHYLAMIDE ◇ LYSERGIC ACID PROPANOLAMIDE ◇ MARGONOVINE ◇ NEOFEMERGEN ◇ SECACORNIN ◇ SECOMETRIN ◇ SYNTOMETRINE

TOXICITY DATA with REFERENCE
ims-inf TDLo:93 mg/kg:CNS,PUL SAMJAF 46,2052,72
ims-inf TDLo:138 mg/kg:CNS,PUL ADCHAK 55,68,72
unr-inf TDLo:176 μg/kg:CNS,PUL JOGBAS 79,764,72
ivn-mus LD50:144 mg/kg JPETAB 105,130,52
ivn-gpg LDLo:80 mg/kg 27ZIAQ -,107,73

SAFETY PROFILE: Poison by intravenous route. Human systemic effects by intramuscular route: convulsions, excitement, motor activity changes, cyanosis, and respiratory depression. When heated to decomposition it emits toxic fumes of NO_x.

LJM000 CAS:2385-87-7 ***HR: 3***
LYSERGIC ACID PYROLIDATE
mf: $C_{20}H_{23}N_3O$ mw: 321.46

SYNS: LSD-25-PYRROLIDATE ◇ d-LYSERGIC ACID PYRROLIDIDE

TOXICITY DATA with REFERENCE
orl-hmn TDLo:10 μg/kg:CNS,CVS PSYPAG 1,20,59
ivn-mus LD50:46 mg/kg 27ZQAG -,99,72
ivn-rbt LD50:400 μg/kg 27ZQAG -,99,72

SAFETY PROFILE: Poison by intravenous route. Human systemic effects by ingestion: hallucinations, distorted perceptions, toxic psychosis, arteriolar or venous dilation. When heated to decomposition it emits toxic fumes of NO_x.

LJM600 CAS:17676-08-3 ***HR: D***
LYSERGIDE TARTRATE
mf: $C_{20}H_{25}N_3O \cdot 1/2C_4H_6O_6$ mw: 398.50

SYNS: ERGOLINE-8-β-CARBOXAMIDE, 9,10-DIDEHYDRO-N,N-DIETHYL-6-METHYL-, TARTRATE (2:1) ◇ LSD TARTRATE ◇ LSD 25 TARTRATE ◇ (+)-LSD TARTRATE ◇ d-LSD TARTRATE

TOXICITY DATA with REFERENCE
orl-rbt TDLo:300 μg/kg (female 7-9D post):TER LANCAO 1,639,68

SAFETY PROFILE: An experimental teratogen. When heated to decomposition it emits toxic fumes of NO_x.

LJM700 CAS:56-87-1 ***HR: D***
LYSINE
mf: $C_6H_{14}N_2O_2$ mw: 146.22

PROP: l-Lysine: Needles from water, hexagonal plates from dil alc. Darkens at 210°; decomp 224.5°. Very freely sol in water; very sltly sol in alc; practically insol in ether.

SYNS: AMINUTRIN ◇ α,epsilon-DIAMINOCAPROIC ACID ◇ 2,6-DIAMINOHEXANOIC ACID ◇ l-LYSINE (9CI) ◇ l-(+)-LYSINE ◇ LYSINE ACID

TOXICITY DATA with REFERENCE
orl-rat TDLo:138 g/kg (5-15D preg):TER JONUAI 74,93,61
orl-rat TDLo:138 g/kg (5-15D preg):REP JONUAI 74,93,61

CONSENSUS REPORTS: EPA Genetic Toxicology Program. Reported in EPA TSCA Inventory.

SAFETY PROFILE: An experimental teratogen. Experimental reproductive effects. When heated to decomposition it emits toxic fumes of NO_x.

LJM800 CAS:57282-49-2 ***HR: 2***
l-LYSINE ACETATE
mf: $C_6H_{14}N_2O_2 \cdot C_2H_4O_2$ mw: 206.28

SYN: l-LYSINE, MONOACETATE

TOXICITY DATA with REFERENCE
orl-rat LD50:14500 mg/kg IYKEDH 12,933,81
ipr-rat LD50:3950 mg/kg IYKEDH 12,933,81
scu-rat LD50:4000 mg/kg IYKEDH 12,933,81
ivn-rat LD50:3650 mg/kg IYKEDH 12,933,81
orl-mus LD50:14400 mg/kg IYKEDH 12,933,81
ipr-mus LD50:5100 mg/kg IYKEDH 12,933,81
scu-mus LD50:5800 mg/kg IYKEDH 12,933,81
ivn-mus LD50:4350 mg/kg IYKEDH 12,933,81

CONSENSUS REPORTS: Reported in EPA TSCA Inventory.

SAFETY PROFILE: Moderately toxic by intraperitoneal, intravenous, and subcutaneous routes. Mildly toxic by ingestion. When heated to decomposition it emits toxic fumes of NO_x.

LJN000 CAS:7274-88-6 ***HR: 1***
d-LYSINE HYDROCHLORIDE
mf: $C_6H_{14}N_2O_2 \cdot ClH$ mw: 182.68

TOXICITY DATA with REFERENCE
ipr-rat LD50:4750 mg/kg ABBIA4 64,319,56

CONSENSUS REPORTS: Reported in EPA TSCA Inventory.

SAFETY PROFILE: When heated to decomposition it emits very toxic fumes of HCl and NO_x.

LJO000 CAS:657-27-2 ***HR: 1***
l-LYSINE MONOHYDROCHLORIDE
mf: $C_6H_{14}N_2O_2{\bullet}ClH$ mw: 182.68

PROP: White powder. Mp: 235-236°. Sol in water; insol in alc and ether. Crystals from dil ethanol. Mp: 263-264° (decomp) when anhyd.

SYNS: 2,6-DIAMINOHEXANOIC ACID HYDROCHLORIDE ◇ l-LYSINE HYDROCHLORIDE ◇ LYSINE MONOHYDROCHLORIDE

TOXICITY DATA with REFERENCE
orl-rat LD50:10 g/kg JPMSAE 62,49,73
ipr-rat LD50:4019 mg/kg ABBIA4 58,253,55

CONSENSUS REPORTS: Reported in EPA TSCA Inventory.

SAFETY PROFILE: Mildly toxic by ingestion. When heated to decomposition it emits very toxic fumes of HCl and NO_x.

LJP000 CAS:55898-33-4 ***HR: 3***
LYSOCELLIN
mf: $C_{34}H_{60}O_{10}$ mw: 628.94

TOXICITY DATA with REFERENCE
orl-mus LD50:350 mg/kg 37ASAA 3,61,78
ipr-mus LD50:65 mg/kg 37ASAA 3,61,78

SAFETY PROFILE: Poison by ingestion and intraperitoneal routes. When heated to decomposition it emits acrid smoke and irritating fumes.

LJP500 CAS:12772-68-8 ***HR: 1***
LYSOL

TOXICITY DATA with REFERENCE
orl-wmn TDLo:5 g/kg:CVS,PUL,KID WILEAR 27,1211,74
orl-wmn TDLo:5 g/kg:KID,MET WILEAR 27,1211,74
orl-wmn LDLo:2 g/kg:CNS,GIT BEGMA5 13,56,35

SAFETY PROFILE: Mildly toxic to humans by ingestion. Human systemic effects by ingestion of large amounts: general anesthetic, coma, vascular relaxation, respiratory system, gastrointestinal system, renal failure, decreased urine volume and metabolic acidosis.

LJQ000 CAS:9011-93-2 ***HR: 2***
LYSOSTAPHIN

PROP: Structure consists of a single polypeptide chain.

TOXICITY DATA with REFERENCE
ivn-rat LD50:530 mg/kg MEIEDD 10,807,83
ivn-mus LD50:820 mg/kg MEIEDD 10,807,83

SAFETY PROFILE: Moderately toxic by intravenous route. Used as an antibacterial enzyme.

LJR000 CAS:147-20-6 ***HR: 3***
LYSSIPOLL
mf: $C_{19}H_{23}NO$ mw: 281.43

SYNS: ALLERGEN ◇ AN 1041 ◇ BELFENE ◇ 4-(BENZHYDRYLOXY)-1-METHYLPIPERIDINE ◇ DAFEN ◇ DAYFEN ◇ DIAFEN ◇ 4-(DIPHENYLMETHOXY)-1-METHYLPIPERIDINE ◇ DIPHENYLPYRALINE ◇ DIPHENYLPYRILENE ◇ HISPRIL ◇ HISTRYL ◇ HISTYN ◇ MEPIBEN ◇ N-METHYLPIPERIDYL-(4)-BENZHYDRYLAETHER SALZSAUREN SALZE (GERMAN) ◇ NEARGAL ◇ P 253

TOXICITY DATA with REFERENCE
eye-rbt 1 mg MLD TXAPA9 50,459,79
orl-mus LD50:250 mg/kg THERAP 26,155,71
ivn-mus LD50:42 mg/kg THERAP 26,1203,71

SAFETY PROFILE: Poison by ingestion and intravenous routes. An eye irritant. When heated to decomposition it emits toxic fumes of NO_x.

LJS000 CAS:80-50-2 ***HR: 3***
LYTISPASM
mf: $C_{17}H_{32}NO_2{\bullet}Br$ mw: 362.41

PROP: Crystals from acetone. Mp: 329°.

SYNS: ANISOTROPINE METHOBROMIDE ◇ ANITSOTROPINE METHYLBROMIDE ◇ endo-8,8-DIMETHYL-3-((1-OXO-2-PROPYLPENTYL)OXY)-8-AZONIABICYCLO(3.2.1)OCTANE BROMIDE ◇ 3-α-HYDROXY-8-METHYL-1-α-H,5-H-TROPANIUM BROMIDE 2-PROPYLVALERATE ◇ 8-METHYL-3-(2-PROPYLPENTANOYLOXY) TROPINIUM BROMIDE ◇ 8-METHYLTROPINIUM BROMIDE 2-PROPYLVALERATE ◇ OCTATROPINE METHYLBROMIDE ◇ 2-PROPYLPENTANOYLTROPINIUM METHYLBROMIDE ◇ VALPIN ◇ VAPIN

TOXICITY DATA with REFERENCE
orl-rat LD50:705 mg/kg NIIRDN 6,24,82
orl-mus LD50:850 mg/kg NIIRDN 6,24,82
ipr-mus LD50:129 mg/kg NIIRDN 6,24,82
scu-mus LD50:133 mg/kg NIIRDN 6,24,82
ivn-mus LD50:6300 μg/kg OYYAA2 2,70,68

SAFETY PROFILE: Poison by subcutaneous, intravenous, and intraperitoneal routes. Moderately toxic by ingestion. When heated to decomposition it emits toxic fumes of Br^- and NO_x.

M

MAB050 CAS:64083-08-5 ***HR: 3***
M-4212
mf: $C_{16}H_{15}N_3O_4$ mw: 313.34

SYN: M 4212 (pesticide) (9CI)

TOXICITY DATA with REFERENCE
mmo-sat 300 ng/plate ENMUDM 3,499,81
dni-mus:ast 500 μg/L CCPHDZ 4,61,80
oms-mus:ast 500 μg/L CCPHDZ 4,61,80
ipr-rat LD50:6500 μg/kg CCPHDZ 4,61,80
ipr-mus LD50:10000 μg/kg CCPHDZ 4,61,80

SAFETY PROFILE: Poison by intraperitoneal route. Mutation data reported. When heated to decomposition it emits toxic fumes of NO_x.

MAB055 CAS:54824-20-3 ***HR: 3***
M-12210
mf: $C_{18}H_{17}N_3O_4$ mw: 339.38

SYN: 2-(2-(PYRROLIDINYL)ETHYL)-5-NITRO-1H-BENZ(de)ISOQUINOLINE-1,3(2H)-DIONE

TOXICITY DATA with REFERENCE
mmo-sat 1 μg/plate ENMUDM 3,499,81
dni-mus:ast 3 mg/L CCPHDZ 4,61,80
oms-mus:ast 3 mg/L CCPHDZ 4,61,80
ipr-rat LD50:4500 μg/kg CCPHDZ 4,61,80
ipr-mus LD50:12600 μg/kg CCPHDZ 4,61,80

SAFETY PROFILE: Poison by intraperitoneal route. Mutation data reported. When heated to decomposition it emits toxic fumes of NO_x.

MAB250 CAS:64431-68-1 ***HR: 3***
MA144 MI
mf: $C_{42}H_{55}NO_{15}$ mw: 813.98

SYN: ANTIBIOTIC MA 144M1

TOXICITY DATA with REFERENCE
dni-mus:leu 470 nmol/L JANTAJ 34,1596,81
oms-mus:leu 43 nmol/L JANTAJ 34,1596,81
ipr-mus LD50:33 mg/kg JANTAJ 31,78-93,78
ivn-mus LD50:30 mg/kg JANTAJ 31,78-93,78

SAFETY PROFILE: Poison by intraperitoneal and intravenous routes. Mutation data reported. When heated to decomposition it emits toxic fumes of NO_x.

MAB300 ***HR: 3***
MABUTEROL
mf: $C_{13}H_{18}ClF_3N_2O \cdot ClH$ mw: 347.24

SYNS: dl-4-AMINO-α-(tert-BUTYLAMINO)-3-CHLORO-5-(TRIFLUOROMETHYL)PHENETHYL ALCOHOL HCl ◇ dl-1-(4-AMINO-3-CHLORO-5-TRIFLUOROMETHYLPHENYL)-2-tert-BUTYLAMINO-ETHANOL HYDROCHLORIDE

TOXICITY DATA with REFERENCE
orl-rat TDLo:2700 μg/kg (female 17-22D post):REP ARZNAD 34,1687,84
orl-hmn TDLo:1143 mg/kg:CNS ARZNAD 34,1697,84
orl-hmn TDLo:60 μg/kg/6W:CNS,CVS ARZNAD 34,1699,84
orl-rat LD50:306 mg/kg ARZNAD 34,1680,84
ipr-rat LD50:76300 μg/kg ARZNAD 34,1680,84
scu-rat LD50:117 mg/kg ARZNAD 34,1680,84
ivn-rat LD50:26400 μg/kg ARZNAD 34,1680,84
orl-mus LD50:200 mg/kg ARZNAD 34,1680,84
ipr-mus LD50:60 mg/kg ARZNAD 34,1680,84
scu-mus LD50:113 mg/kg ARZNAD 34,1680,84
ivn-mus LD50:41500 μg/kg ARZNAD 34,1680,84
orl-dog LD50:229 mg/kg ARZNAD 34,1680,84
ivn-dog LD50:27500 μg/kg ARZNAD 34,1680,84
orl-rbt LD50:243 mg/kg ARZNAD 34,1680,84
ivn-rbt LD50:18300 μg/kg ARZNAD 34,1680,84

SAFETY PROFILE: Poison by ingestion, subcutaneous, intravenous, and intraperitoneal routes. Human systemic effects by ingestion: tremors, headache and pulse rate increase without fall in blood pressure. Experimental reproductive effects. A bronchodilator. When heated to decomposition it emits toxic fumes of F^-, NO_x, and HCl. See also ALCOHOLS.

MAB400 CAS:73341-73-8 ***HR: 3***
MACBECIN II
mf: $C_{30}H_{44}N_2O_8 \cdot H_2O$ mw: 578.78

SYNS: C-14919 E-2 ◇ NSC 330500

TOXICITY DATA with REFERENCE
dnd-mus:leu 60 μmol/L PAACA3 24,321,83
dni-mus:leu 100 nmol/L PAACA3 24,321,83
oms-mus:leu 10 μmol/L PAACA3 24,321,83
ipr-mus LD50:25 mg/kg JANTAJ 33,205,80

SAFETY PROFILE: Poison by intraperitoneal route. Mutation data reported. When heated to decomposition it emits toxic fumes of NO_x.

MAB750 CAS:12634-34-3 ***HR: 3***
MACROMOMYCIN

PROP: Polypeptide antitumor antibiotic produced by *Streptomyces macromomyceticus* (JANTAJ 29,415,76).

SYN: MCR

TOXICITY DATA with REFERENCE
pic-esc 50 ng/plate CNREA8 43,2819,83
dnd-omi 2500 μg/L MOPMA3 17,388,80
dnd-omi 16 μmol/L CNREA8 42,1555,82
dnd-hmn:hla 10 ng/L BBRCA9 83,908,78
dni-hmn:hla 740 ng/L JANTAJ 31,875,78
ipr-mus LD50:5100 μg/kg JANTAJ 29,415,76
ivn-mus LD50:35 mg/kg JANTAJ 32,330,79

SAFETY PROFILE: Poison by intraperitoneal and intravenous routes. Mutation data reported. When heated to decomposition it emits acrid smoke and irritating fumes.

MAC000 CAS:13009-99-9 ***HR: 2***
MAFENIDE ACETATE
mf: $C_7H_{10}N_2O_2S•C_2H_4O_2$ mw: 246.31

SYNS: α-AMINO-p-TOLUENESULFONAMIDE,MONOACETATE ◇ SULFAMYLON ACETATE

TOXICITY DATA with REFERENCE
scu-rat TDLo:7 g/kg (8-14D preg):TER NICHAS 32,973,73
scu-rat TDLo:7 g/kg (8-14D preg):REP NICHAS 32,973,73
orl-rat LD50:9212 mg/kg GTPZAB 31(8),50,87
ipr-rat LD50:1560 mg/kg IYKEDH 10,884,79
ivn-rat LD50:1730 mg/kg IYKEDH 10,884,79
orl-mus LD50:10183 mg/kg GTPZAB 31(8),50,87
ipr-mus LD50:2322 mg/kg GTPZAB 31(8),50,87
ivn-mus LD50:1580 mg/kg LIFSAK 5,2279,66

SAFETY PROFILE: Moderately toxic by intravenous and intraperitoneal routes. An experimental teratogen. Experimental reproductive effects. Used as an antibacterial agent. When heated to decomposition it emits very toxic fumes of NO_x and SO_x.

MAC250 CAS:632-99-5 ***HR: D***
MAGENTA
mf: $C_{20}H_{19}N_3•ClH$ mw: 337.88

PROP: Green powder or greenish crystals with a bronze luster, faint odor. D: 1.22, mp: decomp >200°. Sol in water, alc, and HCl; insol in ether.

SYNS: FUCHSIN ◇ ORIENT BASIC MAGENTA ◇ ROSANILINE CHLORIDE ◇ ROSANILINE HYDROCHLORIDE ◇ ROSANILINIUM CHLORIDE

TOXICITY DATA with REFERENCE
mma-sat 32 μg/plate MUREAV 89,21,81

CONSENSUS REPORTS: IARC Cancer Review: Group 3 IMEMDT 7,238,87; Animal Inadequate Evidence IMEMDT 4,57,74; Human Inadequate Evidence IMEMDT 4,57,74. Reported in EPA TSCA Inventory.

SAFETY PROFILE: Questionable carcinogen. Mutation data reported. When heated to decomposition it emits very toxic fumes of HCl and NO_x.

MAC500 CAS:3248-93-9 ***HR: 3***
MAGENTA BASE
mf: $C_{20}H_{19}N_3$ mw: 301.42

SYNS: C.I. SOLVENT RED 41 ◇ FUCHSINE BASE ◇ ROSANILINE BASE ◇ WAXOLINE RED A

TOXICITY DATA with REFERENCE
skn-hmn 3 mg/3D-I MLD 85DKA8 -,127,77
mmo-esc 5 g/L MUREAV 130,97,84
otr-ham:emb 2 mg/L NCIMAV 58,243,81
orl-rbt LDLo:150 mg/kg CRSBAW 138,201,44

CONSENSUS REPORTS: Reported in EPA TSCA Inventory.

SAFETY PROFILE: Poison by ingestion. A human skin irritant. Mutation data reported. When heated to decomposition it emits toxic fumes of NO_x.

MAC600 CAS:102629-86-7 ***HR: 2***
MAGNAMYCIN HYDROCHLORIDE
mf: $C_{42}H_{67}NO_{16}•ClH$ mw: 878.56

SYN: (12R,13S)-9-DEOXY-12,13-EPOXY-12,13-DIHYDRO-9-OXO-3-ACETATE-4B-(3-METHYLBUTANOATE) LEUCOMYCIN V HYDROCHLORIDE

TOXICITY DATA with REFERENCE
ivn-rat LD50:700 mg/kg 85ERAY 1,64,78
scu-mus LD50:2950 mg/kg 85ERAY 1,64,78
ivn-mus LD50:550 mg/kg 85ERAY 1,64,78
ims-mus LD50:900 mg/kg 85ERAY 1,64,78

SAFETY PROFILE: Moderately toxic by subcutaneous, intravenous, and intramuscular routes. When heated to decomposition it emits toxic fumes of NO_x and HCl.

MAC650 CAS:546-93-0 ***HR: 1***
MAGNESITE
mf: $CO_3•Mg$ mw: 84.32

PROP: Very light, white powder; odorless. D: 3.04; decomp @ 350°. Sol in acids; insol in water and alc.

SYNS: CARBONATE MAGNESIUM ◇ CARBONIC ACID, MAGNESIUM SALT ◇ C.I. 77713 ◇ DCI LIGHT MAGNESIUM CARBONATE ◇ HYDROMAGNESITE ◇ MAGMASTER ◇ MAGNESIA ALBA ◇ MAGNESIUM CARBONATE ◇ MAGNESIUM(II) CARBONATE (1 : 1) ◇ MAGNESIUM CARBONATE, PRECIPITATED ◇ STAN-MAG MAGNESIUM CARBONATE

CONSENSUS REPORTS: Reported in EPA TSCA Inventory.

OSHA PEL: TWA Total Dust: 15 mg/m^3; Respirable Fraction: 5 mg/m^3
ACGIH TLV: TWA (nuisance particulate) 10 mg/m^3 of total dust (when toxic impurities are not present, e.g., quartz < 1%).

SAFETY PROFILE: Incompatible with formaldehyde. When heated to decomposition it emits acrid smoke and irritating fumes. See also MAGNESIUM COMPOUNDS.

MAC750 CAS:7439-95-4 ***HR: 3***
MAGNESIUM
DOT: UN 1418/UN 1869/UN 2950
af: Mg aw: 24.31

PROP: Hexagonal, silvery-white crystals. Mp: 651°, bp: 1100°, d: 1.74 @ 5°, d: 1.738 @ 20°, vap press: 1 mm @ 621°.

SYNS: MAGNESIO (ITALIAN) ◇ MAGNESIUM BORINGS ◇ MAGNESIUM CLIPPINGS (DOT) ◇ MAGNESIUM GRANULES COATED, particle size not less than 149 microns (DOT) ◇ MAGNESIUM METAL (DOT) ◇ MAGNESIUM PELLETS ◇ MAGNESIUM POWDER (DOT) ◇ MAGNESIUM RIBBONS ◇ MAGNESIUM SCALPINGS (DOT) ◇ MAGNESIUM SCRAP (DOT) ◇ MAGNESIUM SHAVINGS (DOT) ◇ MAGNESIUM SHEET ◇ MAGNESIUM TURNINGS (DOT)

TOXICITY DATA with REFERENCE
orl-dog LDLo:230 mg/kg 14CYAT 2,1077,63

CONSENSUS REPORTS: Reported in EPA TSCA Inventory.

DOT Classification: Label: Flammable Solid and Dangerous When Wet.

SAFETY PROFILE: Poison by ingestion. Inhalation of dust and fumes can cause metal fume fever. The powdered metal ignites readily on the skin causing burns. Particles embedded in the skin can produce gaseous blebs which heal slowly.

A dangerous fire hazard in the form of dust or flakes when exposed to flame or oxidizing agents. In solid form, magnesium is difficult to ignite because heat is conducted rapidly away from the source of ignition; it must be heated above its melting point before it will burn. However, in finely divided form, it may be ignited by a spark or the flame of a match. Magnesium fires do not flare up violently unless there is moisture present. Therefore, it must be kept away from water, moisture, etc. It may be ignited by a spark, match flame, or even spontaneously when the material is finely divided and damp, particularly with water-oil emulsion. Moderately explosive in the form of dust when exposed to flame. Also, magnesium reacts with moisture, acids, etc., to evolve hydrogen, a highly dangerous fire and explosion hazard.

Explosive reaction or ignition with calcium carbonate + hydrogen + heat; gold cyanide + heat; mercury cyanide + heat; silver oxide + heat; fused nitrates; phosphates; or sulfates (e.g., ammonium nitrate, metal nitrates); chloroformamidinium nitrate + water (when ignited with powder). The powder may explode on contact with halocarbons (e.g., chloromethane; chloroform; or carbon tetrachloride); and explodes when sparked in dichlorodifluoromethane. Hypergolic reaction with nitric acid + 2-nitroaniline. Mixtures of powdered magnesium and methanol are more powerful than some military explosives. Mixtures of magnesium powder + water can be detonated. Reacts with acetylenic compounds including traces of acetylene found in ethylene gas to form explosive magnesium acetylide.

Violent reactions with ammonium salts; chlorate salts; beryllium fluoride; boron diiodophosphide; carbon tetrachloride + methanol; 1,1,1-trichloroethane; 1,2-dibromoethane; halogens or interhalogens (e.g., fluorine; chlorine; bromine; iodine vapor; chlorine trifluoride; iodine heptafluoride); hydrogen iodide; metal oxides + heat (e.g., beryllium oxide; cadmium oxide; copper oxide; mercury oxide; molybdenum oxide; tin oxide; zinc oxide); nitrogen (when ignited); silicon dioxide powder + heat; polytetrafluoroethylene powder + heat; sulfur + heat; tellurium + heat; barium peroxide; nitric acid vapor; hydrogen peroxide; ammonium nitrate; sodium iodate + heat; sodium nitrate + heat; dinitrogen tetraoxide (when ignited); lead dioxide. Ignites in carbon dioxide at 780°C; molten barium carbonate + water; fluorocarbon polymers + heat; carbon tetrachloride or trichloroethylene (on impact); dichlorodifluoromethane + heat;

Incompatible with ethylene oxide; metal oxosalts; oxidants; potassium carbonate; Al + $KClO_4$; [$Ba(NO_3)_2$ + BaO_2 + Zn]; bromobenzyl trifluoride; CaC; carbonates; $CHCl_3$; [$CuSO_4$ (anhydrous) + NH_4NO_3 + $KClO_3$ + H_2O]; $CuSO_4$; (H_2 + $CaCO_3$); CH_3Cl; NO_2; liquid oxygen; metal cyanides (e.g., cadmium cyanide; cobalt cyanide; copper cyanide; lead cyanide; nickel cyanide; zinc cyanide); performic acid; phosphates; $KClO_3$; $KClO_4$; $AgNO_3$; $NaClO_4$; (Na_2O_2 + CO_2); sulfates; trichloroethylene; Na_2O_2.

To fight fire, operators and fire fighters can approach a magnesium fire to within a few feet if no moisture is present. Water and ordinary extinguishers, such as CO_2, carbon tetrachloride, etc., should not be used on magnesium fires. G-1 powder or powdered talc should be used on open fires. Dangerous when heated; burns violently in air and emits fumes; will react with water or steam to produce hydrogen. See also MAGNESIUM COMPOUNDS.

MAD000 CAS:142-72-3 ***HR: 3***
MAGNESIUM ACETATE
mf: $C_4H_6O_4$•Mg mw: 142.41

PROP: Tetrahydrate, colorless or white, deliquescent crystals. D: 1.45, mp: approx 80°. Very sol in water and alc. Keep container well closed.

SYNS: ACETIC ACID, MAGNESIUM SALT ◇ MAGNESIUM DIACETATE

TOXICITY DATA with REFERENCE
ivn-mus LD50:16 mg/kg JLCMAK 29,809,44

CONSENSUS REPORTS: Reported in EPA TSCA Inventory.

SAFETY PROFILE: Poison by intravenous route. When heated to decomposition it emits acrid smoke and irritating fumes.

MAD050 **_HR: 3_**
MAGNESIUM AUREOLATE
mf: $C_{116}H_{200}O_{58}{\cdot}Mg$ mw: 2546.98

SYNS: MAGNESIUM SALT of AUREOLIC ACID ◇ MITHRAMYCIN, MAGNESIUM SALT

TOXICITY DATA with REFERENCE
ivn-mus LD50:2500 μg/kg ANTCAO 3,1218,53
ivn-dog LDLo:250 μg/kg ANTCAO 3,1218,53
ivn-rbt LDLo:250 μg/kg ANTCAO 3,1218,53

SAFETY PROFILE: A deadly poison by intravenous route. When heated to decomposition it emits acrid smoke and irritating fumes. See also MITHRAMYCIN.

MAD100 CAS:36711-31-6 **_HR: D_**
MAGNESIUM BIS(2,3-DIBROMOPROPYL)PHOSPHATE
mf: $C_6H_{10}O_4P{\cdot}Mg$ mw: 201.44

SYNS: BIS(2,3-DIBROMOPROPYL)PHOSPHATE, MAGNESIUM SALT ◇ 2,3-DIBROMO-1-PROPANOL HYDROGEN PHOSPHATE, MAGNESIUM SALT

TOXICITY DATA with REFERENCE
mmo-sat 3 μmol/plate MUREAV 66,373,79
mma-sat 3 μmol/plate MUREAV 66,373,79
orl-rat TDLo:4860 mg/kg (8-16D preg):REP ESKHA5 100,85,82

SAFETY PROFILE: Experimental reproductive effects. Mutation data reported. When heated to decomposition it emits toxic fumes of PO_x. See also PHOSPHATES.

MAD250 CAS:12007-25-9 **_HR: 3_**
MAGNESIUM BORIDE
mf: B_2Mg_3 mw: 94.56

SAFETY PROFILE: Poison. Reacts with water or acids to form a spontaneous flammable borane gas. See also BORANES and MAGNESIUM COMPOUNDS.

MAE000 CAS:10326-21-3 **_HR: 3_**
MAGNESIUM CHLORATE
DOT: UN 2723
mf: $Cl_2O_6{\cdot}Mg$ mw: 191.21

PROP: White, deliq crystals or powder; bitter taste. Mp: 35°, bp: decomp @ 120°, d: 1.80 @ 25°. Sltly sol in alc.

SYNS: CHLORATE SALT of MAGNESIUM ◇ DE-FOL-ATE ◇ E-Z-OFF ◇ KRMD 58 ◇ MAGNESIUM DICHLORATE ◇ MAGRON ◇ MC DEFOLIANT ◇ ORTHO MC

TOXICITY DATA with REFERENCE
orl-rat TDLo:508 mg/kg (1-20D preg):REP GISAAA 48(4),68,83
orl-rat LD50:6348 mg/kg GISAAA 48(4),68,83
ipr-rat LDLo:1100 mg/kg JPETAB 35,1,29
orl-mus LD50:5235 mg/kg GISAAA 48(4),68,83
orl-rbt LD50:8660 mg/kg GISAAA 48(4),68,83

CONSENSUS REPORTS: Reported in EPA TSCA Inventory.

DOT Classification: Oxidizer; Label: Oxidizer.

SAFETY PROFILE: Moderately toxic by intraperitoneal route. Mildly toxic by ingestion. Probably an eye, skin, and mucous membrane irritant. Experimental reproductive effects. A defoliant. A powerful oxidizer. Explosive reaction with copper(I) sulfide. Incandescent reaction with antimony(III) sulfide; arsenic(III) sulfide; tin(II) sulfide; tin(IV) sulfide. Incompatible with Al; As; C; charcoal; Cu; MnO_2; metal sulfides; dibasic organic acids; organic matter; P; S. When heated to decomposition it emits toxic fumes of Cl^-. See also MAGNESIUM COMPOUNDS and CHLORATES.

MAE250 CAS:7786-30-3 **_HR: 3_**
MAGNESIUM CHLORIDE
mf: Cl_2Mg mw: 95.21

PROP: Thin, white to opaque, gray granules and/or flakes, deliq. Mp: 708° (712° with rapid heating), bp: 1412°, d: 2.325. Sol in water (evolving much heat) and alc.

SYN: DUS-TOP

TOXICITY DATA with REFERENCE
mmo-omi 8000 ppm APMBAY 6,45,58
cyt-hmn:hla 2 mmol/L JCLLAX 78,217,71
dns-rat-ipr 2500 μmol/kg JOENAK 65,45,75
dni-rat:lvr 3300 μmol/L BIJOAK 146,697,75
oms-mus:ast 4 mmol/L AMOKAG 33,141,79
orl-rat LD50:2800 mg/kg JPETAB 35,1,29
ipr-rat LDLo:225 mg/kg JPETAB 35,1,29
scu-rat LDLo:900 mg/kg ENDOAO 24,523,39
ipr-mus LD50:1338 mg/kg COREAF 256,1043,63
ivn-mus LD50:14 mg/kg TXAPA9 22,150,72
ivn-dog LDLo:229 mg/kg JPETAB 1,1,09

CONSENSUS REPORTS: Reported in EPA TSCA Inventory. EPA Genetic Toxicology Program.

SAFETY PROFILE: Poison by intraperitoneal and intravenous routes. Moderately toxic by ingestion and subcutaneous routes. Human mutation data reported. In humid environments it causes steel to rust very rapidly. When heated to decomposition it emits toxic fumes of Cl^-. See also MAGNESIUM.

MAE500 CAS:7791-18-6 ***HR: 3***
MAGNESIUM CHLORIDE HEXAHYDRATE
mf: $Cl_2Mg \cdot 6H_2O$ mw: 203.33

PROP: Deliq crystals. D: 1.59, mp: when rapidly heated approx 118° with decomp. Keep well closed.

SYNS: CHLORURE de MAGNESIUM HYDRATE (FRENCH) ◇ CMH

TOXICITY DATA with REFERENCE
orl-rat LD50:8100 mg/kg AIHAAP 30,470,69
ivn-rat LDLo:176 mg/kg JLCMAK 15,35,29
orl-mus LD50:7600 mg/kg THERAP 31,471,76
ipr-mus LD50:775 mg/kg THERAP 31,471,76

SAFETY PROFILE: Poison by intravenous route. Moderately toxic by intraperitoneal route. Mildly toxic by ingestion. Used in disinfectants and fire extinguishers. When heated to decomposition it emits toxic fumes of Cl^-. See also MAGNESIUM.

MAE750 ***HR: D***
MAGNESIUM COMPOUNDS

SAFETY PROFILE: Variable toxicity. The inhalation of fumes of freshly sublimed magnesium oxide may cause metal fume fever. There is no evidence that magnesium produces true systemic poisoning. Particles of metallic magnesium or magnesium alloy which perforate the skin or gain entry through cuts and scratches may produce a severe local lesion characterized by the evolution of gas and acute inflammatory reaction, frequently with necrosis. The condition has been called a "chemical gas gangrene." Gaseous blebs may develop within 24 hours of the injury. The inflammatory response is marked at the site of injury and there may be signs of lymphangitis. The lesion is very slow to heal. The most serious hazard presented by magnesium is the danger from burns. Protection necessary for personnel handling and processing magnesium is usually no different from that which is necessary for other metals. The toxicity of magnesium compounds is usually that of the anion. When heated to decomposition it emits toxic fumes of MgO. See also MAGNESIUM and specific compounds.

MAF000 CAS:69011-63-8 ***HR: 3***
MAGNESIUM DROSS (HOT)

SYN: MAGNESIUM DROSS, WET (DOT)

CONSENSUS REPORTS: Reported in EPA TSCA Inventory.

DOT Classification: Forbidden (wet or hot)

SAFETY PROFILE: See MAGNESIUM COMPOUNDS.

MAF500 CAS:7783-40-6 ***HR: 2***
MAGNESIUM FLUORIDE
mf: F_2Mg mw: 62.31

PROP: Colorless substance; faint violet, luminous crystals. Mp: 1396°, bp: 2239°, d: 2.9-3.2. Very sol in water; sltly sol in dil acids.

SYNS: AFLUON ◇ IRTRAN 1 ◇ MAGNESIUM FLUORURE (FRENCH) ◇ SELLAITE

TOXICITY DATA with REFERENCE
orl-gpg LDLo:1 g/kg MEIEDD 11,892,89

CONSENSUS REPORTS: Reported in EPA TSCA Inventory.

OSHA PEL: TWA 2.5 mg(F)/m^3
ACGIH TLV: TWA 2.5 mg(F)/m^3
NIOSH REL: TWA 2.5 mg(F)/m^3

SAFETY PROFILE: Moderately toxic by ingestion. When heated to decomposition it emits toxic fumes of F^-. See also MAGNESIUM and FLUORIDES.

MAF750 CAS:7704-71-4 ***HR: 2***
MAGNESIUM FUMARATE
mf: $C_4H_3O_4 \cdot Mg$ mw: 139.38

TOXICITY DATA with REFERENCE
orl-mus LDLo:1394 mg/kg JAPMA8 31,12,42

CONSENSUS REPORTS: Reported in EPA TSCA Inventory.

SAFETY PROFILE: Moderately toxic by ingestion. When heated to decomposition it emits acrid smoke and irritating fumes. See also MAGNESIUM COMPOUNDS.

MAG000 CAS:3632-91-5 ***HR: 3***
MAGNESIUM GLUCONATE
mf: $C_{12}H_{24}O_{14} \cdot Mg$ mw: 416.67

SYN: ALMORA

TOXICITY DATA with REFERENCE
ivn-mus LD50:321 mg/kg RPOBAR 2,299,70

CONSENSUS REPORTS: Reported in EPA TSCA Inventory.

SAFETY PROFILE: Poison by intravenous route.

When heated to decomposition it emits acrid smoke and irritating fumes.

MAG050 CAS:53459-38-4 ***HR: 2***
MAGNESIUM GLUTAMATE HYDROBROMIDE
mf: $C_5H_7NO_4{\cdot}BrH{\cdot}Mg$ mw: 250.36

SYNS: l-GLUTAMIC ACID, MAGNESIUM SALT (1:1), HYDROBROMIDE ◇ PSYCHO-SOMA

TOXICITY DATA with REFERENCE
orl-mus TDLo:600 mg/kg (female 7-12D post):REP TOIZAG 15,73,68
orl-mus TDLo:2250 mg/kg (female 7-12D post):TER TOIZAG 14,132,67
orl-rat LD50:10795 mg/kg TOIZAG 15,60,68
scu-rat LD50:2829 mg/kg TOIZAG 15,60,68
ivn-rat LD50:325 mg/kg TOIZAG 15,60,68
orl-mus LD50:6864 mg/kg TOIZAG 15,60,68
scu-mus LD50:1565 mg/kg TOIZAG 15,60,68
ivn-mus LD50:279 mg/kg TOIZAG 15,60,68

SAFETY PROFILE: Poison by intravenous route. Moderately toxic by subcutaneous route. An experimental teratogen. Experimental reproductive effects. When heated to decomposition it emits toxic fumes of NO_x, Mg, and Br^-.

MAG250 CAS:18972-56-0 ***HR: 3***
MAGNESIUM HEXAFLUOROSILICATE
DOT: UN 2853
mf: $F_6Si{\cdot}Mg$ mw: 166.40

SYNS: FLUOSILICATE de MAGNESIUM (FRENCH) ◇ MAGNESIUM FLUOSILICATE ◇ MAGNESIUM SILICOFLUORIDE

TOXICITY DATA with REFERENCE
orl-gpg LD50:200 mg/kg 28ZEAL 4,265,69
scu-frg LDLo:420 mg/kg CRSBAW 124,133,37

OSHA PEL: TWA 2.5 mg(F)/m^3
NIOSH REL: TWA 2.5 mg(F)/m^3
DOT Classification: Poison B; Label: St. Andrews Cross.

SAFETY PROFILE: Poison by ingestion. Moderately toxic by subcutaneous route. When heated to decomposition it emits toxic fumes of F^-. See also FLUOIDES and MAGNESIUM COMPOUNDS.

MAG500 CAS:60616-74-2 ***HR: 3***
MAGNESIUM HYDRIDE
DOT: UN 2010
mf: H_2Mg mw: 26.33

PROP: White crystals. D: 1.419, mp: decomp >200°. Sol in isopropylamine.

DOT Classification: Flammable Solid; Label: Dangerous When Wet.

SAFETY PROFILE: The powder may ignite spontaneously in air and react violently with water. Incompatible with oxygen and water. See also MAGNESIUM COMPOUNDS and HYDRIDES.

MAG750 CAS:1309-42-8 ***HR: 1***
MAGNESIUM HYDROXIDE
mf: H_2MgO_2 mw: 58.33

PROP: White powder, odorless. D: 2.36, mp: decomp @ 350°. Sol in solns of ammonium salts and dilute acids; almost insol in water and alc.

SYNS: MAGNESIA MAGMA ◇ MAGNESIUM HYDRATE ◇ MILK of MAGNESIA

CONSENSUS REPORTS: Reported in EPA TSCA Inventory.

SAFETY PROFILE: Incompatible with maleic anhydride; phosphorous. See also MAGNESIUM COMPOUNDS.

MAH000 CAS:10377-60-3 ***HR: 3***
MAGNESIUM(II) NITRATE (1:2)
DOT: UN 1474
mf: $N_2O_6{\cdot}Mg$ mw: 148.33

PROP: The dihydrate, [$Mg(NO_3)_2{\cdot}2H_2O$] mw: 184.37, forms white crystals (prisms). D: 2.0256 @ 25°, mp: 129.0°. The hexahydrate, [$Mg(NO_3)_2{\cdot}6H_2O$] mw: 256.43, forms monoclinic, colorless, deliq crystals. D: 1.464, mp: 95°, bp: $-5H_2O$ @ 330°.

SYNS: MAGNESIUM NITRATE (DOT) ◇ NITRIC ACID, MAGNESIUM SALT (2:1)

CONSENSUS REPORTS: Reported in EPA TSCA Inventory.

DOT Classification: Label: Oxidizer.

SAFETY PROFILE: Probably a severe irritant to the eyes, skin, and mucous membranes. A powerful oxidizer. Violent decomposition on contact with dimethylformamide. When heated to decomposition it emits toxic fumes of NO_x. See also NITRATES and MAGNESIUM COMPOUNDS.

MAH250 CAS:10213-15-7 ***HR: 1***
MAGNESIUM(II) NITRATE (1:2), HEXAHYDRATE
mf: $N_2O_6{\cdot}Mg{\cdot}6H_2O$ mw: 256.45

PROP: Colorless, clear, deliq crystals. D: 1.464, mp, approx 95°, very sol in alc. Keep well closed.

SYN: DUSICNAN HORECHATY (CZECH)

TOXICITY DATA with REFERENCE
skn-rbt 500 mg/24H MLD 28ZPAK -,9,72

eye-rbt 500 mg/24H MLD 28ZPAK -,9,72
orl-rat LD50:5440 mg/kg 28ZPAK -,9,72

SAFETY PROFILE: Mildly toxic by ingestion. A skin and eye irritant. When heated to decomposition it emits toxic fumes of NO_x. See also MAGNESIUM and NITRATES.

MAH500 CAS:1309-48-4 ***HR: 3***
MAGNESIUM OXIDE
mf: MgO mw: 40.31

PROP: White, bulky, very fine powder; odorless. Mp: 2500-2800°, d: 3.65-3.75. Very sltly sol in water; sol in dil acids; insol in alc.

SYNS: CALCINED BRUCITE ◇ CALCINED MAGNESIA ◇ CALCINED MAGNESITE ◇ MAGNESIA ◇ MAGNESIA USTA ◇ MAGNEZU TLENEK (POLISH) ◇ SEAWATER MAGNESIA

TOXICITY DATA with REFERENCE
itr-ham TDLo:480 mg/kg/30W-I:ETA CNREA8 33,2209,73
ihl-hmn TCLo:400 mg/m^3 DTLVS* 3,147,71

CONSENSUS REPORTS: Reported in EPA TSCA Inventory.

OSHA PEL: Fume: (Transitional: TWA Total Dust: 15 mg/m^3; Respirable Fraction: 5 mg/m^3) Total Dust: 10 mg/m^3; Respirable Fraction: 5 mg/m^3
ACGIH TLV: TWA 10 mg/m^3 (fume)
DFG MAK: 6 mg/m^3 (fume)

SAFETY PROFILE: Inhalation of the fumes can produce a febrile reaction and leukocytosis in humans. Questionable carcinogen with experimental tumorigenic data. Violent reaction or ignition in contact with interhalogens (e.g., bromine pentafluoride; chlorine trifluoride). Incandescent reaction with phosphorus pentachloride. See also MAGNESIUM COMPOUNDS.

MAH750 CAS:14452-57-4 ***HR: 2***
MAGNESIUM PEROXIDE
DOT: UN 1476
mf: MgO_2 mw: 56.31

PROP: White powder; tasteless and odorless. Insol in water and slowly decomp evolving O_2; sol in dil acids. Keep container closed.

SYNS: IXPER 25M ◇ MAGNESIUM PEROXIDE, solid (DOT)

CONSENSUS REPORTS: Reported in EPA TSCA Inventory.

DOT Classification: Oxidizer; Label: Oxidizer.

SAFETY PROFILE: A powerful oxidizer. Probably a severe irritant to the eyes, skin, and mucous membranes. Flammable by chemical reaction with acidic materials and moisture; an oxidizing agent. Dangerous; reacts vigorously with reducing agents; will decompose violently in or near a fire. See also PEROXIDES, INORGANIC and MAGNESIUM COMPOUNDS.

MAH775 CAS:7782-75-4 ***HR: 1***
MAGNESIUM PHOSPHATE, DIBASIC
mf: $MgPHO_4 \cdot 3H_2O$ mw: 174.33

PROP: White crystalline powder. Sltly sol in water; insol in alc; sol in dil acid.

SYN: DIMAGNESIUM PHOSPHATE

SAFETY PROFILE: A nuisance dust.

MAH780 ***HR: 1***
MAGNESIUM PHOSPHATE, TRIBASIC
mf: $Mg_3(PO_4)_2 \cdot xH_2O$ mw: 262.86

PROP: White crystalline powder; odorless. Sol in dil mineral acids; insol in water.

SYN: TRIMAGNESIUM PHOSPHATE

SAFETY PROFILE: A nuisance dust.

MAI000 CAS:12057-74-8 ***HR: 3***
MAGNESIUM PHOSPHIDE
DOT: UN 2011
mf: Mg_3P_2 mw: 134.87

SYNS: FOSFURI di MAGNESIO (ITALIAN) ◇ MAGNESIUMFOSFIDE (DUTCH) ◇ PHOSPHURE de MAGNESIUM (FRENCH)

TOXICITY DATA with REFERENCE
ihl-rat LCLo:580 ppm/1H ZGSHAM 25,279,33
ihl-cat LCLo:173 ppm/2H ZGSHAM 25,279,33
ihl-gpg LCLo:288 ppm/2H ZGSHAM 25,279,33

CONSENSUS REPORTS: Reported in EPA TSCA Inventory.

DOT Classification: Flammable Solid; Label: Dangerous When Wet, Poison.

SAFETY PROFILE: A poison. Moderately toxic by inhalation. Flammable when exposed to heat, flame, or oxidizing materials. Ignites when heated in chlorine, bromine, or iodine vapors. Incandescent reaction with nitric acid. Reacts with water to evolve flammable phosphine gas. When heated to decomposition it emits toxic fumes of PO_x and phosphine. See also MAGNESIUM and PHOSPHIDES.

MAI250 CAS:1661-03-6 ***HR: 3***
MAGNESIUM PHTHALOCYANINE
mf: $C_{32}H_{16}MgN_8$ mw: 536.87

SYN: (PHTHALOCYANINATO(2−))MAGNESIUM

TOXICITY DATA with REFERENCE
scu-mus TDLo:1000 mg/kg/I:ETA BECCAN 40,30,62

CONSENSUS REPORTS: Reported in EPA TSCA Inventory.

SAFETY PROFILE: Questionable carcinogen with experimental tumorigenic data. When heated to decomposition it emits toxic fumes of NO_x.

MAI500 ***HR: 3***
MAGNESIUM POLYOXYETHYLENE ALKYL ETHER SULFATE

SYNS: AES-MG ◇ AES-2Mg

TOXICITY DATA with REFERENCE
orl-rat LD50:4220 mg/kg TOIZAG 25,876,78
scu-rat LD50:2400 mg/kg TOIZAG 25,876,78
ivn-rat LD50:189 mg/kg TOIZAG 25,876,78
orl-mus LD50:3900 mg/kg TOIZAG 25,876,78
scu-mus LD50:2050 mg/kg TOIZAG 25,876,78
ivn-mus LD50:209 mg/kg TOIZAG 25,876,78

SAFETY PROFILE: Poison by intravenous route. Moderately toxic by ingestion and subcutaneous routes. When heated to decomposition it emits toxic fumes of SO_x. See also ETHERS, MAGNESIUM COMPOUNDS, and SULFATES.

MAI600 CAS:14842-81-0 ***HR: 2***
MAGNESIUM POTASSIUM ASPARTATE
mf: $C_8H_{10}MgN_2O_8 \cdot C_4H_7NO_4 \cdot 2H \cdot K$ mw: 460.75

SYNS: ASPARA ◇ ASPARTAT ◇ MAGNESIUM POTASSIUM-l-ASPARTATE ◇ SPARTASE ◇ TROPHICARD

TOXICITY DATA with REFERENCE
scu-rat LD50:6902 mg/kg NIIRDN 6,11,82
ivn-rat LD50:619 mg/kg NIIRDN 6,11,82
scu-mus LD50:4226 mg/kg NIIRDN 6,11,82
ivn-mus LD50:817 mg/kg NIIRDN 6,11,82

SAFETY PROFILE: Moderately toxic by intravenous route. When heated to decomposition it emits toxic fumes of NO_x and K_2O.

MAJ000 CAS:1343-90-4 ***HR: 1***
MAGNESIUM SILICATE HYDRATE
mf: $Mg_2O_8Si_3 \cdot H_2O$ mw: 278.91

PROP: Fine white powder; odorless and tasteless. Insol in water and alc.

TOXICITY DATA with REFERENCE
skn-hmn 300 μg/3D-I MLD 85DKA8 -,127,77

SAFETY PROFILE: A human skin irritant. See also MAGNESIUM and SILICATES.

MAJ250 CAS:7487-88-9 ***HR: 2***
MAGNESIUM SULFATE (1:1)
mf: $O_4S \cdot Mg$ mw: 120.37

PROP: Opaque needles or granular crystalline powder; odorless with cooling, bitter, salt taste. Sol in water; slowly sol in glycerin; sltly sol in alc.

SYNS: EPSOM SALTS ◇ MAGNESIUM SULPHATE

TOXICITY DATA with REFERENCE
mrc-esc 5 pph JGMIAN 8,45,53
pic-esc 800 umol/L ENMUDM 6,59,84
ipr-rat TDLo:750 mg/kg (17-21D preg):TER GEPHDP 12,25,81
ivn-wmn LDLo:80 mg/kg/2M-I:CVS,PUL SAMJAF 67,145,85
isp-wmn TDLo:20 mg/kg:PNS SAMJAF 68,367,85
orl-mus LDLo:5000 mg/kg HBAMAK 4,1364,35
scu-mus LD50:645 mg/kg CYLPDN 7,178,86
ipr-dog LDLo:1200 mg/kg HBAMAK 4,1364,35
scu-dog LDLo:1500 mg/kg HBAMAK 4,1364,35
scu-cat LDLo:1000 mg/kg AJPHAP 14,366,1905
orl-rbt LDLo:3000 mg/kg HBAMAK 4,1364,35
scu-gpg LDLo:1800 mg/kg AJPHAP 14,366,05

CONSENSUS REPORTS: Reported in EPA TSCA Inventory.

SAFETY PROFILE: Moderately toxic by ingestion, intraperitoneal, and subcutaneous routes. Human systemic effects: heart changes, cyanosis, flaccid paralysis with appropriate anesthesia. An experimental teratogen. Mutation data reported. Potentially explosive reaction when heated with ethoxyethynyl alcohols (e.g., 1-ethoxy-3-methyl-1-butyn-3-ol). When heated to decomposition it emits toxic fumes of SO_x. See also SULFATES.

MAJ500 ***HR: 2***
MAGNESIUM SULFATE HEPTAHYDRATE
mf: $MgO_4S \cdot 7H_2O$ mw: 246.48

PROP: Efflorescent crystals or powder, bitter taste. Mp: $-7H_2O$ @ 200°, d: 1.68. Sltly sol in alc.

SYNS: BITTER SALTS ◇ EPSOM SALTS ◇ SULFURIC ACID, MAGNESIUM SALT (1:1) HEPTAHYDRATE

TOXICITY DATA with REFERENCE
idu-wmn LDLo:5344 mg/kg ATXKA8 27,129,71

SAFETY PROFILE: Moderately toxic by several routes. Parenteral use or use in presence of renal insufficiency may lead to magnesium intoxication. An anticonvulsant and purgative. See also MAGNESIUM SULFATE.

MAJ775 CAS:17300-62-8 ***HR: 3***
MAGNESIUM TETRAHYDROALUMINATE
mf: Al_2H_8Mg mw: 86.33

SAFETY PROFILE: An explosive sensitive to pressure, heating to 150°C. A powerful reducing agent. Potentially explosive reactions with bis(2-methoxy)ethyl ether;

boron trifluoride diethyl etherate; 3,5-dibromocyclopentene; 1,2-dimethoxyethane; dioxane + heat; ethyl acetate; fluoroamides (e.g., N-ethylhelptafluorobutyramide; trifluoroacetamide; tetrafluorsuccinamide; trifluoroacetic acid; heptafluorobutyramide; octafluoroadipamide); and hydrogen peroxide. Violent reaction with alkyl benzoates; pyridine; tetrahydrofuran; and water. See also MAGNESIUM COMPOUNDS.

MAK000 CAS:10124-53-5 ***HR: 3***
MAGNESIUM THIOSULFATE
mf: $O_3S_2{\cdot}Mg$ mw: 136.43

SYN: MAGNOSULF

TOXICITY DATA with REFERENCE
ipr-rat LDLo:805 mg/kg NATWAY 50,479,63
ivn-rat LDLo:103 mg/kg NATWAY 50,479,63

CONSENSUS REPORTS: Reported in EPA TSCA Inventory.

SAFETY PROFILE: Poison by intravenous route. Moderately toxic by intraperitoneal route. When heated to decomposition it emits toxic fumes of SO_x. See also THIOSULFATES.

MAK250 CAS:13446-30-5 ***HR: 3***
MAGNESIUM THIOSULFATE HEXAHYDRATE
mf: $MgO_3S_2{\cdot}6H_2O$ mw: 244.55

PROP: Colorless or white crystals. Loses $3H_2O$ @ 170°, d: 1.82. Insol in alc.

TOXICITY DATA with REFERENCE
scu-mus LD50:850 mg/kg ARZNAD 5,141,55
ivn-mus LD50:400 mg/kg ARZNAD 5,141,55

SAFETY PROFILE: Poison by intravenous route. Moderately toxic by subcutaneous route. When heated to decomposition it emits toxic fumes of SO_x. See also MAGNESIUM THIOSULFATE.

MAK300 ***HR: 2***
MAHOGANY

PROP: A large tree with a dark brown, rough bark and compound leaves. The winged seeds are about 1 inch long. It is native to southern Florida and the Caribbean islands.

SYNS: ACAJOU (HAITI) ◇ CAOBA (CUBA, DOMINICAN REPUBLIC, PUERTO RICO) ◇ SWIETENIA MAHAGONI

SAFETY PROFILE: The seeds contain an unknown toxin. Ingestion of two chewed or crushed seeds may cause unconsciousness, persistent vomiting, low blood pressure, slowed heartbeat, and other heart effects.

MAK500 CAS:510-13-4 ***HR: 3***
MALACHITE GREEN CARBINOL
mf: $C_{23}H_{26}N_2O$ mw: 346.51

SYNS: BIS(p-(DIMETHYLAMINO)PHENYL)PHENYLMETHANOL ◇ CARBINOLBASE des MALACHITGRUEN (GERMAN) ◇ 4-(DIMETHYLAMINO)-α-(4-(DIMETHYLAMINO)PHENYL)-α-PHENYL-BENZENEMETHANOL

TOXICITY DATA with REFERENCE
orl-mus LD50:470 mg/kg ARZNAD 1,5,51
ipr-mus LD50:10 mg/kg ARZNAD 1,5,51

CONSENSUS REPORTS: Reported in EPA TSCA Inventory.

SAFETY PROFILE: Poison by intraperitoneal route. Moderately toxic by ingestion. When heated to decomposition it emits toxic fumes of NO_x.

MAK600 CAS:2437-29-8 ***HR: 3***
MALACHITE GREEN OXALATE
mf: $C_{46}H_{50}N_4{\cdot}C_2H_2O_4{\cdot}2C_2HO_4$ mw: 927.10

TOXICITY DATA with REFERENCE
eye-rbt 76 mg/kg SEV ARTODN 56,43,84
mma-sat 40 μg/plate ARTODN 56,43,84
orl-rat LD50:275 mg/kg ARTODN 56,43,84
orl-mus LD50:50 mg/kg ARTODN 56,43,84

SAFETY PROFILE: Poison by ingestion. A severe eye irritant. Mutation data reported. When heated to decomposition it emits toxic fumes of NO_x. See also OXALATES.

MAK700 CAS:121-75-5 ***HR: 3***
MALATHION
DOT: NA 2783
mf: $C_{10}H_{19}O_6PS_2$ mw: 330.38

PROP: Brown to yellow liquid; characteristic odor. D: 1.23 @ 25°/4°, mp: 2.9°, bp: 156° @ 0.7 mm. Misc in organic solvents; sltly water-sol.

SYNS: AMERICAN CYANAMID 4,049 ◇ S-(1,2-BIS(AETHOXY-CARBONYL)-AETHYL)-O,O-DIMETHYL-DITHIOPHASPHAT(GERMAN) ◇ S-(1,2-BIS(CARBETHOXY)ETHYL)-O,O-DIMETHYL DITHIOPHOSPHATE ◇ S-(1,2-BIS(ETHOXY-CARBONYL)-ETHYL)-O,O-DIMETHYL-DITHIOFOSFAAT (DUTCH) ◇ S-(1,2-BIS(ETHOXYCARBONYL)ETHYL)-O,O-DIMETHYL PHOSPHORODITHIOATE ◇ S-1,2-BIS(ETHOXYCARBONYL)ETHYL-O,O-DIMETHYL THIOPHOSPHATE ◇ S-(1,2-BIS(ETOSSI-CARBONIL)-ETIL)-O,O-DIMETIL-DITIOFOSFATO(ITALIAN) ◇ CALMATHION ◇ CARBETHOXY MALATHION ◇ CARBETOVUR ◇ CARBETOX ◇ CARBOFOS ◇ CARBOPHOS ◇ CELTHIGN ◇ CHEMATHION ◇ CIMEXAN ◇ COMPOUND 4049 ◇ CYTHION ◇ DETMOL MA ◇ DETMOL MA 96% ◇ S-(1,2-DICARBETHOXYETHYL)-O,O-DIMETHYLDITHIOPHOSPHATE ◇ DICARBOETHOXYETHYL-O,O-DIMETHYL PHOSPHORODITHIOATE ◇ 1,2-DI(ETHOXYCARBONYL)ETHYL-O,O-DIMETHYLPHOSPHORODITHIOATE ◇ S-(1,2-DI(ETHOXYCARBONYL)ETHYL DIMETHYL PHOSPHOROTHIOLOTHIONATE ◇ DIETHYL (DIMETHOXYPHOSPHINOTHIOYLTHIO) BUTANEDIOATE ◇ DI-

ETHYL (DIMETHOXYPHOSPHINOTHIOYLTHIO)SUCCINATE ◇ DIETHYL MERCAPTOSUCCINATE-O,O-DIMETHYL DITHIOPHOSPHATE, S-ESTER ◇ DIETHYL MERCAPTOSUCCINATE-O,O-DIMETHYL PHOSPHORODITHIOATE ◇ DIETHYL MERCAPTOSUCCINATE-O,O-DIMETHYL THIOPHOSPHATE ◇ DIETHYL MERCAPTOSUCCINATE-S-ESTER with O,O-DIMETHYLPHOSPHORODITHIOATE ◇ DIETHYL MERCAPTOSUCCINIC ACID O,O-DIMETHYL PHOSPHORODITHIOATE ◇ (DIMETHOXYPHOSPHINOTHIOYL)THIO)BUTANEDIOIC ACID DIETHYL ESTER ◇ O,O-DIMETHYL-S-(1,2-BIS(ETHOXYCARBONYL)ETHYL)DITHIOPHOSPHATE ◇ O,O-DIMETHYL-S-1,2-(DICARBAETHOXYAETHYL)-DITHIOPHOSPHAT(GERMAN) ◇ O,O-DIMETHYL-S-(1,2-DICARBETHOXYETHYL) DITHIOPHOSPHATE ◇ O,O-DIMETHYL-S-(1,2-DICARBETHOXYETHYL)PHOSPHORODITHIOATE ◇ O,O-DIMETHYL-S-(1,2-DICARBETHOXYETHYL) THIOTHIONOPHOSPHATE ◇ O,O-DIMETHYL-S-1,2-DI(ETHOXYCARBAMYL)ETHYL PHOSPHORODITHIOATE ◇ O,O-DIMETHYL-S-1,2-DIKARBETOXYLETHYLDITIOFOSFAT(CZECH) ◇ O,O-DIMETHYLDITHIOPHOSPHATE DIETHYLMERCAPTOSUCCINATE ◇ DITHIOPHOSPHATE de O,O-DIMETHYLE et de S-(1,2-DICARBOETHOXYETHYLE) (FRENCH) ◇ EL 4049 ◇ EMMATOS ◇ EMMATOS EXTRA ◇ ENT 17,034 ◇ S-ESTER with O,O-DIMETHYL PHOSPHOROTHIOATE ◇ ETHIOLACAR ◇ ETIOL ◇ EXPERIMENTAL INSECTICIDE 4049 ◇ EXTERMATHION ◇ FORMAL ◇ FORTHION ◇ FOSFOTHION ◇ FOSFOTION ◇ FOUR THOUSAND FORTY-NINE ◇ FYFANON ◇ HILTHION ◇ HILTHION 25WDP ◇ INSECTICIDE No. 4049 ◇ KARBOFOS ◇ KOP-THION ◇ KYPFOS ◇ MALACIDE ◇ MALAFOR ◇ MALAGRAN ◇ MALAKILL ◇ MALAMAR ◇ MALAMAR 50 ◇ MALAPHELE ◇ MALAPHOS ◇ MALASOL ◇ MALASPRAY ◇ MALATHION ULV CONCENTRATE ◇ MALATHIOZOO ◇ MALATHON ◇ MALATHYL LV CONCENTRATE & ULV CONCENTRATE ◇ MALATION (POLISH) ◇ MALATOL ◇ MALATOX ◇ MALDISON ◇ MALMED ◇ MALPHOS ◇ MALTOX ◇ MALTOX MLT ◇ MERCAPTOSUCCINIC ACID DIETHYL ESTER ◇ MERCAPTOTHION ◇ MERCAPTOTION (SPANISH) ◇ MLT ◇ MOSCARDA ◇ NCI-C00215 ◇ OLEOPHOSPHOTHION ◇ ORTHO MALATHION ◇ PHOSPHORODITHIOIC ACID-O,O-DIMETHYL ESTER-S-ESTER with DIETHYL MERCAPTOSUCCINATE ◇ PHOSPHOTHION ◇ PRIODERM ◇ SADOFOS ◇ SADOPHOS ◇ SF 60 ◇ SIPTOX I ◇ SUMITOX ◇ TAK ◇ TM-4049 ◇ VEGFRU MALATOX ◇ VETIOL ◇ ZITHIOL

TOXICITY DATA with REFERENCE
mmo-sat 10 mg/L TGANAK 15(3),68,81
sce-hmn:lym 40 mg/L MUREAV 88,307,81
orl-rat TDLo:43920 mg/kg (multi) :REP NATUAS 192,464,61
orl-rat TDLo:5550 mg/kg (91D pre/1-20D preg):TER JTEHD6 14,267,84
orl-man LDLo:471 mg/kg:CNS,CVS,PUL ATXKA8 23,11,67
orl-wmn LDLo:246 mg/kg AEHLAU 33,240,78
orl-rat LD50:290 mg/kg 85GMAT -,56,82
ihl-rat LC50:84600 μg/m³/4H GISAAA 51(3),73,86
skn-rat LD50:4444 mg/kg CMEP** -,1,56
ipr-rat LD50:250 mg/kg ARZNAD 22,1926,72
scu-rat LD50:1000 mg/kg JEENAI 50,356,57
ivn-rat LD50:50 mg/kg ARZNAD 5,626,55
orl-mus LD50:190 mg/kg 85GMAT -,56,82
skn-mus LD50:2330 mg/kg ABCHA6 27,684,63
ipr-mus LD50:193 mg/kg PSEBAA 129,699,68
scu-mus LD50:221 mg/kg OIZAAV 71,6099,59
ivn-mus LD50:184 mg/kg CHABA8 52,16618,58
ipr-dog LD50:1857 mg/kg TXAPA9 15,244,69
ihl-cat LCLo:10 mg/m³/4H 85GMAT -,56,82
iat-cat LDLo:1820 μg/kg 14KTAK -,693,64
orl-rbt LDLo:1200 mg/kg AEHA** 99-002-74/76
orl-rbt LD50:250 mg/kg JHEMA2 22,115,78
skn-rbt LD50:4100 mg/kg 28ZEAL 5,142,76

CONSENSUS REPORTS: IARC Cancer Review: Group 3 IMEMDT 7,56,87; Animal No Evidence IMEMDT 30,103,83; NCI Carcinogenesis Bioassay (feed); No Evidence: mouse, rat NCITR* NCI-CG-TR-24,78; No Evidence: rat NCITR* NCI-CG-TR-192,79. EPA Genetic Toxicology Program.

OSHA PEL: (Transitional: TWA Total Dust: 15 mg/m³; Respirable Fraction: 5 mg/m³ (skin)) TWA Total Dust:10 mg/m³' Respirable Fraction: 5 mg/m³ (skin)
ACGIH TLV: TWA 10 mg/m³ (skin)
DFG MAK: 15 mg/m³
NIOSH REL: (Malathion) TWA 15 mg/m³
DOT Classification: ORM-A; Label: None.

SAFETY PROFILE: A human poison by ingestion. An experimental poison by ingestion, inhalation, intraperitoneal, intravenous, intraarterial, and subcutaneous routes. Human systemic effects by ingestion: coma, blood pressure depression, and difficulty in breathing. Questionable carcinogen. An experimental teratogen. Other experimental reproductive effects. Human mutation data reported. Has caused allergic sensitization of the skin. An organic phosphate cholinesterase inhibitor. When heated to decomposition it emits toxic fumes of PO_x and SO_x. See also PHOSPHATES and PARATHION.

MAK900 CAS:110-16-7 ***HR: 2***
MALEIC ACID
DOT: NA 2215
mf: $C_4H_4O_4$ mw: 116.08

PROP: White crystals, faint acidulous odor. Mp: 130.5°, bp: 135° decomp, d: 1.590 @ 20°/4°, vap d: 4.0.

SYNS: cis-BUTENEDIOIC ACID ◇ (Z)-BUTENEDIOIC ACID ◇ cis-1,2-ETHYLENEDICARBOXYLIC ACID ◇ MALEINIC ACID ◇ MALENIC ACID ◇ TOXILIC ACID

TOXICITY DATA with REFERENCE
skn-rbt 500 mg/24H MLD BIOFX* 7-4/70
eye-rbt 100 mg SEV BIOFX* 7-4/70
eye-rbt 1%/2M SEV AJOPAA 33,387,50
orl-rat LD50:708 mg/kg BIOFX* 7-4/70
orl-mus LD50:2400 mg/kg BIJOAK 34,1196,40
skn-rbt LD50:1560 mg/kg BIOFX* 7-4/70

CONSENSUS REPORTS: Reported in EPA TSCA Inventory.

DOT Classification: ORM-A; Label:None

SAFETY PROFILE: Moderately toxic by ingestion and

skin contact. A skin and severe eye irritant. Believed to be more toxic than its isomer, fumeric acid. Combustible when exposed to heat or flame. When heated to decomposition it emits acrid smoke and irritating fumes.

MAL000 CAS:10099-71-5 ***HR: 1***
MALEIC ACID, DIPENTYL ESTER
mf: $C_{14}H_{24}O_4$ mw: 256.38

TOXICITY DATA with REFERENCE
skn-rbt 10 mg/24H open MLD AIHAAP 23,95,62
orl-rat LD50:4920 mg/kg AIHAAP 23,95,62

CONSENSUS REPORTS: Reported in EPA TSCA Inventory.

SAFETY PROFILE: Mildly toxic by ingestion. A skin irritant. When heated to decomposition it emits acrid smoke and irritating fumes. See also ESTERS.

MAL250 CAS:128-53-0 ***HR: 3***
MALEIC ACID-N-ETHYLIMIDE
mf: $C_6H_7NO_2$ mw: 125.14

PROP: Crystals; irritating odor. Mp: 45°.

SYNS: N-ETHYLMALEIMIDE ◇ USAF B-121

TOXICITY DATA with REFERENCE
dni-hmn:oth 6 μmol/L CNREA8 40,1414,80
dni-mus:oth 300 μmol/L BBRCA9 106,1448,82
dni-mus:oth 6 μmol/L CNREA8 40,1414,80
dni-ham:lng 750 nmol/L 32YWA5 -,742,75
ipr-rat LDLo:17 mg/kg JMCMAR 15,534,72
ipr-mus LD50:25 mg/kg NTIS** AD277-689

CONSENSUS REPORTS: Reported in EPA TSCA Inventory.

SAFETY PROFILE: Poison by intraperitoneal route. Human mutation data reported. Vapors are highly irritating. When heated to decomposition it emits toxic fumes of NO_x.

MAL500 CAS:10099-72-6 ***HR: 2***
MALEIC ACID, MONO(HYDROXYETHOXYETHYL) ESTER
mf: $C_8H_{12}O_6$ mw: 204.20

SYNS: 2-BUTENEDIOIC ACID, MONO(2-(2-HYDROXYETHOXY)ETHYL) ESTER ◇ DIETHYLENE GLYCOL, MONO(HYDROGEN MALEATE)

TOXICITY DATA with REFERENCE
eye-rbt 20 mg open SEV AMIHBC 10,61,54
orl-rat LD50:2830 mg/kg AMIHBC 10,61,54

SAFETY PROFILE: Moderately toxic by ingestion. A severe eye irritant. When heated to decomposition it emits acrid smoke and irritating fumes. See also ESTERS.

MAL750 CAS:10099-73-7 ***HR: 2***
MALEIC ACID, MONO(2-HYDROXYPROPYL) ESTER
mf: $C_7H_{10}O_5$ mw: 174.17

SYN: 2-BUTENEDIOIC ACID, MONO(2-HYDROXYPROPYL) ESTER

TOXICITY DATA with REFERENCE
skn-rbt 10 mg/24H open MLD AMIHBC 10,61,54
eye-rbt 2 mg open SEV AMIHBC 10,61,54
orl-rat LD50:3730 mg/kg AMIHBC 10,61,54
skn-rbt LD50:8480 mg/kg AMIHBC 10,61,54

SAFETY PROFILE: Moderately toxic by ingestion. Slightly toxic by skin contact. A skin and severe eye irritant. When heated to decomposition it emits acrid smoke and irritating fumes. See also ESTERS.

MAM000 CAS:108-31-6 ***HR: 3***
MALEIC ANHYDRIDE
DOT: UN 2215
mf: $C_4H_2O_3$ mw: 98.06

OCCH=CHCO•O (ring: bracket joining the first O and the last O)

PROP: Fused black or white crystals. Mp: 52.8°, bp: 202°, flash p: 215°F (CC), d: 1.48 @ 20°/4°, autoign temp: 890°F, vap press: 1 mm @ 44.0°, vap d: 3.4, lel: 1.4%, uel: 7.1%. Sol in dioxane, water @ 30° forming maleic acid; very sltly sol in alc.

SYNS: cis-BUTENEDIOIC ANHYDRIDE ◇ 2,5-DIHYDROFURAN-2,5-DIONE ◇ 2,5-FURANDIONE ◇ MALEIC ACID ANHYDRIDE (MAK) ◇ RCRA WASTE NUMBER U147 ◇ TOXILIC ANHYDRIDE

TOXICITY DATA with REFERENCE
eye-rbt 1% SEV AJOPAA 29,1363,46
cyt-ham:lng 230 mg/L GANMAX 27,95,81
orl-rat TDLo:4060 mg/kg (multi):REP FAATDF 7,359,86
scu-rat TDLo:1220 mg/kg/61W-I:ETA BJCAAI 17,100,63
orl-rat LD50:400 mg/kg IAEC** 17JUN74
ipr-rat LD50:97 mg/kg 85GMAT -,79,82
orl-mus LD50:465 mg/kg GTPZAB 13,42,69
orl-rbt LD50:875 mg/kg 85GMAT -,79,82
skn-rbt LD50:2620 mg/kg TXAPA9 42,417,77
orl-gpg LD50:390 mg/kg 85GMAT -,79,82

CONSENSUS REPORTS: Community Right-To-Know List. Reported in EPA TSCA Inventory.

OSHA PEL: TWA 0.25 ppm
ACGIH TLV: TWA 0.25 ppm
DFG MAK: 0.2 ppm (0.8 mg/m^3)
DOT Classification: ORM-A; Label: None; IMO: Corrosive Material; Label: None.

SAFETY PROFILE: Poison by ingestion and intraperitoneal routes. Moderately toxic by skin contact. A corrosive irritant to eyes, skin, and mucous membranes. Can cause pulmonary edema. Questionable carcinogen with experimental tumorigenic data. Mutation data reported. A pesticide. Combustible when exposed to heat or flame; can react vigorously on contact with oxidizing materials. Explosive in the form of vapor when exposed to heat or flame. Reacts with water or steam to produce heat. Violent reaction with bases (e.g., sodium hydroxide; potassium hydroxide; calcium hydroxide); alkali metals (e.g., sodium; potassium); amines (e.g., dimethylamine; triethylamine); lithium; pyridine. To fight fire, use alcohol foam. Incompatible with cations. When heated to decomposition (above 150°C) it emits acrid smoke and irritating fumes. See also ANHYDRIDES.

MAM250 ***HR: 3***
MALEIC ANHYDRIDE OZONIDE
mf: $C_4H_2O_6$ mw: 146.06

SAFETY PROFILE: Explodes at −40°C. When heated to decomposition it emits acrid smoke and irritating fumes. See also ANHYDRIDES.

MAM750 CAS:541-59-3 ***HR: 3***
MALEIMIDE
mf: $C_4H_3NO_2$ mw: 97.08

SYNS: MALEINIMIDE ◇ PYRROLE-2,5-DIONE ◇ 3-PYRROLINE-2,5-DIONE

TOXICITY DATA with REFERENCE
ipr-mus TDLo:3100 μg/kg (female 9D post):REP ARTODN 37,15,76
ipr-mus TDLo:750 μg/kg (9D preg):TER ARTODN 37,15,76
orl-mus LD50:80 mg/kg GISAAA 40(11),109,75
ipr-mus LD50:7750 μg/kg ARTODN 37,15,76
ivn-mus LD50:18 mg/kg CSLNX* NX#01729

CONSENSUS REPORTS: Reported in EPA TSCA Inventory.

SAFETY PROFILE: Poison by intraperitoneal and intravenous routes. An experimental teratogen. Experimental reproductive effects. When heated to decomposition it emits toxic fumes of NO_x.

MAN000 CAS:6915-15-7 ***HR: 2***
MALIC ACID
mf: $C_4H_6O_5$ mw: 134.10

PROP: White or colorless crystals; acid taste. Exhibits isomeric forms (dl, l and d). D (dl): 1.601, d (d or l): 1.595 @ 20°/40; mp (dl): 128°, mp (d or l): 100°; bp (dl): 150°, bp (d or l): 140° (decomp). Very sol in water and alc; sltly sol in ether.

SYNS: HYDROXYSUCCINIC ACID ◇ KYSELINA JABLECNA

TOXICITY DATA with REFERENCE
skn-rbt 500 mg/24H MOD 28ZPAK -,105,72
eye-rbt 750 μg/24H SEV 28ZPAK -,105,72
orl-rat LDLo:1600 mg/kg 14CYAT 2,1813,63

CONSENSUS REPORTS: Reported in EPA TSCA Inventory.

SAFETY PROFILE: Moderately toxic by ingestion. A skin and severe eye irritant. When heated to decomposition it emits acrid smoke and irritating fumes.

MAN250 CAS:676-46-0 ***HR: 2***
MALIC ACID, SODIUM SALT
mf: $C_4H_4O_5 \cdot 2Na$ mw: 178.06

SYN: NATRIUMMALAT (GERMAN)

TOXICITY DATA with REFERENCE
scu-rat LD50:3500 mg/kg JPETAB 25,467,25
scu-rbt LDLo:1500 mg/kg JPETAB 25,467,25

SAFETY PROFILE: Moderately toxic by subcutaneous route. When heated to decomposition it emits toxic fumes of Na_2O. See also MALIC ACID.

MAN700 CAS:24382-04-5 ***HR: 2***
MALONALDEHYDE SODIUM SALT
mf: $C_3H_3O_2 \cdot Na$ mw: 94.05

SYNS: 3-HYDROXY-2-PROPENAL SODIUM SALT ◇ MALONALDEHYDE, ION(1-), SODIUM ◇ PROPANEDIAL, ION(1-), SODIUM (9CI) ◇ SODIUM MALONDIALDEHYDE

TOXICITY DATA with REFERENCE
mnt-rat:fbr 100 μmol/L MUREAV 101,237,82
cyt-rat:fbr 100 μmol/L MUREAV 101,237,82
orl-rat TDLo:51500 mg/kg/2Y-I:CAR NTPTR* NTP-TR-331,88

CONSENSUS REPORTS: NTP Carcinogenesis Studies (gavage): Clear Evidence: rat NTPTR* NTP-TR-331,88; No Evidence: mouse NTPTR* NTP-TR-331,88

SAFETY PROFILE: Questionable carcinogen with experimental carcinogenic data. Mutation data reported. When heated to decomposition it emits acrid smoke and irritating fumes.

MAN750 CAS:122-31-6 ***HR: 3***
MALONALDEHYDE DIETHYL ACETAL
mf: $C_{11}H_{24}O_4$ mw: 220.35

PROP: Liquid. Mp: −90°, bp: 219.9°, flash p: 190°F (OC), d: 0.9197 @ 20°/20°, vap d: 7.58.

SYNS: MALONALDEHYDE TETRAETHYL DIACETAL ◇ TETRAETHOXY PROPANE ◇ 1,1,3,3-TETRAETHOXYPROPANE ◇ USAF KF-26

TOXICITY DATA with REFERENCE
eye-rbt 500 mg AMIHBC 4,119,51
mmo-sat 4 μmol/plate CNREA8 40,276,80
orl-rat LD50:1610 mg/kg TXAPA9 7,826,65
ipr-mus LD50:200 mg/kg NTIS** AD277-689

SAFETY PROFILE: Poison by intraperitoneal route. Moderately toxic by ingestion. Mutation data reported. An eye irritant. Flammable when exposed to heat or flame; can react with oxidizing materials. To fight fire, use foam, CO_2, dry chemical. When heated to decomposition it emits acrid smoke and irritating fumes. See also ALDEHYDES.

MAO000 CAS:108-13-4 ***HR: 1***
MALONAMIDE
mf: $C_3H_6N_2O_2$ mw: 102.11

PROP: Tetragonal or monoclinic. Mp: 170°, sol in water @ 8°, insol in alc abs, insol in ether.

SYNS: CARBOXAMIDOACETAMIDE ◇ MALONDIAMIDE ◇ MALONIC ACID DIAMIDE ◇ MALONYLDIAMIDE

TOXICITY DATA with REFERENCE
orl-mus LD50:6000 mg/kg BIJOAK 34,1196,40

CONSENSUS REPORTS: Reported in EPA TSCA Inventory.

SAFETY PROFILE: Mildly toxic by ingestion. When heated to decomposition it emits toxic fumes of NO_x.

MAO250 CAS:109-77-3 ***HR: 3***
MALONONITRILE
DOT: UN 2647
mf: $C_3H_2N_2$ mw: 66.07

PROP: White powder. D: 1.049 @ 34°/4°, mp: 30.5°, bp: 220°, flash p: 266°F (TOC). Sol in water, alc, ether.

SYNS: CYANOACETONITRILE ◇ DICYANOMETHANE ◇ DWUMETYLOSULFOTLENKU (POLISH) ◇ MALONIC DINITRILE ◇ METHYLENE CYANIDE ◇ NITRIL KYSELINY MALONOVE (CZECH) ◇ PROPANEDINITRILE ◇ RCRA WASTE NUMBER U149 ◇ USAF A-4600

TOXICITY DATA with REFERENCE
eye-rbt 5 mg/24H SEV 28ZPAK -,158,72
orl-rat LD50:60800 μg/kg 28ZPAK -,158,72
ipr-rat LDLo:10 mg/kg BCPCA6 13,285,64
scu-rat LDLo:7 mg/kg AIPTAK 3,77,1897
orl-mus LD50:19 mg/kg KHZDAN 9,50,66
ipr-mus LD50:13 mg/kg NATUAS 228,1315,70
scu-mus LDLo:8 mg/kg AIPTAK 3,77,1897
ivn-mus LD50:32 mg/kg CSLNX* NX#07576
scu-dog LDLo:6500 μg/kg AIPTAK 3,77,1897
scu-rbt LDLo:6 mg/kg CRSBAW 96,202,27
ivn-rbt LD50:28 mg/kg PJPPAA 31,563,79
scu-frg LDLo:95 mg/kg AIPTAK 3,77,1897

CONSENSUS REPORTS: Cyanide and its compounds are on the Community Right-To-Know List. EPA Extremely Hazardous Substances List. Reported in EPA TSCA Inventory.

NIOSH REL: (Nitriles) TWA 8 mg/m^3
DOT Classification: Poison B; Label: Poison.

SAFETY PROFILE: Poison by ingestion, subcutaneous, intravenous, and intraperitoneal routes. A severe eye irritant. Combustible when exposed to heat or flame. Polymerizes violently when heated to 130°C or on contact with strong base. May spontaneously explode when stored at 70-80°C. To fight fire, use water, fog, spray, foam. When heated to decomposition it emits toxic fumes of NO_x and CN^-. See also NITRILES.

MAO275 CAS:59937-28-9 ***HR: 2***
MALOTILATE
mf: $C_{12}H_{16}O_4S_2$ mw: 288.40

PROP: Pale yellow crystals. Mp: 60.5°. Sol in benzene, cyclohexane, n-hexane, and ether.

SYNS: DIISOPROPYL-1,3-DITHIOL-2-YLIDENEMALONATE ◇ 1,3-DITHIOL-2-YLIDENE-PROPANEDIOIC ACID BIS(1-METHYLETHYL) ESTER ◇ HEPATION ◇ KANTEC ◇ NKK 105

TOXICITY DATA with REFERENCE
orl-rbt TDLo:1920 mg/kg (female 7-18D post):TER TOIZAG 28,880,81
orl-rat TDLo:63 g/kg (35D pre):REP JZKEDZ 8,131,82
orl-rat LD50:2065 mg/kg TOIZAG 25,387,78
ipr-rat LD50:750 mg/kg TOIZAG 25,387,78
orl-mus LD50:3120 mg/kg TOIZAG 25,387,78
ipr-mus LD50:1220 mg/kg TOIZAG 25,387,78
scu-mus LD50:1732 mg/kg SYXUE3 2,269,85
orl-rbt LD50:706 mg/kg TOIZAG 25,387,78
ipr-rbt LD50:656 mg/kg TOIZAG 25,387,78

SAFETY PROFILE: Moderately toxic by ingestion and intraperitoneal routes. An experimental teratogen. Experimental reproductive effects. When heated to decomposition it emits toxic fumes of SO_x. See also ESTERS.

MAO350 CAS:118-71-8 ***HR: 2***
MALTOL
mf: $C_6H_6O_3$ mw: 126.12

PROP: White crystalline powder; caramel-butterscotch odor. Sol in water, alc, glycerin, propylene glycol.

SYNS: CORPS PRALINE ◇ 3-HYDROXY-2-METHYL-4H-PYRAN-4-ONE ◇ 3-HYDROXY-2-METHYL-Γ-PYRONE ◇ 3-HYDROXY-2-METHYL-4-PYRONE ◇ LARIXIC ACID ◇ LARIXINIC ACID ◇ 2-METHYL-3-HYDROXY-4-PYRONE ◇ 2-METHYL-3-OXY-Γ-PYRONE ◇ 2-METHYL PYROMECONIC ACID ◇ PALATONE ◇ TALMON ◇ VETOL

TOXICITY DATA with REFERENCE
skn-rbt 500 mg/24H MOD FCTXAV 13,841,75
mmo-sat 1 mg/plate MUREAV 67,367,79
mma-sat 3333 μg/plate ENMUDM 8(Suppl 7),1,86
sce-hmn:lym 500 μmol/L MUREAV 169,129,86
orl-rat LD50:2330 mg/kg FCTXAV 13,681,75

orl-mus LD50:848 mg/kg TXAPA9 15,604,69
ipr-mus LD16:1400 mg/kg RPTOAN 38,213,75
scu-mus LD50:820 mg/kg CPBTAL 22,1008,74
orl-rbt LD50:1620 mg/kg DOWCC* -,-,67
orl-gpg LD50:1410 mg/kg DOWCC* -,-,67
orl-ckn LD50:3720 mg/kg TXAPA9 15,604,69

CONSENSUS REPORTS: Reported in EPA TSCA Inventory.

SAFETY PROFILE: Moderately toxic by ingestion, intraperitoneal, and subcutaneous routes. A skin irritant. Human mutation data reported. When heated to decomposition it emits acrid smoke and irritating fumes.

MAO500 CAS:69-79-4 ***HR: 3***
MALTOSE
mf: $C_{12}H_{22}O_{11}$ mw: 342.31

PROP: Colorless needles. D: 1.540 @ 17°, mp: decomp. Very sol in water; very sltly sol in cold alc; insol in ether.

SYNS: 4-(α-d-GLUCOPYRANOSIDO)-α-GLUCOPYRANOSE ◇ 4-(α-d-GLUCOSIDO)-d-GLUCOSE ◇ MALTOBIOSE ◇ d-MALTOSE ◇ MALT SUGAR ◇ α-MALT SUGAR

TOXICITY DATA with REFERENCE
ivn-rbt TDLo:105 mg/kg (female 18-31D post):REP OYYAA2 16,971,78
ivn-rat TDLo:15 g/kg (female 9-14D post):TER OYYAA2 6,751,72
scu-mus TDLo:1750 mg/kg/50W-C:ETA GANNA2 48,556,57
orl-rat LD50:34800 mg/kg YACHDS 7,53,79
ipr-rat LD50:30600 mg/kg OYYAA2 6,251,72
ivn-rat LD50:15300 mg/kg OYYAA2 6,251,72
scu-mus LD50:38600 mg/kg YACHDS 7,53,79
ivn-mus LD50:26800 mg/kg YACHDS 7,53,79
ivn-rbt LD50:25200 g/kg NIIRDN 6,805,82

CONSENSUS REPORTS: Reported in EPA TSCA Inventory.

SAFETY PROFILE: Experimental teratogenic and reproductive effects. Questionable carcinogen with experimental tumorigenic data. When heated to decomposition it emits acrid smoke and irritating fumes.

MAO600 CAS:643-84-5 ***HR: 3***
MALVIDIN CHLORIDE
mf: $C_{17}H_{15}O_7$•Cl mw: 366.77

SYN: FLAVYLIUM, 3,4′,5,7-TETRAHYDROXY-3′,5′-DIMETHOXY-, CHLORIDE

TOXICITY DATA with REFERENCE
orl-mky TDLo:3 g/kg (male 60D pre):REP CUSCAM 57,1354,88
ivn-mus LD50:18 mg/kg CSLNX* NX#01634

SAFETY PROFILE: Poison by intravenous route. Experimental reproductive effects. When heated to decomposition it emits toxic fumes of Cl^-.

MAO750 ***HR: 3***
MALVIDOL
mf: $C_{17}H_{15}O_7$ mw: 331.32

SYNS: 3′,5′-DIMETHOXY-3,4′,5,7-TETRAHYDROXYFLAVYLIUM ACID ANION ◇ 3,5,7-TRIHYDROXY-2-(4-HYDROXY-3,5-DIMETHOXYHPENYL)-BENZOPYRYLIUM ACID ANION

TOXICITY DATA with REFERENCE
ipr-rat LD50:2350 mg/kg CHTPBA 2,33,67
ivn-rat LD50:240 mg/kg CHTPBA 2,33,67
ipr-mus LD50:4110 mg/kg CHTPBA 2,33,67
ivn-mus LD50:840 mg/kg CHTPBA 2,33,67

SAFETY PROFILE: Poison by intravenous route. Moderately toxic by intraperitoneal route. When heated to decomposition it emits acrid smoke and irritating fumes.

MAO875 ***HR: 2***
MANCHINEEL

PROP: A deciduous tree which has thick, grey bark and may grow to 30 feet. It produces a small, crabapple-like fruit. The sap changes from white to black when exposed to the air. It grows wild in the Florida everglades and the West Indies.

SYNS: BEACH APPLE ◇ HIPPOMANE MANCINELLA ◇ MANCENILLIER (HAITI) ◇ MANZANILLO (CUBA, PUERTO RICO, DOMINICAN REPUBLIC)

SAFETY PROFILE: The latex contains the poisons hippomane A and B. It can cause direct and allergic dermatitis and conjunctivitis. Inhalation of the sawdust can cause irritation of the nose, throat and lungs. Chewing the fruit causes intense pain and blistering of the lips, mouth, and throat. Ingestion causes vomiting, extreme abdominal pain, and bloody diarrhea.

MAP000 CAS:90-64-2 ***HR: 3***
MANDELIC ACID
mf: $C_8H_8O_3$ mw: 152.16

PROP: Large, white crystals or powder; faint odor. Bp: decomp. D: 1.30, mp: 117-119°. Sol in water, alc, and ether. Darkens and decomp on prolonged exposure to light.

SYNS: AMYGDALIC ACID ◇ AMYGDALINIC ACID ◇ α-HYDROXYPHENYLACETIC ACID ◇ α-HYDROXY-α-TOLUIC ACID ◇ racemic MANDELIC ACID ◇ PARAMANDELIC ACID ◇ PHENYLGLYCOLIC ACID ◇ PHENYLHYDROXYACETIC ACID

TOXICITY DATA with REFERENCE
orl-rat LDLo:3000 mg/kg AIPTAK 64,79,40

ims-rat LD50:300 mg/kg EMSUA8 4,223,46
orl-rbt LDLo:2000 mg/kg AIPTAK 64,79,40

CONSENSUS REPORTS: Reported in EPA TSCA Inventory.

SAFETY PROFILE: Poison by intramuscular route. Moderately toxic by ingestion. Continued absorption can cause kidney irritation. When heated to decomposition it emits acrid smoke and irritating fumes.

MAP250 CAS:532-28-5 ***HR: 3***
MANDELIC ACID NITRILE
mf: C_8H_7NO mw: 133.16

PROP: Yellow, visc liquid. Mp: −10°, bp: 170° decomp, d: 1.124.

SYNS: AMYGDALONITRILE ◇ BENZALDEHYDE CYANOHYDRIN ◇ BENZALDEHYDKYANHYDRIN (CZECH) ◇ HYDROXYPHENYLACETONITRILE ◇ NITRIL KYSELINY MANDLOVE (CZECH) ◇ PHENYLGLYCOLONITRILE

TOXICITY DATA with REFERENCE
eye-rbt 250 μg/24H SEV 28ZPAK -,161,72
mmo-sat 225 nmol/plate SCIEAS 198,625,77
mma-sat 225 nmol/plate SCIEAS 198,625,77
orl-rat LD50:116 mg/kg 28ZPAK -,161,72
scu-mus LDLo:23 mg/kg AIPTAK 12,447,04
ivn-mus LD50:5600 μg/kg CSLNX* NX#07767
scu-rbt LDLo:6 mg/kg AIPTAK 5,161,1899
scu-frg LDLo:600 μg/kg AIPTAK 5,161,1899

CONSENSUS REPORTS: Cyanide and its compounds are on the Community Right-To-Know List. Reported in EPA TSCA Inventory.

SAFETY PROFILE: Poison by ingestion, intravenous, and subcutaneous routes. Mutation data reported. A severe eye irritant. When heated to decomposition it emits toxic fumes of NO_x and CN^-. See also NITRILES.

MAP300 CAS:52623-88-8 ***HR: 1***
MANEB-METHYLTHIOPHANATE mixture
mf: $C_{12}H_{14}N_4O_4S_2 \cdot C_4H_6MnN_2S_4$ mw: 607.72

SYNS: BAS 36801F ◇ CALIGRAN M ◇ DUOSAN ◇ DUOSAN (pesticide) ◇ LABILITE ◇ MANGANESE, ((1,2-ETHANEDIYLBIS(CARBAMODITHIOATO))(2-))-, mixed with DIMETHYL (1,2-PHENYLENEBIS(IMINOCARBONOTHIOYL))BIS(CARBAMATE) ◇ METHYLTHIOPHANATE-MANEB mixture ◇ MF 565 ◇ MF 598 ◇ ORGANIL 644 ◇ PELTAR ◇ THIOPHANATE METHYL-MANEB mixture ◇ TMM

TOXICITY DATA with REFERENCE
orl-rat LD50:10200 mg/kg FMCHA2-,C114,89
skn-rbt LD50:8 g/kg FMCHA2-,C114,89

OSHA PEL: CL 5 mg(Mn)/m^3
ACGIH TLV: TWA 5 mg(Mn)/m^3

SAFETY PROFILE: Mildly toxic by ingestion and skin contact. When heated to decomposition it emits toxic fumes of NO_x, SO_x, and Mn.

MAP600 ***HR: 2***
MANETOL

PROP: Extracted from animal spinal marrow and has blood coagulation properties (KSRNAM 8,7,74).

TOXICITY DATA with REFERENCE
ims-rat TDLo:11235 mg/kg (35D pre):REP KSRNAM 8,7,74
ipr-rat LD50:3082 mg/kg KSRNAM 8,7,74
ivn-rat LD50:788 mg/kg KSRNAM 8,7,74
ipr-mus LD50:3840 mg/kg KSRNAM 8,7,74
ivn-mus LD50:628 mg/kg KSRNAM 8,7,74

SAFETY PROFILE: Moderately toxic by intraperitoneal and intravenous routes. Experimental reproductive effects. When heated to decomposition it emits acrid smoke and irritating fumes.

MAP750 CAS:7439-96-5 ***HR: 3***
MANGANESE
af: Mn aw: 54.94

PROP: Reddish-grey or silvery, brittle, metallic element. Mp: 1260°, bp: 1900°, d: 7.20, vap press: 1 mm @ 1292°.

SYNS: COLLOIDAL MANGANESE ◇ MAGNACAT ◇ MANGAN (POLISH) ◇ MANGAN NITRIDOVANY (CZECH) ◇ TRONAMANG

TOXICITY DATA with REFERENCE
skn-rbt 500 mg/24H MLD 28ZPAK -,21,72
eye-rbt 500 mg/24H MLD 28ZPAK -,21,72
mrc-smc 8 mmol/L/18H MUREAV 42,343,77
ims-rat TDLo:400 mg/kg/1Y-I:ETA NCIUS* PH 43-64-886,SEPT,71
ihl-man TCLo:2300 μg/m^3:BRN,CNS AIHAAP 27,454,66

CONSENSUS REPORTS: Manganese and its compounds are on the Community Right-To-Know List. Reported in EPA TSCA Inventory.

OSHA PEL: Fume: (Transitional: CL 5 mg/m^3) TWA 1 mg/m^3; STEL 3 mg/m^3; Compounds: CL 5 mg/m^3
ACGIH TLV: Fume: 1 mg/m^3; STEL 3 mg/m^3; Dust and Compounds: TWA 5 mg/m^3
DFG MAK: 5 mg/m^3

SAFETY PROFILE: Human systemic effects by inhalation: degenerative brain changes, change in motor activity, muscle weakness. A skin and eye irritant. Questionable carcinogen with experimental tumorigenic data. Mutation data reported. Flammable and moderately explosive in the form of dust or powder when exposed to flame. The dust may be pyrophoric in air and may explode when heated in carbon dioxide. Mixtures of aluminum dust and manganese dust may explode in air. Mixtures with ammonium nitrate may explode when heated. The powdered metal ignites on contact with fluorine; chlorine + heat; hydrogen peroxide; bromine pentafluoride; sulfur dioxide + heat. Violent reaction with NO_2 and oxidants. Incandescent reaction with phosphorus; nitryl fluoride; nitric acid. Will react with water or steam to produce hydrogen; can react with oxidizing materials. To fire fight, use special dry chemical. See also MANGANESE COMPOUNDS.

MAQ000 CAS:638-38-0 ***HR: 2***
MANGANESE ACETATE
mf: $C_4H_6O_4$•Mn mw: 173.04

PROP: Pale red crystals, very sol in water and alc.

SYNS: ACETIC ACID MANGANESE(II) SALT (2:1) ◇ DIACETYLMANGANESE ◇ MANGANESE(2+) ACETATE ◇ MANGANESE(II) ACETATE ◇ MANGANESE DIACETATE ◇ MANGANOUS ACETATE ◇ OCTAN MANGANATY (CZECH)

TOXICITY DATA with REFERENCE
dnr-bcs 50 mmol/L MUREAV 31,185,75
orl-rat LD50:2940 mg/kg MarJV# 29MAR77

CONSENSUS REPORTS: Manganese and its compounds are on the Community Right-To-Know List. Reported in EPA TSCA Inventory.

OSHA PEL: CL 5 mg(Mn)/m^3
ACGIH TLV: TWA 5 mg(Mn)/m^3

SAFETY PROFILE: Moderately toxic by ingestion. Mutation data reported. Used in food packaging. When heated to decomposition it emits acrid smoke and irritating fumes. See also MANGANESE COMPOUNDS.

MAQ250 CAS:6156-78-1 ***HR: 2***
MANGANESE ACETATE TETRAHYDRATE
mf: $C_4H_6O_4$•Mn•$4H_2O$ mw: 245.12

PROP: Pale red, transparent, monoclinic crystals. D: 1.59. Sol in water.

SYNS: MANGANESE(II) ACETATE TETRAHYDRATE ◇ MANGANESE DIACETATE TETRAHYDRATE ◇ MANGANOUS ACETATE TETRAHYDRATE

TOXICITY DATA with REFERENCE
orl-rat LD50:3730 mg/kg AIHAAP 30,470,69

CONSENSUS REPORTS: Manganese and its compounds are on the Community Right-To-Know List.

OSHA PEL: CL 5 mg(Mn)/m^3
ACGIH TLV: TWA 5 mg(Mn)/m^3

SAFETY PROFILE: Moderately toxic by ingestion. When heated to decomposition it emits acrid smoke and irritating fumes. See also MANGANESE COMPOUNDS.

MAQ500 CAS:14024-58-9 ***HR: 3***
MANGANESE ACETYLACETONATE
mf: $C_{10}H_{14}O_4Mn$ mw: 253.18

SYN: MANGANOUS ACETYLACETONATE

TOXICITY DATA with REFERENCE
ims-rat TDLo:1200 mg/kg/26W-I:NEO JNCIAM 60,1171,78
ims-rat TD:1350 mg/kg/21W-I:ETA NCIUS* PH 43-64-886,SEPT,71

CONSENSUS REPORTS: Manganese and its compounds are on the Community Right-To-Know List. Reported in EPA TSCA Inventory.

OSHA PEL: CL 5 mg(Mn)/m^3
ACGIH TLV: TWA 5 mg(Mn)/m^3

SAFETY PROFILE: Questionable carcinogen with experimental neoplastigenic and tumorigenic data. When heated to decomposition it emits acrid smoke and irritating fumes. See also MANGANESE COMPOUNDS.

MAQ600 CAS:85625-90-7 ***HR: 1***
MANGANESE γ-AMINOBUTYRATOPANTOTHENATE
mf: $C_{13}H_{24}MnN_2O_7$ mw: 375.33

SYN: MANGANESE,(4-AMINOBUTANOATO-N,O)(N-(2,4-DIHYDROXY-3,3-DIMETHYL-1-OXOBUTYL)-β-ALANINATO)-

TOXICITY DATA with REFERENCE
ipr-mus LD50:1 g/kg PCJOAU 17,32,83

OSHA PEL: CL 5 mg(Mn)/m^3
ACGIH TLV: TWA 5 mg(Mn)/m^3

SAFETY PROFILE: Mildly toxic by intraperitoneal route. When heated to decomposition it emits toxic fumes of NO_x and Mn.

MAQ780 ***HR: 3***
MANGANESE(II) BIS(ACETYLIDE)
mf: C_4H_2Mn mw: 105.00

$Mn(C\equiv CH)_2$

CONSENSUS REPORTS: Manganese and its compounds are on the Community Right-To-Know List.

SAFETY PROFILE: Highly explosive. When heated to decomposition it emits acrid smoke and irritating fumes. See also MANGANESE COMPOUNDS and ACETYLIDES.

MAR000 CAS:7773-01-5 ***HR: 3***
MANGANESE(II) CHLORIDE (1:2)
mf: Cl_2Mn mw: 125.84

PROP: Cubic, deliquescent, pink crystals. Mp: 650°, bp: 1190°, d: 2.977 @ 25°. Sol in water.

SYNS: MANGANESE DICHLORIDE ◇ MANGANOUS CHLORIDE

TOXICITY DATA with REFERENCE
mmo-esc 5 μmol/L MUREAV 126,9,84
dlt-rat-orl 106 mg/kg/30W-C GISAAA 49(11),80,84
orl-rat TDLo:148 g/kg (female 1-22D post):REP TJADAB 32,1,85
ipr-pig TDLo:4581 mg/kg (female 12-16W post):TER DABBBA 33,2872,72
ipr-mus TDLo:2080 mg/kg/26W-I:CAR FEPRA7 23,393,64
scu-mus TDLo:2080 mg/kg/26W-I:CAR FEPRA7 23,393,64
orl-mus LD50:1715 mg/kg TOLED5 7,221,81
ims-rat LD50:700 mg/kg RPTOAN 38,221,75
ipr-mus LD50:121 mg/kg AEPPAE 244,17,62
scu-mus LDLo:210 mg/kg EQSSDX 1,1,75
ims-mus LD50:255 mg/kg RPTOAN 38,221,75
ivn-dog LD50:202 mg/kg EQSSDX 1,1,75
par-dog LDLo:56 mg/kg CRSBAW 102,262,29
par-rbt LDLo:18 mg/kg CRSBAW 102,262,29

CONSENSUS REPORTS: Manganese and its compounds are on the Community Right-To-Know List. Reported in EPA TSCA Inventory. EPA Genetic Toxicology Program.

OSHA PEL: CL 5 mg(Mn)/m^3
ACGIH TLV: TWA 5 mg(Mn)/m^3

SAFETY PROFILE: Poison by intraperitoneal, subcutaneous, intramuscular, intravenous, and parenteral routes. Moderately toxic by ingestion. Experimental teratogenic and reproductive effects. Questionable carcinogen with experimental carcinogenic data. Mutation data reported. Explosive reaction when heated with zinc foil. Reacts violently with potassium or sodium. When heated to decomposition it emits toxic fumes of Cl^-. See also MANGANESE COMPOUNDS and CHLORIDES.

MAR250 CAS:13446-34-9 ***HR: 3***
MANGANESE(II) CHLORIDE TETRAHYDRATE
mf: $Cl_2Mn•4H_2O$ mw: 197.92

PROP: Reddish, sltly deliquescent, monoclinic crystals. D: 2.01, mp: 58°. Sol in alc; insol in ether. Keep well closed.

SYN: MANGANOUS CHLORIDE TETRAHYDRATE

TOXICITY DATA with REFERENCE
orl-rat TDLo:141 g/kg (90D pre-21D post):REP NTOTDY 5,377,83
orl-rat LD50:1484 mg/kg EVHPAZ 10,95,75
ipr-rat LD50:138 mg/kg EVHPAZ 10,95,75
par-rat LD50:225 mg/kg JINCAO 41,1507,79
ipr-mus LD50:144 mg/kg TXAPA9 63,461,82

CONSENSUS REPORTS: Manganese and its compounds are on the Community Right-To-Know List. EPA Genetic Toxicology Program.

OSHA PEL: CL 5 mg(Mn)/m^3
ACGIH TLV: TWA 5 mg(Mn)/m^3

SAFETY PROFILE: Poison by intraperitoneal and parenteral routes. Moderately toxic by ingestion. Experimental reproductive effects. When heated to decomposition it emits toxic fumes of Cl^-. See also MANGANESE COMPOUNDS.

MAR500 ***HR: 3***
MANGANESE COMPOUNDS

CONSENSUS REPORTS: Manganese and its compounds are on the Community Right-To-Know List.

SAFETY PROFILE: Some are experimental tumorigens. Can cause central nervous system and pulmonary system damage by inhalation of fumes and dust. Very few poisonings have occurred from ingestion. Chronic manganese poisoning is a clearly characterized disease which results from inhalation of fumes or dusts of manganese. Exposure to heavy concentrations of dusts or fumes for as little as three months may produce the condition, but usually cases develop after 1-3 years of exposure. The central nervous system is the chief site of damage. If cases are removed from exposure shortly after appearance of symptoms, some improvement in the patient's condition frequently occurs, though there may be some residual disturbances in gait and speech. When well established, however, the disease results in permanent disability. Exposure to dusts and fumes can possibly increase the incidence of upper respiratory infections and pneumonia. Chronic manganese poisoning usually begins with complaints of languor and sleepiness. This is followed by weakness in the legs and the development of stolid, mask-like faces. The patient speaks with a slow monotonous voice. Then muscular twitching appears, varying from a fine tremor of the hands to coarse, rhythmical movements of the arms, legs, and trunk. Nocturnal cramps of the legs appear about the same time. There is a slight increase in tendon reflexes, ankle and patellar clonus, and a typical Parkinsonian slapping gait. The handwriting may be quite minute. The symptoms may simulate progressive bulbar paralysis, postencephalitic Parkinsonism, multiple sclerosis, amyotrophic lateral sclerosis and progressive lenticular degeneration (Wilson's Disease). Often a history of exposure is the only aid in establishing the diagnosis. Manganese compounds are common air contaminants.

MAR750 CAS:15339-36-3 ***HR: 3***
MANGANESE DIMETHYL DITHIOCARBAMATE
mf: $C_3H_7NS_2•1/2Mn$ mw: 148.69

SYN: MANGANOUSDIMETHYLDITHIOCARBAMATE

TOXICITY DATA with REFERENCE
ivn-mus LD50:32 mg/kg CSLNX* NX#03752

CONSENSUS REPORTS: Manganese and its compounds are on the Community Right-To-Know List.

OSHA PEL: CL 5 mg(Mn)/m^3
ACGIH TLV: TWA 5 mg(Mn)/m^3

SAFETY PROFILE: Poison by intravenous route. A pesticide. When heated to decomposition it emits very toxic fumes of NO_x and SO_x. See also MANGANESE COMPOUNDS and CARBAMATES.

MAS000 CAS:1313-13-9 ***HR: 3***
MANGANESE DIOXIDE
mf: MnO_2 mw: 86.94

PROP: Tetragonal crystals. Mp: $-O_2$ @ 535°, d: 5.0. Insol in water, nitric or cold sulfuric acid.

SYNS: BLACK MANGANESE OXIDE ◇ BOG MANGANESE ◇ BRAUNSTEIN (GERMAN) ◇ BRUINSTEEN (DUTCH) ◇ CEMENT BLACK ◇ C.I. 77728 ◇ C.I. PIGMENT BLACK 14 ◇ C.I. PIGMENT BROWN 8 ◇ MANGAANBIOXYDE (DUTCH) ◇ MANGAANDIOXYDE (DUTCH) ◇ MANGANDIOXID (GERMAN) ◇ MANGANESE BINOXIDE ◇ MANGANESE (BIOSSIDO di) (ITALIAN) ◇ MANGANESE (BIOXYD de) (FRENCH) ◇ MANGANESE BLACK ◇ MANGANESE (DIOSSIDO di) (ITALIAN) ◇ MANGANESE (DIOXYDE de) (FRENCH) ◇ MANGANESE OXIDE ◇ MANGANESE(IV) OXIDE ◇ MANGANESE PEROXIDE ◇ MANGENESE SUPEROXIDE ◇ PYROLUSITE BROWN

TOXICITY DATA with REFERENCE
ihl-mus TCLo:49 mg/m^3/7H (75D pre/1-18D preg):REP FEPRA7 39,623,80
scu-mus LD50:422 mg/kg ZVKOA6 19,186,74
ivn-rbt LDLo:45 mg/kg MEIEDD 10,817,83

CONSENSUS REPORTS: Manganese and its compounds are on the Community Right-To-Know List. Reported in EPA TSCA Inventory.

OSHA PEL: CL 5 mg(Mn)/m^3
ACGIH TLV: TWA 5 mg(Mn)/m^3

SAFETY PROFILE: Poison by intravenous route. Moderately toxic by subcutaneous route. Experimental reproductive effects. A powerful oxidizer. Flammable by chemical reaction. It must not be heated or rubbed in contact with easily oxidizable matter. Violent thermite reaction when heated with aluminum. Potentially explosive reaction with hydrogen peroxide; peroxomonosulfuric acid; chlorates + heat; anilinium perchlorate. Ignition on contact with hydrogen sulfide. Violent reaction with oxidizers; potassium azide (when warmed); diboron tetrafluoride; Incandescent reaction with calcium hydride; chlorine trifluoride; rubidium acetylide (at 350°C). Vigorous reaction with hydroxylaminium chloride. Incompatible with H_2O_2; H_2SO_5; Na_2O_2. Keep away from heat and flammable materials. See also MANGANESE COMPOUNDS.

MAS250 CAS:55448-20-9 ***HR: 3***
MANGANESE EDTA COMPLEX
mf: $C_{10}H_{12}MnN_2O_8$•2H mw: 345.20

TOXICITY DATA with REFERENCE
ipr-rat LD50:1930 mg/kg AMIHAB 21,24,60
ipr-mus LD50:330 mg(Mn)/kg PABIAQ 11,853,63

CONSENSUS REPORTS: Manganese and its compounds are on the Community Right-To-Know List.

OSHA PEL: CL 5 mg(Mn)/m^3
ACGIH TLV: TWA 5 mg(Mn)/m^3

SAFETY PROFILE: Poison by intraperitoneal route. When heated to decomposition it emits toxic fumes of NO_x. See also MANGANESE COMPOUNDS.

MAS500 CAS:12427-38-2 ***HR: 3***
MANGANESE(II) ETHYLENEBIS(DITHIOCARBAMATE)
DOT: UN 2210/UN 2968
mf: $C_4H_6N_2S_4$•Mn mw: 265.30

PROP: Yellow powder or crystals. Sol in water.

SYNS: AAMANGAN ◇ AKZO CHEMIE MANEB ◇ BASF-MANEB SPRITZPULVER ◇ CHEM NEB ◇ CHLOROBLE M ◇ CR 3029 ◇ DITHANE M 22 SPECIAL ◇ ENT 14,875 ◇ 1,2-ETHANEDIYLBIS(CARBAMODITHIOATO)(2−)-MANGANESE ◇ 1,2-ETHANEDIYLBISCARBAMODITHIOIC ACID MANGANESE COMPLEX ◇ 1,2-ETHANEDIYLBISCARBAMODITHIOIC ACID, MANGANESE(2+) SALT (1:1) ◇ 1,2-ETHANEDIYLBISMANEB, MANGANESE (2+) SALT (1:1) ◇ ETHYLENEBISDITHIOCARBAMATE MANGANESE ◇ N,N'-ETHYLENE BIS(DITHIOCARBAMATE MANGANEUX) (FRENCH) ◇ ETHYLENEBIS(DITHIOCARBAMATO) MANGANESE ◇ ETHYLENEBIS (DITHIOCARBAMIC ACID) MANGANESE SALT ◇ ETHYLENEBIS (DITHIOCARBAMIC ACID) MANGANOUS SALT ◇ 1,2-ETHYL- ENEDIYLBIS(CARBAMODITHIOATO)MANGANESE ◇ N,N'-ETILENBIS(DITIOCARBAMMATO) di MANGANESE (ITALIAN) ◇ F 10 (pesticide) ◇ GRIFFIN MANEX ◇ KYPMAN 80 ◇ LONOCOL M ◇ MANAM ◇ MANEB ◇ MANEB, stabilized against self-heating (DOT) ◇ MANEB, with not less than 60% maneb (DOT) ◇ MANEBE (FRENCH) ◇ MANGAAN(II)-(N,N'-ETHYLEEN-BIS(DITHIOCARBAMAAT)) (DUTCH) ◇ MANGAN(II)-(N,N'-AETHYLEN-BIS(DITHIOCARBAMATE))(GERMAN) ◇ MANGANESE ETHYLENE-1,2-BISDITHIOCARBAMATE ◇ MANGANESE(II) ETHYLENE DI(DITHIOCARBAMATE) ◇ MANZATE ◇ MANZATE MANEB FUNGICIDE ◇ MEB ◇ NESPOR ◇ PLANTIFOG 160M ◇ POLYRAM M ◇ REMASAN CHLOROBLE M ◇ TRIMANGOL ◇ UNICROP MANEB ◇ VANCIDE

TOXICITY DATA with REFERENCE
mmo-omi 1000 ppm MMAPAP 50,233,73
mmo-smc 5 ppm RSTUDV 6,161,76
cyt-ham:lng 31 mg/L GMCRDC 27,95,81
orl-mus TDLo:10 g/kg (female 8-12D post):REP TCMUD8 7,7,87
orl-rat TDLo:1420 mg/kg (11D preg):TER TJADAB 14,171,76

orl-rat TDLo:62980 mg/kg/94W-I:CAR VPITAR 29,71,70
imp-rat TDLo:50 mg/kg:ETA VPITAR 29,71,70
orl-rat LD50:3 g/kg GISAAA 36(5),22,71
orl-mus LD50:2600 mg/kg GISAAA 36(5),22,71
orl-gpg LDLo:6400 mg/kg PCOC** -,675,66

CONSENSUS REPORTS: IARC Cancer Review: Group 3 IMEMDT 7,56,87; Animal Inadequate Evidence IMEMDT 12,137,76. Community Right-To-Know List. EPA Genetic Toxicology Program.

OSHA PEL: CL 5 mg(Mn)/m^3
ACGIH TLV: TWA 5 mg(Mn)/m^3
DOT Classification: Flammable Solid; Label: Spontaneously Combustible, Danger When Wet.

SAFETY PROFILE: Moderately toxic by ingestion. Experimental teratogenic and reproductive effects. Questionable carcinogen with experimental carcinogenic and tumorigenic data. Mutation data reported. A fungicide. May ignite spontaneously in air. Dangerous; when heated to decomposition it emits highly toxic fumes of NO_x and SO_x. See also MANGANESE COMPOUNDS and CARBAMATES.

MAS750 CAS:7782-64-1 ***HR: 3***
MANGANESE(II) FLUORIDE
mf: F_2Mn mw: 92.94

PROP: Red, tetragonal crystals or reddish powder. Mp: 856°; d: 3.98. Insol in alc; sol in dilute hydrofluoric acid, concentrated hydrochloric or nitric acid.

SYNS: MANGANESE FLUORIDE ◇ MANGANESE FLUORURE (FRENCH)

TOXICITY DATA with REFERENCE
scu-frg LDLo:224 mg/kg CRSBAW 124,133,37

CONSENSUS REPORTS: Manganese and its compounds are on the Community Right-To-Know List. Reported in EPA TSCA Inventory.

OSHA PEL: TWA 2.5 mg(F)/m^3; Cl 5 mg(Mn)/m^3
ACGIH TLV: TWA 2.5 mg(F)/m^3; TWA 5 mg(Mn)/m^3
NIOSH REL: TWA 2.5 mg(F)/m^3

SAFETY PROFILE: Poison by subcutaneous route. When heated to decomposition it emits toxic fumes of F^-. See also MANGANESE COMPOUNDS and FLUORIDES.

MAT250 CAS:1344-43-0 ***HR: 2***
MANGANESE(II) OXIDE
mf: MnO mw: 70.94

PROP: Grass-green powder, sol in acids, insol in water. D: 5.45, mp: 1650°, converted to Mn_3O_4 if heated in air.

SYNS: CASSEL GREEN ◇ C.I. 77726 ◇ MANGANESE GREEN ◇ MANGANESE MONOXIDE ◇ MANGANOUS OXIDE ◇ NU-MANESE ◇ ROSENSTHIEL

TOXICITY DATA with REFERENCE
scu-mus LD50:1000 mg/kg ZVKOA6 19,186,74

CONSENSUS REPORTS: Manganese and its compounds are on the Community Right-To-Know List. Reported in EPA TSCA Inventory.

OSHA PEL: CL 5 mg(Mn)/m^3
ACGIH TLV: TWA 5 mg(Mn)/m^3

SAFETY PROFILE: Moderately toxic by subcutaneous route. Violent reaction with hydrogen peroxide, $Ca(OCl)_2$, F_2, H_2O_2. See also MANGANESE COMPOUNDS.

MAT500 CAS:1317-34-6 ***HR: 2***
MANGANESE(III) OXIDE
mf: Mn_2O_3 mw: 157.88

PROP: Fine, black powder. D: 4.50. Insol in water; sol in HCl evolving chlorine.

SYNS: CASSEL BROWN ◇ C.I. 77727 ◇ C.I. NATURAL BROWN 8 ◇ COLOGNE EARTH ◇ COLOGNE UMBER ◇ CULLEN EARTH ◇ DIMANGANESE TRIOXIDE ◇ MANGANESE MANGANATE ◇ MANGANESE SISQUIOXIDE ◇ MANGANESE TRIOXIDE ◇ MANGANIC OXIDE ◇ RUBENS BROWN ◇ SOLUBLE VANDYKE BROWN ◇ VANDYKE BROWN ◇ WALNUT STAIN

TOXICITY DATA with REFERENCE
scu-mus LD50:616 mg/kg ZVKOA6 19,186,74

CONSENSUS REPORTS: Manganese and its compounds are on the Community Right-To-Know List. Reported in EPA TSCA Inventory.

OSHA PEL: CL 5 mg(Mn)/m^3
ACGIH TLV: TWA 5 mg(Mn)/m^3

SAFETY PROFILE: Moderately toxic by subcutaneous route. See also MANGANESE COMPOUNDS.

MAT750 CAS:12057-92-0 ***HR: 3***
MANGANESE(VII) OXIDE
mf: Mn_2O_7 mw: 221.88

CONSENSUS REPORTS: Manganese and its compounds are on the Community Right-To-Know List.

SAFETY PROFILE: An unstable explosive sensitive to friction, impact or heating above 40°C. As sensitive as mercury fulminate. Explodes on contact with organic materials (e.g., solvents, oils, fats, fibers, grease). A powerful oxidizer. See also MANGANESE COMPOUNDS.

MAT899 CAS:13770-16-6 ***HR: 3***
MANGANESE(II) PERCHLORATE
mf: Cl_2O_8Mn mw: 253.84

SAFETY PROFILE: Explodes when heated to 195°C. Explosive reaction with 2,2-dimethoxypropane above 65°C. See also MANGANESE COMPOUNDS and PERCHLORATES.

MAU000 CAS:15364-94-0 ***HR: 3***
MANGANESE PERCHLORATE HEXAHYDRATE
mf: $Cl_2O_8•Mn•6H_2O$ mw: 361.96

TOXICITY DATA with REFERENCE
ipr-mus LD50:410 mg/kg JAFCAU 14,512,66

CONSENSUS REPORTS: Manganese and its compounds are on the Community Right-To-Know List.

OSHA PEL: CL 5 mg(Mn)/m^3
ACGIH TLV: TWA 5 mg(Mn)/m^3

SAFETY PROFILE: Moderately toxic by intraperitoneal route. The perchlorate is a powerful oxidizer. Explodes. Incompatible with 2,2-dimethoxypropane. When heated to decomposition it emits toxic fumes of Cl^-. See also MANGANESE COMPOUNDS.

MAU250 CAS:7785-87-7 ***HR: 3***
MANGANESE(II) SULFATE (1:1)
mf: $O_4S•Mn$ mw: 151.00

PROP: Pink granular powder; odorless. Mp: 700°, bp: decomp @ 850°. d: 3.25. Very sol in water, more so in boiling water; insol in alc.

SYNS: MANGANOUS SULFATE ◇ MAN-GRO ◇ NCI-C61143 ◇ SORBA-SPRAY Mn ◇ SULFURIC ACID, MANGANESE(2+) SALT

TOXICITY DATA with REFERENCE
mmo-omi 10 mmol/L JMOBAK 14,453,65
mrc-smc 40 μmol/L MUREAV 137,47,84
ipr-mus TDLo:34356 μg/kg (female 10D post):REP NRTXDN 8,437,87
ipr-mus TDLo:68711 μg/kg (female 8D post):TER NRTXDN 8,437,87
ipr-mus TDLo:660 mg/kg/8W-I:NEO CNREA8 36,1744,76
ipr-mus LD50:332 mg/kg COREAF 256,1043,63

CONSENSUS REPORTS: Manganese and its compounds are on the Community Right-To-Know List. Reported in EPA TSCA Inventory. EPA Genetic Toxicology Program.

OSHA PEL: CL 5 mg(Mn)/m^3
ACGIH TLV: TWA 5 mg(Mn)/m^3

SAFETY PROFILE: Poison by intraperitoneal route. Questionable carcinogen with experimental neoplastigenic data. An experimental teratogen. Experimental reproductive data. Mutation data reported. When heated to decomposition it emits toxic fumes of SO_x and manganese. See also MANGANESE COMPOUNDS and SULFATES.

MAU750 CAS:10101-68-5 ***HR: 2***
MANGANESE(II) SULFATE TETRAHYDRATE (1:1:4)
mf: $O_4S•Mn•4H_2O$ mw: 223.08

SYN: SULFURIC ACID, MANGANESE (2+) SALT (1:1) TETRAHYDRATE

TOXICITY DATA with REFERENCE
dni-mus-ipr 20 g/kg ARGEAR 51,605,81
ipr-mus LD50:534 mg/kg BCPCA6 15,1691,66

CONSENSUS REPORTS: Manganese and its compounds are on the Community Right-To-Know List.

OSHA PEL: CL 5 mg(Mn)/m^3
ACGIH TLV: TWA 5 mg(Mn)/m^3

SAFETY PROFILE: Moderately toxic by intraperitoneal route. Mutation data reported. When heated to decomposition it emits toxic fumes of SO_x. See also MANGANESE COMPOUNDS.

MAU800 CAS:1317-35-7 ***HR: 2***
MANGANESE TETROXIDE
mf: Mn_3O_4 mw: 228.82

PROP: Brownish-black powder. D: 4.7. Insol in water; sol in HCl, liberating chlorine.

SYNS: MANGANESE OXIDE ◇ MANGANOMANGANIC OXIDE ◇ TRIMANGANESE TETRAOXIDE ◇ TRIMANGANESE TETROXIDE

TOXICITY DATA with REFERENCE
orl-mus TDLo:23 g/kg (44D male):REP JTDHD6 6,861,80

CONSENSUS REPORTS: Manganese and its compounds are on the Community Right-To-Know List. Reported in EPA TSCA Inventory.

OSHA PEL: TWA 1 mg(Mn)/m^3
ACGIH TLV: TWA 1 mg(Mn)/m^3
DFG MAK: 1 mg/m^3

SAFETY PROFILE: Experimental reproductive effects. Reacts violently @ <100°. See also MANGANESE COMPOUNDS.

MAV000 CAS:18820-29-6 ***HR: 2***
MANGANESE(II) SULFIDE
mf: MnS mw: 87.00

CONSENSUS REPORTS: Manganese and its compounds are on the Community Right-To-Know List.

SAFETY PROFILE: The dry red sulfide becomes red-hot on exposure to air. When heated to decomposition it emits toxic fumes of SO_x. See also MANGANESE COMPOUNDS and SULFIDES.

MAV250 CAS:12032-88-1 ***HR: 2***
MANGANESE(II) TELLURIDE
mf: MnTe mw: 182.54

CONSENSUS REPORTS: Manganese and its compounds are on the Community Right-To-Know List.

SAFETY PROFILE: Reacts violently with lithium when heated to 230°C. When heated to decomposition it emits toxic fumes of Te. See also TELLURIUM and MANGANESE COMPOUNDS.

MAV500 ***HR: 3***
MANGANESE(II) TETRAHYDROALUMINATE
mf: Al_2H_8Mn mw: 116.96

PROP: Decomp @ −80°C.

CONSENSUS REPORTS: Manganese and its compounds are on the Community Right-To-Know List.

SAFETY PROFILE: Highly unstable; ignites in moist air. See also MANGANESE COMPOUNDS.

MAV750 CAS:12108-13-3 ***HR: 3***
MANGANESE TRICARBONYL METHYLCYCLOPENTADIENYL
mf: $C_9H_7MnO_3$ mw: 218.10

SYNS: AK-33X ◇ ANTIKNOCK-33 ◇ CI-2 ◇ COMBUSTION IMPROVER -2 ◇ METHYLCYCLOPENTADIENYL MANGANESE TRICARBONYL (OSHA) ◇ MMT

TOXICITY DATA with REFERENCE
skn-rbt 100 mg/24H MLD AEHLAU 30,168,75
orl-rat LD50:50 mg/kg TXAPA9 56,353,80
ihl-rat LC50:76 mg/m^3/4H AIHAAP 40,164,79
ipr-rat LD50:23 mg/kg TXAPA9 56,353,80
orl-mus LD50:230 mg/kg AIHAAP 40,164,79
ihl-mus LC50:58600 μg/m^3/4H SAIGBL 20,553,78
ipr-mus LD50:152 mg/kg SKIZAB 34,183,78
orl-dog LDLo:620 mg/kg TMMT** -,-,76
ihl-dog LCLo:489 mg/m^3/2H TMMT** -,-,76
ivn-dog LDLo:10 mg/kg TMMT** -,-,76
skn-rbt LD50:140 mg/kg AIHAAP 40,164,79

CONSENSUS REPORTS: EPA Extremely Hazardous Substances List. Manganese and its compounds are on the Community Right-To-Know List. Reported in EPA TSCA Inventory.

OSHA PEL: TWA 0.2 mg(Mn)/m^3 (skin)
ACGIH TLV: TWA 5 mg(Mn)/m^3

SAFETY PROFILE: Poison by ingestion, inhalation, skin contact, intravenous, and intraperitoneal routes. A skin irritant. When heated to decomposition it emits toxic fumes of CO. See also MANGANESE COMPOUNDS and CARBONYLS.

MAW000 CAS:7783-53-1 ***HR: 3***
MANGANESE TRIFLUORIDE
mf: F_3Mn mw: 111.93

PROP: Red mass; monoclinic crystals. D: 3.54. Stable @ 600°.

CONSENSUS REPORTS: Manganese and its compounds are on the Community Right-To-Know List.

SAFETY PROFILE: Poison irritant. A powerful fluorinating agent. Violent reaction when heated with glass. When heated to decomposition it emits toxic fumes of F^-. See also FLUORIDES and MANGANESE COMPOUNDS.

MAW250 CAS:15825-70-4 ***HR: 3***
MANNITOL HEXANITRATE
DOT: UN 0133
mf: $C_6H_8N_6O_{18}$ mw: 452.17

PROP: Colorless crystals. Bp: explodes @ 120°, d: 1.603 @ O°. Mp: 106-108°. Long needles in regular clusters from alc. Sol in alc and ether; insol in water.

SYNS: DILANGIL ◇ HEXANITROL ◇ HYPERTENAIN ◇ INITIATING EXPLOSIVE NITRO MANNITE (DOT) ◇ MANEXIN ◇ MANICOLE ◇ MANITE ◇ MANNITOL HEXANITRATE, containing, by weight, at least 40% water (DOT) ◇ d-MANNITOL HEXANITRATE ◇ MAXITATE ◇ NITROMANNITE (DOT) ◇ NITROMANNITE (DRY) (DOT) ◇ NITROMANNITOL ◇ SDM No. 5

CONSENSUS REPORTS: Reported in EPA TSCA Inventory.

DOT Classification: Class A Explosive; Label: Explosive A; Forbidden, Dry.

SAFETY PROFILE: Moderately toxic by ingestion and inhalation yielding a fall in blood pressure which may result in weakness, headache, and dizziness. Chronic exposure may produce methemoglobinemia with cyanosis. A powerful explosive sensitive to shock or heat. Upon decomposition it emits toxic fumes of NO_x. See also NITRATES and EXPLOSIVES, HIGH

MAW500 CAS:576-68-1 ***HR: 3***
MANNOMUSTINE
mf: $C_{10}H_{22}Cl_2N_2O_4$ mw: 305.24

SYNS: 1,6-BIS(CHLOROETHYLAMINO)-1,6-BIS-DEOXY-d-MANNITOL ◇ 1,6-BIS(CHLOROETHYLAMINO)-1,6-DIDEOXY-d-MANNITE ◇ 1,6-BIS((β-CHLOROETHYL)AMINO)-1,6-DIDEOXY-d-MANNITOL ◇ 1,6-BIS((2-CHLOROETHYL)AMINO)-1,6-DIDEOXY-d-MANNITOL ◇ DEGRANOL ◇ MANNIT-LOST (GERMAN) ◇ MANNIT-MUSTARD (GERMAN) ◇ MANNITOL NITROGEN MUSTARD

TOXICITY DATA with REFERENCE
dni-hmn:lym 100 μmol/L AGACBH 4,117,74
ipr-rat LD50:56 mg/kg ARZNAD 11,143,61
ivn-rat LD50:56 mg/kg ARZNAD 20,1467,70
ivn-mus LD50:90 mg/kg ANYAA9 68,879,58

ivn-dog LD50:50 mg/kg ANYAA9 68,879,58
ivn-rbt LD50:50 mg/kg ANYAA9 68,879,58
ivn-gpg LDLo:20 mg/kg ANYAA9 68,879,58

SAFETY PROFILE: Poison by intraperitoneal and intravenous routes. Human mutation data reported. An antineoplastic agent. When heated to decomposition it emits very toxic fumes of Cl^- and NO_x.

MAW750 CAS:551-74-6 ***HR: 3***
MANNOMUSTINE DIHYDROCHLORIDE
mf: $C_{10}H_{23}Cl_2N_2O_4$•2ClH mw: 378.13

PROP: Crystals from 80% ethanol. Decomp @ 239-241°. Sol in water; sltly sol in ethanol.

SYNS: 1,6-BIS-(CHLOROETHYLAMINO)-1,6-DESOXY-d-MANNITOLDIHYDROCHLORIDE ◇ 1,6-BIS-(CHLOROETHYLAMINO)-1,6-DIDEOXY-d-MANNITEDIHYDROCHLORIDE ◇ 1,6-DIDEOXY-1,6-DI(2-CHLOROETHYLAMINO)-d-MANNITOLDIHYDROCHLORIDE ◇ MANNITOL MUSTARD DIHYDROCHLORIDE ◇ NSC-9698

TOXICITY DATA with REFERENCE
cyt-hmn:lym 10 μmol/L IPPABX 17,131,81
sce-hmn:lym 3 μmol/L IPPABX 17,131,81
dnd-mus-ipr 200 mg/kg FOBLAN 25,380,79
hma-mus/esc 1 mg/kg MUREAV 21,190,73
orl-rat TDLo:50 mg/kg (13D preg):REP CYGEDX 8(6),44,74
ipr-mus TDLo:23 mg/kg/4W:CAR JNCIAM 36,915,66
ipr-mus TDLo:7265 μg/kg/4W-I:NEO JNCIAM 36,915,66
par-rat LD50:56 mg/kg RRCRBU 52,76,75
scu-mus LD50:120 mg/kg RFECAC 7,296,62
ivn-dog LDLo:8 mg/kg CCSUBJ 2,201,65
ivn-mky LDLo:15 mg/kg CCSUBJ 2,201,65

CONSENSUS REPORTS: IARC Cancer Review: Group 3 IMEMDT 7,56,87; Animal Sufficient Evidence IMEMDT 9,157,75.

SAFETY PROFILE: Poison by intravenous, subcutaneous, and parenteral routes. Experimental reproductive effects. Questionable carcinogen with experimental carcinogenic and neoplastigenic data. Human mutation data reported. A drug used for the treatment of malignant neoplasms. When heated to decomposition it emits very toxic fumes of HCl^- and NO_x. See also MANNOMUSTINE.

MAW800 CAS:7518-35-6 ***HR: D***
MANNOSULFAN
mf: $C_{10}H_{22}O_{14}S_4$ mw: 494.56

SYNS: 1,2,5,6-TETRAMETHANESULFONATE-d-MANNITOL(9CI) ◇ 1,2,5,6-TETRAMESYL-d-MANNITOL ◇ 1,2,5,6-TETRAMETHANESULFONYL-d-MANNITOL ◇ R-52 ◇ TETRA-o-METHYL-SULPHONYL-d-MANNITOL ◇ TMSM ◇ ZITOSTOP

TOXICITY DATA with REFERENCE
mmo-sat 1 mg/plate CNREA8 43,4530,83
oms-hmn:leu 500 μg/L AMSHAR 16,463,68
sce-ham:oth 420 μg/L CNREA8 43,4530,83

SAFETY PROFILE: Human mutation data reported. When heated to decomposition it emits toxic fumes of SO_x. See also SULFONATES.

MAW850 CAS:10347-81-6 ***HR: 3***
MAPROTILINE HYDROCHLORIDE
mf: $C_{20}H_{23}N$•ClH mw: 313.90

SYNS: BA 34276 ◇ CIBA 34 ◇ CIBA 34276 BA ◇ CIBA 34276 HYDROCHLORIDE ◇ LUDIOMIL ◇ 1-(3-METHYLAMINOPROPYL) DIBENZO(b,e)BICYCLO(2.2.2)OCTADIENEHYDROCHLORIDE ◇ 9-(Γ-METHYLAMINOPROPYL)-9,10-DIHYDRO-9,10-ETHANOANTHRACENE HYDROCHLORIDE ◇ N-METHYL-9,10-ETHANOANTHRACENE-9(10H)-PROPANAMINEHYDROCHLORIDE

TOXICITY DATA with REFERENCE
orl-wmn TDLo:15 mg/kg/13D-I:REP AJPSAO 140,641,83
orl-rat TDLo:210 mg/kg (female 9-15D post):TER JZKEDZ 2,69,76
orl-wmn TDLo:7 mg/kg/1W-I JCPYDR 3,264,83
orl-man TDLo:490 μg/kg JCLPDE 47,210,86
orl-hmn TDLo:17 mg/kg:CNS BMJOAE 1,1573,77
orl-rat LD50:760 mg/kg BCFAAI 112,601,73
ipr-rat LD50:72 mg/kg BCFAAI 112,601,73
scu-rat LD50:170 mg/kg BCFAAI 112,601,73
ivn-rat LD50:35 mg/kg JZKEDZ 1,207,75
orl-mus LD50:480 mg/kg JZKEDZ 1,207,75
scu-mus LD50:310 mg/kg JZKEDZ 1,207,75
ivn-mus LD50:31 mg/kg BCFAAI 112,601,73
orl-dog LDLo:20 mg/kg BCFAAI 112,601,73
ivn-dog LD50:20 mg/kg BCFAAI 112,782,73
ivn-cat LDLo:30 mg/kg BCFAAI 112,601,73
ivn-rbt LDLo:20 mg/kg BCFAAI 112,601,73

SAFETY PROFILE: Poison by ingestion, subcutaneous, intravenous, and intraperitoneal routes. Human systemic effects by ingestion, general anesthesia, hallucinations, distorted perceptions, and convulsions. An experimental teratogen. Experimental reproductive effects. An antidepressant. When heated to decomposition it emits toxic fumes of NO_x and HCl.

MAW875 CAS:2212-99-9 ***HR: 3***
MARASMIC ACID
mf: $C_{16}H_{20}O_4$ mw: 276.36

TOXICITY DATA with REFERENCE
mmo-sat 50 μg/disc JANTAJ 36,155,83
dni-omi 100 mg/L JANTAJ 36,155,83
dni-mus:ast 3 mg/L JANTAJ 36,155,83
oms-mus:ast 3 mg/L JANTAJ 36,155,83
ivn-mus LD50:25 mg/kg 85GDA2 6,113,81

SAFETY PROFILE: Poison by intravenous route. Mutation data reported. When heated to decomposition it emits acrid smoke and irritating fumes.

MAX000 CAS:63710-10-1 ***HR: 3***
MARCELLOMYCIN
mf: $C_{42}H_{55}NO_{17}$ mw: 845.98

TOXICITY DATA with REFERENCE
dnd-rat:lvr 11300 nmol/L MOPMA3 14,290,78
dni-mus:leu 950 nmol/L JANTAJ 34,1596,81
oms-mus:leu 50 nmol/L JANTAJ 34,1596,81
ivn-rat TDLo:15 mg/kg:CAR CNREA8 43,5248,83
ipr-mus LD50:10506 μg/kg JANTAJ 30,519,77
ivn-mus LD50:15734 μg/kg DCTODJ 6,21,83
ivn-dog LDLo:2952 μg/kg DCTODJ 6,21,83

SAFETY PROFILE: Questionable carcinogen with experimental carcinogenic data. Poison by intraperitoneal and intravenous routes. Mutation data reported. When heated to decomposition it emits toxic fumes of NO_x.

MAX275 CAS:303-25-3 ***HR: 3***
MAREZINE HYDROCHLORIDE
mf: $C_{18}H_{22}N_2 \cdot ClH$ mw: 302.88

SYNS: N-BENZHYDRYL-N'-METHYLPIPERAZINE HYDROCHLORIDE ◇ N-BENZHYDRYL-N'-METHYLPIPERAZINE MONOHYDROCHLORIDE ◇ CYCLIZINE CHLORIDE ◇ CYCLIZINE HYDROCHLORIDE ◇ (±)-1-DIPHENYLMETHYL-4-METHYLPIPERAZINE HYDROCHLORIDE

TOXICITY DATA with REFERENCE
par-rat TDLo:150 mg/kg (female 7-9D post):REP COREAF 256,3359,63
par-rbt TDLo:350 mg/kg (female 8-14D post):TER COREAF 256,3359,63
orl-mus LD50:165 mg/kg JPETAB 112,297,54
ipr-mus LD50:58 mg/kg PHMGBN 13,241,75
ims-pgn LD50:106 mg/kg JAPMA8 46,140,57

SAFETY PROFILE: Poison by ingestion, intramuscular, and intraperitoneal routes. An experimental teratogen. Experimental reproductive effects. When heated to decomposition it emits toxic fumes of NO_x and HCl.

MBU500 CAS:8015-01-8 ***HR: 1***
MARJORAM OIL, SPANISH

PROP: Main constituent is cineole. From steam distillation of the flowering plant material from the shrub *Thymus mastichina* L. (Fam. *Labiatae)* (FCTXAV 14,443,76). Faintly yellow liquid. D: 0.904-0.920, refr index: 1.463 @ 20°. Sol in fixed oils. Insol in glycerin, propylene glycol, mineral oil.

SYNS: OIL of MARJORAM, SPANISH ◇ SPANISH MARJORAM OIL

TOXICITY DATA with REFERENCE
skn-rbt 500 mg/24H MLD FCTXAV 14,467,76

CONSENSUS REPORTS: Reported in EPA TSCA Inventory.

SAFETY PROFILE: A skin irritant. When heated to decomposition it emits acrid smoke and irritating fumes.

MBU550 ***HR: 2***
MARSH MARIGOLD

PROP: Perennial herbs with large kidney-shaped leaves and long hollow stems. It produces yellow or white flowers in the spring. Various species grow wild in the region bounded by North Carolina, Washington and Alaska.

SYNS: BULL FLOWER ◇ CALTHA (VARIOUS SPECIES) ◇ C. LEPTOSEPALA ◇ COWSLIP ◇ C. PALUSTRIS ◇ GOOLS ◇ HORSE BLOB ◇ KINGCUP ◇ MAY BLOB ◇ MEADOW BRIGHT ◇ POPULAGE ◇ SOLDIER'S BUTTONS ◇ SOUCI D'EAU (CANADA) ◇ WATER GOGGLES

SAFETY PROFILE: All parts of the mature plant contain the direct irritant protoanemonin. The immature plant is edible if boiled. Chewing any part of the plant may causes inflammation and blistering of the mouth and throat. Ingestion may cause vomiting and diarrhea with blood.

MBU750 ***HR: 3***
MARSH ROSEMARY EXTRACT

PROP: Tannin containing extract of root (JNCIAM 57,207,76).

SYNS: LIMONIUM NASHII ◇ TANNIN from LIMONIUM NASHII ◇ TANNIN from MARSH ROSEMARY

TOXICITY DATA with REFERENCE
scu-rat TDLo:530 mg/kg/66W-I:NEO JNCIAM 57,207,76

SAFETY PROFILE: Questionable carcinogen with experimental neoplastigenic data. When heated to decomposition it emits acrid smoke and irritating fumes.

MBU775 CAS:78354-52-6 ***HR: 1***
MARZULENE S
mf: $C_{15}H_{18}O_3S \cdot C_5H_{10}N_2O_3 \cdot Na$ mw: 447.55

TOXICITY DATA with REFERENCE
orl-rat TDLo:195 mg/kg (30D pre):REP KSRNAM 11,510,77
orl-rat LD50:14730 mg/kg KSRNAM 11,510,77
orl-mus LD50:10180 mg/kg KSRNAM 11,510,77

SAFETY PROFILE: Slightly toxic by ingestion. Experimental reproductive effects. When heated to decomposition it emits toxic fumes of SO_x, NO_x, and Na_2O.

MBU780 ***HR: 1***
MASTWOOD

PROP: A tall tree with paired 3- to 8-inch oval leaves with a notch at the tip. It has a light grey outer bark and a pink inner bark. It bears a round fruit about 1.5 inches in diameter containing a large shelled seed. The tree is native to India and Malasia. It is a common ornamental in south Florida, Hawaii, the West Indies and Guam.

SYNS: ALEXANDRIAN LAUREL ◇ BEAUTYLEAF ◇ CALOPHYLLUM INOPHYLLUM ◇ INDIAN LAUREL ◇ KAMANI (HAWAII) ◇ LAURELWOOD ◇ MARIA GRANDE (PUERTO RICO)

SAFETY PROFILE: The seed is toxic and its ingestion may result in nausea and persistent vomiting. The sap is used as a home remedy and may cause inflammation on contact with the cornea.

MBU800 ***HR: 3***
MAY APPLE

PROP: An herb which grows to 1.5 feet with 2 large (1-foot diameter) umbrella-shaped leaves. A single, white, 2-inch flower grows between the leaf and stem. The yellow-green fruit is the size and shape of an egg. Sterile plants have only one leaf.

SYNS: AMERICAN MANDRAKE ◇ BEHEN ◇ DEVIL'S APPLE ◇ HOG APPLE ◇ INDIAN APPLE ◇ PODOPHYLLUM PELTATUM ◇ RACCOON BERRY ◇ UMBRELLA LEAF ◇ WILD JALAP ◇ WILD LEMON

SAFETY PROFILE: The whole plant contains the poisons podophylloresin (a purgative), its glucoside, and α- and β-peltatin (an antimitotic). The fruit is less poisonous. Ingestion of plant parts or extracts causes vomiting and diarrhea. Ingestion of large amounts or repeated skin contact has caused: kidney failure, intestinal blockages, blood abnormalities, coma and death. Industrial workers processing the roots have experienced eye and skin irritation.

MBU820 CAS:35846-53-8 ***HR: 3***
MAYTANSINE
mf: $C_{34}H_{46}ClN_2O_{10}$ mw: 678.27

PROP: Mp: 171-172°. Active principle found in *Maytenus serrata* (BIHAA2 43,495,76).

SYNS: MAITANSINE ◇ MAYSANINE ◇ MAYT ◇ NSC-153858

TOXICITY DATA with REFERENCE
dni-nml:emb 60 nmol/L SCIEAS 189,1002,75
dni-mus:leu 100 nmol/L BCPCA6 24,751,75
oms-mus/ast 100 μg/kg JNCIAM 60,649,78
cyt-mus/ast 100 μg/kg JNCIAM 60,649,78
ipr-mus TDLo:250 μg/kg (female 8D post):TER TJADAB 18,31,78
ivn-hmn TDLo:190 μg/kg/5D:GIT,CNS JNCIAM 60,93,78
scu-rat LD50:480 mg/kg CTRRDO 61,1333,77
ipr-mus LD50:245 μg/kg NCISP* JAN86
ivn-mus LD50:1530 μg/kg NTIS** PB82-165507

SAFETY PROFILE: A deadly poison by intraperitoneal and intravenous routes. Moderately toxic by subcutaneous route. Human systemic effects by intravenous route, hallucinations, distorted perceptions, change in motor activity, and nausea or vomiting. An experimental teratogen. Mutation data reported. Used as an antineoplastic agent. When heated to decomposition it emits very toxic fumes of Cl^- and NO_x.

MBU825 ***HR: 2***
MAY THORN

PROP: Small trees that may grow to 12 feet. *R. californica* is an evergreen with finely toothed leaves. It produces a fruit which contains green, black, or red seeds. It grows wild in California. *R. cathartica* has toothed leaves and scaley buds. It produces clusters of green-white flowers and a red to black fruit with 4 seeds. *R. frangula* has smooth leaves and buds. It produces flat clusters of flowers and a fruit with 3 seeds. *R. cathartica* and *R. frangula* grow wild in the northeastern United States and Canada. Other species are found throughout temperate North America.

SYNS: ALDER BUCKTHORN ◇ ARROW WOOD ◇ BERRY ALDER ◇ BLACK DOGWOOD ◇ BUCKTHORN ◇ CASCARA ◇ COFFEBERRY ◇ HART'S HORN ◇ NERPRUN (CANADA) ◇ PERSIAN BERRY ◇ PURGING BUCKTHORN ◇ RHAMNUS CALIFORNICA ◇ RHAMNUS CATHARTICA ◇ RHAMNUS FRANGULA ◇ RHINE BERRY

SAFETY PROFILE: The fruit and bark contain poisonous hydroxymethylanthraquinones. Ingestion of these plant parts may cause nausea, vomiting, and diarrhea.

MBV100 CAS:32891-29-5 ***HR: 3***
MAZATICOL HYDROCHLORIDE
mf: $C_{21}H_{27}NO_3S_2 \cdot ClH$ mw: 442.07

TOXICITY DATA with REFERENCE
orl-rat LD50:1182 mg/kg IYKEDH 7,614,76
ipr-rat LD50:105 mg/kg IYKEDH 7,614,76
orl-mus LD50:858 mg/kg IYKEDH 7,614,76
ipr-mus LD50:168 mg/kg IYKEDH 7,614,76
scu-mus LD50:521 mg/kg IYKEDH 7,614,76
ivn-mus LD50:20200 μg/kg IYKEDH 7,614,76

SAFETY PROFILE: Poison by intravenous and intraperitoneal routes. Moderately toxic by ingestion and subcutaneous routes. When heated to decomposition it emits toxic fumes of SO_x, NO_x, and HCl. See also ESTERS.

MBV250 CAS:22232-71-9 ***HR: 3***
MAZINDOL
mf: $C_{16}H_{13}ClN_2O$ mw: 284.76

PROP: Crystals from acetone-hexane. Mp: 198-199°.

SYNS: 5-p-CHLOROPHENYL-2,3-DIHYDRO-5H-IMIDAZO(2,1-A) ISOINDOL-5-OL ◇ SA 42-548

TOXICITY DATA with REFERENCE
dni-hmn:fbr 600 mg/L MUREAV 169,171,86
msc-ham:lng 1 g/L MUREAV 169,171,86
orl-man TDLo:13 μg/kg (male 1D pre):REP BMJOAE 287,1763,83
orl-rat TDLo:330 mg/kg (female 7-17D post):TER KSRNAM 20,2279,86
orl-wmn TDLo:140 μg/kg/1W-I AJPSAO 141,1497,84
orl-rat LDLo:180 mg/kg FEPRA7 27,598,68
orl-mus LD50:106 mg/kg FEPRA7 27,598,68
orl-dog LDLo:9 mg/kg FEPRA7 27,598,68
orl-rbt LD50:98 mg/kg FEPRA7 27,598,68

SAFETY PROFILE: Poison by ingestion. An experimental teratogen. Other experimental reproductive effects. Human mutation data reported. An FDA proprietary drug. An anorectic drug. When heated to decomposition it emits very toxic fumes of Cl^- and NO_x.

MBV500 CAS:64521-14-8 ***HR: 3***
7-MBA-3,4-DIHYDRODIOL
mf: $C_{18}H_{16}O_2$ mw: 264.34

SYN: trans-3,4-DIHYDRO-3,4-DIHYDROXY-7-METHYLBENZ(a)ANTHRACENE

TOXICITY DATA with REFERENCE
msc-ham:lng 1 mg/L IJCNAW 19,828,77
msc-ham:ovr 10 mg/L IJCNAW 19,828,77
mma-sat 10 μmol/L BBRCA9 75,427,77
otr-mus:fbr 1 mg/L IJCNAW 19,828,77
sce-ham:ovr 4 mg/L MUREAV 129,365,84
skn-mus TDLo:1000 μg/kg:NEO CALEDQ 3,247,77

CONSENSUS REPORTS: EPA Genetic Toxicology Program.

SAFETY PROFILE: Questionable carcinogen with experimental neoplastigenic data by skin contact. Mutation data reported. When heated to decomposition it emits acrid smoke and irritating fumes.

MBV700 ***HR: 2***
MB PYRETHROID
mf: $C_{20}H_{22}O_3$ mw: 310.42

SYNS: 2,2-DIMETHYL-3-(2-METHYLPROPYL)CYCLOPROPANECARBOXYLIC ACID-p-(METHOXYMETHYL)BENZYL ESTER ◇ 4-(METHYLOXYMETHYL)BENZYL CHRYSANTHEMUMMONOCARBOXYLATE

TOXICITY DATA with REFERENCE
eye-rbt 100 mg MLD CHYCDW 17,8,83
orl-rat LD50:900 mg/kg CHYCDW 17,8,83
orl-mus LD50:1747 mg/kg CHYCDW 17,8,83

SAFETY PROFILE: Moderately toxic by ingestion. An eye irritant. When heated to decomposition, it emits acrid smoke and irritating fumes.

MBV710 CAS:25061-59-0 ***HR: 3***
MBR 3092-42
mf: $C_{13}H_7F_{15}N{\bullet}CF_3O_3S$ mw: 611.28

SYNS: N-(1,1-DIHYDROPERFLUOROOCTYL)PYRIDINIUM TRIFLUOROMETHANESULFONATE ◇ 1-(2,2,3,3,4,4,5,5,6,6,7,7,8,8,8-PENTADECAFLUOROOCTYL)PYRIDINIUM TRIFLUOROMETHANESULFONATE

TOXICITY DATA with REFERENCE
eye-rbt:100 mg JPMSAE 59,188,70
orl-rat LD50:1159 mg/kg JPMSAE 59,188,70
ipr-rat LD50:18 mg/kg JPMSAE 59,188,70
orl-mus LD50:925 mg/kg JPMSAE 59,188,70
ipr-mus LD50:30 mg/kg JPMSAE 59,188,70

SAFETY PROFILE: Poison by intraperitoneal route. Moderately toxic by ingestion. An eye irritant. When heated to decomposition it emits toxic fumes of F^-, NO_x, and SO_x. See also SULFONATES.

MBV720 ***HR: 2***
MDBCP
mf: $C_4H_7Br_2Cl$ mw: 250.38

SYNS: 1,2-DIBROMO-3-CHLORO-2-METHYLPROPANE ◇ METHYLDBCP ◇ 2-METHYL-1,2-DIBROMO-3-CHLOROPROPANE

TOXICITY DATA with REFERENCE
orl-rat LD50:780 mg/kg TOXID9 4,20,84
ihl-rat LC50:225 ppm/6H TOXID9 4,67,84
ihl-mus LC50:227 ppm/6H TOXID9 4,67,84

SAFETY PROFILE: Moderately toxic by inhalation and ingestion. When heated to decomposition it emits toxic fumes of Cl^- and Br^-.

MBV735 CAS:75841-84-8 ***HR: 3***
MDL-899
mf: $C_{14}H_{19}N_5O{\bullet}ClH$ mw: 309.84

SYN: N-(2,5-DIMETHYL-1H-PYRROL-1-YL)-6-(4-MORPHOLINYL)-3-PYRIDAZINAMINE HYDROCHLORIDE

TOXICITY DATA with REFERENCE
orl-rat LD50:2250 mg/kg ARZNAD 35,818,85
ivn-rat LD50:124 mg/kg ARZNAD 35,818,85
orl-dog LD50:2200 mg/kg ARZNAD 35,818,85
ivn-dog LD50:275 mg/kg ARZNAD 35,818,85

SAFETY PROFILE: Poison by intravenous route. Moderately toxic by ingestion. When heated to decomposition it emits toxic fumes of NO_x and HCl.

MBV750 CAS:2126-84-3 ***HR: 3***
MEBANAZINE OXALATE
mf: $C_8H_{12}N_2{\bullet}C_2H_2O_4$ mw: 226.26

SYN: α-PHENYLETHYLHYDRAZINE OXALATE

TOXICITY DATA with REFERENCE
mmo-sat 4400 nmol/plate CNREA8 41,1469,81

mma-sat 4400 nmol/plate CNREA8 41,1469,81
dnr-esc 2200 nmol/plate JTEHD6 9,287,82
dnd-mus-orl 750 mg/kg/5D-C JTEHD6 9,287,82
orl-rat LD50:458 mg/kg IJNEAQ 5,125,66
orl-mus LD50:254 mg/kg IJNEAQ 5,125,66
ipr-mus LD50:176 mg/kg EJPHAZ 6,115,69
scu-mus LD50:121 mg/kg IJNEAQ 5,125,66
ivn-mus LD50:85 mg/kg IJNEAQ 5,125,66

SAFETY PROFILE: Poison by ingestion, intraperitoneal, subcutaneous, and intravenous routes. Mutation data reported. When heated to decomposition it emits toxic fumes of NO_x. See also OXALATES.

MBV775 CAS:16550-39-3 ***HR: D***
MEBANE SODIUM SALT
mf: $C_{16}H_{19}O_2$•Na mw: 266.34

SYNS: 3-ETHYL-2-METHYL-4-PHENYL-4-CYCLOHEXENECARBOXYLIC ACID SODIUM SALT ◇ 5-ETHYL-6-METHYL-4-PHENYL-3-CYCLOHEXENE-1-CARBOXYLIC ACID SODIUM SALT ◇ 2-METHYL-3-ETHYL-4-PHENYL-Δ^4-CYCLOHEXENE CARBOXYLIC ACID SODIUM SALT ◇ ORF 4563

TOXICITY DATA with REFERENCE
orl-rat TDLo:25 μg/kg (female 1D post):REP JPETAB 167,105,69

SAFETY PROFILE: Experimental reproductive effects. When heated to decomposition it emits toxic fumes of Na_2O.

MBW100 CAS:6153-33-9 ***HR: 3***
MEBHYDROLIN NAPADISYLATE
mf: $C_{19}H_{20}N_2$ mw: 276.41

SYN: 2,3,4,5-TETRAHYDRO-2-METHYL-5-(PHENYLMETHYL)-1H-PYRIDO(4,3-b)INDOLE

TOXICITY DATA with REFERENCE
ipr-rat LD50:300 mg/kg NIIRDN 6,850,82
scu-mus LD50:150 mg/kg NIIRDN 6,850,82
ivn-mus LD50:40 mg/kg NIIRDN 6,850,82

SAFETY PROFILE: Poison by subcutaneous, intravenous, and intraperitoneal routes. When heated to decomposition it emits toxic fumes of NO_x.

MBW250 CAS:101809-59-0 ***HR: 2***
MEBICAR

PROP: A compound in the group of N-alkylbiscyclobisurea group; a psychotropic agent with properties occupying the middle position between tranquilizers and neuroleptics (RPTOAN 40,206,77).

TOXICITY DATA with REFERENCE
ipr-rat LD50:3450 mg/kg RPTOAN 40,206,77
ipr-mus LD50:3800 mg/kg RPTOAN 40,206,77

SAFETY PROFILE: Moderately toxic by intraperitoneal route. Human psychotropic effects. When heated to decomposition it emits acrid smoke and irritating fumes.

MBW500 CAS:101809-59-0 ***HR: 2***
MEBICAR-A

PROP: A derivative of bicyclic bis-ureas (FATOAO 40,684,77).

TOXICITY DATA with REFERENCE
ipr-rat LD50:3450 mg/kg FATOAO 40,684,77
ipr-mus LD50:3800 mg/kg FATOAO 40,684,77

SAFETY PROFILE: Moderately toxic by intraperitoneal route. When heated to decomposition it emits acrid smoke and irritating fumes. See also MEBICAR.

MBW750 CAS:64-55-1 ***HR: 3***
MEBUTAMATE
mf: $C_{10}H_{20}N_2O_4$ mw: 232.28

PROP: Crystals. Mp: 77-79°. Sol in most organic solvents. Very sltly sol in water.

SYNS: BUTATENSIN ◇ 2-sec-BUTYL-2-METHYL-1,3-PROPANEDIOL DICARBAMATE ◇ 2-sec-BUTYL-2-METHYLTRIMETHYLENE DICARBAMATE ◇ CARBAMIC ACID-2-sec-BUTYL-2-METHYLTRIMETHYLENE ESTER ◇ CARBUTEN ◇ DICAMOYLMETHTANE ◇ 2,2-DICARBAMYLOXYMETHYL-3-METHYLPENTANE ◇ DORMATE ◇ IPOTENSIVO ◇ MEBUTINA ◇ 2-METHYL-2-sec-BUTYL-1,3-PROPANEDIOL DICARBAMATE ◇ NO-PRESS ◇ PREAN ◇ SIGMAFON ◇ VALLENE

TOXICITY DATA with REFERENCE
orl-rat LD50:1160 mg/kg JPETAB 134,356,61
ipr-rat LD50:410 mg/kg JPETAB 134,356,61
orl-mus LD50:550 mg/kg JPETAB 134,356,61
ipr-mus LD50:460 mg/kg JMCMAR 12,462,69
orl-bwd LD50:100 mg/kg TXAPA9 21,315,72

SAFETY PROFILE: Poison by ingestion. Moderately toxic by intraperitoneal route. An FDA proprietary drug. When heated to decomposition it emits toxic fumes of NO_x.

MBW775 CAS:53-44-1 ***HR: D***
MEC
mf: $C_{23}H_{32}NO_2$ mw: 354.56

SYNS: 3-METHOXY-17-β-CYANETHOXYESTRA-1,3,5(10)-TRIEN ◇ 3-METHOXY-17-β-CYANETHOXYOESTRA-1,3,5(10)-TRIEN ◇ 3-((3-METHOXYESTRA-1,3,5(10)-TRIEN-17-β-YL)OXY)PROPIONITRILE ◇ RS-2196

TOXICITY DATA with REFERENCE
scu-rbt TDLo:300 μg/kg (female 1-3D post):TER FESTAS 20,211,69
scu-rat TDLo:150 μg/kg (1D pre):REP ACENA7 49,193,65

CONSENSUS REPORTS: Cyanide and its compounds are on the Community Right-To-Know List.

SAFETY PROFILE: An experimental teratogen. Experimental reproductive effects. When heated to decomposition it emits toxic fumes of NO_x and CN^-. See also NITRILES.

MBW780 CAS:73561-96-3 ***HR: 3***
MECARBENIL
mf: $C_6H_{10}N_2O_2$ mw: 142.18

SYN: o-METHYLCARBAMOYL-2-METHYLPROPENEALDOXIME

TOXICITY DATA with REFERENCE
orl-rat LD50:79 mg/kg GISAAA 49(4),90,84
orl-mus LD50:79 mg/kg GISAAA 49(4),90,84
orl-rbt LD50:79 mg/kg GISAAA 49(4),90,84

SAFETY PROFILE: Poison by ingestion. When heated to decomposition it emits toxic fumes of NO_x.

MBX000 CAS:26225-59-2 ***HR: 3***
MECINARONE
mf: $C_{24}H_{27}NO_6$ mw: 425.52

SYN: 5-(p-METHOXY-CINNAMOYL)-4,7-DIMETHOXY-6-DIMETHYLAMINOETHOXYBENZOFURAN

TOXICITY DATA with REFERENCE
orl-mus LD50:244 mg/kg ARZNAD 25,782,75
ivn-mus LD50:25 mg/kg DRFUD4 1,133,76
ims-mus LD50:63 mg/kg DRFUD4 1,133,76

SAFETY PROFILE: Poison by ingestion, intravenous, and intramuscular routes. When heated to decomposition it emits toxic fumes of NO_x.

MBX250 CAS:1104-22-9 ***HR: 2***
MECLIZINE DIHYDROCHLORIDE
mf: $C_{25}H_{27}ClN_2•2ClH$ mw: 463.91

PROP: Very sol in chloroform and pyridine. Sltly sol in dilute acids, alc. Insol in H_2O, ether.

SYNS: ANCOLAN DIHYDROCHLORIDE ◇ 1-p-CHLORBENZHYDRYL-m-METHYLBENZYLPIPERAZINEDIHYDROCHLORIDE

TOXICITY DATA with REFERENCE
orl-rat TDLo:1 g/kg (8-15D preg):REP MPHEAE 15,375,66
orl-rat TDLo:800 mg/kg (8-15D preg):TER MPHEAE 15,375,66
orl-mus LD50:1600 mg/kg 29ZVAB -,67,69
ipr-mus LD50:625 mg/kg JAPMA8 43,653,54

CONSENSUS REPORTS: Reported in EPA TSCA Inventory.

SAFETY PROFILE: Moderately toxic by ingestion and intraperitoneal routes. An experimental teratogen. Experimental reproductive effects. When heated to decomposition it emits very toxic fumes of Cl^- and NO_x.

MBX500 CAS:36236-67-6 ***HR: 2***
MECLIZINE HYDROCHLORIDE
mf: $C_{25}H_{27}ClN_2•ClH$ mw: 427.45

SYNS: BONINE ◇ 1-(p-CHLORO-α-PHENYLBENZYL)-4-(m-METHYLBENZYL)PIPERAZINE HYDROCHLORIDE ◇ MECLOZINE HYDROCHLORIDE

TOXICITY DATA with REFERENCE
ipr-mus TDLo:25 mg/kg (female 10D post):TER CAJPBD 3,2,63
orl-rat TDLo:100 mg/kg (11-14D preg):REP SCIEAS 141,353,63
orl-mus LD50:1600 mg/kg NIIRDN 6,814,82
ipr-mus LD50:625 mg/kg NIIRDN 6,814,82

SAFETY PROFILE: Moderately toxic by ingestion and intraperitoneal routes. An experimental teratogen. Experimental reproductive effects. An FDA over-the-counter drug. When heated to decomposition it emits very toxic fumes of NO_x, and Cl^-.

MBX800 CAS:18598-63-5 ***HR: 2***
MECYSTEINE HYDROCHLORIDE
mf: $C_4H_{10}ClNO_2S$ mw: 171.66

PROP: Crystals from methanol. Mp: 140-141°.

SYNS: ACDRILE ◇ ACTIOL ◇ ETHYL ESTER l-CYSTEINE HYDROCHLORIDE (9CI) ◇ L.J. 48 ◇ METHYL CYSTEINE HYDROCHLORIDE ◇ PECTITE ◇ VISCLAIR ◇ ZEOTIN

TOXICITY DATA with REFERENCE
orl-rat TDLo:30 g/kg (30D male):REP KSRNAM 14,1407,80
orl-mus LD50:2333 mg/kg NIIRDN 6,829,82
ipr-mus LD50:1340 mg/kg NIIRDN 6,829,82

SAFETY PROFILE: Moderately toxic by ingestion and intraperitoneal routes. Experimental reproductive effects. When heated to decomposition it emits toxic fumes of SO_x, NO_x, and HCl. See also ESTERS.

MBY000 CAS:2898-11-5 ***HR: 3***
MEDAZEPAM HYDROCHLORIDE
mf: $C_{16}H_{15}ClN_2•ClH$ mw: 307.24

PROP: Orange-red, crystalline powder. Very sol in H_2O, alc.

SYNS: 7-CHLOR-2,3-DIHYDRO-1-METHYL-5-PHENYL-1H-1,4-BENZODIAZEPIN HYDROCHLORID (GERMAN) ◇ 7-CHLORO-2,3-DIHYDRO-1-METHYL-5-PHENYL-1H-1,4-BENZODIAZEPINE,HYDROCHLORIDE

TOXICITY DATA with REFERENCE
orl-rat TDLo:1456 mg/kg (26W pre):REP KSRNAM 4,833,70
orl-rat LD50:900 mg/kg TXAPA9 18,185,71
ipr-rat LD50:210 mg/kg ARZNAD 18,1542,68
scu-rat LD50:1700 mg/kg ARZNAD 18,1542,68

ivn-rat LD50:91 mg/kg KSRNAM 4,833,70
orl-mus LD50:710 mg/kg ARZNAD 18,1542,68
ipr-mus LD50:314 mg/kg KSRNAM 4,833,70
scu-mus LD50:720 mg/kg ARZNAD 18,1542,68
ivn-mus LD50:47 mg/kg KSRNAM 4,833,70
orl-rbt LD50:530 mg/kg ARZNAD 18,1542,68

SAFETY PROFILE: Poison by intraperitoneal and intravenous routes. Moderately toxic by ingestion and subcutaneous routes. Experimental reproductive effects. When heated to decomposition it emits very toxic fumes of HCl and NO_x.

MBY150 CAS:35457-80-8 ***HR: 2***
MEDEMYCIN
mf: $C_{41}H_{67}NO_{15}$ mw: 814.09

SYNS: ANTIBIOTIC SF 837 ◇ ANTIBIOTIC SF 837 A1 ◇ ANTIBIOTIC SF 837 A_1 ◇ ANTIBIOTIC YL 704 B_1 ◇ ANTIBIOTIC YL 704 B1 ◇ ESPINOMYCIN A ◇ MIDECAMYCIN ◇ MIDECAMYCIN A_1 ◇ MYDECAMYCIN ◇ PLATENOMYCIN B1 ◇ SF 837 ◇ TURIMYCIN P_3 ◇ YL 704 B_1

TOXICITY DATA with REFERENCE
orl-rat TDLo:6 g/kg (female 10-15D post):REP JJANAX 25,187,72
orl-rat TDLo:18 g/kg (10-15D preg):TER JJANAX 25,187,72
orl-rat TDLo:9600 mg/kg NIIRDN 6,810,82
orl-mus TDLo:5800 mg/kg NIIRDN 6,810,82
ivn-mus LD50:1000 mg/kg 85GDA2 2,83,80

SAFETY PROFILE: Moderately toxic by intravenous route. Mildly toxic by ingestion. An experimental teratogen. Other experimental reproductive effects. When heated to decomposition it emits toxic fumes of NO_x.

MBY500 CAS:70441-83-7 ***HR: 3***
2,4-MEDP
mf: $C_{14}H_{17}N_3O_2{\cdot}2Cl$ mw: 330.24

SYN: PYRIDINIUM, 2-FORMYL-4'-METHYL-1,1'-(OXYDIMETHYLENE)DI-, DICHLORIDE, OXIME

TOXICITY DATA with REFERENCE
ipr-rat LD50:178 mg/kg ARTODN 41,301,79
ivn-rat LD50:170 mg/kg ARTODN 41,301,79

SAFETY PROFILE: Poison by intraperitoneal and intravenous routes. When heated to decomposition it emits very toxic fumes of NO_x and Cl^-.

MBZ000 CAS:70441-82-6 ***HR: 3***
4,4-MEDP
mf: $C_{14}H_{17}N_3O_2{\cdot}2Cl$ mw: 330.24

SYN: 4-FORMYL-4'-METHYL-1,1'-(OXYDIMETHYLENE)DIPYRIDINIUM, DICHLORIDE OXIME

TOXICITY DATA with REFERENCE
ipr-rat LD50:119 mg/kg ARTODN 41,301,79
ivn-rat LD50:100 mg/kg ARTODN 41,301,79

SAFETY PROFILE: Poison by intraperitoneal and intravenous routes. When heated to decomposition it emits very toxic fumes of NO_x and Cl^-.

MBZ100 CAS:977-79-7 ***HR: 2***
MEDROGESTONE
mf: $C_{23}H_{32}O_2$ mw: 340.55

PROP: Crystals from ether. Mp: 144-146°.

SYNS: AY-62022 ◇ AY 136155 ◇ COLPRO ◇ COLPRONE ◇ 6,17-DIMETHYLPREGNA-4,6-DIENE-3,20-DIONE ◇ ETOGYN ◇ MEDROGESTERONE ◇ 6-METHYL-6-DEHYDRO-17-METHYLPROGESTERONE ◇ METROGESTONE ◇ PROTHIL

TOXICITY DATA with REFERENCE
scu-rat TDLo:26250 μg/kg (female 21D pre):REP JRPFA4 12,473,66
scu-rat TDLo:300 mg/kg (female 15-20D post):TER JRPFA4 12,473,66
orl-gpg LD50:850 mg/kg USXXAM #4230702

SAFETY PROFILE: Moderately toxic by ingestion. An experimental teratogen. Experimental reproductive effects. A steroid. When heated to decomposition it emits acrid smoke and irritating fumes.

MBZ150 CAS:520-85-4 ***HR: D***
MEDROXYPROGESTERONE
mf: $C_{22}H_{32}O_3$ mw: 344.54

PROP: Crystals from chloroform. Mp: 220-223.5°.

SYNS: FARLUTAL ◇ MEDROXYPROGESTERON ◇ 6-α-METHYL-17-α-HYDROXYPROGESTERONE ◇ U 8840

TOXICITY DATA with REFERENCE
scu-rbt TDLo:5 mg/kg (female 1D pre):REP 85GRAA -,57,65
orl-rat TDLo:20 mg/kg (17-20D preg):TER ECJPAE 24,77,77

SAFETY PROFILE: An experimental teratogen. Experimental reproductive effects. A steroid. When heated to decomposition it emits acrid smoke and irritating fumes.

MCA000 CAS:71-58-9 ***HR: 3***
MEDROXYPROGESTERONE ACETATE
mf: $C_{24}H_{34}O_4$ mw: 386.58

PROP: White to off-white, odorless, crystalline powder. Melting range 207-209°. Insol in water; freely sol in chloroform; sparingly sol in alc.

SYNS: 17-α-ACETOXY-6-α-METHYLPREGN-4-ENE-3,20-DIONE ◇ 17-ACETOXY-6-α-METHYLPROGESTERONE ◇ (6-α)-17-(ACETYLOXY)-6-METHYLPREG-4-ENE-3,20-DIONE ◇ DEPO-PROVERA ◇ FARLUTIN ◇ 17-α-HYDROXY-6-α-METHYLPREGN-4-ENE-3,20-DIONE ACETATE

◇ 17-HYDROXY-6-α-METHYLPREGN-4-ENE-3,20-DIONE ACETATE ◇ 17-α-HYDROXY-6-α-METHYLPROGESTERONE ACETATE ◇ 6-α-METHYL-17-α-ACETOXYPREGN-4-ENE-3,20-DIONE ◇ 6-α-METHYL-17-α-ACETOXYPROGESTERONE ◇ 6-α-METHYL-17-α-HYDROXYPROGESTERONE ACETATE ◇ 6-α-METHYL-4-PREGNENE-3,20-DION-17-α-OL ACETATE ◇ METIPREGNONE ◇ NOGEST ◇ ORAGEST ◇ PERLUTEX ◇ REPROMIX

TOXICITY DATA with REFERENCE
dni-hmn:lym 50 μmol/L PSEBAA 146,401,74
unr-wmn TDLo:12 mg/kg (female 6-20W post):REP SCIEAS 211,1171,81
unr-wmn TDLo:3500 μg/kg (female 6W post):TER OBGNAS 23,931,64
imp-rat TDLo:25 mg/kg:NEO BJCAAI 59,210,89
scu-mus TDLo:9600 mg/kg/1Y-I:CAR CALEDQ 33,215,86
ims-dog TD:324 mg/kg/4Y-I:ETA FESTAS 31,340,79
ivn-wmn TDLo:10 mg/kg:EYE ADVPB4 8,103,72
ivn-wmn TDLo:21 mg/kg:EYE ADVPB4 8,103,72

CONSENSUS REPORTS: IARC Cancer Review: Group 2B IMEMDT 7,289,87; Animal Limited Evidence IMEMDT 21,417,79; IMEMDT 6,157,74; Human Inadequate Evidence IMEMDT 21,417,79. Reported in EPA TSCA Inventory. EPA Genetic Toxicology Program.

SAFETY PROFILE: Suspected carcinogen with experimental carcinogenic, neoplastigenic, tumorigenic, and teratogenic data. Human systemic effects by intravenous route: increased intraocular pressure. Human teratogenic effects by an unspecified route: developmental abnormalities of the urogenital system. Human reproductive effects by multiple routes: spermatogenesis, menstrual cycle changes or disorders, postpartum effects, female fertility effects, abortion, newborn behavioral effects. Human mutation data reported. An experimental teratogen. Experimental reproductive effects. A drug for the treatment of secondary amenorrhoea and dysfunctional uterine bleeding. When heated to decomposition it emits acrid smoke and irritating fumes.

MCA025 CAS:13345-50-1 ***HR: 3***
MEDULLIN
mf: $C_{20}H_{30}O_4$ mw: 334.50

SYNS: 7-(2-(3-HYDROXY-1-OCTENYL)-5-OXO-3-CYCLOPENTEN-1-YL)-5-HEPTENOIC ACID ◇ 15-HYDROXY-9-OXO-PROSTA-5,10-13-TRIEN-1-OIC ACID, (5Z,13E,15S)- (9CI) ◇ PGA2 ◇ PGA^2 ◇ 5,6-cis-PGA^2 ◇ (155)-PGA^2 ◇ PROSTAGLANDIN A2 ◇ (+)-PROSTAGLANDIN A^2

TOXICITY DATA with REFERENCE
scu-mus TDLo:5 mg/kg (16D preg):TER PRGLBA 1,191,72
scu-mus TDLo:10 mg/kg (16D preg):REP PRGLBA 1,191,72
ipr-mus LD50:93 mg/kg TXAPA9 25,460,73

SAFETY PROFILE: Poison by intraperitoneal route. An experimental teratogen. Experimental reproductive effects. When heated to decomposition it emits acrid smoke and irritating fumes.

MCA100 CAS:7195-27-9 ***HR: 1***
MEFRUSIDE
mf: $C_{13}H_{19}ClN_2O_5S_2$ mw: 382.91

PROP: dl-Form: Crystals. Mp: 149-150°. d-Form: Crystals. Mp: 146°. l-Form: Crystals. Mp: 146°.

SYNS: B 1500 ◇ BAYCARON ◇ N-(4'-CHLORO-3'-SULFAMOYLBENZENESULFONYL)-N-METHYL-2-AMINOMETHYL-2-METHYLTETRAHYDROFURAN ◇ FBA 1500 ◇ FDA 1902 ◇ MEFRUSID

TOXICITY DATA with REFERENCE
orl-rat TDLo:35 g/kg (28D pre):REP KSRNAM 7,1003,73
ipr-rat LD50:9800 mg/kg KSRNAM 7,1003,73
ipr-mus LD50:5200 mg/kg KSRNAM 7,1003,73

SAFETY PROFILE: Mildly toxic by intraperitoneal route. Experimental reproductive effects. When heated to decomposition it emits toxic fumes of Cl^-, SO_x, and NO_x.

MCA250 CAS:28022-11-9 ***HR: 3***
MEGALOMICIN A
mf: $C_{44}H_{80}N_2O_{15}$ mw: 877.26

SYNS: ANTIBIOTIC W-847-A ◇ ANTIBIOTIC XK 41C ◇ MEGALOMYCIN-A

TOXICITY DATA with REFERENCE
orl-mus LD50:7500 mg/kg 85ERAY 1,52,78
ipr-mus LD50:350 mg/kg 85ERAY 1,52,78
scu-mus LD50:7000 mg/kg 85ERAY 1,52,78

SAFETY PROFILE: Poison by intraperitoneal route. Mildly toxic by ingestion and subcutaneous routes. When heated to decomposition it emits toxic fumes of NO_x.

MCA500 CAS:8064-66-2 ***HR: 3***
MEGESTROL ACETATE + ETHINYLOESTRADIOL

SYNS: 17-HYDROXY-6-METHYLPREGNA-4,6-DIENE-3,20-DIONEACETATE mixed with 19-NOR-17-α-PREGNA-1,3,5(10)-TRIEN-2-YNE-3,17-DIOL ◇ MEGESTROL ACETATE 4 MG., ETHINYLOESTRADIOL 50 μg ◇ MENOQUENS ◇ NEODELPREGNIN ◇ ORACONAL ◇ SERIAL ◇ TRIERVONUM ◇ VOLDYS ◇ VOLIDAN

TOXICITY DATA with REFERENCE
orl-wmn TDLo:2430 μg/kg (30D pre):REP BMJOAE 1,1318,63
orl-wmn TDLo:10584 μg/kg/2Y-I:CAR,LIV,BLD LANCAO 1,273,80
orl-wmn TD:41 mg/kg/2Y-I:CAR,LIV BMJOAE 4,496,75

SAFETY PROFILE: Human reproductive effects by ingestion: female fertility effects. Questionable human carcinogen producing normocytic anemia and liver tumors. An oral contraceptive. When heated to decomposition it emits acrid smoke and irritating fumes.

MCA775 CAS:58001-89-1 ***HR: 1***
MEGLUMINE SODIUM IODAMIDE
mf: $C_{12}H_{11}I_3N_2O_4 \cdot C_{12}H_{11}I_3N_2O_4 \cdot C_7H_{17}NO_5 \cdot Na$ mw: 1474.14

SYN: UROMIRO 380

TOXICITY DATA with REFERENCE
ipr-rat LD50:17900 mg/kg NIIRDN 6,871,82
ivn-mus LD50:9000 mg/kg NIIRDN 6,871,82
ivn-rbt LD50:13200 mg/kg NIIRDN 6,871,82

SAFETY PROFILE: Mildly toxic by several routes. When heated to decomposition it emits toxic fumes of NO_x, I^-, and Na_2O.

MCB000 CAS:108-78-1 ***HR: 3***
MELAMINE
mf: $C_3H_6N_6$ mw: 126.15

PROP: Monoclinic, colorless prisms. Mp: <250°, bp: sublimes, d: 1.573 @ 250°, vap press: 50 mm @ 315°, vap d: 4.34. Sltly sol in water. Very sltly sol in hot alc; insol in ether.

SYNS: AERO ◇ CYANURAMIDE ◇ CYANUROTRIAMIDE ◇ CYANUROTRIAMINE ◇ CYMEL ◇ NCI-C50715 ◇ 2,4,6-TRIAMINO-s-TRIAZINE

TOXICITY DATA with REFERENCE
eye-rbt 500 mg/24H MLD 28ZPAK -,153,72
mnt-mus-orl 1 g/kg ENMUDM 4,342,82
orl-rat TDLo:195 g/kg/2Y-C:CAR TXAPA9 72,292,84
orl-rat TD:197 g/kg/2Y-C:CAR NTPTR* NTP-TR-245,83
orl-rat TD:162 g/kg/2Y-C:ETA TXAPA9 72,292,84
orl-rat LD50:3161 mg/kg TXAPA9 72,292,84
ipr-rat LDLo:3200 mg/kg 14CYAT 3,2769,82
orl-mus LD50:3296 mg/kg 14CYAT 3,2769,82
ipr-mus LDLo:800 mg/kg 14CYAT 3,2769,82

CONSENSUS REPORTS: IARC Cancer Review: Group 3 IMEMDT 7,56,87; Animal Inadequate Evidence IMEMDT 39,333,86. NTP Carcinogenesis Bioassay (feed); No Evidence: mouse NTPTR* NTP-TR-245,83; (feed); Clear Evidence: rat NTPTR* NTP-TR-245,83. Community Right-To-Know List. Reported in EPA TSCA Inventory.

SAFETY PROFILE: Moderately toxic by ingestion and intraperitoneal routes. An eye, skin, and mucous membrane irritant. Causes dermatitis in humans. Questionable carcinogen with experimental carcinogenic and tumorigenic data. Mutation data reported. When heated to decomposition it emits toxic fumes of NO_x and CN^-.

MCB100 CAS:65454-27-5 ***HR: 3***
MELANOMYCIN

TOXICITY DATA with REFERENCE
ipr-mus LD50:125 mg/kg 85GDA2 4(2),238,80
scu-mus LD50:250 mg/kg 85GDA2 4(2),238,80
ivn-mus LD50:50 mg/kg 85GDA2 4(2),238,80

SAFETY PROFILE: Poison by subcutaneous, intravenous and intraperitoneal routes.

MCB250 CAS:102418-04-2 ***HR: 3***
MELANOSPORIN

PROP: Produced by *Streptomyces melanosporus var. melanosporofaciens* (85ERAY 1,256,78).

TOXICITY DATA with REFERENCE
orl-mus LD50:350 mg/kg 85ERAY 1,257,78
ipr-mus LD50:15 mg/kg 85ERAY 1,257,78

SAFETY PROFILE: Poison by ingestion and intraperitoneal routes.

MCB350 CAS:73-31-4 ***HR: 3***
MELATONIN
mf: $C_{13}H_{16}N_2O_2$ mw: 232.31

PROP: A hormone of the pineal gland, also produced by extra-pineal tissues, that lightens skin color in amphibians by reversing the darkening effect of MSH (melanotropin). Pale yellow leaflets from benzene. Mp: 116-118°.

SYNS: N-ACETYL-5-METHOXYTRYPTAMINE ◇ MELATONINE ◇ 5-METHOXY-N-ACETYLTRYPTAMINE

TOXICITY DATA with REFERENCE
scu-rat TDLo:900 mg/kg (male 30D pre):REP TOLED5 31(Suppl),68,86
scu-mus TDLo:4200 mg/kg/20W-I:CAR GETRE8 28(2),47,83
ivn-mus LD50:180 mg/kg CSLNX* NX#02739

SAFETY PROFILE: Questionable carcinogen with experimental carcinogenic data by skin contact. Poison by intravenous route. Experimental reproductive effects. When heated to decomposition it emits toxic fumes of NO_x.

MCB375 CAS:425-51-4 ***HR: D***
MELENGESTRO ACETATE
mf: $C_{23}H_{30}O_4$ mw: 370.53

SYNS: Δ^6-DEHYDRO-17-ACETOXYPROGESTERONE ◇ Δ^6-DEHYDRO-17-α-ACETOXYPROGESTERONE ◇ 17-HYDROXYPREGNA-4,6-DIENE-3,20-DIONE ACETATE (ESTER) ◇ 17-HYDROXYPREGNA-4,6-DIENE-3,20-DIONE ACETATE

TOXICITY DATA with REFERENCE
orl-rbt TDLo:25 μg/kg (1D pre):REP ACEDAB 73,17,63

SAFETY PROFILE: Experimental reproductive effects. When heated to decomposition it yields acrid smoke and irritating fumes.

MCB380 CAS:2919-66-6 ***HR: D***
MELENGESTROL ACETATE
mf: $C_{25}H_{32}O_4$ mw: 396.57

SYNS: 17-(ACETYLOXY)-6-METHYL-16-METHYLENEPREGNA-4,6-DIENE-3,20-DIONE (9CI) ◇ 6-DEHYDRO-16-METHYLENE-6-METHYL-17-ACETOXYPROGESTERONE ◇ 17-HYDROXY-6-METHYL-16-METHYLENEPREGNA-4,6-DIENE-3,20-DIONE, ACETATE ◇ MGA ◇ MGA 100 (STEROID)

TOXICITY DATA with REFERENCE
orl-ctl TDLo:366 μg/kg (female 14-40W post):REP JANSAG 30,433,70
scu-rat TDLo:490 μg/kg (female 7D pre):TER JDSCAE 61,1778,78

SAFETY PROFILE: An experimental teratogen. Experimental reproductive effects. When heated to decomposition it yields acrid smoke and irritating fumes.

MCB500 CAS:3771-19-5 ***HR: 3***
MELIPAN
mf: $C_{20}H_{22}O_3$ mw: 310.42

SYNS: 2-METHYL-2-(4-(1,2,3,4-TETRAHYDRO-1-NAPHTHALENYL)PHENOXY)PROPANOIC ACID ◇ α-METHYL-α-(p-1,2,3,4-TETRAHYDRONAPHTH-1-YLPHENOXY)PROPIONICACID ◇ 2-METHYL-2-(4-(1,2,3,4-TETRAHYDRO-1-NAPHTHYL)PHENOXY) PROPANOIC ACID ◇ 2-METHYL-2-(p-(1,2,3,4-TETRAHYDRO-1-NAPHTHYL)PHENOXY)PROPIONIC ACID ◇ NAFENOIC ACID ◇ NAFENOPIN ◇ SU-13437

TOXICITY DATA with REFERENCE
dns-rat:lvr 10 mg/L CRNGDP 5,1033,84
dni-mus:oth 100 μmol/L CNREA8 40,36,80
orl-rat TDLo:39 g/kg/92W-C:CAR JNCIAM 59,1645,77
orl-mus TDLo:56 g/kg/81W-C:CAR CNREA8 36,1211,76
orl-rat TD:33 g/kg/78W-C:ETA CNREA8 36,1211,76

CONSENSUS REPORTS: IARC Cancer Review: Group 2B IMEMDT 7,56,87; Human Limited Evidence IMEMDT 24,125,80; Animal Sufficient Evidence IMEMDT 24,125,80.

SAFETY PROFILE: Suspected carcinogen with experimental carcinogenic and tumorigenic data. Mutation data reported. A drug for the treatment of hypercholesterolemia or hypertriglyceridemia. When heated to decomposition it emits acrid smoke and irritating fumes.

MCB525 CAS:37231-28-0 ***HR: 3***
MELITTIN
mf: $C_{131}H_{229}N_{39}O_{31}$ mw: 2846.99

PROP: Cream white, water-sol powder. Strongly basic polypeptide comprising 40-50% of the dried venom of the honey bee, *Apis mellifica* (*mellifera*).

SYNS: FORAPIN ◇ MELITTIN-I

TOXICITY DATA with REFERENCE
dns-rat:lvr 100 pmol/L CRNGDP 5,1547,84
ipr-mus LD50:5 mg/kg ARZNAD 22,1921,72
ivn-mus LD50:4 mg/kg 85GDA2 9,170,82
ivn-mus LD50:860 μg/kg TOXIA6 22,308,84

SAFETY PROFILE: Poison by intravenous and intraperitoneal routes. Mutation data reported. When heated to decomposition it emits toxic fumes of NO_x. A major component of bee venom.

MCB535 ***HR: 3***
MELSMON

PROP: Placenta preparations (NIIRDN 6,854,82).

TOXICITY DATA with REFERENCE
scu-rat LD50:77 mg/kg NIIRDN 6,854,82
ipr-mus LD50:66400 mg/kg YACHDS 8,75,80
scu-mus LD50:75200 mg/kg YACHDS 8,75,80

SAFETY PROFILE: Poison by subcutaneous route.

MCB550 CAS:32887-03-9 ***HR: 2***
MELYSIN
mf: $C_{21}H_{33}N_3O_5S \cdot ClH$ mw: 476.09

SYNS: FL-1039 ◇ (+)-6-(((HEXAHYDRO-1H-AZEPIN-1-YL)METHYLENE)AMINO)-3,3-DIMETHYL-7-OXO-4-THIA-1-AZABICYCLO(3.2.0) HEPTANE-2-CARBOXYLIC ACID HYDROXYMETHYL ESTER, PIVALATE (ester), MONOHYDROCHLORIDE ◇ PIVMECILLINAM HYDROCHLORIDE ◇ SELEXID

TOXICITY DATA with REFERENCE
orl-mky TDLo:57330 mg/kg (91D male):REP JZKEDZ 2,171,76
orl-rat LD50:9500 mg/kg NIIRDN 6,632,82
scu-rat LD50:2100 mg/kg IYKEDH 9,1066,78
ivn-rat LD50:465 mg/kg IYKEDH 9,1066,78
orl-mus LD50:3020 mg/kg NIIRDN 6,632,82
scu-mus LD50:1930 mg/kg IYKEDH 9,1066,78
ivn-mus LD50:475 g/kg IYKEDH 9,1066,78

SAFETY PROFILE: Moderately toxic by ingestion, subcutaneous, and intravenous routes. Experimental reproductive effects. When heated to decomposition it emits toxic fumes of SO_x, NO_x, and HCl.

MCB600 CAS:71628-96-1 ***HR: 3***
MENOGAROL
mf: $C_{28}H_{31}NO_{10}$ mw: 541.60

SYNS: 7-CON-o-METHYLNOGAROL ◇ 7-o-METHYLNOGAROL ◇ 7(R)-o-METHYLNORGAROL ◇ NSC-269148 ◇ 7-OMEN ◇ U-52047

TOXICITY DATA with REFERENCE
mnt-rat-par 1560 μg/kg/2D-I CNREA8 43,5293,83
msc-ham:lng 100 μg/L CNREA8 43,5293,83
ivn-rat LD50:77400 μg/kg NTIS** PB84-148410
ipr-mus LD50:83500 μg/kg NTIS** PB84-148410

SAFETY PROFILE: Poison by intravenous and intraperitoneal routes. Mutation data reported. When heated to decomposition it emits toxic fumes of NO_x.

MCB625 ***HR: D***
MENTHA ARVENSIS, OIL

PROP: From *Mentha arvensis var. piperascens* Holmes (forma piperascens Malinvaud) (Fam. *Cabiatae*) (CCPTAY 24,559,81). Colorless to yellow liquid, minty odor. D: 0.888-0.908, refr index: 1.458 @ 20°.Sol in fixed oils, mineral oil, propylene glycol; insol in glycerin.

SYNS: CORNMINT OIL, PARTIALLY DEMENTHOLIZED ◇ MENTHA ARVENSIS OIL, PARTIALLY DEMENTHOLIZED (FCC)

TOXICITY DATA with REFERENCE
scu-rat TDLo:20 mg/kg (9-10D preg):TER CCPTAY 24,559,81
scu-rat TDLo:50 mg/kg (5D male):REP CCPTAY 24,559,81

SAFETY PROFILE: An experimental teratogen. Experimental reproductive effects. When heated to decomposition it emits acrid smoke and irritating fumes.

MCB700 ***HR: 2***
o-1,4-MENTHADIENE
mf: $C_{10}H_{16}$ mw: 136.26

SYN: 1-METHYL-2-(1-METHYLETHYL)-1,4-CYCLOHEXADIENE

TOXICITY DATA with REFERENCE
orl-rat LD50:4270 mg/kg ZDBEA9 (12),39,83
ihl-rat LC50:3550 mg/m^3 ZDBEA9 (12),39,83
orl-mus LD50:2950 mg/kg ZDBEA9 (12),39,83
ihl-mus LC50:25 g/m^3 ZDBEA9 (12),39,83
ipr-mus LD50:2502 mg/kg ZDBEA9 (12),39,83

SAFETY PROFILE: Moderately toxic by ingestion, inhalation, and intraperitoneal routes. When heated to decomposition it emits acrid smoke and irritating fumes.

MCB750 CAS:99-85-4 ***HR: 2***
p-MENTHA-1,4-DIENE
mf: $C_{10}H_{16}$ mw: 136.26

PROP: Colorless liquid; citrus odor. D: 0.841, refr index: 1.4731.477. Sol in alc, fixed oils; insol in water.

SYNS: FEMA No. 3559 ◇ 1-METHYL-4-ISOPROPYLCYCLOHEXADIENE-1,4 ◇ Γ-TERPINENE (FCC)

TOXICITY DATA with REFERENCE
skn-rbt 500 mg MOD FCTXAV 14,659,76
orl-rat LD50:3650 mg/kg FCTXAV 14,875,76

CONSENSUS REPORTS: Reported in EPA TSCA Inventory.

SAFETY PROFILE: Moderately toxic by ingestion. A skin irritant. When heated to decomposition it emits acrid smoke and irritating fumes.

MCC000 CAS:99-83-2 ***HR: 2***
p-MENTHA-1,5-DIENE
mf: $C_{10}H_{16}$ mw: 136.26

PROP: Colorless to sltly yellow liquid; mint odor. D: 0.835-0.865, refr index: 1.471-1.477, flash p: 120°F. Sol in alc; insol in water.

SYNS: α-FELLANDRENE ◇ FEMA No. 2856 ◇ 4-ISOPROPYL-1-METHYL-1,5-CYCLOHEXADIENE ◇ 5-ISOPROPYL-2-METHYL-1,3-CYCLOHEXADIENE ◇ 2-METHYL-5-ISOPROPYL-1,3-CYCLOHEXADIENE ◇ α-PHELLANDRENE (FCC)

TOXICITY DATA with REFERENCE
skn-man 100% SEV FCTXAV 16,843,78
orl-rat LD50:5700 mg/kg FCTXAV 16,843,78

CONSENSUS REPORTS: Reported in EPA TSCA Inventory.

SAFETY PROFILE: Mildly toxic by ingestion. A severe human skin irritant. Incompatible with air. Combustible liquid. When heated to decomposition it emits acrid smoke and irritating fumes.

MCC250 CAS:138-86-3 ***HR: 1***
p-MENTHA-1,8-DIENE
DOT: UN 2052
mf: $C_{10}H_{16}$ mw: 136.26

PROP: Liquid. D: 0.842 @ 20°/4°, mp: −96.9°, bp: 177°. Insol in water; misc in alc and ether.

SYNS: ACINTENE DP ◇ ACINTENE DP DIPENTENE ◇ CAJEPUTENE ◇ CINENE ◇ DIPANOL ◇ DIPENTENE ◇ INACTIVE LIMONENE ◇ KAUTSCHIN ◇ LIMONENE ◇ dl-LIMONENE ◇ 1,8(9)-p-MENTHADIENE ◇ 1-METHYL-4-ISOPROPENYL-1-CYCLOHEXENE ◇ NESOL ◇ Δ-1,8-TERPODIENE ◇ UNITENE

TOXICITY DATA with REFERENCE
skn-rbt 500 mg/24H MOD FCTXAV 12,703,74

CONSENSUS REPORTS: Reported in EPA TSCA Inventory.

DOT Classification: Flammable or Combustible Liquid; Label: Flammable Liquid.

SAFETY PROFILE: A skin irritant. Flammable when exposed to heat or flame; can react vigorously with oxidizing materials. When heated to decomposition it emits acrid smoke and irritating fumes.

MCC500 CAS:5989-54-8 ***HR: 1***
(S)-(−)-p-MENTHA-1,8-DIENE
mf: $C_{10}H_{16}$ mw: 136.26

PROP: Colorless liquid; light odor. D: 0.837-0.841, refr index:.469-1.473. Misc in alc, fixed oils; insol in water.

SYNS: 1-LIMONENE ◇ (−)-LIMONENE (FCC) ◇ 1-METHYL-4-(1-METHYLETHENYL)-(S)-CYCLOHEXENE

TOXICITY DATA with REFERENCE
skn-rbt 500 mg/24H MOD FCTXAV 16,809,78

CONSENSUS REPORTS: Reported in EPA TSCA Inventory.

SAFETY PROFILE: A skin irritant. When heated to decomposition it emits acrid smoke and irritating fumes.

MCD000 CAS:97-45-0 ***HR: 1***
p-MENTHA-6,8-DIEN-2-OL, PROPIONATE
mf: $C_{13}H_{20}O_2$ mw: 208.33

SYNS: 1-CARVYL PROPIONATE ◇ 1-p-MENTHA-6,8(9)-DIEN-2-YL PROPIONATE ◇ 2-METHYL-5-(1-METHYLETHENYL)-2-CYCLOHEXEN-1-OL PROPIONATE

TOXICITY DATA with REFERENCE
skn-rbt 500 mg/24H MLD FCTXAV 16,677,78

CONSENSUS REPORTS: Reported in EPA TSCA Inventory.

SAFETY PROFILE: A skin irritant. When heated to decomposition it emits acrid smoke and irritating fumes.

MCD250 CAS:99-49-0 ***HR: 2***
p-MENTHA-6,8-DIEN-2-ONE
mf: $C_{10}H_{14}O$ mw: 150.24

PROP: Colorless liquid. D: 0.921 @ 20°/4°, bp: 230° @ 755 mm. Insol in water; misc in alc and ether. Found in a score of essential oils and the main constituent of spearmint oil, prepared by isolation from oil of spearmint (FCTXAV 11,1011,73).

SYNS: CARVONE ◇ 6,8(9)-p-MENTHADIEN-2-ONE ◇ Δ-1-METHYL-4-ISOPROPENYL-6-CYCLOHEXEN-2-ONE ◇ $\Delta^{6,8}$-(9)-TERPADIENONE-2 ◇ NCI-C55867

TOXICITY DATA with REFERENCE
scu-mus LD50:2675 mg/kg JAPMA8 46,77,57
orl-rat LD50:1640 mg/kg FCTXAV 2,327,64
orl-gpg LD50:766 mg/kg FCTXAV 2,327,64

CONSENSUS REPORTS: Reported in EPA TSCA Inventory.

SAFETY PROFILE: Moderately toxic by ingestion and subcutaneous routes. Used for flavoring liquors, in perfumes, and soaps. When heated to decomposition it emits acrid smoke and irritating fumes.

MCD750 CAS:80-52-4 ***HR: 3***
p-MENTHANE-1,8-DIAMINE
mf: $C_{10}H_{22}N_2$ mw: 170.34

SYNS: 4-AMINO-a,a,4-TRIMETHYLCYCLOHEXANEMETHAMINE ◇ 1,8-DIAMINO-p-MENTHANE ◇ MENTHANE DIAMINE ◇ USAF RH-4

TOXICITY DATA with REFERENCE
skn-rbt 100 μg/24H open AIHAAP 23,95,62
orl-rat LD50:770 mg/kg AIHAAP 30,470,69
ipr-mus LD50:50 mg/kg NTIS** AD277-689
skn-rbt LD50:292 mg/kg AIHAAP 30,470,69

CONSENSUS REPORTS: Reported in EPA TSCA Inventory.

SAFETY PROFILE: Poison by skin contact and intraperitoneal routes. Moderately toxic by ingestion. A skin irritant. When heated to decomposition it emits toxic fumes of NO_x. See also AMINES.

MCE000 CAS:80-47-7 ***HR: 3***
p-MENTHANE-8-HYDROPEROXIDE
mf: $C_{10}H_{20}O_2$ mw: 172.30

PROP: Clear, pale yellow liquid. D: 0.910-0.925 @ 15.5°/4°.

SYN: p-MENTHANE HYDROPEROXIDE

TOXICITY DATA with REFERENCE
unr-mus TDLo:620 mg/kg:ETA RARSAM 3,193,63

SAFETY PROFILE: Questionable carcinogen with experimental tumorigenic data. When heated to decomposition it emits acrid smoke and irritating fumes. An irritant and powerful oxidizer. See also PEROXIDES, ORGANIC.

MCE250 CAS:1074-95-9 ***HR: 2***
p-MENTHAN-3-ONE racemic
mf: $C_{10}H_{18}O$ mw: 154.28

PROP: Several stereoisomers found in nature; 1-menthone found in essential oils of Russian and American peppermint, Geranium, *Andropogon fragrans*, *Mentha timija*, *Mentha arvensis* and others; d-menthone found in essential oils of *Barosma pulchellum*, *Nepeta japonica maxim* and others; d-isomenthone isolated from *Micromeriabissinica benth.*, *Pelargonium tometosum jacquin,* and others; 1-isomenthone identified in *Reunion geranium*, *Pelargonium capitatum* and others (FCTXAV 14,443,76). Flash p: 156°F.

SYNS: FEMA No. 2667 ◇ 2-ISOPROPYL-5-METHYL-CYCLOHEXAN-1-ONE, racemic ◇ MENTHONE, racemic

TOXICITY DATA with REFERENCE
skn-rbt 500 mg/24H MLD FCTXAV 14,443,76
orl-rat LD50:2180 mg/kg FCTXAV 14(5),443,76

CONSENSUS REPORTS: Reported in EPA TSCA Inventory.

SAFETY PROFILE: Moderately toxic by ingestion. A skin irritant. Combustible liquid. When heated to decomposition it emits acrid smoke and irritating fumes.

MCE275 CAS:11028-39-0 ***HR: 2***
o-1-MENTHENE
mf: $C_{10}H_{18}$ mw: 138.28

SYNS: HEXAHYDROCARQUEJENE ◇ 1-METHYL-2-(1-METHYLETHYL)-1-CYCLOHEXENE, DIDEHYDRO deriv.

TOXICITY DATA with REFERENCE
orl-rat LD50:7200 mg/kg ZDBEA9 (12)39,83
ihl-rat LC50:10600 mg/m^3 ZDBEA9 (12),39,83
orl-mus LD50:4350 mg/kg ZDBEA9 (12),39,83
ihl-mus LC50:76 g/m^3 ZDBEA9 (12),39,83
ipr-mus LD50:2811 mg/kg ZDBEA9 (12),39,83

SAFETY PROFILE: Moderately toxic by intraperitoneal route. Mildly toxic by ingestion. When heated to decomposition it emits acrid smoke and irritating fumes.

MCE500 CAS:500-00-5 ***HR: 1***
p-MENTH-3-ENE
mf: $C_{10}H_{18}$ mw: 138.28

SYN: METHANOMETHENE

TOXICITY DATA with REFERENCE
orl-rat LD50:21 g/kg AIHAAP 30,470,69
ihl-rat LCLo:2000 ppm/4H AIHAAP 30,470,69
skn-rbt LD50:5660 mg/kg AIHAAP 30,470,69

CONSENSUS REPORTS: Reported in EPA TSCA Inventory.

SAFETY PROFILE: Mildly toxic by inhalation, ingestion, and skin contact. When heated to decomposition it emits acrid smoke and irritating fumes.

MCE750 CAS:7786-67-6 ***HR: 2***
p-MENTH-8-EN-3-OL
mf: $C_{10}H_{18}O$ mw: 154.28

PROP: Colorless liquid; mint odor. D: 0.904-0.913, refr index: 1.470-1.475. Misc in alc, ether, fixed oils; sltly sol in water.

SYNS: FEMA No. 2962 ◇ ISOPULEGOL (FCC) ◇ 8(9)-p-MENTHEN-3-OL ◇ 1-METHYL-4-ISOPROPENYLCYCLOHEXAN-3-OL

TOXICITY DATA with REFERENCE
orl-rat LD50:1030 mg/kg FCTXAV 13,681,75
skn-rbt LD50:3000 mg/kg FCTXAV 13,681,75

CONSENSUS REPORTS: Reported in EPA TSCA Inventory.

SAFETY PROFILE: Moderately toxic by ingestion and skin contact. When heated to decomposition it emits acrid smoke and irritating fumes.

MCF250 CAS:89-81-6 ***HR: 2***
p-MENTH-1-EN-3-ONE
mf: $C_{10}H_{16}O$ mw: 152.26

PROP: Liquid. D: 0.926 @ 20°/4°, bp: 233°. Insol in water.

SYNS: 3-CARVOMENTHENONE ◇ 1-METHYL-4-ISOPROPYL-1-CYCLOHEXEN-3-ONE ◇ 3-METHYL-6-(1-METHYLETHYL)-2-CYCLOHEXEN-1-ONE ◇ PIPERITONE

TOXICITY DATA with REFERENCE
skn-rbt 500 mg/24H MOD FCTXAV 16,637,78
orl-rat LD50:2450 mg/kg FCTXAV 16,637,78
scu-mus LD50:1420 mg/kg FCTXAV 16,637,78

CONSENSUS REPORTS: Reported in EPA TSCA Inventory.

SAFETY PROFILE: Moderately toxic by ingestion and subcutaneous routes. A skin irritant. When heated to decomposition it emits acrid smoke and irritating fumes.

MCF500 CAS:15932-80-6 ***HR: 3***
p-MENTH-4(8)-EN-3-ONE
mf: $C_{10}H_{16}O$ mw: 152.26

SYNS: 1-ISOPROPYLIDENE-4-METHYL-2-CYCLOHEXANONE ◇ d-p-MENTH-4(8)-EN-3-ONE ◇ 4(8)-p-MENTHEN-3-ONE ◇ 1-METHYL-4-ISOPROPYLIDENE-3-CYCLOHEXANONE ◇ PULEGONE ◇ d-PULEGONE

TOXICITY DATA with REFERENCE
skn-rbt 500 mg/24H MOD FCTXAV 16,637,78
orl-rat LD50:470 mg/kg FCTXAV 16,637,78
ipr-mus LD50:150 mg/kg NTIS** AD691-490
scu-mus LD50:1709 mg/kg FCTXAV 16,637,78
skn-rbt LD50:3090 mg/kg FCTXAV 16,637,78

CONSENSUS REPORTS: Reported in EPA TSCA Inventory.

SAFETY PROFILE: Poison by intraperitoneal route. Moderately toxic by ingestion, skin contact, and subcutaneous routes. A skin irritant. When heated to decomposition it emits acrid smoke and irritating fumes.

MCF525 CAS:31375-17-4 ***HR: 1***
MENTHENYL KETONE
mf: $C_{13}H_{22}O$ mw: 194.35

SYNS: 1-(p-MENTHEN-6-YL)-1-PROPANONE ◇ NERONE ◇ 1-PROPANONE, 1-p-MENTH-6-EN-2-YL-

TOXICITY DATA with REFERENCE
skn-rbt 500 mg/24H MOD FCTXAV 17,853,79

CONSENSUS REPORTS: Reported in EPA TSCA Inventory.

SAFETY PROFILE: A skin irritant. When heated to decomposition it emits acrid smoke and irritating fumes.

MCF750 CAS:89-78-1 ***HR: 3***
MENTHOL
mf: $C_{10}H_{20}O$ mw: 156.26

PROP: Hexagonal crystals or granules; peppermint taste and odor. D: 0.890 @ 15°/15°, vap press: 1 mm @ 56.0°, vap d: 5.38, mp: 41-43°, bp: 212°, flash p: +199°F. Very sol in alc, chloroform, ether, petr ether, glacial acetic acid, liquid petrolatum; sltly sol in water.

SYNS: FEMA No. 2665 ◇ HEXAHYDROTHYMOL ◇ 2-ISOPROPYL-5-METHYL-CYCLOHEXANOL ◇ p-MENTHAN-3-OL ◇ 1-MENTHOL ◇ 5-METHYL-2-(1-METHYLETHYL)CYCLOHEXANOL ◇ PEPPERMINT CAMPHOR

TOXICITY DATA with REFERENCE
eye-rbt 750 μg SEV AJOPAA 29,1363,46
orl-rat LD50:3180 mg/kg FCTXAV 2,327,64
ipr-rat LDLo:1500 mg/kg APFRAD 10,481,52
ims-rat LD50:10 g/kg AEPPAE 222,244,54
ipr-mus LDLo:1800 mg/kg AIPTAK 63,43,39
orl-cat LDLo:1500 mg/kg HBAMAK 4,1289,35
ivn-cat LDLo:37 mg/kg AIPTAK 63,43,39

CONSENSUS REPORTS: Reported in EPA TSCA Inventory.

SAFETY PROFILE: Poison by intravenous route. Moderately toxic by ingestion and intraperitoneal routes. A severe eye irritant. Incompatible with phenol; β-naphthol; resorcinol or thymol in trituration; potassium permanganate; chromium trioxide; pyrogallol. Combustible liquid. When heated to decomposition it emits acrid smoke and irritating fumes.

MCG000 CAS:15356-70-4 ***HR: 2***
dl-MENTHOL
mf: $C_{10}H_{20}O$ mw: 156.30

SYNS: FEMA No. 2665 ◇ 4-ISOPROPYL-1-METHYLCYCLOHEXAN-3-OL ◇ dl-3-p-MENTHANOL ◇ 3-p-MENTHOL ◇ MENTHOL racemic ◇ MENTHOL racemique (FRENCH) ◇ 5-METHYL-2-(1-METHYLETHYL)-CYCLOHEXANOL (1-α,2-β,5-α)

TOXICITY DATA with REFERENCE
skn-rbt 500 mg/24H MLD FCTXAV 14,443,76
orl-rat LD50:2900 mg/kg FAONAU 44A,59,67
scu-rat LDLo:1 g/kg MMWOAU 73,2011,26
orl-mus LD50:3100 mg/kg QJPPAL 5,233,32
scu-mus LDLo:14 g/kg MMWOAU 73,2011,26
orl-cat LDLo:1500 mg/kg MMWOAU 73,2011,26
ipr-cat LDLo:1500 mg/kg MMWOAU 73,2011,26
ipr-rbt LDLo:2000 mg/kg FAONAU 44A,59,67
ipr-gpg LD50:865 mg/kg APFRAD 10,481,52

CONSENSUS REPORTS: NCI Carcinogenesis Bioassay (feed); No Evidence: mouse, rat NCITR* NCI-GC-TR-98,79. Reported in EPA TSCA Inventory.

SAFETY PROFILE: Moderately toxic by ingestion, intraperitoneal, and subcutaneous routes. A skin irritant. When heated to decomposition it emits acrid smoke and irritating fumes. See also MENTHOL and l-MENTHOL.

MCG250 CAS:2216-51-5 ***HR: 3***
l-MENTHOL
mf: $C_{10}H_{20}O$ mw: 156.30

PROP: Found in high concentrations in oils of Peppermint (*Mentha Piperita)* and Japanese Mint Oil (*Mentha Arvensis)*, and in lower concentrations in Reunion Geranium Oil, and in a large number of essential oils; prepared by isolation from *Mentha arvensis* Oils (FCTXAV 14,443,76).

SYNS: FEMA No. 2665 ◇ (−)-MENTHYL ALCOHOL ◇ (1R-(1-α,2-β,5-α))-5-METHYL-2-(1-METHYLETHYL)CYCLOHEXANOL◇ U.S.P. MENTHOL

TOXICITY DATA with REFERENCE
orl-rat LD50:3300 mg/kg FAONAU 44A,59,67
ipr-rat LD50:700 mg/kg JPPMAB 35,110,83
scu-rat LDLo:1000 mg/kg HBAMAK 4,1365,35
orl-mus LD50:3400 mg/kg QJPPAL 5,233,32
ipr-mus LDLo:2000 mg/kg AIPTAK 63,43,39
scu-mus LDLo:5000 mg/kg HBAMAK 4,1365,35
orl-cat LDLo:800 mg/kg FAONAU 40,59,67
ipr-cat LDLo:800 mg/kg HBAMAK 4,1365,35
ivn-cat LDLo:34 mg/kg AIPTAK 63,43,39
ipr-rbt LDLo:2000 mg/kg FAONAU 44A,59,67
ipr-gpg LDLo:4000 mg/kg AIPTAK 63,43,39

CONSENSUS REPORTS: Reported in EPA TSCA Inventory.

SAFETY PROFILE: Poison by intravenous route. Moderately toxic by ingestion, intraperitoneal, and subcutaneous routes. When heated to decomposition it emits acrid smoke and irritating fumes.

MCG275 CAS:89-80-5 ***HR: 2***
MENTHONE
mf: $C_{10}H_{18}O$ mw: 154.28

PROP: Colorless liquid; mint odor. D: 0.888-0.895, refr index: 1.448-1.453. Sol in alc, fixed oils; very sltly sol in water.

SYNS: FEMA No. 2667 ◇ l-p-MENTHAN-3-ONE ◇ l-MENTHONE (FCC) ◇ p-MENTHONE ◇ trans-MENTHONE ◇ trans-5-METHYL-2-(1-METHYLETHYL)-CYCLOHEXANONE

TOXICITY DATA with REFERENCE
mmo-sat 6400 ng/plate MUREAV 138,17,84
mma-sat 32 μg/plate MUREAV 138,17,84

orl-rat LD50:500 mg/kg FRXXBL #2448856
scu-mus LD50:2180 mg/kg JAPMA8 46,77,57
ivn-dog LDLo:600 mg/kg COREAF 236,633,53

CONSENSUS REPORTS: Reported in EPA TSCA Inventory.

SAFETY PROFILE: Moderately toxic by ingestion, intravenous, and subcutaneous routes. Mutation data reported. When heated to decomposition it emits acrid smoke and irritating fumes. See also KETONES.

MCG500 CAS:16409-45-3 ***HR: 1***
dl-MENTHYL ACETATE
mf: $C_{12}H_{22}O_2$ mw: 198.34

PROP: Colorless liquid; characteristic minty odor. D: 0.919 @ 20°/4°, refr index: 0.443-1.450, bp: 227°, flash p: 197°F. Sltly sol in water, glycerin; misc with alc, ether, propylene glycol, fixed oils.

SYNS: FEMA No. 2668 ◇ MENTHOL, ACETATE (8CI) ◇ MENTHYL ACETATE ◇ MENTHYL ACETATE racemic ◇ p-MENTH-3-YL ESTER-dl-ACETIC ACID

TOXICITY DATA with REFERENCE
skn-rbt 500 mg/24H MLD FCTXAV 14,479,76
orl-rat LD50:7620 mg/kg FCTXAV 14,477,76

SAFETY PROFILE: Mildly toxic by ingestion. A skin irritant. Combustible liquid. When heated to decomposition it emits acrid smoke and irritating fumes.

MCG750 CAS:2623-23-6 ***HR: 1***
1-p-MENTH-3-YL ACETATE
mf: $C_{12}H_{22}O$ mw: 182.34

PROP: Colorless liquid; minty odor. D: 0.919-0.924, refr index: 1.443-1.447. Sol in alc, propylene glycol, fixed oils; sltly sol in water, glycerin.

SYNS: FEMA No. 2668 ◇ l-2-ISOPROPYL-5-METHYL-CYCLOHEXAN-1-OL ACETATE ◇ (−)-MENTHYL ACETATE ◇ l-MENTHYL ACETATE (FCC) ◇ l-p-MENTH-3-YL ACETATE ◇ (R-(1α,2β,5α))-5-METHYL-2-(1-METHYLETHYL)-CYCLOHEXANOL ACETATE (9CI)

TOXICITY DATA with REFERENCE
skn-rbt 500 mg/24H MLD FCTXAV 14,477,76

CONSENSUS REPORTS: Reported in EPA TSCA Inventory.

SAFETY PROFILE: A skin irritant. When heated to decomposition it emits acrid smoke and irritating fumes.

MCG850 CAS:59557-05-0 ***HR: 1***
MENTHYL ACETOACETATE
mf: $C_{14}H_{24}O_3$ mw: 240.38

SYNS: BUTANOIC ACID, 3-OXO-, 5-METHYL-2-(1-METHYLETHYL) CYCLOHEXYL ESTER, (1R-(1-α-2-β, 5α-)- ◇ MENTHOL ACETOACETATE ◇ (-)-MENTHYL ACETOACETATE

TOXICITY DATA with REFERENCE
skn-rbt 500 mg/24H MLD FCTOD7 20,733,82

SAFETY PROFILE: A skin irritant. When heated to decomposition it emits acrid smoke and irritating fumes.

MCH250 CAS:1227-61-8 ***HR: 3***
MEPHEXAMIDE
mf: $C_{15}H_{24}N_2O_3$ mw: 280.41

SYNS: N-(2-(DIETHYLAMINO)ETHYL)-2-(p-METHOXYPHENOXY) ACETAMIDE ◇ 2-(p-METHOXYPHENOXY)-N-(2-(DIETHYLAMINO) ETHYL)ACETAMIDE ◇ MEXEPHENAMIDE

TOXICITY DATA with REFERENCE
ivn-mus LD50:168 mg/kg 27ZQAG -,396,72
ivn-rbt LD50:135 mg/kg AMTYAT 123,141,65
orl-mus LD50:1500 μg/kg GATPBA 1,444,66

SAFETY PROFILE: Poison by ingestion and intravenous routes. When heated to decomposition it emits toxic fumes of NO_x.

MCH525 CAS:4599-60-4 ***HR: 3***
MEPICYCLINE PENICILLINATE
mf: $C_{29}H_{38}N_4O_9 \cdot C_{16}H_{18}N_2O_5S$ mw: 937.13

PROP: Yellowish-white, crystalline powder; sltly bitter taste. Decomp above 143°. Sensitive to light, heat, air. Solubility in water at 20° = 1 g/0.7 mL.

SYNS: CRISEOCIL ◇ DUAMINE ◇ GEOTRICYN ◇ HYDROCYCLINE ◇ MEPENICYCLINE ◇ OLIMPEN ◇ PENETRACYNE ◇ PENILTETRA ◇ PENIMEPICYCLINE ◇ PRESTOCICLINA

TOXICITY DATA with REFERENCE
orl-rat TDLo:12 g/kg (9-14D preg):REP KSRNAM 4,516,70
orl-rat LD50:3990 mg/kg GNRIDX 2,26,68
scu-rat LD50:1550 mg/kg GNRIDX 2,26,68
ivn-rat LD50:345 mg/kg GNRIDX 2,26,68
orl-mus LD50:3 g/kg GNRIDX 2,26,68
scu-mus LD50:1096 mg/kg GNRIDX 2,26,68
ivn-mus LD50:342 mg/kg GNRIDX 2,26,68

SAFETY PROFILE: Poison by intravenous route. Moderately toxic by ingestion and subcutaneous routes. Experimental reproductive effects. When heated to decomposition it emits toxic fumes of NO_x and SO_x.

MCH535 CAS:20344-15-4 ***HR: 3***
MEPIPRAZOLE DIHYDROCHLORIDE
mf: $C_{16}H_{21}ClN_4 \cdot 2ClH$ mw: 377.78

SYNS: 1-(3-CHLOROPHENYL)-4-(2-(5-METHYL-1H-PYRAZOL-3-YL) ETHYL)PIPERAZINE DIHYDROCHLORIDE ◇ H 4007 ◇ MEPIPRAZOLE HYDROCHLORIDE ◇ QUIADON

TOXICITY DATA with REFERENCE
orl-rat TDLo:21 mg/kg (7-13D preg):REP IYKEDH 4,360,73

orl-mus TDLo:210 mg/kg (7-13D preg):TER IYKEDH 4,360,73
orl-rat LD50:570 mg/kg IYKEDH 4,336,73
scu-rat LD50:435 mg/kg IYKEDH 4,336,73
ivn-rat LD50:93 mg/kg IYKEDH 4,336,73
orl-mus LD50:310 mg/kg IYKEDH 4,336,73
scu-mus LD50:440 mg/kg IYKEDH 4,336,73
ivn-mus LD50:110 mg/kg IYKEDH 4,336,73

SAFETY PROFILE: Poison by ingestion and intravenous routes. Moderately toxic by subcutaneous route. An experimental teratogen. Experimental reproductive effects. When heated to decomposition it emits toxic fumes of NO_x and HCl.

MCH550 CAS:18694-40-1 ***HR: 3***
MEPIRIZOL
mf: $C_{11}H_{14}N_4O_2$ mw: 234.29

PROP: Minute, white or cream-colored crystals from isopropyl ether; characteristic odor, bitter taste. Mp: 90-92°. Sparingly sol in water. Sol in dil acids; freely sol in ethanol, benzene, dichloroethane.

SYNS: DA-398 ◇ EPIRIZOLE ◇ MEBRON ◇ 4-METHOXY-2-(5-METHOXY-3-METHYL-1H-PYRAZOL-1-YL)-6-METHYLPYRIMIDINE ◇ 2-(3-METHOXY-5-METHYLPYRAZOL-2-YL)-4-METHOXY-6-METHYLPYRIMIDINE ◇ 1-(4-METHOXY-6-METHYL-2-PYRIMIDINYL)-3-METHYL-5-METHOXYPYRAZOLE ◇ 2-(3-METHYL-5-METHOXY-1-PYRAZOLYL)-4-METHOXY-6-METHYLPYRIMIDINE

TOXICITY DATA with REFERENCE
orl-rat TDLo:44500 mg/kg (26W male):REP IYKEDH 8,524,77
orl-rat LD50:445 mg/kg JMGZAI 7(9),5,70
ipr-rat LD50:214 mg/kg JMGZAI 7(9),5,70
scu-rat LD50:208 mg/kg IYKEDH 8,494,77
ivn-rat LD50:214 mg/kg 85IPAE -,87,75
orl-mus LD50:740 mg/kg OYYAA2 16,1011,78
ipr-mus LD50:540 mg/kg OYYAA2 16,1011,78
scu-mus LD50:550 mg/kg IYKEDH 8,494,77
ivn-mus LD50:550 mg/kg JMGZAI 7(9),5,70

SAFETY PROFILE: Poison by subcutaneous, intravenous and intraperitoneal routes. Moderately toxic by ingestion. Experimental reproductive effects. When heated to decomposition it emits toxic fumes of NO_x.

MCH600 CAS:21362-69-6 ***HR: 2***
MEPITIOSTANE
mf: $C_{25}H_{40}O_2S$ mw: 404.71

PROP: Crystals. Mp: 98-101°.

SYNS: CYCLOPENTANONE-2-α,3-α-EPITHIO-5-α-ANDROSTAN-17-β-YL METHYL ACETAL ◇ 2,2-EPITHIO-17-((1-METHOXYCYCLOPENTYL)OXY)-ANDROSTANE(2-α,3-α,5-α,17-β) ◇ 2-α,3-α-EPITHIO-17-β-YL 1-METHOXYCYCLOPENTYL ETHER ◇ 10364S ◇ THIODELONE ◇ THIODERON

TOXICITY DATA with REFERENCE
orl-rat TDLo:175 mg/kg (35D pre):REP OYYAA2 16,779,78
orl-hmn TDLo:400 μg/kg:SKN,LIV,BIO CANCAR 41,758,78

SAFETY PROFILE: Human systemic effects by ingestion: dermatitis, liver changes, and transaminase activity. Experimental reproductive effects. When heated to decomposition it emits toxic fumes of SO_x.

MCI375 CAS:1953-02-2 ***HR: 2***
MEPRIN
mf: $C_5H_9NO_3S$ mw: 163.21

PROP: Crystals from ethyl acetate. Mp: 95-97°.

SYNS: CAPEN ◇ EPATIOL ◇ MEPRIN (DETOXICANT) ◇ MERCAPTOPROPIONYLGLYCINE ◇ α-MERCAPTOPROPIONYLGLYCINE ◇ (2-MERCAPTOPROPIONYL)GLYCINE ◇ N-(2-MERCAPTOPROPIONYL)GLYCINE ◇ MUCOLYSIN ◇ THIOLA ◇ THIOPRONIN ◇ THIOPRONINE ◇ THIOSOL ◇ TIOPRONIN

TOXICITY DATA with REFERENCE
ims-rat TDLo:600 mg/kg (male 30D pre):REP IJEBA6 24,34,86
orl-rat LD50:1300 mg/kg NIIRDN 6,456,82
orl-mus LD50:2330 mg/kg NIIRDN 6,456,82
ipr-mus LD50:1305 mg/kg YKKZAJ 94,1419,74
ivn-mus LD50:1733 mg/kg JJANAX 38,137,85

SAFETY PROFILE: Moderately toxic by ingestion, intravenous and intraperitoneal routes. When heated to decomposition it emits toxic fumes of SO_x and NO_x. See also MERCAPTANS.

MCI500 CAS:851-68-3 ***HR: 3***
MEPROMAZINE
mf: $C_{19}H_{24}N_2OS$ mw: 328.51

SYNS: DEDORAN ◇ 10-(3-(DIMETHYLAMINO)-2-METHYLPROPYL)-2-METHOXYPHENOTHIAZINE ◇ LEVOMEPROMAZINE ◇ LEVOPROMAZINE ◇ LEVOTOMIN ◇ METHOTRIMEPRAZINE ◇ METHOXYPHENOTHIAZINE ◇ METHOXYTRIMEPRAZINE ◇ MILEZIN ◇ MINOZINAN ◇ NAUROCTIL ◇ NEOTONZIL ◇ NEOZINE ◇ NEURACTIL ◇ NEUROCIL ◇ SINOGAN ◇ SKF 5116 ◇ TISERCIN ◇ VERACTIL

TOXICITY DATA with REFERENCE
orl-rat LD50:1100 mg/kg 27ZQAG -,31,72
scu-rat LD50:45 mg/kg 27ZQAG -,31,72
orl-mus LD50:375 mg/kg PSCBAY 2,17,63
ipr-mus LD50:110 mg/kg CRSBAW 155,1029,61
scu-mus LD50:300 mg/kg PSCBAY 2,17,63
ivn-mus LD50:39 mg/kg PSDTAP 9,159,68
orl-bwd LD50:100 mg/kg TXAPA9 21,315,72

SAFETY PROFILE: Poison by ingestion, intravenous, subcutaneous, and intraperitoneal routes. When heated to decomposition it emits very toxic fumes of SO_x and NO_x.

MCI750 CAS:33396-37-1 *HR: 3*
MEPROSCILLARIN
mf: $C_{31}H_{44}O_8$ mw: 544.75

SYN: CLIFT

TOXICITY DATA with REFERENCE
orl-rat TDLo:145 mg/kg (15-22D preg/21D post):REP ARZNAD 28,506,78
orl-rat LD50:79 mg/kg ARZNAD 28,506,78
ivn-rat LD50:5800 μg/kg ARZNAD 28,506,78
orl-mus LD50:12500 μg/kg ARZNAD 28,506,78
ivn-mus LD50:2800 μg/kg ARZNAD 28,506,78
ivn-cat LDLo:137 μg/kg ARZNAD 28,495,78
ivn-gpg LDLo:678 μg/kg ARZNAD 28,495,78

SAFETY PROFILE: Poison by ingestion and intravenous routes. Experimental reproductive effects. When heated to decomposition it emits acrid smoke and irritating fumes.

MCJ250 CAS:6036-95-9 *HR: 3*
MEPYRAMINE HYDROCHLORIDE
mf: $C_{17}H_{23}N_3O{\cdot}ClH$ mw: 321.89

SYNS: 2-((2-DIMETHYLAMINOETHYL)(p-METHOXYBENZYL) AMINO)PYRIDINE HYDROCHLORIDE ◇ NEOANTERGAN HYDROCHLORIDE ◇ PYRILAMINE HYDROCHLORIDE

TOXICITY DATA with REFERENCE
orl-mus LD50:325 mg/kg JPETAB 93,210,48
scu-rat LD50:115 mg/kg JPETAB 93,210,48
ivn-mus LD50:25 mg/kg JPETAB 93,210,48

CONSENSUS REPORTS: Reported in EPA TSCA Inventory.

SAFETY PROFILE: Poison by ingestion, intravenous and subcutaneous routes. When heated to decomposition it emits very toxic fumes of HCl and NO_x.

MCJ300 CAS:57383-74-1 *HR: 3*
MEPYRAMINE 7-THEOPHYLLINE ACETATE
mf: $C_{17}H_{23}N_3O{\cdot}C_9H_{10}N_4O_4$ mw: 523.66

SYN: 1,2,3,6-TETRAHYDRO-1,3-DIMETHYL-2,6-DIOXO-7H-PURINE-7-ACETIC ACID compounded with N-((4-METHOXYPHENYL)METHYL)-N′,N′-DIMETHYL-N-2-PYRIDINYL-1,2-ETHANEDIAMINE(1:1)

TOXICITY DATA with REFERENCE
orl-rat LD50:234 mg/kg FRXXBL #2244460
ipr-rat LD50:116 mg/kg FRXXBL #2244460
ivn-rat LD50:42700 μg/kg FRXXBL #2244460

SAFETY PROFILE: Poison by ingestion, intravenous, and intraperitoneal routes. When heated to decomposition it emits toxic fumes of NO_x.

MCJ370 CAS:54-36-4 *HR: 3*
MEPYRAPONE
mf: $C_{14}H_{14}N_2O$ mw: 226.30

PROP: Crystals from ether and pentane. Mp: 50-51°.

SYNS: 1,2-DI-3-PYRIDYL-2-METHYL-1-PROPANONE ◇ METHAPYRAPONE ◇ METHOPIRAPONE ◇ METHOPYRAPONE ◇ METHOPYRININE ◇ METHOPYRONE ◇ 2-METHYL-1,2-DI-3-PYRIDINYL-1-PROPANONE (9CI) ◇ METOPIRON ◇ METOPIRONE ◇ METOPYRONE ◇ METYRAPON ◇ METYRAPONE ◇ SU-4885

TOXICITY DATA with REFERENCE
ipr-rat TDLo:270 mg/kg (female 19-21D post):REP PEREBL 20,672,86
ipr-rat TDLo:270 mg/kg (female 19-21D post):TER PEREBL 20,672,86
orl-rat LD50:521 mg/kg NIIRDN 6,827,82
ipr-mus LDLo:300 mg/kg JPETAB 146,395,64

SAFETY PROFILE: Poison by intraperitoneal route. Moderately toxic by ingestion. An experimental teratogen. Experimental reproductive effects. When heated to decomposition it emits toxic fumes of NO_x.

MCJ400 CAS:29216-28-2 *HR: 3*
MEQUITAZINE
mf: $C_{20}H_{22}N_2S$ mw: 322.50

PROP: Crystals from acetonitrile. Mp: 130-131°.

SYNS: LM-209 ◇ METAPLEXAN ◇ MIRCOL ◇ PRIMALAN ◇ 10-(3-QUINCULIDINYLMETHYL)PHENOTHIAZINE

TOXICITY DATA with REFERENCE
orl-rat TDLo:13750 μg/kg (7-17D preg):REP OYYAA2 21,867,81
orl-rat TDLo:220 mg/kg (female 7-17D post):TER OYYAA2 21,867,81
orl-rat LD50:245 mg/kg OYYAA2 22,491,81
ipr-rat LD50:54 mg/kg OYYAA2 22,491,81
scu-rat LD50:690 mg/kg OYYAA2 22,491,81
orl-mus LD50:210 mg/kg OYYAA2 22,491,81
ipr-mus LD50:54 mg/kg OYYAA2 22,491,81
scu-mus LD50:278 mg/kg OYYAA2 22,491,81

SAFETY PROFILE: Poison by ingestion, subcutaneous, and intraperitoneal routes. An experimental teratogen. Experimental reproductive effects. An antihistamine. When heated to decomposition it emits toxic fumes of SO_x and NO_x.

MCJ500 *HR: 2*
MERCAPTANS

PROP: Compounds containing the -SH group bound to carbon. Also called thiols.

SAFETY PROFILE: Generally they have a very offensive odor which may cause nausea and headache. High concentrations of vapor can produce unconsciousness with cyanosis, cold extremities, and rapid pulse. A common air contaminant. Dangerous; when heated to decomposition they almost always emit highly toxic fumes of SO_x. They may react with water, steam or acids to produce toxic and flammable vapors. Aliphatic mercaptans are flammable. They can react violently with powerful oxidizers such as $Ca(OCl)_2$.

MCK000 CAS:4822-44-0 ***HR: 3***
α-MERCAPTOACETANILIDE
mf: C_8H_8NOS mw: 166.23

SYNS: 2-MERCAPTOACETANILIDE ◇ THIOGLYCOLANILIDE ◇ THIOGLYCOLIC ACID ANILIDE ◇ USAF EK-6583

TOXICITY DATA with REFERENCE
ipr-mus LD50:200 mg/kg NTIS** AD277-689

CONSENSUS REPORTS: Reported in EPA TSCA Inventory.

SAFETY PROFILE: Poison by intraperitoneal route. When heated to decomposition it emits very toxic fumes of NO_x and SO_x. See also MERCAPTANS.

MCK300 CAS:54524-31-1 ***HR: 2***
MERCAPTOACETONITRILE
mf: C_2H_3NS mw: 73.11

SYN: ACETONITRILETHIOL

CONSENSUS REPORTS: Cyanide and its compounds are on the Community Right-To-Know List.

SAFETY PROFILE: Concentrated solutions or the solvent-free nitrile may polymerize vigorously. When heated to decomposition it emits toxic fumes of CN^-, SO_x, and NO_x. See also MERCAPTANS and NITRILES.

MCK500 CAS:504-17-6 ***HR: 2***
2-MERCAPTOBARBITURIC ACID
mf: $C_4H_4N_2O_2S$ mw: 144.16
SYNS: 2-THIOBARBITURIC ACID ◇ USAF EK-660

TOXICITY DATA with REFERENCE
ipr-mus LD50:600 mg/kg NTIS** AD277-689

CONSENSUS REPORTS: Reported in EPA TSCA Inventory.

SAFETY PROFILE: Moderately toxic by intraperitoneal route. When heated to decomposition it emits very toxic fumes of SO_x and NO_x. See also MERCAPTANS.

MCK750 CAS:147-93-3 ***HR: 3***
o-MERCAPTOBENZOIC ACID
mf: $C_7H_6O_2S$ mw: 154.19

PROP: Light yellow needles. Mp: 164°, bp: subl. Sltly sol in hot water; sol in alc.

SYNS: THIOSALICYLIC ACID ◇ USAF EK-T-2805 ◇ USAF KF-2 ◇ USAF XR-35

TOXICITY DATA with REFERENCE
ipr-mus LD50:50 mg/kg NTIS** AD277-689

CONSENSUS REPORTS: Reported in EPA TSCA Inventory.

SAFETY PROFILE: Poison by intraperitoneal route. When heated to decomposition it emits toxic fumes of SO_x. See also MERCAPTANS.

MCL500 CAS:5428-95-5 ***HR: 3***
(2-MERCAPTOCARBAMOYL)DI-ACETANILIDE
mf: $C_{18}H_{18}N_4O_4S_2$ mw: 418.52

SYNS: p,p′-BIS(α-THIOLCARBAMYLACETAMIDO)BIPHENYL ◇ CARBAMIC ACID, THIO, S-ESTER with 2-MERCAPTOACETANILIDE ◇ CARBAMINOTHIOGLYCOLIC ACID ANILIDE ◇ USAF UCTL-1766

TOXICITY DATA with REFERENCE
ipr-mus LD50:200 mg/kg NTIS** AD277-689

CONSENSUS REPORTS: Reported in EPA TSCA Inventory.

SAFETY PROFILE: Poison by intraperitoneal route. When heated to decomposition it emits very toxic fumes of NO_x and SO_x. See also MERCAPTANS.

MCM750 CAS:123-93-3 ***HR: 3***
MERCAPTODIACETIC ACID
mf: $C_4H_6O_4S$ mw: 150.16

PROP: A white powder. Mp: 128°.

SYNS: (CARBOXYMETHYLTHIO)ACETIC ACID ◇ DIMETHYLSULFIDE-α,α′-DICARBOXYLIC ACID ◇ THIODIGLYCOLIC ACID ◇ β,β′-THIODIGLYCOLIC ACID ◇ 2,2′-THIODIGLYCOLIC ACID ◇ THIODIGLYCOLLIC ACID ◇ USAF CB-36 ◇ USAF E-2

TOXICITY DATA with REFERENCE
ipr-mus LD50:300 mg/kg NTIS** AD277-689

CONSENSUS REPORTS: Reported in EPA TSCA Inventory.

SAFETY PROFILE: Poison by intraperitoneal route. When heated to decomposition or on contact with acid or acid fumes it emits toxic fumes of SO_x. See also 2-MERCAPTOACETIC ACID.

MCN000 CAS:123-81-9 ***HR: 3***
MERCAPTO DI-ACETIC ACID, ETHYLENE ESTER
mf: $C_6H_{10}O_4S_2$ mw: 210.28

TOXICITY DATA with REFERENCE
orl-rat LD50:330 mg/kg TRIPA7 -,1,73

CONSENSUS REPORTS: Reported in EPA TSCA Inventory.

SAFETY PROFILE: Poison by ingestion. When heated to decomposition it emits toxic fumes of SO_x. See also MERCAPTANS.

MCN250 CAS:60-24-2 ***HR: 3***
2-MERCAPTOETHANOL
DOT: UN 2966
mf: C_2H_6OS mw: 78.14

PROP: Water-white, mobile liquid. Bp: 157-158° (decomp) @ 742 mm, flash p: 165°F (COC), d: 1.1168 @ 20°/20°, vap press: 1.0 mm @ 20°, vap d: 2.69. Pure liquid is misc with water, alc, ether, and benzene.

SYNS: EMERY 5791 ◇ 1-ETHANOL-2-THIOL ◇ 2-HYDROXY-1-ETHANETHIOL ◇ 2-HYDROXYETHYL MERCAPTAN ◇ 2-ME ◇ MERCAPTOETHANOL ◇ β-MERCAPTOETHANOL ◇ MONOTHIOETHYLENEGLYCOL ◇ 2-THIOETHANOL ◇ THIOGLYCOL (DOT) ◇ THIOMONOGLYCOL ◇ USAF EK-4196

TOXICITY DATA with REFERENCE
skn-rbt 10 mg/24H open JIHTAB 26,269,44
eye-rbt 2280 mg SEV AJOPAA 29,1363,46
dnd-omi 10 mmol/L BBRCA9 77,1150,77
mnt-mus:oth 100 mg/L JNCIAM 56,357,76
orl-rat LD50:244 mg/kg GTPZAB 15(2),56,71
orl-mus LD50:190 mg/kg GTPZAB 15(2),56,71
ihl-mus LC50:13200 mg/m³/15M GTPZAB 15(2),56,71
ipr-mus LD50:200 mg/kg NTIS** AD277-689
ivn-mus LD50:480 mg/kg JPMSAE 62,237,73
skn-rbt LD50:150 mg/kg UCDS** 3/23/73
skn-gpg LD50:300 mg/kg JIHTAB 26,269,44

CONSENSUS REPORTS: Reported in EPA TSCA Inventory.

DOT Classification: Poison B; Label: Poison.

SAFETY PROFILE: Poison by ingestion, skin contact, and intraperitoneal routes. Moderately toxic by intravenous route. A skin and severe eye irritant. Human mutation data reported. Flammable when exposed to heat, flame or oxidizers. To fight fire, use alcohol foam, CO_2, dry chemical. When heated to decomposition it emits highly toxic fumes SO_x. See also MERCAPTANS.

MCN500 CAS:51-85-4 ***HR: 3***
β-MERCAPTOETHYLAMINE DISULFIDE
mf: $C_4H_{12}N_2S_2$ mw: 152.30

SYNS: BECAPTAN DISULFURE (FRENCH) ◇ BIS(β-AMINOETHYL) DISULFIDE ◇ CYSTAMINE ◇ CYSTEINAMINE DISULFIDE ◇ CYSTINAMIN (GERMAN) ◇ CYSTINEAMINE ◇ DECARBOXYCYSTINE ◇ β,β'-DIAMINODIETHYL DISULFIDE ◇ 2,2'-DITHIOBIS(ETHYLAMINE) ◇ MERCAMINE DISULFIDE ◇ 2-MERCAPTOETHYLAMINE (OXIDIZED)

TOXICITY DATA with REFERENCE
dnd-mam:lym 200 mmol/L IJRBA3 9,185,65
ipr-rat LD50:99 mg/kg SVLKAO 24,541,81
scu-rat LDLo:150 mg/kg AEPPAE 185,461,37
ims-rat LD50:96 mg/kg SVLKAO 24,541,81
ipr-mus LD50:220 mg/kg RDBGAT 19,593,78
scu-mus LD50:214 mg/kg AIPTAK 109,108,57
scu-cat LDLo:200 mg/kg AEPPAE 185,461,37
scu-gpg LDLo:300 mg/kg AEPPAE 185,461,37

SAFETY PROFILE: Poison by intraperitoneal, intramuscular, and subcutaneous routes. Mutation data reported. When heated to decomposition it emits very toxic fumes of NO_x and SO_x. See also MERCAPTANS and SULFIDES.

MCN750 CAS:156-57-0 ***HR: 3***
2-MERCAPTOETHYLAMINE HYDROCHLORIDE
mf: $C_2H_7NS{\cdot}ClH$ mw: 113.62

PROP: White, sltly hygroscopic crystals. Mp: 70.2-70.7°.

SYNS: CYSTEAMINE HYDROCHLORIDE ◇ CYSTEAMINHYDROCHLORID (GERMAN) ◇ MEA ◇ β-MERCAPTOETHYLAMINE HYDROCHLORIDE ◇ 2-MERCAPTOETHYLAMINE HYDROCHLORIDE ◇ USAF EE-3

TOXICITY DATA with REFERENCE
cyt-rat:lvr 5 mg/L MUREAV 153,57,85
ipr-rat LD50:232 mg/kg ARZNAD 5,421,55
orl-mus LD50:1352 mg/kg CPBTAL 20,721,72
ipr-mus LD50:250 mg/kg JMCMAR 12,510,69
scu-mus LDLo:900 mg/kg ABPAAG 12,142,38
ivn-rbt LD50:150 mg/kg ARZNAD 5,421,55

CONSENSUS REPORTS: Reported in EPA TSCA Inventory.

SAFETY PROFILE: Poison by intravenous and intraperitoneal routes. Moderately toxic by ingestion and subcutaneous routes. Mutation data reported. Used as an antidote to acetaminophen poisoning and as an experimental radioprotective agent. Can react with water or steam to produce toxic fumes. When heated to decomposition it emits highly toxic fumes of SO_x, NO_x, and HCl. See also MERCAPTANS and AMINES.

MCO000 CAS:4542-46-5 ***HR: 3***
4-MERCAPTOETHYLMORPHOLINE
mf: $C_6H_{13}NOS$ mw: 147.26

SYN: N-MORPHOLYLCYSTEAMIN (GERMAN)

TOXICITY DATA with REFERENCE
scu-mus LD50:316 mg/kg AIPTAK 109,108,57

CONSENSUS REPORTS: Reported in EPA TSCA Inventory.

SAFETY PROFILE: Poison by subcutaneous route. When heated to decomposition it emits very toxic fumes of NO_x and SO_x. See also MERCAPTANS.

MCO250 CAS:7538-45-6 ***HR: 3***
2-MERCAPTOETHYL TRIMETHOXY SILANE
mf: $C_5H_{14}O_3SSi$ mw: 182.34

SYN: 2-(TRIMETHOXYSILYL)ETHANETHIOL

TOXICITY DATA with REFERENCE
orl-rat LD50:2460 mg/kg AIHAAP 30,470,69
ipr-mus LD50:17 mg/kg DANKAS 229(4),1011,76

CONSENSUS REPORTS: Reported in EPA TSCA Inventory.

SAFETY PROFILE: Poison by intraperitoneal route. Moderately toxic by ingestion. When heated to decomposition it emits toxic fumes of SO_x. See also MERCAPTANS and SILANES.

MCO500 CAS:60-56-0 ***HR: 3***
2-MERCAPTO-1-METHYLIMIDAZOLE
mf: $C_4H_6N_2S$ mw: 114.18

SYNS: BASOLAN ◇ DANANTIZOL ◇ FAVISTAN ◇ FRENTIROX ◇ MERCAPTAZOLE ◇ MERCAZOLYL ◇ METAZOLO ◇ METHIAMAZOLE ◇ 1-METHYLIMIDAZOLE-2-THIOL ◇ 1-METHYL-2-MERCAPTOIMIDAZOLE ◇ METIZOL ◇ METOTHYRINE ◇ 1-METYLO-2-MERKAPTOIMIDAZOLEM (POLISH) ◇ STRUMAZOLE ◇ TAPAZOLE ◇ THACAPZOL ◇ THIAMAZOLE ◇ THYCAPSOL ◇ USAF EL-30

TOXICITY DATA with REFERENCE
cyt-mus:mmr 3200 μmol/L/24H-C JTSCDR 5,141,80
unr-wmn TDLo:145 mg/kg (1-34W preg):TER ZEGYAX 88,218,66
orl-mus TDLo:320 mg/kg (female 16-21D post):REP TJADAB 27,71A,83
orl-rat TDLo:1100 mg/kg/2Y-C:NEO FCTXAV 11,649,73
orl-rat LD50:2250 mg/kg NIIRDN 6,447,82
scu-rat LD50:1050 mg/kg FRPSAX 14,54,59
orl-mus LD50:860 mg/kg NIIRDN 6,447,82
ipr-mus LD50:500 mg/kg NTIS** AD277-689
scu-mus LD50:345 mg/kg NIIRDN 6,447,82

CONSENSUS REPORTS: Reported in EPA TSCA Inventory.

SAFETY PROFILE: Poison by subcutaneous route. Moderately toxic by ingestion and intraperitoneal routes. Human teratogenic effects. An experimental teratogen. Experimental reproductive effects. Questionable carcinogen with experimental neoplastigenic data. Mutation data reported. An antithyroid drug. When heated to decomposition it emits very toxic fumes of NO_x and SO_x. See also MERCAPTANS.

MCO750 CAS:62571-86-2 ***HR: 3***
1-(d-3-MERCAPTO-2-METHYL-1-OXOPROPYL)-l-PROLINE (S,S)
mf: $C_9H_{12}NO_3S$ mw: 214.28

SYNS: CAPOTEN ◇ LOPIRIN ◇ 1-(3-MERCAPTO-2-METHYL-1-OXOPROPYL)-l-PROLINE ◇ 1-((2S)-3-MERCAPTO-2-METHYLPROPIONYL)-l-PROLINE ◇ d-2-METHYL-3-MERCAPTOPROPANOYL-l-PROLINE ◇ SQ 14,225

TOXICITY DATA with REFERENCE
orl-rbt TDLo:40 mg/kg (15-30D preg):TER PSEBAA 170,378,82
orl-rbt TDLo:8250 μg/kg (female 24-28D post):REP LANCAO 1,1256,80
orl-man TDLo:4 mg/kg/8D-I AJPSAO 142,270,85
orl-man TDLo:471 mg/kg/48W-I:SKN ARDEAC 123,20,87
orl-man LDLo:2500 μg/kg/3D-I:KID PGMJAO 60,561,84
orl-man TDLo:2679 μg/kg/5D-I:KID IJMDAI 21,892,85
orl-man TDLo:7143 μg/kg/2D-I:SKN BMJOAE 294,91,87
orl-man TDLo:12500 μg/kg/25D-I:BLD CMAJAX 129,525,83
orl-wmn TDLo:14 mg/kg/2W-I:KID,BIO AIMEAS 104,126,86
orl-wmn LDLo:1500 μg/kg/7W:CNS,CVS,PUL AIMEAS 94,58,81
orl-wmn TDLo:10 mg/kg/10D-I:SKN BMJOAE 294,91,87
orl-rat LD50:4245 mg/kg IYKEDH 14,297,83
ivn-rat LD50:554 mg/kg IYKEDH 14,297,83
orl-mus LD50:2500 mg/kg PCJOAU 22,212,88
ivn-mus LD50:663 mg/kg IYKEDH 14,297,83

SAFETY PROFILE: Moderately toxic by intravenous route. Mildly toxic by ingestion. Human systemic effects: blood pressure decrease, changes in kidney, decreased urine volume or anuria, dermatitis, dyspnea, hemolysis with or without anemia, metabolic changes, somnolence, ureter or bladder tubules failure. An experimental teratogen. Experimental reproductive effects. Used to treat refractory systemic hypertension and as an experimental drug in heart failure. When heated to decomposition it emits very toxic fumes of NO_x and SO_x. See also MERCAPTANS.

MCO775 CAS:65002-17-7 ***HR: 3***
N-(2-MERCAPTO-2-METHYLPROPANOYL)-l-CYSTEINE
mf: $C_7H_{13}NO_3S_2$ mw: 223.33

SYNS: N-(2-MERCAPTO-2-METHYL-1-OXOPROPYL)-l-CYSTEINE ◇ SA96

TOXICITY DATA with REFERENCE
orl-mus TDLo:840 mg/kg (female 15-21D post):REP IYKEDH 16,665,85
orl-rbt TDLo:3900 mg/kg (female 6-18D post):TER IYKEDH 16,654,85
orl-rat LD50:3900 mg/kg IYKEDH 14,346,83
ipr-rat LD50:353 mg/kg IYKEDH 14,346,83
scu-rat LD50:1021 mg/kg IYKEDH 14,346,83
ivn-rat LD50:1006 mg/kg IYKEDH 14,346,83
orl-mus LD50:4500 mg/kg IYKEDH 14,346,83
ipr-mus LD50:420 mg/kg IYKEDH 14,346,83

scu-mus LD50:1090 mg/kg IYKEDH 14,346,83
ivn-mus LD50:1100 mg/kg IYKEDH 14,346,83
ivn-dog LDLo:200 mg/kg OYYAA2 29,429,85

SAFETY PROFILE: Poison by intraperitoneal and intravenous routes. Moderately toxic by ingestion and subcutaneous routes. An experimental teratogen. Experimental reproductive effects. When heated to decomposition it emits toxic fumes of SO_x and NO_x. An anti-rheumatic agent. See also CYSTEINE and MERCAPTANS.

MCP000 CAS:60764-83-2 ***HR: 3***
MERCAPTOMETHYLTRIETHOXYSILANE
mf: $C_7H_{18}O_3SSi$ mw: 210.40

TOXICITY DATA with REFERENCE
orl-rat LD50:2550 mg/kg MarJV# 29MAR77
ipr-mus LD50:270 mg/kg DANKAS 229(4),1011,76

SAFETY PROFILE: Poison by intraperitoneal route. Moderately toxic by ingestion. When heated to decomposition it emits toxic fumes of SO_x. See also MERCAPTANS and SILANES.

MCP250 CAS:4845-58-3 ***HR: 3***
2-MERCAPTO-6-NITROBENZOTHIAZOLE
mf: $C_7H_4N_2O_2S_2$ mw: 212.25

SYN: USAF EK-3991

TOXICITY DATA with REFERENCE
ipr-mus LD50:25 mg/kg NTIS** AD277-689

CONSENSUS REPORTS: Reported in EPA TSCA Inventory.

SAFETY PROFILE: Poison by intraperitoneal route. When heated to decomposition it emits very toxic fumes of SO_x and NO_x. See also MERCAPTANS.

MCP500 CAS:17654-88-5 ***HR: 2***
5-MERCAPTO-3-PHENYL-2H-1,3,4-THIADIAZOLE-2-THIONE
mf: $C_8H_6N_2S_3$ mw: 226.34

SYNS: BIZMUTHIOL II (CZECH) ◇ 5-MERKAPTO-3-FENYL-1,3,4-THIADIAZOL-2-THION DRASELNY (CZECH)

TOXICITY DATA with REFERENCE
eye-rbt 20 mg/24H SEV 28ZPAK -,202,72
orl-rat LD50:763 mg/kg 28ZPAK -,202,72

SAFETY PROFILE: Moderately toxic by ingestion. A severe eye irritant. When heated to decomposition it emits very toxic fumes of NO_x and SO_x. See also SULFIDES.

MCQ000 CAS:107-96-0 ***HR: 3***
3-MERCAPTOPROPIONIC ACID
mf: $C_3H_6O_2S$ mw: 106.15

SYN: 3MPA

TOXICITY DATA with REFERENCE
ipr-mus LD50:500 mg/kg BCFAAI 119,600,80
orl-rat LD50:96 mg/kg TRIPA7 -,1,73
ipr-rat LD50:66 mg/kg AGACBH 1(5),231,70

CONSENSUS REPORTS: Reported in EPA TSCA Inventory.

SAFETY PROFILE: Poison by ingestion and intraperitoneal routes. When heated to decomposition it emits toxic fumes of SO_x. See also MERCAPTANS.

MCQ100 CAS:6112-76-1 ***HR: D***
6-MERCAPTOPURINE MONOHYDRATE
mf: $C_5H_4N_4S{\cdot}H_2O$ mw: 170.21

SYNS: 1,7-DIHYDRO-6H-PURINE-6-THIONE MONOHYDRATE (9CI) ◇ PURINE-6-THIOL, MONOHYDRATE

TOXICITY DATA with REFERENCE
dlt-mus-ipr 500 mg/kg/40D-I DCTODJ 6,83,83
ipr-ham TDLo:4 mg/kg (female 6-9D post):TER TOXID9 1,27,81
ipr-ham TDLo:4 mg/kg (female 6-9D post):REP TOXID9 1,27,81

SAFETY PROFILE: An experimental teratogen. Experimental reproductive effects. Mutation data reported. When heated to decomposition it emits toxic fumes of SO_x and NO_x. See also MERCAPTANS.

MCQ250 CAS:145-95-9 ***HR: 3***
6-MERCAPTOPURINE 3-N-OXIDE
mf: $C_5H_4N_4OS$ mw: 168.19

SYN: MERCAPTOPURINE-3-N-OXIDE

TOXICITY DATA with REFERENCE
ipr-rat TDLo:100 mg/kg (female 11D post):TER ARPAAQ 86,395,68
ipr-rat LD50:250 mg/kg ARPAAQ 86,395,68

SAFETY PROFILE: Poison by intraperitoneal route. An experimental teratogen. When heated to decomposition it emits very toxic fumes of SO_x and NO_x. See also MERCAPTANS.

MCQ500 CAS:4988-64-1 ***HR: 3***
MERCAPTOPURINE RIBONUCLEOSIDE
mf: $C_{10}H_{12}N_4O_4S$ mw: 284.32

SYNS: 6-MERCAPTOPURINE RIBOSIDE ◇ NSC 4911 ◇ RIBOFURANOSIDE, 9H-PURINE-6-THIOL-9 ◇ RIBOSYL-6-THIOPURINE ◇ THIONOSINE ◇ 6-THIOPURINE RIBONUCLEOSIDE ◇ 6-THIOPURINE RIBOSIDE ◇ TIOINOSINE

TOXICITY DATA with REFERENCE
sce-hmn:lym 100 nmol/L CTRRDO 69,505,85

ipr-rat TDLo:62500 μg/kg (11D preg):TER ARPAAQ 86,395,68
orl-rat LD50:900 mg/kg NIIRDN 6,853,82
ipr-rat LD50:1240 mg/kg NIIRDN 6,853,82
scu-rat LD50:1180 mg/kg NIIRDN 6,853,82
orl-mus LD50:5 g/kg OYYAA2 6,1275,72
ipr-mus LD50:490 mg/kg CPBTAL 16,2080,68
scu-mus LD50:1840 mg/kg NIIRDN 6,853,82
unr-mus LD10:250 mg/kg PMDCAY 7,69,70

SAFETY PROFILE: Poison by an unspecified route. Moderately toxic by ingestion, subcutaneous, and intraperitoneal routes. An experimental teratogen. Human mutation data reported. When heated to decomposition it emits very toxic fumes of SO_x and NO_x. See also MERCAPTANS.

MCQ750 CAS:3811-73-2 ***HR: 3***
2-MERCAPTOPYRIDINE-N-OXIDE SODIUM SALT
mf: $C_5H_5NOS•Na$ mw: 150.16

SYNS: (1-HYDROXY-2-PYRIDINETHIONE), SODIUM SALT, TECH ◇ 2-PYRIDINETHIOL-1-OXIDE SODIUM SALT ◇ SODIUM PYRITHIONE

TOXICITY DATA with REFERENCE
orl-mus LD50:870 mg/kg OYYAA2 4,883,70
ipr-mus LD50:370 mg/kg OYYAA2 4,883,70
scu-mus LD50:428 mg/kg OYYAA2 4,883,70
ivn-mus LD50:320 mg/kg CSLNX* NX#04703
par-mus LDLo:800 mg/kg CBCCT* 7,693,55

CONSENSUS REPORTS: Reported in EPA TSCA Inventory.

SAFETY PROFILE: Poison by intraperitoneal and intravenous routes. Moderately toxic by ingestion, subcutaneous and parenteral routes. Used in preservation of cosmetics. When heated to decomposition it emits very toxic fumes of Na_2O, NO_x, and SO_x. See also MERCAPTANS.

MCR000 CAS:70-49-5 ***HR: 3***
MERCAPTOSUCCINIC ACID
mf: $C_4H_6O_4S$ mw: 150.16

PROP: Off-white powder. Mp: 150°. Sol in water, alc, acetone; sltly sol in ether and benzene.

SYNS: THIOMALIC ACID ◇ 2-THIO-MALIC ACID ◇ USAF EK-P-6297 ◇ USAF M-2

TOXICITY DATA with REFERENCE
orl-rat LD50:800 mg/kg 14CYAT 2,1813,63
ipr-mus LD50:200 mg/kg NTIS** AD277-689

CONSENSUS REPORTS: Reported in EPA TSCA Inventory.

SAFETY PROFILE: Poison by intraperitoneal route. Moderately toxic by ingestion. Has been proposed as an antidote for heavy metal poisoning. Allergic dermatitis in humans has been reported. When heated to decomposition, or on contact with acid or acid fumes, it emits toxic fumes of SO_x. See also MERCAPTANS.

MCR250 CAS:67479-03-2 ***HR: 3***
p-MERCAPTO SULFADIAZINE

SYNS: TSD ◇ USAF LO-3

TOXICITY DATA with REFERENCE
ipr-mus LD50:100 mg/kg NTIS** AD277-689

SAFETY PROFILE: Poison by intraperitoneal route. When heated to decomposition it emits very toxic fumes of SO_x and NO_x. See also MERCAPTANS.

MCR750 CAS:52-67-5 ***HR: 3***
d,3-MERCAPTOVALINE
mf: $C_5H_{11}NO_2S$ mw: 149.23

SYNS: CUPRENIL ◇ CUPRIMINE ◇ DEPEN ◇ DIMETHYLCYSTEINE ◇ β,β-DIMETHYLCYSTEINE ◇ d-MERCAPTOVALINE ◇ METALCAPTASE ◇ PCA ◇ d-PENAMINE ◇ PENICILLAMIN ◇ (S)-PENICILLAMIN ◇ PENICILLAMINE ◇ d-PENICILLAMINE ◇ REDUCED-d-PENICILLAMINE ◇ TROLOVOL

TOXICITY DATA with REFERENCE
mmo-sat 60 μmol/plate BCPCA6 34,3725,85
mma-sat 1 mg/plate ABCHA6 45,2157,81
unr-wmn TDLo:6480 mg/kg (1-39W preg):TER TJADAB 29(3),25A,84
orl-mus TDLo:20160 mg/kg (female 1-21D post):REP TXAPA9 65,273,82
orl-chd TDLo:122 g/kg/3Y-C:CAR,BLD JAMAAP 248,467,82
ipr-mus TDLo:2 g/kg/4W-I:ETA LANCAO 2,1356,76
orl-wmn TDLo:900 mg/kg/26W-I:SKN JRHUA9 12,583,85
orl-wmn TDLo:650 g/kg/81W-I ARHEAW 29,560,86
orl-wmn TDLo:112 mg/kg/1W-I: AIMDAP 145,2271,85
orl-wmn TDLo:105 mg/kg/6W-I:SKN BMJOAE 294,1101,87
orl-man TDLo:400 mg/kg/4W-I:PUL JRHUA9 13,963,86
orl-wmn LDLo:150 mg/kg/30D-I AIMEAS 98,327,83
orl-hmn TDLo:21 mg/kg/D:KID,BLD JAMAAP 240,1870,78
orl-hmn TDLo:893 mg/kg/30W-I JRHUA9 11,251,84
orl-cld TDLo:40 mg/kg/1W-I:BLD,SKN AIMDAP 145,2271,85
orl-man TDLo:482 mg/kg/19W-I:BLD,SKN AIMDAP 143,1487,83
orl-rat LD50:6170 mg/kg NIIRDN 6,758,82
ipr-rat LD50:2080 mg/kg NIIRDN 6,758,82
scu-rat LD50:4020 mg/kg NIIRDN 6,758,82
ivn-rat LD50:2 g/kg ARZNAD 22,1434,72

orl-mus LD50:8419 mg/kg ARZNAD 25,162,75
ipr-mus LD50:298 mg/kg YKKZAJ 94,1419,74
scu-mus LD50:3810 mg/kg NIIRDN 6,758,82
ivn-mus LD50:3840 mg/kg ARZNAD 25,162,75

SAFETY PROFILE: Poison by intraperitoneal route. Moderately toxic by subcutaneous and intravenous routes. Mildly toxic by ingestion. An experimental teratogen. Human systemic effects by ingestion: agranulocytosis, dermatitis, fever, hemorrhage, increased body temperature, dermatitis, leukopenia, proteinuria, thrombocytopenia. Human teratogenic effects by an unspecified route: developmental abnormalities of the craniofacial areas, skin, and skin appendages, and body wall. Experimental reproductive effects. Questionable human carcinogen producing leukemia. Mutation data reported. Used in the treatment of rheumatoid arthritis, metal poisonings, and cystinuria. When heated to decomposition it emits very toxic fumes of NO_x and SO_x. See also MERCAPTANS.

MCS000 CAS:8018-15-3 ***HR: 3***
MERCUMATILIN SODIUM
mf: $C_{14}H_{13}HgO_6 \cdot C_7H_8N_4O_2 \cdot Na$ mw: 681.04

SYNS: CUMERTILIN SODIUM ◇ 8-(Γ-HYDROXYMERCURI-β-METHOXYPROPYL)-3-COUMARINCARBOXYLICACIDTHEOPHYLLINE SODIUM ◇ 8-(3-(HYDROXYMERCURI)-2-METHOXYPROPYL)-2-OXO-2H-1-BENZOPYRAN-3-CARBOXYLIC ACID SODIUM SALT COMPOUND with THEOPHYLLINE (1:1)

TOXICITY DATA with REFERENCE
orl-rat LD50:809 mg/kg JPETAB 105,336,52
ivn-rat LD50:33 mg/kg JPETAB 105,336,52
ims-rat LD50:42 mg/kg JPETAB 105,336,52
scu-mus LD50:282 mg/kg JPETAB 105,336,52
ivn-mus LD50:139 mg/kg JPETAB 105,336,52
ivn-rbt LD50:25 mg/kg JPETAB 105,336,52
ims-rbt LD50:44 mg/kg JPETAB 105,336,52

CONSENSUS REPORTS: Mercury and its compounds are on the Community Right-To-Know List.

ACGIH TLV: TWA 0.1 mg(Hg)/m^3 (skin)
NIOSH REL: (Inorganic Mercury) TWA 0.05 mg(Hg)/m^3

SAFETY PROFILE: A poison by intravenous, intramuscular, and subcutaneous routes. Moderately toxic by ingestion. When heated to decomposition it emits very toxic fumes of Hg, Na_2O and NO_x. See also MERCURY COMPOUNDS.

MCS250 CAS:64049-28-1 ***HR: 3***
2,2'-MERCURIBIS(6-ACETOXYMERCURI-4-NITRO)ANILINE
mf: $C_{16}H_{14}Hg_3N_2O_8$ mw: 964.09

TOXICITY DATA with REFERENCE
ipr-rat LDLo:250 mg/kg NCNSA6 5,12,53

CONSENSUS REPORTS: Mercury and its compounds are on the Community Right-To-Know List.

OSHA PEL: (Transitional: CL 1 mg/10m^3) CL 0.1 mg(Hg)/m^3 (skin)
ACGIH TLV: TWA 0.1 mg(Hg)/m^3 (skin)
NIOSH REL: (Inorganic Mercury) TWA 0.05 mg(Hg)/m^3

SAFETY PROFILE: Poison by intraperitoneal route. When heated to decomposition it emits very toxic fumes of Hg and NO_x. See also MERCURY COMPOUNDS.

MCS500 CAS:64047-26-3 ***HR: 3***
MERCURIBIS(DIETHYL(2,2-DIMETHYL-4-DITHIOCARBOXYAMINO)BUTYLAMMONIUM DICHLORIDE
mf: $C_{22}H_{46}HgN_4S_4 \cdot 2Cl$ mw: 766.45

TOXICITY DATA with REFERENCE
ivn-rbt LDLo:10 mg/kg JPETAB 41,21,31

CONSENSUS REPORTS: Mercury and its compounds are on the Community Right-To-Know List.

OSHA PEL: (Transitional: CL 1 mg/10m^3) CL 0.1 mg(Hg)/m^3 (skin)
ACGIH TLV: TWA 0.1 mg(Hg)/m^3 (skin)
NIOSH REL: (Inorganic Mercury) TWA 0.05 mg (Hg)/m^3

SAFETY PROFILE: Poison by intravenous route. When heated to decomposition it emits very toxic fumes of Hg, NO_x, SO_x, and Cl^-. See also MERCURY COMPOUNDS.

MCS600 CAS:66499-61-4 ***HR: 3***
MERCURIBIS-o-NITROPHENOL
mf: $C_{12}H_8HgN_2O_6$ mw: 476.81

SYNS: BIS(4-HYDROXY-3-NITROPHENYL)MERCURY ◇ MERCURY, BIS(4-HYDROXY-3-NITROPHENYL)-

TOXICITY DATA with REFERENCE
ipr-mus LDLo:105 mg/kg JPETAB 31,87,27
ACGIH TLV: TWA 0.1 mg(Hg)/m^3 (skin)

SAFETY PROFILE: Poison by intraperitoneal route. When heated to decomposition it emits toxic fumes of NO_x and Hg.

MCS750 CAS:1600-27-7 ***HR: 3***
MERCURIC ACETATE
DOT: UN 1629
mf: $C_4H_6O_4 \cdot Hg$ mw: 318.69

PROP: White crystals or powder; slt acetic odor. D:

3.280, mp: 178-180° (overheating causes decomp). Sol in alc. Keep stoppered and protected from light.

SYNS: ACETIC ACID, MERCURY(2+) SALT ◇ BIS(ACETYLOXY) MERCURY ◇ DIACETOXYMERCURY ◇ MERCURIACETATE ◇ MERCURIC DIACETATE ◇ MERCURY ACETATE ◇ MERCURY(2+) ACETATE ◇ MERCURY(II) ACETATE ◇ MERCURY DIACETATE ◇ MERCURYL ACETATE

TOXICITY DATA with REFERENCE
oth-mus:oth 50 mg/L MUREAV 17,93,73
ivn-rat TDLo:3 mg/kg (female 10D post):TER ARTODN 7,382,84
orl-ham TDLo:35 mg/kg (female 8D post):REP ENVRAL 8,207,74
orl-rat LD50:40900 μg/kg GISAAA 46(8),12,81
skn-rat LD50:570 mg/kg GTPZAB 25(7),27,81
orl-mus LD50:23900 μg/kg GISAAA 46(8),12,81
ipr-mus LD50:6500 μg/kg GTPZAB 25(7),27,81
scu-mus LDLo:20 mg/kg MOLAAF 73,751,39
ivn-mus LD50:4390 μg/kg NYKZAU 57,219,61
orl-mam LD50:65 mg/kg GISAAA 49(9),11,84

CONSENSUS REPORTS: EPA Extremely Hazardous Substances List. Mercury and its compounds are on the Community Right-To-Know List. EPA Genetic Toxicology Program. Reported in EPA TSCA Inventory.

OSHA PEL: (Transitional: CL 1 mg/10m^3) CL 0.1 mg (Hg)/m^3 (skin)
ACGIH TLV: TWA 0.1 mg(Hg)/m^3 (skin)
NIOSH REL: (Inorganic Mercury) TWA 0.05 mg(Hg)/m^3
DOT Classification: Poison B; Label: Poison.

SAFETY PROFILE: Poison by ingestion, intravenous, intraperitoneal, and subcutaneous routes. Moderately toxic by skin contact. An experimental teratogen. Experimental reproductive effects. Mutation data reported. When heated to decomposition it emits toxic fumes of Hg. See also MERCURY COMPOUNDS.

MCT000 CAS:6937-66-2 ***HR: 3***
MERCURIC-8,8-DICAFFEINE
mf: $C_{16}H_{18}HgN_8O_4$ mw: 587.01

SYN: BIS(1,3,7-TRIMETHYL-8-XANTHINYL)MERCURY

TOXICITY DATA with REFERENCE
ivn-mus LD50:32 mg/kg CSLNX* NX#04388

CONSENSUS REPORTS: Mercury and its compounds are on the Community Right-To-Know List.

OSHA PEL: (Transitional: CL 1 mg/10m^3) CL 0.1 mg(Hg)/m^3 (skin)
ACGIH TLV: TWA 0.1 mg(Hg)/m^3 (skin)
NIOSH REL: (Inorganic Mercury) TWA 0.05 mg(Hg)/m^3

SAFETY PROFILE: Poison by intravenous route. When heated to decomposition it emits very toxic fumes of Hg and NO_x. See also MERCURY COMPOUNDS.

MCT250 CAS:63766-15-4 ***HR: 3***
MERCURIC DINAPHTHYLMETHANE DISULPHONATE
mf: $C_{21}H_{16}O_6S_2 \cdot xHg$ mw: 1832.62

SYN: 3,3'-METHYLENEDI-2-NAPHTHALENESULFONIC ACID, MERCURY SALT

TOXICITY DATA with REFERENCE
orl-mus LD50:30 mg/kg JPPMAB 2,20,50

CONSENSUS REPORTS: Mercury and its compounds are on the Community Right-To-Know List.

NIOSH REL: (Inorganic Mercury) TWA 0.05 mg (Hg)/m^3

SAFETY PROFILE: Poison by ingestion. When heated to decomposition it emits very toxic fumes of SO_x and Hg. See also MERCURY COMPOUNDS and SULFONATES.

MCT500 CAS:21908-53-2 ***HR: 3***
MERCURIC OXIDE
DOT: UN 1641
mf: HgO mw: 216.59

PROP: Heavy, bright, orange-red or orange-yellow powder. Mp: decomp @ 500°, d: 11.14. Practically insol in water; sol in dil HCl or HNO_3. Protect from light.

SYNS: C.I. 77760 ◇ MERCURIC OXIDE, RED ◇ MERCURIC OXIDE, solid (DOT) ◇ MERCURIC OXIDE, YELLOW ◇ MERCURY(II) OXIDE ◇ OXYDE de MERCURE (FRENCH) ◇ QUECKSILBEROXID (GERMAN) ◇ RED OXIDE of MERCURY ◇ RED PRECIPITATE ◇ SANTAR ◇ YELLOW MERCURIC OXIDE ◇ YELLOW OXIDE of MERCURY ◇ YELLOW PRECIPITATE

TOXICITY DATA with REFERENCE
orl-mus TDLo:34 mg/kg (female 10D post):TER APTOD9 19,A126,80
orl-rat TDLo:10800 μg/kg (5D preg):REP PWPSA8 15,52,72
orl-rat LD50:18 mg/kg NTIS** PB214-270
skn-rat LD50:315 mg/kg GTPZAB 25(7),27,81
ims-rat LDLo:22 mg/kg NCIUS* PH 43-64-886,SEPT,71
orl-mus LD50:16 mg/kg GTPZAB 25(7),27,81
ipr-mus LD50:4500 μg/kg GTPZAB 25(7),27,81

CONSENSUS REPORTS: EPA Extremely Hazardous Substances List. Mercury and its compounds are on the Community Right-To-Know List. Reported in EPA TSCA Inventory.

OSHA PEL: (Transitional: CL 1 mg/10m^3) CL 0.1 mg(Hg)/m^3 (skin)
ACGIH TLV: TWA 0.1 mg(Hg)/m^3 (skin)
NIOSH REL: (Inorganic Mercury) TWA 0.05 mg(Hg)/m^3
DOT Classification: Poison B; Label: Poison.

SAFETY PROFILE: Poison by ingestion, skin contact, intraperitoneal, and intramuscular routes. An experimental teratogen. Experimental reproductive effects. An FDA over-the-counter drug. Used for treating fruit trees. Flammable by chemical reactions. A powerful oxidizer. Explosive reaction with acetyl nitrate; butadiene + ethanol + iodine (at 35°C); chlorine + hydrocarbons (e.g., methane, ethylene); diboron tetrafluoride; hydrogen peroxide + traces of nitric acid; reducing agents (e.g., hydrazine hydrate, phosphinic acid). Forms heat- or impact-sensitive explosive mixtures with non-metals (e.g., phosphorus, sulfur); metals (e.g., magnesium, potassium, sodium-potassium alloy). Reacts violently with hydrogen trisulfide (or ignition); hydrazine hydrate; hydrogen peroxide; hypophosphorus acid; iodine + methanol or ethanol; phospham; acetyl nitrate; S_2Cl_2; reductants. Incandescent reaction with phospham. When heated to decomposition it emits highly toxic fumes of Hg. See also MERCURY COMPOUNDS, INORGANIC.

MCT750 ***HR: 3***
MERCURIC PEROXYBENZOATE
mf: $C_{14}H_{10}HgO_6$ mw: 474.83

SYN: MERCURY(II)PEROXYBENZOATE

CONSENSUS REPORTS: Mercury and its compounds are on the Community Right-To-Know List.

SAFETY PROFILE: Poison. Explodes when heated above 100°C. Upon decomposition it emits toxic fumes of Hg. See also MERCURY COMPOUNDS.

MCU000 CAS:5970-32-1 ***HR: 3***
MERCURIC SALICYLATE
DOT: UN 1644
mf: $C_7H_4HgO_3$ mw: 336.70

PROP: White-yellow or pinkish, odorless powder. Insol in water or alc, sol in warm solns of alkali halides.

SYNS: MERCURIC SALICYLATE, solid (DOT) ◇ MERCURISALICYLIC ACID ◇ MERCURY SALICYLATE ◇ MERCURY SUBSALICYLATE

TOXICITY DATA with REFERENCE
scu-mus LDLo:10 mg/kg MOLAAF 73,751,39

CONSENSUS REPORTS: Mercury and its compounds are on the Community Right-To-Know List.

OSHA PEL: (Transitional: CL 1 mg/10m^3) CL 0.1 mg(Hg)/m^3 (skin)
ACGIH TLV: TWA 0.1 mg(Hg)/m^3 (skin)
NIOSH REL: (Inorganic Mercury) TWA 0.05 mg(Hg)/m^3
DOT Classification: Poison B; Label: Poison.

SAFETY PROFILE: Poison by subcutaneous route. An FDA over-the-counter drug. Incompatible with alkali iodides. When heated to decomposition it emits toxic fumes of Hg. See also MERCURY COMPOUNDS.

MCU250 CAS:592-85-8 ***HR: 3***
MERCURIC SULFOCYANATE
DOT: UN 1646
mf: $C_2HgN_2S_2$ mw: 316.79

PROP: White, odorless powder; sltly sol in cold water; more sol in boiling water (decomp); sol in dil HCl. Protect from light.

SYNS: BIS(THIOCYANATO)-MERCURY ◇ MERCURIC SULFOCYANTE, solid (DOT) ◇ MERCURIC SULFOCYANIDE ◇ MERCURIC THIOCYANATE ◇ MERCURIC THIOCYANATE, solid (DOT) ◇ MERCURY DITHIOCYANATE ◇ MERCURY THIOCYANATE (DOT) ◇ MERCURY(II) THIOCYANATE ◇ THIOCYANIC ACID, MERCURY(2^+) SALT

TOXICITY DATA with REFERENCE
orl-rat LD50:46 mg/kg GTPZAB 25(7),27,81
skn-rat LD50:685 mg/kg GTPZAB 25(7),27,81
orl-mus LD50:24500 μg/kg GTPZAB 25(7),27,81
ipr-mus LD50:3500 μg/kg GTPZAB 25(7),27,81

CONSENSUS REPORTS: Mercury and its compounds are on the Community Right-To-Know List. Reported in EPA TSCA Inventory.

OSHA PEL: (Transitional: CL 1 mg/10m^3) CL 0.1 mg(Hg)/m^3 (skin)
ACGIH TLV: TWA 0.1 mg(Hg)/m^3 (skin)
NIOSH REL: (Inorganic Mercury) TWA 0.05 mg(Hg)/m^3
DOT Classification: Poison B; Label: Poison.

SAFETY PROFILE: A poison by ingestion and intraperitoneal routes. Moderately toxic by skin contact. Thermally unstable and decomposition may be vigorous. When heated to decomposition it emits very toxic fumes of Hg, NO_x, SO_x, and CN^-. See also MERCURY COMPOUNDS and CYANATES.

MCU500 CAS:535-55-7 ***HR: 3***
MERCURIPHENOLDISULFONATE SODIUM
mf: $C_{12}H_8O_8S_2 \cdot Hg \cdot 2Na$ mw: 590.89

SYNS: HERMOPHENYL ◇ p-HYDROXY-BENZENESULFONIC ACID MERCURY DERIVATIVE, DISODIUM SALT ◇ MERCURY and SODIUM PHENOLSULFONATE

TOXICITY DATA with REFERENCE
ivn-rbt LDLo:24 mg/kg JPETAB 41,21,31

CONSENSUS REPORTS: Mercury and its compounds are on the Community Right-To-Know List.

OSHA PEL: (Transitional: CL 1 mg/10m^3) CL 0.1 mg(Hg)/m^3 (skin)
ACGIH TLV: TWA 0.1 mg(Hg)/m^3 (skin)
NIOSH REL: (Inorganic Mercury) TWA 0.05 mg (Hg)/m^3

SAFETY PROFILE: Poison by intravenous route. When heated to decomposition it emits very toxic fumes of Hg, Na_2O and SO_x. See also MERCURY COMPOUNDS and SULFONATES.

MCU750 CAS:55-68-5 ***HR: 3***
MERCURIPHENYL NITRATE
DOT: UN 1895
mf: $C_6H_5HgNO_3$ mw: 339.71

PROP: Crystals. Mp: 176-186°. Insol in cold water.

SYNS: MERPHENYL NITRATE ◇ MERSOLITE 7 ◇ NITRIC ACID, PHENYLMERCURY SALT ◇ PHENALCO ◇ PHENITOL ◇ PHENMERZYL NITRATE ◇ PHENYLMERCURIC NITRATE ◇ PHENYLMERCURY NITRATE ◇ PHERMERNITE

TOXICITY DATA with REFERENCE
scu-rat LDLo:38 mg/kg QJPPAL 12,212,39
scu-mus LD50:45 mg/kg QJPPAL 12,212,39
ivn-mus LD50:27 mg/kg QJPPAL 12,212,39
ivn-rbt LDLo:5 mg/kg JAMAAP 117,1784,41

CONSENSUS REPORTS: Mercury and its compounds are on the Community Right-To-Know List. Reported in EPA TSCA Inventory. EPA Genetic Toxicology Program.

OSHA PEL: (Transitional: CL 1 mg/10m^3) CL 0.1 mg(Hg)/m^3 (skin)
ACGIH TLV: TWA 0.1 mg(Hg)/m^3 (skin)
NIOSH REL: (Inorganic Mercury) TWA 0.05 mg(Hg)/m^3
DOT Classification: Poison B; Label: Poison.

SAFETY PROFILE: Poison by intravenous and subcutaneous routes. FDA over-the-counter drug. When heated to decomposition it emits very toxic fumes of Hg and NO_x. See also MERCURY COMPOUNDS and NITRATES.

MCV000 CAS:129-16-8 ***HR: 2***
MERCUROCHROME
mf: $C_{20}H_{10}Br_2HgO_6 \cdot 2Na$ mw: 752.69

SYNS: ASCEPTICHROME ◇ ASEPTICHROME ◇ CHROMARGYRE ◇ 2,7-DIBROMO-4-HYDROXYMERCURIFLUORESCEINE DISODIUM SALT ◇ DISODIUM-2,7-DIBROM-4-HYDROXY-MERCURI-FLUORESCEIN ◇ DISODIUM-2′,7′-DIBROMO-4′-(HYDROXYMERCURY)FLUORESCEIN ◇ DOMF ◇ FLAVUROL ◇ FLUOROCHROME ◇ GALLOCHROME ◇ GYNOCHROME ◇ MERBROMIN ◇ MERCURANINE ◇ MERCUROCHROME-220 SOLUBLE ◇ MERCUROCOL ◇ MERCUROME ◇ MERCUROPHAGE ◇ PLANOCHROME

TOXICITY DATA with REFERENCE
mrc-smc 10 mg/L EVHPAZ 31,97,79
sln-smc 50 mg/L EVHPAZ 31,97,79
scu-mus LDLo:20 mg/kg MOLAAF 73,751,39
ivn-mus LDLo:105 mg/kg QJPPAL 5,1,32
ivn-rbt LDLo:15 mg/kg JPETAB 35,343,29

CONSENSUS REPORTS: Mercury and its compounds are on the Community Right-To-Know List.

NIOSH REL: (Inorganic Mercury) TWA 0.05 mg (Hg)/m^3

SAFETY PROFILE: Poison by intravenous and subcutaneous routes. Mutation data reported. Relatively nonirritating and nontoxic to damaged skin or tissue. A topical antiseptic. An FDA over-the-counter drug. When heated to decomposition it emits very toxic fumes including fumes of Na_2O, Br^-, and Hg. See also MERCURY COMPOUNDS, ORGANIC.

MCV250 CAS:12002-19-6 ***HR: 3***
MERCUROL
DOT: UN 1639

PROP: Colorless to brownish powder. Contains 20% mercury.

SYN: MERCURY NUCLEATE, solid (DOT)

CONSENSUS REPORTS: Mercury and its compounds are on the Community Right-To-Know List.

NIOSH REL: (Inorganic Mercury) TWA 0.05 mg (Hg)/m^3
DOT Classification: Poison B; Label: Poison.

SAFETY PROFILE: A poison. When heated to decomposition it emits toxic fumes of Hg. See also MERCURY COMPOUNDS.

MCV500 CAS:52486-78-9 ***HR: 3***
MERCUROPHEN
mf: $C_6H_5HgNO_4 \cdot Na$ mw: 378.70

PROP: Brick-red, odorless powder. Sol in hot H_2O.

TOXICITY DATA with REFERENCE
ipr-mus LDLo:30 mg/kg JPETAB 43,71,31

CONSENSUS REPORTS: Mercury and its compounds are on the Community Right-To-Know List.

SAFETY PROFILE: Poison by intraperitoneal route. When heated to decomposition it emits very toxic fumes of NO_x, Na_2O, and Hg vapors. See also MERCURY COMPOUNDS.

MCV750 CAS:8012-34-8 *HR: 3*
MERCUROPHYLLINE
mf: $C_{14}H_{24}HgNO_5 \cdot C_7H_8N_4O_2 \cdot Na$ mw: 690.16

SYNS: MERCUPURIN ◇ MERCUZANTHIN

TOXICITY DATA with REFERENCE
ivn-hmn TDLo:28 mg/kg:CVS JAMAAP 117,1806,41
ipr-rat LD50:121 mg/kg THERAP 10,936,55
scu-mus LD50:163 mg(Hg)/kg JPETAB 105,336,52
ivn-mus LD50:1410 mg/kg JPETAB 99,149,50
ivn-cat LDLo:250 mg/kg JPETAB 99,149,50
ivn-rbt LDLo:177 mg/kg JPETAB 99,149,50

CONSENSUS REPORTS: Mercury and its compounds are on the Community Right-To-Know List.

OSHA PEL: (Transitional: CL 1 mg/10m³) CL 0.1 mg(Hg)/m³ (skin)
ACGIH TLV: TWA 0.1 mg(Hg)/m³ (skin)
NIOSH REL: (Inorganic Mercury) TWA 0.05 mg (Hg)/m³

SAFETY PROFILE: Poison by subcutaneous, intraperitoneal, and intravenous routes. Human systemic effects by intravenous route: cardiac arrythmias. When heated to decomposition it emits toxic fumes of Hg. See also MERCURY COMPOUNDS.

MCW000 CAS:7546-30-7 *HR: 3*
MERCUROUS CHLORIDE
mf: ClHg mw: 236.04

PROP: White, odorless, tasteless, heavy powder or crystals. Subl @ 400°, d: 7.150. Insol in water, alc, and ether. Protect from light. Sunlight causes it to decomp into mercuric chloride and metallic Hg.

SYNS: CALOGREEN ◇ CALOMEL ◇ CALOMELANO (ITALIAN) ◇ CALOSAN ◇ CHLORURE MERCUREUX (FRENCH) ◇ C.I. 77764 ◇ CLORURO MERCUROSO (ITALIAN) ◇ CYCLOSAN ◇ KALOMEL (GERMAN) ◇ MERCUROCHLORIDE (DUTCH) ◇ MERCURY(I) CHLORIDE ◇ MERCURY MONOCHLORIDE ◇ MERCURY PROTOCHLORIDE ◇ MILD MERCURY CHLORIDE ◇ PRECIPITE BLANC ◇ QUECKSILBER(I)-CHLORID (GERMAN) ◇ QUECKSILBER CHLORUER (GERMAN) ◇ SUBCHLORIDE of MERCURY

TOXICITY DATA with REFERENCE
mrc-bcs 50 mmol/L MUREAV 77,109,80
sce-ham:ovr 3200 nmol/L ENMUDM 7,381,85
orl-rat LD50:166 mg/kg GTPZAB 25(7),27,81
skn-rat LD50:1500 mg/kg GTPZAB 25(7),27,81
orl-mus LD50:180 mg/kg GTPZAB 25(7),27,81
ipr-mus LD50:10 mg/kg GTPZAB 25(7),27,81

CONSENSUS REPORTS: Mercury and its compounds are on the Community Right-To-Know List. EPA Genetic Toxicology Program. Reported in EPA TSCA Inventory.

OSHA PEL: (Transitional: CL 1 mg/10m³) CL 0.1 mg(Hg)/m³ (skin)
ACGIH TLV: TWA 0.1 mg(Hg)/m³ (skin)
NIOSH REL: (Inorganic Mercury) TWA 0.05 mg(Hg)/m³

SAFETY PROFILE: Poison by ingestion and intraperitoneal routes. Moderately toxic by skin contact. Mutation data reported. A fungicide. An FDA over-the-counter drug. Incompatible with bromides; iodides; alkali chlorides; sulfates; sulfites; carbonates; hydroxides; lime water; ammonia; golden antimony sulfide; cyanides; copper salts; hydrogen peroxide; iodine; iodoform; lead salts; silver salts; sulfides. When heated to decomposition it emits very toxic fumes of Cl^- and Hg. See also MERCURY COMPOUNDS.

MCW250 CAS:7439-97-6 *HR: 3*
MERCURY
DOT: NA 2809
af: Hg aw: 200.59

PROP: Silvery, heavy, mobile liquid. A liquid metallic element. Mp: −38.89°, bp: 356.9°, d: 13.534 @ 25°, vap press: 2×10^{-3} mm @ 25°. Solid: tin-white, ductile, malleable mass which can be cut with a knife.

SYNS: COLLOIDAL MERCURY ◇ KWIK (DUTCH) ◇ MERCURE (FRENCH) ◇ MERCURIO (ITALIAN) ◇ MERCURY, METALLIC (DOT) ◇ NCI-C60399 ◇ QUECKSILBER (GERMAN) ◇ QUICK SILVER ◇ RCRA WASTE NUMBER U151 ◇ RTEC (POLISH)

TOXICITY DATA with REFERENCE
cyt-man-unr 150 µg/m³ AEHLAU 34,461,79
ihl-rat TCLo:1 mg/m³/24H (female 1-20D post):TER TJADAB 35,59A,87
ihl-rat TCLo:7440 ng/m³/24H (16W male):REP GISAAA 45(3),72,80
ipr-rat TDLo:400 mg/kg/14D-I:ETA ZEKBAI 61,511,57
ihl-man TDLo:44300 µg/m³/8H:CNS,LIV,MET JOCMA7 20,532,78
ihl-wmn TCLo:150 µg/m³/46D:CNS,GIT AEHLAU 33,186,78
skn-man TDLo:129 mg/kg/5H-C:EAR,CNS,SKN DERAAC 172,48,86
ihl-rbt LCLo:29 mg/m³/30H AMIHBC 7,19,53

CONSENSUS REPORTS: Mercury and its compounds are on the Community Right-To-Know List.

OSHA PEL: Vapor: (Transitional: CL 1 mg/10m³) 0.05 mg/m³ (skin)
ACGIH TLV: TWA 0.05 mg(Hg)/m³ (vapor, skin); (Proposed: BEI 35 µg/g creatinine total inorganic mercury in urine, preshift.)
DFG MAK: 0.1 mg/m³; BAT: 5 µg/dL in blood.
NIOSH REL: (Inorganic Mercury) TWA 0.05 mg(Hg)/m³
DOT Classification: Corrosive Material; Label: Corrosive.

SAFETY PROFILE: Poison by inhalation. Corrosive to skin, eyes, and mucous membranes. Human systemic effects by inhalation: wakefulness, muscle weakness, anorexia, headache, tinnitus, hypermotility, diarrhea, liver changes, dermatitis, fever. An experimental teratogen. Experimental reproductive effects. Questionable carcinogen with experimental tumorigenic data. Human mutation data reported. Used in dental applications, electronics, and chemical synthesis.

May explode on contact with 3-bromopropyne; alkynes + silver perchlorate; ethylene oxide; lithium; methylsilane + oxygen (explodes when shaken); peroxyformic acid; chlorine dioxide; tetracarbonylnickel + oxygen. May react with ammonia to form an explosive product. Mixtures with methyl azide are shock- and spark-sensitive explosives. The vapor ignites on contact with boron diiodophosphide. Reacts violently with acetylenic compounds (e.g., acetylene, sodium acetylide, 2-butyne-1,4-diol + acid); metals (e.g., aluminum; calcium; potassium; sodium; rubidium; exothermic formation of amalgams); Cl_2; ClO_2; CH_3N_3; Na_2C_2; nitromethane. Incompatible with methyl azide; oxidants. When heated to decomposition it emits toxic fumes of Hg. See also MERCURY COMPOUNDS.

MCW349 CAS:68833-55-6 ***HR: 3***
MERCURY ACETYLIDE (DOT)
mf: C_2HHg mw: 225.62

SYN: MERCURY ACETYLIDE
ACGIH TLV: TWA 0.01 mg(Hg)/m^3; STEL 0.03 mg(Hg)/m^3

DOT Classification: Forbidden

SAFETY PROFILE: Extremely reactive. When heated to decomposition it emits toxic fumes of Hg.

MCW350 CAS:37297-87-3 ***HR: 3***
MERCURY(II) ACETYLIDE
mf: C_2Hg mw: 224.61

CONSENSUS REPORTS: Mercury and its compounds are on the Community Right-To-Know List.

SAFETY PROFILE: A shock- and heat-sensitive explosive. Upon decomposition it emits toxic fumes of Hg. See also MERCURY COMPOUNDS and ACETYLIDES.

MCW500 CAS:10124-48-8 ***HR: 3***
MERCURY AMIDE CHLORIDE
DOT: UN 1630
mf: ClH_2HgN mw: 252.07

H_2NHgCl

PROP: White, pulverized lumps or powder.

SYNS: AMINOMERCURIC CHLORIDE ◇ AMMONIATED MERCURY ◇ MERCURIC AMMONIUM CHLORIDE, solid ◇ MERCURIC CHLORIDE, AMMONIATED ◇ MERCURY AMINE CHLORIDE ◇ MERCURY AMMONIATED ◇ WHITE MERCURY PRECIPITATED ◇ WHITE PRECIPITATE

CONSENSUS REPORTS: Mercury and its compounds are on the Community Right-To-Know List. Reported in EPA TSCA Inventory.

OSHA PEL: (Transitional: CL 1 mg/10m^3) CL 0.1 mg(Hg)/m^3 (skin)
ACGIH TLV: TWA 0.1 mg(Hg)/m^3 (skin)
NIOSH REL: (Inorganic Mercury) TWA 0.05 mg(Hg)/m^3
DOT Classification: Poison B; Label: Poison.

SAFETY PROFILE: A poison. Explosive reaction with halogens or amine metal salts. When heated to decomposition it emits very toxic fumes of Cl^-, NO_x, and Hg. See also MERCURY COMPOUNDS.

MCX000 CAS:38232-63-2 ***HR: 3***
MERCURY(I) AZIDE
mf: Hg_2N_6 mw: 485.22

SYNS: MERCUROUS AZIDE (DOT) ◇ MERCURY AZIDE

CONSENSUS REPORTS: Mercury and its compounds are on the Community Right-To-Know List.

OSHA PEL: (Transitional: CL 1 mg/10m^3) CL 0.1 mg(Hg)/m^3 (skin)
ACGIH TLV: TWA 0.1 mg(Hg)/m^3 (skin)
NIOSH REL: TWA 0.05 mg(Hg)/m^3
DOT Classification: Forbidden.

SAFETY PROFILE: Poison. Explodes on heating in air. When heated to decomposition it emits very toxic fumes of NO_x and Hg. See also AZIDES and MERCURY COMPOUNDS.

MCX250 CAS:14215-33-9 ***HR: 3***
MERCURY(II) AZIDE
mf: HgN_6 mw: 284.65

CONSENSUS REPORTS: Mercury and its compounds are on the Community Right-To-Know List.

SAFETY PROFILE: Poison. A friction-sensitive explosive with high brisance (shattering power). When heated to decomposition it emits very toxic fumes of Hg and NO_x. See also MERCURY COMPOUNDS and AZIDES.

MCX500 CAS:583-15-3 ***HR: 3***
MERCURY(II) BENZOATE
DOT: UN 1631
mf: $C_{14}H_{10}O_4 \cdot Hg$ mw: 442.83

PROP: White, crystalline, odorless powder. Mp: 165°. Very sol in NaCl soln; sltly sol in alc. Protect from light.

SYNS: MERCURIC BENZOATE ◇ MERCURIC BENZOATE, solid (DOT)

CONSENSUS REPORTS: Mercury and its compounds are on the Community Right-To-Know List.

OSHA PEL: (Transitional: CL 1 mg/10m^3) CL 0.1 mg(Hg)/m^3 (skin)
ACGIH TLV: TWA 0.1 mg(Hg)/m^3 (skin)
NIOSH REL: (Inorganic Mercury) TWA 0.05 mg(Hg)/m^3
DOT Classification: Poison B; Label: Poison.

SAFETY PROFILE: A poison. When heated to decomposition it emits toxic fumes of Hg. See also MERCURY COMPOUNDS.

MCX600 CAS:64771-59-1 ***HR: 3***
MERCURY BIS(CHLOROACETYLIDE)
mf: C_4Cl_2Hg mw: 319.54

$(ClC{\equiv}C)_2Hg$

CONSENSUS REPORTS: Mercury and its compounds are on the Community Right-To-Know List.

SAFETY PROFILE: Explodes violently when heated above its mp of 185°C. Upon decomposition it emits toxic fumes of Hg and Cl^-. See also MERCURY COMPOUNDS and ACETYLIDES.

MCX700 CAS:13465-33-3 ***HR: 3***
MERCURY(I) BROMATE
mf: $Br_2Hg_2O_6$ mw: 656.98

CONSENSUS REPORTS: Mercury and its compounds are on the Community Right-To-Know List.

SAFETY PROFILE: A poison. Ignites on contact with hydrogen sulfide. When heated to decomposition it emits toxic fumes of Br^- and Hg. See also MERCURY COMPOUNDS and BROMATES.

MCX750 CAS:10031-18-2 ***HR: 3***
MERCURY(I) BROMIDE (1:1)
DOT: UN 1634
mf: BrHg mw: 280.50

PROP: White-yellow, odorless, tetragonal crystals or powder. Darkens on exposure to light. D: 7.307, vap d: 19.3. Sublimes @ approx 390° (decomp). Insol in water, alc, and ether; decomp by hot HCl or alkali bromides. Protect from light.

SYN: MERCUROUS BROMIDE, solid (DOT)

CONSENSUS REPORTS: Mercury and its compounds are on the Community Right-To-Know List.

OSHA PEL: (Transitional: CL 1 mg/10m^3) CL 0.1 mg(Hg)/m^3 (skin)
ACGIH TLV: TWA 0.1 mg(Hg)/m^3 (skin)
NIOSH REL: (Inorganic Mercury) TWA 0.05 mg(Hg)/m^3
DOT Classification: Poison B; Label: Poison.

SAFETY PROFILE: A poison. When heated to decomposition it emits very toxic fumes of Br^- and Hg. See also MERCURY COMPOUNDS and BROMIDES.

MCY000 CAS:7789-47-1 ***HR: 3***
MERCURY(II) BROMIDE (1:2)
DOT: UN 1634
mf: Br_2Hg mw: 360.41

PROP: White crystals or crystalline powder. Sensitive to light. Mp: 237°, bp: 322° (subl), d: 6.109 @ 25°, vap press: 1 mm @ 136.5°. Very sol in hot alc, methanol, HCl, HBr, alkali bromide solns; sltly sol in chloroform.

SYNS: MERCURIC BROMIDE ◇ MERCURIC BROMIDE, solid (DOT)

TOXICITY DATA with REFERENCE
orl-rat LD50:40 mg/kg GTPZAB 25(7),27,81
skn-rat LD50:100 mg/kg GTPZAB 25(7),27,81
orl-mus LD50:35 mg/kg GTPZAB 25(7),27,81
ipr-mus LD50:5 mg/kg GTPZAB 25(7),27,81

CONSENSUS REPORTS: Mercury and its compounds are on the Community Right-To-Know List. Reported in EPA TSCA Inventory.

OSHA PEL: (Transitional: CL 1 mg/10m^3) CL 0.1 mg(Hg)/m^3 (skin)
ACGIH TLV: TWA 0.1 mg(Hg)/m^3 (skin)
NIOSH REL: (Inorganic Mercury) TWA 0.05 mg(Hg)/m^3
DOT Classification: Poison B; Label: Poison.

SAFETY PROFILE: A poison by ingestion, skin contact, and intraperitoneal routes. Vigorous reaction with indium at 350°C. Incompatible with sodium and potassium. When heated to decomposition it emits very toxic fumes of Br^- and Hg. See also MERCURY COMPOUNDS and BROMIDES.

MCY250 CAS:64011-37-6 ***HR: 3***
MERCURY(II) BROMIDE COMPLEX with TRIS(2-ETHYLHEXYL) PHOSPHITE
mf: $C_{24}H_{51}O_3P{\cdot}Br_2Hg$ mw: 779.13

SYN: PHOSPHOROUS ACID, TRIS(2-ETHYLHEXYL) ESTER, COMPLEX with MERCURY(II) BROMIDE (1:1)

TOXICITY DATA with REFERENCE
ipr-mus LDLo:31 mg/kg CBCCT* 7,790,55

CONSENSUS REPORTS: Mercury and its compounds are on the Community Right-To-Know List.

OSHA PEL: (Transitional: CL 1 mg/10m^3) CL 0.1 mg(Hg)/m^3 (skin)
ACGIH TLV: TWA 0.1 mg(Hg)/m^3 (skin)
NIOSH REL: (Inorganic Mercury) TWA 0.05 mg(Hg)/m^3

SAFETY PROFILE: Poison by intraperitoneal route. When heated to decomposition it emits very toxic fumes of PO_x, Br^-, and Hg. See also individual components.

MCY475 CAS:7487-94-7 ***HR: 3***
MERCURY(II) CHLORIDE
DOT: UN 1624
mf: Cl_2Hg mw: 271.50

PROP: White crystals or powder. Mp: 276°, bp: 302°, d: 5.440 @ 25°, vap press: 1 mm @ 136.2°.

SYNS: BICHLORIDE of MERCURY ◇ BICHLORURE de MERCURE (FRENCH) ◇ CALOCHLOR ◇ CHLORID RTUTNATY (CZECH) ◇ CHLORURE MERCURIQUE (FRENCH) ◇ CLORURO di MERCURIO (ITALIAN) ◇ CORROSIVE MERCURY CHLORIDE ◇ CORROSIVE SUBLIMATE ◇ MERCURIC CHLORIDE (DOT) ◇ MERCURY BICHLORIDE ◇ MERCURY PERCHLORIDE ◇ NCI-C60173 ◇ QUECKSILBER CHLORID (GERMAN) ◇ PERCHLORIDE of MERCURY ◇ SULEMA (RUSSIAN) ◇ SUBLIMAT (CZECH) ◇ TL 898

TOXICITY DATA with REFERENCE
skn-rbt 500 mg/24H SEV 28ZPAK -,12,72
eye-rbt 50 μg/24H SEV 28ZPAK -,12,72
cyt-ham-scu 6400 μg/kg TJADAB 25,381,82
cyt-hmn:lym 5 μmol/L MUREAV 157,221,85
orl-wmn TDLo:50 μg/kg (10W preg):REP AJOGAH 80,145,60
orl-rat TDLo:2470 μg/kg (female 7D post):TER CUSCAM 55,734,86
orl-wmn TDLo:50 mg/kg:GIT,KID AJOGAH 80,145,60
orl-hmn LDLo:29 mg/kg NEJMAG 244,459,51
orl-man TDLo:29 mg/kg:PUL,KID MJAUAJ 2,125,78
unr-man LDLo:7253 μg/kg 85DCAI 2,73,70
orl-rat LD50:1 mg/kg PEMNDP 8,530,87
skn-rat LD50:41 mg/kg GTPZAB 25(7),27,81
ipr-rat LD50:3210 μg/kg PSDTAP 12,247,71
scu-mus LD50:4500 μg/kg NEZAAQ 34,193,79
skn-rat LD50:41 mg/kg GTPZAB 25(7),27,81
orl-mus LD50:6 mg/kg GISAAA 51(1),76,86
ihl-mus LCLo:300 mg/m^3/10M NDRC** No.9-4-1-9,43

CONSENSUS REPORTS: Mercury and its compounds are on the Community Right-To-Know List. EPA Genetic Toxicology Program. Reported in EPA TSCA Inventory.

OSHA PEL: (Transitional: CL 1 mg/10m^3) CL 0.1 mg(Hg)/m^3 (skin)
ACGIH TLV: TWA 0.1 mg(Hg)/m^3 (skin)
NIOSH REL: (Inorganic Mercury) TWA 0.05 mg(Hg)/m^3
DOT Classification: Poison B; Label: Poison.

SAFETY PROFILE: A human poison by ingestion. Poison experimentally by ingestion, skin contact, and subcutaneous routes. Human systemic effects by ingestion: respiratory obstruction, nausea or vomiting, urine volume decreased or anuria. Human reproductive effects by ingestion: terminates pregnancy. Experimental teratogenic and reproductive effects. Human mutation data reported. A severe eye and skin irritant. Reaction with sodium aci-nitromethanide + acids forms the explosive mercury fulminate. Reacts violently with K; Na. When heated to decomposition it emits toxic fumes of Hg. See also MERCURY COMPOUNDS and CHLORIDES.

MCY500 CAS:63981-49-7 ***HR: 3***
MERCURY(II) CHLORIDE COMPLEX with TRIS(2-ETHYLHEXYL) PHOSPHITE
mf: $C_{24}H_{51}O_3P•Cl_2Hg$ mw: 690.21

SYN: PHOSPHOROUS ACID, TRIS(2-ETHYLHEXYL) ESTER, COMPLEX with MERCURY(II) CHLORIDE (1:1)

TOXICITY DATA with REFERENCE
ipr-mus LDLo:63 mg/kg CBCCT* 7,791,55

CONSENSUS REPORTS: Mercury and its compounds are on the Community Right-To-Know List.

NIOSH REL: (Inorganic Mercury) TWA 0.05 mg(Hg)/m^3

SAFETY PROFILE: Poison by intraperitoneal route. When heated to decomposition it emits very toxic fumes of Hg, Cl^-, and PO_x. See also individual components.

MCY750 ***HR: 3***
MERCURY(I) CHLORITE
mf: $Cl_2Hg_2O_4$ mw: 536.08

CONSENSUS REPORTS: Mercury and its compounds are on the Community Right-To-Know List.

SAFETY PROFILE: A poison. The dry chlorite is spontaneously explosive. Upon decomposition it emits very toxic fumes of Cl^- and Hg. See also MERCURY COMPOUNDS and CHLORITES.

MCY755 CAS:7616-83-3 ***HR: 3***
MERCURY(II) CHLORITE
mf: Cl_2HgO_4 mw: 335.49

SYN: MERCURY DICHLORITE

CONSENSUS REPORTS: Mercury and its compounds are on the Community Right-To-Know List.

SAFETY PROFILE: A poison. The dry chlorite is spontaneously explosive. Upon decomposition it emits toxic fumes of Cl^- and Hg. See also MERCURY COMPOUNDS and CHLORITES.

MCZ000 ***HR: 3***
MERCURY COMPOUNDS, INORGANIC

CONSENSUS REPORTS: Mercury and its compounds are on the Community Right-To-Know List.

SAFETY PROFILE: Mercury is a general protoplasmic poison; after absorption it circulates in the blood and is stored in the liver, kidneys, spleen, and bone. In industrial poisoning, the principal effect is upon the central nervous system, the mouth, and gums. The cardinal symptoms of industrial mercury poisoning are stomatitis, tremors, and psychic disturbances. Usually the first complaints are of excessive salivation and painful chewing. In severe cases there may be gingivitis with loosening of the teeth, and a dark line on the gum margins resembling the "lead line." The psychic disturbance (so called "erethism") includes loss of memory, insomnia, lack of confidence, irritability, vague fears and depression. The dermatitis produced by fulminate of mercury takes the form of small, discrete ulcers on the exposed parts, and is usually accompanied by conjunctivitis and inflammation of the mucous membranes of the nose and throat. In humans, it is readily absorbed by the respiratory tract (elemental mercury vapor, dusts of mercury compounds), intact skin, and the gastrointestinal tract. Occasional incidental swallowing of metallic mercury may be without harm. Spilled and heated elemental mercury is particularly hazardous. A number of mercury compounds, in addition to the fulminate, can cause skin irritation and be absorbed through the skin. They are strong allergens and common air contaminants. Acute Toxicity: Soluble salts have violent corrosive effects on skin and mucous membranes, cause severe nausea, vomiting, abdominal pain, bloody diarrhea, kidney damage, and death usually within 10 days. Many mercury compounds are explosively unstable or undergo hazardous reactions. When heated to decomposition they emit toxic fumes of Hg.

MDA000 ***HR: D***
MERCURY COMPOUNDS, ORGANIC

CONSENSUS REPORTS: Mercury and its compounds are on the Community Right-To-Know List.

DFG MAK: 0.01 mg/m^3

SAFETY PROFILE: The customary grouping of all organic mercurials in a single category is not fully justified by the toxicity of the compounds. Alkyl mercurials have very high toxicity; aryl compounds, particularly the phenyls, are much less toxic, and the organomercurials used in therapeutics are less toxic. The alkyls and aryls commonly cause skin burns and other forms of irritation, and both can be absorbed through the skin. Fatal poisoning has occurred due to exposure to alkyl mercurials and permanent damage to the brain has been reported. Phenyl mercurials appear to be no more toxic than metallic mercury. Organic mercury compounds, like organic lead compounds, seem to have an affinity for lipoid-containing organs, resulting in central nervous system disturbances such as from tetraethyl lead. These are common air contaminants. Many mercury compounds are explosively unstable or undergo hazardous reactions. When heated to decomposition they emit highly toxic fumes of Hg.

MDA100 CAS:72044-13-4 ***HR: 3***
MERCURY(I) CYANAMIDE
mf: CHg_2N_2 mw: 240.61

CONSENSUS REPORTS: Mercury and its compounds are on the Community Right-To-Know List.

SAFETY PROFILE: A poison. Explodes when heated rapidly to 325°C or when exposed to intense light while in a sealed container. When heated to decomposition it emits toxic fumes of NO_x, CN^-, and Hg. See also MERCURY COMPOUNDS.

MDA150 CAS:3021-39-4 ***HR: 3***
MERCURY(II) CYANATE
mf: $C_2HgN_2O_2$ mw: 284.62

CONSENSUS REPORTS: Mercury and its compounds, as well as cyanide and its compounds, are on the Community Right-To-Know List.

SAFETY PROFILE: A poison. A pressure-sensitive explosive. When heated to decomposition it emits toxic fumes of NO_x, CN^-, and Hg. See also MERCURY COMPOUNDS, CYANIDE, and CYANATES.

MDA250 CAS:592-04-1 ***HR: 3***
MERCURY(II) CYANIDE
DOT: UN 1636
mf: C_2HgN_2 mw: 252.63

PROP: Colorless, odorless, transparent prisms; darkened by light. Decomp @ 320°, d: 3.996. Sltly sol in ether.

SYNS: CYANURE de MERCURE (FRENCH) ◇ MERCURIC CYANIDE, solid (DOT)

TOXICITY DATA with REFERENCE
orl-hmn TDLo:27 mg/kg:GIT,KID CTOXAO 11,301,77
orl-wmn TDLo:10 mg/kg:CNS,GIT JAMAAP 66,1694,16
orl-rat LDLo:25 mg/kg NCNSA6 5,28,53
ipr-rat LDLo:7500 μg/kg NCNSA6 5,28,53
orl-mus LD50:33 mg/kg NTIS** PB214-270
scu-dog LD50:2710 μg/kg PSEBAA 116,371,64
ivn-rbt LDLo:2 mg/kg JPETAB 41,21,31

CONSENSUS REPORTS: Reported in EPA TSCA Inventory. Mercury and its compounds, as well as cyanide and its compounds, are on the Community Right-To-Know List.

OSHA PEL: (Transitional: CL 1 mg/10m^3) CL 0.1 mg(Hg)/m^3 (skin)
ACGIH TLV: TWA 0.1 mg(Hg)/m^3 (skin)
NIOSH REL: (Inorganic Mercury) TWA 0.05 mg(Hg)/m^3
DOT Classification: Poison B; Label: Poison.

SAFETY PROFILE: Poison by ingestion, subcutaneous, intravenous, and intraperitoneal routes. Human systemic effects by ingestion: nausea or vomiting, hypermotility, diarrhea, kidney changes, somnolence. Hydrolyzes to toxic fumes. A friction- and impact-sensitive explosive. It may initiate detonation of liquid hydrogen cyanide. Incompatible with fluorine; magnesium; sodium nitrite. When heated to decomposition it emits very toxic fumes of Hg, NO_x, and CN^-. See also CYANIDES and MERCURY COMPOUNDS.

MDA500 CAS:1335-31-5 ***HR: 3***
MERCURY CYANIDE OXIDE
DOT: UN 1642
mf: $C_2Hg_2N_2O$ mw: 469.22

PROP: White, orthorhombic crystals or crystalline powder. D: 4.44.

SYNS: MERCURIC OXYCYANIDE ◇ MERCURIC OXYCYANIDE, solid (desensitized) (DOT) ◇ MERCURY OXYCYANIDE

TOXICITY DATA with REFERENCE
ivn-rbt LDLo:2500 μg/kg JPETAB 41,21,31

CONSENSUS REPORTS: Mercury and its compounds, as well as cyanide and its compounds, are on the Community Right-To-Know List.

OSHA PEL: (Transitional: CL 1 mg/10m^3) CL 0.1 mg(Hg)/m^3 (skin)
ACGIH TLV: TWA 0.1 mg(Hg)/m^3 (skin)
NIOSH REL: (Inorganic Mercury) TWA 0.05 mg(Hg)/m^3
DOT Classification: Poison B; Label: Poison; Forbidden

SAFETY PROFILE: Poison by intravenous route. An explosive sensitive to friction, impact, or heat. The commercial product is stabilized by excess mercury(II) cyanide. When heated to decomposition it emits very toxic fumes of Hg, CN^-, and NO_x. See also MERCURY COMPOUNDS and CYANIDE.

MDA750 CAS:30366-55-3 ***HR: 3***
MERCURY-O,O-DI-n-BUTYL PHOSPHORODITHIOATE
mf: $C_{16}H_{36}HgO_4P_2S_4$ mw: 683.29

SYN: BIS(O,O-DIBUTYLPHOSPHORODITHIOATO-S)MERCURY

TOXICITY DATA with REFERENCE
ivn-mus LD50:180 mg/kg CSLNX* NX#05638

CONSENSUS REPORTS: Mercury and its compounds are on the Community Right-To-Know List.

OSHA PEL: (Transitional: CL 1 mg/10m^3) CL 0.1 mg(Hg)/m^3 (skin)
ACGIH TLV: TWA 0.1 mg(Hg)/m^3 (skin)
NIOSH REL: (Inorganic Mercury) TWA 0.05 mg(Hg)/m^3

SAFETY PROFILE: Poison by intravenous route. When heated to decomposition it emits very toxic fumes of SO_x, PO_x, and Hg. See also MERCURY COMPOUNDS.

MDA800 ***HR: 3***
MERCURY(II) aci-DINITROMETHANIDE
mf: $C_2H_2HgN_4O_8$ mw: 410.65

CONSENSUS REPORTS: Mercury and its compounds are on the Community Right-To-Know List.

SAFETY PROFILE: An explosive detonator. Upon decomposition it emits toxic fumes of Hg and NO_x. See also MERCURY COMPOUNDS.

MDB250 CAS:12558-92-8 ***HR: 3***
MERCURY(II) EDTA COMPLEX
mf: $C_{10}H_{14}HgN_2O_8$ mw: 490.8

SYN: (ETHYLENEDINITRILO)TETRA ACETIC ACID, MERCURY(II) COMPLEX

TOXICITY DATA with REFERENCE
orl-mus LD50:268 mg/kg JEPTDQ 2(6),1529,79
ipr-mus LD50:2700 μg(Hg)/kg PABIAQ 11,853,63
ivn-mus LD50:9500 μg/kg JEPTDQ 2(6),1529,79

CONSENSUS REPORTS: Mercury and its compounds are on the Community Right-To-Know List.

OSHA PEL: (Transitional: CL 1 mg/10m^3) CL 0.1 mg(Hg)/m^3 (skin)
ACGIH TLV: TWA 0.1 mg(Hg)/m^3 (skin)
NIOSH REL: (Inorganic Mercury) TWA 0.05 mg(Hg)/m^3

SAFETY PROFILE: Poison by ingestion, intravenous and intraperitoneal routes. When heated to decomposition it emits very toxic fumes of Hg and NO_x. See also MERCURY COMPOUNDS and EDTA.

MDB500 CAS:63905-89-5 ***HR: 3***
MERCURY(II) FLUOROACETATE
mf: $C_4H_4FO_4 \cdot Hg$ mw: 335.67

SYN: FLUOROACETIC ACID, MERCURY(II) SALT

TOXICITY DATA with REFERENCE
orl-rat LDLo:10 mg/kg NCNSA6 5,7,53

CONSENSUS REPORTS: Mercury and its compounds are on the Community Right-To-Know List.

NIOSH REL: (Inorganic Mercury) TWA 0.05 mg $(Hg)/m^3$

SAFETY PROFILE: Poison by ingestion. When heated to decomposition it emits very toxic fumes of F^- and Hg. See also MERCURY COMPOUNDS and FLUOROACETATES.

MDB775 ***HR: 3***
MERCURY(II) FORMOHYDROXAMATE
mf: $C_2H_4HgN_2O_4$ mw: 320.65

CONSENSUS REPORTS: Mercury and its compounds are on the Community Right-To-Know List.

SAFETY PROFILE: A poison. An explosive. Upon decomposition it emits toxic fumes of Hg and NO_x. See also MERCURY COMPOUNDS.

MDC000 CAS:628-86-4 ***HR: 3***
MERCURY(II) FULMINATE (dry)
DOT: UN 0135
mf: $C_2HgN_2O_2$ mw: 284.63

PROP: White solid. Mp: explodes, d: 4.42.

SYNS: FULMINATE of MERCURY, DRY (DOT) ◇ MERCURY FULMINATE (DOT) ◇ RCRA WASTE NUMBER P065

CONSENSUS REPORTS: Mercury and its compounds are on the Community Right-To-Know List. Reported in EPA TSCA Inventory.

OSHA PEL: (Transitional: CL 1 $mg/10m^3$) CL 0.1 $mg(Hg)/m^3$ (skin)
ACGIH TLV: TWA 0.1 $mg(Hg)/m^3$ (skin)
NIOSH REL: (Inorganic Mercury) TWA 0.05 $mg(Hg)/m^3$
DOT Classification: Forbidden.

SAFETY PROFILE: An explosive sensitive to flame, heat, impact, friction, intense radiation, or contact with sulfuric acid. Self-explodes. Dangerously flammable; should be kept moist until used. Incompatible with sulfuric acid. When heated to decomposition it emits very toxic fumes of Hg and NO_x. See also MERCURY COMPOUNDS and FULMINATES.

MDC250 CAS:628-86-4 ***HR: 3***
MERCURY FULMINATE (wet)
DOT: UN 0135
mf: $C_2HgN_2O_2$ mw: 284.63

SYNS: FULMINATE of MERCURY, WET (DOT) ◇ INITIATING EXPLOSIVE FULMINATE of MERCURY (DOT) ◇ RCRA WASTE NUMBER P065

CONSENSUS REPORTS: Mercury and its compounds are on the Community Right-To-Know List.

OSHA PEL: (Transitional: CL 1 $mg/10m^3$) CL 0.1 mg $(Hg)/m^3$ (skin)
ACGIH TLV: TWA 0.1 $mg(Hg)/m^3$ (skin)
NIOSH REL: (Inorganic Mercury) TWA 0.05 $mg(Hg)/m^3$
DOT Classification: Label: Explosive A.

SAFETY PROFILE: An explosive. It can be kept more safely in wet form. When heated to decomposition it emits very toxic fumes of Hg and NO_x. See also MERCURY FULMINATE (dry).

MDC500 CAS:63937-14-4 ***HR: 3***
MERCURY(I) GLUCONATE
DOT: UN 1637
mf: $C_6H_{11}O_7•Hg$ mw: 395.76

PROP: White solid.

SYNS: MERCUROUS GLUCONATE ◇ MERCUROUS GLUCONATE, solid (DOT)

CONSENSUS REPORTS: Mercury and its compounds are on the Community Right-To-Know List.

OSHA PEL: (Transitional: CL 1 $mg/10m^3$) CL 0.1 $mg(Hg)/m^3$ (skin)
ACGIH TLV: TWA 0.1 $mg(Hg)/m^3$ (skin)
NIOSH REL: (Inorganic Mercury) TWA 0.05 mg $(Hg)/m^3$
DOT Classification: Poison B; Label: Poison.

SAFETY PROFILE: A poison. When heated to decomposition it emits toxic fumes of Hg. See also MERCURY COMPOUNDS, ORGANIC.

MDC750 CAS:7783-30-4 ***HR: 3***
MERCURY(I) IODIDE
DOT: UN 1638
mf: HgI mw: 327.49

PROP: Heavy, odorless, yellow, tetragonal crystals or amorphous powder. D: 7.70, mp: 290° when rapidly heated (partial decomp). Insol in water, alc, and ether; sol in solns of mercurous or mercuric nitrates. Protect from light.

SYNS: IODURE de MERCURE (FRENCH) ◇ MERCUROUS IODIDE ◇ MERCUROUS IODIDE, solid (DOT) ◇ MERCURY PROTOIODIDE ◇ YELLOW MERCURY IODIDE

TOXICITY DATA with REFERENCE
orl-mus LD50:110 mg/kg ATXKA8 20,226,64
ipr-mus LD50:50 mg/kg ATXKA8 20,226,64

CONSENSUS REPORTS: Mercury and its compounds are on the Community Right-To-Know List.

NIOSH REL: (Inorganic Mercury) TWA 0.05 $mg(Hg)/m^3$
DOT Classification: Poison B; Label: Poison.

SAFETY PROFILE: Poison by ingestion and intraperitoneal routes. When heated to decomposition it emits very toxic fumes of Hg and I^-. See also MERCURY and IODIDES.

MDD000 CAS:7774-29-0 ***HR: 3***
MERCURY(II) IODIDE
DOT: UN 1638
mf: HgI_2 mw: 454.39

PROP: Scarlet red, heavy, odorless, almost tasteless powder. Sensitive to light. D: 6.28, mp: 259°, bp: approx 350° (subl); very sol in alkali iodides, $HgCl_2$, $Na_2S_2O_3$.

SYNS: HYDRARGYRUM BIJODATUM (GERMAN) ◇ MERCURIC IODIDE ◇ MERCURIC IODIDE, solid (DOT) ◇ MERCURIC IODIDE, solution (DOT) ◇ MERCURIC IODIDE, RED ◇ MERCURY BINIODIDE ◇ RED MERCURIC IODIDE

TOXICITY DATA with REFERENCE
ihl-rat TCLo:4870 ng/m^3/24H (1-22D preg):TER GISAAA 46(5),73,81
ihl-rat TCLo:4870 ng/m^3/24H (1-22D preg):REP GISAAA 46(5),73,81
orl-man LDLo:357 mg/kg ZKMEAB 106,783,27
orl-rat LD50:18 mg/kg GTPZAB 25(7),27,81
skn-rat LD50:75 mg/kg GTPZAB 25(7),27,81
orl-mus LD50:17 mg/kg GTPZAB 25(7),27,81
ipr-mus LD50:4200 μg/kg GTPZAB 25(7),27,81

CONSENSUS REPORTS: Mercury and its compounds are on the Community Right-To-Know List. Reported in EPA TSCA Inventory.

OSHA PEL: (Transitional: CL 1 $mg/10m^3$) CL 0.1 $mg(Hg)/m^3$ (skin)
ACGIH TLV: TWA 0.1 $mg(Hg)/m^3$ (skin)
NIOSH REL: (Inorganic Mercury) TWA 0.05 $mg(Hg)/m^3$
DOT Classification: Poison B; Label: Poison, solid; Poison B, Label: Poison, solution.

SAFETY PROFILE: A human poison by ingestion. Poison experimentally by ingestion, skin contact, and intraperitoneal routes. An experimental teratogen. Violent reaction with chlorine trifluoride. When heated to decomposition it emits very toxic fumes of Hg and I^-. See also MERCURY COMPOUNDS and IODIDES.

MDD250 CAS:7774-29-0 ***HR: 3***
MERCURY(II) IODIDE (solution)
mf: HgI_2 mw: 454.39

CONSENSUS REPORTS: Reported in EPA TSCA Inventory. Mercury and its compounds are on the Community Right-To-Know List.

OSHA PEL: (Transitional: CL 1 $mg/10m^3$) CL 0.1 $mg(Hg)/m^3$ (skin)
ACGIH TLV: TWA 0.1 $mg(Hg)/m^3$ (skin)
NIOSH REL: (Inorganic Mercury) TWA 0.05 $mg(Hg)/m^3$

SAFETY PROFILE: A poison. When heated to decomposition it emits very toxic fumes of Hg and I^-. See also MERCURY(II) IODIDE.

MDD500 CAS:814-82-4 ***HR: 3***
MERCURY(2+) LACTATE
mf: $C_6H_{12}O_6 \cdot Hg$ mw: 380.77

PROP: White, crystalline powder.

SYNS: MERCURIC LACTATE ◇ PURATIZED B-2

TOXICITY DATA with REFERENCE
orl-rat LD50:200 mg/kg 28ZEAL 4,269,69

CONSENSUS REPORTS: Mercury and its compounds are on the Community Right-To-Know List.

OSHA PEL: (Transitional: CL 1 $mg/10m^3$) CL 0.1 $mg(Hg)/m^3$ (skin)
ACGIH TLV: TWA 0.1 $mg(Hg)/m^3$ (skin)
NIOSH REL: (Inorganic Mercury) TWA 0.05 $mg(Hg)/m^3$

SAFETY PROFILE: Poison by ingestion. When heated to decomposition it emits toxic fumes of Hg. See also MERCURY COMPOUNDS.

MDD750 CAS:115-09-3 ***HR: 3***
MERCURY METHYLCHLORIDE
mf: CH_3ClHg mw: 251.08

PROP: White crystals, characteristic odor. D: 4.063, mp: 170°.

SYNS: CASPAN ◇ CHLOROMETHYLMERCURY ◇ METHYLMERCURIC CHLORIDE ◇ METHYLMERCURY CHLORIDE ◇ MMC ◇ MONOMETHYL MERCURY CHLORIDE

TOXICITY DATA with REFERENCE
cyt-hmn:lym 1 μmol/L ESKGA2 26,99,80
dni-mus:lym 10 nmol/L TXCYAC 36,297,85
orl-rat TDLo:200 μg/kg (female 6-9D post):REP ARTODN 40,103,78
orl-mus TDLo:10 mg/kg (female 12D post):TER NETOD7 6,379,84
orl-mus TDLo:402 mg/kg/58W-C:CAR JTSCDR 8,329,83

orl-mus TD:731 mg/kg/58W-C:CAR TOLED5 18(Suppl 1),114,83
orl-rat LD50:29915 μg/kg BECTA6 14,140,75
ipr-rat LD50:11 mg/kg TXAPA9 22,313,72
ims-rat LDLo:23 mg/kg NCIUS* PH 43-64-886,SEPT,71
orl-mus LD50:57600 μg/kg ACATA5 104,356,79
ihl-mus LC50:80 mg/m^3/4H 85JCAE -,1198,86
ipr-mus LD50:10 mg/kg TXAPA9 42,445,77
ipr-mky LD50:5600 μg/kg ENVRAL 15,5,78
ivn-rbt LDLo:15 mg/kg JPETAB 35,343,29
orl-gpg LD50:21 mg/kg TXAPA9 24,545,73

CONSENSUS REPORTS: Mercury and its compounds are on the Community Right-To-Know List. EPA Genetic Toxicology Program.

OSHA PEL: (Transitional: CL 1 mg/10m^3) TWA 0.01 mg(Hg)/m^3; STEL 0.03 mg/m^3 (skin)
ACGIH TLV: TWA 0.01 mg(Hg)/m^3; STEL 0.03 mg(Hg)/m^3
NIOSH REL: TWA 0.05 mg(Hg)/m^3

SAFETY PROFILE: Poison by ingestion, intramuscular, intravenous, and intraperitoneal routes. Questionable carcinogen with experimental carcinogenic and teratogenic data. Human mutation data reported. Experimental reproductive effects. When heated to decomposition it emits very toxic fumes of Cl^- and Hg. See also MERCURY COMPOUNDS.

MDE000 ***HR: 3***
MERCURY(II) METHYLNITROLATE
mf: $C_2H_2HgN_4O_6$ mw: 340.64

CONSENSUS REPORTS: Mercury and its compounds are on the Community Right-To-Know List.

SAFETY PROFILE: Can explode. When heated to decomposition it emits toxic fumes of Hg and NO_x. See also MERCURY COMPOUNDS.

MDE250 CAS:631-60-7 ***HR: 3***
MERCURY MONOACETATE
DOT: UN 1629
mf: $C_2H_3O_2$•Hg mw: 259.64

PROP: Colorless scales or plates. Mp: decomp. Sol in dil acetic acid; insol in alc, ether.

SYNS: MERCUROUS ACETATE ◇ MERCUROUS ACETATE, solid (DOT) ◇ MERCURY ACETATE

TOXICITY DATA with REFERENCE
orl-rat LD50:175 mg/kg GTPZAB 25(7),27,81
skn-rat LD50:960 mg/kg GTPZAB 25(7),27,81
orl-mus LD50:150 mg/kg GTPZAB 25(7),27,81
ipr-mus LD50:10200 μg/kg GTPZAB 25(7),27,81

CONSENSUS REPORTS: Mercury and its compounds are on the Community Right-To-Know List.

NIOSH REL: (Inorganic Mercury) TWA 0.05 mg(Hg)/m^3
DOT Classification: Poison B; Label: Poison.

SAFETY PROFILE: A poison by ingestion and intraperitoneal routes. Moderately toxic by skin contact. When heated to decomposition it emits toxic fumes of Hg. See also MERCURY COMPOUNDS.

MDE500 CAS:68448-47-5 ***HR: 3***
MERCURY-2-NAPHTHALENEDIAZONIUM TRICHLORIDE
mf: $C_{10}H_7Cl_3HgN_2$ mw: 448.12

SYN: 2-NAPHTHALENEDIAZONIUMTRICHLOROMERCURATE

CONSENSUS REPORTS: Mercury and its compounds are on the Community Right-To-Know List.

SAFETY PROFILE: Explodes violently if heated during drying. When heated to decomposition it emits toxic fumes of Hg. See also MERCURY COMPOUNDS.

MDE750 CAS:10415-75-5 ***HR: 3***
MERCURY(I) NITRATE (1:1)
DOT: UN 1627
mf: NO_3•Hg mw: 262.60

SYNS: MERCUROUS NITRATE, solid (DOT) ◇ NITRATE MERCUREUX (FRENCH) ◇ NITRIC ACID, MERCURY(I) SALT

TOXICITY DATA with REFERENCE
orl-rat LD50:170 mg/kg GISAAA 46(8),12,81
skn-rat LD50:2330 mg/kg GTPZAB 25(7),27,81
orl-mus LD50:49300 μg/kg GISAAA 46(8),12,81
ipr-mus LD50:5 mg/kg ATXKA8 20,226,64
orl-mam LD50:238 mg/kg GISAAA 49(9),11,84

CONSENSUS REPORTS: Mercury and its compounds are on the Community Right-To-Know List. Reported in EPA TSCA Inventory.

OSHA PEL: (Transitional: CL 1 mg/10m^3) CL 0.1 mg(Hg)/m^3 (skin)
ACGIH TLV: TWA 0.1 mg(Hg)/m^3 (skin)
NIOSH REL: (Inorganic Mercury) TWA 0.05 mg(Hg)/m^3
DOT Classification: Oxidizer; Label: Oxidizer; Poison B; Label: Poison.

SAFETY PROFILE: Poison by ingestion and intraperitoneal routes. Moderately toxic by skin contact. A powerful oxidizer. Explodes on contact with red-hot carbon. Mixtures with phosphorus are impact-sensitive explosives. When heated to decomposition it emits very toxic fumes of Hg and NO_x. See also MERCURY COMPOUNDS.

MDF000 CAS:10045-94-0 **HR: 3**
MERCURY(II) NITRATE (1:2)
DOT: UN 1625
mf: $N_2O_6 \cdot Hg$ mw: 324.61

PROP: White-yellowish, deliq powder. Mp: 79°, bp: decomp, d: 4.39.

SYNS: MERCURIC NITRATE ◇ MERCURY NITRATE ◇ MERCURY PERNITRATE ◇ NITRATE MERCURIQUE (FRENCH) ◇ NITRIC ACID, MERCURY(II) SALT

TOXICITY DATA with REFERENCE
orl-rat LD50:26 mg/kg GTPZAB 25(7),27,81
skn-rat LD50:75 mg/kg GTPZAB 25(7),27,81
orl-mus LD50:25 mg/kg GTPZAB 25(7),27,81
ipr-mus LD50:7200 μg/kg GTPZAB 25(7),27,81
scu-mus LDLo:20 mg/kg MOLAAF 73,751,39
orl-mam LD50:87800 μg/kg GISAAA 49(9),11,84

CONSENSUS REPORTS: Mercury and its compounds are on the Community Right-To-Know List. Reported in EPA TSCA Inventory.

OSHA PEL: (Transitional: CL 1 mg/10m^3) CL 0.1 mg(Hg)/m^3 (skin)
ACGIH TLV: TWA 0.1 mg(Hg)/m^3 (skin)
NIOSH REL: (Mercury, Inorganic) TWA 0.05 mg(Hg)/m^3
DOT Classification: Poison B; Label: Poison.

SAFETY PROFILE: Poison by ingestion, skin contact, intraperitoneal, and subcutaneous routes. A powerful oxidizer. Probably an eye, skin, and mucous membrane irritant. Reacts with acetylene to form the explosive mercury acetylide which is sensitive to heat, friction, or contact with sulfuric acid. Reaction with ethanol forms the explosive mercury fulminate. Reaction with isobutene forms an unstable explosive product. Forms explosive mixtures with phosphine (heat- and impact- sensitive); potassium cyanide (heat-sensitive); and sulfur. Violent reaction with phosphinic acid; hypophosphoric acid; unsaturated hydrocarbons; aromatics. Vigorous reaction with petroleum hydrocarbons. When heated to decomposition it emits very toxic fumes of Hg and NO_x. See also MERCURY COMPOUNDS, INORGANIC; and NITRATES.

MDF100 CAS:60345-95-1 **HR: 3**
MERCURY(II) 5-NITROTETRAZOLIDE
mf: $C_2HgN_{10}O_4$ mw: 428.68

CONSENSUS REPORTS: Mercury and its compounds are on the Community Right-To-Know List.

SAFETY PROFILE: A poison. An explosive. When heated to decomposition it emits toxic fumes of Hg and NO_x. See also MERCURY COMPOUNDS.

MDF250 CAS:1191-80-6 **HR: 3**
MERCURY OLEATE
DOT: UN 1640
mf: $C_{36}H_{66}O_4 \cdot Hg$ mw: 763.61

PROP: Yellowish-brown, somewhat transparent, ointment-like mass; odor of oleic acid. Practically insol in water; sltly sol in alc and ether; very sol in oils. Protect from light.

SYNS: MERCURIC OLEATE, solid (DOT) ◇ OLEATE of MERCURY

CONSENSUS REPORTS: Mercury and its compounds are on the Community Right-To-Know List. Reported in EPA TSCA Inventory.

OSHA PEL: (Transitional: CL 1 mg/10m^3) CL 0.1 mg(Hg)/m^3 (skin)
ACGIH TLV: TWA 0.1 mg(Hg)/m^3 (skin)
NIOSH REL: TWA 0.05 mg(Hg)/m^3
DOT Classification: Poison B; Label: Poison.

SAFETY PROFILE: A poison. An FDA over-the-counter drug. When heated to decomposition it emits toxic fumes of Hg. See also MERCURY COMPOUNDS.

MDF350 CAS:7784-37-4 **HR: 3**
MERCURY(II) ORTHOARSENATE
DOT: UN 1623
mf: $AsHO_4 \cdot Hg$ mw: 340.52

PROP: Yellow powder. Mp: decomp. Insol in water; sol in HCl or HNO_3.

SYN: MERCURIC ARSENATE

CONSENSUS REPORTS: Mercury and its compounds, as well as arsenic and its compounds, are on the Community Right-To-Know List.

OSHA PEL: 0.01 mg(As)/m^3; Cancer Hazard; (Transitional: CL 1 mg(Hg)/10m^3) CL 0.1 mg(Hg)/m^3 (skin)
ACGIH TLV: TWA 0.1 mg(Hg)/m^3 (skin)
NIOSH REL: TWA 0.05 mg(Hg)/m^3; CL 2 μg/m^3/15M
DOT Classification: Poison B; Label: Poison.

SAFETY PROFILE: Confirmed human carcinogen. A poison. When heated to decomposition it emits very toxic fumes of Hg and As. See also MERCURY and ARSENIC COMPOUNDS.

MDF500 CAS:3444-13-1 **HR: 3**
MERCURY(II) OXALATE
mf: C_2HgO_4 mw: 288.61

CONSENSUS REPORTS: Mercury and its compounds are on the Community Right-To-Know List.

SAFETY PROFILE: An explosive sensitive to percussion, grinding or heating to 105°C. A storage hazard. When heated to decomposition it emits toxic fumes of

Hg. See also MERCURY COMPOUNDS and OXALATES.

MDF750 CAS:15829-53-5 ***HR: 3***
MERCURY(I) OXIDE
DOT: UN 1641
mf: Hg_2O mw: 417.22

PROP: Black to grayish-black powder. Mp: decomp @ 100°; d: 9.8. Insol in water; sol in HNO_3. Protect from light.

SYNS: MERCUROUS OXIDE, BLACK, solid (DOT) ◇ QUECKSILBEROXID (GERMAN)

CONSENSUS REPORTS: Mercury and its compounds are on the Community Right-To-Know List. Reported in EPA TSCA Inventory.

OSHA PEL: (Transitional: CL 1 mg/10m^3) CL 0.1 mg(Hg)/m^3 (skin)
ACGIH TLV: TWA 0.1 mg(Hg)/m^3 (skin)
NIOSH REL: (Mercury, Inorganic) TWA 0.05 mg(Hg)/m^3
DOT Classification: Poison B; Label: Poison.

SAFETY PROFILE: A poison. Flammable by chemical reaction; an oxidizer. Explosive reaction with hydrogen peroxide; chlorine + ethylene. Reacts violently with molten potassium; molten sodium; S; (H_2S + BaO + air). Forms explosive mixtures with non-metals [e.g., phosphorus (impact-sensitive); sulfur (friction-sensitive)]. Incompatible with alkali metals; reducing materials. Dangerous; when heated to decomposition it emits highly toxic fumes of Hg. See also MERCURY COMPOUNDS, INORGANIC.

MDG000 CAS:1312-03-4 ***HR: 3***
MERCURY OXIDE SULFATE
DOT: NA 2025
mf: Hg_3O_6S mw: 729.83

PROP: Lemon yellow powder; odorless. Bp: volatilizes, d: 6.44, vap d: 25.2. Practically insol in water; sol in acids.

SYNS: BASIC MERCURIC SULFATE ◇ MERCURIC BASIC SULFATE ◇ MERCURIC SUBSULFATE, solid (DOT) ◇ TURPETH MINERAL

CONSENSUS REPORTS: Mercury and its compounds are on the Community Right-To-Know List.

OSHA PEL: (Transitional: CL 1 mg/10m^3) CL 0.1 mg(Hg)/m^3 (skin)
ACGIH TLV: TWA 0.1 mg(Hg)/m^3 (skin)
NIOSH REL: TWA 0.05 mg(Hg)/m^3
DOT Classification: Poison B; Label: Poison.

SAFETY PROFILE: A poison. When heated to decomposition it emits very toxic fumes of Hg and SO_x. See also MERCURY COMPOUNDS.

MDG200 CAS:7616-83-3 ***HR: 3***
MERCURY(II) PERCHLORATE
mf: Cl_2HgO_8 mw: 399.49

CONSENSUS REPORTS: Mercury and its compounds are on the Community Right-To-Know List.

SAFETY PROFILE: A poison. A storage hazard. Solutions may form an explosive precipitate. When heated to decomposition it emits toxic fumes of Cl^- and Hg. See also MERCURY COMPOUNDS.

MDG250 CAS:7783-36-0 ***HR: 3***
MERCURY(I) SULFATE
DOT: UN 1628
mf: $O_4S \cdot 2Hg$ mw: 497.24

PROP: White, crystalline powder. Mp: decomp, d: 7.56. Sltly sol in water; sol in dil HNO_3.

SYN: MERCUROUS SULFATE, solid (DOT)

TOXICITY DATA with REFERENCE
orl-rat LD50:205 mg/kg GTPZAB 25(7),27,81
skn-rat LD50:1175 mg/kg GTPZAB 25(7),27,81
orl-mus LD50:152 mg/kg GTPZAB 25(7),27,81
ipr-mus LD50:11500 μg/kg GTPZAB 25(7),27,81
ivn-mus LD50:5600 mg/kg CSLNX* NX#04883

CONSENSUS REPORTS: Mercury and its compounds are on the Community Right-To-Know List.

OSHA PEL: (Transitional: CL 1 mg/10m^3) CL 0.1 mg(Hg)/m^3 (skin)
ACGIH TLV: TWA 0.1 mg(Hg)/m^3 (skin)
NIOSH REL: (Mercury, Inorganic) TWA 0.05 mg(Hg)/m^3
DOT Classification: Poison B; Label: Poison.

SAFETY PROFILE: A poison by ingestion and intraperitoneal routes. Moderately toxic by skin contact. When heated to decomposition it emits very toxic fumes of Hg and SO_x. See also MERCURY COMPOUNDS.

MDG500 CAS:7783-35-9 ***HR: 3***
MERCURY(II) SULFATE (1:1)
DOT: UN 1633/UN 1645
mf: $O_4S \cdot Hg$ mw: 296.65

PROP: White, crystalline powder; odorless. Mp: decomp, d: 6.47. Sol in HCl, hot dilute H_2SO_4, concentrated solns of NaCl. Protect from light.

SYNS: MERCURIC SULFATE, solid (DOT) ◇ MERCURY BISULFATE ◇ MERCURY PERSULFATE ◇ SULFATE MERCURIQUE (FRENCH) ◇ SULFURIC ACID, MERCURY(2+) SALT (1:1)

TOXICITY DATA with REFERENCE
orl-rat LD50:57 mg/kg NTIS** PB214-270
skn-rat LD50:625 mg/kg GTPZAB 25(7),27,81
orl-mus LD50:25 mg/kg GTPZAB 25(7),27,81
ipr-mus LD50:6300 μg/kg GTPZAB 25(7),27,81

CONSENSUS REPORTS: Mercury and its compounds are on the Community Right-To-Know List. Reported in EPA TSCA Inventory.

OSHA PEL: (Transitional: CL 1 mg/10m^3) CL 0.1 mg (Hg)/m^3 (skin)
ACGIH TLV: TWA 0.1 mg(Hg)/m^3 (skin)
NIOSH REL: (Mercury, Inorganic) TWA 0.05 mg(Hg)/m^3
DOT Classification: Poison B; Label: Poison.

SAFETY PROFILE: Poison by ingestion and intraperitoneal routes. Moderately toxic by skin contact. When heated to decomposition it emits very toxic fumes of Hg and SO_x. See also MERCURY COMPOUNDS.

MDG750 CAS:1344-48-5 ***HR: 3***
MERCURY(II) SULFIDE
mf: HgS mw: 232.65

PROP: *Black:* Black or grayish-black, heavy, odorless, tasteless, amorph powder. Insol in water, alc, dil mineral acids. *Red:* Bright scarlet red powder, lumps, hexagonal crystals. Insol in water; sol in aqua regia.

CONSENSUS REPORTS: Mercury and its compounds are on the Community Right-To-Know List.

SAFETY PROFILE: Poison. Explosive reaction with dichlorine oxide. Mixtures with silver oxide ignite when ground. Incandescent reaction with chlorine. Incompatible with oxidants. When heated to decomposition it emits highly toxic fumes of Hg, SO_x, and H_2S. See also MERCURY COMPOUNDS and SULFIDES.

MDH000 CAS:70224-81-6 ***HR: 3***
MERCURY TETRAVANADATE
mf: $O_{11}V_4$•Hg mw: 580.35

TOXICITY DATA with REFERENCE
scu-mus LDLo:15 mg/kg AJSNAO 1,347,17

CONSENSUS REPORTS: Mercury and its compounds are on the Community Right-To-Know List.

OSHA PEL: (Transitional: CL 1 mg/10m^3) CL 0.1 mg(Hg)/m^3 (skin)
ACGIH TLV: TWA 0.1 mg(Hg)/m^3 (skin)
NIOSH REL: TWA 0.05 mg(Hg)/m^3; 1.0 mg(V)/m^3

SAFETY PROFILE: Poison by subcutaneous route. When heated to decomposition it emits toxic fumes of Hg. See also MERCURY COMPOUNDS and VANADIUM COMPOUNDS.

MDH250 ***HR: 3***
MERCURY(I) THIONITROSYLATE
mf: $(Hg_2N_2S_2)_n$

CONSENSUS REPORTS: Mercury and its compounds are on the Community Right-To-Know List.

SAFETY PROFILE: A poison. Explodes on heating in flame. When heated to decomposition it emits very toxic fumes of Hg, SO_x and NO_x. See also MERCURY COMPOUNDS, SULFIDES, and OXIDES of NITROGEN.

MDH500 CAS:8003-05-2 ***HR: 3***
MERPHENYL NITRATE
mf: C_6H_6HgO•$C_6H_5HgNO_3$ mw: 634.42

SYNS: GYNE-MERFEN ◇ HYDROXYPHENYL-MERCURY COMPOUNDED with NITRATOPHENYLMERCURY (1:1) ◇ MERFEN-STYLI ◇ NITRATE PHENYL MERCURIQUE (FRENCH) ◇ PHEMERNITE ◇ PHENYLMERCURIC NITRATE, BASIC

TOXICITY DATA with REFERENCE
scu-rat LD50:63 mg/kg 28ZEAL 4,317,69

CONSENSUS REPORTS: Mercury and its compounds are on the Community Right-To-Know List.

OSHA PEL: (Transitional: CL 1 mg/10m^3) CL 0.1 mg (Hg)/m^3 (skin)
ACGIH TLV: TWA 0.1 mg(Hg)/m^3 (skin)
NIOSH REL: TWA 0.05 mg(Hg)/m^3

SAFETY PROFILE: Poison by subcutaneous route. When heated to decomposition it emits very toxic fumes of Hg and NO_x. See also MERCURY COMPOUNDS and NITRATES.

MDH750 CAS:63869-00-1 ***HR: 3***
MERSALYL THEOPHYLLINE
mf: $C_{13}H_{17}HgNO_6$•$C_7H_8N_4O_2$•Na mw: 687.08

SYNS: MERCURY, (3-(α-CARBOXYMETHOXYPROPYL)HYDROXY MONOSODIUM SALT, COMPOUNDED with THEOPHYLLINE (1:1) ◇ SALYRGAN THEOPHYLLINE ◇ SODIUM-o-((3-HYDROXYMERCURI-2-METHOXYPROPYL)CARBAMYL)PHENOXYACETATE and THEOPHYLLINE

TOXICITY DATA with REFERENCE
ims-rat LD50:37 mg/kg JPETAB 105,336,52
orl-mus LDLo:530 mg/kg CLDND* 105,336,52

CONSENSUS REPORTS: Mercury and its compounds are on the Community Right-To-Know List.

OSHA PEL: (Transitional: CL 1 mg/10m^3) CL 0.1 mg(Hg)/m^3 (skin)
ACGIH TLV: TWA 0.1 mg(Hg)/m^3 (skin)
NIOSH REL: (Inorganic Mercury) TWA 0.05 mg (Hg)/m^3

SAFETY PROFILE: Poison by intramuscular route. Moderately toxic by ingestion. When heated to decomposition it emits very toxic fumes of Hg, Na_2O, and NO_x. See also MERCURY COMPOUNDS and THEOPHYLLINE.

MDI000 CAS:54-64-8 ***HR: 3***
MERTHIOLATE SODIUM
mf: $C_9H_9HgO_2S{\cdot}Na$ mw: 404.82

SYNS: ((o-CARBOXYPHENYL)THIO)ETHYLMERCURYSODIUM SALT ◇ ELCIDE 75 ◇ ELICIDE ◇ o-(ETHYLMERCURITHIO)BENZOIC ACID SODIUM SALT ◇ ETHYLMERCURITHIOSALICYLIC ACID SODIUM SALT ◇ MERCUROTHIOLATE ◇ MERFAMIN ◇ MERTHIOLATE ◇ MERTHIOLATE SALT ◇ MERTORGAN ◇ MERZONIN SODIUM ◇ SET ◇ SODIUM ETHYLMERCURIC THIOSALICYLATE ◇ SODIUM-o-(ETHYLMERCURITHIO)BENZOATE ◇ SODIUM ETHYLMERCURITHIOSALICYLATE ◇ SODIUM MERTHIOLATE ◇ THIMEROSALATE ◇ THIMEROSOL ◇ THIOMERSALATE

TOXICITY DATA with REFERENCE
eye-rbt 8 μg MLD AJOPAA 78,98,74
ocu-rbt TDLo:112 mg/kg (6-18D preg):TER AROPAW 93,52,75
ipr-rat TDLo:130 mg/kg (6-18D preg):REP AROPAW 93,52,75
scu-rat TDLo:104 mg/kg/1Y-I:NEO CTOXAO 4,185,71
ial-cld LDLo:60 mg/kg/4W-I JOPDAB 104,311,84
orl-rat LD50:75 mg/kg PCOC** -,1130,66
scu-rat LD50:98 mg/kg CTOXAO 4,185,71
orl-mus LD50:91 mg/kg NYKZAU 58,235,62
ivn-mus LDLo:30 mg/kg QJPPAL 12,212,39

CONSENSUS REPORTS: Mercury and its compounds are on the Community Right-To-Know List. EPA Genetic Toxicology Program.

OSHA PEL: (Transitional: CL 1 mg/10m^3) CL 0.1 mg(Hg)/m^3 (skin)
ACGIH TLV: TWA 0.1 mg(Hg)/m^3 (skin)
NIOSH REL: TWA 0.05 mg(Hg)/m^3

SAFETY PROFILE: Poison by ingestion, subcutaneous, and intravenous routes. Experimental teratogenic and reproductive effects. An eye irritant. Questionable carcinogen with experimental neoplastigenic data. An ophthalmic preservative, a topical anti-infective, topical veterinary antibacterial and antifungal agent. An FDA over-the-counter drug. When heated to decomposition it emits very toxic fumes of Hg, Na_2O, and SO_x. See also MERCURY COMPOUNDS.

MDI200 ***HR: 3***
MERURAN

PROP: Composed of a mixture of ethylmercurochloride and the gamma isomer of hexachlorocyclohexane.

TOXICITY DATA with REFERENCE
orl-rat LD50:207 mg/kg 85GMAT -,79,82
orl-mus LD50:137 mg/kg 85GMAT -,79,82
orl-rbt LD50:95 mg/kg 85GMAT -,79,82

CONSENSUS REPORTS: Mercury and its compounds are on the Community Right-To-Know List.

SAFETY PROFILE: Poison by ingestion. When heated to decomposition it emits toxic fumes of Hg. See also CHLOROETHYL MERCURY and 1,2,3,4,5,6-HEXACHLOROCYCLOHEXANE, GAMMA ISOMER.

MDI225 ***HR: 2***
MERVAN ETHANOLAMINE SALT
mf: $C_{11}H_{10}ClO_3{\cdot}C_2H_7NO$ mw: 286.76

SYN: ACETIC ACID, 4-ALLYLOXY-3-CHLOROPHENYL-, compounded with 2-AMINOETHANOL

TOXICITY DATA with REFERENCE
ipr-rat TDLo:1 g/kg (female 6-15D post):TER ARZNAD 20,618,70
ipr-rat LD50:530 mg/kg ARZNAD 20,618,70
scu-rat LD50:630 mg/kg ARZNAD 20,618,70
ipr-mus LD50:555 mg/kg ARZNAD 20,618,70
scu-mus LD50:600 mg/kg ARZNAD 20,618,70
ivn-mus LD50:585 mg/kg ARZNAD 20,618,70

SAFETY PROFILE: Moderately toxic by subcutaneous and intraperitoneal routes. An experimental teratogen. When heated to decomposition it emits toxic fumes of NO_x and Cl^-.

MDI250 CAS:498-24-8 ***HR: 2***
MESACONIC ACID
mf: $C_5H_6O_4$ mw: 130.11

PROP: Orthorhombic needles from alc, monoclinic tablets from ethyl acetate. Mp: 204-205°, d: 1.466, subl, bp: 250° (decomp). Sol in ether; very sltly sol in chloroform, carbon disulfide, and ligroin.

SYNS: MESACONATE ◇ trans-2-METHYL-2-BUTENEDIOIC ACID ◇ METHYLFUMARIC ACID ◇ trans-1-PROPENE-1,2-DICARBOXYLIC ACID

TOXICITY DATA with REFERENCE
ipr-mus LDLo:500 mg/kg CBCCT* 6,147,54

CONSENSUS REPORTS: Reported in EPA TSCA Inventory.

SAFETY PROFILE: Moderately toxic by intraperitoneal route. When heated to decomposition it emits acrid smoke and fumes.

MDI500 CAS:54-04-6 ***HR: 3***
MESCALINE
mf: $C_{11}H_{17}NO_3$ mw: 211.29

PROP: Crystals. Mp: 35-36°, bp: 180° @ 11 mm. Mod sol in water; sol in alc, chloroform, and benzene; practically insol in ether and petr ether.

SYNS: MEZCALINE ◇ MEZCLINE ◇ 3,4,5-TRIMETHOXYBENZENE-ETHANAMINE ◇ 3,4,5-TRIMETHOXYPHENETHYLAMINE

TOXICITY DATA with REFERENCE
scu-ham TDLo:480 mg/kg (female 8-9D post):REP AJOGAH 120,1105,74
scu-ham TDLo:450 μg/kg (female 8D post):TER SCIEAS 158,265,67
ims-hmn TDLo:2500 μg/kg:CNS PSYPAG 3,219,62
ipr-rat LD50:370 mg/kg JPETAB 119,78,57
orl-mus LD50:880 mg/kg JPETAB 131,115,61
ipr-mus LD50:315 mg/kg JMCMAR 13,26,70
scu-mus LD50:534 mg/kg NYKZAU 58,261,62
ivn-mus LD50:157 mg/kg NYKZAU 64,159,68

SAFETY PROFILE: Poison by intraperitoneal and intravenous routes. Moderately toxic by ingestion route. An experimental teratogen. Other experimental reproductive effects. Human systemic effects by intramuscular route: euphoria and hallucinations, distorted perceptions. A psychotoimetic agent (a drug of abuse). When heated to decomposition it emits toxic fumes of NO_x.

MDI750 CAS:832-92-8 ***HR: 3***
MESCALINE HYDROCHLORIDE
mf: $C_{11}H_{17}NO_3 \cdot ClH$ mw: 247.75

PROP: Needles. Mp: 181°. Sol in water and alc.

SYNS: 3,4,5-TRIMETHOXYPHENETHYLAMINEHYDROCHLORIDE ◇ 3,4,5-TRIMETHOXY-β-PHENYLETHYLAMINE HYDROCHLORIDE

TOXICITY DATA with REFERENCE
ipr-rat LD50:132 mg/kg TXAPA9 25,299,73
scu-rat LDLo:320 mg/kg JPETAB 71,62,41
orl-mus LD50:912 mg/kg TXAPA9 45(1),49,78
ipr-mus LD50:212 mg/kg TXAPA9 25,299,73
ivn-mus LD50:110 mg/kg TXAPA9 45(1),49,78
ivn-dog LD50:54 mg/kg TXAPA9 25,299,73
ivn-mky LD50:130 mg/kg TXAPA9 25,299,73
ipr-gpg LD50:328 mg/kg TXAPA9 25,299,73

SAFETY PROFILE: Poison by intravenous, intraperitoneal, and subcutaneous routes. Moderately toxic by ingestion. When heated to decomposition it emits very toxic fumes of HCl and NO_x. See also MESCALINE.

MDJ000 CAS:5967-42-0 ***HR: 3***
MESCALINE SULFATE
mf: $C_{11}H_{17}NO_3 \cdot H_2O_4S$ mw: 309.37

PROP: Crystals. Mp: 158°.

SYNS: MESCALINE ACID SULFATE ◇ MEZCALINE SULFATE ◇ 3,4,5-TRIMETHOXPHENETHYLAMINE SULFATE

TOXICITY DATA with REFERENCE
ipr-rat LD50:370 mg/kg JPETAB 119,78,57
scu-rat LD50:534 mg/kg JPMSAE 59,1699,70
ivn-rat LD50:157 mg/kg JPMSAE 59,1699,70
ipr-mus LD50:240 mg/kg AEPPAE 237,171,59
ivn-mus LD50:157 mg/kg 27ZQAG -,346,72

SAFETY PROFILE: Poison by intraperitoneal and intravenous routes. Moderately toxic by subcutaneous route. When heated to decomposition it emits very toxic fumes of SO_x and NO_x. See also MESCALINE.

MDJ250 CAS:34807-41-5 ***HR: 3***
MESEREIN
mf: $C_{38}H_{38}O_{10}$ mw: 654.76

SYNS: MEZEREIN ◇ (12-β(E,E))-12-((1-OXO-5-PHENYL-2,4-PENTADIENYL)OXY)-DAPHNETOXIN

TOXICITY DATA with REFERENCE
mrc-smc 20 mg/L NATUAS 294,263,81
otr-mus:fbr 100 μg/L FACOEB 1,179,84
otr-mus:oth 16 nmol/L CRNGDP 4,1507,83
dns-mus-skn 340 nmol/kg CNREA8 43,4126,83
skn-mus TDLo:3560 μg/kg/20W-I:NEO CNREA8 39,4791,79

SAFETY PROFILE: Questionable carcinogen with experimental neoplastigenic data. Mutation data reported. When heated to decomposition it emits acrid smoke and irritating fumes.

MDJ750 CAS:141-79-7 ***HR: 3***
MESITYL OXIDE
DOT: UN 1229
mf: $C_6H_{10}O$ mw: 98.16

PROP: Oily, colorless liquid; strong odor. Mp: −59°, bp: 130.0°, flash p: 87°F (CC), d: 0.8539 @ 20°/4°, autoign temp: 652°F, vap press: 10 mm @ 26.0°, vap d: 3.38. Solidifies @ 41.5°; somewhat sol in water @ 20°. Misc in alc and ether and with most organic liquids.

SYNS: ISOBUTENYL METHYL KETONE ◇ ISOPROPYLIDENEACETONE ◇ MESITYLOXID (GERMAN) ◇ MESITYLOXYDE (DUTCH) ◇ METHYL ISOBUTENYL KETONE ◇ 4-METHYL-3-PENTENE-2-ONE ◇ 4-METHYL-3-PENTEN-2-ON (DUTCH, GERMAN) ◇ 2-METHYL-2-PENTEN-4-ONE ◇ 4-METHYL-3-PENTEN-2-ONE ◇ 4-METIL-3-PENTEN-2-ONE (ITALIAN) ◇ OSSIDO di MESITILE (ITALIAN) ◇ OXYDE de MESITYLE (FRENCH)

TOXICITY DATA with REFERENCE
eye-hmn 25 ppm/15M JIHTAB 28,262,46
skn-rbt 430 mg open MLD UCDS** 11/3/71
eye-rbt 4325 μg SEV AJOPAA 29,1363,46
ihl-hmn TCLo:25 ppm:EYE JIHTAB 28,262,46
orl-rat LD50:1120 mg/kg WQCHM* 4,-,74
ihl-rat LC50:9 g/m³/4H 85GMAT -,80,82
orl-mus LD50:710 mg/kg 85GMAT -,80,82
ihl-mus LC50:10 g/m³/2H 85GMAT -,80,82
ipr-mus LD50:354 mg/kg SCCUR* -,6,61
orl-rbt LD50:1000 mg/kg SHELL* -,-,71
skn-rbt LD50:5150 mg/kg NPIRI* 1,71,74

scu-rbt LDLo:840 mg/kg AEXPBL 56,346,1906
scu-frg LDLo:1400 mg/kg AEXPBL 56,346,1906

CONSENSUS REPORTS: Reported in EPA TSCA Inventory.

OSHA PEL: (Transitional: TWA 25 ppm) TWA 15 ppm; STEL 25 ppm
ACGIH TLV: TWA 15 ppm; STEL 25 ppm
DFG MAK: 25 ppm (100 mg/m^3)
NIOSH REL: (Ketones) TWA 40 mg/m^3
DOT Classification: Label: Flammable Liquid.

SAFETY PROFILE: Poison by intraperitoneal route. Moderately toxic by ingestion. Mildly toxic by inhalation and skin contact. Human systemic effects by inhalation: conjunctiva irritation. This compound is highly irritating to all tissues on contact; its vapors also are irritating. High concentrations are narcotic. It is readily absorbed through intact skin. Single exposures tend to indicate that this ketone has greater acute and narcotic action than isophorone. It can have harmful effects upon the kidneys and liver, and may damage the eyes and lungs to a serious degree. Prolonged exposure can injure liver, kidneys, and lungs. It can cause opaque cornea, keratoconus, and extensive necrosis of cornea. Dangerous fire hazard when exposed to heat, sparks, or flame; can react with oxidizing materials. Reacts violently with 2-amino ethanol; chlorosulfonic acid; ethylene diamine; HNO_3; oleum; H_2SO_4. An insect repellent. To fight fire, use alcohol foam, CO_2, dry chemical. When heated to decomposition it emits acrid smoke and irritating fumes. See also KETONES.

MDK000 CAS:78110-23-3 ***HR: 3***
2-MESITYLOXYDIISOPROPYLAMINE HYDROCHLORIDE
mf: $C_{15}H_{25}NO{\cdot}ClH$ mw: 271.87

SYN: C 1686

TOXICITY DATA with REFERENCE
ipr-rat LD50:100 mg/kg ARZNAD 8,708,58
scu-mus LD50:215 mg/kg ARZNAD 8,708,58

SAFETY PROFILE: Poison by intraperitoneal and subcutaneous routes. When heated to decomposition it emits very toxic fumes of NO_x and HCl.

MDK250 CAS:77791-40-3 ***HR: 2***
N-(2-MESITYLOXYETHYL)-N-METHYL-2-(2-METHYLPIPERIDINO)ACETAMIDE HYDROCHLORIDE
mf: $C_{20}H_{32}N_2O_2{\cdot}ClH$ mw: 369.00

SYN: C 2061

TOXICITY DATA with REFERENCE
eye-rbt 2% MLD ARZNAD 8,761,58
scu-mus LD50:560 mg/kg ARZNAD 8,761,58

SAFETY PROFILE: Moderately toxic by subcutaneous route. An eye irritant. When heated to decomposition it emits very toxic fumes of HCl and NO_x.

MDK500 CAS:100836-53-1 ***HR: 3***
(1-MESITYLOXY-2-PROPYL)-N-METHYLCARBAMIC ACID-2-(DIETHYLAMINO)ETHYL ESTER, HYDROCHLORIDE
mf: $C_{20}H_{34}N_2O_3{\cdot}ClH$ mw: 387.02

SYN: C 2136

TOXICITY DATA with REFERENCE
eye-rbt 2% SEV ARZNAD 9,113,59
scu-mus LD50:140 mg/kg ARZNAD 9,113,59

SAFETY PROFILE: Poison by subcutaneous route. Severe eye irritant. When heated to decomposition it emits very toxic fumes of HCl and NO_x. See also CARBAMATES.

MDK750 CAS:101651-36-9 ***HR: 3***
N-(1-MESITYLOXY-2-PROPYL)-N-METHYL-2-(2-METHYLPIPERIDINO) ACETAMIDE HYDROCHLORIDE
mf: $C_{21}H_{34}N_2O_2{\cdot}ClH$ mw: 383.03

SYN: C 2048

TOXICITY DATA with REFERENCE
eye-rbt 2% MLD ARZNAD 9,70,59
scu-mus LD50:150 mg/kg ARZNAD 9,70,59

SAFETY PROFILE: Poison by subcutaneous route. An eye irritant. When heated to decomposition it emits very toxic fumes of HCl and NO_x.

MDK875 CAS:19767-45-4 ***HR: 3***
MESNA
mf: $C_2H_5O_3S_2{\cdot}Na$ mw: 164.18

SYNS: 2-MERCAPTOETHANESULFONIC ACID MONOSODIUM SALT ◇ MESNUM ◇ MISTABRON ◇ MISTABRONCO ◇ MITEXAN ◇ MUCOFLUID ◇ SODIUM-2-MERCAPTOETHANESULFONATE ◇ UROMITEXAN

TOXICITY DATA with REFERENCE
orl-rat LD50:4440 mg/kg EJCODS 18,1377,82
ipr-rat LD50:1251 mg/kg EJCODS 18,1377,82
ivn-rat LD50:1683 mg/kg EJCODS 18,1377,82
orl-mus LD50:6102 mg/kg EJCODS 18,1377,82
ipr-mus LD50:2005 mg/kg EJCODS 18,1377,82
ivn-mus LD50:1887 mg/kg EJCODS 18,1377,82
inv-dog LDLo:400 mg/kg EJCODS 18,1377,82

SAFETY PROFILE: Poison by intravenous route. Moderately toxic by intraperitoneal route. Mildly toxic by ingestion. When heated to decomposition it emits toxic fumes of SO_x and Na_2O. See also SULFONATES and MERCAPTANS.

MDL000 CAS:80-49-9 ***HR: 3***
MESOPIN
mf: $C_{17}H_{24}NO_3{\cdot}Br$ mw: 370.33

SYNS: ARKITROPIN ◇ CAMATROPINE ◇ ESOPIN ◇ HOMAPIN ◇ HOMATROMIDE ◇ HOMATROPINE METHYLBROMIDE ◇ 3-α-HYDROXY-8-METHYL-1-α-H,5-α-H-TROPANIUM BROMIDE MANDELATE ◇ MALCOTRAN ◇ dl-METHYLBROMIDE ◇ METHYLHOMATROPINE BROMIDE ◇ 8-METHYLHOMOTROPINIUM BROMIDE ◇ NOVATRIN ◇ NOVATROPINE ◇ SED-TEMS ◇ SETHYL ◇ TROPINIUM METHOBROMIDE MANDELATE

TOXICITY DATA with REFERENCE
orl-rat LD50:1200 mg/kg JPETAB 105,166,52
ipr-rat LD50:82 mg/kg JPETAB 105,166,52
scu-rat LD50:800 mg/kg JPETAB 105,166,52
orl-mus LD50:1400 mg/kg JPETAB 105,166,52
ipr-mus LD50:60 mg/kg JPETAB 105,166,52
scu-mus LD50:650 mg/kg JPETAB 105,166,52
orl-gpg LD50:1000 mg/kg JPETAB 105,166,52
ipr-gpg LD50:120 mg/kg JPETAB 105,166,52
scu-gpg LD50:75 mg/kg AIPTAK 137,375,62

CONSENSUS REPORTS: Reported in EPA TSCA Inventory.

SAFETY PROFILE: Poison by intraperitoneal and subcutaneous routes. Moderately toxic by ingestion. An anticholinergic agent. An FDA over-the-counter and proprietary drug. When heated to decomposition it emits very toxic fumes of Br^- and NO_x.

MDL250 CAS:1115-12-4 ***HR: 3***
MESOXALONITRILE
mf: C_3N_2O mw: 80.05

SYN: OXOPROPANEDINITRILE (CARBONYL DICYANIDE)

CONSENSUS REPORTS: Cyanide and its compounds are on the Community Right-To-Know List.

SAFETY PROFILE: A poison. Explosive reaction with water. When heated to decomposition it emits highly toxic fumes of CN^- and NO_x. See also NITRILES.

MDL500 CAS:2244-11-3 ***HR: 3***
MESOXALYLUREA MONOHYDRATE
mf: $C_4H_2N_2O_4{\cdot}H_2O$ mw: 160.10

PROP: White crystals, become pink on exposure to air. Colorless, aqueous solution imparts pink color to skin. Mp: 170° (decomp); sol in water and alc.

SYNS: ALLOXAN MONOHYDRATE ◇ MESOXALYLCARBAMIDE MONOHYDRATE ◇ 2,4,5,6(1H,3H)-PYRIMIDINETETRONE HYDRATE ◇ 2,4,5,6-TETRAOXOHEXAHYDROPYRIMIDINE HYDRATE

TOXICITY DATA with REFERENCE
sln-dmg-orl 1 pph ENMUDM 7,325,85
ivn-rat TDLo:60 mg/kg (9D preg):REP JEEMAF 41,93,77
ivn-rat TDLo:60 mg/kg (9D preg):TER JEEMAF 41,93,77
ivn-pig LDLo:200 mg/kg METAAJ 29,40,80

SAFETY PROFILE: Poison by intravenous route. An experimental teratogen. Experimental reproductive effects. Mutation data reported. When heated to decomposition it emits toxic fumes of NO_x.

MDL600 CAS:101-26-8 ***HR: 3***
MESTINON
mf: $C_9H_{13}N_2O_2{\cdot}Br$ mw: 261.15

SYNS: 3-((DIMETHYLAMINO)CARBONYL)OXY)-1-METHYLPYRIDINIUM BROMIDE ◇ DIMETHYLCARBAMIC ACID ESTER of 3-HYDROXY-1-METHYLPYRIDINIUM BROMIDE ◇ 3-HYDROXY-1-METHYLPYRIDINIUM BROMIDE DIMETHYLCARBAMATE (ESTER) ◇ PYRIDOSTIGMINE BROMIDE ◇ REGONAL ◇ RO 1-5130

TOXICITY DATA with REFERENCE
orl-rat LD50:37500 μg/kg GNRIDX 2,828,68
ipr-rat LD50:2699 μg/kg FAATDF 4(2,Pt 2),S195,84
scu-rat LD50:3100 μg/kg JMCMAR 26,145,83
ims-rat LD50:2790 μg/kg DCTODJ 7,507,84
orl-mus LD50:16 mg/kg GNRIDX 2,828,68
ipr-mus LD50:1 mg/kg NIIRDN 6,354,82
scu-mus LD50:1500 μg/kg GNRIDX 2,828,68
ivn-mus LD50:1500 μg/kg MECHAN 3,329,56

SAFETY PROFILE: Poison by ingestion, intramuscular, subcutaneous, intravenous and intraperitoneal routes. When heated to decomposition it emits toxic fumes of Br^- and NO_x. See also CARBAMATES and ESTERS.

MDL750 CAS:8015-29-0 ***HR: 3***
MESTRANOL mixed with NORETHINDRONE

SYNS: ETHYNYLESTRADIOL-3-METHYL ETHER and 17-α-ETHYNYL-17-HYDROXYESTREN-3-ONE ◇ ETHYNYLESTRADIOL-3-METHYL ETHER and 17-α-ETHYNYL-19-NORTESTOSTERONE ◇ 17-α-ETHYNYL-17-HYDROXYESTREN-3-ONE and ETHYNYLESTRADIOL 3-METHYL ETHER ◇ 17-α-ETHYNYL-19-NORTESTOSTERONE and ETHYNYLESTRADIOL 3-METHYL ETHER ◇ MESTRANOL mixed with NORETHISTERONE ◇ NORETHINDRONE mixed with MESTRANOL ◇ NORETHISTERONE mixed with MESTRANOL ◇ NORINYL-1 ◇ ORTHO-NOVUM ◇ SOPHIA

TOXICITY DATA with REFERENCE
orl-wmn TDLo:12 mg/kg (female 60D pre):REP FESTAS 16,158,65
orl-wmn TDLo:37 mg/kg (female 6-26W post):TER JOPDAB 71,128,67
orl-wmn TDLo:26460 μg/kg/5Y-I:CAR,LIV LANCAO 1,310,80
orl-rat TDLo:5639 mg/kg/2Y-C:NEO NAIZAM 25,684,74
orl-wmn TDLo:40 mg/kg/4Y-I:NEO,LIV JAMAAP 235,730,76

orl-wmn TD:71 mg/kg/7Y-I:NEO,LIV ANSUA5 183,239,76
orl-wmn TD:53 mg/kg/7Y-I:CAR,LIV,BLD LANCAO 1,365,80
orl-wmn TDLo:10 mg/kg/Y-I:PUL,GIT,MET LANCAO 1,1479,73

SAFETY PROFILE: Human systemic effects by ingestion: thrombocytopenia (decrease in the number of blood platelets), dyspnea, nausea or vomiting, fever. Human teratogenic and reproductive effects by ingestion: developmental abnormalities of the urogenital system; spermatogenesis; impotence; breast development in males; changes in the uterus, cervix, or vagina; female fertility effects. Experimental reproductive effects. Questionable human carcinogen producing liver tumors.

MDM000 CAS:7660-71-1 ***HR: 3***
MESUPRINE HYDROCHLORIDE
mf: $C_{19}H_{26}N_2O_5S{\cdot}ClH$ mw: 430.99

SYN: 2'-HYDROXY-5'-(1-HYDROXY-2-(p-METHOXYPHENETHYL) AMINO)PROPYL)METHANESULFONANILIDE HCl

TOXICITY DATA with REFERENCE
orl-rat TDLo:22 g/kg/78W-C:ETA TXAPA9 22,276,72

SAFETY PROFILE: Questionable carcinogen with experimental tumorigenic data. When heated to decomposition it emits very toxic fumes of NO_x, SO_x, and HCl.

MDM350 CAS:514-61-4 ***HR: D***
METALUTIN
mf: $C_{19}H_{28}O_2$ mw: 288.47

PROP: Crystals from ether-hexane. Mp: 156-158°.

SYNS: 17-β-HYDROXY-17-METHYLESTR-4-EN-3-ONE ◇ METHALUTIN ◇ METHYLESTRENOLONE ◇ 17-α-METHYL-17-β-HYDROXY-4-ESTREN-3-ONE ◇ METHYLNORTESTOSTERONE ◇ 17-METHYL-19-NORTESTOSTERONE ◇ 17-α-METHYL-19-NORTESTOSTERONE ◇ METHYLOESTRENOLONE ◇ NORMETANDRONE ◇ NORMETHANDROLONE ◇ NORMETHANDRONE ◇ 19-NOR-17-α-METHYLTESTOSTERONE ◇ ORGASTERON

TOXICITY DATA with REFERENCE
orl-wmn TDLo:18 mg/kg (12-24W preg):TER BRGOAY 28,137,58
orl-rbt TDLo:200 μg/kg (female 1D pre):REP 85GRAA -,57,65

SAFETY PROFILE: Human teratogenic effects by ingestion: extra embryonic structures and developmental abnormalities of the urogenital system. An experimental teratogen. Experimental reproductive effects. When heated to decomposition it emits acrid smoke and irritating fumes. See also TESTOSTERONE.

MDM750 CAS:538-79-4 ***HR: 3***
METANICOTINE
mf: $C_{10}H_{14}N_2$ mw: 162.26

TOXICITY DATA with REFERENCE
itr-rat TDLo:75 mg/kg:ETA BJCAAI 16,453,62

SAFETY PROFILE: Questionable carcinogen with experimental tumorigenic data. When heated to decomposition it emits toxic fumes of NO_x.

MDM775 CAS:587-98-4 ***HR: 2***
METANIL YELLOW
mf: $C_{18}H_{15}N_3O_3S{\cdot}Na$ mw: 376.41

PROP: Brownish-yellow powder. Sol in water, alc; moderately sol in benzene and ether; sltly sol in acetone.

SYNS: ACIDIC METANIL YELLOW ◇ ACID LEATHER YELLOW PRW ◇ ACID LEATHER YELLOW R ◇ ACID METANIL YELLOW ◇ ACID YELLOW 36 ◇ AIZEN METANIL YELLOW ◇ AMACID YELLOW M ◇ BRASILAN METANIL YELLOW ◇ BUCACID METANIL YELLOW ◇ CALCOCID YELLOW MXXX ◇ C.I. 13065 ◇ C.I. ACID YELLOW 36 ◇ C.I. ACID YELLOW 36 MONOSODIUM SALT ◇ DIACID METANIL YELLOW ◇ ENIACID METANIL YELLOW GN ◇ EXT D&C YELLOW No. 1 ◇ FENAZO YELLOW M ◇ HIDACID METANIL YELLOW ◇ HISPACID YELLOW MG ◇ JAVA METANIL YELLOW G ◇ KITON ORANGE MNO ◇ KITON YELLOW MS ◇ METANILE YELLOW O ◇ METANIL YELLOW 1955 ◇ METANIL YELLOW C ◇ METANIL YELLOW E ◇ METANIL YELLOW EXTRA ◇ METANIL YELLOW F ◇ METANIL YELLOW G ◇ METANIL YELLOW GRIESBACH ◇ METANIL YELLOW K ◇ METANIL YELLOW KRSU ◇ METANIL YELLOW M3X ◇ METANIL YELLOW O ◇ METANIL YELLOW PL ◇ METANIL YELLOW S ◇ METANIL YELLOW SUPRA P ◇ METANIL YELLOW VS ◇ METANIL YELLOW WS ◇ METANIL YELLOW Y ◇ METANIL YELLOW YK ◇ MITSUI METANIL YELLOW ◇ MONOAZO ◇ REMADERM YELLOW HPR ◇ SHIKISO METANIL YELLOW ◇ SYMULON METANIL YELLOW ◇ TAKAOKA METANIL YELLOW ◇ TERTRACID YELLOW M ◇ TROPAEOLIN G ◇ VONDACID METANIL YELLOW G ◇ 11363 YELLOW ◇ YODOCHROME METANIL YELLOW

TOXICITY DATA with REFERENCE
cyt-hmn:leu 10 μmol/L IJEBA6 16,820,78
orl-rat TDLo:162 g/kg (90D male):REP ENVRAL 15,227,78
orl-rat LD50:5 g/kg 85JCAE -,1310,86
ipr-mus LD50:1000 mg/kg SRTCDF -,15,77

CONSENSUS REPORTS: Reported in EPA TSCA Inventory.

SAFETY PROFILE: Moderately toxic by intraperitoneal route. Experimental reproductive effects. Human mutation data reported. When heated to decomposition it emits toxic fumes of SO_x, NO_x, and Na_2O. See also AROMATIC AMINES and SULFONATES.

MDM800 CAS:5874-97-5 ***HR: 3***
METAPROTERENOL SULFATE
mf: $C_{22}H_{34}N_2O_6{\cdot}H_2O_4S$ mw: 520.66

SYNS: ALOTEC ◇ ALUPENT ◇ 3,5-DIHYDROXY-α-((ISOPROPYL-

AMINO)METHYL)BENZYL ALCOHOL SULFATE ◇ 1-(3,5-DIHYDROXYPHENYL)-2-(ISOPROPYLAMINO)ETHANOLSULFATE ◇ 5-(1-HYDROXY-2-((1-METHYLETHYL)AMINO)ETHYL)-1,3-BENZENEDIOL SULFATE (2:1) SALT ◇ METAPREL ◇ NOVASMASOL ◇ ORCIPRENALINE SULFATE ◇ TH-152

TOXICITY DATA with REFERENCE
orl-mus TDLo:500 mg/kg (female 6-15D post):TER TJADAB 38,15,88
orl-rat LD50:5538 mg/kg TXAPA9 18,185,71
idu-rat LD50:2230 mg/kg AIPTAK 180,155,69
orl-mus LD50:4800 mg/kg NIIRDN 6,163,82
scu-mus LD50:290 mg/kg NIIRDN 6,163,82
ivn-mus LD50:114 mg/kg NIIRDN 6,163,82

SAFETY PROFILE: Poison by subcutaneous and intravenous routes. Moderately toxic by intraduodenal route. Mildly toxic by ingestion. An experimental teratogen. When heated to decomposition it emits toxic fumes of NO_x and SO_x. See also SULFATES.

MDN100 ***HR: 2***
METET
mf: $C_{10}H_{14}N_4O_2S$ mw: 254.34

SYNS: 3,7-DIHYDRO-1,3-DIMETHYL-7-(2-(METHYLTHIO)ETHYL)-1H-PURINE-2,6-DIONE (9CI) ◇ 7-(2-METHYLTHIO)ETHYL)-THEOPHYLLINE ◇ 7-(β-METHYLTHIOETHYL)THEOPHYLLINE ◇ 7-(β-METYLOTIOETYLO)-TEOFILINA(POLISH)

TOXICITY DATA with REFERENCE
orl-mus LD50:550 mg/kg DPHFAK 22,199,70
ipr-mus LD50:422 mg/kg DPHFAK 22,199,70
scu-mus LD50:1780 mg/kg DPHFAK 22,199,70

SAFETY PROFILE: Moderately toxic by ingestion, subcutaneous, and intraperitoneal routes. When heated to decomposition it emits toxic fumes of SO_x and NO_x.

MDN250 CAS:79-41-4 ***HR: 3***
METHACRYLIC ACID
DOT: UN 2531
mf: $C_4H_6O_2$ mw: 86.10

$$H_2C{=}C(CH_3)CO{\bullet}OH$$

PROP: Corrosive liquid or colorless crystals; repulsive odor. Mp: 16°, bp: 163°, flash p: 171°F (COC), d: 1.014 @ 25° (glacial), vap press: 1 mm @ 25.5° Sol in warm water; misc with alc, ether.

SYNS: METHACRYLIC ACID, INHIBITED (DOT) ◇ α-METHYLACRYLIC ACID ◇ 2-METHYLPROPENOIC ACID

TOXICITY DATA with REFERENCE
dnd-esc 50 μmol/L MUREAV 89,95,81
orl-rat LD33:60 mg/kg 85GMAT -,80,82
ipr-mus LD50:48 mg/kg JPMSAE 62,778,73
skn-rbt LD50:500 mg/kg DTLVS* 4,257,80

CONSENSUS REPORTS: Reported in EPA TSCA Inventory.

OSHA PEL: TWA 20 ppm (skin)
ACGIH TLV: TWA 20 ppm
DOT Classification: Corrosive Material; Label: Corrosive.

SAFETY PROFILE: Poison by ingestion and intraperitoneal routes. Moderately toxic by skin contact. Corrosive to skin, eyes, and mucous membranes. Mutation data reported. Flammable when exposed to heat, flame, or oxidizers. A storage hazard; exothermic polymerization may occur spontaneously. To fight fire, use alcohol foam, spray, mist, dry chemical. When heated to decomposition it emits acrid smoke and irritating fumes.

MDN500 CAS:79-39-0 ***HR: 3***
METHACRYLIC ACID AMIDE
mf: C_4H_7NO mw: 85.12

SYNS: METHACRYLIC AMIDE ◇ 2-METHYLACRYLAMIDE ◇ α-METHYL ACRYLIC AMIDE ◇ 2-METHYLPROPENAMIDE ◇ USAF RH-1

TOXICITY DATA with REFERENCE
ihl-man TCLo:3 mg/m^3:BRN,LIV,KID GISAAA 45(10),74,80
orl-rat LD50:459 mg/kg GISAAA 45(10),74,80
orl-mus LD50:451 mg/kg ARTODN 47,179,81
ipr-mus LD50:200 mg/kg NTIS** AD277-689

CONSENSUS REPORTS: Reported in EPA TSCA Inventory.

SAFETY PROFILE: Poison by intraperitoneal route. Moderately toxic by ingestion. Human systemic effects by inhalation: degenerative brain changes and liver and kidney changes. When heated to decomposition it emits toxic fumes of NO_x.

MDN699 CAS:760-93-0 ***HR: 3***
METHACRYLIC ANHYDRIDE
mf: $C_8H_{10}O_3$ mw: 154.18

SYNS: METHACRYLIC ACID ANHYDRIDE ◇ METHACRYLOYL ANHYDRIDE ◇ 2-METHYL-2-PROPENOIC ACID ANHYDRIDE (9CI)

TOXICITY DATA with REFERENCE
ihl-mus LC50:450 mg/m^3/2H 85GMAT -,80,82

CONSENSUS REPORTS: EPA Extremely Hazardous Substances List. Reported in EPA TSCA Inventory.

SAFETY PROFILE: Poison by inhalation. When heated to decomposition it emits acrid smoke and irritating fumes. See also ANHYDRIDES.

MDN899 CAS:920-46-7 ***HR: 3***
METHACRYLOYL CHLORIDE
mf: C_4H_5ClO mw: 104.54

SYNS: METHACRYL CHLORIDE ◇ METHACRYLIC ACID CHLORIDE ◇ METHACRYLIC CHLORIDE ◇ α-METHACRYLOYL CHLORIDE ◇ METHACRYLYL CHLORIDE ◇ 2-METHYLPROPENOIC ACID CHLORIDE ◇ 2-METHYL-2-PROPENOYL CHLORIDE ◇ 2-METHYLPROPENYL CHLORIDE

TOXICITY DATA with REFERENCE
ihl-mus LC50:115 mg/m^3/2H 85GMAT -,80,82
ihl-rat LC50:60 mg/m^3/4H 85GMAT -,80,82

CONSENSUS REPORTS: EPA Extremely Hazardous Substances List. Reported in EPA TSCA Inventory.

SAFETY PROFILE: Poison by inhalation. When heated to decomposition it emits toxic fumes of Cl^-. See also CHLORIDES.

MDO250 CAS:3963-95-9 ***HR: 3***
METHACYCLINE HYDROCHLORIDE
mf: $C_{22}H_{22}N_2O_8$•ClH mw: 478.92

PROP: Yellow, crystalline powder; bitter taste. Decomp @ approx 205°. Sol in water; sltly sol in alc; practically insol in ether, chloroform.

SYNS: ADRIAMICINA ◇ CICLOBIOTIC ◇ GERMICICLIN ◇ GLOBOCICLINA ◇ LONDOMYCIN ◇ MEGAMYCINE ◇ METADOMUS ◇ METHACYCLINE MONOHYDROCHLORIDE ◇ METILENBIOTIC ◇ OPTIMYCIN ◇ PHYSIOMYCINE ◇ RINDEX ◇ RONDOMYCIN

TOXICITY DATA with REFERENCE
ipr-rat LD50:252 mg/kg TXAPA9 18,185,71
ivn-rat LD50:202 mg/kg NIIRDN 6,818,82
orl-mus LD50:3450 mg/kg NIIRDN 6,818,82
ipr-mus LD50:288 mg/kg TXAPA9 18,185,71
ivn-mus LD50:193 mg/kg NIIRDN 6,818,82

SAFETY PROFILE: Poison by intraperitoneal and intravenous routes. Moderately toxic by ingestion. An antibacterial agent. When heated to decomposition it emits very toxic fumes of NO_x and HCl.

MDO750 CAS:76-99-3 ***HR: 3***
METHADONE
mf: $C_{21}H_{27}NO$ mw: 309.49

SYNS: ADANON ◇ AMIDONE ◇ DIAMINON ◇ DOLOPHINE ◇ HEPTADONE ◇ HEPTANON ◇ KETALGIN ◇ MECODIN ◇ PHENADONE ◇ PHYSEPTONE ◇ POLAMIDONE

TOXICITY DATA with REFERENCE
par-rat TDLo:88 mg/kg (female 1-22D post):REP TJADAB 27,37A,83
orl-rat TDLo:770 mg/kg (female 10D pre):TER JPETAB 192,549,75
ivn-man TDLo:571 μg/kg/5H-I:CNS,PUL,GIT FEPRA7 11,346,52
orl-rat LD50:86 mg/kg AIPTAK 180,155,69
ipr-rat LD50:18 mg/kg JAPMA8 47,323,58
scu-rat LD50:30 mg/kg ARZNAD 3,238,53
ivn-rat LD50:11 mg/kg JPPMAB 25,929,73
idu-rat LD50:38 mg/kg AIPTAK 180,155,69
orl-mus LD50:70 mg/kg AIPTAK 135,376,62
ipr-mus LD50:35 mg/kg AIPTAK 135,376,62
scu-mus LD50:35 mg/kg AIPTAK 135,376,62
ivn-mus LD50:20 mg/kg JPETAB 99,163,50
ivn-dog LDLo:26 mg/kg 27ZIAQ -,157,73

CONSENSUS REPORTS: EPA Genetic Toxicology Program.

SAFETY PROFILE: Poison by ingestion, intraperitoneal, intravenous, subcutaneous, and intraduodenal routes. Human systemic effects by intravenous route: nausea or vomiting, respiratory depression, and coma. An experimental teratogen. Experimental reproductive effects. *Caution:* Abuse leads to habituation or addiction. When heated to decomposition it emits toxic fumes of NO_x. See also METHADONE HYDROCHLORIDE.

MDO760 CAS:297-88-1 ***HR: 3***
dl-METHADONE
mf: $C_{21}H_{27}NO$ mw: 309.49

SYNS: 6-(DIMETHYLAMINO)-4,4-DIPHENYL-3-HEPTANONEdl-MIXTURE ◇ 3-HEPTANONE, 6-(DIMETHYLAMINO)-4,4-DIPHENYL-, (±)- ◇ METHADONE ◇ METHADON ◇ (±)-METHADONE

TOXICITY DATA with REFERENCE
dni-rat:tst 10 umol/L BCPCA6 27,123,78
oth-rat:tst 10 umol/L BCPCA6 27,123,78
ipr-rat TDLo:240 mg/kg (female 7D pre-21D post):REP JPETAB 203,340,77
orl-mus LD50:95 mg/kg JPETAB 110,135,54
ipr-mus LD50:28 mg/kg BJPCAL 9,280,54
scu-mus LD50:28 mg/kg YKKZAJ 81,740,61
ivn-dog LD50:26 mg/kg CPBTAL 7,372,59

SAFETY PROFILE: Poison by ingestion, intraperitoneal, subcutaneous, and intravenous routes. Experimental reproductive effects. Mutation data reported. When heated to decomposition it emits toxic fumes of NO_x.

MDO775 CAS:125-58-6 ***HR: 3***
l-METHADONE
mf: $C_{21}H_{27}NO$ mw: 309.49

SYNS: l-6-(DIMETHYLAMINO)-4,4-DIPHENYL-3-HEPTANONE ◇ LEVOMETHADONE ◇ LEVOTHYL ◇ (−)-METHADONE

TOXICITY DATA with REFERENCE
dni-rat:tes 50 μmol/L BCPCA6 27,123,78
oms-rat:tes 50 μmol/L BCPCA6 27,123,78
scu-rat TDLo:1600 μg/kg (1D male):REP JPETAB 198,340,76

orl-mus LD50:97 mg/kg JPETAB 198,340,76
ipr-mus LD50:32 mg/kg BJPCAL 9,280,54
scu-mus LD50:26500 μg/kg MEXPAG 2,323,60

SAFETY PROFILE: Poison by ingestion, subcutaneous, and intraperitoneal routes. Experimental reproductive effects. Mutation data reported. *Caution:* Abuse leads to habituation or addiction. When heated to decomposition it emits toxic fumes of NO_x. See also METHADONE HYDROCHLORIDE.

MDP000 CAS:1095-90-5 ***HR: 3***
METHADONE HYDROCHLORIDE
mf: $C_{21}H_{27}NO•ClH$ mw: 345.95

SYNS: ADANON HYDROCHLORIDE ◇ ALTHOSE HYDROCHLORIDE ◇ AMIDONE HYDROCHLORIDE ◇ DIAMINON HYDROCHLORIDE ◇ DIASONE HYDROCHLORIDE ◇ 6-DIMETHYLAMINO-4,4-DIPHENYL-3-HEPTANONE HYDROCHLORIDE ◇ 1,1-DIPHENYL-1-(β-DIMETHYLAMINOPROPYL)BUTANONE-2HYDROCHLORIDE ◇ DOLOPHINE ◇ DOLOPHINE HYDROCHLORIDE ◇ DOLPHINE ◇ HOECHST 1082

TOXICITY DATA with REFERENCE
dnr-esc 800 ppm DCTODJ 4,1,81
mmo-nsc 300 mg/L DCTODJ 4,1,81
orl-rat TDLo:130 mg/kg (8-22D preg):REP JPETAB 208,106,79
orl-rat TDLo:130 mg/kg (8-22D preg):TER JPETAB 208,106,79
orl-chd LDLo:5 mg/kg JAMAAP 238(23),2516,77
orl-rat LD50:30 mg/kg AIPTAK 123,48,59
ipr-rat LD50:11 mg/kg ANYAA9 51,83,48
scu-rat LD50:12 mg/kg ANYAA9 51,83,48
ivn-rat LD50:9200 μg/kg JPETAB 92,269,48
orl-mus LD50:124 mg/kg AIPTAK 122,434,59
ipr-mus LD50:8300 μg/kg NYKZAU 57,585,61
scu-mus LD50:34 mg/kg ANYAA9 51,83,48
ivn-mus LD50:16 mg/kg TXAPA9 6,334,64
scu-mky LDLo:10 mg/kg PSEBAA 65,113,47
par-frg LD50:56 mg/kg AIPTAK 79,282,49

SAFETY PROFILE: A human poison by ingestion. Poison experimentally by ingestion, subcutaneous, intravenous, parenteral, and intraperitoneal routes. Human systemic effects by ingestion: antipsychotic effects and analgesia. An experimental teratogen. Experimental reproductive effects. Mutation data reported. An analgesic and FDA proprietary drug. A synthetic drug whose action is similar to morphine and heroin, and is almost as addictive. *Caution:* Abuse leads to habituation or addiction. When heated to decomposition it emits very toxic fumes of Cl^- and NO_x. See also other methadone entries.

MDP250 CAS:5967-73-7 ***HR: 3***
l-METHADONE HYDROCHLORIDE
mf: $C_{21}H_{27}NO•ClH$ mw: 345.95

PROP: dl Form: Crystals. Mp: 241°.

SYNS: l-6-DIMETHYLAMINO-4,4-DIPHENYL-3-HEPTANONEHYDROCHLORIDE ◇ LEVADONE ◇ LEVOTHYL ◇ POLAMIDON

TOXICITY DATA with REFERENCE
orl-hmn TDLo:44 μg/kg:CNS JPETAB 93,282,48
ipr-rat LD50:24 mg/kg JPETAB 98,305,50
scu-rat LD50:44 mg/kg JPETAB 98,305,50
ipr-mus LD50:30 mg/kg JPETAB 98,305,50
scu-mus LD50:19 mg/kg JPETAB 98,305,50
ivn-mus LD50:29 mg/kg JPETAB 93,282,48

SAFETY PROFILE: Poison by intravenous, subcutaneous, and intraperitoneal routes. Human systemic effects by ingestion: hallucinations, distorted perceptions, and analgesia. *Caution:* Abuse leads to habituation or addiction. When heated to decomposition it emits very toxic fumes of Cl^- and NO_x. See also METHADONE HYDROCHLORIDE.

MDP500 CAS:15284-15-8 ***HR: 3***
d-METHADONE HYDROCHLORIDE
mf: $C_{21}H_{27}NO•ClH$ mw: 345.95

SYN: d-DOLOPHINE HYDROCHLORIDE

TOXICITY DATA with REFERENCE
orl-hmn TDLo:2380 μg/kg:CNS JPETAB 93,282,48
ipr-rat LD50:72 mg/kg JPETAB 98,305,50
ipr-mus LD50:65 mg/kg JPETAB 98,305,50
ivn-mus LD50:31 mg/kg JPETAB 93,282,48

SAFETY PROFILE: Poison by intravenous and intraperitoneal routes. Human systemic effects by ingestion: euphoria, somnolence (general depressed activity), and analgesia. *Caution:* Abuse leads to habituation or addiction. When heated to decomposition it emits toxic fumes of Cl^- and NO_x. See also METHADONE HYDROCHLORIDE.

MDP750 CAS:125-56-4 ***HR: 3***
dl-METHADONE HYDROCHLORIDE
mf: $C_{21}H_{27}NO•ClH$ mw: 345.95

SYNS: ADANON HYDROCHLORIDE ◇ ALGIDON ◇ ALGOLYSIN ◇ AMIDON HYDROCHLORIDE ◇ AN-148 ◇ BUTALGIN ◇ DEPRIDOL ◇ DIAMINON HYDROCHLORIDE ◇ dl-6-DIMETHYLAMINO-4,4-DIPHENYL-3-HEPTANONE HYDROCHLORIDE ◇ DOLOPHINE HYDROCHLORIDE ◇ DOLOPHIN HYDROCHLORIDE ◇ FENADONE ◇ HEPTADON HYDROCHLORIDE ◇ HOECHST 10,820 ◇ KETALGIN HYDROCHLORIDE ◇ MACODIN ◇ MEPHENON ◇ (±)-METHADONE HYDROCHLORIDE ◇ racemic METHADONE HYDROCHLORIDE ◇ MIADONE ◇ MOHEPTAN ◇ PHENADONE HYDROCHLORIDE ◇ PHYSEPTONE HYDROCHLORIDE ◇ TUSSAL

TOXICITY DATA with REFERENCE
ipr-rat TDLo:105 mg/kg (lactating female 21D post):REP NETOD7 4,455,82
scu-rat TDLo:15 mg/kg (female 6-8D post):TER TXCYAC 42,195,86

orl-rat LD50:95 mg/kg JPETAB 92,269,48
ipr-rat LD50:33 mg/kg JPETAB 98,305,50
ipr-mus LD50:31 mg/kg JPETAB 98,305,50
scu-mus LD50:27 mg/kg JPETAB 98,305,50
ivn-mus LD50:24240 μg/kg BJPCAL 4,98,49
scu-dog LDLo:50 mg/kg JPETAB 98,305,50
ivn-dog LD50:29 mg/kg CPBTAL 7,372,59

SAFETY PROFILE: Poison by ingestion, subcutaneous, intravenous, and intraperitoneal routes. An experimental teratogen. Experimental reproductive effects. *Caution:* Abuse leads to habituation or addiction. When heated to decomposition it emits toxic fumes of NO_x and HCl. See also METHADONE HYDROCHLORIDE.

MDP800 ***HR: 3***
METHAFURYLENE FUMARATE
mf: $C_{14}H_{19}N_3O{\bullet}C_4H_4O_4$ mw: 361.44

SYNS: 2-((2-(DIMETHYLAMINO)ETHYL)FURFURYLAMINO)PYRIDINE FUMARATE ◇ FORALAMIN FUMARATE ◇ FORALMINE FUMARATE ◇ F 151 FUMARATE ◇ N-(2-FURANYLMETHYL)-N',N'-DIMETHYL-N-2-PYRIDINYL-1,2-ETHANEDIAMINEFUMARATE ◇ N-(2-FURFURYL)-N-(2-PYRIDYL)-N',N'-DIMETHYLETHYLENEDIAMINE FUMARATE ◇ METHAFURILEN FUMARATE

TOXICITY DATA with REFERENCE
ipr-rat LDLo:100 mg/kg JPETAB 99,488,50
orl-mus LD50:264 mg/kg JPETAB 99,488,50
ipr-mus LD50:136 mg/kg JPETAB 99,488,50
ipr-cat LDLo:60 mg/kg JPETAB 99,488,50
ipr-gpg LDLo:80 mg/kg JPETAB 99,488,50

SAFETY PROFILE: Poison by ingestion and intraperitoneal routes. When heated to decomposition it emits toxic fumes of NO_x.

MDQ075 CAS:2529-46-6 ***HR: D***
METHALLYL-19-NORTESTOSTERONE
mf: $C_{22}H_{32}O_2$ mw: 328.54

SYNS: 17-β-HYDROXY-17-α-(2-METHYLALLYL)ESTR-4-EN-3-ONE ◇ 17-β-HYDROXY-17-(2-METHYLALLYL)ESTR-4-EN-3-ONE ◇ 17-α-(2-METHYLALLYL)-19-NORTESTOSTERONE ◇ MNT ◇ SC 9022

TOXICITY DATA with REFERENCE
scu-rat TDLo:500 μg/kg (female 16D post):TER CRSBAW 155,967,61
scu-rat TDLo:17500 μg/kg (female 35D pre):REP ENDOAO 63,561,58

SAFETY PROFILE: An experimental teratogen. Experimental reproductive effects. When heated to decomposition it emits acrid smoke and irritating fumes. See also TESTOSTERONE.

MDQ100 CAS:7359-80-0 ***HR: D***
17-α-(1-METHALLYL)-19-NORTESTOSTERONE
mf: $C_{22}H_{32}O_2$ mw: 328.54

SYNS: 17-β-HYDROXY-17-α-(1-METHYLALLYL)ESTR-4-EN-3-ONE ◇ SC-8117

TOXICITY DATA with REFERENCE
scu-rat TDLo:17500 μg/kg (35D pre):REP ENDOAO 63,561,58

SAFETY PROFILE: Experimental reproductive effects. When heated to decomposition it emits acrid smoke and irritating fumes. See also TESTOSTERONE, ALLYL COMPOUNDS, and ESTERS.

MDQ250 CAS:438-41-5 ***HR: 3***
METHAMINODIAZEPOXIDE HYDROCHLORIDE
mf: $C_{16}H_{14}ClN_3O{\bullet}ClH$ mw: 336.24

SYNS: ANSIACAL ◇ A-POXIDE ◇ BENT ◇ BENZODIAPIN ◇ CALMODEN ◇ CEBRUM ◇ CHLORDIAZACHEL ◇ CHLORDIAZEPOXIDE HYDROCHLORIDE ◇ CHLORDIAZEPOXIDE MONOHYDROCHLORIDE ◇ CHLORIDEAZEPOXIDE HYDROCHLORIDE ◇ 7-CHLORO-2-METHYLAMINO-5-PHENYL-3H-1,4-BENZODIAZEPIN, 4-OXIDE, HYDROCHLORIDE ◇ 7-CHLORO-N-METHYL-5-PHENYL-EH-1,4-BENZODIAZEPIN-2-AMINE-4-OXIDE, MONOHYDROCHLORIDE ◇ CORAX ◇ DIAZACHEL (OBS.) ◇ DROXOL ◇ ELENIUM ◇ EQUIBRAL ◇ J-LIBERTY ◇ KALMOCAPS ◇ LABICAN ◇ LENTOTRAN ◇ LIBRIUM ◇ LIBRIUM HYDROCHLORIDE ◇ METHAMINODIAZEPINE HYDROCHLORIDE ◇ MILDMEN ◇ MURCIL ◇ NAPOTON ◇ NOVOSED ◇ PSICHIAL ◇ PSICOSAN ◇ RELIBERAN ◇ RO 5-0690 ◇ SEREN VITA ◇ SK-LYGEN ◇ SOPHIAMIN ◇ TENSINYL ◇ TIMOSIN ◇ TRAKIPEAL ◇ VIANSIN ◇ VIOPSICOL

TOXICITY DATA with REFERENCE
sln-dmg-orl 500 mg/L IJMRAQ 76,348,82
mnt-mus-orl 188 mg/kg TOLED5 18,45,83
mnt-mus-ipr 67 mg/kg TOLED5 18,45,83
dlt-mus-orl 96 g/kg/30D-C IJMRAQ 66(5),847,77
hma-mus/sat 188 mg/kg TOLED5 16,347,83
orl-rat TDLo:525 mg/kg (1-21D preg):REP TXCYAC 17,311,80
orl-mus TDLo:48 g/kg (male 30D pre):TER IJMRAQ 66,847,77
orl-rat LD50:537 mg/kg TXAPA9 16,556,70
ipr-rat LD50:276 mg/kg TXAPA9 16,556,70
scu-rat LD50:800 mg/kg 27ZIAQ -,71,73
ivn-rat LD50:165 mg/kg 27ZQAG -,159,72
orl-mus LD50:530 mg/kg 26RAAN -,39,73
ipr-mus LD50:200 mg/kg JMCMAR 14,1106,71
scu-mus LD50:530 mg/kg 27ZQAG -,159,72
ivn-mus LD50:95 mg/kg JPETAB 129,163,60
orl-dog LD50:1000 mg/kg 27ZIAQ -,71,73
orl-rbt LD50:590 mg/kg AIPTAK 178,216,69
ivn-rbt LD50:36 mg/kg IJNEAQ 5,305,66

CONSENSUS REPORTS: EPA Genetic Toxicology Program.

SAFETY PROFILE: Poison by intraperitoneal and intravenous routes. Moderately toxic by ingestion and subcutaneous routes. An experimental teratogen. Experimental reproductive effects. Mutation data reported. A minor tranquilizer. When heated to decomposition it emits very toxic fumes of HCl and NO_x.

MDQ500 CAS:826-10-8 ***HR: 3***
l-METHAMPHETAMINE HYDROCHLORIDE
mf: $C_{10}H_{15}N{\cdot}ClH$ mw: 185.72

PROP: Crystals; bitter taste. Mp: 170-175°. Sol in water, alc, and chloroform; almost insol in ether.

SYNS: ADIPEX ◇ l-DESOXYEPHEDRINE HYDROCHLORIDE ◇ (−)-N-α-DIMETHYLPHENETHYLAMINE HYDROCHLORIDE ◇ "METH" ◇ l-N-METHYL-β-PHENYLISOPROPYLAMINE HYDROCHLORIDE ◇ "SPEED" ◇ SYNDROX

TOXICITY DATA with REFERENCE
ivn-mus TDLo:15 mg/kg (9-11D preg):TER TJADAB 4,131,71
ivn-mus TDLo:15 mg/kg (9-11D preg):REP TJADAB 4,131,71
ipr-rat LD50:25 mg/kg 27ZQAG -,346,72
scu-rat LD50:30 mg/kg 27ZQAG -,346,72
ipr-mus LD50:70 mg/kg JPETAB 89,382,47
scu-mus LD50:180 mg/kg 27ZQAG -,346,72
ivn-mus LD50:33 mg/kg 27ZQAG -,346,72
orl-dog LD50:10 mg/kg 27ZQAG -,346,72
ivn-dog LD50:2700 μg/kg PSEBAA 118,557,65
scu-cat LD50:50 mg/kg 27ZQAG -,346,72

SAFETY PROFILE: Poison by ingestion, intravenous, intraperitoneal, and subcutaneous routes. An experimental teratogen. Experimental reproductive effects. A powerful central nervous system stimulant. *Caution:* Excessive use may lead to tolerance and habituation. When heated to decomposition it emits very toxic fumes of HCl and NO_x. See also BENZEDRINE.

MDQ750 CAS:74-82-8 ***HR: 3***
METHANE
DOT: UN 1971/UN 1972
mf: CH_4 mw: 16.05

PROP: Colorless, odorless, tasteless gas. Mp: −182.6°. Bp: −161.5°, lel: 5.3%, uel: 15%, fp: −183.2°. D: 0.554 @ 0°/4° (air = 1) or 0.7168 g/L, autoign temp: 650°, vap d: 0.6, flash p: −368.6°F. Sol in water, alc, and ether.

SYNS: FIRE DAMP ◇ MARSH GAS ◇ METHANE, compressed (DOT) ◇ METHANE, refrigerated liquid (DOT) ◇ METHYL HYDRIDE

CONSENSUS REPORTS: Reported in EPA TSCA Inventory.

DOT Classification: Flammable Gas; Label: Flammable Gas

SAFETY PROFILE: A simple asphyxiant. Very dangerous fire and explosion hazard when exposed to heat or flame. Reacts violently with powerful oxidizers (e.g., bromine pentafluoride; chlorine trifluoride; chlorine; fluorine; iodine heptafluoride; dioxygenyl tetrafluoroborate; dioxygen difluoride; trioxygen difluoride; liquid oxygen; ClO_2; NF_3; OF_2). Incompatible with halogens or interhalogens; air (forms explosive mixtures). Explosive in the form of vapor when exposed to heat or flame. To fight fire, stop flow of gas, CO_2 or dry chemical. See also ARGON for a description of asphyxiants.

MDQ800 ***HR: 3***
METHANE BORONIC ANHYDRIDE-PYRIDINE COMPLEX
mf: $CH_3BO{\cdot}C_5H_5N$ mw: 120.95

SAFETY PROFILE: Ignites spontaneously in air. When heated to decomposition it emits toxic fumes of NO_x. See also ANHYDRIDES, PYRIDINE, and BORON COMPOUNDS.

MDR250 CAS:75-75-2 ***HR: 3***
METHANESULFONIC ACID
mf: CH_4O_3S mw: 96.11

PROP: Solid. D: 1.4812 @ 18°/4°, mp: 20°, bp: 167° @ 10 mm. Sol in water, alc, and ether. Corrosive to iron, steel, brass, copper, and lead.

SYN: WSQ 1

TOXICITY DATA with REFERENCE
orl-rat LDLo:200 mg/kg KODAK* 21MAY71
ipr-rat LDLo:50 mg/kg KODAK* 21MAY71
orl-qal LD50:1000 mg/kg JRPFA4 48,371,76

CONSENSUS REPORTS: Reported in EPA TSCA Inventory.

SAFETY PROFILE: Poison by ingestion and intraperitoneal routes. May be corrosive to skin, eyes, and mucous membranes. Explosive reaction with ethyl vinyl ether. Incompatible with hydrogen fluoride. When heated to decomposition it emits toxic fumes of SO_x. See also SULFONATES.

MDR750 CAS:558-25-8 ***HR: 3***
METHANESULFONYL FLUORIDE
mf: CH_3FO_2S mw: 98.10

SYNS: FUMETTE ◇ METHANESULPHONYL FLUORIDE ◇ MSF

TOXICITY DATA with REFERENCE
orl-rat LD50:2 mg/kg IAEC** 17JUN74
ihl-rat LCLo:140 mg/m^3 31ZOAD 1,287,68
ipr-rat LD50:3 mg/kg NATUAS 173,33,54
scu-rat LD50:3500 μg/kg 28ZEAL 4,271,69
scu-mus LDLo:3500 μg/kg 31ZOAD 1,287,68
ivn-mus LD50:1 mg/kg IAEC** 17JUN74
scu-dog LDLo:3500 μg/kg 31ZOAD 1,287,68
ivn-dog LD50:5620 μg/kg IAEC** 17JUN74

scu-rbt LDLo:3500 μg/kg 31ZOAD 1,287,68
ivn-rbt LD50:3370 μg/kg IAEC** 17JUN74

CONSENSUS REPORTS: EPA Extremely Hazardous Substances List. Reported in EPA TSCA Inventory.

SAFETY PROFILE: Poison by ingestion, inhalation, intraperitoneal, intravenous, and subcutaneous routes. When heated to decomposition it emits very toxic fumes of F^- and SO_x. See also FLUORIDES and SULFONATES.

MDR775 CAS:25284-83-7 ***HR: 3***
METHANETELLUROL
mf: CH_4Te mw: 143.64

SAFETY PROFILE: A poison. Ignites spontaneously in air. Explodes on contact with oxygen at room temperature. When heated to decomposition it emits toxic fumes of Te. See also TELLURIUM COMPOUNDS.

MDS500 CAS:100-72-1 ***HR: 2***
2-METHANOL TETRAHYDROPYRAN
mf: $C_6H_{12}O_2$ mw: 116.18

PROP: Liquid. Fp: −70°, bp: 187°, d: 1.0272 @ 20°/20°, vap d: 4.02, vap press: 0.4 mm @ 20°, flash p: 200°F (CC).

SYNS: 2-HYDROXYMETHYLTETRAHYDROPYRAN ◇ TETRAHYDROPYRAN-2-METHANOL ◇ TETRAHYDROPYRANYL-2-METHANOL

TOXICITY DATA with REFERENCE
skn-rbt 515 mg open MLD UCDS** 7/28/66
skn-rbt 500 mg SEV SCCUR* -,6,61
eye-rbt 16450 μg SEV UCDS** 7/28/66
orl-rat LD50:3730 mg/kg AMIHBC 10,61,54
orl-mus LD50:2870 mg/kg SCCUR* -,6,61
skn-rbt LD50:4000 mg/kg UCDS** 7/28/66

SAFETY PROFILE: Moderately toxic by ingestion and skin contact. A severe skin and eye irritant. Combustible when exposed to heat or flame; can react with oxidizing materials. To fight fire, use alcohol foam, spray, mist, dry chemical. When heated to decomposition it emits acrid smoke and irritating fumes.

MDT000 ***HR: 3***
METHAPYRILENE mixed with SODIUM NITRITE (1:2)

SYN: SODIUM NITRITE mixed with METHAPYRILENE (2:1)

TOXICITY DATA with REFERENCE
orl-rat TDLo:121 g/kg/90W-I:CAR FCTXAV 15,269,77

SAFETY PROFILE: Questionable carcinogen with experimental carcinogenic data. When heated to decomposition it emits toxic fumes of Na_2O and NO_x. See also SODIUM NITRITE.

MDT250 CAS:340-56-7 ***HR: 3***
METHAQUALONE HYDROCHLORIDE
mf: $C_{16}H_{14}N_2O{\cdot}ClH$ mw: 286.78

PROP: Crystals. Mp: 255-265°. Sol in ether, ethanol; almost insol in water.

SYNS: METHYLQUINAZOLONE HYDROCHLORIDE ◇ 2-METHYL-3-TOLYLCHINAZOLON-4 HYDROCHLORIDE (GERMAN) ◇ 2-METHYL-3-o-TOLYL-4(3H)-QUINAZOLINONE HYDROCHLORIDE ◇ 2-METHYL-3-(o-TOLYL)-4-QUINAZOLONEHYDROCHLORIDE

TOXICITY DATA with REFERENCE
orl-rat TDLo:1500 mg/kg (1-20D preg):TER NATUAS 200,656,63
orl-rat TDLo:1500 mg/kg (1-20D preg):REP NATUAS 200,656,63
orl-rat LD50:410 mg/kg BCFAAI 111,472,72
ipr-rat LD50:124 mg/kg 27ZQAG -,262,72
ivn-rat LD50:120 mg/kg 27ZQAG -,262,72
orl-mus LD50:400 mg/kg ABMGAJ 13,591,64
ipr-mus LD50:390 mg/kg TXAPA9 37,185,76
ivn-mus LD50:120 mg/kg 27ZQAG -,262,72
ivn-rbt LD50:120 mg/kg 27ZQAG -,262,72
orl-gpg LD50:360 mg/kg 27ZQAG -,262,72

SAFETY PROFILE: Poison by ingestion, intraperitoneal, and intravenous routes. An experimental teratogen. Experimental reproductive effects. When heated to decomposition it emits very toxic fumes of NO_x and HCl.

MDT500 CAS:1229-35-2 ***HR: 3***
METHDILAZINE HYDROCHLORIDE
mf: $C_{18}H_{20}N_2S{\cdot}ClH$ mw: 332.92

PROP: Crystals from isopropyl alc. Mp: 187.5-189°.

SYNS: DILOSYN ◇ DISYNCRAN ◇ 10-((1-METHYL-3-PYRROLIDINYL)METHYL)PHENOTHIAZINE, HYDROCHLORIDE ◇ TACARYL

TOXICITY DATA with REFERENCE
otr-ham:emb 10 mg/L ENMUDM 8(Suppl 6),4,86
ipr-rat TDLo:150 mg/kg (2-4D preg):REP JRPFA4 9,359,65
orl-rat LD50:260 mg/kg TXAPA9 2,68,60
orl-mus LD50:190 mg/kg TXAPA9 2,68,60
ipr-mus LD50:100 mg/kg TXAPA9 2,68,60
ivn-rbt LDLo:17 mg/kg TXAPA9 2,68,60
orl-gpg LD50:263 mg/kg TXAPA9 2,68,60

SAFETY PROFILE: Poison by ingestion, intravenous, and intraperitoneal routes. Mutation data reported. Experimental reproductive effects. An antihistamine. When heated to decomposition it emits very toxic fumes of NO_x and HCl.

MDT600 CAS:51-57-0 ***HR: 3***
METHEDRINE
mf: $C_{10}H_{15}N \cdot ClH$ mw: 185.72

SYNS: ADIPEX ◇ DEOFED ◇ d-DEOXYEPHEDRINE HYDROCHLORIDE ◇ DESOXO-5 ◇ d-DESOXYEPHEDRINE HYDROCHLORIDE ◇ DESOXYFED ◇ DESOXYN ◇ DESOXYNE ◇ DESTIM ◇ DESYPHED ◇ DEXOVAL ◇ DEXTIM ◇ DOXYFED ◇ DRINALFA ◇ EFROXINE ◇ EUFODRIANL ◇ GERVOT ◇ ISOPHEN ◇ METAMPHETAMINE HYDROCHLORIDE ◇ (+)-METHAMPHETAMINE CHLORIDE ◇ (+)-METHAMPHETAMINE HYDROCHLORIDE ◇ METHAMPHETAMINE HYDROCHLORIDE ◇ d-METHAMPHETAMINE HYDROCHLORIDE ◇ METHAMPHETAMINIUM CHLORIDE ◇ METHEDRINE HYDROCHLORIDE ◇ METHYLAMPHETAMINE HYDROCHLORIDE ◇ d-METHYLAMPHETAMINE HYDROCHLORIDE ◇ N-METHYLAMPHETAMINE HYDROCHLORIDE ◇ METHYLISOMYN ◇ NORODIN HYDROCHLORIDE ◇ PERVITIN ◇ PHILOPON ◇ SOXYSYMPAMINE ◇ SYNDROX ◇ TONEDRON

TOXICITY DATA with REFERENCE
orl-rat TDLo:792 mg/kg (4W male/4W pre-3W post):REP JONRA9 20,625,73
scu-rat TDLo:400 mg/kg (1-21D preg/19D post):NEO,TER EAGRDS 5,509,79
ipr-mus LD50:15 mg/kg 27ZQAG -,346,72
scu-mus LD50:7560 μg/kg JPETAB 87,214,46
ivn-mus LD50:6300 μg/kg CSLNX* NX#02170
orl-gpg LD50:90 mg/kg SMWOAS 84,351,54

CONSENSUS REPORTS: EPA Genetic Toxicology Program. Reported in EPA TSCA Inventory.

SAFETY PROFILE: Poison by ingestion, subcutaneous, intravenous, and intraperitoneal routes. An experimental teratogen. Experimental reproductive effects. Questionable carcinogen with experimental neoplastigenic data. When heated to decomposition it emits toxic fumes of NO_x and HCl. See also BENZEDRINE and various amphetamines.

MDT730 CAS:348-67-4 ***HR: 1***
d-METHIONINE
mf: $C_5H_{11}NO_2S$ mw: 149.23

PROP: Loses ammonia.

SYN: d-METIONIEN (AUSTRALIAN)

TOXICITY DATA with REFERENCE
ipr-rat LD50:5223 mg/kg ABBIA4 64,319,56

CONSENSUS REPORTS: Reported in EPA TSCA Inventory.

SAFETY PROFILE: Mildly toxic by intraperitoneal route. When heated to decomposition it emits very toxic fumes of NO_x and SO_x. See also METHIONINE.

MDT740 CAS:59-51-8 ***HR: 2***
dl-METHIONINE
mf: $C_5H_{11}NO_2S$ mw: 149.23

PROP: White crystalline platelets; characteristic odor. Sol in water, dil acids, and alkalies; very sltly sol in alc; insol in ether.

SYNS: ACIMETION ◇ BANTHIONINE ◇ CYNARON ◇ DYPRIN ◇ LOBAMINE ◇ MEONINE ◇ MERTIONIN ◇ METHILANIN ◇ (±)-METHIONINE ◇ METIONE ◇ NESTON

TOXICITY DATA with REFERENCE
orl-rat TDLo:40 g/kg (1-20D preg):TER JRPFA4 33,109,73
orl-rat TDLo:40 g/kg (1-20D preg):REP JRPFA4 33,109,73
ipr-rat LDLo:2 g/kg YACHDS 5,2041,77
orl-mus LDLo:4 g/kg YACHDS 5,2041,77
ipr-mus LDLo:1500 mg/kg YACHDS 5,2041,77
ivn-mus LDLo:300 mg/kg YACHDS 5,2041,77

CONSENSUS REPORTS: EPA Genetic Toxicology Program. Reported in EPA TSCA Inventory.

SAFETY PROFILE: Moderately toxic by ingestion and other routes. An experimental teratogen. Experimental reproductive effects. When heated to decomposition it emits toxic fumes of SO_x and NO_x. See also l-METHIONINE.

MDT750 CAS:63-68-3 ***HR: 1***
l-METHIONINE
mf: $C_5H_{11}NO_2S$ mw: 149.23

PROP: White, crystalline powder or platelets; faint odor. Mp: 281° (decomp), d: 1.340. Sol in water, dil acids, and alkalies; insol in abs alc, alc, benzene, acetone, ether.

SYNS: l-α-AMINO-Γ-METHYLMERCAPTOBUTYRIC ACID ◇ 2-AMINO-4-(METHYLTHIO)BUTYRIC ACID ◇ l(−)-AMINO-Γ-METHYLTHIOBUTYRIC ACID ◇ CYMETHION ◇ LIQUIMETH ◇ METHIONINE ◇ l-(−)-METHIONINE ◇ l-Γ-METHYLTHIO-α-AMINOBUTYRIC ACID

TOXICITY DATA with REFERENCE
mmo-esc 100 mg/L PMRSDJ 1,376,81
dnr-smc 500 mg/L PMRSDJ 1,502,81
orl-rat TDLo:26100 mg/kg (10-20D preg):TER DTTIAF 82,457,75
orl-rat TDLo:14720 mg/kg (10-20D preg):REP DTTIAF 82,457,75
orl-rat LD50:36 g/kg GISAAA 48(6),20,83
ipr-rat LD50:4328 mg/kg ABBIA4 58,253,55

CONSENSUS REPORTS: Reported in EPA TSCA Inventory. EPA Genetic Toxicology Program.

SAFETY PROFILE: Mildly toxic by ingestion and intraperitoneal routes. Human mutation data reported. An experimental teratogen. Experimental reproductive effects. An essential sulfur-containing amino acid.

When heated to decomposition it emits very toxic fumes of NO_x and SO_x.

MDU100 CAS:1982-67-8 ***HR: 3***
METHIONINE SULFOXIMINE
mf: $C_5H_{12}N_2O_3S$ mw: 180.25

SYNS: 2-AMINO-4-(S-METHYLSULFONIMIDOYL)-BUTANOIC ACID (9CI) ◇ dl-METHIONINE-dl-SULFOXIMINE

TOXICITY DATA with REFERENCE
ipr-mus TDLo:350 mg/kg (9-15D preg):TER ASMUAA 38,193,62
ipr-mus LD50:218 mg/kg PSEBAA 94,12,57
ivn-mus LD50:100 mg/kg CSLNX* NX#03652

SAFETY PROFILE: Poison by intraperitoneal and intravenous route. An experimental teratogen. When heated to decomposition it emits toxic fumes of SO_x and NO_x.

MDU300 CAS:3772-76-7 ***HR: 3***
METHOFADIN
mf: $C_{12}H_{14}N_4O_3S$ mw: 294.36

PROP: Crystals. Mp: 146°. Also obtained as the monohydrate.

SYNS: (p-AMINOBENZOLSULFONYL)-4-AMINO-2-METHYL-6-METHOXY-PYRIMIDIN (GERMAN) ◇ 4-AMINO-N-(6-METHOXY-2-METHYL-4-PYRIMIDINYL)BENZENESULFONAMIDE ◇ DUROPROCIN ◇ METHOFADIN ◇ METHOFAZINE ◇ SULFAMETHOMIDINE ◇ SULFAMETOMIDINE ◇ TANASUL ◇ TELEMID

TOXICITY DATA with REFERENCE
orl-rat TDLo:6 g/kg (9-14D preg):REP SEIJBO 13,17,73
orl-rat TDLo:3 g/kg (9-14D preg):TER SEIJBO 13,17,73
ivn-mus LD50:16 mg/kg ARZNAD 3,66,53

SAFETY PROFILE: Poison by intravenous route. Experimental reproductive effects. An experimental teratogen. When heated to decomposition it emits toxic fumes of SO_x and NO_x.

MDU500 CAS:309-36-4 ***HR: 3***
METHOHEXITAL SODIUM
mf: $C_{14}H_{17}N_2O_3$•Na mw: 284.32

PROP: Minute crystals. Sol in water.

SYNS: 5-ALLYL-1-METHYL-5-(1-METHYL-2-PENTYNYL)BARBITURIC ACID SODIUM SALT ◇ BREVIMYTAL ◇ BREVITAL SODIUM ◇ BRIETAL SODIUM ◇ ENALLYNYMAL SODIUM ◇ LILLY 22451 ◇ METHOHEXITONE SODIUM ◇ 1-METHYL-5-ALLYL-5-(1-METHYL-2-PENTYNYL)BARBITURIC ACID SODIUM SALT ◇ SODIUM-dl-5-ALLYL-1-METHYL-5-(1-METHYL-2-PENTYNYL)BARBITURATE ◇ SODIUM METHOHEXITAL ◇ SODIUM METHOHEXITONE ◇ SODIUM A-dl-1-METHYL-5-ALLYL-5-(1-METHYL-2-PENTYNYL)BARBITURATE

TOXICITY DATA with REFERENCE
ivn-wmn TDLo:25 μg/kg:CVS,GIT,SKN JOSUA9 30,906,72
ivn-rat LD50:24890 μg/kg PSEBAA 89,292,55
imp-rat LD50:33 mg/kg 29ZVAB -,75,69
ivn-mus LD50:33600 μg/kg OYYAA2 12,247,76
ivn-dog LD50:21500 μg/kg PSEBAA 89,292,55
imp-dog LD50:25 mg/kg 29ZVAB -,75,69
ivn-rbt LD50:8640 μg/kg PSEBAA 89,292,55
imp-rbt LD50:10 mg/kg 29ZVAB -,75,69

SAFETY PROFILE: Poison by intravenous and implant routes. Human systemic effects by intravenous route: blood pressure lowering, gastrointestinal effects, and allergic dermatitis. An FDA proprietary drug. *Caution:* Excessive use may lead to addiction or habituation. Allergenic effects by intravenous route. When heated to decomposition it emits toxic fumes of Na_2O and NO_x. See also BARBITURATES.

MDU600 CAS:16752-77-5 ***HR: 3***
METHOMYL
mf: $C_5H_{10}N_2O_2S$ mw: 162.23

PROP: White, crystalline solid; slt sulfurous odor. Moderately water-sol, mp: 79°.

SYNS: DU PONT INSECTICIDE 1179 ◇ ENT 27,341 ◇ INSECTICIDE 1,179 ◇ LANNATE ◇ MESOMILE ◇ METHYL N-((METHYLAMINO) CARBONYL)OXY)ETHANIMIDO)THIOATE ◇ METHYL-N-((METHYLCARBAMOYL)OXY)THIOACETIMIDATE ◇ S-METHYL N-[(METHYLCARBAMOYLOXY]THIOACETIMIDATE ◇ 2-METHYLTHIO-ACETALDEHYD-O-(METHYLCARBAMOYL)-OXIM (GERMAN) ◇ 2-METHYLTHIO-PROPIONALDEHYD-O-(METHYLCARBAMOYL)-OXIM(GERMAN) ◇ METOMIL (ITALIAN) ◇ NU-BAIT II ◇ NUDRIN ◇ RCRA WASTE NUMBER P066 ◇ 3-THIABUTAN-2-ONE, O-(METHYLCARBAMOYL)OXIME ◇ WL 18236

TOXICITY DATA with REFERENCE
orl-rat LD50:17 mg/kg GUCHAZ 6,336,73
ihl-rat LC50:77 ppm TXAPA9 40,1,77
scu-rat LD50:9 mg/kg TXAPA9 25,569,73
orl-mus LD50:10 mg/kg JAFCAU 26,550,78
orl-dog LDLo:30 mg/kg TXAPA9 40,1,77
orl-mky LDLo:40 mg/kg TXAPA9 40,1,77
skn-rbt LD50:5880 mg/kg FMCHA2 -,D197,80

CONSENSUS REPORTS: EPA Genetic Toxicology Program. EPA Extremely Hazardous Substances List.

OSHA PEL: TWA 2.5 mg/m^3
ACGIH TLV: TWA 2.5 mg/m^3

SAFETY PROFILE: Poison by ingestion, inhalation, and subcutaneous routes. Mildly toxic by skin contact. When heated to decomposition it emits very toxic fumes of NO_x and SO_x.

MDU750 CAS:522-23-6 ***HR: 3***
METHOPHENAZINE DIFUMARATE
mf: $C_{31}H_{36}ClN_3O_5S$•$C_8H_4O_8$ mw: 826.33

SYNS: FRENOLON DIFUMARATE ◇ METHOPHENAZATE ACID FUMARATE ◇ PHRENOLAN ◇ T-82 DIFUMARATE ◇ 3,4,5-

TRIMETHOXY-BENZOIC ACID 2-(4-(3-(2-CHLOROPHENOTHIAZIN-10-YL)PROPYL)-1-PIPERAZINYL)ETHYL ESTER, DIFUMARATE

TOXICITY DATA with REFERENCE
orl-rat TDLo:80 mg/kg (7-14D preg):TER TJADAB 11,325,75
orl-rat LD50:1635 mg/kg KSRNAM 4,503,70
ipr-rat LD50:939 mg/kg KSRNAM 4,503,70
scu-rat LDLo:2560 mg/kg KSRNAM 4,503,70
orl-mus LD50:580 mg/kg KSRNAM 4,503,70
ipr-mus LD50:150 mg/kg 27ZQAG -,30,72
scu-mus LD50:142 mg/kg 27ZQAG -,30,72
ivn-mus LD50:90 mg/kg 27ZQAG -,30,72

SAFETY PROFILE: Poison by intraperitoneal, subcutaneous, and intravenous routes. Moderately toxic by ingestion. An experimental teratogen. When heated to decomposition it emits very toxic fumes of Cl^-, SO_x, and NO_x. See also ESTERS.

MDV000 CAS:2154-02-1 ***HR: 3***
METHOPHOLINE
mf: $C_{20}H_{24}ClNO_2$ mw: 345.90

SYNS: 1-(p-CHLOROPHENETHYL)-6,7-DIMETHOXY-2-METHYL-1,2,3,4-TETRAHYDROISOQUINOLINE ◇ 1-(p-CHLOROPHENETHYL)-2-METHYL-6,7-DIMETHOXY-1,2,3,4-TETRAHYDROISOQUINOLINE ◇ 1-(p-CHLOROPHENETHYL)-1, 2,3,4-TETRAHYDRO-6,7-DIMETHOXY-2-METHYLISOQUINOLINE ◇ MESOFOLIN ◇ METOFOLINE ◇ NIH 7672 ◇ RO 4-1778 ◇ VERSIDYNE

TOXICITY DATA with REFERENCE
orl-rat LD50:400 mg/kg MDCHAG 5,318,65
ipr-rat LD50:100 mg/kg MDCHAG 5,318,65
scu-rat LD50:400 mg/kg EXPEAM 18,446,62
orl-mus LD50:180 mg/kg EXPEAM 18,446,62
ipr-mus LD50:70 mg/kg MDCHAG 5,318,65
scu-mus LD50:180 mg/kg MEIEDD 11,966,89
ivn-mus LD50:25 mg/kg TXAPA9 6,334,64
orl-dog LD50:295 mg/kg MDCHAG 5,318,65
ivn-rbt LD50:30 mg/kg EXPEAM 18,446,62

SAFETY PROFILE: Poison by ingestion, intraperitoneal, subcutaneous, and intravenous routes. When heated to decomposition it emits very toxic fumes of HCl and NO_x.

MDV250 CAS:5985-35-3 ***HR: 3***
METHORPHINAN HYDROBROMIDE
mf: $C_{17}H_{23}NO•BrH$ mw: 338.33

SYNS: DROMORAN HYDROBROMIDE ◇ dl-3-HYDROXY-N-METHYLMORPHINAN HYDROBROMIDE ◇ NU 2206 ◇ RACEMORPHAN HYDROBROMIDE ◇ RO 1-5431

TOXICITY DATA with REFERENCE
orl-rat LD50:350 mg/kg JPETAB 109,189,53
scu-rat LD50:108 mg/kg AIPTAK 85,387,51
orl-mus LD50:375 mg/kg JPETAB 109,189,53
ipr-mus LD50:120 mg/kg JPETAB 99,163,50
ivn-mus LD50:33 mg/kg AIPTAK 85,387,51
ivn-rbt LD50:19 mg/kg AIPTAK 85,387,51

SAFETY PROFILE: Poison by subcutaneous, intraperitoneal, and intravenous routes. When heated to decomposition it emits very toxic fumes of NO_x and HBr.

MDV500 CAS:59-05-2 ***HR: 3***
METHOTREXATE
mf: $C_{20}H_{22}N_8O_5$ mw: 454.50

SYNS: AMETHOPTERIN ◇ 4-AMINO-4-DEOXY-N^{10}-METHYLPTEROYLGLUTAMATE ◇ 4-AMINO-4-DEOXY-N^{10}-METHYLPTEROYLGLUTAMIC ACID ◇ 4-AMINO-10-METHYLFOLIC ACID ◇ 4-AMINO-N^{10}-METHYLPTEROYLGLUTAMIC ACID ◇ ANTIFOLAN ◇ N-BISMETHYLPTEROYLGLUTAMIC ACID ◇ CL-14377 ◇ l-(+)-N-(p-(((2,4-DIAMINO-6-PTERIDINYL)METHYL)METHYLAMINO)BENZOYL) GLUTAMIC ACID ◇ EMT 25,299 ◇ EMTEXATE ◇ HDMTX ◇ METHOPTERIN ◇ METHOTEXTRATE ◇ METHYLAMINOPTERIN ◇ MTX ◇ NCI-C04671 ◇ NSC-740 ◇ R 9985

TOXICITY DATA with REFERENCE
eye-hmn 150 mg/kg nse CANCAR 48,2158,81
cyt-ham:ovr 5 mg/L ENMUDM 2,455,80
orl-wmn TDLo:250 μg/kg (9W preg):TER JOPDAB 72,790,68
ipr-mus TDLo:50 mg/kg (female 9D post):REP TCMUD8 7,7,87
orl-man TDLo:7 mg/kg/12W-C:CAR,BLD ONCOBS 40,268,83
orl-cld TDLo:125 mg/kg/6Y-I:CAR JAMAAP 238,2631,77
orl-mus TDLo:51 mg/kg/4Y-C:CAR ARDEAC 103,505,71
orl-ham TDLo:210 mg/kg/50W-C:ETA TXAPA9 26,392,73
orl-man TD:74 mg/kg/48W-I:CAR,BLD,SKN ARDEAC 103,501,71
orl-man TD:8260 μg/kg/44W-I:CAR,BLD SJHAAQ 24,234,80
orl-man TDLo:643 μg/kg/6W-I JRHUA9 14,74,87
orl-wmn TDLo:2 mg/kg/17W-I:PUL JRHUA9 14,74,87
isp-wmn LDLo:36 mg/kg/15D NEJMAG 289,770,73
par-wmn TDLo:2600 μg/kg:BRN CANCAR 38,1529,76
unr-wmn TDLo:150 mg/kg:EYE CANCAR 48,2158,81
orl-chd TDLo:2 mg/kg/12D:MET,PUL JAMAAP 209,1861,69
ivn-chd TDLo:100 mg/kg/4H:BLD,BIO CANCAR 33,1151,74
orl-hmn TDLo:43 mg/kg/5Y:LIV ARDEAC 100,523,69
ivn-hmn TDLo:4650 μg/kg/4W-I:LIV PAACA3 5,26,64
ims-hmn TDLo:200 mg/kg/5Y:LIV,PUL ARDEAC 100,531,69
ims-hmn TDLo:35 mg/kg/28W:BPR,PUL BMJOAE 2,156,70
orl-rat LD50:135 mg/kg NIIRDN 6,841,82
ipr-rat LD50:6 mg/kg NIIRDN 6,841,82
ivn-rat LD50:14 mg/kg ARZNAD 20,1467,70
orl-mus LD50:146 mg/kg NIIRDN 6,841,82
ipr-mus LD50:50 mg/kg ANREAK 178,465,74

scu-mus LD50:250 mg/kg NCISP* JAN86
ivn-mus LD50:65 mg/kg NIIRDN 6,841,82

CONSENSUS REPORTS: IARC Cancer Review: Group 3 IMEMDT 7,241,87; Animal Inadequate Evidence IMEMDT 26,267,81; Human Inadequate Evidence IMEMDT 26,267,81. NCI Carcinogenesis Studies (ipr); No Evidence: mouse, rat CANCAR 40,1935,77. Reported in EPA TSCA Inventory.

SAFETY PROFILE: A human poison by intraspinal route. Poison experimentally by ingestion, intravenous, subcutaneous, and intraperitoneal routes. Human teratogenic effects by ingestion: developmental abnormalities of the craniofacial area and the musculoskeletal system. Human systemic effects by multiple routes: thrombocytopenia (decrease in the number of blood platelets), bone marrow changes, other blood changes, cerebral spinal fluid effects, eye effects, blood pressure lowering, cough, dyspnea, fibrosis (pneumoconiosis), cyanosis, gastrointestinal effects, fatty liver degeneration, hepatitis, liver function tests impaired, other liver changes, fever, effects on inflammation or mediation of inflammation, leukopenia. Human mutation data reported. Experimental reproductive effects. A human eye irritant. Questionable human carcinogen producing leukemia, Hodgkin's disease, and skin tumors. An FDA proprietary drug. A chemotherapeutic agent. When heated to decomposition it emits toxic fumes including NO_x.

MDV600 CAS:7413-34-5 ***HR: 3***
METHOTREXATE DISODIUM SALT
mf: $C_{20}H_{20}N_8O_5 \cdot 2Na$ mw: 498.46

SYNS: AMETHOPTERIN SODIUM ◇ 4-AMINO-N^{10}-METHYLPTEROYLGLUTAMIC ACID DISODIUM SALT ◇ DISODIUM METHOTREXATE ◇ GLUTAMIC ACID, N-(p-(((2,4-DIAMINO-6-PTERIDINYL) METHYL)METHYLAMINO)BENZOYL)-, DISODIUM SALT, L-(+)- ◇ MTX DISODIUM ◇ SODIUM METHOTREXATE

TOXICITY DATA with REFERENCE
ipr-mus TDLo:120 mg/kg (male 60D pre):REP JJIND8 58,735,77
ipr-mus LD50:284 mg/kg JJIND8 58,735,77

SAFETY PROFILE: Poison by intraperitoneal route. Experimental reproductive effects. When heated to decomposition it emits toxic fumes of NO_x.

MDV750 CAS:15475-56-6 ***HR: 3***
METHOTREXATE SODIUM
mf: $C_{20}H_{21}N_8O_5 \cdot Na$ mw: 476.48

SYN: MIX SODIUM

TOXICITY DATA with REFERENCE
cyt-hmn:lym 100 nmol/L HEREAY 96,317,82
ivn-rbt TDLo:19200 μg/kg (12D preg):TER TJADAB 15,199,77
ivn-rbt TDLo:19200 μg/kg (12D preg):REP TJADAB 15,199,77
ipr-mus LD50:27 mg/kg PJPPAA 25,327,73

SAFETY PROFILE: Poison by intraperitoneal route. An experimental teratogen. Experimental reproductive effects. Human mutation data reported. When heated to decomposition it emits very toxic fumes of Na_2O and NO_x. See also METHOTREXATE.

MDW000 CAS:61-16-5 ***HR: 3***
METHOXAMINE HYDROCHLORIDE
mf: $C_{11}H_{17}NO_3 \cdot ClH$ mw: 247.71

PROP: Crystals. Mp: 212-216°. Very sol in water; practically insol in ether, benzene, and chloroform.

SYNS: 2-AMINO-1-(2,5-DIMETHOXYPHENYL)-1-PROPANOLHYDROCHLORIDE ◇ α-(1-AMINOETHYL)-2,5-DIMETHOXYBENZYL ALCOHOL HYDROCHLORIDE ◇ β-(2,5-DIMETHOXYPHENYL)-β-HYDROXYISOPROPYLAMINE HYDROCHLORIDE ◇ β-HYDROXY-β-(2,5-DIMETHOXYPHENYL)-ISOPROPYLAMINEHYDROCHLORIDE ◇ PRESSOMIN HYDROCHLORIDE ◇ VASOXINE ◇ VASOXINE HYDROCHLORIDE ◇ VASOXYL HYDROCHLORIDE

TOXICITY DATA with REFERENCE
itt-rat TDLo:400 mg/kg (21D male):REP JRPFA4 58,19,80
orl-mus LDLo:135 mg/kg 27ZIAQ -,160,73
ipr-mus LD50:92 mg/kg NIIRDN 6,836,82
ivn-mus LD50:5030 μg/kg EJPHAZ 9,289,70

SAFETY PROFILE: Poison by ingestion, intravenous, and intraperitoneal routes. Experimental reproductive effects. An FDA proprietary drug. When heated to decomposition it emits very toxic fumes of HCl and NO_x.

MDW100 ***HR: 3***
METHOXERPATE HYDROCHLORIDE
mf: $C_{24}H_{32}N_2O_5 \cdot ClH$ mw: 465.04

SYNS: MEPIRESERPATE HYDROCHLORIDE ◇ MEPISERATE HYDROCHLORIDE ◇ METOSERPATE HYDROCHLORIDE ◇ SU 9064

TOXICITY DATA with REFERENCE
orl-rat LD50:182 mg/kg 27ZQAG -,105,72
ivn-rat LD50:24 mg/kg 27ZQAG -,105,72
ivn-mus LD50:24 mg/kg 27ZQAG -,105,72

SAFETY PROFILE: Poison by ingestion and intravenous routes. When heated to decomposition it emits toxic fumes of NO_x and HCl.

MDW250 CAS:10312-83-1 ***HR: 2***
METHOXYACETALDEHYDE
mf: $C_3H_6O_2$ mw: 74.09

SYNS: α-METHOXYACETALDEHYDE ◇ METHOXYETHANAL

TOXICITY DATA with REFERENCE
orl-rat TDLo:974 mg/kg (male 1D pre):REP TOLED5 32,73,86
orl-rat LD50:2330 mg/kg 34ZIAG -,382,69
ipr-rat LDLo:487 mg/kg TOLED5 32,73,86
skn-rbt LD50:1170 mg/kg 34ZIAG -,382,69

SAFETY PROFILE: Moderately toxic by ingestion, skin contact, and intraperitoneal routes. Experimental reproductive effects. When heated to decomposition it emits acrid smoke and irritating fumes. See also ALDEHYDES.

MDW275 CAS:625-45-6 ***HR: 2***
METHOXYACETIC ACID
mf: $C_3H_6O_3$ mw: 90.09

SYN: 2-METHOXYACETIC ACID

TOXICITY DATA with REFERENCE
dni-mus:emb25 mmol/L TOLED5 45,111,89
ipr-rat TDLo:225 mg/kg (female 8D post):TER TOLED5 22,93,84
ipr-rat TDLo:225 mg/kg (female 8D post):REP TOLED5 22,93,84
orl-rat LDLo:2 g/kg FAATDF 2,158,82

CONSENSUS REPORTS: Reported in EPA TSCA Inventory.

SAFETY PROFILE: Moderately toxic by ingestion. An experimental teratogen. Experimental reproductive effects. Mutation data reported. When heated to decomposition it emits acrid smoke and irritating fumes.

MDW750 CAS:100-06-1 ***HR: 2***
4'-METHOXYACETOPHENONE
mf: $C_9H_{10}O_2$ mw: 150.19

PROP: Colorless to pale yellow fused solid; hawthorn odor. Flash p: +212°F. Sol in fixed oils, propylene glycol; misc in glycerin.

SYNS: ACETANISOLE (FCC) ◇ p-ACETYLANISOLE ◇ 4-ACETYLANISOLE ◇ BANANOTE ◇ FEMA No. 2005 ◇ LINARODIN ◇ p-METHOXYACETOPHENONE ◇ p-METHOXYPHENYL METHYL KETONE ◇ 4-METHOXYPHENYL METHYL KETONE ◇ NOVATONE

TOXICITY DATA with REFERENCE
skn-rbt 500 mg/24H MOD FCTXAV 12,807,74
ihl-hmn TCLo:1700 μg/m³/39W-I:CVS GISAAA 50(4),86,85
orl-rat LD50:1720 mg/kg FCTXAV 12,807,74
orl-mus LD50:820 mg/kg GISAAA 50(4),86,85

CONSENSUS REPORTS: Reported in EPA TSCA Inventory.

SAFETY PROFILE: Moderately toxic by ingestion. Human systemic effects by inhalation: pulse rate increased without fall in blood pressure and blood pressure elevation. A skin irritant. Combustible liquid. When heated to decomposition it emits acrid smoke and irritating fumes. See also KETONES.

MDW780 CAS:38870-89-2 ***HR: 3***
METHOXYACETYL CHLORIDE
mf: $C_3H_5ClO_2$ mw: 108.52

$CH_3OCH_2CO{\cdot}Cl$

SYN: 2-METHOXYETHANOYL CHLORIDE

SAFETY PROFILE: A storage hazard. It evolves HCl gas which can burst a sealed container. When heated to decomposition it emits toxic fumes of Cl^-. See also CHLORIDES.

MDX000 CAS:6443-91-0 ***HR: 3***
METHOXYACETYLENE
mf: C_3H_4O mw: 56.07

$CH_3OC{\equiv}CH$

SYN: ETHYNYLMETHYL ETHER

SAFETY PROFILE: Potentially explosive. A dangerous fire hazard when exposed to heat or flame. When heated to decomposition it emits acrid smoke and irritating fumes. See also ACETYLENE COMPOUNDS.

MDX250 CAS:17959-11-4 ***HR: 3***
(METHOXYACETYL)METHYLCARBAMIC ACID-o-ISOPROPOXYPHENYL ESTER
mf: $C_{14}H_{19}NO_5$ mw: 281.34

SYNS: ENT 27,350 ◇ 2-(1-METHYLETHOXY)PHENYL (METHOXYACETYL)METHYLCARBAMATE ◇ NSC 190948 ◇ UPJOHN U-18120

TOXICITY DATA with REFERENCE
orl-rat LD50:70 mg/kg ARSIM* 20,26,66
orl-mus LD50:200 mg/kg ARSIM* 20,26,66

SAFETY PROFILE: Poison by ingestion. When heated to decomposition it emits toxic fumes of NO_x. See also CARBAMATES.

MDX500 CAS:61417-04-7 ***HR: 3***
4'-(1-METHOXY-9-ACRIDINYLAMINO)METHANESULFONANILIDE
mf: $C_{21}H_{19}N_3O_3S$ mw: 393.49

TOXICITY DATA with REFERENCE
mmo-sat 23 μmol/L JMCMAR 23,269,80
ipr-mus LD10:200 mg/kg JMCMAR 23,269,80

SAFETY PROFILE: Poison by intraperitoneal route. Mutation data reported. When heated to decomposition it emits very toxic fumes of NO_x and SO_x. See also SULFONATES.

MDY000 CAS:59748-95-7 ***HR: 3***
4'-(3-METHOXY-9-ACRIDINYLAMINO) METHANESULFONALIDE
mf: $C_{21}H_{19}N_3O_3S$ mw: 393.49

SYN: N-(4-((3-METHOXY-9-ACRIDINYL)AMINO)PHENYL)-METHANESULFONAMIDE

TOXICITY DATA with REFERENCE
mmo-sat 35 μmol/L JMCMAR 23,269,80
ipr-mus LD10:40 mg/kg JMCMAR 23,269,80

SAFETY PROFILE: Poison by intraperitoneal route. Mutation data reported. When heated to decomposition it emits very toxic fumes of NO_x and SO_x. See also SULFONATES.

MDY250 CAS:61417-05-8 ***HR: 3***
4'-(4-METHOXY-9-ACRIDINYLAMINO) METHANESULFONANILIDE
mf: $C_{21}H_{19}N_3O_3S$ mw: 393.49

TOXICITY DATA with REFERENCE
mmo-sat 43 μmol/L JMCMAR 23,269,80
ipr-mus LD10:120 mg/kg JMCMAR 23,269,80

SAFETY PROFILE: Poison by intraperitoneal route. Mutation data reported. When heated to decomposition it emits very toxic fumes of NO_x and SO_x. See also SULFONATES.

MDY300 CAS:89022-12-8 ***HR: 2***
2'-METHOXY-4'-ALLYLPHENYL 4-GUANIDINOBENZOATE
mf: $C_{18}H_{19}N_3O_3$ mw: 325.40

SYNS: 4-((AMINOIMINOMETHYL)AMINO)BENZOIC ACID 2-METHOXY-4-(2-PROPENYL)PHENYL ESTER ◇ BENZOIC ACID, 4-((AMINOIMINOMETHYL)AMINO)-,2-METHOXY-4-(2-PROPENYL)PHENYL ESTER ◇ BENZOIC ACID, p-GUANIDINO-, 4-ALLYL-2-METHOXYPHENYL ESTER ◇ p-GUANIDINOBENZOIC ACID 4-ALLYL-2-METHOXYPHENYL ESTER

TOXICITY DATA with REFERENCE
ivg-rbt TDLo:100 μg/kg (female 1D pre):REP CCPTAY 32,183,85
ipr-mus LD50:870 mg/kg JMCMAR 29,514,86

SAFETY PROFILE: Moderately toxic by intraperitoneal route. Experimental reproductive effects. When heated to decomposition it emits toxic fumes of NO_x.

MDY750 CAS:1747-60-0 ***HR: 3***
6-METHOXY-2-AMINOBENZOTHIAZOLE
mf: $C_8H_8N_2OS$ mw: 180.24

TOXICITY DATA with REFERENCE
mma-sat 600 nmol/L ENMUDM 3,11,81
dns-rat:lvr 10 μmol/L ENMUDM 3,11,81
orl-mus LD50:241 mg/kg TXAPA9 27,70,74
ivn-mus LD50:140 mg/kg JPETAB 105,486,52

CONSENSUS REPORTS: Reported in EPA TSCA Inventory.

SAFETY PROFILE: Poison by ingestion and intravenous routes. Mutation data reported. When heated to decomposition it emits very toxic fumes of SO_x and NO_x.

MDZ000 CAS:5834-17-3 ***HR: 3***
2-METHOXY-3-AMINODIBENZOFURAN
mf: $C_{13}H_{11}NO_2$ mw: 213.25

SYN: 2-AMINO-3-METHOXYDIPHENYLENOXYD(GERMAN)

TOXICITY DATA with REFERENCE
orl-rat TDLo:13 g/kg/56W-C:CAR ZEKBAI 61,45,56

SAFETY PROFILE: Questionable carcinogen with experimental carcinogenic data. When heated to decomposition it emits toxic fumes of NO_x.

MEA000 CAS:56970-24-2 ***HR: 3***
3-METHOXY-4-AMINODIPHENYL
mf: $C_{13}H_{13}NO$ mw: 199.27

SYN: 3-METHOXYBIPHENYLAMINE

TOXICITY DATA with REFERENCE
scu-rat TDLo:4400 mg/kg/W-I:ETA BMBUAQ 14,141,58

SAFETY PROFILE: Questionable carcinogen with experimental tumorigenic data. When heated to decomposition it emits toxic fumes of NO_x.

MEA250 CAS:64011-44-5 ***HR: 2***
4-METHOXY-4-AMINO-2-PENTANOL
mf: $C_6H_{15}NO_2$ mw: 133.22

TOXICITY DATA with REFERENCE
skn-rbt 500 mg SEV SCCUR* -,6,61
orl-rat LDLo:750 mg/kg SCCUR* -,6,61
orl-mus LD50:1290 mg/kg SCCUR* -,6,61
ihl-mus LCLo:835 ppm/8H SCCUR* -,6,61
skn-rbt LDLo:1380 mg/kg SCCUR* -,6,61

SAFETY PROFILE: Moderately toxic by ingestion and skin contact. Mildly toxic by inhalation. A severe skin irritant. When heated to decomposition it emits toxic fumes of NO_x.

MEA500 CAS:52740-56-4 ***HR: 3***
dl,4-METHOXYAMPHETAMINE HYDROCHLORIDE
mf: $C_{10}H_{15}NO \cdot ClH$ mw: 201.72

SYNS: 4-METHOXYAMPHETAMINE HYDROCHLORIDE ◇ dl-p-METHOXY-α-METHYL-PHENETHYLAMINE HYDROCHLORIDE

TOXICITY DATA with REFERENCE
ipr-rat LD50:46 mg/kg TXAPA9 45,49,78
orl-mus LD50:284 mg/kg TXAPA9 45,49,78
ipr-mus LD50:40 mg/kg JMCMAR 8,100,65
ivn-mus LD50:49 mg/kg TXAPA9 45,49,78
ivn-dog LD50:7 mg/kg TXAPA9 45,49,78

SAFETY PROFILE: Poison by ingestion, intravenous, and intraperitoneal routes. When heated to decomposition it emits very toxic fumes of HCl and NO_x. See also BENZEDRINE.

MEA600 ***HR: 3***
2-METHOXYANILINIUM NITRATE
mf: $C_7H_{10}N_2O_4$ mw: 186.17

SYN: 2-ANISIDINE NITRATE

SAFETY PROFILE: The pure nitrate decomposes exothermically at 146°C. The crude material may decompose as low as 46°C. May ignite with friction. Reacts exothermically with sulfuric acid. When heated to decomposition it emits toxic fumes of NO_x. See also NITRATES.

MEA750 CAS:67293-86-1 ***HR: 3***
METHOXYAZOXYMETHANOLACETATE
mf: $C_4H_8N_2O_4$ mw: 148.14

TOXICITY DATA with REFERENCE
sln-dmg-orl 47 ng PSEBAA 125,988,67
par-rat TDLo:1800 μg/kg/9W-I:ETA PAACA3 14,55,73

SAFETY PROFILE: Questionable carcinogen with experimental tumorigenic data. Mutation data reported. When heated to decomposition it emits toxic fumes of NO_x.

MEA800 CAS:65654-08-2 ***HR: D***
2-(p-METHOXYBENZAMIDO)ACETOHYDROXAMIC ACID
mf: $C_{10}H_{12}N_2O_4$ mw: 224.24

TOXICITY DATA with REFERENCE
mmo-sat 1 μmol/plate JOPHDQ 3,557,80
mma-sat 1 μmol/plate JOPHDQ 3,557,80
dnr-bcs 10 μmol/disc JOPHDQ 3,557,80

SAFETY PROFILE: Mutation data reported. When heated to decomposition it emits toxic fumes of NO_x.

MEB000 CAS:56183-20-1 ***HR: 3***
3-METHOXY-1,2-BENZANTHRACENE
mf: $C_{19}H_{14}O$ mw: 258.33

SYN: 5-METHOXY-BENZ(a)ANTHRACENE

TOXICITY DATA with REFERENCE
scu-mus TDLo:400 mg/kg:ETA JNCIAM 1,303,40

SAFETY PROFILE: Questionable carcinogen with experimental tumorigenic data. When heated to decomposition it emits acrid smoke and irritating fumes.

MEB250 CAS:63019-69-2 ***HR: 3***
5-METHOXY-1,2-BENZANTHRACENE
mf: $C_{19}H_{14}O$ mw: 258.33

SYNS: 5-METHOXY-1,2-BENZ(a)ANTHRACENE ◇ 8-METHOXY-BENZ(a)ANTHRACENE

TOXICITY DATA with REFERENCE
scu-mus TDLo:60 mg/kg:ETA BJCAAI 9,457,55

SAFETY PROFILE: Questionable carcinogen with experimental tumorigenic data. When heated to decomposition it emits acrid smoke and irritating fumes.

MEB500 CAS:6366-20-7 ***HR: 3***
10-METHOXY-1,2-BENZANTHRACENE
mf: $C_{19}H_{14}O$ mw: 258.33

SYN: 7-METHOXY-BENZ(a)ANTHRACENE

TOXICITY DATA with REFERENCE
scu-mus TDLo:600 mg/kg:ETA JNCIAM 1,303,40

SAFETY PROFILE: Questionable carcinogen with experimental tumorigenic data. When heated to decomposition it emits acrid smoke and irritating fumes.

MEB750 CAS:3688-79-7 ***HR: 2***
3-METHOXYBENZANTHRONE
mf: $C_{18}H_{12}O_2$ mw: 260.30

SYNS: ACETATE YELLOW 6G ◇ CELLITON BRILLIANT YELLOW 8G ◇ C.I. 58900 ◇ C.I. DISPERSE YELLOW 13 ◇ DISPERSE YELLOW 6Z ◇ DURANOL BRILLIANT YELLOW G ◇ 3-METHOXY-7H-BENZ(de)ANTHRACEN-7-ONE

TOXICITY DATA with REFERENCE
mma-sat 100 μg/plate MUREAV 66,9,79
ipr-rat LD50:1740 mg/kg RPTOAN 40,137,77
ipr-mus LD50:1200 mg/kg RPTOAN 40,137,77

CONSENSUS REPORTS: Reported in EPA TSCA Inventory. EPA Genetic Toxicology Program.

SAFETY PROFILE: Moderately toxic by intraperitoneal route. When heated to decomposition it emits acrid smoke and irritating fumes.

MEC250 CAS:3811-49-2 ***HR: 3***
2-METHOXY-4H-1,2,3-BENZODIOXAPHOSPHORINE-2-SULFIDE
mf: $C_8H_9O_3PS$ mw: 216.20

SYNS: K-9 ◇ PHOSPHOROTHIOIC ACID, CYCLIC O,O-(METHYLENE-O-PHENYLENE) O-METHYL ESTER ◇ SALITHION ◇ SALITHION-SUMITOMO

TOXICITY DATA with REFERENCE
mmo-sat 500 μg/plate MUREAV 116,185,83
mma-sat 500 μg/plate MUREAV 116,185,83
orl-rat LD50:102 mg/kg PSTDAN 15,3,81
orl-mus LD50:91 mg/kg 30ZDA9 -,336,71
scu-mus LD50:81600 μg/kg FMCHA2 -,C210,83

SAFETY PROFILE: Poison by ingestion and subcutaneous routes. Mutation data reported. An insecticide. When heated to decomposition it emits very toxic fumes of SO_x and PO_x. See also PARATHION.

MEC500 CAS:52351-96-9 ***HR: 3***
6-METHOXYBENZO(a)PYRENE
mf: $C_{21}H_{14}O$ mw: 282.35

TOXICITY DATA with REFERENCE
mma-sat 18 nmol/plate BBRCA9 85,351,78
skn-mus TDLo:180 mg/kg/40W-I:ETA CBINA8 22(1),53,78

SAFETY PROFILE: Questionable carcinogen with experimental tumorigenic data. Mutation data reported. When heated to decomposition it emits acrid smoke and irritating fumes.

MED000 CAS:38860-48-9 ***HR: 3***
N-(4-METHOXY)BENZOYLOXYPIPERIDINE
mf: $C_{13}H_{17}NO_3$ mw: 235.31

TOXICITY DATA with REFERENCE
skn-mus TDLo:19 mg/kg:NEO JNCIAM 54,491,75

SAFETY PROFILE: Questionable carcinogen with experimental neoplastigenic data. When heated to decomposition it emits toxic fumes of NO_x.

MED250 CAS:63059-68-7 ***HR: 3***
8-METHOXY-3,4-BENZPYRENE
mf: $C_{21}H_{14}O$ mw: 282.35

TOXICITY DATA with REFERENCE
scu-rat TDLo:16 mg/kg/13W-I:ETA BJCAAI 6,400,52

SAFETY PROFILE: Questionable carcinogen with experimental tumorigenic data. When heated to decomposition it emits acrid smoke and irritating fumes.

MED500 CAS:105-13-5 ***HR: 2***
p-METHOXYBENZYL ALCOHOL
mf: $C_8H_{10}O_2$ mw: 138.18

PROP: Needles or colorless liquid; floral odor. D: 1.113 @ 15°/15°, refr index: 1.543, mp: 25°, bp: 258.8°, flash p: +210°F. Insol in water; sol in alc and ether fixed oils; sltly sol glycerin.

SYNS: ANISE ALCOHOL ◇ ANISIC ALCOHOL ◇ p-ANISOL ALCOHOL ◇ ANISYL ALCOHOL (FCC) ◇ FEMA No. 2099 ◇ 4-METHOXYBENZENEMETHANOL ◇ 4-METHOXYBENZYL ALCOHOL

TOXICITY DATA with REFERENCE
skn-rbt 500 mg/24H MOD FCTXAV 12,825,74
orl-rat LD50:1200 mg/kg JPETAB 93,26,48
orl-mus LD50:1600 mg/kg JPETAB 93,26,48

CONSENSUS REPORTS: Reported in EPA TSCA Inventory.

SAFETY PROFILE: Moderately toxic by ingestion. A skin irritant. Combustible liquid. When heated to decomposition it emits acrid smoke and irritating fumes. See also ALCOHOLS.

MED750 CAS:14617-95-9 ***HR: 2***
p-METHOXYBENZYL BUTYRATE
mf: $C_{12}H_{16}O_3$ mw: 208.2

SYN: ANISYL-N-BUTYRATE

TOXICITY DATA with REFERENCE
skn-rbt 500 mg MLD FCTXAV 14,659,76
orl-rat LD50:3400 mg/kg FCTXAV 14,683,76

SAFETY PROFILE: Moderately toxic by ingestion. A skin irritant. When heated to decomposition it emits acrid smoke and irritating fumes.

MEE000 ***HR: 3***
p-METHOXYBENZYL CHLORIDE
mf: C_8H_9ClO mw: 156.61

SAFETY PROFILE: Unstable. Has exploded while being stored. When heated to decomposition it emits toxic fumes of Cl^-.

MEF400 CAS:66839-98-3 ***HR: 3***
1-((4-METHOXY(1,1'-BIPHENYL)-3-YL)METHYL) PYRROLIDINE
mf: $C_{18}H_{21}NO$ mw: 267.40

SYNS: PYRROLIDINE, 1-((4-METHOXY(1,1'-BIPHENYL)-3-YL) METHYL)- ◇ 3-PYRROLIDINO-N-METHYL-4-METHOXYBIPHENYL

TOXICITY DATA with REFERENCE
orl-rat TDLo:35 mg/kg (female 1-7D post):REP IJMRAQ 67,392,78
orl-rat LD50:28 mg/kg IJMRAQ 67,392,78

SAFETY PROFILE: Poison by ingestion. Experimental reproductive effects. When heated to decomposition it emits toxic fumes of NO_x.

MEF500 CAS:10024-70-1 ***HR: 2***
3-METHOXY BUTANOIC ACID
mf: $C_5H_{10}O_3$ mw: 118.15

PROP: Liquid. Mp: 12°, d: 1.053 @ 20°/20°, bp: 139° @ 50 mm.

SYN: 3-METHOXYBUTYRIC ACID

TOXICITY DATA with REFERENCE
skn-rbt 10 mg/24H open MLD AMIHBC 10,61,54
eye-rbt 750 μg open SEV AMIHBC 10,61,54
orl-rat LD50:3030 mg/kg AMIHBC 10,61,54

SAFETY PROFILE: Moderately toxic by ingestion. A skin and severe eye irritant. When heated to decomposition it emits acrid smoke and irritating fumes.

MEF750 CAS:5281-76-5 ***HR: 3***
3-METHOXY BUTYRALDEHYDE
mf: $C_5H_{10}O_2$ mw: 102.15

PROP: Flash p: 140°F.

TOXICITY DATA with REFERENCE
orl-rat LD50:540 mg/kg AIHAAP 23,95,62
ihl-rat LCLo:1000 ppm/4H AIHAAP 23,95,62
skn-rbt LD50:310 mg/kg AIHAAP 23,95,62

SAFETY PROFILE: Poison by skin contact. Moderately toxic by ingestion. Mildly toxic by inhalation. Flammable when exposed to heat, open flame, or oxidizers. To fight fire use foam, water spray, mist, CO_2, dry chemical. When heated to decomposition it emits acrid smoke and irritating fumes. See also ALDEHYDES.

MEG000 CAS:55936-76-0 ***HR: D***
4'-METHOXYCARBONYL-N-ACETOXY-N-METHYL-4-AMINOAZOBENZENE
mf: $C_{17}H_{17}N_3O_4$ mw: 327.37

SYNS: p-((p-(ACETOXYMETHYLAMINO)PHENYL)AZO)-BENZOIC ACID METHYL ESTER ◇ 4-((4-((ACETYLOXY)METHYLAMINO)PHENYL)AZO)BENZOIC ACID METHYL ESTER (9CI)

TOXICITY DATA with REFERENCE
mmo-sat 100 nmol/plate CALEDQ 1,91,75
mma-sat 100 nmol/plate CALEDQ 1,91,75

CONSENSUS REPORTS: EPA Genetic Toxicology Program.

SAFETY PROFILE: Mutation data reported. When heated to decomposition it emits toxic fumes of NO_x.

MEG250 CAS:13684-63-4 ***HR: 2***
3-METHOXYCARBONYLAMINOPHENYL-N-3'-METHYLPHENYLCARBAMATE
mf: $C_{16}H_{16}N_2O_4$ mw: 300.34

SYNS: BETANAL ◇ CARBAMIC ACID, (3-METHYLPHENYL)-3-((METHOXYCARBONYL)AMINO)PHENYL ESTER (9CI) ◇ EP-452 ◇ m-HYDROXYCARBANILIC ACID METHYL ESTER m-METHYLCARBANILATE ◇ 3-METHOXYCARBONYL-N-(3'-METHYLPHENYL)-CARBAMAT (GERMAN) ◇ METHYL 3-(m-TOLYLCARBAMOYLOXY) PHENYLCARBAMATE ◇ PHENMEDIPHAM ◇ SCHERING-38584

TOXICITY DATA with REFERENCE
sln-asn 40 mg/L EVHPAZ 31,81,79
cyt-mus-unr 100 mg/kg TGANAK 14(6),41,80
cyt-mus-orl 100 mg/kg CYGEDX 14(6),38,80
orl-rat TDLo:880 mg/kg (9-22D preg):TER GISAAA 49(4),16,84
orl-rat TDLo:2 g/kg (9D preg):REP GISAAA 49(4),16,84
orl-rat LD50:4 g/kg GISAAA 49(4),16,84

CONSENSUS REPORTS: EPA Genetic Toxicology Program.

SAFETY PROFILE: Moderately toxic by ingestion route. An experimental teratogen. Experimental reproductive effects. Mutation data reported. An herbicide. When heated to decomposition it emits toxic fumes of NO_x. See also CARBAMATES.

MEG500 CAS:55936-75-9 ***HR: D***
4'-METHOXYCARBONYL-N-BENZOYLOXY-N-METHYL-4-AMINOAZOBENZENE
mf: $C_{22}H_{19}N_3O_4$ mw: 389.44

SYN: p-((p-BENZOYLOXYMETHYLAMINO)PHENYL)AZO)-BENZOIC ACID METHYL ESTER

TOXICITY DATA with REFERENCE
mmo-sat 100 nmol/plate CALEDQ 1,91,75
mma-sat 100 nmol/plate CALEDQ 1,91,75

SAFETY PROFILE: Mutation data reported. When heated to decomposition it emits toxic fumes of NO_x. See also ESTERS.

MEG750 CAS:79448-03-6 ***HR: D***
N-(2-METHOXYCARBONYLETHYL)-N-(1-ACETOXYBUTYL)NITROSAMINE
mf: $C_{10}H_{18}N_2O_5$ mw: 246.2

SYNS: N-(4-HYDROXYBUTYL)-N-NITROSO-β-ALANINE, METHYL ESTER, ACETATE ◇ MCEABN

TOXICITY DATA with REFERENCE
mmo-sat 1 μmol/plate GANNA2 71,124,80
mmo-esc 1 μmol/plate GANNA2 71,124,80
mrc-bcs 500 nmol/plate GANNA2 71,124,80

SAFETY PROFILE: Mutation data reported. Many N-nitroso compounds are carcinogens. When heated to decomposition it emits toxic fumes of NO_x. See also NITROSAMINES and ESTERS.

MEH000 CAS:70103-81-0 ***HR: D***
N-(2-METHOXYCARBONYLETHYL)-N-(ACETOXYMETHYL)NITROSAMINE
mf: $C_7H_{12}N_2O_5$ mw: 204.15

SYNS: N-(HYDROXYMETHYL)-N-NITROSO-β-ALANINE METHYL ESTER, ACETATE ◇ MCEAMN

TOXICITY DATA with REFERENCE
mmo-sat 1 μmol/plate GANNA2 71,124,80

mmo-esc 1 μmol/plate GANNA2 71,124,80
mrc-bcs 100 nmol/plate GANNA2 71,124,80

SAFETY PROFILE: Mutation data reported. Many N-nitroso compounds are carcinogens. When heated to decomposition it emits toxic fumes of NO_x. See also ESTERS and NITROSAMINES.

MEH250 CAS:55936-78-2 ***HR: D***
4'-METHOXYCARBONYL-N-HYDROXY-N-METHYL-4-AMINOAZOBENZENE
mf: $C_{15}H_{15}N_3O_3$ mw: 285.33

TOXICITY DATA with REFERENCE
mmo-sat 100 nmol/plate CALEDQ 1,91,75
mma-sat 100 nmol/plate GALEDQ 1,91,75

SAFETY PROFILE: Mutation data reported. When heated to decomposition it emits toxic fumes of NO_x.

MEH500 ***HR: D***
N-(METHOXYCARBONYLMETHYL)-N-(1-ACETOXYBUTYL)NITROSAMINE
mf: $C_9H_{16}N_2O_5$ mw: 232.2

TOXICITY DATA with REFERENCE
mmo-sat-1 μmol/plate GANNA2 71,124,80
mmo-esc-1 μmol/plate GANNA2 71,124,80
mrc-bcs-500 μmol/plate GANNA2 71,124,80

SAFETY PROFILE: Mutation data reported. Many N-nitroso compounds are carcinogens. When heated to decomposition it emits toxic fumes of NO_x. See also NITROSAMINES.

MEH750 CAS:70103-80-9 ***HR: D***
N-(METHOXYCARBONYLMETHYL)-N-(ACETOXYMETHYL)NITROSAMINE
mf: $C_6H_{10}N_2O_5$ mw: 190.2

TOXICITY DATA with REFERENCE
mmo-sat 1 μmol/plate GANNA2 71,124,80
mmo-esc 1 μmol/plate GANNA2 71,124,80
mrc-bcs 1 μmol/plate GANNA2 71,124,80

SAFETY PROFILE: Mutation data reported. Many N-nitroso compounds are carcinogens. When heated to decomposition it emits toxic fumes of NO_x. See also NITROSAMINES.

MEH775 CAS:62861-57-8 ***HR: 3***
3-METHOXYCARBONYL PROPEN-2-YL TRIFLUOROMETHANE SULFONATE
mf: $C_6H_7F_3O_5S$ mw: 248.17

$CH_3OCO{\cdot}CH{=}CHCH_2OSO_2CF_3$

SAFETY PROFILE: May explode at room temperature. Reacts violently with aprotic solvents (e.g., DMF, DMSO). When heated to decomposition it emits toxic fumes of F^- and SO_x. See also SULFONATES and FLUORIDES.

MEI000 CAS:100700-29-6 ***HR: 3***
N-(3-METHOXYCARBONYLPROPYL)-N-(1-ACETOXYBUTYL)NITROSAMINE
mf: $C_{11}H_{20}N_2O_5$ mw: 260.2

SYNS: CMPABN ◇ N-(3-CARBOMETHOXYPROPYL)-N-(1-ACETOXYBUTYL)NITROSAMINE ◇ 4-((4-HYDROXYBUTYL) NITROSAMINO)BUTYRIC ACID METHYL ESTER ACETATE (ESTER)

TOXICITY DATA with REFERENCE
mmo-sat 1 μmol/plate GANNA2 71,124,80
mmo-esc 1 μmol/plate GANNA2 71,124,80
mrc-bcs 1 μmol/plate GANNA2 71,124,80
scu-rat TDLo:99 mg/kg 10W-I:CAR IAPUDO 41,619,82
scu-rat TD:70500 μg/kg/10W-I:ETA GANNA2 73,687,82

SAFETY PROFILE: Questionable carcinogen with experimental carcinogenic and tumorigenic data. Mutation data reported. When heated to decomposition it emits toxic fumes of NO_x. See also NITROSAMINES.

MEI250 CAS:70103-82-1 ***HR: D***
N-(3-METHOXYCARBONYLPROPYL)-N-(ACETOXYMETHYL)NITROSAMINE
mf: $C_8H_{14}N_2O_5$ mw: 218.2

SYNS: ((HYDROXYMETHYL)NITROSAMINO)-BUTYRICACID METHYL ESTER, ACETATE ◇ MCPAMN

TOXICITY DATA with REFERENCE
mmo-sat 1 μmol/plate GANNA2 71,124,80
mmo-esc 1 μmol/plate GANNA2 71,124,80
mrc-bcs 100 nmol/plate GANNA2 71,124,80

SAFETY PROFILE: Mutation data reported. Many N-nitroso compounds are carcinogens. When heated to decomposition it emits toxic fumes of NO_x. See also NITROSAMINES.

MEI450 CAS:72-43-5 ***HR: 3***
METHOXYCHLOR
mf: $C_{16}H_{15}Cl_3O_2$ mw: 345.66

PROP: Crystals. Mp: 78°, vap d: 12.

SYNS: 2,2-BIS(p-ANISYL)-1,1,1-TRICHLOROETHANE ◇ 1,1-BIS(p-METHOXYPHENYL)-2,2,2-TRICHLOROETHANE ◇ 2,2-BIS(p-METHOXYPHENYL)-1,1,1-TRICHLOROETHANE ◇ CHEMFORM ◇ DIANISYLTRICHLORETHANE ◇ 2,2-DI-p-ANISYL-1,1,1-TRICHLOROETHANE ◇ p,p'-DIMETHOXYDIPHENYLTRICHLOROETHANE ◇ DIMETHOXY-DT ◇ DIMETHOXY-DDT ◇ 2,2-DI-(p-METHOXYPHENYL)-1,1,1-TRICHLOROETHANE ◇ DI(p-METHOXYPHENYL)-TRICHLOROMETHYL METHANE ◇ DMDT ◇ p,p'-DMDT ◇ ENT 1,716 ◇ MARALATE ◇ MARLATE ◇ METHOXCIDE ◇ METHOXO ◇ p,p'-METHOXYCHLOR ◇ METHOXY-DDT ◇ METOKSYCHLOR (POLISH) ◇ METOX ◇ MOXIE ◇ NCI-C00497 ◇ RCRA WASTE NUMBER U247 ◇ 1,1,1-TRICHLOR-2,2-BIS(4-METHOXY-PHENYL)-AETHAN (GERMAN) ◇ 1,1,1-TRICHLORO-2,2-BIS (p-ANISYL)ETHANE ◇ 1,1,1-TRICHLORO-2,2-BIS(p-METHOXY-PHENOL)ETHANOL ◇ 1,1,1-TRICHLORO-2,2-BIS(p-METHOXY-PHENYL)ETHANE ◇ 1,1,1-TRICHLORO-2,2-DI(4-METHOXYPHENYL) ETHANE ◇ 1,1'-(2,2,2-TRICHLOROETHYLIDENE)BIS(4-METHOXY-BENZENE)

TOXICITY DATA with REFERENCE
spm-rat-orl 28 g/kg/10W-C PSEBAA 176,187,84
cyt-ham-ipr 50 mg/kg ARTODN 58,152,85
orl-rat TDLo:4250 mg/kg (female 42D pre-21D post):REP JAFCAU 22,969,74
orl-rat TDLo:2 g/kg (6-15D preg):TER TXAPA9 45,435,78
orl-rat TDLo:18200 mg/kg/2Y-C:CAR,TER EVHPAZ 36,205,80
orl-mus TDLo:56700 mg/kg/90W-C:CAR,TER JCROD7 93,173,79
orl-dog TDLo:383 g/kg/3Y-C:ETA EVHPAZ 36,205,80
orl-rat TD:80 g/kg/2Y-C:CAR,TER LIFSAK 24,1367,79
orl-rat TD:72800 mg/kg/2Y-C:CAR EVHPAZ 36,205,80
orl-rat TD:87360 mg/kg/2Y-C:CAR EVHPAZ 36,205,80
orl-hmn LDLo:6430 mg/kg PCOC** -,705,66
skn-hmn TDLo:2414 mg/kg:CNS PCOC** -,705,66
orl-rat LD50:5000 mg/kg JPETAB 99,140,50
orl-mus LD50:1 g/kg JAFCAU 25,859,77
ipr-ham LD50:500 mg/kg ARTODN 58,152,85

CONSENSUS REPORTS: IARC Cancer Review: Group 3 IMEMDT 7,56,87; Animal No Evidence IMEMDT 20,259,79; Animal Inadequate Evidence IMEMDT 5,193,74. NCI Carcinogenesis Bioassay (feed); No Evidence: mouse, rat NCITR* NCI-CG-TR-35,78. EPA Genetic Toxicology Program. Community Right-To-Know List.

OSHA PEL: (Transitional: TWA Total Dust: 10 mg/m^3; Respirable Fraction: 5 mg/m^3 TWA Total Dust: 10 mg/m^3; 5 mg/m^3
ACGIH TLV: TWA 10 mg/m^3
DFG MAK: 15 mg/m^3

SAFETY PROFILE: Suspected carcinogen with experimental carcinogenic, tumorigenic, and teratogenic data. Moderately toxic by intraperitoneal and skin contact routes. Human systemic effects by skin contact: somnolence. Experimental reproductive effects. Mutation data reported. When heated to decomposition emits highly toxic fumes of Cl^-. See also DDT and CHLOROPHENOLS.

MEI500 CAS:59928-80-2 ***HR: 2***
METHOXYCHLOR mixed with DIAZINON

PROP: Contains 0.8 lb diazinon and 1.6 lb methoxychlor per gallon of formulation (FMCHA2 -,D10,77).

SYNS: ALFA-TOX ◇ DIAZINON mixed with METHOXYCHLOR

TOXICITY DATA with REFERENCE
skn-rbt 150 mg open MOD CIGET* -,-,77
eye-rbt 30 mg MOD CIGET* -,-,77
orl-rat LD50:2000 mg/kg FMCHA2 -,C9,83
skn-rbt LD50:8 g/kg CIGET* -,-,77

SAFETY PROFILE: Moderately toxic by ingestion. A skin and eye irritant. An insecticide. When heated to decomposition it emits very toxic fumes of Cl^- and NO_x. See also individual components.

MEJ250 CAS:17070-44-9 ***HR: D***
2-METHOXY-6-CHLORO-9-(3-(2-CHLOROETHYL)AMINOPROPYLAMINO) ACRIDINE DIHYDROCHLORIDE

SYNS: N-(2-CHLOROETHYL)-N'-(6-CHLORO-2-METHOXY-9-ACRIDINYL)-1,3-PROPANEDIAMINE DIHYDROCHLORIDE ◇ ICR-191 ◇ ICR DIHYDROCHLORIDE

TOXICITY DATA with REFERENCE
msc-hmn:lym 4 mg/L/24H IMNGBK 4,437,77
dni-hmi 250 μmol/L MUREAV 80,249,81

SAFETY PROFILE: Human mutation data reported. When heated to decomposition it emits very toxic fumes of Cl^- and NO_x.

MEJ500 CAS:61413-39-6 ***HR: 3***
5-METHOXYCHRYSENE
mf: $C_{19}H_{14}O$ mw: 258.33

TOXICITY DATA with REFERENCE
skn-mus TDLo:1200 μg/kg/I:NEO CALEDQ 1,147,76
skn-mus TD:40 mg/kg/3W-I:ETA CCSUDL 1,325,76

SAFETY PROFILE: Questionable carcinogen with experimental neoplastigenic and tumorigenic data by skin contact. When heated to decomposition it emits acrid smoke and irritating fumes.

MEJ750 CAS:1504-74-1 ***HR: 1***
o-METHOXY CINNAMALDEHYDE
mf: $C_{10}H_{10}O_2$ mw: 162.20

SYNS: o-METHOXYCINNAMIC ALDEHYDE ◇ β-(o-METHOXY-PHENYL)ACROLEIN ◇ 3-(2-METHOXYPHENYL)-2-PROPENAL

TOXICITY DATA with REFERENCE
skn-rbt 500 mg/24H MLD FCTXAV 13,681,75
mma-sat 333 μg/plate ENMUDM 8(Suppl 7),1,86

CONSENSUS REPORTS: Reported in EPA TSCA Inventory.

SAFETY PROFILE: A skin irritant. Mutation data reported. When heated to decomposition it emits acrid smoke and irritating fumes. See also ALDEHYDES.

MEJ775 CAS:14737-91-8 ***HR: 2***
cis-2-METHOXYCINNAMIC ACID
mf: $C_{10}H_{10}O_3$ mw: 178.20

SYNS: ACIDE o-METHOXYCINNAMIQUE (FRENCH) ◇ cis-o-METHOXYCINNAMIC ACID ◇ (Z)-3-(2-METHOXYPHENYL)-2-PROPENOIC ACID (9CI) ◇ SUBSTANCE H 36

TOXICITY DATA with REFERENCE
orl-mus LD50:1750 mg/kg AIPTAK 119,443,59
ipr-mus LD50:1500 mg/kg AIPTAK 119,443,59
scu-mus LD50:1100 mg/kg AIPTAK 119,443,59

SAFETY PROFILE: Moderately toxic by ingestion, intraperitoneal, and subcutaneous routes. When heated to decomposition it emits acrid smoke and irritating fumes.

MEK250 CAS:93-51-6 ***HR: 2***
2-METHOXY-p-CRESOL
mf: $C_8H_{10}O_2$ mw: 138.18

PROP: Colorless to yellow liquid. D: 1.092 @ 25°/4°, mp: 5.5°, bp: 220°. Sltly sol in water; sol in alc, benzene, chloroform, ether and acetic acid.

SYNS: p-CREOSOL ◇ HOMOGUAIACOL ◇ 4-HYDROXY-3-METHOXY-1-METHYLBENZENE ◇ 4-HYDROXY-3-METHOXYTOLUENE ◇ 3-METHOXY-4-HYDROXYTOLUENE ◇ 2-METHOXY-4-METHYLPHENOL ◇ p-METHYLGUAIACOL ◇ 4-METHYLGUAIACOL

CONSENSUS REPORTS: Reported in EPA TSCA Inventory.

SAFETY PROFILE: Appears to be at least moderately toxic. An irritant to skin and eyes. When heated to decomposition it emits acrid smoke and irritating fumes.

MEK350 ***HR: 3***
β-(2-METHOXY-5-CYCLOHEXYLBENZOYL)PROPIONIC ACID SODIUM SALT
mf: $C_{17}H_{21}O_4$•Na mw: 312.33

SYNS: 3-(5-CYCLOHEXYL-o-ANISOYL)-PROPIONIC ACID SODIUM SALT ◇ SC-2292

TOXICITY DATA with REFERENCE
orl-mus LD50:1180 mg/kg JPETAB 100,421,50
ipr-mus LD50:185 mg/kg JPETAB 100,421,50
ivn-dog LDLo:23 mg/kg JPETAB 100,421,50

SAFETY PROFILE: Poison by intravenous and intraperitoneal routes. Moderately toxic by ingestion. When heated to decomposition it emits toxic fumes of Na_2O.

MEK700 CAS:865-04-3 ***HR: 3***
10-METHOXY-11-DESMETHOXYRESERPINE
mf: $C_{33}H_{40}N_2O_9$ mw: 608.75

SYNS: CANESCINE 10-METHOXYDERIVATIVE ◇ DEASERPYL ◇ DECASERPIL ◇ DECASERPINE ◇ DECASERPYL ◇ DECASERPYL PLUS ◇ DECOSERPYL ◇ DESERPIDINE, 10-METHOXY- ◇ 10-MD ◇ METHOSERPEDINE ◇ METHOSERPIDINE ◇ 10-METHOXYDESERPIDINE ◇ MINORAN ◇ R 694 ◇ 3-β,20-α-YOHIMBAN-16-β-CARBOXYLIC ACID, 18-β-HYDROXY-10,17-α-DIMETHOXY-, METHYL ESTER, 3,4,5-TRIMETHOXYBENZOATE (ester)

TOXICITY DATA with REFERENCE
orl-wmn TDLo:438 mg/kg/3Y-C:NEO LANCAO 2,672,74
ipr-mus LD50:82 mg/kg JPPMAB 12,677,60
ivn-mus LD50:8750 μg/kg ARZNAD 14,1040,64
ivn-gpg LDLo:52 mg/kg THERAP 15,663,60

SAFETY PROFILE: Poison by intraperitoneal and intravenous routes. Questionable human carcinogen producing colon, nose, and skin tumors. When heated to decomposition it emits toxic fumes of NO_x.

MEK750 CAS:63019-72-7 ***HR: 3***
5-METHOXYDIBENZ(a,h)ANTHRACENE
mf: $C_{23}H_{16}O$ mw: 308.39

SYNS: 3-METHOXY-DBA ◇ 3-METHOXY-1,2:5,6-DIBENZANTHRACENE

TOXICITY DATA with REFERENCE
skn-mus TDLo:300 mg/kg/30W-I:ETA CNREA8 22,78,62

SAFETY PROFILE: Questionable carcinogen with experimental tumorigenic data. When heated to decomposition it emits acrid smoke and irritating fumes.

MEL000 CAS:63041-72-5 ***HR: 3***
7-METHOXYDIBENZ(a,h)ANTHRACENE
mf: $C_{23}H_{16}O$ mw: 308.39

SYNS: 9-METHOXY-DBA ◇ 9-METHOXY-1,2,5,6-DIBENZANTHRACENE

TOXICITY DATA with REFERENCE
skn-mus TDLo:400 mg/kg/40W-I:NEO CNREA8 22,78,62
scu-mus TDLo:80 mg/kg:ETA CNREA8 22,78,62

SAFETY PROFILE: Questionable carcinogen with experimental neoplastigenic and tumorigenic data. When heated to decomposition it emits acrid smoke and irritating fumes.

MEL500 CAS:1918-00-9 ***HR: 2***
2-METHOXY-3,6-DICHLOROBENZOIC ACID
mf: $C_8H_6Cl_2O_3$ mw: 221.04

SYNS: ACIDO(3,6-DICLORO-2-METOSSI)-BENZOICO(ITALIAN) ◇ BANEX ◇ BANLEN ◇ BANVEL ◇ BANVEL HERBICIDE ◇ BRUSH BUSTER ◇ COMPOUND B DICAMBA ◇ DIANAT (RUSSIAN) ◇ DIANATE ◇ DICAMBA (DOT) ◇ 3,6-DICHLOOR-2-METHOXY-BENZOEIZUUR (DUTCH) ◇ 3,6-DICHLOR-3-METHOXY-BENZOESAEURE (GERMAN) ◇ 3,6-DICHLORO-o-ANISIC ACID ◇ 2,5-DICHLORO-6-METHOXYBENZOIC ACID ◇ 3,6-DICHLORO-2-METHOXYBENZOIC ACID ◇ MDBA ◇ MEDIBEN ◇ VELSICOL COMPOUND "R" ◇ VELSICOL 58-CS-11

TOXICITY DATA with REFERENCE
mma-sat 53500 nmol/L MUREAV 136,233,84
dnr-esc 5 mg/disc NTIS** PB80-133226
dnr-bcs 5 mg/disc NTIS** PB80-133226
cyt-mus-unr 500 mg/kg TGANAK 16(1),45,82
orl-rat LD50:1040 mg/kg RREVAH 10,97,65

orl-mus LD50:1190 mg/kg HYSAAV 35(7-9),14,70
orl-rbt LD50:2000 mg/kg HYSAAV 35,14,70
orl-gpg LD50:3000 mg/kg HYSAAV 35,14,70

CONSENSUS REPORTS: EPA Genetic Toxicology Program.

SAFETY PROFILE: Moderately toxic by ingestion. Mutation data reported. When heated to decomposition it emits toxic fumes of Cl^-.

MEL570 ***HR: 2***
2-METHOXY-7,12-DIMETHYLBENZ(a)ANTHRACENE
mf: $C_{21}H_{18}O$ mw: 286.39

SYN: 2-METHOXY-DMBA

TOXICITY DATA with REFERENCE
mma-sat 5 μg/plate CRNGDP 4,1221,83
skn-mus TDLo:440 mg/kg/50W-I:CAR CRNGDP 4,1221,83

SAFETY PROFILE: Questionable carcinogen with experimental carcinogenic data. Mutation data reported. When heated to decomposition it emits acrid smoke and irritating fumes.

MEL580 ***HR: 2***
3-METHOXY-7,12-DIMETHYLBENZ(a)ANTHRACENE
mf: $C_{21}H_{18}O$ mw: 286.39

SYN: 3-METHOXY-DMBA

TOXICITY DATA with REFERENCE
mma-sat 2500 ng/plate CRNGDP 4,1221,83
skn-mus TDLo:440 mg/kg/50W-I:ETA CRNGDP 4,1221,83

SAFETY PROFILE: Questionable carcinogen with experimental tumorigenic data. Mutation data reported. When heated to decomposition it emits acrid smoke and irritating fumes.

MEL600 ***HR: 2***
4-METHOXY-7,12-DIMETHYLBENZ(a)ANTHRACENE
mf: $C_{21}H_{18}O$ mw: 286.39

SYN: 4-METHOXY-DMBA

TOXICITY DATA with REFERENCE
mma-sat 5 μg/plate CRNGDP 4,1221,83
skn-mus TDLo:440 mg/kg/50W-I:CAR CRNGDP 4,1221,83

SAFETY PROFILE: Questionable carcinogen with experimental carcinogenic data. Mutation data reported. When heated to decomposition it emits acrid smoke and irritating fumes.

MEL775 CAS:10371-86-5 ***HR: 3***
METHOXYELLIPTICINE
mf: $C_{18}H_{16}N_2O$ mw: 276.36

SYNS: ICIG 772 ◇ 9-METHOXY-5,11-DIMETHYL-6H-PYRIDO(4,3-b)CARBAZOLE ◇ 9-METHOXY-ELLIPTICINE ◇ 9-METHOXYELLIPTICIN ◇ 9-METHOXYELLIPTICINE ◇ METHOXYELLIPTIONE ◇ NSC 69187

TOXICITY DATA with REFERENCE
mmo-sat 500 ng/plate CNREA8 43,3544,83
mma-sat 500 ng/plate CNREA8 43,3544,83
msc-ham:ovr 100 μg/L CNREA8 43,3544,83
ipr-mus LD50:150 mg/kg BIMDB3 21,101,74

SAFETY PROFILE: Poison by intraperitoneal route. Mutation data reported. When heated to decomposition it emits toxic fumes of NO_x.

MEM250 CAS:3121-61-7 ***HR: 3***
METHOXYETHYL ACRYLATE
mf: $C_6H_{10}O_4$ mw: 146.16

PROP: Liquid. Bp: 61° @ 17 mm, flash p: 180°F (OC), d: 1.0134 @ 20°, vap d: 4.49.

SYNS: ACRYLIC ACID 2-METHOXYETHOXY ESTER ◇ 2-METHOXYETHOXY ACRYLATE ◇ 2-METHOXYETHYL ACRYLATE ◇ 2-PROPENOIC ACID-2-METHOXYETHYL ESTER

TOXICITY DATA with REFERENCE
orl-rat LD50:810 mg/kg AIHAAP 30,470,69
ihl-rat LCLo:500 ppm/4H AIHAAP 30,470,69
skn-rbt LD50:250 mg/kg AIHAAP 30,470,69

CONSENSUS REPORTS: Reported in EPA TSCA Inventory.

SAFETY PROFILE: Poison by skin contact. Moderately toxic by ingestion and inhalation. Flammable when exposed to heat, flame, or sparks. To fight fire, use foam, CO_2, dry chemical. When heated to decomposition it emits acrid smoke and irritating fumes. See also ESTERS.

MEM500 CAS:109-85-3 ***HR: 3***
2-METHOXYETHYLAMINE
mf: C_3H_9NO mw: 75.13

TOXICITY DATA with REFERENCE
ipr-mus LD50:400 mg/kg NTIS** AD691-490

CONSENSUS REPORTS: Reported in EPA TSCA Inventory.

SAFETY PROFILE: Poison by intraperitoneal route. When heated to decomposition it emits toxic fumes of NO_x. See also AMINES.

MEM750 CAS:67262-60-6 ***HR: 3***
2-(2-METHOXYETHYLAMINO)PROPIONANILIDE
mf: $C_{12}H_{18}N_2O_2$ mw: 222.32

TOXICITY DATA with REFERENCE
ipr-mus LD50:275 mg/kg JPMSAE 67,595,78
ivn-mus LD50:70 mg/kg JPMSAE 67,595,78

SAFETY PROFILE: Poison by intraperitoneal and intravenous routes. When heated to decomposition it emits toxic fumes of NO_x.

MEN000 CAS:67262-61-7 ***HR: 3***
2-(2-METHOXYETHYLAMINO)-o-PROPIONOTOLUIDIDE
mf: $C_{13}H_{20}N_2O_2$ mw: 236.35

SYN: 2-(2-METHOXYETHYLAMINO)-2'-METHYL-PROPIONANILIDE

TOXICITY DATA with REFERENCE
ipr-mus LD50:275 mg/kg JPMSAE 67,595,78
ivn-mus LD50:50 mg/kg JPMSAE 67,595,78

SAFETY PROFILE: Poison by intraperitoneal and intravenous routes. When heated to decomposition it emits toxic fumes of NO_x.

MEN250 CAS:67262-80-0 ***HR: 3***
2-(2-METHOXYETHYLAMINO)-2',6'-PROPIONOXYLIDIDE
mf: $C_{14}H_{22}N_2O_2$ mw: 250.38

SYN: 2',6'-DIMETHYL-2-(2-METHOXYETHYLAMINO)-PROPIONANILIDE

TOXICITY DATA with REFERENCE
ipr-mus LD50:100 mg/kg JPMSAE 67,595,78
ivn-mus LD50:22 mg/kg JPMSAE 67,595,78

SAFETY PROFILE: Poison by intraperitoneal and intravenous routes. When heated to decomposition it emits toxic fumes of NO_x.

MEN500 CAS:67262-82-2 ***HR: 3***
2-(2-METHOXYETHYLAMINO)-2',6'-PROPIONOXYLIDIDE PERCHLORATE
mf: $C_{15}H_{24}N_2O_2 \cdot ClHO_4$ mw: 364.87

SYN: 2',6-DIMETHYL-2-(2-METHOXYPROPYLAMINO)-PROPIONANILIDE PERCHLORATE

TOXICITY DATA with REFERENCE
ipr-mus LD50:130 mg/kg JPMSAE 67,595,78
ivn-mus LD50:25 mg/kg JPMSAE 67,595,78

SAFETY PROFILE: Poison by intraperitoneal and intravenous routes. When heated to decomposition it emits very toxic fumes of NO_x and Cl^-. See also PERCHLORATES.

MEN750 CAS:63020-60-0 ***HR: 3***
3-METHOXY-10-ETHYL-1,2-BENZANTHRACENE
mf: $C_{21}H_8O$ mw: 276.29

SYN: 7-ETHYL-5-METHOXY-BENZ(a)ANTHRACENE

TOXICITY DATA with REFERENCE
scu-mus TDLo:900 mg/kg:NEO JNCIAM 1,303,40

SAFETY PROFILE: Questionable carcinogen with experimental neoplastigenic data. When heated to decomposition it emits acrid smoke and irritating fumes.

MEN775 CAS:27807-62-1 ***HR: 3***
2-METHOXYETHYL-BIS(2-CHLOROETHYL) AMINE HYDROCHLORIDE
mf: $C_7H_{15}Cl_2NO \cdot ClH$ mw: 236.59

SYNS: BIS (β-CHLOROETHYL)-α-METHOXYETHYLAMINE HYDROCHLORIDE ◇ TL 783

TOXICITY DATA with REFERENCE
ipr-rat LD50:300 μg/kg CPBTAL 8,99,60
ipr-mus LD50:2410 μg/kg CANCAR 2,1055,49
scu-mus LD50:3 mg/kg NTIS** PB158-507

SAFETY PROFILE: Poison by subcutaneous and intraperitoneal routes. When heated to decomposition it emits toxic fumes of NO_x and HCl.

MEO000 CAS:1616-88-2 ***HR: 3***
METHOXYETHYL CARBAMATE
mf: $C_4H_9NO_3$ mw: 119.24

SYNS: 2-METHOXYETHYL ESTER CARBAMIC ACID ◇ N-METHOXYURETHANE

TOXICITY DATA with REFERENCE
skn-rbt 500 mg open MLD UCDS** 5/19/66
ipr-mus TDLo:6600 mg/kg/10W-I:ETA IJCNAW 4,318,69
orl-rat LD50:11 g/kg UCDS** 5/19/66

CONSENSUS REPORTS: Reported in EPA TSCA Inventory.

SAFETY PROFILE: Questionable carcinogen with experimental tumorigenic data. Mildly toxic by ingestion. A skin irritant. When heated to decomposition it emits toxic fumes of NO_x. See also CARBAMATES.

MEO500 CAS:61738-03-2 ***HR: 3***
1-METHOXY ETHYL ETHYLNITROSAMINE
mf: $C_5H_{12}N_2O_2$ mw: 132.19

SYN: 1-METHOXY-AETHYL-AETHYLNITROSAMIN (GERMAN)

TOXICITY DATA with REFERENCE
orl-rat TDLo:400 mg/kg/23W-I:CAR ZKKOBW 88,25,76
orl-rat TD:800 mg/kg/23W-I:CAR ZKKOBW 88,25,76
orl-rat LD50:1000 mg/kg ZKKOBW 88,25,76

SAFETY PROFILE: Moderately toxic by ingestion.

Questionable carcinogen with experimental carcinogenic data. When heated to decomposition it emits toxic fumes of NO_x. See also NITROSAMINES.

MEO750 CAS:151-38-2 ***HR: 3***
METHOXYETHYL MERCURIC ACETATE
mf: $C_5H_{10}HgO_3$ mw: 318.74

PROP: Crystals. Water-sol.

SYNS: ACETATO(2-METHOXYETHYL)MERCURY ◇ CEKUSIL UNIVERSAL A ◇ LANDISAN ◇ MEMA ◇ MERCURAN ◇ PANOGEN ◇ RADOSAN

TOXICITY DATA with REFERENCE
cyt-dmg-orl 15900 μg/L HEREAY 74,89,73
orl-rat LD50:25 mg/kg OCHRAI 15,5,63

CONSENSUS REPORTS: Mercury and its compounds are on the Community Right-To-Know List.

OSHA PEL: (Transitional: CL 1 mg/10m³) CL 0.1 mg(Hg)/m³ (skin)
ACGIH TLV: TWA 0.1 mg(Hg)/m³; STEL 0.03 mg(Hg)/m³
NIOSH REL: TWA 0.05 mg(Hg)/m³

SAFETY PROFILE: Poison by ingestion. Mutation data reported. A fungicide. When heated to decomposition it emits toxic fumes of Hg. See also MERCURY COMPOUNDS.

MEP000 CAS:19367-79-4 ***HR: 3***
METHOXYETHYL MERCURIC SILICATE
mf: $C_6H_{14}Hg_2O_5Si$ mw: 595.47

SYNS: CRESAN UNIVERSAL TROCKENBEIZE ◇ METHOXYAETHYLQUECKSILBERSILIKAT (GERMAN)

TOXICITY DATA with REFERENCE
orl-rat LD50:1140 mg/kg GUCHAZ 6,340,73
ipr-mus LD50:50 mg/kg OCHRAI 15,5,63

CONSENSUS REPORTS: Mercury and its compounds are on the Community Right-To-Know List.

OSHA PEL: (Transitional: CL 1 mg/10m³) CL 0.1 mg(Hg)/m³ (skin)
ACGIH TLV: TWA 0.1 mg(Hg)/m³ (skin)
NIOSH REL: TWA 0.05 mg(Hg)/m³

SAFETY PROFILE: Poison by intraperitoneal route. Moderately toxic by ingestion. When heated to decomposition it emits toxic fumes of Hg. See also MERCURY COMPOUNDS.

MEP250 CAS:123-88-6 ***HR: 3***
2-METHOXYETHYLMERCURY CHLORIDE
mf: C_3H_7ClHgO mw: 295.14

PROP: Crystals.

SYNS: AGALLOL ◇ AGALLOLAT ◇ ARATAN ◇ ATIRAN ◇ BAYTAN ◇ CEKUSIL UNIVERSAL C ◇ CELMER ◇ CERESAN UNIVERSAL NAZBEIZE ◇ CHLORO(2-METHOXYETHYL)MERCURY ◇ EMISAN 6 ◇ FALISAN ◇ GRAMISAN ◇ HIGOSAN ◇ MEMC ◇ MERCHLORATE ◇ METHOXYAETHYLQUECKSILBERCHLORID (GERMAN) ◇ (β-METHOXYETHYL)MERCURIC CHLORIDE ◇ METHOXYETHYL MERCURIC CHLORIDE ◇ 2-METHOXYETHYLMERCURIC CHLORIDE ◇ β-METHOXYETHYLMERCURY CHLORIDE ◇ METHOXYETHYLMERCURY CHLORIDE ◇ SEDRESAN ◇ TAFASAN ◇ TAYSSATAO ◇ TRIADIMENOL

TOXICITY DATA with REFERENCE
spm-mus-ipr 13 mg/kg BIZNAT 97,173,78
msc-ham:lng 100 ppb HEREAY 90,103,79
orl-rat TDLo:1800 μg/kg (female 7-15D post):REP NRTXDN 10,471,88
orl-wmn TDLo:114 mg/kg JTCTDW 19,391,82
orl-rat LD50:22 mg/kg FMCHA2 -,D192,80
orl-mus LD50:47 mg/kg 85JCAE -,1200,86
scu-mus LD50:88 mg/kg KUMJAX 14,65,61

CONSENSUS REPORTS: Mercury and its compounds are on the Community Right-To-Know List. EPA Genetic Toxicology Program.

OSHA PEL: (Transitional: CL 1 mg/10m³) TWA 0.01 mg(Hg)/m³; STEL 0.03 mg/m³ (skin)
ACGIH TLV: TWA 0.01 mg(Hg)/m³; STEL 0.03 mg(Hg)/m³
NIOSH REL: TWA 0.05 mg(Hg)/m³

SAFETY PROFILE: Poison by ingestion and subcutaneous routes. Experimental reproductive effects. Human mutation data reported. Used to control pineapple disease of sugarcane. When heated to decomposition it emits very toxic fumes of Hg and Cl^-. See also MERCURY COMPOUNDS and CHLORIDES.

MEP500 CAS:61738-05-4 ***HR: 3***
1-METHOXY ETHYL METHYLNITROSAMINE
mf: $C_4H_{10}N_2O_2$ mw: 118.16

SYN: 1-METHOXY-AETHYL-METHYLNITROSAMIN (GERMAN)

TOXICITY DATA with REFERENCE
orl-rat TDLo:920 mg/kg/60W-I:CAR ZKKOBW 88,25,76
orl-rat LD50:240 mg/kg ZKKOBW 88,25,76

SAFETY PROFILE: Poison by ingestion. Questionable carcinogen with experimental carcinogenic data. When heated to decomposition it emits toxic fumes of NO_x. See also NITROSAMINES.

MEP750 CAS:111-10-4 ***HR: 1***
METHOXYETHYL OLEATE
mf: $C_{21}H_{40}O_3$ mw: 340.61

PROP: Light-colored liquid. Mp: −20°, bp: 188-225° @ 4 mm, flash p: 386°F, d: 0.902 @ 20°/20°, vap press: 0.04 mm @ 150°, vap d: 11.8.

SYNS: ETHYLENE GLYCOL MONOMETHYL ETHER OLEATE ◇ OLEIC ACID, 2-METHOXYETHYL ESTER

TOXICITY DATA with REFERENCE
orl-rat LDLo:16000 mg/kg HBTXAC 1,188,56

CONSENSUS REPORTS: Glycol ether compounds are on the Community Right-To-Know List. Reported in EPA TSCA Inventory.

SAFETY PROFILE: Very mildly toxic by ingestion. Combustible when exposed to heat or flame; can react with oxidizing materials. To fight fire, use foam, CO_2, dry chemical. When heated to decomposition it emits acrid smoke and irritating fumes. See also ESTERS.

MEQ000 CAS:16501-01-2 ***HR: 3***
METHOXYETHYL PHTHALATE
mf: $C_{11}H_{12}O_5$ mw: 224.23

PROP: Liquid. Bp: 190-220°, flash p: 275°F (CC), d: 1.17, vap d: 9.75.

SYN: USAF KE-3

TOXICITY DATA with REFERENCE
ipr-mus LD50:300 mg/kg NTIS** AD277-689
orl-gpg LD50:1600 mg/kg 14CYAT 2,1904,63

CONSENSUS REPORTS: Reported in EPA TSCA Inventory.

SAFETY PROFILE: Poison by intraperitoneal route. Moderately toxic by ingestion. Combustible when exposed to heat or flame; can react with oxidizing materials. To fight fire, use CO_2, dry chemical. When heated to decomposition it emits acrid smoke and irritating fumes.

MEQ750 CAS:6893-24-9 ***HR: 3***
1-METHOXY-2-FLUORENAMINE HYDROCHLORIDE
mf: $C_{14}H_{13}NO{\bullet}ClH$ mw: 247.74

SYN: 1-METHOXYFLUOREN-2-AMINE HYDROCHLORIDE

TOXICITY DATA with REFERENCE
orl-rat TDLo:3370 mg/kg/23W-C:CAR CNREA8 28,234,68
ipr-rat TDLo:450 mg/kg/4W-I:ETA CNREA8 28,234,68

SAFETY PROFILE: Questionable carcinogen with experimental carcinogenic and tumorigenic data. When heated to decomposition it emits very toxic fumes of Cl^- and NO_x.

MER000 CAS:6893-20-5 ***HR: 3***
N-(1-METHOXYFLUOREN-2-YL)ACETAMIDE
mf: $C_{16}H_{15}NO_2$ mw: 253.32

SYNS: 1-METHOXY-2-ACETAMIDOFLUORENE ◇ 1-METHOXY-2-FAA ◇ N-(1-METHOXY-2-FLUORENYL)ACETAMIDE

TOXICITY DATA with REFERENCE
orl-rat TDLo:3300 mg/kg/23W-C:CAR CNREA8 28,234,68

SAFETY PROFILE: Questionable carcinogen with experimental carcinogenic data. When heated to decomposition it emits toxic fumes of NO_x.

MER250 CAS:16690-44-1 ***HR: 3***
N-(7-METHOXY-2-FLUORENYL)ACETAMIDE
mf: $C_{16}H_{15}NO_2$ mw: 253.32

SYNS: 7-METHOXY-2-FAA ◇ N-(7-METHOXYFLUOREN-2-YL)ACETAMIDE ◇ 7-METHOXY-N-2-FLUORENYLACETAMIDE

TOXICITY DATA with REFERENCE
orl-rat TDLo:2740 mg/kg/22W-C:CAR CNREA8 28,234,68
orl-rat TD:3940 mg/kg/43W-C:CAR JNCIAM 24,149,60

SAFETY PROFILE: Questionable carcinogen with experimental carcinogenic data. When heated to decomposition it emits toxic fumes of NO_x.

MES000 CAS:131-57-7 ***HR: 3***
4-METHOXY-2-HYDROXYBENZOPHENONE
mf: $C_{14}H_{12}O_3$ mw: 228.26

SYNS: BENZOPHENONE-3 ◇ CYASORB UV 9 ◇ 2-HYDROXY-4-METHOXYBENZOPHENONE ◇ (2-HYDROXY-4-METHOXYPHENYL) PHENYLMETHANONE ◇ MOB ◇ NCI-C60957 ◇ NSC-7778 ◇ OXYBENZONE ◇ SPECTRA-SORB UV 9 ◇ SYNTASE 62 ◇ USAF CY-9 ◇ UVINUL M 40

TOXICITY DATA with REFERENCE
mma-sat 100 μg/plate ENMUDM 9(Suppl 9),1,87
orl-rat TDLo:45 g/kg (90D pre):REP FCTXAV 10,41,72
orl-rat LD50:7400 mg/kg JACTDZ 2(5),35,83
ipr-mus LD50:300 mg/kg NTIS** AD277-689

CONSENSUS REPORTS: Reported in EPA TSCA Inventory.

SAFETY PROFILE: Poison by intraperitoneal route. Mildly toxic by ingestion. Experimental reproductive effects. Mutation data reported. When heated to decomposition it emits acrid smoke and irritating fumes.

MES250 CAS:67465-44-5 ***HR: 3***
N-(2-METHOXY-3-HYDROXYMERCURIPROPYL) BARBITAL
mf: $C_{12}H_{20}HgN_2O_5$ mw: 472.93

SYN: HYDROXY(3-(5,5-DIETHYL-2,4,6-TRIOXO-(1H,3H,5H)-PYRIMIDINO)-2-METHOXYPROPYL)MERCURY

TOXICITY DATA with REFERENCE
ivn-rat LDLo:23200 μg/kg JAPMA8 37,333,48

CONSENSUS REPORTS: Mercury and its compounds are on the Community Right-To-Know List.

OSHA PEL: (Transitional: CL 1 mg/10m³) CL 0.1 mg(Hg)/m³ (skin)
ACGIH TLV: TWA 0.1 mg(Hg)/m³ (skin)
NIOSH REL: (Mercury, Inorganic) TWA 0.05 mg(Hg)/m³

SAFETY PROFILE: Poison by intravenous route. When heated to decomposition it emits very toxic fumes of Hg and NO_x. See also MERCURY COMPOUNDS.

MES300 ***HR: 3***
1-METHOXY IMIDAZOLE-N-OXIDE
mf: $C_4H_6N_2O_2$ mw: 114.10

$CH_3ONCH{=}CHN(O){=}CH$

SAFETY PROFILE: Exothermic decomposition above 140°C may become explosive. When heated to decomposition it emits toxic fumes of NO_x.

MES500 ***HR: 1***
3-METHOXY-5,4'-IMINOBIS(1-BENZAMIDO-ANTHRAQUINONE)
mf: $C_{43}H_{27}O_7$ mw: 655.70

SYN: 5,4'-BIS-BENZOYLAMINO-8-METHOXY-1,1'- DIANTHRIMID (CZECH)

TOXICITY DATA with REFERENCE
skn-rbt 500 mg/24H MOD 28ZPAK -,114,72
eye-rbt 100 mg/24H MOD 28ZPAK -,114,72
orl-rat LD50:9120 mg/kg 28ZPAK -,114,72

SAFETY PROFILE: Mildly toxic by ingestion. A skin and eye irritant. When heated to decomposition it emits acrid smoke and irritating fumes.

MES850 CAS:3471-31-6 ***HR: 3***
5-METHOXYINDOLEACETIC ACID
mf: $C_{11}H_{11}NO_3$ mw: 205.23

SYNS: METHOXYINDOLEACETIC ACID ◇ 5-METHOXYINDOLE-3-ACETIC ACID

TOXICITY DATA with REFERENCE
scu-mus TDLo:4 g/kg/13W-I:NEO BEBMAE 101,605,86
scu-mus TDLo:2 g/kg (10-21D post):NEO,TER BEBMAE 101,605,86
ipr-mus LD50:98 mg/kg JTEHD6 1,515,76

SAFETY PROFILE: Poison by intraperitoneal route. An experimental teratogen. Questionable carcinogen with experimental neoplastigenic and tumorigenic data. When heated to decomposition it emits toxic fumes of NO_x.

MET000 CAS:4484-61-1 ***HR: 1***
2-(METHOXYMETHOXY)ETHANOL
mf: $C_4H_{10}O_3$ mw: 106.14

SYN: (2-HYDROXYETHOXY)METHOXYMETHANE

TOXICITY DATA with REFERENCE
skn-rbt 10 mg/24H open MLD AMIHBC 10,61,54
eye-rbt 500 mg open AMIHBC 10,61,54
orl-rat LD50:6500 mg/kg AMIHBC 10,61,54
skn-rbt LD50:4230 mg/kg AMIHBC 10,61,54

SAFETY PROFILE: Mildly toxic by ingestion and skin contact. A skin and eye irritant. When heated to decomposition it emits acrid smoke and irritating fumes.

MET875 CAS:83876-56-6 ***HR: 2***
3-METHOXY-7-METHYLBENZ(c)ACRIDINE
mf: $C_{19}H_{15}NO$ mw: 273.35

SYN: BENZ(c)ACRIDINE, 3-METHOXY-7-METHYL-

TOXICITY DATA with REFERENCE
scu-mus TDLo:72 mg/kg/12W-I:ETA JMCMAR 26,303,83

SAFETY PROFILE: Questionable carcinogen with experimental tumorigenic data. When heated to decomposition it emits toxic fumes of NO_x.

MEU000 CAS:966-48-3 ***HR: 3***
3-METHOXY-10-METHYL-1,2-BENZANTHRACENE
mf: $C_{20}H_{16}O$ mw: 272.36

SYN: 5-METHOXY-7-METHYL-BENZ(a)ANTHRACENE

TOXICITY DATA with REFERENCE
scu-mus TDLo:140 mg/kg:ETA JNCIAM 1,303,40

SAFETY PROFILE: Questionable carcinogen with experimental tumorigenic data. When heated to decomposition it emits acrid smoke and irritating fumes.

MEU250 CAS:63020-61-1 ***HR: 3***
5-METHOXY-10-METHYL-1,2-BENZANTHRACENE
mf: $C_{20}H_{16}O$ mw: 272.36

SYN: 8-METHOXY-7-METHYL-BENZ(a)ANTHRACENE

TOXICITY DATA with REFERENCE
scu-mus TDLo:80 mg/kg:ETA CNREA8 6,454,46

SAFETY PROFILE: Questionable carcinogen with experimental tumorigenic data. When heated to decomposition it emits acrid smoke and irritating fumes.

MEU500 CAS:16354-47-5 ***HR: 3***
7-METHOXY-12-METHYLBENZ(a)ANTHRACENE
mf: $C_{20}H_{16}O$ mw: 272.36

TOXICITY DATA with REFERENCE
ims-rat TDLo:50 mg/kg:NEO CNREA8 29,506,69

SAFETY PROFILE: Questionable carcinogen with ex-

perimental neoplastigenic data. When heated to decomposition it emits acrid smoke and irritating fumes.

MEU750 CAS:5831-08-3 ***HR: 3***
3-METHOXY-17-METHYL-15H-CYCLOPENTAPHENANTHRENE
mf: $C_{19}H_{16}O$ mw: 260.35

SYN: 3-METHOXY-17-METHYL15H-CYCLOPENTA(a)PHENANTHRENE

TOXICITY DATA with REFERENCE
mma-sat 50 μg/plate CNREA8 36,4525,76
skn-mus TDLo:108 mg/kg/1Y-I:ETA PEXTAR 11,69,69

CONSENSUS REPORTS: EPA Genetic Toxicology Program.

SAFETY PROFILE: Questionable carcinogen with experimental tumorigenic data. Mutation data reported. When heated to decomposition it emits acrid smoke and irritating fumes.

MEV000 CAS:5831-12-9 ***HR: 3***
11-METHOXY-17-METHYL-15H-CYCLOPENTA(a)PHENANTHRENE
mf: $C_{19}H_{16}O$ mw: 260.35

TOXICITY DATA with REFERENCE
mma-sat 50 μg/plate CNREA8 36,4525,76
skn-mus TDLo:108 mg/kg/1Y-I:CAR PEXTAR 11,69,69

CONSENSUS REPORTS: EPA Genetic Toxicology Program.

SAFETY PROFILE: Questionable carcinogen with experimental carcinogenic data. Mutation data reported. When heated to decomposition it emits acrid smoke and irritating fumes.

MEV250 CAS:24684-49-9 ***HR: 3***
6-METHOXY-11-METHYL-15,16-DIHYDRO-17H-CYCLOPENTA(a) PHENANTHREN-17-ONE
mf: $C_{19}H_{16}O_2$ mw: 276.35

SYN: 15,16-DIHYDRO-11-METHYL-6-METHOXY-17H-CYCLOPENTA (a)PHENANTHREN-17-ONE

TOXICITY DATA with REFERENCE
mma-sat 50 μg/plate CNREA8 36,4525,76
skn-mus TDLo:120 mg/kg/50W-I:NEO CNREA8 33,832,73

CONSENSUS REPORTS: EPA Genetic Toxicology Program.

SAFETY PROFILE: Questionable carcinogen with experimental neoplastigenic data. Mutation data reported. When heated to decomposition it emits acrid smoke and irritating fumes.

MEV500 CAS:25498-49-1 ***HR: 2***
(2-(2-METHOXY METHYL ETHOXY)METHYL ETHOXY)PROPANOL
mf: $C_{10}H_{22}O_4$ mw: 206.32

SYN: DOWANOL TPM

TOXICITY DATA with REFERENCE
orl-rat LD50:3300 mg/kg NPIRI* 1,120,74

CONSENSUS REPORTS: Reported in EPA TSCA Inventory.

SAFETY PROFILE: Moderately toxic by ingestion. When heated to decomposition, it emits acrid smoke and irritating fumes. See also GLYCOL ETHERS.

MEV750 CAS:61738-04-3 ***HR: 3***
METHOXYMETHYL ETHYL NITROSAMINE
mf: $C_4H_{10}N_2O_2$ mw: 118.16

SYN: METHOXYMETHYL-AETHYLNITROSAMINE (GERMAN)

TOXICITY DATA with REFERENCE
orl-rat TDLo:1240 mg/kg/23W-I:CAR ZKKOBW 88,25,76
orl-rat TD:1890 mg/kg/17W-I:CAR ZKKOBW 88,25,76
orl-rat LD50:540 mg/kg ZKKOBW 88,25,76

SAFETY PROFILE: Moderately toxic by ingestion. Questionable carcinogen with experimental carcinogenic data. When heated to decomposition it emits toxic fumes of NO_x. See also NITROSAMINES.

MEW000 CAS:13345-60-3 ***HR: 3***
7-METHOXYMETHYL-12-METHYLBENZ(a)ANTHRACENE
mf: $C_{21}H_{18}O$ mw: 286.39

SYN: 9-METHYL-10-METHOXYMETHYL-1,2-BENZANTHRACENE

TOXICITY DATA with REFERENCE
orl-rat TDLo:100 mg/kg:ETA JMCMAR 10,932,67

SAFETY PROFILE: Questionable carcinogen with experimental tumorigenic data. When heated to decomposition it emits acrid smoke and irritating fumes.

MEW250 CAS:39885-14-8 ***HR: 3***
METHOXYMETHYL METHYLNITROSAMINE
mf: $C_3H_8N_2O_2$ mw: 104.13

SYNS: METHOXYMETHYL-METHYLNITROSAMIN (GERMAN) ◇ METHYL(METHOXYMETHYL)NITROSAMINE ◇ N-NITROSO-N-METHOXYMETHYLMETHYLAMINE

TOXICITY DATA with REFERENCE
orl-rat TDLo:1470 mg/kg/50W-I:CAR ZKKOBW 88,25,76
ipr-rat TDLo:417 mg/kg:ETA MCEBD4 6,2716,86
orl-rat LD50:700 mg/kg ZKKOBW 88,25,76
ipr-rat LD50:895 mg/kg PAACA3 16,32,75

SAFETY PROFILE: Moderately toxic by ingestion and

intraperitoneal routes. Questionable carcinogen with experimental carcinogenic and tumorigenic data. When heated to decomposition it emits toxic fumes of NO_x. See also NITROSAMINES.

MEW500 CAS:50308-89-9 ***HR: 3***
8-METHOXY-1-METHYL-4-(p-((p-((1-METHYLPYRIDINIUM-4-YL)AMINO)PHENYL)CARBAMOYL)ANILINO)QUINOLINIUM), DIBROMIDE
mf: $C_{30}H_{29}N_5O_2 \cdot 2Br$ mw: 651.46

TOXICITY DATA with REFERENCE
dnd-mus:lym 760 nmol/L JMCMAR 22,134,79
ipr-mus LD10:6 mg/kg JMCMAR 22,134,79

SAFETY PROFILE: Poison by intraperitoneal route. Mutation data reported. When heated to decomposition it emits very toxic fumes of Br^- and NO_x. See also BROMIDES.

MEW750 CAS:50308-88-8 ***HR: 3***
6-METHOXY-1-METHYL-4-(p-((p-((1-METHYLPYRIDINIUM-4-YL)AMINO)PHENYL)CARBAMOYL)ANILINO)QUINOLINIUM), DI-p-TOLUENESULFONATE
mf: $C_{30}H_{29}N_5O_2 \cdot 2C_7H_7O_3S$ mw: 834.04

TOXICITY DATA with REFERENCE
dnd-mus:lym 680 nmol/L JMCMAR 22,134,79
ipr-mus LD10:50 mg/kg JMCMAR 22,134,79

SAFETY PROFILE: Poison by intraperitoneal route. Mutation data reported. When heated to decomposition it emits very toxic fumes of NO_x and SO_x. See also SULFONATES.

MEW775 CAS:86539-71-1 ***HR: 2***
7-METHOXY-1-METHYL-2-NITRONAPHTHO(2,1-b)FURAN
mf: $C_{14}H_{11}NO_4$ mw: 257.26

SYN: R 7372

TOXICITY DATA with REFERENCE
mmo-sat 1 nmol/plate MUTAEX 1,217,86
oth-esc 1 nmol/tube MUTAEX 1,217,86
msc-ham:ovr 1 µmol/L MUTAEX 1,217,86
scu-rat TDLo:168 mg/kg/42W-I:CAR CALEDQ 35,59,87

SAFETY PROFILE: Questionable carcinogen with experimental carcinogenic data. Mutation data reported. When heated to decomposition it emits toxic fumes of NO_x.

MEW800 CAS:73815-11-9 ***HR: 2***
3-((4-(5-(METHOXYMETHYL)-2-OXO-3-OXAZOLIDINYL)PHENOXY)METHYL)BENZONITRILE
mf: $C_{19}H_{18}N_2O_4$ mw: 338.39

SYN: MD 780515

TOXICITY DATA with REFERENCE
ipr-rat LD50:3000 mg/kg AIPTAK 259,194,82
orl-mus LDLo:3000 mg/kg AIPTAK 259,194,82
ipr-mus LD50:3000 mg/kg AIPTAK 259,194,82

CONSENSUS REPORTS: Cyanide and its compounds are on the Community Right-To-Know List.

SAFETY PROFILE: Moderately toxic by ingestion and intraperitoneal routes. When heated to decomposition it emits toxic fumes of NO_x and CN^-. See also NITRILES.

MEX250 CAS:107-70-0 ***HR: 2***
4-METHOXY-4-METHYL-2-PENTANONE
DOT: UN 2293
mf: $C_7H_{14}O_2$ mw: 130.21

SYN: 4-METHOXY-4-METHYLPENTAN-2-ONE(DOT)

TOXICITY DATA with REFERENCE
skn-rbt 500 mg MOD SCCUR* -,6,61
orl-rat LDLo:3000 mg/kg SCCUR* -,6,61
orl-mus LD50:2050 mg/kg SCCUR* -,6,61
ihl-mus LCLo:2280 ppm/15H SCCUR* -,6,61
skn-rbt LDLo:3 g/kg SCCUR* -,6,61

CONSENSUS REPORTS: Reported in EPA TSCA Inventory.

DOT Classification: Flammable or Combustible Liquid; Label: Flammable Liquid.

SAFETY PROFILE: Moderately toxic by ingestion and skin contact. Mildly toxic by inhalation. A skin irritant. Flammable when exposed to heat or flame, can react vigorously with oxidizing materials. When heated to decomposition it emits acrid smoke and irritating fumes.

MEX300 CAS:3644-11-9 ***HR: 3***
N-(METHOXYMETHYL)-2-PROPENAMIDE
mf: $C_5H_9NO_2$ mw: 115.15

SYNS: ACRYLAMIDE, N-(METHOXYMETHYL)- ◇ METHOXYMETHYLACRYLAMIDE ◇ N-(METHOXYMETHYL)ACRYLAMIDE ◇ 2-PROPENAMIDE, N-(METHOXYMETHYL)-

TOXICITY DATA with REFERENCE
skn-rbt 500 mg/kg MOD JACTDZ 1,41,90
orl-rat LD50:192 mg/kg JACTDZ 1,41,90
skn-rbt LD50:312 mg/kg JACTDZ 1,41,90

SAFETY PROFILE: Poison by ingestion and skin contact. A skin irritant. When heated to decomposition it emits toxic fumes of NO_x.

MEX350 ***HR: 2***
2-METHOXY-3(5)-METHYLPYRAZINE
mf: $C_6H_8N_2O$ mw: 124.14

PROP: Colorless liquid; roasted, hazelnut odor. D: 1.000-1.090 @ 20°, refr index: 1.506, flash p: 131°F. Sol in water, organic solvents.

SYN: FEMA No. 3183

SAFETY PROFILE: Combustible liquid. When heated to decomposition emits toxic fumes of NO_x.

MEY000 CAS:3131-27-9 ***HR: 3***
4-METHOXYMETHYLPYRIDOXINE HYDROCHLORIDE
mf: $C_9H_{13}NO_3$•ClH mw: 219.69

SYNS: 4-METHOXYMETHYL-5-HYDROXY-6-METHYL-3-PYRIDINEMETHANOL HYDROCHLORIDE ◇ 4-METHOXYMETHYLPYRIDOXOL HYDROCHLORIDE

TOXICITY DATA with REFERENCE
orl-rat LD50:2150 mg/kg ARZNAD 11,922,61
scu-rat LD50:705 mg/kg ARZNAD 11,922,61
ivn-rat LD50:420 mg/kg ARZNAD 11,922,61
orl-mus LD50:75 mg/kg ARZNAD 11,922,61
scu-mus LD50:21 mg/kg ARZNAD 11,922,61
ivn-mus LD50:22 mg/kg ARZNAD 11,922,61

SAFETY PROFILE: Poison by ingestion, subcutaneous, and intravenous routes. When heated to decomposition it emits very toxic fumes of HCl and NO_x.

MEY750 CAS:7213-59-4 ***HR: D***
6-METHOXY-3-METHYL-1,7,8-TRIHYDROXYANTHRAQUINONE
mf: $C_{16}H_{12}O_6$ mw: 300.28

SYN: DERMOGLANCIN

TOXICITY DATA with REFERENCE
mmo-sat 100 μg/plate BCSTB5 5,1489,77
mma-sat 100 μg/plate BCSTB5 5,1489,77

CONSENSUS REPORTS: EPA Genetic Toxicology Program.

SAFETY PROFILE: Mutation data reported. When heated to decomposition it emits acrid smoke and irritating fumes.

MEZ300 ***HR: 3***
3-(4-METHOXY-1-NAPHTHOYL)PROPIONIC ACID
mf: $C_{15}H_{14}O_4$ mw: 258.29

SYNS: ACIDE β-(1-METHOXY-4-NAPHTHOYL)PROPIONIQUE (FRENCH) ◇ β-(1-METHOXY-4-NAPHTHOYL)-PROPIONSAEURE (GERMAN)

TOXICITY DATA with REFERENCE
orl-rat LD50:1800 mg/kg AIPTAK 116,154,58
ivn-rat LD50:750 mg/kg AIPTAK 116,154,58
ims-rat LD50:500 mg/kg AEPPAE 222,244,54
orl-mus LD50:700 mg/kg AIPTAK 116,154,58
scu-mus LD50:380 mg/kg AIPTAK 116,154,58

SAFETY PROFILE: Poison by subcutaneous route. Moderately toxic by ingestion, intravenous, and intramuscular routes. When heated to decomposition it emits acrid smoke and irritating fumes.

MFA000 CAS:3178-03-8 ***HR: 3***
1-METHOXY-2-NAPHTHYLAMINE
mf: $C_{11}H_{11}NO$ mw: 173.23

SYN: 2-AMINO-1-METHOXYNAPHTHALENE

TOXICITY DATA with REFERENCE
imp-mus TDLo:100 mg/kg:CAR BMBUAQ 14,147,58

SAFETY PROFILE: Questionable carcinogen with experimental carcinogenic data. When heated to decomposition it emits toxic fumes of NO_x.

MFA250 CAS:63020-03-1 ***HR: 3***
1-METHOXY-2-NAPHTHYLAMINE HYDROCHLORIDE
mf: $C_{11}H_{11}NO$•ClH mw: 209.69

SYNS: 1-METHOXY-2-AMINONAPHTHALENE ◇ o-METHYL-2-AMINO-1-NAPHTHOL HYDROCHLORIDE ◇ NEOSONE D ◇ PHENYL-β-NAPHTHALAMINE

TOXICITY DATA with REFERENCE
scu-mus TDLo:12500 mg/kg/52W:NEO BJCAAI 10,653,56
imp-mus TDLo:50 mg/kg:CAR BJCAAI 12,222,58

SAFETY PROFILE: Questionable carcinogen with experimental neoplastigenic and carcinogenic data. When heated to decomposition it emits very toxic fumes of HCl and NO_x.

MFA300 CAS:42924-53-8 ***HR: 2***
4-(6-METHOXY-2-NAPHTHYL)-2-BUTANONE
mf: $C_{15}H_{16}O_2$ mw: 228.31

SYNS: BRL 147777 ◇ 2-BUTANONE, 4-(6-METHOXY-2-NAPHTHALENYL)- ◇ 4-(6-METHOXY-2-NAPHTHALENYL)-2-BUTANONE ◇ NABUMETONE

TOXICITY DATA with REFERENCE
orl-rat TDLo:1760 mg/kg (female 7-17D post):REP KSRNAM 22,2975,88
orl-rat TDLo:1760 mg/kg (female 7-17D post):TER KSRNAM 22,2975,88
orl-rat LD50:3880 mg/kg KSRNAM 22,2939,88

ipr-rat LD50:1520 mg/kg KSRNAM 22,2939,88
orl-mus LD50:4290 mg/kg KSRNAM 22,2939,88
ipr-mus LD50:2380 mg/kg KSRNAM 22,2939,88
orl-mky LD50:3200 mg/kg KSRNAM 22,2939,88

SAFETY PROFILE: Moderately toxic by ingestion and intraperitoneal routes. An experimental teratogen. Experimental reproductive effects. When heated to decomposition it emits acrid smoke and irritating fumes.

MFA500 CAS:22204-53-1 ***HR: 3***
(+)-2-(METHOXY-2-NAPHTHYL)-PROPIONIC ACID
mf: $C_{14}H_{14}O_3$ mw: 230.28

SYNS: EQUIPROXEN ◇ FLOGINAX ◇ (+)-2-(METHOXY-2-NAPHTHYL)-PROPIONSAEURE (GERMAN) ◇ NAPROSINE ◇ NAPROSYN ◇ NAPRUX ◇ NAXEN ◇ PROXEN

TOXICITY DATA with REFERENCE
dni-hmn:lym 60 ppm ARZNAD 25,288,75
dni-mus:oth 120 ppm ARZNAD 25,288,75
orl-wmn TDLo:20 mg/kg (30W preg):REP ADCHAK 54,942,79
orl-wmn TDLo:120 mg/kg (28W preg):TER ADCHAK 54,942,79
orl-cld TDLo:2250 mg/kg/26W-I:SKN BJRHDF 26,210,87
orl-man TDLo:61 mg/kg/6D-I:KID AIMEAS 105,144,86
orl-wmn TDLo:40 mg/kg/2D-I:KID SJRHAT 15,401,86
orl-wmn TDLo:70 mg/kg/W-I:LIV NEJMAG 295,1201,76
orl-hmn TDLo:50 mg/kg/7D-I ARZNAD 25,281,75
orl-rat LD50:338 mg/kg NIIRDN 6,539,82
ipr-rat LD50:354 mg/kg IYKEDH 9,829,78
scu-rat LD50:928 mg/kg KSRNAM 17,1272,83
orl-mus LD50:360 mg/kg FRPSAX 40,334,85
ipr-mus LD50:500 mg/kg BCFAAI 119,600,80
scu-mus LD50:475 mg/kg OYYAA2 6,1039,72
ivn-mus LD50:435 mg/kg JPETAB 179,114,71

SAFETY PROFILE: Poison by ingestion and intraperitoneal routes. Moderately toxic by subcutaneous, and intravenous routes. Human systemic effects: acute renal failure, acute tubular necrosis, cholestatic jaundice. dermatitis, diarrhea, fever, hypermotility, interstitial nephritis, nausea or vomiting. Human teratogenic effects by ingestion: developmental abnormalities of the cardiovascular (circulatory) system, respiratory system, gastrointestinal system, and other developmental abnormalities. Human reproductive effects by ingestion (effects on newborn): change in condition of newborn at birth, biochemical and metabolic effects, and other neonatal measures or effects. Human mutation data reported. An anti-inflammatory, analgesic and antipyretic. An FDA proprietary drug. When heated to decomposition it emits acrid smoke and irritating fumes.

MFB000 CAS:96-96-8 ***HR: 1***
4-METHOXY-2-NITROANILINE
mf: $C_7H_8N_2O_3$ mw: 168.17

SYNS: 4-AMINO-3-METHOXYAZOBENZENE ◇ 4-METHOXY-2-NITROANILIN (CZECH)

TOXICITY DATA with REFERENCE
orl-rat LD50:14 g/kg 28ZPAK -,118,72

CONSENSUS REPORTS: Reported in EPA TSCA Inventory.

SAFETY PROFILE: Mildly toxic by ingestion. When heated to decomposition it emits toxic fumes of NO_x. See also NITRO COMPOUNDS of AROMATIC HYDROCARBONS.

MFB250 CAS:58683-84-4 ***HR: D***
3-METHOXY-4-NITROAZOBENZENE
mf: $C_{13}H_{11}N_3O_3$ mw: 257.27

SYN: (3-METHOXY-4-NITROPHENYL)PHENYLDIAZENE

TOXICITY DATA with REFERENCE
mmo-sat 100 nmol/plate GANNA2 68,373,77
mma-sat 500 nmol/plate GANNA2 72,921,82

CONSENSUS REPORTS: EPA Genetic Toxicology Program.

SAFETY PROFILE: Mutation data reported. When heated to decomposition it emits toxic fumes of NO_x. See also NITRO COMPOUNDS of AROMATIC HYDROCARBONS.

MFB350 CAS:30335-72-9 ***HR: D***
5-METHOXY-2-NITROBENZOFURAN
mf: $C_9H_7NO_4$ mw: 193.17

SYN: 2-NITRO-5-METHOXYBENZOFURAN

TOXICITY DATA with REFERENCE
mnt-ham:lng 500 μg/L CNREA8 44,1969,84
sce-ham:lng 500 μg/L CNREA8 44,1969,84

SAFETY PROFILE: Mutation data reported. When heated to decomposition it emits toxic fumes of NO_x.

MFB400 CAS:75965-74-1 ***HR: 3***
7-METHOXY-2-NITRONAPHTHO(2,1-b)FURAN
mf: $C_{13}H_9NO_4$ mw: 243.23

SYNS: 2-NITRO-7-METHOXYNAPHTHO(2,1-b)FURAN ◇ R7000

TOXICITY DATA with REFERENCE
oms-ham:lng 4 mg/L MUREAV 157,53,85
cyt-ham:lng 1 mg/L MUREAV 157,53,85
orl-rat TDLo:210 mg/kg/91W-C:CAR CRNGDP 7,1447,86
scu-rat TDLo:24 mg/kg/36W-I:CAR CALEDQ 35,59,87
scu-rat TD:20 mg/kg/10W-I:ETA PAACA3 25,96,84

SAFETY PROFILE: Suspected carcinogen with experimental experimental carcinogenic and tumorigenic data. Mutation data reported. When heated to decomposition it emits toxic fumes of NO_x.

MFB410 CAS:75965-75-2 ***HR: D***
8-METHOXY-2-NITRONAPHTHO(2,1-b)FURAN
mf: $C_{13}H_9NO_4$ mw: 243.23

SYNS: 2-NITRO-8-METHOXYNAPHTHOL(2,1-b)-FURAN ◇ R6998

TOXICITY DATA with REFERENCE
mnt-ham:lng 500 μg/L CNREA8 44,1969,84
oms-ham:lng 4 mg/L MUREAV 157,53,85

SAFETY PROFILE: Mutation data reported. When heated to decomposition it emits toxic fumes of NO_x.

MFB775 CAS:1035-77-4 ***HR: D***
3-METHOXYOESTRADIOL
mf: $C_{19}H_{26}O_2$ mw: 286.45

SYNS: ESTRADIOL 3-METHYL ETHER ◇ 17-β-ESTRADIOL 3-METHYL ETHER ◇ 3-METHOXY-ESTRA-1,3,5(10)-TRIENE-17-β-OL ◇ (17-β)-3-METHOXY-ESTRA-1,3,5(10)-TRIEN-17-OL (9CI) ◇ 3-METHOXYESTRA-1,3,5(10)-TRIEN-17-β-OL ◇ OESTRADIOL 3-METHYL ETHER

TOXICITY DATA with REFERENCE
scu-rat TDLo:2500 μg/kg (female 1D pre):REP ACENA7 49,193,65

SAFETY PROFILE: Experimental reproductive effects. A steroid. When heated to decomposition it emits acrid smoke and irritating fumes.

MFC000 CAS:140-20-5 ***HR: 3***
METHOXYOXIMERCURIPROPYLSUCCINYL UREA
mf: $C_9H_{16}HgN_2O_6$ mw: 448.86

SYNS: N-((3-(HYDROXYMERCURI)-2-METHOXYPROPYL)-CARBAMOYL)SUCCINAMIC ACID ◇ MERALLURIDE ◇ METHOXYHYDROXYMERCURIPROPYLSUCCINYLUREA

TOXICITY DATA with REFERENCE
ims-hmn LDLo:314 mg/kg/18D JAMAAP 147,377,51
scu-rat LD50:28 mg/kg JOPDAB 69,663,66

CONSENSUS REPORTS: Mercury and its compounds are on the Community Right-To-Know List.

OSHA PEL: (Transitional: CL 1 mg/10m^3) CL 0.1 mg(Hg)/m^3 (skin)
ACGIH TLV: TWA 0.1 mg(Hg)/m^3 (skin)
NIOSH REL: TWA 0.05 mg(Hg)/m^3

SAFETY PROFILE: A human poison by intramuscular route. Poison experimentally by subcutaneous route. When heated to decomposition it emits very toxic fumes of Hg and NO_x. See also MERCURY COMPOUNDS.

MFC500 CAS:55-81-2 ***HR: 3***
p-METHOXYPHENETHYLAMINE
mf: $C_9H_{13}NO$ mw: 151.23

SYNS: p-METHOXYPHENYLETHYLAMINE ◇ USAF EL-52

TOXICITY DATA with REFERENCE
ipr-mus LD50:100 mg/kg NTIS** AD277-689

CONSENSUS REPORTS: Reported in EPA TSCA Inventory.

SAFETY PROFILE: Poison by intraperitoneal route. When heated to decomposition it emits toxic fumes of NO_x.

MFC600 CAS:2771-13-3 ***HR: 3***
1-(p-METHOXYPHENETHYL)HYDRAZINE HYDROGEN SULFATE
mf: $C_9H_{14}N_2O•H_2O_4S$ mw: 264.33

SYNS: p-METHOXY-β-PHENYLETHYLHYDRAZINE DIHYDROGEN SULFATE ◇ 1-(p-METHOXYPHENETHYL)-HYDRAZINE SULFATE (1:1)

TOXICITY DATA with REFERENCE
scu-mus TDLo:240 mg/kg (1-6D preg):REP JOENAK 30,205,64
orl-mus LD50:225 mg/kg JMPCAS 5,221,62
scu-mus LD50:182 mg/kg JOENAK 30,205,64

SAFETY PROFILE: Poison by ingestion and subcutaneous routes. Experimental reproductive effects. When heated to decomposition it emits toxic fumes of SO_x and NO_x. See also SULFATES.

MFC700 CAS:150-76-5 ***HR: 3***
4-METHOXYPHENOL
mf: $C_7H_8O_2$ mw: 124.15

PROP: White, waxy solid. Mp: 52.5°, bp: 246°, d: 1.55 @ 20°/20°.

SYNS: HYDROQUINONE MONOMETHYL ETHER ◇ MEQUINOL ◇ p-METHOXYPHENOL ◇ MME ◇ MONO METHYL ETHER HYDROQUINONE ◇ USAF AN-7

TOXICITY DATA with REFERENCE
skn-rbt 6 g/12D-I MLD JIHTAB 31,79,49
orl-rat LD50:1600 mg/kg JIHTAB 31,79,49
ipr-rat LD50:725 mg/kg JIHTAB 31,79,49
ipr-mus LD50:250 mg/kg NTIS** AD691-490
ipr-rbt LD50:970 mg/kg JIHTAB 31,79,49

CONSENSUS REPORTS: Reported in EPA TSCA Inventory. EPA Genetic Toxicology Program.

OSHA PEL: TWA 5 mg/m^3
ACGIH TLV: TWA 5 mg/m^3

SAFETY PROFILE: Poison by intraperitoneal route. Moderately toxic by ingestion. A skin irritant. When heated to decomposition it emits acrid smoke and fumes. See also ETHERS.

MFD250 CAS:70145-83-4 ***HR: 3***
(2-(p-METHOXYPHENOXY)ETHYL)HYDRAZINE HYDROCHLORIDE
mf: $C_9H_{14}N_2O_2 \cdot ClH$ mw: 218.71

TOXICITY DATA with REFERENCE
orl-mus LD50:250 mg/kg JMCMAR 6,63,63
ipr-mus LD50:250 mg/kg JMCMAR 6,63,63

SAFETY PROFILE: Poison by ingestion and intraperitoneal routes. When heated to decomposition it emits very toxic fumes of Cl^- and NO_x.

MFD500 CAS:70-07-5 ***HR: 2***
5-(o-METHOXYPHENOXYMETHYL)-2-OXAZOLIDONE
mf: $C_{11}H_{13}NO_4$ mw: 223.25

SYNS: AHR 233 ◇ ALKAPOL PEG-400 ◇ CL 27,319 ◇ CONTROL ◇ DORSIFLEX ◇ DORSILON ◇ EKILAN ◇ LENETRAN ◇ LENETRAN TAB ◇ LENETRANAT ◇ MEFENOXALONA ◇ MEFENOXALONE ◇ MEPHENOXALONE ◇ METHOXADONE ◇ METHOXYDON(E) ◇ 5-(o-METHOXYPHENOXYMETHYL)-2-OXAZOLIDINONE ◇ METOXADONE ◇ MODERAMIN ◇ OM 518 ◇ OXAZOLIDINONE ◇ PLACIDEX ◇ REPOISE ◇ RISELF ◇ TRANSPOISE ◇ TREPIDONE ◇ VALANAS ◇ XERENE

TOXICITY DATA with REFERENCE
orl-rat LD50:3820 mg/kg TXAPA9 6,642,64
ipr-mus LD50:800 mg/kg RAMAAB 74,82,60
orl-dog LDLo:480 mg/kg TXAPA9 4,220,62

SAFETY PROFILE: Moderately toxic by ingestion and intraperitoneal routes. A tranquilizer. When heated to decomposition it emits toxic fumes of NO_x.

MFE250 CAS:104-01-8 ***HR: 3***
p-METHOXYPHENYLACETIC ACID
mf: $C_9H_{10}O_3$ mw: 166.19

SYNS: ANISYL FORMATE ◇ 2-(p-ANISYL)ACETIC ACID ◇ HOMOANISIC ACID ◇ 4-METHOXYBENZENEACETIC ACID ◇ p-METHOXYBENZYL FORMATE ◇ 4-METHOXYPHENYLACETIC ACID ◇ MOPA

TOXICITY DATA with REFERENCE
orl-mus TDLo:63 g/kg/78W-I:NEO NTIS** PB223-159
orl-rat LD50:1550 mg/kg FCTXAV 14,659,76
ipr-mus LD50:504 mg/kg JMCMAR 20,709,77

CONSENSUS REPORTS: Reported in EPA TSCA Inventory.

SAFETY PROFILE: Moderately toxic by ingestion and intraperitoneal routes. Questionable carcinogen with experimental neoplastigenic data. When heated to decomposition it emits acrid smoke and irritating fumes.

MFF000 CAS:104-47-2 ***HR: 3***
p-METHOXYPHENYLACETONITRILE
mf: C_9H_9NO mw: 147.19

SYNS: ANISYLACETONITRILE ◇ p-METHOXYBENZENEACETONITRILE ◇ p-METHOXYBENZYL CYANIDE ◇ 4-METHOXYPHENYLACETONITRILE

TOXICITY DATA with REFERENCE
ivn-mus LD50:56 mg/kg CSLNX* NX#07882

CONSENSUS REPORTS: Cyanide and its compounds are on the Community Right-To-Know List. Reported in EPA TSCA Inventory.

SAFETY PROFILE: Poison by intravenous route. When heated to decomposition it emits toxic fumes of NO_x and CN^-. See also NITRILES.

MFF250 CAS:3647-17-4 ***HR: 3***
N-(p-METHOXYPHENYL)-1-AZIRIDINECARBOXAMIDE
mf: $C_{10}H_{12}N_2O_2$ mw: 192.24

SYNS: 1-(1-AZIRIDINYL)-N-(p-METHOXYPHENYL)FORMAMIDE ◇ p-METHOXYPHENYL-N-CARBAMOYLAZIRIDINE

TOXICITY DATA with REFERENCE
ipr-mus TDLo:240 mg/kg/4W-I:NEO CNREA8 29,2184,69
ivn-mus LD50:180 mg/kg CSLNX* NX#03949

SAFETY PROFILE: Poison by intravenous route. Questionable carcinogen with experimental neoplastigenic data. When heated to decomposition it emits toxic fumes of NO_x.

MFF500 CAS:3544-23-8 ***HR: 3***
2-METHOXY-4-PHENYLAZOANILINE
mf: $C_{13}H_{13}N_3O$ mw: 227.29

SYNS: 4-AMINO-3-METHOXYAZOBENZENE ◇ 4-(PHENYLAZO)-o-ANISIDINE

TOXICITY DATA with REFERENCE
mma-sat 50 nmol/plate CALEDQ 8,71,79
hma-mus/sat 50 mg/kg JNCIAM 62,911,79
dns-mus:lvr 1 μmol/L GANNA2 72,930,81
orl-rat TDLo:7950 mg/kg/34W-C:CAR CNREA8 21,1068,61
skn-rat TDLo:1490 mg/kg/93W-I:NEO CNREA8 26,2406,66
orl-rat TD:18 g/kg/66W-I:ETA CNREA8 24,1279,64
orl-rat TD:19 g/kg/49W-C:CAR GANNA2 59,131,68

CONSENSUS REPORTS: EPA Genetic Toxicology Program.

SAFETY PROFILE: Questionable carcinogen with experimental carcinogenic, neoplastigenic, and tumorigenic data. Mutation data reported. When heated to decomposition it emits toxic fumes of NO_x.

MFF550 CAS:2592-28-1 ***HR: D***
p-((p-METHOXYPHENYL)AZO)ANILINE
mf: $C_{13}H_{13}NO$ mw: 227.29

TOXICITY DATA with REFERENCE
mma-sat 250 nmol/plate GANNA2 72,921,81
dns-rat:lvr 1 μmol/L GANNA2 72,930,81
dns-mus:lvr 1 μmol/L GANNA2 72,930,81

SAFETY PROFILE: Mutation data reported. When heated to decomposition it emits toxic fumes of NO_x. See also ANILINE DYES.

MFF575 CAS:34289-01-5 ***HR: D***
1-(2-(4-(6-METHOXY-2-PHENYLBENZO(b)THIEN-3-YL)PHENOXY)ETHYL)PYRROLIDINE HYDROCHLORIDE
mf: $C_{27}H_{27}NO_2S•ClH$ mw: 466.07

SYN: 6-METHOXY-3-(p-2-(1-PYRROLIDYL)ETHOXY)PHENYL)-2-PHENYLBENZO(b)THIOPHENE HYDROCHLORIDE

TOXICITY DATA with REFERENCE
orl-rat TDLo:900 μg/kg (4D pre/1-5D preg):REP JMCMAR 14,1185,71

SAFETY PROFILE: Experimental reproductive effects. When heated to decomposition it emits toxic fumes of NO_x, SO_x, and HCl.

MFF580 ***HR: 1***
4-p-METHOXYPHENYL-2-BUTANONE

PROP: Colorless to pale yellow liquid; sweet, floral odor. D: 1.042-1.048, refr index: 1.517-1.521, flash p: +212°F.

SYNS: ANISYLACETONE ◇ FEMA No. 2672

SAFETY PROFILE: Combustible liquid. When heated to decomposition it emits acrid smoke and irritating fumes.

MFF625 CAS:1178-99-0 ***HR: 3***
2-(p-(6-METHOXY-2-PHENYL-3,4-DIHYDRO-1-NAPHTHYL)PHENOXY)TRIETHYLAMINE HYDROCHLORIDE
mf: $C_{29}H_{33}NO_2•ClH$ mw: 464.09

SYN: 1-((p-(2-DIETHYLAMINO)ETHOXY)PHENYL)-3,4-DIHYDRO-6-METHOXY-2-PHENYLNAPHTHALENE HYDROCHLORIDE

TOXICITY DATA with REFERENCE
orl-rat TDLo:1240 μg/kg (female 3-4D post):REP PSEBAA 112,439,63
orl-rat LD50:547 mg/kg PSEBAA 112,439,63
ipr-mus LD50:195 mg/kg PSEBAA 112,439,63

SAFETY PROFILE: Poison by intraperitoneal route. Moderately toxic by ingestion. Experimental reproductive effects. When heated to decomposition it emits toxic fumes of NO_x and HCl.

MFF650 CAS:55308-37-7 ***HR: D***
2-(3-METHOXYPHENYL)-5,6-DIHYDRO-s-TRIAZOLO(5,1-a)ISOQUINOLINE
mf: $C_{17}H_{15}N_3O$ mw: 277.35

SYNS: L-10503 ◇ 2-(m-METHOXYPHENYL)-5,6-DIHYDRO-s-TRIAZOLO(5,1-a)ISOQUINOLINE

TOXICITY DATA with REFERENCE
scu-rat TDLo:100 mg/kg (female 9D post):TER JSTBBK 8,395,77
scu-dog TDLo:100 mg/kg (female 45D post):REP AJVRAH 37,263,76

SAFETY PROFILE: An experimental teratogen. Experimental reproductive effects. When heated to decomposition it emits toxic fumes of NO_x.

MFF750 CAS:16143-89-8 ***HR: 3***
2-(p-METHOXYPHENYL)-3,3-DIPHENYLACRYLONITRILE
mf: $C_{22}H_{17}NO$ mw: 311.40

SYN: α-(p-METHOXYPHENYL)-β,β-DIPHENYLACRYLONITRILE

TOXICITY DATA with REFERENCE
scu-mus TDLo:94 mg/kg/26W-I:CAR MMJJAI 11,95,61
par-mus TDLo:320 mg/kg/1Y-I:ETA NNGZAZ 33,53,57

CONSENSUS REPORTS: Cyanide and its compounds are on the Community Right-To-Know List.

SAFETY PROFILE: Questionable carcinogen with experimental carcinogenic and tumorigenic data. When heated to decomposition it emits toxic fumes of NO_x and CN^-. See also NITRILES.

MFG000 CAS:102-51-2 ***HR: D***
4-METHOXY-m-PHENYLENEDIAMINE
mf: $C_7H_{10}N_2O$ mw: 138.19

SYNS: 3,4-DIAMINOANISOLE ◇ 4-METHOXY-1,2-BENZENEDIAMINE (9CI)

TOXICITY DATA with REFERENCE
dnd-hmn:fbr 50 μmol/L MUREAV 127,107,84

CONSENSUS REPORTS: Reported in EPA TSCA Inventory. Community Right-To-Know List.

SAFETY PROFILE: Human mutation data reported. When heated to decomposition it emits toxic fumes of NO_x.

MFG200 CAS:2598-71-2 ***HR: 3***
o-METHOXY-β-PHENYLETHYLHYDRAZINE DIHYDROGEN SULFATE
mf: $C_9H_{14}N_2O•H_2O_4S$ mw: 264.33

SYNS: HYDRAZINE, 1-(o-METHOXYPHENETHYL)-, SULFATE (1:1) ◇ WL 29

TOXICITY DATA with REFERENCE
scu-mus TDLo:180 mg/kg (female 1-6D post):REP JOENAK 30,205,64
scu-mus LD50:150 mg/kg JOENAK 30,205,64

SAFETY PROFILE: Poison by subcutaneous route. Experimental reproductive effects. When heated to decomposition it emits toxic fumes of NO_x.

MFG250 CAS:34758-84-4 ***HR: 3***
1-(2-METHOXY-2-PHENYL)ETHYL-4-(2-HYDROXY-3-METHOXY-3-PHENYL)PROPYLPIPERAZINE DIHYDROCHLORIDE
mf: $C_{23}H_{31}N_2O_2 \cdot 2HCl$ mw: 440.46

SYNS: 3024 CERM ◇ 1-(2-HYDROXY-3-METHOXY-3-PHENYLPROPYL)-4-(2-METHOXY-2-PHENYLETHYL)PIPERAZINEDIHYDROCHLORIDE ◇ RESPILENE ◇ ZIPEPROL ◇ ZIPEPROL DIHYDROCHLORIDE ◇ ZITOXIL

TOXICITY DATA with REFERENCE
orl-hmn TDLo:11 mg/kg:CNS LANCAO 1,45,84
orl-chd TDLo:154 mg/kg/3D-I:CNS LANCAO 1,45,84
orl-wmn TDLo:18 mg/kg:CNS LANCAO 1,45,84
orl-man TDLo:10714 μg/kg:CNS LANCAO 1,45,84
orl-rat LD50:435 mg/kg OYYAA2 22,355,81
ipr-rat LD50:77600 μg/kg OYYAA2 22,355,81
scu-rat LD50:139 mg/kg OYYAA2 22,355,81
ivn-rat LD50:32700 μg/kg OYYAA2 22,355,81
orl-mus LD50:300 mg/kg ARZNAD 26,523,76
ipr-mus LD50:116 mg/kg OYYAA2 22,355,81
scu-mus LD50:158 mg/kg OYYAA2 22,355,81
ivn-mus LD50:44300 mg/kg OYYAA2 22,355,81
orl-dog LD50:228 mg/kg OYYAA2 22,355,81
par-dog LDLo:54 mg/kg ARZNAD 26,523,76
orl-rbt LD50:173 mg/kg OYYAA2 22,355,81

SAFETY PROFILE: Poison by ingestion, intravenous, intraperitoneal, subcutaneous, and parenteral routes. Human systemic effects by ingestion: convulsions, coma. An antitussive agent. When heated to decomposition it emits toxic fumes of NO_x and ClH.

MFG260 CAS:64-96-0 ***HR: 3***
2-(p-(6-METHOXY-2-PHENYL-3-INDENYL)PHENOXY)TRIETHYLAMINE HYDROCHLORIDE
mf: $C_{28}H_{31}NO_2 \cdot ClH$ mw: 450.06

SYNS: N,N-DIETHYL-2-(4-(6-METHOXY-2-PHENYL-1H-INDEN-3-YL)PHENOXY)-ETHANAMINE HYDROCHLORIDE ◇ 2-(p-(6-METHOXY-2-PHENYLINDEN-3-YL)PHENOXY)TRIETHYLAMINEHYDROCHLORIDE ◇ U-11555A

TOXICITY DATA with REFERENCE
scu-rbt TDLo:30 mg/kg (female 1-3D post):TER FESTAS 20,211,69
orl-rat TDLo:1500 μg/kg (female 1-5D post):REP JRPFA4 13,373,67
orl-rat LD50:547 mg/kg PSEBAA 112,439,63
ipr-mus LD50:247 mg/kg PSEBAA 112,439,63

SAFETY PROFILE: Poison by intraperitoneal route. Moderately toxic by ingestion. An experimental teratogen. Experimental reproductive effects. When heated to decomposition it emits toxic fumes of NO_x and HCl.

MFG275 ***HR: D***
2-(3-METHOXYPHENYL)-8-METHOXY-5H-s-TRIAZOLO(5,1-a)ISOINDOLE
mf: $C_{17}H_{15}N_3O_2$ mw: 293.35

SYN: L 11752

TOXICITY DATA with REFERENCE
scu-rat TDLo:10 mg/kg (1-5D preg):REP JAPRAN 23,295,82

SAFETY PROFILE: Experimental reproductive effects. When heated to decomposition it emits toxic fumes of NO_x.

MFG400 CAS:25355-59-3 ***HR: 2***
1-(p-METHOXYPHENYL)-3-METHYL-3-NITROSOUREA
mf: $C_9H_{11}N_3O_3$ mw: 209.23

SYN: N-METHYL-N′-(p-METHOXYPHENYL)-N-NITROSOUREA

TOXICITY DATA with REFERENCE
mmo-sat 33500 pmol/plate CNREA8 39,5147,79
skn-mus TDLo:621 mg/kg/7W-I:CAR CNREA8 44,1027,84

SAFETY PROFILE: Questionable carcinogen with experimental carcinogenic data. Mutation data reported. When heated to decomposition it emits toxic fumes of NO_x.

MFG510 CAS:53477-43-3 ***HR: 3***
1-(4-METHOXYPHENYL)-3-METHYL TRIAZENE
mf: $C_8H_{11}N_3O$ mw: 165.19

$CH_3OC_6H_4N{=}NNHCH_3$

SAFETY PROFILE: Explodes during vacuum distillation below 1 mbar. Upon decomposition it emits toxic fumes of NO_x.

MFG525 CAS:6732-77-0 ***HR: D***
2-(p-(p-METHOXY-α-PHENYLPHENETHYL) PHENOXY)TRIETHYLAMINE
mf: $C_{27}H_{33}NO_2$ mw: 403.61

SYNS: N,N-DIETHYL-2-(4-(2-(4-METHOXYPHENYL)-1-PHENYLETHYL)PHENOXY)-ETHANAMINE (9CI) ◇ MRL-37

TOXICITY DATA with REFERENCE
scu-mus TDLo:96 mg/kg (1-3D preg):REP JRPFA4 9,277,65

SAFETY PROFILE: Experimental reproductive effects. When heated to decomposition it emits toxic fumes of NO_x.

MFG530 CAS:3063-72-7 *HR: D*
2-(p-(p-METHOXY-α-PHENYLPHENETHYL) PHENOXY)TRIETHYLAMINE HYDROCHLORIDE
mf: $C_{27}H_{33}NO_2 \cdot ClH$ mw: 440.07

SYNS: 1-(p-(β-DIETHYLAMINOETHOXY)PHENYL)-1-PHENYL-2-(p-METHOXYPHENYL)-ETHANE HYDROCHLORIDE ◇ MRL-37 HYDROCHLORIDE

TOXICITY DATA with REFERENCE
orl-rat TDLo:420 mg/kg (21D pre):REP FESTAS 13,472,62
orl-rat TDLo:100 mg/kg (female 18-22D post):TER FESTAS 13,472,62

SAFETY PROFILE: An experimental teratogen. Experimental reproductive effects. When heated to decomposition it emits toxic fumes of NO_x.

MFG600 CAS:21140-85-2 *HR: 3*
trans-3-(o-METHOXYPHENYL)-2-PHENYLACRYLIC ACID
mf: $C_{16}H_{14}O_3$ mw: 254.30

SYNS: ACIDE-α-PHENYL-o-METHOXYCINNAMIQUE (FRENCH) ◇ (E)-o-METHOXY-α-PHENYL-CINNAMIC ACID ◇ SUBSTANCE H 20

TOXICITY DATA with REFERENCE
orl-mus LD50:1100 mg/kg AIPTAK 119,443,59
ipr-mus LD50:400 mg/kg AIPTAK 119,443,59
scu-mus LD50:440 mg/kg AIPTAK 119,443,59
orl-gpg LD50:1000 mg/kg AIPTAK 119,443,59

SAFETY PROFILE: Poison by intraperitoneal route. Moderately toxic by ingestion and subcutaneous routes. When heated to decomposition it emits acrid smoke and irritating fumes.

MFH000 CAS:69103-91-9 *HR: 2*
2-(3-o-METHOXYPHENYLPIPERAZINO)-PROPYL)-3-METHYL-7-METHOXYCHROMONE
mf: $C_{25}H_{30}N_2O_4$ mw: 422.57

TOXICITY DATA with REFERENCE
ipr-mus LD50:500 mg/kg EJMCA5 13,387,78
scu-mus LD50:600 mg/kg EJMCA5 13,387,78

SAFETY PROFILE: Moderately toxic by intraperitoneal and subcutaneous routes. When heated to decomposition it emits toxic fumes of NO_x.

MFH750 CAS:61785-72-6 *HR: D*
7-(4-(3-METHOXYPHENYL)-1-PIPERAZINYL)-4-NITROBENZOFURAZAN-1-OXIDE
mf: $C_{17}H_{17}N_5O_5$ mw: 371.39

SYN: B2772

TOXICITY DATA with REFERENCE
mmo-sat 100 μg/plate MUREAV 48,145,77
mma-sat 50 μg/well CBINA8 19,77,77

SAFETY PROFILE: Mutation data reported. When heated to decomposition it emits toxic fumes of NO_x.

MFH800 CAS:61001-40-9 *HR: D*
2-(m-METHOXYPHENYL)-PYRAZOLO(5,1-a) ISOQUINOLINE
mf: $C_{18}H_{14}N_2O$ mw: 274.34

SYN: 2-(3-METHOXYPHENYL)PYRAZOLO(5,1-a)ISOQUINOLINE

TOXICITY DATA with REFERENCE
scu-rat TDLo:125 mg/kg (6-10D preg):REP EJMCA5 19,215,84

SAFETY PROFILE: Experimental reproductive effects. When heated to decomposition it emits toxic fumes of NO_x.

MFJ000 CAS:70145-82-3 *HR: 3*
(2-(p-METHOXYPHENYLTHIO)ETHYL)HYDRAZINE MALEATE
mf: $C_9H_{14}N_2OS \cdot C_4H_4O_4$ mw: 314.39

TOXICITY DATA with REFERENCE
orl-mus LD50:250 mg/kg JMCMAR 6,63,63
ipr-mus LD50:250 mg/kg JMCMAR 6,63,63

SAFETY PROFILE: Poison by ingestion and intraperitoneal routes. When heated to decomposition it emits very toxic fumes of SO_x and NO_x.

MFJ100 CAS:69095-72-3 *HR: D*
5-(m-METHOXYPHENYL-3-(o-TOLYL)-s-TRIAZOLE
mf: $C_{16}H_{15}N_3O$ mw: 265.34

TOXICITY DATA with REFERENCE
scu-rat TDLo:5 mg/kg (female 1-5D post):REP JAPRAN 23,295,82

SAFETY PROFILE: Experimental reproductive effects. When heated to decomposition it emits toxic fumes of NO_x.

MFJ105 CAS:57170-08-8 *HR: D*
2-(3-METHOXYPHENYL)-5H-s-TRIAZOLO(5,1-a) ISOINDOLE
mf: $C_{16}H_{13}N_3O$ mw: 263.32

SYN: L-10492

TOXICITY DATA with REFERENCE
scu-rat TDLo:12500 μg/kg (6-10D preg):REP NATUAS 256,130,75

SAFETY PROFILE: Experimental reproductive effects. When heated to decomposition it emits toxic fumes of NO_x.

MFJ110 CAS:55309-14-3 ***HR: D***
2-(m-METHOXYPHENYL)-s-TRIAZOLO(5,1-a) ISOQUINOLINE
mf: $C_{17}H_{13}N_3O$ mw: 275.33

TOXICITY DATA with REFERENCE
scu-rat TDLo:15 mg/kg (6-10D preg):REP ARZNAD 33,1222,83

SAFETY PROFILE: Experimental reproductive effects. When heated to decomposition it emits toxic fumes of NO_x.

MFJ115 CAS:85303-91-9 ***HR: D***
5-(m-METHOXYPHENYL)-3-(2,4-XYLYL)-s-TRIAZOLE
mf: $C_{17}H_{17}N_3O$ mw: 279.37

TOXICITY DATA with REFERENCE
orl-ham TDLo:50 mg/kg (4-8D preg):REP JMCMAR 26,1187,83

SAFETY PROFILE: Experimental reproductive effects. When heated to decomposition it emits toxic fumes of NO_x.

MFJ750 CAS:9004-74-4 ***HR: 1***
METHOXY POLYETHYLENE GLYCOL 350
mf: $(C_2H_4O)_n•CH_4O$

TOXICITY DATA with REFERENCE
skn-rbt 500 mg open MLD UCDS** 7/8/71
orl-rat LD50:22 g/kg UCDS** 7/8/71

CONSENSUS REPORTS: Reported in EPA TSCA Inventory.

SAFETY PROFILE: Mildly toxic by ingestion. A skin irritant. When heated to decomposition it emits acrid smoke and irritating fumes.

MFK000 CAS:9004-74-4 ***HR: 1***
METHOXY POLYETHYLENE GLYCOL 550
mf: $(C_2H_4O)_n•CH_4O$

TOXICITY DATA with REFERENCE
skn-rbt 500 mg open MLD UCDS** 4/17/67
eye-rbt 100 mg MLD 34ZIAG -,747,69
orl-rat LD50:40 g/kg UCDS** 4/17/67

CONSENSUS REPORTS: Reported in EPA TSCA Inventory.

SAFETY PROFILE: Mildly toxic by ingestion. A skin and eye irritant. When heated to decomposition it emits acrid smoke and irritating fumes.

MFK250 CAS:9004-74-4 ***HR: 1***
METHOXY POLYETHYLENE GLYCOL 750
mf: $(C_2H_4O)_n•CH_4O$

TOXICITY DATA with REFERENCE
skn-rbt 500 mg open MLD UCDS** 4/25/58
eye-rbt 100 mg MLD 34ZIAG -,747,69
orl-rat LD50:39800 mg/kg 34ZIAG -,747,69
orl-rat LD50:40 mg/kg UCDS** 4/25/58

CONSENSUS REPORTS: Reported in EPA TSCA Inventory.

SAFETY PROFILE: Mildly toxic by ingestion. A skin and eye irritant. When heated to decomposition it emits acrid smoke and irritating fumes.

MFK500 CAS:61-01-8 ***HR: 3***
2-METHOXYPROMAZINE
mf: $C_{18}H_{22}N_2OS$ mw: 314.48

PROP: Crystals. Mp: 44-48°.

SYNS: 10-(3-DIMETHYLAMINOPROPYL)-2-METHOXYPHENOTHIAZINE ◇ 2-METHOXY-10-(3′-DIMETHYLAMINOPROPYL)PHENOTHIAZINE ◇ MOPAZIN ◇ MOPAZINE ◇ NEOPROMA ◇ RP 4632 ◇ TENTON ◇ TENTONE

TOXICITY DATA with REFERENCE
orl-rat LD50:730 mg/kg AIPTAK 125,101,60
ipr-rat LD50:95 mg/kg AIPTAK 125,101,60
orl-mus LD50:408 mg/kg AIPTAK 125,101,60
ipr-mus LD50:112 mg/kg AIPTAK 125,101,60

SAFETY PROFILE: Poison by intraperitoneal route. Moderately toxic by ingestion. When heated to decomposition it emits very toxic fumes of NO_x and SO_x.

MFK750 CAS:3403-42-7 ***HR: 3***
METHOXYPROMAZINE MALEATE
mf: $C_{18}H_{22}N_2OS•C_4H_4O_4$ mw: 430.56

SYNS: 10-(3-(DIMETHYLAMINO)PROPYL)-2-METHOXY)PHENOTHIAZINE, MALEATE ◇ METHOPROMAZINE MALEATE ◇ TENTONE MALEATE

TOXICITY DATA with REFERENCE
orl-rat LD50:730 mg/kg 27ZQAG -,32,72
ipr-rat LD50:130 mg/kg AIPTAK 125,101,60
orl-mus LD50:420 mg/kg 27ZQAG -,32,72
ipr-mus LD50:120 mg/kg 27ZQAG -,32,72
ivn-mus LD50:50 mg/kg 27ZQAG -,32,72

SAFETY PROFILE: Poison by intravenous and intraperitoneal routes. Moderately toxic by ingestion. When heated to decomposition it emits very toxic fumes of NO_x and SO_x.

MFL000 CAS:1589-49-7 ***HR: 1***
3-METHOXY-1-PROPANOL
mf: $C_4H_{10}O_2$ mw: 90.14

SYN: β-PROPYLENE GLYCOL MONOMETHYL ETHER

TOXICITY DATA with REFERENCE
skn-rbt 10 mg/24H open MLD AIHAAP 23,95,62
orl-rat LD50:5710 mg/kg JIHTAB 23,259,41
skn-rbt LD50:5660 mg/kg AIHAAP 30,470,69

CONSENSUS REPORTS: Glycol ether compounds are on the Community Right-To-Know List.

SAFETY PROFILE: Mildly toxic by ingestion and skin contact. A skin irritant. When heated to decomposition it emits acrid smoke and irritating fumes. See also GLYCOL ETHERS.

MFL250 CAS:1610-18-0 ***HR: 2***
METHOXYPROPAZINE
mf: $C_{10}H_{19}N_5O$ mw: 225.34

SYNS: 2,4-BIS(ISOPROPYLAMINO)-6-METHOXY-s-TRIAZINE ◇ 2,6-DIISOPROPYLAMINO-4-METHOXYTRIAZINE ◇ GESAFRAM ◇ 2-METHOXY-4,6-BIS(ISOPROPYLAMINO)-s-TRIAZINE ◇ 2-METHOXY-4,6-BIS(ISOPROPYLAMINO)-1,3,5-TRIAZINE◇ ONTRACK ◇ PRAMITOL ◇ PROMETON

TOXICITY DATA with REFERENCE
skn-rbt 105 mg open MLD CIGET* -,-,77
eye-rbt 21 mg MOD CIGET* -,-,77
orl-rat LD50:1750 mg/kg WRPCA2 9,119,70
ihl-rat LC50:3260 mg/m³/4H FMCHA2 -,C193,83
orl-mus LD50:2160 mg/kg PCOC** -,931,66
skn-rbt LD50:2200 mg/kg GUCHAZ 6,425,73

CONSENSUS REPORTS: Reported in EPA TSCA Inventory.

SAFETY PROFILE: Moderately toxic by inhalation, ingestion, and skin contact. A skin and eye irritant. An herbicide. When heated to decomposition it emits toxic fumes of NO_x.

MFL750 CAS:110-67-8 ***HR: 1***
3-METHOXYPROPIONITRILE
mf: C_4H_7NO mw: 85.12

PROP: Liquid. Mp: −63°, bp: 160°, flash p: 149°F (OC), vap d: 2.94.

TOXICITY DATA with REFERENCE
orl-rat LD50:4390 mg/kg AIHAAP 30,470,69

CONSENSUS REPORTS: Reported in EPA TSCA Inventory. Cyanide and its compounds are on the Community Right-To-Know List.

SAFETY PROFILE: Mildly toxic by ingestion. Flammable when exposed to heat, flame, or oxidizers. Reacts with water, steam, or acids to produce toxic and flammable vapors. To fight fire, use CO_2, dry chemical. When heated to decomposition it emits highly toxic fumes of CN^-. See also NITRILES.

MFM000 CAS:5332-73-0 ***HR: 3***
3-METHOXYPROPYLAMINE
mf: $C_4H_{11}NO$ mw: 89.16

PROP: Colorless liquid. Mp: −75.7°, bp: 116°, flash p: 90°F (TOC), d: 0.8615 @ 30°, vap press: 20 mm @ 30°, vap d: 3.07.

SYN: 3-MPA

TOXICITY DATA with REFERENCE
ivn-mus LD50:180 mg/kg CSLNX* NX#03063

CONSENSUS REPORTS: Reported in EPA TSCA Inventory.

SAFETY PROFILE: Poison by intravenous route. Irritating to skin, eyes, and mucous membranes. Dangerous fire hazard when exposed to heat or flame; can react with oxidizing materials. To fight fire, use CO_2, dry chemical. When heated to decomposition it emits toxic fumes of NO_x. See also AMINES.

MFM750 CAS:2785-87-7 ***HR: 3***
2-METHOXY-4-PROPYLPHENOL
mf: $C_{10}H_{14}O_2$ mw: 166.24

SYNS: CERULIGNOL ◇ DIHYDROEUGENOL ◇ GUAIACYLPROPANE ◇ 4-HYDROXY-3-METHOXYPROPYLBENZENE ◇ p-PROPYLGUAIACOL ◇ p-n-PROPYLGUAIACOL ◇ 4-PROPYLGUAIACOL ◇ 1-PROPYL-3-METHOXY-4-HYDROXYBENZENE

TOXICITY DATA with REFERENCE
skn-rbt 500 mg/24H MOD FCTOD7 20(Suppl),671,82
orl-rat LD50:2600 mg/kg FCTOD7 20(Suppl),671,82
orl-mus LD50:2 g/kg FCTOD7 20(Suppl),671,82
ipr-mus LD50:150 mg/kg NTIS** AD691-490
skn-rbt LD50:310 mg/kg FCTOD7 20(Suppl),671,82

CONSENSUS REPORTS: Reported in EPA TSCA Inventory.

SAFETY PROFILE: Poison by skin contact and intraperitoneal routes. Moderately toxic by ingestion. A skin irritant. When heated to decomposition it emits acrid smoke and irritating fumes.

MFN000 CAS:69242-96-2 ***HR: 3***
9-(2-METHOXY-4-(PROPYLSULFONAMIDO)ANILINO)-4-ACRIDINECARBOXAMIDE HYDROCHLORIDE
mf: $C_{24}H_{24}N_4O_4S{\cdot}ClH$ mw: 501.04

TOXICITY DATA with REFERENCE
mma-sat 642 μmol/L JMCMAR 22,251,79
ipr-mus LD10:35 mg/kg JMCMAR 22,251,79

SAFETY PROFILE: Poison by intraperitoneal route. Mutation data reported. When heated to decomposition it emits very toxic fumes of NO_x, SO_x and HCl.

MFN250 CAS:627-41-8 ***HR: 3***
3-METHOXYPROPYNE
mf: C_4H_6O mw: 70.09

PROP: Bp: 61°.

SYN: METHYL PROPARGYL ETHER

SAFETY PROFILE: Explodes when heated to its boiling point. Ignites spontaneously in air. When heated to decomposition it emits acrid smoke and irritating fumes. See also ETHERS and ACETYLENE COMPOUNDS.

MFN275 CAS:484-20-8 ***HR: 2***
5-METHOXY PSORALEN
mf: $C_{12}H_8O_4$ mw: 216.20

PROP: Naturally occurring analog of psoralen and isomer of methoxsalen. Found in a wide variety of plants. Needles from alc. Mp: 188° (subl). Practically insol in boiling water; sltly sol in glacial acetic acid, chloroform, benzene, warm phenol. Sol in abs alc: 1 part in 60.

SYNS: BERGAPTEN ◇ 4-METHOXY-7H-FURO(3,2-g)(1)BENZOPYRAN-7-ONE ◇ PSORADERM

TOXICITY DATA with REFERENCE
dnd-esc 20 μmol/L CBINA8 21,103,78
dnd-omi 20 μmol/L CBINA8 21,103,78
dnd-omi 20 μmol/L CBINA8 21,103,78
dnd-sal:spr 20 μmol/L CBINA8 21,103,78
dnd-mam:lym 20 μmol/L CBINA8 21,103,78

CONSENSUS REPORTS: IARC Cancer Review: Group 2A IMEMDT 7,242,87; Animal Inadequate Evidence IMEMDT 40,327,86; Human Inadequate Evidence IMEMDT 40,327,86.

SAFETY PROFILE: Suspected carcinogen. Mutation data reported. When heated to decomposition it emits acrid smoke and irritating fumes.

MFN285 CAS:3149-28-8 ***HR: 1***
2-METHOXYPYRAZINE
mf: $C_5H_6N_2O$ mw: 110.12

PROP: Colorless to yellow liquid; nutty, cocoalike odor. D: 1.110-1.140 @ 20°, refr index: 1.508. Sol in alc; insol in water @ 61°.

SYN: FEMA No. 3302

SAFETY PROFILE: Skin and eye irritant. When heated to decomposition it emits toxic fumes of NO_x.

MFN500 CAS:152-47-6 ***HR: 2***
N[1]-(3-METHOXY-2-PYRAZINYL)SULFANILAMIDE
mf: $C_{11}H_{12}N_4O_3S$ mw: 280.33

SYNS: 2-(p-AMINOBENZENESULFANAMIDE)-3-METHOXYPYRAZINE ◇ DALYSEP ◇ FARMITALIA 204/122 ◇ KELFIZIN ◇ LONGUM ◇ 3-METHOXYPYRAZINE SULFANILAMIDE ◇ 3-METHOXY-2-SULFAPYRAZINE ◇ POLYCIDAL ◇ SULFALENE ◇ SULFAMETHOPYRAZINE ◇ SULFAMETHOXYPYRAZINE ◇ SULFAMETOPYRAZINE ◇ SULFAMETOSSIPIRIDAZINA (ITALIAN) ◇ SULFAMETOXYPYRIDAZIN (GERMAN) ◇ 2-SULFANILAMIDO-3-METHOXYPYRAZINE ◇ SULFAPYRAZINEMETHOXINE ◇ SULFAPYRAZINEMETHOXYNE

TOXICITY DATA with REFERENCE
sln-asn 1 g/L MUREAV 26,159,74
orl-mus TDLo:6 g/kg (female 7-12D post):REP OYYAA2 7,1005,73
orl-mus TDLo:3 g/kg (female 7-12D post):TER OYYAA2 7,1005,73
orl-rat LD50:2739 mg/kg ARZNAD 11,459,61
scu-rat LD50:2120 mg/kg NIIRDN 6,390,82
ivn-rat LD50:1790 mg/kg NIIRDN 6,390,82
orl-mus LD50:1292 mg/kg AIPTAK 127,58,60
scu-mus LD50:1590 mg/kg NIIRDN 6,390,82
ivn-mus LD50:893 mg/kg AIPTAK 127,58,60

SAFETY PROFILE: Moderately toxic by ingestion, subcutaneous, and intravenous routes. An experimental teratogen. Experimental reproductive effects. Mutation data reported. An antibacterial agent. When heated to decomposition it emits very toxic fumes of NO_x and SO_x. See also SULFONATES.

MFN600 CAS:1464-33-1 ***HR: 3***
4-METHOXYPYRIDOXINE
mf: $C_9H_{13}NO_3$ mw: 183.23

SYNS: 5-HYDROXY-4-(METHOXYMETHYL)-6-METHYL-3-PYRIDINEMETHANOL ◇ α^4-O-METHYLPYRIDOXOL

TOXICITY DATA with REFERENCE
ipr-rat LD50:540 mg/kg PSDTAP 4,179,64
orl-mus LD50:32 mg/kg PSDTAP 4,179,64
ipr-mus LD50:23 mg/kg PSDTAP 4,179,64
ivn-mus LD50:20 mg/kg PSDTAP 4,179,64
ipr-dog LD50:8 mg/kg PSDTAP 4,179,64
ipr-cat LD50:9 mg/kg PSDTAP 4,179,64
ipr-rbt LD50:17 mg/kg PSDTAP 4,179,64

SAFETY PROFILE: Poison by ingestion, intravenous, and intraperitoneal routes. When heated to decomposition it emits toxic fumes of NO_x.

MFO000 CAS:18179-67-4 ***HR: 2***
N[1]-(5-METHOXY-2-PYRIMIDINYL)SULFANILAMIDE, SODIUM SALT
mf: $C_{11}H_{11}N_4O_3S\cdot Na$ mw: 302.31

SYNS: 2-(p-AMINOBENZENESULFONAMIDO)-5-METHOXYPYRI-

MIDINE SODIUM SALT ◇ 4-AMINO-N-(5-METHOXY-2-PYRIMIDINYL) BENZENESULFONAMIDE SODIUM SALT ◇ METHOXYPYRIMAL SODIUM ◇ 5-METHOXYSULFADIAZINE SODIUM ◇ SULFAMETER SODIUM ◇ SULFAMETHOXYDIAZINE SODIUM ◇ SULFA-5-METHOXYPYRIMIDINE SODIUM SALT ◇ SULFAMETORINE SODIUM ◇ 2-SULFANILAMIDO-5-METHOXYPYRIMIDINE SODIUM SALT

TOXICITY DATA with REFERENCE
orl-rat LD50:1000 mg/kg ARZNAD 11,695,61
ipr-rat LD50:1100 mg/kg ARZNAD 11,695,61
ivn-rat LD50:1200 mg/kg ARZNAD 11,695,61
orl-mus LD50:3000 mg/kg ARZNAD 11,695,61
ipr-mus LD50:1500 mg/kg ARZNAD 11,695,61
ivn-mus LD50:1100 mg/kg ARZNAD 11,695,61
ivn-rbt LD50:1000 mg/kg ARZNAD 11,695,61

SAFETY PROFILE: Moderately toxic by ingestion, intraperitoneal, and intravenous routes. An antibacterial agent. When heated to decomposition it emits very toxic fumes of NO_x, Na_2O, and SO_x. See also SULFONATES.

MFO250 CAS:3949-14-2 ***HR: 3***
5-METHOXY-3-(2-PYRROLIDINOETHYL)INDOLE
mf: $C_{15}H_{20}N_2O$ mw: 244.37

SYNS: CT 4436 ◇ METHOXY-5-PYRROLIDINO-2'-ETHYL-3-INDOLE

TOXICITY DATA with REFERENCE
ipr-rat LD50:4 mg/kg BSCFAS 5,1411,65
ipr-mus LD50:77 mg/kg BSCFAS 5,1411,65
ivn-mus LD50:32 mg/kg BSCFAS 5,1411,65
ivn-rbt LD50:300 μg/kg BSCFAS 5,1411,65

SAFETY PROFILE: Poison by intraperitoneal and intravenous routes. When heated to decomposition it emits toxic fumes of NO_x.

MFO500 CAS:42840-17-5 ***HR: 3***
3-METHOXY-4-PYRROLIDINYLMETHYLDIBENZOFURAN
mf: $C_{18}H_{19}NO_2$ mw: 281.38

TOXICITY DATA with REFERENCE
orl-mus LD50:838 mg/kg TAKHAA 31,247,72
ipr-mus LD50:90 mg/kg CHTPBA 8,57,73
orl-mky LD50:1000 μg/kg ANYAA9 320,151,78
orl-gpg LD50:10 μg/kg ANYAA9 320,151,78

SAFETY PROFILE: Poison by ingestion and intraperitoneal routes. When heated to decomposition it emits toxic fumes of NO_x.

MFP000 CAS:5263-87-6 ***HR: 3***
6-METHOXYQUINOLINE
mf: $C_{10}H_9NO$ mw: 159.20

PROP: Liquid. D: 1.154 @ 20°, mp: 26-28°, bp: 254° @ 310 mm, sol in alc.

TOXICITY DATA with REFERENCE
ipr-mus LDLo:256 mg/kg CBCCT* 3,55,51

CONSENSUS REPORTS: Reported in EPA TSCA Inventory.

SAFETY PROFILE: Poison by intraperitoneal route. When heated to decomposition it emits toxic fumes of NO_x.

MFP500 CAS:73928-02-6 ***HR: 3***
3-METHOXY-4-STILBENAMINE
mf: $C_{15}H_{15}NO$ mw: 225.31

SYN: 3-METHOXY-4-AMINOSTILBENE

TOXICITY DATA with REFERENCE
scu-rat TDLo:250 mg/kg/W-I:ETA BMBUAQ 14,141,58

SAFETY PROFILE: Questionable carcinogen with experimental tumorigenic data. When heated to decomposition it emits toxic fumes of NO_x. See also AMINES.

MFQ250 CAS:19155-52-3 ***HR: 3***
5-METHOXY-1,2,3,4-THIATRIAZOLE
mf: $C_2H_3N_3OS$ mw: 117.13

SAFETY PROFILE: Explodes at room temperature. Upon decomposition it emits toxic fumes of SO_x and NO_x.

MFQ500 CAS:136-90-3 ***HR: 2***
4-METHOXY-m-TOLUIDINE
mf: $C_8H_{11}NO$ mw: 137.20

SYN: 3-AMINO-4-METHOXYTOLUEN (CZECH)

TOXICITY DATA with REFERENCE
skn-rbt 500 mg/24H MOD 28ZPAK -,118,72
eye-rbt 100 mg/24H SEV 28ZPAK -,118,72
orl-rat LD50:2210 mg/kg 28ZPAK -,118,72

SAFETY PROFILE: Moderately toxic by ingestion. A skin and severe eye irritant. When heated to decomposition it emits toxic fumes of NO_x. See also AMINES.

MFQ750 CAS:313-96-2 ***HR: 3***
2-METHOXYTRICYCLOQUINAZOLINE
mf: $C_{22}H_{14}N_4O$ mw: 350.40

TOXICITY DATA with REFERENCE
skn-mus TDLo:1240 mg/kg/1Y-I:ETA BJCAAI 16,275,62

SAFETY PROFILE: Questionable carcinogen with experimental tumorigenic data. When heated to decomposition it emits toxic fumes of NO_x.

MFR000 CAS:2642-50-4 ***HR: 3***
3-METHOXYTRICYCLOQUINAZOLINE
mf: $C_{22}H_{14}N_4O$ mw: 350.40

TOXICITY DATA with REFERENCE
skn-mus TDLo:1200 mg/kg/50W-I:NEO BCPCA6 14,323,65

SAFETY PROFILE: Questionable carcinogen with experimental neoplastigenic data. When heated to decomposition it emits toxic fumes of NO_x.

MFR250 CAS:66967-60-0 ***HR: 1***
METHOXY TRIETHYLENE GLYCOL VINYL ETHER
mf: $C_9H_{16}O_5$ mw: 204.25

TOXICITY DATA with REFERENCE
orl-rat LD50:11300 mg/kg AIHAAP 30,470,69
skn-rbt LD50:10 g/kg AIHAAP 30,470,69

CONSENSUS REPORTS: Glycol ethers are on the Community Right-To-Know List.

SAFETY PROFILE: Mildly toxic by ingestion and skin contact. Many glycol ether compounds have dangerous reproductive effects. When heated to decomposition it emits acrid smoke and irritating fumes. See also GLYCOL ETHERS.

MFR500 CAS:431-46-9 ***HR: 2***
1-METHOXY-2,2,2-TRIFLUOROETHANOL
mf: $C_3H_5F_3O_2$ mw: 130.08

TOXICITY DATA with REFERENCE
orl-mus LD50:750 mg/kg JMCMAR 13,1212,70
ipr-mus LD50:550 mg/kg JMCMAR 13,1212,70

CONSENSUS REPORTS: Reported in EPA TSCA Inventory.

SAFETY PROFILE: Moderately toxic by ingestion and intraperitoneal routes. When heated to decomposition it emits toxic fumes of F^-.

MFR775 CAS:39753-42-9 ***HR: 3***
1-METHOXY-3,4,5-TRIMETHYL PYRAZOLE-N-OXIDE
mf: $C_7H_{12}N_2O_2$ mw: 156.18

$CH_3ONC(CH_3)=C(CH_3)C(CH_3)=N:O$

SAFETY PROFILE: May explode when heated. Upon decomposition it emits toxic fumes of NO_x.

MFS400 CAS:608-07-1 ***HR: 3***
5-METHOXYTRYPTAMINE
mf: $C_{11}H_{14}N_2O$ mw: 190.27

SYNS: 3-(2-AMINOETHYL)-5-METHOXYINDOLE ◇ INDOLE, 3-(2-AMINOETHYL)-5-METHOXY- ◇ METHOXYTRYPTAMINE ◇ 5-MOT

TOXICITY DATA with REFERENCE
otr-mus-scu 3 g/kg EKSODD 7(4),26,85
scu-rat TDLo:8 mg/kg (male 20D pre):REP BIREBV 33,618,85
ipr-mus LD50:176 mg/kg JTEHD6 1,515,76
ivn-mus LD50:106 mg/kg FATOAO 27,681,64

SAFETY PROFILE: Poison by intravenous and intraperitoneal routes. Experimental reproductive effects. Mutation data reported. When heated to decomposition it emits toxic fumes of NO_x.

MFS500 CAS:2736-21-2 ***HR: 3***
6-METHOXYTRYPTAMINE
mf: $C_{11}H_{14}N_2O•HCl$ mw: 226.73

SYN: 3-(2-AMINOETHYL)-6-METHOXYINDOLEHYDROCHLORIDE

TOXICITY DATA with REFERENCE
scu-mus LD50:645 mg/kg RPTOAN 35(1),2,72
ivn-mus LD50:118 mg/kg FATOAO 35,16,72

SAFETY PROFILE: Poison by intravenous route. Moderately toxic by subcutaneous route. When heated to decomposition it emits very toxic fumes of HCl and NO_x.

MFT000 CAS:66-83-1 ***HR: 3***
5-METHOXYTRYPTAMINE HYDROCHLORIDE
mf: $C_{11}H_{14}N_2O•ClH$ mw: 226.73

PROP: Crystals. Decomp @ 248°.

SYNS: 3-(2-AMINOETHYL)-5-METHOXYINDOLEHYDROCHLORIDE ◇ MEXAMINE HYDROCHLORIDE

TOXICITY DATA with REFERENCE
ivn-rat LDLo:6 mg/kg RPTOAN 33,246,70
orl-mus LD50:580 mg/kg FEPRA7 23,5125,64
ipr-mus LD50:227 mg/kg YKKZAJ 94,1620,74
scu-mus LD50:620 mg/kg RPTOAN 35(1),2,72
ivn-mus LD50:103 mg/kg FEPRA7 23,T125,64

SAFETY PROFILE: Poison by intraperitoneal and intravenous routes. Moderately toxic by ingestion and subcutaneous routes. When heated to decomposition it emits very toxic fumes of HCl and NO_x.

MFT250 CAS:35764-54-6 ***HR: 3***
6-METHOXYTRYPTOLINE HYDROCHLORIDE
mf: $C_{12}H_{14}N_2O•ClH$ mw: 238.75

SYNS: 6-METHOXY-1,2,3,4-TETRAHYDRO-β-CARBOLINE HYDROCHLORIDE ◇ 6-METHOXY-1,2,3,4-TETRAHYDRO-9H-PYRIDO(3,4-B)INDOLE HYDROCHLORIDE

TOXICITY DATA with REFERENCE
orl-mus LD50:595 mg/kg ARZNAD 28,42,78
ipr-mus LD50:235 mg/kg ARZNAD 28,42,78
ivn-mus LD50:112 mg/kg ARZNAD 28,42,78

SAFETY PROFILE: Poison by intraperitoneal and intravenous route. Moderately toxic by ingestion. When heated to decomposition it emits very toxic fumes of HCl and NO_x.

MFT300 CAS:712-09-4 *HR: D*
5-METHOXYTRYPTOPHOL
mf: $C_{11}H_{13}NO_2$ mw: 191.25

SYNS: 5-METHOXY-1H-INDOLE-3-ETHANOL (9CI) ◇ 5-METHOXYINDOLE-3-ETHANOL

TOXICITY DATA with REFERENCE
scu-rbt TDLo:750 μg/kg (female 1D pre):REP ENDOAO 83,599,68

SAFETY PROFILE: Experimental reproductive effects. When heated to decomposition it emits toxic fumes of NO_x.

MFT500 CAS:127-25-3 *HR: 1*
METHYL ABIETATE
mf: $C_{21}H_{32}O_2$ mw: 316.47

PROP: Colorless to thick yellow liquid; almost odorless. Flash p: 356°F (OC), vap d: 10.9. D: 1.040 @ 20°/20°, bp: 360-365° with decomp. Insol in water, misc in alc and ether, the usual organic solvents, and with aliphatic hydrocarbons. From the esterification of the resinous residue of turpentine (FCTXAV 12,807,74).

SYNS: ABIETIC ACID, METHYL ESTER ◇ METHYL ESTER of WOOD ROSIN ◇ METHYL ESTER of WOOD ROSIN, partially hydrogenated (FCC)

TOXICITY DATA with REFERENCE
skn-rbt 500 mg/24H MOD FCTXAV 12,807,74

CONSENSUS REPORTS: Reported in EPA TSCA Inventory.

SAFETY PROFILE: A skin irritant. Probably slightly toxic. Combustible liquid when exposed to heat or flame; can react with oxidizing materials. To fight fire, use CO_2, dry chemical. When heated to decomposition it emits acrid smoke and irritating fumes.

MFT750 CAS:79-16-3 *HR: 2*
METHYLACETAMIDE
mf: C_3H_7NO mw: 73.11

SYNS: N-METHYLACETAMIDE ◇ MONOMETHYLACETAMIDE

TOXICITY DATA with REFERENCE
mmo-esc 10 g/L CRSUBM 3,69,55
skn-rat TDLo:1200 mg/kg (12-13D preg):TER TXAPA9 41,35,77
orl-rat TDLo:2 g/kg (7D preg):REP 85DJA5 -,95,71
orl-rat LD50:5000 mg/kg JRPFA4 4,219,62
ipr-rat LD50:2750 mg/kg JRPFA4 4,219,62
scu-rat LD50:3600 mg/kg COREAF 251,1937,60
ipr-mus LD50:4380 mg/kg JPPMAB 16,472,64
ivn-mus LD50:4015 mg/kg JPPMAB 16,472,64
ivn-rbt LDLo:16940 mg/kg ARZNAD 20,1242,70

CONSENSUS REPORTS: Reported in EPA TSCA Inventory.

SAFETY PROFILE: Moderately toxic by intraperitoneal, and subcutaneous routes. Mildly toxic by ingestion and intravenous routes. An experimental teratogen. Experimental reproductive effects. Mutation data reported. When heated to decomposition it emits toxic fumes of NO_x.

MFU000 CAS:102585-60-4 *HR: 2*
2-(2-(3-(N-METHYLACETAMIDO)-2,4,6-TRIIODOPHENOXY)ETHOXY)ACETIC ACID SODIUM SALT
mf: $C_{13}H_{14}I_3NO_5$•Na mw: 667.97

TOXICITY DATA with REFERENCE
orl-mus LD50:1200 mg/kg FRPSAX 31,349,76
ivn-mus LD50:670 mg/kg FRPSAX 31,349,76

SAFETY PROFILE: Moderately toxic by ingestion and intravenous routes. When heated to decomposition it emits very toxic fumes of I^-, Na_2O, and NO_x. See also IODIDES.

MFU250 CAS:100700-34-3 *HR: 3*
2-(2-(3-(N-METHYLACETAMIDO)-2,4,6-TRIIODOPHENOXY)ETHOXY)BUTYRIC ACID SODIUM SALT
mf: $C_{15}H_{18}I_3NO_5$•Na mw: 696.03

TOXICITY DATA with REFERENCE
orl-mus LD50:1850 mg/kg FRPSAX 31,349,76
ivn-mus LD50:340 mg/kg FRPSAX 31,349,76

SAFETY PROFILE: Poison by intravenous route. Moderately toxic by ingestion. When heated to decomposition it emits very toxic fumes of I^-, Na_2O, and NO_x. See also IODIDES.

MFU500 CAS:102585-59-1 *HR: 2*
2-(2-(3-(N-METHYLACETAMIDO)-2,4,6-TRIIODOPHENOXY)ETHOXY)-2-PHENYLACETIC ACID SODIUM SALT
mf: $C_{19}H_{18}I_3NO_5$•Na mw: 744.07

TOXICITY DATA with REFERENCE
orl-mus LD50:1660 mg/kg FRPSAX 31,349,76
ivn-mus LD50:525 mg/kg FRPSAX 31,349,76

SAFETY PROFILE: Moderately toxic by ingestion and intravenous routes. When heated to decomposition it emits very toxic fumes of I^-, Na_2O, and NO_x. See also IODIDES.

MFW000 CAS:579-10-2 ***HR: 3***
N-METHYLACETANILIDE
mf: $C_9H_{11}NO$ mw: 149.21

SYNS: ACETOMETHYLANILIDE ◇ N-ACETYL-METHYLANILINE ◇ METHYLANTIFEBRIN ◇ PHENYLMETHYLACETAMIDE

TOXICITY DATA with REFERENCE
orl-mus LD50:155 mg/kg TXAPA9 19,20,71
ipr-mus LDLo:125 mg/kg CBCCT* 6,213,54

CONSENSUS REPORTS: Reported in EPA TSCA Inventory.

SAFETY PROFILE: Poison by ingestion and intraperitoneal routes. When heated to decomposition it emits toxic fumes of NO_x.

MFW100 CAS:79-20-9 ***HR: 3***
METHYL ACETATE
DOT: UN 1231
mf: $C_3H_6O_2$ mw: 74.09

PROP: Colorless, volatile liquid. Mp: −98.7°, lel: 3.1%, uel: 16%, bp: 57.8°, ULC: 85-90, flash p: 14°F, d: 0.92438, autoign temp: 935°F, vap press: 100 mm @ 9.4°, vap d: 2.55. Moderately sol in water; misc in alc, ether.

SYNS: ACETATE de METHYLE (FRENCH) ◇ ACETIC ACID METHYL ESTER ◇ DEVOTON ◇ METHYL ETHANOATE ◇ METHYLACETAAT (DUTCH) ◇ METHYLACETAT (GERMAN) ◇ METHYLE (ACETATE de) (FRENCH) ◇ METHYLESTER KISELINY OCTOVE (CZECH) ◇ METILE (ACETATO di) (ITALIAN) ◇ OCTAN METYLU (POLISH) ◇ TERETON

TOXICITY DATA with REFERENCE
sln-smc 33800 ppm MUREAV 149,339,85
skn-rbt 500 mg/24H MOD 28ZPAK -,44,72
eye-rbt 100 mg/24H SEV 28ZPAK -,44,72
skn-rbt 500 mg/24H MLD FCTXAV 17(Suppl.),695,79
idu-rbt LD50:3700 mg/kg FCTXAV 17(Suppl.),695,79
ihl-mus LCLo:34 g/m³/4H AGGHAR 5,1,33
idu-rbt LD50:3700 mg/kg FCTXAV 17,859,79
ihl-hmn TCLo:15000 mg/m³:IRR AGGHAR 5,1,33
scu-rat LDLo:8000 mg/kg BSIBAC 18,45,43
ihl-cat LCLo:67000 mg/m³/1H AGGHAR 5,1,33
scu-cat LDLo:3000 mg/kg AGGHAR 5,1,33
orl-rbt LD50:3705 mg/kg IMSUAI 41,31,72
scu-gpg LDLo:3000 mg/kg AGGHAR 5,1,33

CONSENSUS REPORTS: Reported in EPA TSCA Inventory.

OSHA PEL: (Transitional: TWA 200 ppm) TWA 200 ppm; STEL 250 ppm
ACGIH TLV: TWA 200 ppm; STEL 250 ppm
DFG MAK: 200 ppm (610 mg/m³)
DOT Classification: Flammable Liquid; Label: Flammable Liquid.

SAFETY PROFILE: Moderately toxic by several routes. A human systemic irritant by inhalation. A moderate skin and severe eye irritant. Mutation data reported. See also ESTERS. Dangerous fire hazard when exposed to heat, flame, or oxidizers. Moderate explosion hazard when exposed to heat or flame. When heated to decomposition it emits acrid smoke and fumes.

MFW250 CAS:122-00-9 ***HR: 2***
4'-METHYL ACETOPHENONE
mf: $C_9H_{10}O$ mw: 134.19

PROP: Colorless liquid; fruity, actophenone odor. D: 0.996-1.004, refr index: 1.530-1.535, flash p: 198°F. Sol in fixed oils, propylene glycol; insol in glycerin.

SYNS: p-ACETYLTOLUENE ◇ FEMA No. 2677 ◇ MELILOTAL ◇ p-METHYL ACETOPHENONE ◇ 1-METHYL-4-ACETYLBENZENE ◇ METHYL-p-TOLYL KETONE

TOXICITY DATA with REFERENCE
skn-hmn 100% FCTXAV 12,933,74
skn-rbt 500 mg/24H MLD FCTXAV 12,807,74
orl-rat LD50:1400 mg/kg FCTXAV 12,807,74

CONSENSUS REPORTS: Reported in EPA TSCA Inventory.

SAFETY PROFILE: Moderately toxic by ingestion. A human skin irritant. Combustible liquid. When heated to decomposition it emits acrid smoke and irritating fumes. See also KETONES.

MFW500 CAS:520-45-6 ***HR: 3***
METHYLACETOPYRONONE
mf: $C_8H_8O_4$ mw: 168.16

PROP: White crystals or crystalline powder. Mp: 109°, bp: 269.0°, vap press: 1 mm @ 91.7°, vap d: 5.8. Moderately sol in water and organic solvents.

SYNS: 2-ACETYL-5-HYDROXY-3-OXO-4-HEXENOIC ACID Δ-LACTONE ◇ 3-ACETYL-6-METHYL-2,4-PYRANDIONE ◇ 3-ACETYL-6-METHYLPYRANDIONE-2,4 ◇ 3-ACETYL-6-METHYL-2H-PYRAN-2,4(3H)-DIONE ◇ DEHYDRACETIC ACID ◇ DEHYDROACETIC ACID (FCC) ◇ DHA ◇ DHS

TOXICITY DATA with REFERENCE
scu-rat TDLo:592 mg/kg/37W-I:ETA BJCAAI 20,134,66
orl-rat LD50:500 mg/kg WRPCA2 9,119,70
ipr-mus LD50:2000 mg/kg APTOA6 2,109,46
orl-dog LDLo:400 mg/kg HBTXAC 5,62,59
ivn-dog LDLo:400 mg/kg HBTXAC 5,62,59

CONSENSUS REPORTS: Reported in EPA TSCA Inventory.

SAFETY PROFILE: Poison by ingestion and intravenous routes. Moderately toxic by intraperitoneal route. Questionable carcinogen with experimental tumorigenic data. Combustible when exposed to heat or flame. When

heated to decomposition it emits acrid smoke and irritating fumes.

MFW750 CAS:36375-30-1 ***HR: 3***
METHYL-β-ACETOXYETHYL-β-CHLOROETHYLAMINE
mf: $C_7H_{14}ClNO_2$ mw: 179.67

SYNS: 2-ACETOXY-2'-CHLORO-N-METHYL-DIETHYLAMINE ◇ N-ACETOXYETHYL-N-CHLOROETHYLMETHYLAMINE ◇ 2-((2-CHLOROETHYL)METHYLAMINO)ETHANOL ACETATE ◇ TL 1428

TOXICITY DATA with REFERENCE
ihl-mus LCLo:500 mg/m³/10M NDRC** 30101,5,45
scu-mus LD50:20 mg/kg NTIS** PB158-507
ivn-mus LD50:36 mg/kg JPETAB 91,224,47

SAFETY PROFILE: Poison by intravenous and subcutaneous routes. Moderately toxic by inhalation. When heated to decomposition it emits very toxic fumes of Cl^- and NO_x.

MFX000 CAS:7790-01-4 ***HR: 3***
METHYLACETOXYMALONONITRILE
mf: $C_6H_6N_2O_2$ mw: 138.14

SYN: 2-ACETOXYISOSUCCINODINITRILE

TOXICITY DATA with REFERENCE
orl-rat LDLo:120 mg/kg AIHAAP 23,95,62
ihl-rat LC50:597 mg/m³/4H AMIHBC 4,573,51
orl-rbt LDLo:18 mg/kg AMIHBC 4,573,51
ihl-rbt LCLo:108 ppm/4H AMIHBC 4,573,51
skn-rbt LD50:110 mg/kg AIHAAP 23,95,62

CONSENSUS REPORTS: Cyanide and its compounds are on the Community Right-To-Know List.

SAFETY PROFILE: Poison by ingestion and skin contact. Moderately toxic by inhalation. When heated to decomposition it emits toxic fumes of NO_x and CN^-. See also NITRILES.

MFX250 CAS:105-45-3 ***HR: 2***
METHYL ACETYLACETATE
mf: $C_5H_8O_3$ mw: 116.13

PROP: Colorless liquid. Mp: 27.5°, bp: 170°, flash p: 170°F, autoign temp: 536°F, d: 1.077, vap d: 4.00.

SYNS: ACETOACETIC METHYL ESTER ◇ METHYLACETOACETATE ◇ METHYL ACETYLACETONATE ◇ METHYL-3-OXOBUTYRATE ◇ 3-OXOBUTANOIC ACID METHYL ESTER

TOXICITY DATA with REFERENCE
skn-rbt 10 mg/24H JIHTAB 30,63,48
skn-rbt 500 mg/24H MLD FCTXAV 16,637,78
eye-rbt 2 mg SEV AJOPAA 29,1363,46
orl-rat LD50:3228 mg/kg JIHTAB 30,63,48

CONSENSUS REPORTS: Reported in EPA TSCA Inventory.

SAFETY PROFILE: Moderately toxic by ingestion. A skin and severe eye irritant. Combustible when exposed to heat or flame. To fight fire, use foam, CO_2, dry chemical. When heated to decomposition it emits acrid smoke and irritating fumes. See also ESTERS.

MFX500 CAS:59665-11-1 ***HR: D***
N-METHYL-N-ACETYLAMINOMETHYLNITROSAMINE
mf: $C_4H_9N_3O_2$ mw: 131.16

TOXICITY DATA with REFERENCE
cyt-ham:fbr 4 g/L/48H MUREAV 48,337,77

SAFETY PROFILE: Mutation data reported. Many N-nitroso compounds are carcinogens. When heated to decomposition it emits toxic fumes of NO_x. See also NITROSAMINES.

MFX590 CAS:74-99-7 ***HR: 3***
METHYL ACETYLENE
mf: C_3H_4 mw: 40.07

$$HC \equiv CCH_3$$

PROP: Gas. Mp: −104°, lel: 1.7%, bp: −23.3°, vap press: 3876 mm @ 20°, d: 1.787 g/L @ 0°, vap d: 1.38.

SYNS: PROPINE ◇ PROPYNE

CONSENSUS REPORTS: Reported in EPA TSCA Inventory.

OSHA PEL: TWA 1000 ppm
ACGIH TLV: TWA 1000 ppm
DFG MAK: 1000 ppm (1650 mg/m³)

SAFETY PROFILE: This compound is a simple anesthetic and in high concentration is an asphyxiant. Dangerous fire hazard when exposed to heat or flame; can react vigorously with oxidizing materials. Explosive in the form of vapor when exposed to heat or flame. Localized heating of liquid containing cylinders to 95°C may cause an explosion. Product of reaction with silver nitrate ignites at 150°C. A commercial mixture containing 30% propyne in MAPP gas is similar to ethylene in potential hazards and handling requirements. To fight fire, stop flow of gas. When heated to decomposition it emits acrid smoke and irritating fumes. See also ACETYLENE COMPOUNDS.

MFX600 CAS:59355-75-8 ***HR: 3***
METHYL ACETYLENE-PROPADIENE MIXTURE
DOT: UN 1060

SYNS: MAPP ◇ METHYLACETYLENE-PROPADIENE, STABILIZED (DOT) ◇ PROPYNE mixed with PROPADIENE

OSHA PEL: (Transitional: TWA 1000 ppm) TWA 1000 ppm; STEL 1250 ppm
ACGIH TLV: TWA 1000 ppm; STEL 1250 ppm
DFG MAK: 1000 ppm (1650 mg/m^3)
DOT Classification: Flammable Gas; Label: Flammable Gas.

SAFETY PROFILE: A flammable gas mixture. To fight fire, stop flow of gas. When heated to decomposition it emits acrid smoke and irritating fumes. See also PROPYNE and PROPADIENE.

MFX725 CAS:72586-67-5 ***HR: 2***
N-METHYL-N'-(p-ACETYLPHENYL)-N-NITROSOUREA
mf: $C_{10}H_{11}N_3O_3$ mw: 221.24

SYN: 1-(p-ACETYLPHENYL)-3-METHYL-3-NITROSOUREA

TOXICITY DATA with REFERENCE
mmo-sat 33500 pmol/plate CNREA8 39,5147,79
skn-mus TDLo:657 mg/kg/7W-I:CAR CNREA8 44,1027,84

SAFETY PROFILE: Questionable carcinogen with experimental carcinogenic data. Mutation data reported. When heated to decomposition it emits toxic fumes of NO_x.

MFX750 CAS:140-03-4 ***HR: 2***
METHYL ACETYL RICINOLEATE
mf: $C_{21}H_{38}NO_4$ mw: 1306.59

PROP: Crystals. Vap d: 11.9.

SYNS: METHYL-12-ACETOXY-9-OCTADECENOATE ◇ METHYL-12-ACETOXYOLEATE

TOXICITY DATA with REFERENCE
eye-rbt 500 mg AJOPAA 29,1363,46

CONSENSUS REPORTS: Reported in EPA TSCA Inventory.

SAFETY PROFILE: Probably moderately toxic by ingestion. An irritant to the eye and probably the skin and mucous membranes. Combustible when exposed to heat or flame; can react with oxidizing materials. When heated to decomposition it emits toxic fumes of NO_x.

MFY000 CAS:623-59-6 ***HR: D***
N-METHYL-N'-ACETYLUREA
mf: $C_4H_8N_2O_2$ mw: 116.14

PROP: Monoclinic crystals from water. Mp: 180°; very sol in hot water; very sltly sol in hot ether.

TOXICITY DATA with REFERENCE
cyt-ham:fbr 8 g/L/48H MUREAV 48,337,77

CONSENSUS REPORTS: Reported in EPA TSCA Inventory.

SAFETY PROFILE: Mutation data reported. When heated to decomposition it emits toxic fumes of NO_x.

MFY750 CAS:53222-10-9 ***HR: 3***
4'-(2-METHYL-9-ACRIDINYLAMINO)METHANESULFONANILIDE
mf: $C_{21}H_{19}N_3O_2S$ mw: 377.49

TOXICITY DATA with REFERENCE
mmo-sat 49 μmol/L JMCMAR 23,269,80
ipr-mus LD10:130 mg/kg JMCMAR 23,269,80

SAFETY PROFILE: Poison by intraperitoneal route. Mutation data reported. When heated to decomposition it emits very toxic fumes of NO_x and SO_x. See also SULFONATES.

MFZ000 CAS:53478-39-0 ***HR: 3***
4'-(3-METHYL-9-ACRIDINYLAMINO)METHANESULFONANILIDE
mf: $C_{21}H_{19}N_3O_2S$ mw: 377.49

TOXICITY DATA with REFERENCE
mmo-sat 22 μmol/L JMCMAR 23,269,80
ipr-mus LD10:30 mg/kg JMCMAR 23,269,80

SAFETY PROFILE: Poison by intraperitoneal route. Mutation data reported. When heated to decomposition it emits very toxic fumes of NO_x and SO_x. See also SULFONATES.

MGA000 CAS:53221-79-7 ***HR: 3***
4'-(4-METHYL-9-ACRIDINYLAMINO)METHANESULFONANILIDE
mf: $C_{21}H_{19}N_3O_2S$ mw: 377.49

TOXICITY DATA with REFERENCE
mmo-sat 25 μmol/L JMCMAR 23,269,80
ipr-mus LD10:45 mg/kg JMCMAR 23,269,80

SAFETY PROFILE: Poison by intraperitoneal route. Mutation data reported. When heated to decomposition it emits very toxic fumes of NO_x and SO_x. See also SULFONATES.

MGA250 CAS:78-85-3 ***HR: 3***
METHYLACRYLALDEHYDE
DOT: UN 2396
mf: C_4H_6O mw: 70.10

PROP: Colorless liquid. Mp: −81°C, bp: 73.5°, flash p:

35°F (OC), d: 0.830 @ 20°/4°, vap press: 120 mm @ 20°, vap d: 2.42. Sol in water.

SYNS: ISOBUTENAL ◇ METHACRALDEHYDE (DOT) ◇ METHACROLEIN ◇ METHACRYLALDEHYDE (DOT) ◇ METHACRYLIC ALDEHYDE ◇ α-METHYLACROLEIN ◇ 2-METHYL- ACROLEIN ◇ 2-METHYLPROPENAL (CZECH)

TOXICITY DATA with REFERENCE
skn-rbt 500 mg/24H SEV 28ZPAK -,41,72
eye-rbt 50 μg/24H SEV 28ZPAK -,41,72
mmo-sat 500 nmol/L MUREAV 148,25,85
mma-sat 500 μmol/L MUREAV 93,305,82
orl-rat LD50:111 mg/kg 28ZPAK -,224,72
ihl-rat LCLo:125 ppm/4H JIHTAB 31,343,49
skn-rbt LD50:364 mg/kg JIHTAB 31,60,49

CONSENSUS REPORTS: Reported in EPA TSCA Inventory.
DOT Classification: Flammable Liquid; Label: Flammable Liquid and Poison.

SAFETY PROFILE: Poison by ingestion and skin contact. Moderately toxic by inhalation. Severe eye and skin irritant. Mutation data reported. Dangerously flammable when exposed to heat, flame or oxidizers. Can react vigorously with oxidizing materials. To fight fire, use CO_2, alcohol foam, foam, dry chemical. When heated to decomposition it emits acrid smoke and irritating fumes. See also ALDEHYDES.

MGA275 CAS:28051-68-5 ***HR: 3***
2-METHYL ACRYLALDEHYDE OXIME
mf: C_4H_7NO mw: 85.11

$H_2C{=}C(CH_3)CH{=}NOH$

SAFETY PROFILE: Can form heat-sensitive explosive polymeric peroxides. When heated to decomposition it emits toxic fumes of NO_x. See also ALDEHYDES.

MGA300 CAS:1187-59-3 ***HR: 2***
N-METHYLACRYLAMIDE
mf: C_4H_7NO mw: 85.12

SYNS: ACRYLAMIDE, N-METHYL- ◇ 2-PROPENAMIDE, N-METHYL- (9CI)

TOXICITY DATA with REFERENCE
orl-mus TDLo:595 g/kg (male 5W pre):REP ARTODN 59,201,86
orl-rat LD50:476 mg/kg NEZAAQ 34,183,79
orl-mus LD50:477 mg/kg ARTODN 47,179,81

CONSENSUS REPORTS: Reported in EPA TSCA Inventory.

SAFETY PROFILE: Moderately toxic by ingestion. Experimental reproductive effects. When heated to decomposition it emits toxic fumes of NO_x.

MGA500 CAS:96-33-3 ***HR: 3***
METHYL ACRYLATE
DOT: UN 1919
mf: $C_4H_6O_2$ mw: 86.10

$CH_3OCO{\cdot}CH{=}CH_2$

PROP: Colorless liquid; acrid odor. D: 0.9561 @ 20°/4°, mp: −76.5°, bp: 70° @ 608 mm, lel: 2.8%, uel: 25%, fp: −75°, flash p: 27°F (OC), vap press: 100 mm @ 28°, vap d: 2.97. Sol in alc and ether.

SYNS: ACRYLATE de METHYLE (FRENCH) ◇ ACRYLIC ACID METHYL ESTER (MAK) ◇ ACRYLSAEUREMETHYLESTER (GERMAN) ◇ CURITHANE 103 ◇ METHOXYCARBONYLETHYLENE ◇ METHYLACRYLAAT (DUTCH) ◇ METHYL-ACRYLAT (GERMAN) ◇ METHYL ACRYLATE, INHIBITED (DOT) ◇ METHYL PROPENATE ◇ METHYL PROPENOATE ◇ METHYL-2-PROPENOATE ◇ METILACRILATO (ITALIAN) ◇ PROPENOIC ACID METHYL ESTER ◇ 2-PROPENOIC ACID METHYL ESTER

TOXICITY DATA with REFERENCE
eye-rat 578 ppm/49H-I JIHTAB 31,317,49
skn-rbt 100%/1M imm JIHTAB 31,317,49
skn-rbt 10 mg/24H JIHTAB 30,63,48
eye-rbt 19 mg AJOPAA 29,1363,46
mma-mus:lym 22 mg/L ENMUDM 8(Suppl 6),4,86
cyt-mus:lym 22 mg/L ENMUDM 8(Suppl 6),4,86
ihl-hmn TCLo:75 ppm:NOSE,EYE,PUL 34ZIAG -,75,69
orl-rat LD50:300 mg/kg JIHTAB 30,63,48
ihl-rat LC50:1350 ppm/4H JTEHD6 16,811,85
ipr-rat LD50:325 mg/kg AMPMAR 36,58,75
orl-mus LD50:827 mg/kg TOLED5 11,125,82
ihl-mus LCLo:9300 mg/m³ GISAAA 20(8),19,55
ipr-mus LD50:254 mg/kg JDREAF 51,526,72
orl-rbt LDLo:280 mg/kg JIHTAB 31,317,49
ihl-rbt LCLo:2522 ppm/1H JIHTAB 31,317,49
skn-rbt LD50:1243 mg/kg JIHTAB 30,63,48

CONSENSUS REPORTS: IARC Cancer Review: Group 3 IMEMDT 7,56,87; Animal Inadequate Evidence IMEMDT 39,99,86; Human Inadequate Evidence IMEMDT 19,47,79. Community Right-To-Know List. Reported in EPA TSCA Inventory.

OSHA PEL: TWA 10 ppm (skin)
ACGIH TLV: TWA 10 ppm (skin)
DFG MAK: 5 ppm (18 mg/m³)
DOT Classification: Flammable Liquid; Label: Flammable Liquid.

SAFETY PROFILE: Poison by ingestion and intraperitoneal routes. Moderately toxic by skin contact. Mildly toxic by inhalation. Human systemic effects by inhalation: olfaction effects, eye effects and respiratory effects. A skin and eye irritant. Mutation data reported. Chronic exposure has produced injury to lungs, liver, and kidneys in experimental animals. Questionable carcinogen. Dangerously flammable when exposed to heat, flame, or oxidizers. Dangerous explosion hazard in the form of vapor when exposed to heat, sparks, or flame. Can react vigorously with oxidizing materials. A storage hazard; it forms peroxides which may initiate exothermic polymerization. To fight fire, use foam, CO_2, dry chemical. When heated to decomposition it emits acrid smoke and irritating fumes. See also ESTERS.

MGA750 CAS:126-98-7 ***HR: 3***
METHYLACRYLONITRILE
mf: C_4H_5N mw: 67.10

PROP: Mp: −36°, bp: 90.3°, d: 0.805, vap press: 40 mm @ 12.8°, flash p: 55°F.

SYNS: 2-CYANOPROPENE-1 ◇ ISOPROPENE CYANIDE ◇ ISOPROPENYLNITRILE ◇ α-METHYLACRYLONITRILE ◇ 2-METHYLPROPENENITRILE ◇ RCRA WASTE NUMBER U152 ◇ USAF ST-40

TOXICITY DATA with REFERENCE
skn-rbt 200 mg MLD JIHTAB 31,113,49
eye-rbt 50 mg JIHTAB 31,113,49
orl-rat LD50:250 mg/kg AIHAAP 23,95,62
ihl-rat LC50:328 ppm/4H AIHAAP 29,202,68
orl-mus LDLo:15 mg/kg SCCUR* -,6,61
ihl-mus LC50:36 ppm/4H AIHAAP 29,202,68
ipr-mus LDLo:15 mg/kg JIHTAB 31,113,49
ihl-rbt LC50:37 ppm/4H AIHAAP 29,202,68
skn-rbt LD50:320 mg/kg AIHAAP 29,202,68
ihl-gpg LC50:88 ppm/4H AIHAAP 29,202,68

CONSENSUS REPORTS: EPA Extremely Hazardous Substances List. Cyanide and its compounds are on the Community Right-To-Know List. Reported in EPA TSCA Inventory.

OSHA PEL: TWA 1 ppm (skin)
ACGIH TLV: TWA 1 ppm (skin)

SAFETY PROFILE: Poison by ingestion, inhalation, skin contact, and intraperitoneal routes. A skin and eye irritant. A dangerous fire hazard when exposed to heat, flame, or sparks. When heated to decomposition it emits toxic fumes of NO_x and CN^-. See also NITRILES.

MGA850 CAS:109-87-5 ***HR: 3***
METHYLAL
DOT: UN 1234
mf: $C_3H_8O_2$ mw: 76.11

PROP: Colorless liquid, pungent odor. Mp: −104.8°, bp: 42.3°, d: 0.864 @ 20°/4°, vap press: 330 mm @ 20°, vap d: 2.63, autoign temp: 459°F. flash p.: −0.4°F.

SYNS: ANESTHENYL ◇ DIMETHOXYMETHANE ◇ DIMETHYL FORMAL ◇ FORMAL ◇ FORMALDEHYDE DIMETHYLACETAL ◇ METHYLENE DIMETHYL ETHER ◇ METYLAL (POLISH)

TOXICITY DATA with REFERENCE
ihl-rat LC50:15000 ppm NPIRI* 1,73,74
orl-rbt LD50:5708 mg/kg PSEBAA 29,730,32
scu-gpg LDLo:3013 mg/kg BJIMAG 8,279,51

OSHA PEL: TWA 1000 ppm
ACGIH TLV: TWA 1000 ppm
DFG MAK: 1000 ppm (3100 mg/m^3)
DOT Classification: Flammable Liquid; Label: Flammable Liquid.

SAFETY PROFILE: Moderately toxic by subcutaneous route. Mildly toxic by ingestion and inhalation. Can cause injury to lungs, liver, kidneys, and the heart. A narcotic and anesthetic in high concentrations. A very dangerous fire hazard when exposed to heat, flame, or oxidizers. Moderately explosive when exposed to heat or flame. May ignite or explode when heated with oxygen. To fight fire, use foam, CO_2, dry chemical. When heated to decomposition it emits acrid smoke and irritating fumes. See also ETHERS.

MGB000 CAS:62-57-7 ***HR: 2***
2-METHYLALANINE
mf: $C_4H_9NO_2$ mw: 103.14

SYNS: AIB ◇ α-AMINOISOBUTANOIC ACID ◇ α-AMINOISOBUTYRIC ACID ◇ 2-AMINOISOBUTYRIC ACID ◇ 2-AMINO-2-METHYLPROPANOIC ACID ◇ α,α-DIMETHYLGLYCINE ◇ α-METHYLALANINE

TOXICITY DATA with REFERENCE
ipr-mus LD50:750 mg/kg NTIS** AD691-490

CONSENSUS REPORTS: Reported in EPA TSCA Inventory.

SAFETY PROFILE: Moderately toxic by intraperitoneal route. When heated to decomposition it emits toxic fumes of NO_x.

MGB150 CAS:67-56-1 ***HR: 3***
METHYL ALCOHOL
DOT: UN 1230
mf: CH_4O mw: 32.05

PROP: Clear, colorless, very mobile liquid; slt alcoholic odor when pure; crude material may have a repulsive pungent odor. Mp: 64.8°, lel: 6.0%, uel: 36.5%, ULC: 70, fp: −97.8°, d: 0.7915 @ 20°/4°, autoign temp: 878°F, vap press: 100 mm @ 21.2°, vap d: 1.11. Misc in water, ethanol, ether, benzene, ketones, and most other organic solvents.

SYNS: ALCOOL METHYLIQUE (FRENCH) ◇ ALCOOL METILICO (ITALIAN) ◇ CARBINOL ◇ COLONIAL SPIRIT ◇ COLUMBIAN SPIRITS (DOT) ◇ METANOLO (ITALIAN) ◇ METHANOL ◇ METHYLALKOHOL (GERMAN) ◇ METHYL HYDROXIDE ◇ METHYLOL ◇ METYLOWY ALKOHOL (POLISH) ◇ MONOHYDROXYMETHANE ◇ PYROXYLIC SPIRIT ◇ RCRA WASTE NUMBER U154 ◇ WOOD ALCOHOL (DOT) ◇ WOOD NAPHTHA ◇ WOOD SPIRIT

TOXICITY DATA with REFERENCE
skn-rbt 500 mg/24H MOD 28ZPAK -,33,72

eye-rbt 40 mg MOD UCDS** 3/24/70
dni-hmn:lym 300 mmol/L PNASA6 79,1171,82
mma-mus:lym 7900 mg/L ENMUDM 7(Suppl 3),10,85
orl-rat TDLo:7500 mg/kg (17-19D preg):REP TOXID9 1,32,81
ihl-rat TCLo:10000 ppm/7H (7-15D preg):TER FAATDF 5,727,85
orl-man LDLo:6422 mg/kg:CNS,PUL,GIT CMAJAX 128,14,83
orl-man TDLo:3429 mg/kg:EYE AMSVAZ 212,5,82
orl-hmn LDLo:428 mg/kg:CNS,PUL NPIRI* 1,74,74
orl-hmn LDLo:143 mg/kg:EYE,PUL,GIT 34ZIAG -,382,69
orl-wmn TDLo:4 g/kg:EYE,PUL,GIT AMSVAZ 212,5,82
ihl-hmn TCLo:86000 mg/m^3:EYE,PUL AGGHAR 5,1,33
ihl-hmn TCLo:300 ppm:EYE,CNS,PUL NPIRI* 1,74,74
orl-rat LD50:5628 mg/kg GTPZAB 19(11),27,75
ihl-rat LC50:64000 ppm/4H NPIRI* 1,74,74
ipr-rat LD50:7529 mg/kg EVHPAZ 61,321,85
ivn-rat LD50:2131 mg/kg EVHPAZ 61,321,85
orl-mus LD50:7300 mg/kg TXCYAC 25,271,82
ipr-mus LD50:10765 mg/kg EVHPAZ 61,321,85
scu-mus LD50:9800 mg/kg TXAPA9 18,185,71
ivn-mus LD50:4710 mg/kg EVHPAZ 61,321,85
orl-mky LDLo:7000 mg/kg TXAPA9 3,202,61
ihl-mky LCLo:1000 ppm IECHAD 23,931,31
skn-mky LDLo:393 mg/kg IECHAD 23,931,31

CONSENSUS REPORTS: Community Right-To-Know List. Reported in EPA TSCA Inventory. EPA Genetic Toxicology Program.

OSHA PEL: (Transitional: TWA 200 ppm) TWA 200 ppm; STEL 250 ppm (skin)
ACGIH TLV: TWA 200 ppm; STEL 250 ppm (skin)
DFG MAK: 200 ppm (260 mg/m^3); BAT: 30 mg/L in urine at end of shift.
NIOSH REL: TWA 200 ppm; CL 800 ppm/15M
DOT Classification: Flammable Liquid; Label: Flammable Liquid, Poison.

SAFETY PROFILE: A human poison by ingestion. Poison experimentally by skin contact. Moderately toxic experimentally by intravenous and intraperitoneal routes. Mildly toxic by inhalation. Human systemic effects: changes in circulation, cough, dyspnea, headache, lacrimation, nausea or vomiting, optic nerve neuropathy, respiratory effects, visual field changes. An experimental teratogen. Experimental reproductive effects. An eye and skin irritant. Human mutation data reported. A narcotic.

Its main toxic effect is exerted upon the nervous system, particularly the optic nerves and possibly the retinae which can progress to permanent blindness. Once absorbed, methanol is only very slowly eliminated. Coma resulting from massive exposures may last as long as 2-4 days. In the body, the products formed by its oxidation are formaldehyde and formic acid, both of which are toxic. Because of the slow elimination, methanol should be regarded as a cumulative poison. Though single exposures to fumes may cause no harmful effect, daily exposure may result in the accumulation of sufficient methanol in the body to cause illness. Death from ingestion of less than 30 mL has been reported. A common air contaminant.

Flammable liquid. Dangerous fire hazard when exposed to heat, flame, or oxidizers. Explosive in the form of vapor when exposed to heat or flame. Explosive reaction with chloroform + sodium methoxide; diethyl zinc. Violent reaction with alkyl aluminum salts; acetyle bromide; chloroform + sodium hydroxide; CrO_3; cyanuric chloride; (I + ethanol + HgO); $Pb(ClO_4)_2$; $HClO_4$; P_2O_3; (KOH + $CHCl_3$); nitric acid. Incompatible with beryllium dihydride; metals (e.g., potassium; magnesium); oxidants (e.g., barium perchlorate; bromine; sodium hypochlorite; chlorine; hydrogen peroxide); potassium tert-butoxide; carbon tetrachloride + metals (e.g., aluminum; magnesium; zinc); dichloromethane. Dangerous; can react vigorously with oxidizing materials. To fight fire, use alcohol foam. When heated to decomposition it emits acrid smoke and irritating fumes.

MGC200 ***HR: 2***
(±)-2-(p-((2-METHYLALLYL)AMINO)PHENYL) PROPIONIC ACID
mf: $C_{13}H_{17}NO_2$ mw: 219.31

SYNS: (±)-ALMINOPROFEN ◇ BENZENEACETIC ACID, α-METHYL-4-((2-METHYL-2-PROPENYL)AMINO)-,(±)- ◇ EB-382

TOXICITY DATA with REFERENCE
orl-rat TDLo:4750 mg/kg (male 74D pre):REP YACHDS 14,2121,86
orl-rat LD50:550 mg/kg YACHDS 14,2093,86
ipr-rat LD50:700 mg/kg YACHDS 14,2093,86
scu-rat LD50:660 mg/kg YACHDS 14,2093,86
orl-mus LD50:1520 mg/kg YACHDS 14,2093,86
ipr-mus LD50:705 mg/kg YACHDS 14,2093,86

SAFETY PROFILE: Moderately toxic by ingestion, subcutaneous, and intraperitoneal routes. Experimental reproductive effects. When heated to decomposition it emits toxic fumes of NO_x.

MGC225 CAS:12263-85-3 ***HR: 3***
METHYL ALUMINUM SESQUIBROMIDE
DOT: UN 1926
mf: $C_3H_9Al_2Br_3$ mw: 338.81

SYNS: ALUMINUM, TRIBROMOTRIMETHYLDI- ◇ METHYL ALUMINIUM SESQUIBROMIDE (DOT) ◇ METHYL ALUMINUM SESQUIBROMIDE ◇ TRIBROMOTRIMETHYLDIALUMINUM

CONSENSUS REPORTS: Reported in EPA TSCA Inventory.

ACGIH TLV: TWA 2 mg(Al)/m^3
DOT Classification: Flammable Solid; Label: Spontaneously Combustible

SAFETY PROFILE: Flammable solid. Danger from spontaneous combustion. When heated to decomposition it emits toxic fumes of Br^-.

MGC230 CAS:12542-85-7 ***HR: 3***
METHYL ALUMINUM SESQUICHLORIDE (DOT)
DOT: UN 1927
mf: $C_3H_9Al_2Cl_3$ mw: 205.43

PROP: D: 0.877, flash p: 1° F.

SYNS: ALUMINUM, TRICHLOROTRIMETHYLDI- ◇ METHYL ALUMINIUM SESQUICHLORIDE (DOT) ◇ TRICHLOROTRIMETHYLDIALUMINUM

CONSENSUS REPORTS: Reported in EPA TSCA Inventory.

ACGIH TLV: TWA 2 mg(Al)/m^3
DOT Classification: Flammable Solid; Label: Spontaneously Combustible

SAFETY PROFILE: Flammable solid. Danger from spontaneous combustion. When heated to decomposition it emits toxic fumes of Cl^-.

MGC250 CAS:74-89-5 ***HR: 3***
METHYLAMINE
DOT: UN 1061/UN 1235
mf: CH_5N mw: 31.07

PROP: Colorless gas or liquid; powerful ammoniacal odor. Bp: 6.3°, lel: 4.95%, uel: 20.75%, mp: −93.5°, flash p: 32°F (CC), d: 0.662 @ 20°/4°, autoign temp: 806°F, vap d: 1.07. Fuming liquid when liquefied: d: 0.699 @ −10.8°/4°. Sol in alc; misc with ether.

SYNS: AMINOMETHANE ◇ CARBINAMINE ◇ MERCURIALIN ◇ METHANAMINE (9CI) ◇ METHYLAMINEN (DUTCH) ◇ METILAMINE (ITALIAN) ◇ METYLOAMINA (POLISH) ◇ MONOMETHYLAMINE

TOXICITY DATA with REFERENCE
skn-gpg 100 mg open SEV CODEDG 6,140,80
dlt-rat-ihl 10 μg/m^3 GISAAA 46(5),7,81
scu-rat LDLo:200 mg/kg HBAMAK 4,1289,35
ihl-mus LC50:2400 mg/m^3/2H 85GMAT -,80,82
scu-mus LDLo:2500 mg/kg MEIEDD 11,949,89
scu-gpg LDLo:200 mg/kg HBAMAK 4,1289,35
scu-frg LDLo:2000 mg/kg 27ZWAY 1,250,23
ihl-mam LC50:2400 mg/m^3 TPKVAL 14,80,75

CONSENSUS REPORTS: Reported in EPA TSCA Inventory.
OSHA PEL: TWA 10 ppm
ACGIH TLV: TWA 10 ppm; (Proposed: TWA 5 ppm; STEL 15 ppm)
DFG MAK: 10 ppm (12 mg/m^3)
DOT Classification: Flammable Gas; Label: Flammable Gas, Flammable Liquid.

SAFETY PROFILE: Poison by subcutaneous route. Moderately toxic by inhalation. A severe skin irritant. Mutation data reported. A strong base. Flammable gas at ordinary temperature and pressure. Very dangerous fire hazard when exposed to heat, flame, or sparks. Explosive when exposed to heat or flame. To fight fire, stop flow of gas. Forms an explosive mixture with nitromethane. When heated to decomposition it emits toxic fumes of NO_x. See also AMINES.

MGC350 CAS:99-45-6 ***HR: 3***
4-METHYLAMINOACETOCATECHOL
mf: $C_9H_{11}NO_3$ mw: 181.21

PROP: Needles. Decomp 235-236°. Sparingly sol in water, alc, ether.

SYNS: ADRENALONE ◇ ADRENON ◇ ADRENONE ◇ CHEMOSAN ◇ 3,4-DIHYDROXY-α-METHYLAMINOACETOPHENONE ◇ 3,4′-DIHYDROXY-2-(METHYLAMINO)ACETOPHENONE ◇ 1-(3,4-DIHYDROXYPHENYL)-2-(METHYLAMINO)-ETHANONE (9CI) ◇ HAEMODAN ◇ KEPHRINE ◇ KETOGAZE ◇ METHAMINOACETOCATECHOL ◇ 4-METHYLAMINOACETOPYROCATECHOL ◇ REMESTYP ◇ STRYPHNON ◇ STRYPHNONE ◇ STYPHNONE ◇ STYPNON ◇ U 2134

TOXICITY DATA with REFERENCE
dnd-omi 250 μmol/L ABCHA6 42,1019,78
dnd-rat:lng 250 μmol/L ABCHA6 42,1019,78
ivn-mus LD50:275 mg/kg AEPPAE 226,493,55

SAFETY PROFILE: Poison by intravenous route. Mutation data reported. When heated to decomposition it emits toxic fumes of NO_x.

MGD000 CAS:37045-40-2 ***HR: 3***
9-(p-(METHYLAMINO)ANILINO)ACRIDINE HYDROBROMIDE
mf: $C_{20}H_{17}N_3$•BrH mw: 380.32

TOXICITY DATA with REFERENCE
mma-sat 82 μmol/L JMCMAR 22,251,79
ipr-mus LD10:70 mg/kg JMCMAR 22,251,79

SAFETY PROFILE: Poison by intraperitoneal route. Mutation data reported. When heated to decomposition it emits very toxic fumes of NO_x and HBr. See also BROMIDES.

MGD100 CAS:87425-02-3 *HR: 3*
4-METHYLAMINOBENZENE-1,3-BIS(SULFONYL AZIDE)
mf: $C_7H_7N_7O_4S_2$ mw: 317.30

$CH_3NHC_6H_3(SO_2N_3)_2$

SAFETY PROFILE: Explodes when heated. Upon decomposition it emits toxic fumes of SO_x and NO_x. See also AZIDES.

MGD200 CAS:700-07-2 *HR: 3*
3-(METHYLAMINO)-2,1-BENZISOTHIAZOLE
mf: $C_8H_8N_2S$ mw: 166.24

SYN: CI 624

TOXICITY DATA with REFERENCE
orl-rat LD50:400 mg/kg JPETAB 153,292,66
ipr-rat LD50:336 mg/kg JPETAB 153,292,66
scu-rat LD50:1445 mg/kg JPETAB 153,292,66

SAFETY PROFILE: Poison by ingestion and intraperitoneal routes. Moderately toxic by subcutaneous route. When heated to decomposition it emits toxic fumes of SO_x and NO_x.

MGD210 CAS:7765-88-0 *HR: 3*
3-(METHYLAMINO)-2,1-BENZISOTHIAZOLE HYDROCHLORIDE
mf: $C_8H_8N_2S•ClH$ mw: 200.70

SYN: CI 624 HYDROCHLORIDE

TOXICITY DATA with REFERENCE
orl-rat LD50:576 mg/kg JPETAB 153,292,66
ipr-rat LD50:346 mg/kg JPETAB 153,292,66
scu-rat LD50:599 mg/kg JPETAB 153,292,66

SAFETY PROFILE: Poison by intraperitoneal route. Moderately toxic by ingestion and subcutaneous routes. When heated to decomposition it emits toxic fumes of SO_x, NO_x and HCl.

MGD500 CAS:2536-91-6 *HR: 3*
6-METHYL-2-AMINOBENZOTHIAZOLE
mf: $C_8H_8N_2S$ mw: 164.24

SYNS: 2-AMINO-6-METHYLBENZOTHIAZOLE ◇ SKF 1045

TOXICITY DATA with REFERENCE
ivn-mus LD50:84 mg/kg JPETAB 105,486,52

CONSENSUS REPORTS: Reported in EPA TSCA Inventory.

SAFETY PROFILE: Poison by intravenous route. When heated to decomposition it emits very toxic fumes of NO_x and SO_x.

MGD750 CAS:64036-72-2 *HR: 2*
4-METHYL-2-AMINOBENZOTHIAZOLE HYDROCHLORIDE
mf: $C_8H_8N_2S•ClH$ mw: 200.70

SYN: 2-AMINO-4-METHYL-BENZOTHIAZOLE,HYDROCHLORIDE

TOXICITY DATA with REFERENCE
orl-rat LD50:500 mg/kg JPETAB 90,260,47

CONSENSUS REPORTS: Reported in EPA TSCA Inventory.

SAFETY PROFILE: Moderately toxic by ingestion. When heated to decomposition it emits very toxic fumes of HCl, SO_x and NO_x.

MGE000 CAS:63019-98-7 *HR: 3*
3-METHYL-4-AMINOBIPHENYL
mf: $C_{13}H_{13}N$ mw: 183.27

SYN: 3-METHYL-4-AMINODIPHENYL

TOXICITY DATA with REFERENCE
scu-rat TDLo:1200 mg/kg/W-I:ETA BMBUAQ 14,141,58

SAFETY PROFILE: Questionable carcinogen with experimental tumorigenic data. When heated to decomposition it emits toxic fumes of NO_x.

MGE100 CAS:2275-61-8 *HR: 3*
METHYLAMINO-BIS(1-AZIRIDINYL)PHOSPHINE OXIDE
mf: $C_5H_{12}N_3OP$ mw: 161.17

SYNS: AI 3-51254 ◇ P,P-BIS(1-AZIRIDINYL)-N-METHYLPHOSPHINIC AMIDE ◇ ENT 51254 ◇ PHOSPHINIC AMIDE, P,P-BIS(1-AZIRIDINYL)-N-METHYL- ◇ PHOSPHINE OXIDE, BIS(1-AZIRIDINYL) METHYLAMINO-

TOXICITY DATA with REFERENCE
cyt-mus-par 6 mmol/kg FOBLAN 20,1,74
dlt-mus-ipr 500 μg/kg FOBLAN 20,1,74
ipr-mus TDLo:500 μg/kg (male 1D pre):REP FOBLAN 20,1,74
ipr-mus LDLo:18 mg/kg FATOAO 28,70,65
orl-qal LD50:237 mg/kg JRPFA4 48,371,76

SAFETY PROFILE: Poison by ingestion and intraperitoneal routes. Experimental reproductive effects. Mutation data reported. When heated to decomposition it emits toxic fumes of NO_x and PO_x.

MGF000 CAS:63917-71-5 *HR: 3*
METHYLAMINOCOLCHICIDE
mf: $C_{22}H_{26}N_2O_5$ mw: 398.50

SYN: METHYLCOLCHAMINONE

TOXICITY DATA with REFERENCE
ipr-mus LD50:400 mg/kg COREAF 241,1889,55
ivn-mus LD50:20 mg/kg COREAF 241,1889,55

SAFETY PROFILE: Poison by intravenous and intraperitoneal routes. When heated to decomposition it emits toxic fumes of NO_x.

MGF250 CAS:35271-57-9 ***HR: 3***
4-METHYLAMINO CRESOL-2-SULFATE
mf: $C_{16}H_{22}N_2O_2 \cdot H_2O_4S$ mw: 372.48

SYN: 4-(METHYLAMINO)-o-CRESOL, HYDROGEN SULFATE (2:1)

TOXICITY DATA with REFERENCE
orl-rat LDLo:400 mg/kg KODAK* -,-,71
ipr-rat LDLo:25 mg/kg KODAK* -,-,71

CONSENSUS REPORTS: Reported in EPA TSCA Inventory.

SAFETY PROFILE: Poison by ingestion and intraperitoneal routes. When heated to decomposition it emits very toxic fumes of NO_x and SO_x.

MGF500 CAS:63019-97-6 ***HR: 3***
2-METHYL-4-AMINODIPHENYL
mf: $C_{13}H_{13}N$ mw: 183.27

SYN: 2-METHYL-4-PHENYLANILINE

TOXICITY DATA with REFERENCE
scu-rat TDLo:2400 mg/kg/W-I:ETA BMBUAQ 14,141,58

SAFETY PROFILE: Questionable carcinogen with experimental tumorigenic data. When heated to decomposition it emits toxic fumes of NO_x.

MGF750 CAS:1204-78-0 ***HR: 3***
4'-METHYL-4-AMINODIPHENYL
mf: $C_{13}H_{13}N$ mw: 183.27

SYN: 4'-METHYLBIPHENYLAMINE

TOXICITY DATA with REFERENCE
scu-rat TDLo:10 g/kg/W-I:ETA BMBUAQ 14,141,58

SAFETY PROFILE: Questionable carcinogen with experimental tumorigenic data. When heated to decomposition it emits toxic fumes of NO_x. See also AMINES.

MGG000 CAS:109-83-1 ***HR: 3***
2-METHYLAMINOETHANOL
mf: C_3H_9NO mw: 75.11

PROP: Viscous liquid; fishy odor. Corrosive to skin, cork, and metals. A strong base. D: 0.9, vap d: 2.9, bp: 156°, flash p: 165°F (OC). Misc with water, alc, and ether.

SYNS: β-(METHYLAMINO)ETHANOL ◇ N-METHYLAMINOETHANOL ◇ N-METHYLETHANOLAMINE ◇ METHYLETHYLOLAMINE ◇ METHYL(β-HYDROXYETHYL)AMINE ◇ MONOMETHYLAMINOAETHANOL (GERMAN) ◇ N-MONOMETHYLAMINOETHANOL ◇ MONOMETHYLAMINOETHANOL ◇ USAF DO-50

TOXICITY DATA with REFERENCE
skn-rbt 10 mg/24H open MLD AMIHBC 10,61,54
skn-rbt 470 mg open MLD UCDS** 1/20/72
eye-rbt 250 μg open SEV AMIHBC 10,61,54
orl-rat LD50:2340 mg/kg AMIHBC 10,61,54
ipr-rat LD50:1330 mg/kg TXAPA9 12,486,68
ipr-mus LD50:125 mg/kg NTIS** AD277-689
scu-mus LD50:1802 mg/kg AEPPAE 225,428,55

CONSENSUS REPORTS: Reported in EPA TSCA Inventory.

SAFETY PROFILE: Poison by intraperitoneal route. Moderately toxic by ingestion and subcutaneous routes. A corrosive irritant to skin, eyes, and mucous membranes. Flammable when exposed to heat, flame, or oxidizers. To fight fire, use alcohol foam. When heated to decomposition it emits toxic fumes such as NO_x. See also AMINES and ALCOHOLS.

MGG250 CAS:2475-46-9 ***HR: 2***
1-METHYLAMINO-4-ETHANOLAMINO-ANTHRAQUINONE
mf: $C_{17}H_{16}N_2O_3$ mw: 296.35

SYNS: ACETATE BRILLIANT BLUE 4B ◇ ACETOQUINONE LIGHT PURE BLUE R ◇ ALTOCYL BRILLIANT BLUE B ◇ AMACEL BLUE BNN ◇ AMACEL BRILLIANT BLUE B ◇ ARTISIL BLUE BSG ◇ CALCOSYN SAPPHIRE BLUE R ◇ CELANTHRENE BRILLIANT BLUE ◇ CELLITON BLUE FFR ◇ CELUTATE BLUE BLT ◇ C.I. 61505 ◇ CIBACET BRILLIANT BLUE BG NEW ◇ C.I. DISPERSE BLUE 3 ◇ CILLA FAST BLUE FFR ◇ DIACELLITON FAST BRILLIANT BLUE B ◇ DISPERSE BLUE K ◇ DURANOL BRILLIANT BLUE B ◇ EASTMAN BLUE BNN ◇ FENACET AST BLUE FF ◇ 4-HYDROXYETHYLAMINO-1-METHYLAMINOANTHRAQUINONE ◇ INTERCHEM ACETATE BLUE B ◇ KAYALON FAST BLUE FN ◇ LURAFIX BLUE FFR ◇ 1-MA-40EAA (RUSSIAN) ◇ 1-METHYLAMINO-4-(β-HYDROXYETHYLAMINO)ANTHRAQUINONE ◇ 1-METHYLAMINO-4-OXYETHYLAMINOANTHRAQUINONE (RUSSIAN) ◇ MICROSETILE BLUE FF ◇ MIKETON BRILLIANT BLUE B ◇ MODR OSTACETOVA P3R (CZECH) ◇ NACELAN BLUE KLT ◇ NYLOQUINONE PURE BLUE ◇ PERLITON BLUE FFR ◇ SERINYL HOSIERY BLUE ◇ SERISOL BRILLIANT BLUE BG ◇ SETACYL BLUE BN ◇ SUPRACET BRILLIANT BLUE BG ◇ TRANSETILE BLUE P-FER

TOXICITY DATA with REFERENCE
eye-rbt 500 mg/24H MLD 28ZPAK -,245,72
orl-rat LD50:3000 mg/kg 28ZPAK -,245,72
ipr-rat LD50:700 mg/kg GTPZAB 21(12),27,77

CONSENSUS REPORTS: Reported in EPA TSCA Inventory.

SAFETY PROFILE: Moderately toxic by ingestion and intraperitoneal routes. An eye irritant. When heated to decomposition it emits toxic fumes of NO_x.

MGH250 CAS:63991-26-4 ***HR: 3***
α-(1-METHYLAMINOETHYL)BENZYL ALCOHOL HYDROCHLORIDE
mf: $C_{10}H_{15}NO \cdot ClH$ mw: 201.72

SYN: EPHEDRINHYDROCHLORID (GERMAN)

TOXICITY DATA with REFERENCE
orl-hmn TDLo:46 mg/kg:CVS,CNS KLWOAZ 17,1580,38
unr-mus LD50:96 mg/kg JPETAB 92,283,48

SAFETY PROFILE: Poison by an unspecified route. Human systemic effects by ingestion: altered sleep time, anorexia, change in cardiac rate. When heated to decomposition it emits very toxic fumes of HCl and NO_x.

MGH750 CAS:28089-05-6 ***HR: 3***
2-METHYL-3-(β-AMINOETHYL)-5-METHOXY-BENZOFURAN
mf: $C_{12}H_{14}NO_2$•ClH mw: 240.73

SYN: 3-(2-AMINOETHYL)-5-METHOXY-2-METHYLBENZO-FURAN, HYDROCHLORIDE

TOXICITY DATA with REFERENCE
ivn-rat LDLo:39 mg/kg RPTOAN 33,246,70
ivn-mus LD50:50 mg/kg RPTOAN 33,246,70

SAFETY PROFILE: Poison by intravenous route. When heated to decomposition it emits very toxic fumes of HCl and NO_x.

MGJ600 ***HR: 3***
N-(α-METHYLAMINOPHENETHYL)PHENOL HYDROCHLORIDE
mf: $C_{15}H_{17}NO$•ClH mw: 263.79

SYNS: 3-(α-METHYLAMINOPHENETHYL)PHENOL HYDROCHLORIDE ◇ N-METHYL-1-(3-HYDROXYPHENYL)-2-PHENYLETHYLAMINE HYDROCHLORIDE ◇ WIN 6703

TOXICITY DATA with REFERENCE
orl-rat LD50:1830 mg/kg AIPTAK 88,482,52
ivn-rat LD50:54 mg/kg AIPTAK 88,482,52
orl-mus LD50:460 mg/kg AIPTAK 88,482,52
scu-mus LDLo:443 mg/kg JAPMA8 39,354,50
ivn-mus LD50:46 mg/kg AIPTAK 88,482,52

SAFETY PROFILE: Poison by intravenous route. Moderately toxic by ingestion and subcutaneous routes. When heated to decomposition it emits toxic fumes of NO_x and HCl.

MGJ750 CAS:55-55-0 ***HR: 3***
p-METHYLAMINOPHENOLSULFATE
mf: $C_{14}H_{20}N_2O_6S$ mw: 344.38

PROP: Crystals. Discolors in air. Mp: approx 260° (decomp). Sltly sol in cold water and alc; insol in ether; mod sol in boiling H_2O. Keep well closed and protected from light.

SYNS: ELON ◇ GENOL ◇ GRAPHOL ◇ METATYL ◇ METHYL-p-AMINOPHENOL SULFATE ◇ p-(METHYLAMINO)PHENOL SULFATE (2:1) (SALT) ◇ PHOTOL ◇ RHODOL ◇ VEROL

TOXICITY DATA with REFERENCE
mma-sat 333 μg/plate NTPTB* JAN 82
orl-rat LDLo:200 mg/kg KODAK* -,-,71
ipr-rat LDLo:50 mg/kg KODAK* -,-,71

CONSENSUS REPORTS: Reported in EPA TSCA Inventory. EPA Genetic Toxicology Program.

SAFETY PROFILE: Poison by ingestion and intraperitoneal routes. Mutation data reported. When heated to decomposition it emits very toxic fumes of SO_x and NO_x.

MGK750 CAS:52777-39-6 ***HR: 3***
1-(3-METHYLAMINOPROPYL)-2-ADAMANTANOL HYDROCHLORIDE
mf: $C_{14}H_{25}NO$•ClH mw: 259.86

SYNS: 3-(2-HYDROXY-1-ADAMANTYL)-N-METHYLPROPLYAMINE HYDROCHLORIDE ◇ 2-HYDROXY-N-METHYL-1-ADAMANTANEPROPANAMINE HYDROCHLORIDE ◇ 2-HYDROXY-1-(3-METHYLAMINOPROPYL)ADAMANTANE HYDROCHLORIDE

TOXICITY DATA with REFERENCE
orl-mus LD50:300 mg/kg JMCMAR 17,602,74
ipr-mus LD50:150 mg/kg JMCMAR 17,602,74

SAFETY PROFILE: Poison by ingestion and intraperitoneal routes. When heated to decomposition it emits very toxic fumes of NO_x and HCl.

MGL500 CAS:10083-53-1 ***HR: 3***
4-(3′-METHYLAMINOPROPYLIDENE)-9,10-DIHYDRO-4H-BENZO(4,5)CYCLOHEPTA(1,2-b)THIOPHEN
mf: $C_{17}H_{19}NS$ mw: 269.43

SYN: IBD 78

TOXICITY DATA with REFERENCE
orl-rat LD50:500 mg/kg 27ZQAG -,321,72
ivn-rat LD50:28 mg/kg 27ZQAG -,321,72
orl-mus LD50:420 mg/kg 27ZQAG -,321,72
ivn-mus LD50:34 mg/kg 27ZQAG -,321,72
orl-rbt LD50:2000 mg/kg 27ZQAG -,321,72
ivn-rbt LD50:14 mg/kg 27ZQAG -,321,72

SAFETY PROFILE: Poison by intravenous route. Moderately toxic by ingestion. When heated to decomposition it emits very toxic fumes of NO_x and SO_x.

MGL600 CAS:7698-91-1 ***HR: 3***
N-METHYL-4-AMINO-1,2,5-SELENADIAZOLE-3-CARBOXAMIDE
mf: $C_4H_6N_4OSe$ mw: 205.10

SYNS: 4-AMINO-N-METHYL-1,2,5-SELENADIAZOLE-3-CARBOXAMIDE ◇ NSC 93169 ◇ 1,2,5-SELENADIAZOLE-3-CARBOXAMIDE, 4-AMINO-N-METHYL-

TOXICITY DATA with REFERENCE
ipr-mus LDLo:4 mg/kg AACHAX-,551,66

OSHA PEL: TWA 0.2 mg(Se)/m^3
ACGIH TLV: TWA 0.2 mg(Se)/m^3

SAFETY PROFILE: Poison by intraperitoneal route. When heated to decomposition it emits toxic fumes of NO_x and Se.

MGN000 ***HR: 3***
METHYLAMMONIUM CHLORITE
mf: CH_6ClNO_2 mw: 99.52

SAFETY PROFILE: Concentrated solutions of the chlorite are explosively unstable. When heated to decomposition it emits toxic fumes of Cl^-, NH_3, and NO_x. See also CHLORITES.

MGN150 CAS:1941-24-8 ***HR: 3***
METHYLAMMONIUM NITRATE
mf: $CH_6N_2O_3$ mw: 94.07

SYN: METHANAMINIUM NITRATE

SAFETY PROFILE: May explode on contact with rust or copper powder. Rail tanks of 86% aqueous solutions or slurries have exploded during pumping operations. Upon decomposition it emits toxic fumes of NO_x and NH_3. See also NITRATES.

MGN250 ***HR: 3***
METHYLAMMONIUM PERCHLORATE
mf: CH_6ClNO_4 mw: 131.52

SAFETY PROFILE: Concentrated solutions are unstable explosives. Incompatible with ammonium; dimethylammonium; piperidinium perchlorates. Upon decomposition it emits toxic fumes of NO_x, Cl^- and NH_3. See also PERCHLORATES.

MGN500 CAS:110-43-0 ***HR: 2***
METHYL n-AMYL KETONE
DOT: UN 1110
mf: $C_7H_{14}O$ mw: 114.21

PROP: Colorless, mobile liquid; penetrating, fruity odor. Bp: 151.5°, flash p: 120°F (OC), autoign temp: 991°F, vap d: 3.94, d: 0.8197 @ 15°/4°. Very sltly sol in water; sol in alc and ether.

SYNS: AMYL-METHYL-CETONE (FRENCH) ◇ n-AMYL METHYL KETONE ◇ AMYL METHYL KETONE (DOT) ◇ FEMA No. 2544 ◇ 2-HEPTANONE ◇ METHYL-AMYL-CETONE (FRENCH) ◇ METHYL AMYL KETONE (DOT) ◇ METHYL PENTYL KETONE

TOXICITY DATA with REFERENCE
skn-rbt 14 mg/24H open MLD AIHAAP 23,95,62
orl-rat LD50:1670 mg/kg UCDS** 8/11/58
ihl-rat LCLo:4000 ppm/4H AIHAAP 23,95,62
orl-mus LD50:730 mg/kg APJUA8 12,79,62
skn-rbt LD50:12600 mg/kg AIHAAP 23,95,62

CONSENSUS REPORTS: Reported in EPA TSCA Inventory.

OSHA PEL: TWA 100 ppm
ACGIH TLV: TWA 50 ppm
NIOSH REL: (Ketones) TWA 465 mg/m^3
DOT Classification: Combustible Liquid; Label: None; IMO: Flammable or Combustible Liquid; Label: Flammable Liquid.

SAFETY PROFILE: Moderately toxic by ingestion. Mildly toxic by inhalation and skin contact. A skin irritant. Combustible liquid when exposed to heat or flame; can react with oxidizing materials. To fight fire, use foam, CO_2, dry chemical. When heated to decomposition it emits acrid smoke and fumes. See also KETONES.

MGN750 CAS:100-61-8 ***HR: 3***
METHYLANILINE
DOT: UN 2294
mf: C_7H_9N mw: 107.17

PROP: Colorless or sltly yellow liquid; becomes brown on exposure to air. Mp: −57°, d: 0.989 @ 20°/4°, bp: 194-197°. Sol in alc, ether; sltly sol in water.

SYNS: ANILINOMETHANE ◇ (METHYLAMINO)BENZENE ◇ N-METHYLAMINOBENZENE ◇ N-METHYL ANILINE (MAK) ◇ N-METHYLBENZENAMINE ◇ METHYLPHENYLAMINE ◇ N-METHYLPHENYLAMINE ◇ MONOMETHYL ANILINE (OSHA) ◇ N-MONOMETHYLANILINE ◇ N-PHENYLMETHYLAMINE

TOXICITY DATA with REFERENCE
ivn-cat LDLo:24 mg/kg JIHTAB 31,1,49
orl-rbt LDLo:280 mg/kg JIHTAB 31,1,49
ivn-rbt LDLo:24 mg/kg JIHTAB 31,1,49
orl-gpg LDLo:1200 mg/kg XPHBAO 271,16,41
scu-gpg LDLo:1200 mg/kg XPHBAO 271,16,41

CONSENSUS REPORTS: Reported in EPA TSCA Inventory.

OSHA PEL: (Transitional: TWA 2 ppm (skin)) TWA 0.5 ppm (skin)
ACGIH TLV: TWA 0.5 ppm (skin)
DFG MAK: 0.5 ppm (2 mg/m^3)
DOT Classification: Poison B; Label: St. Andrews Cross.

SAFETY PROFILE: Poison by ingestion and intravenous routes. Moderately toxic by subcutaneous route. When heated to decomposition it emits toxic fumes of NO_x.

MGO000 ***HR: 3***
METHYLANILINE and SODIUM NITRITE (1.2:1)

SYN: SODIUM NITRITE and METHYLANILINE (1:1.2)

TOXICITY DATA with REFERENCE
orl-mus TDLo:81 g/kg/28W-C:CAR JNCIAM 46,1029,71

SAFETY PROFILE: Questionable carcinogen with experimental carcinogenic data. When heated to decomposition it emits toxic fumes of NO_x. See also SODIUM NITRITE and METHYL ANILINE.

MGO250 ***HR: 3***
N-METHYLANILINE mixed with SODIUM NITRITE (1:35)

SYNS: N-METHYLANILIN UND NATRIUMNITRIT (GERMAN) ◇ NATRIUMNITRAT UND N-METHYLANILIN (GERMAN) ◇ SODIUM NITRITE mixed with N-METHYLANILINE (35:1)

TOXICITY DATA with REFERENCE
orl-rat TDLo:124 g/kg/16W-C:CAR ARZNAD 21,1572,71

SAFETY PROFILE: Questionable carcinogen with experimental carcinogenic data. When heated to decomposition it emits toxic fumes of NO_x. See also SODIUM NITRITE and METHYLANILINE.

MGO500 CAS:102-50-1 ***HR: 3***
2-METHYL-p-ANISIDINE
mf: $C_8H_{11}NO$ mw: 137.20

SYNS: m-CRESIDINE ◇ 4-METHOXY-2-METHYLANILINE ◇ 4-METHOXY-2-METHYLBENZENAMINE ◇ 2-METHYL-4-METHOXYANILINE ◇ NCI-C02993

TOXICITY DATA with REFERENCE
otr-rat:emb 51500 ng/plate JJATDK 1,190,81
orl-rat TDLo:62 g/kg/77W-I:CAR NCITR* NCI-CG-TR-105,78
orl-mus TDLo:29800 mg/kg/53W-I:CAR IARC** 27,91,82
orl-rat TD:31 g/kg/77W-I:ETA NCITR* NCI-CG-TR-105,78

CONSENSUS REPORTS: IARC Cancer Review: Group 3 IMEMDT 7,56,87; Animal Inadequate Evidence IMEMDT 27,91,82. NCI Carcinogenesis Bioassay (gavage); Clear Evidence: rat NCITR* NCI-CG-TR-105,78; (gavage); Inadequate Studies: mouse NCITR* NCI-CG-TR-105,78. Reported in EPA TSCA Inventory.

SAFETY PROFILE: Suspected carcinogen with experimental carcinogenic and tumorigenic data. Mutation data reported. When heated to decomposition it emits toxic fumes of NO_x.

MGO750 CAS:120-71-8 ***HR: 3***
5-METHYL-o-ANISIDINE
mf: $C_8H_{11}NO$ mw: 137.20

SYNS: m-AMINO-p-CRESOL, METHYL ESTER ◇ 3-AMINO-p-CRESOL METHYL ESTER ◇ 1-AMINO-2-METHOXY-5-METHYLBENZENE ◇ 3-AMINO-4-METHOXYTOLUENE ◇ 2-AMINO-4-METHYLANISOLE ◇ AZOIC RED 36 ◇ C.I. AZOIC RED 83 ◇ CRESIDINE ◇ p-CRESIDINE ◇ KRESIDIN ◇ KREZIDINE ◇ 2-METHOXY-5-METHYLANILINE ◇ 2-METHOXY-5-METHYL-BENZENAMINE (9CI) ◇ 4-METHOXY-m-TOLUIDINE ◇ 4-METHYL-2-AMINOANISOLE ◇ NCI-C02982

TOXICITY DATA with REFERENCE
mmo-sat 62500 ng/plate ENMUDM 7(Suppl 5),1,85
mma-sat 3330 ng/plate ENMUDM 7(Suppl 5),1,85
mma-esc 2 mg/plate ENMUDM 7(Suppl 5),1,85
otr-rat:emb 31 μg/plate JJATDK 1,190,81
orl-rat TDLo:364 g/kg/2Y-C:CAR NCITR* NCI-CG-TR-142,79
orl-mus TDLo:355 g/kg/92W-C:CAR NCITR* NCI-CG-TR-142,79
orl-rat TD:182 g/kg/2Y-C:NEO NCITR* NCI-CG-TR-142,79
orl-rat LD50:1450 mg/kg HURC** -,-,72

CONSENSUS REPORTS: NTP Fifth Annual Report on Carcinogens. IARC Cancer Review: Group 2B IMEMDT 7,56,87; Human Limited Evidence IMEMDT 27,91,82; Animal Sufficient Evidence IMEMDT 27,91,82. NCI Carcinogenesis Bioassay (feed); Clear Evidence: mouse, rat NCITR* NCI-CG-TR-142,79. Reported in EPA TSCA Inventory. Community Right-To-Know List.

SAFETY PROFILE: Confirmed carcinogen with experimental carcinogenic and neoplastigenic data. Moderately toxic by ingestion. Mutation data reported. When heated to decomposition it emits toxic fumes of NO_x. See also ESTERS.

MGP000 CAS:104-93-8 ***HR: 2***
p-METHYL ANISOLE
mf: $C_8H_{10}O$ mw: 122.18

PROP: Found in oil of Ylang-Ylang, Cananga, and others (FCTXAV 12,385,74). Colorless liquid; ylang-ylang odor. D: 0.996-0.970, refr index: 1.510-1.513, flash p: 144°F. Sol in fixed oils; insol in glycerin, propylene glycol.

SYNS: p-CRESOL METHYL ETHER ◇ p-CRESYL METHYL ETHER ◇ FEMA No. 2681 ◇ p-METHOXYTOLUENE ◇ 4-METHOXYTOLUENE ◇ 4-METHYL-1-METHOXYBENZENE ◇ 4-METHYLPHENOL METHYL ETHER ◇ METHYL-p-TOLYL ETHER ◇ p-TOLYL METHYL ETHER

TOXICITY DATA with REFERENCE
skn-rbt 500 mg/24H closed MOD FCTXAV 12,385,74
orl-rat LD50:1920 mg/kg FCTXAV 12,385,74

CONSENSUS REPORTS: Reported in EPA TSCA Inventory.

SAFETY PROFILE: Moderately toxic by ingestion. A skin irritant. Combustible liquid. When heated to decomposition it emits acrid smoke and irritating fumes.

MGP250 CAS:31927-64-7 **HR: 3**
6-METHYLANTHANTHRENE
mf: $C_{23}H_{14}$ mw: 290.37

SYN: 12-METHYLDIBENZO(def,mno)CHRYSENE

TOXICITY DATA with REFERENCE
scu-mus TDLo:72 mg/kg/9W-I:ETA COREAF 246,1477,58

SAFETY PROFILE: Questionable carcinogen with experimental tumorigenic data. When heated to decomposition it emits acrid smoke and irritating fumes.

MGP500 CAS:613-12-7 **HR: D**
2-METHYLANTHRACENE
mf: $C_{15}H_{12}$ mw: 192.27

TOXICITY DATA with REFERENCE
mma-sat 80 μmol/L/2H CNREA8 39,4152,79
msc-hmn:lym 60 μmol/L DTESD7 10,277,82

SAFETY PROFILE: Human mutation data reported. When heated to decomposition it emits acrid smoke and irritating fumes.

MGP750 CAS:779-02-2 **HR: 3**
9-METHYLANTHRACENE
mf: $C_{15}H_{12}$ mw: 192.27

TOXICITY DATA with REFERENCE
mma-sat 5 μg/plate MUREAV 156,61,85
dnd-esc 10 μmol/L PNCCA2 5,39,65
msc-hmn:lym 9 μmol/L DTESD7 10,277,82
dnd-mam:lym 100 μmol BIPMAA 9,689,70
ipr-mus TDLo:11 mg/kg:ETA CANCAR 14,308,61

CONSENSUS REPORTS: Reported in EPA TSCA Inventory.

SAFETY PROFILE: Questionable carcinogen with experimental tumorigenic data. Human mutation data reported. When heated to decomposition it emits acrid smoke and irritating fumes.

MGQ000 CAS:119-68-6 **HR: 1**
N-METHYLANTHRANILIC ACID
mf: $C_8H_9NO_2$ mw: 151.18

SYNS: KYSELINA N-METHYLANTHRANILOVA (CZECH) ◇ o-(METHYLAMINO)BENZOIC ACID ◇ 2-(METHYLAMINO)BENZOIC ACID ◇ N-METHYL-o-AMINOBENZOIC ACID ◇ N-METHYL-2-AMINOBENZOIC ACID

TOXICITY DATA with REFERENCE
eye-rbt 100 mg/24H MOD 28ZPAK -,129,72
orl-rat LD50:11 g/kg 28ZPAK -,129,72

CONSENSUS REPORTS: Reported in EPA TSCA Inventory.

SAFETY PROFILE: Mildly toxic by ingestion. An eye irritant. When heated to decomposition it emits toxic fumes of NO_x.

MGQ250 CAS:85-91-6 **HR: 3**
N-METHYLANTHRANILIC ACID, METHYL ESTER
mf: $C_9H_{11}NO_2$ mw: 165.21

PROP: Pale yellow liquid; grape-like odor. D: 1.126-1.132, refr index: 1.578-1.581, flash p: 196°F. Sol in fixed oils; sltly sol in propylene glycol; insol in water, glycerin.

SYNS: DIMETHYL ANTHRANILATE (FCC) ◇ FEMA No. 2718 ◇ 2-METHYLAMINO METHYL BENZOATE ◇ METHYL METHYLAMINOBENZOATE ◇ METHYL-N-METHYL ANTHRANILATE ◇ MMA

TOXICITY DATA with REFERENCE
orl-rat LDLo:3380 mg/kg FCTXAV 8,359,70
ivn-mus LD50:180 mg/kg CSLNX* NX#07000

CONSENSUS REPORTS: Reported in EPA TSCA Inventory.

SAFETY PROFILE: Poison by intravenous route. Moderately toxic by ingestion. Combustible liquid. When heated to decomposition it emits toxic fumes of NO_x. See also ESTERS.

MGQ525 CAS:35142-06-4 **HR: D**
METHYL ARISTOLATE
mf: $C_{18}H_{14}O_5$ mw: 310.32

SYNS: ARISTOLIC ACID METHYL ESTER ◇ 8-METHOXYPHENANTHRO(3,4-d)-1,3-DIOXOLE-5-CARBOXYLIC ACID METHYL ESTER

TOXICITY DATA with REFERENCE
orl-mus TDLo:60 mg/kg (female 7D post):REP EXPEAM 34,1192,78

SAFETY PROFILE: Experimental reproductive effects. When heated to decomposition it emits acrid smoke and irritating fumes.

MGQ530 CAS:124-58-3 **HR: 2**
METHYLARSENIC ACID
mf: CH_5AsO_3 mw: 139.98

SYNS: KYSELINA METHYLARSONOVA ◇ MAA ◇ METHANEARSONIC ACID ◇ METHYLARSINIC ACID ◇ METHYLARSONIC ACID ◇ MONOMETHYLARSINIC ACID

TOXICITY DATA with REFERENCE
orl-rat LD50:961 mg/kg FAATDF 7,299,86
scu-mus LD50:794 mg/kg 85JCAE-,1265,86

OSHA PEL: TWA 0.5 mg(As)/m^3
ACGIH TLV: TWA 0.2 mg(As)/m^3

SAFETY PROFILE: Moderately toxic by ingestion and subcutaneous routes. When heated to decomposition it emits toxic fumes of As.

MGQ750 CAS:2533-82-6 ***HR: 3***
METHYLARSENIC SULFIDE
mf: CH_3AsS mw: 122.02

SYNS: ASOZIN ◇ BAY 4934 ◇ MAS ◇ METHYLARSINE SULFIDE ◇ METHYLARSINIC SULFIDE ◇ METHYLARSINIC SULPHIDE ◇ METHYLTHIOXOARSINE ◇ MONKIL WP ◇ RHIZOCTOL ◇ (THIOARSENOSO)METHANE ◇ URBASULF

TOXICITY DATA with REFERENCE
mrc-bcs 1 μg/disc/24H MUREAV 40,19,76
orl-rat LD50:100 mg/kg FMCHA2 -,C206,83
skn-rbt LD50:1400 mg/kg GUCHAZ 6,341,73
orl-mam LD50:100 mg/kg 28ZEAL 4,276,69

CONSENSUS REPORTS: Arsenic and its compounds are on the Community Right-To-Know List.

SAFETY PROFILE: Poison by ingestion. Moderately toxic by skin contact. Mutation data reported. When heated to decomposition it emits very toxic fumes of As and SO_x. See also ARSENIC COMPOUNDS and SULFIDES.

MGQ775 CAS:7207-97-8 ***HR: 3***
METHYLARSINE DIIODIDE
mf: CH_3AsI_2 mw: 343.76

PROP: Crystals. Mp: 28°.

SYNS: ARSINE, DIIODOMETHYL- ◇ ARSONOUS DIIODIDE, METHYL-(9CI) ◇ DIIODOMETHYLARSINE ◇ METHYLDIIODOARSINE

TOXICITY DATA with REFERENCE
ivn-mus LD50:18 mg/kg CSLNX* NX#03795

OSHA PEL: TWA 0.5 mg(As)/m^3

SAFETY PROFILE: Poison by intravenous route. When heated to decomposition it emits toxic fumes of As and I^-.

MGR250 CAS:2870-71-5 ***HR: 3***
8-METHYLATROPINIUM BROMIDE
mf: $C_{18}H_{26}NO_3$•Br mw: 384.36

SYNS: ATROPINE METHOBROMIDE ◇ ATROPINE METHYLBROMIDE ◇ 3-α-HYDROXY-8-METHYL-1-α,5-α-H-TROPANIUM BROMIDE (±)TROPATE ◇ HYOSCYAMINE METHYLBROMIDE ◇ METHYLATROPINE BROMIDE ◇ METHYLATROPINIUM BROMIDE ◇ MINTUSSIN ◇ MYDRIASIN ◇ TROPIN

TOXICITY DATA with REFERENCE
orl-rat LD50:1050 mg/kg AIPTAK 180,155,69
scu-rat LD50:1800 mg/kg AIPTAK 180,155,69
idu-rat LD50:312 mg/kg AIPTAK 180,155,69
orl-mus LD50:1640 mg/kg NIIRDN 6,358,82
ipr-mus LD50:75 mg/kg ARZNAD 21,1727,71
scu-mus LD50:242 mg/kg ARZNAD 18,1132,68
ivn-mus LD50:11200 μg/kg CSLNX* NX#03165

CONSENSUS REPORTS: Reported in EPA TSCA Inventory.

SAFETY PROFILE: Poison by subcutaneous, intraduodenal, intraperitoneal, and intravenous routes. Moderately toxic by ingestion. When heated to decomposition it emits very toxic fumes of NO_x and Br^-. See also BROMIDES.

MGR500 CAS:52-88-0 ***HR: 3***
8-METHYLATROPINIUM NITRATE
mf: $C_{18}H_{26}NO_3$•NO_3 mw: 366.46

SYNS: ATROPINE METHONITRATE ◇ ATROPINE METHYL NITRATE ◇ EKOMINE ◇ EUMIDRINA ◇ EUMYDRIN ◇ EUROPEN ◇ HARVATRATE ◇ 3-α-HYDROXY-8-METHYL-1-α-H,5-α-H-TROPANIUM NITRATE (±)-TROPATE (ESTER) ◇ dl-HYOSCYAMINE METHYLNITRATE ◇ dl-HYOSYAMINE METHYLNITRATE ◇ METANITE ◇ METHYL ATROPINE NITRATE ◇ N-METHYLATROPINE NITRATE ◇ N-METHYLATROPINIUM NITRATE ◇ METROPINE ◇ PYLOSTROPIN

TOXICITY DATA with REFERENCE
scu-rat TDLo:5 mg/kg (1D male):REP PSYPAG 10,44,66
ims-hmn TDLo:2 μg/kg:EYE 85IVAW 1,L1,82
orl-rat LD50:1580 mg/kg TXAPA9 1,42,59
orl-mus LD50:1320 mg/kg PSEBAA 120,511,65
ipr-mus LD50:9 mg/kg ATXKA8 29,39,72
ivn-mus LD50:9300 μg/kg JPETAB 110,282,54
scu-gpg LD50:95 mg/kg AIPTAK 137,375,62

CONSENSUS REPORTS: Reported in EPA TSCA Inventory.

SAFETY PROFILE: Poison by intraperitoneal, intravenous and subcutaneous routes. Moderately toxic by ingestion. Human systemic effects: mydriasis. Experimental reproductive effects. When heated to decomposition it emits toxic fumes of NO_x. See also ATROPINE and NITRATES.

MGR750 CAS:624-90-8 ***HR: 3***
METHYL AZIDE
mf: CH_3N_3 mw: 57.05

SAFETY PROFILE: May explode when heated. Presence of mercury increases the sensitivity to shock and spark. Incompatible with (dimethyl malonate + sodium methylate); Hg; methanol; sodium azide; dimethyl sulfate; sodium hydroxide; hydrogen azide. When heated to decomposition it emits toxic fumes of NO_x. See also AZIDES.

MGR800 CAS:16714-23-1 ***HR: 3***
METHYL-2-AZIDOBENZOATE
mf: $C_8H_7N_3O_2$ mw: 177.16

CH_3OCO•$C_6H_4N_3$

SYN: 1-PHENYL-2-METHOXYDIAZENE

SAFETY PROFILE: May explode during distillation. When heated to decomposition it emits toxic fumes of NO_x. See also AZIDES.

MGS500 CAS:11069-34-4 **HR: 3**
METHYL-AZOXY-BUTANE
mf: $C_5H_{12}N_2O$ mw: 116.19

TOXICITY DATA with REFERENCE
scu-rat TDLo:2130 mg/kg/71W-I:ETA PPTCBY 2,73,72
unr-rat LD50:285 mg/kg 23HZAR -,267,70

SAFETY PROFILE: Poison by an unspecified route. Questionable carcinogen with experimental tumorigenic data. When heated to decomposition it emits toxic fumes of NO_x.

MGS700 CAS:71856-48-9 **HR: 2**
METHYLAZOXYMETHANOL-β-D-GLUCOSIDURONIC ACID
mf: $C_8H_{14}N_2O_8$ mw: 266.24

SYNS: β-D-GLUCOPYRANOSIDURONIC ACID, (METHYL-ONN-AZOXY)METHYL- ◇ (METHYL-ONN-AZOXY)METHYL-β-D-GLUCOPYRANOSIDURONIC ACID

TOXICITY DATA with REFERENCE
mma-sat 20 nmol/plate CNREA8 39,3780,79
orl-rat TDLo:71 mg/kg:CAR JJIND8 67,1053,81
orl-rat TD:282 mg/kg/4W-I:CAR JJIND8 67,1053,81

SAFETY PROFILE: Questionable carcinogen with experimental carcinogenic data. Mutation data reported. When heated to decomposition it emits toxic fumes of NO_x.

MGS750 CAS:592-62-1 **HR: 3**
METHYL AZOXYMETHYL ACETATE
mf: $C_4H_8N_2O_3$ mw: 132.14

SYNS: MAM AC ◇ MAM ACETATE ◇ METHYLAZOXYMETHANOL ACETATE ◇ (METHYL-ONN-AZOXY)METHANOL, ACETATE (ESTER)

TOXICITY DATA with REFERENCE
mma-esc 100 μg/plate ENMUDM 6(Suppl 2),1,84
otr-hmn:fbr 7 μmol/L PNASA6 80,7219,83
dnd-hmn:oth 1500 μmol/L/4H PAACA3 21,69,80
ipr-rat TDLo:14 mg/kg (female 15D post):REP NETOD7 7,221,85
scu-rat TDLo:40 mg/kg (female 13D post):TER ESKHA5 (104),73,86
orl-rat TDLo:50 mg/kg/21D-C:ETA JNCIAM 39,355,67
ipr-rat TDLo:35 mg/kg:CAR CBINA8 17,291,77
scu-rat TDLo:62 mg/kg/21D-I:CAR JNCIAM 39,355,67
scu-rat TDLo:30 mg/kg:NEO JJIND8 63,1089,79
orl-mus LDLo:35 mg/kg CNREA8 30,801,70
ipr-mus LD50:105 mg/kg JJIND8 62,911,79
ivn-mus LD50:10 mg/kg CSLNX* NX#04566

CONSENSUS REPORTS: IARC Cancer Review: Group 2B IMEMDT 7,56,87; Animal Sufficient Evidence IMEMDT 10,121,76.

CONSENSUS REPORTS: EPA Genetic Toxicology Program.

SAFETY PROFILE: Suspected carcinogen with experimental carcinogenic, neoplastigenic, tumorigenic, and teratogenic data. Poison by ingestion, intraperitoneal, and intravenous routes. An experimental teratogen. Experimental reproductive effects. Human mutation data reported. When heated to decomposition it emits toxic fumes of NO_x. See also ESTERS.

MGS925 CAS:3527-05-7 **HR: 2**
METHYLAZOXYMETHYL BENZOATE
mf: $C_9H_{10}N_2O_3$ mw: 194.21

SYN: METHANOL, (METHYL-ONN-AZOXY)-, BENZOATE (ester) (9CI)

TOXICITY DATA with REFERENCE
orl-rat TDLo:250 mg/kg/4W-I:ETA IGKEAO 44,211,74

SAFETY PROFILE: Questionable carcinogen with experimental tumorigenic data. When heated to decomposition it emits toxic fumes of NO_x.

MGT000 CAS:54405-61-7 **HR: 3**
METHYLAZOXYOCTANE
mf: $C_9H_{20}N_2O$ mw: 172.31

SYNS: METHYLOCTYLDIAZENE 1-OXIDE ◇ OCTANE-1-NNO-AZOXYMETHANE

TOXICITY DATA with REFERENCE
ipr-mus TDLo:2450 mg/kg/14W-I:CAR JNCIAM 53,1181,74

SAFETY PROFILE: Questionable carcinogen with experimental carcinogenic data. When heated to decomposition it emits toxic fumes of NO_x.

MGT250 CAS:28390-42-3 **HR: 3**
METHYL-AZULENO(5,6,7-c,d)PHENALENE
mf: $C_{21}H_{14}$ mw: 266.35

TOXICITY DATA with REFERENCE
scu-mus TDLo:80 mg/kg/4W-I:ETA NATWAY 57,499,70

SAFETY PROFILE: Questionable carcinogen with experimental tumorigenic data. When heated to decomposition it emits acrid smoke and irritating fumes.

MGT400 **HR: 2**
3-METHYLBENZ(e)ACEPHENANTHRYLENE
mf: $C_{21}H_{14}$ mw: 266.35

SYN: 3-METHYLBENZO(b)FLUORANTHENE

TOXICITY DATA with REFERENCE
mma-sat 125 nmol/plate CRNGDP 6,1023,85
skn-mus TDLo:4262 μg/kg/20D-I:ETA CRNGDP 6,1023,85

SAFETY PROFILE: Questionable carcinogen with experimental tumorigenic data. Mutation data reported. When heated to decomposition it emits acrid smoke and irritating fumes.

MGT410 ***HR: 2***
7-METHYLBENZ(e)ACEPHENANTHRYLENE
mf: $C_{21}H_{14}$ mw: 266.35

SYN: 7-METHYLBENZO(b)FLUORANTHENE

TOXICITY DATA with REFERENCE
skn-mus TDLo:426 μg/kg/20D-I:ETA CRNGDP 6,1023,85

SAFETY PROFILE: Questionable carcinogen with experimental tumorigenic data. When heated to decomposition it emits acrid smoke and irritating fumes.

MGT415 ***HR: 2***
8-METHYLBENZ(e)ACEPHENANTHRYLENE
mf: $C_{21}H_{14}$ mw: 266.35

SYN: 8-METHYLBENZO(b)FLUORANTHENE

TOXICITY DATA with REFERENCE
skn-mus TDLo:426 μg/kg/20D-I:ETA CRNGDP 6,1023,85

SAFETY PROFILE: Questionable carcinogen with experimental tumorigenic data. When heated to decomposition it emits acrid smoke and irritating fumes.

MGT420 ***HR: 2***
12-METHYLBENZ(e)ACEPHENANTHRYLENE
mf: $C_{21}H_{14}$ mw: 266.35

SYN: 12-METHYLBENZO(b)FLUORANTHENE

TOXICITY DATA with REFERENCE
skn-mus TDLo:1065 μg/kg/20D-I:ETA CRNGDP 6,1023,85

SAFETY PROFILE: Questionable carcinogen with experimental tumorigenic data. When heated to decomposition it emits acrid smoke and irritating fumes.

MGT500 CAS:3340-94-1 ***HR: 3***
7-METHYLBENZ(c)ACRIDINE
mf: $C_{18}H_{13}N$ mw: 243.32

SYNS: 9-METHYL-3,4-BENZACRIDINE ◇ 10-METHYL-7,8-BENZACRIDINE (FRENCH) ◇ 7-METHYLBENZO(c)ACRIDINE

TOXICITY DATA with REFERENCE
mma-sat 100 μg/plate MUREAV 66,307,79
mma-ham:lng 1 μmol/L CRNGDP 7,23,86
sce-ham:lng 5 μmol/L CRNGDP 7,23,86
skn-mus TDLo:240 mg/kg/20W-I:ETA ACRSAJ 4,315,56

SAFETY PROFILE: Questionable carcinogen with experimental tumorigenic data. Mutation data reported. When heated to decomposition it emits toxic fumes of NO_x.

MGU500 CAS:3340-93-0 ***HR: 3***
12-METHYLBENZ(a)ACRIDINE
mf: $C_{18}H_{13}N$ mw: 243.32

SYNS: 9-METHYL-1,2-BENZACRIDINE ◇ 10-METHYL-5,6-BENZACRIDINE

TOXICITY DATA with REFERENCE
scu-mus TDLo:200 mg/kg:ETA VOONAW 1,52,55

SAFETY PROFILE: Questionable carcinogen with experimental tumorigenic data. When heated to decomposition it emits toxic fumes of NO_x.

MGU550 CAS:92145-26-1 ***HR: 2***
7-METHYLBENZ(c)ACRIDINE 3,4-DIHYDRODIOL
mf: $C_{18}H_{15}NO_2$ mw: 277.34

SYNS: trans-3,4-DIHYDRO-3,4-DIHYDROXY-7-METHYLBENZ(c)ACRIDINE ◇ trans-3,4-DIHYDROXY-3,4-DIHYDRO-7-METHYLBENZ(c)ACRIDINE

TOXICITY DATA with REFERENCE
mma-sat nmol/plate CRNGDP 7,23,86
sce-ham:lng 5 μmol/L CRNGDP 7,23,86
ipr-mus TDLo:3883 μg/kg:NEO CNREA8 46,4552,86

SAFETY PROFILE: Questionable carcinogen with experimental neoplastigenic data. Mutation data reported. When heated to decomposition it emits acrid smoke and irritating fumes.

MGU750 CAS:2498-77-3 ***HR: 3***
1-METHYLBENZ(a)ANTHRACENE
mf: $C_{19}H_{14}$ mw: 242.33

SYN: 1′-METHYL-1,2-BENZANTHRACENE

TOXICITY DATA with REFERENCE
mma-sat 20 μg/plate CNREA8 36,4525,76
scu-rat TDLo:18 mg/kg:ETA JNCIAM 25,387,60

CONSENSUS REPORTS: EPA Genetic Toxicology Program.

SAFETY PROFILE: Questionable carcinogen with experimental tumorigenic data. Mutation data reported. When heated to decomposition it emits acrid smoke and irritating fumes.

MGV000 CAS:2498-76-2 ***HR: 3***
2-METHYLBENZ(a)ANTHRACENE
mf: $C_{19}H_{14}$ mw: 242.33

SYN: 2'-METHYL-1,2-BENZANTHRACENE

TOXICITY DATA with REFERENCE
mma-sat 10 μg/plate CNREA8 36,4525,76
dnd-omi 1800 μmol/L ZKKOBW 90,37,77
skn-mus TDLo:210 mg/kg/33W-I:ETA CNREA8 11,892,51

SAFETY PROFILE: Questionable carcinogen with experimental tumorigenic data. Mutation data reported. When heated to decomposition it emits acrid smoke and irritating fumes.

MGV250 CAS:2498-75-1 ***HR: 3***
3-METHYLBENZ(a)ANTHRACENE
mf: $C_{19}H_{14}$ mw: 242.33

TOXICITY DATA with REFERENCE
mma-sat 5 μg/plate CNREA8 36,4525,76
scu-mus TDLo:200 mg/kg:ETA AIHAAP 26,475,65

CONSENSUS REPORTS: EPA Genetic Toxicology Program.

SAFETY PROFILE: Questionable carcinogen with experimental tumorigenic data. Mutation data reported. When heated to decomposition it emits acrid smoke and irritating fumes.

MGV500 CAS:316-49-4 ***HR: 3***
4-METHYLBENZ(a)ANTHRACENE
mf: $C_{19}H_{14}$ mw: 242.33

SYN: 4'-METHYL-1:2-BENZANTHRACENE

TOXICITY DATA with REFERENCE
mma-sat 20 μg/plate CNREA8 36,4525,76
scu-rat TDLo:18 mg/kg:ETA JNCIAM 25,387,60

CONSENSUS REPORTS: EPA Genetic Toxicology Program.

SAFETY PROFILE: Questionable carcinogen with experimental tumorigenic data. Mutation data reported. When heated to decomposition it emits acrid smoke and irritating fumes.

MGV750 CAS:2319-96-2 ***HR: 3***
5-METHYLBENZ(a)ANTHRACENE
mf: $C_{19}H_{14}$ mw: 242.33

SYN: 3-METHYL-1,2-BENZANTHRACENE

TOXICITY DATA with REFERENCE
mma-sat 10 μg/plate CNREA8 36,4525,76
scu-rat TDLo:18 mg/kg:ETA JNCIAM 25,387,60

CONSENSUS REPORTS: EPA Genetic Toxicology Program.

SAFETY PROFILE: Questionable carcinogen with experimental tumorigenic data. Mutation data reported. When heated to decomposition it emits acrid smoke and irritating fumes.

MGW000 CAS:316-14-3 ***HR: 3***
6-METHYLBENZ(a)ANTHRACENE
mf: $C_{19}H_{14}$ mw: 242.33

SYN: 4-METHYL-1,2-BENZANTHRACENE

TOXICITY DATA with REFERENCE
mma-sat 10 μg/plate CNREA8 36,4525,76
scu-rat TDLo:18 mg/kg:ETA JNCIAM 25,387,60
scu-mus TDLo:200 mg/kg:CAR AIHAAP 26,475,65

CONSENSUS REPORTS: EPA Genetic Toxicology Program.

SAFETY PROFILE: Questionable carcinogen with experimental carcinogenic and tumorigenic data. Mutation data reported. When heated to decomposition it emits acrid smoke and irritating fumes.

MGW250 CAS:2381-31-9 ***HR: 3***
8-METHYLBENZ(a)ANTHRACENE
mf: $C_{19}H_{14}$ mw: 242.33

SYN: 5-METHYL-1,2-BENZANTHRACENE

TOXICITY DATA with REFERENCE
mma-sat 20 μg/plate CNREA8 36,4525,76
scu-rat TDLo:18 mg/kg:ETA JNCIAM 25,387,60
scu-mus TDLo:200 mg/kg:CAR AIHAAP 26,475,65

CONSENSUS REPORTS: EPA Genetic Toxicology Program.

SAFETY PROFILE: Questionable carcinogen with experimental carcinogenic and tumorigenic data. Mutation data reported. When heated to decomposition it emits acrid smoke and irritating fumes.

MGW500 CAS:2381-16-0 ***HR: 3***
9-METHYLBENZ(a)ANTHRACENE
mf: $C_{19}H_{14}$ mw: 242.33

SYN: 6-METHYL-1,2-BENZANTHRACENE

TOXICITY DATA with REFERENCE
mma-sat 20 μg/plate CNREA8 36,4525,76
scu-rat TDLo:18 mg/kg:ETA JNCIAM 25,387,60

CONSENSUS REPORTS: EPA Genetic Toxicology Program.

SAFETY PROFILE: Questionable carcinogen with experimental tumorigenic data. Mutation data reported. When heated to decomposition it emits acrid smoke and irritating fumes.

MGW750 CAS:2541-69-7 ***HR: 3***
10-METHYL-1,2-BENZANTHRACENE
mf: $C_{19}H_{14}$ mw: 242.33

SYNS: 7-MBA ◇ 10-METHYL-1,2-BENZANTHRACEN (GERMAN) ◇ 7-METHYLBENZ(a)ANTHRACENE

TOXICITY DATA with REFERENCE
dnd-mus-skn 40 μmol/kg IJCNAW 23,201,69
sce-ham:ovr 8 mg/L MUREAV 50,367,78
msc-ham:fbr 1 mg/L DTESD7 8,121,80
orl-rat TDLo:1000 mg/kg:ETA SCIEAS 137,257,62
scu-rat TDLo:8 mg/kg:NEO PSEBAA 124,915,67
scu-mus TDLo:200 mg/kg:CAR AIHAAP 26,475,65

CONSENSUS REPORTS: EPA Genetic Toxicology Program.

SAFETY PROFILE: Questionable carcinogen with experimental carcinogenic, neoplastigenic, and tumorigenic data. Mutation data reported. When heated to decomposition it emits acrid smoke and irritating fumes.

MGX000 CAS:2381-15-9 ***HR: 3***
10-METHYLBENZ(a)ANTHRACENE
mf: $C_{19}H_{14}$ mw: 242.33

SYN: 7-METHYL-1,2-BENZANTHRACENE

TOXICITY DATA with REFERENCE
mma-sat 10 μg/plate CNREA8 36,4525,76
otr-ham:emb 10 mg/L JNCIAM 35,641,65
scu-rat TDLo:18 mg/kg:ETA JNCIAM 25,387,60

CONSENSUS REPORTS: EPA Genetic Toxicology Program.

SAFETY PROFILE: Questionable carcinogen with experimental tumorigenic data. Mutation data reported. When heated to decomposition it emits acrid smoke and irritating fumes.

MGX250 CAS:6111-78-0 ***HR: 3***
11-METHYLBENZ(a)ANTHRACENE
mf: $C_{19}H_{14}$ mw: 242.33

SYN: 8-METHYL-1:2-BENZANTHRACENE

TOXICITY DATA with REFERENCE
mma-sat 50 μg/plate CNREA8 36,4525,76
skn-mus TDLo:1180 mg/kg/49W-I:ETA PRLBA4 129,439,40
scu-mus TDLo:200 mg/kg:CAR AIHAAP 26,475,65

CONSENSUS REPORTS: EPA Genetic Toxicology Program.

SAFETY PROFILE: Questionable carcinogen with carcinogenic and tumorigenic data. Mutation data reported. When heated to decomposition it emits acrid smoke and irritating fumes.

MGX500 CAS:2422-79-9 ***HR: 3***
12-METHYLBENZ(a)ANTHRACENE
mf: $C_{19}H_{14}$ mw: 242.33

SYN: 9-METHYL-1,2-BENZANTHRACENE

TOXICITY DATA with REFERENCE
mma-sat 20 μg/plate CNREA8 36,4525,76
orl-rat TDLo:1000 mg/kg:ETA SCIEAS 137,257,62
scu-mus TDLo:200 mg/kg:CAR AIHAAP 26,475,65

CONSENSUS REPORTS: EPA Genetic Toxicology Program.

SAFETY PROFILE: Questionable carcinogen with experimental carcinogenic and tumorigenic data. Mutation data reported. When heated to decomposition it emits acrid smoke and irritating fumes.

MGX750 CAS:64082-43-5 ***HR: 3***
10-METHYL-1,2-BENZANTHRACENE-5-CARBONAMIDE
mf: $C_{20}H_{15}NO$ mw: 285.36

SYN: 7-METHYLBENZ(a)ANTHRACEN-8-YLCARBAMIDE

TOXICITY DATA with REFERENCE
scu-mus TDLo:4 mg/kg:ETA JNCIAM 1,45,40

SAFETY PROFILE: Questionable carcinogen with experimental tumorigenic data. When heated to decomposition it emits toxic fumes of NO_x.

MGY000 CAS:63018-70-2 ***HR: 3***
7-METHYLBENZ(a)ANTHRACENE-8-CARBONITRILE
mf: $NC_{20}H_{13}$ mw: 267.34

SYNS: 5-CYANO-10-METHYL-1,2-BENZANTHRACENE ◇ 8-CYANO-7-METHYLBENZ(a)ANTHRACINE

TOXICITY DATA with REFERENCE
scu-mus TDLo:200 mg/kg:ETA JNCIAM 1,303,40

CONSENSUS REPORTS: Cyanide and its compounds are on the Community Right-To-Know List.

SAFETY PROFILE: Questionable carcinogen with experimental tumorigenic data. When heated to decomposition it emits toxic fumes of NO_x and CN^-. See also NITRILES.

MGY250 CAS:6366-23-0 ***HR: 3***
7-METHYLBENZ(a)ANTHRACENE-10-CARBONITRILE
mf: $C_{20}H_{13}N$ mw: 267.34

SYN: 7-CYANO-10-METHYL-1,2-BENZANTHRACENE

TOXICITY DATA with REFERENCE
scu-mus TDLo:400 mg/kg:ETA JNCIAM 1,303,40

CONSENSUS REPORTS: Cyanide and its compounds are on the Community Right-To-Know List.

SAFETY PROFILE: Questionable carcinogen with experimental tumorigenic data. When heated to decomposition it emits toxic fumes of NO_x and CN^-. See also NITRILES.

MGY500 CAS:17513-40-5 ***HR: 3***
7-METHYLBENZ(a)ANTHRACENE-12-CARBOXALDEHYDE
mf: $C_{20}H_{14}O$ mw: 270.34

SYN: 12-FORMYL-7-METHYLBENZ(a)ANTHRACENE

TOXICITY DATA with REFERENCE
dni-omi 200 µg/L PNASA6 74,1378,77
skn-mus TDLo:8000 mg/kg:NEO JJIND8 61,135,78

SAFETY PROFILE: Questionable carcinogen with experimental neoplastigenic data. Mutation data reported. When heated to decomposition it emits acrid smoke and irritating fumes. See also ALDEHYDES.

MGZ000 CAS:1155-38-0 ***HR: 3***
7-METHYLBENZ(a)ANTHRACENE-5,6-OXIDE
mf: $C_{19}H_{14}O$ mw: 258.33

SYN: 5,6-EPOXY-5,6-DIHYDRO-7-METHYLBENZ(A)ANTHRACENE

TOXICITY DATA with REFERENCE
mma-sat 500 ng/plate CNREA8 45,2600,85
dns-esc 1 mmol/L ZKKOBW 92,157,78
skn-mus TDLo:120 mg/kg/20W-I:ETA PSEBAA 124,915,67
scu-mus TDLo:400 mg/kg/10W-I:NEO IJCNAW 2,500,67

CONSENSUS REPORTS: EPA Genetic Toxicology Program.

SAFETY PROFILE: Questionable carcinogen with experimental neoplastigenic and tumorigenic data. Mutation data reported. When heated to decomposition it emits acrid smoke and irritating fumes.

MHA000 CAS:66964-37-2 ***HR: 3***
S-(12-METHYL-7-BENZ(a)ANTHRYLMETHYL) HOMOCYSTEINE
mf: $C_{24}H_{23}NO_2S$ mw: 389.54

TOXICITY DATA with REFERENCE
ivn-mus TDLo:5800 mg/kg:ETA JMCMAR 19,1422,76

SAFETY PROFILE: Questionable carcinogen with experimental tumorigenic data. When heated to decomposition it emits very toxic fumes of NO_x and SO_x.

MHA250 CAS:613-93-4 ***HR: 2***
N-METHYLBENZENAMIDE
mf: C_8H_9NO mw: 135.18

SYN: N-METHYLBENZAMIDE

TOXICITY DATA with REFERENCE
orl-mus LD50:840 mg/kg TXAPA9 19,20,71

CONSENSUS REPORTS: Reported in EPA TSCA Inventory.

SAFETY PROFILE: Moderately toxic by ingestion. When heated to decomposition it emits toxic fumes of NO_x.

MHA500 CAS:101-41-7 ***HR: 2***
METHYL BENZENEACETATE
mf: $C_9H_{10}O_2$ mw: 150.19

PROP: Colorless liquid; honey, jasmine odor. D: 1.062, refr index: 1.503-1.509, vap d: 5.18, flash p: 192°F. Sol in alc, fixed oils; insol in glycerin, propylene glycol, water @ 215°.

SYNS: BENZENEACETIC ACID, METHYL ESTER ◇ FEMA No. 2733 ◇ METHYL PHENYLACETATE (FCC) ◇ METHYL-α-TOLUATE ◇ PHENYLACETIC ACID, METHYL ESTER

TOXICITY DATA with REFERENCE
skn-rbt 500 mg/24H FCTXAV 12,807,74
orl-rat LD50:2550 mg/kg FCTXAV 12,807,74
skn-rbt LD50:2400 mg/kg FCTXAV 12,807,74

CONSENSUS REPORTS: Reported in EPA TSCA Inventory.

SAFETY PROFILE: Moderately toxic by ingestion and skin contact. A skin irritant. Combustible liquid. When heated to decomposition it emits acrid smoke and irritating fumes. See also ESTERS.

MHA750 CAS:93-58-3 ***HR: 2***
METHYL BENZENECARBOXYLATE
DOT: UN 2938
mf: $C_8H_8O_2$ mw: 136.16

PROP: Colorless liquid; fragrant odor. Mp: −12.5°, bp: 199.6°, flash p: 181°F, d: 1.082-1.088, refr index: 1.515, vap press: 1 mm @ 39.0°, vap d: 4.69. Sol in alc, fixed oils, propylene glycol, water @ 30°; misc in alc, ether; insol in glycerin.

SYNS: FEMA No. 2683 ◇ METHYL BENZOATE (FCC) ◇ NIOBE OIL ◇ OIL of NIOBE

TOXICITY DATA with REFERENCE
skn-rbt 10 mg/24H MLD AMIHBC 10,61,54
eye-rbt 500 mg AMIHBC 10,61,54
orl-rat LD50:1350 mg/kg FCTXAV 2,327,64
orl-mus LD50:3330 mg/kg FCTXAV 2,327,64
skn-cat LDLo:10 g/kg JPETAB 84,358,45
orl-rbt LD50:2170 mg/kg JPETAB 84,358,45

CONSENSUS REPORTS: Reported in EPA TSCA Inventory.

DOT Classification: Poison B; Label: St. Andrews Cross

SAFETY PROFILE: Moderately toxic by ingestion. Mildly toxic by skin contact. A skin and eye irritant. Combustible liquid when exposed to heat or flame; can react with oxidizing materials. To fight fire, use foam, CO_2, dry chemical, water to blanket fire. When heated to decomposition it emits acrid smoke and irritating fumes.

MHB000 CAS:66217-76-3 ***HR: 3***
METHYL BENZENEDIAZOATE
mf: $C_7H_8N_2O$ mw: 136.16

$C_6H_5N{=}NOCH_3$

SYN: 1-PHENYL-2-METHOXYDIAZENE

SAFETY PROFILE: Explodes on heating or after storage. When heated to decomposition it emits toxic fumes of NO_x.

MHB250 CAS:589-18-4 ***HR: 2***
4-METHYL-BENZENEMETHANOL
mf: $C_8H_{10}O$ mw: 122.18

SYNS: p-METHYLBENZYLALCOHOL ◇ 4-METHYLBENZYL ALCOHOL ◇ p-METHYLBENZYLALKOHOL (GERMAN) ◇ p-TOLYLCARBINOL ◇ 4-TOLYLCARBINOL

TOXICITY DATA with REFERENCE
skn-rbt 500 mg/24H SEV FCTOD7 20(Suppl),839,82
orl-rat LD50:1400 mg/kg ARZNAD 12,347,62

CONSENSUS REPORTS: Reported in EPA TSCA Inventory.

SAFETY PROFILE: Moderately toxic by ingestion. A severe skin irritant. Used as a fragrance and flavor. When heated to decomposition it emits acrid smoke and irritating fumes. See also ALCOHOLS.

MHB500 CAS:1320-44-1 ***HR: 2***
METHYL BENZETHONIUM CHLORIDE MONOHYDRATE
mf: $C_{28}H_{44}NO_2{\cdot}Cl{\cdot}H_2O$ mw: 480.20

PROP: Crystals; bitter taste. Mp: 161-163°. Very sol in water, alc, cellusolve, chloroform, hot benzene.

SYNS: BACTINE ◇ DELAVAN ◇ DIAPARENE ◇ DIAPARENE CHLORIDE ◇ p-DIISOBUTYLCRESOXYETHYLDIMETHYLBENZYLAMMONIUM CHLORIDE MONOHYDRATE ◇ HYAMINE 10X ◇ METHYLBENZETHONIUM CHLORIDE

TOXICITY DATA with REFERENCE
orl-rat LD50:800 mg/kg PCOC** -,590,66
orl-mus LD50:750 mg/kg PCOC** -,590,66

SAFETY PROFILE: Moderately toxic by ingestion. An FDA over-the-counter drug. When heated to decomposition it emits very toxic fumes of NO_x and Cl^-.

MHC000 CAS:5504-68-7 ***HR: 1***
10-METHYL-7H-BENZIMIDAZOL(2,1-a)BENZ(de)ISOQUINOLIN-7-ONE
mf: $C_{19}H_{12}N_2O$ mw: 284.33

SYN: 5-METHYLNAFTOYLBENZIMIDAZOL (CZECH)

TOXICITY DATA with REFERENCE
eye-rbt 500 mg/24H MLD 28ZPAK -,146,72
orl-rat LD50:9490 mg/kg 28ZPAK -,146,72

SAFETY PROFILE: Mildly toxic by ingestion. Moderately toxic by eye irritant. When heated to decomposition it emits toxic fumes of NO_x.

MHC250 CAS:615-15-6 ***HR: 3***
METHYL-2-BENZIMIDAZOLE
mf: $C_8H_8N_2$ mw: 132.18

PROP: Needles from water. Mp: 175-176°. Sol in hot water, NaOH; sltly sol in alc and ether.

SYN: 2-METHYLBENZIMIDAZOLE

TOXICITY DATA with REFERENCE
mmo-sat 250 μg/plate CHIMAD 27,68,73
mma-sat 500 nmol/plate JMCMAR 22,981,79
orl-rat TDLo:46 g/kg (MGN):REP APTOD9 19,A91,80
orl-rat LDLo:500 mg/kg NCNSA6 5,22,53
ivn-mus LD50:200 mg/kg JPETAB 105,486,52

CONSENSUS REPORTS: Reported in EPA TSCA Inventory.

SAFETY PROFILE: Poison by intravenous route. Moderately toxic by ingestion. Experimental reproductive effects. Mutation data reported. When heated to decomposition it emits toxic fumes of NO_x.

MHC500 CAS:614-97-1 ***HR: 2***
5-METHYLBENZIMIDAZOLE
mf: $C_8H_8N_2$ mw: 132.18

PROP: Crystals from water. Mp: 114°.

TOXICITY DATA with REFERENCE
orl-rat LDLo:500 mg/kg NCNSA6 5,22,53

CONSENSUS REPORTS: Reported in EPA TSCA Inventory.

SAFETY PROFILE: Moderately toxic by ingestion. When heated to decomposition it emits toxic fumes of NO_x.

MHC750 CAS:10605-21-7 ***HR: 2***
METHYL BENZIMIDAZOLE-2-YL CARBAMATE
mf: $C_9H_9N_3O_2$ mw: 191.21

SYNS: BAS-3460 ◇ BAVISTIN ◇ BCM ◇ BENZIMIDAZOLE-2-CARBAMIC ACID, METHYL ESTER ◇ 1H-BENZIMIDAZOL-2-YLCARBAMIC ACID METHYL ESTER ◇ BMC ◇ CARBENDAZIM ◇ CAR-

BENDAZOLE ◇ CARBENDAZYM ◇ CTR 6669 ◇ DEROSAL ◇ HOE 17411 ◇ KEMDAZIN ◇ MBC ◇ 2-(METHOXY-CARBONYLAMINO)-BENZIMIDAZOL (GERMAN) ◇ 2-(METHOXYCARBONYLAMINO)-BENZIMIDAZOLE ◇ METHYL-1H-BENZEMEDAZOL-2-YLCARBAMATE ◇ METHYL 2-BENZIMIDAZOLECARBAMATE

TOXICITY DATA with REFERENCE
mnt-hmn:lym 1 μmol/L MUREAV 156,199,85
cyt-ham-orl 100 mg/kg TRENAG 36,396,85
orl-mus TDLo:100 mg/kg (female 10D post):TER CBINA8 26,115,79
orl-rat TDLo:4 g/kg (male 10D pre):REP BIREBV 37,709,87
orl-rat LD50:6400 mg/kg 85AREA 4,131,76/77
skn-rat LD50:2000 mg/kg 85DPAN -,-,71/76
orl-mus LD50:11 g/kg MGONAD 20,163,76

CONSENSUS REPORTS: Reported in EPA TSCA Inventory. EPA Genetic Toxicology Program.

SAFETY PROFILE: Moderately toxic by skin contact. Mildly toxic by ingestion. An experimental teratogen. Experimental reproductive effects. Human mutation data reported. An agricultural chemical and pesticide. When heated to decomposition it emits toxic fumes of NO_x. See also CARBAMATES.

MHD000 CAS:21064-50-6 ***HR: 3***
6-METHYL-3,4-BENZOCARBAZOLE
mf: $C_{17}H_{13}N$ mw: 231.31

SYN: 10-METHYL-7H-BENZO(c)CARBAZOLE

TOXICITY DATA with REFERENCE
scu-mus TDLo:120 mg/kg/9W-I:ETA BAFEAG 42,3,55

SAFETY PROFILE: Questionable carcinogen with experimental tumorigenic data. When heated to decomposition it emits toxic fumes of NO_x.

MHD250 CAS:13127-50-9 ***HR: 3***
9-METHYL-1:2-BENZOCARBAZOLE
mf: $C_{17}H_{13}N$ mw: 231.31

SYN: 11-METHYL-11H-BENZO(a)CARBAZOLE

TOXICITY DATA with REFERENCE
skn-mus TDLo:400 mg/kg/17W-I:ETA CRSBAW 141,635,47

SAFETY PROFILE: Questionable carcinogen with experimental tumorigenic data. When heated to decomposition it emits toxic fumes of NO_x.

MHD300 CAS:102128-78-9 ***HR: 3***
N-(2-METHYLBENZODIOXAN)-N'-ETHYL-β-ALANINAMIDE
mf: $C_{13}H_{18}N_2O_3$ mw: 250.33

SYNS: 3-(((1,4-BENZODIOXAN-2-YL)METHYL)AMINO)-N-METHYLPROsPIONAMIDE ◇ 1205 I.S. ◇ N-(2-METHYL-1,4-BENZODIOXAN)-N'-METHYL-β-ALANINAMIDE

TOXICITY DATA with REFERENCE
ims-rat TDLo:100 mg/kg (5D preg):REP PSEBAA 100,555,59
ipr-mus LD50:350 mg/kg AIPTAK 105,317,56
scu-mus LD50:600 mg/kg AIPTAK 105,317,56
ivn-mus LD50:75 mg/kg AIPTAK 105,317,56
ivn-rbt LD50:90 mg/kg AIPTAK 105,317,56

SAFETY PROFILE: Poison by intravenous and intraperitoneal routes. Moderately toxic by subcutaneous route. Experimental reproductive effects. When heated to decomposition it emits toxic fumes of NO_x.

MHD750 CAS:52400-66-5 ***HR: 3***
4-(2-(2-METHYL-1,3-BENZODIOXOL-2-YL) ETHYL)PIPERAZIN-1-YL-2-ETHANOL DIHYDROCHLORIDE
mf: $C_{16}H_{24}N_2O_3•2ClH$ mw: 365.34

TOXICITY DATA with REFERENCE
ivn-rat LD50:18500 μg/kg EJMCA5 12,413,77
ipr-mus LD50:90 mg/kg EJMCA5 12,413,77

SAFETY PROFILE: Poison by intravenous and intraperitoneal routes. When heated to decomposition it emits very toxic fumes of HCl and NO_x.

MHE000 CAS:3524-62-7 ***HR: 3***
METHYL BENZOIN
mf: $C_{15}H_{14}O_2$ mw: 226.29

SYNS: BENZOIN METHYL ETHER ◇ 2-METHOXY-2-PHENYLACETOPHENONE

TOXICITY DATA with REFERENCE
orl-mus LD50:300 mg/kg TeiD## 16JUN75

CONSENSUS REPORTS: Reported in EPA TSCA Inventory.

SAFETY PROFILE: Poison by ingestion. When heated to decomposition it emits acrid smoke and irritating fumes. See also ETHERS and KETONES.

MHE250 CAS:33942-88-0 ***HR: 3***
5-METHYLBENZO(rat)PENTAPHENE
mf: $C_{25}H_{16}$ mw: 316.41

SYN: 5-METHYL-3,4,9,10-DIBENZPYRENE(FRENCH)

TOXICITY DATA with REFERENCE
scu-mus TDLo:72 mg/kg/9W-I:ETA COREAF 244,273,57

SAFETY PROFILE: Questionable carcinogen with experimental tumorigenic data. When heated to decomposition it emits acrid smoke and irritating fumes.

MHE500 CAS:41699-09-6 *HR: 3*
METHYL-1,12-BENZOPERYLENE
mf: $C_{23}H_{14}$ mw: 290.37

TOXICITY DATA with REFERENCE
scu-mus TDLo:24 mg/kg:ETA COREAF 245,991,57

SAFETY PROFILE: Questionable carcinogen with experimental tumorigenic data. When heated to decomposition it emits acrid smoke and irritating fumes.

MHE750 CAS:1492-55-3 *HR: 3*
7-METHYLBENZO(a)PHENALENO(1,9-hi)ACRIDINE
mf: $C_{28}H_{17}N$ mw: 367.46

TOXICITY DATA with REFERENCE
scu-mus TDLo:72 mg/kg/9W-I:ETA BAFEAG 52,49,65

SAFETY PROFILE: Questionable carcinogen with experimental tumorigenic data. When heated to decomposition it emits toxic fumes of NO_x.

MHF000 CAS:1492-54-2 *HR: 3*
7-METHYLBENZO(h)PHENALENO(1,9-bc)ACRIDINE
mf: $C_{28}H_{17}N$ mw: 367.46

TOXICITY DATA with REFERENCE
scu-mus TDLo:72 mg/kg/9W-I:ETA BAFEAG 52,49,65

SAFETY PROFILE: Questionable carcinogen with experimental tumorigenic data. When heated to decomposition it emits toxic fumes of NO_x.

MHF250 CAS:652-04-0 *HR: 3*
5-METHYLBENZO(c)PHENANTHRENE
mf: $C_{19}H_{14}$ mw: 242.33

SYN: 2-METHYL-3,4-BENZPHENANTHRENE

TOXICITY DATA with REFERENCE
orl-mus TDLo:10 g/kg/51W-I:ETA PRLBA4 129,439,40
ivn-mus TDLo:10 mg/kg:NEO JNCIAM 1,225,40

SAFETY PROFILE: Questionable carcinogen with experimental neoplastigenic and tumorigenic data. When heated to decomposition it emits acrid smoke and irritating fumes.

MHF500 CAS:2381-34-2 *HR: 3*
6-METHYLBENZO(c)PHENANTHRENE
mf: $C_{19}H_{14}$ mw: 242.33

SYN: 1-METHYL-3,4-BENZPHENANTHRENE

TOXICITY DATA with REFERENCE
skn-mus TDLo:720 mg/kg/30W-I:ETA PRLBA4 129,439,40

SAFETY PROFILE: Questionable carcinogen with experimental tumorigenic data. When heated to decomposition it emits acrid smoke and irritating fumes.

MHF750 CAS:134-84-9 *HR: 3*
4-METHYL-p-BENZOPHENONE
mf: $C_{14}H_{12}O$ mw: 196.26

SYNS: PHENYL-p-TOLYL KETONE ◇ USAF DO-54

TOXICITY DATA with REFERENCE
ipr-mus LDLo:250 mg/kg CURL** -,21,62
ipr-mus LD50:250 mg/kg NTIS** AD277-689

CONSENSUS REPORTS: Reported in EPA TSCA Inventory.

SAFETY PROFILE: Poison by intraperitoneal route. When heated to decomposition it emits acrid smoke and irritating fumes.

MHG250 CAS:40568-90-9 *HR: 3*
1-METHYLBENZO(a)PYRENE
mf: $C_{21}H_{14}$ mw: 266.35

SYN: 1-METHYL-BP

TOXICITY DATA with REFERENCE
skn-mus TDLo:2130 μg/kg:NEO CNREA8 40,1073,80

SAFETY PROFILE: Questionable carcinogen with experimental neoplastigenic data. When heated to decomposition it emits acrid smoke and irritating fumes.

MHG500 CAS:16757-82-7 *HR: 3*
2-METHYLBENZO(a)PYRENE
mf: $C_{21}H_{14}$ mw: 266.35

SYN: 9-METHYL-3,4-BENZPYRENE

TOXICITY DATA with REFERENCE
scu-mus TDLo:72 mg/kg/13W-I:ETA IJCNAW 3,238,68

SAFETY PROFILE: Questionable carcinogen with experimental tumorigenic data. When heated to decomposition it emits acrid smoke and irritating fumes.

MHG750 CAS:16757-83-8 *HR: 3*
4-METHYLBENZO(a)PYRENE
mf: $C_{21}H_{14}$ mw: 266.35

TOXICITY DATA with REFERENCE
scu-mus TDLo:72 mg/kg/13W-I:ETA IJCNAW 3,238,68

SAFETY PROFILE: estionable carcinogen with experimental tumorigenic data. When heated to decomposition it emits acrid smoke and irritating fumes.

MHH000 CAS:63041-77-0 *HR: 3*
4'-METHYLBENZO(a)PYRENE
mf: $C_{21}H_{14}$ mw: 266.35

SYNS: 7-METHYLBENZO(a)PYRENE ◇ 4'-METHYL-3:4-BENZ-PYRENE

TOXICITY DATA with REFERENCE
msc-ham:lng 500 nmol/L CRNGDP 4,321,83
ims-rat TDLo:4 mg/kg:NEO NATUAS 273,566,78
imp-mus TDLo:400 mg/kg:ETA AJCAA7 36,211,39

SAFETY PROFILE: Questionable carcinogen with experimental neoplastigenic and tumorigenic data. Mutation data reported. When heated to decomposition it emits acrid smoke and irritating fumes.

MHH200 CAS:31647-36-6 ***HR: 2***
5-METHYLBENZO(a)PYRENE
mf: $C_{21}H_{14}$ mw: 266.35

SYN: 5-METHYL-BP

TOXICITY DATA with REFERENCE
mma-sat 25 μg/plate CNREA8 47,1509,87
skn-mus TDLo:2130 μg/kg:NEO CNREA8 40,1073,80

SAFETY PROFILE: Questionable carcinogen with experimental neoplastigenic data. Mutation data reported. When heated to decomposition it emits acrid smoke and irritating fumes.

MHH500 CAS:63104-32-5 ***HR: 3***
10-METHYLBENZO(a)PYRENE
mf: $C_{21}H_{14}$ mw: 266.35

SYN: 10-MONOMETHYLBENZO(a)PYRENE

TOXICITY DATA with REFERENCE
mma-sat 2900 pmol/plate BBRCA9 85,351,78
ims-rat TDLo:10 mg/kg:NEO NATUAS 273,566,78
skn-mus TDLo:2000 μg/kg/20D-I:ETA CALEDQ 5,179,78

SAFETY PROFILE: Questionable carcinogen with experimental neoplastigenic and tumorigenic data. Mutation data reported. When heated to decomposition it emits acrid smoke and irritating fumes.

MHH750 CAS:16757-80-5 ***HR: 3***
11-METHYLBENZO(a)PYRENE
mf: $C_{21}H_{14}$ mw: 266.35

SYN: 6-METHYL-3,4-BENZPYRENE

TOXICITY DATA with REFERENCE
msc-ham:lng 500 nmol/L CRNGDP 4,321,83
scu-mus TDLo:72 mg/kg/13W-I:ETA IJCNAW 3,238,68

SAFETY PROFILE: Questionable carcinogen with experimental tumorigenic data. Mutation data reported. When heated to decomposition it emits acrid smoke and irritating fumes.

MHI000 CAS:4514-19-6 ***HR: 3***
12-METHYLBENZO(a)PYRENE
mf: $C_{21}H_{14}$ mw: 266.35

TOXICITY DATA with REFERENCE
scu-mus TDLo:72 mg/kg/13W-I:ETA IJCNAW 3,238,68

SAFETY PROFILE: Questionable carcinogen with experimental tumorigenic data. When heated to decomposition it emits acrid smoke and irritating fumes.

MHI250 CAS:553-97-9 ***HR: 3***
2-METHYL-p-BENZOQUINONE
mf: $C_7H_6O_2$ mw: 122.13

SYNS: METHYL-p-BENZOQUINONE ◇ METHYL-1,4-BENZOQUINONE ◇ 2-METHYLBENZOQUINONE-1,4 ◇ 2-METHYL-1,4-QUINONE ◇ p-TOLUQUINONE ◇ 1,4-TOLUQUINONE

TOXICITY DATA with REFERENCE
orl-rat LDLo:250 mg/kg NCNSA6 5,39,53

CONSENSUS REPORTS: Reported in EPA TSCA Inventory.

SAFETY PROFILE: Poison by ingestion. When heated to decomposition it emits acrid smoke and irritating fumes.

MHI300 CAS:1123-91-7 ***HR: 3***
5-METHYL-2,1,3-BENZOSELENADIAZOLE
mf: $C_7H_6N_2Se$ mw: 197.11

SYN: 2,1,3-BENZOSELENADIAZOLE,5-METHYL-

TOXICITY DATA with REFERENCE
ivn-mus LD50:56 mg/kg CSLNX* NX#02914

OSHA PEL: TWA 0.2 mg(Se)/m^3
ACGIH TLV: TWA 0.2 mg(Se)/m^3

SAFETY PROFILE: Poison by intravenous route. When heated to decomposition it emits toxic fumes of NO_x and Se.

MHI500 CAS:2818-88-4 ***HR: 3***
2-METHYLBENZOSELENAZOLE
mf: C_8H_7NSe mw: 196.12

SYN: 2-METHYLBENZSELENAZOL (CZECH)

TOXICITY DATA with REFERENCE
skn-rbt 500 mg/24H MLD 28ZPAK -,222,72
eye-rbt 500 mg/24H MLD 28ZPAK -,222,72
orl-rat LD50:471 mg/kg 28ZPAK -,222,72
ivn-mus LD50:140 mg/kg CSLNX* NX#05958

CONSENSUS REPORTS: Selenium and its compounds are on the Community Right-To-Know List. Reported in EPA TSCA Inventory.

OSHA PEL: TWA 0.2 mg(Se)/m^3
ACGIH TLV: TWA 0.2 mg(Se)/m^3

SAFETY PROFILE: Poison by intravenous route. Moderately toxic by ingestion. An eye and skin irritant. When heated to decomposition it emits very toxic fumes of NO_x and Se. See also SELENIUM COMPOUNDS.

MHI600 CAS:33082-92-7 ***HR: D***
METHYLBENZOTHIADIAZINE CARBAMATE
mf: $C_9H_9N_3O_2S$ mw: 223.27

SYNS: 1H-2,1,4-BENZOTHIADIAZIN-3-YL-CARBAMICACID METHYL ESTER (9CI) ◇ PP010

TOXICITY DATA with REFERENCE
cyt-hmn:lvr 100 μmol/L MUREAV 26,177,74
cyt-ham:lng 100 μmol/L JRIHDC 11(4),84,76

SAFETY PROFILE: Human mutation data reported. When heated to decomposition it emits toxic fumes of SO_x and NO_x. See also CARBAMATES and ESTERS.

MHI750 CAS:120-75-2 ***HR: 3***
2-METHYLBENZOTHIAZOLE
mf: C_8H_7NS mw: 149.22

SYN: USAF EK-1853

TOXICITY DATA with REFERENCE
ivn-mus LD50:105 mg/kg JPETAB 105,486,52
ipr-mus LD50:300 mg/kg NTIS** AD277-689

CONSENSUS REPORTS: Reported in EPA TSCA Inventory.

SAFETY PROFILE: Poison by intravenous and intraperitoneal routes. When heated to decomposition it emits very toxic fumes of NO_x and SO_x.

MHJ000 CAS:6112-39-6 ***HR: 2***
3-METHYLBENZOTHIAZOLIUM-p-TOLUENE SULFONATE

SYN: 3-METHYL-BENZOTHIAZOLIUM SALT with 4-METHYLBENZENESULFONIC ACID (1:1)

TOXICITY DATA with REFERENCE
orl-rat LDLo:1600 mg/kg KODAK* -,-,71
ipr-rat LDLo:800 mg/kg KODAK* -,-,71

CONSENSUS REPORTS: Reported in EPA TSCA Inventory.

SAFETY PROFILE: Moderately toxic by ingestion and intraperitoneal routes. When heated to decomposition it emits very toxic fumes of NO_x and SO_x.

MHJ250 CAS:1128-67-2 ***HR: 3***
3-METHYL-2-BENZOTHIAZOLONE HYDRAZONE
mf: $C_8H_9N_3S$ mw: 179.26

SYN: MBTH

TOXICITY DATA with REFERENCE
ipr-rat LD50:135 mg/kg TXAPA9 36,201,76
ipr-mus LD50:127 mg/kg TXAPA9 33,186,75

CONSENSUS REPORTS: Reported in EPA TSCA Inventory.

SAFETY PROFILE: Poison by intraperitoneal route. When heated to decomposition it emits very toxic fumes of NO_x and SO_x.

MHJ500 CAS:5090-37-9 ***HR: 3***
2-((2-METHYLBENZO(b)THIEN-3-YL)METHYL)-2-IMIDAZOLINE HYDROCHLORIDE
mf: $C_{13}H_{14}N_2S{\bullet}ClH$ mw: 266.81

SYNS: 4,5-DIHYDRO-2-((2-METHYLBENZO(b)THIEN-3-YL)METHYL)-1H-IMIDAZOLE HYDROCHLORIDE ◇ ELLSYL ◇ ELSYL ◇ EX 10-781 ◇ H 1032 ◇ 2-METHYL-3-(Δ^2)-IMIDAZOLINYLMETHYL)BENZO(b) THIOPHENE HYDROCHLORIDE ◇ α-METIL-β-(2-METILENE-4,5-DIIDROIMIDAZOLIL)BENZOTIOFANE CLORIDRATO (ITALIAN) ◇ METIZOLINE HYDROCHLORIDE ◇ RMI 10,482A

TOXICITY DATA with REFERENCE
orl-rat LD50:74 mg/kg FRPPAO 21,204,66
orl-mus LD50:155 mg/kg FRPPAO 21,204,66
ipr-mus LD50:49 mg/kg FRPPAO 21,204,66
ivn-mus LD50:9100 μg/kg FRPPAO 21,204,66

SAFETY PROFILE: Poison by ingestion, intravenous, and intraperitoneal routes. When heated to decomposition it emits very toxic fumes of NO_x, SO_x, and HCl.

MHJ750 CAS:1541-60-2 ***HR: 3***
7-METHYL-6H-(1)BENZOTHIOPYRANO (4,3-b)QUINOLINE
mf: $C_{17}H_{13}NS$ mw: 263.37

TOXICITY DATA with REFERENCE
mma-sat 30 μg/plate MUREAV 66,307,79
scu-mus TDLo:72 mg/kg/9W-I:NEO MUREAV 66,307,79

SAFETY PROFILE: Questionable carcinogen with experimental neoplastigenic data. Mutation data reported. When heated to decomposition it emits very toxic fumes of NO_x and SO_x.

MHK000 CAS:29385-43-1 ***HR: 2***
METHYL-1H-BENZOTRIAZOLE
mf: $C_7H_7N_3$ mw: 133.17

SYN: COBRATEC TT 100

TOXICITY DATA with REFERENCE
orl-rat LD50:675 mg/kg HURC** -,-,72

CONSENSUS REPORTS: Reported in EPA TSCA Inventory.

SAFETY PROFILE: Moderately toxic by ingestion. When heated to decomposition it emits toxic fumes of NO_x.

MHK250 CAS:136-85-6 ***HR: 2***
5-METHYLBENZOTRIAZOLE
mf: $C_7H_7N_3$ mw: 133.17

SYN: 5-METHYL-1,2,3-BENZOTRIAZOLE

TOXICITY DATA with REFERENCE
orl-rat LD50:1600 mg/kg KODAK* -,-,71

CONSENSUS REPORTS: Reported in EPA TSCA Inventory.

SAFETY PROFILE: Moderately toxic by ingestion. When heated to decomposition it emits toxic fumes of NO_x.

MHK500 CAS:95-21-6 ***HR: 3***
2-METHYLBENZOXAZOLE
mf: C_8H_7NO mw: 133.16

SYNS: 2-METHYLBENZOXAZOL (CZECH) ◇ USAF EK-982

TOXICITY DATA with REFERENCE
skn-rbt 500 mg/24H SEV 28ZPAK -,157,72
eye-rbt 250 μg/24H SEV 28ZPAK -,157,72
orl-rat LD50:717 mg/kg 28ZPAK -,157,72
orl-mus LD40:1100 mg/kg JACSAT 67,905,45
ipr-mus LD50:400 mg/kg NTIS** AD277-689

CONSENSUS REPORTS: Reported in EPA TSCA Inventory.

SAFETY PROFILE: Poison by intraperitonel route. Moderately toxic by ingestion. A severe skin and eye irritant. When heated to decomposition it emits toxic fumes of NO_x.

MHL000 CAS:31431-39-7 ***HR: 2***
METHYL-5-BENZOYL BENZIMIDAZOLE-2-CARBAMATE
mf: $C_{16}H_{13}N_3O_3$ mw: 295.32

SYNS: N-2(5-BENZOYL-BENZIMIDAZOLE) CARBAMATE de METHYLE (FRENCH) ◇ 5-BENZOYL-2-BENZIMIDAZOLECARBAMIC ACID METHYL ESTER ◇ N-(BENZOYL-5-BENZIMIDAZOLYL)-2, CARBAMATE de METHYLE (FRENCH) ◇ MBDZ ◇ MEBENDAZOLE (USDA) ◇ OVITELMIN ◇ PANTELMIN ◇ R 17635 ◇ TELMIN ◇ VERMIRAX ◇ VERMOX

TOXICITY DATA with REFERENCE
mma-sat 600 nmol/plate CNREA8 38,4478,78
oms-hmn:leu 1 mg/L THERAP 31,505,76
orl-rat TDLo:78400 μg/kg (8-15D preg):TER THERAP 31,505,76
orl-rat TDLo:78400 μg/kg (female 8-15D post):REP BSVMA8 76,147,74
orl-rat LD50:714 mg/kg IYKEDH 19,735,88
orl-mus LD50:620 mg/kg MPPBAB 47,48,78
ipr-mus LD50:712 mg/kg MPPBAB 48,29,79
orl-gpg LDLo:1260 mg/kg TXAPA9 24,371,73

CONSENSUS REPORTS: EPA Genetic Toxicology Program.

SAFETY PROFILE: Moderately toxic by ingestion and intraperitoneal route. Human mutation data reported. An experimental teratogen. Experimental reproductive effects. When heated to decomposition it emits toxic fumes of NO_x. See also CARBAMATES.

MHL250 CAS:2606-85-1 ***HR: 3***
6-METHYL-3:4-BENZPHENANTHRENE
mf: $C_{19}H_{14}$ mw: 242.33

TOXICITY DATA with REFERENCE
orl-mus TDLo:11 g/kg/56W-I:ETA PRLBA4 129,439,40
scu-mus TDLo:4400 mg/kg/67W-I:ETA,REP PRLBA4 129,439,40

SAFETY PROFILE: Experimental reproductive effects. Questionable carcinogen with experimental tumorigenic data. Experimental reproductive effects. When heated to decomposition it emits acrid smoke and irritating fumes.

MHL500 CAS:2381-19-3 ***HR: 3***
7-METHYL-3,4-BENZPHENANTHRENE
mf: $C_{19}H_{14}$ mw: 242.33

TOXICITY DATA with REFERENCE
skn-mus TDLo:1850 mg/kg/77W-I:ETA PRLBA4 129,439,40

SAFETY PROFILE: Questionable carcinogen with experimental tumorigenic data. When heated to decomposition it emits acrid smoke and irritating fumes.

MHL750 CAS:4076-40-8 ***HR: 3***
8-METHYL-3:4-BENZPHENANTHRENE
mf: $C_{19}H_{14}$ mw: 242.33

SYN: 4-METHYLBENZO(c)PHENANTHRENE

TOXICITY DATA with REFERENCE
orl-mus TDLo:14 g/kg/71W-I:ETA PRLBA4 129,439,40

SAFETY PROFILE: Questionable carcinogen with experimental tumorigenic data. When heated to decomposition it emits acrid smoke and irritating fumes.

MHM000 CAS:2381-39-7 *HR: 3*
5-METHYL-3,4-BENZPYRENE
mf: $C_{21}H_{14}$ mw: 266.35

SYNS: 6-METHYLBENZO(a)PYRENE ◇ 5-METHYL-3,4-BENZO-PYRENE

TOXICITY DATA with REFERENCE
mma-sat 6250 ng/plate CNREA8 47,1509,87
dnd-mus-skn 8 μmol/kg CBINA8 47,111,83
mma-ham:lng 3800 nmol/L PNASA6 73,607,76
dnd-uns:lyms 30 μmol/L CBINA8 47,87,83
scu-rat TDLo:479 μg/kg/60D-I:NEO CBINA8 29,159,80
skn-mus TDLo:43 mg/kg/20W-I:CAR CBINA8 22,53,78
unr-mus TDLo:80 mg/kg/8D-I:ETA BEBMAE 88(11),592,79

CONSENSUS REPORTS: EPA Genetic Toxicology Program.

SAFETY PROFILE: Questionable carcinogen with experimental carcinogenic, neoplastigenic, and tumorigenic data by skin contact. Mutation data reported. When heated to decomposition it emits toxic fumes of NO_x.

MHM250 CAS:16757-81-6 *HR: 3*
8-METHYL-3,4-BENZPYRENE
mf: $C_{21}H_{14}$ mw: 266.35

SYN: 3-METHYLBENZO(a)PYRENE

TOXICITY DATA with REFERENCE
sln-dmg-par 5 mmol/L CNREA8 33,302,73
scu-mus TDLo:40 mg/kg:ETA BJCAAI 6,400,52

CONSENSUS REPORTS: EPA Genetic Toxicology Program.

SAFETY PROFILE: Questionable carcinogen with experimental tumorigenic data. Mutation data reported. When heated to decomposition it emits acrid smoke and irritating fumes.

MHM500 CAS:1929-88-0 *HR: 2*
1-METHYL-3-(2-BENZTHIAZOLYL)UREA
mf: $C_9H_9N_3OS$ mw: 207.27

SYNS: N-(2-BENZOTHIAZOLYL)-N'-METHYLUREA ◇ N-(2-BENZTHIAZOLYL)-N'-METHYLHARNSTOFF (GERMAN) ◇ BENZTHIAZURON ◇ GATINON

TOXICITY DATA with REFERENCE
orl-rat LD50:1280 mg/kg FMCHA2 -,D150,80

CONSENSUS REPORTS: EPA Genetic Toxicology Program.

SAFETY PROFILE: Moderately toxic by ingestion. When heated to decomposition it emits very toxic fumes of SO_x and NO_x.

MHN000 *HR: 3*
N-METHYLBENZYLAMINE mixed with SODIUM NITRITE (1:1)

SYN: SODIUM NITRITE mixed with N-METHYLBENZYLAMINE (1:1)

TOXICITY DATA with REFERENCE
orl-rat TDLo:28 g/kg/8W-C:ETA ZEKBAI 73,54,69

SAFETY PROFILE: Questionable carcinogen with experimental tumorigenic data. When heated to decomposition it emits very toxic fumes of NO_x and Na_2O. See also SODIUM NITRITE.

MHN250 *HR: 3*
METHYLBENZYLAMINE mixed with SODIUM NITRITE (2:3)

SYNS: N-METHYLBENZYLAMINE mixed with SODIUM NITRITE (2:3) ◇ SODIUM NITRITE mixed with METHYLBENZYLAMINE (3:2)

TOXICITY DATA with REFERENCE
orl-rat TDLo:25 g/kg/22W-I:ETA IARCCD 4,159,73

SAFETY PROFILE: Questionable carcinogen with experimental tumorigenic data. When heated to decomposition it emits very toxic fumes of NO_x and Na_2O. See also SODIUM NITRITE.

MHN300 CAS:104-82-5 *HR: 3*
4-METHYLBENZYL CHLORIDE
mf: C_8H_9Cl mw: 140.61

$$CH_3C_6H_4CH_2Cl$$

SAFETY PROFILE: Exothermic decomposition at 55°C is catalyzed by traces of iron. When heated to decomposition it emits toxic fumes of Cl^-. See also CHLORINATED HYDROCARBONS, AROMATIC.

MHN500 CAS:93-96-9 *HR: 1*
α-METHYL BENZYL ETHER
mf: $C_{16}H_{18}O$ mw: 226.34

PROP: Liquid. Mp: −30°, bp: 286.3°, flash p: 275°F (OC), d: 1.0017 @ 20°/20°, vap press: <0.01 mm @ 20°, vap. d: 7.82.

SYN: ETHER-BIS(α-METHYLBENZYL)

TOXICITY DATA with REFERENCE
skn-rbt 500 mg open MLD UCDS** 7/3/67
orl-rat LD50:9800 mg/kg UCDS** 7/3/67

SAFETY PROFILE: Mildly toxic by ingestion. A skin irritant. Combustible when exposed to heat or flame; can react with oxidizing materials. To fight fire, use alcohol foam, CO_2, dry chemical. See also ETHERS.

MHN750 CAS:10309-79-2 ***HR: 3***
1-METHYL-2-BENZYLHYDRAZINE
mf: $C_8H_{12}N_2$ mw: 136.22

SYN: 1-BENZYL-2-METHYLHYDRAZINE

TOXICITY DATA with REFERENCE
scu-rat TDLo:20 mg/kg (10D preg):TER IARCCD 4,45,73
scu-rat TDLo:520 mg/kg/27W-I:ETA 23HZAR -,267,70
scu-rat LD50:270 mg/kg IARCCD 4,45,73

SAFETY PROFILE: Poison by subcutaneous route. Questionable carcinogen with experimental carcinogenic, tumorigenic, and teratogenic data. Mutation data reported. When heated to decomposition it emits toxic fumes of NO_x.

MHO000 CAS:3979-76-8 ***HR: 3***
(α-METHYLBENZYL)HYDRAZINE SULFATE
mf: $C_8H_{12}N_2 \cdot H_2O_4S$ mw: 234.30

SYN: MEBANAZINE SULPHATE

TOXICITY DATA with REFERENCE
orl-mus LD50:271 mg/kg IJNEAQ 5,125,66
scu-mus LD50:175 mg/kg IJNEAQ 5,125,66

SAFETY PROFILE: Poison by ingestion and subcutaneous routes. When heated to decomposition it emits very toxic fumes of SO_x and NO_x.

MHP250 CAS:937-40-6 ***HR: 3***
N-METHYL-N-BENZYLNITROSAMINE
mf: $C_8H_{10}N_2O$ mw: 150.20

SYNS: METHYL-BENZYL-NITROSOAMIN (GERMAN) ◇ N-METHYL-N-NITROSOBENZYLAMINE ◇ N-NITROSOBENZYLMETHYLAMINE ◇ N-NITROSOMETHYLBENZYLAMINE

TOXICITY DATA with REFERENCE
mma-sat 50 μg/plate JMCMAR 29,40,86
mma-esc 5 μmol/plate GANNA2 75,8,84
mma-ham:lng 200 μmol/L CRNGDP 6,1731,85
sce-ham:lng 1 mmol/L CRNGDP 6,1731,85
orl-rat TDLo:16 mg/kg/4W-I:ETA JJIND8 61,145,78
orl-rat LD50:18 mg/kg NATWAY 50,100,63

CONSENSUS REPORTS: EPA Genetic Toxicology Program.

SAFETY PROFILE: Poison by ingestion. Questionable carcinogen with experimental tumorigenic data. Mutation data reported. When heated to decomposition it emits toxic fumes of NO_x. See also NITROSAMINES.

MHP400 CAS:1674-62-0 ***HR: 3***
1-METHYLBIGUANIDE HYDROCHLORIDE
mf: $C_3H_9N_5 \cdot ClH$ mw: 151.63

SYNS: N-METHYLIMIDODICARBONIMIDIC DIAMIDE MONOHYDROCHLORIDE ◇ 1-METILBIGUANIDE CLORIDRATO (ITALIAN)

TOXICITY DATA with REFERENCE
orl-rat LD50:1754 mg/kg FRPSAX 15,521,60
ipr-rat LD50:325 mg/kg FRPSAX 15,521,60
orl-mus LD50:1750 mg/kg ARZNAD 12,314,62
ipr-mus LD50:562 mg/kg JAJAAA 18,196,65

SAFETY PROFILE: Poison by intraperitoneal route. Moderately toxic by ingestion. When heated to decomposition it emits toxic fumes of NO_x and HCl.

MHQ500 CAS:63915-54-8 ***HR: 3***
METHYL-BIS(2-CHLOROETHYL-MERCAPTOETHYL)AMINE HYDROCHLORIDE
mf: $C_9H_{19}Cl_2NS_2 \cdot ClH$ mw: 312.77

SYNS: 2,2′-BIS(2-CHLOROETHYLMERCAPTO)-N-METHYLDIETHYLAMINE HYDROCHLORIDE ◇ METHYLBIS(β-CHLOROETHYLTHIOETHYL)AMINE HYDROCHLORIDE ◇ TL 1002

TOXICITY DATA with REFERENCE
ihl-mus LCLo:220 mg/m³/10M NDRC** No.9-4-1-19,44
ipr-mus LD50:8 mg/kg CANCAR 2,1055,49
scu-mus LDLo:25 mg/kg NTIS** PB158-507

SAFETY PROFILE: Poison by inhalation, subcutaneous, and intraperitoneal routes. When heated to decomposition it emits very toxic fumes of Cl^-, NO_x, and SO_x.

MHQ750 CAS:1555-58-4 ***HR: 2***
METHYL BIS(β-CYANOETHYL)AMINE
mf: $C_7H_{11}N_3$ mw: 137.21

SYN: N-METHYL-3,3′-IMINODIPROPIONITRILE

TOXICITY DATA with REFERENCE
eye-rbt 500 mg open AMIHBC 10,61,54
orl-rat LD50:890 mg/kg AMIHBC 10,61,54
ipr-mus LD50:500 mg/kg FRPSAX 17,753,62
skn-rbt LD50:800 mg/kg AMIHBC 10,61,54

CONSENSUS REPORTS: Cyanide and its compounds are on the Community Right-To-Know List. Reported in EPA TSCA Inventory.

SAFETY PROFILE: Moderately toxic by ingestion, skin contact, and intraperitoneal routes. An eye irritant. When heated to decomposition it emits toxic fumes of NO_x and CN^-. See also NITRILES.

MHQ775 CAS:36148-80-8 ***HR: 3***
N-METHYL-N,N-BIS(3-METHYLSULFONYLOXYPROPYL)AMINE 4,4′-BIPHENYLDISULFONATE
mf: $C_9H_{21}NO_6S_2 \cdot C_{12}H_{10}O_6S_2$ mw: 617.77

SYNS: 838-D ◇ 3,3′-(METHYLIMINO)BIS-1-PROPANOL DIMETHANESULFONATE (ESTER), (1,1′-BIPHENYL)-4,4′-DISULFOANTE (1:1) (SALT)

TOXICITY DATA with REFERENCE
orl-mus LD50:230 mg/kg YKKZAJ 93,47,73
ipr-mus LD50:190 mg/kg YKKZAJ 93,47,73

scu-mus LD50:205 mg/kg YKKZAJ 93,47,73
ivn-mus LD50:190 mg/kg YKKZAJ 93,47,73

SAFETY PROFILE: Poison by ingestion, subcutaneous, intravenous, and intraperitoneal routes. When heated to decomposition it emits toxic fumes of NO_x and SO_x.

MHR000 ***HR: 3***
METHYLBISMUTH OXIDE
mf: CH_3BiO mw: 240.01

SAFETY PROFILE: Ignites spontaneously in air. When heated to decomposition it emits toxic fumes of Bi. See also BISMUTH COMPOUNDS.

MHR100 CAS:81910-05-6 ***HR: D***
METHYL BOTRYODIPLODIN

SYNS: 2-METHOXY-3-METHYL-4-ACETYLTETRAHYDROFURAN ◇ TETRAHYDRO-2-ACETYL-4-METHOXY-3-METHYLFURAN

TOXICITY DATA with REFERENCE
dnd-hmn:oth 10 μmol/L CRNGDP 3,211,82
dnd-rat:oth 10 μmol/L CRNGDP 3,211,82

SAFETY PROFILE: Human mutation data reported. When heated to decomposition it emits toxic fumes of NO_x.

MHR200 CAS:74-83-9 ***HR: 3***
METHYL BROMIDE
DOT: UN 1062
mf: CH_3Br mw: 94.95

PROP: Colorless, transparent, volatile liquid or gas; burning taste, chloroform-like odor. Bp: 3.56°, lel: 13.5%, uel: 14.5%, fp: −93°, flash p: none, d: 1.732 @ 0°/0°, autoign temp: 998°F, vap d: 3.27, vap press: 1824 mm @ 25°.

SYNS: BROM-METHAN (GERMAN) ◇ BROMO METHANE ◇ BROMO-O-GAS ◇ BROMOMETANO (ITALIAN) ◇ BROMURE de METHYLE (FRENCH) ◇ BROMURO di METILE (ITALIAN) ◇ BROOMMETHAAN (DUTCH) ◇ DAWSON 100 ◇ DOWFUME ◇ DOWFUME MC-2 SOIL FUMIGANT ◇ EDCO ◇ EMBAFUME ◇ FUMIGANT-1 (OBS.) ◇ HALON 1001 ◇ ISCOBROME ◇ KAYAFUME ◇ MB ◇ MBX ◇ MEBR ◇ METAFUME ◇ METHOGAS ◇ METHYLBROMID (GERMAN) ◇ METYLU BROMEK (POLISH) ◇ MONOBROMOMETHANE ◇ PESTsMASTER (OBS.) ◇ PROFUME (OBS.) ◇ R 40B1 ◇ RCRA WASTE NUMBER U029 ◇ ROTOX ◇ TERABOL ◇ TERR-O-GAS 100 ◇ ZYTOX

TOXICITY DATA with REFERENCE
mma-sat 5 g/m³ MUREAV 116,185,83
mmo-klp 4750 mg/m³ MUREAV 155,41,85
orl-rat TDLo:3250 mg/kg/13W-I:CAR TXAPA9 72,262,84
ihl-man LCLo:60000 ppm/2H BJIMAG 2,24,45
ihl-chd LCLo:1 mg/m³/2H NHOZAX 23,241,69
ihl-hmn TCLo:35 ppm:GIT INMEAF 11,575,42
orl-rat LD50:214 mg/kg TXAPA9 72,262,84
ihl-rat LC50:302 ppm/8H TXAPA9 81,183,85
ihl-mus LC50:1540 mg/m³/2H 85GMAT -,81,82
ihl-rbt LCLo:2000 mg/m³/11H JIHTAB 22,218,40
ihl-gpg LCLo:300 ppm/9H XPHBAO 185,1,29

CONSENSUS REPORTS: IARC Cancer Review: Group 3 IMEMDT 7,245,87; Human Inadequate Evidence IMEMDT 41,187,86; Animal Limited Evidence IMEMDT 41,187,86. Reported in EPA TSCA Inventory. Community Right-To-Know List. EPA Extremely Hazardous Substances List.

OSHA PEL: (Transitional: CL 20 ppm (skin)) TWA 5 ppm (skin)
ACGIH TLV: TWA 5 ppm (skin)
DFG MAK: 5 ppm (20 mg/m³), Suspected Carcinogen.
NIOSH REL: (Monohalomethanes) Reduce to lowest level
DOT Classification: Poison A; Label: Poison Gas.

SAFETY PROFILE: Suspected carcinogen with experimental carcinogenic data. A human poison by inhalation. Human systemic effects by inhalation: anorexia, nausea or vomiting. Corrosive to skin; can produce severe burns. Human mutation data reported. A powerful fumigant gas which is one of the most toxic of the common organic halides. It is hemotoxic and narcotic with delayed action. The effects are cumulative and damaging to nervous system, kidneys, and lung. Central nervous system effects include blurred vision, mental confusion, numbness, tremors, and speech defects.

Methyl bromide is reported to be eight times more toxic on inhalation than ethyl bromide. Moreover, because of its greater volatility, it is a much more frequent cause of poisoning. Death following acute poisoning is usually caused by its irritant effect on the lungs. In chronic poisoning, death is due to injury to the central nervous system. Fatal poisoning has always resulted from exposure to relatively high concentrations of methyl bromide vapors (from 8,600 to 60,000 ppm). Nonfatal poisoning has resulted from exposure to concentrations as low as 100-500 ppm. In addition to injury to the lung and central nervous system, the kidneys may be damaged with development of albuminuria and, in fatal cases, cloudy swelling and/or tubular degeneration. The liver may be enlarged. There are no characteristic blood changes.

Mixtures of 10-15 percent with air may be ignited with difficulty. Moderately explosive when exposed to sparks or flame. Forms explosive mixtures with air within narrow limits at atmospheric pressure, with wider limits at higher pressure. The explosive sensitivity of mixtures with air may be increased by the presence of aluminum; magnesium; zinc; or their alloys. Incompatible with metals; dimethyl sulfoxide; ethylene oxide. To fight fire, use foam, water, CO_2, dry chemical. When heated to decomposition it emits toxic fumes of Br^-. See also BROMIDES.

MHR250 CAS:96-32-2 ***HR: 3***
METHYL BROMOACETATE
DOT: UN 2643
mf: $C_3H_5BrO_2$ mw: 152.99

PROP: Liquid. Bp: 51-52° @ 15 mm.

SYNS: BROMOACETIC ACID METHYL ESTER ◇ METHYL α-BROMOACETATE ◇ METHYL MONOBROMOACETATE

TOXICITY DATA with REFERENCE
ivn-mus LDLo:16 mg/kg CBCCT* 6,138,54

CONSENSUS REPORTS: Reported in EPA TSCA Inventory.

DOT Classification: Poison B; Label: Poison.

SAFETY PROFILE: Poison by intravenous route. When heated to decomposition it emits toxic fumes of Br^-. See also ESTERS.

MHR500 CAS:583-75-5 ***HR: D***
2-METHYL-4-BROMOANILINE
mf: C_7H_8BrN mw: 186.07

TOXICITY DATA with REFERENCE
mma-sat 1 μmol/plate MUREAV 77,317,80
dnd-ham:lng 3 mmol/L/4H MUREAV 77,317,80

SAFETY PROFILE: Mutation data reported. When heated to decomposition it emits very toxic fumes of NO_x and Br^-. See also BROMIDES.

MHR750 CAS:67880-26-6 ***HR: 3***
METHYL-4-BROMOBENZENEDIAZOATE
mf: $C_7H_7BrN_2O$ mw: 215.05

$BrC_6H_4N{=}NOCH_3$

SAFETY PROFILE: Explodes on heating. When heated to decomposition it emits toxic fumes of Br^- and NO_x.

MHS250 CAS:23471-23-0 ***HR: 3***
METHYL-(BROMOMERCURI)FORMATE
mf: $C_2H_3BrHgO_2$ mw: 339.55

SYN: BROMO(METHOXYCARBONYL) MERCURY

TOXICITY DATA with REFERENCE
ivn-mus LD50:56 mg/kg CSLNX* NX#05824

CONSENSUS REPORTS: Mercury and its compounds are on the Community Right-To-Know List.

OSHA PEL: (Transitional: CL 1 mg/10m³) CL 0.1 mg(Hg)/m³ (skin)
ACGIH TLV: TWA 0.1 mg(Hg)/m³ (skin)
NIOSH REL: (Mercury, Inorganic) TWA 0.05 mg(Hg)/m³

SAFETY PROFILE: Poison by intravenous route. When heated to decomposition it emits very toxic fumes of Hg and Br^-. See also MERCURY COMPOUNDS and BROMIDES.

MHS375 CAS:20680-07-3 ***HR: 2***
1-METHYL-3-(p-BROMOPHENYL)UREA
mf: $C_8H_9BrN_2O$ mw: 229.10

SYNS: BROMDEFENURON ◇ 1-(p-BROMOPHENYL)-3-METHYLUREA ◇ 1-METHYL-3-(p-BROMPHENYL)HARNSTOFF ◇ UREA, N-(4-BROMOPHENYL)-N'-METHYL-(9CI)

TOXICITY DATA with REFERENCE
orl-rat TDLo:722 mg/kg/87W-:ETA ARGEAR 53,329,83

SAFETY PROFILE: Questionable carcinogen with experimental tumorigenic data. When heated to decomposition it emits toxic fumes of NO_x and Br^-.

MHS500 CAS:2938-98-9 ***HR: 2***
2-METHYL-1,4-BUTANEDIOL
mf: $C_5H_{12}O_2$ mw: 104.17

TOXICITY DATA with REFERENCE
eye-rbt 20 mg open SEV AMIHBC 4,119,51
orl-rat LD50:5460 mg/kg AMIHBC 4,119,51
skn-rbt LD50:2620 mg/kg AMIHBC 4,119,51

SAFETY PROFILE: Moderately toxic by skin contact. Mildly toxic by ingestion. A severe eye irritant. When heated to decomposition it emits acrid smoke and irritating fumes.

MHS750 CAS:137-32-6 ***HR: 3***
2-METHYL BUTANOL-1
mf: $C_5H_{12}O$ mw: 88.15

PROP: Colorless liquid. D: 0.81-0.82 @ 20°, fp: < −70°, bp: 128°, flash p: 122°F (OC), vap d: 3.0, lel: 1.4%, uel: 9.0%. Sltly sol in water; misc with alc and ether.

SYNS: dl-sec-BUTYLCARBINOL ◇ 2-METHYLBUTANOL

TOXICITY DATA with REFERENCE
skn-rbt 8193 μg/24H open MLD AIHAAP 23,95,62
orl-rat LD50:4920 mg/kg AIHAAP 23,95,62
ipr-rat LDLo:1900 mg/kg JIHTAB 27,1,45
skn-rbt LDLo:3540 mg/kg AIHAAP 23,95,62

CONSENSUS REPORTS: Reported in EPA TSCA Inventory.

SAFETY PROFILE: Moderately toxic by skin contact and intraperitoneal routes. Mildly toxic by ingestion. An eye, skin, and mucous membrane irritant. Can cause deafness, delerium, headache, nausea, and vomiting. Flammable when exposed to heat, flame, or oxidizers. Explosive in the form of vapor when exposed to heat or flame. Incompatible with H_2S_3. To fight fire, use alcohol foam, spray, mist, dry chemical. When heated to decomposition it emits acrid smoke and irritating fumes. See also ALCOHOLS.

MHT000 CAS:563-46-2 ***HR: 3***
2-METHYL-1-BUTENE
DOT: UN 2371/UN 2459
mf: C_5H_{10} mw: 70.14

PROP: Colorless; extremely volatile liquid or gas. D: 0.7, vap d: 2.4, bp: 38°, flash p: −4°F. Insol in water.

CONSENSUS REPORTS: Reported in EPA TSCA Inventory.

DOT Classification: Flammable Liquid; Label: Flammable Liquid.

SAFETY PROFILE: A simple asphyxiant. Very dangerous fire hazard when exposed to heat, flame or oxidizers. To fight fire, use dry chemical, CO_2, foam. When heated to decomposition it emits acrid smoke and irritating fumes.

MHT250 CAS:563-45-1 ***HR: 3***
3-METHYL-1-BUTENE
DOT: UN 2371/UN 2561
mf: C_5H_{10} mw: 70.14

PROP: Colorless, very volatile, flammable liquid; disagreeable odor. Bp: 31.11°, d: 0.65 @ 20°/20°, fp: −137.5°, flash p: 19.4°F, vap d: 2.4, lel: 1.5%, uel: 9.1%. Insol in water; sol in alc.

CONSENSUS REPORTS: Reported in EPA TSCA Inventory.

DOT Classification: Flammable liquid; Label: Flammable Liquid.

SAFETY PROFILE: Very dangerous fire hazard when exposed to heat, flame, or oxidizers. Explosive in the form of vapor when exposed to heat or flame. To fight fire, use alcohol foam, mist, spray, dry chemical, CO_2. When heated to decomposition it emits acrid smoke and irritating fumes. See also 2-METHYL-1-BUTENE.

MHT500 CAS:541-47-9 ***HR: 2***
3-METHYL-2-BUTENOIC ACID
mf: $C_5H_8O_2$ mw: 100.13

SYNS: β,β-DIMETHYLACRYLIC ACID ◇ 3,3-DIMETHYLACRYLIC ACID ◇ β-METHYLCROTONIC ACID ◇ 3-METHYLCROTONIC ACID ◇ SENECIOIC ACID

TOXICITY DATA with REFERENCE
skn-rbt 500 mg open MLD UCDS** 8/23/67
eye-rbt 1 mg SEV UCDS** 8/23/67
orl-rat LD50:3560 mg/kg TXAPA9 28,313,74
orl-mus LD50:2580 mg/kg GTPZAB 29(4),52,85
orl-gpg LD50:3 g/kg GTPZAB 29(4),52,85

CONSENSUS REPORTS: Reported in EPA TSCA Inventory.

SAFETY PROFILE: Moderately toxic by ingestion. A skin and severe eye irritant. When heated to decomposition it emits acrid smoke and irritating fumes.

MHU100 CAS:115-18-4 ***HR: 2***
3-METHYL-1-BUTEN-3-OL
mf: $C_5H_{10}O$ mw: 86.15

SYNS: METHYLBUTENOL ◇ 2-METHYL-3-BUTEN-2-OL ◇ 3-METHYL-BUTEN-(1)-OL-(3) (GERMAN)

TOXICITY DATA with REFERENCE
ipr-rat LD50:1315 mg/kg PLMEAA 48,120,83
ipr-mus LD50:1117 mg/kg PLMEAA 48,120,83
scu-mus LD50:1680 mg/kg ARZNAD 5,161,55

SAFETY PROFILE: Moderately toxic by intraperitoneal and subcutaneous routes. When heated to decomposition it emits acrid smoke and irritating fumes.

MHU110 CAS:556-82-1 ***HR: 2***
3-METHYL-2-BUTEN-1-OL
mf: $C_5H_{10}O$ mw: 86.15

SYNS: DIMETHYLALLYL ALCOHOL ◇ Γ,Γ-DIMETHYLALLYL ALCOHOL ◇ 3,3-DIMETHYLALLYL ALCOHOL ◇ PRENOL ◇ PRENYL ALCOHOL

TOXICITY DATA with REFERENCE
skn-rbt 500 mg/24H MOD FCTXAV 17,895,79
orl-rat LD50:810 mg/kg FCTXAV 17,895,79
skn-rbt LD50:3900 mg/kg FCTXAV 17,895,79

CONSENSUS REPORTS: Reported in EPA TSCA Inventory.

SAFETY PROFILE: Moderately toxic by skin contact and ingestion. A skin irritant. When heated to decomposition it emits acrid smoke and irritating fumes. See also ALLYL COMPOUNDS and ALCOHOLS.

MHU150 CAS:5205-11-8 ***HR: 2***
3-METHYL-2-BUTENYL BENZOATE
mf: $C_{12}H_{14}O_2$ mw: 190.26

SYNS: BENZOIC ACID, 3-METHYL-2-BUTENYL ESTER ◇ 2-BUTEN-1-OL, 3-METHYL-, BENZOATE ◇ PRENYL BENZOATE

TOXICITY DATA with REFERENCE
skn-rbt 500 mg/24H MLD FCTOD7 20,819,82
orl-rat LD50:4700 mg/kg FCTOD7 20,819,82

CONSENSUS REPORTS: Reported in EPA TSCA Inventory.

SAFETY PROFILE: Moderately toxic by ingestion. A skin irritant. When heated to decomposition it emits acrid smoke and irritating fumes.

MHU200 CAS:6966-40-1 ***HR: 3***
5-(1-METHYL-1-BUTENYL)-5-PROPYLBARBITURIC ACID
mf: $C_{12}H_{18}N_2O_3$ mw: 238.32

TOXICITY DATA with REFERENCE
orl-mus LD50:320 mg/kg JACSAT 61,776,39
ipr-mus LD50:270 mg/kg JACSAT 61,776,39

SAFETY PROFILE: Poison by ingestion and intraperitoneal routes. When heated to decomposition it emits toxic fumes of NO_x. See also BARBITURATES.

MHU250 CAS:78-80-8 ***HR: 3***
2-METHYL-1-BUTEN-3-YNE
mf: C_5H_6 mw: 66.11

PROP: Flash p: <19.4°F.

TOXICITY DATA with REFERENCE
skn-rbt 500 mg MOD SCCUR* -,6,61
orl-rat LDLo:639 mg/kg SCCUR* -,6,61
orl-mus LDLo:350 mg/kg SCCUR* -,6,61
ihl-mus LC50:14 pph/6M SCCUR* -,6,61

SAFETY PROFILE: Poison by ingestion. Mildly toxic by inhalation. A skin irritant. A very dangerous fire hazard when exposed to heat or flame. When heated to decomposition it emits acrid smoke and irritating fumes. See also ACETYLENE COMPOUNDS.

MHU500 ***HR: 3***
3-METHYL-3-BUTEN-1-YNYLTRIETHYLLEAD
mf: $C_{11}H_{20}Pb$ mw: 359.47

CONSENSUS REPORTS: Lead and its compounds are on the Community Right-To-Know List.

SAFETY PROFILE: Explodes on rapid heating. When heated to decomposition it emits toxic fumes of Pb. See also LEAD COMPOUNDS.

MHU750 CAS:97-88-1 ***HR: 2***
2-METHYL BUTYLACRYLATE
DOT: UN 2227
mf: $C_8H_{14}O_2$ mw: 142.22

PROP: Colorless liquid; ester odor. Bp: 163°, flash p: 126°F (TOC), lel: 2%, uel: 8%, autoign temp: 562°F, vap press: 4.9 mm @ 20°, d: 0.895 @ 20°/4°, vap d: 4.8.

SYNS: BUTIL METACRILATO (ITALIAN) ◇ BUTYLMETHACRYLAAT (DUTCH) ◇ N-BUTYL METHACRYLATE ◇ BUTYL-2-METHACRYLATE ◇ BUTYL-2-METHYL-2-PROPENOATE ◇ METHACRYLATE de BUTYLE (FRENCH) ◇ METHACRYLSAEUREBUTYLESTER (GERMAN) ◇ 2-METHYL-BUTYLACRYLAAT (DUTCH) ◇ 2-METHYL-BUTYLACRYLAT (GERMAN)

TOXICITY DATA with REFERENCE
skn-rbt 10 g/kg open JIHTAB 23,343,41
ipr-rat TDLo:690 mg/kg (5-15D preg):TER JDREAF 51,1632,72
ipr-rat TDLo:2304 mg/kg (5-15D preg):REP JDREAF 51,1632,72
orl-rat LD50:22600 mg/kg AIHAAP 30,470,69
ihl-rat LC50:4910 ppm/4H JTEHD6 16,811,85
ipr-rat LD50:2304 mg/kg JDREAF 51,1632,72
orl-mus LD50:13500 mg/kg GTPZAB 19(9),57,75
ipr-mus LD50:1490 mg/kg JPMSAE 62,778,73
orl-rbt LDLo:6270 mg/kg JIHTAB 23,343,41
skn-rbt LD50:11300 mg/kg AIHAAP 30,470,69

CONSENSUS REPORTS: Reported in EPA TSCA Inventory.

DOT Classification: Flammable or Combustible Liquid; Label: Flammable Liquid.

SAFETY PROFILE: Moderately toxic by intraperitoneal route. Mildly toxic by ingestion, inhalation, and skin contact. An experimental teratogen. Experimental reproductive effects. A skin irritant. Flammable when exposed to heat or flame. Explosive in the form of vapor when exposed to heat or flame. Violent polymerization can be caused by heat, moisture, oxidizers. To fight fire, use foam, dry chemical, CO_2. When heated to decomposition it emits acrid smoke and irritating fumes.

MHV000 CAS:110-68-9 ***HR: 3***
N-METHYL-n-BUTYLAMINE
DOT: UN 2945
mf: $C_5H_{13}N$ mw: 87.19

PROP: Liquid. D: 0.7335, bp: 91.1°, vap d: 3.0, flash p: 35.6°. Sol in water.

SYNS: METHYLBUTYLAMINE ◇ N-(METHYL) BUTYL AMINE

TOXICITY DATA with REFERENCE
skn-rbt 100 μg/24H open AIHAAP 23,95,62
eye-rbt 74 mg SEV UCDS** 7/6/70
orl-rat LD50:420 mg/kg UCDS** 7/6/70
ihl-rat LCLo:2000 ppm/4H UCDS** 7/6/70
ipr-mus LD50:471 mg/kg JPETAB 88,82,46
ivn-mus LD50:122 mg/kg JPETAB 88,82,46
skn-rbt LD50:1260 mg/kg AIHAAP 23,95,62

CONSENSUS REPORTS: Reported in EPA TSCA Inventory.

DOT Classification: Flammable Liquid; Label: Flammable Liquid.

SAFETY PROFILE: Poison by intravenous route. Moderately toxic by ingestion, skin contact, and intraperitoneal routes. Mildly toxic by inhalation. A skin and severe eye irritant. A very dangerous fire hazard when exposed to heat, flame, or oxidizers. To fight fire, use alcohol foam. When heated to decomposition it emits toxic fumes of NO_x. See also AMINES.

MHV750 CAS:4435-53-4 ***HR: 2***
METHYL-1,3-BUTYLENE GLYCOL ACETATE
DOT: UN 2708
mf: $C_7H_{14}O_3$ mw: 146.21

PROP: Liquid; bitter taste and acrid odor. Bp: 135°, flash p: 170°F, d: 0.952-0.958 @ 20°/20°, vap d: 5.05.

SYNS: ACETIC ACID-3-METHOXYBUTYL ESTER ◇ BUTOXYL ◇ 3-METHOXYBUTYL ACETATE

TOXICITY DATA with REFERENCE
skn-rbt 10 mg/24H open MLD AMIHBC 10,61,54
eye-rbt 20 mg open AMIHBC 10,61,54
orl-rat LD50:4210 mg/kg AMIHBC 10,61,54

CONSENSUS REPORTS: Reported in EPA TSCA Inventory.

DOT Classification: Flammable or Combustible Liquid; Label: Flammable Liquid.

SAFETY PROFILE: Mildly toxic by ingestion. A skin and eye irritant. Combustible when exposed to heat or flame; can react with oxidizing materials. To fight fire, use alcohol foam, CO_2, dry chemical. When heated to decomposition it emits acrid smoke and irritating fumes.

MHV859 CAS:1634-04-4 ***HR: 3***
METHYL tert-BUTYL ETHER
DOT: UN 2398
mf: $C_5H_{12}O$ mw: 88.17

SYNS: METHYL 1,1-DIMETHYLETHYL ETHER ◇ PROPANE, 2-METHOXY-2-METHYL (9CI)

CONSENSUS REPORTS: Community Right-To-Know List. Reported in EPA TSCA Inventory.

DOT Classification: Flammable Liquid; Label: Flammable Liquid.

SAFETY PROFILE: Flammable when exposed to heat or flame. When heated to decomposition it emits acrid smoke and irritating fumes. See also ETHERS.

MHW000 CAS:20240-62-4 ***HR: 3***
METHYLBUTYL HYDRAZINE
mf: $C_5H_{14}N_2$ mw: 102.21

TOXICITY DATA with REFERENCE
orl-rat TDLo:1425 mg/kg/57W-I:ETA PPTCBY 2,73,72

SAFETY PROFILE: Questionable carcinogen with experimental tumorigenic data. When heated to decomposition it emits toxic fumes of NO_x. See also HYDRAZINE.

MHW250 CAS:73454-79-2 ***HR: 3***
1-METHYL-2-BUTYL-HYDRAZINE DIHYDROCHLORIDE
mf: $C_5H_{14}N_2$•2ClH mw: 175.13

SYN: 1-BUTYL-2-METHYL-HYDRAZINE DIHYDROCHLORIDE

TOXICITY DATA with REFERENCE
orl-rat TDLo:1425 mg/kg/57W-I:ETA 23HZAR -,267,70
unr-rat LD50:600 mg/kg 23HZAR -,267,70

SAFETY PROFILE: Moderately toxic by an unspecified route. Questionable carcinogen with experimental tumorigenic data. When heated to decomposition it emits very toxic fumes of Cl^- and NO_x.

MHW350 CAS:71016-15-4 ***HR: 3***
N-3-METHYLBUTYL-N-1-METHYL ACETONYL-NITROSAMINE
mf: $C_9H_{18}N_2O_2$ mw: 186.29

SYNS: 3-((ISOPENTYL)NITROSOAMINO)-2-BUTANONE ◇ MAMBNA

TOXICITY DATA with REFERENCE
mma-sat 2 g/L CRNGDP 1,867,80
otr-ham:lng 500 mg/L SSBSEF 25,738,82
orl-rat TDLo:21600 mg/kg/74W-C:CAR CCLCDY 7,329,85
orl-mus TDLo:8400 mg/kg/19W-C:CAR CCLCDY 7,329,85
orl-mus TD:19600 mg/kg/32W-C:NEO CCLCDY 7,329,85
orl-rat TD:27 g/kg/74W-C:ETA CMJODS 97,311,84

SAFETY PROFILE: Suspected carcinogen with experimental carcinogenic, neoplastigenic, and tumorigenic data. Mutation data reported. When heated to decomposition it emits toxic fumes of NO_x. See also NITROSAMINES.

MHW500 CAS:7068-83-9 ***HR: 3***
METHYLBUTYLNITROSAMINE
mf: $C_5H_{12}N_2O$ mw: 116.19

SYNS: MBNA ◇ METHYL-BUTYL-NITROSAMIN (GERMAN) ◇ METHYL-N-BUTYLNITROSAMINE ◇ N-METHYL-N-NITROSOBUTYLAMINE ◇ N-NITROSO-N-BUTYLMETHYLAMINE ◇ N-NITROSOMETHYL-N-BUTYLAMINE ◇ NMBA

TOXICITY DATA with REFERENCE
mmo-sat 1 mg/plate TCMUD8 1,295,80
mma-sat 10 μmol/plate TCMUE9 1,13,84
mma-esc 100 μmol/L MUREAV 26,361,74
pic-esc 100 mg/L TCMUE9 1,91,84
orl-rat TDLo:128 mg/kg/20W-I:ETA CRNGDP 1,157,80
orl-mus TDLo:182 mg/kg/1Y-C:CAR 85DUA4 -,129,70

orl-rat LD50:130 mg/kg BJIMAG 19,276,62
ihl-rat LD50:90 mg/kg ZEKBAI 71,135,68
ipr-rat LD50:120 mg/kg BIJOAK 85,72,62
scu-rat LD50:90 mg/kg ZEKBAI 71,135,68
orl-mus LD50:25 mg/kg 85DUA4 -,129,70
scu-mus LD50:10 mg/kg 85DUA4 -,129,70

CONSENSUS REPORTS: EPA Genetic Toxicology Program.

SAFETY PROFILE: Poison by ingestion, inhalation, intraperitoneal, and subcutaneous routes. Questionable carcinogen with experimental carcinogenic and tumorigenic data. Mutation data reported. When heated to decomposition it emits toxic fumes of NO_x. See also NITROSAMINES.

MHW750 CAS:2504-18-9 ***HR: 2***
METHYL-tert-BUTYLNITROSAMINE
mf: $C_5H_{12}N_2O$ mw: 116.19

TOXICITY DATA with REFERENCE
ipr-rat LD50:700 mg/kg BJIMAG 19,276,62
scu-rat LD50:630 mg/kg ZKKOBW 80,17,73

SAFETY PROFILE: Moderately toxic by intraperitoneal and subcutaneous routes. Many nitrosamines are carcinogens. When heated to decomposition it emits toxic fumes of NO_x. See also NITROSAMINES.

MHX000 CAS:38285-49-3 ***HR: 1***
5-METHYL-3-BUTYLTETRAHYDROPYRAN-4-YL ACETATE
mf: $C_{12}H_{22}O_3$ mw: 214.34

SYNS: ACETIC ACID, 3-BUTYL-5-METHYL-TETRAHYDRO-2H-PYRAN-4-YL ESTER ◇ 3-BUTYL-5-METHYL-TETRAHYDRO-2H-PYRAN-4-YL ACETATE

TOXICITY DATA with REFERENCE
skn-rbt 500 mg/24H MLD FCTXAV 14,601,76

CONSENSUS REPORTS: Reported in EPA TSCA Inventory.

SAFETY PROFILE: A skin irritant. When heated to decomposition it emits acrid smoke and irritating fumes.

MHX250 CAS:115-19-5 ***HR: 2***
2-METHYL-3-BUTYN-2-OL
mf: C_5H_8O mw: 84.13

PROP: Colorless liquid. Mp: 2.6°, bp: 104-105°, vap d: 2.49, d: 0.8672 @ 20°/20°, flash p: <69.8°F. Misc with water, acetone, benzene, carbon tetrachloride, cellosolve, etc.

SYNS: DIMETHYLACETYLENECARBINOL ◇ α,α-DIMETHYLPROPARGYL ALCOHOL ◇ 1,1-DIMETHYLPROPARGYL ALCOHOL ◇ ETHYNYLDIMETHYLCARBINOL ◇ 2-HYDROXY-2-METHYL-3-BUTYNE ◇ 3-METHYL-BUTIN-(1)-OL-(3) (GERMAN)

TOXICITY DATA with REFERENCE
orl-rat LD50:1950 mg/kg JPETAB 115,230,55
orl-mus LD50:1800 mg/kg ARZNAD 4,477,54
ipr-mus LD50:3600 mg/kg MEIEDD 10,866,83
scu-mus LD50:2340 mg/kg ARZNAD 5,161,55

CONSENSUS REPORTS: Reported in EPA TSCA Inventory.

SAFETY PROFILE: Moderately toxic by ingestion, intraperitoneal and subcutaneous routes. A very dangerous fire hazard when exposed to heat or flame; can react with oxidizing materials, heat, flames. To fight fire, use alcohol foam, mist, spray, CO_2. When heated to decomposition it emits acrid smoke and irritating fumes. See also ACETYLENE COMPOUNDS.

MHX500 CAS:590-86-3 ***HR: 2***
3-METHYLBUTYRALDEHYDE
mf: $C_5H_{10}O$ mw: 86.15

PROP: Colorless liquid, apple-like odor. Mp: −51°, bp: 92.5°, d: 0.803 @ 17°/4°, vap d: 2.96, flash p: 23°F. Sltly sol in water; sol in alc and ether.

SYNS: ISOPENTALDEHYDE ◇ ISOVALERAL ◇ ISOVALERALDEHYDE ◇ ISOVALERIC ALDEHYDE ◇ 3-METHYLBUTANAL

TOXICITY DATA with REFERENCE
orl-rat LD50:8910 mg/kg TXAPA9 28,313,74
ihl-rat LCLo:16000 ppm/4H TXAPA9 28,313,74
skn-rbt LD50:3180 mg/kg TXAPA9 28,313,74

CONSENSUS REPORTS: Reported in EPA TSCA Inventory.

SAFETY PROFILE: Moderately toxic by skin contact. Mildly toxic by ingestion and inhalation. A very dangerous fire hazard when exposed to heat or flame. Avoid sparks, heat, open flame. When heated to decomposition it emits acrid smoke and irritating fumes. See also ALDEHYDES.

MHY000 CAS:623-42-7 ***HR: 3***
METHYL-n-BUTYRATE
DOT: UN 1237
mf: $C_5H_{10}O_2$ mw: 102.13

PROP: Colorless liquid. Mp: < −97°, bp: 102.3°, flash p: 57°F (CC), d: 0.898, vap press: 40 mm @ 29.6°, vap d: 3.53. Sltly sol in water; misc in alc and ether.

SYNS: METHYL n-BUTANOATE ◇ METHYL BUTYRATE

TOXICITY DATA with REFERENCE
skn-rbt 500 mg/24H MOD FCTOD7 20(Suppl),741,82
ihl-mus LC50:18 g/m^3/2H 85GMAT -,81,82
orl-rbt LD50:3380 mg/kg IMSUAI 41,31,72
skn-rbt LD50:3560 mg/kg TXAPA9 42,417,77

CONSENSUS REPORTS: Reported in EPA TSCA Inventory.

DOT Classification: Flammable Liquid; Label: Flammable Liquid.

SAFETY PROFILE: Moderately toxic by ingestion and skin contact. A skin irritant. A very dangerous fire hazard when exposed to heat, flame or oxidizers. Can react vigorously with oxidizing materials. To fight fire, use alcohol foam, CO_2, dry chemical. When heated to decomposition it emits acrid smoke and irritating fumes. See also ESTERS.

MHY550 CAS:7568-37-8 ***HR: 3***
METHYL CADMIUM AZIDE
mf: CH_3CdN_3 mw: 97.13

CONSENSUS REPORTS: Cadmium and its compounds are on the Community Right-To-Know List.

OSHA PEL: TWA 0.1 mg(Cd)/m^3; CL 0.6 mg(Cd)/m^3 (fume)
ACGIH TLV: TWA 0.05 mg(Cd)/m^3 (Proposed: TWA 0.01 mg(Cd)/m^3 (dust), Suspected Human Carcinogen; 0.002 mg(Cd)/m^3 (respirable dust), Suspected Human Carcinogen); BEI: 10 μg/g creatinine in urine; 10 μg/L in blood.
DFG BAT: Blood 1.5 μg/dL; Urine 15 μg/dL, Suspected Carcinogen.
NIOSH REL: (Cadmium) Reduce to lowest feasible level

SAFETY PROFILE: Confirmed human carcinogen. Hydrolysis reaction in the presence of moisture forms the explosive hydrogen azide gas. When heated to decomposition it emits toxic fumes of Cd and NO_x. See also CADMIUM COMPOUNDS and AZIDES.

MHY600 CAS:52557-97-8 ***HR: 3***
METHYLCAMPHENOATE
mf: $C_{11}H_{18}O_2$ mw: 182.29

SYNS: 2,2-DIMETHYLBICYCLO(2.2.1)HEPTANE-3-CARBOXYLIC ACID, METHYL ESTER ◇ METHYL-2,2-DIMETHYLBICYCLO(2.2.1) HEPTANE-3-CARBOXYLATE

TOXICITY DATA with REFERENCE
skn-rbt 500 mg/24H MOD PESTC* 9(45),4,81
eye-rbt 100 mg MLD PESTC* 9(45),4,81
orl-mam LD50:5 mg/kg PESTC* 9(45),4,81

SAFETY PROFILE: Poison by ingestion. An eye and skin irritant. When heated to decomposition it emits acrid smoke and irritating fumes. See also ESTERS.

MHY750 CAS:2556-73-2 ***HR: 2***
N-METHYL-ε-CAPROLACTAM
mf: $C_7H_{13}NO$ mw: 127.21

SYN: HEXAHYDRO-1-METHYL-2H-AZEPIN-2-ONE

TOXICITY DATA with REFERENCE
orl-rat LD50:1620 mg/kg AIHAAP 30,470,69
skn-rbt LD50:1410 mg/kg AIHAAP 30,470,69

SAFETY PROFILE: Moderately toxic by ingestion and skin contact. When heated to decomposition it emits toxic fumes of NO_x.

MHZ000 CAS:598-55-0 ***HR: 3***
METHYL CARBAMATE
mf: $C_2H_5NO_2$ mw: 75.07

PROP: Needles. Bp: 177°, mp: 52-54°. Very sol in water, alc.

SYNS: BENDIOCARB ◇ METHYLURETHAN ◇ METHYLURETHANE ◇ NCI-C55594 ◇ URETHYLANE

TOXICITY DATA with REFERENCE
mmo-esc 50 g/L/3H CRSUBM 3,69,55
otr-rat:emb 120 μg/L JJIND8 67,1303,81
oms-mus-par 375 mg/kg ZKKOBW 76,69,71
orl-rat TDLo:102 g/kg/2Y:CAR NTPTR* NTP-TR-328,87
skn-mus TDLo:45 mg/kg/14W-I:ETA BJCAAI 9,177,55
ipr-mus LDLo:200 mg/kg TXAPA9 23,288,72
scu-mus LD50:4450 mg/kg AJEBAK 45,507,67
orl-qal LD50:21 mg/kg EESADV 8,551,84

CONSENSUS REPORTS: IARC Cancer Review: Group 3 IMEMDT 7,56,87; Animal Inadequate Evidence IMEMDT 12,151,76. EPA Genetic Toxicology Program. Reported in EPA TSCA Inventory.

SAFETY PROFILE: Poison by ingestion and intraperitoneal routes. Questionable carcinogen with experimental carcinogenic and tumorigenic data. Mutation data reported. When heated to decomposition it emits toxic fumes of NO_x. See also CARBAMATES.

MIA000 CAS:15942-48-0 ***HR: 3***
METHYLCARBAMIC ACID-2-CHLORO-5-tert-PENTYLPHENYL ESTER
mf: $C_{13}H_{18}ClNO_2$ mw: 255.77

SYN: RE 5454

TOXICITY DATA with REFERENCE
orl-rat LD50:75 mg/kg TXAPA9 21,315,72
orl-mus LDLo:42 mg/kg AECTCV 14,111,85
orl-bwd LD50:9 mg/kg TXAPA9 21,315,72

SAFETY PROFILE: Poison by ingestion. When heated to decomposition it emits very toxic fumes of Cl^- and NO_x. See also CARBAMATES.

MIA250 CAS:2631-40-5 ***HR: 3***
METHYLCARBAMIC ACID-o-CUMENYL ESTER
mf: $C_{11}H_{15}NO_2$ mw: 193.27

SYNS: BAY 105807 ◇ ENT 25,670 ◇ ETROFOLAN ◇ HYTOX

◇ ISOPROCARB ◇ ISOPROPYLPHENOL METHYLCARBAMATE ◇ o-ISOPROPYLPHENYL-N-METHYLCARBAMATE ◇ 2-ISOPROPYL-PHENYL-N-METHYLCARBAMATE ◇ KHE 0145 ◇ 2-(1-METHYL-ETHYL)PHENYL METHYLCARBAMATE ◇ MIPC ◇ MIPCIN ◇ MIPSIN

TOXICITY DATA with REFERENCE
orl-rat LD50:178 mg/kg SPEADM 78-1,57,78
ipr-rat LD50:142 mg/kg BWHOA6 44(1-3),241,71
ivn-rat LD50:66000 μg/kg BWHOA6 44(1-3),241,71
orl-mus LD50:150 mg/kg GUCHAZ 6,358,73
skn-mus LD50:7600 mg/kg GUCHAZ 6,358,73
orl-bwd LD50:56200 μg/kg AECTCV 12,355,83

CONSENSUS REPORTS: Reported in EPA TSCA Inventory. EPA Genetic Toxicology Program.

SAFETY PROFILE: Poison by ingestion, intravenous, and intraperitoneal routes. Mildly toxic by skin contact. Used for controlling leafhoppers, planthoppers, and bugs in rice and cacao. When heated to decomposition it emits toxic fumes of NO_x. See also CARBAMATES.

MIA500 CAS:672-06-0 ***HR: 3***
METHYLCARBAMIC ACID-2,4-DICHLORO-5-ETHYL-m-TOLYL ESTER
mf: $C_{11}H_{13}Cl_2NO_2$ mw: 262.15

SYN: U 17556

TOXICITY DATA with REFERENCE
orl-mus LDLo:94 mg/kg AECTCV 14,111,85
orl-bwd LD50:13 mg/kg TXAPA9 21,315,72

SAFETY PROFILE: Poison by ingestion. When heated to decomposition it emits very toxic fumes of Cl^- and NO_x. See also CARBAMATES and ESTERS.

MIA775 CAS:63982-49-0 ***HR: 3***
METHYLCARBAMIC ACID-5-DIMETHYLAMINO-2,4-XYLYL ESTER, HYDROCHLORIDE
mf: $C_{12}H_{18}N_2O_2 \cdot ClH$ mw: 258.78

SYNS: N-METHYL-CARBAMIC ACID 2,4-DIMETHYL-5-DIMETHYL-AMINOPHENYL ESTER, HYDROCHLORIDE ◇ N-METHYLURETH-ANE HYDROCHLORIDE of 6-DIMETHYLAMINO-o-4-XYLENOL ◇ T-1770

TOXICITY DATA with REFERENCE
orl-mus LD50:50 mg/kg JCSOA9 -,182,47
scu-mus LD50:10 mg/kg JCSOA9 -,182,47

SAFETY PROFILE: Poison by ingestion and subcutaneous routes. When heated to decomposition it emits toxic fumes of NO_x and HCl. See also CARBAMATES and ESTERS.

MIB000 CAS:14285-43-9 ***HR: 3***
METHYLCARBAMIC ACID-4-METHYLTHIO-m-CUMENYL ESTER
mf: $C_{12}H_{17}NO_2S$ mw: 239.36

SYN: ACD 7029

TOXICITY DATA with REFERENCE
orl-pgn LD50:13 mg/kg TXAPA9 21,315,72
orl-dck LD50:7500 μg/kg TXAPA9 21,315,72
orl-brd LD50:1800 μg/kg TXAPA9 21,315,72

SAFETY PROFILE: Poison by ingestion. When heated to decomposition it emits very toxic fumes of NO_x and SO_x. See also CARBAMATES.

MIB250 CAS:3566-00-5 ***HR: 3***
METHYLCARBAMIC ACID-4-METHYLTHIO-m-TOLYL ESTER
mf: $C_{10}H_{13}NO_2S$ mw: 211.30

SYNS: BAY S 2758 ◇ BAY 32651 ◇ ENT 25,777

TOXICITY DATA with REFERENCE
orl-rat LD50:50 mg/kg TXAPA9 21,315,72
orl-bwd LD50:40 mg/kg JAFCAU 15,287,67

SAFETY PROFILE: Poison by ingestion. When heated to decomposition it emits very toxic fumes of SO_x and NO_x. See also CARBAMATES.

MIB500 CAS:3279-46-7 ***HR: 3***
METHYLCARBAMIC ACID-o-(2-PROPYNYL-OXY)PHENYL ESTER
mf: $C_{11}H_{11}NO_3$ mw: 205.23

SYNS: ENT 25,810 ◇ HERCULES 9699 ◇ 2-(2-PROPYNYLOXY)PHE-NYL METHYLCARBAMATE

TOXICITY DATA with REFERENCE
orl-rat LD50:80 mg/kg TXAPA9 21,315,72
orl-bwd LD50:45 mg/kg JAFCAU 15,287,67

SAFETY PROFILE: Poison by ingestion. When heated to decomposition it emits toxic fumes of NO_x. See also CARBAMATES.

MIB750 CAS:1129-41-5 ***HR: 3***
METHYLCARBAMIC ACID-m-TOLYL ESTER
mf: $C_9H_{11}NO_2$ mw: 165.21

SYNS: m-CRESYL ESTER of N-METHYLCARBAMIC ACID ◇ m-CRESYL METHYLCARBAMATE ◇ DICRESYL ◇ DRC 3341 ◇ KUMIAI ◇ METACRATE ◇ 3-METHYLPHENYL-N-METHYLCARBA-MATE ◇ MTMC ◇ m-TOLYL-N-METHYLCARBAMATE ◇ 3-TOLYL-N-METHYLCARBAMATE ◇ TSUMACIDE

TOXICITY DATA with REFERENCE
cyt-ham:fbr 125 mg/L/48H MUREAV 48,337,77
cyt-ham:lng 47 mg/L GMCRDC 27,95,81
orl-mus LD50:268 mg/kg FMCHA2 -,D322,80
ihl-rat LD50:475 mg/kg EQSFAP 3,618,75
skn-rat LD50:6000 mg/kg FMCHA2 -,D214,75
orl-mus LD50:109 mg/kg FMCHA2 -,C247,83
skn-mus LD50:6 mg/kg GUCHAZ 6,527,73

CONSENSUS REPORTS: EPA Extremely Hazardous

Substances List. Reported in EPA TSCA Inventory. EPA Genetic Toxicology Program.

SAFETY PROFILE: Poison by ingestion and skin contact. Moderately toxic by inhalation. Mutation data reported. When heated to decomposition it emits toxic fumes of NO_x. See also CARBAMATES.

MIC250 CAS:63982-32-1 ***HR: 3***
METHYLCARBAMIC ESTER of α-3-HYDROXYPHENYLETHYLDIMETHYLAMINE HYDROCHLORIDE
mf: $C_{12}H_{18}N_2O_2$•ClH mw: 258.78

SYNS: N-METHYLCARBAMIC ACID, m-(α-DIMETHYLAMINOETHYL)PHENYL ESTER, HYDROCHLORIDE ◇ MIOTINE ◇ T-1843

TOXICITY DATA with REFERENCE
scu-rat LD50:1 mg/kg NTIS** PB158-508
orl-mus LDLo:2 mg/kg JPETAB 43,413,31
scu-mus LD50:1 mg/kg NTIS** PB158-508
ivn-mus LD50:500 μg/kg NTIS** PB158-508
scu-rbt LD50:1 mg/kg NTIS** PB158-508
scu-gpg LD50:1 mg/kg NTIS** PB158-508

SAFETY PROFILE: Poison by ingestion, intravenous and subcutaneous routes. When heated to decomposition it emits very toxic fumes of NO_x and HCl. See also CARBAMATES and ESTERS.

MID000 CAS:63982-40-1 ***HR: 3***
METHYLCARBAMIC ESTER of 3-OXYPHENYLDIMETHYLAMINE HYDROCHLORIDE
mf: $C_{10}H_{14}N_2O_2$•ClH mw: 230.72

SYNS: AR-12 ◇ N-METHYL-CARBAMIC ACID-3-DIMETHYLAMINOPHENYL ESTER, HYDROCHLORIDE ◇ N-METHYLURETHANE of HYDROCHLORIDE of 3-DIMETHYLAMINOPHENOL

TOXICITY DATA with REFERENCE
orl-mus LD50:100 mg/kg JCSOA9 -,182,47
scu-mus LD50:25 mg/kg JCSOA9 -,182,47
ivn-mus LD80:15 mg/kg NTIS** PB158-508

SAFETY PROFILE: Poison by ingestion, subcutaneous, and intravenous routes. When heated to decomposition it emits very toxic fumes of HCl and NO_x. See also CARBAMATES.

MID250 CAS:60398-22-3 ***HR: 3***
METHYLCARBAMIC ESTER of OXYPHENYLMETHYLDIETHYLAMMONIUM IODIDE
mf: $C_{13}H_{21}N_2O_2$•I mw: 364.26

SYNS: METHIODIDE of N-METHYLURETHANE of 3-DIETHYLAMINOPHENOL ◇ N-METHYLCARBAMIC ACID-3-DIETHYLAMINOPHENYL ESTER, METHIODIDE ◇ N-METHYLCARBAMIC ACID-3-(DIETHYLMETHYLAMMONIO)PHENYL ESTER, IODIDE ◇ (3-(N-METHYLCARBAMOYLOXY)PHENYL)DIETHYLMETHYL-AMMONIUM IODIDE ◇ TL 1217

TOXICITY DATA with REFERENCE
scu-rat LD50:400 μg/kg NTIS** PB158-508
orl-mus LDLo:20 mg/kg JPETAB 43,413,31
scu-mus LD50:129 μg/kg NTIS** PB158-508
ivn-mus LDLo:100 μg/kg JPETAB 43,413,31
scu-dog LD50:75 μg/kg NTIS** PB158-508
scu-mky LD50:200 μg/kg NTIS** PB158-508
scu-cat LD50:75 μg/kg NTIS** PB158-508
scu-rbt LD50:150 μg/kg NTIS** PB158-508
scu-gpg LD50:97 μg/kg NTIS** PB158-508

SAFETY PROFILE: A deadly poison by ingestion, intravenous, and subcutaneous routes. When heated to decomposition it emits very toxic fumes of NO_x, NH_3, and I^-. See also ESTERS and CARBAMATES.

MID500 CAS:63680-78-4 ***HR: 3***
METHYLCARBAMIC ESTER of 8-OXYQUINOLINE METHIODIDE
mf: $C_{12}H_{13}N_2O_2$•I mw: 344.17

SYNS: 8-HYDROXY-1-METHYL QUINOLINIUM IODIDE, METHYLCARBAMATE ◇ N-METHYL CARBAMIC ACID-8-QUINOLINYL ESTER, METHIODIDE

TOXICITY DATA with REFERENCE
orl-mus LDLo:200 mg/kg JPETAB 43,413,31
ipr-mus LD50:89 mg/kg PHARAT 34,142,79
ivn-mus LDLo:100 μg/kg JPETAB 43,413,31

SAFETY PROFILE: Poison by ingestion, intraperitoneal, and intravenous routes. When heated to decomposition it emits very toxic fumes of NO_x and I^-. See also ESTERS and CARBAMATES.

MID750 CAS:59163-97-2 ***HR: 3***
METHYLCARBAMOYLETHYL ACRYLATE
mf: $C_7H_{11}NO_3$ mw: 157.19

SYN: ACRYLIC ACID METHYLCARBAMYLETHYL ESTER

TOXICITY DATA with REFERENCE
skn-rbt 500 mg open MLD UCDS** 3/23/73
eye-rbt 5 mg SEV UCDS** 3/23/73
orl-rat LD50:840 mg/kg UCDS** 3/23/73
skn-rbt LD50:200 mg/kg TXAPA9 28,313,74

SAFETY PROFILE: Poison by skin contact. Moderately toxic by ingestion. A skin and severe eye irritant. When heated to decomposition it emits toxic fumes of NO_x. See also ESTERS and CARBAMATES.

MID800 CAS:69946-37-8 ***HR: 3***
2-METHYL-4-CARBAMOYL-5-HYDROXYIMIDAZOLE
mf: $C_5H_7N_3O_2$ mw: 141.15

SYN: 5-HYDROXY-2-METHYL-1H-IMIDAZOLE-4-CARBOXAMIDE

TOXICITY DATA with REFERENCE
orl-mus LD50:2000 mg/kg GWXXBX #2825738
ipr-mus LD50:500 mg/kg GWXXBX #2825738
ivn-mus LD50:130 mg/kg GWXXBX #2825738

SAFETY PROFILE: Poison by intravenous route. Moderately toxic by ingestion and intraperitoneal routes. When heated to decomposition it emits toxic fumes of NO_x.

MID900 CAS:64049-00-9 ***HR: 3***
(3-(N-METHYLCARBAMOYLOXY)PHENYL) TRIMETHYL-ARSONIUM IODIDE
mf: $C_{13}H_{21}AsNO_2$•I mw: 425.17

SYNS: ARSONIUM, (3-HYDROXYPHENYL)DIETHYLMETHYL-, IODIDE, METHYLCARBAMATE ◇ CARBAMIC ACID, N-METHYL-, 3-DIETHYLARSINOPHENYL ESTER, METHIODIDE ◇ TL-1504

TOXICITY DATA with REFERENCE
scu-mus LD50:500 μg/kg NTIS** PB158-508

OSHA PEL: TWA 0.5 mg(As)/m^3

SAFETY PROFILE: Poison by subcutaneous route. When heated to decomposition it emits toxic fumes of NO_x, As, and I^-.

MIE250 CAS:33531-59-8 ***HR: 3***
o-METHYLCARBANILIC ACID-N-ETHYL-3-PIPERDINYL ESTER
mf: $C_{15}H_{22}N_2O_2$ mw: 262.39

SYN: FUNGIFOS

TOXICITY DATA with REFERENCE
scu-mus LD50:277 mg/kg JMCMAR 14,710,71
ivn-mus LD50:34 mg/kg JMCMAR 14,710,71
orl-dog LD50:5000 mg/kg ARZNAD 26,769,76

SAFETY PROFILE: Poison by subcutaneous and intravenous routes. Mildly toxic by ingestion. An antifungal agent. When heated to decomposition it emits toxic fumes of NO_x.

MIE600 CAS:6700-56-7 ***HR: 3***
1-METHYL-1-CARBETHOXY-4-PHENYL HEXAMETHYLENIMINE CITRATE
mf: $C_{16}H_{23}NO_2$•$7C_6H_8O_7$ mw: 1606.38

SYNS: ETHOHEPTAZINE CITRATE ◇ ZACTANE CITRATE

TOXICITY DATA with REFERENCE
orl-rat LD50:580 mg/kg JPETAB 134,332,61
ipr-rat LD50:156 mg/kg PLRCAT 2,39,70
ipr-mus LD50:217 mg/kg PLRCAT 2,39,70

SAFETY PROFILE: Poison by intraperitoneal route. Moderately toxic by ingestion. When heated to decomposition it emits toxic fumes of NO_x.

MIE750 CAS:629-38-9 ***HR: 2***
METHYL CARBITOL ACETATE
mf: $C_7H_{14}O_4$ mw: 162.21

PROP: Colorless liquid. Bp: 209.1°, flash p: 180°F (OC), d: 1.0396 @ 20°/20°, vap press: 0.12 mm @ 20°.

SYN: DIETHYLENE GLYCOL MONOMETHYL ETHER ACETATE

TOXICITY DATA with REFERENCE
eye-rbt 20 mg AJOPAA 29,1363,46
orl-gpg LD50:3460 mg/kg JIHTAB 23,259,41

CONSENSUS REPORTS: Glycol ether compounds are on the Community Right-To-Know List.

SAFETY PROFILE: Moderately toxic by ingestion. An eye irritant. Combustible when exposed to heat or flame; can react with oxidizing materials. To fight fire, use foam, CO_2, dry chemical, mist, spray. When heated to decomposition it emits acrid smoke and irritating fumes. See also GLYCOL ETHERS.

MIF000 CAS:616-38-6 ***HR: 3***
METHYL CARBONATE
DOT: UN 1161
mf: $C_3H_6O_3$ mw: 90.09

PROP: Colorless liquid; pleasant odor. Mp: 0.5°, d: 1.065 @ 17°/4°, flash p: 66°F (OC). Bp: 90.91°. Misc with acids and alkalies; sol in most organic solvents; insol in water.

SYN: DIMETHYL CARBONATE

TOXICITY DATA with REFERENCE
orl-rat LD50:13 g/kg FCTXAV 17,357,79
ipr-rat LD50:1600 mg/kg FCTXAV 17,357,79
orl-mus LD50:6 g/kg FCTXAV 17,357,79
ipr-mus LD50:800 mg/kg FCTXAV 17,357,79

CONSENSUS REPORTS: Reported in EPA TSCA Inventory.

DOT Classification: Flammable Liquid; Label: Flammable Liquid.

SAFETY PROFILE: Moderately toxic by intraperitoneal route. Mildly toxic by ingestion. An irritant. Violent reaction or ignition on contact with potassium-tert-butoxide. A very dangerous fire hazard when exposed to heat, open flames (sparks), or oxidizers. To fight fire, use alcohol foam. When heated to decomposition it emits acrid smoke and irritating fumes.

MIF250 CAS:61445-55-4 ***HR: 3***
N-METHYL-N-(3-CARBOXYPROPYL)NITROSAMINE
mf: $C_5H_{10}N_2O_3$ mw: 146.17

SYNS: 4-(METHYLNITROSOAMINO)BUTYRIC ACID ◇ N-NITROSOMETHYL-3-CARBOXYPROPYLAMINE

TOXICITY DATA with REFERENCE
mma-sat 48 μmol/plate CNREA8 37,399,77
mmo-smc 16260 mg/L IAPUDO 57,721,84
orl-rat TDLo:6800 mg/kg/57W-I:ETA JJIND8 70,959,83
par-rat TDLo:4500 mg/kg/30W-I:NEO JJCREP 79,309,88

SAFETY PROFILE: Questionable carcinogen with experimental neoplastigenic and tumorigenic data. Mutation data reported. When heated to decomposition it emits toxic fumes of NO_x. See also NITROSAMINES.

MIF500 CAS:140-05-6 ***HR: 1***
METHYL CELLOSOLVE ACETYLRICINOLEATE
mf: $C_{23}H_{42}O_5$ mw: 398.65

PROP: Light, straw-colored liquid. Mp: −60° (gels), bp: 200-260° @ 4 mm, flash p: 446°F, d: 0.966 @ 20°/20°, vap press: <0.01 mm @ 150°, vap d: 13.8.

SYNS: ETHYLENE GLYCOL MONOMETHYL ETHER ACETYLRICINOLEATE ◇ GLYCOL MONOMETHYL ETHER ACETYLRICINOLEATE ◇ 2-METHOXYETHYL-12-ACETOXY-9-OCTADECENOATE ◇ 2-METHOXYETHYL ACETYL RICINOLEATE

TOXICITY DATA with REFERENCE
eye-rbt 500 mg AJOPAA 29,1363,46
orl-rat LD50:20 g/kg JIHTAB 30,63,48
orl-gpg LD50:12 g/kg JIHTAB 23,259,41

CONSENSUS REPORTS: Glycol ether compounds are on the Community Right-To-Know List.

SAFETY PROFILE: Very mildly toxic by ingestion. An eye irritant. Combustible when exposed to heat or flame. To fight fire, use foam, CO_2, dry chemical. When heated to decomposition it emits acrid smoke and irritating fumes. See also GLYCOL ETHERS.

MIF750 CAS:3121-61-7 ***HR: 3***
METHYL CELLOSOLVE ACRYLATE
mf: $C_6H_{10}O_3$ mw: 130.16

PROP: Liquid. Bp: 61° @ 17 mm, flash p: 180°F (OC), d: 1.0134 @ 20°, vap d: 4.49.

SYNS: ACRYLIC ACID-2-METHOXYETHYL ESTER ◇ ETHYLENE GLYCOL MONOMETHYL ETHER ACRYLATE ◇ GLYCOL MONOMETHYL ETHER ACRYLATE ◇ 2-METHOXYETHANOL, ACRYLATE

TOXICITY DATA with REFERENCE
skn-rbt 500 mg open MLD UCDS** 9/17/69
orl-mus TDLo:5200 mg/kg (female 6-13D post):REP TCMUD8 7,29,87
orl-rat LD50:810 mg/kg UCDS** 9/17/69
ihl-rat LC50:500 ppm/4H UCDS** 9/17/69
skn-rbt LD50:250 mg/kg UCDS** 9/17/69

CONSENSUS REPORTS: Glycol ether compounds are on the Community Right-To-Know List. Reported in EPA TSCA Inventory.

SAFETY PROFILE: Poison by skin contact. Moderately toxic by ingestion and inhalation. Experimental reproductive effects. A skin irritant. Flammable when exposed to heat or flame; can react with oxidizing materials. To fight fire, use foam, CO_2, dry chemical. When heated to decomposition it emits acrid smoke and irritating fumes. See also GLYCOL ETHERS.

MIF765 CAS:74-87-3 ***HR: 3***
METHYL CHLORIDE
DOT: UN 1063
mf: CH_3Cl mw: 50.49

PROP: Colorless gas; ethereal odor and sweet taste. D: 0.918 @ 20°/4°, mp: −97°, bp: −23.7°, flash p: <32°F, lel: 8.1%, uel: 17%, autoign temp: 1170°F, vap d: 1.78. Sltly sol in water; misc with chloroform, ether, glacial acetic acid; sol in alc.

SYNS: ARTIC ◇ CHLOOR-METHAAN (DUTCH) ◇ CHLOR-METHAN (GERMAN) ◇ CHLOROMETHANE ◇ CHLORURE de METHYLE (FRENCH) ◇ CLOROMETANO (ITALIAN) ◇ CLORURO di METILE (ITALIAN) ◇ METHYLCHLORID (GERMAN) ◇ METYLU CHLOREK (POLISH) ◇ MONOCHLOROMETHANE ◇ RCRA WASTE NUMBER U045

TOXICITY DATA with REFERENCE
oms-hmn:lym 3 pph MUREAV 155,75,85
sce-hmn:lym 3 pph MUREAV 155,75,85
ihl-mus TCLo:750 ppm/6H (female 6-17D post):TER TJADAB 27,197,83
ihl-rat TCLo:3000 ppm/6H (male 5D pre):REP TXAPA9 86,124,86
ihl-hmn LCLo:20000 ppm/2H:EYE,CNS,GIT 34ZIAG -,386,69
orl-rat LD50:1800 mg/kg 85JCAE -,86,86
ihl-rat LC50:5300 mg/m^3/4H 85GMAT -,82,82
ihl-mus LC50:2200 ppm/6H TXAPA9 86,93,86
ihl-dog LCLo:14661 ppm/6H NIHBAZ 191,1,49
ihl-gpg LCLo:20000 ppm/2H FLCRAP 1,197,67

CONSENSUS REPORTS: IARC Cancer Review: Group 3 IMEMDT 7,246,87; Human Inadequate Evidence IMEMDT 41,161,86; Animal Inadequate Evidence IMEMDT 41,161,86. Reported in EPA TSCA Inventory. EPA Genetic Toxicology Program.

OSHA PEL: (Transitional: TWA 100; CL 200 ppm; Pk 300 ppm/5M)TWA 50 ppm; STEL 100 ppm
ACGIH TLV: TWA 50 ppm; STEL 100 ppm
DFG MAK: 50 ppm (105 mg/m^3); Suspected Carcinogen.
NIOSH REL: (Monohalomethanes) TWA Reduce to lowest level.
DOT Classification: Flammable Gas; Label: Flammable Gas; IMO: Poison A; Label: Poison Gas and Flammable Gas.

SAFETY PROFILE: Suspected carcinogen. Very mildly toxic by inhalation. An experimental teratogen. Other experimental reproductive effects. Human mutation data reported. Human systemic effects by inhalation: convulsions, nausea or vomiting, and unspecified effects on the eye.

Chloromethane has slight irritant properties and may be inhaled without noticeable discomfort. It has some narcotic action, but this effect is weaker than that of chloroform. Acute poisoning, characterized by the narcotic effect, is rare in industry. In exposures to high concentrations, dizziness, drowsiness, incoordination, confusion, nausea and vomiting, abdominal pains, hiccoughs, diplopia, and dimness of vision are followed by delirium, convulsions, and coma. Death may be immediate; however, if the exposure is not fatal, recovery is usually slow. Degenerative changes in the central nervous system are not uncommon. The liver, kidneys, and bone marrow may be affected, with resulting acute nephritis and anemia. Death may occur several days after exposure resulting from degenerative changes in the heart, liver, and especially the kidneys. Repeated exposure to low concentrations causes damage to the central nervous system and, less frequently, to the liver, kidneys, bone marrow and cardiovascular system. Hemorrhages into the lungs, intestinal tract, and dura have been reported. Sprayed on the skin, chloromethane produces anesthesia through freezing of the tissues as it evaporates.

Flammable gas. Very dangerous fire hazard when exposed to heat, flame, or powerful oxidizers. Moderate explosion hazard when exposed to flame and sparks. Explodes on contact with interhalogens (e.g., bromine trifluoride; bromine pentafluoride); magnesium and alloys; potassium and alloys; sodium and alloys; zinc. Potentially explosive reaction with aluminum when heated to 152° in a sealed container. Mixtures with aluminum chloride + ethylene react exothermically and then explode when pressurized to above 30 bar. May ignite on contact with aluminum chloride or powdered aluminum. To fight fire, stop flow of gas and use CO_2, dry chemical, or water spray. When heated to decomposition it emits highly toxic fumes of Cl^-. See also CHLORINATED HYDROCARBONS, ALIPHATIC.

MIF775 CAS:96-34-4 ***HR: 3***
METHYL CHLOROACETATE
DOT: UN 2295
mf: $C_3H_5ClO_2$ mw: 108.53

PROP: Colorless liquid. D: 1.238, mp: −33°, bp: 130-132°. Insol in water; misc with alc, ether.

SYNS: METHYL CHLOROACETATE (DOT) ◇ METHYL MONOCHLORACETATE ◇ METHYL MONOCHLOROACETATE ◇ MONOCHLOROACETIC ACID METHYL ESTER

TOXICITY DATA with REFERENCE
scu-rat LD16:560 mg/kg 85GMAT -,82,82
orl-mus LD50:240 mg/kg 85GMAT -,82,82
ihl-mus LC50:1 g/m³/2H 85GMAT -,82,82

CONSENSUS REPORTS: Reported in EPA TSCA Inventory.

DOT Classification: DOT-IMO: Flammable or Combustible Liquid; Label: Flammable Liquid.

SAFETY PROFILE: Poison by ingestion. Moderately toxic by inhalation and subcutaneous routes. Flammable when exposed to heat or flame; can react vigorously with oxidizing materials. When heated to decomposition it emits toxic fumes of Cl^-.

MIF800 CAS:80-63-7 ***HR: 3***
METHYL-2-CHLOROACRYLATE
mf: $C_4H_5ClO_2$ mw: 120.54

SYNS: 2-CHLOROACRYLIC ACID, METHYL ESTER ◇ 2-CHLORO-2-PROPENOIC ACID METHYL ESTER (9CI) ◇ METHYL-α-CHLOROACRYLATE

TOXICITY DATA with REFERENCE
skn-rbt 500 mg SEV SCCUR* -,6,61
mmo-sat 1 nmol/plate MUREAV 78,113,80
mma-sat 1 nmol/plate MUREAV 78,113,80
ihl-hmn TCLo:5 ppm:EYE 29ZWAE -,90,68
ihl-rat LC50:500 mg/m³/2H FATOAO 19,60,56
ihl-mus LC50:500 mg/m³/2H FATOAO 19,60,56
ihl-cat LC50:500 mg/m³/2H FATOAO 19,60,56
ihl-rbt LC50:500 mg/m³/2H FATOAO 19,60,56
ihl-gpg LC50:500 mg/m³/2H FATOAO 19,60,56

CONSENSUS REPORTS: EPA Extremely Hazardous Substances List. Reported in EPA TSCA Inventory.

SAFETY PROFILE: Moderately toxic by inhalation. Human systemic effects by inhalation: conjunctiva irritation. A severe skin irritant. Mutation data reported. When heated to decomposition it emits toxic fumes of Cl^-. See also ESTERS.

MIG000 CAS:79-22-1 ***HR: 3***
METHYL CHLOROCARBONATE
DOT: UN 1238
mf: $C_2H_3ClO_2$ mw: 94.50

PROP: Colorless liquid. Bp: 71.4°, d: 1.223 @ 20°/4°, vap d: 3.26, flash p: 54°F, autoign temp: 940°F. Sltly sol in water with gradual decomp; misc with alc, benzene, chloroform, and ether.

SYNS: CHLORAMEISENSAEURE METHYLESTER (GERMAN) ◇ CHLOROCARBONATE de METHYLE (FRENCH) ◇ CHLOROCARBONIC ACID METHYL ESTER ◇ CHLOROFORMIC ACID METHYL ESTER ◇ MCF ◇ METHOXYCARBONYL CHLORIDE ◇ METHYLCHLOORFORMIAT (DUTCH) ◇ METHYL CHLOROFORMATE (DOT) ◇ METILCLOROFORMIATO (ITALIAN) ◇ RCRA WASTE NUMBER U156

TOXICITY DATA with REFERENCE
ihl-man TCLo:5 mg/m^3:EYE,PUL GISAAA 42(5),97,77
orl-rat LD50:60 mg/kg GISAAA 42(5),97,77
ihl-rat LC50:88 ppm/1H TXAPA9 42,417,77
orl-mus LD50:67 mg/kg GISAAA 42(5),97,77
ihl-mus LC50:185 mg/m^3/2H GISAAA 42(5),97,77
skn-mus LD50:1750 mg/kg GISAAA 42(5),97,77
ipr-mus LD50:40 mg/kg NTIS** AD691-490
ihl-cat LCLo:1500 mg/m^3/30M GISAAA 42(5),97,77
skn-rbt LD50:7120 mg/kg TXAPA9 42,417,77
orl-gpg LD50:140 mg/kg GISAAA 42(5),97,77

CONSENSUS REPORTS: EPA Extremely Hazardous Substances List. Reported in EPA TSCA Inventory.

DOT Classification: Flammable Liquid; Label: Flammable Liquid, Poison, Corrosive.

SAFETY PROFILE: Poison by ingestion, inhalation, and intraperitoneal routes. Moderately toxic by skin contact. Human systemic effects by inhalation: conjunctiva irritation and respiratory effects. Corrosive to skin, eyes, and mucous membranes. Very dangerous fire hazard when exposed to heat sources, sparks, flame, or oxidizers. Reacts with water or steam to produce toxic and corrosive fumes. When heated to decomposition it emits toxic fumes of Cl^-, methyl chloroformate and phosgene.

MIG250 CAS:4535-90-4 ***HR: 3***
METHYL-β-CHLOROETHYLAMINE HYDROCHLORIDE
mf: $C_3H_8ClN{\cdot}ClH$ mw: 130.03

SYN: 2-CHLORO-N-METHYL-ETHYLAMINEHYDROCHLORIDE

TOXICITY DATA with REFERENCE
ipr-mus LD50:2301 mg/kg JPETAB 94,249,48
ivn-mus LD50:100 mg/kg JPETAB 91,224,47

CONSENSUS REPORTS: Reported in EPA TSCA Inventory.

SAFETY PROFILE: Poison by intravenous route. Moderately toxic by intraperitoneal route. When heated to decomposition it emits very toxic fumes of Cl^- and NO_x. See also AMINES.

MIG500 CAS:62037-49-4 ***HR: D***
N'-METHYL-N'-β-CHLOROETHYLBENZALDEHYDE HYDRAZONE
mf: $C_{10}H_{13}ClN_2$ mw: 196.70

SYN: 2-(2-CHLOROETHYL)-2-METHYLHYDRAZONEBENZALDEHYDE

TOXICITY DATA with REFERENCE
mmo-sat 100 μg/plate ARTODN 42,179,79
mma-sat 100 μg/plate ARTODN 42,179,79

SAFETY PROFILE: Mutation data reported. When heated to decomposition it emits very toxic fumes of Cl^- and NO_x. See also ALDEHYDES.

MIG750 CAS:62258-26-8 ***HR: D***
N'-METHYL-N'-β-CHLOROETHYL-(p-DIMETHYLAMINO)-BENZALDEHYDE HYDRAZONE
mf: $C_{12}H_{18}ClN_3$ mw: 239.78

SYN: p-DIMETHYLAMINOBENZALDEHYDE(2-(2-CHLOROETHYL)-2-METHYL)HYDRAZONE

TOXICITY DATA with REFERENCE
mmo-sat 100 μg/plate ARTODN 42,179,79
mma-sat 100 μg/plate ARTODN 42,179,79

SAFETY PROFILE: Mutation data reported. When heated to decomposition it emits very toxic fumes of Cl^- and NO_x. See also ALDEHYDES.

MIG800 ***HR: 3***
1-METHYL-1-(β-CHLOROETHYL) ETHYLENIMONIUM
mf: $C_5H_{11}ClN$ mw: 120.62

SYN: 1-(2-CHLOROETHYL)-1-METHYLAZIRIDINIUM

TOXICITY DATA with REFERENCE
ivn-dog LD50:500 μg/kg NTIS** PB158-507
iat-dog LD50:250 μg/kg NTIS** PB158-507
scu-rbt LD50:1 mg/kg NTIS** PB158-507
ivn-rbt LD50:1 mg/kg NTIS** PB158-507

SAFETY PROFILE: Poison by subcutaneous, intravenous, and intraarterial routes. When heated to decomposition it emits toxic fumes of Cl^- and NO_x. See also CHLORINATED HYDROCARBONS, ALIPHATIC.

MIG850 CAS:547-95-5 ***HR: 3***
METHYL-β-CHLOROETHYL-ETHYLENIMONIUM PICRYLSULFONATE
mf: $C_5H_{11}ClN{\cdot}C_6H_2N_3O_9S$ mw: 412.79

SYNS: CPS ◇ 1-METHYL-1-(β-CHLOROETHYL)ETHYLENIMONIUM PICRYLSULFONATE

TOXICITY DATA with REFERENCE
ivn-hmn TDLo:800 μg/kg:GIT,BLD CLPTAT 6,50,65
ivn-rat LDLo:2 mg/kg CLPTAT 6,50,65
scu-mus LD50:2400 μg/kg JPETAB 91,224,47
ivn-mus LD50:1500 μg/kg JPETAB 91,224,47
ivn-rbt LDLo:2 mg/kg CLPTAT 6,50,65

SAFETY PROFILE: Poison by subcutaneous and intravenous routes. Human systemic effects by intravenous route: nausea or vomiting, leukopenia and thrombocytopenia. When heated to decomposition it emits toxic fumes of Cl^-, SO_x, and NO_x. See also SULFONATES.

MIH000 CAS:63905-05-5 ***HR: 3***
METHYL-β-CHLOROETHYL-β-HYDROXYETHYLAMINE HYDROCHLORIDE
mf: $C_5H_{12}ClNO•ClH$ mw: 174.09

TOXICITY DATA with REFERENCE
orl-rat LD50:80 mg/kg NTIS** PB158-507
scu-rat LD50:20 mg/kg NTIS** PB158-507
orl-mus LD50:25 mg/kg NTIS** PB158-507
ipr-mus LD50:34 mg/kg JPETAB 91,224,47
scu-mus LD50:10 mg/kg JPETAB 91,224,47
ivn-mus LD50:2250 μg/kg JPETAB 91,224,47
scu-rbt LD50:10 mg/kg NTIS** PB158-507
ivn-rbt LD50:12 mg/kg JPETAB 91,224,47

SAFETY PROFILE: Poison by ingestion, subcutaneous, intraperitoneal, and intravenous routes. When heated to decomposition it emits very toxic fumes of Cl^- and NO_x.

MIH250 CAS:13589-15-6 ***HR: 3***
METHYL-N-(β-CHLOROETHYL)-N-NITROSOCARBAMATE
mf: $C_4H_7ClN_2O_3$ mw: 166.58

SYNS: N-2-CHLOROETHYL-N-NITROSO-CARBAMICACID METHYLESTER ◇ KB-16 ◇ TL 186

TOXICITY DATA with REFERENCE
skn-hmn 200 μg NTIS** PB158-508
eye-mam 500 μg SEV NTIS** PB158-508
orl-rat LD50:20 mg/kg NTIS** PB158-508
ihl-rat LC50:35 mg/m³/10M NTIS** PB158-508
scu-rat LD50:8 mg/kg NTIS** PB158-508
ivn-rat LD50:1100 μg/kg NTIS** PB158-508
ihl-mus LC50:36 mg/m³/10M NTIS** PB158-508
skn-mus LD50:62 mg/kg NTIS** PB158-508
scu-mus LD50:9 mg/kg NTIS** PB158-508
ivn-mus LD50:450 μg/kg NTIS** PB158-508
ihl-dog LC50:100 mg/m³/10M NTIS** PB158-508
ihl-mky LC50:200 mg/m³/10M NTIS** PB158-508

SAFETY PROFILE: Poison by ingestion, skin contact, inhalation, subcutaneous, and intravenous routes. A human skin and severe eye irritant. Many N-nitroso compounds are carcinogens. When heated to decomposition it emits very toxic fumes of Cl^- and NO_x. See also N-NITROSO COMPOUNDS and CARBAMATES.

MIH275 CAS:71-55-6 ***HR: 3***
METHYL CHLOROFORM
DOT: UN 2831
mf: $C_2H_3Cl_3$ mw: 133.40

PROP: Colorless liquid. Bp: 74.1°, fp: −32.5°, flash p: none, d: 1.3376 @ 20°/4°, vap press: 100 mm @ 20.0°. Insol in water; sol in acetone, benzene, carbon tetrachloride, methanol, ether.

SYNS: AEROTHENE TT ◇ CHLOROETENE ◇ CHLOROETHENE ◇ CHLOROTHANE NU ◇ CHLOROTHENE ◇ CHLOROTHENE (inhibited) ◇ CHLOROTHENE NU ◇ CHLOROTHENE VG ◇ CHLORTEN ◇ INHIBISOL ◇ METHYLTRICHLOROMETHANE ◇ NCI-C04626 ◇ RCRA WASTE NUMBER U226 ◇ SOLVENT 111 ◇ STROBANE ◇ α-T ◇ 1,1,1-TCE ◇ 1,1,1-TRICHLOORETHAAN (DUTCH) ◇ 1,1,1-TRICHLORAETHAN (GERMAN) ◇ TRICHLORO-1,1,1-ETHANE (FRENCH) ◇ 1,1,1-TRICHLOROETHANE ◇ α-TRICHLOROETHANE ◇ 1,1,1-TRICLOROETANO (ITALIAN) ◇ TRI-ETHANE

TOXICITY DATA with REFERENCE
eye-man 450 ppm/8H BJIMAG 28,286,71
skn-rbt 5 g/12D-I MLD AIHAAP 19,353,58
skn-rbt 500 mg/24H MOD 28ZPAK -,28,72
eye-rbt 100 mg MLD AIHAAP 19,353,58
eye-rbt 2 mg/24H SEV 28ZPAK -,28,72
dnr-esc 500 mg/L PMRSDJ 1,195,81
otr-mus:emb 20 mg/L CALEDQ 28,85,85
ihl-rat TCLo:2100 ppm/24H (14D pre/1-20D preg):TER TOXID9 1,28,81
ihl-man LCLo:27 g/m³/10M JOCMA7 8,358,66
ihl-man TCLo:350 ppm:CNS WEHSAL 10,82,73
orl-hmn TDLo:670 mg/kg:GIT NTIS** PB257-185
ihl-hmn TCLo:920 ppm/70M:EYE,CNS AIHAAP 19,353,58
ihl-man TCLo:200 ppm/4H:CNS ATSUDG 5,96,82
orl-rat LD50:10300 mg/kg NTIS** PB257-185
ihl-rat LC50:18000 ppm/4H 28ZPAK -,28,72
ipr-rat LD50:3593 mg/kg ENVRAL 40,411,86
orl-mus LD50:11240 mg/kg NTIS** PB257-185
ihl-mus LC50:3911 ppm/2H SAIGBL 13,226,71
ipr-mus LD50:3636 mg/kg SAIGBL 13,290,71
orl-dog LD50:750 mg/kg FMCHA2 -,C242,83
ipr-dog LD50:3100 mg/kg TXAPA9 10,119,67
ivn-dog LDLo:95 mg/kg HBTXAC 5,72,59

CONSENSUS REPORTS: IARC Cancer Review: Group 3 IMEMDT 7,56,87; Animal Inadequate Evidence IMEMDT 20,515,79. NCI Carcinogenesis Bioassay (gavage); Inadequate Studies: mouse, rat NCITR* NCI-CG-TR-3,77. Community Right-To-Know List. Reported in EPA TSCA Inventory. EPA Genetic Toxicology Program.

OSHA PEL: (Transitional: TWA 350 ppm) TWA 350 ppm; STEL 450 ppm
ACGIH TLV: TWA 350 ppm; STEL 450 ppm; BEI: 10 mg/L trichloroacetic acid in urine at end of workweek.
DFG MAK: 200 ppm (1080 mg/m³); BAT: 55 μg/dL in blood after several shifts.
NIOSH REL: (1,1,1-Trichloroethane) CL 350 ppm/15M
DOT Classification: ORM-A; Label: None; Poison B; Label: St. Andrews Cross.

SAFETY PROFILE: Poison by intravenous route. Moderately toxic by ingestion, inhalation, skin contact, subcutaneous, and intraperitoneal routes. An experimental teratogen. Human systemic effects by ingestion and inhalation: conjunctiva irritation, hallucinations or distorted perceptions, motor activity changes, irritability, aggression, hypermotility, diarrhea, nausea or vomiting and other gastrointestinal changes. Experimental reproductive effects. Questionable carcinogen. Mutation data reported. A human skin irritant. An experimental skin and severe eye irritant. Narcotic in high concentrations. Causes a proarrhythmic activity which sensitizes the heart to epinephrine-induced arrhythmias. This sometimes will cause cardiac arrest, particularly when this material is massively inhaled as in drug abuse for euphoria.

Under the proper conditions it can undergo hazardous reactions with aluminum oxide + heavy metals; dinitrogen tetraoxide; inhibitors; metals (e.g., magnesium; aluminum; potassium; potassium-sodium alloy); sodium hydroxide; N_2O_4; oxygen. When heated to decomposition it emits toxic fumes of Cl^-. Used as a cleaning solvent, a chemical intermediate to produce vinylidene chloride, and as a propellant in aerosol cans. See also CHLORINATED HYDROCARBONS, ALIPHATIC.

MIH300 CAS:32179-45-6 ***HR: 3***
1-METHYL-5-CHLOROINDOLINE METHYLBROMIDE
mf: $C_{10}H_{13}ClN \cdot Br$ mw: 262.60

SYNS: S6 (pharmaceutical) ◇ S-6 ◇ 5-CHLORO-2,3-DIHYDRO-1,1-DIMETHYL-1H-INDOLIUM BROMIDE (9CI)

TOXICITY DATA with REFERENCE
orl-mus TDLo:70 mg/kg (female 7-13D post):REP OYYAA2 7,1171,73
orl-mus TDLo:210 mg/kg (female 7-13D post):TER OYYAA2 7,1171,73
orl-rat LD50:447 mg/kg OYYAA2 7,991,73
scu-rat LD50:213 mg/kg OYYAA2 7,991,73
ivn-rat LD50:9100 μg/kg OYYAA2 7,991,73
orl-mus LD50:536 mg/kg OYYAA2 7,991,73
scu-mus LD50:128 mg/kg OYYAA2 7,991,73
ivn-mus LD50:12 mg/kg OYYAA2 7,991,73

SAFETY PROFILE: Poison by subcutaneous and intravenous routes. Moderately toxic by ingestion. An experimental teratogen. Other experimental reproductive effects. A cholinergic agent which stimulates the parasympathetic nerves. When heated to decomposition it emits toxic fumes of Cl^-, Br^-. and NO_x. See also BROMIDES.

MIH500 CAS:4274-06-0 ***HR: 3***
4-METHYL-6-(((2-CHLORO-4-NITRO)PHENYL)AZO)-m-ANISIDINE
mf: $C_{14}H_{13}ClN_4O_3$ mw: 320.76

SYNS: AZO DYE No. 6945 ◇ BROWN SALT NV ◇ 2-CHLORO-4-NITROBENZENEAZO-2'-AMINO-4'-METHOXY-5'-METHYLBENZENE ◇ C.I. 37200 ◇ C.I. AZOIC DIAZO COMPONENT 21 ◇ FAST BROWN SALT RR

TOXICITY DATA with REFERENCE
orl-rat TDLo:49 g/kg/119W-C:ETA ZEKBAI 57,530,51

SAFETY PROFILE: Questionable carcinogen with experimental tumorigenic data. When heated to decomposition it emits very toxic fumes of Cl^- and NO_x.

MIH750 CAS:20405-19-0 ***HR: 3***
(2-METHYL-4-CHLOROPHENOXY)ACETIC ACID, DIETHANOLAMINE SALT
mf: $C_9H_9ClO_3 \cdot C_4H_{11}NO_2$ mw: 305.79

SYNS: ((4-CHLORO-O-TOLYL)OXY)-ACETIC ACID with 2,2'-IMINODIETHANOL (1:1) ◇ MCPA DIETHANOLAMINE SALT

TOXICITY DATA with REFERENCE
orl-rat LD50:800 mg/kg FCTXAV 3,883,65
ipr-rat LD50:300 mg/kg FCTXAV 3,883,65
orl-mus LD50:550 mg/kg FCTXAV 3,883,65
ipr-mus LD50:350 mg/kg FCTXAV 3,883,65

SAFETY PROFILE: Poison by intraperitoneal route. Moderately toxic by ingestion. When heated to decomposition it emits very toxic fumes of Cl^- and NO_x.

MII250 CAS:1788-93-8 ***HR: 3***
2-METHYL-3-(4-CHLOROPHENYL)-4(3H)-QUINAZOLINONE
mf: $C_{15}H_{11}ClN_2O$ mw: 270.73

SYN: 2-METHYL-3-(4'-CHLOROPHENYL)CHINAZOLON-(4)(GERMAN)

TOXICITY DATA with REFERENCE
orl-mus LD50:580 mg/kg ARZNAD 13,688,63
unr-mus LD50:400 mg/kg PCJOAU 7,626,73

SAFETY PROFILE: Poison by an unspecified route. Moderately toxic by ingestion. When heated to decomposition it emits very toxic fumes of Cl^- and NO_x.

MII750 CAS:7142-09-8 ***HR: 2***
2-METHYL-6-CHLORO-4-QUINAZOLINONE
mf: $C_9H_7ClN_2O$ mw: 194.63

SYN: 6-CHLORO-2-METHYL-4-QUINAZOLINONE

TOXICITY DATA with REFERENCE
orl-mus LD50:751 mg/kg ARZNAD 12,1204,62
ipr-mus LD50:514 mg/kg ARZNAD 12,1204,62

SAFETY PROFILE: Moderately toxic by ingestion and intraperitoneal routes. When heated to decomposition it emits very toxic fumes of Cl^- and NO_x.

MIJ000 ***HR: 3***
METHYL CHLOROSULFONATE
mf: CH_3SO_3Cl mw: 130.55

PROP: Colorless liquid, pungent odor. Mp: −70°, bp: 135° (decomp), d: 1.492 @ 10°, vap d: 4.51.

SAFETY PROFILE: Poison irritant to skin, eyes and mucous membranes. Will react with water, steam or acids to produce toxic and corrosive fumes. When heated to decomposition it emits highly toxic fumes of Cl^- and SO_x.

MIJ250 CAS:500-28-7 ***HR: 2***
METHYLCHLOROTHION
mf: $C_8H_9ClNO_5PS$ mw: 297.66

$(CH_3O)_2P(S)C_6H_3(Cl)NO_2$

SYNS: O-(3-CHLOOR-4-NITRO-FENYL)-O,O-DIMETHYL-MONO-THIOFOSFAAT (DUTCH) ◇ CHLOORTHION (DUTCH) ◇ O-(3-CHLOR-4-NITRO-PHENYL)-O,O-DIMETHYL-MONOTHIOPHOSPHAT(GERMAN) ◇ O-(3-CHLORO-4-NITROPHENYL) O,O-DIMETHYL PHOSPHOROTHIOATE ◇ CHLORTHION METHYL ◇ CHLORTION (CZECH) ◇ O-(3-CLORO-4-NITRO-FENIL)-O,O-DIMETIL-MONOTIOFOSFATO (ITALIAN) ◇ O,O-DIMETHYL-O-3-CHLOR-4-NITROFENYLTIOFOSFAT (CZECH) ◇ O,O-DIMETHYL-O-(3-CHLOR-4-NITROPHENYL)-MONOTHIOPHOSPHAT (GERMAN) ◇ O,O-DIMETHYL-O-(3-CHLORO-4-NITROPHENYL) PHOSPHOROTHIOATE ◇ DIMETHYL-3-CHLORO-4-NITROPHENYL THIONOPHOSPHATE ◇ O,O-DIMETHYL-O-(3-CHLORO-4-NITROPHENYL) THIOPHOSPHATE ◇ O,O-DIMETHYL-p-NITRO-m-CHLOROPHENYL THIOPHOSPHATE ◇ O,O-DIMETHYL-O-4-NITRO-3-CHLOROPHENYL THIOPHOSPHATE ◇ ENT 18,861 ◇ p-NITRO-m-CHLOROPHENYL DIMETHYL THIONOPHOSPHATE ◇ THIOPHOSPHATE de O,O-DIMETHYLE et de O-3-CHLORO-4-NITROPHENYLE (FRENCH)

TOXICITY DATA with REFERENCE
orl-rat LD50:625 mg/kg 85GYAZ -,19,71
skn-rat LD50:1500 mg/kg WRPCA2 9,119,70
ipr-rat LD50:750 mg/kg AMIHBC 8,350,53
orl-mus LD50:794 mg/kg ARTODN 41,111,78
orl-rbt LDLo:1 g/kg JEENAI 48,139,55
ipr-gpg LD50:525 mg/kg ARZNAD 5,626,55

SAFETY PROFILE: Moderately toxic by ingestion, skin contact, and intraperitoneal routes. An insecticide. Decomposes and then ignites when heated above 270°C. When heated to decomposition it emits very toxic fumes of Cl^-, SO_x, PO_x, and NO_x.

MIJ500 CAS:127-33-3 ***HR: 3***
METHYLCHLORTETRACYCLINE
mf: $C_{21}H_{21}ClN_2O_8$ mw: 464.89

SYNS: 7-CHLORO-6-DEMETHYLTETRACYCLINE DEMETHYLCHLOROTETRACYCLINE ◇ DECLOMYCIN ◇ DEMECLOCYCLINE ◇ DEMETHYLCHLOROTETRACYCLIN ◇ DEMETHYLCHLOROTETRACYCLINE ◇ 6-DEMETHYLCHLOROTETRACYCLINE ◇ DEMETHYLCHLORTETRACYCLINE ◇ 6-DEMETHYL-7-CHLOROTETRACYCLINEDEMETHYLCHLORTETRACYCLINE ◇ 6-DEMETHYL-7-CHLORTETRACYCLINE ◇ 6-DEMETHYLCHLORTETRACYCLINE ◇ DEMETHYLCHLORTETRACYCLINE, BASE ◇ DMCT ◇ LEDERMYCIN ◇ MEXOCINE ◇ RP 10192

TOXICITY DATA with REFERENCE
dni-hmn:lym 3750 μg/L BCPHBM 16,127,83
unr-wmn TDLo:240 mg/kg (1-39W preg):REP JAMAAP 188,178,64
orl-rat TDLo:5 g/kg (1-22D preg/21D post):TER AJOMAZ 10,89,73
orl-hmn TDLo:420 mg/kg/6W:END,BIO AIMEAS 79,679,73
orl-hmn TDLo:10 mg/kg:IMM,SKN BMJOAE 2,96,77
orl-hmn TDLo:68 mg/kg/8D:KID CTCEA9 15,734,73
ipr-rat LD50:358 mg/kg NIIRDN 6,497,82
ipr-mus LD50:454 mg/kg NIIRDN 6,497,82
ivn-mus LD50:79 mg/kg RPOBAR 2,281,70

SAFETY PROFILE: Poison by intravenous and intraperitoneal routes. Human systemic effects by ingestion: diabetes insipidus, urine volume increase, other changes in urine composition, dermatitis, changes in the nails, allergic rhinitis, serum sickness, effects on cyclic nucleotides. Human reproductive effects by an unspecified route: postnatal measures or effects on newborn. An experimental teratogen. Experimental reproductive effects. Human mutation data reported. When heated to decomposition it emits very toxic fumes of Cl^- and NO_x.

MIJ750 CAS:56-49-5 ***HR: 3***
3-METHYLCHOLANTHRENE
mf: $C_{21}H_{16}$ mw: 268.37

PROP: Pale yellow needles from benzene. Mp: 176.5°, bp: 280° @ 80 mm, d: 1.28 @ 20°. Sol in benzene, xylene, toluene; sltly sol in amyl alc; insol in water.

SYNS: 1,2-DIHYDRO-3-METHYL-BENZ(j)ACEANTHRYLENE ◇ 3-MCA ◇ METHYLCHOLANTHRENE ◇ 20-METHYLCHOLANTHRENE ◇ RCRA WASTE NUMBER U157

TOXICITY DATA with REFERENCE
otr-hmn:lng 500 μg/L/2W GANNA2 74,615,83
sce-hmn:fbr 1 mmol/L MUREAV 117,47,83
ipr-rat TDLo:64 mg/kg (9-21D preg):REP BCPCA6 26,567,77
ipr-mus TDLo:40 mg/kg (female 1D post):TER JJIND8 8,79,47
orl-rat TDLo:600 mg/kg (MGN):CAR,TER CNREA8 12,296,52
orl-rat TDLo:200 mg/kg:ETA JJIND8 48,185,72
skn-rat TDLo:480 mg/kg/20W-I:CAR GANNA2 61,367,70
scu-rat TDLo:8 mg/kg:CAR HOIZAK 57,53,82
itr-rat TDLo:40 mg/kg:NEO JJIND8 49,541,72
orl-mus TDLo:21 mg/kg (15-17D post):CAR,TER TXAPA9 72,427,84
skn-mus TDLo:120 mg/kg (MGN):CAR,TER BEXBAN 71,677,71

ipr-mus TDLo:5 mg/kg (17D post):CAR,TER CRNGDP 6,1389,85
ipr-mus LDLo:100 mg/kg TXAPA9 23,288,72
ivn-frg LDLo:11 mg/kg CNREA8 24,1969,64

CONSENSUS REPORTS: Reported in EPA TSCA Inventory. EPA Genetic Toxicology Program.

SAFETY PROFILE: Suspected carcinogen with experimental carcinogenic, neoplastigenic, and tumorigenic data. Poison by intravenous and intraperitoneal routes. Experimental teratogenic and reproductive effects. Human mutation data reported. When heated to decomposition it emits acrid smoke and irritating fumes.

MIK000 CAS:63041-78-1 ***HR: 3***
5-METHYLCHOLANTHRENE
mf: $C_{21}H_{16}$ mw: 268.37

PROP: Yellow needles from benzene. Mp: 176.5-177.5°.

TOXICITY DATA with REFERENCE
dnd-mus:oth 400 nmol CNREA8 42,1239,82
scu-mus TDLo:40 mg/kg:ETA CNREA8 1,695,41

SAFETY PROFILE: Questionable carcinogen with experimental tumorigenic data. Mutation data reported. When heated to decomposition it emits acrid smoke and irritating fumes.

MIK250 CAS:17012-89-4 ***HR: 3***
22-METHYLCHOLANTHRENE
mf: $C_{21}H_{16}$ mw: 268.37

SYN: 4-METHYLCHOLANTHRENE

TOXICITY DATA with REFERENCE
scu-mus TDLo:40 mg/kg:CAR CNREA8 1,695,41
scu-mus TD:80 mg/kg:CAR CNREA8 1,685,41

SAFETY PROFILE: Questionable carcinogen with experimental carcinogenic data. When heated to decomposition it emits acrid smoke and irritating fumes.

MIK500 ***HR: 3***
20-METHYLCHOLANTHRENE CHOLEIC ACID
mf: $C_{96}H_{160}O_{16}•C_{21}H_{16}$ mw: 1838.93

TOXICITY DATA with REFERENCE
cyt-mus:fbr 1 mg/L AJCAA7 39,149,40
scu-mus TDLo:400 mg/kg:ETA JNCIAM 2,99,41

SAFETY PROFILE: Questionable carcinogen with experimental tumorigenic data. Mutation data reported. When heated to decomposition it emits acrid smoke and irritating fumes.

MIK750 CAS:3342-99-2 ***HR: 3***
cis-3-METHYLCHOLANTHRENE-1,2-DIOL
mf: $C_{21}H_{16}O_2$ mw: 300.37

SYN: cis-1,2-DIHYDROXY-3-METHYLCHOLANTHRENE

TOXICITY DATA with REFERENCE
scu-mus TDLo:120 mg/kg/6W-I:NEO IJCNAW 2,505,67

SAFETY PROFILE: Questionable carcinogen with experimental neoplastigenic data. When heated to decomposition it emits acrid smoke and irritating fumes.

MIL250 CAS:3343-08-6 ***HR: 3***
3-METHYLCHOLANTHRENE-2-ONE
mf: $C_{21}H_{14}O$ mw: 282.35

SYN: 3-METHYLCHOLANTHREN-2-ONE

TOXICITY DATA with REFERENCE
mma-sat 20 nmol/plate CNREA8 38,3398,78
mma-ham:lng 15 nmol/plate CNREA8 38,3398,78
skn-mus TDLo:90 mg/kg/20W-I:CAR CBINA8 22(1),69,78
scu-mus TDLo:120 mg/kg/6W-I:NEO IJCNAW 2,505,67

CONSENSUS REPORTS: EPA Genetic Toxicology Program.

SAFETY PROFILE: Questionable carcinogen with experimental carcinogenic and neoplastigenic data. Mutation data reported. When heated to decomposition it emits acrid smoke and irritating fumes.

MIL500 CAS:3416-21-5 ***HR: 3***
3-METHYLCHOLANTHRENE-11,12-OXIDE
mf: $C_{21}H_{16}O$ mw: 284.37

SYNS: 11,12-DIHYDRO-11,12-EPOXY-3-METHYLCHOLANTHRENE ◇ 11,12-EPOXY-11,12-DIHYDRO-3-METHYLCHOLANTHRENE

TOXICITY DATA with REFERENCE
slt-dmg-par 5 mmol/L CNREA8 33,2354,73
sln-dmg-par 5 mmol/L CNREA8 33,2354,73
dnd-hmn:oth 10 μmol/L CNREA8 36,272,76
otr-mus:oth 750 μg/L PNASA6 68,1098,71
scu-mus TDLo:40 mg/kg/6W-I:NEO IJCNAW 2,505,67

CONSENSUS REPORTS: EPA Genetic Toxicology Program.

SAFETY PROFILE: Questionable carcinogen with experimental neoplastigenic data. Human mutation data reported. When heated to decomposition it emits acrid smoke and irritating fumes.

MIL750 CAS:63041-80-5 ***HR: 3***
20-METHYLCHOLANTHRENE PICRATE
mf: $C_{21}H_{16}•C_6H_3N_3O_7$ mw: 497.49

SYNS: 1,2-DIHYDRO-3-METHYLBENZ(j)ACEANTHRYLENECOMPOUND with 2,4,6-TRINITROPHENOL (1:1) ◇ 3-METHYLCHOLANTHRENE COMPOUND with PICRIC ACID (1:1) ◇ 2,4,6-TRINITROPHENOL COMPOUND with 1,2-DIHYDRO-3-METHYLBENZ(j) ACEANTHRYLENE

TOXICITY DATA with REFERENCE
scu-mus TDLo:200 mg/kg:ETA XPHPAW 149,319,51

SAFETY PROFILE: Questionable carcinogen with experimental tumorigenic data. When heated to decomposition it emits toxic fumes of NO_x. See also PICRIC ACID.

MIM000 CAS:63040-09-5 ***HR: 3***
20-METHYLCHOLANTHRENE-TRINITROBENZENE

SYN: 3-METHYLCHOLANTHRENE COMPOUND with 1,3,5-TRINITROBENZENE (1:1)

TOXICITY DATA with REFERENCE
scu-mus TDLo:60 mg/kg:NEO XPHPAW 149,319,51

SAFETY PROFILE: Questionable carcinogen with experimental neoplastigenic data. When heated to decomposition it emits toxic fumes of NO_x. See also NITRO COMPOUNDS of AROMATIC HYDROCARBONS.

MIM250 CAS:3343-07-5 ***HR: 3***
20-METHYLCHOLANTHREN-15-ONE
mf: $C_{21}H_{14}O$ mw: 282.35

SYNS: 15-KETO-20-METHYLCHOLANTHRENE ◇ 3-METHYLCHOLANTHREN-1-ONE

TOXICITY DATA with REFERENCE
mma-sat 20 nmol/plate CNREA8 38,3398,78
mma-ham:lng 15 nmol/plate CNREA8 38,3398,78
skn-mus TDLo:340 mg/kg/42W-I:ETA PRLBA4 129,439,40
scu-mus TDLo:120 mg/kg/6W-I:NEO IJCNAW 2,505,67

CONSENSUS REPORTS: EPA Genetic Toxicology Program.

SAFETY PROFILE: Questionable carcinogen with experimental neoplastigenic and tumorigenic data. Mutation data reported. When heated to decomposition it emits acrid smoke and irritating fumes.

MIM500 CAS:3351-28-8 ***HR: 3***
1-METHYLCHRYSENE
mf: $C_{19}H_{14}$ mw: 242.33

TOXICITY DATA with REFERENCE
mma-sat 10 μg/plate CNREA8 36,4525,76
skn-mus TDLo:40 mg/kg/3W-I:ETA CCSUDL 1,325,76

CONSENSUS REPORTS: IARC Cancer Review: Group 3 IMEMDT 7,56,87; Animal Inadequate Evidence IMEMDT 32,379,83. EPA Genetic Toxicology Program.

SAFETY PROFILE: Questionable carcinogen with experimental tumorigenic data. Mutation data reported. When heated to decomposition it emits acrid smoke and irritating fumes. See also other methylchrysene entries.

MIM750 CAS:3351-32-4 ***HR: 3***
2-METHYLCHRYSENE
mf: $C_{19}H_{14}$ mw: 242.33

TOXICITY DATA with REFERENCE
mma-sat 10 μg/plate CNREA8 36,4525,76
skn-mus TDLo:40 mg/kg/3W-I:ETA CCSUDL 1,325,76

CONSENSUS REPORTS: IARC Cancer Review: Group 3 IMEMDT 7,56,87; Animal Limited Evidence IMEMDT 32,379,83. EPA Genetic Toxicology Program.

SAFETY PROFILE: Questionable carcinogen with experimental tumorigenic and possible carcinogenic data. Mutation data reported. When heated to decomposition it emits acrid smoke and irritating fumes. See also other methylchrysene entries.

MIN000 CAS:3351-31-3 ***HR: 3***
3-METHYLCHRYSENE
mf: $C_{19}H_{14}$ mw: 242.33

TOXICITY DATA with REFERENCE
mma-sat 20 μg/plate CNREA8 36,4525,76
skn-mus TDLo:40 mg/kg/3W-I:NEO CCSUDL 1,325,76
skn-mus TD:180 mg/kg/15W-I:ETA JNCIAM 53,1121,74

CONSENSUS REPORTS: IARC Cancer Review: Group 3 IMEMDT 7,56,87; Animal Limited Evidence IMEMDT 32,379,83. EPA Genetic Toxicology Program.

SAFETY PROFILE: Questionable carcinogen with experimental neoplastigenic and tumorigenic data. Mutation data reported. When heated to decomposition it emits acrid smoke and irritating fumes. See also other methylchrysene entries.

MIN250 CAS:3351-30-2 ***HR: 3***
4-METHYLCHRYSENE
mf: $C_{19}H_{14}$ mw: 242.33

TOXICITY DATA with REFERENCE
mma-sat 10 μg/plate CNREA8 36,4525,76
skn-mus TDLo:40 mg/kg/3W-I:ETA CCSUDL 1,325,76

CONSENSUS REPORTS: IARC Cancer Review: Group 3 IMEMDT 7,56,87; Animal Limited Evidence IMEMDT 32,379,83. EPA Genetic Toxicology Program.

SAFETY PROFILE: Questionable carcinogen with experimental tumorigenic data. Mutation data reported. When heated to decomposition it emits acrid smoke and irritating fumes. See also other methylchrysene entries.

MIN500 CAS:3697-24-3 ***HR: 3***
5-METHYLCHRYSENE
mf: $C_{19}H_{14}$ mw: 242.33

TOXICITY DATA with REFERENCE
mma-sat 3 μg/plate CRNGDP 7,673,86
dnd-mus-skn 467 μmol/L CRNGDP 4,843,83
skn-mus TDLo:32 μg/kg:CAR CNREA8 45,6406,85
scu-mus TDLo:20 mg/kg/20W-I:CAR CCSUDL 1,325,76
skn-mus TD:40 mg/kg/30W-I:NEO CCSUDL 1,325,76
scu-mus TD:80 mg/kg:ETA CNREA8 3,606,43

CONSENSUS REPORTS: NTP Fifth Annual Report on Carcinogens. IARC Cancer Review: Group 3 IMEMDT 7,56,87; Animal Sufficient Evidence IMEMDT 32,379, 83. EPA Genetic Toxicology Program.

SAFETY PROFILE: Confirmed carcinogen with experimental carcinogenic, neoplastigenic, and tumorigenic data. Mutation data reported. When heated to decomposition it emits acrid smoke and irritating fumes. See also other methylchrysene entries.

MIN750 CAS:1705-85-7 ***HR: 3***
6-METHYLCHRYSENE
mf: $C_{19}H_{14}$ mw: 242.33

TOXICITY DATA with REFERENCE
mma-sat 10 μg/plate CNREA8 36,4525,76
skn-mus TDLo:40 mg/kg/3W-I:ETA CCSUDL 1,325,76

CONSENSUS REPORTS: IARC Cancer Review: Group 3 IMEMDT 7,56,87; Animal Limited Evidence IMEMDT 32,379,83. EPA Genetic Toxicology Program.

SAFETY PROFILE: Questionable carcinogen with experimental tumorigenic data. Mutation data reported. When heated to decomposition it emits acrid smoke and irritating fumes. See also other methylchrysene entries.

MIO000 CAS:101-39-3 ***HR: 2***
α-METHYLCINNAMALDEHYDE
mf: $C_{10}H_{10}O$ mw: 146.20

PROP: Yellow liquid; cinnamon odor. D: 1.035-1.039, refr index: 1.602-1.607, flash p: 174°F. Sol in fixed oils, propylene glycol; insol in glycerin.

SYNS: FEMA No. 2697 ◇ METHYL CINNAMIC ALDEHYDE ◇ α-METHYLCINNAMIC ALDEHYDE ◇ α-METHYLCINNIMAL ◇ 2-METHYL-3-PHENYL-2-PROPENAL

TOXICITY DATA with REFERENCE
skn-gpg 5%/2W MLD ADVEA4 58,121,78
orl-rat LD50:2050 mg/kg FCTXAV 13,681,75

CONSENSUS REPORTS: Reported in EPA TSCA Inventory.

SAFETY PROFILE: Moderately toxic by ingestion. A skin irritant. Combustible liquid. When heated to decomposition it emits acrid smoke and irritating fumes. See also ALDEHYDES.

MIO500 CAS:103-26-4 ***HR: 2***
METHYL CINNAMATE
mf: $C_{10}H_{10}O_2$ mw: 162.20

PROP: White to sltly yellow crystals; fruity odor. D: 1.042 @ 36/0°, mp: 33.4°, bp: 263°, flash p: +212°F. Very sol in alc, ether; sol in fixed oils, glycerin, propylene glycol; insol in water.

SYNS: FEMA No. 2698 ◇ METHYL CINNAMYLATE ◇ METHYL-3-PHENYLPROPENOATE ◇ 3-PHENYL-2-PROPENOIC ACID METHYL ESTER (9CI)

TOXICITY DATA with REFERENCE
orl-rat LD50:2610 mg/kg FCTXAV 13,681,75

CONSENSUS REPORTS: Reported in EPA TSCA Inventory.

SAFETY PROFILE: Moderately toxic by ingestion. Combustible liquid. When heated to decomposition it emits acrid smoke and irritating fumes.

MIO750 CAS:1504-55-8 ***HR: 2***
METHYL CINNAMIC ALCOHOL
mf: $C_{10}H_{12}O$ mw: 148.22

SYNS: α-METHYLCINNAMYL ALCOHOL ◇ 3-PHENYL-2-METHYL-PROPEN-2-OL-1

TOXICITY DATA with REFERENCE
skn-rbt 500 mg/24H MLD FCTXAV 13,681,75
orl-rat LD50:2400 mg/kg FCTXAV 13,681,75

CONSENSUS REPORTS: Reported in EPA TSCA Inventory.

SAFETY PROFILE: Moderately toxic by ingestion. A skin irritant. When heated to decomposition it emits acrid smoke and irritating fumes. See also ALCOHOLS.

MIO975 CAS:21340-68-1 ***HR: 2***
METHYL CLOFENAPATE
mf: $C_{17}H_{17}ClO_3$ mw: 304.79

SYNS: ICI 54856 METHYL ESTER ◇ METHYL-2-(4-(p-CHLOROPHENYL)PHENOXY)-2-METHYLPROPIONATE◇ PROPANOIC ACID, 2-((4'-CHLORO(1,1'-BIPHENYL)-4-YL)OXY)-2-METHYL-,METHYL ESTER (9CI)

TOXICITY DATA with REFERENCE
orl-rat TDLo:31500 mg/kg/75W-C:CAR CNREA8 42,259,82

SAFETY PROFILE: Questionable carcinogen with experimental carcinogenic data. When heated to decomposition it emits toxic fumes of Cl^-.

MIP250 CAS:1184-53-8 ***HR: 3***
METHYL COPPER
mf: CH_3Cu mw: 78.58

CONSENSUS REPORTS: Copper and its compounds are on the Community Right-To-Know List.

SAFETY PROFILE: Explodes violently in air when dry. See also COPPER COMPOUNDS and ORGANOMETALS.

MIP500 CAS:607-71-6 ***HR: 2***
4-METHYLCOUMARIN
mf: $C_{10}H_8O_2$ mw: 160.18

PROP: Needles from benzene. Mp: 82°. Sol in alc and benzene.

TOXICITY DATA with REFERENCE
orl-mus LD50:1691 mg/kg YKKZAJ 83,1124,63
scu-mus LD50:1088 mg/kg YKKZAJ 83,1124,63

SAFETY PROFILE: Moderately toxic by ingestion and subcutaneous routes. When heated to decomposition it emits acrid smoke and irritating fumes.

MIP750 CAS:92-48-8 ***HR: 2***
6-METHYLCOUMARIN
mf: $C_{10}H_8O_2$ mw: 160.18

PROP: White needles from benzene; coconut odor. Mp: 73-76, flash p: +153°F. Sol in alc and benzene.

SYNS: FEMA No. 2690 ◇ 6-MC ◇ 6-METHYL-2H-1-BENZOPYRAN-2-ONE ◇ 6-METHYLBENZOPYRONE ◇ 6-METHYL-1,2-BENZOPYRONE ◇ 6-METHYLCOUMARINIC ANHYDRIDE ◇ NCI-C55812 ◇ TONCARINE

TOXICITY DATA with REFERENCE
skn-rbt 500 mg/24H MLD FCTXAV 14,605,76
mma-sat 3 μmol/plate FCTOD7 21,707,83
orl-rat LD50:1680 mg/kg FCTXAV 14,605,76
scu-mus LD50:253 mg/kg YKKZAJ 76,186,56

CONSENSUS REPORTS: EPA Genetic Toxicology Program. Reported in EPA TSCA Inventory.

SAFETY PROFILE: Poison by subcutaneous route. Moderately toxic by ingestion. A skin irritant. Mutation data reported. Combustible liquid. When heated to decomposition it emits acrid smoke and irritating fumes.

MIP775 CAS:2445-83-2 ***HR: 3***
7-METHYLCOUMARIN
mf: $C_{10}H_8O_2$ mw: 160.18

SYN: 7-METHYL-2H-1-BENZOPYRAN-2-ONE(9CI)

TOXICITY DATA with REFERENCE
skn-rbt 500 mg/24H MLD FCTOD7 20(Suppl),747,82
orl-rat LD50:3800 mg/kg FCTOD7 20(Suppl),747,82
scu-mus LD50:258 mg/kg YKKZAJ 76,168,56

CONSENSUS REPORTS: Reported in EPA TSCA Inventory.

SAFETY PROFILE: Poison by subcutaneous route. Moderately toxic by ingestion. A skin irritant. When heated to decomposition it emits acrid smoke and irritating fumes.

MIP800 CAS:638-10-8 ***HR: 2***
3-METHYLCROTONIC ACID, ETHYL ESTER
mf: $C_7H_{12}O_2$ mw: 128.19

SYNS: ETHYL DIMETHYLACRYLATE ◇ ETHYL β,β-DIMETHYLACRYLATE ◇ ETHYL 3,3-DIMETHYLACRYLATE ◇ ETHYL ISOBUTENOATE ◇ ETHYL ISOPROPYLIDENE ACETATE ◇ ETHYL α-METHYLCROTONATE ◇ ETHYL 3-METHYLCROTONATE ◇ ETHYL SENECIOATE ◇ 3-METHYL-2-BUTENOIC ACID ETHYL ESTER

TOXICITY DATA with REFERENCE
orl-rat LD50:11600 mg/kg GTPZAB 29(4),52,85
orl-mus LD50:2450 mg/kg GTPZAB 29(4),52,85
orl-gpg LD50:500 mg/kg GTPZAB 29(4),52,85

SAFETY PROFILE: Moderately toxic by ingestion. When heated to decomposition it emits acrid smoke and irritating fumes.

MIQ000 CAS:105-34-0 ***HR: 3***
METHYL CYANOACETATE
mf: $C_4H_5NO_2$ mw: 99.10

PROP: Liquid. Mp: −22.5°, bp: 203°, vap d: 3.41, d: 1.123 @ 15/4°. Insol in water; misc in alc and ether.

SYNS: CYANOACETIC ACID METHYL ESTER ◇ METHYL 2-CYANOACETATE ◇ METHYL CYANOETHANOATE ◇ USAF KF-22

TOXICITY DATA with REFERENCE
ipr-mus LD50:200 mg/kg NTIS** AD277-689

CONSENSUS REPORTS: Reported in EPA TSCA Inventory. Cyanide and its compounds are on the Community Right-To-Know List.

SAFETY PROFILE: Poison by intraperitoneal route. When heated to decomposition it emits toxic fumes of NO_x and CN^-. See also ESTERS.

MIQ075 CAS:137-05-3 ***HR: 3***
METHYL 2-CYANOACRYLATE
mf: $C_9H_{13}NO_2$ mw: 111.11

SYNS: ADHERE ◇ COAPT ◇ α-CYANOACRYLATE ACID METHYL ESTER ◇ 2-CYANOACRYLATE ACID METHYL ESTER ◇ CYANOLYT ◇ EASTMAN 910 ◇ MECRLAT

TOXICITY DATA with REFERENCE
eye-hmn TD50:4 ppm AIHAAP 29,558,68

itt-mky TDLo:100 mg/kg (1D male):REP FESTAS 39(Suppl),441,83

CONSENSUS REPORTS: Reported in EPA TSCA Inventory.

OSHA PEL: TWA 2 ppm; STEL 4 ppm
ACGIH TLV: TWA 2 ppm; STEL 4 ppm
DFG MAK: 2 ppm (8 mg/m^3)

SAFETY PROFILE: Experimental reproductive effects. A human eye irritant. When heated to decomposition it emits toxic fumes of NO_x.

MIQ250 CAS:63020-25-7 ***HR: 3***
9-METHYL-10-CYANO-1,2-BENZANTHRACENE
mf: $C_{20}H_{14}N$ mw: 268.35

SYN: 7-CYANO-12-METHYL-BENZ(a)ANTHRACENE

TOXICITY DATA with REFERENCE
skn-mus TDLo:620 mg/kg/26W-I:ETA PRLBA4 129,439,40

CONSENSUS REPORTS: Cyanide and its compounds are on the Community Right-To-Know List.

SAFETY PROFILE: Questionable carcinogen with experimental tumorigenic data. When heated to decomposition it emits toxic fumes of NO_x and CN^-. See also NITRILES.

MIQ350 ***HR: 2***
METHYL CYANOCARBAMATE DIMER
mf: $C_6H_8N_4O_4$ mw: 200.15

SYN: CYANO-CARBAMIC ACID METHYL ESTER, DIMER

TOXICITY DATA with REFERENCE
unr-rat LD50:1600 mg/kg GISAAA 50(6),78,85
unr-mus LD50:500 mg/kg GISAAA 50(6),78,85
unr-gpg LD50:820 mg/kg GISAAA 50(6),78,85

CONSENSUS REPORTS: Cyanide and its compounds are on the Community Right-To-Know List.

SAFETY PROFILE: Moderately toxic by unspecified routes. When heated to decomposition it emits toxic fumes of NO_x and CN^-. See also CARBAMATES, NITRILES, and ESTERS.

MIQ400 CAS:18051-18-8 ***HR: D***
d-6-METHYL-8-CYANOMETHYLERGOLINE
mf: $C_{17}H_{19}N_2$ mw: 251.38

SYNS: d-6-METHYLERGOLINE-8-ACETONITRILE ◇ 6605 VUFB

TOXICITY DATA with REFERENCE
orl-mus TDLo:20 mg/kg (6-7D preg):REP JRPFA4 24,441,71

CONSENSUS REPORTS: Cyanide and its compounds are on the Community Right-To-Know List.

SAFETY PROFILE: Experimental reproductive effects. When heated to decomposition it emits toxic fumes of CN^- and NO_x. See also NITRILES.

MIQ725 CAS:24342-56-1 ***HR: 3***
o-METHYLCYCLIZINE DIHYDROCHLORIDE
mf: $C_{19}H_{24}N_2$•2ClH mw: 353.37

SYNS: 1-METHYL-4-(o-METHYL-α-PHENYLBENZYL)PIPERAZINE DIHYDROCHLORIDE ◇ 1-METHYL-4-((2-METHYLPHENYL)PHENYLMETHYL)PIPERAZINE DIHYDROCHLORIDE (9CI) ◇ 1-METHYL-4-(α-o-TOLYLBENZYL)PIPERAZINEDIHYDROCHLORIDE

TOXICITY DATA with REFERENCE
orl-mus LD50:142 mg/kg AIPTAK 116,17,58
ipr-mus LD50:70 mg/kg AIPTAK 116,17,58
ims-mus LD50:90 mg/kg AIPTAK 116,17,58

SAFETY PROFILE: Poison by ingestion, intramuscular, and intraperitoneal routes. When heated to decomposition it emits toxic fumes of NO_x and HCl.

MIQ740 CAS:108-87-2 ***HR: 3***
METHYLCYCLOHEXANE
DOT: UN 2296
mf: C_7H_{14} mw: 98.21

PROP: Colorless liquid. Mp: −126.4°, lel: 1.2%, uel: 6.7%, bp: 100.3°, flash p: 25°F (CC), d: 0.7864 @ 0°/4°, 0.769 @ 20°/4°, vap press: 40 mm @ 22.0°, vap d: 3.39, autoign temp: 482°F.

SYNS: CYCLOHEXYLMETHANE ◇ HEXAHYDROTOLUENE ◇ METYLOCYKLOHEKSAN (POLISH) ◇ SEXTONE B ◇ TOLUENE HEXAHYDRIDE

TOXICITY DATA with REFERENCE
orl-mus LD50:2250 mg/kg 85GMAT -,82,82
ihl-mus LC50:41500 mg/m^3/2H 85GMAT -,82,82
orl-rbt LDLo:4000 mg/kg JIHTAB 25,199,43

CONSENSUS REPORTS: Reported in EPA TSCA Inventory.

OSHA PEL: (Transitional: TWA 500 ppm) TWA 400 ppm
ACGIH TLV: TWA 400 ppm
DFG MAK: 500 ppm (2000 mg/m^3)
DOT Classification: Flammable Liquid; Label: Flammable Liquid.

SAFETY PROFILE: Moderately toxic by ingestion. Mildly toxic by inhalation. This material does not cause irritation to the eyes and nose, and even at the level of 500 ppm, exhibits only a very faint odor. Therefore, it cannot be said to have any warning properties. It is believed to be about three times as toxic as hexane, and has caused death by tetanic spasm in animals. In sublethal concentrations, it causes narcosis and anesthesia. Dangerous fire hazard and moderate explosion hazard when exposed to heat, flame, or oxidizers. To fight fire, use foam, CO_2, dry chemical. When heated to decomposition it emits acrid smoke and fumes.

MIQ745 CAS:25639-42-3 ***HR: 2***
METHYLCYCLOHEXANOL
DOT: UN 2617
mf: $C_7H_{14}O$ mw: 114.21

PROP: Colorless, viscous liquid; aromatic, menthol-like odor. Bp: 155-180°, flash p: 154°F (CC), autoign temp: 565°F, d: 0.924 @ 15.5°/15.5°, vap d: 3.93.

SYNS: HEXAHYDROCRESOL ◇ HEXAHYDROMETHYLPHENOL ◇ METYLOCYKLOHEKSANOL (POLISH)

TOXICITY DATA with REFERENCE
ihl-hmn TCLo:500 ppm:CNS,LIV,KID TGNCDL 2,55,61
orl-rat LD50:1660 mg/kg JIHTAB 25,415,43
scu-rat LD50:2900 mg/kg JIHTAB 25,415,43
orl-rbt LDLo:1750 mg/kg HBTXAC 1,194,56
skn-rbt LDLo:6800 mg/kg HBTXAC 1,194,56

OSHA PEL: (Transitional: TWA 100 ppm) TWA 50 ppm
ACGIH TLV: TWA 50 ppm
DFG MAK: 50 ppm (235 mg/m^3)
DOT Classification: Flammable or Combustible Liquid; Label: Flammable Liquid.

SAFETY PROFILE: Moderately toxic by ingestion and subcutaneous routes. Mildly toxic by skin contact. Human system effects by inhalation: antipsychotic, unspecified liver and kidney effects. Combustible when exposed to heat, flame or oxidizers. On heating it emits acrid fumes; can react with oxidizing materials. To fight fire, use alcohol foam, CO_2, dry chemical.

MIQ750 CAS:591-23-1 ***HR: 2***
m-METHYLCYCLOHEXANOL
mf: $C_7H_{14}O$ mw: 114.21

TOXICITY DATA with REFERENCE
ims-mus LD50:1000 mg/kg JSICAZ 21,342,62

CONSENSUS REPORTS: Reported in EPA TSCA Inventory.

SAFETY PROFILE: Moderately toxic by intramuscular route. When heated to decomposition it emits acrid smoke and irritating fumes. See also ALCOHOLS.

MIR000 CAS:583-59-5 ***HR: 2***
o-METHYLCYCLOHEXANOL
mf: $C_7H_{14}O$ mw: 114.21

PROP: Colorless liquid. D: 0.934 @ 20°/4°, mp: −9.5°, bp: 165°. Very sltly sol in water, misc in alc and ether.

TOXICITY DATA with REFERENCE
ims-mus LD50:1000 mg/kg JSICAZ 21,342,62

SAFETY PROFILE: Moderately toxic by intramuscular route. When heated to decomposition it emits acrid smoke and irritating fumes. See also ALCOHOLS.

MIR250 CAS:1331-22-2 ***HR: 2***
METHYLCYCLOHEXANONE
DOT: UN 2297
mf: $C_7H_{12}O$ mw: 112.19

PROP: Water-white to pale yellow liquid, acetone-like odor. Mp: -14°C, bp: 160-170°, flash p: 118°F (CC), d: 0.925 @ 15°/5°, vap d: 3.86. Insol in water; sol in ether and alc.

SYN: METYLOCYKLOHEKSANON (POLISH)

TOXICITY DATA with REFERENCE
orl-rbt LDLo:1000 mg/kg JIHTAB 25,199,43
skn-rbt LDLo:4900 mg/kg JIHTAB 25,199,43

CONSENSUS REPORTS: Reported in EPA TSCA Inventory.

DOT Classification: Flammable or Combustible Liquid; Label: Flammable Liquid.

SAFETY PROFILE: Moderately toxic by ingestion. Mildly toxic by skin contact. A toxic compound which can damage the kidneys and the liver. It is similar to cyclohexanol in its toxic action, although it is somewhat less active. Harmful exposure in industry is rare. Experimental animals can withstand prolonged exposures of 0.02-0.05% by volume in air. Combustible when exposed to heat or flame. Can react violently with HNO_3 and other oxidizers. To fight fire, use foam, CO_2, dry chemical. When heated to decomposition it emits acrid smoke and irritating fumes. See also KETONES.

MIR500 CAS:583-60-8 ***HR: 3***
2-METHYLCYCLOHEXANONE
mf: $C_7H_{12}O$ mw: 112.19

PROP: Liquid. D: 0.925 @ 20/4°, mp: −14°, bp: 165.1°. Insol in water; sol in alc and ether.

SYNS: 2-METHYL-CYCLOHEXANON (GERMAN, DUTCH) ◇ 1-METHYLCYCLOHEXAN-2-ONE ◇ o-METHYLCYCLOHEXANONE ◇ 2-METILCICLOESANONE (ITALIAN)

TOXICITY DATA with REFERENCE
orl-rat LD50:2140 mg/kg AIHAAP 30,470,69
ipr-mus LD50:200 mg/kg NTIS** AD691-490
ivn-mus LDLo:270 mg/kg COREAF 236,633,53
orl-rbt LD50:1 g/kg DTLVS* 4,272,80
skn-rbt LD50:1635 mg/kg AIHAAP 30,470,69

CONSENSUS REPORTS: Reported in EPA TSCA Inventory.

OSHA PEL: (Transitional: TWA 100 ppm (skin)) TWA 50 ppm; STEL 75 ppm (skin)
ACGIH TLV: TWA 50 ppm (skin)
DFG MAK: 50 ppm (230 mg/m^3)

SAFETY PROFILE: Poison by intravenous and intraperitoneal routes. Moderately toxic by ingestion and skin contact. When heated to decomposition it emits acrid smoke and irritating fumes. See also KETONES and other methylcyclohexanone entries.

MIR600 CAS:591-24-2 ***HR: 3***
3-METHYLCYCLOHEXANONE
mf: $C_7H_{12}O$ mw: 112.17

$O:C(CH_2)_3CH(CH_3)CH_2$

SYN: METHYL-3-CYCLO-HEXANONE-1 (FRENCH)
ivn-dog LDLo:310 mg/kg COREAF 236,633,53

CONSENSUS REPORTS: Reported in EPA TSCA Inventory.

SAFETY PROFILE: Poison by intravenous route. Reaction with hydrogen peroxide + nitric acid forms an explosive peroxide. When heated to decomposition it emits acrid smoke and irritating fumes. See also KETONES and other methylcyclohexanone entries.

MIR625 CAS:589-92-4 ***HR: 3***
4-METHYLCYCLOHEXANONE
mf: $C_7H_{12}O$ mw: 112.17

$O:CC_2H_4CH(CH_3)CH_2CH_2$

SYN: METHYL-4-CYCLO-HEXANONE-1 (FRENCH)
ivn-dog LDLo:370 mg/kg COREAF 236,633,53
orl-mus LD50:1600 mg/kg 38MKAJ 2C,4783,82
orl-rat LD50:800 mg/kg 38MKAJ 2C,4783,82

CONSENSUS REPORTS: Reported in EPA TSCA Inventory.

SAFETY PROFILE: Poison by intravenous route. Moderately toxic by ingestion. Mixture with nitric acid explodes when heated to 75°C. When heated to decomposition it emits acrid smoke and irritating fumes. See also KETONES and other methylcyclohexanone entries.

MIR750 CAS:591-47-9 ***HR: 2***
4-METHYLCYCLOHEXENE
mf: C_7H_{12} mw: 96.174

$CH{=}CHCH_2CH(CH_3)CH_2CH_2$

PROP: A clear liquid. Bp: 102.5°, flash p: 30.2°F, d: 0.804 @ 15.5°/15.5°, vap press: 10.3 mm @ 38°, vap d: 3.34.

SAFETY PROFILE: Probably an irritant and narcotic in high concentration. Very dangerous fire hazard when exposed to heat or flame; can react vigorously with oxidizing materials. To fight fire, use foam, CO_2, dry chemical. When heated to decomposition it emits acrid smoke and irritating fumes.

MIS250 CAS:2021-21-8 ***HR: D***
N-METHYL-4-CYCLOHEXENE-1,2-DICARBOXIMIDE
mf: $C_9H_{11}NO_2$ mw: 165.21

SYN: N-METHYL-1,2,3,6-TETRAHYDROPHTHALIMIDE

TOXICITY DATA with REFERENCE
ipr-mus TDLo:200 mg/kg (9D preg):REP MOPMA3 13,133,77
ipr-mus TDLo:200 mg/kg (9D preg):TER MOPMA3 13,133,77

SAFETY PROFILE: An experimental teratogen. Other experimental reproductive effects. When heated to decomposition it emits toxic fumes of NO_x.

MIS500 CAS:72299-02-6 ***HR: 1***
1-((6-METHYL-3-CYCLOHEXEN-1-YL)CARBONYL)PIPERIDINE
mf: $C_{13}H_{21}NO$ mw: 207.35

SYN: AI3-36329-A

TOXICITY DATA with REFERENCE
skn-rbt 500 mg/24H MLD NTIS** AD-A075-205
eye-rbt 100 mg/24H MLD NTIS** AD-A075-205

SAFETY PROFILE: A skin and eye irritant. When heated to decomposition it emits toxic fumes of NO_x.

MIS750 CAS:17264-01-6 ***HR: 3***
(METHYL-3-CYCLOHEXENYL)METHANOL
mf: $C_8H_{14}O$ mw: 126.22

TOXICITY DATA with REFERENCE
orl-rat LD50:1410 μg/kg AIHAAP 30,470,69
skn-rbt LD50:620 mg/kg AIHAAP 30,470,69

SAFETY PROFILE: A deadly poison by ingestion. Moderately toxic by skin contact. When heated to decomposition it emits acrid smoke and irritating fumes.

MIT000 CAS:100-60-7 ***HR: 3***
N-METHYL CYCLOHEXYL AMINE
mf: $C_7H_{15}N$ mw: 113.2

$CH_3(C_6H_{11})NH$

SYN: CYCLOHEXYL METHYL AMINE

TOXICITY DATA with REFERENCE
eye-rbt 100 mg SEV 34ZIAG -,388,69
ihl-gpg LC50:7000 mg/m^3/1H 34ZIAG -,388,69
orl-rat LD50:400 mg/kg 34ZIAG -,388,69
skn-rbt LDLo:2 g/kg 34ZIAG -,388,69

CONSENSUS REPORTS: Reported in EPA TSCA Inventory.

SAFETY PROFILE: Poison by ingestion. Moderately toxic by skin contact. Mildly toxic by inhalation. A corrosive irritant to skin, eyes and mucous membranes. Contact can cause severe eye damage. Can cause ptosis, lacrimation, gasping, irregular respiration, nose-bleeding, prostration, convulsions. At a skin contact dose of 2 g/kg rabbits died in 24 hours and skin was badly burned. When heated to decomposition it emits toxic fumes of NO_x. See also AMINES.

MIT250 CAS:21209-02-9 ***HR: 3***
S-2-((4-(4-METHYLCYCLOHEXYL)BUTYL) AMINO)ETHYL THIOSULFATE
mf: $C_{13}H_{27}NO_3S_2$ mw: 309.53

TOXICITY DATA with REFERENCE
orl-mus LD50:900 mg/kg JMCMAR 11,1190,68
ipr-mus LD50:7 mg/kg JMCMAR 11,1190,68

SAFETY PROFILE: Poison by intraperitoneal route. Moderately toxic by ingestion. When heated to decomposition it emits very toxic fumes of NO_x and SO_x. See also THIOSULFATES.

MIT600 CAS:329-99-7 ***HR: 3***
METHYL CYCLOHEXYLFLUOROPHOSPHONATE
mf: $C_7H_{14}FO_2P$ mw: 180.18

SYNS: CMPF ◇ CYCLOHEXYL METHYLPHOSPHONOFLUORIDATE

TOXICITY DATA with REFERENCE
scu-rat LD50:225 μg/kg CJPPA3 44,745,66
scu-mus LD50:400 μg/kg SCJUAD 4,33,67
scu-rbt LD50:100 μg/kg SCJUAD 4,33,67
scu-gpg LD50:100 μg/kg CJPPA3 44,745,66
scu-ham LD50:130 μg/kg CJPPA3 46,109,68

SAFETY PROFILE: A deadly poison by subcutaneous route. When heated to decomposition it emits toxic fumes of PO_x and F^-.

MIT625 CAS:541-91-3 ***HR: 1***
3-METHYL-1-CYCLOPENTADECANONE
mf: $C_{16}H_{30}O$ mw: 238.46

SYNS: CYCLOPENTADECANONE, 3-METHYL- ◇ 3-METHYLCYCLOPENTADECANONE ◇ MOSCHUS KETONE ◇ MUSCONE ◇ MUSKONE

TOXICITY DATA with REFERENCE
skn-rbt 500 mg/24H MOD FCTOD7 20,749,82

CONSENSUS REPORTS: Reported in EPA TSCA Inventory.

SAFETY PROFILE: A skin irritant. When heated to decomposition it emits acrid smoke and irritating fumes.

MIU500 CAS:96-37-7 ***HR: 3***
METHYLCYCLOPENTANE
DOT: UN 2298
mf: C_6H_{12} mw: 84.18

PROP: Colorless liquid or solid. Mp: −142.5°, bp: 71.8°, flash p: <20°F, d: 0.750 @ 20°/4°, vap press: 100 mm @ 17.9°, vap d: 2.9. Insol in water; sol in ether.

TOXICITY DATA with REFERENCE
ihl-mus LCLo:9500 mg/m^3 AEPPAE 149,116,30

CONSENSUS REPORTS: Reported in EPA TSCA Inventory.

DOT Classification: Flammable Liquid; Label: Flammable Liquid.

SAFETY PROFILE: Mildly toxic by inhalation. Probably irritating and narcotic in high concentration. Very dangerous fire hazard when exposed to heat, flame, or oxidizers. Can react vigorously with oxidizing materials. To fight fire, use foam, CO_2, dry chemical. When heated to decomposition it emits acrid smoke and irritating fumes.

MIU750 CAS:3353-08-0 ***HR: 3***
17-METHYL-15H-CYCLOPENTA(a)PHENANTHRENE
mf: $C_{18}H_{14}$ mw: 230.32

TOXICITY DATA with REFERENCE
mma-sat 50 μg/plate CNREA8 36,4525,76
skn-mus TDLo:125 mg/kg/52W-I:ETA NATUAS 210,1281,66

CONSENSUS REPORTS: EPA Genetic Toxicology Program.

SAFETY PROFILE: Questionable carcinogen with experimental tumorigenic data. Mutation data reported. When heated to decomposition it emits acrid smoke and irritating fumes.

MIV250 CAS:63020-76-8 ***HR: 3***
10-METHYL-1,2-CYCLOPENTENOPHENANTHRENE
mf: $C_{18}H_{16}$ mw: 232.34

SYN: 16,17-DIHYDRO-7-METHYL-15H-CYCLOPENTA(a)PHENANTHRENE

TOXICITY DATA with REFERENCE
skn-mus TDLo:1200 mg/kg/37W-I:ETA ARGEAR 6,1,53

SAFETY PROFILE: Questionable carcinogen with ex-

perimental tumorigenic data. When heated to decomposition it emits acrid smoke and irritating fumes.

MIV500 CAS:135-07-9 ***HR: 3***
METHYLCYCLOTHIAZIDE
mf: $C_9H_{11}Cl_2N_3O_4S_2$ mw: 360.25

SYNS: AQUATENSEN ◇ ENDURON ◇ METHYLCHLOROTHIAZIDE ◇ METHYLCLOTHIAZIDE ◇ NSC-110431

TOXICITY DATA with REFERENCE
cyt-ham:fbr 250 mg/L ESKHA5 96,55,78
cyt-ham:lng 140 mg/L GMCRDC 27,95,81
ipr-rat LD50:2000 mg/kg 29ZVAB -,77,69
ipr-mus LD50:870 mg/kg 29ZVAB -,77,69
ivn-mus LD50:400 mg/kg 29ZVAB -,77,69

SAFETY PROFILE: Poison by intravenous route. Moderately toxic by intraperitoneal route. Mutation data reported. An FDA proprietary drug. When heated to decomposition it emits very toxic fumes of SO_x, NO_x, and Cl^-.

MIW000 CAS:19009-56-4 ***HR: 1***
2-METHYL-1-DECANAL
mf: $C_{11}H_{22}O$ mw: 170.33

SYN: METHYL OCTYL ACETALDEHYDE

TOXICITY DATA with REFERENCE
skn-rbt 500 mg/24H MOD FCTXAV 14,601,76

CONSENSUS REPORTS: Reported in EPA TSCA Inventory.

SAFETY PROFILE: A skin irritant. When heated to decomposition it emits acrid smoke and irritating fumes. See also ALDEHYDES.

MIW050 CAS:7011-83-8 ***HR: 1***
4-METHYLDECANOLIDE
mf: $C_{11}H_{20}O_2$ mw: 184.31

SYNS: DECANOIC ACID, 4-HYDROXY-4-METHYL-, Γ-LACTONE ◇ α-METHYL DECALACTONE

TOXICITY DATA with REFERENCE
skn-rbt 500 mg/24H MOD FCTXAV 17,867,79

CONSENSUS REPORTS: Reported in EPA TSCA Inventory.

SAFETY PROFILE: A skin irritant. When heated to decomposition it emits acrid smoke and irritating fumes.

MIW075 CAS:7289-52-3 ***HR: 1***
METHYL n-DECYL ETHER
mf: $C_{11}H_{24}O$ mw: 172.35

SYNS: DECANE, 1-METHOXY- ◇ DECYL METHYL ETHER ◇ ETHER, DECYL METHYL ◇ 1-METHOXYDECANE ◇ METHYL DECYL ETHER

TOXICITY DATA with REFERENCE
skn-rbt 500 mg/24H MOD FCTOD7 20,667,82

CONSENSUS REPORTS: Reported in EPA TSCA Inventory.

SAFETY PROFILE: A skin irritant. When heated to decomposition it emits acrid smoke and irritating fumes.

MIW100 CAS:8022-00-2 ***HR: 3***
METHYL DEMETON

PROP: An oily liquid. D: 1.20. Sltly sol in water.

SYNS: BAY 15203 ◇ BAYER 21/116 ◇ DEMETON METHYL ◇ DEMETHON-METHYL (MAK) ◇ DURATOX ◇ ENT 18,862 ◇ S(and O)-2-(ETHYLTHIO)ETHYL-O,O-DIMETHYLPHOSPHOROTHIOATE ◇ METASYSTOX ◇ METHYL-MERCAPTOPHOS ◇ METHYL SYSTOX

TOXICITY DATA with REFERENCE
orl-rat LD50:65 mg/kg 31ZOAD 1,125,68
skn-rat LD50:300 mg/kg WRPCA2 9,119,70
orl-mus LD50:46 mg/kg 85GMAT -,85,82
orl-cat LDLo:30 mg/kg 85GMAT -,85,82
ihl-cat LC50:20 mg/m^3/4H 85GMAT -,85,82
orl-rbt LDLo:150 mg/kg JEENAI 48,139,55

OSHA PEL: TWA 0.5 mg/m^3 (skin)
ACGIH TLV: TWA 0.5 mg/m^3 (skin)
DFG MAK: 0.5 ppm (5 mg/m^3)

SAFETY PROFILE: Poison by ingestion, skin contact, and inhalation routes. A cholinesterase inhibitor. An insecticide and acaricide. See also PARATHION and various demeton entries.

MIW250 CAS:2587-90-8 ***HR: 3***
METHYL DEMETON METHYL
mf: $C_5H_{13}O_3PS_2$ mw: 216.27

PROP: Pale yellow oil. Bp: 89° @ 0.15 mm, d: 1.207 @ 20°/4°. Sol in water at room temp, sol in organic solvents.

SYNS: CEBETOX ◇ CYMETOX ◇ DEMEPHION ◇ ISONITOX ◇ 2-(METHYLTHIO)-ETHANETHIOL-O,O-DIMETHYL PHOSPHOROTHIOATE ◇ 2-(METHYLTHIO)-ETHANETHIOL-S-ESTER with O,O-DIMETHYL PHOSPHOROTHIOATE ◇ TINOX

TOXICITY DATA with REFERENCE
dnr-omi:50 μL/plate BIZNAT 95,463,76
orl-rat LD50:20 mg/kg PESTD5 16,273,75
skn-rat LD50:68 mg/kg WRPCA2 9,119,70
orl-mus LD50:23 mg/kg PESTD5 16,273,75
orl-dog LD50:37 mg/kg PESTD5 16,273,75

CONSENSUS REPORTS: EPA Extremely Hazardous Substances List.

SAFETY PROFILE: Poison by ingestion and skin contact. Mutation data reported. *Caution:* It is a cholinesterase inhibitor. When heated to decomposition it emits very toxic fumes of PO_x and SO_x. See also PARATHION and various demeton entries.

MIW500 CAS:477-30-5 ***HR: 3***
N-METHYL-N-DESACETYLCOLCHICINE
mf: $C_{21}H_{25}NO_5$ mw: 371.47

SYNS: ALKALOID H 3, from COLCHICUM ANTUMNALE ◇ CIBA 12669A ◇ COLCEMIDE ◇ COLCHAMINE ◇ COLCHINE, N-DEACETYL-N-METHYL ◇ COLEMID ◇ DEACETYLMETHYLCOLCHICINE ◇ DEACETYL-N-METHYLCOLCHICINE ◇ N-DEACETYL-N-METHYLCOLCHICINE ◇ DEMECOLCINE ◇ N-DESACETYL-N-METHYLCOLCHICINE ◇ DESMECOLCINE ◇ 6,7-DIHYDRO-1,2,3,10-TETRAMETHOXY-7-(METHYLAMINO)-BENZO(α)HEPTALEN-9(5H)-ONE ◇ (S)-6,7-DIHYDRO-1,2,3,10-TETRAMETHOXY-7-(METHYLAMINO)-BENZO(a) HEPTALEN-9(5H)-ONE ◇ KOLCHAMIN ◇ METHYLCOLCHICINE ◇ N-METHYL-N-DEACETYLCOLCHICINE ◇ NSC 3096 ◇ OMAINE ◇ REICHSTEIN'S F ◇ SANTAVY'S SUBSTANCE F

TOXICITY DATA with REFERENCE
cyt-hmn:hla 100 nmol/L JJEMAG 40,409,70
cyt-hmn:oth 10 μg/L TSITAQ 12,382,70
sln-mus-ipr 37 mg/kg ENMUDM 6,422,84
otr-ham:emb 10 μg/L CRNGDP 5,89,84
scu-rat TDLo:3 mg/kg (7-9D preg):TER COREAF 247,152,58
ipr-mus TDLo:1800 μg/kg (female 8D post):REP AMZOAF 6,551,66
orl-hmn TDLo:200 μg/kg:SKN 34ZIAG -,184,69
ivn-rat LD50:1700 μg/kg ARZNAD 20,1467,70
par-rat LD50:1700 μg/kg RRCRBU 52,76,75
orl-mus LD50:25530 μg/kg NCISP* JAN86
ipr-mus LD50:35 mg/kg AEPPAE 230,559,57
ims-mus LD50:87 mg/kg JMCMAR 24,257,81

CONSENSUS REPORTS: EPA Genetic Toxicology Program.

SAFETY PROFILE: Poison by ingestion, intraperitoneal, parenteral, intravenous, and intramuscular routes. Human systemic effects by ingestion: (skin and appendages) hair effects. Human mutation data reported. An experimental teratogen. Experimental reproductive effects. When heated to decomposition it emits toxic fumes of NO_x.

MIW750 CAS:4619-66-3 ***HR: 2***
METHYL DIACETOACETATE
mf: $C_7H_{10}O_4$ mw: 158.17

PROP: Colorless liquid. Vap d: 5.45.

TOXICITY DATA with REFERENCE
skn-rbt 10 mg/24H open JIHTAB 30,63,48
eye-rbt 5 mg SEV AJOPAA 29,1363,46
orl-rat LD50:1700 mg/kg JIHTAB 30,63,48

SAFETY PROFILE: Moderately toxic by ingestion. A skin and severe eye irritant. When heated to decomposition it emits acrid smoke and irritating fumes. See also ESTERS.

MIX000 CAS:63991-70-8 ***HR: 3***
2-METHYLDIACETYLBENZIDINE
mf: $C_{17}H_{18}N_2O_2$ mw: 282.37

SYN: 2-METHYL-N,N'-DIACETYLBENZIDINE

TOXICITY DATA with REFERENCE
orl-rat TDLo:5600 mg/kg/35W-C:CAR CNREA8 16,525,56

SAFETY PROFILE: Questionable carcinogen with experimental carcinogenic data. When heated to decomposition it emits toxic fumes of NO_x.

MIX250 CAS:26981-93-1 ***HR: 3***
METHYLDIAZENE
mf: CH_4N_2 mw: 44.06

SYN: METHYL DIAZINE

SAFETY PROFILE: May explode on rapid heating (a mixture exploded when rapidly heated from −196°C to ambient temperature). Incompatible with oxygen. When heated to decomposition it emits toxic fumes of NO_x.

MIX500 CAS:765-31-1 ***HR: 3***
3-METHYLDIAZIRINE
mf: $C_2H_4N_2$ mw: 56.07

SAFETY PROFILE: The gas explodes when heated. When heated to decomposition it emits toxic fumes of NO_x.

MIX750 CAS:6832-16-2 ***HR: 3***
METHYL DIAZOACETATE
mf: $C_3H_4N_2O_2$ mw: 100.08

$CH_3OCO{\bullet}CHN_2$

SAFETY PROFILE: Explodes violently when heated. Upon decomposition it emits toxic fumes of NO_x.

MIY000 CAS:59652-21-0 ***HR: 3***
7-METHYLDIBENZ(c,h)ACRIDINE
mf: $C_{22}H_{15}N$ mw: 293.38

SYN: 9-METHYL-3,4,5,6-DIBENZACRIDINE

TOXICITY DATA with REFERENCE
scu-mus TDLo:72 mg/kg/9W-I:ETA COREAF 251,1322,60

SAFETY PROFILE: Questionable carcinogen with experimental tumorigenic data. When heated to decomposition it emits toxic fumes of NO_x.

MIY200 CAS:79543-29-6 ***HR: 2***
14-METHYLDIBENZ(a,h)ACRIDINE
mf: $C_{22}H_{15}N$ mw: 293.38

SYN: 10-METHYL-1,2:5,6-DIBENZACRIDINE

TOXICITY DATA with REFERENCE
scu-mus TDLo:48 mg/kg/4W-I:ETA COREAF 251,1322,60

SAFETY PROFILE: Questionable carcinogen with experimental tumorigenic data. When heated to decomposition it emits toxic fumes of NO_x.

MIY250 CAS:59652-20-9 ***HR: 3***
14-METHYLDIBENZ(a,j)ACRIDINE
mf: $C_{22}H_{15}N$ mw: 293.38

SYN: 10-METHYL-3,4,5,6-DIBENZACRIDINE

TOXICITY DATA with REFERENCE
skn-mus TDLo:280 mg/kg/23W-I:ETA ACRSAJ 4,315,56

SAFETY PROFILE: Questionable carcinogen with experimental tumorigenic data. When heated to decomposition it emits toxic fumes of NO_x.

MIY500 CAS:63041-83-8 ***HR: 3***
2-METHYLDIBENZ(a,h)ANTHRACENE
mf: $C_{23}H_{16}$ mw: 292.39

SYN: 2'-METHYL-1:2:5:6-DIBENZANTHRACENE

TOXICITY DATA with REFERENCE
skn-mus TDLo:1250 mg/kg/52W-I:ETA PRLBA4 111,485,32

SAFETY PROFILE: Questionable carcinogen with experimental tumorigenic data. When heated to decomposition it emits acrid smoke and irritating fumes.

MIY750 CAS:63041-84-9 ***HR: 3***
3-METHYLDIBENZ(a,h)ANTHRACENE
mf: $C_{23}H_{16}$ mw: 292.39

SYN: 3'-METHYL-1:2:5:6-DIBENZANTHRACENE

TOXICITY DATA with REFERENCE
skn-mus TDLo:552 mg/kg/23W-I:ETA PRLBA4 111,455,32

SAFETY PROFILE: Questionable carcinogen with experimental tumorigenic data. When heated to decomposition it emits acrid smoke and irritating fumes.

MIZ000 CAS:63041-85-0 ***HR: 3***
4-METHYL-1,2,5,6-DIBENZANTHRACENE
mf: $C_{23}H_{16}$ mw: 292.39

SYN: 6-METHYL DIBENZ(a,h)ANTHRACENE

TOXICITY DATA with REFERENCE
skn-mus TDLo:1010 mg/kg/42W-I:ETA PRLBA4 117,318,35

SAFETY PROFILE: Questionable carcinogen with experimental tumorigenic data. When heated to decomposition it emits acrid smoke and irritating fumes.

MJA000 CAS:17278-93-2 ***HR: 3***
10-METHYLDIBENZ(a,c)ANTHRACENE
mf: $C_{23}H_{16}$ mw: 292.39

TOXICITY DATA with REFERENCE
scu-mus TDLo:72 mg/kg/9W-I:ETA EJCAAH 4,123,68

SAFETY PROFILE: Questionable carcinogen with experimental tumorigenic data. When heated to decomposition it emits acrid smoke and irritating fumes.

MJA250 CAS:27093-62-5 ***HR: 3***
N-METHYL-3:4:5:6-DIBENZCARBAZOLE
mf: $C_{21}H_{15}N$ mw: 281.37

SYN: N-METHYL-7H-DIBENZO(c,g)CARBAZOLE

TOXICITY DATA with REFERENCE
skn-mus TDLo:1350 mg/kg/33W-C:ETA BJEPA5 27,179,46

SAFETY PROFILE: Questionable carcinogen with experimental tumorigenic data. When heated to decomposition it emits toxic fumes of NO_x.

MJA500 CAS:33942-87-9 ***HR: 3***
5-METHYL-DIBENZO(b,def)CHRYSENE
mf: $C_{25}H_{16}$ mw: 316.41

SYN: 5-METHYL-3,4:8,9-DIBENZOPYRENE (FRENCH)

TOXICITY DATA with REFERENCE
scu-mus TDLo:72 mg/kg/9W-I:ETA COREAF 246,1477,58

SAFETY PROFILE: Questionable carcinogen with experimental tumorigenic data. When heated to decomposition it emits acrid smoke and irritating fumes.

MJA750 CAS:2869-12-7 ***HR: 3***
7-METHYLDIBENZO(h,rst)PENTAPHENE
mf: $C_{29}H_{18}$ mw: 366.47

SYN: 2'-METHYL-1,2:4,5:8,9-TRIBENZOPYRENE

TOXICITY DATA with REFERENCE
scu-mus TDLo:72 mg/kg/9W-I:ETA COREAF 259,3899,64

SAFETY PROFILE: Questionable carcinogen with experimental tumorigenic data. When heated to decomposition it emits acrid smoke and irritating fumes.

MJB000 CAS:2869-60-5 ***HR: 3***
5-METHYL-1,2,3,4-DIBENZOPYRENE
mf: $C_{25}H_{16}$ mw: 316.41

SYN: 10-METHYLDIBENZO(def,p)CHRYSENE

TOXICITY DATA with REFERENCE
scu-mus TDLo:72 mg/kg/9W-I:ETA COREAF 259,3899,64

SAFETY PROFILE: Questionable carcinogen with ex-

perimental tumorigenic data. When heated to decomposition it emits acrid smoke and irritating fumes.

MJB250 CAS:63041-95-2 ***HR: 3***
7-METHYL-1:2:3:4-DIBENZPYRENE
mf: $C_{25}H_{16}$ mw: 316.41

TOXICITY DATA with REFERENCE
skn-mus TDLo:980 mg/kg/41W-I:ETA PRLBA4 123,343,37

SAFETY PROFILE: Questionable carcinogen with experimental tumorigenic data. When heated to decomposition it emits acrid smoke and irritating fumes.

MJB300 CAS:23777-55-1 ***HR: 3***
METHYLDIBORANE
mf: CH_8B_2 mw: 41.69

SAFETY PROFILE: Ignites spontaneously and then explodes in air. See also BORANES and BORON COMPOUNDS.

MJB500 CAS:3005-27-4 ***HR: 2***
N-METHYL-DIBROMOMALEINIMIDE
mf: $C_5H_3Br_2NO_2$ mw: 268.91

TOXICITY DATA with REFERENCE
ipr-mus TDLo:50 mg/kg (female 9D post):REP ARTODN 37,15,76
ipr-mus TDLo:12500 μg/kg (female 9D post):TER ARTODN 37,15,76
ipr-mus LD50:624 mg/kg ARTODN 37,15,76

SAFETY PROFILE: Moderately toxic by intraperitoneal route. An experimental teratogen. Experimental reproductive effects. When heated to decomposition it emits very toxic fumes of Br^- and NO_x.

MJC750 CAS:7560-83-0 ***HR: 2***
N-METHYLDICYCLOHEXYLAMINE
mf: $C_{13}H_{25}N$ mw: 195.39

TOXICITY DATA with REFERENCE
eye-rbt 100 mg MLD 34ZIAG -,388,69
orl-rat LD50:446 mg/kg 34ZIAG -,388,69
skn-rbt LDLo:2 g/kg 34ZIAG -,388,69

CONSENSUS REPORTS: Reported in EPA TSCA Inventory.

SAFETY PROFILE: Moderately toxic by ingestion and skin contact. An irritant to skin, eyes, and mucous membranes. When heated to decomposition it emits toxic fumes of NO_x. See also AMINES.

MJC775 CAS:4269-88-9 ***HR: 3***
N-METHYL-N-(β-DICYCLOHEXYLAMINOETHYL)PIPERIDINE BROMIDE
mf: $C_{20}H_{39}N_2\cdot Br$ mw: 387.52

SYNS: 1-(2-(DICYCLOHEXYLAMINO)ETHYL)-1-METHYLPIPERIDINIUM BROMIDE ◇ N-METIL-N-(β-DICICLOESILAMINOETIL)PIPERIDINIO BROMURO (ITALIAN)

TOXICITY DATA with REFERENCE
orl-rat LD50:650 mg/kg FRPSAX 15,821,60
ivn-rat LD50:30 mg/kg FRPSAX 15,821,60
ims-rat LD50:100 mg/kg FRPSAX 15,821,60
ipr-rat LD50:120 mg/kg FRPSAX 15,821,60
orl-mus LD50:1008 mg/kg FRPSAX 16,773,61
ipr-mus LD50:120 mg/kg FRPSAX 18,3,63
scu-mus LD50:160 mg/kg FRPSAX 15,821,60
ipr-gpg LD50:100 mg/kg FRPSAX 15,821,60
ims-gpg LD50:100 mg/kg FRPSAX 15,821,60

SAFETY PROFILE: Poison by subcutaneous, intramuscular, intravenous, and intraperitoneal routes. Moderately toxic by ingestion. When heated to decomposition it emits toxic fumes of NO_x and Br^-. See also BROMIDES.

MJD000 CAS:63041-05-4 ***HR: 3***
METHYL DIEPOXYDIALLYLACETATE
mf: $C_9H_{14}O_4$ mw: 186.23

SYN: 4,5-EPOXY-2-(2,3-EPOXYPROPYL)VALERIC ACID, METHYL ESTER

TOXICITY DATA with REFERENCE
unr-mus TDLo:3700 mg/kg:ETA RARSAM 3,193,63

SAFETY PROFILE: Questionable carcinogen with experimental tumorigenic data. When heated to decomposition it emits acrid smoke and irritating fumes. See also ESTERS.

MJD275 CAS:676-99-3 ***HR: 3***
METHYL DIFLUOROPHOSPHITE
mf: CH_3F_2OP mw: 100.01

SYNS: DIFLUORO ◇ DIFLUOROMETHYLPHOSPHINE OXIDE ◇ METHYLPHOSPHONIC DIFLUORIDE ◇ METHYLPHOSPHONYLDIFLUORIDE

TOXICITY DATA with REFERENCE
ihl-rat LCLo:1842 mg/m³/30M TXAPA9 15,131,69
ivn-rat LD50:13700 μg/kg IAEC** 17JUN74
ihl-mus LD50:114 mg/kg TXAPA9 15,131,69
ivn-mus LD50:114 mg/kg IAEC** 17JUN74
ihl-dog LCLo:1842 mg/m³/30M TXAPA9 15,131,69
ivn-dog LD50:25800 μg/kg IAEC** 17JUN74
ihl-mky LCLo:2608 mg/m³/30M TXAPA9 15,131,69
ivn-mky LD50:26900 μg/kg IAEC** 17JUN74
ihl-gpg LCLo:1600 mg/m³ TOXID9 4,22,84

CONSENSUS REPORTS: Reported in EPA TSCA Inventory.

SAFETY PROFILE: Poison by inhalation and intravenous routes. When heated to decomposition it emits toxic fumes of PO_x and F^-. See also FLUORIDES.

MJD300 CAS:30685-43-9 ***HR: 3***
β-METHYLDIGOXIN
mf: $C_{42}H_{66}O_{14}$ mw: 795.08

SYNS: BETA METHYL DIGOXIN ◇ LANIRAPID ◇ LANITOP ◇ MEDIGOXIN ◇ METHYLDIGOXIN ◇ 4'''-METHYLDIGOXIN ◇ 4'''-β-METHYLDIGOXIN ◇ 4'''-o-METHYLDIGOXIN ◇ METILDIGOXIN ◇ METILDIGOXINA (SPANISH) ◇ 3β,12β,14β-TRIHYDROXY-5-β-CARD-20(22)-ENOLIDE-3-(4')-o-METHYL-TRIDIGITOXOSIDE)

TOXICITY DATA with REFERENCE
orl-rbt TDLo:130 mg/kg (female 6-18D post):TER KSRNAM 10,405,76
orl-rbt TDLo:130 mg/kg (female 6-18D post):REP KSRNAM 10,405,76
ivn-man LDLo:160 μg/kg ARZNAD 34,769,84
orl-rat LD50:8300 μg/kg ARZNAD 21,231,71
ipr-rat LD50:6200 μg/kg ARZNAD 21,231,71
scu-rat LD50:5930 μg/kg YKYUA6 31,1127,80
ivn-rat LD50:4800 μg/kg ARZNAD 21,231,71
orl-mus LD50:7800 μg/kg ARZNAD 21,231,71
ipr-mus LD50:4800 μg/kg ARZNAD 21,231,71
scu-mus LD50:9390 μg/kg NIIRDN 6,828,82
ivn-mus LD50:4900 μg/kg ARZNAD 21,231,71
ivn-cat LDLo:190 μg/kg ARZNAD 28,495,78
orl-gpg LD50:2100 μg/kg OYYAA2 7,373,73
ipr-gpg LD50:800 μg/kg OYYAA2 7,373,73
ivn-gpg LDLo:653 μg/kg ARZNAD 28,495,78

SAFETY PROFILE: A deadly poison by ingestion, subcutaneous, intravenous, and intraperitoneal routes. An experimental teratogen. Experimental reproductive effects. A cardiotonic agent. When heated to decomposition it emits acrid smoke and irritating fumes.

MJD500 ***HR: 3***
4'''-o-METHYLDIGOXIN ACETONE (2:1)
mf: $C_{42}H_{66}O_{14}$•$1/2C_3H_6O$ mw: 824.12

SYNS: METILDIGOXIN ◇ 3-β,12-β,14-β-TRIHYDROXY-5-β-CARD-20(22)-ENOLIDE-3-(R'''-O-METHYLTRIDIGITOXOSIDE), ACETONE (2:1)

TOXICITY DATA with REFERENCE
orl-rat LD50:13060 μg/kg IYKEDH 10,710,79
scu-rat LD50:5930 μg/kg IYKEDH 10,710,79
ivn-rat LD50:5450 μg/kg IYKEDH 10,710,79
orl-mus LD50:11100 μg/kg YKYUA6 30,1393,79
ivn-mus LD50:5780 μg/kg TOIZAG 23,198,76
scu-mus LD50:9390 μg/kg TOIZAG 23,198,76
orl-gpg LD50:1800 μg/kg IYKEDH 10,710,79
ipr-gpg LD50:650 μg/kg IYKEDH 10,710,79

SAFETY PROFILE: A deadly poison by ingestion, subcutaneous, intravenous, and intraperitoneal routes. When heated to decomposition it emits acrid smoke and irritating fumes.

MJD600 CAS:68688-86-8 ***HR: D***
trans-3-METHYL-7,8-DIHYDROCHOLANTHRENE-7,8-DIOL
mf: $C_{21}H_{18}O_2$ mw: 302.39

SYN: trans-7,8-DIHYDRO-7,8-DIHYDROXY-3-METHYLCHOLANTHRENE

TOXICITY DATA with REFERENCE
mma-sat 10 μmol/L BBRCA9 85,1568,78
sce-ham:ovr 500 μg/L CALEDQ 7,45,79

SAFETY PROFILE: Mutation data reported. When heated to decomposition it emits acrid smoke and irritating fumes.

MJD610 CAS:68688-87-9 ***HR: 2***
trans-3-METHYL-9,10-DIHYDROCHOLANTHRENE-9,10-DIOL
mf: $C_{21}H_{18}O_2$ mw: 302.39

SYN: trans-9,10-DIHYDRO-9,10-DIHYDROXY-3-METHYLCHOLANTHRENE

TOXICITY DATA with REFERENCE
mma-sat 10 μmol/L BBRCA9 85,1568,78
otr-mus:fbr 250 μg/L BBRCA9 85,1568,78
sce-ham:ovr 500 μg/L CALEDQ 7,45,79
msc-ham:lng 2 mg/L BBRCA9 85,1568,78
skn-mus TDLo:378 ng/kg:ETA CALEDQ 28,223,85

SAFETY PROFILE: Questionable carcinogen with experimental tumorigenic data. Mutation data reported. When heated to decomposition it emits acrid smoke and irritating fumes.

MJD750 CAS:24684-41-1 ***HR: 3***
11-METHYL-15,16-DIHYDRO-17H-CYCLOPENTA(a)PHENANTHRENE
mf: $C_{18}H_{16}$ mw: 232.34

PROP: Crystals from acetic acid. Mp: 126-127°.

SYN: 16,17-DIHYDRO-11-METHYL-15H-CYCLOPENTA(a)PHENANTHRENE

TOXICITY DATA with REFERENCE
mma-sat 20 ng/plate CNREA8 36,4525,76
skn-mus TDLo:1440 mg/kg/45W-I:ETA ARGEAR 6,1,53

SAFETY PROFILE: Questionable carcinogen with experimental tumorigenic data. Mutation data reported. When heated to decomposition it emits acrid smoke and irritating fumes.

MJE250 CAS:37795-71-4 ***HR: 3***
3-METHYL-2,3-DIHYDRO-9H-ISOXAZOLO(3,2-b) QUINAZOLIN-9-ONE
mf: $C_{11}H_{10}N_2O_2$ mw: 202.23

SYN: W 2451

TOXICITY DATA with REFERENCE
orl-rat LD50:765 mg/kg DRFUD4 2,553,77
ipr-rat LD50:670 mg/kg DRFUD4 2,553,77
ivn-rat LD50:290 mg/kg DRFUD4 2,553,77
orl-mus LD50:860 mg/kg DRFUD4 2,553,77
ipr-mus LD50:66 mg/kg DRFUD4 2,553,77
ivn-mus LD50:243 mg/kg DRFUD4 2,553,77

CONSENSUS REPORTS: Reported in EPA TSCA Inventory.

SAFETY PROFILE: Poison by intravenous and intraperitoneal routes. Moderately toxic by ingestion. When heated to decomposition it emits toxic fumes of NO_x.

MJE500 CAS:892-17-1 ***HR: 3***
11-METHYL-15,16-DIHYDRO-17-OXOCYCLOPENTA(a)PHENANTHRENE
mf: $C_{18}H_{14}O$ mw: 246.32

SYNS: 15,16-DIHYDRO-11-METHYLCYCLOPENTA(a)PHENANTHREN-17-ONE ◇ 15,16-DIHYDRO-11-METHYL-17H-CYCLOPENTA(a) PHENANTHREN-17-ONE ◇ 11-METHYL-15,16-DIHYDRO-17H-CYCLOPENTA(a)PHENANTHREN-17-ONE

TOXICITY DATA with REFERENCE
mma-sat 1 μg/plate CNREA8 40,882,80
dnd-mus-ims 120 mg/kg CNREA8 41,4115,81
dnd-mus:oth 10 μg/L CNREA8 43,2261,83
orl-rat TDLo:150 mg/kg:CAR BJCAAI 40,914,79
skn-mus TDLo:16 mg/kg:CAR BJCAAI 40,914,79
skn-mus TD:125 mg/kg/52W-I:ETA NATUAS 210,1281,66
skn-mus TD:46 mg/kg/19W-I:NEO CNREA8 33,832,73

CONSENSUS REPORTS: EPA Genetic Toxicology Program.

SAFETY PROFILE: Suspected carcinogen with experimental carcinogenic, neoplastigenic, and tumorigenic data. Mutation data reported. When heated to decomposition it emits acrid smoke and irritating fumes.

MJE750 CAS:27156-32-7 ***HR: 2***
METHYLDIHYDROPYRAN
mf: $C_6H_{10}O$ mw: 98.16

SYN: 3,4-DIHYDROMETHYL-2H-PYRAN

TOXICITY DATA with REFERENCE
orl-mus LD50:1950 mg/kg GTPZAB 15,49,71
ihl-mus LC50:6500 mg/m^3/2H 85GMAT -,82,82
scu-mus LD50:1250 mg/kg GTPZAB 15,49,71

SAFETY PROFILE: Moderately toxic by ingestion and subcutaneous routes. Mildly toxic by inhalation. When heated to decomposition it emits acrid smoke and irritating fumes.

MJE760 CAS:521-11-9 ***HR: 2***
17-α-METHYLDIHYDROTESTOSTERONE
mf: $C_{20}H_{32}O_2$ mw: 304.52

PROP: Crystals. Mp: 192-193°. Sol in water, acetone, alc, ether, and ethyl acetate.

SYNS: ANDROSTALONE ◇ 5-α-ANDROSTANE-17-α-METHYL-17-β-OL-3-ONE ◇ ANDROSTAN-3-ONE, 17-HYDROXY-17-METHYL-, (5-α-17-β)-(9CI) ◇ ASSIMIL ◇ ERMALONE ◇ 17-β-HYDROXY-17-METHYL-5-α-ANDROSTAN-3-ONE ◇ MESANOLON ◇ MESTALONE ◇ MESTANOLONE ◇ METHYBOL ◇ METHYLANTALON ◇ METHYLDIHYDROTESTOSTERONE ◇ PREROIDE ◇ TANTARONE

TOXICITY DATA with REFERENCE
scu-rat TDLo:160 mg/kg (female 14-21D post):TER ENDOAO 94,979,74
ipr-mus LD50:1240 mg/kg DRUGAY 6,817,82

SAFETY PROFILE: Moderately toxic by intraperitoneal route. An experimental teratogen. When heated to decomposition it emits acrid smoke and irritating fumes.

MJE775 CAS:59122-46-2 ***HR: 3***
METHYL (±)-11-α-16-DIHYDROXY-16-METHYL-9-OXOPROST-13-EN-1-OATE
mf: $C_{22}H_{38}O_5$ mw: 382.60

SYNS: (11-α-13E)-(±)-11,16-DIHYDROXY-16-METHYL-9-OXOPROST-13-EN-1-OIC ACID METHYL ESTER ◇ MISOPROSTOL ◇ PROST-13-EN-1-OIC ACID, 11,16-DIHYDROXY-16-METHYL-9-OXO-, METHYL ESTER, (11-α-13E)-(±)- ◇ SC-29333

TOXICITY DATA with REFERENCE
orl-rat TDLo:320 mg/kg (female 17-22D post):REP JZKEDZ 11,213,85
orl-rbt TDLo:1320 mg/kg (female 6-18D post):TER JZKEDZ 11,237,85
orl-rat LD50:81 mg/kg JZKEDZ 11,33,85
ipr-rat LD50:40 mg/kg DDSCDJ 30,142S,85
ims-rat LD50:19 mg/kg JZKEDZ 11,33,85
orl-mus LD50:27 mg/kg DDSCDJ 30,142S,85
ipr-mus LD50:70 mg/kg DDSCDJ 30,142S,85
ims-mus LD50:16 mg/kg JZKEDZ 11,33,85

SAFETY PROFILE: Poison by ingestion, intramuscular, and intraperitoneal routes. An experimental teratogen. Experimental reproductive effects. When heated to decomposition it emits acrid smoke and irritating fumes.

MJE780 CAS:41372-08-1 ***HR: 2***
α-METHYL-l-3,4-DIHYDROXYPHENYLALANINE
mf: $C_{10}H_{13}NO_4 \cdot 3/2H_2O$ mw: 238.27

SYNS: ALDOMET ◇ ALDOMETIL ◇ ALDOMIN ◇ AMD ◇ BAYER 1440 L ◇ BAYPRESOL ◇ DOPAMET ◇ DOPATEC ◇ DOPEGYT

◇ HYPERPAX ◇ METHYL DOPA SESQUIHYDRATE ◇ MEDOMET ◇ MEDOPREN ◇ METHOPLAIN ◇ MK.B51 ◇ MK-351 ◇ α-MEDOPA ◇ PRESINOL ◇ PRESOLISIN ◇ SEDOMETIL ◇ SEMBRINA ◇ l-TYROSINE, 3-HYDROXY-α-METHYL-, SESQUIHYDRATE

TOXICITY DATA with REFERENCE
orl-rat TDLo:2980 mg/kg (male 65D pre):REP NTPTR* NTP-TR-348,89
orl-mus TDLo:604 g/kg/2Y-C:ETA NTPTR* NTP-TR-348,89

CONSENSUS REPORTS: NTP Carcinogenesis Studies (feed): Equivocal Evidence, mouse NTPTR* NTP-TR-348,89; NTP Carcinogenesis Studies (feed): No Evidence, rat NTPTR* NTP-TR-348,89

SAFETY PROFILE: Experimental reproductive effects. Questionable carcinogen with experimental tumorigenic data. When heated to decomposition it emits toxic fumes of NO_x.

MJE800 ***HR: 3***
1-METHYL-2,6-DI-(p-METHOXYPHENETHYL)PIPERIDINE ETHANESULFONATE
mf: $C_{24}H_{33}NO_2 \cdot C_2H_5O_3S$ mw: 476.71

SYN: 2,6-BIS(p-METHOXYPHENETHYL)-1-METHYLPIPERIDINE ETHANESULFONATE

TOXICITY DATA with REFERENCE
scu-rat LD50:975 mg/kg JPETAB 87,73,46
scu-mus LD50:450 mg/kg JPETAB 87,73,46
ivn-mus LD50:17 mg/kg JPETAB 87,73,46
orl-rbt LD50:112 mg/kg JPETAB 87,73,46
scu-rbt LD50:17500 μg/kg JPETAB 87,73,46
ivn-rbt LD50:5 mg/kg JPETAB 87,73,46

SAFETY PROFILE: Poison by ingestion, subcutaneous, and intravenous routes. When heated to decomposition it emits toxic fumes of NO_x and SO_x.

MJF000 CAS:3732-90-9 ***HR: 3***
3-METHYL-4-DIMETHYLAMINOAZOBENZENE
mf: $C_{15}H_{17}N_3$ mw: 239.35

SYNS: N,N-DIMETHYL-4-(PHENYLAZO)-o-TOLUIDINE ◇ 3-METHYL-4-DAB

TOXICITY DATA with REFERENCE
dni-rat:lvr 10 μmol/L CNREA8 45,337,85
orl-rat TDLo:749 mg/kg/2Y-C:CAR TOPADD 13,257,85
orl-rat TDLo:3024 mg/kg/12W-C:ETA VAAZA2 46,21,84

SAFETY PROFILE: Questionable carcinogen with experimental carcinogenic and tumorigenic data. Mutation data reported. When heated to decomposition it emits toxic fumes of NO_x.

MJF250 CAS:78128-83-3 ***HR: 3***
3-METHYL-4-(N-(2-DIMETHYLAMINOETHYL)-N-PHENYLAMINO)-2(5H) FURANONE
mf: $C_{15}H_{20}N_2O_2$ mw: 260.37

SYN: α-METHYL-β-(N-(2-DIMETHYLAMINOETHYL)-NPHENYL) AMINO-Δ^{α},βBUTENOLID

TOXICITY DATA with REFERENCE
scu-mus LD50:50 mg/kg ARZNAD 11,277,61
ivn-mus LD50:28 mg/kg ARZNAD 11,277,61

SAFETY PROFILE: Poison by subcutaneous and intravenous routes. When heated to decomposition it emits toxic fumes of NO_x.

MJF500 CAS:17400-69-0 ***HR: 3***
3'-METHYL-5'-(p-DIMETHYLAMINOPHENYLAZO)QUINOLINE
mf: $C_{18}H_{18}N_4$ mw: 290.40

SYNS: 5-((p-(DIMETHYLAMINO)PHENYL)AZO)-3-METHYLQUINOLINE ◇ N,N-DIMETHYL-4-(5-(3'-METHYLQUINOLYL)AZO)ANILINE

TOXICITY DATA with REFERENCE
orl-rat TDLo:3276 mg/kg/26W-C:CAR JNCIAM 40,891,68

SAFETY PROFILE: Questionable carcinogen with experimental carcinogenic data. When heated to decomposition it emits toxic fumes of NO_x.

MJF750 CAS:17400-70-3 ***HR: 3***
6'-METHYL-5'-(p-DIMETHYLAMINOPHENYLAZO)QUINOLINE
mf: $C_{18}H_{18}N_4$ mw: 290.40

SYNS: 5-((p-(DIMETHYLAMINO)PHENYL)AZO)-6-METHYLQUINOLINE ◇ N,N-DIMETHYL-4-(5'-(6'-METHYLQUINOLYL)AZO)ANILINE

TOXICITY DATA with REFERENCE
orl-rat TDLo:4914 mg/kg/39W-C:ETA JNCIAM 40,891,68

SAFETY PROFILE: Questionable carcinogen with experimental tumorigenic data. When heated to decomposition it emits toxic fumes of NO_x.

MJG000 CAS:17416-20-5 ***HR: 3***
8'-METHYL-5'-(p-DIMETHYLAMINOPHENYLAZO)QUINOLINE
mf: $C_{18}H_{18}N_4$ mw: 290.40

SYNS: 5-((p-(DIMETHYLAMINO)PHENYL)AZO)-8-METHYLQUINOLINE ◇ N,N-DIMETHYL-4-(5'-(8'-METHYLQUINOLYL)AZO)ANILINE

TOXICITY DATA with REFERENCE
orl-rat TDLo:378 mg/kg/9W-C:CAR JNCIAM 40,891,68

SAFETY PROFILE: Questionable carcinogen with experimental carcinogenic data. When heated to decomposition it emits toxic fumes of NO_x.

MJG500 CAS:2275-23-2 ***HR: 3***
N-METHYL-O,O-DIMETHYLTHIOLOPHOSPHORYL-5-THIA-3-METHYL-2-VALERAMIDE
mf: $C_8H_{18}NO_4PS_2$ mw: 287.36

SYNS: AMERICAN CYANAMID-43073 ◇ O,O-DIMETHYL-S-2-(1-N-METHYLCARBAMOYLETHYLMERCAPTO)ETHYLTHIOPHOSPHATE ◇ O,O-DIMETHYL-S-(2-(1-METHYLCARBAMOYLETHYLTHIO)ETHYL) PHOSPHOROTHIOATE ◇ DIMETHYL-S-(2-(1-METHYLCARBAMOYLETHYLTHIO ETHYL) PHOSPHOROTHIOLATE ◇ ENT 26,613 ◇ KILVAL ◇ N-METHYL-3-THIA-2-METHYL-VALERAMID DER O,O-DIMETHYLTHIOLPHOSPHORSAEURE (GERMAN) ◇ NPH 83 ◇ TRUCIDOR ◇ VAMIDOATE ◇ VAMIDOTHION

TOXICITY DATA with REFERENCE
mma-sat 25 pph MUREAV 40,19,76
mma-esc 25 pph MUREAV 40,19,76
orl-rat LD50:64 mg/kg 28ZEAL 5,233,76
orl-mus LD50:40 mg/kg MZUZA8 8,65,72
skn-mus LD50:1500 mg/kg GUCHAZ 6,530,73
orl-dog LD50:110 mg/kg GUCHAZ 6,530,73
skn-rbt LD50:160 mg/kg WRPCA2 9,119,70
orl-gpg LD50:85 mg/kg GUCHAZ 6,530,73

CONSENSUS REPORTS: EPA Genetic Toxicology Program.

SAFETY PROFILE: Poison by ingestion and skin contact. Mutation data reported. When heated to decomposition it emits very toxic fumes of PO_x, SO_x, and NO_x.

MJG750 CAS:99-80-9 ***HR: 3***
N-METHYL-N,p-DINITROSOANILINE
mf: $C_7H_7N_3O_2$ mw: 165.17.

PROP: Mp: 101°. Contains 30% N-methyl-N,p-dinitrosoaniline (JNCIAM 41,985,68).

SYNS: N,4-DINITROSO-N-METHYLANILINE ◇ ELASTOPAR ◇ ELASTOPAX ◇ HEAT PRE ◇ N-METHYL-N,4-DINITROSOANILINE ◇ N-METHYL-N,4-DINITROSOBENZENAMINE ◇ METHYL-(4-NITROSOPHENYL)NITROSAMINE ◇ N-NITROSO-N-METHYL-4-NITROSO-ANILINE ◇ NITROZAN K

TOXICITY DATA with REFERENCE
mma-sat 5 μg/plate PCBRD2 141,407,84
dni-mus-ipr 20 g/kg ARGEAR 51,605,81
ipr-rat TDLo:520 mg/kg/26W-I:ETA EJCAAH 4,233,68

CONSENSUS REPORTS: IARC Cancer Review: Group 3 IMEMDT 7,56,87; Animal Inadequate Evidence IMEMDT 1,141,72.

SAFETY PROFILE: Questionable carcinogen with experimental tumorigenic data. Mutation data reported. When heated to decomposition it emits toxic fumes of NO_x. See also N-METHYL-N,p-DINITROSO-ANILINE and N-NITROSO COMPOUNDS.

MJH000 CAS:55556-94-0 ***HR: 3***
2-METHYLDINITROSOPIPERAZINE
mf: $C_5H_{11}N_4O_2$ mw: 159.20

SYN: 2-METHYL-DNPZ

TOXICITY DATA with REFERENCE
mma-sat 50 μg/plate TCMUE9 1,13,84
mma-smc 50 μmol/plate MUREAV 77,143,80
orl-rat TDLo:1650 mg/kg/33W-I:ETA CNREA8 35,1270,75

CONSENSUS REPORTS: EPA Genetic Toxicology Program.

SAFETY PROFILE: Questionable carcinogen with experimental tumorigenic data. Mutation data reported. When heated to decomposition it emits toxic fumes of NO_x. See also N-NITROSO COMPOUNDS.

MJH250 CAS:4386-79-2 ***HR: 3***
((2-METHYL-1,3-DIOXALAN-4-YL)METHYL) TRIMETHYLAMMONIUM IODIDE
mf: $C_8H_{18}NO_2 \cdot I$ mw: 287.17

SYNS: ETHYL-Γ-TRIMETHYLAMMONIUM PROPANEDIOL IODIDE ◇ FOURNEAU 2268 ◇ N,N,N,2-TETRAMETHYL-1,3-DIOXOLANE-4-METHANAMINIUM IODIDE (9CI) ◇ TRIMETHYL((2-METHYL-1,3-DIOXOLAN-4-YL)METHYL)AMMONIUM IODIDE (8CI)

TOXICITY DATA with REFERENCE
orl-rat LDLo:50 mg/kg JPETAB 93,287,48
scu-rat LD50:3 mg/kg BSCIA3 26,516,44
orl-mus LD50:5 mg/kg BSCIA3 26,516,44
scu-mus LD50:1200 μg/kg BSCIA3 26,516,44
ivn-mus LD50:100 μg/kg CSLNX* NX#01810
ivn-dog LDLo:1 μg/kg BSCIA3 26,516,44
scu-rbt LDLo:20 μg/kg BSCIA3 26,516,44
scu-gpg LDLo:75 μg/kg BSCIA3 26,516,44

SAFETY PROFILE: Poison by ingestion, intravenous, and subcutaneous routes. When heated to decomposition it emits very toxic fumes of NH_3, NO_x, and I^-. See also IODIDES.

MJH500 CAS:4722-68-3 ***HR: 3***
2-METHYL-1,4-DIOXASPIRO(4.5)DECANE
mf: $C_9H_{16}O_2$ mw: 156.25

TOXICITY DATA with REFERENCE
ipr-mus LDLo:250 mg/kg CBCCT* 8,741,56

CONSENSUS REPORTS: Reported in EPA TSCA Inventory.

SAFETY PROFILE: Poison by intraperitoneal route. When heated to decomposition it emits acrid smoke and irritating fumes.

MJH750 CAS:1331-09-5 ***HR: 2***
METHYL DIOXOLANE
mf: $C_5H_{10}O_2$ mw: 102.15

TOXICITY DATA with REFERENCE
eye-rbt 5 mg SEV AJOPAA 29,1363,46
orl-rat LD50:3000 mg/kg JIHTAB 31,60,49

SAFETY PROFILE: Moderately toxic by ingestion. A severe eye irritant. When heated to decomposition it emits acrid smoke and irritating fumes.

MJH775 CAS:497-26-7 ***HR: 2***
2-METHYL-1,3-DIOXOLANE
mf: $C_4H_8O_2$ mw: 88.12

SYN: METHYLDIOXOLANE

TOXICITY DATA with REFERENCE
orl-rat LD50:2900 mg/kg GISAAA 47(8),89,82
ihl-rat LC50:80 g/m^3 GISAAA 39(11),94,74
orl-mus LD50:3500 mg/kg GISAAA 39(11),94,74
ihl-mus LC50:59 g/m^3 GISAAA 39(11),94,74

CONSENSUS REPORTS: Reported in EPA TSCA Inventory.

SAFETY PROFILE: Moderately toxic by ingestion. Mildly toxic by inhalation. When heated to decomposition it emits acrid smoke and irritating fumes.

MJH800 ***HR: 3***
((2-METHYL-1,3-DIOXOLAN-4-YL)METHYL) TRIMETHYLAMMONIUM CHLORIDE
mf: $C_8H_{18}NO_2{\bullet}Cl$ mw: 195.72

SYN: CHLORURE de l'ETHYLAL TRIMETHYLAMMONIUM PROPANEDIOL (FRENCH)

TOXICITY DATA with REFERENCE
ivn-mus LD50:220 μg/kg AIPTAK 98,399,54
ihl-dog LCLo:10 mg/m^3/30M AIPTAK 98,399,54
ihl-gpg LCLo:10 mg/m^3/30M AIPTAK 98,399,54
scu-gpg LDLo:50 μg/kg AIPTAK 98,399,54

SAFETY PROFILE: A deadly poison by inhalation, subcutaneous and intravenous routes. When heated to decomposition it emits toxic fumes of Cl^-, NH_3, and NO_x. See also CHLORIDES.

MJH900 CAS:83-98-7 ***HR: 3***
o-METHYLDIPHENHYDRAMINE
mf: $C_{18}H_{23}NO$ mw: 269.42

PROP: Liquid. Bp: 195°.

SYNS: BIORPHEN ◇ BROCADISIPAL ◇ BROCASIPAL ◇ DISIPAL ◇ 2-METHYLDIPHENHYDRAMINE ◇ ORPHENADINE ◇ ORPHENADRIN ◇ ORPHENADRINE ◇ WS 2434

TOXICITY DATA with REFERENCE
orl-hmn TDLo:14 mg/kg:CNS,CVS APTSAI 41(2),137,77
orl-hmn TDLo:7143 mg/kg:CNS,CVS ATSUDG 4,425,80
orl-mus LD50:125 mg/kg AIPTAK 138,62,62
ipr-mus LD50:80 mg/kg AIPTAK 138,62,62
scu-mus LD50:150 mg/kg AIPTAK 138,62,62
ivn-mus LD50:33 mg/kg AIPTAK 138,62,62

SAFETY PROFILE: Poison by ingestion, subcutaneous, intravenous, and intraperitoneal routes. Human systemic effects by ingestion: convulsions, coma, arrhythmias, pulse rate increase and blood pressure elevation. When heated to decomposition it emits toxic fumes of NO_x. See also AMINES.

MJJ000 CAS:908-35-0 ***HR: 3***
2-METHYL-1,2-DI-3-PYRIDYL-1-PROPANONE TARTRATE (1:2)
mf: $C_{14}H_{14}N_2O{\bullet}2C_4H_6O_6$ mw: 526.50

SYNS: 2-METHYL-1,2-DIPYRIDYL-(3'-1-OXOPROPANE DITARTRATE ◇ METOPIRON ◇ METOPIRONE DITARTRATE ◇ METYPRAPONE BITARTRATE ◇ METYRAPONE DITARTRATE ◇ SU 4885 DITARTRATE

TOXICITY DATA with REFERENCE
scu-rbt TDLo:200 mg/kg (30-31D preg):REP BNEOBV 27,1,75
ipr-mus LD50:760 mg/kg TXCYAC 17,73,80
ivn-mus LD50:261 mg/kg TXAPA9 23,537,72

SAFETY PROFILE: Poison by intravenous route. Moderately toxic by intraperitoneal route. Experimental reproductive effects. When heated to decomposition it emits toxic fumes of NO_x.

MJJ250 CAS:34419-05-1 ***HR: 3***
N-METHYL-3,6-DITHIA-3,4,5,6-TETRAHYDROPHTHALIMIDE
mf: $C_7H_7NO_2S_2$ mw: 201.27

SYN: 6-METHYL-2,3,5,7-TETRAHYDRO-6H-p-DITHIINO-(2,3-C)PYRROLE-5,7-DIONE

TOXICITY DATA with REFERENCE
ipr-mus TDLo:6250 μg/kg (9D preg):REP MOPMA3 13,133,77
ipr-mus TDLo:6250 μg/kg (9D preg):TER MOPMA3 13,133,77
ipr-mus LD50:210 mg/kg DIPHAH 23,113,71

SAFETY PROFILE: Poison by intraperitoneal route. An experimental teratogen. Experimental reproductive effects. When heated to decomposition it emits very toxic fumes of NO_x and SO_x.

MJJ500 CAS:64037-51-0 ***HR: 3***
1-(1-METHYL-3,3-DI-2-THIENYL-2-PROPENYL) PIPERIDINE HYDROCHLORIDE
mf: $C_{17}H_{15}NS_2{\bullet}ClH$ mw: 333.91

TOXICITY DATA with REFERENCE
scu-rat LD50:95 mg/kg BJPCAL 8,2,53
orl-mus LD50:190 mg/kg JPETAB 107,385,53
scu-mus LD50:119 mg/kg JPETAB 107,385,53
ivn-mus LD50:15 mg/kg BJPCAL 8,2,53
ivn-dog LDLo:22 mg/kg CPBTAL 7,372,59

SAFETY PROFILE: Poison by ingestion, intravenous, and subcutaneous routes. When heated to decomposition it emits very toxic fumes of HCl, SO_x, and NO_x.

MJK500 CAS:701-73-5 ***HR: 3***
METHYL DITHIOCARBANILATE
mf: $C_8H_9NS_2$ mw: 183.30

SYNS: ENT 31,472 ◇ METHYL PHENYLDITHIOCARBAMATE ◇ PHENYLCARBAMODITHIOIC ACID METHYL ESTER (9CI)

TOXICITY DATA with REFERENCE
ivn-mus LD50:180 mg/kg CSLNX* NX#04529

SAFETY PROFILE: Poison by intravenous route. When heated to decomposition it emits very toxic fumes of NO_x and SO_x. See also CARBAMATES.

MJK750 CAS:53384-39-7 ***HR: 3***
METHYLDITHIOCYANATOARSINE HOMOPOLYMER
mf: $(C_3H_3AsN_2S_2)_n$

SYNS: DTAS ◇ MONGARE ◇ POLY-(METHYLBIS(THIOCYANATO) ARSINE)

TOXICITY DATA with REFERENCE
mrc-bcs 5 μg/disc/24H MUREAV 40,19,76

CONSENSUS REPORTS: EPA Genetic Toxicology Program. Arsenic and its compounds are on the Community Right-To-Know List.

SAFETY PROFILE: Arsenic compounds are poisons. Mutation data reported. When heated to decomposition it emits very toxic fumes of As, SO_x, and CN^-. See also ARSENIC COMPOUNDS and THIOCYANATES.

MJL000 CAS:820-54-2 ***HR: 2***
METHYL DIVINYL ACETYLENE
mf: C_7H_8 mw: 92.15

SYN: 2-METHYL-1,5-HEXADIENE-3-YNE

TOXICITY DATA with REFERENCE
orl-rat LDLo:1600 mg/kg WADTAA 55-250,2,55
ihl-rat LC50:24500 mg/m^3 GISAAA 48(10,)89,83
ipr-rat LDLo:2000 mg/kg WADTAA 55-250,2,55

SAFETY PROFILE: Moderately toxic by ingestion and intraperitoneal routes. Mildly toxic by inhalation. When heated to decomposition it emits acrid smoke and irritating fumes. See also ACETYLENE COMPOUNDS.

MJL250 CAS:33089-61-1 ***HR: 3***
2-METHYL-1,3-DI(2,4-XYLYLIMINO)-2-AZAPROPANE
mf: $C_{19}H_{23}N_3$ mw: 293.45

SYNS: AMITRAZ ◇ AMITRAZ ESTRELLA ◇ AZADIENO ◇ BAAM ◇ BOOTS BTS 27419 ◇ N'-(2,4-DIMETHYLPHENYL)-N-(((2,4-DIMETHYLPHENYL)IMINO)METHYL)-N-METHYLMETHANIMIDAMIDE ◇ 1,5-DI(2,4-DIMETHYLPHENYL-3-METHYL-1,3,5-TRIAZAPENTA-1,4-DIENE ◇ ENT 27,967 ◇ N-METHYL-N'-2,4-XYLYL-N-(N-2,4-XYLYLFORMIMIDOYL)FORMAMIDINE ◇ MITABAN ◇ MITAC ◇ TAKTIC ◇ TRIATOX ◇ UPJOHN U-36059

TOXICITY DATA with REFERENCE
orl-rat LD50:400 mg/kg MEIEDD 10,73,83
orl-dog LD50:100 mg/kg SPEADM 78-1,22,78

SAFETY PROFILE: Poison by ingestion. Used to control pear psylla on pears and tetranychid and eriophyid mites on fruit, etc. When heated to decomposition it emits toxic fumes of NO_x.

MJL500 CAS:110-26-9 ***HR: 3***
N,N'-METHYLENEBIS(ACRYLAMIDE)
mf: $C_7H_{10}N_2O_2$ mw: 154.19

PROP: Colorless, crystalline, stable, white powder. Mp: 185° (with decomp), d: 1.235 @ 30°, vap d: 5.31.

SYNS: METHYLENEBISACRYLAMIDE ◇ N,N'-METHYLENEDIACRYLAMIDE ◇ N,N'-METHYLIDENEBISACRYLAMIDE

TOXICITY DATA with REFERENCE
mma-sat 1 mg/plate EMMUEG 11(Suppl 12),1,88
orl-mus TDLo:3208 mg/kg (16D male):REP ARTODN 47,179,81
orl-rat LD50:390 mg/kg 37ASAA 1,306,78
orl-mus LD50:401 mg/kg ARTODN 47,179,81

CONSENSUS REPORTS: Reported in EPA TSCA Inventory.

SAFETY PROFILE: Poison by ingestion. Experimental reproductive effects. Mutation data reported. When heated to decomposition it emits toxic fumes of NO_x. See also AMIDES.

MJL750 CAS:26907-37-9 ***HR: 3***
N,N'-METHYLENEBIS(2-AMINO-1,3,4-THIADIAZOLE)
mf: $C_5H_6N_6S_2$ mw: 214.29

SYNS: BIS-A-TDA ◇ DI-KU-SHUANG ◇ NSC-143019 ◇ TK-5477

TOXICITY DATA with REFERENCE
mnt-mus-orl 200 mg/kg ENVRAL 43,359,87
orl-mus TDLo:60 mg/kg (female 10D post):TER YHHPAL 16,654,81
orl-rat TDLo:5 mg/kg (female 10D post):REP SURR** 2/3,105,78
orl-rat LD50:260 mg/kg CIYPDA 12,345,81

ipr-rat LD50:82 mg/kg TAKHAA 35,68,76
scu-rat LD50:125 mg/kg TAKHAA 35,68,76
orl-mus LD50:2250 mg/kg CIYPDA 12,345,81
ipr-mus LD50:430 mg/kg TAKHAA 35,68,76
scu-mus LD50:145 mg/kg CIYPDA 12,345,81
orl-dog LD50:125 mg/kg TAKHAA 35,68,76
scu-dog LDLo:500 mg/kg CIYPDA 12,345,81
orl-rbt LD50:500 mg/kg TAKHAA 35,68,76
skn-rbt LD50:2000 mg/kg CIYPDA 12,345,81
scu-rbt LDLo:500 mg/kg CIYPDA 12,345,81

SAFETY PROFILE: Poison by ingestion, subcutaneous, and intraperitoneal routes. Moderately toxic by skin contact. An experimental teratogen. Experimental reproductive effects. Mutation data reported. When heated to decomposition it emits toxic fumes of SO_x and NO_x.

MJM200 CAS:101-14-4 ***HR: 3***
4,4'-METHYLENE BIS(2-CHLOROANILINE)
mf: $C_{13}H_{12}Cl_2N_2$ mw: 267.17

SYNS: BIS AMINE ◇ CURALIN M ◇ CURENE 442 ◇ CYANASET ◇ DI(-4-AMINO-3-CHLOROPHENYL)METHANE ◇ DI-(4-AMINO-3-CLOROFENIL)METANO (ITALIAN) ◇ 4,4'-DIAMINO-3,3'-DICHLORODIPHENYLMETHANE ◇ 3,3'-DICHLOR-4,4'-DIAMINODIPHENYLMETHAN (GERMAN) ◇ 3,3'-DICHLORO-4,4'-DIAMINODIPHENYLMETHANE ◇ 3,3'-DICLORO-4,4'-DIAMINODIFENILMETANO (ITALIAN) ◇ MBOCA ◇ 4,4'-METHYLENE(BIS)-CHLOROANILINE ◇ METHYLENE-4,4'-BIS (o-CHLOROANILINE) ◇ p,p'-METHYLENEBIS(α-CHLOROANILINE) ◇ 4,4'-METHYLENEBIS(o-CHLOROANILINE) ◇ p,p'-METHYLENEBIS(o-CHLOROANILINE) ◇ 4,4'-METHYLENEBIS-2-CHLOROBENZENAMINE ◇ METHYLENE-BIS-ORTHOCHLOROANILINE ◇ 4,4-METILENE-BIS-o-CLOROANILINA (ITALIAN) ◇ MOCA ◇ RCRA WASTE NUMBER U158

TOXICITY DATA with REFERENCE
otr-mus:fbr 10 μg/L JJIND8 67,1303,81
sce-ham:ovr 500 μg/L ENMUDM 7,1,85
orl-rat TDLo:4050 mg/kg/77W-C:CAR JEPTDQ 2(1),149,78
scu-rat TDLo:25 g/kg/89W-C:CAR NATWAY 58,578,71
orl-rat TD:27 g/kg/79W-C:ETA NATWAY 58,578,71
orl-rat LD50:2100 mg/kg KCRZAE 26(9),28,67
orl-mus LD50:880 mg/kg KCRZAE 26(9),28,67
ipr-mus LD50:64 mg/kg PMRSDJ 1,682,81

CONSENSUS REPORTS: NTP Fifth Annual Report on Carcinogens. IARC Cancer Review: Group 2A IMEMDT 7,246,87; Animal Sufficient Evidence IMEMDT 4,65,74. EPA Genetic Toxicology Program. Community Right-To-Know List. Reported in EPA TSCA Inventory.

OSHA PEL: TWA 0.02 ppm (skin)
ACGIH TLV: TWA 0.02 ppm (skin); Suspected Human Carcinogen; (Proposed: Confirmed Human Carcinogen)
DFG MAK: Animal Carcinogen, Suspected Human Carcinogen
NIOSH REL: (MOCA) Lowest detectable limit.

SAFETY PROFILE: Confirmed carcinogen with experimental carcinogenic and tumorigenic data. Poison by intraperitoneal route. Moderately toxic by ingestion. Mutation data reported. When heated to decomposition it emits very toxic fumes of Cl^- and NO_x. See also AMINES.

MJM250 CAS:64049-29-2 ***HR: 3***
4,4'-METHYLENE-BIS(2-CHLOROANILINE) HYDROCHLORIDE
mf: $C_{13}H_{12}Cl_2N_2{\bullet}ClH$ mw: 303.63

TOXICITY DATA with REFERENCE
orl-rat TDLo:14 g/kg/78W-C:ETA TXAPA9 31,47,75
orl-mus TDLo:66 g/kg/78W-C:CAR TXAPA9 31,47,75

SAFETY PROFILE: Questionable carcinogen with experimental carcinogenic and tumorigenic data. When heated to decomposition it emits very toxic fumes of Cl^- and NO_x.

MJM500 CAS:97-23-4 ***HR: 3***
2,2'-METHYLENEBIS(4-CHLOROPHENOL)
mf: $C_{13}H_{10}Cl_2O_2$ mw: 269.13

PROP: Crystals, nearly insol in water. Mp: 178°, vap press: 10^{-4} mm @ 100°.

SYNS: ANTIPHEN ◇ BIS(5-CHLOR-2-HYDROXYPHENYL)-METHAN (GERMAN) ◇ BIS(5-CHLORO-2-HYDROXYPHENYL)METHANE ◇ BIS-2-HYDROXY-5-CHLORFENYLMETHAN (CZECH) ◇ BIS(2-HYDROXY-5-CHLOROPHENYL)METHANE ◇ DICESTAL ◇ DICHLOORFEEN (DUTCH) ◇ 5,5'-DICHLORO-2,2'-DIHYDROXYDIPHENYLMETHANE ◇ DI-(5-CHLORO-2-HYDROXYPHENYL)METHANE ◇ 2,2'-DIHYDROXY-5,5'-DICHLORODIPHENYLMETHANE ◇ HYOSAN ◇ KORIUM ◇ O,O-METHYLEEN-BIS(4-CHLOORFENOL) (DUTCH) ◇ O,O-METILEN-BIS(4-CLOROFENOLO) (ITALIAN) ◇ PANACIDE ◇ PREVENTOL ◇ TENIATHANE ◇ WESPURIL

TOXICITY DATA with REFERENCE
skn-rbt 500 mg/24H MLD 28ZPAK -,82,72
eye-rbt 50 μg/24H SEV 28ZPAK -,82,72
mmo-sat 50 nmol/plate MUREAV 90,91,81
orl-rat LD50:2690 mg/kg 28ZPAK -,82,72
ivn-rat LD50:17 mg/kg CRTXB2 2,445,74
orl-mus LD50:1 g/kg JPETAB 96,238,49
orl-dog LD50:2000 mg/kg PCOC** -,361,66
orl-gpg LD50:1250 mg/kg 28ZEAL 5,75,76

CONSENSUS REPORTS: Chlorophenol compounds are on the Community Right-To-Know List. Reported in EPA TSCA Inventory.

SAFETY PROFILE: Poison by intravenous route. Moderately toxic by ingestion. A skin and severe eye irritant. Mutation data reported. Can cause cramps and diarrhea. Possibly similar to DDT. An FDA over-the-counter drug. An anthelmintic. When heated to decomposition it emits toxic fumes of Cl^-. See also DDT and CHLOROPHENOLS.

MJM600 CAS:5124-30-1 ***HR: 3***
METHYLENE BIS(4-CYCLOHEXYLISOCYANATE
mf: $C_{15}H_{22}NO_2$ mw: 262.39

SYNS: BIS(4-ISOCYANATOCYCLOHEXYL)METHANE ◇ NACCONATE H 12

TOXICITY DATA with REFERENCE
ihl-rbt LC50:20 ppm/5H 85INA8 5,392,5(86),86
orl-rat LD50:9900 mg/kg 85INA8 5,392,5(86),86

CONSENSUS REPORTS: Reported in EPA TSCA Inventory.

OSHA PEL: CL 0.01
ACGIH TLV: TWA 0.005 ppm
NIOSH REL: (Diisocyanates): 10H TWA 0.005 ppm; CL 0.02 ppm/10M

SAFETY PROFILE: Poison by inhalation. Mildly toxic by ingestion. When heated to decomposition it emits very toxic fumes of NO_x.

MJN000 CAS:101-61-1 ***HR: 3***
4,4'-METHYLENE BIS(N,N'-DIMETHYLANILINE)
mf: $C_{17}H_{22}N_2$ mw: 254.41

SYNS: p,p'-BIS(DIMETHYLAMINO)DIPHENYLMETHANE ◇ 4,4'-BIS-(DIMETHYLAMINO)DIPHENYLMETHANE ◇ BIS(p-DIMETHYLAMINOPHENYL)METHANE ◇ BIS(p-(N,N-DIMETHYLAMINO)PHENYL)METHANE ◇ p,p'-BIS(N,N-DIMETHYLAMINOPHENYL)METHANE ◇ p,p-DIMETHYLAMINODIPHENYLMETHANE ◇ METHANE BASE ◇ 4,4'-METHYLENEBIS(N,N-DIMETHYL)BENZENAMINE ◇ MICHLER'S BASE ◇ MICHLER'S HYDRIDE ◇ MICHLER'S METHANE ◇ NCI-C01990 ◇ TETRA-BASE ◇ TETRAMETHYLDIAMINODIPHENYLMETHANE ◇ 4,4'-TETRAMETHYLDIAMINODIPHENYLMETHANE ◇ p,p-TETRAMETHYLDIAMINODIPHENYLMETHANE

TOXICITY DATA with REFERENCE
mma-sat 10 μg/plate IARCCD 27,283,80
dnr-esc 20 mg/L JNCIAM 62,873,79
dns-rat:lvr 5 mg/L MUREAV 97,359,82
hma-mus/sat 125 mg/kg JNCIAM 62,911,79
sce-rbt:lym 50 mg/L MUREAV 89,197,81
orl-rat TDLo:8500 mg/kg/59W-C:CAR NCITR* NCI-CG-TR-186,79
orl-mus TDLo:82 g/kg/78W-C:NEO NCITR* NCI-CG-TR-186,79
orl-rat TD:27 g/kg/2Y-C:CAR TOLED5 6,391,80
orl-rat TD:14 g/kg/2Y-C:ETA TOLED5 6,391,80
orl-rat LDLo:500 mg/kg JPETAB 90,260,47
orl-mus LD50:3160 mg/kg NCILB* NIH-NCI-E-C-72-3252,73

CONSENSUS REPORTS: NTP Fifth Annual Report on Carcinogens. IARC Cancer Review: Group 3 IMEMDT 7,56,87; Animal Limited Evidence IMEMDT 27,119,82. NCI Carcinogenesis Bioassay (feed); Clear Evidence: mouse, rat NCITR* NCI-CG-TR-186,79. EPA Genetic Toxicology Program. Reported in EPA TSCA Inventory. Community Right-To-Know List.

DFG MAK: Suspected Carcinogen.

SAFETY PROFILE: Confirmed carcinogen with experimental carcinogenic, neoplastigenic, and tumorigenic data. Moderately toxic by ingestion. Mutation data reported. When heated to decomposition it emits toxic fumes of NO_x.

MJN250 CAS:88-24-4 ***HR: 3***
2,2'-METHYLENEBIS(4-ETHYL-6-tert-BUTYLPHENOL)
mf: $C_{25}H_{36}O_2$ mw: 368.61

SYNS: ANTIOXIDANT 425 ◇ BIS(2-HYDROXY-3-tert-BUTYL-5-ETHYLPHENYL) METHANE ◇ 2,2'-METHYLENEBIS(6-tert-BUTYL-4-ETHYLPHENOL) ◇ PLASTANOX 425 ANTIOXIDANT ◇ USAF CY-6

TOXICITY DATA with REFERENCE
orl-rat TDLo:1250 mg/kg (1-22D preg):REP AJANA2 110,29,62
ipr-mus LD50:50 mg/kg NTIS** AD277-689

CONSENSUS REPORTS: Reported in EPA TSCA Inventory.

SAFETY PROFILE: Poison by intraperitoneal route. Experimental reproductive effects. When heated to decomposition it emits acrid smoke and irritating fumes.

MJN750 CAS:139-25-3 ***HR: 3***
5,5'-METHYLENEBIS(2-ISOCYANATO)TOLUENE
mf: $C_{17}H_{14}N_2O_2$ mw: 278.33

SYNS: 3,3'-DIMETHYLDIPHENYLMETHANE-4,4'-DIISOCYANATE ◇ ISOCYANIC ACID, ESTER with DI-o-TOLUENEMETHANE

TOXICITY DATA with REFERENCE
ivn-mus LD50:320 mg/kg CSLNX* NX#00946

CONSENSUS REPORTS: Reported in EPA TSCA Inventory.

NIOSH REL: (Diisocyanates) TWA 0.005 ppm; CL 0.02 ppm/10M

SAFETY PROFILE: Poison by intravenous route. When heated to decomposition it emits toxic fumes of NO_x.

MJO000 CAS:1807-55-2 ***HR: 3***
4,4'-METHYLENEBIS(N-METHYLANILINE)
mf: $C_{15}H_{18}N_2$ mw: 226.35

SYNS: BIS(N-METHYLANILINE)METHANE ◇ BIS(N-METHYLANILINO)METHAN (GERMAN) ◇ DIMETHYLDIAMINODIPHENYLMETHANE

TOXICITY DATA with REFERENCE
scu-rat TDLo:850 mg/kg:ETA ZEKBAI 71,105,68

CONSENSUS REPORTS: Reported in EPA TSCA Inventory.

SAFETY PROFILE: Questionable carcinogen with ex-

perimental tumorigenic data. When heated to decomposition it emits toxic fumes of NO_x.

MJO250 CAS:838-88-0 ***HR: 3***
4,4'-METHYLENEBIS(2-METHYLANILINE)
mf: $C_{15}H_{18}N_2$ mw: 226.35

PROP: Mp: 149°.

SYNS: BIS-4-AMINO-3-METHYLFENYLMETHAN (CZECH) ◇ 3,3'-DIMETHYL-4,4'-DIAMINODIPHENYLMETHANE ◇ MBOT ◇ ME-MDA ◇ 4,4'-METHYLENEBIS(2-METHYLBENZENAMINE) ◇ 4,4'-METHYLENE DI-o-TOLUIDINE

TOXICITY DATA with REFERENCE
eye-rbt 100 mg/24H MOD 28ZPAK -,72,72
mma-sat 1 mg/plate ARTODN 49,185,82
orl-rat TDLo:4656 mg/kg/55W-C:CAR TXAPA9 31,159,75
orl-dog TDLo:4900 mg/kg/4Y-I:CAR JEPTDQ 1,339,78

CONSENSUS REPORTS: IARC Cancer Review: Group 2B IMEMDT 7,248,87; Animal Limited Evidence IMEMDT 4,73,74. Reported in EPA TSCA Inventory.

DFG MAK: Animal Carcinogen; Suspected Human Carcinogen.

SAFETY PROFILE: Confirmed carcinogen with experimental carcinogenic data. Moderately toxic by ingestion. An eye irritant. Mutation data reported. When heated to decomposition it emits toxic fumes of NO_x.

MJO500 CAS:119-47-1 ***HR: 1***
2,2''-METHYLENEBIS(4-METHYL-6-tert-BUTYLPHENOL)
mf: $C_{23}H_{32}O_2$ mw: 340.55

PROP: Pale cream to white crystals.

SYNS: ADVASTAB 405 ◇ ANTAGE W 400 ◇ ANTI OX ◇ ANTIOXIDANT 1 ◇ BISAKLOFEN BP ◇ 2,2'-BIS-6-TERC.BUTYL-p-KRESYLMETHAN (CZECH) ◇ BKF ◇ CALCO 2246 ◇ CATOLIN 14 ◇ CHEMANOX 21 ◇ 2,2'-METHYLENEBIS(6-tert-BUTYL-p-CRESOL) ◇ NOCRAC NS 6 ◇ OXY CHEK 114 ◇ PLASTANOX 2246 ◇ SYNOX 5LT ◇ VULKANOX BKF

TOXICITY DATA with REFERENCE
eye-rbt 100 mg/24H MOD 28ZPAK -,58,72
orl-rat LD50:4880 mg/kg 28ZPAK -,58,72

CONSENSUS REPORTS: Reported in EPA TSCA Inventory.

SAFETY PROFILE: Mildly toxic by ingestion. An eye irritant. When heated to decomposition it emits acrid smoke and irritating fumes.

MJO750 CAS:7786-17-6 ***HR: 1***
2,2'-METHYLENE-BIS(4-METHYL-6-NONYLPHENOL)
mf: $C_{33}H_{52}O$ mw: 480.85

SYNS: 2,2'-METHYLENEBIS(6-NONYL)-p-CRESOL ◇ NAUGA WHITE

TOXICITY DATA with REFERENCE
orl-rat LD50:33 g/kg RCTEA4 45(3),627,72

CONSENSUS REPORTS: Reported in EPA TSCA Inventory.

SAFETY PROFILE: Mildly toxic by ingestion. When heated to decomposition it emits acrid smoke and irritating fumes.

MJO775 CAS:14168-44-6 ***HR: 3***
METHYLENE BIS(NITRAMINE)
mf: $CH_4N_4O_4$ mw: 136.07

SAFETY PROFILE: A powerful and sensitive explosive. It explodes when heated to 217°C. The lead salt explodes at 195°C. Upon decomposition it emits toxic fumes of NO_x.

MJP400 CAS:101-68-8 ***HR: 3***
METHYLENE BISPHENYL ISOCYANATE
DOT: UN 2489
mf: $C_{15}H_{10}N_2O_2$ mw: 250.27

PROP: Crystals or yellow fused solid. Mp: 37.2°, bp: 194-199° @ 5 mm, d: 1.19 @ 50°, vap press: 0.001 mm @ 40°.

SYNS: BIS(p-ISOCYANATOPHENYL)METHANE ◇ BIS(1,4-ISOCYANATOPHENYL)METHANE ◇ BIS(4-ISOCYANATOPHENYL)METHANE ◇ CARADATE 30 ◇ DESMODUR 44 ◇ DIFENIL-METAN-DIISOCIANATO (ITALIAN) ◇ DIFENYLMETHAAN-DISSOCYANAAT (DUTCH) ◇ 4-4'-DIISOCYANATE de DIPHENYLMETHANE (FRENCH) ◇ 4,4'-DIISOCYANATODIPHENYLMETHANE ◇ DIPHENYLMETHAN-4,4'-DIISOCYANAT (GERMAN) ◇ DIPHENYL METHANE DIISOCYANATE ◇ p,p'-DIPHENYLMETHANE DIISOCYANATE ◇ 4,4'-DIPHENYLMETHANE DIISOCYANATE ◇ DIPHENYLMETHANE 4,4'-DIISOCYANATE (DOT) ◇ HYLENE M50 ◇ ISONATE ◇ MDI ◇ METHYLENEBIS(4-ISOCYANATOBENZENE) ◇ 1,1-METHYLENEBIS(4-ISOCYANATO- BENZENE) ◇ METHYLENEBIS(4-PHENYLENE ISOCYANATE) ◇ METHYLENEBIS(p-PHENYLENE ISOCYANATE) ◇ p,p'-METHYLENEBIS(PHENYL ISOCYANATE) ◇ METHYLENEBIS(p-PHENYL ISOCYANATE) ◇ METHYLENEBIS(4-PHENYL ISOCYANATE) ◇ 4,4'-METHYLENEBIS (PHENYL ISOCYANATE) ◇ 4,4'-METHYLENEDIPHENYL DIISOCYANATE ◇ METHYLENEDI-p-PHENYLENE DIISOCYANATE ◇ METHYLENEDI-p-PHENYLENE ISOCYANATE ◇ 4,4'-METHYLENEDIPHENYLENE ISOCYANATE ◇ METHYLENE DI(PHENYLENE ISOCYANATE) (DOT) ◇ 4,4'-METHYLENEDIPHENYL ISOCYANATE ◇ NACCONATE 300 ◇ NCI-C50668 ◇ RUBINATE 44

TOXICITY DATA with REFERENCE
skn-rbt 500 mg/24H JETOAS 9,41,76
eye-rbt 100 μg MLD AIHAAP 43,89,82
mma-sat 50 μg/plate SWEHDO 6,221,80
ihl-hmn TCLo:130 ppb/30M:IMM,MET AIHAAP 27,121,66
orl-rat LDLo:31690 mg/kg AIHAAP 43,89,82
ihl-rat LC50:178 mg/m³ AIHAAP 43,89,82
orl-mus LD16:10700 mg/kg TPKVAL 15,128,79

CONSENSUS REPORTS: IARC Cancer Review: Group 3 IMEMDT 7,56,87. Reported in EPA TSCA Inventory. Community Right-To-Know List.

OSHA PEL: CL 0.02 ppm
ACGIH TLV: 0.005 ppm
DFG MAK: 0.01 ppm (0.1 mg/m^3)
NIOSH REL: (Diisocyanates) TWA 0.005 ppm; CL 0.02 ppm/10M
DOT Classification: Poison B; Label: St. Andrews Cross.

SAFETY PROFILE: Poison by inhalation. Mildly toxic by ingestion. Human systemic effects by inhalation: increased immune response and body temperature. A skin and eye irritant. An allergic sensitizer. Questionable carcinogen. Mutation data reported. When heated to decomposition it emits toxic fumes of NO_x and SO_x. See also CYANATES.

MJP450 CAS:75-09-2 ***HR: 3***
METHYLENE CHLORIDE
DOT: UN 1593
mf: CH_2Cl_2 mw: 84.93

PROP: Colorless, volatile liquid; odor of chloroform. Bp: 39.8°, lel: 15.5% in O_2, uel: 66.4% in O_2, fp: −96.7°, d: 1.326 @ 20°/4°, autoign temp: 1139°F, vap press: 380 mm @ 22°, vap d: 2.93, refr index: 1.424 @ 20L. Sol in water; misc with alc, acetone, chloroform, ether, and carbon tetrachloride.

SYNS: AEROTHENE MM ◇ CHLORURE de METHYLENE (FRENCH) ◇ DCM ◇ DICHLOROMETHANE (MAK, DOT) ◇ FREON 30 ◇ METHANE DICHLORIDE ◇ METHYLENE BICHLORIDE ◇ METHYLENE DICHLORIDE ◇ METYLENU CHLOREK (POLISH) ◇ NCI-C50102 ◇ RCRA WASTE NUMBER U080 ◇ SOLMETHINE

TOXICITY DATA with REFERENCE
skn-rbt 810 mg/24H SEV JETOAS 9,171,76
eye-rbt 162 mg MOD JETOAS 9,171,76
eye-rbt 10 mg MLD TXCYAC 6,173,76
eye-rbt 17500 mg/m^3/10M TXCYAC 6,173,76
dni-hmn:fbr 5000 ppm/1H-C MUREAV 81,203,81
cyt-ham:ovr 5 g/L MUREAV 116,361,83
dni-ham:lng 5000 ppm/1H-C MUREAV 81,203,81
sce-ham:lng 5000 ppm/1H-C MUREAV 81,203,81
ihl-rat TCLo:4500 ppm/24H (1-17D preg):REP TXAPA9 52,29,80
ihl-mus TCLo:1250 ppm/7H (6-15D preg):TER TXAPA9 32,84,75
ihl-rat TCLo:3500 ppm/6H/2Y-I:CAR FAATDF 4,30,84
ihl-mus TCLo:2000 ppm/5H/2Y-C:CAR NTPTR* NTP-TR-306,86
ihl-rat TCLo:500 ppm/6H/2Y:ETA TXAPA9 48,A185,79
orl-hmn LDLo:357 mg/kg:PNS,CNS 34ZIAG -,390,69
ihl-hmn TCLo:500 ppm/1Y-I:CNS,CVS ABHYAE 43,1123,68
ihl-hmn TCLo:500 ppm/8H:CNS SCIEAS 176,295,72
orl-rat LD50:1600 mg/kg FAONAU 48A,94,70
ihl-rat LC50:88000 mg/m^3/30M FAVUAI 7,35,75
ihl-mus LC50:14400 ppm/7H NIHBAZ 191,1,49
ipr-mus LD50:437 mg/kg AGGHAR 18,109,60
scu-mus LD50:6460 mg/kg TXAPA9 4,354,62
orl-dog LDLo:3 g/kg QJPPAL 7,205,34
ihl-dog LCLo:14108 ppm/7H NIHBAZ 191,1,49
ipr-dog LDLo:950 mg/kg TXAPA9 10,119,67
scu-dog LDLo:2700 mg/kg QJPPAL 7,205,34
ivn-dog LDLo:200 mg/kg QJPPAL 7,205,34
ihl-cat LCLo:43400 mg/m^3/4.5H AHBAAM 116,131,36
orl-rab LDLo:1900 mg/kg HBTXAC 1,94,56
ihl-rbt LCLo:10000 ppm/7H JIHTAB 26,8,44
scu-rbt LDLo:2700 mg/kg QJPPAL 7,205,34
ihl-gpg LCLo:5000 ppm/2H FLCRAP 1,197,67

CONSENSUS REPORTS: NTP Fifth Annual Report on Carcinogens. IARC Cancer Review: Group 2B IMEMDT 7,194,87; Human Inadequate Evidence IMEMDT 41,43,86; Animal Sufficient Evidence IMEMDT 41,43,86; Animal Inadequate Evidence IMEMDT 20,449,79. NTP Carcinogenesis Studies (inhalation); Clear Evidence: mouse, rat NTPTR* NTP-TR-306,86. Reported in EPA TSCA Inventory. EPA Genetic Toxicology Program. Community Right-To-Know List.

OSHA PEL: (Transitional: TWA 500 ppm; CL 1000 ppm; Pk 2000/5M/2H); (Proposed : STEL 126 ppm, 15 min)
ACGIH TLV: TWA 50 ppm, Suspected Human Carcinogen
DFG MAK: 100 ppm (360 mg/m^3); BAT: 5% CO-Hb in blood at end of shift; Suspected Carcinogen.
NIOSH REL: (Methylene Chloride) Reduce to lowest feasible level.
DOT Classification: Poison B; Label: St. Andrews Cross.

SAFETY PROFILE: Confirmed carcinogen with experimental carcinogenic and tumorigenic data. Poison by intravenous route. Moderately toxic by ingestion, subcutaneous, and intraperitoneal routes. Mildly toxic by inhalation. Human systemic effects by ingestion and inhalation: paresthesia, somnolence, altered sleep time, convulsions, euphoria, and change in cardiac rate. An experimental teratogen. Experimental reproductive effects. An eye and severe skin irritant. Human mutation data reported. It is flammable in the range of 12-19 percent in air but ignition is difficult. It will not form explosive mixtures with air at ordinary temperatures. Mixtures in air with methanol vapor are flammable. It will form explosive mixtures with an atmosphere having a high oxygen content; in liquid O_2; N_2O_4; K; Na; NaK. Explosive in the form of vapor when exposed to heat or flame. Reacts violently with Li; NaK; potassium-tert-butoxide;

(KOH + n-methyl-n-nitrosourea). It can be decomposed by contact with hot surfaces and open flame, and then yield toxic fumes which are irritating and give warning of their presence. When heated to decomposition it emits highly toxic fumes of phosgene and Cl^-. See also CHLORINATED HYDROCARBONS, ALIPHATIC.

MJP500 CAS:156-56-9 ***HR: 3***
2-METHYLENECYCLOPROPANYLALANINE
mf: $C_7H_{11}NO_2$ mw: 141.17

PROP: Hypoglycemic principle from the akee plant, *Blighia sapida Kon., Sapindaceae* in the West Indies (LSPPAT 9,1305,70). Yellow plates from methanol + water. Mp: 280-284°.

SYNS: α-AMINOMETHYLENECYCLOPROPANEPROPIONICACID ◇ l-α-AMINO-β-METHYLENECYCLOPROPANEPROPIONIC ACID ◇ α-AMINO-2-METHYLENE-CYCLOPROPANEPROPANOIC ACID (9CI) ◇ α-AMINO-β-(2-METHYLENECYCLOPROPYL)PROPIONIC ACID ◇ 2-AMINO-4,5-METHYLENEHEX-5-ENOIC ACID ◇ HYPOGLYCIN ◇ HYPOGLYCIN A ◇ HYPOGLYCINE A ◇ 2-METHYLENECYCLO-PROPANEALANINE ◇ β-(METHYLENECYCLOPROPYL)ALANINE

TOXICITY DATA with REFERENCE
ipr-rat TDLo:30 mg/kg (3-6D preg):TER NATWAY 55,39,68
ipr-rat TDLo:150 mg/kg (1-5D pre):REP FCTXAV 6,813,68
orl-rat LD50:98 mg/kg: BJPCAL 13,125,58
ipr-rat LD50:97 mg/kg WIMJAD 16,193,67
ivn-rat LDLo:30 mg/kg JPETAB 121,272,57

SAFETY PROFILE: Poison by ingestion, intravenous, and intraperitoneal route. An experimental teratogen. Other experimental reproductive effects. When heated to decomposition it emits toxic fumes of NO_x.

MJP750 CAS:1208-52-2 ***HR: 3***
2,4'-METHYLENEDIANILINE
mf: $C_{13}H_{14}N_2$ mw: 198.29

SYNS: 2',4-BIS(AMINOPHENYL)METHANE ◇ 2,4'-DIAMINO-DIPHENYLMETHAN (GERMAN) ◇ o,p'-DIAMINODIPHENYLMETH-ANE ◇ 2,4'-DIAMINODIPHENYLMETHANE ◇ 2,4'-DIPHENYL-METHANEDIAMINE ◇ 2,4'-METHYLENEBIS(ANILINE)

TOXICITY DATA with REFERENCE
scu-rat LD50:3300 mg/kg NATWAY 57,247,70

CONSENSUS REPORTS: Reported in EPA TSCA Inventory.

DFG MAK: Animal Carcinogen, Suspected Human Carcinogen.

SAFETY PROFILE: Suspected carcinogen. Moderately toxic by subcutaneous route. When heated to decomposition it emits toxic fumes of NO_x.

MJQ000 CAS:101-77-9 ***HR: 3***
4,4'-METHYLENEDIANILINE
DOT: UN 2651
mf: $C_{13}H_{14}N_2$ mw: 198.29

PROP: Tan flakes or lumps; faint amine-like odor. Mp: 90°, flash p: 440°F.

SYNS: 4-(4-AMINOBENZYL)ANILINE ◇ BIS-p-AMINOFENYL-METHAN (CZECH) ◇ BIS(p-AMINOPHENYL)METHANE ◇ BIS(4-AMINOPHENYL)METHANE ◇ CURITHANE ◇ DDM ◇ p,p'-DI-AMINODIFENYLMETHAN (CZECH) ◇ 4,4'-DIAMINODIPHE-NYLMETHAN (GERMAN) ◇ DIAMINODIPHENYLMETHANE ◇ p,p'-DIAMINODIPHENYLMETHANE ◇ 4,4'-DIAMINODIPHENYLMETH-ANE ◇ DI-(4-AMINOPHENYL)METHANE ◇ DIANALINEMETHANE ◇ 4,4'-DIPHENYLMETHANEDIAMINE ◇ EPICURE DDM ◇ MDA ◇ METHYLENEBIS(ANILINE) ◇ 4,4'-METHYLENEBISANILINE ◇ METHYLENEDIANILINE ◇ p,p'-METHYLENEDIANILINE ◇ NCI-C54604 ◇ TONOX

TOXICITY DATA with REFERENCE
eye-rbt 100 mg/24H MOD 28ZPAK -,71,72
mmo-sat 250 μg/plate MUREAV 67,123,79
mma-sat 50 μg/plate MUREAV 67,123,79
dnd-rat-ipr 370 μmol/kg CRNGDP 2,1317,81
sce-mus-ipr 9 mg/kg MUREAV 108,225,83
orl-rat TDLo:320 mg/kg/I:ETA NATUAS 219,1162,68
orl-man TDLo:8420 μg/kg:CNS,LIV BMJOAE 1,514,66
orl-rat LD50:347 mg/kg 28ZPAK -,71,72
ipr-rat LD50:193 mg/kg ZHUGAM 20,393,74
scu-rat LD50:200 mg/kg NATWAY 57,247,70
orl-mus LD50:745 mg/kg ZHYGAM 20,393,74
ipr-mus LD50:74 mg/kg RCOCB8 14,677,76
orl-dog LDLo:300 mg/kg TXCYAC 11,185,78
scu-dog LDLo:400 mg/kg AEXPBL 58,167,1907
orl-rbt LD50:620 mg/kg ZHYGAM 20,393,74
orl-gpg LD50:260 mg/kg ZHYGAM 20,393,74

CONSENSUS REPORTS: NTP Fifth Annual Report on Carcinogens. IARC Cancer Review: Group 2B IMEMDT 7,56,87; Animal Sufficient Evidence IMEMDT 39,347,86; Animal Inadequate Evidence IMEMDT 4,79,74. Community Right-To-Know List. Reported in EPA TSCA Inventory.

ACGIH TLV: TWA 0.1 ppm (skin); Suspected Human Carcinogen.
DFG MAK: Animal Carcinogen, Suspected Human Carcinogen.
DOT Classification: Poison B; Label: St. Andrews Cross.

SAFETY PROFILE: Confirmed carcinogen with experimental tumorigenic data. Poison by ingestion, subcutaneous, and intraperitoneal routes. Human systemic effects by ingestion: rigidity, jaundice, other liver changes. An eye irritant. Mutation data reported. It is not rapidly absorbed through the skin. Combustible when exposed to heat or flame. When heated to decomposition it emits highly toxic fumes of aniline and NO_x.

MJQ100 CAS:13552-44-8 ***HR: 3***
4,4'-METHYLENEDIANILINE DIHYDROCHLORIDE
mf: $C_{13}H_{14}N_2 \cdot 2ClH$ mw: 271.21

SYN: NCI-C54604

TOXICITY DATA with REFERENCE
orl-rat TDLo:10950 mg/kg/2Y-C:CAR NTPTR* NTP-TR-248,83
orl-mus TDLo:21900 mg/kg/2Y-C:CAR NTPTR* NTP-TR-248,83
orl-rat TD:10815 mg/kg/2Y-C:NEO JJIND8 72,1457,84

CONSENSUS REPORTS: NTP Fifth Annual Report on Carcinogens. IARC Cancer Review: Animal Sufficient Evidence IMEMDT 39,347,86. NTP Carcinogenesis Studies (oral); Clear Evidence: mouse, rat NTPTR* NTP-TR-248,83. Reported in EPA TSCA Inventory.

DOT Classification: Poison B; Label: St. Andrews Cross
NIOSH REL: (4,4'-Methylenedianiline): TWA reduce to lowest level.

SAFETY PROFILE: Confirmed carcinogen with experimental carcinogenic and neoplastigenic data. A poison. When heated to decomposition it emits toxic fumes of NO_x and HCl. See also ANILINE DYES.

MJQ250 CAS:34481-84-0 ***HR: 3***
METHYLENEDIANTHRANILIC ACID DIMETHYL ESTER
mf: $C_{17}H_{18}N_2O_4$ mw: 314.37

SYNS: MBMA ◇ METHYLENEBIS(2-AMINO-BENZOIC ACID) DIMETHYL ESTER (9CI) ◇ 4,4'-METHYLENEBIS(2-CARBOMETHOXYANILINE) ◇ METHYLENE-BIS(METHYL ANTHRANILATE)

TOXICITY DATA with REFERENCE
orl-rat TDLo:490 g/kg/78W-C:CAR JEPTDQ 1(3),199,78
orl-rat TD:410 g/kg/78W-C:ETA JEPTDQ 1(3),199,78

SAFETY PROFILE: Questionable carcinogen with experimental carcinogenic and tumorigenic data. When heated to decomposition it emits toxic fumes of NO_x. See also ESTERS.

MJQ325 ***HR: 3***
METHYLENEDILITHIUM
mf: CH_2Li_2 mw: 27.91

SAFETY PROFILE: Ignites spontaneously in air. See also LITHIUM COMPOUNDS.

MJQ500 CAS:156-72-9 ***HR: 3***
METHYLENE DIMETHANESULFONATE
mf: $C_3H_8O_6S_2$ mw: 204.23

SYNS: ENT 51,799 ◇ METHANESULFONIC ACID, METHYLENE ESTER ◇ METHYLENE BIS(METHANESULFONATE)

TOXICITY DATA with REFERENCE
dnd-rat-unr 25 mg/kg CBINA8 7,265,73
dlt-rat-ipr 15 mg/kg CCPTAY 13,639,76
dni-mus:oth 5 mg/L CBINA8 11,501,75
msc-mus:lym 50 μmol/L MUREAV 25,107,74
dnd-mam:oth 250 μmol/L CBINA8 38,119,81
ipr-rat TDLo:15 mg/kg (13D preg):REP JRPFA4 17,325,68
ipr-rat TDLo:15 mg/kg (13D preg):TER JRPFA4 17,325,68
ipr-rat LD50:25 mg/kg BJPCAL 24,24,65

CONSENSUS REPORTS: EPA Genetic Toxicology Program.

SAFETY PROFILE: Poison by intraperitoneal route. An experimental teratogen. Experimental reproductive effects. Mutation data reported. When heated to decomposition it emits toxic fumes of SO_x. See also SULFONATES.

MJQ750 CAS:5625-90-1 ***HR: 3***
4,4'-METHYLENEDIMORPHOLINE
mf: $C_9H_{18}N_2O_2$ mw: 186.29

SYNS: BIS(MORPHOLINO-)METHAN (GERMAN) ◇ BISMORPHOLINO METHANE

TOXICITY DATA with REFERENCE
scu-rat TDLo:50 mg/kg:ETA FCTXAV 6,576,68
scu-rat LD50:1000 mg/kg ZEKBAI 71,105,68

SAFETY PROFILE: Moderately toxic by subcutaneous route. Questionable carcinogen with experimental tumorigenic data. When heated to decomposition it emits toxic fumes of NO_x.

MJQ775 CAS:4764-17-4 ***HR: 3***
METHYLENEDIOXYAMPHETAMINE
mf: $C_{10}H_{13}NO_2$ mw: 179.24

SYNS: MDA ◇ 3,4-METHYLENEDIOXY-AMPHETAMINE ◇ α-METHYL-3,4-(METHYLENEDIOXY)PHENETHYLAMINE

TOXICITY DATA with REFERENCE
orl-mus LD50:13300 μg/kg CTOXAO 6,193,73
ipr-mus LD50:82300 μg/kg CTOXAO 6,193,73
ivn-mus LD50:31100 μg/kg TRBMAV 33,610,75

SAFETY PROFILE: Poison by ingestion, intravenous, and intraperitoneal routes. When heated to decomposition it emits toxic fumes of NO_x. See also AMINES and BENZEDRINE.

MJR000 CAS:274-09-9 ***HR: 2***
1,2-METHYLENEDIOXYBENZENE
mf: $C_7H_6O_2$ mw: 122.13

SYN: METHYLENEDIOXYBENZENE

TOXICITY DATA with REFERENCE
orl-rat LD50:580 mg/kg TXAPA9 7,18,65
orl-mus LD50:1220 mg/kg FCTXAV 2,327,64

CONSENSUS REPORTS: Reported in EPA TSCA Inventory.

SAFETY PROFILE: Moderately toxic by ingestion. When heated to decomposition it emits acrid smoke and irritating fumes.

MJR250 CAS:3160-37-0 ***HR: 2***
3,4-METHYLENEDIOXYBENZYL ACETONE
mf: $C_{11}H_{12}O_3$ mw: 192.23

SYNS: HELIOTROPYL ACETONE ◇ 4-(3,4-METHYLENEDIOXYPHENYL)-2-BUTANONE ◇ PIPERONYL ACETONE

TOXICITY DATA with REFERENCE
orl-rat LD50:4 g/kg FCTXAV 14,659,76

CONSENSUS REPORTS: Reported in EPA TSCA Inventory.

SAFETY PROFILE: Moderately toxic by ingestion. When heated to decomposition it emits acrid smoke and irritating fumes.

MJR500 CAS:64057-70-1 ***HR: 3***
3,4-METHYLENEDIOXY-N,α-DIMETHYL-β-PHENYLETHYLAMINE HYDROCHLORIDE
mf: $C_{11}H_{15}NO_2$•ClH mw: 229.73

TOXICITY DATA with REFERENCE
ipr-rat LD50:49 mg/kg TXAPA9 25,299,73
ipr-mus LD50:97 mg/kg TXAPA9 25,299,73
ivn-dog LD50:14 mg/kg TXAPA9 25,299,73
ivn-mky LD50:22 mg/kg TXAPA9 25,299,73
ipr-gpg LD50:98 mg/kg TXAPA9 25,299,73

SAFETY PROFILE: Poison by intraperitoneal and intravenous routes. When heated to decomposition it emits very toxic fumes of HCl and NO_x. See also AMINES.

MJR750 CAS:42542-07-4 ***HR: 3***
3,4-METHYLENEDIOXY-α-ETHYL-β-PHENYLETHYLAMINE
mf: $C_{11}H_{15}NO_2$•ClH mw: 229.73

TOXICITY DATA with REFERENCE
ipr-rat LD50:95 mg/kg TXAPA9 25,299,73
r-mus LD50:82 mg/kg TXAPA9 25,299,73
ivn-dog LD50:16 mg/kg TXAPA9 25,299,73
ivn-mky LD50:20 mg/kg TXAPA9 25,299,73
ipr-gpg LD50:88 mg/kg TXAPA9 25,299,73

SAFETY PROFILE: Poison by intravenous and intraperitoneal routes. When heated to decomposition it emits very toxic fumes of HCl and NO_x.

MJS250 CAS:64245-99-4 ***HR: 3***
2-(3,4-(METHYLENEDIOXY)PHENOXY)-1-((3,4-(METHYLENEDIOXY)PHENOXY)METHYL) ETHYLAMINE
mf: $C_{17}H_{17}NO_6$ mw: 331.35

SYN: 1,3-BIS(3,4-METILENDIOSSIFENOSSI)-2-AMINOPROPANO (ITALIAN)

TOXICITY DATA with REFERENCE
orl-mus LD50:700 mg/kg FRPSAX 32,502,77
ivn-mus LD50:49 mg/kg FRPSAX 32,502,77

SAFETY PROFILE: Poison by intravenous route. Moderately toxic by ingestion. When heated to decomposition it emits toxic fumes of NO_x.

MJS500 CAS:64246-17-9 ***HR: 3***
1-(5-(3,4-(METHYLENEDIOXY)PHENOXY)-4-((3,4-(METHYLENEDIOXY)PHENOXY)METHYL) PENTYLPIPERIDINE CITRATE
mf: $C_{25}H_{31}NO_6$•$C_6H_8O_7$ mw: 633.71

TOXICITY DATA with REFERENCE
orl-mus LD50:635 mg/kg FRPSAX 32,502,77
ivn-mus LD50:47 mg/kg FRPSAX 32,502,77

SAFETY PROFILE: Poison by intravenous route. Moderately toxic by ingestion. When heated to decomposition it emits toxic fumes of NO_x.

MJS750 CAS:6292-91-7 ***HR: 3***
1-(3,4-METHYLENEDIOXYPHENYL)-2-AMINOPROPANE
mf: $C_{10}H_{13}NO_2$•ClH mw: 215.70

SYNS: 3,4-METHYLENEDIOXY-α-METHYL-β-PHENYLETHYLAMINE HYDROCHLORIDE ◇ α-METHYL-3,4-METHYLENEDIOXYPHENETHYLAMINE HYDROCHLORIDE

TOXICITY DATA with REFERENCE
mmo-sat 5 mg/plate MUREAV 56,199,77
ipr-rat LD50:27 mg/kg TXAPA9 25,299,73
ipr-mus LD50:68 mg/kg TXAPA9 25,299,73
ivn-dog LD50:7 mg/kg TXAPA9 25,299,73
ivn-mky LD50:6 mg/kg TXAPA9 25,299,73
ipr-gpg LD50:28 mg/kg TXAPA9 25,299,73

SAFETY PROFILE: Poison by intravenous and intraperitoneal routes. Mutagenic data reported. When heated to decomposition it emits very toxic fumes of HCl and NO_x.

MJT000 CAS:1653-64-1 ***HR: 3***
3,4-METHYLENEDIOXY-β-PHENYLETHYLAMINE HYDROCHLORIDE
mf: $C_9H_{11}NO_2$•ClH mw: 201.67

TOXICITY DATA with REFERENCE
ipr-rat LD50:55 mg/kg TXAPA9 25,299,73

ipr-mus LD50:176 mg/kg TXAPA9 25,299,73
ivn-dog LD50:28 mg/kg TXAPA9 25,299,73
ivn-mky LD50:45 mg/kg TXAPA9 25,299,73
ipr-gpg LD50:245 mg/kg TXAPA9 25,299,73

SAFETY PROFILE: Poison by intravenous and intraperitoneal routes. When heated to decomposition it emits very toxic fumes of HCl and NO_x.

MJT500 CAS:6317-18-6 ***HR: 3***
METHYLENE DITHIOCYANATE
mf: $C_3H_2N_2S_2$ mw: 130.19

SYN: METHYLENDIRHODANID (CZECH, GERMAN)

TOXICITY DATA with REFERENCE
orl-rat LD50:161 mg/kg MarJV# 29MAR77
ivn-mus LD50:3600 μg/kg CSLNX* NX#03787
scu-rbt LDLo:20 mg/kg AEPPAE 150,257,30

CONSENSUS REPORTS: Reported in EPA TSCA Inventory.

SAFETY PROFILE: Poison by ingestion, intravenous, and subcutaneous routes. When heated to decomposition it emits very toxic fumes of NO_x and SO_x. See also THIOCYANATES.

MJT750 CAS:3693-53-6 ***HR: 3***
METHYLENE DIURETHAN
mf: $C_7H_{14}N_2O_4$ mw: 190.23

SYN: N,N-METHYLENE-BIS(ETHYL CARBAMATE)

TOXICITY DATA with REFERENCE
ipr-mus TDLo:6500 mg/kg/13W-I:ETA JNCIAM 9,35,48

SAFETY PROFILE: Questionable carcinogen with experimental tumorigenic data. When heated to decomposition it emits toxic fumes of NO_x. See also CARBAMATES.

MJU000 CAS:533-31-3 ***HR: 3***
METHYLENE ETHER of OXYHYDROQUINONE
mf: $C_7H_6O_3$ mw: 138.13

SYN: SESAMOL

TOXICITY DATA with REFERENCE
orl-rat TDLo:4608 mg/kg/82W-C:ETA JAFCAU 6,600,58

CONSENSUS REPORTS: Reported in EPA TSCA Inventory.

SAFETY PROFILE: Questionable carcinogen with experimental tumorigenic data. When heated to decomposition it emits acrid smoke and irritating fumes.

MJU250 CAS:2679-01-8 ***HR: 3***
METHYLENE GREEN
mf: $C_{17}H_{17}N_4O_2S{\cdot}Cl$ mw: 364.88

SYN: 3,7-BIS(DIMETHYLAMINO)-4-NITRO-PHENOTHIAZIN-5-IUM, CHLORIDE

TOXICITY DATA with REFERENCE
orl-rat LDLo:500 mg/kg IJLEAG 2,257,34
ipr-rat LDLo:100 mg/kg IJLEAG 2,257,34
ivn-cat LDLo:50 mg/kg IJLEAG 2,257,34
ivn-rbt LDLo:150 mg/kg IJLEAG 2,257,34

SAFETY PROFILE: Poison by intravenous and intraperitoneal routes. Moderately toxic by ingestion. When heated to decomposition it emits very toxic fumes of Cl^-, SO_x, and NO_x.

MJU350 CAS:25382-52-9 ***HR: 3***
METHYLENEMAGNESIUM
mf: CH_2Mg mw: 38.33

SAFETY PROFILE: The polymeric form ignites spontaneously in air. When heated to decomposition it emits toxic fumes of MgO. See also MAGNESIUM COMPOUNDS.

MJU500 CAS:52775-76-5 ***HR: 3***
METHYLENOMYCIN A
mf: $C_9H_{10}O_4$ mw: 182.19

PROP: An antibiotic produced by *Streptomyces violaceoruber* strain No. 2416 (85ERAY 2,1267,78).

SYN: 2-METHYLENE-CYCLOPENTAENE-3-ONO-4,5-EPOXY-4,5-DIMETHYL-1-CARBOXYLIC ACID

TOXICITY DATA with REFERENCE
orl-mus LD50:1500 mg/kg 85ERAY 2,1267,78
ipr-mus LD50:75 mg/kg 85ERAY 2,1267,78
ivn-mus LD50:156 mg/kg 85GDA2 6,45,81

SAFETY PROFILE: Poison by intravenous and intraperitoneal routes. Moderately toxic by ingestion. When heated to decomposition it emits acrid smoke and irritating fumes.

MJU750 CAS:17605-71-9 ***HR: 3***
METHYLEPHEDRINE
mf: $C_{11}H_{17}NO$ mw: 179.25

PROP: dl Form: Crystals from petr ether or methanol. Mp: 63.5-64.5°, very sol in usual solvents. d form: Crystals. Mp: 87-87.5°. l Form: Crystals from petr ether. Mp: 87-88°.

SYNS: METHYLEPHEDRIN (GERMAN) ◇ 1-PHENYL-2-DIMETHYLAMINOPROPANOL

TOXICITY DATA with REFERENCE
orl-rat LDLo:1000 mg/kg AEPPAE 195,647,40
ipr-rat LDLo:210 mg/kg AEPPAE 195,647,40
ipr-mus LDLo:210 mg/kg AEPPAE 195,647,40

SAFETY PROFILE: A poison by intraperitoneal route.

Moderately toxic by ingestion. When heated to decomposition it emits toxic fumes of NO_x.

MJV000 CAS:554-99-4 ***HR: 3***
N-METHYLEPINEPHRINE
mf: $C_{10}H_{15}NO_3$ mw: 197.23

SYNS: 3,4-DIHYDROXY-α-(DIMETHYLAMINOMETHYL)BENZYL ALCOHOL ◇ α-(3,4-DIHYDROXYPHENYL)-β-DIMETHYLAMINO-ETHANOL ◇ α-(3,4-DIHYDROXYPHENYL)-α-HYDROXY-β-DIMETHYLAMINOETHANE ◇ α-(DIMETHYLAMINOMETHYL)PROTOCATECHUYL ALCOHOL ◇ METHADRENE ◇ N-METHYLADRENALINE

TOXICITY DATA with REFERENCE
ipr-rat LD50:50 mg/kg JPETAB 69,1,40
scu-rat LD50:105 mg/kg JPETAB 69,1,40
ivn-rat LD50:5 mg/kg JPETAB 69,1,40
ivn-mus LD50:6750 μg/kg AEPPAE 226,493,55
ivn-dog LD50:8 mg/kg JPETAB 69,1,40
ipr-rbt LD50:20 mg/kg JPETAB 69,1,40
scu-rbt LD50:25 mg/kg JPETAB 69,1,40
ivn-rbt LD50:3 mg/kg JPETAB 69,1,40

SAFETY PROFILE: Poison by intravenous, intraperitoneal, and subcutaneous routes. When heated to decomposition it emits toxic fumes of NO_x.

MJV500 ***HR: 3***
(8-β)-6-METHYLERGOLINE-8-ACETAMIDE TARTRATE (2:1)
mf: $C_{34}H_{42}N_6O_2 \cdot C_4H_6O_6$ mw: 716.92

SYN: ERGOLINE-8-ACETAMIDE, 6-METHYL-, (8-β)-, (R-(R*,R*))-2,3-DIHYDROXYBUTANEDIOATE (2:1)

TOXICITY DATA with REFERENCE
orl-rat TDLo:400 μg/kg (female 6D post):REP CCCCAK 36,2200,71
orl-mus LD50:1 g/kg CCCCAK 36,2200,71
ivn-mus LD50:93 mg/kg CCCCAK 36,2200,71

SAFETY PROFILE: Poison by intravenous route. Experimental reproductive effects. When heated to decomposition it emits toxic fumes of NO_x.

MJV750 CAS:57432-61-8 ***HR: 3***
METHYLERGONOVINE MALEATE
mf: $C_{24}H_{29}N_3O_6$ mw: 455.56

PROP: White to pinkish-tan microcrystalline powder; odorless, bitter taste. Sltly sol in water and alc; very sltly sol in chloroform and ether.

SYNS: BASOFORTINA ◇ MALEIC ACID, METHYL ERGONOVINE ◇ METHEGRIN ◇ METHYLERGOBASINE MALEATE ◇ METHYLERGOMETRINE MALEATE ◇ USAF UCTL-8

TOXICITY DATA with REFERENCE
sce-ham:ovr 10 nmol/L TCMUD8 8,169,88
ims-wmn TDLo:4 μg/kg (female 36W post):REP JRPMAP 33,771,88
ims-wmn TDLo:4 μg/kg (female 36W post):TER JRPMAP 33,771,88
orl-rat LD50:93 mg/kg NIIRDN 6,828,82
ivn-rat LD50:23 mg/kg NIIRDN 6,828,82
orl-mus LD50:187 mg/kg NIIRDN 6,828,82
ipr-mus LD50:6 mg/kg NTIS** AD277-689
ivn-mus LD50:85 mg/kg NIIRDN 6,828,82
ivn-rbt LD50:2600 μg/kg NIIRDN 6,828,82

SAFETY PROFILE: Poison by ingestion, intravenous, and intraperitoneal routes. Human teratogenic and reproductive effects. Mutation data reported. When heated to decomposition it emits toxic fumes of NO_x.

MJW000 CAS:112-61-8 ***HR: 3***
METHYL ESTER STEARIC ACID
mf: $C_{19}H_{38}O_2$ mw: 298.57

PROP: Liquid to semi-solid. Mp: 38°, bp: 215° @ 15 mm, flash p: 307°F (CC), d: 0.860. Sol in water and ether.

SYNS: EMERY 2218 ◇ METHOLENE 2218 ◇ METHYL OCTADECANOATE ◇ METHYL STEARATE ◇ OCTADECANOIC ACID, METHYL ESTER

TOXICITY DATA with REFERENCE
scu-mus TDLo:5200 mg/kg/26W-I:ETA CNREA8 32,880,72

CONSENSUS REPORTS: Reported in EPA TSCA Inventory.

SAFETY PROFILE: Questionable carcinogen with experimental tumorigenic data. Combustible when exposed to heat or flame; can react with oxidizing materials. To fight fire, use CO_2, dry chemical. When heated to decomposition it emits acrid smoke and irritating fumes. See also ESTERS.

MJW250 CAS:1912-28-3 ***HR: D***
METHYL ETHANE SULPHONATE
mf: $C_3H_8O_3S$ mw: 124.17

SYNS: MES ◇ METHYL ETHANE SULFONATE

TOXICITY DATA with REFERENCE
mmo-clr 20 mmol/L MUREAV 7,25,69
dlt-rat-ipr 50 mg/kg GENRA8 4,333,63
orl-rat TDLo:1 g/kg (25D male):REP 15QWAW -,62,65

CONSENSUS REPORTS: Reported in EPA TSCA Inventory.

SAFETY PROFILE: Experimental reproductive effects. Mutation data reported. When heated to decomposition it emits toxic fumes of SO_x. See also SULFONATES.

MJW500 CAS:115-10-6 ***HR: 3***
METHYL ETHER
DOT: UN 1033
mf: C_2H_6O mw: 46.08

PROP: Colorless gas, ether odor. Mp: −138.5°, bp: −23.7°, lel: 3.4%, uel: 27%, flash p: −42°F (CC), autoign temp: 662°F, vap d: 1.617, d: 0.661 (air = 1). Sol in alc, water, ether.

SYNS: DIMETHYL ETHER (DOT) ◇ OXYBISMETHANE ◇ WOOD ETHER

TOXICITY DATA with REFERENCE
ihl-rat LC50:308 g/m^3 TOXID9 1,79,81
ihl-mus LC50:386 ppm/15M EJTXAZ 8,287,75

CONSENSUS REPORTS: Reported in EPA TSCA Inventory.

DOT Classification: Flammable Gas; Label: Flammable Gas.

SAFETY PROFILE: Moderately toxic by inhalation. Very dangerous fire hazard when exposed to heat, flame, or oxidizers. Dangerous explosion hazard when exposed to flame, sparks, etc. Violent reaction with AlH_3 and $LiAlH_2$. Keep in closed container away from heat and open flame. To fight fire, stop flow of gas. When heated to decomposition it emits acrid smoke and irritating fumes. See also ETHERS and ETHYL ETHER.

MJW750 CAS:1320-67-8 ***HR: 1***
METHYL ETHER of PROPYLENE GLYCOL (α)
mf: $C_4H_{10}O_2$ mw: 90.14

PROP: Colorless liquid. Mp: −96.7°, bp: 120°, flash p: 335°F, d: 0.919 @ 25°/25°.

TOXICITY DATA with REFERENCE
eye-rbt 20 mg AJOPAA 29,1363,46

CONSENSUS REPORTS: Glycol ethers are on the Community Right-To-Know List.

SAFETY PROFILE: An eye irritant. Many glycol ether compounds have dangerous human reproductive effects. Combustible when exposed to heat or flame; can react with oxidizing materials. To fight fire, use foam, CO_2, dry chemical. When heated to decomposition it emits acrid smoke and irritating fumes. See also ETHYLENE GLYCOL METHYL ETHER and GLYCOL ETHERS.

MJW875 CAS:13655-95-3 ***HR: 2***
11-β-METHYL-17-α-ETHINYLESTRADIOL
mf: $C_{21}H_{26}O_2$ mw: 310.47

SYN: 11-β-METHYL-19-NOR-17-α-PREGNA-1,3,5(10)-TRIEN-20-YNE-3,17-DIOL

TOXICITY DATA with REFERENCE
imp-ham TDLo:400 mg/kg/39W-C:CAR CNREA8 47,2583,87

SAFETY PROFILE: Questionable carcinogen with experimental carcinogenic data. When heated to decomposition it emits acrid smoke and irritating fumes.

MJX000 CAS:16354-48-6 ***HR: 3***
9-METHYL-10-ETHOXYMETHYL-1,2-BENZANTHRACENE
mf: $C_{21}H_{18}O$ mw: 286.39

SYN: 7-ETHOXY-12-METHYLBENZ(a)ANTHRACENE

TOXICITY DATA with REFERENCE
ims-rat TDLo:50 mg/kg:NEO CNREA8 29,506,69

SAFETY PROFILE: Questionable carcinogen with experimental neoplastigenic data. When heated to decomposition it emits acrid smoke and irritating fumes.

MJX500 CAS:96-17-3 ***HR: 2***
METHYLETHYLACETALDEHYDE
mf: $C_5H_{10}O$ mw: 86.15

SYNS: α-METHYLBUTANAL ◇ 2-METHYLBUTANAL ◇ 2-METHYL-1-BUTANAL ◇ α-METHYLBUTYRALDEHYDE ◇ 2-METHYLBUTYRALDEHYDE

TOXICITY DATA with REFERENCE
skn-rbt 500 mg open MLD UCDS** 10/5/72
eye-rbt 500 mg/24H SEV FCTOD7 20(Suppl),739,82
skn-gpg 100%/24H MOD FCTOD7 20(Suppl),739,82
orl-rat LD50:6400 mg/kg FCTOD7 20(Suppl),739,82
ihl-rat LC50:14000 ppm/4H UCDS** 10/5/72
orl-mus LD50:3200 mg/kg 14CYAT 2,1968,63
skn-rbt LD50:5730 mg/kg UCDS** 10/5/72

CONSENSUS REPORTS: Reported in EPA TSCA Inventory.

SAFETY PROFILE: Mildly toxic by ingestion, inhalation, and skin contact. A skin and severe eye irritant. When heated to decomposition it emits acrid smoke and irritating fumes. See also ALDEHYDES.

MJY000 CAS:24549-06-2 ***HR: 2***
2-METHYL-6-ETHYL ANILINE
mf: $C_9H_{13}N$ mw: 135.23

PROP: Liquid. Bp: 201°. Insol in water; misc in alc and ether.

SYN: 2-ETHYL-6-METHYL-BENZENAMINE

TOXICITY DATA with REFERENCE
orl-rat LD50:885 mg/kg FAATDF 3,285,83

CONSENSUS REPORTS: Reported in EPA TSCA Inventory.

SAFETY PROFILE: Moderately toxic by ingestion. When heated to decomposition it emits toxic fumes of NO_x.

MJY250 CAS:56961-65-0 ***HR: 3***
7-METHYL-9-ETHYLBENZ(c)ACRIDINE
mf: $C_{20}H_{17}N$ mw: 271.38

SYN: 10-METHYL-3-ETHYL-7,8-BENZACRIDINE(FRENCH)

TOXICITY DATA with REFERENCE
skn-mus TDLo:275 mg/kg/22W-I:ETA ACRSAJ 4,315,56

SAFETY PROFILE: Questionable carcinogen with experimental tumorigenic data. When heated to decomposition it emits toxic fumes of NO_x.

MJY500 CAS:25057-89-0 ***HR: 2***
3-(1-METHYLETHYL)-1H-2,1,3-BENZOTHIAZAIN-4(3H)-ONE-2,2-DIOXIDE
mf: $C_{10}H_{12}N_2O_3S$ mw: 240.30

SYNS: BAS 351-H ◇ BASAGRAN ◇ BENDIOXIDE ◇ BENTAZON ◇ 3-ISOPROPYL-2,1,3-BENZOTHIADIAZINON-(4)-2,2-DIOXID (GERMAN) ◇ 3-ISOPROPYL-1H-2,1,3-BENZOTHIADIAZIN-4(3H)-ONE-2,2-DIOXIDE

TOXICITY DATA with REFERENCE
orl-rat TDLo:25 mg/kg (female 6D post):REP AXVMAW 42,261,88
orl-rat TDLo:25 mg/kg (female 6D post):TER AXVMAW 42,261,88
orl-rat LD50:1100 mg/kg GUCHAZ 6,36,73
skn-rat LD50:2500 mg/kg 85DPAN -,-,71/76
orl-dog LD50:450 mg/kg PSSCBG 3,242a,72
orl-cat LD50:500 mg/kg GUCHAZ 6,36,73
orl-rbt LD50:750 mg/kg 85DPAN -,-,71/76
orl-qal LD50:720 mg/kg GUCHAZ 6,36,73

CONSENSUS REPORTS: Reported in EPA TSCA Inventory. EPA Genetic Toxicology Program.

SAFETY PROFILE: Moderately toxic by ingestion and skin contact. An experimental teratogen. Other experimental reproductive effects. When heated to decomposition it emits very toxic fumes of SO_x and NO_x.

MJZ000 CAS:2122-19-2 ***HR: 3***
4-METHYLETHYLENETHIOUREA
mf: $C_4H_8N_2S$ mw: 116.20

SYN: 4-METHYL-2-IMIDAZOLIDINETHIONE ◇ PLTU ◇ PROPILENTIOUREA ◇ PROPYLENE THIOUREA ◇ PROPYLENTHIOHARNSTOFF

TOXICITY DATA with REFERENCE
orl-rat TDLo:100 mg/kg (female 13D post):TER MVMZA8 33,137,78
ipr-mus LD50:50 mg/kg EJMCA5 17,235,82

SAFETY PROFILE: Poison by intraperitoneal route. An experimental teratogen. When heated to decomposition it emits very toxic fumes of SO_x and NO_x.

MKA000 CAS:31218-83-4 ***HR: 3***
(E)-1-METHYLETHYL-3-(((ETHYLAMINO) METHOXYPHOSPHINOTHIOYL)OXY-2-BUTENOATE
mf: $C_{10}H_{20}NO_4PS$ mw: 281.34

SYNS: BLOTIC ◇ ENT 27,989 ◇ (3)-O-2-ISOPROPOXY-CARBONYL-1-METHYLVINYL-O-METHYLETHYLPHOSPHORAMIDOTHIOATE ◇ PROPETAMPHOS ◇ SAFROTIN ◇ SAN 52 139 I ◇ SANDOZ 52139 ◇ VEL 4283

TOXICITY DATA with REFERENCE
orl-rat LD50:75 mg/kg 85ARAE 1,118,77
skn-rat LD50:2300 mg/kg FMCHA2 -,D256,80

SAFETY PROFILE: Poison by ingestion. Moderately toxic by skin contact. When heated to decomposition it emits very toxic fumes of PO_x, SO_x, and NO_x.

MKA250 CAS:64-65-3 ***HR: 3***
3-METHYL-3-ETHYLGLUTARIMIDE
mf: $C_8H_{13}NO_2$ mw: 155.22

SYNS: AHYPNON ◇ BEMEGRIDE ◇ 2,6-DIOXO-4-METHYL-4-ETHYLPIPERIDINE ◇ 4-ETHYL-4-METHYL-2,6-DIOXOPIPERIDINE ◇ β-ETHYL-β-METHYLGLUTARIMIDE ◇ 3-ETHYL-3-METHYLGLUTARIMIDE ◇ 4-ETHYL-4-METHYL-2,6-PIPERIDINEDIONE ◇ EUKRATON ◇ MALYSOL ◇ MEGIMIDE ◇ 4-METHYL-4-ETHYL-2,6-DIOXOPIPERIDINE ◇ β-METHYL-β-ETHYLGLUTARIMIDE ◇ MIKEDIMIDE

TOXICITY DATA with REFERENCE
orl-wmn TDLo:100 mg/kg:CNS LANCAO 2,967,56
orl-man TDLo:20 mg/kg:CNS LANCAO 2,967,56
ipr-rat LD50:24 mg/kg AIPTAK 135,9,62
scu-rat LD50:31 mg/kg JPPMAB 13,244,61
ivn-rat LD50:16 mg/kg AIPTAK 135,9,62
orl-mus LD50:41 mg/kg JPETAB 128,176,60
ipr-mus LD50:25 mg/kg JPETAB 128,176,60
scu-mus LD50:27 mg/kg JPETAB 128,176,60
ivn-mus LD50:16 mg/kg AIPTAK 135,9,62
ims-mus LD50:33 mg/kg NIIRDN 6,763,82
par-mus LD50:32 mg/kg ARZNAD 6,583,56
ivn-rbt LD50:23 mg/kg AIPTAK 135,9,62

CONSENSUS REPORTS: Reported in EPA TSCA Inventory.

SAFETY PROFILE: Poison by ingestion, intravenous, intraperitoneal, intramuscular, subcutaneous, and parenteral routes. Human systemic effects by ingestion: wakefulness, hallucinations, distorted perceptions, toxic psychosis. An analeptic, central nervous system stimulant; used to counteract barbiturate poisoning. When heated to decomposition it emits toxic fumes of NO_x.

MKA400 CAS:78-93-3 ***HR: 3***
METHYL ETHYL KETONE
DOT: UN 1193/UN 1232
mf: C_4H_8O mw: 72.12

$$CH_3CO \cdot CH_2CH_3$$

PROP: Colorless liquid; acetone-like odor. Bp: 79.57°, fp: −85.9°, lel: 1.8%, uel: 11.5%, flash p: 22°F (TOC), d: 0.80615 @ 20°/20°, vap press: 71.2 mm @ 20°, autoign temp: 960°F, vap d: 2.42, ULC: 85-90. Misc with alc, ether, fixed oils, and water.

SYNS: AETHYLMETHYLKETON (GERMAN) ◇ 2-BUTANONE (OSHA) ◇ BUTANONE 2 (FRENCH) ◇ ETHYL METHYL CETONE (FRENCH) ◇ ETHYLMETHYLKETON (DUTCH) ◇ ETHYL METHYL KETONE (DOT) ◇ FEMA No. 2170 ◇ MEK ◇ METHYL ACETONE (DOT) ◇ METILETILCHETONE (ITALIAN) ◇ METYLOETYLOKETON (POLISH) ◇ RCRA WASTE NUMBER U159

TOXICITY DATA with REFERENCE
eye-hmn 350 ppm JIHTAB 25,282,43
skn-rbt 500 mg/24H MOD JIHTAB 25,282,43
skn-rbt 402 mg/24H MLD TXAPA9 19,276,71
skn-rbt 13780 μg/24H open MLD AIHAAP 23,95,62
eye-rbt 80 mg TXAPA9 19,276,71
sln-smc 33800 ppm MUREAV 149,339,85
ihl-rat TCLo:1000 ppm/(6-15D preg):TER TXAPA9 28,452,74
ihl-hmn TCLo:100 ppm/5M:IRR JIHTAB 25,282,43
orl-rat LD50:2737 mg/kg TXAPA9 19,699,71
ihl-rat LC50:23500 mg/m³/8H AIHAAP 20,364,59
orl-mus LD50:4050 mg/kg TOLED5 30,13,86
ihl-mus LC50:40 g/m³/2H 85GMAT -,83,82
ipr-mus LD50:616 mg/kg SCCUR* -,6,61
skn-rbt LD50:6480 mg/kg SCCUR* MSDS-5390-4

CONSENSUS REPORTS: Community Right-To-Know List. EPA Genetic Toxicology Program. Reported in EPA TSCA Inventory.

OSHA PEL: (Transitional: TWA 200 ppm) TWA 200 ppm; STEL 300 ppm
ACGIH TLV: TWA 200 ppm; STEL 300 ppm; BEI: 2 mg(MEK)/L in urine at end of shift.
DFG MAK: 200 ppm (590 mg/m³)
NIOSH REL: (Ketones) TWA 590 mg/m³
DOT Classification: Flammable Liquid; Label: Flammable Liquid.

SAFETY PROFILE: Moderately toxic by ingestion, skin contact, and intraperitoneal routes. Human systemic effects by inhalation: conjunctiva irritation and unspecified effects on the nose and respiratory system. An experimental teratogen. A strong irritant. Human eye irritation @ 350 ppm. Affects peripheral nervous system and central nervous system. See also KETONES. Highly flammable liquid. Reaction with hydrogen peroxide + nitric acid forms a heat- and shock-sensitive explosive product. Ignition on contact with potassium tert-butoxide. Mixture with 2-propanol will produce explosive peroxides during storage. Vigorous reaction with chloroform + alkali. Incompatible with chlorosulfonic acid; oleum. To fight fire, use alcohol foam, CO_2, dry chemical. Used in production of drugs of abuse. When heated to decomposition it emits acrid smoke and fumes.

MKA500 CAS:1338-23-4 ***HR: 3***
METHYL ETHYL KETONE PEROXIDE
DOT: UN 2127/UN 2550
mf: $C_8H_{16}O_4$ mw: 176.24

SYNS: HI-POINT 90 ◇ LUPERSOL ◇ MEKP ◇ MEK PEROXIDE ◇ METHYLETHYLKETONHYDROPEROXIDE ◇ NCI-C55447 ◇ QUICK-SET EXTRA ◇ RCRA WASTE NUMBER U160 ◇ SPRAYSET MEKP ◇ THERMACURE

TOXICITY DATA with REFERENCE
skn-rbt 500 mg AIHAAP 19,205,58
eye-rbt 3 mg AIHAAP 19,205,58
unr-mus TDLo:282 mg/kg:ETA RARSAM 3,193,63
orl-hmn TDLo:480 mg/kg:GIT NCPBBY Jan/Feb,69
orl-rat LD50:484 mg/kg AIHAAP 19,205,58
ihl-rat LC50:200 ppm/4H AIHAAP 19,205,58
ipr-rat LD50:65 mg/kg AIHAAP 19,205,58
orl-mus LD50:470 mg/kg JAMAAP 165,201,57
ihl-mus LC50:170 ppm/4H AIHAAP 19,205,58

CONSENSUS REPORTS: Reported in EPA TSCA Inventory.

OSHA PEL: CL 0.7 ppm
ACGIH TLV: CL 0.2 ppm
DFG MAK: Organic Peroxide, moderate skin irritant.
DOT Classification: Forbidden (in solution with > 9% active oxygen); Organic Peroxide; Label: Organic Peroxide.

SAFETY PROFILE: Poison by intraperitoneal route. Moderately toxic by ingestion and inhalation. Human systemic effects by ingestion: changes in structure or function of esophagus, nausea or vomiting, other gastrointestinal effects. A moderate skin and eye irritant. Questionable carcinogen with experimental tumorigenic data. A shock-sensitive explosive. When heated to decomposition it emits acrid smoke and irritating fumes. See also KETONES and PEROXIDES.

MKA750 CAS:624-46-4 ***HR: 3***
METHYL ETHYL KETONE SEMICARBAZONE
mf: $C_5H_{11}N_3O$ mw: 129.19

SYN: 2-BUTANONE, SEMICARBAZONE

TOXICITY DATA with REFERENCE
ipr-mus LDLo:250 mg/kg CBCCT* 9,129,57
ivn-mus LD50:180 mg/kg CSLNX* NX#05085

SAFETY PROFILE: Poison by intravenous and intraperitoneal routes. When heated to decomposition it emits toxic fumes of NO_x. See also KETONES.

MKB000 CAS:10595-95-6 **HR: 3**
N,N-METHYLETHYLNITROSAMINE
mf: $C_3H_8N_2O$ mw: 88.13

SYNS: ETHYLMETHYLNITROSAMINE ◇ METHYLAETHYLNITROSAMIN (GERMAN) ◇ METHYLETHYLNITROSAMINE ◇ N-METHYL-N-NITROSO-ETHAMINE ◇ N-METHYL-N-NITROSOETHYLAMINE ◇ NEMA ◇ N-NITROSOETHYLMETHYLAMINE ◇ N-NITROSOMETHYLETHYLAMINE (MAK) ◇ NMEA

TOXICITY DATA with REFERENCE
mma-sat 500 μg/plate CRNGDP 5,1091.84
pic-esc 100 mg/L TCMUE9 1,91,84
orl-rat TDLo:600 mg/kg/15W-I:CAR CNREA8 46,2252,86
orl-rat TD:420 mg/kg/71W-C:ETA,REP ZEKBAI 69,103,67
orl-rat TD:2250 mg/kg/30W-C:ETA CALEDQ 14,297,81
orl-rat LD50:90 mg/kg ZEKBAI 69,103,67

CONSENSUS REPORTS: IARC Cancer Review: Group 2B IMEMDT 7,56,87; Animal Limited Evidence IMEMDT 17,221,78. EPA Genetic Toxicology Program.

DFG MAK: Animal Carcinogen, Suspected Human Carcinogen.

SAFETY PROFILE: Confirmed carcinogen with experimental carcinogenic and tumorigenic data. Poison by ingestion. Experimental reproductive effects. Mutation data reported. When heated to decomposition it emits toxic fumes of NO_x. See also NITROSAMINES.

MKB250 CAS:50-12-4 **HR: 3**
3-METHYL-5-ETHYL-5-PHENYLHYDANTOIN
mf: $C_{12}H_{14}N_2O_2$ mw: 218.28

SYNS: EPILAN ◇ 5-ETHYL-3-METHYL-5-PHENYLHYDANTOIN ◇ 5-ETHYL-3-METHYL-5-PHENYL-2,4(3H,5H)-IMIDAZOLEDIONE ◇ 5-ETHYL-3-METHYL-5-PHENYLIMIDAZOLIDIN-2,4-DIONE ◇ 3-ETHYLNIRVANOL ◇ GEROT-EPILAN ◇ INSULTON ◇ MEPHENYTOIN ◇ MESANTOIN ◇ METHOIN ◇ METHYL HYDANTOIN ◇ 3-METHYL-5,5-PHENYLETHYLHYDANTOIN ◇ NSC-34652 ◇ PHENANTOIN ◇ PHENYLETHYLMETHYLHYDANTOIN ◇ SACERNO ◇ SEDANTOINAL ◇ TRIANTOIN

TOXICITY DATA with REFERENCE
cyt-hmn:leu 1 mg/L AJOGAH 116,867,73
ipr-mus TDLo:786 mg/kg (8-10D preg):TER TXAPA9 64,271,82
orl-wmn TDLo:1100 mg/kg:CNS,BLD,MSK JAMAAP 138,498,49
orl-rat LD50:850 mg/kg 27ZQAG -,263,72
ipr-rat LDLo:270 mg/kg MEIEDD 11,919,89
orl-mus LD50:440 mg/kg CKFRAY 4,333,55
ipr-mus LD50:317 mg/kg AIPTAK 156,261,65
orl-cat LD50:190 mg/kg 27ZQAG -,263,72
orl-rbt LD50:430 mg/kg 27ZIAQ-,151,73
orl-gpg LD50:380 mg/kg 27ZQAG -,263,72

SAFETY PROFILE: Poison by ingestion and intraperitoneal routes. Human systemic effects by ingestion: somnolence, hemorrhage, changes in teeth and supporting structures. Human mutation data reported. An experimental teratogen. An FDA proprietary drug used as an anticonvulsant. When heated to decomposition it emits toxic fumes of NO_x.

MKB500 CAS:14551-09-8 **HR: 3**
N-METHYL-N-ETHYL-4-(4'-(PYRIDYL-1'OXIDE) AZO)ANILINE
mf: $C_{14}H_{16}N_4O$ mw: 256.34

SYN: 4-((((4-ETHYL-4-METHYL)AMINO)PHENYL)AZO)PYRIDINE1-OXIDE

TOXICITY DATA with REFERENCE
orl-rat TDLo:2142 mg/kg/17W-C:NEO JNCIAM 37,365,66

SAFETY PROFILE: Questionable carcinogen with experimental neoplastigenic data. When heated to decomposition it emits toxic fumes of NO_x.

MKB750 CAS:72-33-3 **HR: 3**
3-METHYLETHYNYLESTRADIOL
mf: $C_{21}H_{26}O_2$ mw: 310.47

SYNS: COMPOUND 33355 ◇ DELTA-MVE ◇ 17-α-ETHINYL ESTRADIOL 3-METHYL ETHER ◇ ETHINYLESTRADIOL-3-METHYL ETHER ◇ 17-α-ETHINYL OESTRADIOL-3-METHYL ETHER ◇ ETHINYLOESTRADIOL-3-METHYL ETHER ◇ ETHYNYLESTRADIOL-3-METHYL ETHER ◇ 17-ETHYNYLESTRADIOL-3-METHYL ETHER ◇ 17-α-ETHYNYLESTRADIOL-3-METHYL ETHER ◇ (+)-17-α-ETHYNYL-17-β-HYDROXY-3-METHOXY-1,3,5(10)-ESTRATRIENE ◇ (+)-17-α-ETHYNYL-17-β-HYDROXY-3-METHOXY-1,3,5(10)-OESTRATRIENE ◇ 17-α-ETHYNYL-3-METHOXY-1,3,5(10)-ESTRATRIEN-17-β-OL ◇ 17-ETHYNYL-3-METHOXY-1,3,5(10)-ESTRATRIEN-17β-OL ◇ 17-α-ETHYNYL-3-METHOXY-17-β-HYDROXY-Δ-1,3,5(10)-ESTRATRIENE ◇ 17-α-ETHYNYL-3-METHOXY-17-β-HYDROXY-Δ-1,3,5(10)-OESTRATRIENE ◇ 17-ETHYNYL-3-METHOXY-1,3,5(10)-OESTRATIEN-17-β-OL ◇ ETHYNYLOESTRADIOL METHYL ETHER ◇ 17-ETHYNYLOESTRADIOL-3-METHYL ETHER ◇ 17-α-ETHYNYLOESTRADIOL-3-METHYL ETHER ◇ 17-α-ETHYNYLOESTRADIOL METHYL ETHER ◇ MESTRANOL ◇ MESTRENOL ◇ 3-METHOXY-17-α-ETHINYLESTRADIOL ◇ 3-METHOXY-17-α-ETHINYLOESTRADIOL ◇ 3-METHOXYETHYNYLESTRADIOL ◇ 3-METHOXY-17-α-ETHYNOESTRADIOL ◇ 3-METHOXY-17-α-ETHYNYLESTRADIOL ◇ 3-METHOXYETHYNYLOESTRADIOL ◇ 3-METHOXY-17-ETHYNYLOESTRADIOL-17-β ◇ 3-METHOXY-17-α-ETHYNYL-1,3,5(10)-ESTRATRIEN-17-β-OL ◇ 3-METHOXY-17-α-ETHYNYL-1,3,5(10)-OESTRATRIEN-17-β-OL ◇ 3-METHOXY-19-NOR-17-α-PREGNA-1,3,5(10)-TRIEN-10-YN-17-OL ◇ 3-METHOXY-17-α-19-NORPREGNA-K,3,5(10)-TRIEN-20-YN-17-OL ◇ (17-α)-3-METHOXY-19-NORPREGN-1,3,5(10)-TRIEN-20-YN-17-OL ◇ 3-METHYLETHYNYLOESTRADIOL

TOXICITY DATA with REFERENCE
oms-mus-scu 120 μg/kg AJOGAH 120,390,74

orl-wmn TDLo:20 μg/kg (20D pre):REP INJFA3 9,57,64
scu-rbt TDLo:15 μg/kg (female 1-3D post):TER FESTAS 20,211,69
orl-rat TDLo:2400 μg/kg/17W-C:ETA PAACA3 21,76,80
orl-mus TDLo:19 mg/kg/30W-C:NEO CRSBAW 168,1190,74

CONSENSUS REPORTS: NTP Fifth Annual Report on Carcinogens. IARC Cancer Review: Human Limited Evidence IMEMDT 21,257,79; Animal Sufficient Evidence IMEMDT 6,87,74; IMEMDT 21,257,79.

SAFETY PROFILE: Confirmed carcinogen with experimental neoplastigenic, tumorigenic, and teratogenic data. Human reproductive effects by ingestion: changes in ovaries and fallopian tubes, fertility effects. Experimental reproductive effects. Mutation data reported. An FDA proprietary drug. A steroid used in oral contraceptives. When heated to decomposition it emits acrid smoke and irritating fumes.

MKC000 CAS:26509-45-5 ***HR: 2***
METHYL EUGENOL GLYCOL
mf: $C_{11}H_{16}O_4$ mw: 212.27

TOXICITY DATA with REFERENCE
ipr-mus LD50:2005 mg/kg AIPTAK 199,226,72
ivn-mus LD50:880 mg/kg AIPTAK 199,226,72

SAFETY PROFILE: Moderately toxic by intraperitoneal and intravenous routes. When heated to decomposition it emits acrid smoke and irritating fumes. See also ALLYL COMPOUNDS.

MKC250 CAS:3344-14-7 ***HR: 3***
S-METHYL FENITROOXON
mf: $C_9H_{12}NO_5PS$ mw: 277.25

SYNS: ISOSUMITHION ◇ METATHION, S-METHYL ISOMER ◇ S-METHYL FENITROTHION ◇ SUMITHION S-ISOMER ◇ THIOPHOSPHATE de O,S-DIMETHYL et de O-(3-METHYL-4-NITROPHENYLE) (FRENCH)

TOXICITY DATA with REFERENCE
dlt-mus-orl 1 mg/kg PCBPBS 6,280,76
orl-rat LD50:315 mg/kg PCBPBS 6,280,76
orl-mus LD50:320 mg/kg PCBPBS 6,280,76
ipr-mus LD50:54 mg/kg NNGADV 3,35,78

SAFETY PROFILE: Poison by ingestion and intraperitoneal routes. Mutation data reported. When heated to decomposition it emits very toxic fumes of SO_x, PO_x, and NO_x.

MKC500 CAS:33543-31-6 ***HR: 3***
2-METHYLFLUORANTHENE
mf: $C_{17}H_{12}$ mw: 216.29

TOXICITY DATA with REFERENCE
mma-sat 50 μg/plate CRNGDP 3,841,82
skn-mus TDLo:624 mg/kg/52W-I:ETA JNCIAM 49,1165,72

CONSENSUS REPORTS: IARC Cancer Review: Group 3 IMEMDT 7,56,87; Animal Limited Evidence IMEMDT 32,399,83.

SAFETY PROFILE: Questionable carcinogen with experimental tumorigenic data. Mutation data reported. When heated to decomposition it emits acrid smoke and irritating fumes.

MKC750 CAS:1706-01-0 ***HR: 3***
3-METHYLFLUORANTHENE
mf: $C_{17}H_{12}$ mw: 216.29

TOXICITY DATA with REFERENCE
mma-sat 10 μg/plate CRNGDP 3,841,82
skn-mus TDLo:40 mg/kg/20D:NEO JNCIAM 49,1165,72

CONSENSUS REPORTS: IARC Cancer Review: Group 3 IMEMDT 7,56,87; Animal Inadequate Evidence IMEMDT 32,399,83.

SAFETY PROFILE: Questionable carcinogen with experimental neoplastigenic data. Mutation data reported. When heated to decomposition it emits acrid smoke and irritating fumes.

MKD000 CAS:453-18-9 ***HR: 3***
METHYL FLUOROACETATE
mf: $C_3H_5FO_2$ mw: 92.08

SYNS: FLUOROACETIC ACID METHYL ESTER ◇ MFA

TOXICITY DATA with REFERENCE
orl-man TDLo:650 μg/kg:GIT NTIS** PB158-508
orl-rat LD50:3500 μg/kg NTIS** PB158-508
ihl-rat LC50:300 mg/m^3/10M NATUAS 160,179,47
scu-rat LD50:5 mg/kg JPETAB 87,90,46
ims-rat LD50:2500 μg/kg NATUAS 160,179,47
orl-mus LD50:5 mg/kg NTIS** PB158-508
ihl-mus LC50:3200 mg/m^3/10M NATUAS 160,179,47
ipr-mus LD50:7500 μg/kg JOCEAH 21,883,56
scu-mus LD50:5 mg/kg NATUAS 160,179,47
ivn-mus LD50:17 mg/kg NTIS** PB158-508
par-mus LD50:15 mg/kg JCSOA9 -,1471,49
orl-dog LD50:100 μg/kg NTIS** PB158-508
ihl-dog LC50:25 mg/m^3/10M NTIS** PB158-508
skn-rbt LD50:20 mg/kg JCSOA9 -,1773,48

SAFETY PROFILE: Poison by ingestion, inhalation, skin contact, subcutaneous, intramuscular, intraperitoneal, parenteral, and intravenous routes. Human systemic effects by ingestion of very small amounts: nausea or vomiting. When heated to decomposition it emits toxic fumes of F^-. See also ESTERS and FLUORIDES.

MKD250 CAS:482-41-7 ***HR: 3***
7-METHYL-9-FLUOROBENZ(c)ACRIDINE
mf: $C_{18}H_{12}FN$ mw: 261.31

SYN: 3-FLUORO-10-METHYL-7,8-BENZACRIDINE(FRENCH)

TOXICITY DATA with REFERENCE
skn-mus TDLo:132 mg/kg/11W-I:ETA ACRSAJ 4,315,56

SAFETY PROFILE: Questionable carcinogen with experimental tumorigenic data. When heated to decomposition it emits very toxic fumes of F^- and NO_x.

MKD500 CAS:439-25-8 ***HR: 3***
7-METHYL-11-FLUOROBENZ(c)ACRIDINE
mf: $C_{18}H_{12}FN$ mw: 261.31

SYN: 1-FLUORO-10-METHYL-7,8-BENZACRIDINE

TOXICITY DATA with REFERENCE
scu-mus TDLo:20 mg/kg:ETA AICCA6 11,736,55

SAFETY PROFILE: Questionable carcinogen with experimental tumorigenic data. When heated to decomposition it emits very toxic fumes of F^- and NO_x.

MKD750 CAS:436-30-6 ***HR: 3***
10-METHYL-3-FLUORO-5,6-BENZACRIDINE
mf: $C_{18}H_{12}FN$ mw: 261.31

SYNS: 3-FLUORO-10-METHYL-5,6-BENZACRIDINE ◇ 10-FLUORO-12-METHYLBENZ(a)ACRIDINE

TOXICITY DATA with REFERENCE
skn-mus TDLo:245 mg/kg/20W-I:ETA AICCA6 11,736,55

SAFETY PROFILE: Questionable carcinogen with experimental tumorigenic data. When heated to decomposition it emits very toxic fumes of F^- and NO_x.

MKE000 CAS:406-20-2 ***HR: 3***
METHYL-4-FLUOROBUTYRATE
mf: $C_5H_9FO_2$ mw: 120.14

SYNS: 4-FLUOROBUTYRIC ACID METHYL ESTER ◇ METHYLESTER KYSELINY 4-FLUORMASELNE ◇ METHYL-Γ-FLUOROBUTYRATE ◇ METHYL-ω-FLUOROBUTYRATE

TOXICITY DATA with REFERENCE
orl-rat LDLo:25 mg/kg NCNSA6 5,17,53
ihl-rat LC50:350 mg/m³/10M NTIS** PB158-508
ihl-mus LC50:120 mg/m³/10M NTIS** PB158-508
scu-mus LDLo:1 mg/kg 11FYAN 3,78,63
ihl-dog LC50:50 mg/m³/10M NTIS** PB158-508
ihl-mky LC50:500 mg/m³/10M NTIS** PB158-508
ivn-mky LDLo:3 mg/kg PAREAQ 1,383,49
ivn-cat LD50:200 μg/kg PAREAQ 1,383,49

SAFETY PROFILE: Poison by ingestion, inhalation, intravenous, and subcutaneous routes. When heated to decomposition it emits toxic fumes of F^-.

MKE250 CAS:2367-25-1 ***HR: 3***
METHYL-Γ-FLUOROCROTONATE
mf: $C_5H_7FO_2$ mw: 118.12

SYNS: 4-FLUORO-CROTONIC ACID METHYL ESTER ◇ TL 1183

TOXICITY DATA with REFERENCE
ihl-mus LC50:89 mg/m³/10M NTIS** PB158-508
ihl-gpg LC50:150 mg/m³/10M NTIS** PB158-508

SAFETY PROFILE: Poison by inhalation. When heated to decomposition it emits toxic fumes of F^-.

MKE750 CAS:63904-99-4 ***HR: 3***
METHYL-Γ-FLUORO-β-HYDROXYBUTYRATE
mf: $C_5H_9FO_3$ mw: 136.14

SYNS: Γ-FLUORO-β-HYDROXY-BUTYRIC ACID METHYL ESTER ◇ TL 1333

TOXICITY DATA with REFERENCE
orl-rat LDLo:1 mg/kg NCNSA6 5,21,53
ihl-mus LD50:23 mg/m³/10M NTIS** PB158-508
ihl-mky LC50:200 mg/m³/10M NTIS** PB158-508
ihl-cat LC50:100 mg/m³/10M NTIS** PB158-508

SAFETY PROFILE: Poison by ingestion and inhalation. When heated to decomposition it emits toxic fumes of F^-. See also ESTERS.

MKF000 CAS:63732-23-0 ***HR: 3***
METHYL-Γ-FLUORO-β-HYDROXYTHIOLBUTYRATE
mf: $C_5H_9FO_2S$ mw: 152.20

SYN: 4-FLUORO-3-HYDROXY-BUTANETHIOIC ACID METHYL ESTER

TOXICITY DATA with REFERENCE
ihl-dog LC50:63 mg/m³/10M NTIS** PB158-508
ihl-mky LC50:200 mg/m³/10M NTIS** PB158-508

SAFETY PROFILE: Poison by inhalation. When heated to decomposition it emits very toxic fumes of SO_x and F^-.

MKF250 ***HR: 3***
METHYL-FLUORO-PHOSPHORYLCHOLINE
mf: $C_6H_{16}FNO_2P$ mw: 184.20

SYNS: CHOLINE, INNER SALT, METHYLPHOSPHONOFLUORIDATE ◇ METHYLFLUORPHOSPHORSAEURECHOLINESTER (GERMAN)

TOXICITY DATA with REFERENCE
ipr-mus LD50:100 μg/kg AIPTAK 115(4),474,58
ipr-rbt LD50:10 μg/kg DEGEA3 15,2179,60
ivn-rbt LD50:10 μg/kg AIPTAK 115(4),474,58

SAFETY PROFILE: A deadly poison by intraperitoneal and intravenous routes. When heated to decomposition it emits very toxic fumes of PO_x, NO_x, and F^-.

MKG250 CAS:421-20-5 ***HR: 3***
METHYL FLUOROSULFATE
mf: CH_3FO_3S mw: 114.10

PROP: Liquid; ethereal odor. Bp: 92°, d: 1.427 @ 16°, vap d: 3.94.

SYNS: MAGIC METHYL ◇ METHYL ESTER FLUOROSULFURIC ACID ◇ METHYL FLUOROSULFONATE ◇ METHYL FLUORSULFONATE

TOXICITY DATA with REFERENCE
eye-rbt 100 mg/4 sec rns SEV AIHAAP 40,600,79
mmo-sat 4 µg/plate MUREAV 51,285,78
ihl-rat LC50:5 ppm/1H AIHAAP 40,600,79
orl-mus LD50:112 mg/kg AIHAAP 40,600,79
skn-rbt LDLo:455 mg/kg AIHAAP 40,600,79

SAFETY PROFILE: Poison by inhalation and ingestion. Moderately toxic by skin contact. A skin, mucous membrane, and severe eye irritant. Mutation data reported. Reacts with water, steam, or acids to produce toxic and corrosive fumes. When heated to decomposition it emits toxic fumes of F^- and SO_x.

MKG275 CAS:70114-87-3 ***HR: D***
x-METHYLFOLIC ACID
mf: $C_{20}H_{21}N_7O_5$ mw: 439.48

SYNS: ACIDE x-METHYLFOLIQUE (FRENCH) ◇ x-METHYLPTEROYLGLUTAMIC ACID

TOXICITY DATA with REFERENCE
orl-mus TDLo:2400 mg/kg (female 6-9D post):REP COREAF 245,1963,57
orl-rat TDLo:500 mg/kg (female 10-11D post):TER AFPEAM 16,509,59

SAFETY PROFILE: An experimental teratogen. Other experimental reproductive effects. When heated to decomposition it emits toxic fumes of NO_x.

MKG500 CAS:123-39-7 ***HR: 2***
N-METHYLFORMAMIDE
mf: C_2H_5NO mw: 59.08

PROP: Flash p: <71.6°F.

SYNS: METHYLFORMAMIDE ◇ MONOMETHYLFORMAMIDE ◇ NSC 3051

TOXICITY DATA with REFERENCE
eye-rbt 100 mg TXAPA9 39,129,77
skn-rat TDLo:600 mg/kg (female 9D post):TER TXAPA9 41,35,77
orl-rat TDLo:1 g/kg (7D preg):REP 85DJA5 -,95,71
orl-rat LD50:4000 mg/kg JRPFA4 4,219,62
ipr-rat LD50:3500 mg/kg JRPFA4 4,219,62
orl-mus LD50:2600 mg/kg TXCYAC 34,173,85
ipr-mus LD50:802 mg/kg NCISP* JAN86
scu-mus LD50:3100 mg/kg DABBBA 40,549,79
ivn-mus LD50:1580 mg/kg TXCYAC 34,173,85
ims-mus LD50:2700 mg/kg TXCYAC 34,173,85
ivn-dog LD10:1262 mg/kg NTIS** PB82-232158

CONSENSUS REPORTS: Reported in EPA TSCA Inventory. EPA Genetic Toxicology Program.

SAFETY PROFILE: Moderately toxic by ingestion, intraperitoneal, intravenous, intramuscular, and subcutaneous routes. An experimental teratogen. Experimental reproductive effects. An eye irritant. A very dangerous fire hazard when exposed to heat or flame. Violent reaction with benzene sulfonyl chloride. When heated to decomposition it emits toxic fumes of NO_x.

MKG750 CAS:107-31-3 ***HR: 3***
METHYL FORMATE
DOT: UN 1243
mf: $C_2H_4O_2$ mw: 60.06

PROP: Colorless liquid; agreeable odor. Mp: −99.8°, bp: 31.5°, lel: 5.9%, uel: 20%, flash p: −2.2°F, d: 0.98149 @ 15°/4°, 0.975 @ 20°/4°, autoign temp: 869°F, vap press: 400 mm @ 16°/0°, vap d: 2.07. Solidifies at about 100°. Moderately sol in water, methyl alcohol; misc in alc.

SYNS: FORMIATE de METHYLE (FRENCH) ◇ METHYLE (FORMIATE de) (FRENCH) ◇ METHYLFORMIAAT (DUTCH) ◇ METHYLFORMIAT (GERMAN) ◇ METHYL METHANOATE ◇ METIL (FORMIATO di) (ITALIAN)

TOXICITY DATA with REFERENCE
orl-rbt LD50:1622 mg/kg IMSUAI 41,31,72
ihl-gpg LCLo:10000 ppm 14CYAT -,1855,63

CONSENSUS REPORTS: Reported in EPA TSCA Inventory.

OSHA PEL: (Transitional: TWA 100 ppm) TWA 100 ppm; STEL 150 ppm
ACGIH TLV: TWA 100 ppm; STEL 150 ppm
DFG MAK: 100 ppm (250 mg/m^3)
DOT Classification: Flammable Liquid; Label: Flammable Liquid.

SAFETY PROFILE: Moderately toxic by ingestion. Inhalation of vapor can cause irritation to nasal passages and conjunctiva, optic neuritis, narcosis, retching, and death from pulmonary irritation. Industrial fatalities have occurred only with exposure to high concentrations. Flammable liquid. Very dangerous fire hazard when exposed to heat or flame; can react vigorously with oxidizing materials. Explosive in the form of vapor when exposed to heat or flame. Reacts with methanol + sodium methoxide to form an explosive product. To fight fire, use alcohol foam, CO_2, dry chemical. When heated to decomposition it emits acrid smoke and irritating fumes.

MKH000 CAS:534-22-5 ***HR: 3***
2-METHYLFURAN
DOT: UN 2301
mf: C_5H_6O mw: 82.11

PROP: Colorless, mobile liquid; ether-like odor. Bp: 63.7°, fp: −88.7°, flash p: −22°F, d: 0.914 @ 20°/4°, vap press: 139 mm @ 20°, vap d: 2.8.

SYNS: METHYLFURAN ◇ SILVAN (CZECH)

TOXICITY DATA with REFERENCE
eye-rat LD50:500 mg/24H MLD 28ZPAK -,138,72
cyt-ham:ovr 75300 μmol/L CALEDQ 13,89,81
orl-rat LD50:167 mg/kg 28ZPAK -,138,72
ihl-rat LCLo:377 ppm/4H 28ZPAK -,138,72
scu-rat LC50:10 g/m³/2H 85GMAT -,84,82

CONSENSUS REPORTS: Reported in EPA TSCA Inventory.

DOT Classification: Flammable Liquid; Label: Flammable Liquid.

SAFETY PROFILE: Poison by ingestion. Moderately toxic by inhalation. An eye irritant. Mutation data reported. Very dangerous fire hazard when exposed to heat or flame; can react vigorously with oxidizing materials. To fight fire, use CO_2, dry chemical. When heated to decomposition it emits acrid smoke and irritating fumes. See also ETHERS for explosion hazard.

MKH250 CAS:591-12-8 ***HR: 2***
5-METHYL-2(3H)-FURANONE
mf: $C_5H_6O_2$ mw: 98.11

SYN: ANGELICA LACTONE ◇ β,Γ-ANGELICA LACTONE ◇ delta(sup 2)-ANGELICA LACTONE ◇ 4-HYDROXYPENT-3-ENOIC ACID LACTONE ◇ Γ-METHYL-β,Γ-CROTONOLACTONE

TOXICITY DATA with REFERENCE
orl-mus LD50:2800 mg/kg DCTODJ 3,249,80
ipr-mus LD50:3000 mg/kg APTOA6 2,109,46

CONSENSUS REPORTS: Reported in EPA TSCA Inventory.

SAFETY PROFILE: Moderately toxic by intraperitoneal and ingestion routes. When heated to decomposition it emits acrid smoke and irritating fumes. See also KETONES.

MKH500 CAS:591-11-7 ***HR: 3***
5-METHYL-2(5H)-FURANONE
mf: $C_5H_6O_2$ mw: 98.11

SYNS: β-ANGELICA LACTONE ◇ Δ^1-ANGELICA LACTONE ◇ 4-HYDROXY-2-PENTENOIC ACID Γ-LACTONE ◇ Γ-METHYL-α,β-CROTONOLACTONE

TOXICITY DATA with REFERENCE
scu-rat TDLo:2600 mg/kg/65W-I:ETA BJCAAI 19,392,65
ipr-mus LD50:750 mg/kg APTOA6 2,109,46
orl-cat LDLo:500 mg/kg JPETAB 36,355,29

SAFETY PROFILE: Moderately toxic by intraperitoneal route. Questionable carcinogen with experimental tumorigenic data. When heated to decomposition it emits acrid smoke and irritating fumes. See also KETONES.

MKH600 CAS:611-13-2 ***HR: 3***
METHYL FUROATE
mf: $C_6H_6O_3$ mw: 126.12

PROP: Mp: 164°, bp: 181°, d: 1.179.

SYNS: FURAN-α-CARBOXYLIC ACID METHYL ESTER ◇ 2-FUROIC ACID, METHYL ESTER ◇ METHYL 2-FURANCARBOXYLATE ◇ METHYL 2-FUROATE ◇ METHYL PYROMUCATE ◇ PYROMUCIC ACID METHYL ESTER

TOXICITY DATA with REFERENCE
skn-rbt 500 mg/24H MOD FCTXAV 17,869,79
ipr-rat LDLo:75 mg/kg JPETAB 58,174,36

CONSENSUS REPORTS: Reported in EPA TSCA Inventory.

SAFETY PROFILE: Poison by intraperitoneal route. A skin irritant and lachrymator. When heated to decomposition it emits acrid smoke and irritating fumes.

MKH750 CAS:102129-33-9 ***HR: 3***
5-(5-METHYL-2-FURYL)-1-PHENYL-3-(2(PIPERIDINO)ETHYL)-1H-PYRAZOLINE HYDROCHLORIDE
mf: $C_{21}H_{27}N_3O{\cdot}ClH$ mw: 373.97

SYN: 1-PHENYL-3-(β-PIPERIDINO-AETHYL)-5-(α'-METHYL-α-)-FURYL-PYRAZOLIN-HCL (GERMAN)

TOXICITY DATA with REFERENCE
ipr-mus LD50:108 mg/kg ARZNAD 10,925,60
scu-mus LD50:263 mg/kg ARZNAD 10,925,60

SAFETY PROFILE: Poison by intraperitoneal and subcutaneous routes. When heated to decomposition it emits very toxic fumes of HCl and NO_x.

MKI000 CAS:31959-87-2 ***HR: 3***
METHYL GAG

SYNS: 1,1'-(METHYLETHANEDILIDENEDINITRILO)BIGUANIDINE DIHYDROCHLORIDE DIHYDRATE ◇ NSC 32946

TOXICITY DATA with REFERENCE
skn-rbt 5 mg/24H rns TXCYAC 14,117,79
skn-rbt 1%/24H MLD NTIS** PB-269-596
orl-rat TDLo:3900 mg/kg/1Y-I:NEO,REP JNCIAM 41,985,68

SAFETY PROFILE: A skin irritant. Experimental reproductive effects. Questionable carcinogen with experimental neoplastigenic data. When heated to decomposition it emits very toxic fumes of HCl and NO_x.

MKI100 CAS:99-24-1 ***HR: 2***
METHYL GALLATE
mf: $C_8H_8O_5$ mw: 184.16

PROP: Monoclinic prisms from methanol, often hydrated or solvated. When dry: mp: 202°. Sol in hot water, alc, methanol, ether.

SYN: METHYL 3,4,5-TRIHYDROXYBENZOATE

TOXICITY DATA with REFERENCE
orl-mus LD50:1700 mg/kg 85GDA2 8(2),286,82
ipr-mus LD50:784 mg/kg 85GDA2 8(2),286,82
ivn-mus LD50:470 mg/kg 85GDA2 8(2),286,82

CONSENSUS REPORTS: Reported in EPA TSCA Inventory.

SAFETY PROFILE: Moderately toxic by ingestion, intraperitoneal, and intravenous routes. An antioxidant. When heated to decomposition it emits acrid smoke and irritating fumes. See also ESTERS.

MKI250 CAS:6280-15-5 ***HR: 3***
3-METHYL GLUTARALDEHYDE
mf: $C_6H_{10}O_2$ mw: 114.16

SYN: 3-METHYL PENTANEDIAL

TOXICITY DATA with REFERENCE
skn-rbt 10 mg/24H SEV AMIHBC 10,61,54
eye-rbt 250 μg SEV AMIHBC 10,61,54
orl-rat LD50:780 mg/kg AMIHBC 10,61,54
skn-rbt LD50:300 mg/kg AMIHBC 10,61,54

SAFETY PROFILE: Poison by skin contact. Moderately toxic by ingestion. A severe skin and eye irritant. When heated to decomposition it emits acrid smoke and irritating fumes. See also ALDEHYDES.

MKI550 CAS:922-68-9 ***HR: 1***
METHYL GLYOXYLATE
mf: $C_3H_4O_3$ mw: 88.07

SYNS: ACETIC ACID, OXO-, METHYL ESTER ◇ GLYOXYLIC ACID, METHYL ESTER ◇ METHYL OXOETHANOATE

TOXICITY DATA with REFERENCE
skn-rbt 500 mg/24H MOD JACTDZ 1,9,90
eye-rbt 10 mg/20S RNS MOD JACTDZ 1,9,90
orl-rat LDLo:5 g/kg JACTDZ 1,9,90

SAFETY PROFILE: A skin and eye irritant. When heated to decomposition it emits acrid smoke and irritating fumes.

MKI750 CAS:471-29-4 ***HR: 3***
METHYLGUANIDINE
mf: $C_2H_7N_3$ mw: 73.10

PROP: Colorless, deliquescent, strongly alkaline mass. Mp: decomp. Very sol in water; sol in alc.

SYNS: METHYLGUANIDIN (GERMAN) ◇ MONOMETHYL GUANIDIN (GERMAN) ◇ MONOMETHYLGUANIDINE

TOXICITY DATA with REFERENCE
cyt-ham:fbr 1 g/L/48H MUREAV 48,337,77
ipr-rat LDLo:18 mg/kg AEXPBL 90,129,21
scu-rat LDLo:250 mg/kg MEIEDD 10,871,83
ipr-mus LDLo:20 mg/kg ZEPHAR 25,441,11
ivn-mus LDLo:20 mg/kg ZEPHAR 25,441,11

CONSENSUS REPORTS: EPA Genetic Toxicology Program.

SAFETY PROFILE: Poison by subcutaneous, intravenous, and intraperitoneal routes. Mutation data reported. When heated to decomposition it emits toxic fumes of NO_x.

MKJ000 CAS:65272-47-1 ***HR: 3***
METHYLGUANIDINE mixed with SODIUM NITRITE (1:1)

SYN: SODIUM NITRITE mixed with METHYLGUANIDINE (1:1)

TOXICITY DATA with REFERENCE
mmo-sat 80 mmol/L GANNA2 65,45,74
orl-rat TDLo:70 g/kg/63W-C:NEO ZKKOBW 91,189,78

SAFETY PROFILE: Questionable carcinogen with experimental neoplastigenic data. Mutation data reported. When heated to decomposition it emits very toxic fumes of NO_x and Na_2O. See also SODIUM NITRITE and METHYLGUANIDINE.

MKJ250 CAS:63834-87-7 ***HR: 3***
METHYL HEPTANETHIOL
mf: $C_8H_{18}S$ mw: 146.32

PROP: Liquid. Bp: 159-166°, d: 0.848 @ 60°/60°F, vap d: 5.0, flash p: 115°F (OC). Insol in water.

SYN: tert-OCTYL MERCAPTAN

TOXICITY DATA with REFERENCE
eye-rbt 85 mg AIHAAP 19,171,58
orl-rat LD50:85 mg/kg AIHAAP 19,171,58
ihl-rat LC50:51 ppm/4H AIHAAP 19,171,58
skn-rat LD50:1954 mg/kg AIHAAP 19,171,58
ipr-rat LDLo:11 mg/kg AIHAAP 19,171,58
ihl-mus LC50:47 ppm/4H AIHAAP 19,171,58

SAFETY PROFILE: Poison by ingestion, inhalation, and intraperitoneal routes. Moderately toxic by skin contact. Irritating to eyes. Flammable when exposed to heat, flame or oxidizers. To fight fire, use foam, alcohol foam. When heated to decomposition it emits very toxic fumes of SO_x. See also MERCAPTANS.

MKJ750 CAS:1335-09-7 ***HR: 1***
6-METHYL-6-HEPTEN-2-OL
mf: $C_8H_{16}O$ mw: 128.24

SYN: METHYLHEPTENOL

TOXICITY DATA with REFERENCE
skn-rbt 500 mg/24H MOD FCTXAV 16,637,78

CONSENSUS REPORTS: Reported in EPA TSCA Inventory.

SAFETY PROFILE: A skin irritant. When heated to decomposition it emits acrid smoke and irritating fumes.

MKK000 CAS:409-02-9 ***HR: 2***
6-METHYL-5-HEPTEN-2-ONE
mf: $C_8H_{14}O$ mw: 126.22

PROP: Sltly yellow liquid; citrus-lemongrass odor. D: 0.846-0.851, refr index:.438-1.442, mp: −67.1, bp: 173-174°, flash p: 122°F. Insol in water; misc in alc, ether, and chloroform.

SYNS: FEMA No. 2707 ◇ METHYL HEPTENONE

TOXICITY DATA with REFERENCE
skn-rbt 500 mg/24H FCTXAV 13,681,75
orl-rat LD50:3500 mg/kg FCTXAV 13,859,75

CONSENSUS REPORTS: Reported in EPA TSCA Inventory.

SAFETY PROFILE: Moderately toxic by ingestion. A skin irritant. Combustible liquid. When heated to decomposition it emits acrid smoke and irritating fumes. See also KETONES.

MKK250 CAS:7535-34-4 ***HR: 3***
6-METHYL-2-HEPTYLHYDRAZINE
mf: $C_8H_{20}N_2$ mw: 144.30

TOXICITY DATA with REFERENCE
orl-rat LD50:108 mg/kg ARZNAD 12,352,62
scu-rat LD50:65 mg/kg ARZNAD 12,352,62
orl-mus LD50:152 mg/kg ARZNAD 12,352,62
scu-mus LD50:72 mg/kg ARZNAD 12,352,62
ivn-mus LD50:59 mg/kg ARZNAD 12,352,62
orl-gpg LD50:30 mg/kg ARZNAD 12,352,62
scu-gpg LD50:16 mg/kg ARZNAD 12,352,62

SAFETY PROFILE: Poison by ingestion, intravenous, and subcutaneous routes. When heated to decomposition it emits toxic fumes of NO_x.

MKK500 CAS:91336-54-8 ***HR: 3***
6-METHYL-2-HEPTYLISOPROPYLIDENHYDRAZINE
mf: $C_{11}H_{24}N_2$ mw: 184.37

SYNS: KR 492 ◇ 6-METHYL-2-HEPTYL-ISOPROPYLIDENHYDRAZIN (GERMAN)

TOXICITY DATA with REFERENCE
orl-mus LD50:160 mg/kg ARZNAD 12,352,62
scu-mus LD50:135 mg/kg ARZNAD 12,352,62
ivn-mus LD50:72 mg/kg ARZNAD 12,352,62

SAFETY PROFILE: Poison by ingestion, subcutaneous, and intravenous routes. When heated to decomposition it emits toxic fumes of NO_x.

MKK600 CAS:11013-97-1 ***HR: 2***
METHYLHESPERIDIN
mf: $C_{29}H_{36}O_{15}$ mw: 624.65

SYNS: 4H-1-BENZOPYRAN-4-ONE,7-((6-O-(6-DEOXY-α-l-MANNOPYRANOSYL)-β-D-GLUCOPYRANOSYL)OXY)-2,3-DIHYDRO-5-HYDROXY-2-(3-HYDROXY-4-METHOXYPHENYL)-,MONOMETHYL ETHER ◇ MH

TOXICITY DATA with REFERENCE
orl-rat TDLo:52650 mg/kg (female 17-22D post):REP YACHDS 9,1395,81
orl-mus TDLo:19500 mg/kg (female 6-15D post):TER YACHDS 9,1369,81
ivn-mus LD50:750 mg/kg NYKZAU 53,237S,57

SAFETY PROFILE: Moderately toxic by intravenous route. An experimental teratogen. Experimental reproductive effects. When heated to decomposition it emits acrid smoke and irritating fumes.

MKK750 CAS:360-54-3 ***HR: 3***
METHYL HEXAFLUOROISOBUTYRATE
mf: $C_5H_4F_6O_2$ mw: 210.09

SYNS: HEXAFLUOROISOBUTYRIC ACID METHYL ESTER ◇ METHYL-3,3,3-TRIFLUORO-2-(TRIFLUOROMETHYL)PROPIONATE

TOXICITY DATA with REFERENCE
orl-mus LD50:300 mg/kg TXAPA9 14,114,69
ipr-mus LD50:17 mg/kg TXAPA9 14,114,69
ivn-mus LD50:15 mg/kg TXAPA9 14,114,69

SAFETY PROFILE: Poison by ingestion, intravenous, and intraperitoneal routes. When heated to decomposition it emits toxic fumes of F^-. See also FLUORIDES.

MKL250 CAS:591-76-4 ***HR: 3***
2-METHYLHEXANE
mf: C_7H_{16} mw: 100.20

PROP: Colorless liquid. Fp: −118.2, bp: 90.0°, d: 0.6789 @ 20°/4°, vap press: 40 mm @ 14.9°, vap d:

3.45, flash p: <0°F, lel: 1.0%, uel: 6.0%, autoign temp: 536°F.

SYN: ETHYLISOBUTYLMETHANE, ISOHEPTANE

SAFETY PROFILE: Probably mildly toxic by inhalation and ingestion. Very dangerous fire hazard when exposed to heat, sparks, or flame. Explosive in the form of vapor when exposed to heat or flame. To fight fire, use foam, CO_2, dry chemical. When heated to decomposition it emits acrid smoke and irritating fumes.

MKL300 CAS:13706-86-0 ***HR: 1***
5-METHYL-2,3-HEXANEDIONE
mf: $C_7H_{12}O_2$ mw: 128.19

SYNS: ACETYLISOPENTANOYL ◇ ACETYL ISOVALERYL ◇ 2,3-HEXANEDIONE, 5-METHYL-

TOXICITY DATA with REFERENCE
skn-rbt 500 mg/24H MOD FCTOD7 20,637,82

CONSENSUS REPORTS: Reported in EPA TSCA Inventory.

SAFETY PROFILE: A skin irritant. When heated to decomposition it emits acrid smoke and irritating fumes.

MKL400 CAS:7379-12-6 ***HR: 3***
2-METHYL-3-HEXANONE
mf: $C_7H_{14}O$ mw: 114.21

PROP: Bp: 131-132°, d: 0.825, flash p: 75° F.

SYN: 3-HEXANONE, 2-METHYL-

TOXICITY DATA with REFERENCE
skn-rbt 500 mg/24H MLD 85JCAE -,287,86
eye-rbt 500 mg/24H MLD 85JCAE -,287,86
orl-rat LD50:4000 mg/kg TXAPA9 28,313,74
skn-rbt LD50:16 g/kg TXAPA9 28,313,74

SAFETY PROFILE: Moderately toxic by ingestion. A skin and eye irritant. Flammable liquid. When heated to decomposition it emits acrid smoke and irritating fumes.

MKM250 CAS:5166-53-0 ***HR: 2***
5-METHYL-3-HEXEN-2-ONE
mf: $C_7H_{12}O$ mw: 112.19

TOXICITY DATA with REFERENCE
orl-rat LD50:1680 mg/kg TXAPA9 28,313,74
ihl-rat LC50:500 ppm/4H TXAPA9 28,313,74
skn-rbt LD50:450 mg/kg TXAPA9 28,313,74

CONSENSUS REPORTS: Reported in EPA TSCA Inventory.

SAFETY PROFILE: Moderately toxic by inhalation, ingestion and skin contact. When heated to decomposition it emits acrid smoke and irritating fumes. See also KETONES.

MKM300 CAS:690-94-8 ***HR: 3***
2-METHYL-5-HEXEN-3-YN-2-OL
mf: $C_7H_{10}O$ mw: 110.17

SYN: DIMETHYL(VINYL)ETHYNYLCARBINOL

TOXICITY DATA with REFERENCE
orl-rat LD50:600 mg/kg 85GMAT -,60,82
orl-mus LD50:590 mg/kg 85GMAT -,60,82
ihl-mus LCLo:30 mg/m^3/2H 85GMAT -,60,82
orl-rbt LD50:800 mg/kg 85GMAT -,60,82
orl-gpg LD50:600 mg/kg 85GMAT -,60,82

SAFETY PROFILE: Poison by inhalation. Moderately toxic by ingestion. When heated to decomposition it emits acrid smoke and irritating fumes.

MKM500 CAS:28292-43-5 ***HR: 3***
5-METHYL-2-HEXYLAMINE
mf: $C_7H_{17}N$ mw: 115.25

TOXICITY DATA with REFERENCE
ipr-mus LD50:90 mg/kg JPETAB 103,325,51

CONSENSUS REPORTS: Reported in EPA TSCA Inventory.

SAFETY PROFILE: Poison by intraperitoneal route. When heated to decomposition it emits toxic fumes of NO_x. See also AMINES.

MKM750 CAS:30956-43-5 ***HR: 3***
1-METHYLHEXYL-β-OXYBUTYRATE
mf: $C_{11}H_{22}O_3$ mw: 202.33

SYN: 3-HYDROXYBUTYRIC ACID-2-HEPTYL ESTER

TOXICITY DATA with REFERENCE
ipr-rat LD50:640 mg/kg OYYAA2 3,373,69
ivn-mus LD50:120 mg/kg OYYAA2 3,373,69

SAFETY PROFILE: Poison by intravenous route. Moderately toxic by intraperitoneal route. When heated to decomposition it emits acrid smoke and irritating fumes.

MKN000 CAS:60-34-4 ***HR: 3***
METHYL HYDRAZINE
DOT: UN 1244
mf: CH_6N_2 mw: 46.09

PROP: Colorless, hydroscopic liquid; ammonia-like odor. D: 0.874 @ 25°, mp: −20.9°, bp: 87.8°, vap d: 1.6, flash p: 73.4°F, fp: −52.4°, autoign temp: 196°, lel: 2.5%, uel: 97 ±2%. Sltly sol in water; sol in alc, hydrocarbons, and ether; misc with water and hydrazine. Strong reducing agent.

SYNS: HYDRAZOMETHANE ◇ 1-METHYL HYDRAZINE ◇ METHYLHYDRAZINE (DOT) ◇ METYLOHYDRAZYNA (POLISH) ◇ MMH ◇ MONOMETHYL HYDRAZINE ◇ RCRA WASTE NUMBER P068

TOXICITY DATA with REFERENCE
mmo-sat 2 μmol/plate/48H MUREAV 54,167,78
dnd-hmn:fbr 116 pmol NTIS** AD-A092-249
ipr-rat TDLo:100 mg/kg (6-15D preg):REP APTOD9 19,A21,80
ipr-rat TDLo:50 mg/kg (6-15D preg):TER APTOD9 19,A21,80
ihl-rat TCLo:20 ppb/6H/1Y-I:CAR NTIS** AD-A154-659
orl-mus TDLo:10 g/kg/1Y-C:NEO IJCNAW 9,109,72
ihl-mus TCLo:2 ppm/6H/1Y-I:CAR NTIS** AD-A154-659
ipr-mus TDLo:72 mg/kg/8W-I:ETA JNCIAM 42,337,69
orl-rat LD50:32 mg/kg XAW0A2 CWL 2-10,58
ihl-rat LC50:34 ppm/4H AMRL** TR-67-137,67
skn-rat LD50:183 mg/kg CTOXAO 4,435,71
ipr-rat LD50:21 mg/kg CTOXAO 4,435,71
scu-rat LD50:35 mg/kg BJCAAI 30,4329,74
ivn-rat LD50:17 mg/kg CTOXAO 4,435,71
orl-mus LD50:29 mg/kg NTIS** AD-A125-539
ihl-mus LC50:56 ppm/4H AMIHAB 12,609,55
ipr-mus LD50:15 mg/kg PSEBAA 124,172,67
scu-rat LD50:25 mg/kg BJCAAI 30,429,74
ivn-mus LD50:33200 μg/kg MEPAAX 24,71,73
ihl-dog LC50:96 ppm/1H AIHAAP 31,667,70
skn-rbt LD50:95 mg/kg PSEBAA 131,226,69

CONSENSUS REPORTS: Community Right-To-Know List. EPA Extremely Hazardous Substances List. Reported in EPA TSCA Inventory. EPA Genetic Toxicology Program.

OSHA PEL: CL 0.2 ppm (skin))
ACGIH TLV: CL 0.2 ppm; Suspected Human Carcinogen; (Proposed: CL 0.01 ppm; Suspected Human Carcinogen)
NIOSH REL: CL 0.08 mg/m^3/2H
DOT Classification: Flammable Liquid; Label: Flammable Liquid, Corrosive.

SAFETY PROFILE: Suspected carcinogen with experimental carcinogenic, neoplastigenic, tumorigenic, and teratogenic data. Poison by inhalation, ingestion, skin contact, intraperitoneal, subcutaneous, and intravenous routes. Experimental reproductive effects. Human mutation data reported. Corrosive to skin, eyes, and mucous membranes. May self-ignite in air. Very dangerous fire hazard when exposed to heat or flame. To fight fire, use alcohol foam, CO_2, dry chemical. Explosive in the form of vapor when exposed to heat or flame. A powerful reducing agent. It is hypergolic with many oxidants (e.g., dinitrogen tetraoxide and hydrogen peroxide). When heated to decomposition it emits toxic fumes of NO_x. See also HYDRAZINE.

MKN250 CAS:7339-53-9 ***HR: 3***
METHYLHYDRAZINE HYDROCHLORIDE
mf: $CH_6N_2 \cdot ClH$ mw: 82.55

TOXICITY DATA with REFERENCE
orl-rat LD50:58 mg/kg AMIHAB 13,34,56
ipr-rat LD50:58 mg/kg AMIHAB 13,34,56
ivn-rat LD50:59 mg/kg AMIHAB 13,34,56
orl-mus LD50:59 mg/kg AMIHAB 13,34,56
ipr-mus LD50:58 mg/kg AMIHAB 13,34,56
ivn-mus LD50:59 mg/kg AMIHAB 13,34,56
ivn-dog LD50:21 mg/kg AMIHAB 13,34,56
NIOSH REL: (Hydrazines): CL 0.08 mg/m^3/2H

SAFETY PROFILE: Poison by ingestion, intraperitoneal, and intravenous routes. When heated to decomposition it emits very toxic fumes of Cl^- and NO_x. See also METHYLHYDRAZINE.

MKN500 CAS:302-15-8 ***HR: 3***
METHYL HYDRAZINE SULFATE
mf: $CH_6N_2 \cdot H_2O_4S$ mw: 144.17

PROP: Crystals from methyl alcohol. Mp: 142°. Very sol in water; very sltly sol in alc.

TOXICITY DATA with REFERENCE
orl-mus TDLo:2092 mg/kg/83W-C:NEO IJCNAW 9,109,72
ipr-mus LD50:160 mg/kg RPTOAN 36,27,73

CONSENSUS REPORTS: Reported in EPA TSCA Inventory.

SAFETY PROFILE: Poison by intraperitoneal route. Questionable carcinogen with experimental neoplastigenic data. When heated to decomposition it emits very toxic fumes of SO_x and NO_x. See also METHYLHYDRAZINE.

MKN750 CAS:366-71-2 ***HR: 3***
(α-(2-METHYLHYDRAZINO)-p-TOLUOYL)UREA, MONOHYDROBROMIDE
mf: $C_{10}H_{14}N_4O_2 \cdot BrH$ mw: 303.20

SYNS: 1-(p-ALLOPHANOYLBENZYL)-2-METHYLHYDRAZINE HYDROBROMIDE ◇ N-AMINOCARBONYL)-4-((2-METHYLHYDRAZINO)METHYL)BENZAMIDE MONOHYDROBROMIDE ◇ 1-METHYL-2-(p-ALLOPHANOYLBENZYL)HYDRAZINE HYDROBROMIDE ◇ NSC 77517

TOXICITY DATA with REFERENCE
orl-rat TDLo:60 mg/kg (1-12D preg):TER CHDDAT 467,4448,68
orl-rat TDLo:75 mg/kg (8-10D preg):REP CHDDAT 467,4448,68
orl-rat TDLo:900 mg/kg/30D-I:CAR CNREA8 28,924,68
orl-mus LD50:818 mg/kg NCISP* JAN86

SAFETY PROFILE: Moderately toxic by ingestion. Questionable carcinogen with experimental carcinogenic and teratogenic data. Experimental reproductive effects. When heated to decomposition it emits very toxic fumes of HBr and NO_x.

MKO000 CAS:3031-73-0 ***HR: 3***
METHYL HYDROPEROXIDE
mf: CH_4O_2 mw: 48.04

SAFETY PROFILE: A powerful oxidizer. Probably an irritant to the eyes, skin, and mucous membranes. A powerful, shock-sensitive explosive, especially when warm. Explodes on contact with phosphorus pentaoxide. Aqueous solutions may explode on warming with catalytic platinum. Reacts with barium to form dangerously explosive barium salts. Incompatible with barium salts. When heated to decomposition it emits acrid smoke and irritating fumes. See also PEROXIDES.

MKO250 CAS:95-71-6 ***HR: 3***
METHYLHYDROQUINONE
mf: $C_7H_8O_2$ mw: 124.15

SYNS: 2,5-DIHYDROXYTOLUENE ◇ METHYL-p-HYDROQUINONE ◇ 2-METHYLHYDROQUINONE ◇ THQ ◇ 2,5-TOLUENEDIOL ◇ p-TOLUHYDROQUINOL ◇ TOLUQUINOL ◇ TOLYLHYDROQUINONE

TOXICITY DATA with REFERENCE
orl-rat LDLo:200 mg/kg KODAK* -,-,71
ipr-rat LDLo:200 mg/kg KODAK* -,-,71

CONSENSUS REPORTS: Reported in EPA TSCA Inventory.

SAFETY PROFILE: Poison by ingestion and intraperitoneal routes. When heated to decomposition it emits acrid smoke and irritating fumes.

MKP000 CAS:51938-16-0 ***HR: 3***
N-METHYL-N-(4-HYDROXYBUTYL)NITROSAMINE
mf: $C_5H_{12}N_2O_2$ mw: 132.19

SYN: N-NITROSO-N-METHYL-(4-HYDROXYBUTYL)AMINE

TOXICITY DATA with REFERENCE
mma-sat 40 μmol/plate CNREA8 37,399,77
orl-rat TDLo:53 g/kg/20W-C:ETA GANNA2 67,825,76

CONSENSUS REPORTS: EPA Genetic Toxicology Program.

SAFETY PROFILE: Questionable carcinogen with experimental tumorigenic data. Mutation data reported. When heated to decomposition it emits toxic fumes of NO_x. See also NITROSAMINES.

MKP250 CAS:1487-49-6 ***HR: 2***
METHYL-3-HYDROXYBUTYRATE
mf: $C_5H_{10}O_3$ mw: 118.15

PROP: Colorless liquid. Bp: 174.9°, flash p: 180°F (OC), d: 1.0559, vap press: 0.85 mm @ 20°, vap d: 4.1.

TOXICITY DATA with REFERENCE
eye-rbt 5 mg SEV AJOPAA 29,1363,46

SAFETY PROFILE: A severe eye irritant. Combustible when exposed to heat or flame; can react with oxidizing materials. To fight fire, use alcohol foam, CO_2, dry chemical. When heated to decomposition it emits acrid smoke and irritating fumes. See also ESTERS.

MKP500 CAS:90-33-5 ***HR: 2***
4-METHYL-7-HYDROXYCOUMARIN
mf: $C_{10}H_8O_3$ mw: 176.18

SYNS: BILCOLIC ◇ BILICANTE ◇ CANTABILINE ◇ COUMARIN 4 ◇ EUROGALE ◇ 7-HYDROXY-4-METHYLCOUMARIN ◇ 7-HYDROXY-4-METHYL-2-OXO-2H-1-BENZOPYRAN ◇ HYMECROMONE ◇ MEDILLA ◇ MENDIAXON ◇ 4-METHYLUMBELLIFERON (CZECH) ◇ β-METHYLUMBELLIFERONE ◇ 4-METHYLUMBELLIFERONE ◇ OMEGA 127 ◇ PILOT 447

TOXICITY DATA with REFERENCE
orl-rat TDLo:6 g/kg (9-14D preg):REP GNRIDX 4,413,70
orl-rat TDLo:6 g/kg (9-14D preg):TER GNRIDX 4,413,70
orl-rat TDLo:3850 mg/kg KSRNAM 5,1619,71
ipr-rat LD50:2550 mg/kg KSRNAM 5,1619,71
scu-rat LD50:7200 mg/kg KSRNAM 5,1619,71
orl-mus LD50:2850 mg/kg KSRNAM 5,1619,71
ipr-mus LD50:1750 mg/kg KSRNAM 5,1619,71

CONSENSUS REPORTS: Reported in EPA TSCA Inventory. EPA Genetic Toxicology Program.

SAFETY PROFILE: Moderately toxic by ingestion and intraperitoneal routes. An experimental teratogen. Experimental reproductive effects. When heated to decomposition it emits acrid smoke and irritating fumes.

MKQ000 CAS:2470-73-7 ***HR: 3***
(2-METHYL-3-(1-HYDROXYETHOXYETHYL-4-PIPERAZINYL)PROPYL)-10-PHENOTHIAZINE
mf: $C_{24}H_{33}N_3O_2S$ mw: 427.66

SYNS: DIXYRAZINE ◇ ESUCOS ◇ 10-(3-(4-HYDROXYETHOXYETHYL-1-PIPERAZINYL)-2-METHYLPROPYL)PHENOTHIAZINE ◇ 2-(2-(4-(2-METHYL-3-PHENOTHIAZIN-10-YLPROPYL)-1-PIPERAZINYL)ETHOXY)ETHANOL ◇ UCB 3412

TOXICITY DATA with REFERENCE
orl-rat LD50:400 mg/kg ANPBAZ 61,669,61
ivn-rat LD50:37500 μg/kg ANPBAZ 61,669,61

SAFETY PROFILE: Poison by ingestion and intravenous routes. When heated to decomposition it emits very toxic fumes of SO_x and NO_x.

MKQ250 CAS:93-90-3 ***HR: 2***
N-METHYL-N-HYDROXYETHYLANILINE
mf: $C_9H_{13}NO$ mw: 151.23

SYNS: 2-(N-METHYLANILINO)ETHANOL ◇ N-METHYL-N-β-HYDROXYETHYLANILINE

TOXICITY DATA with REFERENCE
skn-rbt 10 mg/24H open MLD AMIHBC 4,119,51
eye-rbt 250 μg open SEV AMIHBC 4,119,51

orl-rat LD50:2830 mg/kg AMIHBC 4,119,51
skn-rbt LD50:3250 mg/kg AMIHBC 4,119,51

CONSENSUS REPORTS: Reported in EPA TSCA Inventory.

SAFETY PROFILE: Moderately toxic by ingestion and skin contact. A skin and severe eye irritant. When heated to decomposition it emits toxic fumes of NO_x.

MKQ500 CAS:63918-39-8 ***HR: 3***
1-METHYL-1-(β-HYDROXYETHYL)ETHYLENAMMONIUM PICRYLSULFONATE
mf: $C_5H_{12}NO•C_6H_2N_3O_9S$ mw: 394.35

SYN: METHYL-β-HYDROXYETHYL-ETHYLENIMONIUM PICRYLSULFONATE

TOXICITY DATA with REFERENCE
ipr-mus LD50:7500 μg/kg JPETAB 91,224,47
ivn-mus LD50:300 mg/kg JPETAB 91,224,47
ivn-rbt LDLo:3 mg/kg NTIS** PB158-507

SAFETY PROFILE: Poison by intraperitoneal and intravenous routes. When heated to decomposition it emits very toxic fumes of SO_x and NO_x.

MKQ875 CAS:593-77-1 ***HR: 3***
METHYLHYDROXYLAMINE
mf: CH_5NO mw: 47.07

SYNS: N-HYDROXY-METHANAMINE ◇ NCI-C60066

TOXICITY DATA with REFERENCE
mmo-bcs 1 mol/L MUREAV 4,517,67
oms-bcs 10 mmol/L MUREAV 4,517,67
cyt-mus:emb 50 μmol/L JNCIAM 32,667,64
cyt-ham:oth 50 μmol/L JNCIAM 32,667,64
ipr-mus LDLo:188 mg/kg JNCIAM 32,667,64

CONSENSUS REPORTS: EPA Genetic Toxicology Program.

SAFETY PROFILE: Poison by intraperitoneal route. Probably a severe irritant to the eyes, skin, and mucous membranes. Mutation data reported. When heated to decomposition it emits toxic fumes of NO_x. See also AMINES.

MKQ880 CAS:67-62-9 ***HR: D***
o-METHYLHYDROXYLAMINE
mf: CH_5NO mw: 47.07

PROP: Free base: mobile liquid; fishy, amine odor. Bp: 49-50°. Misc with water, alc, ether.

SYNS: METHOXYAMINE ◇ NCI-C60060

TOXICITY DATA with REFERENCE
dnd-esc 20 μmol/L MUREAV 89,95,81
mmo-omi 50 mmol/L MUREAV 49,163,78

CONSENSUS REPORTS: EPA Genetic Toxicology Program.

SAFETY PROFILE: Mutation data reported. A strong irritant. When heated to decomposition it emits toxic fumes of NO_x. See also AMINES.

MKR000 CAS:10482-16-3 ***HR: 3***
2-METHYL-4-HYDROXYLAMINOQUINOLINE 1-OXIDE
mf: $C_{10}H_{10}N_2O_2$ mw: 190.22

SYN: 4-(HYDROXYLAMINO)-2-METHYL-QUINOLINE,1-OXIDE

TOXICITY DATA with REFERENCE
scu-mus TDLo:120 mg/kg/50D-I:ETA BCPCA6 16,631,67

SAFETY PROFILE: Questionable carcinogen with experimental tumorigenic data. When heated to decomposition it emits toxic fumes of NO_x.

MKR100 CAS:35440-49-4 ***HR: 3***
METHYL-o-(4-HYDROXY-3-METHOXYCINNAMOYL)RESERPATE
mf: $C_{33}H_{38}N_2O_8$ mw: 590.73

SYN: CD-3400

TOXICITY DATA with REFERENCE
orl-rat TDLo:2 g/kg (female 8-17D post):REP OYYAA2 18,105,79
orl-rat TDLo:1600 mg/kg (female 1-8D post):TER OYYAA2 18,105,79
ipr-rat LD50:210 mg/kg KSRNAM 11,1669,77
ipr-mus LD50:189 mg/kg KSRNAM 11,1669,77

SAFETY PROFILE: Poison by intraperitoneal route. An experimental teratogen. Experimental reproductive effects. When heated to decomposition it emits toxic fumes of NO_x.

MKR150 ***HR: 3***
1-METHYL-3-(HYDROXYMETHYL)IMIDAZOLIUM CHLORIDE LAURATE
mf: $C_{17}H_{31}N_2O_2•Cl$ mw: 330.95

SYN: 3-(DODECANOYLOXYMETHY)-1-METHYL-1H-IMIDAZOLIUM CHLORIDE

TOXICITY DATA with REFERENCE
orl-mus LD50:4300 mg/kg USXXAM #4204065
ipr-mus LD50:140 mg/kg USXXAM #4204065
ivn-mus LD50:100 mg/kg USXXAM #4204065

SAFETY PROFILE: Poison by intravenous and intraperitoneal routes. Mildly toxic by ingestion. When heated to decomposition it emits toxic fumes of NO_x and Cl^-.

MKR250 CAS:297-90-5 ***HR: 3***
N-METHYL-3-HYDROXYMORPHINAN
mf: $C_{17}H_{23}NO$ mw: 257.41

SYNS: CETARIN ◇ DROMORAN ◇ racemic DROMORAN ◇ dl-1,3,4,9,10,10A-HEXAHYDRO-11-METHYL-2H-10,4A-IMINOETHANO PHENANTHREN-6-OL ◇ dl-3-HYDROXY-N-METHYLMORPHINAN ◇ (±)-3-HYDROXY-N-METHYLMORPHINAN ◇ METHORPHINAN ◇ NU 2206 ◇ RACEMORPHAN ◇ RO 1-5431

TOXICITY DATA with REFERENCE
orl-rat LD50:350 mg/kg CLDND* 216,48,52

SAFETY PROFILE: Poison by ingestion. When heated to decomposition it emits toxic fumes of NO_x.

MKR500 CAS:23324-72-3 ***HR: 3***
METHYL HYDROXYOCTADECADIENOATE
mf: $C_{19}H_{34}O_3$ mw: 310.53

SYN: 13-HYDROXY-9,11-OCTADECADIENOIC METHYL ESTER

TOXICITY DATA with REFERENCE
skn-mus TDLo:64 g/kg/53W-I:ETA AMBPBZ 82,127,74

SAFETY PROFILE: Questionable carcinogen with experimental tumorigenic data. When heated to decomposition it emits acrid smoke and irritating fumes. See also ESTERS.

MKS000 CAS:63991-31-1 ***HR: 3***
p-METHYL-m-HYDROXY-PHENYL-PROPANOLAMINE HYDROCHLORIDE
mf: $C_{10}H_{15}NO_2 \cdot ClH$ mw: 217.72

SYNS: α-(1-AMINOETHYL)-3-HYDROXY-4-METHYLBENZYLALCOHOL HYDROCHLORIDE ◇ 1-(3-HYDROXY-4-METHYLPHENYL)-2-AMINO-1-PROPANOL HYDROCHLORIDE

TOXICITY DATA with REFERENCE
scu-rat LDLo:5 mg/kg JPETAB 71,62,41
ivn-rbt LDLo:90 mg/kg JACSAT 53,4149,31

SAFETY PROFILE: Poison by subcutaneous and intravenous routes. When heated to decomposition it emits very toxic fumes of HCl and NO_x.

MKS250 CAS:37286-64-9 ***HR: 1***
α-METHYL-ω-HYDROXYPOLY(OXY(METHYL-1,2-ETHANEDIYL))
mf: $(C_3H_6O)_n \cdot CH_4O$

SYNS: DOWFROTH 250 ◇ JEFFOX OL 2700 ◇ POLYPROPYLENE GLYCOL METHYL ETHER ◇ POLYPROPYLENE GLYCOL MONOMETHYLETHER ◇ PROPYLENE OXIDE-METHANOL ADDUCT ◇ UCON LB-1715

TOXICITY DATA with REFERENCE
orl-rat LD50:49 g/kg UCDS** 6/13/60

CONSENSUS REPORTS: Glycol ether compounds are on the Community Right-To-Know List. Reported in EPA TSCA Inventory.

SAFETY PROFILE: Mildly toxic by ingestion. When heated to decomposition it emits acrid smoke and irritating fumes. See also GLYCOL ETHERS.

MKS750 CAS:78308-53-9 ***HR: 2***
3-METHYL-4-HYDROXY-3-PYRROLIN-2-ONE AMMONIUM SALT
mf: $C_5H_6NO_2 \cdot H_4N$ mw: 130.17

SYN: AMMONIUMSALZ des α-METHYL-β-HYDROXY-Δ^{α},β-BUTYCOLACTAM (GERMAN)

TOXICITY DATA with REFERENCE
scu-mus LD50:2000 mg/kg ARZNAD 11,277,61
ivn-mus LD50:1150 mg/kg ARZNAD 11,277,61

SAFETY PROFILE: Moderately toxic by subcutaneous and intravenous routes. When heated to decomposition it emits toxic fumes of NO_x and NH_3.

MKT000 CAS:593-78-2 ***HR: 3***
METHYL HYPOCHLORITE
mf: CH_3ClO mw: 66.49

SAFETY PROFILE: Explodes violently when exposed to heat, spark, or flame. Upon decomposition it emits toxic fumes of Cl^-. See also HYPOCHLORITES.

MKT500 CAS:616-47-7 ***HR: 3***
1-METHYLIMIDAZOLE
mf: $C_4H_6N_2$ mw: 82.12

TOXICITY DATA with REFERENCE
orl-mus LD50:1400 mg/kg TXAPA9 14,301,69
ipr-mus LD50:380 mg/kg TXAPA9 14,301,69

CONSENSUS REPORTS: Reported in EPA TSCA Inventory.

SAFETY PROFILE: Poison by intraperitoneal route. Moderately toxic by ingestion. Explosive reaction with osmium(VIII) oxide. When heated to decomposition it emits toxic fumes of NO_x.

MKT750 CAS:693-98-1 ***HR: 2***
2-METHYLIMIDAZOLE
mf: $C_4H_6N_2$ mw: 82.12

TOXICITY DATA with REFERENCE
orl-mus LD50:1400 mg/kg TXAPA9 14,301,69
ipr-mus LD50:480 mg/kg TXAPA9 14,301,69

CONSENSUS REPORTS: Reported in EPA TSCA Inventory.

SAFETY PROFILE: Moderately toxic by ingestion and intraperitoneal routes. When heated to decomposition it emits toxic fumes of NO_x.

MKU000 CAS:822-36-6 ***HR: 3***
4-METHYLIMIDAZOLE
mf: $C_4H_6N_2$ mw: 82.12

SYN: 4-ME-I

TOXICITY DATA with REFERENCE
orl-mus LD50:370 mg/kg TXAPA9 14,301,69
ipr-mus LD50:165 mg/kg TXAPA9 14,301,69

CONSENSUS REPORTS: Reported in EPA TSCA Inventory.

SAFETY PROFILE: Poison by ingestion and intraperitoneal routes. When heated to decomposition it emits toxic fumes of NO_x.

MKU250 CAS:105-59-9 ***HR: 2***
N-METHYL-2,2'-IMINODIETHANOL
mf: $C_5H_{13}NO_2$ mw: 119.19

SYNS: N-METHYLDIETHANOLIMINE ◇ 2,2'-(METHYLIMINO) BISETHANOL ◇ N-METHYLIMINODIETHANOL ◇ 2,2'-(METHYLIMINO)DIETHANOL ◇ USAF DO-52

TOXICITY DATA with REFERENCE
skn-rbt 10 mg/24H open MLD AMIHBC 10,61,54
skn-rbt 502 mg open MLD UCDS** 7/13/71
eye-rbt 20 mg open AMIHBC 10,61,54
orl-rat LD50:4780 mg/kg AMIHBC 10,61,54
ipr-mus LD50:500 mg/kg NTIS** AD277-689

CONSENSUS REPORTS: Reported in EPA TSCA Inventory.

SAFETY PROFILE: Moderately toxic by intraperitoneal route. Mildly toxic by ingestion. A skin and eye irritant. When heated to decomposition it emits toxic fumes of NO_x.

MKU750 CAS:306-53-6 ***HR: 3***
((METHYLIMINO)DIETHYLENE)BIS(ETHYLDIMETHYLAMMONIUM BROMIDE)
mf: $C_{13}H_{33}N_3\cdot 2Br$ mw: 391.31

SYNS: AZAMETHONE ◇ AZAMETHONIUM BROMIDE ◇ AZAMETON ◇ C 9295 ◇ CIBA 9295 ◇ GANLION ◇ 3-METHYL-3-AZAPENTANE-1,5-BIS(ETHYLDIMETHYLAMMONIUM)BROMIDE ◇ 2,2'-(METHYLIMINO)BIS(N-ETHYL-N,N-DIMETHYLETHANAMINIUM) DIBROMIDE ◇ PENDIOMID ◇ PENDIOMIDE BROMIDE ◇ PENDIOMIDE DIBROMIDE ◇ PENTAMETHAZENE DIBROMIDE ◇ PENTAMETHAZINE ◇ N,N,N',N',3-PENTAMETHYL-N,N'-DIETHYL-3-AZAPENTYLENE-1,5-DIAMMONIUM DIBROMIDE ◇ PENTAMINE ◇ PENTAMON ◇ PENTOMID ◇ PRAPAR ◇ PRAPARAT 9295

TOXICITY DATA with REFERENCE
orl-mus LDLo:2500 mg/kg MEIEDD 10,129,83
ipr-mus LDLo:155 mg/kg CLDND* 21,8,58
orl-rbt LD50:3000 mg/kg SMWOAS 81,446,51
scu-rbt LD50:160 mg/kg SMWOAS 81,446,51
ivn-rbt LD50:75 mg/kg SMWOAS 81,446,51

SAFETY PROFILE: Poison by intraperitoneal, subcutaneous, and intravenous routes. Moderately toxic by ingestion. When heated to decomposition it emits very toxic fumes of NO_x, NH_3, and Br^-. See also BROMIDES.

MKV500 CAS:876-83-5 ***HR: 3***
2-METHYL-1,3-INDANDIONE
mf: $C_{10}H_8O_2$ mw: 160.18

TOXICITY DATA with REFERENCE
ipr-mus TDLo:25 mg/kg (9D preg):REP ARTODN 33,191,75
ipr-mus TDLo:25 mg/kg (9D preg):TER ARTODN 33,191,75
ipr-mus LDLo:100 mg/kg ARTODN 33,191,75

SAFETY PROFILE: Poison by intraperitoneal route. An experimental teratogen. Experimental reproductive effects. When heated to decomposition it emits acrid smoke and irritating fumes.

MKV750 CAS:83-34-1 ***HR: 3***
β-METHYLINDOLE
mf: C_9H_9N mw: 131.19

PROP: Leaves from ligroin. Mp: 95°, bp: 265° @ 755 mm. Sol in cold water, alc, chloroform, ether, and benzene.

SYNS: 3-METHYLINDOLE ◇ 3-METHYL-1H-INDOLE ◇ 3-MI ◇ SCATOLE ◇ SKATOL ◇ SKATOLE

TOXICITY DATA with REFERENCE
orl-rat LD50:3450 mg/kg FCTXAV 14,863,76
orl-mus LDLo:470 mg/kg AECTCV 14,111,85
ipr-mus LD50:175 mg/kg FCTXAV 14,863,76
scu-frg LDLo:1 g/kg MEIEDD 10,1227,83
orl-dom LDLo:300 mg/kg FCTXAV 14,863,76
orl-ctl LDLo:200 mg/kg FCTXAV 14,863,76
ivn-ctl LDLo:60 mg/kg FCTXAV 14,863,76

CONSENSUS REPORTS: Reported in EPA TSCA Inventory.

SAFETY PROFILE: Poison by ingestion, intravenous, and intraperitoneal routes. Moderately toxic by subcutaneous route. When heated to decomposition it emits toxic fumes of NO_x.

MKW000 CAS:7770-47-0 ***HR: 3***
4-(1-METHYL-3-INDOLYLETHYL)PYRIDINE HYDROCHLORIDE
mf: $C_{16}H_{16}N_2\cdot ClH$ mw: 272.80

SYNS: IN 399 ◇ 4-(N-METHYL-3-INDOLYETHYL)PYRIDINIUM HYDROCHLORIDE ◇ USAF IN-399

TOXICITY DATA with REFERENCE
ipr-mus LD50:200 mg/kg NTIS** AD277-689
ivn-mus LD50:94 mg/kg JPETAB 125,122,59

SAFETY PROFILE: Poison by intraperitoneal and intravenous routes. When heated to decomposition it emits very toxic fumes of HCl and NO_x.

MKW100 ***HR: 3***
4-(2-(1-METHYL-3-INDOLYL)ETHYL)-1-(3-(TRIMETHYLAMMONIO)PROPYL)PYRIDINIUM DICHLORIDE
mf: $C_{22}H_{31}N_3 \cdot 2Cl$ mw: 408.46

SYN: DICHLORURE de TRIMETHYLAMMONIUM-1-(β-N-METHYLINDOYL-3'') ETHYL-4'-PYRIDINIUM)-3 PROPANE

TOXICITY DATA with REFERENCE
orl-rat LD50:200 mg/kg APFRAD 22,523,64
ipr-rat LD50:25 mg/kg APFRAD 22,523,64
scu-rat LD50:140 mg/kg APFRAD 22,523,64
orl-mus LD50:475 mg/kg APFRAD 22,523,64
ipr-mus LD50:40 mg/kg APFRAD 22,523,64
ivn-mus LD50:12 mg/kg APFRAD 22,523,64
ipr-rbt LD50:90 mg/kg APFRAD 22,523,64
rec-rbt LD50:150 mg/kg APFRAD 22,523,64

SAFETY PROFILE: Poison by ingestion, subcutaneous, intravenous, intraperitoneal, and rectal routes. When heated to decomposition it emits toxic fumes of Cl^-, NH_3, and NO_x.

MKW200 CAS:74-88-4 ***HR: 3***
METHYL IODIDE
DOT: UN 2644
mf: CH_3I mw: 141.94

PROP: Colorless liquid, turns brown on exposure to light. Mp: −66.4°, bp: 42.5°, d: 2.279 @ 20°/4°, vap press: 400 mm @ 25.3°, vap d: 4.89. Sol in water @ 15°, misc in alc and ether.

SYNS: IODOMETANO (ITALIAN) ◇ IODOMETHANE ◇ IODURE de METHYLE (FRENCH) ◇ JOD-METHAN (GERMAN) ◇ JOODMETHAAN (DUTCH) ◇ METHYLJODID (GERMAN) ◇ METHYLJODIDE (DUTCH) ◇ METYLU JODEK (POLISH) ◇ MONOIODURO di METILE (ITALIAN) ◇ RCRA WASTE NUMBER U138

TOXICITY DATA with REFERENCE
skn-hmn 1 g/10M MLD BJIMAG 7,122,50
skn-rat 1 g/30M MLD BJIMAG 7,122,50
mmo-esc 20 μmol/L ARTODN 46,277,80
msc-mus:lym 3600 μg/L ENMUDM 7,523,85
scu-rat TDLo:50 mg/kg:ETA ZEKBAI 74,241,70
ipr-mus TDLo:44 mg/kg/8W-I:NEO CNREA8 35,1411,75
orl-rat LDLo:150 mg/kg BJIMAG 7,122,50
ihl-rat LC50:1300 mg/m^3/4H 34ZIAG -,756,69
skn-rat LDLo:800 mg/kg KODAK* 21MAY71
ipr-rat LD50:101 mg/kg 85GMAT -,84,82
scu-rat LDLo:150 mg/kg BJIMAG 7,122,50
ihl-mus LCLo:5000 mg/m^3/1H BJIMAG 7,122,50
ipr-mus LD50:172 mg/kg 85GMAT -,84,82
scu-mus LD50:110 mg/kg TXAPA9 4,354,62
ipr-gpg LD50:51 mg/kg 85GMAT -,84,82

CONSENSUS REPORTS: IARC Cancer Review: Group 3 IMEMDT 7,56,87; Animal Limited Evidence IMEMDT 41,213,86. Deleted from Fifth Annual Report Because of the Re-Evaluation By IARC. Community Right-To-Know List. EPA Genetic Toxicology Program. Reported in EPA TSCA Inventory.

OSHA PEL: (Transitional: TWA 5 ppm (skin)) TWA 2 ppm (skin)
ACGIH TLV: TWA 2 ppm (skin); Suspected Human Carcinogen.
DFG MAK: Animal Carcinogen, Suspected Human Carcinogen.
NIOSH REL: (Monohalomethanes) Reduce to lowest level.
DOT Classification: Poison B; Label: Poison.

SAFETY PROFILE: Suspected carcinogen with experimental neoplastigenic, and tumorigenic data. A poison by ingestion, intraperitoneal, and subcutaneous routes. Moderately toxic by inhalation and skin contact. A human skin irritant. Human mutation data reported. A strong narcotic and anesthetic. Explosive reaction with trialkylphosphines; silver chlorite. Violent reaction with oxygen (at 300-500°C); sodium even in solution. When heated to decomposition it emits toxic fumes of I^-. See also IODIDES.

MKW250 CAS:1910-68-5 ***HR: 3***
N-METHYLISATIN-3-(THIOSEMICARBAZONE)
mf: $C_{10}H_{10}N_4OS$ mw: 234.30

SYNS: BW 33-T-57 ◇ COMPOUND 33T57 ◇ 2-(1,2-DIHYDRO-1-METHYL-2-OXO-3H-INDOLE-3-YLIDENE)HYDRZAINECARBOTHIOAMIDE ◇ KEMOVIRAN ◇ MARBORAN ◇ METHISAZONE ◇ N-METHYLINDOLE-2,3-DIONE THIOSEMICARBAZONE ◇ 1-METHYLINDOLE-2,3-DIONE-3-(THIOSEMICARBAZONE) ◇ N-METHYLISATIN THIOSEMICARBAZONE ◇ METISAZONUM ◇ MIBT ◇ NSC-69811 ◇ VIRUZONA

TOXICITY DATA with REFERENCE
oms-omi 50 μmol/L BBACAQ 519,65,78
dnd-hmn:hla 30 μmol/L BCPCA6 25,821,76
orl-mus LD50:4 g/kg ARZNAD 22,1704,72
ivn-mus LD50:56 mg/kg CSLNX* NX#02255

SAFETY PROFILE: Poison by intravenous route. Moderately toxic by ingestion. Human mutation data reported. When heated to decomposition it emits very toxic fumes of SO_x and NO_x.

MKW450 CAS:110-12-3 ***HR: 2***
METHYL ISOAMYL KETONE
DOT: UN 2302
mf: $C_7H_{14}O$ mw: 114.21

PROP: Colorless, stable liquid; pleasant odor. Bp: 144°, d: 0.8132 @ 20°/20°, fp: −73.9°, flash p: 110°F (OC). Sltly sol in water; misc with most organic solvents.

SYNS: ISOAMYL METHYL KETONE ◇ ISOPENTYL METHYL KETONE ◇ 2-METHYL-5-HEXANONE ◇ 5-METHYL-2-HEXANONE ◇ MIAK

TOXICITY DATA with REFERENCE
orl-rat LD50:4760 mg/kg AIHAAP 23,95,62
ihl-rat LCLo:4000 ppm/4H AIHAAP 23,95,62
orl-mus LDLo:3200 mg/kg KODAK* 21MAY71
skn-rbt LD50:10 g/kg UCDS** 8/7/63

CONSENSUS REPORTS: Reported in EPA TSCA Inventory.

OSHA PEL: TWA 50 ppm
ACGIH TLV: TWA 50 ppm
NIOSH REL: (Ketones (Methyl Isoamyl Ketone): TWA 230 mg/m^3
DOT Classification: Flammable or Combustible Liquid; Label: Flammable Liquid.

SAFETY PROFILE: Moderately toxic by ingestion. Mildly toxic by inhalation and skin contact. Combustible when exposed to heat, flame, or oxidizers. To fight fire, use dry chemical, CO_2, foam, fog. When heated to decomposition it emits acrid smoke and irritating fumes. See also KETONES.

MKW500 CAS:624-44-2 ***HR: 3***
METHYL ISOAMYL KETOXIME
mf: $C_7H_{15}NO$ mw: 129.23

SYN: USAF AM-7

TOXICITY DATA with REFERENCE
ipr-mus LD50:100 mg/kg NTIS** AD277-689

CONSENSUS REPORTS: Reported in EPA TSCA Inventory.

SAFETY PROFILE: Poison by intraperitoneal route. When heated to decomposition it emits toxic fumes of NO_x.

MKW600 CAS:108-11-2 ***HR: 3***
METHYL ISOBUTYL CARBINOL
DOT: UN 2053
mf: $C_6H_{14}O$ mw: 102.20

PROP: Clear liquid. Bp: 131.8°, fp: < −90° (sets to a glass), flash p: 106°F, d: 0.8079 @ 20°/20°, vap press: 2.8 mm @ 20°, vap d: 3.53, lel: 1.0%, uel: 5.5%.

SYNS: ALCOOL METHYL AMYLIQUE (FRENCH) ◇ ISOBUTYL METHYL CARBINOL ◇ ISOBUTYLMETHYLMETHANOL ◇ MAOH ◇ METHYL AMYL ALCOHOL ◇ METHYLISOBUTYL CARBINOL ◇ 2-METHYL-4-PENTANOL ◇ 4-METHYLPENTANOL-2 ◇ 4-METHYL-2-PENTANOL (MAK) ◇ METILAMIL ALCOHOL (ITALIAN) ◇ 4-METILPENTAN-2-OLO (ITALIAN) ◇ MIBC ◇ MIC ◇ 3-MIC

TOXICITY DATA with REFERENCE
skn-rbt 10 mg/24H open MLD AMIHBC 4,119,51
eye-rbt 20 mg open SEV AMIHBC 4,119,51
orl-rat LD50:2590 mg/kg AMIHBC 4,119,51
ihl-rat LCLo:2000 ppm/4H JIHTAB 31,343,49
orl-mus LDLo:1000 mg/kg UCPHAQ 2,217,49
ipr-mus LD50:812 mg/kg SCCUR* -,7,61
skn-rbt LD50:3560 mg/kg AMIHBC 4,119,51

CONSENSUS REPORTS: Reported in EPA TSCA Inventory.

OSHA PEL: (Transitional: TWA 25 ppm (skin)) TWA 25 ppm; STEL 40 ppm (skin)
ACGIH TLV: TWA 25 ppm; STEL 40 ppm (skin)
DFG MAK: 25 ppm (100 mg/m^3)
DOT Classification: Flammable or Combustible Liquid; Label: Flammable Liquid.

SAFETY PROFILE: Moderately toxic by ingestion, skin contact, and intraperitoneal routes. Mildly toxic by inhalation. A skin and severe eye irritant. Inhalation of high concentrations can cause anesthesia. Flammable when exposed to heat or flame; can react with oxidizing materials. A moderate explosion hazard when exposed to heat or flame. To fight fire, use alcohol foam. When heated to decomposition it emits acrid smoke and fumes. See also ALCOHOLS.

MKW750 CAS:105-44-2 ***HR: 3***
METHYL ISOBUTYL KETOXIME
mf: $C_6H_{13}NO$ mw: 115.20

SYN: USAF AM-4

TOXICITY DATA with REFERENCE
ipr-mus LD50:200 mg/kg NTIS** AD277-689

CONSENSUS REPORTS: Reported in EPA TSCA Inventory.

SAFETY PROFILE: Poison by intraperitoneal route. When heated to decomposition it emits toxic fumes of NO_x.

MKX000 CAS:547-63-7 ***HR: 2***
METHYL ISOBUTYRATE
mf: $C_5H_{10}O_2$ mw: 102.15

PROP: Colorless liquid. D: 0.889 @ 20°/4°, mp: −84 to −85°, bp: 93°, flash p: 55.4°F. Sltly sol in water; misc in alc and ether.

SYN: ISOBUTYRIC ACID, METHYL ESTER

TOXICITY DATA with REFERENCE
orl-rat LD50:16000 mg/kg 14CYAT 2,1879,63
ipr-rat LDLo:3200 mg/kg 14CYAT 2,1867,63
ihl-mus LC50:25500 mg/kg 85GMAT -,84,82

CONSENSUS REPORTS: Reported in EPA TSCA Inventory.

SAFETY PROFILE: Moderately toxic by intraperitoneal route. Mildly toxic by ingestion and inhalation. A very dangerous fire hazard when exposed to heat or flame. When heated to decomposition it emits acrid smoke and irritating fumes.

MKX250 CAS:624-83-9 ***HR: 3***
METHYL ISOCYANATE
DOT: UN 2480
mf: C_2H_3NO mw: 57.06

PROP: Liquid. D: 0.9599 @ 20°/20°, bp: 39.1°, flash p: <5°F.

SYNS: ISOCYANATE de METHYLE (FRENCH) ◇ ISOCYANATOMETHANE ◇ ISOCYANIC ACID, METHYL ESTER ◇ METHYLISOCYANAAT (DUTCH) ◇ METHYL ISOCYANAT (GERMAN) ◇ METHYL ISOCYANATE, solutions (DOT) ◇ METIL ISOCIANATO (ITALIAN) ◇ RCRA WASTE NUMBER P064 ◇ TL 1450

TOXICITY DATA with REFERENCE
sce-mus-ihl 3 ppm/6H/4D-C ENMUDM 8(Suppl 6),41,86
ihl-mus TCLo:1 ppm/6H (female 14-17D post):REP EVHPAZ 72,149,87
ihl-mus TCLo:9 ppm/3H (female 8D post):TER JTEHD6 21,265,87
ihl-hmn TCLo:2 ppm:NOSE,EYE,PUL ATXKA8 20,235,64
orl-rat LD50:51500 μg/kg IJEBA6 25,531,87
ihl-rat LC50:6100 ppb/6H FAATDF 6,747,86
orl-mus LD50:120 mg/kg TXAPA9 42,417,77
ihl-mus LC50:12200 ppb/6H FAATDF 6,747,86
skn-rbt LD50:213 mg/kg AIHAAP 30,470,69
ihl-gpg LC50:5400 ppb/6H FAATDF 6,747,86

CONSENSUS REPORTS: Reported in EPA TSCA Inventory.

OSHA PEL: TWA 0.02 ppm (skin)
ACGIH TLV: TWA 0.02 ppm (skin)
DFG MAK: 0.01 ppm (0.025 mg/m^3)
DOT Classification: Flammable Liquid; Label: Flammable Liquid and Poison.

SAFETY PROFILE: Poison by inhalation, ingestion, and skin contact. Human systemic effects by inhalation: conjuctiva irritation, olfactory and pulmonary changes. An experimental teratogen. Other experimental reproductive effects. Mutation data reported. A severe eye, skin, and mucous membrane irritant and a sensitizer. It can be absorbed through the skin. Exposure to high concentrations of the vapor can cause blindness; lung damage, including edema, permanent fibrosis, emphysema, and bronchitis; and gynecological effects. Most deaths are a result of lung tissue damage. Effects of cyanide poisoning have been noted but this may be due to impurities. A very dangerous fire hazard when exposed to heat, flame, or oxidizers. To fight fire, use spray, foam, CO_2, dry chemical. Exothermic reaction with water. When heated to decomposition it emits toxic fumes of NO_x. See also ISOCYANATES.

MKX500 CAS:593-75-9 ***HR: 3***
METHYL ISOCYANIDE
mf: C_2H_3N mw: 41.05

PROP: Colorless liquid. Mp: −45°, bp: 59.6°, d: 0.7464 @ 20°/4°.

SYN: METHYL ISONITRILE

CONSENSUS REPORTS: Community Right-To-Know List. EPA Extremely Hazardous Substances List.

SAFETY PROFILE: Poison. A shock- and heat-sensitive explosive. Explodes when heated in a sealed container. Has exploded on distillation. When heated to decomposition it emits highly toxic fumes of CN^- and NO_x. See also NITRILES.

MKX575 CAS:39687-95-1 ***HR: 3***
METHYL ISOCYANOACETATE
mf: $C_4H_5NO_2$ mw: 99.09

SAFETY PROFILE: Potentially explosive decomposition on contact with traces of heavy metals (e.g., iron; copper; chromium). When heated to decomposition it emits toxic fumes of NO_x and CN^-. See also ISOCYANIDES.

MKY250 CAS:99-48-9 ***HR: 2***
1-METHYL-4-ISOPROPENYL-6-CYCLOHEXEN-2-OL
mf: $C_{10}H_{16}O$ mw: 152.26

SYNS: 1-CARVEOL ◇ l-p-MENTHA-6,8-DIEN-2-OL

TOXICITY DATA with REFERENCE
skn-rbt 500 mg/24H FCTXAV 11,1055,73
orl-rat LD50:3000 mg/kg FCTXAV 11,1055,73

CONSENSUS REPORTS: Reported in EPA TSCA Inventory.

SAFETY PROFILE: Moderately toxic by ingestion. A skin irritant. When heated to decomposition it emits acrid smoke and irritating fumes.

MKY500 CAS:814-78-8 ***HR: 3***
METHYL ISOPROPENYL KETONE
DOT: UN 1246
mf: C_5H_8O mw: 84.119

PROP: Flash p: 69.8°F, lel: 1.8%, uel: 9.0%, vap d: 2.9, bp: 98°.

SYNS: 3-METHYL-3-BUTEN-2-ON (GERMAN) ◇ 2-METHYL-1-BUTEN-3-ONE ◇ METHYL ISOPROPENYL KETONE INHIBITED (DOT)

TOXICITY DATA with REFERENCE
skn-rbt 10 mg/24H open MLD AMIHBC 4,119,51
eye-rbt 50 μg open SEV AMIHBC 4,119,51
ihl-rat LCLo:125 ppm/4H AMIHBC 4,119,51
ipr-mus LD50:490 mg/kg ZoLH## 23OCT75
orl-gpg LDLo:60 mg/kg 14CYAT 2,1754,63
orl-rat LD50:180 mg/kg AMIHBC 4,119,51
skn-rbt LD50:230 mg/kg AMIHBC 4,119,51

CONSENSUS REPORTS: Reported in EPA TSCA Inventory.

DOT Classification: Flammable Liquid; Label: Flammable Liquid.

SAFETY PROFILE: Poison by ingestion and skin contact. Moderately toxic by inhalation and intraperitoneal route. A skin and severe eye irritant. A dangerous fire hazard when exposed to heat or flame. Explosive in the form of vapor when exposed to heat or flame. When heated to decomposition it emits acrid smoke and irritating fumes. See also KETONES.

MLA250 CAS:99-86-5 ***HR: 2***
1-METHYL-4-ISOPROPYLCYCLOHEXADIENE-1,3
mf: $C_{10}H_{16}$ mw: 136.26

PROP: Colorless liquid; lemon odor. D: 0.834 @ 20°/4°, refr index: 1.475-1.480, bp: 181.5°. Insol in water; misc in alc, ether, fixed oils.

SYNS: FEMA No. 3558 ◇ p-MENTHA-1,3-DIENE ◇ 1-METHYL-4-ISOPROPYL-1,3-CYCLOHEXADIENE ◇ α-TERPINENE (FCC)

TOXICITY DATA with REFERENCE
orl-rat LD50:1680 mg/kg FCTXAV 14,873,76

CONSENSUS REPORTS: Reported in EPA TSCA Inventory.

SAFETY PROFILE: Moderately toxic by ingestion. When heated to decomposition it emits acrid smoke and irritating fumes.

MLA750 CAS:563-80-4 ***HR: 3***
METHYL ISOPROPYL KETONE
DOT: UN 2397
mf: $C_5H_{10}O$ mw: 86.15

SYNS: ISOPROPYL METHYL KETONE ◇ 3-METHYL-2-BUTANONE ◇ 3-METHYL BUTAN 2-ONE (DOT) ◇ MIPK

TOXICITY DATA with REFERENCE
skn-rbt 500 mg/24H MOD FCTXAV 16,819,78
skn-rbt 500 mg open MLD FCTXAV 16,637,78
eye-rbt 100 mg/24H MLD FCTXAV 16,819,78
mrc-smc 12300 ppm MUREAV 149,339,85
orl-rat LD50:148 mg/kg SCCUR* -,7,61
ihl-rat LCLo:5700 ppm/4H TXAPA9 28,313,74
skn-rbt LD50:6350 mg/kg FCTXAV 16,819,78

CONSENSUS REPORTS: Reported in EPA TSCA Inventory.

OSHA PEL: TWA 200 ppm
ACGIH TLV: TWA 200 ppm
DOT Classification: Flammable Liquid; Label: Flammable Liquid.

SAFETY PROFILE: Poison by ingestion. Mildly toxic by inhalation and skin contact. Mutation data reported. A skin and eye irritant. Flammable when exposed to heat or flame; can react vigorously with oxidizing materials. When heated to decomposition it emits acrid smoke and irritating fumes. See also KETONES.

MLB000 CAS:63869-07-8 ***HR: 3***
METHYL(5-ISOPROPYL-N-(p-TOLYL)-o-TOLUENESULFONAMIDO)MERCURY
mf: $C_{18}H_{23}HgNO_2S$ mw: 518.07

SYN: N-p-TOLYL-N-METHYLMERCURIAMID KYSELINY 2-METHYL-5-ISOPROPYLBENZENSULFONOVE (CZECH)

TOXICITY DATA with REFERENCE
skn-rbt 500 mg/24H MLD 28ZPAK -,223,72
eye-rbt 20 mg/24H MOD 28ZPAK -,223,72
orl-rat LD50:115 mg/kg 28ZPAK -,223,72

CONSENSUS REPORTS: Mercury and its compounds are on the Community Right-To-Know List.

OSHA PEL: (Transitional: CL 1 mg/10m^3) CL 0.1 mg(Hg)/m^3 (skin)
ACGIH TLV: TWA 0.1 mg(Hg)/m^3 (skin)

SAFETY PROFILE: Poison by ingestion. A skin and eye irritant. When heated to decomposition it emits very toxic fumes of Hg, NO_x and SO_x. See also MERCURY COMPOUNDS.

MLB600 CAS:52547-00-9 ***HR: D***
3-METHYL-5-ISOTHIAZOLAMINE HYDROCHLORIDE
mf: $C_4H_6N_2S{\cdot}ClH$ mw: 150.64

SYN: 3-METHYL-5-ISOTHIAZOLAMINEMONOHYDROCHLORIDE

TOXICITY DATA with REFERENCE
mma-sat 6666 μg/plate MUREAV 155,17,85
mma-mus:lym 486 mg/L MUREAV 155,17,85

SAFETY PROFILE: Mutation data reported. When

heated to decomposition it emits toxic fumes of SO_x, NO_x, and HCl.

MLC000 CAS:8066-01-1 ***HR: 3***
d-d-METHYLISOTHIOCYANATE
mf: $C_3H_6Cl_2 \cdot C_3H_4Cl \cdot C_2H_3NS$ mw: 297.08

SYNS: dd-MENCS ◇ DD-METHYL ISOTHIOCYANATE MIXTURE ◇ 1,2-DICHLOROPROPANE mixed with 1,3-DICHLOROPROPENE and ISOTHIOCYANATOMETHANE ◇ DI-TRAPEX ◇ FORLEX ◇ VORLEX

TOXICITY DATA with REFERENCE
orl-rat LD50:72 mg/kg KONODE 22,100,78
skn-rat LD50:961 mg/kg FMCHA2 -,C254,83
orl-mus LD50:90 mg/kg KONODE 22,100,78
skn-rbt LD50:470 mg/kg FMCHA2 -,C254,83

SAFETY PROFILE: Poison by ingestion. Moderately toxic by skin contact. When heated to decomposition it emits very toxic fumes of Cl^-, SO_x, and NO_x. See also THIOCYANATES.

MLC250 CAS:5707-69-7 ***HR: 3***
3-METHYL-4,5-ISOXAZOLEDIONE-4-((2-CHLOROPHENYL)HYDRAZONE)
mf: $C_{10}H_8ClN_3O_2$ mw: 237.66

SYNS: 4-(2-CHLOROPHENYLHYDRAZONE)-3-METHYL-5-ISOXAZOLONE ◇ 4-(2-CHLOROPHENYLHYDRAZONE)-3-METHYL-5(4H)-ISOXAZOLONE ◇ DRAZOXOLON ◇ DRAZOXOLONE ◇ GANOCIDE ◇ 3-METHYL-4-((o-CHLOROPHENYL)HYDRAZONE)-4,5-ISOXAZOLEDIONE ◇ 3-METHYL-4-(o-CHLOROPHENYLHYDRAZONO)-5-ISOXAZOLONE ◇ MIL-COL ◇ PP781 ◇ SAISAN ◇ SOPRACOL ◇ SOPRACOL 781

TOXICITY DATA with REFERENCE
orl-rat LD50:126 mg/kg FMCHA2 -,D117,80
ipr-rat LD50:20 mg/kg FCTXAV 7,481,69
orl-mus LD50:129 mg/kg FCTXAV 7,481,69
orl-dog LDLo:20 mg/kg FCTXAV 7,481,69
orl-rbt LDLo:100 mg/kg FCTXAV 7,481,69
orl-cat LD50:50 mg/kg 31ZOAD 1,195,68

SAFETY PROFILE: Poison by ingestion and intraperitoneal routes. When heated to decomposition it emits very toxic fumes of Cl^- and NO_x.

MLC500 CAS:64552-25-6 ***HR: D***
N-METHYLJERVINE
mf: $C_{28}H_{41}NO_3$ mw: 439.70

TOXICITY DATA with REFERENCE
orl-ham TDLo:170 mg/kg (7D preg):TER JAFCAU 26,564,78

SAFETY PROFILE: Experimental teratogenic effects. When heated to decomposition it emits toxic fumes of NO_x.

MLC750 CAS:75-86-5 ***HR: 3***
2-METHYLLACTONITRILE
DOT: UN 1541
mf: C_4H_7NO mw: 85.12

PROP: Mp: −20°, bp: 82° @ 23 mm, d: 0.932 @ 19°, autoign temp: 1270°F, flash p: 165°F, vap d: 2.93.

SYNS: ACETONCIANHIDRINEI (ROUMANIAN) ◇ ACETONCIANIDRINA (ITALIAN) ◇ ACETONCYAANHYDRINE (DUTCH) ◇ ACETONCYANHYDRIN (GERMAN) ◇ ACETONECYANHYDRINE (FRENCH) ◇ ACETONE CYANOHYDRIN (DOT) ◇ ACETONKYANHYDRIN (CZECH) ◇ CYANHYDRINE d'ACETONE (FRENCH) ◇ α-HYDROXYISOBUTYRONITRILE ◇ 2-HYDROXY-2-METHYLPROPIONITRILE ◇ RCRA WASTE NUMBER P069 ◇ USAF RH-8

TOXICITY DATA with REFERENCE
orl-rat LD50:17800 μg/kg 28ZPAK -,61,72
ihl-rat LCLo:63 ppm/4H AIHAAP 23,95,62
scu-rat LDLo:8500 μg/kg BJIMAG 19,283,62
orl-mus LD50:14 mg/kg 28ZPAK -,161,72
ihl-mus LCLo:500 mg/m³/2H IGIBA5 11,27,62
ipr-mus LD50:1 mg/kg NTIS** AD277-689
orl-rbt LD50:13500 μg/kg NTIS** PB214-270
skn-rbt LD50:17 mg/kg AIHAAP 23,95,62
orl-gpg LD50:9 mg/kg NTIS** PB214-270

CONSENSUS REPORTS: Cyanide and its compounds are on the Community Right-To-Know List. Reported in EPA TSCA Inventory. EPA Extremely Hazardous Substances List.
NIOSH REL: (Nitriles) CL 4 mg/m³/15M
DOT Classification: Poison B; Label: Poison.

SAFETY PROFILE: Poison by ingestion, skin contact, inhalation, intraperitoneal, and subcutaneous routes. Readily decomposes to HCN and acetone. Keep cool and do not store for long periods. Combustible when exposed to heat or flame. To fight fire, use CO_2, dry chemical, alcohol foam. Vigorous reaction with H_2SO_4. When heated to decomposition it emits toxic fumes of CN^-. See also NITRILES, HYDROCYANIC ACID, and ACETONE.

MLD000 CAS:917-54-4 ***HR: 3***
METHYLLITHIUM
mf: CH_3Li mw: 21.98

SAFETY PROFILE: Ignites spontaneously in air. See also LITHIUM COMPOUNDS.

MLD100 CAS:848-75-9 ***HR: 2***
N-METHYLLORAZEPAM
mf: $C_{16}H_{12}Cl_2N_2O_2$ mw: 335.20

PROP: Crystals from ethanol/tetrahydrofuran. Mp: 205-207°.

SYNS: 7-CHLORO-5-(2-CHLOROPHENYL)-1,3-DIHYDRO-3-HYDROXY-1-METHYL-2H-1,4-BENZODIAZEPIN-2-ONE◇ 7-CHLORO-

5-(2-CHLOROPHENYL)-3-HYDROXY-1-METHYL-2,3-DIHYDRO-1H-1,4-BENZODIAZEPIN-2-ONE ◇ LORAMET ◇ LORMETAZEPAM ◇ METHYLLORAZEPAM ◇ NOCTAMID ◇ RO 5-5516

TOXICITY DATA with REFERENCE
orl-rat TDLo:270 mg/kg (7-22D preg/21D post):REP YACHDS 13(Suppl 3),605,85
orl-rat TDLo:27 mg/kg (7-22D preg/21D post):TER YACHDS 13(Suppl 3),605,85
ipr-rat LD50:6250 mg/kg YACHDS 13(Suppl 3),545,85
orl-mus LD50:1790 mg/kg YACHDS 13(Suppl 3),545,85
ipr-mus LD50:780 mg/kg YACHDS 13(Suppl 3),545,85

SAFETY PROFILE: Moderately toxic by ingestion and intraperitoneal routes. An experimental teratogen. Experimental reproductive effects. When heated to decomposition it emits toxic fumes of Cl^- and NO_x. Note: This is a controlled substance (depressant) listed in the U.S. Code of Federal Regulations, Title 21 Part 1308.14 (1985).

MLD250 CAS:361-37-5 ***HR: 2***
1-METHYLLYSERGIC ACID BUTANOLAMIDE
mf: $C_{21}H_{27}N_3O_2$ mw: 353.51

SYNS: DESERIL ◇ DESERNYL ◇ DESERYL ◇ 9,10-DIDEHYDRO-N-(1-HYDROXYMETHYL)PROPYL)-1,6-DIMETHYLERGOLINE-8β-CARBOXAMIDE ◇ METHYLLYSERGIC ACID BUTANOLAMIDE ◇ METHYSERGID ◇ METHYSERGIDE ◇ UML 491

TOXICITY DATA with REFERENCE
cyt-hmn:lym 100 μg/L/24H MUREAV 48,205,77
orl-hmn TDLo:28 μg/kg:CNS,GIT ARZNAD 16,220,66

SAFETY PROFILE: Human systemic effects by ingestion: anorexia, headache, nausea or vomiting. Human mutation data reported. When heated to decomposition it emits toxic fumes of NO_x.

MLD500 CAS:7240-57-5 ***HR: 3***
1-METHYLLYSERGIC ACID ETHYLAMIDE
mf: $C_{19}H_{23}N_3O$ mw: 309.45

SYNS: 9,10-DIDEHYDRO-N-ETHYL-1,6-DIMETHYLERGOLINE-8β-CARBOXAMIDE ◇ d-1-METHYL LYSERGIC ACID MONOETHYLAMIDE ◇ MLA-74

TOXICITY DATA with REFERENCE
orl-hmn TLDo:25 μg/kg:PSY PSYPAG 1,20,59
ivn-rbt LD50:9 mg/kg 27ZQAG -,101,72

SAFETY PROFILE: Poison by intravenous route. Human systemic effects by ingestion: psychotropic effects. When heated to decomposition it emits toxic fumes of NO_x.

MLE000 CAS:75-16-1 ***HR: 3***
METHYLMAGNESIUM BROMIDE (ethyl ether solution)
mf: CH_3BrMg mw: 119.26
DOT: UN 1928

PROP: Concentration of ethyl ether is not over 40% (FEREAC 41,15972,76).

SYN: METHYL MAGNESIUM BROMIDE in ETHYL ETHER (DOT)

CONSENSUS REPORTS: Reported in EPA TSCA Inventory.

DOT Classification: Flammable Liquid; Label: Flammable Liquid, Spontaneously Combustible.

SAFETY PROFILE: May ignite spontaneously in air. A very dangerous fire hazard when exposed to heat or flame; can react vigorously with oxidizing materials. When heated to decomposition it emits acrid smoke and irritating fumes. See also ETHERS, MAGNESIUM COMPOUNDS, and BROMIDES.

MLE250 CAS:917-64-6 ***HR: 3***
METHYLMAGNESIUM IODIDE
mf: CH_3IMg mw: 166.25

SYN: IODOMETHYLMAGNESIUM

SAFETY PROFILE: Toxic. Explosive reaction with thiophosphoryl chloride or vanadium trichloride. A Grignard reagent. When heated to decomposition it emits toxic fumes of I^-. See also IODIDES and MAGNESIUM COMPOUNDS.

MLE600 CAS:2213-00-5 ***HR: D***
METHYL MARASMATE
mf: $C_{16}H_{20}O_4$ mw: 276.36

SYN: MARASMIC ACID METHYL ESTER

TOXICITY DATA with REFERENCE
mmo-sat 50 μg/disc JANTAJ 36,155,83
dni-mus:ast 3 mg/L JANTAJ 36,155,83

SAFETY PROFILE: Mutation data reported. When heated to decomposition it emits acrid smoke and irritating fumes. See also ESTERS.

MLE650 CAS:74-93-1 ***HR: 3***
METHYL MERCAPTAN
DOT: UN 1064
mf: CH_4S mw: 48.11

PROP: Flammable gas; odor of rotten cabbage. Mp: −123.1°, vap d: 1.66, lel: 3.9%, uel: 21.8%. Bp: 5.95°, d: 0.8665 @ 20°/4°, solidifies @ −123°, flash p: −0.4°F.

SYNS: MERCAPTAN METHYLIQUE (FRENCH) ◇ METHAANTHIOL (DUTCH) ◇ METHANETHIOL ◇ METHANTHIOL (GERMAN) ◇ METHVTIOLO (ITALIAN) ◇ METHYLMERCAPTAAN (DUTCH) ◇ METILMERCAPTANO (ITALIAN) ◇ RCRA WASTE NUMBER U153

TOXICITY DATA with REFERENCE
sln-dmg-ihl 99 pph/6M-C ENVRAL 7,286,74
ihl-rat LC50:675 ppm LacHB# 09JUN78
ihl-mus LC50:6530 μg/m³/2H GTPZAB 16(6),46,72

CONSENSUS REPORTS: EPA Extremely Hazardous Substances List. Reported in EPA TSCA Inventory.

OSHA PEL: (Transitional: CL 10 ppm) TWA 0.5 ppm
ACGIH TLV: TWA 0.5 ppm
DFG MAK: 0.5 ppm (1 mg/m^3)
NIOSH REL: (n-Alkane Monothiols) CL 0.5 ppm/15M
DOT Classification: Flammable Gas; Label: Flammable Gas.

SAFETY PROFILE: Poison by inhalation. Mutation data reported. A common air contaminant. Very dangerous fire hazard when exposed to heat or flame; can react vigorously with oxidizing materials. Explosive in the form of vapor when exposed to heat or flame. Reacts with water, steam, or acids to produce toxic and flammable vapors. Violent reaction with mercury(II) oxide. To fight fire, use alcohol foam, CO_2, dry chemical. Upon decomposition it emits highly toxic fumes of SO_x. See also MERCAPTANS.

MLE750 CAS:2365-48-2 ***HR: 3***
METHYLMERCAPTOACETATE
mf: $C_3H_6O_2S$ mw: 106.15

SYNS: MERCAPTOACETIC ACID METHYL ESTER ◇ METHYL-2-MERCAPTOACETATE ◇ METHYLTHIOGLYCOLATE ◇ THIOGLYCOLIC ACID METHYL ESTER ◇ THIOGLYKOLSAEURE-METHYLESTER (GERMAN) ◇ USAF EK-7119

TOXICITY DATA with REFERENCE
orl-rat LD50:209 mg/kg ZHYGAM 20,575,74
ipr-rat LD50:252 mg/kg ZHYGAM 20,575,74
ipr-mus LD50:100 mg/kg NTIS** AD277-689

CONSENSUS REPORTS: Reported in EPA TSCA Inventory.

SAFETY PROFILE: Poison by ingestion and intraperitoneal routes. When heated to decomposition it emits toxic fumes of SO_x. See also MERCAPTANS and ESTERS.

MLF000 CAS:64038-57-9 ***HR: 3***
1-METHYL-2-MERCAPTO-5-IMIDAZOLE CARBOXYLIC ACID
mf: $C_5H_6N_2O_2S$ mw: 158.19

SYN: USAF EL-97

TOXICITY DATA with REFERENCE
ipr-mus LD50:200 mg/kg NTIS** AD277-689

CONSENSUS REPORTS: Reported in EPA TSCA Inventory.

SAFETY PROFILE: Poison by intraperitoneal route. When heated to decomposition it emits very toxic fumes of SO_x and NO_x.

MLF250 CAS:502-39-6 ***HR: 3***
METHYLMERCURIC DICYANDIAMIDE
mf: $C_3H_6HgN_4$ mw: 298.72

SYNS: AGROSOL ◇ CYANOGUANIDINE METHYLMERCURY DERIV. ◇ CYANO(METHYLMERCURI)GUANIDINE ◇ MEMA ◇ METHYLMERCURIC CYANOGUANIDINE ◇ METHYLMERCURY DICYANDIAMIDE ◇ MMD ◇ MORSODREN ◇ MORTON SOIL DRENCH ◇ PANDRINOX ◇ PANO-DRENCH 4 ◇ PANOGEN ◇ PANOGEN TURF FUNGICIDE ◇ PANOGEN TURF SPRAY ◇ PANOSPRAY 30

TOXICITY DATA with REFERENCE
ipr-mus TDLo:8 mg/kg (9-13D preg):REP TJADAB 5,181,72
ipr-mus TDLo:4 mg/kg (female 10D post):TER ARINAU 19,61,72
orl-rat LD50:68 mg/kg APPHAX 29,623,72
ipr-rat LD50:13 mg/kg TXAPA9 23,197,72
orl-mus LD50:20 mg/kg PCOC** -,735,66
ipr-mus LD50:20 mg/kg AMSVAZ 143,365,52

CONSENSUS REPORTS: Mercury and its compounds are on the Community Right-To-Know List. EPA Extremely Hazardous Substances List.

OSHA PEL: (Transitional: CL 1 mg/10m^3) TWA 0.01 mg(Hg)/m^3; STEL 0.03 mg/m^3 (skin)
ACGIH TLV: TWA 0.01 mg(Hg)/m^3 (skin); 0.01 mg(Hg)/m^3; STEL 0.03 mg(Hg)/m^3
NIOSH REL: TWA 0.05 mg(Hg)/m^3

SAFETY PROFILE: Poison by ingestion and intraperitoneal routes. An experimental teratogen. Experimental reproductive effects. Mutation data reported. When heated to decomposition it emits very toxic fumes of Hg and NO_x. See also MERCURY COMPOUNDS.

MLF500 CAS:5902-79-4 ***HR: 3***
METHYLMERCURICHLORENDIMIDE
mf: $C_{10}H_5Cl_6HgNO_2$ mw: 584.45

SYNS: MEMMI ◇ N-(METHYLMERCURI)-1,4,5,6,7,7-HEXACHLOROBICYCLO(2.2.1)HEPT-5-ENE-2,3-DICARBOXIMIDE◇ N-METHYLMERCURI-1,2,3,6-TETRAHYDRO-3,6-ENDOMETHANO-3,4,5,6,7,7-HEXACHLOROPHTHALIMIDE ◇ N-METHYLMERCURI-1,2,3,6-TETRAHYDRO-3,6-METHANO-3,4,5,6,7,7-HEXACHLOROPHTHALIMIDE

TOXICITY DATA with REFERENCE
orl-rat LD50:155 mg/kg 28ZEAL 4,282,69

CONSENSUS REPORTS: Mercury and its compounds are on the Community Right-To-Know List.

OSHA PEL: (Transitional: CL 1 mg/10m^3) TWA 0.01 mg(Hg)/m^3; STEL 0.03 mg/m^3 (skin)
ACGIH TLV: TWA 0.01 mg(Hg)/m^3; STEL 0.03 mg(Hg)/m^3

SAFETY PROFILE: Poison by ingestion. When heated to decomposition it emits very toxic fumes of Cl^-, Hg, and NO_x. See also MERCURY COMPOUNDS and CHLORIDES.

MLF520 CAS:32787-44-3 ***HR: D***
METHYLMERCURIC PHOSPHATE
mf: CH_5HgO_4P mw: 312.62

SYNS: MERCURY, (DIHYDROGEN PHOSPHATO)METHYL- ◇ MERCURY, METHYL(PHOSPHATO(1-)-O)-(9CI) ◇ METHYLMERCURY PHOSPHATE

TOXICITY DATA with REFERENCE
scu-mus TDLo:40 mg/kg (female 10D post):TER ARINAU 19,61,72

OSHA PEL: (Transitional: CL 1 mg/10m^3) TWA 0.01 mg(Hg)/m^3; STEL 0.03 mg/m^3 (skin)
ACGIH TLV: TWA 0.01 mg(Hg)/m^3; STEL 0.03 mg(Hg)/m^3
NIOSH REL: TWA 0.05 mg(Hg)/m^3

SAFETY PROFILE: An experimental teratogen. When heated to decomposition it emits toxic fumes of PO_x and Hg.

MLF550 CAS:22967-92-6 ***HR: 3***
METHYLMERCURY
mf: CH_3Hg mw: 215.63

SYNS: METHYL-MERCURY(1+) (9CI) ◇ METHYLMERCURY(II) CATION ◇ METHYLMERCURY ION ◇ METHYLMERCURY ION(1+)

TOXICITY DATA with REFERENCE
dnr-smc 1 nmol/L CNJGA8 24,771,82
cyt-ham-ipr 5 mg/kg JACTDZ 3(4),295,84
orl-rat TDLo:8550 mg/kg (female 14D pre-21D post):REP NETOD7 8,585,86
ivn-mus TDLo:2700 μg/kg (female 1D post):TER TJADAB 34,471,86

CONSENSUS REPORTS: Mercury and its compounds are on the Community Right-To-Know List.

OSHA PEL: (Transitional: CL 1 mg/10m^3) TWA 0.01 mg(Hg)/m^3; STEL 0.03 mg/m^3 (skin)
ACGIH TLV: TWA 0.01 mg(Hg)/m^3; STEL 0.03 mg(Hg)/m^3
DFG MAK: 0.01 mg/m^3
NIOSH REL: TWA 0.05 mg(Hg)/m^3

SAFETY PROFILE: A poison. An experimental teratogen. Experimental reproductive effects. Mutation data reported. Used as a fungicide. When heated to decomposition it emits toxic fumes of Hg. See also MERCURY COMPOUNDS, ORGANIC.

MLF750 CAS:72066-32-1 ***HR: 3***
METHYLMERCURY DIMERCAPTOPROPANOL
mf: $C_{15}H_{17}HgNO_4S_2$ mw: 540.04

TOXICITY DATA with REFERENCE
ipr-mus LD50:37 mg/kg OCHRAI 15,5,63

CONSENSUS REPORTS: Mercury and its compounds are on the Community Right-To-Know List.

OSHA PEL: (Transitional: CL 1 mg/10m^3) TWA 0.01 mg(Hg)/m^3; STEL 0.03 mg/m^3 (skin)
ACGIH TLV: TWA 0.01 mg(Hg)/m^3; STEL 0.03 mg(Hg)/m^3
NIOSH REL: TWA 0.05 mg(Hg)/m^3

SAFETY PROFILE: Poison by intraperitoneal route. When heated to decomposition it emits very toxic fumes of Hg, NO_x, and SO_x. See also MERCURY COMPOUNDS and MERCAPTANS.

MLG000 CAS:1184-57-2 ***HR: 3***
METHYLMERCURY HYDROXIDE
mf: CH_4HgO mw: 232.64

SYNS: HYDROXYMETHYLMERCURY ◇ METHYLMERCURIC HYDROXIDE

TOXICITY DATA with REFERENCE
dni-hmn:hla 10 μmol/L AEMBAP 177,229,84
dni-mus-ipr 1 mg/kg JPETAB 194,171,75
orl-mus TDLo:3237 μg/kg (female 8D post):REP PBBHAU 4,385,76
orl-rat TDLo:48 mg/kg (16-22D preg/17D post):TER TJADAB 25(2),34A,82
ipr-mus LD50:20 mg/kg OCHRAI 15,5,63

CONSENSUS REPORTS: Mercury and its compounds are on the Community Right-To-Know List. EPA Genetic Toxicology Program.

OSHA PEL: (Transitional: CL 1 mg/10m^3) TWA 0.01 mg(Hg)/m^3; STEL 0.03 mg/m^3 (skin)
ACGIH TLV: TWA 0.01 mg(Hg)/m^3; STEL 0.03 mg(Hg)/m^3
NIOSH REL: TWA 0.05 mg(Hg)/m^3

SAFETY PROFILE: Poison by intraperitoneal route. An experimental teratogen. Experimental reproductive effects. Human mutation data reported. When heated to decomposition it emits toxic fumes of Hg. See also MERCURY COMPOUNDS.

MLG250 CAS:5902-76-1 ***HR: 3***
METHYLMERCURY PENTACHLOROPHENATE
mf: $C_7H_3Cl_5HgO$ mw: 480.94

SYNS: METHYLMERCURIPENTACHLORFENOLAT(CZECH) ◇ METHYL(PENTACHLOROPHENOXY)MERCURY ◇ STAUFFER MV242

TOXICITY DATA with REFERENCE
skn-rbt 500 mg/24H MLD 28ZPAK -,223,72

eye-rbt 50 μg/24H SEV 28ZPAK -,223,72
orl-rat LD50:56 mg/kg 28ZEAL 4,283,69

CONSENSUS REPORTS: Mercury and its compounds are on the Community Right-To-Know List.

OSHA PEL: (Transitional: CL 1 mg/10m^3) TWA 0.01 mg(Hg)/m^3; STEL 0.03 mg/m^3 (skin)
ACGIH TLV: TWA 0.01 mg(Hg)/m^3; STEL 0.03 mg(Hg)/m^3

SAFETY PROFILE: Poison by ingestion. A skin and severe eye irritant. When heated to decomposition it emits very toxic fumes of Cl^- and Hg. See also CHLORINATED AROMATIC HYDROCARBONS and MERCURY COMPOUNDS.

MLG500 CAS:40661-97-0 ***HR: 3***
METHYLMERCURY PERCHLORATE
mf: CH_3ClHgO_4 mw: 315.08

CONSENSUS REPORTS: Mercury and its compounds are on the Community Right-To-Know List.

SAFETY PROFILE: A spark-sensitive explosive which may be initiated by static discharge. Upon decomposition it emits toxic fumes of Hg and Cl^-. See also MERCURY COMPOUNDS and PERCHLORATES.

MLG750 CAS:2597-95-7 ***HR: 3***
METHYLMERCURY PROPANEDIOLMERCAPTIDE
mf: $C_4H_{10}HgO_2S$ mw: 322.79

SYN: ((DIHDYROXYPROPYL)THIO)METHYLMERCURY

TOXICITY DATA with REFERENCE
ipr-mus LD50:29 mg/kg OCHRAI 15,5,63
ipr-mus LD50:47 mg/kg OCHRAI 15,5,63

CONSENSUS REPORTS: Mercury and its compounds are on the Community Right-To-Know List.

OSHA PEL: (Transitional: CL 1 mg/10m^3) TWA 0.01 mg(Hg)/m^3; STEL 0.03 mg/m^3 (skin)
ACGIH TLV: TWA 0.01 mg(Hg)/m^3; STEL 0,03 mg(Hg)/m^3

SAFETY PROFILE: Poison by intraperitoneal route. When heated to decomposition it emits very toxic fumes of Hg and SO_x. See also MERCURY COMPOUNDS and SULFIDES.

MLH000 CAS:86-85-1 ***HR: 3***
METHYLMERCURY QUINOLINOLATE
mf: $C_{10}H_9HgNO$ mw: 359.79

SYNS: ARTHO LM ◇ LIQUI-SAN ◇ LM SEED PROTECTANT ◇ METASOL ◇ METAZOL ◇ 8-(METHYLMERCURIOXY)QUINOLINE ◇ METHYLMERCURY β-HYDROXYQUINOLATE ◇ METHYLMERCURY 8-HYDROXYQUINOLINATE ◇ METHYLMERCURY OXINATE ◇ METHYLMERCURY OXYQUINOLINATE ◇ ORTHO-LM APPLE SPRAY ◇ ORTHO LM CONCENTRATE ◇ ORTHO LM SEED PROTECTANT ◇ 8-(QUINOLINOLATO)METHYL MERCURY ◇ 8-QUINOLINOL, MERCURY COMPLEX

TOXICITY DATA with REFERENCE
orl-rat LD50:72 mg/kg PCOC** -,739,66
orl-mus LD50:72 mg/kg GUCHAZ 6,347,73

CONSENSUS REPORTS: Mercury and its compounds are on the Community Right-To-Know List.

OSHA PEL: (Transitional: CL 1 mg/10m^3) TWA 0.01 mg(Hg)/m^3; STEL 0.03 mg/m^3 (skin)
ACGIH TLV: TWA 0.01 mg(Hg)/m^3; STEL 0.03 mg(Hg)/m^3
NIOSH REL: TWA 0.05 mg(Hg)/m^3

SAFETY PROFILE: Poison by ingestion. A pesticide. When heated to decomposition it emits very toxic fumes of Hg and NO_x. See also MERCURY COMPOUNDS.

MLH100 CAS:102280-93-3 ***HR: 3***
METHYL-MERCURY TOLUENESULPHAMIDE
mf: $C_9H_{14}Hg_2N_2O_2S$ mw: 615.49

SYNS: N,N-BIS(METHYLQUECKSILBER)-p-TOLUOL-SULFAMID ◇ MERCURY, (((p-TOLYL)SULFAMOYL)IMINO)BIS(METHYL- ◇ METHYL-QUECKSILBER-TOLUOLSULFAMID ◇ (((p-TOLYL)SULFAMOYL)IMINO)BIS(METHYLMERCURY)

TOXICITY DATA with REFERENCE
orl-rat LD50:54 mg/kg AXVMAW 34,383,80
orl-qal LD50:25 mg/kg AXVMAW 34,383,80

ACGIH TLV: TWA 0.01 mg(Hg)/m^3; STEL 0.03 mg(Hg)/m^3

SAFETY PROFILE: Poison by ingestion. When heated to decomposition it emits toxic fumes of NO_x, SO_x, and Hg.

MLH250 CAS:1082-88-8 ***HR: 3***
α-METHYLMESCALINE
mf: $C_{12}H_{19}NO_3$ mw: 225.32

SYNS: 3,4,5-TRIMETHOXYAMPHETAMINE ◇ TRIMETHOXYPHENYL-β-AMINOPROPANE ◇ 3,4,5-TRIMETHOXYPHENYL-β-AMINOPROPANE

TOXICITY DATA with REFERENCE
orl-hmn TDLo:800 μg/kg:PSY JMSCA9 101,317,55
ipr-mus LDLo:200 mg/kg EXPEAM 19,127,63

SAFETY PROFILE: Poison by intraperitoneal. Human systemic effects by ingestion: psychotropic effects. When heated to decomposition it emits toxic fumes of NO_x. See also MESCALINE.

MLH500 CAS:66-27-3 ***HR: 3***
METHYL MESYLATE
mf: $C_2H_6O_3S$ mw: 110.14

PROP: Liquid. D: 1.046 @ 16°/4° bp: 126.5° @ 756 mm. Decomp in water; sol in alc and ether.

SYNS: as-DIMETHYL SULPHATE ◇ METHANESULPHONIC ACID METHYL ESTER ◇ METHYL ESTER of METHANESULFONIC ACID ◇ METHYL ESTER of METHANESULPHONIC ACID ◇ METHYLMETHANSULFONAT (GERMAN) ◇ METHYL METHANESULFONATE ◇ METHYL METHANESULPHONATE ◇ METHYL METHANSULFONATE ◇ METHYL METHANSULPHONATE ◇ MMS ◇ NSC-50256

TOXICITY DATA with REFERENCE
mmo-sat 1 μL/plate MUREAV 130,79,84
dns-hmn:fbr 2400 μmol/L ENMUDM 7,267,85
ipr-mus TDLo:50 mg/kg (male 1D pre):REP BJPCAL 23,521,64
ipr-rat TDLo:100 mg/kg (female 15D post):TER JRPFA4 18,15,69
orl-rat TDLo:420 mg/kg/2W-C:ETA CRNGDP 6,1529,85
ihl-rat TCLo:50 ppm/6H/6W-I:CAR CALEDQ 33,175,86
ivn-rat TDLo:20 mg/kg/(15D preg):ETA,TER EJCAAH 8,641,72
unr-rat TDLo:20 mg/kg (female 15D post):CAR,TER IARCCD 4,143,73
orl-mus TDLo:300 mg/kg/8W-I:NEO TXAPA9 82,19,86
scu-mus TDLo:5640 mg/kg/64W-I:CAR CNREA8 47,3402,87
ivn-rat LD:30 mg/kg (female 21D post):ETA,TER EXPTAX 16,157,78
orl-rat LD50:225 mg/kg FCTXAV 19,347,81
ipr-rat LD50:114 mg/kg FCTOD7 22,665,84
scu-rat LD50:125 mg/kg ZEKBAI 74,241,70
ivn-rat LD50:175 mg/kg ZEKBAI 74,241,70
ipr-mus LDLo:130 mg/kg YOMJA9 20,105,79
ivn-dog LDLo:22 mg/kg CCSUBJ 2,203,65

CONSENSUS REPORTS: IARC Cancer Review: Group 3 IMEMDT 7,56,87; Animal Sufficient Evidence IMEMDT 7,253,74. Reported in EPA TSCA Inventory. EPA Genetic Toxicology Program.

SAFETY PROFILE: Poison by ingestion, intraperitoneal, intravenous, and subcutaneous routes. Human mutation data reported. Experimental teratogenic and reproductive effects. Questionable carcinogen with experimental carcinogenic, neoplastigenic, and tumorigenic data. When heated to decomposition it emits toxic fumes of SO_x. See also SULFONATES and ESTERS.

MLH750 CAS:80-62-6 ***HR: 3***
METHYL METHACRYLATE
DOT: NA 1247
mf: $C_5H_8O_2$ mw: 100.13

PROP: Colorless liquid, very sltly sol in water. Mp: −50°, bp: 101.0°, flash p: 50°F (OC), d: 0.936 @ 20°/4°, vap press: 40 mm @ 25.5°, vap d: 3.45, lel: 2.1%, uel: 12.5%.

SYNS: DIAKON ◇ METAKRYLAN METYLU (POLISH) ◇ METHACRYLATE de METHYLE (FRENCH) ◇ METHACRYLIC ACID, METHYL ESTER (MAK) ◇ METHACRYLSAEUREMETHYL ESTER (GERMAN) ◇ METHYLMETHACRYLAAT (DUTCH) ◇ METHYL-METHACRYLAT (GERMAN) ◇ METHYL METHACRYLATE MONOMER, INHIBITED (DOT) ◇ METHYL-α-METHYLACRYLATE ◇ METHYL-2-METHYL-2-PROPENOATE ◇ 2-METHYL-2-PROPENOIC ACID METHYL ESTER ◇ METIL METACRILATO (ITALIAN) ◇ MME ◇ "MONOCITE" METHACRYLATE MONOMER ◇ NCI-C50680 ◇ RCRA WASTE NUMBER U162

TOXICITY DATA with REFERENCE
skn-rbt 10 g/kg open JIHTAB 23,343,41
eye-rbt 150 mg INMEAF 14,292,45
mma-sat 34 mmol/L JBJSA3 61-A,1203,79
mma-mus:lym 500 mg/L ENMUDM 8(Suppl 6),4,86
ihl-rat TCLo:109 g/m^3/54M (female 6-15D post):TER TXAPA9 50,451,79
ihl-rat TCLo:4480 mg/m^3/2H (female 6-18D post):REP TOLED5 31(Suppl),80,86
imp-rat TDLo:1620 mg/kg:ETA CORTBR 88,223,72
ihl-hmn TCLo:125 ppm:CNS GISAAA 19(10),25,54
ihl-hmn TCLo:60 mg/m^3:CNS,CVS GTPZAB 1,56,57
orl-rat LD50:7872 mg/kg JIHTAB 23,343,41
ihl-rat LC50:3750 ppm 14CYAT 2,1880,63
ipr-rat LD50:1328 mg/kg JDREAF 51,1632,72
scu-rat LD50:7500 mg/kg INMEAF 14,292,45
orl-mus LD50:5204 mg/kg TOLED5 11,125,82
ihl-mus LCLo:13 g/m^3 GISAAA 20(8),19,55
ipr-mus LD50:1000 mg/kg INMEAF 14,292,45
scu-mus LD50:6300 mg/kg INMEAF 14,292,45
orl-dog LDLo:5000 mg/kg INMEAF 14,292,45
scu-dog LD50:4500 mg/kg INMEAF 14,292,45

CONSENSUS REPORTS: IARC Cancer Review: Group 3 IMEMDT 7,56,87; Human Inadequate Evidence IMEMDT 19,187,79; Animal Inadequate Evidence IMEMDT 19,187,79. NTP Carcinogenesis Studies (inhalation); No Evidence: mouse, rat NTPTR* NTP-TR-314,86. Reported in EPA TSCA Inventory. Community Right-To-Know List.

OSHA PEL: TWA 100 ppm
ACGIH TLV: TWA 100 ppm
DFG MAK: 50 ppm (210 mg/m^3)
DOT Classification: Flammable Liquid; Label: Flammable Liquid.

SAFETY PROFILE: Moderately toxic by inhalation and intraperitoneal routes. Mildly toxic by ingestion. Human systemic effects by inhalation: sleep effects, excitement, anorexia, and blood pressure decrease. Experimental teratogenic and reproductive effects. Mutation data reported. A skin and eye irritant. Questionable carcinogen with experimental tumorigenic data. A common air contaminant.

A very dangerous fire hazard when exposed to heat or flame; can react with oxidizing materials. Explosive in the form of vapor when exposed to heat or flame. The monomer may undergo spontaneous, explosive polymerization. Reacts in air to form a heat-sensitive explosive product (explodes on evaporation at 60°C). May ignite on contact with benzoyl peroxide. Potentially violent reaction with the polymerization initiators azoisobutyronitrile; dibenzoyl peroxide; di-tert-butyl peroxide; propionaldehyde. To fight fire, use foam, CO_2, dry chemical. When heated to decomposition it emits acrid smoke and irritating fumes. See also ESTERS.

MLI350 CAS:63160-33-8 ***HR: 3***
METHYL-3-METHOXY CARBONYLAZO-CROTONATE
mf: $C_7H_{10}N_2O_4$ mw: 186.17

$CH_3OCO{\bullet}N{=}NC(CH_3){=}CHCO{\bullet}OCH_3$

SAFETY PROFILE: Explodes during vacuum distillation. Potentially dangerous polymerization reaction during storage at room temperature. When heated to decomposition it emits toxic fumes of NO_x.

MLI750 CAS:67292-68-6 ***HR: 3***
2-METHYL-5-METHOXY-N-DIMETHYLTRYPTAMINE
mf: $C_{14}H_{20}N_2O$ mw: 232.36

SYNS: 4365 CT ◇ 3-(2-(DIMETHYLAMINO)ETHYL)-5-METHOXY-2-METHYLINDOLE ◇ METHYL-2-METHOXY-5-N-DIMETHYLTRYPTAMINE

TOXICITY DATA with REFERENCE
ipr-rat LD50:74 mg/kg BSCFAS (5),1411,65
ipr-mus LD50:100 mg/kg BSCFAS (5),1411,65
ivn-mus LD50:48 mg/kg BSCFAS (5),1411,65
ivn-rbt LD50:22 mg/kg BSCFAS (5),1411,65

SAFETY PROFILE: Poison by intraperitoneal and intravenous routes. When heated to decomposition it emits toxic fumes of NO_x.

MLJ000 CAS:29173-31-7 ***HR: 3***
METHYL(((METHOXYMETHYLPHOSPHINOTHIOYL)THIO)ACETYL) METHYLCARBAMATE
mf: $C_7H_{14}NO_4PS_2$ mw: 271.31

SYNS: MC 2420 ◇ MECARPHON

TOXICITY DATA with REFERENCE
orl-rat LD50:57 mg/kg GUCHAZ 6,324,73
skn-rat LD50:720 mg/kg FMCHA2 -,D191,80

SAFETY PROFILE: Poison by ingestion. Moderately toxic by skin contact. When heated to decomposition it emits very toxic fumes of SO_x, PO_x, and NO_x. See also CARBAMATES.

MLJ500 CAS:926-93-2 ***HR: 3***
1-METHYL-6-(1-METHYLALLYL)-2,5-DITHIOBIUREA
mf: $C_7H_{14}N_4S_2$ mw: 218.37

SYNS: AIMAX ◇ AY-61122 ◇ COMPOUND 33,828 ◇ DITHIOCARBAMOYLHYDRAZINE ◇ I.C.I. 33,828 ◇ MATCH ◇ METALLIBURE ◇ METHALLIBURE ◇ 1-α-METHYLALLYLTHIOCARBAMOYL-2-METHYLTHIOCARBAMOYLHYDRAZINE ◇ N-((1-METHYLALLYL)THIOCARBAMOYL)-N'-(METHYLTHIOCARBAMOYL)HYDRAZINE ◇ 1-METHYL-6-(1-METHYLALLYL)DITHIOBIUREA ◇ NSC-69536 ◇ SUISYNCHRON ◇ TURISYNCHRON

TOXICITY DATA with REFERENCE
cyt-rat-unr 80 mg/kg AXVMAW 36,759,82
sln-rat-unr 80 mg/kg AXVMAW 36,759,82
unr-wmn TDLo:30 mg/kg (30D pre):REP JRPFA4 5,459,63
orl-rat TDLo:300 mg/kg (female 4-6D post):TER JRPFA4 7,211,64
orl-rat LD50:1 g/kg NATUAS 192,1191,61
ivn-mus LD50:180 mg/kg CSLNX* NX#03991

SAFETY PROFILE: Poison by intravenous route. Slightly toxic by ingestion. An experimental teratogen. Human reproductive effects by unspecified route: menstrual cycle changes or disorders. Experimental reproductive effects. Mutation data reported. When heated to decomposition it emits very toxic fumes of SO_x and NO_x.

MLJ750 CAS:78186-61-5 ***HR: 3***
2-METHYL-2-(METHYLAMINO)-1,3-BENZODIOXOLE HYDROCHLORIDE
mf: $C_9H_{11}NO_2{\bullet}ClH$ mw: 201.67

TOXICITY DATA with REFERENCE
ivn-rat LD50:45 mg/kg EJMCA5 12,413,77
ipr-mus LD50:100 mg/kg EJMCA5 12,413,77

SAFETY PROFILE: Poison by intravenous and intraperitoneal routes. When heated to decomposition it emits very toxic fumes of HCl and NO_x.

MLK750 CAS:53499-68-6 ***HR: 3***
N-METHYL-4'-(p-METHYLAMINOPHENYLAZO) ACETANILIDE
mf: $C_{16}H_{18}N_4O$ mw: 282.38

SYNS: 4-(N-ACETYL-N-METHYL)AMINO-4'-N-METHYLAMINO-AZOBENZENE ◇ N'-ACETYL-N'-MONOMETHYL-4'-AMINO-N-MONOMETHYL-4-AMINOAZOBENZENE ◇ N-METHYL-N-(4-((4-(METHYLAMINO)PHENYL)AZO)PHENYLACETAMIDE)

TOXICITY DATA with REFERENCE
mma-sat 250 nmol/plate CNREA8 46,1654,86
dns-rat:lvr 1 μmol/L CNREA8 46,1654,86
orl-rat TDLo:2500 mg/kg/17W-C:ETA CNREA8 34,2274,74

SAFETY PROFILE: Questionable carcinogen with ex-

perimental tumorigenic data. Mutation data reported. When heated to decomposition it emits toxic fumes of NO_x.

MLK800 ***HR: 3***
6-METHYL-8-METHYLAMINO-s-TRIAZOLO(4,3-b)PYRIDAZINE
mf: $C_7H_9N_5$ mw: 163.21

SYNS: COMPOSE 134 P (FRENCH) ◇ METHYL-6-METHYL-AMINO-8-s-TRIAZOLO(4,3b)PYRIDAZINE (FRENCH) ◇ 134 P

TOXICITY DATA with REFERENCE
orl-rat LDLo:75 mg/kg AIPTAK 121,154,59
orl-mus LDLo:54 mg/kg AIPTAK 121,154,59
scu-mus LDLo:65 mg/kg AIPTAK 121,154,59
ivn-mus LDLo:40 mg/kg AIPTAK 121,154,59
ivn-rbt LDLo:70 mg/kg AIPTAK 121,154,59

SAFETY PROFILE: Poison by ingestion, subcutaneous, and intravenous routes. When heated to decomposition it emits toxic fumes of NO_x.

MLL000 CAS:13984-07-1 ***HR: 3***
N-METHYL-N-(5-(N'-METHYLANILINO)-2,4-PENTADIENYLIDENE) ANILINIUM CHLORIDE
mf: $C_{19}H_{21}N_2{\cdot}Cl$ mw: 312.87

TOXICITY DATA with REFERENCE
orl-mus LD50:500 mg/kg JMCMAR 12,806,69
ipr-mus LD50:50 mg/kg JMCMAR 12,806,69

SAFETY PROFILE: Poison by intraperitoneal route. Moderately toxic by ingestion. When heated to decomposition it emits very toxic fumes of NO_x and Cl^-.

MLL250 CAS:80-48-8 ***HR: 3***
METHYL-p-METHYLBENZENESULFONATE
mf: $C_8H_{10}O_3S$ mw: 186.24

PROP: Light brown crystals; crystals of ethyl ligroin. D: 1.230-1.238 @ 25°/25°, vap d: 6.45, mp: 28°. Insol in water; sol in benzene; very sol in alc and ether.

SYNS: METHYLESTER KYSELINY p-TOLUENSULFONOVE (CZECH) ◇ METHYL-4-METHYLBENZENESULFONATE ◇ METHYL-p-TOLUENESULFONATE ◇ METHYL TOLUENE-4-SULFONATE ◇ METHYL TOSYLATE ◇ METHYL-p-TOSYLATE ◇ p-TOLUOLSULFONSAEURE METHYL ESTER (GERMAN)

TOXICITY DATA with REFERENCE
skn-rbt 500 mg/24H SEV 28ZPAK -,197,72
eye-rbt 500 mg/24H MLD 28ZPAK -,197,72
orl-rat TDLo:1400 mg/kg/2Y:ETA ZEKBAI 74,241,70
orl-rat LD50:341 mg/kg 28ZPAK -,197,72
scu-rat LD50:250 mg/kg ZEKBAI 74,241,70

CONSENSUS REPORTS: Reported in EPA TSCA Inventory.

SAFETY PROFILE: Poison by ingestion and subcutaneous route. An eye and severe skin irritant. A vesicant and skin sensitizer. Questionable carcinogen with experimental tumorigenic data. When heated to decomposition it emits toxic fumes of SO_x. See also ESTERS and SULFONATES.

MLL500 CAS:52400-65-4 ***HR: 3***
4-METHYL-1-(2-(2-METHYL-1,3-BENZODIOXOL-2-YL)ETHYL) PIPERAZINE HYDROCHLORIDE
mf: $C_{15}H_{22}N_2O_2{\cdot}ClH$ mw: 298.85

SYN: 2-METHYL-2-(2-(4-METHYL-1-PIPERAZINYL)ETHYL)-1,3-BENZODIOXOLE HYDROCHLORIDE

TOXICITY DATA with REFERENCE
ivn-rat LD50:20 mg/kg EJMCA5 12,413,77
ipr-mus LD50:110 mg/kg EJMCA5 12,413,77

SAFETY PROFILE: Poison by intravenous and intraperitoneal routes. When heated to decomposition it emits very toxic fumes of HCl and NO_x.

MLL600 CAS:53955-81-0 ***HR: 3***
METHYL 2-METHYLBUTYRATE
mf: $C_6H_{12}O_2$ mw: 116.16

PROP: Colorless liquid; sweet, fruity, apple-like odor. D: 0.879, refr index: 1.393-1.397, flash p: 91°F. Sol in alc, fixed oils; insol in water.

SYNS: FEMA No. 2719 ◇ METHYL 2-METHYLBUTANOATE

SAFETY PROFILE: Flammable liquid. When heated to decomposition it emits acrid smoke and irritating fumes.

MLL650 CAS:69462-51-7 ***HR: 1***
2-METHYL-1-((6-METHYL-3-CYCLOHEXEN-1-YL)CARBONYL)PIPERIDINE
mf: $C_{14}H_{23}NO$ mw: 221.38

SYNS: A13-36564 ◇ PIPERIDINE, 2-METHYL-1-((6-METHYL-3-CYCLOHEXEN-1-YL)CARBONYL)-

TOXICITY DATA with REFERENCE
skn-rbt 500 mg/24H MLD AEHA** 51-029-76
eye-rbt 100 mg/24H MOD AEHA** 51-029-76

SAFETY PROFILE: A skin and eye irritant. When heated to decomposition it emits toxic fumes of NO_x.

MLL655 CAS:69462-52-8 ***HR: 1***
3-METHYL-1-((6-METHYL-3-CYCLOHEXEN-1-YL)CARBONYL)PIPERIDINE
mf: $C_{14}H_{23}NO$ mw: 221.38

SYNS: A13-36565 ◇ PIPERIDINE, 3-METHYL-1-((6-METHYL-3-CYCLOHEXEN-1-YL)CARBONYL)-

TOXICITY DATA with REFERENCE
skn-rbt 500 mg/24H MLD AEHA** 51-029-76
eye-rbt 100 mg/24H MLD AEHA** 51-029-76

SAFETY PROFILE: A skin and eye irritant. When heated to decomposition it emits toxic fumes of NO_x.

MLL660 CAS:69462-53-9 ***HR: 1***
4-METHYL-1-((6-METHYL-3-CYCLOHEXEN-1-YL)CARBONYL)PIPERIDINE
mf: $C_{14}H_{23}NO$ mw: 221.38

SYNS: AI3-36566 ◇ PIPERIDINE, 4-METHYL-1-((6-METHYL-3-CYCLOHEXEN-1-YL)CARBONYL)-

TOXICITY DATA with REFERENCE
skn-rbt 500 mg/24H MLD AEHA** 51-029-76
eye-rbt 100 mg/24H MLD AEHA** 51-029-76

SAFETY PROFILE: A skin and eye irritant. When heated to decomposition it emits toxic fumes of NO_x.

MLM500 CAS:68162-93-6 ***HR: 1***
2-METHYL-1-((2-METHYLCYCLOHEXYL)CARBONYL)PIPERIDINE
mf: $C_{14}H_{25}NO$ mw: 223.40

SYN: AI3-36543

TOXICITY DATA with REFERENCE
skn-rbt 500 mg/24H MLD NTIS** AD-A053-882
eye-rbt 100 mg MOD NTIS** AD-A053-882

SAFETY PROFILE: A skin and eye irritant. When heated to decomposition it emits toxic fumes of NO_x.

MLM600 CAS:64387-78-6 ***HR: 2***
3-METHYL-1-((2-METHYLCYCLOHEXYL)CARBONYL)PIPERIDINE
mf: $C_{14}H_{25}NO$ mw: 223.40

SYN: AI3-36558

TOXICITY DATA with REFERENCE
skn-rbt 500 mg/24H MLD AEHA** 51-029-76
eye-rbt 100 mg/24H MLD AEHA** 51-029-76
orl-rat LDLo:3300 mg/kg AEHA** 51-0820-77

SAFETY PROFILE: Moderately toxic by ingestion. A skin and eye irritant. When heated to decomposition it emits toxic fumes of NO_x.

MLN500 CAS:63041-88-3 ***HR: 3***
10-METHYL-1',9-METHYLENE-1,2-BENZANTHRACENE
mf: $C_{20}H_{14}$ mw: 254.34

SYN: 6-METHYL-11H-BENZ(bc)ACEANTHRYLENE

TOXICITY DATA with REFERENCE
scu-mus TDLo:40 mg/kg:ETA CNREA8 6,454,46

SAFETY PROFILE: Questionable carcinogen with experimental tumorigenic data. When heated to decomposition it emits acrid smoke and irritating fumes.

MLO250 CAS:543-39-5 ***HR: 1***
2-METHYL-6-METHYLENE-7-OCTEN-2-OL
mf: $C_{10}H_{18}O$ mw: 154.28

PROP: Found in leaf oils of *Barosma venusta* and in Oil of Hops (FCTXAV 14,601,76).

SYNS: 3-METHYLENE-7-METHYL-1-OCTEN-7-OL ◇ MYRCENOL

TOXICITY DATA with REFERENCE
skn-rbt 500 mg/24H MOD FCTXAV 14,617,76
orl-rat LD50:5300 mg/kg FCTXAV 14,617,76

CONSENSUS REPORTS: Reported in EPA TSCA Inventory.

SAFETY PROFILE: Mildly toxic by ingestion. A skin irritant. When heated to decomposition it emits acrid smoke and irritating fumes.

MLO300 CAS:35700-21-1 ***HR: 3***
15-METHYL-PGF2-α-METHYL ESTER
mf: $C_{22}H_{38}O_5$ mw: 382.60

SYNS: 15(s)15-METHYL-PGF2-α-METHYL ESTER ◇ 15-METHYL-PROSTAGLANDIN-F2-α-METHYL ESTER ◇ 15(s)15-METHYL-PROSTAGLANDIN-F2-α-METHYL ESTER ◇ 15M-PGF2-α

TOXICITY DATA with REFERENCE
ivg-wmn TDLo:10 μg/kg (female 10W post):REP CCPTAY 27,51,83

SAFETY PROFILE: Human reproductive effects by intravaginal route: terminates pregnancy, fertility changes. Other experimental animal reproductive effects. When heated to decomposition it emits acrid smoke and irritating fumes. See also ESTERS and various prostaglandins.

MLO750 CAS:13912-77-1 ***HR: 3***
2-METHYL-2-(α-METHYLHEXYLAMINO)PROPYL-p-AMINOBENZOATE HYDROCHLORIDE
mf: $C_{19}H_{32}N_2O_2 \cdot ClH$ mw: 356.99

SYNS: p-AMINOBENZOICACID-(2-METHYL-2-(1-METHYLHEPTYLAMINO)PROPYL ESTER HYDROCHLORIDE ◇ OCTACAINE HYDROCHLORIDE

TOXICITY DATA with REFERENCE
ipr-mus LD50:81 mg/kg JACSAT 65,1222,43
scu-mus LDLo:240 mg/kg JACSAT 65,1222,43

SAFETY PROFILE: Poison by intraperitoneal and subcutaneous routes. When heated to decomposition it emits very toxic fumes of HCl and NO_x.

MLP250 CAS:36304-84-4 ***HR: 3***
d-3-METHYL-N-METHYLMORPHINAN PHOSPHATE
mf: $C_{18}H_{25}N \cdot H_3O_4P$ mw: 353.44

SYNS: ASTOMIN ◇ AT-17 PHOSPHATE ◇ DIMEMORFAN PHOS-

PHATE ◇ (9-α,13-α,14-α)-3,17-DIMETHYLMORPHINAN PHOSPHATE ◇ 3,17-DIMETHYL-9-α,13-α,14-α-MORPHINAN PHOSPHATE

TOXICITY DATA with REFERENCE
orl-rat TDLo:700 mg/kg (8-14D preg):TER KSRNAM 6,2089,72
orl-rat TDLo:700 mg/kg (8-14D preg):REP KSRNAM 6,2089,72
orl-rat LD50:690 mg/kg MEIEDD 10,466,83
ipr-rat LD50:124 mg/kg OYYAA2 6,1207,72
scu-rat LD50:556 mg/kg NIIRDN 6,345,82
ivn-rat LD50:57 mg/kg MEIEDD 10,466,83
orl-mus LD50:475 mg/kg ARZNAD 26,353,76
scu-mus LD50:223 mg/kg ARZNAD 26,353,76
ivn-mus LD50:33900 μg/kg OYYAA2 6,1207,72
ivn-dog LD50:35 mg/kg ARZNAD 26,353,76

SAFETY PROFILE: Poison by intravenous, intraperitoneal, and subcutaneous routes. Moderately toxic by ingestion. An experimental teratogen. Experimental reproductive effects. Used as an antitussive agent. When heated to decomposition it emits very toxic fumes of NO_x and PO_x. See also PHOSPHATES.

MLP500 CAS:66843-04-7 ***HR: 3***
5-METHYL-5-(1-METHYL-1-PENTENYL)BARBITURIC ACID
mf: $C_{11}H_{16}N_2O_3$ mw: 224.29

TOXICITY DATA with REFERENCE
orl-mus LD50:620 mg/kg JACSAT 61,776,39
ipr-mus LD50:375 mg/kg JACSAT 61,776,39

SAFETY PROFILE: Poison by intraperitoneal route. Moderately toxic by ingestion. When heated to decomposition it emits toxic fumes of NO_x. See also BARBITURATES.

MLP750 CAS:65210-32-4 ***HR: 3***
2-METHYL-2-(2-(N-METHYL-N-PHENETHYLAMINO)ETHYL)-1,3-BENZODIOXOLE HYDROCHLORIDE
mf: $C_{19}H_{23}NO_2$•ClH mw: 333.89

TOXICITY DATA with REFERENCE
ivn-rat LD50:15 mg/kg EJMCA5 12,413,77
ipr-mus LD50:125 mg/kg EJMCA5 12,413,77

SAFETY PROFILE: Poison by intravenous and intraperitoneal routes. When heated to decomposition it emits very toxic fumes of HCl and NO_x.

MLP800 CAS:77-41-8 ***HR: 2***
N-METHYL-α-METHYL-α-PHENYLSUCCINIMIDE
mf: $C_{12}H_{13}NO_2$ mw: 203.26

SYNS: CELONTIN ◇ 1,3-DIMETHYL-3-PHENYL-2,5-DIOXOPYRROLIDINE ◇ 1,3-DIMETHYL-3-PHENYL-PYRROLIDIN-2,5-DIONE ◇ N,2-DIMETHYL-2-PHENYLSUCCINIMIDE ◇ MESUXIMIDE ◇ METHSUXIMIDE ◇ N-METHYL-α-α-METHYLPHENYLSUCCINIMIDE ◇ α-METHYLPHENSUXIMIDE ◇ α-METHYL-α-PHENYL N-METHYL SUCCINIMIDE ◇ METSUCCIMIDE ◇ PETINUTIN ◇ PM 396 ◇ SUCCINIMIDE, N,2-DIMETHYL-2-PHENYL-

TOXICITY DATA with REFERENCE
unr-mus TDLo:1006 mg/kg (female 8-10D post):TER FEPRA7 38,438,79
orl-mus LD50:900 mg/kg JPPMAB 6,740,54

SAFETY PROFILE: Moderately toxic by ingestion. An experimental teratogen. When heated to decomposition it emits toxic fumes of NO_x.

MLR400 ***HR: 3***
6-METHYL-α-(4-METHYL-1-PIPERAZINYLCARBONYL)ERGOLINE-8-β-PROPIONITRILE
mf: $C_{24}H_{31}N_5O$ mw: 405.60

SYN: ERGOLINE-8-β-PROPIONITRILE, 6-METHYL-α-(4-METHYL-1-PIPERAZINYLCARBONYL)-

TOXICITY DATA with REFERENCE
orl-rat TDLo:16 mg/kg (female 5D post):REP ARZNAD 33,1094,83
orl-mus LD50:400 mg/kg ARZNAD 33,1094,83

SAFETY PROFILE: Poison by ingestion. Experimental reproductive effects. When heated to decomposition it emits toxic fumes of NO_x.

MLR500 CAS:37724-45-1 ***HR: 3***
5-METHYL-4-(4-METHYL-1-PIPERAZINYL)THIENO(2,3-d)PYRIMIDINE HYDROCHLORIDE
mf: $C_{12}H_{16}N_4S$•ClH mw: 284.84

SYN: QM-1148

TOXICITY DATA with REFERENCE
ipr-mus LD50:41 mg/kg CHTPBA 7,224,72
ivn-mus LD50:12 mg/kg CHTPBA 7,224,72

SAFETY PROFILE: Poison by intraperitoneal and intravenous routes. When heated to decomposition it emits very toxic fumes of NO_x, SO_x, and HCl.

MLS250 CAS:77791-42-5 ***HR: 2***
N-METHYL-2-(2-METHYLPIPERIDINO)-N-(2-PHENOXYETHYL)ACETAMIDE HYDROCHLORIDE
mf: $C_{17}H_{26}N_2O_2$•ClH mw: 326.91

SYN: C 6575

TOXICITY DATA with REFERENCE
eye-rbt 2% MLD ARZNAD 8,761,58
scu-mus LD50:570 mg/kg ARZNAD 8,761,58

SAFETY PROFILE: Moderately toxic by subcutaneous

route. An eye irritant. When heated to decomposition it emits very toxic fumes of HCl and NO_x.

MLS750 CAS:77791-43-6 ***HR: 3***
N-METHYL-2-(2-METHYLPIPERIDINO)-N-(2-(o-TOLYLOXY)ETHYL)-ACETAMIDE HYDROCHLORIDE
mf: $C_{18}H_{28}N_2O_2 \cdot ClH$ mw: 340.94

SYN: C 6583

TOXICITY DATA with REFERENCE
eye-rbt 2% MLD ARZNAD 8,761,58
scu-mus LD50:350 mg/kg ARZNAD 8,761,58

SAFETY PROFILE: Poison by subcutaneous route. An eye irritant. When heated to decomposition it emits very toxic fumes of NO_x and HCl.

MLT250 CAS:50309-11-0 ***HR: 3***
1-METHYL-4-(p-(p-((1-METHYLPYRIDINIUM-4-YL)AMINO)BENZAMIDO)ANILINO)QUINOLINIUM), DI-p-TOLUENESULFONATE
mf: $C_{29}H_{27}N_5O \cdot 2C_7H_7O_3S$ mw: 804.01

TOXICITY DATA with REFERENCE
dnd-mus:lym 1600 nmol/L JMCMAR 22,134,79
ipr-mus LD10:7500 μg/kg JMCMAR 22,134,79

SAFETY PROFILE: Poison by intraperitoneal route. Mutation data reported. When heated to decomposition it emits very toxic fumes of SO_x and NO_x. See also SULFONATES.

MLT500 CAS:68772-19-0 ***HR: 3***
1-METHYL-4-((4-(3-((4-((1-METHYLPYRIDINIUM-4-YL)AMINO)PHENYL)AMINO)3-OXO-1-PROPENYL)PHENYL)AMINO)QUINOLINIUM), DIBROMIDE
mf: $C_{31}H_{29}N_5O \cdot 2Br$ mw: 647.47

TOXICITY DATA with REFERENCE
dnd-mus:lym 380 nmol/L JMCMAR 22,134,79
ipr-mus LD10:22 mg/kg JMCMAR 22,134,79

SAFETY PROFILE: Poison by intraperitoneal route. Mutation data reported. When heated to decomposition it emits very toxic fumes of NO_x and Br^-.

MLT750 CAS:50309-17-6 ***HR: 3***
1-METHYL-4-(p-((p-((1-METHYLPYRIDINIUM-4-YL)AMINO)PHENYL)CARBAMOYL)ANILINO)-7-NITROQUINOLINIUM), DI-p-TOLUENE SULFONATE
mf: $C_{29}H_{26}N_6O_3 \cdot 2C_7H_7O_3S$ mw: 849.01

TOXICITY DATA with REFERENCE
dnd-mus:lym 690 nmol/L JMCMAR 22,134,79
ipr-mus LD10:27 mg/kg JMCMAR 22,134,79

SAFETY PROFILE: Poison by intraperitoneal route. Mutation data reported. When heated to decomposition it emits very toxic fumes of SO_x and NO_x.

MLU000 CAS:68772-09-8 ***HR: 3***
1-METHYL-4-(p-((p-((1-METHYLPYRIDINIUM-4-YL)AMINO)PHENYL)CARBAMOYL)ANILINO)QUINOLINIUM), DIBROMIDE
mf: $C_{29}H_{27}N_5O \cdot 2Br$ mw: 621.43

TOXICITY DATA with REFERENCE
dnd-mus:lym 1400 nmol/L JMCMAR 22,134,79
ipr-mus LD10:13 mg/kg JMCMAR 22,134,79

SAFETY PROFILE: Poison by intraperitoneal route. Mutation data reported. When heated to decomposition it emits very toxic fumes of NO_x and Br^-.

MLU250 CAS:68772-18-9 ***HR: 3***
1-METHYL-4-(p-(p-((1-METHYLPYRIDINIUM-4-YL)AMINO)STYRYL)ANILINO)QUINOLIUM DIBROMIDE
mf: $C_{30}H_{28}N_4 \cdot 2Br$ mw: 604.44

TOXICITY DATA with REFERENCE
dnd-mus:lym 340 nmol/L JMCMAR 22,134,79
ipr-mus LD10:23 mg/kg JMCMAR 22,134,79

SAFETY PROFILE: Poison by intraperitoneal route. Mutation data reported. When heated to decomposition it emits very toxic fumes of NO_x and Br^-.

MLU750 CAS:50425-35-9 ***HR: 3***
1-METHYL-4-(p-((p-(1-METHYLPYRIDINIUM-4-YL)PHENYL)CARBAMOYL)ANILINO) QUINOLINIUM), DI-p-TOLUENE SULFONATE
mf: $C_{29}H_{26}N_4O \cdot 2C_7H_7O_3S$ mw: 788.99

TOXICITY DATA with REFERENCE
dnd-mus:lym 890 nmol/L JMCMAR 22,134,79
ipr-mus LD10:15 mg/kg JMCMAR 22,134,79

SAFETY PROFILE: Poison by intraperitoneal route. Mutation data reported. When heated to decomposition it emits very toxic fumes of NO_x and SO_x. See also SULFONATES.

MLW250 CAS:19056-01-0 ***HR: 3***
1-METHYL-6-(p-(p-((1-METHYLQUINOLINIUM-6-YL)CARBAMOYL)BENZAMIDO) BENZAMIDO)QUINOLINIUM), DI-p-TOLUENE SULFONATE
mf: $C_{35}H_{29}N_5O_3 \cdot 2C_7H_7O_3S$ mw: 910.09

TOXICITY DATA with REFERENCE
dnd-mus:lym 370 nmol/L JMCMAR 22,134,79
ipr-mus LD10:100 mg/kg JMCMAR 22,134,79

SAFETY PROFILE: Poison by intraperitoneal route.

Mutation data reported. When heated to decomposition it emits very toxic fumes of NO_x and SO_x.

MLW500 CAS:19056-06-5 ***HR: 3***
1-METHYL-6-((p-(p-((1-METHYLQUINOLINIUM-6-YL)CARBAMOYL)BENZAMIDO)PHENYL) CARBAMOYL)QUINOLINIUM), DI-p-TOLUENE SULFONATE
mf: $C_{35}H_{29}N_5O_3 \cdot 2C_7H_7O_3S$ mw: 910.09

TOXICITY DATA with REFERENCE
dnd-mus:lym 300 nmol/L JMCMAR 22,134,79
ipr-mus LD10:90 mg/kg JMCMAR 22,134,79

SAFETY PROFILE: Poison by intraperitoneal route. Mutation data reported. When heated to decomposition it emits very toxic fumes of SO_x and NO_x.

MLX000 CAS:17959-12-5 ***HR: 2***
METHYL ((METHYLTHIO)ACETYL)CARBAMIC ACID-o-ISOPROPOXYPHENYL ESTER
mf: $C_{14}H_{19}NO_4S$ mw: 297.40

SYNS: ENT 27,351 ◇ 2-(1-METHYLETHOXY)PHENYL METHYL (METHYLTHIO)ACETYL)CARBAMATE ◇ NSC 190949 ◇ UPJOHN U-22023

TOXICITY DATA with REFERENCE
orl-rat LD50:650 mg/kg ARSIM* 20,26,66
orl-mus LD50:430 mg/kg ARSIM* 20,26,66

SAFETY PROFILE: Moderately toxic by ingestion. When heated to decomposition it emits very toxic fumes of SO_x and NO_x. See also CARBAMATES.

MLX250 CAS:55514-14-2 ***HR: 3***
3-METHYL-2-(METHYLTHIO)-BENZOTHIAZOLIUM-p-TOLUENESULFONATE
mf: $C_9H_{10}NS_2 \cdot C_7H_7O_3S$ mw: 367.52

SYN: BENZOTHIAZOLIUM, 3-METHYL-2-METHYLTHIO-, p-TOLUENESULFONATE

TOXICITY DATA with REFERENCE
ivn-mus LD50:180 mg/kg CSLNX* NX#02000

CONSENSUS REPORTS: Reported in EPA TSCA Inventory.

SAFETY PROFILE: Poison by intravenous route. When heated to decomposition it emits very toxic fumes of NO_x and SO_x.

MLX300 CAS:25310-48-9 ***HR: D***
METHYL(METHYLTHIO)MERCURY
mf: C_2H_6HgS mw: 262.73

SYNS: MERCURY, (METHANETHIOLATO)METHYL- ◇ MERCURY METHYLMERCURY SULFIDE ◇ MERCURY, METHYL(METHYLTHIO)- ◇ (METHANETHIOLATO)METHYLMERCURY ◇ METHYL METHYLMERCURIC SULFIDE

TOXICITY DATA with REFERENCE
unr-rat TDLo:16 mg/kg (female 1-8D post):TER KUIZAR 41,506,67

ACGIH TLV: TWA 0.01 mg(Hg)/m^3; STEL 0.03 mg(Hg)/m^3

SAFETY PROFILE: An experimental teratogen. When heated to decomposition it emits toxic fumes of SO_x and Hg.

MLX750 CAS:3120-74-9 ***HR: 3***
3-METHYL-4-METHYLTHIOPHENOL
mf: $C_8H_{10}OS$ mw: 154.24

SYNS: 4-(METHYLTHIO)-m-CRESOL ◇ MMTP ◇ USAF MA-17

TOXICITY DATA with REFERENCE
orl-rat LD50:3400 mg/kg GISAAA 43(8),10,78
orl-mus LD50:1000 mg/kg GISAAA 43(8),10,78
orl-gpg LD50:3500 mg/kg GISAAA 43(8),10,78
ipr-mus LDLo:100 mg/kg NTIS** AD438-895

CONSENSUS REPORTS: Reported in EPA TSCA Inventory.

SAFETY PROFILE: Poison by intraperitoneal route. Moderately toxic by ingestion. When heated to decomposition it emits toxic fumes of SO_x.

MLX850 CAS:8064-35-5 ***HR: 3***
METHYL METIRAM

SYNS: BASFUNGIN ◇ BASFUNGINE ◇ PROPYLENE BISDITHIOCARBAMATE

TOXICITY DATA with REFERENCE
mrc-smc 1000 ppm MUREAV 10,533,70
orl-rat TDLo:312 mg/kg (1-20D preg):REP DBANAD 29,1227,76
orl-cat LD50:5200 mg/kg KHZDAN 20,515,77
ipr-mus LD50:200 mg/kg 85JFAN A275,83

CONSENSUS REPORTS: EPA Genetic Toxicology Program.

SAFETY PROFILE: Poison by intraperitoneal route. Mildly toxic by ingestion. Experimental reproductive effects. Mutation data reported. Used as a pesticide and fungicide. When heated to decomposition it emits toxic fumes of SO_x and NO_x. See also CARBAMATES.

MLY000 CAS:801-52-5 ***HR: 3***
N-METHYLMITOMYCIN C
mf: $C_{16}H_{20}N_4O_5$ mw: 348.40

PROP: Antibiotic from *Streptomyces ardus* and *Streptomyces verticillatus* (CCROBU 56,615,72).

SYNS: 8-AZATHIOXANTHINE ◇ ENT 50,825 ◇ METHYLMITOMYCIN ◇ NSC-56410 ◇ PORFIROMYCIN ◇ PORFIROMYCINE ◇ PORPHYROMYCIN ◇ PROFIROMYCIN

◇ 5-THIO-1H-υ-TRIAZOLO(4,5-d)PYRIMIDINE-5,7(4H,6H)-DIONE ◇ U-14743

TOXICITY DATA with REFERENCE
mmo-omi 1 mg/plate JGAMA9 11,129,65
dni-hmn:oth 100 μg/L TUMOAB 53,517,67
ivn-hmn TDLo:1500 μg/kg:BLD CCROBU 56,615,72
orl-rat LD50:68 mg/kg UPJOH* 2(6),-,71
orl-mus LD50:88660 μg/kg NCISP* JAN86
ipr-mus LD50:44 mg/kg CNCRA6 30,9,63
scu-mus LD50:53500 μg/kg NCISP* JAN86
ivn-mus LD50:55 mg/kg JMCMAR 14,103,71

SAFETY PROFILE: Poison by ingestion, intravenous, subcutaneous, and intraperitoneal routes. Human systemic effects by intravenous route: blood effects. Human mutation data reported. When heated to decomposition it emits toxic fumes of NO_x.

MLY250 CAS:64-01-7 ***HR: 3***
3-METHYL-4-MONOMETHYLAMINOAZOBENZENE
mf: $C_{14}H_{14}N_3$ mw: 224.31

SYN: N-METHYL-3-METHYL-p-AMINOAZOBENZENE

TOXICITY DATA with REFERENCE
orl-rat TDLo:9864 mg/kg/39W-C:NEO JEMEAV 87,139,48

SAFETY PROFILE: Questionable carcinogen with experimental neoplastigenic data. When heated to decomposition it emits toxic fumes of NO_x.

MLY500 CAS:2043-24-5 ***HR: 3***
N-METHYLMONOTHIOSUCCINIMIDE
mf: C_5H_7NOS mw: 129.19

SYN: USAF WI-3

TOXICITY DATA with REFERENCE
ipr-mus LD50:250 mg/kg NTIS** AD607-952

CONSENSUS REPORTS: Reported in EPA TSCA Inventory.

SAFETY PROFILE: Poison by intraperitoneal route. When heated to decomposition it emits very toxic fumes of NO_x and SO_x.

MLZ000 CAS:125-72-4 ***HR: 3***
(−)-17-METHYLMORPHINAN-3-OL TARTRATE DIHYDRATE
mf: $C_{17}H_{23}NO{\cdot}C_4H_6O_6{\cdot}2H_2O$ mw: 443.55

SYN: (−)-3-HYDROXY-N-METHYLMORPHINAN TARTRATE DIHYDRATE

TOXICITY DATA with REFERENCE
scu-rat LD50:135 mg/kg AIPTAK 85,387,51
ivn-mus LD50:45 mg/kg AIPTAK 85,387,51
ivn-rbt LD50:23 mg/kg AIPTAK 85,387,51

SAFETY PROFILE: Poison by subcutaneous and intravenous routes. When heated to decomposition it emits toxic fumes of NO_x.

MMA000 CAS:143-98-6 ***HR: 3***
(+)-17-METHYLMORPHINAN-3-OL TARTRATE HYDRATE
mf: $C_{17}H_{23}NO{\cdot}C_4H_6O_6{\cdot}H_2O$ mw: 425.53

SYN: (+)-3-HYDROXY-N-METHYLMORPHINAN TARTRATE HYDRATE

TOXICITY DATA with REFERENCE
scu-rat LD50:500 mg/kg AIPTAK 85,387,51
ivn-mus LD50:75 mg/kg AIPTAK 85,387,51
ivn-rbt LD50:23 mg/kg AIPTAK 85,387,51

SAFETY PROFILE: Poison by intravenous route. Moderately toxic by subcutaneous route. When heated to decomposition it emits toxic fumes of NO_x.

MMA250 CAS:109-02-4 ***HR: 3***
N-METHYL MORPHOLINE
DOT: UN 2535

PROP: Liquid. Flash p: 75.2°F, d: 0.9, vap d: 3.5, bp: 115°.
mf: $C_5H_{11}NO$ mw: 101.17

SYNS: METHYLMORPHOLINE (DOT) ◇ 4-METHYLMORPHOLINE

TOXICITY DATA with REFERENCE
skn-rbt 460 mg open MLD UCDS** 6/30/59
eye-rbt 920 μg SEV UCDS** 6/30/59
orl-rat LD50:2720 mg/kg JIHTAB 31,60,49
ihl-rat LCLo:2000 ppm/4H JIHTAB 30,60,49
orl-rat LD50:1970 mg/kg TPKVAL 15,116,79
ihl-mus LC50:25200 mg/m³/2H TPKVAL 15,116,79
skn-rbt LD50:1242 mg/kg JIHTAB 31,60,49

CONSENSUS REPORTS: Reported in EPA TSCA Inventory.

DOT Classification: Flammable Liquid; Label: Flammable Liquid, Corrosive; Flammable or Combustible Liquid; Label: Flammable Liquid, Corrosive.

SAFETY PROFILE: Moderately toxic by ingestion and skin contact. Mildly toxic by inhalation. A corrosive irritant to skin, eyes, and mucous membranes. Flammable when exposed to heat or flame, can react vigorously with oxidizing materials. When heated to decomposition it emits toxic fumes of NO_x.

MMA525 CAS:2014-29-1 ***HR: 2***
α-METHYL-4-MORPHOLINEACETIC ACID-2,6-XYLYL ESTER HYDROCHLORIDE
mf: $C_{15}H_{21}NO_3 \cdot ClH$ mw: 299.80

SYN: FC 448

TOXICITY DATA with REFERENCE
skn-rbt 200 mg MLD BCFAAI 107,310,68
eye-rbt 1 g MLD BCFAAI 107,310,68
scu-mus LD50:1500 mg/kg BCFAAI 107,310,68

SAFETY PROFILE: Moderately toxic by subcutaneous route. A skin and eye irritant. When heated to decomposition it emits toxic fumes of NO_x and HCl. See also ESTERS.

MMB500 CAS:1321-94-4 ***HR: 1***
METHYLNAPHTHALENE
mf: $C_{11}H_{10}$ mw: 142.21

SYN: METHYLNAFTALEN

TOXICITY DATA with REFERENCE
skn-rbt 500 mg/24H MOD 28ZPAK -,26,72
orl-rat LD50:4360 mg/kg 28ZPAK -,26,72

CONSENSUS REPORTS: Reported in EPA TSCA Inventory.

SAFETY PROFILE: Mildly toxic by ingestion. A skin irritant. When heated to decomposition it emits acrid smoke and irritating fumes. See also other methylnaphthalene entries.

MMB750 CAS:90-12-0 ***HR: 2***
1-METHYLNAPHTHALENE
mf: $C_{11}H_{10}$ mw: 142.21

PROP: Colorless liquid. D: 1.0202 @ 20°/4°, mp: −22°, bp: 244.6°, autoign temp: 984°F. Insol in water, sol in alc and ether.

SYN: α-METHYLNAPHTHALENE

TOXICITY DATA with REFERENCE
mma-sat 6 mmol/L/2H CNREA8 39,4152,79
orl-rat LD50:1840 mg/kg 85GMAT -,85,82

CONSENSUS REPORTS: Reported in EPA TSCA Inventory.

SAFETY PROFILE: Moderately toxic by ingestion. Mutation data reported. Combustible when exposed to heat, flames or oxidizers. To fight fire, use dry chemical, CO_2, water spray or mist, foam. When heated to decomposition it emits acrid smoke and irritating fumes. See also other methylnaphthalene entries.

MMC000 CAS:91-57-6 ***HR: 2***
2-METHYLNAPHTHALENE
mf: $C_{11}H_{10}$ mw: 142.21

PROP: Solid. D: 1.0058 @ 20°/4°, bp: 241.1°, mp: 34.58°. Insol in water; sol in alc and ether.

SYN: β-METHYLNAPHTHALENE

TOXICITY DATA with REFERENCE
orl-rat LD50:1630 mg/kg 85GMAT -,85,82
ipr-mus LDLo:1000 mg/kg TXAPA9 61,185,81

CONSENSUS REPORTS: Reported in EPA TSCA Inventory.

SAFETY PROFILE: Moderately toxic by ingestion and intraperitoneal routes. When heated to decomposition it emits acrid smoke and irritating fumes. See also other methylnaphthalene entries.

MMC250 CAS:481-85-6 ***HR: 3***
2-METHYL-1,4-NAPHTHALENEDIOL
mf: $C_{11}H_{10}O_2$ mw: 174.21

SYNS: MENADIOL ◇ METHYLNAPHTHOHYDROQUINONE ◇ 2-METHYL-1,4-NAPHTHOHYDROQUINONE ◇ 2-METHYL-1,4-NAPHTHOQUINOL

TOXICITY DATA with REFERENCE
orl-mus LD50:30 mg/kg JPETAB 75,111,42
ipr-mus LDLo:400 mg/kg CRSBAW 143,585,49
scu-mus LDLo:80 mg/kg JPETAB 71,210,41

SAFETY PROFILE: Poison by ingestion, subcutaneous, and intraperitoneal routes. When heated to decomposition it emits acrid smoke and irritating fumes.

MMD000 CAS:2869-09-2 ***HR: 3***
5-METHYLNAPHTHO(1,2,3,4-def)CHRYSENE
mf: $C_{25}H_{16}$ mw: 316.41

SYN: 2′-METHYL-1,2:4,5-DIBENZOPYRENE

TOXICITY DATA with REFERENCE
scu-mus TDLo:72 mg/kg/9W-I:ETA COREAF 259,3899,64

SAFETY PROFILE: Questionable carcinogen with experimental tumorigenic data. When heated to decomposition it emits acrid smoke and irritating fumes.

MMD250 CAS:2869-10-5 ***HR: 3***
6-METHYLNAPHTHO(1,2,3,4-def)CHRYSENE
mf: $C_{25}H_{16}$ mw: 316.41

SYN: 3′-METHYL-1,2:4,5-DIBENZOPYRENE

TOXICITY DATA with REFERENCE
scu-mus TDLo:72 mg/kg/9W-I:ETA COREAF 259,3899,64

SAFETY PROFILE: Questionable carcinogen with experimental tumorigenic data. When heated to decomposition it emits acrid smoke and irritating fumes.

MMD500 CAS:58-27-5 ***HR: 3***
2-METHYL-1,4-NAPHTHOQUINONE
mf: $C_{11}H_8O_2$ mw: 172.19

SYNS: AQUAKAY ◇ AQUINONE ◇ HEMODAL ◇ KAERGONA ◇ KANONE ◇ KAPPAXAN ◇ KARCON ◇ KAREON ◇ KATIV-G ◇ KAYKLOT ◇ KAYQUINONE ◇ KIPCA ◇ KLOTTONE ◇ KOAXIN ◇ KOLKLOT ◇ K-THROMBYL ◇ K-VITAN ◇ MENADION ◇ MENADIONE ◇ MENAPHTHON ◇ MENAPHTONE ◇ 2-METHYL-1,4-NAPHTHALENDIONE ◇ 2-METHYL-1,4-NAPHTHALENEDIONE ◇ 2-METHYL-1,4-NAPHTHOCHINON (GERMAN) ◇ 3-METHYL-1,4-NAPHTHOQUINONE ◇ MITENON ◇ MNQ ◇ NSC 4170 ◇ PANOSINE ◇ PROKAYVIT ◇ SYNKAY ◇ THYLOQUINONE ◇ USAF EK-5185 ◇ VITAMIN K2(O) ◇ VITAMIN K3

TOXICITY DATA with REFERENCE
mmo-sat 4 μg/plate ABCHA6 45,327,81
dnd-hmn:fbr 20 μmol/L TOLED5 28,37,85
ivn-mus TDLo:20 mg/kg (8D preg):TER TOIZAG 8,175,61
skn-mus TDLo:1860 mg/kg/27W-I:ETA BJCAAI 7,482,53
ipr-rat LD50:75 mg/kg TXAPA9 18,185,71
orl-mus LD50:500 mg/kg MEIEDD 10,831,83
ipr-mus LD50:50 mg/kg NTIS** AD691-490
scu-mus LD50:138 mg/kg JPETAB 75,111,42
orl-rbt LDLo:230 mg/kg JPETAB 75,111,42

CONSENSUS REPORTS: Reported in EPA TSCA Inventory.

SAFETY PROFILE: Poison by ingestion, intraperitoneal, and subcutaneous routes. Experimental teratogenic effects. Questionable carcinogen with experimental tumorigenic data. Human mutation data reported. When heated to decomposition it emits acrid smoke and irritating fumes.

MMD750 CAS:57414-02-5 ***HR: 3***
2-METHYL-1,4-NAPHTHOQUINONE, SODIUM BISULFITE, TRIHYDRATE
mf: $C_{11}H_9O_2$•$HNaO_3S$•$3H_2O$ mw: 331.32

SYNS: MENADION-NATRIUM-BISULFIT TRIHYDRAT (GERMAN) ◇ 2-METHYL-1,4-NAPHTHOCHINON-NATRIUM-BISULFIT TRIHYDRAT (GERMAN) ◇ VITAMIN K3-NATRIUM-BISULFIT TRIHYDRAT (GERMAN)

TOXICITY DATA with REFERENCE
orl-mus LD50:2500 mg/kg ARZNAD 17,1339,67
ipr-mus LD50:500 mg/kg ARZNAD 17,1339,67
ivn-mus LD50:400 mg/kg ARZNAD 17,1339,67

SAFETY PROFILE: Poison by intravenous route. Moderately toxic by ingestion and intraperitoneal routes. When heated to decomposition it emits toxic fumes of SO_x and Na_2O. See also SULFITES.

MME250 CAS:2216-68-4 ***HR: 2***
N-METHYL-1-NAPHTHYLAMINE
mf: $C_{11}H_{11}N$ mw: 157.23

TOXICITY DATA with REFERENCE
orl-rat LD50:1410 mg/kg AIHAAP 30,470,69

CONSENSUS REPORTS: Reported in EPA TSCA Inventory.

SAFETY PROFILE: Moderately toxic by ingestion. When heated to decomposition it emits toxic fumes of NO_x. See also AMINES.

MME500 CAS:10546-24-4 ***HR: 3***
3-METHYL-2-NAPHTHYLAMINE
mf: $C_{11}H_{11}N$ mw: 157.23

TOXICITY DATA with REFERENCE
scu-ham TDLo:4440 mg/kg/94W-I:CAR JJIND8 67,481,81
scu-rat TDLo:1800 mg/kg/18W-I:ETA CUSCAM 33,45,64

SAFETY PROFILE: Suspected carcinogen with experimental carcinogenic and tumorigenic data. When heated to decomposition it emits toxic fumes of NO_x. See also AMINES.

MME750 CAS:5096-18-4 ***HR: 3***
3-METHYL-2-NAPHTHYLAMINE HYDROCHLORIDE
mf: $C_{11}H_{11}N$•ClH mw: 193.69

TOXICITY DATA with REFERENCE
orl-rat TDLo:1500 mg/kg:CAR CNREA8 26,619,66
orl-rat TD:5250 mg/kg/21W-I:CAR JNCIAM 41,985,68
orl-rat LDLo:1500 mg/kg CNREA8 26,619,66

SAFETY PROFILE: Questionable carcinogen with experimental carcinogenic data. Moderately toxic by ingestion. When heated to decomposition it emits very toxic fumes of NO_x and HCl.

MME800 CAS:27636-33-5 ***HR: 2***
N-METHYL NAPHTHYLCARBAMATE
mf: $C_{12}H_{11}NO_2$ mw: 201.24

SYN: CARBAMIC ACID, METHYL-, NAPHTHALENYL ESTER

TOXICITY DATA with REFERENCE
ipr-mus TDLo:240 mg/kg/4W-I:ETA CNREA8 29,2184,69

SAFETY PROFILE: Questionable carcinogen with experimental tumorigenic data. When heated to decomposition it emits toxic fumes of NO_x.

MME809 CAS:5903-13-9 ***HR: 3***
N-METHYL-N-(1-NAPHTHYL)FLUOROACETAMIDE
mf: $C_{13}H_{12}FNO$ mw: 217.26

SYNS: 1-(N-ACETAMIDOFLUOROMETHYL)-NAPHTHALENE ◇ DP X 1410 ◇ FAM ◇ 2-FLUORO-N-METHYL-N-1-NAPHTHALENYLACETAMIDE ◇ 2-FLUORO-N-METHYL-N-1-NAPHTHYLACETAMIDE ◇ N-METHYL-N-(1-NAPHTHYL)MONOFLUOROACETAMIDE ◇ MFNA ◇ MNFA ◇ NISSOL EC

TOXICITY DATA with REFERENCE
orl-mus TDLo:240 mg/kg (1-12D preg):REP OYYAA2 4,463,70
orl-rat LD50:67 mg/kg SAIGBL 9,563,67
skn-rat LD50:213 mg/kg WRPCA2 9,119,70
scu-rat LD50:41 mg/kg TXAPA9 12,536,68
orl-mus LD50:200 mg/kg OYYAA2 4,463,70
skn-mus LD50:372 mg/kg SPEADM 74-1,-,74
ipr-mus LD50:164 mg/kg TXAPA9 12,536,68
scu-mus LD50:216 mg/kg TXAPA9 12,536,68
orl-dog LD50:2 mg/kg JIDOAA 21,88,72
skn-dog LD50:2750 μg/kg NYKZAU 65,182,69
ipr-dog LDLo:2 mg/kg TXAPA9 12,536,68
orl-mky LD50:300 mg/kg 85DPAN -,-,71/76
skn-mky LDLo:800 mg/kg TXAPA9 12,536,68

SAFETY PROFILE: Poison by ingestion, skin contact, subcutaneous, intravenous, and intraperitoneal routes. Experimental reproductive effects. When heated to decomposition it emits very toxic fumes of F^- and NO_x.

MMF500 CAS:598-58-3 ***HR: 3***
METHYL NITRATE
mf: CH_3NO_3 mw: 77.05

PROP: Colorless liquid. Bp: 65° (explodes), d: 1.208 @ 20°/4°, vap d: 2.66, mp: −83°. Sltly sol in water; sol in alc, ether.

SYN: NITRIC ACID METHYL ESTER

TOXICITY DATA with REFERENCE
orl-rat LD50:344 mg/kg AMRL** TR-77-25,77
ihl-rat LC50:1275 ppm/4H AMRL** TR-77-25,77
orl-mus LD50:1820 mg/kg AMRL** TR-77-25,77
ihl-mus LC50:5942 ppm/4H AMRL** TR-77-25,77
orl-gpg LD50:548 mg/kg AMRL** TR-77-25,77

DOT Classification: Forbidden.

SAFETY PROFILE: Poison by ingestion. Moderately toxic by inhalation. A dangerous fire and explosion hazard by spontaneous chemical reaction. A very shock- and heat-sensitive explosive. Explodes when heated to 65°C. It does not require external O_2 for combustion. A rocket fuel. When heated to decomposition it emits toxic fumes of NO_x. See also NITRATES.

MMF600 ***HR: 3***
2-METHYL-1-NITRATODIMERCURIO-2-NITRATOMERCURIO PROPANE
mf: $C_4H_8Hg_3N_2O_6$ mw: 781.89

$O_3NHgHgCH_2C(CH_3)_2HgNO_3$

CONSENSUS REPORTS: Mercury and its compounds are on the Community Right-To-Know List.

SAFETY PROFILE: Explodes on impact at 80°C. When heated to decomposition it emits toxic fumes of Hg and NO_x. See also MERCURY COMPOUNDS.

MMF750 CAS:624-91-9 ***HR: 3***
METHYL NITRITE
mf: CH_3NO_2 mw: 61.05

PROP: Gas above 10.4°F. Mp: −17°, bp: −12°, d: 0.991 @ 15°. Sol in alc, ether.

SYN: NITROUS ACID, METHYL ESTER

TOXICITY DATA with REFERENCE
mmo-sat 100 ppm MUREAV 117,47,83
mma-sat 100 ppm MUREAV 117,47,83
ihl-rat LCLo:250 ppm/4H BJIMAG 27,1,70

CONSENSUS REPORTS: Reported in EPA TSCA Inventory.

DOT Classification: Forbidden.

SAFETY PROFILE: Moderately toxic by inhalation. Mutation data reported. Narcotic in high concentration. A very dangerous fire and explosion hazard when exposed to heat or flame. A heat-sensitive explosive more powerful than ethyl nitrite. When heated to decomposition it emits toxic fumes of NO_x. See also NITRITES and AMYL NITRITE.

MMF800 CAS:100-15-2 ***HR: 2***
N-METHYL-4-NITROANILINE
mf: $C_7H_8N_2O_2$ mw: 152.15

$O_2NC_6H_4NHCH_3$

SAFETY PROFILE: Vigorous exothermic reaction with carbyl sulfate when heated above 75°C. When heated to decomposition it emits toxic fumes of NO_x. See also NITRO COMPOUNDS of AROMATIC HYDROCARBONS.

MMG000 CAS:129-15-7 ***HR: 3***
2-METHYL-1-NITROANTHRAQUINONE
mf: $C_{15}H_9NO_4$ mw: 267.25

SYNS: 2-METHYL-1-NITRO-9,10-ANTHRACENEDIONE ◇ NCI-C01923 ◇ 1-NITRO-2-METHYLANTHRAQUINONE ◇ 1-N-2-MA (RUSSIAN)

TOXICITY DATA with REFERENCE
mmo-sat 33 μg/plate ENMUDM 8(Suppl 7),1,86
mma-sat 3 μg/plate ENMUDM 8(Suppl 7),1,86
orl-rat TDLo:45 g/kg/2Y-C:CAR NCITR* NCI-CG-TR-29,78
orl-mus TDLo:10 g/kg/41W-C:CAR IJCNAW 19,117,77

orl-rat TD:20 g/kg/78W-C:NEO NCITR* NCI-CG-TR-29,78
ipr-rat LD50:1100 mg/kg GTPZAB 21(12),27,77

CONSENSUS REPORTS: IARC Cancer Review: Group 2B IMEMDT 7,56,87; Animal Sufficient Evidence IMEMDT 27,205,82. NCI Carcinogenesis Bioassay (feed); Clear Evidence: rat NCITR* NCI-CG-TR-29,78; (feed); Clear Evidence: mouse IJCNAW 19,117,77.

SAFETY PROFILE: Suspected carcinogen with experimental carcinogenic and neoplastigenic data. Moderately toxic by intraperitoneal route. Mutation data reported. When heated to decomposition it emits toxic fumes of NO_x. See also NITRO COMPOUNDS of AROMATIC HYDROCARBONS.

MMH000 CAS:62375-91-1 ***HR: 3***
METHYL-2-NITROBENZENE DIAZOATE
mf: $C_7H_7N_3O_3$ mw: 181.15

$O_2NC_6H_4N{=}NOCH_3$

SAFETY PROFILE: Explodes violently when heated. May explode in storage. Upon decomposition it emits toxic fumes of NO_x. See also NITRO COMPOUNDS of AROMATIC HYDROCARBONS.

MMH250 CAS:121-03-9 ***HR: 2***
2-METHYL-5-NITROBENZENESULFONIC ACID
mf: $C_7H_7NO_5S$ mw: 217.21

SYN: KYSELINA-4-NITROTOLUEN-2-SULFONOVA(CZECH)

TOXICITY DATA with REFERENCE
skn-rbt 500 mg/24H MOD 28ZPAK -,182,72
eye-rbt 2 mg/24H SEV 28ZPAK -,182,72
orl-rat LD50:3710 mg/kg 28ZPAK -,182,72

CONSENSUS REPORTS: Reported in EPA TSCA Inventory.

SAFETY PROFILE: Moderately toxic by ingestion. A skin and severe eye irritant. When heated to decomposition it emits very toxic fumes of SO_x and NO_x. See also NITRO COMPOUNDS of AROMATIC HYDROCARBONS and SULFONATES.

MMH400 CAS:97-06-3 ***HR: 2***
4-METHYL-3-NITROBENZENE SULFONIC ACID
mf: $C_7H_7NO_5S$ mw: 217.20

$CH_3(O_2N)C_6H_3SO_2OH$

SAFETY PROFILE: Decomposes exothermically at 170°C. When heated to decomposition it emits toxic fumes of NO_x and SO_x. See also NITRO COMPOUNDS of AROMATIC HYDROCARBONS and SULFONATES.

MMH500 CAS:121-02-8 ***HR: 2***
2-METHYL-5-NITROBENZENESULFONYL CHLORIDE
mf: $C_7H_6ClNO_4S$ mw: 235.65

SYNS: 4-NITROTOLUEN-2-SULFOCHLORID (CZECH) ◇ 5-NITRO-o-TOLUENESULFONYL CHLORIDE

TOXICITY DATA with REFERENCE
skn-rbt 500 mg/24H MOD 28ZPAK -,199,72
eye-rbt 20 mg/24H SEV 28ZPAK -,199,72
orl-rat LD50:7470 mg/kg 28ZPAK -,199,72

SAFETY PROFILE: Mildly toxic by ingestion. A skin and severe eye irritant. When heated to decomposition it emits very toxic fumes of Cl^-, SO_x and NO_x. See also SULFONATES and NITRO COMPOUNDS of AROMATIC HYDROCARBONS.

MMH740 CAS:5709-68-2 ***HR: 3***
1-METHYL-2-NITROBENZIMIDAZOLE
mf: $C_8H_7N_3O_2$ mw: 177.18

TOXICITY DATA with REFERENCE
mmo-sat 1 nmol/plate MUREAV 173,169,86
orl-mus LD50:500 mg/kg AACHAX -,478,65
ipr-mus LD50:199 mg/kg AACHAX -,478,65
scu-mus LD50:375 mg/kg AACHAX -,478,65

SAFETY PROFILE: Poison by subcutaneous and intraperitoneal routes. Moderately toxic by ingestion. Mutation data reported. When heated to decomposition it emits toxic fumes of NO_x. See also AROMATIC AMINES.

MMI000 CAS:619-50-1 ***HR: 3***
METHYL-p-NITROBENZOATE
mf: $C_8H_7NO_4$ mw: 181.16

PROP: Monoclinic crystals. Mp: 96°. Insol in water; sol in alc and ether.

TOXICITY DATA with REFERENCE
ipr-mus LD50:200 mg/kg NTIS** AD691-490

CONSENSUS REPORTS: Reported in EPA TSCA Inventory.

SAFETY PROFILE: Poison by intraperitoneal route. When heated to decomposition it emits toxic fumes of NO_x.

MMI250 CAS:50424-93-6 ***HR: 3***
3-METHYL-2-NITROBENZOYL CHLORIDE
mf: $C_8H_6ClNO_3$ mw: 199.59

$CH_3(O_2N)C_6H_3CO{\bullet}Cl$

SAFETY PROFILE: May explode when heated above 120°C in vacuum. When heated to decomposition it emits toxic fumes of Cl^- and NO_x. See also CHLORIDES and NITRO COMPOUNDS of AROMATIC HYDROCARBONS.

MMI640 CAS:61447-07-2 ***HR: 2***
3-METHYL-4-NITRO-1-BUTEN-3-YL ACETATE
mf: $C_7H_{11}NO_4$ mw: 173.17

$H_2C{=}CHC(CH_3)(OCO{\cdot}CH_3)CH_2NO_2$

SAFETY PROFILE: Decomposes when heated above 100°C. Upon decomposition it emits toxic fumes of NO_x.

MMI650 ***HR: 2***
3-METHYL-4-NITRO-2-BUTEN-1-YL ACETATE
mf: $C_7H_{11}NO_4$ mw: 173.17

$CH_3OCO{\cdot}CH_2CH{=}C(CH_3)CH_2NO_2$

SAFETY PROFILE: Decomposes above 100°C. When heated to decomposition it emits toxic fumes of NO_x.

MMJ000 CAS:21638-36-8 ***HR: 3***
4-METHYL-1-((5-NITROFURFURYLIDENE) AMINO-2-IMIDAZOLIDINONE
mf: $C_8H_{10}N_4O_4$ mw: 226.22

TOXICITY DATA with REFERENCE
mma-sat 100 ng/plate MUREAV 48,295,77
mmo-esc 500 nmol/well CNREA8 34,2266,74
orl-rat TDLo:14 g/kg/46W-C:NEO JNCIAM 51,403,73

CONSENSUS REPORTS: EPA Genetic Toxicology Program.

SAFETY PROFILE: Questionable carcinogen with experimental neoplastigenic data. Mutation data reported. When heated to decomposition it emits toxic fumes of NO_x.

MMJ950 CAS:7194-19-6 ***HR: 2***
5-METHYL-3-(5-NITRO-2-FURYL)ISOXAZOLE
mf: $C_8H_6N_2O_4$ mw: 194.16

TOXICITY DATA with REFERENCE
orl-rat TDLo:37674 mg/kg/46W-C:ETA PAACA3 21,75,80

SAFETY PROFILE: Questionable carcinogen with experimental tumorigenic data. When heated to decomposition it emits toxic fumes of NO_x.

MMJ960 CAS:5052-75-5 ***HR: 2***
5-METHYL-3-(5-NITRO-2-FURYL)PYRAZOLE
mf: $C_8H_7N_3O_3$ mw: 193.18

TOXICITY DATA with REFERENCE
dnd-mus:fbr 300 μmol/L BJCAAI 3,124,78
orl-rat TDLo:37352 mg/kg/46W-C:ETA PAACA3 21,75,80

SAFETY PROFILE: Questionable carcinogen with experimental tumorigenic data. Mutation data reported. When heated to decomposition it emits toxic fumes of NO_x.

MMJ975 CAS:53757-29-2 ***HR: 2***
2-METHYL-4-(5-NITRO-2-FURYL)THIAZOLE
mf: $C_8H_6N_2O_3S$ mw: 210.22

TOXICITY DATA with REFERENCE
mmo-sat 100 ng/plate MUREAV 40,9,76
dnr-esc 500 nmol/well CNREA8 34,2266,74
orl-rat TDLo:40572 mg/kg/46W-C:ETA PAACA3 21,75,80

SAFETY PROFILE: Questionable carcinogen with experimental tumorigenic data. Mutation data reported. When heated to decomposition it emits toxic fumes of SO_x.

MMK000 CAS:10187-79-8 ***HR: 3***
N-(1-METHYL-3-(5-NITRO-2-FURYL)-s-TRIAZOL-5-YL-ACETAMIDE
mf: $C_9H_9N_5O_4$ mw: 251.23

SYNS: 5-ACETAMIDO-1-METHYL-3-(5-NITRO-2-FURYL)-s-TRIAZOLE ◇ N-(1-METHYL-3-(5-NITRO-2-FURYL)-1H-1,2,4-TRIAZOL-5-YL)ACETAMIDE

TOXICITY DATA with REFERENCE
orl-mus LD50:600 mg/kg JMCMAR 16,312,73
ipr-mus LD50:300 mg/kg JMCMAR 16,312,73

SAFETY PROFILE: Poison by intraperitoneal route. Moderately toxic by ingestion. When heated to decomposition it emits toxic fumes of NO_x.

MML250 CAS:4245-76-5 ***HR: D***
N-METHYL-N'-NITROGUANIDINE
mf: $C_2H_6N_4O_2$ mw: 118.12

SYN: 1-METHYL-3-NITROGUANIDINE

TOXICITY DATA with REFERENCE
cyt-ham:lng 3300 mg/L GMCRDC 27,95,81
eye-ham:lng 3300 mg/L GMCRDC 27,95,81

CONSENSUS REPORTS: EPA Genetic Toxicology Program.

SAFETY PROFILE: Mutation data reported. When heated to decomposition it emits toxic fumes of NO_x.

MML500 ***HR: 3***
1-METHYL-3-NITROGUANIDINE mixed with SODIUM NITRITE (1:1)

SYN: SODIUM NITRITE mixed with 1-METHYL-3-NITROGUANIDINE (1:1)

TOXICITY DATA with REFERENCE
dns-hmn:fbr 100 μmol/L/3H IJCNAW 16,284,75
orl-rat TDLo:90 g/kg/67W-C:CAR VOONAW 23(8),54,77

SAFETY PROFILE: Questionable carcinogen with experimental carcinogenic data. Human mutation data reported. When heated to decomposition it emits very toxic fumes of NO_x and Na_2O. See also N-METHYL-N′-NITROGUANIDINE and SODIUM NITRITE.

MML550 ***HR: 3***
1-METHYL-3-NITROGUANIDINIUM NITRATE
mf: $C_2H_7N_5O_5$ mw: 181.11

SAFETY PROFILE: Explodes on impact. When heated to decomposition it emits toxic fumes of NO_x. See also NITRATES.

MML575 ***HR: 3***
1-METHYL-3-NITROGUANIDINIUM PERCHLORATE
mf: $C_2H_7ClN_4O_6$ mw: 218.55

SAFETY PROFILE: Explodes violently upon impact. When heated to decomposition it emits toxic fumes of Cl^- and NO_x. See also PERCHLORATES.

MML750 CAS:1671-82-5 ***HR: 3***
1-METHYL-2-NITROIMIDAZOLE
mf: $C_4H_5N_3O_2$ mw: 127.12

TOXICITY DATA with REFERENCE
mmo-esc 5 μmol/L BJCAAI 37,60,78
orl-mus LD50:126 mg/kg JMCMAR 12,775,69
ipr-mus LD50:126 mg/kg AACHAX -,478,65
scu-mus LD50:158 mg/kg AACHAX -,478,65

SAFETY PROFILE: Poison by ingestion, intraperitoneal, and subcutaneous routes. Mutation data reported. When heated to decomposition it emits toxic fumes of NO_x.

MMM000 CAS:3034-41-1 ***HR: D***
1-METHYL-4-NITRO-1H-IMIDAZOLE
mf: $C_4H_5N_3O_2$ mw: 127.12

SYN: 1-METHYL-4-NITROIMIDAZOLE

TOXICITY DATA with REFERENCE
mmo-sat 1 mmol/L MUREAV 66,207,79
mmo-klp 2 mmol/L/20H MUREAV 66,207,79

SAFETY PROFILE: Mutation data reported. When heated to decomposition it emits toxic fumes of NO_x.

MMM500 CAS:696-23-1 ***HR: D***
2-METHYL-5-NITROIMIDAZOLE
mf: $C_4H_5N_3O_2$ mw: 127.12

SYN: 2-METHYL-5-NITRO-1H-IMIDAZOLE

TOXICITY DATA with REFERENCE
mmo-sat 3 mmol/L MUREAV 66,207,79
mmo-klp 5 mmol/L/20H MUREAV 66,207,79

SAFETY PROFILE: Mutation data reported. When heated to decomposition it emits toxic fumes of NO_x.

MMM750 CAS:14003-66-8 ***HR: D***
4-METHYL-5-NITROIMIDAZOLE
mf: $C_4H_5N_3O_2$ mw: 127.12

SYN: 4-METHYL-5-NITRO-1H-IMIDAZOLE

TOXICITY DATA with REFERENCE
mmo-klp 10 μmol/L/20H MUREAV 66,207,79

CONSENSUS REPORTS: Reported in EPA TSCA Inventory.

SAFETY PROFILE: Mutation data reported. When heated to decomposition it emits toxic fumes of NO_x.

MMN250 CAS:443-48-1 ***HR: 3***
2-METHYL-5-NITROIMIDAZOLE-1-ETHANOL
mf: $C_6H_9N_3O_3$ mw: 171.18

SYNS: ACROMONA ◇ ANAGIARDIL ◇ ATRIVYL ◇ BAYER 5360 ◇ BEXON ◇ CLONT ◇ CONT ◇ DANIZOL ◇ DEFLAMON-WIRKSTOFF ◇ EFLORAN ◇ ELYZOL ◇ ENTIZOL ◇ 1-(β-ETHYLOL)-2-METHYL-5-NITRO-3-AZAPYRROLE ◇ EUMIN ◇ FLAGEMONA ◇ FLAGESOL ◇ FLAGIL ◇ FLAGYL ◇ GIATRICOL ◇ GINEFLAVIR ◇ 1-(β-HYDROXYETHYL)-2-METHYL-5-NITROIMIDAZOLE ◇ 1-(2-HYDROXYETHYL)-2-METHYL-5-NITROIMIDAZOLE ◇ 1-HYDROXYETHYL-2-METHYL-5-NITROIMIDAZOLE ◇ 1-(2-HYDROXY-1-ETHYL)-2-METHYL-5-NITROIMIDAZOLE ◇ KLION ◇ MERONID AL ◇ 2-METHYL-1-(2-HYDROXYETHYL)-5-NITROIMIDAZOLE ◇ 2-METHYL-3-(2-HYDROXYETHYL)-4-NITROIMIDAZOLE ◇ METRONIDAZ ◇ METRONIDAZOL ◇ METRONIDAZOLO ◇ MONAGYL ◇ NALOX ◇ NEO-TRIC ◇ NIDA ◇ NOVONIDAZOL ◇ NSC-50364 ◇ ORVAGIL ◇ 1-(β-OXYETHYL)-2-METHYL-5-NITROIMIDAZOLE ◇ RP 8823 ◇ SANATRICHOM ◇ SC 10295 ◇ TRICHAZOL ◇ TRICHOCIDE ◇ TRICHOMOL ◇ TRICHOMONACID "PHARMACHIM" ◇ TRICHOPOL ◇ TRICOM ◇ TRICOWAS B ◇ TRIKOJOL ◇ TRIMEKS ◇ TRIVAZOL ◇ VAGILEN ◇ VAGIMID ◇ VERTISAL

TOXICITY DATA with REFERENCE
dnr-esc 2 mg/L MUREAV 164,9,86
cyt-hmn:lym 500 mg/L ENMUDM 6,467,84
ipr-mus TDLo:60 mg/kg (female 8-14D post):TER TOLED5 19,37,83
unr-rat TDLo:750 mg/kg (female 30D pre):REP RPHRA6 20,395,64
orl-rat TDLo:219 g/kg/35W-C:CAR JJIND8 63,863,79

orl-mus TDLo:181 g/kg/72W-C:CAR JNCIAM 48,721,72
orl-rat TDLo:27 g/kg/35W-C:ETA,TER JNCIAM 51,403,73
orl-mus TD:8 g/kg/14W-C:NEO TUMOAB 69,379,83
orl-mus TD:1680 mg/kg:CAR,REP JCREA8 112,135,86
orl-wmn TDLo:40 mg/kg SMJOAV 78,627,85
orl-man TDLo:3570 μg/kg/D:LIV,MET JAMAAP 193,1128,65
orl-wmn TDLo:12 mg/kg:EYE,CNS BMJOAE 292,174,86
orl-man TDLo:1030 mg/kg/8W:PNS BMJOAE 2(6087),610,77
ivn-wmn TDLo:30 mg/kg/I:EAR,CNS NZMJAX 97,128,84
orl-rat LD50:3 g/kg OYYAA2 8,1089,74
orl-mus LD50:3800 mg/kg JMCMAR 20,1522,77
ipr-mus LD50:2980 mg/kg OYYAA2 8,1089,74
scu-mus LD50:3640 mg/kg OYYAA2 8,1089,74

CONSENSUS REPORTS: NTP Fifth Annual Report on Carcinogens. IARC Cancer Review: Group 2B IMEMDT 7,250,87; Animal Sufficient Evidence IMEMDT 13,113,77. EPA Genetic Toxicology Program.

SAFETY PROFILE: Confirmed carcinogen with experimental carcinogenic, neoplastigenic, tumorigenic, and teratogenic data. Moderately toxic by ingestion, intraperitoneal, and subcutaneous routes. Human systemic effects by ingestion: paresthesia, nerve or sheath structural changes, eye changes, tremors, fever, jaundice and other liver changes, hearing acuity changes, somnolence, and ataxia. Experimental reproductive effects. Human mutation data reported. When heated to decomposition it emits toxic fumes of NO_x.

MMN500 CAS:936-05-0 ***HR: D***
1-METHYL-5-NITROIMIDAZOLE-2-METHANOL
mf: $C_5H_7N_3O_3$ mw: 157.15

SYN: 1-METHYL-2-HYDROXYMETHYL-5-NITROIMIDAZOLE

TOXICITY DATA with REFERENCE
mmo-sat 1 μmol/L TCMUD8 3,429,83
mmo-klp 10 μmol/L/20H MUREAV 66,207,79

SAFETY PROFILE: Mutation data reported. When heated to decomposition it emits toxic fumes of NO_x.

MMN750 CAS:7681-76-7 ***HR: D***
1-METHYL-5-NITROIMIDAZOLE-2-METHANOL CARBAMATE (ESTER)
mf: $C_6H_8N_4O_4$ mw: 200.18

SYNS: ◇ DUGRO ◇ 1-METHYL-5-NITRO-1H-IMIDAZOLE-2-METHANOL CARBAMATE ESTER ◇ RIDAZOLE ◇ RIDZOL P ◇ RONIDAZOLE

TOXICITY DATA with REFERENCE
mmo-sat 3 μg/plate CBINA8 49,27,84
mmo-smc 500 ppm MUREAV 86,243,81

CONSENSUS REPORTS: EPA Genetic Toxicology Program.

SAFETY PROFILE: Mutation data reported. When heated to decomposition it emits toxic fumes of NO_x. See also CARBAMATES.

MMP000 CAS:70-25-7 ***HR: 3***
N-METHYL-N'-NITRO-N-NITROSOGUANIDINE
DOT: NA 1325
mf: $C_2H_5N_5O_3$ mw: 147.12

PROP: Crystals.

SYNS: METHYLNITRONITROSOGUANIDINE ◇ 1-METHYL-3-NITRO-1-NITROSOGUANIDINE ◇ N-METHYL-N'-NITRO-N-NITROSOGUANIDINE, not exceeding 25 grams in one outside packaging (DOT) ◇ N-METHYL-N-NITROSONITROGUANIDIN (GERMAN) ◇ 1-METHYL-1-NITROSO-3-NITROGUANIDINE ◇ METHYLNITROSOGUANIDINE ◇ N-METHYL-N-NITROSO-N'-NITROGUANIDINE ◇ N-METYLO-N'-NITRO-N-NITROZOGOUANIDYNY (POLISH) ◇ MNG ◇ MNNG ◇ N'-NITRO-N-NITROSO-N-METHYLGUANIDINE ◇ N-NITROSO-N-METHYLNITROGUANIDINE ◇ NSC 9369 ◇ RCRA WASTE NUMBER U163

TOXICITY DATA with REFERENCE
dnd-hmn:fbr 20 μmol/L ENMUDM 7,267,85
sce-hmn:hla 300 nmol/L CRNGDP 5,593,84
ipr-mus TDLo:50 mg/kg (female 11D post):TER TJADAB 25(2),60A,82
ipr-mus TDLo:50 mg/kg (11D preg):REP TJADAB 25(2),60A,82
orl-rat TDLo:500 μg/kg:CAR GANNA2 69,805,78
skn-rat TDLo:720 mg/kg/60W-I:CAR ARGEAR 55,117,85
ivn-rat TDLo:25 mg/kg (15D preg):ETA,TER KFIZAO 84,23,75
ivn-rat TDLo:60 mg/kg:ETA EJCAAH 16,395,80
orl-rat LD50:90 mg/kg ZKKOBW 91,183,78
scu-rat LD50:420 mg/kg ZEKBAI 69,103,67
ivn-rat LD50:80 mg/kg EJCAAH 16,395,80
ipr-mus LD50:66 mg/kg JJIND8 62,911,79
ivn-mus LD50:37300 μg/kg NCISP* JAN86
orl-ham LD50:1070 mg/kg CALEDQ 4,241,78

CONSENSUS REPORTS: IARC Cancer Review: Group 2A IMEMDT 7,248,87; Animal Sufficient Evidence IMEMDT 4,183,74. Reported in EPA TSCA Inventory. EPA Genetic Toxicology Program.

DOT Classification: Flammable Solid; Label: Flammable Solid.

SAFETY PROFILE: Suspected carcinogen with experimental carcinogenic, tumorigenic, and teratogenic data. Poison by ingestion, intraperitoneal, and intravenous routes. Moderately toxic by subcutaneous route. Experimental reproductive effects. Human mutation data reported. An explosive sensitive to heat or impact. Flammable when exposed to heat or flame; can react vigorously with oxidizing materials. When heated to decomposition it emits very toxic fumes of NO_x. See also N-NITROSO COMPOUNDS.

MMP100 CAS:2425-85-6 ***HR: D***
1-((4-METHYL-2-NITROPHENYL)AZO)-2-NAPHTHALENOL
mf: $C_{17}H_{13}N_3O_3$ mw: 307.33

SYNS: ACCOSPERSE TOLUIDINE RED XL ◇ ADC TOLUIDINE RED B ◇ CALCOTONE TOLUIDINE RED YP ◇ CARNELIO HELIO RED ◇ CHROMATEX RED J ◇ C.I. 12120 ◇ C.I. PIGMENT RED 3 ◇ C.P. TOLUIDINE TONER A-2989 ◇ DAINICHI PERMANENT RED 4 R ◇ D&C RED No. 35 ◇ DEEP FASTONA RED ◇ DUPLEX TOLUIDINE RED L 20-3140 ◇ ELJON FAST SCARLET RN ◇ ENIALIT LIGHT RED RL ◇ FASTONA RED B ◇ FAST RED A ◇ GRAPHTOL RED A-4RL ◇ HANSA RED B ◇ HELIO FAST RED BN ◇ HISPALIT FAST SCARLET RN ◇ INDEPENDENCE RED ◇ IRGALITE SCARLET RB ◇ ISOL FAST RED HB ◇ KROMON HELIO FAST RED ◇ LAKE RED 4R ◇ LITHOL FAST SCARLET RN ◇ LUTETIA FAST SCARLET RF ◇ MONOLITE FAST SCARLET CA ◇ NCI-C60366 ◇ No. 2 FORTHFAST SCARLET ◇ ORALITH RED P4R ◇ PERMANENT RED 4R ◇ PIGMENT RUBY ◇ PIGMENT SCARLET (RUSSIAN) ◇ POLYMO RED FON ◇ RECOLITE FAST RED RL ◇ SANYO SCARLET PURE ◇ SCARLET PIGMENT RN ◇ SEGNALE LIGHT RED B ◇ SIEGLE RED BB ◇ SILOGOMMA RED RLL ◇ SILOSOL RED RN ◇ SILOTON RED RLL ◇ SYMULER FAST SCARLET 4R ◇ SYTON FAST SCARLET RB ◇ TERTROPIGMENT RED HAB ◇ TOLUIDINE RED ◇ VERSAL SCARLET PRNL ◇ VULCAFOR SCARLET A

TOXICITY DATA with REFERENCE
mmo-sat 3333 μg/plate ENMUDM 8(Suppl 7),1,86
mma-sat 2500 μg/plate ENMUDM 8(Suppl 7),1,86

CONSENSUS REPORTS: Reported in EPA TSCA Inventory.

SAFETY PROFILE: Mutation data reported. When heated to decomposition it emits toxic fumes of NO_x.

MMP500 CAS:5470-66-6 ***HR: 3***
2-METHYL-4-NITROPYRIDINE-1-OXIDE
mf: $C_6H_6N_2O_3$ mw: 154.14

TOXICITY DATA with REFERENCE
mmo-sat 1 μmol/plate GANNA2 70,799,79
mmo-esc 500 μmol/L GANNA2 70,799,79
dnd-mus:fbr 50 μmol/L CNREA8 35,521,75
scu-mus TDLo:1800 mg/kg/15W-I:ETA GANNA2 70,799,79

SAFETY PROFILE: Questionable carcinogen with experimental tumorigenic data. Mutation data reported. When heated to decomposition it emits toxic fumes of NO_x.

MMP750 CAS:1074-98-2 ***HR: 3***
3-METHYL-4-NITROPYRIDINE-1-OXIDE
mf: $C_6H_6N_2O_3$ mw: 154.14

TOXICITY DATA with REFERENCE
mmo-sat 100 nmol/plate GANNA2 70,799,79
dnd-mus:fbr 50 μmol/L CNREA8 35,521,75
scu-mus TDLo:1680 mg/kg/28W-I:ETA GANNA2 62,325,71

SAFETY PROFILE: Questionable carcinogen with experimental tumorigenic data. Mutation data reported. When heated to decomposition it emits toxic fumes of NO_x.

MMQ250 CAS:4831-62-3 ***HR: 3***
2-METHYL-4-NITROQUINOLINE-1-OXIDE
mf: $C_{10}H_8N_2O_3$ mw: 204.20

SYN: 4-NITROQUINALDINE-N-OXIDE

TOXICITY DATA with REFERENCE
dns-ham:oth 500 μmol/L NATUAS 229,416,71
dns-hmn:fbr 1 nmol/L/90M IJCNAW 16,284,75
cyt-hmn:lvr 5260 nmol/L JNCIAM 47,367,71
skn-mus TDLo:60 mg/kg/10W-I:ETA GANNA2 49,33,58

SAFETY PROFILE: Questionable carcinogen with experimental tumorigenic data. Human mutation data reported. When heated to decomposition it emits toxic fumes of NO_x.

MMQ500 CAS:14073-00-8 ***HR: 3***
3-METHYL-4-NITROQUINOLINE-1-OXIDE
mf: $C_{10}H_8N_2O_3$ mw: 204.20

TOXICITY DATA with REFERENCE
dnd-hmn: nd-hmn:fbr 250 μmol/L CRNGDP 3,1463,82
dns-rat:lvr 50 μmol/L ENMUDM 3,11,81
skn-mus TDLo:600 mg/kg/10W-I:ETA GANNA2 60,523,69
ipr-mus LD50:318 mg/kg PMRSDJ 1,682,81

CONSENSUS REPORTS: EPA Genetic Toxicology Program.

SAFETY PROFILE: Poison by intraperitoneal route. Questionable carcinogen with experimental tumorigenic data. Human mutation data reported. When heated to decomposition it emits toxic fumes of NO_x.

MMQ750 CAS:14094-43-0 ***HR: 3***
5-METHYL-4-NITROQUINOLINE-1-OXIDE
mf: $C_{10}H_8N_2O_3$ mw: 204.20

TOXICITY DATA with REFERENCE
mmo-sat 800 μmol/L CPBTAL 31,959,83
dns-ham:oth 500 nmol/L NATUAS 229,416,71
scu-mus TDLo:120 mg/kg/50D-I:ETA BCPCA6 16,631,67

CONSENSUS REPORTS: EPA Genetic Toxicology Program.

SAFETY PROFILE: Questionable carcinogen with experimental tumorigenic data. Mutation data reported.

When heated to decomposition it emits toxic fumes of NO_x.

MMR000 CAS:715-48-0 ***HR: 3***
6-METHYL-4-NITROQUINOLINE-1-OXIDE
mf: $C_{10}H_8N_2O_3$ mw: 204.20

TOXICITY DATA with REFERENCE
mmo-sat 8 μmol/L CPBTAL 31,959,83
cyt-hmn:lvr 5260 nmol/L JNCIAM 47,367,71
dnd-mus-orl 10 mg/kg IJCNAW 17,765,76
scu-mus TDLo:120 mg/kg/50D-I:ETA BCPCA6 16,631,67
ipr-rat LDLo:60 mg/kg BCPCA6 13,285,64

CONSENSUS REPORTS: EPA Genetic Toxicology Program.

SAFETY PROFILE: Poison by intraperitoneal route. Questionable carcinogen with experimental tumorigenic data. Human mutation data reported. When heated to decomposition it emits toxic fumes of NO_x.

MMR250 CAS:14753-13-0 ***HR: 3***
7-METHYL-4-NITROQUINOLINE-1-OXIDE
mf: $C_{10}H_8N_2O_3$ mw: 204.20

TOXICITY DATA with REFERENCE
dns-ham:oth 500 μmol/L NATUAS 229,416,71
mmo-smc 200 mg/L IGSBAL 85,127,72
cyt-hmn:lvr 5260 nmol/L JNCIAM 47,367,71
dnd-mus:fbr 10 μmol/L CNREA8 35,521,75
scu-mus TDLo:120 mg/kg/50D-I:ETA BCPCA6 16,631,67

SAFETY PROFILE: Questionable carcinogen with experimental tumorigenic data. Human mutation data reported. When heated to decomposition it emits toxic fumes of NO_x.

MMR500 CAS:14094-45-2 ***HR: 3***
8-METHYL-4-NITROQUINOLINE-1-OXIDE
mf: $C_{10}H_8N_2O_3$ mw: 204.20

TOXICITY DATA with REFERENCE
dns-ham:oth 8 μmol/L NATUAS 229,416,71
mmo-smc 100 μg/L TXAPA9 15,451,69
scu-mus TDLo:120 mg/kg/50D-I:ETA BCPCA6 16,631,67

SAFETY PROFILE: Questionable carcinogen with experimental tumorigenic data. Mutation data reported. When heated to decomposition it emits toxic fumes of NO_x.

MMR750 CAS:16699-07-3 ***HR: 3***
1-(4-N-METHYL-N-NITROSAMINOBENZYLIDENE)INDENE
mf: $C_{17}H_{14}N_2O$ mw: 262.33

SYNS: 4-(1-H-INDEN-1-YLIDENEMETHYL)-N-METHYL-N-NITROSOBENZENAMINE ◇ NSC-101983

TOXICITY DATA with REFERENCE
orl-rat TDLo:238 g/kg/34W-C:CAR JNCIAM 51,1313,73

SAFETY PROFILE: Questionable carcinogen with experimental carcinogenic data. When heated to decomposition it emits toxic fumes of NO_x. See also NITROSAMINES.

MMR800 ***HR: 3***
1-(METHYLNITROSAMINO)-2-BUTANONE
mf: $C_5H_{10}N_2O_2$ mw: 130.17

SYNS: M-2-OB ◇ N-NITROSOMETHYL(2-OXOBUTYL)AMINE

TOXICITY DATA with REFERENCE
msc-ham:lng 700 μmol/L 50EYAN -,241,83
scu-ham TDLo:67 mg/kg/29W-I:CAR CNREA8 43,4885,83
scu-ham TD:100 mg/kg:CAR CNREA8 43,4885,83
scu-ham LD50:92 mg/kg CNREA8 43,4885,83

SAFETY PROFILE: Poison by subcutaneous route. Questionable carcinogen with experimental carcinogenic data. Mutation data reported. When heated to decomposition it emits toxic fumes of NO_x. See also NITROSAMINES.

MMR810 ***HR: 3***
4-(METHYLNITROSAMINO)-2-BUTANONE
mf: $C_5H_{10}N_2O_2$ mw: 130.17

SYNS: M-3-OB ◇ N-NITROSOMETHYL(3-OXOBUTYL)AMINE

TOXICITY DATA with REFERENCE
msc-ham:lng 700 μmol/L 50EYAN -,241,83
mma-ham:lng 600 μmol/L MUREAV 163,303,86
scu-ham TDLo:770 mg/kg/38W-I:CAR CNREA8 43,4885,83
scu-ham TD:1377 mg/kg/34W-I:CAR CNREA8 43,4885,83
scu-ham LD50:705 mg/kg CNREA8 43,4885,83

SAFETY PROFILE: Moderately toxic by subcutaneous route. Questionable carcinogen with experimental carcinogenic data. Mutation data reported. When heated to decomposition it emits toxic fumes of NO_x. See also NITROSAMINES.

MMS000 CAS:67557-57-7 ***HR: 3***
METHYLNITROSAMINOMETHYL-d3 ESTER ACETIC ACID
mf: $C_4H_5D_3N_2O_3$ mw: 135.14

SYNS: ACETOXYMETHYLTRIDEUTEROMETHYLNITROSAMINE ◇ 1-ACETOXY-N-NITROSO-N-TRIDEUTEROMETHYLMETHYLAMINE ◇ NITROSO-N-(1-ACETOXYMETHYL)TRIDEUTEROMETHYLAMINE ◇ TRIDEUTEROMETHYL ACETOXYMETHYLNITROSAMINE

TOXICITY DATA with REFERENCE
orl-rat TDLo:51 mg/kg/90D-I:ETA ZKKOBW 91,317,78
orl-rat LD50:120 mg/kg ZKKOBW 91,317,78

SAFETY PROFILE: Poison by ingestion. Questionable carcinogen with experimental tumorigenic data. When heated to decomposition it emits toxic fumes of NO_x. See also ESTERS and NITROSAMINES.

MMS200 CAS:60153-49-3 ***HR: 3***
3-METHYLNITROSAMINOPROPIONITRILE
mf: $C_4H_7N_3O$ mw: 113.14

SYNS: MNPN ◇ PROPANENITRILE, 3-(METHYLNITROSOAMINO)-

TOXICITY DATA with REFERENCE
scu-rat TDLo:103 mg/kg/20W-I:CAR CNREA8 47,467,87
scu-rat TD:622 mg/kg:ETA PAACA3 25,99,84

CONSENSUS REPORTS: IARC Cancer Review: Group 2B IMEMDT 7,56,87, Animal Sufficient Evidence IMEMDT 37,263,85

SAFETY PROFILE: Suspected carcinogen with experimental carcinogenic and tumorigenic data. When heated to decomposition it emits toxic fumes of NO_x.

MMS250 CAS:64091-90-3 ***HR: 3***
4-(N-METHYL-N-NITROSAMINO)-4-(3-PYRIDYL) BUTANAL
mf: $C_{10}H_{13}N_3O_2$ mw: 207.26

SYNS: Γ-(METHYLNITROSAMINO)-3-PYRIDINEBUTYRALDEHYDE ◇ Γ-(METHYLNITROSOAMINO)-3-PYRIDINEBUTANAL ◇ 4-(N-NITROSO-N-METHYLAMINO)-4-(3-PYRIDYL)BUTANAL ◇ NNA

TOXICITY DATA with REFERENCE
ipr-mus TDLo:880 mg/kg/I:NEO IARCCD 19,395,78
ipr-mus TD:880 mg/kg/7W-I:ETA JNCIAM 60,819,78

CONSENSUS REPORTS: IARC Cancer Review: Group 3 IMEMDT 7,56,87; Animal Inadequate Evidence IMEMDT 37,205,85.

SAFETY PROFILE: Questionable carcinogen with experimental neoplastigenic and tumorigenic data. When heated to decomposition it emits toxic fumes of NO_x. See also N-NITROSO COMPOUNDS.

MMS500 CAS:64091-91-4 ***HR: 3***
4-(N-METHYL-N-NITROSAMINO)-1-(3-PYRIDYL)-1-BUTANONE
mf: $C_{10}H_{13}N_3O_2$ mw: 207.26

SYNS: 4-(N-METHYL-N-NITROSOAMINO)-4-(3-PYRIDYL)-1-BUTANONE ◇ N-METHYL-N-NITROSO-4-OXO-4-(3-PYRIDYL)BUTYL AMINE ◇ 4-(NITROSOAMINO-N-METHYL)-1-(3-PYRIDYL)-1-BUTANONE ◇ 4-(N-NITROSO-N-METHYLAMINO)-1-(3-PYRIDYL)-1-BUTANONE ◇ NNK

TOXICITY DATA with REFERENCE
mma-sat 1 μmol/plate CRNGDP 4,305,83
dnd-rat-ipr 1200 mg/kg/12D-C CNREA8 46,1280,86
scu-rat TDLo:68396 μg/kg/20W-I:CAR CRNGDP 8,291,87
ipr-mus TDLo:300 mg/kg/3D-I:NEO,TER CNREA8 49,3770,89
ipr-mus TDLo:880 mg/kg/I:NEO IARCCD 19,395,78
scu-ham TDLo:1504 mg/kg/25W-I:CAR IAPUDO 41,309,82

CONSENSUS REPORTS: IARC Cancer Review: Group 2B IMEMDT 7,56,87; Animal Sufficient Evidence IMEMDT 37,209,85. EPA Genetic Toxicology Program.

SAFETY PROFILE: Suspected carcinogen with experimental carcinogenic and neoplastigenic data. An experimental teratogen. Mutation data reported. When heated to decomposition it emits toxic fumes of NO_x. See also N-NITROSO COMPOUNDS.

MMS750 CAS:16699-10-8 ***HR: 3***
4-(4-N-METHYL-N-NITROSAMINOSTYRYL)QUINOLINE
mf: $C_{18}H_{15}N_3O$ mw: 289.36

SYNS: N-METHYL-N-NITROSO-4-(2-(4-QUINOLINYL)ETHENYL) BENZENAMINE ◇ NSC-101984

TOXICITY DATA with REFERENCE
orl-rat TDLo:12 g/kg/34W-C:CAR JNCIAM 51,1313,73

SAFETY PROFILE: Questionable carcinogen with experimental carcinogenic data. When heated to decomposition it emits toxic fumes of NO_x. See also N-NITROSO COMPOUNDS.

MMT000 CAS:7417-67-6 ***HR: 3***
METHYLNITROSOACETAMIDE
mf: $C_3H_6N_2O_2$ mw: 102.11

SYNS: METHYLNITROSOACETAMID (GERMAN) ◇ N-METHYL-N-NITROSOACETAMIDE ◇ N-NITROSO-N-METHYLACETAMIDE

TOXICITY DATA with REFERENCE
mma-sat 2 nmol/plate CNREA8 39,1328,79
sln-dmg-par 3000 ppm BPYKAU 4,90,67
mmo-smc 100 μmol/L ZEVBA5 95,55,65
orl-rat TDLo:375 mg/kg/75W-I:ETA PPTCBY 2,73,72
orl-rat LD50:20 mg/kg ZEKBAI 69,103,67

CONSENSUS REPORTS: EPA Genetic Toxicology Program.

SAFETY PROFILE: Poison by ingestion. Questionable carcinogen with experimental tumorigenic data. Mutation data reported. When heated to decomposition it emits toxic fumes of NO_x. See also N-NITROSO COMPOUNDS.

MMT250 CAS:21928-82-5 ***HR: 3***
N-METHYL-N-NITROSOADENINE
mf: $C_6H_6N_6O$ mw: 178.18

SYNS: 6-(METHYLNITROSAMINO)PURINE ◇ N-METHYL-N-NITROSO-1H-PURIN-6-AMINE ◇ 6-MNA ◇ N^6-METHYL-N^6-NITROSO-1H-PURIN-6-AMINE

TOXICITY DATA with REFERENCE
orl-mus TDLo:10263 mg/kg/72W-I:CAR IJCNAW 24,319,79
orl-mus TD:485 g/kg/34W-I:NEO PAACA3 18,79,77

SAFETY PROFILE: Questionable carcinogen with experimental carcinogenic and neoplastigenic data. When heated to decomposition it emits toxic fumes of NO_x. See also N-NITROSO COMPOUNDS.

MMT300 ***HR: 3***
N^6-(METHYLNITROSO)ADENOSINE
mf: $C_{11}H_{14}N_6O_5$ mw: 310.31

SYN: N^6-METHYL-N^6-NITROSO-9b-d-RIBOFURANOSYL-9H-PURIN-6-AMINE

TOXICITY DATA with REFERENCE
orl-mus TDLo:16368 mg/kg/66W-I:CAR IJCNAW 24,319,79
orl-mus TD:17856 mg/kg/72W-I:CAR IJCNAW 24,319,79
orl-mus TD:26 g/kg/2Y-I:NEO IAPUDO 31,787,80

SAFETY PROFILE: Questionable carcinogen with experimental carcinogenic and neoplastigenic data. When heated to decomposition it emits toxic fumes of NO_x. See also N-NITROSO COMPOUNDS.

MMT500 CAS:4549-43-3 ***HR: 3***
N-METHYL-N-NITROSOALLYLAMINE
mf: $C_4H_8N_2O$ mw: 100.14

SYNS: METHYLALLYLNITROSAMIN (GERMAN) ◇ METHYLALLYLNITROSAMINE ◇ N-METHYL-N-NITROSO-2-PROPEN-1-AMINE ◇ N-NITROSOALLYLMETHYLAMINE ◇ NITROSOMETHYLALLYLAMINE ◇ N-NITROSOMETHYLALLYLAMINE

TOXICITY DATA with REFERENCE
mmo-sat 40 μmol/L ENMUDM 3,11,81
mmo-esc 40 μmol/L ENMUDM 3,11,81
dns-rat:lvr 500 μmol/L ENMUDM 3,11,81
orl-rat TDLo:800 mg/kg/76W-C:ETA ARZNAD 19,1077,69
orl-rat LD50:340 mg/kg ZEKBAI 69,103,67
ivn-rat LD50:320 mg/kg ZEKBAI 69,103,67

SAFETY PROFILE: Poison by ingestion and intravenous routes. Questionable carcinogen with experimental tumorigenic data. Mutation data reported. When heated to decomposition it emits toxic fumes of NO_x. See also N-NITROSO COMPOUNDS.

MMT750 CAS:3684-97-7 ***HR: 3***
2-(N-METHYL-N-NITROSO)AMINOACETONITRILE
mf: $C_3H_5N_3O$ mw: 99.11

SYNS: N-NITROSOMETHYLAMINACETONITRIL (GERMAN) ◇ N-NITROSOMETHYLAMINOACETONITRILE

TOXICITY DATA with REFERENCE
orl-rat TDLo:620 mg/kg/89W-C:ETA ZEKBAI 69,103,67
orl-rat LD50:45 mg/kg ZEKBAI 69,103,67

CONSENSUS REPORTS: Cyanide and its compounds are on the Community Right-To-Know List.

SAFETY PROFILE: Poison by ingestion. Questionable carcinogen with experimental tumorigenic data. When heated to decomposition it emits toxic fumes of NO_x and CN^-. See also N-NITROSO COMPOUNDS and NITRILES.

MMU000 CAS:41735-28-8 ***HR: 3***
5-(N-METHYL-N-NITROSO)AMINO-3-(5-NITRO-2-FURYL)-s-TRIAZOLE
mf: $C_7H_6N_6O_4$ mw: 238.19

SYN: N-METHYL-3-(5-NITRO-2-FURYL)-N-NITROSO-1H-1,2,4-TRIAZOL-5-AMINE

TOXICITY DATA with REFERENCE
orl-mus LD50:400 mg/kg JMCMAR 16,312,73
ipr-mus LD50:300 mg/kg JMCMAR 16,312,73

SAFETY PROFILE: Poison by ingestion and intraperitoneal routes. Many N-nitroso compounds are carcinogens. When heated to decomposition it emits toxic fumes of NO_x. See also N-NITROSO COMPOUNDS.

MMU250 CAS:614-00-6 ***HR: 3***
N-METHYL-N-NITROSOANILINE
mf: $C_7H_8N_2O$ mw: 136.17

SYNS: N-METHYL-N-NITROSOBENZENAMINE ◇ METHYLPHENYLNITROSAMINE ◇ MNA ◇ NITROSOMETHYLANILINE ◇ N-NITROSO-N-METHYLANILINE ◇ N-NITROSOMETHYLPHENYLAMINE (MAK) ◇ NMA ◇ PHENYLMETHYLNITROSAMINE

TOXICITY DATA with REFERENCE
mma-sat 735 nmol/plate CALEDQ 15,289,82
mma-esc 50 nmol/plate GANNA2 75,8,84
mrc-smc 2500 mg/L IAPUDO 57,721,84
dni-mus-ipr 20 g/kg ARGEAR 51,605,81
ipr-rat TDLo:140 mg/kg (9D preg):TER IARCCD 4,112,73
orl-rat TDLo:61 mg/kg/29W-C:CAR CALEDQ 1,215,76
scu-rat TDLo:78 mg/kg/39W-I:CAR CALEDQ 1,215,76
orl-mus TDLo:2744 mg/kg/28W-C:NEO JNCIAM 46,1029,71
orl-rat TD:3400 mg/kg/35W-C:ETA ZEKBAI 69,103,67
orl-ham TDLo:1960 mg/kg/50W-I:ETA tumors CNREA8 48,6648,88
orl-rat LD50:225 mg/kg TXAPA9 17,426,70
ipr-rat LD50:180 mg/kg ZEKBAI 71,32,68
orl-ham LD50:150 mg/kg TXAPA9 17,426,70

CONSENSUS REPORTS: Reported in EPA TSCA Inventory. EPA Genetic Toxicology Program.

DFG MAK: Animal Carcinogen, Suspected Human Carcinogen.

SAFETY PROFILE: Confirmed carcinogen with experimental carcinogenic, neoplastigenic, tumorigenic, and teratogenic data. Poison by ingestion and intraperitoneal routes. Mutation data reported. When heated to decomposition it emits toxic fumes of NO_x. See also N-NITROSO COMPOUNDS.

MMU500 CAS:63412-06-6 ***HR: 3***
N-METHYL-N-NITROSOBENZAMIDE
mf: $C_8H_8N_2O_2$ mw: 164.18

SYN: MNB

TOXICITY DATA with REFERENCE
mmo-sat 340 nmol/L MUREAV 48,131,77
orl-rat TDLo:2800 mg/kg/52W-I:CAR JJIND8 62,1523,79
ipr-rat LD50:70 mg/kg JJIND8 62,1523,79

CONSENSUS REPORTS: EPA Genetic Toxicology Program.

SAFETY PROFILE: Poison by intraperitoneal route. Questionable carcinogen with experimental carcinogenic data. Mutation data reported. When heated to decomposition it emits toxic fumes of NO_x. See also N-NITROSO COMPOUNDS.

MMV000 CAS:13860-69-0 ***HR: 3***
N-METHYL-N-NITROSOBIURET
mf: $C_3H_6N_4O_3$ mw: 146.13

SYNS: 1-METHYL-1-NITROSOBIURET ◇ N-METHYL-N-NITROSO-N'-CARBAMOYLUREA ◇ N-NITROSO-N-METHYLBIURET

TOXICITY DATA with REFERENCE
mmo-omi 2500 ppm/2H-C ANTBAL 24,168,79
mmo-omi 5000 ppm/4H-C ANTBAL 24,168,79
mmo-omi 500 ppm/2H-C ANTBAL 24,168,79
mmo-omi 5000 ppm/2H ANTBAL 21,795,76
mmo-omi 1 pph ANTBAL 27,738,82
orl-rat TDLo:25 mg/kg:ETA ZEKBAI 75,229,71
orl-rat LD50:450 mg/kg PPTCBY 2,73,72

CONSENSUS REPORTS: EPA Genetic Toxicology Program.

SAFETY PROFILE: Moderately toxic by ingestion. Questionable carcinogen with experimental tumorigenic data. Mutation data reported. When heated to decomposition it emits toxic fumes of NO_x. See also N-NITROSO COMPOUNDS.

MMV250 CAS:58169-97-4 ***HR: D***
METHYLNITROSOCARBAMIC ACID-o-CHLOROPHENYL ESTER
mf: $C_8H_7ClN_2O_3$ mw: 214.62

SYNS: 2-CHLOROPHENYL METHYLNITROSOCARBAMATE ◇ HOPCIDE, NITROSATED (JAPANESE)

TOXICITY DATA with REFERENCE
mmo-esc 500 ng/plate BECTA6 14,389,75
mrc-bcs 500 ng/plate BECTA6 14,389,75

CONSENSUS REPORTS: EPA Genetic Toxicology Program.

SAFETY PROFILE: Mutation data reported. Many N-nitroso compounds are carcinogens. When heated to decomposition it emits very toxic fumes of Cl^- and NO_x. See also N-NITROSO COMPOUNDS and CARBAMATES.

MMV500 CAS:100836-61-1 ***HR: D***
METHYLNITROSOCARBAMIC ACID-o-(1,3-DIOXOLAN-2-YL)PHENYL ESTER
mf: $C_{11}H_{12}N_2O_3$ mw: 220.25

SYN: NITROSO DIOXACARB

TOXICITY DATA with REFERENCE
mmo-sat 1/µL/plate MUREAV 48,225,77

SAFETY PROFILE: Mutation data reported. Many N-nitroso compounds are carcinogens. When heated to decomposition it emits toxic fumes of NO_x. See also N-NITROSO COMPOUNDS and CARBAMATES.

MMV750 CAS:100836-62-2 ***HR: 2***
METHYLNITROSOCARBAMIC ACID-α-(ETHYLTHIO)-o-TOLYL ESTER
mf: $C_{11}H_{14}N_2O_3S$ mw: 254.33

SYN: NITROSOETHIOFENCARB

TOXICITY DATA with REFERENCE
mmo-sat 1 µL/plate MUREAV 48,225,77

SAFETY PROFILE: Mutation data reported. Many N-nitroso compounds are carcinogens. When heated to decomposition it emits very toxic fumes of SO_x and NO_x. See also N-NITROSO COMPOUNDS and CARBAMATES.

MMW250 CAS:58139-32-5 ***HR: D***
METHYLNITROSOCARBAMIC ACID-o-ISOPROPYLPHENYL ESTER
mf: $C_{11}H_{14}N_2O_3$ mw: 222.27

SYNS: 2-ISOPROPYLPHENYL N-METHYLCARBAMATE, nitrosated ◇ MIPSIN, nitrosated (JAPANESE)

TOXICITY DATA with REFERENCE
mmo-esc 50 μg/plate BECTA6 14,389,75
mrc-bcs 10 ng/plate BECTA6 14,389,75

CONSENSUS REPORTS: EPA Genetic Toxicology Program.

SAFETY PROFILE: Mutation data reported. Many N-nitroso compounds are carcinogens. When heated to decomposition it emits toxic fumes of NO_x. See also N-NITROSO COMPOUNDS and CARBAMATES.

MMW750 CAS:58139-34-7 ***HR: D***
METHYLNITROSOCARBAMIC ACID-3,5-XYLYL ESTER
mf: $C_{10}H_{12}N_2O_3$ mw: 208.24

SYNS: MAQBARL, nitrosated (JAPANESE) ◇ 3,5-XYLYL-N-METHYLCARBAMATE, NITROSATED

TOXICITY DATA with REFERENCE
mmo-esc 5 μg/plate BECTA6 14,389,75
mrc-bcs 500 ng/plate BECTA6 14,389,75

SAFETY PROFILE: Mutation data reported. Many N-nitroso compounds are carcinogens. When heated to decomposition it emits toxic fumes of NO_x. See also N-NITROSO COMPOUNDS and CARBAMATES.

MMW775 CAS:25355-61-7 ***HR: 3***
1-METHYL-1-NITROSO-3-(p-CHLOROPHENYL) UREA
mf: $C_8H_8ClN_3O_2$ mw: 213.64

SYNS: 1-(p-CHLOROPHENYL)-3-METHYL-3-NITROSOUREA ◇ 3-(p-CHLOROPHENYL)-1-METHYL-1-NITROSOUREA ◇ N-METHYL-N'-(p-CHLOROPHENYL)-N-NITROSOUREA

TOXICITY DATA with REFERENCE
mmo-sat 66 μmol/L CNREA8 39,5147,79
cyt-ham:lng 10 μmol/L IAPUDO 31,797,80
sce-ham:lng 10 μmol/L IAPUDO 31,685,80
orl-rat TDLo:598 mg/kg/20W-I:CAR ARGEAR 56,9,86
skn-mus TDLo:634 mg/kg/7W-I:CAR CNREA8 44,1027,84

SAFETY PROFILE: Suspected carcinogen with experimental carcinogenic data. Mutation data reported. When heated to decomposition it emits toxic fumes of Cl^- and NO_x. See also N-NITROSO COMPOUNDS.

MMX000 CAS:33868-17-6 ***HR: 3***
METHYLNITROSOCYANAMIDE
mf: $C_2H_3N_3O$ mw: 85.08

SYN: MNC

TOXICITY DATA with REFERENCE
mmo-sat 2500 μmol/L GMCRDC 17,17,75
mmo-esc 5 μmol/L CNREA8 38,4630,78
cyt-ham:emb 2 mmol/L/24H MUREAV 43,429,77
orl-rat TDLo:490 mg/kg/28W-C:ETA GANNA2 68,813,77
orl-mus LD50:69 mg/kg PJACAW 50,497,74

CONSENSUS REPORTS: EPA Extremely Hazardous Substances List.

SAFETY PROFILE: Poison by ingestion. Questionable carcinogen with experimental tumorigenic data. Mutation data reported. When heated to decomposition it emits toxic fumes of NO_x. See also N-NITROSO COMPOUNDS.

MMX200 CAS:75881-22-0 ***HR: 2***
N-METHYL-N-NITROSODECYLAMINE
mf: $C_{11}H_{24}N_2O$ mw: 200.37

SYN: NITROSOMETHYL-n-DECYLAMINE

TOXICITY DATA with REFERENCE
mma-sat 25 μg/plate TCMUD8 1,295,80
orl-rat TDLo:5200 mg/kg/2Y-C:ETA CNREA8 41,1288,81

SAFETY PROFILE: Questionable carcinogen with experimental tumorigenic data. Mutation data reported. When heated to decomposition it emits toxic fumes of NO_x.

MMX250 CAS:615-53-2 ***HR: 3***
N-METHYL-N-NITROSOETHYLCARBAMATE
mf: $C_4H_8N_2O_3$ mw: 132.14

$CH_3CH_2OCO \cdot N(NO)CH_3$

SYNS: ETHYL ESTER of METHYLNITROSO-CARBAMIC ACID ◇ N-METHYL-N-NITROSOCARBAMIC ACID, ETHYL ESTER ◇ METHYLNITROSOURETHAN (GERMAN) ◇ METHYLNITROSOURETHANE ◇ N-METHYL-N-NITROSO-URETHANE ◇ MNU ◇ NITROSOMETHYLURETHAN (GERMAN) ◇ NITROSOMETHYLURETHANE ◇ N-NITROSO-N-METHYLURETHANE ◇ NMUM ◇ NMUT ◇ RCRA WASTE NUMBER U178

TOXICITY DATA with REFERENCE
pic-esc 1 mg/L TCMUE9 1,91,84
dnd-gpg:oth 20 mmol/L CBINA8 20,77,78
ivn-rat TDLo:40 mg/kg (8D preg):REP IARCCD 4,100,73
ivn-rat TDLo:80 mg/kg (female 13D post):TER IARCCD 4,100,73
orl-rat TDLo:23 mg/kg/4W-I:ETA NATUAS 199,190,63
ivn-rat TDLo:20 mg/kg (15D preg):ETA,TER IARCCD 4,100,73
ipc-rat TDLo:100 mg/kg (21D preg):ETA,TER IARCCD 4,1,73
scu-mus TDLo:5 mg/kg (9D preg):CAR,TER CNREA8 34,3373,74
orl-ham TDLo:220 mg/kg/28W-I:CAR,TER JNCIAM 37,389,66
orl-rat LD50:180 mg/kg NATWAY 48,165,61
ipr-rat LDLo:50 mg/kg BJCAAI 22,316,68
scu-rat LDLo:125 mg/kg BJCAAI 16,92,62
ivn-rat LD50:4 mg/kg ZEKBAI 69,103,67

scu-mus LDLo:7 mg/kg BJCAAI 23,167,69
orl-mus TDLo:50 mg/kg/I GANMAX 3,61,66
ihl-mus LCLo:600 mg/m^3/10M NDRC** NDCrc-132,Mar,42
ipr-mus LD50:37 mg/kg CNREA8 30,11,70
scu-ham LDLo:21 mg/kg AJPAA4 50,639,67

CONSENSUS REPORTS: IARC Cancer Review: Group 2B IMEMDT 7,56,87; Animal Sufficient Evidence IMEMDT 4,211,74. Reported in EPA TSCA Inventory. EPA Genetic Toxicology Program.

SAFETY PROFILE: Suspected carcinogen with experimental carcinogenic, tumorigenic, and teratogenic data. Poison by ingestion, intraperitoneal, subcutaneous, and intravenous routes. Moderately toxic by inhalation. Experimental reproductive effects. Mutation data reported. Has been implicated as a transplacental brain carcinogen. Combustible when exposed to heat, sparks, open flame, and powerful oxidizers. Explodes when heated. A storage hazard. When heated to decomposition it emits toxic fumes of NO_x. See also NITROSAMINES and CARBAMATES.

MMX500 CAS:31364-55-3 ***HR: 3***
N-METHYL-N-NITROSO-β-d-GLUCOSAMINE
mf: $C_7H_{14}N_2O_6$ mw: 222.23

SYNS: N-METHYL-N-NITROSO-β-d-GLUCOSYLAMIN (GERMAN) ◇ N-METHYL-N-NITROSO-β-d-GLUCOSYLAMINE

TOXICITY DATA with REFERENCE
orl-rat TDLo:2500 mg/kg/25W-I:ETA ZEKBAI 75,296,71

SAFETY PROFILE: Questionable carcinogen with experimental tumorigenic data. When heated to decomposition it emits toxic fumes of NO_x. See also N-NITROSO COMPOUNDS.

MMX750 CAS:16339-21-2 ***HR: 3***
4-METHYL-4-N-NITROSOMETHYLAMINO)-2-PENTANONE
mf: $C_7H_{14}N_2O_2$ mw: 158.23

SYNS: METHYL-1,1-DIMETHYLBUTANON(3)-NITROSAMIN(GERMAN) ◇ 2-METHYLNITROSAMINO-2-METHYLPENTANON(4) (GERMAN)

TOXICITY DATA with REFERENCE
orl-rat TDLo:16 g/kg/89W-C:ETA ZEKBAI 69,103,67
orl-rat LD50:2100 mg/kg ZEKBAI 69,103,67

SAFETY PROFILE: Moderately toxic by ingestion. Questionable carcinogen with experimental tumorigenic data. When heated to decomposition it emits toxic fumes of NO_x. See also N-NITROSO COMPOUNDS.

MMY000 CAS:15567-46-1 ***HR: D***
N-METHYL-N-NITROSOOCTANAMIDE
mf: $C_9H_{18}N_2O_2$ mw: 186.29

SYN: N-NITROSO-N-METHYLCAPRYLAMIDE

TOXICITY DATA with REFERENCE
mrc-smc 10 μmol/L ZEVBA5 98,230,66

CONSENSUS REPORTS: EPA Genetic Toxicology Program.

SAFETY PROFILE: Mutation data reported. Many N-nitroso compounds are carcinogens. When heated to decomposition it emits toxic fumes of NO_x. See also N-NITROSO COMPOUNDS.

MMY250 CAS:16339-01-8 ***HR: 3***
N-METHYL-N-NITROSO-4-(PHENYLAZO)ANILINE
mf: $C_{13}H_{12}N_4O$ mw: 240.29

SYNS: 4-METHYLAMINO-N-NITROSOAZOBENZENE ◇ N-NITROSO-4-METHYLAMINOAZOBENZENE ◇ N-NITROSO-4-METHYLAMINOAZOBENZOL (GERMAN)

TOXICITY DATA with REFERENCE
dni-mus-orl 20 g/kg ARGEAR 51,605,81
orl-rat TDLo:100 mg/kg:ETA ZEKBAI 69,103,67
orl-rat LD50:1200 mg/kg ZEKBAI 69,103,67

SAFETY PROFILE: Moderately toxic by ingestion. Questionable carcinogen with experimental tumorigenic data. Mutation data reported. When heated to decomposition it emits toxic fumes of NO_x. See also N-NITROSO COMPOUNDS.

MMY500 CAS:21561-99-9 ***HR: 3***
1-METHYL-1-NITROSO-3-PHENYLUREA
mf: $C_8H_9N_3O_2$ mw: 179.20

SYNS: N-METHYL-N-NITROSO-N'-PHENYLUREA ◇ METHYLPHENYLNITROSOUREA ◇ N-METHYL-N'-PHENYL-N-NITROSOUREA ◇ MPNU ◇ NITROSOMETHYLPHENYLUREA ◇ 3-PHENYL-1-METHYL-1-NITROSOHARNSTOFF(GERMAN)

TOXICITY DATA with REFERENCE
mmo-sat 66 μmol/L CNREA8 39,5147,79
cyt-ham:lng 10 μmol/L IAPUDO 31,797,80
sce-ham:lng 10 μmol/L IAPUDO 31,797,80
orl-rat TDLo:750 mg/kg/32W-I:CAR CALEDQ 4,299,78
scu-rat TDLo:250 mg/kg:ETA MVMZA8 33,128,78
skn-mus TDLo:532 mg/kg/7W-I:CAR CNREA8 44,1027,84

SAFETY PROFILE: Suspected carcinogen with experimental carcinogenic and tumorigenic data. Mutation data reported. When heated to decomposition it emits toxic fumes of NO_x. See also N-NITROSO COMPOUNDS.

MMY750 CAS:13603-07-1 ***HR: 3***
3-METHYLNITROSOPIPERIDINE
mf: $C_6H_{12}N_2O$ mw: 128.20

SYN: 1-NITROSO-3-PIPECOLINE

TOXICITY DATA with REFERENCE
mma-sat 1 μmol/plate MUREAV 56,131,77
sln-dmg-orl 5 mmol/L/24H MUREAV 67,27,79
mma-smc 25 mmol/L/24H MUREAV 57,155,78
sce-hmn:lym 10 mmol/L TCMUE9 1,129,84
orl-rat TDLo:2520 mg/kg/50W-I:ETA IJCNAW 16,318,75

CONSENSUS REPORTS: EPA Genetic Toxicology Program.

SAFETY PROFILE: Questionable carcinogen with experimental tumorigenic data. Human mutation data reported. When heated to decomposition it emits toxic fumes of NO_x. See also N-NITROSO COMPOUNDS.

MMZ000 CAS:15104-03-7 ***HR: 3***
4-METHYLNITROSOPIPERIDINE
mf: $C_6H_{12}N_2O$ mw: 128.20

SYN: 1-NITROSO-4-PIPECOLINE

TOXICITY DATA with REFERENCE
mma-sat 200 μg/plate MUREAV 56,131,77
sln-dmg-orl 5 mmol/L/24H MUREAV 67,27,79
mma-smc 25 mmol/L/24H MUREAV 57,155,78
orl-rat TDLo:2016 mg/kg/40W-I:ETA IJCNAW 16,318,75

CONSENSUS REPORTS: EPA Genetic Toxicology Program.

SAFETY PROFILE: Questionable carcinogen with experimental tumorigenic data. Mutation data reported. When heated to decomposition it emits toxic fumes of NO_x. See also N-NITROSO COMPOUNDS.

MMZ800 CAS:71677-48-0 ***HR: 2***
3-METHYL-1-NITROSO-4-PIPERIDONE
mf: $C_6H_{10}N_2O_2$ mw: 142.18

SYN: NITROSO-3-METHYL-4-PIPERIDONE

TOXICITY DATA with REFERENCE
mma-sat 250 μg/plate TCMUD8 1,295,80
orl-rat TDLo:1640 mg/kg/26W-I:ETA CRNGDP 5,1351,84

SAFETY PROFILE: Questionable carcinogen with experimental tumorigenic data. Mutation data reported. When heated to decomposition it emits toxic fumes of NO_x.

MNA000 CAS:924-46-9 ***HR: 3***
N-METHYL-N-NITROSO-1-PROPANAMINE
mf: $C_4H_{10}N_2O$ mw: 102.16

SYNS: METHYL-N-PROPYLNITROSAMINE ◇ METHYLPROPYLNITROSOAMINE ◇ MPN ◇ NITROSOMETHYL-N-PROPYLAMINE ◇ NITROSOMETHYLPROPYLAMINE

TOXICITY DATA with REFERENCE
pic-esc 50 mg/L TCMUE9 1,91,84
mma-ham:lng 10 mmol/L CNREA8 37,1044,77
orl-rat TDLo:160 mg/kg/23W-I:ETA JJIND8 70,959,83
scu-rat TDLo:240 mg/kg/45W-I:NEO JNCIAM 54,937,75
scu-mus TDLo:185 mg/kg/48W-I:CAR ZKKOBW 90,253,77
scu-ham TDLo:1250 mg/kg/24W-I:CAR ZKKOBW 90,141,77
scu-ham TDLo:100 mg/kg/(14D preg):NEO,TER ZKKOBW 90,119,77
scu-rat LD50:106 mg/kg JNCIAM 54,937,75
scu-mus LD50:77 mg/kg ZKKOBW 90,253,77
scu-ham LD50:493 mg/kg JNCIAM 52,457,74

CONSENSUS REPORTS: EPA Genetic Toxicology Program.

SAFETY PROFILE: Suspected carcinogen with experimental carcinogenic, neoplastigenic, and tumorigenic data. Poison by subcutaneous route. An experimental teratogen. Mutation data reported. When heated to decomposition it emits toxic fumes of NO_x. See also N-NITROSO COMPOUNDS.

MNA250 CAS:16395-80-5 ***HR: 3***
N-METHYL-N-NITROSOPROPIONAMIDE
mf: $C_4H_8N_2O_2$ mw: 116.14

SYNS: METHYLNITROSO-PROPIONAMIDE ◇ METHYL-NITROSOPROPIONSAEUREAMID (GERMAN) ◇ METHYLNITROSOPROPIONYLAMIDE

TOXICITY DATA with REFERENCE
orl-rat TDLo:445 mg/kg/89W-I:ETA PPTCBY 2,73,72

SAFETY PROFILE: Questionable carcinogen with experimental tumorigenic data. When heated to decomposition it emits toxic fumes of NO_x. See also N-NITROSO COMPOUNDS.

MNA500 ***HR: D***
2-METHYL-N-NITROSOTHIAZOLIDINE
mf: $C_4H_8N_2OS$ mw: 132.2

SYN: MNT

TOXICITY DATA with REFERENCE
mmo-sat 200 nmol/plate JAFCAU 28,781,80
mma-sat 200 nmol/plate JAFCAU 28,781,80

SAFETY PROFILE: Mutation data reported. Many N-nitroso compounds are carcinogens. When heated to decomposition it emits very toxic fumes of NO_x and SO_x. See also N-NITROSO COMPOUNDS.

MNA650 CAS:23139-00-6 ***HR: 2***
1-METHYL-1-NITROSO-3-(p-TOLYL)UREA
mf: $C_9H_{11}N_3O_2$ mw: 193.23

SYN: N-METHYL-N′-(p-METHYLPHENYL)-N-NITROSOUREA

TOXICITY DATA with REFERENCE
mmo-sat 33500 pmol/plate CNREA8 39,5147,79
skn-mus TDLo:574 mg/kg/7W-I:CAR CNREA8 44,1027,84

SAFETY PROFILE: Questionable carcinogen with experimental carcinogenic data. Mutation data reported. When heated to decomposition it emits toxic fumes of NO_x.

MNA750 CAS:684-93-5 ***HR: 3***
N-METHYL-N-NITROSOUREA
mf: $C_2H_5N_3O_2$ mw: 103.10

SYNS: METHYLNITROSO-HARNSTOFF (GERMAN) ◇ N-METHYL-N-NITROSO-HARNSTOFF (GERMAN) ◇ METHYLNITROSOUREA ◇ 1-METHYL-1-NITROSOUREA ◇ METHYLNITROSOUREE (FRENCH) ◇ MNU ◇ N-NITROSO-N-METHYLCARBAMIDE ◇ N-NITROSO-N-METHYL-HARNSTOFF (GERMAN) ◇ NITROSOMETHYLUREA ◇ N-NITROSO-N-METHYLUREA ◇ 1-NITROSO-1-METHYLUREA ◇ NMH ◇ NMU ◇ NSC 23909 ◇ RCRA WASTE NUMBER U177 ◇ SKI 24464 ◇ SRI 859

TOXICITY DATA with REFERENCE
sln-dmg-par 1 mmol/L MUREAV 149,193,85
oms-mam:lym 1 mmol/L CBINA8 46,179,83
ipr-rat TDLo:5 mg/kg (female 13D post):REP TOLED5 31(Suppl),81,86
ipr-rat TDLo:5 mg/kg (female 15D post):TER BINKBT 20(9),65,77
orl-rat TDLo:6 mg/kg:CAR CNREA8 45,4827,85
skn-rat TDLo:576 mg/kg/24W-I:NEO,TER ARGEAR 55,117,85
ipr-rat TDLo:20 mg/kg (18D post):CAR IJCNAW 15,385,75
ivn-rat TDLo:20 mg/kg (22D post):NEO,TER IARCCD 4,127,73
ivn-rat TDLo:10 mg/kg (15D post):NOO,TER KFIZAO 84,23,75
ocu-rat TDLo:800 μg/kg:ETA EXERA6 42,83,86
par-rat TDLo:5 mg/kg (21D post):NEO,TER IARCCD 4,1,73
unr-rat TDLo:10 mg/kg:NEo,TER 40RMA7 -,64,78
unr-rat TDLo:20 mg/kg (21D post):NEO,TER VOONAW 20(12),76,74
orl-mus TDLo:300 mg/kg/8W-I:NEO TXAPA9 82,19,86
scu-mus TDLo:75 mg/kg (14-19D post):NEO,TER VOONAW 17(8),75,71
ivn-mus TDLo:50 mg/kg (15D post):NEO,TER SEIJBO 21,261,81
orl-rat LD50:110 mg/kg ZEKBAI 69,103,67
ipr-rat LD50:110 mg/kg NATUAS 222,1064,69
ivn-rat LD50:108 mg/kg APEPA2 257,296,67
imp-rat LDLo:30 mg/kg CBINA8 5,139,72
ipr-mus LD50:144 mg/kg CNREA8 30,11,70
scu-ham LD50:113 mg/kg BJCAAI 29,359,74
ivn-ham LDLo:42 mg/kg JPBAA7 92,35,66
orl-mam LD50:110 mg/kg GMCRDC 17,107,75
ivn-mam LD50:110 mg/kg GMCRDC 17,107,75

CONSENSUS REPORTS: NTP Fifth Annual Report on Carcinogens. IARC Cancer Review: Group 2A IMEMDT 7,56,87; Human Limited Evidence IMEMDT 17,227,78; Animal Sufficient Evidence IMEMDT 17,227,78, IMEMDT 1,125,72. EPA Genetic Toxicology Program. Community Right-To-Know List. Reported in EPA TSCA Inventory.

SAFETY PROFILE: Confirmed carcinogen with experimental carcinogenic, neoplastigenic, tumorigenic, and teratogenic data. Poison by ingestion, implant, intraperitoneal, subcutaneous, and intravenous routes. Experimental reproductive effects. Human mutation data reported. Explodes at room temperature. Can detonate with (KOH + CH_2Cl_2). When heated to decomposition it emits toxic fumes of NO_x. See also N-NITROSO COMPOUNDS.

MNB000 CAS:37793-22-9 ***HR: D***
d-1-(3-METHYL-3-NITROSOUREIDO)-1-DEOXYGALACTOPYRANOSE
mf: $C_8H_{15}N_3O_7$ mw: 265.26

SYNS: 1-DEOXY-1-(3-METHYL-3-NITROSOUREIDO)-d-GALACTOPYRANOSE ◇ 3-β-d-GALACTOPYRANOSYL-l-METHYL-l-NITROSOUREA ◇ SZAZ

TOXICITY DATA with REFERENCE
mma-sat 400 mg/L/1H MUREAV 40,281,76
ham-lng 100 mg/L CBINA8 13,173,76

CONSENSUS REPORTS: EPA Genetic Toxicology Program.

SAFETY PROFILE: Mutation data reported. Many N-nitroso compounds are carcinogens. When heated to decomposition it emits toxic fumes of NO_x. See also N-NITROSO COMPOUNDS.

MNB250 CAS:39070-08-1 ***HR: 2***
1-METHYL-2-NITRO-5-VINYL-1H-IMIDAZOLE
mf: $C_6H_7N_3O_2$ mw: 153.16

SYNS: L 8580 ◇ MEV

TOXICITY DATA with REFERENCE
mmo-esc 800 μmol/L JGMIAN 100,283,77
dnd-esc 30 μmol/L JGMIAN 100,283,77
dni-esc 170 μmol/L JGMIAN 100,271,77
pic-esc 33 μmol/L JGMIAN 100,271,77
dnr-bcs 5 μg/disc AEMIDF 43,177,82
orl-mus LD50:480 mg/kg JMCMAR 20,656,77

SAFETY PROFILE: Moderately toxic by ingestion. Mutation data reported. When heated to decomposition it emits toxic fumes of NO_x.

MNB300 CAS:62421-98-1 ***HR: D***
o-METHYL NOGALAROL
mf: $C_{30}H_{33}NO_{12}$ mw: 599.64

SYN: 7-o-METHYL NOGALAROL

TOXICITY DATA with REFERENCE
dni-hmn:oth 850 nmol/L HXPHAU 38(Pt 2),623,75
dnd-mam:lym 12 µmol/L CBINA8 36,1,81

SAFETY PROFILE: Human mutation data reported. When heated to decomposition it emits toxic fumes of NO_x.

MNB500 CAS:53153-66-5 ***HR: 2***
3-METHYL-2(3)-NONENENITRILE
mf: $C_{10}H_{17}N$ mw: 151.28

SYNS: CITGRENILE ◇ 2-NONENENITRILE, 3-METHYL-

TOXICITY DATA with REFERENCE
skn-rbt 500 mg/24H MOD FCTOD7 20,757,82
orl-rat LD50:1720 mg/kg FCTOD7 20,757,82

CONSENSUS REPORTS: Reported in EPA TSCA Inventory.

SAFETY PROFILE: Moderately toxic by ingestion. A skin irritant. When heated to decomposition it emits toxic fumes of NO_x.

MNB750 CAS:111-79-5 ***HR: 1***
METHYL NONYLENATE
mf: $C_{10}H_{18}O_2$ mw: 170.28

SYNS: METHYL-2-NONENOATE ◇ 2-NONENOIC ACID, METHYL ESTER

TOXICITY DATA with REFERENCE
skn-rbt 500 mg/24H MOD FCTXAV 14,811,76

CONSENSUS REPORTS: Reported in EPA TSCA Inventory.

SAFETY PROFILE: A skin irritant. When heated to decomposition it emits acrid smoke and irritating fumes. See also ESTERS.

MNC000 CAS:111-80-8 ***HR: 2***
METHYL-2-NONYNOATE
mf: $C_{10}H_{16}O_2$ mw: 168.26

SYNS: METHYL OCTINE CARBONATE ◇ METHYL OCTYNE CARBONATE ◇ 2-NONYNOIC ACID, METHYL ESTER ◇ OCTYNECARBOXYLIC ACID, METHYL ESTER

TOXICITY DATA with REFERENCE
orl-rat LD50:2220 mg/kg FCTXAV 13,871,75
skn-rbt LD50:5000 mg/kg FCTXAV 13,871,75

CONSENSUS REPORTS: Reported in EPA TSCA Inventory.

SAFETY PROFILE: Moderately toxic by ingestion. Mildly toxic by skin contact. When heated to decomposition it emits acrid smoke and irritating fumes. See also ESTERS.

MNC100 CAS:7359-79-7 ***HR: D***
21-METHYLNORETHISTERONE
mf: $C_{21}H_{28}O_2$ mw: 312.49

SYNS: 17-β-HYDROXY-17-(1-PROPYNYL)ESTR-4-EN-3-ONE ◇ 17-β-HYDROXY-17-α-(1-PROPYNYL)ESTR-4-EN-3-ONE ◇ 17-α-METHYLETHYNYL-19-NORTESTOSTERONE

TOXICITY DATA with REFERENCE
orl-wmn TDLo:6 mg/kg (20D pre):REP PRSMA4 53,435,60

SAFETY PROFILE: Human female reproductive effects by ingestion: menstrual cycle changes or disorders. A steroid. When heated to decomposition it emits acrid smoke and irritating fumes.

MNC125 CAS:53-38-3 ***HR: D***
2-α-METHYL-A-NOR-17-α-PREGN-20-YNE-2-β,17-β-DIOL
mf: $C_{21}H_{28}O_2$ mw: 312.49

TOXICITY DATA with REFERENCE
par-rat TDLo:2500 µg/kg (1D preg):REP STEDAM 4,657,64

SAFETY PROFILE: Experimental reproductive effects. A steroid. When heated to decomposition it emits acrid smoke and irritating fumes.

MNC150 CAS:3570-10-3 ***HR: D***
17-α-METHYL-B-NORTESTOSTERONE
mf: $C_{19}H_{28}O_2$ mw: 288.47

SYNS: BENORTERONE ◇ BENOTERONE ◇ 17-β-HYDROXY-17-METHYL-B-NORANDROST-4-EN-3-ONE ◇ SKF 7690

TOXICITY DATA with REFERENCE
orl-mus TDLo:17500 mg/kg (35D male):REP ENDOAO 85,960,69

SAFETY PROFILE: Experimental reproductive effects. A steroid. When heated to decomposition it emits acrid smoke and irritating fumes. See also TESTOSTERONE.

MNC175 CAS:7786-29-0 ***HR: 1***
2-METHYLOCTANAL
mf: $C_9H_{18}O$ mw: 142.27

SYNS: METHYLHEXYLACETALDEHYDE ◇ α-METHYLOCTANAL ◇ OCTANAL, 2-METHYL-

TOXICITY DATA with REFERENCE
skn-rbt 500 mg/24H MOD FCTOD7 20,753,82

CONSENSUS REPORTS: Reported in EPA TSCA Inventory.

SAFETY PROFILE: A skin irritant. When heated to decomposition it emits acrid smoke and irritating fumes.

MNC250 CAS:3221-61-2 ***HR: 2***
2-METHYLOCTANE
mf: C_9H_{20} mw: 128.26

PROP: Autoign temp: 428°F, d: 0.71, vap d: 4.43, bp: 143.3°.

SAFETY PROFILE: A very dangerous fire hazard when exposed to heat, flames, oxidizers. To fight fire, use water spray, foam, fog, dry chemical, CO_2. When heated to decomposition it emits acrid smoke and irritating fumes.

MNC500 CAS:2216-33-3 ***HR: 2***
3-METHYLOCTANE
mf: C_9H_{20} mw: 128.26

PROP: Autoign temp: 428°, d: 0.72, vap d: 4.43, bp: 143.8°.

SAFETY PROFILE: A very dangerous fire hazard when exposed to heat, flames, oxidizers. To fight fire, use water spray, foam, fog, dry chemical, CO_2. When heated to decomposition it emits acrid smoke and irritating fumes.

MNC750 CAS:2216-34-4 ***HR: 2***
4-METHYLOCTANE
mf: C_9H_{20} mw: 128.26

PROP: Autoign temp: 437°F, bp: 142.4°.

SAFETY PROFILE: A very dangerous fire hazard when exposed to heat, flames, oxidizers. To fight fire, use water spray, foam, fog, dry chemical, CO_2. When heated to decomposition it emits acrid smoke and irritating fumes.

MND000 CAS:37205-85-9 ***HR: 1***
METHYL OCTANOATE and METHYL DECANOATE

PROP: Methyl ester of fatty acids with 8-12 alkyl carbons (85ARAE 3,8176).

SYNS: OCTANOIC ACID, METHYL ESTER mixed with METHYL DECANOATE ◇ OFF-SHOOT-O

TOXICITY DATA with REFERENCE
orl-rat LD50:20500 mg/kg 85ARAE 3,81,76/77

SAFETY PROFILE: Mildly toxic by ingestion. When heated to decomposition it emits acrid smoke and irritating fumes. See also ESTERS.

MND050 CAS:5340-36-3 ***HR: 2***
3-METHYL-3-OCTANOL
mf: $C_9H_{20}O$ mw: 144.29

SYNS: AMYLETHYLMETHYLCARBINOL ◇ APROL 161 ◇ 3-METHYLOCTAN-3-OL ◇ 3-OCTANOL, 3-METHYL-

TOXICITY DATA with REFERENCE
skn-rbt 500 mg/24H MOD FCTOD7 20,759,82
orl-rat LD50:3400 mg/kg FCTOD7 20,759,82

CONSENSUS REPORTS: Reported in EPA TSCA Inventory.

SAFETY PROFILE: Moderately toxic by ingestion. A skin irritant. When heated to decomposition it emits acrid smoke and irritating fumes.

MND100 CAS:24089-00-7 ***HR: 2***
3-METHYL-1-OCTEN-OL
mf: $C_9H_{18}O$ mw: 142.27

SYNS: APROL 160 ◇ 1-OCTEN-3-OL, 3-METHYL-

TOXICITY DATA with REFERENCE
skn-rbt 500 mg/24H SEV FCTOD7 20,761,82
orl-rat LD50:1800 mg/kg FCTOD7 20,761,82

CONSENSUS REPORTS: Reported in EPA TSCA Inventory.

SAFETY PROFILE: Moderately toxic by ingestion. A severe skin irritant. When heated to decomposition it emits acrid smoke and irritating fumes.

MND275 CAS:111-12-6 ***HR: 2***
METHYL 2-OCTYNOATE
mf: $C_9H_{14}O_2$ mw: 154.23

PROP: Colorless to sltly yellow liquid; powerful, unpleasant odor; violet odor when diluted. D: 0.919, refr index: 1.446, flash p: +212°F. Sol in fixed oils; sltly sol in propylene glycol; insol in glycerin

SYNS: FEMA No. 2729 ◇ FOLIONE ◇ METHYL HEPTINE CARBONATE ◇ METHYL 2-OCTINATE

TOXICITY DATA with REFERENCE
skn-rbt 500 mg/24H MOD FCTXAV 17,375,79
orl-rat LD50:1530 mg/kg FCTXAV 17,375,79
skn-rbt LD50:3300 mg/kg FCTXAV 17,375,79

CONSENSUS REPORTS: Reported in EPA TSCA Inventory.

SAFETY PROFILE: Moderately toxic by ingestion and skin contact. A moderate skin and eye irritant. A combustible liquid. When heated to decomposition it emits acrid smoke and irritating fumes.

MND500 CAS:1181-54-0 ***HR: 3***
N'-METHYLOL-o-CHLORTETRACYCLINE
mf: $C_{23}H_{27}ClN_2O_9$ mw: 510.97

TOXICITY DATA with REFERENCE
orl-mus LD50:2830 mg/kg 85ERAY 1,501,78
ipr-mus LD50:273 mg/kg 85ERAY 1,501,78
ivn-mus LD50:115 mg/kg 85ERAY 1,501,78

SAFETY PROFILE: Poison by intraperitoneal and intravenous routes. Moderately toxic by ingestion. When heated to decomposition it emits very toxic fumes of Cl^- and NO_x.

MND600 CAS:547-58-0 ***HR: 3***
METHYL ORANGE
mf: $C_{14}H_{14}N_3O_3S{\cdot}Na$ mw: 327.36

PROP: Orange-yellow powder or crystalline scales. Sol in 500 parts water; more sol in hot water; practically insol in alc.

SYNS: C.I. 13025 ◇ C.I. ACID ORANGE 52 ◇ DIAZOBEN ◇ 4-DIMETHYLAMINOAZOBENZENE-4'-SULPHONIC ACID SODIUM SALT ◇ p-((p-(DIMETHYLAMINO)PHENYL)AZO)BENZENESULFONIC ACID SODIUM SALT ◇ ENIAMETHYL ORANGE ◇ GOLD ORANGE ◇ HELIANTHINE ◇ HELIANTHINE B ◇ KCA METHYL ORANGE ◇ METHYL ORANGE B ◇ ORANGE 3 ◇ ORANGE III ◇ TROPAEOLIN

TOXICITY DATA with REFERENCE
mma-ssp 100 mg/L PMRSDJ 1,424,81
dns-hmn:fbr 4 mg/L PMRSDJ 1,528,81
ipr-mus LD50:101 mg/kg PMRSDJ 1,682,81

CONSENSUS REPORTS: EPA Genetic Toxicology Program. Reported in EPA TSCA Inventory.

SAFETY PROFILE: Poison by intraperitoneal route. Human mutation data reported. When heated to decomposition it emits toxic fumes of SO_x, NO_x and Na_2O. See also SULFONATES.

MND750 CAS:19836-78-3 ***HR: 1***
3-METHYL-2-OXAZOLIDONE
mf: $C_4H_7NO_2$ mw: 101.12

TOXICITY DATA with REFERENCE
orl-rat LD50:7130 mg/kg AIHAAP 30,470,69
skn-rbt LD50:10800 mg/kg AIHAAP 30,470,69

CONSENSUS REPORTS: Reported in EPA TSCA Inventory.

SAFETY PROFILE: Mildly toxic by ingestion and skin contact. When heated to decomposition it emits toxic fumes of NO_x.

MNE250 CAS:9038-95-3 ***HR: 1***
METHYLOXIRANE, POLYMER with OXIRANE, MONOBUTYL ESTER

SYN: UCON FLUID 50-HB 260

TOXICITY DATA with REFERENCE
orl-rat LD50:5000 mg/kg TXAPA9 17,498,70

CONSENSUS REPORTS: Reported in EPA TSCA Inventory.

SAFETY PROFILE: Mildly toxic by ingestion. When heated to decomposition it emits acrid smoke and irritating fumes. See also ESTERS.

MNF250 CAS:21308-79-2 ***HR: 3***
METHYL-12-OXO-trans-10-OCTADECENOATE
mf: $C_{19}H_{34}O_3$ mw: 310.53

SYNS: 12-OXO-trans-10-OCTADECENOIC ACID, METHYL ESTER ◇ (E)-12-OXO-10-OCTADECENOIC ACID, METHYL ESTER

TOXICITY DATA with REFERENCE
skn-mus TDLo:55 g/kg/46W-I:ETA AMBPBZ 82,127,74

SAFETY PROFILE: Questionable carcinogen with experimental tumorigenic data. When heated to decomposition it emits acrid smoke and irritating fumes.

MNG000 CAS:73-09-6 ***HR: 3***
(3-METHYL-4-OXO-5-PIPERIDINO-2-THIAZOLIDINYLIDENE)ACETIC ACID ETHYL ESTER
mf: $C_{13}H_{20}N_2O_3S$ mw: 284.41

SYNS: 2-CARBETHOXYMETHYLENE-3-METHYL-5-PIPERIDINO-4-THIAZOLIDONE ◇ ELKAPIN ◇ ETHYL (Z)-(3-METHYL-4-OXO-5-PIPERIDINO-THIAZOLIDIN-2-YLIDENE)ACETATE ◇ ETOZOLINE ◇ GOE 687 (GERMAN) ◇ 3-METHYL-4-OXO-5-PIPERIDINO-$\Delta^{2,\alpha}$-THIAZOLIDINEACETIC ACID ETHYL ESTER ◇ Z-(3-METHYL-4-OXO-5-PIPERIDINO-THIAZOLIDIN-2-YLIDEN)-ESSIGSAEUREAETHYLESTER (GERMAN) ◇ (3-METHYL-4-OXO-5-(1-PIPERIDINYL)-2-THIAZOLIDINYLIDENE)ACETIC ACID ETHYL ESTER ◇ W 2900 A

TOXICITY DATA with REFERENCE
orl-rat TDLo:67200 mg/kg (female 16-22D post):REP ARZNAD 27,1758,77
orl-rat TDLo:10 g/kg (6-15D preg):TER ARZNAD 27,1758,77
orl-rat LD50:10250 mg/kg ARZNAD 27,1745,77
ipr-rat LD50:1575 mg/kg ARZNAD 27,1745,77
ivn-rat LDLo:110 mg/kg ARZNAD 27,1745,77
orl-mus LD50:8670 mg/kg ARZNAD 27,1745,77
ipr-mus LD50:1210 mg/kg ARZNAD 27,1745,77
ivn-mus LDLo:57 mg/kg ARZNAD 27,1745,77
orl-dog LDLo:3200 mg/kg ARZNAD 27,1745,77

SAFETY PROFILE: Poison by intravenous route. Moderately toxic by ingestion and intraperitoneal routes. An experimental teratogen. Experimental reproductive effects. Used as a diuretic. When heated to decomposition it emits very toxic fumes of NO_x and SO_x. See also ESTERS.

MNG250 CAS:27343-29-9 ***HR: 3***
11-METHYL-1-OXO-1,2,3,4-TETRAHYDRO-CHRYSENE
mf: $C_{19}H_{16}O$ mw: 260.35

SYN: 1,2,3,4-TETRAHYDRO-11-METHYLCHRYSEN-1-ONE

TOXICITY DATA with REFERENCE
mma-sat 50 µg/plate CNREA8 36,4525,76
skn-mus TDLo:62 mg/kg/26W-I:NEO CNREA8 33,832,73
skn-mus TD:16 mg/kg/4D-C:ETA BJCAAI 38,148,78

CONSENSUS REPORTS: EPA Genetic Toxicology Program.

SAFETY PROFILE: Questionable carcinogen with experimental neoplastigenic and tumorigenic data. Mutation data reported. When heated to decomposition it emits acrid smoke and irritating fumes.

MNG500 CAS:593-56-6 ***HR: D***
METHYLOXYLAMMONIUM CHLORIDE
mf: $CH_5NO•ClH$ mw: 83.53

SYNS: METHOXYAMINE HYDROCHLORIDE ◇ METHOXYLAMINE HYDROCHLORIDE ◇ o-METHYLHYDROXYLAMINE ◇ o-METHYL-HYDROXYLAMINE HYDROCHLORIDE

TOXICITY DATA with REFERENCE
oms-bcs 900 mmol/L DKBSAS 257,154,81
msc-ham:lng 2 g/L SOGEBZ 11,475,75

CONSENSUS REPORTS: Reported in EPA TSCA Inventory. EPA Genetic Toxicology Program.

SAFETY PROFILE: Mutation data reported. When heated to decomposition it emits very toxic fumes of HCl, NH_3 and NO_x. See also AMINES.

MNG525 CAS:34388-29-9 ***HR: 2***
4-(METHYLOXYMETHYL)BENZYL CHRYSAN-THEMUM MONOCARBOXYLATE
mf: $C_{19}H_{26}O_3$ mw: 302.45

SYNS: CYCLOPROPANECARBOXYLIC ACID, 2,2-DIMETHYL-3-(2-METHYLPROPENYL)-, p-(METHOXYMETHYL) BENZYL ESTER ◇ 2,2-DIMETHYL-3-(2-METHYLPROPYL)CYCLOPROPANECARBOXYLIC ACIDp-(METHOXYMETHYL)BENZYL ESTER ◇ MB PYRETHROID

TOXICITY DATA with REFERENCE
eye-rbt 100 mg MLD CHYCDW 17,8,83
orl-rat LD50:900 mg/kg CHYCDW 17,8,83
orl-mus LD50:1747 mg/kg CHYCDW 17,8,83

SAFETY PROFILE: Moderately toxic by ingestion. An eye irritant. When heated to decomposition it emits acrid smoke and irritating fumes.

MNH000 CAS:298-00-0 ***HR: 3***
METHYL PARATHION
DOT: NA 2783
mf: $C_8H_{10}NO_5PS$ mw: 263.22

PROP: Crystals. Vap d: 9.1, mp: 37-38°, d: 1.358 @ 20°/4°. Sol in most organic solvents.

SYNS: A-GRO ◇ AZOFOS ◇ AZOPHOS ◇ BAY E-601 ◇ BAY 11405 ◇ BLADAN-M ◇ CEKUMETHION ◇ DALF ◇ DEVITHION ◇ O,O-DIMETHYL-O-p-NITROFENYLESTER KYSELINY THIOFOSFORECNE (CZECH) ◇ O,O-DIMETHYL-O-(4-NITROFENYL)-MONOTHIOFOSFA-AT (DUTCH) ◇ DIMETHYL p-NITROPHENYL MONOTHIOPHOS-PHATE ◇ O,O-DIMETHYL-O-(4-NITRO-PHENYL)-MONOTHIOPHOS-PHAT (GERMAN) ◇ O,O-DIMETHYL-O-(p-NITROPHENYL) PHOS-PHOROTHIOATE ◇ O,O-DIMETHYL-O-(4-NITROPHENYL) PHOS-PHOROTHIOATE ◇ DIMETHYL 4-NITROPHENYL PHOS-PHOROTHIONATE ◇ O,O-DIMETHYL-O-(p-NITROPHENYL)-THIONOPHOSPHAT (GERMAN) ◇ O,O-DIMETHYL-O-(4-NITROPHENYL)-THIONOPHOSPHAT (GERMAN) ◇ DIMETHYL-p-NITROPHENYL THIONPHOSPHATE ◇ DIMETHYL p-NITROPHENYL THIOPHOSPHATE ◇ O,O-DIMETHYL-O-p-NITROPHENYL THIOPHOS-PHATE ◇ DIMETHYL PARATHION ◇ O,O-DIMETIL-O-(4-NITRO-FENIL)-MONOTIOFOSFATO (ITALIAN) ◇ DREXEL METHYL PARA-THION 4E ◇ ENT 17,292 ◇ FOLIDOL M ◇ GEARPHOS ◇ ME-PARATHION ◇ MEPATON ◇ MEPTOX ◇ METACIDE ◇ METAFOS ◇ METAPHOR ◇ METAPHOS ◇ METHYL-E 605 ◇ METHYL FOSFERNO ◇ METHYL NIRAN ◇ METHYLTHIOPHOS ◇ METILPARATION (HUNGARIAN) ◇ METRON ◇ METYLOPARA-TION (POLISH) ◇ METYLPARATION (CZECH) ◇ NCI-C02971 ◇ p-NITROPHENYLDIMETHYLTHIONO- PHOSPHATE ◇ NITROX ◇ OLEOVOFOTOX ◇ PARAPEST M-50 ◇ PARATAF ◇ M-PARATHION ◇ PARATHION METHYL ◇ PARATHION-METILE (ITALIAN) ◇ PARATOX ◇ PENNCAP-M ◇ RCRA WASTE NUMBER P071 ◇ SINAFID M-48 ◇ SIXTY-THREE SPECIAL E.C. INSECTICIDE ◇ TEKWAISA ◇ THIOPHENIT ◇ THIOPHOSPHATE de O,O-DIMETHYLE et de O-(4-NITROPHENYLE) (FRENCH) ◇ THYLFAR M-50 ◇ TOLL ◇ VERTAC METHYL PARATHION TECHNISCH 80% ◇ VOFATOX ◇ WOFATOS ◇ WOFATOX ◇ WOFOTOX

TOXICITY DATA with REFERENCE
mmo-sat 667 µg/plate ENMUDM 5(Suppl 1),3,83
sce-hmn:lym 10 mg/L MUREAV 88,307,81
orl-rat TDLo:9 mg/kg (7-15D preg):REP JPFCD2 15,365,80
ipr-mus TDLo:60 mg/kg (female 11D post):TER AEHLAU 15,609,67
orl-rat LD50:6 mg/kg 28ZPAK -,208,72
ihl-rat LC50:34 mg/m³/4H EGESAQ 34,173,80
skn-rat LD50:63 mg/kg SPEADM 78-1,38,78
ipr-rat LD50:2800 µg/kg APCRAW 4,117,61
scu-rat LD50:6 mg/kg JEENAI 50,356,57
ivn-rat LD50:9 mg/kg NTIS** PB277-077
orl-mus LD50:18 mg/kg 85GMAT -,59,82
ihl-mus LC50:120 mg/m³/4H NTIS** PB241-840
skn-mus LD50:1200 mg/kg JTEHD6 9,491,82
ipr-mus LD50:5400 µg/kg PCBPBS 8,302,78
scu-mus LD50:18 mg/kg BLLIAX 38,151,58
ivn-mus LD50:9800 µg/kg NTIS** PB241-840
skn-rbt LD50:300 mg/kg AFDOAQ 16,3,52

CONSENSUS REPORTS: IARC Cancer Review:

Group 3 IMEMDT 7,56,87; Animal No Evidence IMEMDT 30,131,83. NCI Carcinogenesis Bioassay (feed); No Evidence: mouse, rat NCITR* NCI-CG-TR-157,79. EPA Genetic Toxicology Program. EPA Extremely Hazardous Substances List.

OSHA PEL: TWA 0.2 mg/m^3 (skin)
ACGIH TLV: TWA 0.2 mg/m^3 (skin)
NIOSH REL: (Methyl Parathion) TWA 0.2 mg/m^3
DOT Classification: Poison B; Label: Poison.

SAFETY PROFILE: Poison by inhalation, ingestion, skin contact, subcutaneous, intravenous, and intraperitoneal routes. Experimental teratogenic and reproductive effects. Questionable carcinogen. Human mutation data reported. A cholinesterase inhibitor type of insecticide. When heated to decomposition it emits very toxic fumes of NO_x, PO_x, and SO_x. See also PARATHION.

MNH010 CAS:298-00-0 ***HR: 3***
METHYL PARATHION (dry)
DOT: NA 2783
mf: $C_8H_{10}NO_5PS$ mw: 263.22

SYN: PHOSPHOROTHIOIC ACID-O,O-DIMETHYL-O-(p-NITROPHENYL) ESTER (dry mixture)

DOT Classification: Poison B; Label: Poison.

SAFETY PROFILE: A poison. When heated to decomposition it emits toxic fumes of NO_x. See also METHYL PARATHION.

MNH020 CAS:298-00-0 ***HR: 3***
METHYL PARATHION MIXTURE, liquid
DOT: NA 2783
mf: $C_8H_{10}NO_5PS$ mw: 263.22

SYN: PHOSPHOROTHIOIC ACID-O,O-DIMETHYL O-(p-NITROPHENYL) ESTER (liquid mixture)

DOT Classification: Poison B; Label: Poison.

SAFETY PROFILE: A poison. When heated to decomposition it emits very toxic fumes of NO_x, PO_x and SO_x. See also METHYL PARATHION.

MNH250 CAS:1825-21-4 ***HR: 3***
METHYL PENTACHLOROPHENATE
mf: $C_7H_3Cl_5O$ mw: 280.35

SYNS: METHYL PENTACHLOROPHENYL ESTER ◇ NCI-C56520 ◇ PENTACHLOROANISOLE ◇ 2,3,4,5,6-PENTACHLOROANISOLE ◇ PENTACHLOROMETHOXYBENZENE ◇ PENTACHLOROPHENYL METHYL ETHER

TOXICITY DATA with REFERENCE
mmo-sat 3333 μg/plate ENMUDM 8(Suppl 7),1,86
mmo-sat 3333 μg/plate ENMUDM 8(Suppl 7),1,86
orl-rat TDLo:724 mg/kg (female 23W pre):TER FCTOD7 25,163,87
orl-rat TDLo:7420 mg/kg (female 23W pre):REP FCTOD7 25,163,87
orl-rat LDLo:500 mg/kg JPETAB 90,260,47
orl-mus LD50:318 mg/kg TECSDY 11,37,86
ipr-mus LD50:281 mg/kg TECSDY 11,37,86

SAFETY PROFILE: Poison by ingestion and intraperitoneal routes. An experimental teratogen. Experimental reproductive effects. Mutation data reported. A pesticide. When heated to decomposition it emits toxic fumes of Cl^-. See also ESTERS, ETHERS, and CHLORINATED HYDROCARBONS, AROMATIC.

MNH500 CAS:54363-49-4 ***HR: 3***
METHYLPENTADIENE
DOT: UN 2461
mf: C_6H_{10} mw: 82.16

PROP: Liquid. Bp: 75-77°, flash p: < −4°F, d: 0.7184 @ 20°/4°, vap d: 2.83.

TOXICITY DATA with REFERENCE
skn-rbt 500 mg MOD SCCUR* -,7,61
ihl-mus LCLo:12200 ppm/4H SCCUR* -,7,61

DOT Classification: Flammable Liquid; Label: Flammable liquid.

SAFETY PROFILE: Mildly toxic by inhalation. A skin irritant. A very dangerous fire hazard when exposed to heat, flame, or oxidizers. Keep away from heat and open flame. To fight fire, use foam, CO_2, dry chemical. When heated to decomposition it emits acrid smoke and irritating fumes.

MNH750 CAS:1118-58-7 ***HR: 2***
2-METHYL-1,3-PENTADIENE
mf: C_6H_{10} mw: 82.15

$$H_2C{=}C(CH_3)CH{=}CHCH_3$$

SYN: ISOPRENE

PROP: Flash p: < −4°F, bp: 75-77°, d: 0.7184 @ 20°/4°, vap d: 2.83.

SAFETY PROFILE: Probably irritating and narcotic in high concentration. A very dangerous fire hazard when exposed to heat, flame or oxidizers. Keep away from heat or open flame. To fight fire, use foam, CO_2, dry chemical. When heated to decomposition it emits acrid smoke and irritating fumes.

MNI000 CAS:926-56-7 ***HR: 2***
4-METHYL-1,3-PENTADIENE
mf: C_6H_{10} mw: 82.15

$$H_2C{=}CHCH{=}C(CH_3)_2$$

PROP: Flash p: −29.2°F.

SAFETY PROFILE: A very dangerous fire hazard when exposed to heat or flame. When heated to decomposition it emits acrid smoke and irritating fumes. See also 2-METHYL-1,3-PENTADIENE.

MNI500 CAS:96-14-0 ***HR: 3***
3-METHYLPENTANE
DOT: UN 1208/UN 2462
mf: C_6H_{14} mw: 86.18

PROP: Flash p: 19.4°F, lel: 1.2%, uel: 7.0%, bp: 63.3°, fp: −118° (sets to a glass), d: 0.664 @ 20°/4°, vap press: 100 mm @ 10.5°, vap d: 2.97.

SYN: DIETHYLMETHYL METHANE

CONSENSUS REPORTS: Reported in EPA TSCA Inventory.

DOT Classification: Flammable Liquid; Label: Flammable Liquid.

SAFETY PROFILE: May have narcotic or anesthetic properties. A very dangerous fire hazard when exposed to heat or flame; can react vigorously with oxidizing materials. Explosive in the form of vapor when exposed to heat or flame. To fight fire, use foam, CO_2, dry chemical. When heated to decomposition it emits acrid smoke and irritating fumes.

MNI525 CAS:21586-21-0 ***HR: 2***
2-METHYL-2,4-PENTANEDIAMINE
mf: $C_6H_{16}N_2$ mw: 116.24

SYNS: 2,4-DIAMINO-2-METHYLPENTANE ◇ 2,4-PENTANEDIAMINE, 2-METHYL-

TOXICITY DATA with REFERENCE
skn-rbt 500 mg/24H SEV ZACTDZ 1,14,90
eye-rbt 100 mg SEV ZACTDZ 1,14,90
orl-rat LD50:431 mg/kg ZACTDZ 1,14,90
skn-rbt LD50:1600 mg/kg ZACTDZ 1,14,90

SAFETY PROFILE: Moderately toxic by ingestion and skin contact. A severe skin and eye irritant. When heated to decomposition it emits toxic fumes of NO_x.

MNI750 CAS:4457-71-0 ***HR: 3***
3-METHYL-1,5-PENTANEDIOL
mf: $C_6H_{14}O_2$ mw: 118.20

PROP: Clear liquid. Bp: 248.4°, fp: −60°, d: 0.9755 @ 20°/20°, vap press: < 0.01 mm @ 20°, vap d: 4.

TOXICITY DATA with REFERENCE
ivn-mus LD50:320 mg/kg CSLNX* NX#01094

CONSENSUS REPORTS: Reported in EPA TSCA Inventory.

SAFETY PROFILE: Poison by intravenous route. Flammable when exposed to heat or flame; can react with oxidizing materials. To fight fire, use foam, dry chemical, CO_2. When heated to decomposition it emits acrid smoke and irritating fumes. See also ALCOHOLS.

MNJ000 CAS:542-54-1 ***HR: 3***
4-METHYLPENTANENITRILE
mf: $C_6H_{11}N$ mw: 97.18

PROP: Liquid. D: 0.804 @ 20°/4°, mp: −51.1°, bp: 155.5°. Insol in water; misc in alc and ether.

SYNS: ISOAMYL CYANIDE ◇ ISOCAPRONITRILE ◇ 4-METHYLVALERONITRILE

TOXICITY DATA with REFERENCE
orl-mus LD50:488 mg/kg NEZAAQ 39,423,84
scu-rbt LDLo:90 mg/kg AIPTAK 5,161,1899
scu-frg LDLo:1600 mg/kg AIPTAK 5,161,1899

CONSENSUS REPORTS: Cyanide and its compounds are on the Community Right-To-Know List. Reported in EPA TSCA Inventory.

SAFETY PROFILE: Poison by subcutaneous route. Moderately toxic by ingestion. When heated to decomposition it emits toxic fumes of NO_x and CN^-. See also NITRILES.

MNJ100 CAS:77-74-7 ***HR: 2***
3-METHYL-3-PENTANOL
mf: $C_6H_{14}O$ mw: 102.20

SYNS: METHYLDIAETHYLCARBINOL (GERMAN) ◇ METHYLDIETHYLCARBINOL ◇ 3-METHYL-PENTANOL-(3) (GERMAN)

TOXICITY DATA with REFERENCE
orl-rat LD50:710 mg/kg JPETAB 115,230,55
orl-mus LDLo:750 mg/kg LDTU** -,-,31
scu-mus LD50:1100 mg/kg ARZNAD 5,161,55

CONSENSUS REPORTS: Reported in EPA TSCA Inventory.

SAFETY PROFILE: Moderately toxic by ingestion and subcutaneous routes. When heated to decomposition it emits acrid smoke and irritating fumes. See also ALCOHOLS.

MNJ750 CAS:623-36-9 ***HR: 2***
2-METHYL-2-PENTEN-1-AL
mf: $C_6H_{10}O$ mw: 98.16

SYNS: 2-METHYL-2-PENTENAL ◇ 2-METHYL-2-PENTENE-1-AL

TOXICITY DATA with REFERENCE
skn-rbt 10 mg/24H open MLD AMIHBC 10,61,54
eye-rbt 20 mg open SEV AMIHBC 10,61,54
orl-rat LD50:4290 mg/kg AMIHBC 10,61,54
ihl-rat LC50:2000 ppm/4H AMIHBC 10,61,54
skn-rbt LD50:4500 mg/kg AMIHBC 10,61,54

CONSENSUS REPORTS: Reported in EPA TSCA Inventory.

SAFETY PROFILE: Mildly toxic by ingestion, skin contact and inhalation. A skin and severe eye irritant. When heated to decomposition it emits acrid smoke and irritating fumes. See also ALDEHYDES.

MNK000 CAS:763-29-1 ***HR: 3***
2-METHYLPENTENE
mf: C_6H_{12} mw: 84.18

PROP: Liquid. Mp: −135.8°, bp: 62°, flash p: −18.4°F, d: 0.684 @ 15.5°/15.5°, vap press: 326 mm @ 37.3°, vap d: 2.9, autoign temp: 572°F.

SYNS: 2-METHYL-1-PENTENE ◇ 2-METHYL-PENTENE-1

TOXICITY DATA with REFERENCE
ihl-rat LC50:115 g/m^3/4H RPTOAN 31,162,68
ihl-mus LC50:127 g/m^3/2H RPTOAN 31,162,68

CONSENSUS REPORTS: Reported in EPA TSCA Inventory.

SAFETY PROFILE: Mildly toxic by inhalation. Probably irritating and narcotic in high concentration. A very dangerous fire hazard when exposed to heat, flame or oxidizers; can react vigorously with oxidizing materials. To fight fire, use CO_2, dry chemical. When heated to decomposition it emits acrid smoke and irritating fumes.

MNK250 ***HR: 3***
4-METHYL-1-PENTENE
mf: C_6H_{12} mw: 84.18

PROP: A liquid. Mp: −153.6°, bp: 54°, flash p: 19.4°F, d: 0.668 @ −15.5°/15.5°, vap press: 424 mm @ 38° vap d: 2.9, autoign temp: 572°F.

SAFETY PROFILE: Irritating and narcotic in high concentration. A very dangerous fire hazard when exposed to heat, flame or oxidizers; can react vigorously with oxidizing materials. To fight fire, use CO_2, dry chemical. When heated to decomposition it emits acrid smoke and irritating fumes.

MNK500 ***HR: 3***
4-METHYL-2-PENTENE
mf: C_6H_{12} mw: 84.18

PROP: Liquid. Mp: −134.4°, bp: 58°, d: 0.670 @ 20°/4°, vap d: 2.90, flash p: <20°F.

SYN: 1-ISOPROPYL-2-METHYLETHYLENE

SAFETY PROFILE: Irritating and narcotic in high concentration. A very dangerous fire hazard when exposed to heat, flame, or oxidizers; can react with oxidizing materials. To fight fire, use dry chemical, CO_2, foam. When heated to decomposition it emits acrid smoke and irritating fumes.

MNK750 CAS:4461-48-7 ***HR: 3***
trans-4-METHYL-2-PENTENE
mf: C_6H_{12} mw: 84.18

$CH_3CH{=}CHCH(CH_3)_2$

PROP: Flash p: −20.2°F.

SAFETY PROFILE: A very dangerous fire hazard when exposed to heat or flame. When heated to decomposition it emits acrid smoke and irritating fumes.

MNL250 CAS:918-85-4 ***HR: 2***
3-METHYL-1-PENTEN-3-OL
mf: $C_6H_{12}O$ mw: 100.18

SYN: 3-METHYL-PENTEN-(1)-OL-(3)(GERMAN)

TOXICITY DATA with REFERENCE
orl-rat LD50:700 mg/kg JPETAB 115,230,55
orl-mus LD50:1152 mg/kg ARZNAD 4,477,54
scu-mus LD50:1160 mg/kg ARZNAD 5,161,55

CONSENSUS REPORTS: Reported in EPA TSCA Inventory.

SAFETY PROFILE: Moderately toxic by ingestion and subcutaneous routes. When heated to decomposition it emits acrid smoke and irritating fumes.

MNL500 CAS:63468-05-3 ***HR: 2***
4-METHYL-2-PENTEN-4-OL
mf: $C_6H_{12}O$ mw: 100.18

TOXICITY DATA with REFERENCE
skn-rbt 500 mg MLD SCCUR* -,7,61
orl-rat LDLo:3750 mg/kg SCCUR* -,7,61
orl-mus LD50:1940 mg/kg SCCUR* -,7,61
ihl-mus LCLo:4390 ppm/1H SCCUR* -,7,61

SAFETY PROFILE: Moderately toxic by ingestion. Mildly toxic by inhalation. A skin irritant. When heated to decomposition it emits acrid smoke and irritating fumes.

MNL775 CAS:105-29-3 ***HR: 3***
3-METHYL-2-PENTEN-4-YN-1-OL
DOT: UN 2705
mf: C_6H_8O mw: 96.13

$HC{\equiv}CC(CH_3){=}CHCH_2OH$

DOT Classification: Corrosive Material; Label:Corrosive

SAFETY PROFILE: Decomposes violently when heated

above 155°C. May polymerize exothermically when heated above 100°C. Polymerization is catalyzed by traces of acid or base. Reaction with sodium hydroxide forms an explosive salt. The cis- isomer readily cyclizes to form the dangerous 2,3-dimethylfuran. When heated to decomposition it emits acrid smoke and irritating fumes. See also ACETYLENE COMPOUNDS.

MNM500 CAS:302-66-9 ***HR: 3***
METHYLPENTYNOL CARBAMATE
mf: $C_7H_{11}NO_2$ mw: 141.19

PROP: Sltly water-sol crystals. Mp: 57°, bp: 121° @ 16 mm.

SYNS: ANANSIOL ◇ CALMINOL ◇ CARBAMATE de METHYL-PENTINOL (FRENCH) ◇ CARBAMIC ACID-1-ETHYL-1-METHYL-2-PROPYNYL ESTER ◇ CARBAMIC ACID-2-ETHYNYL-2-BUTYL ESTER ◇ 3-CARBAMOYLOXY-3-METHYL-4-PENTYNE ◇ COMESA ◇ 1-ETHYL-1-METHYL-2-PROPYNYL CARBAMATE ◇ 2-ETHYNYL-2-BUTYL CARBAMATE ◇ FORMARIN ◇ MEPARFYNOL CARBAMATE ◇ MEPENTAMATE ◇ MEPENTAMATO ◇ METHYLPARAFYNOL CARBAMATE ◇ 3-METHYL-PENTIN-(1)-OL-(3) (GERMAN) ◇ 3-METHYL-1-PENTYN-3-OL CARBAMATE ◇ 3-MPC ◇ OBLIVON C ◇ OBLIVON CARBAMATE ◇ OLOSED ◇ OVETTEN ◇ PENTIN C ◇ PLACIDAL ◇ PLACIDAS ◇ PSICOPLEGIL ◇ PSICOSEDINA ◇ TRUSONO ◇ USAF EL-108 ◇ VEREDEN

TOXICITY DATA with REFERENCE
orl-mus LD50:337 mg/kg JPPMAB 10,315,58
ipr-mus LD50:100 mg/kg NTIS** AD277-689
scu-mus LD50:450 mg/kg ARZNAD 5,161,55

SAFETY PROFILE: Poison by ingestion and intraperitoneal routes. Moderately toxic by subcutaneous route. A sedative and tranquilizer which can cause central nervous system depression and death by overdose. Toxic effects are enhanced with the use of alcohol and barbiturates. When heated to decomposition it emits toxic fumes of NO_x. See also CARBAMATES.

MNM750 CAS:17043-56-0 ***HR: 3***
METHYL PERCHLORATE
mf: CH_3ClO_4 mw: 114.49

SAFETY PROFILE: An unstable explosive very sensitive to heat, shock and friction. Incompatible with alkyl perchlorates; oxygen. Upon decomposition it emits toxic fumes of Cl^-. See also PERCHLORATES.

MNN000 CAS:685-09-6 ***HR: 3***
METHYL PERFLUOROMETHACRYLATE
mf: $C_5H_3F_5O_2$ mw: 190.08

SYNS: 3,3-DIFLUORO-2-(TRIFLUOROMETHYL)ACRYLICACID, METHYL ESTER ◇ 3,3-DIFLUORO-2-(TRIFLUOROMETHYL)-2-PROPENOIC ACID, METHYL ESTER

TOXICITY DATA with REFERENCE
orl-mus LD50:220 mg/kg TXAPA9 14,114,69
ipr-mus LD50:17 mg/kg TXAPA9 14,114,69
ivn-mus LD50:20 mg/kg TXAPA9 14,114,69

SAFETY PROFILE: Poison by ingestion, intravenous, and intraperitoneal routes. When heated to decomposition it emits toxic fumes of F^-. See also FLUORIDES.

MNN250 CAS:3871-82-7 ***HR: 3***
METHYLPERIDOL HYDROCHLORIDE
mf: $C_{22}H_{26}FNO_2{\cdot}ClH$ mw: 391.95

SYNS: 4'-FLUORO-4-(4-HYDROXY-4-p-TOLYLPIPERIDINO)BUTYROPHENONE, HYDROCHLORIDE ◇ 1-(4-FLUOROPHENYL)-4-(4-HYDROXY-4-(4-METHYLPHENYL)-1-PIPERIDINYL)-1-BUTANONEHYDROCHLORIDE ◇ LUVATRENE ◇ METHYLPERIDOL ◇ MOPERONE CHLORHYDRATE ◇ MOPERONE HYDROCHLORIDE ◇ R 1658

TOXICITY DATA with REFERENCE
orl-rat LD50:152 mg/kg KSRNAM 4,869,70
scu-rat LD50:24500 μg/kg KSRNAM 4,869,70
ivn-rat LD50:12100 μg/kg KSRNAM 4,869,70
orl-mus LD50:218 mg/kg KSRNAM 4,869,70
orl-mus LD50:28100 μg/kg KSRNAM 4,869,70
ivn-mus LD50:15500 μg/kg KSRNAM 4,869,70

SAFETY PROFILE: Poison by ingestion, subcutaneous, and intravenous routes. When heated to decomposition it emits very toxic fumes of F^-, HCl, and NO_x.

MNN350 CAS:18760-80-0 ***HR: 3***
dl-N-METHYLPHEDRINE HYDROCHLORIDE
mf: $C_{11}H_{17}NO{\cdot}ClH$ mw: 215.75

SYN: dl-METHYLEPHENDRINE HYDROCHLORIDE

TOXICITY DATA with REFERENCE
orl-mus LD50:758 mg/kg NIIRDN 6,827,82
scu-mus LD50:484 mg/kg NIIRDN 6,827,82
ivn-mus LD50:134 mg/kg NIIRDN 6,827,82

SAFETY PROFILE: Poison by intravenous route. Moderately toxic by ingestion and subcutaneous routes. When heated to decomposition it emits toxic fumes of NO_x and HCl.

MNN500 CAS:832-69-9 ***HR: D***
1-METHYLPHENANTHRENE
mf: $C_{15}H_{12}$ mw: 192.27

TOXICITY DATA with REFERENCE
mma-sat 5 μg/plate MUREAV 156,61,85
msc-hmn:lym 25 μmol/L MUREAV 128,221,84

CONSENSUS REPORTS: IARC Cancer Review: Group 3 IMEMDT 7,56,87; Animal Inadequate Evidence IMEMDT 32,405,83.

SAFETY PROFILE: Questionable carcinogen. Human

mutation data reported. When heated to decomposition it emits acrid smoke and irritating fumes.

MNO250 CAS:21917-91-9 ***HR: 3***
2-METHYLPHENANTHRO(2,1-d)THIAZOLE
mf: $C_{16}H_{11}NS$ mw: 249.34

TOXICITY DATA with REFERENCE
skn-mus TDLo:800 mg/kg/31W-I:ETA VOONAW 15(8),54,69

SAFETY PROFILE: Questionable carcinogen with experimental tumorigenic data. When heated to decomposition it emits very toxic fumes of NO_x and SO_x.

MNO500 CAS:299-11-6 ***HR: D***
5-METHYLPHENAZINE METHYLSULFATE
mf: $C_{13}H_{11}N_2$•CH_3O_4S mw: 306.36

SYNS: N-METHYLPHENAZIUM METHOSULFATE ◇ 5-METHYLPHENAZIUM METHYL SULFATE ◇ N-METHYLPHENAZONIUM METHOSULFATE ◇ 5-N-METHYLPHENAZONIUM METHOSULFATE ◇ N-METHYLPHENAZONIUM METHOSULPHATE

TOXICITY DATA with REFERENCE
mmo-sat 100 μg/plate MUREAV 40,203,76
mma-sat 100 μg/plate MUREAV 40,203,76

CONSENSUS REPORTS: Reported in EPA TSCA Inventory.

SAFETY PROFILE: Mutation data reported. When heated to decomposition it emits very toxic fumes of NO_x and SO_x.

MNO750 CAS:3735-23-7 ***HR: 3***
METHYL PHENCAPTON
mf: $C_9H_{11}Cl_2O_2PS_3$ mw: 349.25

SYNS: ((2,5-DICHLOROPHENYLTHIO)METHANETHIOL-S-ESTER with O,O-DIMETHYL PHOSPHORODITHIOATE ◇ S-(((2,5-DICHLOROPHENYL)THIO)METHYL) O,O-DIMETHYL PHOSPHORODITHIOATE ◇ O,O-DIMETHYL S-(2,5-DICHLOROPHENYLTHIO)METHYL PHOSPHORODITHIOATE ◇ ENT 25,554-X ◇ GEIGY 30494

TOXICITY DATA with REFERENCE
orl-rat LD50:220 mg/kg ARSIM* 20,10,66
orl-mus LD50:11 mg/kg ARSIM* 20,10,66

CONSENSUS REPORTS: EPA Extremely Hazardous Substances List.

SAFETY PROFILE: Poison by ingestion. When heated to decomposition it emits very toxic fumes of Cl^-, SO_x and PO_x.

MNO775 CAS:21085-56-3 ***HR: 3***
2-METHYLPHENELZINE
mf: $C_9H_{14}N_2$ mw: 150.25

SYNS: (o-METHYLPHENETHYL)HYDRAZINE ◇ (2-(2-METHYLPHENYL)ETHYL)-HYDRAZINE ◇ SL 31

TOXICITY DATA with REFERENCE
scu-mus TDLo:300 mg/kg (1-6D preg):REP JOENAK 49,635,71
scu-mus LD50:250 mg/kg JOENAK 49,635,71

SAFETY PROFILE: Poison by subcutaneous route. Experimental reproductive effects. When heated to decomposition it emits toxic fumes of NO_x.

MNP250 CAS:65210-29-9 ***HR: 3***
2-METHYL-2-(2-(PHENETHYLAMINO)ETHYL)-1,3-BENZODIOXOLE HYDROCHLORIDE
mf: $C_{18}H_{21}NO_2$•ClH mw: 319.86

TOXICITY DATA with REFERENCE
ivn-rat LD50:20 mg/kg EJMCA5 12,413,77
ipr-mus LD50:100 mg/kg EJMCA5 12,413,77

SAFETY PROFILE: Poison by intravenous and intraperitoneal routes. When heated to decomposition it emits very toxic fumes of HCl and NO_x.

MNP300 CAS:60789-89-1 ***HR: 3***
1-(α-METHYLPHENETHYL)-2-(5-METHYL-3-ISOXAZOLYLCARBONYL)HYDRAZINE
mf: $C_{14}H_{17}N_3O_2$ mw: 259.34

SYN: RO 5-1226

TOXICITY DATA with REFERENCE
orl-rat LD50:89 mg/kg 27ZQAG -,296,72
ipr-rat LD50:94 mg/kg 27ZQAG -,296,72
ipr-mus LD50:150 mg/kg 27ZQAG -,296,72

SAFETY PROFILE: Poison by ingestion and intraperitoneal routes. When heated to decomposition it emits toxic fumes of NO_x.

MNP400 CAS:2598-76-7 ***HR: 2***
1-(α-METHYLPHENETHYL)-2-PHENETHYLHYDRAZINE
mf: $C_{17}H_{22}N_2$ mw: 254.41

SYNS: HYDRAZINE, 1-(α-METHYLPHENETHYL)-2-PHENETHYL- ◇ WL 23

TOXICITY DATA with REFERENCE
scu-mus TDLo:600 mg/kg (female 1-6D post):REP JOENAK 30,205,64
scu-mus LD50:526 mg/kg JOENAK 30,205,64

SAFETY PROFILE: Moderately toxic by subcutaneous route. Experimental reproductive effects. When heated to decomposition it emits toxic fumes of NO_x.

MNP500 CAS:2598-74-5 ***HR: 2***
2-(p-METHYLPHENETHYL)-3-THIOSEMICARBAZIDE
mf: $C_{10}H_{15}N_3S$ mw: 209.34

SYNS: SEMICARBAZIDE,2-(p-METHYLPHENETHYL)-3-THIO- ◇ WL 39

TOXICITY DATA with REFERENCE
scu-mus TDLo:120 mg/kg (female 1-6D post):REP JOENAK 30,205,64
scu-mus LD50:500 mg/kg JOENAK 30,205,64

SAFETY PROFILE: Moderately toxic by subcutaneous route. Experimental reproductive effects. When heated to decomposition it emits toxic fumes of NO_x and SO_x

MNQ000 CAS:113-45-1 ***HR: 3***
METHYL PHENIDYL ACETATE
mf: $C_{14}H_{19}NO_2$ mw: 233.34

SYNS: CALOCAIN ◇ CENTEDEIN ◇ CENTREDIN ◇ MERIDIL ◇ METHYLPHENIDAN ◇ METHYL PHENIDATE ◇ METHYL α-PHENYL-α-(2-PIPERIDYL)ACETATE ◇ NCI-C56280 ◇ PHENIDYLATE ◇ α-PHENYL-2-PIPERIDINEACETIC ACID METHYL ESTER ◇ PLIMASINE ◇ RITALIN ◇ RITALINE ◇ RITCHER WORKS

TOXICITY DATA with REFERENCE
ipr-mus TDLo:200 mg/kg (10D preg):TER CAJPBD 3,2,63
orl-cld TDLo:180 mg/kg/26W-I AJPSAO 143,1176,85
ivn-man LDLo:445 mg/kg/34W-I:PUL HPCQA4 3,67,72
ipr-rat LD50:430 mg/kg APTOA6 17,121,60
orl-mus LD50:190 mg/kg JPETAB 131,115,61
ipr-mus LD50:32 mg/kg JNPHAG 17,37,86
scu-mus LD50:218 mg/kg AIPTAK 184,34,70
ivn-mus LD50:41 mg/kg BCPCA6 8,263,61

SAFETY PROFILE: Poison experimentally by ingestion, intraperitoneal, intravenous, and subcutaneous routes. Moderately toxic to humans by intravenous route. Human systemic effects by intravenous route: dyspnea. An experimental teratogen. When heated to decomposition it emits toxic fumes of NO_x.

MNQ250 CAS:64057-75-6 ***HR: 3***
2-METHYLPHENISOPROPYLAMINE SULFATE
mf: $C_{10}H_{15}N \cdot 1/2H_2O_4S$ mw: 198.30

TOXICITY DATA with REFERENCE
orl-man TDLo:1500 μg/kg:GIT JPETAB 100,298,50
ipr-mus LD50:152 mg/kg JPETAB 100,298,50

SAFETY PROFILE: Poison by intraperitoneal route. Human systemic effects by ingestion: gastrointestinal tract effects. When heated to decomposition it emits very toxic fumes of NO_x and SO_x.

MNQ500 CAS:13993-65-2 ***HR: 3***
10-METHYLPHENOTHIAZINE-2-ACETIC ACID
mf: $C_{15}H_{13}NO_2S$ mw: 271.35

SYNS: ACIDE METIAZINIQUE (FRENCH) ◇ AMBRUNATE ◇ MA ◇ METHIAZIC ACID ◇ METHIAZINIC ACID ◇ (10-METHYL-2-PHENOTHIAZINYL)ACETIC ACID ◇ N-METHYL-3-PHENOTHIAZINYL-ACETIC ACID ◇ METIAZIC ACID ◇ METIAZINIC ACID ◇ RP 16,091 ◇ SORIDERMAL ◇ SORIPAL

TOXICITY DATA with REFERENCE
orl-rat TDLo:140 mg/kg (MGN):REP OYYAA2 8,1587,74
orl-rat TDLo:660 mg/kg (female 9-14D post):TER KSRNAM 8,3436,74
orl-rat LD50:495 mg/kg NIIRDN 6,822,82
ipr-rat LD50:365 mg/kg OYYAA2 9,429,75
scu-rat LD50:405 mg/kg IYKEDH 8,107,77
orl-mus LD50:800 mg/kg ARZNAD 19,1198,69
ipr-mus LD50:350 mg/kg OYYAA2 9,429,75
scu-mus LD50:490 mg/kg ARZNAD 19,1207,69
ivn-mus LD50:350 mg/kg ARZNAD 19,1198,69
orl-dog LD50:2000 mg/kg ARZNAD 19,1207,69
orl-gpg LD50:225 mg/kg ARZNAD 19,1207,69

SAFETY PROFILE: Poison by ingestion, intraperitoneal, and intravenous routes. Moderately toxic by subcutaneous route. An experimental teratogen. Experimental reproductive effects. An anti-inflammatory agent. When heated to decomposition it emits very toxic fumes of SO_x and NO_x.

MNR100 CAS:2598-72-3 ***HR: 3***
1-(2-(o-METHYLPHENOXY)ETHYL)HYDRAZINE HYDROGEN SULFATE
mf: $C_9H_{14}N_2O \cdot H_2O_4S$ mw: 264.33

SYNS: HYDRAZINE, 1-(2-(o-METHYLPHENOXY)ETHYL)-, HYDROGEN SULFATE (1:1) ◇ WEG 148

TOXICITY DATA with REFERENCE
scu-mus TDLo:180 mg/kg (female 1-6D post):REP JOENAK 30,205,64
scu-mus LD50:150 mg/kg JOENAK 30,205,64

SAFETY PROFILE: Poison by subcutaneous route. Experimental reproductive effects. When heated to decomposition it emits toxic fumes of NO_x and SO_x

MNR250 CAS:140-39-6 ***HR: 2***
4-METHYLPHENYL ACETATE
mf: $C_9H_{10}O_2$ mw: 150.19

PROP: Colorless liquid; strong floral odor. D: 1.044 @ 16°, refr index: 1.499-1.502, bp: decomp @ 360°, mp: 220°, vap d: 5.18, flash p: 203°F. Sol in fixed oils, propylene glycol, misc in alc and ether; insol in water, glycerin.

SYNS: ACETIC ACID-4-METHYLPHENYL ESTER ◇ p-ACETOXYTOLUENE ◇ 4-ACETOXYTOLUENE ◇ p-CRESOL ACETATE ◇ p-CRESYL ACETATE (FCC) ◇ FEMA No. 3073 ◇ 4-METHYLBENZOIC ACID METHYL ESTER ◇ p-METHYLPHENYL ACETATE ◇ NARCEOL ◇ PARACRESYL ACETATE ◇ p-TOLYL ACETATE ◇ p-TOLYL ETHANOATE

TOXICITY DATA with REFERENCE
orl-rat LD50:1900 mg/kg FCTXAV 12,391,74
skn-rbt LD50:2100 mg/kg FCTXAV 12,391,74

CONSENSUS REPORTS: Reported in EPA TSCA Inventory.

SAFETY PROFILE: Moderately toxic by ingestion and skin contact. Combustible liquid. When heated to decomposition it emits toxic smoke and irritating fumes. See also ESTERS.

MNR500 CAS:621-90-9 ***HR: 3***
N-METHYL-p-(PHENYLAZO)ANILINE
mf: $C_{13}H_{13}N_3$ mw: 211.29

SYNS: MAB ◇ 4-(METHYLAMINO)AZOBENZENE ◇ N-METHYL-4-AMINOAZOBENZENE ◇ N-METHYL-p-AMINOAZOBENZENE ◇ p-MONOMETHYLAMINOAZOBENZENE ◇ 4-MONOMETHYL-AMINOAZOBENZENE

TOXICITY DATA with REFERENCE
dnd-rat-orl 200 μmol/kg CBINA8 31,1,80
dns-rat:lvr 1 μmol/L GANNA2 72,930,81
ipr-mus TDLo:400 mg/kg (8-9D preg):TER KAIZAN 37,179,62
ipr-mus TDLo:400 mg/kg (8-9D preg):REP KAIZAN 37,179,62
orl-rat TDLo:634 mg/kg/5W-I:NEO CNREA8 39,3411,79
ipr-rat TDLo:317 mg/kg/3W-I:CAR CNREA8 44,2540,84
scu-rat TDLo:480 mg/kg/12W-I:ETA CNREA8 27,1600,67
orl-rat TD:2100 mg/kg/5W-C:CAR CRNGDP 8,577,87
scu-mus LDLo:600 mg/kg OFAJAE 36,195,60

CONSENSUS REPORTS: EPA Genetic Toxicology Program.

SAFETY PROFILE: Moderately toxic by subcutaneous route. Experimental teratogenic and reproductive effects. Questionable carcinogen with experimental carcinogenic, neoplastigenic, and tumorigenic data. Mutation data reported. When heated to decomposition it emits toxic fumes of NO_x.

MNR750 CAS:722-23-6 ***HR: D***
p-(3-METHYLPHENYLAZO)ANILINE
mf: $C_{13}H_{13}N_3$ mw: 211.29

SYN: 3'-METHYL-4-AMINOAZOBENZENE

TOXICITY DATA with REFERENCE
mmo-sat 1 μmol/plate CRNGDP 4,1487,83
dns-rat:lvr 10 μmol/L CRNEA8 46,1654,86

SAFETY PROFILE: Mutation data reported. When heated to decomposition it emits toxic fumes of NO_x.

MNS000 CAS:10121-94-5 ***HR: 3***
N-METHYL-4-(PHENYLAZO)-o-ANISIDINE
mf: $C_{14}H_{15}N_3O$ mw: 241.32

SYNS: 3-METHOXYMETHYLAMINOAZOBENZENE ◇ 3-METHOXY-4-MONOMETHYLAMINOAZOBENZENE

TOXICITY DATA with REFERENCE
orl-rat TDLo:9300 mg/kg/34W-C:ETA CNREA8 21,1068,61
skn-rat TDLo:660 mg/kg/41W-I:NEO CNREA8 26,2406,66

SAFETY PROFILE: Questionable carcinogen with experimental tumorigenic and neoplastigenic data. When heated to decomposition it emits toxic fumes of NO_x.

MNS250 CAS:21075-41-2 ***HR: 3***
5-METHYL-7-PHENYL-1:2-BENZACRIDINE
mf: $C_{24}H_{17}N$ mw: 319.42

SYNS: 7-METHYL-9-PHENYLBENZ(c)ACRIDINE ◇ 3-PHENYL-10-METHYL-7:8 BENZACRIDINE (FRENCH)

TOXICITY DATA with REFERENCE
skn-mus TDLo:390 mg/kg/32W-I:ETA AICCA6 7,184,50

SAFETY PROFILE: Questionable carcinogen with experimental tumorigenic data. When heated to decomposition it emits toxic fumes of NO_x.

MNS500 CAS:7055-03-0 ***HR: 3***
2-METHYL-N-PHENYLBENZAMIDE
mf: $C_{14}H_{13}NO$ mw: 211.28

SYN: BAS-3050

TOXICITY DATA with REFERENCE
dlt-rat-orl 1350 mg/kg/90D MUREAV 46,240,77
orl-rat LD50:6 g/kg 28ZEAL 5,144,76
ipr-mus LD50:200 mg/kg 85DPAN -,-,71/76

SAFETY PROFILE: Poison by intraperitoneal route. Mildly toxic by ingestion. Mutation data reported. When heated to decomposition it emits toxic fumes of NO_x. See also AMIDES.

MNT000 CAS:103-07-1 ***HR: 1***
2-METHYL-4-PHENYL-2-BUTYL ACETATE
mf: $C_{13}H_{18}O_2$ mw: 206.31

SYNS: α,α-DIMETHYLBENZENEPROPANOL ACETATE ◇ DI-METHYLPHENYLETHYLCARBINYL ACETATE ◇ (1,1-DIMETHYL-3-PHENYLPROPYL)ESTER ACETIC ACID ◇ 2-METHYL-4-PHENYL-2-BUTANOL ACETATE

TOXICITY DATA with REFERENCE
skn-rbt 500 mg/24H MLD FCTXAV 16,637,78
orl-rat LD50:4850 mg/kg FCTXAV 16,637,78

CONSENSUS REPORTS: Reported in EPA TSCA Inventory.

SAFETY PROFILE: Mildly toxic by ingestion. A skin irritant. When heated to decomposition it emits acrid smoke and irritating fumes.

MNT075 ***HR: 2***
METHYL PHENYLCARBINYL ACETATE
mf: $C_{10}H_{12}O_2$ mw: 164.20

PROP: Colorless liquid; gardenia odor. D: 1.023, refr index: 1.493-1.497, flash p: 176°F. Sol in fixed oils, glycerin; insol in water.

SYNS: FEMA No. 2684 ◇ α-PHENYL ETHYL ACETATE

SAFETY PROFILE: Combustible liquid. When heated to decomposition it emits acrid smoke and irritating fumes.

MNT100 ***HR: 3***
1-METHYL-3-PHENYL-5-CHLOROIMIDAZO (4,5-b)PYRIDIN-2-ONE
mf: $C_{13}H_{10}ClN_3O$ mw: 259.71

SYN: 5-CHLORO-1-METHYL-3-PHENYL-1H-IMIDAZO(4,5-b) PYRIDIN-2(3H)-ONE

TOXICITY DATA with REFERENCE
orl-rat LD50:1750 mg/kg EJMCA5 18,501,83
par-rat LD50:210 mg/kg EJMCA5 18,501,83
orl-mus LD50:1000 mg/kg EJMCA5 18,501,83

SAFETY PROFILE: Poison by parenteral route. Moderately toxic by ingestion. When heated to decomposition it emits toxic fumes of Cl^- and NO_x.

MNT500 CAS:20240-98-6 ***HR: 3***
1-(2-METHYLPHENYL)-3,3-DIMETHYLTRIAZENE
mf: $C_9H_{13}N_3$ mw: 163.25

SYNS: 3,3-DIMETHYL-1-(o-METHYLPHENYL)TRIAZENE ◇ 3,3-DIMETHYL-1-(o-TOLYL)TRIAZENE ◇ 1-(o-METHYLPHENYL)-3,3-DIMETHYL-TRIAZEN (GERMAN) ◇ 1-(o-METHYLPHENYL)-3,3-DIMETHYL-TRIAZENE

TOXICITY DATA with REFERENCE
orl-rat TDLo:300 mg/kg:CAR,REP ZKKOBW 81,285,74
scu-rat TDLo:500 mg/kg:CAR ZKKOBW 81,285,74
scu-rat TD:1620 mg/kg/45W-I:CAR ZKKOBW 81,285,74
orl-rat LD50:350 mg/kg ZKKOBW 81,285,74
scu-rat LD50:500 mg/kg ZKKOBW 81,285,74

SAFETY PROFILE: Poison by ingestion. Moderately toxic by subcutaneous route. Experimental reproductive effects. Questionable carcinogen with experimental carcinogenic data. When heated to decomposition it emits toxic fumes of NO_x.

MNT750 CAS:56713-63-4 ***HR: 3***
1,1'-(4-METHYL-1,3-PHENYLENE)BIS(3-(2-CHLOROETHYL)-3-NITROSOUREA)
mf: $C_{13}H_{16}Cl_2N_6O_4$ mw: 391.25

TOXICITY DATA with REFERENCE
mrc-smc 100 μmol/L/16H MUREAV 42,45,77
ivn-rat LD50:20 mg/kg EJCAAH 13,937,77

SAFETY PROFILE: Poison by intravenous route. Mutation data reported. Many N-nitroso compounds are carcinogens. When heated to decomposition it emits very toxic fumes of NO_x and Cl^-. See also N-NITROSO COMPOUNDS.

MNU000 CAS:4760-34-3 ***HR: D***
N-METHYL-o-PHENYLENEDIAMINE
mf: $C_7H_{10}N_2$ mw: 122.19

SYN: 2-AMINO-N-METHYLANILINE

TOXICITY DATA with REFERENCE
mma-sat 2500 μg/plate FCTOD7 23,695,85

CONSENSUS REPORTS: Reported in EPA TSCA Inventory.

SAFETY PROFILE: Mutation data reported. When heated to decomposition it emits toxic fumes of NO_x. See also AMINES.

MNU050 CAS:73791-44-3 ***HR: 3***
METHYL(2-PHENYLETHYL)ARSINIC ACID
mf: $C_9H_{13}AsO_2$ mw: 228.14

SYN: ARSINE OXIDE, HYDROXYMETHYLPHENETHYL-

TOXICITY DATA with REFERENCE
ivn-mus LD50:56 mg/kg CSLNX* NX#01208

OSHA PEL: TWA 0.5 mg(As)/m^3

SAFETY PROFILE: Poison by intravenous route. When heated to decomposition it emits toxic fumes of As.

MNU100 CAS:2598-70-1 ***HR: 3***
o-METHYL-β-PHENYLETHYLHYDRAZINE DIHYDROGEN SULFATE
mf: $C_9H_{14}N_2 \cdot H_2O_4S$ mw: 248.33

SYNS: HYDRAZINE, 1-(o-METHYLPHENETHYL)-, SULFATE (1:1) ◇ WL 31

TOXICITY DATA with REFERENCE
scu-mus TDLo:300 mg/kg (female 1-6D post):REP JOENAK 30,205,64
scu-mus LD50:217 mg/kg JOENAK 30,205,64

SAFETY PROFILE: Poison by subcutaneous route. Experimental reproductive effects. When heated to decomposition it emits toxic fumes of NO_x and SO_x.

MNU150 CAS:156-48-9 ***HR: 3***
p-METHYL-β-PHENYLETHYLHYDRAZINE DIHYDROGEN SULFATE
mf: $C_9H_{14}N_2{\cdot}H_2O_4S$ mw: 248.33

SYNS: HYDRAZINE, 1-(p-METHYLPHENETHYL)-, SULFATE (1:1) ◇ HYDRAZINE, (2-(4-METHYLPHENYL)ETHYL)-, SULFATE (1:1) (9CI) ◇ WL 32

TOXICITY DATA with REFERENCE
scu-mus TDLo:240 mg/kg (female 1-6D post):REP JOENAK 30,205,64
scu-mus LD50:182 mg/kg JOENAK 30,205,64

SAFETY PROFILE: Poison by subcutaneous route. Experimental reproductive effects. When heated to decomposition it emits toxic fumes of NO_x and SO_x.

MNU250 CAS:13256-11-6 ***HR: 3***
METHYL-PHENYLETHYL-NITROSAMINE
mf: $C_9H_{12}N_2O$ mw: 164.23

SYNS: N-METHYL-N-NITROSOPHENETHYLAMINE ◇ METHYL(2-PHENYLAETHYL)NITROSAMIN (GERMAN) ◇ N-NITROSO-N-METHYL-2-PHENYLETHYLAMINE

TOXICITY DATA with REFERENCE
mma-sat 10 μg/plate MUREAV 66,1,79
orl-rat TDLo:5200 μg/kg/33W-I:CAR FCTOD7 20,393,82
orl-rat TD:165 mg/kg/33W-I:ETA XENOBH 3,271,73
orl-rat TD:14400 μg/kg/33W-I:CAR FCTOD7 20,393,82
orl-rat LD50:48 mg/kg ZEKBAI 69,103,67

SAFETY PROFILE: Poison by ingestion. Questionable carcinogen with experimental carcinogenic and tumorigenic data. Mutation data reported. When heated to decomposition it emits toxic fumes of NO_x. See also NITROSAMINES.

MNU500 CAS:637-60-5 ***HR: 3***
4-METHYLPHENYLHYDRAZINE HYDROCHLORIDE
mf: $C_7H_{10}N_2{\cdot}ClH$ mw: 158.65

TOXICITY DATA with REFERENCE
mmo-sat 800 μg/plate NEZAAQ 33,474,78
mma-sat 800 μg/plate NEZAAQ 33,474,78
orl-mus TDLo:1750 mg/kg/7W-I:NEO ZKKOBW 89,245,77
scu-mus TDLo:3640 mg/kg/26W-I:CAR ZKKOBW 89,245,77

CONSENSUS REPORTS: Reported in EPA TSCA Inventory.

SAFETY PROFILE: Questionable carcinogen with experimental carcinogenic and neoplastigenic data. Mutation data reported. When heated to decomposition it emits very toxic fumes of Cl^- and NO_x.

MNU750 CAS:74764-93-5 ***HR: 2***
N′-(5-METHYL-3-PHENYL-1-INDOLYL)-N,N,N′-TRIMETHYLETHYLENEDIAMINE HYDROCHLORIDE
mf: $C_{20}H_{25}N_3{\cdot}ClH$ mw: 343.65

SYN: N,N,N′-TRIMETHYL-N′-(5-METHYL-3-PHENYL-1H-INDOL-1-YL)-1,2-ETHANEDIAMINE HYDROCHLORIDE

TOXICITY DATA with REFERENCE
orl-rat LD50:2000 mg/kg ARZNAD 30,919,80
orl-mus LD50:1750 mg/kg ARZNAD 30,919,80

SAFETY PROFILE: Moderately toxic by ingestion. When heated to decomposition it emits very toxic fumes of NO_x and HCl.

MNV000 CAS:1136-45-4 ***HR: 3***
5-METHYL-3-PHENYLISOXAZOLE-4-CARBOXYLIC ACID
mf: $C_{11}H_9NO_3$ mw: 203.21

SYN: 3-PHENYL-5-METHYLISOXAZOL-4-CARBONSAEURE(GERMAN)

TOXICITY DATA with REFERENCE
ivn-mus LD50:300 mg/kg ARZNAD 15,322,65

CONSENSUS REPORTS: Reported in EPA TSCA Inventory.

SAFETY PROFILE: Poison by intravenous route. When heated to decomposition it emits toxic fumes of NO_x.

MNV250 CAS:1173-88-2 ***HR: 3***
5-METHYL-3-PHENYL-4-ISOXAZOLYL PENICILLIN, SODIUM
mf: $C_{19}H_{18}N_3O_5S{\cdot}Na$ mw: 423.45

SYNS: BACTOCILL ◇ BRISTOPHEN ◇ BRL 1400 ◇ CRYPTOCILLIN ◇ MICROPENIN ◇ OXABEL ◇ OXACILLIN SODIUM SALT ◇ P 12 ◇ PENICILLIN P-12 ◇ PENSTAPHOCID ◇ PROSTAPHILIN ◇ RESISTOPHEN ◇ SODIUM OXACILLIN ◇ SQ 16423 ◇ STAPENOR ◇ STAPHCILLIN V ◇ V-CILLIN K

TOXICITY DATA with REFERENCE
orl-wmn TDLo:4560 mg/kg/11W-I:LIV,MSK RMMJAK 62,34,65
ivn-hmn TDLo:1200 mg/kg/7D:LIV AIMDAP 138,915,78
scu-rat LD50:3900 μg/kg NIIRDN 6,147,82
ivn-rat LD50:2150 μg/kg NIIRDN 6,147,82
scu-mus LD50:2600 μg/kg NIIRDN 6,147,82
ivn-mus LD50:1399 mg/kg FATOAO 31,232,68

SAFETY PROFILE: Poison by subcutaneous and intravenous routes. Human systemic effects by ingestion and intravenous routes: blood effects, liver function impaired and other liver changes. When heated to decomposition it emits very toxic fumes of NO_x, Na_2O, and SO_x.

MNV500 CAS:58139-35-8 ***HR: D***
3-METHYLPHENYL-N-METHYL-N-NITROSOCARBAMATE
mf: $C_9H_{10}N_2O_3$ mw: 194.21

SYNS: METHYLNITROSOCARBAMIC ACID-3-METHYLPHENYL ESTER ◇ 3-METHYLPHENYL-N-NITROSO-N-METHYLCARBAMATE ◇ NITROSO-MTMC

TOXICITY DATA with REFERENCE
mmo-esc 1 μg/plate MUREAV 54,283,78
cyt-ham:lng 5800 μg/L GMCRDC 27,95,81

CONSENSUS REPORTS: EPA Genetic Toxicology Program.

SAFETY PROFILE: Mutation data reported. Many N-nitroso compounds are carcinogens. When heated to decomposition it emits toxic fumes of NO_x. See also N-NITROSO COMPOUNDS and CARBAMATES.

MNV750 CAS:1707-14-8 ***HR: 3***
3-METHYL-2-PHENYLMORPHOLINE HYDROCHLORIDE
mf: $C_{11}H_{15}NO•ClH$ mw: 213.73

SYNS: A 66 HYDROCHLORIDE ◇ MARSIN ◇ 3-METHYL-2-PHENYLTETRAHYDRO-2H-1,4-OXAZINE HYDROCHLORIDE ◇ NEOZINE ◇ PHENMETRAZINE HYDROCHLORIDE ◇ 2-PHENYL-3-METHYLTETRAHYDRO-1,4-OXAZINE HYDROCHLORIDE ◇ PRELUDIN HYDROCHLORIDE ◇ PROBESE-P HYDROCHLORIDE ◇ PSYCHAMINE A 66 HYDROCHLORIDE ◇ USAF GE-1

TOXICITY DATA with REFERENCE
orl-wmn TDLo:84 mg/kg (28-84D preg):REP BMJOAE 2,1327,62
orl-wmn TDLo:84 mg/kg (28-84D preg):TER BMJOAE 2,1327,62
orl-rat LD50:165 mg/kg ARZNAD 13,711,63
ipr-rat LD50:175 mg/kg TXAPA9 2,589,60
scu-rat LD50:350 mg/kg TXAPA9 2,589,60
orl-mus LD50:165 mg/kg ARZNAD 13,711,63
ipr-mus LD50:50 mg/kg NTIS** AD277-689
scu-mus LD50:240 mg/kg APSXAS 4,37,67
ivn-mus LD50:71 mg/kg TXAPA9 2,589,60
ivn-rbt LD50:41 mg/kg 27ZQAG -,286,72

SAFETY PROFILE: Poison by ingestion, intravenous, intraperitoneal, and subcutaneous routes. Human reproductive effects. Human teratogenic effects by ingestion: developmental abnormalities of the respiratory and gastrointestinal systems, and effects on newborn including neonatal measures or effects. When heated to decomposition it emits very toxic fumes of NO_x and HCl.

MNW100 CAS:52670-78-7 ***HR: 3***
METHYLPHENYLPHOSPHORAMIDIC ACID DIETHYL ESTER
mf: $C_{11}H_{18}NO_3P$ mw: 243.27

TOXICITY DATA with REFERENCE
orl-rat LD50: 41 mg/kg JACTDZ 3(2),162,84
ihl-rat LCLo: 11600 mg/m^3/15M JACTDZ 3(2),162,84
skn-rbt LD50: 840 mg/kg JACTDZ 3(2),162,84
ocu-rbt LDLo: 50 mg/kg JACTDZ 3(2),162,84

SAFETY PROFILE: Poison by ingestion and ocular routes. Moderately toxic by skin contact. Mildly toxic by inhalation. When heated to decomposition it emits toxic fumes of NO_x and PO_x. See also ESTERS.

MNW150 CAS:3074-43-9 ***HR: 3***
1-METHYL-4-PHENYLPIPERAZINE
mf: $C_{11}H_{16}N_2$ mw: 176.29

SYN: A 1390

TOXICITY DATA with REFERENCE
orl-mus LD50:420 mg/kg JPETAB 110,157,54
ipr-mus LD50:140 mg/kg JPETAB 110,157,54
ivn-mus LD50:36 mg/kg JPETAB 110,157,54
ipr-dog LDLo:125 mg/kg JPETAB 110,157,54

SAFETY PROFILE: Poison by intravenous and intraperitoneal routes. Moderately toxic by ingestion. When heated to decomposition it emits toxic fumes of NO_x.

MNW775 ***HR: 3***
1-METHYL-4-PHENYL-4-PROPIONOXYPIPERIDINE HYDROCHLORIDE
mf: $C_{15}H_{21}NO_2•ClH$ mw: 283.83

SYN: 1-METHYL-4-PHENYL-4-PIPERIDINOL PROPIONATE (ESTER) HYDROCHLORIDE

TOXICITY DATA with REFERENCE
orl-mus LD50:78 mg/kg AITEAT 15,290,67
ipr-mus LD50:46 mg/kg AITEAT 15,290,67
scu-mus LD50:50 mg/kg AITEAT 15,290,67
ivn-mus LD50:22 mg/kg AITEAT 15,290,67

SAFETY PROFILE: Poison by ingestion, subcutaneous, intravenous, and intraperitoneal routes. When heated to decomposition it emits toxic fumes of NO_x and HCl.

MNW780 CAS:4956-14-3 ***HR: 3***
1-METHYL-4-PHENYL-4-PROPIONOXYPIPERIDINE N-OXIDE HYDROCHLORIDE
mf: $C_{15}H_{21}NO_3•ClH$ mw: 299.83

SYN: 1-METHYL-4-PHENYL-4-PIPERIDINOL PROPIONATE (ester) 1-OXIDE HYDROCHLORIDE

TOXICITY DATA with REFERENCE
orl-mus LD50:625 mg/kg AITEAT 15,290,67
ipr-mus LD50:550 mg/kg AITEAT 15,290,67
scu-mus LD50:600 mg/kg AITEAT 15,290,67
ivn-mus LD50:350 mg/kg AITEAT 15,290,67

SAFETY PROFILE: Poison by intravenous route. Moderately toxic by ingestion and intraperitoneal routes. When heated to decomposition it emits toxic fumes of NO_x and HCl.

MNX000 CAS:1944-83-8 ***HR: 1***
2-METHYL-1-PHENYL-2-PROPYL HYDROPEROXIDE
mf: $C_{10}H_{14}O_2$ mw: 164.18

SYN: 2-PHENYL-1,1-DIMETHYLETHYLHYDROPEROXIDE

SAFETY PROFILE: An irritant to the eyes, skin and mucous membranes. A powerful oxidant. Preparative hazard. When heated to decomposition it emits acrid smoke and irritating fumes. See also PEROXIDES.

MNX250 ***HR: 3***
N-METHYL-N-(1-PHENYL-2-PROPYL)-2-(PYRROLIDINYL)ACETAMIDE HYDROCHLORIDE
mf: $C_{16}H_{24}N_2O{\bullet}ClH$ mw: 296.88

SYN: C 2094

TOXICITY DATA with REFERENCE
ipr-rat LD50:220 mg/kg ARZNAD 9,113,59
scu-mus LD50:562 mg/kg ARZNAD 9,113,59

SAFETY PROFILE: Poison by intraperitoneal route. Moderately toxic by subcutaneous route. When heated to decomposition it emits very toxic fumes of HCl and NO_x.

MNX260 ***HR: 3***
N-METHYL-N-(3-PHENYLPROPYL)-2-(PYRROLIDINYL)ACETAMIDE HYDROCHLORIDE
mf: $C_{16}H_{24}N_2O{\bullet}ClH$ mw: 296.88

SYN: C 661

TOXICITY DATA with REFERENCE
eye-rbt 2% MLD ARZNAD 9,113,59
ipr-rat LD50:115 mg/kg ARZNAD 9,113,59
scu-mus LD50:510 mg/kg ARZNAD 9,113,59

SAFETY PROFILE: Poison by intraperitoneal route. Moderately toxic by subcutaneous route. An eye irritant. When heated to decomposition it emits very toxic fumes of HCl and NO_x.

MNX850 CAS:20921-41-9 ***HR: 2***
1-METHYL-1-PHENYL-2-PROPYNYL CYCLOHEXANECARBAMATE
mf: $C_{17}H_{21}NO_2$ mw: 271.39

SYNS: CYCLOHEXANECARBAMIC ACID, 1-METHYL-1-PHENYL-2-PROPYNYL ESTER ◇ 1-METHYL-1-PHENYL-2-PROPYNYL-N-CYCLOHEXYLCARBAMATE

TOXICITY DATA with REFERENCE
orl-rat TDLo:38 g/kg/96W-C:ETA JJIND8 71,211,83

SAFETY PROFILE: Questionable carcinogen with experimental tumorigenic data. When heated to decomposition it emits toxic fumes of NO_x.

MNY750 CAS:57962-60-4 ***HR: 3***
5-METHYL-1-PHENYL-2-(PYRROLIDINYL)IMIDAZOLE
mf: $C_{14}H_{17}N_3$ mw: 227.34

SYN: METHYL-5 PHENYL-1 (PYRROLIDINYL-1)-2 IMIDAZOLE (FRENCH)

TOXICITY DATA with REFERENCE
orl-mus LD50:92120 μg/kg EJMCA5 13,469,78
ipr-mus LD50:36690 μg/kg EJMCA5 13,469,78
ivn-mus LD50:6870 μg/kg EJMCA5 13,469,78

SAFETY PROFILE: Poison by ingestion, intravenous, and intraperitoneal routes. When heated to decomposition it emits toxic fumes of NO_x.

MNZ000 CAS:86-34-0 ***HR: 2***
N-METHYL-2-PHENYL-SUCCINIMIDE
mf: $C_{11}H_{11}NO_2$ mw: 189.23

SYNS: EPIMID ◇ FENOSUCCIMIDE ◇ LIFENE ◇ 1-METHYL-3-PHENYLPYRROLIDIN-2,5-DIONE ◇ 1-METHYL-3-PHENYL-2,5-PYRROLIDINEDIONE ◇ METHYLPHENYLSUCCINIMIDE ◇ N-METHYL-α-PHENYLSUCCINIMIDE ◇ MILONTIN ◇ MIROTIN ◇ PHENOSUCCIMIDE ◇ PHENYLSUXIMIDE ◇ PM 334 ◇ SUCCITIMAL

TOXICITY DATA with REFERENCE
cyt-hmn:leu 100 μg/L AJOGAH 116,867,73
unr-mus TDLo:1533 mg/kg (8-10D preg):TER FEPRA7 38,438,79
orl-mus LD50:700 mg/kg ARZNAD 23,377,73
ipr-mus LD50:402 mg/kg EJMCA5 13,465,78

SAFETY PROFILE: Moderately toxic by ingestion and intraperitoneal routes. An experimental teratogen. Human mutation data reported. An anticonvulsant. When heated to decomposition it emits toxic fumes of NO_x.

MOA000 CAS:52968-02-2 ***HR: D***
1-METHYL-3-(4-PHENYL-2-THIAZOLYL)UREA
mf: $C_{11}H_{11}N_3OS$ mw: 233.31

SYNS: N-METHYL-N'-(4'-PHENYL-THIAZOLYL(2'))-HARNSTOFF (GERMAN) ◇ N-METHYL-N'-(4'-PHENYL-THIAZOLYL(2'))-UREA ◇ N-METHYL-N'-(4-PHENYL-2-THIAZOLYL)UREA

TOXICITY DATA with REFERENCE
cyt-rat-ipr 200 mg/kg/24H ARZNAD 25,1716,75
hma-mus/srm 30 mg/kg ARZNAD 25,1716,75

CONSENSUS REPORTS: EPA Genetic Toxicology Program.

SAFETY PROFILE: Mutation data reported. When

heated to decomposition it emits very toxic fumes of NO_x and SO_x.

MOA250 CAS:73840-42-3 ***HR: 3***
1-METHYL-4-(PHENYLTHIO)PYRIDINIUM IODIDE
mf: $C_{12}H_{12}NS \cdot I$ mw: 329.21

TOXICITY DATA with REFERENCE
ipr-mus LD50:57 mg/kg PHARAT 33,120,78
ivn-mus LD50:56200 μg/kg CSLNX* NX#02330

SAFETY PROFILE: Poison by intravenous and intraperitoneal routes. When heated to decomposition it emits very toxic fumes of NO_x, SO_x, and I^-. See also IODIDES.

MOA500 CAS:2724-69-8 ***HR: 3***
N-METHYL-N'-PHENYL THIOUREA
mf: $C_8H_{10}N_2S$ mw: 166.26

PROP: A powder.

SYN: NITROSOMETHYL UREA

TOXICITY DATA with REFERENCE
dlt-mus-ipr 50 mg/kg MGGEAE 117,197,72
orl-rat LDLo:50 mg/kg JPETAB 93,287,48

CONSENSUS REPORTS: Reported in EPA TSCA Inventory.

SAFETY PROFILE: Poison by ingestion. Mutation data reported. When heated to decomposition it emits very toxic fumes of SO_x and NO_x.

MOA600 CAS:2156-27-6 ***HR: 2***
1-(1-METHYL-2-((α-PHENYL-o-TOLYL)OXY) ETHYL)PIPERIDINE
mf: $C_{21}H_{27}NO$ mw: 309.49

SYNS: 1-(2-(2-BENZILFENOSSI)-1-METILETIL)-PIPERIDINA ◇ BLASCORID ◇ PIPERIDINE, 1-(1-METHYL-2-(2-(PHENYLMETHYL) PHENOXY)ETHYL)-(9CI) ◇ PIPERIDINE, 1-(1-METHYL-2-((α-PHENYL-o-TOLYL)OXY)ETHYL)- ◇ PIREXYL

TOXICITY DATA with REFERENCE
orl-rat TDLo:6 g/kg (female 60D pre):REP BCFAAI 109,476,70
orl-mus LD50:1087 mg/kg BCFAAI 109,476,70

SAFETY PROFILE: Moderately toxic by ingestion. Experimental reproductive effects. When heated to decomposition it emits toxic fumes of NO_x.

MOA725 CAS:16033-21-9 ***HR: 3***
3-METHYL-1-PHENYLTRIAZENE
mf: $C_7H_9N_3$ mw: 135.19

SYNS: 1-PHENYL-3-METHYLTRIAZINE ◇ 1-PHENYL-3-MONOMETHYLTRIAZENE ◇ PMT

TOXICITY DATA with REFERENCE
mmo-sat 100 μmol/L CNREA8 42,1446,82
mma-sat 1 mmol/L CNREA8 42,1446,82
mmo-nsc 600 μmol/L MUREAV 13,276,71
dnd-ofs-sal:spr 250 g/L BCPCA6 19,1505,70
skn-mus TDLo:284 mg/kg/8W-I:NEO CNREA8 34,1671,74
orl-rat LD50:420 mg/kg NEOLA4 25,153,78
scu-mus LDLo:45 mg/kg CNREA8 34,1671,74

SAFETY PROFILE: Poison by ingestion and subcutaneous routes. Questionable carcinogen with experimental neoplastigenic data. Mutation data reported. When heated to decomposition it emits toxic fumes of NO_x.

MOA750 CAS:31185-58-7 ***HR: 3***
4-METHYL-5-PHENYL-2-TRIFLUOROMETHOXAZOLIDINE
mf: $C_{11}H_{12}F_3NO$ mw: 231.24

SYN: 2-(METHOXY(METHYLTHIO)PHOSPHINYLIMINO)-3-ETHYL-5-METHYL-1,3-OXAZOLIDINE

TOXICITY DATA with REFERENCE
orl-rat LDLo:120 μg/kg RREVAH 53,19,74
orl-mus LD50:800 mg/kg JMCMAR 13,1212,70
ipr-mus LD50:600 mg/kg JMCMAR 13,1212,70

SAFETY PROFILE: A deadly poison by ingestion. Moderately toxic by intraperitoneal route. When heated to decomposition it emits very toxic fumes of F^- and NO_x. See also FLUORIDES.

MOB000 CAS:593-54-4 ***HR: 3***
METHYL PHOSPHINE
mf: CH_5P mw: 48.02

PROP: Colorless gas. Bp: −14°.

SAFETY PROFILE: A poison. Ignites spontaneously in air. Dangerous fire hazard when exposed to heat or flame. Incompatible with primary lower alkylphosphones. Can react vigorously with oxidizing materials. When heated to decomposition it emits toxic fumes of PO_x and phosphine. See also PHOSPHINE.

MOB250 CAS:18466-11-0 ***HR: 3***
METHYLPHOSPHODITHIOIC ACID-S-(((p-CHLOROPHENYL)THIO)METHYL)-O-METHYL ESTER
mf: $C_9H_{12}ClOPS_2$ mw: 266.75

SYNS: ENT 27,180 ◇ N 4548 ◇ STAUFFER N-4548

TOXICITY DATA with REFERENCE
orl-rat LD50:31 mg/kg ARSIM* 20,22,66
orl-gpg LDLo:50 mg/kg JEENAI 61(5),1261,68
scu-gpg LDLo:100 mg/kg JEENAI 61(5),1261,68

SAFETY PROFILE: Poison by ingestion and subcutaneous routes. When heated to decomposition it emits very toxic fumes of Cl^-, PO_x, and SO_x.

MOB275 CAS:71840-26-1 ***HR: 3***
METHYLPHOSPHONIC ACID, (2-(BIS(1-METHYLETHYL)AMINO)ETHYL) ETHYL ESTER
mf: $C_{11}H_{26}NO_3P$ mw: 251.35

TOXICITY DATA with REFERENCE
skn-rbt 1 g/15M MOD IAEC** 17JUN74
eye-rbt 200 mg/15M MLD IAEC** 17JUN74
ivn-mus LD50:204 mg/kg IAEC** 17JUN74
ivn-rbt LD50:164 mg/kg IAEC** 17JUN74

SAFETY PROFILE: Poison by intravenous route. An eye and skin irritant. When heated to decomposition it emits toxic fumes of PO_x and NO_x. See also ESTERS.

MOB399 CAS:676-97-1 ***HR: 3***
METHYL PHOSPHONIC DICHLORIDE
DOT: NA 9206
mf: CH_3Cl_2OP mw: 132.91

TOXICITY DATA with REFERENCE
ihl-rat LC50:26 ppm/4H AIHAAP 25,470,64

CONSENSUS REPORTS: EPA Extremely Hazardous Substances List. Reported in EPA TSCA Inventory.

DOT Classification: Corrosive Material; Label: Corrosive and Poison.

SAFETY PROFILE: Poison by inhalation. A corrosive irritant to the eyes, skin, and mucous membranes. When heated to decomposition it emits toxic fumes of Cl^- and PO_x. See also CHLORIDES.

MOB500 CAS:18278-44-9 ***HR: 3***
METHYLPHOSPHONODITHIOIC ACID O-METHYL ESTER, S-ESTER with 2-MERCAPTO-N-METHYLACETAMIDE
mf: $C_5H_{12}NO_2PS_3$ mw: 245.33

SYNS: ENT 25,977 ◇ MONSANTO CP-19203

TOXICITY DATA with REFERENCE
orl-rat LD50:10 mg/kg ARSIM* 20,15,66
orl-gpg LDLo:10 mg/kg JEENAI 60(3),733,67
scu-gpg LDLo:25 mg/kg JEENAI 60(3),733,67

SAFETY PROFILE: Poison by ingestion and subcutaneous routes. When heated to decomposition it emits very toxic fumes of NO_x, PO_x, and SO_x.

MOB599 CAS:2703-13-1 ***HR: 3***
METHYLPHOSPHONOTHIOIC ACID-O-ETHYL O-(p-(METHYLTHIO)PHENYL)ESTER
mf: $C_{10}H_{15}O_2PS_2$ mw: 262.34

SYNS: BAYER 29952 ◇ ENT 25,612 ◇ METHYLPHOSPHONOTHIOIC ACID-O-ETHYL O-(4-(METHYLTHIO)PHENYL)ESTER (9CI)

TOXICITY DATA with REFERENCE
orl-rat LDLo:1 mg/kg ARSIM* 20,4,66

CONSENSUS REPORTS: EPA Extremely Hazardous Substances List.

SAFETY PROFILE: Deadly poison by ingestion. When heated to decomposition it emits toxic fumes of PO_x and SO_x. See also ESTERS.

MOB699 CAS:2665-30-7 ***HR: 3***
METHYLPHOSPHONOTHIOIC ACID-O-(4-NITROPHENYL)-O-PHENYL ESTER
mf: $C_{13}H_{12}NO_4PS$ mw: 309.29

SYNS: COLEP ◇ CP 40294 ◇ ENT 25,787 ◇ METHYLPHOSPHONOTHIOIC ACID-O-(p-NITROPHENYL)-O-PHENYL ESTER ◇ MONSANTO CP-40294 ◇ O-(4-NITROPHENYL) O-PHENYLMETHYL PHOSPHONOTHIOATE

TOXICITY DATA with REFERENCE
orl-rat LD50:8 mg/kg ARSIM* 20,15,66

CONSENSUS REPORTS: EPA Extremely Hazardous Substances List.

SAFETY PROFILE: Poison by ingestion. When heated to decomposition it emits toxic fumes of PO_x, SO_x, and NO_x. See also ESTERS.

MOB750 CAS:5954-90-5 ***HR: 3***
METHYLPHOSPHONOTHIOIC ACID-O-PHENYL ESTER, O-ESTER with p-HYDROXYBENZONITRILE
mf: $C_{14}H_{12}NO_2PS$ mw: 289.30

SYNS: CP-40507 ◇ ENT 25,870 ◇ MONSANTO CP-40507

TOXICITY DATA with REFERENCE
orl-rat LD50:79 mg/kg ARSIM* 20,15,66
orl-gpg LDLo:1 mg/kg JEENAI 61(5),1261,68
scu-gpg LDLo:5 mg/kg JEENAI 61(5),1261,68

CONSENSUS REPORTS: Cyanide and its compounds are on The Community Right-To-Know List.

SAFETY PROFILE: Poison by ingestion and subcutaneous routes. When heated to decomposition it emits very toxic fumes of SO_x, PO_x, NO_x, and CN^-. See also NITRILES.

MOC000 CAS:676-98-2 ***HR: 2***
METHYL PHOSPHONOTHIOIC DICHLORIDE anhydrous
DOT: NA 1760
mf: CH_3Cl_2PS mw: 148.97

DOT Classification: Corrosive Material; Label: Corrosive.

SAFETY PROFILE: A corrosive irritant to skin, eyes and mucous membranes. When heated to decomposition it emits very toxic fumes of Cl^-, PO_x and SO_x. See also PHOSPHATES, CHLORIDES, and SULFIDES.

MOC250 CAS:676-83-5 ***HR: 3***
METHYLPHOSPHONOUS DICHLORIDE
DOT: UN 2845

CONSENSUS REPORTS: Reported in EPA TSCA Inventory.

DOT Classification: Flammable Liquid; Label: Flammable Liquid.

SAFETY PROFILE: A poison. A corrosive irritant to the skin, eyes, and mucous membranes. Flammable when exposed to heat or flame; can react vigorously with oxidizing materials. When heated to decomposition it emits very toxic fumes of Cl^- and PO_x. See also HYDROCHLORIC ACID.

MOC500 CAS:19143-28-3 ***HR: D***
METHYL-4-PHTHALIMIDO-dl-GLUTARAMATE
mf: $C_{14}H_{14}N_2O_5$ mw: 290.30

TOXICITY DATA with REFERENCE
orl-rbt TDLo:900 mg/kg (6-11D preg):TER TJADAB 2,265,69

SAFETY PROFILE: An experimental teratogen. When heated to decomposition it emits toxic fumes of NO_x.

MOC750 CAS:42472-93-5 ***HR: 2***
N-METHYL-2-PHTHALIMIDOGLUTARIMIDE
mf: $C_{14}H_{12}N_2O_4$ mw: 272.28

TOXICITY DATA with REFERENCE
orl-rbt TDLo:650 mg/kg (4-16D preg):TER LIFSAK 3,721,64
orl-mus LD50:670 mg/kg LIFSAK 3,721,64

SAFETY PROFILE: Moderately toxic by ingestion. An experimental teratogen. When heated to decomposition it emits toxic fumes of NO_x.

MOD000 CAS:85-71-2 ***HR: 2***
METHYL PHTHALYL ETHYL GLYCOLATE
mf: $C_{13}H_{14}O_6$ mw: 266.27

PROP: Liquid. Bp: 310°, flash p: 380°F (CC), d: 1.220, vap d: 9.16.

SYNS: ETHYLo-(o-(METHOXYCARBONYL)BENZOYL)GLYCOLATE ◇ ETHYL o-(METHOXYCARBONYL)BENZOYLOXYACETATE ◇ GLYCOLIC ACID, ETHYL ESTER, METHYL PHTHALATE

TOXICITY DATA with REFERENCE
eye-rbt 500 mg AJOPAA 29,1363,46
orl-rat LD50:3200 mg/kg 14CYAT 2,1906,63

SAFETY PROFILE: Moderately toxic by ingestion. An eye irritant. Combustible when exposed to heat or flame; can react with oxidizing materials. To fight fire, use CO_2, dry chemical. When heated to decomposition it emits acrid smoke and irritating fumes. See also ESTERS.

MOD250 CAS:109-01-3 ***HR: 3***
N-METHYLPIPERAZINE
mf: $C_5H_{12}N_2$ mw: 100.19

PROP: A hygroscopic solid; typical amine-like odor. D: 0.9031 20°/20°, mp: 65.5°, bp: 139°, flash p: 108°F (OC), vap d: 3.5.

SYN: 1-METHYLPIPERAZINE

TOXICITY DATA with REFERENCE
skn-rbt 100 μg/24H open AIHAAP 23,95,62
eye-rbt 100 mg SEV 34ZIAG -,689,69
orl-rat LD50:2830 mg/kg AIHAAP 23,95,62
orl-mus LD50:1450 mg/kg TPKVAL 15,116,79
ihl-mus LC50:2740 mg/m³/2H TPKVAL 11,123,69
ipr-mus LD50:150 mg/kg NTIS** AD691-490
skn-rbt LD50:1490 mg/kg AIHAAP 23,95,62

CONSENSUS REPORTS: Reported in EPA TSCA Inventory.

SAFETY PROFILE: Poison by intraperitoneal route. Moderately toxic by inhalation, ingestion, and skin contact. A skin and severe eye irritant. Flammable when exposed to heat or flame; can react with oxidizing materials. To fight fire, use alcohol foam, CO_2, dry chemical. When heated to decomposition it emits toxic fumes of NO_x.

MOD500 CAS:23491-45-4 ***HR: 3***
p-(5-(5-(4-METHYL-1-PIPERAZINYL)-2-BENZIMIDAZOLYL)-2-BENZIMIDAZOLYL)-PHENOL TRIHYDROCHLORIDE
mf: $C_{25}H_{24}N_6O{\bullet}3ClH$ mw: 533.93

SYNS: BISBENZIMIDAZOLE ◇ HOE 33258 ◇ 33258 HOECHST ◇ HOECHST DYE 33258 ◇ 4-(5-(4-METHYL-1-PIPERAZINYL)(2,5'-BI-1H-BENZIMIDAZOL)-2'-YL)-PHENOL TRIHYDROCHLORIDE ◇ NSC-322921

TOXICITY DATA with REFERENCE
mmo-sat 100 μmol/L AMACCQ 9,77,76
dnd-esc 10 mg/L MUREAV 89,95,81
sce-ham:lng 15 μmol/L HEREAY 96,295,82
sce-ham:ovr 10 μmol/L NATUAS 258,121,75
ivn-rat LD50:32200 μg/kg NTIS** PB84-171206
ivn-mus LD50:36900 μg/kg NTIS** PB84-171206

SAFETY PROFILE: Poison by intravenous route. Mutation data reported. When heated to decomposition it emits very toxic fumes of HCl and NO_x.

MOD750 CAS:74203-59-1 ***HR: 3***
4-(4-METHYL-1-PIPERAZINYLCARBONYL)-1-PHENYL-2-PYRROLIDINONE HYDROCHLORIDE
mf: $C_{16}H_{21}N_3O_2 \cdot ClH$ mw: 323.86

TOXICITY DATA with REFERENCE
ipr-mus LD50:380 mg/kg CHTPBA 7,398,72
ivn-mus LD50:300 mg/kg CHTPBA 7,398,72

SAFETY PROFILE: Poison by intraperitoneal and intravenous routes. When heated to decomposition it emits very toxic fumes of HCl and NO_x.

MOE250 CAS:60706-43-6 ***HR: 3***
10-(2-(4-METHYL-1-PIPERAZINYL)ETHYL)PHENOTHIAZINE
mf: $C_{19}H_{23}N_3S$ mw: 325.51

SYNS: N-METHYL-PIPERAZINYL-N'-AETHYL-PHENOTHIAZIN (GERMAN) ◇ P 527

TOXICITY DATA with REFERENCE
ipr-mus LD50:220 mg/kg ARZNAD 7,106,57
scu-mus LD50:640 mg/kg ARZNAD 7,106,57
ivn-mus LD50:135 mg/kg ARZNAD 7,106,57
ivn-rbt LD50:36 mg/kg ARZNAD 7,106,57

SAFETY PROFILE: Poison by intraperitoneal and intravenous routes. Moderately toxic by subcutaneous route. When heated to decomposition it emits very toxic fumes of NO_x and SO_x.

MOF750 CAS:37724-43-9 ***HR: 3***
4-(4-METHYL-1-PIPERAZINYL)-5,6,7,8-TETRAHYDRO-(1)-BENZOTHIENO(2,3-d) PYRIMIDINE HYDROCHLORIDE
mf: $C_{15}H_{20}N_4S \cdot ClH$ mw: 324.91

SYN: QM-1143

TOXICITY DATA with REFERENCE
orl-mus LD50:160 mg/kg CHTPBA 7,224,72
ipr-mus LD50:67 mg/kg CHTPBA 7,224,72
ivn-mus LD50:29 mg/kg CHTPBA 7,224,72

SAFETY PROFILE: Poison by ingestion, intraperitoneal and intravenous routes. When heated to decomposition it emits very toxic fumes of HCl, SO_x, and NO_x.

MOG000 CAS:37724-47-3 ***HR: 3***
4-(4-METHYL-1-PIPERAZINYL)THIENO(2,3-d) PYRIMIDINE HYDROCHLORIDE
mf: $C_{11}H_{14}N_4S \cdot ClH$ mw: 270.81

SYN: QM-1149

TOXICITY DATA with REFERENCE
ipr-mus LD50:58 mg/kg CHTPBA 7,224,72
ivn-mus LD50:24 mg/kg CHTPBA 7,224,72

SAFETY PROFILE: Poison by intraperitoneal and intravenous routes. When heated to decomposition it emits very toxic fumes of SO_x, NO_x, and HCl.

MOG250 CAS:70301-64-3 ***HR: 3***
2-(4-METHYL-1-PIPERAZINYL)-11-(p-TOLYL)-10,11-DIHYDROPYRIDAZINO(3,4-b)(1,4) BENZOXAZEPINE
mf: $C_{28}H_{25}N_5O$ mw: 447.58

TOXICITY DATA with REFERENCE
scu-rat LD50:255 mg/kg PCJOAU 13,256,79
scu-mus LD50:255 mg/kg PCJOAU 13,256,79

SAFETY PROFILE: Poison by subcutaneous route. When heated to decomposition it emits toxic fumes of NO_x.

MOG500 CAS:626-67-5 ***HR: 3***
N-METHYLPIPERIDINE
DOT: UN 2399
mf: $C_6H_{13}N$ mw: 99.20

PROP: Liquid. D: 0.821 @ 15°, bp: 107°, flash p: <73.4°F.

SYN: 1-METHYLPIPERIDINE (DOT)

TOXICITY DATA with REFERENCE
scu-rbt LDLo:400 mg/kg BDCGAS 34,2408,01

CONSENSUS REPORTS: Reported in EPA TSCA Inventory.

DOT Classification: Flammable Liquid; Label: Flammable Liquid.

SAFETY PROFILE: Poison by subcutaneous route. A very dangerous fire hazard when exposed to heat or flame. When heated to decomposition it emits toxic fumes of NO_x.

MOG750 CAS:109-05-7 ***HR: 3***
2-METHYLPIPERIDINE
mf: $C_6H_{13}N$ mw: 99.20

PROP: Liquid. D: 0.862 @ 0°, bp: 118-119° @ 753 mm, flash p: 50°F. Sol in water; insol in aqueous KOH.

SYN: α-PIPECOLIN

TOXICITY DATA with REFERENCE
scu-rbt LDLo:300 mg/kg BDCGAS 34,2408,01

CONSENSUS REPORTS: Reported in EPA TSCA Inventory.

SAFETY PROFILE: Poison by subcutaneous route. A very dangerous fire hazard when exposed to heat or flame. When heated to decomposition it emits toxic fumes of NO_x.

MOH000 CAS:626-56-2 ***HR: 2***
3-METHYLPIPERIDINE
mf: $C_6H_{13}N$ mw: 99.18

PROP: Flash p: 36.4°F.

SAFETY PROFILE: A very dangerous fire hazard when exposed to heat or flame. When heated to decomposition it emits toxic fumes of NO_x. See also other methylpiperidine entries.

MOH250 CAS:626-58-4 ***HR: 2***
4-METHYLPIPERIDINE
mf: $C_6H_{13}N$ mw: 99.18

PROP: Flash p: 48.2°F.

SAFETY PROFILE: A very dangerous fire hazard when exposed to heat or flame. When heated to decomposition it emits toxic fumes of NO_x. See also other methylpiperidine entries.

MOI250 CAS:78219-35-9 ***HR: 3***
β-4-METHYLPIPERIDINOETHYL BENZOATE HYDROCHLORIDE
mf: $C_{15}H_{21}NO_2•ClH$ mw: 283.83

SYN: BENZOIC ACID-2-(4-METHYLPIPERIDINO)ETHYL ESTER HYDROCHLORIDE

TOXICITY DATA with REFERENCE
scu-mus LDLo:1400 mg/kg JACSAT 52,1633,30
ivn-mus LDLo:43 mg/kg JACSAT 52,1633,30

SAFETY PROFILE: Poison by intravenous route. Moderately toxic by subcutaneous route. When heated to decomposition it emits very toxic fumes of NO_x and HCl.

MOI500 CAS:65210-33-5 ***HR: 3***
2-METHYL-2-(2-PIPERIDINOETHYL)-1,3-BENZODIOXOLE HYDROCHLORIDE
mf: $C_{15}H_{21}NO_2•ClH$ mw: 283.83

TOXICITY DATA with REFERENCE
ivn-rat LD50:40 mg/kg EJMCA5 12,413,77
ipr-mus LD50:150 mg/kg EJMCA5 12,413,77

SAFETY PROFILE: Poison by intravenous and intraperitoneal routes. When heated to decomposition it emits very toxic fumes of HCl and NO_x.

MOJ500 CAS:38589-14-9 ***HR: 3***
β-METHYL-4-PIPERIDINOPHENETHYLAMINE DIHYDROCHLORIDE
mf: $C_{14}H_{22}N_2•2ClH$ mw: 291.30

SYN: 2-METHYL-2-(4-PIPERIDINOPHENYL)ETHYLAMINEDIHYDROCHLORIDE

TOXICITY DATA with REFERENCE
orl-mus LD50:600 mg/kg JMCMAR 22,1460,79
unr-dog LDLo:150 mg/kg JMCMAR 22,1460,79

SAFETY PROFILE: Poison by an unspecified route. Moderately toxic by ingestion. When heated to decomposition it emits very toxic fumes of NO_x and HCl.

MOK000 CAS:73790-27-9 ***HR: 3***
METHYL-4-(3-PIPERIDINOPROPIONYLAMINO) SALICYLATE, METHIODIDE
mf: $C_{17}H_{25}N_2O_4•I$ mw: 448.34

SYN: 4-(3-PIPERIDINOPROPIONAMIDO)SALICYCLIC ACID METHYL ESTER, METHIODIDE

TOXICITY DATA with REFERENCE
ipr-mus LD50:54 mg/kg JMCMAR 10,235,67
ivn-mus LD50:18 mg/kg JMCMAR 10,235,67

SAFETY PROFILE: Poison by intraperitoneal and intravenous routes. When heated to decomposition it emits very toxic fumes of NO_x and I^-.

MOK500 CAS:69766-22-9 ***HR: 3***
Γ-3-METHYLPIPERIDINOPROPYL-p-AMINOBENZOATE HYDROCHLORIDE
mf: $C_{16}H_{24}N_2O_2•ClH$ mw: 312.88

SYN: p-AMINOBENZOIC ACID-3-(3-METHYLPIPERIDINO) PROPYL ESTER HYDROCHLORIDE

TOXICITY DATA with REFERENCE
ivn-rat LDLo:30 mg/kg JACSAT 49,2835,27
scu-mus LDLo:250 mg/kg JACSAT 49,2835,27

SAFETY PROFILE: Poison by intravenous and subcutaneous routes. When heated to decomposition it emits very toxic fumes of HCl and NO_x.

MOL250 CAS:3478-94-2 ***HR: 2***
3-(2-METHYLPIPERIDINO)PROPYL-3,4-DICHLOROBENZOATE
mf: $C_{16}H_{21}Cl_2NO_2$ mw: 330.28

SYNS: Γ-(2-METHYLPIPERIDINO)PROPYL-3,4-DICHLOROBENZOATE ◇ PIPERALIN

TOXICITY DATA with REFERENCE
orl-rat LD50:2500 mg/kg FMCHA2 -,D245,80
orl-gpg LD50:1560 mg/kg PCOC** -,911,66

SAFETY PROFILE: Moderately toxic by ingestion. When heated to decomposition it emits very toxic fumes of Cl^- and NO_x. See also ESTERS.

MOM750 CAS:2856-75-9 ***HR: 3***
2-METHYL-3-PIPERIDINOPYRAZINE MONOSULFATE
mf: $C_{10}H_{15}N_3•H_2O_4S$ mw: 275.36

SYNS: 2-METHYL-3-PIPERIDINOPYRAZINE SULFATE ◇ MODALINE SULFATE ◇ W3207B

TOXICITY DATA with REFERENCE
orl-hmn TDLo:51 μg/kg:CNS JNDRAK 4,86,64
orl-rat LD50:730 mg/kg 27ZQAG -,265,72
ipr-rat LD50:190 mg/kg 27ZQAG -,265,72
orl-mus LD50:780 mg/kg 27ZQAG -,265,72
ipr-mus LD50:275 mg/kg 27ZQAG -,265,72
ivn-mus LD50:100 mg/kg CSLNX* NX#01635

SAFETY PROFILE: Poison by intravenous and intraperitoneal routes. Moderately toxic by ingestion. Human systemic effects by ingestion: changes in motor activity. When heated to decomposition it emits very toxic fumes of NO_x and SO_x.

MON000 CAS:78219-61-1 ***HR: 3***
1-METHYL-4-PIPERIDYL-p-AMINOBENZOATE HYDROCHLORIDE
mf: $C_{13}H_{18}N_2O_2 \cdot ClH$ mw: 270.79

SYN: p-AMINO-BENZOIC ACID-1-METHYL-4-PIPERIDYL ESTER, HYDROCHLORIDE

TOXICITY DATA with REFERENCE
ivn-rat LDLo:10 mg/kg JACSAT 51,922,29
scu-mus LDLo:15 mg/kg JACSAT 51,922,29

SAFETY PROFILE: Poison by intravenous and subcutaneous routes. When heated to decomposition it emits very toxic fumes of NO_x and HCl.

MON250 CAS:3321-80-0 ***HR: 3***
N-METHYL-3-PIPERIDYL BENZILATE
mf: $C_{20}H_{23}NO_3$ mw: 325.44

SYNS: BENZILIC ACID-1-METHYL-3-PIPERIDYL ESTER ◇ JB 336

TOXICITY DATA with REFERENCE
ipr-rat LD50:150 mg/kg IPPABX 4,179,68
ivn-rat LD50:22 mg/kg AIPTAK 120,186,59
ivn-mus LD50:40 mg/kg AIPTAK 120,186,59
ivn-gpg LD50:17 mg/kg AIPTAK 120,186,59

SAFETY PROFILE: Poison by intravenous and intraperitoneal routes. When heated to decomposition it emits toxic fumes of NO_x. See also ESTERS.

MON500 CAS:40378-58-3 ***HR: 3***
1-METHYL-4-PIPERIDYL BENZOATE HYDROCHLORIDE
mf: $C_{13}H_{17}NO_2 \cdot ClH$ mw: 255.77

SYN: BENZOIC ACID-1-METHYL-4-PIPERIDYL ESTER HYDROCHLORIDE

TOXICITY DATA with REFERENCE
ivn-rat LDLo:18 mg/kg JACSAT 51,922,29
scu-mus LDLo:125 mg/kg JACSAT 51,922,29

SAFETY PROFILE: Poison by intravenous and subcutaneous routes. When heated to decomposition it emits very toxic fumes of NO_x and HCl.

MON750 CAS:5588-33-0 ***HR: 3***
10-(2-(1-METHYL-2-PIPERIDYL)ETHYL)-2-METHYLSULFINYL PHENOTHIAZINE
mf: $C_{21}H_{26}N_2OS_2$ mw: 386.61

SYNS: MESORIDAZINE ◇ THIORIDAZIEN THIOMETHYL SULFOXIDE

TOXICITY DATA with REFERENCE
orl-hmn LDLo:42 mg/kg ARGPAQ 34,955,77
orl-rat LD50:664 mg/kg 27ZQAG -,29,72
orl-mus LD50:560 mg/kg 27ZQAG -,29,72
ivn-mus LD50:26 mg/kg 27ZQAG -,29,72
orl-rbt LD50:7800 mg/kg 27ZQAG -,29,72

SAFETY PROFILE: Human poison by ingestion. Experimental poison by intravenous route. When heated to decomposition it emits very toxic fumes of SO_x and NO_x.

MOO250 CAS:50-52-2 ***HR: 3***
10-(2-(1-METHYL-2-PIPERIDYL)ETHYL)-2-(METHYLTHIO)PHENOTHIAZINE
mf: $C_{21}H_{26}N_2S_2$ mw: 370.61

SYNS: MALLOROL ◇ MELERIL ◇ MELLARIL ◇ MELLERETTE ◇ MELLERETTEN ◇ MELLERIL ◇ 2-METHYLMERCAPTO-10-(2-N-METHYL-2-PIPERIDYL)ETHYL)PHENOTHIAZINE ◇ SONAPAX ◇ THIORIDAZIN ◇ THIORIDAZINE ◇ TP-21

TOXICITY DATA with REFERENCE
sln-dmg-orl 100 mg/L IJEBA6 11,403,73
orl-man TDLo:10 mg/kg (7D male):REP AJPSAO 118,171,61
orl-cld TDLo:28 mg/kg/2W-I AJPSAO 143,1176,85
orl-wmn TDLo:14500 mg/kg/15Y-I:EYE CMAJAZ 132,737,85
orl-hmn LDLo:43 mg/kg ARGPAQ 34,955,77
orl-chd TDLo:25 mg/kg:CNS,CVS AJDCAI 130,507,76
orl-hmn TDLo:24 mg/kg:ANS AACRAT 52,938,73
orl-rat LD50:995 mg/kg TXAPA9 18,185,71
ipr-rat LD50:150 mg/kg PCJOAU 10,1001,76
scu-rat LD50:640 mg/kg MDCHAG 4,199,67
orl-mus LD50:385 mg/kg ARZNAD 15,841,65
ipr-mus LD50:65 mg/kg JMCMAR 13,23,70
scu-mus LD50:310 mg/kg YKKZAJ 90,800,70

CONSENSUS REPORTS: EPA Genetic Toxicology Program.

SAFETY PROFILE: Human poison by ingestion. Experimental poison by ingestion, subcutaneous, and intraperitoneal routes. Human systemic effects by ingestion: visual field and retinal changes, toxic psychosis, parasympatholytic, and heart rate change. Human reproductive effects. Experimental reproductive effects.

Mutation data reported. An antipsychotic and sedative. When heated to decomposition it emits very toxic fumes of NO_x and SO_x.

MOO500 CAS:130-61-0 ***HR: 3***
10-(2-(1-METHYL-2-PIPERIDYL)ETHYL)-2-METHYLTHIOPHENOTHIAZINE HYDROCHLORIDE
mf: $C_{21}H_{26}N_2S_2 \cdot ClH$ mw: 407.07

SYNS: MELLARIL HYDROCHLORIDE ◇ 2-METHYLMERCAPTO-10-(2-(N-METHYL-2-PIPERIDYL)ETHYLPHENOTHIAZINEHYDROCHLORIDE ◇ THIORIDAZINE HYDROCHLORIDE ◇ THORIDAZINE HYDROCHLORIDE ◇ TIORIDAZIN ◇ TP-21 ◇ USAF SZ-3 ◇ USAF SZ-B

TOXICITY DATA with REFERENCE
dlt-mus-orl 1200 mg/kg/30D IJEBA6 11,403,73
orl-rat LD50:1060 mg/kg NIIRDN 6,458,82
ipr-rat LD50:160 mg/kg TXAPA9 24,37,73
ivn-rat LD50:71 mg/kg 27ZQAG -,53,72
orl-mus LD50:360 mg/kg 27ZQAG -,53,72
ipr-mus LD50:100 mg/kg NTIS** AD277-689
ivn-mus LD50:51 mg/kg 27ZQAG -,53,72
orl-dog LD50:160 mg/kg 27ZQAG -,53,72
orl-rbt LD50:1100 mg/kg 27ZQAG -,53,72
ivn-rbt LD50:26 mg/kg 27ZQAG -,53,72

CONSENSUS REPORTS: EPA Genetic Toxicology Program.

SAFETY PROFILE: Poison by ingestion, intravenous and intraperitoneal routes. Mutation data reported. When heated to decomposition it emits very toxic fumes of NO_x, SO_x, and HCl.

MOO750 CAS:314-03-4 ***HR: 3***
9-(1-METHYL-4-PIPERIDYLIDENE)THIOXANTHENE
mf: $C_{19}H_{19}NS$ mw: 293.45

SYNS: BP 400 ◇ CALMIXENE

TOXICITY DATA with REFERENCE
orl-rat LD50:550 mg/kg 27ZQAG -,62,72
ivn-rat LD50:13 mg/kg 27ZQAG -,62,72
orl-mus LD50:400 mg/kg 27ZQAG -,62,72
ivn-mus LD50:23 mg/kg 27ZQAG -,62,72
orl-rbt LD50:460 mg/kg 27ZQAG -,62,72
ivn-rbt LD50:18 mg/kg 27ZQAG -,62,72

SAFETY PROFILE: Poison by ingestion and intravenous routes. When heated to decomposition it emits very toxic fumes of NO_x and SO_x.

MOP000 ***HR: 3***
9-(N-METHYL-PIPERIDYLIDEN-4)THIOXANE MALEATE
mf: $C_{19}H_{19}NS \cdot C_4H_4O_4$ mw: 409.53

SYNS: BP-400 ◇ 9-(4'-(N-METHYLPIPERIDYLENE)THIOXANTHENE MALEATE

TOXICITY DATA with REFERENCE
orl-rat LD50:850 mg/kg ANAEA3 21,233,63
ivn-rat LD50:13 mg/kg ANAEA3 21,233,63
orl-mus LD50:310 mg/kg ANAEA3 21,233,63
ivn-mus LD50:22.5 mg/kg ANAEA3 21,233,63
ivn-rbt LD50:18 mg/kg ANAEA3 21,233,63

SAFETY PROFILE: Poison by ingestion and intravenous routes. When heated to decomposition it emits very toxic fumes of NO_x and SO_x.

MOP500 CAS:60706-49-2 ***HR: 3***
9-(METHYL-2-PIPERIDYL)METHYLCARBAZOLE
mf: $C_{19}H_{22}N_2$ mw: 278.43

SYN: 9-(1-METHYL-PIPERIDYL-(2)-METHYL)-CARBAZOL(GERMAN)

TOXICITY DATA with REFERENCE
ivn-rat LD50:30 mg/kg ARZNAD 9,219,59
orl-mus LD50:780 mg/kg ARZNAD 9,219,59
ipr-mus LD50:125 mg/kg ARZNAD 9,219,59
scu-mus LD50:1250 mg/kg ARZNAD 9,219,59
ivn-mus LD50:50 mg/kg ARZNAD 9,219,59
ivn-rbt LDLo:12500 μg/kg ARZNAD 9,219,59

SAFETY PROFILE: Poison by intravenous and intraperitoneal routes. Moderately toxic by ingestion and subcutaneous routes. When heated to decomposition it emits toxic fumes of NO_x.

MOQ000 CAS:60706-52-7 ***HR: 3***
1-(1-METHYL-2-PIPERIDYL)METHYLPHENOTHIAZINE
mf: $C_{19}H_{22}N_2S$ mw: 310.49

SYNS: P 892 ◇ PROMONTA

TOXICITY DATA with REFERENCE
ivn-rat LD50:30 mg/kg 27ZQAG -,35,72
orl-mus LD50:780 mg/kg 27ZQAG -,35,72
ipr-mus LD50:125 mg/kg 27ZQAG -,35,72
scu-mus LD50:1250 mg/kg 27ZQAG -,35,72
ivn-mus LD50:50 mg/kg 27ZQAG -,35,72
ivn-dog LD50:13 mg/kg 27ZQAG -,35,72

SAFETY PROFILE: Poison by intravenous and intraperitoneal routes. Moderately toxic by ingestion and subcutaneous routes. When heated to decomposition it emits very toxic fumes of NO_x and SO_x.

MOQ250 CAS:60-89-9 ***HR: 3***
(N-METHYL-3-PIPERIDYL)METHYLPHENOTHIAZINE
mf: $C_{19}H_{22}N_2S$ mw: 310.49

SYNS: LACUMIN ◇ MEPAZIN ◇ MEPAZINE BASE ◇ 10-(1-METHYL-PIPERIDYL-3-METHYL)PHENOTHIAZINE ◇ 10-(1-METHYL-3-

PIPERIDYL)METHYL PHENOTHIAZINE ◇ P 391 ◇ PACATAL ◇ PACATAL BASE ◇ PAXITAL ◇ PECATAL ◇ PECAZINE

TOXICITY DATA with REFERENCE
ipr-mus LD50:140 mg/kg ARZNAD 8,489,58
scu-mus LD50:750 mg/kg CANJAE 3,224,56
ivn-mus LD50:70 mg/kg ARZNAD 4,232,54
ivn-rbt LD50:20 mg/kg CANJAE 3,224,56

SAFETY PROFILE: Poison by intraperitoneal and intravenous routes. Moderately toxic by subcutaneous route. When heated to decomposition it emits very toxic fumes of NO_x and SO_x.

MOQ500 CAS:4354-45-4 ***HR: 3***
1-METHYL-3-PIPERIDYL-α-PHENYLCYCLOHEXANEGLYCOLATE
mf: $C_{20}H_{29}NO_3$ mw: 331.50

SYNS: N-METHYL-3-PIPERIDYL-α-CYCLOHEXYL MANDELATE ◇ OXYCLIPINE

TOXICITY DATA with REFERENCE
ivn-rat LD50:18 mg/kg AIPTAK 120,186,59
ivn-mus LD50:32 mg/kg AIPTAK 120,186,59

SAFETY PROFILE: Poison by intravenous route. When heated to decomposition it emits toxic fumes of NO_x.

MOR250 CAS:17814-73-2 ***HR: 3***
METHYL POTASSIUM
mf: CH_3K mw: 54.13

SAFETY PROFILE: Since it is incompatible with moisture (as in all living tissue), it must be considered a poison. When dry it ignites spontaneously in air. When heated to decomposition it emits toxic fumes of K_2O. See also POTASSIUM COMPOUNDS.

MOR500 CAS:83-43-2 ***HR: 2***
METHYLPREDNISOLONE
mf: $C_{22}H_{30}O_5$ mw: 374.52

PROP: Crystals. Mp: 228-237°.

SYNS: MEDROL ◇ MEDROL DOSEPAK ◇ MEDRONE ◇ Δ^1-6-α-METHYLHYDROCORTISONE ◇ 6-α-METHYLPREDNISOLONE ◇ METRISONE ◇ NSC-19987 ◇ 11-β,17,21-TRIHYDROXY-6-α-METHYLPREGNA-1,4-DIENE-3,20-DIONE ◇ 11-β,17-α,21-TRIHYDROXY-6-α-METHYL-1,4-PREGNADIENE-3,20-DIONE ◇ URBASON ◇ URBASONE ◇ WYACORT

TOXICITY DATA with REFERENCE
ipr-mus LD50:2292 mg/kg NIIRDN 6,832,82

CONSENSUS REPORTS: Reported in EPA TSCA Inventory.

SAFETY PROFILE: Moderately toxic by intraperitoneal route. A steroid hormone. When heated to decomposition it emits acrid smoke and irritating fumes.

MOR750 CAS:75-28-5 ***HR: 3***
2-METHYLPROPANE
DOT: UN 1075/UN 1969
mf: C_4H_{10} mw: 58.14

PROP: Colorless gas. Bp: −11.7°, lel: 1.9%, uel: 8.5%, fp: −160°, d: 0.5572 @ 20°, autoign temp: 864°F, vap d: 2.01.

SYN: ISOBUTANE

CONSENSUS REPORTS: Reported in EPA TSCA Inventory.

DOT Classification: Label: Flammable Gas.

SAFETY PROFILE: An asphyxiant. A common air contaminant. A very dangerous fire and explosion hazard when exposed to heat, flame, or oxidizers. To fight fire, stop flow of gas. When heated to decomposition it emits acrid smoke and irritating fumes.

MOS000 CAS:75-66-1 ***HR: 3***
2-METHYL-2-PROPANETHIOL
mf: $C_4H_{10}S$ mw: 90.20

PROP: Mobile liquid; heavy skunk odor. Mp: −0.5°, bp: 63.7-64.2°, d: 0.79-0.82 @ 15.5°/15.5°, flash p: < −20°F, vap d: 3.1, n (25/D) 1.41984. Sltly sol in water; very sol in alc, ether, and liquid H_2S.

SYNS: tert-BUTANETHIOL ◇ tert-BUTYL MERCAPTAN

TOXICITY DATA with REFERENCE
eye-rbt 84 mg AIHAAP 19,171,58
orl-rat LD50:4729 mg/kg AIHAAP 19,171,58
ipr-rat LD50:590 mg/kg AIHAAP 19,171,58

CONSENSUS REPORTS: Reported in EPA TSCA Inventory.

SAFETY PROFILE: Moderately toxic by intraperitoneal route. Mildly toxic by ingestion. An eye irritant. A very dangerous fire hazard when exposed to heat or flame. Can react vigorously with oxidizing materials. To fight fire, use alcohol foam, dry chemical, mist, fog. When heated to decomposition or on contact with acid or acid fumes it emits highly toxic fumes of SO_x. See also MERCAPTANS.

MOS250 CAS:555-57-7 ***HR: 3***
N-METHYL-N-PROPARGYLBENZYLAMINE
mf: $C_{11}H_{13}N$ mw: 159.25

SYNS: N-BENZYL-N-METHYL-2-PROPYNYLAMINE ◇ EUTONYL ◇ N-METHYL-N-BENZYLPROPYNYLAMINE ◇ N-METHYL-N-2-PROPYNYLBENZYLAMINE ◇ PARAGLYINE ◇ PARGLYAMINE ◇ PARGYLINE

TOXICITY DATA with REFERENCE
ipr-rat TDLo:180 mg/kg (12D male):REP FESTAS 27,1326,76

orl-chd TDLo:8750 μg/kg:NOSE,EYE,PUL 34ZIAG -,452,69
orl-rat LD50:300 mg/kg JMCMAR 14,913,71
ipr-rat LD50:142 mg/kg ANYAA9 107,1068,63
orl-mus LD50:680 mg/kg JMCMAR 14,913,71
ipr-mus LD50:290 mg/kg THERAP 22,367,67
scu-mus LD50:380 mg/kg BCPCA6 17,369,68
ivn-mus LD50:56 mg/kg CSLNX* NX#02420
orl-dog LD50:175 mg/kg ANYAA9 107,1068,63
ipr-mky LD50:200 mg/kg ANYAA9 107,1068,63
ipr-cat LD50:200 mg/kg ANYAA9 107,1068,63

SAFETY PROFILE: Poison by ingestion, intraperitoneal, subcutaneous, and intravenous routes. Human systemic effects by ingestion: eye lacrimation, olfactory and pulmonary effects. Experimental reproductive effects. When heated to decomposition it emits toxic fumes of NO_x. See also AMINES.

MOS875 CAS:922-67-8 ***HR: 3***
METHYL PROPIOLATE
mf: $C_4H_4O_2$ mw: 84.08

$CH_3OCO{\cdot}C{\equiv}CH$

SYN: ACETYLENECARBOXYLIC ACID METHYL ESTER ◇ METHYL PROPYNOATE

TOXICITY DATA with REFERENCE
ivn-mus LD50:18 mg/kg CSLNX* NX#08364

SAFETY PROFILE: Poison by intravenous route. Octakis(trifluorophosphine)rhodium catalyzes the violent polymerization of methyl propiolate. When heated to decomposition it emits acrid smoke and irritating fumes.

MOT000 CAS:554-12-1 ***HR: 3***
METHYL PROPIONATE
DOT: UN 1248
mf: $C_4H_8O_2$ mw: 88.12

PROP: Colorless liquid. Mp: −87.0°, bp: 79.8°, flash p: 28°F (CC) −2°C, d: 0.937 @ 4°, autoign temp: 876°F, vap press: 40 mm @ 11.0°, vap d: 3.03, lel: 2.50%, uel: 13%, d: 0.915 @ 20°/4°. Sol in water @ 20°; misc in alc and ether.

SYNS: METHYL PROPANOATE ◇ METHYL PROPYLATE ◇ PROPANOIC ACID, METHYL ESTER ◇ PROPIONATE de METHYLE (FRENCH)

TOXICITY DATA with REFERENCE
skn-rbt 500 mg/24H MOD FCTOD7 20(Suppl),765,82
orl-rat LD50:5 g/kg FCTOD7 20(Suppl),765,82
orl-mus LD50:3460 mg/kg GTPZAB 18(3),48,74
ihl-mus LC50:27 g/m^3 GTPZAB 18(3),48,74
orl-rbt LDLo:2550 mg/kg AMIHAB 21,100,60

CONSENSUS REPORTS: Reported in EPA TSCA Inventory.

DOT Classification: Flammable Liquid; Label: Flammable Liquid.

SAFETY PROFILE: Moderately toxic by ingestion. Mildly toxic by inhalation. A skin irritant. A very dangerous fire hazard when exposed to heat, flame, or oxidizers. Explosive in the form of vapor when exposed to heat or flame. To fight fire, use foam, CO_2, dry chemical. When heated to decomposition it emits acrid smoke and irritating fumes.

MOT750 CAS:94-14-4 ***HR: 3***
(2-METHYLPROPYL)-p-AMINOBENZOATE
mf: $C_{11}H_{15}NO_2$ mw: 193.27

SYN: p-AMINOBENZOIC ACID ISOBUTYL ESTER

TOXICITY DATA with REFERENCE
ipr-mus LD50:48 mg/kg JMCMAR 17,900,74

CONSENSUS REPORTS: Reported in EPA TSCA Inventory.

SAFETY PROFILE: Poison by intraperitoneal route. When heated to decomposition it emits toxic fumes of NO_x.

MOT800 CAS:73791-45-4 ***HR: 3***
METHYLPROPYLARSINIC ACID
mf: $C_4H_{11}AsO_2$ mw: 166.07

SYN: ARSINE OXIDE, HYDROXYMETHYLPROPYL-

TOXICITY DATA with REFERENCE
ivn-mus LD50:180 mg/kg CSLNX* NX#01191

OSHA PEL: TWA 0.5 mg(As)/m^3

SAFETY PROFILE: Poison by intravenous route. When heated to decomposition it emits toxic fumes of As.

MOU250 CAS:16354-54-4 ***HR: 3***
12-METHYL-7-PROPYLBENZ(a)ANTHRACENE
mf: $C_{22}H_{20}$ mw: 284.42

TOXICITY DATA with REFERENCE
ims-rat TDLo:50 mg/kg:NEO PNASA6 58,2253,67

SAFETY PROFILE: Questionable carcinogen with experimental neoplastigenic data. When heated to decomposition it emits acrid smoke and irritating fumes.

MOU500 CAS:541-95-7 ***HR: 3***
METHYLPROPYLCARBINOL CARBAMATE
mf: $C_6H_{13}NO_2$ mw: 131.20

SYNS: HEDONAL ◇ 1-METHYLBUTYL ESTER CARBAMIC ACID ◇ 2-PENTANOL CARBAMATE

TOXICITY DATA with REFERENCE
orl-rat LD50:421 mg/kg TXAPA9 1,150,59

orl-mus LDLo:1 g/kg LDTU** -,-,31
ipr-mus LD50:343 mg/kg TXAPA9 1,150,59

SAFETY PROFILE: Poison by intraperitoneal route. Moderately toxic by ingestion. When heated to decomposition it emits toxic fumes of NO_x. See also CARBAMATES.

MOU750 CAS:2917-19-3 ***HR: 3***
3-(1-METHYLPROPYL)-6-CHLOROPHENYL METHYLCARBAMATE
mf: $C_{12}H_{16}ClNO_2$ mw: 241.74

SYNS: CAL CHEM 5655 ◇ CHEVRON RE 5655 ◇ 2-CHLORO-5-(1-METHYLPROPYL)PHENYL METHYLCARBAMATE ◇ ENT 27,128 ◇ METHYLCARBAMIC ACID-3-sec-BUTYL-6-CHLOROPHENYL ESTER ◇ METHYLCARBAMIC ACID-2-CHLORO-5-(1-METHYLPROPYL)PHENYL ESTER ◇ ORTHO-5655 ◇ RE 5655

TOXICITY DATA with REFERENCE
orl-rat LD50:50 mg/kg ARSIM* 20,7,66
orl-ckn LD50:19 mg/kg TXAPA9 11,49,67
orl-bwd LD50:2400 μg/kg TXAPA9 21,315,72

SAFETY PROFILE: Poison by ingestion. When heated to decomposition it emits very toxic fumes of Cl^- and NO_x. See also CARBAMATES.

MOV000 CAS:3766-81-2 ***HR: 3***
2-(1-METHYLPROPYL)PHENYL METHYLCARBAMATE
mf: $C_{12}H_{17}NO_2$ mw: 207.30

SYNS: BASSA ◇ BAYCARD ◇ BPMC ◇ o-sec-BUTYLPHENYL METHYLCARBAMATE ◇ 2-sec-BUTYLPHENYL-N-METHYLCARBAMATE ◇ CARVIL ◇ HOPCIN ◇ OSBAC

TOXICITY DATA with REFERENCE
orl-rat LD50:410 mg/kg FMCHA2 -,D227,80
orl-mus LD50:173 mg/kg SPEADM 78-1,57,78
skn-mus LD50:340 mg/kg BESAAT 15,132,69
ipr-mus LD50:140 mg/kg FAATDF 4,724,84
ivn-mus LD50:42 mg/kg FAATDF 4,724,84

CONSENSUS REPORTS: Reported in EPA TSCA Inventory. EPA Genetic Toxicology Program.

SAFETY PROFILE: Poison by ingestion, skin contact, intravenous, and intraperitoneal routes. Used as an insecticide. When heated to decomposition it emits toxic fumes of NO_x. See also CARBAMATES.

MOV500 CAS:4268-36-4 ***HR: 3***
2-METHYL-2-PROPYLTRIMETHYLENE BUTYLCARBAMATE CARBAMATE
mf: $C_{13}H_{26}N_2O_4$ mw: 274.41

SYNS: 2-(((AMINOCARBONYL)OXY)METHYL)-2-METHYLPENTYL ESTER BUTYL CARBAMIC ACID ◇ BENVIL ◇ N-N-BUTYL-2-METHYL-2-PROPYL-1,3-PROPANEDIOL DICARBAMATE ◇ N-BUTYL-2-METHYL-2-PROPYL-1,3-PROPANEDIOL DICARBAMATE ◇ CARBAMIC ACID, ESTER with 2-(HYDROXYMETHYL)-2-METHYLPENTYL BUTYLCARBAMATE ◇ CARBAMIC ACID, ESTER with 2-METHYL-2-PROPYL-1,3-PROPANEDIOL BUTYLCARBAMATE ◇ EFFISAX ◇ 2-(HYDROXYMETHYL)-2-(METHYLPENTYL) BUTYLCARBAMATE CARBAMATE ◇ 2-(HYDROXYMETHYL)-2-METHYLPENTYL ESTER, CARBAMATE, BUTYL CARBAMIC ACID ◇ IDALENE ◇ 2-METHYL-2-PROPYL-1,3-PROPANEDIOL BUTYLCARBAMATE CARBAMATE ◇ NOSPAN ◇ SOLACEN ◇ SOLACIN ◇ TIBAMATO ◇ TYBAMATE ◇ TYBATRAN ◇ W 713

TOXICITY DATA with REFERENCE
orl-hmn TDLo:18 mg/kg/D:CNS,GIT JAGSAF 12,1066,64
orl-rat LD50:1040 mg/kg 27ZQAG -,418,72
ipr-rat LD50:465 mg/kg 27ZQAG -,418,72
orl-mus LD50:830 mg/kg 27ZQAG -,418,72
ipr-mus LD50:514 mg/kg JMCMAR 12,462,69
ivn-mus LD50:254 mg/kg 27ZQAG -,418,72
ivn-rbt LD50:105 mg/kg IJNEAQ 5,305,66

SAFETY PROFILE: Poison by intravenous route. Moderately toxic by ingestion and intraperitoneal routes. Human systemic effects by ingestion: somnolence, hallucinations or distorted perceptions, and nausea or vomiting. When heated to decomposition it emits toxic fumes of NO_x. See also CARBAMATES.

MOV800 CAS:35700-27-7 ***HR: 3***
15(s)-15-METHYL-PROSTAGLANDIN E2
mf: $C_{21}H_{34}O_5$ mw: 366.55

SYNS: (5Z,11-α,13E,15S,17Z)-11,15-DIHYDROXY-15-METHYL-9-OXO-PROSTA-5,13-DIEN-1-OIC ACID ◇ 15(s)-15-METHYL-PGE2

TOXICITY DATA with REFERENCE
ims-wmn TDLo:500 ng/kg (13W preg):REP JOGBAS 79,737,72

SAFETY PROFILE: Human reproductive effects by intramuscular and intravaginal routes: terminates pregnancy. A steroid. When heated to decomposition it emits acrid smoke and irritating fumes.

MOW000 CAS:59177-70-7 ***HR: 3***
N-METHYL-4-PROTOADAMANTANEAMINE HYDROCHLORIDE
mf: $C_{11}H_{19}N \cdot ClH$ mw: 201.77

TOXICITY DATA with REFERENCE
orl-mus LD50:403 mg/kg JMCMAR 19,967,76
ipr-mus LD50:140 mg/kg JMCMAR 19,967,76

SAFETY PROFILE: Poison by intraperitoneal route. Moderately toxic by ingestion. When heated to decomposition it emits very toxic fumes of NO_x and HCl.

MOW250 CAS:59177-76-3 ***HR: 3***
N-METHYL-4-PROTOADAMANTANEMETHANAMINE MALEATE
mf: $C_{12}H_{21}N \cdot C_4H_4O_4$ mw: 295.42

TOXICITY DATA with REFERENCE
orl-mus LD50:295 mg/kg JMCMAR 19,967,76
ipr-mus LD50:148 mg/kg JMCMAR 19,967,76

SAFETY PROFILE: Poison by ingestion and intraperitoneal routes. When heated to decomposition it emits toxic fumes of NO_x.

MOW500 CAS:3690-50-4 ***HR: 3***
METHYL PROTOANEMONIN
mf: $C_6H_6O_2$ mw: 110.12

SYNS: ETHYLIDENE-2(5H)-FURANONE ◇ 4-HYDROXYHEXA-2,4-DIENOIC ACID LACTONE

TOXICITY DATA with REFERENCE
scu-rat TDLo:2440 mg/kg/61W-I:ETA BJCAAI 15,85,61

SAFETY PROFILE: Questionable carcinogen with experimental tumorigenic data. When heated to decomposition it emits acrid smoke and irritating fumes.

MOW750 CAS:109-08-0 ***HR: 2***
2-METHYLPYRAZINE
mf: $C_5H_6N_2$ mw: 94.13

PROP: Liquid; nutty, cocoa odor. Mp: −29°, bp: 133° @ 737 mm, flash p: 122°F (COC), d: 1.0224 @ 25°/25°, refr index: 1.504, vap d: 3.2. Misc with water, alc, acetone, fixed oils.

SYN: FEMA No. 3309

TOXICITY DATA with REFERENCE
mmo-smc 8500 μg/L FCTXAV 18,581,80
cyt-ham:ovr 2500 μg/L FCTXAV 18,581,80
orl-rat LD50:1800 mg/kg DCTODJ 3,249,80
ipr-mus LD50:1820 mg/kg TXAPA9 17,244,70

CONSENSUS REPORTS: Reported in EPA TSCA Inventory.

SAFETY PROFILE: Moderately toxic by ingestion and intraperitoneal routes. Mutation data reported. Combustible liquid when exposed to heat, flame, or oxidizers. Can react with oxidizing materials. To fight fire, use water spray, foam, dry chemical, CO_2. When heated to decomposition it emits highly toxic fumes of NO_x.

MOX000 CAS:7554-65-6 ***HR: 3***
4-METHYLPYRAZOLE
mf: $C_4H_6N_2$ mw: 82.12

TOXICITY DATA with REFERENCE
orl-rat LD50:650 mg/kg EXPEAM 28,1198,72
ivn-rat LD50:310 mg/kg EXPEAM 28,1198,72
orl-mus LD50:640 mg/kg EXPEAM 28,1198,72
ivn-mus LD50:310 mg/kg EXPEAM 28,1198,72

SAFETY PROFILE: Poison by intravenous route. Moderately toxic by ingestion. When heated to decomposition it emits toxic fumes of NO_x.

MOX250 CAS:108-34-9 ***HR: 3***
METHYLPYRAZOLYL DIETHYLPHOSPHATE
mf: $C_8H_{15}N_2O_4P$ mw: 234.22

SYNS: O,O-DIAETHYL-O-(3-METHYL-1H-PYRAZOL-5-YL)-PHOSPHAT (GERMAN) ◇ O,O-DIETHYL-O-(3-METHYL-1H-PYRAZOL-5-YL)-FOSFAAT (DUTCH) ◇ DIETHYL-3-METHYL-5-PYRAZOLYL PHOSPHATE ◇ O,O-DIETHYL-O-(3-METHYL-5-PYRAZOLYL) PHOSPHATE ◇ O,O-DIETIL-O-(3-METIL-1H-PIRAZOL-5-IL)-FOSFATO (ITALIAN) ◇ ENT 24,723 ◇ 3-METHYLPYRAZOLYL-5-DIETHYLPHOSPHATE ◇ PHOSPHATE de DIETHYLE et de 3-METHYL-5-PYRAZOLYLE (FRENCH) ◇ PHOSPHORIC ACID-DIETHYL-(3-METHYL-5-PYRAZOLYL) ESTER ◇ PIRAZOXON (ITALIAN)

TOXICITY DATA with REFERENCE
scu-rat LD50:7 mg/kg JEENAI 50(3),356,57
orl-mus LD50:4 mg/kg ARSIM* 20,17,66
orl-bwd LD50:40 mg/kg TXAPA9 21,315,72

SAFETY PROFILE: Poison by ingestion and subcutaneous routes. When heated to decomposition it emits very toxic fumes of NO_x and PO_x.

MOX875 CAS:2381-21-7 ***HR: 2***
3-METHYLPYRENE
mf: $C_{17}H_{12}$ mw: 216.29

SYN: 1-METHYLPYRENE

TOXICITY DATA with REFERENCE
mma-sat 180 μmol/L/2H CNREA8 39,4152,79
ipr-mus TDLo:42 mg/kg/3D-I:NEO JTEHD6 21,525,87

SAFETY PROFILE: Questionable carcinogen with experimental neoplastigenic data. Mutation data reported. When heated to decomposition it emits acrid smoke and irritating fumes.

MOY000 CAS:109-06-8 ***HR: 3***
2-METHYLPYRIDINE
DOT: UN 2313
mf: C_6H_7N mw: 93.14

N=CHCH=CHCH=CCH$_3$ (ring: N bonded to C)

PROP: Colorless liquid; strong unpleasant odor. Mp: −70°, bp: 129°, flash p: 102°F (OC), d: 0.95 @ 15°/4°, autoign temp: 1000°F, vap press: 10 mm @ 24.4°, vap d: 3.2. Very sol in water; misc in alc and ether.

SYNS: α-METHYLPYRIDINE ◇ 2-PICOLINE ◇ α-PICOLINE ◇ o-PICOLINE (DOT) ◇ RCRA WASTE NUMBER U191

TOXICITY DATA with REFERENCE
skn-rbt 10 mg/24H open MLD AMIHBC 4,119,51
skn-rbt 470 mg open MLD UCDS** 2/21/58
eye-rbt 750 μg open SEV AMIHBC 4,119,51
orl-rat LD50:790 mg/kg HYSAAV 33,341,68

ihl-rat LCLo:4000 ppm/4H AMIHBC 4,119,51
ipr-rat LD50:200 mg/kg FAATDF 5,920,85
orl-mus LD50:674 mg/kg HYSAAV 33,341,68
skn-rbt LD50:410 mg/kg AMIHBC 4,119,51
orl-gpg LD50:900 mg/kg HYSAAV 33,341,68

CONSENSUS REPORTS: Reported in EPA TSCA Inventory.

DOT Classification: Flammable or Combustible Liquid; Label: Flammable Liquid.

SAFETY PROFILE: Poison by intraperitoneal route. Moderately toxic by ingestion and skin contact. Mildly toxic by inhalation. A skin and severe eye irritant. Flammable when exposed to heat and flame. To fight fire, use CO_2, dry chemical. Mixtures with hydrogen peroxide + iron(II) sulfate + sulfuric acid may ignite and then explode. When heated to decomposition it emits toxic fumes of NO_x. See also 4-METHYLPYRIDINE.

MOY250 CAS:108-89-4 ***HR: 3***
4-METHYLPYRIDINE
DOT: UN 2313
mf: C_6H_7N mw: 93.14

PROP: Colorless liquid; disagreeable odor. Bp: 145°, fp: 3.7°, d: 0.9571 @ 15°/4°, vap d: 3.21, flash p: 134°F (OC).

SYNS: Γ-PICOLINE ◇ 4-PICOLINE ◇ p-PICOLINE (DOT)

TOXICITY DATA with REFERENCE
skn-rbt 10 mg/24H open SEV AMIHBC 10,61,54
skn-rbt 480 mg open MOD UCDS** 2/21/58
eye-rbt 750 μg open SEV AMIHBC 10,61,54
orl-rat LD50:1290 mg/kg UCDS** 2/21/58
ihl-rat LCLo:1000 ppm/4H AMIHBC 10,61,54
ipr-rat LD50:163 mg/kg FAATDF 5,920,85
skn-rbt LD50:270 mg/kg AMIHBC 10,61,54
orl-bwd LD50:422 mg/kg AECTCV 12,355,83

CONSENSUS REPORTS: Reported in EPA TSCA Inventory.

DOT Classification: Flammable or Combustible Liquid; Label: Flammable Liquid.

SAFETY PROFILE: Poison by skin contact and intraperitoneal routes. Moderately toxic by ingestion. Mildly toxic by inhalation. A severe skin and eye irritant. Flammable when exposed to heat, flames, oxidizers. To fight fire, use alcohol foam. When heated to decomposition it emits toxic fumes of NO_x. See also 2-METHYLPYRIDINE.

MOY500 CAS:63019-78-3 ***HR: 3***
2-METHYLPYRIDINE-4-AZO-p-DIMETHYLANILINE
mf: $C_{14}H_{16}N_4$ mw: 240.34

TOXICITY DATA with REFERENCE
orl-rat TDLo:5550 mg/kg/22W-C:ETA CNREA8 14,715,54

SAFETY PROFILE: Questionable carcinogen with experimental tumorigenic data. When heated to decomposition it emits toxic fumes of NO_x.

MOY750 CAS:31932-35-1 ***HR: 3***
3-METHYLPYRIDINE-1-OXIDE-4-AZO-p-DIMETHYL-ANILINE
mf: $C_{14}H_{16}N_4O$ mw: 256.34

SYN: N,N-DIMETHYL-4-(4'-(3'-METHYLPYRIDYL-1'-OXIDE)AZO) ANILINE

TOXICITY DATA with REFERENCE
orl-rat TDLo:3700 mg/kg/35W-C:NEO JNCIAM 41,855,68
orl-rat TD:4284 mg/kg/17W-C:ETA JNCIAM 41,855,68

SAFETY PROFILE: Questionable carcinogen with experimental neoplastigenic and tumorigenic data. When heated to decomposition it emits toxic fumes of NO_x.

MOY875 CAS:7680-73-1 ***HR: 3***
METHYL PYRIDINIUM CHLORIDE
mf: $C_6H_8N{\cdot}Cl$ mw: 129.60

SYNS: N-METHYLPYRIDINIUM CHLORIDE ◇ 1-METHYLPYRIDINIUM CHLORIDE

TOXICITY DATA with REFERENCE
orl-rat LD50:285 mg/kg GISAAA 49(1),74,84
scu-rat LD50:280 mg/kg APFRAD 8,773,50
orl-mus LD50:286 mg/kg GISAAA 49(1),74,84
ipr-mus LDLo:220 mg/kg JCINAO 25,908,46

SAFETY PROFILE: Poison by ingestion, subcutaneous, and intraperitoneal routes. When heated to decomposition it emits toxic fumes of Cl^- and NO_x.

MOZ000 CAS:930-73-4 ***HR: 2***
1-METHYLPYRIDINIUM IODIDE
mf: $C_6H_8N{\cdot}I$ mw: 221.05

TOXICITY DATA with REFERENCE
scu-mus LD50:1360 mg/kg NDRC** -,105,43
ipr-mus LD50:491 mg/kg PHARAT 33,120,78

CONSENSUS REPORTS: Reported in EPA TSCA Inventory.

SAFETY PROFILE: Moderately toxic by subcutaneous and intraperitoneal routes. When heated to decomposition it emits very toxic fumes of NO_x and I^-.

MPA050 CAS:486-84-0 **HR: 3**
1-METHYL-9H-PYRIDO(3,4-b)INDOLE
mf: $C_{12}H_{10}N_2$ mw: 182.24

SYNS: ARIBINE ◇ HARMAN ◇ HARMANE ◇ LOCUTURINE ◇ LOTURINE ◇ 2-METHYL-β-CARBOLINE ◇ 3-METHYL-4-CARBOLINE ◇ 1-METHYLNORHARMAN ◇ PASSIFLORIN

TOXICITY DATA with REFERENCE
dni-hmn:oth 200 μmol/L BBRCA9 86,124,79
msc-ham:lng 100 mg/L CALEDQ 17,249,83
ipr-mus LD50:50 mg/kg AIPTAK 149,164,64

SAFETY PROFILE: Poison by intraperitoneal route. Human mutation data reported. When heated to decomposition it emits toxic fumes of NO_x.

MPA100 **HR: 3**
α-((4-METHYL-2-PYRIDYLAMINO)METHYL) BENZYL ALCOHOL HYDROCHLORIDE
mf: $C_{14}H_{16}N_2O•ClH$ mw: 264.78

TOXICITY DATA with REFERENCE
orl-mus LD50:685 mg/kg JPETAB 128,65,60
ipr-mus LD50:230 mg/kg JPETAB 128,65,60
ivn-mus LD50:83 mg/kg JPETAB 128,65,60

SAFETY PROFILE: Poison by intravenous and intraperitoneal routes. Moderately toxic by ingestion. When heated to decomposition it emits toxic fumes of NO_x and HCl.

MPA250 CAS:41288-00-0 **HR: 3**
3-(6-(5-METHYL-2-PYRIDYLOXY)HEXYL) THIAZOLIDINE DIHYDROCHLORIDE
mf: $C_{15}H_{24}N_2OS•2ClH$ mw: 353.39

SYN: 5-METHYL-2-(6-(3-THIAZOLIDINYL)HEXYLOXY)PYRIDINEDIHYDROCHLORIDE

TOXICITY DATA with REFERENCE
orl-mus LD50:500 mg/kg JMCMAR 16,319,73
ipr-mus LD50:200 mg/kg JMCMAR 16,319,73

SAFETY PROFILE: Poison by intraperitoneal route. Moderately toxic by ingestion. When heated to decomposition it emits very toxic fumes of NO_x, SO_x, and HCl.

MPB000 CAS:96-54-8 **HR: 2**
1-METHYLPYRROLE
mf: C_5H_7N mw: 81.12

PROP: Liquid. D: 0.9, vap d: 2.8, bp: 112°, fp: −57°, flash p: 60.8°F. Insol in water.

SAFETY PROFILE: A very dangerous fire hazard when exposed to heat, flame, or oxidizers. To fight fire, use dry chemical, CO_2, foam. When heated to decomposition it emits acrid smoke and irritating fumes. See also PYRROLE.

MPB175 CAS:53365-77-8 **HR: 2**
1-METHYLPYRROLE-2,3-DIMETHANOL
mf: $C_7H_{11}NO_2$ mw: 141.19

SYNS: 1,2-BISHYDROXYMETHYL-1-METHYLPYRROLE ◇ 2,3-BISHYDROXYMETHYL-1-METHYLPYRROLE

TOXICITY DATA with REFERENCE
oth-hmn:hlas 20 μmol/L CBINA8 30,325,80
sce-hmn:lyms 5 μmol/L MUREAV 149,485,85
skn-mus TDLo:1325 mg/kg/47W-I:CAR CALEDQ 17,61,82

SAFETY PROFILE: Questionable carcinogen with experimental carcinogenic data. Mutation data reported. When heated to decomposition it emits toxic fumes of NO_x.

MPB250 CAS:120-94-5 **HR: 3**
1-METHYLPYRROLIDINE
mf: $C_5H_{11}N$ mw: 85.15

PROP: Colorless to yellow liquid; penetrating amine-like odor. Bp: 80.5°, fp: −90°, d: 0.8054 @ 20°/20°, flash p: 37.4°F, vap d: 2.9.

SYN: N-METHYLTETRAHYDROPYRROLE

TOXICITY DATA with REFERENCE
ipr-mus LD50:178 mg/kg JPETAB 88,82,46
ivn-mus LD50:47 mg/kg JPETAB 88,82,46

CONSENSUS REPORTS: Reported in EPA TSCA Inventory.

SAFETY PROFILE: Poison by intraperitoneal and intravenous routes. This material is strongly alkaline. Liquid and vapors are corrosive to the skin, eyes, or mucous membranes. A very dangerous fire hazard; keep away from sparks, heat sources, and powerful oxidizers. Keep in closed containers. To fight fire, use alcohol foam. When heated to decomposition it emits highly toxic fumes of NO_x. See also AMMONIA.

MPB500 CAS:3690-18-4 **HR: 3**
β-METHYL-1-PYRROLIDINEPROPIONANILIDE
mf: $C_{14}H_{20}N_2O$ mw: 232.36

SYNS: ANILIDE of (PYRROLIDINO-N)-3-N-BUTYRIC ACID ◇ l'ANILIDE de l'ACIDE (PYRROLIDINO-N)-3-N-BUTYRIQUE (FRENCH)

TOXICITY DATA with REFERENCE
orl-mus LD50:1350 mg/kg AIPTAK 130,235,60
ipr-mus LD50:200 mg/kg AIPTAK 130,235,60
scu-mus LD50:660 mg/kg AIPTAK 130,235,60
ivn-gpg LDLo:92 mg/kg AIPTAK 130,235,60

SAFETY PROFILE: Poison by intraperitoneal and intravenous routes. Moderately toxic by ingestion and subcutaneous routes. When heated to decomposition it emits toxic fumes of NO_x.

MPC250 CAS:22966-83-2 ***HR: 3***
1-(2-METHYL-5-PYRROLIDINO-2,4-PENTADIENYLIDENE)PYRROLIDINIUM PERCHLORATE
mf: $C_{14}H_{23}N_2{\cdot}ClO_4$ mw: 318.84

TOXICITY DATA with REFERENCE
orl-mus LD50:200 mg/kg JMCMAR 12,806,69
ipr-mus LD50:10 mg/kg JMCMAR 12,806,69

SAFETY PROFILE: Poison by ingestion and intraperitoneal routes. When heated to decomposition it emits very toxic fumes of NO_x and Cl^-. See also PERCHLORATES.

MPC300 ***HR: 3***
6-METHYL-α-(1-PYRROLIDINYLCARBONYL) ERGOLINE-8-β-PROPIONITRILE
mf: $C_{23}H_{28}N_4O$ mw: 376.55

SYN: ERGOLINE-8-β-PROPIONITRILE, 6-METHYL-α-(1-PYRROLIDINYLCARBONYL)-

TOXICITY DATA with REFERENCE
orl-rat TDLo:8 mg/kg (female 5D post):REP ARZNAD 33,1094,83
orl-mus LD50:200 mg/kg ARZNAD 33,1094,83

SAFETY PROFILE: Poison by ingestion. Experimental reproductive effects. When heated to decomposition it emits toxic fumes of NO_x.

MPD000 CAS:7236-83-1 ***HR: 3***
3-(1-METHYL-2-PYRROLIDINYL)INDOLE
mf: $C_{13}H_{16}N_2$ mw: 200.31

TOXICITY DATA with REFERENCE
ipr-rat LD50:176 mg/kg JMCMAR 7,415,64
ipr-mus LD50:65 mg/kg JMCMAR 7,415,64

SAFETY PROFILE: Poison by intraperitoneal route. When heated to decomposition it emits toxic fumes of NO_x.

MPD250 CAS:3671-00-9 ***HR: 3***
3-(1-METHYL-3-PYRROLIDINYL)INDOLE
mf: $C_{13}H_{16}N_2$ mw: 200.31

TOXICITY DATA with REFERENCE
orl-rat LD50:413 mg/kg JMCMAR 9,136,66
ipr-rat LD50:107 mg/kg JMCMAR 9,136,66
orl-mus LD50:244 mg/kg JMCMAR 9,136,66
ipr-mus LD50:100 mg/kg JMCMAR 9,136,66

SAFETY PROFILE: Poison by ingestion and intraperitoneal routes. When heated to decomposition it emits toxic fumes of NO_x.

MPE250 CAS:1982-37-2 ***HR: 3***
10-((1-METHYL-3-PYRROLIDINYL)METHYL)-PHENOTHIAZINE
mf: $C_{18}H_{20}N_2S$ mw: 296.46

SYNS: DILOSYN ◇ DISYNCRAM ◇ DISYNCRAN ◇ METHDILAZINE ◇ MJ 5022 ◇ NCI-C60720 ◇ PRODUCT 5022 ◇ TACARYL ◇ TACAZYL ◇ TACRYL

TOXICITY DATA with REFERENCE
orl-hmn TDLo:4 mg/kg:CNS,GIT,PUL 34ZIAG -,379,69
orl-rat LD50:162 mg/kg TXAPA9 18,185,71
orl-mus LD50:225 mg/kg 27ZQAG -,29,72
ipr-mus LD50:183 mg/kg 27ZQAG -,29,72
orl-gpg LD50:263 mg/kg 27ZQAG -,29,72

SAFETY PROFILE: Poison by ingestion and intraperitoneal routes. Human systemic effects by ingestion: somnolence, dyspnea and gastrointestinal changes. When heated to decomposition it emits very toxic fumes of SO_x and NO_x.

MPF200 CAS:872-50-4 ***HR: 3***
N-METHYLPYRROLIDONE
mf: C_5H_9NO mw: 99.15

PROP: Colorless liquid; mild odor. Bp: 202°, fp: −24°, flash p: 204°F (OC), d: 1.027 @ 25°/4°, vap d: 3.4.

SYNS: N-METHYL-2-PYRROLIDINONE ◇ 1-METHYL-2-PYRROLIDINONE ◇ 1-METHYL-5-PYRROLIDINONE ◇ N-METHYLPYRROLIDINONE ◇ METHYLPYRROLIDONE ◇ N-METHYL-2-PYRROLIDONE ◇ 1-METHYL-2-PYRROLIDONE ◇ M-PYROL ◇ NMP

TOXICITY DATA with REFERENCE
eye-rbt 100 mg MOD FCTOD7 26,475,88
sln-smc 154 mmol/L EMMUEG 11,31,88
orl-rat TDLo:9700 mg/kg (female 6-15D post):TER EPASR* 8EHQ-1087-0695
skn-rat TDLo:7500 mg/kg (female 6-15D post):REP FAATDF 2,73,82
orl-rat LD50:3914 mg/kg ARZNAD 26,1581,76
ipr-rat LD50:2472 mg/kg ARZNAD 26,1581,76
ivn-rat LD50:80500 μg/kg IYKEDH 18,922,87
orl-mus LD50:5130 mg/kg EPASR* 8EHQ-1087-0695
ipr-mus LD50:3050 mg/kg EPASR* 8EHQ-1087-0695
ivn-mus LD50:54500 μg/kg IYKEDH 18,922,87
skn-rbt LD50:8000 mg/kg NPIRI* 1,84,74

CONSENSUS REPORTS: Reported in EPA TSCA Inventory.

DFG MAK: 100 ppm (400 mg/m^3)

SAFETY PROFILE: Poison by intravenous route. Moderately toxic by ingestion, intraperitoneal, and intravenous routes. Mildly toxic by skin contact. An experimental teratogen. Experimental reproductive effects. Mutation data reported. Combustible when exposed to heat, open flame, or powerful oxidizers. To fight fire, use foam, CO_2, dry chemical. When heated to decomposition it emits toxic fumes of NO_x.

MPF500 CAS:1769-24-0 ***HR: 2***
N-METHYLPYRROLIDONE
mf: $C_9H_8N_2O$ mw: 160.19

SYN: 2-METHYL-4(3H)-QUINAZOLINONE

TOXICITY DATA with REFERENCE
orl-mus LD50:859 mg/kg ARZNAD 12,1204,62
ipr-mus LD50:592 mg/kg ARZNAD 12,1204,62
par-mus LD50:500 mg/kg PCJOAU 7,626,73

SAFETY PROFILE: Moderately toxic by ingestion, intraperitoneal and parenteral routes. When heated to decomposition it emits toxic fumes of NO_x.

MPF750 CAS:2436-66-0 ***HR: 2***
3-METHYL-4-QUINAZOLINONE
mf: $C_9H_8N_2O$ mw: 160.19

TOXICITY DATA with REFERENCE
orl-mus LD50:517 mg/kg ARZNAD 12,1204,62
ipr-mus LD50:414 mg/kg ARZNAD 12,1204,62

SAFETY PROFILE: Moderately toxic by ingestion and intraperitoneal routes. When heated to decomposition it emits toxic fumes of NO_x.

MPF800 CAS:91-62-3 ***HR: 3***
6-METHYLQUINOLINE
mf: $C_{10}H_9N$ mw: 143.20

SYNS: p-METHYLQUINOLINE ◇ p-TOLUQUINOLINE

TOXICITY DATA with REFERENCE
skn-rbt 500 mg/24H MOD FCTXAV 17,871,79
mma-sat 25 μg/plate APTOD9 19,A105,80
orl-rat LD50:1260 mg/kg FCTXAV 17,871,79
ipr-mus LD50:386 mg/kg FCTXAV 17,871,79
skn-rbt LD50:5000 mg/kg FCTXAV 17,871,79

CONSENSUS REPORTS: EPA Genetic Toxicology Program. Reported in EPA TSCA Inventory.

SAFETY PROFILE: Poison by intraperitoneal route. Moderately toxic by ingestion. Mildly toxic by skin contact. Mutation data reported. A skin irritant. When heated to decomposition it emits toxic fumes of NO_x.

MPG000 CAS:2525-21-5 ***HR: 3***
1-METHYLQUINOLINIUM CHLORIDE
mf: $C_{10}H_{10}N•Cl$ mw: 179.66

TOXICITY DATA with REFERENCE
ivn-mus LD50:56 mg/kg CSLNX* NX#03581

CONSENSUS REPORTS: Reported in EPA TSCA Inventory.

SAFETY PROFILE: Poison by intravenous route. When heated to decomposition it emits very toxic fumes of NO_x and Cl^-.

MPG250 CAS:14628-06-9 ***HR: 3***
8-(METHYLQUINOLYL)-N-METHYL CARBAMATE
mf: $C_{12}H_{12}N_2O_2$ mw: 216.26

SYNS: CIBA C-7824 ◇ ENT 27,407 ◇ GIEGY GS-13798 ◇ GS-13,798 ◇ NSC 190997

TOXICITY DATA with REFERENCE
orl-rat LD50:120 mg/kg 28ZEAL 4,286,69
ipr-mus LD50:27 mg/kg PHARAT 34,142,79
scu-gpg LDLo:100 mg/kg JEENAI 62,934,69

SAFETY PROFILE: Poison by ingestion, intraperitoneal, and subcutaneous routes. When heated to decomposition it emits toxic fumes of NO_x. See also CARBAMATES.

MPH300 CAS:2901-66-8 ***HR: 3***
METHYL RESERPATE
mf: $C_{23}H_{30}N_2O_5$ mw: 414.55

SYNS: METHYL RESERPINOLATE ◇ MR ◇ RESERPATE de METHYLE (FRENCH)

TOXICITY DATA with REFERENCE
orl-rat LD50:479 mg/kg FRXXBL #2390163
orl-mus LD50:210 mg/kg FRXXBL #23901643
ivn-mus LD50:48 mg/kg FRXXBL #2390163

SAFETY PROFILE: Poison by ingestion and intravenous routes. When heated to decomposition it emits toxic fumes of NO_x.

MPH500 CAS:504-15-4 ***HR: 3***
5-METHYLRESORCINOL
mf: $C_7H_8O_2$ mw: 124.15

SYNS: 1,3-DIHYDROXY-5-METHYLBENZENE ◇ 3,5-DIHYDROXYTOLUENE ◇ 5-METHYL-1,3-BENZENDIOL ◇ 5-METHYLRESORCINOL ORCINOL ◇ ORCIN ◇ ORCINOL

TOXICITY DATA with REFERENCE
orl-rat LD50:844 mg/kg GTPPAF 8,145,72
scu-rat LDLo:1000 mg/kg RMSRA6 15,561,1895
orl-mus LD50:772 mg/kg HYSAAV 34(4-6),16,69
ipr-mus LD50:405 mg/kg CTYAD8 12,410,81
ivn-mus LD50:290 mg/kg CTYAD8 12,410,81
orl-rbt LD50:2400 mg/kg HYSAAV 34(4-6),16,69
orl-gpg LD50:1687 mg/kg HYSAAV 34(4-6),16,69
scu-gpg LDLo:600 mg/kg RMSRA6 15,561,1895
scu-frg LDLo:50 mg/kg RMSRA6 15,561,1895
skn-mam LD50:7800 mg/kg GISAAA 45(10),16,80

CONSENSUS REPORTS: EPA Genetic Toxicology Program. Reported in EPA TSCA Inventory.

SAFETY PROFILE: Poison by subcutaneous and intravenous routes. Moderately toxic by ingestion and intraperitoneal routes. Mildly toxic by skin contact. When heated to decomposition it emits acrid smoke and irritating fumes.

MPH750 CAS:4807-55-0 ***HR: 3***
3-METHYLRHODANINE
mf: $C_4H_5NOS_2$ mw: 147.22

SYN: USAF T-2

TOXICITY DATA with REFERENCE
ipr-mus LD50:400 mg/kg NTIS** AD277-689
ivn-mus LD50:180 mg/kg CSLNX* NX#03766

CONSENSUS REPORTS: Reported in EPA TSCA Inventory.

SAFETY PROFILE: Poison by intraperitoneal and intravenous routes. When heated to decomposition it emits very toxic fumes of NO_x and SO_x.

MPI000 CAS:119-36-8 ***HR: 3***
METHYL SALICYLATE
mf: $C_8H_8O_3$ mw: 152.16

PROP: From steam distillation of leaves from *Gaultheria procumbens* L. (Fam. *Ericacaae*) or from the bark of *Betula lenta* L. (Fam. *Betulaceae*). Colorless, yellowish or reddish oily liquid; odor and taste of wintergreen. Mp: −8.6°, bp: 223.3°, ULC: 20-25, flash p: 214°F (CC), fp: −1.2°, d: 1.1840 @ 25°/25°, refr index: 1.535, autoign temp: 850°F, vap press: 1 mm @ 54.0°, vap d: 5.24. Sltly sol in water @ 222° (decomp); sol in chloroform, ether, alc, glacial acetic acid.

SYNS: ACIDE ANISIQUE (FRENCH) ◇ ACIDE METHYL-o-BENZOIQUE (FRENCH) ◇ o-ANISIC ACID ◇ BETULA OIL ◇ FEMA No. 2745 ◇ GAULTHERIA OIL, ARTIFICIAL ◇ o-HYDROXYBENZOIC ACID, METHYL ESTER ◇ 2-HYDROXYBENZOIC ACID METHYL ESTER ◇ o-METHOXYBENZOIC ACID ◇ 2-METHOXYBENZOIC ACID ◇ METHYL-o-HYDROXYBENZOATE ◇ METYLESTER KYSELINY SALICYLOVE (CZECH) ◇ NATURAL WINTERGREEN OIL ◇ OIL of WINTERGREEN ◇ SALICYLIC ACID, METHYL ESTER ◇ SWEET BIRCH OIL ◇ SYNTHETIC WINTERGREEN OIL ◇ TEABERRY OIL ◇ WINTERGREEN OIL (FCC) ◇ WINTERGREEN OIL, SYNTHETIC

TOXICITY DATA with REFERENCE
skn-rbt 500 mg/24H MOD FCTXAV 16,637,78
eye-rbt 500 mg/24H MLD 28ZPAK -,106,72
eye-rbt 500 mg/24H SEV 28ZPAK -,106,72
skn-gpg 100% SEV FCTXAV 16,821,78
eye-gpg 100% SEV FCTXAV 16,821,78
orl-rat TDLo:36450 mg/kg (MGN):REP TXAPA9 18,755,71
scu-rat TDLo:500 mg/kg (10D preg):TER AJPAA4 35,315,59
orl-man LDLo:101 mg/kg AJMSA9 193,772,37
orl-chd LDLo:228 mg/kg:PUL,GIT AJDCAI 69,37,45
orl-chd LDLo:700 mg/kg:PNS,CNS,PUL ADCHAK 28,475,53
orl-wmn LDLo:355 mg/kg AJMSA9 193,772,37
orl-inf LDLo:1480 mg/kg:PUL,GIT AJMSA9 193,772,37
orl-hmn LDLo:506 mg/kg MEIEDD 10,876,83
unr-man LDLo:522 mg/kg 85DCAI 2,73,70
orl-rat LD50:887 mg/kg FCTXAV 2,327,64
orl-mus LD50:1110 mg/kg JPETAB 132,207,61
orl-dog LD50:2100 mg/kg FCTXAV 16,821,78
scu-dog LDLo:2250 mg/kg FCTXAV 16,821,78
orl-rbt LD50:1300 mg/kg FCTXAV 16,821,78
scu-rbt LDLo:4250 mg/kg FCTXAV 16,821,78
orl-gpg LD50:1060 mg/kg FCTXAV 2,327,64
scu-gpg LDLo:1500 mg/kg FAONAU 44A,63,67

CONSENSUS REPORTS: Reported in EPA TSCA Inventory.

SAFETY PROFILE: Human poison by ingestion. Moderately toxic to humans by an unspecified route. Moderately toxic experimentally by ingestion, intraperitoneal, intravenous, and subcutaneous routes. An experimental teratogen. Human systemic effects by ingestion: flaccid paralysis without anesthesia, general anesthesia, dyspnea, and nausea or vomiting, respiratory stimulation. Experimental reproductive effects. A severe skin and eye irritant. Ingestion of relatively small amounts has caused severe poisoning and death. Combustible liquid when exposed to heat or flame; can react with oxidizing materials. To fight fire, use CO_2, dry chemical. When heated to decomposition it emits acrid smoke and irritating fumes. See also ESTERS.

MPI200 CAS:6547-08-6 ***HR: 2***
METHYLSELENO-2-BENZOIC ACID
mf: $C_8H_8O_2Se$ mw: 215.12

SYNS: ACIDE METHYL SELENO-2-BENZOIQUE ◇ BENZOIC ACID, o-(METHYLSELENO)- ◇ BENZOIC ACID, 2-(METHYLSELENO)- ◇ o-(METHYLSELENO)BENZOIC ACID

TOXICITY DATA with REFERENCE
ipr-rat LDLo:1100 mg/kg CRSBAW 172,383,78
ivn-rat LDLo:500 mg/kg CRSBAW 172,383,78

OSHA PEL: TWA 0.2 mg(Se)/m^3
ACGIH TLV: TWA 0.2 mg(Se)/m^3

SAFETY PROFILE: Moderately toxic by intraperitoneal and intravenous routes. When heated to decomposition it emits toxic fumes of Se.

MPI205 CAS:22262-18-6 ***HR: 2***
o-(METHYLSELENO)BENZOIC ACID
mf: $C_8H_7O_2Se \cdot Na$ mw: 237.10

SYN: BENZOIC ACID, o-(METHYLSELENO)-, SODIUM SALT

TOXICITY DATA with REFERENCE
ipr-mus LD50:644 mg/kg CHDDAT 268,2807,69

OSHA PEL: TWA 0.2 mg(Se)/m^3
ACGIH TLV: TWA 0.2 mg(Se)/m^3

SAFETY PROFILE: Moderately toxic by intraperitoneal route. When heated to decomposition it emits toxic fumes of Se.

MPI225 CAS:21992-92-7 ***HR: 3***
(METHYLSELENO)TRIS(DIMETHYLAMINO) PHOSPHONIUM IODIDE
mf: $C_7H_{21}N_3PSe \cdot I$ mw: 384.14

SYN: PHOSPHONIUM,(METHYLSELENO)TRIS(DIMETHYLAMINO)-, IODIDE

TOXICITY DATA with REFERENCE
ivn-mus LD50:8 mg/kg CSLNX* NX#05351

OSHA PEL: TWA 0.2 mg(Se)/m^3
ACGIH TLV: TWA 0.2 mg(Se)/m^3

SAFETY PROFILE: Poison by intravenous route. When heated to decomposition it emits toxic fumes of PO_x, Se, and I^-.

MPI600 ***HR: 3***
1-METHYLSILACYCLOPENTA-2,4-DIENE
mf: C_5H_8Si mw: 96.20

CH=CHSi(CH_3)CH=CH (ring)

SAFETY PROFILE: Violently explosive reaction with dienophiles (e.g., maleic anhydride; tetracyanoethylene; acetylenedicarboxylate). When heated to decomposition it emits acrid smoke and irritating fumes.

MPI625 CAS:992-94-9 ***HR: 3***
METHYLSILANE
mf: CH_6Si mw: 46.14

SAFETY PROFILE: Mixtures with mercury explode when shaken in the air. When heated to decomposition it emits acrid smoke and irritating fumes. See also SILANE.

MPI750 CAS:681-84-5 ***HR: 3***
METHYL SILICATE
DOT: UN 2606
mf: $C_4H_{12}O_4Si$ mw: 152.25

PROP: Clear liquid. Vap d: 5.25.

SYNS: METHYL ESTER of o-SILICIC ACID ◇ METHYL ORTHOSILICATE ◇ TETRAMETHOXY SILANE ◇ TETRAMETHYLSILICATE ◇ TL 199

TOXICITY DATA with REFERENCE
eye-rbt 250 μg open SEV AMIHBC 4,119,51
orl-rat LDLo:700 mg/kg 30ZNA5 3,935,68
ihl-rat LCLo:250 ppm/4H AMIHBC 4,119,51
ihl-mus LCLo:1000 mg/m^3/10M NDRC** NDCrc-132,May,42
ipr-mus LD50:250 mg/kg CBCCT* 2,56,50
skn-rbt LD50:17 g/kg AMIHBC 4,119,51
ivn-rbt LDLo:100 mg/kg 30ZNA5 3,935,68

CONSENSUS REPORTS: Reported in EPA TSCA Inventory.

OSHA PEL: TWA 1 ppm
ACGIH TLV: TWA 1 ppm
DOT Classification: Flammable Liquid; Label: Flammable Liquid and Poison.

SAFETY PROFILE: Poison by intravenous and intraperitoneal routes. Moderately toxic by ingestion and inhalation. Mildly toxic by skin contact. A severe eye irritant. This material can cause extensive necrosis (experimentally), keratoconus, and opaque cornea. It also causes severe human eye injuries, as well as necrosis of corneal cells, which progresses long after exposure has ceased. It is destructive and its effects resist treatment. Permanent blindness is possible from exposure to it. The kidney seems to be most subject to injury regardless of the mode of exposure. Pulmonary edema has also occurred. This material is more toxic than either ethyl silicate or silicic acid, although it has been thought that the injury caused is largely due to the action of the silicic acid. Flammable when exposed to heat or flame; can react vigorously with oxidizing materials. Potentially violent reaction with metal hexafluorides (e.g., rhenium; molybdenum; tungsten). When heated to decomposition it emits acrid smoke and irritating fumes. See also SILICATES, and SILICA, AMORPHOUS HYDRATED.

MPI800 CAS:75993-65-6 ***HR: 3***
METHYLSILVER
mf: CH_3Ag mw: 122.90

CONSENSUS REPORTS: Silver and its compounds are on the Community Right-To-Know List.

SAFETY PROFILE: Explodes at −20°C. When heated to decomposition it emits acrid smoke and irritating fumes. See also SILVER COMPOUNDS.

MPJ000 CAS:18356-02-0 ***HR: 3***
METHYL SODIUM
mf: CH_3Na mw: 38.02

SAFETY PROFILE: Ignites spontaneously in air. Corrosive and irritating material. Incompatible with p-chloronitrobenzene. When heated to decomposition it emits toxic fumes of Na_2O. See also SODIUM COMPOUNDS and METHANOL.

MPJ100 CAS:23362-09-6 ***HR: 3***
METHYL STIBINE
mf: CH_5Sb mw: 138.80

CONSENSUS REPORTS: Antimony and its compounds are on the Community Right-To-Know List.

SAFETY PROFILE: A shock- and heat-sensitive explosive. When heated to decomposition it emits toxic fumes of Sb. See also ANTIMONY COMPOUNDS.

MPJ250 CAS:73928-03-7 ***HR: 3***
2-METHYL-4-STILBENAMINE
mf: $C_{15}H_{15}N$ mw: 209.31

SYN: 2-METHYL-4-AMINOSTILBENE

TOXICITY DATA with REFERENCE
scu-rat TDLo:240 mg/kg/W-I:ETA BMBUAQ 14,141,58

SAFETY PROFILE: Questionable carcinogen with experimental tumorigenic data. When heated to decomposition it emits toxic fumes of NO_x.

MPJ500 CAS:73928-04-8 ***HR: 3***
3-METHYL-4-STILBENAMINE
mf: $C_{15}H_{15}N$ mw: 209.31

SYN: 3-METHYL-4-AMINOSTILBENE

TOXICITY DATA with REFERENCE
orl-rat TDLo:350 mg/kg/W:ETA BMBUAQ 14,141,58

SAFETY PROFILE: Questionable carcinogen with experimental tumorigenic data. When heated to decomposition it emits toxic fumes of NO_x.

MPJ800 CAS:29847-17-4 ***HR: 3***
β-METHYLSTREPTOZOTOCIN
mf: $C_9H_{17}N_3O_7$ mw: 279.29

SYN: β-d-METHYL-2-DEOXY-2-(3-METHYL-3-NITROSOUREIDO) GLUCOPYRANOSIDE

TOXICITY DATA with REFERENCE
ipr-mus LD50:1700 mg/kg JMCMAR 19,918,76
ivn-mus LD50:1620 mg/kg JMCMAR 19,918,76
ivn-dog LD50:400 mg/kg JMCMAR 19,918,76

SAFETY PROFILE: Poison by intravenous route. Moderately toxic by intraperitoneal route. When heated to decomposition it emits toxic fumes of NO_x.

MPK250 CAS:98-83-9 ***HR: 3***
α-METHYL STYRENE
DOT: UN 2303
mf: C_9H_{10} mw: 118.19

PROP: Colorless liquid. D: 0.862 @ 20°/4°, mp: −96.0°, bp: 152.4°. Insol in water; misc in alc and ether.

SYNS: ISOPROPENIL-BENZOLO (ITALIAN) ◇ ISOPROPENYLBENZEEN (DUTCH) ◇ ISOPROPENYLBENZENE ◇ ISOPROPENYL-BENZOL (GERMAN) ◇ α-METIL-STIROLO (ITALIAN) ◇ as-METHYL- PHENYLETHYLENE ◇ α-METHYLSTYREEN (DUTCH) ◇ α-METHYL-STYROL (GERMAN) ◇ 2-PHENYLPROPENE ◇ β-PHENYLPROPENE ◇ 2-PHENYLPROPYLENE ◇ β-PHENYLPROPYLENE

TOXICITY DATA with REFERENCE
skn-rbt 100% MOD AMIHAB 14,387,56
eye-rbt 91 mg MLD AMIHAB 14,387,56
ihl-hmn TCLo:600 ppm:IRR AMIHAB 14,387,56
ihl-rat LCLo:3000 ppm 28ZRAQ -,131,60
ihl-gpg LCLo:3000 ppm 28ZRAQ -,131,60

CONSENSUS REPORTS: Reported in EPA TSCA Inventory.

OSHA PEL: (Transitional: TWA CL 100 ppm) TWA 50 ppm; STEL 100 ppm
ACGIH TLV: TWA 50 ppm; STEL 100 ppm
DFG MAK: 100 ppm (480 mg/m^3)
DOT Classification: Flammable Liquid.

SAFETY PROFILE: Mildly toxic by inhalation. Human systemic effects by inhalation: irritant effects. A skin and eye irritant. Flammable when exposed to heat or flame; can react vigorously with oxidizing materials. When heated to decomposition it emits acrid smoke and irritating fumes.

MPK500 CAS:25013-15-4 ***HR: 2***
METHYL STYRENE (mixed isomers)
DOT: UN 2618
mf: C_9H_{10} mw: 118.19

PROP: A mixture containing 55-70% m-vinyltoluene and 30-45% p-vinyltoluene (AMIHAB 14,387,56).

SYN: VINYLTOLUENE (mixed isomers) (OSHA)

TOXICITY DATA with REFERENCE
ihl-hmn TCLo:400 ppm:IRR AMIHAB 14,387,56
orl-rat LD50:4000 mg/kg AMIHAB 14,387,56

CONSENSUS REPORTS: Reported in EPA TSCA Inventory.

OSHA PEL: TWA 100 ppm
ACGIH TLV: TWA 50 ppm; STEL 100 ppm
DFG MAK: 100 ppm (480 mg/m^3)
DOT Classification: Flammable or Combustible Liquid; Label:Flammable Liquid

SAFETY PROFILE: Moderately toxic by ingestion. Human systemic effects by inhalation: irritant effects. When heated to decomposition it emits acrid smoke and irritating fumes. See also other methyl styrene entries.

MPK750 CAS:13107-39-6 ***HR: D***
p-METHYLSTYRENE OXIDE
mf: $C_9H_{10}O$ mw: 134.19

SYNS: p-(EPOXYETHYL)TOLUENE ◇ (4-METHYLPHENYL)OXIRANE ◇ 2-(4-METHYLPHENYL)OXIRANE ◇ 4-METHYLSTYRENE OXIDE ◇ p-TOLYLOXIRANE

TOXICITY DATA with REFERENCE
mmo-sat 340 μmol/L MUREAV 58,159,78
sce-hmn:lym 1 mmol/L MUREAV 116,379,83

CONSENSUS REPORTS: EPA Genetic Toxicology Program.

SAFETY PROFILE: Human mutation data reported. When heated to decomposition it emits acrid smoke and irritating fumes.

MPL000 CAS:1322-90-3 ***HR: 2***
METHYL STYRYLPHENYL KETONE
mf: $C_{16}H_{14}O$ mw: 222.30

PROP: Liquid. Mp: −30°. Bp: 246° @ 50 mm, flash p: 350°F (OC), d: 1.093 @ 20°/20°, vap press: <0.01 @ 20°, vap d: 7.67.

SYNS: 1,3-DIPHENYL-2-BUTEN-1-ONE ◇ DYPNONE ◇ β-METHYLCHALCONE ◇ 1-((2-PHENYLETHENYL)PHENYL)ETHANONE ◇ ar-STYRYLACETOPHENONE

TOXICITY DATA with REFERENCE
skn-rbt 10 mg/24H open MLD JIHTAB 31,60,49
eye-rbt 500 mg AJOPAA 29,1363,46
orl-rat LD50:3600 mg/kg JIHTAB 31,60,49
skn-rbt LD50:6300 mg/kg JIHTAB 31,60,49

SAFETY PROFILE: Moderately toxic by ingestion. Mildly toxic by skin contact. A skin and eye irritant. Combustible when exposed to heat or flame; can react with oxidizing materials. To fight fire, use alcohol foam; CO_2, dry chemical. Used as a sunscreen. When heated to decomposition it emits acrid smoke and irritating fumes. See also KETONES.

MPL250 CAS:13292-87-0 ***HR: 3***
METHYL SULFIDE compound with BORANE (1:1)
mf: $C_2H_6S{\cdot}BH_3$ mw: 75.98

SYNS: BORANE, compound with DIMETHYLSULFIDE ◇ DI-METHYLSULFIDE BORANE

TOXICITY DATA with REFERENCE
eye-rbt 100 mg SEV PESTC* 9,4,80
skn-rbt LDLo:200 mg/kg PESTC* 9,4,80

CONSENSUS REPORTS: Reported in EPA TSCA Inventory.

SAFETY PROFILE: Poison by skin contact. A severe eye irritant. When heated to decomposition it emits toxic fumes of SO_x. See also BORANES, BORON COMPOUNDS, and SULFIDES.

MPL500 CAS:27302-90-5 ***HR: 3***
2-((METHYLSULFINYL)ACETYL)PYRIDINE
mf: $C_8H_9NO_2S$ mw: 183.24

SYN: OXISURAN

TOXICITY DATA with REFERENCE
orl-rat TDLo:437 g/kg/2Y-C:NEO TXCYAC 28,17,83
orl-mus TDLo:336 g/kg/80W-C:CAR TXCYAC 28,17,83
orl-rat LD50:6550 mg/kg TXAPA9 36,49,76
ivn-rat LD50:1800 mg/kg TXAPA9 36,49,76
orl-mus LD50:5280 mg/kg TXAPA9 36,49,76
ivn-mus LD50:2510 mg/kg TXAPA9 36,49,76
orl-rbt LD50:6610 mg/kg TXAPA9 36,49,76

SAFETY PROFILE: Moderately toxic by intravenous route. Mildly toxic by ingestion. Questionable carcinogen with experimental carcinogenic and neoplastigenic data. When heated to decomposition it emits very toxic fumes of SO_x and NO_x.

MPL600 CAS:49575-13-5 ***HR: 3***
METHYLSULFINYL ETHYLTHIAMINE DISULFIDE
mf: $C_{26}H_{38}N_8O_4S_4$ mw: 654.96

TOXICITY DATA with REFERENCE
scu-rat LD50:690 mg/kg SKNEA7 (24),13,74
ivn-rat LD50:330 mg/kg SKNEA7 (24),13,74
scu-mus LD50:430 mg/kg SKNEA7 (24),13,74
ivn-mus LD50:310 mg/kg SKNEA7 (24),13,74

SAFETY PROFILE: Poison by intravenous route. Moderately toxic by subcutaneous route. When heated to decomposition it emits toxic fumes of SO_x and NO_x. See also SULFIDES.

MPL750 CAS:72738-90-0 ***HR: 3***
9-(p-(METHYLSULFONAMIDO)ANILINO)-3-ACRIDINE CARBAMIC ACID METHYL ESTER
mf: $C_{22}H_{20}N_4O_4S$ mw: 436.52

TOXICITY DATA with REFERENCE
mmo-sat 85 μmol/L JMCMAR 23,269,80
ipr-mus LD50:170 mg/kg JMCMAR 23,269,80

SAFETY PROFILE: Poison by intraperitoneal route. Mutation data reported. When heated to decomposition it emits very toxic fumes of NO_x and SO_x. See also ESTERS and CARBAMATES.

MPM000 CAS:53222-15-4 ***HR: 3***
N-(9-(p-(METHYLSULFONAMIDO)ANILINO)ACRIDIN-3-YL)ACETAMIDE METHANESULFONATE
mf: $C_{22}H_{20}N_4O_3S \cdot CH_5O_3S$ mw: 517.64

SYN: N-(9-((4-((METHYLSULFONYL)AMINO)PHENYL)AMINO)-3-ACRIDINYL)ACETAMIDEMETHANESULFONATE

TOXICITY DATA with REFERENCE
mma-sat 282 μmol/L JMCMAR 23,251,79
ipr-mus LD10:19 mg/kg JMCMAR 22,251,79

SAFETY PROFILE: Poison by intraperitoneal route. Mutation data reported. When heated to decomposition it emits very toxic fumes of NO_x and SO_x. See also SULFONATES.

MPM750 ***HR: 3***
2-METHYLSULFONYL-10-(3-(4′-CARBAMOYLPIPERIDINO)PROPYL)-PHENOTHIAZINE
mf: $C_{22}H_{26}N_3O_3S_2$ mw: 444.63

SYNS: 1-(3-(2-(METHANESULFONYL)PHENOTHIAZIN-10-YL)PROPYL-4-PIPERIDINECARBOXAMIDE ◇ 9965 RP

TOXICITY DATA with REFERENCE
orl-mus LD50:1154 mg/kg OYYAA2 1,97,67
ipr-mus LD50:78 mg/kg OYYAA2 1,97,67
scu-mus LD50:712 mg/kg OYYAA2 1,97,67
ivn-mus LD50:64 mg/kg OYYAA2 1,97,67

SAFETY PROFILE: Poison by intravenous and intraperitoneal routes. Moderately toxic by ingestion and subcutaneous routes. When heated to decomposition it emits very toxic fumes of SO_x and NO_x.

MPN000 CAS:15318-45-3 ***HR: 3***
METHYLSULFONYL CHLORAMPHENICOL
mf: $C_{12}H_{15}Cl_2NO_5S$ mw: 356.24

SYNS: 8065 C.B. ◇ DEXTROSULPHENIDOL ◇ TAP ◇ THIAMPHENICOL ◇ THIOCYMETIN ◇ THIOPHENICOL ◇ WIN-5063-2

TOXICITY DATA with REFERENCE
orl-mus TDLo:6 g/kg (female 7-12D post):TER OYYAA2 7,41,73
ipr-rat TDLo:400 mg/kg (female 1-7D post):REP FESTAS 21,431,70
unr-hmn TDLo:214 mg/kg/10D:CNS,GIT,SKN ARZNAD 24,944,74
orl-rat LD50:7 g/kg OYYAA2 3,390,69
ipr-rat LD50:5 g/kg OYYAA2 3,390,69
scu-rat LD50:4 g/kg OYYAA2 3,390,69
ivn-rat LD50:339 mg/kg OYYAA2 3,390,69
orl-mus LD50:7 g/kg OYYAA2 3,390,69
ipr-mus LD50:5 g/kg OYYAA2 3,390,69
scu-mus LD50:4190 mg/kg OYYAA2 3,390,69
ivn-mus LD50:368 mg/kg OYYAA2 3,390,69

SAFETY PROFILE: Poison by intravenous route. Moderately toxic by subcutaneous route. Mildly toxic by ingestion. Human systemic effects by an unspecified route: sleep disorders, dermatitis, nausea or vomiting. An experimental teratogen. Experimental reproductive effects. When heated to decomposition it emits very toxic fumes of SO_x, NO_x, and Cl^-. See also CHLORAMPHENICOL.

MPN100 CAS:505-44-2 ***HR: 3***
3-METHYLSULPHINYLPROPYLISOTHIOCYANATE
mf: $C_5H_9OS_2$ mw: 149.26

SYNS: IBERIN ◇ ISOTHIOCYANIC ACID, 3-(METHYLSULFINYL) PROPYL ESTER ◇ PROPANE, 1-ISOTHIOCYANATO-3-(METHYLSULFINYL)-(9CI)

TOXICITY DATA with REFERENCE
scu-rat TDLo:100 mg/kg (female 8D post):REP FCTXAV 18,159,80
scu-rat LD50:90 mg/kg FCTXAV 18,159,80

SAFETY PROFILE: Poison by subcutaneous route. Experimental reproductive effects. When heated to decomposition it emits toxic fumes of SO_x.

MPN250 CAS:595-48-2 ***HR: 2***
METHYL TARTRONIC ACID
mf: $C_4H_6O_5$ mw: 134.10

SYNS: 2-HYDROXY-2-METHYLMALONATE ◇ HYDROXYMETHYLPROPANEDIOIC ACID (9CI) ◇ ISOMALIC ACID

TOXICITY DATA with REFERENCE
orl-mus LD50:2450 mg/kg TXAPA9 9,274,66
ipr-mus LD50:1400 mg/kg TXAPA9 9,274,66
orl-rbt LDLo:7000 mg/kg IECHAD 15,628,23

SAFETY PROFILE: Moderately toxic by ingestion and intraperitoneal routes. When heated to decomposition it emits acrid smoke and irritating fumes.

MPN275 CAS:22262-19-7 ***HR: 3***
o-(METHYLTELLURO)BENZOIC ACID
mf: $C_8H_7O_2Te \cdot Na$ mw: 285.74

SYN: BENZOIC ACID, o-(METHYLTELLURO)-, SODIUM SALT

TOXICITY DATA with REFERENCE
ipr-mus LD50:382 mg/kg CHDDAT 268,2807,69

ACGIH TLV: TWA 0.1 mg(Te)/m^3

SAFETY PROFILE: Poison by intraperitoneal route. When heated to decomposition it emits toxic fumes of Te.

MPN500 CAS:58-18-4 ***HR: 3***
17-METHYLTESTOSTERONE
mf: $C_{20}H_{30}O_2$ mw: 302.50

SYNS: ANDROMETH ◇ ANDROSAN ◇ ANDROSAN (tablets) ◇ ANDROSTEN ◇ 4-ANDROSTENE-17-α-METHYL-17-β-OL-3-ONE ◇ ANERTAN ◇ ANERTAN (tablets) ◇ DELATESTRYL ◇ DIANABOL ◇ DUMOGRAN ◇ GLOSSO STERANDRYL ◇ HOMANDREN ◇ HORMALE ◇ 17-β-HYDROXY-17-METHYLANDROST-4-EN-3-ONE ◇ MALESTRONE ◇ MALOGEN ◇ MASENONE ◇ MASTESTONA ◇ MESTERONE ◇ METANDREN ◇ 17-METHYLTESTOSTERON ◇ METHYLTESTOSTERONE ◇ 17-α-METHYLTESTOSTERONE ◇ METRONE ◇ M.T. MUCORETTES ◇ NABOLIN ◇ NEO-HOMBREOL-M ◇ NSC-9701 ◇ NU MAN ◇ ORAVIRON ◇ ORETON-M ◇ ORETON METHYL ◇ STENOLON ◇ STERONYL ◇ SYNANDRETS ◇ SYNANDROTABS ◇ TESTHORMONE ◇ TESTORA ◇ TESTOVIRON ◇ TESTRED

TOXICITY DATA with REFERENCE
orl-wmn TDLo:30 mg/kg (6-39W preg):TER MACPAJ 32,41,57
skn-rat TDLo:1900 μg/kg (female 1-19D post):REP GTPZAB 22(6),25,78
orl-man TDLo:5366 mg/kg/7Y-C:CAR,LIV BJSUAM 66,212,79
orl-man TDLo:420 mg/kg/4Y-C:CAR,LIV LANCAO 2,1273,72
orl-man TDLo:39 mg/kg/33W-I:LIV,SYS AJGAAR 82,461,87
ipr-rat LD50:1050 mg/kg NIIRDN 6,830,82
orl-mus LD50:1860 mg/kg NIIRDN 6,830,82
ipr-mus LD50:400 mg/kg NIIRDN 6,830,82

CONSENSUS REPORTS: Reported in EPA TSCA Inventory. EPA Genetic Toxicology Program.

SAFETY PROFILE: Poison by intraperitoneal route. Moderately toxic by ingestion. Human teratogenic effects by ingestion: developmental abnormalities of the urogenital system. Experimental teratogenic and reproductive effects. Human systemic effects: cholestatic jaundice, weight loss or decreased weight gain. Questionable human carcinogen producing liver tumors. A synthetic androgenic steroid. When heated to decomposition it emits acrid smoke and irritating fumes. See also TESTOSTERONE.

MPN600 CAS:57716-89-9 ***HR: D***
4-o-METHYL-12-o-TETRADECANOYLPHORBOL-13-ACETATE
mf: $C_{37}H_{58}O_8$ mw: 402.74

SYNS: 4-o-METHYL-TPA ◇ Me-TPA

TOXICITY DATA with REFERENCE
dns-mus-skn 4 μmol/kg RCOCB8 24,533,79
dni-gpg:oth 4 μmol/L CNREA8 42,1975,82

CONSENSUS REPORTS: EPA Genetic Toxicology Program.

SAFETY PROFILE: Human mutation data reported. When heated to decomposition it emits acrid smoke and irritating fumes.

MPN750 CAS:29553-26-2 ***HR: 2***
2-METHYL-3,3,4,5-TETRAFLUORO-2-BUTANOL
mf: $C_5H_8F_4O$ mw: 160.13

TOXICITY DATA with REFERENCE
orl-rat LDLo:670 mg/kg JOCMA7 4,262,62
ihl-rat LCLo:1000 ppm/4H JOCMA7 4,262,62

CONSENSUS REPORTS: Reported in EPA TSCA Inventory.

SAFETY PROFILE: Moderately toxic by ingestion and inhalation. When heated to decomposition it emits toxic fumes of F^-. See also FLUORIDES and ALCOHOLS.

MPO000 CAS:89-94-1 ***HR: 2***
2-METHYL-1,2,3,6-TETRAHYDROBENZ-ALDEHYDE
mf: $C_8H_{12}O$ mw: 124.20

SYNS: 2-METHYL-4-CYCLOHEXENE-1-CARBOXALDEHYDE ◇ 6-METHYL-3-CYCLOHEXENE-1-CARBOXALDEHYDE

TOXICITY DATA with REFERENCE
skn-rbt 10 mg/24H open MLD AIHAAP 23,95,62
orl-rat LD50:5660 mg/kg AIHAAP 23,95,62
skn-rbt LD50:3150 mg/kg AIHAAP 23,95,62

CONSENSUS REPORTS: Reported in EPA TSCA Inventory.

SAFETY PROFILE: Moderately toxic by skin contact. Mildly toxic by ingestion. A skin irritant. When heated to decomposition it emits acrid smoke and irritating fumes. See also ALDEHYDES.

MPO250 CAS:63020-37-1 ***HR: 3***
6-METHYL-1,2,3,4-TETRAHYDROBENZ(a)ANTHRACENE
mf: $C_{19}H_{18}$ mw: 246.37

SYN: 4-METHYL-1′,2′,3′,4′-TETRAHYDRO-1,2-BENZANTHRACENE

TOXICITY DATA with REFERENCE
imp-mus TDLo:80 mg/kg:ETA JNCIAM 2,241,41

SAFETY PROFILE: Questionable carcinogen with experimental tumorigenic data. When heated to decomposition it emits acrid smoke and irritating fumes.

MPO390 CAS:101607-49-2 ***HR: 2***
7-METHYL-1,2,3,4-TETRAHYDRODIBENZ(c,h)ACRIDINE
mf: $C_{22}H_{19}N$ mw: 297.42

SYN: 10-METHYL-1,2-TETRAHYDRO-1,2:7,8-BENZACRIDINE

TOXICITY DATA with REFERENCE
scu-mus TDLo:72 mg/kg/9W-I:ETA COREAF 251,1322,60

SAFETY PROFILE: Questionable carcinogen with ex-

perimental tumorigenic data. When heated to decomposition it emits toxic fumes of NO_x.

MPO400 CAS:101607-48-1 ***HR: 2***
14-METHYL-8,9,10,11-TETRAHYDRODIBENZ(a,h)ACRIDINE
mf: $C_{22}H_{19}N$ mw: 297.42

SYN: 10-METHYL-1,2-TETRAHYDRO-1,2:5,6-BENZACRIDINE

TOXICITY DATA with REFERENCE
scu-mus TDLo:72 mg/kg/9W-I:ETA COREAF 251,1322,60

SAFETY PROFILE: Questionable carcinogen with experimental tumorigenic data. When heated to decomposition it emits toxic fumes of NO_x.

MPO500 CAS:96-47-9 ***HR: 3***
METHYLTETRAHYDROFURAN
mf: $C_5H_{10}O$ mw: 86.15

PROP: Liquid; ether-like odor. Bp: 80°, flash p: 12°F, d: 0.853 @ 20°/4°, vap d: 2.97.

SYN: 2-METHYLTETRAHYDROFURAN (CZECH)

TOXICITY DATA with REFERENCE
eye-rbt 500 mg/24H MLD 28ZPAK -,138,72
orl-rat LDLo:5720 mg/kg 28ZPAK -,138,72
ihl-rat LC50:6000 ppm/4H 34ZIAG -,395,69
skn-rbt LD50:4500 mg/kg 34ZIAG -,395,69

CONSENSUS REPORTS: Reported in EPA TSCA Inventory.

SAFETY PROFILE: Mildly toxic by ingestion, inhalation, and skin contact. An eye irritant. A very dangerous fire hazard when exposed to heat, flame, or oxidizers. Can react vigorously with oxidizing materials. To fight fire, use alcohol foam, CO_2, dry chemical. When heated to decomposition it emits acrid smoke and irritating fumes.

MPO750 CAS:53112-33-7 ***HR: 3***
N-METHYL-1,2,3,4-TETRAHYDROISOQUINOLINE HYDROCHLORIDE
mf: $C_{10}H_{13}N \cdot ClH$ mw: 183.70

SYN: MTIQ

TOXICITY DATA with REFERENCE
ipr-mus LD50:131 μg/kg JPETAB 209,79,78
par-mus LDLo:145 mg/kg JPETAB 62,165,38

SAFETY PROFILE: Poison by intraperitoneal and parenteral routes. When heated to decomposition it emits very toxic fumes of HCl and NO_x.

MPP000 CAS:26590-20-5 ***HR: 2***
METHYLTETRAHYDROPHTHALIC ANHYDRIDE
mf: $C_9H_{10}O_3$ mw: 166.19

SYN: 3a,4,7,7a-TETRAHYDROMETHYL-1,3-ISOBENZOFURANDIONE

TOXICITY DATA with REFERENCE
orl-rat LD50:2140 mg/kg AIHAAP 30,470,69
skn-rbt LDLo:1410 mg/kg AIHAAP 30,470,69

CONSENSUS REPORTS: Reported in EPA TSCA Inventory.

SAFETY PROFILE: Moderately toxic by ingestion and skin contact. When heated to decomposition it emits acrid smoke and irritating fumes. See also ANHYDRIDES.

MPP250 CAS:28839-49-8 ***HR: D***
N-METHYL-3,4,5,6-TETRAHYDROPHTHALIMIDE
mf: $C_9H_{11}NO_2$ mw: 165.21

SYNS: N-METHYL-1-CYCLOPHENE-1,2-DICARBOXIMIDE ◇ 3,4,5,6-TETRAHYDRO-N-METHYLPHTHALIMIDE

TOXICITY DATA with REFERENCE
ipr-mus TDLo:200 mg/kg (9D preg):REP MOPMA3 13,133,77
ipr-mus TDLo:50 mg/kg (9D preg):TER MOPMA3 13,133,77

SAFETY PROFILE: An experimental teratogen. Experimental reproductive effects. When heated to decomposition it emits toxic fumes of NO_x.

MPP750 CAS:3655-88-7 ***HR: 3***
N-METHYL-TETRAHYDROTHIAMIDINTHIONE ACETIC ACID
mf: $C_6H_{10}N_2O_2S_2$ mw: 206.30

SYNS: 5-CARBOXYMETHYL-3-METHYL-2H-1,3,5-THIADIAZINE-2-THIONE ◇ 5-METHYL-6-THIOXOTETRAHYDRO-3-THIADIAZINE-ACETIC ACID

TOXICITY DATA with REFERENCE
orl-rat LD50:1000 mg/kg FMCHA2 -,D303,80
ipr-rat LD50:300 mg/kg 31ZOAD 1,71,68

SAFETY PROFILE: Poison by intraperitoneal route. Moderately toxic by ingestion. When heated to decomposition it emits very toxic fumes of NO_x and SO_x.

MPQ250 CAS:13183-79-4 ***HR: 3***
1-METHYL-1H-TETRAZOLE-5-THIOL
mf: $C_2H_4N_4S$ mw: 116.16

SYN: 1-METHYL-5-MERCAPTO-1,2,3,4-TETRAZOLE

TOXICITY DATA with REFERENCE
scu-rat TDLo:7 g/kg (male 14D pre):REP TJADAB 36,467,87
ipr-mus LD50:400 mg/kg NTIS** AD691-490

CONSENSUS REPORTS: Reported in EPA TSCA Inventory.

SAFETY PROFILE: Poison by intraperitoneal route. Experimental reproductive effects. When heated to decomposition it emits very toxic fumes of NO_x and SO_x. See also SULFIDES.

MPQ500 CAS:22885-98-9 ***HR: 3***
α-METHYL TETRONIC ACID
mf: $C_5H_6O_3$ mw: 114.11

SYN: 4-HYDROXY-5-METHYL-2(5H)-FURANONE

TOXICITY DATA with REFERENCE
scu-rat TDLo:2600 mg/kg/65W-I:ETA BJCAAI 19,392,65

SAFETY PROFILE: Questionable carcinogen with experimental tumorigenic data. When heated to decomposition it emits acrid smoke and irritating fumes.

MPQ750 CAS:144-82-1 ***HR: 3***
N^1-(5-METHYL-1,3,4-THIADIAZOL-2-YL)-SULFANILAMIDE
mf: $C_9H_{10}N_4O_2S_2$ mw: 270.35

SYNS: 2-(p-AMINOBENZENESULFONAMIDO)-5-METHYLTHIA-DIAZOLE ◇ 5-METHYL-2-SULFANILAMIDO-1,3,4-THIADIAZOLE ◇ SULFAMETHIZOLE ◇ SULFAMETHYLTHIADIAZOLE ◇ 2-SULFANILAMIDO-5-METHYL-1,3,4-THIADIAZOLE

TOXICITY DATA with REFERENCE
par-mus TDLo:6000 mg/kg/4W-I:ETA ACRAAX 37,258,52
scu-mus LD50:1210 mg/kg 29ZVAB -,111,69

CONSENSUS REPORTS: Reported in EPA TSCA Inventory.

SAFETY PROFILE: Moderately toxic by subcutaneous route. Questionable carcinogen with experimental tumorigenic data. When heated to decomposition it emits very toxic fumes of NO_x and SO_x. See also SULFONATES.

MPR000 CAS:24050-16-6 ***HR: 3***
2-METHYLTHIAZOLIDINE
mf: C_4H_9NS mw: 103.20

SYN: METHYL-2-THIAZOLIDINE (FRENCH)

TOXICITY DATA with REFERENCE
mmo-sat 100 mg/L JAFCAU 28,62,80
mma-sat 1 mg/L JAFCAU 28,62,80
ipr-mus LD50:250 mg/kg CHTPBA 5,312,70

SAFETY PROFILE: Poison by intraperitoneal route. Mutation data reported. When heated to decomposition it emits very toxic fumes of NO_x and SO_x.

MPR250 CAS:65400-79-5 ***HR: 3***
N-(3-METHYL-2-THIAZOLIDINYLIDENE)NICOTINAMIDE
mf: $C_{10}H_{11}N_3OS$ mw: 221.30

TOXICITY DATA with REFERENCE
orl-mus LD50:570 mg/kg JMCMAR 23,773,80
ivn-mus LD50:279 mg/kg JMCMAR 23,773,80

SAFETY PROFILE: Poison by intravenous route. Moderately toxic by ingestion. When heated to decomposition it emits very toxic fumes of SO_x and NO_x.

MPS250 CAS:16960-39-7 ***HR: 3***
METHYLTHIOACETALDEHYDE-o-(CARBAMOYL)OXIME
mf: $C_4H_8N_2O_2S$ mw: 148.20

SYNS: N-((AMINOCARBONYL)OXY)ETHANIMIDOTHIOIC ACID, METHYL ESTER ◇ DUPONT INSECTICIDE 1642 ◇ EI-1642 ◇ ENT 27,411 ◇ METHYL N-(CARBAMOYLOXY)THIOACETIMIDATE ◇ NSC 191001

TOXICITY DATA with REFERENCE
orl-rat LDLo:60 mg/kg 28ZEAL 5,104,76
skn-rbt LDLo:3400 mg/kg 28ZEAL 5,104,76

SAFETY PROFILE: Poison by ingestion. Moderately toxic by skin contact. Used as an insecticide. When heated to decomposition it emits very toxic fumes of SO_x and NO_x. See also ALDEHYDES.

MPS300 CAS:77430-23-0 ***HR: 2***
METHYL(THIOACETAMIDO) MERCURY
mf: C_3H_7HgNS mw: 289.76

TOXICITY DATA with REFERENCE
skn-hmn TDLo:23 mg/kg NEURAI 14,69,64

OSHA PEL: (Transitional: CL 1 mg/10m^3) TWA 0.01 mg(Hg)/m^3; STEL 0.03 mg/m^3 (skin)
ACGIH TLV: TWA 0.01 mg(Hg)/m^3; STEL 0.03 mg(Hg)/m^3
NIOSH REL: TWA 0.05 mg(Hg)/m^3

SAFETY PROFILE: Human systemic effects by skin contact. When heated to decomposition it emits toxic fumes of NO_x, SO_x, and Hg.

MPS500 CAS:7152-24-1 ***HR: 3***
2-METHYLTHIOBENZIMIDAZOLE
mf: $C_8H_8N_2S$ mw: 164.24

TOXICITY DATA with REFERENCE
ivn-mus LD50:100 mg/kg CSLNX* NX#04989

CONSENSUS REPORTS: Reported in EPA TSCA Inventory.

SAFETY PROFILE: Poison by intravenous route. When heated to decomposition it emits very toxic fumes of SO_x and NO_x.

MPT000 CAS:556-64-9 ***HR: 3***
METHYL THIOCYANATE
mf: C_2H_3NS mw: 73.12

PROP: Liquid. D: 1.068 @ 20°, mp: −51°, bp: 130-133°. Very sltly sol in water; misc in alc and ether.

SYNS: METHYLRHODANID (GERMAN) ◇ METHYL SULFOCYANATE ◇ METHYLTHIOKYANAT

TOXICITY DATA with REFERENCE
orl-rat LD50:60 mg/kg 85JCAE-,1036,86
ipr-mus LD50:23 mg/kg PCBPBS 2,95,72
ivn-mus LD50:18 mg/kg CSLNX* NX#02864
orl-cat LDLo:8500 μg/kg MEIEDD 11,963,89
scu-rbt LDLo:20 mg/kg AEPPAE 150,257,30

CONSENSUS REPORTS: Reported in EPA TSCA Inventory. EPA Extremely Hazardous Substances List.

SAFETY PROFILE: Poison by ingestion, intravenous, and subcutaneous routes. When heated to decomposition it emits very toxic fumes of NO_x and SO_x. See also THIOCYANATES.

MPT100 ***HR: D***
17-β-(METHYLTHIO)ESTRA-1,3,5(10)-TRIEN-3-OL
mf: $C_{19}H_{26}OS$ mw: 302.51

TOXICITY DATA with REFERENCE
unr-rat TDLo:7500 μg/kg (female 1D post):REP 85GRAA -,111,65

SAFETY PROFILE: Experimental reproductive effects. When heated to decomposition it emits toxic fumes of SO_x.

MPT250 CAS:4836-09-3 ***HR: 2***
2-METHYLTHIOETHYL ACRYLATE
mf: $C_6H_{10}O_2S$ mw: 146.22

SYNS: ACRYLIC ACID METHYLTHIOETHYL ESTER ◇ ACRYLIC ACID-2-(METHYLTHIO)ETHYL ESTER ◇ 2-(METHYLTHIO)ETHANOL ACRYLATE ◇ 2-PROPENOIC ACID-2-(METHYLTHIO)ETHYL ESTER

TOXICITY DATA with REFERENCE
skn-rbt 100 μg/24H open AIHAAP 23,95,62
orl-rat LD50:1230 mg/kg AIHAAP 23,95,62
orl-mus LD50:3730 mg/kg 85GMAT -,87,82
skn-rbt LD50:1490 mg/kg AIHAAP 23,95,62

SAFETY PROFILE: Moderately toxic by ingestion and skin contact. A skin irritant. When heated to decomposition it emits toxic fumes of SO_x. See also ESTERS and ACRYLIC ACID.

MPT300 CAS:17719-22-1 ***HR: 3***
2-(2-METHYLTHIOETHYLAMINO) ETHYLGUANIDINE SULFATE
mf: $C_6H_{16}N_4S \cdot 1/2H_2O_4S$ mw: 225.33

TOXICITY DATA with REFERENCE
orl-rat LD50:1850 mg/kg JMCMAR 11,848,68
orl-mus LD50:1200 mg/kg JMCMAR 11,848,68
ivn-mus LD50:93 mg/kg JMCMAR 11,848,68

SAFETY PROFILE: Poison by intravenous route. Moderately toxic by ingestion. When heated to decomposition it emits toxic fumes of SO_x and NO_x. See also SULFATES.

MPT500 CAS:834-12-8 ***HR: 2***
2-METHYLTHIO-4-ETHYLAMINO-6-ISOPROPYLAMINO-s-TRIAZINE
mf: $C_9H_{17}N_5S$ mw: 227.37

SYNS: 2-ETHYLAMINO-4-ISOPROPYLAMINO-6-METHYLMERCARPO-s-TRIAZINE ◇ 2-ETHYLAMINO-4-ISOPROPYLAMINO-6-METHYLTHIO-s-TRIAZINE ◇ 2-ETHYLAMINO-4-ISOPROPYLAMINO-6-METHYLTHIO-1,3,5-TRIAZINE ◇ 2-METHYLMERCAPTO-4-ETHYLAMINO-6-ISOPROPYLAMINO-s-TRIAZINE ◇ 2-METHYLMERCAPTO-4-ISOPROPYLAMINO-6-ETHYLAMINO-s-TRIAZINE

TOXICITY DATA with REFERENCE
eye-rbt 76 mg MLD CIGET* -,-,77
orl-rat LD50:1750 mg/kg FMCHA2 -,D133,80
orl-mus LD50:965 mg/kg PCOC** -,29,66
skn-rbt LD50:8160 mg/kg 85DPAN -,-,71/76

SAFETY PROFILE: Moderately toxic by ingestion. Mildly toxic by skin contact. An eye irritant. When heated to decomposition it emits very toxic fumes of NO_x and SO_x.

MPU000 CAS:342-69-8 ***HR: 3***
METHYLTHIOINOSINE
mf: $C_{11}H_{14}N_4O_4S$ mw: 298.35

SYNS: 6-METHYLMERCAPTOPURINERIBONUCLEOSIDE ◇ 6-METHYLMERCAPTOPURINE RIBOSIDE ◇ 6-METHYL-9-RIBOFURANOSYLPURINE-6-THIOL ◇ 6-METHYL-MP-RIBOSIDE ◇ 6-METHYLTHIOINOSINE ◇ 6-(METHYLTHIO)PURINE RIBONUCLEOSIDE ◇ 6-METHYLTHIOPURINE RIBOSIDE ◇ NCI-C04784 ◇ NSC 40774 ◇ β-d-RIBOSYL-6-METHYLTHIOPURINE ◇ SQ 21977

TOXICITY DATA with REFERENCE
dni-mus:lym 2 μmol/L CNREA8 44,2272,84
oms-mus:lym 3 μmol/L CNREA8 43,1587,83
ipr-rat TDLo:65 mg/kg (11D preg):TER ARPAAQ 86,395,68
ipr-rat TDLo:44 mg/kg/7W-I:ETA CANCAR 40(Suppl 4),1935,77

ipr-rat LD50:65 mg/kg ARPAAQ 86,395,68
ipr-rat LD50:137 mg/kg NCISP* JAN86

CONSENSUS REPORTS: NCI Carcinogenesis Studies (ipr); Equivocal Evidence: mouse CANCAR 40,1935,77; (ipr); No Evidence: rat CANCAR 40,1935,77.

SAFETY PROFILE: Poison by intraperitoneal route. Experimental teratogenic effects. Questionable carcinogen with experimental tumorigenic data. Mutation data reported. When heated to decomposition it emits very toxic fumes of SO_x and NO_x.

MPU250 CAS:34681-10-2 ***HR: 3***
3-(METHYLTHIO)-o-((METHYLAMINO)CARBONYL)OXIME-2-BUTANONE
mf: $C_7H_{14}N_2O_2S$ mw: 190.29

SYNS: AFILINE ◇ BUTOCARBOXIM (GERMAN) ◇ DRAWIN 755 ◇ 2-METHYLTHIO-o-(N-METHYLCARBAMOYL)-BUTANONOXIM-3 (GERMAN)

TOXICITY DATA with REFERENCE
orl-rat LD50:153 mg/kg 85DPAN -,-,71/76
skn-rbt LD50:360 mg/kg FMCHA2 -,C39,83

SAFETY PROFILE: Poison by ingestion and skin contact. Used as an insecticide. When heated to decomposition it emits very toxic fumes of SO_x and NO_x.

MPU500 CAS:66104-23-2 ***HR: 3***
8-β-((METHYLTHIO)METHYL)-6-PROPYLERGOLINE METHANESULFONATE
mf: $C_{19}H_{25}N_2S \cdot CH_4O_3S$ mw: 409.63

SYNS: MPE ◇ PERGOLIDE MESYLATE

TOXICITY DATA with REFERENCE
orl-rat LD50:15 mg/kg TXAPA9 48,A95,79
orl-mus LD50:54 mg/kg TXAPA9 48,A95,79
ipr-mus LD50:100 mg/kg JMCMAR 26,522,83

SAFETY PROFILE: Poison by ingestion and intraperitoneal routes. When heated to decomposition it emits very toxic fumes of SO_x and NO_x. See also SULFONATES.

MPV000 CAS:554-14-3 ***HR: 3***
2-METHYLTHIOPHENE
mf: C_5H_6S mw: 98.17

$SCH{=}CHCH{=}CCH_3$ (ring)

PROP: Oil. D: 1.022 @ 20°/4°, bp: 115.4°, flash p.: 46.4°F.

TOXICITY DATA with REFERENCE
orl-rat LD50:3200 mg/kg 85GMAT -,87,82
ipr-rat LD50:1 g/kg 85GMAT -,87,82
orl-mus LD50:1460 mg/kg 85GMAT -,87,82
ihl-mus LC50:11500 mg/m³/2H 85GMAT -,87,82
ipr-mus LDLo:500 mg/kg CBCCT* 1,47,49

CONSENSUS REPORTS: Reported in EPA TSCA Inventory.

SAFETY PROFILE: Moderately toxic by ingestion and intraperitoneal routes. Mildly toxic by inhalation. A very dangerous fire hazard when exposed to heat or flame. Ignites on contact with nitric acid. When heated to decomposition it emits toxic fumes of SO_x. See also SULFIDES.

MPV250 CAS:616-44-4 ***HR: 2***
3-METHYLTHIOPHENE
mf: C_5H_6S mw: 98.17

SYN: 3-THIOTOLENE

TOXICITY DATA with REFERENCE
orl-mus LD50:1800 mg/kg 85GMAT -,87,82
ihl-mus LC50:18 g/m³/2H 85GMAT -,87,82
ipr-mus LDLo:512 mg/kg CBCCT* 1,47,49

CONSENSUS REPORTS: Reported in EPA TSCA Inventory.

SAFETY PROFILE: Moderately toxic by ingestion and intraperitoneal routes. Mildly toxic by inhalation. When heated to decomposition it emits toxic fumes of SO_x. See also SULFIDES.

MPV750 CAS:867-44-7 ***HR: 3***
METHYL THIOPSEUDOUREA SULFATE
mf: $C_2H_6N_2S \cdot H_2O_4S$ mw: 188.24

SYNS: 2-METHYL-2-THIOPSEUDOUREA SULFATE ◇ USAF EK-1231

TOXICITY DATA with REFERENCE
orl-rat LD50:800 mg/kg JPETAB 90,260,47
ipr-mus LD50:400 mg/kg NTIS** AD438-895
ivn-mus LD50:180 mg/kg CSLNX* NX#05644

CONSENSUS REPORTS: Reported in EPA TSCA Inventory.

SAFETY PROFILE: Poison by intravenous and intraperitoneal routes. Moderately toxic by ingestion. When heated to decomposition it emits very toxic fumes of NO_x and SO_x. See also SULFATES.

MPV800 ***HR: 3***
2-METHYL-2-THIOPSEUDOUREA SULFATE (1:1) SODIUM SALT
mf: $C_2H_6N_2S \cdot O_4S \cdot 7Na$ mw: 347.15

TOXICITY DATA with REFERENCE
orl-rat LDLo:400 mg/kg ARPAAQ 37,253,44
ivn-rat LDLo:200 mg/kg ARPAAQ 37,253,44
ivn-rbt LDLo:50 mg/kg ARPAAQ 37,253,44

SAFETY PROFILE: Poison by ingestion and intravenous routes. When heated to decomposition it emits toxic fumes of NO_x, SO_x, and Na_2O. See also METHYL THIOPSEUDOUREA SULFATE.

MPW500 CAS:56-04-2 ***HR: 3***
6-METHYLTHIOURACIL
mf: $C_5H_6N_2OS$ mw: 142.19

PROP: Bitter crystals or colorless liquid; odor of onions. Decomp @ 326-331°, sublimes readily. Very sltly sol in ether, cold water, alkaline hydroxides, NH_3; sltly sol in alc, acetone, almost insol in benzene, chloroform.

SYNS: ALKIRON ◇ ANTIBASON ◇ BASECIL ◇ BASETHYRIN ◇ 2,3-DIHYDRO-6-METHYL-2-THIOXO-4(1H)-PYRIMIDINONE ◇ 2-MERCAPTO-4-HYDROXY-6-METHYLPYRIMIDINE ◇ 2-MERCAPTO-6-METHYLPYRIMID-4-ONE ◇ 2-MERCAPTO-6-METHYL-4-PYRIMIDONE ◇ METACIL ◇ METHIACIL ◇ METHIOCIL ◇ 6-METHYL-2-THIO-2,4-(1H3H)PYRIMIDINEDIONE ◇ METHYLTHIOURACIL ◇ 4-METHYL-2-THIOURACIL ◇ 6-METHYL-2-THIOURACIL ◇ 4-METHYLURACIL ◇ 6-METIL-TIOURACILE (ITALIAN) ◇ MTU ◇ MURACIL ◇ ORCANON ◇ PROSTRUMYL ◇ RCRA WASTE NUMBER U164 ◇ STRUMACIL ◇ THIMECIL ◇ THIOMECIL ◇ 2-THIO-6-METHYL-1,3-PYRIMIDIN-4-ONE ◇ 6-THIO-4-METHYLURACIL ◇ THIOMIDIL ◇ 2-THIO-4-OXO-6-METHYL-1,3-PYRIMIDINE ◇ THIORYL ◇ THIOTHYMIN ◇ THIOTHYRON ◇ THIURYL ◇ THYREONORM ◇ THYREOSTAT ◇ THYRIL ◇ TIOMERACIL ◇ TIORALE M ◇ TIOTIRON ◇ USAF EK-6454

TOXICITY DATA with REFERENCE
unr-wmn TDLo:1344 mg/kg (19-42W preg):REP LANCAO 1,1281,53
unr-wmn TDLo:1344 mg/kg (19-42W preg):TER LANCAO 1,1281,53
orl-rat TDLo:9100 mg/kg/2Y-C:CAR CNREA8 19,870,59
orl-mus TDLo:196 g/kg/1Y-C:NEO CANCAR 40,2188,77
orl-rat TD:53750 mg/kg/65W-I:ETA BJCAAI 7,181,53
orl-rat LD50:2790 mg/kg FSTEAI 7,313,52
ipr-rat LD50:920 mg/kg FRPSAX 13,882,58
ipr-mus LD50:200 mg/kg NTIS** AD277-689
orl-rbt LDLo:2500 mg/kg MEIEDD 10,877,83

CONSENSUS REPORTS: IARC Cancer Review: Group 2B IMEMDT 7,56,87; Animal Sufficient Evidence IMEMDT 7,53,74. Reported in EPA TSCA Inventory.

SAFETY PROFILE: Suspected carcinogen with experimental carcinogenic, neoplastigenic, tumorigenic, and teratogenic data. Poison by intraperitoneal route. Moderately toxic by ingestion. Human teratogenic and reproductive effects by an unspecified route: developmental abnormalities of the endocrine system and effects on newborn including neonatal measures or effects. Experimental reproductive effects. Used to treat hyperthyroidism. When heated to decomposition it emits very toxic fumes of NO_x and SO_x.

MPW600 CAS:598-52-7 ***HR: 3***
METHYL THIOUREA
mf: $C_2H_6N_2S$ mw: 90.16

SYNS: 1-METHYLTHIOUREA ◇ UREA, 1-METHYL-2-THIO-

TOXICITY DATA with REFERENCE
orl-rat TDLo:250 mg/kg (female 14D post):TER TJADAB 23,335,81
orl-rat LD50:50 mg/kg JPETAB 90,260,47
ipr-rat LD50:1850 mg/kg THERAP 35,409,80
orl-mus LDLo:2000 mg/kg TJADAB 23,335,81

SAFETY PROFILE: Poison by ingestion. Moderately toxic by intraperitoneal route. An experimental teratogen. When heated to decomposition it emits toxic fumes of SO_x.

MPX850 CAS:99-75-2 ***HR: 2***
METHYL-4-TOLUATE
mf: $C_9H_{10}O_2$ mw: 150.19

SYNS: p-CARBOMETHOXYTOLUENE ◇ METHYL-p-METHYLBENZOATE ◇ METHYL-4-METHYLBENZOATE ◇ METHYL-p-TOLUATE ◇ MPT

TOXICITY DATA with REFERENCE
skn-rbt 500 mg/24H MOD FCTXAV 17,873,79
orl-rat LD50:3300 mg/kg FCTXAV 17,873,79
orl-mus LD50:3800 mg/kg 85GMAT -,87,82
ipr-mus LD50:1250 mg/kg 85GMAT -,87,82

CONSENSUS REPORTS: Reported in EPA TSCA Inventory.

SAFETY PROFILE: Moderately toxic by ingestion and intraperitoneal routes. A skin irritant. When heated to decomposition it emits acrid smoke and irritating fumes.

MPY000 CAS:2058-62-0 ***HR: 3***
N-METHYL-p-(m-TOLYLAZO)ANILINE
mf: $C_{14}H_{15}N_3$ mw: 225.32

SYNS: N-METHYL-3′-METHYL-p-AMINOAZOBENZENE ◇ N-METHYL-3′-METHYL-4-AMINOAZOBENZENE ◇ 3′-METHYL-4-MONOMETHYLAMINOAZOBENZENE

TOXICITY DATA with REFERENCE
mmo-sat 1 μmol/plate CRNGDP 1,121,80
dns-rat:lvr 1 μmol/L CNREA8 46,1654,86
orl-rat TDLo:2520 mg/kg/20W-C:ETA JNCIAM 15,67,54

SAFETY PROFILE: Questionable carcinogen with experimental tumorigenic data. Mutation data reported. When heated to decomposition it emits toxic fumes of NO_x.

MPY250 CAS:17018-24-5 ***HR: 3***
N-METHYL-p-(o-TOLYLAZO)ANILINE
mf: $C_{14}H_{15}N_3$ mw: 225.32

SYNS: N-METHYL-2'-METHYL-p-AMINOAZOBENZENE ◇ N-METHYL-2'-METHYL-4-AMINOAZOBENZENE

TOXICITY DATA with REFERENCE
orl-rat TDLo:2350 mg/kg/14W-C:ETA CNREA8 8,141,48

SAFETY PROFILE: Questionable carcinogen with experimental tumorigenic data. When heated to decomposition it emits toxic fumes of NO_x.

MPY500 CAS:28149-22-6 ***HR: 3***
N-METHYL-p-(p-TOLYLAZO)ANILINE
mf: $C_{14}H_{15}N_3$ mw: 225.32

SYNS: N-METHYL-4'-METHYL-p-AMINOAZOBENZENE ◇ N-METHYL-4'-METHYL-4-AMINOAZOBENZENE

TOXICITY DATA with REFERENCE
orl-rat TDLo:2400 mg/kg/14W-C:ETA CNREA8 8,141,48

SAFETY PROFILE: Questionable carcinogen with experimental tumorigenic data. When heated to decomposition it emits toxic fumes of NO_x.

MPY750 CAS:2298-13-7 ***HR: D***
2-METHYL-4-((o-TOLYL)AZOANILINE HYDROCHLORIDE
mf: $C_{14}H_{15}N_3$•ClH mw: 261.78

SYNS: C.I. 37210 ◇ C.I. SOLVENT YELLOW 3 MONOHYDROCHLORIDE

TOXICITY DATA with REFERENCE
mma-sat 500 μg/plate MUREAV 56,249,78

CONSENSUS REPORTS: Reported in EPA TSCA Inventory.

SAFETY PROFILE: Mutation data reported. When heated to decomposition it emits very toxic fumes of NO_x and HCl.

MPZ000 CAS:100733-36-6 ***HR: 3***
8-METHYL-3-(α-(o-TOLYL)BENZYLOXY) TROPANIUM IODIDE
mf: $C_{23}H_{30}NO$•I mw: 463.44

TOXICITY DATA with REFERENCE
orl-mus LD50:750 mg/kg ARZNAD 14,964,64
ivn-mus LD50:3500 μg/kg ARZNAD 14,964,64

SAFETY PROFILE: Poison by intravenous route. Moderately toxic by ingestion. When heated to decomposition it emits very toxic fumes of NO_x and I^-. See also IODIDES.

MQA000 CAS:50838-36-3 ***HR: 3***
N-METHYL-N-(m-TOLYL)CARBAMOTHIOIC ACID-(1,2,3,4-TETRAHYDRO-1,4-METHANONAPHTHALEN-6-YL) ESTER
mf: $C_{20}H_{21}NOS$ mw: 323.48

SYNS: KC 9147 ◇ o-(1,4-METHANO-1,2,3,4-TETRAHYDRO-6-NAPHTHYL)-N-METHYL-N-(m-TOLYL)-THIOCARBAMATE ◇ o-(1,2,3,4-TETRAHYDRO-1,4-METHANONAPHTHALEN-6-YL)-m,N-DIMETHYLTHIO-CARBANILATE ◇ 1,2,3,4-TETRAHYDRO-1,4-METHANONAPHTHALEN-6-YLN-METHYL-N-(m-TOLYL)CARBAMOTHIOATE ◇ o-(5,6,7,8-TETRAHYDRO-5,8-METHANO-2-NAPHTHYL)-N-METHYL-N-(m-METHYLPHENYL)THIOCARBAMATE ◇ TOLCICLATE

TOXICITY DATA with REFERENCE
orl-rat LD50:6000 mg/kg DRFUD4 1,543,76
ivn-rat LD50:151 mg/kg KSRNAM 15,2399,81
orl-mus LD50:4000 mg/kg DRFUD4 1,543,76
ivn-mus LD50:80200 μg/kg KSRNAM 15,2399,81
ivn-dog LD50:124 mg/kg KSRNAM 15,2399,81
ivn-rbt LD50:94800 μg/kg KSRNAM 15,2399,81

SAFETY PROFILE: Poison by intravenous route. Moderately toxic by ingestion. When heated to decomposition it emits very toxic fumes of NO_x and SO_x. See also ESTERS and CARBAMATES.

MQB100 CAS:20064-41-9 ***HR: 3***
3-METHYL-5-(p-TOLYLSULFONYL)-1,2,4-THIADIAZOLE
mf: $C_{10}H_{10}N_2O_2S_2$ mw: 254.34

TOXICITY DATA with REFERENCE
eye-rbt 200 mg/1M YKKZAJ 88,1437,68
ivn-rat LD50:340 mg/kg YKKZAJ 88,1437,68
orl-mus LD50:1622 mg/kg YKKZAJ 88,1437,68
ipr-mus LD50:97900 μg/kg YKKZAJ 88,1437,68
scu-mus LD50:107 mg/kg YKKZAJ 88,1437,68
ivn-mus LD50:229 mg/kg YKKZAJ 88,1437,68

SAFETY PROFILE: Poison by subcutaneous, intravenous, and intraperitoneal routes. Moderately toxic by ingestion. An eye irritant. When heated to decomposition it emits toxic fumes of SO_x and NO_x.

MQB250 CAS:21124-13-0 ***HR: 3***
3-METHYL-1-(p-TOLYL)-TRIAZENE
mf: $C_7H_{11}N_3$ mw: 137.21

SYN: 3-METHYL-1-(4-METHYLPHENYL)TRIAZENE

TOXICITY DATA with REFERENCE
mmo-sat 100 μmol/L CNREA8 42,1446,82
mma-sat 1 mmol/L CNREA8 42,1446,82
orl-rat LD50:380900 μg/kg NEOLA4 25,153,78

CONSENSUS REPORTS: Reported in EPA TSCA Inventory.

SAFETY PROFILE: Poison by ingestion. Mutation data reported. When heated to decomposition it emits toxic fumes of NO_x.

MQB500 CAS:4253-34-3 ***HR: 2***
METHYLTRIACETOXYSILANE
mf: $C_7H_{12}O_6Si$ mw: 220.28

TOXICITY DATA with REFERENCE
orl-rat LD50:2060 mg/kg MarJV# 29MAR77

CONSENSUS REPORTS: Reported in EPA TSCA Inventory.

SAFETY PROFILE: Moderately toxic by ingestion. When heated to decomposition it emits acrid smoke and irritating fumes. See also SILANE.

MQB750 CAS:3413-72-7 ***HR: 3***
5-(3-METHYL-1-TRIAZENO)IMIDAZOLE-4-CARBOXAMIDE
mf: $C_5H_8N_6O$ mw: 168.19

SYNS: MTIC ◇ NSC-407347

TOXICITY DATA with REFERENCE
mmo-sat 100 μg/plate CRNGDP 3,467,82
dnd-hmn:oth 100 μmol/L CRNGDP 7,259,86
ipr-rat TDLo:10 mg/kg:ETA,REP JNCIAM 54,951,75
orl-rat TDLo:4450 mg/kg/14W-C:CAR,REP JNCIAM 54,951,75
ipr-mus LD50:775 mg/kg GANNA2 59,207,68

SAFETY PROFILE: Questionable carcinogen with experimental carcinogenic, and tumorigenic data. Experimental reproductive effects. Human mutation data reported. When heated to decomposition it emits toxic fumes of NO_x.

MQC000 CAS:41814-78-2 ***HR: 3***
5-METHYL-1,2,4-TRIAZOLE(3,4-b)BENZOTHIAZOLE
mf: $C_9H_7N_3S$ mw: 189.25

SYNS: BEAM ◇ BIM ◇ EL-291 ◇ 5-METHYL-s-TRIAZOLO(3,4-b) BENZOTHIAZOLE ◇ TRICYCLAZOLE ◇ TRICYCLAZONE

TOXICITY DATA with REFERENCE
orl-rat LD50:250 mg/kg 85ARAE 4,120,76
orl-mus LD50:250 mg/kg FMCHA2 -,C243,83

CONSENSUS REPORTS: Reported in EPA TSCA Inventory.

SAFETY PROFILE: Poison by ingestion. A fungicide. When heated to decomposition it emits very toxic fumes of NO_x and SO_x.

MQC150 CAS:598-99-2 ***HR: 3***
METHYL TRICHLOROACETATE
mf: $C_3H_3Cl_3O_2$ mw: 177.41

SAFETY PROFILE: Reacts violently with trimethylamine when heated. When heated to decomposition it emits toxic fumes of Cl^-. See also CHLORIDES.

MQC250 CAS:63991-43-5 ***HR: 1***
N-METHYL-2,4,5-TRICHLOROBENZENESULFONAMIDE
mf: $C_7H_6Cl_3NO_2S$ mw: 274.55

SYN: METHYLAMIDKYSELINY-2,4-5-TRICHLORBENZEN-SULFONOVE (CZECH)

TOXICITY DATA with REFERENCE
skn-rbt 500 mg/24H MLD 28ZPAK -,199,72
eye-rbt 500 mg/24H MLD 28ZPAK -,199,72
orl-rat LD50:5100 mg/kg 28ZPAK -,199,72

SAFETY PROFILE: Mildly toxic by ingestion. A skin and eye irritant. When heated to decomposition it emits very toxic fumes of Cl^-, NO_x, and SO_x.

MQC500 CAS:75-79-6 ***HR: 3***
METHYLTRICHLOROSILANE
DOT: UN 1250
mf: CH_3Cl_3Si mw: 149.48

SYNS: METHYL-TRICHLORSILAN (CZECH) ◇ TRICHLOROMETHYLSILANE

TOXICITY DATA with REFERENCE
skn-rbt 500 mg/24H MLD 28ZPAK -,216,72
eye-rbt 5 mg/24H SEV 28ZPAK -,216,72
orl-rat LDLo:1000 mg/kg JIHTAB 30,332,48
ihl-rat LC50:450 ppm/4H 28ZPAK -,216,72
ipr-rat LDLo:30 mg/kg JIHTAB 30,332,48
ihl-mus LC50:180 mg/m^3/2H TPKVAL (3),23,61

CONSENSUS REPORTS: Reported in EPA TSCA Inventory. EPA Extremely Hazardous Substances List.

DOT Classification: Flammable Liquid; Label: Flammable Liquid, Corrosive.

SAFETY PROFILE: Poison by inhalation and intraperitoneal routes. Moderately toxic by ingestion. A corrosive irritant to skin, eyes, and mucous membranes. Flammable when exposed to heat or flame; can react vigorously with oxidizing materials. When heated to decomposition it emits toxic fumes of Cl^-. See also CHLOROSILANES.

MQC750 CAS:993-16-8 ***HR: 2***
METHYLTRICHLOROSTANNANE
mf: CH_3Cl_3Sn mw: 240.08

SYNS: METHYLTIN TRICHLORIDE ◇ METHYLTRICHLOROTIN ◇ MONOMETHYLTIN TRICHLORIDE ◇ TRICHLOROMETHYLSTANNANE ◇ TRICHLOROMETHYLTIN

TOXICITY DATA with REFERENCE
orl-rat TDLo:211 mg/kg (14D pre-21D post):REP NTOTDY 4,539,82
unr-rat LD50:1370 mg/kg TIUSAD 107,1,76

CONSENSUS REPORTS: Reported in EPA TSCA Inventory.

OSHA PEL: TWA 0.1 mg(Sn)/m^3 (skin)
ACGIH TLV: TWA 0.1 mg(Sn)/m^3 (skin) (Proposed: TWA 0.1 mg(Sn)/m^3; STEL 0.2 mg(Sn)/m^3 (skin))
NIOSH REL: (Organotin Compounds) TWA 0.1 mg(Sn)/m^3

SAFETY PROFILE: Moderately toxic by an unspecified route. Experimental reproductive effects. When heated to decomposition it emits toxic fumes of Cl^-. See also TIN COMPOUNDS and CHLORIDES.

MQD000 CAS:63041-14-5 ***HR: 3***
1-METHYLTRICYCLOQUINAZOLINE
mf: $C_{22}H_{14}N_4$ mw: 334.40

TOXICITY DATA with REFERENCE
skn-mus TDLo:1240 mg/kg/1Y-I:NEO BJCAAI 16,275,62

SAFETY PROFILE: Questionable carcinogen with experimental neoplastigenic data. When heated to decomposition it emits toxic fumes of NO_x.

MQD250 CAS:28522-57-8 ***HR: 3***
3-METHYLTRICYCLOQUINAZOLINE
mf: $C_{22}H_{14}N_4$ mw: 334.40

TOXICITY DATA with REFERENCE
skn-mus TDLo:1240 mg/kg/1Y-I:NEO BJCAAI 16,275,62

SAFETY PROFILE: Questionable carcinogen with experimental neoplastigenic data. When heated to decomposition it emits toxic fumes of NO_x.

MQD500 CAS:63041-15-6 ***HR: 3***
4-METHYLTRICYCLOQUINAZOLINE
mf: $C_{22}H_{14}N_4$ mw: 334.40

TOXICITY DATA with REFERENCE
skn-mus TDLo:1240 mg/kg/1Y-I:NEO BJCAAI 16,275,62

SAFETY PROFILE: Questionable carcinogen with experimental neoplastigenic data. When heated to decomposition it emits toxic fumes of NO_x.

MQD750 CAS:2031-67-6 ***HR: 1***
METHYLTRIETHOXYSILANE
mf: $C_7H_{18}O_3Si$ mw: 178.34

SYNS: TRIETHOXYMETHYLSILANE ◇ UNION CARBIDE A-162

TOXICITY DATA with REFERENCE
skn-rbt 500 mg/24H MLD 28ZPAK -,219,72
eye-rbt 89 mg JIHTAB 30,332,48
eye-rbt 500 mg/24H MLD 28ZPAK -,219,72
orl-rat LDLo:15700 mg/kg 28ZPAK -,219,72
ihl-rat LCLo:4000 ppm/8H JIHTAB 30,332,48

CONSENSUS REPORTS: Reported in EPA TSCA Inventory.

SAFETY PROFILE: Mildly toxic by ingestion and inhalation. A skin and eye irritant. When heated to decomposition it emits acrid smoke and irritating fumes. See also SILANE.

MQE000 CAS:31185-56-5 ***HR: 2***
5-METHYL-2-TRIFLUOROMETHYLOXAZOLIDINE
mf: $C_5H_8F_3NO$ mw: 155.14

TOXICITY DATA with REFERENCE
ipr-mus LD50:1000 mg/kg JMCMAR 13,1212,70

SAFETY PROFILE: Moderately toxic by intraperitoneal route. When heated to decomposition it emits very toxic fumes of F^- and NO_x. See also FLUORIDES.

MQF000 CAS:3216-06-6 ***HR: 3***
8-METHYL-3-(α-(α,α,α-TRIFLUORO-o-TOLYL) BENZYLOXY)TROPANIUM IODIDE
mf: $C_{23}H_{27}F_3NO\cdot I$ mw: 517.41

TOXICITY DATA with REFERENCE
orl-mus LDLo:75 mg/kg ARZNAD 14,964,64
ivn-mus LDLo:2500 μg/kg ARZNAD 14,964,64

SAFETY PROFILE: Poison by ingestion and intravenous routes. When heated to decomposition it emits very toxic fumes of F^-, NO_x, and I^-. See also IODIDES and FLUORIDES.

MQF200 CAS:3823-94-7 ***HR: 3***
METHYL TRIFLUOROVINYL ETHER
mf: $C_3H_3F_3O$ mw: 112.05

SAFETY PROFILE: Very explosive. Upon ignition it decomposes more violently than acetylene. When heated to decomposition it emits toxic fumes of Cl^-. See also ETHERS and FLUORIDES.

MQF250 CAS:518-82-1 ***HR: 3***
6-METHYL-1,3,8-TRIHYDROXYANTHRAQUINONE
mf: $C_{15}H_{10}O_5$ mw: 270.25

PROP: Red needles from acetic acid. Mp: 255-257°. Insol in water; sol in alkali, alc, benzene, chloroform, ether, acetic acid.

SYNS: C.I. 75440 ◇ C.I. NATURAL YELLOW 14 ◇ EMODIN ◇ EMODOL ◇ FRANGULA EMODIN ◇ PERSIAN BERRY LAKE ◇ RHEUM EMODIN ◇ SCHUTTGELB ◇ 1,3,8-TRIHYDROXY-6-METHYL-9,10-ANTHRACENEDIONE ◇ 1,3,8-TRIHYDROXY-6-METHYLANTHRAQUINONE

TOXICITY DATA with REFERENCE
mmo-sat 50 μg/plate BCSTB5 5,1489,77

mma-sat 10 μg/plate MUREAV 125,135,84
ipr-mus LD50:35 mg/kg JAFCAU 27,1342,79

CONSENSUS REPORTS: EPA Genetic Toxicology Program.

SAFETY PROFILE: Poison by intraperitoneal route. Mutation data reported. When heated to decomposition it emits acrid smoke and irritating fumes.

MQF500 CAS:1185-55-3 ***HR: 1***
METHYLTRIMETHOXYSILANE
mf: $C_4H_{12}O_3Si$ mw: 136.25

SYN: UNION CARBIDE A-163

TOXICITY DATA with REFERENCE
skn-rbt 500 mg open MLD UCDS** 1/17/72
skn-rbt 500 mg/24H MLD 28ZPAK -,217,72
eye-rbt 500 mg/24H MOD 28ZPAK -,217,72
orl-rat LD50:12500 mg/kg AIHAAP 30,470,69

CONSENSUS REPORTS: Reported in EPA TSCA Inventory.

SAFETY PROFILE: Mildly toxic by ingestion. A skin and eye irritant. When heated to decomposition it emits acrid smoke and fumes. See also SILANE.

MQF750 CAS:53597-29-8 ***HR: 3***
(2-METHYL-2-(2-(TRIMETHYLAMMONIO) ETHOXY)ETHYL)DIETHYLMETHYL AMMONIUM DIIODIDE
mf: $C_{13}H_{32}N_2O•2I$ mw: 486.27

SYNS: (((2-(DIETHYLMETHYLAMMONIO)-1-METHYL)ETHOXY) ETHYL)TRIMETHYLAMMONIUM DIIODIDE ◇ N,N-DIETHYL-N-METHYL-2-(2-(TRIMETHYLAMMONIO)ETHOXY)-1-PROPANAMINIUM DIIODIDE ◇ N,N-DIETHYL-N,N',N',N'-TETRAMETHYL-N,N'-(2-METHYL-3-OXAPENTAMETHYLENE)BIS(AMMONIUM IODIDE) ◇ 2-(2-DIMETHYLAMINOETHOXY)-N,N-DIETHYLPROPYLAMINE DIMETHIODIDE ◇ 2-DIMETHYLAMINOETHYL 2'-DIETHYLAMINOISOPROPYL ETHER BISMETHIODIDE ◇ β-DIMETHYLAMINOETHYL β'-DIETHYLAMINO-α'-METHYLETHYL ETHER DIMETHIODIDE ◇ HL-8731 ◇ PLEGATIL

TOXICITY DATA with REFERENCE
orl-mus LD50:1125 mg/kg ARZNAD 7,616,57
ivn-mus LD50:39 mg/kg ARZNAD 7,616,57

SAFETY PROFILE: Poison by intravenous route. Moderately toxic by ingestion. When heated to decomposition it emits very toxic fumes of NO_x, NH_3, and I^-. See also IODIDES.

MQG500 CAS:4426-51-1 ***HR: 3***
4-METHYL TRIMETHYLENE SULFITE
mf: $C_4H_8O_3S$ mw: 136.18

SYNS: 1,3-BUTANEDIOL, CYCLIC SULFITE ◇ NSC-60195

TOXICITY DATA with REFERENCE
ipr-rat LD50:300 mg/kg NCISS* -,57,64
ivn-rat LD50:300 mg/kg NCISS* -,57,64
ipr-mus LD50:300 mg/kg NCISS* -,57,64
ivn-mus LD50:300 mg/kg NCISS* -,57,64
ivn-cat LD50:175 mg/kg NCISS* -,57,64

SAFETY PROFILE: Poison by intravenous and intraperitoneal routes. When heated to decomposition it emits toxic fumes of SO_x. See also SULFITES.

MQG750 CAS:24589-78-4 ***HR: 2***
N-METHYL-N-TRIMETHYLSILYLTRIFLUOROACETAMIDE
mf: $C_6H_{12}F_3NOSi$ mw: 199.28

SYN: MSTFA

TOXICITY DATA with REFERENCE
ipr-mus LDLo:1000 mg/kg StoGD# 27May75

CONSENSUS REPORTS: Reported in EPA TSCA Inventory.

SAFETY PROFILE: Moderately toxic by intraperitoneal route. When heated to decomposition it emits very toxic fumes of F^- and NO_x.

MQH000 CAS:5137-55-3 ***HR: 3***
METHYLTRIOCTYLAMMONIUM CHLORIDE
mf: $C_{25}H_{54}N•Cl$ mw: 404.25

SYNS: ALIQUAT 336 ◇ N-METHYL-N,N-DIOCTYL-1-OCTANAMINIUM CHLORIDE ◇ METHYLTRICAPRYLYLAMMONIUMCHLORIDE ◇ TRICAPRYLMETHYLAMMONIUM CHLORIDE ◇ TRICAPRYLYLMETHYLAMMONIUM CHLORIDE ◇ TRIOCTYLMETHYLAMMONIUM CHLORIDE

TOXICITY DATA with REFERENCE
orl-rat LD50:223 mg/kg AEHA** 33-3-68/71
ipr-rat LD50:6600 μg/kg AEHA** 33-3-68/71
orl-mus LD50:280 mg/kg AEHA** 33-3-68/71
ipr-mus LD50:3300 μg/kg AEHA** 33-3-68/71
ipr-gpg LDLo:10 mg/kg AEHA** 33-3-68/71

CONSENSUS REPORTS: Reported in EPA TSCA Inventory.

SAFETY PROFILE: Poison by ingestion and intraperitoneal routes. When heated to decomposition it emits very toxic fumes of NO_x, NH_3, and Cl^-. See also CHLORIDES.

MQH100 CAS:22223-55-8 ***HR: 3***
4-METHYL-2,6,7-TRIOXA-1-ARSABICYCLO (2.2.2)OCTANE
mf: $C_5H_9AsO_3$ mw: 192.06

SYN: 2,6,7-TRIOXA-1-ARSABICYCLO(2.2.2)OCTANE,4-METHYL-

TOXICITY DATA with REFERENCE
ivn-mus LD50:24 mg/kg EJMCA5 13,207,78

OSHA PEL: TWA 0.5 mg(As)/m^3

SAFETY PROFILE: Poison by intravenous route. When heated to decomposition it emits toxic fumes of As.

MQH250 CAS:1949-07-1 ***HR: 3***
1-METHYL-4-(3,3,3-TRIS(p-CHLOROPHENYL) PROPIONYLPIPERAZINE MONOHYDRO-CHLORIDE
mf: $C_{26}H_{25}Cl_3N_2O{\cdot}ClH$ mw: 524.34

SYNS: DICRODEN ◇ HETOLIN ◇ 1-(β,β,β-TRIS(p-CHLOROPHENYL) PROPIONYL)-4-METHYLPIPERAZINEHYDROCHLORIDE

TOXICITY DATA with REFERENCE
orl-mus LD50:610 mg/kg MEIEDD 11,739,89
orl-dom LDLo:200 mg/kg FAZMAE 17,108,73

SAFETY PROFILE: Poison by ingestion. When heated to decomposition it emits very toxic fumes of Cl^- and NO_x.

MQH500 CAS:57583-34-3 ***HR: 2***
METHYLTRIS(2-ETHYLHEXYLOXYCAR-BONYLMETHYLTHIO)STANNANE
mf: $C_{31}H_{60}O_6S_3Sn$ mw: 743.78

TOXICITY DATA with REFERENCE
unr-rat LD50:920 mg/kg TIUSAD 107,1,76

CONSENSUS REPORTS: Reported in EPA TSCA Inventory.

OSHA PEL: TWA 0.1 mg(Sn)/m^3 (skin)
ACGIH TLV: TWA 0.1 mg(Sn)/m^3 (skin) (Proposed: TWA 0.1 mg(Sn)/m^3; STEL 0.2 mg(Sn)/m^3 (skin))
NIOSH REL: (Organotin Compounds) TWA 0.1 mg(Sn)/m^3

SAFETY PROFILE: Moderately toxic by an unspecified route. When heated to decomposition it emits toxic fumes of SO_x. See also TIN COMPOUNDS and CARBONYLS.

MQH750 CAS:953-17-3 ***HR: 3***
METHYL TRITHION
mf: $C_9H_{12}ClO_2PS_3$ mw: 314.81

SYNS: ((p-CHLOROPHENYL)THIO)METHANETHIOL-S-ESTERwith O,O-DIMETHYL PHOSPHORODITHIOATE ◇ S-(((p-CHLOROPHENYL) THIO)METHYL) O,O-DIMETHYL PHOSPHORODITHIOATE ◇ S-(((4-CHLOROPHENYL)THIO)METHYL)O,O-DIMETHYLPHOSPHORO-DITHIOATE ◇ DIMETHYL-p-CHLOROPHENYLTHIOMETHYL DITHIOPHOSPHATE ◇ O,O-DIMETHYL-S-(p-CHLOROPHENYL-THIOMETHYL)PHOSPHORODITHIOATE ◇ O,O-DIMETHYLTHIO-PHOSPHORIC ACID, p-CHLOROPHENYL ESTER ◇ ENT 25,599 ◇ G-29288 ◇ GEIGY G-29288 ◇ METHYLCARBOPHENOTHION ◇ R-1492 ◇ STAUFFER R-1492 ◇ TRI-ME

TOXICITY DATA with REFERENCE
orl-rat LD50:48 mg/kg ARSIM* 20,10,66
skn-rat LD50:190 mg/kg WRPCA2 9,119,70
orl-mus LD50:112 mg/kg ARSIM* 20,10,66
skn-rbt LD50:2420 mg/kg BESAAT 12,161,66

SAFETY PROFILE: Poison by ingestion and skin contact route. A cholinesterase inhibitor type of insecticide. When heated to decomposition it emits very toxic fumes of Cl^-, PO_x, and SO_x. See also PARATHION.

MQI000 CAS:583-80-2 ***HR: 3***
4-METHYL-TROPOLONE
mf: $C_8H_8O_2$ mw: 136.16

SYN: 2-HYDROXY-4-METHYL-2,4,6-CYCLOHEPTATRIEN-1-ONE

TOXICITY DATA with REFERENCE
ipr-mus LD50:535 mg/kg JMCMAR 6,755,63
scu-mus LD50:189 mg/kg NYKZAU 53,918,57
ivn-mus LD50:51 mg/kg NYKZAU 53,918,57

SAFETY PROFILE: Poison by subcutaneous and intravenous routes. Moderately toxic by intraperitoneal route. When heated to decomposition it emits acrid smoke and irritating fumes. See also KETONES.

MQI250 CAS:154-06-3 ***HR: 3***
l-5-METHYLTRYPTOPHAN
mf: $C_{12}H_{14}N_2O_2$ mw: 218.28

TOXICITY DATA with REFERENCE
scu-mus TDLo:2000 mg/kg/20W-I:ETA AICCA6 19,660,63

SAFETY PROFILE: Questionable carcinogen with experimental tumorigenic data. When heated to decomposition it emits toxic fumes of NO_x.

MQI500 CAS:7361-31-1 ***HR: 3***
dl-α-METHYL-p-TYROSINE METHYL ESTER HYDROCHLORIDE
mf: $C_{11}H_{15}NO_3{\cdot}ClH$ mw: 245.73

SYN: (±)-α-METHYL-p-TYROSINE METHYL ESTER HYDROCHLORIDE

TOXICITY DATA with REFERENCE
ipr-mus LD50:400 mg/kg TXAPA9 28,227,74

CONSENSUS REPORTS: Reported in EPA TSCA Inventory.

SAFETY PROFILE: Poison by intraperitoneal route. When heated to decomposition it emits very toxic fumes of NO_x and HCl. See also ESTERS.

MQI750 CAS:626-48-2 ***HR: 3***
6-METHYLURACIL
mf: $C_5H_6N_2O_2$ mw: 126.13

PROP: Crystals from glacial acetic acid. Decomp above 300°.

SYNS: 6-METHYL-2,4(1H,3H)-PYRIMIDINEDIONE ◇ 4-METHYLURACIL ◇ PSEUDOTHYMINE

TOXICITY DATA with REFERENCE
orl-rat TDLo:2 g/kg (10D preg):TER BEXBAN 66,1382,68
scu-rat TDLo:20 mg/kg (4D pre):REP JSICAZ 19,264,60
orl-rat TDLo:18 g/kg/52W-C:NEO VOONAW 8(10),49,62
orl-rat LD50:64500 mg/kg LONZA# 03FEB81
orl-mam LD50:2 g/kg RPTOAN 48,186,85

CONSENSUS REPORTS: Reported in EPA TSCA Inventory.

SAFETY PROFILE: Moderately toxic by ingestion. Questionable carcinogen with experimental neoplastigenic and teratogenic data. Experimental reproductive effects. When heated to decomposition it emits toxic fumes of NO_x.

MQJ000 CAS:598-50-5 ***HR: 2***
N-METHYLUREA
mf: $C_2H_6N_2O$ mw: 74.10

PROP: Crystals. Mp: 101°, bp: decomp, d: 1.205 @ 20°/20°. Very sol in water and alc; insol in ether.

SYNS: METHYLUREA ◇ 1-METHYLUREA

TOXICITY DATA with REFERENCE
mmo-bcs 5 g/L MUREAV 42,19,77
mmo-omi 500 ppm/1H SOGEBZ 12,772,76
mmo-clr 400 mmol/L FOMIAZ 20,452,75
orl-rat LDLo:500 mg/kg NCNSA6 5,47,53
par-mus LDLo:10189 mg/kg JPETAB 52,216,34

CONSENSUS REPORTS: Reported in EPA TSCA Inventory. EPA Genetic Toxicology Program.

SAFETY PROFILE: Moderately toxic by ingestion. Mutation data reported. Combustible when exposed to heat or flame; can react with oxidizing materials. When heated to decomposition it emits toxic fumes of NO_x.

MQJ250 ***HR: 3***
METHYL UREA and SODIUM NITRITE

SYNS: METHYLHARNSTOFF and NATRIUMNITRIT (GERMAN) ◇ SODIUM NITRITE and METHYL UREA

TOXICITY DATA with REFERENCE
orl-rat TDLo:80 mg/kg (9D preg):TER BEXBAN 83,88,77
orl-rat TDLo:15 g/kg/50D-C:NEO ZAPPAN 121,61,77
orl-rat TDLo:12 g/kg/17D-C:ETA ARZNAD 21,1707,71

SAFETY PROFILE: Experimental reproductive effects. Questionable carcinogen with experimental neoplastigenic, tumorigenic, and teratogenic data. When heated to decomposition it emits toxic fumes of Na_2O. See also SODIUM NITRITE and N-METHYLUREA.

MQJ500 CAS:1119-16-0 ***HR: 1***
4-METHYL VALERALDEHYDE
mf: $C_6H_{12}O$ mw: 100.18

SYN: ISOCAPROALDEHYDE

TOXICITY DATA with REFERENCE
skn-rbt 10 mg/24H open MLD AIHAAP 23,95,62
skn-rbt LD50:4460 mg/kg AIHAAP 23,95,62

CONSENSUS REPORTS: Reported in EPA TSCA Inventory.

SAFETY PROFILE: Mildly toxic by skin contact. A skin irritant. When heated to decomposition it emits acrid smoke and irritating fumes. See also ALDEHYDES.

MQJ750 CAS:97-61-0 ***HR: 2***
2-METHYLVALERIC ACID
mf: $C_6H_{12}O_2$ mw: 116.18

PROP: Water-white liquid. D: 0.8947 @ 20°/4°, bp: 126.5° @ 750 mm, vap press: 0.02 mm @ 20°, fp: sets to glass $< -85°$, flash p: 225°F (OC). Very sltly sol in water; misc in alc and ether.

SYNS: 2-METHYLPENTANOIC ACID ◇ METHYLPROPYLACETIC ACID ◇ α-METHYLVALERIC ACID ◇ 2-PENTANECARBOXYLIC ACID

TOXICITY DATA with REFERENCE
skn-rbt 500 mg open MLD UCDS** 2/4/59
skn-rbt 500 mg/24H MOD FCTOD7 20(Suppl),763,82
eye-rbt 250 μg open SEV AMIHBC 10,61,54
orl-rat LD50:2040 mg/kg AMIHBC 10,61,54
skn-rbt LD50:2500 mg/kg FCTOD7 20(Suppl),763,82

CONSENSUS REPORTS: Reported in EPA TSCA Inventory.

SAFETY PROFILE: Moderately toxic by ingestion and skin contact. A skin and severe eye irritant. Combustible when exposed to heat, flame, or oxidizers. To fight fire, use dry chemical, CO_2, foam. When heated to decomposition it emits acrid smoke and irritating fumes.

MQK000 CAS:4553-62-2 ***HR: 3***
2-METHYL-1,5-VALERODINITRILE
mf: $C_6H_7N_2$ mw: 107.15

TOXICITY DATA with REFERENCE
ivn-mus LD50:56 mg/kg CSLNX* NX#02814

CONSENSUS REPORTS: Cyanide and its compounds are on the Community Right-To-Know List. Reported in EPA TSCA Inventory.

SAFETY PROFILE: Poison by intravenous route.

When heated to decomposition it emits toxic fumes of NO_x and CN^-. See also NITRILES.

MQK500 CAS:3195-78-6 ***HR: 2***
N-METHYL-N-VINYLACETAMIDE
mf: C_5H_9NO mw: 99.15

SYNS: N-VINYLMETHYLACETAMIDE ◇ N-VINYL-N-METHYL-ACETAMIDE

TOXICITY DATA with REFERENCE
orl-rat LDLo:2830 mg/kg AIHAAP 23,95,62
skn-rbt LDLo:1410 mg/kg AIHAAP 23,95,62

CONSENSUS REPORTS: Reported in EPA TSCA Inventory.

SAFETY PROFILE: Moderately toxic by ingestion and skin contact. When heated to decomposition it emits toxic fumes of NO_x.

MQK750 CAS:108-22-5 ***HR: 3***
METHYLVINYL ACETATE
DOT: UN 2403
mf: $C_5H_8O_2$ mw: 100.13

PROP: Water-white liquid. Mp: −92.9°, bp: 96.6° @ 746 mm, flash p: 60°F (OC), d: 0.9226 @ 20°/20°, vap d: 3.45.

SYN: ISOPROPENYL ACETATE

TOXICITY DATA with REFERENCE
eye-rbt 500 mg AJOPAA 29,1363,46
orl-rat LD50:3000 mg/kg JIHTAB 31,60,49

CONSENSUS REPORTS: Reported in EPA TSCA Inventory.

DOT Classification: Flammable Liquid; Label:Flammable Liquid

SAFETY PROFILE: Moderately toxic by ingestion. An irritant to eyes and mucous membranes. A very dangerous fire hazard when exposed to heat or flame; can react vigorously with oxidizing materials. To fight fire, use alcohol foam, CO_2, dry chemical. When heated to decomposition it emits acrid smoke nd irritating fumes.

MQL000 CAS:2969-87-1 ***HR: 1***
METHYL VINYL ADIPATE
mf: $C_9H_{14}O_4$ mw: 186.23

SYNS: ADIPIC ACID, METHYL VINYL ESTER ◇ HEXANEDIOIC ACID ETHENYL METHYL ESTER ◇ VINYL METHYLADIPATE

TOXICITY DATA with REFERENCE
orl-rat LD50:6200 mg/kg GISAAA 35(10),88,70
orl-mus LD50:6200 mg/kg GISAAA 35(10),88,70
orl-rbt LD50:6200 mg/kg GISAAA 35(10),88,70

SAFETY PROFILE: Mildly toxic by ingestion. When heated to decomposition it emits acrid smoke and irritating fumes. See also ESTERS.

MQL250 CAS:598-32-3 ***HR: 1***
METHYL VINYL CARBINOL
mf: C_4H_8O mw: 72.12

PROP: Liquid. D: 0.831 20°/4° mp: < −80°, bp: 97°. Misc in water.

SYNS: 3-BUTEN-2-OL ◇ 1-METHYL PROPENOL

TOXICITY DATA with REFERENCE
eye-hmn 50 ppm/15M JIHTAB 28,262,46
mmo-sat 50 mmol/L MUREAV 93,305,82
mma-sat 50 mmol/L MUREAV 93,305,82
ihl-hmn TCLo:50 ppm:NOSE,EYE JIHTAB 28,262,46

CONSENSUS REPORTS: Reported in EPA TSCA Inventory.

SAFETY PROFILE: Human systemic effects by inhalation: eye problems. A human eye irritant. Human mutation data reported. When heated to decomposition it emits acrid smoke and irritating fumes. See also ALCOHOLS.

MQL750 CAS:107-25-5 ***HR: 3***
METHYL VINYL ETHER
DOT: UN 1087
mf: C_3H_6O mw: 58.09

$$CH_3OCH{=}CH_2$$

PROP: Colorless, easily liquefied gas or colorless liquid. Bp: 6.0°, d: 0.7500, vap d: 2.0, fp: −121.6°, vap press: 1052 mm @ 20°, flash p: −68.8°F, lel: 2.6%, uel: 39.0%.

SYNS: METHOXYETHENE ◇ VINYL METHYL ETHER (DOT)

TOXICITY DATA with REFERENCE
orl-rat LD50:4900 mg/kg 34ZIAG -,395,69

CONSENSUS REPORTS: Reported in EPA TSCA Inventory.

DOT Classification: Flammable Gas; Label: Flammable Gas.

SAFETY PROFILE: Mildly toxic by ingestion. A very dangerous fire hazard when exposed to heat, flame, or oxidizers. Explosive in the form of vapor when exposed to heat or flame. Can react vigorously with oxidizing materials. To fight fire, stop flow of gas. Potentially explosive reaction with halogens (e.g., bromine; chlorine) or hydrogen halides (e.g., hydrogen bromide; hydrogen chloride). Reaction with acids forms acetaldehyde. Weak acids catalyze the exothermic polymerization of the ether. The unstabilized ether can form dangerous peroxides. When heated to decomposition it emits acrid smoke and irritating fumes. See also ETHERS and PEROXIDES.

MQM000 CAS:107-25-5 ***HR: 3***
METHYL VINYL ETHER (inhibited)

SYN: VINYL METHYL ETHER, INHIBITED (DOT)

CONSENSUS REPORTS: Reported in EPA TSCA Inventory.

DOT Classification: Label: Flammable Gas.

SAFETY PROFILE: Flammable gas. Can explode in air. See also METHYL VINYL ETHER.

MQM100 ***HR: 3***
METHYL VINYL KETONE
mf: C_4H_6O mw: 70.09

$H_2C{=}CHCO{\bullet}CH_3$

SYN: 1-BUTEN-3-ONE

SAFETY PROFILE: May polymerize violently on exposure to heat or light. When heated to decomposition it emits acrid smoke and irritating fumes. See also KETONES.

MQM500 CAS:140-76-1 ***HR: 3***
2-METHYL-5-VINYLPYRIDINE
mf: C_8H_9N mw: 119.18

SYNS: 5-ETHENYL-2-METHYLPYRIDINE ◇ 5-VINYL-2-PICOLINE

TOXICITY DATA with REFERENCE
orl-rat LD50:1167 mg/kg AIHAAP 30,470,69
ihl-rat LC50:189 mg/m^3/2H 85GMAT -,88,82
scu-rat LD50:1290 mg/kg 85GMAT -,88,82
orl-mus LD50:775 mg/kg 85GMAT -,88,82
ihl-mus LC50:213 mg/m^3/2H 85GMAT -,88,82
scu-mus LD50:532 mg/kg 85GMAT -,88,82
skn-rbt LD50:718 mg/kg AIHAAP 30,470,69

CONSENSUS REPORTS: Reported in EPA TSCA Inventory. EPA Extremely Hazardous Substances List.

SAFETY PROFILE: Poison by inhalation. Moderately toxic by ingestion, skin contact and subcutaneous routes. When heated to decomposition it emits toxic fumes of NO_x.

MQM750 CAS:3680-02-2 ***HR: 3***
METHYL VINYL SULFONE
mf: $C_3H_6O_2S$ mw: 106.15

TOXICITY DATA with REFERENCEbt 10 mg/24H open MLD AIHAAP 23,95,62
orl-rat LD50:570 mg/kg AIHAAP 23,95,62
orl-rat LDLo:570 mg/kg AIHAAP 23,95,62
skn-rbt LD50:32 mg/kg AIHAAP 23,95,62

SAFETY PROFILE: Poison by skin contact. Moderately toxic by ingestion. A skin irritant. When heated to decomposition it emits toxic fumes of SO_x. See also SULFIDES.

MQM775 CAS:15284-39-6 ***HR: 3***
2-METHYL-5-VINYL TETRAZOLE
mf: $C_4H_6N_4$ mw: 110.12

$N{=}N(CH_3)N{=}NCCH{=}CH_2$ (ring)

SAFETY PROFILE: Forms an explosive complex with aluminum hydride. When heated to decomposition it emits toxic fumes of NO_x.

MQN000 CAS:5974-19-6 ***HR: 3***
METHYL VIOLET 6B
mf: $C_{31}H_{34}N_3{\bullet}Cl$ mw: 484.13

SYN: PENTAMETHYLBENZYL-p-ROSANILINECHLORIDE

TOXICITY DATA with REFERENCE
orl-mus LDLo:25 mg/kg CRSBAW 138,838,44
orl-rbt LDLo:75 mg/kg CRSBAW 138,838,44

SAFETY PROFILE: Poison by ingestion. When heated to decomposition it emits very toxic fumes of NO_x and Cl^-.

MQN025 CAS:8004-87-3 ***HR: 3***
METHYL VIOLET BB

SYNS: BASIC VIOLET K ◇ C.I. 42535 ◇ C.I. BASIC VIOLET 1 ◇ METHYL VIOLET ◇ METHYL VIOLET 2B ◇ METHYL VIOLET FN ◇ METHYL VIOLET N ◇ METHYL-VIOLETT (GERMAN) ◇ PARIS VIOLET R ◇ VIOLET POWDER H 2503

TOXICITY DATA with REFERENCE
mmo-sat 320 ng/plate MUREAV 89,21,81
orl-mus LD50:105 mg/kg ARZNAD 1,5,51
ipr-mus LD50:6 mg/kg ARZNAD 1,5,51

CONSENSUS REPORTS: EPA Genetic Toxicology Program. Reported in EPA TSCA Inventory.

SAFETY PROFILE: Poison by ingestion and intraperitoneal routes. Mutation data reported. When heated to decomposition it emits acrid smoke and irritating fumes.

MQN250 CAS:39425-26-8 ***HR: 3***
METHYL VIOLET CARBINOL

SYNS: CARBINOLBASE des METHYLVIOLETT (GERMAN) ◇ METHYLVIOLET CARBINOL BASE

TOXICITY DATA with REFERENCE
orl-mus LD50:880 mg/kg ARZNAD 1,5,51
ipr-mus LD50:36 mg/kg ARZNAD 1,5,51

SAFETY PROFILE: Poison by intraperitoneal route.

Moderately toxic by ingestion. When heated to decomposition it emits acrid smoke and irritating fumes.

MQN500 CAS:1076-22-8 ***HR: 3***
3-METHYLXANTHINE
mf: $C_6H_6N_4O_2$ mw: 166.16

TOXICITY DATA with REFERENCE
ipr-mus LD50:1300 mg/kg FATOAO 46(5),104,83
ivn-dog LDLo:300 mg/kg AEXPBL 43,305,1900
ivn-rbt LDLo:500 mg/kg AEXPBL 43,305,1900

SAFETY PROFILE: Poison by intravenous route. Moderately toxic by intraperitoneal route. When heated to decomposition it emits toxic fumes of NO_x.

MQO250 CAS:78186-37-5 ***HR: 3***
1-METHYL-N-(2,6-XYLYL)ISONIPECOTAMIDE HYDROCHLORIDE
mf: $C_{15}H_{22}N_2O•ClH$ mw: 282.85

SYNS: C 3134 ◇ 1-METHYL-PIPERIDINE-4-CARBONSAURE-O,O-XYLIDID HYDROCHLORID (GERMAN)

TOXICITY DATA with REFERENCE
ipr-rat LD50:270 mg/kg ARZNAD 8,609,58
scu-mus LD50:595 mg/kg ARZNAD 8,609,58

SAFETY PROFILE: Poison by intraperitoneal route. Moderately toxic by subcutaneous route. When heated to decomposition it emits very toxic fumes of HCl and NO_x.

MQO500 CAS:100836-65-5 ***HR: 3***
N-METHYL-N-(1-(3,5-XYLYLOXY)-2-PROPYL)CARBAMIC ACID-2-DIETHYLAMINO) ETHYL ESTER, HYDROCHLORIDE
mf: $C_{19}H_{32}N_2O_3•ClH$ mw: 372.99

SYN: C 2140

TOXICITY DATA with REFERENCE
eye-rbt 2% MLD ARZNAD 9,113,59
scu-mus LD50:175 mg/kg ARZNAD 9,113,59

SAFETY PROFILE: Poison by subcutaneous route. An eye irritant. When heated to decomposition it emits very toxic fumes of NO_x and HCl. See also CARBAMATES.

MQO750 CAS:100836-66-6 ***HR: 3***
N-METHYL-N-(1-(3,5-XYLYLOXY)-2-PROPYL)CARBAMIC ACID-2(2-METHYLPIPERIDINO)ETHYL ESTER HYDROCHLORIDE
mf: $C_{21}H_{34}N_2O_3•ClH$ mw: 399.03

SYN: C 2141

TOXICITY DATA with REFERENCE
eye-rbt 2% MOD ARZNAD 9,113,59
scu-mus LD50:260 mg/kg ARZNAD 9,113,59

SAFETY PROFILE: Poison by subcutaneous route. An eye irritant. When heated to decomposition it emits very toxic fumes of NO_x and HCl. See also ESTERS and CARBAMATES.

MQP000 CAS:100836-67-7 ***HR: 3***
N-METHYL-N-(1-(3,5-XYLYLOXY)-2-PROPYL)CARBAMIC ACID-2-(PYRROLIDINYL) ETHYL ESTER, HCl
mf: $C_{19}H_{30}N_2O_3•ClH$ mw: 370.97

SYN: C 6005

TOXICITY DATA with REFERENCE
eye-rbt 2% MOD ARZNAD 9,113,59
scu-mus LD50:160 mg/kg ARZNAD 9,113,59

SAFETY PROFILE: Poison by subcutaneous route. An eye irritant. When heated to decomposition it emits very toxic fumes of NO_x and HCl. See also ESTERS and CARBAMATES.

MQP250 CAS:18815-73-1 ***HR: 3***
METHYLZINC IODIDE
mf: CH_3IZn mw: 207.31

SYN: IODOMETHYLZINC

CONSENSUS REPORTS: Zinc and its compounds are on the Community Right-To-Know List.

SAFETY PROFILE: Ignites on contact with nitromethane. When heated to decomposition it emits toxic fumes of I^- and ZnO. See also ZINC COMPOUNDS and IODIDES.

MQP500 CAS:129-49-7 ***HR: 3***
METHYSERGIDE DIMALEATE
mf: $C_{21}H_{27}N_3O_2•C_4H_4O_4$ mw: 469.59

PROP: Decomp above 165°. Sol in methanol; less sol in water; insol in abs ethanol.

SYNS: 1-(HYDROXYMETHYL)PROPYLAMIDE of 1-METHYL-(+)-LYSERGIC ACID HYDROGEN MALEATE ◇ METHYLSERGIDE BIMALEATE ◇ SANSERT

TOXICITY DATA with REFERENCE
cyt-hmn:lyms 100 μg/L MUREAV 48,205,77
dlt-mus-ipr 100 mg/kg MUREAV 50,317,78
ipr-mus TDLo:400 μg/kg (1D pre):REP JOENAK 37,327,67
orl-rat LD50:200 mg/kg 27ZQAG -,101,72
ivn-rat LD50:125 mg/kg 27ZQAG -,101,72
orl-mus LD50:581 mg/kg 27ZQAG -,101,72
ivn-mus LD50:185 mg/kg 27ZQAG -,101,72
ivn-rbt LD50:28 mg/kg 27ZQAG -,101,72

SAFETY PROFILE: Poison by ingestion and intravenous routes. Experimental reproductive effects. Human mutation effects reported. When heated to decomposition it emits toxic fumes of NO_x.

MQQ000 CAS:5800-19-1 ***HR: 3***
METIAPINE
mf: $C_{19}H_{21}N_3S$ mw: 323.49

SYN: 2-METHYL-11-(4-METHYL-1-PIPERAZINYL)-DIBENZO(b,f)(1,4) THIAZEPINE

TOXICITY DATA with REFERENCE
orl-rat TDLo:42 mg/kg (14D pre):REP TXAPA9 25,212,73
orl-rbt TDLo:24 mg/kg (9-16D preg):TER TXAPA9 25,212,73
orl-rat LD50:934 mg/kg TXAPA9 22,335,72
ipr-rat LD50:487 mg/kg TXAPA9 25,220,73
orl-mus LD50:680 mg/kg TXAPA9 22,335,72
ipr-mus LD50:320 mg/kg TXAPA9 25,220,73

SAFETY PROFILE: Poison by intraperitoneal route. Moderately toxic by ingestion. An experimental teratogen. Experimental reproductive effects. When heated to decomposition it emits very toxic fumes of NO_x and SO_x.

MQQ050 CAS:1084-65-7 ***HR: 3***
METICRANE
mf: $C_{10}H_{13}NO_4S_2$ mw: 275.36

PROP: Crystals from methyl cellosolve. Mp: 236-237°.

SYNS: ARRESTEN ◇ FONTILIX ◇ E-103-E ◇ FONTILIZ ◇ 6-METHYL-3,4-DIHYDRO-2H-1-BENZOTHIOPYRAN-7-SULFONAMIDE 1,1-DIOXIDE ◇ 6-METHYL-7-SULFAMIDO-THIOCHROMAN-1,1-DIOXIDE ◇ 6-METHYLTHIOCHROMAN-7-SULFONAMIDE 1,1-DIOXIDE ◇ METICRAN ◇ SD 171-02

TOXICITY DATA with REFERENCE
orl-mus TDLo:120 mg/kg (female 7-12D post):REP OYYAA2 8,217,74
orl-rat TDLo:60 mg/kg (9-14D preg):TER OYYAA2 8,217,74
ivn-rat LD50:445 mg/kg OYYAA2 8,217,74
ipr-mus LD50:10 g/kg OYYAA2 8,217,74
ivn-mus LD50:325 mg/kg OYYAA2 8,217,74

SAFETY PROFILE: Poison by intravenous route. An experimental teratogen. Experimental reproductive effects. When heated to decomposition it emits toxic fumes of SO_x and NO_x.

MQQ250 CAS:9006-42-2 ***HR: 2***
METIRAM

PROP: A mixture of 5.2 parts by weight of ammoniates of ethylenebis(dithiocarbamate)zinc with 1 part by weight ethylenebisdithiocarbamic acid bimolecular and trimolecular cyclic anhydrosulfides and disulfides (85ARAE 4,51,76).

SYNS: AMAREX ◇ CARBATENE ◇ FMC-9102 ◇ NIA 9102 ◇ POLIKARBATSIN (RUSSIAN) ◇ POLYCARBACIN ◇ POLYCARBACINE ◇ POLYCARBAZIN ◇ POLYCARBAZINE ◇ POLYMARCIN ◇ POLYMARCINE ◇ POLYMARSIN ◇ POLYMARZIN ◇ POLYMARZINE ◇ POLYMAT ◇ POLYRAM ◇ POLYRAM 80 ◇ POLYRAM COMBI ◇ POLYRAM 80WP ◇ ZINC METIRAM ◇ ZINEB-ETHYLENE THIURAM DISULFIDE ADDUCT

TOXICITY DATA with REFERENCE
mmo-smc 5 ppm RSTUDV 6,161,76
mrc-smc 50 ppm MUREAV 10,533,70
orl-rat LD50:2850 mg/kg VRDEA5 (9).130,75
orl-mus LD50:2630 mg/kg VETNAL 56(6),59,80
orl-rbt LD50:620 mg/kg VETNAL 56(6),59,80

CONSENSUS REPORTS: Zinc and its compounds are on the Community Right-To-Know List. EPA Genetic Toxicology Program.

SAFETY PROFILE: Moderately toxic by ingestion. Mutation data reported. When heated to decomposition it emits toxic fumes of SO_x, NO_x, ZnO, and NH_3. See also ETHYLENEBIS(DITHIOCARBAMATO)ZINC, CARBAMATES, SULFIDES, and ZINC COMPOUNDS.

MQQ300 ***HR: 3***
METOCLOPRAMIDE DIHYDROCHLORIDE MONOHYDRATE
mf: $C_{14}H_{22}ClN_3O_2•2ClH•H_2O$ mw: 390.78

SYNS: 4-AMINO-5-CHLORO-N-(2-(DIETHYLAMINO)ETHYL)-o-ANISAMIDE DIHYDROCHLORIDE MONOHYDRATE ◇ 2-METHOXY-4-AMINO-5-CHLORO-N-β-DIETHYLAMINOETHYL)BENZAMIDE DIHYDROCHLORIDE MONOHYDRATE

TOXICITY DATA with REFERENCE
orl-rat TDLo:2100 mg/kg (30D pre):REP YACHDS 8,3991,80
orl-rat LD50:647 mg/kg YACHDS 8,3991,80
ipr-rat LD50:130 mg/kg AAREAV 23,663,66
scu-rat LD50:633 mg/kg AAREAV 23,663,66
ivn-rat LD50:42100 μg/kg YACHDS 8,3991,80
orl-mus LD50:335 mg/kg YACHDS 8,3991,80
ipr-mus LD50:138 mg/kg AAREAV 23,663,66
scu-mus LD50:223 mg/kg AAREAV 23,663,66
ivn-mus LD50:37500 μg/kg YACHDS 8,3991,80
ivn-rbt LD50:23 mg/kg AAREAV 23,663,66

SAFETY PROFILE: Poison by ingestion, subcutaneous, intravenous, and intraperitoneal routes. Experimental reproductive effects. When heated to decomposition it emits toxic fumes of NO_x and HCl.

MQQ450 CAS:51218-45-2 ***HR: 2***
METOLACHLOR
mf: $C_{15}H_{22}ClNO_2$ mw: 283.83

PROP: Clear, odorless liquid. Bp: 100°, n (20/D) 1.5301, vap press @ 20°: 0.000013 mm Hg. Solubility in water at 20°: 530 ppm. Sol in most organic solvents.

SYNS: 2-AETHYL-6-METHYL-N-(1-METHYL-2-METHOXYAETHYL)-CHLORACETANILID (GERMAN) ◇ BICEP ◇ CGA-24705 ◇ α-CHLOR-6'-AETHYL-n-(2-METHOXY-1-METHYLAETHYL)-ACET-o-TOLUIDIN (GERMAN) ◇ 2-CHLORO-6'-ETHYL-N-(2-METHOXY-1-METHYL-ETHYL)ACET-o-TOLUIDIDE ◇ α-CHLORO-2'-ETHYL-6'-METHYL-N-(1-METHYL-2-METHOXYETHYL)-ACETANILIDE ◇ 2-CHLORO-N-(2-ETHYL-6-METHYLPHENYL)-N-(2-METHOXY-1-METHYLETHYL) ACETAMIDE ◇ 2-CHLORO-N-(6-ETHYL-o-TOLYL)-N-(2-METHOXY-1-METHYLETHYL)-ACETAMIDE ◇ CODAL ◇ COTORAN MULTI ◇ DUAL ◇ 2-ETHYL-6-METHYL-1-N-(2-METHOXY-1-EMTHYLETHYL) CHLOROACETANILIDE ◇ METELILACHLOR ◇ MILOCEP ◇ ONTRACK 8E ◇ PRIMAGRAM ◇ PRIMEXTRA

TOXICITY DATA with REFERENCE
skn-rbt 334 mg/kg open MLD CIGET* -,-,77
mma-sat 10 μg/plate MUREAV 136,233,84
mrc-smc 1 μg/plate MUREAV 136,233,84
orl-rat LD50:2534 mg/kg FMCHA2 -,C91,83
skn-rat LD50:3710 mg/kg LTIRC* TR-78-0310

CONSENSUS REPORTS: EPA Genetic Toxicology Program.

SAFETY PROFILE: Moderately toxic by skin contact and ingestion. Mutation data reported. A skin irritant. When heated to decomposition it emits toxic fumes of Cl^- and NO_x.

MQQ500 CAS:5377-20-8 ***HR: 3***
METOMIDATE
mf: $C_{13}H_{14}N_2O_2$ mw: 230.29

SYN: METHYL 1-(α-METHYLBENZYL)IMIDAZOLE-5-CARBOXYLATE

TOXICITY DATA with REFERENCE
orl-pgn LD50:56 mg/kg TXAPA9 21,315,72
orl-dck LD50:56 mg/kg TXAPA9 21,315,72
orl-brd LD50:32 mg/kg TXAPA9 21,315,72

SAFETY PROFILE: Poison by ingestion. When heated to decomposition it emits toxic fumes of NO_x.

MQQ750 CAS:35944-74-2 ***HR: 3***
METOMIDATE HYDROCHLORIDE
mf: $C_{13}H_{14}N_2O_2 \cdot ClH$ mw: 266.75

SYNS: 1-(α-METHYLBENZYL)IMIDAZOLE-5-CARBOXYLIC ACID METHYL ESTER, HYDROCHLORIDE ◇ 1-(1-PHENETHYL)-IMIDAZOLE-5-CARBOXYLIC ACID, METHYL ESTER, HYDROCHLORIDE

TOXICITY DATA with REFERENCE
ivn-mus LD50:18 mg/kg CSLNX* NX#09250
orl-pgn LD50:42 mg/kg TXAPA9 21,315,72
orl-dck LD50:133 mg/kg TXAPA9 21,315,72
orl-bwd LD50:32 mg/kg TXAPA9 21,315,72

SAFETY PROFILE: Poison by ingestion and intravenous routes. When heated to decomposition it emits very toxic fumes of NO_x and HCl.

MQR000 CAS:14008-44-7 ***HR: 3***
METOPIMAZINE
mf: $C_{22}H_{27}N_3O_3S_2$ mw: 445.64

PROP: Solid. Mp: 170-171°.

SYNS: 10-(3-(4-CARBAMOYLPIPERIDINE)PROPYL)-2-(METHANESULFONYL)PHENOTHIAZINE ◇ EXP 999 ◇ 2-METHYLSULFONYL-10-(3-(4-CARBAMOYLPIPERIDINO)PROPYL)PHENOTHIAZINE ◇ 1-(3-(2-(METHYLSULFONYL)PHENOTHIAZIN-10-YL)PROPYL)ISONIPECOTAMIDE ◇ 1-(3-(2-METHYLSULFONYL)PHENOTHIAZIN-10-YL)PROPYL)-4-PIPERIDINE CARBOXAMIDE ◇ 1-(3-(2-(METHYLSULFONYL)-10H-PHENOTHIAZIN-10-YL)PROPYL)-4-PIPERIDINECARBOXAMIDE ◇ RP 9965 ◇ VOGALENE

TOXICITY DATA with REFERENCE
orl-rat LD50:976 mg/kg TXAPA9 18,185,71
scu-rat LD50:1080 mg/kg TXAPA9 18,185,71
orl-mus LD50:800 mg/kg 27ZQAG -,32,72
ipr-mus LD50:95 mg/kg 27ZQAG -,32,72
scu-mus LD50:315 mg/kg 27ZQAG -,32,72
ivn-mus LD50:90 mg/kg 27ZQAG -,32,72

SAFETY PROFILE: Poison by intravenous, intraperitoneal, and subcutaneous routes. Moderately toxic by ingestion. An anti-emetic agent. When heated to decomposition it emits very toxic fumes of SO_x and NO_x.

MQR100 CAS:7761-45-7 ***HR: 3***
METOPRINE
mf: $C_{11}H_{10}Cl_2N_4$ mw: 269.15

SYNS: BW 50-197 ◇ BW-197U ◇ DDMP ◇ 2,4-DIAMINO-5-(3,4-DICHLOROPHENYL)-6-METHYLPYRIMIDINE ◇ 2,4-DIAMINO-5-(3',4'-DICHLOROPHENYL)-6-METHYLPYRIMIDINE ◇ 5-(3,4-DICHLOROPHENYL)-6-METHYL-2,4-PYRIMIDINEDIAMINE ◇ METHODICHLOROPHEN ◇ NSC-19494 ◇ PYRIMETHAMINE ◇ U-197

TOXICITY DATA with REFERENCE
cyt-hmn-unr 2800 μg/kg RBBIAL 25,145,65
orl-rat TDLo:2500 μg/kg (7D preg):TER PSEBAA 87,571,54
orl-rat TDLo:5 mg/kg (8-9D preg):REP PSEBAA 87,571,54
ipr-rat LD50:7 mg/kg 14XBAV -,367,64
scu-mus LD50:50 mg/kg CTRRDO 60,547,76

SAFETY PROFILE: Poison by subcutaneous and intraperitoneal routes. An experimental teratogen. Experimental reproductive effects. Human mutation data reported. When heated to decomposition it emits toxic fumes of Cl^- and NO_x.

MQR144 CAS:37350-58-6 ***HR: 3***
METOPROLOL
mf: $C_{15}H_{25}NO_3$ mw: 267.41

SYNS: CGP 2175 ◇ H 93/26 ◇ (±)-1-(4-(2-METHOXYETHYL) PHENOXY)-3-((1-METHYLTHYL)AMINO)-2-PROPANOL ◇ (±)-METOPROLOL

TOXICITY DATA with REFERENCE
orl-rat TDLo:60 g/kg (9W male/2W pre-3W post):REP APTSAI 36,96,75
orl-wmn TDLo:17 mg/kg/17D-I AIMEAS 108,67,88
orl-wmn TDLo:150 mg/kg:EYE,CVS,CNS AMSVAZ 218,525,85
orl-mus LD50:1050 mg/kg USXXAM #4252984
ivn-mus LD50:62 mg/kg USXXAM #4252984

SAFETY PROFILE: Poison by intravenous route. Moderately toxic by ingestion. Human systemic effects by ingestion: cardiac, eye, and behavioral effects. Experimental reproductive effects. When heated to decomposition it emits toxic fumes of NO_x.

MQR150 ***HR: 3***
METOPROLOL TARTRATE
mf: $C_{15}H_{25}NO_3•1/2C_4H_6O_6$ mw: 342.41

SYNS: (±)-1-(ISOPROPYLAMINO)-3-(p-(2-METHOXYETHYL) PHENOXY)-2-PROPANOL HEMI-l-TARTRATE ◇ 1-(ISOPROPYL-AMINO)-3-(p-(2-METHOXYETHYL)PHENOXY)-2-PROPANOLTARTRATE (2:1) ◇ METOPROLOL HEMITARTRATE

TOXICITY DATA with REFERENCE
orl-rat TDLo:5500 mg/kg (female 7-17D post):REP KSRNAM 13,3216,79
orl-rat TDLo:5500 mg/kg (female 7-17D post):TER KSRNAM 13,3216,79
orl-rat LD50:5500 mg/kg IYKEDH 14,297,83
ipr-rat LD50:219 mg/kg YAKUD5 24,2439,82
scu-rat LD50:1150 mg/kg KSRNAM 13,2839,79
ivn-rat LD50:90 mg/kg IYKEDH 14,297,83
orl-mus LD50:1480 mg/kg IYKEDH 14,297,83
ipr-mus LD50:204 mg/kg IYKEDH 14,297,83
scu-mus LD50:510 mg/kg IYKEDH 14,297,83
ivn-mus LD50:84 mg/kg IYKEDH 14,297,83
orl-dog LD50:1090 mg/kg IYKEDH 14,297,83
orl-rbt LD50:604 mg/kg IYKEDH 14,297,83
ivn-rbt LD50:28700 μg/kg IYKEDH 14,297,83

SAFETY PROFILE: Poison by intravenous and intraperitoneal routes. Moderately toxic by ingestion and subcutaneous routes. An experimental teratogen. Experimental reproductive effects. When heated to decomposition it emits toxic fumes of NO_x.

MQR200 CAS:1178-29-6 ***HR: 3***
METOSERPATE HYDROCHLORIDE
mf: $C_{24}H_{32}N_2O_5•ClH$ mw: 465.04

SYNS: METHYL-18-EPIRESERPATE METHYL ETHER HYDROCHLORIDE ◇ PACITRAN ◇ SU-9064 ◇ SU 8842 HYDROCHLORIDE

TOXICITY DATA with REFERENCE
orl-rat LD50:182 mg/kg JPETAB 138,78,62
ivn-rat LD50:21 mg/kg 27ZQAG -,112,72
ivn-mus LD50:24 mg/kg JPETAB 138,78,62
orl-bwd LD50:100 mg/kg AECTCV 12,355,83

SAFETY PROFILE: Poison by ingestion and intravenous routes. When heated to decomposition it emits toxic fumes of NO_x and HCl.

MQR225 CAS:19937-59-8 ***HR: 2***
METOXURON
mf: $C_{10}H_{13}ClN_2O_2$ mw: 228.70

SYNS: N'-(3-CHLOR-4-METHOXY-PHENYL)-N,N-DIMETHYL-HARNSTOFF (GERMAN) ◇ 3-(3-CHLOR-4-METHOXYPHENYL)-1,1-DIMETHYLHARNSTOFF (GERMAN) ◇ N-(3-CHLORO-4-METHOXY-PHENYL)-N',N'-DIMETHYLUREA ◇ 3-(3-CHLORO-4-METHOXY-PHENYL)-1,1-DIMETHYL-UREA ◇ DEFTOR ◇ N,N-DIMETHYL-N'-(4-METHOXY-3-CHLOROPHENYL)UREA ◇ DOSAFLO ◇ DOSAGRAN ◇ DOSANEX ◇ DOSANEX FL ◇ DOSANEX MG ◇ FL ◇ HERBICIDE 6602 ◇ PURIVEL

TOXICITY DATA with REFERENCE
mma-sat 5 μg/plate MUREAV 58,353,78
dni-mus-orl 500 mg/kg MUREAV 58,353,78
orl-rat LD50:1600 mg/kg WRPCA2 9,119,70
orl-mus LD50:2542 mg/kg GISAAA 47(2),13,82
orl-rbt LD50:2300 mg/kg GISAAA 48(2),85,83

SAFETY PROFILE: Moderately toxic by ingestion. Mutation data reported. When heated to decomposition it emits toxic fumes of Cl^- and NO_x.

MQR250 ***HR: 3***
4-o-METPA
mf: $C_{37}H_{58}O_8$ mw: 630.95

TOXICITY DATA with REFERENCE
skn-mus 1450 ng MLD CCSUDL 2,11,78
skn-mus TDLo:966 mg/kg/48W-I:ETA CNREA8 39,4183,79

SAFETY PROFILE: Questionable carcinogen with experimental tumorigenic data by skin contact. A skin irritant. When heated to decomposition it emits acrid smoke and irritating fumes.

MQR275 CAS:21087-64-9 ***HR: 3***
METRIBUZIN
mf: $C_8H_{14}N_4OS$ mw: 214.32

SYNS: 4-AMINO-6-tert-BUTYL-3-(METHYLTHIO)-1,2,4-TRIAZIN-5-ONE ◇ 4-AMINO-6-tert-BUTYL-3-METHYLTHIO-as-TRIAZIN-5-ONE ◇ 4-AMINO-6-(1,1-DIMETHYLETHYL)-3-(METHYLTHIO)-1,2,4-TRIAZIN-5(4H)-ONE ◇ BAY 61597 ◇ BAY DIC 1468 ◇ BAYER 94337 ◇ BAYER 6159H ◇ BAYER 6443H ◇ DIC 1468 ◇ LEXONE ◇ SENCOR ◇ SENCORAL ◇ SENCORER ◇ SENCOREX

TOXICITY DATA with REFERENCE
orl-rat LD50:1100 mg/kg FMCHA2 -,C156,83
orl-mus LD50:698 mg/kg 28ZEAL 5,154,76
orl-gpg LD50:250 mg/kg 85DPAN -,-,71/76

CONSENSUS REPORTS: EPA Genetic Toxicology Program.

OSHA PEL: TWA 5 mg/m^3
ACGIH TLV: TWA 5 mg/m^3

SAFETY PROFILE: Poison by ingestion. When heated to decomposition it emits very toxic fumes of NO_x and SO_x.

MQR300 CAS:31112-62-6 ***HR: 3***
METRIZAMIDE
mf: $C_{18}H_{22}I_3N_3O_8$ mw: 789.13

PROP: White crystals from isopropyl alcohol. Mp: 230° (decomp). Very sol in water (50% w/v) at room temp.

SYNS: 2-(3-ACETAMIDO-5-N-METHYL-ACETAMIDO-2,4,6-TRIIODOBENZAMIDO)2-DEOXY-d-GLUCOSE ◇ 2-(3-ACETAMIDO-2,4,6-TRIIODO-5-(N-METHYLACETAMIDO)BENAZMIDO)-2-DEOXY-d-GLUCOPYRANOSE ◇ 2-(3-ACETAMIDO-2,4,6-TRIIODO-5-(N-METHYLACETAMIDO)BENZAMIDO)-2-DEOXY-d-GLUCOSE ◇ 2-((3-(ACETYLAMINO)-5-(ACETYLMETHYLAMINO)-2,4,6-TRIIODOBENZOYL)AMINO)-2-DEOXY-d-GLUCOSE ◇ 3-ACETYL-AMINO-5-N-METHYL-ACETYLAMINO-2,4,6-TRIIODOBENZOYL GLUCOSAMINE ◇ AMIPAQUE ◇ TELEBRIX 300 ◇ WIN 39103

TOXICITY DATA with REFERENCE
ipr-rat TDLo:34369 mg/kg (female 1-7D post):TER NIIHAO 24,277,79
ivn-rat TDLo:54008 mg/kg (female 7-17D post):REP NIIHAO 24,287,79
isp-man TDLo:143 mg/kg:CNS SMJOAV 77,88,84
ipr-rat LD50:9300 mg/kg NIIRDN 6,842,82
scu-rat LD50:17 mg/kg YKYUA6 32,865,81
ivn-rat LD50:13100 mg/kg NIIRDN 6,842,82
ice-rat LD50:135 mg/kg RADLAX 140,713,81
ipr-mus LD50:9950 mg/kg USXXAM #4001323
scu-mus LD50:7500 mg/kg NIIRDN 6,842,82
ivn-mus LD50:1534 mg/kg INVRAV 15(Suppl),248,80
ice-mus LD50:2910 mg/kg USXXAM #4001323
par-rbt LD50:207 mg/kg USXXAM #4001323

SAFETY PROFILE: Poison by subcutaneous, intracerebral, and parenteral routes. Moderately toxic by intraperitoneal and intravenous routes. Human systemic effects: meningeal changes to spinal cord. An experimental teratogen. An experimental teratogen. Experimental reproductive effects. When heated to decomposition it emits toxic fumes of I^- and NO_x.

MQR350 CAS:1949-45-7 ***HR: 3***
METRIZOIC ACID
mf: $C_{12}H_{11}I_3N_2O_4$ mw: 627.95

PROP: Crystals. Mp: 281-282°.

SYNS: 3-(ACETYLAMINO)-5-(ACETYLMETHYLAMINO)-2,4,6-TRIIODO-BENZOIC ACID ◇ METRIZOATE

TOXICITY DATA with REFERENCE
orl-rat LD50:38100 mg/kg NIIRDN 6,843,82
scu-rat LD50:29600 mg/kg NIIRDN 6,843,82
ivn-rat LD50:14300 mg/kg NIIRDN 6,843,82
scu-mus LD50:36 g/kg NIIRDN 6,843,82
ivn-mus LD50:12200 mg/kg INVRAV 15,502,80
ice-mus LD50:234 mg/kg NIIRDN 6,843,82

SAFETY PROFILE: Poison by intracerebral route. Mildly toxic by ingestion. When heated to decomposition it emits toxic fumes of I^- and NO_x.

MQR500 CAS:826-39-1 ***HR: 3***
MEVASIN HYDROCHLORIDE
mf: $C_{11}H_{21}N{\bullet}ClH$ mw: 203.79

SYNS: INVERSINE HYDROCHLORIDE ◇ MECAMINE HYDROCHLORIDE ◇ MECAMYLAMINE HYDROCHLORIDE ◇ MEKAMIN HYDROCHLORIDE ◇ 3-METHYLAMINOISOCAMPHANE HYDROCHLORIDE ◇ N-METHYL-dl-ISOBORNYLAMINE HYDROCHLORIDE ◇ N,2,3,3-TETRAMETHYL-2-NORBORNANAMINEHYDROCHLORIDE

TOXICITY DATA with REFERENCE
orl-rat LD50:208 mg/kg JPETAB 117,169,56
ipr-rat LD50:53 mg/kg RPOBAR 2,301,70
scu-rat LD50:177 mg/kg JPETAB 117,169,56
ivn-rat LD50:21 mg/kg RPOBAR 2,300,70
orl-mus LD50:92 mg/kg NIIRDN 6,814,82
ipr-mus LD50:17 mg/kg RPOBAR 2,300,70
scu-mus LD50:81 mg/kg RPOBAR 2,300,70
ivn-mus LD50:14 mg/kg RPOBAR 2,300,70
orl-gpg LD50:175 mg/kg JPETAB 117,169,56
ipr-gpg LD50:63 mg/kg JPETAB 117,169,56
scu-gpg LD50:155 mg/kg JPETAB 117,169,56

SAFETY PROFILE: Poison by ingestion, intraperitoneal, intravenous, and subcutaneous routes. When heated to decomposition it emits very toxic fumes of HCl and NO_x.

MQR750 CAS:7786-34-7 ***HR: 3***
MEVINPHOS
mf: $C_7H_{13}O_6P$ mw: 224.17

SYNS: APAVINPHOS ◇ α-2-CARBOMETHOXY-1-METHYLVINYL DIMETHYL PHOSPHATE ◇ 2-CARBOMETHOXY-1-PROPEN-2-YL DIMETHYL PHOSPHATE ◇ CMDP ◇ COMPOUND 2046 ◇ 3-((DIMETHOXYPHOSPHINYL)OXY)-2-BUTENOIC ACID METHYL ESTER ◇ O,O-DIMETHYL-O-(2-CARBOMETHOXY-1-METHYLVINYL) PHOSPHATE ◇ DIMETHYL-1-CARBOMETHOXY-1-PROPEN-2-YL PHOSPHATE ◇ DIMETHYL ESTER PHOSPHORIC ACID ESTER with METHYL 3-HYDROXYCROTONATE ◇ O,O-DIMETHYL-O-2-METHOXYCARBONYL-1-METHYL-VINYL-PHOSPHAT(GERMAN) ◇ DIMETHYL 2-METHOXYCARBONYL-1-METHYLVINYL PHOSPHATE ◇ DIMETHYL METHOXYCARBONYLPROPENYL PHOSPHATE ◇ DIMETHYL (1-METHOXYCARBOXYPROPEN-2-YL)PHOSPHATE ◇ O,O-DIMETHYL O-(1-METHYL-2-CARBOXYVINYL) PHOSPHATE ◇ DURAPHOS ◇ ENT 22,374 ◇ FOSDRIN ◇ GESFID ◇ GESTID ◇ 3-HYDROXYCROTONIC ACID METHYL ESTER DIMETHYL PHOSPHATE ◇ MENIPHOS ◇ MENITE ◇ (2-METHOXY-CARBONYL-1-METHYL-VINYL)-DIMETHYL-PHOSPHAT(GERMAN) ◇ (2-METHOXYCARBONYL-1-METHYL-VINYL)-DIMETHYL-FOSFAAT (DUTCH) ◇ 2-METHOXYCARBONYL-1-METHYLVINYL DIMETHYL-PHOSPHATE ◇ 1-METHOXYCARBONYL-1-PROPEN-2-YL DIMETHYL

PHOSPHATE ◇ (1-METHOXYCARBOXYPROPEN-2-YL)PHOSPHORIC ACID, DIMETHYL ESTER ◇ METHYL-3-(DIMETHOXYPHOSPHINYL-OXY)CROTONATE ◇ (2-METOSSICARBONIL-1-METIL-VINIL)-DIMETIL-FOSFATO (ITALIAN) ◇ MEVINFOS (DUTCH) ◇ OS 2046 ◇ PHOSDRIN (OSHA) ◇ PHOSFENE ◇ PHOSPHATE de DIMETHYLE et de 2-METHOXYCARBONYL-1 METHYLVINYLE (FRENCH) ◇ PHOS-PHENE (FRENCH)

TOXICITY DATA with REFERENCE
orl-man TDLo:700 μg/kg/28D-I:PNS TXAPA9 42,351,77
orl-rat LD50:3 mg/kg DOEAAH 36,25,79
ihl-rat LC50:14 ppm/1H AMIHBC 9,45,54
skn-rat LD50:4200 μg/kg TXAPA9 2,88,60
ipr-rat LD50:1350 μg/kg PSEBAA 114,509,63
orl-mus LD50:4 mg/kg AMIHBC 9,45,54
skn-mus LD50:12 mg/kg JTEHD6 9,491,82
ipr-mus LD50:2 mg/kg CJBPAZ 39,1790,61
scu-mus LD50:1180 μg/kg JPPMAB 19,612,67
ivn-mus LD50:680 μg/kg JPPMAB 19,612,67
skn-rbt LD50:4700 μg/kg GUCHAZ 6,353,73

CONSENSUS REPORTS: EPA Genetic Toxicology Program. EPA Extremely Hazardous Substances List.

OSHA PEL: (Transitional: TWA 0.1 mg/m^3 (skin)) TWA 0.01 ppm; STEL 0.03 ppm (skin)
ACGIH TLV: TWA 0.01 ppm; STEL 0.03 ppm (skin)
DFG MAK: 0.01 ppm (0.1 mg/m^3)
DOT Classification: Poison B; Label: Poison.

SAFETY PROFILE: Poison by ingestion, inhalation, skin contact, subcutaneous, intravenous, and intraperitoneal routes. Human systemic effects by ingestion: peripheral motor nerve recording changes. An insecticide. When heated to decomposition it emits toxic fumes of PO_x.

MQR760 CAS:31868-18-5 ***HR: 2***
MEXAZOLAM
mf: $C_{18}H_{16}Cl_2N_2O_2$ mw: 363.26

PROP: Crystals. Mp: 172-175°.

SYNS: CS 386 ◇ MELEX

TOXICITY DATA with REFERENCE
orl-rat LD50:4500 mg/kg SKKNAJ 30,175,78
ipr-rat LD50:4000 mg/kg SKKNAJ 30,175,78
scu-rat LD50:4000 mg/kg SKKNAJ 30,175,78
orl-mus LD50:4571 mg/kg SKKNAJ 30,175,78
ipr-mus LD50:6000 mg/kg SKKNAJ 30,175,78
scu-mus LD50:6000 mg/kg SKKNAJ 30,175,78

SAFETY PROFILE: Moderately toxic by intraperitoneal and subcutaneous routes. Mildly toxic by ingestion. When heated to decomposition it emits toxic fumes of Cl^- and NO_x.

MQR775 CAS:5370-01-4 ***HR: 3***
MEXILETINE HYDROCHLORIDE
mf: $C_{11}H_{17}NO{\cdot}ClH$ mw: 215.75

SYNS: KO 1173 ◇ KOE 1173 HYDROCHLORIDE ◇ 1-METHYL-2-(2,6-XYLYLOXY)-ETHYLAMINE HYDROCHLORIDE ◇ MEXITIL

TOXICITY DATA with REFERENCE
orl-rat TDLo:1350 mg/kg (female 17-22D post):REP IYKEDH 14,527,83
orl-rat TDLo:825 mg/kg (female 7-17D post):TER IYKEDH 14,527,83
orl-rat LD50:330 mg/kg IYKEDH 13,922,82
scu-rat LD50:500 mg/kg IYKEDH 13,922,82
ivn-rat LD50:27 mg/kg IYKEDH 13,922,82
ims-rat LD50:190 mg/kg IYKEDH 13,922,82
orl-mus LD50:272 mg/kg ARZNAD 38,1398,88
scu-mus LD50:235 mg/kg IYKEDH 13,922,82
ivn-mus LD50:21 mg/kg BLLIAX 85,152,86
ims-mus LD50:128 mg/kg IYKEDH 13,922,82
orl-dog LD50:356 mg/kg IYKEDH 16,590,85

SAFETY PROFILE: Poison by ingestion, subcutaneous, intramuscular, and intravenous routes. Experimental reproductive effects. When heated to decomposition it emits toxic fumes of NO_x and HCl. See also AMINES.

MQS100 CAS:30578-37-1 ***HR: 3***
MEZINIUM METHYL SULFATE
mf: $C_{11}H_{12}N_3O{\cdot}CH_3O_4S$ mw: 313.36

PROP: Crystals from water. Mp: 176° (decomp).

SYNS: AMEZINIUMMETILSULFAT (GERMAN) ◇ AMEZINIUM METILSULFATE ◇ 4-AMINO-6-METHOXY-1-PHENYLPYRIDAZINIUM-METHYLSULFAT (GERMAN) ◇ 4-AMINO-6-METHOXY-1-PHENYL-PYRIDAZINIUM METHYL SULFATE ◇ LU 1631 ◇ REGULTON ◇ SUPRATONIN

TOXICITY DATA with REFERENCE
orl-rat TDLo:11340 mg/kg (male 9W pre):TER YACHDS 16,1543,88
orl-rat TDLo:11340 mg/kg (male 9W pre):REP YACHDS 16,1543,88
orl-rat LD50:1410 mg/kg ARZNAD 31,1580,81
ivn-rat LD50:45 mg/kg ARZNAD 31,1580,81
orl-mus LD50:1330 mg/kg YACHDS 16,1471,88
ivn-mus LD50:40 mg/kg ARZNAD 31,1580,81

SAFETY PROFILE: Poison by intravenous route. Moderately toxic by ingestion. An experimental teratogen. Other experimental reproductive effects. When heated to decomposition it emits toxic fumes of NO_x and SO_x.

MQS200 CAS:51481-65-3 ***HR: 2***
MEZLOCILLIN
mf: $C_{21}H_{24}N_5O_8S_2{\cdot}Na$ mw: 561.61

SYN: ANTIBIOTIC BAY-f 1353

TOXICITY DATA with REFERENCE
ivn-rat TDLo:5500 mg/kg (female 7-17D post):REP IYKEDH 9,986,78
ivn-rat TDLo:81 g/kg (60D male/14D pre-7D pre):TER IYKEDH 9,981,78
ivn-rat LD50:2636 mg/kg OYYAA2 17,513,79
ivn-mus LD50:6329 mg/kg OYYAA2 17,513,79

SAFETY PROFILE: Moderately toxic by intravenous route. An experimental teratogen. Experimental reproductive effects. When heated to decomposition it emits toxic fumes of SO_x, NO_x, and Na_2O.

MQS215 ***HR: D***
MIANG TEA LEAF EXTRACT

TOXICITY DATA with REFERENCE
dns-hmn:lym 5000 ppm CNREA8 39,4802,79
dni-rat:mmr 25000 ppm CNREA8 39,4802,79

SAFETY PROFILE: Human mutation data reported. When heated to decomposition it emits acrid smoke and irritating fumes.

MQS220 CAS:24219-97-4 ***HR: 3***
MIANSERINE
mf: $C_{18}H_{20}N_2$ mw: 264.40

SYNS: DIBENZO(c,f)PYRAZINO(1,2-a)AZEPINE,1,2,3,4,10,14b-HEXAHYDRO-2-METHYL- ◇ 1,2,3,4,10,14b-HEXAHYDRO-2-METHYLDIBENZO(c,f)PYRAZINO(1,2-a)AZEPINE ◇ MIANSERIN ◇ MIANSERYNA

TOXICITY DATA with REFERENCE
scu-rat TDLo:65 mg/kg (female 8-20D post):REP ARTODN 7,504,84
orl-wmn TDLo:25 mg/kg/6W-I:BLD BMJOAE 291,1375,85
orl-mus LD50:365 mg/kg DDREDK 3,357,83
ipr-mus LD50:130 mg/kg APPHAX 40,235,83

SAFETY PROFILE: Poison by ingestion and intraperitoneal routes. Human systemic effects by ingestion: agranulocytosis. Experimental reproductive effects. When heated to decomposition it emits toxic fumes of NO_x.

MQS225 CAS:3704-09-4 ***HR: 3***
MIBOLERONE
mf: $C_{20}H_{30}O_2$ mw: 302.50

PROP: Crystalline solid. Solubility in deionized water: 0.0454 mg/mL @ 37°.

SYNS: CHEQUE ◇ (7-α,17-β)-17-HYDROXY-7,17-DIMETHYL-ESTR-4-EN-3-ONE (9CI) ◇ 17-β-HYDROXY-7-α,17-DIMETHYLESTR-4-EN-3-ONE ◇ MATENON ◇ MIBOLERON ◇ U 10997

TOXICITY DATA with REFERENCE
scu-mus TDLo:800 μg/kg (10D pre):REP STEDAM 9,235,67
orl-dog TDLo:196 μg/kg (6-61D preg/42D post):TER AJVRAH 39,837,78
orl-dog TDLo:8985 μg/kg/9.6Y-I:NEO TOPADD 13,177,85

SAFETY PROFILE: Experimental teratogenic and reproductive effects. Questionable carcinogen with experimental neoplastigenic data. When heated to decomposition it emits acrid smoke and irritating fumes.

MQS250 CAS:12001-26-2 ***HR: 2***
MICA

PROP: Containing less than 1% crystalline silica (FEREAC 39,23540,74).

SYNS: MICA SILICATE ◇ SUZORITE MICA

OSHA PEL: (Transitional: TWA 20 mppcf) TWA Respirable Fraction: 3 mg/m^3
ACGIH TLV: TWA Respirable Fraction: 3 mg/m^3

SAFETY PROFILE: The dust is injurious to lungs. See SILICATES.

MQS500 CAS:90-94-8 ***HR: 3***
MICHLER'S KETONE
mf: $C_{17}H_{20}N_2O$ mw: 268.39

PROP: Leaves from ethanol. Mp: 172°, bp: >360° decomp. Insol in water; very sol in benzene; sol in alc; very sltly sol in ether.

SYNS: p,p'-BIS(N,N-DIMETHYLAMINO)BENZOPHENONE ◇ 4,4'-BIS(DIMETHYLAMINO)BENZOPHENONE ◇ BIS(p-(N,N-DIMETHYLAMINO)PHENYL)KETONE ◇ BIS(4-(DIMETHYLAMINO)PHENYL) METHANONE ◇ p,p'-MICHLER'S KETONE ◇ NCI-C02006 ◇ TETRA- nn-METHYLDIAMINOBENZOPHENONE

TOXICITY DATA with REFERENCE
mma-sat 33300 ng/plate ENMUDM 7(Suppl 5),1,85
cyt-ham:fbr 1500 μg/L MUTAEX 1,17,86
orl-rat TDLo:15 g/kg/78W-C:CAR NCITR* NCI-TR-181,79
orl-mus TDLo:82 g/kg/78W-C:CAR NCITR* NCI-TR-181,79
orl-rat TD:8 g/kg/78W-C:NEO NCITR* NCI-TR-181,79

CONSENSUS REPORTS: NTP Fifth Annual Report on Carcinogens. NCI Carcinogenesis Bioassay (feed); Clear Evidence: mouse, rat NCITR* NCI-CG-TR-181,79. Reported in EPA TSCA Inventory. EPA Genetic Toxicology Program.

DFG MAK: Suspected Carcinogen.

SAFETY PROFILE: Confirmed human carcinogen with experimental carcinogenic and neoplastigenic data. Mutation data reported. When heated to decomposition it emits toxic fumes of NO_x. See also KETONES.

MQS550 CAS:22916-47-8 ***HR: 3***
MICONAZOLE
mf: $C_{18}H_{14}Cl_4N_2O$ mw: 416.14

PROP: (+)-Form: Mp: 135.3°. (−)-Form: Mp: 135°.

SYNS: 1-(2-(2,4-DICHLOROPHENYL)-2-((2,4-DICHLOROPHENYL) METHOXY)ETHYL-IMIDAZOLE (9CI) ◇ R 18134

TOXICITY DATA with REFERENCE
sln-asn 410 μg/L MUREAV 79,169,80
orl-rbt TDLo:3100 mg/kg (female 1-31D post):REP DRUGAY 19,7,80
ivn-inf TDLo:104 mg/kg/4D-I MJAUAJ 146,57,87
ipr-rat LD50:349 mg/kg ARZNAD 31,2145,81
orl-mus LD50:872 mg/kg ARZNAD 31,2145,81
ipr-mus LD50:451 mg/kg ARZNAD 31,2145,81
ivn-dog LD50:60 mg/kg DRUGAY 19,7,80
ivn-mam LD50:100 mg/kg DRUGAY 19,7,80

SAFETY PROFILE: Poison by intravenous and intraperitoneal routes. Moderately toxic by ingestion. Experimental reproductive effects. Mutation data reported. When heated to decomposition it emits toxic fumes of Cl^- and NO_x.

MQS560 CAS:22832-87-7 ***HR: 3***
MICONAZOLE NITRATE
mf: $C_{18}H_{14}Cl_4N_2O•HNO_3$ mw: 479.16

SYNS: 1-(2,4-DICHLORO-β-((2,4-DICHLOROBENZYL)OXY) PHENETHYL)-IMIDAZOLE NITRATE ◇ MCZ NITRATE ◇ R 14889

TOXICITY DATA with REFERENCE
orl-rat TDLo:270 mg/kg (female 17-22D post):REP IYKEDH 7,535,76
orl-rat TDLo:1100 mg/kg (7-17D preg):TER IYKEDH 7,367,76
orl-rat LD50:920 mg/kg NIIRDN 6,809,82
ipr-rat LD50:1060 mg/kg NIIRDN 6,809,82
ivn-rat LD50:14700 μg/kg IYKEDH 11,181,80
orl-mus LD50:578 mg/kg CHTHBK 17(6),392,72
ipr-mus LD50:480 mg/kg NIIRDN 6,809,82
ivn-mus LD50:28 mg/kg IYKEDH 11,181,80
orl-gpg LD50:276 mg/kg CHTHBK 17(6),392,72

SAFETY PROFILE: Poison by ingestion and intravenous routes. Moderately toxic by intraperitoneal route. An experimental teratogen. Experimental reproductive effects. An antimycotic. When heated to decomposition it emits toxic fumes of Cl^- and NO_x. See also NITRATES.

MQS579 CAS:52093-21-7 ***HR: 3***
MICROMYCIN
mf: $C_{20}H_{41}N_5O_7$ mw: 463.66

SYNS: ANTIBIOTIC KW 1062 ◇ ANTIBIOTIC XK 62-2 ◇ GENTAMICIN C2b ◇ GENTAMICIN C(^{2}b) ◇ KW-1062 ◇ 6'-N-METHYLGENTAMICIN C(^{1}a) ◇ MICROMICIN ◇ SAGAMICIN ◇ SAGAMICIN (OBS.) ◇ XK-62-2

TOXICITY DATA with REFERENCE
unr-rat TDLo:3250 mg/kg (female 19-22D post):REP JJANAX 30,432,77
unr-rat TDLo:2250 mg/kg (8-17D preg):TER JJANAX 30,432,77
scu-rat LD50:1223 mg/kg JJANAX 30,408,77
ivn-rat LD50:104 mg/kg JJANAX 30,408,77
ims-rat LD50:625 mg/kg JJANAX 30,408,77
orl-mus LD50:15600 mg/kg JJANAX 30,408,77
ipr-mus LD50:274 mg/kg JJANAX 30,408,77
scu-mus LD50:350 mg/kg JJANAX 30,408,77
ivn-mus LD50:75 mg/kg JJANAX 30,408,77
ims-mus LD50:245 mg/kg JJANAX 30,408,77
ims-dog LD50:459 mg/kg JJANAX 30,408,77

SAFETY PROFILE: Poison by intravenous, intraperitoneal, subcutaneous, and intramuscular routes. An experimental teratogen. Experimental reproductive effects. An antibiotic. When heated to decomposition it emits toxic fumes of NO_x.

MQS600 CAS:82643-25-2 ***HR: 3***
MICRONOMYCIN SULFATE
mf: $C_{21}H_{43}N_5O_6•H_2O_4S$ mw: 559.77

SYNS: d-o-2-AMINO-2,3,4,6-TETRADEOXY-6-(METHYLAMINO)-α-d-erythro-HEXOPYRANOSYL-(1-4)-o-(3-DEOXY-4-C-METHYL-3-(METHYLAMINO)-β-l-ARABINOPYRANOSYL-(1-6))-2-DEOXY-STREPTAMINE SULFATE ◇ MICRONOMICIN SULFATE

TOXICITY DATA with REFERENCE
scu-rat LD50:1223 mg/kg NIIRDN 6,APP-18,82
ivn-rat LD50:104 mg/kg JJANAX 36,3204,83
ims-rat LD50:625 mg/kg JJANAX 36,3204,83
scu-mus LD50:350 mg/kg NIIRDN 6,APP-18,82
ivn-mus LD50:245 mg/kg NIIRDN 6,APP-18,82
ivn-dog LD50:459 mg/kg NIIRDN 6,APP-18,82
ims-dog LD50:459 mg/kg JJANAX 36,3204,83

SAFETY PROFILE: Poison by subcutaneous and intravenous routes. Moderately toxic by intramuscular route. When heated to decomposition it emits toxic fumes of NO_x and SO_x.

MQS750 ***HR: 3***
MICRURUS ALLENI YATESI VENOM

SYN: VENOM, COSTA RICAN SNAKE, MICRURUS ALLENI YATESI

TOXICITY DATA with REFERENCE
ipr-mus LD50:113 μg/kg AJTHAB 21,360,72
ivn-mus LD50:750 μg/kg AJTHAB 21,360,72

SAFETY PROFILE: A deadly poison by intraperitoneal and intravenous routes.

MQT000 ***HR: 3***
MICRURUS CARINICAUDUS DUMERILII VENOM

SYNS: M. CARINICAUDUS DUMERILII VENOM ◇ M.C. DUMERILLI VENOM

TOXICITY DATA with REFERENCE
ipr-mus LD50:1075 μg/kg TOXIA6 13,139,75
ivn-mus LD50:1075 μg/kg TOXIA6 13,139,75

SAFETY PROFILE: A deadly poison by intraperitoneal and intravenous routes.

MQT100 ***HR: 3***
MICRURUS FULVIUS VENOM

SYN: VENOM, SNAKE, NICRURUS FULVIUS

TOXICITY DATA with REFERENCE
ipr-rat LDLo:8 mg/kg TOXIA6 9,219,71
ipr-mus LD50:330 μg/kg 14FHAR -,409,63
scu-mus LD50:1300 μg/kg TOXIA6 5,47,67
ivn-mus LD50:240 μg/kg 14FHAR -,409,63
ivn-dog LDLo:500 μg/kg 19DDA6 1,269,67
ivn-cat LDLo:2 mg/kg TOXIA6 9,219,71
ipr-mam LD50:970 μg/kg CLPTAT 8,849,67

SAFETY PROFILE: A deadly poison by subcutaneous, intravenous, and intraperitoneal routes.

MQT250 ***HR: 3***
MICRURUS MIPARTIUS HERTWIGI VENOM

SYN: VENOM, COSTA RICAN SNAKE, MICRURUS MIPARTITUS HERTWIGI

TOXICITY DATA with REFERENCE
ipr-mus LD50:238 μg/kg AJTHAB 21,360,72
ivn-mus LD50:1875 μg/kg AJTHAB 21,360,72

SAFETY PROFILE: A deadly poison by intraperitoneal and intravenous routes.

MQT500 ***HR: 3***
MICRURUS NIGROCINCTUS VENOM

SYN: VENOM, COSTA RICAN SNAKE, MICRURUS NIGROCINCTUS

TOXICITY DATA with REFERENCE
ipr-mus LD50:444 μg/kg AJTHAB 21,360,72
ivn-mus LD50:419 μg/kg AJTHAB 21,360,72

SAFETY PROFILE: A deadly poison by intraperitoneal and intravenous routes.

MQT525 CAS:59467-70-8 ***HR: 3***
MIDAZOLAM
mf: $C_{18}H_{13}ClFN_3$ mw: 325.79

SYN: 8-CHLORO-6-(2-FLUOROPHENYL)-1-METHYL-4H-IMIDAZO (1,5-a)(1,4)BENZODIAZEPINE

TOXICITY DATA with REFERENCE
orl-rat LD50:1600 mg/kg BCPHBM 16(Suppl 1),375,83
ivn-rat LD50:75 mg/kg BCPHBM 16(Suppl 1),375,83
orl-mus LD50:1600 mg/kg BCPHBM 16(Suppl 1),375,83
ivn-mus LD50:50 mg/kg BCPHBM 16(Suppl 1),375,83

SAFETY PROFILE: Poison by intravenous route. Moderately toxic by ingestion. When heated to decomposition it emits toxic fumes of F^-, Cl^-, and NO_x.

MQT530 CAS:3092-17-9 ***HR: 3***
MIDODRINE
mf: $C_{12}H_{18}N_2O_4 \cdot ClH$ mw: 290.78

SYNS: ACETAMIDE,2-AMINO-N-(2-(2,5-DIMETHOXYPHENYL)-2-HYDROXYETHYL)-, MONOHYDROCHLORIDE, (±)- (9CI) ◇ ACETAMIDE, 2-AMINO-N-(β-HYDROXY-2,5-DIMETHOXYPHENETHYL)-, MONOHYDROCHLORIDE, (±)- ◇ 1-(2′,5′-DIMETHOXYPHENYL)-2-GLYCINAMIDOETHANOL HYDROCHLORIDE ◇ GUTRON ◇ (±)-MIDODRINE HYDROCHLORIDE ◇ ST 1085

TOXICITY DATA with REFERENCE
orl-rat TDLo:560 mg/kg (female 17-22D post):REP KSRNAM 20,8789,86
orl-rbt TDLo:130 mg/kg (female 6-18D post):TER KSRNAM 20,8781,86
orl-rat LD50:68800 μg/kg KSRNAM 21,49,87
ipr-rat LD50:31300 μg/kg KSRNAM 21,49,87
scu-rat LD50:51 mg/kg KSRNAM 21,49,87
ivn-rat LD50:18200 μg/kg KSRNAM 21,49,87
orl-mus LD50:246 mg/kg KSRNAM 21,49,87
ipr-mus LD50:171 mg/kg KSRNAM 21,49,87
scu-mus LD50:170 mg/kg KSRNAM 21,49,87
orl-dog LD50:150 mg/kg KSRNAM 21,49,87

SAFETY PROFILE: Poison by ingestion, subcutaneous, intravenous, and intraperitoneal routes. An experimental teratogen. Experimental reproductive effects. When heated to decomposition it emits toxic fumes of NO_x and HCl.

MQT550 CAS:51781-21-6 ***HR: 3***
MIKELAN
mf: $C_{16}H_{24}N_2O_3 \cdot ClH$ mw: 328.88

SYNS: ABBOTT-43326 ◇ 5-(3-tert-BUTYLAMINO-2-HYDROXY)PROPOXY-3,4-DIHYDROCARBOSTYRIL HYDROCHLORIDE ◇ 5-(3-tert-BUTYLAMINO-2-HYDROXY-PROPOXY)-3,4-DIHYDRO-(2(1H)-CHINOLINON-HYDROCHLORID (GERMAN) ◇ CARTEOLOL HYDROCHLORIDE ◇ ENDAK ◇ ENDAK MITE ◇ OC 1085 ◇ OPC 1085

TOXICITY DATA with REFERENCE
orl-mus TDLo:4350 mg/kg (female 15-22D post):REP JTSCDR 4,59,79
ivn-rat TDLo:700 μg/kg (female 9-15D post):TER OYYAA2 11,197,76
orl-rat LD50:1330 mg/kg OYYAA2 11,159,76
ipr-rat LD50:390 mg/kg OYYAA2 11,159,76
scu-rat LD50:1950 mg/kg OYYAA2 11,159,76

ivn-rat LD50:153 mg/kg OYYAA2 11,159,76
orl-mus LD50:810 mg/kg ARZNAD 33,290,83
ipr-mus LD50:375 mg/kg ARZNAD 33,290,83
scu-mus LD50:600 mg/kg IYKEDH 12,668,81
ivn-mus LD50:54500 μg/kg IYKEDH 12,668,81
orl-dog LD50:830 mg/kg ARZNAD 33,290,83
orl-rbt LD50:740 mg/kg OYYAA2 11,165,76
ivn-rbt LD50:112 mg/kg OYYAA2 11,165,76

CONSENSUS REPORTS: EPA Genetic Toxicology Program.

SAFETY PROFILE: Poison by intravenous and intraperitoneal routes. Moderately toxic by ingestion and subcutaneous routes. An experimental teratogen. Experimental reproductive effects. When heated to decomposition it emits toxic fumes of NO_x and HCl.

MQT600 CAS:77855-81-3 ***HR: 2***
MILBEMYCIN D
mf: $C_{22}H_{48}O_7$ mw: 424.06

SYNS: ANTIBIOTIC B 41D ◇ B 41D ◇ (6R,25)-5-o-DEMETHYL-28-DEOXY-6,28-EPOXY-25-(1-METHYLETHYL)MILBEMYCINB

TOXICITY DATA with REFERENCE
orl-rat LD50:2467 mg/kg SKKNAJ 35,71,83
orl-mus LD50:1610 mg/kg SKKNAJ 35,71,83
ipr-mus LD50:668 mg/kg SKKNAJ 35,71,83

SAFETY PROFILE: Moderately toxic by ingestion and intraperitoneal routes. When heated to decomposition it emits acrid smoke and irritating fumes.

MQT750 ***HR: 3***
MILBEX mixed with PHOSALON (2:1)

SYNS: MILBEX mixed with PHOSALONE (2:1) ◇ PHOSALON mixed with MILBEX (1:2) ◇ PHOSPHORODITHIOIC ACID, O,O-DIETHYL ESTER, S-ESTER with 6-CHLORO-3-(MERCAPTOMETHYL)-2-BENZOXAZOLINONE mixed with MILBEX (1:2)

TOXICITY DATA with REFERENCE
orl-rat LD50:2000 mg/kg GTPZAB 19(9),55,75
orl-mus LD50:335 mg/kg GTPZAB 19(9),55,75

SAFETY PROFILE: Poison by ingestion. When heated to decomposition it emits very toxic fumes of NO_x and PO_x.

MQU000 CAS:67527-71-3 ***HR: 2***
MILDIOMYCIN
mf: $C_{19}H_{30}N_8O_9 \cdot H_2O$ mw: 532.59

PROP: Isolated from culture filtrate of *Streptoverticillium rimofaciens* B-98891 (JANTAJ 31,519,78).

SYNS: ANTIBIOTIC B-98891 ◇ B-98891 ◇ MIL

TOXICITY DATA with REFERENCE
orl-rat LD50:2500 mg/kg JANTAJ 31,519,78
scu-rat LD50:500 mg/kg JANTAJ 31,519,78
ivn-rat LD50:500 mg/kg JANTAJ 31,519,78
orl-mus LD50:2500 mg/kg JANTAJ 31,519,78
scu-mus LD50:500 mg/kg JANTAJ 31,519,78
ivn-mus LD50:500 mg/kg JANTAJ 31,519,78

SAFETY PROFILE: Moderately toxic by ingestion, subcutaneous, and intravenous routes. When heated to decomposition it emits toxic fumes of NO_x.

MQU250 ***HR: 2***
MIL L 7808

PROP: Composition consisting of bis(2-ethylhexyl) sebacate, tricresyl phosphate, and phenothiazine (94.5%, 5.0%, 0.5% by wt) (MRLR** No. 256,54).

TOXICITY DATA with REFERENCE
orl-rat LD50:4740 mg/kg MRLR** #256,54
ivn-rat LD50:1260 mg/kg MRLR** #256,54
ivn-rbt LD50:760 mg/kg MRLR** #256,54

SAFETY PROFILE: Moderately toxic by intravenous route. Mildly toxic by ingestion. When heated to decomposition it emits very toxic fumes of NO_x, SO_x, and PO_x. See also BIS(2-ETHYLHEXYL)SEBACATE, TRITOLYL PHOSPHATE, and PEHNOTHIAZINE.

MQU500 ***HR: 3***
MIL L 17535

PROP: Composition consisting of diisoamyl adipate, barium petroleum sulfonate, phenothiazine (96.65%, 3.0%; 0.35% by wt) (MRLR** No. 256,54).

TOXICITY DATA with REFERENCE
ivn-rat LD50:160 mg/kg MRLR** No.256,54
ivn-rbt LD50:190 mg/kg MRLR** No.256,54

CONSENSUS REPORTS: Barium and its compounds are on the Community Right-To-Know List.

SAFETY PROFILE: Poison by intravenous route. When heated to decomposition it emits very toxic fumes of SO_x and NO_x. See also ADIPIC ACID DIISOPENTYL ESTER and PHENOTHIAZINE.

MQU525 CAS:37065-29-5 ***HR: D***
MILOXACIN
mf: $C_{12}H_9NO_6$ mw: 263.22

PROP: Colorless prisms from DMF. Mp: 264° (decomp).

SYNS: AB 206 ◇ ANTIBIOTIC AB 206 ◇ 5,8-DIHYDRO-5-METHOXY-8-OXO-2H-1,3-DIOXOLO(4,5-g)QUINOLINE-7-CARBOXYLICACID ◇ FULDAZIN

TOXICITY DATA with REFERENCE
dni-esc 100 μg/L AMACCQ 17,763,80
dni-srm 270 mg/L AMACCQ 17,763,80

dni-omi 290 mg/L AMACCQ 17,763,80
dni-omi 140 mg/L AMACCQ 17,763,80
orl-rat TDLo:3300 mg/kg (female 7-17D post):REP OYYAA2 19,815,80
orl-rat TDLo:3300 mg/kg (female 7-17D post):TER OYYAA2 19,815,80

SAFETY PROFILE: An experimental teratogen. Experimental reproductive effects. Mutation data reported. When heated to decomposition it emits toxic fumes of NO_x.

MQU750 CAS:57-53-4 ***HR: 3***
MILTOWN
mf: $C_9H_{18}N_2O_4$ mw: 218.29

SYNS: AMEPROMAT ◇ AMOSENE ◇ ANASTRESS ◇ ANATHYLMON ◇ ANDAKSIN ◇ ANDAXIN ◇ ANEURAL ◇ ANEUXRAL ◇ ANSIATAN ◇ ANSIL ◇ ANSIOWAS ◇ ANURAL ◇ ANXIETIL ◇ APASCIL ◇ ARCOBAN ◇ ARTOLON ◇ ATRAXINE ◇ AYERMATE ◇ BAMD 400 ◇ BIOBAMAT ◇ BROBAMATE ◇ CALMADIN ◇ CALMAX ◇ CALMIREN ◇ CANQUIL-400 ◇ CAP-O-TRAN ◇ CIRPONYL ◇ CRESTANIL ◇ CYPRON ◇ DAPAZ ◇ DICANDIOL ◇ 2,2-DI(CARBAMOYLOXYMETHYL)PENTANE ◇ DIVERON ◇ DORMABROL ◇ ECUANIL ◇ EDENAL ◇ ENORDEN ◇ EPICUR ◇ EQUANIL SUSPENSION ◇ EQUILIUM ◇ EQUINIL ◇ ERINA ◇ ESTASIL ◇ FAS-CILE ◇ GADEXYL ◇ HARMONIN ◇ HARTOL ◇ HOLBAMATE ◇ IPSOTIAN ◇ KESSOBAMATE ◇ KLORT ◇ LARTEN ◇ LEPENIL ◇ LEPETOWN ◇ LETYL ◇ LIBIOLAN ◇ MADIOL ◇ MAR BATE ◇ MARGONIL ◇ MENDEL ◇ MEPAMTIN ◇ MEPAVLON ◇ MEPIOSINE ◇ MEPOSED ◇ MEPRANIL ◇ MEPROBAM ◇ MEPROBAMAT (GERMAN) ◇ MEPROBAMATE ◇ MEPROBAMATO (ITALIAN) ◇ MEPROCOMPREN ◇ MEPROCON CMC ◇ MEPRODIL ◇ MEPROLEAF ◇ MEPROSAN ◇ MEPROTABS ◇ MEPROZINE ◇ MEPTRAN ◇ 2-METHYL-2-N-PROPYL-1,3-PROPANEDIOL DICARBAMATE ◇ 2-METHYL-2-PROPYLTRIMETHYLENE CARBAMATE ◇ METRACTYL ◇ MILPREM ◇ MILTANN ◇ NEO-TRAN ◇ NEPHENTINE ◇ OROLEVOL ◇ PANCALMA ◇ PAN-TRANQUIL ◇ PEREQUIL ◇ PLACIDON ◇ PROCALMIDOL ◇ PROQUANIL ◇ QUIETIDON ◇ RESTENIL ◇ ROBAMATE ◇ SEDABAMATE ◇ SERIL ◇ SPANTRAN ◇ TRANQUILAN ◇ TRELMAR ◇ URBIL ◇ VISTABAMATE ◇ WARDAMATE ◇ WYSEALS ◇ ZIRPON

TOXICITY DATA with REFERENCE
cyt-mus-orl 26667 μg/kg PISCAD 59(Pt. 3),547,72
orl-rat TDLo:5160 mg/kg (female 1-22D post):REP PSYPAG 41,113,75
scu-mus TDLo:500 mg/kg (female 10D post):TER KAIZAN 38,258,63
orl-cld TDLo:80 mg/kg HUTODJ 4,215,85
orl-wmn LDLo:760 mg/kg JAMAAP 207,361,69
orl-hmn TDLo:280 mg/kg:CNS OSMJAT 52,1304,56
orl-man TDLo:5700 μg/kg:PUL,GIT,SKN JIMSAX 41,119,57
orl-wmn TDLo:384 mg/kg:CVS,PUL NOMDA6 56,321,57
unr-man LDLo:441 mg/kg 85DCAI 2,73,70
unr-man LDLo:441 mg/kg 85DCAI 2,73,70
orl-rat LD50:1000 mg/kg 29QHAQ -,232,74
ipr-rat LD50:410 mg/kg JPETAB 129,75,60
ivn-rat LD50:350 mg/kg PMARAU 44,915,57
orl-mus LD50:750 mg/kg JMCMAR 15,998,72
ipr-mus LD50:331 mg/kg DPHFAK 23,281,71
scu-mus LD50:520 mg/kg JPPMAB 26,109,74
ivn-mus LD50:230 mg/kg EJMCA5 12,447,77
ivn-rbt LD50:260 mg/kg IJNEAQ 5,305,66
scu-gpg LD50:380 mg/kg AIPTAK 137,375,62
orl-ham LD50:1410 mg/kg JPETAB 129,75,60
ipr-ham LD50:625 mg/kg JPETAB 129,75,60

CONSENSUS REPORTS: Reported in EPA TSCA Inventory. EPA Genetic Toxicology Program.

SAFETY PROFILE: Human poison by unspecified routes. Moderately toxic to humans and experimentally by ingestion. Experimental poison by intravenous, intraperitoneal, and subcutaneous routes. An experimental teratogen. Human systemic effects by ingestion: coma, blood pressure decrease, regional or general arteriolar constriction, dyspnea, cyanosis, respiratory depression, nausea or vomiting and allergic skin dermatitis. Experimental reproductive effects. Mutation data reported. Implicated in aplastic anemia. Used as a tranquilizer. When heated to decomposition it emits toxic fumes of NO_x. See also CARBAMATES.

MQV000 CAS:8031-03-6 ***HR: 1***
MIMOSA ABSOLUTE

PROP: From the flowers of *Acacia decurrens var. dealbata* (FCTXAV 13,681,75).

SYN: ABSOLUTE MIMOSA

TOXICITY DATA with REFERENCE
skn-gpg 100% MLD FCTXAV 13,873,75

CONSENSUS REPORTS: Reported in EPA TSCA Inventory.

SAFETY PROFILE: A skin irritant.

MQV250 CAS:1401-55-4 ***HR: 3***
MIMOSA TANNIN

SYNS: ACACIA MOLLISSIMA TANNIN ◇ TANNIN from MIMOSA

TOXICITY DATA with REFERENCE
scu-rat TDLo:350 mg/kg/12W-I:ETA BJCAAI 14,147,60
ipr-mus LD50:320 mg/kg JPPMAB 9,98,57
ivn-mus LD50:130 mg/kg JPPMAB 9,98,57

SAFETY PROFILE: Poison by intravenous and intraperitoneal routes. Questionable carcinogen with experimental tumorigenic data. When heated to decomposition it emits acrid smoke and irritating fumes. See also TANNIN.

MQV500 *HR: D*
MINERAL DUSTS

SAFETY PROFILE: Variable toxicity. From the economic and toxicity standpoints, the most important are those containing free silica which can cause silicosis upon inhalation of sufficient quantity. These include sand, sandstone, quartz, and flint. They consist mainly of silica in the form of quartz; diatomaceous earth, which is essentially amorphous silica; and granite, which contains 20-40% quartz. Minerals that contain combined silica in the form of silicates but no free silica are generally less capable of causing silicosis. Asbestos, however, can cause a fibrotic lung condition of its own, known as asbestosis, and lung cancer. (See also various asbestos entries.) Mica and talc dust are also considered somewhat hazardous. Non-siliceous minerals, like limestone, marble, dolomite, etc., which do not contain toxic elements, do not ordinarily present any significant dust hazard. Minerals containing toxic elements, such as cryolite which contains fluorine, and pyrolusite, which contains manganese, may cause systemic poisoning upon inhalation or ingestion of sufficient quantity. In any event, the minerals are usually less reactive than synthetic compounds of the same elements and, in fact, may be relatively inert by comparison. These are common air contaminants. See also specific materials.

MQV750 CAS:8012-95-1 *HR: 2*
MINERAL OIL

PROP: Colorless, oily liquid; practically tasteless and odorless. D: 0.83-0.86 (light), 0.875-0.905 (heavy), flash p: 444°F (OC), ULC: 10-20. Insol in water and alc; sol in benzene, chloroform, and ether. A mixture of liquid hydrocarbons from petroleum.

SYNS: ADEPSINE OIL ◇ ALBOLINE ◇ BAYOL F ◇ BLANDLUBE ◇ CRYSTOSOL ◇ DRAKEOL ◇ FONOLINE ◇ GLYMOL ◇ KAYDOL ◇ KONDREMUL ◇ MINERAL OIL, WHITE (FCC) ◇ MOLOL ◇ NEOCULTOL ◇ NUJOL ◇ OIL MIST, MINERAL (OSHA, ACGIH) ◇ PAROL ◇ PAROLEINE ◇ PARRAFIN OIL ◇ PENETECK ◇ PENRECO ◇ PERFECTA ◇ PETROGALAR ◇ PETROLATUM, liquid ◇ PRIMOL 335 ◇ PROTOPET ◇ SAXOL ◇ TECH PET F ◇ WHITE MINERAL OIL

TOXICITY DATA with REFERENCE
skn-rbt 100 mg/24H MLD CTOIDG 94(8),41,79
eye-rbt 250 mg/5D MLD AMIHAB 14,265,56
skn-gpg 100 mg/24H MLD CTOIDG 94(8),41,79
ihl-man TCLo:5 mg/m^3/5Y-I:CAR,GIT,TER JOCMA7 23,333,81
skn-mus TDLo:332 g/mg/20W-I:ETA ANYAA9 132,439,65
orl-mus LD50:22 g/kg ATXKA8 30,243,73

CONSENSUS REPORTS: Reported in EPA TSCA Inventory.
OSHA PEL: Oil Mist: TWA 5 mg/m^3
ACGIH TLV: TWA 5 mg/m^3; STEL 10 mg/m^3

SAFETY PROFILE: A human teratogen by inhalation which causes testicular tumors in the fetus. Inhalation of vapor or particulates can cause aspiration pneumonia. A skin and eye irritant. Questionable human carcinogen producing gastrointestinal tumors. Combustible liquid when exposed to heat or flame. To fight fire, use dry chemical, CO_2, foam. When heated to decomposition it emits acrid smoke and fumes.

MQV755 CAS:64741-49-7 *HR: 3*
MINERAL OIL, PETROLEUM CONDENSATES, VACUUM TOWER

SYNS: CONDENSATES (PETROLEUM), VACUUM TOWER (9CI) ◇ VACUUM RESIDUUM

CONSENSUS REPORTS: IARC Cancer Review: Group 1 IMEMDT 7,252,87; Animal Sufficient Evidence IMEMDT 33,87,84. Reported in EPA TSCA Inventory.

SAFETY PROFILE: Confirmed carcinogen. When heated to decomposition it emits acrid smoke and irritating fumes.

MQV760 CAS:64742-18-3 *HR: 3*
MINERAL OIL, PETROLEUM DISTILLATES, ACID-TREATED HEAVY NAPHTHENIC

SYNS: ACID-TREATED HEAVY NAPHTHENIC DISTILLATE ◇ DISTILLATES (PETROLEUM), ACID-TREATED HEAVY NAPHTHENIC (9CI)

CONSENSUS REPORTS: IARC Cancer Review: Group 1 IMEMDT 7,252,87; Animal Sufficient Evidence IMEMDT 33,87,84. Reported in EPA TSCA Inventory.

SAFETY PROFILE: Confirmed carcinogen. When heated to decomposition it emits acrid smoke and irritating fumes.

MQV765 CAS:64742-20-7 *HR: 3*
MINERAL OIL, PETROLEUM DISTILLATES, ACID-TREATED HEAVY PARAFFINIC

SYNS: ACID-TREATED HEAVY PARAFFINIC DISTILLATE ◇ DISTILLATES (PETROLEUM), ACID-TREATED HEAVY PARAFFINIC (9CI)

CONSENSUS REPORTS: IARC Cancer Review: Group 1 IMEMDT 7,252,87; Animal Sufficient Evidence IMEMDT 33,87,84. Reported in EPA TSCA Inventory.

SAFETY PROFILE: Confirmed carcinogen. When heated to decomposition it emits acrid smoke and irritating fumes.

MQV770 CAS:64742-19-4 ***HR: 3***
MINERAL OIL, PETROLEUM DISTILLATES, ACID-TREATED LIGHT NAPHTHENIC

SYNS: ACID-TREATED LIGHT NAPHTHENIC DISTILLATE ◇ DISTILLATES (PETROLEUM), ACID-TREATED LIGHT NAPHTHENIC (9CI)

CONSENSUS REPORTS: IARC Cancer Review: Group 1 IMEMDT 7,252,87; Animal Sufficient Evidence IMEMDT 33,87,84. Reported in EPA TSCA Inventory.

SAFETY PROFILE: Confirmed carcinogen. When heated to decomposition it emits acrid smoke and irritating fumes.

MQV775 CAS:64742-21-8 ***HR: 3***
MINERAL OIL, PETROLEUM DISTILLATES, ACID-TREATED LIGHT PARAFFINIC

SYNS: ACID-TREATED LIGHT PARAFFINIC DISTILLATE ◇ DISTILLATES (PETROLEUM), ACID-TREATED LIGHT PARAFFINIC (9CI)

CONSENSUS REPORTS: IARC Cancer Review: Group 1 IMEMDT 7,252,87; Animal Sufficient Evidence IMEMDT 33,87,84. Reported in EPA TSCA Inventory.

SAFETY PROFILE: Confirmed carcinogen. When heated to decomposition it emits acrid smoke and irritating fumes.

MQV780 CAS:64741-53-3 ***HR: 3***
MINERAL OIL, PETROLEUM DISTILLATES, HEAVY NAPHTHENIC

SYNS: DISTILLATES (PETROLEUM), HEAVY NAPHTHENIC (9CI) ◇ HEAVY NAPHTHENIC DISTILLATE ◇ HEAVY NAPHTHENIC DISTILLATES (PETROLEUM)

TOXICITY DATA with REFERENCE
mmo-sat 5 μL/plate CBTOE2 2,63,86
skn-mus TDLo:216 g/kg/36W-I:NEO EPASR* 8EHQ-0887-0691S

CONSENSUS REPORTS: IARC Cancer Review: Group 1 IMEMDT 7,252,87; Animal Sufficient Evidence IMEMDT 33,87,84. Reported in EPA TSCA Inventory.

SAFETY PROFILE: Confirmed carcinogen with experimental neoplastigenic data. Mutation data reported. When heated to decomposition it emits acrid smoke and irritating fumes.

MQV785 CAS:64741-51-1 ***HR: 3***
MINERAL OIL, PETROLEUM DISTILLATES, HEAVY PARAFFINIC

SYNS: DISTILLATES (PETROLEUM), HEAVY PARAFFINIC (9CI) ◇ HEAVY PARAFFINIC DISTILLATE

TOXICITY DATA with REFERENCE
mmo-sat 3 uL/plate CBTOE2 2,63,86

CONSENSUS REPORTS: IARC Cancer Review: Group 1 IMEMDT 7,252,87; Animal Sufficient Evidence IMEMDT 33,87,84. Reported in EPA TSCA Inventory.

SAFETY PROFILE: Confirmed carcinogen. Mutation data reported.

MQV790 CAS:64742-52-5 ***HR: 3***
MINERAL OIL, PETROLEUM DISTILLATES, HYDROTREATED HEAVY NAPHTHENIC

SYNS: DISTILLATES (PETROLEUM), HYDROTREATED HEAVY NAPHTHENIC (9CI) ◇ HYDROTREATED HEAVY NAPHTHENIC DISTILLATE ◇ HYDROTREATED HEAVY NAPHTHENIC DISTILLATES (PETROLEUM) ◇ PETROLEUM DISTILLATES, HYDROTREATED HEAVY NAPHTHENIC

TOXICITY DATA with REFERENCE
mmo-sat 10 μL/plate CBTOE2 2,63,86
skn-mus TDLo:480 g/kg/80W-I:NEO EPASR* 8EHQ-0887-0691S
skn-mus LD:402 g/kg/78W-I:ETA BJCAAI 48,429,83

CONSENSUS REPORTS: IARC Cancer Review: Group 1 IMEMDT 7,252,87; Animal Inadequate Evidence IMEMDT 33,87,84. Reported in EPA TSCA Inventory.

SAFETY PROFILE: Confirmed carcinogen with experimental tumorigenic data. Mutation data reported. When heated to decomposition it emits acrid smoke and irritating fumes.

MQV795 CAS:64742-54-7 ***HR: 3***
MINERAL OIL, PETROLEUM DISTILLATES, HYDROTREATED HEAVY PARAFFINIC

SYNS: DISTILLATES (PETROLEUM), HYDROTREATED HEAVY PARAFFINIC (9CI) ◇ HYDROTREATED HEAVY PARAFFINIC DISTILLATE

CONSENSUS REPORTS: IARC Cancer Review: Group 1 IMEMDT 7,252,87; Animal Inadequate Evidence IMEMDT 33,87,84. Reported in EPA TSCA Inventory.

SAFETY PROFILE: Confirmed carcinogen. When heated to decomposition it emits acrid smoke and irritating fumes.

MQV800 CAS:64742-53-6 ***HR: 3***
MINERAL OIL, PETROLEUM DISTILLATES, HYDROTREATED LIGHT NAPHTHENIC

SYNS: DISTILLATES (PETROLEUM), HYDROTREATED LIGHT NAPHTHENIC (9CI) ◇ HYDROTREATED LIGHT NAPHTHENIC DISTILLATE ◇ HYDROTREATED LIGHT NAPHTHENIC DISTILLATES (PETROLEUM)

TOXICITY DATA with REFERENCE
skn-mus TDLo:480 g/kg/80W-I:NEO EPASR* 8EHQ-0887-0691S

CONSENSUS REPORTS: IARC Cancer Review: Group 1 IMEMDT 7,252,87; Animal Sufficient Evidence IMEMDT 33,87,84. Reported in EPA TSCA Inventory.

SAFETY PROFILE: Confirmed carcinogen with experimental neoplastigenic data. When heated to decomposition it emits acrid smoke and irritating fumes.

MQV805 CAS:64742-55-8 ***HR: 3***
MINERAL OIL, PETROLEUM DISTILLATES, HYDROTREATED LIGHT PARAFFINIC

SYNS: DISTILLATES (PETROLEUM), HYDROTREATED LIGHT PARAFFINIC (9CI) ◇ HYDROTREATED LIGHT PARAFFINIC DISTILLATE

CONSENSUS REPORTS: IARC Cancer Review: Group 1 IMEMDT 7,252,87; Animal Sufficient Evidence IMEMDT 33,87,84 Reported in EPA TSCA Inventory.

SAFETY PROFILE: Confirmed carcinogen. When heated to decomposition it emits acrid smoke and irritating fumes.

MQV810 CAS:64741-52-2 ***HR: 3***
MINERAL OIL, PETROLEUM DISTILLATES, LIGHT NAPHTHENIC

SYNS: DISTILLATES (PETROLEUM), LIGHT NAPHTHENIC (9CI) ◇ LIGHT NAPHTHENIC DISTILLATE ◇ LIGHT NAPHTHENIC DISTILLATES (PETROLEUM)

TOXICITY DATA with REFERENCE
skn-mus TDLo:480 g/kg/80W-I:NEO EPASR* 8EHQ-0887-0691S

CONSENSUS REPORTS: IARC Cancer Review: Group 1 IMEMDT 7,252,87; Animal Sufficient Evidence IMEMDT 33,87,84. Reported in EPA TSCA Inventory.

SAFETY PROFILE: Confirmed carcinogen with experimental neoplastigenic data. When heated to decomposition it emits acrid smoke and irritating fumes.

MQV815 CAS:64741-50-0 ***HR: 3***
MINERAL OIL, PETROLEUM DISTILLATES, LIGHT PARAFFINIC

SYNS: DISTILLATES (PETROLEUM), LIGHT PARAFFINIC (9CI) ◇ LIGHT PARAFFINIC DISTILLATE

TOXICITY DATA with REFERENCE
mmo-sat 5 uL/plate CBTOE2 2,63,86

CONSENSUS REPORTS: IARC Cancer Review: Group 1 IMEMDT 7,252,87; Animal Sufficient Evidence IMEMDT 33,87,84. Reported in EPA TSCA Inventory.

SAFETY PROFILE: Confirmed carcinogen. Mutation data reported. When heated to decomposition it emits acrid smoke and irritating fumes.

MQV820 CAS:64742-63-8 ***HR: 2***
MINERAL OIL, PETROLEUM DISTILLATES, SOLVENT-DEWAXED HEAVY NAPHTHENIC

SYNS: DISTILLATES (PETROLEUM), SOLVENT-DEWAXED HEAVY NAPHTHENIC (9CI) ◇ SOLVENT-DEWAXED HEAVY NAPHTHENIC DISTILLATE

CONSENSUS REPORTS: IARC Cancer Review: Group 3 IMEMDT 7,252,87; Animal No Evidence IMEMDT 33,87,84. Reported in EPA TSCA Inventory.

SAFETY PROFILE: Questionable carcinogen. When heated to decomposition it emits acrid smoke and irritating fumes.

MQV825 CAS:64742-65-0 ***HR: 2***
MINERAL OIL, PETROLEUM DISTILLATES, SOLVENT-DEWAXED HEAVY PARAFFINIC

SYNS: DISTILLATES (PETROLEUM), SOLVENT-DEWAXED HEAVY PARAFFINIC (9CI) ◇ PETROLEUM DISTILLATES, SOLVENT-DEWAXED HEAVY PARAFFINIC ◇ SOLVENT-DEWAXED HEAVY PARAFFINIC DISTILLATE

TOXICITY DATA with REFERENCE
skn-mus TDLo:386 g/kg/22W-I:ETA BJCAAI 48,429,83

CONSENSUS REPORTS: IARC Cancer Review: Group 3 IMEMDT 7,252,87; Animal No Evidence IMEMDT 33,87,84. Reported in EPA TSCA Inventory.

SAFETY PROFILE: Questionable carcinogen with experimental tumorigenic data. When heated to decomposition it emits acrid smoke and irritating fumes.

MQV835 CAS:64742-64-9 ***HR: 3***
MINERAL OIL, PETROLEUM DISTILLATES, SOLVENT-DEWAXED LIGHT NAPHTHENIC

SYNS: DISTILLATES (PETROLEUM), SOLVENT-DEWAXED LIGHT NAPHTHENIC (9CI) ◇ SOLVENT-DEWAXED LIGHT NAPHTHENIC DISTILLATE

CONSENSUS REPORTS: IARC Cancer Review: Group 1 IMEMDT 7,252,87; Animal Sufficient Evidence IMEMDT 33,87,84. Reported in EPA TSCA Inventory.

SAFETY PROFILE: Confirmed carcinogen. When heated to decomposition it emits acrid smoke and irritating fumes.

MQV840 CAS:64742-56-9 ***HR: 3***
MINERAL OIL, PETROLEUM DISTILLATES, SOLVENT-DEWAXED LIGHT PARAFFINIC

SYNS: DISTILLATES (PETROLEUM), SOLVENT-DEWAXED LIGHT PARAFFINIC (9CI) ◇ SOLVENT-DEWAXED LIGHT PARAFFINIC DISTILLATE

CONSENSUS REPORTS: IARC Cancer Review: Group 1 IMEMDT 7,252,87; Animal Sufficient Evidence IMEMDT 33,87,84. Reported in EPA TSCA Inventory.

SAFETY PROFILE: Confirmed carcinogen. When heated to decomposition it emits acrid smoke and irritating fumes.

MQV845 CAS:64741-96-4 ***HR: 2***
MINERAL OIL, PETROLEUM DISTILLATES, SOLVENT-REFINED HEAVY NAPHTHENIC

SYNS: DISTILLATES (PETROLEUM), SOLVENT-REFINED HEAVY NAPHTHENIC (9CI) ◇ SOLVENT-REFINED HEAVY NAPHTHENIC DISTILLATE

CONSENSUS REPORTS: IARC Cancer Review: Group 3 IMEMDT 7,252,87; Animal No Evidence IMEMDT 33,87,84. Reported in EPA TSCA Inventory.

SAFETY PROFILE: Questionable carcinogen. When heated to decomposition it emits acrid smoke and irritating fumes.

MQV850 CAS:64741-88-4 ***HR: 2***
MINERAL OIL, PETROLEUM DISTILLATES, SOLVENT-REFINED HEAVY PARAFFINIC

SYNS: DISTILLATES (PETROLEUM), SOLVENT-REFINED HEAVY PARAFFINIC (9CI) ◇ SOLVENT-REFINED HEAVY PARAFFINIC DISTILLATE

CONSENSUS REPORTS: IARC Cancer Review: Group 3 IMEMDT 7,252,87; Animal No Evidence IMEMDT 33,87,84. Reported in EPA TSCA Inventory.

SAFETY PROFILE: Questionable carcinogen. When heated to decomposition it emits acrid smoke and irritating fumes.

MQV852 CAS:64741-97-5 ***HR: 3***
MINERAL OIL, PETROLEUM DISTILLATES, SOLVENT-REFINED LIGHT NAPHTHENIC

SYNS: DISTILLATES (PETROLEUM), SOLVENT-REFINED LIGHT NAPHTHENIC (9CI) ◇ SOLVENT-REFINED LIGHT NAPHTHENIC DISTILLATE

CONSENSUS REPORTS: IARC Cancer Review: Group 1 IMEMDT 7,252,87; Animal Sufficient Evidence IMEMDT 33,87,84. Reported in EPA TSCA Inventory.

SAFETY PROFILE: Confirmed carcinogen. When heated to decomposition it emits acrid smoke and irritating fumes.

MQV855 CAS:64741-89-5 ***HR: 3***
MINERAL OIL, PETROLEUM DISTILLATES, SOLVENT-REFINED LIGHT PARAFFINIC

SYNS: DISTILLATES (PETROLEUM), SOLVENT-REFINED LIGHT PARAFFINIC (9CI) ◇ SOLVENT-REFINED LIGHT PARAFFINIC DISTILLATE

CONSENSUS REPORTS: IARC Cancer Review: Group 1 IMEMDT 7,252,87; Animal Sufficient Evidence IMEMDT 33,87,84. Reported in EPA TSCA Inventory.

SAFETY PROFILE: Confirmed carcinogen. When heated to decomposition it emits acrid smoke and irritating fumes.

MQV857 CAS:64742-11-6 ***HR: 3***
MINERAL OIL, PETROLEUM EXTRACTS, HEAVY NAPHTHENIC DISTILLATE SOLVENT

SYNS: EXTRACTS (PETROLEUM), HEAVY NAPHTHENIC DISTILLATE SOLVENT (9CI) ◇ HEAVY NAPHTHENIC DISTILLATE SOLVENT EXTRACT

TOXICITY DATA with REFERENCE
ihl-rat TCLo:660 mg/m^3/6H (12-16D post):TER
JJATDK 2,260,82
ihl-rat TCLo:660 mg/m^3/6H (12-16D post):REP
JJATDK 2,260,82

CONSENSUS REPORTS: IARC Cancer Review: Group 1 IMEMDT 7,252,87; Animal Sufficient Evidence IMEMDT 33,87,84. Reported in EPA TSCA Inventory.

SAFETY PROFILE: Confirmed carcinogen. An experimental teratogen. Experimental reproductive data. When heated to decomposition it emits acrid smoke and irritating fumes.

MQV859 CAS:64742-04-7 ***HR: 3***
MINERAL OIL, PETROLEUM EXTRACTS, HEAVY PARAFFINIC DISTILLATE SOLVENT

SYNS: EXTRACTS (PETROLEUM), HEAVY PARAFFINIC DISTILLATE SOLVENT (9CI) ◇ HEAVY PARAFFINIC DISTILLATE, SOLVENT EXTRACT

CONSENSUS REPORTS: IARC Cancer Review: Group 1 IMEMDT 7,252,87; Animal Sufficient Evidence IMEMDT 33,87,84. Reported in EPA TSCA Inventory.

SAFETY PROFILE: Confirmed carcinogen. When heated to decomposition it emits acrid smoke and irritating fumes.

MQV860 CAS:64742-03-6 ***HR: 3***
MINERAL OIL, PETROLEUM EXTRACTS, LIGHT NAPHTHENIC DISTILLATE SOLVENT

SYNS: EXTRACTS (PETROLEUM), LIGHT NAPHTHENIC DISTILLATE SOLVENT (9CI) ◇ LIGHT NAPHTHENIC DISTILLATE, SOLVENT EXTRACT

CONSENSUS REPORTS: IARC Cancer Review: Group 1 IMEMDT 7,252,87; Animal Sufficient Evidence 33,87,84. Reported in EPA TSCA Inventory.

SAFETY PROFILE: Confirmed carcinogen. When heated to decomposition it emits acrid smoke and irritating fumes.

MQV862 CAS:64742-05-8 ***HR: 3***
MINERAL OIL, PETROLEUM EXTRACTS, LIGHT PARAFFINIC DISTILLATE SOLVENT

SYNS: EXTRACTS (PETROLEUM), LIGHT PARAFFINIC DISTILLATE SOLVENT (9CI) ◇ LIGHT PARAFFINIC DISTILLATE, SOLVENT EXTRACT

CONSENSUS REPORTS: IARC Cancer Review: Group 1 IMEMDT 7,252,87; Animal Sufficient Evidence IMEMDT 33,87,84. Reported in EPA TSCA Inventory.

SAFETY PROFILE: Confirmed carcinogen. When heated to decomposition it emits acrid smoke and irritating fumes.

MQV863 CAS:64742-10-5 ***HR: 3***
MINERAL OIL, PETROLEUM EXTRACTS, RESIDUAL OIL SOLVENT

SYNS: EXTRACTS (PETROLEUM), RESIDUAL OIL SOLVENT (9CI) ◇ RESIDUAL OIL SOLVENT EXTRACT

CONSENSUS REPORTS: IARC Cancer Review: Group 1 IMEMDT 7,252,87; Animal Sufficient Evidence IMEMDT 33,87,84. Reported in EPA TSCA Inventory.

SAFETY PROFILE: Confirmed carcinogen. When heated to decomposition it emits acrid smoke and irritating fumes.

MQV865 CAS:64742-68-3 ***HR: 3***
MINERAL OIL, PETROLEUM NAPHTHENIC OILS, CATALYTIC DEWAXED HEAVY

SYNS: CATALYTIC-DEWAXED HEAVY NAPHTHENIC DISTILLATE ◇ NAPHTHENIC OILS (PETROLEUM), CATALYTIC DEWAXED HEAVY(9CI)

CONSENSUS REPORTS: IARC Cancer Review: Group 1 IMEMDT 7,252,87; Animal Sufficient Evidence IMEMDT 33,87,84. Reported in EPA TSCA Inventory.

SAFETY PROFILE: Confirmed carcinogen. When heated to decomposition it emits acrid smoke and irritating fumes.

MQV867 CAS:64742-69-4 ***HR: 3***
MINERAL OIL, PETROLEUM NAPHTHENIC OILS, CATALYTIC DEWAXED LIGHT

SYNS: CATALYTIC-DEWAXED LIGHT NAPHTHENIC DISTILLATE ◇ NAPHTHENIC OILS (PETROLEUM), CATALYTIC DEWAXED LIGHT (9CI)

CONSENSUS REPORTS: IARC Cancer Review: Group 1 IMEMDT 7,252,87; Animal Sufficient Evidence IMEMDT 33,87,84. Reported in EPA TSCA Inventory.

SAFETY PROFILE: Confirmed carcinogen. When heated to decomposition it emits acrid smoke and irritating fumes.

MQV868 CAS:64742-70-7 ***HR: 3***
MINERAL OIL, PETROLEUM PARAFFIN OILS, CATALYTIC DEWAXED HEAVY

SYNS: CATALYTIC-DEWAXED HEAVY PARAFFINIC DISTILLATE ◇ PARAFFIN OILS (PETROLEUM), CATALYTIC DEWAXED HEAVY (9CI)

CONSENSUS REPORTS: IARC Cancer Review: Group 1 IMEMDT 7,252,87; Animal Sufficient Evidence IMEMDT 33,87,84. Reported in EPA TSCA Inventory.

SAFETY PROFILE: Confirmed carcinogen. When heated to decomposition it emits acrid smoke and irritating fumes.

MQV870 CAS:64742-71-8 ***HR: 3***
MINERAL OIL, PETROLEUM PARAFFIN OILS, CATALYTIC DEWAXED LIGHT

SYNS: CATALYTIC-DEWAXED LIGHT PARAFFINIC DISTILLATE ◇ PARAFFIN OILS (PETROLEUM), CATALYTIC DEWAXED LIGHT (9CI)

CONSENSUS REPORTS: IARC Cancer Review: Group 1 IMEMDT 7,252,87; Animal Sufficient Evidence IMEMDT 33,87,84. Reported in EPA TSCA Inventory.

SAFETY PROFILE: Confirmed carcinogen. When heated to decomposition it emits acrid smoke and irritating fumes.

MQV872 CAS:64742-17-2 ***HR: 3***
MINERAL OIL, PETROLEUM RESIDUAL OILS, ACID-TREATED

SYNS: ACID-TREATED RESIDUAL OIL ◇ RESIDUAL OILS (PETROLEUM), ACID-TREATED (9CI)

CONSENSUS REPORTS: IARC Cancer Review: Group 1 IMEMDT 7,252,87; Animal Sufficient Evidence IMEMDT 33,87,84.

SAFETY PROFILE: Confirmed carcinogen. When heated to decomposition it emits acrid smoke and irritating fumes.

MQV875 CAS:8042-47-5 ***HR: 3***
MINERAL OIL, SLAB OIL

SYNS: SLAB OIL (9CI) ◇ WHITE MINERAL OIL

CONSENSUS REPORTS: IARC Cancer Review: Group 1 IMEMDT 7,252,87; Animal Inadequate Evidence IMEMDT 33,87,84. Reported in EPA TSCA Inventory.

SAFETY PROFILE: Confirmed carcinogen. When heated to decomposition it emits acrid smoke and irritating fumes.

MQW000 ***HR: 2***
MINING REAGENT
DOT: NA 2022

PROP: Solution contains 20% or more cresylic acid (FEREAC 41,15972,76).

SYN: MINING REAGENT, liquid (containing 20% or more cresylic acid) (DOT)

DOT Classification: Corrosive Material; Label: Corrosive.

SAFETY PROFILE: A corrosive irritant to skin, eyes and mucous membranes. See also CRESYLIC ACID.

MQW100 CAS:13614-98-7 ***HR: 3***
MINOCIN
mf: $C_{23}H_{27}N_3O_7$•ClH mw: 493.99

SYNS: MINOCYCLINE CHLORIDE ◇ MINOCYCLINE HYDROCHLORIDE ◇ TRI-MINO ◇ TRI-MINOCYCLINE

TOXICITY DATA with REFERENCE
orl-rat LD50:2380 mg/kg NIIRDN 6,811,82
ipr-rat LD50:331 mg/kg NIIRDN 6,811,82
ivn-rat LD50:164 mg/kg NIIRDN 6,811,82
orl-mus LD50:3600 mg/kg NIIRDN 6,811,82
ipr-mus LD50:299 mg/kg NIIRDN 6,811,82
ivn-mus LD50:154 mg/kg NIIRDN 6,811,82
par-mus LDLo:620 mg/kg CHTHBK 24,17,78

SAFETY PROFILE: Poison by intravenous and intraperitoneal routes. Moderately toxic by ingestion and parenteral routes. When heated to decomposition it emits toxic fumes of NO_x and HCl.

MQW250 CAS:10118-90-8 ***HR: 3***
MINOCYCLINE
mf: $C_{23}H_{27}N_3O_7$ mw: 457.53

PROP: Bright yellow-orange, amorphous solid.

SYNS: CL 59806 ◇ 7-DIMETHYLAMINO-6-DEMETHYL-6-DEOXYTETRACYCLINE ◇ MINOCYCLIN

TOXICITY DATA with REFERENCE
dni-hmn:lym 3750 μg/L BCPHBM 16,127,83
orl-wmn TDLo:100 mg/kg:KID BMJOAE 1,524,79
orl-mus LD50:3100 mg/kg 85ERAY 1,501,78
ipr-mus LD50:310 mg/kg 85ERAY 1,501,78
ivn-mus LD50:140 mg/kg 85ERAY 1,501,78
ice-mus LD50:38 mg/kg NKRZAZ 26,196,80

SAFETY PROFILE: Poison by intraperitoneal, intravenous, and intracerebral routes. Moderately toxic by ingestion. Human systemic effects by ingestion: interstitial nephritis, proteinuris and hematuris. Human mutation data reported. When heated to decomposition it emits toxic fumes of NO_x.

MQW500 CAS:2385-85-5 ***HR: 3***
MIREX
mf: $C_{10}Cl_{12}$ mw: 545.50

PROP: Very white, odorless crystals. Decomp @ 485°. Water-insol; sol in dioxane and benzene.

SYNS: BICHLORENDO ◇ CG-1283 ◇ DECHLORANE 4070 ◇ DODECACHLOROOCTAHYDRO-1,3,4-METHENO-2H-CYCLOBUTA(c,d)PENTALENE ◇ 1,1a,2,2,3,3a,4,5,5,5a,5b,6-DODECACHLOROOCTAHYDRO-1,3,4-METHENO-1H-CYCLOBUTA(c,d)PENTALENE◇ DODECACHLOROPENTACYCLODECANE ◇ DODECACHLOROPENTACYCLO(3,2,2,$0^{2,6}$,$0^{3,9}$,$0^{5,10}$)DECANE ◇ ENT 25,719 ◇ FERRIAMICIDE ◇ HEXACHLOROCYCLOPENTADIENEDIMER ◇ 1,2,3,4,5,5-HEXACHLORO-1,3-CYCLOPENTADIENE DIMER ◇ HRS 1276 ◇ NCI-C06428 ◇ PERCHLORODIHOMOCUBANE ◇ PERCHLOROPENTACYCLODECANE ◇ PERCHLOROPENTACYCLO (5.2.1.$0^{2,6}$.$0^{3,9}$.$0^{5,8}$)DECANE

TOXICITY DATA with REFERENCE
oms-rat-orl 100 mg/kg TOLED5 23,127,84
orl-mus TDLo:22 mg/kg (female 1-21D post):REP FEPRA7 37,938,78
orl-rat TDLo:50 mg/kg (5-9D preg):TER TJADAB 27,401,83
orl-rat TDLo:1635 mg/kg/78W-C:NEO JJIND8 58,133,77
orl-mus TDLo:2222 mg/kg/58W-C:CAR JNCIAM 42,1101,69
orl-rat TDLo:2340 mg/kg/56W-C:ETA EXMPA6 38,271,83
orl-rat LD50:235 mg/kg SPEADM 78-1,14,78
skn-rbt LD50:800 mg/kg FMCHA2 -,C159,83
orl-ham LD50:125 mg/kg TXAPA9 48,A192,79
orl-dck LD50:2400 mg/kg ENVRAL 14,212,77
ihl-brd LC50:1400 ppm ENVRAL 14,212,77

CONSENSUS REPORTS: NTP Fifth Annual Report on

Carcinogens. IARC Cancer Review: Group 2B IMEMDT 7,56,87; Human Limited Evidence IMEMDT 20,283,79; Animal Sufficient Evidence IMEMDT 20,283,79; IMEMDT 5,203,74. EPA Genetic Toxicology Program.

SAFETY PROFILE: Confirmed carcinogen with experimental carcinogenic, tumorigenic, and teratogenic data. Poison by ingestion. Moderately toxic by inhalation and skin contact. An experimental teratogen. Experimental reproductive effects. Mutation data reported. A persistent insecticide which is toxic to non-target species. It can bioaccumulate. See also CHLORINATED HYDROCARBONS, ALIPHATIC.

MQW525 ***HR: 3***
MISTLETOE (AMERICAN)

PROP: Parasitic plants which grow on the trunks and branches of trees. They have thick, leathery leaves and white or pink berries. *P. rubrum* grows only on mahogany trees in southern Florida and the West Indies. *P. serotinum* grows on deciduous trees in the region bounded by New Jersey, Florida, Texas, and Illinois. It is commonly sold as a mistletoe plant at Christmas. *P. tomentosum* grows in the region bounded by Kansas, Louisiana and Mexico.

SYNS: CEPA CABALLERO (CUBA) ◇ FALSE MISTLETOE ◇ INJERTO (TEXAS, MEXICO) ◇ PHORADENDRON RUBRUM ◇ PHORADENDRON SEROTINUM ◇ PHORADENDRON TOMENTOSUM

SAFETY PROFILE: The leaves, stems, and berries contain the poisonous lectin phoratoxin (a toxalbumin). Ingestion of these plant parts may cause after a delay period of several hours: severe vomiting, abdominal cramps, and diarrhea. Deaths have been reported from ingestion of the berries. See also ABRIN as an example toxalbumin.

MQW750 CAS:18378-89-7 ***HR: 3***
MITHRAMYCIN
mf: $C_{52}H_{76}O_{24}$ mw: 1085.28

PROP: Antibiotic substance isolated from the fermentation broth of three strains of an unidentified *Streptomyces* species (ANTCAO 3,1218,53).

SYNS: A-2371 ◇ ANTIBIOTIC LA 7017 ◇ AUREOLIC ACID ◇ AURLELIC ACID ◇ MITHRACIN ◇ MITHRAMYCIN A ◇ MITRAMYCIN ◇ NSC 24559 ◇ PA 144

TOXICITY DATA with REFERENCE
dnd-hmn:hla 1 mg/L CNREA8 45,2813,85
msc-hmn:hla 40 µg/L CNREA8 45,2813,85
unr-rat TDLo:2 mg/kg (11D preg):TER 85DJA5 -,95,71
unr-rat TDLo:2 mg/kg (11D preg):REP 85DJA5 -,95,71
orl-hmn TDLo:50 µg/kg/5D:BLD CORTBR 127,106,77
ipr-rat LD50:2500 µg/kg ADTEAS 3,181,68
ivn-rat LD50:1741 µg/kg NCICP* -,215,66
orl-mus LD50:500 mg/kg 85GDA2 1,342,80
ipr-mus LD50:1090 µg/kg ANTBAL 21,258,76
scu-mus LD50:2810 µg/kg ANTBAL 21,258,76
ivn-mus LD50:350 µg/kg 85ERAY 2,1405,78
ivn-dog LD50:250 µg/kg 85ERAY 2,1405,78
ivn-rbt LD50:250 µg/kg 85ERAY 2,1405,78

SAFETY PROFILE: A deadly poison by intravenous, intraperitoneal, and subcutaneous routes. Moderately toxic by ingestion. Human systemic effects by ingestion: blood thrombocytopenia. An experimental teratogen. Experimental reproductive effects. Human mutation data reported. When heated to decomposition it emits acrid smoke and irritating fumes.

MQX000 CAS:11043-98-4 ***HR: 3***
MITOCROMIN

SYNS: MITOCHROMIN ◇ NSC 77471

TOXICITY DATA with REFERENCE
orl-rat LD50:13 mg/kg NCICP* -,143,65
ivn-rat LD50:700 µg/kg NCICP* -,143,65
orl-mus LD50:20 mg/kg NCICP* -,143,65
ivn-mus LD50:97 µg/kg NCICP* -,143,65
ivn-dog LD50:300 µg/kg NCICP* -,143,65

SAFETY PROFILE: A deadly poison by ingestion and intravenous routes.

MQX250 CAS:11043-99-5 ***HR: 3***
MITOMALCIN

SYNS: AMRITAMYCIN ◇ B 2992 ACTIVE PRINCIPLE ◇ NSC B-2992 ◇ NSC-113233

TOXICITY DATA with REFERENCE
ipr-mus LD50:2640 µg/kg NCICP*
scu-mus LD50:7053 µg/kg NCISP* JAN86
ivn-mus LD50:4952 µg/kg NCISP* JAN86
ims-mus LD50:7521 µg/kg NCISP* JAN86

SAFETY PROFILE: Poison by intraperitoneal, intravenous, subcutaneous, and intramuscular routes.

MQX500 CAS:4055-39-4 ***HR: 3***
MITOMYCIN A
mf: $C_{16}H_{19}N_3O_6$ mw: 349.38

PROP: Red-violet crystals from (acetone + carbon tetrachloride). Decomp 159-161°. Sol in water, benzene, toluene, trichloroethylene, nitrobenzene and many organic solvents. Insol in xylene, carbon tetrachloride, carbon disulfide, petr ether, ligroin, cyclohexane.

TOXICITY DATA with REFERENCE
pic-esc 3 µg/L JMCMAR 20,767,77
ivn-mus LD50:1 mg/kg 85FZAT -,421,67

SAFETY PROFILE: Poison by intravenous route. Mutation data reported. When heated to decomposition it emits toxic fumes of NO_x.

MQX750 CAS:4055-40-7 ***HR: 3***
MITOMYCIN B
mf: $C_{16}H_{19}N_3O_6$ mw: 349.38

PROP: Violet crystals from (acetone + carbon tetrachloride. Decomp @ 182-184°. Sol in water and many organic solvents; insol in xylene, carbon tetrachloride, carbon sulfide, petr ether, ligroin, cyclohexane, benzene, toluene, trichloroethylene, and nitrobenzene.

TOXICITY DATA with REFERENCE
dnd-esc 60 μmol/L CJBIAE 56,296,78
dnd-mam:lym 60 μmol/L CJBIAE 56,296,78
ivn-mus LD50:3 mg/kg JMCMAR 14,103,71

SAFETY PROFILE: Poison by intravenous route. Mutation data reported. When heated to decomposition it emits toxic fumes of NO_x.

MQX775 CAS:54824-17-8 ***HR: D***
MITONAFIDE
mf: $C_{16}H_{15}N_3O_4$•ClH mw: 349.80

SYNS: M 4212 (PHARMACEUTICAL) ◇ 5-NITRO-2-(2-DIMETHYLAMINOETHYL)BENZO(de)ISOQUINOLINE-1,3-DIONEHYDROCHLORIDE

TOXICITY DATA with REFERENCE
dnd-ham:ovr 10 μmol/L JJIND8 70,1097,83
cyt-ham:ovr 1 μmol/L JJIND8 70,1097,83

SAFETY PROFILE: Mutation data reported. When heated to decomposition it emits toxic fumes of NO_x and HCl. See also AROMATIC AMINES.

MQY090 CAS:65271-80-9 ***HR: 3***
MITOXANTRONE
mf: $C_{22}H_{28}N_4O_6$ mw: 444.54

PROP: Crystals from ethanol/hexane. Mp: 160-162°.

SYNS: 5,8-BIS((2-((HYDROXYETHYL)AMINO)ETHYL)AMINO)-1,4-DIHYDROXYANTHRAQUINONE ◇ DHAQ ◇ DIHYDROXYANTHRAQUINONE ◇ 1,4-DIHYDROXY-5,8-BIS((2-((HYDROXYETHYL)AMINO)ETHYL)AMINO)-9,10-ANTHRACENEDIONE (9CI) ◇ MITOXANTHRONE ◇ NSC 279836

TOXICITY DATA with REFERENCE
mmo-sat 20 μg/plate TCMUD8 5,319,85
cyt-hmn:lym 100 μmol/L MUREAV 93,185,82
ipr-mus LD50:24120 μg/kg NCISP* JAN86
scu-mus LD50:20910 μg/kg NCISP* JAN86
ivn-mus LD50:17640 μg/kg NCISP* JAN86

SAFETY PROFILE: Poison by subcutaneous, intravenous, and intraperitoneal routes. Human mutation data reported. When heated to decomposition it emits toxic fumes of NO_x.

MQY100 CAS:70476-82-3 ***HR: D***
MITOXANTRONE HYDROCHLORIDE
mf: $C_{22}H_{28}N_4O_6$•2ClH mw: 517.46

SYNS: CL 232315 ◇ NSC 301739

TOXICITY DATA with REFERENCE
dni-hmn:oth 1 nmol/L PAACA3 24,252,83
cyt-hmn:hla 100 μg/L CNREA8 43,3270,83
cyt-ham:lng 8 μg/L CNREA8 45,3593,85

SAFETY PROFILE: Human mutation data reported. When heated to decomposition it emits toxic fumes of NO_x and HCl.

MQY110 CAS:85622-95-3 ***HR: 3***
MITOZOLOMIDE
mf: $C_7H_7ClN_6O_2$ mw: 242.65

SYNS: AZOLASTONE ◇ 8-CARBAMOYL-3-(2-CHLOROETHYL)IMIDAZO(5,1-d)-1,2,3,5-TETRAZIN-4(3H)-ONE◇ CCRG 81010 ◇ 3-(2-CHLOROETHYL)-3,4-DIHYDRO-4-OXOIMIDAZO(5,1-d)-1,2,3,5-TETRAZINE-8-CA RBOXAMIDE ◇ IMIDAZO(5,1-d)-1,2,3,5-TETRAZINE-8-CARBOXAMIDE,3-(2-CHLOROETHYL)-3,4-DIHYDRO-4-OXO-◇ M & B 39565 ◇ NSC 353451

TOXICITY DATA with REFERENCE
dnd-hmn:emb 50 umol/L CNREA8 44,1772,84
dnd-mus:leu 50 umol/L CNREA8 44,1767,84
ipr-mus TDLo:40 mg/kg (male 1D pre):REP ARTODN 11,273,87
orl-hmn TDLo:3108 μg/kg:GIT,BLD BJCAAI 55,433,87
ipr-mus LD10: 45 mg/kg BJCAAI 58,139,88

SAFETY PROFILE: Poison by intraperitoneal route. Human systemic effects by ingestion: nausea or vomiting and blood changes. Experimental reproductive effects. Human mutation data reported. When heated to decomposition it emits toxic fumes of NO_x and Cl^-.

MQY125 CAS:42794-63-8 ***HR: 3***
MK-142 DIMETHANESULFONATE
mf: $C_{22}H_{32}N_2O_6$•$2CH_4O_3S$ mw: 612.78

SYNS: N,N′-DI(3-(p-METHOXYPHENOXY)-2-HYDROXYPROPYL)ETHYLENEDIAMINEDIMETHANESULPHOANTE ◇ 1,1′-(1,2-ETHANEDIYLDIIMINO)BIS(3-(4-METHOXYPHENOXY)-2-PROPANOL, DIMETHANESULFONATE (salt) ◇ MK 142

TOXICITY DATA with REFERENCE
orl-rat LD50:3100 mg/kg MMDPA6 9,216,77
orl-mus LD50:1570 mg/kg PJPPAA 27,365,75
ivn-mus LD50:135 mg/kg PJPPAA 27,365,75

SAFETY PROFILE: Poison by intravenous route. Moderately toxic by ingestion. When heated to decomposition it emits toxic fumes of NO_x and SO_x. See also SULFONATES.

MQY250 ***HR: 2***
MLO-5277 (ORGANO-SILICATE)

TOXICITY DATA with REFERENCE
ihl-rat LCLo:500 mg/m^3 XAWPA2 CWL 2-10,58
ivn-rat LD50:900 mg/kg MRLR** #256,54
ivn-rbt LD50:540 mg/kg MRLR** #256,54

SAFETY PROFILE: Moderately toxic by inhalation and intravenous routes. See also SILICATES.

MQY300 CAS:84504-69-8 ***HR: 2***
MN-1695
mf: $C_9H_7Cl_2N_5 \cdot C_4H_4O_4$ mw: 372.19

SYN: 2,4-DIAMINO-6-(2,5-DICHLOROPHENYL)-s-TRIAZINEMALEATE

TOXICITY DATA with REFERENCE
orl-rat TDLo:810 mg/kg (female 17-22D post):REP OYYAA2 32,657,86
orl-rbt TDLo:390 mg/kg (female 6-18D post):TER OYYAA2 33,121,87
orl-rat LD50:2898 mg/kg EPXXDW #114907
ipr-rat LD50:545 mg/kg ARZNAD 34,474,84
scu-rat LD50:1524 mg/kg ARZNAD 34,474,84
orl-mus LD50:5697 mg/kg ARZNAD 34,474,84
ipr-mus LD50:775 mg/kg ARZNAD 34,474,84
scu-mus LD50:2841 mg/kg ARZNAD 34,474,84

SAFETY PROFILE: Moderately toxic by ingestion, intraperitoneal and subcutaneous routes. An experimental teratogen. Other experimental reproductive effects. When heated to decomposition it emits toxic fumes of Cl^- and NO_x.

MQY325 CAS:63642-17-1 ***HR: 3***
MNCO
mf: $C_7H_{14}N_4O_4$ mw: 218.25

SYNS: N^{Δ}-(N-METHYL-N-NITROSOCARBAMOYL)-l-ORNITHINE ◇ N^5-(METHYLNITROSOCARBAMOYL)-l-ORNITHINE ◇ N^5-(N-METHYL-N-NITROSOCARBAMOYL)-l-ORNITHINE

TOXICITY DATA with REFERENCE
dnd-rat-ipr 1 mmol/kg CRNGDP 5,555,84
otr-ham-par 2616 mg/kg/12W-I JJIND8 71,1327,83
dnd-ham:oth 218 mg/L JJIND8 71,1327,83
ipr-rat TDLo:1870 mg/kg/26W-I:CAR JEPTDQ 4(1),117,80
ipr-ham TDLo:2616 mg/kg/12W-I:CAR JJIND8 71,1327,83
ipr-ham TDLo:1308 mg/kg/6W-I:ETA CALEDQ 8,163,79

SAFETY PROFILE: Suspected carcinogen with experimental carcinogenic and tumorigenic data. Mutation data reported. When heated to decomposition it emits toxic fumes of NO_x. See also N-NITROSO COMPOUNDS.

MQY350 CAS:87209-80-1 ***HR: 2***
MOBILAT

TOXICITY DATA with REFERENCE
scu-rat TDLo:27500 mg/kg (female 7-17D post):REP KSRNAM 11,2175,77
scu-rat TDLo:5500 mg/kg (female 7-17D post):TER KSRNAM 11,2175,77
skn-rat LDLo:51230 mg/kg ARZNAD 14,1309,64
scu-rat LD50:1184 mg/kg NIIRDN 6,856,82
scu-mus LD50:1016 mg/kg NIIRDN 6,856,82

SAFETY PROFILE: Moderately toxic by subcutaneous route. An experimental teratogen. Experimental reproductive effects.

MQY400 CAS:2210-63-1 ***HR: 2***
MOBUTAZON
mf: $C_{13}H_{16}N_2O_2$ mw: 232.28

PROP: Crystals from ethanol + water. Mp: 102-103°.

SYNS: ARCOMONOL TABLETS ◇ 4-BUTYL-1-PHENYL-3,5-DIOXOPYRAZOLIDINE ◇ 4-BUTYL-1-PHENYL-3,5-PYRAZOLIDINEDIONE ◇ 2 FDBP ◇ MOBUZON ◇ MOFEBUTAZONE ◇ MONAZAN ◇ MONOBUTYL ◇ MONOPHENYLBUTAZONE ◇ MONORHEUMETTEN ◇ 2-PHENYL-3,5-DIHYDROXY-4-BUTYLPYRAZOLIDINE ◇ REUMATOX

TOXICITY DATA with REFERENCE
ipr-rat LD50:848 mg/kg FRPSAX 14,347,59
ipr-mus LD50:937 mg/kg FRPSAX 14,347,59
ivn-mus LD50:600 mg/kg AEPPAE 233,365,58

SAFETY PROFILE: Moderately toxic by intravenous and intraperitoneal routes. When heated to decomposition it emits toxic fumes of NO_x.

MRA000 ***HR: 2***
MODECCIN TOXIN

SYNS: MODECCIN ◇ TOXIN, ADENIA DIGITATA (MODECCA DIGITATA), MODECCIN

TOXICITY DATA with REFERENCE
dni-hmn:hla 30 ug/L BBRCA9 79,1176,77
ipr-rat LD50:1300 ng/kg BIJOAK 174,491,78
ipr-mus LD50:2300 mg/kg 85GDA2 8(1),109,82

SAFETY PROFILE: Moderately toxic by intraperitoneal route. Human mutation data reported.

MRA100 ***HR: 2***
MODIFIER 113-63

SYN: SILICON ORGANIC LACQUER MODIFICATOR 113-63

TOXICITY DATA with REFERENCE
orl-rat LD50:1 g/kg GISAAA 50(3),81,85
orl-mus LD50:1 g/kg GISAAA 50(3),81,85
orl-gpg LD50:1 g/kg GISAAA 50(3),81,85

SAFETY PROFILE: Moderately toxic by ingestion.

MRA250 CAS:11015-37-5 ***HR: 3***
MOENOMYCIN

PROP: Produced by *Streptomyces roseoflavus* (85ERAY 1,740,78).

SYNS: BAMBERMYCIN ◇ FLAVOMYCIN ◇ FLAVOPHOSPHOLIPOL ◇ MENOMYCIN ◇ MOENOMYCIN A

TOXICITY DATA with REFERENCE
scu-mus LD50:500 mg/kg 85ERAY 1,740,78
ivn-mus LD50:200 mg/kg 85ERAY 1,740,78
ivn-dog LDLo:600 mg/kg AACHAX -,743,65

SAFETY PROFILE: Poison by intravenous route. Moderately toxic by subcutaneous route.

MRA275 CAS:11052-70-3 ***HR: D***
MOGALAROL
mf: $C_{29}H_{31}NO_{12}$ mw: 585.61

TOXICITY DATA with REFERENCE
dni-hmn:oth 3180 nmol/L HXPHAU 39(Pt 2),623,75
dnd-mam:lym 12 μmol/L CBINA8 36,1,81

SAFETY PROFILE: Human mutation data reported. When heated to decomposition it emits toxic fumes of NO_x.

MRA300 CAS:53681-76-8 ***HR: 3***
MOLANTIN P

TOXICITY DATA with REFERENCE
skn-rbt 500 mg/24H MOD 28ZPAK -,284,72
eye-rbt 20 mg/24H MOD 28ZPAK -,284,72
orl-rat LD50:278 mg/kg GISAAA 43(7),70,78
orl-mus LD50:315 mg/kg GISAAA 43(7),70,78

SAFETY PROFILE: Poison by ingestion. An eye and skin irritant.

MRA750 ***HR: 2***
MOLECULAR SIEVE 13X with 14.6% DI-n-BUTYLAMINE

TOXICITY DATA with REFERENCE
eye-rbt 26 mg SEV UCDS** 6/14/60
orl-rat LD50:2000 mg/kg UCDS** 6/14/60
ihl-pig LCLo:1650 mg/m^3/2.5H UCDS** 6/14/60

SAFETY PROFILE: Moderately toxic by ingestion and inhalation. A severe eye irritant. Potenially hazardous reactions with ethylene; triaryl phosphates; nitromethane; tert-butyl hydroperoxide; benzyl bromide; mercury(II) perchlorate. See also DI-N-BUTYLAMINE.

MRB250 CAS:15622-65-8 ***HR: 3***
MOLINDONE HYDROCHLORIDE
mf: $C_{16}H_{24}N_2O_2$•ClH mw: 312.88

SYNS: 3-ETHYL-6,7-DIHYDRO-2-METHYL-5-MORPHOLINOMETHYLINDOLE-4(5H)-ONE HYDROCHLORIDE ◇ 3-ETHYL-6,7-DIHYDRO-2-METHYL-5-MORPHOLINOMETHYLINDOL-4(5H)-ONE HYDROCHLORIDE

TOXICITY DATA with REFERENCE
orl-man TDLo:2500 μg/kg (male 2D pre):REP TXMDAX 81(9),47,85
orl-man TDLo:4762 μg/kg/3D-I:KID JCLPDE 47,607,86
orl-rat LD50:261 mg/kg TXAPA9 18,185,71
orl-mus LD50:670 mg/kg FEPRA7 26,738,67
ipr-mus LD50:243 mg/kg 27ZQAG -,134,72

SAFETY PROFILE: Poison by ingestion and intraperitoneal routes. Human reproductive effects. Human systemic effects by changes in kidney tubules. When heated to decomposition it emits very toxic fumes of NO_x and HCl.

MRC000 CAS:12656-85-8 ***HR: 3***
MOLYBDATE ORANGE

SYNS: CHROME VERMILION ◇ C.I. 77605 ◇ C.I. PIGMENT RED 104 ◇ MINERAL FIRE RED 5GS ◇ MOLYBDATE RED ◇ MOLYBDEN RED ◇ MOLYBDENUM RED ◇ NCI-C54626

CONSENSUS REPORTS: IARC Cancer Review: Group 1 IMEMDT 49,49,90; Human Sufficient Evidence IMEMDT 49,49,90. Chromium and its compounds are on the Community Right-To-Know List. Reported in EPA TSCA Inventory.

OSHA PEL: TWA 5 mg(Mo)/m^3
ACGIH TLV: TWA 5 mg(Mo)/m^3

SAFETY PROFILE: Confirmed carcinogen. Dusts are poison by inhalation. See MOLYBDENUM COMPOUNDS and CHROMIUM COMPOUNDS.

MRC250 CAS:7439-98-7 ***HR: 3***
MOLYBDENUM
af: Mo aw: 95.94

PROP: Cubic, silver-white metallic crystals or gray-black powder. Mp: 2622°, bp: approx 4825°, d: 10.2, vap press: 1 mm @ 3102°.

SYN: MOLYBDATE

TOXICITY DATA with REFERENCE
cyt-rat-ihl 19500 μg/m^3 GTPZAB 24(9),33,80
orl-mus TDLo:448 mg/kg (multi):TER AEHLAU 23,102,71
orl-rat TDLo:6050 μg/kg (female 35W pre):REP GISAAA 42(8),30,77
ipr-rat LDLo:114 mg/kg 28ZLA8 -,214,61
itr-rbt LDLo:70 mg/kg NTIS** PB249-458

CONSENSUS REPORTS: Reported in EPA TSCA Inventory.

OSHA PEL: Soluble Compounds: TWA 5 mg(Mo)/m^3; Insoluble Compounds: (Transitional: TWA Total Dust: 15 mg/m^3; Respirable Fraction: 5 mg/m^3) TWA Total Dust: 10 mg/m^3; Respirable Fraction: 5 mg/m^3
ACGIH TLV: Soluble Compounds: TWA 5 mg(Mo)/m^3; Insoluble Compounds: TWA 10 mg(Mo)/m^3
DFG MAK: (insoluble compounds) 15 mg/m^3; (soluble compounds) 5 mg/m^3

SAFETY PROFILE: Poison by intraperitoneal and intratracheal routes. Mutation data reported. An experimental teratogen. Experimental reproductive effects. Flammable or explosive in the form of dust; when exposed to heat or flame. Violent reaction with oxidants (e.g., bromine trifluoride; bromine pentafluoride; chlorine trifluoride; potassium perchlorate; nitryl fluoride; fluorine; iodine pentafluoride; sodium peroxide; lead dioxide). When heated to decomposition it emits toxic fumes of Mo. See also POWDERED METALS and MOLYBDENUM COMPOUNDS.

MRC500 ***HR: 3***
MOLYBDENUM AZIDE PENTACHLORIDE
mf: Cl_5MoN_3 mw: 315.22

SAFETY PROFILE: Extremely explosive. When heated to decomposition it emits very toxic fumes of Mo, NO_x, and Cl^-. See also MOLYBDENUM COMPOUNDS, AZIDES, and CHLORIDES.

MRC600 CAS:68825-98-9 ***HR: 3***
MOLYBDENUM AZIDE TRIBROMIDE
mf: Br_3MoN_3 mw: 377.64

SAFETY PROFILE: Highly explosive. When heated to decomposition it emits toxic fumes of Mo, Br^- and NO_x. See also AZIDES, BROMIDES, and MOLYBDENUM COMPOUNDS.

MRC650 CAS:12007-97-5 ***HR: 3***
MOLYBDENUM BORIDE
mf: B_5Mo_2 mw: 245.93

PROP: Mp: 2200°, d: 7.48.

TOXICITY DATA with REFERENCE
itr-rat LDLo:125 mg/kg GTPZAB 9(6),40,65
ipr-mus LD50:1377 mg/kg GTPZAB 9(6),40,65
orl-rbt LD50:1380 mg/kg 85AEA9 -,115,75

ACGIH TLV: TWA 10 mg(Mo)/m^3
OSHA PEL: TWA 15 mg(Mo)/m^3

CONSENSUS REPORTS: Reported in EPA TSCA Inventory.

SAFETY PROFILE: Poison by intratracheal route. Moderately toxic by ingestion and intraperitoneal routes. When heated to decomposition it emits toxic fumes of boron and Mo.

MRC750 ***HR: 3***
MOLYBDENUM COMPOUNDS

SAFETY PROFILE: Poison by subcutaneous and intraperitoneal routes. Molybdenum and its compounds are highly toxic based upon animal experiments. Symptoms of acute poisoning include severe gastrointestinal irritation with diarrhea, coma, and deaths from heart failure. Experimental animals exposed to high levels (57 mg Mo/m^3)of molybdenum dust for 120 days (4 hours/day) accumulated Mo in the lungs, spleen, and heart, and showed a decrease of DNA and RNA in the liver, kidneys, and spleen. Workers exposed to Mo or MoO_3 (concentrations of 1-19 mg Mo/m^3) over a period of 3-7 years have suffered from pneumoconiosis. Inhalation of molybdenum dust from alloys or carbides can cause "hard-metal lung disease". It is suggested that suitable precautions should be taken against human inhalation of significant amounts of the more soluble molybdenum compounds. MoO_3 and Na_2MoO_4 are soluble. $CaMoO_4$, MoO, and MoS_2 are insoluble. Hexavalent molybdenum compounds are readily absorbed through the gastrointestinal tract. Coal-fired electrical power plants can be significant sources of molybdenum. Application of some fertilizers may raise molybdenum concentrations in ground water. Molybdenum is rapidly excreted by the body. Molybdenum is an important trace element in the normal growth and development of plants. It is found also in animal tissue, although its precise function is unknown. It is a common air contaminant. See also specific compounds.

MRD000 CAS:14259-66-6 ***HR: 3***
MOLYBDENUM DIAZIDE TETRACHLORIDE
mf: Cl_4MoN_6 mw: 321.79

SAFETY PROFILE: Highly explosive. When heated to decomposition it emits very toxic fumes of Mo, NO_x, and Cl^-. See also MOLYBDENUM COMPOUNDS, AZIDES, and CHLORIDES.

MRD250 CAS:18868-43-4 ***HR: 3***
MOLYBDENUM OXIDE
mf: MoO_2 mw: 127.94

SYNS: MOLYBDENUM DIOXIDE ◇ MOLYBDENUM(VI) OXIDE

TOXICITY DATA with REFERENCE
scu-mus LD50:318 mg/kg ZVKOA6 19,186,74

CONSENSUS REPORTS: Reported in EPA TSCA Inventory.

OSHA PEL: TWA 10 mg(Mo)/m^3, total dust; TWA 5 mg(Mo)/m^3, respirable fraction
ACGIH TLV: TWA 10 mg(Mo)/m^3, total dust; TWA 5 mg(Mo)/m^3, respirable fraction

SAFETY PROFILE: Poison by subcutaneous route. Incandescent reaction with air. When heated to decomposition it emits toxic fumes of Mo. See also MOLYBDENUM COMPOUNDS.

MRD500 CAS:10241-05-1 ***HR: 3***
MOLYBDENUM PENTACHLORIDE
DOT: UN 2508
mf: Cl_5Mo mw: 273.19

PROP: Green-black solid, dark-red as liquid or vapor. Hygroscopic, reacting with water and air. Mp: 194°, bp: 268°, d: 2.9. Sol in dry ether, dry alc, and other anhydrous organic solvents.

CONSENSUS REPORTS: Reported in EPA TSCA Inventory. EPA Genetic Toxicology Program.

OSHA PEL: TWA 5 mg(Mo)/m^3
ACGIH TLV: TWA 5 mg(Mo)/m^3
DOT Classification: ORM-B; Label: None; Corrosive Material; Label: Corrosive.

SAFETY PROFILE: A poison. A corrosive irritant to skin, eyes, and mucous membranes. Reacts with moisture to form hydrochloric acid. When heated to decomposition it emits toxic fumes of Mo and Cl^-. See also MOLYBDENUM COMPOUNDS and HYDROCHLORIC ACID.

MRD750 CAS:1317-33-5 ***HR: 3***
MOLYBDENUM(IV) SULFIDE
mf: MoS_2 mw: 160.07

PROP: Black, lustrous powder. Mp: 1185°, bp: decomp in air, d: 4.80 @ 14°.

SYNS: MOLYBDENITE ◇ MOLYBDIC SULFIDE ◇ MOLYKOTE ◇ MOPOLM

CONSENSUS REPORTS: Reported in EPA TSCA Inventory. EPA Genetic Toxicology Program.

SAFETY PROFILE: Mixtures with potassium nitrate are explosive. Violent reaction with H_2O_2. When heated to decomposition it emits toxic fumes of SO_x and Mo. See also SULFIDES and MOLYBDENUM COMPOUNDS.

MRE000 CAS:1313-27-5 ***HR: 3***
MOLYBDENUM TRIOXIDE
mf: MoO_3 mw: 143.94

PROP: White or yellow to sltly bluish powder or granules. Mp: 795°; bp: 1155°; d: 4.696 @ 26°/4°. Sol in 1000 parts water, in concentrated mineral acids, solutions of alkali hydroxides. Sol in ammonia or potassium bitartrate, solidifying to a yellowish-white mass.

SYNS: MOLYBDENUM(VI) OXIDE ◇ MOLYBDIC ANHYDRIDE ◇ MOLYBDIC TRIOXIDE

TOXICITY DATA with REFERENCE
ipr-mus TDLo:4750 mg/kg/7W-I:NEO CNREA8 36,1744,76
ihl-man TCLo:6 mg/m^3/4Y:PUL NTIS** AEC-TR-6710
orl-rat LD50:125 mg/kg 28ZLA8 -,214,61
scu-mus LD50:94 mg/kg ZVKOA6 19,186,74
ipr-gpg LDLo:400 mg/kg EQSSDX 1,1,75

CONSENSUS REPORTS: Reported in EPA TSCA Inventory. EPA Extremely Hazardous Substances List.

OSHA PEL: TWA 5 mg(Mo)/m^3
ACGIH TLV: TWA 5 mg(Mo)/m^3

SAFETY PROFILE: Poison by ingestion, subcutaneous, and intraperitoneal routes. Human systemic effects by inhalation: pulmonary fibrosis and cough. Questionable carcinogen with experimental neoplastigenic data. A powerful irritant. Explodes on contact with molten magnesium. Violent reaction with interhalogens (e.g., bromine pentafluoride; chlorine trifluoride). Incandescent reaction with hot sodium, potassium, or lithium. When heated to decomposition it emits toxic fumes of Mo. See also MOLYBDENUM COMPOUNDS.

MRE225 CAS:17090-79-8 ***HR: 3***
MONENSIC ACID
mf: $C_{36}H_{62}O_{11}$ mw: 670.98

PROP: Crystals. Mp: 103-105° (monohydrate). Very stable under alkaline conditions. Sltly sol in water; more sol in hydrocarbons; very sol in other organic solvents.

SYNS: A 3823A ◇ ELANCOBAN ◇ MONELAN ◇ MONENSIN (USDA) ◇ MONENSIN A

TOXICITY DATA with REFERENCE
skn-rbt 100 mg/24H MLD DCTODJ 8,451,85
eye-rbt 50 mg MOD DCTODJ 8,451,85
orl-rat LD50:100 mg/kg DCTODJ 8,451,85
ipr-rat LD50:15 mg/kg DCTODJ 8,451,85
orl-mus LD50:43800 μg/kg 85GDA2 5,472,81
ipr-mus LD50:10 mg/kg DCTODJ 8,451,85
orl-hor LD50:2 mg/kg AJVRAH 42,456,81
orl-dom LD50:12 mg/kg AJVRAH 45,1142,84

SAFETY PROFILE: Poison by ingestion and intraperitoneal routes. An eye and skin irritant. When heated to decomposition it emits acrid smoke and irritating fumes.

MRE230 CAS:22373-78-0 ***HR: 3***
MONENSIN SODIUM
mf: $C_{36}H_{62}O_{11}$•Na mw: 693.97

SYNS: COBAN ◇ MONENSIN, MONOSODIUM SALT (9CI) ◇ MONENSIN SODIUM SALT ◇ RUMENSIN ◇ SODIUM MONENSIN

TOXICITY DATA with REFERENCE
orl-rat LD50:29 mg/kg JANSAG 58,1512,84
orl-mus LD50:70 mg/kg JANSAG 58,1512,84
orl-rbt LD50:42 mg/kg JANSAG 58,1512,84
orl-pig LD50:17 mg/kg JANSAG 58,1512,84
orl-ckn LD50:200 mg/kg JANSAG 58,1512,84
orl-ctl LD50:26 mg/kg JANSAG 58,1512,84
orl-hor LD50:2 mg/kg JANSAG 58,1512,84

SAFETY PROFILE: Poison by ingestion. When heated to decomposition it emits toxic fumes of Na_2O. See also MONENSIC ACID.

MRE250 CAS:31876-38-7 ***HR: 3***
MONILIFORMIN
mf: $C_4H_2O_3$ mw:98.06

PROP: Isolated from *Fusarium moniliforme sheldon*, a fungus found on maize (FCTXAV 15,579,77).

SYNS: 3-HYDROXY-CYCLOBUT-3-ENE-1,2-DIONE ◇ 1-HYDROXY-CYCLOBUT-1-ENE-3,4-DIONE

TOXICITY DATA with REFERENCE
orl-rat LD50:41 mg/kg FCTXAV 15,579,77
ipr-mus LD50:21 mg/kg POSCAL 60,1415,81
orl-ckn LD50:4 mg/kg FCTXAV 15,579,77
orl-dck LD50:3680 mg/kg FCTXAV 15,579,77

CONSENSUS REPORTS: EPA Genetic Toxicology Program.

SAFETY PROFILE: Poison by ingestion and intraperitoneal routes. When heated to decomposition it emits acrid smoke and irritating fumes.

MRE275 ***HR: 3***
MONKSHOOD

PROP: An erect perennial 2 to 6 feet tall resembling a delphinium. The flowers are usually blue and helmet-shaped, and grow at the top of stalk in summer or autumn. The roots are tuberous. Various species are native to the temperate zones of North America including Canada and Alaska. Some are cultivated in the same areas. They are not found in the warmer areas of North America.

SYNS: ACONITUM (VARIOUS SPECIES) ◇ A. COLUMBIANUM ◇ A. NAPELLUS ◇ ACONITE ◇ FRIAR'S CAP ◇ HELMET FLOWER ◇ SOLDIER'S CAP ◇ WOLFSBANE

SAFETY PROFILE: The whole plant contains the poison aconitine and some related alkaloids with the greatest amounts found in the leaves and roots. The immediate effects of ingestion are a burning feeling in the lips, mouth, and throat, followed by numbness. Later effects may include difficulty with speech, nausea, vomiting, blurred and yellow-green vision, weakness, dizziness, poor coordination, abnormal sensations, and cardiac arrhythmias which may result in death. See also ACONITINE.

MRE500 CAS:101809-53-4 ***HR: 3***
MONOACETYLISOSPIRAMYCIN

TOXICITY DATA with REFERENCE
ipr-rat LD50:664 mg/kg JJANAX 23,429,70
ipr-mus LD50:600 mg/kg JJANAX 23,429,70
scu-mus LD50:1500 mg/kg JJANAX 23,429,70
ivn-mus LD50:313 mg/kg JJANAX 23,429,70
ivn-rbt LD50:198 mg/kg JJANAX 23,429,70

SAFETY PROFILE: Poison by intravenous route. Moderately toxic by intraperitoneal and subcutaneous routes.

MRE750 CAS:71251-30-4 ***HR: 3***
MONOACETYLSPIRAMYCIN

SYN: ACETYL KITASAMYCIN

TOXICITY DATA with REFERENCE
ipr-rat LD50:613 mg/kg TXAPA9 18,185,71
ivn-mus LD50:206 mg/kg JJANAX 23,429,70

SAFETY PROFILE: Poison by intravenous route. Moderately toxic by intraperitoneal route.

MRF000 CAS:7558-63-6 ***HR: 2***
MONOAMMONIUM GLUTAMATE
mf: $C_5H_9NO_4 \cdot H_3N$ mw: 164.19

PROP: White crystalline powder; odorless. Sol in water; insol in common organic solvents.

SYNS: AMMONIUMGLUTAMINAT (GERMAN) ◇ MAG ◇ MONOAMMONIUM l-GLUTAMATE

TOXICITY DATA with REFERENCE
ipr-rat LD50:1000 mg/kg HSZPAZ 300,97,55

CONSENSUS REPORTS: Reported in EPA TSCA Inventory.

SAFETY PROFILE: Moderately toxic by intraperitoneal route. When heated to decomposition it emits toxic fumes including NO_x and NH_3. See also SODIUM GLUTAMATE.

MRF200 CAS:2425-41-4 ***HR: 3***
MONOBENZALPENTAERYTHRITOL
mf: $C_{12}H_{16}O_4$ mw: 224.28

SYN: 2-PHENYL-m-DIOXANE-5,5-DIMETHANOL

TOXICITY DATA with REFERENCE
ipr-rat LD50:103 mg/kg AITEAT 10,925,62

ipr-mus LD50:98 mg/kg AITEAT 10,925,62
scu-mus LD50:127 mg/kg AITEAT 10,925,62

SAFETY PROFILE: Poison by intraperitoneal and subcutaneous routes. When heated to decomposition it emits acrid smoke and irritating fumes.

MRF250 CAS:621-92-1 ***HR: 3***
MONOBENZYL-p-AMINOPHENOL HYDROCHLORIDE
mf: $C_{13}H_{13}NO•ClH$ mw: 235.73

TOXICITY DATA with REFERENCE
orl-rat LDLo:400 mg/kg KODAK* -,-,71
ipr-rat LD50:100 mg/kg KODAK* -,-,71

CONSENSUS REPORTS: Reported in EPA TSCA Inventory.

SAFETY PROFILE: Poison by ingestion and intraperitoneal routes. When heated to decomposition it emits very toxic fumes of NO_x and HCl.

MRF275 CAS:4704-77-2 ***HR: 3***
MONOBROMOGLYCEROL
mf: $C_3H_7BrO_2$ mw: 155.01

SYNS: BROMODEOXYGLYCEROL ◇ α-BROMOHYDRIN ◇ 3-BROMO-1,2-PROPANEDIOL

TOXICITY DATA with REFERENCE
mmo-sat 5 μmol/plate JTEHD6 5,1149,79
mma-sat 5 μmol/plate JTEHD6 5,1149,79
ipr-mus LD50:212 mg/kg JMCMAR 20,644,77

SAFETY PROFILE: Poison by intraperitoneal route. Mutation data reported. When heated to decomposition it emits toxic fumes of Br^-.

MRF500 CAS:16456-56-7 ***HR: 2***
MONOBUTYL PHOSPHITE
mf: $C_4H_{11}O_3P$ mw: 138.12

SYN: BUTYL PHOSPHITE

TOXICITY DATA with REFERENCE
unr-mus LD50:1640 mg/kg AMIHAB 11,487,55

CONSENSUS REPORTS: Reported in EPA TSCA Inventory.

SAFETY PROFILE: Moderately toxic by an unspecified route. When heated to decomposition it emits toxic fumes of PO_x.

MRF525 CAS:131-70-4 ***HR: D***
MONOBUTYL PHTHALATE
mf: $C_{12}H_{14}O_4$ mw: 222.26

SYNS: BUTYL HYDROGEN PHTHALATE ◇ MBP ◇ MONO-n-BUTYL PHTHALATE

TOXICITY DATA with REFERENCE
orl-rat TDLo:5600 mg/kg (male 7D pre):REP NFGZAD 28,159,82
ipr-mus LD50:1 g/kg CHDDAT 273,2165,71

CONSENSUS REPORTS: Reported in EPA TSCA Inventory.

SAFETY PROFILE: Experimental reproductive effects. When heated to decomposition it emits acrid smoke and irritating fumes.

MRG000 CAS:55398-86-2 ***HR: 2***
MONOCHLORO DIPHENYL OXIDE
mf: $C_{12}H_9ClO$ mw: 204.66

SYNS: MONOCHLOROPHENYLETHER ◇ PHENYL ETHER MONOCHLORO

TOXICITY DATA with REFERENCE
orl-gpg LDLo:600 mg/kg 14CYAT 2,1707,63

OSHA PEL: TWA 500 $\mu g/m^3$

SAFETY PROFILE: Moderately toxic by ingestion. When heated to decomposition it emits toxic fumes of Cl^-. See also ETHERS and CHLORINATED HYDROCARBONS, ALIPHATIC.

MRG100 ***HR: 2***
MONOCHORIA VAGINALIS (Burm. f.) Presl., extract

PROP: Indian plant belonging to the family *Pontederiaceae* IJEBA6 22,487,84

TOXICITY DATA with REFERENCE
orl-rat TDLo:150 mg/kg (female 12-14D post):REP IJEBA6 22,487,84
ipr-mus LD50:562 mg/kg IJEBA6 22,487,84

SAFETY PROFILE: Moderately toxic by intraperitoneal route. Experimental reproductive effects. When heated to decomposition it emits acrid smoke and irritating fumes.

MRH000 CAS:315-22-0 ***HR: 3***
MONOCROTALINE
mf: $C_{16}H_{23}NO_6$ mw: 325.40

PROP: Prisms from abs ethanol. Decomp @ 197-198°.

SYNS: CROTALINE ◇ 14,19-DIHYDRO-12,13-DIHYDROXY(13-α,14-α)-20-NORCROTALANAN-11,15-DIONE ◇ MONOCRATILIN ◇ NCI-C56462 ◇ NSC 28693

TOXICITY DATA with REFERENCE
cyt-hmn:lvr 50 mg/L BJCAAI 14,637,60
dns-rat:lvr 2 μmol/L CNREA8 45,3125,85
scu-rat TDLo:130 mg/kg/1Y-I:CAR JNCIAM 56,787,76
orl-rat LD50:66 mg/kg TXAPA9 18,387,71

ipr-rat LDLo:130 mg/kg CBINA8 12,299,76
scu-rat LDLo:60 mg/kg TXAPA9 23,470,72
ivn-rat LD50:92 mg/kg JPETAB 83,265,45
ipr-mus LD50:259 mg/kg TXAPA9 59,424,81
ivn-mus LD50:261 mg/kg JPETAB 75,78,42

CONSENSUS REPORTS: IARC Cancer Review: Group 2B IMEMDT 7,56,87; Animal Limited Evidence IMEMDT 10,291,76. EPA Genetic Toxicology Program.

SAFETY PROFILE: Suspected carcinogen with experimental carcinogenic data. Poison by ingestion, intravenous, intraperitoneal, and subcutaneous routes. Human mutation data reported. When heated to decomposition it emits toxic fumes of NO_x.

MRH209 CAS:6923-22-4 ***HR: 3***
MONOCROTOPHOS
mf: $C_7H_{14}NO_5P$ mw: 223.19

PROP: A reddish-brown solid; mild ester odor. Bp: 125°.

SYNS: APADRIN ◇ AZODRIN ◇ BILOBRAN ◇ CRISODRIN ◇ CRISODIN ◇ 3-(DIMETHOXYPHOSPHINYLOXY)N-METHYL-cis-CROTONAMIDE ◇ O,O-DIMETHYL-O-(2-N-METHYLCARBAMOYL-1-METHYL-VINYL)-FOSFAAT (DUTCH) ◇ O,O-DIMETHYL-O-(2-N-METHYLCARBAMOYL-1-METHYL)-VINYL-PHOSPHAT(GERMAN) ◇ O,O-DIMETHYL-O-(2-N-METHYLCARBAMOYL-1-METHYL-VINYL) PHOSPHATE ◇ DIMETHYL-1-METHYL-2-(METHYLCARBAMOYL) VINYLPHOSPHATE, cis ◇ (E)-DIMETHYL 1-METHYL-3-(METHYL-AMINO)-3-OXO-1-PROPENYL PHOSPHATE ◇ DIMETHYL PHOSPHATE ESTER OF 3-HYDROXY-N-METHYL-cis-CROTONAMIDE ◇ DIMETHYL PHOSPHATE OF 3-HYDROXY-N-METHYL-cis-CROTONAMINE ◇ O,O-DIMETIL-O-(2-N-METILCARBAMOIL-1-METIL-VINIL)-FOSFATO (ITALIAN) ◇ ENT 27,129 ◇ HAZODRIN ◇ 3-HYDROXY-N-METHYL-cis-CROTONAMIDE DIMETHYL PHOSPHATE ◇ cis-1-METHYL-2-METHYL CARBAMOYL VINYL PHOSPHATE ◇ MONOCIL 40 ◇ MONOCRON ◇ NUVACRON ◇ PHOSPHATE de DIMETHYLE et de 2-METHYLCARBAMOYL-1-METHYL VINYLE (FRENCH) ◇ PHOSPHORIC ACID, DIMETHYL ESTER, ESTER with cis-3-HYDROXY-N-METHYLCROTONAMIDE ◇ PILLARDRIN ◇ PLANTDRIN ◇ SHELL SD 9129 ◇ SUSVIN ◇ ULVAIR

TOXICITY DATA with REFERENCE
mmo-sat 500 μg/plate NTIS** PB84-138973
mmo-esc 5 μL/plate MUREAV 28,405,75
skn-rat LD50:112 mg/kg WRPCA2 9,119,70
orl-rat LD50:8 mg/kg FMCHA2 -,C161,83
ihl-rat LC50:63 mg/m^3/4H EGESAQ 24,173,80
ipr-rat LDLo:20 mg/kg BECTA6 19,47,78
scu-rat LD50:6964 μg/kg BJPCBM 40,124,70
ivn-rat LD50:9200 μg/kg NTIS** PB277-077
orl-mus LD50:15 mg/kg ARSIM* 20,21,66
ipr-mus LD50:3800 μg/kg TXAPA9 13,37,68
scu-mus LD50:8710 μg/kg JPPMAB 19,612,67
ivn-mus LD50:9200 μg/kg JPPMAB 19,612,67
skn-rbt LD50:354 mg/kg SHELL*
orl-dck LD50:3360 μg/kg TXAPA9 22,556,72
skn-dck LD50:30 mg/kg TXAPA9 47,451,79

CONSENSUS REPORTS: EPA Genetic Toxicology Program. EPA Extremely Hazardous Substances List.

OSHA PEL: TWA 0.25 mg/m^3
ACGIH TLV: TWA 0.25 mg/m^3

SAFETY PROFILE: Poison by ingestion, inhalation, skin contact, intraperitoneal, subcutaneous, and intravenous routes. Mutation data reported. Use may be restricted. When heated to decomposition it emits very toxic NO_x and PO_x.

MRH212 CAS:22771-18-2 ***HR: 2***
MONOCYCLOHEXYLTIN ACID
mf: $C_6H_{12}O_2Sn$ mw: 234.87

SYN: STANNANE, CYCLOHEXYLHYDROXYOXO-

TOXICITY DATA with REFERENCE
orl-rat LD50:3600 mg/kg GISAAA 48(3),55,83

OSHA PEL: TWA 0.1 mg(Sn)/m^3 (skin)
ACGIH TLV: TWA 0.1 mg(Sn)/m^3 (skin)

SAFETY PROFILE: Moderately toxic by ingestion. When heated to decomposition it emits toxic fumes of Sn.

MRH250 CAS:151-41-7 ***HR: 2***
MONODODECYL ESTER SULFURIC ACID
mf: $C_{12}H_{26}O_4S$ mw: 266.4

SYNS: DODECYL SULFATE ◇ DODECYLSULFURIC ACID ◇ LAURYL SULFATE ◇ LAURYL SULFURIC ACID

TOXICITY DATA with REFERENCE
orl-rat LD50:1300 mg/kg JSCCA5 13,469,62

CONSENSUS REPORTS: Reported in EPA TSCA Inventory.

SAFETY PROFILE: Moderately toxic by ingestion. When heated to decomposition it emits toxic fumes of SO_x. See also SULFATES and ESTERS.

MRI000 CAS:1498-64-2 ***HR: 3***
MONOETHYLDICHLOROTHIOPHOSPHATE
mf: $C_2H_5Cl_2OPS$ mw: 179.00

SYNS: ETHYLDICHLORTHIOFOSFAT (CZECH) ◇ O-ETHYLESTER KYSELINY DICHLORTHIOFOSFORECNE (CZECH) ◇ TL 429

TOXICITY DATA with REFERENCE
orl-rat LD50:900 mg/kg HYSAAV 33(12),334,68
ihl-rat LCLo:32 ppm/4H 28ZPAK -,214,72
orl-mus LD50:720 mg/kg HYSAAV 33(12),334,68
ihl-mus LCLo:3060 mg/m^3/10M NDRC** NDCrc-132,Nov,42
orl-rbt LD50:850 mg/kg HYSAAV 33(12),334,68
orl-gpg LD50:750 mg/kg HYSAAV 33(12),334,68

SAFETY PROFILE: Poison by inhalation. Moderately toxic by ingestion. A powerful irritant. When heated to decomposition it emits very toxic fumes of Cl^-, PO_x, and SO_x.

MRI100 CAS:4376-20-9 **HR: 3**
MONOETHYLHEXYL PHTHALATE
mf: $C_{16}H_{22}O_4$ mw: 278.38

SYNS: MEHP ◇ MONO(2-ETHYLHEXYL)PHTHALATE

TOXICITY DATA with REFERENCE
skn-rat 100 mg NTIS** PB-250-102
mmo-sat 1250 μg/plate EVHPAZ 45,119,82
cyt-ham-orl 375 mg/kg EVHPAZ 45,119,82
orl-rat TDLo:4500 mg/kg (6-15D preg):REP BECTA6 27,181,81
orl-mus TDLo:100 mg/kg (female 7D post):TER JEPTDQ 4(2-3),533,80
orl-rat LD50:1340 mg/kg TXAPA9 45,250,78
ipr-rat LD50:415 mg/kg NTIS** PB250-102
ivn-rat LD50:150 mg/kg NTIS** PB250-102
ipr-mus LD50:240 mg/kg NTIS** PB250-102
ivn-mus LD50:208 mg/kg NTIS** PB250-102

SAFETY PROFILE: Poison by intravenous and intraperitoneal routes. Moderately toxic by ingestion. An experimental teratogen. Experimental reproductive effects. Mutation data reported. A skin irritant. When heated to decomposition it emits acrid smoke and irritating fumes.

MRI250 CAS:21124-09-4 **HR: D**
MONOETHYLPHENYLTRIAZENE
mf: $C_8H_{11}N_3$ mw: 149.22

SYN: 1-ETHYL-3-PHENYLTRIAZENE

TOXICITY DATA with REFERENCE
unr-rat TDLo:200 mg/kg (10D preg):TER XENOBH 3,271,73

SAFETY PROFILE: An experimental teratogen. When heated to decomposition it emits toxic fumes of NO_x.

MRI500 CAS:63041-07-6 **HR: 3**
MONOGLYCIDYL ETHER of N-PHENYLDIETHANOLAMINE
mf: $C_{13}H_{19}NO_3$ mw: 237.33

SYN: 2-(N-(2-(2,3-EPOXYPROPOXY)ETHYL)ANILINO)ETHANOL

TOXICITY DATA with REFERENCE
scu-mus TDLo:17 g/kg/43W-I:ETA FCTXAV 4,365,66

SAFETY PROFILE: Questionable carcinogen with experimental tumorigenic data. When heated to decomposition it emits toxic fumes of NO_x. See also ETHERS and AMINES.

MRI750 CAS:39801-14-4 **HR: 3**
8-MONOHYDRO MIREX
mf: $C_{10}HCl_{11}$ mw: 511.06

SYNS: HYDROMIREX ◇ PHOTOMIREX ◇ 1,2,3,4,5,5,6,7,9,10,10-UNDECACHLOROPENTACYCLO(5.3.O.$O^{2,6}$.$0^{3,9}$.$0^{4,8}$)DECANE

TOXICITY DATA with REFERENCE
spm-mus-ipr 9 mg/kg/5D-C ENMUDM 8(Suppl 6),39,86
orl-rbt TDLo:130 mg/kg (6-18D preg):TER TXAPA9 45,332,78
orl-rat TDLo:1276 mg/kg/91W-C:ETA TXAPA9 59,268,81
orl-rat LD50:200 mg/kg JAFCAU 26,388,78

SAFETY PROFILE: Poison by ingestion. Experimental teratogenic effects. Questionable carcinogen with experimental tumorigenic data. Mutation data reported. When heated to decomposition it emits toxic fumes of Cl^-. See also MIREX.

MRI775 CAS:30833-53-5 **HR: D**
MONO-ISO-BUTYL PHTHALATE
mf: $C_{12}H_{14}O_4$ mw: 222.26

SYNS: 1,2-BENZENEDICARBOXYLIC ACID, MONO(2-METHYLPROPYL) ESTER (9CI) ◇ MIBP

TOXICITY DATA with REFERENCE
orl-rat TDLo:9200 mg/kg (7D male):REP TXCYAC 15,197,80

SAFETY PROFILE: Experimental reproductive effects. When heated to decomposition it emits acrid smoke and irritating fumes.

MRJ000 CAS:142-18-7 **HR: 1**
1-MONOLAURIN
mf: $C_{15}H_{30}O_4$ mw: 274.45

SYNS: DODECANOIC ACID-2,3-DIHYDROXYPROPYL ESTER ◇ GLYCERYL MONOLAURATE

TOXICITY DATA with REFERENCE
orl-rat LD50:53 g/kg FOREAE 21,348,56

CONSENSUS REPORTS: Reported in EPA TSCA Inventory.

SAFETY PROFILE: Mildly toxic by ingestion. When heated to decomposition it emits acrid smoke and irritating fumes. See also ESTERS.

MRJ125 **HR: 3**
MONOLITHIUM ACETYLIDE-AMMONIA
mf: $C_2HLi \cdot H_3N$ mw: 49.00

SAFETY PROFILE: The complex ignites on contact with carbon dioxide, sulfur dioxide, chlorine, or water. When heated to decomposition it emits toxic fumes of NH_3 and NO_x. See also ACETYLIDES and LITHIUM COMPOUNDS.

MRJ250 CAS:29674-96-2 *HR: 3*
MONOMETHYLHYDRAZINE NITRATE
mf: CH_6N_2•HNO_3 mw: 109.11

SYN: METHYLHYDRAZINIUM NITRATE

TOXICITY DATA with REFERENCE
orl-rat LD50:133 mg/kg CTOXAO 4,435,71
skn-rat LD50:285 mg/kg CTOXAO 4,435,71
ipr-rat LD50:43 mg/kg CTOXAO 4,435,71
ivn-rat LD50:38 mg/kg CTOXAO 4,435,71
orl-ham LD50:72 mg/kg CTOXAO 4,435,71
skn-ham LD50:326 mg/kg CTOXAO 4,435,71
ipr-ham LD50:45 mg/kg CTOXAO 4,435,71

SAFETY PROFILE: Poison by ingestion, skin contact, intravenous, and intraperitoneal routes. An impact-sensitive explosive. When heated to decomposition it emits toxic fumes of NO_x.

MRJ600 CAS:54597-56-7 *HR: 3*
MONOMYCIN

TOXICITY DATA with REFERENCE
ipr-mus LD50:510 mg/kg 85FZAT -,433,67
scu-mus LD50:615 mg/kg 85FZAT -,433,67
ivn-mus LD50:77 mg/kg AITEAT 10,947,62

SAFETY PROFILE: Poison by intravenous route. Moderately toxic by intraperitoneal and subcutaneous routes.

MRJ700 CAS:77893-24-4 *HR: D*
MONONITROSOCIMETIDINE
mf: $C_{10}H_{17}N_7OS$ mw: 283.40

TOXICITY DATA with REFERENCE
mmo-sat 10 mg/plate CRNGDP 2,261,81
cyt-ham:ovr 260 μmol/L/2H CRNGDP 2,261,81

SAFETY PROFILE: Mutation data reported. When heated to decomposition it emits toxic fumes of SO_x and NO_x.

MRJ750 CAS:5632-47-3 *HR: 3*
MONONITROSOPIPERAZINE
mf: $C_4H_9N_3O$ mw: 115.16

SYNS: N-NITROSOPIPERAZINE ◇ 1-NITROSOPIPERAZINE

TOXICITY DATA with REFERENCE
mma-sat 5 μmol/plate MUREAV 57,1,78
hma-mus/sat 8 mg/kg CNREA8 37,4572,77
orl-rat TDLo:5400 mg/kg/60W-I:CAR,REP ZEKBAI 74,179,70
orl-mus TDLo:720 mg/kg/20W-I:ETA JNCIAM 55,633,75
orl-rat TD:1400 mg/kg/60W-I:NEO ZEKBAI 74,179,70
orl-rat LD50:2260 mg/kg ZEKBAI 74,179,70

SAFETY PROFILE: Moderately toxic by ingestion. Questionable carcinogen with experimental carcinogenic, neoplastigenic, and tumorigenic data. Experimental reproductive effects. Mutation data reported. When heated to decomposition it emits toxic fumes of NO_x. See also N-NITROSO COMPOUNDS.

MRK000 CAS:3504-13-0 *HR: 3*
MONOPEROXY SUCCINIC ACID
mf: $C_4H_6O_5$ mw: 134.09

HOCO•C_2H_4CO•OOH

SAFETY PROFILE: Explodes on exposure to flame. It is slightly shock sensitive. A powerful oxidant. Upon decomposition it emits acrid smoke and irritating fumes. See also PEROXIDES, ORGANIC.

MRK250 CAS:3454-11-3 *HR: 3*
MONOPOTASSIUM aci-1-DINITROETHANE
mf: $C_2H_3KN_2O_3$ mw: 142.16

$O_2NC(CH_3)$=NOK

SYN: POTASSIUM 1-NITROETHANE-1-OXIMATE

SAFETY PROFILE: An explosive. When heated to decomposition it emits very toxic fumes of K_2O and NO_x.

MRK500 CAS:19473-49-5 *HR: 1*
MONOPOTASSIUM GLUTAMATE
mf: $C_5H_8NO_4$•K mw: 185.24

PROP: White, free-flowing, hygroscopic crystalline powder; practically odorless. Freely sol in water; sltly sol in alc.

SYNS: l-GLUTAMIC ACID, MONOPOTASSIUM SALT ◇ MONO- POTASSIUM l-GLUTAMATE (FCC) ◇ MPG ◇ POTASSIUM GLUTAMATE ◇ POTASSIUM GLUTAMINATE

TOXICITY DATA with REFERENCE
orl-hmn TDLo:57 mg/kg:CNS SCIEAS 163,826,69
orl-mus LD50:4500 mg/kg FATOAO 42,274,79

CONSENSUS REPORTS: Reported in EPA TSCA Inventory. EPA Genetic Toxicology Program.

SAFETY PROFILE: Mildly toxic by ingestion. Human systemic effects by ingestion: headache. When heated to decomposition it emits toxic fumes of K_2O and NO_x.

MRK609 CAS:1066-26-8 *HR: 3*
MONOSODIUM ACETYLIDE
mf: C_2HNa mw: 48.02

SAFETY PROFILE: When heated to 150°C it decomposes to form a gas which ignites spontaneously in air. The dry powder can ignite spontaneously if a large surface area is exposed to air (e.g. on a filter paper). When heated to decomposition it emits toxic fumes of Na_2O. See also ACETYLIDES.

MRK750 CAS:4390-16-3 ***HR: 3***
MONOSODIUM BARBITURATE
mf: $C_4H_4N_2O_3 \cdot Na$ mw: 151.09

SYNS: 2,4,6(1H,3H,5H)-PYRIMIDINETRIONE, MONOSODIUM SALT (9CI) ◇ SODIUM BARBITURATE

TOXICITY DATA with REFERENCE
orl-rat TDLo:14 g/kg/20W-C:NEO JJIND8 63,1089,79

CONSENSUS REPORTS: Reported in EPA TSCA Inventory.

SAFETY PROFILE: Questionable carcinogen with experimental neoplastigenic data. When heated to decomposition it emits toxic fumes of NO_x and Na_2O.

MRL000 CAS:18996-35-5 ***HR: 3***
MONOSODIUM CITRATE
mf: $C_6H_7O_7 \cdot Na$ mw: 214.12

SYNS: CITRIC ACID, MONOSODIUM SALT ◇ CITRIC ACID, SODIUM SALT ◇ CITROFLUYL ◇ 2-HYDROXY-1,2,3-PROPANETRICARBOXYLIC ACID MONOSODIUM SALT ◇ MONOSODIUM DIHYDROGEN CITRATE ◇ SODIUM CITRATE ◇ SODIUM DIHYDROGEN CITRATE

TOXICITY DATA with REFERENCE
ipr-rat LD50:1348 mg/kg JPETAB 94,65,48
ipr-mus LD50:1635 mg/kg JPETAB 94,65,48
ivn-mus LD50:49 mg/kg JPETAB 94,65,48
ivn-rbt LD50:379 mg/kg JPETAB 94,65,48

CONSENSUS REPORTS: Reported in EPA TSCA Inventory.

SAFETY PROFILE: Poison by intravenous route. Moderately toxic by intraperitoneal route. When heated to decomposition it emits toxic fumes of Na_2O.

MRL500 CAS:142-47-2 ***HR: 2***
MONOSODIUM GLUTAMATE
mf: $C_5H_8NO_4 \cdot Na$ mw: 169.13

PROP: White or almost white crystals or powder; slt peptone-like odor, meal-like taste. Very sol in water; sltly sol in alc.

SYNS: ACCENT ◇ AJINOMOTO ◇ CHINESE SEASONING ◇ GLUTACYL ◇ GLUTAMIC ACID, SODIUM SALT ◇ GLUTAMMATO MONOSODICO (ITALIAN) ◇ GLUTAVENE ◇ MONOSODIOGLUTAMMATO (ITALIAN) ◇ MONOSODIUM-l-GLUTAMATE (FCC) ◇ α-MONOSODIUM GLUTAMATE ◇ MSG ◇ NATRIUMGLUTAMINAT (GERMAN) ◇ RL-50 ◇ SODIUM GLUTAMATE ◇ SODIUM l-GLUTAMATE ◇ l(+) SODIUM GLUTAMATE ◇ VETSIN ◇ ZEST

TOXICITY DATA with REFERENCE
orl-rat TDLo:48 g/kg (14D pre-21D post):REP TXAPA9 50,267,79
orl-rat TDLo:19500 mg/kg (multi) :TER QUNUAZ 31,85,71
orl-wmn TDLo:50 mg/kg:PUL NEJMAG 305,1154,81
orl-hmn TDLo:43 mg/kg:CNS,GIT HYSAAV 36(9),364,71
orl-man TDLo:3571 μg/kg:SKN LANCAO 1,988,87
ivn-hmn TDLo:714 μg/kg:CNS SCIEAS 163,826,69
orl-rat LD50:16600 mg/kg FRPPAO 27,19,72
ipr-rat LD50:4253 mg/kg HSZPAZ 300,97,55
scu-rat LD50:5580 mg/kg OYYAA2 15,433,78
ivn-rat LD50:3300 mg/kg OYYAA2 15,433,78
orl-mus LD50:11400 mg/kg AEPPAE 227,214,55
ipr-mus LD50:3800 mg/kg AEPPAE 227,214,55
scu-mus LD50:8200 mg/kg OYYAA2 15,433,78
ivn-mus LD50:30 g/kg FAONAU 53A,437,74
scu-cat LD50:8000 mg/kg EbeAG# 19NOV73
ipr-gpg LD50:15 g/kg FAONAU 53A,437,74

CONSENSUS REPORTS: Reported in EPA TSCA Inventory. EPA Genetic Toxicology Program.

SAFETY PROFILE: Moderately toxic by intravenous route. Mildly toxic by ingestion and other routes. An experimental teratogen. Other experimental reproductive effects. Human systemic effects by ingestion and intravenous routes: somnolence, hallucinations and distorted perceptions, headache, dyspnea, nausea or vomiting, dermatitis. The cause of "Chinese restaurant syndrome." When heated to decomposition it emits toxic fumes of NO_x and Na_2O.

MRL750 CAS:2163-80-6 ***HR: 3***
MONOSODIUM METHYLARSONATE
mf: $CH_4AsO_3 \cdot Na$ mw: 161.96

SYNS: ANSAR 170 ◇ ARSONATE liquid ◇ ASAZOL ◇ BUENO ◇ DACONATE 6 ◇ DAL-E-RAD ◇ HERB-ALL ◇ HERBAN M ◇ MERGE ◇ MESAMATE ◇ MESAMATE CONCENTRATE ◇ METHYLARSENIC ACID, SODIUM SALT ◇ MONATE ◇ MONOSODIUM ACID METHANEARSONATE ◇ MONOSODIUM ACID METHARSONATE ◇ MONOSODIUM METHANEARSONATE ◇ MONOSODIUM METHANEARSONIC ACID ◇ MSMA ◇ NCI-C60071 ◇ PHYBAN ◇ SILVISAR 550 ◇ SODIUM ACID METHANEARSONATE ◇ SODIUM METHANEARSONATE ◇ TARGET MSMA ◇ TRANS-VERT ◇ WEED 108 ◇ WEED-E-RAD ◇ WEED-HOE

TOXICITY DATA with REFERENCE
skn-rbt 54 mg open MLD CIGET* -,-,77
eye-rbt 34 mg MLD CIGET* -,-,77
orl-rat LD50:700 mg/kg FMCHA2 -,C163,83
unr-mam LD50:50 mg/kg 30ZDA9 -,260,71

CONSENSUS REPORTS: Arsenic and its compounds are on the Community Right-To-Know List. EPA Genetic Toxicology Program.

OSHA PEL: TWA 0.5 mg(As)/m^3
ACGIH TLV: TWA 0.2 mg(As)/m^3

SAFETY PROFILE: Poison by unspecified route. Moderately toxic by ingestion. A skin and eye irritant. When heated to decomposition it emits toxic fumes of As and Na_2O. See also ARSENIC COMPOUNDS.

MRM000 CAS:5736-15-2 ***HR: 2***
MONOSODIUM SALT of 2,2'-METHYLENE BIS(3,4,6-TRICHLOROPHENOL)
mf: $C_{13}H_5Cl_6O_2$•Na mw: 428.87

SYNS: HEXACHLOROPHENE ◇ ISOBAC 20 ◇ SERIBAK

TOXICITY DATA with REFERENCE
orl-rat LD50:560 mg/kg FMCHA2 -,C133,83
orl-qal LD50:575 mg/kg FMCHA2 -,C133,83
orl-dck LD50:1450 mg/kg FMCHA2 -,C133,83

CONSENSUS REPORTS: Chlorophenol compounds are on the Community Right-To-Know List.

SAFETY PROFILE: Moderately toxic by ingestion. A fungicide and disinfectant. When heated to decomposition it emits toxic fumes of Cl^- and Na_2O. See also CHLOROPHENOLS.

MRM250 CAS:547-32-0 ***HR: 2***
MONOSODIUM-2-SULFANILAMIDOPYRIMIDINE
mf: $C_{10}H_{10}N_4O_2S$•Na mw: 273.29

SYNS: N^1-2-PYRIMIDINYLSULFANILAMIDE MONOSODIUM SALT ◇ SODIUM SULFADIAZINE ◇ SODIUM SULFAPYRIMIDINE ◇ SOLUBLE SULFADIAZINE ◇ 2-SULFANILAMIDOPYRIMIDINE SODIUM SALT

TOXICITY DATA with REFERENCE
orl-mus TDLo:1440 mg/kg (1D preg):REP JPETAB 101,362,51
orl-mus LD50:1700 mg/kg AIPTAK 94,338,53
scu-mus LD50:1408 mg/kg AMSSAQ 142,64,43
ivn-mus LD50:540 mg/kg AIPTAK 94,338,53

SAFETY PROFILE: Moderately toxic by ingestion and subcutaneous routes. Experimental reproductive effects. When heated to decomposition it emits very toxic fumes of NO_x, Na_2O and SO_x.

MRM750 CAS:96-27-5 ***HR: 3***
MONOTHIOGLYCEROL
mf: $C_3H_8O_2S$ mw: 108.17

PROP: Liquid. Bp: 118° @ 5 mm, d: 1.248 @ 20°/20°.

SYNS: 1-MERCAPTOGLYCEROL ◇ 1-MERCAPTO-2,3-PROPANEDIOL ◇ 3-MERCAPTO-1,2-PROPANEDIOL ◇ α-MONOTHIOGLYCEROL ◇ THIOGLYCERIN ◇ α-THIOGLYCEROL ◇ 1-THIOGLYCEROL ◇ THIOVANOL ◇ USAF B-40 ◇ USAF CB-37

TOXICITY DATA with REFERENCE
oms-esc 2000 ppm AMACCQ 19,556,81
scu-ham TDLo:1925 mg/kg (35D male):REP CCPTAY 17,413,78
ipr-rat LD50:390 mg/kg JPETAB 97,349,49
ipr-mus LD50:340 mg/kg JPETAB 97,349,49
ivn-cat LD50:220 mg/kg JPETAB 97,349,49
ivn-rbt LD50:250 mg/kg JPETAB 97,349,49

CONSENSUS REPORTS: Reported in EPA TSCA Inventory.

SAFETY PROFILE: Poison by intraperitoneal and intravenous routes. Experimental reproductive effects. Mutation data reported. Flammable when exposed to heat or flame; can react with oxidizing materials. When heated to decomposition it emits highly toxic fumes of SO_x. See also MERCAPTANS.

MRN000 CAS:4166-00-1 ***HR: 3***
MONOTHIOSUCCINIMIDE
mf: C_4H_5NOS mw: 115.16

SYNS: THIOSUCCINIMIDE ◇ USAF WI-1

TOXICITY DATA with REFERENCE
orl-rat TDLo:29 g/kg/2W-C:ETA GANNA2 68,397,77
ipr-mus LD50:100 mg/kg NTIS** AD414-344

SAFETY PROFILE: Poison by intraperitoneal route. Questionable carcinogen with experimental tumorigenic data. When heated to decomposition it emits very toxic fumes of NO_x and SO_x.

MRN100 ***HR: 2***
MOONSEED

PROP: A woody vine which grows to 12 feet. The leaves are about 8 inches long and broad. The blue-black fruit looks and forms clusters like grapes and has a crescent-shaped seed. The vine has been mistaken for wild grape. It grows wild in moist wooded areas in the region bounded by Georgia, Oklahoma, Manitoba, and Quebec.

SYNS: CANADA MOONSEED ◇ MENISPERMUM CANADENSE ◇ RAISIN de COULEUVRE (CANADA) ◇ TEXAS SARSAPARILLA ◇ YELLOW PARILLA (CANADA) ◇ YELLOW SARSAPARILLA

SAFETY PROFILE: The berries contain poisonous alkaloids with picrotoxin-like effects. Ingestion of the berries can cause convulsions. See also PICROTOXIN.

MRN250 CAS:19395-58-5 ***HR: 3***
MOQUIZONE
mf: $C_{20}H_{21}N_3O_3$ mw: 351.41

PROP: Crystals from isopropanol. Mp: 135-137°. Crystals from ethyl acetate or benzene, petr ether. Mp: 128-130°.

SYN: 2,3-DIHYDRO-1-(MORPHOLINOACETYL)-3-PHENYL-4(1H)-QUINAZOLINONE

TOXICITY DATA with REFERENCE
orl-rat LD50:2220 mg/kg TXAPA9 16,571,70
ivn-rat LD50:146 mg/kg TXAPA9 16,571,70
orl-mus LD50:930 mg/kg JJPAAZ 22,235,72
ipr-mus LD50:559 mg/kg TXAPA9 16,571,70
scu-mus LD50:774 mg/kg TXAPA9 16,571,70

ivn-mus LD50:237 mg/kg TXAPA9 16,571,70
orl-rbt LD50:1730 mg/kg TXAPA9 16,571,70
ivn-rbt LD50:180 mg/kg TXAPA9 16,571,70

SAFETY PROFILE: Poison by intravenous route. Moderately toxic by ingestion, intraperitoneal, and subcutaneous routes. Experimental reproductive effects. When heated to decomposition it emits toxic fumes of NO_x.

MRN275 CAS:25717-80-0 ***HR: 2***
MORIAL
mf: $C_9H_{14}N_4O_4$ mw: 242.27

PROP: Colorless crystals or white, crystalline powder; practically tasteless and odorless. Mp: 140-141°. Freely sol in $CHCl_3$; sol in dil HCl, ethanol, ethyl acetate, methanol; sparingly sol in water, acetone, benzene; very sltly sol in ether, petr ether.

SYNS: N-CARBOXY-3-MORPHOLINOSYDNONIMINE ETHYL ESTER ◇ CORVASAL ◇ CORVATON ◇ N-(ETHOXYCARBONYL)-3-(4-MORPHOLINYL)SYDNONE IMINE ◇ MOLSIDOLAT ◇ MOLSIDOMINE ◇ MORSYDOMINE ◇ MOTAZOMIN ◇ SIN-10

TOXICITY DATA with REFERENCE
orl-rat LD50:1050 mg/kg NIIRDN 6,858,82
ipr-rat LD50:1250 mg/kg MEIEDD 10,892,83
scu-rat LD50:1360 mg/kg MEIEDD 10,892,83
ivn-rat LD50:800 mg/kg MEIEDD 10,892,83
orl-mus LD50:830 mg/kg NIIRDN 6,858,82

SAFETY PROFILE: Moderately toxic by ingestion, intravenous, intraperitoneal, and subcutaneous routes. When heated to decomposition it emits toxic fumes of NO_x.

MRN500 CAS:480-16-0 ***HR: 2***
MORIN
mf: $C_{15}H_{10}O_7$ mw: 302.25

PROP: Anhydr needles from abs alc. Decomp @ 285-290°. Crystalized with 1 or 2 mols water. Sltly sol in water, ether, and acetic acid; very sol in alc.

SYNS: A1-MORIN ◇ AURANTICA ◇ BOIS D'ARC (FRENCH) ◇ CALICO YELLOW ◇ C.I. NATURAL YELLOW 8 ◇ 2-2-(2,4-DIHYDROXYPHENYL)-3,5,7-TRIHYDROXY-4H-1-BENZOPYRAN-4-ONE◇ 2'-HYDROXYPELARGIDENOLON 1522 ◇ OSAGE ORANGE ◇ OSAGE ORANGE CRYSTALS ◇ OSAGE ORANGE EXTRACT ◇ 2',3,4',5,7-PENTAHYDROXYFLAVONE ◇ TOXYLON POMIFERUM

TOXICITY DATA with REFERENCE
mmo-sat 250 μg/plate BCSTB5 5,1489,77
mma-sat 166 nmol/plate MUREAV 54,297,78
dnr-bcs 2 mg/disc TRENAF 27,153,76
ipr-mus LD50:555 mg/kg AIPTAK 123,395,60

CONSENSUS REPORTS: Reported in EPA TSCA Inventory.

SAFETY PROFILE: Moderately toxic by intraperitoneal route. Mutation data reported. When heated to decomposition it emits acrid smoke and irritating fumes.

MRN550 ***HR: 3***
MORINGA OLEIFERA Lamk., extract excluding roots

PROP: Indian plant belonging to the family *Moringaceae* IJEBA6 18,594,80

TOXICITY DATA with REFERENCE
orl-ham TDLo:500 mg/kg (female 1-5D post):REP IJEBA6 18,594,80
ipr-mus LD50:8 mg/kg IJEBA6 18,594,80

SAFETY PROFILE: Poison by intraperitoneal route. Experimental reproductive effects. When heated to decomposition it emits acrid smoke and irritating fumes.

MRN600 CAS:26162-66-3 ***HR: 3***
MORISYLYTE CITRATE
mf: $C_{16}H_{25}NO_3 \cdot C_6H_8O_7$ mw: 471.56

SYNS: CITRATE de ACETOXY-THYMOXY-ETHYL-DIMETHYLAMINE (FRENCH) ◇ 5-(2-(N,N-DIMETHYLAMINO)ETHOXY)CARVACROL ACETATE CITRATE

TOXICITY DATA with REFERENCE
orl-rat LD50:950 mg/kg THERAP 26,775,71
scu-rat LD50:220 mg/kg THERAP 26,775,71
orl-mus LD50:550 mg/kg THERAP 26,775,71
scu-mus LD50:225 mg/kg THERAP 26,775,71

SAFETY PROFILE: Poison by subcutaneous route. Moderately toxic by ingestion. When heated to decomposition it emits toxic fumes of NO_x. See also ESTERS.

MRN675 CAS:469-81-8 ***HR: 3***
MORPHERIDINE DIHYDROCHLORIDE
mf: $C_{20}H_{30}N_2O_3 \cdot 2ClH$ mw: 419.44

PROP: Liquid. Bp: 188-192°, n (18/D) 1.5276.

SYNS: ETHYL 1-(2'-MORPHOLINOETHYL)-4-PHENYLPIPERIDINE-4-CARBOXYLATE DIHYDROCHLORIDE ◇ MORPHERIDINE ◇ MORPHOLINOETHYL NORPETHIDINE DIHYDROCHLORIDE ◇ 1-(2-MORPHOLINOETHYL)-4-PHENYLISONIPECOTIC ACID ETHYL ESTER DIHYDROCHLORIDE ◇ 1-(2-MORPHOLINYL)ETHYL)-4-PHENYL-4-PIPERIDINECARBOXYLIC ACID ETHYL ESTER DIHYDROCHLORIDE ◇ TA 1

TOXICITY DATA with REFERENCE
scu-rat LD50:70 mg/kg BJPCAL 11,27,56
ipr-mus LD50:118 mg/kg BJPCAL 11,27,56
ivn-mus LD50:45 mg/kg MEIEDD 10,898,83

SAFETY PROFILE: Poison by subcutaneous, intravenous, and intraperitoneal routes. Caution: May be habit forming. This is a controlled substance (opiate) listed in the U.S. Code of Federal Regulations, Title 21 Part

1308.11 (1985). When heated to decomposition it emits toxic fumes of NO_x and HCl. See also ESTERS.

MRN700 CAS:1109-85-9 ***HR: 3***
MORPHINAN, 3-METHOXY-17-PHENETHYL-, TARTRATE, (−)-
mf: $C_{25}H_{31}NO•C_4H_6O_6$ mw: 511.67

TOXICITY DATA with REFERENCE
scu-mus LD50:388 mg/kg 31ZPAG 2,83,66
ivn-mus LD50:17 mg/kg 31ZPAG 2,83,66

SAFETY PROFILE: Poison by subcutaneous and intravenous routes. When heated to decomposition it emits toxic fumes of NO_x.

MRO500 CAS:57-27-2 ***HR: 3***
(−)-MORPHINE
mf: $C_{17}H_{19}NO_3$ mw: 285.37

PROP: White, crystalline alkaloid. Short, orthorhombic, columnar prisms from anisole. Decomp @ 254°, mp: 197°, subl @ 190-200°.

SYNS: MORFINA (ITALIAN) ◇ MORPHIA ◇ MORPHINA ◇ MORPHINE ◇ MORPHINISM ◇ MORPHINUM ◇ MORPHIUM ◇ 4a,5,7a,8-TETRAHYDRO-12-METHYL-9H-9,9c-IMINOETHANOPHENANTHRO(4,5-bcd)FURAN-3,5-DIOL

TOXICITY DATA with REFERENCE
unr-wmn TDLo:648 mg/kg (1-39W preg):REP PEDIAU 23,288,59
orl-man TDLo:3857 μg/kg/1W-I LANCAO 2,98,87
orl-rat LD50:335 mg/kg DRFUD4 2,39,77
ipr-rat LD50:160 mg/kg PHARAT 31,655,76
scu-rat LD50:122 mg/kg ARZNAD 24,600,74
ivn-rat LD50:140 mg/kg BJPCAL 7,196,52
orl-mus LD50:524 mg/kg ARZNAD 24,600,74
ipr-mus LD50:140 mg/kg NYKZAU 53,56S,57
scu-mus LD50:220 mg/kg ARZNAD 8,25,58
ivn-mus LD50:135 mg/kg THERAP 7,21,52
ice-mus LD50:6900 μg/kg SCIEAS 134,1078,61

SAFETY PROFILE: Poison experimentally by ingestion, intracerebral, intraperitoneal, subcutaneous, and intravenous routes. Human reproductive effects by an unspecified route: effects on newborn including drug dependence. Experimental reproductive effects. Morphine is the constituent of opium most responsible for its toxic effects. When taken orally, the effects of morphine poisoning begin to appear in 20-40 minutes; if taken hypodermically, the symptoms appear much earlier and narcotism is more likely to follow the early symptoms. Abuse leads to habituation or addiction. Individual susceptibility varies greatly and children are more susceptible than adults. When heated to decomposition it emits toxic fumes of NO_x.

MRO750 CAS:52-26-6 ***HR: 3***
MORPHINE HYDROCHLORIDE
mf: $C_{17}H_{19}NO_3•ClH$ mw: 321.83

PROP: Trihydrate: White flakes or crystalline powder; bitter taste. Mp: approx 200° (decomp). Dissolves slowly in glycerol; insol in chloroform, ether.

SYNS: 7,8-DIDEHYDRO-4,5-α-EPOXY-17-METHYLMORPHINAN-3,6-α-DIOL HYDROCHLORIDE ◇ 7,8-DIDEHYDRO-4,5-α-EPOXY-17-METHYLMORPHINE HYDROCHLORIDE ◇ MORPHINE CHLORHYDRATE ◇ MORPHINE CHLORIDE

TOXICITY DATA with REFERENCE
scu-rat TDLo:840 mg/kg (4-10D preg):TER KAEMAW 51,41,78
scu-rat TDLo:480 mg/kg (7-10D preg):REP KAEMAW 51,41,78
orl-rat LD50:335 mg/kg AIPTAK 115,213,58
scu-rat LD50:480 mg/kg AIPTAK 115,213,58
ivn-rat LD50:265 mg/kg AIPTAK 115,213,58
orl-mus LD50:745 mg/kg JJPAAZ 18,406,68
ipr-mus LD50:148 mg/kg BJPCBM 73,887,81
scu-mus LD50:354 mg/kg KSRNAM 5,1787,71
ivn-mus LD50:180 mg/kg ATSUDG 7,90,84
par-mus LD50:440 mg/kg WKMRAH 24(4),65,83
ivn-dog LD50:175 mg/kg CPBTAL 7,372,59

CONSENSUS REPORTS: EPA Genetic Toxicology Program.

SAFETY PROFILE: Poison by ingestion, intraperitoneal, intravenous, parenteral, and subcutaneous routes. An experimental teratogen. Experimental reproductive effects. Abuse leads to habituation or addiction. When heated to decomposition it emits very toxic fumes of NO_x and HCl. See also MORPHINE.

MRP000 CAS:14075-02-6 ***HR: 3***
MORPHINE METHOCHLORIDE
mf: $C_{18}H_{22}NO_3•Cl$ mw: 335.86

SYNS: 7,8-DIDEHYDRO-4,5-α-EPOXY-17-METHYLMORPHINAN-3,6-α-DIOL ◇ N-METHYL MORPHINE CHLORIDE ◇ N-METHYLMORPHINIUM CHLORIDE ◇ MORPHINE METHYLCHLORIDE

TOXICITY DATA with REFERENCE
scu-mus LDLo:271 mg/kg JPETAB 49,319,33
ivn-mus LD50:32700 μg/kg JPETAB 157,185,67
scu-rbt LDLo:172 mg/kg JPETAB 49,319,33
par-frg LDLo:2500 mg/kg JPETAB 49,319,33

SAFETY PROFILE: Poison by intravenous and subcutaneous routes. Moderately toxic by parenteral route. When heated to decomposition it emits very toxic fumes of NO_x and Cl^-. See also MORPHINE.

MRP100 CAS:6009-81-0 ***HR: D***
MORPHINE MONOHYDRATE
mf: $C_{17}H_{19}NO_3•H_2O$ mw: 303.39

SYN: MORPHINAN-3,6-DIOL,7,8-DIDEHYDRO-4,5-EPOXY-17-METHYL-(5-α-6-α)-, MONOHYDRATE

TOXICITY DATA with REFERENCE
imp-rat TDLo:300 mg/kg (male 3D pre):REP JPETAB 192,542,75

SAFETY PROFILE: Experimental reproductive effects. When heated to decomposition it emits toxic fumes of NO_x.

MRP250 CAS:64-31-3 ***HR: 3***
MORPHINE SULFATE
mf: $C_{34}H_{38}N_2O_6 \cdot H_2O_4S$ mw: 668.82

SYN: MORPHINE SULPHATE

TOXICITY DATA with REFERENCE
mnt-mus-ipr 6400 mg/kg IJMRAQ 75,112,82
orl-mus TDLo:2800 mg/kg (female 8-21D post):REP NATUAS 247,376,74
ivn-rat TDLo:20 mg/kg (female 21D post):TER APTOA6 37,265,75
orl-wmn TDLo:8 mg/kg LANCAO 1,573,87
orl-man TDLo:12 mg/kg/4W-I DICPBB 22,397,88
orl-rat LD50:461 mg/kg BJPCAL 30,11,67
ipr-rat LD50:235 mg/kg APTOA6 22,241,65
scu-rat LD50:108 mg/k BJPCAL 10,260,55
ivn-rat LD50:70 mg/kg TXAPA9 17,250,70
ims-rat LD50:78 mg/kg AIPTAK 190,124,71
orl-mus LD50:600 mg/kg JPETAB 133,400,61
ipr-mus LD50:275 mg/kg JMCMAR 10,627,67
scu-mus LD50:180 mg/kg AUVJA2 49,525,73
ivn-mus LD50:156 mg/kg JPETAB 157,185,67
ivn-dog LD50:316 mg/kg AIPTAK 149,571,64
par-frg LD50:678 mg/kg AIPTAK 79,282,49

SAFETY PROFILE: Poison by subcutaneous, intravenous, intraperitoneal, and intramuscular routes. Moderately toxic by ingestion and parenteral routes. An experimental teratogen. Experimental reproductive effects. Mutation data reported. Used as a narcotic. Abuse leads to habituation or addiction. When heated to decomposition it emits very toxic fumes of NO_x and SO_x. See also MORPHINE and SULFATES.

MRP500 CAS:67238-91-9 ***HR: D***
MORPHOCYCLINE
mf: $C_{27}H_{33}N_3O_9$ mw: 543.63

TOXICITY DATA with REFERENCE
ivn-rat TDLo:20 mg/kg (9D preg):TER ANTBAL 20(2),158,75

SAFETY PROFILE: An experimental teratogen. When heated to decomposition it emits toxic fumes of NO_x.

MRP750 CAS:110-91-8 ***HR: 3***
MORPHOLINE
DOT: UN 1760/UN 2054
mf: C_4H_9NO mw: 87.14

$HNC_2H_4OCH_2CH_2$ (ring)

PROP: Colorless, hygroscopic oil; amine odor. Bp: 128.9°, fp: −7.5°, flash p: 100°F (OC), autoign temp: 590°F, vap press: 10 mm @ 23°, vap d: 3.00, mp: −4.9°, d: 1.007 @ 20°/4°. Volatile with steam; misc with water evolving some heat; misc with acetone, benzene, ether, castor oil, methanol, ethanol, ethylene, glycol, linseed oil, turpentine, pine oil. Immiscible with concentrated NaOH solns.

SYNS: DIETHYLENEIMIDE OXIDE ◇ DIETHYLENE IMIDOXIDE ◇ DIETHYLENE OXIMIDE ◇ DIETHYLENIMIDE OXIDE ◇ MORPHOLINE, AQUEOUS MIXTURE (DOT) ◇ 1-OXA-4-AZACYCLOHEXANE ◇ TETRAHYDRO-p-ISOXAZINE ◇ TETRAHYDRO-1,4-ISOXAZINE ◇ TETRAHYDRO-1,4-OXAZINE ◇ TETRAHYDRO-2H-1,4-OXAZINE

TOXICITY DATA with REFERENCE
skn-rbt 995 mg/24H SEV BIOFX* 10-4/70
skn-rbt 500 mg open MOD UCDS** 4/21/67
eye-rbt 2 mg SEV AJOPAA 29,1363,46
otr-mus:lym 1 μL/L ENMUDM 4,390,82
orl-mus TDLo:2560 mg/kg/Y-C:NEO GISAAA 44(8),15,79
orl-rat LD50:1050 mg/kg UCDS** 4/21/67
ihl-rat LC50:8000 ppm/8H NPIRI* 1,85,74
ihl-mus LC50:1320 mg/m^3/2H TPKVAL 8,60,66
ipr-mus LD50:413 mg/kg CANCAR 2,1055,49
skn-rbt LD50:500 mg/kg AMIHBC 10,61,54

CONSENSUS REPORTS: Reported in EPA TSCA Inventory. EPA Genetic Toxicology Program.

OSHA PEL: (Transitional: TWA 20 ppm (skin) TWA 20 ppm (skin); STEL 30 ppm (skin)
ACGIH TLV: TWA 20 ppm (skin)
DFG MAK: 20 ppm (70 mg/m^3)
DOT Classification: Flammable Liquid; Label: FLAM. Liquid (NA2054, UN2054); Corrosive Material; Label: Corrosive (NA1760); Flammable or Combustible Liquid; Label: Flammable Liquid (UN2054).

SAFETY PROFILE: Moderately toxic by ingestion, inhalation, skin contact, and intraperitoneal routes. Mutation data reported. A corrosive irritant to skin, eyes, and mucous membranes. Can cause kidney damage. Questionable carcinogen with experimental neoplastigenic data. Flammable liquid. A very dangerous fire hazard when exposed to flame, heat, or oxidizers; can react with oxidizing materials. To fight fire, use alcohol foam, CO_2, dry chemical. Mixtures with nitromethane are explosive. May ignite spontaneously in contact with cellulose nitrate of high surface area. When heated to decomposition it emits highly toxic fumes of NO_x.

MRQ250 CAS:4856-95-5 ***HR: 3***
MORPHOLINE, compounded with BORANE (1 : 1)
mf: $C_4H_9NO•BH_3$ mw: 100.98

SYNS: BORANE, compounded with MORPHOLINE ◇ MORPHOLINEBORANE

TOXICITY DATA with REFERENCE
ipr-mus LD50:475 mg/kg MPMSAE 19,1025,80
ivn-mus LD50:320 mg/kg CSLNX* NX#03238

CONSENSUS REPORTS: Reported in EPA TSCA Inventory.

SAFETY PROFILE: Poison by intravenous route. Moderately toxic by intraperitoneal route. When heated to decomposition it emits toxic fumes of NO_x. See also BORANES, BORON COMPOUNDS, and MORPHOLINE.

MRQ500 CAS:622-40-2 ***HR: 2***
MORPHOLINE ETHANOL
mf: $C_6H_{13}NO_2$ mw: 131.20

PROP: Colorless liquid. Bp: 225.5°, mp: 1.6°, vap press: 0.1 mm @ 20°, flash p: 210°F (OC), vap d: 4.54, d: 1.07.

SYNS: N-β-HYDROXYETHYLMORPHOLINE ◇ N-(2-HYDROXYETHYL)MORPHOLINE ◇ β-OXYAETHYL-MORPHOLIN (GERMAN)

TOXICITY DATA with REFERENCE
skn-rbt 10 mg/24H open JIHTAB 30,63,48
eye-rbt 5 mg SEV AJOPAA 29,1363,46
dns-rat:lvr 5 g/L MUREAV 136,153,84
orl-rat LD50:12 g/kg JIHIAB 30,63,48
ipr-mus LD50:3600 mg/kg CANCAR 2,1055,49
scu-mus LD50:2650 mg/kg AIPTAK 109,108,57
par-mus LDLo:4800 mg/kg CBCCT* 7,691,55
skn-gpg LD50:2500 mg/kg JIHTAB 30,63,48

CONSENSUS REPORTS: Reported in EPA TSCA Inventory.

SAFETY PROFILE: Moderately toxic by skin contact, subcutaneous, and intraperitoneal routes. Mildly toxic by ingestion and parenteral routes. Mutation data reported. A skin and severe eye irritant. Combustible when exposed to heat or flame; can react with oxidizing materials. To fight fire, use alcohol foam, CO_2, dry chemical. When heated to decomposition it emits toxic fumes of NO_x.

MRQ600 CAS:10024-89-2 ***HR: 2***
MORPHOLINE HYDROCHLORIDE
mf: $C_4H_9NO•ClH$ mw: 123.60

SYN: MORPHOLINE, HYDROCHLORIDE

TOXICITY DATA with REFERENCE
orl-rat LD50:4580 mg/kg JIHTAB 23,259,41
ipr-mus LD50:1350 mg/kg JJPAAZ 17,475,67
scu-mus LD50:1640 mg/kg AIPTAK 112,36,57
ivn-mus LD50:900 mg/kg AIPTAK 112,36,57

SAFETY PROFILE: Moderately toxic by intraperitoneal, subcutaneous, and intravenous routes. Mildly toxic by ingestion. When heated to decomposition it emits toxic fumes of NO_x and HCl. See also MORPHOLINE.

MRQ750 CAS:5299-64-9 ***HR: 3***
4-MORPHOLINENONYLIC ACID
mf: $C_{13}H_{25}NO_3$ mw: 227.39

SYNS: AI 318284 ◇ N-MORPHOLINO NONANAMIDE ◇ 4-NONANOYLMORPHOLINE ◇ PELARGONIC MORPHOLIDE

TOXICITY DATA with REFERENCE
eye-rbt 500 μg AIHAAP 23,194,62
ihl-hmn TCLo:21 mg/m³/3M:PUL,EYE,NOSE AIHAAP 23,199,62
ihl-rat LD50:23 mg/kg AIHAAP 23,194,62
ihl-mus LD50:104 mg/kg AIHAAP 23,194,62
ivn-mus LD50:18 mg/kg CSLNX* NX#08842
ivn-rbt LD50:21 mg/kg AIHAAP 23,194,62
ihl-gpg LD50:6 mg/kg AIHAAP 23,194,62

SAFETY PROFILE: Poison by inhalation and intravenous routes. Human systemic effects by inhalation: lacrimation, deviated nasal septum, and cough. An eye and intense mucous membrane irritant. When heated to decomposition it emits toxic fumes of NO_x.

MRR075 CAS:2958-89-6 ***HR: 3***
4-MORPHOLINE SULFENYL CHLORIDE
mf: C_4H_8ClNOS mw: 153.63

$OC_2H_4N(SCl)CH_2CH_2$

SAFETY PROFILE: An unstable explosive. When heated to decomposition it emits toxic fumes of Cl^- SO_x, and NO_x. See also CHLORIDES.

MRR100 CAS:35175-75-8 ***HR: 2***
MORPHOLINIUM PERCHLORATE
mf: $C_4H_{10}ClNO_5$ mw: 187.58

$OC_2H_4N^-H_2CH_2CH_2ClO_4^-$

SAFETY PROFILE: Decomposes exothermically at 230°C. When heated to decomposition it emits toxic fumes of Cl^- and NO_x. See also PERCHLORATES.

MRR125 CAS:102071-88-5 ***HR: 3***
N-MORPHOLINO-β-(2-AMINOMETHYLBENZO-DIOXAN)-PROPIONAMIDE
mf: $C_{16}H_{22}N_2O_4$ mw: 306.40

SYNS: 3-(((1,4-BENZODIOXAN-2-YL)METHYL)AMINO)-1-MORPHOLINO-1-PROPANONE ◇ 1530 I.S.

TOXICITY DATA with REFERENCE
ipr-mus LD50:400 mg/kg AIPTAK 105,317,56
ivn-mus LD50:80 mg/kg AIPTAK 105,317,56
ivn-rbt LD50:70 mg/kg AIPTAK 105,317,56

SAFETY PROFILE: Poison by intravenous and intraperitoneal routes. When heated to decomposition it emits toxic fumes of NO_x.

MRR750 CAS:15029-32-0 ***HR: 3***
MORPHOLINOCARBONYLACETONITRILE
mf: $C_7H_{10}N_2O_2$ mw: 154.19

SYNS: 1-(CYANOACETYL)MORPHOLINE ◇ 4-CYANOACETYL-MORPHOLINE

TOXICITY DATA with REFERENCE
ivn-mus LD50:180 mg/kg CSLNX* NX#04762

CONSENSUS REPORTS: Cyanide and its compounds are on the Community Right-To-Know List. Reported in EPA TSCA Inventory.

SAFETY PROFILE: Poison by intravenous route. When heated to decomposition it emits toxic fumes of NO_x and CN^-. See also NITRILES.

MRR775 CAS:72122-60-2 ***HR: 3***
MORPHOLINO-CNU
mf: $C_7H_{14}ClN_4O_3$ mw: 237.70

SYN: 1-(2-CHLOROETHYL)-3-MORPHOLINO-1-NITROSOUREA

TOXICITY DATA with REFERENCE
ivn-rat TDLo:32 mg/kg/60W-I:ETA DTESD7 8,273,80

SAFETY PROFILE: Questionable carcinogen with experimental tumorigenic data. When heated to decomposition it emits toxic fumes of Cl^- and NO_x. See also N-NITROSO COMPOUNDS.

MRR850 CAS:79867-78-0 ***HR: 2***
MORPHOLINODAUNOMYCIN
mf: $C_{31}H_{35}NO_{11}$ mw: 597.67

SYN: 5,12-NAPHTHACENEDIONE,7,8,9,10-TETRAHYDRO-8-ACETYL-1-METHOXY-10-((2,3,6-TRIDEOXY-3-(4-MORPHOLINYL)-α-L-lyxo-HEXOPYRANOSYL)OXY)-6,8,11-TRIHYDROXY-,(8S-cis)-

TOXICITY DATA with REFERENCE
dns-rat:lvr 2 mg/L CNREA8 44,5599,84
otr-mus:fbr 5 μg/L CBTOE2 3,17,87
msc-ham:lng 10 μg/L CNREA8 44,5599,84
ivn-rat TDLo:250 μg/kg:CAR CBTOE2 3,17,87

SAFETY PROFILE: Questionable carcinogen with experimental carcinogenic data. Mutation data reported. When heated to decomposition it emits toxic fumes of NO_x.

MRT100 CAS:80790-68-7 ***HR: 3***
3′-MORPHOLINO-3′-DEAMINODAUNORUBICIN
mf: $C_{31}H_{38}NO_{12}$ mw: 616.70

SYNS: 3′-DEAMINO-3′-MORPHOLINO-ADRIAMYCIN ◇ 3′-DEAMINO-3′-(4-MORPHOLINYL)DAUNORUBICIN ◇ 5,12-NAPHTHACENEDIONE,7,8,9,10-TETRAHYDRO-8-(HYDROXYACETYL)-1-METHOXY-10-((2,3,6-TRIDEOXY-MORPHOLINO-α-L-lyxo-HEXOPYRANOSYL)OXY)-6,8,11-TRIHYDROXY-

TOXICITY DATA with REFERENCE
dni-hmn:oth 50 nmol/L CNREA8 43,1044,83
oth-hmn:oth 50 nmol/L CNREA8 43,1044,83
dnd-uns:lyms 2500 nmol/L CNREA8 43,1044,83
ivn-rat TDLo:250 μg/kg:ETA PAACA3 27,238,86
orl-mus LD50:500 μg/kg JANTAJ 35,117,82
ipr-mus LD50:250 μg/kg JANTAJ 35,117,82

SAFETY PROFILE: Poison by ingestion and intraperitoneal routes. Questionable carcinogen with experimental tumorigenic data. Human mutation data reported. When heated to decomposition it emits toxic fumes of NO_x.

MRT250 CAS:77791-41-4 ***HR: 2***
2-(MORPHOLINO)-N-METHYL-N-(2-MESITYLOXYETHYL)ACETAMIDE HYDROCHLORIDE
mf: $C_{18}H_{28}N_2O_3{\cdot}ClH$ mw: 356.94

SYNS: C 2085 ◇ N-(2-MESITYLOXYETHYL)-N-METHYL-2-(MORPHOLINE)ACETAMIDE HYDROCHLORIDE

TOXICITY DATA with REFERENCE
eye-rbt 2% MOD ARZNAD 8,761,58
scu-mus LD50:655 mg/kg ARZNAD 8,761,58

SAFETY PROFILE: Moderately toxic by subcutaneous route. An eye irritant. When heated to decomposition it emits very toxic fumes of HCl and NO_x.

MRU000 CAS:58139-48-3 ***HR: 3***
4-MORPHOLINO-2-(5-NITRO-2-THIENYL)QUINAZOLINE
mf: $C_{16}H_{14}N_4O_3S$ mw: 342.40

TOXICITY DATA with REFERENCE
mmo-sat 1250 μg/plate CNREA8 35,3611,75
orl-rat TDLo:14 g/kg/60W-C:NEO JNCIAM 57,277,76

SAFETY PROFILE: Questionable carcinogen with experimental neoplastigenic data. Mutation data reported.

When heated to decomposition it emits very toxic fumes of SO_x and NO_x.

MRU075 CAS:16142-27-1 ***HR: 3***
3-MORPHOLINOSYDNONE IMINE HYDROCHLORIDE
mf: $C_6H_{10}N_4O_2 \cdot ClH$ mw: 206.66

TOXICITY DATA with REFERENCE
orl-mus LD50:480 mg/kg OYYAA2 2,280,68
ipr-mus LD50:315 mg/kg OYYAA2 2,280,68
ivn-mus LD50:260 mg/kg OYYAA2 2,280,68

SAFETY PROFILE: Poison by intravenous and intraperitoneal routes. Moderately toxic by ingestion. When heated to decomposition it emits toxic fumes of NO_x and HCl.

MRU080 CAS:10329-95-0 ***HR: 2***
MORPHOLINO-THALIDOMIDE
mf: $C_{18}H_{19}N_3O_5$ mw: 357.40

SYNS: CG 601 ◇ E-298 ◇ 1-METHYL-MORPHOLINO-3-PHTHALIMIDO-GLUTARIMIDE ◇ N-(1-(MORPHOLINOMETHYL)-2,6-DIOXO-3-PIPERIDYL)PHTHALIMIDE ◇ 1-MORPHOLINOMETHYLTHALIDOMIDE

TOXICITY DATA with REFERENCE
ipr-rat LD50:1300 mg/kg AIPTAK 194,39,71
orl-mus LD50:3000 mg/kg AIPTAK 194,39,71
ipr-mus LD50:840 mg/kg AIPTAK 194,39,71

SAFETY PROFILE: Moderately toxic by ingestion and intraperitoneal routes. When heated to decomposition it emits toxic fumes of NO_x.

MRU100 ***HR: 3***
4-(2-MORPHOLINYL)PYROCATECHOL
mf: $C_{10}H_{13}NO_3$ mw: 195.24

SYN: 2-(3,4-DIOXYPHENYL)TETRAHYDRO-1,4-OXAZIN (GERMAN)

TOXICITY DATA with REFERENCE
orl-mus LDLo:1600 mg/kg AEPPAE 222,540,54
scu-mus LD50:65 mg/kg AEPPAE 222,540,54
ivn-mus LD50:13 mg/kg AEPPAE 226,493,55

SAFETY PROFILE: Poison by subcutaneous and intravenous routes. Moderately toxic by ingestion. When heated to decomposition it emits toxic fumes of NO_x.

MRU250 CAS:144-41-2 ***HR: 3***
MORPHOTHION
mf: $C_8H_{16}NO_4PS_2$ mw: 285.34

PROP: Colorless solid. Mp: 65°. Sol in acetone, dioxane, acetonitrile.

SYNS: O,O-DIMETHYL-S-((MORFOLINO-CARBONYL)-METHYL)-DITHIOFOSFAAT (DUTCH) ◇ O,O-DIMETHYL-S-(MORPHOLINOCARBAMOYLMETHYL) DITHIOPHOSPHATE ◇ O,O-DIMETHYL-S-((MORPHOLÌNO-CARBONYL)-METHYL)-DITHIOPHOSPHAT(GERMAN) ◇ O,O-DIMETHYL MORPHOLINOCARBONYLMETHYL PHOSPHORODITHIOATE ◇ O,O-DIMETHYL-S-(MORPHOLINOCARBONYLMETHYL) PHOSPHORODITHIOATE ◇ DIMETHYL S-(MORPHOLINOCARBONYLMETHYL)PHOSPHOROTHIOLOTHIONATE ◇ O,O-DIMETIL-S-((MORFOLINO-CARBONIL)-METIL)-DITIOFOSFATO (ITALIAN) ◇ DITHIOPHOSPHATE de O,O-DIMETHYLE et de S-((MORPHOLINOCARBONYLE)-METHYLE) (FRENCH) ◇ 4-(MERCAPTOACETYL)MORPHOLINE O,O-DIMETHYL PHOSPHORODITHIOATE ◇ MORFOTHION (DUTCH) ◇ PHOSPHORODITHIOIC ACID, O,O-DIMETHYL S-(MORPHOLINOCARBONYLMETHYL) ESTER

TOXICITY DATA with REFERENCE
orl-rat LD50:190 mg/kg 28ZEAL 5,159,76
skn-rat LD50:283 mg/kg WRPCA2 9,119,70
orl-mus LD50:130 mg/kg BESAAT 12,161,66
orl-rbt LD50:190 mg/kg BESAAT 12,161,66

SAFETY PROFILE: Poison by ingestion and skin contact. A cholinesterase inhibitor. When heated to decomposition it emits very toxic fumes of PO_x, SO_x, and NO_x. See also PARATHION.

MRU255 ***HR: 3***
MOSE
mf: $C_{12}H_{24}N_2O_2Se \cdot 2ClH$ mw: 380.26

SYNS: BIS(β-(N,N-DIMORPHOLINO)ETHYL)SELENIDEDIHYDROCHLORIDE ◇ 4,4'-(SELENODI-2,1-ETHANEDIYL)BISMORPHOLINE DIHYDROCHLORIDE

TOXICITY DATA with REFERENCE
ivn-rat LD50:400 mg/kg DCTODJ 7,41,84
ipr-mus LD50:1250 mg/kg DCTODJ 7,41,84
ivn-rbt LD50:80 mg/kg DCTODJ 7,41,84

CONSENSUS REPORTS: Selenium and its compounds are on the Community Right-To-Know List.

OSHA PEL: TWA 0.2 mg(Se)/m^3
ACGIH TLV: TWA 0.2 mg(Se)/m^3

SAFETY PROFILE: Poison by intravenous route. Moderately toxic by intraperitoneal route. When heated to decomposition it emits toxic fumes of NO_x, HCl, and Se. See also SELENIUM COMPOUNDS.

MRU359 ***HR: 3***
MOUNTAIN LAUREL

PROP: Evergreen shrubs with thick leaves and white, pink, or purple flowers. The various species grow along the Pacific coast from Alaska to California and in the region bounded by Nova Scotia, Georgia, and Michigan.

SYNS: AMERICAN LAUREL ◇ BIG LEAF IVY ◇ CALF KILL ◇ CALICO BUSH ◇ DWARF LAUREL ◇ IVY BUSH ◇ KALMIA ANGUSTIFOLIA ◇ KALMIA LATIFOLIA ◇ KALMIA MICROPHYLLA ◇ KID KILL ◇ LAMB KILL ◇ SHEEP LAUREL ◇ SPOONWOOD IVY ◇ WICKY ◇ WOOD LAUREL

SAFETY PROFILE: The leaves and nectar contain poisonous grayanotoxins (andromedotoxins). Ingestion of these plant parts results in immediate pain in the mouth and may be followed several hours later by vomiting, diarrhea, headache, impaired vision, irregular heartbeat, severe low blood pressure, coma, convulsions, and death.

MRU600 CAS:34816-55-2 ***HR: 2***
MOXESTROL
mf: $C_{21}H_{26}O_3$ mw: 326.47

PROP: Crystals. Mp: 280°.

SYNS: 11-β-METHOXY-19-NOR-17-α-PREGNA-1,3,5(10)-TRIEN-20-YNE-3,17-DIOL ◇ R 2858 ◇ RU 2858 ◇ SURESTRYL

TOXICITY DATA with REFERENCE
scu-rat TDLo:50 μg/kg (16-20D preg):REP JRPFA4 59,43,80
scu-rat TDLo:50 μg/kg (16-20D preg):TER JRPFA4 59,43,80
imp-ham TDLo:400 mg/kg/39W-C:CAR CNREA8 47,2583,87

SAFETY PROFILE: An experimental teratogen. Experimental reproductive effects. Questionable carcinogen with experimental carcinogenic data. A steroid. When heated to decomposition it emits acrid smoke and irritating fumes.

MRU750 CAS:52279-59-1 ***HR: 2***
MOXNIDAZOLE
mf: $C_{13}H_{18}N_6O_5{\cdot}ClH$ mw: 374.83

SYN: 3-(5-NITRO-1-METHYL-2-IMIDAZOLYL)-METHYLENE-AMINO-5-MORPHOLINO-METHYL-2-OXAZOLIDONEHCl

TOXICITY DATA with REFERENCE
mmo-esc 28 μmol/L JEPTDQ 2(3),657,79
sln-dmg-orl 10 mmol/L JEPTDQ 2(3),657,79
slt-mus-orl 3 g/kg EXPEAM 34,500,78
orl-mus TDLo:3 g/kg (8-10D preg):TER EXPEAM 34,500,78
orl-mus LD50:3500 mg/kg ARZNAD 28,1665,78

CONSENSUS REPORTS: EPA Genetic Toxicology Program.

SAFETY PROFILE: Moderately toxic by ingestion. An experimental teratogen. Mutation data reported. When heated to decomposition it emits toxic fumes of NO_x and HCl.

MRU755 ***HR: 2***
MS-ANTIGEN 40

TOXICITY DATA with REFERENCE
orl-rat LD50:6504 mg/kg NIIRDN 6,125,82
ipr-rat LD50:1643 mg/kg NIIRDN 6,125,82
scu-rat LD50:4044 mg/kg NIIRDN 6,125,82
orl-mus LD50:5643 mg/kg NIIRDN 6,125,82
ipr-mus LD50:2342 mg/kg NIIRDN 6,125,82
scu-mus LD50:2018 mg/kg NIIRDN 6,125,82
ivn-mus LD50:841 mg/kg NIIRDN 6,125,82

SAFETY PROFILE: Moderately toxic by intraperitoneal, subcutaneous, and intravenous routes. Mildly toxic by ingestion.

MRU757 CAS:57314-55-3 ***HR: 3***
MT-45
mf: $C_{24}H_{32}N_2{\cdot}2ClH$ mw: 421.50

SYN: (±)-1-CYCLOHEXYL-4-(1,2-DIPHENYLETHYL)PIPERAZINEDIHYDROCHLORIDE

TOXICITY DATA with REFERENCE
orl-rat LD50:150 mg/kg DRFUD4 2,39,77
scu-rat LD50:136 mg/kg DRFUD4 2,39,77
ivn-rat LD50:7800 μg/kg DRFUD4 2,39,77
orl-mus LD50:329 mg/kg DRFUD4 2,39,77
ipr-mus LD50:58400 μg/kg DRFUD4 2,39,77
scu-mus LD50:743 mg/kg DRFUD4 2,39,77
ivn-mus LD50:17800 μg/kg DRFUD4 2,39,77

SAFETY PROFILE: Poison by ingestion, subcutaneous, intravenous and intraperitoneal routes. When heated to decomposition it emits toxic fumes of NO_x and HCl.

MRU760 CAS:41208-07-5 ***HR: 2***
MTDQ
mf: $C_{25}H_{30}N_2$ mw: 358.57

SYNS: 6,6′-METHYLENEBIS(1,2-DIHYDRO-2,2,4-TRIMETHYL-QUINOLINE) ◇ NSC 217697

TOXICITY DATA with REFERENCE
orl-rat LD50:5 g/kg NEOLA4 24,253,77
orl-mus LD50:6250 mg/kg NCISP* JAN86
ipr-mus LD50:1433 mg/kg NCISP* JAN86

SAFETY PROFILE: Moderately toxic by intraperitoneal route. Mildly toxic by ingestion. When heated to decomposition it emits toxic fumes of NO_x.

MRU775 CAS:34521-15-8 ***HR: 3***
MTEAI
mf: $C_{10}H_{24}N_3S{\cdot}Br{\cdot}BrH$ mw: 379.26

SYNS: (2-((IMINO(METHYLAMINO)METHYL)THIO)ETHYL) TRIETHYLAMMONIUM BROMIDE HYDROBROMIDE ◇ S-(2-TRIETHYLAMINOETHYL)-1′-METHYLISOTHIURONIUM BROMIDE HYDROBROMIDE ◇ N,N,N- TRIETHYL-2-((IMINO(METHYLAMINO) METHYL)THIO)-ETHANAMINIUM BROMIDE, MONOHYDROBROMIDE

TOXICITY DATA with REFERENCE
ipr-mus LD50:148 mg/kg CPBTAL 23,1639,75
scu-mus LD50:172 mg/kg CPBTAL 23,1639,75
ivn-mus LD50:99300 μg/kg CPBTAL 23,1639,75

SAFETY PROFILE: Poison by subcutaneous, intravenous, and intraperitoneal routes. When heated to decomposition it emits toxic fumes of SO_x, NO_x, NH_3, and Br^-.

MRU900 CAS:87-56-9 ***HR: 3***
MUCOCHLORIC ACID
mf: $C_4H_2Cl_2O_3$ mw: 168.96

SYNS: ALDEHYDODICHLOROMALEIC ACID ◇ 2-BUTENOIC ACID, 2,3-DICHLOR-4-OXO-, (Z)-(9CI) ◇ α-β-DICHLORO-β-FORMYL ACRYLIC ACID ◇ 3,4-DICHLORO-2-HYDROXYCROTONOLACTONE ◇ 3,4-DICHLORO-2-HYDROXYCROTONOLACTONIC ACID ◇ DICHLOROMALEALDEHYDIC ACID ◇ 2,3-DICHLOROMALEIC ALDEHYDE ACID ◇ 2,3-DICHLORO-4-OXO-2-BUTENOIC ACID ◇ KYSELINA MUKOCHLOROVA ◇ MALEALDEHYDIC ACID, DICHLORO-4-OXO-, (Z)-(9CI)

TOXICITY DATA with REFERENCE
skn-rbt 20 mg/24H MOD 85JCAE-,558,86
eye-rbt 50 μg/24H SEV 85JCAE-,558,86
mmo-sat 100 ng/plate BECTA6 24,590,80
orl-mus TDLo:6100 mg/kg/78W-I:ETA NTIS** PB223-159
orl-mus LD50:84 mg/kg 85GMAT-,46,82

CONSENSUS REPORTS: Reported in EPA TSCA Inventory.

SAFETY PROFILE: Poison by ingestion. Moderate skin and severe eye irritant. Questionable carcinogen with experimental tumorigenic data. Mutation data reported. When heated to decomposition it emits toxic fumes of Cl^-.

MRV000 CAS:4412-09-3 ***HR: 2***
MUCOCHLORIC ANHYDRIDE
mf: $C_8H_2Cl_4O_5$ mw: 319.90

PROP: α Isomer: mp: 141-143°; β isomer: mp: 180°. Insol in water; sol in many organic solvents, as acetone, xylene, cyclohexanone, methylnapthehalones. Monoclinic prisms: sltly sol in cold water; sol in hot water, hot benzene and alc.

SYNS: BIS(3,4-DICHLOROFURANON-5-YL-2) ETHER ◇ BIS(3,4-DICHLORO-2(5)-FURANONYL) ETHER ◇ GC-2466 ◇ 5,5′-OXYBIS(3,4-DICHLORO-2(5H)-FURANONE ◇ 4,4′-OXYBIS(2,3-DICHLORO-4-HYDROXYCROTONIC ACID-DI-Γ-LACTONE

TOXICITY DATA with REFERENCE
orl-rat LD50:2000 mg/kg FMCHA2 -,D211,80

SAFETY PROFILE: Moderately toxic by ingestion. When heated to decomposition it emits very toxic fumes of Cl^-.

MRV250 CAS:992-21-2 ***HR: 3***
MUCOMYCIN
mf: $C_{29}H_{33}N_4O_{10}$ mw: 597.66

SYNS: N^2-(((+)-5-AMINO-5-CARBOXYPENTYLAMINO)METHYL) TETRACYCLINE ◇ ARMYL ◇ LYMECYCLINE ◇ N-LYSINOMETHYL-TETRACYCLINE ◇ TETRACICLINA-l-METILENLISINA (ITALIAN) ◇ TETRACYCLINE-l-METHYLENE LYSINE ◇ TETRALISAL ◇ TETRALYSAL

TOXICITY DATA with REFERENCE
ivn-mus LD50:181 mg/kg KPOBAR 2,333,70

SAFETY PROFILE: Poison by intravenous route. When heated to decomposition it emits toxic fumes of NO_x.

MRV500 CAS:3148-09-2 ***HR: 3***
MUCONOMYCIN A
mf: $C_{27}H_{34}O_9$ mw: 502.61

PROP: Plates from acetone + H_2O. Decomp above 240°. Derived from cultures of *Myrothecium verrucaria* (ARZNAD 15,893,65).

SYNS: VER A ◇ VERRUCARIN A

TOXICITY DATA with REFERENCE
skn-gpg 402 ng MLD FAATDF 4(2, Pt 2)S124,84
dni-mus:lym 1 nmol/L PLMEAA 34,231,78
ivn-rat LD50:870 μg/kg ARZNAD 15,893,65
ipr-mus LD50:500 μg/kg TXAPA0 15,262,69
ivn-mus LD50:1500 μg/kg ARZNAD 15,893,65
ivn-rbt LD50:540 μg/kg ARZNAD 15,893,65

SAFETY PROFILE: A poison by intravenous and intraperitoneal routes. Mutation data reported. A skin irritant. When heated to decomposition it emits acrid smoke and irritating fumes.

MRV525 ***HR: 2***
MUCOPOLYSACCHARIDE, POLYSULFURIC ACID ESTER

SYN: MUCOPOLYSACCHARIDIPOLYSCHWEFELSAEUREESTER (GERMAN)

TOXICITY DATA with REFERENCE
ivn-rbt TDLo:450 mg/kg (female 8-16D post):TER KSRNAM 9,251,75
scu-rbt TDLo:320 mg/kg (female 32D pre):REP KSRNAM 16,5119,82
ipr-rat LD50:2400 mg/kg KSRNAM 9,241,75
scu-rat LD50:1550 mg/kg KSRNAM 9,241,75
ivn-rat LD50:3260 mg/kg KSRNAM 9,241,75

scu-mus LD50:4500 mg/kg KSRNAM 9,241,75
ivn-mus LD50:3480 mg/kg KSRNAM 9,241,75
scu-dog LDLo:500 mg/kg KSRNAM 9,241,75
ivn-rbt LD50:4020 mg/kg KSRNAM 16,5119,82
scu-rbt LD50:2020 mg/kg KSRNAM 16,5119,82

SAFETY PROFILE: Moderately toxic by intraperitoneal, subcutaneous, and intravenous routes. An experimental teratogen. Experimental reproductive effects. See also ESTERS. When heated to decomposition it emits toxic fumes of SO_x.

MRV600 ***HR: 3***
MUCUNA MONOSPERMA DC. ex Wight (extract excluding roots)

PROP: Indian plant belonging to the family *Leguminosae* IJEBA6 18,594,80

TOXICITY DATA with REFERENCE
orl-rat TDLo:150 mg/kg (female 12-14D post):REP IJEBA6 18,594,80
ipr-mus LD50:250 mg/kg IJEBA6 18,594,80

SAFETY PROFILE: Poison by intraperitoneal route. Experimental reproductive effects. When heated to decomposition it emits acrid smoke and irritating fumes.

MRV750 CAS:36069-45-1 ***HR: 2***
MULDAMINE
mf: $C_{29}H_{48}NO_3$ mw: 458.78

TOXICITY DATA with REFERENCE
orl-ham TDLo:600 mg/kg (7D preg):REP JAFCAU 26,561,78
orl-ham TDLo:600 mg/kg (7D preg):TER JAFCAU 26,561,78
orl-ham LDLo:600 mg/kg JAFCAU 26,561,78

SAFETY PROFILE: Moderately toxic by ingestion. An experimental teratogen. Experimental reproductive effects. When heated to decomposition it emits toxic fumes of NO_x.

MRW000 CAS:58-34-4 ***HR: 3***
MULTERGAN METHYL SULFATE
mf: $C_{18}H_{23}N_2S•CH_3O_4S$ mw: 410.59

SYNS: METHYLPHENAZONIUM METHOSULFATE ◇ MULTERGAN ◇ MULTEZIN ◇ PADISAL ◇ N-(β-(10-PHENOTHIAZINYL)PROPYL)TRIMETHYLAMMONIUM METHYL SULFATE ◇ PMS ◇ PROTHAZIN METHOSULFATE ◇ RP 3554 ◇ N,N,N-α-TETRAMETHYL-10H-PHENOTHIAZINE-10-ETHANAMINIUM METHYL SULFATE ◇ THIAZINAMIUM METHYL SULFATE ◇ TRIMETHYL (1-METHYL-2-PHENOTHIAZIN-10-YLETHYL)AMMONIUM METHYL SULFATE ◇ TRIMETHYL(1-METHYL-2-(10-PHENOTHIAZINYL)ETHYL)AMMONIUM METHYL SULFATE ◇ VALAN

TOXICITY DATA with REFERENCE
orl-mus LD50:375 mg/kg MEIEDD 10,1333,83
ipr-mus LDLo:61 mg/kg CLDND*

SAFETY PROFILE: Poison by ingestion and intraperitoneal routes. An antihistamine and anticholinergic agent. When heated to decomposition it emits very toxic fumes of SO_x, NH_3, and NO_x.

MRW125 CAS:3644-61-9 ***HR: 3***
MUSCALM
mf: $C_{16}H_{23}NO•ClH$ mw: 281.86

SYNS: 2,4′-DIMETHYL-3-PIPERIDINOPROPIOPHENONEHYDROCHLORIDE ◇ 2-METHYL-3-PIPERIDINO-1-p-TOLYLPROPAN-1-ONE HYDROCHLORIDE ◇ MYDOCALM ◇ N-553 ◇ 1-PIPERIDINO-2-METHYL-3-(p-TOLYL)-3-PROPANONEHYDROCHLORIDE ◇ TOLPERISONE HYDROCHLORIDE

TOXICITY DATA with REFERENCE
orl-rat LD50:1450 mg/kg NIIRDN 6,531,82
ipr-rat LD50:170 mg/kg NIIRDN 6,531,82
scu-rat LD50:645 mg/kg NIIRDN 6,531,82
ivn-dog LD50:45 mg/kg OYYAA2 9,809,75
ivn-cat LD50:40 mg/kg OYYAA2 9,809,75

SAFETY PROFILE: Poison by intravenous and intraperitoneal routes. Moderately toxic by ingestion and subcutaneous routes. When heated to decomposition it emits toxic fumes of NO_x and HCl.

MRW250 CAS:300-54-9 ***HR: 3***
MUSCARINE
mf: $C_9H_{20}NO_2$ mw: 174.30

SYNS: MUSCARIN ◇ dl-MUSCARINE ◇ MUSK ◇ MUSKARIN ◇ TRIMETHYL(TETRAHYDRO-4-HYDROXY-5-METHYLFURFURYL) AMMONIUM

TOXICITY DATA with REFERENCE
unr-man LDLo:735 μg/kg 85DCAI 2,73,70
orl-mus LDLo:750 mg/kg 27ZIAQ -,167,73
ipr-mus LD50:5 mg/kg AIPTAK 192,88,71
ivn-mus LD50:230 μg/kg TOLED5 1000 (Sp 1),42,80
orl-cat LDLo:7 mg/kg HBAMAK 4,1289,35
scu-cat LDLo:10 mg/kg AEXPBL 61,283,09
ivn-cat LDLo:1100 μg/kg SCIEAS 144,1100,64
orl-rbt LDLo:200 mg/kg AEXPBL 61,283,09
scu-rbt LDLo:27 mg/kg AEXPBL 61,283,09

SAFETY PROFILE: Human poison by an unspecified route. Experimental poison by ingestion, intravenous, intraperitoneal, and subcutaneous routes. When heated to decomposition it emits toxic fumes of NO_x and NH_3.

MRW269 *HR: D*
MUSHROOMS

SAFETY PROFILE: Variable toxicity. Symptoms developing within 2 hours of ingestion are seldom dangerous. Symptoms which develop longer than 6 hours after ingestion tend to be more severe and possibly life threatening.

Amanita phaloides and its relatives cause most of the fatalities due to mushroom poisoning. They contain amatoxins, cyclic octapeptides which inhibit the liver enzyme RNA polymerase II and thus cause liver failure. First symptoms appear 12 hours after ingestion and include persistent nausea, vomiting, intestinal pain, and profuse watery diarrhea. A latent period of up to 5 days follows and is then succeeded by signs of liver failure. Even with intensive care, the fatality rate is 10-15%.

Ingestion of some species of the genus *Cortinarius* is followed within 3-17 days by polydipsia, polyuria and then kidney failure, fatty degeneration of the liver and severe inflammation of the intestine.

Gyromitra esculenta contains gyromitrin and related hydrazones which are hydrolyzed to the toxic monomethylhydrazine. This is a volatile and water-soluble material, and the mushrooms may be detoxified by boiling in water or air drying. Ingestion of the mushrooms or breathing the vapor from cooking can cause very sudden emesis which usually recedes in 2-6 days, although deaths have occurred.

Mushrooms of the genera *Clitocybe* and *Inocybe* produce the parasympathetic nervous system stimulant muscarine which causes sweating, abdominal pain, emesis, blurred vision and other parasympathetic responses. Symptoms usually recede after 2 hours.

Mushrooms of the genera *Psilocybe*, *Panaeolus*, *Copelandia*, *Gymnopilus*, *Conocybe* and *Pluteus* contain the toxic psilocybin which has central nervous system effects including hallucinations and hyperthermia. Symptoms usually recede within 2 hours. Children ingesting large amounts may experience tonic-clonic convulsions, coma and death. Symptoms lasting more than 24 hours may be due to the addition of phencyclidine (PCP) to the mushroom.

Amanita muscaria and *Amanita pantherina* contain the toxins ibotenic acid and muscimol. Muscimol acts as a GABA agonist and effects the central nervous system. Symptoms appear within 2 hours and may include abdominal discomfort, drowsiness, dizziness, sleep followed by increased motor activity, hallucinations, delirium and manic excitement. Children ingesting large amounts may experience convulsions, coma, and central nervous system effects lasting up to 12 hours.

The mushroom *Coprinus atramentarius* is edible but contains the amino acid coprine which is metabolized into an acetaldehyde dehydrogenase inhibitor. The resultant disturbance of alcohol metabolism may cause headache, nausea and vomiting, flushing and cardiovascular effects if alcohol is consumed within 72 hours after the mushrooms are eaten.

See also 5-AMINOMETHYL-3-ISOXYZOLE (muscimol), α-AMINO-3-HYDROXY-5-ISOOXAZOLE-ACETIC ACID HYDRATE (ibotenic acid), O-PHOSPHORYL-4-HYDROXY-N,N-DIMETHYLTRYPTAMINE (psilocybin), MUSCARINE, and MONOMETHYLHYDRAZINE.

MRW275 CAS:34513-77-4 *HR: 3*
MYBORIN
mf: $C_{17}H_{27}BN_2$ mw: 270.27

SYNS: (T-4)-DIETHYL(2-(1-ETHYL-1-(2H-PYRROL-5-YL)PROPYL)-1H-PYRROLATO-N[1],N[2])-BORON ◇ 4,4,8,8-TETRAETHYL-3,3a,4,8-TETRAHYDRO-3a,4a,4-DIAZABORA-5-INDACENE

TOXICITY DATA with REFERENCE
orl-mus LD50:180 mg/kg PSEBAA 150,434,75
ipr-mus LD50:70 mg/kg PSEBAA 150,434,75
scu-mus LD50:420 mg/kg PSEBAA 150,434,75

SAFETY PROFILE: Poison by ingestion and intraperitoneal routes. Moderately toxic by subcutaneous route. When heated to decomposition it emits toxic fumes of Na_2O. See also BORON COMPOUNDS.

MRW500 CAS:73665-15-3 *HR: 3*
MYCINAMICIN 1
mf: $C_{36}H_{61}NO_{12}$ mw: 699.7

PROP: Produced from culture broth of *Micromonospora griseorubida* (JANTAJ 33,364,80).

TOXICITY DATA with REFERENCE
ipr-mus LD50:177 mg/kg JANTAJ 33,364,80
scu-mus LD50:310 mg/kg JANTAJ 33,364,80

SAFETY PROFILE: Poison by intraperitoneal and subcutaneous routes. When heated to decomposition it emits toxic fumes of NO_x.

MRW750 CAS:73684-69-2 *HR: 3*
MYCINAMICINS II
mf: $C_{36}H_{61}NO_{13}$ mw: 715.98

PROP: Produced from culture broth of *Micromonospora grisedrubida* (JANTAJ 33,364,80).

SYN: A 11725 II

TOXICITY DATA with REFERENCE
ipr-mus LD50:363 mg/kg JANTAJ 33,364,80
scu-mus LD50:465 mg/kg JANTAJ 33,364,80

SAFETY PROFILE: Poison by intraperitoneal route. Moderately toxic by subcutaneous route. When heated to decomposition it emits toxic fumes of NO_x.

MRW800 CAS:12609-89-1 ***HR: 3***
MYCOHEPTYNE
mf: $C_{47}H_{71}NO_{17}$ mw: 922.19

SYNS: ANTIBIOTIC 281471 ◇ ANTIBIOTIC 44 VI ◇ MYCOHEPTIN

TOXICITY DATA with REFERENCE
orl-rat TDLo:100 mg/kg (9D preg):REP ANTBAL 20,45,75
orl-mus LD50:2560 mg/kg KHFZAN 11(2),140,88
ipr-mus LD50:20 mg/kg 85GDA2 2,289,80

SAFETY PROFILE: Poison by intraperitoneal route. Moderately toxic by ingestion. Experimental reproductive effects. When heated to decomposition it emits toxic fumes of NO_x.

MRX000 CAS:24280-93-1 ***HR: 2***
MYCOPHENOLIC ACID
mf: $C_{17}H_{20}O_6$ mw: 320.37

PROP: Needles from hot water. Mp: 141°, a weak dibasic acid. Practically insol in cold water; very sol in alc; mod sol in ether, chloroform; sltly sol in benzene, toluene.

SYNS: 6-(5-CARBOXY-3-METHYL-2-PENTENYL)-7-HYDROXY-5-METHOXY-4-METHYLPHTHALIDE ◇ (E)-6-(4-HYDROXY-6-METHOXY-7-METHYL-3-OXO-5-PHTHALANYL)-4-METHYL-4-HEXENOICACID ◇ LILLY-68618 ◇ MELBEX ◇ (E)-4-METHYL-5-METHOXY-7-HYDROXY-6-(5-CARBOXY-3METHYLPENT-2-EN-1-YL)PHTHALIDE ◇ MICOFENOLICO ACIDO (SPANISH) ◇ NSC-129185

TOXICITY DATA with REFERENCE
dni-mus:lym 200 nmol/L CNREA8 45,5512,85
oms-mus:lym 400 nmol/L CNREA8 45,5512,85
orl-rat TDLo:900 mg/kg (male 30D pre):REP TOIZAG 29,400,82
orl-rat LD50:352 mg/kg TOIZAG 29,400,82
ivn-rat LD50:450 mg/kg PMDCAY 9,1,73
orl-mus LDLo:1000 mg/kg 85GDA2 5,445,81
ipr-mus LD50:505 mg/kg TOIZAG 29,400,82
ivn-mus LDLo:500 mg/kg 85ERAY 3,1740,78

CONSENSUS REPORTS: EPA Genetic Toxicology Program.

SAFETY PROFILE: Moderately toxic by ingestion, intraperitoneal, and intravenous routes. Experimental reproductive effects. Mutation data reported. When heated to decomposition it emits acrid smoke and irritating fumes.

MRX500 CAS:23593-75-1 ***HR: 3***
MYCOSPORIN
mf: $C_{22}H_{17}ClN_2$ mw: 344.86

SYNS: BAY 5097 ◇ BISPHENYL-(2-CHLORPHENYL)-1-IMIDAZOLYL-METHAN (GERMAN) ◇ CANESTEN ◇ 1-(o-CHLORO-α,α-DIPHENYL-BENZYL)IMIDAZOLE ◇ 1-(α-(2-CHLOROPHENYL)BENZHYDRYL)IMIDAZOLE ◇ 1-((2-CHLOROPHENYL)DIPHENYLMETHYL)-1H-IMIDAZOLE ◇ (CHLOROTRITYL)IMIDAZOLE ◇ 1-(o-CHLOROTRITYL)IMIDAZOLE ◇ CHLOTRIMAZOLE ◇ CLOTRIMAZOL ◇ DIPHENYL-(2-CHLOROPHENYL)-1-IMIDAZOYLMETHANE ◇ EMPECID ◇ FB 5097 ◇ GYNE-LOTRIMIN ◇ LOTRIMIN ◇ MYCELAX ◇ MYCELEX ◇ TRIMYSTEN

TOXICITY DATA with REFERENCE
scu-rat TDLo:600 mg/kg (9-14D preg):REP KSRNAM 7,359,73
scu-mus TDLo:600 mg/kg (female 7-12D post):TER KSRNAM 7,359,73
ivg-wmn TDLo:28 mg/kg/7D:SKN CTOXAO 18,41,81
orl-rat LD50:708 mg/kg ARZNAD 22,1272,72
ipr-rat LD50:445 mg/kg KSRNAM 7,1333,73
orl-mus LD50:761 mg/kg ARZNAD 22,1272,72
ipr-mus LD50:347 mg/kg KSRNAM 7,1333,73
orl-rbt LDLo:2000 mg/kg ARZNAD 22,1272,72
orl-mam LD50:750 mg/kg AACHAX -,271,69

SAFETY PROFILE: Poison by intraperitoneal route. Moderately toxic by ingestion. An experimental teratogen. Human systemic effects by intrvaginal route: primary skin irritations. Experimental reproductive effects. A fungicide. When heated to decomposition it emits very toxic fumes of Cl^- and NO_x.

MRY000 CAS:1404-01-9 ***HR: 3***
MYCOTICIN (1:1)
mf: $C_{18}H_{30}O_5$ mw: 326.48

SYN: MYCOTICIN

TOXICITY DATA with REFERENCE
ipr-mus LD50:10 mg/kg JIDEAE 23,163,54
scu-mus LD50:100 mg/kg 85ERAY 2,997,78

SAFETY PROFILE: Poison by intraperitoneal and subcutaneous routes. When heated to decomposition it emits acrid smoke and irritating fumes.

MRY100 CAS:52955-41-6 ***HR: 3***
MYOMYCIN B
mf: $C_{27}H_{51}N_9O_{14}$ mw: 725.87

TOXICITY DATA with REFERENCE
orl-mus LD50:9000 mg/kg 85GDA2 1,235,80
ipr-mus LD50:800 mg/kg 85GDA2 1,235,80
scu-mus LD50:1100 mg/kg 85GDA2 1,235,80
ivn-mus LD50:165 mg/kg 85GDA2 1,235,80

SAFETY PROFILE: Poison by intravenous route.

Moderately toxic by intraperitoneal and subcutaneous routes. Mildly toxic by ingestion. When heated to decomposition it emits toxic fumes of NO_x.

MRY250 ***HR: 3***
MYOMYCIN SULFATE

PROP: Isolated from beer filtrates (JANTAJ 26,273, 73).

TOXICITY DATA with REFERENCE
ipr-mus LD50:800 mg/kg JANTAJ 26,272,73
scu-mus LD50:1100 mg/kg JANTAJ 26,272,73
ivn-mus LD50:165 mg/kg JANTAJ 26,272,73

SAFETY PROFILE: Poison by intravenous route. Moderately toxic by intraperitoneal and subcutaneous routes. When heated to decomposition it emits very toxic fumes of NO_x and SO_x. See also SULFATES.

MRY600 ***HR: 3***
MYOPORUM LAETUM

PROP: A large, thick-trunked tree which may have buttress for the limbs. The bark is a light grey-white color. The elliptical leaves are about 3 inches long with serrated edges. When held up to a light, the leaves show many distinctive dark glands. The seed pod is a small ball and contains one seed. It is native to New Zealand, but is commonly planted along streets in northern California.

SAFETY PROFILE: The leaves and fruit contain (−) ngaione, a furanoid sesquiterpene ketone. Ingestion of these plant parts can cause persistent vomiting, convulsions, coma, and death.

MRZ100 ***HR: 3***
MYRABOLAM TANNIN

SYN: TANNIN from MYRABOLAM

TOXICITY DATA with REFERENCE
ipr-mus LD50:150 mg/kg JPPMAB 9,98,57
scu-mus LD50:150 mg/kg JPPMAB 9,98,57
ivn-mus LD50:150 mg/kg JPPMAB 9,98,57
ims-mus LD50:150 mg/kg JPPMAB 9,98,57

SAFETY PROFILE: Poison by subcutaneous, intramuscular, intravenous and intraperitoneal routes. See also TANNIN.

MRZ150 CAS:123-35-3 ***HR: 1***
MYRCENE
mf: $C_{10}H_{16}$ mw: 136.26

PROP: Colorless to pale yellow liquid; sweet, balsamic odor. D: 0.789, refr index: 1.466-1.471, flash p: 99°F. Sol in alc, fixed oils; insol in water.

SYNS: FEMA No. 2762 ◇ 3-METHYLENE-7-METHYL-1,6-OCTADIENE ◇ 7-METHYL-3-METHYLENE-1,6-OCTADIENE

TOXICITY DATA with REFERENCE
skn-rbt 500 mg/24H MOD FCTXAV 14,615,76

CONSENSUS REPORTS: Reported in EPA TSCA Inventory.

SAFETY PROFILE: A moderate skin and eye irritant. Flammable liquid. When heated to decomposition it emits acrid smoke and irritating fumes.

MSA250 CAS:544-63-8 ***HR: 3***
MYRISTIC ACID
mf: $C_{14}H_{28}O_2$ mw: 228.36

PROP: White or faintly yellow crystals from methanol. Mp: 58.5°, bp: 250.5° @ 100 mm, d: 0.8622 @ 54°/4°. Sol in abs alc, methanol, ether, petr ether, benzene, chloroform; insol in water.

SYNS: CRODACID ◇ EMERY 655 ◇ HYDROFOL ACID 1495 ◇ HYSTRENE 9014 ◇ TETRADECANOIC ACID ◇ n-TETRADECOIC ACID ◇ 1-TRIDECANECARBOXYLIC ACID ◇ UNIVOL U 316S

TOXICITY DATA with REFERENCE
skn-hmn 75 mg/3D-I MOD 85DKA8 -,127,77
sln-smc 2500 ppb ANYAA9 407,186,83
ivn-mus LD50:43 mg/kg APTOA6 18,141,61

CONSENSUS REPORTS: Reported in EPA TSCA Inventory.

SAFETY PROFILE: Poison by intravenous route. Mutation data reported. A human skin irritant. When heated to decomposition it emits acrid smoke and irritating fumes.

MSA500 CAS:607-91-0 ***HR: 3***
MYRISTICIN
mf: $C_{11}H_{12}O_3$ mw: 192.23

SYN: 5-ALLYL-1-METHOXY-2,3-(METHYLENEDIOXY)BENZENE

TOXICITY DATA with REFERENCE
dnd-mus-ipr 400 mg/kg CRNGDP 5,1613,84
orl-hmn TDLo:5700 μg/kg:CNS JNEUAY 2,205,61
orl-rat LD50:4260 mg/kg JAFCAU 30,563,82
orl-cat LDLo:400 mg/kg AMJPA6 80,563,09
scu-rbt LDLo:900 mg/kg AMJPA6 80,563,09
scu-gpg LDLo:2000 mg/kg AMJPA6 80,563,09

SAFETY PROFILE: Poison by ingestion. Moderately toxic by subcutaneous route. Human systemic effects by ingestion: wakefulness, euphoria, hallucinations and distorted perceptions. Mutation data reported. When heated to decomposition it emits acrid smoke and irritating fumes.

MSA750 CAS:64817-78-3 ***HR: 3***
9-MYRISTOYL-1,7,8-ANTHRACENETRIOL
mf: $C_{28}H_{36}O_4$ mw: 436.64

SYNS: 1,8-DIHYDROXY-10-MYRISTOYL-9-ANTHRONE ◇ 1,8-DIHYDROXY-10-(1-OXOTETRADECYL)-9(10H)-ANTHRACENONE ◇ 10-MYRISTOYL-1,8,9-ANTHRACENETRIOL

TOXICITY DATA with REFERENCE
skn-mus TDLo:2621 mg/kg/73W-I:NEO JMCMAR 21,26,78

SAFETY PROFILE: Questionable carcinogen with experimental neoplastigenic data. When heated to decomposition it emits acrid smoke and irritating fumes.

MSB000 CAS:63021-43-2 ***HR: 3***
1-MYRISTOYLAZIRIDINE
mf: $C_{16}H_{31}NO$ mw: 253.48

SYNS: MYRISTOYLETHYLENEIMINE ◇ TETRADECANOYLETHYLENEIMINE

TOXICITY DATA with REFERENCE
scu-mus TDLo:4 g/kg/32W-I:NEO BJPCAL 9,306,54

SAFETY PROFILE: Questionable carcinogen with experimental neoplastigenic data. When heated to decomposition it emits toxic fumes of NO_x.

MSB100 CAS:79127-47-2 ***HR: 2***
N-MYRISTOYLOXY-N-ACETYL-2-AMINO-7-IODOFLUORENE
mf: $C_{29}H_{38}INO_3$ mw: 575.58

SYNS: ACETOHYDROXAMIC ACID, N-(7-IODOFLUOREN-2-YL)-O-MYRISTOYL- ◇ HYDROXYLAMINE, N-ACETYL-N-(7-IODO-2-FLUORENYL)-O-MYRISTOYL- ◇ N-MYRISTOYLOXY-AAIF

TOXICITY DATA with REFERENCE
scu-rat TDLo:147 mg/kg/6W-I:CAR CRNGDP 2,655,81

SAFETY PROFILE: Questionable carcinogen with experimental carcinogenic data. When heated to decomposition it emits toxic fumes of NO_x.

MSB250 CAS:63224-46-4 ***HR: 3***
N-MYRISTOYLOXY-N-MYRISTOYL-2-AMINOFLUORENE
mf: $C_{41}H_{63}NO_3$ mw: 618.05

SYN: N-TETRADECANOYL-N-TETRADECANOYLOXY-2-AMINOFLUORENE

TOXICITY DATA with REFERENCE
dnr-hmn:fbr 100 μmol/L CNREA8 37,1461,77
dns-hmn:fbr 100 μmol/L/5H IJCNAW 16,284,75
scu-rat TDLo:157 mg/kg/5W-I:NEO CNREA8 37,1461,77

SAFETY PROFILE: Questionable carcinogen with experimental neoplastigenic data. Human mutation data reported. When heated to decomposition it emits toxic fumes of NO_x.

MSB500 CAS:2748-88-1 ***HR: 3***
MYRISTYL-Γ-PICOLINIUM CHLORIDE
mf: $C_{20}H_{36}N\cdot Cl$ mw: 326.02

SYNS: QUATRESIN ◇ WET-TONE B

TOXICITY DATA with REFERENCE
orl-rat LD50:250 mg/kg ARTODN 32,245,74
ipr-rat LD50:7500 μg/kg JAPMA8 35,89,46
scu-rat LD50:200 mg/kg JAPMA8 35,89,46
ivn-rat LD50:30 mg/kg JAPMA8 35,89,46

CONSENSUS REPORTS: Reported in EPA TSCA Inventory.

SAFETY PROFILE: Poison by ingestion, intraperitoneal, intravenous, and subcutaneous routes. When heated to decomposition it emits very toxic fumes of NO_x and Cl^-.

MSB750 ***HR: 3***
MYROBALANS TANNIN

SYNS: TANNIN from MYROBALANS ◇ TERMINALIA CHEBULA RETZ TANNING

TOXICITY DATA with REFERENCE
scu-mus TDLo:750 mg/kg/12W-I:ETA BJCAAI 14,147,60

SAFETY PROFILE: Questionable carcinogen with experimental tumorigenic data. See also TANNIN.

MSC000 ***HR: 3***
MYRTAN TANNIN

SYNS: EUCALYPTUS REDUNCA TANNIN ◇ TANNIN from MYRTAN

TOXICITY DATA with REFERENCE
scu-mus TDLo:750 mg/kg/12W-I:ETA BJCAAI 14,147,60

SAFETY PROFILE: Questionable carcinogen with experimental tumorigenic data. See also TANNIN.

MSC100 CAS:115-79-7 ***HR: 3***
MYSURAN CHLORIDE
mf: $C_{28}H_{42}Cl_2N_4O_2\cdot 2Cl$ mw: 608.54

PROP: Crystals. Mp: 196-199°. Freely sol in water.

SYNS: AMBENONIUM CHLORIDE ◇ AMBENONIUM DICHLORIDE ◇ AMBESTIGMIN CHLORIDE ◇ N,N'-BIS(2-DIETHYLAMINOETHYL) OXAMIDE BIS(2-CHLOROBENZYL CHLORIDE) ◇ MISURAN ◇ MYSURAN ◇ MYTELASE ◇ MYTELASE CHLORIDE ◇ OKSAZIL ◇ (OXALYBIS(IMINOETHYLENE)BIS((o-CHLOROBENZYL)DIETHYLAMMONIUM) DICHLORIDE ◇ OXAMIZIL ◇ OXAZIL ◇ OXAZYL ◇ WIN 8077

TOXICITY DATA with REFERENCE
orl-rat LD50:18500 μg/kg NIIRDN 6,132,82
scu-rat LD50:4500 μg/kg NIIRDN 6,132,82

ivn-rat LD50:2720 μg/kg NIIRDN 6,132,82
orl-mus LD50:145 mg/kg NIIRDN 6,132,82
scu-mus LD50:3700 μg/kg NIIRDN 6,132,82
ivn-mus LD50:1510 μg/kg NIIRDN 6,132,82

SAFETY PROFILE: Poison by ingestion, subcutaneous, and intravenous routes. When heated to decomposition it emits toxic fumes of Cl^-, NH_3, and NO_x. See also CHLORIDES.

N

NAC000 CAS:25134-21-8 ***HR: 3***
NADIC METHYL ANHYDRIDE
mf: $C_{10}H_{10}O_3$ mw: 178.20

SYNS: METHYLBICYCLO(2.2.1)HEPTENE-2,3-DICARBOXYLICANHYDRIDE ISOMERS ◇ NMA

TOXICITY DATA with REFERENCE
ipr-mus LD50:100 mg/kg NTIS** AD441-640

CONSENSUS REPORTS: Reported in EPA TSCA Inventory.

SAFETY PROFILE: Poison by intraperitoneal route. An irritant. When heated to decomposition it emits acrid smoke and irritating fumes. See also ANHYDRIDES.

NAC500 CAS:35991-93-6 ***HR: 3***
NADOXOLOL HYDROCHLORIDE
mf: $C_{14}H_{16}N_2O_3$•ClH mw: 296.78

PROP: White, crystalline powder from ethyl ether-methanol. Mp: 188°. Sol in water, methanol, ethanol; insol in ethyl ether.

SYNS: BRADYL ◇ CHLORHYDRATE de (NAPHTHYLOXY-1)-4 HYDROXY-3 BUTYRAMIDOXIME (FRENCH) ◇ N,3-DIHYDROXY-4-(1-NAPHTHALENYLOXY)BUTANINIDAMIDE HYDROCHLORIDE ◇ 3-HYDROXY-4-(1-NAPHTHYLOXY)BUTYRAMIDOXIMEHYDROCHLORIDE ◇ LL 1530 ◇ 4-α-NAPHTHYLOXY-3-HYDROXY-BUTYRAMIDOXIM-HYDROCHLORID (GERMAN) ◇ 4-(α-NAPHTHYLOXY)-3-HYDROXYBUTYRAMIDOXIMEHYDROCHLORIDE

TOXICITY DATA with REFERENCE
orl-hmn TDLo:80 mg/kg:CVS TOERD9 4,163,82
orl-chd TDLo:100 mg/kg:CVS TOERD9 4,163,82
orl-mus LD50:1 g/kg GWXXBX #2132113
ivn-mus LD50:180 mg/kg GWXXBX #2132113

SAFETY PROFILE: Poison by intravenous route. Moderately toxic by ingestion. Human systemic effects by ingestion: decreased pulse rate and cardiac changes. Used to treat cardiac arrhythmia. When heated to decomposition it emits very toxic fumes of NO_x and HCl.

NAD000 CAS:5051-16-1 ***HR: 3***
NAFIVERINE DIHYDROCHLORIDE
mf: $C_{34}H_{38}N_2O_4$•2ClH mw: 611.66

PROP: Crystals. Mp: 220-221°.

SYN: N,N'-DI(α-(1-NAPHTHYL)PROPIONYLOXY-2-ETHYL)PIPERAZINE DIHYDROCHLORIDE

TOXICITY DATA with REFERENCE
orl-rat LD50:5550 mg/kg ARZNAD 16,386,66
ipr-mus LD50:331 mg/kg ARZNAD 16,386,66
ivn-mus LD50:82 mg/kg ARZNAD 16,386,66

SAFETY PROFILE: Poison by intraperitoneal and intravenous routes. Mildly toxic by ingestion. When heated to decomposition it emits very toxic fumes of HCl and NO_x.

NAD500 CAS:1845-11-0 ***HR: 3***
NAFOXIDINE
mf: $C_{29}H_{31}NO_2$ mw: 425.61

SYNS: 1-(2-(4-(3,4-DIHYDRO-6-METHOXY-2-PHENYL-1-NAPHTHALENYL)PHENOXY)ETHYL)PYRROLIDENE ◇ 1-(2-(p-(3,4-DIHYDRO-6-METHOXY-2-PHENYL-1-NAPHTHYL)PHENOXY)ETHYL) PYRROLIDINE

TOXICITY DATA with REFERENCE
dni-hmn:mmr 1 μmol/L CNREA8 45,1611,85
scu-rat TDLo:250 μg/kg (1D pre):REP JMCMAR 25,323,82
scu-rat TDLo:500 μg/kg:ETA JSTBBK 12,47,80

SAFETY PROFILE: Experimental reproductive effects. Questionable carcinogen with experimental tumorigenic data. Human mutation data reported. When heated to decomposition it emits toxic fumes of NO_x.

NAD750 CAS:1847-63-8 ***HR: 3***
NAFOXIDINE HYDROCHLORIDE
mf: $C_{29}H_{31}NO_2$•ClH mw: 462.07

SYNS: 1-(2-(3,4-DIHYDRO-6-METHOXY-2-PHENYL-1-NAPHTHALENYL)PHENOXY)ETHYL)-PYRROLIDINE HCl ◇ 1-(2-(p-3,4-DIHYDRO-6-METHOXY-2-PHENYL-1-NAPHTHYL)PHENOXY)ETHYL) PYRROLIDINE HYDROCHLORIDE ◇ U 11100 ◇ U 11100A

TOXICITY DATA with REFERENCE
spm-rbt-scu 45 mg/kg JRPFA4 11,107,66
spm-rbt-unr 45 mg/kg/45D-I ADRPBI 4,65,69
orl-rbt TDLo:12300 μg/kg (female 6-8D post):TER FESTAS 16,281,65
orl-rat TDLo:175 μg/kg (female 7D pre):REP JMCMAR 10,78,67
orl-rat LD50:302 mg/kg PSEBAA 112,439,63
ipr-mus LD50:143 mg/kg PSEBAA 112,439,63

CONSENSUS REPORTS: EPA Genetic Toxicology Program.

SAFETY PROFILE: Poison by ingestion and intraperitoneal routes. An experimental teratogen. Experimental reproductive effects. Mutation data reported. When heated to decomposition it emits toxic fumes of NO_x and HCl.

NAE000 CAS:31329-57-4 ***HR: 3***
NAFTIDROFURYL
mf: $C_{24}H_{33}NO_3$ mw:383.58

SYN: N-DIETHYLAMINOETHYL-β-(1-NAPHTHYL)-β-TETRAHYDROFURYL ISOBUTYRATE ◇ 2-(DIETHYLAMINO)ETHYL TETRAHYDRO-α-(1-NAPHTHYLMETHYL)-2-FURANPROPIONATE ◇ DUBIMAX ◇ NAFRONYL ◇ 3-(1-NAPHTHYL)-2-TETRAHYDROFURFURYLPROPIONIC ACID 2-(DIETHYLAMINO)ETHYL ESTER ◇ α-TETRAHYDROFURFURYL-1-NAPHTHALENEPROPIONIC ACID 2-(DIETHYLAMINO) ETHYL ESTER ◇ TETRAHYDRO-α-(1-NAPHTALENYLMETHYL)-2-FURANPROPANOIC ACID 2-(DIETHYLAMINO)ETHYL ESTER ◇ TETRAHYDRO-α-(1-NAPHTHYLMETHYL)-2-FURANPROPANOIC ACID 2-(DIETHYLAMINO)ETHYL ESTER

TOXICITY DATA with REFERENCE
orl-rat LD50:1890 mg/kg GNRIDX 5,1,71
ipr-rat LD50:120 mg/kg GNRIDX 5,1,71
scu-rat LD50:2500 mg/kg GNRIDX 5,1,71
orl-mus LD50:365 mg/kg EJMCA5 10,291,75
ipr-mus LD50:129 mg/kg GNRIDX 5,1,71
scu-mus LD50:272 mg/kg GNRIDX 5,1,71
ivn-mus LD50:23 mg/kg JETOAS 8(3),188,75

SAFETY PROFILE: Poison by ingestion, subcutaneous, intraperitoneal, and intravenous routes. When heated to decomposition it emits toxic fumes of NO_x.

NAE100 CAS:3200-06-4 ***HR: 3***
NAFTIDROFURYL OXALATE
mf: $C_{24}H_{33}NO_3 \cdot C_2H_2O_4$ mw: 473.62

SYNS: 2-(DIETHYLAMINO)ETHYL TETRAHYDRO-α-(1-NAPHTHYLMETHYL)-2-FURANPROPIONATE OXALATE (1:1) ◇ EU-1806 ◇ 2-FURANPROPIONIC ACID, TETRAHYDRO-α-(1-NAPHTHYLMETHYL)-, 2-(DIETHYLAMINO)ETHYL ESTER, OXALATE (1:1) ◇ LS 121 ◇ NAFRONYL OXALATE ◇ PRAXILENE

TOXICITY DATA with REFERENCE
orl-rat TDLo:1680 mg/kg (female 6-19D post):REP JETOAS 2,40,69
orl-mus TDLo:4680 mg/kg (female 6-18D post):TER JETOAS 2,40,69
orl-rat LD50:711 mg/kg YACHDS 14,1237,86
ipr-rat LD50:98 mg/kg YACHDS 14,1237,86
scu-rat LD50:896 mg/kg YACHDS 14,1237,86
ivn-rat LD50:16500 μg/kg CHTPBA 3,463,68
orl-mus LD50:567 mg/kg YACHDS 14,1237,86
ipr-mus LD50:159 mg/kg YACHDS 14,1237,86
ivn-mus LD50:22 mg/kg YACHDS 14,1237,86
ivn-dog LDLo:71 mg/kg CHTPBA 3,463,68

SAFETY PROFILE: Poison by intravenous and intraperitoneal routes. Moderately toxic by ingestion and subcutaneous route. An experimental teratogen. Experimental reproductive effects. When heated to decomposition it emits toxic fumes of NO_x.

NAE505 ***HR: 3***
NAGARMOTHA OIL

PROP: The oil is a mixture of at least 21 compounds, mostly sesquiterpenes (IJEBA6 10,41,72).

SYNS: CYPERUS SCARIOSUS OIL ◇ NAGARMUSTA OIL

TOXICITY DATA with REFERENCE
mmo-sat 4830 μg/plate BECTA6 32,179,84
mma-sat 483 μg/plate BECTA6 32,179,84
orl-mus LD50:1400 mg/kg IJEBA6 10,41,72
ipr-mus LD50:240 mg/kg IJEBA6 10,41,72

SAFETY PROFILE: Poison by intraperitoneal route. Moderately toxic by ingestion. Mutation data reported. When heated to decomposition it emits acrid smoke and irritating fumes.

NAE510 ***HR: 3***
NAJA FLAVA VENOM

SYN: VENOM, SNAKE, NAJA FLAVA

TOXICITY DATA with REFERENCE
scu-rat LDLo:400 μg/kg BIJOAK 70,293,58
ivn-mus LD50:371 μg/kg TOXIA6 2,5,64
scu-dog LDLo:100 μg/kg BIJOAK 70,293,58
ivn-rbt LD50:75 μg/kg SCIEAS 117,47,53

SAFETY PROFILE: A deadly poison by subcutaneous and intravenous routes.

NAE512 ***HR: 3***
NAJA HAJE ANNULIFERA VENOM

SYN: VENOM, EGYPTIAN COBRA, NAJA HAJE ANNULIFERA

TOXICITY DATA with REFERENCE
ipr-mus LD50:160 μg/kg AIMJA9 26,199,75
scu-mus LD50:1750 μg/kg TOXIA6 5,47,67
ivn-mus LD50:4500 μg/kg BBACAQ 746,18,83
ice-mus LD50:95 μg/kg TOXIA6 22,308,84
ipr-gpg LD50:500 μg/kg TOXIA6 18,374,80

SAFETY PROFILE: A deadly poison by subcutaneous, intravenous, intracerebral, and intraperitoneal routes.

NAE515 ***HR: 3***
NAJA HAJE VENOM

SYN: VENOM, SNAKE, NAJA HAJE

TOXICITY DATA with REFERENCE
ipr-mus LD50:500 μg/kg HSZPAZ 297,92,54
ivn-mus LD50:287 μg/kg TOXIA6 8,51,70
ivn-rbt LDLo:600 μg/kg TOXIA6 2,5,64

SAFETY PROFILE: A deadly poison by intravenous and intraperitoneal routes.

NAE875 **HR: 3**
NAJA MELANOLEUCA VENOM

SYN: VENOM, SNAKE, NAJA MELANOLEUCA

TOXICITY DATA with REFERENCE
ipr-mus LD50:324 μg/kg TOXIA6 9,131,71
scu-mus LD50:225 μg/kg TOXIA6 5,47,67
ivn-mus LD50:93 μg/kg CHDDAT 280,1757,75

SAFETY PROFILE: A deadly poison by subcutaneous, intravenous, and intraperitoneal routes.

NAF000 **HR: 3**
NAJA NAJA ATRA VENOM

SYN: VENOM, SEA SNAKE, NAJA NAJA ATRA

TOXICITY DATA with REFERENCE
eye-rbt 5 μg SEV KTHYAC 1,200,51
ice-rbt LDLo:12500 ng/kg KTHYAC 2,170,52
ipr-gpg LDLo:150 mg/kg 30FDAB 1,21,70
ivn-rat LDLo:200 μg/kg TOXIA6 24,233,86
par-rat LD50:38 μg/kg TOXIA6 19,61,81
ipr-mus LD50:370 μg/kg JBCHA3 240,1616,65
scu-mus LD50:280 μg/kg TOXIA6 5,47,67
ivn-mus LD50:290 μg/kg BCPCA6 20,1549,71

SAFETY PROFILE: A deadly poison by intracerebral, intraperitoneal, subcutaneous, parenteral, and intravenous routes. A severe eye irritant.

NAF200 **HR: 3**
NAJA NAJA KAOUTHIA VENOM

SYN: VENOM, SNAKE, NAJA NAJA KAOUTHIA

TOXICITY DATA with REFERENCE
ipr-mus LD50:329 μg/kg TOXIA6 9,131,71
scu-mus LD50:240 μg/kg TOXIA6 5,47,67
ivn-mus LD50:250 μg/kg TOXIA6 7,239,69

SAFETY PROFILE: A deadly poison by subcutaneous, intravenous, and intraperitoneal routes.

NAF250 **HR: 3**
NAJA NAJA SIAMENSIS VENOM

SYN: VENOM, THAILAND COBRA, NAJA NAJA SIAMENSIS

TOXICITY DATA with REFERENCE
ipr-mus LD50:360 μg/kg 19DDA6 1,283,67
scu-mus LDLo:250 μg/kg TOXIA6 5,207,68
ivn-mus LDLo:275 μg/kg EJBCAI 21,1,71

SAFETY PROFILE: A deadly poison by subcutaneous, intravenous, and intraperitoneal routes.

NAG000 **HR: 3**
NAJA NIGRICOLLIS VENOM

SYNS: N. NIGRICOLLIS VENOM ◇ VENOM, SNAKE, NAJA NIGRICOLLIS

TOXICITY DATA with REFERENCE
ims-mus TDLo:10 μg/kg (10D preg):REP TOXIA6 13,475,75
ims-mus TDLo:10 μg/kg (10D preg):TER TOXIA6 13,475,75
scu-rat LDLo:1500 μg/kg 30FDAB 1,21,71
par-rat LD50:2250 ng/kg TOXIA6 19,61,81
ipr-mus LD50:352 μg/kg TOXIA6 18,384,80
scu-mus LD50:2 mg/kg HSZPAZ 297,104,54
ivn-mus LD50:30 μg/kg TOXIA6 20,111,82
ims-mus LD50:440 μg/kg TOXIA6 13,475,75
ivn-rbt LDLo:300 μg/kg TOXIA6 2,5,64

SAFETY PROFILE: A deadly poison by subcutaneous, intramuscular, parenteral, intravenous, and intraperitoneal routes. An experimental teratogen. Other experimental reproductive effects.

NAG200 **HR: 3**
NAJA NIVEA VENOM

SYN: VENOM, SOUTH AFRICAN CAPE COBRA, NAJA NIVEA

TOXICITY DATA with REFERENCE
ipr-mus LD50:600 μg/kg 19DDA6 1,283,67
scu-mus LD50:400 μg/kg TOXIA6 5,47,67
ivn-rat LDLo:900 μg/kg TOXIA6 23,63,85
ivn-mus LD50:389 μg/kg 23EIAT 1,437,68

SAFETY PROFILE: A deadly poison by subcutaneous, intravenous, and intraperitoneal routes.

NAG400 CAS:300-76-5 **HR: 3**
NALED

mf: $C_4H_7Br_2Cl_2O_4P$ mw: 380.80

PROP: Mp: 27.0°. Slightly sol in aliphatic hydrocarbons; very sol in aromatic hydrocarbons.

SYNS: ARTHODIBROM ◇ BROMCHLOPHOS ◇ BROMEX ◇ DIBROM ◇ O-(1,2-DIBROM-2,2-DICHLORAETHYL)-O,O-DIMETHYL-PHOSPHAT (GERMAN) ◇ 1,2-DIBROMO-2,2-DICHLOROETHYL DIMETHYL PHOSPHATE ◇ O-(1,2-DIBROMO-2,2-DICLORO-ETIL)-O,O-DIMETIL-FOSTATO (ITALIAN) ◇ O-(1,2-DIBROOM-2,2-DICHLOOR-ETHYL)-O,O-DIMETHYL-FOSFAAT (DUTCH) ◇ O,O-DIMETHYL-O-(1,2-DIBROMO-2,2-DICHLOROETHYL)PHOSPHATE ◇ DIMETHYL-1,2-DIBROMO-2,2-DICHLOROETHYL PHOSPHATE (OSHA) ◇ O,O-DIMETHYL-O-2,2-DICHLORO-1,2-DIBROMOETHYLPHOSPHATE ◇ ENT 24,988 ◇ HIBROM ◇ ORTHO 4355 ◇ ORTHODIBROM ◇ ORTHODIBROMO ◇ PHOSPHATE de O,O-DIMETHLE et de O-(1,2-DIBROMO-2,2-DICHLORETHYLE) (FRENCH) ◇ RE-4355

TOXICITY DATA with REFERENCE
skn-man 42 mg/21D-I open TXAPA9 21,369,72
skn-rbt 500 mg/24H SEV TXAPA9 21,369,72
mma-sat 50 μg/plate JAFCAU 29,268,81

mmo-bcs 50 μg/plate JAFCAU 29,268,81
mma-bcs 50 μg/plate JAFCAU 29,268,81
mmo-sat 5 μL/plate MUREAV 28,405,75
orl-rat LD50:250 mg/kg TXAPA9 14,515,69
ihl-rat LD50:7700 μg/kg BECTA6 19,113,78
skn-rat LD50:800 mg/kg WRPCA2 9,119,70
orl-mus LD50:330 mg/kg ARTODN 34,103,75
ihl-mus LD50:156 mg/kg BECTA6 19,113,78
skn-rbt LD50:1100 mg/kg 28ZEAL 5,161,76

OSHA PEL: (Transitional: TWA 3 mg/m^3) TWA 3 mg/m^3 (skin)
ACGIH TLV: TWA 3 mg/m^3 (skin)

SAFETY PROFILE: Poison by ingestion and inhalation. Moderately toxic by skin contact. A human skin irritant and a severe experimental skin irritant. Mutation data reported. A cholinesterase inhibitor. When heated to decomposition it emits very toxic fumes of Br^-, Cl^-, and PO_x. See also PARATHION.

NAG500 CAS:57-29-4 ***HR: 3***
NALORPHINE HYDROCHLORIDE
mf: $C_{19}H_{21}NO_3{\cdot}ClH$ mw: 347.87

PROP: Crystals from alc. Mp: 260-263°. Sol in water; mod sol in alc.

SYNS: N-ALLYL-2,12-DIHYDROXY-1,11-EPOXYMORPHINENE-13 HYDROCHLORIDE ◇ ALLYLMORPHINE HYDROCHLORIDE ◇ N-ALLYLNORMORPHINE HYDROCHLORIDE ◇ MIROMORFALIL ◇ NALINE HYDROCHLORIDE

TOXICITY DATA with REFERENCE
scu-rat TDLo:350 mg/kg (14D male):REP JRPFA4 14,481,67
ivn-hmn TDLo:143 μg/kg:CNS AJPSAO 113,887,57
orl-rat LD50:1150 mg/kg AIPTAK 123,48,59
scu-rat LD50:1460 mg/kg TXAPA9 18,185,71
ipr-mus LD50:447 mg/kg AIPTAK 103,489,55
scu-mus LD50:670 mg/kg CLDND*
ivn-dog LD50:120 mg/kg AIPTAK 149,571,64

SAFETY PROFILE: Poison by intravenous route. Moderately toxic by ingestion, subcutaneous, and intraperitoneal routes. Human systemic effects by intravenous route: euphoria, hallucinations and distorted perceptions, and toxic psychosis. Experimental reproductive effects. Caution: Violent withdrawal symptoms may occur in narcotic addicts. When heated to decomposition it emits very toxic fumes of NO_x and HCl. See also MORPHINE.

NAG550 CAS:465-65-6 ***HR: 3***
l-NALOXONE
mf: $C_{19}H_{21}NO_4$ mw: 327.41 P2U1

SYNS: l-N-ALLYL-7,8-DIHYDRO-14-HYDROXYNORMORPHINONE ◇ 17-ALLYL-4,5-α-EPOXY-3,14-DIHYDROXYMORPHINAN-6-ONE ◇ l-N-ALLYL-14-HYDROXYNORDIHYDROMORPHINONE ◇ 12-ALLYL-7,7a,8,9-TETRAHYDRO-3,7a-DIHYDROXY-4aH-8,9c-IMINOETHANOPHENANTHRO(4,5-bcd)FURANONE ◇ MORPHINAN-6-ONE, 4,5-α-EPOXY-3,14-DIHYDROXY-17-(2-PROPENYL)- ◇ NALOXONE ◇ NORMORPHINONE, N-ALLYL-7,8-DIHYDRO-14-HYDROXY-, (-)-

TOXICITY DATA with REFERENCE
dni-hmn:oth 30 umol/48H JJCREP 78,519,87
oth-hmn:oth 120 umol/L/24H JJCREP 78,519,87
imp-mus TDLo:500 μg/kg (female 17-21D post):REP PBBHAU 11,235,79
scu-mus LD50:260 mg/kg ANYAA9 281,321,76

SAFETY PROFILE: Poison by intraperitoneal route. Experimental reproductive effects. Mutation data reported. When heated to decomposition it emits toxic fumes of NO_x.

NAH000 CAS:357-08-4 ***HR: 3***
NALOXONE HYDROCHLORIDE
mf: $C_{19}H_{21}NO_4{\cdot}ClH$ mw: 363.87

PROP: Crystals from (ethanol + ether). Mp: 200-205°. Sol in water and alc; insol in ether.

SYNS: N-ALLYL-7,8-DIHYDRO-14-HYDROXYNORMORPHINONE, HYDROCHLORIDE ◇ 17-ALLYL-4,5-α-EPOXY-3,14-DIHYDROXYMORPHINAN-6-ONE HYDROCHLORIDE ◇ 1-N-ALLYL-14-HYDROXYNORDIHYDROMORPHINONE HYDROCHLORIDE ◇ EN-1530 ◇ EN-15304 ◇ NARCAN

TOXICITY DATA with REFERENCE
ipr-rat TDLo:560 mg/kg (7-20D preg):REP NTOTDY 3,295,81
ivn-dom TDLo:3 mg/kg (lactating female 2D post):TER BIREBV 39,532,88
ivn-wmn TDLo:32 μg/kg/3M-C:CVS,PUL,SKN AEMED3 16,1294,87
ipr-rat LD50:239 mg/kg IYKEDH 16,357,85
scu-rat LD50:500 mg/kg YKYUA6 36,655,85
ivn-rat LD50:107 mg/kg YKYUA6 36,655,85
ipr-mus LDLo:80 mg/kg RCOCB8 17,255,77
scu-mus LD50:286 mg/kg JMCMAR 20,844,77
ivn-mus LD50:90 mg/kg YKYUA6 36,655,85

SAFETY PROFILE: Poison by intraperitoneal, intravenous, and subcutaneous routes. Human systemic effects by intravenous route: pulse rate increase, acute pulmonary edema, sweating. An experimental teratogen. Experimental reproductive effects. When heated to decomposition it emits very toxic fumes of NO_x and HCl.

NAH500 CAS:835-31-4 ***HR: 3***
NAPHAZOLINE
mf: $C_{14}H_{14}N_2$ mw: 210.30

SYNS: ANTAN ◇ IMIDIN ◇ NAPHTHIZINE ◇ α-NAPHTHYLMETHYL IMIDAZOLINE ◇ 2-(α-NAPHTHYLMETHYL)-IMIDAZOLINE ◇ 2-(1-NAPHTHYLMETHYL)-2-IMIDAZOLINE ◇ PRIVINE ◇ RHINAZINE ◇ SANORIN

TOXICITY DATA with REFERENCE
scu-rat LD50:385 mg/kg FEPRA7 9,280,50
ipr-mus LD50:92 mg/kg JPETAB 160,22,68
scu-mus LD50:514 mg/kg RPTOAN 37,198,74
ivn-mus LD50:170 mg/kg FEPRA7 9,280,50
scu-rbt LD50:950 μg/kg FEPRA7 9,280,50
ivn-rbt LD50:800 μg/kg FEPRA7 9,280,50

SAFETY PROFILE: Poison by subcutaneous, intraperitoneal, and intravenous routes. When heated to decomposition it emits toxic fumes of NO_x.

NAH550 CAS:5144-52-5 ***HR: 3***
NAPHAZOLINE NITRATE
mf: $C_{14}H_{14}N_2 \cdot HNO_3$ mw: 273.32

SYNS: 4,5-DIHYDRO-2-(1-NAPHTHALENYLMETHYL)-1H-IMIDAZOLE MONONITRATE (9CI) ◇ NAFTIZIN ◇ NAPHTHISEN ◇ NAPHTHIZEN ◇ NAPHTYZIN ◇ PRIVINE ◇ PRIVINE NITRATE ◇ RINAZIN

TOXICITY DATA with REFERENCE
orl-rat LD50:1260 mg/kg FRPPAO 21,204,66
scu-rat LD50:385 mg/kg NIIRDN 6,539,82
orl-mus LD50:265 mg/kg FRPPAO 21,204,66
ipr-mus LD50:66 mg/kg FRPPAO 21,204,66
scu-mus LD50:170 mg/kg NIIRDN 6,539,82
ivn-mus LD50:13200 μg/kg FRPPAO 21,204,66
orl-rbt LD50:50 mg/kg NIIRDN 6,539,82
scu-rbt LD50:950 μg/kg NIIRDN 6,539,82
ivn-rbt LD50:800 μg/kg NIIRDN 6,539,82

SAFETY PROFILE: Poison by ingestion, subcutaneous, intravenous, and intraperitoneal routes. When heated to decomposition it emits toxic fumes of NO_x. See also NITRATES.

NAH800 CAS:550-34-5 ***HR: 3***
NAPHTHACAINE HYDROCHLORIDE
mf: $C_{17}H_{22}N_2O_2 \cdot ClH$ mw: 322.87

SYNS: 4-AMINO-1-NAPHTHOIC ACID 2-(DIETHYLAMINO)ETHYL ESTER HYDROCHLORIDE ◇ NAPHTHOCAINE HYDROCHLORIDE

TOXICITY DATA with REFERENCE
ipr-mus LD50:118 mg/kg JPETAB 94,299,48
scu-mus LD50:107 mg/kg JPETAB 70,315,40
scu-rbt LD50:54 mg/kg JPETAB 70,315,40

SAFETY PROFILE: Poison by subcutaneous and intraperitoneal routes. When heated to decomposition it emits toxic fumes of NO_x and HCl. See also ESTERS.

NAI000 CAS:92-24-0 ***HR: 3***
NAPHTHACENE
mf: $C_{18}H_{12}$ mw: 228.30

PROP: Orange crystals. Subl in vacuo. Mp: 341°, d: 1.35.

SYNS: BENZ(b)ANTHRACENE ◇ 2,3-BENZANTHRACENE ◇ 2,3-BENZANTHRENE ◇ TETRACENE

TOXICITY DATA with REFERENCE
mma-sat 5 μmol/L ENMUDM 3,11,81
dns-rat:lvr 500 μmol/L ENMUDM 3,11,81
dnd-mam:lym 20 mg BIPMAA 4,409,66

CONSENSUS REPORTS: Reported in EPA TSCA Inventory.

SAFETY PROFILE: Mutation data reported. Explosion hazard; shock will explode it. Can react on contact with oxidizing materials. When heated to decomposition it emits acrid smoke and irritating fumes.

NAI500 CAS:8030-30-6 ***HR: 3***
NAPHTHA
DOT: UN 1255/UN 1256/UN 2553

PROP: Dark straw-colored to colorless liquid. Bp: 149-216°, flash p: 107°F (CC), d: 0.862-0.892, autoign temp: 531°F. Sol in benzene, toluene, xylene, etc. Made from American coal oil and consists chiefly of pentane, hexane, and heptane (XPHPAW 255,43,40).

SYNS: AROMATIC SOLVENT ◇ BENZIN ◇ COAL TAR NAPHTHA ◇ HI-FLASH NAPHTHAETHYLEN ◇ NAPHTA (DOT) ◇ NAPHTHA DISTILLATE (DOT) ◇ NAPHTHA PETROLEUM (DOT) ◇ NAPHTHA, SOLVENT (DOT) ◇ PETROLEUM BENZIN ◇ PETROLEUM DISTILLATES (NAPHTHA) ◇ PETROLEUM ETHER (DOT) ◇ PETROLEUM NAPHTHA (DOT) ◇ PETROLEUM SPIRIT (DOT) ◇ SKELLY-SOLVE-F ◇ VM & P NAPHTHA (ACGIH)

TOXICITY DATA with REFERENCE
ihl-hmn LCLo:3 pph/5M TABIA2 3,231,33
ivn-man LDLo:27 mg/kg:PUL CTOXAO 16,335,80
ihl-rat LCLo:1600 ppm/6H CHINAG (17),1078,39
ipr-mam LDLo:2500 mg/kg AJHYA2 7,276,27

CONSENSUS REPORTS: Reported in EPA TSCA Inventory.

OSHA PEL: TWA 100 ppm
ACGIH TLV: TWA 300 ppm
NIOSH REL: (Refined Petroleum Solvents): 10H TWA 350 mg/m^3; CL 1800 mg/m^3/15M
DOT Classification: Flammable Liquid; Label: Flammable Liquid.

SAFETY PROFILE: A human poison via intravenous route. Human systemic effects by intravenous route: dyspnea, respiratory stimulation, and other unspecified respiratory effects. Mildly toxic by inhalation. Can cause unconsciousness which may go into coma, stentorious breathing, and bluish tint to the skin. Recovery follows removal from exposure. In mild form, intoxication resembles drunkenness. On a chronic basis, no true poisoning; sometimes headache, lack of appetite, dizziness, sleeplessness, indigestion, and nausea. A common air contaminant. Flammable when exposed to heat or flame; can react with oxidizing materials. Keep containers tightly closed. Slight explosion hazard. To fight fire, use foam, CO_2, dry chemical. See also NAPHTHA, PENTANE, HEXANE, and HEPTANE.

NAJ000 CAS:66-77-3 ***HR: 3***
α-NAPHTHAL

TOXICITY DATA with REFERENCE
scu-dog LDLo:330 mg/kg ZMWIAJ 19,545,1881

CONSENSUS REPORTS: Reported in EPA TSCA Inventory.

SAFETY PROFILE: Poison by subcutaneous route. When heated to decomposition it emits acrid smoke and irritating fumes.

NAJ500 CAS:91-20-3 ***HR: 3***
NAPHTHALENE
DOT: UN 1334/UN 2304
mf: $C_{10}H_8$ mw: 128.18

PROP: Aromatic odor; white, crystalline, volatile flakes. Mp: 80.1°, bp: 217.9°, flash p: 174°F (OC), d: 1.162, lel: 0.9%, uel: 5.9%, vap press: 1 mm @ 52.6°, vap d: 4.42, autoign temp: 1053°F (567°C). Sol in alc, benzene; insol in water; very sol in ether, CCl_4, CS_2, hydronaphthalenes, in fixed and volatile oils.

SYNS: CAMPHOR TAR ◇ MIGHTY 150 ◇ MOTH BALLS (DOT) ◇ MOTH FLAKES ◇ NAFTALEN (POLISH) ◇ NAPHTHALENE, crude or refined (DOT) ◇ NAPHTHALENE, molten (DOT) ◇ NAPHTHALIN (DOT) ◇ NAPHTHALINE ◇ NAPHTHENE ◇ NCI-C52904 ◇ RCRA WASTE NUMBER U165 ◇ TAR CAMPHOR ◇ WHITE TAR

TOXICITY DATA with REFERENCE
skn-rbt 495 mg open MLD UCDS** 1/11/68
eye-rbt 100 mg MLD BIOFX* 16-4/70
orl-mus TDLo:2400 mg/kg (7-14D preg):REP JTEHD6 15,25,85
ipr-rat TDLo:5925 mg/kg (1-15D preg):TER TXAPA9 48,A35,79
scu-rat TDLo:3500 mg/kg/12W-I:ETA APAVAY 329,141,56
orl-chd LDLo:100 mg/kg 28ZRAQ -,228,60
unr-hmn LDLo:29 mg/kg YKYUA6 31,1499,80
unr-man LDLo:74 mg/kg 85DCAI 2,73,70
orl-rat LD50:490 mg/kg 85GMAT-,89,82
orl-mus LD50:533 mg/kg FAATDF 4(3, Pt 1),406,84
ipr-mus LD50:150 mg/kg NTIS** AD691-490
scu-mus LD50:969 mg/kg TOIZAG 20,772,73
ivn-mus LD50:100 mg/kg CSLNX* NX#00203
orl-dog LDLo:400 mg/kg HBAMAK 4,1289,35
orl-cat LDLo:1000 mg/kg HBAMAK 4,1289,35
orl-rbt LDLo:3 g/kg HBAMAK 4,1289,35
orl-gpg LD50:1200 mg/kg GISAAA 47(11),78,82

CONSENSUS REPORTS: Reported in EPA TSCA Inventory. EPA Genetic Toxicology Program. Community Right-To-Know List.

OSHA PEL: (Transitional: TWA 10 ppm) TWA 10 ppm; STEL 15 ppm
ACGIH TLV: TWA 10 ppm; STEL 15 ppm
DFG MAK: 10 ppm (50 mg/m^3)
DOT Classification: ORM-A; Label: None; Flammable Solid; Label: Flammable Solid.

SAFETY PROFILE: Human poison by ingestion. Experimental poison by ingestion, intravenous, and intraperitoneal routes. Moderately toxic by subcutaneous route. An experimental teratogen. Experimental reproductive effects. An eye and skin irritant. Can cause nausea, headache, diaphoresis, hematuria, fever, anemia, liver damage, vomiting, convulsions, and coma. Poisoning may occur by ingestion of large doses, inhalation, or skin absorption. Questionable carcinogen with experimental tumorigenic data. Flammable when exposed to heat or flame; reacts with oxidizing materials. Explosive reaction with dinitrogen pentaoxide. Reacts violently with CrO_3; aluminum chloride + benzoyl chloride. Fires in the benzene scrubbers of coke oven gas plants have been attributed to oxidation of naphthalene. Explosive in the form of vapor or dust when exposed to heat or flame. To fight fire, use water, CO_2, dry chemical. When heated to decomposition it emits acrid smoke and irritating fumes.

NAK000 CAS:86-86-2 ***HR: 2***
1-NAPHTHALENEACETAMIDE
mf: $C_{12}H_{11}NO$ mw: 185.24

SYNS: N-ACETYL-1-NAPHTHYLAMINE ◇ AMID-THIN ◇ FRUITONE ◇ NAAM ◇ NAD ◇ NAPHTHALENE ACETAMIDE ◇ α-NAPHTHALENEACETAMIDE ◇ α-NAPHTHYLACETAMIDE ◇ 1-NAPHTHYLACETAMIDE ◇ ROOTONE ◇ ROSETONE

TOXICITY DATA with REFERENCE
orl-mam LD50:1000 mg/kg FMCHA2 -,C165,83

CONSENSUS REPORTS: Reported in EPA TSCA Inventory.

SAFETY PROFILE: Moderately toxic by ingestion. When heated to decomposition it emits toxic fumes of NO_x.

NAK500 CAS:86-87-3 ***HR: 3***
1-NAPHTHALENEACETIC ACID
mf: $C_{12}H_{10}O_2$ mw: 186.22

PROP: Needles from water; white, odorless crystals. Mp: 134.5-135.5°. Only sltly water-sol; sol in approx 30 parts alc; very sol in acetone, ether, chloroform.

SYNS: AGRONAA ◇ ALPHASPRA ◇ ANA ◇ APPL-SET ◇ CELMONE ◇ FRUITONE ◇ KLINGTITE ◇ LIQUI-STIK ◇ NAA 800

◇ NAFUSAKU ◇ NAPHTHALENE-1-ACETIC ACID ◇ α-NAPHTHALENEACETIC ACID ◇ α-NAPHTHYLACETIC ◇ NAPHTHYLACETIC ACID ◇ α-NAPHTHYLACETIC ACID ◇ 1-NAPHTHYLACETIC ACID ◇ α-NAPHTHYLENEACETIC ACID ◇ α-NAPHTHYLESSIGSAEURE (GERMAN) ◇ NAPHYL-1-ESSIGSAEURE (GERMAN) ◇ NIAGARA-STIK ◇ NU-TONE ◇ PARMONE ◇ PHYMONE ◇ PIMACOL-SOL ◇ PLANOFIX ◇ PLUCKER ◇ PRIMACOL ◇ ROOTONE ◇ STAFAST ◇ STIK ◇ STOP-DROP ◇ TEKKAM ◇ TIP-OFF ◇ TRANSPLANTONE ◇ TRE-HOLD ◇ VARDHAK

TOXICITY DATA with REFERENCE
eye-rbt 100 mg SEV PESTC* 9,10,80
mmo-smc 500 mg/L IDZAAW 48,185,73
orl-rat TDLo:27 g/kg (90D male):REP TXCYAC 5,371,76
orl-rat LD50:1000 mg/kg 85DPAN -,-,71/76
ipr-rat LD50:100 mg/kg PESTC* 9,10,80
orl-mus LD50:743 mg/kg TOIZAG 14,132,67
ipr-mus LD50:609 mg/kg TOIZAG 15,443,68

CONSENSUS REPORTS: Reported in EPA TSCA Inventory.

SAFETY PROFILE: Poison by intraperitoneal route. Moderately toxic by ingestion. Experimental reproductive effects. Mutation data reported. A skin, mucous membrane and severe eye irritant. Can cause depression. A pesticide. When heated to decomposition it emits acrid smoke and irritating fumes.

NAM000 CAS:2243-62-1 ***HR: 3***
1,5-NAPHTHALENEDIAMINE
mf: $C_{10}H_{10}N_2$ mw: 158.22

SYNS: 1,5-DIAMINONAPHTHALENE ◇ 1,5-NAPHTHYLENEDIAMINE ◇ NCI-C03021

TOXICITY DATA with REFERENCE
mma-sat 33300 ng/plate ENMUDM 7(Suppl 5),1,85
mma-sat 33300 ng/plate ENMUDM 7(Suppl 5),1,85
otr-rat:emb 5200 ng/plate JJATDK 1,190,81
orl-rat TDLo:721 g/kg/1Y-C:CAR IARC** 27,127,82
orl-mus TDLo:120 g/kg/2Y-C:CAR NCITR* NCI-CG-TR-143,78
orl-rat TD:18 g/kg/88W-C:ETA,REP NCITR* NCI-CG-TR-143,78
orl-rat TDLo:54 g/kg/2Y-C:NEO,REP NCITR* NCI-CG-TR-143,78
orl-rat TDLo:361 g/kg/1Y-C:ETA IARC** 27,127,82

CONSENSUS REPORTS: IARC Cancer Review: Group 3 IMEMDT 7,56,87; Animal Limited Evidence IMEMDT 27,127,82. NCI Carcinogenesis Bioassay (feed); Clear Evidence: mouse, rat NCITR* NCI-CG-TR-143,78. EPA Genetic Toxicology Program.

SAFETY PROFILE: Suspected carcinogen with experimental carcinogenic, neoplastigenic, tumorigenic data. Experimental reproductive effects. Mutation data reported. When heated to decomposition it emits toxic fumes of NO_x. See also AMINES.

NAM500 CAS:3173-72-6 ***HR: 2***
1,5-NAPHTHALENE DIISOCYANATE
mf: $C_{12}H_6N_2O_2$ mw: 210.20

PROP: White to light yellow crystals.

SYNS: 1,5-DIISOCYANATONAPHTHALENE ◇ ISOCYANIC ACID-1,5-NAPHTHYLENE ESTER

CONSENSUS REPORTS: IARC Cancer Review: Group 3 IMEMDT 7,56,87. Reported in EPA TSCA Inventory.

DFG MAK: 0.01 ppm (0.09 mg/m^3)

SAFETY PROFILE: A powerful allergen. An irritant. Questionable carcinogen. When heated to decomposition it emits toxic fumes of NO_x. See also ISOCYANATES.

NAN000 CAS:132-86-5 ***HR: 3***
1,3-NAPHTHALENEDIOL
mf: $C_{10}H_8O_2$ mw: 160.18

PROP: White, crystalline leaves. Mp: 125°.

SYN: NAPHTHORESORCINOL

TOXICITY DATA with REFERENCE
mmo-sat 250 μmol/L ENMUDM 3,11,81
ipr-mus LD50:200 mg/kg NTIS** AD691-490

CONSENSUS REPORTS: Reported in EPA TSCA Inventory.

SAFETY PROFILE: Poison by intraperitoneal route. Mutation data reported. When heated to decomposition it emits acrid smoke and irritating fumes.

NAN500 CAS:571-60-8 ***HR: 3***
1,4-NAPHTHALENEDIOL
mf: $C_{10}H_8O_2$ mw: 160.18

TOXICITY DATA with REFERENCE
orl-rat LDLo:100 mg/kg NCNSA6 5,35,53

CONSENSUS REPORTS: Reported in EPA TSCA Inventory.

SAFETY PROFILE: Poison by ingestion. When heated to decomposition it emits acrid smoke and irritating fumes.

NAO500 CAS:582-17-2 ***HR: 3***
2,7-NAPHTHALENEDIOL
mf: $C_{10}H_8O_2$ mw: 160.18

SYNS: C.I. 76645 ◇ 2,7-DIHYDROXYNAPHTHALENE ◇ NAPHTHALENE-2,7-DIOL ◇ NAPHTHALENEDIOL-2,7 (FRENCH)

TOXICITY DATA with REFERENCE
ipr-mus LD50:102 mg/kg EJMCA5 13,381,78

CONSENSUS REPORTS: Reported in EPA TSCA Inventory.

SAFETY PROFILE: Poison by intraperitoneal route. When heated to decomposition it emits acrid smoke and irritating fumes.

NAP000 CAS:120-18-3 ***HR: 3***
2-NAPHTHALENESULFONIC ACID
mf: $C_{10}H_8O_3S$ mw: 208.24

SYNS: β-NAPHTHALENESULFONIC ACID ◇ NAPHTHALENE-2-SULFONIC ACID ◇ β-NAPHTHYLSULFONIC ACID

TOXICITY DATA with REFERENCE
orl-rat LD50:400 mg/kg 14CYAT 2,1842,63

CONSENSUS REPORTS: Reported in EPA TSCA Inventory.

SAFETY PROFILE: Poison by ingestion. When heated to decomposition it emits toxic fumes of SO_x. See also SULFONATES.

NAP300 CAS:81-30-1 ***HR: 3***
1,4,5,8-NAPHTHALENETETRACARBOXYLIC ACID-1,8:4,5-DIANHYDRIDE
mf: $C_{14}H_4O_6$ mw: 268.18

TOXICITY DATA with REFERENCE
orl-rat LD50:7400 mg/kg 85GMAT -,90,82
itr-rat LDLo:50 mg/kg 85GMAT -,90,82
orl-mus LD50:7100 mg/kg 85GMAT -,90,82

SAFETY PROFILE: Poison by intratracheal route. Mildly toxic by ingestion. When heated to decomposition it emits acrid smoke and irritating fumes. See also ANHYDRIDES.

NAP500 CAS:91-60-1 ***HR: 3***
2-NAPHTHALENETHIOL
mf: $C_{10}H_8S$ mw: 160.24

PROP: Crystals from ethanol, disagreeable odor. Mp: 81°, bp: 286°. Very sol in ethanol, ether, petr ether; sltly sol in water; sltly volatile with steam.

SYNS: 2-MERCAPTONAPHTHALENE ◇ NAPHTHALENE-2-THIOL ◇ β-NAPHTHYL MERCAPTAN ◇ 2-NAPHTHYL MERCAPTAN ◇ 2-NAPHTHYL THIOL ◇ THIO-β-NAPHTHOL ◇ β-THIONAPHTHOL ◇ USAF CY-4

TOXICITY DATA with REFERENCE
orl-mus LD50:385 mg/kg DCTODJ 3,249,80
ipr-mus LD50:200 mg/kg NTIS** AD277-689

CONSENSUS REPORTS: Reported in EPA TSCA Inventory.

SAFETY PROFILE: Poison by ingestion and intraperitoneal routes. A mosquito larvicide. When heated to decomposition it emits highly toxic fumes of SO_x. See also MERCAPTANS.

NAQ000 CAS:81-84-5 ***HR: 1***
1,8-NAPHTHALIC ANHYDRIDE
mf: $C_{12}H_6O_3$ mw: 198.18

PROP: Needles in alcohol. Mp: 273-274°. Sltly sol in acetic acid; very sol in alc and ether.

SYN: PROTECT

TOXICITY DATA with REFERENCE
orl-rat LD50:12340 mg/kg FMCHA2 -,C199,83

CONSENSUS REPORTS: Reported in EPA TSCA Inventory.

SAFETY PROFILE: Mildly toxic by ingestion. When heated to decomposition it emits acrid smoke and irritating fumes. See also ANHYDRIDES.

NAQ500 CAS:2668-92-0 ***HR: 3***
NAPHTHALOXIMIDODIETHYL THIOPHOSPHATE
mf: $C_{16}H_{16}NO_5PS$ mw: 365.36

SYNS: O,O-DIETHYL-o-NAPHTHALIMIDEPHOSPHOROTHIOATE ◇ O,O-DIETHYL-o-NAPHTHALOXIMIDO PHOSPHOROTHIOATE ◇ O,O-DIETHYL-o-NAPHTHALOXIMIDOPHOSPHOROTHIONATE ◇ O,O-DIETHYL-o-NAPHTHYLAMIDOPHOSPHOROTHIOATE ◇ ENT 24,970 ◇ N-HYDROXYNAPHTHALIMIDE-O,O-DIETHYL PHOSPHOROTHIOATE ◇ NAPHTHALOXIMIDE-O,O-DIETHYL PHOSPHOROTHIOATE ◇ PHOSPHOROTHIOIC ACID-O,O-DIETHYL ESTER, -o-NAPHTHALIMIDO DERIVATIVE ◇ PHOSPHOROTHIOIC ACID-O,O-DIETHYL-o-NAPHTHYLAMIDOESTER

TOXICITY DATA with REFERENCE
orl-rat LDLo:500 mg/kg TXAPA9 21,315,72
orl-ckn LD50:3169 mg/kg TXAPA9 3,521,61
orl-bwd LD50:24 mg/kg TXAPA9 21,315,72

SAFETY PROFILE: Poison by ingestion. When heated to decomposition it emits very toxic SO_x, PO_x, and NO_x. See also ESTERS.

NAR000 CAS:1338-24-5 ***HR: 2***
NAPHTHENIC ACID

PROP: Odorless crystals. D: 1.034, mp: 31°, bp: 233°. Sltly water-sol.

SYNS: AGENAP ◇ NAPHID ◇ SUNAPTIC ACID B ◇ SUNAPTIC ACID C

TOXICITY DATA with REFERENCE
orl-rat LD50:3000 mg/kg AMIHAB 12,477,55
ipr-rat LD50:640 mg/kg AMIHAB 12,477,55

CONSENSUS REPORTS: Reported in EPA TSCA Inventory.

SAFETY PROFILE: Moderately toxic by ingestion and

intraperitoneal routes. When heated to decomposition it emits acrid smoke and irritating fumes.

NAR500 CAS:61789-51-3 *HR: 3*
NAPHTHENIC ACID, COBALT SALT
DOT: UN 2001

PROP: Brown, amorph powder or bluish-red solid. Flash p: 120°F, d: 0.9, autoign temp: 529°F. Water-insol; sol in oil, alc, ether. Contains 6% cobalt (AMIHAB 12,477,55).

SYNS: COBALT NAPHTHENATE, POWDER (DOT) ◇ NAPHTHENATE de COBALT (FRENCH)

TOXICITY DATA with REFERENCE
orl-rat LD50:3900 mg/kg AMIHAB 12,477,55

CONSENSUS REPORTS: Cobalt and its compounds are on the Community Right-To-Know List. Reported in EPA TSCA Inventory.

DOT Classification: Flammable Solid; Label: Flammable Solid.

SAFETY PROFILE: Moderately toxic by ingestion. Flammable when exposed to heat or flame. When heated to decomposition it emits acrid smoke and irritating fumes. See also COBALT COMPOUNDS.

NAS000 CAS:1338-02-9 *HR: 3*
NAPHTHENIC ACID, COPPER SALT

PROP: A solid. Flash p: 100°F, d: 1.055. Contains 8% copper (AMIHAB 12,477,55).

SYNS: COPPER NAPHTHENATE ◇ COPPER UVERSOL ◇ CUPRINOL ◇ WITTOX C

TOXICITY DATA with REFERENCE
orl-mus LDLo:110 mg/kg SCCUR* -,3,61

CONSENSUS REPORTS: Copper and its compounds are on the Community Right-To-Know List. Reported in EPA TSCA Inventory.

SAFETY PROFILE: Poison by ingestion. A pesticide. A dangerous fire hazard when exposed to heat or flame; can react with oxidizing materials. To fight fire, use foam, CO_2, dry chemical. See also COPPER COMPOUNDS.

NAS500 CAS:61790-14-5 *HR: 3*
NAPHTHENIC ACID, LEAD SALT
mf: $C_7H_{12}O_2 \cdot xPb$ mw: 1578.52

PROP: Contains 24% lead (AMIHAB 12,477,55).

SYNS: CYCLOHEXANECARBOXYLIC ACID, LEAD SALT ◇ LEAD NAPHTHENATE

TOXICITY DATA with REFERENCE
skn-mus TDLo:50 g/kg/29W-I:ETA BECCAN 39,420,61
orl-rat LD50:5100 mg/kg AMIHAB 12,477,55
ipr-rat LD50:520 mg/kg AMIHAB 12,477,55

CONSENSUS REPORTS: IARC Cancer Review: Animal Inadequate Evidence IMEMDT 23,325,80. Lead and its compounds are on the Community Right-To-Know List. Reported in EPA TSCA Inventory.

SAFETY PROFILE: A poison. Moderately toxic by intraperitoneal route. Mildly toxic by ingestion. Questionable carcinogen with experimental tumorigenic data. When heated to decomposition it emits toxic fumes of lead. See also LEAD COMPOUNDS.

NAT000 CAS:12001-85-3 *HR: 1*
NAPHTHENIC ACID, ZINC SALT

PROP: A solid. Contains 8% zinc (AMIHAB 12,477,55).

SYNS: ZINC NAPHTHENATE ◇ ZINC UVERSOL

TOXICITY DATA with REFERENCE
orl-rat LD50:4920 mg/kg AIHAAP 30,470,69

CONSENSUS REPORTS: Zinc and its compounds are on the Community Right-To-Know List. Reported in EPA TSCA Inventory.

SAFETY PROFILE: Mildly toxic by ingestion. Combustible when exposed to heat or flame. When heated to decomposition it emits toxic fumes of ZnO. A fungicide and mildew preventive. See also ZINC COMPOUNDS.

NAT500 CAS:192-65-4 *HR: 3*
NAPHTHO(1,2,3,4-def)CHRYSENE
mf: $C_{24}H_{14}$ mw: 302.38

SYNS: DB(a,e)P ◇ DIBENZO(a,e)PYRENE ◇ 1,2,4,5-DIBENZOPYRENE

TOXICITY DATA with REFERENCE
skn-mus TDLo:312 mg/kg/52W-I:NEO ZEKBAI 68,137,66
scu-mus TDLo:72 mg/kg/9W-I:ETA COREAF 259,3899,64

CONSENSUS REPORTS: NTP Fifth Annual Report on Carcinogens. IARC Cancer Review: Group 2B IMEMDT 7,56,87; Animal Sufficient Evidence IMEMDT 32,327,83; IMEMDT 3,201,73.

SAFETY PROFILE: Confirmed carcinogen with experimental neoplastigenic and tumorigenic data. When heated to decomposition it emits acrid smoke and irritating fumes.

NAU000 CAS:16566-64-6 *HR: 3*
NAPHTHO(1,8-gh:4,5-g'h')DIQUINOLINE
mf: $C_{22}H_{12}N_2$ mw: 304.36

SYNS: 1,12-DIAZADIBENZO(a,i)PYRENE ◇ DIPYRIDO(2,3-d,2,3-1)PYRENE

TOXICITY DATA with REFERENCE
scu-mus TDLo:72 mg/kg/9W-I:ETA BJCAAI 26,262,72

SAFETY PROFILE: Questionable carcinogen with experimental tumorigenic data. When heated to decomposition it emits toxic fumes of NO_x.

NAU500 CAS:16566-62-4 ***HR: 3***
NAPHTHO(1,8-gh:5,4-g'h')DIQUINOLINE
mf: $C_{22}H_{12}N_2$ mw: 304.36

SYNS: 4,11-DIAZADIBENZO(a,h)PYRENE ◇ DIPYRIDO(2,3-D,2,3-K)PYRENE

TOXICITY DATA with REFERENCE
ipr-rat TDLo:100 mg/kg:ETA NEOLA4 26,23,79

SAFETY PROFILE: Questionable carcinogen with experimental tumorigenic data. When heated to decomposition it emits toxic fumes of NO_x.

NAU525 CAS:6051-87-2 ***HR: 3***
β-NAPHTHOFLAVONE
mf: $C_{19}H_{12}O_2$ mw: 272.31

SYNS: 1H-NAPHTHO(2,1-b)PYRAN-1-ONE, 3-PHENYL- ◇ β-NF ◇ 3-PHENYL-1H-NAPHTHO(2,1-b)PYRAN-1-ONE

TOXICITY DATA with REFERENCE
ipr-rat TDLo:120 mg/kg (female 8-10D post):TER TJADAB 24,1,81
ipr-rat TDLo:120 mg/kg (female 8-10D post):REP TJADAB 24,1,81
ivn-mus LD50:180 mg/kg CSLNX* NX#03239

SAFETY PROFILE: Poison by intraperitoneal route. An experimental teratogen. Experimental reproductive effects. When heated to decomposition it emits acrid smoke and irritating fumes.

NAV000 CAS:10335-79-2 ***HR: D***
2-NAPHTHOHYDROXAMIC ACID
mf: $C_{11}H_9NO_2$ mw: 187.21

SYNS: N-HYDROXY-2-NAPHTHALENECARBOXAMIDE ◇ 2-NAPHTHOHYDROXIMIC ACID ◇ 2-NAPHTHOYLHYDROXAMIC ACID ◇ 2-NAPHTHYLHYDROXAMIC ACID

TOXICITY DATA with REFERENCE
mmo-sat 250 nmol/plate MUREAV 135,139,84
sce-ham:ovr 20 μmol/L PAACA3 21,126,80

CONSENSUS REPORTS: EPA Genetic Toxicology Program.

SAFETY PROFILE: Mutation data reported. When heated to decomposition it emits toxic fumes of NO_x.

NAV500 CAS:93-09-4 ***HR: 2***
2-NAPHTHOIC ACID
mf: $C_{11}H_8O_2$ mw: 172.19

PROP: Two forms: α and β. α: Needles, sltly sol in hot water; sol in hot alc and ether. β: Plates or needles, sltly sol in hot water; sol in alc and ether. Mp (α): 160-161°, mp (β): 184-185°, bp (α): 300°, bp (β): >300°.

SYNS: ISONAPHTHOIC ACID ◇ 2-MAYTHIC ACID ◇ 2-NAPHTHALENECARBOXYLIC ACID ◇ NAPHTHALENE-β-CARBOXYLIC ACID ◇ β-NAPHTHOIC ACID

TOXICITY DATA with REFERENCE
orl-rat LD50:4500 mg/kg 85GMAT -,90,82
orl-mus LD50:4700 mg/kg 85GMAT -,90,82
ipr-mus LDLo:500 mg/kg CBCCT* 4,379,52

CONSENSUS REPORTS: Reported in EPA TSCA Inventory.

SAFETY PROFILE: Moderately toxic by intraperitoneal route. Mildly toxic by ingestion. A skin and mucous membrane irritant. When heated to decomposition it emits acrid smoke and irritating fumes.

NAW000 CAS:1321-67-1 ***HR: 3***
NAPHTHOL
mf: $C_{10}H_8O$ mw: 144.18

PROP: Monoclinic. D: 1.217 @ 4°, mp: 122-123°, bp: 285-286°. Sol in cold and hot water and in chloroform; very sol in alc and benzene.

TOXICITY DATA with REFERENCE
scu-dog LDLo:333 mg/kg ZMWIAJ 19,545,1881
scu-rbt LDLo:1000 mg/kg ZMWIAJ 19,545,1881

SAFETY PROFILE: Poison by subcutaneous route. When heated to decomposition it emits acrid smoke and irritating fumes.

NAW500 CAS:90-15-3 ***HR: 3***
1-NAPHTHOL
mf: $C_{10}H_8O$ mw: 144.18

PROP: Colorless crystals; odor of phenol, disagreeable taste. Mp: 96°, bp: 282.5°, d: 1.0954 @ 98.7°/4°, vap press: 1 mm @ 94.0°. Very sltly sol in water; sol in alc and ether.

SYNS: BASF URSOL ERN ◇ C.I. 76605 ◇ C.I. OXIDATION BASE 33 ◇ DURAFUR DEVELOPER D ◇ FOURAMINE ERN ◇ FOURRINE 99 ◇ FOURRINE ERN ◇ FURRO ER ◇ α-HYDROXYNAPHTHALENE ◇ 1-HYDROXYNAPHTHALENE ◇ NAKO TRB ◇ 1-NAPHTHALENOL ◇ α-NAPHTHOL ◇ TERTRAL ERN ◇ URSOL ERN ◇ ZOBA ERN

TOXICITY DATA with REFERENCE
skn-rbt 500 mg/24H SEV BIOFX* 25-4/73
skn-rbt 550 mg open MOD UCDS** 11/10/64
eye-rbt 1 mg SEV UCDS** 11/10/64
mma-sat 500 μg/plate BJCAAI 37,873,78
dnr-esc 20 μL/disc MUREAV 97,1,82
scu-mus TDLo:90 mg/kg (6-14D preg):TER NTIS** PB223-160

scu-mus TDLo:90 mg/kg (6-14D preg):REP NTIS** PB223-160
orl-mus LD50:275 mg/kg GISAAA 30(9),22,65
orl-cat LD50:134 mg/kg GISAAA 30(9),22,65
orl-rbt LD50:9 g/kg GISAAA 30(9),22,65
skn-rbt LD50:880 mg/kg AIHAAP 23,95,62

CONSENSUS REPORTS: Reported in EPA TSCA Inventory. EPA Genetic Toxicology Program.

SAFETY PROFILE: Poison by ingestion. Moderately toxic by skin contact. An experimental teratogen. Experimental reproductive effects. A severe eye and skin irritant. Mutation data reported. Ingestion of large amounts can cause nephritis, vomiting, diarrhea, circulatory collapse, anemia, convulsions, and death. Can cause kidney irritation and injury to cornea and lens of the eye. Combustible when exposed to heat or flame. When heated to decomposition it emits acrid smoke and irritating fumes.

NAX000 CAS:135-19-3 ***HR: 3***
2-NAPHTHOL
mf: $C_{10}H_8O$ mw: 144.18

PROP: White to yellowish-white crystals; slt phenolic odor. Flash p: 307°F, vap press: 10 mm @ 145.5°, vap d: 4.97 mp: 121-123°, bp: 285-286°, d: 1.22. Darkens with age or exposure to light. Subl when heated; distills in vacuum. Sltly sol in water; more sol in boiling water, glycerol, olive oil, solns of alkali hydroxides. Very sol in alc, ether; sol in chloroform.

SYNS: AZOGEN DEVELOPER A ◇ C.I. 37500 ◇ C.I. AZOIC COUPLING COMPONENT 1 ◇ C.I. DEVELOPER 5 ◇ DEVELOPER A ◇ DEVELOPER AMS ◇ DEVELOPER BN ◇ DEVELOPER SODIUM ◇ β-HYDROXYNAPHTHALENE ◇ 2-HYDROXYNAPHTHALENE ◇ ISONAPHTHOL ◇ β-MONOXYNAPHTHALENE ◇ β-NAFTOL (DUTCH) ◇ 2-NAFTOL (DUTCH) ◇ β-NAFTOLO (ITALIAN) ◇ 2-NAFTOLO (ITALIAN) ◇ 2-NAPHTHALENOL ◇ NAPHTHOL B ◇ β-NAPHTHOL ◇ β-NAPHTHYL ALCOHOL ◇ β-NAPHTHYL HYDROXIDE ◇ 2-NAPHTOL (FRENCH) ◇ β-NAPHTOL (GERMAN)

TOXICITY DATA with REFERENCE
skn-rbt 500 mg/24H MLD BIOFX* 26-4/73
eye-rbt 100 mg MOD BIOFX* 26-4/73
dnr-esc 20 μL/disc MUREAV 97,1,82
scu-rat LDLo:2940 mg/kg AEPPAE 186,195,37
orl-rat LD50:1960 mg/kg GTPPAF 8,145,72
orl-mus LDLo:100 mg/kg HBAMAK 4,1289,35
ipr-mus LD50:97500 mg/kg EJMCA5 13,381,78
scu-mus LDLo:100 mg/kg ZHINAV 64,113,1909
orl-cat LDLo:100 mg/kg HBAMAK 4,1289,35
scu-rbt LDLo:3 g/kg HBAMAK 4,1289,35
scu-gpg LDLo:2670 mg/kg AEPPAE 186,195,37

CONSENSUS REPORTS: Reported in EPA TSCA Inventory. EPA Genetic Toxicology Program.

SAFETY PROFILE: Poison by ingestion and subcutaneous routes. Mutation data reported. A skin and eye irritant. Combustible when exposed to heat or flame. To fight fire, use CO_2 dry chemical. Incompatible with antipyrine; camphor; phenol; ferric salts; menthol; potassium permanganate and other oxidizing materials; urethane. When heated to decomposition it emits toxic fumes of Na_2O. See also α-NAPHTHOL.

NAX500 CAS:19381-50-1 ***HR: 2***
NAPHTHOL GREEN B
mf: $C_{30}H_{15}FeN_3O_{15}S_3 \cdot 3Na$ mw: 878.48

SYNS: C.I. ACID GREEN 1 ◇ D&C GREEN 1 ◇ EXT D&C GREEN No. 1

TOXICITY DATA with REFERENCE
ivn-mus LD50:460 mg/kg TXAPA9 44,225,78

CONSENSUS REPORTS: Reported in EPA TSCA Inventory.

SAFETY PROFILE: Moderately toxic by intravenous route. When heated to decomposition it emits very toxic fumes of SO_x and NO_x.

NAY000 CAS:6471-49-4 ***HR: D***
NAPHTHOL RED B
mf: $C_{24}H_{17}N_5O_7$ mw: 487.46

SYNS: ALKALI RESISTANT RED DARK ◇ CALCOTONE RED 3B ◇ CARNATION RED TONER B ◇ C.I. 12355 ◇ C.I. PIGMENT RED 23 ◇ CONGO RED R-138 ◇ FENALAC RED FKB EXTRA ◇ MALTA RED X 2284 ◇ NAPHTHOL RED B 20-7575 ◇ NAPHTHOL RED DEEP 10459 ◇ NAPHTHOL RED D TONER 35-6001 ◇ NCI-C60377 ◇ PIGMENT RED 23 ◇ PIGMENT RED BH ◇ RUBESCENCE RED MT-21 ◇ SANYO FAST RED 10B ◇ SAPONA RED LAKE RL-6280 ◇ SEGNALE LIGHT RUBINE RG ◇ TEXTILE RED WD-263

TOXICITY DATA with REFERENCE
mmo-sat 33 μg/plate ENMUDM 8(Suppl 7),1,86
mma-sat 33 μg/plate ENMUDM 8(Suppl 7),1,86

CONSENSUS REPORTS: Reported in EPA TSCA Inventory.

SAFETY PROFILE: Mutation data reported. When heated to decomposition it emits toxic fumes of NO_x.

NAY500 CAS:72017-28-8 ***HR: 2***
NAPHTHOPYRIN
mf: $C_{21}H_{18}N_3$ mw: 312.42

TOXICITY DATA with REFERENCE
scu-rat LDLo:3600 mg/kg AEPPAE 186,195,37
scu-gpg LDLo:4100 mg/kg AEPPAE 186,195,37

SAFETY PROFILE: Moderately toxic by subcutaneous route. When heated to decomposition it emits toxic fumes of NO_x.

NAZ000 CAS:224-98-6 ***HR: 3***
NAPHTHO(2,3-f)QUINOLINE
mf: $C_{17}H_{11}N$ mw: 229.29

PROP: Colorless leaflets. Mp: 170°, bp: 446°. Insol in water; sol in alc and ether.

SYN: β-ANTHRAQUINOLINE

TOXICITY DATA with REFERENCE
mma-sat 500 μg/plate CRNGDP 3,947,82
scu-rat LDLo:40 mg/kg/4W-I:ETA AJCAA7 35,534.39

SAFETY PROFILE: Questionable carcinogen with experimental tumorigenic data. Mutation data reported. When heated to decomposition it emits toxic fumes of NO_x.

NBA000 CAS:524-42-5 ***HR: 3***
1,2-NAPHTHOQUINONE
mf: $C_{10}H_6O_2$ mw: 158.16

PROP: Red needles from ether. Mp: decomp @ 115-120°. Sol in water, H_2SO_4, benzene, and ether. Golden yellow needles decomp @ 145-147°. Sol in alc, benzene, ether; insol in water.

SYNS: 1,2-NAPHTHALENEDIONE ◇ 1,2 NAPHTHAQUINONE ◇ β-NAPHTHOQUINONE

TOXICITY DATA with REFERENCE
orl-rat LDLo:250 mg/kg NCNSA6 5,39,53
orl-mus LDLo:1070 mg/kg AECTCV 14,111,85
orl-bwd LD50:75 mg/kg AECTCV 12,355,83

CONSENSUS REPORTS: Reported in EPA TSCA Inventory.

SAFETY PROFILE: Poison by ingestion. When heated to decomposition it emits acrid smoke and irritating fumes.

NBA500 CAS:130-15-4 ***HR: 3***
1,4-NAPHTHOQUINONE
mf: $C_{10}H_6O_2$ mw: 158.16

PROP: Yellow triclinic; odor of benzoquinone. Mp: 125-126°, D: 1.422. Very sltly sol in cold water; very sol in hot alc; sol in ether, benzene, chloroform, carbon bisulfide, acetic acid, alkali hydroxide solns. Volatile with steam.

SYNS: 1,4-DIHYDRO-1,4-DIKETONAPHTHALENE ◇ 1,4-NAPHTHALENEDIONE ◇ α-NAPHTHOQUINONE ◇ RCRA WASTE NUMBER U166 ◇ USAF CY-10

TOXICITY DATA with REFERENCE
scu-rat TDLo:5 mg/kg (1D pre):REP ENDOAO 57,466,55
skn-mus TDLo:800 mg/kg/29W-C:ETA PIATA8 16,309,40
orl-rat LD50:190 mg/kg VKMGA7 6,37,66
ipr-rat LDLo:50 mg/kg 85GMAT -,91,82
scu-rat LD50:202 mg/kg GTPZAB 16(5),52,72
orl-mus LDLo:80 mg/kg JPETAB 71,210,41
ipr-mus LD50:5500 μg/kg JMCMAR 26,570,83
scu-mus LDLo:40 mg/kg JPETAB 71,210,41
ivn-mus LDLo:10 mg/kg JPETAB 71,210,41
orl-gpg LD50:400 mg/kg VKMGA7 6,37,66
orl-bwd LD50:133 mg/kg AECTCV 12,355,83

CONSENSUS REPORTS: Reported in EPA TSCA Inventory.

SAFETY PROFILE: Poison by ingestion, intravenous, subcutaneous, and intraperitoneal routes. Experimental reproductive effects. Questionable carcinogen with experimental tumorigenic data. When heated to decomposition it emits acrid smoke and irritating fumes.

NBC000 CAS:63021-45-4 ***HR: 3***
1-(2-NAPHTHOYL)-AZIRIDINE
mf: $C_{13}H_{11}NO$ mw: 197.25

SYN: β-NAPHTHOYLETHYLENEIMINE

TOXICITY DATA with REFERENCE
scu-rat TDLo:975 mg/kg/20W-I:NEO BJPCAL 9,306,54
ipr-mus LD50:122 mg/kg NCISA* PH-43-63-1132

SAFETY PROFILE: Poison by intraperitoneal route. Questionable carcinogen with experimental neoplastigenic data. When heated to decomposition it emits toxic fumes of NO_x.

NBC500 CAS:76790-18-6 ***HR: D***
N-(2-NAPHTHOYL)-o-PROPIONYLHYDROXYLAMINE
mf: $C_{14}H_{13}NO_3$ mw: 243.28

SYN: 2-NAPHTHOHYDROXAMIC ACID-o-PROPIONATE ESTER

TOXICITY DATA with REFERENCE
mmo-sat 1 μmol/plate PAACA3 21,126,80
mma-sat 1 μmol/plate CBINA8 34,267,81

SAFETY PROFILE: Mutation data reported. When heated to decomposition it emits toxic fumes of NO_x. See also AMINES.

NBD000 CAS:2508-23-8 ***HR: 3***
N-2-NAPHTHYLACETOHYDROXAMIC ACID
mf: $C_{12}H_{11}NO_2$ mw: 201.24

SYNS: N-ACETYL-2-NAPHTHYLHYDROXYLAMINE ◇ 2-(N-HYDROXYACETAMIDO)NAPHTHALENE ◇ N-HYDROXY-2-ACETYLAMINONAPHTHALENE ◇ N-HYDROXY-N-2-NAPHTHALENYLACETAMIDE

TOXICITY DATA with REFERENCE
dns-hmn:oth 10 μmol/L JJIND8 72,847,84
imp-mus TDLo:160 mg/kg:NEO ANYAA9 108,924,63

SAFETY PROFILE: Human mutation data reported.

Questionable carcinogen with experimental neoplastigenic data. When heated to decomposition it emits toxic fumes of NO_x.

NBD500 CAS:132-75-2 ***HR: 3***
α-(1-NAPHTHYL)ACETONITRILE
mf: $C_{12}H_9N$ mw: 167.22

SYN: α-NAPHTHYL ACETONITRILE

TOXICITY DATA with REFERENCE
ivn-mus LD50:180 mg/kg CSLNX* NX#03835

CONSENSUS REPORTS: Cyanide and its compounds are on the Community Right-To-Know List. Reported in EPA TSCA Inventory.

SAFETY PROFILE: Poison by intravenous route. When heated to decomposition it emits toxic fumes of NO_x and CN^-. See also NITRILES.

NBE000 CAS:134-32-7 ***HR: 3***
1-NAPHTHYLAMINE
DOT: UN 2077
mf: $C_{10}H_9N$ mw: 143.20

PROP: White crystals, reddening on exposure to air; unpleasant odor. Mp: 50°, bp: 300.8°, flash p: 315°F, d: 1.131, vap press: 1 mm @ 104.3°, vap d: 4.93. Sublimes, volatile with steam. Sol in 590 parts water; very sol in alc, ether. Keep well closed and away from light.

SYNS: ALFANAFTILAMINA (ITALIAN) ◇ ALFA-NAFTYLOAMINA (POLISH) ◇ 1-AMINONAFTALEN (CZECH) ◇ 1-AMINONAPHTHALENE ◇ C.I. AZOIC DIAZO COMPONENT 114 ◇ 1-NAFTILAMINA (SPANISH) ◇ α-NAFTYLAMIN (CZECH) ◇ 1-NAFTYLAMINE (DUTCH) ◇ NAPHTHALIDINE ◇ 1-NAPHTHYLAMIN (GERMAN) ◇ α-NAPHTHYLAMINE ◇ RCRA WASTE NUMBER U167

TOXICITY DATA with REFERENCE
mma-sat 10 μg/plate ENMUDM 6,497,84
dns-ham:lvr 10 μmol/L ENMUDM 6,1,84
ipr-rat TDLo:1300 mg/kg/13W-I:ETA CNREA8 28,535,68
orl-rat LD50:779 mg/kg 28ZPAK -,67,72
ipr-mus LD50:96 mg/kg PMRSDJ 1,682,81
scu-rbt LDLo:300 mg/kg XPHBAO 271,174,41
orl-mam LDLo:4000 mg/kg JIDHAN 13,87,31

CONSENSUS REPORTS: IARC Cancer Review: Group 3 IMEMDT 7,260,87; Animal Inadequate Evidence IMEMDT 4,87,74; Human Limited Evidence IMEMDT 4,87,74. EPA Genetic Toxicology Program. Community Right-To-Know List. Reported in EPA TSCA Inventory.

OSHA PEL: Cancer Suspect Agent
NIOSH REL: (α-Naphthylamine): TWA use 29 CFR 1910.1004
DOT Classification: Poison B; Label: St. Andrews Cross.

SAFETY PROFILE: Confirmed carcinogen with experimental tumorigenic data. Along with β-naphthylamine and benzidine, it has been incriminated as a cause of urinary bladder cancer. Poison by subcutaneous and intraperitoneal routes. Moderately toxic by ingestion. Mutation data reported. Combustible when exposed to heat or flame. Incompatible with nitrous acid. To fight fire, use dry chemical, CO_2, mist, spray. When heated to decomposition it emits toxic fumes of NO_x. See also 2-NAPHTHYLAMINE and AROMATIC AMINES.

NBE500 CAS:91-59-8 ***HR: 3***
β-NAPHTHYLAMINE
DOT: UN 1650
mf: $C_{10}H_9N$ mw: 143.20

PROP: White to faint pink, lustrous leaflets; faint aromatic odor. Mp: 111.5°, d: 1.061 @ 98°/4°, vap press: 1 mm @ 108.0°. Bp: 306° (also listed as 294°). Sol in hot water, alc, and ether.

SYNS: 2-AMINONAFTALEN (CZECH) ◇ 2-AMINONAPHTHALENE ◇ BETA-NAFTYLOAMINA (POLISH) ◇ C.I. 37270 ◇ FAST SCARLET BASE B ◇ NA ◇ β-NAFTYLAMIN (CZECH) ◇ β-NAFTILAMINA (ITALIAN) ◇ 2-NAFTYLAMINE (DUTCH) ◇ β-NAFTYLOAMINA (POLISH) ◇ 2-NAPHTHALAMINE ◇ 2-NAPHTHALENAMINE ◇ β-NAPHTHYLAMIN (GERMAN) ◇ 2-NAPHTHYLAMIN (GERMAN) ◇ 6-NAPHTHYLAMINE ◇ 2-NAPHTHYLAMINE ◇ 2-NAPHTHYLAMINE MUSTARD ◇ RCRA WASTE NUMBER U168 ◇ USAF CB-22

TOXICITY DATA with REFERENCE
mmo-sat 2500 ng/plate PMRSDJ 1,261,81
dnd-hmn:fbr 50 μmol/L MUREAV 127,107,84
orl-rat TDLo:17100 mg/kg/57W-I:CAR BJCAAI 46,646,82
ipr-rat TDLo:1300 mg/kg/13W-I:ETA CNREA8 28,535,68
orl-mus TDLo:600 mg/kg/8W-I:NEO TXAPA9 82,19,86
par-mus TDLo:18 mg/kg:CAR BECCAN 40,42,62
mul-dog TDLo:3077 mg/kg/78W-C:EPA AJPAA4 13,656,37
orl-rat TD:13 g/kg/1Y-I:ETA,REP JNCIAM 41,985,68
orl-rat LD50:727 mg/kg 28ZPAK -,67,72
ipr-mus LD50:200 mg/kg NTIS** AD277-689

CONSENSUS REPORTS: NTP Fifth Annual Report on Carcinogens. IARC Cancer Review: Group 1 IMEMDT 7,261,87; Animal Sufficient Evidence IMEMDT 4,97,74; Human Sufficient Evidence IMEMDT 4,97,74. Community Right-To-Know List. EPA Genetic Toxicology Program. Reported in EPA TSCA Inventory.

OSHA PEL: Cancer Suspect Agent
ACGIH TLV: Confirmed Human Carcinogen.
DFG MAK: Human Carcinogen.
NIOSH REL: (β-Naphthylamine): TWA use 29 CFR 1910.1009
DOT Classification: Poison B; Label: Poison.

SAFETY PROFILE: Confirmed human carcinogen with experimental neoplastigenic, and tumorigenic data. Long and continued exposure to even small amounts may produce tumors and cancers of the bladder. Poison by intraperitoneal route. Moderately toxic by ingestion. Experimental reproductive effects. Human mutation data reported. A very toxic chemical in any of its physical forms, such as flake, lump, dust, liquid, or vapor. It can be absorbed into the body through the lungs, the gastrointestinal tract, or the skin. Combustible when exposed to heat or flame. At elevated temperatures it evolves a vapor which is flammable and explosive. Incompatible with nitrous acid. When heated to decomposition it emits toxic fumes of NO_x. See also AROMATIC AMINES and 1-NAPHTHYLAMINE.

NBE850 CAS:86-65-7 ***HR: 2***
2-NAPHTHYLAMINE-6,8-DISULFONIC ACID
mf: $C_{10}H_9NO_6S_2$ mw: 303.32

SYNS: AMIDO-G-ACID ◇ 7-AMINO-1,3-NAPHTHALENEDISULFONIC ACID

TOXICITY DATA with REFERENCE
ipr-mus TDLo:7500 mg/kg/8W-I:NEO JJIND8 67,1299,81

CONSENSUS REPORTS: Reported in EPA TSCA Inventory.

SAFETY PROFILE: Questionable carcinogen with experimental neoplastigenic data. When heated to decomposition it emits toxic fumes of NO_x and SO_x.

NBF000 CAS:552-46-5 ***HR: 2***
1-NAPHTHYLAMINE HYDROCHLORIDE
mf: $C_{10}H_9N{\bullet}ClH$ mw: 179.66

PROP: Needles. Bp: subl. Sol in water, alc, and ether.

SYN: α-NAPHTHYLAMINE HYDROCHLORIDE

TOXICITY DATA with REFERENCE
orl-rat TDLo:7 g/kg/50W-C:ETA AICCA6 19,539,63

CONSENSUS REPORTS: Reported in EPA TSCA Inventory.

SAFETY PROFILE: Questionable carcinogen. When heated to decomposition it emits very toxic fumes of NO_x and HCl. See also 1-NAPHTHYLAMINE.

NBF500 CAS:93-45-8 ***HR: 2***
p-(2-NAPHTHYLAMINO)PHENOL
mf: $C_{16}H_{13}NO$ mw: 235.30

SYNS: N-p-HYDROXYPHENYL-β-NAPHTHYLAMINE ◇ N-p-HYDROXYPHENYL-2-NAPHTHYLAMINE ◇ p-HYDROXYPHENYL-β-NAPHTHYLAMINE ◇ p-HYDROXYPHENYL-2-NAPHTHYLAMINE

TOXICITY DATA with REFERENCE
orl-rat LD50:5200 mg/kg HYSAAV 31,183,66
orl-mus LD50:2200 mg/kg HYSAAV 31,183,66

CONSENSUS REPORTS: Reported in EPA TSCA Inventory.

SAFETY PROFILE: Moderately toxic by ingestion. When heated to decomposition it emits toxic fumes of NO_x. See also AROMATIC AMINES.

NBG000 CAS:63978-93-8 ***HR: 3***
4-(1-NAPHTHYLAZO)-2-NAPHTHYLAMINE
mf: $C_{20}H_{15}N_3$ mw: 297.38

TOXICITY DATA with REFERENCE
orl-rat TDLo:23 g/kg/64W-C:ETA ZEKBAI 57,530,51

SAFETY PROFILE: Questionable carcinogen with experimental tumorigenic data. When heated to decomposition it emits toxic fumes of NO_x.

NBG500 CAS:6416-57-5 ***HR: 3***
4-(1-NAPHTHYLAZO)-m-PHENYLENEDIAMINE
mf: $C_{16}H_{14}N_4$ mw: 262.34

SYNS: C.I. 11285 ◇ FAT BROWN ◇ 1-NAPHTHALENAZO-2′,4′-DIAMINOBENZENE ◇ 4-(1-NAPHTHALENYLAZO)-1,3-PHENYLENEDIAMINE ◇ ORGANOL BROWN 2R ◇ RESINOL BROWN RRN ◇ SUDAN BROWN RR

TOXICITY DATA with REFERENCE
orl-rat TDLo:34 g/kg/89W-C:ETA ZEKBAI 57,530,51

CONSENSUS REPORTS: IARC Cancer Review: Group 3 IMEMDT 7,56,87; Animal Inadequate Evidence IMEMDT 8,249,75. Reported in EPA TSCA Inventory.

SAFETY PROFILE: Questionable carcinogen with experimental tumorigenic data. When heated to decomposition it emits toxic fumes of NO_x. See also AROMATIC AMINES.

NBH200 CAS:3759-61-3 ***HR: 3***
1-NAPHTHYL CHLOROFORMATE
mf: $C_{11}H_7ClO_2$ mw: 206.63

SYNS: CARBANOCHLORIDIC ACID, NAPHTHYL ESTER ◇ CARBONOCHLORIDIC ACID, 1-NAPHTHALENYL ESTER ◇ CHLOROFORMIC ACID, ESTER with 1-NAPHTHOL ◇ CHLOROFORMIC ACID 1-MAPHTHYL ESTER ◇ CHLOROMROWCZAN 1-NAFTYLU (CZECH) ◇ 1-NAPHTHYL CHLOROCARBONATE ◇ α-NAPHTHYL CHLOROFORMATE

TOXICITY DATA with REFERENCE
skn-rbt 500 mg SEV MEPAAX 29,293,78
eye-rbt 100 mg SEV MEPAAX 29,293,78
orl-rat LD50:2550 mg/kg MEPAAX 29,293,78
ipr-rat LD50:300 mg/kg MEPAAX 29,293,78
ivn-mus LD50:56 mg/kg CSLNX* NX#07351

SAFETY PROFILE: Poison by intravenous and intraperitoneal routes. Moderately toxic by ingestion. A severe eye and skin irritant. When heated to decomposition it emits toxic fumes of Cl^-. See also ESTERS.

NBH500 CAS:1465-25-4 ***HR: 3***
N-(1-NAPHTHYL)ETHYLENEDIAMINE DIHYDROCHLORIDE
mf: $C_{12}H_{14}N_2 \cdot 2ClH$ mw: 259.20

PROP: Long, hexagonal prisms. Mp: 188-190°. Very sol in 95% alc, dil HCl, hot water; sltly sol in cold water, acetone, abs alc.

SYNS: N-1-NAPHTHALENYL-1,2-ETHANEDIAMINEDIHYDROCHLORIDE ◇ NCI-C03281

TOXICITY DATA with REFERENCE
ipr-mus LD50:150 mg/kg NTIS** AD691-490

CONSENSUS REPORTS: NCI Carcinogenesis Bioassay (feed); No Evidence: mouse, rat NCITR* NCI-CG-TR-168,79. Reported in EPA TSCA Inventory.

SAFETY PROFILE: Poison by intraperitoneal route. When heated to decomposition it emits very toxic fumes of NO_x and HCl. See also AMINES.

NBI000 CAS:76206-36-5 ***HR: D***
1-NAPHTHYL-N-ETHYL-N-NITROSOCARBAMATE
mf: $C_{13}H_{12}N_2O_3$ mw: 244.27

SYN: N-ETHYL-N-NITROSOCARBAMIC ACID 1-NAPHTHYL ESTER

TOXICITY DATA with REFERENCE
mmo-sat 10 nmol/plate ENMUDM 2,395,80
mrc-smc 5 nmol/plate ENMUDM 2,395,80

SAFETY PROFILE: Mutation data reported. Many N-nitroso compounds are carcinogens. When heated to decomposition it emits toxic fumes of NO_x. See also N-NITROSO COMPOUNDS and CARBAMATES.

NBI500 CAS:613-47-8 ***HR: 3***
2-NAPHTHYLHYDROXYLAMINE
mf: $C_{10}H_9NO$ mw: 159.20

SYNS: N-HYDROXY-2-AMINONAPHTHALENE ◇ N-HYDROXY-2-NAPHTHYLAMINE ◇ NHA

TOXICITY DATA with REFERENCE
dnd-hmn:fbr 10 μmol/L GANNA2 75,349,84
dns-hmn:oth 10 μmol/L JJIND8 72,847,84
skn-rat TDLo:624 mg/kg 52W-I:ETA RCOCB8 18,353,77
ipr-rat TDLo:1300 mg/kg/13W-I:NEO CNREA8 28,535,68
imp-mus TDLo:80 mg/kg:CAR BJCAAI 17,127,63

CONSENSUS REPORTS: EPA Genetic Toxicology Program.

SAFETY PROFILE: Questionable carcinogen with experimental carcinogenic, neoplastigenic, and tumorigenic data. Human mutation data reported. When heated to decomposition it emits toxic fumes of NO_x. See also AMINES.

NBJ000 CAS:2173-57-1 ***HR: 1***
β-NAPHTHYL ISOBUTYL ETHER
mf: $C_{14}H_{16}O$ mw: 200.30

SYN: ISOBUTYL(2-NAPHTHYL)ETHER

TOXICITY DATA with REFERENCE
orl-rat LDLo:4750 mg/kg FCTXAV 2,327,64

CONSENSUS REPORTS: Reported in EPA TSCA Inventory.

SAFETY PROFILE: Mildly toxic by ingestion. When heated to decomposition it emits acrid smoke and irritating fumes. See also ETHERS.

NBJ500 CAS:7090-25-7 ***HR: 3***
1-NAPHTHYL METHYLNITROSOCARBAMATE
mf: $C_{12}H_{10}N_2O_3$ mw: 230.24

SYNS: DENAPON, NITROSATED (JAPANESE) ◇ METHYL-NITROSOCARBAMIC ACID-1-NAPHTHYL ESTER ◇ 1-NAPHTHYL-N-METHYL-N-NITROSOCARBAMATE ◇ N-NITROSOCARBARYL ◇ NITROSO-NAC

TOXICITY DATA with REFERENCE
mmo-asn 1 ug/plate MUREAV 97,293,82
dnr-hmn:fbr 10 μmol/L MUREAV 44,1,77
orl-rat TDLo:600 mg/kg/10W-I:CAR EESADV 2,413,78
scu-rat TDLo:1000 mg/kg:NEO FCTXAV 13,365,75
skn-mus TDLo:921 mg/kg/50W-I:CAR JCREA8 102,13,81
orl-rat TD:2500 mg/kg/10W-I:ETA CALEDQ 1,275,76

CONSENSUS REPORTS: IARC Cancer Review: Animal Sufficient Evidence IMEMDT 12,37,76. EPA Genetic Toxicology Program.

SAFETY PROFILE: Confirmed carcinogen with experimental carcinogenic, neoplastigenic, and tumorigenic data. Human mutation data reported. When heated to decomposition it emits toxic fumes of NO_x. See also N-NITROSO COMPOUNDS and CARBAMATES.

NBK000 CAS:69782-11-2 ***HR: 3***
(2-(α-NAPHTHYLOXY)ETHYL)HYDRAZINE MALEATE
mf: $C_{12}H_{14}N_2O \cdot C_4H_4O_4$ mw: 318.36

TOXICITY DATA with REFERENCE
orl-mus LD50:625 mg/kg JMCMAR 6,63,63
ipr-mus LD50:375 mg/kg JMCMAR 6,63,63

SAFETY PROFILE: Poison by intraperitoneal route. Moderately toxic by ingestion. When heated to decomposition it emits toxic fumes of NO_x.

NBL000 CAS:93-46-9 ***HR: 3***
2-NAPHTHYL-p-PHENYLENEDIAMINE
mf: $C_{26}H_{20}N_2$ mw: 360.48

SYNS: ACETO DIPP ◇ AGERITE WHITE ◇ DI-β-NAPHTHYL-p-PHENYLDIAMINE ◇ DI-β-NAPHTHYL-p-PHENYLENEDIAMINE ◇ N,N'-DI-β-NAPHTHYL-p-PHENYLENEDIAMINE ◇ sym-DI-β-NAPHTHYL-p-PHENYLENEDIAMINE ◇ DNPD ◇ DWU-β-NAFTYLO-p-FENYLODWUAMINA (POLISH) ◇ NONOX CL ◇ TISPERSE MB-2X

TOXICITY DATA with REFERENCE
skn-hmn 500 mg/48H MLD AMIHBC 5,311,52
skn-rbt 500 mg/24H MLD 28ZPAK -,74,72
eye-rbt 500 mg/24H MLD 28ZPAK -,74,72
pic-esc 100 mmol/L MDMIAZ 31,11,79
orl-mus TDLo:31 g/kg/78W-I:ETA NTIS** PB223-159
ipr-mam LD50:4500 mg/kg AMIHBC 5,311,52

CONSENSUS REPORTS: Reported in EPA TSCA Inventory.

SAFETY PROFILE: A human skin irritant. An experimental skin and eye irritant. Mutation data reported. Questionable carcinogen with experimental tumorigenic data. When heated to decomposition it emits toxic fumes of NO_x. See also AMINES.

NBM500 CAS:76206-37-6 ***HR: D***
1-NAPHTHYL-N-PROPYL-N-NITROSOCARBAMATE
mf: $C_{14}H_{14}N_2O_3$ mw: 258.30

SYN: N-NITROSO-N-PROPYLCARBAMIC ACID-1-NAPHTHYL ESTER

TOXICITY DATA with REFERENCE
mmo-sat 10 nmol/plate ENMUDM 2,395,80
mrc-smc 5 nmol/plate ENMUDM 2,395,80

SAFETY PROFILE: Mutation data reported. Many N-nitroso compounds are carcinogens. When heated to decomposition it emits toxic fumes of NO_x. See also N-NITROSO COMPOUNDS and CARBAMATES.

NBN000 CAS:17518-47-7 ***HR: 3***
1-NAPHTHYL THIOACETAMIDE
mf: $C_{12}H_{11}NOS$ mw: 217.30

TOXICITY DATA with REFERENCE
orl-rat LD50:60 mg/kg JPETAB 90,260,47

CONSENSUS REPORTS: Reported in EPA TSCA Inventory.

SAFETY PROFILE: Poison by ingestion. When heated to decomposition it emits very toxic fumes of SO_x and NO_x.

NBO500 CAS:32524-44-0 ***HR: 3***
5-β-NAPHTHYL-2:4:6-TRIAMINOAZOPYRIMIDINE
mf: $C_{14}H_{13}N_7$ mw: 279.34

SYN: 5-(β-NAPHTHYLAZO)-2,4,6-TRIAMINOPYRIMIDINE

TOXICITY DATA with REFERENCE
ipr-rat TDLo:270 mg/kg/14D-I:ETA JRMSAS 74,59,54

SAFETY PROFILE: Questionable carcinogen with experimental tumorigenic data. When heated to decomposition it emits toxic fumes of NO_x.

NBO515 ***HR: 3***
N-1-NAPHTHYL-N,N',N'-TRIETHYLETHYLENEDIAMINE
mf: $C_{18}H_{26}N_2$ mw: 270.46

TOXICITY DATA with REFERENCE
ipr-rat LDLo:45 mg/kg BJPCAL 11,1,56
ipr-mus LD50:65 mg/kg BJPCAL 11,1,56
scu-mus LD50:306 mg/kg BJPCAL 11,1,56

SAFETY PROFILE: Poison by subcutaneous and intraperitoneal routes. When heated to decomposition it emits toxic fumes of NO_x. See also AMINES.

NBO525 CAS:25796-01-4 ***HR: 3***
4-(1-NAPHTHYLVINYL)-PYRIDINE HYDROCHLORIDE
mf: $C_{17}H_{13}N \cdot ClH$ mw: 267.77

SYNS: 4-(2-(1-NAPHTHALENYL)ETHENYL)-PYRIDINEHYDROCHLORIDE (9CI) ◇ 4-(2-(1-NAPHTHYL)VINYL)PYRIDINE HYDROCHLORIDE ◇ NVP ◇ YuB-25

TOXICITY DATA with REFERENCE
ipr-mus LD50:300 mg/kg EJPHAZ 11,187,70
scu-mus LD50:150 mg/kg EJPHAZ 11,187,70
ivn-mus LD50:75 mg/kg EJPHAZ 11,187,70

SAFETY PROFILE: Poison by subcutaneous, intravenous, and intraperitoneal routes. When heated to decomposition it emits toxic fumes of NO_x and HCl.

NBO550 CAS:26159-34-2 ***HR: D***
NAPROXEN SODIUM
mf: $C_{14}H_{14}O_3 \cdot Na$ mw: 253.27

SYNS: l-(-)-6-METHOXY-α-METHYL-2-NAPHTHALENEACETIC ACID SODIUM SALT ◇ 2-NAPHTHALENEACETIC ACID, 6-METHOXY-α-METHYL-, SODIUM SALT, l-(-)- ◇ 2-NAPHTHALENEACETIC ACID, 6-METHOXY-α-METHYL-, SODIUM SALT, (R)- (9CI)

TOXICITY DATA with REFERENCE
orl-mky TDLo:20 mg/kg (female 2D pre):REP CCPTAY 27,505,83

SAFETY PROFILE: Experimental reproductive effects. When heated to decomposition it emits acrid smoke and irritating fumes.

NBP000 CAS:8023-75-4 ***HR: 1***
NARCISSUS ABSOLUTE

PROP: The components include linalool, benzyl acetate, benzyl alcohol, terpineol, cineol, phenylpropyl alcohol and its acetate, phenylethyl alcohol and its acetate, n-heptanol, n-nonanal, indole and traces of undecalactone (FCTXAV 16,637,78).

TOXICITY DATA with REFERENCE
skn-gpg 100% MLD FCTXAV 16,827,78

CONSENSUS REPORTS: Reported in EPA TSCA Inventory.

SAFETY PROFILE: A skin irritant. See also individual components.

NBP275 CAS:6035-40-1 ***HR: 3***
NARCOTINE
mf: $C_{22}H_{23}NO_7$ mw: 413.46

PROP: Long needles from methanol. Mp: 232° (decomp). Freely sol in carbon disulfide, hot chloroform. Sol in about 1500 parts alc; sparingly sol in benzene, water.

SYNS: COSCOPIN ◇ GNOSCOPINE ◇ NOSCAPINE ◇ NSC-5366 ◇ OPIAN

TOXICITY DATA with REFERENCE
ipr-rat LD50:750 mg/kg APTOA6 6,235,50
orl-mus LD50:2784 mg/kg APFRAD 15,640,57
scu-mus LD50:491 mg/kg APFRAD 15,640,57
ivn-mus LD50:83 mg/kg 85GDA2 8(1),206,82
ivn-dog LDLo:73 mg/kg CPBTAL 7,372,59

SAFETY PROFILE: Poison by intravenous route. Moderately toxic by ingestion, intraperitoneal, and subcutaneous routes. When heated to decomposition it emits toxic fumes of NO_x.

NBP300 ***HR: 2***
NARCOTINE N-OXIDE HYDROCHLORIDE
mf: $C_{22}H_{23}NO_8 \cdot ClH$ mw: 465.92

SYN: (s-(4*,S*))-6,7-DIMETHOXY-3-(5,6,7,8-TETRAHYDRO-4-METHOXY-6-METHYL-1,3-DIOXOLO(4,5-g)ISOQUINOLIN-5-YL)-1(3H)-ISOBENZOFURANONE, N-OXIDE, HYDROCHLORIDE

TOXICITY DATA with REFERENCE
orl-mus LD50:1740 mg/kg YKKZAJ 81,266,61
scu-mus LD50:1450 mg/kg YKKZAJ 81,266,61
ivn-mus LD50:647 mg/kg YKKZAJ 81,266,61

SAFETY PROFILE: Moderately toxic by ingestion, subcutaneous and intravenous routes. When heated to decomposition it emits toxic fumes of NO_x and HCl.

NBP500 CAS:5591-45-7 ***HR: 3***
NAVARON
mf: $C_{23}H_{29}N_3O_2S_2$ mw: 443.67

SYNS: CP-12,252-1 ◇ N,N-DIMETHYL-9-(3-(4-METHYL-1-PIPERANIZYL)PROPYLIDENE)-9H-THIOXANTHENE-2-SULFONAMIDE, (Z)- ◇ N,N-DIMETHYL-9-(3-(4-METHYL-1-PIPERAZINYL)PROPYLIDENE)THIAXANTHENE-2-SULFONAMIDE ◇ N,N-DIMETHYL- 9-(3-(4-METHYL-1-PIPERAZINYL)PROPYLIDENE)THIOXANTHENE-2-SULFONAMIDE ◇ 2-(DIMETHYLSULFAMOYL)-(9-(4-METHYL-1-PIPERAZINYL)PROPYLIDENE)THIOXANTHENE ◇ NAVANE ◇ ORBINAMON ◇ P-4657-B ◇ THIOTHIXENE ◇ THIOTHIXINE ◇ TIOTIXENE

TOXICITY DATA with REFERENCE
orl-man TDLo:3 mg/kg (male 1W pre):REP JCLPDE 48,216,87
orl-mus TDLo:180 mg/kg (female 7-12D post):TER OYYAA2 3,315,69
orl-wmn TDLo:300 μg/kg/2D-I JCLPDE 44,77,83
orl-rat LD50:720 mg/kg NIIRDN 6,455,82
ipr-rat LD50:55 mg/kg PSYPAG 12,142,68
scu-rat LD50:2 g/kg KSRNAM 3,709,69
orl-mus LD50:400 mg/kg NIIRDN 6,455,82
ipr-mus LD50:100 mg/kg PSYPAG 12,142,68
scu-mus LD50:4 g/kg KSRNAM 3,709,69

SAFETY PROFILE: Poison by ingestion and intraperitoneal routes. Moderately toxic by subcutaneous route. An experimental teratogen. Experimental reproductive effects. When heated to decomposition it emits very toxic fumes of SO_x and NO_x.

NBR000 CAS:500-38-9 ***HR: 1***
NDGA
mf: $C_{18}H_{22}O_4$ mw: 302.40

PROP: Crystals from acetic acid. Mp: 184-185°. Sol in methanol, ethanol, and ether; sltly sol in hot water and chloroform; nearly insol in benzene and petroleum ether.

SYNS: 1,4-BIS(3,4-DIHYDROXYPHENYL)-2,3-DIMETHYLBUTANE ◇ DIHYDRONORGUAIARETIC ACID ◇ β,Γ-DIMETHYL-α,Δ-BIS(3,4-DIHYDROXYPHENYL)BUTANE ◇ 4,4′-(2,3-DIMETHYLTETRAMETHYLENE)DIPYROCATECHOL ◇ NORDIHYDROGUAIARETIC ACID ◇ NORDIHYDROGUAIRARETIC ACID

TOXICITY DATA with REFERENCE
orl-rat LD50:2 g/kg JAOCA7 54,239,77
orl-mus LD50:2 g/kg JAOCA7 54,239,77
ipr-mus LD50:550 mg/kg AFREAW 3,197,51
orl-gpg LD50:830 mg/kg AFREAW 3,197,51

SAFETY PROFILE: Moderately toxic by intraperitoneal route. An antioxidant food additive. When heated to decomposition it emits acrid smoke and irritating fumes.

NBR100 ***HR: 3***
NDPEA
mf: $C_5H_{12}N_2O_4$ mw: 164.19

SYNS: 3-((2-HYDROXYETHYL)NITROSAMINO)-1,2-PROPANEDIOL ◇ N-NITRISO-N-(2,3-DIHYDROXYPROPYL)-N-(2-HYDROXYETHYL)AMINE ◇ N-NITROSODIHYDROXYPROPYLETHANOLAMINE ◇ N-NITROSO-2,3-DIHYDROXYPROPYL-2-HYDROXYETHYLAMINE

TOXICITY DATA with REFERENCE
mmo-sat 1 mg/plate MUREAV 111,135,83
mma-sat 50 µg/plate MUREAV 111,135,83
orl-rat TDLo:2750 mg/kg/2Y-C:ETA CRNGDP 5,167,84

SAFETY PROFILE: Questionable carcinogen with experimental tumorigenic data. Mutation data reported. Many N-nitroso compounds are carcinogens. When heated to decomposition it emits toxic fumes of NO_x. See also N-NITROSO COMPOUNDS and AMINES.

NBR500 CAS:11048-13-8 ***HR: 3***
NEBRAMYCIN

PROP: Produced by *Streptomyces tenebrarius* ATCC 17920 (85ERAY 1,725,78).

SYNS: TENEBRIMYCIN ◇ TENEMYCIN

TOXICITY DATA with REFERENCE
ipr-mus LD50:1250 mg/kg 85ERAY 1,725,78
scu-mus LD50:1500 mg/kg 85ERAY 1,725,78
ivn-mus LD50:300 mg/kg 85ERAY 1,725,78

SAFETY PROFILE: Poison by intravenous route. Moderately toxic by subcutaneous and intraperitoneal routes.

NBR800 ***HR: 3***
NECKLACEPOD SOPHORA

PROP: Shrubs or small trees with compound leaves. *S. secundiflora* produces clusters of purple flowers and seed pods containing bright red seeds. It grows wild in Texas, New Mexico, Mexico and on the coastal dunes of Baja California, Hawaii, and Guam. *S. tomentosa* produces clusters of bright yellow flowers and seed pods containing yellow seeds. It grows wild in Florida, Bermuda, and the West Indies.

SYNS: BURN BEAN ◇ COLORINES (MEXICO) ◇ CORAL BEAN ◇ FRIJOLILLO (MEXICO) ◇ MESCAL BEAN ◇ PAGODA TREE ◇ RED BEAN ◇ RED HOTS ◇ SILVER BUSH ◇ SOPHORA SECUNDIFLORA ◇ SOPHORA TOMENTOSA ◇ TAMBALISA (CUBA) ◇ TEXAS MOUNTAIN LAUREL

SAFETY PROFILE: The seeds contain poisonous cytisine and related alkaloids whose action is similar to nicotine. Ingestion of the seeds may cause nausea, vomiting, headache, dizziness, diarrhea, convulsions and respiratory paralysis. See also CYTISINE and NICOTINE.

NBS300 CAS:8002-65-1 ***HR: 1***
NEEM OIL

SYNS: MARGOSA OIL ◇ NIM OIL ◇ OILS, NIM

TOXICITY DATA with REFERENCE
orl-rat TDLo:20 g/kg (female 1-10D post):TER IJMRAQ 83,89,86
ivg-mky TDLo:800 mg/kg (female 4D pre):REP IJMRAQ 79,131,84
orl-rat LD50:14 g/kg JOETD7 23,39,88
orl-rbt LD50:24 g/kg JOETD7 23,39,88

SAFETY PROFILE: Slightly toxic by ingestion. An experimental teratogen. Experimental reproductive effects. When heated to decomposition it emits acrid smoke and irritating fumes.

NBS500 CAS:23327-57-3 ***HR: 3***
NEFOPAM HYDROCHLORIDE
mf: $C_{17}H_{19}NO \cdot ClH$ mw: 289.83

SYNS: ACUPAN ◇ AJAN ◇ FENAZOXINE HYDROCHLORIDE ◇ 5-METHYL-1-PHENYL-3,4,5,6-TETRAHYDRO-1H-2,5-BENZOXAZOCINE HYDROCHLORIDE ◇ SINALGICO

TOXICITY DATA with REFERENCE
orl-rat LD50:80 mg/kg TXAPA9 33,46,75
ivn-rat LD50:28 mg/kg 27ZQAG -,269,72
orl-mus LD50:119 mg/kg 27ZQAG -,269,72
ipr-mus LD50:70 mg/kg 27ZQAG -,269,72
ivn-mus LD50:45 mg/kg 27ZQAG -,269,72
ivn-dog LDLo:20 mg/kg TXAPA9 33,46,75
ims-dog LDLo:30 mg/kg TXAPA9 33,46,75

SAFETY PROFILE: Poison by ingestion, intravenous, intraperitoneal, and intramuscular routes. When heated to decomposition it emits very toxic fumes of HCl and NO_x.

NBS700 CAS:25417-20-3 ***HR: 2***
NEKAL
mf: $C_{18}H_{24}O_3S \cdot Na$ mw: 343.47

SYNS: DIBUTYL-NAPHTHALENE SULFATE, SODIUM SALT ◇ NAFTALIN-BUTIL-SOLFONATO (ITALIAN) ◇ SODIUM DIBUTYLNAPHTHALENE SULFATE ◇ SODIUM DIBUTYLNAPHTHALENESULFONATE ◇ SODIUM DIBUTYLNAPHTHYLSULFONATE

TOXICITY DATA with REFERENCE
orl-rat LD50:1250 mg/kg GISAAA 45(4),39,80
orl-mus LD50:1134 mg/kg GISAAA 45(4),39,80
orl-gpg LD50:1500 mg/kg GISAAA 45(4),39,80
scu-gpg LDLo:800 mg/kg MELAAD 37,140,46

CONSENSUS REPORTS: Reported in EPA TSCA Inventory.

SAFETY PROFILE: Moderately toxic by ingestion and subcutaneous routes. When heated to decomposition it emits toxic fumes of SO_x and Na_2O. See also SULFONATES.

NBT000 CAS:1317-43-7 ***HR: 3***
NEMALITE
mf: H_2MgO_2 mw: 58.33

PROP: Dust used was fibrous (NATWAY 59,318,72).

SYNS: BRUCITE ◇ NEMALITH (GERMAN)

TOXICITY DATA with REFERENCE
ipr-rat TDLo:450 mg/kg/3W-I:CAR ZHPMAT 162,467,76
ipl-rat TDLo:200 mg/kg:NEO BJCAAI 28,173,73

SAFETY PROFILE: Questionable carcinogen with experimental carcinogenic and neoplastigenic data.

NBT500 CAS:76-74-4 ***HR: 3***
NEMBUTAL
mf: $C_{11}H_{18}N_2O_3$ mw: 226.31

SYNS: DORSITAL ◇ ETHAMINAL ◇ 5-ETHYL-5-(1-METHYLBUTYL)BARBITURIC ACID ◇ 5-ETHYL-5-(1-METHYLBUTYL)MALONYLUREA ◇ 5-ETHYL-5-(1-METHYLBUTYL)-2,4,6(1H,3H,5H)-PYRIMIDINETRIONE (9CI) ◇ MEBUBARBITAL ◇ NEODORM (NEW) ◇ PENTABARBITONE ◇ PENTOBARBITAL ◇ PENTOBARBITURATE ◇ PENTOBARBITURIC ACID ◇ RIVADORM

TOXICITY DATA with REFERENCE
scu-ham TDLo:1200 μg/kg (male 3D pre):REP DEPBA5 12,49,79
orl-hmn LDLo:36 mg/kg CTOXAO 10,327,77
orl-rat LD50:125 mg/kg FEPRA7 7,262,48
ipr-rat LD50:108 mg/kg PHMGBN 19,182,79
scu-rat LD50:144 mg/kg JPETAB 118,139,56
orl-mus LD50:170 mg/kg JPETAB 118,139,56
ipr-mus LD50:123 mg/kg RCOCB8 50,209,85
ivn-mus LD50:65 mg/kg AIPTAK 141,83,63
ims-mus LD50:124 mg/kg NIIRDN 6,779,82
ivn-dog LD50:50 mg/kg AIPTAK 141,83,63
orl-rbt LDLo:175 mg/kg JPETAB 44,325,32
ipr-rbt LDLo:65 mg/kg JPETAB 44,325,32
ivn-rbt LD50:33 mg/kg JPETAB 96,209,49

SAFETY PROFILE: Human poison by ingestion. Experimental poison by ingestion, intraperitoneal, intravenous, intramuscular, and subcutaneous routes. Experimental reproductive effects. A sedative, hypnotic and anticonvulsant. When heated to decomposition it emits toxic fumes of NO_x. See also BARBITURATES.

NBU000 CAS:57-33-0 ***HR: 3***
NEMBUTAL SODIUM
mf: $C_{11}H_{18}N_2O_3$•Na mw: 249.30

PROP: White, crystalline powder. Sol in water and alc; insol in ether.

SYNS: AUROPAN ◇ BARPENTAL ◇ BIOSEDAN ◇ BUTYLONE ◇ CARBRITAL ◇ CONTINAL ◇ DIABUTAL ◇ EMBUTAL ◇ ETAMINAL SODIUM ◇ ETHAMINAL SODIUM ◇ 5-ETHYL-5-(1-METHYLBUTYL)BARBITURIC ACID SODIUM SALT ◇ 5-ETHYL-5-(1-METHYLBUTYL)-2,4,6(1H,3H,5H)-PYRIMIDINETRIONE MONOSODIUM SALT (9CI) ◇ EUTHATAL ◇ IPRAL SODIUM ◇ ISOBARB ◇ MEBUBARBITAL SODIUM ◇ MEBUMAL NATRIUM ◇ MEBUMAL SODIUM ◇ MINTAL ◇ NAPENTAL ◇ PACIFAN ◇ PALAPENT ◇ PENBAR ◇ PENTABARBITAL SODIUM ◇ PENTAL ◇ PENTOBARBITONE SODIUM ◇ PENTONAL ◇ PENTYL ◇ PROPYLMETHYLCARBINYLETHYL BARBITURIC ACID SODIUM SALT ◇ RIVADORN ◇ SAGATAL ◇ SODITAL ◇ SODIUM ETHAMINAL ◇ SODIUM-5-ETHYL-5-(1-METHYLBUTYL)BARBITURATE◇ SODIUM NEMBUTAL ◇ SODIUM-PENT ◇ SODIUM PENTABARBITAL ◇ SODIUM PENTABARBITONE ◇ SODIUM PENTOBARBITAL ◇ SODIUM PENTOBARBITONE ◇ SODIUM PENTOBARBITURATE ◇ SOLUBLE PENTOBARBITAL ◇ SOMNOPENTYL ◇ SONISTAN ◇ SONTOBARBITAL NABITONE ◇ SOPENTAL ◇ SOTYL ◇ VETBUTAL

TOXICITY DATA with REFERENCE
cyt-ham-scu 100 mg/kg AEHLAU 36,316,81
dlt-ham-scu 100 mg/kg AEHLAU 36.316.81
scu-rat TDLo:520 mg/kg (female 9-21D post):REP TJADAB 25(2),60A,82
scu-ham TDLo:100 mg/kg (female 1D post):TER AEHLAU 36,316,81
orl-wmn TDLo:60 mg/kg:CNS JLCMAK 29,265,44
orl-man TDLo:6430 μg/kg:CNS JPETAB 197,488,76
orl-rat LD50:118 mg/kg TXAPA9 21,315,72
ipr-rat LD50:36 mg/kg FCTXAV 3,597,65
scu-rat LD50:47 mg/kg AIPTAK 180,155,69
ivn-rat LD50:65 mg/kg JPETAB 135,213,62
idu-rat LD50:39 mg/kg AIPTAK 180,155,69
orl-mus LD50:239 mg/kg ARZNAD 31,2180,81
ipr-mus LD50:126 mg/kg JMCMAR 12,989,69
scu-mus LD50:122 mg/kg TXAPA9 18,185,71
ims-mus LD50:124 mg/kg JPETAB 102,138,51
ice-mus LD50:6500 μg/kg JPETAB 197,479,76

SAFETY PROFILE: Poison by ingestion, intraperitoneal, subcutaneous, intravenous, intraduodenal, intramuscular, intracerebral, parenteral, and rectal routes. An experimental teratogen. Other experimental reproductive effects. Human systemic effects by ingestion: wakefulness, change in motor activity, ataxia and antipsychotic effects. Mutation data reported. When heated to decomposition it emits toxic fumes of NO_x and Na_2O. See also BARBITURATES.

NBU800 CAS:471-77-2 ***HR: D***
NEOABIETIC ACID
mf: $C_{20}H_{30}O_2$ mw: 302.50

SYN: 13-ISOPROPYLIDENEPODOCARP-8(14)-EN-15-OICACID

TOXICITY DATA with REFERENCE
mmo-sat 500 μg/plate ENMUDM 1,361,79
mmo-smc 750 mg/L MUREAV 119,273,83

SAFETY PROFILE: Mutation data reported. When heated to decomposition it emits acrid smoke and fumes.

NBV000 CAS:63681-05-0 ***HR: 3***
NEOANTERGAN PHOSPHATE
mf: $C_{17}H_{23}N_3O{\cdot}H_3O_4P$ mw: 383.43

SYN: N-α-PYRIDYL-N-p-METHOXYBENZYL-N',N'-DIMETHYLETHYLENEDIAMINE PHOSPHATE

TOXICITY DATA with REFERENCE
ipr-mus LD50:90 mg/kg JPETAB 89,247,47
scu-mus LD50:150 mg/kg PHREA7 27,542,47
ivn-mus LD50:30 mg/kg PHREA7 27,542,47
scu-gpg LD50:70 mg/kg PHREA7 27,542,47

SAFETY PROFILE: Poison by intravenous, intraperitoneal, and subcutaneous routes. When heated to decomposition it emits very toxic fumes of PO_x and NO_x.

NBV500 CAS:9014-02-2 ***HR: 3***
NEOCARZINOSTATIN

PROP: An acidic, single-chain polypeptide with a molecular weight of approximately 10,700 isolated from the culture filtrate of *Streptomyces carzinostaticus* Var F41 (MUREAV 56,91,77).

SYNS: NCS ◇ NEOCARCINOSTATIN ◇ NEOCARZINOSTATIN K ◇ NSC 69856 ◇ ZINOSTATIN

TOXICITY DATA with REFERENCE
mmo-esc 5 μg/plate TAKHAA 44,96,85
dni-hmn:fbr 200 μg/L CRNGDP 4,917,83
dns-ham:ovr 300 μg/L CNREA8 44,1748,84
ipr-rat TDLo:800 μg/kg (8-15D preg):TER OYYAA2 11,329,76
ipr-rat TDLo:800 μg/kg (8-15D preg):REP OYYAA2 11,329,76
orl-rat LDLo:300 mg/kg OYYAA2 11,309,76
ipr-rat LD50:940 μg/kg OYYAA2 11,309,76
scu-rat LD50:4980 μg/kg OYYAA2 11,309,76
ivn-rat LD50:830 μg/kg JANTAJ 28,64,75
orl-mus LD50:1050 mg/kg JANTAJ 28,64,75
ipr-mus LD50:1720 μg/kg JANTAJ 31,468,78
scu-mus LD50:7200 μg/kg OYYAA2 11,309,76
ivn-mus LD50:960 μg/kg JANTAJ 28,64,75
ivn-rbt LD50:900 μg/kg JANTAJ 28,64,75

CONSENSUS REPORTS: EPA Genetic Toxicology Program.

SAFETY PROFILE: A deadly poison by ingestion, intraperitoneal, subcutaneous, and intravenous routes. An experimental teratogen. Experimental reproductive effects. Human mutation data reported. When heated to decomposition it emits acrid smoke and irritating fumes.

NBW000 CAS:64093-79-4 ***HR: 3***
NEOCHROMIUM
mf: $CrHO_5S$ mw: 165.07

SYNS: BASIC CHROMIC SULFATE ◇ BASIC CHROMIC SULPHATE ◇ BASIC CHROMIUM SULFATE ◇ BASIC CHROMIUM SULPHATE ◇ CHROMIUM HYDROXIDE SULFATE ◇ CHROMIUM SULFATE ◇ CHROMIUM SULFATE, BASIC ◇ CHROMIUM SULPHATE ◇ KOREON ◇ MONOBASIC CHROMIUM SULFATE ◇ MONOBASIC CHROMIUM SULPHATE ◇ SULFURIC ACID, CHROMIUM SALT, BASIC

TOXICITY DATA with REFERENCE
oms-hmn:oth 500 mg/L BJCAAI 44,219,81
dni-ham:kdy 500 mg/L BJCAAI 44,219,81
scu-rat TDLo:135 mg/kg:NEO ANYAA9 271,431,76
scu-rat TD:135 mg/kg:ETA PBPHAW 14,47,78

CONSENSUS REPORTS: Chromium and its compounds are on the Community Right-To-Know List.

OSHA PEL: TWA 0.5 mg(Cr)/m^3
ACGIH TLV: TWA 0.5 mg(Cr)/m^3

SAFETY PROFILE: Questionable carcinogen with experimental neoplastigenic and tumorigenic data. Human mutation data reported. When heated to decomposition it emits toxic fumes of SO_x. See also CHROMIUM COMPOUNDS.

NBW500 CAS:26896-20-8 ***HR: 2***
NEODECANOIC ACID
mf: $C_{10}H_{20}O_2$ mw: 172.30

PROP: A mixture of isomeric 10-carbon, saturated monocarboxilic acids (FMCHA2 -,D334,80).

SYN: WILTZ-65

TOXICITY DATA with REFERENCE
orl-mam LD50:3400 mg/kg FMCHA2 -,C256,83

CONSENSUS REPORTS: Reported in EPA TSCA Inventory.

SAFETY PROFILE: Moderately toxic by ingestion. When heated to decomposition it emits acrid smoke and irritating fumes.

NBX000 CAS:7440-00-8 ***HR: 2***
NEODYMIUM
af: Nd aw: 144.24

PROP: It is a bright, silvery, lustrous, very reactive rare earth metal. Bp: 3127°, d: 7.003, mp: approx 1024°.

TOXICITY DATA with REFERENCE
ice-hmn TDLo:17 μg/kg:BLD JCINAO 21,447,42

CONSENSUS REPORTS: Reported in EPA TSCA Inventory.

SAFETY PROFILE: Human systemic effects by intracerebral route: blood changes. It may be an anticoagulant lanthanon. Care in handling is advised. Flammable in the form of dust when exposed to heat or flame. Slight explosion hazard in the form of dust when exposed to flame. Can react violently with air; halogens; N_2. Violent reaction with phosphorus above 400°C. Many of its compounds are poisons. See also RARE EARTHS and various neodymium compounds.

NBX500 CAS:6192-13-8 ***HR: 3***
NEODYMIUM ACETATE
mf: $C_6H_9O_6$•Nd mw: 321.39

SYNS: NEODYMACETAT (GERMAN) ◇ NEODYMIUM TRIACETATE

TOXICITY DATA with REFERENCE
ivn-cat LDLo:50 mg/kg ZGEMAZ 113,536,44

CONSENSUS REPORTS: Reported in EPA TSCA Inventory.

SAFETY PROFILE: Poison by intravenous route. When heated to decomposition it emits acrid smoke and irritating fumes. See also NEODYMIUM and RARE EARTHS.

NBY000 CAS:10024-93-8 ***HR: 3***
NEODYMIUM CHLORIDE
mf: Cl_3Nd mw: 250.59

PROP: Large, purple prisms. Sol in water, alc.

TOXICITY DATA with REFERENCE
skn-rbt 500 mg/24H MLD TXAPA9 6,614,64
eye-rbt 50 mg TXAPA9 6,614,64
ipr-rat LD50:150 mg/kg AIPTAK 37,305,30
scu-rat LDLo:150 mg/kg AIPTAK 37,305,30
ivn-rat LD50:3500 μg/kg AMIHAB 16,475,57
orl-mus LD50:3692 mg/kg EQSSDX 1,1,75
ipr-mus LD50:347 mg/kg AEHLAU 5,437,62
scu-mus LD50:278 mg/kg EQSSDZ 1,1,75
ivn-rbt LD50:139 mg/kg EQSSDX 1,1,75
ipr-gpg LD50:140 mg/kg AMIHAB 15,9,57
scu-gpg LD50:243 mg/kg EQSSDX 1,1,75

CONSENSUS REPORTS: Reported in EPA TSCA Inventory.

SAFETY PROFILE: Poison by intravenous, intraperitoneal, and subcutaneous routes. Moderately toxic by ingestion. A skin and eye irritant. When heated to decomposition it emits very toxic fumes of Cl^-. See also NEODYMIUM, CHLORIDES, and RARE EARTHS.

NCB000 CAS:10045-95-1 ***HR: 3***
NEODYMIUM(III) NITRATE (1:3)
mf: N_3O_9•Nd mw: 330.27

PROP: Triclinic crystals.

SYN: NITRIC ACID, NEODYMIUM SALT

TOXICITY DATA with REFERENCE
orl-rat LD50:2072 mg/kg EQSSDX 1,1,75
ipr-rat LD50:204 mg/kg EQSSDX 1,1,75
ivn-rat LD50:4800 μg/kg EQSSDX 1,1,75
ipr-mus LD50:183 mg/kg EQSSDX 1,1,75
ivn-rbt LDLo:50 mg/kg JCINAO 21,447,42

CONSENSUS REPORTS: Reported in EPA TSCA Inventory.

SAFETY PROFILE: Poison by intraperitoneal and intravenous routes. Moderately toxic by ingestion. When heated to decomposition it emits very toxic fumes of NO_x. See also NITRATES and NEODYMIUM.

NCB500 CAS:16454-60-7 ***HR: 3***
NEODYMIUM(III) NITRATE, HEXAHYDRATE (1:3:6)
mf: N_3O_9•Nd•$6H_2O$ mw: 438.39

SYN: NITRIC ACID, NEODYMIUM (3+) SALT, HEXAHYDRATE

TOXICITY DATA with REFERENCE
orl-rat LD50:2750 mg/kg TXAPA9 5,750,63
ipr-rat LD50:270 mg/kg TXAPA9 5,750,63
ivn-rat LD50:6 mg/kg TXAPA9 5,750,63
ipr-mus LD50:270 mg/kg TXAPA9 5,750,63

SAFETY PROFILE: Poison by intraperitoneal and intravenous routes. Moderately toxic by ingestion. When heated to decomposition it emits toxic fumes of NO_x. See also NEODYMIUM(III) NITRATE.

NCC000 CAS:1313-97-9 ***HR: 2***
NEODYMIUM OXIDE
mf: Nd_2O_3 mw: 336.48

PROP: Blue powder, hexagonal structure. Very stable; sol in dil acids.

TOXICITY DATA with REFERENCE
ipr-rat LDLo:800 mg/kg CURL** 35,117,60

CONSENSUS REPORTS: Reported in EPA TSCA Inventory.

SAFETY PROFILE: Moderately toxic by intraperitoneal route. See also NEODYMIUM and RARE EARTHS.

NCD500 CAS:91-85-0 ***HR: 3***
NEOHETRAMINE
mf: $C_{16}H_{22}N_4O$ mw: 286.42

SYNS: 2-((2-DIMETHYLAMINO)ETHYL)(p-METHOXYBENZYL) AMINO)PYRIMIDINE ◇ N,N-DIMETHYL-N'-(p-METHOXYBENZYL)-N'-(2-PYRIMIDYL)ETHYLENEDIAMINE ◇ NCI-C60708

TOXICITY DATA with REFERENCE
ims-rat TDLo:50 mg (5D preg):REP PSEBAA 100,555,59
orl-wmn TDLo:2 mg/kg JOALAS 19,313,48
orl-mus LD50:245 mg/kg PSEBAA 66,512,47
ipr-mus LD50:119 mg/kg PSEBAA 66,512,47
orl-gpg LD50:493 mg/kg PSEBAA 66,512,47

SAFETY PROFILE: Poison by ingestion and intraperitoneal routes. Experimental reproductive effects. When heated to decomposition it emits toxic fumes of NO_x.

NCD550 CAS:28002-70-2 ***HR: 3***
NEOMIX
mf: $C_{23}H_{46}N_6O_{13}\cdot xH_2O_4S$ mw: 1301.31

SYNS: BYKOMYCIN ◇ o-2,6-DIAMINO-2,6-DIDEOXY-β-l-IDOPYRANOSYL-(1-3)-o-(β-d-RIBOFURANOSYL-(1-5)-o-(2,6-DIAMINO-2,6-DIDEOXY-α-d-GLUCOPYRANOSYL-(1-4))-2-DEOXYD-STREPTAMINE SULFATE (SALT) ◇ FRAMYCETIN SULFATE ◇ FRAMYCIN SULFATE ◇ FRAQUINOL ◇ MYACINE ◇ NEOBRETTIN ◇ NEOMYCIN B SULFATE (SALT) ◇ NEOSULF ◇ TUTTOMYCIN

TOXICITY DATA with REFERENCE
ipr-rat LD50:250 mg/kg THERAP 13,945,58
scu-rat LD50:700 mg/kg THERAP 13,945,58
ims-rat LD50:550 mg/kg THERAP 13,945,58
ipr-mus LD50:270 mg/kg THERAP 13,945,58
scu-mus LD50:450 mg/kg THERAP 13,945,58
ivn-mus LD50:55 mg/kg THERAP 13,945,58
ims-mus LD50:450 mg/kg THERAP 13,945,58
scu-gpg LD50:450 mg/kg THERAP 13,945,58

SAFETY PROFILE: Poison by intravenous and intraperitoneal routes. Moderately toxic by subcutaneous and intramuscular routes. When heated to decomposition it emits toxic fumes of NO_x and SO_x.

NCE000 CAS:1404-04-2 ***HR: 3***
NEOMYCIN

PROP: An antibiotic.

SYNS: MYACYNE ◇ MYCIFRADIN ◇ NEOMCIN ◇ NIVEMYCIN ◇ ONAMYCIN POWDER V

TOXICITY DATA with REFERENCE
dnd-esc 20 mg/L MUREAV 89,95,81
cyt-mus-par 50 mg/kg CHRTBC 4,277,72
orl-wmn TDLo:220 mg/kg/2D:EAR,KID LARYA8 78,1734,68
unr-hmn TDLo:57 mg/kg:EAR,KID AIMEAS 66,1052,67
orl-rat LD50:2750 mg/kg TXAPA9 18,185,71
scu-rat LD50:200 mg/kg ARZNAD 12,597,62
orl-mus LD50:2880 mg/kg 85ERAY 1,664,78
ipr-mus LD50:116 mg/kg 85ERAY 1,664,78
scu-mus LD50:275 mg/kg ANIBAL 17,522,72
ivn-mus LD50:35 mg/kg ARZNAD 13,391,63

SAFETY PROFILE: Poison by intraperitoneal, intravenous, and subcutaneous routes. Moderately toxic by ingestion. Human systemic effects: changes in acuity, liver tubule changes and decreased urine volume or anuria. Mutation data reported. When heated to decomposition it emits acrid smoke and irritating fumes.

NCE500 CAS:3947-65-7 ***HR: 3***
NEOMYCIN A
mf: $C_{12}H_{26}N_4O_6$ mw: 322.42

SYNS: NEAMIN ◇ NEBRAMYCIN X ◇ NEGAMICIN

TOXICITY DATA with REFERENCE
scu-mus LD50:1250 mg/kg 85ERAY 1,664,78
ivn-mus LD50:320 mg/kg 85ERAY 1,664,78
ivn-mus LD50:125 mg/kg JANTAJ 27,677,74

SAFETY PROFILE: Poison by intravenous route. Moderately toxic by subcutaneous route. When heated to decomposition it emits toxic fumes of NO_x.

NCF000 CAS:119-04-0 ***HR: 3***
NEOMYCIN B
mf: $C_{23}H_{46}N_6O_{13}$ mw: 614.75

PROP: Water-sol.

SYNS: ACTILIN ◇ ANTIBIOTIQUE ◇ FRAMYCETIN ◇ SOFRAMYCIN

TOXICITY DATA with REFERENCE
ivn-mus LD50:24 mg/kg JANTAJ 27,677,74
ipr-mus LD50:250 mg/kg 85ERAY 1,741,78
scu-mus LD50:220 mg/kg HBTXAC 2,121,57
ivn-mus LD50:65 mg/kg APFRAD 11,44,53

SAFETY PROFILE: Poison by intraperitoneal, subcutaneous, and intravenous routes. When heated to decomposition it emits toxic fumes of NO_x.

NCF300 CAS:66-86-4 ***HR: 3***
NEOMYCIN C
mf: $C_{23}H_{46}N_6O_{13}$ mw: 614.75

TOXICITY DATA with REFERENCE
ipr-mus LD50:116 mg/kg 85GDA2 1,126,80
scu-mus LD50:290 mg/kg HBTXAC 2,121,57
ivn-mus LD50:44 mg/kg JANTAJ 27,677,74

SAFETY PROFILE: Poison by subcutaneous, intravenous, and intraperitoneal routes. When heated to decomposition it emits toxic fumes of NO_x.

NCF500 CAS:7542-37-2 ***HR: 3***
NEOMYCIN E
mf: $C_{23}H_{45}N_5O_{14}$ mw: 615.73

SYNS: AMINOSIDIN ◇ ANTIBIOTIC SF 767B ◇ CATENULIN

◇ CRESTOMYCIN ◇ ESTOMYCIN ◇ GABBROMYCIN ◇ HUMATIN ◇ HUMYCIN ◇ HYDROXYMYCIN ◇ MONOMYCIN A ◇ PARGONYL ◇ PAROMOMYCIN ◇ PAUCIMYCIN ◇ QUINTOMYCIN C ◇ R 400 ◇ ZYGOMYCIN A1

TOXICITY DATA with REFERENCE
orl-rat LD50:21620 mg/kg NKRZAZ 16,124,68
scu-rat LD50:1010 mg/kg ANTCAO 12,243,62
ivn-rat LD50:156 mg/kg ACHTA6 9,730,59
ims-rat LD50:1200 mg/kg NKRZAZ 16,124,68
orl-mus LD50:2275 mg/kg 85ERAY 1,674,78
ipr-mus LD50:930 mg/kg 85FZAT -,332,67
scu-mus LD50:423 mg/kg ACHTA6 9,730,59
ivn-mus LD50:90 mg/kg 85ERAY 1,674,78
ivn-mky LDLo:80 mg/kg ACHTA6 9,730,59

SAFETY PROFILE: Poison by intravenous route. Moderately toxic by ingestion, intramuscular, intraperitoneal, and subcutaneous routes. When heated to decomposition it emits toxic fumes of NO_x.

NCG000 CAS:1405-10-3 ***HR: 3***
NEOMYCIN SULFATE

SYNS: BIOSOL VETERINARY ◇ FRADIOMYCIN SULFATE ◇ LIDAMYCIN CREME ◇ MYCAIFRADIN SULFATE ◇ MYCIFRADIN-N ◇ MYCIGIENT ◇ NEOBIOTIC ◇ NEO-MANTLE CREME ◇ NEOMIX ◇ NEOMYCINE SULFATE ◇ NEOMYCIN SULPHATE ◇ OTOBIOTIC ◇ QUINTESS-N ◇ USAF CB-19

TOXICITY DATA with REFERENCE
skn-hmn 6 mg/3D-I MLD 85DKA8 -,127,77
orl-wmn TDLo:12600 mg/kg/7D:CNS ARSUAX 111,822,76
scu-rat TDLo:534 mg/kg KSRNAM 13,7,79
ipr-mus LD50:300 mg/kg NTIS** AD277-689
scu-mus LD50:190 mg/kg ANTBAL 18,444,73
ivn-mus LD50:17400 μg/kg AITEAT 10,947,62
ims-mus LD50:142 mg/kg AIMDAP 119,493,67

CONSENSUS REPORTS: Reported in EPA TSCA Inventory. EPA Genetic Toxicology Program.

SAFETY PROFILE: Poison by intraperitoneal, intramuscular, intravenous, and subcutaneous routes. Human systemic effects by ingestion: somnolence, hallucinations and distorted perceptions and anorexia. A human skin irritant. When heated to decomposition it emits very toxic fumes of SO_x. See also NEOMYCIN.

NCG500 CAS:7440-01-9 ***HR: 1***
NEON
DOT: UN 1065/UN 1913
af: Ne aw: 20.18

PROP: Colorless, inert, gaseous element; odorless. Mp: −248.67°, bp: −245.9°. D: (liquid): 1.204 @ −245.9°; (gas) 0.89994 g/L @ 0°.

SYNS: NEON, compressed (DOT) ◇ NEON, refrigerated liquid (DOT)

CONSENSUS REPORTS: Reported in EPA TSCA Inventory.

ACGIH TLV: Simple Asphyxiant
DOT Classification: Nonflammable Gas; Label: Nonflammable Gas.

SAFETY PROFILE: An inert asphyxiant gas. See also ARGON.

NCG700 CAS:63035-21-2 ***HR: 3***
NEONOL

TOXICITY DATA with REFERENCE
orl-rat LD50:6875 mg/kg GISAAA 47(2),85,82
orl-rbt LDLo:135 mg/kg JPHAA3 25,597,36
ipr-rbt LDLo:110 mg/kg JPHAA3 25,597,36

SAFETY PROFILE: Poison by ingestion and intraperitoneal routes.

NCG715 CAS:81296-95-9 ***HR: 2***
NEONOL AF-14

TOXICITY DATA with REFERENCE
orl-rat LD50:1800 mg/kg GISAAA 48(1),84,83
orl-mus LD50:1800 mg/kg GISAAA 48(1),84,83
orl-rbt LD50:1800 mg/kg GISAAA 48(1),84,83
orl-gpg LD50:1800 mg/kg GISAAA 48(1),84,83

SAFETY PROFILE: Moderately toxic by ingestion.

NCG725 ***HR: 2***
NEONOL 2B1317-12

TOXICITY DATA with REFERENCE
orl-rat LD50:1800 mg/kg GISAAA 48(1),84,83
orl-mus LD50:1800 mg/kg GISAAA 48(1),84,83
orl-rbt LD50:1800 mg/kg GISAAA 48(1),84,83
orl-gpg LD50:1800 mg/kg GISAAA 48(1),84,83

SAFETY PROFILE: Moderately toxic by ingestion.

NCH000 CAS:463-82-1 ***HR: 3***
NEOPENTANE
DOT: UN 1265/UN 2044
mf: C_5H_{12} mw: 72.17

PROP: Liquid or gas. Solidifies @ −19.8°, bp: 9.5°, d: 0.613° @ 0°/0° (liquid), flash p: <19.4°F, lel: 1.4%, uel: 7.5%. Insol in water.

SYNS: 2,2-DIMETHYLPROPANE (DOT) ◇ tert-PENTANE (DOT)

TOXICITY DATA with REFERENCE
ipr-mus LD50:100 mg/kg HINAAU 10,206,68

CONSENSUS REPORTS: Reported in EPA TSCA Inventory.

NIOSH REL: TWA 120 ppm; CL 610 ppm/15M
DOT Classification: Flammable Liquid; Label: Flammable Liquid; Flammable Gas; Label: Flammable Gas.

SAFETY PROFILE: Poison by intraperitoneal route. Both the gas and the liquid are flammable when exposed to heat or flame; can react vigorously with oxidizing materials. When heated to decomposition it emits acrid smoke and irritating fumes.

NCH500 CAS:78-12-6 ***HR: 2***
1,1′,1″,1‴-(NEOPENTANE TETRAYLTETRAOXY) TETRAKIS(2,2,2-TRICHLOROETHANOL)
mf: $C_{13}H_{16}Cl_{12}O_8$ mw: 725.69

SYN: PENTAERYTHRITOL CHLOROL ◇ PERICHLOR ◇ PERICHLORAL ◇ PERICLOR ◇ PETRICHLORAL

TOXICITY DATA with REFERENCE
orl-mus LD50:1030 mg/kg 27ZQAG -,422,72

SAFETY PROFILE: Moderately toxic by ingestion. When heated to decomposition it emits toxic fumes of Cl^-. See also CHLORINATED HYDROCARBONS, ALIPHATIC.

NCI300 CAS:17557-23-2 ***HR: 3***
NEOPENTYL GLYCOL DIGLYCIDYL ETHER
mf: $C_{11}H_{20}O_4$ mw: 216.31

SYNS: 1,3-BIS(2,3-EPOXYPROPOXY)-2,2-DIMETHYLPROPANE ◇ DIGLYCIDYL ETHER of NEOPENTYL GLYCOL ◇ 2,2′-((2,2-DIMETHYL-1,3-PROPANEDIYL)BIS(OXYMETHYLENE))BISOXIRANE ◇ HELOXY WC68

TOXICITY DATA with REFERENCE
skn-mus TDLo:9393 mg/kg/2Y-I:ETA NTIS** ORNL-5762

CONSENSUS REPORTS: Glycol ether compounds are on the Community Right-To-Know List. Reported in EPA TSCA Inventory.

SAFETY PROFILE: Questionable carcinogen with experimental tumorigenic data. When heated to decomposition it emits acrid smoke and irritating fumes. See also GLYCOL ETHERS.

NCI500 CAS:126-99-8 ***HR: 3***
NEOPRENE
DOT: UN 1991
mf: C_4H_5Cl mw: 88.54

$H_2C{=}CClCH{=}CH_2$

PROP: Colorless liquid. An oil-resistant synthetic rubber made by the polymerization of chloroprene. D: 0.958 @ 20°/20°, bp: 59.4°, flash p: −4°F, lel: 4.0%, uel: 20%, vap d: 3.0, brittle point: −35°, softens @ approx 80°. Sltly sol in water; misc in alc and ether.

SYNS: 2-CHLOOR-1,3-BUTADIEEN (DUTCH) ◇ 2-CHLOR-1,3-BUTADIEN (GERMAN) ◇ CHLOROBUTADIENE ◇ 2-CHLOROBUTA-1,3-DIENE ◇ 2-CHLORO-1,3-BUTADIENE ◇ CHLOROPREEN (DUTCH) ◇ CHLOROPREN (GERMAN, POLISH) ◇ CHLOROPRENE ◇ β-CHLOROPRENE (OSHA, MAK) ◇ CHLOROPRENE, inhibited (DOT) ◇ CHLOROPRENE, uninhibited (DOT) ◇ 2-CLORO-1,3-BUTADIENE (ITALIAN) ◇ CLOROPRENE (ITALIAN)

TOXICITY DATA with REFERENCE
mma-sat 2 pph/4H ARTODN 41,249,79
cyt-hmn-unr 1 mg/m^3 MUREAV 147,301,85
ihl-rat TCLo:4 mg/m^3/24H (female 3-4D post):TER GTPZAB 19(3),30,75
ihl-rat TCLo:10 ppm/4H (female 3-20D post):REP TXAPA9 44,81,78
orl-rat LD50:450 mg/kg 85GMAT -,38,82
ihl-rat LC50:11800 mg/m^3/4H 85GMAT -,38,82
scu-rat LDLo:500 mg/kg JIHTAB 18,240,36
orl-mus LD50:146 mg/kg 85GMAT -,38,82
ihl-mus LC50:2300 mg/m^3 ZKMAAX (6),66,69
scu-mus LDLo:1000 mg/kg JIHTAB 18,240,36
ihl-cat LCLo:1290 mg/m^3/8H JIDHAN 18,240,36
scu-cat LDLo:100 mg/kg JIDHAN 18,240,36
ivn-rbt LDLo:96 mg/kg SBLEA2 44,63,42

CONSENSUS REPORTS: IARC Cancer Review: Group 3 IMEMDT 7,160,87; Animal Inadequate Evidence IMEMDT 19,131,79; Human Inadequate Evidence IMEMDT 19,131,79. Reported in EPA TSCA Inventory. Community Right-To-Know List.

OSHA PEL: (Transitional: TWA 25 ppm (skin)) TWA 10 ppm (skin)
ACGIH TLV: TWA 10 ppm (skin)
DFG MAK: 10 ppm (36 mg/m^3)
NIOSH REL: CL (Chloroprene) 1 ppm/15M
DOT Classification: Flammable Liquid; Label: Flammable Liquid (UN1991); Forbidden (uninhibited); Flammable Liquid; Label: Flammable Liquid, Poison (UN1991).

SAFETY PROFILE: Poison by ingestion, intravenous, and subcutaneous routes. Moderately toxic by inhalation. An experimental teratogen. Experimental reproductive effects. Human mutation data reported. Human exposure has caused dermatitis, conjunctivitis, corneal necrosis, anemia, temporary loss of hair, nervousness, and irritability. Exposure to the vapor can cause respiratory tract irritation leading to asphyxia. Other effects are central nervous system depression, drop in blood pressure, severe degenerative changes in the liver, kidneys, lungs, and other vital organs. Questionable carcinogen. A very dangerous fire hazard when exposed to heat or flame. Explosive in the form of vapor when exposed to heat or flame. To fight fire, use alcohol foam. Auto-oxidizes in air to form an unstable peroxide which catalyzes exothermic polymerization of the monomer. Incompatible with liquid or gaseous fluorine. When heated to decomposition it emits toxic fumes of Cl^-. See also CHLORINATED HYDROCARBONS, ALIPHATIC.

NCI525 CAS:6795-60-4 ***HR: D***
NEOPROGESTIN
mf: $C_{20}H_{28}O_2$ mw: 300.48

PROP: Crystals from ethyl acetate + petr ether. Mp: 169-171°.

SYNS: 17-HYDROXY-19-NOR-17-α-PREGNA-4,20-DIEN-3-ONE ◇ 17-HYDROXY-17-α-VINYL-4-ESTREN-3-ONE ◇ NOR-PROGESTELEA ◇ NORVINISTERONE ◇ 19-NOR-17-α-VINYLTESTOSTERONE ◇ SC 4641 ◇ 17-VINYL-19-NORTESTOSTERONE ◇ 17-α-VINYL-19-NORTESTOSTERONE

TOXICITY DATA with REFERENCE
scu-rbt TDLo:250 μg/kg (female 5D pre):REP PSEBAA 94,717,57

SAFETY PROFILE: Experimental reproductive effects. A steroid. When heated to decomposition it emits acrid smoke and irritating fumes.

NCI550 ***HR: 2***
NEOPROSERINE

PROP: An anterior pituitary hormone (NIIRDN 6,559,82).

TOXICITY DATA with REFERENCE
scu-rat TDLo:8 mg/kg (8-15D preg):TER OYYAA2 12,789,76
scu-rat TDLo:800 mg/kg (8-15D preg):REP OYYAA2 12,789,76
ipr-rat LD50:1330 mg/kg NIIRDN 6,559,82

SAFETY PROFILE: Moderately toxic by intraperitoneal route. An experimental teratogen. Other experimental reproductive effects.

NCI600 CAS:124-97-0 ***HR: 3***
NEOPROTOVERATRINE
mf: $C_{41}H_{63}NO_{15}$ mw: 810.05

SYNS: NEOPROTOVERATRIN ◇ PROTOVERATRINE B ◇ VERATRIN ◇ VERATRINE

TOXICITY DATA with REFERENCE
scu-mus LD50:210 μg/kg JPETAB 113,89,55
ivn-mus LD50:65 μg/kg APFRAD 14,783,56
scu-rbt LD50:170 μg/kg APFRAD 14,783,56

SAFETY PROFILE: A deadly poison by subcutaneous and intravenous routes. When heated to decomposition it emits toxic fumes of NO_x.

NCJ000 CAS:69343-45-9 ***HR: 3***
NEOPSICAINE HYDROCHLORIDE
mf: $C_{19}H_{25}NO_4 \cdot ClH$ mw: 367.91

SYNS: 3-(BENZOYLOXY)-8-METHYL-8-AZABICYCLO(3.2.1)OCTANE-2-CARBOXYLIC ACID PROPYL ESTER, HYDROCHLORIDE (1R-(2-endo,3-exo)) ◇ PSICAINE-NEU HYDROCHLORIDE ◇ PSICAIN-NEW HYDROCHLORIDE

TOXICITY DATA with REFERENCE
eye-rbt 2% MLD ARZNAD 8,181,58
ipr-rat LD50:41 mg/kg ARZNAD 8,181,58
ipr-mus LD50:57 mg/kg ARZNAD 8,181,58
scu-mus LD50:200 mg/kg ARZNAD 8,181,58
ivn-mus LDLo:20 mg/kg WDMU** -,-,36
scu-gpg LDLo:50 mg/kg FDHU** -,-,33

SAFETY PROFILE: Poison by intraperitoneal, intravenous, and subcutaneous routes. An eye irritant. When heated to decomposition it emits very toxic fumes of NO_x and HCl.

NCJ500 CAS:457-60-3 ***HR: 3***
NEOSALVARSAN
mf: $C_{13}H_{13}As_2N_2H_2O_4S \cdot Na$ mw: 466.17

PROP: Yellow powder.

SYNS: ((5-(3-AMINO-4-HYDROXYPHENYL)ARSENO)-2-HYDROXYANILINO)METHANOL SULFOXYLATE SODIUM ◇ ARSEVAN ◇ ARSPHENAMINE METHYLENESULFOXYLIC ACID SODIUM SALT ◇ COLLUNOVAR ◇ COLLUNOVER ◇ 3,3′-DIAMINO-4,4′-DIHYDROXY ARSENOBENZENE METHYLENESULFOXYLATE SODIUM ◇ MIARSENOL ◇ NEOARSOLUIN ◇ NEOARSPHENAMINE ◇ NOVARSAN ◇ NOVARSENOBENZOL ◇ NOVARSENOBILLON ◇ VETARSENOBILLON

TOXICITY DATA with REFERENCE
scu-rat LDLo:100 mg/kg BIZEA2 184,360,27
ivn-rat LDLo:280 mg/kg MADCAJ 6,195,37
ivn-mus LDLo:250 mg/kg HBAMAK 4,1289,35
ivn-rbt LDLo:250 mg/kg HBAMAK 4,1289,35

CONSENSUS REPORTS: Arsenic and its compounds are on the Community Right-To-Know List.

OSHA PEL: TWA 0.5 mg(As)m^3
ACGIH TLV: TWA 0.2 mg(As)/m^3

SAFETY PROFILE: Poison by intravenous and subcutaneous routes. When heated to decomposition it emits very toxic fumes of As, NO_x, Na_2O, and SO_x. See also ARSENIC COMPOUNDS.

NCK000 CAS:36519-25-2 ***HR: 3***
NEOSOLANIOL
mf: $C_{18}H_{26}O_8$ mw: 370.44

SYNS: 4-β,15-DIACETOXY-3-α,8-α-DIHYDROXY-12,13-EPOXYTRICHOTHEC-9-ENE ◇ 12,13-EPOXY-4-β,15-DIACETOXY-3-α,8-α-DIHYDROXYTRICHOTHEC-9-ENE

TOXICITY DATA with REFERENCE
skn-gpg 370 ng MLD FAATDF 4(2, Pt 2),S124,84
ipr-mus LD50:14500 μg/kg 85GDA2 6,190,81
orl-ckn LD50:24870 μg/kg AEMIDF 35,636,78

SAFETY PROFILE: Poison by ingestion and intraperitoneal routes. A skin irritant. When heated to decomposition it emits acrid smoke and irritating fumes.

NCK500 CAS:102418-06-4 ***HR: 3***
NEOSPIRAMYCIN

TOXICITY DATA with REFERENCE
scu-mus LD50:1187 mg/kg JJANAX 23,429,70
ivn-mus LD50:168 mg/kg JJANAX 23,429,70

SAFETY PROFILE: Poison by intravenous route. Moderately toxic by subcutaneous route. When heated to decomposition it emits toxic fumes of NO_x.

NCL000 CAS:1344-34-9 ***HR: 3***
NEOSTAM
mf: $C_{36}H_{49}N_3O_{22}Sb_3 \cdot Na$ mw: 1264.12

SYNS: GLUCOSTIBAMINE SODIUM ◇ GLUCOSTIMIDINE SODIUM ◇ NEOSTAM STIBAMINE GLUCOSIDE ◇ NITROGEN GLUCOSIDE of SODIUM-p-AMINOPHENYLSTIBONATE ◇ SODIUM-p-AMINOBENZENESTIBONATE GLUCOSIDE ◇ SODIUM STIBANILATE GLUCOSIDE ◇ STIBAMINE GLUCOSIDE

TOXICITY DATA with REFERENCE
ivn-mus LD50:1475 mg/kg JPETAB 81,224,44
ims-rbt LD50:91 mg/kg JPETAB 87,119,46

CONSENSUS REPORTS: Antimony and its compounds are on The Community Right-To-Know List.

OSHA PEL: TWA 0.5 mg(Sb)/m^3
ACGIH TLV: TWA 0.5 mg(Sb)/m^3
NIOSH REL: (Antimony) TWA 0.5 mg(Sb)/m^3

SAFETY PROFILE: Poison by intramuscular route. Moderately toxic by intravenous route. When heated to decomposition it emits toxic fumes of Sb, NO_x, and Na_2O. See also ANTIMONY COMPOUNDS.

NCL100 CAS:59-99-4 ***HR: 3***
NEOSTIGMINE
mf: $C_{12}H_{19}N_2O_2$ mw: 223.33

SYNS: EUSTIGMIN ◇ EUSTIGMINE ◇ (m-HYDROXYPHENYL) TRIMETHYLAMMONIUM DIMETHYLCARBAMATE (ester) ◇ INTRASTIGMINA ◇ JUVASTIGMIN ◇ NORMASTIGMIN ◇ PROSTIGMIN ◇ PROSTIGMINE ◇ VAGOSTIGMINE

TOXICITY DATA with REFERENCE
ivn-rat LD50:225 μg/kg AIPTAK 101,205,55
orl-mus LD50:12340 μg/kg JPETAB 117,75,56
ipr-mus LD50:380 μg/kg AIPTAK 192,88,71
scu-mus LD50:280 μg/kg JPPMAB 34,603,82
ivn-mus LD50:200 μg/kg JPETAB 121,113,57
scu-dog LDLo:769 μg/kg FEPRA7 5,184,46
ims-rbt LD50:340 μg/kg DCTODJ 3,319,80

SAFETY PROFILE: A deadly poison by ingestion, subcutaneous, intramuscular, intravenous, and intraperitoneal routes. When heated to decomposition it emits toxic fumes of NO_x and NH_3. See also CARBAMATES.

NCL300 CAS:5787-73-5 ***HR: 3***
NEOSUPRANOL HYDROCHLORIDE
mf: $C_8H_{19}N \cdot ClH$ mw: 165.74

SYNS: EA-1 HYDROCHLORIDE ◇ N,1-DIMETHYLHEXYLAMINE HYDROCHLORIDE ◇ 2-METHYLAMINO-HEPTANE, HYDROCHLORIDE ◇ OENETHYL HYDROCHLORIDE ◇ PACAMINE HYDROCHLORIDE

TOXICITY DATA with REFERENCE
orl-man TDLo:2857 μg/kg:CNS JAPMA8 39,12,50
scu-man TDLo:1428 μg/kg:CNS,SKN JAPMA8 39,12,50
orl-rat LD50:320 mg/kg JAPMA8 39,12,50
ipr-rat LD50:60 mg/kg JPETAB 81,235,44
scu-rat LD50:60 mg/kg JAPMA8 39,12,50
scu-mus LD50:88 mg/kg THERAP 6,123,51
ivn-mus LD50:36 mg/kg JAPMA8 39,12,50
ivn-dog LD50:26 mg/kg JAPMA8 39,12,50

SAFETY PROFILE: Poison by ingestion, subcutaneous, intravenous, and intraperitoneal routes. Human systemic effects by ingestion or subcutaneous routes: sweating and headache. When heated to decomposition it emits toxic fumes of NO_x and HCl. See also AMINES.

NCL500 CAS:59-42-7 ***HR: 3***
NEOSYNEPHRINE
mf: $C_9H_{13}NO_2$ mw: 167.23

SYNS: 1-α-HYDROXY-β-METHYLAMINO-3-HYDROXY-1-ETHYLBENZENE ◇ (R)-3-HYDROXY-α-((METHYLAMINO)METHYL) BENZENEMETHANOOL ◇ 1-m-HYDROXY-α-((METHYLAMINO) METHYL)-BENZYL ALCOHOL ◇ (−)-m-HYDROXY-α-(METHYLAMINOMETHYL)BENZYL ALCOHOL ◇ 1-1-(m-HYDROXYPHENYL)-2-METHYLAMINOETHANOL ◇ 1-(3-HYDROXYPHENYL)-N-METHYLETHANOLAMINE ◇ ISOPHRIN ◇ MESATON ◇ METAOXEDRIN ◇ METASYMPATOL ◇ METASYNEPHRINE ◇ m-METHYLAMINOETHANOLPHENOL ◇ MEZATON ◇ R(−)-MEZATON ◇ m-OXEDRINE ◇ (−)-m-OXEDRINE ◇ PHENYLEPHRINE ◇ (−)-PHENYLEPHRINE ◇ R(−)-PHENYLEPHRINE ◇ m-SYMPATHOL ◇ m-SYMPATOL ◇ m-SYNEPHRINE ◇ VISADRON

TOXICITY DATA with REFERENCE
ivn-dom TDLo:20 μg/kg (16W preg):REP AJOGAH 143,170,82
ivn-dom TDLo:40 μg/kg (16W preg):TER AJOGAH 143,170,82
ocu-wmn TDLo:188 μg/kg:CVS ANESAV 36,187,72
ocu-inf TDLo:5500 μg/kg:CVS ANESAV 36,187,72
orl-rat LD50:350 mg/kg AIPTAK 180,155,69
scu-rat LD50:28 mg/kg JPETAB 85,119,45
ipr-mus LDLo:60 mg/kg JPETAB 160,22,68
scu-mus LD50:875 mg/kg FATOAO 28,46,65
ivn-mus LD50:38 mg/kg APTOA6 38,474,76

SAFETY PROFILE: Poison by ingestion, subcutaneous, intravenous, intraperitoneal, and intraduodenal routes. Human systemic effects by ocular route: blood pressure increase. An experimental teratogen. Other experimental reproductive effects. A nasal decongestant. When heated to decomposition it emits toxic fumes of NO_x.

NCM275 CAS:67298-49-1 ***HR: 3***
NEOTHRAMYCIN
mf: $C_{13}H_{14}N_2O_4$ mw: 262.29

TOXICITY DATA with REFERENCE
orl-mus LD50:100 mg/kg JANTAJ 35,1093,82
ipr-mus LD50:22 mg/kg JANTAJ 35,1093,82
ivn-mus LD50:23100 μg/kg JANTAJ 35,1093,82

SAFETY PROFILE: Poison by ingestion, intravenous, and intraperitoneal routes. When heated to decomposition it emits toxic fumes of NO_x.

NCM700 CAS:555-89-5 ***HR: 1***
NEOTRAN
mf: $C_{13}H_{10}Cl_{12}O_2$ mw: 269.13

SYNS: BIS(p-CHLOROPHENOXY)METHANE ◇ DCPM ◇ DI-p-CHLORODIPHENOXYMETHANE ◇ DI-(p-CHLOROPHENOXY)METHANE ◇ DI-(4-CHLOROPHENOXY)METHANE ◇ ENT 15,208 ◇ K 1875 ◇ 1,1'-(METHYLENEBIS(OXY))BIS(4-CHLORO)BENZENE ◇ OXYTHANE

TOXICITY DATA with REFERENCE
orl-rat LD50:5800 mg/kg AIHOAX 1,341,50
ihl-rat LCl0:18000 ppm MEIEDD 10,176,83
orl-mus LD50:5800 mg/kg FMCHA2 -,C167,83

SAFETY PROFILE: Mildly toxic by ingestion and inhalation. When heated to decomposition it emits toxic fumes of Cl^-. See also CHLORINATED HYDROCARBONS, AROMATIC.

NCM800 CAS:26552-50-1 ***HR: 3***
NEPTAL
mf: $C_{12}H_{15}HgNO_6$ mw: 469.87

SYNS: ACETIC ACID, (o-((2-HYDROXY-3-HYDROXYMERCURI)PROPYL)CARBAMOYL)PHENOXY- ◇ (o-((2-HYDROXY-3-(HYDROXY MERCURE)PROPYL)CARBAMOYL)PHENOXY)ACETIC ACID ◇ HYDROXYMERCURIPROPANOLAMIDE of o-CARBOXYPHENOXYACETIC ACID ◇ MERCURY, (3-(α-CARBOXY-o-ANISAMIDO)-2-HYDROXYPROPYL)HYDROXY-

TOXICITY DATA with REFERENCE
ivn-rbt LDLo:12 mg/kg JPETAB 41,21,31
ACGIH TLV: TWA 0.1 $mg(Hg)/m^3$ (skin)
NIOSH REL: (Mercury, Inorganic): TWA 0.05 $mg(Hg)/m^3$

SAFETY PROFILE: Poison by intravenous route. When heated to decomposition it emits toxic fumes of NO_x and Hg.

NCN500 ***HR: 3***
NEPTUNIUM
af: Np aw: 237.048

PROP: Exists in α, β, and Γ forms. Mp: 640°, bp: 3902°, d: 20.45 @ 20°. The first synthetic transuranium element discovered. It is a silvery, radioactive, chemically active metal. Emits α particles and has a half-life of $2.2x10^6$ years.

SAFETY PROFILE: Strong radiotoxicity. The soluble compounds, including the nitrates and organic complexes (e.g., tributyl phosphate), are the most important industrially.

NCN800 CAS:56001-43-5 ***HR: 1***
NEROLIDYL ACETATE
mf: $C_{17}H_{28}O_2$ mw: 264.45

SYNS: 1,6,10-DODECATRIEN-3-OL, 3,7,11-TRIMETHYL-, ACETATE, (S-(Z))- ◇ 3,7,11-TRIMETHYL-1,6,10-DODECATRIEN-3-YL ACETATE

TOXICITY DATA with REFERENCE
skn-rbt 500 mg/24H MOD FCTXAV 17,875,79

CONSENSUS REPORTS: Reported in EPA TSCA Inventory.

SAFETY PROFILE: A skin irritant. When heated to decomposition it emits acrid smoke and irritating fumes.

NCO000 CAS:8016-38-4 ***HR: 1***
NEROLI OIL, TUNISIAN

PROP: Distilled from the blossoms of *Citrus aurantium* (FCTXAV 14,659,76).

SYN: NEROLI BIGARADE OIL, TUNISIAN

TOXICITY DATA with REFERENCE
orl-rat LD50:4550 mg/kg FCTXAV 14,813,76

CONSENSUS REPORTS: Reported in EPA TSCA Inventory.

SAFETY PROFILE: Mildly toxic by ingestion. When heated to decomposition it emits acrid smoke and irritating fumes.

NCO500 CAS:3915-83-1 ***HR: 1***
NERYL ISOVALERIANATE
mf: $C_{15}H_{26}O_2$ mw: 238.41

SYNS: (Z)-3,7-DIMETHYL-2,6-OCTADIENYL ESTER ISOVALERIC ACID ◇ 3,7-DIMETHYL-2-cis-6-OCTADIEN-1-YL ISOVALERATE ◇ NERYL-β-METHYL BUTYRATE

TOXICITY DATA with REFERENCE
skn-rbt 500 mg/24H MOD FCTXAV 14,815,76

CONSENSUS REPORTS: Reported in EPA TSCA Inventory.

SAFETY PROFILE: A skin irritant. When heated to decomposition it emits acrid smoke and irritating fumes. See also ESTERS.

NCP000 CAS:105-91-9 ***HR: 1***
NERYL PROPIONATE
mf: $C_{13}H_{22}O_2$ mw: 210.35

SYNS: cis-3,7-DIMETHYL-2,6-OCTADIEN-1-OL PROPIONATE ◇ (Z)-3,7-DIMETHYL-2,6-OCTADIEN-1-OL PROPIONATE ◇ PROPIONIC ACID-3,7-DIMETHYL-2,6-OCTADIEN-1-YL ESTER ◇ PROPIONIC ACID, NERYL ESTER

TOXICITY DATA with REFERENCE
skn-rbt 500 mg/24H MOD FCTXAV 14,629,76

CONSENSUS REPORTS: Reported in EPA TSCA Inventory.

SAFETY PROFILE: A skin irritant. When heated to decomposition it emits acrid smoke and irritating fumes. See also ESTERS.

NCP500 CAS:7783-33-7 ***HR: 3***
NESSLER REAGENT
DOT: UN 1643
mf: $HgI_4 2K$ mw: 786.39

SYNS: CHANNING'S SOLUTION ◇ MERCURIC POTASSIUM IODIDE ◇ MERCURIC POTASSIUM IODIDE, solid (DOT) ◇ MERCURY(II) POTASSIUM IODIDE ◇ POTASSIUM IODOHYDRAGYRATE ◇ POTASSIUM MERCURIC IODIDE ◇ POTASSIUM TETRAIODOMERCURATE(II) ◇ TETRAIODOMERCURATE(2-), DIPOTASSIUM

TOXICITY DATA with REFERENCE
skn-gpg LDLo:1000 mg/kg AEHLAU 11,201,65
ipr-gpg LDLo:1000 mg/kg AEHLAU 11,201,65

CONSENSUS REPORTS: Mercury and its compounds are on the Community Right-To-Know List. Reported in EPA TSCA Inventory.

OSHA PEL: (Transitional: CL 1 mg/10m^3) CL 0.1 mg(Hg)/m^3 (skin)
ACGIH TLV: TWA 0.1 mg(Hg)/m^3 (skin)
NIOSH REL: (Inorganic Mercury) TWA 0.05 mg(Hg)/m^3
DOT Classification: Poison B; Label: Poison.

SAFETY PROFILE: A poison. Moderately toxic by skin contact and intraperitoneal routes. When heated to decomposition it emits very toxic fumes of Hg, K_2O, and I^-. See also MERCURY and IODINE COMPOUNDS.

NCP550 CAS:56391-57-2 ***HR: 3***
NETILMICIN SULFATE
mf: $C_{21}H_{41}N_5O_7 \cdot 5/2H_2O_4S$ mw: 720.80

SYNS: CERTOMYCIN ◇ o-3-DEOXY-4-C-METHYL-3-(METHYLAMINO)-β-l-ARABINOPYRANOSYL-(1-6)-o-(2,b-DIAMINO-2,3,4,6-TETRADEOXY-α-d-glycero-HEX-4-ENOPYRANOSYL-(1-4))-2-DEOXY-N^1-ETHYL-d-STREPTAMINE SULFATE (2:5) (salt) ◇ NETILLIN ◇ NETROMYCIN ◇ NTL SULFATE ◇ SCH 20569

TOXICITY DATA with REFERENCE
ims-rat TDLo:138 mg/kg (7-17D preg):REP JJANAX 35,614,82
ims-rat TDLo:550 mg/kg (7-17D preg):TER JJANAX 35,614,82
ipr-rat LD50:270 mg/kg IYKEDH 16,866,85
scu-rat LD50:200 mg/kg IYKEDH 16,866,85
ivn-rat LD50:40700 μg/kg IYKEDH 16,866,85
ims-rat LD50:197 mg/kg IYKEDH 16,866,85
ipr-mus LD50:189 mg/kg AMACCQ 10,827,76
scu-mus LD50:265 mg/kg AACHAX 10,827,76
ivn-mus LD50:60613 μg/kg AMACCQ 10,827,76
ims-mus LD50:63100 μg/kg IYKEDH 16,866,85

SAFETY PROFILE: Poison by subcutaneous, intramuscular, intravenous, and intraperitoneal routes. An experimental teratogen. Other experimental reproductive effects. When heated to decomposition it emits toxic fumes of NO_x and SO_x.

NCP875 CAS:1438-30-8 ***HR: D***
NETROPSIN

TOXICITY DATA with REFERENCE
dnd-omi 200 μmol EJBCAI 26,81,72
dnd-omi 200 μmol EJBCAI 26,81,72

SAFETY PROFILE: Mutation data reported. When heated to decomposition it emits toxic fumes of NO_x and HCl. See also NETROPSIN HYDROCHLORIDE.

NCQ000 CAS:63770-20-7 ***HR: 3***
NETROSPIN HYDROCHLORIDE
mf: $C_{18}H_{26}N_{10}O_3 \cdot ClH$ mw: 467.00

SYN: CONGOCIDINE DIHYDROCHLORIDE

TOXICITY DATA with REFERENCE
cyt-mus:ast 10 μmol/L PHARAT 32,178,77
sce-ham:lng 100 μmol/L HEREAY 96,295,82
scu-mus LD50:70 mg/kg JACSAT 73,341,51
ivn-mus LD50:17 mg/kg JACSAT 73,341,51

SAFETY PROFILE: Poison by subcutaneous and intravenous routes. Mutation data reported. An antibacterial agent. When heated to decomposition it emits very toxic fumes of HCl and NO_x.

NCQ100 CAS:7654-03-7 ***HR: 3***
NEURALEX
mf: $C_{15}H_{16}N_2O$ mw: 240.33

PROP: Crystals from isopropyl ether or petr ether + benzene. Mp: 93-94°.

SYNS: BENMOXINE ◇ 1-(BENZOYL)-2-(α-METHYLBENZYL)HYDRA-

ZINE ◇ 75-20 C ◇ (α-METHYLBENZYL)-1-BENZOYL-2-HYDRAZINE ◇ NERUSIL ◇ 2-(1-PHENYLETHYL)HYDRAZIDE BENZOIC ACID

TOXICITY DATA with REFERENCE
orl-rat LD50:675 mg/kg 27ZQAG -,376,72
ipr-rat LD50:475 mg/kg 27ZQAG -,376,72
orl-mus LD50:250 mg/kg 27ZQAG -,376,72
ipr-mus LD50:130 mg/kg PPRPAS 21,621,66
orl-gpg LD50:460 mg/kg 27ZQAG -,376,72
ipr-gpg LD50:325 mg/kg 27ZQAG -,376,72

SAFETY PROFILE: Poison by ingestion and intraperitoneal routes. An antidepressant. When heated to decomposition it emits toxic fumes of NO_x.

NCQ500 CAS:102395-05-1 ***HR: 3***
NEUROSTAIN

PROP: Contains methylene blue, vitamin C and sodium bicarbonate (NIIRDN 6,552,82).

TOXICITY DATA with REFERENCE
scu-rat LD50:190 mg/kg NIIRDN 6,552,82
ivn-rat LD50:36 mg/kg NIIRDN 6,552,82
scu-mus LD50:164 mg/kg NIIRDN 6,552,82
ivn-mus LD50:39 mg/kg NIIRDN 6,552,82

SAFETY PROFILE: Poison by subcutaneous and intravenous routes. When heated to decomposition it emits toxic fumes of Na_2O. See also individual components.

NCQ600 ***HR: 2***
NEWLASE

TOXICITY DATA with REFERENCE
ipr-rat LD50:3280 mg/kg KSRNAM 8,3080,74
scu-rat LD50:10700 mg/kg KSRNAM 8,3080,74
ipr-mus LD50:3170 mg/kg KSRNAM 8,3080,74
scu-mus LD50:6350 mg/kg KSRNAM 8,3080,74

SAFETY PROFILE: Moderately toxic by intraperitoneal route.

NCQ820 CAS:464-45-9 ***HR: 1***
NGAI CAMPHOR
mf: $C_{10}H_{18}O$ mw: 154.28

SYNS: 1-2-BORNANOL ◇ (−)-BORNEOL ◇ (1S,2R,4S)-(−)-1-BORNEOL ◇ 1-BORNYL ALCOHOL ◇ 1-2-CAMPHANOL ◇ LINDEROL

TOXICITY DATA with REFERENCE
skn-rbt 500 mg/24H MLD FCTXAV 16,637,78
orl-rat LD50:5800 mg/kg FCTXAV 16,637,78

CONSENSUS REPORTS: Reported in EPA TSCA Inventory. EPA Genetic Toxicology Program.

SAFETY PROFILE: Mildly toxic by ingestion. A skin irritant. When heated to decomposition it emits acrid smoke and irritating fumes. See also ALCOHOLS.

NCQ900 CAS:59-67-6 ***HR: 3***
NIACIN
mf: $C_6H_5NO_2$ mw: 123.12

PROP: The anti-pellagra vitamin. Colorless needles or white crystalline powder; slt odor. Mp: 236°, subl above mp, d: 1.473. Sol in water and boiling alc; insol in most lipid solvents. Nonhygroscopic and stable in air.

SYNS: ACIDE NICOTINIQUE (FRENCH) ◇ ACIDUM NICOTINICUM ◇ AKOTIN ◇ ANTI-PELLAGRA VITAMIN ◇ APELAGRIN ◇ BIONIC ◇ 3-CARBOXYPYRIDINE ◇ DASKIL ◇ DAVITAMON PP ◇ DIREKTAN ◇ EFACIN ◇ NAH ◇ NAOTIN ◇ NICACID ◇ NICAMIN ◇ NICANGIN ◇ NICO ◇ NICO-400 ◇ NICOBID ◇ NICOCAP ◇ NICOCIDIN ◇ NICOCRISINA ◇ NICODAN ◇ NICODELMINE ◇ NICOLAR ◇ NICONACID ◇ NICONAT ◇ NICONAZID ◇ NICOROL ◇ NICOSIDE ◇ NICO-SPAN ◇ NICOSYL ◇ NICOTAMIN ◇ NICOTENE ◇ NICOTIL ◇ NICOTINE ACID ◇ NICOTINIC ACID ◇ NICOTINIPCA ◇ NICOTINOYL HYDRAZINE ◇ NICOTINSAURE (GERMAN) ◇ NICOVASAN ◇ NICOVASEN ◇ NICOVEL ◇ NICYL ◇ NIPELLEN ◇ PELLAGRAMIN ◇ PELLAGRA PREVENTIVE FACTOR ◇ PELLAGRIN ◇ PELONIN ◇ PEVITON ◇ PP FACTOR ◇ P.P. FACTOR-PELLAGRA PREVENTIVE FACTOR ◇ PYRIDINE-3-CARBONIC ACID ◇ PYRIDINE-β-CARBOXYLIC ACID ◇ PYRIDINE-3-CARBOXYLIC ACID ◇ 3-PYRIDINECARBOXYLIC ACID ◇ PYRIDINE-CARBOXYLIQUE-3 (FRENCH) ◇ S115 ◇ SK-NIACIN ◇ TINIC ◇ VITAPLEX N ◇ WAMPOCAP

TOXICITY DATA with REFERENCE
orl-mus TDLo:174 g/kg/94W-C:CAR ONCOBS 38,106,81
orl-rat LD50:7000 mg/kg NIIRDN 6,544,82
ipr-rat LD50:730 mg/kg OYYAA2 14,741,77
scu-rat LD50:5000 mg/kg NIIRDN 6,544,82
ivn-rat LDLo:3500 mg/kg PSEBAA 38,241,38
orl-mus LD50:3720 mg/kg OYYAA2 7,149,73
ipr-mus LD50:358 mg/kg OYYAA2 14,741,77
scu-mus LD50:3500 mg/kg OYYAA2 14,741,77
ivn-mus LD50:5000 mg/kg THERAP 26,831,71
orl-rbt LD50:4550 mg/kg OYYAA2 14,741,77
ivn-gpg LDLo:3500 mg/kg PSEBAA 38,241,38

CONSENSUS REPORTS: Reported in EPA TSCA Inventory.

SAFETY PROFILE: Poison by intraperitoneal route. Moderately toxic by ingestion, intravenous, and subcutaneous routes. Questionable carcinogen with experimental carcinogenic data. When heated to decomposition it emits toxic fumes of NO_x.

NCR000 CAS:98-92-0 ***HR: 2***
NIACINAMIDE
mf: $C_6H_6N_2O$ mw: 122.14

PROP: Colorless needles or white crystalline powder; odorless with a bitter taste. Mp: 129°, d: 1.40. Very sol in water, ether, glycerin.

SYNS: ACID AMIDE ◇ AMIDE PP ◇ AMINICOTIN ◇ AMIXICOTYN ◇ AMNICOTIN ◇ AUSTROVIT PP ◇ BENICOT ◇ DELONIN AMIDE ◇ DIPEGYL ◇ DIPIGYL ◇ ENDOBION ◇ FACTOR PP ◇ HANSAMID ◇ INOVITAN PP ◇ NAM ◇ NANDERVIT-N ◇ NIACEVIT ◇ NIAMIDE

◇ NICAMIDE ◇ NICAMINA ◇ NICAMINDON ◇ NICASIR ◇ NICOBION ◇ NICOFORT ◇ NICOGEN ◇ NICOMIDOL ◇ NICOSAN 2 ◇ NICOTA ◇ NICOTAMIDE ◇ NICOTILAMIDE ◇ NICOTILILAMIDO ◇ NICOTINE ACID AMIDE ◇ NICOTINIC ACID AMIDE ◇ NICOTINIC AMIDE ◇ NICOTINSAUREAMID (GERMAN) ◇ NICOTOL ◇ NICOTYLAMIDE ◇ NICOVEL ◇ NICOVIT ◇ NICOVITOL ◇ NICOZYMIN ◇ NIKO-TAMIN ◇ NIKOTINSAEUREAMID (GERMAN) ◇ NIOCINAMIDE ◇ NIOZYMIN ◇ PELMIN ◇ PELMINE ◇ PELONIN AMIDE ◇ PP-FACTOR ◇ PYRIDINE-3-CARBOXYLIC ACID AMIDE ◇ 3-PYRIDINECARBOXYLIC ACID AMIDE ◇ VI-NICOTYL ◇ VI-NICTYL ◇ VITAMIN B3 ◇ VITAMIN PP ◇ WITAMINA PP

TOXICITY DATA with REFERENCE
dni-rat:lvr 20 mmol/L JJIND8 69,1353,82
sce-ham:ovr 5 mmol/L MUREAV 126,63,84
scu-rat LD50:1680 mg/kg PSEBAA 62,19,46
ivn-rat LD50:2200 mg/kg JPETAB 76,75,42
orl-mus LD50:2500 mg/kg NIIRDN 6,545,82
ipr-mus LD50:2300 mg/kg CYLPDN 5,173,84
scu-mus LD50:2 g/kg QJPPAL 19,48,46

CONSENSUS REPORTS: Reported in EPA TSCA Inventory.

SAFETY PROFILE: Moderately toxic by ingestion, intravenous, intraperitoneal, and subcutaneous routes. Mutation data reported. When heated to decomposition it emits toxic fumes of NO_x.

NCS000 CAS:62765-93-9 ***HR: 3***
NIAX CATALYST ESN

PROP: Mixture of 95% dimethylaminopropionitrile and 5% bis-dimethylaminoethyl ether (DCTODJ 2,223,79).

TOXICITY DATA with REFERENCE
orl-rat LD50:2460 mg/kg DCTODJ 2,223,79
ipr-rat LDLo:2000 mg/kg JEPTDQ 4(2-3),555,80
skn-rbt LD50:445 mg/kg DCTODJ 2,223,79

CONSENSUS REPORTS: Cyanide and its compounds are on the Community Right-To-Know List.
NIOSH REL: (Niax ESN): Exposure to be minimized

SAFETY PROFILE: Moderately toxic by ingestion, skin contact, and intraperitoneal routes. When heated to decomposition it emits toxic fumes of NO_x and CN^-. See also 3-(DIMETHYLAMINO)PROPIONITRILE and BIS(2-DIMETHYLAMINOETHYL) ETHER.

NCT000 CAS:25791-96-2 ***HR: 2***
NIAX POLYOL LG-168

TOXICITY DATA with REFERENCE
skn-rbt 500 mg open MLD UCDS** 1/7/71
orl-rat LD50:2830 mg/kg UCDS** 1/7/71

CONSENSUS REPORTS: Reported in EPA TSCA Inventory.

SAFETY PROFILE: Moderately toxic by ingestion. A skin irritant.

NCT500 CAS:100215-34-7 ***HR: 1***
NIAX POLYOL LHT-42

TOXICITY DATA with REFERENCE
skn-rbt 500 mg open MLD UCDS** 4/29/69
orl-rat LD50:20 g/kg UCDS** 4/29/69

CONSENSUS REPORTS: Reported in EPA TSCA Inventory.

SAFETY PROFILE: Mildly toxic by ingestion. A skin irritant.

NCV500 CAS:102395-10-8 ***HR: 1***
NIAX TRIOL 6000

TOXICITY DATA with REFERENCE
skn-rbt 500 mg open MLD UCDS** 6/15/71
orl-rat LD50:57 g/kg UCDS** 6/15/71

SAFETY PROFILE: Mildly toxic by ingestion. A skin irritant.

NCW000 CAS:550-99-2 ***HR: 3***
NIAZOL
mf: $C_{14}H_{14}N_2 \cdot ClH$ mw: 246.76

SYNS: ALBALON LIQUIFILM ◇ CLERA ◇ COLDAN ◇ 4,5-DIHYDRO-2-(1-NAPHTHALENYLMETHYL)-1H-IMIDAZOLE MONOHYDROCHLORIDE ◇ NAPHAZOLINE HYDROCHLORIDE ◇ NAPHCON ◇ NAPHCON FORTE ◇ 2-(1-NAPHTHYLMETHYL)IMIDAZOLINE HYDROCHLORIDE ◇ 2-(1-NAPHTHYLMETHYL)-2-IMIDAZOLINE HYDROCHLORIDE ◇ PRIVINE HYDROCHLORIDE ◇ PRIZOLE HYDROCHLORIDE ◇ RHINANTIN ◇ RHINOPERD ◇ SANORIN-SPOFA ◇ STRICYLON ◇ VASOCON

TOXICITY DATA with REFERENCE
ipr-rat LD50:50 mg/kg JPETAB 86,284,46
scu-rat LD50:325 mg/kg JPETAB 86,284,46
ivn-rat LD50:6 mg/kg JPETAB 113,341,55
scu-mus LD50:170 mg/kg JPETAB 86,280,46
ivn-mus LD50:16500 μg/kg JPETAB 113,341,55
scu-rbt LD50:950 mg/kg JPETAB 86,284,46
ivn-rbt LD50:800 μg/kg JPETAB 86,284,46
ims-rbt LD50:950 μg/kg JPETAB 86,284,46

SAFETY PROFILE: Poison by intraperitoneal, subcutaneous, intravenous, and intramuscular routes. An FDA over the counter and proprietary drug. When heated to decomposition it emits very toxic fumes of NO_x and HCl.

NCW300 CAS:5868-05-3 ***HR: 1***
NICERITROL
mf: $C_{29}H_{24}N_4O_8$ mw: 556.57

PROP: Crystals. Mp: 160-164°.

SYNS: 2,2-BIS(((3-PYRIDINYLCARBONYL)OXY)METHYL)-1,3-PROPANEDIYL ESTER of 3-PYRIDINECARBOXYLIC ACID ◇ BURFOR ◇ CARDIOLIPOL ◇ PENTAERYTHRITOL TETRANICOTINATE ◇ PERYCIT ◇ SK 1

TOXICITY DATA with REFERENCE
orl-rat TDLo:42 g/kg (35D pre):REP OYYAA2 14,755,77
orl-rat LD50:20 g/kg NIIRDN 6,546,82
ipr-rat LD50:5000 mg/kg NIIRDN 6,546,82
scu-rat LD50:5000 mg/kg NIIRDN 6,546,82
orl-mus LD50:20 g/kg NIIRDN 6,546,82
ipr-mus LD50:5000 mg/kg NIIRDN 6,546,82
scu-mus LD50:5000 mg/kg NIIRDN 6,546,82
ipr-rbt LD50:5000 mg/kg NIIRDN 6,546,82

SAFETY PROFILE: Mildly toxic by ingestion. Experimental reproductive effects. When heated to decomposition it emits toxic fumes of NO_x. See also ESTERS.

NCW500 CAS:7440-02-0 ***HR: 3***
NICKEL
af: Ni aw: 58.71

PROP: A silvery-white, hard, malleable and ductile metal. D: 8.90 @ 25°, vap press: 1 mm @ 1810°. Crystallizes as metallic cubes. Mp: 1455°, bp: 2730°. Stable in air at room temp.

SYNS: C.I. 77775 ◇ Ni 270 ◇ NICKEL 270 ◇ NICKEL (DUST) ◇ NICKEL (ITALIAN) ◇ NICKEL PARTICLES ◇ NICKEL SPONGE ◇ Ni 0901-S ◇ Ni 4303T ◇ NP 2 ◇ RANEY ALLOY ◇ RANEY NICKEL

TOXICITY DATA with REFERENCE
otr-ham:kdy 400 mg/L IAPUDO 53,193,84
otr-ham:emb 5 μmol/L TOXID9 1,132,81
orl-rat TDLo:158 mg/kg (MGN):TER AEHLAU 23,102,71
scu-rat TDLo:3000 mg/kg/6W-I:ETA JNCIAM 16,55,55
ims-rat TDLo:56 mg/kg:CAR IAPUDO 53,127,84
par-rat TDLo:40 mg/kg/52W-I:ETA,TER AEHLAU 5,445,62
imp-rat TDLo:250 mg/kg:CAR JNCIAM 16,55,55
ims-mus TDLo:200 mg/kg:NEO NCIUS* PH 43-64-886,SEPT,70
imp-rbt TDLo:165 mg/kg/2Y-I:NEO,TER JNCIAM 16,55,55
orl-rat LDLo:5 g/kg FDRLI* 7684D,83
itr-rat LDLo:12 mg/kg NTIS** AEC-TR-6710
ivn-mus LDLo:50 mg/kg FATOAO 23,549,60
ivn-dog LDLo:10 mg/kg 14CYAT 2,1120,63
scu-rat LDLo:12500 μg/kg NTIS** PB158-508
ipr-rbt LDLo:7 mg/kg NTIS** PB158-508
scu-rbt LDLo:7500 μg/kg NTIS** PB158-508
orl-gpg LDLo:5 mg/kg AMPMAR 25,247,64

CONSENSUS REPORTS: NTP Fifth Annual Report on Carcinogens. IARC Cancer Review: Group 1 IMEMDT 7,264,87; Animal Sufficient Evidence IMEMDT 11,75,76; Animal Inadequate Evidence IMEMDT 2,126,73. Community Right-To-Know List. Reported in EPA TSCA Inventory.

OSHA PEL: TWA Soluble: Compounds: 0.1 mg(Ni)/m^3; Insoluble Compounds: 1 mg(Ni)/m^3
ACGIH TLV: TWA 1 mg(Ni)/m^3; (Proposed: TWA 0.05 mg(Ni)/m^3; Human Carcinogen)
DFG TRK: 0.5 mg/m^3; Human Carcinogen.
NIOSH REL: (Inorganic Nickel) TWA 0.015 mg(Ni)/m^3

SAFETY PROFILE: Confirmed carcinogen with experimental carcinogenic, neoplastigenic, tumorigenic, and teratogenic data. Poison by ingestion, intratracheal, intraperitoneal, subcutaneous, and intravenous routes. An experimental teratogen. Ingestion of soluble salts causes nausea, vomiting, diarrhea. Mutation data reported. Hypersensitivity to nickel is common and can cause allergic contact dermatitis, pulmonary asthma, conjunctivitis, and inflammatory reactions around nickel-containing medical implants and prostheses. Powders may ignite spontaneously in air. Reacts violently with F_2; NH_4NO_3; hydrazine; NH_3; (H_2 + dioxane); performic acid; P; Se; S; (Ti + $KClO_3$). Incompatible with oxidants (e.g., bromine pentafluoride; peroxyformic acid; potassium perchlorate; chlorine; nitryl fluoride; ammonium nitrate). Raney-nickel catalysts may initiate hazardous reactions with ethylene + aluminum chloride; p-dioxane; hydrogen; hydrogen + oxygen; magnesium silicate; methanol; organic solvents + heat; sulfur compounds. Nickel catalysts have caused many industrial accidents.

NCX000 CAS:373-02-4 ***HR: 3***
NICKEL(II) ACETATE (1:2)
mf: $C_4H_6O_4$•Ni mw: 176.81

PROP: Green prisms. Mp: decomp, d: 1.798.

SYNS: ACETIC ACID, NICKEL(2+) SALT ◇ NICKELOUS ACETATE

TOXICITY DATA with REFERENCE
pic-esc 160 μmol/L ENMUDM 6,59,84
dns-rat-ipr 129 μmol/kg/5D-I CRNGDP 6,1819,85
otr-ham:kdy 225 mg/L IAPUDO 53,193,84
ivn-ham TDLo:10 mg/kg (8D preg):REP ADTEAS 5,51,72
ims-rat TDLo:420 mg/kg/47W-I:NEO NCIUS* PH 43-64-886,JUL,68
imp-rat TDLo:95 mg/kg/78W-C:ETA PAACA3 5,50,64
orl-rat LD50:350 mg/kg PWPSA8 11,39,68
ipr-rat LD50:23 mg/kg PWPSA8 11,39,68
orl-mus LD50:410 mg/kg PWPSA8 11,39,68
ipr-mus LD50:32 mg/kg PWPSA8 11,39,68
scu-gpg LDLo:20 mg/kg JOHYAY 8,565,08

CONSENSUS REPORTS: NTP Fifth Annual Report on Carcinogens. IARC Cancer Review: Group 1 IMEMDT 7,264,87. Nickel and its compounds are on the Community Right-To-Know List. Reported in EPA TSCA Inventory.

OSHA PEL: (Transitional: TWA 1 mg/m^3) TWA 0.1 mg (Ni)/m^3
ACGIH TLV: TWA 0.1 mg(Ni)/m^3; (Proposed: TWA 0.05 mg(Ni)/m^3; Human Carcinogen)
NIOSH REL: (Inorganic Nickel) TWA 0.015 mg(Ni)/m^3

SAFETY PROFILE: Confirmed carcinogen with experimental neoplastigenic and tumorigenic data. Poison by ingestion, intraperitoneal, and subcutaneous routes. Experimental reproductive effects. Mutation data reported. When heated to decomposition it emits irritating fumes. See also NICKEL COMPOUNDS.

NCX500 CAS:6018-89-9 ***HR: 3***
NICKEL ACETATE TETRAHYDRATE
mf: $C_4H_6O_4$•Ni•$4H_2O$ mw: 248.89

TOXICITY DATA with REFERENCE
cyt-mus:mmr 100 μmol/L MUREAV 68,337,79
ipr-mus LD50:45700 μg/kg RCOCB8 30,133,80

CONSENSUS REPORTS: Nickel and its compounds are on the Community Right-To-Know List.

OSHA PEL: (Transitional: TWA 1 mg/m^3) TWA 0.1 mg (Ni)/m^3
ACGIH TLV: TWA 0.1 mg(Ni)/m^3; (Proposed: TWA 0.05 mg(Ni)/m^3; Human Carcinogen)
NIOSH REL: (Inorganic Nickel) TWA 0.015 mg(Ni)/m^3

SAFETY PROFILE: Suspected carcinogen. Poison by intraperitoneal route. Mutation data reported. When heated to decomposition it emits acrid smoke and irritating fumes. See also NICKEL(II) ACETATE and NICKEL COMPOUNDS.

NCY000 CAS:37227-61-5 ***HR: D***
NICKEL ALLOY, Ni,Be

SYN: BERYLLIUM-NICKEL ALLOY

CONSENSUS REPORTS: IARC Cancer Review: Animal Inadequate Evidence IMEMDT 23,143,80. Nickel and its compounds as well as beryllium and its compounds are on the Community Right-To-Know List.

SAFETY PROFILE: Questionable carcinogen. See also NICKEL COMPOUNDS and BERYLLIUM COMPOUNDS.

NCY050 CAS:15699-18-0 ***HR: 3***
NICKEL AMMONIUM SULFATE
mf: O_8S_2•Ni•$2H_4N$ mw: 286.93

PROP: Crystals.

SYNS: AMMONIUM DISULFATONICKELATE(II) ◇ AMMONIUM NICKEL SULFATE ◇ SULFURIC ACID, AMMONIUM NICKEL(2+) SALT (2:2:1)

TOXICITY DATA with REFERENCE
orl-rat LD50:400 mg/kg FDRLI* 8005A,84

CONSENSUS REPORTS: IARC Cancer Review: Animal Limited Evidence IMEMDT 49,257,90. Reported in EPA TSCA Inventory.

OSHA PEL: TWA 1 mg(Ni)/m^3
ACGIH TLV: TWA 1 mg(Ni)/m^3; (Proposed: TWA 0.05 mg(Ni)/m^3; Human Carcinogen)

SAFETY PROFILE: Poison by ingestion. When heated to decomposition it emits toxic fumes of NO_x, SO_x, and Ni.

NCY100 CAS:12035-52-8 ***HR: 2***
NICKEL ANTIMONIDE
mf: NiSb mw: 180.46

SYNS: ANTIMONY, compounded with NICKEL (1:1) ◇ NICKEL MONOANTIMONIDE

TOXICITY DATA with REFERENCE
ims-rat TDLo:172 mg/kg:CAR IAPUDO 53,127,84

CONSENSUS REPORTS: Reported in EPA TSCA Inventory.

SAFETY PROFILE: Questionable carcinogen with experimental carcinogenic data. When heated to decomposition it emits toxic fumes of Ni and Sb.

NCY125 CAS:12255-10-6 ***HR: 3***
NICKEL ARSENIDE SULFIDE
mf: AsNiS mw: 165.69

SYN: NICKEL SULFARSENIDE

TOXICITY DATA with REFERENCE
ims-rat TDLo:158 mg/kg:CAR IAPUDO 53,127,84
irn-rat TDLo:79022 μg/kg:NEO CRNGDP 5,1511,84

OSHA: Cancer Hazard
ACGIH TLV: TWA 1 mg(Ni)/m^3; (Proposed: TWA 0.05 mg(Ni)/m^3; Human Carcinogen)

SAFETY PROFILE: Confirmed human carcinogen with experimental carcinogenic and neoplastigenic data. When heated to decomposition it emits toxic fumes of Ni, As, and SO_x.

NCY500 CAS:3333-67-3 ***HR: 3***
NICKEL(II) CARBONATE (1:1)
mf: $CNiO_3$ mw: 118.72

PROP: Rhombic, light green crystals. Mp: decomp.

SYNS: BASIC NICKEL CARBONATE ◇ CARBONIC ACID, NICKEL SALT (1:1) ◇ C.I. 77779 ◇ NICKELOUS CARBONATE

TOXICITY DATA with REFERENCE
dnd-rat-ipr 10 mg/kg/3H CNREA8 42,3544,82

oms-rat-ipr 40 mg/kg CNREA8 44,3892,84
sce-ham:ovr 400 nmol/L ENMUDM 7,381,85
imp-rat TDLo:95 mg/kg/78W-C:ETA PAACA3 5,50,64
scu-gpg LDLo:32 mg/kg JOHYAY 8,565,08

CONSENSUS REPORTS: NTP Fifth Annual Report on Carcinogens. IARC Cancer Review: Group 1 IMEMDT 7,264,87; Animal Sufficient Evidence IMEMDT 11,75,76. Nickel and its compounds are on the Community Right-To-Know List. Reported in EPA TSCA Inventory.

OSHA PEL: TWA 1 mg(Ni)/m^3
ACGIH TLV: TWA 1 mg(Ni)/m^3; (Proposed: TWA 0.05 mg(Ni)/m^3; Human Carcinogen)
DFG TRK: 0.5 mg/m^3; Human Carcinogen.
NIOSH REL: (Inorganic Nickel) TWA 0.015 mg(Ni)/m^3

SAFETY PROFILE: Confirmed carcinogen with experimental carcinogenic and tumorigenic data. Poison by subcutaneous route. Mutation data reported. See also NICKEL COMPOUNDS.

NCZ000 CAS:13463-39-3 ***HR: 3***
NICKEL CARBONYL
DOT: UN 1259
mf: C_4NiO_4 mw: 170.75

$Ni(CO)_4$

PROP: Colorless, volatile liquid or needles. Bp: 43°, mp: −19.3°, lel: 2% @ 20°, d: 1.3185 @ 17°, vap press: 400 mm @ 25.8°, flash p: < −4°. Oxidizes in air. Sol in alc, benzene, chloroform, acetone, and carbon tetrachloride.

SYNS: NICKEL CARBONYLE (FRENCH) ◇ NICHEL TETRACARBONILE (ITALIAN) ◇ NICKEL TETRACARBONYL ◇ NICKEL TETRACARBONYLE (FRENCH) ◇ NIKKELTETRACARBONYL (DUTCH) ◇ RCRA WASTE NUMBER P073

TOXICITY DATA with REFERENCE
ihl-rat TCLo:160 mg/m^3/15M (female 7D post):REP SCIEAS 203,550,79
ihl-ham TCLo:60 mg/m^3/15M (female 4D post):TER 48ECAY 3,301,80
ihl-rat TCLo:250 mg/m^3/30M/1Y:ETA AJCPAI 35,203,61
ivn-rat TDLo:54 mg/kg/20W-I:CAR CNREA8 32,2253,72
ihl-hmn TCLo:7 mg/m^3:CNS GISAAA 22(11),30,57
ihl-hmn LCLo:30 ppm/30M AEHLAU 23,373,71
ihl-rat LC50:35 ppm/30M AJCPAI 26,107,56
ipr-rat LD50:39 mg/kg AEHLAU 14,604,67
scu-rat LD50:63 mg/kg AEHLAU 14,604,67
ivn-rat LD50:66 mg/kg AEHLAU 14,604,67
ihl-mus LC50:67 mg/m^3/30M AMIHBC 8,48,53
ihl-dog LCLo:360 ppm/90M JOHYAY 7,525,07
ihl-cat LC50:1890 mg/m^3/30M EQSSDX 1,1,75
ihl-rbt LCLo:300 mg/m^3/30M YAKUD5 22,455,80

CONSENSUS REPORTS: NTP Fifth Annual Report on Carcinogens. IARC Cancer Review: Group 1 IMEMDT 7,264,87; Animal Limited Evidence IMEMDT 2,126,73; IMEMDT 11,75,76. EPA Extremely Hazardous Substances List. Nickel and its compounds are on the Community Right-To-Know List. Reported in EPA TSCA Inventory.

OSHA PEL: TWA 0.001 ppm (Ni)
ACGIH TLV: TWA 0.05 ppm (Ni); (Proposed: TWA 0.05 mg(Ni)/m^3; Human Carcinogen)
DFG TRK: 0.1 ppm; Animal Carcinogen, Suspected Human Carcinogen.
NIOSH REL: (Nickel Carbonyl) TWA 0.001 ppm
DOT Classification: Poison B; Label: Poison, Flammable Liquid; Flammable Liquid; Label: Flammable Liquid and Poison.

SAFETY PROFILE: Confirmed carcinogen with experimental carcinogenic, tumorigenic, and teratogenic data. A human poison by inhalation. Poison experimentally by inhalation, intravenous, subcutaneous, and intraperitoneal routes. An experimental teratogen. Other experimental reproductive effects. Human systemic effects by inhalation: somnolence, fever and other pulmonary changes. Vapors may cause coughing, dyspnea (difficult breathing), irritation, congestion and edema of the lungs, tachycardia (rapid pulse), cyanosis, headache, dizziness, weakness. Toxicity by inhalation is believed to be caused by both the nickel and carbon monoxide liberated in the lungs. Chronic exposure may cause cancer of lungs, nasal sinuses. Sensitization dermatitis is fairly common. Probably the most hazardous compound of nickel in the workplace. A common air contaminant. It is lipid soluble and can cross biological membranes (e.g., lung alveolus, blood-brain barrier, placental barrier).

A very dangerous fire hazard when exposed to heat, flame, or oxidizers. Moderate explosion hazard when exposed to heat or flame. Explodes when heated to about 60°. Explosive reaction with liquid bromine; mercury + oxygen; oxygen + butane. Violent reaction with dinitrogen tetraoxide; air; oxygen. Reacts with tetrachloropropadiene to form the extremely sensitive explosive dicarbonyl trichloropropenyl dinickel chloride dimer. Can react with oxidizing materials. To fight fire, use water, foam, CO_2, dry chemical. When heated to decomposition or on contact with acid or acid fumes it emits highly toxic fumes of carbon monoxide. See also NICKEL COMPOUNDS and CARBONYLS.

NDA000 CAS:7791-20-0 ***HR: 3***
NICKEL(II) CHLORIDE HEXAHYDRATE (1:2:6)
mf: $Cl_2Ni{\cdot}6H_2O$ mw: 237.73

PROP: A: Yellow, deliq scales; b: monoclinic, green crystals. A: $NiCl_2$, b: $NiCl_2{\cdot}6H_2O$; mw (a): 129.60, mw (b): 237.70, mp (a): subl, bp (a): 987°, d (a): 3.55, vap press: 1 mm @ 671°.

TOXICITY DATA with REFERENCE
cyt-mus:mmr 100 µmol/L MUREAV 68,337,79
sce-ham:fbr 32 mg/L MUREAV 104,141,82
ipr-mus LD50:48 mg/kg AEPPAE 244,17,62
ivn-dog LD50:40 mg/kg EQSSDX 1,1,75

CONSENSUS REPORTS: Nickel and its compounds are on the Community Right-To-Know List.

OSHA PEL: (Transitional: TWA 1 mg/m^3) TWA 0.1 mg (Ni)/m^3
ACGIH TLV: TWA 0.1 mg(Ni)/m^3; (Proposed: TWA 0.05 mg(Ni)/m^3; Human Carcinogen)
NIOSH REL: (Inorganic Nickel) TWA 0.015 mg(Ni)/m^3

SAFETY PROFILE: Suspected carcinogen. Poison by intraperitoneal and intravenous routes. Mutation data reported. Violent reaction with potassium. When heated to decomposition it emits very toxic fumes of Cl^-. See also NICKEL COMPOUNDS and CHLORIDES.

NDA500 CAS:1271-28-9 ***HR: 3***
NICKEL, COMPOUND with pi-CYCLOPENTADIENYL (1:2)
mf: $C_{10}H_{10}$•Ni mw: 188.91

SYNS: pi-CYCLOPENTADIENYL COMPOUND with NICKEL ◇ DI-pi-CYCLOPENTADIENYLNICKEL ◇ NICKEL BISCYCLOPENTADIENE ◇ NICKELOCENE

TOXICITY DATA with REFERENCE
ims-rat TDLo:600 mg/kg/52W-I:NEO NCIBR* PH43-64-886,DEC65
ims-rat TD:50 mg/kg:ETA NCIUS* PH-43-64-886,SEPT,65
orl-rat LD50:490 mg/kg PWPSA8 11,39,68
ipr-rat LD50:50 mg/kg NCIBR* PH43-64-886,JAN65
ims-rat LD50:82 mg/kg NCIBR* PH43-64-886,AUG64
orl-mus LD50:600 mg/kg PWPSA8 11,39,68
ipr-mus LD50:86 mg/kg PWPSA8 11,39,68

CONSENSUS REPORTS: NTP Fifth Annual Report on Carcinogens. IARC Cancer Review: Group 1 IMEMDT 7,264,87; Animal Sufficient Evidence IMEMDT 11,75,76; Animal Inadequate Evidence IMEMDT 2,126,73. Nickel and its compounds are on the Community Right-To-Know List. Reported in EPA TSCA Inventory.

SAFETY PROFILE: Confirmed carcinogen with experimental carcinogenic, neoplastigenic, and tumorigenic data. Poison by intraperitoneal and intramuscular routes. Moderately toxic by ingestion. When heated to decomposition it emits acrid smoke and irritating fumes. See also NICKEL COMPOUNDS.

NDB000 ***HR: 3***
NICKEL COMPOUNDS

CONSENSUS REPORTS: Nickel and its compounds are on the Community Right-To-Know List.

OSHA PEL: TWA Soluble: Compounds: 0.1 mg(Ni)/m^3; Insoluble Compounds: 1 mg(Ni)/m^3
ACGIH TLV: TWA 1 mg/m^3; (Proposed: TWA 0.05 mg(Ni)/m^3; Human Carcinogen)
DFG MAK: Human Carcinogen.
NIOSH REL: (Inorganic Nickel) TWA 0.015 mg(Ni)/m^3

SAFETY PROFILE: Many are human carcinogens by inhalation. Nickel and many of its compounds are poisons and carcinogens. All airborne nickel contaminating dusts are regarded as carcinogenic by inhalation. Nickel carbonyl is probably the most hazardous compound of nickel in the workplace. It is carcinogenic and highly irritating to the lungs and can produce asphyxia by decomposing to form carbon monoxide. Nickel chloride ($NiCl_2$), sulfate ($NiSO_4 \cdot 6H_2O$), nitrate [$Ni(NO_3)_2 \cdot 6H_2O$], carbonate ($NiCO_3$), hydroxide [$Ni(OH)_2$] and acetate [$Ni(COOCH_3)_2$] are the salts of greatest commercial importance.

Ingestion of large doses of nickel compounds (1-3 mg/kg) has been shown to cause intestinal disorders, convulsions, and asphyxia. Hypersensitivity to nickel is common and can cause allergic contact dermatitis, pulmonary asthma, conjunctivitis, and inflammatory reactions around nickel-containing medical implants and prostheses. The most common effect resulting from exposure to nickel compounds is the development of "nickel itch". It occurs primarily in persons doing nickel-plating and is most frequent under conditions of high temperature and humidity, when the skin is moist, and mainly affects the hands and arms. There is marked variation in individual susceptibility to the dermatitis.

Nickel refinery workers experience increased mortality rates from cancer of the lungs and nasal cavities attributed to inhalation of airborne nickel compounds. Cancer develops in rodents after administration of Ni_3S_2, NiO, and $Ni(CO)_4$. Nickel chloride, sulfate, carbonate, and carbonyl are experimental teratogens.

Pulmonary damage develops in rodents chronically exposed to aerosols of nickel dust, $NiCl_2$, or NiO. Divalent nickel salts cause hyperglycemia, immune system effects, kidney damage, liver damage, and heart effects in experimental animals by parenteral administration. These compounds are common air contaminants. See also NICKEL and specific compounds.

NDB500 CAS:557-19-7 ***HR: 3***
NICKEL CYANIDE (solid)
DOT: UN 1653
mf: C_2N_2Ni mw: 110.75

PROP: Apple-green plates or powder.

SYNS: NICKEL CYANIDE (DOT) ◇ RCRA WASTE NUMBER P074

CONSENSUS REPORTS: Cyanide and its compounds, as well as nickel and its compounds, are on the Community Right-To-Know List. Reported in EPA TSCA Inventory.

OSHA PEL: (Transitional: TWA 1 mg/m^3) TWA 0.1 mg $(Ni)/m^3$
ACGIH TLV: TWA 0.1 $mg(Ni)/m^3$; (Proposed: TWA 0.05 $mg(Ni)/m^3$; Human Carcinogen)
DOT Classification: Poison B; Label: Poison.

SAFETY PROFILE: Suspected carcinogen. A poison. Incandescent reaction when heated with magnesium. When heated to decomposition it emits very toxic fumes of CN^-. See also CYANIDES and NICKEL COMPOUNDS.

NDB875 CAS:12035-51-7 ***HR: 2***
NICKEL DISULFIDE
mf: NiS_2 mw: 122.83

SYN: NICKEL SULFIDE

TOXICITY DATA with REFERENCE
ims-rat TDLo:117 mg/kg:CAR IAPUDO 53,127,84
irn-rat TDLo:58580 μg/kg:NEO CRNGDP 5,1511,84

SAFETY PROFILE: Questionable carcinogen with experimental carcinogenic and neoplastigenic data. When heated to decomposition it emits toxic fumes of SO_x.

NDC000 CAS:14708-14-6 ***HR: 2***
NICKEL(II) FLUOBORATE
mf: B_2F_8•Ni mw: 232.33

SYNS: NICKEL BOROFLUORIDE ◇ NICKEL FLUOROBORATE ◇ NICKELOUS TETRAFLUOROBORATE ◇ NICKEL(II) TETRAFLUOROBORATE ◇ TL 1091

TOXICITY DATA with REFERENCE
orl-rat LDLo:500 mg/kg NCNSA6 5,28,53
ihl-mus LCLo:530 mg/m^3/10M NDRC** No.9-4-1-19,44

CONSENSUS REPORTS: Nickel and its compounds are on the Community Right-To-Know List. Reported in EPA TSCA Inventory.

OSHA PEL: (Transitional: TWA 1 mg/m^3) TWA 0.1 mg $(Ni)/m^3$; 2.5 $mg(F)/m^3$
ACGIH TLV: TWA 0.1 $mg(Ni)/m^3$; (Proposed: TWA 0.05 $mg(Ni)/m^3$; Human Carcinogen)
NIOSH REL: (Inorganic Nickel) TWA 0.015 $mg(Ni)/m^3$

SAFETY PROFILE: Suspected carcinogen. Moderately toxic by ingestion and inhalation. When heated to decomposition it emits very toxic fumes of F^-. See also FLUORIDES, BORON, and NICKEL COMPOUNDS.

NDC500 CAS:10028-18-9 ***HR: 3***
NICKEL(II) FLUORIDE (1:2)
mf: F_2Ni mw: 96.71

PROP: Green crystals. D: 4.63. Sltly water-sol; decomp by boiling water; insol in alc, ether.

SYNS: NICKEL DIFLUORIDE ◇ NICKELOUS FLUORIDE

TOXICITY DATA with REFERENCE
ivn-mus LD50:130 mg/kg 19UQAS -,30,65

CONSENSUS REPORTS: Nickel and its compounds are on the Community Right-To-Know List. Reported in EPA TSCA Inventory.

OSHA PEL: (Transitional: TWA 1 mg/m^3) TWA 0.1 mg $(Ni)/m^3$; 2.5 $mg(F)/m^3$
ACGIH TLV: TWA 0.1 $mg(Ni)/m^3$; (Proposed: TWA 0.05 $mg(Ni)/m^3$; Human Carcinogen)
NIOSH REL: (Inorganic Nickel) TWA 0.015 $mg(Ni)/m^3$

SAFETY PROFILE: Suspected carcinogen. Poison by intravenous route. Reacts violently with potassium. Chronic exposure may cause mottling of teeth, changes in bones. When heated to decomposition it emits toxic fumes of F^-. See also FLUORIDES and NICKEL COMPOUNDS.

NDD000 CAS:26043-11-8 ***HR: 3***
NICKEL(II) FLUOSILICATE (1:1)
mf: F_6Si•Ni mw: 200.80

SYN: HEXAFLUOROSILICATE(2–), NICKEL

TOXICITY DATA with REFERENCE
orl-rat LDLo:100 mg/kg NCNSA6 5,28,53

CONSENSUS REPORTS: Nickel and its compounds are on the Community Right-To-Know List. Reported in EPA TSCA Inventory.

OSHA PEL: (Transitional: TWA 1 mg/m^3) TWA 0.1 mg $(Ni)/m^3$
ACGIH TLV: TWA 1 $mg(Ni)/m^3$; (Proposed: TWA 0.05 $mg(Ni)/m^3$; Human Carcinogen)
NIOSH REL: (Inorganic Nickel) TWA 0.015 $mg(Ni)/m^3$

SAFETY PROFILE: Suspected carcinogen. Poison by ingestion. When heated to decomposition it emits toxic fumes of F^-. See also NICKEL COMPOUNDS.

NDD500 CAS:56668-59-8 ***HR: 3***
NICKEL-GALLIUM ALLOY

PROP: Nickel (60%) - gallium (40%) alloy (JDREAF 29,1023,60).

SYN: GALLIUM-NICKEL ALLOY

TOXICITY DATA with REFERENCE
imp-rat TDLo:86 mg/kg/30W-C:ETA JDREAF 39,1023,60

CONSENSUS REPORTS: Nickel and its compounds are on the Community Right-To-Know List.

OSHA PEL: TWA 1 mg(Ni)/m^3
ACGIH TLV: TWA 1 mg(Ni)/m^3; (Proposed: TWA 0.05 mg(Ni)/m^3; Human Carcinogen)
NIOSH REL: (Inorganic Nickel) TWA 0.015 mg(Ni)/m^3

SAFETY PROFILE: Suspected carcinogen with experimental tumorigenic data. See also NICKEL COMPOUNDS and GALLIUM COMPOUNDS.

NDE000 CAS:12125-56-3 ***HR: 3***
NICKEL(II) HYDROXIDE
mf: H_2NiO_2 mw: 92.73

PROP: Light green crystals or amorphous.

SYNS: NICKEL HYDROXIDE (DOT) ◇ NICKELOUS HYDROXIDE

TOXICITY DATA with REFERENCE
ims-rat TDLo:480 μg/kg:CAR CRNGDP 4,275,83
ims-rat TD:60 mg/kg:ETA 45ICAX -,59,80
orl-rat LD50:1500 mg/kg FDRLI* 7702C,83
scu-mus LD50:50 mg/kg ZVKOA6 19,186,74
orl-rat LD50:1500 mg/kg FDRLI* 7702C,83
scu-mus LD50:50 mg/kg ZVKOA6 19,186,74

CONSENSUS REPORTS: NTP Fifth Annual Report on Carcinogens. IARC Cancer Review: Group 1 IMEMDT 7,264,87; Animal Sufficient Evidence IMEMDT 11,75,76. Nickel and its compounds are on the Community Right-To-Know List. Reported in EPA TSCA Inventory.

OSHA PEL: (Transitional: TWA 1 mg/m^3) TWA 0.1 mg (Ni)/m^3
ACGIH TLV: TWA 0.1 mg(Ni)/m^3; (Proposed: TWA 0.05 mg(Ni)/m^3; Human Carcinogen)
NIOSH REL: (Inorganic Nickel) TWA 0.015 mg(Ni)/m^3

SAFETY PROFILE: Confirmed carcinogen with experimental carcinogenic and tumorigenic data. Poison by subcutaneous route. See also NICKEL COMPOUNDS.

NDE010 CAS:12125-56-3 ***HR: 3***
NICKEL(III) HYDROXIDE
mf: H_3NiO_3 mw: 109.74

SYNS: NICKEL BLACK ◇ NICKELIC HYDROXIDE

TOXICITY DATA with REFERENCE
scu-mus LD50:25 mg/kg ZVKOA6 19,186,74

CONSENSUS REPORTS: NTP Fifth Annual Report On Carcinogens. IARC Cancer Review: Animal Sufficient Evidence IMEMDT 11,75,76

OSHA PEL: TWA 1 mg(Ni)/m^3
ACGIH TLV: TWA 0.05 mg(Ni)/m^3; Confirmed Carcinogen

SAFETY PROFILE: Poison by intravenous route. When heated to decomposition it emits toxic fumes of Ni.

NDE500 CAS:59978-65-3 ***HR: 3***
NICKEL IRON SULFIDE
mf: $FeNi_4S_4$ mw: 418.93

SYNS: IRON NICKEL SULFIDE ◇ NICKEL-IRON SULFIDE MATTE

TOXICITY DATA with REFERENCE
ims-rat TDLo:147 mg/kg:NEO RCOCB8 14,319,76
ims-rat TD:400 mg/kg:CAR IAPUDO 53,127,84

CONSENSUS REPORTS: Nickel and its compounds are on the Community Right-To-Know List.

OSHA PEL: TWA 1 mg(Ni)/m^3
ACGIH TLV: TWA 1 mg(Ni)/m^3; (Proposed: TWA 0.05 mg(Ni)/m^3; Human Carcinogen)
NIOSH REL: (Inorganic Nickel) TWA 0.015 mg(Ni)/m^3

SAFETY PROFILE: Suspected carcinogen with experimental carcinogenic and neoplastigenic data. When heated to decomposition it emits toxic fumes of SO_x. See also NICKEL COMPOUNDS, SULFIDES, and IRON COMPOUNDS.

NDF000 CAS:74203-45-5 ***HR: 3***
NICKEL(II) ISODECYL ORTHOPHOSPHATE (3:2)

SYN: PHOSPHORIC ACID, ISODECYL NICKEL(2+) SALT (2:3)

TOXICITY DATA with REFERENCE
ims-rat TDLo:320 mg/kg/35W-I:NEO NCIUS* PH 43-64-886,DEC,68
ims-rat TD:1260 mg/kg/30W-I:ETA NCIUS* PH 43-64-886,AUG,69
ipr-rat LD50:400 mg/kg NCIUS* PH 43-64-886,JUL,68

CONSENSUS REPORTS: Nickel and its compounds are on the Community Right-To-Know List.

NIOSH REL: (Inorganic Nickel) TWA 0.015 mg(Ni)/m^3

SAFETY PROFILE: Poison by intraperitoneal route. Questionable carcinogen with experimental neoplastigenic and tumorigenic data. When heated to decomposition it emits very toxic fumes of PO_x. See also NICKEL COMPOUNDS.

NDF400 CAS:1314-05-2 ***HR: 2***
NICKEL MONOSELENIDE
mf: NiSe mw: 137.67

SYN: NICKEL SELENIDE

TOXICITY DATA with REFERENCE
ims-rat TDLo:131 mg/kg:CAR IAPUDO 53,127,84

SAFETY PROFILE: Questionable carcinogen with experimental carcinogenic data.

NDF500 CAS:1313-99-1 ***HR: 3***
NICKEL MONOXIDE
mf: NiO mw: 74.71

PROP: Cubic, green-black crystals; yellow when hot. Mp: 1900°, d: 7.45. Insol in water; sol in acids.

SYNS: BUNSENITE ◇ C.I. 77777 ◇ GREEN NICKEL OXIDE ◇ NICKELOUS OXIDE ◇ NICKEL OXIDE (MAK) ◇ NICKEL(II) OXIDE (1:1) ◇ NICKEL PROTOXIDE

TOXICITY DATA with REFERENCE
otr-mus:emb 500 μmol/L PAACA3 24,100,83
otr-ham:kdy 100 mg/L IAPUDO 53,193,84
ims-rat TDLo:71264 μg/kg:CAR IAPUDO 53,127,84
ipl-rat TDLo:40 mg/kg:ETA 55DXAE -,37,85
itr-rat LDLo:20 mg/kg NTIS** AEC-TR-6710
scu-mus LD50:50 mg/kg ZVKOA6 19,186,74
ivn-dog LDLo:9 mg/kg EQSSDX 1,1,75
ivn-cat LDLo:12700 μg/kg EQSSDX 1,1,75

CONSENSUS REPORTS: NTP Fifth Annual Report on Carcinogens. IARC Cancer Review: Group 1 IMEMDT 7,264,87; Animal Inadequate Evidence IMEMDT 2,126,73; Animal Sufficient Evidence IMEMDT 11,75,76. Nickel and its compounds are on the Community Right-To-Know List. Reported in EPA TSCA Inventory.

OSHA PEL: TWA 1 mg(Ni)/m^3
ACGIH TLV: TWA 1 mg(Ni)/m^3; (Proposed: TWA 0.05 mg(Ni)/m^3; Human Carcinogen)
DFG TRK: 0.5 mg/m^3; Human Carcinogen.
NIOSH REL: (Inorganic Nickel) TWA 0.015 mg(Ni)/m^3

SAFETY PROFILE: Confirmed carcinogen with experimental carcinogenic and tumorigenic data. Poison by intratracheal, intravenous, and subcutaneous routes. Mutation data reported. Can react violently with fluorine; hydrogen peroxide; hydrogen sulfide, iodine; barium oxide + air. See also NICKEL COMPOUNDS.

NDG000 CAS:13138-45-9 ***HR: 3***
NICKEL(II) NITRATE (1:2)
DOT: UN 2725
mf: N_2O_6•Ni mw: 182.73

PROP: Green, deliquescent crystals. Mp: 56.7°, bp: 136.7°, d: 2.05.

SYNS: NICKEL NITRATE (DOT) ◇ NITRIC ACID, NICKEL(II) SALT

TOXICITY DATA with REFERENCE
dlt-mus-unr 56 mg/kg MUREAV 97,180,82
unr-mus TDLo:1960 mg/kg (male 35D pre):REP MUREAV 97,180,82
ivn-mus LDLo:9 mg/kg FATOAO 23,549,60

CONSENSUS REPORTS: Nickel and its compounds are on the Community Right-To-Know List. Reported in EPA TSCA Inventory.

OSHA PEL: (Transitional: TWA 1 mg/m^3) TWA 0.1 mg (Ni)/m^3
ACGIH TLV: TWA 0.1 mg(Ni)/m^3; (Proposed: TWA 0.05 mg(Ni)/m^3; Human Carcinogen)
NIOSH REL: (Inorganic Nickel) TWA 0.015 mg(Ni)/m^3
DOT Classification: Oxidizer; Label:Oxidizer.

SAFETY PROFILE: Suspected carcinogen. Poison by intravenous route. Experimental reproductive effects. Mutation data reported. A powerful oxidizer. When heated to decomposition it emits very toxic fumes of NO_x. See also NICKEL COMPOUNDS and NITRATES.

NDG500 CAS:13478-00-7 ***HR: 2***
NICKEL(II) NITRATE, HEXAHYDRATE (1:2:6)
mf: N_2O_6•Ni•$6H_2O$ mw: 290.85

PROP: Green, deliquescent crystals. Mp: 56.7°, bp: 136.7°, d: 2.05. Sol in 0.4 parts water or alc. Keep well closed.

SYNS: NICKEL(2+) NITRATE, HEXAHYDRATE ◇ NITRIC ACID, NICKEL(2+) SALT, HEXAHYDRATE

TOXICITY DATA with REFERENCE
orl-rat LD50:1620 mg/kg AIHAAP 30,470,69

CONSENSUS REPORTS: Nickel and its compounds are on the Community Right-To-Know List.

OSHA PEL: (Transitional: TWA 1 mg/m^3) TWA 0.1 mg (Ni)/m^3
ACGIH TLV: TWA 0.1 mg(Ni)/m^3; (Proposed: TWA 0.05 mg(Ni)/m^3; Human Carcinogen)
NIOSH REL: (Inorganic Nickel) TWA 0.015 mg(Ni)/m^3

SAFETY PROFILE: Suspected carcinogen. Moderately toxic by ingestion. When heated to decomposition it emits toxic fumes of NO_x. See also NICKEL(II) NITRATE.

NDG550 CAS:17861-62-0 ***HR: 3***
NICKEL NITRITE
DOT: UN 2726
mf: N_2O_4•Ni mw: 150.73

SYNS: NICKEL DINITRITE ◇ NICKEL NITRITE ◇ NITROUS ACID, NICKEL (2+) SALT

OSHA PEL: TWA 1 mg(Ni)/m^3
ACGIH TLV: TWA 0.05 mg(Ni)/m^3; Confirmed Carcinogen
DOT Classification: Oxidizer; Label: Oxidizer

SAFETY PROFILE: Confirmed carcinogen. When heated to decomposition it emits toxic fumes of NO_x and Ni.

NDH000 CAS:7718-54-9 ***HR: 3***
NICKELOUS CHLORIDE
DOT: NA 9139
mf: Cl_2Ni mw: 129.61

SYNS: NICKEL CHLORIDE (DOT) ◇ NICKEL(II) CHLORIDE (1:2)

TOXICITY DATA with REFERENCE
dnr-esc 200 mg/L CRNGDP 2,189,81
cyt-ham:ovr 3200 nmol/L CNREA8 45,2320,85
orl-rat TDLo:10 g/kg (male 11W pre):REP TJADAB 33,90C,86
ipr-mus TDLo:5078 μg/kg (female 11D post):TER TJADAB 19,137,79
orl-rat LD50:105 mg/kg RPTOAN 32,102,69
ipr-rat LD50:20597 μg/kg TXCYAC 35,47,85
ipr-mus LD50:26 mg/kg AEPPAE 244,17,62
ivn-dog LDLo:10 mg/kg 14CYAT 2,1120,63
ims-rbt LD50:27 mg/kg JPETAB 87,119,46

CONSENSUS REPORTS: Nickel and its compounds are on the Community Right-To-Know List. Reported in EPA TSCA Inventory. EPA Genetic Toxicology Program.

OSHA PEL: (Transitional: TWA 1 mg/m^3) TWA 0.1 mg (Ni)/m^3
ACGIH TLV: TWA 0.1 mg(Ni)/m^3; (Proposed: TWA 0.05 mg(Ni)/m^3; Human Carcinogen)
NIOSH REL: (Inorganic Nickel) TWA 0.015 mg(Ni)/m^3
DOT Classification: ORM-E; Label: None.

SAFETY PROFILE: Suspected carcinogen. Poison by ingestion, intravenous, intramuscular, and intraperitoneal routes. An experimental teratogen. Experimental reproductive effects. Mutation data reported. When heated to decomposition it emits very toxic fumes of Cl^-. See also NICKEL COMPOUNDS.

NDH500 CAS:1314-06-3 ***HR: 3***
NICKEL PEROXIDE
mf: Ni_2O_3 mw: 165.42

PROP: Gray-black powder. Mp: $-O_2$ @ 600°, d: 4.83. Decomp about 600° into NiO and O_2. Insol in water; very sltly sol in cold acid, dissolved by hot HCl evolving Cl_2; dissolved by hot H_2SO_4 or HNO_3 evolving O_2.

SYNS: DINICKEL TRIOXIDE ◇ NICKELIC OXIDE ◇ NICKEL OXIDE ◇ NICKEL OXIDE PEROXIDE ◇ NICKEL SISQUIOXIDE ◇ NICKEL TRIOXIDE

TOXICITY DATA with REFERENCE
otr-mus:emb 500 μmol/L PAACA3 24,100,83
scu-mus LD50:50 mg/kg ZVKOA6 19,186,74

CONSENSUS REPORTS: Nickel and its compounds are on the Community Right-To-Know List. Reported in EPA TSCA Inventory.

OSHA PEL: (Transitional: TWA 1 mg/m^3) TWA 0.1 mg (Ni)/m^3
ACGIH TLV: TWA 1 mg(Ni)/m^3; (Proposed: TWA 0.05 mg(Ni)/m^3; Human Carcinogen)
NIOSH REL: (Inorganic Nickel) TWA 0.015 mg(Ni)/m^3

SAFETY PROFILE: Suspected carcinogen. Poison by subcutaneous route. Mutation data reported. Hazardous reaction with hydrogen peroxide. Presence of the oxide increases the sensitivity of nitroalkanes (e.g., nitromethane, nitroethane, 1-nitropropane) to heat. See also NICKEL COMPOUNDS and PEROXIDES.

NDI000 CAS:14220-17-8 ***HR: 3***
NICKEL POTASSIUM CYANIDE
mf: $C_4N_4Ni{\cdot}2K$ mw: 240.99

SYNS: DIPOTASSIUM NICKEL TETRACYANIDE ◇ DIPOTASSIUM TETRACYANONICKELATE ◇ POTASSIUM TETRACYANONICKELATE ◇ POTASSIUM TETRACYANONICKELATE(II)

TOXICITY DATA with REFERENCE
cyt-mus:mmr 1 mmol/L/48H MUREAV 67,221,79
orl-mus LD50:275 mg/kg JPMSAE 52,59,63

CONSENSUS REPORTS: Nickel and its compounds, as well as cyanide and its compounds, are on the Community Right-To-Know List. Reported in EPA TSCA Inventory.

OSHA PEL: (Transitional: TWA 1 mg/m^3) TWA 0.1 mg (Ni)/m^3
ACGIH TLV: TWA 0.1 mg(Ni)/m^3; (Proposed: TWA 0.05 mg(Ni)/m^3; Human Carcinogen)
NIOSH REL: (Inorganic Nickel) TWA 0.015 mg(Ni)/m^3

SAFETY PROFILE: Suspected carcinogen. Poison by ingestion. Mutation data reported. When heated to decomposition it emits very toxic fumes of NO_x, K_2O, and CN^-. See also NICKEL COMPOUNDS and CYANIDE.

NDI500 ***HR: 3***
NICKEL REFINERY DUST

PROP: *Analysis:* Cupric oxide (3.4%), nickel sulfate (20.0%), nickel sulfide (57.0%), nickel oxide (6.3%), cobalt oxide (1.0%), ferric oxide (1.8%), silicon dioxide

(1.2%), misc. (2.0%), water (7.3%) (CNREA8 22,158,62).

TOXICITY DATA with REFERENCE
ims-rat TDLo:90 mg/kg:CAR CNREA8 22,152,62
ims-mus TDLo:135 mg/kg:NEO CNREA8 22,152,62
ims-mus LDLo:800 mg/kg CNREA8 22,152,62

CONSENSUS REPORTS: IARC Cancer Review: Human Sufficient Evidence IMEMDT 2,126,73. Nickel and its compounds are on the Community Right-To-Know List.

OSHA PEL: TWA 1 mg(Ni)/m^3
ACGIH TLV: TWA 1 mg(Ni)/m^3; (Proposed: TWA 0.05 mg(Ni)/m^3; Human Carcinogen)
NIOSH REL: (Inorganic Nickel) TWA 0.015 mg(Ni)/m^3

SAFETY PROFILE: Confirmed carcinogen with experimental carcinogenic and neoplastigenic data. A human carcinogenic. Moderately toxic by intramuscular route. When heated to decomposition it emits toxic fumes of SO_x. See also NICKEL, NICKEL COMPOUNDS, and individual components.

NDJ000 CAS:13520-61-1 ***HR: 3***
NICKEL(2+) SALT PERCHLORIC ACID HEXAHYDRATE
mf: Cl_2O_8•Ni•$6H_2O$ mw: 365.73

SYN: NICKEL(2+) PERCHLORATE, HEXAHYDRATE

TOXICITY DATA with REFERENCE
ipr-mus LDLo:100 mg/kg JAFCAU 14,512,66

CONSENSUS REPORTS: Nickel and its compounds are on the Community Right-To-Know List.

OSHA PEL: (Transitional: TWA 1 mg/m^3) TWA 0.1 mg (Ni)/m^3
ACGIH TLV: TWA 0.1 mg(Ni)/m^3; (Proposed: TWA 0.05 mg(Ni)/m^3; Human Carcinogen)
NIOSH REL: TWA 15 μg(Ni)/m^3

SAFETY PROFILE: Suspected carcinogen. Poison by intraperitoneal route. A powerful oxidizer. Mixtures with 2,2-dimethoxypropane explode when heated above 65°C. When heated to decomposition it emits toxic fumes of Cl^-. See also NICKEL COMPOUNDS and PERCHLORATES.

NDJ399 CAS:12255-80-0 ***HR: 3***
NICKEL SUBARSENIDE
mf: As_2Ni_5 mw: 443.39

SYN: NICKEL ARSENIDE (As_2-Ni_5)

TOXICITY DATA with REFERENCE
ims-rat TDLo:443 mg/kg:CAR IAPUDO 53,127,84
orl-rat LD50:5840 mg/kg FDRLI* 8005F,84

CONSENSUS REPORTS: IARC Cancer Review: Animal Limited Evidence

OSHA: Cancer Hazard
ACGIH TLV: TWA 0.1 mg(Ni)/m^3; (Proposed: TWA 0.05 mg(Ni)/m^3; Human Carcinogen)

SAFETY PROFILE: Confirmed human carcinogen with experimental carcinogenic data. Moderately toxic by ingestion. When heated to decomposition it emits toxic fumes of As.

NDJ400 CAS:12256-33-6 ***HR: 3***
NICKEL SUBARSENIDE
mf: As_8Ni_{11} mw: 1245.17

SYN: NICKEL ARSENIDE (As_8-Ni_{11})

TOXICITY DATA with REFERENCE
ims-rat TDLo:108 mg/kg:CAR IAPUDO 53,127,84

CONSENSUS REPORTS: IARC Cancer Review: Animal Limited Evidence IMEMDT 49,257,90

OSHA: Cancer Hazard
ACGIH TLV: TWA 0.05 mg(Ni)/m^3; Confirmed Confirmed Carcinogen

SAFETY PROFILE: Confirmed human carcinogen with experimental carcinogenic data. When heated to decomposition it emits toxic fumes of As.

NDJ475 CAS:12137-13-2 ***HR: 3***
NICKEL SUBSELENIDE
mf: Ni_3Se_2 mw: 334.05

SYNS: NICKEL SELENIDE ◇ NICKEL SELENIDE (3:2) CRYSTALLINE

TOXICITY DATA with REFERENCE
otr-ham:emb 5 μmol/plate TOXID9 1,132,81
ims-rat TDLo:319 mg/kg:CAR IAPUDO 53,127,84

OSHA PEL: TWA Soluble: Compounds: 0.1 mg(Ni)/m^3; Insoluble Compounds: 1 mg(Ni)/m^3; TWA 0.2 mg(Se)/m^3
ACGIH TLV: TWA 1 mg(Ni)/m^3; (Proposed: TWA 0.05 mg(Ni)/m^3; Human Carcinogen); TWA 0.2 mg (Se)/m^3
DFG TRK: 0.5 mg/m^3; Human Carcinogen.
NIOSH REL: (Inorganic Nickel) TWA 0.015 mg(Ni)/m^3

SAFETY PROFILE: Confirmed carcinogen with experimental carcinogenic data. Mutation data reported.

NDJ500 CAS:12035-72-2 ***HR: 3***
NICKEL SUBSULFIDE
mf: Ni_3S_2 mw: 240.25

SYNS: HEAZLEWOODITE ◇ NICKEL SUBSULPHIDE ◇ NICKEL SULFIDE ◇ α-NICKEL SULFIDE (3:2) CRYSTALLINE ◇ NICKEL SULPHIDE ◇ NICKEL TRITADISULPHIDE

TOXICITY DATA with REFERENCE
sce-hmn:lym 1 mg/L CNREA8 41,4136,81
otr-ham:emb 5 μmol/plate TOXID9 1,132,81
irn-rat TDLo:50 mg/kg (1D pre):REP 50WKA3 -,399,83
ims-rat TDLo:80 mg/kg (6D preg):TER TXAPA9 43,381,78
ihl-rat TCLo:970 μg/m^3/6H/76W-I:CAR JNCIAM 54,1165,75
scu-rat TDLo:25 mg/kg:ETA JJCREP 79,212,88
ivn-rat TDLo:10 mg/kg:CAR FEPRA7 39(3,Pt.1),330
mul-ham TDLo:272 g/kg/18W-I:NEO AEMBAP 91,57,78
itt-rat TD:40 mg/kg:ETA,TER PAACA3 18,52,77
ipr-gpg LDLo:102 μg/kg JOHYAY 8,565,08

CONSENSUS REPORTS: NTP Fifth Annual Report on Carcinogens. IARC Cancer Review: Group 1 IMEMDT 7,264,87; Animal Sufficient Evidence IMEMDT 2,126,73, IMEMDT 11,75,76. Nickel and its compounds are on the Community Right-To-Know List. Reported in EPA TSCA Inventory.

OSHA PEL: TWA 1 mg(Ni)/m^3
ACGIH TLV: TWA 1 mg(Ni)/m^3; (Proposed: TWA 0.05 mg(Ni)/m^3; Human Carcinogen)
NIOSH REL: TWA 0.015 mg(Ni)/m^3

SAFETY PROFILE: Confirmed carcinogen with experimental carcinogenic, neoplastigenic, tumorigenic, and teratogenic data. Poison by intraperitoneal route. An experimental teratogen. Other experimental reproductive effects. Human mutation data reported. When heated to decomposition it emits toxic fumes of SO_x. See also NICKEL COMPOUNDS and SULFIDES.

NDK000 CAS:13770-89-3 ***HR: 3***
NICKEL (II) SULFAMATE
mf: $H_4N_2NiO_6S_2$ mw: 250.89

TOXICITY DATA with REFERENCE
ipr-mus LDLo:250 mg/kg CBCCT* 4,320,52

CONSENSUS REPORTS: Nickel and its compounds are on the Community Right-To-Know List. Reported in EPA TSCA Inventory.

OSHA PEL: (Transitional: TWA 1 mg/m^3) TWA 0.1 mg (Ni)/m^3
ACGIH TLV: TWA 0.1 mg(Ni)/m^3; (Proposed: TWA 0.05 mg(Ni)/m^3; Human Carcinogen)
NIOSH REL: (Inorganic Nickel) TWA 0.015 mg(Ni)/m^3

SAFETY PROFILE: Suspected carcinogen. Poison by intraperitoneal route. When heated to decomposition it emits very toxic fumes of SO_x and NO_x. See also NICKEL COMPOUNDS.

NDK500 CAS:7786-81-4 ***HR: 3***
NICKEL SULFATE
mf: $O_4S{\bullet}Ni$ mw: 154.77

PROP: Cubic yellow crystals. Mp: $-SO_3$ @ 840°, d: 3.68.

SYNS: NCI-C60344 ◇ NICKELOUS SULFATE ◇ NICKEL(II) SULFATE (1:1) ◇ NICKEL(II) SULFATE ◇ NICKEL SULFATE(1:1) ◇ NICKEL(2+) SULFATE(1:1) ◇ SULFURIC ACID, NICKEL(2+)SALT ◇ SULFURIC ACID, NICKEL(2+) SALT (1:1)

TOXICITY DATA with REFERENCE
skn-chd 5%/48H ARDEAC 87,378,63
mmo-smc 100 mmol/L MUREAV 117,149,83
sce-hmn:leu 23 μmol/L DMBUAE 27,40,80
imp-rat TDLo:95 mg/kg/78W-C:ETA PAACA3 5,50,64
ipr-rat LD50:500 mg/kg GTPZAB 23(6),48,79
ipr-mus LD50:55 mg/kg COREAF 256,1043,63
ivn-mus LDLo:7640 μg/kg FATOAO 23,549,60
scu-dog LDLo:38 mg/kg JOHYAY 8,565,08
ivn-dog LDLo:38 mg/kg JOHYAY 8,565,08
scu-cat LDLo:24 mg/kg JOHYAY 8,565,08
scu-rbt LDLo:33 mg/kg JOHYAY 8,565,08
ivn-rbt LDLo:33 mg/kg JOHYAY 8,565,08

CONSENSUS REPORTS: Nickel and its compounds are on the Community Right-To-Know List. Reported in EPA TSCA Inventory. EPA Genetic Toxicology Program.

OSHA PEL: (Transitional: TWA 1 mg/m^3) TWA 0.1 mg (Ni)/m^3
ACGIH TLV: TWA 0.1 mg(Ni)/m^3; (Proposed: TWA 0.05 mg(Ni)/m^3; Human Carcinogen)
NIOSH REL: (Inorganic Nickel) TWA 0.015 mg(Ni)/m^3
DOT Classification: ORM-E; Label: None.

SAFETY PROFILE: Suspected carcinogen with experimental tumorigenic data. Poison by intravenous, intraperitoneal, and subcutaneous routes. Human mutation data reported. A human skin irritant. When heated to decomposition it emits very toxic fumes of SO_x. See also NICKEL COMPOUNDS and SULFATES.

NDL000 CAS:10101-97-0 ***HR: 3***
NICKEL(II) SULFATE HEXAHYDRATE (1:1:6)
mf: $NiO_4S{\bullet}_6H_2O$ mw: 262.89

SYNS: NICKEL MONOSULFATE HEXAHYDRATE ◇ NICKEL SULFATE HEXAHYDRATE ◇ NICKEL (II) SULFATE HEXAHYDRATE ◇ NICKEL SULPHATE HEXAHYDRATE ◇ SULFURIC ACID, NICKEL (2+) SALT, HEXAHYDRATE

TOXICITY DATA with REFERENCE
cyt-hmn:lym 5 mg/L ENMUDM 3,597,81
otr-ham:emb 5 mg/L DTESD7 8,259,80
ipr-rat TDLo:403 mg/kg (30D male):REP ARTODN 37,159,77
orl-rat LD50:275 mg/kg FDRLI* 7702A,83

scu-dog LDLo:500 mg/kg 27ZWAY 3.2,1446,
ivn-dog LDLo:89500 μg/kg EQSSDX 1,1,75
scu-cat LDLo:500 mg/kg EQSSDX 1,1,75
ivn-cat LDLo:71600 μg/kg EQSSDX 1,1,75
scu-rbt LDLo:500 mg/kg EQSSDX 1,1,75
scu-gpg LDLo:62 mg/kg EQSSDX 1,1,75

CONSENSUS REPORTS: Nickel and its compounds are on the Community Right-To-Know List. EPA Genetic Toxicology Program.

OSHA PEL: (Transitional: TWA 1 mg/m^3) TWA 0.1 mg (Ni)/m^3
ACGIH TLV: TWA 0.1 mg(Ni)/m^3; (Proposed: TWA 0.05 mg(Ni)/m^3; Human Carcinogen)
NIOSH REL: (Inorganic Nickel) TWA 0.015 mg(Ni)/m^3

SAFETY PROFILE: Suspected carcinogen. Poison by ingestion, intravenous, and subcutaneous routes. Experimental reproductive effects. Human mutation data reported. When heated to decomposition it emits toxic fumes of SO_x. See also NICKEL SULFATE.

NDL100 CAS:11113-75-0 ***HR: 3***
NICKEL SULFIDE
mf: NiS mw: 90.77

SYNS: NICKEL MONOSULFIDE ◇ NICKELOUS SULFIDE ◇ NICKEL(II) SULFIDE ◇ α-NICKEL SULFIDE (1:1) CRYSTALLINE

TOXICITY DATA with REFERENCE
cyt-mus:mmr 1 mmol/L/48H MUREAV 67,221,79
dnd-ham:ovr 1 mg/L CALEDQ 15,35,82
ims-rat TDLo:86585 μg/kg:CAR IAPUDO 53,127,84
irn-rat TDLo:43290 μg/kg:NEO CRNGDP 5,1511,84

CONSENSUS REPORTS: Nickel and its compounds are on the Community Right-To-Know List.

OSHA PEL: TWA 1 mg(Ni)/m^3
ACGIH TLV: TWA 1 mg(Ni)/m^3; (Proposed: TWA 0.05 mg(Ni)/m^3; Human Carcinogen)
DFG TRK: 0.5 mg/m^3; Human Carcinogen.
NIOSH REL: TWA 0.015 mg(Ni)/m^3

SAFETY PROFILE: Confirmed carcinogen with experimental carcinogenic and neoplastigenic data. Mutation data reported. When heated to decomposition it emits toxic fumes of SO_x. See also NICKEL COMPOUNDS and SULFIDES.

NDL425 CAS:12142-88-0 ***HR: 3***
NICKEL TELLURIDE
mf: NiTe mw: 186.31

TOXICITY DATA with REFERENCE
ims-rat TDLo:178 mg/kg:CAR IAPUDO 53,127,84
ims-rat TD:56 mg/kg:ETA NTIS** DOE/EV/03140-5

CONSENSUS REPORTS: Reported in EPA TSCA Inventory.

OSHA PEL: TWA 1 mg(Ni)/m^3; Confirmed Carcinogen; 1 mg(Te)/m^3
ACGIH TLV: TWA 0.1 mg(Te)/m^3
NIOSH REL: (Nickel, Inorganic) TWA 0.015 mg(Ni)/m^3

SAFETY PROFILE: Confirmed carcinogen with experimental carcinogenic and tumorigenic data.

NDL500 CAS:12035-39-1 ***HR: 3***
NICKEL TITANIUM OXIDE
mf: NiO_3Ti mw:154.61

SYNS: NICKEL-TITANATE ◇ TITANIUM NICKEL OXIDE

TOXICITY DATA with REFERENCE
ims-rat TDLo:210 mg/kg/91W-I:CAR NCIUS* PH 43-69-886,DEC,68
imp-rat TDLo:200 mg/kg:ETA JJIND8 67,965,81

CONSENSUS REPORTS: Nickel and its compounds are on the Community Right-To-Know List. Reported in EPA TSCA Inventory.

OSHA PEL: TWA 1 mg(Ni)/m^3
ACGIH TLV: TWA 0.1 mg(Ni)/m^3; (Proposed: TWA 0.05 mg(Ni)/m^3; Human Carcinogen)
NIOSH REL: (Inorganic Nickel) TWA 0.015 mg(Ni)/m^3

SAFETY PROFILE: Suspected carcinogen with experimental carcinogenic and tumorigenic data. See also NICKEL COMPOUNDS and TITANIUM COMPOUNDS.

NDL800 CAS:65141-46-0 ***HR: 3***
NICORANDIL
mf: $C_8H_9N_3O_4$ mw: 211.20

PROP: Colorless needles from ether/ethanol. Mp: 92-93°.

SYNS: N-(2-HYDROXYETHYL)NICOTINAMIDE NITRATE (ESTER) ◇ 2-NICOTINAMIDOETHYL NITRATE ◇ PERISOLOL ◇ SIGMART

TOXICITY DATA with REFERENCE
orl-rat LD50:1220 mg/kg YKYUA6 35,1627,84
ipr-rat LD50:1100 mg/kg YKYUA6 35,1627,84
scu-rat LD50:1200 mg/kg YKYUA6 35,1627,84
orl-mus LD50:1150 mg/kg YKYUA6 35,1627,84
ipr-mus LD50:990 mg/kg YKYUA6 35,1627,84
scu-mus LD50:1350 mg/kg YKYUA6 35,1627,84
orl-dog LD50:62500 μg/kg YKYUA6 35,1627,84

SAFETY PROFILE: Poison by ingestion. Moderately toxic by intraperitoneal and subcutaneous routes. When heated to decomposition it emits toxic fumes of NO_x. See also NITRATES.

NDM000 CAS:27848-84-6 ***HR: 3***
NICOTERGOLINE
mf: $C_{24}H_{26}BrN_3O_3$ mw: 484.44

SYNS: 8-β-((5-BROMONICOTINOYLOXY)METHYL)-1,6-DIMETHYL-10-α-METHOXYERGOLINE ◇ FI 6714 ◇ 10-METHOXY-1,6-DIMETHYL-ERGOLINE-8-METHANOL-5-BROMO-3-PYRIDINECARBOXYLATE (ESTER) ◇ 10-METHOXY-1,6-DIMETHYL-ERGOLIN-8-β-METHANOL-(5-BROMNICOTINAT) (GERMAN) ◇ 1-METHYL-LUMILYSERGOL-8-(5-BROMONICOTINATE)-10-METHYL ETHER ◇ MNE ◇ NARGOLINE ◇ NICERGOLIN (GERMAN) ◇ NICERGOLINE ◇ NIMERGOLINE ◇ SERMION

TOXICITY DATA with REFERENCE
orl-rat TDLo:2600 mg/kg (female 17-22D post):REP KSRNAM 20,6375,86
orl-rat TDLo:2475 mg/kg (female 7-17D post):TER KSRNAM 20,6375,86
orl-rat LD50:1193 mg/kg IYKEDH 19,735,88
scu-rat LD50:778 mg/kg ARZNAD 29,1223,79
ivn-rat LD50:42 mg/kg ARZNAD 29,1206,79
orl-mus LD50:633 mg/kg ARZNAD 29,1223,79
ivn-mus LD50:46 mg/kg ARZNAD 29,1206,79
ivn-dog LD50:20 mg/kg ARZNAD 29,1206,79

SAFETY PROFILE: Poison by intravenous route. Moderately toxic by ingestion and subcutaneous routes. An experimental teratogen. Other experimental reproductive effects. A vasodilator. When heated to decomposition it emits very toxic fumes of Br^- and NO_x.

NDN000 CAS:54-11-5 ***HR: 3***
NICOTINE
DOT: UN 1654
mf: $C_{10}H_{14}N_2$ mw: 162.26

PROP: An alkaloid from tobacco. In its pure state, a colorless and almost odorless oil; sharp burning taste. Mp: < −80°, bp: 247.3° (partial decomp), lel: 0.75%, uel: 4.0%, d: 1.0092 @ 20°, autoign temp: 471°F, vap press: 1 mm @ 61.8°, vap d: 5.61. Volatile with steam; misc with water below 60°; very sol in alc, chloroform ether, petr ether, and kerosene oils.

SYNS: BLACK LEAF ◇ DESTRUXOL ORCHID SPRAY ◇ EMO-NIK ◇ ENT 3,424 ◇ FLUX MAAG ◇ FUMETOBAC ◇ MACH-NIC ◇ 1-METHYL-2-(3-PYRIDYL)PYRROLIDINE ◇ 3-(N-METHYLPYRROLIDINO)PYRIDINE ◇ (S)-3-(1-METHYL-2-PYRROLIDINYL)PYRIDINE (9CI) ◇ 3-(1-METHYL-2-PYRROLIDINYL)PYRIDINE ◇ l-3-(1-METHYL-2-PYRROLIDYL)PYRIDINE ◇ (−)-3-(1-METHYL-2-PYRROLIDYL)PYRIDINE ◇ NIAGARA P.A. DUST ◇ NICOCIDE ◇ NICO-DUST ◇ NICO-FUME ◇ NICOTINA (ITALIAN) ◇ (−)-NICOTINE ◇ l-NICOTINE ◇ NICOTINE, liquid (DOT) ◇ NICOTINE, solid (DOT) ◇ NICOTINE ALKALOID ◇ NIKOTIN (GERMAN) ◇ NIKOTYNA (POLISH) ◇ ORTHO N-4 DUST ◇ ORTHO N-5 DUST ◇ PYRIDINE, 3-(TETRAHYDRO-1-METHYLPYRROL-2-YL) ◇ β-PYRIDYL-α-N-METHYLPYRROLIDINE ◇ RCRA WASTE NUMBER P075 ◇ TENDUST ◇ dl-TETRAHYDRONICOTYRINE ◇ XL ALL INSECTICIDE

TOXICITY DATA with REFERENCE
mmo-smc 100 ppm RSTUDV 6,161,76
dlt-mus-ipr 10 mg/kg/2D-I IRLCDZ 6,461,78
dni-rbt:oth 1 mmol/L TXCYAC 34,309,85
orl-wmn TDLo:40 μg/kg (24W preg):TER BJOGAS 90,710,83
orl-rat TDLo:594 mg/kg (male 4W pre):REP NETOD7 1,221,79
unr-man LDLo:882 μg/kg 85DCAI 2,73,70
rec-hmn TDLo:1430 μg/kg:CNS,GIT CTOXAO 10,391,77
orl-rat LD50:50 mg/kg FMCHA2 -,D218,80
skn-rat LD50:140 mg/kg WRPCA2 9,119,70
ipr-rat LD50:14560 μg/kg JPETAB 124,350,58
scu-rat LD50:25 mg/kg FATOAO 47(5),85,84
ivn-rat LDLo:1 mg/kg 27ZIAQ -,171,73
ims-rat LDLo:15 mg/kg SCIEAS 127,1054,58
par-rat LDLo:34 mg/kg JPETAB 48,317,33
itr-rat LD50:19300 μg/kg ARDSBL 129,112,84
idu-rat LDLo:30 mg/kg ARZNAD 25,1037,75
orl-mus LD50:3340 μg/kg HYSAAV 34(4-6),187,69
ipr-mus LD50:5900 μg/kg TXAPA9 28,227,74
scu-mus LDLo:16 mg/kg AEPPAE 188,605,38
ivn-mus LD50:300 μg/kg BJPCBM 35,161,69 SCIEAS 127,1054,58
skn-rbt LD50:50 mg/kg AFDOAQ 16,3,52

CONSENSUS REPORTS: EPA Extremely Hazardous Substances List. Reported in EPA TSCA Inventory. EPA Genetic Toxicology Program.

OSHA PEL: TWA 0.5 mg/m^3 (skin)
ACGIH TLV: TWA 0.5 mg/m^3 (skin)
DFG MAK: 0.07 ppm; 0.5 mg/m^3
DOT Classification: Poison B; Label: St. Andrews Cross; Poison B; Label: Poison.

SAFETY PROFILE: A deadly human poison by unspecified route. Experimental poison by ingestion, skin contact, intraperitoneal, subcutaneous, intravenous, intramuscular, parenteral, intratracheal, and intraduodenal routes. Human systemic effects by rectal route: hallucinations, distorted perceptions, nausea or vomiting. Human teratogenic effects by ingestion: developmental abnormalities of the cardiovascular system. Human blood pressure effects. Can be absorbed by intact skin. An experimental teratogen. Experimental reproductive effects. "Nicotinism," poisoning by nicotine, is characterized by stimulation and subsequent depression of the central and autonomic nervous systems. Death can result from respiratory paralysis. Mutation data reported. Used as a pesticide and in veterinary medicine as an external parasiticide. Combustible when exposed to heat or flame. Moderately explosive in the form of vapor when exposed to heat or flame. Can react with oxidizing materials. To fight fire, use alcohol foam, dry chemical, CO_2. When heated to decomposition it emits toxic fumes of NO_x, CO, and other highly toxic fumes. See also SMOKELESS TOBACCO.

NDP400 CAS:2820-51-1 ***HR: 3***
NICOTINE HYDROCHLORIDE
DOT: UN 1656
mf: $C_{10}H_{14}N_2{\cdot}xClH$ mw: 417.48

SYNS: CHLORHYDRATE de NICOTINE (FRENCH) ◇ NICOTINE HYDROCHLORIDE (d,l) ◇ NICOTINE HYDROCHLORIDE, solution (DOT)

TOXICITY DATA with REFERENCE
ipr-mus LD50:13500 mg/kg AIPTAK 223,132,76
scu-mus LD50:31900 μg/kg BAFEAG 44,426,57
ivn-rbt LDLo:6500 μg/kg JPETAB 77,343,43
ipr-gpg LDLo:32 mg/kg AJEBAK 25,83,47

DOT Classification: Poison B; Label: Poison.

SAFETY PROFILE: Poison by intravenous, subcutaneous, and intraperitoneal routes. When heated to decomposition it emits very toxic fumes of Cl^-, NO_x, and CO. See also NICOTINE.

NDP500 CAS:21361-93-3 ***HR: 3***
l-NICOTINE HYDROCHLORIDE
mf: $C_{10}H_{14}N_2{\cdot}ClH$ mw: 198.72

PROP: Deliquescent crystals.

TOXICITY DATA with REFERENCE
ipr-rat LDLo:20 mg/kg AJEBAK 25,83,47
ipr-gpg LDLo:20 mg/kg AJEBAK 25,83,47

SAFETY PROFILE: Poison by intraperitoneal route. When heated to decomposition it emits very toxic fumes of Cl^- and NO_x. See also NICOTINE.

NDQ000 CAS:69782-38-3 ***HR: 3***
(+)-NICOTINE HYDROCHLORIDE
mf: $C_{10}H_{14}N_2{\cdot}ClH$ mw: 198.72

SYN: d-NICOTINE HYDROCHLORIDE

TOXICITY DATA with REFERENCE
ipr-rat LDLo:24 mg/kg AJEBAK 25,83,47
ipr-gpg LDLo:33 mg/kg AJEBAK 25,83,47

SAFETY PROFILE: Poison by intraperitoneal route. When heated to decomposition it emits very toxic fumes of Cl^-, CO, and NO_x. See also NICOTINE.

NDR000 CAS:29790-52-1 ***HR: 3***
NICOTINE MONOSALICYLATE
DOT: UN 1657
mf: $C_{10}H_{14}N_2{\cdot}C_7H_6O_3$ mw: 300.39

PROP: Mp: 118°. Very sol in water or alc.

SYN: NICOTINE SALICYLATE (DOT)

DOT Classification: Poison B; Label: Poison.

SAFETY PROFILE: A poison. Symptoms of exposure: Extreme nausea, vomiting, evacuation of bowel and bladder, mental confusion, twitching, convulsions. Base is readily absorbed through mucous membranes and intact skin. Institute treatment immediately. When heated to decomposition it emits toxic fumes of NO_x and CO. See also NICOTINE.

NDR500 CAS:65-30-5 ***HR: 3***
NICOTINE SULFATE
DOT: UN 1658
mf: $C_{20}H_{26}N_4{\cdot}O_4S$ mw: 418.56

SYNS: 1-1-METHYL-2-(3-PYRIDYL)-PYRROLIDINE SULFATE ◇ (S)-3-(1-METHYL-2-PYRROLIDINYL)PYRIDINE SULFATE (2:1) ◇ 1-3-(1-METHYL-2-PYRROLIDYL)PYRIDINE SULFATE ◇ NICOTINE SULFATE (2:1) ◇ NICOTINE SULFATE, liquid (DOT) ◇ NICOTINE SULFATE, solid (DOT) ◇ SULFATE de NICOTINE (FRENCH)

TOXICITY DATA with REFERENCE
orl-rat LD50:55 mg/kg TXAPA9 21,315,72
skn-rat LD50:285 mg/kg TXAPA9 2,88,60
orl-mus LD50:16 mg/kg TXAPA9 42,417,77

CONSENSUS REPORTS: Reported in EPA TSCA Inventory.

DOT Classification: Poison B; Label: Poison.

SAFETY PROFILE: Poison by ingestion and skin contact. When heated to decomposition it emits very toxic fumes of SO_x and organic fumes. See also NICOTINE and SULFATES.

NDS500 CAS:65-31-6 ***HR: 3***
NICOTINE TARTRATE (1:2)
DOT: UN 1659
mf: $C_{10}H_{14}N_2{\cdot}2C_4H_6O_6$ mw:462.46

SYNS: (S)-3-(1-METHYL-2-PYRROLIDINYL-PYRIDINE(R-(R,R))-2,3-DIHYDROXYBUTANEDIOATE (1:2) ◇ NICOTINE ACID TARTRATE ◇ NICOTINE BITARTRATE ◇ NICOTINE HYDROGEN TARTRATE ◇ (−)-NICOTINE HYDROGEN TARTRATE ◇ NICOTINE TARTRATE ◇ NICOTINE TARTRATE (DOT) ◇ TARTRATE de NICOTINE (FRENCH)

TOXICITY DATA with REFERENCE
scu-rat TDLo:1 mg/kg (female 21D post):REP JPETAB 248,786,89
scu-rat TDLo:1 mg/kg (female 21D post):TER JPETAB 248,786,89
orl-rat LD50:65 mg/kg PLRCAT 5,341,73
ipr-rat LD50:83 mg/kg TXAPA9 22,325,72
ivn-rat LD50:600 mg/kg PLRCAT 5,341,73
orl-mus LD50:65 mg/kg PLRCAT 5,341,73
ipr-mus LD50:9100 μg/kg TXAPA9 65,366,82
scu-mus LDLo:59 mg/kg AEPPAE 188,605,38
ivn-mus LD50:300 μg/kg BJPCBM 35,161,69

DOT Classification: Poison B; Label: Poison

SAFETY PROFILE: Poison by ingestion, intravenous, intraperitoneal, and subcutaneous routes. An experimental teratogen. Experimental reproductive effects.

When heated to decomposition it emits toxic fumes of NO_x and CO. See also NICOTINE.

NDT500 CAS:6938-06-3 ***HR: 3***
NICOTINIC ACID, BUTYL ESTER
mf: $C_{10}H_{13}NO_2$ mw: 179.24

TOXICITY DATA with REFERENCE
ipr-mus LDLo:63 mg/kg CBCCT* 4,46,52

CONSENSUS REPORTS: Reported in EPA TSCA Inventory.

SAFETY PROFILE: Poison by intraperitoneal route. When heated to decomposition it emits toxic fumes of NO_x. See also ESTERS.

NDU000 CAS:69782-43-0 ***HR: 3***
NICOTINIC ACID, 3-(2,6-DIMETHYLPIPERIDINO) PROPYL ESTER, HYDROCHLORIDE
mf: $C_{16}H_{24}N_2O_2{\cdot}ClH$ mw: 312.88

SYN: 3-PYRIDINECARBOXYLIC ACID, 3-(2,6-DIMETHYLPIPERIDINO)PROPYL ESTER, HYDROCHLORIDE

TOXICITY DATA with REFERENCE
scu-mus LD50:330 mg/kg JACSAT 68,2592,46
ivn-mus LD50:115 mg/kg JACSAT 68,2592,46

SAFETY PROFILE: Poison by subcutaneous and intravenous routes. When heated to decomposition it emits very toxic fumes of HCl and NO_x.

NDU500 CAS:553-53-7 ***HR: 3***
NICOTINIC ACID, HYDRAZIDE
mf: $C_6H_7N_3O$ mw: 137.16

PROP: Needles from dil alc or benzene. Mp: 158-159°. Very sol in water and alc; sltly sol in benzene.

SYNS: NICOTINOYL HYDRAZINE ◇ NICOTINYLHYDRAZIDE ◇ 3-PYRIDOYL HYDRAZINE ◇ WS 102

TOXICITY DATA with REFERENCE
orl-mus TDLo:124 g/kg/71W-C:CAR ONCOBS 38,106,81
ivn-mus LD50:180 mg/kg CSLNX* NX#04178
par-mus LDLo:100 mg/kg CBCCT* 7,691,55

SAFETY PROFILE: Poison by intravenous and parenteral routes. Questionable carcinogen with experimental carcinogenic data. When heated to decomposition it emits toxic fumes of NO_x.

NDV000 CAS:93-60-7 ***HR: 2***
NICOTINIC ACID, METHYL ESTER
mf: $C_7H_7NO_2$ mw: 137.15

SYNS: METHYL NICOTINATE ◇ NICOMETH

TOXICITY DATA with REFERENCE
par-mus LDLo:2000 mg/kg CBCCT* 7,691,55

CONSENSUS REPORTS: Reported in EPA TSCA Inventory.

SAFETY PROFILE: Moderately toxic by parenteral route. When heated to decomposition it emits toxic fumes of NO_x.

NDW000 CAS:69782-42-9 ***HR: 3***
NICOTINIC ACID, 3-(2-METHYLPIPERIDINO) PROPYL ESTER, HYDROCHLORIDE
mf: $C_{15}H_{22}N_2O_2{\cdot}ClH$ mw: 298.85

SYN: 3-PYRIDINECARBOXYLIC ACID, 3-(2-METHYLPIPERIDINO)PROPYL ESTER, HYDROCHLORIDE

TOXICITY DATA with REFERENCE
scu-mus LD50:500 mg/kg JACSAT 68,2592,46
ivn-mus LD50:100 mg/kg JACSAT 68,2592,46

SAFETY PROFILE: Poison by intravenous route. Moderately toxic by subcutaneous route. When heated to decomposition it emits very toxic fumes of HCl and NO_x. See also ESTERS.

NDW500 CAS:54-86-4 ***HR: 2***
NICOTINIC ACID, SODIUM SALT
mf: $C_6H_4O_2{\cdot}Na$ mw: 131.09

SYNS: NATRIUMNICOTINAT (GERMAN) ◇ SODIUM NICOTINATE

TOXICITY DATA with REFERENCE
scu-rat LD50:5000 mg/kg JPETAB 73,85,41
scu-mus LDLo:3000 mg/kg JPETAB 65,95,39
ivn-mus LD50:2900 mg/kg YKKZAJ 81,1748,61

CONSENSUS REPORTS: Reported in EPA TSCA Inventory.

SAFETY PROFILE: Moderately toxic by intravenous and subcutaneous routes. When heated to decomposition it emits toxic fumes of Na_2O.

NDW525 CAS:29876-14-0 ***HR: 2***
N-NICOTINOYLTRYPTAMIDE
mf: $C_{16}H_{15}N_3O$ mw: 265.34

SYNS: N-(2-(3-INDOLYL)ETHYL)-NICOTINAMIDE ◇ N-(2-INDOL-3-YLETHYL)NICOTINAMIDE ◇ NICOTINAMIDE, N-(2-INDOL-3-YLETHYL)- ◇ 3-PYRIDINECARBOXAMIDE, N-(2-(1H-INDOL-3-YL)ETHYL)-(9CI) ◇ N-3-PYRIDOYLTRYPTAMINE ◇ TRYPTAMIDE

TOXICITY DATA with REFERENCE
orl-mus TDLo:4500 mg/kg (female 6-14D post):TER PJPPAA 39,779,87
ipr-mus LD50:1250 mg/kg DIPHAH 22,313,70

SAFETY PROFILE: Moderately toxic by intraperitoneal route. An experimental teratogen. When heated to decomposition it emits toxic fumes of NO_x.

NDX300 CAS:487-19-4 ***HR: 3***
β-NICOTYRINE
mf: $C_{10}H_{10}N_2$ mw: 158.22

PROP: Oily liquid; characteristic odor, darkens on standing. Bp: 280-281°, d: 1.241; n (20/D) 1.6057. Sol in alc, dil acids; sparingly sol in water.

SYNS: 1-METHYL-2-(3-PYRIDYL)PYRROLE ◇ 3-(1-METHYL-2-PYRROLYL)PYRIDINE ◇ 3-(1-METHYL-1H-PYRROL-2-YL)-PYRIDINE (9CI) ◇ NICOTYRINE ◇ 3,2'-NICOTYRINE

TOXICITY DATA with REFERENCE
ivn-rat LD50:750 mg/kg USXXAM #4065561
orl-mus LD50:575 mg/kg USXXAM #4065561
ipr-mus LD50:191 mg/kg

SAFETY PROFILE: Poison by intraperitoneal route. Moderately toxic by ingestion and intravenous routes. When heated to decomposition it emits toxic fumes of NO_x.

NDX500 CAS:4394-00-7 ***HR: 3***
NIFLURIL
mf: $C_{13}H_9F_3N_2O_2$ mw: 282.24

PROP: Crystals from ethanol. Mp: 204°.

SYNS: ACIDE NIFLUMIQUE (FRENCH) ◇ ACIDO NIFLUMICO (ITALIAN) ◇ ACTOL ◇ FORENOL ◇ LANDRUMA ◇ NIFLUMIC ACID ◇ 2-(3-(TRIFLUOROMETHYL)ANILINO)NICOTINIC ACID ◇ 2-(3-(TRIFLUOROMETHYL)-PHENYL)AMINONICOTINIC ACID ◇ UP 83

TOXICITY DATA with REFERENCE
orl-rbt TDLo:100 mg/kg (1D pre):REP FESTAS 38,238,82
orl-rat LD50:250 mg/kg JMCMAR 16,780,73
ipr-rat LD50:155 mg/kg JMCMAR 19,135,76
orl-mus LD50:375 mg/kg AGACBH 14,247,84
ipr-mus LD50:196 mg/kg NYKZAU 68,683,72

SAFETY PROFILE: Poison by ingestion and intraperitoneal routes. Experimental reproductive effects. An anti-inflammatory agent. When heated to decomposition it emits very toxic fumes of F^- and NO_x.

NDY000 CAS:555-84-0 ***HR: 3***
NIFURADENE
mf: $C_8H_8N_4O_4$ mw: 224.20

PROP: Lemon-yellow solid from nitromethane. Mp: 261.5-263° (decomp).

SYNS: NF 246 ◇ 1-(((5-NITRO-2-FURANYL)METHYLENE)AMINO)-2-IMIDAZOLIDINONE ◇ N-(5-NITRO-2-FURFURYLIDENE)-1-AMINO-2-IMIDAZOLIDONE ◇ 1-((5-NITROFURFURYLIDENE)AMINO)-2-IMIDAZOLIDINONE ◇ N-(5-NITRO-2-FURFURYLIDENEAMINO)-2-IMIDAZOAIDINONE ◇ NSC-6470 ◇ OXAFURADENE ◇ OXYFURADENE ◇ RENAFUR

TOXICITY DATA with REFERENCE
orl-rat TDLo:34 g/kg/46W-C:CAR JNCIAM 51,403,73
orl-rat LD50:1680 mg/kg TXAPA9 18,185,71
ipr-mus LD50:1000 mg/kg JPPMAB 16,663,64
orl-rbt LD50:540 mg/kg TXAPA9 18,185,71

CONSENSUS REPORTS: IARC Cancer Review: Group 2B IMEMDT 7,56,87; Animal Limited Evidence IMEMDT 7,181,74.

SAFETY PROFILE: Suspected carcinogen with experimental carcinogenic data. Moderately toxic by ingestion and intraperitoneal routes. When heated to decomposition it emits toxic fumes of NO_x.

NDY350 CAS:24632-47-1 ***HR: 3***
NIFURPIPONE
mf: $C_{12}H_{17}N_5O_4$ mw: 295.34

SYNS: 4-METHYL-1-PIPERAZINEACETIC ACID (5-NITROFURFURYLIDENE)HYDRAZIDE ◇ 4-METHYL-PIPERAZINO-ACETOHYDRAZONE-5-NITROFURFUROL ◇ (4-METHYLPIPER- AZINYLACETYL)HYDRAZONE-5-NITRO-2-FURALDEHYDE

TOXICITY DATA with REFERENCE
orl-rat LD50:990 mg/kg BCFAAI 114,73,74
ipr-rat LD50:212 mg/kg BCFAAI 114,73,74
orl-mus LD50:880 mg/kg BCFAAI 114,73,74
ipr-mus LD50:315 mg/kg JMCMAR 14,633,71

CONSENSUS REPORTS: EPA Genetic Toxicology Program.

SAFETY PROFILE: Poison by intraperitoneal route. Moderately toxic by ingestion. When heated to decomposition it emits toxic fumes of NO_x.

NDY360 CAS:51799-29-2 ***HR: 3***
NIFURPIPONE ACETATE
mf: $C_{12}H_{17}N_5O_4 \cdot C_2H_4O_2$ mw: 355.40

SYNS: 4-METHYL-1-PIPERAZINEACETIC, ACID ((5-NITRO-2-FURANYL)METHYLENE)HYDRAZIDE, ACETATE ◇ 4-METHYL-1-PIPERAZINEACETIC, ACID (5-NITROFURFURYLIDENE)HYDRAZIDE ACETATE

TOXICITY DATA with REFERENCE
orl-rat LD50:833 mg/kg BCFAAI 114,73,75
orl-mus LD50:1125 mg/kg BCFAAI 114,73,75
ipr-mus LD50:228 mg/kg BCFAAI 114,73,75
scu-mus LD50:288 mg/kg BCFAAI 114,73,75
ivn-mus LD50:295 mg/kg BCFAAI 114,73,75
orl-rbt LD50:400 mg/kg BCFAAI 114,73,75
ivn-rbt LD50:127 mg/kg BCFAAI 114,73,75

SAFETY PROFILE: Poison by ingestion, subcutaneous, intravenous, and intraperitoneal routes. When heated to decomposition it emits toxic fumes of NO_x.

NDY370 CAS:24632-48-2 ***HR: 3***
NIFURPIPONE DIHYDROCHLORIDE
mf: $C_{12}H_{17}N_5O_4 \cdot 2ClH$ mw: 368.26

SYN: 4-METHYL-1-PIPERAZINEACETIC, ACID(5-NITROFURFURYLIDENE)HYDRAZIDE DIHYDROCHLORIDE

TOXICITY DATA with REFERENCE
orl-rat LD50:960 mg/kg BCFAAI 114,73,75
orl-mus LD50:688 mg/kg BCFAAI 114,73,75
ipr-mus LD50:252 mg/kg BCFAAI 114,73,75
ivn-mus LD50:222 mg/kg BCFAAI 114,73,75

SAFETY PROFILE: Poison by intravenous and intraperitoneal routes. Moderately toxic by ingestion. When heated to decomposition it emits toxic fumes of NO_x and HCl.

NDY380 CAS:56217-89-1 ***HR: 3***
NIFURPIPONE HYDROCHLORIDE
mf: $C_{12}H_{17}N_5O_4 \cdot ClH$ mw: 331.80

SYN: 4-METHYL-1-PIPERAZINEACETIC, ACID (5-NITROFURFURYLIDENE)HYDRAZIDE HYDROCHLORIDE

TOXICITY DATA with REFERENCE
ivn-rat LD50:140 mg/kg BCFAAI 114,73,75
orl-mus LD50:594 mg/kg BCFAAI 114,73,75
ivn-mus LD50:300 mg/kg BCFAAI 114,73,75
ivn-mky LD50:240 mg/kg BCFAAI 114,73,75

SAFETY PROFILE: Poison by intravenous route. Moderately toxic by ingestion. When heated to decomposition it emits toxic fumes of NO_x and HCl.

NDY390 CAS:56217-90-4 ***HR: 3***
NIFURPIPONE MALEATE (1:1)
mf: $C_{12}H_{17}N_5O_4 \cdot C_4H_4O_4$ mw: 411.42

SYN: 4-METHYL-1-PIPERAZINEACETIC ACID, (5-NITROFURFURYLIDENE)HYDRAZIDE MALEATE (1:1)

TOXICITY DATA with REFERENCE
orl-rat LD50:1350 mg/kg BCFAAI 114,73,75
ivn-rat LD50:145 mg/kg BCFAAI 114,73,75
orl-mus LD50:750 mg/kg BCFAAI 114,73,75
ivn-mus LD50:365 mg/kg BCFAAI 114,73,75
ivn-rbt LD50:150 mg/kg BCFAAI 114,73,75

SAFETY PROFILE: Poison by intravenous route. Moderately toxic by ingestion. When heated to decomposition it emits toxic fumes of NO_x.

NDY400 CAS:13411-16-0 ***HR: 2***
NIFURPIRINOL
mf: $C_{12}H_{10}N_2O_4$ mw: 246.24

PROP: Yellow needles from acetone or methanol. Mp: 170-171°C.

SYNS: FURANACE ◇ FURANACE-10 ◇ FURPIRINOL ◇ FURPYRINOL ◇ 6-HYDROXYMETHYL-2-(2-(5-NITRO-2-FURYL)VINYL)-PYRIDINE ◇ NF 323 ◇ 6-(2-(5-NITRO-2-FURANYL)ETHENYL)-2-PYRIDINEMETHANOL (9CI) ◇ 6-(2-(5-NITRO-2-FURYL)VINYL-2-PYRIDINE-METHANOL ◇ P-7138

TOXICITY DATA with REFERENCE
mmo-sat 1 mg/L MUREAV 104,61,82
dns-hmn:fbr 5 μmol/L IDZAAW 48,291,73
orl-rat LD50:4050 mg/kg MEIEDD 10,938,83
orl-mus LD50:3720 mg/kg MEIEDD 10,938,83

SAFETY PROFILE: Moderately toxic by ingestion. Human mutation data reported. When heated to decomposition it emits toxic fumes of NO_x.

NDY500 CAS:3570-75-0 ***HR: 3***
NIFURTHIAZOLE
mf: $C_8H_6N_4O_4S$ mw: 254.24

PROP: Bright yellow plates. Mp: 215.5°.

SYNS: AS-17665 ◇ FNT ◇ FORMIC 2-(4-(5-NITROFURYL)-2-THIAZOLYL)HYDRAZIDE ◇ 2-(2-FORMYLHYDRAZINO)-4-(5-NITRO-2-FURYL)THIAZOLE ◇ NEFURTHIAZOLE ◇ 2-(4-(5-NITRO-2-FURANYL)-2-THIAZOLYL)-HYDRAZINECARBOXALDEHYDE ◇ NSC-525334

TOXICITY DATA with REFERENCE
mmo-sat 200 ng/plate CNJGA8 21,319,79
spm-mus-ipr 1575 mg/kg/5D-I CMMUAO 5,257,78
orl-rat TDLo:31 g/kg/44W-C:CAR PAACA3 10,15,69
orl-mus TDLo:30 g/kg/33W-C:CAR CNREA8 38,1398,78
orl-rat TD:28 g/kg/46W-C:NEO JNCIAM 47,437,71

CONSENSUS REPORTS: IARC Cancer Review: Group 2B IMEMDT 7,56,87; Animal Sufficient Evidence IMEMDT 7,151,74. EPA Genetic Toxicology Program.

SAFETY PROFILE: Suspected carcinogen with experimental carcinogenic and neoplastigenic data. Mutation data reported. When heated to decomposition it emits very toxic fumes of NO_x and SO_x.

NDY600 CAS:22609-73-0 ***HR: 3***
NILUDIPINE
mf: $C_{25}H_{34}N_2O_8$ mw: 490.61

SYNS: BAY-a 7168 ◇ BIS(2-PROPOXYETHYL)-1,4-DIHYDRO-2,6-DIMETHYL-4-(3-NITROPHENYL)-3,5-PYRIDINEDICARBOXYLATE ◇ 1,4-DIHYDRO-2,6-DIMETHYL-4-(m-NITROPHENYL)-3,5-PYRIDINEDICARBOXYLIC ACID BIS(2-PROPOXYETHYL) ESTER

TOXICITY DATA with REFERENCE
orl-rat TDLo:270 mg/kg (female 17-21D post):REP IYKEDH 12,1100,81
orl-rat LD50:1469 mg/kg IYKEDH 12,714,81
scu-rat LD50:830 mg/kg IYKEDH 12,714,81
ivn-rat LD50:40 mg/kg IYKEDH 12,714,81
orl-mus LD50:1415 mg/kg IYKEDH 12,714,81
scu-mus LD50:642 mg/kg IYKEDH 12,714,81
ivn-mus LD50:90 mg/kg IYKEDH 12,714,81

SAFETY PROFILE: Poison by intravenous route. Moderately toxic by ingestion and subcutaneous routes.

Experimental reproductive effects. When heated to decomposition it emits toxic fumes of NO_x.

NDY650 CAS:75530-68-6 ***HR: 2***
NILVADIPINE
mf: $C_{19}H_{19}N_3O_6$ mw: 385.41

SYNS: FK 235 ◇ FR 34235 ◇ 3,5-PYRIDINEDICARBOXYLIC ACID, 2-CYANO-1,4-DIHYDRO-6-METHYL-4-(3-NITROPHENYL)-,3-METHYL-5-(1-METHYLETHYL) ESTER

TOXICITY DATA with REFERENCE
orl-rat TDLo:8352 mg/kg (male 9W pre):REP KSRNAM 21,1765,87
orl-rat LD50:1560 mg/kg KSRNAM 21,1723,87
ivn-rat LD50:9650 μg/kg KSRNAM 21,1723,87
orl-mus LD50:1300 mg/kg KSRNAM 21,1723,87
ivn-mus LD50:9150 μg/kg KSRNAM 21,1723,87
orl-dog LD50:510 mg/kg KSRNAM 21,1723,87
ivn-dog LD50:3850 μg/kg KSRNAM 21,1723,87

SAFETY PROFILE: Moderately toxic by ingestion and intravenous routes. Experimental reproductive effects. When heated to decomposition it emits toxic fumes of NO_x.

NDY700 CAS:66085-59-4 ***HR: 3***
NIMODIPINE
mf: $C_{21}H_{26}N_2O_7$ mw: 418.45

PROP: Crystals from petr ether/acetic ester. Mp: 125°.

SYNS: BAY-E 9736 ◇ 1,4-DIHYDRO-2,6-DIMETHYL-4-(3-NITROPHENYL)- 3,5-PYRIDINEDICARBOXYLIC ACID 2-METHOXYETHYL-1-METHYLETHYL ESTER ◇ NIMOTOP

TOXICITY DATA with REFERENCE
orl-dog TDLo:2730 mg/kg (13W male):REP KSRNAM 19,6933,85
orl-rat LD50:2738 mg/kg KSRNAM 19,68899,85
scu-rat LD50:4234 mg/kg KSRNAM 19,68899,85
ivn-rat LD50:5 mg/kg ARTODN 54,275,83
orl-mus LD50:940 mg/kg KSRNAM 19,68899,85
scu-mus LD50:9500 mg/kg KSRNAM 19,68899,85
ivn-mus LD50:26200 μg/kg KSRNAM 19,68899,85
orl-dog LD50:1 g/kg KSRNAM 19,68899,85

SAFETY PROFILE: Poison by intravenous route. Moderately toxic by ingestion. Experimental reproductive effects. When heated to decomposition it emits toxic fumes of NO_x.

NDY800 CAS:42471-28-3 ***HR: D***
NIMUSTINE
mf: $C_9H_{13}ClN_6O_2$ mw: 272.73

SYN: 3-(4-AMINO-2-METHYL-5-PYRIMIDINYL)METHYL-1-(2-CHLOROETHYL)-1-NITROSOUREA

TOXICITY DATA with REFERENCE
mmo-sat 75 μg/plate TAKHAA 44,96,85
dnr-bcs 100 μg/plate TAKHAA 44,96,85

SAFETY PROFILE: Human mutation data reported. When heated to decomposition it emits toxic fumes of NO_x.

NDZ000 ***HR: 1***
NIOBIUM
af: Nb aw: 92.906

PROP: Steel gray, cubic crystals. Mp: 2468 ± 10°; bp: 4742°; d: 8.57. Sol in aqua regia, fused alkali. An element which occurs throughout nature.

SAFETY PROFILE: An eye and severe skin irritant. No reports of human intoxication. Can cause kidney damage. Experimentally, there is a moderate fibrogenic effect on the lungs after intratracheal administration. Some niobium is found in all parts of the body. Flammable in the form of dust when exposed to flame or by chemical reaction. Moderately explosive in the form of dust when exposed to flame. Ignites in fluorine; chlorine (at 205°C). Incandescent reaction with bromine trifluoride.

NEA000 CAS:10026-12-7 ***HR: 3***
NIOBIUM CHLORIDE
mf: Cl_5Nb mw: 270.16

PROP: Yellow, very deliq, monoclinic crystals. Decomp in moist air evolving HCl. D: 2.75, mp: 204.7-209.5°, bp: approx 250°, subl @ 125°. Sol in HCl, carbon tetrachloride.

SYNS: COLUMBIUM PENTACHLORIDE ◇ NIOBIUM PENTACHLORIDE

TOXICITY DATA with REFERENCE
orl-rat LD50:1400 mg/kg HYSAAV 31,328,66
ipr-rat LD50:40 mg/kg EQSSDX 1,1,75
orl-mus LD50:829 mg/kg HYSAAV 31,328,66
ipr-mus LD50:61 mg/kg 34ZIAG -,422,69

CONSENSUS REPORTS: Reported in EPA TSCA Inventory. EPA Genetic Toxicology Program.

SAFETY PROFILE: Poison by intraperitoneal route. Moderately toxic by ingestion. May cause kidney injury. When heated to decomposition it emits very toxic fumes of Cl^-. See also NIOBIUM and CHLORIDES.

NEA100 CAS:81486-22-8 ***HR: 3***
NIPRADILOL
mf: $C_{15}H_{22}N_2O_6$ mw: 326.39

SYNS: 3,4-DIHYDRO-8-(2-HYDROXY-3-(ISOPROPYLAMINO)PROPOXY)-2H-1-BENZOPYRAN-3-OL-3-NITRATE ◇ 3,4-DIHYDRO-8-(2-HYDROXY-3-ISOPROPYLAMINO)PROPOXY-3-NITROXY-2H-1-

BENZOPYRAN ◇ 3,4-DIHYDRO-8-(2-HYDROXY-3-((1-METHYLETHYL)AMINO)PROPOXY)-2H-1-BENZOPYRAN-3-OL-3-NITRATE ◇ K 351

TOXICITY DATA with REFERENCE
orl-rat TDLo:5400 mg/kg (17-22D preg/21D post):REP OYYAA2 30,1,85
orl-rat TDLo:2200 mg/kg (female 7-17D post):TER OYYAA2 29,747,85
orl-rat LD50:1040 mg/kg OYYAA2 29,725,85
ipr-rat LD50:144 mg/kg OYYAA2 29,725,85
scu-rat LD50:850 mg/kg OYYAA2 29,725,85
ivn-rat LD50:78 mg/kg OYYAA2 29,725,85
orl-mus LD50:461 mg/kg OYYAA2 29,725,85
ipr-mus LD50:147 mg/kg OYYAA2 29,725,85
scu-mus LD50:416 mg/kg OYYAA2 29,725,85
ivn-mus LD50:68 mg/kg OYYAA2 29,725,85
ivn-dog LD50:20 mg/kg OYYAA2 29,725,85
orl-rbt LD50:300 mg/kg OYYAA2 29,725,85

SAFETY PROFILE: Poison by ingestion, intravenous, and intraperitoneal routes. Moderately toxic by subcutaneous route. An experimental teratogen. Experimental reproductive effects. When heated to decomposition it emits toxic fumes of NO_x. See also NITRATES.

NEA500 CAS:77-20-3 ***HR: 3***
NISENTIL
mf: $C_{16}H_{23}NO_2$ mw: 261.40

SYNS: ALPHAPRODINE ◇ 1,3-DIMETHYL-4-PHENYL-4-PIPERIDINOL PROPIONATE (ESTER) ◇ α-1,3-DIMETHYL-4-PHENYL-4-PIPERIDINYL PROPIONATE ◇ α-1,3-DIMETHYL-4-PHENYL-4-PROPIONOXYPIPERIDINE ◇ 1,3-DIMETHYL-4-PHENYL-4-PROPIONOXYPIPERIDINE ◇ NU-1196 ◇ PRISILIDINE ◇ α-PRODINE ◇ PROPIONIC ACID, α-1,3-DIMETHYL-4-PHENYL-4-PIPERIDYL ESTER

TOXICITY DATA with REFERENCE
ivn-wmn TDLo:400 μg/kg (38W preg):TER JRPMAP 27,439,82
orl-rat LD50:90 mg/kg JPETAB 99,16,50
ipr-rat LD50:22 mg/kg JMCMAR 19,852,76
scu-mus LD50:23 mg/kg JPETAB 99,312,50
ipr-mus LD50:73 mg/kg JPETAB 99,312,50
scu-mus LD50:98 mg/kg JPETAB 99,312,50
ivn-mus LD50:54 mg/kg JPETAB 99,312,50

SAFETY PROFILE: Experimental poison by ingestion, intraperitoneal, subcutaneous, and intravenous routes. Human teratogenic effects by intravenous route: developmental abnormalities of the cardiovascular system. A narcotic analgesic. When heated to decomposition it emits toxic fumes of NO_x.

NEB000 CAS:561-78-4 ***HR: 3***
NISENTIL HYDROCHLORIDE
mf: $C_{16}H_{23}NO_2{\bullet}ClH$ mw: 297.86

SYNS: ALPHAPRODINE HYDROCHLORIDE ◇ 1,3-DIMETHYL-4-PHENYL-4-PIPERIDINOL, PROPIONATE, HYDROCHLORIDE ◇ 1,3-DIMETHYL-4-PHENYL-4-PIPERIDYL PROPIONATE HYDROCHLORIDE ◇ dl-1,3-DIMETHYL-4-PHENYL-4-PIPERIDINOL PROPIONATE HYDROCHLORIDE ◇ (±)-1,3-DIMETHYL-4-PHENYL-4-PIPERIDYL ESTER PROPIONIC ACID HYDROCHLORIDE ◇ (±)-α-1,3-DIMETHYL-4-PHENYL-4-PROPIONOXYPIPERIDINEHYDROCHLORIDE ◇ NISENTIL ◇ NU-1196 ◇ PRISILIDENE HYDROCHLORIDE ◇ α-PRODINE HYDROCHLORIDE

TOXICITY DATA with REFERENCE
orl-hmn LDLo:1440 μg/kg CTOXAO 5,307,72
orl-rat LD50:90 mg/kg JPETAB 93,314,48
ipr-rat LD50:22 mg/kg JPETAB 99,312,50
scu-rat LD50:23 mg/kg JPETAB 99,312,50
ivn-rat LD50:25 mg/kg AIPTAK 109,171,57
ipr-mus LD50:73 mg/kg JPETAB 99,312,50
scu-mus LD50:98 mg/kg JPETAB 99,312,50
ivn-mus LD50:32 mg/kg JPETAB 99,314,48
ivn-dog LD50:36200 μg/kg AIPTAK 109,171,57
ivn-rbt LD50:19 mg/kg JPETAB 99,312,50

SAFETY PROFILE: A deadly human poison by ingestion. Poison experimentally by ingestion, subcutaneous, intravenous, and intraperitoneal routes. When heated to decomposition it emits toxic fumes of NO_x and HCl. See also NISENTIL.

NEC000 CAS:1432-75-3 ***HR: 3***
NITRALAMINE HYDROCHLORIDE
mf: $C_{10}H_{13}ClN_2O_2S{\bullet}ClH$ mw: 297.22

SYNS: 2-((o-CHLORO-α-(NITROMETHYL)BENZYL)THIO)ETHYLAMINE HYDROCHLORIDE ◇ ((α-NITROMETHYL)-o-CHLOROBENZYLTHIO)ETHYLAMINEHYDROCHLORIDE

TOXICITY DATA with REFERENCE
ivn-mus LD50:100 mg/kg CSLNX* NX#08976
ipr-rat LD50:124 mg/kg 28ZEAL 4,301,69
ipr-mus LD50:130 mg/kg 28ZEAL 4,301,69

SAFETY PROFILE: Poison by intravenous and intraperitoneal routes. When heated to decomposition it emits very toxic fumes of SO_x, NO_x, and Cl^-.

NEC500 CAS:101976-64-1 ***HR: 3***
NITRAMINOACETIC ACID
mf: $C_2H_4N_2O_4$ mw: 120.08

SYNS: O-ACETYL NITROHYDROXYAMINE ◇ O-ACETYL-N-NITROHYDROXYLAMINE

TOXICITY DATA with REFERENCE
orl-mus LD50:40 mg/kg 85ERAY 1,902,78
ipr-mus LD50:43 mg/kg 85ERAY 1,902,78
ivn-mus LD50:32 mg/kg 85ERAY 1,902,78

SAFETY PROFILE: Poison by ingestion, intraperitoneal, and intravenous routes. When heated to decomposition it emits toxic fumes of NO_x. See also AMINES.

NED000 ***HR: 3***
NITRATES

PROP: Organic nitrates are usually termed nitro compounds. These compounds are a combination of the nitro ($-NO_2$) group and an organic radical. However, this term is often used to denote nitric acid esters of an organic material. Inorganic nitrates are compounds of metals which are combined with the mono-valent NO_3 radical.

SAFETY PROFILE: Large amounts taken by mouth may have serious or even fatal effects. The symptoms are dizziness, abdominal cramps, vomiting, bloody diarrhea, weakness, convulsions, and collapse. Small, repeated doses may lead to weakness, general depression, headache and mental impairment. Also, there is some implication of increased cancer incidence among those exposed.

Flammable by spontaneous chemical reaction; practically all nitrates are powerful oxidizing agents. Some nitrates may explode when shocked, exposed to heat or flame or by spontaneous chemical reaction (see also EXPLOSIVES, HIGH). All the inorganic nitrates act as oxygen carriers; under proper conditions, these can give up their oxygen to other materials which may in turn detonate. Ammonium nitrate has all the properties of the other nitrates, but it is also able to detonate by itself under certain conditions. It is therefore a high explosive, although very insensitive to impact and difficult to detonate. In the pure state, it requires a combination of an initiator and a high explosive. It is a relatively safe high explosive which, however, must be stored in a cool, ventilated place away from acute fire hazards and easily oxidized materials. Ammonium nitrate must not be confined because, if a fire should start, confinement can cause detonation with extremely violent results.

Violent reaction with Al; BP; cyanide; esters; PN_2H; P; NaCN; $SnCl_2$; sodium hypophosphite; thiocyanates. Dangerous disaster hazard due to fire and explosion hazard. When heated to decomposition it emits toxic fumes of NO_x. They are powerful oxidizing agents which may cause violent reaction with reducing materials. Nitrates should be protected carefully in storage.

NED500 CAS:7697-37-2 ***HR: 3***
NITRIC ACID
DOT: NA 1760/UN 2031
mf: HNO_3 mw: 63.02

PROP: Transparent, colorless or yellowish, fuming, suffocating, caustic and corrosive liquid. Mp: −42°, bp: 86°, d: 1.50269 @ 25°/4°.

SYNS: ACIDE NITRIQUE (FRENCH) ◇ ACIDO NITRICO (ITALIAN) ◇ AQUA FORTIS ◇ AZOTIC ACID ◇ AZOTOWY KWAS (POLISH) ◇ HYDROGEN NITRATE ◇ NITRIC ACID, over 40% (DOT) ◇ SALPETERSAURE (GERMAN) ◇ SALPETERZUUROPLOSSINGEN (DUTCH)

TOXICITY DATA with REFERENCE
orl-rat TDLo:2345 mg/kg (female 18D post):REP ZHYGAM 29,667,83
orl-rat TDLo:21150 mg/kg (female 1-21D post):TER ZHYGAM 29,667,83
orl-hmn LDLo:430 mg/kg YAKUD5 22,651,80
unr-man LDLo:110 mg/kg 85DCAI 2,73,70

CONSENSUS REPORTS: Reported in EPA TSCA Inventory. EPA Genetic Toxicology Program. Community Right-To-Know List.

OSHA PEL: (Transitional: TWA 2 ppm) TWA 2 ppm; STEL 4 ppm
ACGIH TLV: TWA 2 ppm; STEL 4 ppm
DFG MAK: 10 ppm (25 mg/m^3)
NIOSH REL: (Nitric Acid) TWA 2 ppm
DOT Classification: Oxidizer; Label: Oxidizer and Corrosive (UN2031); Corrosive Material; Label: Corrosive (NA1760, UN2031).

SAFETY PROFILE: Human poison by ingestion. An experimental teratogen. Experimental reproductive effects. Corrosive to eyes, skin, mucous membranes, and teeth. Causes upper respiratory irritation which may seem to clear up only to return in a few hours and more severely. Depending on environmental factors the vapor will consist of a mixture of the various oxides of nitrogen and nitric acid. Flammable by chemical reaction with reducing agents. It is a powerful oxidizing agent.

Explosive reaction with acetic anhydride, acetone + acetic acid (in storage), acetone + hydrogen peroxide, acetone + sulfuric acid (if confined), alcohols, alkane thiols, 2-aminothiazole, 2-aminothiazole + sulfuric acid, dinitrogen tetraoxide or sulfuric acid + aromatic amines (e.g., aniline, n-ethylamine, o-toluidine, xylidine, p-phenylenediamine), benzidine (hypergolic), benzonitrile + sulfuric acid, 5-acetylamino-3-bromobenzo(b)thiophene, 1,4-bis(methoxymethyl)-2,3,5,5-tetramethylbenzene (at 80°C), 1,3-bis(trifluoromethyl)benzene + sulfuric acid (vapors are initiated by spark), tert-butyl-m-xylene, cadmium phosphide, chlorobenzene, cotton + rubber + sulfuric acid + water (mixture has caused industrial explosions), crotonaldehyde (hypergolic), cyclohexylamine, 1,2-diaminoethanebis(trimethylgold), diethyl ether, diethyl ether + sulfuric acid, 1,1-dimethyl hydrazine (hypergolic), dimethyl sulfide, dimethyl sulfide + 1,4-dioxane, dimethyl sulfoxide + water, dinitrogen tetraoxide + nitrogenous fuels (hypergolic with triethylamine, dimethylhydrazine, mixo-xylidine), dioxane + perchloric acid, diphenyldistibene, divinyl ether (hypergolic), ethane sulfonamide, 5-ethyl-2-methylpyridine, fat + sulfuric acid (when confined), 2-formylamino-1-phenyl-1,3-propanediol, fluorine, furfurylidene ketones (hypergolic), glycerol + sulfuric acid, hexalithium disilicide, 2,2,4,4,6,6-hexamethyl-trithiane, hydrazine (hypergolic), hydrocarbons, hydrocarbons + 1,1-dimethylhydrazine, hydrogen peroxide + soils, ion exchange resins, magnesium + 2-nitroaniline (hypergolic), metal acetylides (e.g., cesium acetylide, rubidium acetylide), metal cyanides, metal hexacyanoferrates (3-) or (4-), metal thiocyanates, 4-methylcyclohexanone (above 75°C), methylthiophene, nitrobenzene + sulfuric acid, 1-nitronaphthalene + sulfuric acid, non-metal hydrides (e.g., arsine, phosphine, tetraborane(10), stibene), phosphorus, organic materials + oxidizers (e.g., perchloric acid, potassium chlorate, sulfuric acid), phenylacetylene + 1,1-dimethylhydrazine, phosphine derivatives (e.g., phosphine, phosphonium iodide, ethyl phosphine, tris(iodomercury) phosphine), tetraphosphorus diiodo triselenide, phosphorus trichloride, polyurethane foam, propiophenone + sulfuric acid, pyrocatechol (hypergolic), potassium phosphinate + heat, resorcinol, rubber, silicone oil, silver buten-3-ynide, sulfur dioxide, 1,3,5-triacetylhexahydro-1,3,5-triazine + trifluoroacetic anhydride, triazine + trifluoroacetic anhydride, zinc ethoxide. Explosive or hypergolic reaction with various hydrocarbons (e.g., acetylene derivatives, benzene, 3-carene, cashew nut shell oil, cyclopentadienes, dicyclopentadiene, dienes, hexamethylbenzene, mesitylene, burning petroleum products, toluene, turpentine, p-xylene).

Ignition on contact with acetone, alcohols + disulfuric acid, alcohols + potassium permanganate, aliphatic amines, O-alkyl ethylene dithiophosphate, ammonia, anilinium nitrate + inorganic materials [e.g., copper(I) chloride, potassium permanganate, sodium pentacyanonitrosylferrate, vanadium(V) oxide, ammonium metavanadate, sodium metavanadate), aromatic amines + metal compounds (e.g., ammonium metavanadate, copper(I) oxide, copper(II) oxide, iron(III) chloride, iron(III) oxide, potassium chromate, potassium dichromate, potassium hexacyanoferrate(I), potassium hexacyanoferrate(III), sodium metavanadate, sodium pentacyanonitrosylferrate(II), vanadium(V) oxide], dichromates + organic fuels (e.g., ammonium dichromate, potassium dichromate, potassium chromate, cyclohexanol, 2-cresol, 3-cresol, furfural), diphenyl tin, lead-containing rubber, metals (e.g., lithium, sodium, magnesium), non-metal hydrides (e.g., hydrogen iodide, hydrogen selenide, hydrogen sulfide, hydrogen telluride), phosphorus vapor, nickel tetraphosphide, tetraphosphorus iodide, polysilylene, turpentine + catalysts (e.g., concentrated sulfuric acid, iron(III) chloride, ammonium metavanadate, copper(II) chloride), wood.

Forms explosive mixtures with acetic acid, acetic acid + sodium hexahydroxyplatinate(IV), acetic anhydride + hexamethylenetetramine acetate, acetoxyethylene glycol, ammonium nitrate, anilinium nitrate, 1,2-dichloroethane, dichloroethylene, dichloromethane, diethylaminoethanol, 3,6-dihydro-1,2,2H-oxazine, dimethyl ether, 1,3-dinitrobenzene, disodium phenyl orthophosphate, 2-hexenal (heat sensitive), hydrofluoric acid + lactic acid, hydrofluoric acid + propylene glycol + silver nitrate, hydrogen peroxide + ketones (e.g., 2-butanone, 3-pentanone, cyclopentanone, cyclohexanone, 3-methylcyclohexanone), hydrogen peroxide + mercury(II) oxide, metals (e.g., titanium, uranium, tin), metal salicylates, nitroaromatics (e.g., mono- and di-nitrobenzenes, di- and tri-nitrotoluenes), nitrobenzene + water, nitromethane, salicylic acid.

Incompatible with 4-acetoxy-3-methoxybenzaldehyde, acetylene, acrolein, acrylonitrile, acrylonitrile + methacrylate copolymer, allyl alcohol, allyl chloride, 2-amino ethanol, ammonium hydroxide, aniline, anion exchange resins, antimony, SbH_3, arsenic hydride, arsine + boron tribromide, benzo[b]thiophene derivatives, N-benzyl-N-ethylaniline + sulfuric acid, bismuth, boron, boron decahydride, B_4H_{10}, boron phosphide, bromine pentafluoride, butanethiol, 2,6-di-tert-butyl phenol, calcium hypophosphite, carbon, $C_2H_5PH_2$, cellulose, Cs_2C_2, chlorine trifluoride, 4-chloro-2-nitroaniline, chlorosulfonic acid, coal, CuN_3, Cu_3N_2, copper(I) nitride, cresol, cumene, cyanides, cyclic ketones, cyclohexanol + cyclohexanone, diborane, di-2-6-butoxyethylether (butex), dichromate + anion exchange resins, diisopropylether, dimethylaminomethylferrocene + water, uns-dimethyl hydrazine, diphenylmercury, epichlorohydrin, ethanol, m-ethylaniline, ethylene diamine, ethylene imine, 5-ethyl-2-picoline, formaldehyde + impurities, formic acid + heat, formic acid + urea, furfural, furfuryl alcohol, germanium, glycerol + hydrofluoric acid, glyoxal, hydrogen iodide, HN_3, hydrogen peroxide, hydrogen selenide, hydrogen sulfide, H_2Te, indane + sulfuric acid, FeO, iron(II) oxide powder, isoprene, ketones + hydrogen peroxide, lactic acid + HF, Li_6Si_2, magnesium silicide, magnesium phosphide, magnesium-titanium alloy, manganese, mesityl oxide, mesitylene, metals (e.g., bismuth powder, germanium powder, uranium powder, molten zinc), 2-methyl-5-ethyl pyridine, NdP, non-metal powders (e.g., boron, silicon, arsenic, carbon), n-butyraldehyde, oleum, phosphorus halides, phthalic acid, phtahlic anhydride, polyalkenes (e.g., polyetheylene, polypropylene), polydibromosilane, polyethylene oxide derivatives, KH_2PO_2, β-propiolactone, propylene oxide, pyridine, Rb_2C_2, reducants, selenium, selenium iodophosphide, silver + ethanol, sodium, sodium hydroxide, NaN_3, sucrose, sulfamic acid, sulfuric acid, sulfuric acid + $C_6H_5CH_3$, sulfuric acid + glycerides, sulfur halides (e.g., sulfur dichloride, sulfur dibromide, disulfur dibromide), sulfuric acid + terephthalic acid, terpenes, thioaldehydes, thiocyanates, thioketones, thiophene, titanium, titanium alloy, titanium-magnesium alloy, toluidine, triazine, tricadmium diphosphide, triethylgallium monoetherate, trimagnesium diphosphide,

2,4,6-trimethyltrioxane, uranium, uranium disulfide, uranium-neodymium alloy, uranium-neodymium-zirconium alloy, vinylacetate, vinylidene chloride, zinc, zirconium-uranium alloys.

Will react with water or steam to produce heat and toxic and corrosive fumes. To fight fire, use water. When heated to decomposition emits highly toxic fumes of NO_x and hydrogen nitrate.

NEE000 CAS:13277-59-3 ***HR: 1***
NITRIC ACID, DODECYL ESTER
mf: $C_{12}H_{25}NO_3$ mw: 231.38

SYNS: DODECYL NITRATE ◇ LAURYLESTER KYSELINY DUSICNE (CZECH) ◇ LAURYLNITRAT (CZECH) ◇ LAURYL NITRATE

TOXICITY DATA with REFERENCE
skn-rbt 500 mg/24H MLD 28ZPAK -,49,72
eye-rbt 500 mg/24H MLD 28ZPAK -,49,72

SAFETY PROFILE: A skin and eye irritant. When heated to decomposition it emits toxic fumes of NO_x. See also ESTERS.

NEE500 CAS:7697-37-2 ***HR: 3***
NITRIC ACID (RED FUMING)
DOT: UN 2032
mf: HNO_3 mw:63.02

PROP: Colorless to yellow to red corrosive liquid. D: > 1.480. Contains from 8 to 17% NO_2 (AMIHBC 10,418,54).

SYNS: NITRIC ACID, FUMING (DOT) ◇ NITRIC ACID, RED FUMING (DOT) ◇ NITROUS FUMES ◇ RED FUMING NITRIC ACID ◇ RFNA

TOXICITY DATA with REFERENCE
ihl-rat LC50:67 ppm (NO_2)/4H AMIHBC 10,418,54

CONSENSUS REPORTS: EPA Genetic Toxicology Program.

DOT Classification: Corrosive Material; Label: Corrosive, Oxidizer and Poison.

SAFETY PROFILE: Poison by inhalation. A corrosive irritant to skin, eyes, and mucous membranes. A very dangerous fire hazard and very powerful oxidizing agent. Can react explosively with many reducing agents. Will react with water or steam to produce heat and toxic, corrosive, and flammable vapors. When heated to decomposition it emits highly toxic fumes of NO_x. See also NITRIC ACID.

NEF000 ***HR: 3***
NITRIC ACID (WHITE FUMING)

PROP: Contains from 0.1 to 0.4% NO_2 (AMIHBC 10,418,54).

SYNS: WFNA ◇ WHITE FUMING NITRIC ACID

TOXICITY DATA with REFERENCE
ihl-rat LC50:244 ppm (NO_2)/30M AMIHBC 10,418,54

SAFETY PROFILE: Moderately toxic by inhalation. A corrosive irritant to skin, eyes, and mucous membranes. A very dangerous fire hazard and a very powerful oxidizing agent. Can react explosively with many reducing agents. Will react with water or steam to produce heat and toxic, corrosive, and flammable vapors. When heated to decomposition it emits highly toxic fumes of NO_x. See also NITRIC ACID.

NEF500 CAS:10196-18-6 ***HR: 2***
NITRIC ACID, ZINC SALT, HEXAHYDRATE
mf: $N_2O_6 \cdot Zn \cdot 6H_2O$ mw: 297.51

PROP: Tetragonal, colorless crystals. D: 2.065 @ 14°, mp: 36.4°, bp: $-H_2O$ @ 105-131°.

SYNS: DUSICNAN ZINECNATY (CZECH) ◇ ZINC(II) NITRATE, HEXAHYDRATE (1:2:6)

TOXICITY DATA with REFERENCE
skn-rbt 500 mg/24H SEV 28ZPAK -,10,72
eye-rbt 20 mg/24H SEV 28ZPAK -,10,72
orl-rat LD50:1190 mg/kg 28ZPAK -,10,72

CONSENSUS REPORTS: Zinc and its compounds are on the Community Right-To-Know List.

SAFETY PROFILE: Moderately toxic by ingestion. A severe skin and eye irritant. Can react violently with carbon; copper; metal sulfides; organic matter; phosphorus; sulfur. When heated to decomposition it emits toxic fumes of NO_x and ZnO. See also ZINC COMPOUNDS and NITRATES.

NEG000 CAS:7782-94-7 ***HR: 3***
NITRIC AMIDE
mf: $H_2N_2O_2$ mw: 62.03

O_2NNH_2

PROP: Shiny, white leaflets from ether + petr ether. Mp: 72-75° (decomp); sol in ether, alc, acetone, water; sltly sol in benzene; insol in petr ether, chloroform.

SYN: NITRAMIDE

SAFETY PROFILE: Unstable. Reactions are violent. Can be explosive at specific temperatures. Explodes on contact with concentrated alkali or concentrated sulfuric acid. When heated to decomposition it emits toxic fumes of NO_x. See also AMMONIA (which is easily evolved) and AMIDES.

NEG100 CAS:10102-43-9 ***HR: 3***
NITRIC OXIDE
DOT: UN 1660
mf: NO mw: 30.01

PROP: Colorless gas, blue liquid and solid. Mp: −161°, bp: −151.18, d: 1.3402 g/L; liquid, 1.269 @ −150°; gas, 1.04.

SYNS: BIOXYDE d'AZOTE (FRENCH) ◇ NITROGEN MONOXIDE ◇ OXYDE NITRIQUE (FRENCH) ◇ RCRA WASTE NUMBER P076 ◇ STICKMONOXYD (GERMAN)

TOXICITY DATA with REFERENCE
mmo-sat 30 ppm MUREAV 136,119,84
msc-rat-ihl 27 ppm/3H-C MUREAV 136,119,84
msc-ham:lng 10 ppm/10M-C MUREAV 136,119,84
ihl-rat LC50:1068 mg/m^3 GTPZAB 19(4),52
ihl-mus LCLo:320 ppm AEPPAE 181,145,36

CONSENSUS REPORTS: Reported in EPA TSCA Inventory. EPA Extremely Hazardous Substances List.

OSHA PEL: TWA 25 ppm
ACGIH TLV: TWA 25 ppm
NIOSH REL: (Oxides of Nitrogen) TWA 25 ppm
DOT Classification: Poison A; Label: Poison Gas.

SAFETY PROFILE: A poison gas. A severe eye, skin, and mucous membrane irritant. A systemic irritant by inhalation. Mutation data reported. Exposure may occur whenever nitric acid acts upon organic material, such as wood, sawdust, and refuse; it occurs when nitric acid is heated, and when organic nitro compounds are burned, for example, celluloid, cellulose nitrate (guncotton), and dynamite. The action of nitric acid upon metals, as in metal etching and pickling, also liberates the fumes. In high temperature welding, as with the oxyacetylene or electric torch, the nitrogen and oxygen of the air unite to form oxides of nitrogen. Automobile exhaust and power plant emissions are also sources of NO_x. Exposure occurs in many manufacturing processes when nitric acid is made or used. Oxides of nitrogen have been implicated as a cause of acid rain.

The oxides of nitrogen are somewhat soluble in water reacting with it in the presence of oxygen to form nitric and nitrous acids. This is the action that takes place deep in the respiratory system. The acids formed are irritating and can cause congestion in the throat and bronchi and edema of the lungs. The acids are neutralized by the alkalies present in the tissues with the formation of nitrates and nitrites. The latter may cause some arterial dilation, fall in blood pressure, headache and dizziness, and there may be some formation of methemoglobin. However, the nitrite effect is of secondary importance.

Because of their relatively low solubility in water, the nitrogen oxides are initially only slightly irritating to the mucous membranes of the upper respiratory tract. Their warning power is therefore low, and dangerous amounts of the fumes may be breathed before the worker notices any real discomfort. Higher concentrations (60-150 ppm) cause immediate irritation of the nose and throat with coughing and burning in the throat and chest. These symptoms often clear upon breathing fresh air, and the worker may feel well for several hours. Some 6-24 hours after exposure, a sensation of tightness and burning in the chest develops, followed by shortness of breath, sleeplessness, and restlessness. Dyspnea and air hunger may increase rapidly with development of cyanosis and loss of consciousness followed by death. In cases which recover from the pulmonary edema, there is usually no permanent disability, but pneumonia may develop later. Concentrations of 100-150 ppm are dangerous for short exposures of 30-60 minutes. Concentrations of 200-700 ppm may be fatal after even very short exposures.

Continued exposure to low concentrations of the fumes, insufficient to cause pulmonary edema, is said to result in chronic irritation of the respiratory tract with cough, headache, loss of appetite, dyspepsia, corrosion of the teeth, and gradual loss of strength.

Exposure to NO_x is always potentially serious, and persons so exposed should be kept under close observation for at least 48 hours.

An oxidizer. The liquid is a sensitive explosive. Explosive reaction with carbon disulfide (when ignited); methanol (when ignited); pentacarbonyl iron (at 50°C); phosphine + oxygen; sodium diphenylketyl; dichlorine oxide; fluorine; nitrogen trichloride; ozone; perchloryl fluoride (at 100-300°C); vinyl chloride. Reacts to form explosive products with dienes (e.g., 1,3-butadiene, cyclopentadiene, propadiene).

Can react violently with acetic anhydride; Al; amorphous boron; BaO; BCl_3; $CsHC_2$; calcium; carbon + potassium hydrogentartrate; charcoal; ClO; pyrophoric chromium; 1,2-dichloroethane; dichloroethylene; ethylene; fuels; hydrocarbons; hydrogen + oxygen; Na_2O; uns-dimethyl hydrazine; NH_3; $CHCl_3$; Fe; Mg; Mn; CH_2Cl_2; olefins; phosphorus; PNH_2; PH_3; potassium; potassium sulfide; propylene; rubidium acetylide; Na; S; WC; trichloroethylene; 1,1,1-trichloroethane; uns-tetrachloroethane; uranium; uranium dicarbide. Will react with water or steam to produce heat and corrosive fumes; can react vigorously with reducing materials.

NEH000 *HR: 2*
NITRIDES

PROP: Compounds of N(3−) as the anion, such as Li_3N, Ca_3N_2, etc.

SAFETY PROFILE: The details of the toxicity of nitrides as a group are unknown. However, many nitrides react with moisture to evolve ammonia. This gas is an irritant to mucous membranes. To the extent that many nitrides evolve flammable ammonia gas upon contact with moisture, nitrides can be fire hazards. A moderate explosion hazard. When heated to decomposition they emit toxic fumes of NH_3. See also AMMONIA.

NEH500 ***HR: 3***
NITRILES

PROP: Nitriles are organic compounds containing the (−C≡N) grouping, e.g., acrylonitrile (CH_2:CHC≡N).

CONSENSUS REPORTS: Cyanides and its compounds are on the Community Right-To-Know List.

SAFETY PROFILE: Nitriles are organic cyanides; acrylonitrile, propionitrile, and some others resemble cyanides in toxicity. Other nitriles, such as cyanamides and cyanates, have no cyanide effect. Can react violently with ($LiAlH_4$ + H_2O). The nitriles may be used as insecticides. Many are flammable. When heated to decomposition they emit highly toxic fumes of CN^-. See also specific compounds and CYANIDES.

NEI000 CAS:18662-53-8 ***HR: 3***
NITRILOTRIACETIC ACID TRISODIUM SALT MONOHYDRATE
mf: $C_6H_6NO_6$•3Na•H_2O mw: 311.16

SYNS: N,N-BIS(CARBOXYMETHYL)GLYCINE TRISODIUM SALT MONOHYDRATE ◇ NCI-C01445 ◇ NITRILOACETIC ACID TRISODIUM SALT MONOHYDRATE ◇ NTA SODIUM HYDRATE ◇ TRISODIUM NITRILOTRIACETATE MONOHYDRATE

TOXICITY DATA with REFERENCE
msc-hmn:oth 11 μmol/L TOLED5 25,137,85
orl-rat TDLo:900 μg/kg (9-14D preg):TER TXAPA9 23,222,72
orl-rat TDLo:830 g/kg/2Y-C:CAR NCITR* NCI-CG-TR-6,77
orl-mus TDLo:315 g/kg/75W-C:CAR NCITR* NCI-CG-TR-6,77
orl-rat TD:876 g/kg/2Y-C:ETA FCTOD7 20,441,82
ipr-mus LD50:500 mg/kg NTIS** AD691-490

CONSENSUS REPORTS: NCI Carcinogenesis Bioassay (feed); Clear Evidence: mouse, rat NCITR* NCI-CG-TR-6,77. Cyanides and its compounds are on the Community Right-To-Know List.

SAFETY PROFILE: Suspected carcinogen with experimental carcinogenic, tumorigenic, and teratogenic data. Moderately toxic by intraperitoneal route. Human mutation data reported. When heated to decomposition it emits toxic fumes of NO_x, CN^-, and Na_2O.

NEI500 CAS:122-20-3 ***HR: 2***
***1,1',1''*-NITRILOTRI-2-PROPANOL**
mf: $C_9H_{21}NO_3$ mw: 191.31

PROP: Pure, crystalline, white solid. Mp: 45°, bp: 305°, flash p: 320°F (OC), d: 1.0200 @ 20°/20°, vap press: <0.01 mm @ 20°.

SYNS: TRIISOPROPANOLAMINE ◇ TRIS(2-HYDROXYPROPYL) AMINE ◇ TRIS(2-HYDROXY-1-PROPYL)AMINE

TOXICITY DATA with REFERENCE
eye-rbt 5 mg SEV AJOPAA 29,1363,46
orl-rat LD50:6500 mg/kg JIHTAB 30,63,48
orl-mus LD50:2520 mg/kg GISAAA 45(3),79,80
orl-rbt LD50:11 g/kg GISAAA 45(3),79,80
skn-rbt LDLo:10 g/kg OCDS** 4/21/67
orl-gpg LD50:1080 mg/kg JIHTAB 23,259,41

CONSENSUS REPORTS: Reported in EPA TSCA Inventory. Cyanide and its compounds are on the Community Right-To-Know List.

SAFETY PROFILE: Moderately toxic by ingestion. Mildly toxic by skin contact. A severe eye irritant. Combustible when exposed to heat or flame. To fight fire, use alcohol foam, water, CO_2, dry chemical. When heated to decomposition it emits toxic fumes of NO_x and CN^-. See also AMINES.

NEJ000 ***HR: 3***
NITRITES

PROP: Salts of nitrous acid.

SAFETY PROFILE: Large amounts taken by mouth may produce nausea, vomiting, cyanosis (due to methemoglobin formation), collapse, and coma. Repeated small doses cause a fall in blood pressure, rapid pulse, headache, and visual disturbances. They have been implicated in an increased incidence of cancer. They may react with organic amines in the body to form carcinogenic nitrosamines. Organic nitrites are used to treat angina pectoris. Fire hazards are variable. They are generally powerful oxidizers. On contact with readily oxidized materials, a violent reaction such as a fire or explosion may ensue. Explosion hazards are also variable. Organic nitrites may decompose violently in contact with NH_4; salts; cyanide; KCN. Dangerous; shock may explode them; can react vigorously with reducing materials. When heated to decomposition they emit highly toxic fumes of NO_x. See also SODIUM NITRITE and specific compounds.

NEJ500 CAS:602-87-9 ***HR: 3***
5-NITROACENAPHTHENE
mf: $C_{12}H_9NO_2$ mw: 199.22

SYNS: 1,2-DIHYDRO-5-NITRO-ACENAPHTHYLENE ◇ 5-NAN ◇ NCI-C01967 ◇ 5-NITROACENAPHTHYLENE ◇ 5-NITROACENAPTHENE ◇ 5-NITRONAPHTHALENE ETHYLENE

TOXICITY DATA with REFERENCE
mmo-sat 30 ng/plate ENMUDM 7(Suppl 5),1,85
mma-sat 300 ng/plate ENMUDM 7(Suppl 5),1,85
mma-esc 10 μg/plate ENMUDM 7(Suppl 5),1,85
orl-rat TDLo:40 g/kg/78W-C:CAR NCITR* NCI-TR-118,78
orl-mus TDLo:33 g/kg/78W-C:CAR,REP NCITR* NCI-TR-118,78

ipr-mus TDLo:3744 mg/kg/78W-I:ETA NEZAAQ 24,263,69
orl-rat TD:120 g/kg/17W-C:CAR TJIDAH 89,475,74
orl-mus TD:80 g/kg/78W-C:CAR,REP NCITR* NCI-TR-118,78

CONSENSUS REPORTS: IARC Cancer Review: Group 2B IMEMDT 7,56,87; Animal Sufficient Evidence IMEMDT 16,319,78. NCI Carcinogenesis Bioassay (feed); Clear Evidence: mouse, rat NCITR* NCI-CG-TR-118,78. Reported in EPA TSCA Inventory. EPA Genetic Toxicology Program.

DFG MAK: Animal Carcinogen, Suspected Human Carcinogen.

SAFETY PROFILE: Confirmed carcinogen with experimental carcinogenic, neoplastigenic data. Experimental reproductive effects. Mutation data reported. When heated to decomposition it emits toxic fumes of NO_x. See also NITRO COMPOUNDS OF AROMATIC HYDROCARBONS.

NEJ600 CAS:5653-21-4 ***HR: 3***
2-NITROACETALDEHYDE OXIME
mf: $C_2H_4N_2O_3$ mw: 104.07

SYN: METHAZONIC ACID

SAFETY PROFILE: Explodes when heated above 110°C. When heated to decomposition it emits toxic fumes of NO_x. See also ALDEHYDES.

NEK000 CAS:104-04-1 ***HR: 2***
p-NITROACETANILIDE
mf: $C_8H_8N_2O_3$ mw: 180.18

$O_2NC_6H_4NHCO{\cdot}CH_3$

PROP: White crystals. Mp: 214-216°. Almost insol in cold water; sol in hot water, alc, ether.

SYNS: p-ACETAMIDONITROBENZENE ◇ 4-NITROACETANILIDE ◇ 4′-NITROACETANILIDE ◇ N-(4-NITROPHENYL)ACETAMIDE

TOXICITY DATA with REFERENCE
ipr-rat LDLo:500 mg/kg NCNSA6 5,10,53

CONSENSUS REPORTS: Reported in EPA TSCA Inventory.

SAFETY PROFILE: Moderately toxic by intraperitoneal route. Reaction with sulfuric acid is vigorous with much gas evolution. When heated to decomposition it emits toxic fumes of NO_x. See also ACETANILIDE and NITRO COMPOUNDS OF AROMATIC HYDROCARBONS.

NEL000 CAS:1777-84-0 ***HR: 3***
3-NITRO-p-ACETOPHENETIDIDE
mf: $C_{10}H_{12}N_2O_4$ mw: 224.24

PROP: Yellow needles in water. M: 103-104°. Sol in abs alc, ether, and chloroform.

SYNS: 4-ACETAMINO-2-NITROPHENETOLE ◇ N-(4-ETHOXY-3-NITRO)PHENYLACETAMIDE ◇ N-(4-ETHOXYPHENYL)-3′-NITROACETAMIDE ◇ NCI C01978 ◇ 2-NITRO-4-ACETAMINOFENETOL (CZECH) ◇ 3-NITRO-p-ACETOPHENETIDE ◇ 5-NITRO-p-ACETOPHENETIDIDE ◇ 3′-NITRO-p-ACETOPHENETIDIN

TOXICITY DATA with REFERENCE
eye-rbt 100 mg/24H MOD 28ZPAK -,115,72
mmo-sat 1 mg/plate ENMUDM 5(Suppl 1),3,83
mma-sat 33 μg/plate ENMUDM 5(Suppl 1),3,83
orl-mus TDLo:957 g/kg/78W-C:NEO NCITR* NCI-CG-TR-133,79
orl-mus TD:478 g/kg/78W-C:ETA NCITR* NCI-CG-TR-133,79
orl-rat LD50:664 mg/kg NCIMR* NIH-71-E-2144

CONSENSUS REPORTS: NCI Carcinogenesis Bioassay (feed); Clear Evidence: mouse NCITR* NCI-CG-TR-133,79; (feed); No Evidence: rat NCITR* NCI-CG-TR-133,79. Reported in EPA TSCA Inventory.

SAFETY PROFILE: Moderately toxic by ingestion. An eye irritant. Mutation data reported. Questionable carcinogen with experimental carcinogenic, neoplastigenic, and tumorigenic data. When heated to decomposition it emits toxic fumes of NO_x. See also NITRO COMPOUNDS OF AROMATIC HYDROCARBONS.

NEL450 CAS:577-59-3 ***HR: 2***
2′-NITROACETOPHENONE
mf: $C_8H_7NO_3$ mw: 165.15

$O_2NC_6H_4CO{\cdot}CH_3$

SYNS: o-NITROACETOPHENONE ◇ 1-(2-NITROPHENYL)ETHANONE

TOXICITY DATA with REFERENCE
dnr-bcs 5 μL/disc MUREAV 170,11,86
orl-mus LD50:800 mg/kg PESTC 9(44),6,81
orl-rat LD50:1600 mg/kg PESTC 9(44),6,81

CONSENSUS REPORTS: Reported in EPA TSCA Inventory.

SAFETY PROFILE: Moderately toxic by ingestion. Mutation data reported. Solutions with potassium methyl selenide in dimethyl formamide solvent undergo an explosive reaction. When heated to decomposition it emits toxic fumes of NO_x. See also NITRO COMPOUNDS OF AROMATIC HYDROCARBONS.

NEL500 CAS:121-89-1 ***HR: 3***
3′-NITROACETOPHENONE
mf: $C_8H_7NO_3$ mw: 165.16

PROP: Needles. Mp: 80-81°, bp: 202°. Insol in water; sol in alc.

SYNS: m-NITROACETOPHENONE ◇ 1-(3-NITROPHENYL)ETHANONE ◇ USAF MA-1

TOXICITY DATA with REFERENCE
skn-rbt 500 mg/2H MLD 85JCAE -,730,86
eye-rbt 500 mg open AMIHBC 10,61,54
mmo-sat 1 mg/plate ENMUDM 9(Suppl 9),1,87
orl-rat LD50:3250 mg/kg AMIHBC 10,61,54
ipr-mus LD50:200 mg/kg NTIS** AD277-689
skn-rbt LD50:3 g/kg AMIHBC 10,61,54

CONSENSUS REPORTS: Reported in EPA TSCA Inventory.

SAFETY PROFILE: Poison by intraperitoneal route. Moderately toxic by ingestion and skin contact. An eye and skin irritant. Mutation data reported. When heated to decomposition it emits toxic fumes of NO_x. See also 2′-NITROACETOPHENONE.

NEM000 CAS:59748-51-5 ***HR: 3***
4′-(3-NITRO-9-ACRIDINYLAMINO) METHANESULFONANILIDE
mf: $C_{20}H_{16}N_4O_4S$ mw: 408.46

SYN: N-(4-((3-NITRO-9-ACRIDINYL)AMINO)PHENYL)METHANESULFONAMIDE

TOXICITY DATA with REFERENCE
mmo-sat 4074 µmol/L JMCMAR 23,269,80
ipr-mus LD10:12500 µg/kg JMCMAR 23,269,80

SAFETY PROFILE: Poison by intraperitoneal route. Mutation data reported. When heated to decomposition it emits very toxic fumes of NO_x and SO_x.

NEM300 ***HR: 2***
NITROALKANES

PROP: Compounds of the form RNO_2 where R is an aliphatic hydrocarbon.

SAFETY PROFILE: Generally mild oxidants. If subjected to high temperatures or pressures, they may decompose violently. Polynitroalkanes are more unstable. Nitroalkanes react with inorganic bases to form explosive products. Metal oxides promote violent decomposition reactions. Nitrate esters have very low stability. Nitroalkenes are generally more reactive and less stable. When heated to decomposition they emit toxic fumes of NO_x.

NEM480 CAS:119-34-6 ***HR: 3***
2-NITRO-4-AMINOPHENOL
mf: $C_6H_6N_2O_3$ mw: 154.14

SYNS: 4-AMINO-2-NITROPHENOL ◇ C.I. 76555 ◇ FOURRINE 57 ◇ FOURRINE BROWN PR ◇ FOURRINE BROWN PROPYL ◇ 4-HYDROXY-3-NITROANILINE ◇ NCI-C03963 ◇ o-NITRO-p-AMINOPHENOL ◇ OXIDATION BASE 25

TOXICITY DATA with REFERENCE
mma-sat 100 µg/plate IAPUDO 27,283,80
eye-rbt 100 mg/24H SEV 28ZPAK -,107,72
mmo-sat 100 µg/plate ENMUDM 8(Suppl 5),1,85
otr-rat:emb 11 µg/plate JJATDK 1,190,81
orl-rat TDLo:108 g/kg/2Y-C:CAR NCITR* NCI-CG-TR-94,78
orl-rat LD50:1470 mg/kg NCILB* NIH-NCI-E-C-72-3252,73
ipr-rat LD50:302 mg/kg JTEHD6 2,657,77
orl-mus LD50:1470 mg/kg NCILB* NIH-NCI-E-C-72-3252,73

CONSENSUS REPORTS: IARC Cancer Review: Group 3 IMEMDT 7,56,87; Animal Inadequate Evidence IMEMDT 16,43,78. NCI Carcinogenesis Bioassay (feed); No Evidence: mouse NCITR* NCI-CG-TR-94,78; Clear Evidence: rat NCITR* NCI-CG-TR-94,78. Reported in EPA TSCA Inventory. EPA Genetic Toxicology Program.

DFG MAK: Suspected Carcinogen.

SAFETY PROFILE: Suspected carcinogen with experimental carcinogenic data. Very poisonous by intraperitoneal route. Moderately toxic by ingestion. A severe eye irritant. Mutation data reported. When heated to decomposition it emits toxic fumes of NO_x.

NEM500 CAS:99-57-0 ***HR: 3***
p-NITRO-o-AMINOPHENOL
mf: $C_6H_6N_2O_3$ mw: 154.14

SYNS: 3-AMINO-4-HYDROXYNITROBENZENE ◇ 2-AMINO-4-NITROPHENOL ◇ 2-HYDROXY-5-NITROANILINE ◇ NCI-C55958 ◇ 4-NITRO-2-AMINOFENOL (CZECH) ◇ p-NITROAMINOFENOL (POLISH)

TOXICITY DATA with REFERENCE
eye-rbt 100 mg/24H MOD 28ZPAK -,107,72
mmo-sat 20 µg/plate CMMUAO 8,151,83
mma-sat 100 µg/plate PNASA6 72,2423,75
pic-esc 100 mmol/L MDMIAZ 31,11,79
orl-rat TDLo:64375 mg/kg/2Y-C:NEO NTPTR* NTP-TR-339,88
ipr-rat LD50:246 mg/kg JTEHD6 2,657,77
orl-mus LD50:850 mg/kg GTPZAB 25(8),50,81
ipr-mus LD50:143 mg/kg GENEA3 26,109,85

CONSENSUS REPORTS: Reported in EPA TSCA Inventory. EPA Genetic Toxicology Program.

SAFETY PROFILE: Poison by intraperitoneal route. Moderately toxic by ingestion. An eye irritant. Mutation data reported. Questionable carcinogen with experimental tumorigenic data. When heated to decomposition it emits toxic fumes of NO_x. See also NITRO COMPOUNDS OF AROMATIC HYDROCARBONS.

NEM600 ***HR: 3***
2-(N-NITROAMINO)PYRIDINE-N-OXIDE
mf: $C_5H_5N_3O_3$ mw: 155.11

$O_2NNHC_5H_4N:O$

SAFETY PROFILE: A powerful explosive. When heated to decomposition it emits toxic fumes of NO_x. See also NITRO COMPOUNDS OF AROMATIC HYDROCARBONS.

NEN000 ***HR: 3***
5-N-NITROAMINOTETRAZOLE
mf: $CH_2N_6O_2$ mw: 130.07

SAFETY PROFILE: Explodes when heated to 140°C. Upon decomposition it emits toxic fumes of NO_x.

NEN300 CAS:52096-16-9 ***HR: 3***
4-NITROAMINO-1,2,4-TRIAZOLE
mf: $C_2H_3N_5O_2$ mw: 129.08

SAFETY PROFILE: Explodes at the melting point of 72°C. When heated to decomposition it emits toxic fumes of NO_x.

NEN500 CAS:99-09-2 ***HR: 3***
m-NITROANILINE
DOT: UN 1661
mf: $C_6H_6N_2O_2$ mw: 138.14

PROP: Yellow, rhombic crystals. D: 0.9011 @ 25°/4°, mp: 114°, bp: 306.4°. Sol in water, alc, and ether.

SYNS: m-AMINONITROBENZENE ◇ 1-AMINO-3-NITROBENZENE ◇ AZOBASE MNA ◇ C.I. 37030 ◇ C.I. AZOIC DIAZO COMPONENT 7 ◇ DAITO ORANGE BASE R ◇ DEVOL ORANGE R ◇ DIAZO FAST ORANGE R ◇ FAST ORANGE R SALT ◇ HILTONIL FAST ORANGE R BASE ◇ MNA ◇ NAPHTOELAN ORANGE R BASE ◇ m-NITROAMINOBENZENE ◇ m-NITRANILINE ◇ 3-NITROANILINE ◇ 3-NITROBENZENAMINE ◇ m-NITROPHENYLAMINE ◇ ORANGE BASE IRGA I

TOXICITY DATA with REFERENCE
mmo-sat 15 mmol/L AEMIDF 44,801,82
mma-sat 10 μmol/plate MUREAV 58,11,78
orl-rat LD50:535 mg/kg AMRL** TR-72-62,72
orl-mus LD50:308 mg/kg AMRL** TR-72-62,72
ipr-dog LDLo:70 mg/kg JIDHAN 2,247,20
orl-gpg LD50:450 mg/kg VKMGA7 6,89,66
orl-qal LD50:562 mg/kg AECTCV 12,355,83

CONSENSUS REPORTS: Reported in EPA TSCA Inventory. EPA Genetic Toxicology Program.

DOT Classification: Poison B; Label: Poison.

SAFETY PROFILE: Poison by ingestion and intraperitoneal routes. Mutation data reported. Absorbed through the skin and by inhalation of the dust. Acute exposure may cause methemoglobinemia cyanosis. Chronic exposure may cause liver damage. Decomposes exothermically at 247°C. Possibly explosive reaction with ethylene oxide at 130°C. When heated to decomposition it emits toxic fumes of NO_x. See also o-NITROANILINE, p-NITROANILINE, and ANILINE DYES.

NEO000 CAS:88-74-4 ***HR: 3***
o-NITROANILINE
DOT: UN 1661
mf: $C_6H_6N_2O_2$ mw: 138.14

PROP: Orange-yellow crystals. Mp: 69-71°, bp: 284.5°, vap press: 1 mm @ 104°, d: 0.9015 @ 25°/4°. Sltly sol in cold water; sol in hot water, alc, and ether.

SYNS: 1-AMINO-2-NITROBENZENE ◇ AZOENE FAST ORANGE GR SALT ◇ BRENTAMINE FAST ORANGE GR BASE ◇ C.I. 37025 ◇ C.I. AZOIC DIAZO COMPONENT 6 ◇ DEVOL ORANGE B ◇ FAST ORANGE BASE GR ◇ HILTONIL FAST ORANGE GR BASE ◇ NATASOL FAST ORANGE GR SALT ◇ o-NITRANILINE ◇ ORANGE BASE CIBA II ◇ ORTHONITROANILINE (DOT)

TOXICITY DATA with REFERENCE
mma-sat 2500 μg/plate FCTOD7 23,695,85
orl-rat LD50:1600 mg/kg VKMGA7 6,89,66
orl-mus LD50:1070 mg/kg GTPZAB 25(8),50,81
skn-rbt LD50:20 g/kg TXAPA9 42,417,77
orl-gpg LD50:2350 mg/kg VKMGA7 6,89,66
orl-qal LD50:750 mg/kg AECTCV 12,355,83

CONSENSUS REPORTS: Reported in EPA TSCA Inventory.

DOT Classification: Poison B; Label: Poison.

SAFETY PROFILE: A poison. Moderately toxic by ingestion. Mildly toxic by skin contact. Mutation data reported. Mixtures with magnesium are hypergolic on contact with nitric acid. Forms extremely explosive addition compounds with hexanitroethane. Vigorous reaction with sulfuric acid above 200°C. When heated to decomposition it emits toxic fumes of NO_x. See also m-NITROANILINE, p-NITROANILINE, and ANILINE DYES.

NEO500 CAS:100-01-6 ***HR: 3***
p-NITROANILINE
DOT: UN 1661
mf: $C_6H_6N_2O_2$ mw: 138.14

PROP: Bright yellow powder. Mp: 148.5°, bp: 332°, flash p: 390°F (CC), d: 1.424, vap press: 1 mm @ 142.4°. Sol in water, alc, ether, benzene, methanol.

SYNS: p-AMINONITROBENZENE ◇ 1-AMINO-4-NITROBENZENE ◇ AZOFIX RED GG SALT ◇ AZOIC DIAZO COMPONENT 37 ◇ C.I. 37035 ◇ C.I. AZOIC DIAZO COMPONENT 37 ◇ C.I. DEVELOPER 17 ◇ DEVELOPER P ◇ DIAZO FAST RED GG ◇ FAST RED BASE GG ◇ FAST RED 2G SALT ◇ NAPHTOELAN RED GG BASE ◇ NCI-C60786 ◇ p-NITRANILINE ◇ 4-NITRANILINE ◇ NITRAZOL CF EXTRA ◇ p-NITROANILINA (POLISH) ◇ 4-NITROANILINE (MAK) ◇ 4-NITROBENZENAMINE ◇ p-NITROPHENYLAMINE ◇ PARANITROANILINE, solid (DOT) ◇ PNA ◇ RCRA WASTE NUMBER P077 ◇ RED 2G BASE ◇ SHINNIPPON FAST RED GG BASE

TOXICITY DATA with REFERENCE
mmo-sat 333 μl/plate ENMUDM 5(Suppl 1),3,83

mma-sat 1 mg/L ENMUDM 5,803,83
mma-esc 10 g/L FCTXAV 18,215,80
dnr-esc 10 g/L FCTXAV 18,215,80
orl-mus TDLo:9600 mg/kg (female 6-13D post):REP TCMUD8 7,29,87
orl-rat LD50:750 mg/kg CEHYAN 23,168,78
ipr-rat LDLo:600 mg/kg 85GMAT -,92,82
orl-mus LD50:810 mg/kg TXAPA9 42,417,77
ipr-mus LD50:250 mg/kg NTIS** AD691-490
ims-mus LD50:800 mg/kg IGIBA5 15,151,66
orl-gpg LD50:450 mg/kg VKMGA7 6,89,66
orl-qal LD50:1 g/kg AECTCV 12,355,83
ivn-mam LDLo:40 mg/kg XPHBAO 271,34,41
orl-bwd LD50:75 mg/kg TXAPA9 21,315,72

CONSENSUS REPORTS: Reported in EPA TSCA Inventory. EPA Genetic Toxicology Program.

OSHA PEL: (Transitional: TWA 6 mg/m^3 (skin)) TWA 3 mg/m^3 (skin)
ACGIH TLV: TWA 3 mg/m^3 (skin)
DFG MAK: 1 ppm (6 mg/m^3)
DOT Classification: Poison B; Label: Poison.

SAFETY PROFILE: Poison by ingestion, intravenous, and intraperitoneal routes. Moderately toxic by intramuscular route. Mutation data reported. Acute symptoms of exposure are headache, nausea, vomiting, weakness and stupor, cyanosis and methemoglobinemia. Chronic exposure can cause liver damage. Experimental reproductive effects. Combustible when exposed to heat or flame. See NITRATES for explosion and disaster hazards. To fight fire, use water spray or mist, foam, dry chemical, CO_2. Vigorous reaction with sulfuric acid above 200°C. Reaction with sodium hydroxide at 130°C under pressure may produce the explosive sodium-4-nitrophenoxide. When heated to decomposition it emits toxic fumes of NO_x. See also m-NITROANILINE, o-NITROANILINE NITRO COMPOUNDS OF AROMATIC HYDROCARBONS, and ANILINE DYES.

NEP000 CAS:66827-74-5 ***HR: 3***
p-NITROANILINE MERCURY(II) derivative

TOXICITY DATA with REFERENCE
orl-rat LDLo:500 mg/kg NCNSA6 5,32,53
ipr-rat LDLo:500 mg/kg NCNSA6 5,32,53

CONSENSUS REPORTS: Mercury and its compounds are on the Community Right-To-Know List.

OSHA PEL: (Transitional: CL 1 mg/10m^3) CL 0.1 mg(Hg)/m^3 (skin)
ACGIH TLV: TWA 0.1 mg(Hg)/m^3 (skin)
NIOSH REL: (Mercury, Inorganic) TWA 0.05 mg(Hg)/m^3

SAFETY PROFILE: Mercury compounds are poisons. Moderately toxic by ingestion and intraperitoneal routes. When heated to decomposition it emits very toxic fumes of NO_x and Hg. See also MERCURY and p-NITROANILINE.

NEP500 CAS:96-75-3 ***HR: 2***
4-NITROANILINE-2-SULFONIC ACID
mf: $C_6H_6N_2O_5S$ mw: 218.20

PROP: Yellow needles. Very sol in water; sol in aqua H_2SO_4, concentrated HCl; sltly sol in alc.

SYN: 2-AMINO-5-NITRO BENZENESULFONIC ACID

TOXICITY DATA with REFERENCE
orl-rat LDLo:500 mg/kg JPETAB 90,260,47

CONSENSUS REPORTS: Reported in EPA TSCA Inventory.

SAFETY PROFILE: Moderately toxic by ingestion. When heated to decomposition it emits very toxic fumes of SO_x and NO_x. See also NITRO COMPOUNDS OF AROMATIC HYDROCARBONS.

NEP600 ***HR: 3***
4-NITROANILINIUM PERCHLORATE
mf: $C_6H_7ClN_2O_6$ mw: 238.58

$$O_2NC_6H_4NH_3^-ClO_4^-$$

SAFETY PROFILE: A shock- and heat-sensitive explosive. When heated to decomposition it emits toxic fumes of Cl^- and NO_x. See also PERCHLORATES and NITRO COMPOUNDS OF AROMATIC HYDROCARBONS.

NEQ000 CAS:97-52-9 ***HR: 2***
4-NITRO-o-ANISIDINE
mf: $C_7H_8N_2O_3$ mw: 168.17

PROP: Pale yellow needles. D: 1.211 @ 156°, mp: 139-140°.

SYNS: 2-AMINO-5-NITROANISOL (CZECH) ◇ C.I. 37125

TOXICITY DATA with REFERENCE
orl-rat LD50:997 mg/kg MarJV# 29MAR77

CONSENSUS REPORTS: Reported in EPA TSCA Inventory.

SAFETY PROFILE: Moderately toxic by ingestion. When heated to decomposition it emits toxic fumes of NO_x. See also NITRO COMPOUNDS OF AROMATIC HYDROCARBONS.

NEQ500 CAS:99-59-2 ***HR: 3***
5-NITRO-o-ANISIDINE
mf: $C_7H_8N_2O_3$ mw: 168.17

PROP: Red needles from alc. D: 1.207 @ 156°, mp:

118°. Sol in hot benzene, alc, and acetic acid; very sltly sol in ligroin.

SYNS: 2-AMINO-1-METHOXY-4-NITROBENZENE ◇ 3-AMINO-4-METHOXYNITROBENZENE ◇ 2-AMINO-4-NITROANISOLE ◇ o-ANISIDINE NITRATE ◇ AZOAMINE SCARLET ◇ AZOGENE ECARLATE R ◇ AZOIC DIAZO COMPONENT 13 BASE ◇ C.I. AZOIC DIAZO COMPONENT 13 ◇ C.I. 37130 ◇ FAST SCARLET R ◇ 2-METHOXY-5-NITROANILINE ◇ 2-METHOXY-5-NITROBENZENAMINE ◇ NCI-C01934 ◇ 3-NITRO-6-METHOXYANILINE ◇ 5-NITRO-2-METHOXYANILINE

TOXICITY DATA with REFERENCE
mmo-sat 10 μmol/plate MUREAV 58,11,78
mma-sat 10 μmol/plate MUREAV 58,11,78
orl-rat TDLo:109 g/kg/78W-C:CAR NCITR* NCI-CG-TR-127,78
orl-mus TDLo:413 g/kg/78W-C:CAR NCITR* NCI-CG-TR-127,78
orl-mus TD:524 g/kg/78W-C:ETA NCITR* NCI-CG-TR-127,78
orl-rat LD50:704 mg/kg NCIMR* NIH-71-E-2144
orl-mus LD50:1060 mg/kg GTPZAB 25(8),50,81

CONSENSUS REPORTS: NTP Fifth Annual Report on Carcinogens. IARC Cancer Review: Group 3 IMEMDT 7,56,87; Animal Limited Evidence IMEMDT 27,133,82. NCI Carcinogenesis Bioassay (feed); Clear Evidence: mouse, rat NCITR* NCI-CG-TR-127,78. Reported in EPA TSCA Inventory. EPA Genetic Toxicology Program. Community Right-To-Know List.

SAFETY PROFILE: Confirmed carcinogen with experimental carcinogenic and tumorigenic data. Moderately toxic by ingestion. Mutation data reported. When heated to decomposition it emits toxic fumes of NO_x. See also NITRO COMPOUNDS OF AROMATIC HYDROCARBONS.

NER000 CAS:91-23-6 ***HR: 3***
o-NITROANISOLE
DOT: UN 2730
mf: $C_7H_7NO_3$ mw: 153.15

$CH_3OC_6H_4NO_2$

PROP: Colorless crystals. D: 1.254 @ 20°/4°, mp: 9.5-10.5°, bp: 277°. Insol in water; sol in alc, ether.

SYNS: 2-METHOXYNITROBENZENE ◇ 1-METHOXY-2-NITROBENZENE ◇ NCI-C60388 ◇ 2-NITROANISOLE ◇ o-NITROPHENYL METHYL ETHER

TOXICITY DATA with REFERENCE
mmo-sat 666 μg/plate ENMUDM 5(Suppl 1),3,83
mma-sat 20 μmol/plate MUREAV 58,11,78
orl-rat LD50:1980 mg/kg GTPZAB 25(8),50,81
orl-mus LD50:1450 mg/kg GTPZAB 25(8),50,81

CONSENSUS REPORTS: Reported in EPA TSCA Inventory. EPA Genetic Toxicology Program.

DOT Classification: Poison B; Label: St. Andrews Cross.

SAFETY PROFILE: A poison. Moderately toxic by ingestion. Mutation data reported. Explosive reaction with sodium hydroxide + zinc. Vigorous reaction with hydrogen + catalyst (at 250°C/34bar). When heated to decomposition it emits toxic fumes of NO_x. See also p-NITROANISOLE.

NER500 CAS:100-17-4 ***HR: 3***
p-NITROANISOLE
DOT: UN 2730
mf: $C_7H_7NO_3$ mw: 153.15

PROP: Prisms from alc. D: 1.233 @ 20°, mp: 54°, bp: 260°. Sltly sol in cold petr ether; sol in water; very sol in alc, boiling petr ether, and ether.

SYNS: 1-METHOXY-4-NITROBENZENE ◇ p-METHOXYNITROBENZENE ◇ 4-METHOXYNITROBENZENE ◇ p-NITROANISOL ◇ 4-NITROANISOLE

TOXICITY DATA with REFERENCE
mmo-sat 6500 μmol/L ENMUDM 3,11,81
mma-sat 20 μmol/plate MUREAV 58,11,78
orl-rat LD50:2600 mg/kg GTPZAB 25(8),50,81
ipr-rat LD50:1400 mg/kg AGGHAR 17,217,59
orl-mus LD50:1710 mg/kg GTPZAB 25(8),50,81
ipr-mus LD50:698 mg/kg TXAPA9 41,216,77

CONSENSUS REPORTS: Reported in EPA TSCA Inventory. EPA Genetic Toxicology Program.

DOT Classification: Poison B; Label: St. Andrews Cross.

SAFETY PROFILE: A poison. Moderately toxic by ingestion and intraperitoneal routes. Mutation data reported. Can explode in presence of Ni. When heated to decomposition it emits toxic fumes of NO_x. See also o-NITROANISOLE and NITRO COMPOUNDS OF AROMATIC HYDROCARBONS.

NES000 CAS:3586-69-4 ***HR: D***
2-NITROANTHRACENE
mf: $C_{14}H_9NO_2$ mw: 223.24

PROP: Yellow needles from alc. Mp: 146°, bp: >360°. Insol in aq alkali; sltly sol in alc; very sol in benzene.

TOXICITY DATA with REFERENCE
mmo-sat 25 ng/plate CBINA8 26,11,79
mma-sat 100 ng/plate CBINA8 26,11,79

SAFETY PROFILE: Mutation data reported. When heated to decomposition it emits toxic fumes of NO_x. See also NITRO COMPOUNDS OF AROMATIC HYDROCARBONS.

NES500 CAS:619-17-0 ***HR: 2***
4-NITROANTHRANILIC ACID
mf: $C_7H_6N_2O_4$ mw: 182.15

SYNS: 2-AMINO-4-NITRO-BENZOIC ACID ◇ NCI-C01945

TOXICITY DATA with REFERENCE
mmo-sat 10 μg/plate ENMUDM 8(Suppl 7),1,86
mma-sat 10 μg/plate ENMUDM 8(Suppl 7),1,86
orl-rat LD50:640 mg/kg NCIMR* NIH-71-E-2144,73

CONSENSUS REPORTS: NCI Carcinogenesis Bioassay (feed); No Evidence: mouse, rat NCITR* NCI-CG-TR-109,78. Reported in EPA TSCA Inventory.

SAFETY PROFILE: Moderately toxic by ingestion. Mutation data reported. When heated to decomposition it emits toxic fumes of NO_x. See also NITRO COMPOUNDS OF AROMATIC HYDROCARBONS.

NET000 CAS:82-34-8 ***HR: 2***
1-NITROANTHRAQUINONE
mf: $C_{14}H_7NO_4$ mw: 253.22

PROP: Needles in acetic acid. Mp: 230°, bp: 270° @ 7 mm. Insol in water; sltly sol in alc; very sltly sol in ether.

SYNS: 1-NITRO-9,10-ANTHRACENEDIONE ◇ 1-NITROANTHRACHINON (CZECH) ◇ α-NITROANTHRAQUINONE

TOXICITY DATA with REFERENCE
skn-rbt 500 mg/24H MLD 28ZPAK -,120,72
eye-rbt 100 mg/24H MOD 28ZPAK -,120,72
ipr-rat LD50:1050 mg/kg GISAAA 49(4),90,84
orl-mus LD50:1540 mg/kg GISAAA 49(4),90,84

CONSENSUS REPORTS: Reported in EPA TSCA Inventory.

SAFETY PROFILE: Moderately toxic by ingestion and intraperitoneal routes. A skin and eye irritant. When heated to decomposition it emits toxic fumes of NO_x. See also NITRO COMPOUNDS OF AROMATIC HYDROCARBONS.

NET500 CAS:2491-52-3 ***HR: D***
p-NITROAZOBENZENE

SYN: 4-NITROAZOBENZENE

TOXICITY DATA with REFERENCE
mmo-sat 100 nmol/plate GANNA2 68,373,77

SAFETY PROFILE: Mutation data reported. When heated to decomposition it emits toxic fumes of NO_x. See also NITRO COMPOUNDS OF AROMATIC HYDROCARBONS.

NEU500 CAS:552-89-6 ***HR: 2***
2-NITROBENZALDEHYDE
mf: $C_7H_5NO_3$ mw: 151.13

PROP: Yellow needles from water. Mp: 42-43.5°, bp: 153° @ 23 mm. Sltly sol in water; sol in alc, ether, benzene; volatile with steam.

SYN: o-NITROBENZALDEHYDE

TOXICITY DATA with REFERENCE
mma-sat 1 μmol/plate MUREAV 58,11,78
orl-mus LD50:600 mg/kg AFECAT 76(815),65,83

CONSENSUS REPORTS: Reported in EPA TSCA Inventory. EPA Genetic Toxicology Program.

SAFETY PROFILE: Moderately toxic by ingestion. Mutation data reported. Violent reaction with pyrrole. Thermal decompositon is very dangerous; pressure increases rapidly. When heated to decomposition it emits toxic fumes of NO_x. See also ALDEHYDES.

NEV000 CAS:99-61-6 ***HR: 3***
3-NITROBENZALDEHYDE
mf: $C_7H_5NO_3$ mw: 151.13

PROP: Yellowish, crystalline powder. Mp: 58°, bp: 164° @ 23 mm. Volatile with steam. Insol in water; sol in alc, chloroform, ether.

SYNS: 3-FORMYLNITROBENZENE ◇ m-NITROBENZALDEHYDE

TOXICITY DATA with REFERENCE
ivn-mus LD50:180 mg/kg CSLNX* NX#02435

CONSENSUS REPORTS: Reported in EPA TSCA Inventory.

SAFETY PROFILE: Poison by intravenous route. Can explode violently during vacuum distillation. Thermal decomposition is very dangerous; pressure increases rapidly. When heated to decomposition it emits toxic fumes of NO_x. See also ALDEHYDES.

NEV500 CAS:555-16-8 ***HR: 2***
4-NITROBENZALDEHYDE
mf: $C_7H_5NO_3$ mw: 151.13

$O_2NC_6H_4CO{\cdot}H$

PROP: White to yellow crystals. Mp: 106-107°, subl. Sltly volatile with steam. Sltly sol in water, ether; sol in alc, benzene, glacial acetic acid.

SYNS: p-FORMYLNITROBENZENE ◇ p-NITROBENZALDEHYDE

TOXICITY DATA with REFERENCE
mma-sat 25 μmol/plate MUREAV 58,11,78
orl-rat LD50:4700 mg/kg GISAAA 24(9),15,59
skn-rat LD50:16000 mg/kg GISAAA 24(9),15,59
ipr-rat LD50:545 mg/kg HINEL* AF33(657)-11756,64

CONSENSUS REPORTS: Reported in EPA TSCA Inventory.

SAFETY PROFILE: Moderately toxic by intraperi-

toneal route. Mildly toxic by ingestion and skin contact. Mutation data reported. Exothermic decomposition is vigorous with a very high rate of pressure increase. Reaction with pyrrole in acetic acid is violent. When heated to decomposition it emits toxic fumes of NO_x. See also ALDEHYDES and NITRO COMPOUNDS OF AROMATIC HYDROCARBONS.

NEW550 CAS:39735-49-4 ***HR: D***
2-(p-NITROBENZAMIDO)ACETOHYDROXAMIC ACID
mf: $C_9H_9N_3O_5$ mw: 239.21

TOXICITY DATA with REFERENCE
mmo-sat 1 μmol/plate JOPHDQ 3,557,80
dnr-bcs 10 μmol/disc JOPHDQ 3,557,80

SAFETY PROFILE: Mutation data reported. When heated to decomposition it emits toxic fumes of NO_x.

NEX000 CAS:98-95-3 ***HR: 3***
NITROBENZENE
DOT: UN 1662
mf: $C_6H_5NO_2$ mw: 123.12

PROP: Bright yellow crystals or yellow, oily liquid; odor of volatile almond oil. Mp: 6°, bp: 210-211°, ULC: 20-30%, lel: 1.8% @ 200°F, flash p: 190°F (CC), d: 1.205 @ 15°/4°, autoign temp: 900°F, vap press: 1 mm @ 44.4°, vap d: 4.25. Volatile with steam; sol in about 500 parts water; very sol in alc, benzene, ether, oils.

SYNS: ESSENCE of MIRBANE ◇ ESSENCE of MYRBANE ◇ MIRBANE OIL ◇ NCI-C60082 ◇ NITROBENZEEN (DUTCH) ◇ NITROBENZEN (POLISH) ◇ NITROBENZENE, liquid (DOT) ◇ NITROBENZOL (DOT) ◇ NITROBENZOL, liquid (DOT) ◇ OIL of MIRBANE (DOT) ◇ OIL of MYRBANE ◇ RCRA WASTE NUMBER U169

TOXICITY DATA with REFERENCE
skn-rbt 500 mg/24H MOD 28ZPAK -,61,72
eye-rbt 500 mg/24H MLD 28ZPAK -,61,72
cyt-smc 10 mmol/tube HEREAY 33,457,47
ihl-rat TCLo:1260 μg/m³/4H (female 1-21D post):TER GTPZAB 31(4),48,87
ihl-rat TCLo:1260 μg/m³/4H (female 1-21D post):REP GTPZAB 31(4),48,87
orl-wmn TDLo:200 mg/kg:CNS,CVS,PUL ATXKA8 28,208,71
unr-man LDLo:35 mg/kg 85DCAI 2,73,70
skn-rat LD50:2100 mg/kg GISAAA 24(9),15,59
ipr-rat LD50:640 mg/kg AGGHAR 17,217,59
scu-rat LDLo:800 mg/kg HBAMAK 4,1375,35
orl-mus LD50:590 mg/kg GTPZAB 25(8),50,81
skn-mus LDLo:480 mg/kg XPHBAO 271,79,41
scu-mus LDLo:286 mg/kg PSEBAA 42,844,39
orl-dog LDLo:750 mg/kg HBAMAK 4,1375,35
ivn-dog LDLo:150 mg/kg XPHBAO 271,79,41
orl-cat LDLo:1 g/kg XPHBAO 271,78,41
skn-cat LDLo:25 g/kg HBAMAK 4,1375,35
orl-rbt LDLo:700 mg/kg PCOC** -,805,66
skn-rbt LDLo:600 mg/kg HBAMAK 4,1375,35

CONSENSUS REPORTS: Community Right-To-Know List. EPA Extremely Hazardous Substances List. Reported in EPA TSCA Inventory.

OSHA PEL: TWA 1 ppm (skin)
ACGIH TLV: TWA 1 ppm (skin)
DFG MAK: 1 ppm (5 mg/m³); BAT: 100 μg/L of aniline in blood after several shifts.
DOT Classification: Poison B; Label: Poison.

SAFETY PROFILE: Human poison by an unspecified route. Poison experimentally by subcutaneous and intravenous routes. Moderately toxic by ingestion, skin contact, and intraperitoneal routes. Human systemic effects by ingestion: general anesthetic, respiratory stimulation, and vascular changes. An experimental teratogen. Experimental reproductive effects. Mutation data reported. An eye and skin irritant. Can cause cyanosis due to formation of methemoglobin. It is absorbed rapidly through the skin. The vapors are hazardous.

An oxidant. Combustible when exposed to heat and flame. Moderate explosion hazard when exposed to heat or flame. Explosive reaction with solid or concentrated alkali + heat (e.g., sodium hydroxide or potassium hydroxide); aluminum chloride + phenol (at 120°C); aniline + glycerol + sulfuric acid; nitric + sulfuric acid + heat. Forms explosive mixtures with aluminum chloride; oxidants (e.g., fluorodinitromethane; uronium perchlorate; tetranitromethane; sodium chlorate; nitric acid; nitric acid + water; peroxodisulfuric acid; dinitrogen tetraoxide); phosphorous pentachloride; potassium; sulfuric acid. Reacts violently with aniline + glycerin; N_2O; $AgClO_4$. To fight fire, use water, foam, CO_2, dry chemical. Incompatible with potassium hydroxide. When heated to decomposition it emits toxic fumes of NO_x. See also NITRO COMPOUNDS OF AROMATIC HYDROCARBONS.

NEX500 CAS:5410-29-7 ***HR: 3***
o-NITROBENZENEARSONIC ACID
mf: $C_6H_6AsNO_5$ mw: 247.05

TOXICITY DATA with REFERENCE
orl-rat LDLo:100 mg/kg NCNSA6 5,13,53

CONSENSUS REPORTS: Arsenic and its compounds are on the Community Right-To-Know List. Reported in EPA TSCA Inventory.

OSHA PEL: TWA 0.5 mg(As)/m³

SAFETY PROFILE: Poison by ingestion. When heated to decomposition it emits very toxic fumes of NO_x and

As. See also ARSENIC COMPOUNDS and NITRO COMPOUNDS OF AROMATIC HYDROCARBONS.

NEY000 CAS:2243-76-7 ***HR: D***
p-NITROBENZENEAZOSALICYLIC ACID
mf: $C_{13}H_9N_3O_5$ mw: 287.25

SYNS: ALIZARINE CHROME ORANGE G ◇ ALIZARINE ORANGE ◇ ALIZARINE ORANGE 2GN ◇ ALIZARINE ORANGE R ◇ ALIZARINE ORANGE RD ◇ ALIZARINE R-CF ◇ ALIZARINE YELLOW P ◇ ALIZARINE YELLOW R ◇ ALIZARINE YELLOW RW ◇ ALIZAROL ORANGE R ◇ ALPHACROIC ORANGE R ◇ ANTHRANOL CHROME YELLOW R ◇ AZOCHROMAL ORANGE R ◇ BRASILIAN CHROME ORANGE R ◇ CALCOCHROME ORANGE R ◇ CHROME FAST ORANGE RW ◇ CHROME ORANGE MR ◇ CHROME ORANGE R ◇ CHROME ORANGE RLE ◇ CHROME YELLOW 3RN ◇ CHROMOL ORANGE R. EXTRA ◇ C.I. 14030 ◇ C.I. MORDANT ORANGE 1 ◇ DIAMOND CHROME YELLOW 3R ◇ DUROCHROME YELLOW 2RN ◇ ENIACROMO ORANGE R ◇ ENRIOCHROME ORANGE AOR ◇ FENAKROM ORANGE R ◇ HIDACHROME ORANGE R ◇ HISPACROM ORANGE R ◇ 2-HYDROXY-5-((4-NITROPHENYL)AZO)BENZOIC ACID ◇ JAVA CHROME YELLOW 3R ◇ JAVA UNICHROME YELLOW 3R ◇ KCA CHROME ORANGE R ◇ KENACHROME ORANGE ◇ LIGHTHOUSE CHROME ORANGE ◇ MAGRACROM ORANGE ◇ METACHROME ORANGE R ◇ MITSUI CHROME ORANGE A ◇ MITSUI CROME ORANGE AN ◇ MONOCHROME ORANGE R ◇ MONOCHROME YELLOW 3R ◇ 5-(p-NITROPHENYL)AZO)SALICYLIC ACID ◇ ORTHOCHROME ORANGE R ◇ PEERACHROME YELLOW R ◇ PNBAS ◇ PONTACHROME YELLOW 3RN ◇ SOLOCHROME ORANGE M ◇ SYNCHROMATE ORANGE AOR ◇ TERRACOTTA 2RN ◇ TERRA COTTA RRN ◇ TERTROCHROME YELLOW 3R ◇ YODOCHROME ORANGE A

TOXICITY DATA with REFERENCE
mma-sat 500 μg/plate MUREAV 56,249,78

CONSENSUS REPORTS: Reported in EPA TSCA Inventory. EPA Genetic Toxicology Program.

SAFETY PROFILE: Mutation data reported. When heated to decomposition it emits toxic fumes of NO_x. See also NITRO COMPOUNDS OF AROMATIC HYDROCARBONS.

NEY500 CAS:13331-27-6 ***HR: 3***
m-NITROBENZENEBORONIC ACID
mf: $C_6H_6BNO_4$ mw: 166.94

TOXICITY DATA with REFERENCE
ivn-mus LD50:180 mg/kg CSLNX* NX#01859

CONSENSUS REPORTS: Reported in EPA TSCA Inventory.

SAFETY PROFILE: Poison by intravenous route. When heated to decomposition it emits toxic fumes of NO_x. See also BORON COMPOUNDS and NITRO COMPOUNDS OF AROMATIC HYDROCARBONS.

NEZ000 ***HR: 3***
4-NITROBENZENEDIAZONIUM AZIDE
mf: $C_6H_4N_6O_2$ mw: 192.14

$O_2NC_6H_4N_2^-N_3^-$

SAFETY PROFILE: The dry salt is explosive. When heated to decomposition it emits toxic fumes of NO_x. See also AZIDES and NITRO COMPOUNDS OF AROMATIC HYDROCARBONS.

NEZ100 CAS:2028-76-4 ***HR: 2***
3-NITROBENZENEDIAZONIUM CHLORIDE
mf: $C_6H_4ClN_3O_2$ mw: 185.57

$O_2NC_6H_4N_2^-Cl^-$

SAFETY PROFILE: Reacts with o-ethyldithiocarbamate at 70°C to explosively evolve nitrogen gas. When heated to decomposition it emits toxic fumes of Cl^- and NO_x. See also NITRO COMPOUNDS OF AROMATIC HYDROCARBONS and CHLORIDES.

NFA000 CAS:42238-29-9 ***HR: 3***
4-NITROBENZENEDIAZONIUM NITRATE
mf: $C_6H_4N_4O_5$ mw: 212.12

$O_2NC_6H_4N_2^-NO_3^-$

SAFETY PROFILE: The dry salt explodes when heated. When heated to decomposition it emits toxic fumes of NO_x. See also NITRATES and NITRO COMPOUNDS OF AROMATIC HYDROCARBONS.

NFA500 CAS:22751-24-2 ***HR: 3***
3-NITROBENZENEDIAZONIUM PERCHLORATE
mf: $C_6H_4ClN_3O_6$ mw: 249.57

$O_2NC_6H_4N_2^-ClO_4^-$

DOT Classification: Forbidden.

SAFETY PROFILE: An extremely shock- and heat-sensitive explosive. Upon decomposition it emits toxic fumes of Cl^- and NO_x. See also PERCHORATES and NITRO COMPOUNDS OF AROMATIC HYDROCARBONS.

NFA600 CAS:25910-37-6 ***HR: 3***
2-NITROBENZENEDIAZONIUM SALTS
mf: $C_6H_4N_3O_2Z$

$O_2NC_6H_4N_2^-Z^-$

SAFETY PROFILE: Reaction with sodium disulfide, sodium polysulfide, or other sulfides forms powerfully explosive products. When heated to decomposition it emits toxic fumes of NO_x. See also NITRO COMPOUNDS OF AROMATIC HYDROCARBONS.

NFA625 CAS:14368-49-1 ***HR: 3***
4-NITROBENZENEDIAZONIUM SALTS
mf: $C_6H_4N_3O_2Z$

$O_2NC_6H_4N_2^-Z^-$

SAFETY PROFILE: Reaction with hydrogen sulfide, sodium sulfide, or potassium O,O-diphenyl phosphorodithioate forms powerfully explosive products. When heated to decomposition it emits toxic fumes of NO_x. See also NITRO COMPOUNDS OF AROMATIC HYDROCARBONS.

NFB000 CAS:6325-93-5 ***HR: 2***
p-NITROBENZENESULFONAMIDE
mf: $C_6H_6N_2O_4S$ mw: 202.20

TOXICITY DATA with REFERENCE
orl-rat LDLo:500 mg/kg NCNSA6 5,34,53

CONSENSUS REPORTS: Reported in EPA TSCA Inventory.

SAFETY PROFILE: Moderately toxic by ingestion. When heated to decomposition it emits very toxic fumes of NO_x and SO_x. See also NITRO COMPOUNDS OF AROMATIC HYDROCARBONS.

NFB500 CAS:98-47-5 ***HR: 3***
3-NITROBENZENESULFONIC ACID
mf: $C_6H_5NO_5S$ mw: 203.18

$O_2NC_6H_4SO_2OH$

PROP: Hygroscopic leaflets. Mp: 70°, bp: decomp. Very sol in water; sol in alc and alkali; insol in ether.

SYNS: KYSELINA NITROBENZEN-m-SULFONOVA (CZECH) ◇ m-NITROBENZENESULFONIC ACID

TOXICITY DATA with REFERENCE
skn-rbt 500 mg/24H MOD 28ZPAK -,178,72
eye-rbt 2 mg/24H SEV 28ZPAK -,178,72

CONSENSUS REPORTS: Reported in EPA TSCA Inventory.

SAFETY PROFILE: A skin and severe eye irritant. Decomposes violently at about 200°. Mixture with sulfuric acid + sulfur trioxide may explode above 150°C. When heated to decomposition it emits toxic fumes of SO_x and NO_x. See also NITRO COMPOUNDS OF AROMATIC HYDROCARBONS and SULFONATES.

NFC500 CAS:127-68-4 ***HR: 1***
m-NITROBENZENESULFONIC ACID, SODIUM SALT
mf: $C_6H_4NO_5S{\cdot}Na$ mw: 225.16

SYNS: LUDIGOL F,60 ◇ NITROBENZEN-m-SULFONAN SODNY (CZECH) ◇ TISKAN (CZECH)

TOXICITY DATA with REFERENCE
skn-rbt 500 mg/24H MLD 28ZPAK -,179,72
eye-rbt 20 mg/24H SEV 28ZPAK -,179,72
orl-rat LD50:11 g/kg 28ZPAK -,179,72

CONSENSUS REPORTS: Reported in EPA TSCA Inventory.

SAFETY PROFILE: Mildly toxic by ingestion. A skin and severe eye irritant. When heated to decomposition it emits very toxic fumes of NO_x, Na_2O and SO_x. See also NITRO COMPOUNDS OF AROMATIC HYDROCARBONS and SULFONATES.

NFC700 CAS:5709-67-1 ***HR: 3***
2-NITROBENZIMIDAZOLE
mf: $C_7H_5N_3O_2$ mw: 163.15

SYN: 2-NITRO-1H-BENZIMIDAZOLE(9CI)

TOXICITY DATA with REFERENCE
mmo-sat 100 nmol/plate MUREAV 173,169,86
ipr-mus LD50:330 mg/kg AACHAX -,478,65
scu-mus LD50:595 mg/kg AACHAX -,478,65

SAFETY PROFILE: Poison by intraperitoneal route. Moderately toxic by subcutaneous route. Mutation data reported. When heated to decomposition it emits toxic fumes of NO_x. See also NITRO COMPOUNDS OF AROMATIC HYDROCARBONS.

NFD500 CAS:94-52-0 ***HR: 3***
6-NITRO-BENZIMIDAZOLE
mf: $C_7H_5N_3O_2$ mw: 163.15

PROP: Needles from water. Mp: 204°. Sltly sol in water, chloroform, ether, benzene; sol in acid, alkali carbonate.

SYNS: NCI-C01912 ◇ 5-NITRO-1H-BENZIMIDAZOLE

TOXICITY DATA with REFERENCE
mmo-sat 33 μg/plate ENMUDM 8(Suppl 7),1,86
mma-sat 33 μg/plate ENMUDM 8(Suppl 7),1,86
orl-mus TDLo:160 g/kg/78W-C:CAR NCITR* NCI-CG-TR-117,79
orl-mus TD:79 g/kg/78W-C:ETA NCITR* NCI-CG-TR-117,79
orl-rat LDLo:500 mg/kg NCNSA6 5,22,53

CONSENSUS REPORTS: NCI Carcinogenesis Bioassay (feed); Clear Evidence: mouse NCITR* NCI-CG-TR-117,78; (feed); Negative: rat NCITR* NCI-CG-TR-117,78. Reported in EPA TSCA Inventory.

SAFETY PROFILE: Moderately toxic by ingestion. Questionable carcinogen with experimental carcinogenic and tumorigenic data. Mutation data reported. When heated to decomposition it emits toxic fumes of NO_x. See also NITRO COMPOUNDS OF AROMATIC HYDROCARBONS.

NFD700 CAS:33094-66-5 **_HR: D_**
2-NITROBENZOFURAN
mf: $C_8H_5NO_3$ mw: 163.14

TOXICITY DATA with REFERENCE
mmo-sat 1 nmol/plate MUREAV 88,355,81
pic-sat 1 nmol/plate MUREAV 104,1,82

SAFETY PROFILE: Mutation data reported. When heated to decomposition it emits toxic fumes of NO_x. See also NITRO COMPOUNDS OF AROMATIC HYDROCARBONS.

NFF000 CAS:58131-55-8 **_HR: D_**
2,2'-((7-NITRO-4-BENZOFURAZAZANYL) IMINO)BISETHANOL OXIDE
mf: $C_{10}H_{12}N_4O_6$ mw: 284.26

SYNS: B2837 ◇ 4-(N,N-BIS(2-HYDROXYETHYL)AMINO)-7-NITROBENZOFURAZAN-1-OXIDE

TOXICITY DATA with REFERENCE
mmo-sat 5 μg/plate MUREAV 48,145,77
dni-mus:leu 2500 μmol/L CNREA8 35,3735,75

CONSENSUS REPORTS: EPA Genetic Toxicology Program.

SAFETY PROFILE: Mutation data reported. When heated to decomposition it emits toxic fumes of NO_x.

NFG000 CAS:121-92-6 **_HR: 2_**
m-NITROBENZOIC ACID
mf: $C_7H_5NO_4$ mw: 167.13

PROP: Monoclinic leaves, bitter taste. D: 1.494, mp: 141-142°. Sol in water, alc, and ether. Melts in hot water; very sol in benzene, carbon disulfide, petr ether.

TOXICITY DATA with REFERENCE
mma-sat 40 μmol/plate MUREAV 58,11,78
ipr-rat LD50:670 mg/kg CRSBAW 160,1097,66
ivn-rat LD50:680 mg/kg CRSBAW 160,1097,66
par-rat LD50:1820 mg/kg CRSBAW 160,1097,66
ipr-mus LD50:610 mg/kg CRSBAW 160,1097,66
ivn-mus LD50:640 mg/kg CRSBAW 160,1097,66
par-mus LD50:1290 mg/kg CRSBAW 160,1097,66

CONSENSUS REPORTS: Reported in EPA TSCA Inventory. EPA Genetic Toxicology Program.

SAFETY PROFILE: Moderately toxic by intraperitoneal, intravenous and parenteral routes. Mutation data reported. When heated to decomposition it emits toxic fumes of NO_x. See also NITRO COMPOUNDS OF AROMATIC HYDROCARBONS.

NFG500 CAS:552-16-9 **_HR: D_**
o-NITROBENZOIC ACID
mf: $C_7H_5NO_4$ mw: 167.13

PROP: Yellowish-white, triclinic crystals from water; intensely sweet. D: 1.575, mp: 147.5°. Very sltly sol in benzene, carbon disulfide, petr ether; sol in water, alc, and ether.

SYN: 2-NITROBENZOIC ACID

TOXICITY DATA with REFERENCE
mma-sat 10 μmol/plate MUREAV 58,11,78
dnr-bcs 5 mg/disc MUREAV 170,11,86

CONSENSUS REPORTS: Reported in EPA TSCA Inventory.

SAFETY PROFILE: Mutation data reported. When heated to decomposition it emits toxic fumes of NO_x. See also NITRO COMPOUNDS OF AROMATIC HYDROCARBONS.

NFH000 CAS:636-97-5 **_HR: 3_**
p-NITROBENZOIC ACID HYDRAZIDE
mf: $C_7H_7N_3O_3$ mw: 181.17

TOXICITY DATA with REFERENCE
ivn-mus LD50:56 mg/kg CSLNX* NX#00861

CONSENSUS REPORTS: Reported in EPA TSCA Inventory.

SAFETY PROFILE: Poison by intravenous route. When heated to decomposition it emits toxic fumes of NO_x. See also NITRO COMPOUNDS OF AROMATIC HYDROCARBONS.

NFH500 CAS:619-24-9 **_HR: 3_**
m-NITROBENZONITRILE
mf: $C_7H_4N_2O_2$ mw: 148.13

PROP: Needles from water. Mp: 117-118°, bp: subl. Sol in hot water and alc, very sol in ether.

SYNS: m-CYANONITROBENZENE ◇ 3-CYANONITROBENZENE ◇ 3-NITROBENZONITRILE

TOXICITY DATA with REFERENCE
mmo-sat 10 μg/plate ENMUDM 4,163,82
mma-sat 10 μg/plate ENMUDM 4,163,82
ipr-mus LDLo:250 mg/kg CBCCT* 6,216,54

CONSENSUS REPORTS: Cyanide and its compounds are on The Community Right-To-Know List. Reported in EPA TSCA Inventory.

SAFETY PROFILE: Poison by intraperitoneal route. Mutation data reported. When heated to decomposition it emits toxic fumes of NO_x and CN^-. See also NI-

TRILES and NITRO COMPOUNDS OF AROMATIC HYDROCARBONS.

NFI000 CAS:612-24-8 *HR: D*
o-NITROBENZONITRILE
mf: $C_7H_4N_2O_2$ mw: 148.12

TOXICITY DATA with REFERENCE
dnr-bcs 50 μg/plate MUREAV 170,11,86
mmo-sat 10 mmol/plate MUREAV 58,11,78

CONSENSUS REPORTS: Cyanide and its compounds are on the Community Right-To-Know List. EPA Genetic Toxicology Program.

SAFETY PROFILE: Mutation data reported. Unstable. Incompatible with nitrogen trichloride. When heated to decomposition it emits very toxic fumes of NO_x and CN^-. See also NITRILES and NITRO COMPOUNDS OF AROMATIC HYDROCARBONS.

NFI100 CAS:70021-99-7 *HR: D*
1-NITROBENZO(a)PYRENE
mf: $C_{20}H_{11}NO_2$ mw: 297.32

TOXICITY DATA with REFERENCE
mmo-sat 1 μg/plate ENMUDM 6,417,84
mma-ham:ovr 5 mg/L CRNGDP 7,681,85

SAFETY PROFILE: Mutation data reported. When heated to decomposition it emits toxic fumes of NO_x. See also NITRO COMPOUNDS OF AROMATIC HYDROCARBONS.

NFJ000 CAS:2338-12-7 *HR: 3*
5-NITROBENZOTRIAZOLE
DOT: UN 0385
mf: $C_6H_4N_4O_2$ mw: 164.14

SYNS: 5-NITROBENZOTRIAZOL (DOT) ◇ 5-NITRO-1H-BENZOTRIAZOLE ◇ 6-NITRO-1H-BENZOTRIAZOLE

TOXICITY DATA with REFERENCE
ipr-mus LDLo:250 mg/kg CSLNX* NX#00275
ivn-mus LD50:178 mg/kg CSLNX* NX#00275

CONSENSUS REPORTS: Reported in EPA TSCA Inventory.

DOT Classification: Class A Explosive; Label: Explosive A.

SAFETY PROFILE: Poison by intraperitoneal and intravenous routes. An explosive. When heated to decomposition it emits toxic fumes of NO_x. See also NITRO COMPOUNDS OF AROMATIC HYDROCARBONS.

NFJ500 CAS:98-46-4 *HR: 3*
3-NITROBENZOTRIFLUORIDE
DOT: UN 2306
mf: $C_7H_4F_3NO_2$ mw: 191.12

PROP: Thin, pale, straw-colored, oily liquid with aromatic odor. Mp: −5°, bp: 202.8°, flash p: 217°F (OC), d: 1.437 @ 15.5°/15.5°, vap press: 0.3 mm @ 25°.

SYNS: m-NITROBENZOTRIFLUORIDE (DOT) ◇ m-NITROTRIFLUOROTOLUENE ◇ m-NITROTRIFLUORTOLUOL(GERMAN) ◇ m-(TRIFLUOROMETHYL)NITROBENZENE ◇ 3-TRIFLUOROMETHYLNITROBENZENE ◇ α,α,α-TRIFLUORO-m-NITROTOLUENE ◇ USAF MA-5

TOXICITY DATA with REFERENCE
orl-rat LD50:610 mg/kg TPKVAL 10,131,68
ihl-rat LC50:870 mg/kg 85GMAT -,92,82
orl-mus LD50:520 mg/kg TPKVAL 10,131,68
ihl-mus LC50:880 mg/m^3/2H 85GMAT -,92,82
ipr-mus LD50:100 mg/kg NTIS** AD277-689
scu-frg LDLo:573 mg/kg AEPPAE 130,250,28

CONSENSUS REPORTS: Reported in EPA TSCA Inventory.

DOT Classification: Poison B; Label: Poison.

SAFETY PROFILE: Poison by intraperitoneal route. Moderately toxic by ingestion, inhalation, and subcutaneous routes. Combustible when exposed to heat or flame. When heated to decomposition it emits very toxic fumes of F^- and NO_x. See also FLUORIDES and NITRO COMPOUNDS OF AROMATIC HYDROCARBONS.

NFK100 CAS:122-04-3 *HR: 2*
4-NITROBENZOYL CHLORIDE
mf: $C_7H_4ClNO_3$ mw: 185.57

PROP: Bright yellow needles from petr ether; pungent odor. Mp: 75°, bp: 205°. Decomposed by water and alc; sol in ether. Keep well closed.

SYNS: p-NITROBENZOIC ACID CHLORIDE ◇ 4-NITROBENZOIC ACID CHLORIDE ◇ p-NITROBENZOYL CHLORIDE

TOXICITY DATA with REFERENCE
mmo-sat 3 μg/plate ENMUDM 9(Suppl 9),1,87
mma-sat 33 μg/plate ENMUDM 9(Suppl 9),1,87
ihl-uns TCLo:11830 $\mu g/m^3$/4H (male 16W pre):REP GTPZAB 28(9),47,84
orl-rat LD50:5600 mg/kg GTPZAB 28(9),47,84
orl-mus LD50:3440 mg/kg GTPZAB 28(9),47,84
orl-rbt LD50:4750 mg/kg GTPZAB 28(9),47,84

CONSENSUS REPORTS: Reported in EPA TSCA Inventory.

SAFETY PROFILE: Moderately toxic by ingestion. Experimental reproductive effects. Mutation data reported. When heated to decomposition it emits toxic fumes of NO_x and Cl^-. See also NITRO COMPOUNDS OF AROMATIC HYDROCARBONS and CHLORIDES.

NFK500 CAS:121-90-4 ***HR: 3***
m-NITROBENZOYL CHLORIDE
mf: $C_7H_4ClNO_3$ mw: 185.57

PROP: Crystals. Mp: 34-35°, bp: 275-278°. Slt decomp; decomp in water and alc; very sol in ether. Yellow to brown liquid, partially crystallized at room temp. Vap d: 6.43.

TOXICITY DATA with REFERENCE
orl-rat LD50:2460 mg/kg AIHAAP 30,470,69
skn-rbt LD50:790 mg/kg AIHAAP 30,470,69

CONSENSUS REPORTS: Reported in EPA TSCA Inventory.

SAFETY PROFILE: Moderately toxic by ingestion and skin contact. Very unstable. Has exploded spontaneously. Will react with water or steam to produce toxic and corrosive fumes. When heated to decomposition it emits highly toxic fumes of Cl^- and NO_x. See also NITRO COMPOUNDS OF AROMATIC HYDROCARBONS and CHLORIDES.

NFL000 CAS:610-14-0 ***HR: 3***
2-NITROBENZOYL CHLORIDE
mf: $C_7H_4ClNO_3$ mw: 185.57

$O_2NC_6H_4CO{\bullet}Cl$

SYN: o-NITROBENZOYL CHLORIDE

SAFETY PROFILE: May explode when heated above 110°C. When heated to decomposition it emits toxic fumes of Cl^- and NO_x. See also CHLORIDES and NITRO COMPOUNDS OF AROMATIC HYDROCARBONS.

NFL100 ***HR: 3***
3-NITROBENZOYL NITRATE
mf: $C_7H_4N_2O_6$ mw: 212.12

$O_2NC_6H_4CO{\bullet}ONO_2$

SAFETY PROFILE: Explodes if heated rapidly. When heated to decomposition it emits toxic fumes of NO_x. See also NITRATES and NITRO COMPOUNDS OF AROMATIC HYDROCARBONS.

NFL500 CAS:38860-52-5 ***HR: 3***
N-(4-NITRO)BENZOYLOXYPIPERIDINE
mf: $C_{12}H_{14}N_2O_4$ mw: 250.28

SYN: p-NITROBENZOIC ACID, PIPERIDINO ESTER

TOXICITY DATA with REFERENCE
skn-mus TDLo:50 mg/kg:NEO JNCIAM 54,491,75

SAFETY PROFILE: Questionable carcinogen with experimental neoplastigenic data. When heated to decomposition it emits toxic fumes of NO_x. See also NITRO COMPOUNDS OF AROMATIC HYDROCARBONS.

NFM000 CAS:70021-98-6 ***HR: D***
3-NITROBENZ(a)PYRENE
mf: $C_{20}H_{11}NO_2$ mw: 297.32

TOXICITY DATA with REFERENCE
mmo-sat 100 ng/plate CRNGDP 6,1235,85
mma-ham:ovr 5 mg/L CRNGDP 7,681,85

SAFETY PROFILE: Mutation data reported. When heated to decomposition it emits toxic fumes of NO_x. See also NITRO COMPOUNDS OF AROMATIC HYDROCARBONS.

NFM500 CAS:63041-90-7 ***HR: D***
6-NITROBENZ(a)PYRENE
mf: $C_{20}H_{11}NO_2$ mw: 297.32

TOXICITY DATA with REFERENCE
mma-sat 3 μg/plate ENMUDM 6,417,84
otr-hmn:fbr 4 μmol/L CRNGDP 4,353,83

CONSENSUS REPORTS: IARC Cancer Review: Group 3 IMEMDT 7,56,87; Animal Inadequate Evidence IMEMDT 33,187,84. EPA Genetic Toxicology Program.

SAFETY PROFILE: Questionable carcinogen. Human mutation data reported. When heated to decomposition it emits toxic fumes of NO_x. See also NITRO COMPOUNDS OF AROMATIC HYDROCARBONS.

NFM550 CAS:612-25-9 ***HR: 2***
2-NITROBENZYL ALCOHOL
mf: $C_7H_7NO_3$ mw: 153.14

$O_2NC_6H_4CH_2OH$

SAFETY PROFILE: Decomposes exothermically at 201°C. When heated to decomposition it emits toxic fumes of NO_x. See also NITRO COMPOUNDS OF AROMATIC HYDROCARBONS and ALCOHOLS.

NFM560 ***HR: 2***
4-NITROBENZYL ALCOHOL
mf: $C_7H_7NO_3$ mw: 153.14

$O_2NC_6H_4CH_2OH$

SAFETY PROFILE: Decomposes exothermically at 236°C. When heated to decomposition it emits toxic fumes of NO_x. See also NITRO COMPOUNDS OF AROMATIC HYDROCARBONS and ALCOHOLS.

NFM700 CAS:3958-60-9 ***HR: 3***
2-NITROBENZYL BROMIDE
mf: $C_7H_6BrNO_2$ mw: 232.03

$O_2NC_6H_4CH_2Br$

SAFETY PROFILE: Decomposes exothermically at 100°C. It has caused plant scale explosions. When heated to decomposition it emits toxic fumes of NO_x and Br^-. See also 4-NITROBENZYL BROMIDE and NITRO COMPOUNDS OF AROMATIC HYDROCARBONS.

NFN000 CAS:19689-86-2 ***HR: 3***
4-NITROBENZYL BROMIDE
mf: $C_7H_6BrNO_2$ mw: 216.05

PROP: Needles from alc. Mp: 99-100°. Insol in water; sol in alc; very sol in ether.

SYNS: 2-BROMOETHYL-3-NITROANISOLE ◇ 1-(BROMOMETHYL)-4-NITROBENZENE ◇ α-BROMO-p-NITROTOLUENE ◇ α-BROMOPARA-NITROTOLUENE ◇ p-NITROBENZYL BROMIDE

TOXICITY DATA with REFERENCE
mmo-sat 5 μmol/L ENMUDM 3,11,81
dns-rat:lvr 10 μmol/L ENMUDM 3,11,81
ivn-mus LD50:56 mg/kg CSLNX* NX#07695

CONSENSUS REPORTS: Reported in EPA TSCA Inventory.

SAFETY PROFILE: Poison by intravenous route. Mutation data reported. When heated to decomposition it emits very toxic fumes of NO_x and Br^-. See also 2-NITROBENZYL BROMIDE and NITRO COMPOUNDS OF AROMATIC HYDROCARBONS.

NFN400 CAS:100-14-1 ***HR: 3***
p-NITROBENZYL CHLORIDE
mf: $C_7H_6ClNO_2$ mw: 171.59

SYN: α-CHLORO-p-NITROTOLUENE

TOXICITY DATA with REFERENCE
mmo-sat 33 μg/plate NTPTB* APR 82
mma-sat 100 μg/plate NTPTB* APR 82
ihl-rat LCLo:280 mg/m^3/4H TOXID9 4,66,84

CONSENSUS REPORTS: EPA Extremely Hazardous Substances List. Reported in EPA TSCA Inventory.

SAFETY PROFILE: Poison by inhalation. Mutation data reported. When heated to decomposition it emits toxic fumes of Cl^- and NO_x. See also 2-NITROBENZYL CHLORIDE and NITRO COMPOUNDS OF AROMATIC HYDROCARBONS.

NFN500 CAS:612-23-7 ***HR: D***
2-NITROBENZYL CHLORIDE
mf: $C_7H_6ClNO_2$ mw: 171.59

PROP: Crystals. Mp: 48-49°. Insol in water, sol in alc, very sltly sol hot ether.

SYNS: 1-(CHLOROMETHYL)-2-NITROBENZENE ◇ α-CHLORO-o-NITROTOLUENE ◇ o-NITROBENZYL CHLORIDE

TOXICITY DATA with REFERENCE
mma-sat 100 nmol/plate MUREAV 58,11,78

SAFETY PROFILE: Mutation data reported. Thermal decompositon is highly exothermic with much gas evolution. When heated to decomposition it emits very toxic fumes of Cl^- and NO_x. See also p-NITROBENZYL CHLORIDE and NITRO COMPOUNDS OF AROMATIC HYDROCARBONS.

NFO700 CAS:10598-82-0 ***HR: 2***
1-(p-NITROBENZYL)-2-NITROIMIDAZOLE
mf: $C_{10}H_8N_4O_4$ mw: 248.22

SYN: 2-NITRO-1-(p-NITROBENZYL)IMIDAZOLE

TOXICITY DATA with REFERENCE
orl-mus LD50:1000 mg/kg AACHAX -,513,67
ipr-mus LD50:707 mg/kg AACHAX -,513,67
scu-mus LD50:1320 mg/kg AACHAX -,513,67

SAFETY PROFILE: Moderately toxic by ingestion, subcutaneous, and intraperitoneal routes. When heated to decomposition it emits toxic fumes of NO_x. See also NITRO COMPOUNDS OF AROMATIC HYDROCARBONS.

NFP500 CAS:86-00-0 ***HR: 2***
o-NITROBIPHENYL
mf: $C_{12}H_9NO_2$ mw: 199.22

PROP: Light yellow- to reddish-colored liquid or crystalline solid. Mp: 35°, bp: 330°, flash p: 354°F, d: 1.189 @ 40°/15.5°, autoign temp: 356°F, vap press: 2 mm @ 140°, vap d: 5.9. Insol in water; sol in methanol, ethanol, tetrahydrofurfuryl alc, acetone, dimethyl-formamide.

SYNS: 2-NITRODIPHENYL ◇ o-NITRODIPHENYL

TOXICITY DATA with REFERENCE
otr-ham:kdy 80 μg/L BJCAAI 37,873,78
mma-sat 20 μg/plate BJCAAI 37,873,78
orl-rat LD50:1230 mg/kg JIHTAB 29,1,47
orl-rbt LD50:1580 mg/kg JIHTAB 29,1,47

SAFETY PROFILE: Moderately toxic by ingestion. Mutation data reported. An irritant. Combustible when exposed to heat or flame. To fight fire, use CO_2, dry chemical. When heated to decomposition it emits toxic fumes of NO_x. See also p-NITROBIPHENYL and NITRO COMPOUNDS OF AROMATIC HYDROCARBONS.

NFQ000 CAS:92-93-3 ***HR: 3***
4-NITROBIPHENYL
mf: $C_{12}H_9NO_2$ mw: 199.22

PROP: Needles from alc. Mp: 113-114°, bp: 340°. Insol in water; sltly sol in cold alc; very sol in ether.

SYNS: p-NITRODIPHENYL ◇ p-PHENYL-NITROBENZENE ◇ 4-PHENYL-NITROBENZENE ◇ PNB

TOXICITY DATA with REFERENCE
mmo-bcs 50 mmol/L EXPEAM 39,530,83
dns-rat:lvr 100 μmol/L ENMUDM 3,11,81
orl-dog TDLo:7000 mg/kg/2Y-I:NEO JAMAAP 172,1611,60
orl-dog TD:7000 mg/kg/3Y-I:ETA IMSUAI 27,634,58
orl-rat LD50:2230 mg/kg JIHIAB 29,1,47
ipr-mus LD50:347 mg/kg JJIND8 62,911,79
orl-rbt LD50:1970 mg/kg JIHTAB 29,1,47

CONSENSUS REPORTS: IARC Cancer Review: Group 3 IMEMDT 7,56,87; Animal Limited Evidence IMEMDT 4,113,74. Reported in EPA TSCA Inventory.

OSHA: Cancer Suspect Agent
ACGIH TLV: Confirmed Human Carcinogen.
DFG MAK: Animal Carcinogen, Suspected Human Carcinogen.
NIOSH REL: (4-Nitrobiphenyl): TWA use 29 CFR 1910.1003

SAFETY PROFILE: Confirmed carcinogen with experimental carcinogenic, neoplastigenic, and tumorigenic data. Poison by intraperitoneal route. Moderately toxic by ingestion. Mutation data reported. When heated to decomposition it emits toxic fumes of NO_x. See also NITRO COMPOUNDS OF AROMATIC HYDROCARBONS.

NFQ500 CAS:594-70-7 ***HR: 3***
tert-NITROBUTANE
mf: $C_4H_9NO_2$ mw: 103.12

$O_2NC(CH_3)_3$

SYN: 2-METHYL-2-NITROPROPANE

SAFETY PROFILE: Potentially explosive when heated. When heated to decomposition it emits toxic fumes of NO_x. See also NITRO COMPOUNDS.

NFR000 ***HR: 3***
1-NITRO-3-BUTENE
mf: $C_4H_7NO_2$ mw: 101.11

CONSENSUS REPORTS: Community Right-To-Know List.

SAFETY PROFILE: An unstable material. When heated to decomposition it emits toxic fumes of NO_x. See also 2-NITRO-2-BUTENE.

NFR500 CAS:4812-23-1 ***HR: 3***
2-NITRO-2-BUTENE
mf: $C_4H_7NO_2$ mw: 101.12

PROP: Liquid. D: 0.988 @ 0°, bp: 138-139° @ 747 mm.

TOXICITY DATA with REFERENCE
eye-hmn 1 μg/L IMSUAI 27,375,58
skn-rbt 500 mg open SEV AMIHAB 18,312,58
eye-rbt 50 mg SEV AMIHAB 18,312,58
orl-rat LDLo:280 mg/kg IMSUAI 34,800,65
ihl-rat LCLo:1400 ppm AMIHAB 18,312,58
ipr-rat LDLo:80 mg/kg TXAPA9 1,156,59
skn-rbt LDLo:620 mg/kg IMSUAI 34,800,65

SAFETY PROFILE: Poison by ingestion and intraperitoneal routes. Moderately toxic by skin contact and inhalation. A human eye irritant and severe experimental eye and skin irritant. An unstable compound. When heated to decomposition it emits toxic fumes of NO_x. See also NITRO COMPOUNDS.

NFS500 ***HR: 3***
NITRO CARBO NITRATE

PROP: A solid explosive made of ammonium nitrate, dinitrocotton, etc.

DOT Classification: Blasting Agent; Label: Blasting Agent.

SAFETY PROFILE: An explosive and oxidizer. When heated to decomposition it emits toxic fumes of NO_x and NH_3. See also EXPLOSIVES, HIGH; AMMONIUM NITRATE, and NITRATES.

NFS525 CAS:100-00-5 ***HR: 3***
p-NITROCHLOROBENZENE
DOT: UN 1578
mf: $C_6H_4ClNO_2$ mw: 157.56

PROP: D: 1.520, mp: 83°, bp: 242°, flash p: 110°. Insol in water; sltly sol in alc; very sol in CS_2 and ether.

SYNS: 1-CHLOOR-4-NITROBENZEEN (DUTCH) ◇ 1-CHLOR-4-NITROBENZOL (GERMAN) ◇ p-CHLORONITROBENZENE ◇ 4-CHLORONITROBENZENE ◇ 1-CHLORO-4-NITROBENZENE ◇ 4-CHLORO-1-NITROBENZENE ◇ 1-CLORO-4-NITROBENZENE (ITALIAN) ◇ p-NITROCHLOORBENZEEN (DUTCH) ◇ p-NITROCHLOROBENZENE solid (DOT) ◇ p-NITROCHLOROBENZOL (GERMAN) ◇ p-NITROCLOROBENZENE (ITALIAN) ◇ PNCB

TOXICITY DATA with REFERENCE
mmo-sat 819 μg/plate MUREAV 116,217,83
mma-sat 100 μg/plate ENMUDM 5(Suppl 1),3,83
dnd-mus-ipr 60 mg/kg BSIBAC 56,1680,80
orl-mus TDLo:194 g/kg/78W-C:CAR JEPTDQ 2(2),325,78
orl-mus TD:390 g/kg/78W-C:CAR JEPTDQ 2(2),325,78
orl-rat LD50:420 mg/kg AGGHAR 17,217,59
skn-rat LD50:16 g/kg AGGHAR 17,217,59
ipr-rat LD50:420 mg/kg AGGHAR 17,217,59
orl-mus LD50:650 mg/kg FATOAO 29,619,66

CONSENSUS REPORTS: Reported in EPA TSCA Inventory.

OSHA PEL: TWA 1 mg/m^3 (skin)
ACGIH TLV: TWA 0.1 ppm (skin)
DFG MAK: 1 mg/m^3
DOT Classification: Poison B; Label: Poison.

SAFETY PROFILE: A poison by ingestion. Questionable carcinogen with experimental carcinogenic data. Mutation data reported. May explode on heating. Potentially violent reaction with sodium methoxide. When heated to decomposition it emits very toxic fumes of NO_x and Cl^-. See also other chloronitrobenzene entries and NITRO COMPOUNDS OF AROMATIC HYDROCARBONS

NFS550 CAS:10199-89-0 ***HR: D***
4-NITRO-7-CHLOROBENZOFURAZAN
mf: $C_6H_2ClN_3O_3$ mw: 199.56

SYNS: 4-CHLORO-7-NITROBENZO-2-OXA-1,3-DIAZOLE ◇ NBD-C 1 ◇ NBD-CHLORIDE

TOXICITY DATA with REFERENCE
mmo-sat 2 μg/plate MUREAV 88,351,81
dni-mus:leu 4 μmol/L CNREA8 35,3735,75

CONSENSUS REPORTS: Reported in EPA TSCA Inventory.

SAFETY PROFILE: Mutation data reported. When heated to decomposition it emits toxic fumes of Cl^- and NO_x. See also NITRO COMPOUNDS OF AROMATIC HYDROCARBONS.

NFS700 CAS:121-17-5 ***HR: 3***
3-NITRO-4-CHLOROBENZOTRIFLUORIDE
DOT: UN 2307
mf: $C_7H_3ClF_3NO_2$ mw: 225.56

SYNS: 4-CHLORO-3-NITRO-α,α,α-TRIFLUOROTOLUENE ◇ 3-NITRO-4-CHLORO-α,α,α-TRIFLUOROTOLUENE

TOXICITY DATA with REFERENCE
dns-hmn:emb 1 g/L AISSAW 18,123,82
orl-rat LD50:1075 mg/kg GISAAA 47(3),88,82
orl-mus LD50:400 mg/kg GISAAA 47(3),88,82
ivn-mus LD50:56 mg/kg CSLNX* NX#07616

DOT Classification: Poison B; Label: Poison.

SAFETY PROFILE: Poison by ingestion and intravenous routes. Human mutation data reported. When heated to decomposition it emits toxic fumes of F^-, Cl^-, and NO_x. See also CHLORINATED HYDROCARBONS, AROMATIC; and NITRO COMPOUNDS OF AROMATIC HYDROCARBONS.

NFT000 CAS:2463-84-5 ***HR: 3***
p-NITRO-o-CHLOROPHENYL DIMETHYL THIONOPHOSPHATE
mf: $C_8H_9ClNO_5PS$ mw: 297.66

PROP: White solid. Mp: 52°. Insol in water; sol in acetone, cyclohexane, xylene, toluene, and ethylacetate.

SYNS: O-(4-CHLOOR-3-NITRO-FENYL)-O,O-DIMETHYLMONOTHIOFOSFAAT (DUTCH) ◇ O-(4-CHLOR-3-NITRO-PHENYL)-O,O-DIMETHYL-MONOTHIOPHOSPHAT (GERMAN) ◇ O-(2-CHLORO-4-NITROPHENYL) O,O-DIMETHYL PHOSPHOROTHIOATE ◇ O-(4-CLORO-3-NITRO-FENIL)-O,O-DIMETIL-MONOIIOFOSFATO(ITALIAN) ◇ O,O-DIMETHYL O-2-CHLORO-4-NITROPHENYL PHOSPHOROTHIOATE ◇ DIMETHYL-2-CHLORONITROPHENYL THIOPHOSPHATE ◇ ENT 17,035 ◇ EXPERIMENTAL INSECTICIDE 4124 ◇ ISOCHLOORTHION (DUTCH) ◇ ISOMERIC CHLORTHION ◇ THIOPHOSPHATE de O,O-DIMETHYLE et de O-4-CHLORO-3-NITROPHENYLE (FRENCH)

TOXICITY DATA with REFERENCE
orl-rat LD50:330 mg/kg TXAPA9 2,88,60
skn-rat LD50:790 mg/kg TXAPA9 2,88,60
orl-mus LD50:331 mg/kg ARTODN 41,111,78
orl-ckn LD50:248 mg/kg TXAPA9 3,521,61

SAFETY PROFILE: Poison by ingestion. Moderately toxic by skin contact. When heated to decomposition it emits very toxic fumes of PO_x, SO_x, NO_x, and Cl^-. See also PARATHION.

NFT400 CAS:7496-02-8 ***HR: 3***
6-NITROCHRYSENE
mf: $C_{18}H_{11}NO_2$ mw: 273.30

TOXICITY DATA with REFERENCE
mmo-sat 5 μg/plate CNREA8 44,3408,84
mma-sat 2500 ng/plate CNREA8 44,3408,84
otr-ham:emb 3700 nmol/L CRNGDP 4,357,83
ipr-mus TDLo:9163 μg/kg/2W-I:CAR CRNGDP 6,801,85

CONSENSUS REPORTS: IARC Cancer Review: Group 3 IMEMDT 7,56,87; Animal Limited Evidence IMEMDT 33,195,84.

SAFETY PROFILE: Questionable carcinogen with experimental carcinogenic data. Mutation data reported. When heated to decomposition it emits toxic fumes of NO_x. See also CHRYSENE and NITRO COMPOUNDS OF AROMATIC HYDROCARBONS.

NFT425 CAS:1734-79-8 ***HR: 2***
p-NITROCINNAMALDEHYDE
mf: $C_9H_7NO_3$ mw: 177.17

TOXICITY DATA with REFERENCE
ipr-mus LDLo:500 mg/kg CBCCT* 6,145,54

CONSENSUS REPORTS: Reported in EPA TSCA Inventory.

SAFETY PROFILE: Moderately toxic by intraperitoneal route. When heated to decomposition it emits toxic fumes of NO_x. See also NITRO COMPOUNDS and ALDEHYDES.

NFT459 ***HR: 3***
NITRO COMPOUNDS

PROP: Compounds of the form C-NO_2 or N-NO_2.

SAFETY PROFILE: The presence of a C– or N– linked nitro group in an organic compound can significantly decrease its reactivity and stability. Nitrate esters, with the nitro group linked to O, are very unstable. Nitro alkanes are mild oxidants but high temperatures and pressures may cause them to react violently. Polynitroalkanes may be explosive. Alkalies react with nitro alkanes to form explosive metal salts. The presence of metal oxides increases the thermal sensitivity of the lower nitro alkanes (e.g., nitromethane, nitroethane, and 1-nitropropane. Many nitro alkenes are highly reactive and may be explosive. Compounds with more than one nitro group (polynitroalkyls, such as trinitromethane and dinitroacetonitrile) are generally explosive. See also specific compounds and NITRO COMPOUNDS OF AROMATIC HYDROCARBONS.

NFT500 ***HR: 3***
NITRO COMPOUNDS of AROMATIC HYDROCARBONS

SAFETY PROFILE: The mono-, di-, and trinitrobenzenes are absorbed chiefly through the skin and through inhalation of the dust or vapor when these materials are heated. The dinitrobenzenes are believed to be somewhat more toxic than the mononitrobenzene and more toxic than aniline. The effect of di- and trinitrobenzene on the body is similar to that of aniline and mononitrobenzene with reduction of the oxygen-carrying power of the blood and depression of the nervous system being responsible for most of the symptoms following acute exposure. Poisoning with the solid nitro compounds is usually slower and less severe than is the case with the liquid nitro and amino benzenes since absorption is less rapid. Thus, chronic poisoning occurs more frequently than acute, the picture observed in the chronic form being one of anemia, moderate cyanosis, fatigue, slight dizziness, headache, insomnia, and loss of weight. Prolonged chronic exposure may result in damage to the liver and kidneys, with production of acute yellow atrophy, toxic hepatitis, and fatty degeneration of the kidneys. The introduction of one or more Cl atoms into the nitrobenzene ring results in the formation of chloronitrobenzene compounds or nitrochlors. The chloro-mono-nitrobenzenes have essentially the same toxic effect as nitrobenzene. The Cl derivatives of dinitrobenzene, on the other hand, while resembling dinitrobenzene in their systemic effects are much more irritating to the skin. They act as direct irritants and, in addition, may cause sensitization.

Dangerous; many of these compounds are highly flammable and some are explosive, especially those with more than one nitro group on the ring (polynitroaryls, such as trinitrobenzene, trinitrotoluene, tetranitro-N-methylaniline, trinitrophenol). The presence of alkali increases the thermal sensitivity of the explosive materials. Industrial explosions have occurred in this manner. When heated to decomposition they evolve highly toxic fumes of NO_x. See specific nitro compounds.

NFU500 CAS:119-33-5 ***HR: 2***
2-NITRO-p-CRESOL
mf: $C_7H_7NO_3$ mw: 153.15

$CH_3(O_2N)C_6H_3OH$

PROP: Yellow in aqueous ethyl alc. D: 1.240 @ 28.6°, mp: 32°, bp: 125° @ 22 mm. Very sltly sol in water; very sol in alc and ether.

SYN: 4-METHYL-2-NITROPHENOL

TOXICITY DATA with REFERENCE
orl-rat LD50:3360 mg/kg HURC** -,-,72

CONSENSUS REPORTS: Reported in EPA TSCA Inventory.

SAFETY PROFILE: Moderately toxic by ingestion. Potentially explosive reaction with sodium hydroxide + sodium carbonate + methanol. When heated to decomposition it emits toxic fumes of NO_x. See also NITRO COMPOUNDS OF AROMATIC HYDROCARBONS.

NFV000 CAS:2581-34-2 ***HR: 2***
4-NITRO-m-CRESOL
mf: $C_7H_7NO_3$ mw: 153.15

TOXICITY DATA with REFERENCE
orl-rat LD50:4410 mg/kg MarJV# 29MAR77
ipr-mus LDLo:500 mg/kg CBCCT* 6,54,54

CONSENSUS REPORTS: Reported in EPA TSCA Inventory.

SAFETY PROFILE: Moderately toxic by intraperitoneal route. Mildly toxic by ingestion. When heated to decomposition it emits toxic fumes of NO_x. See also NITRO COMPOUNDS OF AROMATIC HYDROCARBONS.

NFV500 CAS:1122-60-7 ***HR: 3***
NITROCYCLOHEXANE
mf: $C_6H_{11}NO_2$ mw: 129.18

TOXICITY DATA with REFERENCE
ihl-rat LC50:150 mg/m^3/4H 85GMAT -,92,82
orl-mus LD50:250 mg/kg 85GMAT -,92,82
ihl-mus LCLo:10 mg/m^3/2H 85GMAT -,92,82

CONSENSUS REPORTS: EPA Extremely Hazardous Substances List.

SAFETY PROFILE: Poison by ingestion and inhalation. When heated to decomposition it emits toxic fumes of NO_x. See also NITRO COMPOUNDS.

NFW000 CAS:4185-47-1 ***HR: 3***
NITRODIETHANOLAMINE DINITRATE
mf: $C_4H_8N_4O_8$ mw: 240.16

$O_2NN(C_2H_4ONO_2)_2$

SYNS: DIETHANOLNITRAMINE DINITRATE ◇ DIETHANOL-N-NITRAMINE DINITRATE ◇ DINA ◇ 2,2'-DINITRATO-N-NITRODIETHYLAMINE ◇ sym-DINITROXYDIETHYLNITRAMINE ◇ N-NITRODIETHANOLAMINE DINITRATE ◇ 2,2'-(NITROIMINO) BISETHANOL, DINITRATE (ESTER) (9CI) ◇ 2,2'-NITROIMINOBIS (ETHYLNITRATE) ◇ 2,2'-NITROIMINODIETHANOL NITRATE ◇ 2,2'-(NITROIMINO)ETHANOL DINITRATE

TOXICITY DATA with REFERENCE
orl-mus LD50:3360 mg/kg KHZDAN 9,50,66
ipr-mus LD50:180 mg/kg PCJOAU 10,1504,76
ivn-gpg LD50:25 mg/kg THERAP 8,565,53

SAFETY PROFILE: Poison by intraperitoneal and intravenous routes. Moderately toxic by ingestion. An explosive. Chromium compounds increase the burning rate. When heated to decomposition it emits toxic fumes of NO_x. See also NITRO COMPOUNDS and NITRATES.

NFW100 CAS:17074-42-9 ***HR: 3***
1-NITRO-9-(DIETHYLAMINOETHYLAMINE)-ACRIDINE DIHYDROCHLORIDE
mf: $C_{19}H_{22}N_4O_2{\bullet}2ClH$ mw: 411.37

SYNS: C-516 ◇ N,N-DIETHYL-N'-(1-NITRO-9-ACRIDINYL)-1,2-ETHANEDIAMINE DIHYDROCHLORIDE (9CI)

TOXICITY DATA with REFERENCE
dnd-hmn:hla 2500 nmol/L CBINA8 49,311,84
dns-hmn:hla 1 μmol/L CBINA8 49,311,84
dni-hmn:hla 70 nmol/L BBACAQ 825,244,85
orl-rat LD50:42200 μg/kg MMDPA6 8,252,76
ivn-rat LD50:3400 μg/kg MMDPA6 8,252,76
orl-mus LD50:56400 μg/kg MMDPA6 8,252,76
ivn-mus LD50:4900 μg/kg MMDPA6 8,252,76
ivn-pgn LD50:4920 mg/kg AITEAT 28,777,80

SAFETY PROFILE: Poison by ingestion and intravenous routes. Human mutation data reported. When heated to decomposition it emits toxic fumes of NO_x and HCl.

NFW200 CAS:17074-44-1 ***HR: 3***
1-NITRO-9-(DIETHYLAMINOPROPYLAMINE)-ACRIDINE DIHYDROCHLORIDE
mf: $C_{20}H_{24}N_4O_2{\bullet}2ClH$ mw: 425.40

SYNS: C-410 ◇ N,N-DIETHYL-N'-(1-NITRO-9-ACRIDINYL)-1,3-PROPANEDIAMINE DIHYDROCHLORIDE (9CI)

TOXICITY DATA with REFERENCE
orl-rat LD50:63200 μg/kg MMDPA6 8,252,76
ivn-rat LD50:1200 μg/kg MMDPA6 8,252,76
orl-mus LD50:27500 μg/kg MMDPA6 8,252,76
ivn-mus LD50:1600 μg/kg MMDPA6 8,252,76
ivn-pgn LD50:1600 μg/kg AITEAT 28,777,80

SAFETY PROFILE: Poison by ingestion and intravenous routes. When heated to decomposition it emits toxic fumes of NO_x and HCl.

NFW350 CAS:77162-70-0 ***HR: 3***
1-NITRO-9-(2-DIHYDROXYETHYLAMINO-ETHYLAMINO)-ACRIDINE HYDROCHLORIDE
mf: $C_{19}H_{22}N_4O_4{\bullet}2ClH$ mw: 443.37

SYNS: C-835 ◇ 2,2'-((2-(8-NITRO-9-ACRIDINYLAMINO)ETHYL) DIETHANOL HYDROCHLORIDE

TOXICITY DATA with REFERENCE
dnd-hmn:hla 2500 nmol/L CBINA8 49,311,84
dns-hmn:hla 5 μmol/L CBINA8 49,311,84
dni-hmn:hla 20 nmol/L BBACAQ 825,244,85
orl-rat LD50:176 mg/kg AITEAT 28,735,80
ivn-rat LD50:1700 μg/kg AITEAT 28,735,80
orl-mus LD50:127 mg/kg AITEAT 28,735,80
ivn-mus LD50:2700 μg/kg AITEAT 28,735,80

SAFETY PROFILE: Poison by ingestion and intravenous routes. Human mutation data reported. When heated to decomposition it emits toxic fumes of NO_x and HCl.

NFW400 CAS:63944-10-5 ***HR: 3***
1-NITRO-9-(DIMETHYLAMINE)-ACRIDINE DIHYDROCHLORIDE
mf: $C_{15}H_{13}N_3O_2{\bullet}2ClH$ mw: 340.23
SYN: C-702

TOXICITY DATA with REFERENCE
orl-rat LD50:23700 μg/kg MMDPA6 8,252,76
ivn-rat LD50:1400 μg/kg MMDPA6 8,252,76
orl-mus LD50:17500 μg/kg MMDPA6 8,252,76
ivn-mus LD50:3700 μg/kg MMDPA6 8,252,76

SAFETY PROFILE: Poison by ingestion and intravenous routes. When heated to decomposition it emits toxic fumes of NO_x and HCl. See also AMINES.

NFW425 CAS:28743-45-5 ***HR: 3***
1-NITRO-9-(DIMETHYLAMINO)-ACRIDINE HYDROCHLORIDE
mf: $C_{15}H_{13}N_3O_2{\bullet}ClH$ mw: 303.77
SYN: C 702

TOXICITY DATA with REFERENCE
orl-rat LD50:24 mg/kg AITEAT 22,823,74
ivn-rat LD50:1450 μg/kg AITEAT 22,823,74
orl-mus LD50:17500 μg/kg AITEAT 22,823,74
ivn-mus LD50:3700 μg/kg AITEAT 22,823,74
ivn-pgn LD50:3700 μg/kg AITEAT 28,777,80

SAFETY PROFILE: Poison by ingestion and intravenous routes. When heated to decomposition it emits toxic fumes of NO_x and HCl.

NFW430 CAS:24268-87-9 ***HR: 3***
1-NITRO-10-(DIMETHYLAMINOETHYL)-9-ACRIDONE HYDROCHLORIDE
mf: $C_{17}H_{17}N_3O_3$•ClH mw: 347.83

SYNS: C-541 ◇ 10-(2-(DIMETHYLAMINO)ETHYL)-1-NITRO-9(10H)-ACRIDINONE MONOHYDROCHLORIDE (9CI)

TOXICITY DATA with REFERENCE
orl-rat LD50:41 mg/kg MMDPA6 8,252,76
ivn-rat LD50:300 μg/kg MMDPA6 8,252,76
orl-mus LD50:33300 μg/kg MMDPA6 8,252,76
ivn-mus LD50:900 μg/kg MMDPA6 8,252,76
ivn-pgn LD50:900 μg/kg AITEAT 28,777,80

SAFETY PROFILE: Poison by ingestion and intravenous routes. When heated to decomposition it emits toxic fumes of NO_x and HCl.

NFW435 CAS:65094-73-7 ***HR: 3***
1-NITRO-9-((2-DIMETHYLAMINO)-1-METHYLETHYLAMINO)-ACRIDINE DIHYDROCHLORIDE
mf: $C_{18}H_{20}N_4O_2$•2ClH mw: 397.34

SYNS: C-829 ◇ DIHYDROCHLORIDE-1-NITRO-9-((2-DIMETHYLAMINO)-1-METHYLETHYLAMINO)-ACRIDINE ◇ 9-((2-(DIMETHYLAMINO)-1-METHYLETHYL)AMINO)-1-NITROACRIDINEDIHYDROCHLORIDE ◇ 9-((2-(DIMETHYLAMINO)-1-METHYLETHYL)AMINO)-1-NITROACRIDINE HYDROCHLORIDE ◇ N-DIMETHYL-1-METHYL-N'-(1-NITRO-9-ACRIDINYL)-1,2-ETHANEDIAMINE DIHYDROCHLORIDE ◇ N^1,N^1-DIMETHYL-N^2-(1-NITRO-9-ACRIDINYL)-1,2-PROPANEDIAMINE DIHYDROCHLORIDE ◇ N'-(1-NITRO-9-ACRIDINYL)-N,N,1-TRIMETHYL-1,2-ETHANEDIAMINE HYDROCHLORIDE

TOXICITY DATA with REFERENCE
sce-hmn:lym 80 nmol/L/69H MUREAV 67,93,79
orl-rat LD50:105 mg/kg AITEAT 27,749,79
ivn-rat LD50:13300 μg/kg AITEAT 27,749,79
orl-mus LD50:99200 μg/kg AITEAT 27,749,79
ivn-mus LD50:21300 μg/kg AITEAT 27,749,79
ivn-pgn LD50:21 mg/kg

SAFETY PROFILE: Poison by ingestion and intravenous routes. Human mutation data reported. When heated to decomposition it emits toxic fumes of NO_x and HCl.

NFW450 ***HR: 3***
1-NITRO-9-(5-DIMETHYLAMINOPENTYLAMINO)-ACRIDINE DIHYDROCHLORIDE
mf: $C_{20}H_{24}N_4O_2$•2ClH mw: 424.41

SYN: C 609

TOXICITY DATA with REFERENCE
dnd-hmn:hla 2500 nmol/L CBINA8 49,311,84
dns-hmn:hla 1 μmol/L CBINA8 49,311,84
dni-hmn:hla 10 nmol/L BBACAQ 825,244,85
orl-rat LD50:63300 μg/kg AITEAT 21,775,73
ivn-rat LD50:500 μg/kg AITEAT 21,775,73
orl-mus LD50:76700 μg/kg AITEAT 21,775,73
ivn-mus LD50:1200 μg/kg AITEAT 21,775,73
ivn-pgn LD50:1200 μg/kg AITEAT 28,777,80

SAFETY PROFILE: Poison by ingestion and intravenous routes. Human mutation data reported. When heated to decomposition it emits toxic fumes of NO_x and HCl.

NFW460 CAS:24268-89-1 ***HR: 3***
1-NITRO-10-(3-DIMETHYLAMINOPROPYL)-ACRIDONE HYDROCHLORIDE
mf: $C_{18}H_{19}N_3O_3$•ClH mw: 361.86

SYNS: C-492 ◇ 10-(3-(DIMETHYLAMINO)PROPYL)-1-NITRO-9-ACRIDANONE HYDROCHLORIDE ◇ 10-(3-(DIMETHYLAMINO)PROPYL)-1-NITRO-9(10H)-ACRIDINONE MONOHYDROCHLORIDE (9CI) ◇ 1-NITRO-14-(DIMETHYLAMINOPROPYL)-ACRIDONE HYDROCHLORIDE ◇ 1-NITRO 10-(3-DIMETHYLAMINOPROPYL)-ACRIDON HYDROCHLORIDE

TOXICITY DATA with REFERENCE
orl-rat LD50:53300 μg/kg AITEAT 21,775,73
ivn-rat LD50:900 μg/kg AITEAT 21,775,73
orl-mus LD50:17500 μg/kg MMDPA6 8,252,76
ivn-mus LD50:1 mg/kg AITEAT 21,775,76
ivn-pgn LD50:1000 mg/kg AITEAT 28,777,80

SAFETY PROFILE: Poison by ingestion and intravenous routes. When heated to decomposition it emits toxic fumes of NO_x and HCl.

NFW470 CAS:20064-00-0 ***HR: 3***
1-NITRO-9-(3-DIMETHYLAMINOPROPYLAMINE)-ACRIDINE-N-OXIDE DIHYDROCHLORIDE
mf: $C_{18}H_{20}N_4O_3$•2ClH mw: 413.34

SYNS: C-684 ◇ N,N-DIMETHYL-N'-(1-NITRO-9-ACRIDINYL)-1,3-PROPANEDIAMINE N-OXIDE, DIHYDROCHLORIDE (9CI)

TOXICITY DATA with REFERENCE
orl-rat LD50:356 mg/kg MMDPA6 8,252,76
ivn-rat LD50:5600 μg/kg MMDPA6 8,252,76
orl-mus LD50:300 mg/kg MMDPA6 8,252,76
ivn-mus LD50:6200 μg/kg MMDPA6 8,252,76
ivn-pgn LD50:6250 μg/kg AITEAT 28,777,80

SAFETY PROFILE: Poison by ingestion and intravenous routes. When heated to decomposition it emits toxic fumes of NO_x and HCl.

NFW500 CAS:4533-39-5 ***HR: 3***
1-NITRO-9-(3'-DIMETHYLAMINOPROPYL-AMINO)-ACRIDINE
mf: $C_{18}H_{20}N_4O_2$ mw: 324.42

SYNS: C 283 ◇ N,N-DIMETHYL-N'-(1-NITRO-9-ACRIDINYL)-1,3-PROPANEDIAMINE ◇ LEDAKRIN

TOXICITY DATA with REFERENCE
mmo-sat 1 pmol/plate MUREAV 78,7,80
dns-hmn:hla 250 nmol/L CBINA8 49,311,84
orl-rat LD50:25700 μg/kg MMDPA6 8,252,76
ivn-rat LD50:700 μg/kg MMDPA6 8,252,76
orl-mus LD50:34 mg/kg MMDPA6 8,252,76
ivn-mus LD50:1 mg/kg MMDPA6 8,252,76

SAFETY PROFILE: Poison by ingestion and intravenous routes. Human mutation data reported. When heated to decomposition it emits toxic fumes of NO_x.

NFX500 CAS:22751-18-4 ***HR: 3***
1-NITRO-3-(2,4-DINITROPHENYL)UREA
mf: $C_7H_5N_5O_7$ mw: 271.15

$O_2NNHCO{\cdot}NHC_6H_3(NO_2)_2$

SAFETY PROFILE: A very impact- and friction-sensitive explosive less powerful than picric acid. Upon decomposition it emits toxic fumes of NO_x. See also NITRO COMPOUNDS OF AROMATIC HYDROCARBONS.

NFY000 CAS:836-30-6 ***HR: 1***
p-NITRODIPHENYLAMINE
mf: $C_{12}H_{10}N_2O_2$ mw: 214.24

PROP: Yellow needles. Mp: 132-133°. Insol in dilute acid; sol in alc, acetic acid.

SYNS: 4-NITRODIFENYLAMIN (CZECH) ◇ 4-NITRODIPHENYLAMINE ◇ 4-NITRO-N-PHENYLBENZENAMINE

TOXICITY DATA with REFERENCE
eye-rbt 500 mg/24H MLD 28ZPAK -,134,72
orl-rat LD50:11800 mg/kg 28ZPAK -,134,72

CONSENSUS REPORTS: Reported in EPA TSCA Inventory.

SAFETY PROFILE: Mildly toxic by ingestion. An eye irritant. When heated to decomposition it emits toxic fumes of NO_x. See also NITRO COMPOUNDS OF AROMATIC HYDROCARBONS and AMINES.

NFY500 CAS:79-24-3 ***HR: 3***
NITROETHANE
DOT: UN 2842
mf: $C_2H_5NO_2$ mw: 75.08

PROP: Oily, colorless liquid. Agreeable odor. Mp: −90°, bp: 114.0°, fp: −50°, d: 1.052 @ 20°/20°, autoign temp: 778°F, flash p: 106°F, decomp @ 335-382°, lel: 4.0%, vap press: 15.6 mm @ 20°, vap d: 2.58. Sol in water, acid, and alkali; misc in alc, chloroform, and ether.

SYN: NITROETAN (POLISH)

TOXICITY DATA with REFERENCE
orl-rat LD50:1100 mg/kg GISAAA 32(9),9,67
orl-mus LD50:860 mg/kg GISAAA 32(9),9,67
ipr-mus LD50:310 mg/kg KHFZAN 10(6),53,76
orl-rbt LDLo:500 mg/kg JIHTAB 22,315,40

CONSENSUS REPORTS: Reported in EPA TSCA Inventory. EPA Genetic Toxicology Program.

OSHA PEL: TWA 100 ppm
ACGIH TLV: TWA 100 ppm
DFG MAK: 100 ppm (310 mg/m^3)
DOT Classification: Flammable or Combustible Liquid; Label: Flammable Liquid.

SAFETY PROFILE: Poison by intraperitoneal route. Moderately toxic by ingestion. Causes injury to liver and kidneys. An eye and mucous membrane irritant. A dangerous fire hazard when exposed to heat, flame, or oxidizers. To fight fire, use alcohol foam, CO_2, dry chemical; water can blanket fire. Incompatible with $Ca(OH)_2$; hydrocarbons; hydroxides; inorganic bases; KOH; NaOH; metal oxides. Explodes when heated. When heated to decomposition it emits toxic fumes of NO_x. See also NITRO COMPOUNDS.

NFY550 CAS:625-48-9 ***HR: 3***
2-NITROETHANOL
mf: $C_2H_5NO_3$ mw: 91.07

SAFETY PROFILE: An explosive. Traces of alkali may promote explosive decomposition. When heated to decomposition it emits toxic fumes of NO_x. See also NITRO COMPOUNDS and ALCOHOLS.

NGA500 CAS:3741-14-8 ***HR: 3***
4-NITRO-2-ETHYLQUINOLINE-N-OXIDE
mf: $C_{11}H_{10}N_2O_3$ mw: 218.23

TOXICITY DATA with REFERENCE
skn-mus TDLo:69 mg/kg/11W-I:ETA GANNA2 49,33,58

SAFETY PROFILE: Questionable carcinogen with experimental tumorigenic data. When heated to decomposition it emits toxic fumes of NO_x.

NGA700 CAS:892-21-7 ***HR: 3***
3-NITROFLUORANTHENE

SYN: 4-NITROFLUORANTHENE

TOXICITY DATA with REFERENCE
mmo-sat 130 ng/plate MUREAV 91,321,81

otr-ham:emb 4100 nmol/L CRNGDP 4,357,83
scu-rat TDLo:120 mg/kg/8W-I:CAR CALEDQ 15,1,82

CONSENSUS REPORTS: IARC Cancer Review: Group 3 IMEMDT 7,56,87; Animal Inadequate Evidence IMEMDT 33,201,84

SAFETY PROFILE: Questionable carcinogen with experimental carcinogenic data. Mutation data reported. When heated to decomposition it emits toxic fumes of NO_x. See also NITRO COMPOUNDS OF AROMATIC HYDROCARBONS.

NGB000 CAS:607-57-8 ***HR: 3***
2-NITROFLUORENE
mf: $C_{13}H_9NO_2$ mw: 211.23

PROP: Needles from 50% acetic acid. Mp: 157-158°.

TOXICITY DATA with REFERENCE
mmo-sat 250 ng/plate MUREAV 143,213,85
otr-hmn:lvr 80 μg/L BJCAAI 37,873,78
orl-rat TDLo:3400 mg/kg/23W-C:ETA JNCIAM 10,1201,50
skn-rat TDLo:310 mg/kg/79W-I:NEO JNCIAM 10,1201,50
ipr-mus LD50:1600 mg/kg JJIND8 62,911,79

CONSENSUS REPORTS: Reported in EPA TSCA Inventory. EPA Genetic Toxicology Program.

SAFETY PROFILE: Moderately toxic by intraperitoneal route. Questionable carcinogen with experimental neoplastigenic and tumorigenic data. Human mutation data reported. When heated to decomposition it emits toxic fumes of NO_x. See also NITRO COMPOUNDS OF AROMATIC HYDROCARBONS.

NGB500 CAS:42135-22-8 ***HR: D***
3-NITRO-9-FLUORENONE
mf: $C_{13}H_7NO_3$ mw: 225.21

SYNS: 3-NITRO-9H-FLUOREN-9-ONE ◇ 3-NITROFLUORENONE

TOXICITY DATA with REFERENCE
mmo-sat 1 μg/plate CALEDQ 15,209,82
mrc-esc 20 μg/well MUREAV 46,53,77

SAFETY PROFILE: Mutation data reported. When heated to decomposition it emits toxic fumes of NO_x. See also NITRO COMPOUNDS OF AROMATIC HYDROCARBONS and KETONES.

NGB700 CAS:67-28-7 ***HR: 2***
5-NITRO-2-FURALDEHYDE ACETYLHYDRAZONE
mf: $C_7H_7N_3O_4$ mw: 197.17

PROP: Yellow crystals from acetic acid + ethanol. Decomp 230-235°. Solubility in water 1:20,000.

SYNS: FURITON ◇ HC-064 ◇ NF 64 ◇ NIDRAFUR ◇ NIDRAFUR P ◇ NIFURHYDRAZONUM ◇ NIHYDRAZON ◇ NIHYDRAZONE ◇ ((5-NITRO-2-FURANYL)METHYLENE)HYDRAZIDEACETIC ACID (9CI) ◇ 5-NITROFURFURALACETYLHYDRAZONE ◇ ZONIFUR

TOXICITY DATA with REFERENCE
pic-esc 8 mg/L CJMIAZ 10,932,64
mmo-eug 10 mg/L JPROAR 17,129,70
orl-mus LD50:1500 mg/kg JAPMA8 39,313,50

SAFETY PROFILE: Moderately toxic by ingestion. Mutation data reported. When heated to decomposition it emits toxic fumes of NO_x. See also ALDEHYDES and NITRO COMPOUNDS.

NGC000 CAS:555-15-7 ***HR: 3***
5-NITRO-2-FURALDEHYDE OXIME
mf: $C_5H_4N_2O_4$ mw: 156.11

SYNS: MICOFUR ◇ NIFUROXIME ◇ 5-NITRO-2-FURALDOXIME ◇ NITROFUROXIME ◇ USAF EA-5

TOXICITY DATA with REFERENCE
mma-sat 10 μg/plate MUREAV 48,295,77
mmo-esc 2 mg/L MUREAV 53,289,78
mma-esc 10 mg/L BJCAAI 37(Suppl 3),124,78
mmo-eug 20 mg/L JPROAR 17,129,70
ipr-mus LD50:100 mg/kg NTIS** AD277-689
ivn-mus LD50:80 mg/kg CSLNX* NX#05399

SAFETY PROFILE: Poison by intravenous and intraperitoneal routes. Mutation data reported. When heated to decomposition it emits toxic fumes of NO_x. See also ALDEHYDES and NITRO COMPOUNDS.

NGC400 CAS:831-71-0 ***HR: 3***
5-NITRO-2-FURALDEHYDE THIOSEMICARBAZONE
mf: $C_6H_6N_4O_3S$ mw: 214.22

SYNS: BENZAZON VII ◇ USAF EA-14

TOXICITY DATA with REFERENCE
mmo-sat 125 μg/L CHABA8 93,108591c,80
dnr-esc 10 mg/L FOMIAZ 25,388,80
orl-mus LD50:1500 mg/kg JAPMA8 39,313,50
ipr-mus LD50:50 mg/kg NTIS** AD277-689

SAFETY PROFILE: Poison by intraperitoneal route. Moderately toxic by ingestion. Mutation data reported. When heated to decomposition it emits toxic fumes of SO_x and NO_x. See also ALDEHYDES.

NGD000 CAS:772-43-0 ***HR: D***
5-NITRO-2-FURAMIDOXIME
mf: $C_5H_4N_2O_5$ mw: 172.11

SYN: N-HYDROXY-5-NITRO-2-FURANCARBOXIMIDAMIDE

TOXICITY DATA with REFERENCE
mmo-sat 100 pmol/plate CRNGDP 2,283,81
dnr-sat 500 nmol/well CNREA8 34,2266,74

CONSENSUS REPORTS: EPA Genetic Toxicology Program.

SAFETY PROFILE: Mutation data reported. When heated to decomposition it emits toxic fumes of NO_x.

NGD400 CAS:609-39-2 ***HR: D***
2-NITROFURAN
mf: $C_4H_3NO_3$ mw: 113.08

SYNS: NITROFURAN ◇ 5-NITROFURAN

TOXICITY DATA with REFERENCE
mmo-esc 100 μg/plate BBIADT 44,485,85
dnr-esc 20 μg/disc MUREAV 81,21,81

SAFETY PROFILE: Mutation data reported. When heated to decomposition it emits toxic fumes of NO_x.

NGD600 CAS:2493-04-1 ***HR: D***
5-NITRO-2-FURANMETHANOL
mf: $C_5H_5NO_4$ mw: 143.11

SYN: 5-NITROFURFURYL ALCOHOL

TOXICITY DATA with REFERENCE
mmo-sat 1 nmol/plate MUREAV 81,21,81
pic-esc 1 μg/plate MUREAV 81,21,81

SAFETY PROFILE: Mutation data reported. When heated to decomposition it emits toxic fumes of NO_x. See also ALCOHOLS.

NGE000 CAS:67-20-9 ***HR: 3***
NITROFURANTOIN
mf: $C_8H_6N_4O_5$ mw: 238.18

SYNS: BENKFURAN ◇ BERKFURIN ◇ CHEMIOFURAN ◇ CYANTIN ◇ DANTAFUR ◇ FURADANTIN ◇ FURADONIN ◇ FURANTOIN ◇ FUROBACTINA ◇ ITURAN ◇ MACRODANTIN ◇ NCI-C55196 ◇ N-(5-NITROFURFURYLIDENE)-1-AMINOHYDANTOIN ◇ N-(5-NITRO-2-FURFURYLIDENE)-1-AMINOHYDANTOIN ◇ 1-((5-NITROFURFURYLIDENE)AMINO)HYDANTOIN ◇ NSC 2107 ◇ N-TOIN ◇ ORAFURAN ◇ PARFURAN ◇ URIZEPT ◇ USAF EA-2 ◇ WELFURIN ◇ ZOOFURIN

TOXICITY DATA with REFERENCE
mmo-sat 500 ng/plate AEMIDF 46,596,83
dnd-hmn:oth 112 g/L CBINA8 45,77,83
orl-man TDLo:140 mg/kg (14D male):REP JOURAA 77,275,57
unr-mus TDLo:750 mg/kg (female 9-11D post):TER TJADAB 14,250,76
orl-rat TDLo:135 g/kg/2Y-C:NEO CALEDQ 21,303,84
orl-hmn TDLo:80 mg/kg:PNS,CNS ARNEAS 18,680,68
orl-wmn TDLo:24 mg/kg:BLD AJCPAI 58,408,72
unr-hmn TDLo:60 mg/kg:KID BMJOAE 2,1044,76
orl-rat LD50:604 mg/kg RPOBAR 2,303,70
ipr-rat LD50:112 mg/kg RPOBAR 2,303,70
orl-mus LD50:360 mg/kg JMCMAR 16,312,73
ipr-mus LD50:100 mg/kg NTIS** AD277-689

CONSENSUS REPORTS: Reported in EPA TSCA Inventory. EPA Genetic Toxicology Program.

SAFETY PROFILE: Poison by ingestion and intraperitoneal routes. Human systemic effects: peripheral motor nerve recording changes, ataxia, changes in urine composition and hemolysis with or without anemia. Human reproductive effects by ingestion: spermatogenesis. An experimental teratogen. Other experimental reproductive effects. Questionable carcinogen with experimental neoplastigenic data. Human mutation data reported. When heated to decomposition it emits toxic fumes of NO_x.

NGE500 CAS:59-87-0 ***HR: 3***
NITROFURAZONE
mf: $C_6H_6N_4O_4$ mw: 198.16

PROP: Odorless, lemon-yellow crystals; bitter aftertaste. Darkens upon prolonged exposure to light. Decomp @ 236-240°. Very sltly sol in water; sltly sol in alc, propylene glycol; sol in alkaline solns; insol in ether.

SYNS: ALDOMYCIN ◇ ALFUCIN ◇ AMIFUR ◇ BABROCID ◇ BIOFUREA ◇ CHEMOFURAN ◇ COCAFURIN ◇ COXISTAT ◇ DERMOFURAL ◇ DYNAZONE ◇ ELDEZOL ◇ FEDACIN ◇ FLAVAZONE ◇ FRACINE ◇ FURACILLIN ◇ FURACINETTEN ◇ FURACOCCID ◇ FURACORT ◇ FURACYCLINE ◇ FURALDON ◇ FURAN-OFTENO ◇ FURAPLAST ◇ FURASEPTYL ◇ FURAZONE ◇ FURESOL ◇ FURFURIN ◇ FUVACILLIN ◇ HEMOFURAN ◇ IBIOFURAL ◇ MAMMEX ◇ MONOFURACIN ◇ NCI-C56064 ◇ NEFCO ◇ NF ◇ NIFUZON ◇ 5-NITROFURALDEHYDE SEMICARBAZIDE ◇ 6-NITROFURALDEHYDE SEMICARBAZIDE ◇ 5-NITRO-2-FURALDEHYDE SEMICARBAZONE ◇ 5-NITROFURAN-2-ALDEHYDE SEMICARBAZONE ◇ 5-NITRO-2-FURANCARBOXALDEHYDE SEMICARBAZONE ◇ 2((5-NITRO-2-FURANYL)METHYLENE)HYDRAZINECARBOXAMIDE ◇ 5-NITROFURFURAL SEMICARBAZONE ◇ (5-NITRO-2-FURFURYLIDENEAMINO)UREA ◇ NITROZONE ◇ NSC-2100 ◇ OTOFURAN ◇ SANFURAN ◇ SPRAY-DERMIS ◇ SPRAY-FORAL ◇ U-6421 ◇ USAF EA-4 ◇ VABROCID ◇ VADROCID ◇ VETERINARY NITROFURAZONE ◇ YATROCIN

TOXICITY DATA with REFERENCE
dnd-esc 1 mg/L MUREAV 107,1,83
dnd-hmn:oth 250 μmol/L CNREA8 35,781,75
orl-mus TDLo:800 mg/kg (female 6-13D post):REP TCMUD8 7,29,87
orl-mus TDLo:817 mg/kg (female 6-15D post):TER NTIS** PB86-145844
orl-rat TDLo:8652 mg/kg/2Y-C:CAR NTPTR* NTP-TR-337,88
orl-mus TDLo:10094 mg/kg/2Y-C:CAR FEPRA7 25,419,66
scu-mus TDLo:300 mg/kg/18D-I:ETA SEIJBO 15,234,75
orl-rat TD:31 g/kg/45W-C:NEO FEPRA7 25,419,66
orl-rat LD50:590 mg/kg JAMAAP 133,299,47

ipr-rat LDLo:150 mg/kg BCPCA6 13,285,64
scu-rat LD50:3000 mg/kg JAMAAP 133,299,47
orl-mus LD50:249 mg/kg TRSTAZ 74,43,80
ipr-mus LD50:96 mg/kg JMCMAR 10,530,67
scu-mus LD50:753 mg/kg JPETAB 102,185,51

CONSENSUS REPORTS: IARC Cancer Review: Group 2B IMEMDT 7,56,87; Animal Inadequate Evidence IMEMDT 7,171,74. Reported in EPA TSCA Inventory. EPA Genetic Toxicology Program.

SAFETY PROFILE: Suspected carcinogen with experimental carcinogenic, neoplastigenic, tumorigenic, and teratogenic data. Poison by ingestion and intraperitoneal routes. Moderately toxic by subcutaneous route. An experimental teratogen. Other experimental reproductive effects. A human sensitizer. Human mutation data reported. When heated to decomposition it emits toxic fumes of NO_x.

NGE775 CAS:698-63-5 ***HR: D***
5-NITROFURFURAL
mf: $C_5H_3NO_4$ mw: 141.09

SYNS: 5-NITRO-5-FURANCARBOXALDEHYDE (9CI) ◇ 5-NITRO-2-FURALDEHYDE ◇ NITROFURFURAL ◇ 5-NITROFURFURALDEHYDE

TOXICITY DATA with REFERENCE
mmo-sat 1 nmol/plate MUREAV 81,21,81
dnr-esc 50 μg/disc MUREAV 81,21,81

CONSENSUS REPORTS: EPA Genetic Toxicology Program.

SAFETY PROFILE: Mutation data reported. When heated to decomposition it emits toxic fumes of NO_x. See also ALDEHYDES.

NGG000 CAS:23256-30-6 ***HR: 3***
4-((5-NITROFURFURYLIDENE)AMINO)-3-METHYLTHIOMORPHOLINE-1,1-DIOXIDE
mf: $C_{10}H_{13}N_3O_5S$ mw: 287.32

SYNS: BAYER 2502 ◇ LAMPIT ◇ 3-METHYL-4-(5'-NITROFURYLIDENE-AMINO)-TETRAHYDRO-4H-1,4-THIAZINE-1,1-DIOXIDE ◇ NIFURTIMOX ◇ 1-((5-NITROFURFURYLIDENE)AMINO)-2-METHYLTETRAHYDRO-1,4-THIAZINE-4,4-DIOXIDE ◇ TETRAHYDRO-3-METHYL-4-((5-NITROFURFURYLIDENE)AMINO)-2H-1,4-THIAZINE-1,1-DIOXIDE

TOXICITY DATA with REFERENCE
mmo-esc 50 mg/L MUREAV 77,241,80
dnr-esc 500 mg/L FOMIAZ 25,388,80
dnd-omi 100 μmol/L BCPCA6 34,1457,84
mnt-mus-orl 600 mg/kg TOLED5 25,259,85
orl-rat TDLo:500 mg/kg (6-15D preg):TER ARZNAD 22,1603,72
orl-rat TDLo:2520 mg/kg (70D male):REP ARZNAD 22,1603,72
orl-rat TDLo:11700 mg/kg/47W-I:ETA ARZNAD 22,1607,72
orl-rat LD50:4050 mg/kg ARZNAD 22,1590,72
orl-mus LD50:2291 mg/kg TRSTAZ 74,43,80
orl-rbt LD50:2 g/kg ARZNAD 22,1590,72

SAFETY PROFILE: Moderately toxic by ingestion. Experimental reproductive effects. Questionable carcinogen with experimental tumorigenic and teratogenic data. Mutation data reported. When heated to decomposition it emits very toxic fumes of NO_x and SO_x. See also NITRO COMPOUNDS OF AROMATIC HYDROCARBONS.

NGG500 CAS:67-45-8 ***HR: 3***
3-((5-NITROFURFURYLIDENE)AMINO)-2-OXAZOLIDONE
mf: $C_8H_7N_3O_5$ mw: 225.18

SYNS: BIFURON ◇ CORIZIUM ◇ DIAFURON ◇ ENTEROTOXON ◇ FURAXONE ◇ FURAZOL ◇ FURAZOLIDON ◇ FURAZOLIDONE (USDA) ◇ FURAZON ◇ FURIDON ◇ FUROVAG ◇ FUROX ◇ FUROXAL ◇ FUROXANE ◇ FUROXONE SWINE MIX ◇ FUROZOLIDINE ◇ GIARDIL ◇ GIARLAM ◇ MEDARON ◇ NEFTIN ◇ NG-180 ◇ NICOLEN ◇ NIFULIDONE ◇ NIFURAN ◇ 3-(((5-NITRO-2-FURANYL)METHYLENE)AMINO)-2-OXAZOLIDINONE ◇ NITROFURAZOLIDONE ◇ NITROFURAZOLIDONUM ◇ 3-(5'-NITROFURFURALAMINO)-2-OXAZOLIDONE ◇ N-(5-NITRO-2-FURFURYLIDENE)-3-AMINOOXAZOLIDINE-2-ONE ◇ N-(5-NITRO-2-FURFURYLIDENE)-3-AMINO-2-OXAZOLIDONE ◇ NITROFUROXON ◇ 3-((5-NITROFURYLIDENE)AMINO)-2-OXAZOLIDONE ◇ 5-NITRO-N-(2-OXO-3-OXAZOLIDINYL)-2-FURANMETHANIMINE ◇ PURADIN ◇ ROPTAZOL ◇ SCLAVENTEROL ◇ TIKOFURAN ◇ TOPAZONE ◇ TRICHOFURON ◇ TRICOFURON ◇ USAF EA-1 ◇ VIOFURAGYN

TOXICITY DATA with REFERENCE
pic-esc 2 mg/L MUREAV 156,69,85
cyt-hmn:lym 2 mg/L/72H MUREAV 59,139,79
orl-mus TDLo:1 g/kg (female 1D post):REP JOENAK 15,355,57
orl-man TDLo:11 mg/kg:PUL,BLD ARDSBL 105,823,72
orl-rat LD50:2336 mg/kg TXAPA9 18,185,71
orl-mus LD50:1782 mg/kg TRSTAZ 74,43,80
ipr-mus LD50:300 mg/kg PMDCAY 5,320,67
orl-ckn LDLo:150 mg/kg CLDND*

CONSENSUS REPORTS: IARC Cancer Review: Group 3 IMEMDT 7,56,87; Animal Inadequate Evidence IMEMDT 31,141,83. Reported in EPA TSCA Inventory. EPA Genetic Toxicology Program.

SAFETY PROFILE: Poison by ingestion and intraperitoneal routes. Human systemic effects by ingestion: dyspnea, respiratory depression and rosinophillis. Experimental reproductive effects. Human mutation data reported. Questionable carcinogen. When heated to decomposition it emits toxic fumes of NO_x.

NGH500 CAS:645-12-5 *HR: D*
5-NITROFUROIC ACID
mf: $C_5H_3NO_5$ mw: 157.09

PROP: Light yellow in water. Mp: 185°, bp: subl. Very sltly sol in water; sol in alc and ether; insol in chloroform.

SYNS: 5-NITRO-2-FURANCARBOXYLIC ACID (9CI) ◇ 5-NITROFURAN-CARBOXYLIC ACID ◇ NITROFURATE ◇ 5-NITROPYROMUCATE

TOXICITY DATA with REFERENCE
mmo-sat 1 mg/L MUREAV 44,139,77
mmo-esc 350 mg/L MUREAV 130,97,84

CONSENSUS REPORTS: EPA Genetic Toxicology Program.

SAFETY PROFILE: Mutation data reported. When heated to decomposition it emits toxic fumes of NO_x.

NGI000 CAS:6281-23-8 *HR: D*
3-(5-NITRO-2-FURYL)ACRYLIC ACID
mf: $C_7H_5NO_5$ mw: 183.13

SYNS: 5-NFA ◇ 5-NITRO-2-FURANACRYLIC ACID ◇ 3-(5-NITRO-2-FURANYL)-2-PROPENOIC ACID ◇ NITROFURYLACRYLAMIDE ◇ 5-NITRO-2-FURYL ACRYLIC ACID

TOXICITY DATA with REFERENCE
dnr-esc 50 μg/disc MUREAV 81,21,81
bfa-hmn/sat 14 mg/kg MUREAV 77,13,80

CONSENSUS REPORTS: EPA Genetic Toxicology Program.

SAFETY PROFILE: Human mutation data reported. When heated to decomposition it emits toxic fumes of NO_x.

NGI500 CAS:712-68-5 *HR: 3*
2-(5-NITRO-2-FURYL)-5-AMINO-1,3,4-THIADIAZOLE
mf: $C_6H_4N_4O_3S$ mw: 212.20

SYNS: 2-AMINO-5-(5-NITRO-2-FURYL)-1,3,4-THIADIAZOLE◇ 5-AMINO-2-(5-NITRO-2-FURYL)-1,3,4-THIADIAZOLE◇ FURIDIAZINE ◇ 5-(5-NITRO-2-FURANYL)-1,3,4-THIADIAZOL-2-AMINE ◇ 5-(5-NITRO-2-FURYL)-2-AMINO-1,3,4-THIADIAZOLE

TOXICITY DATA with REFERENCE
orl-rat TDLo:6 g/kg/46W-I:CAR JNCIAM 54,841,75
orl-rat TD:2240 mg/kg/32W-C:NEO,REP FEPRA7 29,817,70

CONSENSUS REPORTS: IARC Cancer Review: Group 2B IMEMDT 7,56,87; Animal Limited Evidence IMEMDT 7,143,74.

SAFETY PROFILE: Suspected carcinogen with experimental carcinogenic and neoplastigenic data. Experimental reproductive effects. When heated to decomposition it emits very toxic fumes of NO_x and SO_x. See also NITRO COMPOUNDS OF AROMATIC HYDROCARBONS.

NGI800 CAS:75198-31-1 *HR: 3*
3-(5-NITRO-2-FURYL)-IMIDAZO(1,2-a)PYRIDINE
mf: $C_{11}H_7N_3O_3$ mw: 229.21

SYN: NFIP

TOXICITY DATA with REFERENCE
orl-rat TDLo:11200 mg/kg/20W-C:CAR TUMOAB 66,131,80
orl-mus TDLo:33600 mg/kg/40W-C:CAR TUMOAB 66,131,80

SAFETY PROFILE: Suspected carcinogen with experimental carcinogenic data. When heated to decomposition it emits toxic fumes of NO_x.

NGK000 CAS:2122-86-3 *HR: 3*
5-(5-NITRO-2-FURYL)-1,3,4-OXADIAZOLE-2-OL
mf: $C_6H_3N_3O_5$ mw: 197.12

TOXICITY DATA with REFERENCE
orl-rat TDLo:12 g/kg/46W-C:NEO JNCIAM 54,841,75

SAFETY PROFILE: Questionable carcinogen with experimental neoplastigenic data. When heated to decomposition it emits toxic fumes of NO_x.

NGK500 CAS:36133-88-7 *HR: 3*
N-((3-(5-NITRO-2-FURYL)-1,2,4-OXADIAZOLE-5-YL)METHYL)ACETAMIDE
mf: $C_9H_8N_4O_5$ mw: 252.21

TOXICITY DATA with REFERENCE
mma-sat 1 μg/plate MUREAV 40,9,76
dnr-sat 500 nmol/well CNREA8 34,2266,74
mmo-esc 300 nmol/well CNREA8 34,2266,74
mrc-esc 500 nmol/well CNREA8 34,2266,74
orl-rat TDLo:11 g/kg/46W-C:CAR JNCIAM 54,841,75

CONSENSUS REPORTS: EPA Genetic Toxicology Program.

SAFETY PROFILE: Questionable carcinogen with experimental carcinogenic data. Mutation data reported. When heated to decomposition it emits toxic fumes of NO_x.

NGL000 CAS:53757-31-6 *HR: 3*
3-(5-NITRO-2-FURYL)-2-PHENYLACRYLAMIDE
mf: $C_{13}H_{10}N_2O_4$ mw: 258.25

SYN: 3-(5-NITRO-2-FURYL)-2-PHENYL-2-PROPENAMIDE

TOXICITY DATA with REFERENCE
mma-sat 1 μg/plate MUREAV 40,9,76
dnr-sat 500 nmol/well CNREA8 34,2266,74
mmo-esc 300 nmol/well CNREA8 34,2266,74

mrc-esc 500 nmol/well CNREA8 34,2266,74
orl-rat TDLo:700 mg/kg (male 7D pre):REP APHGBP 14,261,64
orl-rat TDLo:474 g/kg/46W-C:ETA FEPRA7 38,1403,79

SAFETY PROFILE: Questionable carcinogen with experimental tumorigenic data. Mutation data reported. When heated to decomposition it emits toxic fumes of NO_x.

NGL500 CAS:710-25-8 *HR: 3*
3-(5-NITRO-2-FURYL)-2-PROPENAMIDE
mf: $C_7H_6N_2O_4$ mw: 182.15

SYNS: 5-NITRO-2-FURANACRYLAMIDE ◇ 3-(5-NITRO-2-FURYL)ACRYLAMIDE ◇ 5-NITRO-2-FURYLACRYLAMIDE

TOXICITY DATA with REFERENCE
mmo-sat 10 μg/plate JOPHDQ 1,15,78
dnr-sat 500 mmol/well CNREA8 34,2266,74
orl-rat TDLo:475 g/kg/46 W-C:ETA FEPRA7 38,1403,79

CONSENSUS REPORTS: EPA Genetic Toxicology Program.

SAFETY PROFILE: Questionable carcinogen with experimental tumorigenic data. Mutation data reported. When heated to decomposition it emits toxic fumes of NO_x.

NGM400 CAS:53757-28-1 *HR: 2*
4-(5-NITRO-2-FURYL)THIAZOLE
mf: $C_7H_4N_2O_3S$ mw: 196.19

TOXICITY DATA with REFERENCE
mmo-sat 10 ng/plate CNREA8 41,2648,81
dnr-sat 500 nmol/well CNREA8 34,2266,74
orl-rat TDLo:37996 mg/kg/46W-C:ETA PAACA3 21,75,80

SAFETY PROFILE: Questionable carcinogen with experimental tumorigenic data. Mutation data reported. When heated to decomposition it emits toxic fumes of NO_x and SO_x.

NGM500 CAS:24554-26-5 *HR: 3*
N-(4-(5-NITRO-2-FURYL)-2-THIAZOLYL)FORMAMIDE
mf: $C_8H_5N_3O_4S$ mw: 239.22

SYNS: FANFT ◇ 2-FORMYLAMINO-4-(5-NITRO-2-FURYL)THIAZOLE ◇ N-(4-(5-NITRO-2-FURYL)-2-THIAZOLYL)FORMAMID (GERMAN)

TOXICITY DATA with REFERENCE
mmo-esc 3 ng/plate ENMUDM 6(Suppl 2),1,84
mma-esc 30 ng/plate ENMUDM 6(Suppl 2),1,84
bfa-rat/sat 658 mg/kg/7D-C MUREAV 135,169,84
otr-ham:emb 1 mg/L JJIND8 67,1303,81
dns-dog:oth 10 μmol/L CNREA8 42,3974,82
orl-rat TDLo:14500 mg/kg/20W-C:CAR CALEDQ 34,249,87
orl-mus TDLo:1008 mg/kg/12W-C:CAR CNREA8 41,1397,81
orl-dog TDLo:26 g/kg/2Y-C:ETA JNCIAM 45,535,70
orl-rat TD:6000 mg/kg/6W-C:CAR,REP CNREA8 39,1207,79

CONSENSUS REPORTS: EPA Genetic Toxicology Program.

SAFETY PROFILE: Suspected carcinogen with experimental carcinogenic, tumorigenic data. Mutation data reported. Experimental reproductive effects. When heated to decomposition it emits very toxic fumes of SO_x and NO_x.

NGN000 CAS:18523-69-8 *HR: 3*
(4-(5-NITRO-2-FURYL)THIAZOL-2-YL)HYDRAZONOACETONE
mf: $C_{10}H_{10}N_4O_3S$ mw: 266.30

SYN: 2-(2-ISOPROPYLIDENEHYDRAZONO)-4-(5-NITRO-2-FURYL)-THIAZOLE

TOXICITY DATA with REFERENCE
orl-rat TDLo:16 g/kg/45W-C:NEO FEPRA7 25,419,66

SAFETY PROFILE: Questionable carcinogen with experimental neoplastigenic data. When heated to decomposition it emits very toxic fumes of NO_x and SO_x. See also NITRO COMPOUNDS OF AROMATIC HYDROCARBONS.

NGN500 CAS:42011-48-3 *HR: 3*
N-(4-(5-NITRO-2-FURYL)-2-THIAZOLYL)-2,2,2-TRIFLUOROACETAMIDE
mf: $C_9H_4F_3N_3O_4S$ mw: 307.22

SYNS: 2-(2,2,2-TRIFLUOROACETAMIDO)-4-(5-NITRO-2-FURYL)THIAZOLE ◇ 2,2,2-TRIFLUORO-N-(4-(5-NITRO-2-FURYL)-2-THIAZOLYL)ACETAMIDE

TOXICITY DATA with REFERENCE
pic-esc 500 μg/L MUREAV 26,3,74
mma-sat 100 ng/plate MUREAV 48,295,77
dnr-sat 500 nmol/well CNREA8 34,2266,74
mmo-esc 300 nmol/well CNREA8 34,2266,74
mrc-esc 500 nmol/well CNREA8 34,2266,74
orl-rat TDLo:18 g/kg/46W-C:CAR,REP JNCIAM 54,841,75
orl-mus TDLo:108 g/kg/46W-C:CAR CNREA8 33,1593,73

SAFETY PROFILE: Suspected carcinogen with experimental carcinogenic data. Experimental reproductive effects. Mutation data reported. When heated to decomposition it emits very toxic fumes of F^-, NO_x, and SO_x.

NGP500 CAS:7727-37-9 ***HR: 1***
NITROGEN
DOT: UN 1066/UN 1977
mf: N_2 mw: 28.02

PROP: Colorless gas, colorless liquid or cubic crystals at low temp. Mp: −210.0°, d: 1.2506 g/L @ 0°, d (liquid): 0.808 @ −195.8°. Condenses to a liquid. Sltly sol in water; sol in liquid ammonia, alc.

SYNS: NITROGEN, compressed (DOT) ◇ NITROGEN, refrigerated liquid (DOT) ◇ NITROGEN GAS

CONSENSUS REPORTS: Reported in EPA TSCA Inventory.

DOT Classification: Nonflammable Gas; Label: Nonflammable Gas.

SAFETY PROFILE: Low toxicity. In high concentrations it is a simple asphyxiant. The release of nitrogen from solution in the blood, with formation of small bubbles, is the cause of most of the symptoms and changes found in compressed air illness (caisson disease). It is a narcotic at high concentration and high pressure. Both the narcotic effects and the bends are hazards of compressed air atmospheres such as found in underwater diving. Nonflammable Gas. Can react violently with lithium, neodymium, titanium under the proper conditions. See also ARGON.

NGP510 CAS:7727-37-9 ***HR: 3***
NITROGEN (cryogenic liquid)

CONSENSUS REPORTS: Reported in EPA TSCA Inventory.

DOT Classification: Nonflammable Gas; Label: Nonflammable Gas.

SAFETY PROFILE: Nitrogen (liquid) can explode during use. See also NITROGEN.

NGQ500 CAS:10025-85-1 ***HR: 3***
NITROGEN CHLORIDE
mf: Cl_3N mw: 120.37

PROP: Volatile, yellowish oil or rhombic crystals; pungent odor. Mp: < −40°, explodes @ 93°, bp: < 71°, d: 1.653, vap press: 150 mm @ 20°.

SYNS: CHLORINE NITRIDE (NITROGEN) TRICHLORIDE ◇ TRICHLORAMINE ◇ TRICHLORINE NITRIDE

DOT Classification: Forbidden.

SAFETY PROFILE: An irritant to the eyes, skin, mucous membranes, and a systemic central nervous system irritant. An explosive sensitive to impact, light, and ultrasound. The solid explodes on melting. The liquid explodes above 60°C. Concentrated solutions are also explosive. Explosive decomposition is initiated by contact with: concentrated ammonia; arsenic; dinitrogen tetraoxide; hydrogen sulfide; hydrogen trisulfide; nitrogen oxide; organic matter; ozone; phosphine; phosphorus; potassium cyanide; potassium hydroxide solutions; selenium; hydrogen chloride; hydrogen fluoride; hydrogen bromide; hydrogen iodide. Mixtures with chlorine + hydrogen are potentially explosive. Upon decomposition it emits toxic fumes of Cl^- and NO_x. See also CHLORIDES.

NGR000 ***HR: 3***
NITROGEN CHLORIDE DIFLUORIDE
mf: ClF_2N mw: 87.46

SAFETY PROFILE: Very unstable. Use caution in handling.

NGR500 CAS:10102-44-0 ***HR: 3***
NITROGEN DIOXIDE
DOT: NA 1067
mf: NO_2 mw: 46.01

PROP: Colorless solid to yellow liquid; irritating odor. Mp: −9.3° (yellow liquid), bp: 21° (red-brown gas with decomp), d: 1.491 @ 0°, vap press: 400 mm @ 80°. Liquid below 21.15°. Sol in concentrated sulfuric acid, nitric acid. Corrosive to steel when wet.

SYNS: AZOTE (FRENCH) ◇ AZOTO (ITALIAN) ◇ NITRITO ◇ NITROGEN PEROXIDE, liquid (DOT) ◇ RCRA WASTE NUMBER P078 ◇ STICKSTOFFDIOXID (GERMAN) ◇ STIKSTOFDIOXYDE (DUTCH)

TOXICITY DATA with REFERENCE
mmo-sat 6 ppm MUREAV 136,119,84
sce-ham:lng 5 ppm/10M-C MUREAV 89,303,81
ihl-mus TDLo:22 ppm (female 7-18D post):REP NRTXDN 9,545,88
ihl-rat TCLo:85 $\mu g/m^3$/24H (female 1-22D post):TER GISAAA 42(12),22,77
ihl-hmn LCLo:200 ppm/1M AOHYA3 17,159,74
ihl-man TCLo:6200 ppb/10M:PUL KEKHA7 17,337,68
ihl-man TCLo:90 ppm/40M:PUL JOCMA7 8,301,66
ihl-rat LC50:88 ppm/4H AMIHBC 10,418,54
ihl-mus LC50:1000 ppm/10M JCTODH 4,246,77
ihl-dog LCLo:123 mg/m^3 TXAPA9 9,160,66
ihl-mky LCLo:123 mg/m^3/8H TXAPA9 9,160,66
ihl-rbt LC50:315 ppm/15M AIHAAP 23,457,62
ihl-gpg LC50:30 ppm/1H AEHLAU 10,220,65

CONSENSUS REPORTS: Reported in EPA TSCA Inventory. EPA Genetic Toxicology Program.

OSHA PEL: (Transitional: CL 5 ppm) STEL 1 ppm
ACGIH TLV: TWA 3 ppm; STEL 5 ppm
DFG MAK: 5 ppm (9 mg/m^3)
NIOSH REL: CL (Oxides of Nitrogen) 1 ppm/15M
DOT Classification: Poison A; Label: Poison Gas and Oxidizer.

SAFETY PROFILE: Experimental poison by inhalation. Moderately toxic to humans by inhalation. An experimental teratogen. Other experimental reproductive effects. Human systemic effects by inhalation: pulmonary vascular resistance changes, cough, dyspnea, and other pulmonary changes. Mutation data reported. Violent reaction with cyclohexane; F_2; formaldehyde and alcohols; nitrobenzene; petroleum; toluene. When heated to decomposition it emits toxic fumes of NO_x. See also NITRIC OXIDE.

NGS000 CAS:10102-44-0 ***HR: 3***
NITROGEN DIOXIDE (liquid)

CONSENSUS REPORTS: Reported in EPA TSCA Inventory. EPA Extremely Hazardous Substances List.

DOT Classification: Poison A; Label: Oxidizer and Poison Gas.

SAFETY PROFILE: A poison. See also NITROGEN DIOXIDE, NITROGEN MONOXIDE, and NITRIC OXIDE.

NGS500 CAS:13847-65-9 ***HR: 3***
NITROGEN FLUORIDE OXIDE
mf: F_3NO mw: 87.01

SYNS: AMOX ◇ TRIFLUOROAMINE OXIDE

TOXICITY DATA with REFERENCE
ihl-rat LC50:24 ppm/4H TXAPA9 13,76,68
ipr-rat LD50:38 mg/kg TXAPA9 13,76,68
ihl-mus LC50:18 ppm/4H TXAPA9 13,76,68
ipr-mus LD50:30 mg/kg TXAPA9 13,76,68

OSHA PEL: TWA 2.5 mg(F)/m^3
ACGIH TLV: TWA 2.5 mg(F)/m^3
NIOSH REL: (Inorganic Fluorides) TWA 2.5 mg(F)/m^3

SAFETY PROFILE: Poison by inhalation and intraperitoneal routes. A skin, eye, and mucous membrane irritant. When heated to decomposition it emits very toxic fumes of F^- and NO_x. See also FLUORIDES.

NGT500 CAS:63907-41-5 ***HR: 3***
NITROGEN MONOXIDE, mixed with NITROGEN TETROXIDE
DOT: UN 1975

PROP: Containing up to 33.2% by weight nitric oxide (FEREAC 41,15972,76).

SYNS: AZOTU TLENKI (POLISH) ◇ NITRIC OXIDE and NITROGEN TETROXIDE MIXTURES (DOT) ◇ NITROGEN TETROXIDE-NITRIC OXIDE MIXTURE (DOT)

TOXICITY DATA with REFERENCE
ihl-rat LC50:115 ppm/1H AIHAAP 23,457,62

DOT Classification: Poison A; Label: Oxidizer and Poison Gas.

SAFETY PROFILE: A poison. Moderately toxic by inhalation. When heated to decomposition it emits toxic fumes of NO_x. See also components as listed.

NGU000 CAS:10024-97-2 ***HR: 2***
NITROGEN OXIDE
DOT: UN 1070/UN 2201
mf: N_2O mw: 44.02

PROP: Colorless gas, liquid or cubic crystals; slt sweet odor. Mp: −90.8°, bp: −88.49°, d: 1.977 g/L (liquid 1.226 @ −89°).

SYNS: DINITROGEN MONOXIDE ◇ FACTITIOUS AIR ◇ HYPONITROUS ACID ANHYDRIDE ◇ LAUGHING GAS ◇ NITROUS OXIDE (DOT) ◇ NITROUS OXIDE, compressed (DOT) ◇ NITROUS OXIDE, refrigerated liquid (DOT)

TOXICITY DATA with REFERENCE
sln-dmg-ihl 99 pph/6M-C ENVRAL 7,286,74
dni-rat-ihl 75000 ppm/24H AACRAT 62,738,83
ihl-mus TCLo:5000 ppm/4H (female 14D post):REP NETOD7 8,189,86
ihl-mus TCLo:75 pph/6H (female 14D post):TER TJADAB 32,26A,85
ihl-hmn TDLo:24 mg/kg/2H:CNS,CVS,MET BJANAD 35,631,63
ihl-rat LC50:1068 mg/m^3/4H TPKVAL 15,53,79

CONSENSUS REPORTS: Reported in EPA TSCA Inventory. EPA Genetic Toxicology Program.
ACGIH TLV: 50 ppm
NIOSH REL: (Waste Anesthetic Gases and Vapors) TWA 25 ppm
DOT Classification: Nonflammable Gas; Label: Nonflammable Gas, Oxidizer, compressed.

SAFETY PROFILE: Moderately toxic by inhalation. Human systemic effects by inhalation: general anesthetic, decreased pulse rate without blood pressure fall and body temperature decrease. An experimental teratogen. Experimental reproductive effects. Mutation data reported. An asphyxiant. Does not burn but is flammable by chemical reaction and supports combustion. Moderate explosion hazard; it can form an explosive mixture with air. Violent reaction with Al; B; hydrazine; LiH; LiC_6H_5; PH_3; Na; WC. Also self-explodes at high temperatures.

NGU500 CAS:10544-72-6 ***HR: 3***
NITROGEN TETROXIDE
mf: N_2O_4 mw: 92.02
DOT: NA 1067

PROP: Nitrogen tetroxide is a dimer of nitrogen dioxide (AIHAAP 23,457,62).

SYNS: DINITROGEN DIOXIDE ◇ DINITROGEN TETROXIDE (DOT) ◇ NITROGEN TETROXIDE, liquid (DOT)

TOXICITY DATA with REFERENCE
ihl-rbt LC50:315 ppm/15M AIHAAP 23,457,62

CONSENSUS REPORTS: Reported in EPA TSCA Inventory.

DOT Classification: Poison A; Label: Poison Gas and Oxidizer.

SAFETY PROFILE: A poison. Moderately toxic by inhalation. When heated to decomposition it emits toxic fumes of NO_x. See also NITROGEN MONOXIDE.

NGV000 CAS:10544-72-6 ***HR: 3***
NITROGEN TETROXIDE (liquid)
mf: N_2O_4 mw: 92.02

SYNS: NITROGEN DI-DIOXIDE ◇ NITROGEN PEROXIDE, liquid (DOT)

CONSENSUS REPORTS: Reported in EPA TSCA Inventory.

DOT Classification: Poison A; Label: Oxidizer and Poison Gas.

SAFETY PROFILE: A poison. See also NITROGEN MONOXIDE and NITROGEN TETROXIDE. When heated to decomposition it emits toxic fumes of NO_x.

NGV500 ***HR: 3***
NITROGEN TRIBROMIDE HEXAAMMONIATE
mf: $Br_3N•6H_3N$ mw: 355.90

SYN: TRIBROMAMINE HEXAAMMONIATE

SAFETY PROFILE: Explodes at $-67°C$ or on contact with phosphorous or arsenic. When heated to decomposition it emits toxic fumes of Br^-, NH_3, and NO_x. Incompatible with phosphorus, arsenic. See also BROMIDES.

NGW000 CAS:7783-54-2 ***HR: 3***
NITROGEN TRIFLUORIDE
DOT: UN 2451
mf: F_3N mw: 71.01

PROP: Colorless gas; odor of mold. Mp: $-208.5°$, bp: $-129°$, d (liquid): 1.537 @ $-129°$; d: (liquid @ bp:) 1.885. Insol in H_2O.

SYN: NITROGEN FLUORIDE

TOXICITY DATA with REFERENCE
ihl-rat LC50:6700 ppm/1H TXAPA9 26,1,73
ihl-mus LC50:2000 ppm/4H 34ZIAG -,427,69
ihl-dog LC50:9600 ppm/1H AMRL** TR-69-84,69
ihl-mky LC50:7500 ppm/1H AMRL** TR-69-84,69

CONSENSUS REPORTS: Reported in EPA TSCA Inventory.

OSHA PEL: TWA 10 ppm
ACGIH TLV: TWA 10 ppm
NIOSH REL: (Inorganic Fluorides) TWA 2.5 mg(F)/m^3
DOT Classification: Poison A; Label: Poison Gas; Nonflammable Gas; Label: Nonflammable Gas.

SAFETY PROFILE: A poison. Mildly toxic by inhalation. Prolonged absorption may cause mottling of teeth, skeletal changes. Severe explosion hazard by chemical reaction with reducing agents, particularly when under pressure. A very dangerous fire hazard; a very powerful oxidizer; otherwise inert at normal temperatures and pressures. Reacts violently when ignited with H_2. When pure (dry) it does not attack glass or mercury at normal temperatures. Can react violently with NH_3; CO; diborane; H_2; H_2S; CH_4; tetrafluorohydrazine. Can react vigorously with reducing materials. Particularly hazardous under pressure. Incompatible with charcoal, hydrogen-containing compounds, tetrafluorohydrazine. When heated to decomposition it emits highly toxic fumes of F^-. See also FLUORIDES.

NGW500 CAS:13444-85-4 ***HR: 3***
NITROGEN TRIIODIDE
mf: NI_3 mw: 394.7

PROP: Black crystals. Mp: explodes, bp: subl in vacuum.

SYN: NITROGEN IODIDE

DOT Classification: Forbidden.

SAFETY PROFILE: A severe explosion hazard when shocked, exposed to heat or flame, or by spontaneous chemical reaction. It has no known uses as an explosive because it is far too sensitive in the dry state to store or handle safely. If this material must be worked with, it should be kept wet. A convenient way of keeping it wet is with ether; when it is needed in the dry state, it simply has to be taken out into the open and the ether will evaporate, leaving it perfectly dry. When dry, it will explode when given the slightest touch, vibration, or rise in temperature. Even a puff of air directed into it can cause it to detonate. It is a high explosive and is very violent. Incompatible with O_3; H_2S; Cl_2; Br_2; acids. See also IODIDES.

NGX000 ***HR: 3***
NITROGEN TRIIODIDE-AMMONIA
mf: $I_3N•H_3N$ mw: 411.74

PROP: Mp: decomp below 20°, bp: explodes, d: 3.5.

SAFETY PROFILE: Severe explosion hazard. This material is extremely unstable when dry. The slightest shock or heat will cause it to decompose explosively. Readily formed from ammonia and iodine. It should be kept moist. Can self-explode. Explodes with friction (even when moist) and heat. When heated to decomposition it emits highly toxic fumes of I^-, NO_x, and NH_3. See also IODINE and AMMONIA.

NGX500 **HR: 3**
NITROGEN TRIIODIDE-SILVER AMIDE
mf: $I_3N•AgH_2N$ mw: 518.60

SAFETY PROFILE: Dry complex may explode. When heated to decomposition it emits highly toxic fumes of NO_x and I^-. See also SILVER COMPOUNDS, AMIDES, and IODIDES.

NGY000 CAS:55-63-0 **HR: 3**
NITROGLYCERIN
DOT: UN 0143/UN 0144/UN 1204
mf: $C_3H_5N_3O_9$ mw: 227.11

PROP: Colorless to yellow liquid; sweet taste. Mp: 13°, bp: explodes @ 218°, d: 1.599 @ 15°/15°, vap press: 1 mm @ 127°, vap d: 7.84, autoign temp: 518°F, decomp @ 50-60°, volatile @ 100°. Misc with ether, acetone, glacial acetic acid, ethyl acetate, benzene, nitrobenzene, pyridine, chloroform, ethylene bromide, dichloroethylene; sltly sol in petr ether, glycerol.

SYNS: ANGININE ◇ BLASTING GELATIN (DOT) ◇ BLASTING OIL ◇ GLONOIN ◇ GLYCERINTRINITRATE (CZECH) ◇ GLYCEROL, NITRIC ACID TRIESTER ◇ GLYCEROL TRINITRATE ◇ GLYCEROL (TRINITRATE de) (FRENCH) ◇ GLYCEROLTRINTRAAT (DUTCH) ◇ GLYCERYL NITRATE ◇ GLYCERYL TRINITRATE ◇ GLYCERYL TRINITRATE, solution up to 1% in alcohol (DOT) ◇ GTN ◇ KLAVI KORDAL ◇ MYOCON ◇ NG ◇ NIGLYCON ◇ NIONG ◇ NITRIC ACID TRIESTER OF GLYCEROL ◇ NITRINE-TDC ◇ NITROGLICERINA (ITALIAN) ◇ NITROGLICERYNA (POLISH) ◇ NITROGLYCERIN, liquid, desensitized (DOT) ◇ NITROGLYCERIN, liquid, not desensitized (DOT) ◇ NITROGLYCERINE ◇ NITROGLYCEROL ◇ NITROGLYN ◇ NITROL ◇ NITROLINGUAL ◇ NITROLOWE ◇ NITRONET ◇ NITRONG ◇ NITRO-SPAN ◇ NITROSTAT ◇ NK-843 ◇ NTG ◇ PERGLOTTAL ◇ 1,2,3-PROPANETRIOL, TRINITRATE ◇ 1,2,3-PROPANETRIYL NITRATE ◇ RCRA WASTE NUMBER P081 ◇ SK-106N ◇ SOUP ◇ TNG ◇ TRINITRIN ◇ TRINITROGLYCERIN ◇ TRINITROGLYCEROL

TOXICITY DATA with REFERENCE
skn-rbt 500 mg/24H MLD NTIS** AD-B011-150
mma-sat 50 μg/well CBINA8 19,77,77
ipr-rat TDLo:11 mg/kg (female 7-17D post):TER YACHDS 13,3807,85
ipr-rat TDLo:11 mg/kg (female 7-17D post):REP YACHDS 13,3807,85
orl-rat TDLo:36500 mg/kg/2Y-C:ETA ATSUDG 4,88,80
ivn-man TDLo:51429 μg/kg/2D-I:CNS,EYE AIMEAS 101,500,84
orl-rat LD50:105 mg/kg YACHDS 13,3649,85
ipr-rat LD50:102 mg/kg IYKEDH 13,90,82
scu-rat LD50:94 mg/kg YACHDS 13,3649,85
ivn-rat LD50:23200 μg/kg JJATDK 3,161,83
orl-mus LD50:115 mg/kg YACHDS 13,3649,85
ipr-mus LD50:104 mg/kg IYKEDH 13,90,82
scu-mus LD50:110 mg/kg YACHDS 13,3649,85
ivn-mus LD50:10600 μg/kg IYKEDH 13,90,82
ivn-dog LD50:19 mg/kg OYYAA2 22,629,81
scu-cat LDLo:150 mg/kg AEPPAE 200,271,42
scu-rbt LDLo:400 mg/kg AEPPAE 200,271,42
ivn-rbt LD50:45 mg/kg NIIRDN 6,547,82

CONSENSUS REPORTS: Reported in EPA TSCA Inventory. Community Right-To-Know List.

OSHA PEL: (Transitional: TWA CL 0.2 ppm (skin)) STEL: 0.1 mg/m^3 (skin)
ACGIH TLV: TWA 0.05 ppm (skin)
DFG MAK: 0.05 ppm (0.5 mg/m^3) (skin)
NIOSH REL: CL (Nitroglycerin or EGDN) 0.1 mg/m^3/20M
DOT Classification: Class A Explosive; Label: EXPLOS. A, desensitized (UN0144); Forbidden, not desensitized; Flammable Liquid; Label: Flammable Liquid (UN1204, NA1204); Class A Explosive; Label: Explosive A, Poison (UN0143).

SAFETY PROFILE: Human poison by an unspecified route. Poison experimentally by ingestion, intraperitoneal, subcutaneous, and intravenous routes. An experimental teratogen. Other experimental reproductive effects. A skin irritant. Questionable carcinogen with experimental tumorigenic data. Mutation data reported. It can cause respiratory difficulties and death due to respiratory paralysis by ingestion. The acute symptoms of nitroglycerin poisoning are headaches, nausea, vomiting, abdominal cramps, convulsions, methemoglobinemia, circulatory collapse and reduced blood pressure, excitement, vertigo, fainting, respiratory rales and cyanosis. Toxic effects may occur by ingestion, inhalation of dust, or absorption through intact skin. Human systemic effects by intravenous route: encephalitis, miosis, corneal damage. Used as a vasodilator and as an explosive.

A very dangerous fire hazard when exposed to heat, flame, or by spontaneous chemical reaction. A severe explosion hazard when shocked or exposed to O_3, heat, or flame. Nitroglycerin is a powerful explosive, very sensitive to mechanical shock, heat, or UV radiation. Small quantities of it can readily be detonated by a hammer blow on a hard surface, particularly when it has been absorbed in filter paper. It explodes when heated to 215°C. Frozen nitroglycerin is somewhat less sensitive than the liquid. However, a half-thawed or partially thawed mixture is more sensitive than either one. When heated to decomposition it emits toxic fumes of NO_x. See also EXPLOSIVES, HIGH; and DYNAMITE.

NGY500 CAS:53569-64-5 **HR: 3**
NITROGLYCERIN mixed with ETHYLENE GLYCOL DINITRATE (1:1)
mf: $C_3H_5N_3O_9•C_2H_4N_2O_6$ mw: 379.19

SYN: ETHYLENE GLYCOL DINITRATE mixed with NITROGLYCERIN (1:1)

TOXICITY DATA with REFERENCE
scu-cat LDLo:75 mg/kg AEPPAE 200,271,42
scu-rbt LDLo:300 mg/kg AEPPAE 200,271,42

OSHA PEL: (Transitional: TWA CL 0.2 ppm (skin)) STEL: 0.1 mg/m^3 (skin)
ACGIH TLV: TWA 0.05 ppm (skin)
NIOSH REL: CL (Nitroglycerin or EGDN) 0.1 mg/m^3/20M

SAFETY PROFILE: A human poison. Poison by subcutaneous route. A high explosive. When heated to decomposition it emits very toxic fumes of NO_x. See also EXPLOSIVES, HIGH; and individual components.

NGY700 CAS:10339-31-8 ***HR: 3***
N-NITROGLYCINE
mf: $C_2H_4N_2O_4$ mw: 120.08

SYNS: NITRAMINOACETIC ACID ◇ N-NITROGLICIN

TOXICITY DATA with REFERENCE
orl-mus LD50:40 mg/kg 85GDA2 4(1),56,80
ipr-mus LD50:43 mg/kg 85GDA2 4(1),56,80
ivn-mus LD50:32 mg/kg 85GDA2 4(1),56,80

SAFETY PROFILE: Poison by ingestion, intravenous, and intraperitoneal routes. When heated to decomposition it emits toxic fumes of NO_x. See also NITRO COMPOUNDS.

NHA000 CAS:3251-56-7 ***HR: 2***
4-NITROGUAIACOL
mf: $C_7H_7NO_4$ mw: 169.15

PROP: Yellow needles from water. Mp: 104-105°.

TOXICITY DATA with REFERENCE
ipr-mus LDLo:500 mg/kg CBCCT* 6,221,54

CONSENSUS REPORTS: Reported in EPA TSCA Inventory.

SAFETY PROFILE: Moderately toxic by intraperitoneal route. When heated to decomposition it emits toxic fumes of NO_x.

NHA500 CAS:556-88-7 ***HR: 3***
α-NITROGUANIDINE
DOT: UN 0282/UN 1336
mf: $CH_4N_4O_2$ mw: 104.09

PROP: Yellow solid, high explosive. Usually stable needles. Mp: 246°, decomp: 225-250°. Sltly sol in alc, concentration acids, cold solns of alkalies; sol in water; very sltly sol in ether.

SYNS: NITROGUANIDINE ◇ 1-NITROGUANIDINE ◇ 2-NITROGUANIDINE ◇ NITROGUANIDINE, containing less than 20% water (DOT) ◇ NITROGUANIDINE DRY (DOT) ◇ PICRITE (THE EXPLOSIVE)

TOXICITY DATA with REFERENCE
cyt-ham:fbr 4 g/L/24H MUREAV 48,337,77
ipr-mus LD50:48 mg/kg PCJOAU 10,1504,76

CONSENSUS REPORTS: Reported in EPA TSCA Inventory.

DOT Classification: Class A Explosive; Label: EXPLOSIVE A (UN0282); Flammable Solid; Label: Flammable Solid (UN1336)

SAFETY PROFILE: Poison by intraperitoneal route. Mutation data reported. A very dangerous fire hazard when exposed to heat, flame or by chemical reaction with oxidizers. A severe explosion hazard when shocked or exposed to heat or flame. It is about as powerful as TNT. It is normally mixed with colloided nitrocellulose or ammonium nitrate and paraffin wax. Can react vigorously with oxidizing materials and the derivatives can be explosive. The mercury and silver salts and other derivatives are much more impact-sensitive. When heated to decomposition it emits highly toxic fumes of NO_x. See also NITRO COMPOUNDS.

NHB000 CAS:6065-14-1 ***HR: 3***
2-NITRO-2-HEPTENE
mf: $C_7H_{13}NO_2$ mw: 143.21

TOXICITY DATA with REFERENCE
skn-rbt 500 mg open SEV AMIHAB 18,312,58
eye-rbt 50 mg SEV AMIHAB 18,312,58
orl-rat LDLo:940 mg/kg IMSUAI 34,800,65
ipr-rat LDLo:280 mg/kg TXAPA9 1,156,59
skn-rbt LDLo:940 mg/kg IMSUAI 34,800,65

SAFETY PROFILE: Poison by intraperitoneal route. Moderately toxic by ingestion and skin contact. A severe eye and skin irritant. When heated to decomposition it emits toxic fumes of NO_x. See also NITRO COMPOUNDS.

NHB500 CAS:6065-13-0 ***HR: 3***
3-NITRO-2-HEPTENE
mf: $C_7H_{13}NO_2$ mw: 143.21

TOXICITY DATA with REFERENCE
skn-rbt 500 mg open SEV AMIHAB 18,312,58
eye-rbt 50 mg SEV AMIHAB 18,312,58
orl-rat LDLo:420 mg/kg IMSUAI 34,800,65
ipr-rat LDLo:280 mg/kg TXAPA9 1,156,59
skn-rbt LDLo:1400 mg/kg IMSUAI 34,800,65

SAFETY PROFILE: Poison by intraperitoneal route. Moderately toxic by ingestion and skin contact. A severe eye and skin irritant. When heated to decomposition it emits toxic fumes of NO_x. See also NITRO COMPOUNDS.

NHD000 CAS:6065-17-4 ***HR: 3***
2-NITRO-2-HEXENE
mf: $C_6H_{11}NO_2$ mw: 129.18

TOXICITY DATA with REFERENCE
eye-hmn 2500 ng/L IMSUAI 27,375,58
skn-rbt 500 mg open SEV AMIHAB 18,312,58
eye-rbt 50 mg SEV AMIHAB 18,312,58
orl-rat LDLo:420 mg/kg IMSUAI 34,800,65
ipr-rat LDLo:120 mg/kg TXAPA9 1,156,59
skn-rbt LDLo:1400 mg/kg IMSUAI 34,800,65

SAFETY PROFILE: Poison by intraperitoneal route. Moderately toxic by ingestion and skin contact. A severe eye and skin irritant. When heated to decomposition it emits toxic fumes of NO_x. See also NITRO COMPOUNDS.

NHE000 CAS:4812-22-0 ***HR: 3***
3-NITRO-3-HEXENE
mf: $C_6H_{11}NO_2$ mw: 129.18

TOXICITY DATA with REFERENCE
skn-rbt 500 mg open SEV AMIHAB 18,312,58
eye-rbt 50 mg SEV AMIHAB 18,312,58
ihl-mus TCLo:200 ppb/63W-I:NEO TXAPA9 5,445,63
orl-rat LDLo:420 mg/kg IMSUAI 34,800,65
ipr-rat LDLo:80 mg/kg TXAPA9 1,156,59
skn-rbt LDLo:940 mg/kg IMSUAI 34,800,65

SAFETY PROFILE: Poison by intraperitoneal route. Moderately toxic by ingestion and skin contact. A severe eye and skin irritant. Questionable carcinogen with experimental neoplastigenic data. When heated to decomposition it emits toxic fumes of NO_x. See also NITRO COMPOUNDS.

NHE500 ***HR: 3***
NITROHYDRENE

PROP: An oil. Composition: nitroglycerin + nitrosucrose.

SAFETY PROFILE: A very dangerous fire hazard when exposed to heat or flame or by chemical reaction. A severe explosion hazard when shocked or exposed to heat. It is a powerful explosive, about as powerful as nitroglycerin, and is used to stretch glycerin supplies. It is made by dissolving up to 25% of sucrose in glycerin and nitrating the resulting mixture to give an explosive oil. This procedure saves considerable quantities of glycerin. The material is almost exactly like nitroglycerin. Shock and heat will explode it. Incompatible with oxidizing materials. When heated to decomposition it emits highly toxic fumes of NO_x. See also NITROGLYCERIN, and EXPLOSIVES, HIGH.

NHE600 CAS:102107-61-9 ***HR: 3***
3-NITRO-4-HYDROXYPHENYLARSENOUS ACID
mf: $C_6H_4AsNO_4$ mw: 229.03

SYNS: 4-ARSENOSO-2-NITROPHENOL ◇ PHENOL, 4-ARSENOSO-2-NITRO-

TOXICITY DATA with REFERENCE
ivn-rat LD50:5 mg/kg JACSAT 70,1762,48
ivn-mus LDLo:7500 μg/kg PHBUA9 2,19,54

OSHA PEL: TWA 0.5 mg(As)/m^3
ACGIH TLV: TWA 0.2 mg(As)/m^3

SAFETY PROFILE: Poison by intravenous route. When heated to decomposition it emits toxic fumes of NO_x and As.

NHF500 CAS:4008-48-4 ***HR: 3***
5-NITRO-8-HYDROXYQUINOLINE
mf: $C_9H_6N_2O_3$ mw: 190.17

SYN: 5-NITRO-8-QUINOLINOL

TOXICITY DATA with REFERENCE
orl-rat TDLo:13 g/kg/52W-C:ETA TXAPA9 8,343,66
orl-rat LD50:510 mg/kg ATXKA8 20,313,65
orl-mus LD50:104 mg/kg ATXKA8 20,313,65
ivn-mus LD50:8300 μg/kg ATXKA8 20,313,65

SAFETY PROFILE: Poison by ingestion. Questionable carcinogen with experimental tumorigenic data. When heated to decomposition it emits toxic fumes of NO_x.

NHG000 CAS:527-73-1 ***HR: 3***
2-NITROIMIDAZOLE
mf: $C_3H_3N_3O_2$ mw: 113.09

SYNS: 11A ◇ AZOMYCIN ◇ 2-NITRO-1H-IMIDAZOLE

TOXICITY DATA with REFERENCE
mmo-sat 1 nmol/plate MUREAV 58,1,78
mmo-omi 1 mg/plate JGAMA9 11,129,65
orl-mus LD50:316 mg/kg AACHAX -,478,65
ipr-mus LD50:80 mg/kg 85ERAY 1,745,78
scu-mus LD50:316 mg/kg AACHAX -,478,65
ivn-mus LD50:80 mg/kg JAJAAA 6,182,53

CONSENSUS REPORTS: Reported in EPA TSCA Inventory.

SAFETY PROFILE: Poison by ingestion, intraperitoneal, subcutaneous, and intravenous routes. Mutation data reported. When heated to decomposition it emits toxic fumes of NO_x.

NHH000 CAS:6506-37-2 ***HR: 2***
4-(2-(5-NITROIMIDAZOL-1-YL)ETHYL)MORPHO-LINE
mf: $C_9H_{14}N_4O_3$ mw: 226.27

SYNS: ACTEROL ◇ ESCLAMA ◇ 1-(N-p-ETHYLMORPHOLINE)-5-NITROIMIDAZOLE ◇ K 1900 ◇ N-2-MORPHOLINOETHYL-5-NITROIMIDAZOLE ◇ 1-(2-N-MORPHOLINYLETHYL)-5-NITROIMIDAZOLE ◇ NAXOFEM ◇ NAXOGIN ◇ NIMORAZOLE ◇ NIMORAZOLO ◇ NITRIMIDAZINE ◇ NULOGYL ◇ SIRLEDI

TOXICITY DATA with REFERENCE
mmo-sat 1 μmol/L TCMUD8 3,429,83
mma-sat 50 μg/plate MUREAV 77,301,80
mmo-esc 1 mmol/L MUREAV 26,483,74
mmo-klp 50 μmol/L/20H MUREAV 66,207,79
mmo-omi 250 μmol/L MUREAV 26,483,74
scu-mus TDLo:700 mg/kg (female 7-13D post):REP KSRNAM 6,729,72
orl-rat LD50:1540 mg/kg KSRNAM 5,1403,71
ipr-rat LD50:1200 mg/kg KSRNAM 5,1403,71
scu-rat LD50:1090 mg/kg KSRNAM 5,1403,71
orl-mus LD50:1530 mg/kg JMCMAR 21,781,78
ipr-mus LD50:1200 mg/kg KSRNAM 5,1408,71
scu-mus LD50:1140 mg/kg KSRNAM 5,1408,71

SAFETY PROFILE: Moderately toxic by ingestion, intraperitoneal and subcutaneous routes. Experimental reproductive effects. Mutation data reported. When heated to decomposition it emits toxic fumes of NO_x. See also NITRO COMPOUNDS of AROMATIC HYDROCARBONS.

NHH500 CAS:13551-87-6 ***HR: 2***
1-(2-NITROIMIDAZOL-1-YL)-3-METHOXYPROPAN-2-OL
mf: $C_7H_{11}N_3O_4$ mw: 201.21

SYNS: 1-(2-HYDROXY-3-METHOXYPROPYL)-2-NITROIMIDAZOLE ◇ α-(METHOXYMETHYL)-2-NITRO-1H-IMIDAZOLE-1-ETHANOL (9CI) ◇ α-(METHOXYMETHYL)-2-NITROIMIDAZOLE-1-ETHANOL ◇ α-(METHOXYMETHYL)-2-NITRO-1H-IMIDAZOLE-1-ETHANOL ◇ MISONIDAZOLE ◇ 1-(2-NITRO-1-IMIDAZOLYL)-3-METHOXY-2-PROPANOL ◇ NSC-261037 ◇ Ro 7-0582 ◇ SR 1354 ◇ SRI 1354

TOXICITY DATA with REFERENCE
cyt-mus-ipr 1 g/kg CNREA8 43,2210,83
sce-mus:emb 3 mmol/L IOBPD3 6,915,80
ipr-mus TDLo:1 g/kg (8D preg):TER BJRAAP 54,154,81
ipr-mus TDLo:1 g/kg (8D preg):REP BJRAAP 54,154,81
unr-man TDLo:228 mg/kg/3W-I:SKN CTRRDO 63,123,79
orl-rat LD50:2131 mg/kg NTIS** PB81-121212
orl-mus LD50:1869 mg/kg NCISP* JAN86
ipr-mus LD50:1340 mg/kg RAREAE 91,186,82
scu-mus LD50:1414 mg/kg AACHAX -,513,67

CONSENSUS REPORTS: EPA Genetic Toxicology Program.

SAFETY PROFILE: Moderately toxic by ingestion, intraperitoneal, and subcutaneous routes. An experimental teratogen. Experimental reproductive effects. Mutation data reported. When heated to decomposition it emits toxic fumes of NO_x.

NHI000 CAS:13551-92-3 ***HR: 2***
3-(2-NITROIMIDAZOL-1-YL)-1,2-PROPANEDIOL
mf: $C_6H_9N_3O_4$ mw: 187.18

SYNS: DEMETHYLMISONIDAZOLE ◇ DESMETHYLMISONIDAZOLE ◇ 3-(2-NITRO-1H-IMIDAZOL-1-YL)-1,2-PROPANEDIOL ◇ NSC 261036 ◇ RO 5-9963

TOXICITY DATA with REFERENCE
mmo-klp 20 μmol/L/20H MUREAV 66,207,79
otr-mus:emb 1 mmol/L/3D-C 45OJAG -,146,80
ipr-mus LD50:3205 mg/kg NCISP* JAN86

SAFETY PROFILE: Moderately toxic by intraperitoneal route. Mutation data reported. When heated to decomposition it emits toxic fumes of NO_x.

NHI500 CAS:7046-61-9 ***HR: 3***
NITROIMINODIETHYLENEDIISOCYANIC ACID
mf: $C_6H_8N_4O_4$ mw: 200.18

SYN: 3-NITRO-3-AZAPENTANE-1,5-DIISOCYANATE

TOXICITY DATA with REFERENCE
ivn-mus LD50:56 mg/kg CSLNX* NX#05191

CONSENSUS REPORTS: Reported in EPA TSCA Inventory.
NIOSH REL: (Diisocyanates) TWA 0.005 ppm; CL 0.02 ppm/10M

SAFETY PROFILE: Poison by intravenous route. When heated to decomposition it emits toxic fumes of NO_x.

NHJ000 CAS:34701-14-9 ***HR: 3***
4-NITROINDANE
mf: $C_9H_9NO_2$ mw: 163.17

SAFETY PROFILE: A heat-sensitive explosive. Residue from distillation explodes with admission of air. Upon decomposition it emits toxic fumes of NO_x.

NHJ009 CAS:7436-07-9 ***HR: 3***
5-NITROINDANE
mf: $C_9H_9NO_2$ mw: 163.17

SAFETY PROFILE: A heat-sensitive explosive. Residue from distillation explodes with admission of air. Upon decomposition it emits toxic fumes of NO_x.

NHK000 CAS:2942-42-9 ***HR: D***
7-NITROINDAZOLE
mf: $C_7H_5N_3O_2$ mw: 163.15

SYN: 7-NITRO-1H-INDAZOLE

TOXICITY DATA with REFERENCE
mmo-sat 100 nmol/plate MUREAV 173,169,86
mma-sat 100 nmol/plate MUREAV 58,11,78

CONSENSUS REPORTS: EPA Genetic Toxicology Program.

SAFETY PROFILE: Mutation data reported. When heated to decomposition it emits toxic fumes of NO_x. See also NITRO COMPOUNDS.

NHK500 CAS:6146-52-7 ***HR: D***
5-NITROINDOLE
mf: $C_8H_6N_2O_2$ mw: 162.16

SYN: 5-NITRO-1H-INDOLE

TOXICITY DATA with REFERENCE
mmo-sat 50 nmol/plate MUREAV 173,169,86
mma-sat 100 nmol/plate MUREAV 58,11,78

CONSENSUS REPORTS: EPA Genetic Toxicology Program.

SAFETY PROFILE: Mutation data reported. When heated to decomposition it emits toxic fumes of NO_x. See also NITRO COMPOUNDS.

NHK800 CAS:24458-48-8 ***HR: 2***
NITROL
mf: $C_{10}H_{13}N_3O_3$ mw: 223.26

SYNS: BENZENAMINE,N-(2-METHYL-2-NITROPROPYL)-p-NITROSO-(9CI) ◇ CP 25017 ◇ N-(2-METHYL-2-NITROPROPYL)-p-NITROSOANILINE ◇ N-(2-METHYL-2-NITROPROPYL)-4-NITROSOBENZAMINE ◇ NITROL (PROMOTER) ◇ N-(p-NITROSOANILINOMETHYL)-2-NITROPROPANE

TOXICITY DATA with REFERENCE
unr-rat TDLo:81800 mg/kg/2Y:CAR EPASR* 8EHQ-0183-0165
orl-rat LD50:2730 mg/kg CHIP** 27DEC83

CONSENSUS REPORTS: Reported in EPA TSCA Inventory.

SAFETY PROFILE: Moderately toxic by ingestion. Questionable carcinogen with experimental carcinogenic data. When heated to decomposition it emits toxic fumes of NO_x.

NHK900 CAS:133-58-4 ***HR: 3***
NITROMERSOL
mf: $C_7H_5HgNO_3$ mw: 351.72

PROP: Powder; yellow, odorless, and tasteless. Sol in boiling glacial acetic acid, insol in water, ether, acetone, or alc.

SYNS: MERCURY,(2-METHYL-5-NITROPHENOLATO(2-)-C^6,O^1)-(9CI) ◇ METAPHEN ◇ 5-METHYL-2-NITRO-7-OXA-8-MERCURABICYCLO(4.2.0)OCTA-1,3,5-TRIENE ◇ 4-NITRO-5-HYDROXYMERCURIORTHOCRESOL ◇ NITROMERSOL SOLUTION ◇ 7-OXA-8-MERCURABICYCLO(4.2.0)OCTA-1,3,5-TRIENE, 5-METHYL-2-NITRO-

TOXICITY DATA with REFERENCE
ipr-rat LD50:3 mg/kg JAPMA8 38,270,49
ivn-dog LDLo:3500 μg/kg JPETAB 44,423,32
ACGIH TLV: TWA 0.1 mg(Hg)/m^3 (skin)

SAFETY PROFILE: Poison by intraperitoneal and intravenous routes. When heated to decomposition it emits toxic fumes of NO_x and Hg.

NHM000 CAS:603-71-4 ***HR: 3***
NITROMESITYLENE
mf: $C_9H_{11}NO_2$ mw: 165.19

SAFETY PROFILE: Can explode in preparation. When heated to decomposition it emits toxic fumes of NO_x. See also NITRO COMPOUNDS OF AROMATIC HYDROCARBONS.

NHM500 CAS:75-52-5 ***HR: 3***
NITROMETHANE
DOT: UN 1261
mf: CH_3NO_2 mw: 61.05

PROP: An oily liquid; moderate to strong, disagreeable odor. Bp: 101°, lel: 7.3%, fp: −29°, flash p: 95°F (CC), d: 1.1322 @ 25°/4°, autoign temp: 785°F, vap press: 27.8 mm @ 20°, vap d: 2.11. Sltly sol in water; sol in alc, ether.

SYNS: NITROCARBOL ◇ NITROMETAN (POLISH)

TOXICITY DATA with REFERENCE
orl-rat LD50:940 mg/kg GISAAA 32(9),9,67
orl-mus LD50:950 mg/kg GISAAA 32(9),9.67
ipr-mus LD50:110 mg/kg KHFZAN 10(6),53,76
orl-dog LDLo:125 mg/kg AMIHAB 11,102,55
ivn-dog LDLo:800 mg/kg AMIHAB 11,102,55
ihl-mky LCLo:1000 ppm INMEAF 16,441,47
orl-rbt LDLo:750 mg/kg JIHTAB 22,315,40
ivn-rbt LDLo:750 mg/kg AMIHAB 11,102,55

CONSENSUS REPORTS: Reported in EPA TSCA Inventory.

OSHA PEL: TWA 100 ppm
ACGIH TLV: TWA 100 ppm; (Proposed: 20 ppm)
DFG MAK: 100 ppm (250 mg/m^3)
DOT Classification: Flammable Liquid; Label: Flammable Liquid; Flammable or Combustible Liquid; Label: Flammable Liquid.

SAFETY PROFILE: Poison by ingestion and intraperitoneal routes. Moderately toxic by intravenous route. Mildly toxic by inhalation. In humans it may cause anorexia, nausea, vomiting, diarrhea, and kidney injury and liver damage.

A very dangerous fire hazard when exposed to heat, oxidizers, or flame. May explode by detonation, heat, or shock. Its sensitivity is increased when mixed with acids; bases; acetone; aluminum powder; ammonium salts + organic solvents; bis(2-aminoethyl)amine; 1,2-diaminoethane + N,2,4,6-tetranitro-N-methyl aniline; haloforms (e.g., chloroform; bromoform); hydrazine + methanol. Ignites when mixed with alkyl metal halides (e.g., diethylaluminum bromide; dimethylaluminum bromide; ethylaluminum bromide iodide; methyl zinc iodide; methylaluminum diiodide). Can react violently with $AlCl_3$ + organic matter; $Ca(OH)_2$; m-methyl aniline; $Ca(OCl)_2$; hexamethylbenzene; hydrocarbons; inorganic bases, hydroxides, organic amines, KOH, formaldehyde, nitric acid, metal oxides, 1,2-diaminomethane, lithium perchlorate; sodium hydride. Reacts with aqueous silver nitrate to form the explosive silver fulminate. When heated to decomposition it emits toxic fumes of NO_x. See also NITROALKANES.

NHN500 CAS:598-57-2 ***HR: 3***
N-NITROMETHYLAMINE
mf: $CH_4N_2O_2$ mw: 76.07

SYNS: METHYLNITRAMINE ◇ N-NITROMETHANAMINE

TOXICITY DATA with REFERENCE
ipr-mus LD50:500 mg/kg PCJOAU 10,1505,76

SAFETY PROFILE: Moderately toxic by intraperitoneal route. Explodes on contact with concentrated sulfuric acid. When heated to decomposition it emits toxic fumes of NO_x. See also NITRO COMPOUNDS and AMINES.

NHN550 CAS:33094-74-5 ***HR: D***
2-NITRO-3-METHYL-5-CHLOROBENZOFURAN
mf: $C_9H_6ClNO_3$ mw: 211.61

SYN: 5-CHLORO-3-METHYL-2-NITROBENZOFURAN

TOXICITY DATA with REFERENCE
mmo-sat 1 nmol/plate MUREAV 88,355,81
pic-sat 1 nmol/plate MUREAV 104,1,82

SAFETY PROFILE: Mutation data reported. When heated to decomposition it emits toxic fumes of Cl^- and NO_x.

NHO500 CAS:77-49-6 ***HR: 2***
2-NITRO-2-METHYL-1,3-PROPANEDIOL
mf: $C_4H_9NO_4$ mw: 135.14

PROP: Crystals. Mp: 148 ± 1°, bp: decomp.

SYNS: 1,1-DIMETHYLOL-1-NITROETHANE ◇ 2-METHYL-2-NITROPROPANE-1,3-DIOL

TOXICITY DATA with REFERENCE
orl-rat LDLo:2000 mg/kg JIHTAB 22,315,40
ipr-mus LD50:1600 mg/kg KHFZAN 11(1),73,77
orl-rbt LDLo:1000 mg/kg JIHTAB 22,315,40

CONSENSUS REPORTS: Reported in EPA TSCA Inventory.

SAFETY PROFILE: Moderately toxic by ingestion and intraperitoneal routes. Flammable when exposed to heat, flame or powerful oxidizers. When heated to decomposition it emits toxic fumes of NO_x.

NHP000 CAS:76-39-1 ***HR: 2***
2-NITRO-2-METHYL-1-PROPANOL
mf: $C_4H_9NO_3$ mw: 119.14

PROP: Crystals. Mp: 91°, bp: 95° @ 10 mm.

TOXICITY DATA with REFERENCE
orl-rbt LDLo:1000 mg/kg JIHTAB 22,315,40

CONSENSUS REPORTS: Reported in EPA TSCA Inventory.

SAFETY PROFILE: Moderately toxic by ingestion. When heated to decomposition it emits toxic fumes of NO_x. See also NITRO COMPOUNDS.

NHP500 CAS:5863-35-4 ***HR: 2***
NITROMIFENE CITRATE
mf: $C_{27}H_{28}N_2O_4 \cdot C_6H_8O_7$ mw: 636.71

SYNS: CI-628 ◇ CI-628 CITRATE ◇ CN-55945-27 ◇ 1-(2-(p-(α-(p-METHOXYPHENYL)-β-NITROSTYRYL)PHENOXY)ETHYL)PYRROLIDINE CITRATE (1:1) ◇ 1-(2-(p-(α-(p-METHOXYPHENYL)-β-NITROSTYRYL)PHENOXY)ETHYL)PYRROLIDINE MONOCITRATE ◇ PARKE DAVIS CI-628

TOXICITY DATA with REFERENCE
orl-dog TDLo:3750 μg/kg (female 1-15D post):REP TJADAB 7,199,73
orl-dog TDLo:3750 μg/kg (female 1-15D post):TER TJADAB 7,199,73
orl-rat LD50:1778 mg/kg NATUAS 211,538,66
orl-mus LD50:832 mg/kg NATUAS 211,538,66

SAFETY PROFILE: Moderately toxic by ingestion. An experimental teratogen. Experimental reproductive effects. When heated to decomposition it emits toxic fumes of NO_x.

NHQ000 CAS:86-57-7 ***HR: 3***
1-NITRONAPHTHALENE
mf: $C_{10}H_7NO_2$ mw: 173.18

PROP: Yellow crystals. Bp: 304°, flash p: 327°F (CC), d: 1.331 @ 4°/4°, vap d: 5.96. Mp: 59-61°. Insol in water; sol in CS_2, alc, chloroform, and ether.

SYNS: NCI-C01956 ◇ α-NITRONAPHTHALENE

TOXICITY DATA with REFERENCE
mmo-sat 1 μg/plate MUREAV 122,243,83
mma-sat 3300 ng/plate ENMUDM 7(Suppl 5),1,85
orl-rat LD50:120 mg/kg NCIMR* NIH-71-E-2144,73
ipr-rat LD50:86 mg/kg JJATDK 4,253,84

CONSENSUS REPORTS: NCI Carcinogenesis Bioassay (feed); No Evidence: mouse, rat NCITR* NCI-CG-TR-64,78. Reported in EPA TSCA Inventory.

DFG MAK: Suspected Carcinogen.

SAFETY PROFILE: Poison by ingestion and intraperitoneal routes. Mutation data reported. A skin, eye, and mucous membrane irritant. Combustible when exposed to heat or flame. To fight fire, use CO_2, dry chemical or water spray. Explosive reaction with nitric acid + sulfuric acid above 60°C. Forms a sensitive explosive mixture with tetranitromethane. When heated to decomposition it emits toxic fumes of NO_x. See also 2-NITRONAPHTHALENE and NITRO COMPOUNDS OF AROMATIC HYDROCARBONS.

NHQ500 CAS:581-89-5 ***HR: 3***
2-NITRONAPHTHALENE
mf: $C_{10}H_7NO_2$ mw: 173.18

PROP: Colorless in ethanol. Mp: 79°, bp: 165° @ 15 mm. Insol in water; very sol in alc and ether.

SYN: β-NITRONAPHTHALENE

TOXICITY DATA with REFERENCE
mmo-sat 1 μg/plate MUREAV 122,243,83
mrc-smc 1 pph JJIND8 62,901,79
orl-dog TDLo:2400 mg/kg/34/W-I:ETA XPHCI* 17DEC76
orl-rat LD50:4400 mg/kg 34ZIAG -,428,69
ipr-mus LD50:1300 mg/kg JJIND8 62,911,79
orl-rbt LD50:2650 mg/kg 34ZIAG -,428,693

CONSENSUS REPORTS: EPA Genetic Toxicology Program.

DFG TRK: 0.035 ppm; Animal Carcinogen, Suspected Human Carcinogen.

SAFETY PROFILE: Confirmed carcinogen with experimental tumorigenic data. Moderately toxic by ingestion and intraperitoneal routes. Mutation data reported. A skin and lung irritant. For explosion and disaster hazards, see NITRATES. Combustible when exposed to heat or flame. When heated to decomposition it emits toxic fumes of NO_x. See also 1-NITRONAPHTHALENE.

NHQ950 CAS:69267-51-2 ***HR: 2***
2-NITRONAPHTHO(2,1-b)FURAN
mf: $C_{12}H_7NO_3$ mw: 213.20

SYN: R6597

TOXICITY DATA with REFERENCE
mmo-sat 1 nmol/plate MUREAV 88,355,81
mma-sat 1 nmol/plate MUREAV 88,355,81
pic-sat 100 pmol/plate MUREAV 104,1,82 CNREA8 44,1969,84
msc-ham:lng 500 μg/L CNREA8 44,1969,84
skn-mus TDLo:10489 μg/kg:ETA JJCREP 78,565,87

SAFETY PROFILE: Questionable carcinogen with experimental tumorigenic data. Mutation data reported. When heated to decomposition it emits toxic fumes of NO_x.

NHR500 CAS:13115-28-1 ***HR: 3***
3-NITRO-2-NAPHTHYLAMINE
mf: $C_{10}H_8N_2O_2$ mw: 188.20

TOXICITY DATA with REFERENCE
orl-rat TDLo:38 g/kg/51W-I:CAR,REP JNCIAM 41,985,68
orl-rat TD:130 mg/kg/1Y-I:NEO JNCIAM 41,985,68

SAFETY PROFILE: Questionable carcinogen with experimental carcinogenic and neoplastigenic data. Experimental reproductive effects. When heated to decomposition it emits toxic fumes of NO_x. See also AMINES and NITRO COMPOUNDS OF AROMATIC HYDROCARBONS.

NHS000 CAS:13010-07-6 ***HR: D***
3-NITRO-1-NITROSO-1-PROPYLGUANIDINE
mf: $C_4H_9N_5O_3$ mw: 175.18

SYN: N-NITROSO-N′-NITRO-N-PROPYLGUANIDINE

TOXICITY DATA with REFERENCE
mmo-sat 100 mmol/plate IDZAAW 50,403,75
pic-esc 500 μg/L ESKHA5 88,118,70

CONSENSUS REPORTS: EPA Genetic Toxicology Program.

SAFETY PROFILE: Mutation data reported. Many N-nitroso compounds are carcinogens. When heated to decomposition it emits toxic fumes of NO_x. See also N-NITROSO COMPOUNDS and NITRO COMPOUNDS.

NHS500 CAS:13826-86-3 ***HR: 3***
NITRONIUM TETRAFLUOROBORATE(1−)
mf: BF_4NO_2 mw: 132.82

TOXICITY DATA with REFERENCE
ivn-mus LD50:180 mg/kg CSLNX* NX#05621

CONSENSUS REPORTS: Reported in EPA TSCA Inventory.

OSHA PEL: TWA 2.5 mg(F)/m^3
NIOSH REL: (Inorganic Fluorides) TWA 2.5 mg(F)/m^3

SAFETY PROFILE: Poison by intravenous route. A powerful nitrating agent when dissolved in sulfolane. Incompatible with tetrahydrothiophene-1,1-dioxide; organics; α-propanol; methane; ethane @ −78° or higher. When heated to decomposition it emits very toxic fumes of F^- and NO_x. See also BORON COMPOUNDS.

NHT000 CAS:4812-25-3 ***HR: 3***
2-NITRO-2-NONENE
mf: $C_9H_{17}NO_2$ mw: 171.27

TOXICITY DATA with REFERENCE
eye-hmn 10 μg/L IMSUAI 27,375,58
skn-rbt 500 mg open SEV AMIHAB 18,312,58
eye-rbt 50 mg SEV AMIHAB 18,312,58
orl-rat LDLo:2100 mg/kg IMSUAI 34,800,65
ipr-rat LDLo:180 mg/kg TXAPA9 1,156,59
skn-rbt LDLo:620 mg/kg IMSUAI 34,800,65
ihl-ckn LCLo:43 ppm AMIHAB 18,312,58

SAFETY PROFILE: Poison by inhalation and intraperitoneal routes. Moderately toxic by ingestion and skin contact. A severe eye and skin irritant. When heated to decomposition it emits toxic fumes of NO_x. See also NITRO COMPOUNDS.

NHU000 CAS:6065-04-9 ***HR: 2***
3-NITRO-3-NONENE
mf: $C_9H_{17}NO_2$ mw: 171.27

TOXICITY DATA with REFERENCE
skn-rbt 500 mg open SEV AMIHAB 18,312,58
eye-rbt 50 mg SEV AMIHAB 18,312,58
orl-rat LDLo:2100 mg/kg IMSUAI 34,800,65
ipr-rat LDLo:420 mg/kg TXAPA9 1,156,59
skn-rbt LDLo:420 mg/kg IMSUAI 34,800,65

SAFETY PROFILE: Moderately toxic by ingestion, skin contact and intraperitoneal routes. A severe eye and skin irritant. When heated to decomposition it emits toxic fumes of NO_x. See also NITRO COMPOUNDS.

NHV500 CAS:6065-01-6 ***HR: 3***
5-NITRO-4-NONENE
mf: $C_9H_{17}NO_2$ mw: 171.27

TOXICITY DATA with REFERENCE
orl-rat LDLo:940 mg/kg IMSUAI 34,800,65
ipr-rat LDLo:280 mg/kg TXAPA9 1,156,59
skn-rbt LDLo:420 mg/kg IMSUAI 34,800,65

SAFETY PROFILE: Poison by intraperitoneal route. Moderately toxic by ingestion and skin contact. When heated to decomposition it emits toxic fumes of NO_x. See also NITRO COMPOUNDS.

NHW000 CAS:6065-11-8 ***HR: 3***
2-NITRO-2-OCTENE
mf: $C_8H_{15}NO_2$ mw: 157.24

TOXICITY DATA with REFERENCE
skn-rbt 500 mg open SEV AMIHAB 18,312,58
eye-rbt 50 mg SEV AMIHAB 18,312,58
orl-rat LDLo:1400 mg/kg IMSUAI 34,800,65
ipr-rat LDLo:280 mg/kg TXAPA9 1,156,59
skn-rbt LDLo:620 mg/kg IMSUAI 34,800,65

SAFETY PROFILE: Poison by intraperitoneal route. Moderately toxic by ingestion and skin contact. A severe skin and eye irritant. When heated to decomposition it emits toxic fumes of NO_x. See also NITRO COMPOUNDS.

NHW500 CAS:6065-10-7 ***HR: 3***
3-NITRO-2-OCTENE
mf: $C_8H_{15}NO_2$ mw: 157.24

TOXICITY DATA with REFERENCE
skn-rbt 500 mg open SEV AMIHAB 18,312,58
eye-rbt 50 mg SEV AMIHAB 18,312,58
orl-rat LDLo:620 mg/kg IMSUAI 34,800,65
ipr-rat LDLo:180 mg/kg TXAPA9 1,156,59
skn-rbt LDLo:620 mg/kg IMSUAI 34,800,65

SAFETY PROFILE: Poison by intraperitoneal route. Moderately toxic by ingestion and skin contact. A severe skin and eye irritant. When heated to decomposition it emits toxic fumes of NO_x. See also NITRO COMPOUNDS.

NHX000 CAS:6065-09-4 ***HR: 3***
3-NITRO-3-OCTENE
mf: $C_8H_{15}NO_2$ mw: 157.24

TOXICITY DATA with REFERENCE
skn-rbt 500 mg open SEV AMIHAB 18,312,58
eye-rbt 50 mg SEV AMIHAB 18,312,58
orl-rat LDLo:620 mg/kg IMSUAI 34,800,65
ipr-rat LDLo:180 mg/kg IMSUAI 34,800,65
skn-rbt LDLo:940 mg/kg IMSUAI 34,800,65

SAFETY PROFILE: Poison by intraperitoneal route. Moderately toxic by ingestion and skin contact. A severe skin and eye irritant. When heated to decomposition it emits toxic fumes of NO_x. See also NITRO COMPOUNDS.

NHY100 CAS:600-26-0 ***HR: 3***
1-NITRO-1-OXIMINOETHANE
mf: $C_2H_4N_2O_3$ mw: 104.07

SYN: ETHYLNITROLIC ACID

SAFETY PROFILE: An explosive solid. When heated

to decomposition it emits toxic fumes of NO_x. See also NITRO COMPOUNDS.

NHY250 CAS:625-49-0 ***HR: 3***
NITROOXIMINOMETHANE
mf: $CH_2N_2O_3$ mw: 90.04

SYN: METHYLNITROLIC ACID

SAFETY PROFILE: An unstable and explosive solid. When heated to decomposition it emits toxic fumes of NO_x. See also NITRO COMPOUNDS.

NHY500 CAS:53663-14-2 ***HR: 3***
7-NITRO-3-OXO-3H-2,1-BENZOXAMERCUROLE
mf: $C_7H_3HgNO_4$ mw: 365.70

SYNS: 2-HYDROXYMERCURI-3-NITROBENZOICACID-1,2-CYCLIC ANHYDRIDE ◇ 7-NITRO-3H-2,1-BENZOXAMERCUROL-3-ONE

TOXICITY DATA with REFERENCE
ivn-mus LD50:56 mg/kg CSLNX* NX#05674

CONSENSUS REPORTS: Mercury and its compounds are on the Community Right-To-Know List.

OSHA PEL: (Transitional: CL 1 mg/10m^3) CL 0.1 mg(Hg)/m^3 (skin)
ACGIH TLV: TWA 0.1 mg(Hg)/m^3 (skin)
NIOSH REL: (Inorganic Mercury) TWA 0.05 mg(Hg)/m^3

SAFETY PROFILE: Poison by intravenous route. When heated to decomposition it emits very toxic fumes of Hg and NO_x. See also NITRO COMPOUNDS and MERCURY COMPOUNDS.

NHZ000 CAS:6065-19-6 ***HR: 3***
2-NITRO-2-PENTENE
mf: $C_5H_9NO_2$ mw: 115.15

TOXICITY DATA with REFERENCE
skn-rbt 500 mg open SEV AMIHAB 18,312,58
eye-rbt 50 mg SEV AMIHAB 18,312,58
orl-rat LDLo:280 mg/kg IMSUAI 34,800,65
ipr-rat LDLo:80 mg/kg TXAPA9 1,156,59
skn-rbt LDLo:940 mg/kg IMSUAI 34,800,65

SAFETY PROFILE: Poison by ingestion and intraperitoneal routes. Moderately toxic by skin contact. A severe skin and eye irritant. When heated to decomposition it emits toxic fumes of NO_x. See also NITRO COMPOUNDS.

NIA000 CAS:6065-18-5 ***HR: 3***
3-NITRO-2-PENTENE
mf: $C_5H_9NO_2$ mw: 115.15

TOXICITY DATA with REFERENCE
skn-rbt 500 mg open SEV AMIHAB 18,312,58
eye-rbt 50 mg SEV AMIHAB 18,312,58
orl-rat LDLo:420 mg/kg IMSUAI 34,800,65
ihl-rat LCLo:268 ppm/5H AMIHAB 18,312,58
ipr-rat LDLo:55 mg/kg TXAPA9 1,156,59
skn-rbt LDLo:620 mg/kg IMSUAI 34,800,65

SAFETY PROFILE: Poison by intraperitoneal route. Moderately toxic by ingestion, inhalation, and skin contact. A severe eye and skin irritant. When heated to decomposition it emits toxic fumes of NO_x. See also NITRO COMPOUNDS.

NIA700 CAS:20731-44-6 ***HR: 3***
3-NITROPERCHLORYLBENZENE
mf: $C_6H_4ClNO_5$ mw: 205.55

$O_2NC_6H_4ClO_3$

SAFETY PROFILE: A shock-sensitive explosive. When heated to decomposition it emits toxic fumes of Cl^- and NO_x. See also PERCHLORATES and NITRO COMPOUNDS OF AROMATIC HYDROCARBONS.

NIB000 CAS:943-39-5 ***HR: 3***
p-NITROPEROXYBENZOIC ACID
mf: $C_7H_5NO_5$ mw: 183.13

TOXICITY DATA with REFERENCE
scu-mus TDLo:228 mg/kg/57W-I:ETA CNREA8 30,1037,70

SAFETY PROFILE: Questionable carcinogen with experimental tumorigenic data. When heated to decomposition it emits toxic fumes of NO_x. See also PEROXIDES and NITRO COMPOUNDS OF AROMATIC HYDROCARBONS.

NIC000 CAS:17024-18-9 ***HR: D***
2-NITROPHENANTHRENE
mf: $C_{14}H_9NO_2$ mw: 223.24

TOXICITY DATA with REFERENCE
mmo-sat 125 mg/plate CBINA8 26,11,79
mma-sat 1 μg/plate CBINA8 26,11,79

SAFETY PROFILE: Mutation data reported. When heated to decomposition it emits toxic fumes of NO_x. See also NITRO COMPOUNDS.

NID000 CAS:100-29-8 ***HR: 2***
p-NITROPHENETOLE
mf: $C_8H_9NO_3$ mw: 167.18

PROP: Colorless monoclinic. D: 1.18 @ 15°, mp: 59-60°, bp: 283° @ 758 mm. Very sltly sol in water; very sol in hot alc, ether.

SYNS: 1-ETHOXY-4-NITROBENZENE ◇ p-NITROPHENETOL (GERMAN)

TOXICITY DATA with REFERENCE
mmo-sat 250 μg/plate
/S CRNGDP 3,167,82 @te = ipr-rat LD50:2100 mg/kg
AGGHAR 17,217,59

CONSENSUS REPORTS: Reported in EPA TSCA Inventory.

SAFETY PROFILE: Moderately toxic by intraperitoneal route. Mutation data reported. When heated to decomposition it emits toxic fumes of NO_x. See also NITRO COMPOUNDS OF AROMATIC HYDROCARBONS.

NIE000 CAS:554-84-7 ***HR: 3***
3-NITROPHENOL
DOT: UN 1663
mf: $C_6H_5NO_3$ mw: 139.12

PROP: Monoclinic crystals. Mp: 97°, bp: 194° @ 70 mm, d: 1.485 @ 20°/4°. Decomposes when distilled at ordinary pressure. Sol in hot water and dil acids, caustic solns; insol in petr ether.

SYN: m-NITROPHENOL (DOT)

TOXICITY DATA with REFERENCE
skn-rbt 500 mg/24H MOD 28ZPAK -,107,72
eye-rbt 5 mg/24H SEV 28ZPAK -,107,72
orl-rat LD50:328 mg/kg GTPPAF 8,145,72
scu-rat LDLo:500 mg/kg RMSRA6 16,449,1896
orl-mus LD50:1070 mg/kg GTPZAB 25(8),50,81
ipr-mus LDLo:70 mg/kg RBPMAZ 22,1,52
scu-gpg LDLo:500 mg/kg RMSRA6 16,449,1896
scu-frg LDLo:160 mg/kg RMSRA6 16,449,1896
orl-mam LD50:250 mg/kg GISAAA 45(10),16,80
skn-mam LD50:543 mg/kg GISAAA 45(10),16,80

CONSENSUS REPORTS: EPA Genetic Toxicology Program. Reported in EPA TSCA Inventory.

SAFETY PROFILE: Poison by ingestion, subcutaneous, and intraperitoneal routes. Moderately toxic by skin contact. A skin and severe eye irritant. When heated to decomposition it emits toxic fumes of NO_x. See also other nitrophenol entries.

NIE500 CAS:88-75-5 ***HR: 3***
o-NITROPHENOL
DOT: UN 1663
mf: $C_6H_5NO_3$ mw: 139.12

PROP: Light yellow crystals; aromatic odor. Mp: 45°, bp: 214.5°, d: 1.495 @ 20°, vap press: 1 mm @ 49.3°. Sol in water; very sol in alc, ether, benzene, CS; volatile with steam.

SYNS: 2-HYDROXYNITROBENZENE ◇ 2-NITROPHENOL

TOXICITY DATA with REFERENCE
orl-rat LD50:334 mg/kg GTPPAF 8,145,72
scu-rat LDLo:1100 mg/kg RMSRA6 16,449,1896
orl-mus LD50:1300 mg/kg TXAPA9 42,417,77
ipr-mus LD50:378 mg/kg JMCMAR 18,868,75
ims-mus LDLo:600 mg/kg HBTXAC 5,120,59
scu-cat LDLo:600 mg/kg HBTXAC 5,120,59
scu-rbt LDLo:1700 mg/kg HBTXAC 5,120,59
scu-gpg LDLo:900 mg/kg RMSRA6 16,449,1896
scu-frg LDLo:300 mg/kg HBTXAC 5,120,59

CONSENSUS REPORTS: EPA Genetic Toxicology Program. Reported in EPA TSCA Inventory. Community Right-To-Know List.

DOT Classification: Poison B; Label: St. Andrews Cross.

SAFETY PROFILE: Poison by ingestion, subcutaneous, and intraperitoneal routes. Moderately toxic by intramuscular route. Can cause liver and kidney damage. The liquid phenol reacts violently with KOH. Product of reaction with chlorosulfuric acid decomposes violently at 24°C. When heated to decomposition it emits toxic fumes of NO_x. See also NITRO COMPOUNDS OF AROMATIC HYDROCARBONS.

NIF000 CAS:100-02-7 ***HR: 3***
4-NITROPHENOL
DOT: UN 1663
mf: $C_6H_5NO_3$ mw: 139.12

PROP: Colorless to sltly yellow; odorless crystals, sweet then burning taste. D: 1.270 @ 120°/4°, mp: 113-114° (subl). Sltly sol in cold water; very sol in alc, chloroform, ether; sol in alkali solns, hydroxides, and carbonates.

SYNS: 4-HYDROXYNITROBENZENE ◇ NCI-C55992 ◇ 4-NITROFENOL (DUTCH) ◇ p-NITROPHENOL (DOT) ◇ PARANITROFENOL (DUTCH) ◇ PARANITROFENOLO (ITALIAN) ◇ PARANITROPHENOL (FRENCH, GERMAN) ◇ RCRA WASTE NUMBER U170

TOXICITY DATA with REFERENCE
dnd-esc 50 μmol/L MUREAV 89,95,81
dni-hmn:fbr 1 mmol/L CNREA8 35,1392,75
scu-rat LDLo:200 mg/kg RMSRA6 16,449,1896
orl-mus LD50:380 mg/kg GTPZAB 25(8),50,81
ipr-mus LD50:75 mg/kg NTIS** AD691-490
ivn-dog LDLo:10 mg/kg XPHBAO 271,131,41
scu-gpg LDLo:200 mg/kg RMSRA6 16,449,1896
ims-pgn LDLo:65 mg/kg JPETAB 49,187,33

CONSENSUS REPORTS: EPA Genetic Toxicology Program. Community Right-To-Know List. Reported in EPA TSCA Inventory.

SAFETY PROFILE: Poison by ingestion, subcutaneous, intraperitoneal, intravenous, and intramuscular routes. Moderately toxic by skin contact. Human mutation data reported. Its exothermic decomposition causes a dangerous fast pressure increase. Mixtures with diethyl phosphite may explode when heated. When heated to decomposition it emits toxic fumes of NO_x. See also other nitrophenol entries and NITRO COMPOUNDS OF AROMATIC HYDROCARBONS.

NIG500 CAS:22098-38-0 ***HR: 3***
p-NITROPHENOL, MERCURY(II) SALT

TOXICITY DATA with REFERENCE
ipr-rat LDLo:50 mg/kg NCNSA6 5,36,53

CONSENSUS REPORTS: Mercury and its compounds are on the Community Right-To-Know List.
NIOSH REL: (Mercury, Inorganic) TWA 0.05 $mg(Hg)/m^3$

SAFETY PROFILE: Poison by intraperitoneal route. When heated to decomposition it emits very toxic fumes of Hg and NO_x. See also MERCURY and 4-NITROPHENOL.

NIH000 CAS:64047-82-1 ***HR: 3***
p-NITROPHENOL TIN(IV) SALT

TOXICITY DATA with REFERENCE
orl-rat LDLo:250 mg/kg NCNSA6 5,36,53
ipr-rat LDLo:100 mg/kg NCNSA6 5,36,53

OSHA PEL: TWA 2 $mg(Sn)/m^3$
ACGIH TLV: TWA 2 $mg(Sn)/m^3$

SAFETY PROFILE: Poison by ingestion and intraperitoneal routes. When heated to decomposition it emits toxic fumes of NO_x. See also TIN COMPOUNDS and 4-NITROPHENOL.

NIH500 CAS:64047-83-2 ***HR: 3***
p-NITROPHENOL ZINC SALT

TOXICITY DATA with REFERENCE
orl-rat LDLo:250 mg/kg NCNSA6 5,36,53
ipr-rat LDLo:250 mg/kg NCNSA6 5,36,53

CONSENSUS REPORTS: Zinc and its compounds are on the Community Right-To-Know List.

SAFETY PROFILE: Poison by ingestion and intraperitoneal routes. When heated to decomposition it emits toxic fumes of NO_x and ZnO. See also ZINC COMPOUNDS and 4-NITROPHENOL.

NII000 CAS:1798-11-4 ***HR: 2***
p-NITROPHENOXYACETIC ACID
mf: $C_8H_7NO_5$ mw: 197.16

SYN: (4-NITROPHENOXY)ACETIC ACID

TOXICITY DATA with REFERENCE
ipr-mus LD50:936 mg/kg CHTPBA 5,211,70

CONSENSUS REPORTS: Reported in EPA TSCA Inventory.

SAFETY PROFILE: Moderately toxic by intraperitoneal route. When heated to decomposition it emits toxic fumes of NO_x. See also NITRO COMPOUNDS OF AROMATIC HYDROCARBONS.

NII200 CAS:3644-32-4 ***HR: 3***
p-NITROPHENOXYTRIBUTYLTIN
mf: $C_{18}H_{31}NO_3Sn$ mw: 428.19

SYN: STANNANE,(p-NITROPHENOXY)TRIBUTYL-

TOXICITY DATA with REFERENCE
ivn-mus LD50:10 mg/kg CSLNX* NX#05585

OSHA PEL: TWA 0.1 $mg(Sn)/m^3$ (skin)
ACGIH TLV: TWA 0.1 $mg(Sn)/m^3$; STEL 0.2 mg/m^3 (skin)
NIOSH REL: (Organotin Compound):10H TWA 0.1 $mg(Sn)/m^3$

SAFETY PROFILE: Poison by intravenous route. When heated to decomposition it emits toxic fumes of NO_x and Sn.

NII500 CAS:3740-52-1 ***HR: D***
o-NITROPHENYLACETIC ACID
mf: $C_8H_7NO_4$ mw: 181.16

SYNS: 2-NITROBENZENEACETIC ACID ◇ (2-NITROPHENYL)ACETIC ACID ◇ 2-(o-NITROPHENYL)ACETIC ACID

TOXICITY DATA with REFERENCE
mma-sat 10 µmol/plate MUREAV 58,11,78

CONSENSUS REPORTS: Reported in EPA TSCA Inventory.

SAFETY PROFILE: Mutation data reported. Forms an explosive mixture with thionyl chloride. When heated to decomposition it emits toxic fumes of NO_x. See also NITRO COMPOUNDS OF AROMATIC HYDROCARBONS.

NIJ000 CAS:555-21-5 ***HR: 3***
p-NITROPHENYLACETONITRILE
mf: $C_8H_6N_2O_2$ mw: 162.16

PROP: Leaves from alc. Mp: 116-117°. Insol in water; sol in alc.

SYNS: p-NITROBENZENEACETONITRILE ◇ 4-NITROBENZENEACETONITRILE ◇ p-NITROBENZYLCYANIDE ◇ 4-NITROBENZYL CYANIDE ◇ 4-NITRO-BENZYL-CYANID (GERMAN) ◇ 4-NITROPHENYLACETONITRILE

TOXICITY DATA with REFERENCE
ivn-mus LD50:32 mg/kg CSLNX* NX#02093

CONSENSUS REPORTS: Cyanide and its compounds are on the Community Right-To-Know List. Reported in EPA TSCA Inventory.

SAFETY PROFILE: Poison by intravenous route. When heated to decomposition it emits toxic fumes of NO_x and CN^-. See also NITRILES and NITRO COMPOUNDS OF AROMATIC HYDROCARBONS.

NIJ200 CAS:22751-23-1 ***HR: 3***
2-NITROPHENYLACETYL CHLORIDE
mf: $C_8H_6ClNO_3$ mw: 199.59

$O_2NC_6H_4CH_2CO{\bullet}Cl$

SAFETY PROFILE: Explodes when heated. When heated to decomposition it emits toxic fumes of Cl^- and NO_x. See also CHLORIDES and NITRO COMPOUNDS OF AROMATIC HYDROCARBONS.

NIJ300 ***HR: 3***
NITROPHENYL ACETYLENE
mf: $C_8H_5NO_6$ mw: 211.13

SAFETY PROFILE: An explosive. When heated to decomposition it emits toxic fumes of NO_x. See also ACETYLENE COMPOUNDS and NITRO COMPOUNDS OF AROMATIC HYDROCARBONS.

NIJ400 CAS:51-58-1 ***HR: 3***
1-p-NITROPHENYL-3-AMIDINOUREA HYDROCHLORIDE
mf: $C_8H_9N_5O_3{\bullet}ClH$ mw: 259.68

SYNS: 3-AMIDINO-1-p-NITROPHENYLUREAHYDROCHLORIDE ◇ N-(AMINOIMINOMETHYL)-N'-(4-NITROPHENYL)UREA MONOHYDROCHLORIDE ◇ NITROGUANIL ◇ UREA, 1-AMIDINO-3-(p-NITROPHENYL)-, MONOHYDROCHLORIDE

TOXICITY DATA with REFERENCE
orl-mus TDLo:35 mg/kg (female 7D pre):REP TXAPA9 13,228,68
ivn-rat LD50:240 mg/kg PJPPAA 38,425,86
ipr-mus LD50:225 mg/kg JMCMAR 10,521,67

SAFETY PROFILE: Poison by intravenous and intraperitoneal routes. Experimental reproductive effects. When heated to decomposition it emits toxic fumes of NO_x and HCl.

NIJ500 CAS:98-72-6 ***HR: 3***
4-NITROPHENYLARSONIC ACID
mf: $C_6H_6AsNO_5$ mw: 247.05

SYNS: NITARSONE ◇ 4-NITROBENZENEARSONIC ACID ◇ p-NITROPHENYLARSONIC ACID ◇ RAS-26

TOXICITY DATA with REFERENCE
orl-rat LDLo:100 mg/kg NCNSA6 5,13,53
ivn-mus LD50:18 mg/kg CSLNX* NX#01192

CONSENSUS REPORTS: Arsenic and its compounds are on the Community Right-To-Know List.

OSHA PEL: TWA 0.5 mg(As)/m^3
ACGIH TLV: TWA 0.2 mg(As)m^3

SAFETY PROFILE: Poison by ingestion and intravenous routes. When heated to decomposition it emits very toxic fumes of NO_x and As. See also ARSENIC COMPOUNDS and NITRO COMPOUNDS OF AROMATIC HYDROCARBONS.

NIK000 CAS:2581-69-3 ***HR: 3***
4-((p-NITROPHENYL)AZO)DIPHENYLAMINE
mf: $C_{18}H_{14}N_4O_2$ mw: 318.36

TOXICITY DATA with REFERENCE
orl-rat TDLo:31 g/kg/88W-C:ETA ZEKBAI 57,530,51

CONSENSUS REPORTS: Reported in EPA TSCA Inventory.

SAFETY PROFILE: Questionable carcinogen with experimental tumorigenic data. When heated to decomposition it emits toxic fumes of NO_x.

NIK500 CAS:63019-77-2 ***HR: 3***
7-((p-NITROPHENYLAZO)METHYLBENZ(c)ACRIDINE
mf: $C_{24}H_{16}N_4O_2$ mw: 392.44

SYN: 9-METHYL-ω-(p-NITROBENZENEAZO)-3,4-BENZACRIDINE

TOXICITY DATA with REFERENCE
scu-mus TDLo:200 mg/kg:ETA VOONAW 1,52,55

SAFETY PROFILE: Questionable carcinogen with experimental tumorigenic data. When heated to decomposition it emits toxic fumes of NO_x.

NIL500 CAS:4752-02-7 ***HR: 3***
1-(p-NITROPHENYL)BIGUANIDE HYDROCHLORIDE
mf: $C_8H_{10}N_6O_2{\bullet}ClH$ mw: 258.70

SYN: N-(4-NITROPHENYL)-IMIDODICARBONIMIDICDIAMIDE MONOHYDROCHLORIDE

TOXICITY DATA with REFERENCE
ipr-mus LD50:165 mg/kg JMCMAR 10,521,67
ivn-mus LD50:180 mg/kg CSLNX* NX#04028

SAFETY PROFILE: Poison by intraperitoneal and intravenous routes. When heated to decomposition it emits very toxic fumes of HCl and NO_x.

NIL625 CAS:51138-16-0 ***HR: 3***
N-(2-NITROPHENYL)-1,3-DIAMINOETHANE
mf: $C_8H_{11}N_3O_2$ mw: 181.19

SAFETY PROFILE: May explode during vacuum distillation. When heated to decomposition it emits toxic fumes of NO_x. See also NITRO COMPOUNDS OF AROMATIC HYDROCARBONS.

NIM000 CAS:1224-64-2 ***HR: 3***
p-NITROPHENYLDI-N-BUTYLPHOSPHINATE
mf: $C_{14}H_{22}NO_4P$ mw: 299.34

SYNS: DIBUTYL-PHOSPHINIC ACID, 4-NITROPHENYL ESTER ◇ NIBUFIN ◇ p-NITROPHENYLDIBUTYLPHOSPHINATE

TOXICITY DATA with REFERENCE
ipr-rat LD50:12300 μg/kg TXAPA9 49,23,79
ims-rat LD50:3360 μg/kg TXAPA9 49,23,79
scu-mus LD50:7200 μg/kg RPTOAN 42,106,79
ivn-mus LD50:4800 μg/kg RPTOAN 42,106,79
ims-mus LD50:8900 μg/kg TXAPA9 49,23,79
ivn-rbt LD50:660 μg/kg FATOAO 14,32,51
ims-rbt LD50:560 μg/kg TXAPA9 49,23,79

SAFETY PROFILE: A deadly poison by intraperitoneal, intramuscular, intravenous, and subcutaneous routes. When heated to decomposition it emits very toxic fumes of PO_x and NO_x. See also NITRO COMPOUNDS OF AROMATIC HYDROCARBONS.

NIM500 CAS:311-45-5 ***HR: 3***
p-NITROPHENYL DIETHYLPHOSPHATE
mf: $C_{10}H_{14}NO_6P$ mw: 275.22

PROP: Oily liquid; slt odor. Bp: 170° @ 1 mm, d: 1.2736 @ 20°/4°. Sltly water-sol; freely sol in organic solvents.

SYNS: CHINORTA ◇ O,O'-DIAETHYL-p-NITROPHENYLPHOSPHAT (GERMAN) ◇ DIAETHYL-p-NITROPHENYLPHOSPHORSAEURE-ESTER (GERMAN) ◇ DIETHYL-p-NITROFENYL ESTER KYSELINY FOSFORECNE (CZECH) ◇ DIETHYL p-NITROPHENYL PHOSPHATE ◇ O,O-DIETHYL O-p-NITROPHENYL PHOSPHATE ◇ DIETHYL PARAOXON ◇ O,O-DIETHYL PHOSPHORIC ACID O-p-NITROPHENYL ESTER ◇ O,O-DIETYL-o-p-NITROFENYLFOSFAT (CZECH) ◇ E 600 ◇ ENT 16,087 ◇ ESTER 25 ◇ ETHYL p-NITROPHENYL ETHYLPHOSPHATE ◇ ETHYL PARAOXON ◇ ETICOL ◇ FOSFAKOL ◇ HC 2072 ◇ MINTACO ◇ MINTACOL ◇ MIOTISAL ◇ MIOTISAL A ◇ P-NITROPHENOL, ESTER with DIETHYL PHOSPHATE ◇ OXY-PARATHION ◇ PARAOXON ◇ PARAOXONE ◇ PAROXAN ◇ PESTOX 101 ◇ PHOSPHACOL ◇ PHOSPHORIC ACID DIETHYL 4-NITROPHENYL ESTER ◇ RCRA WASTE NUMBER P401 ◇ SOLUGLACIT ◇ TS 219

TOXICITY DATA with REFERENCE
mmo-ssp 5 mmol/L MUREAV 117,139,83
sce-ham:ovr 100 μmol/L JTEHD6 8,939,81
orl-rat LD50:1800 μg/kg BCPCA6 16,1183,67
ipr-rat LD50:930 μg/kg ARZNAD 14,85,64
scu-rat LD50:230 μg/kg NIIRDN 6,113,82
ivn-rat LD50:240 μg/kg CJBPAZ 34,197,56
ims-rat LD50:446 μg/kg BJIMAG 22,317,65
orl-mus LD50:760 μg/kg HPFSDS FDA-80-1076,212,80
ipr-mus LD50:330 μg/kg FATOAO 35,494,72
scu-mus LD50:270 μg/kg JPPMAB 34,603,82
ivn-mus LD50:520 μg/kg TXAPA9 22,684,72
ims-mus LD50:710 μg/kg AEPPAE 237,211,59
par-mus LDLo:600 μg/kg GUCHAZ 6,388,73
scu-rbt LD50:230 μg/kg NIIRDN 6,113,82
ivn-rbt LDLo:100 μg/kg BJPCAL 8,208,53

SAFETY PROFILE: A deadly poison by ingestion, intraperitoneal, intravenous, subcutaneous, intramuscular, and parenteral routes. Mutation data reported. A cholinesterase inhibitor. An insecticide. When heated to decomposition it emits toxic fumes of PO_x and NO_x. See also PARATHION.

NIN000 CAS:3015-74-5 ***HR: 3***
p-NITROPHENYL ETHYLBUTYLPHOSPHONATE
mf: $C_{12}H_{18}NO_5P$ mw: 287.28

TOXICITY DATA with REFERENCE
scu-mus LD50:1200 μg/kg RPTOAN 42,106,79
ivn-mus LD50:1000 μg/kg RPTOAN 42,106,79
ivn-rbt LD50:72 μg/kg BCPCA6 13,1229,64
scu-rat LD50:1200 μg/kg FATOAO 42(3),299,79
ivn-rat LD50:1 mg/kg FATOAO 42(3),299,79

SAFETY PROFILE: A deadly poison by intravenous and subcutaneous routes. When heated to decomposition it emits very toxic fumes of PO_x and NO_x.

NIQ500 CAS:21658-26-4 ***HR: 3***
p-NITROPHENYL-p'-GUANIDINOBENZOATE
mf: $C_{14}H_{12}N_4O_4$ mw: 300.30

SYNS: p-GUANIDINOBENZOIC ACID p-NITROPHENYL ESTER ◇ p-NITROPHENYL p-GUANIDINO-BENZOATE ◇ 4-NITROPHENYL 4-GUANIDINOBENZOATE ◇ NPGB

TOXICITY DATA with REFERENCE
mmo-sat 100 μg/plate MUREAV 67,81,79
mma-sat 100 μg/plate MUREAV 67,81,79
imp-mus TDLo:1600 μg/kg (2D pre):REP CCPTAY 26,137,82
ipr-mus LD50:180 mg/kg BIREBV 20,1045,79

SAFETY PROFILE: Poison by intraperitoneal route. Experimental reproductive effects. Mutation data reported. When heated to decomposition it emits toxic fumes of NO_x.

NIQ800 CAS:3034-19-3 ***HR: 3***
2-NITROPHENYLHYDRAZINE
mf: $C_6H_7N_3O_2$ mw: 153.16

SYN: (o-NITROPHENYL)HYDRAZINE

TOXICITY DATA with REFERENCE
mmo-sat 4 mmol/L AEMIDF 44,801,82
mmo-omi 7 mg/L MUREAV 173,233,86
ivn-mus LD50:178 mg/kg CSLNX* NX#00632

SAFETY PROFILE: Poison by intravenous route. Mutation data reported. When heated to decomposition it emits toxic fumes of NO_x. See also NITRO COMPOUNDS OF AROMATIC HYDROCARBONS.

NIR000 CAS:100-16-3 ***HR: 3***
p-NITROPHENYLHYDRAZINE
mf: $C_6H_7N_3O_2$ mw: 153.16

PROP: Orange-red leaflets or needles. Mp: approx 157° (decomp). Sltly sol in cold water; sol in hot water, hot benzene, alc, ether, chloroform, ethyl acetate.

SYNS: p-HYDRAZINONITROBENZENE ◇ 4-NITROPHENYLHYDRAZINE

TOXICITY DATA with REFERENCE
mmo-sat 2 mmol/L AEMIDF 44,801,82
mmo-omi 780 μg/L MUREAV 173,233,86
ipr-mus LDLo:250 mg/kg CBCCT* 6,57,54

CONSENSUS REPORTS: Reported in EPA TSCA Inventory.

SAFETY PROFILE: Poison by intraperitoneal route. Mutation data reported. When heated to decomposition it emits toxic fumes of NO_x. See also NITRO COMPOUNDS OF AROMATIC HYDROCARBONS.

NIR550 CAS:38753-50-3 ***HR: 2***
o-NITROPHENYL ISOPROPYL ETHER
mf: $C_9H_{11}NO_3$ mw: 181.21

SYNS: o-ISOPROPOXYNITROBENZENE ◇ 2-ISOPROPOXYNITROBENZENE ◇ ISOPROPYL o-NITROPHENYL ETHER ◇ o-IZOPROPOKSYNITROBENZEN (CZECH)

TOXICITY DATA with REFERENCE
skn-rbt 500 mg MLD MEPAAX 29,293,78
eye-rbt 100 mg MLD MEPAAX 29,293,78
orl-rat LD50:2500 mg/kg MEPAAX 29,293,78
ipr-rat LD50:800 mg/kg MEPAAX 29,293,78

SAFETY PROFILE: Moderately toxic by ingestion and intraperitoneal routes. A skin and eye irritant. When heated to decomposition it emits toxic fumes of NO_x. See also NITRO COMPOUNDS OF AROMATIC HYDROCARBONS.

NIS000 CAS:63868-95-1 ***HR: 3***
o-NITROPHENYLMERCURY ACETATE
mf: $C_8H_7HgNO_4$ mw: 381.75

SYN: o-NITROPHENYLMERCURY (ACETATO)

TOXICITY DATA with REFERENCE
orl-rat LDLo:100 mg/kg NCNSA6 5,30,53
ipr-rat LDLo:25 mg/kg NCNSA6 5,30,53

CONSENSUS REPORTS: Mercury and its compounds are on the Community Right-To-Know List.

OSHA PEL: (Transitional: CL 1 mg/10m^3) CL 0.1 mg(Hg)/m^3 (skin)
ACGIH TLV: TWA 0.1 mg(Hg)/m^3 (skin)
NIOSH REL: (Inorganic Mercury) TWA 0.05 mg(Hg)/m^3

SAFETY PROFILE: Poison by ingestion and intraperitoneal routes. When heated to decomposition it emits very toxic fumes of Hg and NO_x. See also MERCURY and NITRO COMPOUNDS OF AROMATIC HYDROCARBONS.

NIT500 CAS:2216-12-8 ***HR: 2***
2-NITROPHENYL PHENYL ETHER
mf: $C_{12}H_9NO_3$ mw: 215.22

PROP: Yellow oil. D: 1.258 @ 15°, mp: < −20°, bp: 235° @ 60 mm. Insol in water; sol in abs alc and benzene.
SYN: 2-NITRODIPHENYL ETHER

TOXICITY DATA with REFERENCE
ivn-mus LD50:838 mg/kg TXCYAC 8,347,77

CONSENSUS REPORTS: Reported in EPA TSCA Inventory.

SAFETY PROFILE: Moderately toxic by intravenous route. When heated to decomposition it emits toxic fumes of NO_x. See also ETHERS and NITRO COMPOUNDS OF AROMATIC HYDROCARBONS.

NIU000 CAS:620-88-2 ***HR: 1***
p-NITROPHENYL PHENYLETHER
mf: $C_{12}H_9NO_3$ mw: 215.22

SYNS: 4-NITROBIPHENYL ESTER ◇ 4-NITRODIFENYLETHER (CZECH) ◇ p-NITRODIPHENYL ETHER ◇ 4-NITRODIPHENYL ETHER ◇ 1-NITRO-4-PHENOXYBENZENE (9CI) ◇ 4-NITROPHENYL PHENYL ETHER

TOXICITY DATA with REFERENCE
eye-rbt 500 mg/24H MLD 28ZPAK -,118,72
mmo-sat 100 μg/plate MUREAV 67,123,79
mma-sat 50 μg/plate MUREAV 67,123,79

CONSENSUS REPORTS: Reported in EPA TSCA Inventory.

SAFETY PROFILE: Mutation data reported. An eye irritant. When heated to decomposition it emits toxic fumes of NO_x. See also ETHERS and NITRO COMPOUNDS OF AROMATIC HYDROCARBONS.

NIV000 CAS:61785-70-4 ***HR: D***
4-NITRO-5-(4-PHENYL-1-PIPERAZINYL) BENZOFURAZAN OXIDE
mf: $C_{16}H_{15}N_5O_4$ mw: 341.36

SYN: B2734

TOXICITY DATA with REFERENCE
mmo-sat 100 μg/plate MUREAV 48,145,77
mma-sat 50 μg/well CBINA8 19,77,77

CONSENSUS REPORTS: EPA Genetic Toxicology Program.

SAFETY PROFILE: Mutation data reported. When heated to decomposition it emits toxic fumes of NO_x.

NIW300 CAS:49558-46-5 ***HR: 3***
2-NITROPHENYL SULFONYL DIAZOMETHANE
mf: $C_7H_5N_3O_4S$ mw: 227.19

$O_2NC_6H_4SO_2CHN_2$

SAFETY PROFILE: Explodes at room temperature. When heated to decomposition it emits toxic fumes of SO_x and NO_x. See also NITRO COMPOUNDS OF AROMATIC HYDROCARBONS.

NIW400 CAS:3704-42-5 ***HR: 2***
4-(4-NITROPHENYL)THIAZOLE
mf: $C_9H_6N_2O_2S$ mw: 206.23

TOXICITY DATA with REFERENCE
orl-rat TDLo:39928 mg/kg/46W-C:ETA PAACA3 21,75,80

SAFETY PROFILE: Questionable carcinogen with experimental tumorigenic data. When heated to decomposition it emits toxic fumes of NO_x and SO_x.

NIW500 CAS:1836-77-7 ***HR: 1***
p-NITROPHENYL-2,4,6-TRICHLOROPHENYL ETHER
mf: $C_{12}H_6Cl_3NO_3$ mw: 318.54

SYNS: CHLORNITROFEN ◇ CNP ◇ CNP 1032 ◇ MC 338 ◇ MC 1478 ◇ MO ◇ MO 338 ◇ 2',4',6'-TRICHLORO-4-NITROBIPHENYL ETHER ◇ 2,4,6-TRICHLORO-4'-NITRODIPHENYL ETHER ◇ 1,3,5-TRICHLORO-2-(4-NITROPHENOXY)BENZENE ◇ 2,4,6-TRICHLOROPHENYL-4-NITROPHENYL ETHER

TOXICITY DATA with REFERENCE
mma-sat 100 μg/plate CBINA8 44,133,83
orl-rat LD50:10800 mg/kg 85ARAE 2,195,77
orl-mus LD50:11800 mg/kg FMCHA2 -,C160,83

CONSENSUS REPORTS: EPA Genetic Toxicology Program.

SAFETY PROFILE: Mildly toxic by ingestion. Mutation data reported. Used as an herbicide. When heated to decomposition it emits very toxic fumes of NO_x and Cl^-. See also ETHERS.

NIX000 CAS:15457-05-3 ***HR: 1***
p-NITROPHENYL-α,α,α-TRIFLUORO-2-NITRO-p-TOLYL ETHER
mf: $C_{13}H_7F_3N_2O_5$ mw: 328.22

SYNS: C-6989 ◇ 2,4'-DINITRO-4-TRIFLUOROMETHYL-DIPHENYL ETHER ◇ FLUORODIFEN ◇ 2-NITRO-1-(4-NITROPHENOXY)-4-(TRIFLUOROMETHYL)BENZENE ◇ p-NITROPHENYL-2-NITRO-4-(TRIFLUOROMETHYL) PHENYL ETHER ◇ PREFORAN ◇ SOYEX

TOXICITY DATA with REFERENCE
eye-rbt 100 mg MOD CIGET* -,-,77
orl-rat LD50:9000 mg/kg FMCHA2 -,C194,83

CONSENSUS REPORTS: EPA Genetic Toxicology Program. Reported in EPA TSCA Inventory.

SAFETY PROFILE: Mildly toxic by ingestion. An eye irritant. Used as an herbicide. When heated to decomposition it emits very toxic fumes of NO_x and F^-.

NIX500 CAS:108-03-2 ***HR: 3***
1-NITROPROPANE
mf: $C_3H_7NO_2$ mw: 89.11

$O_2NCH_2CH_2CH_3$

PROP: Colorless liquid. Bp: 132°, fp: −108°, flash p: 93°F (TCC), d: 1.003 @ 20°/20°, autoign temp: 789°F, vap press: 7.5 mm @ 20°, vap d: 3.06, lel: 2.2%. Sltly sol in water; misc with alc, ether, and many organic solvents.

SYN: 1-NP

TOXICITY DATA with REFERENCE
eye-hmn 150 ppm/15M JIHTAB 28,262,46
ihl-hmn TCLo:150 ppm:EYE JIHTAB 28,262,46
orl-rat LD50:455 mg/kg NPIRI* 1,91,74
ihl-rat LC50:3100 ppm/8H NPIRI* 1,91,74
orl-mus LD50:800 mg/kg GISAAA 32(9),9,67
ipr-mus LD50:250 mg/kg KHFZAN 10(6),53,76
orl-rbt LDLo:250 mg/kg JIHTAB 22,315,40

CONSENSUS REPORTS: Reported in EPA TSCA Inventory.

OSHA PEL: TWA 25 ppm
ACGIH TLV: TWA 25 ppm
DFG MAK: 25 ppm (90 mg/m^3)

SAFETY PROFILE: Poison by ingestion and intraperitoneal routes. Mildly toxic by inhalation. A human eye irritant. Human systemic effects by inhalation: conjunctiva irritation. Very dangerous fire hazard when exposed to heat, open flame, or oxidizers. Reacts violently with $Ca(OH)_2$; hydrocarbons; hydroxides; inorganic bases. May explode on heating. Metal oxides increase its sensitivity to thermal ignition. To fight fire, use alcohol foam, CO_2, dry chemical, water spray. When heated to decomposition it emits toxic fumes of NO_x. See also 2-NITROPROPANE, NITROALKANES, and NITRO COMPOUNDS.

NIY000 CAS:79-46-9 ***HR: 3***
2-NITROPROPANE
mf: $C_3H_7NO_2$ mw: 89.11

$O_2NCH(CH_3)_2$

PROP: Colorless liquid. Bp: 120°, fp: −93°, flash p: 82°F (TCC), d: 0.992 @ 20°/20°, autoign temp: 802°F, vap press: 10 mm @ 15.8°, vap d: 3.06, lel: 2.6%. Misc with organic solvents; sol in water, alc, and ether.

SYNS: DIMETHYLNITROMETHANE ◇ ISONITROPROPANE ◇ NIPAR S-20 ◇ NIPAR S-20 SOLVENT ◇ NIPAR S-30 SOLVENT ◇ NITROISOPROPANE ◇ β-NITROPROPANE ◇ 2-NP ◇ RCRA WASTE NUMBER U171

TOXICITY DATA with REFERENCE
mmo-sat 1 mg/plate ENMUDM 5(Suppl 1),3,83
mma-sat 1 mg/plate ENMUDM 5(Suppl 1),3,83
ipr-rat TDLo:2550 mg/kg (1-15D preg):TER SWEHDO 7(Suppl 4),66,81
ipr-rat TDLo:2550 mg/kg (1-15D preg):REP SWEHDO 7(Suppl 4),66,81
orl-rat TDLo:4277 mg/kg/16W-I:CAR CRNGDP 8,1947,87
ihl-rat TCLo:207 ppm/7H/26W-I:CAR JEPTDQ 2(5),233,79
ihl-rat TC:207 ppm/26W-I:ETA XPHCI* 25APR77
ihl-man TCLo:20 ppm:CNS,GIT INMEAF 16,441,47
orl-rat LD50:720 mg/kg 34ZIAG -,430,69
ihl-rat LC50:400 ppm/6H JEPTDQ 2(2),233,78
ihl-mus LC50:10 g/m^3/2H 85GMAT -,94,82
ipr-mus LD50:75 mg/kg NTIS** AD691-490
ihl-cat LCLo:714 ppm/5H AMIHBC 5,52,52
orl-rbt LDLo:500 mg/kg JIHTAB 22,315,40
ihl-rbt LCLo:2381 ppm/5H AMIHBC 5,52,52
ihl-gpg LCLo:4622 ppm/5H AMIHBC 5,52,52

CONSENSUS REPORTS: NTP Fifth Annual Report on Carcinogens. IARC Cancer Review: Group 2B IMEMDT 7,56,87; Human Inadequate Evidence IMEMDT 29,331,82; Animal Sufficient Evidence IMEMDT 29,331,82. Community Right-To-Know List. EPA Genetic Toxicology Program. Reported in EPA TSCA Inventory.

OSHA PEL: (Transitional: TWA 25 ppm) TWA 10 ppm
ACGIH TLV: TWA 10 ppm; Suspected Human Carcinogen.
DFG TRK: 5 ppm; Animal Carcinogen, Suspected Human Carcinogen.
NIOSH REL: (2-Nitropropane): TWA reduce to lowest feasible level

SAFETY PROFILE: Confirmed carcinogen with experimental carcinogenic, tumorigenic, and teratogenic data. Poison by intraperitoneal route. Moderately toxic by ingestion and inhalation. Human systemic effects by inhalation: anorexia, hypermotility, diarrhea, nausea or vomiting. An experimental teratogen. Other experimental reproductive effects. Mutation data reported. Can cause liver and kidney injury, methemoglobinemia, and cyanosis. Very dangerous fire hazard when exposed to heat, open flame, or oxidizers. May explode on heating. Violent reactions with chlorosulfonic acid; oleum. May react with amines + heavy metal oxides (e.g., mercury oxide or silver oxide) to form explosive salts. May ignite on contact with mixtures of carbon + hopcalite which are used in some respirators. Hopcalite is a catalyst consisting of coprecipitated copper(II) oxide and manganese(IV) oxide. To fight fire, use alcohol foam, CO_2, dry chemical, water spray. When heated to decomposition it emits toxic fumes of NO_x. See also NITRO ALKANES.

NIY200 CAS:4749-28-4 ***HR: 2***
2-NITROPROPENE
mf: $C_3H_5NO_2$ mw: 87.08

$H_2C{=}C(NO_2)CH_3$

SAFETY PROFILE: A lachrimator. Potentially dangerous polymerization reaction on contact with alkalies. When heated to decomposition it emits toxic fumes of NO_x. See also NITROALKANES.

NIY500 CAS:504-88-1 ***HR: 3***
3-NITROPROPIONIC ACID
mf: $C_3H_5NO_4$ mw: 119.09

SYNS: BNP ◇ BOVINOCIDIN ◇ HIPTAGENIC ACID ◇ NCI-C03076 ◇ β-NITROPROPIONIC ACID

TOXICITY DATA with REFERENCE
mmo-sat 100 μg/plate IAPUDO 27,283,80
mma-sat 100 μg/plate ENMUDM 7(Suppl 5),1,85
otr-rat:emb 970 ng/plate JJATDK 1,190,81
dns-rat:lvr 100 mg/L MUREAV 97,359,82
orl-rat TDLo:1870 mg/kg/2Y-I:NEO NCITR* NCI-CG-TR-52,78
ipr-rat LD50:67 mg/kg TXAPA9 78,310,85
ivn-mus LD50:50 mg/kg 85ERAY 2,925,78

CONSENSUS REPORTS: NCI Carcinogenesis Bioassay (gavage); Some Evidence: rat NCITR* NCI-CG-TR-52,78; (gavage); No Evidence: mouse NCITR* NCI-CG-TR-52,78. EPA Genetic Toxicology Program.

SAFETY PROFILE: Poison by intravenous and intraperitoneal routes. Questionable carcinogen with experimental neoplastigenic data. Mutation data reported. When heated to decomposition it emits toxic fumes of NO_x.

NJA000 CAS:5522-43-0 ***HR: 3***
3-NITROPYRENE
mf: $C_{16}H_9NO_2$ mw: 247.26

SYN: 1-NITROPYRENE

TOXICITY DATA with REFERENCE
mmo-esc 3 μg/plate MUREAV 142,163,85
otr-hmn:fbr 3 μmol/L CRNGDP 4,353,83
orl-rat TDLo:396 mg/kg/16W-I:CAR CNREA8 48,4256,88
ipr-rat TDLo:20 mg/kg/4W-I:CAR DTESD7 13,279,86
scu-rat TD:99 mg/kg/8W-I:ETA CNREA8 44,1158,84
ipr-mus TD:1780 mg/kg:NEO PAACA3 25,122,84
scu-rat TD:160 mg/kg/10W-I:ETA,REP CALEDQ 25,239,85

CONSENSUS REPORTS: IARC Cancer Review: Group 3 IMEMDT 7,56,87; Animal Limited Evidence IMEMDT 33,209,84. Reported in EPA TSCA Inventory.

DFG MAK: Suspected Carcinogen.

SAFETY PROFILE: Suspected carcinogen with experimental carcinogenic, neoplastigenic, and tumorigenic data. Experimental reproductive effects. Human mutation data reported. When heated to decomposition it emits toxic fumes of NO_x.

NJA100 CAS:57835-92-4 ***HR: 2***
4-NITROPYRENE
mf: $C_{16}H_9NO_2$ mw: 247.26

TOXICITY DATA with REFERENCE
mmo-sat 1 nmol/plate MUREAV 143,173,85
dnr-bcs 100 ng/disc MUREAV 174,89,86
ipr-mus TDLo:27700 μg/kg/15D-I:CAR CRNGDP 7,1317,86

SAFETY PROFILE: Questionable carcinogen with experimental carcinogenic data. Mutation data reported. When heated to decomposition it emits toxic fumes of NO_x.

NJA500 CAS:1124-33-0 ***HR: 3***
4-NITROPYRIDINE-N-OXIDE
mf: $C_5H_4N_2O_3$ mw: 140.11

$O_2NC_5H_4N{:}O$

SYN: 4-NITROPYRIDINE-1-OXIDE

TOXICITY DATA with REFERENCE
mmo-esc 500 μmol/L GANNA2 70,799,79
dnd-mus:fbr 50 μmol/L CNREA8 35,521,75
scu-mus TDLo:960 mg/kg/15W-I:ETA GANNA2 70,799,79
orl-rat LDLo:8 mg/kg 34ZIAG -,430,69
orl-dog LDLo:34 mg/kg 34ZIAG -,430,69
skn-rbt LDLo:360 mg/kg 34ZIAG -,430,69
orl-ckn LDLo:23 mg/kg 34ZIAG -,430,69

CONSENSUS REPORTS: EPA Extremely Hazardous Substances List. EPA Genetic Toxicology Program. Reported in EPA TSCA Inventory.

SAFETY PROFILE: Poison by ingestion and skin contact. Questionable carcinogen with experimental tumorigenic data. Mutation data reported. Mixtures with diethyl-1,4-dihydro-2,6-dimethylpyridine-3,5-dicarboxyl ate explode when heated above 130°C. When heated to decomposition it emits toxic fumes of NO_x.

NJB500 CAS:18714-34-6 ***HR: 3***
2-NITROQUINOLINE
mf: $C_9H_6N_2O_2$ mw: 174.17

TOXICITY DATA with REFERENCE
scu-mus TDLo:680 mg/kg/31W-I:ETA GANNA2 60,609,69

SAFETY PROFILE: Questionable carcinogen with experimental tumorigenic data. When heated to decomposition it emits toxic fumes of NO_x. See also other nitroquinoline entries.

NJC000 CAS:607-34-1 ***HR: D***
5-NITROQUINOLINE
mf: $C_9H_6N_2O_2$ mw: 174.17

TOXICITY DATA with REFERENCE
mmo-sat 10 μg/plate MUREAV 92,29,82
mmo-esc 500 μg/plate CNREA8 32,2369,72

CONSENSUS REPORTS: EPA Genetic Toxicology Program.

SAFETY PROFILE: Mutation data reported. When heated to decomposition it emits toxic fumes of NO_x. See also other nitroquinoline entries.

NJC500 CAS:613-50-3 ***HR: D***
6-NITROQUINOLINE
mf: $C_9H_6N_2O_2$ mw: 174.17

PROP: Needles. Mp: 149-150°; bp: subl. Sltly sol in cold water; cold alc and ether; sol in benzene.

TOXICITY DATA with REFERENCE
mma-sat 100 μmol/plate MUREAV 58,11,78
mmo-esc 5700 nmol/L ENMUDM 3,11,81

SAFETY PROFILE: Mutation data reported. When heated to decomposition it emits toxic fumes of NO_x. See also other nitroquinoline entries.

NJD500 CAS:607-35-2 ***HR: 3***
8-NITROQUINOLINE
mf: $C_9H_6N_2O_2$ mw: 174.17

PROP: Monoclinic crystals from alcohol. Mp: 88-89°. Sol in hot water, alc, ether, and benzene.

TOXICITY DATA with REFERENCE
mmo-sat 50 μg/plate MUREAV 39,285,77
mma-sat 100 nmol/plate MUREAV 58,11,78
dnr-sat 350 μg/disc MUREAV 39,285,77
orl-rat TDLo:17 g/kg/48W-C:CAR CALEDQ 4,265,78
orl-rat TD:42 g/kg/48W-C:CAR CALEDQ 4,265,78
orl-rat TD:36400 mg/kg/2Y-C:CAR,REP CALEDQ 14,115,81
ipr-mus LDLo:125 mg/kg CBCCT* 6,149,54

CONSENSUS REPORTS: EPA Genetic Toxicology Program. Reported in EPA TSCA Inventory.

SAFETY PROFILE: Poison by intraperitoneal route. Experimental reproductive effects. Questionable carcinogen with experimental carcinogenic data. Mutation data reported. When heated to decomposition it emits toxic fumes of NO_x. See also other nitroquinoline entries.

NJF000 CAS:56-57-5 ***HR: 3***
4-NITROQUINOLINE-N-OXIDE
mf: $C_9H_6N_2O_3$ mw: 190.17

SYNS: 4-NITROCHINOLIN N-OXID (SWEDISH) ◇ 4-NITROQUINOLINE-1-OXIDE ◇ 4-NQO

TOXICITY DATA with REFERENCE
mmo-smc 1 mg/L GNKAA5 21,914,85
dns-hmn:oth 1 μmol/L JJIND8 69,557,82
iut-rat TDLo:50 mg/kg (12D preg):REP 49VWAG -,71,83
orl-rat TDLo:13500 μg/kg/18W-I:ETA ACPADQ 92,437,84
scu-rat TDLo:50 mg/kg/10W-I:CAR GANNA2 57,559,66
orl-mus TDLo:100 mg/kg/12W-I:CAR GANNA2 58,389,67
scu-mus TDLo:80 mg/kg/20W-I:CAR application DTESD7 13,253,86
scu-mus TDLo:25 mg/kg (17D preg):CAR,TER CNREA8 34,3373,74
orl-mus TD:115 mg/kg/23W-I:NEO GANNA2 56,429,65
ipr-rat LDLo:40 mg/kg BCPCA6 13,285,64
scu-rat LD50:12600 μg/kg JJPAAZ 20,264,70
ipr-mus LD50:190 mg/kg JJIND8 62,911,79

CONSENSUS REPORTS: EPA Genetic Toxicology Program.

SAFETY PROFILE: Suspected carcinogen with experimental carcinogenic, neoplastigenic, and tumorigenic data. Poison by intraperitoneal and subcutaneous routes. An experimental teratogen. Other experimental reproductive effects. Human mutation data reported. When heated to decomposition it emits toxic fumes of NO_x.

NJH000 ***HR: 3***
NITROSAMINES

PROP: Compounds which have the chemical group =N N=O attached to an alkyl or aryl group. They are formed by reaction between an amine and NO_x or nitrites.

SAFETY PROFILE: Confirmed carcinogen of the lung, nasal sinus, brain, esophagus, stomach, liver, bladder, and kidney. They are often produced in food as by-products from processing and preparation. They are found in whisky, herbicides, and cosmetics as well as in tanneries, rubber factories, and iron foundries. They can be formed within the body by reaction of amine-containing foods or drugs with the nitrites resulting from bacterial conversion of nitrates. See also N-NITROSO COMPOUNDS.

NJH500 CAS:20661-60-3 ***HR: 3***
N-NITROSARCOSINE
mf: $C_3H_6N_2O_4$ mw: 134.11

SYN: N-METHYL-N-NITROGLYCINE

TOXICITY DATA with REFERENCE
mma-sat 300 μmol/plate JMCMAR 20,1588,77
ipr-mus LD50:247 mg/kg JMCMAR 20,1588,77

SAFETY PROFILE: Poison by intraperitoneal route. Mutation data reported. Many N-nitroso compounds are carcinogens. When heated to decomposition it emits toxic fumes of NO_x. See also N-NITROSO COMPOUNDS.

NJH750 ***HR: D***
NITROSATED COAL DUST EXTRACT

SYN: COAL DUST, EXTRACT, NITROSATED

TOXICITY DATA with REFERENCE
sce-hmn:lym 10 g/L ENMUDM 5,440,83
sce-ham:ovr 3300 mg/L ENMUDM 5,440,83

SAFETY PROFILE: Human mutation data reported.

NJI700 CAS:938-81-8 ***HR: 3***
N-NITROSOACETANILIDE
mf: $C_8H_8N_2O_2$ mw: 164.16

$C_6H_5N(NO)CO{\bullet}CH_3$

SAFETY PROFILE: Many N-nitroso compounds are carcinogens. Explodes on contact with piperidine. Mixture with thiophene explodes at room temperature. When heated to decomposition it emits toxic fumes of NO_x. See also N-NITROSO COMPOUNDS.

NJI850 CAS:75881-17-3 ***HR: 2***
1-NITROSO-4-ACETYL-3,5-DIMETHYLPIPERAZINE
mf: $C_8H_{15}N_3O_2$ mw: 185.26

SYN: 1-ACETYL-4-NITROSO-2,6-DIMETHYLPIPERAZINE

TOXICITY DATA with REFERENCE
orl-rat TDLo:1950 mg/kg/30W-I:CAR CNREA8 41,1034,81

SAFETY PROFILE: Questionable carcinogen with experimental carcinogenic data. When heated to decomposition it emits toxic fumes of NO_x.

NJJ500 CAS:57644-85-6 ***HR: 3***
NITROSOALDICARB
mf: $C_7H_{13}N_3O_3S$ mw: 219.29

SYNS: 2-METHYL-2-(METHYLTHIO)PROPANAL-o-((METHYLNITROSOAMINO)CARBONYL)OXIME ◇ 2-METHYL-2-(METHYLTHIO)PROPIONALDEHYDE-o-((METHYLNITROSO)CARBAMOYL) OXIME

TOXICITY DATA with REFERENCE
mmo-sat 1 μg/plate TCMUE9 1,13,84
dnd-hmn:fbr 10 μmol/L MUREAV 44,1,77
orl-rat TDLo:40 mg/kg/2W-I:ETA EESADV 2,413,78

SAFETY PROFILE: Questionable carcinogen with experimental tumorigenic data. Human mutation data reported. Many N-nitroso compounds are carcinogens. When heated to decomposition it emits very toxic fumes of SO_x and NO_x. See also N-NITROSO COMPOUNDS and CARBANOLATE.

NJJ875 ***HR: 3***
N-NITROSOALLYLETHANOLAMINE
mf: $C_5H_{10}N_2O_2$ mw: 130.17

TOXICITY DATA with REFERENCE
mmo-sat 10 μg/plate TCMUE9 1,13,84
mma-sat 10 μg/plate TCMUE9 1,13,84
orl-rat TDLo:4375 mg/kg/50W-I:ETA CALEDQ 22,281,84

SAFETY PROFILE: Questionable carcinogen with experimental tumorigenic data. Mutation data reported. Many N-nitroso compounds are carcinogens. When heated to decomposition it emits toxic fumes of NO_x. See also N-NITROSO COMPOUNDS and AMINES.

NJJ950 ***HR: 2***
N-NITROSOALLYL-2-HYDROXYPROPYLAMINE
mf: $C_6H_{12}N_2O_2$ mw: 144.20

SYNS: NAHP ◇ NITROSO-ALLYL-2-HYDROXYPROPYLAMINE ◇ N-NITROSO-N-ALLYL-N-(2-HYDROXYPROPYL)AMINE

TOXICITY DATA with REFERENCE
orl-rat TDLo:4950 mg/kg/50W-I:ETA CALEDQ 22,281,84

SAFETY PROFILE: Questionable carcinogen with experimental tumorigenic data. When heated to decomposition it emits toxic fumes of NO_x.

NJK000 CAS:760-56-5 ***HR: D***
NITROSOALLYLUREA
mf: $C_4H_7N_3O_2$ mw: 129.14

SYNS: 1-ALLYL-1-NITROSOUREA ◇ N-NITROSO-N-2-PROPENYLUREA

TOXICITY DATA with REFERENCE
mma-sat 25 μg/plate TCMUE9 1,13,84
pic-esc 4 mg/L TCMUE9 1,91,84

CONSENSUS REPORTS: EPA Genetic Toxicology Program. @thr = SAFETY PROFILE: Mutation data reported. Many N-nitroso compounds are carcinogens. When heated to decomposition it emits toxic fumes of NO_x. See also N-NITROSO COMPOUNDS.

NJK150 CAS:1133-64-8 ***HR: 2***
1-NITROSOANABASINE
mf: $C_{10}H_{13}N_3O$ mw: 191.26

SYNS: NAB ◇ N-NITROSOANABASINE ◇ N′-NITROSOANABASINE ◇ N-NITROSO-2-(3′-PYRIDYL)PIPERIDINE ◇ PYRIDINE, 3-(1-NITROSO-2-PIPERIDINYL)-,(S)-(9CI)

TOXICITY DATA with REFERENCE
orl-rat TDLo:6000 mg/kg/50W-I:CAR BJCAAI 18,265,64

SAFETY PROFILE: Questionable carcinogen with experimental carcinogenic data. When heated to decomposition it emits toxic fumes of NO_x.

NJK500 CAS:40580-89-0 ***HR: 3***
1-NITROSOAZACYCLOTRIDECANE
mf: $C_{12}H_{24}N_2O$ mw: 212.38

SYNS: NDMI ◇ N-NITROSODODECAMETHYLENEIMINE ◇ NITROSODODECAMETHYLENIMINE ◇ N-NITROSODODECAMETHYLENIMINE

TOXICITY DATA with REFERENCE
mma-sat 10 μg/plate TCMUE9 1,13,84
mma-esc 7500 μmol/L IAPUDO 41,543,82
orl-rat TDLo:1040 mg/kg/1Y-I:CAR CRNGDP 5,537,84
orl-mus TDLo:2500 mg/kg/52W-I:ETA IJCNAW 11,369,73
orl-rat TD:1300 mg/kg/1Y-I:CAR CRNGDP 5,537,84

SAFETY PROFILE: Questionable carcinogen with experimental carcinogenic and tumorigenic data. Mutation data reported. When heated to decomposition it emits toxic fumes of NO_x. See also N-NITROSO COMPOUNDS.

NJL000 CAS:15216-10-1 ***HR: 3***
1-NITROSOAZETIDINE
mf: $C_3H_6N_2O$ mw: 86.11

SYNS: N-NITROSAZETIDINE ◇ NITROSO-AZETIDIN (GERMAN) ◇ NITROSOAZETIDINE ◇ N-NITROSOAZETIDINE ◇ NITROSOTRIMETHYLENEIMINE ◇ N-NITROSOTRIMETHYLENEIMINE

TOXICITY DATA with REFERENCE
mmo-sat 1 μg/plate MUREAV 51,319,78
mma-sat 250 μg/plate MUREAV 67,21,79
orl-rat TDLo:198 mg/kg/17W-I:ETA ZKKOBW 89,215,77
orl-rat LDLo:1600 mg/kg NATWAY 54,518,67

SAFETY PROFILE: Moderately toxic by ingestion. Questionable carcinogen with experimental tumorigenic data. Mutation data reported. Many N-nitroso compounds are carcinogens. When heated to decomposition it emits toxic fumes of NO_x. See also N-NITROSO COMPOUNDS.

NJL850 CAS:61034-40-0 ***HR: 2***
1-NITROSO-4-BENZOYL-3,5-DIMETHYLPIPERAZINE
mf: $C_{13}H_{18}N_3O_2$ mw: 248.34

SYNS: 4-BENZOYL-3,5-DIMETHYL-N-NITROSOPIPERAZINE◇ N-NITROSO-4-BENZOYL-3,5-DIMETHYLPIPERAZINE◇ PIPERAZINE, 1-BENZOYL-2,6-DIMETHYL-4-NITROSO-(9CI)

TOXICITY DATA with REFERENCE
mma-sat 250 μg/plate MUREAV 111,135,83
orl-rat TDLo:4300 mg/kg/50W-I:NEO CNREA8 41,1034,81

SAFETY PROFILE: Questionable carcinogen with experimental neoplastigenic data. Mutation data reported. When heated to decomposition it emits toxic fumes of NO_x.

NJM000 CAS:775-11-1 ***HR: 3***
NITROSOBENZYLUREA
mf: $C_8H_9N_3O_2$ mw: 179.20

SYNS: 1-BENZYL-1-NITROSOUREA ◇ N-NITROSO-N-(PHENYLMETHYL)UREA

TOXICITY DATA with REFERENCE
mmo-sat 1 μg/plate MUREAV 68,1,79
mma-sat 250 μg/plate TCMUE9 1,13,84
orl-rat TDLo:1600 mg/kg/92W-I:ETA ZKKOBW 91,63,78

SAFETY PROFILE: Questionable carcinogen with experimental tumorigenic data. Mutation data reported. Many N-nitroso compounds are carcinogens. When heated to decomposition it emits toxic fumes of NO_x. See also N-NITROSO COMPOUNDS.

NJM400 CAS:10125-76-5 ***HR: D***
4-NITROSOBIPHENYL
mf: $C_{12}H_9NO$ mw: 183.22

SYN: 4-NITROSO-1,1'-BIPHENYL

TOXICITY DATA with REFERENCE
mmo-sat 5500 nmol/L ENMUDM 3,11,81
dns-rat:lvr 5 μmol/L ENMUDM 3,11,81

CONSENSUS REPORTS: EPA Genetic Toxicology Program.

SAFETY PROFILE: Mutation data reported. Many N-nitroso compounds are carcinogens. When heated to decomposition it emits toxic fumes of NO_x. See also N-NITROSO COMPOUNDS.

NJM500 CAS:60414-81-5 ***HR: 3***
N-NITROSOBIS(2-ACETOXYPROPYL)AMINE
mf: $C_{10}H_{18}N_2O_5$ mw: 246.30

SYNS: BAP ◇ 1,1-(N-NITROSOIMINO)DI-2-PROPANOL, DIACETATE

TOXICITY DATA with REFERENCE
mma-sat 1 mg/palte CRNGDP 6,415,85
dns-rat:lvr 1 mmol/L MUREAV 144,197,85
scu-mus TDLo:6370 mg/kg/26W-I:CAR CALEDQ 9,257,80
skn-ham TDLo:14 g/kg/41W-I:ETA CALEDQ 3,109,77
scu-ham TDLo:3500 mg/kg/20W-I:NEO CNREA8 36,2877,76
scu-mus TD:7350 mg/kg/15W-I:CAR CALEDQ 9,257,80
scu-mus LD50:4898 mg/kg CALEDQ 9,257,80
scu-ham LD50:6500 mg/kg CALEDQ 1,197,76

SAFETY PROFILE: Questionable carcinogen with experimental carcinogenic, neoplastigenic, and tumorigenic data. Mutation data reported. When heated to decomposition it emits toxic fumes of NO_x. See also N-NITROSO COMPOUNDS.

NJN000 CAS:60599-38-4 ***HR: 3***
N-NITROSOBIS(2-OXOPROPYL)AMINE
mf: $C_6H_{10}N_2O_3$ mw: 158.18

SYNS: BIS-(2-OXOPROPYL)-N-NITROSAMINE◇ BOP ◇ DI-OXO-DI-N-PROPYLNITROSAMINE ◇ 2,2'-DIOXO-DI-N-PROPYLNITROSAMINE ◇ 2,2'-DIOXO-N-NITROSODIPROPYLAMINE ◇ N,N-DI(2-OXOPROPYL)NITROSAMINE ◇ 2,2'-DIOXOPROPYL-N-PROPYLNITROSAMINE ◇ DOPN ◇ N-NITROSO-N,N-DI(2-OXYPROPYL)AMINE ◇ (NITROSOIMINO)DIACETONE

TOXICITY DATA with REFERENCE
mma-sat 250 μg/plate MUREAV 111,135,83
otr-hmn:oth 200 μg/L BANRDU 12,15,82
dns-rat:lvr 1 μmol/L MUREAV 144,197,85
mma-ham:lng 20 μmol/L CNREA8 45,5219,85
orl-rat TDLo:500 mg/kg/1Y-I:CAR,REP CALEDQ 13,303,81

scu-rat TDLo:20 mg/kg (18D post):CAR,REP CNREA8 46,4135,86
scu-gpg TDLo:180 mg/kg/5W-I:CAR CALEDQ 5,31,78
orl-ham TDLo:57 mg/kg/13W-C:CAR CALEDQ 2,323,77
scu-ham TDLo:40 mg/kg (8-14D post):NEO,TER CNREA8 46,3663,86
scu-rat TD:100 mg/kg:NEO CALEDQ 5,13,78
scu-ham TD:10 mg/kg:ETA JJIND8 71,355,83
scu-rat LD50:100 mg/kg CALEDQ 5,13,78
scu-gpg LD50:205 mg/kg CALEDQ 5,31,78
orl-ham LD50:142 mg/kg CALEDQ 2,323,77
scu-ham LD50:89 mg/kg JNCIAM 58,1449,77

SAFETY PROFILE: Suspected carcinogen with experimental carcinogenic, neoplastigenic, tumorigenic data. Poison by ingestion and subcutaneous routes. An experimental teratogen. Human mutation data reported. When heated to decomposition it emits toxic fumes of NO_x. See also N-NITROSO COMPOUNDS.

NJN300 CAS:83335-32-4 ***HR: 3***
N-NITROSO-BIS-(4,4,4-TRIFLUORO-n-BUTYL) AMINE
mf: $C_8H_{12}F_6N_2O$ mw: 266.22

SYNS: F-6-NDBA ◇ 4,4,4,4',4',4'-HEXAFLUORO-N-NITROSODIBUTYLAMINE ◇ N-NITROSO-4,4,4,4',4',4'-HEXAFLUORODIBUTYLAMINE

TOXICITY DATA with REFERENCE
mma-sat 1 μmol/plate CRNGDP 3,781,82
dnd-esc 100 nmol/tube CRNGDP 3,781,82
orl-rat TDLo:960 g/kg/32W-I:CAR CRNGDP 3,1219,82
orl-rat TD:3150 mg/kg/32W-I:CAR CRNGDP 3,1219,82
orl-rat LD50:750 mg/kg CRNGDP 3,1219,82

SAFETY PROFILE: Moderately toxic by ingestion. Questionable carcinogen with experimental carcinogenic data. Mutation data reported. When heated to decomposition it emits toxic fumes of F^- and NO_x. See also N-NITROSO COMPOUNDS.

NJN500 CAS:69113-02-6 ***HR: D***
NITROSOBROMOETHYLUREA
mf: $C_3H_6BrN_3O_2$ mw: 196

SYN: 1-(2-BROMOETHYL)-1-NITROSOUREA

TOXICITY DATA with REFERENCE
mmo-sat 1 μg/plate MUREAV 68,1,79

SAFETY PROFILE: Mutation data reported. Many N-nitroso compounds are carcinogens. When heated to decomposition it emits very toxic fumes of Br^- and NO_x. See also N-NITROSO COMPOUNDS.

NJO000 CAS:56375-33-8 ***HR: D***
N-NITROSOBUTYLAMINE
mf: $C_4H_{10}N_2O$ mw: 102.16

SYNS: BUTYLNITROSAMINE ◇ N-NITROSOBUTANAMINE

TOXICITY DATA with REFERENCE
cyt-rat-orl 350 mg/kg ZKKOBW 86,47,76

SAFETY PROFILE: Mutation data reported. Many N-nitroso compounds are carcinogens. When heated to decomposition it emits toxic fumes of NO_x. See also N-NITROSO COMPOUNDS.

NJO150 ***HR: 2***
N-NITROSO-N-BUTYLBUTYRAMIDE
mf: $C_8H_{16}N_2O_2$ mw: 172.26

SYNS: N-BUTYL-N-NITROSOBUTYRAMIDE ◇ N-BUTYL-N-(1-OXOBUTYL)NITROSAMINE ◇ NBBA ◇ N-NITROSO-N-BUTYROXYBUTYLAMINE

TOXICITY DATA with REFERENCE
scu-rat TDLo:65 mg/kg/10W-I:CAR IAPUDO 41,619,82
scu-rat TD:70500 μg/kg/10W-I:ETA GANNA2 73,687,82

SAFETY PROFILE: Questionable carcinogen with experimental carcinogenic and tumorigenic data. When heated to decomposition it emits toxic fumes of NO_x.

NJO200 CAS:73487-24-8 ***HR: 3***
N-NITROSO-N-(BUTYL-N-BUTYROLACTONE) AMINE
mf: $C_8H_{14}N_2O_3$ mw: 186.24

SYNS: BBAL ◇ N-n-BUTYL-N-4-(1,4-BUTYROLACTONE)NITROSAMINE ◇ 4-(N-BUTYLNITROSAMINO)-4-HYDROXYBUTYRIC ACID LACTONE

TOXICITY DATA with REFERENCE
mmo-sat 50 nmol/plate CNREA8 40,162,80
mmo-esc 500 nmol/plate CNREA8 40,162,80
dnr-bcs 100 nmol/plate CNREA8 40,162,80
dns-rat:oth 1 mmol/L CBINA8 53,99,85
scu-rat TDLo:70500 μg/kg/10W-I:CAR GANNA2 73,687,82
scu-rat TD:70 mg/kg/10W-I:CAR IAPUDO 41,619,82

SAFETY PROFILE: Questionable carcinogen with experimental carcinogenic data. Mutation data reported. When heated to decomposition it emits toxic fumes of NO_x. See also N-NITROSO COMPOUNDS.

NJO300 CAS:46061-25-0 ***HR: 2***
N-NITROSO-4-tert-BUTYLPIPERIDINE
mf: $C_9H_{18}N_2O$ mw: 170.29

SYNS: 4-tert-BUTYL-1-NITROSOPIPERIDINE ◇ PIPERIDINE, 4-tert-BUTYL-1-NITROSO

TOXICITY DATA with REFERENCE
mma-sat 250 μg/plate MUREAV 111,135,83
orl-rat TDLo:4500 mg/kg/2Y-I:CAR CRNGDP 2,1045,81

SAFETY PROFILE: Questionable carcinogen with experimental carcinogenic data. Mutation data reported.

When heated to decomposition it emits toxic fumes of NO_x.

NJO500 CAS:71752-66-4 **HR: 3**
NITROSO-sec-BUTYLUREA
mf: $C_5H_{11}N_3O_2$ mw: 145.19

SYN: 1-sec-BUTYL-1-NITROSOUREA

TOXICITY DATA with REFERENCE
mmo-sat 100 µg/plate MUREAV 68,1,79
mma-sat 500 µg/plate TCMUE9 1,13,84
skn-mus TDLo:585 mg/kg/50W-I:ETA JCROD7 102,13,81

SAFETY PROFILE: Questionable carcinogen with experimental tumorigenic data. Mutation data reported. Many N-nitroso compounds are carcinogens. When heated to decomposition it emits toxic fumes of NO_x. See also N-NITROSO COMPOUNDS.

NJP000 CAS:62573-59-5 **HR: 3**
NITROSO-BUX-TEN

SYN: 3-(1-ETHYLPROPYL)PHENOLMETHYLNITROSOCARBAMATE mixed with 3-(1-METHYLBUTYL)PHENYL METHYLNITROSOCARBAMATE

TOXICITY DATA with REFERENCE
orl-rat TDLo:660 mg/kg/10W-I:ETA IARCCD 19,495,78

CONSENSUS REPORTS: EPA Genetic Toxicology Program. EPA Genetic Toxicology Program.

SAFETY PROFILE: Questionable carcinogen with experimental tumorigenic data. Many N-nitroso compounds are carcinogens. When heated to decomposition it emits toxic fumes of NO_x. See also N-NITROSO COMPOUNDS and CARBAMATES.

NJP500 CAS:101418-02-4 **HR: D**
N-NITROSOCARBANILIC ACID ISOPROPYLESTER
mf: $C_{10}H_{12}N_2O_3$ mw: 208.24

SYN: NITROSOPROPHAM

TOXICITY DATA with REFERENCE
mmo-sat 1 µL/plate MUREAV 48,225,77

SAFETY PROFILE: Mutation data reported. Many N-nitroso compounds are carcinogens. When heated to decomposition it emits toxic fumes of NO_x. See also N-NITROSO COMPOUNDS.

NJQ000 CAS:13256-15-0 **HR: 3**
N-NITROSO-N'-CARBETHOXYPIPERAZINE
mf: $C_7H_{13}N_3O_3$ mw: 187.23

SYNS: ETHYL 4-NITROSO-1-PIPERAZINECARBOXYLATE ◇ N-NITROSO-N'-CARBAETHOXYPIPERAZIN (GERMAN)

TOXICITY DATA with REFERENCE
scu-rat TDLo:2300 mg/kg/66W-I:ETA ZEKBAI 69,103,67
scu-rat LD50:400 mg/kg ZEKBAI 69,103,67

SAFETY PROFILE: Poison by subcutaneous route. Questionable carcinogen with experimental tumorigenic data. Many N-nitroso compounds are carcinogens. When heated to decomposition it emits toxic fumes of NO_x. See also N-NITROSO COMPOUNDS.

NJQ500 CAS:62593-23-1 **HR: 3**
NITROSOCARBOFURAN
mf: $C_{12}H_{14}N_2O_4$ mw: 250.28

SYNS: 2,3-DIHYDRO-2,2-DIMETHYL-7-BENZOFURANYLNITROSOMETHYL CARBAMATE ◇ METHYLNITROSOCARBAMIC ACID-2,3-DIHYDRO-2,3-DIMETHYL-7-BENZOFURANYLESTER

TOXICITY DATA with REFERENCE
mmo-sat 2500 ng/plate TCMUE9 1,13,84
dnd-hmn:fbr 10 µmol/L MUREAV 44,1,77
orl-rat TDLo:380 mg/kg/23W-I:CAR EESADV 2,413,78

SAFETY PROFILE: Questionable carcinogen with experimental carcinogenic data. Human mutation data reported. When heated to decomposition it emits toxic fumes of NO_x. See also N-NITROSO COMPOUNDS and CARBAMATES.

NJR000 CAS:69113-00-4 **HR: 3**
NITROSOCHLOROETHYLDIETHYLUREA
mf: $C_7H_{14}ClN_3O_2$ mw: 207.69

TOXICITY DATA with REFERENCE
orl-rat TDLo:176 mg/kg/46W-I:ETA JCROD7 94,131,79

SAFETY PROFILE: Questionable carcinogen with experimental tumorigenic data. Many N-nitroso compounds are carcinogens. When heated to decomposition it emits very toxic fumes of Cl^- and NO_x. See also N-NITROSO COMPOUNDS.

NJR500 CAS:59960-30-4 **HR: 3**
NITROSOCHLOROETHYLDIMETHYLUREA
mf: $C_5H_{10}ClN_3O_2$ mw: 179.63

SYN: 1-NITROSO-1-(2-CHLOROETHYL)-3,3-DIMETHYLUREA

TOXICITY DATA with REFERENCE
orl-rat TDLo:168 mg/kg/50W-I:ETA JCROD7 94,131,79

SAFETY PROFILE: Questionable carcinogen with experimental tumorigenic data. Many N-nitroso compounds are carcinogens. When heated to decomposition it emits very toxic fumes of Cl^- and NO_x. See also N-NITROSO COMPOUNDS.

NJS300 CAS:73785-40-7 **HR: 2**
NITROSOCIMETIDINE
mf: $C_{10}H_{15}N_7OS$ mw: 281.38

SYN: N-NITROSOCIMETIDINE

TOXICITY DATA with REFERENCE
mma-sat 1 mg/plate TOLED5 12,281,82
sce-hmn:lym 260 µmol/L MUREAV 156,117,85
orl-mus TDLo:1316 mg/(pre-post):CAR,TER CNREA8 45,3561,85

SAFETY PROFILE: Questionable carcinogen with experimental carcinogenic data. An experimental teratogen. Human mutation data reported. Many N-nitroso compounds are carcinogens. When heated to decomposition it emits toxic fumes of SO_x and NO_x. See also N-NITROSO COMPOUNDS.

NJS500 CAS:33904-55-1 ***HR: D***
NITROSO-dl-CITRULLINE
mf: $C_6H_{12}N_4O_4$ mw: 204.22

SYNS: dl-N^5-(AMINOCARBONYL)-N^5-NITROSOORNITHINE ◇ dl-N^5-CARBAMOYL-N^5-NITROSOORNITHINE

TOXICITY DATA with REFERENCE
mmo-sat 1500 µmol/L MUREAV 48,131,77

SAFETY PROFILE: Mutation data reported. Many N-nitroso compounds are carcinogens. When heated to decomposition it emits toxic fumes of NO_x. See also N-NITROSO COMPOUNDS.

NJT000 CAS:42713-66-6 ***HR: D***
NITROSO-l-CITRULLINE
mf: $C_6H_{12}N_4O_4$ mw: 204.22

SYNS: l-N^5-(AMINOCARBONYL)-N^5-NITROSOORNITHINE ◇ l-N^5-CARBAMOYL-N^5-NITROSOORNITHINE

TOXICITY DATA with REFERENCE
mmo-sat 1500 µmol/L MUREAV 48,131,77

SAFETY PROFILE: Mutation data reported. Many N-nitroso compounds are carcinogens. When heated to decomposition it emits toxic fumes of NO_x. See also N-NITROSO COMPOUNDS.

NJT500 ***HR: 3***
NITROSO COMPOUNDS

PROP: Compounds of the form C−N=O or N−N=O. Organic nitrogen compounds.

SAFETY PROFILE: Usually highly toxic carcinogens, teratogens, and mutagens by almost all routes of exposure. Some of these compounds may have hazardous instabilities under the appropriate conditions. When heated to decomposition they emit very toxic fumes of NO_x. See also specific compounds.

NJT550 ***HR: 3***
N-NITROSO COMPOUNDS

PROP: A class of organic compounds of the form R_2−N N=O or R=N−N=O.

SAFETY PROFILE: Many members of this class are toxins, carcinogens, teratogens, and mutagens. Sources of exposure to N-nitroso compounds are: formation in the environment and absorption from food, water, air, or industrial and consumer products; formation in the body from precursors in food, water, or air; from tobacco; and from naturally occurring compounds. Some are used in the production of rubber and they may be formed as by-products in industrial processes. Nitrosamines have been found in food and cosmetics. N-nitroso compounds can be formed from the reaction of nitrates with nitrite under acidic conditions. These conditions can occur in the environment, mouth, and stomach. Nitrites are formed in the mouth by the action of bacteria on nitrates. Nitrosatable substances in the environment include secondary and tertiary amines, quaternary ammonium compounds, ureas, carbamates, and guanidines. Many of the resulting N-nitroso compounds are experimental carcinogens and mutagens. See also individual compounds, NITROSAMINES, CARBAMATES, and SMOKELESS TOBACCO.

NJU000 CAS:100700-20-7 ***HR: D***
4-(NITROSOCYANAMIDO)BUTYRAMIDE
mf: $C_5H_8N_4O_2$ mw: 156.17

SYN: Γ-GUANIDINOBUTYRIC ACID AMIDE, NITROSATED

TOXICITY DATA with REFERENCE
mmo-sat 156 µmol/L GANNA2 65,45,74

SAFETY PROFILE: Mutation data reported. Many N-nitroso compounds are carcinogens. When heated to decomposition it emits toxic fumes of NO_x. See also N-NITROSO COMPOUNDS.

NJU500 CAS:102584-88-3 ***HR: D***
5-(NITROSOCYANAMIDO)-2-HYDROXYVALERAMIDE
mf: $C_6H_{10}N_4O_3$ mw: 186.20

SYN: l-ARGININE, nitrosated

TOXICITY DATA with REFERENCE
mmo-sat 10 mmol/L GANNA2 65,45,74

SAFETY PROFILE: Mutation data reported. Many N-nitroso compounds are carcinogens. When heated to decomposition it emits toxic fumes of NO_x. See also N-NITROSO COMPOUNDS.

NJV000 CAS:877-31-6 ***HR: 3***
NITROSOCYCLOHEXYLUREA
mf: $C_7H_{13}N_3O_2$ mw: 171.23

SYN: 1-CYCLOHEXYL-1-NITROSOUREA

TOXICITY DATA with REFERENCE
mmo-sat 100 μg/plate MUREAV 68,1,79
mma-sat 250 μg/plate TCMUE9 1,13,84
pic-esc 10 mg/L TCMUE9 1,91,84
skn-mus TDLo:685 mg/kg/50W-I:ETA JCROD7 102,13,81

SAFETY PROFILE: Questionable carcinogen with experimental tumorigenic data. Mutation data reported. Many N-nitroso compounds are carcinogens. When heated to decomposition it emits toxic fumes of NO_x. See also N-NITROSO COMPOUNDS.

NJV500 CAS:16338-97-9 ***HR: 3***
N-NITROSODIALLYL AMINE
mf: $C_6H_{10}N_2O$ mw: 126.18

SYNS: DIALLYINITROSAMIN (GERMAN) ◇ DIALLYLNITROSAMINE

TOXICITY DATA with REFERENCE
mma-sat 10 μg/plate MUREAV 66,1,79
mma-esc 5 mmol/L MUREAV 89,209,81
mmo-omi 1 pph/48H-C SOGEBZ 10,522,74
dni-mus-ipr 20 g/kg ARGEAR 51,605,81
orl-rat TDLo:11760 mg/kg/42W-C:ETA JCROD7 109,5,85
orl-ham TDLo:805 g/kg/37W-I:CAR JJIND8 74,1043,85
scu-ham TDLo:500 mg/kg:NEO JNCIAM 59,1569,77
orl-rat LD50:800 mg/kg ZEKBAI 69,103,67
scu-ham LD50:1230 mg/kg JNCIAM 59,1569,77

CONSENSUS REPORTS: EPA Genetic Toxicology Program.

SAFETY PROFILE: Moderately toxic by ingestion and subcutaneous routes. Questionable carcinogen with experimental carcinogenic, neoplastigenic, and tumorigenic data. Mutation data reported. When heated to decomposition it emits toxic fumes of NO_x. See also N-NITROSO COMPOUNDS and AMINES.

NJW000 CAS:5350-17-4 ***HR: 3***
NITROSODI-sec-BUTYLAMINE
mf: $C_8H_{18}N_2O$ mw: 158.28

SYN: N-NITROSO-DI-sec-BUTYLAMINE

TOXICITY DATA with REFERENCE
orl-rat TDLo:2750 mg/kg/50W-I:ETA JJIND8 62,407,79

SAFETY PROFILE: Questionable carcinogen with experimental tumorigenic data. Many N-nitroso compounds are carcinogens. When heated to decomposition it emits toxic fumes of NO_x. See also N-NITROSO COMPOUNDS and AMINES.

NJW500 CAS:55-18-5 ***HR: 3***
N-NITROSODIETHYLAMINE
mf: $C_4H_{10}N_2O$ mw: 102.16

PROP: Yellow oil. D: 0.9422 @ 20°/4°, bp: 47° @ 5 mm, bp: 176.9°. Sol in water, alc, and ether.

SYNS: DANA ◇ DEN ◇ DENA ◇ DIAETHYLNITROSAMIN (GERMAN) ◇ DIETHYLNITROSAMINE ◇ N,N-DIETHYLNITROSAMINE ◇ DIETHYLNITROSOAMINE ◇ N-ETHYL-N-NITROSO-ETHANAMINE ◇ NDEA ◇ N-NITROSODIAETHYLAMINE (GERMAN) ◇ NITROSODIETHYLAMINE ◇ RCRA WASTE NUMBER U174

TOXICITY DATA with REFERENCE
mma-esc 2 μmol/plate MUREAV 135,87,84
otr-hmn:oth 1 g/L/9W-I PAACA3 25,135,84
orl-ham TDLo:150 mg/kg (lactating female 30D post):REP PSEBAA 136,1007,71
orl-rat TDLo:158 mg/kg (female 22D post):TER IARCCD 4,45,73
orl-rat TDLo:119 mg/kg/3.3Y-C:CAR CRNGDP 8,1635,87
orl-rat TDLo:150 mg/kg (22D post):CAR,TER IARCCD 4,92,73
ipr-rat LD:75 mg/kg:NEO CNREA8 48,2492,88
scu-rat TDLo:480 mg/kg/12D-I:ETA,TER NATWAY 54,47,67
scu-rat TDLo:480 mg/kg (10-21D post):ETA NATWAY 54,47,67
ivn-rat TDLo:80 mg/kg:CAR JNCIAM 49,1729,72
ivn-rat TDLo:150 mg/kg (22D post):CAR,TER IARCCD 4,45,73
ipr-mus TDLo:120 mg/kg (21D post):NEO,REP VOONAW 17(1),45,71
scu-mus TDLo:80 mg/kg (15-20D post):NEO,REP ZEKBAI 67,152,65
orl-ham TDLo:150 mg/kg (post):ETA,REP PSEBAA 136,1007,71
ipr-ham TDLo:30 mg/kg (15D post):ETA,TER ZAPPAN 121,82,77
scu-ham TDLo:1250 μg/kg (15D post):NEO,TER CNREA8 47,5112,87
orl-rat LD50:280 mg/kg ARZNAD 13,841,63
ipr-rat LD50:216 mg/kg BIJOAK 85,72,62
scu-rat LD50:195 mg/kg EXPADD 27,171,85
ivn-rat LD50:280 mg/kg IARCCD 4,45,73
ipr-mus LD50:132 mg/kg JJIND8 62,911,79
orl-gpg LD50:250 mg/kg ZEKBAI 69,103,67

CONSENSUS REPORTS: NTP Fifth Annual Report on Carcinogens. IARC Cancer Review: Group 2A IMEMDT 7,56,87; Animal Sufficient Evidence IMEMDT 1,107,72, IMEMDT 17,83,78, IMEMDT 28,151,82; Human Limited Evidence IMEMDT 17,83,78. NCI Carcinogenesis Studies (ipr); Clear Evidence: mouse, rat

RRCRBU 52,1,75. Reported in EPA TSCA Inventory. Community Right-To-Know List.

DFG MAK: Animal Carcinogen, Suspected Human Carcinogen.

SAFETY PROFILE: Confirmed carcinogen with experimental carcinogenic, neoplastigenic, and tumorigenic data. Poison by ingestion, intravenous, intraperitoneal, and subcutaneous routes. An experimental teratogen. Other experimental reproductive effects. Human mutation data reported. A transplacental carcinogen. When heated to decomposition it emits toxic fumes of NO_x. See also N-NITROSO COMPOUNDS and AMINES.

NJX000 CAS:3276-41-3 ***HR: 3***
N-NITROSO-3,6-DIHYDRO-1,2-OXAZINE
mf: $C_4H_6N_2O_2$ mw: 114.12

SYNS: 3,6-DIHYDRO-2-NITROSO-2H-1,2-OXAZINE ◇ N-NITROSO-3,6-DIHYDROOXAZIN-1,2(GERMAN)

TOXICITY DATA with REFERENCE
orl-rat TDLo:22 g/kg/94W-I:NEO ZKKOBW 79,114,73
orl-rat LD50:900 mg/kg ZKKOBW 79,114,73

SAFETY PROFILE: Moderately toxic by ingestion. Questionable carcinogen with experimental neoplastigenic data. Many N-nitroso compounds are carcinogens. When heated to decomposition it emits toxic fumes of NO_x. See also N-NITROSO COMPOUNDS.

NJX500 CAS:62641-67-2 ***HR: D***
1-NITROSO-5,6-DIHYDROTHYMINE
mf: $C_5H_7N_3O_3$ mw: 157.15

SYNS: DIHYDRO-5-METHYL-1-NITROSO-2,4(1H,3H)-PYRIMIDINEDIONE ◇ 5,6-DIHYDRO-1-NITROSOTHYMINE

TOXICITY DATA with REFERENCE
mmo-sat 38 μmol/L MUREAV 48,131,77
dnd-rat-ipr 700 μmol/kg PAACA3 24,73,83

CONSENSUS REPORTS: EPA Genetic Toxicology Program.

SAFETY PROFILE: Mutation data reported. Many N-nitroso compounds are carcinogens. When heated to decomposition it emits toxic fumes of NO_x. See also N-NITROSO COMPOUNDS.

NJY000 CAS:16813-36-8 ***HR: 3***
1-NITROSO-5,6-DIHYDROURACIL
mf: $C_4H_5N_3O_3$ mw: 143.12

SYNS: DIHYDRO-1-NITROSO-2,4(1H,3H)-PYRIMIDINEDIONE ◇ 5,6-DIHYDRO-1-NITROSOURACIL ◇ NDHU ◇ NO-DHU

TOXICITY DATA with REFERENCE
mmo-sat 2900 nmol/L MUREAV 48,131,77
dnd-rat-ipr 1 mg/kg BBRCA9 53,773,73
dns-rat:lvr 7 mmol/L JJIND8 73,515,84
orl-rat TDLo:440 mg/kg/41W-I:CAR JJIND8 62,1523,79
ipr-rat TDLo:1200 mg/kg/8D-I:ETA TUMOAB 61,509,75
orl-rat TD: 3000 mg/kg/32W-I:CAR ZKKOBW 79,304,73
ipr-rat LD50:850 mg/kg ZKKOBW 79,304,73

CONSENSUS REPORTS: EPA Genetic Toxicology Program.

SAFETY PROFILE: Moderately toxic by intraperitoneal route. Questionable carcinogen with experimental carcinogenic and tumorigenic data. Mutation data reported. When heated to decomposition it emits toxic fumes of NO_x. See also N-NITROSO COMPOUNDS.

NJY500 CAS:88208-16-6 ***HR: 2***
N-NITROSO-2,3-DIHYDROXYPROPYLALLYL-AMINE
mf: $C_6H_{12}N_2O_3$ mw: 160.20

SYNS: 3-(ALLYLNITROSAMINO)-1,2-PROPANEDIOL ◇ N-NITROSOALLYL-2,3-DIHYDROXYPROPYLAMINE ◇ 1,2-PROPANEDIOL, 3-(NITROSO-2-PROPENYLAMINO)-

TOXICITY DATA with REFERENCE
mma-sat 250 μg/plate MUREAV 111,135,83
orl-rat TDLo:5500 mg/kg/50W-I:ETA CALEDQ 22,281,84

SAFETY PROFILE: Questionable carcinogen with experimental tumorigenic data. Mutation data reported. When heated to decomposition it emits toxic fumes of NO_x.

NJY550 ***HR: 3***
NITROSO-DIHYDROXYPROPYLOXOPRO-PYLAMINE
mf: $C_6H_{12}N_2O_4$ mw: 176.20

SYNS: 1-((2,3-DIHYDROXYPROPYL)NITROSAMINO)-2-PROPANONE ◇ N-NITROSODIHYDROXYPROPYL-2-OXOPROPYLAMINE ◇ N-NITROSO-2-OXOPROPYL-2,3-DIHYDROXYPROPYLAMINE

TOXICITY DATA with REFERENCE
mma-sat 50 μg/plate TCMUE9 1,13,84
orl-rat TDLo:528 mg/kg/26W-I:ETA IAPUDO 57,617,84

SAFETY PROFILE: Questionable carcinogen with experimental tumorigenic data. Mutation data reported. Many N-nitroso compounds are carcinogens. When heated to decomposition it emits toxic fumes of NO_x. See also N-NITROSO COMPOUNDS and AMINES.

NKA000 CAS:601-77-4 ***HR: 3***
N-NITROSODIISOPROPYLAMINE
mf: $C_6H_{14}N_2O$ mw: 130.22

SYNS: DIISOPROPYLNITROSAMIN (GERMAN) ◇ N-NITROSODI-i-PROPYLAMINE (MAK)

TOXICITY DATA with REFERENCE
orl-rat TDLo:14 g/kg/110W-C:CAR ZEKBAI 69,103,67

orl-rat TD:1800 mg/kg/50W-I:ETA JJIND8 62,407,79
orl-rat LD50:850 mg/kg ZEKBAI 69,103,67

DFG MAK: Animal Carcinogen, Suspected Human Carcinogen.

SAFETY PROFILE: Confirmed carcinogen with experimental carcinogenic and tumorigenic data. Moderately toxic by ingestion. When heated to decomposition it emits toxic fumes of NO_x. See also N-NITROSO COMPOUNDS and AMINES.

NKA500 CAS:63124-33-4 ***HR: D***
NITROSODIMETHOATE
mf: $C_5H_{11}N_2O_4PS_2$ mw: 258.27

SYN: PHOSPHORODITHIOIC ACID-O,O-DIMETHYL ESTER, S-ESTER with 2-MERCAPTO-N-METHYL-N-NITROSO ACETAMIDE

TOXICITY DATA with REFERENCE
mmo-sat 1 μL/plate MUREAV 48,225,77

SAFETY PROFILE: Mutation data reported. Many N-nitroso compounds are carcinogens. When heated to decomposition it emits very toxic fumes of SO_x, PO_x and NO_x. See also N-NITROSO COMPOUNDS and MERCAPTANS.

NKA600 CAS:62-75-9 ***HR: 3***
N-NITROSODIMETHYLAMINE
mf: $C_2H_6N_2O$ mw: 74.10

PROP: Yellow liquid; sol in water, alc, and ether. Bp: 152°, d: 1.005 @ 20°/4°.

SYNS: DIMETHYLNITROSAMIN (GERMAN) ◇ DIMETHYLNITROSAMINE ◇ N,N-DIMETHYLNITROSAMINE ◇ DIMETHYLNITROSOAMINE ◇ DMN ◇ DMNA ◇ N-METHYL-N-NITROSOMETHANAMINE ◇ NDMA ◇ NITROSODIMETHYLAMINE ◇ RCRA WASTE NUMBER P082

TOXICITY DATA with REFERENCE
mma-sat 100 μg/plate ENMUDM 6(Suppl 2),1,84
msc-hmn:lym 14 mmol/L MUREAV 128,221,84
ipr-mus TDLo:10 mg/kg (female 18D post):REP TXCYAC 24,251,82
orl-rat TDLo:30 mg/kg (10D preg):TER VOONAW 13(5),87,67
orl-rat TDLo:30 mg/kg (female 21D post):NEO,TER BEXBAN 78,1308,74
orl-rat TDLo:23 mg/kg/2Y-I:CAR CNREA8 41,4997,81
ihl-rat TCLo:200 μg/m³/45W-C:CAR VOONAW 21(6),107,75
scu-rat TDLo:24466 μg/kg/20W-I:ETA CNREA8 46,498,86
ihl-mus TCLo:200 μg/m³/45W-C:NEO VOONAW 21(6),107,75
orl-wmn LDLo:20 mg/kg/2.5Y:GIT,MET ONCOBS 37,273,80
orl-rat LD50:58 mg/kg TXAPA9 27,380,74
ihl-rat LC50:78 ppm/4H AMIHAB 12,617,55
ipr-rat LD50:26500 μg/kg BJIMAG 11,167,54
scu-rat LD50:45 mg/kg ZKKOBW 80,17,73
ivn-rat LDLo:40 mg/kg BJIMAG 11,167,54
ihl-mus LC50:57 ppm/4H AMIHAB 12,617,55
ipr-mus LD50:19 mg/kg EXPEAM 32,495,76
orl-dog LDLo:20 mg/kg BJIMAG 11,167,54
ihl-dog LCLo:16 ppm/4H AMIHAB 12,617,55

CONSENSUS REPORTS: NTP Fifth Annual Report on Carcinogens. IARC Cancer Review: Group 2A IMEMDT 7,56,87; Animal Sufficient Evidence IMEMDT 17,125,78, IMEMDT 1,95,72; Human Limited Evidence IMEMDT 17,125,78; Human Inadequate Evidence IMEMDT 1,95,72. Reported in EPA TSCA Inventory. EPA Genetic Toxicology Program. Community Right-To-Know List. EPA Extremely Hazardous Substances List.

OSHA PEL: Cancer Suspect Agent
ACGIH TLV: Suspected Human Carcinogen
DFG MAK: Animal Carcinogen, Suspected Human Carcinogen.

SAFETY PROFILE: Confirmed carcinogen with experimental carcinogenic, neoplastigenic, tumorigenic, and teratogenic data. A transplacental carcinogen. Human poison by ingestion. Experimental poison by ingestion, inhalation, intraperitoneal, subcutaneous, and intravenous routes. Human systemic effects by ingestion: ulceration or bleeding from small intestine, nausea or vomiting, and fever. Experimental reproductive effects. Human mutation data reported. Has caused fatal liver disease in humans. When heated to decomposition it emits toxic fumes of NO_x. See also NITROSAMINES.

NKA695 CAS:69091-16-3 ***HR: 2***
cis-NITROSO-2,6-DIMETHYLMORPHOLINE
mf: $C_6H_{12}N_2O_2$ mw: 144.20

SYNS: MORPHOLINE, 2,6-DIMETHYL-N-NITROSO-, (cis)- ◇ MORPHOLINE, 2,6-DIMETHYL-4-NITROSO-, (Z)- ◇ cis-N-NITROSO-2,6-DIMETHYLMORPHOLINE

TOXICITY DATA with REFERENCE
mma-sat 100 nmol/plate MUREAV 77,215,80
orl-rat TDLo:190 mg/kg/30W-I:ETA CRNGDP 1,501,80

SAFETY PROFILE: Questionable carcinogen with experimental tumorigenic data. Mutation data reported. When heated to decomposition it emits toxic fumes of NO_x.

NKA700 CAS:69091-15-2 ***HR: 3***
trans-NITROSO-2,6-DIMETHYLMORPHOLINE
mf: $C_6H_{12}N_2O_2$ mw: 144.20

SYN: trans-N-NITROSO-2,6-DIMETHYLMORPHOLINE

TOXICITY DATA with REFERENCE
mma-sat 1 mg/plate TCMUD8 1,295,80
pic-esc 100 mg/L TCMUE9 1,91,84
orl-rat TDLo:100 mg/kg/30W-I:ETA CRNGDP 1,501,80

SAFETY PROFILE: Questionable carcinogen with experimental tumorigenic data. Mutation data reported. Many N-nitroso compounds are carcinogens. When heated to decomposition it emits toxic fumes of NO_x. See also N-NITROSO COMPOUNDS.

NKA850 CAS:67774-31-6 ***HR: 2***
1-NITROSO-3,5-DIMETHYLPIPERAZINE
mf: $C_6H_{13}N_3O$ mw: 143.22

SYNS: 3,5-DIMETHYL-1-NITROSOPIPERAZINE ◇ NITROSO-3,5-DIMETHYLPIPERAZINE ◇ N-NITROSO-3,5-DIMETHYLPIPERAZINE

TOXICITY DATA with REFERENCE
mma-sat 10 μg/plate TCMUE9 1,13,84
orl-rat TDLo:1450 mg/kg/29W-I:CAR CNREA8 41,1034,81

SAFETY PROFILE: Questionable carcinogen with experimental carcinogenic data. Mutation data reported. When heated to decomposition it emits toxic fumes of NO_x.

NKB000 ***HR: D***
NITROSODIOCTYLAMINE
mf: $C_{16}H_{34}N_2O$ mw: 270.52

SYN: N-NITROSO-DI-N-OCTYLAMINE

TOXICITY DATA with REFERENCE
mma-sat 250 μg/plate MUREAV 51,319,78

SAFETY PROFILE: Mutation data reported. Many N-nitroso compounds are carcinogens. When heated to decomposition it emits toxic fumes of NO_x. See also N-NITROSO COMPOUNDS and AMINES.

NKB500 CAS:156-10-5 ***HR: 3***
p-NITROSODIPHENYLAMINE
mf: $C_{12}H_{10}N_2O$ mw: 198.24

PROP: Green plates with bluish luster (from benzene) or steel blue prisms or plates (from ether + H_2O). Mp: 144-145°. Sltly sol in water or petr ether; very sol in alc, ether, benzene, chloroform.

SYNS: NAUGARD TKB ◇ NCI-C02244 ◇ p-NITROSODIFENYLAMIN (CZECH) ◇ 4-NITROSODIPHENYLAMINE ◇ p-NITROSO-N-PHENYLANILINE ◇ 4-NITROSO-N-PHENYLANILINE ◇ 4-NITROSO-N-PHENYLBENZENAMINE ◇ N-PHENYL-p-NITROSOANILINE ◇ TKB

TOXICITY DATA with REFERENCE
eye-rbt 100 mg/24H MOD 28ZPAK -,133,72
mma-sat 200 μg/plate ENMUDM 5(Suppl 1),3,83
dns-mus:ast 50 μmol/L MUREAV 89,95,81
orl-rat TDLo:1365 g/kg/78W-C:CAR IARC** 27,227,82
orl-mus TDLo:204 g/kg/57W-C:CAR NCITR* NCI-CG-TR-190,79
orl-rat TD:164 g/kg/78W-C:NEO NCITR* NCI-CG-TR-190,79
orl-rat LD50:2140 mg/kg 28ZPAK -,133,72
ivn-mus LD50:178 mg/kg CSLNX* NX#00267

CONSENSUS REPORTS: NTP Fifth Annual Report on Carcinogens. IARC Cancer Review: Group 3 IMEMDT 7,56,87; Animal Inadequate Evidence IMEMDT 27,227,82. NCI Carcinogenesis Bioassay (feed); Clear Evidence: mouse, rat NCITR* NCI-CG-TR-190,79. Community Right-To-Know List. Reported in EPA TSCA Inventory.

SAFETY PROFILE: Confirmed carcinogen with experimental carcinogenic and neoplastigenic data. Poison by intravenous route. Moderately toxic by ingestion. Mutation data reported. An eye irritant. When heated to decomposition it emits toxic fumes of NO_x. See also N-NITROSO COMPOUNDS and AMINES.

NKB700 CAS:621-64-7 ***HR: 3***
N-NITROSODI-N-PROPYLAMINE
mf: $C_6H_{14}N_2O$ mw: 130.22

SYNS: DI-n-PROPYLNITROSAMINE ◇ DIPROPYLNITROSOAMINE ◇ DPN ◇ DPNA ◇ NDPA ◇ N-NITROSODIPROPYLAMINE ◇ N-NITROSO-N-DIPROPYLAMINE ◇ N-NITROSO-N-PROPYL-1-PROPANAMINE ◇ N-NITROSO-N-PROPYLPROPANAMINE ◇ RCRA WASTE NUMBER U111

TOXICITY DATA with REFERENCE
mma-sat 1 μmol/plate GANNA2 75,8,84
dns-hmn:hla 100 μmol/L CNREA8 38,2621,78
orl-rat TDLo:660 mg/kg/60W-I:CAR CALEDQ 19,207,83
scu-rat TDLo:926 mg/kg/38W-I:NEO JNCIAM 54,937,75
orl-mus TDLo:120 mg/kg/50W-I:ETA IAPUDO 41,643,82
scu-mus TDLo:1516 mg/kg/44W-I:CAR ZKKOBW 90,253,77
scu-ham TDLo:100 mg/kg (14D preg):NEO,TER ZKKOBW 90,79,77
orl-rat LD50:480 mg/kg ZEKBAI 69,103,67
scu-rat LD50:487 mg/kg JNCIAM 54,937,75
scu-mus LD50:689 mg/kg ZKKOBW 90,253,77
scu-ham LD50:600 mg/kg JNCIAM 51,1019,73

CONSENSUS REPORTS: NTP Fifth Annual Report on Carcinogens. IARC Cancer Review: Group 2B IMEMDT 7,56,87; Animal Sufficient Evidence IMEMDT 17,177,78; Human Limited Evidence IMEMDT 17,177,78. EPA Genetic Toxicology Program. Community Right-To-Know List. Reported in EPA TSCA Inventory.

DFG MAK: Animal Carcinogen, Suspected Human Carcinogen.

SAFETY PROFILE: Confirmed carcinogen with experimental carcinogenic, neoplastigenic, tumorigenic data.

Moderately toxic by ingestion and subcutaneous routes. An experimental teratogen. Human mutation data reported. When heated to decomposition it emits toxic fumes of NO_x. See also NITROSAMINES.

NKC000 CAS:17608-59-2 ***HR: 3***
N-NITROSOEPHEDRINE
mf: $C_{10}H_{14}N_2O_2$ mw: 194.26

SYNS: α-(1-(N-METHYL-N-NITROSOAMINO)ETHYL)BENZYLALCOHOL ◇ 2-(N-METHYL-N-NITROSOAMINO)-1-PHENYL-1-PROPANOL

TOXICITY DATA with REFERENCE
mma-sat 50 μg/plate CALEDQ 21,63,83
otr-ham:kdy 80 μg/L BJCAAI 37,873,78
orl-rat TDLo:17 g/kg/71W-I:CAR CALEDQ 5,103,78
ipr-mus TDLo:600 mg/kg/7D-I:CAR CNREA8 35,1981,75
ipr-mus LD50:392 mg/kg CNREA8 35,1981,75

SAFETY PROFILE: Suspected carcinogen with experimental carcinogenic data. Poison by intraperitoneal route. Mutation data reported. When heated to decomposition it emits toxic fumes of NO_x. See also N-NITROSO COMPOUNDS.

NKC300 CAS:71785-87-0 ***HR: 3***
N-NITROSO-3,4-EPOXYPIPERIDINE
mf: $C_5H_8N_2O_2$ mw: 128.15

SYN: 3-NITROSO-7-OXA-3-AZABICYCLO(4.1.0)HEPTANE

TOXICITY DATA with REFERENCE
mmo-sat 100 μg/plate TCMUE9 1,13,84
mma-sat 50 μg/plate TCMUE9 1,13,84
orl-rat TDLo:700 mg/kg/30W-I:ETA CRNGDP 1,753,80

SAFETY PROFILE: Questionable carcinogen with experimental tumorigenic data. An experimental carcinogenic. Mutation data reported. Many N-nitroso compounds are carcinogens. When heated to decomposition it emits toxic fumes of NO_x. See also N-NITROSO COMPOUNDS.

NKC500 CAS:56235-95-1 ***HR: 3***
NITROSOETHOXYETHYLAMINE
mf: $C_4H_{10}N_2O_2$ mw: 118.16

SYNS: O,N-DIETHYL-N-NITROSOHYDROXYLAMINE ◇ N-NITROSO-O,N-DIAETHYLHYDROXYLAMIN (GERMAN) ◇ N-NITROSO-O,N-DIETHYLHYDROXYLAMINE

TOXICITY DATA with REFERENCE
orl-rat TDLo:7800 mg/kg/39W-I:CAR ZKKOBW 83,205,75
orl-rat LD50:1000 mg/kg ZKKOBW 83,205,75

SAFETY PROFILE: Moderately toxic by ingestion. When heated to decomposition it emits toxic fumes of NO_x. See also N-NITROSO COMPOUNDS and AMINES.

NKD000 CAS:612-64-6 ***HR: 3***
N-NITROSO-N-ETHYL ANILINE
mf: $C_8H_{10}N_2O$ mw: 150.20

PROP: Yellow oil. D: 1.087 @ 20°/4°, bp: 119-120° @ 15 mm. Insol in water.

SYNS: ETHYLNITROSOANILINE ◇ N-ETHYL-N-NITROSOBENZENAMINE ◇ NEA ◇ NITROSOETHYLANILINE ◇ N-NITROSOETHYLPHENYLAMINE (MAK)

TOXICITY DATA with REFERENCE
otr-mus:fbr 400 μg/L JJIND8 67,1303,81
otr-ham:emb 6300 μg/L JJIND8 67,1303,81
ipr-rat TDLo:180 mg/kg (9D preg):TER IARCCD 4,112,73
orl-rat LD50:180 mg/kg VOONAW 14(3),37,68
ipr-rat LD50:180 mg/kg VOONAW 14(3),37,68

CONSENSUS REPORTS: EPA Genetic Toxicology Program

DFG MAK: Animal Carcinogen, Suspected Human Carcinogen.

SAFETY PROFILE: Confirmed carcinogen. Poison by ingestion and intraperitoneal routes data. An experimental teratogen. Mutation data reported. Many N-nitroso compounds are carcinogens. When heated to decomposition it emits toxic fumes of NO_x. See also N-NITROSO COMPOUNDS.

NKD500 CAS:3398-69-4 ***HR: 3***
N-NITROSOETHYL-tert-BUTYLAMINE
mf: $C_6H_{14}N_2O$ mw: 130.22

SYNS: AETHYL-tert-BUTYL-NITROSOAMIN (GERMAN) ◇ EBNA ◇ ETHYL-tert-BUTYLNITROSAMINE ◇ N-ETHYL-N-NITROSO-tert-BUTANAMINE ◇ N-NITROSO-tert-BUTYLETHYLAMINE

TOXICITY DATA with REFERENCE
orl-rat TDLo:370 mg/kg:ETA NATWAY 50,735,63
orl-rat LD50:1600 mg/kg NATWAY 50,100,63

CONSENSUS REPORTS: EPA Genetic Toxicology Program.

SAFETY PROFILE: Moderately toxic by ingestion. Questionable carcinogen with experimental tumorigenic data. Many N-nitroso compounds are carcinogens. When heated to decomposition it emits toxic fumes of NO_x. See also N-NITROSO COMPOUNDS.

NKE000 CAS:54897-63-1 ***HR: 3***
N-NITROSO-ETHYL(3-CARBOXYPROPYL) AMINE
mf: $C_6H_{12}N_2O_3$ mw: 160.20

SYNS: ECPN ◇ N-ETHYL-N-(3-CARBOXYPROPYL)NITROSOAMINE ◇ 4-(ETHYLNITROSOAMINO)BUTYRIC ACID ◇ N-NITROSO-N-(3-CARBOXYPROPYL)ETHYLAMINE

TOXICITY DATA with REFERENCE
mma-sat 10 μmol/plate CNREA8 37,399,77
orl-rat TDLo:5800 mg/kg/20W-I:ETA GANNA2 65,565,74

CONSENSUS REPORTS: EPA Genetic Toxicology Program.

SAFETY PROFILE: Questionable carcinogen with experimental tumorigenic data. Mutation data reported. Many N-nitroso compounds are carcinogens. When heated to decomposition it emits toxic fumes of NO_x. See also N-NITROSO COMPOUNDS and AMINES.

NKE500 CAS:614-95-9 ***HR: 3***
N-NITROSO-N-ETHYLURETHAN
mf: $C_5H_{10}N_2O_3$ mw: 146.17

SYNS: AETHYLNITROSOURETHAN (GERMAN) ◇ ENU ◇ ETHYLNITROSOCARBAMIC ACID, ETHYL ESTER ◇ N-ETHYL-N-NITROSOCARBAMIC ACID ETHYL ESTER ◇ N-ETHYL-N-NITROSOURETHANE ◇ NEU ◇ NITROSOETHYLURETHAN

TOXICITY DATA with REFERENCE
mma-sat 5 μg/plate TCMUE9 1,13,84
slt-mus-ipr 150 mg/kg MUREAV 92,193,82
ivn-rat TDLo:160 mg/kg (15D preg):TER IARCCD 4,45,73
orl-rat TDLo:630 mg/kg/48W-C:CAR GANNA2 73,48,82
skn-rat TDLo:8100 mg/kg/27W-I:ETA GANNA2 70,653,79
orl-rat TD:1680 mg/kg/32W-C:CAR GANNA2 73,48,82
ivn-rat LD50:160 mg/kg IARCCD 4,45,73

CONSENSUS REPORTS: EPA Genetic Toxicology Program.

SAFETY PROFILE: Poison by intravenous route. Experimental teratogenic effects. Mutation data reported. Questionable carcinogen with experimental carcinogenic data. When heated to decomposition it emits toxic fumes of NO_x. See also N-NITROSO COMPOUNDS and CARBAMATES.

NKF000 CAS:13256-13-8 ***HR: 3***
N-NITROSO-N-ETHYLVINYLAMINE
mf: $C_4H_8N_2O$ mw: 100.14

SYNS: AETHYL-VINYL-NITROSOAMIN (GERMAN) ◇ N-ETHYL-N-NITROSOETHENAMINE ◇ N-ETHYL-N-NITROSOETHENYLAMINE ◇ N-ETHYL-N-NITROSOVINYLAMINE ◇ ETHYLVINYLNITROSAMINE ◇ N-NITROSOETHYLVINYLAMINE ◇ VINYLETHYLNITROSAMIN (GERMAN) ◇ VINYLETHYLNITROSAMINE

TOXICITY DATA with REFERENCE
scu-ham TDLo:5 mg/kg (15D preg):TER EXPADD 22,67,82
orl-rat TDLo:230 mg/kg/25W-C:ETA ARZNAD 19,1077,69
scu-ham TDLo:31 mg/kg/50W-I:CAR JCROD7 102,227,82
scu-ham TD:55 mg/kg/44W-I:CAR JCROD7 102,227,82
orl-rat LD50:88 mg/kg NATWAY 50,100,63
ivn-rat LD50:88 mg/kg ZEKBAI 69,103,67
scu-ham LD50:109 mg/kg JNCIAM 58,439,77

CONSENSUS REPORTS: EPA Genetic Toxicology Program.

SAFETY PROFILE: Poison by ingestion, intravenous, and subcutaneous routes. Experimental teratogenic effects. Questionable carcinogen with experimental carcinogenic and tumorigenic data. When heated to decomposition it emits toxic fumes of NO_x. See also N-NITROSO COMPOUNDS and AMINES.

NKF500 CAS:2508-20-5 ***HR: 3***
2-NITROSOFLUORENE
mf: $C_{13}H_9NO$ mw: 195.23

TOXICITY DATA with REFERENCE
mmo-sat 50 ng/plate CBINA8 54,71,85
dnd-esc 10 μmol/L MUREAV 89,95,81
ipr-rat TDLo:160 mg/kg/4W-I:NEO CNREA8 30,1485,70
scu-rat TDLo:200 mg/kg/8W-I:CAR PSEBAA 120,538,65

CONSENSUS REPORTS: EPA Genetic Toxicology Program.

SAFETY PROFILE: Questionable carcinogen with experimental carcinogenic and neoplastigenic data. Mutation data reported. When heated to decomposition it emits toxic fumes of NO_x. See also NITROSO COMPOUNDS.

NKG000 CAS:69112-98-7 ***HR: 3***
NITROSOFLUOROETHYLUREA
mf: $C_3H_6FN_3O_2$ mw: 135.12

SYN: 1-(2-FLUOROETHYL)-1-NITROSO-UREA

TOXICITY DATA with REFERENCE
mmo-sat 1 μg/plate MUREAV 68,1,79
mma-sat 2500 ng/plate TCMUE9 1,13,84
pic-esc 2 mg/L TCMUE9 1,91,84
skn-mus TDLo:108 mg/kg/50W-I:CAR JCROD7 102,13,81
skn-mus TD:216 mg/kg/50W-I:ETA JCROD7 102,13,81

SAFETY PROFILE: Questionable carcinogen with experimental carcinogenic and tumorigenic data. Mutation data reported. When heated to decomposition it emits very toxic fumes of F^- and NO_x. See also N-NITROSO COMPOUNDS.

NKG450 CAS:29291-35-8 ***HR: 2***
NITROSOFOLIC ACID
mf: $C_{19}H_{18}N_8O_7$ mw: 470.45

SYN: GLUTAMIC ACID, N-NITROSO-N-PTEROYL-, l-

TOXICITY DATA with REFERENCE
mma-sat 100 μg/plate BJCAAI 37,873,78

otr-ham:kdy 10 mg/L BJCAAI 37,873,78
ipr-mus TDLo:375 mg/kg/7D-I:NEO CNREA8 35,1981,75

CONSENSUS REPORTS: IARC Cancer Review: Group 3 IMEMDT 7,56,87; Animal Inadequate Evidence IMEMDT 7,217,78.

SAFETY PROFILE: Questionable carcinogen with experimental neoplastigenic data. Mutation data reported. When heated to decomposition it emits toxic fumes of NO_x.

NKG500 CAS:56516-72-4 ***HR: D***
NITROSOGLYPHOSATE
mf: $C_3H_7N_2O_6P$ mw: 198.09

SYN: N-NITROSO-N-(PHOSPHONOMETHYL)GLYCINE

TOXICITY DATA with REFERENCE
mmo-sat 1 μL/plate MUREAV 48,225,77

SAFETY PROFILE: Mutation data reported. Many N-nitroso compounds are carcinogens. When heated to decomposition it emits very toxic fumes of PO_x and NO_x. See also N-NITROSO COMPOUNDS.

NKH000 CAS:674-81-7 ***HR: 3***
NITROSOGUANIDINE
mf: CH_4N_4O mw: 88.09

PROP: A solid.

SYNS: INITIATING EXPLOSIVE NITROSOGUANIDINE (DOT) ◇ NITROSOGUANIDIN (GERMAN) ◇ N-NITROSOGUANIDINE

TOXICITY DATA with REFERENCE
ipr-mus LDLo:21 mg/kg TXAPA9 23,288,72

DOT Classification: Class A Explosive; Label: Explosive A.

SAFETY PROFILE: Poison by intraperitoneal route. Many N-nitroso compounds are carcinogens. An explosive. Dangerous when stored in sealed containers as it decomposes to release nitrogen. When heated to decomposition it emits toxic fumes of NO_x. See also N-NITROSO COMPOUNDS, and EXPLOSIVES, HIGH.

NKH500 CAS:55557-02-3 ***HR: 2***
NITROSOGUVACOLINE
mf: $C_7H_{10}N_2O_3$ mw: 170.19

SYNS: N-NITROSOGUVACINE ◇ N-NITROSOGUVACOLINE ◇ 1-NITROSO-1,2,5,6-TETRAHYDRONICOTINIC ACID METHYL ESTER ◇ 1,2,5,6-TETRAHYDRO-1-NITROSO-3-PYRIDINECARBOXYLIC ACID METHYL ESTER

TOXICITY DATA with REFERENCE
mma-sat 10 μg/plate MUREAV 111,135,83
sce-hmn:lym 1 mmol/L TCMUE9 1,129,84

CONSENSUS REPORTS: IARC Cancer Review: Group 3 IMEMDT 7,56,87; Animal Inadequate Evidence IMEMDT 37,263,85. EPA Genetic Toxicology Program.

SAFETY PROFILE: Questionable carcinogen. Human mutation data reported. When heated to decomposition it emits toxic fumes of NO_x. See also N-NITROSO COMPOUNDS.

NKI000 CAS:932-83-2 ***HR: 3***
N-NITROSOHEXAHYDROAZEPINE
mf: $C_6H_{12}N_2O$ mw: 128.20

SYNS: HEXAHYDRO-1-NITROSO-1H-AZEPINE ◇ N-6-MI ◇ N-NITROSOAZACYCLOHEPTANE ◇ NITROSOHEXAMETHYLENIMINE ◇ N-NITROSOHEXAMETHYLENEIMINE ◇ N-NITROSOPERHYDROAZEPINE

TOXICITY DATA with REFERENCE
mma-sat 10 μg/plate TCMUE9 1,13,84
mma-esc 7500 μmol/L IAPUDO 41,543,82
pic-esc 100 mg/L TCMUE9 1,91,84
scu-ham TDLo:10 mg/kg (15D preg):REP ZKKOBW 86,69,76
orl-rat TDLo:1570 mg/kg/28W-I:CAR CALEDQ 12,99,81
orl-mus TDLo:700 mg/kg/8W-I:CAR JJIND8 73,1215,84
scu-mus TDLo:4 mg/kg:ETA ZKKOBW 78,78,72
scu-ham TDLo:80 mg/kg (8-15D preg):CAR,TER ZKKOBW 86,69,76
orl-mus LD:2400 mg/kg/30W-I:NEO CALEDQ 41,139,88
orl-rat LD50:336 mg/kg CNREA8 28,1217,68
ipr-rat LDLo:300 mg/kg CNREA8 28,1217,68
scu-mus LD50:65 mg/kg ZKKOBW 78,78,72
scu-ham LD50:142 mg/kg JNCIAM 50,323,73

SAFETY PROFILE: Suspected carcinogen with experimental carcinogenic, tumorigenic, and teratogenic data. Poison by ingestion, intraperitoneal, and subcutaneous routes. Experimental reproductive effects. Mutation data reported. When heated to decomposition it emits toxic fumes of NO_x. See also N-NITROSO COMPOUNDS.

NKI500 CAS:60391-92-6 ***HR: 3***
NITROSO HYDANTOIC ACID
mf: $C_3H_5N_3O_4$ mw: 147.11

SYNS: N-(AMINOCARBONYL)-N-NITROGLYCINE ◇ N-CARBAMOYL-N-NITROSOGLYCINE ◇ CARBOXYMETHYLNITROSOUREA ◇ 1-(CARBOXYMETHYL)-1-NITROSOUREA

TOXICITY DATA with REFERENCE
mmo-esc 5 mmol/L GANNA2 70,705,79
cyt-ham:fbr 125 mg/L/48H MUREAV 48,337,77
cyt-ham:lng 50 mg/L GMCRDC 27,95,81
orl-rat TDLo:6720 mg/kg/64W-C:CAR JCROD7 106,12,83
orl-rat TD:12390 mg/kg/59W-C:CAR JCROD7 106,12,83

orl-rat TD:4 g/kg/74W-I:NEO JJIND8 62,1523,79
ipr-rat LD50:210 mg/kg JJIND8 62,1523,79

CONSENSUS REPORTS: EPA Genetic Toxicology Program.

SAFETY PROFILE: Poison by intraperitoneal route. Questionable carcinogen with experimental carcinogenic and neoplastigenic data. Mutation data reported. When heated to decomposition it emits toxic fumes of NO_x. See also N-NITROSO COMPOUNDS.

NKJ000 CAS:42579-28-2 ***HR: 3***
1-NITROSOHYDANTOIN
mf: $C_3H_3N_3O_3$ mw: 129.09

SYN: NHYD

TOXICITY DATA with REFERENCE
mmo-sat 12 μmol/L/48H MUREAV 48,131,77
orl-rat TDLo:5900 mg/kg/52W-I:NEO JJIND8 62,1523,79
ipr-rat LD50:470 mg/kg JJIND8 62,1523,79

CONSENSUS REPORTS: EPA Genetic Toxicology Program.

SAFETY PROFILE: Moderately toxic by intraperitoneal route. Questionable carcinogen with experimental neoplastigenic data. Mutation data reported. Many nitroso compounds are carcinogens. When heated to decomposition it emits toxic fumes of NO_x. See also NITROSO COMPOUNDS.

NKJ050 CAS:96806-34-7 ***HR: 2***
1-NITROSO-1-HYDROXYETHYL-3-CHLOROETHYLUREA
mf: $C_5H_{10}ClN_3O_3$ mw: 195.63

SYNS: CHNU-I ◇ 1-NITROSO-1-(2-HYDROXYETHYL)-3-(2-CHLOROETHYL)UREA

TOXICITY DATA with REFERENCE
mmo-sat 50 μg/plate MUREAV 178,157,87
orl-rat TDLo:656 mg/kg/40W-I:CAR JCREA8 112,221,86

SAFETY PROFILE: Questionable carcinogen with experimental carcinogenic data. Mutation data reported. When heated to decomposition it emits toxic fumes of NO_x and Cl^-.

NKJ100 CAS:96806-35-8 ***HR: 2***
1-NITROSO-1-HYDROXYPROPYL-3-CHLOROETHYLUREA
mf: $C_6H_{12}ClN_3O_3$ mw: 209.66

SYNS: CPNU-I ◇ 1-NITROSO-1-(2-HYDROXYPROPYL)-3-(2-CHLOROETHYL)UREA

TOXICITY DATA with REFERENCE
mmo-sat 50 μg/plate MUREAV 178,157,87
orl-rat TDLo:1104 mg/kg/30W-I:CAR JCREA8 112,221,86

SAFETY PROFILE: Questionable carcinogen with experimental carcinogenic data. Mutation data reported. When heated to decomposition it emits toxic fumes of NO_x and Cl^-.

NKK000 CAS:71752-70-0 ***HR: 3***
NITROSO-3-HYDROXYPROPYLUREA
mf: $C_4H_9N_3O_3$ mw: 147.16

SYNS: 1-(3-HYDROXYPROPYL)-1-NITROSOUREA ◇ NITROSO-3-HYDROXY-N-PROPYLUREA

TOXICITY DATA with REFERENCE
mmo-sat 25 μg/plate MUREAV 68,1,79
skn-mus TDLo:530 mg/kg/45W-I:CAR CNREA8 43,214,83
orl-ham TDLo:589 mg/kg/37W-I:ETA PAACA3 24,92,83

SAFETY PROFILE: Questionable carcinogen with experimental carcinogenic and tumorigenic data. Mutation data reported. When heated to decomposition it emits toxic fumes of NO_x. See also N-NITROSO COMPOUNDS.

NKK500 CAS:3715-92-2 ***HR: 3***
N-NITROSOIMIDAZOLIDINETHIONE
mf: $C_3H_5ON_3S$ mw: 131.17

SYNS: N-NITROSOETHYLENETHIOUREA ◇ NO-ETU

TOXICITY DATA with REFERENCE
mnt-mus-ipr 120 mg/kg MUREAV 48,225,77
dlt-mus-orl 500 mg/kg/5D-I MUREAV 54,258,78
sce-ham:lng 50 μmol/L MUREAV 78,177,80
orl-rat TDLo:120 mg/kg (female 15D post):REP TJADAB 21,367,80
orl-rat TDLo:120 mg/kg (13D preg):TER TJADAB 21,367,80
orl-mus TDLo:530 mg/kg/9W-I:NEO CALEDQ 7,339,79
orl-mus LDLo:400 mg/kg MUREAV 56,335,78

CONSENSUS REPORTS: EPA Genetic Toxicology Program.

SAFETY PROFILE: Poison by ingestion. Experimental teratogenic and reproductive effects. Questionable carcinogen with experimental neoplastigenic data. Mutation data reported. Many N-nitroso compounds are carcinogens. When heated to decomposition it emits very toxic fumes of NO_x and SO_x. See also N-NITROSO COMPOUNDS.

NKL000 CAS:3844-63-1 ***HR: 3***
1-NITROSOIMIDAZOLIDINONE
mf: $C_3H_5N_3O_2$ mw: 115.11

SYNS: ETHYLENENITROSOUREA ◇ N-NITRO-2-IMIDAZOLIDONE ◇ 1-NITRO-2-IMIDAZOLIDONE ◇ N-NITROSO-IMIDAZOLIDON (GERMAN) ◇ 1-NITROSO-2-IMIDAZOLIDINONE ◇ N-NITROSOIMIDAZOLIDONE ◇ NSC 73438 ◇ SRI 1869

TOXICITY DATA with REFERENCE
mmo-sat 16 μmol/L/48H MUREAV 48,131,77
mmo-smc 400 mg/L/30M GENTAE 78,1101,74
mrc-smc 5 mmol/L ZEVBA5 98,230,66
scu-rat TDLo:600 mg/kg/20W-I:ETA ZEKBAI 69,103,67
scu-rat LD50:250 mg/kg ZEKBAI 69,103,67
ipr-mus LD50:28200 μg/kg NCISP* JAN86

CONSENSUS REPORTS: EPA Genetic Toxicology Program.

SAFETY PROFILE: Poison by intraperitoneal and subcutaneous routes. Mutation data reported. Questionable carcinogen with experimental tumorigenic data. Many N-nitroso compounds are carcinogens. When heated to decomposition it emits toxic fumes of NO_x. See also N-NITROSO COMPOUNDS.

NKL300 CAS:77698-19-2 ***HR: 3***
1,1'-(NITROSOIMINO)BIS-2-BUTANONE
mf: $C_8H_{14}N_2O_3$ mw: 186.24

SYNS: BOB ◇ N-NITROSOBIS(2-OXOBUTYL)AMINE

TOXICITY DATA with REFERENCE
mma-sat 5 mg/plate JJCREP 77,107,86
scu-ham TDLo:200 mg/kg:ETA CALEDQ 12,223,81
scu-ham LD50:282 mg/kg CALEDQ 12,223,81

SAFETY PROFILE: Poison by subcutaneous route. Questionable carcinogen with experimental tumorigenic data. Mutation data reported. Many N-nitroso compounds are carcinogens. When heated to decomposition it emits toxic fumes of NO_x. See also N-NITROSO COMPOUNDS and KETONES.

NKL500 CAS:16339-18-7 ***HR: 3***
2,2'-(N-NITROSOIMINO)DIACETONITRILE
mf: $C_4H_4N_4O$ mw: 124.12

SYNS: N-NITROSAMINO DIACETONITRIL (GERMAN) ◇ NITROSIMINODIACETONITRILE ◇ N-NITROSODIACETONITRILE ◇ N-NITROSODI(CYANOMETHYL)AMINE ◇ 2,2'-(NITROSOIMINO) BISACETONITRILE

TOXICITY DATA with REFERENCE
orl-rat TDLo:1584 mg/kg/2Y-C:ETA,REP ZEKBAI 69,103,67
orl-rat LD50:163 mg/kg ZEKBAI 69,103,67

CONSENSUS REPORTS: Cyanide and its compounds are on The Community Right-To-Know List.

SAFETY PROFILE: Poison by ingestion. Questionable carcinogen with experimental tumorigenic data. Experimental reproductive effects. Many N-nitroso compounds are carcinogens. When heated to decomposition it emits toxic fumes of NO_x and CN^-. See also N-NITROSO COMPOUNDS and NITRILES.

NKM000 CAS:1116-54-7 ***HR: 3***
NITROSOIMINO DIETHANOL
mf: $C_4H_{10}N_2O_3$ mw: 134.16

SYNS: BIS(β-HYDROXYAETHYL)NITROSAMIN (GERMAN) ◇ BIS(β-HYDROXYETHYL)NITROSAMINE ◇ DIAETHANOLNITROSAMIN (GERMAN) ◇ DIETHANOLNITROSOAMINE ◇ 2,2'-DIHYDROXY-N-NITROSODIETHYLAMINE ◇ 2,2'-IMINODI-N-NITROSOETHANOL ◇ NCI-C55583 ◇ NDELA ◇ N-NITROSOAMINODIETHANOL ◇ N-NITROSOBIS(2-HYDROXYETHYL)AMINE ◇ N-NITROSODIAETHANOLAMIN (GERMAN) ◇ N-NITROSODIETHANOLAMINE (MAK) ◇ 2,2'-(NITROSOIMINO)BISETHANOL ◇ RCRA WASTE NUMBER U173

TOXICITY DATA with REFERENCE
mmo-sat 100 μmol/plate MUREAV 158,141,85
sln-dmg-skn 2 mmol/L TCMUD8 4,437,84
orl-rat TDLo:855 mg/kg/2Y-I:CAR CNREA8 42,5167,82
skn-ham TDLo:21600 mg/kg/36W-I:NEO IAPUDO 41,299,82
scu-ham TDLo:6080 mg/kg/27W-I:CAR CNREA8 43,2521,83
orl-rat TD:1202 mg/kg/2Y-C:ETA IAPUDO 41,591,82
orl-rat LD50:7500 mg/kg GISAAA 40(1),81,75
scu-ham LDLo:11 g/kg CALEDQ 4,55,77

CONSENSUS REPORTS: NTP Fifth Annual Report on Carcinogens. IARC Cancer Review: Group 2B IMEMDT 7,56,87; Animal Sufficient Evidence IMEMDT 17,77,78; Human Limited Evidence IMEMDT 17,77,78. Reported in EPA TSCA Inventory.

DFG MAK: Animal Carcinogen, Suspected Human Carcinogen.

SAFETY PROFILE: Confirmed carcinogen with experimental carcinogenic, neoplastigenic, and tumorigenic data. Mildly toxic by ingestion. Mutation data reported. When heated to decomposition it emits toxic fumes of NO_x. See also N-NITROSO COMPOUNDS and ALCOHOLS.

NKM500 CAS:13256-19-4 ***HR: 3***
N-NITROSO-2,2'-IMINODIETHANOLDIACETATE
mf: $C_8H_{14}N_2O_5$ mw: 218.24

SYNS: ACETIC ACID, ESTER with N-NITROSO-2,2'-IMINODIETHANOL ◇ BIS-(ACETOXYAETHYL)NITROSAMIN (GERMAN) ◇ N-NITROSOBIS (ACETOXYETHYL)AMINE

TOXICITY DATA with REFERENCE
mmo-sat 100 μg/plate MUREAV 89,217,81
orl-rat TDLo:40 g/kg/58W-C:ETA ZEKBAI 69,103,67
orl-rat LD50:5000 mg/kg ZEKBAI 69,103,67

SAFETY PROFILE: Mildly toxic by ingestion. Questionable carcinogen with experimental tumorigenic data. Mutation data reported. Many N-nitroso compounds are carcinogens. When heated to decomposition it emits toxic fumes of NO_x. See also N-NITROSO COMPOUNDS and ESTERS.

NKN000 CAS:7633-57-0 ***HR: 3***
N-NITROSOINDOLINE
mf: $C_8H_8N_2O$ mw: 148.18

SYNS: N-NITROSOINDOLIN (GERMAN) ◇ 1-NITROSOINDOLINE

TOXICITY DATA with REFERENCE
orl-rat TDLo:20 g/kg/2Y-C:ETA ZEKBAI 69,103,67
orl-rat LD50:320 mg/kg ZEKBAI 69,103,67

SAFETY PROFILE: Poison by ingestion. Questionable carcinogen with experimental tumorigenic data. Many N-nitroso compounds are carcinogens. When heated to decomposition it emits toxic fumes of NO_x. See also N-NITROSO COMPOUNDS.

NKN500 CAS:72505-67-0 ***HR: D***
N-NITROSOISOBUTYLTHIAZOLIDINE
mf: $C_7H_{14}N_2OS$ mw: 174.2

SYN: 2-ISOBUTYL-3-NITROSOTHIAZOLIDINE

TOXICITY DATA with REFERENCE
mmo-sat 1 mg/L JAFCAU 28,62,80
mma-sat 1 mg/L JAFCAU 28,62,80

SAFETY PROFILE: Mutation data reported. Many N-nitroso compounds are carcinogens. When heated to decomposition it emits very toxic fumes of NO_x and SO_x. See also N-NITROSO COMPOUNDS.

NKO000 CAS:6238-69-3 ***HR: D***
N-NITROSOISONIPECTOIC ACID
mf: $C_6H_{10}N_2O_3$ mw: 158.18

SYN: 1-NITROSO-4-PIPERIDINECARBOXYLICACID

TOXICITY DATA with REFERENCE
mma-sat 5 mmol/L MUREAV 57,85,78

SAFETY PROFILE: Mutation data reported. Many N-nitroso compounds are carcinogens. When heated to decomposition it emits toxic fumes of NO_x. See also N-NITROSO COMPOUNDS.

NKO400 CAS:71752-69-7 ***HR: 3***
NITROSOISOPROPANOLUREA
mf: $C_4H_9N_3O_3$ mw: 147.16

SYNS: 1-(2-HYDROXYPROPYL)-1-NITROSOUREA ◇ NITROSO-2-HYDROXY-N-PROPYLUREA ◇ N-NITROSO-2-HYDROXY-N-PROPYLUREA

TOXICITY DATA with REFERENCE
mmo-sat 1 μg/plate MUREAV 68,1,79
mma-sat 1 μg/plate TCMUE9 1,13,84
dnd-mam:lym 10 mmol/L CBINA8 48,169,84
orl-rat TDLo:768 mg/kg/30W-C:CAR CNREA8 43,214,83
skn-mus TDLo:471 mg/kg/40W-I:CAR CNREA8 43,214,83
orl-ham TDLo:662 mg/kg/36W-I:ETA IAPUDO 57,617,84

SAFETY PROFILE: Suspected carcinogen with experimental carcinogenic and tumorigenic data. Mutation data reported. When heated to decomposition it emits toxic fumes of NO_x. See also N-NITROSO COMPOUNDS.

NKO425 CAS:16830-14-1 ***HR: 3***
NITROSOISOPROPYLUREA
mf: $C_4H_9N_3O_2$ mw: 131.16

SYNS: 1-ISOPROPYL-1-NITROSOUREA ◇ N-(1-METHYLETHYL)-N-NITROSO-UREA (9CI)

TOXICITY DATA with REFERENCE
mmo-sat 100 μg/plate TCMUE9 1,13,84
mma-sat 100 μg/plate TCMUE9 1,13,84
pic-esc 4 mg/L TCMUE9 1,91,84
skn-mus TDLo:525 mg/kg/50W-I:CAR JCROD7 102,13,81

SAFETY PROFILE: Questionable carcinogen with experimental carcinogenic data. Mutation data reported. When heated to decomposition it emits toxic fumes of NO_x. See also N-NITROSO COMPOUNDS.

NKO500 ***HR: D***
NITROSO LINURON
mf: $C_9H_9Cl_2N_3O_2$ mw: 262.11

SYN: 3-(3,4-DICHLOROPHENYL)-1,1-DIMETHYL-3-NITROSOUREA

TOXICITY DATA with REFERENCE
mmo-sat 1 μL/plate MUREAV 48,225,77

SAFETY PROFILE: Mutation data reported. Many N-nitroso compounds are carcinogens. When heated to decomposition it emits very toxic fumes of Cl^- and NO_x. See also N-NITROSO COMPOUNDS.

NKO600 CAS:73239-98-2 ***HR: 3***
N-NITROSO-2-METHOXY-2,6-DIMETHYL-MORPHOLINE
mf: $C_7H_{14}N_2O_3$ mw: 174.23

SYN: 2,6-DIMETHYL-2-METHOXY-4-NITROSO-MORPHOLINE

TOXICITY DATA with REFERENCE
mma-sat 200 nmol/plate JJIND8 64,157,80
bfa-ham/sat 100 mg/kg JJIND8 64,157,80
par-ham TDLo:50 mg/kg:ETA CALEDQ 13,233,81
par-ham LDLo:1000 mg/kg CALEDQ 13,233,81

SAFETY PROFILE: Moderately toxic by parenteral route. Questionable carcinogen with experimental tumorigenic data. Mutation data reported. Many N-nitroso compounds are carcinogens. When heated to decomposition it emits toxic fumes of NO_x. See also N-NITROSO COMPOUNDS.

NKO900 CAS:108278-70-2 ***HR: 2***
1-NITROSOMETHOXYETHYLUREA
mf: $C_4H_9N_3O_3$ mw: 147.16

SYNS: N-(2-METHOXYETHYL)-N-NITROSOUREA ◇ 1-(2-METHOXYETHYL)-1-NITROSOUREA ◇ NITROSO-2-METHOXYETHYLUREA ◇ UREA, N-(2-METHOXYETHYL)-N-NITROSO-

TOXICITY DATA with REFERENCE
mmo-sat 10 μg/plate MUREAV 178,157,87
pic-esc 250 mg/L MUREAV 178,157,87
orl-rat TDLo:480 mg/kg/30W-I:CAR JJCREP 79,181,88

SAFETY PROFILE: Questionable carcinogen with experimental carcinogenic data. Mutation data reported. When heated to decomposition it emits toxic fumes of NO_x.

NKP000 CAS:53198-46-2 ***HR: D***
N-NITROSO-N-METHYL-N-α-ACETOXYBENZYLAMINE
mf: $C_{10}H_{12}N_2O_3$ mw: 208.24

SYNS: N-METHYL-N-(α-ACETOXYBENZYL)NITROSAMINE ◇ α-METHYLNITROSOAMINOBENZYL ALCOHOL ACETATE (ESTER)

TOXICITY DATA with REFERENCE
mma-sat 250 ng/plate CALEDQ 2,305,77
msc-ham:lng 1 μmol/L GANNA2 73,517,82

CONSENSUS REPORTS: EPA Genetic Toxicology Program.

SAFETY PROFILE: Mutation data reported. Many N-nitroso compounds are carcinogens. When heated to decomposition it emits toxic fumes of NO_x. See also N-NITROSO COMPOUNDS.

NKP500 CAS:52977-61-4 ***HR: D***
o-(N-NITROSO-N-METHYL-β-ALANYL)-l-SERINE
mf: $C_7H_{13}N_3O_5$ mw: 219.23

TOXICITY DATA with REFERENCE
dnr-esc 200 μg/disc CNREA8 34,1658,74

SAFETY PROFILE: Mutation data reported. Many N-nitroso compounds are carcinogens. When heated to decomposition it emits toxic fumes of NO_x. See also N-NITROSO COMPOUNDS.

NKQ000 CAS:16219-98-0 ***HR: 3***
2-NITROSOMETHYLAMINOPYRIDINE
mf: $C_6H_7N_3O$ mw: 137.16

SYNS: 2-(METHYLNITROSAMINO)PYRIDINE ◇ N-METHYL-N-NITROSO-2-AMINOPYRIDINE ◇ N-NITROSO-N-METHYL-2-AMINOPYRIDINE

TOXICITY DATA with REFERENCE
mma-sat 20 mmol/L CRNGDP 3,415,82
hma-mus/esc 60 mg/kg CRNGDP 3,415,82
orl-rat TDLo:250 mg/kg/50W-I:CAR JJIND8 62,153,79
orl-rat LD50:60 mg/kg JJIND8 62,153,79

SAFETY PROFILE: Poison by ingestion. Questionable carcinogen with experimental carcinogenic data. Mutation data reported. When heated to decomposition it emits toxic fumes of NO_x. See also N-NITROSO COMPOUNDS.

NKQ500 CAS:13256-21-8 ***HR: 3***
N-NITROSOMETHYLAMINOSULFOLANE
mf: $C_5H_{10}N_2O_3S$ mw: 178.23

SYNS: 3-(N-NITROSOMETHYLAMINO)SULFOLAN(GERMAN) ◇ TETRAHYDRO-N-METHYL-N-NITROSO-3-THIOPHENAMINE 1,1-DIOXIDE

TOXICITY DATA with REFERENCE
orl-rat TDLo:2700 mg/kg/48W-C:ETA ARZNAD 19,1077,69
orl-rat LD50:750 mg/kg ZEKBAI 69,103,67

SAFETY PROFILE: Moderately toxic by ingestion. Questionable carcinogen with experimental tumorigenic data. Many N-nitroso compounds are carcinogens. When heated to decomposition it emits very toxic fumes of NO_x and SO_x. See also N-NITROSO COMPOUNDS.

NKR000 CAS:51542-33-7 ***HR: 3***
N-NITROSO-N-METHYL-N'-(2-BENZOTHIAZOLYL)UREA
mf: $C_9H_8N_4OS$ mw: 220.27

SYNS: N-(2-BENZOTHIAZOLYL)-N'-METHYL-N'NITROSOUREA ◇ N-METHYL-N-NITROSO-N'-(2-BENZOTHIAZOLYL)-HARNSTOFF(GERMAN) ◇ N-METHYL-N-NITROSO-N'-(2-BENZOTHIAZOLYL)-UREA ◇ N-NITROSOBENZTHIAZURON

TOXICITY DATA with REFERENCE
slt-smc 4 mmol/L/16H MUREAV 22,121,74
orl-rat TDLo:1156 mg/kg/64W-I:CAR IARCCD 14,429,76
orl-rat TDLo:400 mg/kg:ETA ZKKOBW 81,217,74
orl-rat LDLo:500 mg/kg ZKKOBW 81,217,74

SAFETY PROFILE: Moderately toxic by ingestion. Questionable carcinogen with experimental carcinogenic and tumorigenic data. Mutation data reported. Many N-nitroso compounds are carcinogens. When heated to decomposition it emits very toxic fumes of SO_x and NO_x. See also N-NITROSO COMPOUNDS.

NKR500 CAS:62783-48-6 ***HR: 3***
N-NITROSO-N-(2-METHYLBENZYL)METHYLAMINE
mf: $C_9H_{12}N_2O$ mw: 164.23

SYNS: N,o-DIMETHYL-N-NITROSOBENZYLAMINE ◇ N-NITROSO-N-(2-METHYLBENZYL)-METHYLAMIN(GERMAN)

TOXICITY DATA with REFERENCE
orl-rat TDLo:300 mg/kg/29W-C:ETA ZKKOBW 88,231,77
orl-rat LD50:90 mg/kg ZKKOBW 88,231,77

SAFETY PROFILE: Poison by ingestion. Questionable carcinogen with experimental tumorigenic data. Many N-nitroso compounds are carcinogens. When heated to decomposition it emits toxic fumes of NO_x. See also N-NITROSO COMPOUNDS and AMINES.

NKS000 CAS:62783-49-7 ***HR: 3***
N-NITROSO-N-(3-METHYLBENZYL)METHYLAMINE
mf: $C_9H_{12}N_2O$ mw: 164.23

SYNS: N,m-DIMETHYL-N-NITROSOBENZYLAMINE ◇ N-NITROSO-N-(3-METHYLBENZYL)-METHYLAMIN(GERMAN)

TOXICITY DATA with REFERENCE
mma-sat 100 μg/plate JMCMAR 29,40,86
orl-rat TDLo:170 mg/kg/49W-C:ETA ZKKOBW 88,231,77
orl-rat LD50:600 mg/kg ZKKOBW 88,231,77

SAFETY PROFILE: Moderately toxic by ingestion. Questionable carcinogen with experimental tumorigenic data. Mutation data reported. Many N-nitroso compounds are carcinogens. When heated to decomposition it emits very toxic fumes of NO_x. See also N-NITROSO COMPOUNDS and AMINES.

NKS500 CAS:62783-50-0 ***HR: 3***
N-NITROSO-N-(4-METHYLBENZYL)METHYLAMINE
mf: $C_9H_{12}N_2O$ mw: 164.23

SYNS: N,p-DIMETHYL-N-NITROSOBENZYLAMINE ◇ N-NITROSO-N-(4-METHYLBENZYL)-METHYLAMIN(GERMAN)

TOXICITY DATA with REFERENCE
mma-sat 500 μg/plate JMCMAR 26,309,83
orl-rat TDLo:155 mg/kg/44W-C:ETA ZKKOBW 88,231,77
orl-rat LD50:400 mg/kg ZKKOBW 88,231,77

SAFETY PROFILE: Poison by ingestion. Questionable carcinogen with experimental tumorigenic data. Mutation data reported. Many N-nitroso compounds are carcinogens. When heated to decomposition it emits toxic fumes of NO_x. See also N-NITROSO COMPOUNDS.

NKT000 CAS:69112-99-8 ***HR: 3***
NITROSOMETHYLBIS(CHLOROETHYL)UREA
mf: $C_6H_{11}Cl_2N_3O_2$ mw: 228.10

SYN: 1-NITROSO-1-METHYL-3,3-BIS-(2-CHLOROETHYL)UREA

TOXICITY DATA with REFERENCE
orl-rat TDLo:76 mg/kg/6W-I:ETA JCROD7 94,131,79

SAFETY PROFILE: Questionable carcinogen with experimental tumorigenic data. Many N-nitroso compounds are carcinogens. When heated to decomposition it emits very toxic fumes of Cl^- and NO_x. See also NITROSO COMPOUNDS.

NKT100 CAS:75016-34-1 ***HR: 2***
N-NITROSO-N-METHYL-N-n-BUTYL-1-d^2-AMINE

SYN: NMBA-d2

TOXICITY DATA with REFERENCE
orl-rat TDLo:132 mg/kg/20W-I:ETA CRNGDP 1,157,80

SAFETY PROFILE: Questionable carcinogen with experimental tumorigenic data. When heated to decomposition it emits toxic fumes of NO_x.

NKT105 CAS:75016-36-3 ***HR: 2***
NITROSOMETHYL-d^3-n-BUTYLAMINE
mf: $C_5H_9D_3N_2O$ mw: 119.19

SYN: NMBA-d3

TOXICITY DATA with REFERENCE
orl-rat TDLo:152 mg/kg/23W-I:ETA CRNGDP 1,157,80

SAFETY PROFILE: Questionable carcinogen with experimental tumorigenic data. When heated to decomposition it emits toxic fumes of NO_x.

NKT500 CAS:5432-28-0 ***HR: 3***
N-NITROSO-N-METHYLCYCLOHEXYLAMINE
mf: $C_7H_{14}N_2O$ mw: 142.23

SYNS: METHYLCYCLOHEXYLNITROSAMIN(GERMAN) ◇ METHYLCYCLOHEXYLNITROSAMINE ◇ N-METHYL-N-NITROSOCYCLOHEXYLAMINE ◇ N-NITROSOMETHYLCYCLOHEXYLAMINE

TOXICITY DATA with REFERENCE
mma-sat 1 μg/plate MUREAV 51,319,78
pic-esc 100 mg/L TCMUE9 1,91,84
orl-rat TDLo:150 mg/kg/54W-C:ETA ARZNAD 19,1077,69
orl-rat LD50:30 mg/kg ZEKBAI 69,103,67
ipr-rat LD50:28 mg/kg TXAPA9 17,426,70
ivn-rat LD50:30 mg/kg ZEKBAI 69,103,67
orl-mus LD50:57 mg/kg TXAPA9 17,426,70
orl-ham LD50:168 mg/kg TXAPA9 17,426,70

SAFETY PROFILE: Poison by ingestion, intravenous, and intraperitoneal routes. Questionable carcinogen with experimental tumorigenic data. Mutation data reported. Many N-nitroso compounds are carcinogens. When heated to decomposition it emits toxic fumes of

NO_x. See also N-NITROSO COMPOUNDS and AMINES.

NKU000 CAS:55090-44-3 ***HR: 3***
NITROSOMETHYL-n-DODECYLAMINE
mf: $C_{13}H_{28}N_2O$ mw: 228.43

SYNS: N-METHYL-N-NITROSOLAURYLAMINE ◇ N-NITROSO-N-METHYL-N-DODECYLAMIN (GERMAN) ◇ N-NITROSO-N-METHYL-N-DODECYLAMINE ◇ NMDDA

TOXICITY DATA with REFERENCE
mma-sat 1 μg/plate MUREAV 51,319,78
orl-rat TDLo:3600 mg/kg/30W-I:ETA FCTOD7 21,601,83
orl-gpg TDLo:8000 mg/kg/40W-I:CAR CNREA8 40,1879,80
orl-ham TDLo:2000 mg/kg/67W-I:CAR ZKKOBW 90,227,77
orl-ham LD50:930 mg/kg ZKKOBW 90,227,77

CONSENSUS REPORTS: EPA Genetic Toxicology Program.

SAFETY PROFILE: Suspected carcinogen with experimental carcinogenic and tumorigenic data. Moderately toxic by ingestion. Mutation data reported. When heated to decomposition it emits toxic fumes of NO_x. See also N-NITROSO COMPOUNDS and AMINES.

NKU350 CAS:26921-68-6 ***HR: 2***
N-NITROSOMETHYLETHANOLAMINE
mf: $C_3H_8N_2O_2$ mw: 104.13

SYNS: METHYL-2-HYDROXYETHYLNITROSAMINE ◇ METHYL-2-HYDROXYETHYLNITROSOAMINE ◇ 2-(METHYLNITROSOAMINO)ETHANOL ◇ N-NITROSOMETHYL-(2-HYDROXYETHYL)AMINE

TOXICITY DATA with REFERENCE
dnd-rat-orl 30 μmol/kg PAACA3 27,105,86
orl-rat TDLo:1602 mg/kg/60W-I:CAR CNREA8 48,1533,88

SAFETY PROFILE: Questionable carcinogen with experimental carcinogenic data. Mutation data reported. When heated to decomposition it emits toxic fumes of NO_x.

NKU400 CAS:28538-70-7 ***HR: 3***
NITROSOMETHYL-n-HEXYLAMINE
mf: $C_7H_{16}N_2O$ mw: 144.25

SYNS: N-METHYL-N-NITROSO-1-HEXANAMINE ◇ N-METHYL-N-NITROSOHEXYLAMINE ◇ NITROSO-n-HEXYLMETHYLAMINE

TOXICITY DATA with REFERENCE
mma-sat 10 μg/plate TCMUE9 1,13,84
pic-esc 50 mg/L TCMUE9 1,91,84
orl-rat TDLo:420 mg/kg/21W-I:ETA JCROD7 106,171,83

SAFETY PROFILE: Questionable carcinogen with experimental tumorigenic data. Mutation data reported. Many N-nitroso compounds are carcinogens. When heated to decomposition it emits toxic fumes of NO_x. See also N-NITROSO COMPOUNDS and AMINES.

NKU500 CAS:75411-83-5 ***HR: 3***
N-NITROSOMETHYL-2-HYDROXYPROPYLAMINE
mf: $C_4H_{10}N_2O_2$ mw: 118.16

SYNS: 2-HYDROXYPROPYLMETHYLNITROSAMINE ◇ 1-(METHYLNITROSOAMINO)-2-PROPANOL ◇ MHP ◇ NMHP

TOXICITY DATA with REFERENCE
mmo-sat 500 μg/plate TCMUE9 1,13,84
dns-rat:lvr 1 mmol/L MUREAV 144,197,85
orl-rat TDLo:300 mg/kg/30W-I:ETA CALEDQ 22,83,84
par-rat TDLo:600 mg/kg/30W-I:NEO JJCREP 79,309,88

SAFETY PROFILE: Questionable carcinogen with experimental neoplastigenic and tumorigenic data. Mutation data reported. Many N-nitroso compounds are carcinogens. When heated to decomposition it emits toxic fumes of NO_x. See also N-NITROSO COMPOUNDS and AMINES.

NKU550 CAS:75881-16-2 ***HR: 3***
NITROSO-2-METHYLMORPHOLINE
mf: $C_5H_{10}N_2O_2$ mw: 130.17

SYNS: 2-METHYL-4-NITROSOMORPHOLINE ◇ N-NITROSO-2-METHYLMORPHOLINE

TOXICITY DATA with REFERENCE
mmo-sat 1 mg/plate TCMUD8 1,295,80
mma-sat 250 μg/plate TCMUD8 1,295,80
orl-rat TDLo:146 g/kg/50W-I:CAR CRNGDP 3,911,82
orl-ham TDLo:1296 mg/kg/30W-I:ETA CRNGDP 5,875,84

SAFETY PROFILE: Questionable carcinogen with experimental carcinogenic and tumorigenic data. Mutation data reported. When heated to decomposition it emits toxic fumes of NO_x. See also N-NITROSO COMPOUNDS.

NKU570 CAS:31820-22-1 ***HR: 2***
NITROSOMETHYLNEOPENTYLAMINE
mf: $C_6H_{14}N_2O$ mw: 130.22

SYNS: NMNA ◇ N,2,2-TRIMETHYL-N-NITROSO-1-PROPYLAMINE

TOXICITY DATA with REFERENCE
orl-rat TDLo:200 mg/kg/22W-I:ETA JJIND8 68,681,82

SAFETY PROFILE: Questionable carcinogen with experimental tumorigenic data. When heated to decomposition it emits toxic fumes of NO_x.

NKU580 CAS:75881-19-5 ***HR: 2***
NITROSO-N-METHYL-n-NONYLAMINE
mf: $C_{10}H_{22}N_2O$ mw: 186.34

SYNS: NITROSOMETHYL-n-NONYLAMINE ◇ N-NITROSOMETHYL-n-NONYLAMINE ◇ 1-NONYLAMINE, N-METHYL-N-NITROSO-

TOXICITY DATA with REFERENCE
mma-sat 25 μg/plate TCMUD8 1,295,80
orl-rat TDLo:4800 mg/kg/51W-C:ETA CNREA8 41,1288,81

SAFETY PROFILE: Questionable carcinogen with experimental tumorigenic data. When heated to decomposition it emits toxic fumes of NO_x.

NKU590 CAS:34423-54-6 ***HR: 2***
NITROSO-N-METHYL-n-OCTYLAMINE
mf: $C_9H_{20}N_2O$ mw: 172.31

SYNS: N-METHYL-N-NITROSOOCTYLAMINE ◇ NITROSOMETHYL-n-OCTYLAMINE ◇ N-NITROSOMETHYL-n-OCTYLAMINE ◇ 1-OCTYLAMINE, N-METHYL-N-NITROSO-

TOXICITY DATA with REFERENCE
mma-sat 50 μg/plate TCMUD8 1,295,80
orl-rat TDLo:4480 mg/kg/43W-C:ETA CNREA8 41,1288,81

SAFETY PROFILE: Questionable carcinogen with experimental tumorigenic data. Mutation data reported. When heated to decomposition it emits toxic fumes of NO_x.

NKU600 CAS:39884-53-2 ***HR: 3***
NITROSO-2-METHYL-1,3-OXAZOLIDINE
mf: $C_4H_8N_2O_2$ mw: 116.14

SYNS: 2-METHYL-3-NITROSO-1,3-OXAZOLIDINE ◇ N-NITROSO-2-METHYL-1,3-OXAZOLIDINE

TOXICITY DATA with REFERENCE
mmo-sat 500 μg/plate TCMUD8 1,295,80
mma-sat 250 μg/plate TCMUD8 1,295,80
orl-rat TDLo:250 mg/kg/50W-I:CAR CRNGDP 3,911,82

SAFETY PROFILE: Questionable carcinogen with experimental carcinogenic data. Mutation data reported. When heated to decomposition it emits toxic fumes of NO_x. See also N-NITROSO COMPOUNDS.

NKU875 CAS:35631-27-7 ***HR: 3***
NITROSO-5-METHYLOXAZOLIDONE
mf: $C_4H_8N_2O_2$ mw: 116.14

SYNS: 5-METHYL-3-NITROSO-1,3-OXAZOLIDINE ◇ NITROSO-5-METHYL-1,3-OXAZOLIDINE ◇ N-NITROSO-5-METHYL-1,3-OXAZOLIDINE

TOXICITY DATA with REFERENCE
mmo-sat 1 μg/plate TCMUE9 1,13,84
mma-sat 5 μg/plate TCMUE9 1,13,84
pic-esc 10 mg/L TCMUE9 1,91,84
orl-rat TDLo:250 g/kg/50W-I:CAR CRNGDP 3,911,82
skn-mus TDLo:465 mg/kg/50W-I:CAR CNREA8 43,214,83
orl-ham TDLo:960 mg/kg/24W-I:ETA CRNGDP 5,875,84

SAFETY PROFILE: Suspected carcinogen with experimental carcinogenic and tumorigenic data. Mutation data reported. When heated to decomposition it emits toxic fumes of NO_x. See also N-NITROSO COMPOUNDS.

NKV000 CAS:55984-51-5 ***HR: 3***
N-NITROSOMETHYL-2-OXOPROPYLAMINE
mf: $C_4H_8N_2O_2$ mw: 116.14

SYNS: 1-(METHYLNITROSOAMINO)2-PROPANONE ◇ MOP ◇ NMOP

TOXICITY DATA with REFERENCE
dnd-rat-ipr 60 mg/kg CRNGDP 5,565,84
dns-rat:lvr 100 μmol/L MUREAV 144,197,85
orl-rat TDLo:300 mg/kg/30W-I:ETA JJIND8 70,959,83
par-rat TDLo:600 mg/kg/30W-I:CAR JJCREP 79,309,88
orl-ham TDLo:176 mg/kg/2ZW-I:CAR JCREA8 109,1,85
scu-ham LD50:35400 μg/kg CNREA8 40,3585,80

SAFETY PROFILE: Suspected carcinogen with experimental carcinogenic and tumorigenic data. Poison by subcutaneous route. Mutation data reported. When heated to decomposition it emits toxic fumes of NO_x. See also N-NITROSO COMPOUNDS.

NKV100 ***HR: 3***
N-NITROSOMETHYLPENTYLNITROSAMINE
mf: $C_6H_{13}N_3O_2$ mw: 159.22

SYNS: MPN ◇ N-NITROSO-N-(NITROSOMETHYL)PENTYLAMINE

TOXICITY DATA with REFERENCE
scu-rat TDLo:200 mg/kg/20W-I:ETA SACAB7 25,66,81
orl-rat LDLo:50 mg/kg SACAB7 25,66,81
orl-mky LDLo:142 mg/kg SACAB7 25,66,81

SAFETY PROFILE: Poison by ingestion. Questionable carcinogen with experimental tumorigenic data. Many N-nitroso compounds are carcinogens. When heated to decomposition it emits toxic fumes of NO_x. See also N-NITROSO COMPOUNDS and AMINES.

NKV500 CAS:68426-46-0 ***HR: 3***
NITROSOMETHYLPHENYLCARBAMATE
mf: $C_8H_8N_2O_3$ mw: 180.18

SYNS: N-METHYL-N-NITROSOCARBAMIC ACID, PHENYL ESTER ◇ N-NITROSO-N-METHYLPHENYLCARBAMATE ◇ PHENYL METHYLNITROSOCARBAMATE

TOXICITY DATA with REFERENCE
mmo-sat 1 μg/plate MUREAV 68,1,79
mma-sat 5 μg/plate TCMUE9 1,13,84

pic-esc 100 µg/L TCMUE9 1,91,84
orl-rat TDLo:470 mg/kg/10W-I:CAR EESADV 2,413,78

SAFETY PROFILE: Questionable carcinogen with experimental carcinogenic data. Mutation data reported. When heated to decomposition it emits toxic fumes of NO_x. See also N-NITROSO COMPOUNDS and CARBAMATES.

NKW000 CAS:68690-89-1 ***HR: 3***
N-NITROSO-N-METHYL-1-(1-PHENYL) ETHYLAMINE
mf: $C_9H_{12}N_2O$ mw: 164.23

SYNS: α,N-DIMETHYL-N-NITROSOBENZYLAMINE ◇ N-METHYL-N-NITROSO-1-PHENYLETHYLAMINE ◇ N-NITROSO-N-METHYL-(1-PHENYL)-ETHYLAMIN (GERMAN)

TOXICITY DATA with REFERENCE
orl-rat TDLo:610 mg/kg/59W-C:ETA ZKKOBW 92,235,78
orl-rat LD50:600 mg/kg ZKKOBW 92,235,78

SAFETY PROFILE: Moderately toxic by ingestion. Questionable carcinogen with experimental tumorigenic data. Many N-nitroso compounds are carcinogens. When heated to decomposition it emits toxic fumes of NO_x. See also N-NITROSO COMPOUNDS and AMINES.

NKW500 CAS:16339-07-4 ***HR: 3***
1-NITROSO-4-METHYLPIPERAZINE
mf: $C_5H_{11}N_3O$ mw: 129.19

SYNS: N'-METHYL-N-NITROSOPIPERAZINE ◇ 1-METHYL-4-NITROSOPIPERAZINE ◇ N-NITROSO-N'-METHYLPIPERAZIN (GERMAN) ◇ N-NITROSO-N'-METHYLPIPERAZINE

TOXICITY DATA with REFERENCE
mma-sat 50 nmol/plate MUREAV 57,1,78
mma-smc 50 µmol/plate MUREAV 77,143,80
orl-rat TDLo:10 g/kg/59W-C:ETA ZEKBAI 69,103,67
orl-rat LD50:100 mg/kg ZEKBAI 69,103,67

SAFETY PROFILE: Poison by ingestion. Questionable carcinogen with experimental tumorigenic data. Mutation data reported. Many N-nitroso compounds are carcinogens. When heated to decomposition it emits toxic fumes of NO_x. See also N-NITROSO COMPOUNDS.

NKW800 CAS:75881-20-8 ***HR: 2***
N-NITROSO-N-METHYL-n-TETRADECYLAMINE
mf: $C_{15}H_{32}N_2O$ mw: 256.49

TOXICITY DATA with REFERENCE
mma-sat 50 µg/plate MUREAV 111,135,83
orl-rat TDLo:6720 mg/kg/2Y-C:ETA CNREA8 41,1288,81

SAFETY PROFILE: Questionable carcinogen with experimental tumorigenic data. When heated to decomposition it emits toxic fumes of NO_x.

NKX000 CAS:57117-24-5 ***HR: 3***
NITROSO-2-METHYLTHIOPROPIONALDEHYDE-o-METHYL CARBAMOYL-OXIME
mf: $C_5H_9N_3O_3S$ mw: 191.23

SYNS: N-((METHYLNITROSOCARBAMOYL)OXY)-2-METHYL-THIOACETIMIDIC ACID ◇ NITROSO-METHOMYL

TOXICITY DATA with REFERENCE
mmo-sat 1 µg/plate TCMUE9 1,13,84
dnd-hmn:fbr 10 µmol/L MUREAV 44,1,77
orl-rat TDLo:500 mg/kg/10W-I:CAR EESADV 2,413,78
orl-rat TD:500 mg/kg/10W-I:ETA IARCCD 19,495,78

CONSENSUS REPORTS: EPA Genetic Toxicology Program.

SAFETY PROFILE: Questionable carcinogen with experimental carcinogenic and tumorigenic data. Human mutation data reported. When heated to decomposition it emits very toxic fumes of NO_x and SO_x. See also NITROSO COMPOUNDS.

NKX300 CAS:819-35-2 ***HR: 3***
NITROSOMETHYL-2-TRIFLUOROETHYLAMINE
mf: $C_3H_5F_3N_2O$ mw: 142.10

TOXICITY DATA with REFERENCE
mmo-sat 250 µg/plate TCMUD8 1,295,80
mma-sat 1 mg/plate TCMUD8 1,295,80
orl-rat TDLo:300 mg/kg/30W-I:ETA JJIND8 68,681,82

SAFETY PROFILE: Questionable carcinogen with experimental tumorigenic data. Mutation data reported. Many nitroso compounds are carcinogens. When heated to decomposition it emits toxic fumes of F^- and NO_x. See also NITROSO COMPOUNDS and AMINES.

NKX500 CAS:68107-26-6 ***HR: 3***
NITROSOMETHYLUNDECYLAMINE
mf: $C_{12}H_{26}N_2O$ mw: 214.40

SYN: N-METHYL-N-NITROSOUNDECYLAMINE

TOXICITY DATA with REFERENCE
mma-sat 1 µg/plate MUREAV 51,319,78
orl-rat TDLo:5600 mg/kg/30W-I:CAR CALEDQ 5,209,78

SAFETY PROFILE: Questionable carcinogen with experimental carcinogenic data. Mutation data reported. When heated to decomposition it emits toxic fumes of NO_x. See also N-NITROSO COMPOUNDS and AMINES.

NKY000 CAS:4549-40-0 ***HR: 3***
N-NITROSOMETHYLVINYLAMINE
mf: $C_3H_6N_2O$ mw: 86.11

SYNS: N-METHYL-N-NITROSO-ETHENYLAMINE ◇ N-METHYL-N-NITROSOVINYLAMINE ◇ METHYLVINYLNITROSAMIN (GERMAN)

◇ METHYLVINYLNITROSAMINE ◇ MVNA ◇ NMVA ◇ RCRA WASTE NUMBER P084

TOXICITY DATA with REFERENCE
orl-rat TDLo:110 mg/kg/52W-C:ETA ARZNAD 19,1077,69
orl-rat LD50:24 mg/kg ZEKBAI 69,103,67
ihl-rat LD50:22 mg/kg ZEKBAI 69,103,67

CONSENSUS REPORTS: NTP Fifth Annual Report on Carcinogens. IARC Cancer Review: Group 2B IMEMDT 7,56,87; Animal Sufficient Evidence IMEMDT 17,257,78; Human Limited Evidence IMEMDT 17,257,78. Community Right-To-Know List. EPA Genetic Toxicology Program.

SAFETY PROFILE: Confirmed carcinogen with experimental tumorigenic data. Poison by ingestion and inhalation. When heated to decomposition it emits toxic fumes of NO_x. See also N-NITROSO COMPOUNDS and AMINES.

NKY500 CAS:102433-74-9 ***HR: D***
NITROSOMETOXURON
mf: $C_{10}H_{12}ClN_3O_3$ mw: 257.70

SYN: 3-(3-CHLORO-4-METHOXYPHENYL)-1,1-DIMETHYL-3-NITROSOUREA

TOXICITY DATA with REFERENCE
mmo-sat 1 μL/plate MUREAV 48,225,77

SAFETY PROFILE: Mutation data reported. Many N-nitroso compounds are carcinogens. When heated to decomposition it emits very toxic fumes of Cl^- and NO_x. See also N-NITROSO COMPOUNDS.

NKZ000 CAS:59-89-2 ***HR: 3***
4-NITROSOMORPHOLINE
mf: $C_4H_8N_2O_2$ mw: 116.14

SYNS: N-NITROSOMORPHOLIN (GERMAN) ◇ NITROSOMORPHOLINE ◇ N-NITROSOMORPHOLINE (MAK) ◇ NMOR

TOXICITY DATA with REFERENCE
slt-dmg-orl 210 μmol/kg MUREAV 144,177,85
dns-hmn:fbr 160 μg/L PMRSDJ 1,528,81
orl-rat TDLo:320 mg/kg:CAR JCROD7 94,233,79
ivn-rat TDLo:286 mg/kg/57W-I:ETA ZEKBAI 66,138,64
orl-mus TDLo:3140 mg/kg/28W-C:NEO JNCIAM 46,1029,71
orl-ham TDLo:365 mg/kg/61W-C:CAR CALEDQ 17,333,83
orl-rat TD:180 mg/kg/I:ETA,REP ZEKBAI 66,1,64
orl-rat LD50:282 mg/kg NAIWAY 48,134,61
ipr-rat LD50:282 mg/kg BJIMAG 19,276,62
scu-rat LD50:170 mg/kg ZKKOBW 80,17,73
ivn-rat LD50:98 mg/kg ZEKBAI 69,103,67
ihl-mus LCLo:1000 mg/m^3/10M NDRC** NDCrc-132,Sept,42
ipr-mus LD50:64 mg/kg PMRSDJ 1,682,81
orl-ham LD50:956 mg/kg CALEDQ 17,333,83
scu-ham LD50:490 mg/kg ZKKOBW 81,251,74

CONSENSUS REPORTS: NTP Fifth Annual Report on Carcinogens. IARC Cancer Review: Group 2B IMEMDT 7,56,87; Animal Sufficient Evidence IMEMDT 17,263,78; Human Limited Evidence IMEMDT 17,263,78. Community Right-To-Know List. EPA Genetic Toxicology Program.

DFG MAK: Animal Carcinogen, Suspected Human Carcinogen.

SAFETY PROFILE: Confirmed carcinogen with experimental carcinogenic, neoplastigenic, and tumorigenic data. Poison by ingestion, intraperitoneal, subcutaneous, and intravenous routes. Moderately toxic by inhalation. Human mutation data reported. Experimental reproductive effects. When heated to decomposition it emits toxic fumes of NO_x. See also N-NITROSO COMPOUNDS.

NLA000 CAS:21711-65-9 ***HR: 3***
1-NITROSONAPHTHALENE
mf: $C_{10}H_7NO$ mw: 157.18

TOXICITY DATA with REFERENCE
mma-sat 1 μg/plate MUREAV 94,315,82
skn-rat TDLo:624 mg/kg/52W-I:ETA RCOCB8 18,353,77

SAFETY PROFILE: Questionable carcinogen with experimental tumorigenic data. Mutation data reported. When heated to decomposition it emits toxic fumes of NO_x. See also NITROSO COMPOUNDS.

NLA500 CAS:6610-08-8 ***HR: 3***
2-NITROSONAPHTHALENE
mf: $C_{10}H_7NO$ mw: 157.18

TOXICITY DATA with REFERENCE
mmo-sat 50 μg/plate PNASA6 69,3128,72
mma-sat 100 μg/plate PNASA6 72,5135,75
dni-sat 100 μmol/L CNREA8 34,3102,74
pic-sat 100 μmol/L/2M CNREA8 34,3102,74
skn-rat TDLo:624 mg/kg/52W-I:ETA RCOCB8 18,353,77
scu-mus TDLo:250 mg/kg:NEO CNREA8 31,1461,71

CONSENSUS REPORTS: EPA Genetic Toxicology Program.

SAFETY PROFILE: Questionable carcinogen with experimental neoplastigenic and tumorigenic data. Mutation data reported. Many nitroso compounds are carcinogens. When heated to decomposition it emits toxic fumes of NO_x. See also NITROSO COMPOUNDS.

NLB000 CAS:131-91-9 ***HR: 2***
1-NITROSO-2-NAPHTHOL
mf: $C_{10}H_7NO_2$ mw: 173.18

SYNS: NITROSO-β-NAPHTHOL ◇ α-NITROSO-β-NAPHTHOL

TOXICITY DATA with REFERENCE
orl-rat LDLo:500 mg/kg NCNSA6 5,35,53

CONSENSUS REPORTS: Reported in EPA TSCA Inventory.

SAFETY PROFILE: Moderately toxic by ingestion. Many nitroso compounds are carcinogens. Ignites when heated to 124°C. When heated to decomposition it emits toxic fumes of NO_x. See also NITROSO COMPOUNDS.

NLB500 CAS:132-53-6 ***HR: 3***
2-NITROSO-1-NAPHTHOL
mf: $C_{10}H_7NO_2$ mw: 173.18

TOXICITY DATA with REFERENCE
orl-rat TDLo:31 g/kg/1Y-I:ETA,REP JNCIAM 41,985,68

CONSENSUS REPORTS: Reported in EPA TSCA Inventory.

SAFETY PROFILE: Many nitroso compounds are carcinogens. Experimental reproductive effects. Questionable carcinogen with experimental tumorigenic data. When heated to decomposition it emits toxic fumes of NO_x. See also NITROSO COMPOUNDS.

NLB700 ***HR: 3***
1-NITROSO-2-NITROAMINO-2-IMIDAZOLINE
mf: $C_3H_5N_5O_3$ mw: 159.13

SYNS: 4,5-DIHYDRO-N-NITRO-1-NITROSO-1H-IMIDAZOL-2-AMINE ◇ NSC 25958

TOXICITY DATA with REFERENCE
orl-mus LD50:624 mg/kg NCISP* JAN86
ipr-mus LD50:249 mg/kg NCISP* JAN86
scu-mus LD50:175 mg/kg NCISP* JAN86

SAFETY PROFILE: Poison by subcutaneous and intraperitoneal routes. Moderately toxic by ingestion. Many N-nitroso compounds are carcinogens. When heated to decomposition it emits toxic fumes of NO_x. See also N-NITROSO COMPOUNDS.

NLC000 CAS:13010-08-7 ***HR: 3***
1-NITROSO-3-NITRO-1-BUTYLGUANIDINE
mf: $C_5H_{11}N_5O_3$ mw: 189.21

SYNS: N-BUTYL-N'-NITRO-N-NITROSOGUANIDINE ◇ N-NITROSO-N'-NITRO-N-BUTYLGUANIDINE

TOXICITY DATA with REFERENCE
mmo-sat 100 nmol/plate IDZAAW 50,403,75
mmo-omi 500 ppm SOGEBZ 11,183,75
scu-rat TDLo:69 mg/kg/5W-I:ETA GANNA2 69,277,78

CONSENSUS REPORTS: EPA Genetic Toxicology Program.

SAFETY PROFILE: Questionable carcinogen with experimental tumorigenic data. Mutation data reported. Many N-nitroso compounds are carcinogens. When heated to decomposition it emits toxic fumes of NO_x. See also N-NITROSO COMPOUNDS.

NLC500 CAS:13010-10-1 ***HR: 3***
1-NITROSO-3-NITRO-1-PENTYLGUANIDINE
mf: $C_6H_{13}N_5O_3$ mw: 203.24

SYNS: 3-NITRO-1-NITROSO-1-PENTYLGUANIDINE ◇ 1-PENTYL-3-NITRO-1-NITROSOGUANIDINE

TOXICITY DATA with REFERENCE
mmo-sat 100 nmol/plate IDZAAW 50,403,75
mmo-esc 5 mg/L ESKHA5 88,118,70
pic-esc 10 mg/L ESKHA5 88,118,70
dnr-smc 2 μmol/well IDZAAW 50,403,75
cyt-ham:fbr 30 mg/L/24H MUREAV 48,337,77
cyt-ham:lng 16 mg/L GMCRDC 27,95,81
scu-rat TDLo:69 mg/kg/5W-I:ETA GANNA2 69,277,78

CONSENSUS REPORTS: EPA Genetic Toxicology Program.

SAFETY PROFILE: Questionable carcinogen with experimental tumorigenic data. Mutation data reported. Many nitroso compounds are carcinogens. When heated to decomposition it emits very toxic fumes of NO_x. See also NITROSO COMPOUNDS.

NLD000 CAS:71598-10-2 ***HR: 3***
3-NITROSO-1-NITRO-1-PROPYLGUANIDINE
mf: $C_4H_9N_5O_3$ mw: 175.18

SYNS: PNNG ◇ N-PROPYL-N'-NITRO-N-NITROSOGUANIDINE

TOXICITY DATA with REFERENCE
mmo-sat 300 nmol/L ENMUDM 3,11,81
dns-rat-orl 1 g/kg BANRDU 13,123,82
orl-rat TDLo:2172 mg/kg/52W-C:NEO GANNA2 70,181,79
scu-rat TDLo:60 mg/kg/5W-I:ETA GANNA2 69,277,78

SAFETY PROFILE: Questionable carcinogen with experimental neoplastigenic and tumorigenic data. Mutation data reported. Many N-nitroso compounds are carcinogens. When heated to decomposition it emits toxic fumes of NO_x. See also N-NITROSO COMPOUNDS and NITRO COMPOUNDS.

NLD500 CAS:16543-55-8 ***HR: 3***
N'-NITROSONORNICOTINE
mf: $C_9H_{11}N_3O$ mw: 177.23

SYNS: 1'-NITROSO-1'-DEMETHYLNICOTINE ◇ 1-NITROSO-2-(3-

PYRIDYL)PYRROLIDINE ◇ 3-(1-NITROSO-2-PYRROLIDINYL)PYRIDINE ◇ (s)-3-(1-NITROSO-2-PYRROLIDINYL)PYRIDINE ◇ NNN

TOXICITY DATA with REFERENCE
mma-sat 50 μg/plate MUREAV 40,19,76
dnd-rat-ivn 360 μg/kg CNREA8 42,2877,82
orl-rat TDLo:2552 mg/kg/36W-C:ETA CALEDQ 20,333,83
scu-rat TDLo:567 mg/kg/22W-I:CAR CNREA8 44,2285,84
ipr-mus TDLo:880 mg/kg/7W-I:NEO JNCIAM 60,819,78
ipr-ham TDLo:1422 mg/kg/25W-I:CAR CNREA8 41,2849,81

CONSENSUS REPORTS: NTP Fifth Annual Report on Carcinogens. IARC Cancer Review: Group 2B IMEMDT 7,56,87; Animal Sufficient Evidence IMEMDT 17,281,78; Human Limited Evidence IMEMDT 17,281,78. Community Right-To-Know List. EPA Genetic Toxicology Program.

SAFETY PROFILE: Confirmed carcinogen with experimental carcinogenic, neoplastigenic, and tumorigenic data. Mutation data reported. When heated to decomposition it emits toxic fumes of NO_x. See also N-NITROSO COMPOUNDS.

NLD525 ***HR: 2***
N'-NITROSONORNICOTINE-1-N-OXIDE
mf: $C_9H_{11}N_3O_2$ mw: 193.23

SYNS: NICOTINE, 1'-DEMETHYL-1'-NITROSO-, 1-OXIDE ◇ NNN-1-N-OXIDE

TOXICITY DATA with REFERENCE
orl-rat TDLo:2799 mg/kg/36W-C:ETA CALEDQ 20,333,83

SAFETY PROFILE: Questionable carcinogen with experimental tumorigenic data. When heated to decomposition it emits toxic fumes of NO_x.

NLD800 CAS:18207-29-9 ***HR: 2***
1-NITROSO-1-OCTYLUREA
mf: $C_9H_{19}N_3O_2$ mw: 201.31

SYN: n-OCTYL NITROSUREA

TOXICITY DATA with REFERENCE
orl-rat TDLo:800 mg/kg:ETA ANYAA9 381,250,82

SAFETY PROFILE: Moderately toxic by subcutaneous route. Questionable carcinogen with experimental tumorigenic data. When heated to decomposition it emits toxic fumes of NO_x.

NLE000 CAS:39884-52-1 ***HR: 3***
N-NITROSOOXAZOLIDINE
mf: $C_3H_6N_2O_2$ mw: 102.11

SYNS: N-NITROSOOXAZOLIDIN (GERMAN) ◇ NITROSOOXAZOLIDONE ◇ N-NITROSO-1,3-OXAZOLIDINE ◇ 3-NITROSOOXAZOLIDINE

TOXICITY DATA with REFERENCE
mmo-sat 2500 ng/plate TCMUE9 1,13,84
mma-sat 1 μg/plate TCMUE9 1,13,84
pic-esc 100 mg/L TCMUE9 1,191,84
orl-rat TDLo:790 mg/kg/36W-I:CAR ZKKOBW 88,25,76
skn-mus TDLo:408 mg/kg/50W-I:CAR CNREA8 43,214,83
orl-ham TDLo:1288 mg/kg/35W-I:ETA CRNGDP 5,875,84
orl-rat LD50:1500 mg/kg ZKKOBW 88,25,76

SAFETY PROFILE: Suspected carcinogen with experimental carcinogenic and tumorigenic data. Moderately toxic by ingestion. Mutation data reported. When heated to decomposition it emits toxic fumes of NO_x. See also N-NITROSO COMPOUNDS.

NLE400 CAS:77698-20-5 ***HR: 2***
N-NITROSO(2-OXOBUTYL)(2-OXOPROPYL) AMINE
mf: $C_7H_{12}N_2O_3$ mw: 172.21

SYN: OBOB

TOXICITY DATA with REFERENCE
scu-ham TDLo:100 mg/kg:ETA CALEDQ 12,223,81
scu-ham LD50:174 mg/kg CALEDQ 12,223,81

SAFETY PROFILE: Poison by subcutaneous route. Questionable carcinogen with experimental tumorigenic data. When heated to decomposition it emits toxic fumes of NO_x.

NLE500 CAS:61734-86-9 ***HR: 3***
N-NITROSO-N-PENTYL-(4-HYDROXYBUTYL)AMINE
mf: $C_9H_{20}N_2O_2$ mw: 188.31

SYNS: N-PENTYL-N-(4-HYDROXYBUTYL)NITROSAMINE ◇ 4-(PENTYLNITROSAMINO)-1-BUTANOL

TOXICITY DATA with REFERENCE
mma-sat 10 μmol/plate CNREA8 37,399,77
orl-rat TDLo:76 g/kg/20W-C:ETA GANNA2 67,825,76

CONSENSUS REPORTS: EPA Genetic Toxicology Program.

SAFETY PROFILE: Questionable carcinogen with experimental tumorigenic data. Mutation data reported. Many N-nitroso compounds are carcinogens. When heated to decomposition it emits toxic fumes of NO_x. See also N-NITROSO COMPOUNDS.

NLE550 ***HR: 3***
N-NITROSOPHENACETIN
mf: $C_{10}H_{12}N_2O_3$ mw: 208.24

TOXICITY DATA with REFERENCE
mmo-sat 100 ng/plate JJIND8 72,863,84
mma-sat 100 ng/plate JJIND8 72,863,84
mmo-omi 5 μg/plate JJIND8 72,863,84
orl-rat LD50:21 mg/kg JJIND8 72,863,84

SAFETY PROFILE: Poison by ingestion. Mutation data reported. Many N-nitroso compounds are carcinogens. When heated to decomposition it emits toxic fumes of NO_x. See also N-NITROSO COMPOUNDS.

NLF000 CAS:38241-20-2 ***HR: D***
2-NITROSOPHENANTHRENE
mf: $C_{14}H_9NO$ mw: 207.24

TOXICITY DATA with REFERENCE
mmo-sat 5 μg/plate PNASA6 69,3128,72

SAFETY PROFILE: Mutation data reported. Many nitroso compounds are carcinogens. When heated to decomposition it emits toxic fumes of NO_x. See also NITROSO COMPOUNDS.

NLF200 CAS:104-91-6 ***HR: 3***
NITROSOPHENOL
mf: $C_6H_5NO_2$ mw: 123.12

PROP: Pale yellow, orthorhombic needles. Mp: 144° (decomp). Sltly sol in water; sol in dilute alkalies, alc, ether, and acetone.

SYNS: p-NITROSOPHENOL ◇ 4-NITROSOPHENOL ◇ QUINONE MONOXIME ◇ QUINONE OXIME

TOXICITY DATA with REFERENCE
mmo-sat 50 μg/plate FCTXAV 18,523,80
cyt-rat :lvr 750 μg/L MUREAV 153,57,85
ipr-mus LDLo:250 mg/kg CBCCT* 6,148,54
par-mus LDLo:200 mg/kg CBCCT* 7,692,55

CONSENSUS REPORTS: Reported in EPA TSCA Inventory.

SAFETY PROFILE: Poison by parenteral and intraperitoneal routes. Mutation data reported. An irritant and sensitizer. Many nitroso compounds are carcinogens. A very dangerous fire and explosion hazard. When exposed to heat or flame, it burns explosively. Contamination by acid or alkali may cause ignition. Can heat spontaneously and cause fire. When heated to decomposition it emits toxic fumes of NO_x. See also NITROSO COMPOUNDS.

NLF300 CAS:13168-78-0 ***HR: 3***
2-NITROSOPHENOL
mf: $C_6H_5NO_2$ mw: 123.11

SYN: 1,2-BENZOQUINONE MONOXIME

SAFETY PROFILE: Many nitroso compounds are carcinogens. Explodes when heated or on contact with concentrated acids. When heated to decomposition it emits toxic fumes of NO_x. See also NITROSO COMPOUNDS.

NLG000 CAS:777-79-7 ***HR: D***
NITROSOPHENYLETHYLUREA
mf: $C_9H_{11}N_3O_2$ mw: 193.23

SYNS: 1-NITROSO-1-PHENETHYLUREA ◇ N-NITROSO-N-(2-PHENYLETHYL)UREA

TOXICITY DATA with REFERENCE
mma-sat 10 μg/plate TCMUE9 1,13,84
pic-esc 2 mg/L TCMUE9 1,91,84

SAFETY PROFILE: Mutation data reported. Many N-nitroso compounds are carcinogens. When heated to decomposition it emits toxic fumes of NO_x. See also N-NITROSO COMPOUNDS.

NLG500 CAS:6268-32-2 ***HR: 3***
NITROSOPHENYLUREA
mf: $C_7H_7N_3O_2$ mw: 165.17

SYNS: N-NITROSO-N-PHENYLUREA ◇ PHENYL-NITROSO-HANRSTOFF (GERMAN) ◇ N-PHENYL-N-NITROSOUREA

TOXICITY DATA with REFERENCE
mmo-sat 10 μg/plate TCMUE9 1,13,84
mma-sat 10 μg/plate TCMUE9 1,13,84
scu-rat TDLo:150 mg/kg:ETA ZEKBAI 71,63,68
scu-rat LD50:150 mg/kg ZEKBAI 71,63,68

SAFETY PROFILE: Poison by subcutaneous route. Questionable carcinogen with experimental tumorigenic data. Mutation data reported. Many N-nitroso compounds are carcinogens. When heated to decomposition it emits toxic fumes of NO_x. See also N-NITROSO COMPOUNDS.

NLH000 CAS:13256-23-0 ***HR: 3***
N-NITROSO-4-PICOLYLETHYLAMINE
mf: $C_8H_{11}N_3O$ mw: 165.22

SYNS: AETHYL-4-PICOLYLNITROSAMIN (GERMAN) ◇ 4-((ETHYLNITROSAMINO)METHYL)PYRIDINE

TOXICITY DATA with REFERENCE
orl-rat TDLo:1200 mg/kg/43W-C:ETA ARZNAD 19,1077,69
orl-rat LD50:40 mg/kg ZEKBAI 69,103,67
ivn-rat LD50:35 mg/kg ZEKBAI 69,103,67

SAFETY PROFILE: Poison by ingestion and intravenous routes. Questionable carcinogen with experimental tumorigenic data. Many N-nitroso compounds are carcinogens. When heated to decomposition it emits toxic fumes of NO_x. See also N-NITROSO COMPOUNDS and AMINES.

NLH500 CAS:4515-18-8 ***HR: 3***
1-NITROSOPIPECOLIC ACID
mf: $C_6H_{10}N_2O_3$ mw: 158.18

SYNS: N-NITROSOPIPECOLIC ACID ◇ 1-NITROSO-2-PIPERIDINE-CARBOXYLIC ACID

TOXICITY DATA with REFERENCE
mma-sat 5 μmol/plate MUREAV 57,85,78
sce-hmn:lym 10 mmol/L TCMUE9 1,129,84
ipr-mus LD50:203 mg/kg JMCMAR 16,583,73

CONSENSUS REPORTS: EPA Genetic Toxicology Program.

SAFETY PROFILE: Poison by intraperitoneal route. Human mutation data reported. Many N-nitroso compounds are carcinogens. When heated to decomposition it emits toxic fumes of NO_x. See also N-NITROSO COMPOUNDS.

NLI000 CAS:7247-89-4 ***HR: 3***
1-NITROSO-2-PIPECOLINE
mf: $C_6H_{12}N_2O$ mw: 128.20

SYN: 2-METHYLNITROSOPIPERIDINE

TOXICITY DATA with REFERENCE
sln-dmg-orl 5 mmol/L/24H MUREAV 67,27,79
mma-smc 50 mmol/L/24H MUREAV 57,155,78
orl-rat TDLo:2520 mg/kg/50W-I:ETA IJCNAW 16,318,75

SAFETY PROFILE: Questionable carcinogen with experimental tumorigenic data. Mutation data reported. Many nitroso compounds are carcinogens. When heated to decomposition it emits toxic fumes of NO_x. See also NITROSO COMPOUNDS.

NLI500 CAS:14026-03-0 ***HR: 3***
R(−)-N-NITROSO-α-PIPECOLINE
mf: $C_6H_{12}N_2O$ mw: 128.20

SYNS: R(−)-N-NITROSO-2-METHYL-PIPERIDIN (GERMAN) ◇ R(−)-N-NITROSO-2-METHYLPIPERIDINE

TOXICITY DATA with REFERENCE
orl-rat TDLo:6300 mg/kg/40W-I:CAR ZKKOBW 79,118,73
orl-rat LD50:600 mg/kg ZKKOBW 79,118,73

SAFETY PROFILE: Moderately toxic by ingestion. Questionable carcinogen with experimental carcinogenic data. When heated to decomposition it emits toxic fumes of NO_x. See also N-NITROSO COMPOUNDS.

NLJ000 CAS:36702-44-0 ***HR: 3***
s(+)-N-NITROSO-α-PIPECOLINE
mf: $C_6H_{12}N_2O$ mw: 128.20

SYNS: s(+)-N-NITROSO-2-METHYL-PIPERIDIN (GERMAN) ◇ s(+)-N-NITROSO-2-METHYLPIPERIDINE

TOXICITY DATA with REFERENCE
orl-rat TDLo:5700 mg/kg/37W-I:CAR ZKKOBW 79,118,73
orl-rat LD50:600 mg/kg ZKKOBW 79,118,73

SAFETY PROFILE: Questionable carcinogen with experimental carcinogenic data. Moderately toxic by ingestion. When heated to decomposition it emits toxic fumes of NO_x. See also N-NITROSO COMPOUNDS.

NLJ500 CAS:100-75-4 ***HR: 3***
N-NITROSOPIPERIDINE
mf: $C_5H_{10}N_2O$ mw: 114.17

PROP: Light yellow oil. D: 1.063 @ 18.5°/4°, bp: 217-218°. Sol in water; very sol in acid solns.

SYNS: HEXAHYDRO-N-NITROSOPYRIDINE ◇ NITROSOPIPERIDIN (GERMAN) ◇ N-NITROSO-PIPERIDIN (GERMAN) ◇ 1-NITROSOPIPERIDINE ◇ N-N-PIP ◇ NPIP ◇ NO-PIP ◇ RCRA WASTE NUMBER U179

TOXICITY DATA with REFERENCE
trn-dmg-orl 425 ppm ENMUDM 7,349,85
sce-hmn:lym 10 mmol/L TCMUE9 1,129,84
orl-rat TDLo:350 mg/kg/2Y-I:CAR IAPUDO 31,657,80
scu-rat TDLo:680 mg/kg/34W-I:ETA XENOBH 3,217,73
ivn-rat TDLo:400 mg/kg/40W-I:CAR NATWAY 50,100,63
scu-ham TDLo:100 mg/kg (14D preg):NEO,TER ZKKOBW 90,71,77
scu-ham TD:100 mg/kg:NEO ZKKOBW 90,71,77
orl-rat LD50:200 mg/kg ZEKBAI 69,103,67
scu-rat LD50:100 mg/kg ZKKOBW 80,17,73
ivn-rat LD50:60 mg/kg ZEKBAI 69,103,67
orl-ham LD50:617 mg/kg CALEDQ 21,219,83
scu-ham LD50:324 mg/kg ZKKOBW 81,251,74

CONSENSUS REPORTS: NTP Fifth Annual Report on Carcinogens. IARC Cancer Review: Group 2B IMEMDT 7,56,87; Human Limited Evidence IMEMDT 17,287,78; Animal Sufficient Evidence IMEMDT 17,287,78; IMEMDT 28,151,82. Community Right-To-Know List. EPA Genetic Toxicology Program. Reported in EPA TSCA Inventory.

DFG MAK: Animal Carcinogen, Suspected Human Carcinogen.

SAFETY PROFILE: Confirmed carcinogen with experimental carcinogenic, neoplastigenic, and tumorigenic data. Poison by ingestion, intravenous, and subcutaneous routes. An experimental teratogen. Human mutation data reported. When heated to decomposition it emits toxic fumes of NO_x. See also N-NITROSO COMPOUNDS.

NLK000 CAS:55556-85-9 ***HR: 3***
NITROSO-3-PIPERIDINOL
mf: $C_5H_{10}N_2O_2$ mw: 130.17

SYNS: N-NITROSO-3-HYDROXYPIPERIDINE ◇ N-NITROSO-3-PIPERIDINOL

TOXICITY DATA with REFERENCE
mma-sat 1 μmol/plate MUREAV 56,131,77
sce-hmn:lym 10 mmol/L TCMUE9 1,129,84
orl-rat TDLo:1666 mg/kg/36W-I:ETA JNCIAM 55,705,75

CONSENSUS REPORTS: EPA Genetic Toxicology Program.

SAFETY PROFILE: Questionable carcinogen with experimental tumorigenic data. Human mutation data reported. Many N-nitroso compounds are carcinogens. When heated to decomposition it emits toxic fumes of NO_x. See also N-NITROSO COMPOUNDS.

NLK500 CAS:55556-93-9 ***HR: 3***
NITROSO-4-PIPERIDINOL
mf: $C_5H_{10}N_2O_2$ mw: 130.17

TOXICITY DATA with REFERENCE
mma-sat 1 μmol/plate MUREAV 56,131,77
orl-rat TDLo:1666 mg/kg/36W-I:ETA JNCIAM 55,705,75

SAFETY PROFILE: Questionable carcinogen with experimental tumorigenic data. Mutation data reported. Many nitroso compounds are carcinogens. When heated to decomposition it emits toxic fumes of NO_x. See also NITROSO COMPOUNDS.

NLL000 CAS:55556-91-7 ***HR: 3***
NITROSO-4-PIPERIDONE
mf: $C_5H_8N_2O_2$ mw: 128.15

SYNS: N-NITROSO-4-PIPERIDINONE ◇ NITROSO-4-PIPERIDINONE ◇ 1-NITROSO-4-PIPERIDONE

TOXICITY DATA with REFERENCE
mma-sat 250 μg/plate TCMUE9 1,13,84
orl-rat TDLo:1640 mg/kg/36W-I:ETA JNCIAM 55,705,75

SAFETY PROFILE: Questionable carcinogen with experimental tumorigenic data. Mutation data reported. Many N-nitroso compounds are carcinogens. When heated to decomposition it emits toxic fumes of NO_x. See also N-NITROSO COMPOUNDS and KETONES.

NLL500 CAS:7519-36-0 ***HR: 3***
1-NITROSO-l-PROLINE
mf: $C_5H_8N_2O_3$ mw: 144.15

SYN: N-NITROSO-l-PROLINE ◇ NO-Pro ◇ NPRO

TOXICITY DATA with REFERENCE
orl-rat TDLo:770 mg/kg/8W-I:ETA JMCMAR 16,583,73
ipr-mus LD50:203 mg/kg JMCMAR 16,583,73

CONSENSUS REPORTS: IARC Cancer Review: Group 3 IMEMDT 7,56,87; Animal Inadequate Evidence IMEMDT 17,303,78.

SAFETY PROFILE: Poison by intraperitoneal route. Questionable carcinogen with experimental tumorigenic data. Many nitroso compounds are carcinogens. When heated to decomposition it emits toxic fumes of NO_x. See also NITROSO COMPOUNDS.

NLM000 CAS:67809-15-8 ***HR: D***
N-NITROSO-N-PROPYLACETAMIDE
mf: $C_5H_{10}N_2O_2$ mw: 130.17

SYN: N-NITROSO-N(N-PROPYL)ACETAMIDE

TOXICITY DATA with REFERENCE
mma-sat 200 nmol/plate/48H CNREA8 39,1328,79

SAFETY PROFILE: Mutation data reported. Many N-nitroso compounds are carcinogens. When heated to decomposition it emits toxic fumes of NO_x. See also N-NITROSO COMPOUNDS.

NLM500 CAS:39603-53-7 ***HR: 3***
1-(NITROSOPROPYLAMINO)-2-PROPANOL
mf: $C_6H_{14}N_2O_2$ mw: 146.22

SYNS: β-HYDROXYPROPYLPROPYLNITROSAMINE ◇ (2-HYDROXYPROPYL)PROPYLNITROSOAMINE ◇ N-NITROSO-2-HYDROXY-N-PROPYL-N-PROPYLAMINE

TOXICITY DATA with REFERENCE
mma-sat 500 nmol/plate ZKKOBW 86,293,76
scu-rat TDLo:3500 mg/kg/55W-I:CAR JNCIAM 54,937,75
scu-mus TDLo:3272 mg/kg/53W-I:CAR ZKKOBW 91,189,78
scu-ham TDLo:100 mg/kg/(14D preg):NEO,TER ZKKOBW 90,119,77
scu-ham TD:1150 mg/kg/23W-I:ETA JNCIAM 51,287,73
scu-rat LD50:1273 mg/kg JNCIAM 54,937,75
scu-mus LD50:1235 mg/kg ZKKOBW 91,189,78
scu-ham LD50:1500 mg/kg JNCIAM 52,1245,74

CONSENSUS REPORTS: IARC Cancer Review: Animal Sufficient Evidence IMEMDT 17,177,78.

SAFETY PROFILE: Confirmed carcinogen with experimental carcinogenic and tumorigenic data. Moderately toxic by subcutaneous route. An experimental teratogen. Mutation data reported. When heated to decomposition it emits toxic fumes of NO_x. See also N-NITROSO COMPOUNDS.

NLN000 CAS:51938-12-6 ***HR: 3***
N-NITROSO-N-PROPYL-(4-HYDROXYBUTYL)AMINE
mf: $C_7H_{16}N_2O_2$ mw: 160.25

SYNS: PHBN ◇ PROPYL(4-HYDROXYBUTYL)NITROSAMINE ◇ 4-(PROPYLNITROSAMINO)-1-BUTANOL

TOXICITY DATA with REFERENCE
mma-sat 20 μmol/plate CNREA8 37,399,77
orl-rat TDLo:5200 mg/kg/20W-C:ETA GANNA2 65,13,74

CONSENSUS REPORTS: EPA Genetic Toxicology Program.

SAFETY PROFILE: Questionable carcinogen with experimental tumorigenic data. Mutation data reported. Many N-nitroso compounds are carcinogens. When heated to decomposition it emits toxic fumes of NO_x. See also N-NITROSO COMPOUNDS.

NLN500 CAS:65792-56-5 ***HR: 3***
N-NITROSO-N-PROPYLPROPIONAMIDE
mf: $C_6H_{12}N_2O_2$ mw: 144.20

SYNS: N-NITROSO-N-PROPYL-PROPRIONAMID (GERMAN) ◇ 1-OXOPROPYLPROPYLNITROSAMINE ◇ 1-OXOPROPYLPROPYL-NITROSAMIN (GERMAN)

TOXICITY DATA with REFERENCE
scu-ham TDLo:125 mg/kg:ETA,REP ZKKOBW 90,221,77
scu-ham LD50:308 mg/kg ZKKOBW 90,221,77

SAFETY PROFILE: Poison by subcutaneous route. Experimental reproductive effects. Questionable carcinogen with experimental tumorigenic data. Many N-nitroso compounds are carcinogens. When heated to decomposition it emits toxic fumes of NO_x. See also N-NITROSO COMPOUNDS.

NLO000 ***HR: D***
3-NITROSO-2-PROPYLTHIAZOLIDINE
mf: $C_6H_{12}N_2OS$ mw: 160.26

SYNS: N-NITROSO-2-N-PROPYLTHIAZOLIDINE ◇ 2-PROPYL-N-NITROSOTHIAZOLIDINE

TOXICITY DATA with REFERENCE
mmo-sat 1 mg/L JAFCAU 28,62,80
mma-sat 1 mg/L JAFCAU 28,62,80

SAFETY PROFILE: Mutation data reported. Many N-nitroso compounds are carcinogens. When heated to decomposition it emits very toxic fumes of NO_x and SO_x. See also N-NITROSO COMPOUNDS.

NLO500 CAS:816-57-9 ***HR: 3***
N-NITROSO-N-PROPYLUREA
mf: $C_4H_9N_3O_2$ mw: 131.16

SYNS: NITROSOPROPYLUREA ◇ NITROSO-N-PROPYLUREA ◇ NPU ◇ PNU ◇ N-PROPYLNITROSOHARNSTOFF (GERMAN) ◇ N-PROPYLNITROSUREA ◇ 1-PROPYL-1-NITROSOUREA

TOXICITY DATA with REFERENCE
mmo-sat 1 μg/plate MUREAV 68,1,79
cyt-ham:fbr 125 mg/L/48H MUREAV 48,337,77
cyt-ham:lng 68 mg/L GMCRDC 27,95,81
sce-ham:lng 1 g/L CNREA8 44,3270,84
par-rat TDLo:80 mg/kg (9D preg):TER IARCCD 4,112,73
orl-rat TDLo:120 mg/kg (19D preg):ETA,TER ARGEAR 40,99,72
orl-rat TDLo:200 mg/kg:ETA GANNA2 67,121,76
skn-mus TDLo:525 mg/kg/50W-I:CAR JCROD7 102,13,81
orl-ham TDLo:3024 mg/kg/36W-C:CAR GANNA2 73,695,82
orl-rat TD:3 g/kg/10W-C:NEO JJIND8 73,757,84
orl-rat LD50:480 mg/kg ARGEAR 40,99,72

CONSENSUS REPORTS: EPA Genetic Toxicology Program.

SAFETY PROFILE: Suspected carcinogen with experimental carcinogenic, neoplastigenic, tumorigenic data. Moderately toxic by ingestion. An experimental teratogen. Mutation data reported. When heated to decomposition it emits toxic fumes of NO_x. See also N-NITROSO COMPOUNDS.

NLP000 CAS:29291-35-0 ***HR: 3***
N-NITROSO-N-PTEROYL-l-GLUTAMIC ACID
mf: $C_{19}H_{18}N_8O_7$ mw: 470.45

SYNS: N-(p-(((2-AMINO-4-OXO-6-PTERIDINYL)METHYL)-N-NITROSOAMINO)BENZOYL)-l-GLUTAMIC ACID ◇ NITROSOFOLIC ACID

TOXICITY DATA with REFERENCE
mma-sat 100 μg/plate BJCAAI 37,873,78
otr-ham:kdy 80 μg/L BJCAAI 37,873,78
ipr-mus TDLo:375 mg/kg/7D-I:NEO CNREA8 35,1981,75
ipr-mus LD50:327 mg/kg CNREA8 35,1981,75

CONSENSUS REPORTS: IARC Cancer Review: Animal Inadequate Evidence IMEMDT 17,217,78.

SAFETY PROFILE: Poison by intraperitoneal route. Questionable carcinogen with experimental neoplastigenic data. Mutation data reported. Many N-nitroso compounds are carcinogens. When heated to decomposition it emits toxic fumes of NO_x. See also N-NITROSO COMPOUNDS.

NLP375 CAS:86674-51-3 ***HR: 2***
1-NITROSOPYRENE
mf: $C_{16}H_{19}NO$ mw: 241.36

TOXICITY DATA with REFERENCE
mmo-sat 500 ng/plate CNREA8 43,3132,83
dnd-hmn:fbr 900 nmol/L CRNGDP 7,1279,86
orl-rat TDLo:386 mg/kg/16W-I:CAR CNREA8 48,4256,88

SAFETY PROFILE: Questionable carcinogen with experimental carcinogenic data. Human mutation data reported. When heated to decomposition it emits toxic fumes of NO_x.

NLP480 CAS:35884-45-8 ***HR: 2***
NITROSOPYRROLIDINE
mf: $C_4H_8N_2O$ mw: 100.14

SYN: PYRROLIDINE, NITROSO-

TOXICITY DATA with REFERENCE
mma-sat 10 μg/plate TCMUE9 1,13,84
pic-esc 20 mg/L TCMUE9 1,91,84
orl-rat TDLo:2804 mg/kg/84W-C:CAR CNREA8 46,1285,86

SAFETY PROFILE: Questionable carcinogen with experimental carcinogenic data. Mutation data reported. When heated to decomposition it emits toxic fumes of NO_x.

NLP500 CAS:930-55-2 ***HR: 3***
N-NITROSOPYRROLIDINE
mf: $C_4H_8N_2O$ mw: 100.14

SYNS: N-NITROSOPYRROLIDIN (GERMAN) ◇ 1-NITROSOPYRROLIDINE ◇ NO-PYR ◇ N-N-PYR ◇ NPYR ◇ RCRA WASTE NUMBER U180 ◇ TETRAHYDRO-N-NITROSOPYRROLE

TOXICITY DATA with REFERENCE
mma-esc 7500 μmol/L IAPUDO 41,543,82
dnd-hmn:lng 100 μmol/plate JNCIAM 59,1401,77
orl-rat TDLo:685 mg/kg/98W-C:CAR ZKKOBW 90,161,77
orl-mus TDLo:3235 mg/kg/92W-C:NEO 85DUA4 -,129,70
orl-ham TDLo:465 mg/kg/52W-C:CAR JCROD7 104,75,82
orl-rat TD:1350 mg/kg/30W-C:ETA PAACA3 25,126,84
orl-rat LD50:900 mg/kg ZEKBAI 69,103,67
orl-mus LD50:125 mg/kg 85DUA4 -,129,70
scu-mus LD50:125 mg/kg 85DUA4 -,129,70
orl-ham LD50:1023 mg/kg JCROD7 104,75,82

CONSENSUS REPORTS: NTP Fifth Annual Report on Carcinogens. IARC Cancer Review: Group 2B IMEMDT 7,56,87; Animal Sufficient Evidence IMEMDT 17,313,78; Human Limited Evidence IMEMDT 17,313,78. EPA Genetic Toxicology Program. Reported in EPA TSCA Inventory.

DFG MAK: Animal Carcinogen, Suspected Human Carcinogen.

SAFETY PROFILE: Confirmed carcinogen with experimental carcinogenic, neoplastigenic, and tumorigenic data. Poison by ingestion and subcutaneous routes. Human mutation data reported. When heated to decomposition it emits toxic fumes of NO_x. See also N-NITROSO COMPOUNDS.

NLP600 CAS:54634-49-0 ***HR: 3***
N-NITROSO-2-PYRROLIDINE
mf: $C_4H_6N_2O_2$ mw: 114.12

SYN: 1-NITROSO-2-PYRROLIDINONE

TOXICITY DATA with REFERENCE
dnr-bcs 1500 μg/disc YKKZAJ 97,320,77
orl-rat TDLo:1460 mg/kg/52W-C:ETA YKKZAJ 97,320,77
orl-mus LD50:291 mg/kg YKKZAJ 97,320,77

CONSENSUS REPORTS: EPA Genetic Toxicology Program.

SAFETY PROFILE: Poison by ingestion. Questionable carcinogen with experimental tumorigenic data. Mutation data reported. Many N-nitroso compounds are carcinogens. When heated to decomposition it emits toxic fumes of NO_x. See also N-NITROSO COMPOUNDS.

NLP700 CAS:56222-35-6 ***HR: 2***
1-NITROSO-3-PYRROLIDINOL
mf: $C_4H_8N_2O_2$ mw: 116.14

SYNS: 3-HYDROXYNITROSOPYRROLIDINE ◇ N-NITROSO-3-HYDROXYPYRROLIDINE

TOXICITY DATA with REFERENCE
mma-sat 500 μg/plate MUREAV 89,35,81
orl-rat TDLo:1778 mg/kg/2Y-I:CAR,REP IAPUDO 31,657,80

SAFETY PROFILE: Experimental reproductive effects. Questionable carcinogen with experimental carcinogenic data. Mutation data reported. When heated to decomposition it emits toxic fumes of NO_x.

NLQ000 CAS:10552-94-0 ***HR: 3***
N-NITROSO-3-PYRROLINE
mf: $C_4H_6N_2O$ mw: 98.12

SYNS: 2,5-DIHYDRO-1-NITROSO-1H-PYRROLE ◇ NITROSO-3-PYRROLIN (GERMAN)

TOXICITY DATA with REFERENCE
mma-sat 1 μg/plate MUREAV 51,319,78
mma-esc 2500 mg/L MUREAV 89,35,81
orl-rat TDLo:2700 mg/kg/60W-I:ETA ZKKOBW 77,257,72

SAFETY PROFILE: Questionable carcinogen with experimental tumorigenic data. Mutation data reported. Many N-nitroso compounds are carcinogens. When heated to decomposition it emits toxic fumes of NO_x. See also N-NITROSO COMPOUNDS.

NLQ500 CAS:1130-69-4 ***HR: 3***
4-NITROSOQUINOLINE-1-OXIDE
mf: $C_9H_6N_2O_2$ mw: 174.17

TOXICITY DATA with REFERENCE
scu-mus TDLo:60 mg/kg/9W-I:ETA GANNA2 69(4),499,78

SAFETY PROFILE: Many nitroso compounds are carcinogens. Questionable carcinogen with experimental tumorigenic data. When heated to decomposition it emits toxic fumes of NO_x. See also NITROSO COMPOUNDS.

NLR000 CAS:3565-26-2 ***HR: 3***
5-NITROSO-8-QUINOLINOL
mf: $C_9H_6N_2O_2$ mw: 174.17

SYN: 8-HYDROXY-5-NITROSOQUINOLINE

TOXICITY DATA with REFERENCE
ivn-mus LD50:56 mg/kg CSLNX* NX#01332

CONSENSUS REPORTS: Reported in EPA TSCA Inventory.

SAFETY PROFILE: Poison by intravenous route. Many nitroso compounds are carcinogens. When heated to decomposition it emits toxic fumes of NO_x. See also NITROSO COMPOUNDS.

NLR500 CAS:13256-22-9 ***HR: 3***
N-NITROSOSARCOSINE
mf: $C_3H_6N_2O_3$ mw: 118.11

SYNS: N-METHYL-N-NITROSOGLYCINE ◇ N-NITROSOMETHYLGLYCINE ◇ NITROSO SARKOSIN (GERMAN)

TOXICITY DATA with REFERENCE
orl-rat TDLo:29 g/kg/41W-C:ETA ARZNAD 19,1077,69
orl-rat LD50:5000 mg/kg ZEKBAI 69,103,67

CONSENSUS REPORTS: NTP Fifth Annual Report on Carcinogens. IARC Cancer Review: Group 2B IMEMDT 7,56,87; Animal Sufficient Evidence IMEMDT 17,327,78; Human Limited Evidence IMEMDT 17,327,78.

SAFETY PROFILE: Confirmed carcinogen with experimental tumorigenic data. Mildly toxic by ingestion. When heated to decomposition it emits toxic fumes of NO_x. See also N-NITROSO COMPOUNDS.

NLS000 CAS:13344-50-8 ***HR: 3***
N-NITROSOSARCOSINE, ETHYL ESTER
mf: $C_5H_{10}N_2O_3$ mw: 146.17

SYN: NITROSO-SARKOSIN-AETHYLESTER

TOXICITY DATA with REFERENCE
orl-rat TDLo:7900 mg/kg/23W-C:ETA NATWAY 50,99,63
orl-rat TD:160 g/kg/8W-I:CAR JJIND8 71,75,83
orl-rat LD50:4000 mg/kg ZEKBAI 69,103,67
ivn-rat LD50:4000 mg/kg ZEKBAI 69,103,67

SAFETY PROFILE: Moderately toxic by ingestion and intravenous routes. Many N-nitroso compounds are carcinogens. Questionable carcinogen with experimental carcinogenic and tumorigenic data. When heated to decomposition it emits toxic fumes of NO_x. See also N-NITROSO COMPOUNDS and ESTERS.

NLS500 CAS:53051-16-4 ***HR: D***
o-(N-NITROSOSARCOSYL)-l-SERINE
mf: $C_6H_{11}N_3O_5$ mw: 205.20

SYN: l-SERINE, ESTER with N-METHYL-N-NITROSOGLYCINE

TOXICITY DATA with REFERENCE
dnr-esc 250 μg/disc CNREA8 34,1658,74

CONSENSUS REPORTS: EPA Genetic Toxicology Program.

SAFETY PROFILE: Mutation data reported. Many N-nitroso compounds are carcinogens. When heated to decomposition it emits toxic fumes of NO_x. See also N-NITROSO COMPOUNDS.

NLT000 CAS:38241-21-3 ***HR: D***
4-NITROSO-trans-STILBENE
mf: $C_{14}H_{11}NO$ mw: 209.26

TOXICITY DATA with REFERENCE
mmo-sat 50 μg/plate PNASA6 69,3128,72

SAFETY PROFILE: Mutation data reported. Many nitroso compounds are carcinogens. When heated to decomposition it emits toxic fumes of NO_x. See also NITROSO COMPOUNDS.

NLT500 CAS:40548-68-3 ***HR: 3***
N-NITROSO-TETRAHYDRO-1,2-OXAZINE
mf: $C_4H_8N_2O_2$ mw: 116.14

SYNS: N-NITROSO-TETRAHYDROOXAZIN-1,2 (GERMAN) ◇ N-NITROSOTETRAHYDRO-1,2-OXAZIN ◇ 2-NITROSOTETRAHYDRO-2H-1,2-OXAZINE ◇ TETRAHYDRO-2-NITROSO-2H-1,2-OXAZINE

TOXICITY DATA with REFERENCE
orl-rat TDLo:18 g/kg/63W-I:CAR ZKKOBW 79,114,73
orl-rat LD50:830 mg/kg ZKKOBW 79,114,73

SAFETY PROFILE: Moderately toxic by ingestion. Questionable carcinogen with experimental carcinogenic data. When heated to decomposition it emits toxic fumes of NO_x. See also N-NITROSO COMPOUNDS.

NLU000 CAS:35627-29-3 ***HR: 3***
N-NITROSO-TETRAHYDRO-1,3-OXAZINE
mf: $C_4H_8N_2O_2$ mw: 116.14

SYNS: N-NITROSOTETRAHYDROOXAZIN-1,3 (GERMAN) ◇ 3-NITROSO-TETRAHYDRO-1,3-OXAZINE

TOXICITY DATA with REFERENCE
mma-sat 250 μg/plate TCMUE9 1,13,84
orl-rat TDLo:250 g/kg/50W-I:CAR CRNGDP 3,911,82
orl-rat TD:2900 mg/kg/24W-I:ETA ZKKOBW 82,257,74
orl-rat LD50:600 mg/kg ZKKOBW 82,257,74

SAFETY PROFILE: Moderately toxic by ingestion. Questionable carcinogen with experimental carcinogenic and tumorigenic data. Mutation data reported. When heated to decomposition it emits toxic fumes of NO_x. See also N-NITROSO COMPOUNDS.

NLU480 CAS:70501-82-5 ***HR: 2***
N-NITROSO-1,2,3,4-TETRAHYDROPYRIDINE
mf: $C_5H_8N_2O$ mw: 112.15

SYNS: N-NITROSO-Δ²-PIPERIDINE ◇ N-NITROSO-Δ²-TETRAHYDROPYRIDINE ◇ 1-NITROSO-1,2,3,4-TETRAHYDROPYRIDINE

TOXICITY DATA with REFERENCE
mmo-sat 25 μg/plate TCMUE9 1,13,84
mma-sat 25 μg/plate TCMUD8 1,295,80
orl-rat TDLo:1250 mg/kg/25W-I:ETA CRNGDP 1,753,80

SAFETY PROFILE: Questionable carcinogen with experimental tumorigenic data. Mutation data reported. When heated to decomposition it emits toxic fumes of NO_x.

NLU500 CAS:55556-92-8 ***HR: 3***
N-NITROSO-1,2,3,6-TETRAHYDROPYRIDINE
mf: $C_5H_8N_2O$ mw: 112.15

SYNS: N-NITROSO-Δ³-PIPERIDINE ◇ N-NITROSO-Δ³-TETRAHYDROPYRIDINE

TOXICITY DATA with REFERENCE
mmo-sat 50 μg/plate TCMUE9 1,13,84
mma-sat 1 μg/plate MUREAV 51,319,78
orl-rat TDLo:33 mg/kg/25W-I:CAR EESADV 6.513,82
orl-rat TD:50 mg/kg/2Y-I:CAR EESADV 6,513,82

CONSENSUS REPORTS: EPA Genetic Toxicology Program.

SAFETY PROFILE: Questionable carcinogen with experimental carcinogenic data. Mutation data reported. When heated to decomposition it emits toxic fumes of NO_x. See also N-NITROSO COMPOUNDS.

NLV500 CAS:73870-33-4 ***HR: 2***
N-NITROSOTHIAZOLIDINE
mf: $C_3H_6N_2OS$ mw: 118.17

SYNS: 3-NITROSOTHIAZOLIDINE ◇ NT

TOXICITY DATA with REFERENCE
mmo-sat 2500 ng/plate FCTOD7 22,253,84
mma-sat 2500 ng/plate FCTOD7 22,253,84
dnr-bcs 2500 μg/disc FCTOD7 22,1013,84
dns-rat:lvr 200 μmol FCTOD7 22,1013,84
orl-rat LD50:1950 mg/kg JTEHD6 13,595,84

SAFETY PROFILE: Moderately toxic by ingestion. Mutation data reported. Many N-nitroso compounds are carcinogens. When heated to decomposition it emits very toxic fumes of NO_x and SO_x. See also N-NITROSO COMPOUNDS.

NLW000 CAS:26541-51-5 ***HR: 3***
N-NITROSOTHIOMORPHOLINE
mf: $C_4H_8N_2OS$ mw: 132.20

SYN: 4-NITROSOTHIOMORPHOLINE

TOXICITY DATA with REFERENCE
mma-sat 1 μmol/plate MUREAV 57,1,78
orl-rat TDLo:860 mg/kg/38W-I:CAR ZEKBAI 74,179,70

SAFETY PROFILE: Questionable carcinogen with experimental carcinogenic data. Mutation data reported. When heated to decomposition it emits very toxic fumes of NO_x and SO_x. See also N-NITROSO COMPOUNDS.

NLW500 CAS:611-23-4 ***HR: 2***
2-NITROSOTOLUENE
mf: C_7H_7NO mw: 121.15

PROP: Needles or prisms. Mp: 72.5°. Very sol in chloroform, alc, and ether.

SYNS: 1-METHYL-2-NITROSO-BENZENE (9CI) ◇ o-METHYLNITROSOBENZENE ◇ o-NITROSOTOLUENE

TOXICITY DATA with REFERENCE
mma-sat 5 μmol/plate JMCMAR 22,981,79
orl-rat TDLo:103 g/kg/73W-C:CAR CALEDQ 16,103,82

SAFETY PROFILE: Questionable carcinogen with experimental carcinogenic data. Mutation data reported. Many nitroso compounds are carcinogens. When heated to decomposition it emits toxic fumes of NO_x. See also NITROSO COMPOUNDS.

NLX000 CAS:71752-68-6 ***HR: 3***
NITROSOTRIDECYLUREA
mf: $C_{14}H_{29}N_3O_2$ mw: 271.46

SYN: 1-NITROSO-1-TRIDECYLUREA

TOXICITY DATA with REFERENCE
mmo-sat 1 μg/plate MUREAV 68,1,79
mma-sat 100 μg/plate TCMUE9 1,13,84
skn-mus TDLo:1085 mg/kg/50W-I:ETA JCROD7 102,13,81

SAFETY PROFILE: Questionable carcinogen with experimental tumorigenic data. Mutation data reported. Many nitroso compounds are carcinogens. When heated to decomposition it emits toxic fumes of NO_x. See also NITROSO COMPOUNDS.

NLX500 CAS:50285-70-6 ***HR: 3***
NITROSOTRIETHYLUREA
mf: $C_7H_{15}N_3O_2$ mw: 173.25

SYNS: NITROSOTRIAETHYLHARNSTOFF (GERMAN) ◇ 1,1,3-TRIETHYL-3-NITROSOUREA

TOXICITY DATA with REFERENCE
mma-sat 1 μg/plate MUREAV 51,319,78
orl-rat TDLo:150 mg/kg:CAR,REP CRNGDP 9,573,88
ivn-rat TDLo:150 mg/kg:NEO,REP CRNGDP 9,573,88
orl-ham TDLo:1733 mg/kg/W-I:ETA PAACA3 24,92,83
orl-rat TD:2550 mg/kg/31W-C:ETA,REP JJIND8 65,451,80

SAFETY PROFILE: Experimental reproductive effects. Questionable carcinogen with experimental carcinogenic, neoplastigenic, and tumorigenic data. Mutation data reported. Many nitroso compounds are carcinogens. When heated to decomposition it emits toxic fumes of NO_x. See also NITROSO COMPOUNDS.

NLX700 CAS:82018-90-4 ***HR: 3***
N-NITROSO-2,2,2-TRIFLUORODIETHYLAMINE
mf: $C_4H_7F_3N_2O$ mw: 156.13

SYN: N-NITROSO-2,2,2-TRIFLUOROETHYL-ETHYLAMINE

TOXICITY DATA with REFERENCE
mma-sat 20 μmol/plate CRNGDP 3,155,82
orl-rat TDLo:945 mg/kg/63W-I:CAR CRNGDP 4,755,83
orl-rat TD:1305 mg/kg/44W-I:CAR CRNGDP 4,755,83
orl-rat LD50:960 mg/kg CRNGDP 4,755,83

SAFETY PROFILE: Moderately toxic by ingestion. Questionable carcinogen with experimental carcinogenic data. Mutation data reported. When heated to decomposition it emits toxic fumes of F^- and NO_x. See also N-NITROSO COMPOUNDS and AMINES.

NLY000 CAS:16339-14-3 ***HR: 3***
1-NITROSO-1,2,2-TRIMETHYLHYDRAZINE
mf: $C_3H_9N_3O$ mw:103.15

SYNS: N-NITROSOTRIMETHYLHYDRAZIN (GERMAN) ◇ N-NITROSO-N,N',N'-TRIMETHYLHYDRAZINE

TOXICITY DATA with REFERENCE
orl-rat TDLo:200 mg/kg/75W-C:ETA ZEKBAI 69,103,67
orl-rat LD50:95 mg/kg ZEKBAI 69,103,67

SAFETY PROFILE: Poison by ingestion. Questionable carcinogen with experimental tumorigenic data. Many N-nitroso compounds are carcinogens. When heated to decomposition it emits very toxic fumes of NO_x. See also N-NITROSO COMPOUNDS.

NLY500 CAS:62178-60-3 ***HR: 3***
NITROSOTRIMETHYLPHENYL-N-METHYLCARBAMATE
mf: $C_{11}H_{14}N_2O_3$ mw: 222.27

SYNS: LANDRIN, NITROSO DERIVATIVE ◇ N-METHYL-N-NITROSOCARBAMIC ACID, TRIMETHYLPHENYL ESTER ◇ NITROSOLANDRIN

TOXICITY DATA with REFERENCE
mmo-sat 1 μg/plate MUREAV 68,1,79
mma-sat 111 ng/plate MUREAV 56,1,77
dnd-hmn:fbr 10 μmol/L MUREAV 44,1,77
orl-rat TDLo:580 mg/kg/10W-I:CAR EESADV 2,413,78

SAFETY PROFILE: Questionable carcinogen with experimental carcinogenic data. Human mutation data reported. When heated to decomposition it emits toxic fumes of NO_x. See also N-NITROSO COMPOUNDS, CARBAMATES, and ESTERS.

NLY750 CAS:88208-15-5 ***HR: 3***
NITROSO-3,4,5-TRIMETHYLPIPERAZINE
mf: $C_7H_{15}N_3O$ mw: 157.25

SYNS: N-NITROSO-3,4,5-TRIMETHYLPIPERAZINE ◇ 1-NITROSO-3,4,5-TRIMETHYLPIPERAZINE

TOXICITY DATA with REFERENCE
mma-sat 10 μg/plate TCMUE9 1,13,84
orl-rat TDLo:270 mg/kg/30W-I:CAR CRNGDP 4,1165,83
orl-ham TDLo:2208 mg/kg/46W-I:CAR CRNGDP 4,1165,83
orl-rat TD:1100 mg/kg/27W-I:ETA IAPUDO 57,617,84

SAFETY PROFILE: Suspected carcinogen with experimental carcinogenic and tumorigenic data. Mutation data reported. When heated to decomposition it emits toxic fumes of NO_x. See also N-NITROSO COMPOUNDS.

NLZ000 CAS:69113-01-5 ***HR: 3***
NITROSOTRIS(CHLOROETHYL)UREA
mf: $C_7H_{12}Cl_3N_3O_2$ mw: 276.57

SYN: 1-NITROSO-1,3,3-TRIS-(2-CHLOROETHYL)UREA

TOXICITY DATA with REFERENCE
orl-rat TDLo:95 mg/kg/17W-I:ETA JCROD7 94,131,79

SAFETY PROFILE: Questionable carcinogen with experimental tumorigenic data. Many nitroso compounds are carcinogens. When heated to decomposition it emits very toxic fumes of Cl^- and NO_x. See also NITROSO COMPOUNDS.

NMA000 CAS:71752-67-5 ***HR: 3***
NITROSOUNDECYLUREA
mf: $C_{12}H_{25}N_3O_2$ mw: 243.40

SYN: N-NITROSO-N-UNDECYLUREA

TOXICITY DATA with REFERENCE
mmo-sat 1 μg/plate MUREAV 68,1,79
mma-sat 100 μg/plate TCMUE9 1,13,84
skn-mus TDLo:974 mg/kg/50W-I:CAR JCROD7 102,13,81

SAFETY PROFILE: Questionable carcinogen with experimental carcinogenic data. Mutation data reported. When heated to decomposition it emits toxic fumes of NO_x. See also N-NITROSO COMPOUNDS.

NMA500 CAS:13010-20-3 ***HR: D***
NITROSOUREA
mf: $CH_3N_3O_2$ mw: 89.07

TOXICITY DATA with REFERENCE
cyt-mus-ipr 3200 μg/kg MUREAV 38,106,76
spm-mus-ipr 3200 μg/kg MUREAV 38,106,76

SAFETY PROFILE: Mutation data reported. Many N-nitroso compounds are carcinogens. When heated to decomposition it emits toxic fumes of NO_x. See also N-NITROSO COMPOUNDS.

NMB000 CAS:9056-38-6 ***HR: 3***
NITROSTARCH
DOT: UN 0146/UN 1337

PROP: Solid.

SYNS: NITROSTARCH, containing less than 20% water (DOT) ◇ NITROSTARCH, DRY (DOT)

CONSENSUS REPORTS: Reported in EPA TSCA Inventory.

DOT Classification: Class A Explosive; Label: Explosive A (UN0146); Flammable Solid; Label: Flammable Solid (UN1337); Flammable Liquid; Label: Flammable Liquid (UN1337).

SAFETY PROFILE: A very dangerous fire and explosion hazard when exposed to heat, flame, shock, or oxidizers. It is a powerful high explosive. Nitrostarch is not a definite compound, but a mixture of various nitric acid esters of starch with different degrees of nitration. When heated to decomposition it emits toxic fumes of NO_x. See also NITRO COMPOUNDS.

NMC000 CAS:4003-94-5 ***HR: 3***
4-NITROSTILBENE
mf: $C_{14}H_{11}NO_2$ mw: 225.26

SYN: 4-NITROSTILBEN (GERMAN)

TOXICITY DATA with REFERENCE
mmo-sat 5 μg/plate CBINA8 26,11,79
mma-sat 5 μg/plate CBINA8 26,11,79
orl-rat TDLo:2700 mg/kg/57W-C:ETA NATWAY 42,128,55

SAFETY PROFILE: Questionable carcinogen with experimental tumorigenic data. Mutation data reported. When heated to decomposition it emits toxic fumes of NO_x. See also NITRO COMPOUNDS OF AROMATIC HYDROCARBONS.

NMD500 CAS:63021-48-7 ***HR: 3***
7-(m-NITROSTYRYL)BENZ(c)ACRIDINE
mf: $C_{25}H_{16}N_2O_2$ mw: 376.43

SYN: m-NITROBENZYLIDENE-3,4-BENZ-9-METHYLACRIDINE

TOXICITY DATA with REFERENCE
scu-mus TDLo:200 mg/kg:ETA VOONAW 1,52,55

SAFETY PROFILE: Questionable carcinogen with experimental tumorigenic data. When heated to decomposition it emits toxic fumes of NO_x.

NME000 CAS:63021-49-8 ***HR: 3***
7-(o-NITROSTYRYL)BENZ(c)ACRIDINE
mf: $C_{25}H_{16}N_2O_2$ mw: 376.43

SYN: o-NITROBENZYLIDENE-3,4-BENZ-9-METHYLACRIDINE

TOXICITY DATA with REFERENCE
scu-mus TDLo:200 mg/kg:ETA VOONAW 1,52,55

SAFETY PROFILE: Questionable carcinogen with experimental tumorigenic data. When heated to decomposition it emits toxic fumes of NO_x.

NME500 CAS:63021-50-1 ***HR: 3***
7-(p-NITROSTYRYL)BENZ(c)ACRIDINE
mf: $C_{25}H_{16}N_2O_2$ mw: 376.43

SYN: p-NITROBENZYLIDENE-3,4-BENZ-9-METHYLACRIDINE

TOXICITY DATA with REFERENCE
scu-mus TDLo:200 mg/kg:ETA VOONAW 1,52,55

SAFETY PROFILE: Questionable carcinogen with experimental tumorigenic data. When heated to decomposition it emits toxic fumes of NO_x.

NMF000 CAS:63021-46-5 ***HR: 3***
12-(m-NITROSTYRYL)BENZ(a)ACRIDINE
mf: $C_{25}H_{16}N_2O_2$ mw: 376.43

SYN: m-NITROBENZYLIDENE-1,2-BENZ-9-METHYLACRIDINE

TOXICITY DATA with REFERENCE
scu-mus TDLo:200 mg/kg:ETA VOONAW 1,52,55

SAFETY PROFILE: Questionable carcinogen with experimental tumorigenic data. When heated to decomposition it emits toxic fumes of NO_x.

NMF500 CAS:63021-47-6 ***HR: 3***
12-(o-NITROSTYRYL)BENZ(a)ACRIDINE
mf: $C_{25}H_{16}N_2O_2$ mw: 376.43

SYN: o-NITROBENZYLIDENE-1,2-BENZ-9-METHYLACRIDINE

TOXICITY DATA with REFERENCE
scu-mus TDLo:200 mg/kg:ETA VOONAW 1,52,55

SAFETY PROFILE: Questionable carcinogen with experimental tumorigenic data. When heated to decomposition it emits toxic fumes of NO_x.

NMG000 CAS:22188-15-4 ***HR: 3***
12-(p-NITROSTYRYL)BENZ(a)ACRIDINE
mf: $C_{25}H_{16}N_2O_2$ mw: 376.43

SYN: p-NITROBENZYLIDENE-1,2-BENZ-9-METHYLACRIDINE

TOXICITY DATA with REFERENCE
scu-mus TDLo:200 mg/kg:ETA VOONAW 1,52,55

SAFETY PROFILE: Questionable carcinogen with experimental tumorigenic data. When heated to decomposition it emits toxic fumes of NO_x.

NMG500 CAS:62316-46-5 ***HR: 3***
NITROSYL AZIDE
mf: N_4O mw: 72.03

$O:NN_3$

PROP: Yellow compound.

SAFETY PROFILE: Probably poison by inhalation and ingestion. Spontaneous decomposition at −50° may be explosive. Has exploded during preparation. Upon decomposition it emits very toxic fumes of NO_x and N_3. See also AZIDES.

NMH000 CAS:2696-92-6 ***HR: 3***
NITROSYL CHLORIDE
DOT: UN 1069
mf: ClNO mw: 65.46

O:NCl

PROP: Yellow gas, irritating odor. Mp: −64.5°, bp: −5.8°, d (liquid): 1.250 @ 30°, vap d: 2.3, vap press: 76 mm @ 50°. Non-explosive, very corrosive. Liquid @ −5.5°, solid @ −61.5°. Sol in fuming H_2SO_4.

SYN: NITROGEN OXYCHLORIDE

CONSENSUS REPORTS: Reported in EPA TSCA Inventory.

DOT Classification: Nonflammable Gas; Label: Nonflammable Gas; Poison A; Label: Poison Gas, Corrosive.

SAFETY PROFILE: Poison by inhalation and ingestion. A corrosive irritant to skin, eyes, and mucous membranes. Inhalation may cause pulmonary edema and hemorrhage. Potentially explosive reaction with acetone + platinum. Mixtures with hydrogen + oxygen ignite spontaneously. When heated to decomposition it emits very toxic fumes of Cl^- and NO_x.

NMH275 CAS:4343-68-4 ***HR: 3***
NITROSYL CYANIDE
mf: CN_2O mw: 56.02

O=NC≡N

CONSENSUS REPORTS: Cyanide and its compounds are on the Community Right-To-Know List.

SAFETY PROFILE: The impure gas explodes below −20°C. Handle as the gas only!! When heated to decomposition it emits toxic fumes of CN^- and NO_x. See also CYANIDE.

NMH500 CAS:7789-25-5 ***HR: 3***
NITROSYL FLUORIDE
mf: FNO mw: 49.01

O:NF

PROP: Colorless gas, often bluish because of impurities. Mp: −132.5°, bp: −59.9°; d: (liquid @ bp) 1.326, (solid) 1.719.

SYN: NITROGEN OXYFLUORIDE

SAFETY PROFILE: A poison. A severe irritant to skin, eyes, and mucous membranes. Reacts vigorously with glass; corrodes quartz. Explosive reaction with alkenes; oxygen difluoride. Incandescent reaction with metals (e.g., antimony, bismuth, tin, sodium); non-metals (e.g., arsenic, boron, red phosphorus, silicon). When heated to decomposition it emits highly toxic fumes of F^- and NO_x. See also FLUORIDES and HYDROFLUORIC ACID.

NMI000 CAS:15605-28-4 ***HR: 3***
NITROSYL PERCHLORATE
mf: $ClNO_5$ mw: 129.46

$O:NOClO_3$

SAFETY PROFILE: Below 100° it slowly decomposes; at about 115-120°C it speeds up and explodes. A powerful oxidizer. Explosive reaction with pentaammineazidocobalt(III) perchlorate + phenyl isocyanate; organic materials (e.g., pinene; acetone; ethanol; ether); primary aromatic amines (e.g., aniline; toluidines; xylidines; mesidine). Ignition on mixing with urea. Incompatible with metal salts. When heated to decomposition it emits very toxic fumes of NO_x and Cl^-. See also PERCHLORATES.

NMI500 CAS:18902-42-6 ***HR: 3***
NITROSYLRUTHENIUM TRICHLORIDE
mf: Cl_3NORu mw: 237.44

$O{:}NRuCl_3$

SAFETY PROFILE: Very unstable. Decomposes violently at 400°C. When heated to decomposition it emits highly toxic fumes of NO_x, Cl^-, and RuO_4. See also RUTHENIUM COMPOUNDS.

NMJ000 CAS:7782-78-7 ***HR: 3***
NITROSYLSULFURIC ACID
DOT: UN 2308
mf: HNO_5S mw: 127.07

$O{:}NOSO_2OH$

PROP: Prisms. Decomp @ 73.5°. Sol in sulfuric acid, decomp in water.

SYNS: CHAMBER CRYSTALS ◇ NITRO ACID SULFITE ◇ NITROSONIUM BISULFITE ◇ NITROSYL HYDROGEN SULFATE ◇ NITROSYL SULFATE ◇ SULFURIC ACID, MONOANHYDRIDE with NITROUS ACID

CONSENSUS REPORTS: Reported in EPA TSCA Inventory.

DOT Classification: Corrosive Material; Label: Corrosive.

SAFETY PROFILE: A poison. A corrosive irritant to skin, eyes, and mucous membranes. Explosive reaction above 50°C with 2-chloro-4,6-dinitroaniline and 4-chloro-2,6-dinitroaniline. Potentially explosive reaction with dinitroaniline. When heated to decomposition it emits toxic fumes of SO_x and NO_x. See also SULFATES and NITRATES.

NMJ400 CAS:13444-89-8 ***HR: 2***
NITROSYL TRIBROMIDE
mf: Br_3NO mw: 269.73

SAFETY PROFILE: Probably very toxic. A severe skin, eye, and mucous membrane irritant. Ignites on contact with sodium-antimony alloy. When heated to decomposition it emits very toxic fumes of Br^- and NO_x. See also BROMIDES.

NMJ500 CAS:603-11-2 ***HR: 2***
NITROTEREPHTHALIC ACID
mf: $C_8H_5NO_6$ mw: 211.14

SAFETY PROFILE: A preparative hazard. When heated to decomposition it emits toxic fumes of NO_x.

NMK000 CAS:55011-46-6 ***HR: 3***
5-NITROTETRAZOL
mf: CHN_5O_2 mw: 115.05

SAFETY PROFILE: The silver and mercury salts are explosive. When heated to decomposition it emits toxic fumes of NO_x.

NML000 CAS:61-57-4 ***HR: 3***
NITROTHIAZOLE
mf: $C_6H_6N_4O_3S$ mw: 214.22

SYNS: AMBILHAR ◇ BA 32644 ◇ BA 32644 CIBA ◇ CIBA 32644 ◇ CIBA 32644-BA ◇ NIRIDAZOLE ◇ NITRIDAZOLE ◇ NITROTHIAMIDAZOL ◇ NITROTHIAMIDAZOLE ◇ 1-(5-NITRO-2-THIAZOLYL)IMIDAZOLIDIN-2-ONE ◇ 1-(5-NITRO-2-THIAZOLYL)-2-IMIDAZOLIDINONE ◇ 1-(5-NITRO-2-THIAZOLYL)-2-IMIDAZOLINONE ◇ 1-(5-NITRO-2-THIAZOLYL)-2-OXOTETRAHYDROIMIDAZOL ◇ 1-(5-NITRO-2-THIAZOLYL)-2-OXOTETRAHYDROIMIDAZOLE ◇ NTOI

TOXICITY DATA with REFERENCE
mma-sat 33 ng/plate ENMUDM 7,429,85
bfa-hmn/sat 25 mg/kg CMMUAO 4,171,76
orl-mus TDLo:175 mg/kg (male 5D pre):REP JRPFA4 46,217,76
orl-rat TDLo:8 g/kg/90W-I:NEO CALEDQ 4,305,78
orl-mus TDLo:15 g/kg/49W-C:CAR CALEDQ 1,69,75
orl-mus TD:16 g/kg/75W-C:CAR JNCIAM 59,1625,77
orl-rat LD50:900 mg/kg FAZMAE 17,108,73
orl-mus LD50:2500 mg/kg YHHPAL 15,456,80
ipr-mus LD50:220 mg/kg JPETAB 228,662,84

CONSENSUS REPORTS: IARC Cancer Review: Group 2B IMEMDT 7,56,87; Animal Sufficient Evidence IMEMDT 13,123,77. EPA Genetic Toxicology Program.

SAFETY PROFILE: Suspected carcinogen with experimental carcinogenic and neoplastigenic data. Poison by intraperitoneal route. Moderately toxic by ingestion. Experimental reproductive effects. Human mutation data reported. Used as an amoebicide and schistosomicidal agent. When heated to decomposition it emits very toxic fumes of SO_x and NO_x.

NMO000 CAS:609-40-5 ***HR: 3***
2-NITROTHIOPHENE
mf: $C_4H_3NO_2S$ mw: 129.14

PROP: Monoclinic. Mp: 46°, bp: 224-225°. Insol in alkali; sol in alc and ether.

TOXICITY DATA with REFERENCE
ipr-mus LDLo:64 mg/kg CBCCT* 2,244,50

CONSENSUS REPORTS: Reported in EPA TSCA Inventory.

SAFETY PROFILE: Poison by intraperitoneal route. When heated to decomposition it emits very toxic fumes of NO_x and SO_x.

NMO400 CAS:40358-04-1 ***HR: 3***
4-NITROTHIOPHENE-2-SULFONYL CHLORIDE
mf: $C_4H_2ClNO_4S_2$ mw: 224.64

$O_2NC{=}CHSC(SO_2Cl){=}CH$

SAFETY PROFILE: Residue left from distillation decomposes vigorously at 147°C. When heated to decomposition it emits toxic fumes of Cl^-, SO_x, and NO_x.

NMO500 CAS:99-08-1 ***HR: 3***
m-NITROTOLUENE
DOT: UN 1664
mf: $C_7H_7NO_2$ mw: 137.15

PROP: Liquid. Mp: 15.1°, flash p: 233°F (CC), d: 1.1630 @ 15°/4°, vap press: 1 mm @ 50.2°, vap d: 4.72, bp: 231.9°. Misc with alc, ether; sol in benzene; sol in water @ 30°.

SYNS: 3-METHYLNITROBENZENE ◇ m-METHYLNITROBENZENE ◇ MNT ◇ 3-NITROTOLUENE ◇ 3-NITROTOLUOL

TOXICITY DATA with REFERENCE
orl-rat LD50:1072 mg/kg NTIS** PB214-270
orl-mus LD50:330 mg/kg CHABA8 73,91015e,70
orl-rbt LD50:2400 mg/kg CHABA8 73,91015e,70
orl-gpg LD50:3600 mg/kg CHABA8 73,91015e,70

CONSENSUS REPORTS: Reported in EPA TSCA Inventory.

OSHA PEL: (Transitional: TWA 5 ppm (skin)) TWA 2 ppm (skin)
ACGIH TLV: TWA 2 ppm (skin)
DFG MAK: 5 ppm (30 mg/m^3)
DOT Classification: Poison B; Label: Poison.

SAFETY PROFILE: Poison by ingestion. Combustible when exposed to heat, flame, or oxidizers. To fight fire, use water, CO_2, dry chemical. Probably an explosive. When heated to decomposition it emits toxic fumes of NO_x. See also other methylnitrobenzene entries and NITRO COMPOUNDS OF AROMATIC HYDROCARBONS.

NMO525 CAS:88-72-2 ***HR: 3***
o-NITROTOLUENE
DOT: UN 1664
mf: $C_7H_7NO_2$ mw: 137.15

PROP: Yellowish liquid. Mp: −10°, bp: 222.3°, flash p: 223°F (CC), d: 1.1622 @ 19°/15°, vap press: 1 mm @ 50°, vap d: 4.72. Insol in water; sol in SO_2 and petr ether; misc in alc, benzene, and ether. Sltly sol in NH_3.

SYNS: 2-METHYLNITROBENZENE ◇ o-METHYLNITROBENZENE ◇ 2-NITROTOLUENE ◇ ONT

TOXICITY DATA with REFERENCE
dns-rat-orl 150 mg/kg CGINA8 52,131,84
dns-rat-orl 200 mg/kg CNREA8 43,2836,83
orl-rat LD50:891 mg/kg NTIS** PB214-270
orl-mus LD50:970 mg/kg GTPZAB 25(8),50,81

CONSENSUS REPORTS: Reported in EPA TSCA Inventory.

OSHA PEL: (Transitional: TWA 5 ppm (skin)) TWA 2 ppm (skin)
ACGIH TLV: TWA 2 ppm (skin)
DFG MAK: 5 ppm (30 mg/m^3)
DOT Classification: Poison B; Label: Poison.

SAFETY PROFILE: A poison. Moderately toxic by ingestion. Mucous membrane effects by inhalation. Mutation data reported. Combustible when exposed to heat or open flame. To fight fire, use water spray, fog, foam, CO_2. Potentially explosive reaction with alkali (e.g., sodium hydroxide). When heated to decomposition it emits toxic fumes of NO_x. See also other methylnitrobenzene entries and NITRO COMPOUNDS OF AROMATIC HYDROCARBONS.

NMO550 CAS:99-99-0 ***HR: 3***
p-NITROTOLUENE
DOT: UN 1664
mf: $C_7H_7NO_2$ mw: 137.15

PROP: Yellowish crystals. Bp: 238.3°, flash p: 223°F (CC), d: 1.286, vap press: 1 mm @ 53.7°, vap d: 4.72. Mp: 53-54°. Insol in water; sol in alc, benzene, ether, chloroform, and acetone.

SYNS: 4-METHYLNITROBENZENE ◇ p-METHYL NITROBENZENE ◇ NCI-C60537 ◇ 4-NITROTOLUENE ◇ 4-NITROTOLUOL ◇ PNT

TOXICITY DATA with REFERENCE
mmo-sat 10 μg/plate ENMUDM 4,163,82
mma-sat 10 μg/plate ENMUDM 4,163,82
orl-rat LD50:1960 mg/kg GTPZAB 25(8),50,81
skn-rat LD50:16000 mg/kg GISAAA 24(9),15,59
ipr-rat LD50:940 mg/kg GISAAA 24(9),15,59
orl-mus LD50:1231 mg/kg NTIS** PB214-270

CONSENSUS REPORTS: Reported in EPA TSCA Inventory.

OSHA PEL: (Transitional: TWA 5 ppm (skin)) TWA 2 ppm (skin)
ACGIH TLV: TWA 2 ppm (skin)
DFG MAK: 5 ppm (30 mg/m^3)
DOT Classification: Poison B; Label: Poison.

SAFETY PROFILE: A poison. Moderately toxic by ingestion and intraperitoneal routes. Mildly toxic by skin contact. Mutation data reported. Combustible when exposed to heat or flame. To fight fire, use CO_2, dry chemical, foam. The residue from vacuum distillation may explode spontaneously. Reacts with sodium to form an ignitable product. Violent reaction with concentrated sulfuric acid (above 160°C); sulfuric acid + sulfur trioxide (above 52°C). Mixtures with tetranitromethane are sensitive high explosives. May explode on standing. It has been involved in plant scale explosions. When heated to decomposition it emits toxic fumes of NO_x. See also other methylnitrobenzene entries and NITRO COMPOUNDS OF AROMATIC HYDROCARBONS.

NMO600 CAS:1321-12-6 ***HR: 3***
mixo-NITROTOLUENE
mf: $C_7H_7NO_2$ mw: 137.14

$O_2NC_6H_4CH_3$
DFG MAK: 5 ppm (30 mg/m^3)

SAFETY PROFILE: A poison. May decompose explosively if heated above 190°C. When heated to decomposition it emits toxic fumes of NO_x. See also NITRO COMPOUNDS OF AROMATIC HYDROCARBONS.

NMP000 CAS:119-32-4 ***HR: 3***
5-NITRO-4-TOLUIDINE
mf: $C_7H_8N_2O_2$ mw: 152.17

PROP: Clear liquid. Mp: 116°, flash p: 315°F (CC), d: 1.312, vap d: 5.80. Sltly sol in hot water, sol in alc.

SYNS: 4-AMINO-2-NITROTOLUENE ◇ 4-METHYL-3-NITROANILINE ◇ 2-NITRO-4-AMINOTOLUENE ◇ 3-NITRO-4-METHYLANILINE ◇ 3-NITRO-4-TOLUIDIN (CZECH) ◇ m-NITRO-p-TOLUIDINE ◇ 3-NITRO-p-TOLUIDINE

TOXICITY DATA with REFERENCE
orl-rat LD50:6860 mg/kg 28ZPAK -,133,72
ivn-mus LD50:180 mg/kg CSLNX* NX#04522
orl-bwd LD50:3 mg/kg TXAPA9 21,315,72

CONSENSUS REPORTS: Reported in EPA TSCA Inventory.

SAFETY PROFILE: Poison by ingestion and intravenous routes. Combustible when exposed to heat or flame. To fight fire, use CO_2, spray, foam. When heated to decomposition it emits toxic fumes of NO_x. See also NITRO COMPOUNDS OF AROMATIC HYDROCARBONS.

NMP500 CAS:99-55-8 ***HR: 3***
5-NITRO-o-TOLUIDINE
mf: $C_7H_8N_2O_2$ mw: 152.17

SYNS: 2-AMINO-4-NITROTOLUENE ◇ AZOFIX SCARLET G SALT ◇ AZOGENE FAST SCARLET G ◇ C.I. 37105 ◇ C.I. AZOIC DIAZO COMPONENT 12 ◇ DAINICHI FAST SCARLET G BASE ◇ DAITO SCARLET BASE G ◇ DEVOL SCARLET B ◇ DIABASE SCARLET G ◇ DIAZO FAST SCARLET G ◇ FAST RED SG BASE ◇ FAST SCARLET G ◇ HILTONIL FAST SCARLET G BASE ◇ KAYAKU SCARLET G BASE ◇ LAKE SCARLET G BASE ◇ LITHOSOL ORANGE R BASE ◇ 6-METHYL-3-NITROANILINE ◇ 2-METHYL-5-NITRO-BENZENEAMINE ◇ MITSUI SCARLET G BASE ◇ NAPHTHANIL SCARLET G BASE ◇ NAPHTOELAN FAST SCARLET G SALT ◇ NCI-C01843 ◇ 4-NITRO-2-AMINOTOLUENE (MAK) ◇ PNOT ◇ RCRA WASTE NUMBER U181 ◇ SCARLET BASE CIBA II ◇ SUGAI FAST SCARLET G BASE ◇ SYMULON SCARLET G BASE

TOXICITY DATA with REFERENCE
mmo-sat 10 μg/plate ENMUDM 4,163,82
mma-sat 33300 ng/plate ENMUDM 7(Suppl 5),1,85
otr-rat:emb 41 μg/plate JJATDK 1,190,81
orl-mus TDLo:150 g/kg/78W-C:CAR NCITR* NCI-CG-TR-107,78
orl-rat LD50:574 mg/kg NCIMR* NIH-71-E-2144,73

CONSENSUS REPORTS: NCI Carcinogenesis Bioassay (feed); Clear Evidence: mouse NCITR* NCI-CG-TR-107,78. Reported in EPA TSCA Inventory.

DFG MAK: Animal Carcinogen, Suspected Human Carcinogen.

SAFETY PROFILE: Confirmed carcinogen with experimental carcinogenic data. Moderately toxic by ingestion. Mutation data reported. Decomposes exothermically when heated to 150°C. When heated to decomposition it emits toxic fumes of NO_x. See also NITRO COMPOUNDS OF AROMATIC HYDROCARBONS.

NMQ000 CAS:464-10-8 ***HR: 3***
NITROTRIBROMOMETHANE
mf: CBr_3NO_2 mw: 297.75

PROP: Crystals. Bp: 127°, mp: 103°, d: 2.79 @ 18°.

SYNS: BROMOPICRIN ◇ NITROBROMOFORM ◇ TRIBROMONITROMETHANE

TOXICITY DATA with REFERENCE
ipr-mus LD50:15 mg/kg KHFZAN 10(6),53,76

CONSENSUS REPORTS: Reported in EPA TSCA Inventory.

SAFETY PROFILE: Poison by intraperitoneal route. In vapor form it is highly toxic by ingestion and inhalation and on contact with skin, eyes, and mucous membranes. An explosive. When heated to decomposition it emits very toxic fumes of Br^- and NO_x. See also NITRO COMPOUNDS and BROMIDES.

NMQ500 CAS:556-89-8 ***HR: 3***
NITROUREA
DOT: UN 0147
mf: $CH_3N_3O_3$ mw: 105.06

PROP: Crystals.

SYN: m-NITROCARBAMIDE

CONSENSUS REPORTS: Reported in EPA TSCA Inventory.

DOT Classification: Class A Explosive; Label: Explosive A.

SAFETY PROFILE: A very dangerous fire hazard when exposed to heat or flame. A severe explosion hazard when shocked or exposed to heat. Can react vigorously with oxidizing materials. It is a high explosive. Incompatible with mercuric and silver salts. When heated to decomposition it emits highly toxic fumes of NO_x. See also EXPLOSIVES, HIGH; and NITRATES.

NMR000 CAS:7782-77-6 ***HR: 2***
NITROUS ACID
mf: HNO_2 mw: 47.02

PROP: Pale blue solution.

SYN: NITROSYL HYDROXIDE

TOXICITY DATA with REFERENCE
mmo-omi 20 mmol/L EJMBA2 16,133,81
dnr-ssp 132 μmol/L CNJGA8 24,771,82

CONSENSUS REPORTS: EPA Genetic Toxicology Program. Reported in EPA TSCA Inventory.

SAFETY PROFILE: Mutation data reported. Flammable by chemical reaction; a powerful oxidizer. Explodes on contact with phosphorus trichloride. Reacts violently with PH_3 and PCl_3. Reactions with 1-amino-5-nitrophenol, ammonium decahydroborate(2−), hydrazine (product is hydrogen azide) may give explosive products. Incompatible with anilines (e.g., 4-bromoaniline, 2-chloroaniline, 3-chloroaniline, 2-nitroaniline, 3-nitroaniline, 4-nitroaniline, aniline), semicarbazone, silver nitrate. When heated to decomposition it emits highly toxic fumes of NO_x. See also NITRIC OXIDE.

NMS000 CAS:25168-04-1 ***HR: 3***
NITROXYLENE
DOT: NA 1665
mf: $C_8H_9NO_2$ mw: 151.18

PROP: Light yellow liquid. Mp: 2°, bp: 244°, d: 1.135 @ 15°/4°.

SYNS: NITRODIMETHYLBENZENE ◇ NITROXYLOL (DOT)

TOXICITY DATA with REFERENCE
orl-rat LD50:2440 mg/kg MarJV# 29MAR77

DOT Classification: Poison B; Label: Poison.

SAFETY PROFILE: A poison. Moderately toxic by ingestion. When heated to decomposition it emits toxic fumes of NO_x. See also NITRO COMPOUNDS OF AROMATIC HYDROCARBONS.

NMS500 CAS:81-20-9 ***HR: 3***
2-NITRO-m-XYLENE
mf: $C_8H_7NO_2$ mw: 149.16

SYN: 2,6-DIMETHYLNITROBENZENE

TOXICITY DATA with REFERENCE
ivn-mus LD50:45 mg/kg CSLNX* NX#00236

CONSENSUS REPORTS: Reported in EPA TSCA Inventory.

SAFETY PROFILE: Poison by intravenous route. When heated to decomposition it emits toxic fumes of NO_x. See also NITRO COMPOUNDS OF AROMATIC HYDROCARBONS.

NMT000 CAS:13444-90-1 ***HR: 3***
NITRYL CHLORIDE
mf: $ClNO_2$ mw: 81.46

PROP: Corrosive, toxic, colorless gas. Decomp > 120°, vap d: 2.81 g/L @ 100°, bp: −14.3°, mp: −145°, d: (liquid): 1.37 @ 0°.

SYN: NITROXYL CHLORIDE

SAFETY PROFILE: A poison by inhalation. A corrosive irritant to skin, eyes, and mucous membranes. A powerful oxidizer. The gas or liquid may attack organic matter with explosive violence. Violent reaction with ammonia or sulfur dioxide. Incompatible with tin(II) bromide or tin(II) iodide. Reacts with water or steam to produce corrosive fumes. When heated to decomposition it emits toxic fumes of Cl^- and NO_x. See also CHLORIDES.

NMT500 CAS:10022-50-1 ***HR: 3***
NITRYL FLUORIDE
mf: FNO_2 mw: 65.01

PROP: Colorless gas; pungent odor. Mp: −166.0°, bp: −72.4°, d (liquid): 1.796 @ bp, d (solid): 1.924.

SAFETY PROFILE: Poison by inhalation. A severe irritant to skin, eyes, and mucous membranes. A powerful oxidizing agent. This gas is intensely reactive. Explosive reaction with hydrogen at 200-300°C. Ignites on contact with antimony; arsenic; boron; iodine; phosphorus; selenium. Ignites when warmed with bismuth; carbon; chromium; lead; sulfur. Incandescent reaction with aluminum; cadmium; cobalt; iron; molybdenum; nickel; potassium; sodium; thorium; titanium; tungsten; uranium; vanadium; zinc; zirconium; lithium (at 200-300°C); manganese (at 200-300°C). Incompatible with metals; non-metals. When heated to decomposition it emits toxic fumes of F^- and NO_x. See also FLUORIDES.

NMU000 CAS:7789-26-6 ***HR: 3***
NITRYL HYPOFLUORITE
mf: FNO_3 mw: 81.00

PROP: Colorless gas, acrid odor. D: (liquid) 1.507 @ −45.9°, (solid) 1.951 @ −193.2°, mp: −175°, bp: −45.9°.

SYN: FLUORINE NITRATE

CONSENSUS REPORTS: EPA Extremely Hazardous Substances List.

SAFETY PROFILE: Probably a poison irritant to skin, eyes, mucous membranes. A powerful oxidant and dangerous explosive. Explodes on contact with organic materials (e.g., alcohol; ether; grease). Ignites on mixing with ammonia; dinitrogen oxide; and hydrogen sulfide. When heated to decomposition it emits toxic fumes of F^- and NO_x. See also FLUORIDES and NITRATES.

NMU500 ***HR: 3***
NITRYL PERCHLORATE
mf: $ClNO_6$ mw: 145.46

SAFETY PROFILE: Explosive reaction with organic solvents (e.g., benzene; acetone; ether). A common impurity in ammonium perchlorate. The mixture is more sensitive than pure ammonium perchlorate. When heated to decomposition it emits highly toxic fumes of NO_x and Cl^-. See also PERCHLORATES.

NMV000 CAS:23282-20-4 ***HR: 3***
NIVALENOL
mf: $C_{15}H_{20}O_7$ mw: 312.35

PROP: A toxic substance from moldy rice infected by *Fusarium nivale* (CNREA8 28,2393,68).

SYNS: (3-α,4-β,7-α)-12,13-EPOXY-3,4,7,15-TETRAHYDROXYTRICHOTHEC-9-EN-8-ONE ◇ 12,13-EPOXY-3,4,7,15-TETRAHYDROXYTRICHOTHEC-9-EN-8-ONE ◇ 3-α,4-β,7-α,15-TETRAHYDROXYSCIRP-9-EN-8-ONE

TOXICITY DATA with REFERENCE
skn-gpg 3124 ng MLD FAATDF 4(2, Pt 2),S124,84
dni-mus:ast 1 mg/L EXPEAM 24,1032,68
ipr-mus LD50:4 mg/kg CNREA8 28,2393,68
scu-mus LD50:5200 μg/kg JJEMAG 42,91,72
ivn-mus LD50:6300 μg/kg FAATDF 4(2, Pt 2),S124,84

SAFETY PROFILE: Poison by intravenous, intraperitoneal, and subcutaneous routes. Mutation data reported. A skin irritant. When heated to decomposition it emits acrid smoke and irritating fumes.

NMV400 ***HR: 3***
NIZOFENONE FUMARATE
mf: $C_{21}H_{21}ClN_4O_3 \cdot C_4H_4O_4$ mw: 528.99

SYN: 1-(2-(2-CHLOROBENZOYL)-4-NITROPHENYL)-2-(DIETHYLAMINOMETHYL)IMIDAZOLEFUMARATE

TOXICITY DATA with REFERENCE
ivn-rat TDLo:5500 μg/kg (female 7-17D post):REP IYKEDH 16,1,85
ivn-rat TDLo:110 mg/kg (female 7-17D post):TER IYKEDH 16,1,85
orl-rat LD50:1580 mg/kg OYYAA2 30,727,85
scu-rat LD50:1629 mg/kg OYYAA2 30,627,85
ivn-rat LD50:63 mg/kg OYYAA2 30,627,85
orl-mus LD50:495 mg/kg OYYAA2 30,627,85
scu-mus LD50:270 mg/kg OYYAA2 30,627,85
ivn-mus LD50:62 mg/kg OYYAA2 30,627,85

SAFETY PROFILE: Poison by subcutaneous and intravenous routes. Moderately toxic by ingestion. An experimental teratogen. Experimental reproductive effects. When heated to decomposition it emits toxic fumes of Cl^- and NO_x.

NMV450 CAS:86451-37-8 ***HR: 3***
NMDHP

SYNS: 3-(METHYLNITROSAMINO)-1,2-PROPANEDIOL ◇ N-NITROSOMETHYL(2,3-DIHYDROXYPROPYL)AMINE

TOXICITY DATA with REFERENCE
mmo-sat 500 μg/plate TCMUE9 1,13,84
mma-sat 100 μg/plate TCMUE9 1,13,84
dns-rat:lvr 5 mmol/L MUREAV 144,197,85
orl-rat TDLo:580 mg/kg/40W-I:ETA CALEDQ 22,83,84

SAFETY PROFILE: Questionable carcinogen with experimental tumorigenic data. Mutation data reported. Many N-nitroso compounds are carcinogens. When heated to decomposition it emits toxic fumes of NO_x. See also N-NITROSO COMPOUNDS.

NMV480 CAS:76631-42-0 ***HR: 3***
NOCARDICIN COMPLEX
mf: $C_{16}H_{17}N_3O_3$ mw: 299.36

SYN: FR-1923

TOXICITY DATA with REFERENCE
orl-mus LD50:10 g/kg 85GDA2 4(1),114,80
ipr-mus LD50:2500 mg/kg 85GDA2 4(1),114,80
ivn-mus LD50:100 mg/kg 85GDA2 4(1),114,80

SAFETY PROFILE: Poison by intravenous route. Moderately toxic by intraperitoneal route. Mildly toxic by ingestion. When heated to decomposition it emits toxic fumes of NO_x.

NMV500 CAS:1404-15-5 ***HR: 3***
NOGALAMYCIN
mf: $C_{39}H_{49}O_{16}$ mw: 773.88

SYNS: ANTIBIOTIC NSC 70845 ◇ ANTIBIOTIC 205T3 ◇ NOGALOMYCIN ◇ NSC-70845 ◇ U-15167

TOXICITY DATA with REFERENCE
pic-esc 250 ng/plate CNREA8 43,2819,83
dni-hmn:oth 410 nmol/L HXPHAU 38(Pt 2),623,75
ipr-rat LD50:500 μg/kg ADTEAS 3,181,68
ivn-rat LDLo:4600 μg/kg NCIRI* -,192,66
ipr-mus LD50:13750 μg/kg IJEBA6 17,595,79
ivn-mus LDLo:8500 μg/kg NCIRI* -,192,66
ivn-dog LDLo:250 μg/kg NCIRI* -,192,66

SAFETY PROFILE: Poison by intravenous and intraperitoneal routes. Human mutation data reported. An antibiotic. When heated to decomposition it emits toxic fumes of NO_x.

NMV600 CAS:64267-46-5 ***HR: D***
NOGAMYCIN
mf: $C_{37}H_{47}NO_{14}$ mw: 729.85

SYN: DISNOGAMYCIN

TOXICITY DATA with REFERENCE
dni-mus:leu 3200 nmol/L JANTAJ 34,1596,81
dnd-mam:lym 12 μmol/L CBINA8 36,1,81

SAFETY PROFILE: Mutation data reported. When heated to decomposition it emits toxic fumes of NO_x.

NMV700 CAS:24526-64-5 ***HR: 3***
NOMIFENSINE
mf: $C_{16}H_{18}N_2$ mw: 238.33

PROP: Mp: 179-181°.

SYNS: 8-AMINO-2-METHYL-4-PHENYL-1,2,3,4-TETRAHYDROISOQUINOLINE ◇ NOMIFENSIN ◇ 1,2,3,4-TETRAHYDRO-2-METHYL-4-PHENYL-8-ISOQUINOLINAMINE(9CI)

TOXICITY DATA with REFERENCE
orl-wmn TDLo:7 mg/kg/7D-I:BLD BMJOAE 288,830,84
orl-wmn TDLo:56 mg/kg/4W-I:LIV BMJOAE 289,1268,84
ivn-rat LD50:72 mg/kg DRUGAY 18,1,79
orl-mus LD50:260 mg/kg ARZNAD 32,873,82
ivn-mus LD50:90 mg/kg DRUGAY 18,1,79
ivn-gpg LD50:264 mg/kg DRUGAY 18,1,79

SAFETY PROFILE: Poison by ingestion and intravenous routes. Human systemic effects by ingestion: diffuse hepatitis, hemorrhage and decrease in the number of blood platelets (thromobcytopenia). When heated to decomposition it emits toxic fumes of NO_x. See also NOMIFENSINE MALEATE.

NMV725 CAS:32795-47-4 ***HR: 3***
NOMIFENSINE MALEATE
mf: $C_{16}H_{18}N_2 \cdot C_4H_4O_4$ mw: 354.44

SYNS: ALIVAL ◇ ALIVAL (ANTIDEPRESSANT) ◇ 8-AMINO-2-METHYL-4-PHENYL-1,2,3,4-TETRAHYDROISOQUINOLINEMALEATE ◇ 8-AMINO-1,2,3,4-TETRAHYDRO-2-METHYL-4-PHENYLISOQUINOLINE MALEATE ◇ (Z)-2-BUTENEDIOATE (1:1) 1,2,3,4-TETRAHYDRO-2-METHYL-4-PHENYL-8-ISOQUINOLINAMINE ◇ HOE 984 ◇ HOSTALIVAL ◇ MERITAL ◇ NOMIFENSIN ◇ NOMIFENSINE HYDROGEN MALEATE ◇ NOMIFENSIN HYDROGEN MALEATE ◇ PSICRONIZER

TOXICITY DATA with REFERENCE
orl-mus TDLo:600 mg/kg (6-15D preg):REP OYYAA2 28,171,84
orl-man TDLo:66429 μg/kg/31D-I CMAJAX 133,207,85
ivn-man TDLo:1429 μg/kg/20M-C:CVS,BPR CLPTAT 41,88,87
orl-rat LD50:430 mg/kg ARZNAD 23,45,73
ipr-rat LD50:112 mg/kg OYYAA2 27,1019,84
scu-rat LD50:193 mg/kg OYYAA2 27,1019,84
ivn-rat LD50:66 mg/kg OYYAA2 27,1019,84
orl-mus LD50:300 mg/kg OYYAA2 27,1019,84
ipr-mus LD50:140 mg/kg OYYAA2 27,1019,84
scu-mus LD50:195 mg/kg OYYAA2 27,1019,84
ivn-mus LD50:68 mg/kg OYYAA2 27,1019,84
orl-rbt LD50:854 mg/kg ARZNAD 23,45,73

SAFETY PROFILE: Poison by ingestion, subcutaneous, intravenous, and intraperitoneal routes. Human systemic effects by intravenous route: heart rate changes, blood pressure elevation. Experimental reproductive effects. An antidepressant. When heated to decomposition it emits toxic fumes of NO_x. See also NOMIFENSINE.

NMV735 CAS:27753-52-2 ***HR: 2***
NONABROMOBIPHENYL
mf: $C_{12}HBr_9$ mw: 864.32

SYNS: 1,1-BIPHENYL, NONABROMO- ◇ BROMKAL 80-9D
orl-mus TDLo:4469 mg/kg/78W-C:CAR YACHDS 14,5541,86

CONSENSUS REPORTS: Reported in EPA TSCA Inventory.

SAFETY PROFILE: Questionable carcinogen with experimental carcinogenic data. When heated to decomposition it emits toxic fumes of Br^-.

NMV740 CAS:20982-74-5 ***HR: 3***
NONACARBONYL DIIRON
mf: $C_9Fe_2O_9$ mw: 363.79

SYN: IRON CARBONYL

SAFETY PROFILE: The commercial product ignites on contact with brass when heated above 93°C. When heated to decomposition it emits acrid smoke and irritating fumes.

NMV750 CAS:49780-10-1 ***HR: 3***
NONACHLAZINE
mf: $C_{22}H_{24}ClN_3OS \cdot 2ClH$ mw: 486.92

SYNS: AY-25,329 ◇ 2-CHLORO-10-(3-(HEXAHYDROPYRROLO(1,2-a)PYRAZIN-2(1H)-YL)-PROPIONYL)PHENOTHIAZINE 2HCl ◇ 10'-(β-(1,4-DIAZABICYCLO(4.3.0)NONANYL-4)-PROPIONYL)-2-CHLOROPHENTHIAZIN DICHLORIDE

TOXICITY DATA with REFERENCE
orl-mus LD50:160 mg/kg FATOAO 42,243,79
scu-mus LD50:210 mg/kg FATOAO 44,342,81
ivn-mus LD50:55 mg/kg FATOAO 38(6),732,75

SAFETY PROFILE: Poison by ingestion, subcutaneous, and intravenous routes. When heated to decomposition it emits toxic fumes of SO_x, NO_x, and HCl.

NMV760 CAS:557-48-2 ***HR: 2***
trans,cis-2,6-NONADIENAL
mf: $C_9H_{14}O$ mw: 138.23

PROP: Sltly yellow liquid; powerful, violet, cucumber odor. D: 0.850-0.870, refr index: 1.470. Sol in alc, fixed oils; insol in water.

SYNS: CUCUMBER ALDEHYDE ◇ FEMA No. 3317 ◇ trans,cis-2,6-NONADIENAL ◇ trans-2,cis-6-NONADIENAL ◇ 2,6-NONADIENAL ◇ VIOLET LEAF ALDEHYDE

TOXICITY DATA with REFERENCE
skn-rbt 500 mg/24H MOD FCTOD7 20,769,82

CONSENSUS REPORTS: Reported in EPA TSCA Inventory.

SAFETY PROFILE: A moderate skin irritant. When heated to decomposition it emits acrid smoke and irritating fumes.

NMV780 CAS:28069-72-9 ***HR: 1***
trans,cis-2,6-NONADIENOL
mf: $C_9H_{16}O$ mw: 140.25

PROP: White to yellow liquid; powerful, vegetable odor. D: 0.860-0.880, refr index: 1.464. Insol in water.

SYNS: CUCUMBER ALCOHOL ◇ FEMA No. 2780 ◇ NONADIENOL ◇ 2-trans-6-cis-NONADIEN-1-OL ◇ VIOLET LEAF ALCOHOL

TOXICITY DATA with REFERENCE
skn-gpg 100%/24H MOD FCTOD7 20,771,82

CONSENSUS REPORTS: Reported in EPA TSCA Inventory.

SAFETY PROFILE: A skin irritant. When heated to decomposition it emits acrid smoke and irritating fumes.

NMW500 CAS:124-19-6 ***HR: 2***
1-NONANAL
mf: $C_9H_{18}O$ mw: 142.27

PROP: Found in at least 20 essential oils, including rose and citrus oils and several species of pine oil (FCTXAV 11, 95,73). Colorless to light yellow liquid; citrus-rose odor. D: 0.820-0.830, refr index: 1.422-1.429, flash p: 162°F. Sol in alc; fixed oils, propylene glycol; insol in glycerin.

SYNS: ALDEHYDE C-9 ◇ C-9 ALDEHYDE ◇ FEMA No. 2782 ◇ NCI-C61018 ◇ 1-NONALDEHYDE ◇ 1-NONYL ALDEHYDE ◇ PELARGONIC ALDEHYDE

TOXICITY DATA with REFERENCE
skn-rbt 500 mg/24H SEV FCTXAV 11,1079,73

CONSENSUS REPORTS: Reported in EPA TSCA Inventory.

SAFETY PROFILE: A severe skin irritant. Combustible liquid. When heated to decomposition it emits acrid smoke and irritating fumes. See also ALDEHYDES.

NMX000 CAS:111-84-2 ***HR: 3***
NONANE
DOT: UN 1920
mf: C_9H_{20} mw: 128.29

PROP: Colorless liquid. Mp: −53.7°, bp: 150.7°, lel: 0.8%, uel: 2.9%, flash p: 88°F (CC), d: 0.718 @ 20°/4°, autoign temp: 374°F, vap press: 10 mm @ 38.0°, vap d: 4.41. Insol in water; sol in abs alc and ether.

SYN: SHELLSOL 140

TOXICITY DATA with REFERENCE
ihl-rat LC50:3200 ppm/4H DTLVS* 4,312,80
ivn-mus LD50:218 mg/kg JPMSAE 67,566,78

CONSENSUS REPORTS: Reported in EPA TSCA Inventory.

OSHA PEL: TWA 200 ppm
ACGIH TLV: TWA 200 ppm

DOT Classification: Flammable or Combustible Liquid; Label: Flammable Liquid.

SAFETY PROFILE: Poison by intravenous route. Mildly toxic by inhalation. Irritating to respiratory tract. Narcotic in high concentrations. A very dangerous fire hazard when exposed to heat or flame; can react with oxidizing materials. Explosive in the form of vapor when exposed to heat or flame. Emitted from modern building materials. (CENEAR 69,22,91) To fight fire, use CO_2, dry chemical. When heated to decomposition it emits acrid smoke and irritating fumes.

NMX500 CAS:31080-39-4 ***HR: 2***
4-NONANECARBOXYLIC ACID
mf: $C_{10}H_{20}O_2$ mw: 172.30

SYNS: 2-PROPYLHEPTANIC ACID ◇ 2-PROPYLHEPTANSAEURE (GERMAN)

TOXICITY DATA with REFERENCE
orl-rat LDLo:2500 mg/kg ARZNAD 3,86,53
ipr-rat LDLo:1 g/kg ARZNAD 3,86,53

SAFETY PROFILE: Moderately toxic by ingestion. When heated to decomposition it emits acrid smoke and irritating fumes.

NMY000 CAS:112-05-0 ***HR: 3***
NONANOIC ACID
mf: $C_9H_{18}O_2$ mw: 158.27

PROP: Oily, colorless liquid. Bp: 254°, mp: 12°, d: 0.9055 @ 20°/4°. Very sltly sol in water.

SYNS: CIRRASOL 185A ◇ EMFAC 1202 ◇ HEXACID C-9 ◇ n-NONOIC ACID ◇ n-NONYLIC ACID ◇ 1-OCTANECARBOXYLIC ACID ◇ PELARGIC ACID ◇ PELARGON (RUSSIAN) ◇ PELARGONIC ACID

TOXICITY DATA with REFERENCE
skn-rbt 500 mg/24H MOD FCTXAV 16,839,78
eye-rbt 91 mg SEV EMERY* S3B,-,64
skn-gpg 100% SEV FCTXAV 16,839,78
orl-rat LDLo:3200 mg/kg 14CYAT 2,1788,63
orl-mus LD50:15 g/kg GISAAA 42(3),99,77
ivn-mus LD50:224 mg/kg APTOA6 18,141,61

CONSENSUS REPORTS: Reported in EPA TSCA Inventory.

SAFETY PROFILE: Poison by intravenous route. Moderately toxic by ingestion. A severe skin and eye irritant. When heated to decomposition it emits acrid smoke and irritating fumes.

NMY500 CAS:821-55-6 ***HR: 2***
2-NONANONE
mf: $C_9H_{18}O$ mw: 142.27

PROP: Colorless liquid. Mp: −9°, bp: 194°, flash p: 160°F (CC), vap d: 4.9, d: 0.832 @ 30°. Insol in water; sol in alc, ether.

SYNS: HEPTYL METHYL KETONE ◇ METHYL HEPTYL KETONE ◇ NONAN-2-ONE

TOXICITY DATA with REFERENCE
orl-rat LD50:3200 mg/kg KODAK* -,-,71

CONSENSUS REPORTS: Reported in EPA TSCA Inventory.

SAFETY PROFILE: Moderately toxic by ingestion. Combustible when exposed to heat or flame; can react with oxidizing materials. To fight fire, use foam, CO_2, dry chemical. When heated to decomposition it emits acrid smoke and irritating fumes.

NMZ000 CAS:502-56-7 ***HR: 2***
5-NONANONE
mf: $C_9H_{18}O$ mw: 142.27

SYNS: BUTYL KETONE ◇ DIBUTYL KETONE ◇ NONAN-5-ONE ◇ 5-OXONONANE

TOXICITY DATA with REFERENCE
orl-rat LDLo:1000 mg/kg CTOXAO 17,271,80
ivn-mus LD50:1379 mg/kg JPMSAE 67,566,78

CONSENSUS REPORTS: Reported in EPA TSCA Inventory.

SAFETY PROFILE: Moderately toxic by ingestion and intravenous routes. When heated to decomposition it emits acrid smoke and irritating fumes. See also KETONES.

NNA000 CAS:63021-51-2 ***HR: 3***
1-NONANOYLAZIRIDINE
mf: $C_{11}H_{21}NO$ mw: 183.33

SYN: NONANOYLETHYLENEIMINE

TOXICITY DATA with REFERENCE
scu-rat TDLo:720 mg/kg/20W-I:NEO BJPCAL 9,306,54
scu-mus TDLo:500 mg/kg/20W-I:ETA BJPCAL 9,306,54

SAFETY PROFILE: Questionable carcinogen with experimental neoplastigenic and tumorigenic data. When heated to decomposition it emits toxic fumes of NO_x.

NNA300 CAS:2463-53-8 ***HR: 2***
2-NONENAL
mf: $C_9H_{16}O$ mw: 140.25

PROP: White to sltly yellow liquid; fatty, violet odor. D: 0.850-0.870, refr index: 1.457. Sol in alc, fixed oils; insol in water.

SYNS: FEMA No. 3213 ◇ HEPTYLIDENE ALDEHYDE ◇ β-HEXYLACROLEIN ◇ 2-NONEN-1-AL ◇ α-NONENYL ALDEHYDE ◇ trans-2-NONENAL (FCC)

TOXICITY DATA with REFERENCE
skn-rbt 500 mg/24H SEV FCTOD7 20(Suppl),775,82
orl-rat LD50:5 g/kg FCTOD7 20(Suppl),775,82
skn-rbt LD50:3700 mg/kg FCTOD7 20(Suppl),775,82

CONSENSUS REPORTS: Reported in EPA TSCA Inventory.

SAFETY PROFILE: Moderately toxic by skin contact. Mildly toxic by ingestion. A severe skin irritant. When heated to decomposition it emits acrid smoke and irritating fumes. See also ALDEHYDES.

NNA325 CAS:2277-19-2 ***HR: 2***
6-NONENAL, (Z)-
mf: $C_9H_{16}O$ mw: 140.25

PROP: Liquid. Flash p: 130° F.

SYNS: cis-6-NONENAL ◇ cis-6-NONEN-1-AL ◇ (Z)-6-NONENAL

TOXICITY DATA with REFERENCE
skn-gpg 100%/24H MLD FCTOD7 20,777,82

CONSENSUS REPORTS: Reported in EPA TSCA Inventory.

SAFETY PROFILE: A skin irritant. Combustible liquid. When heated to decomposition it emits acrid smoke and irritating fumes.

NNB000 ***HR: 3***
2-NONEN-4,6,8-TRIYN-1-AL
mf: C_9H_4O mw: 128.13

$$H(C\equiv C)_3CH{=}CHCO{\bullet}H$$

SAFETY PROFILE: Very unstable. Can explode at room temperature. Upon decomposition it emits acrid smoke and irritating fumes. See also ALDEHYDES and ACETYLENE COMPOUNDS.

NNB300 CAS:26027-38-3 ***HR: 3***
NONOXYNOL-9
mf: $(C_2H_4O)_n{\bullet}C_{15}H_{24}O$

PROP: Very stable. Compounds with n < 15 are yellow to almost colorless liquids; n > 20 are pale yellow to off-white pastes or waxes. Compounds with n < 6 are sol in oil; higher n's are sol in water. Almost colorless liquid. D: 1.06; fp 26°F, pour point 37°F, flash p: 535-555°F, visc (25°) 175-250 cp. Sol in water, ethanol, ethylene glycol, ethylene dichloride, xylene, corn oil. Insol in Stoddard solvent, deodorized kerosene, low viscosity white mineral oil.

SYNS: N-9 ◇ NONYLPHENOXYPOLY(ETHYLENEOXY)ETHANOL ◇ NP-9

TOXICITY DATA with REFERENCE
otr-mus:fbr 10 ppm CRNGDP 3,553,82
ivg-rbt TDLo:1 mg/kg (female 1D pre):REP CCPTAY 32,183,85
ipr-mus LD50:150 mg/kg CCPTAY 33,1,86

CONSENSUS REPORTS: Reported in EPA TSCA Inventory.

SAFETY PROFILE: Poison by intraperitoneal route. Experimental reproductive effects. Mutation data reported. An active ingredient in contraceptive jellies, foams, and creams. Combustible when exposed to heat or flames. When heated to decomposition it emits acrid smoke and irritating fumes. See also ALCOHOLS.

NNB400 CAS:143-13-5 ***HR: 2***
NONYL ACETATE
mf: $C_{11}H_{22}O_2$ mw: 186.29

PROP: Colorless liquid; fruity odor. D: 0.864, refr index: 1.422, flash p: +153°F. Sol in alc, ether; insol in water.

SYN: FEMA No. 2788

SAFETY PROFILE: Combustible liquid. When heated to decomposition it emits acrid smoke and irritating fumes.

NNB500 CAS:143-08-8 ***HR: 2***
n-NONYL ALCOHOL
mf: $C_9H_{20}O$ mw: 144.29

PROP: Colorless liquid; rose-citrus odor. D: 0.827 @ 20°/4°, refr index: 1.43-1.435, mp: −5°, bp: 213.5°, flash p: 169°F. Insol in water; misc in alc, ether, chloroform.

SYNS: ALCOHOL C-9 ◇ FEMA No. 2789 ◇ NONALOL ◇ 1-NONANOL ◇ NONAN-1-OL ◇ NONYL ALCOHOL ◇ OCTYL CARBINOL ◇ PELARGONIC ALCOHOL

TOXICITY DATA with REFERENCE
orl-rat LDLo:1400 mg/kg 14CYAT 2,1466,63
orl-mus LDLo:10 g/kg TPKVAL 5,51,63
ihl-mus LC50:5500 mg/m^3/2H 85GMAT -,94,82
skn-rbt LD50:5660 mg/kg FCTXAV 11,95,73

CONSENSUS REPORTS: Reported in EPA TSCA Inventory.

SAFETY PROFILE: Moderately toxic by ingestion. Mildly toxic by skin contact and inhalation. Combustible liquid. When heated to decomposition it emits acrid smoke and irritating fumes. See also ALCOHOLS.

NNC100 CAS:63885-67-6 ***HR: 3***
o-NONYL-HARMOL HYDROCHLORIDE
mf: $C_{21}H_{27}N_2O{\bullet}ClH$ mw: 359.92

SYN: 1-METHYL-7-(NONYLOXY)-9H-PYRIDO(3,4-b)INDOLEHYDROCHLORIDE

TOXICITY DATA with REFERENCE
ipr-mus LDLo:150 mg/kg QJPPAL 5,63,32
ipr-gpg LDLo:200 mg/kg QJPPAL 5,63,32
scu-frg LDLo:210 mg/kg QJPPAL 5,63,32

SAFETY PROFILE: Poison by subcutaneous and intraperitoneal routes. When heated to decomposition it emits toxic fumes of NO_x and HCl.

NNC500 CAS:25154-52-3 ***HR: 2***
NONYL PHENOL (mixed isomers)
mf: $C_{15}H_{24}O$ mw: 220.39

PROP: Clear, straw-colored, viscous liquid; slt phenolic odor. Bp: 293-297°, pour p: 2°, vap d: 7.59, flash p: 285°F, d: 0.949 @ 20°/4°. Insol in water, dil aq NaOH; sol in benzene, chlorinated solvents, aniline, heptane, aliphatic alc, ethylene glycol.

SYNS: 2,6-DIMETHYL-4-HEPTYLPHENOL, (o and p) ◇ HYDROXY No. 253

TOXICITY DATA with REFERENCE
skn-rbt 10 mg/24H SEV AMIHBC 4,119,51
skn-rbt 500 mg open MOD UCDS** 6/9/59
eye-rbt 50 μg SEV AMIHBC 4,119,51
orl-rat LD50:1620 mg/kg UCDS** 6/9/59
skn-rbt LD50:2140 mg/kg AIHAAP 30,470,69

CONSENSUS REPORTS: Reported in EPA TSCA Inventory.

SAFETY PROFILE: Moderately toxic by ingestion and skin contact. A severe skin and eye irritant. Combustible when exposed to heat or flame. When heated to decomposition it emits acrid smoke and irritating fumes.

NND000 CAS:11096-42-7 ***HR: 2***
NONYLPHENOXYPOLYETHYLENEOXY ETHANOL-IODINE COMPLEX
mf: $(C_2H_4O)_n{\cdot}C_{15}H_{24}O{\cdot}xI_2$

SYNS: BIOPAL CVL-10 ◇ BIOPAL NR-20 ◇ BIOPAL VRO 10 ◇ BIOPAL VRO 20

TOXICITY DATA with REFERENCE
orl-rat LD50:2100 mg/kg 28ZEAL 4,301,69

CONSENSUS REPORTS: Reported in EPA TSCA Inventory.

SAFETY PROFILE: Moderately toxic by ingestion. A disinfecting cleaner. When heated to decomposition it emits toxic fumes of I^-. See also IODIDES.

NND500 CAS:9016-45-9 ***HR: 2***
NONYL PHENYL POLYETHYLENE GLYCOL ETHER
mf: $(C_2H_4O)_n{\cdot}C_{15}H_{24}O$

SYNS: CHEMAX NP SERIES ◇ GLYCOLS, POLYETHYLENE, MONO(NONYLPHENYL) ETHER ◇ NONOXYNOL ◇ NONYLPHENOL, POLYOXYETHYLENE ETHER ◇ NONYL PHENYL POLYETHYLENE GLYCOL ◇ POLYOXYETHYLENE NONYLPHENOL ◇ TERGITOL NPX ◇ TRITON N-100 ◇ TRYCOL NP-1

TOXICITY DATA with REFERENCE
skn-rbt 500 mg open MLD UCDS** 6/16/65
eye-rbt 5 mg SEV UCDS** 6/16/65
orl-rat LD50:1310 mg/kg UCDS** 6/16/65
skn-rbt LD50:2000 mg/kg UCDS** 6/16/65

CONSENSUS REPORTS: Glycol ether compounds are on the Community Right-To-Know List. Reported in EPA TSCA Inventory.

SAFETY PROFILE: Moderately toxic by ingestion and skin contact. A skin and severe eye irritant. When heated to decomposition it emits acrid smoke and irritating fumes. See also GLYCOL ETHERS.

NNE000 CAS:5283-67-0 ***HR: 2***
NONYLTRICHLOROSILANE
DOT: UN 1799
mf: $C_9H_{19}Cl_3Si$ mw: 261.72

CONSENSUS REPORTS: Reported in EPA TSCA Inventory.

DOT Classification: Corrosive Material; Label: Corrosive.

SAFETY PROFILE: A corrosive irritant to skin, eyes, and mucous membranes. When heated to decomposition it emits toxic fumes of Cl^-. See also CHLOROSILANES.

NNE100 CAS:13257-44-8 ***HR: 1***
2-NONYNAL DIMETHYLACETAL
mf: $C_{11}H_{20}O_2$ mw: 184.31

SYNS: 2-NONYN-1-AL DIMETHYLACETAL ◇ 2-NONYNAL, DIMETHYL ACETAL ◇ PARMAVERT

TOXICITY DATA with REFERENCE
skn-rbt 500 mg/24H MOD FCTOD7 20,779,82

CONSENSUS REPORTS: Reported in EPA TSCA Inventory.

SAFETY PROFILE: A skin irritant. When heated to decomposition it emits acrid smoke and irritating fumes.

NNE400 CAS:7491-74-9 ***HR: 1***
NOOTROPYL
mf: $C_6H_{10}N_2O_2$ mw: 142.18

PROP: Crystals from isopropanol. Mp: 151.5-152.5°

SYNS: 1-ACETAMIDO-2-PYRROLIDINONE ◇ CEREBROFORTE ◇ ENCETROP ◇ EUVIFOR ◇ GABACET ◇ GENOGRIS ◇ 2-KETOPYRROLIDINE-1-YLACETAMIDE ◇ NOOTRON ◇ NOOTROPIL ◇ NORMABRAIN ◇ NORZETAM ◇ 2-OXO-PYRROLIDINE ACETAMIDE ◇ 2-OXO-1-PYRROLIDINEACETAMIDE ◇ 2-OXOPYRROLIDIN-1-YLACETAMIDE ◇ PIRACETAM ◇ PIRAZETAM ◇ PIRROXIL ◇ PYRACETAM ◇ PYRAMEM ◇ 2-PYRROLIDINONEACETAMIDE ◇ 2-PYRROLIDONEACETAMIDE ◇ UCB 6215

TOXICITY DATA with REFERENCE
orl-rat LDLo:5600 mg/kg FRPPAO 32,47,77
orl-mus LDLo:10 g/kg FRPPAO 32,47,77
scu-mus LD50:12 g/kg KHFZAN 11(8),132,77
ivn-mus LD50:10 g/kg KHFZAN 11(8),132,77

SAFETY PROFILE: Mildly toxic by ingestion. When heated to decomposition it emits toxic fumes of NO_x.

NNE500 CAS:101975-70-6 ***HR: 3***
NOPCO 2272-R

PROP: Sulfated butyl oleate (JAPMA8 38,428,49).

TOXICITY DATA with REFERENCE
eye-rbt 1% MLD JAPMA8 38,428,49
ivn-mus LD50:63 mg/kg JAPMA8 38,428,49

SAFETY PROFILE: Poison by intravenous route. An eye irritant. When heated to decomposition it emits toxic fumes of SO_x.

NNE550 CAS:360-70-3 *HR: D*
NORANDROSTENOLONE DECANOATE
mf: $C_{28}H_{44}O_3$ mw: 428.72

SYNS: DECA-DURABOL ◇ DECA-DURABOLIN ◇ ESTR-4-EN-3-ONE, 17-β-HYDROXY-, DECANOATE ◇ ESTR-4-EN-3-ONE, 17-((1-OXODECYL)OXY)-, (17-β)- (9CI) ◇ 17-β-HYDROXYESTR-4-EN-3-ONE DECANOATE ◇ NANDROLONE DECANOATE ◇ 19-NORANDROSTENOLONE DECANOATE ◇ NORTESTOSTERONE DECANOATE ◇ 19-NORTESTOSTERONE DECANOATE ◇ RETABOLIL

TOXICITY DATA with REFERENCE
par-rat TDLo:5 mg/kg (female 1D pre):REP ARGYAJ 221,103,76

SAFETY PROFILE: Experimental reproductive effects. When heated to decomposition it emits acrid smoke and irritating fumes.

NNE600 CAS:797-58-0 *HR: 1*
NORBOLETHONE
mf: $C_{21}H_{32}O_2$ mw: 316.53

SYN: 18,19-DINOR-17-α-PREGN-4-EN-3-ONE, 13-ETHYL-17-HYDROXY-

TOXICITY DATA with REFERENCE
orl-rat TDLo:1800 mg/kg (female 2-19D post):REP FESTAS 22,735,71
orl-mus LD50:5010 mg/kg TXAPA9 18,185,71

SAFETY PROFILE: Mildly toxic by ingestion. Experimental reproductive effects. When heated to decomposition it emits acrid smoke and irritating fumes.

NNF000 CAS:991-42-4 *HR: 3*
NORBORMIDE
mf: $C_{33}H_{25}N_3O_3$ mw: 511.61

PROP: Crystals. Mp: 190-198°. Insol in water.

SYNS: COMPOUND S-6,999 ◇ ENT 51,762 ◇ 5-(α-HYDROXY-α-2-PYRIDYLBENZYL)-7-(α-2-PYRIDYLBENZYLIDENE)-5-NORBORNENE-2,3-DICARBOXIMIDE ◇ MCN 1025 ◇ RATICATE ◇ S-6,999 ◇ SHOXIN

TOXICITY DATA with REFERENCE
orl-rat LD50:3800 μg/kg NYKZAU 66,224,70
ivn-rat LDLo:250 μg/kg EJPHAZ 16,117,71
orl-mus LD50:2250 mg/kg 28ZEAL 5,167,76
orl-dog LD50:1 g/kg YKYUA6 31,1385,80
orl-mky LD50:1 g/kg YKYUA6 31,1385,80
orl-cat LD50:1 g/kg YKYUA6 31,1385,80
orl-rbt LD50:1 g/kg YKYUA6 31,1385,80
orl-gpg LD50:300 mg/kg YKYUA6 31,1385,80
orl-ham LD50:140 mg/kg 31ZOAD 1,323,68

SAFETY PROFILE: Poison by ingestion and intravenous routes. A rodenticide. When heated to decomposition it emits toxic fumes of NO_x.

NNG000 CAS:121-46-0 *HR: 3*
NORBORNADIENE
mf: C_7H_8 mw: 92.15

SYNS: BICYCLO(2.2.1)HEPTADIENE ◇ 2,5-NORBORNADIENE

TOXICITY DATA with REFERENCE
orl-rat LD50:890 mg/kg SHELL*
ihl-rat LC50:14100 ppm/8H SCCUR* -,1,61
ipr-rat LD50:890 mg/kg SCCUR* -,1,61
orl-mus LD50:3850 mg/kg SCCUR* -,1,61
ihl-mus LC50:27700 ppm/30M SCCUR* -,1,61
ivn-mus LD50:56 mg/kg CSLNX* NX#04158

CONSENSUS REPORTS: Reported in EPA TSCA Inventory.

SAFETY PROFILE: Poison by intravenous route. Moderately toxic by ingestion and intraperitoneal routes. Mildly toxic by inhalation. When heated to decomposition it emits acrid smoke and irritating fumes.

NNG500 CAS:61242-71-5 *HR: 2*
endo-2-NORBORNANECARBOXYLIC ACID ETHYL ESTER
mf: $C_{10}H_{16}O_2$ mw: 168.26

SYN: 2,5-ENDOMETHYLENE CYCLOHEXANECARBOXYLIC ACID, ETHYL ESTER (mixed formyl isomers)

TOXICITY DATA with REFERENCE
skn-rbt 10 mg/24H open MLD AIHAAP 23,95,62
orl-rat LD50:7460 mg/kg AIHAAP 23,95,62
skn-rbt LD50:10 g/kg AIHAAP 23,95,62

SAFETY PROFILE: Mildly toxic by ingestion and skin contact. A skin irritant. Can react violently with HNO_3; 2-aminoethanol; chlorosulfonic acid; ethylene diamine; ethylene imine; oleum; H_2SO_4; K-tert-butoxide. Can react with oxidizing materials. To fight fire, use foam, alcohol foam, CO_2, dry chemical. When heated to decomposition emits acrid smoke and irritating fumes. See also ESTERS.

NNH000 CAS:5240-72-2 *HR: 2*
2-NORBORNANEMETHANOL
mf: $C_8H_{14}O$ mw: 126.22

SYNS: EXPERIMENTAL CHEMOTHERAPEUTANT 1,207 ◇ 2-HYDROXYMETHYLBICYCLO(2.2.1)HEPTANE ◇ 2-HYDROXYMETHYLNORCAMPHANE ◇ 2-NORCAMPHANE METHANOL

TOXICITY DATA with REFERENCE
skn-rbt 10 mg/24H open MLD AIHAAP 23,95,62
orl-rat LD50:1600 mg/kg 28ZEAL 4,302,69
skn-rbt LDLo:800 mg/kg AIHAAP 30,470,69

CONSENSUS REPORTS: Reported in EPA TSCA Inventory.

SAFETY PROFILE: Moderately toxic by ingestion and skin contact. A skin irritant. When heated to decomposition it emits acrid smoke and irritating fumes.

NNH500 CAS:498-66-8 ***HR: 1***
2-NORBORNENE
mf: C_7H_{10} mw: 94.17

SYNS: NORBORNYLENE ◇ NORCAMPHENE

TOXICITY DATA with REFERENCE
orl-rat LD50:11300 mg/kg AIHAAP 30,470,69

CONSENSUS REPORTS: Reported in EPA TSCA Inventory.

SAFETY PROFILE: Mildly toxic by ingestion. When heated to decomposition it emits acrid smoke and irritating fumes.

NNI000 CAS:4582-21-2 ***HR: 2***
trans-5-NORBORNENE-2,3-DICARBONYL CHLORIDE
mf: $C_9H_8Cl_2O_2$ mw: 219.07

SYN: trans-BICYCLO(2.2.1)HEPT-5-ENYL-2,3-DICARBONYLCHLORIDE

TOXICITY DATA with REFERENCE
orl-rat LD50:1830 mg/kg TXAPA9 28,313,74
skn-rbt LD50:1000 mg/kg TXAPA9 28,313,74

SAFETY PROFILE: Moderately toxic by ingestion and skin contact. When heated to decomposition it emits toxic fumes of Cl^-.

NNJ600 ***HR: 3***
NORBUTORPHANOL TARTRATE
mf: $C_{16}H_{21}NO_2 \cdot 1/2C_4H_6O_6$ mw: 334.39

SYN: 1-6,10a-β-DIHYDROXY-1,2,3,9,10,10a-HEXAHYDRO-4H(10),4a-IMINOETHANOPHENANTHRENETARTRATE

TOXICITY DATA with REFERENCE
scu-rat TDLo:4500 mg/kg (30D male):REP IYKEDH 13,792,82
scu-rat LD50:315 mg/kg IYKEDH 13,792,82
ivn-rat LD50:53900 μg/kg IYKEDH 13,792,82
scu-mus LD50:182 mg/kg IYKEDH 13,792,82
ivn-mus LD50:29300 μg/kg IYKEDH 13,792,82

SAFETY PROFILE: Poison by subcutaneous and intravenous routes. Experimental reproductive effects. When heated to decomposition it emits toxic fumes of NO_x.

NNK000 CAS:497-38-1 ***HR: 3***
NORCAMPHOR
mf: $C_7H_{10}O$ mw: 110.17

SYN: 2-NORBORNANONE

TOXICITY DATA with REFERENCE
ivn-mus LD50:180 mg/kg CSLNX* NX#04039

CONSENSUS REPORTS: Reported in EPA TSCA Inventory.

SAFETY PROFILE: Poison by intravenous route. When heated to decomposition it emits acrid smoke and irritating fumes.

NNK500 CAS:303-26-4 ***HR: D***
NORCHLORCYCLIZINE
mf: $C_{17}H_{19}ClN_2$ mw: 286.83

SYNS: N-(p-CHLOROBENZHYDRYL)PIPERAZINE ◇ 4-(4-CHLOROBENZHYDRYL)PIPERAZINE ◇ 1-(4-CHLOROBENZHYDRYL) PIPERAZINE ◇ 1-(4-CHLORO-α-PHENYLBENZYL)PIPERAZINE ◇ 1-(α-PHENYL-4-CHLOROBENZYL)PIPERAZINE

TOXICITY DATA with REFERENCE
orl-rat TDLo:500 mg/kg (12-15D preg):REP JPETAB 147,391,65
orl-rat TDLo:100 mg/kg (12-15D preg):TER JPETAB 147,391,65

SAFETY PROFILE: An experimental teratogen. Experimental reproductive effects. When heated to decomposition it emits very toxic fumes of NO_x and Cl^-.

NNL000 CAS:18719-22-7 ***HR: D***
NORCHLORCYCLIZINE HYDROCHLORIDE
mf: $C_{17}H_{19}ClN_2 \cdot 7ClH$ mw: 542.05

SYNS: 1-(p-CHLORO-α-PHENYLBENZYL)-PIPERAZINE HYDROCHLORIDE ◇ 1-(α-(4-CHLOROPHENYL)BENZYL)-PIPERAZINE HYDROCHLORIDE

TOXICITY DATA with REFERENCE
orl-rat TDLo:192 mg/kg (12-15D preg):TER TJADAB 18,199,78

SAFETY PROFILE: An experimental teratogen. Experimental reproductive effects. When heated to decomposition it emits toxic fumes of NO_x and Cl^-.

NNL400 CAS:16111-27-6 ***HR: 3***
NORDIMAPRIT
mf: $C_5H_{13}N_3S \cdot 2ClH$ mw: 220.19

SYNS: CARBAMIMIDOTHIOICACID-2-(DIMETHYLAMINO)ETHYL ESTER DIHYDROCHLORIDE ◇ N,N-DIMETHYL-β-AMINOAETHYL-ISOTHIURONIUM DIHYDROCHLORID (GERMAN) ◇ 2-DIMETHYL-AMINOETHYLISOTHIURONIUM CHLORIDE HYDROCHLORIDE

◇ s-(2-(DIMETHYLAMINO)ETHYL)PSEUDOTHIOUREA DIHYDROCHLORIDE ◇ 2-(2-(DIMETHYLAMINO)ETHYL)-2-THIOPSEUDOUREA DIHYDROCHLORIDE ◇ SKF 91487

TOXICITY DATA with REFERENCE
ipr-mus LD50:150 mg/kg YKKZAJ 92,442,72
scu-mus LD50:63 mg/kg ARZNAD 7,576,57
ivn-mus LD50:49100 μg/kg YKKZAJ 88,156,68

SAFETY PROFILE: Poison by subcutaneous, intravenous, and intraperitoneal routes. When heated to decomposition it emits toxic fumes of NO_x, SO_x, and HCl.

NNL500 CAS:8056-51-7 ***HR: 3***
NORDIOL-28
mf: $C_{21}H_{28}O_2.C_{20}H_{24}O_2$ mw:608.93

SYNS: DUOLUTON ◇ EDIWAL ◇ ETHINYL ESTRADIOL mixed with NORGESTREL ◇ ETHINYLESTRADIOL mixed with dl-NORGESTREL ◇ ETHINYLOESTRADIOL mixed with NORGESTREL ◇ EUGYON ◇ FOLLINYL ◇ GRAVISTAT ◇ LO/OVRAL ◇ MICROGYNON ◇ MINIDRIL ◇ NEOGYNON ◇ NEOVLETTA ◇ NORDETTE ◇ NORDIOL ◇ NORGESTREL mixed with ETHINYL ESTRADIOL ◇ dl-NORGESTREL mixed with ETHINYLESTRADIOL ◇ ORASECRON ◇ OVIDON ◇ OVRAL ◇ OVRAL 21 ◇ OVRAL 28 ◇ OVRAN ◇ OVRANETT ◇ PRIMOVLAR ◇ PRO-DUOSTERONE ◇ RIGEVIDON ◇ SEQUILAR ◇ SEQUOSTAT ◇ SH 71121 ◇ SHB 261AB ◇ SHB 264AB ◇ STEDIRIL ◇ STEDIRIL D ◇ TRIQUILAR ◇ WL 20 ◇ WL 33 ◇ WY-E 104

TOXICITY DATA with REFERENCE
orl-wmn TDLo:44 μg/kg (1D preg):REP FESTAS 37,508,82
orl-wmn TDLo:18144 μg/kg/12Y-I:CAR,LIV LANCAO 1,273,80
orl-wmn TD:5 mg/kg/26W-I:NEO,LIV LANCAO 2,926,73
orl-wmn TD:756 μg/kg/26W-I:NEO,LIV MJAUAJ 2,223,78
orl-wmn TD:50 mg/kg/5Y-I:CAR,LIV JAMAAP 235,730,76

SAFETY PROFILE: Human reproductive effects by ingestion: menstrual cycle changes or disorders, postpartum changes, mating performance, female fertility index and other fertility changes. Experimental reproductive effects. Questionable human carcinogen producing liver tumors. A steroid used as a pharmaceutical and veterinary drug. When heated to decomposition it emits acrid smoke and irritating fumes.

NNM000 CAS:492-41-1 ***HR: 3***
(−)-NOREPHEDRINE
mf: $C_9H_{13}NO$ mw: 151.23

SYNS: α-(1-AMINOETHYL)-BENZYL ALCOHOL ◇ 2-AMINO-1-PHENYL-1-PROPANOL ◇ FENILPROPANOLAMINA (ITALIAN) ◇ MYDRIATIN ◇ PHENYLPROPANOLAMINE ◇ PPA ◇ PROPADRINE ◇ USAF CS-6

TOXICITY DATA with REFERENCE
orl-wmn LDLo:8 mg/kg:CNS,PUL BMJOAE 289,591,84
orl-cld TDLo:2500 μg/kg:PSY AJDCAI 138,683,84
orl-hmn TDLo:714 μg/kg:BPR LANCAO 1,60,80
orl-wmn TDLo:500 μg/kg:CNS NEURAI 36,593,86
orl-rat LD50:1538 mg/kg BCFAAI 109,132,70
ipr-rat LDLo:60 mg/kg TXAPA9 22,138,72
orl-mus LD50:1060 mg/kg BCFAAI 109,132,70
ipr-mus LD50:200 mg/kg NTIS** AD277-689
scu-mus LD50:124 mg/kg FEPRA7 4,139,45
ivn-mus LDLo:114 mg/kg CLDND* 153,161,30
ivn-rbt LDLo:75 mg/kg JPETAB 36,363,29

CONSENSUS REPORTS: Reported in EPA TSCA Inventory.

SAFETY PROFILE: A human poison by ingestion. Poison experimentally by intravenous, subcutaneous, and intraperitoneal routes. Moderately toxic by an unspecified route. Human systemic effects by ingestion: sleep, increased pulse rate without blood pressure decrease and chronic pulmonary edema or congestion, convulsions, headache, and blood pressure elevation. Used in production of drugs of abuse. When heated to decomposition it emits toxic fumes of NO_x.

NNM500 CAS:14838-15-4 ***HR: 2***
dl-NOREPHEDRINE
mf: $C_9H_{13}NO$ mw: 151.23

SYNS: dl-α-(1-AMINOETHYL)BENZYL ALCOHOL ◇ dl-2-AMINO-1-HYDROXY-1-PHENYLPROPANE ◇ dl-α-HYDROXY-β-AMINOPROPYLBENZENE ◇ dl-1-PHENYL-2-AMINOPROPANOL-1

TOXICITY DATA with REFERENCE
scu-rat LD50:850 mg/kg JPETAB 85,199,45

CONSENSUS REPORTS: Reported in EPA TSCA Inventory.

SAFETY PROFILE: Moderately toxic by subcutaneous route. When heated to decomposition it emits very toxic fumes of NO_x.

NNN000 CAS:4345-16-8 ***HR: 3***
NOREPHEDRINE HYDROCHLORIDE
mf: $C_9H_{13}NO{\cdot}ClH$ mw: 187.69

SYNS: α-(1-AMINOETHYL)BENZYL ALCOHOL HYDROCHLORIDE ◇ 2-AMINO-1-PHENYL-1-PROPANOL HYDROCHLORIDE ◇ MYDRIATINE ◇ PHENYLPROPANOLAMINE HYDROCHLORIDE ◇ PROPADRINE HYDROCHLORIDE

TOXICITY DATA with REFERENCE
orl-wmn TDLo:1 mg/kg:CVS,PUL JAMAAP 245,601,81
orl-rat LD50:1490 mg/kg TXAPA9 18,185,71
ipr-rat LDLo:175 mg/kg JACSAT 51,2262,29
scu-rat LDLo:350 mg/kg CHREAY 9,389,31
orl-mus LD50:1443 mg/kg JPMSAE 60,1810,71
ipr-mus LD50:300 mg/kg JPETAB 93,114,48
scu-mus LDLo:700 mg/kg AEPPAE 153,161,30

ivn-mus LD50:180 mg/kg CSLNX* NX#02746
scu-gpg LDLo:350 mg/kg CRSBAW 106,552,31

CONSENSUS REPORTS: Reported in EPA TSCA Inventory.

SAFETY PROFILE: Poison by intraperitoneal, intravenous, and subcutaneous routes. Moderately toxic by ingestion. Human systemic effects by ingestion: change in heart rate and respiratory stimulation. Used in cold remedies and diet tablets. When heated to decomposition it emits very toxic fumes of HCl and NO_x. See also (−)-NOREPHEDRINE.

NNN500 CAS:3198-15-0 ***HR: 3***
l-NOREPHEDRINE HYDROCHLORIDE
mf: $C_9H_{13}NO•ClH$ mw: 187.69

SYN: (−)-NOREPHEDRINE HYDROCHLORIDE

TOXICITY DATA with REFERENCE
ipr-mus LD50:176 mg/kg JPETAB 158,135,67

CONSENSUS REPORTS: Reported in EPA TSCA Inventory.

SAFETY PROFILE: Poison by intraperitoneal route. When heated to decomposition it emits very toxic fumes of HCl and NO_x. See also (−)-NOREPHEDRINE.

NNO000 CAS:2153-98-2 ***HR: 3***
(−)-NOREPHEDRINE HYDROCHLORIDE
mf: $C_9H_{13}NO•ClH$ mw: 187.69

SYN: d-NOREPHEDRINE HYDROCHLORIDE

TOXICITY DATA with REFERENCE
ipr-mus LD50:377 mg/kg JPETAB 158,135,67

CONSENSUS REPORTS: Reported in EPA TSCA Inventory.

SAFETY PROFILE: Poison by intraperitoneal route. When heated to decomposition it emits very toxic fumes of HCl and NO_x. See also (−)-NOREPHEDRINE.

NNO500 CAS:51-41-2 ***HR: 3***
l-NOREPINEPHRINE
mf: $C_8H_{11}NO_3$ mw: 169.20

PROP: Microcrystals. Decomp @ 216.5-218°.

SYNS: ADRENOR ◇ AKTAMIN ◇ l-2-AMINO-1-(3,4-DIHYDROXYPHENYL)ETHANOL ◇ (R)-4-(2-AMINO-1-HYDROXYETHYL)-1,2-BENZENEDIOL ◇ l-α-(AMINOMETHYL)-3,4-DIHYDROXYBENZYL ALCOHOL ◇ (−)-α-(AMINOMETHYL)PROTOCATECHUYL ALCOHOL ◇ ARTERENOL ◇ l-ARTERENOL ◇ l-1-(3,4-DIHYDROXYPHENYL)-2-AMINOETHANOL ◇ l-3,4-DIHYDROXYPHENYLETHANOLAMINE ◇ LEVARTERENOL ◇ LEVOARTERENOL ◇ LEVONORADRENALINE ◇ LEVONOREPINEPHRINE ◇ LEVOPHED ◇ (−)-NORADREC ◇ NORADRENALIN ◇ NORADRENALINA (ITALIAN) ◇ NORADRENALINE ◇ (−)-NORADRENALINE ◇ d-(−)-NORADRENALINE ◇ l-NORADRENALINE ◇ NORADRENLINE ◇ NORARTRINAL ◇ NOREPINEPHRINE ◇ (−)-NOREPINEPHRINE ◇ NOREPIRENAMINE ◇ SYMPATHIN E

TOXICITY DATA with REFERENCE
dni-hmn:oth 5800 nmol/L CNREA8 40,1414,80
dni-mus:oth 5800 nmol/L CNREA8 40,1414,80
ipr-rat TDLo:8 mg/kg (7-14D preg):REP JCPPAV 58,309,64
ivn-gpg TDLo:10 μg/kg (female 63D post):TER JRPFA4 66,23,82
ivn-rat LD50:100 μg/kg JPETAB 95,502,49
orl-mus LD50:20 mg/kg APTOA6 31,49,72
ipr-mus LD50:6 mg/kg NIIRDN 6,567,82
scu-mus LD50:5 mg/kg NATWAY 56,615,69
ivn-mus LD50:550 μg/kg AEPPAE 226,493,55
ivn-rbt LDLo:250 μg/kg AIPTAK 41,365,31

SAFETY PROFILE: Poison by ingestion, intraperitoneal, subcutaneous, and intravenous routes. An experimental teratogen. Experimental reproductive effects. Human mutation data reported. A sympathomimetic vasopressor. When heated to decomposition it emits toxic fumes of NO_x.

NNO699 CAS:51-40-1 ***HR: 3***
l-NOREPINEPHRINE BITARTRATE
mf: $C_8H_{11}NO_3•C_4H_6O_6$ mw: 319.30

SYNS: l-ARTERENOL BITARTRATE ◇ LEVARTERENOL BITARTRATE ◇ (−)-NORADRENALINE ACID TARTRATE ◇ (−)-NORADRENALINE BITARTRATE ◇ l-NORADRENALINE BITARTRATE ◇ (−)-NORADRENALINE BITARTRATE MONOHYDRATE ◇ NORADRENALINE HYDROGEN TARTRATE ◇ NORADRENALINE TARTRATE (1:1) ◇ (−)-NORADRENALINE TARTRATE ◇ l-NORADRENALINE TARTRATE ◇ (−)-NOREPINEPHRINE BITARTRATE ◇ (−)-NOREPINEPHRINE TARTRATE

TOXICITY DATA with REFERENCE
ivn-rat LD50:210 μg/kg YACHDS 7,627,79
ipr-mus LD50:26800 μg/kg DPHFAK 23,435,71
ivn-mus LD50:1025 μg/kg EJPHAZ 9,289,70

SAFETY PROFILE: Poison by intravenous and intraperitoneal routes. When heated to decomposition it emits toxic fumes of NO_x. See also NOREPINEPHRINE.

NNP000 CAS:329-56-6 ***HR: 3***
NOREPINEPHRINE HYDROCHLORIDE
mf: $C_8H_{11}NO_3•ClH$ mw: 205.66

PROP: X-form-crystals. Mp: 145.2-146.4°. Very sol in water.

SYNS: AKTAMIN HYDROCHLORIDE ◇ LEVOPHED HYDROCHLORIDE ◇ NORADRENALINE HYDROCHLORIDE ◇ NOR-ADRENALIN HYDROCHLORIDE

TOXICITY DATA with REFERENCE
orl-rat LD50:132 mg/kg AIPTAK 180,155,69

scu-rat LD50:29 mg/kg AIPTAK 180,155,69
idu-rat LD50:26 mg/kg AIPTAK 180,155,69

SAFETY PROFILE: Poison by ingestion, subcutaneous, and intraduodenal routes. When heated to decomposition it emits very toxic fumes of HCl and NO_x. See also NOREPINEPHRINE.

NNP050 CAS:55-27-6 ***HR: 3***
dl-NOREPINEPHRINE HYDROCHLORIDE
mf: $C_8H_{11}NO_3$•ClH mw: 205.66

SYNS: dl-ARTERENOL HYDROCHLORIDE ◇ 3,4-DIHYDROXYPHENYL)-1-AMINO-2-ETHANOL-1-HYDROCHLORIDE ◇ (±)-NORADRENALINE HYDROCHLORIDE ◇ dl-NORADRENALINE HYDROCHLORIDE ◇ (±)-NOREPINEPHRINE HYDROCHLORIDE

TOXICITY DATA with REFERENCE
scu-rat TDLo:13 mg/kg (7-19D preg):REP NYKZAU 85,79,85
scu-rat LDLo:2 mg/kg JPETAB 71,62,41
ipr-mus LD50:15600 μg/kg JPETAB 92,369,48

SAFETY PROFILE: Poison by intraperitoneal and subcutaneous routes. Experimental reproductive effects. When heated to decomposition it emits toxic fumes of NO_x and HCl. See also other norepinephrine entries.

NNP500 CAS:68-22-4 ***HR: 3***
19-NORETHISTERONE
mf: $C_{20}H_{26}O_2$ mw: 298.46

SYNS: 17-α-ETHINYLESTRA-4-EN-17-β-OL-3-ONE ◇ 17-α-ETHINYL-17-β-HYDROXY-Δ:4-ESTREN-3-ONE ◇ 17-α-ETHINYL-19-NORTESTOSTERONE ◇ 17-α-ETHYNYL-4-ESTREN-17-OL-3-ONE ◇ 17-α-ETHYNYL-17-HYDROXY-4-ESTREN-3-ONE ◇ 17-α-ETHYNYL-17-β-HYDROXY-19-NORANDROST-4-EN-3- ONE ◇ 17-α-ETHYNYL-19-NORANDROST-4-EN-17-β-OL-3-ONE ◇ 17-α-ETHYNYL-19-NOR-4-ANDROSTEN-17-β-OL-3-ONE ◇ 17-α-ETHYNYL-19-NORTESTOSTERONE ◇ (17-α)-17-HYDROXY-19-NORPREGN- 4-EN-20-YN-3-ONE ◇ 17-β-HYDROXY-19-NORPREGN-4-EN-20-YN-3-ONE ◇ 17-HYDROXY-19-NOR-17-α-PREGN-4-EN-20-YN-3-ONE ◇ 19-NOR-ETHINYL-4,5-TESTOSTERONE ◇ 19-NOR-17-α-ETHYNYLANDRO- STEN-17-β-OL-3-ONE ◇ 19-NOR-17-α-ETHYNYL-17-β-HYDROXY-4-ANDROSTEN-3-ONE ◇ 19-NOR-17-α-ETHYNYLTESTOSTERONE ◇ NORLUTIN

TOXICITY DATA with REFERENCE
dni-hmn:lym 50 μmol/L PSEBAA 146,401,74
oms-mus:oth 10 μg/L AJOGAH 120,390,74
oms-dom:oth 100 μg/L AJOGAH 120,390,74
unr-wmn TDLo:12 mg/kg (female 6-20W post):REP SCIEAS 211,1171,81
unr-wmn TDLo:4200 μg/kg (female 6-8W post):TER NEJMAG 268,255,63
scu-mus TDLo:163 mg/kg/78W-C:ETA BJCAAI 21,153,67
orl-wmn TDLo:42 mg/kg:SKN,END AJOGAH 84,962,62
orl-mus LD50:12 g/kg NIIRDN 6,566,82

CONSENSUS REPORTS: NTP Fifth Annual Report on Carcinogens. IARC Cancer Review: Animal Limited Evidence IMEMDT 21,441,79; Animal Sufficient Evidence IMEMDT 6,179,74. EPA Genetic Toxicology Program.

SAFETY PROFILE: Confirmed carcinogen with experimental carcinogenic, tumorigenic, and teratogenic data. Mildly toxic by ingestion. Human systemic effects by ingestion: dermatitis and androgenic effects. Human teratogenic effects: developmental abnormalities of the musculoskeletal system and urogenital system; and behavioral effects in the newborn. Human reproductive effects: spermatogenesis; testes, epididymis, sperm duct changes; impotence; male breast development; other male effects; ovaries, fallopian tube changes; menstrual cycle changes or disorders uterus, cervix, vagina effects; postpartum effects; changes in female fertility. Experimental reproductive effects. Human mutation data reported. When heated to decomposition it emits acrid smoke and irritating fumes.

NNQ000 CAS:3836-23-5 ***HR: 3***
NORETHISTERONE ENANTHATE
mf: $C_{27}H_{38}O_3$ mw: 410.65

SYNS: 17-α-ETHINYL-19-NORTESTOSTERONE ENANTHATE ◇ 17-α-ETHYNYL-17-β-HEPTANOYLOXY-4-ESTREN-3-ONE ◇ 17-β-HEPTANOYLOXY-19-NOR-17-α-PREGNEN-20-YNONE ◇ NORETHISTERONE OENANTHATE ◇ NORIGEST ◇ NORLUTIN ENANTHATE ◇ SH 393

TOXICITY DATA with REFERENCE
ims-wmn TDLo:4 mg/kg (1D pre):REP CCPTAY 5,21,72
scu-rat TDLo:150 mg/kg (14-19D preg):TER ENDOAO 71,448,62
ims-rat TDLo:7000 mg/kg/70W:NEO 36PYAS-,163,77
ivn-wmn TDLo:4 mg/kg:EYE ADVPB4 8,103,72

SAFETY PROFILE: An experimental teratogen. Questionable carcinogen with experimental tumorigenic data. Human systemic effects by intravenous route: increased intraocular pressure. Human reproductive effects by intramuscular route: menstrual cycle changes or disorders, changes in female fertility. Experimental reproductive effects. When heated to decomposition it emits acrid smoke and irritating fumes.

NNQ500 CAS:6533-00-2 ***HR: 3***
NORGESTREL
mf: $C_{21}H_{28}O_2$ mw: 312.49

PROP: Crystals from diethyl ether-hexane. Mp: 142-143°.

SYNS: 13-ETHYL-17-α-ETHYNYLGON-4-EN-17-β-OL-3-ONE ◇ 13-ETHYL-17-α-ETHYNYL-17-β-HYDROXY-4-GONEN-3-ONE ◇ dl-13-β-ETHYL-17-α-ETHYNYL-17-β-HYDROXYGON-4-EN-3-ONE ◇ (±)-13-ETHYL-17-α-ETHYNYL-17-HYDROXYGON-4-EN-3-ONE ◇ dl-13-β-ETHYL-17-α-ETHYNYL-19-NORTESTOSTERONE ◇ (±)-13-ETHYL-17-HYDROXY-18,19-DINOR-17-α-PREGN-4-EN-20-YN-3-ONE ◇ 17-ETHYNYL-18-METHYL-19-NORTESTOSTERONE ◇ FH 122-A ◇ 17-β-

HYDROXY-18-METHYL-19-NOR-17-α-PREGN-4-EN-20-YN-3-ONE ◇ 18-METHYL-17-α-ETHYNYL-19-NORTESTOSTERONE ◇ MONOVAR ◇ d(−)-NORGESTREL ◇ (±)-NORGESTREL ◇ α-NORGESTREL ◇ d-NORGESTREL ◇ dl-NORGESTREL ◇ LD NORGESTREL (FRENCH) ◇ POSTINOR ◇ SH 850 ◇ SH 70850 ◇ WY 3707

TOXICITY DATA with REFERENCE
orl-wmn TDLo:30 μg/kg (female 30D pre):REP BMJOAE 4,489,68
scu-rat TDLo:60 mg/kg (female 17-20D post):TER AVBIB9 13,71,74
orl-mus TDLo:57 mg/kg/68W-C:NEO CRSBAW 168,1190,74
orl-rat LD50:5010 mg/kg TXAPA9 18,185,71
ipr-rat LD50:11200 mg/kg YKYUA6 29,1479,78
orl-mus LD50:5010 mg/kg TXAPA9 18,185,71
ipr-mus LD50:7300 mg/kg YKYUA6 29,1479,78

SAFETY PROFILE: Human reproductive effects by ingestion and implant: menstrual cycle changes or disorders and female fertility index changes. An experimental teratogen. Experimental reproductive effects. Questionable carcinogen with experimental neoplastigenic data. An oral contraceptive. When heated to decomposition it emits acrid smoke and irritating fumes.

NNQ520 CAS:797-63-7 ***HR: 2***
d(−)-NORGESTREL
mf: $C_{21}H_{28}O_2$ mw: 312.49

SYNS: 17-α-ETHINYL-13-β-ETHYL-17-β-HYDROXY-4-ESTREN-3-ONE ◇ 13-ETHYL-17-α-ETHYNYLGON-4-EN-17-β-OL-3-ONE ◇ 13-ETHYL-17-α-ETHYNYL-17-β-HYDROXY-4-GONEN-3-ONE ◇ 17-ETHYNYL-18-METHYL-19-NORTESTOSTERONE ◇ 17-β-HYDROXY-18-METHYL-19-NOR-17-α-PREGN-4-EN-20-YN-3-ONE ◇ 18-METHYL-17-α-ETHYNYL-19-NORTESTOSTERONE ◇ d-NORGESTREL ◇ POSTINOR

TOXICITY DATA with REFERENCE
imp-wmn TDLo:1600 μg/kg (52 W pre):REP CCPTAY 12,615,75
orl-mus TDLo:29 mg/kg/69W-C:NEO CRSBAW 168,1190,74

CONSENSUS REPORTS: IARC Cancer Review: Animal Inadequate Evidence IMEMDT 6,201,74; IMEMDT 21,479,79; Human Inadequate Evidence IMEMDT 21,479,79

SAFETY PROFILE: Human reproductive effects by ingestion: menstrual cycle changes, fertility index. Questionable carcinogen with experimental neoplastigenic data. When heated to decomposition it emits acrid smoke and irritating fumes.

NNQ525 CAS:797-64-8 ***HR: 2***
l-NORGESTREL
mf: $C_{21}H_{28}O_2$ mw: 312.49

SYNS: LEVONORGESTREL ◇ NORPLANT

TOXICITY DATA with REFERENCE
orl-wmn TDLo:219 μg/kg (52W pre):REP CCPTAY 25,243,82
orl-mus TDLo:500 mg/kg (female 8-12D post):TER EXCEDS 81,197,83
ipr-mus LDLo:2 g/kg EXCEDS 81,175,83

CONSENSUS REPORTS: EPA Genetic Toxicology Program.

SAFETY PROFILE: Moderately toxic by intraperitoneal route. Human male reproductive effects by ingestion: disorders of spermatogenesis and the androgenic endocrine system, impotence and other effects. Human female reproductive effects by several routes: changes in the uterus, cervix, vagina, menstrual cycle and fertility. An experimental teratogen. Experimental reproductive effects. When heated to decomposition it emits acrid smoke and irritating fumes.

NNR000 ***HR: 3***
NORGESTREL mixed with MESTRANOL

SYN: MESTRANOL mixed with NORGESTREL

TOXICITY DATA with REFERENCE
orl-wmn TDLo:11 mg/kg/Y-I:CAR JAMAAP 235,730,76

SAFETY PROFILE: Questionable human carcinogen producing liver tumors.

NNR125 CAS:848-21-5 ***HR: D***
NORGESTRIENONE
mf: $C_{20}H_{22}O_2$ mw: 294.42

PROP: Pale yellow needles from diisopropyl ether. Mp: 169°. Sol in alc, ether, acetone, benzene, chloroform; practically insol in water, dil aq acids and alkalies.

SYNS: 17-α-ETHYNYL-17-β-HYDROXYESTRA-4,9,11-TRIEN-3-ONE ◇ 17-HYDROXY-19-NOR-17-α-PREGNA-4,9,11-TRIEN-20-YN-3-ONE ◇ OGYLINE ◇ R 2010 ◇ RU 2010

TOXICITY DATA with REFERENCE
ivg-wmn TDLo:1 mg/kg (36W pre):REP CCPTAY 20,511,79

SAFETY PROFILE: Human female reproductive effects by implant and intravaginal routes: changes in menstrual cycle and fertility. When heated to decomposition it emits acrid smoke and irritating fumes.

NNR200 CAS:21848-62-4 ***HR: 3***
NORGLAUCINE
mf: $C_{20}H_{23}NO_4$ mw: 341.44

SYNS: NORGLAUCIN ◇ d-N-NORGLAUCINE ◇ 1,2,9,10-TETRAMETHOXY-6a-α-NORAPORPHINE

TOXICITY DATA with REFERENCE
orl-mus LD50:450 mg/kg APFRAD 38,537,80

ipr-mus LD50:170 mg/kg APFRAD 38,537,80
ivn-mus LD50:90 mg/kg APFRAD 38,537,80

SAFETY PROFILE: Poison by intravenous and intraperitoneal routes. Moderately toxic by ingestion. When heated to decomposition it emits toxic fumes of NO_x.

NNR300 CAS:244-63-3 ***HR: 3***
NORHARMAN
mf: $C_{11}H_8N_2$ mw: 168.21

SYN: 9H-PYRIDO(3,4-b)INDOLE

TOXICITY DATA with REFERENCE
dnd-hmn:hla 50 μmol/L MUREAV 89,95,81
msc-ham:lng 150 mg/L CALEDQ 17,249,83
orl-rat TDLo:2520 mg/kg (28D male):REP TOLED5 3,157,79
ivn-mus LD50:100 mg/kg CSLNX* NX#02316

SAFETY PROFILE: Poison by intravenous route. Experimental reproductive effects. Human mutation data reported. When heated to decomposition it emits toxic fumes of NO_x.

NNR500 CAS:494-97-3 ***HR: 3***
NORNICOTINE
mf: $C_9H_{12}N_2$ mw: 148.23

PROP: Hygroscopic, visc liquid; slt amine odor. Bp: 270°, d: 1.0737 @ 20°/4°. Misc with water; very sol in chloroform, alc, ether, petr ether, oils, kerosine.

SYNS: 1'-DEMETHYL-(s)-NICOTINE ◇ NORNICOTIN ◇ 2-(3-PYRIDYL)PYRROLIDINE ◇ 3-(2-PYRROLIDINYL)-(s)-PYRIDINE ◇ 3-(2-PYRROLIDINYL)PYRIDINE

TOXICITY DATA with REFERENCE
ipr-rat LDLo:23500 μg/kg MEIEDD 10,963,83
ipr-mus LD50:14660 μg/kg NAIZAM 8,69,57
ivn-mus LD50:3409 μg/kg JPETAB 88,82,46
ivn-rbt LD50:3 mg/kg PSEBAA 58,231,45

SAFETY PROFILE: Poison by intraperitoneal and intravenous routes. About 1/3 the toxicity of nicotine. Causes faintness, prostration, muscular weakness, severe nausea, vomiting, diarrhea, and collapse without convulsions. An insecticide. When heated to decomposition emits highly toxic fumes of NO_x.

NNS500 CAS:5746-86-1 ***HR: 3***
NORNICOTINE (+)-HYDROCHLORIDE
mf: $C_9H_{12}N_2 \cdot ClH$ mw: 184.69

SYN: d-NORNICOTINE HYDROCHLORIDE

TOXICITY DATA with REFERENCE
ipr-rat LDLo:6 mg/kg AJEBAK 25,83,47
ipr-gpg LDLo:10 mg/kg AJEBAK 25,83,47

SAFETY PROFILE: Poison by intraperitoneal route. When heated to decomposition it emits very toxic fumes of NO_x and HCl. See also NORNICOTINE.

NNT000 CAS:89-25-8 ***HR: 2***
NORPHENAZONE
mf: $C_{10}H_{10}N_2O$ mw: 174.22

PROP: White powder. Bp: 191° @ 7 mm, mp: 128.9°.

SYNS: 3-METHYL-1-PHENYL-2-PYRAZOLIN-5-ONE ◇ 3-METHYL-1-PHENYL-5-PYRAZOLONE ◇ NCI-C03952 ◇ 1-PHENYL-3-METHYL-5-PYRAZOLONE ◇ 1-PHENYL-3-METHYLPYRAZOLONE-5

TOXICITY DATA with REFERENCE
eye-rbt 500 mg/24H MOD 28ZPAK -,144,72
orl-rat LD50:3500 mg/kg LONZA# 08FEB79
ipr-mus LD50:2012 mg/kg RPTOAN 36,27,73

CONSENSUS REPORTS: NCI Carcinogenesis Bioassay (feed); No Evidence: mouse, rat NCITR* NCI-CG-TR-141,78. Reported in EPA TSCA Inventory.

SAFETY PROFILE: Moderately toxic by ingestion and intraperitoneal routes. An eye irritant. When heated to decomposition it emits toxic fumes of NO_x.

NNT100 CAS:15308-34-6 ***HR: 3***
(±)-NORPHENYLEPHRINE HYDROCHLORIDE
mf: $C_8H_{11}NO_2 \cdot ClH$ mw: 189.66

SYNS: (±)-α-(AMINOMETHYL)-m-HYDROXYBENZYLALCOHOL HYDROCHLORIDE ◇ NORFENEFRINE HYDROCHLORIDE ◇ NOVADRAL ◇ dl-m-OCTOPAMINE HYDROCHLORIDE ◇ ZONDEL

TOXICITY DATA with REFERENCE
orl-rat LD50:390 mg/kg NIIRDN 6,569,82
ipr-rat LD50:32 mg/kg NIIRDN 6,569,82
scu-rat LD50:28 mg/kg NIIRND 6,569,82
ivn-rat LD50:17400 μg/kg NIIRDN 6,569,82
orl-mus LD50:3300 mg/kg NIIRDN 6,569,82
ipr-mus LD50:645 mg/kg NIIRDN 6,569,82
scu-mus LD50:700 mg/kg NIIRDN 6,569,82
ivn-mus LD50:148 mg/kg NIIRDN 6,569,82

SAFETY PROFILE: Poison by ingestion, subcutaneous, intravenous and intraperitoneal routes. When heated to decomposition it emits toxic fumes of NO_x and HCl.

NNT500 CAS:472-54-8 ***HR: 3***
19-NORPREGN-4-ENE-3,20-DIONE
mf: $C_{20}H_{28}O_2$ mw: 300.48

SYNS: 19-NOR-P ◇ 19-NORPROGESTERONE

TOXICITY DATA with REFERENCE
orl-rat TDLo:350 mg/kg (7D pre):REP JRPFA4 10,105,65

imp-mus TDLo:234 mg/kg/56W-C:ETA NATUAS 212,686,66
imp-mus TD:400 mg/kg/56W-C:ETA,TER JRPFA4 6,99,63

SAFETY PROFILE: Experimental reproductive effects. Questionable carcinogen with experimental tumorigenic and teratogenic data. When heated to decomposition it emits acrid smoke and irritating fumes.

NNU000 CAS:21466-08-0 ***HR: D***
19-NOR-17-α-PREGN-5(10)-EN-20-YNE-3-α,17-DIOL
mf: $C_{20}H_{28}O_2$ mw: 300.48

SYNS: 17-α-ETHYNYL-ESTR-5(10)-ENE-3-α,17-β-DIOL ◇ 3-α-HYDROXYNORETHYNODREL

TOXICITY DATA with REFERENCE
orl-mus TDLo:450 μg/kg/(8-10D preg):TER TJADAB 3,339,70
orl-mus TDLo:450 μg/kg/(8-10D preg):REP TJADAB 3,339,70

SAFETY PROFILE: An experimental teratogen. Experimental reproductive effects. When heated to decomposition it emits acrid smoke and irritating fumes.

NNU500 CAS:2307-97-3 ***HR: D***
19-NOR-17-α-PREGN-5(10)-EN-20-YNE-3-β, 17-DIOL
mf: $C_{20}H_{28}O_2$ mw: 300.48

SYNS: 17-α-ETHYNYL-ESTR-5(10)-ENE-3-β,17-β-DIOL ◇ 3-β-HYDROXYNORETHYNODREL

TOXICITY DATA with REFERENCE
orl-mus TDLo:450 μg/kg/(8-10D preg):TER TJADAB 3,339,70
orl-mus TDLo:450 μg/kg/(8-10D preg):REP TJADAB 3,339,70

SAFETY PROFILE: An experimental teratogen. Experimental reproductive effects. When heated to decomposition it emits acrid smoke and irritating fumes.

NNV000 CAS:52-76-6 ***HR: 2***
19-NOR-17-α-PREGN-4-EN-20-YN-17-OL
mf: $C_{20}H_{28}O$ mw: 284.48

SYNS: 3-DESOXYNORLUTIN ◇ ETHINYLESTRENOL ◇ Δ^4-17-α-ETHINYLESTREN-17-β-OL ◇ 17-α-ETHINYL-17-β-HYDROXYESTR-4-ENE ◇ 17-α-ETHINYL-17-β-HYDROXYOESTR-4-ENE ◇ ETHINYLOESTRANOL ◇ ETHINYL OESTRENOL ◇ Δ^4-17-α-ETHINYLOESTREN-17-β-OL ◇ 17-α-ETHYNIL-Δ^4-ESTRENE-17-β-OL ◇ ETHYNYLESTRENOL ◇ 17-α-ETHYNYLESTRENOL ◇ 17-α-ETHYNYLESTR-4-EN-17-β-OL ◇ ETHYNLOESTRENOL ◇ 17-α-ETHYNYLOESTRENOL ◇ 17-α-ETHYNYLOESTR-4-EN-17-β-OL ◇ EXLUTEN ◇ EXLUTION ◇ EXLUTON ◇ EXLUTONA ◇ LINESTRENOL ◇ LYNENOL ◇ LYNESTRENOL ◇ LYNOESTRENOL ◇ (17-α)-19-NORPREGN-4-EN-20-YN-17-OL ◇ NSC-37725 ◇ ORG 485-50 ◇ ORGAMETIL ◇ ORGAMETRIL ◇ ORGAMETROL

TOXICITY DATA with REFERENCE
orl-wmn TDLo:5600 μg/kg (80W pre):REP CCPTAY 8,315,73
orl-rbt TDLo:1300 μg/kg (female 6-18D post):TER ARTODN 52,23,83

CONSENSUS REPORTS: IARC Cancer Review: Animal Inadequate Evidence IMEMDT 21,407,79.

SAFETY PROFILE: An experimental teratogen. Experimental reproductive effects. Human reproductive effects by ingestion and implant routes: menstrual cycle changes or disorders, female fertility index and other fertility changes and effects on females. Questionable carcinogen. Used in oral contraceptives and in the treatment of dysfunctional uterine bleeding. When heated to decomposition it emits acrid smoke and irritating fumes.

NNV500 CAS:36393-56-3 ***HR: 2***
(−)-NORPSEUDOEPHEDRINE

SYN: 1-NOR-PSI-EPHEDRIN(GERMAN)

TOXICITY DATA with REFERENCE
scu-mus LD50:450 mg/kg ARZNAD 5,367,55

CONSENSUS REPORTS: Reported in EPA TSCA Inventory.

SAFETY PROFILE: Moderately toxic by subcutaneous route. Used in production of drugs of abuse. When heated to decomposition it emits toxic fumes of NO_x.

NNW500 CAS:2153-98-2 ***HR: 3***
(+)-NORPSEUDOEPHEDRINE HYDROCHLORIDE
mf: $C_9H_{13}NO{\bullet}ClH$ mw: 187.69

PROP: Prisms. Mp: 180-181°. Sol in water.

SYNS: ADIPOSETTIN ◇ AMORPHAN ◇ EXPONCIT ◇ FUGOA ◇ MINUSIN ◇ d-NORPSEUDOEPHEDRINE HYDROCHLORIDE ◇ REDUFORM

TOXICITY DATA with REFERENCE
ipr-mus LD50:161 mg/kg JPETAB 158,135,67
scu-mus LD50:275 mg/kg ARZNAD 5,367,55
ivn-rbt LDLo:75 mg/kg JPETAB 36,363,29

CONSENSUS REPORTS: Reported in EPA TSCA Inventory.

SAFETY PROFILE: Poison by intraperitoneal, subcutaneous and intravenous routes. When heated to decomposition it emits very toxic fumes of HCl and NO_x.

NNX300 CAS:1235-13-8 ***HR: D***
19-NORSPIROXENONE
mf: $C_{21}H_{30}O_2$ mw: 314.51

SYNS: L 586.153 ◇ 2′,3′-α-TETRAHYDROFURAN-2′-SPIRO-17-(ESTER-4-EN-3-ONE)

TOXICITY DATA with REFERENCE
spm-rat-scu 40 mg/kg PSEBAA 150,226,75
scu-rat TDLo:1680 mg/kg (42D male):REP PSEBAA 150,226,75

SAFETY PROFILE: Experimental reproductive effects. Mutation data reported. When heated to decomposition it emits acrid smoke and irritating fumes.

NNX400 CAS:434-22-0 ***HR: D***
NORTESTONATE
mf: $C_{18}H_{26}O_2$ mw: 274.44

PROP: Dimorphic crystals. Mp: 112 and 124°. Sol in alc, ether, chloroform.

SYNS: 17-β-HYDROESTR-4-EN-3-ONE ◇ (17-β)-17-HYDROXY-ESTR-4-EN-3-ONE (9CI) ◇ MENIDRABOL ◇ NANDROLON ◇ NANDROLONE ◇ NORANDROSTENOLON ◇ NORANDROSTENOLONE ◇ 19-NORANDROSTENOLONE ◇ NORTESTOSTERONE ◇ 19-NORTESTOSTERONE ◇ (+)-19-NORTESTOSTERONE ◇ OESTRENOLON

TOXICITY DATA with REFERENCE
scu-rat TDLo:1200 μg/kg (female 3D pre):REP ACENA7 47,444,64

SAFETY PROFILE: Experimental reproductive effects. When heated to decomposition it emits acrid smoke and irritating fumes.

NNX600 CAS:54025-36-4 ***HR: 3***
19-NORTESTOSTERONE-17-N-(2-CHLOROETHYL)-N-NITROSOCARBAMATE
mf: $C_{21}H_{29}ClN_2O_4$ mw: 408.97

SYNS: (17-β)-17-((((2-CHLOROETHYL)NITROSOAMINO)CARBONYL)OXY)ESTR-4-EN-3-ONE ◇ LEO 1727 ◇ LS 1727

TOXICITY DATA with REFERENCE
orl-rat LD50:2190 mg/kg APTOA6 49,290,81
ipr-rat LD50:155 mg/kg APTOA6 49,290,81
ivn-rat LD50:4 mg/kg APTOA6 49,290,81
orl-mus LD10:1630 mg/kg APTOA6 49,290,81
ipr-mus LD50:98 mg/kg APTOA6 49,290,81
ivn-mus LD50:8 mg/kg APTOA6 49,290,81

SAFETY PROFILE: Poison by intravenous and intraperitoneal routes. Moderately toxic by ingestion. Many N-nitroso compounds are carcinogens. When heated to decomposition it emits toxic fumes of Cl^- and NO_x. See also N-NITROSO COMPOUNDS and CARBAMATES.

NNX650 CAS:38911-59-0 ***HR: D***
19-NORTESTOSTERONE HOMOFARNESATE
mf: $C_{34}H_{50}O_3$ mw: 506.84

SYN: (17-β)-17-((1-OXO-4,8,12-TRIMETHYL-3,7,11-TRIDECATRIENYL)OXY)ESTR-4-EN-3-ONE

TOXICITY DATA with REFERENCE
ims-rat TDLo:175 mg/kg (14D male):REP ARZNAD 23,693,73

SAFETY PROFILE: Experimental reproductive effects. When heated to decomposition it emits acrid smoke and irritating fumes.

NNX700 ***HR: 3***
NORTHERN COPPERHEAD VENOM

SYNS: A. CONTORTRIX MOKASEN VENOM ◇ AGKISTRODON CONTORTRIX MOKASEN VENOM ◇ VENOM, SNAKE, AGKISTRODON CONTORTRIX MOKASEN

TOXICITY DATA with REFERENCE
ipr-mus LD50:220 μg/kg CBPBB8 64,201,79
scu-mus LD50:600 μg/kg CBPBB8 64,201,79
ivn-mus LD50:2711 μg/kg TOXIA6 9,131,71
ims-mus LD50:24 μg/kg JTASAG 55,96,80
ivn-dog LDLo:750 μg/kg 19DDA6 1,269,67

SAFETY PROFILE: A deadly poison by subcutaneous, intramuscular, intravenous and intraperitoneal routes.

NNY000 CAS:72-69-5 ***HR: 3***
NORTRIPTYLINE
mf: $C_{19}H_{21}N$ mw: 263.41

SYNS: DEMETHYLAMITRIPTYLENE ◇ DESMETHYLAMITRIPTYLINE ◇ 10,11-DIHYDRO-N-METHYL-5H-DIBENZO(a,d)CYCLOHEPTANE-Δ,Γ-PROPYLAMINE ◇ 5-(3-(METHYLAMINO)PROPYLIDENE)DIBENZO(a,e)CYCLOHEPTA(1,5)DIENE ◇ 5-(3-METHYLAMINOPROPYLIDENE)-10,11-DIHYDRO-5H-DIBENZO(a,d)CYCLOHEPTENE ◇ NORAMITRIPTYLINE

TOXICITY DATA with REFERENCE
orl-rat LD50:502 mg/kg HEPHD2 55,527,80
ivn-rat LD50:22 mg/kg HEPHD2 55,527,80
orl-mus LD50:387 mg/kg BJPCBM 34,236,68
ipr-mus LD50:70 mg/kg JMCMAR 6,338,63
ivn-mus LD50:17 mg/kg JMCMAR 17,65,74

SAFETY PROFILE: Poison by ingestion, intraperitoneal, and intravenous routes. When heated to decomposition it emits toxic fumes of NO_x.

NOA000 CAS:128-62-1 ***HR: 2***
NOSCAPINE
mf: $C_{22}H_{23}NO_7$ mw: 413.46

PROP: Orthorhombic prisms. D: 1.395, mp: 175°. Sublimes @ 150-160° under 11 mm pressure. Insol in vegetable oils; sltly sol in NH_4OH, hot solns of KOH, NaOH.

SYNS: CAPVAL ◇ COSCOPIN ◇ COSCOTABS ◇ KEY-TUSSCAPINE ◇ LONGATIN ◇ LYOBEX ◇ METHOXYHYDRASTINE ◇ NARCOMPREN ◇ NARCOSINE ◇ NARCOTINE ◇ 1-α-NARCOTINE ◇ NARCOTUSSIN ◇ NECTADON ◇ NICOLANE ◇ NIPAXON ◇ NOSCAPAL ◇ NOSCAPALIN ◇ NSC 5366 ◇ OPIAN ◇ OPIANINE ◇ TERBENOL ◇ TUSSCAPINE ◇ VADEBEX

TOXICITY DATA with REFERENCE
cyt-ham:lng 100 mg/L ATSUDG (4),41,80
orl-mus LD50:960 mg/kg NIIRDN 6,565,82
scu-mus LD50:725 mg/kg RPTOAN 31,338,68

SAFETY PROFILE: Moderately toxic by ingestion and subcutaneous routes. Mutation data reported. An antitussive. When heated to decomposition it emits toxic fumes of NO_x.

NOA500 CAS:912-60-7 ***HR: 3***
NOSCAPINE HYDROCHLORIDE
mf: $C_{22}H_{23}NO_7$•ClH mw: 449.92

PROP: Crystals. Very sol in water.

SYNS: CAPVAL HYDROCHLORIDE ◇ COSCOPIN HYDROCHLORIDE ◇ COSCOTABS HYDROCHLORIDE ◇ KEY-TUSSCAPINE HYDROCHLORIDE ◇ LONGATIN HYDROCHLORIDE ◇ LYOBEX HYDROCHLORIDE ◇ METHOXYHYDRASTINE HYDROCHLORIDE ◇ NARCOMPREN HYDROCHLORIDE ◇ NARCOSINE HYDROCHLORIDE ◇ NARCOSINE, HYDROCHLORIDE (8CI) ◇ 1-α-NARCOTINE HYDROCHLORIDE ◇ NECTADON HYDROCHLORIDE ◇ NSC 5366 HYDROCHLORIDE ◇ OPIAN HYDROCHLORIDE ◇ OPIANINE HYDROCHLORIDE ◇ TERBENOL HYDROCHLORIDE ◇ TUSSCAPINE HYDROCHLORIDE ◇ VADEBEX HYDROCHLORIDE

TOXICITY DATA with REFERENCE
cyt-ham:fbr 60 mg/L ESKHA5 96,55,78
cyt-ham:lng 60 mg/L GMCRDC 27,95,81
orl-mus LD50:853 mg/kg YKKZAJ 81,266,61
scu-mus LD50:1350 mg/kg ARZNAD 19,1916,69
ivn-dog LD50:52 mg/kg JPETAB 123,287,58
ivn-gpg LDLo:83 mg/kg AIPTAK 119,205,59

SAFETY PROFILE: Poison by intravenous route. Moderately toxic by ingestion and subcutaneous routes. Mutation data reported. When heated to decomposition it emits very toxic fumes of HCl and NO_x.

NOA600 CAS:10540-29-1 ***HR: 2***
NOVADEX
mf: $C_{26}H_{29}NO$ mw: 371.56

PROP: Crystals from petr ether. Mp: 96-98°. cis-Form base: mp: 72-74° from methanol.

SYNS: cis-1-(p-(2-(N,N-DIMETHYLAMINO)ETHOXY)PHENYL)-1,2-DIPHENYLBUT-1-ENE ◇ cis-N,N-DIMETHYL-2-(p-(1,2-DIPHENYL-1-BUTENYL)PHENOXY)ETHYLAMINE ◇ (Z)-2-(p-(1,2-DIPHENYL-1-BUTENYL)-PHENOXY)-N,N-DIMETHYLETHYLAMINE ◇ ICI 46,474 ◇ TAMOXIFEN

TOXICITY DATA with REFERENCE
dni-hmn:mmr 1 μmol/L CNREA8 42,1727,82
dni-rat-scu 500 μg/kg BBACAQ 656,45,81
orl-rbt TDLo:3250 μg/kg (female 6-18D post):TER JZKEDZ 6,217,80
orl-mus TDLo:2400 μg/kg (female 1-3D post):REP JMCMAR 14,952,71
orl-wmn TDLo:5600 μg/kg/1W-I:SKN BMJOAE 291,1172,85
unr-wmn TDLo:200 μg/kg/D:BLD,GIT CTRRDO 60,1431,76

CONSENSUS REPORTS: EPA Genetic Toxicology Program.

SAFETY PROFILE: Human systemic effects by an unspecified route: nausea or vomiting, leukopenia, thrombocytopenia, and skin changes. An experimental teratogen. Other experimental reproductive effects. Human mutation data reported. When heated to decomposition it emits toxic fumes of NO_x.

NOA700 CAS:51109-89-8 ***HR: 3***
NOVALICHIN

TOXICITY DATA with REFERENCE
orl-mus LD50:260 mg/kg AMNIB6 20,37,81
ipr-mus LD50:260 mg/kg AMNIB6 20,37,81
ivn-mus LD50:30 mg/kg AMNIB6 20,37,81

SAFETY PROFILE: Poison by ingestion, intravenous, and intraperitoneal routes. An antifungal antibiotic.

NOB000 CAS:1476-53-5 ***HR: 3***
NOVOBIOCIN, MONOSODIUM SALT
mf: $C_{31}H_{35}N_2O_{11}$•Na mw: 634.67

SYNS: ALBAMYCIN ◇ ALBAMYCIN SODIUM ◇ CATHOMYCIN SODIUM ◇ CATHOMYCIN SODIUM LYOVAC ◇ INAMYCIN ◇ MONOSODIUM NOVOBIOCIN ◇ NOVOBIOCIN MONOSODIUM ◇ NOVOBIOCIN, SODIUM derivative ◇ SODIUM ALBAMYCIN ◇ SODIUM NOVOBIOCIN ◇ U-6591

TOXICITY DATA with REFERENCE
ipr-rat LD50:400 mg/kg TXAPA9 24,37,73
scu-rat LD50:700 mg/kg TXAPA9 9,445,66
orl-mus LD50:962 mg/kg ANTCAO 6,226,56
ipr-mus LD50:260 mg/kg TXAPA9 24,37,73
ivn-mus LD50:407 mg/kg ANTCAO 6,226,56
ipr-gpg LD50:12 mg/kg UPJOH* 2(6),-,71
scu-gpg LD50:28 mg/kg UPJOH* 2(6),-,71

SAFETY PROFILE: Poison by intraperitoneal and subcutaneous routes. Moderately toxic by ingestion and subcutaneous routes. An antibiotic with serious side effects which include liver and blood disease. It may also promote the development of resistant strains of staphylococcus. When heated to decomposition it emits toxic fumes of NO_x and Na_2O.

NOB700 CAS:1936-40-9 ***HR: 3***
NOVOEMBICHIN
mf: $C_7H_{14}Cl_3N$•ClH mw: 255.03

PROP: Colorless oil. Bp: 114°.

SYNS: BIS(β-CHLOROETHYL)-β-CHLOROPROPYLAMINE HYDRO-

CHLORIDE ◇ N,N-BIS(2-CHLOROETHYL)-2-CHLOROPROPYLAMINE HYDROCHLORIDE ◇ 2-CHLORO-N,N-BIS(2-CHLOROETHYL)-1-PROPYLAMINE HYDROCHLORIDE ◇ EMBICHIN 7 ◇ NOVEMBICHIN ◇ NOVEMBIKHIN

TOXICITY DATA with REFERENCE
dni-mus:ast 30 mg/L NEOLA4 22,105,75
oms-mus:ast 30 mg/L NEOLA4 22,105,75
ipr-rat LD50:300 μg/kg CPBTAL 8,807,60
scu-mus LDLo:25 mg/kg NTIS** PB158-507

SAFETY PROFILE: Poison by intraperitoneal and subcutaneous routes. Mutation data reported. An anticancer agent. When heated to decomposition it emits toxic fumes of NO_x and HCl.

NOC000 CAS:54-30-8 ***HR: 3***
NOVOSPASMIN
mf: $C_{19}H_{32}N_2O_2$ mw: 320.53

SYNS: ACAMYLOPHENINE ◇ ADOPON ◇ AVADYL ◇ AVOCAN ◇ BELOSIN ◇ CAMYLOFINE ◇ CAMYLOPIN ◇ α-(2-(DIETHYLAMINOETHYL)AMINO)-BENZENEACETICACID-3-METHYLBUTYL ESTER ◇ α-(N,β-DIETHYLAMINOETHYL) AMINOPHENYLACETIC ACID ISOAMYL ESTER ◇ B-DIETHYLAMINOETHYLAMINOPHENYLACETIC ACID ISOAMYL ESTER ◇ N-(2-DIETHYLAMINOETHYL)- 2-PHENYLGLYCINE ISOPENTYL ESTER ◇ ISOAMYL N-(β-DIETHYLAMINOETHYL)-α-AMINOPHENYLACETATE ◇ ISOAMYL α-(N-(β-DIETHYLAMINOETHYL))-AMINOPHENYLACETATE ◇ ISOPENTYL ALCHOL, ESTER with N-(2-(DIETHYLAMINO)ETHYL)-2-PHENYLGLYCINE ◇ NAVADYL ◇ PHENYLDIETHYLAMINOETHYL-1'-AMINOACETIC ACID, ISOPENTYL ESTER ◇ SINTESPASMIL ◇ SPASMOCAN

TOXICITY DATA with REFERENCE
orl-mus LD50:760 mg/kg DMWOAX 76,479,51
ipr-mus LD50:175 mg/kg YKKZAJ 76,1175,56

SAFETY PROFILE: Poison by intraperitoneal route. Moderately toxic by ingestion. When heated to decomposition it emits toxic fumes of NO_x.

NOC400 CAS:89911-79-5 ***HR: 2***
NTPA
mf: $C_6H_{14}N_2O_4$ mw: 178.22

SYNS: 3-((2-HYDROXYPROPYL)NITROSAMINO)-1,2-PROPANEDIOL ◇ N-NITROSO-2,3-DIHYDROXYPROPYL-2-HYDROXYPROPYLAMINE ◇ NITROSOTRIHYDROXY-DIPROPYLAMINE

TOXICITY DATA with REFERENCE
mma-sat 25 μg/plate TCMUE9 1,13,84
orl-rat TDLo:280 mg/kg/37W-C:CAR CRNGDP 5,167,84
orl-rat TD:720 mg/kg/24W-C:CAR CRNGDP 5,167,84
orl-rat TD:3150 mg/kg/26W-C:ETA CRNGDP 5,167,84

SAFETY PROFILE: Questionable carcinogen with experimental carcinogenic and tumorigenic data. Mutation data reported. When heated to decomposition it emits toxic fumes of NO_x. See also N-NITROSO COMPOUNDS.

NOE500 CAS:475-83-2 ***HR: 3***
1-NUCIFERINE
mf: $C_{19}H_{21}NO_2$ mw: 295.41

PROP: Alkaloid obtained from the lotus *Nelumbo nucifera* and *Nelumbo lutea* (AIPTAK 197,261,72).

SYNS: 1-5,6-DIMETHOXYAPORPHINE ◇ (R)-1,2-DIMETHOXYAPORPHINE ◇ 1,2-DIMETHOXY-6a-β-APORPHINE ◇ NUCIFERIN ◇ NUCIFERINE ◇ (−)-NUCIFERINE ◇ (R)-5,6,6a,7-TETRAHYDRO-1,2-DIMETHOXY-6-METHYL-4H-DIBENZO(de,g)QUINOLINE

TOXICITY DATA with REFERENCE
orl-rat LD50:280 mg/kg AIPTAK 197,261,72
orl-mus LD50:240 mg/kg AIPTAK 197,261,72
ipr-mus LD50:83 mg/kg JPMSAE 61,813,72

SAFETY PROFILE: Poison by ingestion and intraperitoneal routes. When heated to decomposition it emits toxic fumes of NO_x.

NOE525 ***HR: 3***
NU-1326 HYDROBROMIDE
mf: $C_{19}H_{17}N{\cdot}BrH$ mw: 340.29

SYNS: 2,3-DIHYDRO-2-METHYL-9-PHENYL-1H-INDENO(2,1-c)PYRIDINE HYDROBROMIDE ◇ 2-METHYL-9-PHENYL-2,3-DIHYDRO-1-PYRIDINDENE HYDROBROMIDE

TOXICITY DATA with REFERENCE
orl-mus LD50:275 mg/kg JPETAB 92,249,48
ipr-mus LD50:150 mg/kg JPETAB 92,249,48
scu-mus LD50:260 mg/kg JPETAB 92,249,48
ivn-mus LD50:40 mg/kg JPETAB 92,249,48

SAFETY PROFILE: Poison by ingestion, subcutaneous, intravenous and intraperitoneal routes. When heated to decomposition it emits toxic fumes of NO_x and HBr.

NOE550 CAS:63039-90-7 ***HR: 3***
NU-1932
mf: $C_{17}H_{25}NO_2{\cdot}ClH$ mw: 311.89

SYNS: 3-β-AETHYL-1-METHYL-4-PHENYL-4α-PIPERIDYLPROPIONAT HYDROCHLORID (GERMAN) ◇ 3-β-AETHYL-1-METHYL-4-PHENYL-4-α-PROPIONYLOXYPIPERIDIN HYDROCHLORID (GERMAN) ◇ MEPRODINE (GERMAN)

TOXICITY DATA with REFERENCE
orl-rat LD50:175 mg/kg JPETAB 93,314,48
scu-rat LD50:90 mg/kg JPETAB 93,314,48
ivn-rbt LD50:18 mg/kg JPETAB 93,314,48
ipr-mus LD50:136 mg/kg JPETAB 93,314,48
scu-mus LD50:150 mg/kg JPETAB 93,314,48
ivn-mus LD50:57 mg/kg JPETAB 93,314,48

SAFETY PROFILE: Poison by ingestion, subcutaneous, intravenous, and intraperitoneal routes. When heated to decomposition it emits very toxic fumes of HCl and NO_x.

NOF000 ***HR: 1***
NUISANCE DUSTS and AEROSOLS

SAFETY PROFILE: Variable toxicity depending upon composition. Causes local irritation of eyes, nose, throat, and lungs. Some may lead to chronic bronchitis, emphysema, and bronchial asthma. Dermatitis may result from short contact. Asthma, angioneurotic edema, hives, etc., may result from short periods of inhalation. A topic eczema, angioneurotic edema, hives, etc., may also result from prolonged contact. A common air contaminant. Nuisance aerosols do evoke some tissue response in the lung upon inhalation of sufficient amounts. However, this reaction is potentially reversible and leaves no scar tissue.

NOF500 CAS:61-12-1 ***HR: 3***
NUPERCAINE HYDROCHLORIDE
mf: $C_{20}H_{29}N_3O_2 \cdot ClH$ mw: 379.98

PROP: Crystals. Mp: decomp @ 90-98°, vap d: 13.1.

SYNS: BENZOLIN ◇ BENZOLIN HYDROCHLORIDE ◇ BUTOXYCINCHONINIC ACID DIETHYLETHYLENEDIAMIDE HYDROCHLORIDE ◇ 2-N-BUTOXY-N-(2-DIETHYLAMINOETHYL)CINCHONINAMIDE HYDROCHLORIDE ◇ 2-BUTOXY-N-(2-DIETHYLAMINOETHYL) CINCHONINAMIDE HYDROCHLORIDE ◇ 2-BUTOXY-N-(2-DIETHYLAMINOETHYL)CINCHONINIC ACID AMIDE HYDROCHLORIDE ◇ CINCAINE HYDROCHLORIDE ◇ CINCHOCAINE HYDROCHLORIDE ◇ CINCHOCAINIUM CHLORIDE ◇ DIBUCAIN ◇ DIBUCAINE HYDROCHLORIDE ◇ PERCAIN ◇ PERCAINE HYDROCHLORIDE ◇ SOVCAIN ◇ SOVCAINE HYDROCHLORIDE ◇ SOVKAIN ◇ SOVOCAINE HYDROCHLORIDE

TOXICITY DATA with REFERENCE
eye-rbt 2% SEV ARZNAD 8,181,58
dnd-esc 15 μmol/L MUREAV 89,95,81
ipr-rat LD50:30 mg/kg JPETAB 103,306,51
ipr-mus LD50:29 mg/kg JPETAB 94,299,48
scu-mus LD50:25400 μg/kg PSEBAA 103,353,60
ivn-mus LD50:5151 μg/kg JAPMA8 44,393,55
scu-rbt LDLo:5 mg/kg ODFU** -,-,31
ivn-rbt LD50:2 mg/kg JPETAB 103,306,51
isp-rbt LDLo:2 mg/kg JPETAB 57,221,36
par-frg LDLo:55 mg/kg AEPPAE 166,437,32

SAFETY PROFILE: Poison by ingestion, subcutaneous, intravenous, intraspinal, parenteral, and intraperitoneal routes. Mutation data reported. A severe eye irritant. A local anesthetic. When heated to decomposition it emits very toxic fumes of HCl and NO_x.

NOG000 CAS:84082-68-8 ***HR: 2***
NUTMEG

SYNS: MYRISTICA ◇ NUCES NUCISTAE ◇ NUX MOSCHATA

TOXICITY DATA with REFERENCE
orl-hmn TDLo:214 mg/kg:CNS,CVS JNEUAY 2,205,61
ipr-rat LD50:500 mg/kg JNEUAY 2,205,61

SAFETY PROFILE: Moderately toxic by intraperitoneal route. Human systemic effects by ingestion: central nervous system and cardiovascular system effects. When heated to decomposition it emits acrid smoke and irritating fumes.

NOG500 CAS:8008-45-5 ***HR: 2***
NUTMEG OIL, EAST INDIAN

PROP: Major components are α- and β-pinene, camphene, myristicin, dipentene and sabanene. Found in fruit of *Myristica fragrans Houttuyn* (Fam. *Myristicaceae)*. Prepared by steam distillation of dried nutmeg (FCTXAV 14,601,76). Colorless to pale yellow liquid; odor and taste of nutmeg. East Indian: d: 0.880-0.910, refr index: 1.474-1.488; West Indian: d: 0.854-0.880, refr index: 1.469-1.476 @20°. Sol in fixed oils, mineral oil; sltly sol in cold alc; very sol in hot alc, chloroform, ether; insol in glycerin, propylene glycol.

SYNS: MYRISTICA OIL ◇ NUTMEG OIL ◇ OIL of MYRISTICA ◇ OIL of NUTMEG

TOXICITY DATA with REFERENCE
skn-rbt 500 mg/24H MOD FCTXAV 14,631,76
dnr-bcs 20 mg/disc TOFOD5 8,91,85
orl-mus TDLo:4 g/kg (40D male):TER TOLED5 7,239,81
orl-mus TDLo:2400 mg/kg (40D male):REP TOLED5 7,239,81
orl-rat LD50:2620 mg/kg FCTXAV 14,631,76

CONSENSUS REPORTS: Reported in EPA TSCA Inventory.

SAFETY PROFILE: Moderately toxic by ingestion. An experimental teratogen. Experimental reproductive effects. Mutation data reported. A skin irritant. When heated to decomposition it emits acrid smoke and irritating fumes.

NOG800 ***HR: 3***
NUX-VOMICA TREE

PROP: A small tree with oval leaves about 3.5 inches long. The hard-shelled fruit looks like a small yellow or orange grapefruit about 1.5 inches in diameter. The grey, velvety seeds are the size and shape of a nickel. The tree is grown in Hawaii.

SYNS: STRYCHNINE ◇ STRYCHNOS NUX-VOMICA

SAFETY PROFILE: The whole plant including the seeds contains the poison strychnine. Ingestion of any part of the plant may cause severe muscle spasms, fever, anoxia, metabolic acidosis, and convulsions. See also STRYCHNINE.

NOH000 CAS:63428-83-1 ***HR: 1***
NYLON
mf: $(C_6H_{11}NO)_n$

PROP: Crystalline solid. Sol in phenol, cresols, xylene, and formic acid. Insol in alc, esters, ketones, hydrocarbons. Film used for implant study. (CNREA8 15,333,55).

SYNS: AMILAN ◇ ASHLENE ◇ CAPROLON ◇ ENKALON ◇ GRILON ◇ KAPRON ◇ MIRLON ◇ PERLON ◇ PHRILON ◇ POLYAMID (GERMAN) ◇ SILON ◇ TROGAMID T ◇ VYDYNE

TOXICITY DATA with REFERENCE
imp-rat TDLo:123 mg/kg:ETA CNREA8 15,333,55

SAFETY PROFILE: Questionable carcinogen with experimental tumorigenic data by implant. Reacts violently with F_2. When heated to decomposition it emits toxic fumes of NO_x.

NOH500 CAS:1400-61-9 ***HR: 3***
NYSTATIN
mf: $C_{46}H_{83}NO_{18}$ mw: 938.30

PROP: Yellow to light-tan powder; odor suggestive of cereals. Mp: decomp > 160°. Sparingly sol in methanol and ethanol; very sltly sol in water; insol in chloroform, ether, and benzene.

SYNS: BIOFANAL ◇ CANDEX ◇ CANDIO-HERMAL ◇ DIASTATIN ◇ MORONAL ◇ MYCOSTATIN ◇ MYCOSTATIN 20 ◇ NILSTAT ◇ NYSTAN ◇ NYSTATINE ◇ NYSTAVESCENT ◇ O-V STATIN

TOXICITY DATA with REFERENCE
cyt-mus-par 50 mg/kg EXPEAM 33,306,77
orl-rat TDLo:100 mg/kg (9D preg):TER ANTBAL 20,45,75
orl-rat LD50:10 g/kg APPHAX 30,81,73
ipr-rat LD50:24305 μg/kg ABANAE 3,697,55/56
orl-mus LD50:8000 mg/kg PMDCAY 14,105,77
ipr-mus LD50:4400 μg/kg JPMSAE 65,905,76
scu-mus LD50:120 mg/kg NYKZAU 53,228S,57
ivn-mus LD50:3 mg/kg JANTAJ 32,1230,79

CONSENSUS REPORTS: EPA Genetic Toxicology Program.

SAFETY PROFILE: Poison by intraperitoneal and intravenous routes. Moderately toxic by subcutaneous route. Mildly toxic by ingestion. An experimental teratogen. Mutation data reported. An antibiotic. When heated to decomposition it emits toxic fumes of NO_x.

O

OAD090 CAS:37203-43-3 ***HR: 3***
OCHRATOXIN

TOXICITY DATA with REFERENCE
orl-pig TDLo:35 μg/kg (female 21-28D post):TER BVJOA9 133,412,77
orl-rat LD50:20 mg/kg 19FKA3 -,153,67

SAFETY PROFILE: Poison by ingestion. An experimental teratogen. When heated to decomposition it emits acrid smoke and irritating fumes.

OAD100 CAS:89930-55-2 ***HR: D***
OCHRATOXIN A SODIUM SALT
mf: $C_{20}H_{18}ClNO_6 \cdot Na$ mw: 426.83

TOXICITY DATA with REFERENCE
orl-rat TDLo:5 mg/kg (8-9D preg):TER ARCVBP 5,167,74
orl-rat TDLo:5 mg/kg (8-9D preg):REP ARCVBP 5,167,74

SAFETY PROFILE: An experimental teratogen. Experimental reproductive effects. When heated to decomposition it emits toxic fumes of Cl^-, NO_x, and Na_2O.

OAD200 ***HR: D***
OCIMUM SANCTUM LINN., LEAF EXTRACT

TOXICITY DATA with REFERENCE
orl-rat TDLo:500 mg/kg (1-5D preg):REP IJMRAQ 59,777,71

SAFETY PROFILE: Experimental reproductive effects.

OAF000 CAS:68917-09-9 ***HR: 2***
OCOTEA CYMBARUM OIL

PROP: Found in the tree *Ocotea cymbarum* containing 93% safrole, 5% sesquiterpenes, 0.7% pinene, 0.6% eugenol, 0.2% cineol, and 0.2% fufural (FCTXAV 16,637,78).

SYN: OIL of SASSAFRAS BRAZILIAN

TOXICITY DATA with REFERENCE
skn-mus 100% FCTXAV 16,831,78
skn-rbt 500 mg/24H MOD FCTXAV 16,831,78
skn-pig 100% FCTXAV 16,831,78
orl-rat LD50:1580 mg/kg FCTXAV 16,831,78

CONSENSUS REPORTS: Reported in EPA TSCA Inventory.

SAFETY PROFILE: Moderately toxic by ingestion. A skin irritant. When heated to decomposition it emits acrid smoke and irritating fumes. See also individual components.

OAH000 CAS:27858-07-7 ***HR: 2***
OCTABROMODIPHENYL
mf: $C_{12}H_2Br_8$ mw: 785.42

SYNS: BB-8 ◇ BROMKAL 80 ◇ OBB ◇ OCTABROMOBIPHENYL ◇ ar,ar,ar,ar,ar',ar',ar',ar' OCTABROMO-1,1'-BIPHENYL

TOXICITY DATA with REFERENCE
orl-rat TDLo:50 mg/kg (6-15D preg):REP AIHAAP 38,307,77
orl-rat TDLo:500 mg/kg (6-15D preg):TER AIHAAP 38,307,77
orl-rat LDLo:2000 mg/kg TXAPA9 22,316,72

CONSENSUS REPORTS: Reported in EPA TSCA Inventory.

SAFETY PROFILE: Moderately toxic by ingestion. An experimental teratogen. Experimental reproductive effects. Irritating to skin, eyes, and mucous membranes. Can cause kidney and liver enlargement and thyroid hyperplasia; it is stored in fatty tissue. When heated to decomposition it emits toxic fumes of Br^-. See also POLYBROMINATED BIPHENYLS.

OAJ000 CAS:3268-87-9 ***HR: 3***
OCTACHLORODIBENZODIOXIN
mf: $C_{12}Cl_8O_2$ mw: 459.72

PROP: Colorless crystals. Mp: 239°.

SYNS: NCI-C03678 ◇ OCDD ◇ OCTACHLORODIBENZO(b,e)(1,4)DIOXIN ◇ OCTACHLORODIBENZO-p-DIOXIN ◇ 1,2,3,4,6,7,8,9-OCTACHLORODIBENZODIOXIN

TOXICITY DATA with REFERENCE
eye-rbt 2 mg MLD EVHPAZ 5,87,73
orl-rat TDLo:5 g/kg (6-15D preg):TER ADCSAJ 120,55,73
skn-mus TDLo:290 mg/kg/60W-I:ETA EVHPAZ 5,163,73
orl-rat LD50:1 mg/kg EXPEAM 38,879,82

CONSENSUS REPORTS: IARC Cancer Review: Animal Inadequate Evidence IMEMDT 15,41,77.

SAFETY PROFILE: Poison by ingestion. An experimental teratogen. An eye irritant. Questionable carcinogen with experimental tumorigenic data. When heated to decomposition it emits toxic fumes of Cl^-.

OAL000 CAS:127-90-2 ***HR: 2***
OCTACHLORODIPROPYLETHER
mf: $C_6H_6Cl_8O$ mw: 377.72

SYNS: BIS(2,3,3,3-TETRACHLOROPROPYL)ETHER ◇ ENT 25,456 ◇ MONSANTO CP-16226 ◇ S 421

TOXICITY DATA with REFERENCE
orl-rat LD50:3630 mg/kg ARSIM* 20,15,66
orl-rbt LD50:2500 mg/kg BESAAT 12,161,66

SAFETY PROFILE: Moderately toxic by ingestion. When heated to decomposition it emits toxic fumes of Cl^-. See also ETHERS and CHLORINATED HYDROCARBONS, ALIPHATIC.

OAN000 CAS:297-78-9 ***HR: 3***
1,3,4,5,6,8,8-OCTACHLORO-1,3,3a,4,7,7a-HEXAHYDRO-4,7-METHANO ISOBENZOFURAN
mf: $C_9H_4Cl_8O$ mw: 411.73

SYNS: CP 14,957 ◇ ENT 25,545 ◇ ENT 25,545-X ◇ ISOBENZAN ◇ OCTACHLORO-HEXAHYDRO-METHANOISOBENZOFURAN◇ 1,3,4,5,6,-7,10,10-OCTACHLORO-4,7-endo-METHYLENE-4,7,8,9-TETRAHYDROPHTHALAN ◇ 1,3,4,5,6,7,8,8-OCTACHLORO-2-OXA-3a,4,7,7a-TETRAHYDRO-4,7-METHANOINDENE◇ OMTAN ◇ R 6700 ◇ SD 4402 ◇ SHELL 4402 ◇ SHELL WL 1650 ◇ TELODRIN ◇ WL 1650

TOXICITY DATA with REFERENCE
orl-mus TDLo:71 g/kg/78W-I:ETA NTIS** PB223-159
orl-rat LD50:4800 μg/kg ARSIM* 20,13,66
skn-rat LD50:5 mg/kg WRPCA2 9,119,70
ipr-rat LD50:3560 μg/kg 32ZDAL -,88,70
ivn-rat LD50:1800 μg/kg AEPPAE 232,227,57
orl-mus LD50:8400 μg/kg ARSIM* 20,13,66
ipr-mus LD50:8170 μg/kg 32ZDAL -,88,70
orl-dog LD50:1 mg/kg 32ZDAL -,88,70
orl-rbt LD50:4 mg/kg 32ZDAL -,88,70
skn-rbt LD50:12 mg/kg PCOC** -,1099,66
skn-gpg LD50:2 mg/kg 32ZDAL -,88,70

SAFETY PROFILE: Poison by ingestion, skin contact, intraperitoneal, and intravenous routes. Questionable carcinogen with experimental tumorigenic data. Used as an insecticide. When heated to decomposition it emits toxic fumes of Cl^-. See also CHLORINATED HYDROCARBONS, ALIPHATIC.

OAP000 CAS:2234-13-1 ***HR: 3***
OCTACHLORONAPHTHALENE
mf: $C_{10}Cl_8$ mw: 403.70

CONSENSUS REPORTS: Community Right-To-Know List. Reported in EPA TSCA Inventory.

OSHA PEL: (Transitional: TWA 0.1 mg/m^3 (skin)) TWA 0.1 mg/m^3 (skin); STEL 0.3 mg/m^3
ACGIH TLV: TWA 0.1 mg/m^3 (skin); STEL 0.3 mg/m^3

SAFETY PROFILE: Poison by inhalation, ingestion, and skin contact. When heated to decomposition it emits highly toxic fumes of Cl^-. See also CHLORINATED HYDROCARBONS, AROMATIC.

OAR000 CAS:124-26-5 ***HR: 3***
OCTADECANNAMIDE
mf: $C_{18}H_{37}NO$ mw: 283.56

SYNS: KEMAMIDE S ◇ STEARAMIDE

TOXICITY DATA with REFERENCE
imp-mus TDLo:1000 mg/kg:ETA CNREA8 26,105,66

CONSENSUS REPORTS: Reported in EPA TSCA Inventory.

SAFETY PROFILE: Questionable carcinogen with experimental tumorigenic data. When heated to decomposition it emits toxic fumes of NO_x. See also AMIDES.

OAT000 CAS:2223-93-0 ***HR: 3***
OCTADECANOIC ACID, CADMIUM SALT
mf: $C_{36}H_{72}O_4$•Cd mw: 681.48

SYNS: CADMIUM STEARATE ◇ KADMIUMSTEARAT (GERMAN) ◇ STEARIC ACID, CADMIUM SALT

TOXICITY DATA with REFERENCE
ihl-wmn TCLo:147 mg/m^3/35M:CNS,GIT ZHYGAM 17,308,71
ihl-hmn TCLo:1800 μg/m^3/2Y:CVS,CNS,MET HYSAAV 33,187,68
orl-rat LD50:1125 mg/kg GISAAA 35(2),98,70
ihl-rat LC50:130 mg/m^3/2H ZHYGAM 17,308,71
orl-mus LD50:590 mg/kg HYSAAV 33,187,68

CONSENSUS REPORTS: EPA Extremely Hazardous Substances List. Cadmium and its compounds are on the Community Right-To-Know List. Reported in EPA TSCA Inventory.

OSHA PEL: TWA 0.1 mg(Cd)/m^3; CL 0.6 mg(Cd)/m^3 (fume)
ACGIH TLV: TWA 0.05 mg(Cd)/m^3 (Proposed: TWA 0.01 mg(Cd)/m^3 (dust), Suspected Human Carcinogen; 0.002 mg(Cd)/m^3 (respirable dust), Suspected Human Carcinogen); BEI: 10 μg/g creatinine in urine; 10 μg/L in blood.
DFG BAT: Blood 1.5 μg/dL; Urine 15 μg/dL, Suspected Carcinogen.
NIOSH REL: (Cadmium) Reduce to lowest feasible level

SAFETY PROFILE: Confirmed carcinogen. Poison by inhalation. Moderately toxic by ingestion. Human systemic effects by inhalation: hallucinations, or distorted perceptions; nausea or vomiting, other gastrointestinal effects; weight loss or decreased weight gain; cardiac effects. When heated to decomposition it emits toxic fumes of Cd. See also CADMIUM COMPOUNDS.

OAV000 CAS:31566-31-1 ***HR: 3***
OCTADECANOIC ACID, MONOESTER with 1,2,3-PROPANETRIOL
mf: $C_{21}H_{42}O_4$ mw: 358.63

PROP: Pure white- or cream-colored, wax-like solid; faint odor. Mp: 58-59°, d: 0.97. Sol in (hot) alc, oils, and hydrocarbons.

SYNS: ABRACOL S.L.G ◇ ADMUL ◇ ADVAWAX 140 ◇ ALDO HMS ◇ ALDO MS ◇ ALDO MSA ◇ ALDO MSLG ◇ ALDO-28 ◇ ALDO-72 ◇ ARLACEL 161 ◇ ARLACEL 169 ◇ ARMOSTAT 801 ◇ ATMOS 150 ◇ ATMUL 67 ◇ ATMUL 84 ◇ ATMUL 124 ◇ CEFATIN ◇ CELINHOL -A ◇ CERASYNT 1000-D ◇ CERASYNT S ◇ CERASYNT SD ◇ CERASYNT SE ◇ CERASYNT WM ◇ CITOMULGAN M ◇ CYCLOCHEM GMS ◇ DERMAGINE ◇ DISTEARIN ◇ DREWMULSE TP ◇ DREWMULSE V ◇ DRUMULSE AA ◇ EMCOL CA ◇ EMEREST 2400 ◇ EMEREST 2401 ◇ EMCOL MSK ◇ EMUL P.7 ◇ ESTOL 603 ◇ GLYCERIN MONOSTEARATE ◇ GLYCEROL MONOSTEARATE ◇ GLYCERYL MONOSTEARATE ◇ GROCOR 5500 ◇ GROCOR 6000 ◇ HODAG GMS ◇ IMWITOR 191 ◇ IMWITOR 900K ◇ KESSCO 40 ◇ LIPO GMS 410 ◇ LIPO GMS 450 ◇ LIPO GMS 600 ◇ MONELGIN ◇ MONOSTEARIN ◇ OGEEN 515 ◇ OGEEN GRB ◇ OGEEN M ◇ ORBON ◇ PROTACHEM GMS ◇ SEDETINE ◇ STARFOL GMS 450 ◇ STARFOL GMS 600 ◇ STARFOL GMS 900 ◇ STEARIC ACID, MONOESTER with GLYCEROL ◇ STEARIC MONOGLYCERIDE ◇ TEGIN ◇ TEGIN 503 ◇ TEGIN 515 ◇ UNIMATE GMS ◇ USAF KE-7 ◇ WITCONOL MS ◇ WITCONOL MST

TOXICITY DATA with REFERENCE
ipr-mus LD50:200 mg/kg NTIS** AD277-689

CONSENSUS REPORTS: Reported in EPA TSCA Inventory.

SAFETY PROFILE: Poison by intraperitoneal route. When heated to decomposition it emits acrid smoke and irritating fumes. See also ESTERS.

OAX000 CAS:112-92-5 ***HR: 3***
1-OCTADECANOL
mf: $C_{18}H_{38}O$ mw: 270.56

PROP: Colorless solid or flakes. Mp: 58°, bp: 202° @ 10 mm, d: 0.8124 @ 59°/4°.

SYNS: ADOL ◇ ADOL 68 ◇ ATALCO S ◇ CO-1895 ◇ CO-1897 ◇ CRODACOL-S ◇ DECYL OCTYL ALCOHOL ◇ DYTOL E-46 ◇ LOROL 28 ◇ OCTADECANOL ◇ n-OCTADECANOL ◇ OCTA DECYL ALCOHOL ◇ n-OCTADECYL ALCOHOL ◇ POLAAX ◇ SIPOL S ◇ SIPONOL S ◇ STEAROL ◇ STEARYL ALCOHOL ◇ STERAFFINE ◇ USP XIII STEARYL ALCOHOL

TOXICITY DATA with REFERENCE
imp-mus TDLo:1000 mg/kg:NEO CNREA8 26,105,66
orl-rat LD50:20 g/kg 37ASAA 1,722,28

CONSENSUS REPORTS: Reported in EPA TSCA Inventory.

SAFETY PROFILE: Mildly toxic by ingestion. Questionable carcinogen with experimental neoplastigenic data. Flammable when exposed to heat or flame; can react with oxidizing materials. To fight fire, use foam, CO_2, dry chemical. When heated to decomposition it emits acrid smoke and irritating fumes. See also ALCOHOLS.

OAX050 CAS:76379-67-4 ***HR: 1***
5,7,11,13-OCTADECATETRAYNE-1,18-DIOL
mf: $C_{18}H_{22}O_2$ mw: 270.40

TOXICITY DATA with REFERENCE
skn-rbt 500 mg/kg/25H MLD JJATDK 8,35,88
mma-ham:ovr 1 g/L JJATDK 8,35,88
sce-ham:ovr 100 mg/L JJATDK 8,35,88
orl-rat TDLo:5100 mg/kg (female 2D pre):REP JJATDK 8,35,88

SAFETY PROFILE: A skin irritant. Experimental reproductive effects. Mutation data reported. When heated to decomposition it emits acrid smoke and irritating fumes.

OAX100 CAS:463-40-1 ***HR: 2***
9,12,15-OCTADECATRIENOIC ACID
mf: $C_{18}H_{30}O_2$ mw: 278.43

$$CH_3(CH_2CH{=}CH)_3(CH_2)_7CO{\bullet}OH$$

SYN: LINOLENIC ACID

SAFETY PROFILE: Mixture with cobalt naphthenate forms explosive peroxides when exposed to air. When heated to decomposition it emits acrid smoke and irritating fumes.

OAZ000 CAS:1912-84-1 ***HR: 3***
(Z)-9-OCTADECENOIC ACID, TIN (2+) SALT
mf: $C_{36}H_{66}O_4Sn$ mw: 581.71

SYN: STANNOUS OLEATE

TOXICITY DATA with REFERENCE
ivn-mus LD50:100 mg/kg CSLNX* NX#02288

CONSENSUS REPORTS: Reported in EPA TSCA Inventory.

SAFETY PROFILE: Poison by intravenous route. When heated to decomposition it emits acrid smoke and irritating fumes. See also TIN COMPOUNDS.

OBA000 CAS:143-28-2 ***HR: 1***
(Z)-9-OCTADECEN-1-OL
mf: $C_{18}H_{36}O$ mw: 268.54

SYNS: ADOL ◇ ADOL 34 ◇ ADOL 80 ◇ ADOL 85 ◇ ADOL 90 ◇ ADOL 320 ◇ ADOL 330 ◇ ADOL 340 ◇ ATALCO O ◇ CACHALOT O-1 ◇ CACHALOT O-3 ◇ CACHALOT O-8 ◇ CACHALOT O-15 ◇ CONDITIONER 1 ◇ CRODACOL A.10 ◇ CRODACOL-O ◇ DERMAFFINE ◇ H.D. EUTANOL ◇ HD OLEYL ALCOHOL 70/75 ◇ HD OLEYL ALCOHOL 80/85 ◇ HD OLEYL ALCOHOL 90/95 ◇ HD OLEYL ALCOHOL CG ◇ LANCOL ◇ LOXANOL 95 ◇ LOXANOL M ◇ NOVOL ◇ OCENOL ◇ OCEOL ◇ cis-9-OCTADECEN-1-OL ◇ OLEOL ◇ OLEYL ALCOHOL ◇ SATOL ◇ SIPOL O ◇ UNJECOL 50 ◇ UNJECOL 70 ◇ UNJECOL 90 ◇ UNJECOL 110

TOXICITY DATA with REFERENCE
skn-hmn 75 mg/3D-I MLD 85DKA8 -,127,77

CONSENSUS REPORTS: Reported in EPA TSCA Inventory.

SAFETY PROFILE: A human skin irritant. An ingredient in cosmetics. When heated to decomposition it emits acrid smoke and irritating fumes. See also ALCOHOLS.

OBC000 CAS:124-30-1 ***HR: 3***
OCTADECYLAMINE
mf: $C_{18}H_{39}N$ mw: 269.58

SYNS: N-OCTADECYLAMINE ◇ OKTADECYLAMIN (CZECH) ◇ STEARYLAMINE

TOXICITY DATA with REFERENCE
skn-rbt 500 mg/24H MOD 28ZPAK -,63,72
ipr-mus LD50:250 mg/kg NTIS** AD691-490

CONSENSUS REPORTS: Reported in EPA TSCA Inventory.

SAFETY PROFILE: Poison by intraperitoneal route. A skin irritant. When heated to decomposition it emits toxic fumes of NO_x. See also AMINES.

OBE000 CAS:1838-08-0 ***HR: 3***
OCTADECYLAMINE HYDROCHLORIDE
mf: $C_{18}H_{39}N•ClH$ mw: 306.04

TOXICITY DATA with REFERENCE
unr-mus LDLo:200 mg/kg ATMPA2 32,177,38

CONSENSUS REPORTS: Reported in EPA TSCA Inventory.

SAFETY PROFILE: Poison by an unspecified route. When heated to decomposition it emits very toxic fumes of NO_x and HCl. See also OCTADECYLAMINE.

OBG000 CAS:112-96-9 ***HR: 3***
OCTADECYL ISOCYANATE
mf: $C_{19}H_{37}NO$ mw: 295.57

PROP: Mp: 15-18°, bp: 150-180° @ 0.75 mm, d: 0.86 @ 25°.

SYN: ISOCYANIC ACID, OCTADECYL ESTER

TOXICITY DATA with REFERENCE
ivn-mus LD50:100 mg/kg CSLNX* NX#04505

CONSENSUS REPORTS: Reported in EPA TSCA Inventory.

SAFETY PROFILE: Poison by intravenous route. When heated to decomposition it emits toxic fumes of NO_x. See also ISOCYANATES.

OBI000 CAS:112-04-9 ***HR: 2***
OCTADECYLTRICHLOROSILANE
DOT: UN 1800
mf: $C_{18}H_{37}Cl_3Si$ mw: 387.99

PROP: Bp: 223° @ 10 mm, d: 0.984 @ 25°.

SYN: TRICHLOROOCTADECYLSILANE

CONSENSUS REPORTS: Reported in EPA TSCA Inventory.

DOT Classification: Corrosive Material; Label: Corrosive.

SAFETY PROFILE: A corrosive irritant to skin, eyes and mucous membranes. Reacts with water or steam to produce toxic and corrosive fumes. When heated to decomposition it emits toxic fumes of Cl^-. See also CHLOROSILANES.

OBK000 CAS:3710-30-3 ***HR: 2***
1,7-OCTADIENE
mf: C_8H_{14} mw: 110.22

$$H_2C=CH(CH_2)_4CH=CH_2$$

PROP: Flash p: 77°F.

TOXICITY DATA with REFERENCE
orl-rat LD50:14696 mg/kg AIHAAP 30,470,69
ihl-rat LCLo:8000 ppm/4H AIHAAP 30,470,69
skn-rbt LD50:10519 mg/kg AIHAAP 30,470,69

CONSENSUS REPORTS: Reported in EPA TSCA Inventory.

SAFETY PROFILE: Mildly toxic by ingestion, inhalation and skin contact. A very dangerous fire hazard when exposed to heat or flame; can react vigorously with oxidizing materials. When heated to decomposition it emits acrid smoke and irritating fumes.

OBK100 CAS:355-66-8 ***HR: 2***
OCTAFLUOROADIPAMIDE
mf: $C_6H_4F_8N_2O_2$ mw: 288.10

$$(-C_2F_4CO•NH_2)_2$$

SAFETY PROFILE: Forms a dangerously unstable complex with lithium tetrahydroaluminate. When heated to decomposition it emits toxic fumes of F^- and NO_x. See also FLUORIDES and AMIDES.

OBM000 CAS:382-21-8 ***HR: 3***
OCTAFLUORO-sec-BUTENE
mf: C_4F_8 mw: 200.04

SYNS: OCTAFLUOROISOBUTENE ◇ OCTAFLUOROISOBUTYLENE ◇ PERFLUOROISOBUTYLENE (ACGIH) ◇ PFIB

TOXICITY DATA with REFERENCE
ihl-rat LC50:500 ppb/6H 34ZIAG -,310,69
ihl-mus LCLo:10 mg/m^3/2H GTPZAB 5(3),3,61
ihl-rbt LC50:1200 ppb/2H BIMEA5 57,247,68
ihl-gpg LC50:1050 ppb/2H BIMEA5 57,247,68
ACGIH TLV: (Proposed: CL 0.01 ppm)

SAFETY PROFILE: A deadly poison by inhalation. A skin, eye, and mucous membrane irritant. Human acute exposure causes marked irritation of conjunctivae, throat, and lungs. When heated to decomposition it emits toxic fumes of F^-. See also FLUORIDES.

OBO000 CAS:360-89-4 ***HR: 1***
1,1,1,2,3,4,4,4-OCTAFLUORO-2-BUTENE
DOT: UN 2422
mf: C_4F_8 mw: 200.04

PROP: Bp: 1.2°, fp: −135°.

SYNS: FC-1318 ◇ OCTAFLUOROBUTENE-2 ◇ OCTAFLUOROBUT-2-ENE (DOT) ◇ PERFLUOROBUT-2-ENE ◇ PERFLUORO-2-BUTENE (DOT)

TOXICITY DATA with REFERENCE
sln-dmg-ihl 99 pph/10M ENVRAL 7,275,74
ihl-rat LCLo:6100 ppm/4H DUPON* 25JUL67

CONSENSUS REPORTS: Reported in EPA TSCA Inventory. EPA Genetic Toxicology Program.

DOT Classification: Nonflammable Gas; Label: Nonflammable Gas.

SAFETY PROFILE: Mildly toxic by inhalation. Mutation data reported. When heated to decomposition it emits toxic fumes of F^-. See also FLUORIDES.

OBS800 CAS:39660-55-4 ***HR: 2***
OCTAFLUOROPENTANOL
mf: $C_5H_4F_8O$ mw: 232.09

SYNS: OCTAFLUOROAMYL ALCOHOL ◇ OCTAFLUORO-1-PENTANOL ◇ OCTAFLUOROPENTYL ALCOHOL ◇ omega-H-OKTAFLUORPENTANOL-1 (GERMAN)

TOXICITY DATA with REFERENCE
orl-rat LD50:1110 mg/kg ZHYGAM 26,9,80
ihl-mus LC50:10500 mg/m^3/2H 85GMAT -,94,82
ipr-mus LD50:410 mg/kg ZHYGAM 26,9,80

SAFETY PROFILE: Moderately toxic by ingestion and intraperitoneal routes. Mildly toxic by inhalation. When heated to decomposition it emits toxic fumes of F^-. See also ALCOHOLS and FLUORIDES.

OBU000 CAS:355-80-6 ***HR: 1***
2,2,3,3,4,4,5,5-OCTAFLUORO-1-PENTANOL
mf: $C_5H_4F_8O$ mw: 232.09

PROP: Liquid. D: 1.6647 @ 20°/4°, bp: 140-141°.

TOXICITY DATA with REFERENCE
ihl-rat LCLo:2500 ppm/4H JOCMA7 4,262,62

CONSENSUS REPORTS: Reported in EPA TSCA Inventory.

SAFETY PROFILE: Mildly toxic by inhalation. When heated to decomposition it emits toxic fumes of F^-. See also FLUORIDES and ALCOHOLS.

OBV200 CAS:4430-31-3 ***HR: 2***
OCTAHYDROCOUMARIN
mf: $C_9H_{14}O_2$ mw: 154.23

SYNS: OCTAHYDRO-1-BENZOPYRAN-2-ONE ◇ OCTAHYDRO-2H-1-BENZOPYRAN-2-ONE (9CI) ◇ BICYCLONONALACTONE

TOXICITY DATA with REFERENCE
skn-rbt 500 mg/24H MLD FCTOD7 20(Suppl),781,82
orl-rat LD50:3900 mg/kg FCTOD7 20(Suppl),781,82
skn-rbt LD50:3500 mg/kg FCTOD7 20(Suppl),781,82

CONSENSUS REPORTS: Reported in EPA TSCA Inventory.

SAFETY PROFILE: Moderately toxic by ingestion and skin contact. A skin irritant. When heated to decomposition it emits acrid smoke and irritating fumes.

OBW000 CAS:63021-67-0 ***HR: 3***
OCTAHYDRO-1:2:5:6-DIBENZANTHRACENE
mf: $C_{22}H_{22}$ mw: 286.44

SYN: OCTAHYDRODIBENZ(a,h)ANTHRACENE

TOXICITY DATA with REFERENCE
skn-mus TDLo:1250 mg/kg/52W-I:ETA PRLBA4 111,485,32

SAFETY PROFILE: Questionable carcinogen with experimental tumorigenic data. When heated to decomposition it emits acrid smoke and irritating fumes.

OBY000 CAS:20917-49-1 ***HR: 3***
OCTAHYDRO-1-NITROSOAZOCINE
mf: $C_7H_{14}N_2O$ mw: 142.23

SYNS: NHMI ◇ N-NITROSOAZACYCLOOCTANE ◇ NITROSOHEPTAMETHYLENEIMINE ◇ N-NITROSOHEPTAMETHYLENEIMINE ◇ NITROSO-HEPTAMETHYLENIMIN (GERMAN)

TOXICITY DATA with REFERENCE
mma-sat 10 μg/plate TCMUE9 1,13,84
pic-esc 50 mg/L TCMUE9 1,91,84
orl-rat TDLo:150 mg/kg/20W-I:ETA IJCNAW 15,301,75
scu-ham TDLo:650 mg/kg/59W-I:CAR JJIND8 61,239,78
scu-ham TD:760 mg/kg/46W-I:CAR JJIND8 61,239,78
scu-ham TD:66 mg/kg/W-I AJPAA4 93,45,78
orl-rat TD:1200 mg/kg/20W-I ANTRD4 2,381,82
orl-rat LD50:283 mg/kg PSEBAA 130,945,69
scu-ham LD50:220 mg/kg JJIND8 61,239,78

SAFETY PROFILE: Poison by ingestion and subcutaneous routes. Questionable carcinogen with experimental carcinogenic and tumorigenic data. Mutation data reported. When heated to decomposition it emits toxic fumes of NO_x. See also N-NITROSO COMPOUNDS.

OCA000 CAS:20917-50-4 ***HR: 3***
OCTAHYDRO-1-NITROSO-1H-AZONINE
mf: $C_8H_{16}N_2O$ mw: 156.26

SYNS: N-NITROSOAZACYCLONONANE ◇ N-NITROSOOCTAMETHYLENEIMINE

TOXICITY DATA with REFERENCE
mma-sat 10 µg/plate TCMUE9 1,13,84
pic-esc 10 mg/L TCMUE9 1,91,84
orl-rat TDLo:742 mg/kg/33W-I:ETA PSEBAA 130,945,69
orl-rat LD50:566 mg/kg PSEBAA 130,945,69

SAFETY PROFILE: Moderately toxic by ingestion. Questionable carcinogen with experimental tumorigenic data. Mutation data reported. Many N-nitroso compounds are carcinogens. When heated to decomposition it emits toxic fumes of NO_x. See also N-NITROSO COMPOUNDS.

OCE000 CAS:104-50-7 ***HR: 1***
Γ-OCTALACTONE
mf: $C_8H_{14}O_2$ mw: 142.22

PROP: Colorless to pale yellow liquid; coconut odor. D: 0.970-0.980, refr index: 1.443-1.447. Sol in alc; sltly sol in water.

SYNS: Γ-n-BUTYL-Γ-BUTYROLACTONE ◇ FEMA No. 2798 ◇ 5-HYDROXYOCTANOIC ACID LACTONE ◇ OCTANOLIDE-1,4 ◇ TETRAHYDRO-6-PROPYL-2H-PYRAN-2-ONE

TOXICITY DATA with REFERENCE
skn-rbt 500 mg/24H MOD FCTXAV 14,821,76
orl-rat LD50:4400 mg/kg FCTXAV 14,821,76

CONSENSUS REPORTS: Reported in EPA TSCA Inventory.

SAFETY PROFILE: Mildly toxic by ingestion. A skin irritant. When heated to decomposition it emits acrid smoke and irritating fumes.

OCI000 CAS:5591-33-3 ***HR: 2***
5,5'-(OCTAMETHYLENEBIS(CARBONYLIMINO)-BIS(N-METHYL-2,4,6-TRIIODOISOPHTHALAMIC ACID)
mf: $C_{28}H_{28}I_6N_4O_8$ mw: 1310.00

TOXICITY DATA with REFERENCE
ivn-rat LD50:6600 mg/kg JMCMAR 9,964,66
ivn-mus LD50:1576 mg/kg INVRAV 15(Suppl),248,80
ice-mus LD50:1120 mg/kg JMCMAR 9,964,66
ivn-dog LD50:10 g/kg JMCMAR 9,964,66

SAFETY PROFILE: Moderately toxic by intracerebral and intravenous routes. When heated to decomposition it emits very toxic fumes of NO_x and I^-.

OCM000 CAS:152-16-9 ***HR: 3***
OCTAMETHYLPYROPHOSPHORAMIDE
mf: $C_8H_{24}N_4O_3P_2$ mw: 286.30

PROP: Viscous liquid. Mp: 20-21°, bp: 154° @ 2.0 mm, d: 1.09 @ 25°/4°. Misc with water; sol in most organic solvents; almost insol in higher aliphatic hydrocarbons.

SYNS: BIS(BISDIMETHYLAMINOPHOSPHONOUS)ANHYDRIDE ◇ BIS(DIMETHYLAMINO)PHOSPHONOUS ANYHYDRIDE ◇ BIS-(DIMETHYLAMINO)PHOSPHORIC ANHYDRIDE ◇ BIS-N,N,N',N'-TETRAMETHYLPHOSPHORODIAMIDIC ANHYDRIDE ◇ ENT 17,291 ◇ LETHALAIRE G-59 ◇ OCTAMETHYL-DIFOSFORZUUR-TETRAMIDE (DUTCH) ◇ OCTAMETHYLDIPHOSPHORAMIDE ◇ OCTAMETHYL-DIPHOSPHORSAEURE-TETRAMID (GERMAN) ◇ OCTAMETHYL PYROPHOSPHORTETRAMIDE ◇ OCTAMETHYL TETRAMIDO PYROPHOSPHATE ◇ OMPA ◇ OMPACIDE ◇ OMPATOX ◇ OMPAX ◇ OTTOMETIL-PIROFOSFORAMMIDE (ITALIAN) ◇ PESTOX ◇ PESTOX 3 ◇ PESTOX III ◇ PYROPHOSPHORIC ACID OCTAMETHYLTETRAAMIDE ◇ PYROPHOSPHORYLTETRAKISDIMETHYLAMIDE ◇ RCRA WASTE NUMBER P085 ◇ SCHRADAN ◇ SCHRADANE (FRENCH) ◇ SYSTAM ◇ SYSTOPHOS ◇ SYTAM ◇ TETRAKISDIMETHYLAMINOPHOSPHONOUS ANHYDRIDE

TOXICITY DATA with REFERENCE
orl-man TDLo:643 µg/kg/30D-I:BIO TXAPA9 14,603,69
orl-rat LD50:5 mg/kg WRPCA2 4,36,65
ihl-rat LCLo:8 mg/m^3/4H 85GMAT -,95,82
skn-rat LD50:15 mg/kg TXAPA9 14,515,69
ipr-rat LD50:4900 µg/kg JJPAAZ 17,509,67
scu-rat LD50:9 mg/kg JEENAI 50(3),356,57
orl-mus LDLo:2 mg/kg BESAAT 12,161,66
ipr-mus LD50:10 mg/kg PSEBAA 129,699,68
scu-mus LD50:14 mg/kg FEPRA7 23,T927,64
ivn-mus LD50:9400 µg/kg BLLIAX 38,151,58
ivn-dog LD50:5 mg/kg JPETAB 99,376,50
orl-rbt LD50:25 mg/kg BESAAT 12,161,66
skn-rbt LDLo:20 mg/kg 85GMAT -,95,82
ocu-rbt LDLo:5 mg/kg 85GMAT -,95,82

CONSENSUS REPORTS: EPA Extremely Hazardous Substances List. Reported in EPA TSCA Inventory.

SAFETY PROFILE: Poison by ingestion, inhalation, skin contact, intraperitoneal, intravenous, subcutaneous, and ocular routes. Human systemic effects by ingestion: a cholinesterase inhibitor. Has been found to inhibit peripheral cholinesterase without pronounced effects on the central nervous system. An insecticide. When heated to decomposition it emits toxic fumes of NO_x and PO_x. See also PARATHION and ANHYDRIDES.

OCO000 CAS:124-13-0 ***HR: 1***
1-OCTANAL
mf: $C_8H_{16}O$ mw: 128.24

PROP: Found in about 20 essential oils, including a number of citrus oils (FCTXAV 11,95,73). Colorless to light yellow liquid; fatty-orange odor. Bp: 163.4, flash p: 125°F (CC), d: 0.821 @ 20°/4°, refr index: 1.417-1.425, vap d: 4.41. Sol in alc, fixed oils, propylene glycol; insol in glycerin.

SYNS: ALDEHYDE C-8 ◇ C-8 ALDEHYDE ◇ FEMA No. 2797 ◇ OCTANALDEHYDE ◇ n-OCTYL ALDEHYDE

TOXICITY DATA with REFERENCE
skn-rbt 500 mg/24H MLD FCTXAV 11,1079,73
eye-rbt 100 mg MLD FCTXAV 11,1079,73
orl-rat LD50:5630 mg/kg FCTXAV 11,95,73
skn-rbt LD50:6350 mg/kg FCTXAV 11,95,73

CONSENSUS REPORTS: Reported in EPA TSCA Inventory.

SAFETY PROFILE: Mildly toxic by ingestion and skin contact. A skin and eye irritant. Combustible when exposed to heat or flame; can react with oxidizing materials. To fight fire, use foam, CO_2, dry chemical. See also ALDEHYDES.

OCU000 CAS:111-65-9 ***HR: 3***
OCTANE
DOT: UN 1262
mf: C_8H_{18} mw: 114.26

PROP: Clear liquid. Bp: 125.8°, lel: 1.0%, uel: 4.7%; fp: −56.5°, flash p: 56°F, d: 0.7036 @ 20°/4°, autoign temp: 428°F, vap press: 10 mm @ 19.2°, vap d: 3.86. Insol in water, sltly sol in alc, ether; misc with benzene.

SYNS: OKTAN (POLISH) ◇ OKTANEN (DUTCH) ◇ OTTANI (ITALIAN)

CONSENSUS REPORTS: Reported in EPA TSCA Inventory.

OSHA PEL: (Transitional: TWA 500 ppm) TWA 300 ppm; STEL 375 ppm
ACGIH TLV: TWA 300 ppm; STEL 375 ppm
DFG MAK: 500 ppm (2350 mg/m^3)
NIOSH REL: (Alkanes) TWA 350 mg/m^3

DOT Classification: Flammable Liquid; Label: Flammable Liquid.

SAFETY PROFILE: May act as a simple asphyxiant. See also ARGON for a description of simple asphyxiants. A narcotic in high concentration. Human dermal exposure to undiluted octane for five hours resulted in blister formation but no anesthesia; one hour caused diffuse burning sensation. A very dangerous fire hazard and severe explosion hazard when exposed to heat, flame, or oxidizers. When heated to decomposition it emits acrid smoke and irritating fumes. See also ALKANES.

OCW000 CAS:7613-16-3 ***HR: 2***
1,8-OCTANEDIAMINE, DIHYDROCHLORIDE
mf: $C_8H_{20}N_2$•2ClH mw: 217.22

SYN: OCTANE-1,8-DIAMINE DIHYDROCHLORIDE

TOXICITY DATA with REFERENCE
ipr-mus LD50:760 mg/kg JPETAB 107,332,53

CONSENSUS REPORTS: Reported in EPA TSCA Inventory.

SAFETY PROFILE: Moderately toxic by intraperitoneal route. When heated to decomposition it emits very toxic fumes of HCl and NO_x. See also AMINES.

OCY000 CAS:124-07-2 ***HR: 2***
OCTANOIC ACID
mf: $C_8H_{16}O_2$ mw: 144.24

PROP: Colorless, oily liquid; unpleasant odor, burning rancid taste. D: 0.91 @ 20°, bp: 240°, mp: 17°. Sltly sol in water; sol in most organic solvents.

SYNS: C-8 ACID ◇ CAPRYLIC ACID ◇ n-CAPRYLIC ACID ◇ 1-HEPTANECARBOXYLIC ACID ◇ HEXACID 898 ◇ NEO-FAT 8 ◇ OCTIC ACID ◇ n-OCTOIC ACID ◇ n-OCTYLIC ACID

TOXICITY DATA with REFERENCE
skn-rbt 500 mg/24H MOD FCTXAV 19,237,81
sln-smc 5 ppm ANYAA9 407,186,83
oms-nml:oth 10 mmol/L CHROAU 40,1,73
cyt-nml:oth 10 mmol/L CHROAU 40,1,73
orl-rat LD50:10080 mg/kg FCTXAV 2,327,64
ivn-mus LD50:600 mg/kg APTOA6 18,141,61

CONSENSUS REPORTS: Reported in EPA TSCA Inventory.

SAFETY PROFILE: Moderately toxic by intravenous route. Mildly toxic by ingestion. Mutation data reported. A skin irritant. Yields irritating vapors which can cause coughing. When heated to decomposition it emits acrid smoke and irritating fumes.

OCY100 CAS:20296-29-1 ***HR: 1***
3-OCTANOL
mf: $C_8H_{18}O$ mw: 130.28

PROP: Colorless liquid; strong, nutty odor. D: 0.816-0.821, refr index: 1.425. Sol in alc, fixed oils; insol in water.

SYNS: AMYLETHYLCARBINOL ◇ ETHYLAMYLCARBINOL ◇ ETHYL-n-AMYLCARBINOL ◇ FEMA No. 3581 ◇ OCTANOL-3

TOXICITY DATA with REFERENCE
skn-rbt 500 mg/24H MOD FCTXAV 17,881,79

SAFETY PROFILE: A moderate skin and eye irritant. When heated to decomposition it emits acrid smoke and irritating fumes.

ODE000 ***HR: 1***
OCTANOL, mixed isomers
mf: $C_8H_{18}O$ mw: 130.26

SYNS: ALFOL-10 ◇ ANTAK ◇ FAIR 85 ◇ ROYALTAC ◇ SPROUT-OFF ◇ SUCKER PLUCKER ◇ TOBACCO SUCKER CONTROL AGENT 148 ◇ TOBACCO SUCKER CONTROL AGENT 504

TOXICITY DATA with REFERENCE
skn-rbt 500 mg open MLD UCDS** 1/11/68
orl-rat LD50:18 g/kg 85ARAE 3,84,76/77
skn-rbt LD50:5660 mg/kg UCDS** 1/11/68

SAFETY PROFILE: Mildly toxic by ingestion and skin contact. A skin irritant. When heated to decomposition it emits acrid smoke and irritating fumes. See also ALCOHOLS.

ODE200 CAS:124-07-2 ***HR: 2***
OCTANOLIC ACID (mixed isomers)
mf: $C_8H_{16}O_2$ mw: 144.24

TOXICITY DATA with REFERENCE
skn-rbt 10 mg/24H open SEV AIHAAP 23,95,62
orl-rat LD50:1410 mg/kg AIHAAP 23,95,62
skn-rbt LDLo:710 mg/kg AIHAAP 23,95,62

CONSENSUS REPORTS: Reported in EPA TSCA Inventory.

SAFETY PROFILE: Moderately toxic by ingestion and skin contact. A severe skin irritant. When heated to decomposition it emits acrid smoke and irritating fumes. See also ESTERS.

ODG000 CAS:111-13-7 ***HR: 2***
2-OCTANONE
mf: $C_8H_{16}O$ mw: 128.24

PROP: Colorless liquid; pleasant apple odor. D: 0.813-0.818, refr index: 1.414-1.418, mp: −20.9°, bp: 173.5°, vap d: 4.4, flash p: 160°F. Sltly sol in water; sol in alc, hydrocarbons, ether, esters.

SYNS: FEMA No. 2802 ◇ METHYL HEXYL KETONE (FCC)

TOXICITY DATA with REFERENCE
skn-rbt 500 mg/24H MLD FCTXAV 13,681,75

CONSENSUS REPORTS: Reported in EPA TSCA Inventory.

SAFETY PROFILE: A skin irritant. Combustible liquid when exposed to heat, flame, or oxidizers. To fight fire, use foam, alcohol foam. When heated to decomposition it emits acrid smoke and irritating fumes. See also ETHER and KETONES.

ODI000 CAS:106-68-3 ***HR: 2***
3-OCTANONE
DOT: UN 2271
mf: $C_8H_{16}O$ mw: 128.24

PROP: Liquid; fruity odor. Bp: 157-162°, d: 0.822 @ 20°/20°, flash p: 138°F.

SYNS: AMYL ETHYL KETONE ◇ EAK ◇ ETHYL AMYL KETONE

TOXICITY DATA with REFERENCE
skn-rbt 500 mg/24H MOD FCTXAV 12,715,74
ipr-mus LD50:406 mg/kg SCCUR* -,5,61

CONSENSUS REPORTS: Reported in EPA TSCA Inventory.

OSHA PEL: TWA 25 ppm

DOT Classification: Flammable or Combustible Liquid; Label: Flammable Liquid

SAFETY PROFILE: Poison by intraperitoneal route. Moderately irritating to skin, eyes, and mucous membranes by inhalation. Narcotic in high concentration. Combustible liquid and dangerously flammable from heat, flame, and oxidizers. To fight fire, use foam, CO_2, dry chemical. When heated to decomposition it emits acrid smoke. See also KETONES.

ODO000 CAS:63937-47-3 ***HR: 1***
2-OCTANOYL-1,2,3,4-TETRAHYDRO-ISOQUINOLINE
mf: $C_{17}H_{25}NO$ mw: 259.43

SYNS: AI3-36420 ◇ 1,2,3,4-TETRAHYDRO-2-OCTANOYLISOQUINOLINE ◇ 1,2,3,4-TETRAHYDRO-2-(1-OXOOCTYL)ISOQUINOLINE

TOXICITY DATA with REFERENCE
skn-rbt 500 mg MOD NTIS** AD-A042-527
eye-rbt 100 mg MLD NTIS** AD-A042-527

SAFETY PROFILE: A skin and eye irritant. When heated to decomposition it emits toxic fumes of NO_x.

ODQ300 CAS:16607-77-5 ***HR: 3***
1,3,7-OCTATRIEN-5-YNE
mf: C_8H_8 mw: 104.15

$$H(HC{=}CH)_3C{\equiv}CH$$

SAFETY PROFILE: Decomposes violently when heated to 156°C. A dangerous storage hazard; it polymerizes to a shock-sensitive explosive solid. When heated to decomposition it emits acrid smoke and irritating fumes. See also ACETYLENE COMPOUNDS.

ODW000 CAS:3391-86-4 ***HR: 3***
1-OCTEN-3-OL
mf: $C_8H_{16}O$ mw: 128.24

SYNS: AMYLVINYLCARBINOL ◇ MATSUTAKE ALCOHOL (JAPANESE)

TOXICITY DATA with REFERENCE
orl-rat LD50:340 mg/kg FCTXAV 14,681,76
skn-rbt LD50:3300 mg/kg FCTXAV 14,681,76
ivn-mus LD50:56 mg/kg CSLNX* NX#02545

CONSENSUS REPORTS: Reported in EPA TSCA Inventory.

SAFETY PROFILE: Poison by ingestion and intravenous routes. Moderately toxic by skin contact. When heated to decomposition it emits acrid smoke and irritating fumes.

ODY000 CAS:6168-86-1 ***HR: 3***
OCTIN HYDROCHLORIDE
mf: $C_9H_{19}N•ClH$ mw: 177.75

SYNS: ISOMETHEPTENE HYDROCHLORIDE ◇ 2-METHYLAMINO-ISOOCTANE HYDROCHLORIDE ◇ METHYLAMINO-METHYLHEPTENE HYDROCHLORIDE

TOXICITY DATA with REFERENCE
ipr-rat LD50:70 mg/kg JPETAB 81,235,44
scu-rat LDLo:160 mg/kg JPETAB 71,62,41
scu-mus LD50:171 mg/kg JPETAB 92,214,48
ivn-mus LD50:18 mg/kg JPETAB 92,214,48
scu-dog LD50:76 mg/kg JPETAB 92,214,48
ivn-dog LD50:26 mg/kg JPETAB 92,214,48
scu-rbt LD50:101 mg/kg JPETAB 92,214,48
ivn-rbt LD50:18 mg/kg JPETAB 92,214,48
par-frg LDLo:200 mg/kg QJPPAL 9,647,36

SAFETY PROFILE: Poison by intraperitoneal, intravenous, parenteral, and subcutaneous routes. An adrenergic agent. When heated to decomposition it emits very toxic fumes of Cl^- and NO_x.

ODY100 CAS:13448-22-1 ***HR: 3***
OCTOCLOTHEPINE
mf: $C_{19}H_{21}ClN_2S$ mw: 344.93

SYNS: 1-(8-CHLORO-10,11-DIHYDRODIBENZO(b,f)THIEPIN-10-YL)-4-METHYLPIPERAZINE ◇ CLOROTEPINE ◇ CLOTEPIN ◇ CLOTHEPIN ◇ OCTOCLOTHEPIN ◇ PIPERAZINE, 1-(8-CHLORO-10,11-DIHYDRODIBENZO(b,f)THIEPIN-10-YL)-4-METHYL-

TOXICITY DATA with REFERENCE
orl-rat TDLo:1080 mg/kg (female 2-19D post):TER THERAP 22,1429,67
orl-mus LD50:78 mg/kg CCCCAK 48,144,83

SAFETY PROFILE: Poison by ingestion. An experimental teratogen. When heated to decomposition it emits toxic fumes of NO_x, SO_x, and Cl^-.

OEE000 CAS:103-09-3 ***HR: 2***
OCTYL ACETATE
mf: $C_{10}H_{20}O_2$ mw: 172.30

PROP: Water-white, stable liquid. D: 0.873 @ 20°/20°, mp: −93°, bp: 199°, flash p: 190°F. Sol in water; misc in alc and ether. (OC).

SYNS: ACETIC ACID α-ETHYLEXYL ESTER ◇ 2-ETHYLHEXANYL ACETATE ◇ β-ETHYLHEXYL ACETATE ◇ 2-ETHYLHEXYL ACETATE ◇ 2-ETHYLHEXYL ETHANOATE

TOXICITY DATA with REFERENCE
skn-rbt 10 mg/24H open JIHTAB 26,269,44
skn-rbt 500 mg open MLD UCDS** 4/21/67
eye-rbt 436 mg AJOPAA 29,1363,46
orl-rat LD50:3000 mg/kg JIHTAB 26,269,44

CONSENSUS REPORTS: Reported in EPA TSCA Inventory.

SAFETY PROFILE: Moderately toxic by ingestion. A skin and eye irritant. Flammable when exposed to heat or flame; can react with oxidizing materials. To fight fire, use foam, CO_2, dry chemical. When heated to decomposition it emits acrid smoke and irritating fumes. See also ESTERS, OCTYL ALCOHOL, and ACETIC ACID.

OEG000 CAS:112-14-1 ***HR: 2***
1-OCTYL ACETATE
mf: $C_{10}H_{20}O_2$ mw: 172.30

PROP: Colorless liquid; orange-jasmine odor. D: 0.865, refr index: 1.418-1.421, mp: −38.5°, bp: 210°, flash p: 190°F. Insol in water; misc with alc, ether, fixed oils.

SYNS: ACETATE C-8 ◇ ACETIC ACID, OCTYL ESTER ◇ CAPRYLYL ACETATE ◇ FEMA No. 2806 ◇ 1-OCTANOL ACETATE ◇ n-OCTANYL ACETATE ◇ OCTYL ACETATE ◇ n-OCTYL ACETATE ◇ OCTYL ALCOHOL ACETATE

TOXICITY DATA with REFERENCE
skn-rbt 500 mg/24H MLD FCTXAV 12,807,74
orl-rat LD50:3000 mg/kg AMIHBC 10,61,54

CONSENSUS REPORTS: Reported in EPA TSCA Inventory.

SAFETY PROFILE: Moderately toxic by ingestion. A skin irritant. Combustible liquid. When heated to decomposition it emits acrid smoke and irritating fumes. See also ESTERS.

OEG100 ***HR: 1***
3-OCTYL ACETATE
mf: $C_{10}H_{20}O_2$ mw: 172.27

PROP: Colorless liquid; rosy, minty odor. D: 0.856-0.860, refr index: 1.414, fp: 190°. Sol in alc, propylene glycol, fixed oils; sltly sol in water.

SYN: FEMA No. 3583

SAFETY PROFILE: Combustible liquid. When heated to decomposition it emits acrid smoke and irritating fumes.

OEI000 CAS:111-87-5 ***HR: 3***
OCTYL ALCOHOL
mf: $C_8H_{18}O$ mw: 130.26

PROP: Colorless liquid. D: 0.827 @ 20° 16/4°, mp: −16.7°, bp: 194.5°, flash p: 178°F. Sol in water; misc in alc, ether, and chloroform. Found in several citrus oils and at least 10 other natural sources. (FCTXAV 11,95,73).

SYNS: ALCOHOL C-8 ◇ ALFOL 8 ◇ CAPRYL ALCOHOL ◇ CAPRYLIC ALCOHOL ◇ DYTOL M-83 ◇ EPAL 8 ◇ FEMA No. 2800 ◇ HEPTYL CARBINOL ◇ 1-HYDROXYOCTANE ◇ LOROL 20 ◇ OCTANOL ◇ n-OCTANOL ◇ 1-OCTANOL (FCC) ◇ OCTILIN ◇ OCTYL ALCOHOL, NORMAL-PRIMARY ◇ PRIMARY OCTYL ALCOHOL ◇ SIPOL L8

TOXICITY DATA with REFERENCE
skn-rbt 500 mg/24H MLD FCTXAV 11,1079,73
cyt-smc 2 mmol/tube HEREAY 33,457,47
orl-mus LD50:1790 mg/kg HYSAAV 31,310,66
ivn-mus LD50:69 mg/kg AIPTAK 135,330,62

CONSENSUS REPORTS: Reported in EPA TSCA Inventory.

SAFETY PROFILE: Poison by intravenous route. Moderately toxic by ingestion. Mutation data reported. A skin irritant. Combustible liquid when exposed to heat or flame; can react with oxidizing materials. To fight fire, use water foam, fog, alcohol foam, dry chemical, CO_2. See also ALCOHOLS.

OEK000 CAS:111-86-4 ***HR: 3***
N-OCTYLAMINE
mf: $C_8H_{19}N$ mw: 129.28

PROP: Bp: 170-179°, flash p: 140°F, d: 0.779 @ 20°/20°, vap d: 4.46.

SYNS: CAPRYLAMINE ◇ CAPRYLYLAMINE ◇ 1-OCTANAMINE

TOXICITY DATA with REFERENCE
ipr-mus LD50:100 mg/kg NTIS** AD691-490

CONSENSUS REPORTS: Reported in EPA TSCA Inventory.

SAFETY PROFILE: Poison by intraperitoneal route. Administration to humans has caused headache, fall in blood pressure, rapid pulse, urticaria, and itching of skin due to release of histamine. Flammable when exposed to heat or flame, can react with oxidizing materials. To fight fire, use alcohol foam, dry chemical. Incompatible with oxidizing materials. See also AMINES and FATTY AMINES.

OEM000 CAS:5843-82-3 ***HR: 3***
N-OCTYLATROPINIUM BROMIDE
mf: $C_{25}H_{40}NO_3 \cdot Br$ mw: 482.57

SYNS: AD 122 ◇ ATROPINE OCTABROMIDE ◇ ATROPINE-N-OCTYL BROMIDE ◇ ATROPIN-N-OCTYLBROMID (GERMAN) ◇ N-n-OCTYLATROPINE BROMIDE ◇ N-n-OCTYL-ATROPINIUMBROMID (GERMAN) ◇ N-n-OCTYLATROPINIUM BROMIDE ◇ 8-OCTYLATROPINIUM BROMIDE ◇ (− +)-TROPATE-3-α-HYDROXY-8-OCTYL-1-α-H,5-α-H-TROPANIUM

TOXICITY DATA with REFERENCE
orl-mus LD50:380 mg/kg ARZNAD 7,217,57
ipr-mus LD50:102 mg/kg OYYAA2 5,813,71
scu-mus LD50:335 mg/kg ARZNAD 7,217,57
ivn-mus LD50:8200 μg/kg OYYAA2 5,813,71

SAFETY PROFILE: Poison by ingestion, intraperitoneal, subcutaneous, and intravenous routes. When heated to decomposition it emits very toxic fumes of NO_x and Br^-. See also ATROPINE and BROMIDES.

OEQ000 CAS:3575-31-3 ***HR: 3***
4-n-OCTYLBENZOIC ACID
mf: $C_{15}H_{22}O_2$ mw: 234.37

SYN: p-OCTYLBENZOIC ACID

TOXICITY DATA with REFERENCE
ivn-mus LD50:56 mg/kg CSLNX* NX#10074

CONSENSUS REPORTS: Reported in EPA TSCA Inventory.

SAFETY PROFILE: Poison by intravenous route. When heated to decomposition it emits acrid smoke and irritating fumes.

OES000 CAS:113-48-4 ***HR: 2***
N-OCTYL BICYCLOHEPTENE DICARBOXIMIDE
mf: $C_{17}H_{25}NO_2$ mw: 275.43

SYNS: BICYCLO(2.2.1)HEPTENE-2-DICARBOXYLIC ACID, 2-ETHYLHEXYLIMIDE ◇ ENDOMETHYLENETETRAHYDROPHTHALIC ACID, N-2-ETHYLHEXYL IMIDE ◇ ENT 8,184 ◇ N-(2-ETHYL-HEXYL)BICYCLO-(2,2,1)-HEPT-5-ENE-2,3-DICARBOXIMIDE ◇ N-2-ETHYLHEXYLIMIDEENDOMETHYLENETETRAHYDROPHTHALIC ACID ◇ N-(2-ETHYLHEXYL)-5-NORBORNENE-2,3-DICARBOXIMIDE ◇ 2-(2-ETHYLHEXYL)-3a,4,7,7a-TETRAHYDRO-4,7-METHANO-1H-ISOINDOLE-1,3(2H)-DIONE ◇ MGK-264 ◇ OCTACIDE 264 ◇ N-OCTYLBICYCLO-(2.2.1)-5-HEPTENE-2,3-DICARBOXIMIDE ◇ PYRODONE ◇ SYNERGIST 264 ◇ VAN DYK 264

TOXICITY DATA with REFERENCE
orl-rat TDLo:346 mg/kg (MGN):REP TXAPA9 9,555,66
orl-rat LD50:2800 mg/kg FMCHA2 -,C157,83
skn-rat LD50:470 mg/kg 30ZDA9 -,139,71
orl-mus LD50:1 g/kg YKYUA6 32,605,81
skn-rbt LD50:470 mg/kg SPEADM 78-1,5,78

SAFETY PROFILE: Moderately toxic by ingestion, skin contact, and intraperitoneal routes. Experimental reproductive effects. Large doses can cause central nervous system stimulation followed by depression. When heated to decomposition it emits toxic fumes of NO_x.

OEU000 CAS:119-07-3 ***HR: 1***
OCTYL DECYL PHTHALATE
mf: $C_{26}H_{42}O_4$ mw: 418.68

PROP: A clear liquid. Bp: 239° @ 4 mm, fp: −50°, flash p: 455°F (COC), d: 0.980 @ 20°/20°.

SYNS: 1,2-BENZENEDICARBOXYLIC ACID, DECYL OCTYL ESTER ◇ DECYL OCTYL PHTHALATE ◇ N-DECYL-N-OCTYL PHTHALATE ◇ N-OCTYL-N-DECYL PHTHALATE

TOXICITY DATA with REFERENCE
orl-rat LD50:45 g/kg AIHAAP 30,470,69

CONSENSUS REPORTS: Reported in EPA TSCA Inventory.

SAFETY PROFILE: Mildly toxic by ingestion. Combustible when exposed to heat or flame; can react vigorously with oxidizing materials. When heated to decomposition it emits acrid smoke and irritating fumes. See also ESTERS.

OEW000 CAS:959-55-7 ***HR: 1***
OCTYL-DIMETHYL-BENZYLAMMONIUM CHLORIDE
mf: $C_{17}H_{30}N{\cdot}Cl$ mw: 283.93

SYNS: BENZYLDIMETHYLOCTYLAMMONIUM CHLORIDE ◇ N,N-DIMETHYL-N-OCTYLBENZENEMETHANAMINIUMCHLORIDE

TOXICITY DATA with REFERENCE
skn-rbt 1 mg/24H OYYAA2 6,329,72
eye-rbt 1 mg OYYAA2 6,329,72

CONSENSUS REPORTS: Reported in EPA TSCA Inventory.

SAFETY PROFILE: A skin and eye irritant. When heated to decomposition it emits very toxic fumes of NO_x, NH_3, and Cl^-.

OEY000 CAS:629-82-3 ***HR: 2***
OCTYL ETHER
mf: $C_{16}H_{34}O$ mw: 242.50

PROP: D: 0.82, vap d: 8.36, bp: 300°, flash p: >212°F, autoign temp: 401°F.

SYNS: CAPRYLIC ETHER ◇ DIOCTYL ETHER

TOXICITY DATA with REFERENCE
ivn-mus LD50:1183 mg/kg JPMSAE 67,566,78

CONSENSUS REPORTS: Reported in EPA TSCA Inventory.

SAFETY PROFILE: Moderately toxic by intravenous route. Combustible when exposed to heat or flame; can react vigorously with oxidizing materials. To fight fire, use dry chemical, CO_2, water spray or mist, foam. When heated to decomposition it emits acrid smoke and irritating fumes. See also ETHERS.

OEY100 CAS:112-32-3 ***HR: 1***
OCTYL FORMATE
mf: $C_9H_{18}O_2$ mw: 158.27

PROP: Colorless liquid; fruity odor. D: 0.869, refr index: 1.418. Sol in fixed oils, propylene glycol; insol in glycerin.

SYN: FEMA No. 2809 ◇ FORMIC ACID, OCTYL ESTER ◇ n-OCTYL FORMATE

TOXICITY DATA with REFERENCE
skn-rbt 500 mg/24H MOD FCTXAV 17,883,79

CONSENSUS REPORTS: Reported in EPA TSCA Inventory.

SAFETY PROFILE: A skin irritant. When heated to decomposition it emits acrid smoke and irritating fumes.

OFA000 CAS:1034-01-1 ***HR: 1***
OCTYL GALLATE
mf: $C_{15}H_{22}O_5$ mw: 282.37

TOXICITY DATA with REFERENCE
orl-rat LD50:4700 mg/kg FOMAAB 26,99,51

CONSENSUS REPORTS: Reported in EPA TSCA Inventory.

SAFETY PROFILE: Mildly toxic by ingestion. When heated to decomposition it emits acrid smoke and irritating fumes.

OFE000 CAS:26530-20-1 ***HR: 2***
2-OCTYL-4-ISOTHIAZOLIN-3-ONE
mf: $C_{11}H_{19}NOS$ mw: 213.37

SYNS: KATHON LP PRESERVATIVE ◇ KATHON SP 70 ◇ MICRO-CHEK 11 ◇ MICRO-CHEK SKANE ◇ OCTHILINONE ◇ 2-OCTYL-3(2H)-ISOTHIAZOLONE ◇ PANCIL ◇ RH 893 ◇ SKANE M8

TOXICITY DATA with REFERENCE
skn-rbt 500 mg/24H MosJN# 15AUG79
eye-rbt 100 mg SEV MosJN# 15AUG79
orl-rat LD50:550 mg/kg MosJN# 15AUG79
skn-rbt LD50:690 mg/kg MosJN# 15AUG79

CONSENSUS REPORTS: Reported in EPA TSCA Inventory.

SAFETY PROFILE: Moderately toxic by ingestion and skin contact. A skin and severe eye irritant. A mildewcide. When heated to decomposition it emits very toxic fumes of SO_x and NO_x. See also KETONES.

OFE100 CAS:29806-73-3 ***HR: 1***
OCTYL PALMITATE
mf: $C_{24}H_{48}O_2$ mw: 368.72

SYNS: CERAPHYL 368 ◇ 2-ETHYLHEXYL PALMITATE

◇ HEXADECANOIC ACID, 2-ETHYLHEXYL ESTER (9CI) ◇ PALMITIC ACID, 2-ETHYLHEXYL ESTER ◇ WICKENOL 155

TOXICITY DATA with REFERENCE
skn-rbt 500 mg/24H MLD FCTXAV 19,251,81

CONSENSUS REPORTS: Reported in EPA TSCA Inventory.

SAFETY PROFILE: A skin irritant. When heated to decomposition it emits acrid smoke and irritating fumes.

OFG000 CAS:5249-88-7 ***HR: 3***
4-OCTYLOXY-β-(1-PIPERIDYL)PROPIOPHENONE HYDROCHLORIDE
mf: $C_{22}H_{35}NO_2 \cdot ClH$ mw: 382.04

SYN: 4'-(OCTYLOXY)-3-PIPERIDINOPROPIOPHENONEHYDROCHLORIDE

TOXICITY DATA with REFERENCE
ipr-mus LD50:35 mg/kg JPETAB 115,419,55
ivn-mus LD50:19 mg/kg JPETAB 115,419,55

SAFETY PROFILE: Poison by intraperitoneal and intravenous routes. When heated to decomposition it emits very toxic fumes of NO_x and HCl.

OFI000 CAS:7530-07-6 ***HR: 2***
OCTYLPEROXIDE
DOT: NA 2129
mf: $C_{16}H_{34}O_2$ mw: 258.50

SYN: CAPRYLYL PEROXIDE, solution (DOT)

DOT Classification: Label: Organic Peroxide.

SAFETY PROFILE: A powerful oxidizer. Probably a severe eye, skin, and mucous membrane irritant. When heated to decomposition it emits acrid smoke and irritating fumes. See also PEROXIDES, ORGANIC.

OFK000 CAS:27193-28-8 ***HR: 3***
OCTYL PHENOL
mf: $C_{14}H_{22}O$ mw: 206.36

PROP: White to light pink flakes. Bp: 280-283°, fp: 72-74°, d: 0.941 @ 24°/24°.

SYN: USAF RH-6

TOXICITY DATA with REFERENCE
ipr-mus LD50:25 mg/kg NTIS** AD277-689

CONSENSUS REPORTS: Reported in EPA TSCA Inventory.

SAFETY PROFILE: Poison by intraperitoneal route. Combustible when exposed to heat or flame; can react vigorously with oxidizing materials. When heated to decomposition it emits acrid smoke and irritating fumes. See also PHENOL.

OFM000 CAS:9036-19-5 ***HR: 2***
OCTYL PHENOL condensed with 3 MOLES ETHYLENE OXIDE
mf: $(C_2H_4O)_n \cdot C_{14}H_{22}O$

SYNS: OCTYLPHENOL EO (3) ◇ OPE-3

TOXICITY DATA with REFERENCE
eye-rbt 15 mg MLD PSTGAW 20,16,53
orl-rat LD50:4000 mg/kg PSTGAW 20,16,53

CONSENSUS REPORTS: Reported in EPA TSCA Inventory.

SAFETY PROFILE: Moderately toxic by ingestion. An eye irritant. When heated to decomposition it emits acrid smoke and irritating fumes.

OFO000 CAS:9036-19-5 ***HR: 2***
OCTYL PHENOL condensed with 5 MOLES ETHYLENE OXIDE

TOXICITY DATA with REFERENCE
orl-rat LD50:3800 mg/kg PSTGAW 20,16,53

CONSENSUS REPORTS: Reported in EPA TSCA Inventory.

SAFETY PROFILE: Moderately toxic by ingestion. When heated to decomposition it emits acrid smoke and irritating fumes.

OFQ000 CAS:9036-19-5 ***HR: 2***
OCTYL PHENOL condensed with 8-10 MOLES ETHYLENE OXIDE
mf: $(C_2H_4O)_n \cdot C_{14}H_{22}O$

SYNS: OCTYLPHENOL EO (10) ◇ TRITON X100

TOXICITY DATA with REFERENCE
orl-rat LD50:1800 mg/kg PSTGAW 20,16,53

CONSENSUS REPORTS: Reported in EPA TSCA Inventory.

SAFETY PROFILE: Moderately toxic by ingestion. When heated to decomposition it emits acrid smoke and irritating fumes.

OFS000 CAS:9036-19-5 ***HR: 2***
OCTYL PHENOL EO (16)
mf: $(C_2H_4O)_n \cdot C_{14}H_{22}O$

SYN: OCTYL PHENOL condensed with 16 MOLES ETHYLENE OXIDE

TOXICITY DATA with REFERENCE
orl-rat LD50:2800 mg/kg TXAPA9 5,782,63

CONSENSUS REPORTS: Reported in EPA TSCA Inventory.

SAFETY PROFILE: Moderately toxic by ingestion.

When heated to decomposition it emits acrid smoke and irritating fumes.

OFU000 CAS:9036-19-5 **HR: 2**
OCTYL PHENOL EO (20)

SYN: OCTYL PHENOL condensed with 20 MOLES ETHYLENE OXIDE

TOXICITY DATA with REFERENCE
orl-rat LD50:3600 mg/kg TXAPA9 5,782,63

CONSENSUS REPORTS: Reported in EPA TSCA Inventory.

SAFETY PROFILE: Moderately toxic by ingestion. When heated to decomposition it emits acrid smoke and irritating fumes.

OFU100 CAS:7425-87-8 **HR: 2**
1-OCTYL-2-PYRROLIDINONE
mf: $C_{12}H_{23}NO$ mw: 197.36

PROP: Mp: -23°, d: 0.920, flash p: > 230° F. /SH,3 SYNS: N-OCTYLPYRROLIDINONE ◇ N-OCTYLPYRROLIDONE ◇ 2-PYRROLIDINONE, 1-OCTYL-

TOXICITY DATA with REFERENCE
skn-rbt 500 mg/24H SEV FCTOD7 26,475,88
eye-rbt 100 mg SEV FCTOD7 26,475,88
orl-rat LD50:2050 mg/kg FCTOD7 26,475,88

SAFETY PROFILE: Moderately toxic by ingestion. A severe skin and eye irritant. A corrosive. Combustible liquid. When heated to decomposition it emits toxic fumes of NO_x.

OGA000 CAS:3547-33-9 **HR: 1**
2-(OCTYLTHIO)ETHANOL
mf: $C_{10}H_{22}OS$ mw: 190.38

SYNS: 2-HYDROXYETHYL-n-OCTYL SULFIDE ◇ MGK REPELLENT 874 ◇ R-874 ◇ R-874 PHILLIPS

TOXICITY DATA with REFERENCE
orl-rat LD50:8530 mg/kg FMCHA2 -,C157,83
skn-rbt LD50:13590 mg/kg 28ZEAL 5,196,76

CONSENSUS REPORTS: Reported in EPA TSCA Inventory.

SAFETY PROFILE: Mildly toxic by ingestion and skin contact. An insect repellent. When heated to decomposition it emits toxic fumes of SO_x.

OGE000 CAS:5283-66-9 **HR: 2**
OCTYLTRICHLOROSILANE
DOT: UN 1801
mf: $C_8H_{17}Cl_3Si$ mw: 247.69

PROP: Fuming liquid.

CONSENSUS REPORTS: Reported in EPA TSCA Inventory.

DOT Classification: Corrosive Material; Label: Corrosive.

SAFETY PROFILE: A corrosive irritant to skin, eyes, and mucous membranes. Will react with water or steam to produce toxic and corrosive fumes. When heated to decomposition it emits toxic fumes of Cl^-. See also SILANES and CHLOROSILANES.

OGG000 CAS:3091-25-6 **HR: 2**
OCTYLTRICHLOROSTANNANE
mf: $C_8H_{17}Cl_3Sn$ mw: 338.29

SYNS: MONO-N-OCTYLTIN TRICHLORIDE ◇ MONO-N-OCTYL-ZINN-TRICHLORID (GERMAN)

TOXICITY DATA with REFERENCE
orl-rat LD50:4600 mg/kg ARZNAD 19,934,69
unr-rat LD50:3800 mg/kg TIUSAD 107,1,76

CONSENSUS REPORTS: Reported in EPA TSCA Inventory.

OSHA PEL: TWA 0.1 mg(Sn)/m^3 (skin)
ACGIH TLV: TWA 0.1 mg(Sn)/m^3 (skin) (Proposed: TWA 0.1 mg(Sn)/m^3; STEL 0.2 mg(Sn)/m^3 (skin))
NIOSH REL: (Organotin Compounds) TWA 0.1 mg (Sn)/m^3

SAFETY PROFILE: Moderately toxic by an unspecified route. Mildly toxic by ingestion. When heated to decomposition it emits toxic fumes of Cl^-. See also TIN COMPOUNDS.

OGI000 CAS:27107-89-7 **HR: 2**
OCTYLTRIS(2-ETHYLHEXYLOXYCARBONYL-METHYLTHIO)STANNANE
mf: $C_{38}H_{74}O_6S_3Sn$ mw: 841.99

TOXICITY DATA with REFERENCE
orl-mus LD50:1500 mg/kg ATXKA8 26,196,70

CONSENSUS REPORTS: Reported in EPA TSCA Inventory.

OSHA PEL: TWA 0.1 mg(Sn)/m^3 (skin)
ACGIH TLV: TWA 0.1 mg(Sn)/m^3 (skin) (Proposed: TWA 0.1 mg(Sn)/m^3; STEL 0.2 mg(Sn)/m^3 (skin))
NIOSH REL: (Organotin Compounds) TWA 0.1 mg(Sn)/m^3

SAFETY PROFILE: Moderately toxic by ingestion. When heated to decomposition it emits toxic fumes of SO_x. See also TIN COMPOUNDS.

OGI100 CAS:20436-27-5 ***HR: 3***
ODODIBORANE
mf: B_2H_5I mw: 153.56

SAFETY PROFILE: Violent ignition occurs spontaneously in air. When heated to decomposition it emits acrid smoke and irritating fumes. See also BORANES and BORON COMPOUNDS.

OGI300 CAS:82419-36-1 ***HR: 3***
OFLOXACIN
mf: $C_{18}H_{20}FN_3O_4$ mw: 361.41

PROP: Colorless needles from ethanol. Mp: 250-257° (decomp).

SYNS: 2,3-DIHYDRO-9-FLUORO-3-METHYL-10-(4-METHYL-1-PIPERAZINYL)-7-OXO-7H-PYRIDO(1,2,3-de)-1,4-BENZOXAZINE-6-CARBOXYLIC ACID ◇ DL-820 ◇ DL-8280 ◇ OXALDIN ◇ TARIVID

TOXICITY DATA with REFERENCE
dnd-hmn:lym 80 mg/L MUREAV 211,171,89
orl-rat TDLo:1620 mg/kg (female 9-10D post):REP ARZNAD 36,1244,86
orl-rbt TDLo:2080 mg/kg (female 6-18D post):TER ARZNAD 36,1244,86
orl-rat LD50:3590 mg/kg NKRZAZ 32(Suppl 1),1084,84
scu-rat LD50:7070 mg/kg NKRZAZ 32(Suppl 1),1084,84
ivn-rat LD50:273 mg/kg NKRZAZ 32(Suppl 1),1084,84
orl-mus LD50:5290 mg/kg NKRZAZ 32(Suppl 1),1084,84
scu-mus LDLo:7690 mg/kg NKRZAZ 32(Suppl 1),1084,84
ivn-mus LD50:208 mg/kg NKRZAZ 32(Suppl 1),1084,84
ivn-dog LDLo:100 mg/kg NKRZAZ 32(Suppl 1),1084,84
orl-mky LD50:500 mg/kg NKRZAZ 32(Suppl 1),1084,84

SAFETY PROFILE: Poison by intravenous route. Moderately toxic by ingestion. An experimental teratogen. Other experimental reproductive effects. Mutation data reported. When heated to decomposition it emits toxic fumes of F^- and NO_x.

OGK000 CAS:8015-79-0 ***HR: 3***
OIL of CALAMUS, GERMAN

PROP: Extract of *Acorus calamus L., araceae*. Containing: asarone, eugenol; esters of acetic and heptylic acids. Volatile oil. Yellow to yellowish-brown liquid (viscid); aromatic odor, bitter taste. D: 0.960-0.9707 @ 20°/20°. Very sltly sol in water; misc with alc. Keep well closed, cool, and protected from light.

SYNS: CALAMUS OIL ◇ KALMUS OEL (GERMAN) ◇ OIL of SWEET FLAG

TOXICITY DATA with REFERENCE
orl-rat TDLo:10 g/kg/59W-C:ETA PAACA3 8,24,67
orl-rat LD50:777 mg/kg FCTXAV 2,327,64
ipr-rat LD50:221 mg/kg FCTXAV 15,623,77

CONSENSUS REPORTS: Reported in EPA TSCA Inventory.

SAFETY PROFILE: Poison by intraperitoneal route. Moderately toxic by ingestion. Questionable carcinogen with experimental tumorigenic data. When heated to decomposition it emits acrid smoke and irritating fumes. See also individual components.

OGL020 ***HR: 3***
OIL of CALAMUS, SPANISH

PROP: Oil extracted from the Indian plant belonging to the family *Araceae* and having an asarone content of nearly 80% (ANTCAO 7,378,57).

SYNS: ACORUS CALAMUS OIL ◇ ACORUS CALAMUS Linn., oil extract ◇ ESSENTIAL OIL of ACORUS CALAMUS Linn. ◇ OIL of ACORUS CALAMUS Linn. ◇ OIL de ACORUS CALAMUS (SPANISH)

TOXICITY DATA with REFERENCE
ipr-rat LD50:221 mg/kg JPETAB 125,353,59
ipr-mus LD50:177 mg/kg JPPMAB 11,163,59
ipr-gpg LD50:297 mg/kg ANTCAO 7,378,57

SAFETY PROFILE: Poison by intraperitoneal route.

OGM000 ***HR: 3***
OIL GAS

PROP: A gas derived from petroleum. Composition: illuminants 4.2%, carbon monoxide 10.4%, hydrogen 47.6%, methane 27.0%, carbon dioxide 4.6%, nitrogen 5.8%, oxygen 0.4%. Lel: 4.8%; uel: 32.5%; autoign temp: 637°F.

SAFETY PROFILE: A poison. A very dangerous fire hazard when exposed to heat or flame; can react vigorously with oxidizing materials. Explosive in the form of vapor when exposed to heat or flame. To fight fire, use CO_2, dry chemical, water spray. See also CARBON MONOXIDE.

OGO000 CAS:8008-26-2 ***HR: 2***
OIL of LIME, distilled

PROP: From distillation of juice or crushed fruit of *Citrus aurantofolia* Swingle. Colorless to green-yellow liquid. Sol in fixed oils, mineral oil; insol glycerin, propylene glycol.

SYNS: DISTILLED LIME OIL ◇ LIME OIL ◇ LIME OIL, distilled (FCC) ◇ OILS, LIME

TOXICITY DATA with REFERENCE
skn-rbt 500 mg/24H MLD FCTXAV 12,729,74
dnr-bcs 20 mg/disc TOFOD5 8,91,85
orl-mus TDLo:67 g/kg/39W-I:ETA JNCIAM 35,771,65

CONSENSUS REPORTS: Reported in EPA TSCA Inventory.

SAFETY PROFILE: A skin irritant. Questionable carcinogen with experimental tumorigenic data. Mutation data reported. When heated to decomposition it emits acrid smoke and irritating fumes.

OGQ100 CAS:8007-12-3 ***HR: 2***
OIL of MACE

PROP: From steam distillation of dried arillode of the ripe seed of *Myristica fragrans* Houtt. (Fam. *Myristicaceae*). Colorless to pale yellow liquid; odor and taste of nutmeg. East Indian: d: 0.880-0.930, refr index: 1.474-1.488; West Indian: d: 0.854-0.880, refr index: 1.469-1.480 @20°. Sol in fixed oils, mineral oil; sltly sol in cold alc; very sol in hot alc, chloroform, ether; insol in glycerin, propylene glycol.

SYNS: NCI-C56484 ◇ MACE OIL ◇ OIL of NUTMEG, expressed

TOXICITY DATA with REFERENCE
skn-rbt 500 mg/24H MOD FCTXAV 17,851,79
orl-rat LD50:3640 mg/kg FCTXAV 2,327,64

CONSENSUS REPORTS: Reported in EPA TSCA Inventory. EPA Genetic Toxicology Program.

SAFETY PROFILE: Moderately toxic by ingestion. A skin irritant. Human ingestion causes symptoms similar to volatile oil of nutmeg. When heated to decomposition it emits acrid smoke and irritating fumes.

OGS000 ***HR: 2***
OIL of MUSTARD, EXPRESSED mixed with OIL of ARGEMONE

PROP: Mustard oil mixed with 0.5% argemone oil (IJMRAQ 61,428,73).

SYNS: ARGEMONE OIL mixed with MUSTARD OIL ◇ OIL of ARGEMONE mixed with OIL of MUSTARD

TOXICITY DATA with REFERENCE
skn-mus TDLo:2564 g/kg/92W-C:ETA IJCNAW 10,652,72

SAFETY PROFILE: Questionable carcinogen with experimental tumorigenic data. When heated to decomposition it emits acrid smoke and irritating fumes.

OGU000 CAS:8008-46-6 ***HR: 2***
OIL of MYRTLE

PROP: Volatile oil from leaves of *Myrtus communis L.* (TXAPA9 45,264,78). Yellow to greenish liquid; fragrant odor. D: 0.890-0.915 @ 15°/15°. Insol in water; sol in alc, chloroform, ether.

SYNS: ESSENTIAL OIL from MYRTLE ◇ MYRTLE OIL

TOXICITY DATA with REFERENCE
skn-rbt 500 mg/24H MOD FCTOD7 21,869,83
orl-rat LD50:3800 mg/kg TXAPA9 45,264,78
orl-mus LD50:2230 mg/kg TXCYAC 12,335,79

CONSENSUS REPORTS: Reported in EPA TSCA Inventory.

SAFETY PROFILE: Moderately toxic by ingestion. A skin irritant. When heated to decomposition it emits acrid smoke and irritating fumes.

OGY000 CAS:8008-57-9 ***HR: 2***
OIL of ORANGE

PROP: Yellow to deep-orange liquid; characteristic orange taste and odor. D: 0.842-0.846 @ 25°/25°, refr index: 1.472 @ 20°. Sol in 2 vols 90% alc, in 1 vol glacial acetic acid; sltly sol in water; misc with abs alc, carbon disulfide. Keep well closed, cool, and protected from light. Oil expressed from the peel of *Citrus sinensis* L. Osbeck (Fam. *Rutaceae*) (BJCAAI 13,92,59).

SYNS: NEAT OIL of SWEET ORANGE ◇ OIL of SWEET ORANGE ◇ ORANGE OIL ◇ ORANGE OIL, coldpressed (FCC) ◇ SWEET ORANGE OIL

TOXICITY DATA with REFERENCE
skn-rbt 500 mg/24H MOD FCTXAV 12,733,74
orl-mus TDLo:67 g/kg/40W-I:NEO JNCIAM 35,771,65

CONSENSUS REPORTS: Reported in EPA TSCA Inventory.

SAFETY PROFILE: A skin irritant. Questionable carcinogen with experimental neoplastigenic data. When heated to decomposition it emits acrid smoke and irritating fumes.

OHA000 CAS:85-86-9 ***HR: 3***
OIL RED
mf: $C_{22}H_{16}N_4O$ mw: 352.42

SYNS: BENZENEAZOBENZENEAZO-β-NAPHTHOL ◇ CERASINROT ◇ C.I. SOLVENT RED 23 ◇ D&C RED No. 17 ◇ FETTSCHARLACH ◇ OIL SCARLET ◇ ORGANOL SCARLET ◇ 1-((4-(PHENYLAZO)PHENYL)AZO)-2-NAPHTHALENOL ◇ 1-(p-PHENYLAZOPHENYLAZO)-2-NAPHTHOL ◇ ROUGE CERASINE ◇ SOMALIA RED III ◇ SUDAN III ◇ TETRAZOBENZENE-β-NAPHTHOL ◇ TONY RED

TOXICITY DATA with REFERENCE
cyt-ham:ovr 20 μmol/L/5H-C ENMUDM 1,27,79
ipr-rbt LDLo:250 mg/kg JPBAA7 87,317,64
scu-rbt LDLo:1000 mg/kg JPBAA7 87,317,64
ipl-rbt LDLo:500 mg/kg JPBAA7 87,317,64

CONSENSUS REPORTS: IARC Cancer Review: Group 3 IMEMDT 7,56,87; Animal Inadequate Evidence IMEMDT 8,241,75. Reported in EPA TSCA Inventory.

SAFETY PROFILE: Poison by intraperitoneal route. Moderately toxic by subcutaneous and intrapleural routes. Questionable carcinogen. Mutation data reported. When heated to decomposition it emits toxic fumes of NO_x.

OHC000 ***HR: 1***
OIL ROSE TURKISH

PROP: From the flowers of *Rose damascena,* includes citronellol and geraniol (FCTXAV 13,681,75).

SYNS: ATTAR ROSE TURKISH ◇ OTTO ROSE TURKISH ◇ TURKISH ROSE OIL

TOXICITY DATA with REFERENCE
skn-rbt 500 mg/24H MOD FCTXAV 13,681,75

SAFETY PROFILE: A skin irritant. When heated to decomposition it emits acrid smoke and irritating fumes.

OHE000 CAS:8014-29-7 ***HR: 2***
OIL of RUE

PROP: Volatile oil from *Ruta graveolens L., Rutacene.* Pale yellow liquid; characteristic, sharp, unpleasant odor. D: 0.832-0.845 @ 15°/15°. Solidifies at 8-10°. Very sltly sol in water; sol in 3 vols 70% alc. Keep well closed, cool, and protected from light.

TOXICITY DATA with REFERENCE
skn-rbt 500 mg/2H MLD FCTXAV 13,455,75
orl-mus LD50:2070 mg/kg FCTXAV 13,455,75

CONSENSUS REPORTS: Reported in EPA TSCA Inventory.

SAFETY PROFILE: Moderately toxic by ingestion. A skin irritant. Frequent skin contact causes erythema, vesication. Ingestion of large quantities produces epigastric pain, nausea, vomiting, confusion, convulsions, death; may terminate pregnancy. When heated to decomposition it emits acrid smoke and irritating fumes.

OHG000 CAS:115-71-9 ***HR: 2***
OIL of SANDALWOOD, EAST INDIAN
mf: $C_{15}H_{24}O$ mw: 220.39

PROP: From steam distillation of the ground dried wood of *Santalus album L.* (FCTXAV 12,807,74). Colorless to sltly yellow viscous liquid; sandalwood odor. D: 0.965-0.973, refr index: 1.505. Very sol in alc, fixed oils, propylene glycol; in in water, glycerin.

SYNS: 5-(2,3-DIMETHYLTRICYCLO(2.2.1.$0^{2,6}$)HEPT-3-YL)-2-METHYL-2-PENTEN-1-OL ◇ FEMA No. 3006 ◇ SANDALWOOD OIL, EAST INDIAN ◇ α-SANTALOL (FCC)

TOXICITY DATA with REFERENCE
skn-mus 500 mg MLD FCTXAV 12,807,74
skn-rbt 500 mg/24H FCTXAV 12,807,74
orl-rat LD50:3800 mg/kg FCTXAV 12,807,74

CONSENSUS REPORTS: Reported in EPA TSCA Inventory.

SAFETY PROFILE: Moderately toxic by ingestion. A skin irritant. When heated to decomposition it emits acrid smoke and irritating fumes.

OHI000 CAS:8006-80-2 ***HR: 3***
OIL of SASSAFRAS

PROP: Yellow to reddish-yellow liquid; characteristic odor and taste of sassafras. D: 1.065-1.077 @ 25°/25°. Very sltly sol in water; sol in 2 vols 90% alc. Keep well closed, cool, and protected from light. 80% Safrol (27ZTAP 3,106,69).

TOXICITY DATA with REFERENCE
orl-man LDLo:83 mg/kg ADCHAK 28,475,53

CONSENSUS REPORTS: Reported in EPA TSCA Inventory.

SAFETY PROFILE: Human poison by ingestion. When heated to decomposition it emits acrid smoke and irritating fumes.

OHI875 CAS:2481-94-9 ***HR: D***
OIL YELLOW DEA
mf: $C_{16}H_{19}N_3$ mw: 253.38

SYNS: C-299 ◇ CERES YELLOW GGN ◇ C.I. 11021 ◇ C.I. SOLVENT YELLOW 56 ◇ p-(DIETHYLAMINO)AZOBENZENE ◇ 4-(DIETHYLAMINO)AZOBENZENE ◇ N,N-DIETHYL-4-AMINOAZOBENZENE ◇ N,N-DIETHYL-p-(PHENYLAZO)ANILINE ◇ N,N-DIETHYL-4-(PHENYLAZO)BENZENAMINE ◇ DIETHYL YELLOW ◇ OIL YELLOW 2635 ◇ OIL YELLOW DE ◇ OIL YELLOW E190 ◇ OIL YELLOW ENC ◇ OIL YELLOW GA ◇ OIL YELLOW NB ◇ SUDAN YELLOW GGN ◇ WAXOLINE YELLOW ED

TOXICITY DATA with REFERENCE
mma-sat 1 μg/plate EPASR* 8EHQ-0982-0455
otr-ham:kdy 250 μg/L BJCAAI 38,34,78

CONSENSUS REPORTS: EPA Genetic Toxicology Program. Reported in EPA TSCA Inventory.

SAFETY PROFILE: Mutation data reported. When heated to decomposition it emits toxic fumes of NO_x.

OHK000 CAS:6370-43-0 ***HR: 2***
OIL YELLOW HA
mf: $C_{14}H_{14}N_2O$ mw: 226.30

SYNS: C.I. 11860 ◇ C.I. SOLVENT YELLOW 12 ◇ OIL YELLOW OPS ◇ OLEAL YELLOW RE

TOXICITY DATA with REFERENCE
scu-mus TDLo:6000 mg/kg/57W-I:ETA GMJOAZ 30,364,49

SAFETY PROFILE: Questionable carcinogen with ex-

perimental tumorigenic data. When heated to decomposition it emits toxic fumes of NO_x.

OHM700 CAS:112-90-3 ***HR: 2***
OLEAMINE
mf: $C_{18}H_{37}N$ mw: 267.56

SYNS: ALAMINE 11 ◇ ARMEEN O ◇ KEMAMINE P 989 ◇ NORAM O ◇ cis-9-OCTADECENYLAMINE ◇ OLEINAMINE ◇ OLEYL AMINE ◇ OLEYLAMIN (GERMAN)

TOXICITY DATA with REFERENCE
orl-mus TDLo:800 mg/kg (9D preg):REP DZZEA7 32,861,77
ipr-mus TDLo:400 mg/kg (9D preg):TER DZZEA7 32,861,77
ipr-mus LD50:888 mg/kg DZZEA7 35,1070,80

CONSENSUS REPORTS: Reported in EPA TSCA Inventory.

SAFETY PROFILE: Moderately toxic by intraperitoneal route. An experimental teratogen. Experimental reproductive effects. When heated to decomposition it emits toxic fumes of NO_x. See also AMINES.

OHM875 ***HR: 3***
OLEANDER

PROP: A shrub which grows to 20 feet. The leaves are narrow, about 10 inches long and are usually in groups of 3. The red, pink or white flowers grow in small clusters. The thin, 5-inch long seed capsules contain seeds with fluffy wings. It is native to the Mediterranean region, but is cultivated outdoors in warm areas and also is grown as a house plant.

SYNS: ADELFA (PUERTO RICO) ◇ ALHELI EXTRANJERO (PUERTO RICO) ◇ LAURIER ROSE (HAITI) ◇ NERIUM OLEANDER ◇ 'OLEANA (HAWAII) ◇ 'OLINANA (HAWAII) ◇ 'OLIWA (HAWAII) ◇ ROSA FRANCESA (CUBA) ◇ ROSE BAY ◇ ROSE LAUREL (MEXICO)

SAFETY PROFILE: Poisonous cardiac glycosides are found in all parts of the plant, smoke from burning plants, and water which has been used to soak the plants. Human systemic effects by ingestion include: mouth pain, nausea, vomiting, abdominal pain and cramps, diarrhea. Cardiac glycosides may cause death by their effect on heart function. See also OLEANDRIN and DIGITALIS.

OHO000 CAS:6696-47-5 ***HR: 3***
OLEANDOMYCIN HYDROCHLORIDE
mf: $C_{35}H_{61}NO_{12}{\bullet}ClH$ mw: 724.43

PROP: Long needles from ethyl acetate. Mp: 134-135°. Very sol in water.

SYN: OLEANDOMYCINMONOHYDROCHLORIDE

TOXICITY DATA with REFERENCE
scu-rat LD50:10000 mg/kg ARZNAD 8,386,58
ivn-rat LD50:376 mg/kg ANTCAO 7,419,57
orl-mus LD50:4000 mg/kg ANTCAO 7,419,57
scu-mus LD50:2500 mg/kg ARZNAD 8,386,58
ivn-mus LD50:550 mg/kg 85ERAY 1,30,78

SAFETY PROFILE: Poison by intravenous route. Moderately toxic by ingestion and subcutaneous routes. When heated to decomposition it emits very toxic fumes of NO_x and HCl.

OHO200 CAS:7060-74-4 ***HR: 3***
OLEANDOMYCIN PHOSPHATE
mf: $C_{35}H_{61}NO_{12}{\bullet}H_3O_4P$ mw: 785.97

SYN: MATROMYCIN

TOXICITY DATA with REFERENCE
ivn-rat LD50:480 mg/kg ANTCAO 7,419,57
orl-mus LD50:4 g/kg NIIRDN 6,164,82
scu-mus LD50:1 g/kg NIIRDN 6,164,82
ivn-mus LD50:400 mg/kg NIIRDN 6,164,82

SAFETY PROFILE: Poison by intravenous route. Moderately toxic by ingestion and subcutaneous routes. When heated to decomposition it emits toxic fumes of PO_x and NO_x. See also PHOSPHATES.

OHQ000 CAS:465-16-7 ***HR: 3***
OLEANDRIN
mf: $C_{32}H_{48}O_9$ mw: 576.80

PROP: From the leaves of *Nerium oleander L., Apocynaceae (Laurier rose)*. Crystals from dil methanol. Mp: 250°. Practically insol in water; sol in alc, chloroform.

SYNS: CORRIGEN ◇ FOLIANDRIN ◇ FOLINERIN ◇ FOLINEVIN ◇ NERIOL ◇ NERIOLIN ◇ NERIOSTENE ◇ OLEANDRINE

TOXICITY DATA with REFERENCE
ivn-cat LD50:300 μg/kg MEIEDD 10,979,83
scu-frg LDLo:2250 μg/kg 27ZWAY E.1,78,-

SAFETY PROFILE: Poison by subcutaneous and intravenous routes. When heated to decomposition it emits acrid smoke and irritating fumes.

OHQ300 CAS:50657-29-9 ***HR: D***
OLEANOGLYCOTOXIN A
mf: $C_{48}H_{78}O_{18}$ mw: 943.26

TOXICITY DATA with REFERENCE
iut-rat TDLo:2500 μg/kg (female 1D post):REP CCPTAY 14,39,76

SAFETY PROFILE: Experimental reproductive effects. When heated to decomposition it emits acrid smoke and irritating fumes.

OHS000 ***HR: 2***
OLEFINS

PROP: Unsaturated aliphatic hydrocarbons having one or more double bonds.

SAFETY PROFILE: Unsaturated aliphatic hydrocarbons do not differ greatly from paraffins, particularly insofar as their toxic effect on working personnel is concerned. Ethylene and some of its homologs occur in manufactured and natural gases. Ethylene can be used as an anesthetic, and on inhalation in sufficient quantity it can be an asphyxiant. However, the greatest hazard from its use is the danger of fire and explosion. Prolonged or repeated exposures to high concentrations of various olefins have caused certain toxic effects in animals, such as liver damage and hyperplasia of the bone marrow (due to butene-2), but no corresponding effects have been discovered in human beings due to industrial exposures. The diolefins, butadiene and isoprene, are more irritating than paraffins or mono-olefins of the same volatility. The α-olefins (e.g., 1-octene, 1-octadecene) are particularly reactive because the double bond is on the first carbon. In general the olefins have comparatively low toxicity, but are fire and explosion hazards.

OHU000 CAS:112-80-1 ***HR: 3***
OLEIC ACID
mf: $C_{18}H_{34}O_2$ mw: 282.52

$$CH_3(CH_2)_7CH{=}CH(CH_2)_7CO{\bullet}OH$$

PROP: Colorless liquid; odorless when pure. Mp: 6°, bp: 360.0°, flash p: 372°F (CC), d: 0.895 @ 25°/25°, autoign temp: 685°F, vap press: 1 mm @ 176.5°, bp: 286° @ 100 mm. Insol in water; misc in alc and ether.

SYNS: CENTURY CD FATTY ACID ◇ EMERSOL 210 ◇ EMERSOL 213 ◇ EMERSOL 6321 ◇ EMERSOL 233LL ◇ EMERSOL 221 LOW TITER WHITE OLEIC ACID ◇ EMERSOL 220 WHITE OLEIC ACID ◇ GLYCON RO ◇ GLYCON WO ◇ GROCO 2 ◇ GROCO 4 ◇ GROCO 5L ◇ HY-PHI 1055 ◇ HY-PHI 1088 ◇ HY-PHI 2066 ◇ HY-PHI 2088 ◇ HY-PHI 2102 ◇ INDUSTRENE 105 ◇ INDUSTRENE 205 ◇ INDUSTRENE 206 ◇ K 52 ◇ l'ACIDE OLEIQUE (FRENCH) ◇ METAUPON ◇ NEO-FAT 90-04 ◇ NEO-FAT 92-04 ◇ cis-Δ^9-OCTADECENOIC ACID ◇ cis-OCTADEC-9-ENOIC ACID ◇ cis-9-OCTADECENOIC ACID ◇ 9,10-OCTADECENOIC ACID ◇ PAMOLYN ◇ RED OIL ◇ TEGO-OLEIC 130 ◇ VOPCOLENE 27 ◇ WECOLINE OO ◇ WOCHEM No. 320

TOXICITY DATA with REFERENCE
skn-hmn 15 mg/3D-I MOD 85DKA8 -,127,77
skn-rbt 500 mg open MLD UCDS** 11/29/63
cyt-smc 100 mg/L NATUAS 294,263,81
dns-mus-rec 35 mg/kg CALEDQ 23,253,84
cyt-ham:fbr 2500 μg/L CRNGDP 3,499,82
scu-rbt TDLo:390 mg/kg/17W-I:ETA CRSBAW 137,760,43
orl-rat LD50:74 g/kg UCDS** 11/29/63
ivn-rat LD50:2400 μg/kg AJPAA4 103,376,81
ivn-mus LD50:230 mg/kg APTOA6 18,141,61

CONSENSUS REPORTS: Reported in EPA TSCA Inventory.

SAFETY PROFILE: Poison by intravenous route. Mildly toxic by ingestion. Mutation data reported. A human and experimental skin irritant. Questionable carcinogen with experimental tumorigenic data. Combustible when exposed to heat or flame. To fight fire, use CO_2, dry chemical. The peroxidized acid explodes on contact with aluminum. Potentially dangerous reaction with perchloric acid + heat. When heated to decomposition it emits acrid smoke and irritating fumes.

OHW000 CAS:112-62-9 ***HR: 2***
cis-OLEIC ACID, METHYL ESTER
mf: $C_{19}H_{36}O_2$ mw: 296.55

PROP: Oil. D: 0.874 @ 20°/4°, bp: 168-170°. Insol in water; misc in alc and ether.

SYNS: EMEREST 2301 ◇ EMEREST 2801 ◇ EMERY 2219 ◇ EMERY 2310 ◇ EMERY OLEIC ACID ESTER 2301 ◇ KEMESTER 105 ◇ KEMESTER 115 ◇ KEMESTER 205 ◇ KEMESTER 213 ◇ METHYL-9-OCTADECENOATE ◇ METHYL cis-9-OCTADECENOATE ◇ METHYL (Z)-9-OCTADECENOATE ◇ METHYL OLEATE ◇ (Z)-9-OCTADECENOIC ACID METHYL ESTER

TOXICITY DATA with REFERENCE
skn-mus TDLo:54 g/kg/45W-I:ETA AMBPBZ 82,127,74

CONSENSUS REPORTS: Reported in EPA TSCA Inventory.

SAFETY PROFILE: Questionable carcinogen with experimental tumorigenic data by skin contact. When heated to decomposition it emits acrid smoke and irritating fumes.

OHY000 CAS:143-18-0 ***HR: 1***
OLEIC ACID, POTASSIUM SALT
mf: $C_{18}H_{34}O_2{\bullet}K$ mw: 321.62

SYNS: POTASSIUM cis-9-OCTADECENOIC ACID ◇ POTASSIUM OLEATE

TOXICITY DATA with REFERENCE
eye-rbt 12 mg/48H JANCA2 56,905,73

CONSENSUS REPORTS: Reported in EPA TSCA Inventory.

SAFETY PROFILE: An eye irritant. When heated to decomposition it emits toxic fumes of K_2O.

OIA000 CAS:143-19-1 ***HR: 3***
OLEIC ACID, SODIUM SALT
mf: $C_{18}H_{33}O_2{\bullet}Na$ mw: 304.50

PROP: White powder; slt tallow odor. Mp: 232-235°.

SYNS: EUNATROL ◇ SODIUM OLEATE

TOXICITY DATA with REFERENCE
ivn-mus LD50:152 mg/kg RPOBAR 2,327,70
ivn-rbt LDLo:150 mg/kg JPETAB 24,221,25

CONSENSUS REPORTS: Reported in EPA TSCA Inventory.

SAFETY PROFILE: Poison by intravenous route. Migrates to food from packaging materials. Combustible when exposed to heat or flame. When heated to decomposition it emits toxic fumes of Na_2O.

OIC000 CAS:63021-11-4 ***HR: 2***
1-OLEOYLAZIRIDINE
mf: $C_{20}H_{37}NO$ mw: 307.58

SYN: OLEOYLETHYLENEIMINE

TOXICITY DATA with REFERENCE
scu-rat TDLo:1700 mg/kg/49D-I:NEO BJPCAL 9,306,54

SAFETY PROFILE: Questionable carcinogen with experimental neoplastigenic data. When heated to decomposition it emits toxic fumes of NO_x.

OIG000 CAS:9004-98-2 ***HR: 1***
OLEYL ALCOHOL condensed with 2 moles ETHYLENE OXIDE
mf: $(C_2H_4O)_n \cdot C_{18}H_{36}O$

PROP: A polyoxyethylene alkyl ether of fatty alcohols (FCTXAV 2,509,64).

SYNS: BRIJ 92 ◇ BRIJ 92((2)-OLEYL) ◇ (Z)-α-9-OCTADECENYL-omega-HYDROXYPOLY(OCY-1,2-ETHANEDIYL)◇ OLEYL ALCOHOL EO (2)

TOXICITY DATA with REFERENCE
orl-rat LD50:25800 mg/kg SPCOAH 38,47,65

CONSENSUS REPORTS: Reported in EPA TSCA Inventory.

SAFETY PROFILE: Mildly toxic by ingestion. When heated to decomposition it emits acrid smoke and irritating fumes. See also ETHERS.

OIG040 CAS:9004-98-2 ***HR: 2***
OLEYL ALCOHOL condensed with 10 moles ETHYLENE OXIDE
mf: $(C_2H_4O)n \cdot C_{18}H_{36}O$

SYNS: BRIJ 96((10) OLEYL) ◇ DECAETHOXY OLEYL ETHER ◇ (Z)-α-9-OCTADECENYL-omega-HYDROXYPOLY(OXY-1,2-ETHANEDIYL) ◇ OLEYL ALCOHOL EO (10) ◇ POLYETHYLENE GLYCOL MONOLEYL ETHER ◇ POLYOXYL 10 OLEYL ETHER

TOXICITY DATA with REFERENCE
skn-rbt 500 mg/24H MOD TXAPA9 19,276,71
eye-rbt 100 mg TXAPA9 19,276,71
orl-rat LD50:2700 mg/kg SPCOAH 38,47,65

CONSENSUS REPORTS: Reported in EPA TSCA Inventory.

SAFETY PROFILE: Moderately toxic by ingestion. A skin and eye irritant. When heated to decomposition it emits acrid smoke. See also ETHERS.

OII200 CAS:7333-84-8 ***HR: 2***
OLEYLAMINE HYDROFLUORIDE
mf: $C_{18}H_{37}N \cdot FH$ mw: 287.57

SYNS: OLEYLAMINE-HF ◇ OLEYLAMINHYDROFLUORID (GERMAN)

TOXICITY DATA with REFERENCE
orl-mus TDLo:800 mg/kg (9D preg):REP DZZEA7 32,861,77
ipr-mus TDLo:200 mg/kg (9D preg):TER DZZEA7 32,861,77
ipr-mus LD50:459 mg/kg DZZEA7 35,1070,80

SAFETY PROFILE: Moderately toxic by intraperitoneal route. An experimental teratogen. Experimental reproductive effects. When heated to decomposition it emits toxic fumes of NO_x and HF. See also FLUORIDES and AMINES.

OIK000 CAS:9004-98-2 ***HR: 3***
OLEYLPOLYOXETHYLENE GLYCOL ETHER
mf: $C_{58}H_{116}O_{21}$ mw: 1149.74

TOXICITY DATA with REFERENCE
ipr-rat LD50:235 mg/kg JAPMA8 42,556,53

CONSENSUS REPORTS: Glycol ether compounds are on the Community Right-To-Know List. Reported in EPA TSCA Inventory.

SAFETY PROFILE: Poison by intraperitoneal route. Some glycol ether compounds may have dangerous human reproductive effects. When heated to decomposition it emits acrid smoke and irritating fumes. See also GLYCOL ETHERS.

OIM000 CAS:8050-07-5 ***HR: 1***
OLIBANUM GUM

PROP: Contains 3-8% volatile oil (pinene, dipentene, etc.), 60% resins, 20% gum (polysaccharide fraction) and 6-8% bassorin (FCTXAV 16,637,78). A gum from the trees *Boswellia carterii* Birdw. and other *Boswellia* species (Fam. *Burseraceae*).

SYN: FRANKINCENSE GUM

TOXICITY DATA with REFERENCE
skn-rbt 500 mg/24H MOD FCTXAV 16,837,78

CONSENSUS REPORTS: Reported in EPA TSCA Inventory.

SAFETY PROFILE: A skin irritant. When heated to decomposition it emits acrid smoke and irritating fumes.

OIM025 CAS:8050-07-5 ***HR: 1***
OLIBANUM OIL

PROP: Distilled from a gum from the trees *Boswellia carterii* Birdw. and other *Boswellia* species (Fam. *Burseraceae*). Pale liquid; pleasant balsamic odor. D: 0.862-0.889, refr index: 1.465-1.482 @ 20°. Sol in fixed oils, mineral oil; insol in glycerin, propylene glycol.

SYN: FRANKINCENSE OIL

TOXICITY DATA with REFERENCE
skn-rbt 500 mg/24H MOD FCTXAV 16,837,78

SAFETY PROFILE: A skin irritant. When heated to decomposition it emits acrid smoke and irritating fumes.

OIQ000 CAS:8001-25-0 ***HR: 2***
OLIVE OIL

PROP: Yellow oil; pleasing, delicate flavor. Mp: −6°, flash p: 437°F (CC), autoign temp: 650°F, d: 0.909-0.915 @ 25°/25°. Becomes rancid on exposure to air. Sltly sol in alc; misc with ether, chloroform, carbon disulfide. From fruit of *Olea europaea* (85DIA2 2,196,77).

TOXICITY DATA with REFERENCE
skn-hmn 300 mg/3D-I MLD 85DKA8 -,127,77
inv-mus LD50:1320 mg/kg KEKHB8 (3),19,73

CONSENSUS REPORTS: Reported in EPA TSCA Inventory. EPA Genetic Toxicology Program.

SAFETY PROFILE: Moderately toxic by intraperitoneal route. A human skin irritant. Combustible when exposed to heat or flame; can react with oxidizing materials. Some spontaneous heating. To fight fire, use CO_2, dry chemical. When heated to decomposition it emits acrid smoke and irritating fumes.

OIS000 CAS:11006-70-5 ***HR: 3***
OLIVOMYCIN

PROP: A mixture of antibiotics produced by *Streptomyces olivoreticuli*. Composed of olivomycins A, B, C, and D with olivomycin A as the major component.

SYNS: ABURAMYCIN ◇ CHROMOMYCIN ◇ OLIGOMYCIN A, mixed with OLIGOMYCIN B ◇ OLIGOMYCIN B, mixed with OLIGOMYCIN A ◇ OLIVOMITSIN

TOXICITY DATA with REFERENCE
dnd-esc 100 μmol/L PMSBA4 2,48,71
dnd-hmn:hla 5 mg/L CNREA8 45,2813,85
orl-mus LD50:20 mg/kg 85FZAT -,85,67
ipr-mus LD50:1700 μg/kg 85FZAT -,214,67
scu-mus LD50:3 mg/kg 85ERAY 2,1401,78
ivn-mus LD50:5 mg/kg 85ERAY 2,1322,78

SAFETY PROFILE: Poison by ingestion, intravenous, intraperitoneal, and subcutaneous routes. Human mutation data reported. See also other olivomycin entries.

OIU000 CAS:6988-58-5 ***HR: 3***
OLIVOMYCIN A
mf: $C_{58}H_{84}O_{26}$ mw: 1197.42

PROP: Yellow crystals. Mp: 160-165°. Sol in alc, ether, chloroform; insol in benzene, water.

SYNS: NSC-76411 ◇ OLIVOMYCIN I
ipr-mus LD50:6600 μg/kg ANTBAL 29,666,84
ivn-mus LD50:4600 μg/kg 85GDA2 1,336,80

SAFETY PROFILE: Poison by intraperitoneal and intravenous routes. See also OLIVOMYCIN.

OIU499 CAS:102647-16-5 ***HR: 3***
OLIVOMYCIN D
mf: $C_{58}H_{84}O_{26}$ mw: 1197.42

SYNS: ANTIBIOTIC A-649 ◇ ANTIBIOTIC A-64922 ◇ BRISTOL A-649 ◇ C-1228 ◇ NSC 38270 ◇ OLIVOMITSIN ◇ OLIVOMYCIN ◇ OLIVOMYCIN A ◇ OLIVOMYCINE

TOXICITY DATA with REFERENCE
cyt-hmn-leu:15 μg/L CNREA8 26,2437,66
ipr-mus LD50:5474 μg/kg NCISP* JAN86
scu-mus LD50:3861 μg/kg NCISP* JAN86

SAFETY PROFILE: Poison by subcutaneous and intraperitoneal routes. Human mutation data reported. When heated to decomposition it emits acrid smoke and irritating fumes. See also other olivomycin entries.

OIU850 CAS:43143-11-9 ***HR: D***
OMADINE MDS
mf: $C_{10}H_8N_2O_2S_2{\bullet}O_4S{\bullet}Mg$ mw: 372.69

SYNS: MAGNESIUM SULFATE adduct of 2,2-DITHIO-BIS-PYRIDINE 1-OXIDE ◇ SULFURIC ACID, MAGNESIUM SALT (1:1), compounded with 2,2′-DITHIOBIS(PYRIDINE) 1,1′-OXIDE

TOXICITY DATA with REFERENCE
orl-rat TDLo:210 mg/kg (15-22D preg/20D post):REP FAATDF 4,81,84

CONSENSUS REPORTS: Reported in EPA TSCA Inventory.

SAFETY PROFILE: Experimental reproductive effects. When heated to decomposition it emits toxic fumes of SO_x and NO_x.

OIW000 CAS:52684-23-8 ***HR: 3***
OMP-1
mf: $(C_{16}H_{32}O_2Sn{\bullet}C_{13}H_{26}O_2Sn{\bullet}C_5H_6O_2)_x$

PROP: Trialkyltin methacrylate polymer (NTIS** AD-A062-138).

SYN: 2-METHYL-2-PROPENOIC ACID METHYL ESTER, POLYMER withTRIBUTYL(92-METHYL-1-OXO-2-PROPENYL)OXY)STANNANE and((2-METHYL-1-OXO-2-PROPENYL)OXY)TRIPOPYLSTANNANE

TOXICITY DATA with REFERENCE
skn-rbt 500 mg/24H NTIS** AD-A062-138
eye-rbt 100 mg NTIS** AD-A062-138
orl-rat LD50:280 mg/kg NTIS** AD-A062-138
ihl-rat LC50:64 mg/m^3/4H NTIS** AD-A062-138

SAFETY PROFILE: Poison by ingestion and inhalation. An eye and skin irritant. When heated to decomposition it emits acrid smoke and irritating fumes. See also TIN COMPOUNDS and ORGANOMETALS.

OIY000 CAS:26354-18-7 ***HR: 3***
OMP 2
mf: $(C_{16}H_{32}O_2Sn \cdot C_5H_8O_2)_x$

PROP: Trialkyltin methacrylate polymer (NTIS** AD-A062-138).

SYNS: 2-METHYL-2-PROPENOIC ACID METHYL ESTER, POLYMER withTRIBUTYL(92-METHYL-1-OXO-2-PROPENYL)OXY)STANNANE ◇ TRIBUTYL(METHACRYLOYLOXY)-STANNANE POLYMER with METHYL METHACRYLATE (8CI)

TOXICITY DATA with REFERENCE
skn-rbt 500 mg/24H NTIS** AD-A062-138
eye-rbt 100 mg NTIS** AD-A062-138
orl-rat LD50:230 mg/kg NTIS** AD-A062-138
ihl-rat LC50:51 mg/m^3/4H NTIS** AD-A062-138

CONSENSUS REPORTS: Reported in EPA TSCA Inventory.

SAFETY PROFILE: Poison by ingestion and inhalation. A skin and eye irritant. When heated to decomposition it emits acrid smoke and irritating fumes. See also TIN COMPOUNDS and ORGANOMETALS.

OJA000 ***HR: 3***
OMP-4

PROP: Trialkyltin methacrylate polymer containing 31.06% tin (NTIS** AD-A062-138).

TOXICITY DATA with REFERENCE
skn-rbt 500 mg/24H MOD NTIS** AD-A062-138
eye-rbt 100 mg MOD NTIS** AD-A062-138
orl-rat LD50:268 mg/kg NTIS** AD-A062-138
ipr-rat LD50:1400 μg/kg NTIS** AD-A062-138
orl-mus LD50:56 mg/kg NTIS** AD-A062-138
ipr-mus LD50:15 mg/kg NTIS** AD-A062-138

SAFETY PROFILE: Poison by ingestion and intraperitoneal routes. A skin and eye irritant. When heated to decomposition it emits acrid smoke and irritating fumes. See also TIN COMPOUNDS and ORGANOMETALS.

OJC000 CAS:96231-64-0 ***HR: 3***
OMP-5

PROP: Trialkyltin methacrylate polymer containing 28% tin (NTIS** AD-A062-138).

TOXICITY DATA with REFERENCE
eye-rbt 100 mg MOD NTIS** AD-A062-138
orl-rat LD50:1427 mg/kg NTIS** AD-A062-138
ipr-rat LD50:19 mg/kg NTIS** AD-A062-138
orl-mus LD50:406 mg/kg NTIS** AD-A062-138
ipr-mus LD50:29 mg/kg NTIS** AD-A062-138

SAFETY PROFILE: Poison by intraperitoneal route. Moderately toxic by ingestion. An eye irritant. When heated to decomposition it emits acrid smoke and irritating fumes. See also TIN COMPOUNDS and ORGANOMETALS.

OJD100 CAS:31430-18-9 ***HR: 3***
ONCODAZOLE
mf: $C_{14}H_{11}N_3O_3S$ mw: 301.34

SYNS: METHYL5-(2-THENOYL)-2-BENZIMIDAZOLECARBAMATE ◇ METHYL (5-(2-THIENYLCARBONYL)-1H-BENZIMIDAZOLE-2-YL)CARBAMATE ◇ NOCODAZOLE ◇ NSC 238159 ◇ R 17934 ◇ N-(5-(2-THIENOYL)-2-BENZIMIDAZOLYL)CARBAMIC ACID METHYL ESTER ◇ 5-(2-THIENOYL)-2-BENZIMIDAZOLECARBAMIC ACID METHYL ESTER ◇ 5-(2-THIENYLCARBONYL)-2-BENZIMIDAZOLECARBAMIC ACID METHYL ESTER ◇ (5-(2-THIENYLCARBONYL)-1H-BENZI-MID-AZOL-2-YL)-CARBAMIC ACID METHYL ESTER

TOXICITY DATA with REFERENCE
sln-smc 1 ppm/16H MUREAV 141,15,84
cyt-hmn:hla 40 μg/L CNREA8 36,905,76
ipr-mus TDLo:10 mg/kg (female 1D post):TER MUREAV 210,313,89
ipr-mus TDLo:10 mg/kg (female 1D post):REP MUREAV 210,313,89
ipr-mus LD50:39350 μg/kg NCISP* JAN86

SAFETY PROFILE: Poison by intraperitoneal route. An experimental teratogen. Other experimental reproductive effects. Human mutation data reported. When heated to decomposition it emits toxic fumes of SO_x and NO_x. See also CARBAMATES.

OJD200 ***HR: 1***
ONION OIL

PROP: From steam distillation of bulbs of *Allium ceoa* L. (Fam. *Lillaceae*). Clear amber liquid; strong pungent odor and taste of onion. Sol in fixed oils, mineral oil, alc; insol in glycerin, propylene glycol.

SYN: OIL of ONION

SAFETY PROFILE: Skin irritant. When heated to decomposition it emits acrid smoke and irritating fumes.

OJD300 CAS:484-23-1 ***HR: 3***
OPHTHAZIN
mf: $C_8H_{10}N_6$ mw: 190.24

SYNS: C 7441 ◇ CASSELLA 532 ◇ DIHYDRALAZIN ◇ DIHYDRALAZINE ◇ DIHYDRALLAZIN ◇ 1,4-DIHYDRAZINONAPHTHALAZINE ◇ DIHYDRAZINOPHTHALAZINE ◇ 1,4-DIHYDRAZINOPHTHALAZINE ◇ HYPOPRESOL ◇ NEPRESOL ◇ NEPRESOLIN ◇ NEPRESSOL ◇ PHTHALAZINE, 1,4-DIHYDRAZINO- ◇ 1,4-PHTHALAZINEDIONE, 2,3-DIHYDRO-, DIHYDRAZONE (9CI) ◇ TONOLYSIN

TOXICITY DATA with REFERENCE
ivn-wmn TDLo:250 μg/kg (female 32W post):TER OBGNAS 55,519,80
ivn-mus LD50:300 mg/kg EJMCA5 11,107,76

SAFETY PROFILE: Poison by intravenous route. An experimental teratogen. When heated to decomposition it emits toxic fumes of NO_x.

OJG000 ***HR: 3***
OPIUM

PROP: Air dried, milky exudation from incised, unripe capsules of *Papaver somniferum L* or *P. album Mill.* Morphine is the most important alkaloid and occurs to the extent of 10-16%.

SYN: GUM OPIUM

TOXICITY DATA with REFERENCE
mma-sat 800 μg/plate LANCAO 2,494,78

SAFETY PROFILE: Poison by ingestion. Mutation data reported. Use may lead to habituation and addiction. A narcotic, sedative, analgesic, and hypnotic. Source of morphine, codeine, papaverine, thebaine, etc. Can cause nausea, vomiting, constipation, and respiratory problems. Combustible when exposed to heat or flame. See also MORPHINE.

OJG509 ***HR: D***
OPIUM, PYROLYZATE

TOXICITY DATA with REFERENCE
mmo-sat 100 μg/plate CRNGDP 3,577,82
sce-hmn:lym 10 mg/L CRNGDP 4,227,83

SAFETY PROFILE: Human mutation data reported. When heated to decomposition it emits acrid smoke and irritating fumes.

OJI600 CAS:956-03-6 ***HR: 3***
ORACAINE HYDROCHLORIDE
mf: $C_{14}H_{21}NO_2 \cdot ClH$ mw: 271.82

PROP: Crystals. Mp: 150-151°. Sol in water, alc.

SYNS: MEPRYLCAINE HYDROCHLORIDE ◇ 2-METHYL-2-(PROPYLAMINO)-1-PROPANOL BENZOATE (ester), HYDROCHLORIDE ◇ 2-METHYL-2-(PROPYLAMINO)-1-PROPANOL BENZOATE HYDROCHLORIDE

TOXICITY DATA with REFERENCE
ipr-mus LD50:225 mg/kg AIPTAK 115,483,58
scu-mus LD50:262 mg/kg AIPTAK 115,483,58
ivn-mus LD50:21 mg/kg AIPTAK 115,483,58

CONSENSUS REPORTS: Reported in EPA TSCA Inventory.

SAFETY PROFILE: Poison by subcutaneous, intravenous, and intraperitoneal routes. When heated to decomposition it emits toxic fumes of NO_x and HCl.

OJI750 CAS:51234-28-7 ***HR: 3***
ORAFLEX
mf: $C_{16}H_{12}ClNO_3$ mw: 301.74

SYNS: BENOXAPROFEN ◇ 2-(4-CHLOROPHENYL)-α-METHYL-5-BENZOXAZOLEACETIC ACID ◇ COMPOUND 90459 ◇ COXIGON ◇ OPREN ◇ UNIPROFEN

TOXICITY DATA with REFERENCE
dnd-hmn:leu 15 mg/L CNREA8 48,3094,88
orl-rat TDLo:525 mg/kg (35D male):REP YACHDS 9,4453,81
orl-wmn TDLo:168 mg/kg/2W-I:KID,BLD NPRNAY 35,279,83
orl-man TDLo:780 mg/kg/13W-I:END BMJOAE 284,1365,82
orl-wmn TDLo:552 mg/kg/46D-C:LIV LANCAO 1,959,82
orl-wmn LDLo:780 mg/kg/65D-C:LIV LANCAO 1,959,82
orl-rat LD50:118 mg/kg YACHDS 9,4445,81
ipr-rat LD50:129 mg/kg YACHDS 9,4445,81
scu-rat LD50:121 mg/kg YACHDS 9,4445,81
orl-mus LD50:800 mg/kg JMCMAR 18,53,75
ipr-mus LD50:398 mg/kg YACHDS 9,4445,81
scu-mus LD50:482 mg/kg YACHDS 9,4445,81

SAFETY PROFILE: Poison by ingestion, subcutaneous, and intraperitoneal routes. Moderately toxic to humans by ingestion. Human systemic effects by ingestion: jaundice and gynecomastia (excessive development of the male mammary glands), changes in kidney tubules, decreased urine volume or anuria, and eosinophilia. Experimental reproductive effects. Mutation data reported. When heated to decomposition it emits toxic fumes of Cl^- and NO_x.

OJK300 ***HR: D***
ORF 13811
mf: $C_{20}H_{34}O_5$ mw: 354.54

SYNS: 4-(4,8-DIMETHYL-5-HYDROXY-7-NONENYL)-4-METHYL-3,8-DIOXABICYCLO(3.2.1)OCTANE-1-ACETICACID

TOXICITY DATA with REFERENCE
orl-rat TDLo:7500 μg/kg (16D preg):REP CCPTAY 30,39,84

SAFETY PROFILE: Experimental reproductive effects.

When heated to decomposition it emits acrid smoke and irritating fumes.

OJM000 ***HR: 3***
ORGANOMETALS

PROP: Compounds containing carbon and a metal. Ordinarily metallic carbonates (calcium carbonate, etc.) are excluded and also metallic salts of common organic acids. Examples of organic metal compounds are Grignard compounds, such as methyl magnesium iodide (CH_3MgI), and metallic alkyls, such as butyllithium (C_4H_9Li), tetraethyllead, triethyl aluminum, tetrabutyl titanate, sodium methylate, copper phthalocyanine, and metallocenes. Also, there are many organotin compounds, such as monoalkyltins, monoaryltins, dialkyltins, diaryltins, trialkyltins, triaryltins, tetraalkyltins and tetraaryltins.

SAFETY PROFILE: Many are highly toxic or flammable. As an example, organotin compounds are poisons by ingestion and intravenous routes. Irritating to skin, eyes, and mucous membranes. Can damage lung tissue and the liver. Trialkyltins are most toxic as a group. Next are the dialkyltins and the monoalkyltins. In each major organotin group the ethyltin derivative is the most toxic, followed by the methyltins. This group of compounds is constantly growing in importance, but there is relatively little toxicity information on most of them. Alkyl compounds of lead, tin, mercury, and aluminum are known to be highly toxic. Less is known about other organometals, but for the most part they are highly reactive chemically and therefore dangerous, if only on direct contact. It is prudent to exercise great caution in handling organometals, particularly the alkyl forms. Many organolithium compounds are explosive. See also individual compounds.

OJM400 CAS:9016-01-7 ***HR: 3***
ORGOTEINS

PROP: Water-soluble protein congeners isolated from red blood cells, liver, and other tissues. Molecular weight is about 34,000.

SYNS: ONTOSEIN ◇ ORGOTEIN ◇ ORMETEIN ◇ PALOSEIN

TOXICITY DATA with REFERENCE
ims-rat TDLo:2200 μg/kg (6-16D preg):TER TXAPA9 26,184,73
ipr-mus LDLo:60 mg/kg TXAPA9 26,184,73
ivn-rbt LDLo:10 mg/kg TXAPA9 26,184,73

SAFETY PROFILE: Poison by intravenous and intraperitoneal routes. An experimental teratogen.

OJO000 CAS:8007-11-2 ***HR: 3***
ORIGANUM OIL

PROP: Main constituent is carvacrol. From steam distillation of the herb *Thymus capitatus* Hoffm. et Link (FCTXAV 12,807,74). Yellow to dark red brown liquid; pungent spicy odor of thyme oil. D: 0.935-0.960, refr index: 1.502 @ 20°. Sol in fixed oil, propylene glycol, mineral oil; insol in glycerin.

SYN: OIL of ORIGANUM

TOXICITY DATA with REFERENCE
skn-mus 100%:SEV FCTXAV 12,945,74
skn-rbt 500 mg/24H MOD FCTXAV 12,945,74
orl-rat LD50:1850 mg/kg FCTXAV 12,945,74
skn-rbt LD50:320 mg/kg FCTXAV 12,945,74

CONSENSUS REPORTS: Reported in EPA TSCA Inventory.

SAFETY PROFILE: Poison by skin contact. Moderately toxic by ingestion. A severe skin irritant. When heated to decomposition it emits acrid smoke and irritating fumes. See also CARVACROL.

OJS000 CAS:16773-42-5 ***HR: 3***
ORNIDAZOLE
mf: $C_7H_{10}ClN_3O_3$ mw: 219.65

SYNS: α-(CHLORMETHYL)-2-METHYL-5-NITRO-IMIDAZOL-1-AETHANOL (GERMAN) ◇ 1-(3-CHLORO-2-HYDROXYPROPYL)-2-METHYL-5-NITROIMIDAZOLE ◇ α-(CHLOROMETHYL)-2-METHYL-5-NITRO-1H-IMIDAZOLE-1-ETHANOL ◇ RO 7-0207 ◇ TIBERAL

TOXICITY DATA with REFERENCE
mmo-sat 1 nmol/plate MUREAV 58,1,78
mmo-esc 20 μmol/L MUREAV 48,155,77
unr-rat TDLo:5600 mg/kg (14D male):REP TOXID9 4,81,84
orl-rat LD50:1780 mg/kg ARZNAD 28,612,78
orl-mus LD50:1139 mg/kg ARZNAD 28,612,78
ipr-mus LD50:1120 mg/kg ARZNAD 28,612,78
ivn-mus LD50:375 mg/kg ARZNAD 28,612,78

SAFETY PROFILE: Poison by intravenous route. Moderately toxic by ingestion and intraperitoneal routes. Experimental reproductive effects. Mutation data reported. When heated to decomposition it emits very toxic fumes of Cl^- and NO_x.

OJU000 CAS:3184-13-2 ***HR: 1***
l-ORNITHINE HYDROCHLORIDE
mf: $C_5H_{12}N_2O_2 \cdot ClH$ mw: 168.65

PROP: Decomp @ 233°. Sol in water.

SYN: l-ORNITHINE MONOHYDROCHLORIDE

TOXICITY DATA with REFERENCE
orl-rat LD50:10 g/kg JPMSAE 62,49,73

CONSENSUS REPORTS: Reported in EPA TSCA Inventory.

SAFETY PROFILE: Mildly toxic by ingestion. When heated to decomposition it emits very toxic fumes of NO_x and HCl.

OJV500 CAS:65-86-1 ***HR: 2***
OROTIC ACID
mf: $C_5H_4N_2O_4$ mw: 156.11

PROP: Crystals from water. Mp: 345-346°.

SYNS: ANIMAL GALACTOSE FACTOR ◇ 6-CARBOXYURACIL ◇ ORIDIN ◇ OROPUR ◇ OROTONIN ◇ OROTSAURE (GERMAN) ◇ OROTURIC ◇ OROTYL ◇ 6-URACILCARBOXYLIC ACID ◇ WHEY FACTOR

TOXICITY DATA with REFERENCE
pic-esc 1 g/L ZAPOAK 12,583,72
dnd-rat-orl 21 g/kg/5W-C CRNGDP 6,765,85
orl-mus LD50:2 g/kg NIIRDN 6,165,82
ipr-mus LD50:841 mg/kg NIIRDN 6,165,82
ivn-mus LD50:770 mg/kg NIIRDN 6,165,82

CONSENSUS REPORTS: Reported in EPA TSCA Inventory.

SAFETY PROFILE: Moderately toxic by ingestion, intraperitoneal, and intravenous routes. Mutation data reported. When heated to decomposition it emits toxic fumes of NO_x.

OJV525 ***HR: 2***
OROTIC ACID mixed with CHOLESTEROL and CHOLIC ACID (2:2:1)

SYNS: CHOLESTEROL mixed with OROTIC ACID mixed with CHOLIC ACID (2:2:1) ◇ CHOLIC ACID mixed with CHOLESTEROL mixed with OROTIC ACID (1:2:2)

TOXICITY DATA with REFERENCE
orl-rat TDLo:585 g/kg/56W-C:ETA DABBBA 41,515,80

SAFETY PROFILE: Questionable carcinogen with experimental tumorigenic data. When heated to decomposition it emits acrid smoke and irritating fumes.

OJW000 CAS:341-69-5 ***HR: 3***
ORPHENADRINE HYDROCHLORIDE
mf: $C_{18}H_{23}NO{\bullet}ClH$ mw: 305.88

PROP: Crystals. Mp: 156-157°. Sol in water, alc, chloroform; sltly sol in acetone, benzene; almost insol in ether.

SYNS: BF 5930 ◇ BG 5930 ◇ BROCADISIPAL ◇ BROCASIPAL ◇ BS 5930 ◇ 2-DIMETHYLAMINOETHYL-2-METHYLBENZHYDRYL ETHERHYDROCHLORIDE ◇ N,N-DIMETHYL-2-(o-METHYL-α-PHENYLBENZYLOXY)ETHYLAMINE HYDROCHLORIDE ◇ DISIPAL HYDROCHLORIDE ◇ MEPHENAMIN HYDROCHLORIDE ◇ MEPHENAMINE HYDROCHLORIDE

TOXICITY DATA with REFERENCE
orl-mus TDLo:150 mg/kg (7-12D preg):REP KSRNAM 6,219,72
orl-rat TDLo:30 mg/kg (7-16D preg):TER TXAPA9 21,230,72
orl-wmn TDLo:150 mg/kg ATXKA8 23,264,68
orl-man TDLo:71 mg/kg:EYE,PSY,CPR HUTODJ 4,331,85
orl-cld LDLo:33 mg/kg:CNS ATXKA8 25,76,69
orl-rat LD50:255 mg/kg GNRIDX 2,311,68
ipr-rat LD50:93 mg/kg AIPTAK 177,28,69
scu-rat LD50:230 mg/kg ARZNAD 5,72,55
ivn-rat LD50:27500 μg/kg ARZNAD 5,72,55
orl-mus LD50:100 mg/kg ARZNAD 5,72,55
ipr-mus LD50:65 mg/kg 27ZQAG -,373,72
scu-mus LD50:88 mg/kg 27ZQAG -,373,72
scu-gpg LD50:74 mg/kg AIPTAK 177,28,69

SAFETY PROFILE: Poison by ingestion, intravenous, intraperitoneal, and subcutaneous routes. Human systemic effects: mydriasis (pupillary dilation), hallucinations, distorted perceptions, pulse rate increase, intracranial pressure increase. An experimental teratogen. Experimental reproductive effects. When heated to decomposition it emits toxic fumes of NO_x and HCl.

OJY000 CAS:5588-10-3 ***HR: 3***
ORTHOXINE HYDROCHLORIDE
mf: $C_{11}H_{17}NO{\bullet}ClH$ mw: 215.75

SYNS: METHOXIPHENADRIN HYDROCHLORIDE ◇ o-METHOXY-N-α-DIMETHYLPHENETHYLAMINEHYDROCHLORIDE ◇ METHOXYPHENAMINE HYDROCHLORIDE ◇ METHOXYPHENAMINIUM CHLORIDE ◇ β-(o-METHOXYPHENYL)ISO-p-TROPYLMETHYLAMINEHYDROCHLORIDE ◇ α-(2-METHOXYPHENYL)-β-METHYLAMINOPROPANEHYDROCHLORIDE ◇ ORTODRINEX HYDROCHLORIDE ◇ PROASMA HYDROCHLORIDE

TOXICITY DATA with REFERENCE
orl-rat LD50:630 mg/kg NIIRDN 6,837,82
scu-rat LD50:573 mg/kg NIIRDN 6,837,82
ivn-rat LD50:50 mg/kg NIIRDN 6,837,82
orl-mus LD50:605 mg/kg NIIRDN 6,837,82
ipr-mus LD50:90 mg/kg 27ZQAG -,348,72
scu-mus LD50:380 mg/kg NIIRDN 6,837,82
orl-rbt LD50:652 mg/kg NIIRDN 6,837,82
scu-rbt LD50:269 mg/kg NIIRDN 6,837,82
ivn-rbt LD50:30 mg/kg NIIRDN 6,837,82

SAFETY PROFILE: Poison by intravenous, intraperitoneal, and subcutaneous routes. Moderately toxic by ingestion. An FDA over the counter drug used as a bronchodilator. When heated to decomposition it emits very toxic fumes of HCl and NO_x.

OKE000 CAS:7440-04-2 ***HR: 3***
OSMIUM
af: Os aw: 190.20

PROP: A lustrous, bluish-white, extremely hard and dense, brittle metal. Bp: 5027°, d: 22.57, mp: approx 2700°.

SYN: METALLIC OSMIUM

TOXICITY DATA with REFERENCE
ivn-dog LDLo:17 mg/kg SMSJAR 26,131,1826

CONSENSUS REPORTS: Reported in EPA TSCA Inventory.

SAFETY PROFILE: Poison by intravenous route. An irritant to eyes and mucous membranes. The principal effects of exposure are ocular disturbances and an asthmatic condition caused by inhalation. Furthermore, it causes dermatitis and ulceration of the skin upon contact. When osmium is heated, it gives off a pungent, poisonous fume of osmium tetraoxide. One case of osmium poisoning reported in the literature resulted from the inhalation of osmium tetraoxide, which gave rise to a capillary bronchitis and dermatitis. The vapor has a pronounced and nauseating odor which should be taken as a warning of a possibly toxic concentration in the atmosphere, and personnel should immediately move to an area of fresh air. The metal itself is not highly toxic. Flammable in the form of dust when exposed to heat or flame. Slight explosion hazard in the form of dust when exposed to heat or flame. Violent reaction or ignition with chlorine trichloride or oxygen difluoride. Ignites when heated to 100°C with fluorine. Incandescent reaction in phosphorus vapor. When heated to decomposition it emits toxic fumes of OsO_4. See also OSMIUM TETROXIDE.

OKG000 CAS:13768-38-2 ***HR: 3***
OSMIUM HEXAFLUORIDE
mf: F_6Os mw: 304.20

PROP: Pale yellow, volatile solid. Mp: 32.1°, bp: 45.9°.

SAFETY PROFILE: Highly poisonous. A very corrosive eye, skin, and mucous membrane irritant. Ignites paraffin oil and other organic materials. Explosive reaction with silicon. When heated to decomposition it emits highly toxic fumes of OsO_4 and F^-. See also OSMIUM and FLUORIDES.

OKI000 CAS:12036-02-1 ***HR: 3***
OSMIUM(IV) OXIDE
mf: O_2Os mw: 80.06

SAFETY PROFILE: Amorphous form can explode spontaneously in air. When heated to decomposition it emits toxic fumes of Os. See also OSMIUM and OSMIUM TETROXIDE.

OKK000 CAS:20816-12-0 ***HR: 3***
OSMIUM TETROXIDE
DOT: UN 2471
mf: O_4Os mw: 254.20

PROP: (A) Monoclinic, colorless crystals; (B) yellow mass; pungent, chlorine-like odor. Mp (A): 39.5°, mp: (B): 41°, bp: 130° (subl), d: 4.906 @ 22°, vap press (A): 10 mm @ 26.0°, vap press (B): 10 mm @ 31.3°. Sol in benzene.

SYNS: OSMIC ACID ◇ OSMIUM(VIII) OXIDE ◇ RCRA WASTE NUMBER P087

TOXICITY DATA with REFERENCE
mrc-bcs 5 mmol/L MUREAV 77,109,80
dns-ham:emb 200 μmol/L MUREAV 131,173,84
itt-rat TDLo:20336 μg/kg (1D male):REP JRPFA4 7,21,64
ihl-man TCLo:133 μg/m³:EYE,PUL BJIMAG 3,183,46
ihl-rat LCLo:40 ppm/4H SCCUR* -,8,61
ipr-rat LD50:14100 μg/kg SCCUR* -,8,61
orl-mus LD50:162 mg/kg SCCUR* -,8,61
ihl-mus LCLo:40 ppm/4H SCCUR* -,8,61
ipr-mus LD50:13500 μg/kg SCCUR* -,8,61
ihl-rbt LCLo:1316 mg/m³/30M JIDHAN 15,136,33

CONSENSUS REPORTS: Community Right-To-Know List. EPA Genetic Toxicology Program. Reported in EPA TSCA Inventory.

OSHA PEL: (Transitional: TWA 0.002 mg/m³ (Os)) TWA 0.0002 ppm; STEL 0.0006 ppm (Os)
ACGIH TLV: TWA 0.0002 ppm; STEL 0.0006 ppm (Os)
DFG MAK: 0.0002 ppm (0.002 mg/m³)
DOT Classification: Poison B; Label: Poison.

SAFETY PROFILE: Poison by ingestion, inhalation, and intraperitoneal routes. Human systemic effects by inhalation: lacrimation and other eye effects and structural or functional changes in trachea or bronchi. Experimental reproductive effects. Mutation data reported. Explodes on contact with 1-methylimidazole. Catalytic decomposition of hydrogen peroxide can be hazardous. See also OSMIUM.

OKK500 CAS:12798-63-9 ***HR: 3***
OS-40 (PHOSPHATE ESTER)

SYN: WF-104

TOXICITY DATA with REFERENCE
orl-rat LD50:2520 mg/kg MRLR** No.256,54
ihl-rat LC50:3900 mg/m³ XAWPA2 CWL 2-10,58
ivn-rat LD50:130 mg/kg MRLR** No.256,54
ivn-rbt LD50:190 mg/kg MRLR** No.256,54

SAFETY PROFILE: Poison by intravenous route. Moderately toxic by ingestion and inhalation. When heated to decomposition it emits toxic fumes of PO_x. See also ESTERS.

OKO200 CAS:61840-39-9 ***HR: 2***
OTAMOL

TOXICITY DATA with REFERENCE
skn-rbt 500 mg/24H SEV 28ZPAK -,288,72
eye-rbt 100 mg/24H MOD 28ZPAK -,288,72
orl-rat LD50:9110 mg/kg 28ZPAK -,288,72

SAFETY PROFILE: Mildly toxic by ingestion. An eye and severe skin irritant.

OKO400 CAS:26095-59-0 ***HR: 3***
OTILONIUM BROMIDE
mf: $C_{29}H_{43}N_2O_4$•Br mw: 563.65

SYN: OTTILONIO BROMURO (ITALIAN)

TOXICITY DATA with REFERENCE
orl-mus LD50:3 g/kg FRPSAX 39,3,84
ipr-mus LD50:85 mg/kg FRPSAX 39,3,84
scu-mus LD50:1 g/kg FRPSAX 39,3,84

SAFETY PROFILE: Poison by intraperitoneal route. Moderately toxic by ingestion and subcutaneous routes. When heated to decomposition it emits toxic fumes of Br^- and NO_x.

OKO500 CAS:1218-35-5 ***HR: 3***
OTRIVINE HYDROCHLORIDE
mf: $C_{16}H_{24}N_2$•ClH mw: 280.88

SYNS: 2-(4-tert-BUTYL-2,6-DIMETHYLBENZYL)-2-IMIDAZOLINEHYDROCHLORIDE ◇ 2-(4-tert-BUTYL-2,6-DIMETHYLBENZYL)-2-IMIDAZOLINE MONOHYDROCHLORIDE ◇ OTRIVIN HYDROCHLORIDE ◇ XYLOMETAZOLINE HYDROCHLORIDE

TOXICITY DATA with REFERENCE
orl-rat LD50:230 mg/kg KSRNAM 5,555,71
ipr-rat LD50:43 mg/kg KSRNAM 5,555,71
scu-rat LD50:90 mg/kg KSRNAM 5,555,71
orl-mus LD50:75 mg/kg KSRNAM 5,555,71
scu-mus LD50:53 mg/kg KSRNAM 5,555,71
ivn-mus LD50:12500 μg/kg KSRNAM 5,555,71

SAFETY PROFILE: Poison by ingestion, subcutaneous, intravenous, and intraperitoneal routes. When heated to decomposition it emits toxic fumes of NO_x and HCl.

OKS000 CAS:630-60-4 ***HR: 3***
OUABAIN
mf: $C_{29}H_{44}O_{12}$ mw: 584.73

PROP: A natural plant product (JPETAB 49,561,33).

SYNS: ACOCANTHERIN ◇ ASTROBAIN ◇ GRATIBAIN ◇ GRATUS STROPHANTHIN ◇ G-STROPHANTHIN ◇ OUABAGENIN-l-RHAMNOSID (GERMAN) ◇ OUABAGENIN-l-RHAMNOSIDE ◇ OUBAIN ◇ OUABAINE ◇ PUROSTROPHAN ◇ STROPHANTHIN G ◇ STROPHOPERM

TOXICITY DATA with REFERENCE
ipr-rat LD50:47 mg/kg TXAPA9 24,37,73
scu-rat LDLo:50 mg/kg ARZNAD 23,1125,73
ivn-rat LD50:14 mg/kg TXAPA9 20,44,71
ipr-mus LD50:11 mg/kg AIPTAK 155,165,65
scu-mus LDLo:8 mg/kg ARZNAD 23,1125,73
ivn-mus LD50:2200 μg/kg PSEBAA 118,756,65
orl-dog LDLo:1500 μg/kg HBAMAK 4,1289,35
scu-dog LDLo:100 μg/kg ARZNAD 23,1125,73
ivn-dog LDLo:54 μg/kg JPETAB 179,447,71
ivn-mky LDLo:102 μg/kg ARZNAD 13,412,63
ipr-cat LD50:100 μg/kg AIPTAK 155,165,65
idu-cat LDLo:3614 mg/kg ARZNAD 19,687,69
ims-rbt LDLo:1 mg/kg COREAF 149,306,09
ims-gpg LDLo:220 μg/kg JPETAB 52,1,34

CONSENSUS REPORTS: EPA Extremely Hazardous Substances List. Reported in EPA TSCA Inventory.

SAFETY PROFILE: Poison by ingestion, intramuscular, intraperitoneal, intravenous, subcutaneous, and parenteral routes. Moderately toxic by intraduodenal route. A cardiac stimulant. When heated to decomposition it emits acrid smoke and irritating fumes.

OKS100 CAS:31323-50-9 ***HR: 3***
OUDENONE
mf: $C_{12}H_{16}O_3$ mw: 208.28

SYN: (S)-2-(DIHYDRO-5-PROPYL-2(3H)-FURYLIDENE)-1,3-CYCLOPENTANEDIONE

TOXICITY DATA with REFERENCE
orl-mus LD50:1100 mg/kg 85ERAY 3,2018,78
ipr-mus LD50:163 mg/kg 85ERAY 3,2018,78
ivn-mus LD50:138 mg/kg 85ERAY 3,2018,78

SAFETY PROFILE: Poison by intravenous and intraperitoneal routes. Moderately toxic by ingestion. When heated to decomposition it emits acrid smoke and irritating fumes.

OKS150 CAS:34200-96-9 ***HR: 2***
OUDENONE SODIUM SALT
mf: $C_{12}H_{18}O_4$•Na mw: 249.29

TOXICITY DATA with REFERENCE
orl-mus LD50:2000 mg/kg 85ERAY 3,2018,78
ipr-mus LD50:1850 mg/kg 85ERAY 3,2018,78
ivn-mus LD50:1000 mg/kg 85ERAY 3,2018,78

SAFETY PROFILE: Moderately toxic by ingestion, intraperitoneal and intravenous routes. When heated to decomposition it emits toxic fumes of Na_2O. See also OUDENONE.

OKU000 CAS:109-29-5 ***HR: 1***
OXACYCLOHEPTADECAN-2-ONE
mf: $C_{16}H_{30}O_2$ mw: 254.46

SYNS: CYCLOHEXADECANOLIDE ◇ DIHYDROAMBRETTOLIDE ◇ 1,16-HEXADECANOLACTONE ◇ HEXADECANOLIDE ◇ 16-HYDROXYHEXADECANOIC ACID LACTONE ◇ JUNIPERIC ACID LACTONE

TOXICITY DATA with REFERENCE
skn-rbt 500 mg/24H MOD FCTXAV 13,452,75

CONSENSUS REPORTS: Reported in EPA TSCA Inventory.

SAFETY PROFILE: A skin irritant. When heated to decomposition it emits acrid smoke and irritating fumes.

OKW000 CAS:7450-97-7 ***HR: 3***
OXAFLUMAZINE DISUCCINATE
mf: $C_{26}H_{32}F_3N_3O_2S•C_8H_{12}O_8$ mw: 743.87

PROP: Crystals from acetonitrile. Mp: 136-138°.

SYNS: OXAFLUMINE ◇ SD 270-31

TOXICITY DATA with REFERENCE
orl-mus LD50:919 mg/kg THERAP 26,481,71
ipr-mus LD50:175 mg/kg THERAP 26,481,71
ivn-mus LD50:94 mg/kg THERAP 26,481,71

SAFETY PROFILE: Poison by intraperitoneal and intravenous routes. Moderately toxic by ingestion. When heated to decomposition it emits very toxic fumes of F^-, NO_x, and SO_x.

OKW100 CAS:3391-83-1 ***HR: 1***
11-OXAHEXADECANOLIDE
mf: $C_{15}H_{28}O_3$ mw: 256.43

SYNS: 1,7-DIOXACYCLOHEPTADECAN-17-ONE ◇ 16-HYDROXY-11-OXAHEXADECANOIC ACID, omega-LACTONE ◇ MUSK R 1

TOXICITY DATA with REFERENCE
skn-gpg 100%/24H MLD FCTOD7 20,787,82

CONSENSUS REPORTS: Reported in EPA TSCA Inventory.

SAFETY PROFILE: A skin irritant. When heated to decomposition it emits acrid smoke and irritating fumes.

OKW110 CAS:6707-60-4 ***HR: 1***
12-OXAHEXADECANOLIDE
mf: $C_{15}H_{28}O_3$ mw: 256.43

SYNS: CERVOLIDE ◇ 1,6-DIOXACYCLOHEPTADECAN-17-ONE ◇ HIBISCOLIDE ◇ 16-HYDROXY-12 OXAHEXADECANOIC ACID, omega-LACTONE

TOXICITY DATA with REFERENCE
skn-rbt 500 mg/24H MLD FCTOD7 20,789,82

CONSENSUS REPORTS: Reported in EPA TSCA Inventory.

SAFETY PROFILE: A skin irritant. When heated to decomposition it emits acrid smoke and irritating fumes.

OKY000 ***HR: 3***
OXALATES

PROP: Salts of oxalic acid.

SAFETY PROFILE: Poisons by ingestion and inhalation. Powerful irritants. Oxalates are corrosive to tissue and produce local irritation. When ingested they have a caustic effect on the mouth, esophagus, and stomach. The soluble oxalates are readily absorbed from the gastrointestinal tract and can cause severe damage to the kidneys. Oxalates are common components of poisonous plants. When heated to decomposition they emit toxic and irritating fumes. See also OXALIC ACID.

OLA000 CAS:144-62-7 ***HR: 3***
OXALIC ACID
mf: $C_2H_2O_4$ mw: 90.04

PROP: Orthorhombic colorless crystals. Mp: 101°, subl @ 150°, d: 1.653. Sol in water, abs alc, and ether

SYNS: ACIDE OXALIQUE (FRENCH) ◇ ACIDO OSSALICO (ITALIAN) ◇ ETHANEDIOIC ACID ◇ ETHANEDIONIC ACID ◇ KYSELINA STAVELOVA (CZECH) ◇ NCI-C55209 ◇ OXAALZUUR (DUTCH) ◇ OXALSAEURE (GERMAN)

TOXICITY DATA with REFERENCE
skn-rbt 500 mg/24H MLD 28ZPAK -,50,72
eye-rbt 250 μg/24H SEV 28ZPAK -,50,72
eye-rbt 100 mg/4S rns SEV FCTOD7 20,573,82
orl-mus TDLo:8400 mg/kg (male 7D pre):REP NTIS** PB86-167053
orl-rat LD50:7500 mg/kg TXAPA9 42,417,77
scu-cat LDLo:112 mg/kg HBAMAK 4,1377,35
scu-frg LDLo:757 mg/kg HBAMAK 4,1377,35

CONSENSUS REPORTS: Reported in EPA TSCA Inventory.

OSHA PEL: (Transitional: TWA 1 mg/m^3) TWA 1 mg/m^3; STEL 2 mg/m^3
ACGIH TLV: TWA 1 mg/m^3; STEL 2 mg/m^3
SAFETY PROFILE: Poison by subcutaneous route. Moderately toxic by ingestion and subcutaneous routes. A skin and severe eye irritant. Acute oxalic poisoning results from ingestion of a solution of the acid. There is marked corrosion of the mouth, esophagus, and stomach with symptoms of vomiting, burning and abdominal pain, collapse and sometimes convulsions. Death may follow quickly. The systemic effects are attributed to the removal by the oxalic acid of the calcium in the blood. The renal tubules become obstructed by the insoluble calcium oxalate, and there is profound kidney disturbance. The chief effects of inhalation of the dusts or

vapor are severe irritation of the eyes and upper respiratory tract, gastrointestinal disturbances, albuminuria, gradual loss of weight, increasing weakness and nervous system complaints, ulceration of the mucous membranes of the nose and throat, epistaxis, headache, irritation, and nervousness. Oxalic acid has a caustic action on the skin and may cause dermatitis; a case of early gangrene of the fingers resembling that caused by phenol has been described. More severe cases may show albuminuria, chronic cough, vomiting, pain in the back and gradual emaciation and weakness. The skin lesions are characterized by cracking and fissuring of the skin and the development of slow-healing ulcers. The skin may be bluish in color, and the nails brittle and yellow. Violent reaction with furfuryl alcohol; Ag; $NaClO_3$; NaOCl. When heated to decomposition it emits acrid smoke and irritating fumes. See also OXALATES.

OLE000 CAS:127-95-7 ***HR: 2***
OXALIC ACID, MONOPOTASSIUM SALT
mf: C_2HO_4•K mw:128.13

PROP: Monoclinic, colorless crystals. Mp: decomp, d: 2.0.

SYNS: KLEESALZ (GERMAN) ◇ POTASSIUM HYDROGEN OXALATE ◇ POTASSIUM SALT of SORREL ◇ SORREL SALT

TOXICITY DATA with REFERENCE
orl-wmn TDLo:100 mg/kg:CNS,CVS,KID MMWOAU 79,1481,32
orl-wmn TDLo:400 mg/kg:CNS,CVS,GIT MMWOAU 79,1481,32
ivn-man TDLo:1071 mg/kg:CNS,GIT MLDCAS 4,178,71
orl-wmn LDLo:660 mg/kg MMWOAU 79,1481,32

CONSENSUS REPORTS: Reported in EPA TSCA Inventory.

SAFETY PROFILE: Moderately toxic to humans by ingestion. Human systemic effects by ingestion and intravenous routes: general anesthetic, somnolence, fluid intake, blood pressure increase or decrease, esophagus changes, nausea or vomiting, and urine volume decrease or anuria. When heated to decomposition it emits toxic fumes of K_2O. See also OXALATES.

OLG000 CAS:63042-11-5 ***HR: 2***
OXALYL-o-AMINOAZOTOLUENE
mf: $C_{16}H_{15}N_3O_3$ mw: 297.34

SYNS: 2'-METHYL-4'-(o-TOLYLAZO)OXANILIC ACID ◇ 4'-OXALYLAMINO-2,3'-DIMETHYLAZOBENZENE

TOXICITY DATA with REFERENCE
orl-rat TDLo:20 g/kg/17W-C:ETA GANNA2 32,232,38

SAFETY PROFILE: Questionable carcinogen with experimental tumorigenic data. When heated to decomposition it emits toxic fumes of NO_x.

OLI000 CAS:79-37-8 ***HR: 3***
OXALYL CHLORIDE
mf: $C_2Cl_2O_2$ mw: 126.93

ClCO•CO•Cl

PROP: Colorless, fuming liquid; penetrating odor. Mp: −12°, bp: 63-64°, d: 1.488 @ 13°/4°.

SAFETY PROFILE: Poison. Violently decomposed by water and alcohol. Severe irritant to skin, eyes, respiratory tract. Explodes on contact with dimethyl sulfoxide. Forms shock-sensitive explosive mixtures with potassium or with K-Na alloy. Will react with water or steam to produce toxic and corrosive fumes. When heated to decomposition it emits toxic fumes of Cl^-. See also OXALIC ACID.

OLK000 CAS:359-40-0 ***HR: 1***
OXALYL FLUORIDE
mf: $C_2F_2O_2$ mw: 94.02

SYN: TL 108

TOXICITY DATA with REFERENCE
ihl-mus LCLo:4500 mg/m³/10M NDRC** NDCrc-132,May,42

CONSENSUS REPORTS: Reported in EPA TSCA Inventory.

SAFETY PROFILE: Mildly toxic by inhalation. When heated to decomposition it emits toxic fumes of F^-. See also FLUORIDES and OXALATES.

OLK200 CAS:15219-97-3 ***HR: 2***
OXALYSINE
mf: $C_5H_{12}N_2O_3$ mw: 148.19

SYNS: (L)-3-(2-AMINOETHOXY)ALANINE ◇ I 677 ◇ 1-4-OXALYSINE

TOXICITY DATA with REFERENCE
orl-mus LD50:1550 mg/kg YHHPAL 15,391,80
ipr-mus LD50:1960 mg/kg YHHPAL 15,391,80
ivn-mus LD50:1790 mg/kg YHHPAL 15,391,80
ims-mus LD50:2330 mg/kg YHHPAL 15,391,80

SAFETY PROFILE: Moderately toxic by ingestion, intraperitoneal, intravenous, and intramuscular routes. When heated to decomposition it emits toxic fumes of NO_x.

OLM000 CAS:6569-69-3 ***HR: 3***
1,4-OXAMERCURANE
mf: C_4H_8HgO mw: 272.71

TOXICITY DATA with REFERENCE
ivn-mus LD50:180 mg/kg CSLNX* NX#05133

CONSENSUS REPORTS: Mercury and its compounds are on the Community Right-To-Know List.

OSHA PEL: (Transitional: CL 1 mg/10m^3) CL 0.1 mg(Hg)/m^3 (skin)
ACGIH TLV: TWA 0.1 mg(Hg)/m^3 (skin)
NIOSH REL: (Inorganic Mercury) TWA 0.05 mg (Hg)/m^3

SAFETY PROFILE: Poison by intravenous route. When heated to decomposition it emits toxic fumes of Hg. See also MERCURY COMPOUNDS.

OLM300 CAS:27035-30-9 ***HR: 3***
OXAMETHACIN
mf: $C_{19}H_{17}ClN_2O_4$ mw: 372.81

PROP: Crystals from dioxane. Mp: 181-182° (decomp). Sol in most organic solvents at elevated temperatures.

SYNS: ABC 8/3 ◇ ACIDO 1-(p-CLOROBENZOIL)-5-METOSSI-2-METIL-3-INDOLILACETOIDROSSAMICO (ITALIAN) ◇ ACIDO INDOXAMICO (ITALIAN) ◇ 1-(4-CHLOROBENZOYL)-N-HYDROXY-5-METHOXY-2-METHYL-1H-INDOLE-3-ACETAMIDE ◇ 1-(p-CHLOROBENZOYL)-5-METHOXY-2-METHYL-3-INDOLYLACETOHYDROXAMIC ACID ◇ 1-(p-CHLOROBENZOYL)-5-METHOXY-2-METHYLINDOLE-3-ACETOHYDROXAMIC ACID ◇ DINULCID ◇ FLOGAR ◇ INDOXAMIC ACID ◇ OXAMETACIN ◇ OXAMETACINE

TOXICITY DATA with REFERENCE
orl-rat LD50:78 mg/kg BCFAAI 114,319,75
ipr-rat LD50:40 mg/kg BCFAAI 114,319,75
orl-mus LD50:92 mg/kg BCFAAI 114,319,75
ipr-mus LD50:32 mg/kg BCFAAI 114,319,75

SAFETY PROFILE: Poison by ingestion and intraperitoneal routes. An anti-inflammatory agent. When heated to decomposition it emits toxic fumes of Cl^- and NO_x.

OLO000 CAS:471-46-5 ***HR: 3***
OXAMIDE
mf: $C_2H_4N_2O_2$ mw: 88.08

PROP: Triclinic needles. Decomp @ 350°, d: 1.667 @ 20°/4°. Sltly sol in hot water, alc.

SYNS: AMID KYSELINY STAVELOVE (CZECH) ◇ 1-CARBAMOYLFORMIMIDIC ACID ◇ ETHANEDIAMIDE ◇ OXALAMIDE ◇ OXALIC ACID DIAMIDE ◇ OXAMID (CZECH) ◇ OXAMIMIDIC ACID

TOXICITY DATA with REFERENCE
eye-rbt 100 mg/24H MOD 28ZPAK -,53,72
orl-rat LD50:447 mg/kg GISAAA 47(7),78,82
orl-mus LD50:235 mg/kg GISAAA 47(7),78,82
ipr-mus LDLo:128 mg/kg CBCCT* 3,127,51

CONSENSUS REPORTS: Reported in EPA TSCA Inventory.

SAFETY PROFILE: Poison by ingestion and intraperitoneal routes. An eye irritant. When heated to decomposition it emits toxic fumes of NO_x. See also AMIDES.

OLS000 CAS:10039-54-0 ***HR: 3***
OXAMMONIUM SULFATE
DOT: UN 2865
mf: $H_6N_2O_2•H_2O_4S$ mw: 164.16

PROP: A crystalline material. Mp: 177°.

SYNS: BIS(HYDROXYLAMINE) SULFATE ◇ HYDROXYLAMINE NEUTRAL SULFATE ◇ HYDROXYLAMINE SULFATE ◇ HYDROXYLAMINE SULFATE (2:1) ◇ HYDROXYLAMMONIUM SULFATE

TOXICITY DATA with REFERENCE
cyt-dmg-orl 5000 ppm CARYAB 31,1,78
ipr-mus LDLo:102 mg/kg TXAPA9 23,288,72

CONSENSUS REPORTS: Reported in EPA TSCA Inventory.

DOT Classification: Corrosive Material; Label: Corrosive.

SAFETY PROFILE: Poison by intraperitoneal route. Mutation data reported. A corrosive irritant to skin, eyes, and mucous membranes. Moderately explosive when exposed to heat or by chemical reaction. In the presence of alkalies at elevated temperatures, free hydroxylamine is liberated and may decompose explosively. When heated to decomposition it emits toxic fumes of SO_x and NO_x. See also AMINES and SULFATES.

OLT000 CAS:21738-42-1 ***HR: 3***
OXAMNIQUINE
mf: $C_{14}H_{21}N_3O_3$ mw: 279.38

SYNS: 6-HYDROXYMETHYL-2-ISOPROPYLAMINOMETHYL-7-NITRO-1,2,3,4-TETRAHYDROQUINOLINE ◇ 2-((ISOPROPYLAMINO)METHYL)-7-NITRO-1,2,3,4-TETRAHYDRO-6-QUINOLINEMETHANOL ◇ MANSIL ◇ 1,2,3,4-TETRAHYDRO-2-((ISOPROPYLAMINO)METHYL)-7-NITRO-6-QUINOLINEMETHANOL ◇ 1,2,3,4-TETRAHYDRO-2-(((1-METHYL-ETHYL)AMINO)METHYL)-7-NITRO-6-QUINOLINEMETHANOL ◇ UK 4261 ◇ UK 4271 ◇ VANSIL

TOXICITY DATA with REFERENCE
mmo-sat 364 nmol/plate JPETAB 200,1,77
msc-ham:lng 1 μmol/L MUREAV 157,1,85
orl-rat LD50:30 mg/kg DDREDK 4,229,84
ipr-rat LD50:20 mg/kg DDREDK 4,229,84
ims-rat LD50:60 mg/kg DDREDK 4,229,84
orl-mus LD50:1300 mg/kg ARZNAD 31,555,81
ipr-mus LD50:650 mg/kg DDREDK 4,229,84
orl-rbt LD50:500 mg/kg DDREDK 4,229,84
orl-ham LD50:950 mg/kg DDREDK 4,229,84

CONSENSUS REPORTS: EPA Genetic Toxicology Program.

SAFETY PROFILE: Poison by ingestion, intraperitoneal and intramuscular routes. Human mutation data reported. An antischistosomal agent. When heated to decomposition it emits toxic fumes of NO_x.

OLU000 CAS:126-93-2 ***HR: 2***
OXANAMIDE
mf: $C_8H_{15}NO_2$ mw: 157.24

PROP: Tasteless, odorless, white crystals from petr ether. Mp: 90-91°. One part sol in 95 parts of water at 30°.

SYNS: 2,3,-EPOXY-2-ETHYLHEXANAMIDE ◇ 2-ETHYL-3-PROPYL-2,3-EPOXYPROPIONAMIDE ◇ 2-ETHYL-3-PROPYLGLYCIDAMIDE ◇ QUIACTIN

TOXICITY DATA with REFERENCE
orl-rat LD50:1250 mg/kg FEPRA7 7,262,48
orl-mus LD50:1220 mg/kg PSEBAA 103,101,60
ipr-mus LD50:720 mg/kg PSEBAA 103,101,60

SAFETY PROFILE: Moderately toxic by ingestion and intraperitoneal routes. When heated to decomposition it emits toxic fumes of NO_x.

OLW000 CAS:500-72-1 ***HR: 2***
OXANILIC ACID
mf: $C_8H_7NO_3$ mw: 165.16

TOXICITY DATA with REFERENCE
ipr-mus LDLo:500 mg/kg CBCCT* 7,395,55

CONSENSUS REPORTS: Reported in EPA TSCA Inventory.

SAFETY PROFILE: Moderately toxic by intraperitoneal route. When heated to decomposition it emits toxic fumes of NO_x.

OLW400 CAS:6577-41-9 ***HR: 3***
OXAPIUM IODIDE
mf: $C_{22}H_{34}NO_2 \cdot I$ mw: 471.47

PROP: α-Form: White crystals from monochlorobenzene or isopropyl alcohol. Mp: 195-197°. Insol in trchloroethylene. β-Form: White crystals from monochlorobenzene. Mp: 150-152°. Sol in hot trichloroethylene. Both sol in methanol, ethanol, chloroform, and tetrachlorethane, hardly sol in benzene, toluene, xylene, and water.

SYNS: ANC 113 ◇ CICLONIUM IODIDE ◇ 2-CYCLOHEXYL-2-PHENYL-4-PIPERIDINOMETHYL-DIOXOLANE-1,3 METHIODIDE ◇ CYCLONIUM IODIDE ◇ ESPERAN ◇ N-METHYL-N-(2-CYCLOHEXYL-2-PHENYL-1,3-DIOXOLAN-4-YL-METHYL)-PIPERIDINIUMIODIDE ◇ SH 100

TOXICITY DATA with REFERENCE
orl-mus TDLo:600 mg/kg (7-14 D preg):REP OYYAA2 4,109,70
orl-rat LD50:494 mg/kg OYYAA2 4,109,70
scu-rat LD50:244 mg/kg NIIRDN 6,864,82
ivn-rat LD50:13900 μg/kg NIIRDN 6,864,82
orl-mus LD50:512 mg/kg MEIEDD 10,393,83
ipr-mus LD50:16 mg/kg PJPPAA 30,493,78
scu-mus LD50:87900 μg/kg MEIEDD 10,393,83
ivn-mus LD50:6970 μg/kg MEIEDD 10,393,83
ivn-dog LD50:31 mg/kg NIIRDN 6,864,82

SAFETY PROFILE: Poison by subcutaneous, intravenous, and intraperitoneal routes. Moderately toxic by ingestion. Experimental reproductive effects. When heated to decomposition it emits toxic fumes of NO_x and I^-.

OLW600 CAS:21256-18-8 ***HR: 3***
OXAPROZIN
mf: $C_{18}H_{15}NO_3$ mw: 293.34

PROP: Crystals from methanol. Mp: 160.5-161.5°.

SYNS: ALVO ◇ 4,5-DIPHENYL-2-OXAZOLEPROPANOIC ACID ◇ 4,5-DIPHENYL-2-OXAZOLEPROPIONIC ACID ◇ DURAPROST ◇ OXAPRO ◇ WY-21743

TOXICITY DATA with REFERENCE
orl-rat TDLo:540 mg/kg (female 17-22D post):REP IYKEDH 15,265,84
orl-rbt TDLo:39 mg/kg (female 6-18D post):TER IYKEDH 15,250,84
orl-rat LD50:4470 mg/kg IYKEDH 15,359,84
ipr-rat LD50:506 mg/kg IYKEDH 15,359,84
scu-rat LD50:2910 mg/kg IYKEDH 15,359,84
ivn-rat LD50:82 mg/kg IYKEDH 15,359,84
orl-mus LD50:1210 mg/kg IYKEDH 15,359,84
ipr-mus LD50:376 mg/kg IYKEDH 15,359,84
scu-mus LD50:556 mg/kg IYKEDH 15,359,84
ivn-mus LD50:93 mg/kg IYKEDH 15,359,84
ipr-dog LD50:200 mg/kg IYKEDH 15,359,84
ivn-dog LD50:124 mg/kg IYKEDH 15,359,84

SAFETY PROFILE: Poison by ingestion, intravenous, and intraperitoneal routes. Moderately toxic by subcutaneous route. An experimental teratogen. Experimental reproductive effects. An anti-inflammatory agent. When heated to decomposition it emits toxic fumes of NO_x.

OLY000 CAS:15980-15-1 ***HR: 2***
1,4-OXATHIANE
mf: C_4H_8OS mw: 104.18

$\overline{OC_2H_4SCH_2CH_2}$

PROP: Water-white, refractive, mobile liquid; characteristic odor. Bp: 148.7°, fp: −17°, flash p: 108°F (CC): d: 1.117 @ 20°.

SYNS: OXATHIANE ◇ p-THIOXANE

TOXICITY DATA with REFERENCE
skn-rbt 10 mg/24H MLD AMIHBC 4,119,51
eye-rbt 20 mg open AMIHBC 4,119,51
orl-rat LD50:2830 mg/kg AMIHBC 4,119,51
ihl-rat LCLo:4000 ppm/4H JIHTAB 31,343,49

SAFETY PROFILE: Moderately toxic by ingestion. Mildly toxic by inhalation. A skin and eye irritant. Combustible when exposed to heat, flame or oxidizers. Forms explosive complexes with silver perchlorate or copper(I) perchlorate. Incompatible with metal perchlorates and oxidizing materials. To fight fire, use water, CO_2, dry chemical. When heated to decomposition it emits toxic fumes of SO_x.

OMA000 CAS:73806-23-2 ***HR: 3***
1,4-OXATHIANE compound with MERCURIC CHLORIDE
mf: $C_4H_8OS{\cdot}Cl_2Hg$ mw: 375.67

SYN: MERCURIC CHLORIDE-1,4-OXATHIANE

TOXICITY DATA with REFERENCE
ivn-mus LD50:18 mg/kg CSLNX* NX#05126

CONSENSUS REPORTS: Mercury and its compounds are on the Community Right-To-Know List.

OSHA PEL: (Transitional: CL 1 mg/10m^3) CL 0.1 mg(Hg)/m^3 (skin)
ACGIH TLV: TWA 0.1 mg(Hg)/m^3 (skin)
NIOSH REL: (Inorganic Mercury) TWA 0.05 mg(Hg)/m^3

SAFETY PROFILE: Poison by intravenous route. When heated to decomposition it emits very toxic fumes of Hg, Cl^-, and SO_x. See also individual components.

OMC000 CAS:4378-73-8 ***HR: D***
1,2-OXATHIETANE-2,2-DIOXIDE
mf: $C_3H_6O_3S$ mw: 122.15

TOXICITY DATA with REFERENCE
mmo-sat 100 nmol/plate CBINA8 19,241,77
hma-mus/sat 10 μmol/kg CBINA8 19,241,77

SAFETY PROFILE: Mutation data reported. When heated to decomposition it emits toxic fumes of SO_x.

OMG000 CAS:60607-34-3 ***HR: 3***
OXATIMIDE
mf: $C_{27}H_{30}N_4O$ mw: 426.61

SYNS: 1-(3-(4-(DIPHENYLMETHYL)-1-PIPERAZINYL)PROPYL)-2-BENZIMIDAZOLINONE ◇ 1-(3-(4-(DIPHENYLMETHYL)-1-PIPERAZINYL)PROPYL)-1,3-DIHYDRO-2H-BENZIMIDAZOL-2-ONE ◇ KW-4354 ◇ OXATOMIDA ◇ OXATOMIDE ◇ R 35443 ◇ TINSET

TOXICITY DATA with REFERENCE
orl-rat TDLo:110 mg/kg (female 7-17D post):REP YACHDS 12,2851,84
orl-rat TDLo:440 mg/kg (female 7-17D post):TER YACHDS 12,2851,84
orl-rat LD50:1410 mg/kg YACHDS 12,2769,84
ipr-rat LD50:63 mg/kg YACHDS 12,2769,84
ivn-rat LD50:29 mg/kg YACHDS 12,2769,84
orl-mus LD50:9596 mg/kg YACHDS 12,2769,84
ipr-mus LD50:7926 mg/kg YACHDS 12,2769,84
ivn-mus LD50:25 mg/kg YACHDS 12,2769,84
orl-gpg LD50:320 mg/kg DRFUD4 3,465,78
ivn-gpg LD50:23 mg/kg DRFUD4 3,465,78

SAFETY PROFILE: Poison by ingestion, intraperitoneal, and intravenous routes. An experimental teratogen. Experimental reproductive effects. Used to treat allergies and asthma. When heated to decomposition it emits toxic fumes of NO_x.

OMK000 CAS:32388-21-9 ***HR: 3***
OXAZINOMYCIN
mf: $C_9H_{11}NO_7$ mw: 245.21

SYNS: MINIMYCIN ◇ 5-β-d-RIBOFURANOSYL-2H-1,3-OXAZINE-2,4(3H)-DIONE ◇ 5-β-d-RIBOFURANOSYL-1,3-OXAZINE-2,4-DIONE

TOXICITY DATA with REFERENCE
ipr-mus LD50:5 mg/kg 85GDA2 5,281,81
scu-mus LD50:20 mg/kg JANTAJ 25,44,72
ivn-mus LD50:50 mg/kg 85GDA2 5,281,81

SAFETY PROFILE: Poison by intraperitoneal, subcutaneous, and intravenous routes. When heated to decomposition it emits toxic fumes of NO_x.

OMK300 CAS:24143-17-7 ***HR: 2***
OXAZOLAZEPAM
mf: $C_{18}H_{17}ClN_2O_2$ mw: 328.82

SYNS: OXAZOLAM ◇ SERENAL

TOXICITY DATA with REFERENCE
orl-mus TDLo:600 mg/kg (female 7-12D post):REP SKKNAJ 21,107,69
orl-rat TDLo:1800 mg/kg (9-14D preg):TER SKKNAJ 21,107,69
orl-mus LD50:5200 mg/kg NYKZAU 66,107,70
ipr-mus LD50:768 mg/kg 85IPAE -,91,72

SAFETY PROFILE: Moderately toxic by intraperitoneal route. Mildly toxic by ingestion. An experimental teratogen. Experimental reproductive effects. When heated to decomposition it emits toxic fumes of Cl^- and NO_x.

OMM000 CAS:497-25-6 ***HR: 2***
2-OXAZOLIDINONE
mf: $C_3H_5NO_2$ mw: 87.09

SYNS: (2-HYDROXYETHYL)CARBAMIC ACID,Γ-LACTONE ◇ OXAZOLIDONE

TOXICITY DATA with REFERENCE
orl-mus TDLo:1000 mg/kg:ETA BCPCA6 2,168,59

SAFETY PROFILE: Questionable carcinogen with experimental tumorigenic data. When heated to decomposition it emits toxic fumes of NO_x. See also CARBAMATES.

OMS400 CAS:16485-39-5 ***HR: 3***
OXELADINE CITRATE
mf: $C_{20}H_{33}NO_3 \cdot xC_6H_8O_7$

SYN: OXELADIN CITRATE

TOXICITY DATA with REFERENCE
orl-rat LD50:183 mg/kg NIIRDN 6,157,82
orl-mus LD50:130 mg/kg NIIRDN 6,157,82
scu-mus LD50:244 mg/kg YKKZAJ 82,1314,62
ivn-mus LD50:13 mg/kg NIIRDN 6,157,82

SAFETY PROFILE: Poison by ingestion, subcutaneous, and intravenous routes. When heated to decomposition it emits toxic fumes of NO_x.

OMU000 CAS:55689-65-1 ***HR: 3***
OXEPINAC
mf: $C_{16}H_{12}O_4$ mw: 268.28

SYNS: 6,11-DIHYDRO-11-OXO-DIBENZ(b,e)OXEPIN-3-ACETICACID ◇ OXEPINACO (SPANISH)

TOXICITY DATA with REFERENCE
orl-rat TDLo:300 mg/kg (6-15D preg):REP ARZNAD 28,451,78
orl-rbt TDLo:390 mg/kg (female 6-18D post):TER ARZNAD 28,451,78
orl-rat LD50:110 mg/kg JMCMAR 19,941,76
ipr-rat LD50:174 mg/kg ARZNAD 28,445,78
scu-rat LD50:113 mg/kg ARZNAD 28,445,78
orl-mus LD50:852 mg/kg ARZNAD 28,445,78
ipr-mus LD50:596 mg/kg ARZNAD 28,445,78
scu-mus LD50:645 mg/kg ARZNAD 28,445,78
orl-dog LD50:600 mg/kg DRFUD4 3,602,78
orl-rbt LD50:308 mg/kg ARZNAD 28,445,78

SAFETY PROFILE: Poison by ingestion, intraperitoneal, and subcutaneous routes. An experimental teratogen. Experimental reproductive effects. When heated to decomposition it emits acrid smoke and irritating fumes.

OMW000 CAS:503-30-0 ***HR: 2***
OXETANE
mf: C_3H_6O mw: 58.09

PROP: Oil; agreeable odor. D: 0.8930 @ 25°/4°, bp: 480 @ 750 mm.

SYNS: 1,3-PROPYLENE OXIDE ◇ TRIMETHYLENE OXIDE ◇ TRIMETHYLENOXID (GERMAN)

TOXICITY DATA with REFERENCE
scu-rat TDLo:2240 mg/kg/56W-I:ETA ZEKBAI 74,241,70
scu-rat LD50:500 mg/kg ZEKBAI 74,241,70

CONSENSUS REPORTS: Reported in EPA TSCA Inventory.

SAFETY PROFILE: Moderately toxic by subcutaneous route. May be narcotic in high concentrations. Questionable carcinogen with experimental tumorigenic data. When heated to decomposition it emits acrid smoke and irritating fumes.

OMY500 CAS:53716-50-0 ***HR: D***
OXFENDAZOLE
mf: $C_{15}H_{13}N_3O_3S$ mw: 315.37

PROP: Crystals from chloroform-methanol. Mp: 253° (decomp).

SYNS: METHYL5-(PHENYLSULFINYL)-2-BENZIMIDAZOLECARBAMATE ◇ METHYL (5-PHENYLSULFINYL)-1H-BENZIMIDAZOL-2-YL CARBAMATE ◇ OFDZ ◇ 5-(PHENYLSULFINYL)-2-BENZIMIDAZOLECARBAMIC ACID METHYL ESTER ◇ (5-(PHENYLSULFINYL)-1H-BENZIMIDAZOL-2-YL)CARBAMIC ACID METHYL ESTER ◇ 5-PHENYLSULFINYL-2-CARBOMETHOXYAMINOBENZIMIDAZOLE ◇ RS-8858 ◇ SYNANTHIC ◇ SYSTAMEX

TOXICITY DATA with REFERENCE
oms-hmn:oth 2 mg/L THERAP 31,505,76
orl-dom TDLo:22500 μg/kg (17D preg):REP RMVEAG 153,639,77
orl-rat TDLo:126 mg/kg (8-15D preg):TER RMVEAG 153,639,77

SAFETY PROFILE: An experimental teratogen. Experimental reproductive effects. Human mutation data reported. When heated to decomposition it emits toxic fumes of SO_x and NO_x. See also CARBAMATES and ESTERS.

OMY700 CAS:67049-95-0 ***HR: 3***
OXIBENDAZOLE

SYNS: LODITAC ◇ METHYL 5-n-PROPOXY-2-BENZIMIDAZOLE CARBAMATE ◇ OBDZ ◇ N-(PROPOXY-5-BENZIMIDAZOLYL)-2, CARBAMATE de METHYLE (FRENCH) ◇ N-(2-(5-PROPOXYBENZIMIDAZOLYL)) METHYL CARBAMATE ◇ SKF 30310

TOXICITY DATA with REFERENCE
orl-dom TDLo:90 mg/kg (female 14-21D post):REP RMVEAG 152,467,76
orl-rat TDLo:900 mg/kg (female 8-15D post):TER RMVEAG 152,467,76
orl-mus LDLo:32 mg/kg AJVRAH 38,809,77

SAFETY PROFILE: Poison by ingestion. An experimental teratogen. Experimental reproductive effects. When heated to decomposition it emits toxic fumes of NO_x. See also CARBAMATES and ESTERS.

OMY800 ***HR: 2***
3-N-OXIDE PURIN-6-THIOL MONOHYDRATE
mf: $C_5H_4N_4OS \cdot H_2O$ mw: 186.21

SYN: 6-MERCAPTOPURINE 3-N-OXIDE MONOHYDRATE

TOXICITY DATA with REFERENCE
scu-rat TDLo:6500 mg/kg/26W-I:NEO CNREA8 27,925,67

SAFETY PROFILE: Questionable carcinogen with experimental neoplastigenic data. When heated to decomposition it emits toxic fumes of NO_x and SO_x.

OMY850 CAS:58-36-6 ***HR: 3***
10,10'-OXIDIPHENOXARSINE
mf: $C_{24}H_{16}As_2O_3$ mw: 502.24

PROP: Colorless, monoclinic prisms. Mp: 184-185°, decomp 380°, sp. gr. 1.40-1.42. Sol in alc, chloroform, methylene chloride. Practically insol in water (5 ppm at 20°) and alkali.

SYNS: BIS(PHENOXARSIN-10-YL) ETHER ◇ BIS(10-PHENOXARSYL) OXIDE ◇ BIS(10-PHENOXYARSINYL) OXIDE ◇ 10,10'-BIS(PHENOXYARSINYL) OXIDE ◇ DID 47 ◇ OBPA ◇ 10-10' OXYBISPHENOXYARSINE ◇ PHENOXAKSINE OXIDE ◇ PXO ◇ SA 546 ◇ VINADINE ◇ VINYZENE ◇ VINYZENE bp 5 ◇ VINYZENE bp 5-2 ◇ VINYZENE (pesticide) ◇ VINYZENE SB 1

TOXICITY DATA with REFERENCE
skn-gpg 250 mg/5D SEV TXCYAC 10,341,78
orl-rat LD50:40 mg/kg TCXYAC 10,341,78
orl-mus LDLo:42 mg/kg AECTCV 14,111,85
orl-gpg LD50:24 mg/kg TXCYAC 10,341,78
ihl-gpg LCLo:141 mg/m^3/2H TXAPA9 10,341,78
orl-bwd LD50:24 mg/kg TXAPA9 21,315,72

CONSENSUS REPORTS: Arsenic and its compounds are on The Community Right-To-Know List.

OSHA PEL: TWA 0.5 mg(As)/m^3

SAFETY PROFILE: Poison by ingestion and inhalation. A severe skin irritant. When heated to decomposition it emits toxic fumes of As. See also ARSENIC COMPOUNDS.

OMY899 ***HR: 3***
OXIMES

PROP: Compounds of the form RC=NOH.

SAFETY PROFILE: May explode when heated. Their instability may be due to the presence of peroxides resulting from autooxidation. The presence of iron(III) chloride increases their sensitivity to heat. Oxime carbamates (C=NOCO•NH–) are heat-sensitive explosives. When heated to decomposition it emits toxic fumes of NO_x.

OMY925 CAS:88321-09-9 ***HR: 2***
OXIRANECARBOXYLIC ACID, 3-(((3-METHYL-1-(((3-METHYLBUTYL)AMINO)CARBONYL) BUTYL)AMINO) CARBONYL)-, ETHYL ESTER, (2S-(2-α-3-β(R*)))-
mf: $C_{17}H_{30}N_2O_5$ mw: 342.49

SYNS: EP 453 ◇ EST

TOXICITY DATA with REFERENCE
cyt-ham:lng 200 mg/L IYKEDH 17,815,86
orl-rat TDLo:2750 mg/kg (female 7-17D post):REP IYKEDH 17,617,86
orl-rbt TDLo:6500 mg/kg (female 6-18D post):TER IYKEDH 17,632,86
ipr-rat LD50:2050 mg/kg IYKEDH 17,736,86
ipr-mus LD50:3270 mg/kg IYKEDH 17,736,86

SAFETY PROFILE: Moderately toxic by intraperitoneal route. An experimental teratogen. Experimental reproductive effects. Mutation data reported. When heated to decomposition it emits toxic fumes of NO_x.

ONC000 CAS:61695-72-5 ***HR: 2***
7-OXIRANYLBENZ(a)ANTHRACENE
mf: $C_{20}H_{14}O$ mw: 270.34

SYNS: BENZ(a)ANTHRACEN-7-YL-OXIRANE ◇ 7-BENZANTHRYL-OXIRANE ◇ 7-(EPOXYETHYL)-BENZ(a)ANTHRACENE

TOXICITY DATA with REFERENCE
mmo-sat 1 nmol/plate CNREA8 38,3247,78
dnd-omi 1100 μmol/L CNREA8 38,3247,78
scu-rat TDLo:11 mg/kg/40D-I:ETA CALEDQ 1,339,76

CONSENSUS REPORTS: EPA Genetic Toxicology Program.

SAFETY PROFILE: Questionable carcinogen with experimental tumorigenic data. Mutation data reported. When heated to decomposition it emits acrid smoke and irritating fumes.

ONE000 CAS:61695-69-0 ***HR: 2***
6-OXIRANYLBENZO(a)PYRENE
mf: $C_{22}H_{14}O$ mw: 294.36

TOXICITY DATA with REFERENCE
scu-rat TDLo:12 mg/kg/40D-I:ETA CALEDQ 1,339,76

SAFETY PROFILE: Questionable carcinogen with experimental tumorigenic data. When heated to decomposition it emits acrid smoke and irritating fumes.

ONG000 CAS:61695-74-7 ***HR: 2***
1-OXIRANYLPYRENE
mf: $C_{18}H_{12}O$ mw: 244.30

SYNS: 1-EPOXYETHYLPYRENE ◇ 1-PYRENYLOXIRANE

TOXICITY DATA with REFERENCE
mmo-sat 100 pmol/plate CNREA8 40,642,80
dnd-omi 27 μmol/L CNREA8 38,3247,78
msc-ham:lng 100 μg/L IJCNAW 24,203,79
dnd-mam:lym 1790 μmol/L BBRCA9 82,929,78
skn-mus TDLo:977 μg/kg:ETA CNREA8 40,642,80

CONSENSUS REPORTS: EPA Genetic Toxicology Program.

SAFETY PROFILE: Questionable carcinogen with experimental tumorigenic data. Mutation data reported.

When heated to decomposition it emits acrid smoke and irritating fumes.

ONI000 CAS:30286-75-0 ***HR: 3***
OXITROPIUM BROMIDE
mf: $C_{19}H_{26}NO_4 \cdot Br$ mw: 412.37

SYNS: BA 253 BR ◇ Ba-253-BR-L ◇ BROMURO de OXITROPIO(SPANISH) ◇ (−)-N-ETHYL-NORHYOSCINE-METHOBROMIDE ◇ N-ETHYLNORSCOPOLAMINEMETHOBROMIDE ◇ OTB ◇ OXYTROPIUM BROMIDE ◇ VENTILAT

TOXICITY DATA with REFERENCE
orl-rat LD50:2250 mg/kg DRFUD4 4,117,79
ivn-rat LD50:44 mg/kg DRFUD4 4,117,79
orl-mus LD50:1600 mg/kg DRFUD4 4,117,79
scu-mus LD50:1150 mg/kg DRFUD4 4,117,79
ivn-mus LD50:25700 μg/kg DRFUD4 4,117,79
orl-dog LD50:3 g/kg DRFUD4 4,117,79
ivn-dog LD50:50 mg/kg DRFUD4 4,117,79
ivn-rbt LD50:34 mg/kg DRFUD4 4,117,79

SAFETY PROFILE: Poison by intravenous route. Moderately toxic by ingestion and subcutaneous routes. An anticholinergic bronchodilator. When heated to decomposition it emits very toxic fumes of NO_x and Br^-. See also BROMIDES.

ONO000 CAS:566-28-9 ***HR: 2***
7-OXOCHOLESTEROL
mf: $C_{27}H_{44}O_2$ mw: 400.71

SYNS: 3-β-HYDROXYCHOLEST-5-EN-7-ONE (8CI) ◇ (3-β)3-HYDROXYCHOLEST-5-EN-7-ONE (9CI) ◇ 7-KETOCHOLESTEROL ◇ SC 4722

TOXICITY DATA with REFERENCE
scu-mus TDLo:600 mg/kg/72W-I:ETA JNCIAM 19,977,57

SAFETY PROFILE: Questionable carcinogen with experimental tumorigenic data. When heated to decomposition it emits acrid smoke and irritating fumes.

ONQ000 CAS:77791-69-6 ***HR: D***
N-(2-OXO-3,5,7-CYCLOHEPTATRIEN-1-YL) AMINOOXOACETIC ACID ETHYL ESTER
mf: $C_{10}H_{13}NO_4$ mw: 211.24

SYN: AY-25,674

TOXICITY DATA with REFERENCE
orl-rat TDLo:5 g/kg (6-15D preg):TER APTOD9 18,A37,79

SAFETY PROFILE: An experimental teratogen. When heated to decomposition it emits toxic fumes of NO_x. See also ESTERS.

ONU000 ***HR: 3***
OXODIPEROXODIPYRIDINECHROMIUM(VI)
mf: $C_{10}H_{10}CrN_2O_5$ mw: 290.20

$(C_5H_5N)_2CrO(O_2)_2$

SYN: PYRIDINE PERCHROMATE

CONSENSUS REPORTS: Chromium and its compounds are on The Community Right-To-Know List.

SAFETY PROFILE: A powerful oxidant. Violently explosive when dry. When heated to decomposition it emits toxic fumes of NO_x. See also CHROMIUM COMPOUNDS.

ONU100 CAS:38293-27-5 ***HR: 3***
OXODIPEROXOPYRIDINE CHROMIUM-N-OXIDE
mf: $C_5H_5CrNO_6$ mw: 227.09

$OCr(O_2)_2C_5H_5N{:}O$

CONSENSUS REPORTS: Chromium and its compounds are on the Community Right-To-Know List.

SAFETY PROFILE: An explosive. When heated to decomposition it emits toxic fumes of NO_x. See also CHROMIUM COMPOUNDS and PEROXIDES.

ONW000 ***HR: 3***
OXODISILANE
mf: H_4OSi_2 mw: 76.20

$H_3SiSi(O)H$

SAFETY PROFILE: Ignites spontaneously in air. See also SILANE.

ONY000 CAS:1707-95-5 ***HR: 3***
2-(3-OXO-1-INDANYLIDENE)-1,3-INDANDIONE
mf: $C_{18}H_{10}O_3$ mw: 274.28

SYN: BINDON

TOXICITY DATA with REFERENCE
ipr-mus TDLo:12500 μg/kg (9D preg):REP ARTODN 33,191,75
ipr-mus TDLo:12500 μg/kg (9D preg):TER ARTODN 33,191,75
ipr-mus LDLo:100 mg/kg ARTODN 33,191,75

SAFETY PROFILE: Poison by intraperitoneal route. An experimental teratogen. Experimental reproductive effects. When heated to decomposition it emits acrid smoke and irritating fumes.

OOA000 CAS:5100-91-4 ***HR: 2***
8-OXO-8H-ISOCHROMENO(4',3':4,5)PYRROLO-(2,3-f)QUINOLINE
mf: $C_{18}H_{10}N_2O_2$ mw: 286.30

TOXICITY DATA with REFERENCE
scu-mus TDLo:72 mg/kg/9W-I:ETA SCIEAS 158,387,67

SAFETY PROFILE: Questionable carcinogen with experimental tumorigenic data. When heated to decomposition it emits toxic fumes of NO_x.

OOC000 CAS:959-14-8 ***HR: 2***
OXOLAMINE
mf: $C_{14}H_{19}N_3O$ mw: 245.36

PROP: Liquid. Bp: 127° @ 0.4 mm.

SYN: N,N-DIETHYL-3-PHENYL-1,2,4-OXADIAZOLE-5-ETHANAMINE ◇ 3-PHENYL-5-(β-DIETHYLAMINOETHYL)-1,2,4-OXODIAZOLE

TOXICITY DATA with REFERENCE
orl-mus LD50:925 mg/kg BCFAAI 109,476,70
scu-mus LD50:672 mg/kg JMCMAR 10,411,67

SAFETY PROFILE: Moderately toxic by ingestion and subcutaneous routes. When heated to decomposition it emits toxic fumes of NO_x. See also AMINES and other oxolamine entries.

OOE000 CAS:1949-20-8 ***HR: 3***
OXOLAMINE CITRATE
mf: $C_{14}H_{19}N_3O \cdot C_6H_8O_7$ mw:437.50

PROP: Crystals. Sltly sol in water and alc.

SYNS: 5-β-DIETHYLAMINOETHYL-3-PHENYL-1,2,4-OXADIAZOLE CITRATE ◇ 3-PHENYL-5-(β-(DIETHYLAMINO)ETHYL)-1,2,4-OXADIAZOLE CITRATE

TOXICITY DATA with REFERENCE
scu-mus TDLo:120 mg/kg (9-11D preg):REP ARZNAD 17,781,67
scu-mus TDLo:120 mg/kg (9-11D preg):TER ARZNAD 17,781,67
orl-rat TDLo:15 g/kg/1Y-I:CAR EMPSAL 2,1,63
orl-rat LD50:1650 mg/kg BJPCAL 16,209,61
orl-mus LD50:650 mg/kg THERAP 19,631,64
ipr-mus LD50:351 mg/kg BJPCAL 16,209,61

SAFETY PROFILE: Poison by intraperitoneal route. Moderately toxic by ingestion. Experimental teratogenic and reproductive effects. Questionable carcinogen with experimental carcinogenic data. When heated to decomposition it emits toxic fumes of NO_x. See also AMINES and other oxolamine entries.

OOE100 CAS:1219-20-1 ***HR: 3***
OXOLAMINE HYDROCHLORIDE
mf: $C_{14}H_{19}N_3O \cdot ClH$ mw: 281.82

SYNS: AF 438 HYDROCHLORIDE ◇ 5-(2-(DIETHYLAMINO)ETHYL)-3-PHENYL-1,2,4-OXADIAZOLE HYDROCHLORIDE ◇ 5-(2-DIETILAMINOETIL)-3-FENIL-1,2,4-OXADIEZOLOCLORIDRATO (ITALIAN) ◇ 683 M HYDROCHLORIDE ◇ OXOLAMINA CLORIDRATO (ITALIAN) ◇ 3-PHENYL-5-(β-DIETHYLAMINOETHYL)-1,2,4-OXADIAZOLE HYDROCHLORIDE

TOXICITY DATA with REFERENCE
orl-rat LD50:1650 mg/kg BJPCAL 16,209,61
ipr-rat LD50:185 mg/kg BJPCAL 16,209,61
orl-mus LD50:1061 mg/kg BCFAAI 109,476,70
ipr-mus LD50:208 mg/kg BJPCAL 16,209,61
scu-mus LD50:465 mg/kg BJPCAL 16,209,61
ivn-mus LD50:63 mg/kg BJPCAL 16,209,61

SAFETY PROFILE: Poison by intravenous and intraperitoneal routes. Moderately toxic by ingestion and subcutaneous routes. When heated to decomposition it emits toxic fumes of NO_x and HCl. See also AMINES and other oxolamine entries.

OOG000 CAS:14698-29-4 ***HR: 2***
OXOLINIC ACID
mf: $C_{13}H_{11}NO_5$ mw: 261.25

PROP: Crystals from DMF. Mp: 314-316° (decomp).

SYNS: EMYRENIL ◇ 1-ETHYL-1,4-DIHYDRO-6,7-METHYLENEDIOXY-4-OXO-3-QUINOLINECARBOXYLIC ACID ◇ 5-ETHYL-5,8-DIHYDRO-8-OXO-1,3-DIOXOLO(4,5-g)QUINOLINE-7-CARBOXYLIC ACID ◇ 1-ETHYL-6,7-METHYLENEDIOXY-4-QUINOLONE-3-CARBOXYLIC ACID ◇ NIDANTIN ◇ NSC-110364 ◇ OSSIAN ◇ OXOBOI ◇ PIETIL ◇ PRODOXOL ◇ URITRATE ◇ URO-ALVAR ◇ UROTRATE ◇ UROXOL ◇ UTIBID ◇ W 4565

TOXICITY DATA with REFERENCE
mmo-esc 130 μg/L AMACCQ 17,763,80
oms-esc 50 mg/L EJBCAI 62,491,76
orl-rat LD50:525 mg/kg TXAPA9 18,185,71
orl-mus LD50:1890 mg/kg TXAPA9 18,185,71

SAFETY PROFILE: Moderately toxic by ingestion. Mutation data reported. When heated to decomposition it emits toxic fumes of NO_x.

OOI000 CAS:125-29-1 ***HR: 3***
6-OXO-3-METHOXY-N-METHYL-4,5-EPOXYMORPHINAN
mf: $C_{18}H_{21}NO_3$ mw: 299.40

SYNS: BEKADID ◇ DICO ◇ DICODID ◇ DIHYDROCODEINONE ◇ 4,5α-EPOXY-3-METHOXY-17-METHYLMORPHINAN-6-ONE ◇ HYDROCODONE

TOXICITY DATA with REFERENCE
scu-rat LD50:150 mg/kg ARZNAD 3,238,53
scu-mus LD50:86 mg/kg JPETAB 52,468,34
ivn-cat LDLo:9 mg/kg AEPPAE 194,296,40
ivn-rbt LDLo:7 mg/kg AEPPAE 194,296,40

SAFETY PROFILE: Poison by intravenous and subcutaneous routes. When heated to decomposition it emits toxic fumes of NO_x.

OOK000 CAS:55764-18-6 ***HR: 2***
9-OXO-8-OXATRICYCLO(5.3.1.0^{2,6})UNDECANE
mf: $C_{11}H_{14}O_2$ mw: 178.25

SYN: 8-OXA-9-KETOTRICYCLO(5.3.1.0^{2,6})UNDECANE

TOXICITY DATA with REFERENCE
orl-rat LD50:840 mg/kg TXAPA9 28,313,74
skn-rbt LD50:1270 mg/kg TXAPA9 28,313,74

CONSENSUS REPORTS: Reported in EPA TSCA Inventory.

SAFETY PROFILE: Moderately toxic by ingestion and skin contact. When heated to decomposition it emits acrid smoke and irritating fumes.

OOK100 CAS:306-12-7 ***HR: 3***
OXOPHENARSINE
mf: $C_6H_6AsNO_2$ mw: 199.05

SYNS: 2-AMINO-4-ARSENOSOPHENOL ◇ (3-AMINO-4-HYDROXY-PHENYL)ARSENOUS ACID ◇ PHENARSEN ◇ PHENOL, 2-AMINO-4-ARSENOSO-

TOXICITY DATA with REFERENCE
orl-mus LDLo:150 mg/kg JPETAB 76,358,42
ivn-mus LDLo:30 mg/kg JPETAB 76,358,42

OSHA PEL: TWA 0.5 mg(As)/m^3

SAFETY PROFILE: Poison by ingestion and intravenous routes. When heated to decomposition it emits toxic fumes of NO_x and As.

OOM300 CAS:42472-96-8 ***HR: 2***
N-(2-OXO-3-PIPERIDYL)PHTHALIMIDE
mf: $C_{13}H_{12}N_2O_3$ mw: 244.27

SYNS: 3-(1,3-DIHYDRO-1,3-DIOXO-2H-ISOINDOL-2-YL)-2-OXOPIPERIDINE ◇ EM 136

TOXICITY DATA with REFERENCE
orl-rbt TDLo:1200 mg/kg (7-18D preg):TER ARZNAD 31,941,81
orl-rbt TDLo:1200 mg/kg (7-18D preg):REP ARZNAD 31,941,81
orl-mus LD50:1350 mg/kg ARZNAD 31,941,81
ipr-mus LD50:747 mg/kg ARZNAD 31,941,81

SAFETY PROFILE: Moderately toxic by ingestion and intraperitoneal routes. An experimental teratogen. Experimental reproductive effects. When heated to decomposition it emits toxic fumes of NO_x.

OOM400 ***HR: 3***
OXOPROPANEDINITRILE
mf: C_3N_2O mw: 80.05

O:C(CN)$_2$

SYN: CARBONYL DICYANIDE

CONSENSUS REPORTS: Cyanide and its compounds are on the Community Right-To-Know List.

SAFETY PROFILE: Explosive reaction on contact with water. When heated to decomposition it emits toxic fumes of NO_x and CN^-. See also NITRILES.

OOO000 CAS:64058-54-4 ***HR: 3***
p-(2-OXOPROPOXY)BENZENEARSONIC ACID
mf: $C_9H_{11}AsO_5$ mw: 274.12

TOXICITY DATA with REFERENCE
orl-rat LDLo:2500 mg/kg JPETAB 63,122,38
ivn-rat LDLo:900 mg/kg JPETAB 63,122,38
ims-rat LDLo:600 mg/kg JPETAB 63,122,38

CONSENSUS REPORTS: Arsenic and its compounds are on The Community Right-To-Know List.

OSHA PEL: TWA 0.5 mg(As)/m^3

SAFETY PROFILE: Arsenic compounds are poisons. Moderately toxic by ingestion, intravenous, and intramuscular routes. When heated to decomposition it emits toxic fumes of As. See also ARSENIC COMPOUNDS.

OOO100 CAS:40942-73-2 ***HR: 1***
3-(2-OXOPROPYL)-2-PENTYLCYCLO-PENTANONE
mf: $C_{13}H_{22}O_2$ mw: 210.35

SYNS: CYCLOPENTANONE, 3-(2-OXOPROPYL)-2-PENTYL- ◇ MAGNOLIONE ◇ PENTYLCYCLOPENTANONEPROPANONE

TOXICITY DATA with REFERENCE
skn-rbt 500 mg/24H MOD FCTOD7 20,795,82

CONSENSUS REPORTS: Reported in EPA TSCA Inventory.

SAFETY PROFILE: A skin irritant. When heated to decomposition it emits acrid smoke and irritating fumes.

OOS000 CAS:22755-01-7 ***HR: 3***
OXOSILANE
mf: H_2OSi mw: 46.10

O:SiH$_2$

SAFETY PROFILE: Spontaneously flammable in air or Cl_2. See also SILANE.

OOY000 CAS:70-22-4 ***HR: 3***
OXOTREMORINE
mf: $C_{12}H_{18}N_2O$ mw: 206.32

PROP: Pale yellow liquid. Bp: 150-151° @ 0.6 mm; 124° @ 0.1 mm.

SYNS: 2′-OXOPYRROLIDINO-1-PYRROLIDINO-4-BUTYNE ◇ OXOTREMORIN ◇ 1-(4-(1-PYRROLIDINYL)-2-BUTYNYL)-2-PYRROLIDINONE

TOXICITY DATA with REFERENCE
ipr-rat LD50:4400 μg/kg TXAPA9 14,67,69
orl-mus LD50:11300 μg/kg HPFSDS FDA-80-1076,212,80

ipr-mus LD50:3 mg/kg ATXKA8 29,39,72
ivn-mus LD50:1400 mg/kg BJPCAL 26,56,66

SAFETY PROFILE: Poison by ingestion and intraperitoneal routes. Moderately toxic by intravenous route. When heated to decomposition it emits toxic fumes of NO_x.

OPC000 CAS:13380-94-4 ***HR: 2***
8-OXOTRICYCLO(5.2.1.0^{2,6})DECANE
mf: $C_{10}H_{14}O$ mw: 150.24

SYNS: CORODANE ◇ 8-KETOTRICYCLO(5.2.1.0^{2,6})DECANE ◇ TETRAHYDRO-4,7-METHANOINDAN-5(4H)-ONE

TOXICITY DATA with REFERENCE
orl-rat LD50:710 mg/kg TXAPA9 28,313,74
skn-rbt LD50:1260 mg/kg TXAPA9 28,313,74

CONSENSUS REPORTS: Reported in EPA TSCA Inventory.

SAFETY PROFILE: Moderately toxic by ingestion and skin contact. When heated to decomposition it emits acrid smoke and irritating fumes.

OPE000 CAS:80-51-3 ***HR: D***
OXYBIS(BENZENESULFONYL HYDRAZIDE)
mf: $C_{12}H_{14}N_4O_5S_2$ mw: 358.42

SYNS: CELLMIC S ◇ CELOGEN OT ◇ CENITRON OB ◇ NITROPORE OBSH ◇ p,p'-OXYBISBENZENE DISULFONYLHYDRAZIDE ◇ OXYBISBENZENESULFONIC ACID DIHYDRAZIDE ◇ p,p'-OXYBIS(BENZENESULFONYL) HYDRAZINE

TOXICITY DATA with REFERENCE
mma-sat 800 μg/plate NEZAAQ 33,474,78
mmo-sat 400 μg/plate NEZAAQ 33,474,78

CONSENSUS REPORTS: Reported in EPA TSCA Inventory.

SAFETY PROFILE: Mutation data reported. When heated to decomposition it emits very toxic fumes of NO_x and SO_x.

OPE100 CAS:74007-80-0 ***HR: 3***
OXYBIS(DIBUTYL(2,4,5-TRICHLOROPHENOXY)TIN)
mf: $C_{28}H_{40}Cl_6O_3Sn_2$ mw: 874.76

SYNS: 1,3-BIS(2,4,5-TRICHLOROPHENOXY)-1,1,3,3-TETRABUTYLDISTANNOXANE ◇ DISTANNOXANE, 1,3-BIS(2,4,5-TRICHLOROPHENOXY)-1,1,3,3-TETRABUTYL- ◇ STANNANE, OXYBIS(DIBUTYL (2,4,5-TRICHLOROPHENOXY)-

TOXICITY DATA with REFERENCE
ivn-mus LD50:100 mg/kg CSLNX* NX#03082

OSHA PEL: TWA 0.1 mg(Sn)/m^3 (skin)
ACGIH TLV: TWA 0.1 mg(Sn)/m^3; STEL 0.2 mg/m^3 (skin)
NIOSH REL: (Organotin Compound):10H TWA 0.1 mg(Sn)/m^3

SAFETY PROFILE: Poison by intravenous route. When heated to decomposition it emits toxic fumes of Sn and Cl^-.

OPG000 ***HR: 3***
OXYBIS(N,N-DIMETHYLACETAMIDE-TRIPHENYLSTIBONIUM) DIPERCHLORATE
mf: $C_{44}H_{50}Cl_2N_2O_{11}Sb$ mw: 940.10

CONSENSUS REPORTS: Antimony and its compounds are on The Community Right-To-Know List.

SAFETY PROFILE: Many antimony compounds are poisons. A very unstable explosive. When heated to decomposition it emits very toxic fumes of Sb, Cl^-, and NO_x. See also PERCHLORATES and ANTIMONY COMPOUNDS.

OPI200 ***HR: 2***
2,2'-OXYBIS(6-OXABICYCLO(3.1.0)HEXANE) mixed with 2,2-BIS(p-(2,3-EPOXYPROPOXY)PHENYL)PROPANE

SYNS: ARALDITE 6010 mixed with ERR 4205 (1:1) ◇ BIS(2,3-EPOXYCYCLOPENTYL) ETHER mixed with DIGLYCIDYL ETHER of BISPHENOL A (1:1) ◇ 2,2-BIS(p-(2,3-EPOXYPROPOXY)PHENYL)PROPANE mixed with 2,2'-OXYBIS(6-OXABICYCLO(3.1.0)HEXANE) ◇ DIGLYCIDYL ETHER of BISPHENOL A mixed with BIS(2,3-EPOXYCYCLOPENTYL) ETHER (1:1) ◇ EPI-REZ 508 mixed with ERR 4205 (1:1) ◇ EPON 828 mixed with ERR 4205 (1:1) ◇ ERR 4205 mixed with ARALDITE 6010 (1:1) ◇ ERR 4205 mixed with EPI-REZ 508 (1:1) ◇ ERR 4205 mixed with EPON 828 (1:1)

TOXICITY DATA with REFERENCE
skn-mus TDLo:72 g/kg/2Y-I:ETA NTIS** ORNL-5762

SAFETY PROFILE: Questionable carcinogen with experimental tumorigenic data. When heated to decomposition it emits acrid smoke and irritating fumes.

OPI300 CAS:5987-82-6 ***HR: 3***
OXYBUPROCAINE HYDROCHLORIDE
mf: $C_{17}H_{28}N_2O_3 \cdot ClH$ mw: 344.93

SYNS: BENOXIL ◇ BENOXINATE HYDROCHLORIDE ◇ CEBESINE ◇ CONJUNCAIN ◇ DORSACAINE ◇ DORSACAINE HYDROCHLORIDE ◇ LACRIMIN ◇ NOVESIN ◇ NOVESINE

TOXICITY DATA with REFERENCE
scu-rat LD50:60 mg/kg NIIRDN 6,154,82
scu-mus LD50:42500 μg/kg NIIRDN 6,154,82

SAFETY PROFILE: Poison by subcutaneous route. A topical anesthetic. When heated to decomposition it emits toxic fumes of NO_x and HCl.

OPK000 CAS:1508-65-2 ***HR: 3***
OXYBUTYNIN CHLORIDE
mf: $C_{22}H_{31}NO_3 \cdot ClH$ mw: 394.00

PROP: Crystals. Mp: 129-130°.

SYNS: 4-DIETHYLAMINO-2-BUTYNYLPHENYL(CYCLOHEXYL)GLYCOLATEHYDROCHLORIDE ◇ DITROPAN ◇ MJ 4309-1 ◇ OXIBUTININA HYDROCHLORIDE ◇ α-PHENYLCYCLOHEXANEGLYCOLIC ACID 4-(DIETHYLAMINO)-2-BUTYNYLESTER HYDROCHLORIDE ◇ TROPAX

TOXICITY DATA with REFERENCE
eye-rbt 1% OYYAA2 27,931,84
orl-rat TDLo:1350 mg/kg (female 17-22D post):REP TXCYAC 40,31,86
orl-rat TDLo:1100 mg/kg (female 7-17D post):TER TXCYAC 40,31,86
orl-rat LD50:460 mg/kg IYKEDH 19,735,88
ipr-rat LD50:223 mg/kg IYKEDH 19,735,88
scu-rat LD50:740 mg/kg IYKEDH 19,735,88
ivn-rat LD50:61 mg/kg IYKEDH 19,735,88
orl-mus LD50:725 mg/kg AIPTAK 156,467,65
ipr-mus LD50:185 mg/kg AIPTAK 156,467,65
ivn-mus LD50:42 mg/kg CSLNX* NX#11090

SAFETY PROFILE: Poison by ingestion, intravenous, and intraperitoneal routes. An experimental teratogen. Experimental reproductive effects. An eye irritant. An anticholinergic agent. When heated to decomposition it emits very toxic fumes of HCl and NO_x.

OPK250 CAS:1634-78-2 ***HR: 3***
OXYCARBOPHOS
mf: $C_{10}H_{19}O_7PS$ mw: 314.32

SYNS: CARBETHOXY MALAOXON ◇ S-(1,2-DIETHOXYCARBNYL) ETHYL O,O-DIMETHYL PHOSPHOROTHIOATE ◇ ((DIMETHOXY-PHOSPHINYL)THIO)-BUTANEDIOIC ACID DIETHYL ESTER (9CI) ◇ O,O-DIMETHYL-S-1,2-BIS(ETHOXYCARBONYL)ETHYL PHOSPHOROTHIOATE ◇ O,O-DIMETHYL-S-(1,2-DICARBETHOXY)ETHYL PHOSPHOROTHIOATE ◇ O,O-DIMETHYL ESTER PHOSPHOROTHIOIC ACID-S-ESTER with 1,2-BIS(METHOXYCARBONYL)ETHANETHIOL ◇ LIROMAT ◇ MALAOXON ◇ MALAOXONE ◇ MALATHION-O-ANALOG ◇ NCI-C08628 ◇ SUCCINIC ACID, MERCAPTO-, DIETHYL ESTER, S-ESTER with O,O-DIMETHYLPHOSPHOROTHIOATE

TOXICITY DATA with REFERENCE
sce-ham:ovr 100 μmol/L JTEHD6 8,939,81
orl-rat LD50:158 mg/kg TXAPA9 9,408,66
ipr-rat LD50:17500 μg/kg CJPPA3 45,621,67

CONSENSUS REPORTS: NCI Carcinogenesis Bioassay (feed); No Evidence: mouse, rat NCITR* NCI-CG-TR-135,79

SAFETY PROFILE: Poison by ingestion and intraperitoneal routes. Mutation data reported. When heated to decomposition it emits toxic fumes of SO_x and PO_x.

OPK300 CAS:485-89-2 ***HR: 3***
OXYCINCHOPHEN
mf: $C_{16}H_{11}NO_3$ mw: 265.28

PROP: Minute, deep yellow prisms from alc. Decomp @ 206-207°. Sol in acetic acid, alkalies, hot alc, benzene; sparingly sol in water, ether.

SYNS: CHINOXONE ◇ FENIDRONE ◇ HPC ◇ 3-HYDROXY-CINCHOPEN ◇ HYDROXYCINCHOPHENE ◇ 3-HYDROXY-2-PHENYLCINCHONINIC ACID ◇ 3-HYDROXY-2-PHENYL-4-QUINOLINECARBOXYLIC ACID ◇ MAGNOFENYL ◇ MAGNOPHENYL ◇ OXINOFEN ◇ REUMALON ◇ REUMARTRIL ◇ SINTOFENE

TOXICITY DATA with REFERENCE
orl-mus LD50:495 mg/kg JPETAB 103,79,51
scu-mus LDLo:400 mg/kg JHHBAI 86,89,50
ivn-mus LD50:144 mg/kg JPETAB 103,79,51

SAFETY PROFILE: Poison by intravenous and subcutaneous routes. Moderately toxic by ingestion. When heated to decomposition it emits toxic fumes of NO_x.

OPM000 CAS:101-80-4 ***HR: 3***
4,4'-OXYDIANILINE
mf: $C_{12}H_{12}N_2O$ mw: 200.26

PROP: Colorless crystals. Mp: 187°, bp: >300°.

SYNS: p-AMINOPHENYL ETHER ◇ 4-AMINOPHENYL ETHER ◇ BIS(4-AMINOPHENYL)ETHER ◇ BIS(p-AMINOPHENYL)ETHER ◇ DADPE ◇ 4,4'-DIAMINOBIPHENYLOXIDE ◇ DIAMINODIPHENYL ETHER ◇ 4,4-DIAMINODIPHENYL ETHER ◇ p,p'-DIAMINODIPHENYL ETHER ◇ 4,4'-DIAMINODIPHENYL OXIDE ◇ 4,4'-DIAMINOPHENYL ETHER ◇ NCI-C50146 ◇ OXYBIS(4-AMINOBENZENE) ◇ 4,4'-OXYBISANILINE ◇ p,p'-OXYBIS(ANILINE) ◇ 4,4'-OXYBISBENZENAMINE ◇ OXYDIANILINE ◇ p,p'-OXYDIANILINE ◇ 4,4'-OXYDIPHENYLAMINE ◇ OXYDI-p-PHENYLENEDIAMINE

TOXICITY DATA with REFERENCE
mmo-sat 250 μg/plate MUREAV 67,123,79
mma-sat 10 μg/plate MUREAV 143,11,85
dnd-rat-ipr 3640 μmol/kg CRNGDP 2,1317,81
orl-rat TDLo:17108 mg/kg (13W male):REP VTPHAK 15,649,78
orl-rat TDLo:8652 mg/kg/2Y-C:CAR NCITR* NCI-CG-TR-205,80
scu-rat TDLo:8550 mg/kg/76W-I:ETA VOONAW 21(3),69,75
orl-mus TDLo:13 g/kg/2Y-C:CAR NCITR* NCI-CG-TR-205,80
orl-mus TD:12978 mg/kg/2Y-C:NEO JJIND8 72,1457,84
orl-rat LD50:725 mg/kg HYSAAV 33,137,68
ipr-rat LD50:365 mg/kg HYSAAV 33,137,68
orl-mus LD50:685 mg/kg HYSAAV 33,137,68
ipr-mus LD50:300 mg/kg HYSAAV 33,137,68
orl-rbt LD50:700 mg/kg HYSAAV 33,137,68
ipr-rbt LD50:650 mg/kg HYSAAV 33,137,68
orl-gpg LD50:650 mg/kg 85GMAT -,43,82

CONSENSUS REPORTS: NTP Fifth Annual Report on

Carcinogens. IARC Cancer Review: Group 2B IMEMDT 7,56,87; Animal Sufficient Evidence IMEMDT 29,203,82; Animal Inadequate Evidence IMEMDT 16,301,78. NCI Carcinogenesis Bioassay (feed); Clear Evidence: mouse, rat NCITR* NCI-CG-TR-205,80. Reported in EPA TSCA Inventory.
DFG MAK: Animal Carcinogen, Suspected Human Carcinogen.

SAFETY PROFILE: Confirmed carcinogen with experimental carcinogenic, neoplastigenic, and tumorigenic data. Poison by intraperitoneal route. Moderately toxic by ingestion. Mutation data reported. Experimental reproductive effects. Mutation data reported. When heated to decomposition it emits toxic fumes of NO_x.

OPO000 CAS:106-75-2 ***HR: 2***
OXYDIETHYLENE BIS(CHLOROFORMATE)
mf: $C_6H_8Cl_2O_5$ mw: 231.04

SYN: OXYDIETHYLENE CHLOROFORMATE

TOXICITY DATA with REFERENCE
orl-mus LD50:813 mg/kg 37ASAA 4,758,78
ihl-mus LD50:169 ppm/1H 37ASAA 4,758,78
skn-mus LD50:2000 mg/kg 37ASAA 4,758,78

CONSENSUS REPORTS: Reported in EPA TSCA Inventory.

SAFETY PROFILE: Moderately toxic by ingestion, inhalation, and skin contact. When heated to decomposition it emits toxic fumes of Cl^-.

OPQ000 CAS:62912-51-0 ***HR: 3***
(N,N'-OXYDIETHYLENEDIOXYDIETHYLENE) BIS(TRIETHYLAMMONIUM IODIDE)
mf: $C_{20}H_{46}N_2O_3 \cdot I_2$ mw: 616.48

TOXICITY DATA with REFERENCE
orl-mus LD50:400 mg/kg FRPSAX 32,129,77
ipr-mus LD50:8 mg/kg FRPSAX 32,129,77

SAFETY PROFILE: Poison by ingestion and intraperitoneal routes. When heated to decomposition it emits very toxic fumes of I^-, NH_3, and NO_x. See also IODIDES.

OPY000 CAS:73873-83-3 ***HR: 3***
1,1'-(OXYDIMETHYLENE)BIS(4-FORMYLPYRIDINIUM) DINITRATE DIOXIME
mf: $C_{14}H_{16}N_4O_3 \cdot N_2O_6$ mw: 412.36

SYN: LUH6-DINITRAT (GERMAN)

TOXICITY DATA with REFERENCE
ipr-mus LD50:180 mg/kg ARZNAD 14,5,64
ivn-mus LD50:139 mg/kg ARZNAD 14,5,64
ims-mus LD50:225 mg/kg ARZNAD 14,5,64

SAFETY PROFILE: Poison by intraperitoneal, intravenous, and intramuscular routes. When heated to decomposition it emits toxic fumes of NO_x. See also NITRATES and OXIMES.

OQI000 CAS:1965-09-9 ***HR: 3***
4,4'-OXYDIPHENOL
mf: $C_{12}H_{10}O_3$ mw: 202.22

SYN: p,p'-OXYDIPHENOL

TOXICITY DATA with REFERENCE
ipr-mus LD50:150 mg/kg NTIS** AD691-490

CONSENSUS REPORTS: Reported in EPA TSCA Inventory.

SAFETY PROFILE: Poison by intraperitoneal route. When heated to decomposition it emits acrid smoke and irritating fumes.

OQM000 CAS:110-98-5 ***HR: 1***
1,1'-OXYDI-2-PROPANOL
mf: $C_6H_{14}O_3$ mw: 134.20

PROP: Colorless, sltly viscous liquid; nearly odorless. Bp: 231.8°, flash p: 280°F (OC), d: 1.0252 @ 20°/20°, vap press: 1 mm @ 73.8°, vap d: 4.63.

SYNS: 2,2'-DIHYDROXYDIPROPYL ETHER ◇ 2,2'-DIHYDROXYISOPROPYL ETHER ◇ DIPROPYLENE GLYCOL

TOXICITY DATA with REFERENCE
skn-rbt 500 mg/24H MLD FCTXAV 16,637,78
eye-rbt 510 mg AJOPAA 29,1363,46
mma-sat 10 mg/plate CALEDQ 18,271,83
orl-rat LD50:14850 mg/kg 34ZIAG -,731,69
ipr-rat LD50:10 g/kg FCTXAV 16,637,78
ivn-rat LD50:5800 mg/kg AMIHBC 3,448,51
ipr-mus LD50:4494 mg/kg FEPRA7 6,342,47
ivn-dog LD50:11500 mg/kg FCTXAV 16,637,78
orl-gpg LD50:17600 mg/kg GWXXBX #2703360

CONSENSUS REPORTS: Reported in EPA TSCA Inventory.

SAFETY PROFILE: Mildly toxic by ingestion. A skin and eye irritant. Mutation data reported. Combustible when exposed to heat or flame, can react vigorously with oxidizing materials. To fight fire, use alcohol foam, CO_2, dry chemical. When heated to decomposition it emits acrid smoke and irritating fumes.

OQO000 CAS:36788-39-3 ***HR: 1***
OXYDIPROPANOL PHOSPHITE (3:1)
mf: $C_{18}H_{39}O_9P$ mw: 430.54

SYN: TRIS(DIPROPYLENE GLYCOL)PHOSPHINE

TOXICITY DATA with REFERENCE
orl-rat LD50:19700 mg/kg AIHAAP 30,470,69

CONSENSUS REPORTS: Reported in EPA TSCA Inventory.

SAFETY PROFILE: Mildly toxic by ingestion. When heated to decomposition it emits toxic fumes of PO_x.

OQQ000 CAS:1656-48-0 ***HR: 2***
3,3′-OXYDIPROPIONITRILE
mf: $C_6H_8N_2O$ mw: 124.16

PROP: Colorless liquid. Mp: −26.3° bp: 172° @ 10 mm, d: 1.041 @ 30°.

SYNS: β,β′-DICYANODIETHYL ETHER ◇ ETHER, BIS(2-CYANOETHYL) ◇ β,β′-OXYDIPROPIONITRILE

TOXICITY DATA with REFERENCE
eye-rbt 500 mg AMIHBC 10,61,54
orl-rat LD50:2830 mg/kg AMIHBC 10,61,54

CONSENSUS REPORTS: Cyanide and its compounds are on The Community Right-To-Know List. Reported in EPA TSCA Inventory.

SAFETY PROFILE: Moderately toxic by ingestion. An eye irritant. When heated to decomposition it emits toxic fumes of NO_x and CN^-. See also NITRILES.

OQS000 CAS:2497-07-6 ***HR: 3***
OXYDISULFOTON
mf: $C_8H_{19}O_3PS_3$ mw: 290.42

SYNS: BAY 23323 ◇ O,O-DIETHYL-S-((ETHYLSULFINYL)ETHYL) PHOSPHORODITHIOATE ◇ O,O-DIETHYL S-(2-(ETHYLSULFINYL) ETHYL) PHOSPHORODITHIOATE ◇ DISULFOTON DISULIDE ◇ DISULFOTON SULFOXIDE ◇ DISYSTON SULFOXIDE ◇ ETHYLTHIOMETON SULFOXIDE

TOXICITY DATA with REFERENCE
orl-rat LD50:3500 μg/kg FMCHA2 -,D111,80
skn-rat LD50:192 mg/kg FMCHA2 -,D111,80
orl-mus LD50:12 mg/kg OYYAA2 2,397,68
skn-mus LD50:263 mg/kg OYYAA2 2,397,68

SAFETY PROFILE: Poison by ingestion and skin contact. When heated to decomposition it emits very toxic fumes of SO_x and PO_x.

OQU000 CAS:16648-69-4 ***HR: 3***
1-OXYFEDRIN
mf: $C_{19}H_{23}NO_3 \cdot ClH$ mw: 349.89

PROP: Crystals from methanol. Mp: 190-193°.

SYNS: 1-(1-HYDROXY-1-PHENYL-2-PROPYLAMINO)-1-(m-METHOXYPHENYL)-1-PROPANONE HYDROCHLORIDE ◇ 1-3-METHOXY-ω-(1-HYDROXY-1-PHENYLISOPROPYLAMINO)PROPIOPHENONEHYDROCHLORIDE

TOXICITY DATA with REFERENCE
orl-rat LD50:500 mg/kg ARZNAD 17,1478,67
ipr-rat LD50:92 mg/kg ARZNAD 17,1478,67
ivn-rat LD50:46 mg/kg ARZNAD 17,1478,67
orl-mus LD50:510 mg/kg ARZNAD 17,1478,67
ipr-mus LD50:95 mg/kg ARZNAD 17,1478,67
ivn-mus LD50:29 mg/kg RPTOAN 38,199,75
ivn-dog LD50:50 mg/kg ARZNAD 17,1478,67
ivn-cat LD50:21 mg/kg ARZNAD 17,1478,67
orl-gpg LD50:340 mg/kg ARZNAD 17,1478,67
ipr-gpg LD50:83 mg/kg ARZNAD 17,1478,67

SAFETY PROFILE: Poison by ingestion, intraperitoneal, and intravenous routes. When heated to decomposition it emits very toxic fumes of NO_x and HCl.

OQW000 CAS:7782-44-7 ***HR: 3***
OXYGEN
DOT: UN 1072/UN 1073
mf: O_2 mw: 32.00

PROP: Colorless, odorless, tasteless gas, liquid, or hexagonal crystals. Supports combustion. D (liquid): 1.14 @ −183.0°, d (solid): 1.426 @ −252.5°, vap d: 1.429 @ 0°. D: (gas) 1.429 g/L @ 0°, mp: −218.4°, bp: −182.96°. One vol gas dissolves in 32 vols water @ 20°, dissolves in 7 vols alc @ 20°. Sol in other organic liquids to a greater extent than water.

SYNS: OXYGEN, compressed (DOT) ◇ OXYGEN, refrigerated liquid (DOT)

TOXICITY DATA with REFERENCE
cyt-ham:lng 80 pph MUREAV 57,27,78
cyt-ham:lng 80 pph ACATA5 94,520,76
ihl-wmn TCLo:12 pph/10M (26-39W preg):TER AJOGAH 82,1055,61
ihl-hmn TCLo:100 pph/14H:PUL JAMAAP 128,710,45

CONSENSUS REPORTS: Reported in EPA TSCA Inventory. EPA Genetic Toxicology Program.

DOT Classification: Nonflammable Gas; Label: Nonflammable Gas, Oxidizer; Nonflammable Gas; Label: Oxidizer.

SAFETY PROFILE: Human systemic effects by inhalation: cough and other pulmonary changes. Human teratogenic effects by inhalation: developmental abnormalities of the fetal cardiovascular system. Mutation data reported. Not toxic as gas. In liquid form it can cause severe "burns" and tissue damage on contact with the skin due to extreme cold.

An oxidant. Though itself nonflammable, it is essential to combustion. Even a slight increase in the oxygen content of the air above the normal 21% greatly increases the oxidation or burning rate (and the hazard) of many materials. Exclusion of O_2 from the neighborhood of a fire is one of the principal methods of extinguishment. Avoid smoking, flames, electric sparks. Liquid O_2 can explode on contact with readily oxidizable materials, especially at high temperatures. Under the proper conditions of temperature, pressure, and reagent concentration it can react violently with acetaldehyde; acetylene; acetone; secondary-alcohols (e.g., 2-propanol, 2-butanol) aluminum; $Al(BH_4)_3$; AlH_3; aluminum-titanium alloys; alkali metals (lithium; cesium; potassium; rubidium; sodium; potassium); ammonia; ammonia + platinum; asphalt; CCl_4; chlorinated hydrocarbons; cyanogen; barium; benzene; 1,4-benzenediol + 1-propanol; benzoic acid; $Be(BH_4)_2$; biological materials + ether; BAs_2Br_3; B_2H_{10}; B_2H_6; boron tribromide; boron trichloride; bromine + chlorotrifluoroethylene; butane + $Ni(CO)_4$; carbon disulfide; carbon disulfide + mercury + anthracene; carbon monoxide; CsH; calcium; calcium phosphide; copper + hydrogen sulfide; $C_{10}H_{14}$; cyclohexane-1,2-dione bis(phenylhydrazone); cyclooctatetraene; diborane; diboron tetrafluoride; dimethoxymethane; dimethylketene; dimethyl sulfide; diphenyl ethylene; disilane; ethers (e.g., diethyl ether; diisopropyl ether; tetrahydrofuran; dioxane; ethyl ether); fibrous fabrics; fluorine + hydrogen; fuels; germanium; glycerol; halocarbons (e.g., 1,1,1-trichloroethane; trichloroethylene; chlorotrifluoroethylene; bromotrifluoroethylene); hydrazine; hydrocarbons (e.g., 1,1-diphenylethylene; gasoline; cyclohexane; ethylene; cumene; p-xylene; buten-3-yne); hydrocarbons + promoters (e.g., methyl nitrate; nitromethane; ethyl nitrate; tetrafluorohydrazine); hydrogen; hydrogen sulfide; lithiated dialkylnitrosoamines; magnesium; metals; metal hydrides (e.g., sodium hydride; uranium hydride; lithium hydride; potassium hydride; rubidium hydride; cesium hydride; magnesium hydride); methane; methoxycyclooctatetraene; 4-methoxytoluene; $Ni(CO)_4$ + butane; non-metal hydrides (e.g., diborane; tetraborane(10); phosphine; pentaborane(11); pentaborane(9); decaborane(14); aluminum tetrahydroborate); oil films; organic matter; (OF_2 + H_2O); phosphorus; phosphorus tribromide; phosphorus trifluoride; phosphorus(III) oxide; polymers [e.g., foam rubber; neoprene; polytetrafluoroethylene (teflon)]; polytetrafluoroethylene + stainless steel; polyurethane; polyvinyl chloride; propylene oxide; K_2O_2; rhenium; trirhenium nonachloride; rubber + ozone; rubberized fabric; selenium; NaH; sodium hydroxide + tetramethyldisiloxane; strontium; tetracarbonylnickel; tetracarbonylnickel + mercury; tetrafluoroethylene; tetrafluorohydrazine; tetrasilane; titanium and alloys; trisilane; CH_2Cl_2; oil; paraformaldehyde; wood; charcoal. Compressed O_2 is shipped in steel cylinders under high pressure. If these containers are broken due to shock or exposed to high temperature, an explosion and fire may result.

ORA000 CAS:7783-41-7 ***HR: 3***
OXYGEN DIFLUORIDE
DOT: UN 2190
mf: F_2O mw: 54.00

PROP: Colorless gas or yellowish-brown liquid. Reacts slowly with water. D: (liquid) 1.90 @ −224°, mp: −223.8°, bp: −144.8°.

SYNS: FLUORINE MONOXIDE ◇ FLUORINE OXIDE ◇ OXYGEN FLUORIDE

TOXICITY DATA with REFERENCE
ihl-hmn TCLo:500 ppb:PUL 34ZIAG -,444,69
ihl-rat LC50:136 ppm/1H AIHAAP 33,661,72
ihl-mus LC50:62 ppm/1H AIHAAP 33,661,72
ihl-dog LC50:128 ppm/1H AIHAAP 33,661,72
ihl-mky LC50:26 ppm/1H AMRL** TR-70-77/70

OSHA PEL: (Transitional: TWA 0.1 mg/m^3) CL 0.05 ppm
ACGIH TLV: CL 0.05 ppm

DOT Classification: Poison A; Label: Poison Gas.

SAFETY PROFILE: Poison by inhalation. Human systemic effects by inhalation: chronic pulmonary edema or congestion. A corrosive skin, eye, and mucous membrane irritant. Attacks lungs with delayed appearance of symptoms. A very powerful oxidizer. Must be kept away from contact with reducing agents. Explosive reaction with adsorbents (e.g., silica gel; alumina; molecular sieve); diborane; halogens + heat; metal halides; aluminum chloride; antimony pentachloride (at 150°C); tungsten + heat; hydrogen sulfide; liquid nitrogen oxide; nitrosyl fluoride; charcoal; sulfur tetrafluoride. Forms spark-sensitive explosive mixtures with water or combustible gases (e.g., carbon monoxide; hydrogen; methane). Ignites on contact with diborane tetrafluoride; non-metals (e.g., red phosphorus; boron powder; silicon); phosphorus(V) oxide; nitrogen oxide gas. Incandescent reaction with metals (e.g., aluminum, barium; cadmium; magnesium; strontium; zinc; zirconium; lithium (above 400°C); potassium (above 400°C); sodium. Incompatible with NH_3; As_2O_3; Cl_2 + Cu; CrO_3; Ir; O_3; O_2 + H_2O; Pd; Pt; Rh; Ru; SiO_2. When heated to decomposition it emits highly toxic fumes of F^-. See also FLUORIDES.

ORA100 CAS:1491-59-4 ***HR: 3***
OXYMETHAZOLINE
mf: $C_{16}H_{24}N_2O$ mw: 260.42

SYNS: 6-t-BUTYL-3-(2-IMIDAZOLIN-2-YLMETHYL)-2,4-DIMETHYLPHENOL ◇ NASIVINE ◇ OXYMETAZOLINE ◇ OXYMETOZOLINE ◇ PHENOL, 6-t-BUTYL-3-(2-IMIDAZOLIN-2-YLMETHYL)-2,4-DIMETHYL-

TOXICITY DATA with REFERENCE
ipr-mus TDLo:20 mg/kg (female 1-5D post):REP CCPTAY 11,563,75

orl-rat LD50:800 μg/kg TXAPA9 18,185,71
scu-rat LD50:1100 μg/kg TXAPA9 18,185,71

SAFETY PROFILE: Poison by ingestion and subcutaneous routes. Experimental reproductive effects. When heated to decomposition it emits toxic fumes of NO_x.

ORE000 CAS:3176-03-2 ***HR: 3***
OXYMETHEBANOL
mf: $C_{19}H_{27}NO_4$ mw: 333.47

SYNS: DIHYDRO-14-HYDROXY-4-o-METHYL-6-β-THEBAINOL ◇ DIHYDRO-14-HYDROXY-6-β-THEBAINOL 4 METHYL ESTER ◇ 6-β,14-DIHYDROXY-3,4-DIMETHOXY-N-METHYLMORPHINAN ◇ 3,4-DIMETHOXY-17-METHYLMORPHINAN-6-β,14-DIOL ◇ DROTEBANOL ◇ 14-HYDROXYDIHYDRO-6-β-THEBAINOL 4-METHYL ETHER ◇ 14-HYDROXY-6-β-THEBAINOL 4-METHYL ETHER ◇ METEBANYL ◇ OXYMETEBANOL ◇ RAM-326

TOXICITY DATA with REFERENCE
scu-rat LD50:120 mg/kg NIIRDN 6,156,82
ivn-rat LD50:122 mg/kg NIIRDN 6,156,82
orl-mus LD50:697 mg/kg NIIRDN 6,156,82
scu-mus LD50:1150 mg/kg CPBTAL 19,1,71
ivn-mus LD50:91 mg/kg ARZNAD 20,43,70
scu-gpg LD50:919 mg/kg ARZNAD 20,43,70

SAFETY PROFILE: Poison by subcutaneous and intravenous routes. Moderately toxic by ingestion. An antitussive agent. When heated to decomposition it emits toxic fumes of NO_x. See also ESTERS.

ORG000 CAS:67-47-0 ***HR: 2***
5-OXYMETHYLFURFUROLE
mf: $C_6H_6O_3$ mw: 126.12

SYNS: HMF ◇ 5-HYDROXYMETHYLFURALDEHYDE ◇ 5-(HYDROXYMETHYL)-2-FURANCARBOXALDEHYDE ◇ 5-(HYDROXYMETHYL) FURFURAL ◇ HYDROXYMETHYLFURFUROLE

TOXICITY DATA with REFERENCE
scu-rat TDLo:200 mg/kg:ETA JNCIAM 47,1037,71

CONSENSUS REPORTS: Reported in EPA TSCA Inventory.

SAFETY PROFILE: Questionable carcinogen with experimental tumorigenic data. When heated to decomposition it emits acrid smoke and irritating fumes. See also ALDEHYDES.

ORG100 CAS:357-07-3 ***HR: 3***
OXYMORPHINONE HYDROCHLORIDE
mf: $C_{17}H_{19}NO_4$•ClH mw: 337.83

SYNS: 4,5-α-EPOXY-3,14-DIHYDROXY-17-METHYLMORPHINAN-6-ONE HYDROCHLORIDE ◇ MORPHINAN-6-ONE, 3,14-DIHYDROXY-4,5-α-EPOXY-17-METHYL-, HYDROCHLORIDE ◇ MORPHINAN-6-ONE, 4,5-EPOXY-3,14-DIHYDROXY-17-METHYL-, HYDROCHLORIDE, (5-α)- (9CI) ◇ NUMORPHAN HYDROCHLORIDE ◇ OXYMORPHONE HYDROCHLORIDE

TOXICITY DATA with REFERENCE
scu-ham TDLo:172 mg/kg (female 8D post):TER AJOGAH 123,705,75
orl-mus LD50:402 mg/kg THERAP 15,1208,60
ipr-mus LD50:301 mg/kg THERAP 15,1208,60
scu-mus LD50:285 mg/kg THERAP 15,1208,60
ivn-mus LD50:185 mg/kg THERAP 15,1208,60

SAFETY PROFILE: Poison by ingestion, intravenous, intraperitoneal, and subcutaneous routes. An experimental teratogen. When heated to decomposition it emits toxic fumes of NO_x and HCl.

ORI300 CAS:11005-02-0 ***HR: 3***
OXYPANAMINE

TOXICITY DATA with REFERENCE
ipr-rat LD50:30 mg/kg JPETAB 125,73,59
orl-mus LD50:600 mg/kg JPETAB 125,73,59
ipr-mus LD50:13900 μg/kg JPETAB 125,85,59
ivn-mus LD50:3300 μg/kg JPETAB 125,73,59
ivn-dog LDLo:1 mg/kg JPETAB 125,73,59

SAFETY PROFILE: Poison by intravenous and intraperitoneal routes. Moderately toxic by ingestion. See also AMINES.

ORQ000 CAS:50-10-2 ***HR: 3***
OXYPHENONIUM BROMIDE
mf: $C_{21}H_{34}NO_3$•Br mw: 428.47

SYNS: ANTRENIL ◇ ANTRENYL ◇ ANTRENYL BROMIDE ◇ DIETHYL(2-HYDROXYETHYL)METHYLAMMONIUM BROMIDE α-PHENYLCYCLOHEXANEGLYCOLATE ◇ ETHANAMINIUM, 2-((CYCLOHEXYLHYDROXYPHENYLACETYL)OXY)-N,N-DIETHYL-N-METHYL-, BROMIDE (9CI) ◇ METACIN ◇ METATSIN ◇ METHACIN ◇ OXIFENON ◇ OXYFENON ◇ OXYPHENON ◇ OXYPHENONIUM ◇ SPASMOPHEN

TOXICITY DATA with REFERENCE
orl-hmn TDLo:357 μg/kg:CVS JPETAB 108,292,53
orl-rat LD50:995 mg/kg JPETAB 108,292,53
scu-rat LD50:786 mg/kg JPETAB 108,292,53
ivn-rat LD50:13200 μg/kg JPETAB 108,292,53
ims-rat LD50:400 mg/kg JPETAB 108,292,53
orl-mus LD50:400 mg/kg CLDND* 9,73,54
scu-mus LD50:350 mg/kg CLDND* 151,614,57

SAFETY PROFILE: Poison by ingestion, intramuscular, intraperitoneal, intravenous, and subcutaneous routes. Human systemic effects by ingestion: change in heart rate. When heated to decomposition it emits very toxic fumes of NO_x, NH_3, and Br^-.

ORS000 CAS:39603-54-8 ***HR: 2***
β-OXYPROPYLPROPYLNITROSAMINE
mf: $C_6H_{12}N_2O_2$ mw: 144.20

SYNS: N-NITROSO-2-OXO-N-PROPYL-N-PROPYLAMINE ◇ 1-(NITROSOPROPYLAMINO)-2-PROPANONE ◇ 2-OXI-PROPYL-PRO-

PYLNITROSAMIN (GERMAN) ◇ 2-OXO-PROPYL-PROPYLNITROSAMINE ◇ (2-OXOPROPYL)PROPYLNITROSOAMINE

TOXICITY DATA with REFERENCE
mma-sat 500 nmol/plate ZKKOBW 86,293,76
scu-rat TDLo:3100 mg/kg/33W-I:CAR ZKKOBW 81,23,74
scu-mus TDLo:1688 mg/kg/56W-I:CAR ZKKOBW 91,189,78
scu-ham TDLo:100 mg/kg/(14D preg):NEO,TER ZKKOBW 90,119,77
scu-ham TD:700 mg/kg/14W-I:ETA JNCIAM 51,287,73
scu-rat LD50:424 mg/kg ZKKOBW 81,23,74
scu-mus LD50:603 mg/kg ZKKOBW 91,189,78

SAFETY PROFILE: Suspected carcinogen with experimental carcinogenic, neoplastigenic, and tumorigenic data. Moderately toxic by subcutaneous route. An experimental teratogen. Mutation data reported. When heated to decomposition it emits toxic fumes of NO_x. See also NITROSAMINES.

ORS200 CAS:136-16-3 ***HR: 2***
OXYTHIAMINE
mf: $C_{12}H_{16}N_3O_2S$ mw: 266.37

SYNS: 5-(2-HYDROXYETHYL)-3-((4-HYDROXY-2-METHYL-5-PYRIMIDINYL)METHYL)-4-METHYL-THIAZOLIUM ◇ OXYTHIAMIN ◇ THIAZOLIUM, 5-(2-HYDROXYETHYL)-3-((4-HYDROXY-2-METHYL-5-PYRIMIDINYL)METHYL)-4-METHYL-

TOXICITY DATA with REFERENCE
scu-rat TDLo:160 mg/kg (female 1-8D post):REP FESTAS 24,962,73
scu-mus LD50:1430 mg/kg PCJOAU 16,848,82

SAFETY PROFILE: Moderately toxic by subcutaneous route. Experimental reproductive effects. When heated to decomposition it emits toxic fumes of NO_x and SO_x.

ORU000 CAS:2439-01-2 ***HR: 2***
OXYTHIOQUINOX
mf: $C_{10}H_6N_2OS_2$ mw: 234.30

SYNS: BAYER 4964 ◇ BAYER 36205 ◇ CHINOMETHIONATE ◇ CYCLIC-S,S-(6-METHYL-2,3-QUINOXALINEDIYL)DITHIOCARBONATE ◇ ENT 25,606 ◇ ERADE ◇ ERAZIDON ◇ FORSTAN ◇ 6-METHYL-CHINOXALIN-2,3-DITHIOL-CYCLO-CARBONAT (GERMAN) ◇ 6-METHYL-1,3-DITHIOLO(4,5-b)QUINOXALIN-2-ONE ◇ 6-METHYL-2-OXO-1,3-DITHIOLO(4,5-b)QUINOXALINE ◇ 6-METHYL-2,3-QUINOXALINE DITHIOCARBONATE ◇ 6-METHYL-2,3-QUINOXALINEDITHIOL CYCLIC CARBONATE ◇ 6-METHYL-2,3-QUINOXALINEDITHIOL CYCLIC DITHIOCARBONATE ◇ 6-METHYL-QUINOXALINE-2,3-DITHIOLCYCLOCARBONATE ◇ MORESTAN ◇ MORESTANE ◇ QUINOMETHIONATE ◇ SS 2074

TOXICITY DATA with REFERENCE
orl-rat LD50:1100 mg/kg WRPCA2 9,119,70
skn-rat LD50:500 mg/kg GUCHAZ 6,387,73
ipr-rat LDLo:95 mg/kg TXAPA9 14,632,69
orl-mus LDLo:1070 mg/kg AECTCV 14,111,85
ipr-mus LD50:473 mg/kg JPETAB 173,60,70
orl-gpg LD50:1500 mg/kg PCOC** -,769,66

CONSENSUS REPORTS: EPA Genetic Toxicology Program.

SAFETY PROFILE: Moderately toxic by intraperitoneal, ingestion, and skin contact routes. A pesticide. When heated to decomposition it emits very toxic fumes of NO_x and SO_x.

ORU500 CAS:50-56-6 ***HR: 3***
OXYTOCIN
mf: $C_{43}H_{66}N_{12}O_{12}S_2$ mw: 1007.33

PROP: White powder. Sol in water, 1-butanol, 2-butanol.

SYNS: ATONIN O ◇ DI-SIPIDIN ◇ ENDOPITUITRINA ◇ α-HYPOPHAMINE ◇ NOBITOCIN S ◇ ORASTHIN ◇ OXYSTIN ◇ PARTOCON ◇ PITOCIN ◇ PITON S ◇ POSTERIOR PITUITARY EXTRACT ◇ PRESOXIN ◇ SYNPITAN ◇ SYNTHETIC OXYTOCIN ◇ SYNTOCIN ◇ SYNTOCINON ◇ SYNTOCINONE ◇ UTEDRIN ◇ UTERACON

TOXICITY DATA with REFERENCE
orl-rat TDLo:20 μg/kg (21D preg):REP OYYAA2 8,181,74
ivn-mus LD50:5800 μg/kg NIIRDN 6,151,82

SAFETY PROFILE: Poison by intravenous route. Experimental reproductive effects. A pituitary hormone which stimulates uterine contraction and milk production. The principal uterus-contracting and lactation-stimulating hormone of the posterior pituitary gland. Note: Unlike vasopressin which occurs in at least two forms, oxytocin from beef and hog sources shows no difference in amino acid composition. When heated to decomposition it emits toxic fumes of SO_x and NO_x.

ORU900 CAS:12198-93-5 ***HR: 1***
OZOKERITE

SYN: OZOCERITE

TOXICITY DATA with REFERENCE
skn-rbt 500 mg/24H MLD JACTDZ 3(3),43,84
eye-rbt 100 mg MLD JACTDZ 3(3),43,84

SAFETY PROFILE: A skin and eye irritant.

ORW000 CAS:10028-15-6 ***HR: 3***
OZONE
mf: O_3 mw: 48.00

PROP: Unstable colorless gas or dark blue liquid; characteristic odor. Mp: −193°, bp: −111.9°, d (gas): 2.144 g/L, 1.71 @ −183°. D: (liquid) 1.614 g/mL @ −195.4°.

SYNS: OZON (POLISH) ◇ TRIATOMIC OXYGEN

TOXICITY DATA with REFERENCE
eye-rbt 2 ppm/4H JPCAAC 10,17,60
mmo-esc 100 ppb/20M MEHYDY 4,165,78

dnr-esc 50 ppm/30M BBRCA9 77,220,77
dnd-omi 3300 nmol/L BBACAQ 655,323,81
dlt-oin-ihl 30 ppm/3H-C ENMUDM 4,657,82
sce-hmn:lng 250 ppb/1H-C ENVRAL 18,336,79
cyt-hmn:leu 7230 ppb/36H ENVRAL 12,188,76
dni-ham:lng 2 ppm/1H JTEHD6 17,119,86
cyt-ham-ihl 200 ppb/5H ENVRAL 4,262,71
cyt-rat-ihl 28 mg/m^3/5D-C GISAAA 44(9),12,79
ihl-rat TCLo:1500 ppb/24H (17-20D preg):REP TOLED5 5,3,80
ihl-rat TCLo:1040 ppt/24H (6-9D preg):TER TXAPA9 48,19,79
ihl-mus TCLo:5 ppm/2H/75D-I:NEO AEHLAU 20,16,70
ihl-mus TC:608 $\mu g/m^3$/24W-I:ETA JJIND8 75,771,85
ihl-hmn LCLo:50 ppm/30M 34ZIAG -,446,69
ihl-hmn TCLo:100 ppm/1M:SKN,PUL NEACA9 19,686,41
ihl-man TCLo:1860 ppb/75M:EYE,CVS,PUL AEHLAU 10,517,65
ihl-hmn TCLo:1 ppm:PUL AEHLAU 10,295,65
ihl-hmn TCLo:600 ppb/2H:PUL ARDSBL 118,287,78
ihl-rat LC50:4800 ppm/4H AMIHAB 15,181,57
ihl-mus LC50:12600 ppb/3H FEPRA7 16,22,57
ihl-cat LC50:34500 ppb/3H IMSUAI 25,301,56
ihl-rbt LC50:36 ppm/3H IMSUAI 25,301,56
ihl-gpg LC50:24800 ppb/3H IMSUAI 26,63,57
ihl-ham LC50:10500 ppb/4H AMIHAB 15,181,57

CONSENSUS REPORTS: Reported in EPA TSCA Inventory. EPA Genetic Toxicology Program.

OSHA PEL: (Transitional: TWA 0.1 ppm) TWA 0.1 ppm; STEL 0.3 ppm
ACGIH TLV: TWA CL 0.1 ppm
DFG MAK: 0.1 ppm (0.2 mg/m^3)

SAFETY PROFILE: A human poison by inhalation. Human systemic effects by inhalation: visual field changes, lacrimation, headache, decreased pulse rate with fall in blood pressure, blood pressure decrease, dermatitis, cough, dyspnea, respiratory stimulation and other pulmonary changes. Experimental teratogenic and reproductive effects. Human mutation data reported. A skin, eye, upper respiratory system and mucous membrane irritant. Questionable carcinogen with experimental neoplastigenic and tumorigenic data. Can be a safe water disinfectant in low concentration. Concentration of 0.015 ppm of ozone in air produces a barely detectable odor. Concentrations of 1 ppm produce a disagreeable sulfur-like odor and may cause headache and irritation of eyes and the upper respiratory tract; symptoms disappear after leaving the exposure.

A powerful oxidizing agent. Dangerous chemical reaction with acetylene; alkenes; alkylmetals (e.g., dimethylzinc; diethylzinc); antimony; aromatic compounds (e.g., benzene, aniline); benzene + oxygen + rubber; bromine; charcoal + potassium iodide; citronellic acid; combustible gases (e.g., carbon monoxide; ethylene; nitrogen oxide; ammonia; phosphine); (diallyl methyl carbinol + acetic acid); trans-2,3-dichloro-2-butene; dicyanogen; dienes + oxygen; diethyl ether; 1,1-difluoroethylene; N_2O_5; ethylene; ethylene + formyl fluoride; fluoroethylene; liquid hydrogen; hydrogen + oxygen difluoride; hydrogen bromide; hydrogen iodide; 4-hydroxy-4-methyl-1,6-heptadiene; 23-hydroxy-2,2,4-trimethyl-3-pentenoic acid lactone; isopropylidene compounds; nitrogen; NO_2; NO; nitrogen trichloride; nitrogen triiodide; nitroglycerin; organic liquids; organic matter; oxygen + rubber powder; oxygen fluorides (e.g., dioxygen difluoride; dioxygen trifluoride); silica gel; stibine; tetrafluorohydrazine; tetramethylammonium hydroxide; trifluoroethylene; unsaturated acetals. A severe explosion hazard in liquid form when shocked, exposed to heat or flame, or in concentrated form by chemical reaction with powerful reducing agents. Incompatible with rubber; dinitrogen tetraoxide. See also OZONIDES and PEROXIDES.

ORY000 ***HR: 3***
OZONE mixed with NITROGEN OXIDES (53% : 47%)

SYN: NITROGEN OXIDES mixed with OZONE (47%:53%)

TOXICITY DATA with REFERENCE
ihl-hmn TCLo:1 ppm/2H:CNS NEACA9 19,686,41
ihl-mus LCLo:10 ppm/5H NEACA9 19,686,41

SAFETY PROFILE: Poison by inhalation. Human systemic effects by inhalation: central nervous system effects. See also OZONE and NITROGEN MONOXIDE.

ORY499 ***HR: 3***
OZONIDES

SYN: TRIOXOLANES

SAFETY PROFILE: Many are unstable explosives. The presence of peroxides is thought to be the cause of instability. Polymeric alkene ozonides (e.g., trans-2-butene ozonide) are shock-sensitive explosives. They are decomposed by the catalytic action of powdered palladium, platinum, silver, or iron(II) salts.

P

PAB000 *HR: 3*
P 271
mf: $C_{17}H_{36}N_2 \cdot 2Br$ mw: 428.37

SYNS: 1,5-PENTAMETHYLEN-BIS-N-AETHYLPYRROLIDINIUM)-DIBROMID (GERMAN) ◇ 1,1'-PENTAMETHYLENEBIS(1-ETHYL-PYRROLIDINIUM DIBROMIDE)

TOXICITY DATA with REFERENCE
orl-mus LD50:510 mg/kg AEPPAE 232,219,57
scu-mus LD50:31 mg/kg AEPPAE 232,219,57
ivn-mus LD50:18250 μg/kg AEPPAE 232,219,57

SAFETY PROFILE: Poison by subcutaneous and intravenous routes. Moderately toxic by ingestion. When heated to decomposition it emits toxic fumes of Br^- and NO_x.

PAB250 CAS:492-08-0 *HR: 3*
PACHYCARPINE
mf: $C_{15}H_{26}N_2$ mw: 234.43

PROP: Isolated from *Sophora pachicarpa* (32ZCAI -, 175,67).

SYNS: d-SPARTEINE ◇ SPARTEINE, d-ISOMER

TOXICITY DATA with REFERENCE
scu-mus LD50:90 mg/kg FATOAO 21(5),46,58
ivn-mus LD50:26 mg/kg 32ZCAI -,175,67

SAFETY PROFILE: Poison by subcutaneous and intravenous routes. When heated to decomposition it emits toxic fumes of NO_x.

PAB500 CAS:23668-11-3 *HR: 3*
PACTAMYCIN
mf: $C_{28}H_{38}N_4O_8$ mw: 558.70

PROP: Sol in ethanol, chloroform, methylene chloride, benzene, ether. Unstable in solution.

SYNS: NSC-52947 ◇ U-15800

TOXICITY DATA with REFERENCE
mmo-eug 150 mg/L NEOLA4 19,579,72
dni-eug 150 mg/L NEOLA4 19,579,72
dni-rat-ipr 6 mg/kg RCOCB8 11,311,75
ivn-rat LD50:1400 μg/kg 85ERAY 2,1420,78
orl-mus LD50:10 mg/kg UPJOH* 2(6),-,71
ipr-mus LD50:32 mg/kg CNCRA6 30,9,63
ivn-mus LD50:16 mg/kg 85ERAY 2,1420,78

SAFETY PROFILE: Poison by ingestion, intravenous, and intraperitoneal routes. Mutation data reported. When heated to decomposition it emits toxic fumes of NO_x.

PAB750 CAS:4630-95-9 *HR: 3*
PADRIN
mf: $C_{22}H_{28}N \cdot Br$ mw: 386.42

SYNS: 1,1-DIETHYL-2-METHYL-3-DIPHENYLMETHYLENE-PYRROLIDINIUM BROMIDE ◇ 3-(DIPHENYLMETHYLENE)-1,1-DI-ETHYL-2-METHYLPYRROLIDINIUM BROMIDE ◇ 3-(DIPHENYL-METHYLENE)-1-ETHYL-2-METHYLPYRROLIDINE ETHYL BROMIDE

TOXICITY DATA with REFERENCE
scu-mus TDLo:120 mg/kg (female 7-12D post):TER ARZNAD 22,706,72
orl-rat TDLo:120 mg/kg (7-12D preg):REP ARZNAD 22,706,72
orl-rat LD50:1090 mg/kg ARZNAD 22,706,72
ipr-rat LD50:46 mg/kg ARZNAD 22,706,72
scu-rat LD50:170 mg/kg ARZNAD 22,706,72
orl-mus LD50:310 mg/kg ARZNAD 22,706,72
ipr-mus LD50:40 mg/kg ARZNAD 22,706,72
scu-mus LD50:30 mg/kg ARZNAD 22,706,72
ivn-mus LD50:11 mg/kg ARZNAD 22,706,72
orl-dog LD50:327 mg/kg ARZNAD 22,706,72
scu-dog LD50:88 mg/kg ARZNAD 22,706,72
ivn-dog LD50:18 mg/kg ARZNAD 22,706,72

SAFETY PROFILE: Poison by ingestion, intraperitoneal, subcutaneous, and intravenous routes. An experimental teratogen. Experimental reproductive effects. When heated to decomposition it emits very toxic fumes of NO_x and Br^-. See also BROMIDES.

PAC000 CAS:23180-57-6 *HR: 2*
PAEONIFLORIN
mf: $C_{23}H_{28}O_{11}$ mw:480.51

PROP: Chief component of herb paeony root which is used in Chinese medicine (YKKZAJ 89(7),879,69).

SYNS: PAEONIA MOUTAN ◇ PAEONY ROOT

TOXICITY DATA with REFERENCE
ipr-mus LD50:9530 mg/kg YKKZAJ 89,879,69
ivn-mus LD50:3530 mg/kg YKKZAJ 89,879,69

SAFETY PROFILE: Moderately toxic by intravenous route. Mildly toxic by intraperitoneal route. When heated to decomposition it emits acrid smoke and irritating fumes.

PAC250 CAS:552-41-0 ***HR: 3***
PAEONOL
mf: $C_9H_{10}O_3$ mw:166.19

PROP: Main component of the root bark of paeony (*paeonia moutan*) (YKKZAJ 89,1205,69).

TOXICITY DATA with REFERENCE
orl-mus LD50:3430 mg/kg YKKZAJ 89,1205,69
ipr-mus LD50:781 mg/kg YKKZAJ 89,1205,69
ivn-mus LD50:196 mg/kg YKKZAJ 89,1205,69

SAFETY PROFILE: Poison by intravenous route. Moderately toxic by ingestion and intraperitoneal routes. When heated to decomposition it emits acrid smoke and irritating fumes.

PAC500 CAS:62229-70-3 ***HR: 2***
PALANILCARRIER A

TOXICITY DATA with REFERENCE
skn-rbt 500 mg/24H SEV 28ZPAK -,289,72
eye-rbt 50 μg/24H SEV 28ZPAK -,289,72
orl-rat LD50:1650 mg/kg 28ZPAK -,289,72
ipr-rat LD50:1010 mg/kg MEPAAX 20,519,69

SAFETY PROFILE: Moderately toxic by ingestion and intraperitoneal routes. A severe skin and eye irritant. When heated to decomposition it emits acrid smoke and irritating fumes.

PAD250 CAS:7440-05-3 ***HR: 1***
PALLADIUM
af: Pd aw: 106.4

PROP: A steely white, stable metal; can be annealed to be soft and ductile. Mp: 1555°, bp: 3167°, d: 12.02 @ 20°/4. Volatile at high temps.

SAFETY PROFILE: May be a skin sensitizer. This metal in the form of palladium chloride has been administered orally in dosage of about 1 grain daily in the treatment of tuberculosis without apparent ill effects. In the laboratory, palladium appears to bind to many cell components; blocks the action of a number of enzymes and interferes with the use of energy by nerves and muscles; induces lung malfunction and produces abnormal fetuses. Lethal intravenous doses cause appetite loss, hemolysis, renal deposition and bone marrow damage. Poorly absorbed by the body when ingested. Palladium dust can be a fire and explosion hazard. Combustible in the form of dust when exposed to heat, or flame. Explosive reaction with hydrogen + hydrogen peroxide. Reaction with formic acid or sodium tetrahydroborate releases explosive hydrogen gas. Violent reaction with isopropyl alcohol; OF_2S. Under the proper conditions it undergoes hazardous reactions with aluminum; arsenic; carbon; methanol; ozonides; sulfur.

PAD500 CAS:7647-10-1 ***HR: 3***
PALLADIUM(2+) CHLORIDE
mf: Cl_2Pd mw: 177.30

PROP: Dark brown, deliq crystals. D: 4.0 @ 18°, mp: 678-680° (decomp). Sol in water, alc, acetone, and hydrochloric acid.

SYNS: NCI-C60184 ◇ PALLADIUM CHLORIDE ◇ PALLADOUS CHLORIDE

TOXICITY DATA with REFERENCE
skn-rbt 100 mg/24H MLD AEHLAU 30,168,75
dni-hmn:lym 600 μmol/L IAAAAM 79,83,86
itt-rat TDLo:3546 μg/kg (1D male):REP JRPFA4 7,21,64
orl-mus TDLo:880 mg/kg/75W-C:CAR JONUAI 101,1431,71
orl-rat LD50:2704 mg/kg GTPZAB 21(7),55,77
ipr-rat LD50:70 mg/kg EVHPAZ 10,63,75
ivn-rat LD50:3 mg/kg EVHPAZ 10,63,75
itr-rat LD50:6 mg/kg EVHPAZ 10,63,75
ipr-mus LD50:174 mg/kg COREAF 256,1043,63
ivn-rbt LDLo:18600 μg/kg MEIEDD 10,1003,83

CONSENSUS REPORTS: Reported in EPA TSCA Inventory. EPA Genetic Toxicology Program.

SAFETY PROFILE: Poison by intraperitoneal, intravenous, and intratracheal routes. Moderately toxic by ingestion. Experimental reproductive effects. A skin irritant. Questionable carcinogen with experimental carcinogenic data. Human mutation data reported. When heated to decomposition it emits highly toxic fumes of Cl^-. See also PALLADIUM.

PAD750 ***HR: 3***
PALLADIUM CHLORIDE, DIHYDRIDE
mf: $Cl_2Pd{\cdot}2H_2O$ mw: 213.34

PROP: Dark brown crystals. Sol in water, alc, acetone.

TOXICITY DATA with REFERENCE
orl-rat LD50:576 mg/kg EVHPAZ 10,95,75
ipr-rat LD50:85 mg/kg EVHPAZ 10,95,75
ivn-rbt LDLo:18 mg/kg EQSSDX 1,1,75

SAFETY PROFILE: Poison by intraperitoneal and intravenous routes. Moderately toxic by ingestion. Used in photography, electroplating, and in the manufacture of indelible ink. When heated to decomposition it emits toxic fumes of Cl^-. See also PALLADIUM(2+) CHLORIDE.

PAE000 CAS:8014-19-5 ***HR: 1***
PALMAROSA OIL

PROP: From steam distillation of the grass *Cymbopogon Martini* Stapf. Var. Motia, mainly *Geraniol* (FCTXAV 12,807,74). Yellow oily liquid. D: 0.879-

0.892, refr index: 1.473 @ 20°. Sol in fixed oils, propylene glycol, mineral oil; insol in glycerin.

SYNS: GERANIUM OIL, EAST INDIAN TYPE ◇ GERANIUM OIL, TURKISH TYPE ◇ OIL of PALMAROSA

TOXICITY DATA with REFERENCE
skn-rbt 500 mg/24H MOD FCTXAV 12,947,74

CONSENSUS REPORTS: Reported in EPA TSCA Inventory.

SAFETY PROFILE: A skin irritant. When heated to decomposition it emits acrid smoke and irritating fumes.

PAE100 CAS:131-04-4 ***HR: D***
PALMATINIUM HYDROXIDE
mf: $C_{21}H_{22}NO_4 \cdot HO$ mw: 369.45

SYNS: CALYSTIGINE ◇ DIBENZO(a,g)QUINOLIZINIUM, 5,6-DIHYDRO-2,3,9,10-TETRAMETHOXY-, HYDROXIDE ◇ 5,6-DIHYDRO-2,3,9,10-TETRAMETHOXYDIBENZO(a,g)QUINOLIZINIUMHYDROXIDE ◇ PALMATINE HYDROXIDE ◇ 7,8,13,13a-TETRADEHYDRO-2,3,9,10-TETRAMETHOXYBERBINIUMHYDROXIDE

TOXICITY DATA with REFERENCE
orl-dog TDLo:1800 mg/kg (male 60D pre):REP JETHDA 25,151,89

SAFETY PROFILE: Experimental reproductive effects. When heated to decomposition it emits toxic fumes of NO_x.

PAE250 CAS:57-10-3 ***HR: 3***
PALMITIC ACID
mf: $C_{17}H_{32}O_2$ mw: 256.48

PROP: Colorless plates or white crystalline powder; slt characteristic odor and taste. D: 0.849 @ 70°/4°, mp: 63-64°, bp: 271.5° @ 100 mm. Insol in water; very sltly sol in petr ether; sol in abs ether, chloroform.

SYNS: CETYLIC ACID ◇ EMERSOL 140 ◇ EMERSOL 143 ◇ HEXADECANOIC ACID ◇ n-HEXADECOIC ACID ◇ HEXADECYLIC ACID ◇ HYDROFOL ◇ HYSTRENE 8016 ◇ INDUSTRENE 4516 ◇ 1-PENTADECANECARBOXYLIC ACID

TOXICITY DATA with REFERENCE
skn-hmn 75 mg/3D-I MLD 85DKA8 -,127,77
imp-mus TDLo:1000 mg/kg:NEO CNREA8 26,105,66
ivn-mus LD50:57 mg/kg APTOA6 18,141,61

CONSENSUS REPORTS: Reported in EPA TSCA Inventory.

SAFETY PROFILE: A poison by intravenous route. A human skin irritant. Questionable carcinogen with experimental neoplastigenic data. When heated to decomposition it emits acrid smoke and irritating fumes.

PAE500 CAS:8002-75-3 ***HR: D***
PALM OIL

SYNS: OILS, PALM ◇ PALM BUTTER

TOXICITY DATA with REFERENCE
orl-rat TDLo:55 g/kg (female 5-15D post):TER SEIJBO 20,139,80
orl-rat TDLo:55 g/kg (female 5-15D post):REP SEIJBO 20,139,80

CONSENSUS REPORTS: Reported in EPA TSCA Inventory.

SAFETY PROFILE: An experimental teratogen. Experimental reproductive effects. When heated to decomposition it emits acrid smoke and irritating fumes.

PAE600 ***HR: 3***
PALUDRINE DIHYDROCHLORIDE
mf: $C_{11}H_{16}ClN_5 \cdot 2ClH$ mw: 326.69

SYN: 1-(p-CHLOROPHENYL)-5-ISOPROPYLBIGUANIDEDIHYDROCHLORIDE

TOXICITY DATA with REFERENCE
orl-rat LDLo:257 mg/kg JPETAB 90,233,47
orl-mus LD50:30 mg/kg JPETAB 90,233,47
ims-mus LD50:26 mg/kg JPETAB 90,233,47
ims-dog LDLo:205 mg/kg JPETAB 90,233,47
ims-mky LDLo:205 mg/kg JPETAB 90,233,47

SAFETY PROFILE: Poison by ingestion and intramuscular routes. When heated to decomposition it emits toxic fumes of NO_x and HCl.

PAE750 CAS:12174-11-7 ***HR: 3***
PALYGORSCITE

SYNS: ATTAPULGITE ◇ PALYGORSKIT (GERMAN)

TOXICITY DATA with REFERENCE
ipr-rat TDLo:338 mg/kg/2W-I:NEO ZHPMAT 162,467,76
imp-rat TDLo:200 mg/kg:ETA JJIND8 67,965,81

SAFETY PROFILE: Questionable carcinogen with experimental neoplastigenic and tumorigenic data by implant route. When heated to decomposition it emits acrid smoke and irritating fumes.

PAE875 CAS:58718-68-6 ***HR: 3***
PALYTHOATOXIN

SYN: PTX

TOXICITY DATA with REFERENCE
ipr-mus LDLo:600 ng/kg 35FUAR 2,379,76
ivn-mus LDLo:530 ng/kg 35FUAR 2,379,76
ivn-cat LDLo:200 ng/kg 35FUAR 2,379,76

SAFETY PROFILE: A deadly poison by intravenous

and intraperitoneal routes. When heated to decomposition it emits acrid smoke and irritating fumes.

PAF000 CAS:11077-03-5 **HR: 3**
PALYTOXIN
mf: $C_{120}H_{200}N_4O_{52}$ mw: 2531.24

PROP: Isolated from the zoanthids "Limu-make-o-Hana" (SCIEAS 172,495,71).

TOXICITY DATA with REFERENCE
skn-rbt 250 ng/4M MOD TOXIA6 12,427,74
eye-rbt 300 mg/24H SEV TOXIA6 12,427,74
ipr-rat LD50:710 ng/kg TXAPA9 34,214,75
scu-rat LD50:400 ng/kg TXAPA9 34,214,75
ivn-rat LD50:89 ng/kg TXAPA9 34,214,75
ims-rat LD50:240 ng/kg TXAPA9 34,214,75
par-rat LDLo:25 ng/kg 34TRAD -,311,76
itr-rat LD50:360 ng/kg TXAPA9 34,214,75
ipr-mus LD50:50 mg/kg PWPSA8 17,294,74
ivn-dog LD50:33 ng/kg TXAPA9 34,214,75
par-dog LDLo:50 ng/kg 34TRAD -,311,76
ivn-mky LD50:78 ng/kg TXAPA9 34,214,75

SAFETY PROFILE: A deadly poison by intraperitoneal, parenteral, subcutaneous, intravenous, intramuscular, and intratracheal routes. A skin and severe eye irritant. When heated to decomposition it emits toxic fumes of NO_x.

PAF100 CAS:130-85-8 **HR: 3**
PAMOIC ACID
mf: $C_{23}H_{16}O_6$ mw: 388.39

PROP: Crystals from dil pyridine. Decomp above 280° without melting. Practically insoluble in water, alc, ether, benzene, acetic acid; sparingly sol in chloroform; sol in nitrobenzene, pyridine.

SYNS: EMBONIC ACID ◇ 4,4′-METHYLENEBIS(3-HYDROXY-2-NAPHTHOIC ACID)

TOXICITY DATA with REFERENCE
eye-rbt 100 mg MOD FCTOD7 20,573,82
eye-rbt 100 mg/45 rns MLD FCTOD7 20,573,82
ipr-mus LD50:390 mg/kg EJMCA5 13,381,78

CONSENSUS REPORTS: Reported in EPA TSCA Inventory.

SAFETY PROFILE: Poison by intraperitoneal route. An eye irritant. When heated to decomposition it emits acrid smoke and irritating fumes.

PAF250 CAS:19562-30-2 **HR: 3**
PANACID
mf: $C_{14}H_{16}N_4O_3$ mw: 288.34

SYNS: BACTRAMYL ◇ 5,8-DIHYDRO-8-ETHYL-5-OXO-2 (1-PYRROLIDINYL) PYRIDO(2,3-D)PYRIMIDINE-6-CARBOXYLIC ACID ◇ PD 93 ◇ PIRODAL ◇ PIROMIDIC ACID ◇ REELON ◇ SEPTURAL

TOXICITY DATA with REFERENCE
ipr-mus LD50:900 mg/kg 37ASAA 2,782,78
scu-mus LD50:200 mg/kg 37ASAA 2,782,78
ivn-mus LD50:100 mg/kg 37ASAA 2,782,78
skn-rbt LD50:1500 mg/kg 37ASAA 2,782,78

SAFETY PROFILE: Poison by subcutaneous and intravenous routes. Moderately toxic by skin contact and intraperitoneal routes. An antibacterial agent. When heated to decomposition it emits toxic fumes of NO_x.

PAF450 CAS:41753-43-9 **HR: 3**
PANAX SAPONIN E
mf: $C_{54}H_{92}O_{23}$ mw: 1109.46

SYNS: GINSENOSIDE RB1 ◇ PSEUDOGINSENOSIDE D ◇ SANCHINOSIDE E1

TOXICITY DATA with REFERENCE
oms-rat-ipr 50 mg/kg CPBTAL 24,2400,76
ipr-mus LD50:1110 mg/kg ARZNAD 25,343,75
ivn-mus LD50:243 mg/kg SYHJAM 10,61,79

SAFETY PROFILE: Poison by intravenous route. Moderately toxic by intraperitoneal route. Mutation data reported. When heated to decomposition it emits acrid smoke and irritating fumes.

PAF500 CAS:804-36-4 **HR: 3**
PANAZONE
mf: $C_{14}H_{12}N_6O_6$ mw: 360.32

SYNS: 1,5-BIS(5-NITRO-2-FURANYL)-1,4-PENTADIEN-3-ONE, (AMINOIMINOMETHYL)HYDRAZONE ◇ BIS(5-NITROFURFURYLIDENE)ACETONE GUANYLHYDRAZONE ◇ sym-BIS(5-NITRO-2-FURFURYLIDENE) ACETONE GUANYLHYDRAZONE ◇ 1,5-BIS(5-NITRO-2-FURYL)-3-PENTADIENONE AMIDINONHYDRAZONE ◇ 1,5-BIS(5-NITRO-2-FURYL)-3-PENTADIENONEGUANYLHYDRAZONE ◇ DIFURAN ◇ DIFURAZONE ◇ ((3-(5-NITRO-2-FURYL)-1-(2-(5-NITRO-2-FURYL)VINYL)ALLYIDENE)AMINO)GUANIDINE ◇ PAYZONE

TOXICITY DATA with REFERENCE
orl-mus LD50:5330 mg/kg PMDCAY 5,320,67
ipr-mus LD50:300 mg/kg PMDCAY 5,320,67
scu-mus LD50:3750 mg/kg PMDCAY 5,320,67

CONSENSUS REPORTS: IARC Cancer Review: Group 3 IMEMDT 7,117,87; Animal Inadequate Evidence IMEMDT 31,185,83.

SAFETY PROFILE: Poison by intraperitoneal route. Moderately toxic by subcutaneous route. Mildly toxic by ingestion. Questionable carcinogen. A growth promoter in chickens. When heated to decomposition it emits very toxic fumes of NO_x.

PAF550 CAS:9087-70-1 ***HR: D***
PANCREATIC BASIC TRYPSIN INHIBITOR
mf: $C_{284}H_{432}N_{84}O_{79}S_7$ mw: 6512.42

SYNS: ANTAGOSAN ◇ ANTIKREIN ◇ ANTILYSIN ◇ ANTILYSINE ◇ APROTININ ◇ BASIC PANCREATIC TRYPSIN INHIBITOR ◇ BAYER A-128 ◇ FOSTEN ◇ INIPROL ◇ KALLIKREIN-TRYPSIN INACTIVATOR ◇ KIR RICHTER ◇ ONQUININ ◇ REPULSON ◇ RIKER 52G ◇ RP-9921 ◇ PANCREATIC TRYPSIN INHIBITOR ◇ PANCREATIC TRYPSIN INHIBITOR (KUNITZ) ◇ TRASYLOL ◇ TRAZININ ◇ TRYPSIN INHIBITOR ◇ TRYPSIN INHIBITOR, PANCREATIC BASIC ◇ TRYPSIN-KALLIKREIN INHIBITOR (KUNITZ) ◇ ZYMOFREN

TOXICITY DATA with REFERENCE
iut-mus TDLo:160 mg/kg (female 2D post):REP FESTAS 25,954,74

SAFETY PROFILE: Experimental reproductive effects. When heated to decomposition it emits toxic fumes of NO_x and SO_x.

PAF600 CAS:8049-47-6 ***HR: 2***
PANCREATIN

TOXICITY DATA with REFERENCE
ipr-rat LD50:2550 mg/kg YACHDS 3,2427,75
ipr-mus LD50:2100 mg/kg YACHDS 3,2427,75
scu-mus LD50:3870 mg/kg YACHDS 3,2427,75

CONSENSUS REPORTS: Reported in EPA TSCA Inventory.

SAFETY PROFILE: Moderately toxic by subcutaneous and intraperitoneal routes.

PAF625 CAS:15500-66-0 ***HR: 3***
PANCURONIUM BROMIDE
mf: $C_{35}H_{60}N_2O_4 \cdot 2Br$ mw: 732.79

PROP: Odorless crystals with bitter taste. Mp: 215°. One gram dissolves in 30 parts chloroform, one part water (20°).

SYNS: 5α-ANDROSTAN-3α,17β-DIOL, 2β,16β-DIPIPECOLINIO-, DIBROMIDE, DIACETATE ◇ 1,1'-(3,17-BIS(ACETYLOXY)ANDROSTANE-2,16-DIYL)BIS(1-METHYLPIPERIDINIUM) DIBROMIDE ◇ 3-α,17-β-DIACETOXY-2-β,16-β-DIPIPERIDINO-5-α-ANDROSTANE DIMETHOBROMIDE ◇ MIOBLOCK ◇ NA 97 ◇ ORG NA 97 ◇ PAVULON

TOXICITY DATA with REFERENCE
ivn-man LDLo:143 μg/kg:BLD,PUL JATOD3 4,275,80
ivn-rat LD50:153 μg/kg NIIRDN 6,353,82
ipr-mus LD50:128 μg/kg OYYAA2 5,715,71
scu-mus LD50:167 μg/kg NIIRDN 6,353,82
ivn-mus LD50:13 μg/kg OYYAA2 5,715,71
ivn-rbt LD50:16 μg/kg NIIRDN 6,353,82
ivn-ckn LD50:63 μg/kg OYYAA2 5,715,71

CONSENSUS REPORTS: EPA Genetic Toxicology Program.

SAFETY PROFILE: A human poison by intravenous route. Poison experimentally by intravenous, subcutaneous, and intraperitoneal routes. Human systemic effects by intravenous route: chronic pulmonary edema, hemorrhage. When heated to decomposition it emits toxic fumes of NO_x and Br^-.

PAF630 CAS:15500-66-0 ***HR: 3***
PANCURONIUM BROMIDE HYDRATE
mf: $C_{35}H_{60}N_2O_4 \cdot 2Br \cdot H_2O$ mw: 768.83

SYN: PANCURONIUM DIBROMIDE HYDRATE ◇ PAVULON

TOXICITY DATA with REFERENCE
ivn-rbt TDLo:36 μg/kg (8-16D preg):REP SKKNAJ 22,187,70
orl-rat LD50:202 mg/kg SKKNAJ 22,187,70
ipr-rat LD50:479 μg/kg SKKNAJ 22,187,70
scu-rat LD50:436 μg/kg SKKNAJ 22,187,70
ivn-rat LD50:129 μg/kg SKKNAJ 22,187,70
orl-mus LD50:21200 μg/kg SKKNAJ 22,187,70
ipr-mus LD50:154 μg/kg SKKNAJ 22,187,70
scu-mus LD50:168 μg/kg SKKNAJ 22,187,70
ivn-mus LD50:36 μg/kg SKKNAJ 22,187,70

SAFETY PROFILE: A deadly poison by ingestion, subcutaneous, intravenous, and intraperitoneal routes. Experimental reproductive effects. When heated to decomposition it emits toxic fumes of NO_x and Br^-.

PAG050 CAS:58580-55-5 ***HR: 3***
PANIMYCIN
mf: $C_{18}H_{37}N_5O_8 \cdot 7H_2O_4S$ mw: 1138.16

SYNS: o-3-AMINO-3-DEOXY-α-d-GLUCOPYRANOSYL-(1-6)-o-(2,6-DIAMINO-2,3,4,6-TETRADEOXY-α-d-erythro-HEXOPYRANOSYL-(1-4))-2-DEOXY-d-STREPTAMINE SULFATE (SALT) ◇ DEBECACIN SULFATE ◇ DIBEKACIN SULFATE ◇ ORBICIN

TOXICITY DATA with REFERENCE
ipr-rat LD50:799 mg/kg NIIRDN 6,340,82
ivn-rat LD50:140 mg/kg NIIRDN 6,340,82
ims-rat LD50:560 mg/kg NIIRDN 6,340,82
ipr-mus LD50:431 mg/kg NIIRDN 6,340,82
ivn-mus LD50:62600 μg/kg NIIRDN 6,340,82
ims-mus LD50:396 mg/kg NIIRDN 6,340,82

SAFETY PROFILE: Poison by intravenous and intramuscular routes. Moderately toxic by intraperitoneal route. When heated to decomposition it emits toxic fumes of SO_x and NO_x.

PAG075 CAS:53994-73-3 ***HR: 2***
PANORAL
mf: $C_{15}H_{14}ClN_3O_4S$ mw: 367.83

SYNS: CCL ◇ CEFACLOR ◇ 3-CHLORO-7-d-(2-PHENYLGLYCINAMIDO)-3-CEPHEM-4-CARBOXYLIC ACID ◇ LILLY 99638

TOXICITY DATA with REFERENCE
orl-mus TDLo:10 g/kg (female 6-15D post):REP NKRZAZ 27(Suppl 7),846,79
orl-mus TDLo:10 g/kg (female 6-15D post):TER NKRZAZ 27(Suppl 7),846,79
orl-cld TDLo:131 mg/kg/7D-I:MSK,SKN CMAJAX 126,1032,82
orl-inf TDLo:525 mg/kg/7D-I:SKN,SYS CMAJAX 126,1032,82
ipr-rat LD50:1259 mg/kg NKRZAZ 27(Suppl 7),765,79
scu-rat LD50:4838 mg/kg NKRZAZ 27(Suppl 7),765,79
ipr-mus LD50:1227 mg/kg NKRZAZ 27(Suppl 7),765,79
scu-mus LD50:4180 mg/kg NKRZAZ 27(Suppl 7),765,79

SAFETY PROFILE: Moderately toxic by intraperitoneal route. Human systemic effects by ingestion: joints, dermatitis, increased body temperature. An experimental teratogen. Experimental reproductive effects. When heated to decomposition it emits toxic fumes of Cl^-, SO_x, and NO_x.

PAG150 CAS:16816-67-4 ***HR: 2***
d-PANTETHINE
mf: $C_{22}H_{42}N_4O_8S_2$ mw: 554.80

SYNS: BIS(PANTOTHENAMIDOETHYL) DISULFIDE ◇ LBF DISULFIDE ◇ PANTETHINE ◇ PANTETINA ◇ PANTOMIN ◇ PANTOSIN

TOXICITY DATA with REFERENCE
ims-rat TDLo:120 mg/kg (9-14D preg):REP OYYAA2 2,55,68
ims-rat TDLo:3 g/kg (9-14D preg):TER OYYAA2 2,55,68
ipr-mus LD50:4800 mg/kg NIIRDN 6,598,82
scu-mus LD50:4870 mg/kg NIIRDN 6,598,82
ivn-mus LD50:3400 mg/kg NIIRDN 6,598,82
ims-mus LD50:5100 mg/kg NIIRDN 6,598,82

SAFETY PROFILE: Moderately toxic by intravenous route. An experimental teratogen. Experimental reproductive effects. When heated to decomposition it emits toxic fumes of SO_x and NO_x. See also AMIDES.

PAG200 CAS:81-13-0 ***HR: 2***
d-PANTHENOL
mf: $C_9H_{19}NO_4$ mw: 205.29

PROP: Viscous, somewhat hygroscopic liquid; sltly bitter taste. D: (20/20) 1.2, bp: 118-120°, easily decomp on distillation. Freely sol in water, alc, methanol, ether; sltly sol in glycerin. Natural pH about 9.5.

SYNS: ALCOPAN-250 ◇ BEPANTHEN ◇ BEPANTHENE ◇ BEPANTOL ◇ COZYME ◇ DEXPANTHENOL (FCC) ◇ d-(+)-2,4-DIHYDROXY-N-(3-HYDROXYPROPYL)-3,3-DIMETHYLBUTYRAMIDE ◇ D-P-A INJECTION ◇ ILOPAN ◇ MOTILYN ◇ PANADON ◇ PANTHENOL ◇ d(+)-PANTHENOL (FCC) ◇ PANTHODERM ◇ PANTOL ◇ PANTOTHENOL ◇ d-PANTOTHENOL ◇ PANTOTHENYL ALCOHOL ◇ d-PANTOTHENYL ALCOHOL ◇ d(+)-PANTOTHENYL ALCOHOL ◇ THENALTON ◇ ZENTINIC

TOXICITY DATA with REFERENCE
ipr-mus LD50:9 g/kg FRPSAX 14,43,59
ivn-mus LD50:7000 mg/kg NIIRDN 6,599,82
ivn-rbt LD50:4000 mg/kg NIIRDN 6,599,82

CONSENSUS REPORTS: Reported in EPA TSCA Inventory.

SAFETY PROFILE: Moderately toxic by intravenous route. When heated to decomposition it emits toxic fumes of NO_x. See also AMIDES.

PAG225 CAS:53025-21-1 ***HR: 1***
PANTOCRIN

TOXICITY DATA with REFERENCE
ims-rat TDLo:63 mg/kg (7-13D preg):TER OYYAA2 6,727,72
ims-rat TDLo:63 mg/kg (7-13D preg):REP OYYAA2 6,727,72
orl-rat LD50:87800 mg/kg OYYAA2 6,739,72
ipr-rat LD50:82 g/kg OYYAA2 6,739,72
scu-rat LD50:112 g/kg OYYAA2 6,739,72
ivn-rat LD50:17800 mg/kg OYYAA2 6,739,72
orl-mus LD50:117 g/kg OYYAA2 6,739,72
ipr-mus LD50:104 g/kg OYYAA2 6,739,72
scu-mus LD50:114 g/kg OYYAA2 6,739,72
ivn-mus LD50:34 g/kg OYYAA2 6,739,72
ims-mus LD50:97800 mg/kg OYYAA2 6,739,72

SAFETY PROFILE: Very mildly toxic by ingestion and several other routes. An experimental teratogen. Experimental reproductive effects. When heated to decomposition it emits acrid smoke and irritating fumes.

PAG500 CAS:9001-73-4 ***HR: 3***
PAPAIN

PROP: White to gray, sltly hygroscopic powder. Sol in water and glycerin; insol in other common organic solvents. The most thermostatic enzyme known, digests protein. Isolated from the latex of the green fruit and leaves of *Carcia papaya L.* (IJMRAQ 67,499,78).

SYNS: ARBUZ ◇ CAROID ◇ NEMATOLYT ◇ PAPAYOTIN ◇ SUMMETRIN ◇ TROMASIN ◇ VEGETABLE PEPSIN ◇ VELARDON ◇ VERMIZYM

TOXICITY DATA with REFERENCE
orl-rbt TDLo:500 mg/kg (female 16-18D post):REP JAINAA 28,6,79
ipr-rat TDLo:375 mg/kg (female 12D post):TER IJMRAQ 67,499,78
orl-man TDLo:71 mg/kg:GIT JCGADC 9,127,87
orl-mus LD50:12500 mg/kg NYKZAU 51,27,55
ipr-mus LDLo:50 mg/kg TOXIA6 19,851,81

CONSENSUS REPORTS: Reported in EPA TSCA Inventory.

SAFETY PROFILE: Poison by intraperitoneal route. Human systemic effects by ingestion: changes in structure or function of esophagus. Experimental teratogenic and reproductive effects. An allergen. When heated to decomposition it emits toxic fumes of NO_x.

PAG750 CAS:63905-64-6 ***HR: 3***
PAPAVERIN CARBOXYLIC ACID, SODIUM SALT
mf: $C_{21}H_{20}NO_6 \cdot Na$ mw: 405.41

SYNS: 1-(3,4)-DIMETHOXYBENZYL-6,7-DIMETHOXYISOQUINOLINE-3-CARBOXYLIC ACID, SODIUM SALT ◇ 6,7-DIMETHOXY-1-VERATRYLISOQUINOLINE-3-CARBOXYLIC ACID SODIUM SALT

TOXICITY DATA with REFERENCE
ivn-man TDLo:8250 μg/kg:PNS,CNS AEPPAE 213,30,51
orl-mus LD50:260 mg/kg AEPPAE 213,30,51
scu-mus LD50:225 mg/kg AEPPAE 213,30,51
ivn-mus LD50:200 mg/kg AEPPAE 213,30,51

SAFETY PROFILE: Poison by ingestion, intravenous, and subcutaneous routes. Human systemic effects by intravenous route: sensory changes in peripheral nerves, spasticity. When heated to decomposition it emits toxic fumes of NO_x and Na_2O.

PAH000 CAS:58-74-2 ***HR: 3***
PAPAVERINE
mf: $C_{20}H_{21}NO_4$ mw: 339.42

PROP: Colorless, rhombic needles. Mp: 147°, bp: decomp, d: 1.337 @ 20°/4°. Insol in water; sol in hot benzene, glacial acetic acid, acetone; sltly sol in chloroform, carbon tetrachloride, petr ether.

SYNS: 1-((3,4-DIMETHOXYPHENYL)METHYL)-6,7-DIMETHOXYISOQUINOLINE ◇ 6,7-DIMETHOXY-1-VERATRYLISOQUINOLINE ◇ PAPANERINE ◇ PAPAVERINA (ITALIAN)

TOXICITY DATA with REFERENCE
oms-mus:oth μmol/L JIDEAE 66,313,76
orl-rat LD50:325 mg/kg ARZNAD 20,1338,70
ipr-rat LD50:64 mg/kg APTOA6 6,235,50
scu-rat LD50:151 mg/kg ARZNAD 20,1338,70
ivn-rat LD50:18 mg/kg ARZNAD 20,1338,70
orl-mus LD50:230 mg/kg ARZNAD 16,1127,66
ipr-mus LD50:117 mg/kg ARZNAD 16,386,66
scu-mus LD50:280 mg/kg ARZNAD 20,1338,70
ivn-mus LD50:25 mg/kg AIPTAK 59,149,38
ims-mus LD50:175 mg/kg AITEAT 15,415,67
idr-mus LD50:150 mg/kg AIPTAK 59,149,38
scu-dog LDLo:120 mg/kg FDWU** -,-,31
scu-cat LDLo:120 mg/kg FDWU** -,-,31
orl-rbt LDLo:190 mg/kg HBAMAK 4,1289,35
scu-rbt LDLo:250 mg/kg HBAMAK 4,1289,35
ivn-rbt LD50:25 mg/kg AIPTAK 59,149,38
scu-pgn LDLo:150 mg/kg FDWU** -,-,31
scu-frg LDLo:140 mg/kg HBAMAK 4,1289,35

SAFETY PROFILE: Poison by ingestion, intramuscular, subcutaneous, intradermal, intraperitoneal, and intravenous routes. Its central nervous system action is about midway between morphine and codeine, and large doses do not produce the amount of excitement caused by codeine or the soporific action of morphine. Mutation data reported. A cerebral vasodilator and smooth muscle relaxant. Combustible when exposed to heat or flame. When heated to decomposition it emits toxic fumes of NO_x. See also MORPHINE.

PAH250 CAS:61-25-6 ***HR: 3***
PAPAVERINE CHLOROHYDRATE
mf: $C_{20}H_{21}NO_4 \cdot ClH$ mw: 375.88

SYNS: CARDOVERINA ◇ CERESPAN ◇ CHLORHYDRATE de PAPAVERINE (FRENCH) ◇ 1-((3,4-DIMETHOXYPHENYL)METHYL)-6,7-DIMETHOXYISOQUINOLINE HYDROCHLORIDE ◇ DISPAMIL ◇ NCI-C56359 ◇ PAPAVARINE CHLORHYDRATE ◇ PAPANERIN-HCL (GERMAN) ◇ PAPAVERINE HYDROCHLORIDE ◇ PAPAVERINE MONOHYDROCHLORIDE ◇ PAP H ◇ 6,7,3',4'-TETRAMETHOXY-1-BENZYLISOQUINOLINE HYDROCHLORIDE ◇ THERAPAV ◇ VASAL

TOXICITY DATA with REFERENCE
cyt-ham:fbr 125 mg/L ESKHA5 96,55,78
cyt-ham:lng 340 mg/L GMCRDC 27,95,81
scu-mus TDLo:140 mg/kg (9D preg):TER DGDFA5 22,61,80
orl-rat LD50:360 mg/kg AIPTAK 149,200,64
ipr-rat LD50:64 mg/kg ARZNAD 6,454,56
scu-rat LD50:368 mg/kg ARZNAD 6,454,56
ivn-rat LD50:20 mg/kg AIPTAK 123,264,60
idu-rat LD50:71 mg/kg AIPTAK 180,155,69
orl-mus LD50:130 mg/kg JOETD7 6,127,82
ipr-mus LD50:108 mg/kg JMCMAR 16,1063,73
ivn-mus LD50:23 mg/kg JPETAB 87,73,46

CONSENSUS REPORTS: Reported in EPA TSCA Inventory.

SAFETY PROFILE: Poison by ingestion, intraperitoneal, intraduodenal, intravenous, and subcutaneous routes. An experimental teratogen. Mutation data reported. When heated to decomposition it emits very toxic fumes of NO_x and HCl. See also PAPAVERINE.

PAH500 CAS:115-67-3 ***HR: 2***
PARADIONE
mf: $C_7H_{11}NO_3$ mw: 157.19

SYNS: A 348 ◇ 3,5-DIMETHYL-5-ETHYLOXAZOLIDINE-2,4-DIONE ◇ 5-ETHYL-3,5-DIMETHYLOXAZOLIDINE-2,4-DIONE ◇ ISOETHADIONE ◇ PARAMETADIONE ◇ PARAMETHADIONE

TOXICITY DATA with REFERENCE
orl-rat TDLo:2640 mg/kg (6-15D preg):TER TXAPA9 37,126,76
orl-rat TDLo:2640 mg/kg (6-15D preg):REP TXAPA9 37,126,76

orl-mus LD50:1000 mg/kg 27ZQAG -,280,72
ipr-mus LD50:750 mg/kg 27ZQAG -,280,72

SAFETY PROFILE: Moderately toxic by ingestion and intraperitoneal routes. Experimental teratogenic effects. Other experimental reproductive effects. When heated to decomposition it emits toxic fumes of NO_x.

PAH750 CAS:8002-74-2 ***HR: 3***
PARAFFIN

PROP: Colorless or white, translucent wax; odorless. D: approx 0.90, mp: 50-57°. Insol in water, alc; sol in benzene, chloroform, ether, carbon disulfide, oils; misc with fats.

SYNS: PARAFFIN WAX ◇ PARAFFIN WAX FUME (ACGIH)

TOXICITY DATA with REFERENCE
imp-rat TDLo:120 mg/kg:ETA CNREA8 33,1225,73

CONSENSUS REPORTS: Reported in EPA TSCA Inventory.

OSHA PEL: Fume: TWA 2 mg/m^3 (fume)
ACGIH TLV: Fume: TWA 2 mg/m^3 (fume)

SAFETY PROFILE: Questionable carcinogen with experimental tumorigenic data by implant route. Many paraffin waxes contain carcinogens. See also PARAFFIN HYDROCARBONS.

PAH770 ***HR: D***
PARAFFIN HYDROCARBONS

SAFETY PROFILE: The effects of the paraffin hydrocarbons vary with the volatility. The gaseous hydrocarbons, such as methane, ethane, etc., have but slight anesthetic effects and are hazardous only when present in sufficient concentration to dilute the oxygen to a point below that which is necessary to sustain life. With the volatile liquid hydrocarbons, or with the next higher fraction, the anesthetic action predominates, and with the higher molecular weights or with the less volatile compounds, the anesthetic increases, but at the same time an irritant action becomes more pronounced. For information concerning toxic and hazardous properties of these materials, see the individual compounds. Paraffins are common air contaminants. Can be a dangerous fire hazard depending on volatility.

PAH800 CAS:63449-39-8 ***HR: 3***
PARAFFIN WAXES and HYDROCARBON WAXES, CHLORINATED (C12, 60% CHLORINE)

SYN: CHLORINATED PARAFFINS (C12, 60% CHLORINE)

TOXICITY DATA with REFERENCE
orl-rat TDLo:227 g/kg/2Y-I:CAR NTPTR* NTP-TR-308,86
orl-mus TDLo:182 g/kg/2Y-I:CAR NTPTR* NTP-TR-308,86
orl-rat TD:227 g/kg/2Y-I:NEO NTPTR* NTP-TR-308,86

CONSENSUS REPORTS: NTP Fifth Annual Report on Carcinogens. IARC Cancer Review: Group 2B IMEMDT 48,55,90; Animal Sufficient Evidence IMEMDT 48,55,90. NTP Carcinogenesis Studies (gavage); Clear Evidence: mouse, rat NTPTR* NTP-TR-308,86. Reported in EPA TSCA Inventory.

SAFETY PROFILE: Confirmed carcinogen with carcinogenic and neoplastigenic data. When heated to decomposition it emits acrid smoke and irritating fumes.

PAH810 CAS:63449-39-8 ***HR: 2***
PARAFFIN WAXES and HYDROCARBON WAXES, CHLORINATED (C23, 43% CHLORINE)

SYN: CHLORINATED PARAFFINS (C23, 43% CHLORINE)

TOXICITY DATA with REFERENCE
orl-rat TDLo:657 g/kg/2Y-I:NEO FAATDF 9,454,87
orl-mus TDLo:2575 g/kg/2Y-I:CAR NTPTR* NTP-TR-305,86
orl-mus TD:3750 g/kg/2Y-I:CAR FAATDF 9,454,87

CONSENSUS REPORTS: NTP Carcinogenesis Studies (gavage): Clear Evidence: mouse NTPTR* NTP-TR-305,86; Equivocal Evidence: rat NTPTR* NTP-TR-305,86. Reported in EPA TSCA Inventory.

SAFETY PROFILE: Questionable carcinogen with experimental carcinogenic and neoplastigenic data. When heated to decomposition it emits acrid smoke and irritating fumes.

PAI000 CAS:30525-89-4 ***HR: 3***
PARAFORMALDEHYDE
DOT: UN 2213
mf: $(CH_2O)_n$

PROP: White crystals; odor of formaldehyde. Flash p: 158°F, autoign temp: 572°F. Sltly sol in cold water; moderately sol in hot water yielding formaldehyde.

SYNS: FLO-MOR ◇ FORMAGENE ◇ PARAFORSN ◇ TRIFORMOL ◇ TRIOXYMETHYLENE

TOXICITY DATA with REFERENCE
skn-rbt 500 mg/2H SEV BIOFX* 28-4/73
eye-rbt 100 mg SEV BIOFX* 28-4/73
otr-rat:emb 2200 ng/plate JJATDK 1,190,81
orl-rat LD50:800 mg/kg 28ZEAL 4,308,69
skn-rbt LDLo:10000 mg/kg BIOFX* 28-4/73

CONSENSUS REPORTS: Reported in EPA TSCA Inventory.

DOT Classification: ORM-A; Label: None; DOT-IMO: Flammable Solid; Label: None.

SAFETY PROFILE: Moderately toxic by ingestion. A severe eye and skin irritant. Mutation data reported. Flammable when exposed to heat or flame; can react with oxidizing materials. To fight fire, use alcohol foam, CO_2, dry chemical. Incompatible with liquid oxygen. Dangerous; when heated to decomposition it emits toxic formaldehyde gas. See also FORMALDEHYDE.

PAI250 CAS:123-63-7 ***HR: 3***
PARALDEHYDE
DOT: UN 1264
mf: $C_6H_{12}O_3$ mw: 132.18

$OCH(CH_3)OCH(CH_3)OCHCH_3$

PROP: Colorless liquid; disagreeable taste, aromatic odor. Mp: 12.6°, lel: 1.3%, bp: 124.4° @ 752 mm, flash p: 62.6°F, d: 0.9943 @ 20°/4°, autoign temp: 460°F, vap d: 4.55. Sol in water; misc with alc, ether, oils, chloroform.

SYNS: ACETALDEHYDE, TRIMER ◇ ELALDEHYDE ◇ PARACETALDEHYDE ◇ PARAL ◇ PARALDEHYD (GERMAN) ◇ PARALDEIDE (ITALIAN) ◇ PCHO ◇ RCRA WASTE NUMBER U182 ◇ TRIACETALDEHYDE (FRENCH) ◇ 2,4,6-TRIMETHYL-1,3,5-TRIOXAAN (DUTCH) ◇ 2,4,6-TRIMETHYL-s-TRIOXANE ◇ 2,4,6-TRIMETHYL-1,3,5-TRIOXANE ◇ s-TRIMETHYLTRIOXYMETHYLENE ◇ 2,4,6-TRIMETIL-1,3,5-TRIOSSANO (ITALIAN)

TOXICITY DATA with REFERENCE
skn-rbt 500 mg open MLD UCDS** 3/24/70
eye-rbt 5 mg SEV UCDS** 3/24/70
unr-man LDLo:1462 mg/kg 85DCAI 2,73,70
rec-man LDLo:333 mg/kg:GIT SAMJAF 29,1021,55
orl-rat LD50:1530 mg/kg UCDS** 24MAR70
ihl-rat LCLo:2000 ppm/4H UCDS** 24MAR70
ipr-rat LDLo:2100 mg/kg TXAPA9 1,156,59
scu-rat LDLo:1650 mg/kg AEPPAE 182,348,36
orl-cat LDLo:3300 mg/kg HBAMAK 4,1289,35
orl-rbt LD50:3304 mg/kg PSEBAA 29,730,32
skn-rbt LD50:14 g/kg UCDS** 24MAR70

CONSENSUS REPORTS: Reported in EPA TSCA Inventory.

DOT Classification: Flammable Liquid; Label: Flammable Liquid.

SAFETY PROFILE: A human poison by rectal route. Moderately toxic to humans by an unspecified route. Moderately toxic experimentally by inhalation, ingestion, intraperitoneal, and subcutaneous routes. Human systemic effects by rectal route: necrotic changes. A skin and severe eye irritant. Low doses produce hypnotic and analgesic effects. Larger doses depress the nervous system with loss of reflexes, coma, and respiratory depression leading to respiratory paralysis and death. Chronic effects include weight loss, muscular weakness, and mental fatigue. However, poisoning is rare. A hypnotic agent. Dangerous fire hazard when exposed to heat, flame, or oxidizers. Slight explosion hazard when exposed to heat or flame. Dangerous; keep away from heat and open flame. To fight fire, use alcohol foam, CO_2, dry chemical. Potentially violent reaction with nitric acid. Incompatible with alkalies; hydrocyanic acid; iodides; oxidizers. When heated to decomposition it emits acrid smoke and irritating fumes. See also ALDEHYDES.

PAI750 CAS:62046-63-3 ***HR: 2***
PARAPLEX RG-2 (60%)

PROP: 60% oil modified sebacic alkyd, 40% toluene (NPIRI* 2,206,75).

TOXICITY DATA with REFERENCE
orl-rat LD50:3000 mg/kg NPIRI* 2,206,75
skn-rbt LD50:4000 mg/kg NPIRI* 2,206,75

SAFETY PROFILE: Moderately toxic by skin contact and ingestion. When heated to decomposition it emits acrid smoke and irritating fumes.

PAI990 CAS:4685-14-7 ***HR: 3***
PARAQUAT
mf: $C_{12}H_{14}N_2$ mw: 186.28

SYNS: DIMETHYL VIOLOGEN ◇ GRAMOXONE S ◇ METHYL VIOLOGEN (2+) ◇ PARAQUAT DICATION

TOXICITY DATA with REFERENCE
mmo-omi 20 ppm MUREAV 138,39,84
orl-rat LD50:150 mg/kg FMCHA2 -,C118,83
ipr-rat LD50:14800 μg/kg VHTODE 22,395,80
orl-mus LD50:120 mg/kg GEPHDP 14,541,83
orl-gpg LD50:40 mg/kg GEPHDP 14,541,83

CONSENSUS REPORTS: EPA Genetic Toxicology Program.

OSHA PEL: Respirable Dust: (Transitional: TWA 0.5 mg/m^3 (skin)) TWA 0.1 mg/m^3 (skin)
ACGIH TLV: TWA 0.1 mg/m^3

SAFETY PROFILE: Poison by ingestion and intraperitoneal routes. Mutation data reported. Causes ulceration of digestive tract, diarrhea, vomiting, renal damage, jaundice, edema, hemorrhage, fibrosis of lung, and death from anoxia may result. When heated to decomposition it emits toxic fumes of NO_x. See also PARAQUAT DICHLORIDE.

PAI995 CAS:4685-14-7 ***HR: 3***
PARAQUAT DIHYDRIDE
mf: $C_{12}H_{14}N_2 \cdot 2H_2O$ mw: 222.32

SYN: 1,1'-DIMETHYL-4,4'-BIPYRIDINIUM DIHYDRATE ◇ DIPYRIDYLDIHYDRATE

TOXICITY DATA with REFERENCE
orl-rat LD50:100 mg/kg IYKEDH 10,520,79
ipr-rat LD50:80 mg/kg IYKEDH 10,520,79
orl-mus LD50:120 mg/kg GEPHDP 17,359,86
orl-dog LD50:25 mg/kg IYKEDH 10,520,79
skn-rbt LD50:236 mg/kg PEMNDP 8,630,87
orl-gpg LD50:30 mg/kg IYKEDH 10,520,79

OSHA PEL: Respirable Dust: (Transitional: TWA 0.5 mg/m^3 (skin)) TWA 0.1 mg/m^3 (skin)
ACGIH TLV: TWA 0.1 mg/m^3

SAFETY PROFILE: Poison by ingestion, skin, and intraperitoneal routes. An herbicide. When heated to decomposition it emits toxic fumes of NO_x.

PAJ000 CAS:1910-42-5 ***HR: 3***
PARAQUAT DICHLORIDE
mf: $C_{12}H_{14}N_2{\cdot}2Cl$ mw: 257.18

PROP: Yellow solid. Sol in water.

SYNS: CEKUQUAT ◇ CRISQUAT ◇ DEXTRONE ◇ N,N'-DIMETHYL-4,4'-BIPYRIDINIUM DICHLORIDE ◇ 1,1'-DIMETHYL-4,4'-DIPYRIDYLIUM CHLORIDE ◇ 1,1'-DIMETHYL-4,4'-DIPYRIDINIUM-DICHLORID (GERMAN) ◇ DIMETHYL VICLOGEN CHLORIDE ◇ ESGRAM ◇ GRAMOZONE ◇ METHYLVIOLOGEN ◇ OK 622 ◇ PARAQUAT CHLORIDE

TOXICITY DATA with REFERENCE
eye-rbt 25 mg MLD BJIMAG 23,126,66
sce-ham:fbr 1 mmol/L PAACA3 25,110,84
sce-ham:lng 100 μmol/L MUREAV 151,263,85
ipr-mus TDLo:30150 μg/kg (8-16D preg):REP TXAPA9 33,450,75
ipr-rat TDLo:6500 μg/kg (6D preg):TER 26UZAB 6,257,68/70
orl-hmn LDLo:214 mg/kg YKYUA6 30,985,79
orl-man TDLo:32 mg/kg:PUL,GIT BMJOAE 3,290,68
orl-man LDLo:43 mg/kg BMJOAE 2,396,78
orl-wmn LDLo:3000 mg/kg:CNS,PUL,GIT FSIZAQ 27,494,77
orl-wmn LDLo:111 mg/kg:PUL,GIT,KID AIMDAP 146,681,86
orl-rat LD50:57 mg/kg RREVAH 10,97,65
skn-rat LD50:80 mg/kg WRPCA2 9,119,70
ipr-rat LD50:26 mg/kg BJIMAG 23,126,66
scu-rat LD50:24 mg/kg CSLNX* NX#00224
orl-mus LD50:120 mg/kg GEPHDP 12,225,81
ipr-mus LD50:20 mg/kg JPTLAS 128,21,79
ivn-mus LD50:180 mg/kg CSLNX* NX#00224
orl-dog LD50:25 mg/kg IYKEDH 10,520,79
orl-mky LD50:50 mg/kg EXMPA6 17,317,72

CONSENSUS REPORTS: EPA Extremely Hazardous Substances List.

ACGIH TLV: TWA 0.1 mg/m^3

SAFETY PROFILE: A human poison by ingestion. Poison experimentally by ingestion, skin contact, intraperitoneal, intravenous, and subcutaneous routes. Human systemic effects by ingestion: acute renal failure, acute tubular necrosis, cough, diarrhea, dyspnea, headache, hypermotility, nausea or vomiting, other gastrointestinal effects, other pulmonary effects, ulceration or bleeding from stomach. An experimental teratogen. Experimental reproductive effects. Has a delayed damaging effect on the lung alveoli. Has caused fatal poisoning in humans with severe injury to lungs. Has been implicated in aplastic anemia. An eye irritant. Mutation data reported. The National Institute of Drug Abuse (NIDA), USA, has concluded that contamination of marihuana with the herbicide paraquat may pose a serious threat to marihuana (cannabis) smokers, and issued a warning that marihuana contaminated with the herbicide paraquat could lead to permanent lung damage for regular and heavy users of marihuana. The maximum level of contamination permitted for domestic uses is 0.05 ppm; but paraquat has been found in marihuana samples at levels ranging from 3 to 2,204 ppm, averaging 452 ppm. It tends to concentrate in lung tissue whether it is ingested or inhaled, and produces a condition called fibrosis which reduces the capacity of the lung to absorb oxygen. *Inhalation* of paraquat creates a greater risk of lung damage than *ingestion* of an identical amount. An herbicide. When heated to decomposition it emits toxic fumes of Cl^- and NO_x. See also PARAQUAT.

PAJ250 CAS:2074-50-2 ***HR: 3***
PARAQUAT DIMETHYL SULPHATE
mf: $C_{12}H_{14}N_2{\cdot}2CH_3O_4S$ mw: 408.48

SYNS: 1,1'-DIMETHYL-4,4'-BIPYRIDYNIUM DIMETHYLSULFATE ◇ 1',1'-DIMETHYL-4,4'-DIPYRIDINIUM DI(METHYLSULFATE) ◇ GRAMOXONE METHYL SULFATE ◇ PARAQUAT BIS(METHYL SULFATE) ◇ PARAQUAT DIMETHYL SULFATE

TOXICITY DATA with REFERENCE
ipr-rat LD50:35 mg/kg BJIMAG 23,126,66
orl-dog LD50:25 mg/kg IYKEDH 10,520,79
orl-cat LD50:35 mg/kg IYKEDH 10,520,79
orl-gpg LD50:30 mg/kg IYKEDH 10,520,79

CONSENSUS REPORTS: EPA Extremely Hazardous Substances List.
ACGIH TLV: TWA 0.1 mg/m^3

SAFETY PROFILE: Poison by ingestion and intraperitoneal routes. An herbicide. When heated to decomposition it emits very toxic fumes of SO_x and NO_x. See also PARAQUAT.

PAJ500 CAS:10048-32-5 ***HR: 3***
PARASCORBIC ACID
mf: $C_6H_8O_2$ mw: 112.14

PROP: Oily liquid; sweet, aromatic odor. Bp: 104-105°

@ 14 mm, 119-123° @ 22 mm; d: 1.079 @ 18°/4°. Sol in water; very sol in alc, ether.

SYNS: (S)-(+)-5,6-DIHYDRO-6-METHYL-2H-PYRAN-2-ONE ◇ Γ-HEXENOLACTONE ◇ 2-HEXEN-5,1-OLIDE ◇ D''-HEXENOLLACTONE ◇ 5-HYDROXY-2-HEXENOIC ACID LACTONE ◇ PARASORBIC ACID ◇ (+)-PARASORBINSAEURE (GERMAN) ◇ SORBIC OIL

TOXICITY DATA with REFERENCE
scu-rat TDLo:1280 mg/kg/32W-I:NEO BJCAAI 17,100,63
ipr-mus LD50:250 mg/kg 85GDA2 8(1),253,82
ivn-mus LD50:195 mg/kg ARZNAD 19,617,69
skn-rbt LD50:5040 mg/kg AIHAAP 30,470,69

CONSENSUS REPORTS: IARC Cancer Review: Group 3 IMEMDT 7,56,87; Animal Limited Evidence IMEMDT 10,199,76.

SAFETY PROFILE: Poison by intraperitoneal and intravenous routes. Mildly toxic by skin contact. Questionable carcinogen with experimental neoplastigenic data. When heated to decomposition it emits acrid smoke and irritating fumes.

PAJ750 CAS:84-08-2 ***HR: 3***
PARATHIAZINE
mf: $C_{18}H_{20}N_2S$ mw: 296.46

SYNS: PYRATHIAZINE ◇ PYRROLAZOATE ◇ PYRROLIDINO-AETHYLPHENTHIAZIN (GERMAN) ◇ 10-(2-(1-PYRROLIDINYL) ETHYL)PHENOTHIAZINE ◇ 10-(2-(1-PYRROLIDYL)ETHYL)PHENOTHIAZINE ◇ ROLAZOTE

TOXICITY DATA with REFERENCE
scu-rat TDLo:450 mg/kg (6-8D preg):REP PSEBAA 89,629,55
orl-mus LD50:707 mg/kg ARZNAD 7,237,57
ipr-mus LD50:190 mg/kg ARZNAD 7,237,57

SAFETY PROFILE: Poison by intraperitoneal route. Moderately toxic by ingestion. Experimental reproductive effects. When heated to decomposition it emits very toxic fumes of NO_x and SO_x.

PAK000 CAS:56-38-2 ***HR: 3***
PARATHION
DOT: NA 2783
mf: $C_{10}H_{14}NO_5PS$ mw: 291.28

PROP: Pale-yellow liquid. Bp: 375°, mp: 6°. Very sol in alcs, esters, ethers, ketones, aromatic hydrocarbons; insol in water, petr ether, kerosene.

SYNS: AAT ◇ AATP ◇ ALLERON ◇ APHAMITE ◇ ARALO ◇ BAY E-605 ◇ BAYER E-605 ◇ BLADAN ◇ COMPOUND 3422 ◇ COROTHION ◇ CORTHION ◇ CORTHIONE ◇ DANTHION ◇ O,O-DIAETHYL-O-(4-NITROPHENYL)-MONOTHIOPHOSPHAT (GERMAN) ◇ O,O-DIETHYL-O-(4-NITRO-FENIL)-MONOTHIOFOSFAAT (DUTCH) ◇ O,O-DIETHYL-O-p-NITROFENYLESTER KYSELINYTHIOFOSFORECNE (CZECH) ◇ O,O-DIETHYL-O-p-NITROFENYLTIOFOSFAT (CZECH) ◇ O,O-DIETHYL-O-4-NITROPHENYLPHOSPHOROTHIOATE ◇ O,O-DIETHYL-O-(p-NITROPHENYL) PHOSPHOROTHIOATE ◇ O,O-DIETHYL-O-(4-NITROPHENYL) PHOSPHOROTHIOATE ◇ DIETHYL-4-NITROPHENYL PHOSPHOROTHIONATE ◇ DIETHYL-p-NITROPHENYLTHIONOPHOSPHATE ◇ DIETHYL-p-NITROPHENYLTHIOPHOSPHATE ◇ O,O-DIETHYL-O-(p-NITROPHENYL)THIONOPHOSPHATE ◇ O,O-DIETHYL-O-p-NITROPHENYL THIOPHOSPHATE ◇ O,O-DIETHYL-O-4-NITROPHENYL THIOPHOSPHATE ◇ DIETHYLPARATHION ◇ O,O-DIETIL-O-(4-NITRO-FENIL)-MONOTIOFOSFATO (ITALIAN) ◇ DNTP ◇ DPP ◇ DREXEL PARATHION 8E ◇ ECATOX ◇ EKATOX ◇ ENT 15,108 ◇ ETHLON ◇ ETHYL PARATHION ◇ FOLIDOL ◇ FOSFERMO ◇ FOSFEX ◇ FOSFIVE ◇ FOSOVA ◇ FOSTERN ◇ FOSTOX ◇ GEARPHOS ◇ GENITHION ◇ KOLPHOS ◇ KYPTHION ◇ LETHALAIRE G-54 ◇ LIROTHION ◇ MURFOS ◇ NCI-C00226 ◇ NIRAN ◇ p-NITROPHENOL, O-ESTER with O,O-DIETHYLPHOSPHOROTHIOATE ◇ NITROSTIGMIN (GERMAN) ◇ NITROSTIGMINE ◇ NOURITHION ◇ OLEOFOS 20 ◇ OLEOPARAPHENE ◇ OLEOPARATHION ◇ ORTHOPHOS ◇ PANTHION ◇ PARADUST ◇ PARAMAR ◇ PARAPHOS ◇ PARATHENE ◇ PARATHION, liquid (DOT) ◇ PARATHION-ETHYL ◇ PARAWET ◇ PESTOX PLUS ◇ PETHION ◇ PHOSKIL ◇ PHOSPHEMOL ◇ PHOSPHENOL ◇ PHOSPHOROTHIOIC ACID, O,O-DIETHYL-O-(4-NITROPHENYL) ESTER ◇ PHOSPHOSTIGMINE ◇ RCRA WASTE NUMBER P089 ◇ RHODIASOL ◇ RHODIATOX ◇ RHODIATROX ◇ SELEPHOS ◇ SIXTY-THREE SPECIAL E.C. INSECTICIDE ◇ SNP ◇ SOPRATHION ◇ STATHION ◇ STRATHION ◇ SULPHOS ◇ SUPER RODIATOX ◇ THIOPHOS ◇ THIOPHOSPHATE de O,O-DIETHYLE et de O-(4-NITROPHENYLE) (FRENCH) ◇ TIOFOS ◇ TOX 47 ◇ VAPOPHOS ◇ VITREX

TOXICITY DATA with REFERENCE
mma-sat 500 μg/plate JTEHD6 16,403,85
dns-hmn:fbr 10 μmol/L NTIS** PB268-647
orl-rat TDLo:360 μg/kg (female 2-22D post):REP TOLED5 3,11,79
scu-rat TDLo:9800 μg/kg (female 7-13D post):TER OYYAA2 6,667,72
orl-rat TDLo:1260 mg/kg/80W-C:ETA NCITR* NCI-CG-TR-70,79
orl-man TDLo:429 μg/kg/4D-I:SYS TXAPA9 14,603,69
orl-hmn LDLo:171 μg/kg CMEP** -,1,56
orl-wmn TDLo:5670 μg/kg:CNS,PUL,KID ANYAA9 160,383,69
orl-hmn LD50:3 mg/kg 85JFAN A311,86
skn-hmn LDLo:7143 μg/kg SKEZAP 9,1,68
itr-hmn LDLo:714 μg/kg SKEZAP 9,1,68
orl-rat LD50:2 mg/kg TXAPA9 11,546,67
ihl-rat LC50:84 mg/m³/4H NTIS** AD-A041-973
skn-rat LD50:6800 μg/kg TXAPA9 2,88,60
ipr-rat LD50:3600 μg/kg PSEBAA 114,509,63
ims-rat LD50:6 mg/kg JCINAO 37,350,58
orl-mus LD50:5 mg/kg ABCHA6 30,181,66
ihl-mus LCLo:15 mg/m³ 85GMAT -,52,82
skn-mus LD50:19 mg/kg ABCHA6 26,257,62

CONSENSUS REPORTS: IARC Cancer Review: Group 3 IMEMDT 7,56,87; Human Inadequate Evidence IMEMDT 30,153,83; Animal Inadequate Evidence IMEMDT 30,153,83. NCI Carcinogenesis Bioassay (feed); Clear Evidence: rat NCITR* NCI-CG-TR-70,79; (feed); No Evidence: mouse NCITR* NCI-CG-TR-70,79. EPA Farm Worker Field Reentry FEREAC 39,16888,74. EPA Extremely Hazardous Substances List. Community

Right-To-Know List. EPA Genetic Toxicology Program.

OSHA PEL: TWA 0.1 mg/m^3 (skin)
ACGIH TLV: TWA 0.1 mg/m^3 (skin); BEI: 0.5 mg/g creatinine in urine at end of shift.
DFG MAK: 0.1 mg/m^3; BAT: 500 μg/L p-nitrophenol in urine after several shifts.
NIOSH REL: (Parathion) TWA 0.05 mg/m^3
DOT Classification: Poison B; Label: Poison.

SAFETY PROFILE: A deadly poison by all routes. Human systemic effects by ingestion: general anesthetic; pulmonary effects; and kidney, ureter, bladder effects, true cholinesterase changes. Experimental teratogenic and reproductive effects. Questionable carcinogen with experimental carcinogenic and tumorigenic data. Human mutation data reported. A cholinesterase inhibitor. Parathion, like the other organic phosphorus poisons, acts as an irreversible inhibitor of the enzyme cholinesterase and thus allows the accumulation of large amounts of acetylcholine. When a critical level of cholinesterase depletion is reached, grave symptoms appear. Whether death is actually caused entirely by cholinesterase depletion or by the disturbance of a number of enzymes is not yet known. Recovery is apparently complete if a poisoned animal or man has time to reform a critical amount of cholinesterase. The organism exposed remains susceptible to relatively low dosages of parathion until the cholinesterase level has regenerated. Small doses at frequent intervals are, therefore, more or less additive. There is not, however, at the present time, any indication that, when recovery from a given exposure is entirely complete, the exposed organism is prejudiced in any way. Combustible when exposed to heat or flame. Violent reaction with endrin. Highly dangerous; shock can shatter the container releasing the contents. A broad spectrum insecticide in agricultural applications. When heated to decomposition it emits highly toxic fumes of NO_x, PO_x, SO_x.

PAK230 ***HR: 3***
PARATHION and compressed gas mixture (DOT)
DOT: NA 1967

SYN: PHOSPHOROTHIOIC ACID-O,O-DIETHYL, O-(p-NITROPHENYL)ESTER, mixed with compressed gas

DOT Classification: Poison A; Label: Poison Gas.

SAFETY PROFILE: A poison gas. Questionable carcinogen. When heated to decomposition it emits very toxic fumes of PO_x, SO_x, and NO_x. See also PARATHION.

PAK250 CAS:56-38-2 ***HR: 3***
PARATHION (mixture, dry)
DOT: NA 2783

SYN: O,O-DIETHYL-O-(p-NITROPHENYL)ESTER (DRY MIXTURE) PHOSPHOROTHIOIC ACID

DOT Classification: Poison B; Label: Poison.

SAFETY PROFILE: A powerful poison. Questionable carcinogen. When heated to decomposition it emits toxic fumes of PO_x, SO_x, and NO_x. See also PARATHION.

PAK260 CAS:56-38-2 ***HR: 3***
PARATHION MIXTURE, liquid (DOT)
DOT: NA 2783

SYN: PHOSPHOROTHIOIC ACID-O,O-DIETHYL, O-(p-NITROPHENYL) ESTER (liquid mixture)

DOT Classification: Poison B; Label: Poison.

SAFETY PROFILE: A deadly poison. Questionable carcinogen. When heated to decomposition it emits very toxic fumes of NO_x, SO_x, and PO_x. See also PARATHION.

PAK300 CAS:611-59-6 ***HR: D***
PARAXANTHINE
mf: $C_7H_8N_4O_2$ mw: 180.19

SYNS: 1,7-DIMETHYLXANTHINE ◇ 1H-PURINE-2,6-DIONE, 3,7-DIHYDRO-1,7-DIMETHYL-(9CI)

TOXICITY DATA with REFERENCE
mmo-esc 150 mg/L CSHSAZ 16,337,52
dns-hmn:oth 1 mmol/L BIOJAU 35,665,81
ipr-mus TDLo:600 mg/kg (female 11-12D post):TER TJADAB 34,279,86

SAFETY PROFILE: An experimental teratogen. Mutation data reported. When heated to decomposition it emits toxic fumes of NO_x.

PAL500 CAS:144-11-6 ***HR: 3***
PARKOPAN
mf: $C_{20}H_{31}NO$ mw: 301.52

SYNS: BENZHEXOL ◇ α-CYCLOHEXYL-α-PHENYL-1-PIPERIDINEPROPANOL ◇ 2-CYCLOHEXYL-2-PHENYL-1-PIPERIDINO-1-PROPANOL ◇ TRIHEXYLPHENIDYL ◇ TRIPHENIDYL

TOXICITY DATA with REFERENCE
ipr-mus LD50:150 mg/kg ARZNAD 21,1727,71
scu-mus LD50:202 mg/kg AEPPAE 226,18,55
ivn-mus LD50:41 mg/kg AEPPAE 226,18,55

SAFETY PROFILE: Poison by intraperitoneal, subcutaneous, and intravenous routes. When heated to decomposition it emits toxic fumes of NO_x.

PAL600 CAS:1597-82-6 ***HR: 3***
PARMATHASONE ACETATE
mf: $C_{24}H_{31}FO_6$ mw: 434.55

SYNS: 21-ACETYL-6-α-FLUORO-16-α-METHYLPREDNISOLONE ◇ 6-

α-FLUORO-16-α-METHYLPREDNISOLONE-21-ACETATE ◇ HALDRONE ◇ 16-α-METHYL-6-α-FLUORO-PREDNISOLONE-21-ACETATE ◇ MONOCORTIN ◇ PARAMETHASONE 21-ACETATE ◇ PARAMETHAZONE ACETATE ◇ SINTECORT ◇ STEMAX

TOXICITY DATA with REFERENCE
unr-rat TDLo:2625 μg/kg (1-21D preg):TER NFGZAD 13,211,68
ipr-rat LD50:392 mg/kg TXAPA9 18,185,71

SAFETY PROFILE: Poison by intraperitoneal route. An experimental teratogen. When heated to decomposition it emits toxic fumes of F^-.

PAL750 CAS:8000-68-8 ***HR: 2***
PARSLEY OIL

PROP: From steam distillation of above ground parts (herb oil) or ripe seed (seed oil) of *Petroselinium sativum* Hoffm. (Fam. *Umbelligerae*). Yellow to light brown liquid; odor of parsley. D (herb oil): 0.908-0.940, (seed oil): 1.040; refr index (herb oil): 1.503-1.530 @ 20°, (seed oil): 1.513-1.522 @ 20°. Sol in fixed oils, mineral oil; sltly sol in propylene glycol; insol in glycerin.

SYNS: OIL of PARSLEY ◇ PARSLEY HERB OIL (FCC) ◇ PARSLEY SEED OIL (FCC) ◇ PETERSILIENSAMEN OEL (GERMAN)

TOXICITY DATA with REFERENCE
skn-hmn 10 mg/48H MLD FCTXAV 13,681,75
skn-rbt 500 mg/24H MLD FCTXAV 13,681,75
skn-gpg 500 mg/24H MLD FCTXAV 13,681,75
skn-gpg 100% MLD FCTXAV 13,897,75
orl-rat LD50:3300 mg/kg FCTOD7 21,871,83
orl-mus LD50:1520 mg/kg FCTXAV 13,897,75

CONSENSUS REPORTS: Reported in EPA TSCA Inventory.

SAFETY PROFILE: Moderately toxic by ingestion. A human skin irritant. When heated to decomposition it emits acrid smoke and irritating fumes.

PAM000 CAS:113-42-8 ***HR: 3***
PARTERGIN
mf: $C_{20}H_{25}N_3O_2$ mw: 339.48

SYNS: BASOFORTINA ◇ 9,10-DIDEHYDRO-N-(α-(HYDROXYMETHYL)PROPYL)-6-METHYL-ERGOLINE-8-β-CARBOXAMIDE ◇ ME 277 ◇ METHERGINE ◇ METHYLERGOBASINE ◇ METHYLERGOBREVIN ◇ METHYLERGOMETRIN ◇ METHYLERGOMETRINE ◇ METHYLERGONOVIN ◇ METHYLERGONOVINE

TOXICITY DATA with REFERENCE
ipr-rat TDLo:10 mg/kg (female 12-21D post):REP AJPHAP 180,296,55
orl-rat LD50:93 mg/kg 27ZIAQ -,163,73
ivn-rat LD50:23 mg/kg 27ZQAG -,100,72
orl-mus LD50:187 mg/kg 27ZIAQ -,163,73
ivn-mus LD50:85 mg/kg 27ZIAQ -,163,73
ivn-rbt LD50:2600 μg/kg 27ZIAQ -,163,73

SAFETY PROFILE: Poison by ingestion and intravenous routes. When heated to decomposition it emits toxic fumes of NO_x.

PAM175 CAS:508-59-8 ***HR: 3***
PARTHENICIN
mf: $C_{15}H_{18}O_4$ mw: 262.33

PROP: Crystals from water. Mp: 163-166°. Practically insol in water; sol in alc, chloroform, ether, ethyl acetate.

SYN: PARTHENIN

TOXICITY DATA with REFERENCE
cyt-hmn:leu 1 nmol/L IJEBA6 16,1117,78
mnt-mus-ipr 200 mg/kg/24H-I IJEBA6 16,1117,78
cyt-ham:ovr 3120 ppb BSECBU 13,365,85
ipr-rat LD50:42 mg/kg JOBSDN 6,729,84

SAFETY PROFILE: Poison by intraperitoneal route. Human mutation data reported. When heated to decomposition it emits acrid smoke and irritating fumes.

PAM500 CAS:20170-20-1 ***HR: 2***
PASALIN
mf: $C_{20}H_{22}N_4O$ mw: 334.46

SYNS: AP-14 ◇ DIFENAMIZOLE ◇ 2-(DIMETHYLAMINO)-N-(1,3-DIPHENYL-1H-PYRAZOL-5-YL) PROPANAMIDE ◇ DIPHENAMIZOLE ◇ 1,3-DIPHENYL-5-(2-DIMETHYLAMINOPROPIONAMIDO)PYRAZOLE

TOXICITY DATA with REFERENCE
orl-rat LD50:815 mg/kg GDIKAN 20,360,72
ipr-rat LD50:1480 mg/kg IYKEDH 4,467,73
orl-mus LD50:960 mg/kg GDIKAN 20,360,72

SAFETY PROFILE: Moderately toxic by intraperitoneal and ingestion routes. An analgesic and anti-inflammatory agent. When heated to decomposition it emits toxic fumes of NO_x.

PAM775 CAS:53760-19-3 ***HR: 3***
PASPALIN
mf: $C_{30}H_{39}NO_5$ mw: 493.70

SYNS: 21-ACETOXY-7,18-DIHYDROXY-16,18-DIMETHYL-10-PHENYL-(11)CYTOCHALASA-6(12),13,19-TRIENE-1-ONE ◇ CYTOCHALASIN-H ◇ KODOCYTOCHALASIN-1 ◇ PASPALIN P I

TOXICITY DATA with REFERENCE
cyt-hmn:lym 50 μg/kg IJEBA6 16,430,78
ivn-mus LD50:2 mg/kg 85GDA2 2,439,80
ipr-mam LD50:2341 μg/kg INDRBA 15,76,78

SAFETY PROFILE: Poison by intravenous and intraperitoneal routes. Human mutation data reported. When heated to decomposition it emits toxic fumes of NO_x.

PAM780 ***HR: 1***
PASQUE FLOWER

PROP: Small perennial herbs bearing yellow, red, white, or purple flowers. Various species are found in Alaska, Canada, and the temperate areas of the United States, and some are native to Europe. They are common in rock gardens.

SYNS: A. CANADENSIS ◇ ANEMONE ◇ ANEMONE (VARIOUS SPECIES) ◇ A. NUTTALLIANA ◇ A. PATENS ◇ APRIL FOOLS ◇ A. PULSATILLA ◇ CATS EYES ◇ GOSLING ◇ HARTSHORN PLANT ◇ LILY OF THE FIELD ◇ LION'S BEARD ◇ NIGHTCAPS ◇ NIMBLE WEED ◇ PRAIRIE CROCUS ◇ PRAIRIE HEN FLOWER ◇ PRIARIE SMOKE ◇ P. VULGARIS ◇ THIMBLEWEED ◇ WILD CROCUS ◇ WINDFLOWER

SAFETY PROFILE: The whole plant contains the poison protoanemonin which is a direct irritant and drying agent. Chewing pieces of the plant causes inflammation and blistering of the mouth and throat. Because of the immediate pain ingestion seldom occurs but can cause vomiting and bloody diarrhea.

PAM782 CAS:8057-62-3 ***HR: 2***
PASSIONFLOWER EXTRACT

SYNS: PASSIFLORAE INCARNATAE EXTRACTUM ◇ PASSIFLORA EXTRACT ◇ PASSIFLORA INCARNATA, EXTRACT

TOXICITY DATA with REFERENCE
orl-rat TDLo:440 mg/kg (7-17D preg):REP KSRNAM 15,3431,81
ipr-rat LD50:3510 mg/kg NIIRDN 6,APP-15,82
ipr-mus LD50:3140 mg/kg NIIRDN 6,APP-15,82
scu-mus LD50:8300 mg/kg KSRNAM 15,3393,81

SAFETY PROFILE: Moderately toxic by intraperitoneal route. Experimental reproductive effects. When heated to decomposition it emits acrid smoke and irritating fumes.

PAM785 CAS:3060-89-7 ***HR: 2***
PATORAN
mf: $C_9H_{11}BrN_2O_2$ mw: 259.13

PROP: Crystals from cyclohexane. Mp: 95-96°. Vap press at 20°: 0.000003 mm Hg. Solubility in water at 20°: 320 ppm. Sol in methanol, ethanol, acetone, chloroform.

SYNS: N'-(4-BROMOPHENYL)-N-METHOXY-N-METHYLUREA ◇ 3-(p-BROMOPHENYL)-1-METHOXY-1-METHYLUREA ◇ 3-(p-BROMOPHENYL)-1-METHYL-1-METHOXYUREA ◇ 3-(4-BROMPHENYL)-1-METHOXYHARNSTOFF (GERMAN) ◇ C-3126 ◇ CIBA-3126 ◇ METOBROMURON ◇ PATTONEX

TOXICITY DATA with REFERENCE
eye-rbt 50 mg MLD CIGET* -,-,77
orl-rat LD50:2000 mg/kg WRPCA2 9,119,70
ipr-rat LD50:430 mg/kg PESTD5 17,351,76
orl-mus LD50:2098 mg/kg PESTD5 17,351,76
ipr-mus LD50:847 mg/kg PESTD5 17,351,76

CONSENSUS REPORTS: EPA Genetic Toxicology Program.

SAFETY PROFILE: Moderately toxic by ingestion and intraperitoneal routes. An eye irritant. When heated to decomposition it emits toxic fumes of Br^- and NO_x.

PAM789 CAS:53962-20-2 ***HR: 3***
PATRINOSIDE
mf: $C_{21}H_{34}O_{11}$ mw: 462.55

TOXICITY DATA with REFERENCE
ipr-mus LD50:555 mg/kg SHZAAY 34,200,80
scu-mus LD50:350 mg/kg SHZAAY 34,200,80
ivn-mus LD50:595 mg/kg SHZAAY 34,200,80

SAFETY PROFILE: Poison by subcutaneous route. Moderately toxic by intravenous and intraperitoneal routes. When heated to decomposition it emits acrid smoke and irritating fumes.

PAM800 CAS:76319-15-8 ***HR: 3***
PATRINOSIDE-AGLYCONE
mf: $C_{15}H_{24}O_6$ mw: 300.39

TOXICITY DATA with REFERENCE
ipr-mus LD50:182 mg/kg SHZAAY 34,200,80
scu-mus LD50:234 mg/kg SHZAAY 34,200,80
ivn-mus LD50:143 mg/kg SHZAAY 34,200,80

SAFETY PROFILE: Poison by subcutaneous, intravenous, and intraperitoneal routes. When heated to decomposition it emits acrid smoke and irritating fumes.

PAN100 CAS:434-07-1 ***HR: 3***
PAVISOID
mf: $C_{21}H_{32}O_3$ mw: 332.53

PROP: Crystals from ethyl acetate. Mp: 178-180°.

SYNS: ADROIDIN ◇ ADROYD ◇ ANADROL ◇ ANADROYD ◇ ANAPOLON ◇ ANASTERON ◇ ANASTERONAL ◇ ANASTERONE ◇ BECOREL ◇ CI-406 ◇ 4,5-DIHYDRO-2-HYDROXYMETHYLENE-17-α-METHYLTESTOSTERONE ◇ DYNASTEN ◇ HMD ◇ 17-β-HYDROXY-2-HYDROXYMETHYLENE-17-α-METHYL-3-ANDROSTANONE ◇ 17-β-HYDROXY-2-(HYDROXYMETHYLENE)-17-α-METHYL-5-α-ANDROSTAN-3-ONE ◇ 17-β-HYDROXY-2-(HYDROXYMETHYLENE)-17-METHYL-5-α-ANDROSTAN-3-ONE ◇ 17-HYDROXY-2-(HYDROXYMETHYLENE)-17-METHYL-5-α-17-β-ANDROST-3-ONE ◇ 2-HYDROXYMETHYLENE-17-α-METHYL-5-α-ANDROSTAN-17-β-OL-3-ONE ◇ 2-HYDROXY-METHYLENE-17-α-METHYL-DIHYDROTESTOSTERONE ◇ 2-(HYDROXYMETHYLENE)-17-α-METHYLDIHYDROTESTOSTERONE ◇ 2-HYDROXYMETHYLENE-17-α-METHYL-17-β-HYDROXY-3-ANDROSTANONE ◇ METHABOL ◇ 17-α-METHYL-2-HYDROXYMETHYLENE-17-HYDROXY-5-α-ANDROSTAN-3-ONE ◇ NASTENON ◇ NSC-26198 ◇ OXIMETHOLONUM ◇ OXIMETOLONA ◇ OXITOSONA-50 ◇ OXYMETHALONE ◇ OXYMETHENOLONE ◇ OXYMETHOLONE ◇ PARDROYD ◇ PLENASTRIL ◇ PROTANABOL ◇ ROBORAL ◇ SYNASTERON ◇ ZENALOSYN

TOXICITY DATA with REFERENCE
scu-rat TDLo:28 mg/kg (female 14D pre):REP CCPTAY 5,489,72
orl-rat TDLo:20 mg/kg (17-20D preg):TER ECJPAE 24,77,77
orl-chd TDLo:270 mg/kg/9W-C:CAR NEJMAG 296,1411,77
orl-man TDLo:2336 mg/kg/2Y-C:CAR CANCAR 43,440,79
orl-hmn TDLo:46 mg/kg/14W-I:LIV JAMAAP 240,243,78

CONSENSUS REPORTS: NTP Fifth Annual Report on Carcinogens. IARC Cancer Review: Human Inadequate Evidence IMEMDT 13,131,77.

SAFETY PROFILE: Confirmed human carcinogen producing liver tumors. Human systemic effects by ingestion: impaired liver function. An experimental teratogen. Experimental reproductive effects. When heated to decomposition it emits acrid smoke and irritating fumes. See also TESTOSTERONE.

PAN250 CAS:25265-19-4 ***HR: 3***
PBAN-560

PROP: Terpolymer of polybutadiene, acrylic acid, and acrylonitrile (NTIS** AD441-640).

TOXICITY DATA with REFERENCE
ipr-mus LDLo:316 mg/kg NTIS** AD441-640

CONSENSUS REPORTS: Reported in EPA TSCA Inventory. Cyanide and its compounds are on The Community Right-To-Know List.

SAFETY PROFILE: Poison by intraperitoneal route. When heated to decomposition it emits toxic fumes of NO_x and CN^-. See also NITRILES and individual components.

PAN500 CAS:25265-19-4 ***HR: 2***
PBAN-560 (degassed)

PROP: Terpolymer of polybutadiene, acrylic acid, and acrylonitrile (NTIS** AD441-640).

TOXICITY DATA with REFERENCE
ipr-mus LD50:750 mg/kg NTIS** AD441-640

CONSENSUS REPORTS: Reported in EPA TSCA Inventory. Cyanide and its compounds are on The Community Right-To-Know List.

SAFETY PROFILE: Moderately toxic by intraperitoneal route. When heated to decomposition it emits toxic fumes of CN^-. See also NITRILES and individual components.

PAN750 CAS:102070-98-4 ***HR: 2***
PC ALCOHOL DF

TOXICITY DATA with REFERENCE
skn-rbt 500 mg/24H SEV 28ZPAK -,36,72
eye-rbt 250 μg/24H SEV 28ZPAK -,36,72
orl-rat LD50:4230 mg/kg 28ZPAK -,36,72

SAFETY PROFILE: Mildly toxic by ingestion. Severe skin and eye irritant. When heated to decomposition it emits acrid smoke and irritating fumes. See also ALCOHOLS.

PAN775 ***HR: 3***
PCMH

mf: $C_{10}H_{12}N_2O_6 \cdot Mg$ mw: 280.55

SYNS: MAGNESIUM-5-OXO-2-PYRROLIDINECARBOXYLATE◇ 5-OXO-2-PYRROLIDINECARBOXYLIC ACID MAGNESIUM SALT (2:1) ◇ PYRROLIDONE CARBOXYLATE de MAGNESIUM HYDRATE (FRENCH)

TOXICITY DATA with REFERENCE
ipr-rat LD50:1460 mg/kg THERAP 31,471,76
orl-mus LD50:11800 mg/kg THERAP 31,471,76
ipr-mus LD50:1350 mg/kg THERAP 31,471,76
scu-mus LD50:1400 mg/kg THERAP 31,471,76
ivn-mus LD50:305 mg/kg THERAP 31,471,76

SAFETY PROFILE: Poison by intravenous route. Moderately toxic by subcutaneous and intraperitoneal routes. Very mildly toxic by ingestion. When heated to decomposition it emits toxic fumes of NO_x. See also MAGNESIUM COMPOUNDS.

PAN800 CAS:102366-69-8 ***HR: 3***
PE-043

TOXICITY DATA with REFERENCE
orl-rat TDLo:600 mg/kg (female 9-14D post):REP TOIZAG 15,491,68
orl-mus TDLo:600 mg/kg (female 7-12D post):TER TOIZAG 15,491,68
scu-rat LD50:648 mg/kg TOIZAG 15,478,68
ivn-rat LD50:41 mg/kg TOIZAG 15,478,68
orl-mus LD50:15560 mg/kg TOIZAG 15,478,68
scu-mus LD50:242 mg/kg TOIZAG 15,478,68
ivn-mus LD50:87 mg/kg TOIZAG 15,478,68

SAFETY PROFILE: Poison by intravenous and subcutaneous routes. Moderately toxic by subcutaneous route. An experimental teratogen. Experimental reproductive effects. When heated to decomposition it emits acrid smoke and irritating fumes.

PAO000 CAS:8002-03-7 ***HR: 1***
PEANUT OIL

PROP: Straw-yellow to greenish-yellow or nearly colorless oil; nutty odor and bland taste. Mp: 2.7°, flash p: 540°F, d: 0.92, autoign temp: 833°F. Misc with ether, petr ether, chloroform, carbon disulfide; sol in benzene, carbon tetrachloride, oils; very sltly sol in alc. From seed of *Arachis hypogaea* (85DIA2 2,201,77).

SYNS: ARACHIS OIL ◇ EARTHNUT OIL ◇ GROUNDNUT OIL ◇ INDIGENOUS PEANUT OIL ◇ KATCHUNG OIL ◇ PECAN SHELL POWDER

TOXICITY DATA with REFERENCE
skn-hmn 300 mg/3D-I MLD 85DKA8 -,127,77
mma-sat 10 μL/plate FCTXAV 18,467,80
orl-mus TDLo:952 g/kg/1Y-I:ETA IJMRAQ 61,422,73

CONSENSUS REPORTS: Reported in EPA TSCA Inventory.

SAFETY PROFILE: A human skin irritant and mild allergen. Questionable carcinogen with experimental tumorigenic data. Mutation data reported. Combustible when exposed to heat or flame; can react with oxidizing materials. Slight spontaneous heating. To fight fire, use CO_2, dry chemical. When heated to decomposition it emits acrid smoke and irritating fumes.

PAO500 CAS:2858-66-4 ***HR: 3***
PELLETIERINE
mf: $C_8H_{15}NO$ mw: 141.24

PROP: Sltly colored, oily liquid. Bp: 195°, d: 0.988 @ 20°/4°. Sol in alc, ether, chloroform.

SYNS: 2-ACETONYLPIPERIDINE ◇ ISOPELLETIERINE ◇ (R)-(−)-PELLETIERINE ◇ 1-(2-PIPERIDYL)-2-PROPANONE ◇ PUNICINE

TOXICITY DATA with REFERENCE
ivn-gpg LDLo:400 mg/kg HBAMAK 4,1289,35
ivn-pgn LDLo:140 mg/kg HBAMAK 4,1289,35

SAFETY PROFILE: Poison by intravenous route. When heated to decomposition it emits toxic fumes of NO_x.

PAP000 CAS:18968-99-5 ***HR: 3***
PEMOLINE MAGNESIUM
mf: $C_9H_{10}MgN_2O_4$ mw: 234.52

PROP: Monohydrate, amorph material. Mp: higher than 300°. Insol in water.

SYNS: ABBOTT 30400 ◇ BIS(AQUO)-N,N'-(2-IMINO-5-PHENYL-4-OXAZOLIDINONE)MAGNESIUM(II) ◇ CYLERT ◇ ECYLERT ◇ (2-IMINO-5-PHENYL-4-OXAZOLIDINONATO(2-))DIAQUOMAGNESIUM ◇ 2-IMINO-5-PHENYL-4-OXAZOLIDINONE MAGNESIUM CHELATE ◇ MAGNESIUM PEMOLINE ◇ PEMOLIN and MAGNESIUM HYDROXIDE ◇ PEMOLINE MAGNESIUM CHELATE ◇ PMH ◇ TAMILAN

TOXICITY DATA with REFERENCE
orl-rat LD50:485 mg/kg ARZNAD 23,810,73
orl-mus LD50:500 mg/kg AIPTAK 181,441,69
ipr-mus LD50:487 mg/kg AIPTAK 181,441,69
orl-mky LD50:400 mg/kg AIPTAK 181,441,69

SAFETY PROFILE: Poison by ingestion. Moderately toxic by intraperitoneal route. When heated to decomposition it emits toxic fumes of NO_x. See also MAGNESIUM COMPOUNDS and ORGANOMETALS.

PAP100 CAS:6152-95-0 ***HR: 3***
PEMPIDINE HYDROCHLORIDE
mf: $C_{10}H_{21}N{\bullet}ClH$ mw: 191.78

SYN: 1,2,2,6,6-PENTAMETHYLPIPERIDINEHYDROCHLORIDE

TOXICITY DATA with REFERENCE
orl-mus LD50:510 mg/kg BJPCAL 13,501,58
ipr-mus LD50:161 mg/kg BJPCAL 13,501,58
ivn-mus LD50:91 mg/kg BJPCAL 13,501,58

SAFETY PROFILE: Poison by intravenous and intraperitoneal routes. Moderately toxic by ingestion. When heated to decomposition it emits toxic fumes of NO_x and HCl.

PAP110 CAS:546-48-5 ***HR: 3***
PEMPIDINE TARTRATE
mf: $C_{10}H_{21}N{\bullet}C_4H_6O_6$ mw: 305.42

SYNS: PEMPIDINA TARTRATO (ITALIAN) ◇ 1,2,2,6,6-PENTAMETHYLPIPERIDINE TARTRATE

TOXICITY DATA with REFERENCE
orl-mus LD50:400 mg/kg BCFAAI 103,490,64
ipr-mus LD50:164 mg/kg BJPCAL 13,501,58
scu-mus LD50:230 mg/kg BCFAAI 103,490,64
ivn-mus LD50:113 mg/kg BJPCAL 13,501,58

SAFETY PROFILE: Poison by ingestion, subcutaneous, intravenous, and intraperitoneal routes. When heated to decomposition it emits toxic fumes of NO_x.

PAP225 CAS:38363-40-5 ***HR: 3***
PENBUTOLOL
mf: $C_{18}H_{29}NO_2$ mw: 291.48

PROP: Crystals. Mp: 68-72°. Sol in methanol, ethanol, chloroform.

SYNS: 1-(2-CYCLOPENTYLPHENOXY)-3-((1,1-DIMETHYLETHYL)AMINO)-2-PROPANOL, (S)- ◇ (−)-PENBUTOLOL ◇ HOE 893 ◇ HOE 893D

TOXICITY DATA with REFERENCE
orl-rat LD50:1265 mg/kg DRUGAY 22,1,81
ivn-rat LD50:22 mg/kg DRUGAY 22,1,81
orl-mus LD50:1230 mg/kg DRUGAY 22,1,81
ivn-mus LD50:18 mg/kg DRUGAY 22,1,81
orl-rbt LD50:300 mg/kg DRUGAY 22,1,81
ivn-rbt LD50:17 mg/kg DRUGAY 22,1,81

SAFETY PROFILE: Poison by ingestion and intrave-

nous routes. When heated to decomposition it emits toxic fumes of NO_x.

PAP230 CAS:38363-32-5 ***HR: 3***
PENBUTOLOL SULFATE
mf: $C_{18}H_{29}NO_2 \cdot 1/2H_2O_4S$ mw: 340.47

SYNS: (−)-1-(tert-BUTYLAMINO)-3-(o-CYCLOPENTYLPHENOXYL)-2-PROPANOL SULFATE ◇ HOE 893D ◇ (−)-TERBUCLOMINE

TOXICITY DATA with REFERENCE
orl-rat LD50:1350 mg/kg YKYUA6 36,657,85
ipr-rat LD50:76 mg/kg OYYAA2 21,1001,81
scu-rat LD50:860 mg/kg OYYAA2 21,1001,81
ivn-rat LD50:26 mg/kg YKYUA6 36,657,85
orl-mus LD50:1230 mg/kg OYYAA2 21,1001,81
ipr-mus LD50:71 mg/kg OYYAA2 21,1001,81
scu-mus LD50:405 mg/kg OYYAA2 21,1001,81
ivn-mus LD50:38 mg/kg OYYAA2 21,1001,81

SAFETY PROFILE: Poison by intravenous and intraperitoneal routes. Moderately toxic by ingestion and subcutaneous route. When heated to decomposition it emits toxic fumes of NO_x and SO_x.

PAP250 CAS:26864-56-2 ***HR: 3***
PENFLURIDOL
mf: $C_{28}H_{27}ClF_5NO$ mw: 524.01

PROP: White microcrystals. Mp: 105-107°. Sltly sol in water.

SYNS: 4-(4-CHLORO-α,α,α-TRIFLUORO-m-TOLYL)-1-(4,4-BIS(p-FLUOROPHENYL)BUTYL)-4-PIPERIDINOL ◇ McN-JR-16,341 ◇ R 16341 ◇ SEMAP ◇ TLP-607

TOXICITY DATA with REFERENCE
orl-rat TDLo:3 mg/kg (lactating female 3D post):REP NSAPCC 279,31,73
orl-mus TDLo:48 mg/kg (7-12D preg):TER OYYAA2 13,581,77
orl-rat LD50:160 mg/kg EJPHAZ 11,139,70
orl-mus LD50:87 mg/kg EJPHAZ 11,139,70
ivn-dog LDLo:5 mg/kg EJPHAZ 11,139,70
orl-gpg LD50:90 mg/kg EJPHAZ 11,139,70

SAFETY PROFILE: Poison by ingestion and intravenous routes. Experimental teratogenic and reproductive effects. A neuroleptic agent. When heated to decomposition it emits very toxic fumes of Cl^-, F^-, and NO_x.

PAP500 CAS:52-66-4 ***HR: 3***
dl-PENICILLAMINE
mf: $C_5H_{11}NO_2S$ mw: 149.23

SYNS: 3-MERCAPTO-dl-VALINE (9CI) ◇ dl-α-MERCAPTOVALINE

TOXICITY DATA with REFERENCE
pic-esc 100 mg/L APMBAY 12,234,64
orl-rat LD50:365 mg/kg ARZNAD 22,1434,72
ipr-mus LD50:358 mg/kg TOLED5 26,95,85

SAFETY PROFILE: Poison by ingestion and intraperitoneal routes. Mutation data reported. When heated to decomposition it emits toxic fumes of SO_x and NO_x.

PAP550 CAS:2219-30-9 ***HR: 3***
PENICILLAMINE HYDROCHLORIDE
mf: $C_5H_{11}NO_2S \cdot ClH$ mw: 185.69

SYNS: DISTAMINE ◇ METALCAPTASE ◇ d-PENICILLAMINE HYDROCHLORIDE ◇ USAF EL-23

TOXICITY DATA with REFERENCE
orl-hmn TDLo:5460 mg/kg:SKN BMJOAE 1,838,77
ipr-mus LD50:200 mg/kg NTIS** AD277-689
orl-mam LD50:3670 mg/kg ANTBAL 14,837,69
ivn-mam LD50:2170 mg/kg ANTBAL 14,837,69

SAFETY PROFILE: A poison by intraperitoneal route. Moderately toxic by ingestion and intravenous route. Human systemic effects by ingestion: dermatitis. When heated to decomposition it emits very toxic fumes of NO_x, SO_x, and HCl.

PAP600 CAS:69388-84-7 ***HR: 3***
PENICILLANIC ACID SULFONE SODIUM SALT
mf: $C_8H_{10}NO_5S \cdot Na$ mw: 207.24

SYNS: CP 45899 SODIUM SALT ◇ PENICILLANIC ACID DIOXIDE SODIUM SALT ◇ PENICILLANIC ACID 1,1-DIOXIDE SODIUM SALT ◇ SODIUM-1,1-DIOXOPENICILLANATE ◇ SODIUM PENICILLANATE 1,1-DIOXIDE ◇ SODIUM SULBACTAM ◇ 4-THIA-1-AZABICYCLO (3.2.0)HEPTANE-2-CARBOXYLIC ACID, 3,3-DIMETHYL-7-OXO-, 4,4-DIOXIDE, SODIUM SALT, (2S-cis)-

TOXICITY DATA with REFERENCE
ivn-rat TDLo:13500 mg/kg (female 17-22D post):REP NKRZAZ 32(Suppl 4),108,84
ivn-rat LD50:6500 mg/kg NKRZAZ 32(Suppl 4),97,84
ivn-mus LD50:8100 μg/kg IYKEDH 17,1106,86

SAFETY PROFILE: Poison by intravenous route. Experimental reproductive effects. When heated to decomposition it emits toxic fumes of NO_x and SO_x.

PAP750 CAS:90-65-3 ***HR: 3***
PENICILLIC ACID
mf: $C_8H_{10}O_4$ mw: 170.18

PROP: Needles from petr ether. Mp: 83-84°. Sltly sol in cold water, hot petr ether; very sol in hot water, alc, ether, benzene chloroform; insol in pentane-hexane.

SYNS: Γ-KETO-β-METHOXY-Δ-METHYLENE-Δ^{α}-HEXENOIC ACID ◇ 3-METHOXY-5-METHYL-4-OXO-2,5-HEXADIENOIC ACID ◇ PA ◇ PENCILLIC ACID

TOXICITY DATA with REFERENCE
mrc-bcs 100 μg/disc CNREA8 36,445,76
dnd-hmn:hla 320 mg/L/1H JJEMAG 42,527,72
ipr-mus TDLo:90 mg/kg (10D preg):TER TOXIA6 16,92,78
ipr-mus TDLo:90 mg/kg (10D preg):REP TOXIA6 16,92,78
scu-rat TDLo:960 mg/kg/48W-I:NEO BJCAAI 15,85,61
ipr-rat LD50:90 mg/kg JTEHD6 7,169,81
orl-mus LD50:600 mg/kg JPETAB 88,119,46
ipr-mus LD50:90 mg/kg TXAPA9 52,1,80
scu-mus LD50:100 mg/kg MEIEDD 10,1017,83
ivn-mus LD50:250 mg/kg JPETAB 88,119,46

CONSENSUS REPORTS: IARC Cancer Review: Group 3 IMEMDT 7,56,87; Animal Sufficient Evidence IMEMDT 10,211,76. EPA Genetic Toxicology Program.

SAFETY PROFILE: Poison by intravenous, subcutaneous, and intraperitoneal routes. Moderately toxic by ingestion. An experimental teratogen. Experimental reproductive effects. Questionable carcinogen with experimental neoplastigenic data. Human mutation data reported. When heated to decomposition it emits acrid smoke and irritating fumes.

PAQ000 CAS:1406-05-9 ***HR: 3***
PENICILLIN
mf: $(CH_3)_2C_5H_3NSO(COOH)NHCOOR$ (bicyclic)

PROP: A group of isomeric and closely related antibiotic compounds with outstanding bacterial activity. An extract from *Penicillium notatum* (JPETAB 77,40,43). Different varieties of penicillin are produced by adding the proper precursors to the nutrient solution.

SYN: PENIZILLIN (GERMAN)

TOXICITY DATA with REFERENCE
orl-wmn TDLo:72 mg/kg (13-15W preg):REP BSFDA3 57,534,50
ims-man TDLo:12385 μg/kg/16D-I:SKN MJAUAJ 1,305,47
scu-mus LDLo:3200 mg/kg JPETAB 77,70,43
ivn-mus LDLo:1000 mg/kg JPETAB 77,70,43
ipr-gpg LD50:30 mg/kg 85ERAY 3,1663,78
scu-gpg LD50:30 mg/kg 85ERAY 3,1663,78

CONSENSUS REPORTS: EPA Genetic Toxicology Program.

SAFETY PROFILE: Poison by intraperitoneal and subcutaneous routes. Moderately toxic by intravenous route. Human reproductive effects by ingestion: abortion. Human systemic effects by intramuscular route: dermatitis. Experimental reproductive effects. Has been implicated in aplastic anemia. When heated to decomposition it emits very toxic fumes of NO_x and SO_x.

PAQ060 ***HR: 3***
PENICILLIN compounded with CHOLINE CHLORIDE

SYNS: C 172 ◇ PENICILLIN-CHOLINESTER CHLORID (GERMAN)

TOXICITY DATA with REFERENCE
ipr-mus LD50:62500 μg/kg ARZNAD 5,698,55
scu-mus LD50:450 mg/kg ARZNAD 5,698,55
ivn-mus LD50:37500 μg/kg ARZNAD 5,698,55

SAFETY PROFILE: Poison by intraperitoneal and intravenous routes. Moderately toxic by subcutaneous route. When heated to decomposition it emits very toxic fumes of NO_x and Cl^-. See also PENICILLIN and CHOLINE.

PAQ100 CAS:751-84-8 ***HR: 3***
PENICILLIN G BENETHAMINE
mf: $C_{31}H_{35}N_3O_4S$ mw: 545.701

PROP: Crystals; slt characteristic amine taste. Mp: 146-147°. Very sltly sol in water (0.1 w/v at 40°).

SYNS: BENEPEN ◇ BENETACIL ◇ BENETHAMINE PENICILLIN ◇ BENETHAMINE PENICILLIN G ◇ BENETOLIN ◇ BETAPEN

TOXICITY DATA with REFERENCE
ivn-rat LD50:122 mg/kg ANTCAO 5, 152,55
ivn-mus LD50:170 mg/kg ANTCAO 5, 152,55
ivn-rbt LD50:55 mg/kg ANTCAO 5, 152,55

SAFETY PROFILE: Poison by intravenous route. When heated to decomposition it emits toxic fumes of SO_x and NO_x. See also PENICILLIN.

PAQ120 CAS:7177-45-9 ***HR: 3***
PENICILLIN G EPHEDRINE SALT
mf: $C_{16}H_{18}N_2O_4S•C_{10}H_{15}NO$ mw: 499.68

SYNS: 3,3-DIMETHYL-7-OXO-6-(2-PHENYLACETAMIDO)-4-THIA-1-AZABICYCLO(3.2.0)HEPTANE-2-CARBOXYLIC ACID compounded with EPHEDRINE (1:1) ◇ EPHEDRINE PENICILLIN ◇ TERSAVIN

TOXICITY DATA with REFERENCE
ipr-rat LD50:680 mg/kg JPETAB 95,336,49
scu-rat LD50:2400 mg/kg JPETAB 95,336,49
ipr-mus LD50:630 mg/kg JPETAB 95,336,49
ivn-mus LD50:175 mg/kg JPETAB 95,336,49
ivn-rat LD50:175 mg/kg JPETAB 95,336,49

SAFETY PROFILE: Poison by intravenous route. Moderately toxic by intraperitoneal and subcutaneous routes. When heated to decomposition it emits toxic fumes of SO_x and NO_x. See also PENICILLIN and EPHEDRINE.

PAQ875 ***HR: 3***
PENICILLIUM ROQUEFORTI TOXIN

SYNS: P. ROQUEFORTI TOXIN ◇ PR TOXIN ◇ TOXIN, PENICILLIUM ROQUEFORTI

TOXICITY DATA with REFERENCE
mmo-sat 625 ng/plate MUREAV 130,79,84
dnd-rat:lvr 25 μmol/L TOERD9 2,273,79
oms-rat-ipr 10 mg/kg CBINA8 14,207,76
oms-rat:lvr 35 mg/L CBINA8 18,153,77
orl-rat TDLo:100 mg/kg/52D-C:ETA MYCPAH 78,125,82
orl-rat LDLo:115 mg/kg APMBAY 25,111,73
ipr-rat LD50:7 mg/kg FEBLAL 88,341,78
ipr-mus LD50:7 mg/kg FEBLALA 88,341,78

SAFETY PROFILE: Poison by ingestion and intraperitoneal routes. Questionable carcinogen with experimental tumorigenic data. Mutation data reported. When heated to decomposition it emits acrid smoke and irritating fumes.

PAR250 CAS:12627-35-9 ***HR: 3***
PENITREM A
mf: $C_{37}H_{44}ClNO_6$ mw: 634.27

SYN: TREMORTIN A

TOXICITY DATA with REFERENCE
ipr-mus TDLo:3 mg/kg (9D preg):TER TOXIA6 16,92,78
orl-mus LD50:10 mg/kg 41KEAL -,108,78
ipr-mus LDLo:1100 μg/kg CTOXAO 17,45,80
ipr-dog LDLo:500 μg/kg TXAPA9 35,311,76

CONSENSUS REPORTS: EPA Genetic Toxicology Program.

SAFETY PROFILE: Poison by ingestion and intraperitoneal routes. An experimental teratogen. When heated to decomposition it emits very toxic fumes of Cl^- and NO_x.

PAR500 ***HR: 3***
PENNYROYAL OIL

PROP: Chief constituent is d-pulegone. From steam distillation of *Mentha pulegium L.* (Fam. *Labiatae*) (FCTXAV 12,807,74). Light yellow liquid; mint odor. Sol in fixed oils, propylene glycol, mineral oil; insol in glycerin.

SYN: AMERICAN PENNYROYAL OIL

TOXICITY DATA with REFERENCE
skn-mus 100% MOD FCTXAV 12,949,74
orl-rat LD50:400 mg/kg FCTXAV 12,949,74

SAFETY PROFILE: Experimental poison by ingestion. A skin irritant. When heated to decomposition it emits acrid smoke and irritating fumes.

PAR600 CAS:17088-72-1 ***HR: 3***
PENOCTONIUM BROMIDE
mf: $C_{26}H_{50}NO_2{\cdot}Br$ mw: 488.68

SYNS: α,α-DICYCLOPENTYL-ACETICACID-DIETHYLAMINO-ETHYLESTER BROMOCTYLATE ◇ N-(2-((DICYCLOPENTYLACETYL)OXY)ETHYL)-N,N-DIETHYL-1-OCTANAMINIUM BROMIDE (9CI) ◇ α,α-DICYCLOPENTYLESSIGSAURE-DIAETHYLAMINO-AETHYLESTER-BROMOCTYLAT (GERMAN) ◇ UG 767

TOXICITY DATA with REFERENCE
eye-rbt 1% ARZNAD 18,137,68
orl-rat LD50:1200 mg/kg ARZNAD 18,137,68
ivn-rat LD50:19 mg/kg ARZNAD 18,137,68
orl-mus LD50:900 mg/kg ARZNAD 18,137,68
ipr-mus LD50:26 mg/kg ARZNAD 18,137,68
scu-mus LD50:60 mg/kg ARZNAD 18,137,68
ivn-mus LD50:20 mg/kg ARZNAD 18,137,68

SAFETY PROFILE: Poison by subcutaneous, intravenous, and intraperitoneal routes. Moderately toxic by ingestion. An eye irritant. When heated to decomposition it emits toxic fumes of Br^- and NO_x. See also BROMIDES.

PAR750 CAS:13820-81-0 ***HR: 3***
PENTAAMMINEAQUACOBALT(III) CHLORATE
mf: $Cl_3CoH_{17}N_5O_{10}$ mw: 412.44

CONSENSUS REPORTS: Cobalt and its compounds are on the Community Right-To-Know List.

SAFETY PROFILE: Explodes on impact or heating to 130°C. When heated to decomposition it emits very toxic fumes of Cl^- and NO_x. See also CHLORATES and COBALT COMPOUNDS.

PAR799 CAS:41481-90-7 ***HR: 3***
PENTAAMMINEPYRAZINERUTHENIUM(II) PERCHLORATE
mf: $C_4H_{19}Cl_2N_7O_8Ru$ mw: 465.21

SAFETY PROFILE: An extremely shock-sensitive explosive. When heated to decomposition it emits very toxic fumes of RuO_4, Cl^- and NO_x. See also PERCHLORATES and RUTHENIUM COMPOUNDS.

PAS829 CAS:19482-31-6 ***HR: 3***
PENTAAMMINEPYRIDINERUTHENIUM(II) PERCHLORATE
mf: $C_5H_{20}Cl_2N_6O_8Ru$ mw: 464.22

SAFETY PROFILE: An extremely touch-sensitive explosive. When heated to decomposition it emits very toxic fumes of RuO_4, Cl^- and NO_x. See also PERCHLORATES and RUTHENIUM COMPOUNDS.

PAS859 CAS:15663-42-0 ***HR: 3***
PENTAAMMINETHIOCYANATOCOBALT(III) PERCHLORATE
mf: $CH_{15}Cl_2CoN_6O_8S$ mw: 401.06

CONSENSUS REPORTS: Cobalt and its compounds are on the Community Right-To-Know List.

SAFETY PROFILE: Explodes when heated to 325°C. It has medium impact-sensitivity. When heated to decomposition it emits very toxic fumes of Cl^- and NO_x. See also PERCHLORATES and COBALT COMPOUNDS.

PAS879 CAS:38139-15-0 ***HR: 3***
PENTAAMMINETHIOCYANATORUTHENIUM(II) PERCHLORATE
mf: $CH_{15}Cl_2N_6O_8RuS$ mw: 443.20

SAFETY PROFILE: The dry material and its solutions in ether are extremely touch-sensitive explosives. When heated to decomposition it emits very toxic fumes of RuO_4, Cl^-, and NO_x. See also PERCHLORATES and RUTHENIUM COMPOUNDS.

PAT750 CAS:19624-22-7 ***HR: 3***
PENTABORANE(9)
DOT: UN 1380
mf: B_5H_9 mw: 63.14

PROP: Colorless gas or liquid; bad odor. Mp: −46.6°, d: 0.61 @ 0°, vap d: 2.2, vap press: 66 mm @ 0°, lel: 0.42%, bp: 60.°

TOXICITY DATA with REFERENCE
ihl-rat LC50:6 ppm/4H AIHAAP 19,46,58
ipr-rat LD50:11100 μg/kg 14KTAK-,693,64
ihl-mus LC50:3400 ppb/4H AIHAAP 19,46,58
ihl-dog LC50:92 mg/m³/15M JPETAB 145,382,64
ihl-mky LC50:640 mg/m³/2M JPETAB 145,382,64

CONSENSUS REPORTS: EPA Extremely Hazardous Substances List. Reported in EPA TSCA Inventory.

OSHA PEL: (Transitional: 0.005 ppm) TWA 0.005 ppm; STEL 0.015 ppm
ACGIH TLV: TWA 0.005 ppm; STEL 0.013 ppm
DFG MAK: 0.005 ppm (0.01 mg/m³)
DOT Classification: Flammable Liquid; Label: Spontaneously Combustible, Poison; Label: Flammable Liquid and Poison.

SAFETY PROFILE: Poison by inhalation and intraperitoneal routes. Dangerous fire hazard by chemical reaction; spontaneously flammable in air. Dangerous explosion hazard. To fight fire, use special fire-fighting materials; water is not effective; reacts violently with halogenated extinguishing agents. Get instructions from supplier. Explosive reaction with oxygen. Forms shock-sensitive solutions in solvents containing carbonyl, ether, or ester functions; or halogens. Incompatible with dimethyl sulfoxide. Upon decomposition it emits toxic fumes of B. See also BORANES and BORON COMPOUNDS.

PAT799 CAS:18433-84-6 ***HR: 3***
PENTABORANE(11)
mf: B_5H_{11} mw: 65.16

SAFETY PROFILE: Ignites spontaneously in air. When heated to decomposition it emits toxic fumes of B. See also BORANES and BORON COMPOUNDS.

PAU250 CAS:608-71-9 ***HR: 3***
PENTABROMOPHENOL
mf: C_6HBr_5O mw: 488.62

PROP: Mp: 229.5°, bp: subl. Insol in water; sltly sol in alc and ether.

TOXICITY DATA with REFERENCE
ipr-mus LDLo:250 mg/kg CBCCT* 5,288,53

CONSENSUS REPORTS: Reported in EPA TSCA Inventory.

SAFETY PROFILE: Poison by intraperitoneal route. When heated to decomposition it emits toxic fumes of Br^-. See also BROMIDES.

PAU500 CAS:1163-19-5 ***HR: 3***
PENTABROMOPHENYL ETHER
mf: $C_{12}Br_{10}O$ mw: 959.22

SYNS: BERKFLAM B 10E ◇ BR 55N ◇ BROMKAL 83-10DE ◇ BROMKAL 82-ODE ◇ DBDPO ◇ DECABROMOBIPHENYL ETHER ◇ DECABROMOBIPHENYL OXIDE ◇ DECABROMODIPHENYL OXIDE ◇ DECABROMOPHENYL ETHER ◇ DE 83R ◇ FR 300 ◇ FRP 53 ◇ NCI-C55287 ◇ 1,1′-OXYBIS(2,3,4,5,6-PENTABROMOBENZENE (9CI) ◇ SAYTEX 102 ◇ SAYTEX 102E ◇ TARDEX 100

TOXICITY DATA with REFERENCE
orl-rat TDLo:100 mg/kg (6-15D preg):REP COTODO 1,52,74
orl-rat TDLo:1092 g/kg/2Y-C:NEO NTPTR* NTP-TR-309,86

CONSENSUS REPORTS: NTP Carcinogenesis Studies (feed); Some Evidence: rat NTPTR* NTP-TR-309,86; (feed); Equivocal Evidence: mouse NTPTR* NTP-TR-309,86. EPA Extremely Hazardous Substances List. Polybrominated biphenyl compounds are on the Community Right-To-Know List. Reported in EPA TSCA Inventory.

SAFETY PROFILE: Questionable carcinogen with experimental neoplastigenic data. Experimental reproductive effects. Used as a flame retardant for thermoplastics. When heated to decomposition it emits toxic fumes of Br^-. See also ETHERS and BROMIDES.

PAV000 CAS:135-48-8 ***HR: D***
PENTACENE
mf: $C_{22}H_{14}$ mw: 278.36

PROP: Deep blue needles. Insol in water; sltly sol in organic solvents.

SYNS: BENZO(B)NAPHTHACENE ◇ 2,3:6,7-DIBENZANTHRACENE ◇ LIN-DIBENZANTHRACENE ◇ LIN-NAPHTHOANTHRACENE

TOXICITY DATA with REFERENCE
dnd-mam:lym 20 mg BIPMAA 4,409,66

CONSENSUS REPORTS: Reported in EPA TSCA Inventory.

SAFETY PROFILE: Mutation data reported. When heated to decomposition it emits acrid smoke and irritating fumes.

PAV225 CAS:1768-31-6 ***HR: 3***
PENTACHLOROACETONE
mf: C_3HCl_5O mw: 230.29

SYNS: 1,1,1,3,3-PENTACHLOROPROPANONE ◇ 1,1,1,3,3-PENTACHLORO-2-PROPANONE

TOXICITY DATA with REFERENCE
mmo-sat 300 μL/plate ENMUDM 7,163,85
mrc-smc 100 μL/L MUREAV 155,53,85
orl-rat LD50:200 mg/kg 85GMAT -,96,82
ihl-mus LC50:450 mg/m³/2H 85GMAT -,96,82

SAFETY PROFILE: Poison by inhalation and ingestion. Mutation data reported. When heated to decomposition it emits toxic fumes of Cl^-. See also CHLORINATED HYDROCARBONS, ALIPHATIC.

PAV250 CAS:25201-35-8 ***HR: 3***
PENTACHLOROACETOPHENONE
mf: $C_8Cl_5H_3O$ mw: 292.36

SYN: 2',3',4',5',6'-PENTACHLOROACETOPHENONE

TOXICITY DATA with REFERENCE
orl-rat TDLo:360 mg/kg (male 72D pre):REP GISAAA 51(7),75,86
orl-rat LD50:1250 mg/kg GTPZAB 24(3),48,80
skn-rat LD50:1800 mg/kg GTPZAB 24(3),48,80
ipr-rat LD50:290 mg/kg GTPZAB 24(3),48,80
orl-mus LD50:700 mg/kg GTPZAB 24(3),48,80
ipr-mus LD50:160 mg/kg GTPZAB 24(3),48,80

SAFETY PROFILE: Poison by intraperitoneal route. Moderately toxic by ingestion and skin contact. Experimental reproductive effects. When heated to decomposition it emits toxic fumes of Cl^-. See also CHLORINATED HYDROCARBONS, ALIPHATIC.

PAV500 CAS:608-93-5 ***HR: 2***
PENTACHLOROBENZENE
mf: C_6HCl_5 mw: 250.32

PROP: Needles from alc. D: 1.834 @ 17°, mp: 85-86°, bp: 275-277°; Insol in water; sol in hot alc; very sol in ether.

SYNS: QCB ◇ RCRA WASTE NUMBER U183

TOXICITY DATA with REFERENCE
orl-rat TDLo:2 g/kg (6-15D preg):TER TXCYAC 5,117,75
orl-rat LD50:1080 mg/kg JEPTDQ 4(5-6),183,80
orl-mus LD50:1175 mg/kg JEPTDQ 4(5-6),183,80

CONSENSUS REPORTS: Reported in EPA TSCA Inventory.

SAFETY PROFILE: Moderately toxic by ingestion. An experimental teratogen. When heated to decomposition it emits toxic fumes of Cl^-. See also CHLORINATED HYDROCARBONS, AROMATIC.

PAV600 CAS:25429-29-2 ***HR: 3***
PENTACHLOROBIPHENYL
mf: $C_{12}H_5Cl_5$ mw: 326.42

SYNS: APIROLIO 1476 C ◇ BIPHENYL, PENTACHLORO- ◇ 1,1'-BIPHENYL, PENTACHLORO-(9CI) ◇ DIPHENYL PENTACHLORIDE ◇ KANEKROL 500 ◇ PENTACHLORODIPHENYL ◇ PYRALENE 1476 ◇ PYROCLOR 5

TOXICITY DATA with REFERENCE
unr-rat LD50:3580 mg/kg GISAAA 53(5),6,88

CONSENSUS REPORTS: NTP Fifth Annual Report On Carcinogens.

SAFETY PROFILE: Confirmed carcinogen. Moderately toxic by unspecified route. When heated to decomposition it emits toxic fumes of Cl^-.

PAV750 CAS:31391-27-2 ***HR: 2***
PENTACHLOROBUTANE
mf: $C_4H_5Cl_5$ mw: 230.34

TOXICITY DATA with REFERENCE
orl-mus LD50:2500 mg/kg GISAAA 28,9,63
orl-gpg LD50:1400 mg/kg GISAAA 28,9,63

SAFETY PROFILE: Moderately toxic by ingestion. When heated to decomposition it emits toxic fumes of Cl^-. See also CHLORINATED HYDROCARBONS, ALIPHATIC.

PAV775 ***HR: D***
2,2,3,4,4-PENTACHLORO-3-BUTENOIC ACID
mf: $C_4HCl_5O_2$ mw: 258.30

SYN: PENTACHLORO-3-BUTENOIC ACID

TOXICITY DATA with REFERENCE
mmo-sat 10 μg/plate MUREAV 137,89,84
otr-ham:emb 800 μg/L CALEDQ 23,297,84

SAFETY PROFILE: Mutation data reported. When heated to decomposition it emits toxic fumes of Cl^-. See also CHLORINATED HYDROCARBONS, ALIPHATIC.

PAW000 CAS:40321-76-4 ***HR: 3***
1,2,3,7,8-PENTACHLORODIBENZO-p-DIOXIN
mf: $C_{12}H_3Cl_5O_2$ mw: 356.40

TOXICITY DATA with REFERENCE
orl-mus LD50:337500 ng/kg TXAPA9 44,335,78
orl-gpg LD50:3100 ng/kg TXAPA9 44,335,78

CONSENSUS REPORTS: IARC Cancer Review: Animal Inadequate Evidence IMEMDT 15,41,77.

SAFETY PROFILE: Poison by ingestion. Questionable carcinogen. When heated to decomposition it emits toxic fumes of Cl^-. See also CHLORINATED HYDROCARBONS, AROMATIC.

PAW100 CAS:57117-31-4 ***HR: 3***
2,3,4,7,8-PENTACHLORODIBENZOFURAN
mf: $C_{12}H_3Cl_5O$ mw: 340.40

SYN: DIBENZOFURAN,2,3,4,7,8-PENTACHLORO-

TOXICITY DATA with REFERENCE
orl-mus TDLo:4 μg/kg (female 10-13D post):TER TXAPA9 90,206,87
orl-rat LD50:916 μg/kg FAATDF 11,236,88
ivn-mky LDLo:34 μg/kg TXAPA9 93,231,88
orl-gpg LDLo:10 μg/kg ANYAA9 320,151,79

SAFETY PROFILE: Poison by ingestion and intravenous route. An experimental teratogen. When heated to decomposition it emits toxic fumes of Cl^-.

PAW250 CAS:42279-29-8 ***HR: 3***
PENTACHLORO DIPHENYL OXIDE
mf: $C_{12}H_5Cl_5O$ mw: 342.42

SYNS: ETHER, PENTACHLOROPHENYL ◇ PHENYL ETHER PENTACHLORO

TOXICITY DATA with REFERENCE
orl-gpg LDLo:100 mg/kg 14CYAT 2,1707,63

OSHA PEL: TWA 0.5 mg/m^3

SAFETY PROFILE: Poison by ingestion. When heated to decomposition it emits toxic fumes of Cl^-.

PAW500 CAS:76-01-7 ***HR: 3***
PENTACHLOROETHANE
DOT: UN 1669
mf: C_2HCl_5 mw: 202.28

PROP: Colorless liquid; chloroform-like odor. Mp: −29°, bp: 161-162°, d: 1.6728 @ 25°/4°. Insol in water; misc in alc and ether.

SYNS: ETHANE PENTACHLORIDE ◇ NCI-C53894 ◇ PENTACHLOROETHAAN (DUTCH) ◇ PENTACHLORAETHAN (GERMAN) ◇ PENTACHLORETHANE (FRENCH) ◇ PENTACLOROETANO (ITALIAN) ◇ PENTALIN ◇ RCRA WASTE NUMBER U184

TOXICITY DATA with REFERENCE
msc-mus:lyms 70 mg/L EMMUEG 12,85,88
sce-ham:ovr 100 mg/L SCIEAS 236,933,87
orl-mus TDLo:129 g/kg/2Y-I:CAR NTPTR* NTP-TR-232,82
orl-mus TD:258 g/kg/2Y-I:CAR NTPTR* NTP-TR-232,82
ihl-rat LC50:4238 ppm/2H 85JCAE-,97,86
ihl-mus LCLo:35 mg/m^3/2H AEPPAE 141,19,29
orl-dog LDLo:500 mg/kg AJHYA2 16,325,32
ivn-dog LDLo:100 mg/kg QJPPAL 7,205,34
scu-rbt LDLo:700 mg/kg QJPPAL 7,205,34

CONSENSUS REPORTS: IARC Cancer Review: Group 3 IMEMDT 7,56,87; Animal Limited Evidence IMEMDT 41,99,86. NTP Carcinogenesis Bioassay (gavage); Clear Evidence: mouse NTPTR* NTP-TR-232,82; (gavage); No Evidence: rat NTPTR* NTP-TR-232,82. Reported in EPA TSCA Inventory.

DFG MAK: 5 ppm (40 mg/m^3)
DOT Classification: Poison B; Label: Poison.

SAFETY PROFILE: Poison by inhalation and intravenous routes. Moderately toxic by ingestion and subcutaneous routes. An irritant. Questionable carcinogen with experimental carcinogenic data. Flammable when exposed to heat or flame. Moderately explosive by spontaneous chemical reaction. To fight fire, use water, CO_2, dry chemical. Dehalogenation by reaction with alkalies; metals, etc. will produce spontaneously explosive chloroacetylenes. Violent reaction with NaK alloy + bromoform. Mixtures with potassium are very shock-sensitive explosives. When heated to decomposition it emits highly toxic fumes of Cl^-. See also CHLORINATED HYDROCARBONS, ALIPHATIC.

PAW600 CAS:102395-72-2 ***HR: 3***
2',2'',4',4'',5-PENTACHLORO-4-HYDROXYISOPHTHALANILIDE
mf: $C_{20}H_{11}Cl_5N_2O_3$ mw: 504.58

SYN: SC. 121

TOXICITY DATA with REFERENCE
orl-mus LD50:450 mg/kg FRPSAX 16,679,61
ipr-mus LD50:66 mg/kg FRPSAX 16,679,61
scu-mus LD50:68 mg/kg FRPSAX 16,679,61

SAFETY PROFILE: Poison by intraperitoneal and subcutaneous routes. Moderately toxic by ingestion. When heated to decomposition it emits toxic fumes of Cl^- and NO_x.

PAW750 CAS:1321-64-8 ***HR: 3***
PENTACHLORONAPHTHALENE
mf: $C_{10}H_3Cl_5$ mw: 300.38

PROP: White solid.

CONSENSUS REPORTS: Reported in EPA TSCA Inventory.

OSHA PEL: TWA 0.5 mg/m^3 (skin)
ACGIH TLV: TWA 0.5 mg/m^3
DFG MAK: 0.5 mg/m^3

SAFETY PROFILE: Poison by ingestion, inhalation, and skin contact. An irritant. Action similar to chlorinated naphthalenes and chlorinated diphenyls. Dangerous; when heated to decomposition it emits highly toxic fumes of Cl^-. See also CHLORINATED HYDROCARBONS, AROMATIC.

PAX000 CAS:82-68-8 ***HR: 3***
PENTACHLORONITROBENZENE
mf: $C_6Cl_5NO_2$ mw: 295.32

PROP: Colorless crystals. Mp: 146°, bp: 328°, vap press: 0.013 mm @ 25°.

SYNS: AVICOL ◇ BATRILEX ◇ BRASSICOL ◇ EARTHCIDE ◇ FARTOX ◇ FOLOSAN ◇ FOMAC 2 ◇ FUNGICLOR ◇ GC 3944-3-4 ◇ KOBU ◇ KOBUTOL ◇ KP 2 ◇ NCI-C00419 ◇ OLPISAN ◇ PCNB ◇ PENTACHLORNITROBENZOL (GERMAN) ◇ PENTAGEN ◇ PKhNB ◇ QUINTOCENE ◇ QUINTOZEN ◇ QUINTOZENE ◇ RCRA WASTE NUMBER U185 ◇ SANICLOR 30 ◇ TERRACHLOR ◇ TERRAFUN ◇ TILCAREX ◇ TRI-PCNB ◇ TRITISAN

TOXICITY DATA with REFERENCE
dnd-esc 20 μmol/L MUREAV 89,95,81
mrc-asn 40 μmol/L MUREAV 147,288,85
orl-mus TDLo:5 g/kg (7-16D preg):TER TXAPA9 35,239,76
orl-mus TDLo:4176 mg/kg (female 6-14D post):REP NTIS** PB223-160
orl-mus TDLo:135 g/kg/77W-C:CAR JNCIAM 42,1101,69
orl-rat LD50:1100 mg/kg FAATDF 7,299,86
ihl-rat LC50:1400 mg/m^3 85GMAT -,96,82
skn-rat LD50:4 g/kg FMCHA2 -,C221,89
orl-mus LC50:1400 mg/kg 85GMAT -,96,82
ihl-mus LC50:2 g/m^3 85GMAT -,96,82
ipr-mus LD50:4500 mg/kg ARTODN 51,329,82

CONSENSUS REPORTS: IARC Cancer Review: Group 3 IMEMDT 7,56,87; Animal Sufficient Evidence IMEMDT 5,211,74. NCI Carcinogenesis Bioassay (feed); No Evidence: mouse, rat NCITR* NCI-CG-TR-61,78. EPA Extremely Hazardous Substances List. Reported in EPA TSCA Inventory. EPA Genetic Toxicology Program.

ACGIG TLV: (Proposed: 0.5 mg/m^3)

SAFETY PROFILE: Moderately toxic by ingestion. An experimental teratogen. Questionable carcinogen with experimental carcinogenic data. Mutation data reported. Used as a fungicide. Dangerous; when heated to decomposition it emits highly toxic fumes of NO_x and Cl^-. See also NITRO COMPOUNDS OF AROMATIC HYDROCARBONS.

PAX250 CAS:87-86-5 ***HR: 3***
PENTACHLOROPHENOL
DOT: UN 2020
mf: C_6HCl_5O mw: 266.32

PROP: Dark-colored flakes and sublimed needle crystals; characteristic odor. Mp: 191°, bp: 310° (decomp), d: 1.978, vap press: 40 mm @ 211.2°. Sol in ether, benzene; very sol in alc; insol in water; sltly sol in cold petr ether.

SYNS: CHEM-TOL ◇ CHLOROPHEN ◇ CRYPTOGIL OL ◇ DOWCIDE 7 ◇ DOWICIDE 7 ◇ DOWICIDE EC-7 ◇ DOWICIDE G ◇ DOW PENTACHLOROPHENOL DP-2 ANTIMICROBIAL ◇ DUROTOX ◇ EP 30 ◇ FUNGIFEN ◇ GLAZD PENTA ◇ GRUNDIER ARBEZOL ◇ LAUXTOL ◇ LAUXTOL A ◇ LIROPREM ◇ NCI-C54933 ◇ NCI-C55378 ◇ NCI-C56655 ◇ PCP ◇ PENCHLOROL ◇ PENTA ◇ PENTACHLOORFENOL (DUTCH) ◇ PENTACHLOROFENOL ◇ PENTACHLOROPHENATE ◇ 2,3,4,5,6-PENTACHLOROPHENOL ◇ PENTACHLOROPHENOL, DOWICIDE EC-7 ◇ PENTACHLOROPHENOL, DP-2 ◇ PENTACHLOROPHENOL (GERMAN) ◇ PENTACHLOROPHENOL, TECHNICAL ◇ PENTACLOROFENOLO (ITALIAN) ◇ PENTACON ◇ PENTA-KIL ◇ PENTASOL ◇ PENWAR ◇ PERATOX ◇ PERMACIDE ◇ PERMAGARD ◇ PERMASAN ◇ PERMATOX DP-2 ◇ PERMATOX PENTA ◇ PERMITE ◇ PRILTOX ◇ RCRA WASTE NUMBER U242 ◇ SANTOBRITE ◇ SANTOPHEN ◇ SANTOPHEN 20 ◇ SINITUHO ◇ TERM-I-TROL ◇ THOMPSON'S WOOD FIX ◇ WEEDONE

TOXICITY DATA with REFERENCE
skn-rbt 10 mg/24H open MLD AIHAAP 23,95,62
mma-sat 40 nmol/plate AIDZAC 10,305,82
orl-rat TDLo:4 g/kg (female 77D pre-28D post):REP TXAPA9 41,138,77
scu-mus TDLo:450 mg/kg (female 6-14D post):TER NTIS** PB223-160
orl-mus TDLo:8736 mg/kg/2Y-C:CAR NTPTR* NTP-TR-349,89
scu-mus TDLo:46 mg/kg:ETA NTIS** PB223-159
orl-man LDLo:401 mg/kg EESADV 1,343,77
orl-rat LD50:27 mg/kg JPETAB 76,104,42
ihl-rat LC50:355 mg/m^3 GTPZAB 13(9),58,69
skn-rat LD50:96 mg/kg GTPZAB 13(9),58,69
ipr-rat LD50:56 mg/kg BJPCAL 13,20,58
scu-rat LD50:100 mg/kg FEPRA7 2,76,43
orl-mus LD50:117 mg/kg TOLED5 29,39,85
ipr-mus LD50:58 mg/kg JTEHD6 10,699,82
scu-dog LDLo:135 mg/kg HBTXAC 5,123,59
orl-rbt LDLo:70 mg/kg JPETAB 76,104,42

skn-rbt LDLo:40 mg/kg JPETAB 76,104,42
ipr-rbt LDLo:135 mg/kg HBTXAC 5,123,59
scu-rbt LDLo:70 mg/kg JPETAB 76,104,42

CONSENSUS REPORTS: IARC Cancer Review: Human Limited Evidence IMEMDT 41,319,86; Animal Inadequate Evidence IMEMDT 20,303,79. Chlorophenol compounds are on The Community Right-To-Know List. Reported in EPA TSCA Inventory. EPA Genetic Toxicology Program.

OSHA PEL: TWA 0.5 mg/m^3 (skin)
ACGIH TLV: TWA 0.5 mg/m^3 (skin); BEI: 2 mg/g creatinine in urine prior to last shift of workweek.
DFG MAK: 0.05 ppm (0.5 mg/m^3); BAT: 1000 μg/L in plasma/serum.
DOT Classification: ORM-E; Label: None.

SAFETY PROFILE: Suspected human carcinogen with experimental tumorigenic data. Human poison by ingestion. Poison experimentally by ingestion, skin contact, intraperitoneal, and subcutaneous routes. An experimental teratogen. Other experimental reproductive effects. A skin irritant. Mutation data reported. Acute poisoning is marked by weakness with changes in respiration, blood pressure, and urinary output. Also causes dermatitis, convulsions, and collapse. Chronic exposure can cause liver and kidney injury. Dangerous; when heated to decomposition it emits highly toxic fumes of Cl^-. See also CHLOROPHENOLS.

PAX750 ***HR: 3***
PENTACHLOROPHENOL, SODIUM derivative mixed with TETRACHLOROPHENOL SODIUM derivative (4:1)

SYN: CRYPTOGYL NA (ITALIAN)

TOXICITY DATA with REFERENCE
orl-rbt LDLo:218 mg/kg FOMDAK No.2,105,56
scu-rbt LDLo:257 mg/kg FOMDAK No.2,105,56
ivn-rbt LDLo:22 mg/kg FOMDAK No.2,105,56

CONSENSUS REPORTS: Chlorophenol compounds are on The Community Right-To-Know List.

SAFETY PROFILE: Poison by ingestion, subcutaneous, and intravenous routes. When heated to decomposition it emits toxic fumes of Cl^- and Na_2O. See also PENTACHLOROPHENOL and CHLOROPHENOLS.

PAY000 CAS:16714-68-4 ***HR: 2***
1,1,2,2,3-PENTACHLOROPROPANE
mf: $C_3H_3Cl_5$ mw: 216.31

TOXICITY DATA with REFERENCE
unr-rat LD50:819 mg/kg CHABA8 89,142,78
unr-mus LD50:551 mg/kg CHABA8 89,142,78
unr-rbt LD50:500 mg/kg CHABA8 89,142,78
unr-gpg LD50:560 mg/kg CHABA8 89,142,78

CONSENSUS REPORTS: Reported in EPA TSCA Inventory.

SAFETY PROFILE: Moderately toxic by unspecified routes. When heated to decomposition it emits toxic fumes of Cl^-. See also CHLORINATED HYDROCARBONS, ALIPHATIC.

PAY200 CAS:1600-37-9 ***HR: D***
1,1,2,3,3-PENTACHLOROPROPENE
mf: C_3HCl_5 mw: 214.29

SYNS: 1,1,2,3,3-PENTACHLORO-1-PROPENE ◇ 1,1,2,3,3-PENTACHLOROPROPYLENE

TOXICITY DATA with REFERENCE
mmo-sat 10 μg/plate MUREAV 79,203,80
mmo-smc 100 μg/L MUREAV 119,273,83

CONSENSUS REPORTS: Reported in EPA TSCA Inventory.

SAFETY PROFILE: Mutation data reported. When heated to decomposition it emits toxic fumes of Cl^-. See also CHLORINATED HYDROCARBONS, ALIPHATIC.

PAY250 CAS:2176-62-7 ***HR: 3***
2,3,4,5,6-PENTACHLOROPYRIDINE
mf: C_5Cl_5N mw: 251.31

TOXICITY DATA with REFERENCE
ipr-mus LD50:235 mg/kg TXAPA9 11,361,67

CONSENSUS REPORTS: Reported in EPA TSCA Inventory.

SAFETY PROFILE: Poison by intraperitoneal route. When heated to decomposition it emits very toxic fumes of Cl^- and NO_x.

PAY500 CAS:133-49-3 ***HR: 3***
PENTACHLOROTHIOPHENOL
mf: C_6HCl_5S mw: 282.38

SYNS: PCTP ◇ PENTACHLORO-BENZENETHIOL ◇ PENTACHLORTHIOFENOL (CZECH) ◇ USAF B-51

TOXICITY DATA with REFERENCE
eye-rbt 500 mg/24H SEV 28ZPAK -,168,72
orl-rat LD50:11900 g/kg 28ZPAK -,168,72
ipr-mus LD50:100 mg/kg NTIS** AD277-689

CONSENSUS REPORTS: Chlorophenol compounds are on The Community Right-To-Know List. Reported in EPA TSCA Inventory.

SAFETY PROFILE: Poison by intraperitoneal route. Mildly toxic by ingestion. Severe eye irritant. When heated to decomposition it emits very toxic fumes of SO_x and Cl^-. See also CHLOROPHENOLS.

PAY600 CAS:24378-32-3 ***HR: 3***
PENTACYANONITROSYLFERRATE BARIUM
mf: $C_5FeN_6O{\cdot}Ba$ mw: 353.30

SYN: BARIUMNITROSYLPENTACYANOFERRATE

TOXICITY DATA with REFERENCE
orl-rat LD50:183 mg/kg ARZNAD 24,308,74
ipr-rat LD50:19400 μg/kg ARZNAD 24,308,74
orl-mus LD50:72 mg/kg ARZNAD 24,308,74
ipr-mus LD50:18200 μg/kg ARZNAD 24,308,74
ivn-rbt LDLo:5600 μg/kg ARZNAD 24,308,74

CONSENSUS REPORTS: Barium and its compounds and cyanide and its compounds are on the Community Right-To-Know List.

SAFETY PROFILE: Poison by ingestion, intravenous and intraperitoneal routes. When heated to decomposition it emits toxic fumes of NO_x and CN^-. See also BARIUM COMPOUNDS and CYANIDE.

PAY610 CAS:26045-95-4 ***HR: 3***
PENTACYANONITROSYLFERRATE COBALT
mf: $C_5FeN_6O{\cdot}Co$ mw: 274.89

SYN: COBALTNITROSYLPENTACYANOFERRATE

TOXICITY DATA with REFERENCE
orl-rat LD50:147 mg/kg ARZNAD 24,308,74
ipr-rat LD50:14900 μg/kg ARZNAD 24,308,74
orl-mus LD50:74 mg/kg ARZNAD 24,308,74
ipr-mus LD50:10700 μg/kg ARZNAD 24,308,74
ivn-rbt LDLo:9400 μg/kg ARZNAD 24,308,74

CONSENSUS REPORTS: Cobalt and its compounds and cyanide and its compounds are on the Community Right-To-Know List.

SAFETY PROFILE: Poison by ingestion, intravenous, and intraperitoneal routes. When heated to decomposition it emits toxic fumes of NO_x and CN^-. See also CYANIDE and COBALT COMPOUNDS.

PAY750 CAS:629-62-9 ***HR: 2***
N-PENTADECANE
mf: $C_{15}H_{32}$ mw: 212.47

PROP: Colorless liquid. D: 0.770 @ 20°/4°, mp: 10°, bp: 270.5°. Insol in water; very sol in alc and ether.

TOXICITY DATA with REFERENCE
ivn-mus LD50:3494 mg/kg JPMSAE 67,566,78

CONSENSUS REPORTS: Reported in EPA TSCA Inventory.

SAFETY PROFILE: Moderately toxic by intravenous route. When heated to decomposition it emits acrid smoke and irritating fumes. See also PARAFFIN HYDROCARBONS.

PAZ000 CAS:1002-84-2 ***HR: 3***
PENTADECANOIC ACID
mf: $C_{15}H_{30}O_2$ mw: 242.45

SYN: PENTADECYCLIC ACID

TOXICITY DATA with REFERENCE
ivn-mus LD50:54 mg/kg APTOA6 18,141,61

CONSENSUS REPORTS: Reported in EPA TSCA Inventory.

SAFETY PROFILE: Poison by intravenous route. When heated to decomposition it emits acrid smoke and irritating fumes.

PBA000 CAS:2570-26-5 ***HR: 3***
PENTADECYLAMINE
mf: $C_{15}H_{33}N$ mw: 227.49

SYNS: 1-PENTADECANAMINE ◇ n-PENTADECYLAMINE ◇ 1-PENTADECYLAMINE

TOXICITY DATA with REFERENCE
orl-rat LD50:660 mg/kg TPKVAL 10,124,68
ihl-rat LC50:900 mg/m³/4H TPKVAL 10,124,68
orl-mus LD50:520 mg/kg TPKVAL 10,124,68
ihl-mus LC50:240 mg/m³/2H TPKVAL 10,124,68
ihl-mam LC50:200 mg/m³ TPKVAL 14,80,75

CONSENSUS REPORTS: EPA Extremely Hazardous Substances List. Reported in EPA TSCA Inventory.

SAFETY PROFILE: Poison by inhalation. Moderately toxic by ingestion. When heated to decomposition it emits toxic fumes of NO_x. See also AMINES.

PBA250 CAS:504-60-9 ***HR: 3***
(E)-1,3-PENTADIENE
mf: C_5H_8 mw: 68.13

$$H_2C{=}CHCH{=}CHCH_3$$

PROP: Flash p.: −45.4°F, lel: 2%, uel: 8.3%.

SYNS: trans-1,3-PENTADIENE ◇ trans-PIPERYLENE

TOXICITY DATA with REFERENCE
ivn-mus LD50:18 mg/kg CSLNX* NX#04179

CONSENSUS REPORTS: Reported in EPA TSCA Inventory.

SAFETY PROFILE: Poison by intravenous route. A very dangerous fire and explosion hazard when exposed to heat or flame; can react vigorously with oxidizing materials. When heated to decomposition it emits acrid smoke and irritating fumes.

PBB000 CAS:4911-55-1 ***HR: 3***
1,3-PENTADIYNE
mf: C_5H_4 mw: 64.09

$$HC{\equiv}CC{\equiv}CCH_3$$

SAFETY PROFILE: Explodes easily on distillation at atmospheric pressure. When heated to decomposition it emits acrid smoke and irritating fumes. See also ACETYLENE COMPOUNDS.

PBB229 ***HR: 3***
1,3-PENTADIYN-1-YL COPPER
mf: C_5H_3Cu mw: 126.62

$$CH_3C{\equiv}CC{\equiv}CCu$$

CONSENSUS REPORTS: Copper and its compounds are on The Community Right-To-Know List.

SAFETY PROFILE: An impact- and friction-sensitive explosive. When heated to decomposition it emits acrid smoke and irritating fumes. See also ACETYLENE COMPOUNDS and COPPER COMPOUNDS.

PBB449 ***HR: 3***
1,3-PENTADIYN-1-YL SILVER
mf: C_5H_3Ag mw: 170.95

$$CH_3C{\equiv}CC{\equiv}CAg$$

CONSENSUS REPORTS: Silver and its compounds are on The Community Right-To-Know List.

SAFETY PROFILE: An impact- and friction-sensitive explosive. Explodes on contact with sulfuric acid. When heated to decomposition it emits acrid smoke and irritating fumes. See also ACETYLENE COMPOUNDS and SILVER COMPOUNDS.

PBB750 CAS:115-77-5 ***HR: 1***
PENTAERYTHRITOL
mf: $C_5H_{12}O_4$ mw: 136.17

$$C(CH_2OH)_4$$

PROP: Crystals. Mp: 262°, d: 1.38 @ 25°/4°.

SYNS: AUXINUTRIL ◇ 2,2-BIS(HYDROXYMETHYL)-1,3-PROPANEDIOL ◇ HERCULES P6 ◇ METHANE TETRAMETHYLOL ◇ MONOPENTEK ◇ PE ◇ PENTAERYTHRITE ◇ PENTEK ◇ TETRAHYDROXYMETHYLMETANE ◇ TETRAKIS (HYDROXYMETHYL)METHANE ◇ TETRAMETHYLOLMETHANE

TOXICITY DATA with REFERENCE
orl-mus LD50:25500 mg/kg TXAPA9 6,351,64
orl-gpg LD50:11300 mg/kg TXAPA9 6,351,64

CONSENSUS REPORTS: Reported in EPA TSCA Inventory.

OSHA PEL: (Transitional: TWA Total Dust: 15 mg/m^3; Respirable Fraction: 5 mg/m^3) TWA Total Dust: 10 mg/m^3; Respirable Fraction: 5 mg/m^3
ACGIH TLV: TWA (nuisance particulate) 10 mg/m^3 of total dust (when toxic impurities are not present, e.g., quartz < 1%).

SAFETY PROFILE: Mildly toxic by ingestion. A nuisance dust. Flammable from heat or flame or oxidizers. Mixtures with thiophosphoryl chloride react when heated to form a product which ignites and then explodes on contact with air. Used in coatings, stabilizers, explosives, P.E.T.N resins, drugs, insecticides, and lubricants. When heated to decomposition it emits acrid smoke and irritating fumes.

PBC000 CAS:4196-86-5 ***HR: 1***
PENTAERYTHRITOL TETRABENZOATE
mf: $C_{33}H_{28}O_8$ mw: 552.61

SYNS: BENZOFLEX S-552 ◇ BENZOIC ACID, TETRAESTER with PENTAERYTHRITOL ◇ 2,2-BIS((BENZOYLOXY)METHYL)-1,3-PROPANEDIOL, DIBENZOATE

TOXICITY DATA with REFERENCE
orl-rat LD50:10 g/kg NPIRI* 2,83,75

CONSENSUS REPORTS: Reported in EPA TSCA Inventory.

SAFETY PROFILE: Mildly toxic by ingestion. When heated to decomposition it emits acrid smoke and irritating fumes.

PBC250 CAS:78-11-5 ***HR: 3***
PENTAERYTHRITOL TETRANITRATE
DOT: UN 0150/UN 0411
mf: $C_5H_8N_4O_{12}$ mw: 316.17

PROP: Crystals. Mp: 138-140°, bp: explodes @ 205-215°, d: 1.773 @ 20°/4°. Sol in acetone; insol in water; sltly sol in alc, ether.

SYNS: ANGICAP ◇ ANGITET ◇ ANTORA ◇ ARCOTRATE ◇ BARITRATE ◇ 2,2-BISDIHYDROXYMETHYL-1,3-PROPANEDIOL TETRANITRATE ◇ 2,2-BIS(HYDROXYMETHYL)-1,3-PROPANEDIOL TETRANITRATE ◇ CHOT ◇ DELTRATE-20 ◇ 1,3-DINITRATO-2,2-BIS(NITRATOMETHYL)PROPANE ◇ DUOTRATE ◇ EL PETN ◇ ERINIT ◇ HASETHROL ◇ INITIATING EXPLOSIVE PENTAERYTHRITE TETRANITRATE (DOT) ◇ KAYTRATE ◇ LOWETRATE ◇ MARTRATE-45 ◇ METRANIL ◇ MYCARDOL ◇ MYOTRATE "10" ◇ NCI-C55743 ◇ NEO-COROVAS ◇ NEOPENTANETETRAYL NITRATE ◇ NIPERYT ◇ NIPERYTH ◇ NITROPENTA ◇ NITROPENTAERYTHRITE ◇ NITROPENTAERYTHRITOL ◇ PENCARD ◇ PENTAERYTHRITE TETRANITRATE ◇ PENTAERYTHRITE TETRANITRATE (DOT) ◇ PENTAERYTHRITE TETRANITRATE, desensitized, wet (DOT) ◇ PENTAERYTHRITE TETRANITRATE, dry (DOT) ◇ PENTAERYTHRITE THERANITRATE, with not less than 7% wax (DOT) ◇ PENTAERYTHRITOL TETRANITRATE, diluted ◇ PENTAFIN ◇ PENTESTAN-80 ◇ PENTETRATE UNICELLES ◇ PENTRATE ◇ PENTRIOL ◇ PENTRYATE 80 ◇ PERGITRAL ◇ PERIDEX-LA ◇ PERITRATE ◇ PERITYL ◇ PREVANGOR ◇ QUINTRATE ◇ RYTHRITOL ◇ SDM No. 23 ◇ SUBICARD

◇ TENTRATE-20 ◇ TETRANITROPENTAERYTHRITE ◇ TETRASULE ◇ TRANITE D-LAY ◇ VASITOL ◇ VASODIATOL ◇ VASO-80 UNICELIES

TOXICITY DATA with REFERENCE
orl-man TDLo:1669 mg/kg/8Y-C:SKN BJDEAZ 87,498,72

CONSENSUS REPORTS: Reported in EPA TSCA Inventory.

DOT Classification: Class A Explosive; Label: Explosive A, desensitized, wet: Forbidden, dry.

SAFETY PROFILE: Human systemic effects by ingestion: dermatitis. Effects are similar to nitroglycerin, i.e., headache, weakness, and fall in blood pressure. Severe explosion hazard when shocked or exposed to heat. It explodes at 215°C. On decomposition it emits highly toxic fumes of NO_x; can react vigorously with oxidizing materials. Used in detonators and explosive specialities. See also NITRATES and EXPLOSIVES, HIGH.

PBC500 CAS:78-11-5 ***HR: 3***
PENTAERYTHRITOL TETRANITRATE (desensitized, wet)

CONSENSUS REPORTS: Reported in EPA TSCA Inventory.

DOT Classification: Class A Explosive; Label: Explosive A.

SAFETY PROFILE: A high explosive. See also PENTAERYTHRITOL TETRANITRATE and EXIPLOSIVES, HIGH.

PBC750 CAS:3524-68-3 ***HR: 2***
PENTAERYTHRITOL TRIACRYLATE
mf: $C_{14}H_{18}O_7$ mw: 298.32

SYNS: ACRYLIC ACID, PENTAERITHRITOL TRIESTER ◇ PETA ◇ 2-PROPENOICACID-2-(HYDROXYMETHYL)-2-(((1-OXO-2-PROPENYL) OXY)METHYL)-1,3-PROPANEDIYLESTER

TOXICITY DATA with REFERENCE
skn-rbt 500 mg open MLD UCDS** 3/28/72
eye-rbt 1 mg SEV UCDS** 3/28/72
skn-mus TDLo:16 g/kg/80W-I:ETA JTEHD6 19,149,86
orl-rat LD50:2460 mg/kg TXAPA9 28,313,74
skn-rbt LD50:4000 mg/kg TXAPA9 28,313,74

CONSENSUS REPORTS: Reported in EPA TSCA Inventory.

SAFETY PROFILE: Moderately toxic by ingestion and skin contact. Skin and severe eye irritant. Questionable carcinogen with experimental tumorigenic data. Used for radiation-cured adhesives, coatings, inks, textiles, photo resists, and coatings. When heated to decomposition it emits acrid smoke and irritating fumes.

PBD000 CAS:4067-16-7 ***HR: 2***
PENTAETHYLENEHEXAMINE
mf: $C_{10}H_{28}N_6$ mw: 124.35

SYNS: PEHA ◇ 3,6,9,12-TETRAAZATETRADECANE-1,14-DIAMINE

TOXICITY DATA with REFERENCE
mmo-sat 3333 µg/plate ENMUDM 8(Suppl 7),1,86
mma-sat 1 mg/plate ENMUDM 8(Suppl 7),1,86
orl-rat LD50:1600 mg/kg 37ASAA 7,580,79

CONSENSUS REPORTS: Reported in EPA TSCA Inventory.

SAFETY PROFILE: Moderately toxic by ingestion. Mutation data reported. When heated to decomposition it emits toxic fumes of NO_x. See also AMINES.

PBD250 CAS:771-60-8 ***HR: 3***
PENTAFLUOROANILINE
mf: $C_6H_2F_5N$ mw: 183.09

SYNS: AMINOPENTAFLUOROBENZENE ◇ 2,3,4,5,6-PENTAFLUOROANILINE ◇ PENTAFLUOROPHENYLAMINE

TOXICITY DATA with REFERENCE
ipr-rat LD50:390 mg/kg IZSBAI 3,91,65
ipr-mus LD50:384 mg/kg IZSBAI 3,91,65

CONSENSUS REPORTS: Reported in EPA TSCA Inventory.

SAFETY PROFILE: Poison by intraperitoneal route. When heated to decomposition it emits very toxic fumes of F^- and NO_x. See also FLUORIDES.

PBD300 CAS:21892-31-9 ***HR: 3***
1,2,3,4,5-PENTAFLUOROBICYCLO(2.2.0)HEXA-2,5-DIENE
mf: C_6HF_5 mw: 168.07

SAFETY PROFILE: The liquid is explosive. When heated to decomposition it emits toxic fumes of F^-. See also FLUORIDES.

PBD500 CAS:10051-06-6 ***HR: 3***
PENTAFLUOROGUANIDINE
mf: CF_5N_3 mw: 149.02

SAFETY PROFILE: Several of the adducts formed with alcohols are powerful explosives sensitive to impact, friction, heat, or phase changes. Used as a liquid rocket fuel oxidant. When heated to decomposition it emits toxic fumes of F^-. See also FLUORIDES.

PBD750 CAS:771-61-9 ***HR: 3***
PENTAFLUOROPHENOL
mf: C_6HF_5O mw: 184.07

TOXICITY DATA with REFERENCE
scu-rat LD50:322 mg/kg IZSBAI 3,91,65

scu-mus LD50:283 mg/kg IZSBAI 3,91,65
orl-mam LD50:330 mg/kg GISAAA 45(10),16,80
skn-mam LD50:1120 mg/kg GISAAA 45(10),16,80

CONSENSUS REPORTS: Reported in EPA TSCA Inventory.

SAFETY PROFILE: Poison by ingestion and subcutaneous routes. Moderately toxic by skin contact. When heated to decomposition it emits toxic fumes of F^-. See also FLUORIDES.

PBE000 CAS:4457-90-3 ***HR: 3***
PENTAFLUOROPHENYLALUMINUM DIBROMIDE
mf: $C_6AlBr_2F_5$ mw: 334.86

SAFETY PROFILE: Ignites spontaneously in air. Explodes on contact with water. When heated to decomposition it emits very toxic fumes of Br^- and F^-. See also ALUMINUM COMPOUNDS, FLUORIDES and BROMIDES.

PBE250 CAS:1076-44-4 ***HR: 3***
PENTAFLUOROPHENYLLITHIUM
mf: C_6F_5Li mw: 174.00

SAFETY PROFILE: Explodes on contact with water or deuterium oxide. When heated to decomposition it emits very toxic fumes of Li_2O and F^-. See also FLUORIDES and LITHIUM COMPOUNDS.

PBE500 CAS:422-63-9 ***HR: 2***
2,2,3,3,3-PENTAFLUORO-1,1-PROPANEDIOL
mf: $C_3H_3F_5O_2$ mw: 166.06

TOXICITY DATA with REFERENCE
orl-mus LD50:600 mg/kg JMCMAR 13,1212,70
ipr-mus LD50:600 mg/kg JMCMAR 13,1212,70

CONSENSUS REPORTS: Reported in EPA TSCA Inventory.

SAFETY PROFILE: Moderately toxic by ingestion and intraperitoneal routes. When heated to decomposition it emits toxic fumes of F^-. See also FLUORIDES.

PBE750 CAS:422-05-9 ***HR: 2***
2,2,3,3,3-PENTAFLUORO-1-PROPANOL
mf: $C_3H_3F_5O$ mw: 150.06

TOXICITY DATA with REFERENCE
orl-rat LDLo:2250 mg/kg JOCMA7 4,262,62

CONSENSUS REPORTS: Reported in EPA TSCA Inventory.

SAFETY PROFILE: Moderately toxic by ingestion. When heated to decomposition it emits toxic fumes of F^-. See also FLUORIDES.

PBF000 CAS:422-64-0 ***HR: 3***
PENTAFLUOROPROPIONIC ACID
mf: $C_3HF_5O_2$ mw: 164.04

TOXICITY DATA with REFERENCE
orl-rat LD10:750 mg/kg GTPZAB 10(3),13,66
ipr-rbt LD50:68 mg/kg CBCCT* 2,299,50

CONSENSUS REPORTS: Reported in EPA TSCA Inventory.

SAFETY PROFILE: Poison by intraperitoneal route. Moderately toxic by ingestion. When heated to decomposition it emits toxic fumes of F^-. See also FLUORIDES.

PBF250 CAS:509-09-1 ***HR: 2***
PENTAFLUOROPROPIONIC ACID SILVER SALT
mf: $C_3F_5O_2 \cdot Ag$ mw: 270.90

SYN: PENTAFLUORPROPIONAN STRIBRNY (CZECH)

TOXICITY DATA with REFERENCE
skn-rbt 500 mg/24H MOD 28ZPAK -,8,72
eye-rbt 5 mg/24H SEV 28ZPAK -,8,72
orl-rat LD50:1770 mg/kg 28ZPAK -,8,72

CONSENSUS REPORTS: Silver and its compounds are on the Community Right-To-Know List. Reported in EPA TSCA Inventory.

OSHA PEL: TWA 0.01 mg(Ag)/m^3
ACGIH TLV: TWA 0.01 mg(Ag)/m^3

SAFETY PROFILE: Moderately toxic by ingestion. Skin and severe eye irritant. When heated to decomposition it emits toxic fumes of F^-. See also FLUORIDES and SILVER COMPOUNDS.

PBF300 CAS:422-61-7 ***HR: 3***
PENTAFLUOROPROPIONYL FLUORIDE
mf: C_3F_6O mw: 166.02

SAFETY PROFILE: May explode on contact with fluorinated catalysts. When heated to decomposition it emits toxic fumes of F^-. See also FLUORIDES.

PBF400 ***HR: 3***
PENTAFLUOROPROPIONYL HYPOCHLORITE
mf: $C_3ClF_5O_2$ mw: 198.48

SAFETY PROFILE: Unstable at 22°C. The gas explodes at pressures above 27-62 mbar. Upon decomposition it emits toxic fumes of F^- and Cl^-. See also HYPOCHLORITES and FLUORIDES.

PBF500 ***HR: 3***
PENTAFLUOROPROPIONYL HYPOFLUORITE
mf: $C_3F_6O_2$ mw: 182.02

PROP: Bp: 2°.

SAFETY PROFILE: Distillation at 2°C/1 bar will cause it to explode, as will sparks or heating. When heated to decomposition it emits toxic fumes of F^-. See also FLUORIDES.

PBG100 CAS:60672-60-8 ***HR: 3***
PENTAFLUOROSULFUR PEROXYACETATE
mf: $C_2H_3F_5O_3S$ mw: 202.10

SAFETY PROFILE: Potentially explosive and may detonate with thermal or mechanical shock. When heated to decomposition it emits toxic fumes of F^- and SO_x. See also PEROXIDES.

PBG500 CAS:3030-47-5 ***HR: 3***
***N,N,N',N',N''*-PENTAMETHYLDIETHYLENETRIAMINE**
mf: $C_9H_{23}N_3$ mw: 173.35

SYNS: N-(2-(DIMETHYLAMINO)ETHYL)-N,N,N'-TRIMETHYL-1,2-ETHANEDIAMINE, (9CI) ◇ PENTAMETHYLDIETHYLENETRIAMINE ◇ PMDT

TOXICITY DATA with REFERENCE
orl-rat LD50:1630 mg/kg AIHAAP 30,470,69
skn-rbt LD50:280 mg/kg AIHAAP 30,470,69

CONSENSUS REPORTS: Reported in EPA TSCA Inventory.

SAFETY PROFILE: Poison by skin contact. Moderately toxic by ingestion. When heated to decomposition it emits toxic fumes of NO_x. See also AMINES.

PBG725 ***HR: 3***
1,1'-PENTAMETHYLENEBIS(sec-BUTYLPIPERIDINIUM)DIBROMIDE
mf: $C_{23}H_{48}N_2$•2Br mw: 512.55

SYN: 1,5-PENTAMETHYLEN-BIS(N-sec-BUTYLPIPERIDINIUM)-DIBROMID (GERMAN)

TOXICITY DATA with REFERENCE
orl-mus LD50:420 mg/kg AEPPAE 232,219,57
scu-mus LD50:125 mg/kg AEPPAE 232,219,57
ivn-mus LD50:60 mg/kg AEPPAE 232,219,57

SAFETY PROFILE: Poison by subcutaneous and intravenous routes. Moderately toxic by ingestion. When heated to decomposition it emits toxic fumes of NO_x and Br^-.

PBG750 CAS:60784-44-3 ***HR: 3***
1,1'-PENTAMETHYLENEBIS(3-(2-CHLOROETHYL)-3-NITROSOUREA
mf: $C_{11}H_{20}Cl_2N_6O_4$ mw: 371.27

TOXICITY DATA with REFERENCE
sln-dmg-orl 100 μmol/L/24H MUREAV 57,293,78
mrc-smc 500 μmol/L/16H MUREAV 42,45,77
ipr-rat LD50:45 mg/kg EJCAAH 13,937,77

SAFETY PROFILE: Poison by intraperitoneal route. Mutation data reported. Many N-nitroso compounds are carcinogens. When heated to decomposition it emits very toxic fumes of Cl^- and NO_x. See also N-NITROSO COMPOUNDS.

PBG850 ***HR: 3***
1,1'-PENTAMETHYLENEBIS(1-ETHYLPIPERIDINIUM) DIBROMIDE
mf: $C_{23}H_{48}N_2$•2Br mw: 512.55

SYN: 1,5-PENTAMETHYLEN-BIS-(N-n-BUTYLPIPERIDINIUM)-DIBROMID (GERMAN)

TOXICITY DATA with REFERENCE
orl-mus LD50:275 mg/kg AEPPAE 232,219,57
scu-mus LD50:38 mg/kg AEPPAE 232,219,57
ivn-mus LD50:14500 μg/kg AEPPAE 232,219,57

SAFETY PROFILE: Poison by ingestion, subcutaneous, and intravenous routes. When heated to decomposition it emits toxic fumes of NO_x and Br^-. See also BROMIDES.

PBH075 CAS:63938-82-9 ***HR: 3***
1,1'-PENTAMETHYLENEBIS(1-METHYLPIPERIDINIUM) DIBROMIDE
mf: $C_{17}H_{36}N_2$•2Br mw: 428.37

TOXICITY DATA with REFERENCE
orl-mus LD50:250 mg/kg AEPPAE 232,219,57
scu-mus LD50:30 mg/kg AEPPAE 232,219,57
ivn-mus LD50:18 mg/kg AEPPAE 232,219,57

SAFETY PROFILE: Poison by ingestion, subcutaneous and intravenous routes. When heated to decomposition it emits toxic fumes of NO_x and Br^-. See also BROMIDES.

PBH100 ***HR: 3***
1,1'-PENTAMETHYLENEBIS(PYRIDINIUM BROMIDE)
mf: $C_{15}H_{20}N_2$•2Br mw: 388.19

SYNS: G.L. 105 ◇ 1,1'-PENTAMETHLENEBIS-PYRIDINIUM DIBROMIDE ◇ P.M. 388

TOXICITY DATA with REFERENCE
orl-mus LD50:1025 mg/kg AEPPAE 232,219,57
scu-mus LD50:126 mg/kg AEPPAE 232,219,57
ivn-mus LD50:32 mg/kg AIPTAK 90,271,52

SAFETY PROFILE: Poison by subcutaneous and intravenous routes. Moderately toxic by ingestion. When heated to decomposition it emits toxic fumes of Br^- and NO_x. See also BROMIDES.

PBH150 CAS:24771-52-6 **HR: 2**
1,1'-(PENTAMETHYLENEDIOXY)BIS(3-CHLORO-2-PROPANOL)
mf: $C_{11}H_{22}Cl_2O_4$ mw: 289.23

SYNS: 2-PROPANOL, 1,1'-(PENTAMETHYLENEDIOXY)BIS(3-CHLORO- ◇ U-15,646

TOXICITY DATA with REFERENCE
orl-rat TDLo:960 mg/kg (male 8D pre):REP JRPFA4 21,263,70
orl-rat LD50:2248 mg/kg JRPFA4 21,267,70

SAFETY PROFILE: Moderately toxic by ingestion. Experimental reproductive effects. When heated to decomposition it emits toxic fumes of Cl^-.

PBH500 CAS:62912-45-2 **HR: 3**
(N,N'-PENTAMETHYLENE DIOXYDIETHYLENE)BIS(DIMETHYLETHYL AMMONIUM IODIDE)
mf: $C_{17}H_{40}N_2O_2 \cdot I_2$ mw: 558.39

TOXICITY DATA with REFERENCE
orl-mus LD50:300 mg/kg FRPSAX 32,129,77
ipr-mus LD50:8 mg/kg FRPSAX 32,129,77

SAFETY PROFILE: Poison by ingestion and intraperitoneal routes. When heated to decomposition it emits very toxic fumes of NH_3, NO_x, and I^-. See also IODIDES.

PBI000 CAS:62912-47-4 **HR: 3**
(N,N'-PENTAMETHYLENE DIOXYDIETHYLENE)BIS(TRIETHYL AMMONIUM IODIDE)
mf: $C_{21}H_{48}N_2O_2 \cdot I_2$ mw: 614.51

TOXICITY DATA with REFERENCE
orl-mus LD50:120 mg/kg FRPSAX 32,129,77
ipr-mus LD50:600 μg/kg FRPSAX 32,129,77

SAFETY PROFILE: Poison by ingestion and intraperitoneal routes. When heated to decomposition it emits very toxic fumes of NH_3, NO_x, and I^-. See also IODIDES.

PBI200 **HR: 3**
1,1'-PENTAMETHYLENEDIPIPERIDINE DIHYDROBROMIDE
mf: $C_{15}H_{30}N_2 \cdot 2BrH$ mw: 400.31

SYN: 1,5-PENTAMETHYLEN-BIS-(PIPERIDINIUM)-DIHYDROBROMID (GERMAN)

TOXICITY DATA with REFERENCE
orl-mus LD50:1300 mg/kg AEPPAE 232,219,57
scu-mus LD50:140 mg/kg AEPPAE 232,219,57
ivn-mus LD50:87500 μg/kg AEPPAE 232,219,57

SAFETY PROFILE: Poison by subcutaneous and intravenous routes. Moderately toxic by ingestion. When heated to decomposition it emits toxic fumes of NO_x and Br^-.

PBI500 CAS:54-95-5 **HR: 3**
1,5-PENTAMETHYLENETETRAZOLE
mf: $C_6H_{10}N_4$ mw: 138.20

PROP: White, crystalline powder. Mp: 57-58°. Sol in water, alc and ether.

SYNS: α,β-CYCLOPENTAMETHYLENETETRAZOLE ◇ PENTAMETHYLENETETRAZOL ◇ PENTAMETHYLENE-1,5-TETRAZOLE ◇ PENTYLENETETRAZOL ◇ 6,7,8,9-TETRAHYDRO-5-AZEPOTETRAZOLE ◇ 6,7,8,9-TETRAHYDRO-5H-TETRAZOLOAZEPINE ◇ 7,8,9,10-TETRAZABICYCLO(5.3.0)-8,10-DECADIENE ◇ 1,2,3,3A-TETRAZACYCLOHEPTA-8A,2-CYCLOPENTADIENE

TOXICITY DATA with REFERENCE
orl-man LDLo:147 mg/kg 85DCAI 2,73,70
ivn-man LDLo:29 mg/kg 85DCAI 2,73,70
orl-rat LD50:140 mg/kg JPPMAB 13,244,61
ipr-rat LD50:62 mg/kg TXAPA9 18,185,71
scu-rat LD50:85 mg/kg TXAPA9 18,185,71
ivn-rat LD50:45 mg/kg AIPTAK 135,9,62
rec-rat LD50:8 mg/kg AACRAT 46,395,67
orl-mus LD50:88 mg/kg JPETAB 128,176,60
ipr-mus LD50:71 mg/kg JPETAB 128,176,60
scu-mus LD50:80 mg/kg ARZNAD 8,190,58
ivn-mus LD50:51 mg/kg JPPMAB 13,244,61
par-mus LD50:72 mg/kg ARZNAD 6,583,56
ivn-dog LDLo:40 mg/kg 27ZIAQ -,184,73
ivn-rbt LD50:69 mg/kg AIPTAK 135,9,62

CONSENSUS REPORTS: Reported in EPA TSCA Inventory.

SAFETY PROFILE: A human poison by ingestion and intravenous routes. Poison experimentally by ingestion, intravenous, intraperitoneal, subcutaneous, rectal, and parenteral routes. When heated to decomposition it emits toxic fumes of NO_x.

PBJ000 CAS:79-55-0 **HR: 3**
1,2,2,6,6-PENTAMETHYLPIPERIDINE
mf: $C_{10}H_{21}N$ mw: 155.32

SYNS: PEROLYSEN ◇ PYRILENE

TOXICITY DATA with REFERENCE
orl-mus LD50:275 mg/kg NATUAS 184,1707,59
ivn-mus LD50:40 mg/kg NATUAS 184,1707,59

CONSENSUS REPORTS: Reported in EPA TSCA Inventory.

SAFETY PROFILE: Poison by ingestion and intrave-

nous routes. When heated to decomposition it emits toxic fumes of NO_x.

PBJ600 CAS:53378-72-6 ***HR: 3***
PENTAMETHYLTANTALUM
mf: $C_5H_{15}Ta$ mw: 256.12

SAFETY PROFILE: Explodes at room temperature. When heated to decomposition it emits acrid smoke and irritating fumes. See also TANTALUM.

PBJ750 CAS:556-72-9 ***HR: 2***
2,2,4,6,6-PENTAMETHYLTETRAHYDRO PYRIMIDINE
mf: $C_9H_{19}N_2$ mw: 155.30

TOXICITY DATA with REFERENCE
orl-rat LDLo:1000 mg/kg SCCUR* -,8,61
orl-mus LDLo:700 mg/kg SCCUR* -,8,61
orl-rbt LDLo:1200 mg/kg SCCUR* -,8,61

CONSENSUS REPORTS: Reported in EPA TSCA Inventory.

SAFETY PROFILE: Moderately toxic by ingestion. When heated to decomposition it emits toxic fumes of NO_x.

PBJ875 CAS:57590-20-2 ***HR: 3***
PENTANAL METHYLFORMYLHYDRAZONE
mf: $C_7H_{14}N_2O$ mw: 142.23

SYNS: FORMIC ACID, METHYLPENTYLIDENEHYDRAZIDE ◇ PENTANAL, N-FORMYL-N-METHYLHYDRAZONE ◇ PENTYLIDENE GYROMITRIN

TOXICITY DATA with REFERENCE
orl-mus TDLo:2600 mg/kg/52W-I:CAR MYCPAH 98,83,87
orl-mus TD:2600 mg/kg/52W-I:NEO MYCPAH 98,83,87
orl-mus LD50:594 mg/kg TXAPA9 45,429,78
orl-rbt LDLo:350 mg/kg NATWAY 62,395,75

SAFETY PROFILE: Poison by ingestion. Questionable carcinogen with experimental carcinogenic and neoplastigenic data. When heated to decomposition it emits toxic fumes of NO_x.

PBK000 CAS:136-25-4 ***HR: 2***
PENTANATE
mf: $C_{11}H_9Cl_5O_3$ mw: 366.45

SYNS: 2,2-DICHLOROPROPIONIC ACID, 2-(2,4,5-TRICHLOROPHENOXY)ETHYL ESTER ◇ ERBN ◇ ERBON ◇ ETHANOL-2-(2,4,5-TRICHLOROPHENOXY)-, 2,2-DICHLOROPROPIONATE ◇ NOVEGE ◇ NOVON ◇ 2-(2,4,5-TRICHLOROPHENOXY)ETHYL-2,2-DICHLOROPROPIONATE ◇ 2,4,5-TRICHLOROPHENOXYETHYL-α,α-DICHLOROPROPIONATE

TOXICITY DATA with REFERENCE
orl-rat LD50:1000 mg/kg FMCHA2 -,D128,80
orl-mus LD50:912 mg/kg HYSAAV 34,174,69
orl-rbt LD50:2193 mg/kg HYSAAV 34,174,69
orl-ckn LD50:3170 mg/kg PCOC** -,484,66

SAFETY PROFILE: Moderately toxic by ingestion. Used as a herbicide to control perennial broadleaf weeds. When heated to decomposition it emits toxic fumes of Cl^-.

PBK250 CAS:109-66-0 ***HR: 3***
PENTANE
DOT: UN 1265
mf: C_5H_{12} mw: 72.17

PROP: Colorless liquid. Bp: 36.1°, flash p: < −40°F, fp: −129.8°, d: 0.626 @ 20°/4°, autoign temp: 588°F, vap press: 400 mm @ 18.5°, vap d: 2.48, lel: 1.5%, uel: 7.8%. Sol in water; misc in alc, ether, organic solvents.

SYNS: AMYL HYDRIDE (DOT) ◇ PENTAN (POLISH) ◇ PENTANEN (DUTCH) ◇ PENTANI (ITALIAN)

TOXICITY DATA with REFERENCE
ivn-mus LD50:446 mg/kg JPMSAE 67,566,78

CONSENSUS REPORTS: Reported in EPA TSCA Inventory.

OSHA PEL: (Transitional: TWA 1000 ppm; STEL 750 ppm) TWA 600 ppm; STEL 750 ppm
ACGIH TLV: TWA 600 ppm; STEL 750 ppm
DFG MAK: 1000 ppm (2950 mg/m^3)
NIOSH REL: (Alkanes) TWA 350 mg/m^3
DOT Classification: Flammable Liquid; Label: Flammable Liquid.

SAFETY PROFILE: Moderately toxic by intravenous route. Narcotic in high concentration. The liquid can cause blisters on contact. Flammable liquid. Highly dangerous fire hazard when exposed to heat, flame, or oxidizers. Severe explosion hazard when exposed to heat or flame. Shock can shatter metal containers and release contents. To fight fire, use foam, CO_2, dry chemical. When heated to decomposition it emits acrid smoke and irritating fumes.

PBK500 CAS:462-94-2 ***HR: 3***
1,5-PENTANEDIAMINE
mf: $C_5H_{14}N_2$ mw: 102.21

PROP: Colorless, thick liquid; characteristic odor. Mp: 9°, bp: 178-180°, d: 0.873 @ 25°/4°. Very sol in water and alc; sltly sol in ether.

SYNS: ANIMAL CONIINE ◇ CADAVERIN ◇ CADAVERINE ◇ 1,5-DIAMINOPENTANE ◇ PENTAMETHYLENEDIAMINE ◇ 1,5-PENTAMETHYLENEDIAMINE

TOXICITY DATA with REFERENCE
dni-mus:ast 10 mmol/L AMOKAG 33,149,79

dni-mus:lvr 20 mmol/L AMOKAG 33,149,79
dnd-mam:lym 200 mmol/L IJRBA3 9,185,65
scu-rat LDLo:250 mg/kg ZEPTAT 17,59,15
orl-mus LDLo:1600 mg/kg AECTCV 14,111,85
ivn-rbt LDLo:100 mg/kg CRSBAW 83,481,20
rec-rbt LDLo:400 mg/kg CRSBAW 83,481,20

SAFETY PROFILE: Poison by intravenous, rectal, and subcutaneous routes. Moderately toxic by skin contact. An irritant, sensitizer, and allergen. Mutgenic data. When heated to decomposition it emits highly toxic fumes of NO_x. See also AMINES.

PBK750 CAS:111-29-5 ***HR: 1***
1,5-PENTANEDIOL
mf: $C_5H_{12}O_2$ mw: 104.17

PROP: Colorless, viscous, odorless liquid; bitter taste. Mp: −18°, bp: 239°, flash p: 275°F, fp: −15.6°, d: 0.994 @ 20°/4°, vap press: <0.01 mm @ 20°, vap d: 3.59, autoign temp: 635°F. Misc with water, methanol, alc, acetone.

SYNS: 1,5-DIHYDROXYPENTANE ◇ PENTAMETHYLENE GLYCOL ◇ PENTANE-1,5-DIOL ◇ 1,5-PENTYLENE GLYCOL

TOXICITY DATA with REFERENCE
skn-rbt 495 mg open MLD UCDS** 5/22/59
eye-rbt 100 mg MLD 34ZIAG -,731,69
orl-rat LD50:5890 mg/kg UCDS** 5/22/50

CONSENSUS REPORTS: Reported in EPA TSCA Inventory.

SAFETY PROFILE: Mildly toxic by ingestion. A skin and eye irritant. Combustible when exposed to heat or flame; can react with oxidizing materials. To fight fire, use foam, CO_2, dry chemical. Used as a plasticizer in cellulose products and adhesives, and in brake fluid. When heated to decomposition it emits acrid smoke and irritating fumes.

PBL000 CAS:625-69-4 ***HR: 1***
2,4-PENTANEDIOL
mf: $C_5H_{12}O_2$ mw: 104.17

PROP: Syrup. D: 0.989 @ 20°, bp: 202-203°.

SYNS: 2,4-AMYLENEGLYCOL ◇ ISOAMYLENE ALCOHOL ◇ PENTANEDIOL-2,4

TOXICITY DATA with REFERENCE
eye-rbt 500 mg open AMIHBC 10,61,54
orl-rat LD50:6860 mg/kg AMIHBC 10,61,54
orl-mus LD50:5792 mg/kg JAPMA8 45,669,56
skn-rbt LD50:14100 mg/kg AMIHBC 10,61,54

SAFETY PROFILE: Mildly toxic by ingestion and skin contact. Eye irritant. When heated to decomposition it emits acrid smoke and irritating fumes.

PBL250 CAS:64025-06-5 ***HR: 3***
(2,4-PENTANEDIONATO-O,O')PHENYLMERCURY
mf: $C_{11}H_{12}HgO_2$ mw: 376.82

SYN: 2,4-PENTANEDIONE, PHENYLMERCURIC SALT

TOXICITY DATA with REFERENCE
ipr-mus LDLo:16 mg/kg CBCCT* 2,241,50

CONSENSUS REPORTS: Mercury and its compounds are on the Community Right-To-Know List.

OSHA PEL: (Transitional: CL 1 mg/10m^3) CL 0.1 mg(Hg)/m^3 (skin)
ACGIH TLV: TWA 0.1 mg(Hg)/m^3 (skin)
NIOSH REL: TWA 0.05 mg(Hg)/m^3

SAFETY PROFILE: Poison by intraperitoneal and probably other routes. When heated to decomposition it emits toxic fumes of Hg. See also MERCURY COMPOUNDS.

PBL350 CAS:600-14-6 ***HR: 3***
2,3-PENTANEDIONE
mf: $C_5H_8O_2$ mw: 100.13

PROP: Mp: -52°, bp: 110-112°, d: 0.957, flash p: 66° F.

SYN: ACETYLPROPIONYL

TOXICITY DATA with REFERENCE
skn-rbt 500 mg/24H MOD FCTXAV 17,699,79
orl-rat LD50:3000 mg/kg FCTXAV 17,699,79

CONSENSUS REPORTS: Reported in EPA TSCA Inventory.

SAFETY PROFILE: Moderately toxic by ingestion. A skin irritant. Flammable liquid. When heated to decomposition it emits acrid smoke and irritating fumes.

PBL500 CAS:3264-82-2 ***HR: 3***
2,4-PENTANEDIONE, NICKEL(II) DERIVATIVE
mf: $C_{10}H_{14}NiO_4$ mw: 256.95

TOXICITY DATA with REFERENCE
ipr-mus LD50:125 mg/kg CBCCT* 4,231,52

CONSENSUS REPORTS: Nickel and its compounds are on the Community Right-To-Know List. Reported in EPA TSCA Inventory.
NIOSH REL: TWA 0.015 mg(Ni)/m^3

SAFETY PROFILE: Poison by intraperitoneal route. When heated to decomposition it emits acrid smoke and irritating fumes. See also NICKEL COMPOUNDS.

PBL750 CAS:17501-44-9 ***HR: 3***
2,4-PENTANEDIONE, ZIRCONIUM COMPLEX
mf: $C_{20}H_{28}O_8Zr$ mw: 487.70

SYN: ACETYLZIRCONIUM, ACETONATE

TOXICITY DATA with REFERENCE
ipr-mus LDLo:316 mg/kg NTIS** AD441-640

CONSENSUS REPORTS: Reported in EPA TSCA Inventory.

OSHA PEL: (Transitional: TWA 5 mg(Zr)/m^3) TWA 5 mg(Zr)/m^3; STEL 10 mg(Zr)/m^3
ACGIH TLV: TWA 5 mg(Zr)/m^3; STEL 10 mg(Zr)/m^3
DFG MAK: 5 mg(Zr)/m^3

SAFETY PROFILE: Poison by intraperitoneal route. When heated to decomposition it emits acrid smoke and irritating fumes. Used as a crosslinking gent for oxygen containing polymers. See also ZIRCONIUM COMPOUNDS.

PBM000 CAS:110-66-7 ***HR: 3***
1-PENTANETHIOL
DOT: UN 1111
mf: $C_5H_{12}S$ mw: 104.23

$CH_3(CH_2)_4SH$

PROP: Water-white to yellow liquid. D: 0.857 @ 20°, bp: 123.64°, flash p: 65°F, vap press: 13.8 mm @ 25°, vap d: 3.59. Insol in water; misc in alc and ether.

SYNS: AMYL HYDROSULFIDE ◇ n-AMYL MERCAPTAN ◇ AMYL MERCAPTAN (DOT) ◇ AMYL SULFHYDRATE ◇ AMYL THIOALCOHOL ◇ MERCAPTAN AMYLIQUE (FRENCH) ◇ PENTYL MERCAPTAN

TOXICITY DATA with REFERENCE
ihl-rat LCLo:2000 ppm/4H JIHTAB 31,343,49

CONSENSUS REPORTS: Reported in EPA TSCA Inventory.

NIOSH REL: (Thiols (n-Alkane Mono)) CL 0.5 ppm/15M
DOT Classification: Label: Flammable Liquid.

SAFETY PROFILE: Moderately toxic by inhalation. A weak sensitizer and allergen. Local contact may cause contact dermatitis. Dangerous fire hazard when exposed to heat or flame; can react vigorously with oxidizing materials. Hypergolic reaction with concentrated nitric acid. To fight fire, use foam, CO_2, dry chemical. See also MERCAPTANS.

PBM100 ***HR: 3***
PENTANITROANILINE
mf: $C_6H_2N_6O_{10}$ mw: 318.12

SAFETY PROFILE: A very sensitive explosive. When heated to decomposition it emits toxic fumes of NO_x. See also NITRO COMPOUNDS OF AROMATIC HYDROCARBONS.

PBM500 CAS:62-68-0 ***HR: 3***
PENTANOIC ACID, 2,2-DIPHENYL, 2-(N,N-DIETHYLAMINO)ETHYL ESTER, HYDROCHLORIDE
mf: $C_{23}H_{31}NO_2$•ClH mw: 390.01

SYNS: DIETHYLAMINOETHANOL ESTER OF DIPHENYLPROPYLACETIC ACID HYDROCHLORIDE ◇ 2'-DIETHYLAMINOETHYL-2,2-DIPHENYLPENTANOATE HYDROCHLORIDE ◇ β-DIETHYLAMINOETHYL DIPHENYLPROPYLACETATE HYDROCHLORIDE ◇ 2-DIETHYLAMINOETHYL-2,2-DIPHENYLVALERATEHYDROCHLORIDE ◇ α-PHENYL-α-PROPYLBENZENEACETIC ACID-2-(DIETHYLAMINO)ETHYL ESTER HYDROCHLORIDE ◇ PROADIFEN HYDROCHLORIDE ◇ PROPYLADIPHENIN ◇ 5171 RP ◇ SKF 525A ◇ U 5446

TOXICITY DATA with REFERENCE
scu-rat TDLo:20 mg/kg (female 4D pre):REP JSICAZ 19,264,60
orl-rat LD50:2140 mg/kg 27ZQAG -,366,72
ipr-rat LD50:163 mg/kg 27ZQAG -,366,72
orl-mus LD50:538 mg/kg 27ZQAG -,366,72
ipr-mus LD50:110 mg/kg PCJOAU 3,508,69
scu-mus LD50:90 mg/kg JPETAB 98,121,50
ivn-mus LD50:60 mg/kg 27ZQAG -,366,72
ivn-cat LDLo:25 mg/kg TXAPA9 8,118,66

SAFETY PROFILE: Poison by intraperitoneal, subcutaneous, and intravenous routes. Moderately toxic by ingestion. Experimental reproductive effects. See also ESTERS. When heated to decomposition it emits very toxic fumes of NO_x and HCl.

PBM750 CAS:6032-29-7 ***HR: 3***
2-PENTANOL
DOT: UN 1105
mf: $C_5H_{12}O$ mw: 88.17

PROP: Colorless liquid. Bp: 119.3°, flash p: 105°F (OC), ULC:40-45, uel: 9.0%, lel: 1.2%, fp: −50°, d: 0.8169 @ 20°/20°, autoign temp: 650-725°F, vap d: 3.04. Sltly sol in water; misc in alc, and ether.

SYNS: sec-AMYL ALCOHOL (DOT) ◇ METHYL PROPYL CARBINOL ◇ PENTANOL-2 ◇ sec-PENTYL ALCOHOL

TOXICITY DATA with REFERENCE
skn-rbt 500 mg/24H MOD 28ZPAK -,36,72
eye-rbt 20 mg/24H SEV 28ZPAK -,36,72
orl-rat LD50:1470 mg/kg 28ZPAK -,36,72
ipr-rat LDLo:2130 mg/kg JIHTAB 27,1,45
orl-rbt LDLo:3500 mg/kg JLCMAK 10,985,25

CONSENSUS REPORTS: Reported in EPA TSCA Inventory.

DOT Classification: Flammable Liquid; Label: Flammable Liquid: Flammable or Combustible Liquid; Label: Flammable Liquid.

SAFETY PROFILE: Moderately toxic by ingestion and

intraperitoneal routes. A narcotic. A skin and severe eye irritant. Flammable when exposed to heat or flame; can react with oxidizing materials. A severe explosion hazard when exposed to heat or flame. To fight fire, use alcohol foam, dry chemical. When heated to decomposition it emits acrid smoke and irritating fumes. See also ALCOHOLS.

PBN250 CAS:107-87-9 ***HR: 3***
2-PENTANONE
DOT: UN 1249
mf: $C_5H_{10}O$ mw: 86.15

PROP: Water-white liquid; fruity, ethereal odor. D: 0.801-0.806, vap d: 3.0, bp: 216°F, flash p: 45°F, autoign temp: 941°F, lel: 1.5%, uel: 8.2%. Sltly sol in water; misc with alc, ether.

SYNS: ETHYL ACETONE ◇ FEMA No. 2842 ◇ METHYL-PROPYL-CETONE (FRENCH) ◇ METHYL-n-PROPYL KETONE ◇ METHYL PROPYL KETONE (ACGIH, DOT) ◇ METYLOPROPYLOKETON (POLISH) ◇ MPK

TOXICITY DATA with REFERENCE
skn-rbt 405 mg open MLD UCDS** 7/20/67
sln-smc 13600 ppm MUREAV 149,339,85
ihl-hmn TCLo:1500 ppm:EYE,CNS,GIT NPIRI* 1,83,74
orl-rat LD50:3730 mg/kg AIHAAP 23,95,62
ihl-rat LCLo:2000 ppm/4H AIHAAP 23,95,62
ipr-rat LD50:1250 mg/kg NPIRI* 1,83,74
skn-rbt LD50:6500 mg/kg NPIRI* 1,83,74

OSHA PEL: (Transitional: TWA 200 ppm; STEL 250 ppm) TWA 200 ppm; STEL 250 ppm
ACGIH TLV: TWA 200 ppm; STEL 250 ppm
DFG MAK: 200 ppm (700 mg/m^3)
NIOSH REL: TWA 530 mg/m^3
DOT Classification: Label: Flammable Liquid.

SAFETY PROFILE: Moderately toxic by ingestion and intraperitoneal routes. Mildly toxic by skin contact and inhalation. Human systemic effects by inhalation: headache, nausea, irritation of the respiratory passages, eyes, and skin. A skin irritant. Mutation data reported. A highly flammable liquid. A very dangerous fire hazard when exposed to heat or flame; can react vigorously with oxidizing materials. An explosion hazard in the form of vapor when exposed to heat or flame. To fight fire, use alcohol foam. Mixtures with bromine trifluoride may explode during evaporation. When heated to decomposition it emits acrid smoke and irritating fumes. See also KETONES.

PBO000 CAS:33100-27-5 ***HR: 2***
1,4,7,10,13-PENTAOXACYLOPENTADECANE
mf: $C_{10}H_{20}O_5$ mw: 220.30

SYN: 15-CROWN-5

TOXICITY DATA with REFERENCE
skn-rbt 100 mg/24H MLD DCTODJ 8,451,85
eye-rbt 50 mg MOD DCTODJ 8,451,85
orl-mus LD50:1020 mg/kg TXAPA9 44,263,78
ipr-mus LD50:514 mg/kg DCTODJ 8,451,85
skn-rbt LD50:2520 mg/kg DCTODJ 8,451,85

CONSENSUS REPORTS: Reported in EPA TSCA Inventory.

SAFETY PROFILE: Moderately toxic by ingestion, skin contact, and intraperitoneal routes. A skin and eye irritant. When heated to decomposition it emits acrid smoke and irritating fumes.

PBO250 CAS:4353-28-0 ***HR: 1***
3,6,9,12,15-PENTAOXAHEPTADECANE
mf: $C_{12}H_{26}O_5$ mw: 250.38

SYNS: DIETHOXYTETRAETHYLENE GLYCOL ◇ ETHER, BIS(2-(2-ETHOXYETHOXY)ETHYL) ◇ GLYCOL, DIETHOXYTETRAETHYLENE ◇ TETRAETHYLENE GLYCOL DIETHYL ETHER

TOXICITY DATA with REFERENCE
orl-rat LD50:4290 mg/kg AIHAAP 30,470,69
skn-rbt LD50:6350 mg/kg AIHAAP 30,470,69

CONSENSUS REPORTS: Glycol ether compounds are on the Community Right-To-Know List. Reported in EPA TSCA Inventory.

SAFETY PROFILE: Mildly toxic by ingestion and skin contact. Many glycol ethers have dangerous reproductive effects. When heated to decomposition it emits acrid smoke and irritating fumes. See also GLYCOL ETHERS.

PBO500 CAS:143-24-8 ***HR: 1***
2,5,8,11,14-PENTAOXAPENTADECANE
mf: $C_{10}H_{22}O_5$ mw: 222.32

SYNS: ANSUL ETHER 181AT ◇ BIS(2-(2-METHOXYETHOXY) ETHYL) ETHER ◇ BIS(2-METHOXYETHYL)ETHER ◇ DIMETHOXYTETRAETHYLENE GLYCOL ◇ DIMETHOXYTETRAGLYCOL ◇ ETHER, BIS(2-(2-METHOXYETHOXY)ETHYL) ◇ TETRAETHYLENE GLYCOL DIMETHYL ETHER ◇ TETRAGLYME

TOXICITY DATA with REFERENCE
eye-rbt 500 mg AJOPAA 29,1363,46
ihl-rat TCLo:1000 ppm/7H (5D male):REP TXAPA9 70,303,83
orl-rat LD50:5140 mg/kg JIHTAB 23,259,41

CONSENSUS REPORTS: Glycol ether compounds are on the Community Right-To-Know List. Reported in EPA TSCA Inventory.

SAFETY PROFILE: Mildly toxic by ingestion. Experimental reproductive effects. An eye irritant. Many glycol ethers are suspected of having dangerous human reproductive effects. When heated to decomposition it emits acrid smoke and irritating fumes. See also GLYCOL ETHERS.

PBO800 ***HR: 3***
PENTASILVER DIAMIDOPHOSPHATE
mf: $Ag_5N_2O_2P$ mw: 630.33

CONSENSUS REPORTS: Silver and its compounds are on the Community Right-To-Know List.

SAFETY PROFILE: An explosive very sensitive to friction or contact with sulfuric acid. When heated to decomposition it emits toxic fumes of PO_x and NO_x. See also SILVER COMPOUNDS.

PBP000 ***HR: 3***
PENTASILVER DIIMIDOTRIPHOSPHATE
mf: $Ag_5H_2N_2O_8P_3$ mw: 790.30

CONSENSUS REPORTS: Silver and its compounds are on the Community Right-To-Know List.

SAFETY PROFILE: An explosive sensitive to friction, heat, or contact with sulfuric acid. When heated to decomposition it emits very toxic fumes of PO_x and NO_x. See also SILVER COMPOUNDS and PHOSPHATES.

PBP100 ***HR: 3***
PENTASILVER ORTHODIAMIDOPHOSPHATE
mf: $Ag_5H_2N_2O_3P$ mw: 617.37

CONSENSUS REPORTS: Silver and its compounds are on the Community Right-To-Know List.

SAFETY PROFILE: Explodes on heating, friction, or on contact with sulfuric acid. When heated to decomposition it emits toxic fumes of PO_x and NO_x. See also SILVER COMPOUNDS.

PBP300 CAS:2276-52-0 ***HR: 3***
PENTAZOCINE HYDROCHLORIDE
mf: $C_{19}H_{27}NO{\cdot}ClH$ mw: 321.93

SYN: 1,2,3,4,5,6-HEXAHYDRO-6,11-DIMETHYL-3-(3-METHYL-2-BUTENYL)-2,6-METHANO-3-BENZAZOCIN-8-OLHCl

TOXICITY DATA with REFERENCE
orl-rat TDLo:616 mg/kg (16-22D preg/21D post):REP NYKZAU 74,111,78
scu-mus TDLo:100 mg/kg (9D preg):TER DGDFA5 22,61,80
orl-wmn TDLo:50 mg/kg IJMDAI 22,385,86
scu-mus LD50:220 mg/kg JMCMAR 18,996,75

SAFETY PROFILE: Poison by subcutaneous route. An experimental teratogen. Experimental reproductive effects. When heated to decomposition it emits toxic fumes of NO_x and HCl.

PBP500 CAS:764-39-6 ***HR: 3***
2-PENTENAL
mf: C_5H_8O mw: 84.13

TOXICITY DATA with REFERENCE
ipr-mus LD50:244 mg/kg ZolH## 23OCT75

CONSENSUS REPORTS: Reported in EPA TSCA Inventory.

SAFETY PROFILE: Poison by intraperitoneal route. When heated to decomposition it emits acrid smoke and irritating fumes. See also ALDEHYDES.

PBP750 CAS:2100-17-6 ***HR: 2***
4-PENTENAL
mf: C_5H_8O mw: 84.13

TOXICITY DATA with REFERENCE
orl-rat LD50:620 mg/kg AIHAAP 23,95,62
ihl-rat LCLo:250 ppm/4H AIHAAP 23,95,62
skn-rbt LD50:1590 mg/kg AIHAAP 23,95,62

CONSENSUS REPORTS: Reported in EPA TSCA Inventory.

SAFETY PROFILE: Moderately toxic by ingestion, inhalation, and skin contact. When heated to decomposition it emits acrid smoke and irritating fumes. See also ALDEHYDES.

PBQ000 ***HR: 3***
1-PENTENE
mf: C_5H_{10} mw: 70.14

$H_2C{=}CHCH_2CH_2CH_3$

PROP: Liquid. Flash p: −0.4°F, lel: 1.5%, uel: 8.7%, d: 0.6429 @ 20°/4°, bp: 30.1°. Insol in water; misc with alc, ether, benzene.

SAFETY PROFILE: A very dangerous fire and explosion hazard when exposed to heat or flames; can react vigorously with oxidizing materials. When heated to decomposition it emits acrid smoke and irritating fumes.

PBQ250 CAS:646-04-8 ***HR: 3***
2-PENTENE
mf: C_5H_{10} mw: 70.14

$CH_3CH{=}CHCH_2CH_3$

PROP: cis Form: Liquid. Flash p: −0.4°F, d: 0.6503 @ 20°/4°, mp: −180° to −178°, bp: 37.0°. trans Form: Liquid. D: 0.6482 @ 20°/4°, mp: −136° to −135°, bp: 35.85°.

SAFETY PROFILE: A dangerous fire and explosion hazard when exposed to heat or flame; can react vigorously with oxidizing materials. When heated to decomposition it emits acrid smoke and irritating fumes.

PBQ300 CAS:16187-03-4 ***HR: 3***
trans-2-PENTENE OZONIDE
mf: $C_5H_{10}O_3$ mw: 118.13

$OOCH(CH_2CH_3)OCHCH_3$

SYN: 3-ETHYL-5-METHYL-1,2,4-TRIOXOLANE

SAFETY PROFILE: Explodes when heated to ignition. When heated to decomposition it emits acrid smoke and irritating fumes. See also OZONIDES.

PBQ750 CAS:591-80-0 ***HR: 3***
4-PENTENOIC ACID
mf: $C_5H_8O_2$ mw: 100.13

SYN: ALLYLACETIC ACID

TOXICITY DATA with REFERENCE
orl-rat LD50:470 mg/kg FCTXAV 2,327,64
orl-mus LD50:610 mg/kg FCTXAV 2,327,64
ipr-mus LD50:315 mg/kg JPPMAB 21,85,69
scu-mus LD50:315 mg/kg JPPMAB 21,85,69

CONSENSUS REPORTS: Reported in EPA TSCA Inventory.

SAFETY PROFILE: Poison by subcutaneous and intraperitoneal routes. Moderately toxic by ingestion. When heated to decomposition it emits acrid smoke and irritating fumes.

PBR000 CAS:821-09-0 ***HR: 3***
4-PENTEN-1-OL
mf: $C_5H_{10}O$ mw: 86.14

PROP: Flash p: <73.4°F.

SAFETY PROFILE: Dangerous fire hazard when exposed to heat or flame; can react vigorously with oxidizing materials. When heated to decomposition it emits acrid smoke and irritating fumes.

PBR250 CAS:1629-58-9 ***HR: 2***
1-PENTEN-3-ONE
mf: C_5H_8O mw: 84.13

SYN: ETHYL VINYL KETONE

TOXICITY DATA with REFERENCE
ivn-mus LD50:56 mg/kg CSLNX* NX#00948

CONSENSUS REPORTS: Reported in EPA TSCA Inventory.

SAFETY PROFILE: Poison by intravenous route. When heated to decomposition it emits acrid smoke and irritating fumes. See also KETONES.

PBR500 CAS:625-33-2 ***HR: 3***
3-PENTEN-2-ONE
mf: C_5H_8O mw: 84.13

PROP: Colorless liquid. Bp: 122-124°, d: 0.856, vap d: 2.89.

SYNS: ETHYLIDENE ACETONE ◇ METHYL PROPENYL KETONE

TOXICITY DATA with REFERENCE
skn-rbt 10 mg/24H JIHTAB 30,63,48
eye-rbt 500 mg JIHTAB 30,63,48
orl-rat LD50:3730 mg/kg UCDS** 7/20/67
ihl-rat LCLo:250 ppm/4H JIHTAB 31,343,49
skn-rbt LD50:500 mg/kg JIHTAB 30,63,48

CONSENSUS REPORTS: Community Right-To-Know List. Reported in EPA TSCA Inventory.

SAFETY PROFILE: Moderately toxic by ingestion, inhalation, and skin contact. A skin and eye irritant. Flammable when exposed to heat or flame; can react with oxidizing materials. To fight fire, use foam, CO_2, dry chemical. When heated to decomposition it emits acrid smoke and irritating fumes. See also KETONES.

PBR750 ***HR: 3***
2-PENTEN-4-YN-3-OL
mf: C_5H_6O mw: 82.103

$CH_3CH{=}C(OH)C{\equiv}CH$

SAFETY PROFILE: An explosion hazard. Distillation residue explodes above 90°C. When heated to decomposition it emits acrid smoke and irritating fumes. See also ACETYLENE COMPOUNDS.

PBS000 CAS:60-44-6 ***HR: 3***
PENTHIENATE BROMIDE
mf: $C_{18}H_{30}NO_3S \cdot Br$ mw: 420.46

SYNS: α-CYCLOPENTYL-2-THIOPHENEGLYCOLATEDIETHYL(2-HYDROXYETHYL)METHYLAMMONIUM BROMIDE ◇ 2-DIETHYLAMINOETHYL-2-CYCLOPENTYL-2-(2-THIENYL)HYDROXYACETATE METHOBROMIDE ◇ 2-DIETHYLAMINOETHYL-α-CYCLOPENTYL-2-THIOPHENEGLYCOLATE METHOBROMIDE ◇ MONODRAL ◇ MONODRAL BROMIDE ◇ WIN 4369

TOXICITY DATA with REFERENCE
orl-mus LD50:2080 mg/kg JPETAB 110,282,54
scu-mus LD50:350 mg/kg CLDND*
ivn-mus LD50:16 mg/kg JPETAB 110,282,54
ivn-gpg LDLo:99 mg/kg CRSBAW 151,614,57

SAFETY PROFILE: Poison by intravenous and subcutaneous routes. Moderately toxic by ingestion. When heated to decomposition it emits very toxic fumes of NH_3, NO_x, SO_x, and Br^-.

PBS250 CAS:115-58-2 ***HR: 3***
PENTOBARBITAL
mf: $C_{11}H_{18}N_2O_3$ mw: 226.31

SYNS: 5-AETHYL-5-PENTYL-(2')-BARBITURSAEURE (GERMAN) ◇ 5-ETHYL-5-PENTYLBARBITURIC ACID

TOXICITY DATA with REFERENCE
ipr-rat LDLo:210 mg/kg JPHAA3 26,317,37
ipr-mus LD50:124 mg/kg ARZNAD 15,688,65

SAFETY PROFILE: Poison by intraperitoneal route. When heated to decomposition it emits toxic fumes of NO_x. See also BARBITURATES.

PBS500 CAS:21642-83-1 ***HR: 3***
R(+)-PENTOBARBITAL SODIUM
mf: $C_{11}H_{17}N_2O_3$•Na mw: 248.29

PROP: Crystal granules or white powder; sltly bitter taste. Decomp at approx 127°. Very sol in water, alc; insol in ether.

SYNS: R(+)-5-ETHYL-5-(1-METHYLBUTYL)BARBITURIC ACID SODIUM SALT ◇ SODIUM (R)(+)-PENTOBARBITAL

TOXICITY DATA with REFERENCE
ipr-mus LD50:185 mg/kg TXAPA9 26,495,73
ivn-mus LD50:147 mg/kg TXAPA9 26,495,73

SAFETY PROFILE: Poison by intraperitoneal and intravenous routes. Abuse may lead to addiction. When heated to decomposition it emits toxic fumes of NO_x and Na_2O. See also BARBITURATES.

PBS750 CAS:21642-82-0 ***HR: 3***
S(−)-PENTOBARBITAL SODIUM
mf: $C_{11}H_{17}N_2O_3$•Na mw: 248.29

SYN: S(−)-5-ETHYL-5-(1-METHYLBUTYL)BARBITURIC ACID SODIUM SALT

TOXICITY DATA with REFERENCE
ipr-mus LD50:92 mg/kg TXAPA9 26,495,73
ivn-mus LD50:78 mg/kg TXAPA9 26,495,73

SAFETY PROFILE: Poison by intraperitoneal and intravenous routes. When heated to decomposition it emits toxic fumes of NO_x and Na_2O. See also BARBITURATES.

PBT000 CAS:52-62-0 ***HR: 3***
PENTOLINIUM TARTRATE
mf: $C_{15}H_{32}N_2$•$2C_4H_5O_6$ mw: 538.67

PROP: Crystals; acid taste. Decomp @ 203°. Very sol in water; insol in ether, chloroform.

SYNS: ANSOLYSEN ◇ ANSOLYSEN BITARTRATE ◇ ANSOLYSEN TARTRATE ◇ MB 2050A ◇ PENTALINIUM TARTRATE ◇ 1,1'-PENTAMETHYLENEBIS(1-METHYLPYRROLIDINIUM HYDROGEN TARTRATE) ◇ PENTAMETHYLENE-1,5-BIS(1'-METHYLPYRROLIDINIUM TARTRATE) ◇ 1,1'-PENTAMETHYLENEBIS(1-METHYLPYRROLIDINIUM TARTRATE) ◇ PENTAPYRROLIDIUM BITARTRATE ◇ PENTILIUM ◇ PENTOLINIUM BITARTRATE ◇ PENTOLINIUM DITARTRATE

TOXICITY DATA with REFERENCE
orl-rat LD50:890 mg/kg PSEBAA 120,511,65
orl-mus LD50:512 mg/kg NIIRDN 6,780,82
ipr-mus LD50:36 mg/kg NIIRDN 6,780,82
scu-mus LDLo:94 mg/kg CLDND*
ivn-mus LD50:29 mg/kg NIIRDN 6,780,82
ivn-dog LD50:10 mg/kg JAPMA8 46,346,57

SAFETY PROFILE: Poison by intraperitoneal, subcutaneous, and intravenous routes. Moderately toxic by ingestion. When heated to decomposition it emits toxic fumes of NO_x.

PBT100 CAS:53910-25-1 ***HR: 3***
PENTOSTATIN
mf: $C_{11}H_{16}N_4O_4$ mw: 268.31

SYNS: 2-CDF ◇ CL 67310465 ◇ CO-VIDARABINE ◇ DEOXYCOFORMYCIN ◇ 2'-DEOXYCOFORMYCIN ◇ (R)-3-(2-DEOXY-β-d-PENTOFURANOSYL)-3,6,7,8-TETRAHYDROIMIDAZO(4,5-d)(1,3)DIAZEPIN-8-OL ◇ NSC-218321

TOXICITY DATA with REFERENCE
dnd-hmn:lym 10 μmol/L JBCHA3 259,9426,84
ipr-mus TDLo:500 μg/kg (10D preg):TER TJADAB 25(2),361,82
ivn-mus LD50:122 mg/kg NTIS** PB84-211424

SAFETY PROFILE: Poison by intravenous route. An experimental teratogen. Human mutation data reported. When heated to decomposition it emits toxic fumes of NO_x.

PBT250 CAS:76-75-5 ***HR: 3***
PENTOTHAL
mf: $C_{11}H_{18}N_2O_2S$ mw: 242.37

SYNS: 5-ETHYLDIHYDRO-5-(1-METHYLBUTYL)-2-THIOXO-4,6,(1H,5H)-PYRIMIDINEDIONE (9CI) ◇ 5-ETHYL-5-(1-METHYLBUTYL)-2-THIOBARBITURIC ACID ◇ FARMOTAL ◇ INTRAVAL ◇ NESDONAL ◇ PENTHIOBARBITAL ◇ PENTOTHIOBARBITAL ◇ THIOMEBUMAL ◇ THIOPENTAL ◇ THIOPENTOBARBITAL ◇ THIOPENTONE ◇ THIOTHAL ◇ TIOPENTALE (ITALIAN)

TOXICITY DATA with REFERENCE
cyt-hmn:leu 1500 mg/L TGANAK 18(1),13,84
ipr-rat TDLo:1020 mg/kg (1-17D preg):TER REANBJ 26,137,79
ipr-rat LD50:120 mg/kg 27ZIAQ -,260,65
ivn-rat LD50:64 mg/kg PSEBAA 89,292,55
orl-mus LD50:600 mg/kg BJPCAL 1,215,46
ipr-mus LD50:110 mg/kg AIPTAK 128,391,60
ivn-mus LD50:70 mg/kg ARZNAD 4,441,54
orl-rbt LDLo:600 mg/kg JPETAB 60,189,37
rec-rbt LDLo:110 mg/kg JPETAB 60,189,37

SAFETY PROFILE: Poison by intraperitoneal, intravenous, and rectal routes. Moderately toxic by ingestion. An experimental teratogen. Human mutation data reported. A short-acting intravenous anesthetic. When heated to decomposition it emits very toxic fumes of NO_x and SO_x. See also PENTOTHAL SODIUM and BARBITURATES.

PBT500 CAS:71-73-8 ***HR: 3***
PENTOTHAL SODIUM
mf: $C_{11}H_{18}N_2O_2S \cdot Na$ mw: 265.36

SYNS: 5-ETHYLDIHYDRO-5-(1-METHYLBUTYL)-2-THIOXO-4,6(1H,5H)-PYRIMIDINEDIONE, MONOSODIUM SALT ◇ 5-ETHYL-5-(1-METHYLBUTYL)-2-THIOBARBITURIC ACID MONOSODIUM ◇ FARMOTAL ◇ HYPNOSTAN ◇ INTRAVAL SODIUM ◇ LEOPENTAL ◇ MONOSODIUM-5-ETHYL-5-(1-METHYLBUTYL) THIOBARBITURATE ◇ NESDONAL SODIUM ◇ PENTHIOBARBITAL SODIUM ◇ RAVONAL ◇ SODIUM-5-ETHYL-5-(1-METHYLBUTYL)-2-THIOBARBITURATE ◇ SODIUM PENTHIOBARBITAL ◇ SODIUM PENTOTHAL ◇ SODIUM PENTOTHIOBARBITAL ◇ SODIUM THIOPENTAL ◇ SODIUM THIOPENTOBARBITAL ◇ SODIUM THIOPENTONE ◇ SOLUBLE THIOPENTONE ◇ THIOMEBUMAL SODIUM ◇ THIONEMBUTAL ◇ THIOPENTAL SODIUM ◇ THIOPENTAL SODIUM SALT ◇ THIOPENTONE SODIUM ◇ THIOTHAL SODIUM ◇ THIPENTAL SODIUM ◇ TIOPENTAL SODIUM ◇ TRAPANAL ◇ TRAPANAL SODIUM

TOXICITY DATA with REFERENCE
ipr-mus TDLo:75 mg/kg (female 11D post):TER OFAJAE 43,219,67
scu-mus TDLo:432 mg/kg (female 3-6D post):REP JOENAK 43,225,69
rec-hmn TDLo:45 mg/kg:CVS,PUL AACRAT 46,395,67
iat-hmn TDLo:7 mg/kg:CVS LANCAO 2,571,43
orl-rat LD50:117 mg/kg JPETAB 115,432,55
ipr-rat LD50:115 mg/kg ANESAV 8,589,47
ivn-rat LD50:43600 μg/kg JPETAB 116,317,56
rec-rat LD50:102 mg/kg AACRAT 46,395,67
orl-mus LD50:208 mg/kg JPETAB 115,432,55
ipr-mus LD50:115 mg/kg NIIRDN 6,457,82
scu-mus LD50:225 mg/kg TXAPA9 27,70,74
ivn-mus LD50:57 mg/kg AIPTAK 105,221,56
orl-dog LDLo:150 mg/kg JPETAB 60,125,37
ivn-dog LD50:36 mg/kg CRAAA7 29,89,50
ipr-rbt LD50:35 mg/kg NIIRDN 6,457,82
ivn-rbt LD50:31 mg/kg CRAAA7 29,89,50
ipr-gpg LD50:57500 μg/kg NIIRDN 6,457,82

SAFETY PROFILE: Poison by ingestion, intraperitoneal, rectal, subcutaneous, and intravenous routes. Human systemic effects by intraarterial route: acute arterial occlusion; by rectal route: respiratory depression, body temperature decrease, general anesthetic. An experimental teratogen. Experimental reproductive effects. An intravenous anesthetic. When heated to decomposition it emits toxic fumes of NO_x and Na_2O. See also PENTOTHAL and BARBITURATES.

PBU100 CAS:6493-05-6 ***HR: 3***
PENTOXYPHYLLINE
mf: $C_{13}H_{18}N_4O_3$ mw: 278.35

PROP: Colorless needles from methanol; bitter tasting. Mp: 105°. Solubility in water: 77 mg/mL at 25°, 191 mg/mL at 37°; in benzene: 11 g/100 mL.

SYNS: BL 191 ◇ 3,7-DIHYDRO-3,7-DIMETHYL-1-(5-OXOHEXYL)-1H-PURINE-2,6-DIONE ◇ 3,7-DIMETHYL-1-(5-OXOHEXYL)-1H,3H-PURIN-2,6-DIONE ◇ DIMETHYLOXOHEXYLXANTHINE ◇ 3,7-DIMETHYL-1-(5-OXOHEXYL)XANTHINE ◇ 1-(5-OXOHEXYL)-3,7-DIMETHYLXANTHINE ◇ 1-(5-OXOHEXYL)THEOBROMINE ◇ OXPENTIFYLLINE ◇ PENTOXIFYLLIN ◇ PENTOXIFYLLINE ◇ PENTOXIPHYLLIUM ◇ RENTYLIN ◇ TRENTAL ◇ TORENTAL ◇ VAZOFIRIN

TOXICITY DATA with REFERENCE
dns-hmn:oth 1 mmol/L BIOJAU 35,665,81
dni-hmn:oth 4 mmol/L BIOJAU 35,665,81
orl-rat TDLo:6270 mg/kg (6-16D preg):REP ARZNAD 21,1446,71
ivn-rat TDLo:300 mg/kg (female 9-14D post):TER OYYAA2 10,773,75
orl-wmn TDLo:80 mg/kg:CVS BMJOAE 288,26,84
orl-rat LD50:1170 mg/kg YACHDS 9,13,81
ipr-rat LD50:230 mg/kg YACHDS 9,13,81
scu-rat LD50:375 mg/kg OYYAA2 15,153,78
ivn-rat LD50:231 mg/kg ARZNAD 21,1446,71
orl-mus LD50:1225 mg/kg NIIRDN 6,779,82
ipr-mus LD50:239 mg/kg ARZNAD 21,1446,71
scu-mus LD50:480 mg/kg IYKEDH 8,267,77
ivn-mus LD50:108 mg/kg IYKEDH 8,267,77
ivn-rbt LD50:100 mg/kg IYKEDH 8,267,77

CONSENSUS REPORTS: EPA Genetic Toxicology Program.

SAFETY PROFILE: Poison by subcutaneous, intravenous, and intraperitoneal routes. Moderately toxic by ingestion. Human systemic effects by ingestion: pulse rate decrease with fall in blood pressure. Human mutation data reported. An experimental teratogen. Experimental reproductive effects. When heated to decomposition it emits toxic fumes of NO_x.

PBV000 CAS:75-85-4 ***HR: 3***
tert-PENTYL ALCOHOL
DOT: UN 1105
mf: $C_5H_{12}O$ mw: 88.17

PROP: Colorless liquid. Mp: −11.9°, bp: 101.8°, flash p: 105°F (CC), d: 0.809, autoign temp: 819°F, vap press: 10 mm @ 17.2°, lel: 1.2%, uel: 9%, vap d: 3.03. Sltly sol in water; sol in alc and ether.

SYNS: tert-AMYL ALCOHOL (DOT) ◇ AMYLENE HYDRATE ◇ DIMETHYLETHYLCARBINOL ◇ 2-METHYL BUTANOL-2 ◇ 2-METHYL-2-BUTANOL ◇ 3-METHYLBUTAN-3-OL ◇ tert-PENTANOL

TOXICITY DATA with REFERENCE
unr-man LDLo:441 mg/kg 85DCAI 2,73,70
orl-rat LD50:1000 mg/kg SCIEAS 116,663,52
ipr-rat LDLo:1530 mg/kg JIHTAB 27,1,45
scu-rat LDLo:1400 mg/kg JPETAB 49,36,33
rec-rat LDLo:1400 mg/kg JPETAB 49,36,33
scu-mus LD50:2100 mg/kg ARZNAD 5,161,55

CONSENSUS REPORTS: Reported in EPA TSCA Inventory.

DOT Classification: Flammable or Combustible Liquid; Label: Flammable Liquid.

SAFETY PROFILE: Moderately toxic to humans by an unspecified route. Moderately toxic experimentally by ingestion, intraperitoneal, subcutaneous and rectal routes. Narcotic in high concentration. Flammable when exposed to heat, flame, or oxidizing materials. Moderately explosive in the form of vapor when exposed to heat or flame. A hypnotic agent. When heated to decomposition it emits acrid smoke and irritating fumes.

PBV500 ***HR: 3***
PENTYLAMINE (mixed isomers)
DOT: UN 1106
mf: $C_5H_{13}N$ mw: 87.17

PROP: Water-white liquid. Mp: −55°, bp: 104°, flash p: 45°F (OC), d: 0.7614 @ 20°/4°, vap d: 3.01, lel: 2.2%, uel: 22%.

SYN: AMYLAMINE (mixed isomers) (DOT)

TOXICITY DATA with REFERENCE
skn-rbt 500 mg open SEV UCDS** 8/12/68
orl-rat LD50:470 mg/kg UCDS** 8/12/68
ihl-rat LCLo:2000 ppm/4H AIHAAP 30,470,69
skn-rbt LD50:1120 mg/kg AIHAAP 30,470,69

DOT Classification: Label: Flammable Liquid.

SAFETY PROFILE: Moderately toxic by ingestion and skin contact. A severe skin irritant. Dangerous fire hazard when exposed to heat or flame; can react with oxidizing materials. Moderately explosive in the form of vapor when exposed to heat or flame. To fight fire, use alcohol foam, dry chemical. When heated to decomposition it emits toxic fumes of NO_x. See also AMINES.

PBV750 CAS:2188-67-2 ***HR: 3***
2,N-PENTYLAMINOETHYL-p-AMINOBENZOATE
mf: $C_{14}H_{22}N_2O_2$ mw: 250.38

SYNS: p-AMINOBENZOIC ACID-2-N-AMYLAMINOETHYL ESTER ◇ 2-N-AMYLAMINOETHYL-p-AMINOBENZOATE ◇ AMYLCAINE ◇ AMYLSINE ◇ NAEPAINE

TOXICITY DATA with REFERENCE
scu-mus LDLo:350 mg/kg JPETAB 62,69,38
ivn-rbt LDLo:18 mg/kg JPETAB 62,69,38
scu-gpg LDLo:100 mg/kg JPETAB 62,69,38

SAFETY PROFILE: Poison by intravenous and subcutaneous routes. When heated to decomposition it emits toxic fumes of NO_x.

PBW000 CAS:26311-45-5 ***HR: 3***
4-n-PENTYLBENZOIC ACID
mf: $C_{12}H_{16}O_2$ mw: 192.28

PROP: Colorless liquid. D: 0.992 @ 14°/14°, bp: 260.7 @ 746 mm. Insol in water; misc in alc and ether.

SYNS: p-AMYLBENZOIC ACID ◇ p-PENTYLBENZOIC ACID

TOXICITY DATA with REFERENCE
ivn-mus LD50:56 mg/kg CSLNX* NX#10073

CONSENSUS REPORTS: Reported in EPA TSCA Inventory.

SAFETY PROFILE: Poison by intravenous route. When heated to decomposition it emits acrid smoke and irritating fumes.

PBW400 CAS:20675-51-8 ***HR: 3***
PENTYLCANNABICHROMENE
mf: $C_{21}H_{30}O_2$ mw: 314.51

SYNS: 2H-1-BENZOPYRAN-5-OL,2-METHYL-2-(4-METHYL-3-PENTENYL)-7-PENTYL- ◇ CANNABICHROME ◇ CANNABICHROMENE ◇ CANNANBICHROMENE

TOXICITY DATA with REFERENCE
dni-hmn:lyms 100 umol/L FEPRA7 36,1748,77
orl-mus TDLo:50 mg/kg (female 12D post):REP TJADAB 33,195,86
ipr-mus LD50:113 mg/kg GEPHDP 12,357,81
ivn-mky LD50:270 mg/kg TXAPA9 58,118,81

SAFETY PROFILE: Poison by intravenous and intraperitoneal routes. Experimental reproductive effects. Mutation data reported. When heated to decomposition it emits acrid smoke and irritating fumes.

PBW500 CAS:543-59-9 ***HR: 3***
PENTYL CHLORIDE
DOT: UN 1107
mf: $C_5H_{11}Cl$ mw: 106.61

PROP: Water-white liquid; sweet odor. Mp: −99°, bp: 108.2°, flash p: 54°F (OC), d: 0.883 @ 20°/4°, autoign temp: 500°F, vap d: 3.67, lel: 1.4%, uel: 8.6%.

SYNS: n-AMYL CHLORIDE ◇ AMYL CHLORIDE (DOT) ◇ 1-CHLOROPENTANE

CONSENSUS REPORTS: Reported in EPA TSCA Inventory.

DOT Classification: Label: Flammable Liquid.

SAFETY PROFILE: Dangerous fire hazard when exposed to heat or flame; can react with oxidizing materials. Moderately explosive in the form of vapor when exposed to heat or flame. To fight fire, use foam, CO_2, dry chemical. Dangerous; when heated to decomposition it emits highly toxic fumes of phosgene and Cl^-. See also CHLORINATED HYDROCARBONS, ALIPHATIC.

PBW750 CAS:12789-46-7 ***HR: 2***
PENTYL ESTER PHOSPHORIC ACID
DOT: UN 2819
mf: $C_5H_{13}O_4P$ mw: 168.15

SYN: AMYL ACID PHOSPHATE (DOT)

DOT Classification: Corrosive Material; Label: Corrosive.

SAFETY PROFILE: Corrosive and irritating to the skin, eyes, and mucous membranes. When heated to decomposition it emits toxic fumes of PO_x. See also ESTERS.

PBX000 CAS:693-65-2 ***HR: 3***
PENTYL ETHER
mf: $C_{10}H_{22}O$ mw: 158.32

PROP: Liquid. Mp: −69.3°, bp: 187°, flash p: 135°F (OC), d: 0.783 @ 20°/4°, vap d: 5.46, autoign temp: 340°F.

SYNS: N-AMYL ETHER ◇ AMYL ETHER ◇ DIPENTYL ETHER ◇ 1,1-OXYBIS PENTANE

TOXICITY DATA with REFERENCE
ivn-mus LD50:164 mg/kg JPMSAE 67,566,78

CONSENSUS REPORTS: Reported in EPA TSCA Inventory.

SAFETY PROFILE: Poison by intravenous routes. Flammable when exposed to heat or flame; reacts with oxidizing materials. To fight fire, use alcohol foam, dry chemical. When heated to decomposition it emits acrid smoke and irritating fumes. See also ETHERS.

PBX250 CAS:1119-68-2 ***HR: 3***
n-PENTYLHYDRAZINE HYDROCHLORIDE
mf: $C_5H_{14}N_2 \cdot ClH$ mw: 138.67

SYN: n-AMYLHYDRAZINE HYDROCHLORIDE

TOXICITY DATA with REFERENCE
orl-mus TDLo:11 g/kg/80W-C:NEO BJCAAI 31,492,75
orl-mus TD:12 mg/kg/W-C:ETA PAACA3 16,61,75

SAFETY PROFILE: Questionable carcinogen with experimental neoplastigenic and tumorigenic data. When heated to decomposition it emits very toxic fumes of NO_x and HCl.

PBX325 CAS:3425-61-4 ***HR: 3***
tert-PENTYL HYDROPEROXIDE
mf: $C_5H_{12}O_2$ mw: 104.17

SYNS: tert-AMYL HYDROPEROXIDE ◇ 1,1-DIMETHYLPROPYL HYDROPEROXIDE

TOXICITY DATA with REFERENCE
orl-rat LD50:863 mg/kg 85GMAT -,21,82
ipr-rat LD50:250 mg/kg FATOAO 40,369,77
orl-mus LD50:450 mg/kg 85GMAT -,21,82
ipr-mus LD50:275 mg/kg 85GMAT -,21,82

CONSENSUS REPORTS: Reported in EPA TSCA Inventory.

SAFETY PROFILE: Poison by intraperitoneal route. Moderately toxic by ingestion. When heated to decomposition it emits acrid smoke and irritating fumes. See also PEROXIDES, ORGANIC.

PBX350 CAS:25677-40-1 ***HR: 2***
2-PENTYLIDENECYCLOHEXANONE
mf: $C_{11}H_{18}O$ mw: 166.29

TOXICITY DATA with REFERENCE
skn-rbt 500 mg/24H SEV FCTOD7 20(Suppl),797,82
orl-rat LD50:5 g/kg FCTOD7 20(Suppl),797,82
skn-rbt LD50:3500 mg/kg FCTOD7 20(Suppl),797,82

SAFETY PROFILE: Moderately toxic by skin contact. Mildly toxic by ingestion. A severe skin irritant. When heated to decomposition it emits acrid smoke and irritating fumes. See also KETONES.

PBX500 CAS:10589-74-9 ***HR: 3***
n-PENTYLNITROSOUREA
mf: $C_6H_{13}N_3O_2$ mw: 159.22

SYNS: 1-AMYL-1-NITROSOUREA ◇ n-AMYLNITROSOUREA ◇ ANU ◇ 1-NITROSO-1-PENTYLUREA

TOXICITY DATA with REFERENCE
mmo-sat 1 μg/plate MUREAV 68,1,79
mma-sat 25 μg/plate TCMUE9 1,13,84
pic-esc 2 mg/L TCMUE9 1,91,84
cyt-ham:fbr 250 mg/L/48H MUREAV 48,337,77
orl-rat TDLo:184 g/kg/49W-I:CAR GANNA2 71,464,80
scu-rat TDLo:510 mg/kg:ETA ANYAA9 381,250,82
skn-mus TDLo:629 mg/kg/50W-I:CAR JCROD7 102,13,81
orl-rat LD50:560 mg/kg PPTCBY 2,85,72

SAFETY PROFILE: Suspected carcinogen with experimental carcinogenic and tumorigenic data. Moderately toxic by ingestion. Mutation data reported. When heated to decomposition it emits toxic fumes of NO_x. See also N-NITROSO COMPOUNDS.

PBX750 CAS:62573-57-3 ***HR: 3***
m-(3-PENTYL)PHENYL-N-METHYL-N-NITROSOCARBAMATE
mf: $C_{13}H_{18}N_2O_3$ mw: 250.33

SYNS: N-METHYL-N-NITROSOCARBAMICACID-m-3-PENTYLPHENYL ESTER ◇ NITROSO-BUX-TEN

TOXICITY DATA with REFERENCE
mmo-esc 100 μmol/L IARCCD 14,425,75
mma-sat 125 ng/plate MUREAV 56,1,77
dnd-hmn:fbr 10 μmol/L MUREAV 44,1,77
orl-rat TDLO:660 mg/kg/10W-I:CAR EESADV 2,413,78

SAFETY PROFILE: Questionable carcinogen with experimental carcinogenic data. Human mutation data reported. When heated to decomposition it emits toxic fumes of NO_x. See also N-NITROSO COMPOUNDS and CARBAMATES.

PBY750 CAS:107-72-2 ***HR: 2***
PENTYLTRICHLOROSILANE
DOT: UN 1728
mf: $C_5H_{11}Cl_3Si$ mw: 205.60

SYN: AMYL TRICHLOROSILANE

TOXICITY DATA with REFERENCE
skn-rbt 100 μg/24H open AIHAAP 23,95,62
orl-rat LD50:2340 mg/kg AIHAAP 23,95,62
ihl-rat LCLo:2000 ppm/4H JIHTAB 31,343,49
skn-rbt LD50:780 mg/kg AIHAAP 23,95,62

CONSENSUS REPORTS: Reported in EPA TSCA Inventory.

DOT Classification: Corrosive Material; Label: Corrosive.

SAFETY PROFILE: Moderately toxic by ingestion and skin contact. Mildly toxic by inhalation. A corrosive irritant to the eyes, skin, and mucous membranes. When heated to decomposition it emits toxic fumes of Cl^-. See also CHLOROSILANES.

PBZ000 CAS:2761-24-2 ***HR: 1***
PENTYLTRIETHOXYSILANE
mf: $C_{11}H_{26}O_3Si$ mw: 234.46

SYN: AMYLTRIETHOXYSILANE

TOXICITY DATA with REFERENCE
skn-rbt 10 mg/24H MLD AMIHBC 10,61,54
eye-rbt 500 mg AMIHBC 10,61,54
orl-rat LD50:20 g/kg AMIHBC 10,61,54
skn-rbt LD50:7130 mg/kg AMIHBC 10,61,54

CONSENSUS REPORTS: Reported in EPA TSCA Inventory.

SAFETY PROFILE: Mildly toxic by ingestion and skin contact. A skin and eye irritant. When heated to decomposition it emits acrid smoke and irritating fumes. See also SILANE.

PCA250 CAS:627-19-0 ***HR: 3***
1-PENTYNE
mf: C_5H_8 mw: 68.119

$$HC{\equiv}CCH_2CH_2CH_3$$

PROP: Flash p: −4°F.

SAFETY PROFILE: Dangerous fire hazard when exposed to heat or flame; can react vigorously with oxidizing materials. When heated to decomposition it emits acrid smoke and irritating fumes. See also ACETYLENE COMPOUNDS.

PCA300 CAS:627-21-4 ***HR: 3***
2-PENTYNE
mf: C_5H_8 mw: 68.12

$$CH_3C{\equiv}CCH_2CH_3$$

SAFETY PROFILE: A very dangerous fire hazard when exposed to heat or flame; can react vigorously with oxidizing materials. Solutions with silver perchlorate explode on contact with mercury. When heated to decomposition it emits acrid smoke and irritating fumes. See also ACETYLENE COMPOUNDS.

PCB000 CAS:70384-29-1 ***HR: 3***
PEPLEOMYCIN SULFATE

SYN: NK 631

TOXICITY DATA with REFERENCE
scu-rat TDLo:23 mg/kg/61W-I:CAR ONCOBS 41,114,84
par-rat TDLo:17 mg/kg/52W-I:CAR PAACA3 24,96,83
ipr-rat LD50:180 mg/kg JJANAX 31,719,78
scu-rat LD50:199 mg/kg JJANAX 31,719,78
ivn-rat LD50:215 mg/kg JJANAX 31,719,78
ipr-mus LD50:77 mg/kg JJANAX 31,719,78
scu-mus LD50:80 mg/kg JJANAX 31,719,78
ivn-mus LD50:45 mg/kg JJANAX 31,719,78

SAFETY PROFILE: Poison by intraperitoneal, subcutaneous, and intravenous routes. Questionable carcinogen with experimental carcinogenic data. When heated to decomposition it emits very toxic fumes of NO_x and SO_x.

PCB250 CAS:8006-90-4 ***HR: 2***
PEPPERMINT OIL

PROP: From steam distillation of *Mentha piperita* L. (Fam. *Labiatae*). Colorless to pale yellow liquid; strong odor and taste of peppermint. D: 0.896-0.908 @ 25°/25°, refr index: 1.459 @ 20°.

SYN: PFEFFERMINZ OEL (GERMAN)

TOXICITY DATA with REFERENCE
dnr-bcs 5 μL/disc TOFOD5 8,91,85
orl-rat LD50:2426 mg/kg JPMSAE 54,1071,65
ipr-rat LD50:819 mg/kg JPMSAE 54,1071,65
orl-mus LD50:2490 mg/kg TOFOD5 8,91,85

CONSENSUS REPORTS: Reported in EPA TSCA Inventory.

SAFETY PROFILE: Moderately toxic by ingestion and intraperitoneal routes. An allergen. Mutation data reported. When heated to decomposition it emits acrid smoke and irritating fumes.

PCB275 ***HR: 1***
PEPPER PLANTS

PROP: About 20 species of perennial herbs native to tropical America. When ripe the fruit is a red, orange or yellow pod containing many small seeds. Most species are cultivated as spices or ornamentals. The bird pepper grows wild in the southern United States and Mexico.

SYNS: AJI CABALLERO (PUERTO RICO) ◇ AJI de GALLINA (PUERTO RICO) ◇ AJI GUAGUAO (CUBA) ◇ AJI PICANTE (PUERTO RICO) ◇ BIRD PEPPER ◇ C. ANNUUM ◇ C. FRUTESCENS ◇ CAPSICUM (VARIOUS SPECIES) ◇ CAYENNE PEPPER ◇ CHERRY PEPPER ◇ CHILE PEPPER ◇ HOT PEPPER ◇ LONG PEPPER ◇ NIOI-PEPA (HAWAII) ◇ PIMENT (HAITI) ◇ PIMENT BOUC (HAITI) ◇ RED PEPPER ◇ TABASCO PEPPER

SAFETY PROFILE: The fruit and seeds contain the irritant capsaicin. Except for the group including the bell pepper, sweet pepper, and green pepper, chewing the fruit can cause a painful but harmless irritation of the mouth and lips. Skin contact may cause redness but not blistering. See also CAPSAICIN.

PCB300 ***HR: 2***
PEPPER TREE

PROP: The pepper tree grows to 35 feet usually with a twisted trunk and hanging branches like a willow. The Brazilian pepper tree grows to about 20 feet. Both species have compound leaves which smell of black pepper when crushed and produce small red berries. The Brazilian pepper is a common nuisance tree in southern Florida, Hawaii, Guam, and the West Indies. The pepper tree is cultivated in California, Hawaii and the West Indies.

SYNS: ARBOL de PERU (MEXICO) ◇ BRAZILIAN PEPPER TREE ◇ CALIFORNIA PEPPER TREE ◇ CHRISTMAS BERRY TREE ◇ COPAL (CUBA) ◇ FLORIDA HOLLY ◇ NANI-O-HILO (HAWAII) ◇ PERUVIAN MASTIC TREE ◇ PIMIENTA del BRAZIL (PUERTO RICO) ◇ PIMIENTO de AMERICA (CUBA) ◇ SCHINUS MOLLE ◇ SCHINUS TEREBINTHIFOLIUS ◇ WILELAIKI (HAWAII)

SAFETY PROFILE: The berries contain poisonous triterpines. Ingestion of the berries can cause throat irritation, vomiting and diarrhea.

PCB750 CAS:26305-03-3 ***HR: 2***
PEPSTATIN A
mf: $C_{34}H_{63}N_5O_9$ mw: 686.02

PROP: Colorless needles. Mp: 228-229° (decomp). Sol in methanol, ethanol, acetic acid, DMSO; insol in benzene, chloroform, ether, water.

SYN: PEPSTATIN

TOXICITY DATA with REFERENCE
ipr-rat LD50:875 mg/kg JANTAJ 23,259,70
ipr-mus LD50:1090 mg/kg JANTAJ 23,259,70
ipr-dog LD50:450 mg/kg JANTAJ 23,259,70
ipr-rbt LD50:820 mg/kg JANTAJ 23,259,70

SAFETY PROFILE: Moderately toxic by intraperitoneal route. When heated to decomposition it emits toxic fumes of NO_x.

PCC000 CAS:9076-25-9 ***HR: 3***
PEPTICHEMIO

PROP: Made up of 6 peptides of m-(di-(2-chloroethyl) amino-1-phenylalanine (WDMBAM 5,303,75).

SYNS: NSC 247516 ◇ PEP ◇ PTC

TOXICITY DATA with REFERENCE
mmo-sat 100 μg/plate GSLNAG 9,10,77
dns-hmn:leu 10 μmol/L EJCODS 17,991,81
ipr-rat LD50:7180 μg/kg WDMBAM 5(3),303,75
ipr-mus LD50:10310 μg/kg WDMBAM 5(3),303,75
scu-mus LD50:13 mg/kg WDMBAM 5(3),303,75
ivn-mus LD50:10560 μg/kg WDMBAM 5(3),303,75

SAFETY PROFILE: Poison by intravenous, intraperitoneal, and subcutaneous routes. Human mutation data reported. When heated to decomposition it emits very toxic fumes of Cl^- and NO_x.

PCC475 CAS:6002-77-3 ***HR: 3***
PERAZINE MALEATE
mf: $C_{20}H_{25}N_3S \cdot xC_4H_4O_4$

SYN: 10H-PHENOTHIAZINE, 10-(3-(4-METHYL-1-PIPERAZINYL)PROPYL)-(Z)-2-BUTENEDIOATE

TOXICITY DATA with REFERENCE
orl-rat LD50:1381 mg/kg NIIRDN 6,765,82
ipr-rat LD50:95200 μg/kg NIIRDN 6,765,82
orl-mus LD50:1196 mg/kg NIIRDN 6,765,82
ipr-mus LD50:189 mg/kg NIIRDN 6,765,82

SAFETY PROFILE: Poison by intraperitoneal route. Moderately toxic by ingestion. When heated to decomposition it emits toxic fumes of SO_x and NO_x.

PCC750 CAS:25251-03-0 ***HR: 3***
PERBROMYL FLUORIDE
mf: $BrFO_3$ mw: 146.91

PROP: Attacks glass, polytetrafluoroethylene, polychlortri-fluoroethylene, halogen oxides.

SAFETY PROFILE: Poison irritant to skin, eyes and mucous membranes. When heated to decomposition it emits very toxic fumes of Br^- and F^-. See also BROMIDES and FLUORIDES.

PCD000 ***HR: 3***
PERCHLORATES

PROP: Composition: combinations with the monovalent $-ClO_4$ radical.

SAFETY PROFILE: Perchlorates are unstable materials, and are irritant to the body wherever they come in contact with it. Avoid skin contact. Flammable by chemical reaction; powerful oxidizers. All perchlorates are potentially hazardous when in contact with reducing materials. Moderate explosion hazard when shocked or exposed to heat or by chemical reaction. Perchlorates, when mixed with carbonaceous material, form explosive mixtures. Many perchlorates of nitrogenous bases (e.g., hydroxylamine, urea, methylamaine, ethylamine, isopropylamine, 4-ethylpyridine, diaminoethane) and organic perchlorates are explosives. Diazonium perchlorates are very dangerous. All perchlorates are considered to be fire and explosive hazards when associated with carbonaceous materials or finely divided metals. This is also true of the presence of calcium hydride; sulfur; powdered magnesium; aluminum; zinc. They react violently with benzene; CaH_2; charcoal; olefins; ethanol; SrH_2; S; H_2SO_4; and reducing materials. To fight fire, use water or foam. When heated to decomposition it emits toxic fumes of Cl^-. See also EXPLOSIVES, HIGH.

PCD250 CAS:7601-90-3 ***HR: 3***
PERCHLORIC ACID
DOT: UN 1802/UN 1873
mf: $ClHO_4$ mw: 100.46

PROP: Colorless, fuming, unstable liquid. Mp: −112°, bp: 19° @ 11 mm, d: 1.768 @ 22°.

TOXICITY DATA with REFERENCE
orl-rat LD50:1100 mg/kg GTPZAB 17(8),33,73
scu-mus LD50:250 mg/kg GTPZAB 17(8),33,73
orl-dog LD50:400 mg/kg GTPZAB 17(8),33,73

CONSENSUS REPORTS: Reported in EPA TSCA Inventory.

DOT Classification: Oxidizer; Label: Oxidizer (UN1802, UN1873); Corrosive Material; Label: Oxidizer and Corrosive (UN1802); Oxidizer; Label: Oxidizer and Corrosive (UN1873); Forbidden (more than 72%).

SAFETY PROFILE: Poison by ingestion and subcutaneous routes. A severe irritant to the eyes, skin, and mucous membranes. A powerful oxidizer. A severe explosion hazard; the anhydrous form can explode spontaneously.

Potentially explosive reaction with acetic anhydride + acetic acid + organic materials; acetic anhydride + organic materials + transition metals (e.g., chromium, iron, nickel); acetonitrile; alcohols; azo dyes + orthoperiodic acid; bis(2-hydroxyethyl)terephthalate + ethanol + ethylene glycol; bismuth (above 110°C); antimony (above 110°C); carbon; charcoal + chromium trioxide + heat; cellulose and derivatives + heat; combustible materials; dehydrating agents; dichloromethane + dimethylsulfoxide; diethyl ether; dimethyl ether; dioxane + nitric acid + heat; fecal material + nitric acid; graphitic carbon + nitric acid; hydrofluoric acid + structural materials; iron(II) sulfate; nitric acid + organic matter + heat; nitric acid + pyridine + sulfuric acid; nitrogenous epoxides; organic materials + sodium hydrogen carbonate (above 200°C); phenyl acetylene (at −78°C); sodium phosphinate + heat; sulfuric acid + organic materials; sulfur trioxide. Reacts to form explosive products with aniline + formaldehyde; ethylbenzene + thallium triacetate (at 65°C); fluorine (forms fluorine perchlorate); glycerol + lead oxide; hydrogen + heat; hydrogen halides; phosphine; pyridine; sulfoxides. Violent reaction or ignition with acetic acid; acetic acid + acetic anhydride; acetic anhydride; acetic anhydride + carbon tetrachloride + 2-methyl cyclohexanone; antimony compounds; azo pigments; bis-1;2-diaminopropane-cis-dichlorochromium(III) perchlorate; carbon; 1,3-bis(di-n-cyclopentadienyl iron)-2-propen-1-one; CH_3OH; CCl_4; copper dichromium tetraoxide (at 120°C); DNA; dibutyl sulfoxide; dimethyl sulfoxide; ethylbenzene; glycol ethers; glycols; HNO_3; HCl; H_2SO_4; hypophosphites; iron sulfate; iodides; ketones; PbO + glycerin; methanol + triglycerides; 2-methylpropene + metal oxides; 2-methyl cyclohexanone; NI_3; nitrogenous epoxides; nitrosophenol; o-periodic acid; oleic acid; organophosphorus compounds; paper; P_2O_5 + $CHCl_3$; P_2O_5; P_2Zn_3; sodium iodide + hydroiodic acid; sodium phosphinate; steel; sulfinyl chloride; SO_3; trichloroethylene; vegetable matter; wood; zinc phosphide. When heated to decomposition it emits toxic fumes of Cl^-. See also PERCHLORATES.

PCD500 CAS:7790-98-9 ***HR: 3***
PERCHLORIC ACID, AMMONIUM SALT
DOT: UN 0402/UN 1442
mf: $ClO_4 \cdot H_4N$ mw: 117.50

PROP: White crystals. Mp: decomp, d: 1.95.

SYN: AMMONIUM PERCHLORATE (DOT)

TOXICITY DATA with REFERENCE
par-rat LDLo:3500 mg/kg RPTOAN 32,159,69
par-mus LDLo:2 g/kg RPTOAN 32,159,69
par-rbt LDLo:750 mg/kg RPTOAN 32,159,69

CONSENSUS REPORTS: Reported in EPA TSCA Inventory.

DOT Classification: Label: Oxidizer; Class A Explosive; Label: EXPLOSIVE A

SAFETY PROFILE: Moderately toxic by parenteral route. Flammable when exposed to heat or flame or by spontaneous chemical reaction with reducing materials. A very powerful oxidizer which has caused explosions in industry. Ignites violently with combustibles. Severe explosion hazard; decomposes at 130° and explodes at 380°. When contaminated by powdered carbon; ferrocene; S; organic matter; powdered metals; nitryl perchlorate; potassium periodate; potassium permanganate it becomes impact sensitive. Potentially explosive reactions with carbon (above 240°C); dichromium trioxide (at 270°C); cadmium oxide (at 260°C); zinc oxide (at 200°C); copper chromite; copper oxide; iron oxide; potassium permanganate; potassium dichromate; mono-, di-, tri-, or tetra-methylammonium perchlorates; metal perchlorates (e.g., lithium perchlorate; zinc perchlorate); nitrophenol-formaldehyde polymer. Mixtures with aluminum or copper burn violently when ignited. Mixtures with ethylene dinitrate ignite when stored at 60°C. When heated to decomposition it emits toxic fumes of NH_3 and Cl^-. See also PERCHLORATES and EXPLOSIVES, HIGH.

PCD750 CAS:13465-95-7 ***HR: 3***
PERCHLORIC ACID, BARIUM SALT•$3H_2O$
DOT: UN 1447
mf: Cl_2O_8•Ba•$3H_2O$ mw: 336.24

PROP: Colorless crystals. Mp: decomp @ 400°, d: 2.74.

SYN: BARIUM PERCHLORATE (DOT)

CONSENSUS REPORTS: Barium and its compounds are on the Community Right-To-Know List. Reported in EPA TSCA Inventory.

OSHA PEL: TWA 0.5 mg(Ba)/m^3
ACGIH TLV: TWA 0.5 mg(Ba)/m^3
DOT Classification: Oxidizer; Label: Oxidizer and Poison.

SAFETY PROFILE: A poison. An unstable material. An oxidizer. When refluxed with an alcohol highly explosive alkyl perchlorates are formed. When heated to decomposition it emits toxic fumes of Cl^-. A dessicant. See also PERCHLORATES and BARIUM COMPOUNDS.

PCE000 CAS:10034-81-8 ***HR: 2***
PERCHLORIC ACID, MAGNESIUM SALT
DOT: UN 1475
mf: Cl_2O_8•Mg mw: 223.21

PROP: White, hygroscopic crystals. Mp: decomp @ 251°, d: 2.60 @ 25°.

SYNS: ANHYDRONE ◇ DEHYDRITE ◇ MAGNESIUM PERCHLORATE ◇ PERCHLORATE de MAGNESIUM (FRENCH)

TOXICITY DATA with REFERENCE
ipr-mus LD50:1500 mg/kg JAFCAU 14,512,66.

CONSENSUS REPORTS: Reported in EPA TSCA Inventory.

DOT Classification: Label: Oxidizer.

SAFETY PROFILE: Moderately toxic by intraperitoneal route. A powerful oxidizer which has caused many explosions in industry. Potentially explosive reactions with alkenes (above 220°C); ammonia; aryl hydrazine + ether; dimethyl sulfoxide + heat; ethylene oxide; fluorobutane + water; organic materials; phosphorus; trimethyl phosphate. Reacts to form explosive products with ethanol (forms ethyl perchlorate); cellulose + dinitrogen tetraoxide + oxygen (forms cellulose nitrate). Avoid contact with mineral acids; butyl fluorides; hydrocarbons. A drying agent. When heated to decomposition it emits toxic fumes of MgO and Cl^-. See also MAGNESIUM COMPOUNDS and PERCHLORATES.

PCE250 CAS:26388-85-2 ***HR: 3***
PERCHLORIC ACID, NICKEL(II) SALT compounded with OCTAMETHYL PYROPHOSPHORAMIDE
mf: $C_{24}H_{72}N_{12}NiO_9P_6$•$2ClO_4$ mw: 1116.51

SYN: TRIS(OCTAMETHYLPYROPHOSPHORAMIDE)NICKEL(2+), DIPERCHLORATE

TOXICITY DATA with REFERENCE
ipr-mus LD50:15 mg/kg JAFCAU 14,512,66

CONSENSUS REPORTS: Nickel and its compounds are on the Community Right-To-Know List.

NIOSH REL: TWA 0.015 mg(Ni)/m^3

SAFETY PROFILE: Poison by intraperitoneal route. When heated to decomposition it emits very toxic fumes of NO_x, PO_x, and Cl^-. See also PERCHLORATES and NICKEL COMPOUNDS.

PCE750 CAS:7601-89-0 ***HR: 3***
PERCHLORIC ACID, SODIUM SALT
DOT: UN 1502
mf: ClO_4•Na mw: 122.44

PROP: Colorless, deliquescent crystals. Mp: 482° (decomp).

SYNS: NATRIUMPERCHLORAAT (DUTCH) ◇ NATRIUMPERCHLORAT (GERMAN) ◇ PERCHLORATE de SODIUM (FRENCH) ◇ SODIO (PERCLORATO DI) (ITALIAN) ◇ SODIUM PERCHLORATE ◇ SODIUM PERCHLORATE (DOT)

TOXICITY DATA with REFERENCE
orl-rat LD50:2100 mg/kg GTPZAB 17(8),33,73
ipr-mus LD50:551 mg/kg COREAF 257,791,63

CONSENSUS REPORTS: Reported in EPA TSCA Inventory.

DOT Classification: Oxidizer; Label: Oxidizer

SAFETY PROFILE: Moderately toxic by ingestion and intraperitoneal routes. A powerful oxidizer. Forms explosive mixture with acetone; 1,3-butylene glycol; 2,3-butylene glycol; CaH_2; charcoal; diaminoethane; dimethyl formamide; ethanolamine; ethylene glycol; formamide; galactose; glycerol; hydrazine; water; NH_4NO_3; Mg; reducing agents; SrH_2; urea. When heated to decomposition it emits toxic fumes of Cl^- and Na_2O. See also PERCHLORATES.

PCF250 CAS:3200-96-2 ***HR: 3***
PERCHLORO-2-CYCLOBUTENE-1-ONE
mf: C_4Cl_4O mw: 205.84

SYN: 2,3,4,5-TETRACHLORO-2-CYCLOBUTEN-1-ONE

TOXICITY DATA with REFERENCE
scu-mus TDLo:280 mg/kg/70W-I:ETA JNCIAM 46,143,71

SAFETY PROFILE: Questionable carcinogen with experimental tumorigenic data. When heated to decomposition it emits toxic fumes of Cl^-.

PCF275 CAS:127-18-4 ***HR: 3***
PERCHLOROETHYLENE
DOT: UN 1897
mf: C_2Cl_4 mw: 165.82

PROP: Colorless liquid; chloroform-like odor. Mp: −23.35°, bp: 121.20°, d: 1.6311 @ 15°/4°, vap press: 15.8 mm @ 22°, vap d: 5.83.

SYNS: ANKILOSTIN ◇ ANTISOL 1 ◇ CARBON BICHLORIDE ◇ CARBON DICHLORIDE ◇ CZTEROCHLOROETYLEN (POLISH) ◇ DIDAKENE ◇ DOW-PER ◇ ENT 1,860 ◇ ETHYLENE TETRACHLORIDE ◇ FEDAL-UN ◇ NCI-C04580 ◇ NEMA ◇ PERAWIN ◇ PERCHLOORETHYLEEN, PER (DUTCH) ◇ PERCHLOR ◇ PERCHLORAETHYLEN, PER (GERMAN) ◇ PERCHLORETHYLENE ◇ PERCHLORETHYLENE, PER (FRENCH) ◇ PERCLENE ◇ PERCLOROETILENE (ITALIAN) ◇ PERCOSOLVE ◇ PERK ◇ PERKLONE ◇ PERSEC ◇ RCRA WASTE NUMBER U210 ◇ TETLEN ◇ TETRACAP ◇ TETRACHLOORETHEEN (DUTCH) ◇ TETRACHLORAETHEN (GERMAN) ◇ TETRACHLOROETHENE ◇ TETRACHLOROETHYLENE (DOT) ◇ 1,1,2,2-TETRACHLOROETHYLENE ◇ TETRACLOROETENE (ITALIAN) ◇ TETRALENO ◇ TETRALEX ◇ TETRAVEC ◇ TETROGUER ◇ TETROPIL

TOXICITY DATA with REFERENCE
skn-rbt 810 mg/24H SEV JETOAS 9,171,76
eye-rbt 162 mg MLD JETOAS 9,171,76
dns-hmn:lng 100 mg/L NTIS** PB82-185075
otr-rat:emb 97 μmol/L ITCSAF 14,290,78
ihl-rat TCLo:900 ppm/7H (7-13D preg):REP TJADAB 19,41A,79
ihl-rat TCLo:1000 ppm/24H (14D pre/1-22D preg):TER APTOD9 19,A21,80
ihl-rat TCLo:200 ppm/6H/2Y-I:CAR NTPTR* NTP-TR-311,86
orl-mus TDLo:195 g/kg/50W-I:CAR NCITR* NCI-TR-13,77
ihl-rat TC:200 ppm/6H/2Y-I:NEO TOLED5 31(Suppl),16,86
ihl-hmn TCLo:96 ppm/7H:PNS,EYE,CNS NTIS** PB257-185
ihl-man TCLo:280 ppm/2H:EYE,CNS AMIHBC 5,566,52
ihl-man TCLo:600 ppm/10M:EYE,CNS AMIHBC 5,566,52
ihl-man LDLo:2857 mg/kg:CNS,PUL MLDCAS 5,152,72
orl-rat LD50:2629 mg/kg AIHAAP 20,364,59
ihl-rat LC50:34200 mg/m³/8H AIHAAP 20,364,59
orl-mus LD50:8100 mg/kg NTIS** PB257-185
ihl-mus LC50:5200 ppm/4H APTOA6 9,303,53
orl-dog LDLo:4000 mg/kg AJHYA2 9,430,29
ipr-dog LD50:2100 mg/kg TXAPA9 10,119,67
ivn-dog LDLo:85 mg/kg QJPPAL 7,205,34
orl-cat LDLo:4000 mg/kg AJHYA2 9,430,29

CONSENSUS REPORTS: NTP Fifth Annual Report on Carcinogens. IARC Cancer Review: Group 2B IMEMDT 7,355,87; Animal Limited Evidence IMEMDT 20,491,79. NCI Carcinogenesis Bioassay (gavage); Clear Evidence: mouse NCITR* NCI-CG-TR-13,77; (inhalation); Clear Evidence: mouse, rat NTPTR* NTP-TR-311,86; (gavage); Inadequate Studies: rat NCITR* NCI-CG-TR-13,77. Reported in EPA TSCA Inventory. EPA Genetic Toxicology Program. Community Right-To-Know List.

OSHA PEL: (Transitional: TWA 100 ppm; CL 200 ppm; Pk 600 ppm/5M)TWA 25 ppm
ACGIH TLV: TWA 50 ppm; STEL 200 ppm; BEI: 7 mg/L trichloroacetic acid in urine at end of workweek.
DFG MAK: 50 ppm (345 mg/m³); BAT: blood 100 μg/dl
NIOSH REL: (Tetrachloroethylene) Minimize workplace exposure.
DOT Classification: Poison B; Label: St. Andrews Cross; ORM-A; Label: None.

SAFETY PROFILE: Confirmed carcinogen with experimental carcinogenic, neoplastigenic, and teratogenic data. Experimental poison by intravenous route. Moderately toxic to humans by inhalation with the following effects: local anesthetic, conjunctiva irritation, general anesthesia, hallucinations, distorted perceptions, coma, and pulmonary changes. Moderately experimentally toxic by ingestion, inhalation, intraperitoneal, and subcutaneous routes. An experimental teratogen. Experimental reproductive effects. Human mutation data reported. An eye and severe skin irritant. The liquid can cause injuries to the eyes; however, with proper precautions it can be handled safely. The symptoms of acute intoxication from this material are the result of its effects upon the nervous system. Can cause dermatitis, particularly after repeated or prolonged contact with the skin. Irritates the gastrointestinal tract upon ingestion. It may be handled in the presence or absence of air, water, and light with any of the common construction materials at temperatures up to 140°. This material is extremely stable and resists hydrolysis. A common air contaminant. Reacts violently under the proper conditions with Ba; Be; Li; N_2O_4; metals; NaOH. When heated to decomposition it emits highly toxic fumes of Cl^-. See also CHLORINATED HYDROCARBONS, ALIPHATIC.

PCF300 CAS:594-42-3 ***HR: 3***
PERCHLOROMETHYL MERCAPTAN
DOT: UN 1670
mf: CCl_4S mw: 185.87

PROP: Yellow, oily liquid. Bp: slt decomp @ 149°, d: 1.700 @ 20°, vap d: 6.414.

SYNS: CLAIRSIT ◇ MERCAPTAN METHYLIQUE PERCHLORE (FRENCH) ◇ PCM ◇ PERCHLORMETHYLMERKAPTAN (CZECH) ◇ RCRA WASTE NUMBER P118 ◇ TRICHLOROMETHANE SULFENYL CHLORIDE ◇ TRICHLOROMETHYLSULFENYL CHLORIDE ◇ TRICHLOROMETHYLSULPHENYL CHLORIDE

TOXICITY DATA with REFERENCE
skn-rbt 500 mg/24H SEV 28ZPAK -,13,72
eye-rbt 50 μg/24H SEV 28ZPAK -,13,72
orl-rat LD50:826OO μg/kg 28ZPAK -,13,72
ihl-rat LCLo:260 mg/m³/4H 85GMAT -,97,82
ihl-mus LC50:296 mg/m³/2H 85GMAT -,97,82
ivn-mus LD50:56 mg/kg CSLNX* NX#06768

CONSENSUS REPORTS: EPA Extremely Hazardous Substances List. Reported in EPA TSCA Inventory.

OSHA PEL: TWA 0.1 ppm
ACGIH TLV: TWA 0.1 ppm
DOT Classification: Poison B; Label: Poison.

SAFETY PROFILE: Poison by ingestion, inhalation, and intravenous routes. A severe skin, eye, and mucous membrane irritant. When heated to decomposition it emits very toxic fumes of Cl^- and SO_x. See also MERCAPTANS.

PCF500 CAS:5390-07-8 ***HR: 2***
PERCHLORYLBENZENE
mf: $C_6H_5ClO_3$ mw: 160.56

SAFETY PROFILE: A powerful oxidant. Probably very toxic. Mixtures with aluminum trichloride may explode spontaneously. When heated to decomposition it emits toxic fumes of Cl^-.

PCF750 CAS:7616-94-6 ***HR: 2***
PERCHLORYL FLUORIDE
mf: $ClFO_3$ mw: 102.45

PROP: Colorless, noncorrosive gas; characteristic sweet odor. Mp: −146°, bp: −46.8°, d: (liquid): 1.434. d: (gas): 0.637.

SYNS: CHLORINE FLUORIDE OXIDE ◇ CHLORINE OXYFLUORIDE

TOXICITY DATA with REFERENCE
ihl-rat LCLo:2000 ppm/40M TXAPA9 27,527,74
ihl-mus LC50:630 ppm/4H PENNS*
ihl-dog LCLo:451 ppm/4H XAWPA2 CWL 2-10,58

CONSENSUS REPORTS: Reported in EPA TSCA Inventory.

OSHA PEL: (Transitional: TWA 3 ppm) TWA 3 ppm; STEL 6 ppm
ACGIH TLV: TWA 3 ppm; STEL 6 ppm

SAFETY PROFILE: Moderately toxic by inhalation route. Forms methemoglobin in the body and destroys red cells causing anemia, anorexia and cyanosis. Recovery is said to be rapid, leaving no permanent physiological damage. Can be absorbed through the skin. Its odor can be detected as low as 10 ppm although this cannot be relied upon as an indication of toxic concentration in air. While nonflammable, it supports combustion. It is a powerful oxidizer. Moderately explosive. Potentially explosive reactions with combustible gases or vapors; benzene + aluminum trichloride; benzocyclobutene + butyllithium + potassium tert-butoxide; calcium acetylide; potassium cyanide; potassium thiocyanate; sodium iodide; charcoal; ethyl-4-fluorobenzoylacetate; hydrocarbons; hydrogen sulfide; nitrogen oxide; sulfur dichloride; vinylidene chloride; 3α-hydroxy-5β-androstane-11,17-dione-17-hydrazone; lithiated compounds; 2-lithio (dimethylaminomethyl)ferroxene; methyl-2-bromo-5,5-ethylene dioxy(2.2.1)bicycloheptane-7-carboxylate; aliphatic heterocyclic amines; sodium methoxide + methanol; vinylidene chloride. Reacts to form explosive products with nitrogenous bases (e.g., isopropylamine, isobutylamine; aniline; phenyl hydrazine; 1,2-diphenyl hydrazine); sawdust; lampblack. Violent reaction with finely divided organic materials. A fluorinating agent in chemical synthesis, and as an oxidant in rocket fuel. When heated to decomposition it emits toxic fumes of F^- and Cl^-. See also FLUORINE and PERCHLORATES.

PCF775 CAS:10294-48-1 ***HR: 3***
PERCHLORYL PERCHLORATE
mf: Cl_2O_7 mw: 182.90

SAFETY PROFILE: An explosive sensitive to impact or rapid heating. Explodes on contact with iodine. Mixtures with bromine pentafluoride are shock-sensitive explosives. When heated to decomposition it emits toxic fumes of Cl^-. See also PERCHLORATES.

PCG000 CAS:768-34-3 ***HR: 3***
1-PERCHLORYLPIPERIDINE
mf: $C_5H_{10}ClNO_3$ mw: 167.60

$O_3CN(CH_2)_4CH_2$

SAFETY PROFILE: May explode in storage or if heated. Explodes on contact with piperidine. Powerful oxidizer. When heated to decomposition it emits very toxic fumes of NO_x and Cl^-.

PCG250 ***HR: 3***
PERCHROMATES

CONSENSUS REPORTS: Chromium and its compounds are on the Community Right-To-Know List.

SAFETY PROFILE: Poisons. Violent reactions with aniline, olefins, pyridine, quinoline. Powerful oxidants. See also CHROMIUM COMPOUNDS.

PCG500 CAS:76-42-6 ***HR: 3***
PERCODAN
mf: $C_{18}H_{21}NO_4$ mw: 315.40

SYNS: DIHYDROHYDROXYCODEINONE ◇ DIHYDRO-14-HYDROXYCODEINONE ◇ 14-HYDROXYDIHYDROCODEINONE ◇ OXYCODEINONE ◇ PERCOBARB

TOXICITY DATA with REFERENCE
ipr-mus LD50:320 mg/kg TXAPA9 3,261,61
scu-mus LD50:426 mg/kg 28ZNAE 138,27,38

SAFETY PROFILE: Poison by intraperitoneal route. Moderately toxic by subcutaneous route. When heated to decomposition it emits toxic fumes of NO_x.

PCG550 CAS:54527-84-3 ***HR: 3***
PERDIPINE
mf: $C_{26}H_{29}N_3O_6{\cdot}ClH$ mw: 516.04

SYNS: 1,4-DIHYDRO-2,6-DIMETHYL-4-(m-NITROPHENYL)-3,5-PYRIDINEDICARBOXYLIC ACID 2-(BENZYLMETHYL AMINO)ETHYL METHYL ESTER, MONOHYDROCHLORIDE ◇ NICARDIPINE HYDROCHLORIDE ◇ NICODEL ◇ RS-79216 ◇ YC 93

TOXICITY DATA with REFERENCE
orl-rat TDLo:1350 mg/kg (female 17-22D post):REP KSRNAM 13,1160,79
orl-rbt TDLo:1650 mg/kg (female 7-17D post):TER KSRNAM 13,1160,79
orl-man TDLo:26 mg/kg/30D-I:KID JAMAAP 258,3388,87
orl-wmn TDLo:1800 µg/kg/36H-I:KID JAMAAP 258,3388,87
orl-rat LD50:187 mg/kg ARZNAD 26,2172,76
ipr-rat LD50:155 mg/kg NIIRDN 6,543,82
scu-rat LD50:683 mg/kg NIIRDN 6,543,82
ivn-rat LD50:15500 µg/kg ARZNAD 26,2172,76
orl-mus LD50:322 mg/kg CYLPDN 4,97,83
ipr-mus LD50:123 mg/kg CYLPDN 4,97,83
scu-mus LD50:540 mg/kg NIIRDN 6,543,82
ivn-mus LD50:19900 µg/kg OYYAA2 18,301,79

SAFETY PROFILE: Poison by ingestion, intravenous, and intraperitoneal routes. Moderately toxic by subcutaneous route. Human systemic effects: decreased urine volume or anuria. Experimental reproductive effects. When heated to decomposition it emits toxic fumes of NO_x and HCl.

PCG600 CAS:376-53-4 ***HR: 3***
PERFLUOROADIPONITRILE
mf: $C_6F_8N_2$ mw: 252.08

SYNS: OCTAFLUOROADIPONITRILE ◇ PERFLUOROADIPIC ACID DINITRILE ◇ PERFLUOROADIPINIC ACID DINITRILE

TOXICITY DATA with REFERENCE
orl-rat LD50:2917 mg/kg 85GMAT -,97,82
ihl-rat LC50:62 mg/m³/4H 85GMAT -,97,82
orl-mus LD50:1955 mg/kg 85GMAT -,97,82
ihl-mus LC50:140 mg/m³/4H 85GMAT -,97,82

CONSENSUS REPORTS: Cyanide and its compounds are on the Community Right-To-Know List.

SAFETY PROFILE: Poison by inhalation. Moderately toxic by ingestion. When heated to decomposition it emits toxic fumes of F^-, CN^-, and NO_x. See also ADIPONITRILE and NITRILES.

PCG650 CAS:66793-67-7 ***HR: 3***
PERFLUORO-tert-BUTYL PEROXYHYPOFLUORITE
mf: $C_4F_{10}O_2$ mw: 270.03

$(F_3C)_3COOF$

SAFETY PROFILE: The liquid explodes at 22°C. The gas phase is stable. When heated to decomposition it emits toxic fumes of F^-. See also PEROXIDES.

PCG725 CAS:335-76-2 **HR: 3**
PERFLUORODECANOIC ACID
mf: $C_{10}HF_{19}O_2$ mw: 514.11

SYNS: NDFDA ◇ NONADECAFLUORODECANOIC ACID ◇ NONADECAFLUORO-n-DECANOIC ACID ◇ PERFLUORO-N-DECANOIC ACID ◇ PFDA

TOXICITY DATA with REFERENCE
orl-rat TDLo:30 mg/kg (12D preg):TER NTIS** AD-A095-370
orl-rat TDLo:30 mg/kg (12D preg):REP NTIS** AD-A095-370
orl-rat LD50:57 mg/kg TXAPA9 85,169,86
ipr-rat LD50:40 mg/kg TOXID9 1,16,81
ipr-mus LD50:150 mg/kg TOXID9 1,16,81

CONSENSUS REPORTS: Reported in EPA TSCA Inventory.

SAFETY PROFILE: Poison by ingestion and intraperitoneal routes. An experimental teratogen. Experimental reproductive effects. When heated to decomposition it emits toxic fumes of F^-.

PCG755 CAS:308-48-5 **HR: 2**
PERFLUORO-n-DIBUTYL ETHER
mf: $C_8F_{18}O$ mw: 454.08

SYNS: DIPERFLUOROBUTYL ETHER ◇ PERFLUORODIBUTYLETHER

TOXICITY DATA with REFERENCE
ihl-rat LC50:80 g/m^3/4H 85GMAT -,97,82
ihl-mus LC50:49500 mg/m^3 TPKVAL 15,101,79
ipr-mus LDLo:512 mg/kg CBCCT* 3,361,51

SAFETY PROFILE: Moderately toxic by intraperitoneal route. Mildly toxic by inhalation. When heated to decomposition it emits toxic fumes of F^-. See also ETHERS and FLUORIDES.

PCG760 CAS:358-21-4 **HR: 1**
PERFLUORO ETHER
mf: $C_4F_{10}O$ mw: 254.04

SYN: BIS(PENTAFLUOROETHYL)ETHER

TOXICITY DATA with REFERENCE
orl-rat LD50:20 g/kg KBAMAJ 12(1),84,78
ihl-mus LC50:177 g/m^3/2H KBAMAJ 12(1),84,78
ivn-mus LD50:29 g/kg KBAMAJ 12(1),84,78

SAFETY PROFILE: Mildly toxic by ingestion, inhalation, and intravenous routes. When heated to decomposition it emits toxic fumes of F^-. See also ETHERS and FLUORIDES.

PCG775 CAS:14362-70-0 **HR: 3**
PERFLUOROFORMAMIDINE
mf: CF_4N_2 mw: 116.02

SAFETY PROFILE: Explodes when shocked or during phase changes. When heated to decomposition it emits toxic fumes of F^- and NO_x. See also FLUORIDES.

PCH000 CAS:335-57-9 **HR: 3**
PERFLUORO-n-HEPTANE
mf: C_7F_{16} mw: 388.07

SYNS: HEXADECAFLUOROHEPTANE ◇ PERFLUOROHEPTANE

TOXICITY DATA with REFERENCE
ipr-mus LDLo:100 mg/kg CBCCT* 4,108,52

CONSENSUS REPORTS: Reported in EPA TSCA Inventory.

SAFETY PROFILE: Poison by intraperitoneal route. When heated to decomposition it emits very toxic fumes such as F^-. See also FLUORIDES.

PCH100 CAS:355-43-1 **HR: 3**
PERFLUOROHEXYL IODIDE
mf: $C_6F_{13}I$ mw: 445.95

$F_3C(CF_2)_4CF_2I$

SAFETY PROFILE: Explodes in contact with sodium when heated to the boiling point of the iodide. When heated to decomposition it emits toxic fumes of F^- and I^-. See also IODIDES and FLUORIDES.

PCH275 CAS:356-69-4 **HR: 2**
PERFLUOROMETHOXYPROPIONIC ACID METHYL ESTER
mf: $C_5H_3F_7O_3$ mw: 244.08
orl-mus LD50:8000 mg/kg GTPZAB 18(3),48,74
ihl-mus LC50:22 g/m^3 GTPZAB 18(3),48,74
ipr-mus LDLo:1000 mg/kg GTPZAB 18(3),48,74

SAFETY PROFILE: Moderately toxic by intraperitoneal route. Mildly toxic by ingestion and inhalation. When heated to decomposition it emits toxic fumes of F^-. See also ESTERS.

PCH290 CAS:355-02-2 **HR: 1**
PERFLUOROMETHYLCYCLOHEXANE
mf: C_7F_{14} mw: 350.07

SYN: 1-TRIFLUOROMETHYL-1,2,2,3,3,4,4,5,5,6,6-UNDECAFLUORO CYCLOHEXANE

TOXICITY DATA with REFERENCE
ihl-rat LCLo:825 ppm/14H 11FYAN 3,90,63

CONSENSUS REPORTS: Reported in EPA TSCA Inventory.

SAFETY PROFILE: Mildly toxic by inhalation. When heated to decomposition it emits toxic fumes of F^-. See also FLUORIDES.

PCH300 CAS:354-93-8 **HR: 3**
PERFLUORO-tert-NITROSOBUTANE
mf: C_4F_9NO mw: 249.04

$(F_3C)_3CN:O$

SYN: TRIS(TRIFLUOROMETHYL)NITROSOMETHANE

SAFETY PROFILE: Potentially explosive reaction with nitrogen oxides. When heated to decomposition it emits toxic fumes of F^- and NO_x. See also NITROSO COMPOUNDS and FLUORIDES.

PCH350 CAS:356-69-4 **HR: 2**
PERFLUOROPROPIONIC ACID, METHYL ESTER
mf: $C_4H_3F_5O_2$ mw: 178.07

SYN: PENTAFLUORO-PROPIONIC ACID METHYL ESTER

TOXICITY DATA with REFERENCE
orl-mus LD50:7000 mg/kg GTPZAB 18(3),48,74
ihl-mus LC50:23 g/m^3 GTPZAB 18(3),48,74
ipr-mus LDLo:1000 mg/kg GTPZAB 18(3),48,74

SAFETY PROFILE: Moderately toxic by intraperitoneal route. Mildly toxic by ingestion and inhalation. When heated to decomposition it emits toxic fumes of F^-. See also ESTERS.

PCH500 CAS:434-64-0 **HR: 1**
PERFLUOROTOLUENE
mf: C_7F_8 mw: 236.07

SYN: OCTAFLUOROTOLUENE

TOXICITY DATA with REFERENCE
ihl-mus LCLo:5000 ppm/10M ANASAB 19,167,64

CONSENSUS REPORTS: Reported in EPA TSCA Inventory.

SAFETY PROFILE: Mildly toxic by inhalation. When heated to decomposition it emits toxic fumes of F^-. See also FLUORIDES.

PCH800 CAS:6724-53-4 **HR: 3**
PERHEXILINE MALEATE
mf: $C_{19}H_{35}N \cdot C_4H_4O_4$ mw: 393.63

SYNS: 2-(2,2-DICYCLOHEXYLETHYL)PIPERIDINEMALEATE ◇ PEXID

TOXICITY DATA with REFERENCE
mma-ham:lng 10 μmol/L MUREAV 157,1,85
orl-man TDLo:857 mg/kg/40W:PNS,CNS, BMJOAE 1,1256,76
orl-man LDLo:4693 mg/kg/3Y:PNS,LIV NZMJAX 96,202,83
orl-rat LD50:2150 mg/kg OYYAA2 14,337,77
ipr-rat LD50:106 mg/kg OYYAA2 21,83,81
orl-mus LD50:2641 mg/kg OYYAA2 14,337,77
ipr-mus LD50:107 mg/kg OYYAA2 21,83,81

SAFETY PROFILE: Poison by intraperitoneal route. Moderately toxic by ingestion. Human systemic effects by ingestion: muscle weakness, paresthesia, and hepatitis. Mutation data reported. When heated to decomposition it emits toxic fumes of NO_x.

PCI250 CAS:969-33-5 **HR: 3**
PERIACTIN HYDROCHLORIDE
mf: $C_{21}H_{21}N \cdot ClH$ mw: 323.89

SYNS: CYPROHEPTADIENE HYDROCHLORIDE ◇ CYPROHEPTADINE HYDROCHLORIDE ◇ 4-(5-DIBENZO(a,e)CYCLOHEPTATRIENYLIDENE)PIPERIDINE HYDROCHLORIDE ◇ 1-METHYL-4-(5-DIBENZO(a,e)CYCLOHEPTATRIENYLIDENE)PIPERIDINEHYDROCHLORIDE

TOXICITY DATA with REFERENCE
cyt-hmn:lyms 2200 umol/L/24H-C ZEKIA5 118,219,74
orl-rat TDLo:60 mg/kg (5-16D preg):TER RCOCB8 7,701,74
orl-rat TDLo:250 mg/kg (6-15D preg):REP TCMUD8 3,439,83
orl-wmn TDLo:7 mg/kg/4W-C:GIT,LIV BMJOAE 1,753,78
orl-rat LD50:295 mg/kg DRUGAY 6,340,82
ipr-rat LD50:52400 μg/kg YAKUD5 22,375,80
orl-mus LD50:69 mg/kg AGACBH 4,264,74
ipr-mus LD50:55300 μg/kg YAKUD5 22,375,80

SAFETY PROFILE: Poison by ingestion and intraperitoneal routes. Human systemic effects by ingestion: jaundice, liver function tests impaired, gastrointestinal effects. An experimental teratogen. Experimental reproductive effects. Human mutation data reported. When heated to decomposition it emits very toxic fumes of NO_x and HCl.

PCI550 CAS:536-59-4 **HR: 2**
PERILLA ALCOHOL
mf: $C_{10}H_{16}O$ mw: 152.26

PROP: Bp: 119-121°/11 mm Hg, d: 0.960, flash p: >230° F.

SYNS: CYCLOHEX-1-ENE-1-METHANOL,4-(1-METHYLETHENYL)- ◇ DIHYDROCUMINYL ALCOHOL ◇ 4-ISOPROPENYL-CYCLOHEX-1-ENE-1-METHANOL ◇ p-MENTHA-1,8-DIEN-7-OL ◇ PERILLOL ◇ PERILLYL ALCOHOL

TOXICITY DATA with REFERENCE
skn-rbt 500 mg/24H SEV FCTXAV 19,253,81
orl-rat LD50:2100 mg/kg FCTXAV 19,253,81

CONSENSUS REPORTS: Reported in EPA TSCA Inventory.

SAFETY PROFILE: Moderately toxic by ingestion. A

severe skin irritant. Combustible liquid. When heated to decomposition it emits acrid smoke and irritating fumes.

PCI500 CAS:129-03-3 ***HR: 3***
PERIACTINOL
mf: $C_{21}H_{21}N$ mw: 287.43

SYNS: CYPROHEPTADINE ◇ 4-(5H-DIBENZO(a,d)CYCLOHEPTEN-5-YLIDENE)-1-METHYLPIPERIDINE ◇ 4-(5-DIBENZO(a,d)CYCLOHEPTEN-5-YLIDINE)-1-METHYLPIPERIDINE ◇ DRONACTIN ◇ MK 141 ◇ PERIACTIN

TOXICITY DATA with REFERENCE
ipr-rat TDLo:20 mg/kg (10D preg):REP AIPTAK 257,168,82
ipr-rat TDLo:10 mg/kg (10D preg):TER AIPTAK 257,168,82
orl-rat LD50:295 mg/kg 27ZQAG -,69,72
ipr-rat LD50:52 mg/kg 27ZQAG -,69,72
orl-mus LD50:106 mg/kg ARZNAD 25,1723,75
scu-mus LD50:107 mg/kg 27ZQAG -,69,72
orl-dog LDLo:50 mg/kg 27ZIAQ -,82,73
ivn-rbt LDLo:3500 μg/kg 27ZQAG -,69,72

CONSENSUS REPORTS: EPA Genetic Toxicology Program.

SAFETY PROFILE: Poison by ingestion, intraperitoneal, subcutaneous and intravenous routes. Experimental teratogenic and reproductive effects. When heated to decomposition it emits toxic fumes of NO_x.

PCI600 ***HR: 1***
PERILLA FRUTESCENS (Linn.) Britt., extract

PROP: Indian plant belonging to the family *Labiatae* IJEBA6 15,208,77

SYN: PERILLA OCIMOIDES Linn., extract

TOXICITY DATA with REFERENCE
orl-rat TDLo:150 mg/kg (female 12-14D post):REP IJEBA6 15,208,77
ipr-mus LD50:1 g/kg IJEBA6 15,208,77

SAFETY PROFILE: Slightly toxic by intraperitoneal route. Experimental reproductive effects. When heated to decomposition it emits acrid smoke and irritating fumes.

PCI750 CAS:553-84-4 ***HR: 3***
PERILLA KETONE
mf: $C_{10}H_{14}O_2$ mw: 166.24

PROP: A potent lung toxin from the mint plant, *Perilla frutescens* (SCIEAS 197,573,77).

SYNS: 1-(3-FURANYL)-4-METHYL-1-PENTANONE ◇ β-FURYL ISOAMYL KETONE ◇ 1-(3-FURYL)-4-METHYL-1-PENTANONE ◇ PURPLE MINT PLANT EXTRACT

TOXICITY DATA with REFERENCE
ipr-rat LD50:10 mg/kg SCIEAS 197,573,77
ipr-mus LD50:2500 μg/kg SCIEAS 197,573,77
ivn-dog LD50:106 mg/kg JANSAG 60,248,85
ivn-rbt LD50:14 mg/kg JANSAG 60,248,85
ipr-ham LD50:13700 μg/kg JANSAG 60,248,85
ivn-ctl LDLo:30 mg/kg TXAPA9 45,300,78
ivn-dom LDLo:10 mg/kg TXAPA9 45,300,78

SAFETY PROFILE: Poison by intravenous and intraperitoneal routes. A potent pulmonary edemagenic agent (experimental). May also be hazardous to humans. When heated to decomposition it emits acrid smoke and irritating fumes.

PCJ000 CAS:30950-27-7 ***HR: 2***
1-PERILLALDEHYDE-α-ANTIOXIME
mf: $C_{10}H_{15}NO$ mw: 165.26

SYNS: 4-ISOPROPENYL-1-CYCLOHEXENE-1-CARBOXALDEHYDE, ANTI-OXIME ◇ PERILLARTINE ◇ PERILLA SUGAR

TOXICITY DATA with REFERENCE
orl-rat LD50:2500 mg/kg AFDOAQ 15,82,51

CONSENSUS REPORTS: Reported in EPA TSCA Inventory.

SAFETY PROFILE: Moderately toxic by ingestion. When heated to decomposition it emits toxic fumes of NO_x.

PCJ250 CAS:13444-71-8 ***HR: 3***
O-PERIODIC ACID
mf: HIO_4 mw: 191.91

SAFETY PROFILE: Powerful oxidizer. Potentially explosive reaction with dimethyl sulfoxide. Reaction with tetraethylammonium hydroxide forms an explosive product. Incompatible with azo-pigments. Used as a sweetening agent in Japan. When heated to decomposition it emits toxic fumes of I^-. See also IODIDES.

PCJ325 CAS:2139-25-5 ***HR: 3***
PERISOXAL CITRATE
mf: $C_{16}H_{20}N_2O_2 \cdot 1/2C_6H_8O_7$ mw: 368.41

SYNS: 3-(1-HYDROXY-2-PIPERIDINOETHYL)-5-PHENYLISOXAZOLE ◇ 3-(1-HYDROXY-2-PIPERIDINOETHYL)-5-PHENYLISOXAZOLE CITRATE (2:1) ◇ ISOXAL ◇ α-(5-PHENYL-3-ISOXAZOLYL)-1-PIPERIDINEETHANOL CITRATE (2:1) ◇ 31252-S

TOXICITY DATA with REFERENCE:1479 mg/kg IYKEDH 10,232,79
scu-rat LD50:878 mg/kg OYYAA2 6,1285,72
ivn-rat LD50:31300 μg/kg IYKEDH 10,232,79
orl-mus LD50:565 mg/kg YACHDS 2,1028,74
scu-mus LD50:373 mg/kg IYKEDH 10,232,79
ivn-mus LD50:35900 μg/kg IYKEDH 10,232,79

SAFETY PROFILE: Poison by subcutaneous and intravenous routes. Moderately toxic by ingestion. When heated to decomposition it emits toxic fumes of NO_x.

PCJ350 CAS:19395-78-9 ***HR: 3***
PERISTIL
mf: $C_{20}H_{21}N_3O_3 \cdot ClH$ mw: 387.90

SYNS: 2,3-DIHYDRO-1-(MORPHOLINOACETYL)-3-PHENYL-4(1H)-QUINAZOLINONE HYDROCHLORIDE ◇ MOQUIZONE HYDROCHLORIDE ◇ 1-MORPHOLINOACETYL-3-PHENYL-2,3-DIHYDRO-4(1H)-QUINAZOLINONE HYDROCHLORIDE

TOXICITY DATA with REFERENCE
orl-rat TDLo:960 mg/kg (female 2-17D post):REP ARZNAD 20,1559,70
orl-rat LD50:2135 mg/kg ARZNAD 22,1894,72
ipr-rat LD50:450 mg/kg ARZNAD 22,1894,72
ivn-rat LD50:146 mg/kg ARZNAD 22,1894,72
orl-mus LD50:1155 mg/kg ARZNAD 22,1894,72
ipr-mus LD50:559 mg/kg ARZNAD 22,1894,72
scu-mus LD50:774 mg/kg ARZNAD 22,1894,72
ivn-mus LD50:237 mg/kg ARZNAD 22,1894,72
ivn-dog LD50:200 mg/kg ARZNAD 22,1894,72

SAFETY PROFILE: Poison by intravenous route. Moderately toxic by ingestion, subcutaneous, and intraperitoneal routes. Experimental reproductive effects. When heated to decomposition it emits toxic fumes of NO_x and HCl.

PCJ400 CAS:93763-70-3 ***HR: 1***
PERLITE

PROP: Average density of 0.13. Expands when finely ground and heated. Natural glass, amorphous mineral consisting of fused sodium potassium aluminum silicate, containing <1% quartz.

TOXICITY DATA with REFERENCE
orl-mus LD50:12960 mg/kg JTSCDR 10,83,85

OSHA PEL: TWA Total Dust: 15 mg/m^3; Respirable Fraction: 5 mg/m^3
ACGIH TLV: TWA (nuisance particulate) 10 mg/m^3 of total dust (when toxic impurities are not present, e.g., quartz < 1%).

SAFETY PROFILE: Slightly toxic by ingestion. A nuisance dust.

PCJ500 ***HR: 3***
PERMANGANATES

PROP: Compounds containing an MnO_4^- radical.

CONSENSUS REPORTS: Manganese and its compounds are on The Community Right-To-Know List.

SAFETY PROFILE: Poisons. Many are strong oxidizing agents, hence irritating. Flammable by chemical reaction with reducing agents. Moderately explosive when shocked or exposed to heat. Silver permanganate and other metallic permanganates may detonate when exposed to high temperatures or when they are involved in fires or severely shocked. Store in a cool, ventilated area, away from acute fire hazards and easily oxidized materials. They may be disposed of by dissolving in water since practically all permanganates are soluble in water. They can react vigorously on contact with reducing materials. Incompatible with acetic acid; acetic anhydride; H_2SO_4 + C_6H_6. See also MANGANESE COMPOUNDS.

PCJ750 CAS:13446-10-1 ***HR: 2***
PERMANGANIC ACID AMMONIUM SALT
DOT: NA 9190
mf: $MnO_4 \cdot H_4N$ mw: 136.99

PROP: Crystalline solid. Mp: explodes; d: 2.208 @ 10°.

SYN: AMMONIUM PERMANGANATE

CONSENSUS REPORTS: Manganese and its compounds are on the Community Right-To-Know List.

OSHA PEL: CL 5 mg(Mn)/m^3
ACGIH TLV: TWA 5 mg(Mn)/m^3
DOT Classification: Label: Oxidizer.

SAFETY PROFILE: Probably an irritant. A powerful oxidizer. Moderately flammable by chemical reaction with reducing agents. Explosive when shocked or warmed to 60°. Can be exploded by percussion. When heated to decomposition it emits toxic fumes of NO_x and NH_3. Incompatible with reducing material, friction. When heated to decomposition it emits toxic fumes of NO_x and NH_3. See also PERMANGANATES and MANGANESE COMPOUNDS.

PCK000 CAS:7787-36-2 ***HR: 2***
PERMANGANIC ACID, BARIUM SALT
DOT: UN 1448
mf: $Mn_2O_8 \cdot Ba$ mw: 375.22

SYN: BARIUM PERMANGANATE

CONSENSUS REPORTS: Barium and its compounds and manganese and its compounds are on the Community Right-To-Know List.

OSHA PEL: 8H TWA 0.5 mg(Ba)/m^3; CL 5 mg(Mn) /m^3
ACGIH TLV: TWA 0.5 mg(Ba)/m^3
DOT Classification: Oxidizer; Label: Oxidizer

SAFETY PROFILE: Probably an irritant. A powerful oxidizer. When heated to decomposition it emits acrid smoke and irritating fumes. See also BARIUM COMPOUNDS.

PCK500 CAS:84-97-9 ***HR: 3***
PERNAZINE
mf: $C_{20}H_{25}N_3S$ mw: 339.54

SYNS: N-METHYL-PIPERAZINYL-N'-PROPYL-PHENOTHIAZIN (GERMAN) ◇ 10-(3-(4-METHYL-1-PIPERAZINYL)PROPYL)-10H-PHENOTHIAZINE (9CI) ◇ N-(3-(4-METHYL-1-PIPERAZINYL)PROPYL)PHENOTHIAZINE ◇ PERAZINE ◇ PSYTOMIN ◇ TAXILAN

TOXICITY DATA with REFERENCE
unr-hmn TDLo:270 mg/kg/28D ARZNAD 32,911,82
orl-rat LD50:500 mg/kg AIPTAK 115,1,58
ipr-rat LD50:300 mg/kg ARZNAD 8,199,58
ivn-rat LD50:80 mg/kg ARZNAD 8,199,58
orl-mus LD50:640 mg/kg ARZNAD 8,199,58
ipr-mus LD50:185 mg/kg ARZNAD 8,199,58
scu-mus LD50:560 mg/kg ARZNAD 8,199,58
ivn-mus LD50:75 mg/kg ARZNAD 8,199,58
ivn-rbt LD50:25 mg/kg ARZNAD 8,199,58

SAFETY PROFILE: Poison by intravenous and intraperitoneal routes. Moderately toxic by ingestion and subcutaneous routes. Human toxic effects by unspecified route. When heated to decomposition it emits very toxic fumes of NO_x and SO_x.

PCK600 ***HR: 3***
PERNETTYA (various species)

PROP: Small evergreen shrubs with thick leaves and white flowers. The fruit is white or some bright color, and stays on the plant all winter. They are native to Central and South America, New Zealand, and Tasmania. They are cultivated in the United States.

SAFETY PROFILE: The leaves and nectar contain poisonous grayanotoxins (andromedotoxins). Ingestion of leaves or honey made from the nectar causes immediate pain in the mouth and may be followed several hours later by vomiting, diarrhea, headache, muscle weakness, impaired vision, slowed heartbeat, severe low blood pressure, convulsions, coma, and death.

PCK639 CAS:28416-66-2 ***HR: 3***
PEROCAN CITRATE
mf: $C_{16}H_{24}N_2 \cdot C_6H_8O_7$ mw: 436.56

SYNS: ISOAMINILE CITRATE ◇ α-(ISOPROPYL)-α-(β-DIMETHYLAMINOPROPYL)PHENYLACETONITRIL CITRATE ◇ PERACON

TOXICITY DATA with REFERENCE
orl-rat LD50:250 mg/kg KSRNAM 5,2212,71
scu-rat LD50:220 mg/kg KSRNAM 5,2212,71
ivn-rat LD50:36 mg/kg JDNRAK 2,43,62
orl-mus LD50:274 mg/kg KSRNAM 5,2212,71
scu-mus LD50:157 mg/kg NIIRDN 6,71,82
ivn-mus LD50:52 mg/kg KSRNAM 5,2212,71
ivn-dog LD50:36 mg/kg JNDRAK 2,43,62
ivn-rbt LD50:36 mg/kg JNDRAK 2,43,62

SAFETY PROFILE: Poison by ingestion, subcutaneous, and intravenous routes. When heated to decomposition it emits toxic fumes of NO_x.

PCK669 ***HR: 3***
PEROCAN CYCLAMATE
mf: $C_{16}H_{25}N_2 \cdot C_6H_{13}NO_3S$ mw: 424.69

SYNS: ISOAMINILE CYCLAMATE ◇ α-(ISOPROPYL)-α-(β-DIMETHYLAMINOPROPYL)PHENYLACETONITRILCYCLAMATE

TOXICITY DATA with REFERENCE
orl-mus LD50:380 mg/kg KSRNAM 5,1787,71
scu-mus LD50:175 mg/kg KSRNAM 5,1787,71
ivn-mus LD50:86 mg/kg KSRNAM 5,1787,71

SAFETY PROFILE: Poison by ingestion, subcutaneous, and intravenous routes. When heated to decomposition it emits toxic fumes of SO_x and NO_x.

PCL000 ***HR: 3***
PEROXIDES, INORGANIC

SAFETY PROFILE: Variable toxicity. They may cause injury on contact with skin or mucous membranes. Moderate to dangerous fire hazard by chemical reaction with reducing agents and contaminants; strong oxidizing agents; contact with moisture may produce much heat. Moderate explosion hazard; heat, shock, or catalysts can cause violent decomposition. Contact with reducing agents may give rise to explosively violent reactions. See also HYDROGEN PEROXIDE and SODIUM PEROXIDE.

PCL250 ***HR: 3***
PEROXIDES, ORGANIC

PROP: Organic compounds containing the –OO– group.

SAFETY PROFILE: Often highly toxic and irritating to the skin, eyes, and mucous membranes. Dangerous fire hazard by chemical reaction with reducing agents or exposure to heat. They readily release oxygen and thus are powerful oxidizers. Severe explosion hazard when shocked, exposed to heat, or by spontaneous chemical reaction. Many peroxides are very unstable. Upon contact with reducing materials, such as organic matter, thiocyanates, an explosion can occur.

Many solvents form dangerous levels of peroxides during storage: e.g.; dipropyl ether; divinylacetylene; vinylidene chloride; potassium amide; sodium amide. Other compounds form peroxides in storage but concentration is required to reach dangerous levels: e.g.; diethyl ether; ethyl vinyl ether; tetrahydrofuran; p-dioxane; 1,1-diethoxyethane; ethylene glycol dimethyl ether; propyne; butadiyne; dicyclopentadiene; cyclohexene; tetrahydronaphthalenes; deca-hydronaphthalenes. Some monomeric materials can form peroxides which catalyze hazardous polymerization reactions: e.g.; acrylic acid; acrylonitrile; butadiene; 2-chlorobutadiene; chlorotrifluoroethylene; methyl methacrylate; styrene; tetrafluoroethylene; vinyl acetate; vinylacetylene; vinyl chloride; vinylidine chloride; vinylpyridine. Compounds which contain one or two ether functions are especially susceptible to peroxide formation.

Peroxyacids (RCO•OOH) are some of the most powerful oxidants of the organic peroxides. Some of the simple peroxyacids are peroxy-formic acid; peroxy-acetic acid; peroxypivalic acid; peroxytrifluoroacetic acid. Traces of transition metals (e.g., cobalt; iron; manganese; nickel; vanadium) can catalyze explosive decomposition of these acids.

Handle all peroxides or peroxide-containing materials with great care. See also specific compounds.

PCL500 CAS:79-21-0 ***HR: 3***
PEROXYACETIC ACID
DOT: NA 2131
mf: $C_2H_4O_3$ mw: 76.06

PROP: Not over 40% peracetic acid and not over 6% hydrogen peroxide (FEREAC 41,15972,76). Colorless liquid; strong odor. Bp: 105°, explodes @ 110°, flash p: 105°F (OC), d: 1.15 @ 20°. Water-sol. Powerful oxidizer.

SYNS: ACETYL HYDROPEROXIDE ◇ ACIDE PERACETIQUE (FRENCH) ◇ ETHANEPEROXOIC ACID ◇ HYDROPEROXIDE, ACETYL ◇ PERACETIC ACID (MAK) ◇ PERACETIC ACID, solution (DOT) ◇ PEROXYACETIC ACID, maximum concentration 43% in acetic acid (DOT)

TOXICITY DATA with REFERENCE
skn-rbt 500 mg open SEV UCDS** 12/12/68
eye-rbt 1 mg SEV UCDS** 12/12/68
skn-mus TDLo:21 g/kg/26W-I:ETA JNCIAM 55,1359,75
orl-rat LD50:1540 mg/kg UCDS** 12/12/68
ihl-rat LC50:450 mg/m^3 GISAAA 48(6),28,83
orl-mus LD50:210 mg/kg GISAAA 48(6),28,83
skn-rbt LD50:1410 mg/kg UCDS** 12/12/68
orl-gpg LD50:10 mg/kg BSPII* 1/75-19B

CONSENSUS REPORTS: EPA Extremely Hazardous Substances List. Community Right-To-Know List. Reported in EPA TSCA Inventory.

DFG MAK: Very strong skin effects.
DOT Classification: Organic Peroxide; Label: Organic Peroxide, Corrosive.

SAFETY PROFILE: Poison by ingestion. Moderately toxic by inhalation and skin contact. A corrosive eye, skin, and mucous membrane irritant. Questionable carcinogen with experimental tumorigenic data by skin contact. Flammable when exposed to heat or flames. Severe explosion hazard when exposed to heat or by spontaneous chemical reaction. Explodes violently at 110°C. A powerful oxidizing agent. Explosive reaction with acetic anhydride; 5-p-chlorophenyl-2,2-dimethyl-3-hexanone. Violent reaction with ether solvents (e.g., tetrahydrofuran; diethyl ether); metal chloride solutions (e.g., calcium chloride; potassium chloride; sodium chloride); olefins; organic matter. Dangerous; keep away from combustible materials. When heated to decomposition it emits acrid smoke and irritating fumes. To fight fire, use water, foam, CO_2. Used as a polymerization initiator, curing agent, and cross-linking agent. See also PEROXIDES, ORGANIC.

PCL750 CAS:2278-22-0 ***HR: 3***
PEROXYACETYL NITRATE
mf: $C_2H_3NO_5$ mw: 121.06

SYNS: ACETYLNITROPEROXIDE ◇ NITRIC ACID, ANHYDRIDE with PEROXYACETIC ACID ◇ PAN

TOXICITY DATA with REFERENCE
eye-hmn 5 ppm/5M IAPWAR 4,79,61
mmo-sat 2200 nmol/plate MUREAV 157,123,85
mma-sat 2200 nmol/plate MUREAV 157,123,85
ihl-mus LC50:106 ppm/2H AEHLAU 15,739,67
ihl-rat LC50:95 ppm/4H TXCYAC 8,231,77

SAFETY PROFILE: Poison by inhalation. A human eye irritant. Mutation data reported. Extremely explosive. When heated to decomposition it emits toxic fumes of NO_x. See also NITRATES and PEROXIDES.

PCL775 CAS:66955-43-9 ***HR: 3***
PEROXYACETYL PERCHLORATE
mf: $C_2H_3ClO_6$ mw: 158.50

SAFETY PROFILE: An explosive sensitive to detonation or friction. When heated to decomposition it emits toxic fumes of Cl^-. See also PERCHLORATES and PEROXIDES.

PCM000 CAS:93-59-4 ***HR: 3***
PEROXYBENZOIC ACID
mf: $C_7H_6O_3$ mw: 138.13

PROP: Leaflets. Mp: 42°, bp: explodes @ 80-100°. Insol in water; sol in alc and ether.

SYNS: BENZENECARBOPEROXOIC ACID (9CI) ◇ BENZOYLHYDROGEN PEROXIDE ◇ BENZOYL HYDROPEROXIDE ◇ PERBENZOIC ACID

TOXICITY DATA with REFERENCE
skn-mus TDLo:1040 mg/kg/26W-I:ETA JNCIAM 55,1359,75

SAFETY PROFILE: Moderately irritating to skin, eyes, and mucous membranes by ingestion and inhalation. Questionable carcinogen with experimental tumorigenic data by skin contact. A dangerous fire hazard when exposed to heat, flame, or reducing materials. A powerful oxidizing agent. Severe explosion hazard when exposed to heat or flame. Violent reaction with olefins. Can react vigorously with reducing materials. Avoid evaporation. When heated to decomposition it emits acrid smoke and irritating fumes. See also PEROXIDES, ORGANIC.

PCM250 CAS:13709-32-5 ***HR: 3***
PEROXYDISULFURYL DIFLUORIDE
mf: $F_2O_6S_2$ mw: 198.13

FSO_2OOSO_2F

SAFETY PROFILE: A powerful irritant and corrosive to skin, eyes, and mucous membranes. A very powerful oxidizer. A dangerous fire hazard. Ignites organic materials immediately on contact. Explosive reaction with carbon monoxide; dichloromethane. Violent reaction with boron nitride. When heated to decomposition it emits very toxic fumes of F^- and SO_x. See also PEROXIDES and FLUORIDES.

PCM500 CAS:107-32-4 ***HR: 3***
PEROXYFORMIC ACID
mf: CH_2O_3 mw: 62.03

SYN: METHANEPEROXOIC ACID

SAFETY PROFILE: A powerful irritant and an oxidizer. A dangerous fire hazard when exposed to heat, flame, or reducing materials. Unstable and shock-sensitive. 80% solution is explosive. Extremely dangerous when moved. Violent reaction with carbon; red phosphorus; silicon; formaldehyde; benzaldehyde; aniline; alkenes. Self reactive. Incompatible with metals; nonmetals; organic materials. When heated to decomposition it emits acrid smoke and irritating fumes.

PCM550 CAS:5797-06-8 ***HR: 3***
PEROXYFUROIC ACID
mf: $C_5H_4O_4$ mw: 128.08

OCH=CHCH=CCO•OOH (ring: O bonded to C)

SAFETY PROFILE: Explodes violently at 40°C. Explodes or decomposes violently on contact with finely divided metal salts; ergosterol; pyrogallol; or animal charcoal. When heated to decomposition it emits acrid smoke and irritating fumes.

PCM750 CAS:5106-46-7 ***HR: 3***
PEROXYHEXANOIC ACID
mf: $C_6H_{12}O_3$ mw: 132.16

SAFETY PROFILE: A powerful oxidizer. Explodes and ignites on rapid heating. When heated to decomposition it emits acrid smoke and irritating fumes.

PCN000 CAS:19356-22-0 ***HR: 3***
PEROXYLINOLENIC ACID
mf: $C_{18}H_{30}O_3$ mw: 294.48

SYNS: LINOLENIC HYDROPEROXIDE ◇ LINOLENATE HYDROPEROXIDE ◇ 9,12,15-OCTADECATRIENEPEROXOIC ACID (Z,Z,Z) (9CI)

TOXICITY DATA with REFERENCE
mma-sat 100 μg/plate ABCHA6 44,1989,80
scu-rat TDLo:1 g/kg/45W:CAR FCTXAV 12,451,74
scu-rat TD:1765 mg/kg/95W:NEO FCTXAV 12,451,74

SAFETY PROFILE: Questionable carcinogen with experimental carcinogenic and neoplastigenic data. Mutation data reported. When heated to decomposition it emits acrid smoke and irritating fumes.

PCN250 CAS:1931-62-0 ***HR: 3***
PEROXYMALEIC ACID, O,O-(tert-BUTYL) ESTER
DOT: UN 2099/UN 2100/UN 2101
mf: $C_8H_{12}O_5$ mw: 188.20

$(CH_3)_3COOCO•CH=CHCO•OH$

SYN: O-O-tert-BUTYLHYDROGEN MONOPEROXY MALEATE

TOXICITY DATA with REFERENCE
ipr-mus LDLo:16 mg/kg CBCCT* 2,302,50

CONSENSUS REPORTS: Reported in EPA TSCA Inventory.

DOT Classification: Organic Peroxide; Label:Organic Peroxide

SAFETY PROFILE: Poison by intraperitoneal route. A slightly shock-sensitive explosive. When heated to decomposition it emits acrid smoke and irritating fumes. See also PEROXIDES and ESTERS.

PCN500 ***HR: 3***
PEROXYMONOPHOSPHORIC ACID
mf: H_3O_5P mw: 87.0

SYN: PEROXOMONOPHOSPHORIC ACID

SAFETY PROFILE: Very irritating and corrosive to tissues of skin, eyes, mucous membranes. Powerful oxidizer. A dangerous fire hazard when exposed to heat, flame, or reducing materials. Causes ignition on contact with organic materials. Hazardous reaction with potassium permanganate + coal. When heated to decomposition it emits toxic fumes of PO_x. See also PHOSPHATES.

PCN750 CAS:7722-86-3 ***HR: 3***
PEROXYMONOSULFURIC ACID
mf: H_2O_5S mw: 114.08

SYN: PEROXOMONOSULFURIC ACID

SAFETY PROFILE: Strong irritant. Powerful oxidizer. An explosive. Explosive reaction acetone; alcohols; aromatics (e.g., aniline; benzene; phenol); platinum; manganese dioxide; silver. Incompatible with acetone; catalysts; fibers. When heated to decomposition it emits toxic fumes of SO_x. See also PEROXIDES.

PCO000 CAS:26604-66-0 ***HR: 3***
PEROXYNITRIC ACID
mf: HNO_4 mw: 79.02

SAFETY PROFILE: A poison. Very irritating and corrosive to tissue. Decomposes explosively. A dangerous fire hazard when exposed to heat, flame, or reducing materials. Upon decomposition it emits toxic fumes of NO_x. See also PEROXIDES and NITRIC ACID.

PCO100 CAS:4212-43-5 ***HR: 3***
PEROXYPROPIONIC ACID
mf: $C_3H_6O_3$ mw: 90.08

SYN: PROPANE PEROXOIC ACID

SAFETY PROFILE: Burns very rapidly (deflagrates) when heated. Mixtures with organic solvents explode when heated. When heated to decomposition it emits acrid smoke and irritating fumes. See also PEROXIDES.

PCO150 CAS:5796-89-4 ***HR: 3***
PEROXYPROPIONYL NITRATE
mf: $C_3H_5NO_5$ mw: 135.08

$CH_3CH_2CO{\bullet}OONO_2$

SAFETY PROFILE: An extremely explosive gas. Handle only if highly diluted by air or nitrogen. When heated to decomposition it emits toxic fumes of NO_x. See also PEROXIDES and NITRATES.

PCO175 CAS:66955-44-0 ***HR: 3***
PEROXYPROPIONYL PERCHORATE
mf: $C_3H_5ClO_6$ mw: 172.52

$CH_3CH_2CO{\bullet}OOClO_3$

SAFETY PROFILE: An explosive sensitive to detonation or friction. When heated to decomposition it emits toxic fumes of Cl^-. See also PERCHLORATES and PEROXIDES.

PCO250 CAS:359-48-8 ***HR: 3***
PEROXYTRIFLUOROACETIC ACID
mf: $C_2HF_3O_3$ mw: 130.03

SAFETY PROFILE: A poison. A powerful oxidizer. Corrosive and irritating to skin, eyes, mucous membranes. Reaction with 4-iodo-3,5-dimethylisoxazole forms an explosive by-product. When heated to decomposition it emits toxic fumes of F^-. See also FLUORIDES and PEROXIDES.

PCO500 CAS:2015-28-3 ***HR: 3***
PERPHENAZINE DIHYDROCHLORIDE
mf: $C_{21}H_{26}ClN_3OS{\bullet}2ClH$ mw: 476.93

PROP: Crystals. Mp: 225-226°.

SYN: 4-(3-(2-CHLOROPHENOTHIAZIN-10-YL)PROPYL)-1-PIPERAZINEETHANOL DIHYDROCHLORIDE

TOXICITY DATA with REFERENCE
orl-rat LD50:318 mg/kg 27ZQAG -,36,72
ipr-rat LD50:124 mg/kg 27ZQAG -,36,72
ivn-rat LD50:34 mg/kg AIPTAK 118,358,59
orl-mus LD50:110 mg/kg AIPTAK 118,358,59
scu-mus LD50:960 mg/kg 27ZQAG -,36,72
ivn-mus LD50:19 mg/kg 27ZQAG -,36,72
ivn-dog LD50:51 mg/kg 27ZQAG -,36,72
ivn-cat LD50:35 mg/kg 27ZQAG -,36,72
ivn-pgn LD50:32 mg/kg 27ZQAG -,36,72
ims-pgn LD50:250 mg/kg 27ZQAG -,36,72

SAFETY PROFILE: Poison by ingestion, intraperitoneal, intravenous, and intramuscular routes. Moderately toxic by subcutaneous route. When heated to decomposition it emits very toxic fumes of Cl^-, NO_x, SO_x, and HCl.

PCO750 CAS:3111-71-5 ***HR: 3***
PERPHENAZINE HYDROCHLORIDE
mf: $C_{21}H_{26}ClN_3OS{\bullet}ClH$ mw: 440.47

TOXICITY DATA with REFERENCE
orl-rat TDLo:70 mg/kg (7-16D preg):TER TXAPA9 21,230,72
orl-rat TDLo:7 mg/kg (7-16D preg):REP TXAPA9 21,230,72
ivn-mus LD50:46 mg/kg CCCCAK 38,2137,73

SAFETY PROFILE: Poison by intravenous route. An experimental teratogen. Experimental reproductive effects. When heated to decomposition it emits very toxic fumes of Cl^-, NO_x, SO_x and HCl.

PCO850 CAS:5352-90-9 ***HR: 3***
PERPHENAZINE MALEATE
mf: $C_{21}H_{26}ClN_3OS{\bullet}C_4H_4O_4$ mw: 520.09

SYN: 4-(3-(2-CHLOROPHENOTHIAZIN-10-YL)PROPYL)-1-PIPERAZINEETHANOLMALEATE

TOXICITY DATA with REFERENCE
orl-rat LD50:317 mg/kg NIIRDN 6,768,82
ipr-rat LD50:133 mg/kg NIIRDN 6,768,82
orl-mus LD50:392 mg/kg NIIRDN 6,768,82
ipr-mus LD50:206 mg/kg NIIRDN 6,768,82

SAFETY PROFILE: Poison by ingestion and intraperitoneal routes. When heated to decomposition it emits toxic fumes of Cl^-, NO_x, and SO_x.

PCP250 CAS:58-32-2 ***HR: 3***
PERSANTIN
mf: $C_{24}H_{40}N_8O_4$ mw: 504.72

SYNS: ANGINAL ◇ 2,6-BIS(BIS(2-HYDROXYETHYL)AMINO)-4,8-DIPIPERIDINOPYRIMIDO(5,4-d)PYRIMIDINE ◇ 2,6-BIS(DIETHANOLAMINO)-4,8-DIPIPERIDINOPYRIMIDO(5,4-d)PYRIMIDINE ◇ CARDOXIN ◇ CLERIDIUM 150 ◇ CORONARINE ◇ CURANTYL ◇ 2,2',2'',2'''-(4,8-DIPIPERIDINOPYRIMIDO(5,4-d)PYRIMIDINE-2,6-DIYLDINITRILO)TETRAETHANOL ◇ DIPYRIDAMINE ◇ DIPYRIDAMOL ◇ DIPYRIDAMOLE ◇ DIPYRIDAN ◇ DIPYUDAMINE ◇ GULLIOSTIN ◇ NATYL ◇ PERIDAMOL ◇ PERSANTINE ◇ PIROAN ◇ PRANDIOL ◇ RA 8 ◇ USAF GE-12

TOXICITY DATA with REFERENCE
dni-rat:lvr 200 nmol/L CNREA8 43,1616,83
oms-rat:lvr 300 nmol/L CNREA8 43,1616,83
oms-mus:ast 60 μmol/L BCPCA6 22,2511,73
orl-rat LD50:8400 mg/kg MEIEDD 10,489,83
scu-rat LD50:1650 mg/kg ARZNAD 22,892,72
ivn-rat LD50:195 mg/kg ARZNAD 22,892,72
orl-mus LD50:2150 mg/kg ARZNAD 11,848,61
ipr-mus LD50:150 mg/kg NTIS** AD691-490
scu-mus LD50:2700 mg/kg NIIRDN 6,330,82
ivn-mus LD50:150 mg/kg ARZNAD 11,848,61

SAFETY PROFILE: Poison by intraperitoneal and intravenous routes. Moderately toxic by ingestion and subcutaneous routes. Mutation data reported. Used as a coronary vasodilator. When heated to decomposition it emits toxic fumes of NO_x.

PCP500 ***HR: 3***
PERSIMMON

PROP: Tannin containing fraction of unripe fruit of *Diospyros virginiana* (JNCIAM 57,207,76).

SYNS: DIOSPYROS VIRGINIANA ◇ TANNIN from PERSIMMON

TOXICITY DATA with REFERENCE
scu-rat TDLo:1800 mg/kg/57W-I:NEO JNCIAM 57,207,76

SAFETY PROFILE: Questionable carcinogen with experimental neoplastigenic data. When heated to decomposition it emits acrid smoke and irritating fumes. See also TANNIN.

PCP750 CAS:8007-00-9 ***HR: 1***
PERU BALSAM

PROP: Peru Balsam is the exudation from the trunk of the tree *Myroxylon pereirae* (FCTXAV 12,807,74).

TOXICITY DATA with REFERENCE
skn-rbt 500 mg/24H MOD FCTXAV 12,807,74

CONSENSUS REPORTS: Reported in EPA TSCA Inventory.

SAFETY PROFILE: A skin irritant. When heated to decomposition it emits acrid smoke and irritating fumes.

PCQ000 CAS:8007-00-9 ***HR: 2***
PERU BALSAM OIL

PROP: Constituents are benzyl benzoate and benzyl cinnamate, found in balsam of *Myroxylon pereirae* (FCTXAV 12,807,74).

TOXICITY DATA with REFERENCE
skn-rbt 500 mg/24H MLD FCTXAV 12,807,74
orl-rat LD50:2360 mg/kg FCTXAV 12,807,74

CONSENSUS REPORTS: Reported in EPA TSCA Inventory.

SAFETY PROFILE: Moderately toxic by ingestion. A skin irritant. When heated to decomposition it emits acrid smoke and irritating fumes.

PCQ250 CAS:198-55-0 ***HR: D***
PERYLENE
mf: $C_{20}H_{12}$ mw: 252.32

SYNS: DIBENZ(de,kl)ANTHRACENE ◇ PERI-DINAPHTHALENE ◇ PERILENE

TOXICITY DATA with REFERENCE
mmo-sat 25 nmol/plate TXCYAC 18,219,80
mma-sat 2 μg/plate CRNGDP 5,925,84

CONSENSUS REPORTS: IARC Cancer Review: Group 3 IMEMDT 7,56,87; Animal Inadequate Evidence IMEMDT 32,411,83. Reported in EPA TSCA Inventory. EPA Genetic Toxicology Program.

SAFETY PROFILE: Questionable carcinogen. Mutation data reported. When heated to decomposition it emits acrid smoke and irritating fumes.

PCQ750 CAS:60102-37-6 ***HR: 3***
PETASITENINE
mf: $C_{19}H_{27}NO_7$ mw: 381.47

PROP: Pyrrolizidine alkaloid isolated from *Petasites japonicus Maxim* (JNCIAM 58,1155,77).

SYN: FUKINOTOXIN

TOXICITY DATA with REFERENCE
mma-sat 1 mg/plate MUREAV 68,211,79
dns-rat:lvr 2 μmol/L CNREA8 45,3125,85
orl-rat TDLo:921 mg/kg/13W-C:CAR JNCIAM 58,1155,77

CONSENSUS REPORTS: IARC Cancer Review: Group 3 IMEMDT 7,56,87; Animal Limited Evidence IMEMDT 31,207,83.

SAFETY PROFILE: Questionable carcinogen with experimental carcinogenic data. Mutation data reported. Used as a food and herbal remedy. When heated to decomposition it emits toxic fumes of NO_x.

PCR000 CAS:91845-41-9 ***HR: 3***
PETASITES JAPONICUS MAXIM

PROP: Dried flower stalk of *Petasites Japonicus Maxim* (GANNA2 64,527,73).

SYNS: COLTS FOOT ◇ FUKI-NO-TOH (JAPANESE)

TOXICITY DATA with REFERENCE
orl-rat TDLo:1060 mg/kg/69W-C:CAR GANNA2 64,527,73
orl-mus TDLo:2300 g/kg/69W-C:CAR TOLED5 1,291,78
orl-rat TD:960 g/kg/68W-C:ETA GANNA2 68(6),841,77

SAFETY PROFILE: Suspected carcinogen with experimental carcinogenic and tumorigenic data. When heated to decomposition it emits acrid smoke and irritating fumes.

PCR250 CAS:8002-05-9 ***HR: 3***
PETROLEUM
DOT: UN 1267/UN 1270/NA 1270

PROP: A thick flammable, dark yellow to brown or green-black liquid. D: 0.780-0.970, flash p: 20-90°F. Insol in water; sol in benzene, chloroform, ether. Consists of a mixture of hydrocarbons from C_2H_6 and up, chiefly of the paraffins, cycloparaffins, or of cyclic aromatic hydrocarbons, with small amounts of benzene hydrocarbons, sulfur, and oxygenated compounds.

SYNS: BASE OIL ◇ COAL LIQUID ◇ COAL OIL ◇ CRUDE OIL ◇ PETROLEUM CRUDE ◇ ROCK OIL ◇ SENECA OIL

TOXICITY DATA with REFERENCE
mma-sat 1 mg/plate CRNGDP 3,21,78
skn-mus TDLo:3744 mg/kg/2Y-I:CAR NTIS** CONF-790334
skn-mus TD:40 g/kg/10W-I:ETA BECCAN 39,420,61
skn-mus TD:21216 mg/kg/2Y-I:CAR NTIS** CONF-801143
skn-mus TD:3744 mg/kg/2Y-I:NEO JOCMA7 21,614,79

CONSENSUS REPORTS: IARC Cancer Review: Group 3 IMEMDT 45,119,89; Animal Limited Evidence IMEMDT 45,119,89; Human Inadequate Evidence IMEMDT 45,119,89. Reported in EPA TSCA Inventory.

DOT Classification: Combustible Liquid; Label: None; Flammable Liquid; Label: Flammable Liquid; Flammable or Combustible Liquid; Label: Flammable Liquid.

SAFETY PROFILE: Questionable carcinogen with experimental carcinogenic, neoplastigenic, and tumorigenic data by skin contact. A dangerous fire hazard when exposed to heat, flame, or powerful oxidizers. To fight fire, use foam, CO_2, dry chemical. When heated to decomposition it emits acrid smoke and irritating fumes. See also MINERAL OILS.

PCR500 CAS:8052-42-4 ***HR: 3***
PETROLEUM ASPHALT

PROP: Steam refined asphalt (IMSUAI 34,255,65).

SYNS: ASPHALT, PETROLEUM ◇ PETROLEUM ROOFING TAR ◇ ROAD ASPHALT

TOXICITY DATA with REFERENCE
ims-rat TDLo:5400 mg/kg/24W-I:NEO ARPAAQ 70,372,60

CONSENSUS REPORTS: Reported in EPA TSCA Inventory.

SAFETY PROFILE: Questionable carcinogen with experimental neoplastigenic data on skin contact. When heated to decomposition it emits acrid smoke and irritating fumes. See also ASPHALT.

PCR750 CAS:64743-05-1 ***HR: 1***
PETROLEUM COKE (calcined)

SYN: PETROLEUM COKE CALCINED (DOT)

CONSENSUS REPORTS: Reported in EPA TSCA Inventory.

DOT Classification: ORM-C; Label: None.

SAFETY PROFILE: Flammable material. When heated to decomposition it emits acrid smoke and irritating fumes.

PCS000 CAS:64741-79-3 ***HR: 1***
PETROLEUM COKE (uncalcined)

SYN: PETROLEUM COKE UNCALCINED (DOT)

CONSENSUS REPORTS: Reported in EPA TSCA Inventory.

DOT Classification: ORM-C; Label: None.

SAFETY PROFILE: When heated to decomposition it emits acrid smoke and irritating fumes.

PCS250 CAS:8002-05-9 ***HR: 2***
PETROLEUM DISTILLATE
DOT: UN 1268
SYN: NAPHTHA

TOXICITY DATA with REFERENCE
par-man TDLo:57 mg/kg DICPBB 15,693,81

CONSENSUS REPORTS: Reported in EPA TSCA Inventory.

OSHA PEL: (Transitional: TWA 500 ppm) TWA 400 ppm
DOT Classification: Combustible Liquid; Label: None; Flammable Liquid; Label: Flammable Liquid.

SAFETY PROFILE: Human systemic effects by parenteral route: cough, dyspnea, nausea or vomiting. Flammable or combustible liquid when exposed to heat or flame. When heated to decomposition it emits acrid smoke and irritating fumes. Used as a vehicle for pesticides. See also PETROLEUM and ASPHALT.

PCS260 CAS:64742-44-5 ***HR: 2***
PETROLEUM DISTILLATES, CLAY-TREATED HEAVY NAPHTHENIC
TOXICITY DATA with REFERENCE
skn-mus TDLo:406 g/kg/22W-I:ETA BJCAAI 48,429,83

CONSENSUS REPORTS: Reported in EPA TSCA Inventory.

SAFETY PROFILE: Questionable carcinogen with experimental tumorigenic data. When heated to decomposition it emits acrid smoke and irritating fumes.

PCS270 CAS:64742-45-6 ***HR: 2***
PETROLEUM DISTILLATES, CLAY-TREATED LIGHT NAPHTHENIC

TOXICITY DATA with REFERENCE
skn-mus TDLo:577 g/kg/78W-I:ETA BJCAAI 48,429,83

CONSENSUS REPORTS: Reported in EPA TSCA Inventory.

SAFETY PROFILE: Questionable carcinogen with experimental tumorigenic data. When heated to decomposition it emits acrid smoke and irritating fumes.

PCS750 ***HR: 2***
PETROLEUM 60 SOLVENT

PROP: Consists of a mixture of hydrocarbons from C_5-C_{10}, chiefly of paraffins, cycloparaffins, alkyl benzenes, and benzene (TXAPA9 34,374,75).

TOXICITY DATA with REFERENCE
ihl-hmn TCLo:550 ppm/15M:EYE TXAPA9 34,374,75
ihl-rat LC50:4900 ppm/4H TXAPA9 34,374,75
ihl-cat LCLo:4100 ppm/4H TXAPA9 34,374,75
NIOSH REL: (Refined Petroleum Solvents) TWA 350 mg/m^3; CL 1800 mg/m^3/15M

SAFETY PROFILE: Mildly toxic by inhalation. Human systemic effects by inhalation: change in sense of smell and taste, conjunctiva irritation. When heated to decomposition it emits acrid smoke and irritating fumes. See also individual components.

PCT000 ***HR: 2***
PETROLEUM 70 SOLVENT

PROP: Consists of a mixture of hydrocarbons from C_5-C_{10}, chiefly of paraffins, monocycloparaffins, and alkyl benzenes (TXAPA9 34,395,75)

TOXICITY DATA with REFERENCE
ihl-hmn TCLo:180 ppm/15M:EYE TXAPA9 34,395,75
ihl-dog LCLo:930 ppm/4H TXAPA9 34,395,75
NIOSH REL: (Refined Petroleum Solvents) TWA 350 mg/m^3; CL 1800 mg/m^3/15M

SAFETY PROFILE: Mildly toxic by inhalation. Human systemic effects by inhalation: lacrimation. When heated to decomposition it emits acrid smoke and irritating fumes. See also OLEFINS.

PCT250 CAS:64475-85-0 ***HR: 3***
PETROLEUM SPIRITS
DOT: UN 1271

PROP: Volatile, clear, colorless and non-fluorescent liquid. Mp: < −73°, bp: 40-80°, ULC: 95-100, lel: 1.1%, uel: 5.9%, flash p: < 0°F, d: 0.635-0.660, autoign temp: 550°F, vap d: 2.50.

SYNS: BENZINE ◇ BENZOLINE ◇ CANADOL ◇ HERBITOX ◇ LIGROIN ◇ MINERAL SPIRITS ◇ MINERAL THINNER ◇ MINERAL TURPENTINE ◇ PAINTERS' NAPHTHA ◇ REFINED SOLVENT NAPHTHA ◇ SKELLY-SOLVE S ◇ SOLVENT NAPHTHA ◇ STODDARD SOLVENT ◇ VARNISH MAKERS' NAPHTHA ◇ VARNISH MAKERS' AND PAINTERS' NAPHTHA ◇ VARSOL ◇ VM&P NAPHTHA ◇ WHITE SPIRITS

TOXICITY DATA with REFERENCE
eye-hmn 880 ppm/15M TXAPA9 32,263,75
ihl-hmn TCLo:600 mg/m^3/8H:EYE,PUL,GIT TPKVAL 10,116,68
unr-man LDLo:1470 mg/kg 85DCAI 2,73,70
ihl-rat LC50:3400 ppm/4H TXAPA9 32,263,75
ipr-rat LDLo:8560 mg/kg TXAPA9 1,156,59
ihl-mus LCLo:50000 mg/m^3 TPKVAL 10,116,68

CONSENSUS REPORTS: Reported in EPA TSCA Inventory.

OSHA PEL: TWA 300 ppm; STEL 400 PPM
ACGIH TLV: TWA 300 ppm
NIOSH REL: TWA 350 mg/m^3; CL 1800 mg/m^3/15M

SAFETY PROFILE: Moderately toxic to humans by an unspecified route. Mildly toxic by inhalation and intraperitoneal routes. Ingestion can cause a burning sensation, vomiting, diarrhea, drowsiness, and, in severe cases, pulmonary edema. Inhalation of concentrated vapors can cause intoxication resembling that from alcohol, headache, nausea, coma, and hemorrhage to various vital organs. Highly dangerous fire hazard when exposed to heat, flame sparks, or oxidizing materials. Explosive in the form of vapor when exposed to heat or flame. Highly dangerous; keep away from heat or flame.! To fight fire, use foam, CO_2, dry chemical. When heated to decomposition it emits acrid smoke and irritating fumes.

PCT500 ***HR: 1***
PETROLEUM 50 THINNER

PROP: A mixture of paraffins, monocycloparaffins, condensed cycloparaffins, benzene, toluene, and C_8 alkyl benzenes (TXAPA9 36,427,76).

SYN: 50 THINNER

TOXICITY DATA with REFERENCE
eye-hmn 530 ppm/30M MLD TXAPA9 36,427,76
ihl-rat LC50:8300 ppm/4H TXAPA9 36,427,76

SAFETY PROFILE: Mildly toxic by inhalation. A human eye irritant. When heated to decomposition it emits acrid smoke and irritating fumes. See also BENZENE, TOLUENE, and OLEFINS.

PCU000 CAS:1429-30-7 ***HR: 3***
PETUNIDOL
mf: $C_{16}H_{13}O_7$ mw: 317.29

SYNS: 2-(3,4-DIHYDROXY-5-METHOXYPHENYL)-3,5,7-TRIHYDROXYBENZOPYRYLIUM, ACID ANION ◇ 3,3′,4′,5,7-PENTAHYDROXY-5′-METHOXYFLAVYLIUM ACID ANION

TOXICITY DATA with REFERENCE
ipr-rat LD50:2350 mg/kg CHTPBA 2,33,67
ivn-rat LD50:240 mg/kg CHTPBA 2,33,67
ipr-mus LD50:4110 mg/kg CHTPBA 2,33,67
ivn-mus LD50:840 mg/kg CHTPBA 2,33,67

SAFETY PROFILE: Poison by intravenous route. Moderately toxic by intraperitoneal route. When heated to decomposition it emits acrid smoke and irritating fumes.

PCU350 CAS:17466-45-4 ***HR: 3***
PHALLOIDIN
mf: $C_{35}H_{48}N_8O_{11}S$ mw: 788.97

PROP: Hexahydrate, needles from water. Mp: 280-282°. Solubility in water 0.5%; much more sol in hot water; freely sol in methanol, ethanol, butanol, pyridine.

SYN: PHALLOIDINE

TOXICITY DATA with REFERENCE
ipr-mus LD50:2 mg/kg NEJMAG 269,223,63
ivn-mus LDLo:6600 μg/kg AEPPAE 190,406,38
orl-mam LDLo:1000 μg/kg CTOXAO 17,45,80

SAFETY PROFILE: Poison by ingestion, intraperitoneal, and intravenous routes. When heated to decomposition it emits toxic fumes of SO_x, and NO_x.

PCU360 CAS:9000-70-8 ***HR: D***
PHARMAGEL A

SYNS: ABSORBABLE GELATIN SPONGE ◇ GELATIN ◇ GELATINE ◇ GELATIN FOAM ◇ GELATINS ◇ GELFOAM ◇ GT ◇ PHARMAGEL AdB ◇ PHARMAGEL B ◇ PURAGEL ◇ SPONGIOFORT ◇ VEE GEE GELATIN

TOXICITY DATA with REFERENCE
ipr-mus TDLo:700 mg/kg (female 7-13D post):REP OYYAA2 8,981,74
ipr-mus TDLo:700 mg/kg (female 7-13D post):TER OYYAA2 8,981,74

CONSENSUS REPORTS: Reported in EPA TSCA Inventory.

SAFETY PROFILE: An experimental teratogen. Experimental reproductive effects. When heated to decomposition it emits acrid smoke and irritating fumes.

PCU375 ***HR: 3***
PHEASANT'S EYE

PROP: Small perennials which have 2- to 3-inch yellow or red flowers in the spring. Native to the alpine areas of southern Europe, Asia Minor, and Asia, they are cultivated in the northern United States, particularly in rock gardens.

SYNS: A. AESTIVALIS ◇ A. AMURENSIS ◇ A. ANNUA ◇ ADONIS (VARIOUS SPECIES) ◇ A. VERNALIS ◇ FLOR de ADONIS (CUBA) ◇ GOTA de SANGRE (CUBA) ◇ RED MOROCCO

SAFETY PROFILE: Most parts of the plant contain a cardiac glycoside similar to digitalis and may also contain a strong irritant. No human poisonings have been reported in the United States. Cardiac glycosides may cause death by their effect on heart function. See also DIGITALIS.

PCU390 ***HR: 1***
PHELLOBERIN A

PROP: Contains 37.5 mg berberine chloride, 37.5 mg

clioquinol in carboxymethylcellulose (NIIRDN 6,666,82).

TOXICITY DATA with REFERENCE
orl-rat TDLo:7500 mg/kg (30D pre):REP KSRNAM 11,464,77
orl-rat LD50:10 g/kg NIIRDN 6,666,82
orl-mus LD50:7120 mg/kg KSRNAM 11,464,77

SAFETY PROFILE: Mildly toxic by ingestion. Experimental reproductive effects. When heated to decomposition it emits toxic fumes of NO_x and Cl^-. A bacteriostatic agent. See also BERBERINE and CARBOXYMETHYLCELLULOSE.

PCU400 CAS:14436-50-1 ***HR: 3***
PHEM
mf: $C_{19}H_{29N}O_4$ mw: 335.49

SYNS: ETHYLPHENYLMALONIC ACID 2-(DIETHYLAMINO)ETHYL ETHYL ESTER ◇ ETHYLPHENYL-PROPANEDIOIC ACID-2-(DIETHYLAMINO)ETHYL ETHYL ESTER (9CI) ◇ PHENYLAETHYLMALONSAEURE-AETHYL-DIAETHYLAMINOAETHYL-DI-ESTER (GERMAN) ◇ Sch 5705

TOXICITY DATA with REFERENCE
orl-rat LD50:3700 mg/kg AEPPAE 237,264,59
ipr-rat LD50:320 mg/kg AEPPAE 237,264,59
orl-mus LD50:440 mg/kg AEPPAE 237,264,59
ipr-mus LD50:210 mg/kg AEPPAE 237,264,59

SAFETY PROFILE: Poison by intraperitoneal route. Moderately toxic by ingestion. When heated to decomposition it emits toxic fumes of NO_x. See also ESTERS.

PCU425 CAS:10477-72-2 ***HR: 3***
PHENACID
mf: $C_{12}H_{15}Cl_2NO_2$ mw: 276.18

SYNS: 4-(BIS(2-CHLOROETHYL)AMINO)-BENZENEACETICACID (9CI) ◇ (p-(BIS(2-CHLOROETHYL)AMINO)PHENYL)ACETIC ACID ◇ 2-(p-(BIS(2-CHLOROETHYL)AMINO)PHENYL)ACETIC ACID ◇ CB 1331 ◇ p-N,N-DI-(2-CHLOROETHYL)AMINOPHENYL ACETIC ACID ◇ NSC 71964

TOXICITY DATA with REFERENCE
sln-dmg-par 3 mmol/L GENRA8 1,173,60
sln-dmg-unr 10 mmol/L ANYAA9 160,228,69
orl-rat TDLo:10 mg/kg (14D pre):REP STEDAM 19,771,72
ipr-rat LDLo:15 mg/kg BCPCA6 5,192,60

CONSENSUS REPORTS: EPA Genetic Toxicology Program.

SAFETY PROFILE: Poison by intraperitoneal route. Experimental reproductive effects. Mutation data reported. When heated to decomposition it emits toxic fumes of Cl^- and NO_x.

PCU500 CAS:551-16-6 ***HR: 3***
PHENACYL-6-AMINOPENICILLINATE
mf: $C_8H_{12}N_2O_3S$ mw: 216.28

PROP: Obtained from *Penicillum chrysogenum* and *Pleurotus ostroeatus*.

SYNS: 6-AMINOPENICILLANIC ACID ◇ PENICIN

TOXICITY DATA with REFERENCE
scu-rat TDLo:2600 mg/kg/65W-I:ETA BJCAAI 19,392,65

CONSENSUS REPORTS: Reported in EPA TSCA Inventory.

SAFETY PROFILE: Questionable carcinogen with experimental tumorigenic data. When heated to decomposition it emits very toxic fumes of NO_x and SO_x.

PCV350 CAS:2522-81-8 ***HR: 2***
PHENACYLPIVALATE
mf: $C_{13}H_{16}O_3$ mw: 220.29

SYNS: BENZOYLCARBINOL TRIMETHYLACETATE ◇ BENZOYLCARBINYL TRIMETHYLACETATE ◇ 2,2-DIMETHYL-PROPANOIC ACID-2-OXO-2-PHENYLETHYL ESTER (9CI) ◇ PIBECARB

TOXICITY DATA with REFERENCE
orl-dog TDLo:91 g/kg (26W male):REP KSRNAM 8,1913,74
ipr-rat LD50:2600 mg/kg GNRIDX 4,1,70
orl-mus LD50:1900 mg/kg KSRNAM 8,1895,74
ipr-mus LD50:1400 mg/kg KSRNAM 8,1895,74
scu-mus LD50:4900 mg/kg KSRNAM 8,1895,74

SAFETY PROFILE: Moderately toxic by ingestion and intraperitoneal routes. Experimental reproductive effects. When heated to decomposition it emits acrid smoke and irritating fumes.

PCV500 CAS:189-92-4 ***HR: 3***
PHENALENO(1,9-gh)QUINOLINE
mf: $C_{19}H_{11}N$ mw: 253.31

SYNS: PYRENOLINE ◇ PYRIDO(2',3':4)PYRENE

TOXICITY DATA with REFERENCE
scu-mus TDLo:72 mg/kg/9W-I:ETA COREAF 258,3387,64

SAFETY PROFILE: Questionable carcinogen with experimental tumorigenic data. When heated to decomposition it emits toxic fumes of NO_x.

PCV750 CAS:31031-74-0 ***HR: 3***
PHENAMACIDE HYDROCHLORIDE
mf: $C_{13}H_{19}NO_2$•ClH mw: 257.79

SYNS: AKLONIN (GERMAN) ◇ AKLONINE ◇ (±)-α-AMINOBENZENEACETIC ACID,3-METHYLBUTYL ESTER HYDROCHLORIDE (9CI) ◇ ISOAMYL PHENYLAMINOACETATE HYDROCHLORIDE ◇ ISOPENTYL-2-PHENYLGLYCINATE HYDROCHLORIDE ◇ 3-METHYLBUTYL α-AMINOBENZENEACETATE HYDROCHLORIDE

(±)- ◇ PHENYLAMINOACETIC ACID ISOAMYL ESTER HYDROCHLORIDE ◇ dl-2-PHENYLGLYCINISOAMYLESTERHYDROCHLORID (GERMAN)

TOXICITY DATA with REFERENCE
orl-mus LD50:2600 mg/kg PHARAT 33,749,78
ipr-mus LD50:415 mg/kg PHARAT 30,765,75
ivn-mus LD50:77 mg/kg PHARAT 33,749,78

SAFETY PROFILE: Poison by intravenous route. Moderately toxic by ingestion and intraperitoneal routes. When heated to decomposition it emits very toxic fumes of Cl^- and NO_x.

PCV775 CAS:40068-20-0 ***HR: D***
PHENAMIDE
mf: $C_{14}H_{20}Cl_2N_2O$ mw: 303.26

SYNS: 2-(p-(BIS(2-CHLOROETHYL)AMINO)-N-ETHYLACETAMIDE ◇ 4-(BIS(2-CHLOROETHYL)AMINO)-N-ETHYL-BENZENEACETAMIDE ◇ NSC 122053

TOXICITY DATA with REFERENCE
orl-rat TDLo:100 mg/kg (14D pre):REP STEDAM 19,771,72

SAFETY PROFILE: Experimental reproductive effects. When heated to decomposition it emits toxic fumes of Cl^- and NO_x.

PCW000 CAS:7258-91-5 ***HR: 3***
PHENANTHRA-ACENAPHTHENE
mf: $C_{24}H_{16}$ mw: 304.40

SYN: 4,5-DIHYDRONAPHTH(1,2-k)ACEPHENANTHRYLENE

TOXICITY DATA with REFERENCE
skn-mus TDLo:1250 mg/kg/52W-I:ETA PRLBA4 117,318,35

SAFETY PROFILE: Questionable carcinogen with experimental tumorigenic data by skin contact. When heated to decomposition it emits acrid smoke and irritating fumes.

PCW250 CAS:85-01-8 ***HR: 3***
PHENANTHRENE
mf: $C_{14}H_{10}$ mw: 178.24

PROP: Solid or monoclinic crystals. Mp: 100°, bp: 339°, d: 1.179 @ 25°, vap press: 1 mm @ 118.3°, vap d: 6.14. Insol in water; sol in CS_2, benzene, and hot alc; very sol in ether.

SYNS: PHENANTHREN (GERMAN) ◇ PHENANTRIN

TOXICITY DATA with REFERENCE
mma-sat 100 μg/plate APSXAS 17,189,80
sce-ham-ipr 900 mg/kg/24H MUREAV 66,65,79
skn-mus TDLo:71 mg/kg:NEO JNCIAM 50,1717,73
skn-mus TD:22 g/kg/10W-I:ETA BJCAAI 10,363,56
orl-mus LD50:700 mg/kg HYSAAV 29,19,64
ivn-mus LD50:56 mg/kg CSLNX* NX#00190

CONSENSUS REPORTS: IARC Cancer Review: Group 3 IMEMDT 7,56,87; Animal Inadequate Evidence IMEMDT 32,419,83. Reported in EPA TSCA Inventory. EPA Genetic Toxicology Program.

OSHA PEL: TWA 0.2 mg/m^3

SAFETY PROFILE: Poison by intravenous route. Moderately toxic by ingestion. Mutation data reported. A human skin photosensitizer. Questionable carcinogen with experimental neoplastigenic and tumorigenic data by skin contact. Combustible when exposed to heat or flame; can react vigorously with oxidizing materials. To fight fire, use water, foam, CO_2, dry chemical. When heated to decomposition it emits acrid smoke and irritating fumes.

PCW500 CAS:20057-09-4 ***HR: 3***
PHENANTHRENE-3,4-DIHYDRODIOL
mf: $C_{14}H_{12}O_2$ mw: 212.26

SYNS: 3,4-DIHYDROMORPHOL ◇ 3,4-DIHYDRO-3,4-PHENANTHRENEDIOL

TOXICITY DATA with REFERENCE
skn-mus TDLo:85 mg/kg:ETA CNREA8 39,4069,79

SAFETY PROFILE: Questionable carcinogen with experimental tumorigenic data by skin contact. When heated to decomposition it emits acrid smoke and irritating fumes.

PCX000 CAS:585-08-0 ***HR: 3***
9,10-PHENANTHRENE OXIDE
mf: $C_{14}H_{10}O$ mw: 194.24

PROP: Colorless needles. Mp: 152-153°. Very sltly sol in water; very sol in alc, ether.

SYNS: 1a,9b-DIHYDROPHENANTHRO(9,10-B)OXIRENE (9CI) ◇ 9,10-EPOXY-9,10-DIHYDROPHENANTHRENE ◇ PHENANTHRENE-9,10-EPOXIDE

TOXICITY DATA with REFERENCE
mmo-sat 50 μg/plate MUTAEX 1,35,86
mma-sat 100 μg/plate MUREAV 66,337,79
dni-omi 100 μg/L PNASA6 74,1378,77
skn-mus TDLo:40 mg/kg:ETA JNCIAM 39,1217,67

CONSENSUS REPORTS: EPA Genetic Toxicology Program.

SAFETY PROFILE: Questionable carcinogen with experimental tumorigenic data by skin contact. Mutation data reported. When heated to decomposition it emits acrid smoke and irritating fumes.

PCX250 CAS:84-11-7 ***HR: 3***
PHENANTHRENEQUINONE
mf: $C_{14}H_8O_2$ mw: 208.22

PROP: Orange needles. D: 1.405 @ 4°, mp: 206.5-207.5°, bp: >300° subl. Very sltly sol in water; sol in hot alc, benzene; sltly sol in ether.

SYNS: 9,10-PHENANTHRAQUINONE ◇ 9,10-PHENANTHRENE-DIONE ◇ 9,10-PHENANTHRENEQUINONE

TOXICITY DATA with REFERENCE
mma-sat 30 μmol/L PNASA6 81,1696,84
skn-mus TDLo:800 mg/kg/29W-C:ETA PIATA8 16,309,40
ipr-mus LDLo:165 mg/kg HBTXAC 5,110,59

CONSENSUS REPORTS: Reported in EPA TSCA Inventory. EPA Genetic Toxicology Program.

SAFETY PROFILE: Poison by acute intraperitoneal route. Questionable carcinogen with experimental tumorigenic data by skin contact. Mutation data reported. When heated to decomposition it emits acrid smoke and irritating fumes.

PCY250 CAS:66-71-7 ***HR: 3***
1,10-PHENANTHROLINE
mf: $C_{12}H_8N_2$ mw: 180.22

PROP: Crystals from benzene. Mp: 93-94°, anhydr. 117°. Sol in water, alc, ether.

SYNS: 4,5-DIAZAPHENANTHRENE ◇ ORTHOPHENANTHROLINE ◇ β-PHENANTHROLINE ◇ o-PHENANTHROLINE ◇ 1,10-o-PHENANTHROLINE

TOXICITY DATA with REFERENCE
ipr-mus TDLo:30 mg/kg (female 8D post):TER JPMSAE 66,1755,77
ipr-mus LD50:75 mg/kg JPETAB 196,478,76
ivn-mus LD50:18 mg/kg JPETAB 137,1,62

CONSENSUS REPORTS: Reported in EPA TSCA Inventory.

SAFETY PROFILE: Poison by ingestion and intraperitoneal routes. An experimental teratogen. Mutation data reported. When heated to decomposition it emits toxic fumes of NO_x.

PCY300 CAS:84-12-8 ***HR: 3***
4,7-PHENANTHROLINE-5,6-DIONE
mf: $C_{12}H_6N_2O_2$ mw: 210.20

PROP: Crystals from methanol. Mp: 295° (decomp). Sparingly sol in water, alc; sol in dil mineral acids.

SYNS: 11925 C ◇ CIBA 11925 ◇ ENTHOHEX ◇ ENTOBEX ◇ ENTRONON ◇ PHANQUINONE ◇ PHANQUINONUM ◇ PHANQUONE ◇ 4,7-PHENANTHROLENE-5,6-QUINONE ◇ PQ

TOXICITY DATA with REFERENCE
sln-asn 1 g/L MUREAV 26,159,74
orl-rat LD50:5 mg/kg ANTCAO 8,297,58
orl-mus LD50:4 mg/kg ANTCAO 8,297,58

SAFETY PROFILE: Poison by ingestion. Mutation data reported. When heated to decomposition it emits toxic fumes of NO_x.

PCY400 CAS:14635-33-7 ***HR: 3***
PHENANTHRO(2,1-d)THIAZOLE
mf: $C_{15}H_9NS$ mw: 235.31

TOXICITY DATA with REFERENCE
skn-mus TDLo:400 mg/kg/13W-I:NEO VOONAW 15(8),54,69

SAFETY PROFILE: Questionable carcinogen with experimental neoplastigenic data. When heated to decomposition it emits toxic fumes of NO_x and SO_x.

PCY500 CAS:4120-78-9 ***HR: 3***
N-3-PHENANTHRYLACETAMIDE
mf: $C_{16}H_{13}NO$ mw: 235.30

SYNS: 3-ACETAMIDOPHENANTHRENE ◇ 3-ACETAMINOPHENANTHRENE ◇ 3-ACETYLAMINOPHENANTHRENE

TOXICITY DATA with REFERENCE
orl-rat TDLo:4572 mg/kg/32W-C:CAR CNREA8 15,188,55
ims-mus TDLo:80 mg/kg:ETA ZEKBAI 72,321,69

SAFETY PROFILE: Questionable carcinogen with experimental carcinogenic and tumorigenic data. When heated to decomposition it emits toxic fumes of NO_x.

PCY750 CAS:4235-09-0 ***HR: 3***
N-9-PHENANTHRYLACETAMIDE
mf: $C_{16}H_{13}NO$ mw: 235.30

SYNS: 9-ACETAMIDOPHENANTHRENE ◇ 9-ACETAMINOPHENANTHRENE ◇ 9-ACETYLAMINOPHENANTHRENE ◇ 9-PHENANTHRYLACETAMIDE

TOXICITY DATA with REFERENCE
ims-mus TDLo:80 mg/kg:ETA ZEKBAI 72,321,69

SAFETY PROFILE: Questionable carcinogen with experimental tumorigenic data. When heated to decomposition it emits toxic fumes of NO_x.

PCZ000 CAS:2438-51-9 ***HR: 3***
N-2-PHENANTHRYLACETOHYDROXAMIC ACID
mf: $C_{16}H_{13}NO_2$ mw: 251.30

SYNS: 2-(N-HYDROXYACETAMIDO)PHENANTHRENE ◇ N-HYDROXY-2-ACETYLAMINOPHENANTHRENE ◇ N-HYDROXY-N-2-PHENANTHRENYL-ACETAMIDE (9CI) ◇ 2-PHENANTHRYLACETHYDROXAMIC ACID

TOXICITY DATA with REFERENCE
oms-bcs 10 g/L CNREA8 30,1473,70
dns-hmn:fbr 10 μmol/L/5H IJCNAW 16,284,75
dnd-rat-ipr 40 mg/kg CRNGDP 5,231,84
scu-rat TDLo:30 mg/kg/4W-I:CAR CNREA8 26,2239,66

SAFETY PROFILE: Questionable carcinogen with experimental carcinogenic data. Human mutation data reported. When heated to decomposition it emits toxic fumes of NO_x.

PDA250 CAS:3366-65-2 ***HR: 3***
2-PHENANTHRYLAMINE
mf: $C_{14}H_{11}N$ mw: 193.26

SYN: 2-AMINOPHENANTHRENE

TOXICITY DATA with REFERENCE
mma-sat 50 ng/plate CBINA8 26,11,79
orl-rat TDLo:450 mg/kg/6D-I:ETA ZEKBAI 72,321,69

SAFETY PROFILE: Questionable carcinogen with experimental tumorigenic data. Mutation data reported. When heated to decomposition it emits toxic fumes of NO_x. See also AROMATIC AMINES.

PDA500 CAS:1892-54-2 ***HR: 3***
3-PHENANTHRYLAMINE
mf: $C_{14}H_{11}N$ mw: 193.26

SYN: 3-AMINOPHENANTHRENE

TOXICITY DATA with REFERENCE
mma-sat 200 ng/plate ENMUDM 6,497,84
orl-rat TDLo:450 mg/kg/6D-I:ETA ZEKBAI 72,321,69

SAFETY PROFILE: Questionable carcinogen with experimental tumorigenic data. Mutation data reported. When heated to decomposition it emits toxic fumes of NO_x. See also AROMATIC AMINES.

PDA750 CAS:947-73-9 ***HR: 3***
9-PHENANTHRYLAMINE
mf: $C_{14}H_{11}N$ mw: 193.26

SYN: 9-AMINOPHENANTHRENE

TOXICITY DATA with REFERENCE
mma-sat 250 ng/plate JJIND8 71,293,83
orl-rat TDLo:450 mg/kg/6D-I:ETA ZEKBAI 72,321,69

SAFETY PROFILE: Questionable carcinogen with experimental tumorigenic data. Mutation data reported. When heated to decomposition it emits toxic fumes of NO_x. See also AROMATIC AMINES.

PDB000 CAS:578-94-9 ***HR: 3***
PHENARSAZINE CHLORIDE
DOT: UN 1698
mf: $C_{12}H_9AsClN$ mw: 277.59

PROP: Light yellow to green granules; irr odor. Mp: 195°, bp: 410° (decomp), d: 1.65, vap press: very low @ 20°, vap d: 9.6. Very sltly sol in water; sltly sol in benzene, brass. Corrodes iron, bronze.

SYNS: ADAMSITE ◇ 5-AZA-10-ARSENAANTHRACENE CHLORIDE ◇ 10-CHLORO-5,10-DIHYDROARSACRIDINE ◇ 10-CHLORO-5,10-DIHYDROPHENARSAZINE ◇ DIPHENYLAMINECHLORARSINE ◇ DIPHENYLAMINECHLOROARSINE (DOT) ◇ DM

TOXICITY DATA with REFERENCE
ihl-hmn LCLo:30 g/m^3 SCJUAD 4,33,67
ihl-hmn TCLo:19 mg/m^3/3M:GIT:PUL AIHAAP 23,199,62
ihl-hmn LCLo:54 ppm/30M NTIS** PB214-270
ivn-mus LD50:35 mg/kg CSLNX* NX#11444
ivn-rbt LD50:6 mg/kg AIHAAP 23,194,62

CONSENSUS REPORTS: Arsenic and its compounds are on the Community Right-To-Know List.

DOT Classification: Irritating Material; Label: Irritant; Poison B; Label: Poison.

OSHA PEL: TWA 0.5 mg(As)/m^3

SAFETY PROFILE: Human poison by inhalation. Poison experimentally by intravenous route. Human systemic effects by inhalation: changes in function or structure of salivary glands, nausea or vomiting, cough. May be irritating to skin, eyes, and mucous membranes. A vomiting type of poison gas (non-persistent). When heated to decomposition it emits very toxic fumes of As and Cl^-. See also ARSENIC COMPOUNDS.

PDB250 CAS:4095-45-8 ***HR: 3***
PHENARSAZINE OXIDE
mf: $C_{24}H_{18}As_2N_2O$ mw: 500.28

SYN: PZO

TOXICITY DATA with REFERENCE
skn-gpg 250 mg/5D SEV TXCYAC 10,341,78
orl-rat LD50:83 mg/kg TXCYAC 10,341,78
orl-gpg LD50:77 mg/kg TXCYAC 10,341,78

CONSENSUS REPORTS: Arsenic and its compounds are on the Community Right-To-Know List.

OSHA PEL: TWA 0.05 mg(As)/m^3

SAFETY PROFILE: Poison by ingestion. Moderately toxic by inhalation. A severe skin irritant. When heated to decomposition it emits very toxic fumes of As and NO_x. See also ARSENIC COMPOUNDS.

PDB300 CAS:73973-02-1 ***HR: 3***
S-(10-PHENASAZINYL)-O,O-DIISOOCTYLPHOSPHORODITHIOATE
mf: $C_{28}H_{43}AsNO_2PS_2$ mw: 595.73

SYNS: 10-PHENASAZINETHIOL, S-ESTER with O,O-

DIISOOCTYLPHOSPHORODITHIOATE ◇ PHOSPHORODITHIOIC ACID, O,O-DIISOOCTYL S-(10-PHENASAZINYL) ESTER

TOXICITY DATA with REFERENCE
ivn-mus LD50:20 mg/kg CSLNX* NX#00779

OSHA PEL: TWA 0.5 mg(As)/m^3

SAFETY PROFILE: Poison by intravenous route. When heated to decomposition it emits toxic fumes of NO_x, SO_x, PO_x, and As.

PDB350 CAS:51753-57-2 ***HR: 2***
PHENAZEPAM
mf: $C_{15}H_{10}BrClN_2O$ mw: 349.63

SYN: FENAZEPAM

TOXICITY DATA with REFERENCE
orl-rat TDLo:70 mg/kg (16-22D preg):REP FATOAO 43,299,80
ipr-rat LD50:720 mg/kg KHFZAN 13(10),108,79
orl-mus LD50:2400 mg/kg KHFZAN 13(10),108,79
ipr-mus LD50:620 mg/kg KHFZAN 13(10),108,79

SAFETY PROFILE: Moderately toxic by ingestion and intraperitoneal routes. Experimental reproductive effects. When heated to decomposition it emits toxic fumes of Cl^-, Br^-, and NO_x.

PDB500 CAS:92-82-0 ***HR: 3***
PHENAZINE
mf: $C_{12}H_8N_2$ mw: 180.22

PROP: Pale yellow crystals. Mp: 171°, bp: > 360° (subl). Very sltly sol in water; sol in cold and hot alc, ether.

SYNS: AZOPHENYLENE ◇ DIBENZOPARADIAZINE ◇ DIBENZOPYRAZINE

TOXICITY DATA with REFERENCE
imp-rat TDLo:7 mg/kg:ETA COREAF 240,1738,55
ipr-mus LD50:400 mg/kg EJMCA5 10,273,75
ivn-mus LD50:180 mg/kg CSLNX* NX#02474

CONSENSUS REPORTS: Reported in EPA TSCA Inventory.

SAFETY PROFILE: Poison by intraperitoneal and intravenous routes. Questionable carcinogen with experimental tumorigenic data. When heated to decomposition it emits toxic fumes of NO_x.

PDB750 CAS:304-81-4 ***HR: 3***
PHENAZIN-5-OXIDE
mf: $C_{12}H_8N_2O$ mw: 196.22

SYNS: N-OXIDE de PHENAZINE (FRENCH) ◇ PHENAZIN ◇ PHENAZINE-N-OXIDE ◇ PHENAZINE-5N-OXIDE ◇ PHENAZINE-9-OXIDE ◇ PHENAZIN OXIDE

TOXICITY DATA with REFERENCE
mma-sat 1 mg/plate MUREAV 116,185,83
orl-mus TDLo:201 mg/kg (91D pre-21D post):REP OYYAA2 17,505,79
orl-mus TDLo:201 mg/kg (91D pre-21D post):TER OYYAA2 17,505,79
orl-rat LD50:6600 mg/kg OYYAA2 12,483,76
orl-mus LD50:12 mg/kg GUCHAZ 6,398,73
ipr-mus LD50:500 mg/kg EJMCA5 10,273,75

SAFETY PROFILE: Poison by ingestion. Moderately toxic by intraperitoneal route. Experimental teratogenic and reproductive effects. Mutation data reported. When heated to decomposition it emits toxic fumes of NO_x.

PDC000 CAS:91-75-8 ***HR: 3***
PHENAZOLINE
mf: $C_{17}H_{19}N_3$ mw: 265.39

SYNS: ANTASTEN ◇ ANTAZOLINE ◇ ANTIHISTAL ◇ ANTISTINE ◇ AZALONE ◇ BEN-A-HIST ◇ 2-(N-BENZYLANILINOMETHYL)-2-IMIDAZOLINE ◇ 4,5-DIHYDRO-N-PHENYL-N-PHENYLMETHYL-1H-IMIDAZOLE-2-METHANAMINE ◇ HISTOSTAB ◇ IMIDAMINE ◇ 5512-M ◇ 2-(N-PHENYL-N-BENZYLAMINOMETHYL)IMIDAZOLINE ◇ 2-PHENYL-BENZYL-AMINO-METHYLIMIDAZOLIN(GERMAN)

TOXICITY DATA with REFERENCE
eye-rbt 1% OPHTAD 143,154,62
orl-mus LD50:398 mg/kg ARZNAD 7,237,57
ipr-mus LD50:100 mg/kg AEPPAE 237.171.59
scu-mus LD50:135 mg/kg ARZNAD 17,214,67

SAFETY PROFILE: Poison by ingestion, subcutaneous, and intraperitoneal routes. An eye irritant. When heated to decomposition it emits toxic fumes of NO_x.

PDC250 CAS:136-40-3 ***HR: 3***
PHENAZOPYRIDINIUM CHLORIDE
mf: $C_{11}H_{11}N_5 \cdot ClH$ mw: 249.73

PROP: Red crystals; sltly bitter taste. Sltly sol in cold water, alc; sol in acetic acid; insol in acetone, benzene, chloroform, ether.

SYNS: AZODINE ◇ AZODIUM ◇ AZODYNE ◇ AZO GANTRISIN ◇ AZO GASTANOL ◇ AZO-MANDELAMINE ◇ AZOMINE ◇ AZO-STANDARD ◇ AZO-STAT ◇ AZOTREX ◇ BARIDIUM ◇ BISTERIL ◇ CYSTAMINE "MCCLUNG" ◇ CYSTOPYRIN ◇ CYSTURAL ◇ 2,6-DIAMINO-3-PHENYLAZOPYRIDINE HYDROCHLORIDE ◇ 2,6-DIAMINO-3-(PHENYLAZO)PYRIDINE MONOHYDROCHLORIDE ◇ DI-AZO ◇ DIRIDONE ◇ DOLONIL ◇ EUCISTIN ◇ GIRACID ◇ MALLOFEEN ◇ MALLOPHENE ◇ NC 150 ◇ NCI-C01672 ◇ NEFRECIL ◇ PAP ◇ PDP ◇ PHENAZO ◇ PHENAZODINE ◇ PHENAZOPYRIDINE HYDROCHLORIDE ◇ PHENYLAZO-DIAMINOPYRIDINE HYDROCHLORIDE ◇ β-PHENYLAZO-α,α'-DIAMINOPYRIDINE HYDROCHLORIDE ◇ 3-PHENYLAZO-2,6-DIAMINOPYRIDINE HYDROCHLORIDE ◇ PHENYLAZO-α,α'-DIAMINOPYRIDINE MONOHYDROCHLORIDE ◇ PHENYLAZO-PYRIDINE HYDROCHLORIDE ◇ 3-(PHENYLAZO)-2,6-PYRIDINEDIAMINE, HYDROCHLORIDE ◇ PHENYLAZO TABLETS ◇ PHENYL-IDIUM ◇ PHENYL-IDIUM 200 ◇ PIRID ◇ PIRIDACIL ◇ PYRAZODINE ◇ PYRAZOFEN ◇ PYREDAL ◇ PYRIDACIL ◇ PYRIDENAL

◇ PYRIDENE ◇ PYRIDIATE ◇ PYRIDIUM ◇ PYRIDIVITE ◇ PYRIPYRIDIUM ◇ PYRIZIN ◇ SEDURAL ◇ SULADYNE ◇ SULODYNE ◇ THIOSULFIL-A FORTE ◇ URAZIUM ◇ URIDINAL ◇ URIPLEX ◇ UROBIOTIC-250 ◇ URODINE ◇ UROFEEN ◇ UROMIDE ◇ UROPHENYL ◇ UROPYRIDIN ◇ UROPYRINE ◇ UTOSTAN ◇ VESTIN ◇ W 1655

TOXICITY DATA with REFERENCE
orl-rat TDLo:225 g/kg/78W-C:CAR NCITR* NCI-CG-TR-99,78
orl-mus TDLo:81 g/kg/80W-C:CAR NCITR* NCI-CG-TR-99,78
orl-rat TD:110 g/kg/78W-C:ETA NCITR* NCI-CG-TR-99,78
orl-chd TDLo:400 mg/kg:CNS,PUL AJDCAI 110,105,65
orl-wmn TDLo:34 mg/kg:GIT,KID,MET AIMEAS 72,89,70
orl-chd TDLo:125 mg/kg:PUL,BLD CPEDAM 10,537,71
orl-wmn TDLo:140 mg/kg:GIT,BLD,MET CMAJAX 91,756,64
orl-rat LD50:403 mg/kg TXAPA9 1,42,59
ipr-rat LDLo:200 mg/kg JPETAB 51,200,34
ivn-mus LD50:180 mg/kg CSLNX* NX#04014

CONSENSUS REPORTS: NTP Fifth Annual Report on Carcinogens. IARC Cancer Review: Group 2B IMEMDT 7,312,87; Animal Sufficient Evidence IMEMDT 24,163,80; Human Limited Evidence IMEMDT 24,163,80; Animal Inadequate Evidence IMEMDT 8,117,75. NCI Carcinogenesis Bioassay (feed); Clear Evidence: mouse, rat NCITR* NCI-CG-TR-99,78.

SAFETY PROFILE: Confirmed carcinogen with experimental carcinogenic and tumorigenic data. A poison by intraperitoneal and intravenous routes. Moderately toxic by ingestion. Human systemic effects by ingestion: somnolence, cyanosis, diarrhea, nausea or vomiting, anuria or decreased urine volume, normocytic anemia, methemoglobinemia-carboxhemoglobinemia, dehydration, changes in blood sodium levels. When heated to decomposition it emits very toxic fumes of NO_x and HCl.

PDC750 CAS:2275-14-1 ***HR: 3***
PHENCAPTON
DOT: NA 2783
mf: $C_{11}H_{13}Cl_2O_2PS_3$ mw: 375.29

SYNS: O,O-DIAETHYL-S((2,5-DICHLOR-PHENYL-THIO)-METHYL)-DITHIOPHOSPHAT (GERMAN) ◇ S-(2,5-DICHLOROPHENYLTHIOMETHYL) O,O-DIETHYL PHOSPHORODITHIOATE ◇ 2,5-DICHLOROPHENYLTHIOMETHYL O,O-DIETHYL PHOSPHORODITHIOATE ◇ S-(2,5-DICHLOROPHENYLTHIOMETHYL) DIETHYL PHOSPHOROTHIOLOTHIONDATE ◇ O,O-DIETHYL-S-(2,5-DICHLOROPHENYLTHIOMETHYL) DITHIOPHOSPHATE ◇ O,O-DIETHYL-S-(2,5-DICHLOROPHENYLTHIOMETHYL) DITHIOPHOSPHORAN ◇ O,O-DIETHYL-S-(2,5-DICHLOROPHENYLTHIOMETHYL) PHOSPHORODITHIOATE ◇ O,O-DIETHYL-S-(2,5-DICHLOROPHENYLTHIOMETHYL)PHOSPHOROTHIOLOTHIONATE ◇ DITHIOPHOSPHATE de-O,O-DIETHYLE et de S(2,5-DICHLORO- PHENYL) THIOMETHYLE (FRENCH) ◇ EENKAPTON (DUTCH) ◇ ENT 25,585 ◇ GEIGY G-28029 ◇ PRZEDZIORKOFOS (POLISH)

TOXICITY DATA with REFERENCE
orl-rat LD50:61 mg/kg PAPOAC 19,415,68
skn-rat LD50:652 mg/kg 85DPAN -,-,71/76
orl-mus LD50:220 mg/kg ARSIM* 20,10,66
orl-ckn LD50:886 mg/kg TXAPA9 11,49,67

DOT Classification: ORM-A; Label: None.

SAFETY PROFILE: Poison by ingestion. Moderately toxic by skin contact. A cholinesterase inhibitor. When heated to decomposition it emits very toxic fumes of Cl^-, PO_x, and SO_x. See also PARATHION.

PDC850 CAS:3735-90-8 ***HR: 3***
PHENCARBAMID
mf: $C_{19}H_{24}N_2OS$ mw: 328.51

PROP: Crystals. Mp: 48-49°, bp: 120-126°. Practically insol in water; freely sol in methanol, ether, chloroform. Sol in petr ether.

SYNS: BAYER 1355 ◇ DIPHENYLTHIOCARBAMIC ACID-S-(2-(DIETHYLAMINO)ETHYL) ESTER ◇ ESCORPAL ◇ FENCARBAMIDE ◇ PHENCARBAMIDE ◇ RISPASULF

TOXICITY DATA with REFERENCE
orl-rat LD50:370 mg/kg AIPTAK 155,393,65
scu-rat LD50:400 mg/kg AIPTAK 155,393,65
ivn-rat LD50:30 mg/kg AIPTAK 155,393,65
scu-mus LD50:93 mg/kg AIPTAK 155,393,65
ivn-mus LD50:32 mg/kg AIPTAK 155,393,65

SAFETY PROFILE: Poison by ingestion, subcutaneous, and intravenous routes. When heated to decomposition it emits toxic fumes of SO_x and NO_x. See also CARBAMATES and ESTERS.

PDC875 CAS:58-13-9 ***HR: 3***
PHENCARBAMIDE HYDROCHLORIDE
mf: $C_{19}H_{24}N_2OS•ClH$ mw: 364.97

SYNS: BA 1355 ◇ S-(2-(DIETHYLAMINO)ETHYL) DIPHENYLTHIOCARBAMATE HYDROCHLORIDE ◇ DIPHENYLAMIN-(β-DIAETHYLAMINOAETHYL)CARBAMIDTHIOESTER(GERMAN) ◇ DIPHENYLCARBAMOTHIOIC ACID-S-(2-(DIETHYLAMINO)ETHYL) ESTER HYDROCHLORIDE ◇ DIPHENYLTHIOCARBAMIC ACID-S-(2-(DIETHYLAMINO)ETHYL ESTER HYDROCHLORIDE ◇ ESCORPAL ◇ FENCARBAMIDE HYDROCHLORIDE ◇ PHENCARBAMID HYDROCHLORIDE (GERMAN)

TOXICITY DATA with REFERENCE
orl-rat LD50:410 mg/kg AIPTAK 155,393,65
scu-rat LD50:400 mg/kg AIPTAK 155,393,65
ivn-rat LD50:33 mg/kg AIPTAK 155,393,65
scu-mus LD50:103 mg/kg AIPTAK 155,393,65
ivn-mus LD50:32 mg/kg AIPTAK 151,515,64

SAFETY PROFILE: Poison by subcutaneous and intravenous routes. Moderately toxic by ingestion. When heated to decomposition it emits toxic fumes of SO_x, NO_x, and HCl. See also CARBAMATES and ESTERS.

PDC890 CAS:77-10-1 ***HR: 3***
PHENCYCLIDINE
mf: $C_{17}H_{25}N$ mw: 243.43

PROP: Colorless crystals. Mp: 46-46.5°, bp: 135-137°.

SYNS: CL-395 ◇ HOG ◇ PCP (anesthetic) ◇ 1-(1-PHENYLCYCLOHEXYL)PIPERIDINE

TOXICITY DATA with REFERENCE
orl-mus TDLo:130 mg/kg (5D pre/1-21D preg):REP TJADAB 28,319,83
ipr-mus LD50:2800 μg/kg TOLED5 12,171,82

SAFETY PROFILE: Poison by intraperitoneal route. Experimental reproductive effects. Caution: This is a controlled substance (depressant) listed in the U.S. Code of Federal Regulations, Title 21 Part 1308.12 (1985). The ethylamine, pyrrolidine and thiophene analogs are listed as hallucinogens, Title 21 Part 1308.11. When heated to decomposition it emits toxic fumes of NO_x.

PDD000 CAS:569-59-5 ***HR: 3***
PHENDIMETRAZINE TARTRATE
mf: $C_{19}H_{19}N•C_4H_6O_6$ mw: 411.49

SYNS: 2-METHYL-9-PHENYL-2,3,4,9-TETRAHYDRO-1H-INDENO(2,1-c)PYRIDINE TARTRATE ◇ 2-METHYL-9-PHENYL-2,3,4,9-TETRAHYDRO-1-PYRIDINDENE TARTRATE ◇ PHENINDAMINE HYDROGEN TARTRATE ◇ PLEGINE ◇ THEOPHORIN TARTRATE ◇ THEPHORIN TARTRATE

TOXICITY DATA with REFERENCE
orl-rat LD50:280 mg/kg 29ZVAB -,91,69
scu-rat LDLo:200 mg/kg CRSBAW 144,887,50
orl-mus LD50:265 mg/kg JPETAB 105,291,52
ipr-mus LD50:88 mg/kg JPETAB 105,291,52
ivn-mus LD50:18 mg/kg CSLNX* NX#03034
orl-rbt LD50:577 mg/kg 29ZVAB -,91,69
ivn-gpg LD50:24500 μg/kg AIPTAK 113,313,58

SAFETY PROFILE: Poison by ingestion, intraperitoneal, subcutaneous, and intravenous routes. An antihistamine. When heated to decomposition it emits toxic fumes of NO_x.

PDD300 CAS:65-29-2 ***HR: 3***
(v-PHENENYLTRIS(OXYETHYLENE))TRIS(TRIETHYLAMMONIUM IODIDE)
mf: $C_{30}H_{60}N_3O_3•3I$ mw: 891.63

SYNS: BENZCURINE IODIDE ◇ F 2559 ◇ FLAXEDIL ◇ GALLAMINE ◇ RELAXAN ◇ RETENSIN ◇ RP 3697 ◇ SYNCURARINE ◇ TRICURAN ◇ 1,2,3-TRI(β-DIETHYLAMINOETHOXY)BENZENE TRIETHIODIDE ◇ TRI(β-DIETHYLAMINOETHOXY)-1,2,3-BENZENE TRI-IODOETHYLATE ◇ TRIIODOETHYLATE de GALLAMINE (FRENCH) ◇ TRIIODOETHYLATE OF TRI(DIETHYLAMINOETHYLOXY)-1,2,3-BENZENE ◇ 1,2,3-TRIS(2-DIETHYLAMINOETHOXY)BENZENE TRIETHIODIDE ◇ 1,2,3-TRIS(2-DIETHYLAMINOETHOXY)BENZENE TRIS(ETHYLIODIDE) ◇ 1,2,3-TRIS(2-TRIETHYLAMMONIUM ETHOXY)BENZENE TRIIODIDE

TOXICITY DATA with REFERENCE
ipr-rat LD50:23200 μg/kg NIIRDN 6,178,82
scu-rat LD50:28500 μg/kg NIIRDN 6,178,82
ivn-rat LD50:5100 μg/kg NIIRDN 6,178,82
idu-rat LD50:380 mg/kg AIPTAK 180,155,69
orl-mus LD50:425 mg/kg AIPTAK 80,172,49
ipr-mus LD50:10 mg/kg AIPTAK 184,75,70
scu-mus LD50:16400 μg/kg OYYAA2 9,117,75
ivn-mus LD50:1800 μg/kg CSLNX* NX#01979
par-mus LD50:1700 μg/kg APFRAD 7,368,49
ivn-dog LD50:800 μg/kg AIPTAK 80,172,49
orl-rbt LD50:100 mg/kg AIPTAK 80,172,49
scu-rbt LDLo:2 mg/kg AIPTAK 80,172,49
ivn-rbt LD50:440 μg/kg IJNEAQ 5,305,66
ims-rbt LDLo:2 mg/kg AIPTAK 80,172,49
ivn-ckn LD50:600 μg/kg AIPTAK 122,152,59

SAFETY PROFILE: Poison by ingestion, subcutaneous, intravenous, parenteral, intraduodenal, intraperitoneal, and intramuscular routes. When heated to decomposition it emits very toxic fumes of NH_3, NO_x, and I^-. See also IODIDES.

PDD350 CAS:132-93-4 ***HR: 3***
PHENETHICILLIN K
mf: $C_{17}H_{19}N_2O_5S•K$ mw: 402.54

PROP: dl-Form: Crystals from acetone. Decomp 230-232°. Much less hygroscopic than benzylpenicillin sodium. Freely sol in water.

SYNS: ALFACILLIN ◇ ALFOCILLIN ◇ ALTICINA ◇ ASTRACILLIN ◇ BENDRALAN ◇ BL P 152 ◇ BRL 152 ◇ BROCSIL ◇ BROXIL ◇ CHEMIPEN ◇ CHEMIPEN-C ◇ CVK ◇ DARCIL ◇ CRAMCILLIN-S ◇ FENETICILLINE ◇ K PHENETHICILLIN ◇ MAXIPEN ◇ OPTIPEN ◇ α-ORACILLIN ◇ ORALOPEN ◇ PEN 200 ◇ PENSIG ◇ PHENETHICILLIN K SALT ◇ PHENETHECILLIN POTASSIUM ◇ PHENETHECILLIN POTASSIUM SALT ◇ PHENETICILLIN POTASSIUM ◇ PHENO-m-PENICILLIN ◇ PHENOXYAETHYLPENICILLIN K-SALZ (GERMAN) ◇ α-PHENOXYETHYLPENICILLIN ◇ α-PHENOXYETHYLPENICILLIN POTASSIUM ◇ α-PHENOXYETHYLPENICILLIN POTASSIUM SALT ◇ POTASSIUM METHYLPHENOXYMETHYLPENICILLIN ◇ POTASSIUM PHENETHICILLIN ◇ POTASSIUM-α-PHENOXYETHYL PENICILLIN ◇ POTASSIUM (1-PHENOXYETHYL)PENICILLIN ◇ POTASSIUM-6-(α-PHENOXYPROPIONAMIDO)PENICILLANATE ◇ PRIOSPEN ◇ RO-CILLIN ◇ SEMOPEN ◇ SYNAPEN ◇ SYNCILLIN ◇ SYNERPENIN ◇ SYNTHECILLIN ◇ SYNTHECILLINE

TOXICITY DATA with REFERENCE
ipr-rat LD50:1750 mg/kg ARZNAD 12,751,62
ipr-mus LD50:1896 mg/kg NIIRDN 6,661,82
ivn-mus LD50:312 mg/kg NIIRDN 6,661,82
ivn-gpg LD50:324 mg/kg RPOBAR 2,311,70

SAFETY PROFILE: Poison by intravenous route. Moderately toxic by intraperitoneal route. When heated to decomposition it emits toxic fumes of K_2O, SO_x, and NO_x.

PDD500 CAS:156-43-4 ***HR: 3***
PHENETHIDINE
DOT: UN 2311
mf: $C_8H_{11}NO$ mw: 137.20

SYNS: p-AMINOPHENETOLE ◇ 4-AMINOPHENETOLE ◇ p-ETHOXYANILINE ◇ 4-ETHOXYANILINE ◇ p-PHENETIDINE (DOT)

TOXICITY DATA with REFERENCE
skn-rbt 500 mg MLD FCTOD7 20,563,82
eye-rbt 100 mg MOD FCTOD7 20,563,82
eye-rbt 100 mg/45 rns MLD FCTOD7 20,563,82
mma-sat 1 μmol/plate CPBTAL 33,2877,85
orl-rat LD50:580 mg/kg GISAAA 35(8),28,70
ihl-rat LCLo:250 mg/m^3 GISAAA 35(8),28,70
orl-mus LD50:530 mg/kg GTPZAB 25(8),50,81
ipr-mus LD50:692 mg/kg JMCMAR 17,900,74

CONSENSUS REPORTS: Reported in EPA TSCA Inventory.

DOT Classification: Poison B; Label: St. Andrews Cross.

SAFETY PROFILE: Poison by inhalation. Moderately toxic by ingestion and intraperitoneal routes. Caution: It can be absorbed through the skin. A skin and eye irritant. Mutation data reported. When heated to decomposition it emits toxic fumes of NO_x.

PDD750 CAS:60-12-8 ***HR: 3***
PHENETHYL ALCOHOL
mf: $C_8H_{10}O$ mw: 122.18

PROP: Colorless liquid; floral odor of roses. Mp: −27°, bp: 220°, flash p: 216°F, d: 1.0245 @ 15°, vap d: 4.21. Misc with alc, ether; sol in fixed oils, glycerin, propylene glycol.

SYNS: BENZYL CARBINOL ◇ FEMA No. 2858 ◇ PHENETHANOL ◇ β-PHENETHYL ALCOHOL ◇ 2-PHENETHYL ALCOHOL ◇ β-PHENYLETHANOL ◇ 2-PHENYLETHANOL ◇ β-PHENYLETHYL ALCOHOL ◇ 2-PHENYLETHYL ALCOHOL

TOXICITY DATA with REFERENCE
eye-rbt 12 g/10M MLD ARZNAD 9,349,59
skn-gpg 100 mg MLD FCTXAV 13,903,75
orl-rat TDLo:43 mg/kg (6-15D preg):REP JTEHD6 12,235,83
orl-rat TDLo:430 mg/kg (6-15D preg):TER JTEHD6 12,235,83
orl-rat LD50:1790 mg/kg FCTXAV 2,327,64
orl-mus LD50:800 mg/kg 14CYAT 2,1496,63
ipr-mus LD50:200 mg/kg 14CYAT 2,1476,63
skn-rbt LD50:790 mg/kg TXAPA9 28,313,74
orl-rbt LDLo:2 g/kg JEENAI 48,139,55
orl-gpg LD50:400 mg/kg 14CYAT 2,1476,63
skn-gpg LD50:5000 mg/kg 14CYAT 2,1476,63

CONSENSUS REPORTS: Reported in EPA TSCA Inventory.

SAFETY PROFILE: Poison by ingestion and intraperitoneal routes. Moderately toxic by skin contact. A skin and eye irritant. Experimental teratogenic effects. Other experimental reproductive effects. Causes severe central nervous system injury to experimental animals. Combustible when exposed to heat or flame; can react with oxidizing materials. To fight fire, use CO_2, dry chemical. When heated to decomposition it emits acrid smoke and irritating fumes.

PDE000 CAS:98-85-1 ***HR: 3***
α-PHENETHYL ALCOHOL
DOT: UN 2937
mf: $C_8H_{10}O$ mw: 122.18

PROP: Colorless liquid; hyacinth odor. Bp: 204°, fp: 21.4°, d: 1.015 @ 20°/20°, refr index: 1.525, vap press: 0.1 mm @ 20°, vap d: 4.21, flash p: 205°F (OC). Sol in fixed oils, propylene glycol; very sol in glycerin.

SYNS: BENZENEMETHANOL, α-METHYL- ◇ ETHANOL, 1-PHENYL- ◇ FEMA No. 2685 ◇ 1-FENYLETHANOL ◇ FENYL-METHYLKARBINOL ◇ α-METHYLBENZYL ALCOHOL (FCC) ◇ METHYLPHENYLCARBINOL ◇ METHYPHENYLMETHANOL ◇ NCI-C55685 ◇ 1-PHENYLETHANOL ◇ PHENYLMETHYLCARBINOL ◇ STYRALLYL ALCOHOL ◇ STYRALYL ALCOHOL

TOXICITY DATA with REFERENCE
skn-rbt 10 mg/24H open JIHTAB 26,269,44
skn-rbt 500 mg/24H MOD FCTXAV 12,995,74
eye-rbt 2 mg SEV AJOPAA 29,1363,46
orl-rat LD50:400 mg/kg JIHTAB 26,269,44
scu-mus LD50:250 mg/kg AIPTAK 116,154,58
ivn-dog LDLo:200 mg/kg JPETAB 15,129,20
skn-rbt LD50:2500 mg/kg FCTXAV 12,995,74

CONSENSUS REPORTS: Reported in EPA TSCA Inventory. EPA Genetic Toxicology Program.

DOT Classification: Poison B; Label: St. Andrews Cross.

SAFETY PROFILE: Poison by ingestion and subcutaneous routes. Moderately toxic by skin contact. A skin and severe eye irritant. Combustible when exposed to heat or flame; can react with oxidizing materials. To fight fire, use alcohol foam, foam, CO_2, dry chemical.

PDE250 CAS:64-04-0 ***HR: 3***
β-PHENETHYLAMINE
mf: $C_8H_{11}N$ mw: 121.20

PROP: Colorless to sltly yellow liquid; fishy odor. Bp:

194.5-195°, d: 0.96 @ 15.5°/15.5°, vap d: 4.18. Sol in water; very sol in alc, ether.

SYNS: β-AMINOETHYLBENZENE ◇ 1-AMINO-2-PHENYLETHANE ◇ β-PHENYLAETHYLAMIN (GERMAN) ◇ 1-PHENYL-2-AMINO-ATHAN (GERMAN) ◇ 1-PHENYL-2-AMINOETHANE ◇ PHENYLETHYLAMINE ◇ ω-PHENYLETHYLAMINE ◇ 2-PHENYLETHYLAMINE

TOXICITY DATA with REFERENCE
orl-rat LDLo:800 mg/kg AEPPAE 195,647,40
ipr-rat LDLo:100 mg/kg AEPPAE 195,647,40
orl-mus LD50:400 mg/kg JACSAT 63,602,41
scu-mus LD50:320 mg/kg ARZNAD 7,620,57
ivn-mus LD50:100 mg/kg JPETAB 106,341,52
icv-mus LD50:39 mg/kg TYKNAQ 27,131,80

CONSENSUS REPORTS: Reported in EPA TSCA Inventory.

SAFETY PROFILE: Poison by ingestion, intraperitoneal, subcutaneous, intracervical, and intravenous routes. A strong base. A skin irritant and possible sensitizer. When heated to decomposition it emits toxic fumes of NO_x. See also AMINES.

PDE500 CAS:156-28-5 ***HR: 3***
β-PHENETHYLAMINE HYDROCHLORIDE
mf: $C_8H_{11}N \cdot ClH$ mw: 157.66

PROP: Orthorhombic platelets. Mp: 217°. Very sol in water; sol in alc; insol in ether.

SYNS: β-PHENYLATHYLAMINHYDROCHLORID (GERMAN) ◇ 1-PHENYL-2-AMINOETHANE HYDROCHLORIDE ◇ β-PHENYLETHYLAMINE HYDROCHLORIDE ◇ 2-PHENYLETHYLAMINE HYDROCHLORIDE ◇ USAF EL-76

TOXICITY DATA with REFERENCE
orl-hmn TDLo:714 mg/kg:CVS AIMDAP 39,404,27
orl-hmn TDLo:45 mg/kg:CNS KLWOAZ 17,1580,38
scu-rat LDLo:160 mg/kg JPETAB 71,62,41
scu-mus LD50:470 mg/kg JPPMAB 15,472,63
ivn-rbt LDLo:50 mg/kg JPETAB 36,363,29
skn-gpg LDLo:200 mg/kg JPETAB 32,121,27

CONSENSUS REPORTS: Reported in EPA TSCA Inventory.

SAFETY PROFILE: Poison by skin contact, subcutaneous, and intravenous routes. A skin irritant and possible sensitizer. Human systemic effects by ingestion: blood pressure lowering, anoxeria. When heated to decomposition it emits very toxic fumes of NO_x and HCl. See also AMINES.

PDF000 CAS:114-86-3 ***HR: 3***
β-PHENETHYLBIGUANIDE
mf: $C_{10}H_{15}N_5$ mw: 205.30

SYNS: DBI ◇ N'-β-FENETILFORMAMIDINILIMINOUREA (ITALIAN) ◇ FENFORMINA ◇ NCI-C01741 ◇ β-PHENETHYBIGUANIDE ◇ 1-PHENETHYLBIGUANIDE ◇ PHENETHYLDIGUANIDE ◇ N'-β-PHENETHYLFORMAMIDINYLLIMINOUREA ◇ PHENFORMINE

TOXICITY DATA with REFERENCE
orl-rat LD50:1650 mg/kg BCFAAI 110,470,71
orl-mus LD50:830 mg/kg BCFAAI 110,470,71
ipr-mus LD50:140 mg/kg BCFAAI 97,396,58
scu-gpg LD50:16 mg/kg MEXPAG 8,237,63
ivn-gpg LDLo:30 mg/kg 43ZRAD 1,337,80

CONSENSUS REPORTS: NCI Carcinogenesis Bioassay (feed); No Evidence: mouse, rat NCITR* NCI-CG-TR-7,77.

SAFETY PROFILE: Poison by intraperitoneal, subcutaneous, and intravenous routes. Moderately toxic by ingestion. When heated to decomposition it emits toxic fumes of NO_x.

PDF250 CAS:834-28-6 ***HR: 3***
PHENETHYLBIGUANIDE HYDROCHLORIDE
mf: $C_{10}H_{15}N_5 \cdot ClH$ mw: 241.76

SYNS: DBI-TD ◇ DIPAR ◇ 1-FENETILBIGUANIDE CLORIDRATO (ITALIAN) ◇ MELTROL ◇ N'-β-PHENETHYLBIGUANIDE HYDROCHLORIDE ◇ 1-PHENETHYLBIGUANIDE HYDROCHLORIDE ◇ PHENOFORMINE HYDROCHLORIDE ◇ 1-PHENYLAETHYLBIGUANID HYDROCHLORID (GERMAN) ◇ USAF VI-6

TOXICITY DATA with REFERENCE
orl-rat LD50:938 mg/kg FRPSAX 15,521,60
ipr-rat LD50:172 mg/kg FRPSAX 15,521,60
ivn-rat LD50:17500 μg/kg ARZNAD 23,1571,73
orl-mus LD50:407 mg/kg TXAPA9 14,393,69
ipr-mus LD50:150 mg/kg NTIS** AD691-490
ivn-mus LD50:17800 μg/kg CSLNX* NX#00094

SAFETY PROFILE: Poison by intraperitoneal and intravenous routes. Moderately toxic by ingestion. When heated to decomposition it emits very toxic fumes of HCl and NO_x.

PDF500 CAS:7476-91-7 ***HR: 2***
PHENETHYL CHLORACETATE
mf: $C_{10}H_{11}ClO_2$ mw: 198.66

SYN: CHLORO-ACETIC ACID, PHENETHYL ESTER

TOXICITY DATA with REFERENCE
ipr-mus LDLo:500 mg/kg CBCCT* 8,99,56

CONSENSUS REPORTS: Reported in EPA TSCA Inventory.

SAFETY PROFILE: Moderately toxic by intraperitoneal route. When heated to decomposition it emits toxic fumes of Cl^-.

PDF750 CAS:103-48-0 ***HR: 1***
PHENETHYL ISOBUTYRATE
mf: $C_{12}H_{16}O_2$ mw: 192.28

PROP: Colorless to light yellow liquid; fruity, rosy odor. D: 0.9871.486-1.490, flash p: +212°F. Sol in alc, fixed oils; insol in water @ 230°.

SYNS: BENZYLCARBINOL ISOBUTYRATE ◇ BENZYLCARBINYL ISOBUTYRATE ◇ FEMA No. 2862 ◇ PHENYLETHYL ISOBUTYRATE ◇ β-PHENYLETHYL ISOBUTYRATE ◇ 2-PHENYLETHYL ISOBUTYRATE ◇ 2-PHENYLETHYL-2-METHYLPROPIONATE

TOXICITY DATA with REFERENCE
orl-rat LD50:5200 mg/kg FCTXAV 16,637,78

CONSENSUS REPORTS: Reported in EPA TSCA Inventory.

SAFETY PROFILE: Mildly toxic by ingestion. Combustible liquid. When heated to decomposition it emits acrid smoke and irritating fumes.

PDF775 CAS:140-26-1 ***HR: 1***
PHENETHYL ISOVALERATE
mf: $C_{13}H_{18}O_2$ mw: 206.31

PROP: Colorless to sltly yellow liquid; fruity, rosy odor. D: 0.973, refr index: 1.484, flash p: +212°F. Sol in alc, fixed oils; insol in water @ 263°.

SYNS: FEMA No. 2871 ◇ 3-METHYL-BUTANOIC ACID 2-PHENYLETHYL ESTER ◇ PHENETHYL ESTER ISOVALERIC ACID ◇ PHENYLETHYL ISOVALERATE ◇ β-PHENYLETHYL ISOVALERATE ◇ 2-PHENYLETHYL-3-METHYLBUTIRATE

TOXICITY DATA with REFERENCE
orl-rat LD50:6220 mg/kg VPITAR 33(5),48,74
orl-mus LD50:6220 mg/kg VPITAR 33(5),48,74
orl-gpg LD50:6220 mg/kg VPITAR 33(5),48,74

CONSENSUS REPORTS: Reported in EPA TSCA Inventory.

SAFETY PROFILE: Mildly toxic by ingestion. Combustible liquid. When heated to decomposition it emits acrid smoke and irritating fumes. See also ESTERS.

PDF790 ***HR: 1***
2-PHENETHYL 2-METHYLBUTYRATE
mf: $C_{13}H_{18}O_2$ mw: 206.28

PROP: Colorless liquid; floral, fruity odor. D: 0.973, refr index: 1.484, flash p: +212°F. Sol in alc, fixed oils; insol in water.

SYN: FEMA No. 3632

SAFETY PROFILE: Combustible liquid. When heated to decomposition it emits acrid smoke and irritating fumes.

PDF800 CAS:2550-26-7 ***HR: 2***
PHENETHYL METHYL KETONE
mf: $C_{10}H_{12}O$ mw: 148.22

SYNS: BENZYLACETONE ◇ METHYL PHENETHYL KETONE ◇ METHYL PHENYLETHYL KETONE ◇ METHYL-2-PHENYLETHYL KETONE ◇ 4-PHENYLBUTAN-2-ONE ◇ 4-PHENYL-2-BUTANONE ◇ β-PHENYLETHYL METHYL KETONE

TOXICITY DATA with REFERENCE
skn-rbt 500 mg/24H SEV FCTOD7 21,647,83
eye-rbt 2000 ppm FCTOD7 21,647,83
orl-rat LD50:3200 mg/kg FCTOD7 21,647,83
orl-mus LD50:1590 mg/kg YHTPAD 15(5),7,80
ipr-mus LD50:583 mg/kg YHTPAD 15(5),7,80

CONSENSUS REPORTS: Reported in EPA TSCA Inventory.

SAFETY PROFILE: Moderately toxic by ingestion and intraperitoneal routes. An eye and severe skin irritant. When heated to decomposition it emits acrid smoke and irritating fumes. See also KETONES.

PDI000 CAS:102-20-5 ***HR: 2***
PHENETHYL PHENYLACETATE
mf: $C_{16}H_{16}O_2$ mw: 240.32

PROP: Colorless to sltly yellow liquid above 26°; rosy, hyacinth odor. D: 1.079-1.082, flash p: +212°F. Sol in alc; insol in water.

SYNS: BENZENEACETIC ACID, 2-PHENYLETHYL ESTER ◇ BENZYLCARBINYL-α-TOLUATE ◇ FEMA No. 2866 ◇ PHENYLACETIC ACID, PHENETHYL ESTER ◇ β-PHENYLETHYL PHENYLACETATE ◇ 2-PHENYLETHYL PHENYLACETATE ◇ 2-PHENYLETHYL-α-TOLUATE

TOXICITY DATA with REFERENCE
orl-mus LD50:3190 mg/kg VPITAR 33(5),48,74
orl-gpg LD50:3190 mg/kg VPITAR 33(5),48,74
orl-rat LD50:15 g/kg FCTXAV 2,327,64

CONSENSUS REPORTS: Reported in EPA TSCA Inventory.

SAFETY PROFILE: Moderately toxic by ingestion. Combustible liquid. When heated to decomposition it emits acrid smoke and irritating fumes. See also ESTERS.

PDI250 CAS:33384-03-1 ***HR: 3***
N-(p-PHENETHYL)PHENYLACETOHYDROXAMIC ACID
mf: $C_{16}H_{17}NO_2$ mw: 255.34

SYNS: N-HYDROXY-4-ACETYLAMINOBIBENZYL ◇ N-HYDROXY-4'-PHENETHYLACETANILIDE

TOXICITY DATA with REFERENCE
dns-hmn:fbr 100 μmol/L/5H IJCNAW 16,284,75
ipr-rat TDLo:380 mg/kg/3W-I:NEO CNREA8 24,128,64

SAFETY PROFILE: Questionable carcinogen with experimental neoplastigenic data. Human mutation data reported. When heated to decomposition it emits toxic fumes of NO_x.

PDI500 CAS:332-14-9 ***HR: 3***
1-PHENETHYLPIPERIDINE
mf: $C_{13}H_{19}N$ mw: 189.33

SYNS: 1-PIPERIDINO-2-PHENYL-AETHAN (GERMAN) ◇ 1-PIPERIDINO-2-PHENYLETHANE

TOXICITY DATA with REFERENCE
scu-mus LD50:150 mg/kg ARZNAD 15,126,65

CONSENSUS REPORTS: Reported in EPA TSCA Inventory.

SAFETY PROFILE: Poison by subcutaneous route. When heated to decomposition it emits toxic fumes of NO_x.

PDJ000 CAS:78219-62-2 ***HR: 3***
1-PHENETHYL-4-PIPERIDYL-p-AMINOBENZOATE HYDROCHLORIDE
mf: $C_{20}H_{22}N_2O_2$•ClH mw: 358.90

SYN: p-AMINO-BENZOIC ACID-1-PHENETHYL-4-PIPERIDYL ESTER HYDROCHLORIDE

TOXICITY DATA with REFERENCE
ivn-rat LDLo:13 mg/kg JACSAT 51,922,29
scu-mus LDLo:100 mg/kg JACSAT 51,922,29

SAFETY PROFILE: Poison by intravenous and subcutaneous routes. When heated to decomposition it emits very toxic fumes of HCl and NO_x. See also ESTERS.

PDJ250 CAS:78219-63-3 ***HR: 3***
1-PHENETHYL-4-PIPERIDYL BENZOATE HYDROCHLORIDE
mf: $C_{20}H_{23}NO_2$•ClH mw: 345.90

SYN: BENZOIC ACID-1-PHENETHYL-4-PIPERIDYL ESTER HYDROCHLORIDE

TOXICITY DATA with REFERENCE
ivn-rat LDLo:25 mg/kg JACSAT 51,922,29
scu-mus LDLo:1000 mg/kg JACSAT 51,922,29

SAFETY PROFILE: Poison by intravenous route. Moderately toxic by subcutaneous route. When heated to decomposition it emits very toxic fumes of HCl and NO_x. See also ESTERS.

PDK000 CAS:122-70-3 ***HR: 2***
PHENETHYL PROPIONATE
mf: $C_{11}H_{14}O_2$ mw: 178.25

SYNS: BENZYLCARBINYL PROPIONATE ◇ ENT 18,544 ◇ 2-PHENYLETHYL PROPIONATE ◇ PROPANOIC ACID-2-PHENYLETHYL ESTER

TOXICITY DATA with REFERENCE
orl-rat LD50:4000 mg/kg FCTXAV 12,807,74

CONSENSUS REPORTS: Reported in EPA TSCA Inventory.

SAFETY PROFILE: Moderately toxic by ingestion. When heated to decomposition it emits acrid smoke and irritating fumes. See also ESTERS.

PDK200 ***HR: 1***
PHENETHYL SALICYLATE
mf: $C_{15}H_{14}O_3$ mw: 242.27

PROP: White crystals; balsamic odor. Solidification point: 41°, flash p: +212°F. Sol in alc; insol in water.

SYN: FEMA No. 2868

SAFETY PROFILE: Combustible liquid. When heated to decomposition it emits acrid smoke and irritating fumes.

PDK300 CAS:3898-45-1 ***HR: 2***
1-PHENETHYLSEMICARBAZIDE
mf: $C_9H_{13}N_3O$ mw: 179.25

SYNS: SEMICARBAZIDE, 1-PHENETHYL- ◇ WL 20

TOXICITY DATA with REFERENCE
scu-mus TDLo:600 mg/kg (female 1-6D post):REP JOENAK 30,205,64
scu-mus LD50:1111 mg/kg JOENAK 30,205,64

SAFETY PROFILE: Moderately toxic by subcutaneous route. Experimental reproductive effects. When heated to decomposition it emits toxic fumes of NO_x.

PDK500 CAS:3473-12-9 ***HR: 2***
2-PHENETHYL-3-THIOSEMICARBAZIDE
mf: $C_9H_{13}N_3S$ mw: 195.31

SYNS: SEMICARBAZIDE, 2-PHENETHYL-3-THIO- ◇ WL 19

TOXICITY DATA with REFERENCE
scu-mus TDLo:900 mg/kg (female 1-6D post):REP JOENAK 30,205,64
scu-mus LD50:750 mg/kg JOENAK 30,205,64

SAFETY PROFILE: Moderately toxic by subcutaneous route. Experimental reproductive effects. When heated to decomposition it emits toxic fumes of NO_x and SO_x.

PDK750 CAS:2158-04-5 ***HR: 3***
PHENETHYLUREA
mf: $C_9H_{12}N_2O$ mw: 164.23

SYN: N-DIMETHYL-N-NITROSOUREA

TOXICITY DATA with REFERENCE
orl-mus LD50:2500 mg/kg JPPMAB 21,366,69
ipr-mus LD50:328 mg/kg JPETAB 52,211,34

SAFETY PROFILE: Poison by intraperitoneal route. Moderately toxic by ingestion. Many N-nitroso compounds are carcinogens. When heated to decomposition it emits toxic fumes of NO_x. See also N-NITROSO COMPOUNDS.

PDK900 CAS:49720-83-4 ***HR: 3***
o-PHENETIDINE ANTIMONYL TARTRATE

TOXICITY DATA with REFERENCE
ipr-mus LD50:21 mg(Sb)/kg AJTMAQ 25,263,45

CONSENSUS REPORTS: Antimony and its compounds are on The Community Right-To-Know List.

OSHA PEL: TWA 0.5 mg(Sb)/m^3
ACGIH TLV: TWA 0.5 mg(Sb)/m^3
NIOSH REL: TWA 0.5 mg/m^3

SAFETY PROFILE: Poison by intraperitoneal route. When heated to decomposition it emits toxic fumes of Sb. See also ANTIMONY COMPOUNDS.

PDL000 CAS:63957-36-8 ***HR: 3***
m-PHENETIDINE ANTIMONYL TARTRATE

TOXICITY DATA with REFERENCE
ipr-mus LD50:16 mg(Sb)/kg AJTMAQ 25,263,45

CONSENSUS REPORTS: Antimony and its compounds are on The Community Right-To-Know List.

OSHA PEL: TWA 0.5 mg(Sb)/m^3
ACGIH TLV: TWA 0.5 mg(Sb)/m^3
NIOSH REL: TWA 0.5 mg/m^3

SAFETY PROFILE: Poison by intraperitoneal route. When heated to decomposition it emits toxic fumes of Sb. See also ANTIMONY COMPOUNDS.

PDL500 CAS:6273-75-2 ***HR: 3***
p-PHENETIDINE ANTIMONYL TARTRATE
mf: $C_{12}H_{16}NO_8Sb \cdot H_2O$ mw: 442.06

SYN: p-PHENETIDINANTIMONYLTARTRAT(GERMAN)

TOXICITY DATA with REFERENCE
ipr-mus LD50:25 mg(Sb)/kg AJTMAQ 25,263,45
ivn-mus LDLo:80 mg/kg HBAMAK 4,1289,35

CONSENSUS REPORTS: Antimony and its compounds are on the Community Right-To-Know List.

OSHA PEL: TWA 0.5 mg(Sb)/m^3
ACGIH TLV: TWA 0.5 mg(Sb)/m^3
NIOSH REL: (Antimony) TWA 0.5 mg(Sb)/m^3

SAFETY PROFILE: Poison by intraperitoneal and intravenous routes. When heated to decomposition it emits toxic fumes of NO_x and Sb. See also ANTIMONY COMPOUNDS.

PDL750 CAS:637-56-9 ***HR: 2***
PHENETIDINE HYDROCHLORIDE
mf: $C_8H_{11}NO \cdot ClH$ mw: 173.66

TOXICITY DATA with REFERENCE
orl-rat LD50:2080 mg/kg GISAAA 35,28,70
orl-mus LD50:1180 mg/kg GISAAA 35,28,70

CONSENSUS REPORTS: Reported in EPA TSCA Inventory.

SAFETY PROFILE: Moderately toxic by ingestion. When heated to decomposition it emits very toxic fumes of HCl and NO_x.

PDM000 CAS:103-73-1 ***HR: 2***
PHENETOLE
mf: $C_8H_{10}O$ mw: 122.18

PROP: Colorless liquid. D: 0.967 @ 20°/4°, mp: −30°, bp: 172°, flash p: 145°F. Very insol in water; very sol in alc, ether.

SYNS: ETHOXYBENZENE ◇ ETHYL PHENYL ETHER

TOXICITY DATA with REFERENCE
scu-rat LDLo:3500 mg/kg RMSRA6 15,561,1895
orl-mus LD50:2200 mg/kg JPETAB 88,400,46

CONSENSUS REPORTS: Reported in EPA TSCA Inventory.

SAFETY PROFILE: Moderately toxic by ingestion and subcutaneous routes. Combustible when exposed to heat or flame, can react vigorously with oxidizing materials. To fight fire, use dry chemical, CO_2, foam, spray or mist. When heated to decomposition it emits acrid smoke and irritating fumes. See also ETHERS.

PDM250 CAS:404-82-0 ***HR: 3***
PHENFLUORAMINE HYDROCHLORIDE
mf: $C_{12}H_{16}F_3N \cdot ClH$ mw: 267.75

SYNS: N-ETHYL-α-METHYL-m-TRIFLUOROMETHYLPHENETHYLAMINE ◇ N-ETHYL-α-METHYL-m-(TRIFLUOROMETHYL)PHENETHYLAMINE HYDROCHLORIDE ◇ FENFLURAMINE HYDROCHLORIDE ◇ PONDERAL ◇ PONDERAX ◇ PONDIMIN ◇ 1-(3-TRIFLUOROMETHYLPHENYL)-2-ETHYLAMINOPROPANEHYDROCHLORIDE

TOXICITY DATA with REFERENCE
orl-rat TDLo:280 mg/kg (female 7-20D post):REP SCIEAS 205,1220,79
orl-rat TDLo:280 mg/kg (female 7-20D post):TER SCIEAS 205,1220,79
orl-hmn TDLo:7 mg/kg:EYE,CNS,GIT BMJOAE 1,740,67
orl-rat LD50:69 mg/kg TXAPA9 14,182,69
ipr-rat LD50:90 mg/kg TXAPA9 19,705,71
orl-mus LD50:170 mg/kg THERAP 20,297,65
ivn-mus LD50:90 mg/kg TXAPA9 19,705,71
orl-dog LD50:100 mg/kg TXAPA9 19,705,71

ivn-dog LD50:23 mg/kg TXAPA9 14,182,69
orl-cat LD50:60 mg/kg TXAPA9 19,705,71
orl-rbt LD50:50 mg/kg TXAPA9 19,705,71
ipr-gpg LD50:100 mg/kg TXAPA9 19,705,71

SAFETY PROFILE: Poison by ingestion, intravenous, and intraperitoneal routes. Human systemic effects by ingestion: mydriasis, change in motor activity, nausea. An experimental teratogen. Other experimental reproductive effects. When heated to decomposition it emits very toxic fumes of F^-, NO_x, and HCl.

PDM500 CAS:92-43-3 ***HR: 3***
PHENIDONE
mf: $C_9H_{10}N_2O$ mw: 162.21

PROP: Mp: 121°.

SYNS: 1-PHENYL-3-OXOPYRAZOLIDINE ◇ 1-PHENYL-3-PYRAZOLIDINONE ◇ 1-PHENYL-3-PYRAZOLIDONE

TOXICITY DATA with REFERENCE
orl-rat LD50:200 mg/kg KODAK* -,-,71
ipr-rat LD50:200 mg/kg KODAK* -,-,71

CONSENSUS REPORTS: Reported in EPA TSCA Inventory.

SAFETY PROFILE: Poison by ingestion and intraperitoneal routes. When heated to decomposition it emits toxic fumes of NO_x.

PDM750 CAS:577-91-3 ***HR: 3***
PHENIODOL
mf: $C_{15}H_{12}I_2O_3$ mw: 494.07

SYNS: BILIOGNOST ◇ BILISELECTAN ◇ β-(3,5-DIIODO-4-HYDROXYPHENYL)-α-PHENYLPROPIONIC ACID ◇ 3,5-DIIODO-α-PHENYLPHLORETIC ACID ◇ DIKOL ◇ β-(4-HYDROXY-3,5-DIIODOPHENYL)-α-PHENYLPROPIONIC ACID ◇ IODOALPHIONIC ACID ◇ IODOBIL ◇ JODOBIL ◇ β-(4-OXY-3,5-DIJOD-PHENYL)-α-PHENYL-PROPIONSAEURE (GERMAN) ◇ PRIODAX ◇ TENICID

TOXICITY DATA with REFERENCE
orl-rat LD50:1100 mg/kg KLWOAZ 20,125,41
ipr-rat LD50:510 mg/kg JAPMA8 42,476,53
scu-rat LD50:540 mg/kg KLWOAZ 20,125,41
ivn-rat LD50:390 mg/kg KLWOAZ 20,125,41
ipr-mus LD50:640 mg/kg JAPMA8 42,476,53
ivn-mus LD50:400 mg/kg JMCMAR 13,997,70
orl-cat LDLo:500 mg/kg JLCMAK 27,1376,42
ivn-cat LDLo:100 mg/kg JLCMAK 27,1376,42
ipr-gpg LD50:930 mg/kg JAPMA8 42,476,53

SAFETY PROFILE: Poison by intravenous route. Moderately toxic by ingestion, intraperitoneal, and subcutaneous routes. When heated to decomposition it emits very toxic fumes of I^-.

PDN000 CAS:55-52-7 ***HR: 3***
PHENIPRAZINE
mf: $C_9H_{14}N_2$ mw: 150.25

SYNS: CASTRON ◇ CATRAL ◇ CATRAN ◇ CATRONIAZIDE ◇ CAVODIL ◇ DICATRON ◇ FENILISOPROPILIDRAZINA ◇ 2-HYDRAZINO-1-PHENYLPROPANE ◇ JB 516 ◇ KATRON ◇ KATRONIAZID ◇ (α-METHYLPHENETHYL)HYDRAZINE ◇ MIRAL ◇ P 1142 ◇ PHENIZINE ◇ 1-PHENYL-2-HYDRAZINOPROPANE ◇ β-PHENYLISOPROPYLHYDRAZINE ◇ PHENYLISOPROPYLHYDRAZINE ◇ PIH ◇ PSICOSTEN ◇ RUN

TOXICITY DATA with REFERENCE
oms-bcs 10 mmol/L MUREAV 5,343,68
unr-man TDLo:117 mg/kg/2.5Y:EYE BMJOAE 1,331,63
orl-rat LD50:34 mg/kg 27ZQAG -,353,72
ipr-rat LD50:40 mg/kg 27ZQAG -,353,72
scu-rat LD50:45 mg/kg 27ZQAG -,353,72
ivn-rat LD50:44 mg/kg 27ZQAG -,353,72
orl-mus LD50:150 mg/kg JJPAAZ 13,186,63
ipr-mus LD50:48 mg/kg JJPAAZ 13,186,63
scu-mus LD50:95 mg/kg ANYAA9 80,568,59
ivn-mus LD50:12 mg/kg ARZNAD 12,352,62
scu-cat LDLo:30 mg/kg JPETAB 128,7,60
ivn-cat LDLo:30 mg/kg JPETAB 128,7,60

SAFETY PROFILE: Poison by ingestion, intraperitoneal, subcutaneous, and intravenous routes. Human systemic effects by an unspecified route: visual field effects. Mutation data reported. When heated to decomposition it emits toxic fumes of NO_x.

PDN250 CAS:66-05-7 ***HR: 3***
PHENIPRAZINE HYDROCHLORIDE
mf: $C_9H_{14}N_2$•ClH mw: 186.71

SYNS: CATRON HYDROCHLORIDE ◇ CATRONIACID ◇ JB 516 ◇ (1-METHYL-2-PHENYLETHYL)-HYDRAZINEIUM CHLORIDE ◇ PHENYLISOPROPYLHYDRAZINE HYDROCHLORIDE

TOXICITY DATA with REFERENCE
orl-mus LD50:59 mg/kg IJNEAQ 5,125,66
ipr-mus LD50:112 mg/kg JMCMAR 18,20,75
scu-mus LD50:87 mg/kg IJNEAQ 5,125,66
ivn-mus LD50:66 mg/kg IJNEAQ 5,125,66

SAFETY PROFILE: Poison by ingestion, intraperitoneal, subcutaneous, and intravenous routes. When heated to decomposition it emits very toxic fumes of Cl^-, NO_x, and HCl.

PDN500 CAS:537-05-3 ***HR: 3***
PHENODIANISYL HYDROCHLORIDE
mf: $C_{23}H_{25}N_3O_3$•ClH mw: 427.97

PROP: Crystals, odorless. Mp: 176°. Very sol in alc; insol in water, oils.

SYNS: ACOINE ◇ AKOIN HYDROCHLORID (GERMAN) ◇ N,N'-BIS(4-METHOXYPHENYL)-N'-(4-ETHOXYPHENYL)GUANIDINE HYDROCHLORIDE ◇ α,Γ-DI-p-ANISYL-β-(ETHOXYPHENYL)GUANIDINE

HYDROCHLORIDE ◇ DIANISYL-MONOPHENETHYLGUANIDINE HYDROCHLORIDE ◇ DIPARAANISYL-MONOPHENETHYL-GUANIDIN-HYDROCHLORID (GERMAN) ◇ 2-(4-ETHOXYPHENYL)-1,3-BIS(4-METHOXYPHENYL)GUANIDINE HYDROCHLORIDE ◇ GUANICAINE ◇ PHENODIANISYL

TOXICITY DATA with REFERENCE
scu-mus LDLo:300 mg/kg HDTU** -,-,33
ivn-mus LDLo:53 mg/kg WDMU** -,-,36
orl-dog LDLo:75 mg/kg HBAMAK 4,1291,35
scu-rbt LDLo:150 mg/kg MEIEDD 10,1043,83
scu-gpg LDLo:150 mg/kg HBAMAK 4,1291,35

SAFETY PROFILE: Poison by ingestion, intravenous, and subcutaneous routes. Solutions are decomposed by light. When heated to decomposition it emits very toxic fumes of HCl and NO_x.

PDN750 CAS:108-95-2 ***HR: 3***
PHENOL
DOT: UN 1671/UN 2312/NA 2821
mf: C_6H_6O mw: 94.12

PROP: White, crystalline mass which turns pink or red if not perfectly pure; burning taste, distinctive odor. Mp: 40.6°, bp: 181.9°, flash p: 175°F (CC), d: 1.072, autoign temp: 1319°F, vap press: 1 mm @ 40.1°, vap d: 3.24. Sol in water; misc in alc and ether.

SYNS: ACIDE CARBOLIQUE (FRENCH) ◇ BAKER'S P AND S LIQUID and OINTMENT ◇ BENZENOL ◇ CARBOLIC ACID ◇ CARBOLSAURE (GERMAN) ◇ FENOL (DUTCH, POLISH) ◇ FENOLO (ITALIAN) ◇ HYDROXYBENZENE ◇ MONOHYDROXYBENZENE ◇ MONOPHENOL ◇ NCI-C50124 ◇ OXYBENZENE ◇ PHENIC ACID ◇ PHENOL, molten (DOT) ◇ PHENOL ALCOHOL ◇ PHENOLE (GERMAN) ◇ PHENYL HYDRATE ◇ PHENYL HYDROXIDE ◇ PHENYLIC ACID ◇ PHENYLIC ALCOHOL ◇ RCRA WASTE NUMBER U188

TOXICITY DATA with REFERENCE
skn-rbt 500 mg/24H SEV BIOFX* 27-4/73
skn-rbt 535 mg open SEV UCDS** 1/6/66
eye-rbt 5 mg SEV UCDS** 1/6/66
oms-hmn:hla 17 mg/L WATRAG 19,577,85
dns-rat-orl 4 g/kg JJIND8 74,1283
orl-mus TDLo:2300 mg/kg (female 6-15D post):TER NTIS** PB85-104461
orl-rat TDLo:300 mg/kg (female 6-15D post):REP NTIS** PB83-247726
skn-mus TDLo:16 g/kg/40W-I:CAR CNREA8 19,413,59
skn-mus TD:4000 mg/kg/24W-I:NEO CNREA8 19,413,59
orl-inf LDLo:10 mg/kg 34ZIAG -,463,69
orl-hmn LDLo:14 g/kg 34ZIAG -,463,69
orl-hmn LDLo:140 mg/kg 29ZWAE -,329,68
orl-rat LD50:317 mg/kg PSEBAA 32,592,35
ihl-rat LC50:316 mg/m^3 GISAAA 41(6),103,76
skn-rat LD50:669 mg/kg BJIMAG 27,155,70
scu-rat LD50:460 mg/kg TOIZAG 10,1,63
orl-mus LD50:270 mg/kg GISAAA 38(8),6,73
ihl-mus LC50:177 mg/m^3 GISAAA 41(6),103,76
ivn-mus LD50:112 mg/kg QJPPAL 12,212,39
orl-dog LDLo:500 mg/kg HBAMAK 4,1319,35
orl-cat LDLo:80 mg/kg HBAMAK 4,1319,35
scu-cat LDLo:80 mg/kg JPETAB 80,233,44
skn-rbt LD50:850 mg/kg AIHAAP 37(10),596,76
par-rbt LDLo:300 mg/kg RMSRA6 15,561,1895
ipr-gpg LDLo:300 mg/kg HBTXAC 1,228,56

CONSENSUS REPORTS: NCI Carcinogenesis Bioassay (oral); No Evidence: mouse, rat NCITR* NCI-CG-TR-203,80. EPA Extremely Hazardous Substances List. Community Right-To-Know List. Reported in EPA TSCA Inventory. EPA Genetic Toxicology Program.

OSHA PEL: TWA 5 ppm (skin)
ACGIH TLV: TWA 5 ppm (skin); BEI: 250 mg(total phenol)/g creatinine in urine at end of shift.
DFG MAK: 5 ppm (19 mg/m^3); BAT: 300 mg/L at end of shift.
NIOSH REL: (Phenol) TWA 20 mg/m^3; CL 60 mg/m^3/15M
DOT Classification: Poison B; Label: Poison.

SAFETY PROFILE: Human poison by ingestion. An experimental poison by ingestion, subcutaneous, intravenous, parenteral, and intraperitoneal routes. Moderately toxic by skin contact. A severe eye and skin irritant. Questionable carcinogen with experimental carcinogenic and neoplastigenic data. Human mutation data reported. An experimental teratogen. Other experimental reproductive effects. Absorption of phenolic solutions through the skin may be very rapid, and can cause death within 30 minutes to several hours by exposure of as little as 64 square inches of skin. Lesser exposures can cause damage to the kidneys, liver, pancreas and spleen, and edema of the lungs. Ingestion can cause corrosion of the lips, mouth, throat, esophagus and stomach, and gangrene. Ingestion of 15 grams has killed. Chronic exposures can cause death from liver and kidney damage. Dermatitis resulting from contact with phenol or phenol-containing products is fairly common in industry. A common air contaminant.

Combustible when exposed to heat, flame, or oxidizers. Potentially explosive reaction with aluminum chloride + nitromethane (at 110°C/100 bar); formaldehyde; peroxydisulfuric acid; peroxymonosulfuric acid; sodium nitrite + heat. Violent reaction with aluminum chloride + nitrobenzene (at 120°C); sodium nitrate + trifluoroacetic acid; butadiene. Can react with oxidizing materials. To fight fire, use alcohol foam, CO_2, dry chemical. When heated to decomposition it emits acrid smoke and irritating fumes.

PDO000 CAS:108-95-2 ***HR: 3***
PHENOL (liquid)
mf: $C_6H_5{\bullet}OH$ mw: 94.11

PROP: A liquid tar acid containing over 50% benzophenol (FEREAC 41,15972,76).

SYN: CARBOLIC ACID, LIQUID (DOT)

CONSENSUS REPORTS: EPA Extremely Hazardous Substances List. Community Right-To-Know List. Reported in EPA TSCA Inventory.

DOT Classification: Poison B; Label: Poison.

SAFETY PROFILE: Poison by ingestion, inhalation and skin contact. See also PHENOL.

PDO250 CAS:98-14-6 ***HR: 3***
PHENOL-p-ARSONIC ACID
mf: $C_6H_7AsO_4$ mw: 218.05

PROP: White powder. Mp: 175-180° decomp. Very sol in water; sol in alc.

SYNS: p-HYDROXYBENZENEARSONIC ACID ◇ p-HYDROXYPHENYLARSONIC ACID ◇ OXARSANILIC ACID

TOXICITY DATA with REFERENCE
orl-rat LDLo:450 mg/kg JPETAB 63,122,38
ivn-rat LDLo:100 mg/kg JPETAB 63,122,38
ims-rat LDLo:90 mg/kg JPETAB 63,122,38
ivn-mus LD50:56 mg/kg CSLNX* NX#04548

CONSENSUS REPORTS: Arsenic and its compounds are on the Community Right-To-Know List.
OSHA PEL: TWA 0.5 mg(As)/m^3

SAFETY PROFILE: Poison by intravenous and intramuscular routes. Moderately toxic by ingestion. When heated to decomposition it emits toxic fumes of As. See also ARSENIC COMPOUNDS.

PDO750 CAS:77-09-8 ***HR: 2***
PHENOLPHTHALEIN
mf: $C_{20}H_{14}O_4$ mw: 318.34

PROP: Small crystals. Mp: 258-262°, d: 1.299. Insol in water; very sol in chloroform.

SYNS: 3,3-BIS(p-HYDROXYPHENOL)-PHTHALIDE ◇ NCI-C55798

TOXICITY DATA with REFERENCE
ipr-rat LDLo:500 mg/kg NCNSA6 5,30,53

CONSENSUS REPORTS: Reported in EPA TSCA Inventory.

SAFETY PROFILE: Moderately toxic by intraperitoneal route. Used in medicine as a laxative; in chemistry as an indicator. When heated to decomposition it emits acrid smoke and irritating fumes.

PDP100 CAS:63496-48-0 ***HR: 2***
PHENOSMOLIN

SYN: FENOSMOLIN

TOXICITY DATA with REFERENCE
orl-rat LD50:1200 mg/kg VETNAL 57(8),29,81
orl-mus LD50:1646 mg/kg VETNAL 57(8),29,81
skn-mus LD50:7500 mg/kg VETNAL 57(8),29,81

SAFETY PROFILE: Moderately toxic by ingestion. Mildly toxic by skin contact. When heated to decomposition it emits acrid smoke and irritating fumes.

PDP250 CAS:92-84-2 ***HR: 3***
PHENOTHIAZINE
mf: $C_{12}H_9NS$ mw: 199.28

PROP: Yellow, rhombic leaflets or diamond-shaped plates from toluene or butanol. Mp: 185.1°, sublimes at 130° at 1 mm, bp: 371°. Freely sol in benzene; sol in ether, hot acetic acid; sltly sol in alc and in mineral oils; practically insol in petr ether, chloroform, water.

SYNS: AFI-TIAZIN ◇ AGRAZINE ◇ ANTIVERM ◇ BIVERM ◇ CONTAVERM ◇ DIBENZOPARATHIAZINE ◇ DIBENZOTHIAZINE ◇ DIBENZO-1,4-THIAZINE ◇ ENT 38 ◇ FEENO ◇ FENOTHIAZINE (DUTCH) ◇ FENOTIAZINA (ITALIAN) ◇ FENOVERM ◇ FENTIAZIN ◇ HELMETINA ◇ LETHELMIN ◇ NEMAZENE ◇ NEMAZINE ◇ ORIMON ◇ PADOPHENE ◇ PENTHAZINE ◇ PHENEGIC ◇ PHENOSAN ◇ PHENOVERM ◇ PHENOVIS ◇ PHENOXUR ◇ PHENTHIAZINE ◇ RECONOX ◇ SOUFRAMINE ◇ THIODIFENYLAMINE (DUTCH) ◇ THIODIPHENYLAMIN (GERMAN) ◇ THIODIPHENYLAMINE ◇ TIODIFENILAMINA (ITALIAN) ◇ VERMITIN ◇ WURM-THIONAL ◇ XL-50

TOXICITY DATA with REFERENCE
orl-rat TDLo:1250 mg/kg (1-22D preg):REP AJANA2 110,29,62
orl-chd LDLo:425 mg/kg/5D LANCAO 242,86,42
orl-mus LD50:5000 mg/kg DNEUD5 7,45,80
ivn-mus LD50:178 mg/kg CSLNX* NX#00586

CONSENSUS REPORTS: EPA Genetic Toxicology Program. Reported in EPA TSCA Inventory.

OSHA PEL: TWA 5 mg/m^3 (skin)
ACGIH TLV: TWA 5 mg/m^3 (skin)

SAFETY PROFILE: Poison by intravenous route. Moderately toxic to humans by ingestion. Experimental reproductive effects. An insecticide. Large doses, i.e., heavy exposure, may cause hemolytic anemia and toxic degeneration of the liver. Can cause skin irritation and photosensitization. Dangerous; when heated to decomposition or on contact with acid or acid fumes it emits highly toxic fumes of SO_x and NO_x.

PDP500 CAS:13764-35-7 ***HR: 3***
PHENOTHIAZINE-10-CARBODITHIOIC ACID-2-(DIETHYLAMINO)ETHYL ESTER
mf: $C_{19}H_{22}N_2S_3$ mw: 374.61

SYNS: DIETHYLAMINOETHYLPHENO-THIAZINYL-10-DITHIOCARBOXYLATE ◇ PHENOTHIAZINYL-10-DITHIACARBOXYLATE de DIETHYLAMINOETHYLE (FRENCH)

TOXICITY DATA with REFERENCE
orl-rat LD50:2000 mg/kg CRSBAW 154,965,60
scu-rat LD50:2000 mg/kg CRSBAW 154,965,60
orl-mus LD50:400 mg/kg CRSBAW 154,965,60
ipr-mus LD50:130 mg/kg CRSBAW 154,965,60
scu-mus LD50:800 mg/kg CRSBAW 154,965,60
ivn-mus LD50:85 mg/kg CRSBAW 154,965,60

SAFETY PROFILE: Poison by ingestion, intravenous, and intraperitoneal routes. Moderately toxic by subcutaneous route. When heated to decomposition it emits very toxic fumes of SO_x and NO_x. See also ESTERS.

PDP600 CAS:15904-73-1 ***HR: 1***
10H-PHENOTHIAZINE-10-PROPANAMINE, 3-METHOXY-N,N,β-TRIMETHYL-
mf: $C_{19}H_{24}N_2OS$ mw: 328.51

SYNS: 10-(3-(DIMETHYLAMINO)-2-METHYLPROPYL)-3-METHOXYPHENOTHIAZINE ◇ 10-(3′-DIMETHYLAMINO-2′-METHYL-1′-PROPYL)-3-METHOXYPHENOTHIAZINE ◇ PHENOTHIAZINE, 10-(3-(DIMETHYLAMINO)-2-METHYLPROPYL)-3-METHOXY-

TOXICITY DATA with REFERENCE
eye-rbt 100 mg MLD FCTOD7 20,573,82
eye-rbt 100 mg/30S RNS MLD FCTOD7 20,573,82

SAFETY PROFILE: An eye irritant. When heated to decomposition it emits toxic fumes of SO_x.

PDQ750 CAS:262-20-4 ***HR: 3***
PHENOXATHRIN
mf: $C_{12}H_8OS$ mw: 200.26

SYNS: DIBENZOTHIOXIN ◇ 1,4-DIBENZOTHIOXINE ◇ PHENOTHIOXIN ◇ PHENOXATHINE ◇ PHENOXTHIN ◇ USAF DO-17

TOXICITY DATA with REFERENCE
ipr-mus LD50:200 mg/kg NTIS** AD277-689

CONSENSUS REPORTS: Reported in EPA TSCA Inventory.

SAFETY PROFILE: Poison by intraperitoneal route. When heated to decomposition it emits toxic fumes of SO_x.

PDR000 CAS:2120-70-9 ***HR: 1***
PHENOXYACETALDEHYDE
mf: $C_8H_8O_2$ mw: 136.16

SYN: CORTEX ALDEHYDE

TOXICITY DATA with REFERENCE
skn-rbt 500 mg/24H MOD FCTXAV 14,659,76

CONSENSUS REPORTS: Reported in EPA TSCA Inventory.

SAFETY PROFILE: A skin irritant. When heated to decomposition it emits acrid smoke and irritating fumes.

PDR100 CAS:122-59-8 ***HR: 3***
PHENOXYACETIC ACID
mf: $C_8H_8O_3$ mw: 152.16

PROP: Needles from water. Mp: 98°; bp: 285° (some decomposition). One gram dissolves in about 75 mL water. Freely sol in alc, ether, benzene, carbon disulfide, glacial acetic acid.

SYNS: GLYCOLIC ACID PHENYL ETHER ◇ GLYCOLLIC ACID PHENYL ETHER ◇ PHENOXYETHANOIC ACID ◇ o-PHENYLGLYCOLIC ACID ◇ PHENYLIUM

TOXICITY DATA with REFERENCE
skn-rbt 500 mg/24H MLD FCTXAV 17,887,79
orl-rat LD50:3700 mg/kg GISAAA 46(1),25,81
ipr-rat LD50:323 mg/kg FCTXAV 17,887,79
orl-mus LD50:3750 mg/kg GISAAA 46(1),25,81

CONSENSUS REPORTS: Reported in EPA TSCA Inventory.

SAFETY PROFILE: Poison by intraperitoneal route. Moderately toxic by ingestion. When heated to decomposition it emits acrid smoke and irritating fumes. See also ETHERS.

PDR250 CAS:4279-76-9 ***HR: 3***
PHENOXYACETYLENE
mf: C_8H_6O mw: 118.14

SAFETY PROFILE: Explodes on rapid heating. When heated to decomposition it emits acrid smoke and irritating fumes. See also ACETYLENE COMPOUNDS.

PDR500 CAS:139-59-3 ***HR: 3***
p-PHENOXYANILINE
mf: $C_{12}H_{11}NO$ mw: 185.24

SYNS: 4-AMINODIPHENYL ETHER ◇ p-AMINOPHENYL PHENYL ETHER ◇ 4-PHENOXYANILINE

TOXICITY DATA with REFERENCE
skn-rbt 500 mg/24H MLD 28ZPAK -,119,72
eye-rbt 100 mg/24H MOD 28ZPAK -,119,72
mmo-sat 25 μg/plate MUREAV 67,123,79
mma-sat 25 μg/plate MUREAV 67,123,79
orl-rat LD50:1100 mg/kg 28ZPAK -,119,72
orl-mus LD50:685 mg/kg HYSAAV 33,137,68
ipr-mus LD50:365 mg/kg HYSAAV 33,137,68

CONSENSUS REPORTS: Reported in EPA TSCA Inventory.

SAFETY PROFILE: Poison by intraperitoneal route. Moderately toxic by ingestion. Mutation data reported. A skin and eye irritant. When heated to decomposition it emits toxic fumes of NO_x. See also ETHERS.

PDS250 CAS:69782-16-7 ***HR: 3***
(4-PHENOXYBUTYL)HYDRAZINE MALEATE
mf: $C_{10}H_{16}N_2O{\bullet}C_4H_4O_4$ mw: 296.36

TOXICITY DATA with REFERENCE
orl-mus LD50:375 mg/kg JMCMAR 6,63,63
ipr-mus LD50:250 mg/kg JMCMAR 6,63,63

SAFETY PROFILE: Poison by ingestion and intraperitoneal routes. When heated to decomposition it emits fumes of NO_x.

PDS500 CAS:46231-41-8 ***HR: 3***
(2-PHENOXYETHYL)GUANIDINE
mf: $C_9H_{13}N_3O$ mw: 179.25

TOXICITY DATA with REFERENCE
orl-mus LD50:750 mg/kg JMCMAR 6,705,63
scu-mus LD50:250 mg/kg JMCMAR 6,705,63
ivn-mus LD50:44096 μg/kg BCPCA6 12,229,63

SAFETY PROFILE: Poison by subcutaneous and intravenous route. Moderately toxic by ingestion. When heated to decomposition it emits very toxic fumes of NO_x.

PDS750 CAS:4230-21-1 ***HR: 3***
(2-PHENOXYETHYL)HYDRAZINE HYDROCHLORIDE
mf: $C_8H_{12}N_2O{\bullet}ClH$ mw: 188.68

TOXICITY DATA with REFERENCE
orl-mus LD50:200 mg/kg JMCMAR 6,63,63
ipr-mus LD50:200 mg/kg JMCMAR 6,63,63

SAFETY PROFILE: Poison by ingestion and intraperitoneal routes. When heated to decomposition it emits very toxic fumes of Cl^-, NO_x, and HCl.

PDS900 ***HR: 1***
PHENOXYETHYL ISOBUTYRATE
mf: $C_{12}H_{16}O_3$ mw: 208.26

PROP: Colorless liquid; honey, roselike odor. D: 1.044, refr index: 1.492, flash p: +212°F. Misc in alc, chloroform, ether; insol in water.

SYN: FEMA No. 2873

SAFETY PROFILE: Combustible liquid. When heated to decomposition it emits acrid smoke and irritating fumes.

PDT250 CAS:59-96-1 ***HR: 3***
N-PHENOXYISOPROPYL-N-BENZYL-β-CHLOROETHYLAMINE
mf: $C_{18}H_{22}ClNO$ mw: 303.86

SYNS: A 688 ◇ BENSYLYTE ◇ 2-(N-BENZYL-2-CHLOROETHYLAMINO)-1-PHENOXYPROPANE ◇ BENZYL(2-CHLOROETHYL)-(1-METHYL-2-PHENOXYETHYL)AMINE ◇ BENZYLT ◇ N-(2-CHLOROETHYL)-N-(1-METHYL-2-PHENOXYETHYL)BENZENEMETHANAMINE ◇ N-(2-CHLOROETHYL)-N-(1-METHYL-2-PHENOXYETHYL)BENZYLAMINE ◇ DIBENYLIN ◇ DIBENYLINE ◇ DIBENZYLINE ◇ NSC 37448 ◇ PHENOXYBENZAMINE

TOXICITY DATA with REFERENCE
orl-man TDLo:2571 μg/kg (18D male):REP CCPTAY 29,479,84
ipr-mus TDLo:100 mg/kg/8W-I:NEO CNREA8 33,3069,73
orl-rat LD50:2500 mg/kg 27ZIAQ -,-,65
ice-rat LD50:3400 μg/kg AITEAT 24,223,76
orl-mus LD50:1535 mg/kg 27ZIAQ -,195,73
ivn-dog LDLo:10 mg/kg 27ZIAQ -,195,73
orl-gpg LD50:500 mg/kg 27ZIAQ -,195,73

CONSENSUS REPORTS: IARC Cancer Review: Animal Sufficient Evidence IMEMDT 24,185,80; Animal Limited Evidence IMEMDT 9,223,75

SAFETY PROFILE: Confirmed carcinogen with experimental carcinogenic and neoplastigenic data. Poison by intravenous and intracerebral routes. Moderately toxic by ingestion. Human reproductive effects by ingestion: spermatogenesis. Experimental reproductive effects. When heated to decomposition it emits very toxic fumes of Cl^- and NO_x.

PDT500 CAS:87-08-1 ***HR: 3***
PHENOXYMETHYLPENICILLIN
mf: $C_{16}H_{18}N_2O_5S$ mw: 350.42

SYNS: ACIPEN V ◇ APOPEN ◇ BEROMYCIN ◇ DISTAQUAINE V ◇ ESKACILLIAN V ◇ FENACILIN ◇ FENOSPEN ◇ FENOXYPEN ◇ MEROPENIN ◇ ORACILLIN ◇ ORATREN ◇ OSPEN ◇ PENICILLIN PHENOXYMETHYL ◇ PENICILLIN V ◇ PEN-ORAL ◇ PEN V ◇ PENVEE ◇ PHENOPENICILLIN ◇ 6-PHENOXYACETAMIDOPENICILLANIC ACID ◇ PHENOXYMETHYLENEPENICILLINIC ACID ◇ STABICILLIN ◇ V-CIL ◇ V-CILLIN ◇ VEBECILLIN

TOXICITY DATA with REFERENCE
dni-mus:ast 200 μg/L NEOLA4 22,105,75
oms-mus:ast 200 μg/L NEOLA4 22,105,75
orl-wmn TDLo:10 mg/kg/2D:LIV,MET,SKN LANCAO 2,1297,76
orl-mus LD50:6578 mg/kg ANTBAL 3(4),37,58
ipr-mus LD50:12 mg/kg ANTBAL 17,616,72
scu-mus LD50:24 mg/kg ANTBAL 17,616,72
ivn-mus LD50:8 mg/kg ANTBAL 17,616,72

SAFETY PROFILE: Poison by intraperitoneal, subcutaneous, and intravenous routes. Human systemic effects by ingestion: impaired liver function, dermatitis, fever. Mutation data reported. When heated to decomposition it emits very toxic fumes of SO_x and NO_x.

PDT750 CAS:132-98-9 ***HR: 2***
d-α-PHENOXYMETHYLPENICILLINATE K SALT
mf: $C_{16}H_{17}N_2O_5S{\bullet}K$ mw: 388.51

SYNS: ANTIBIOCIN ◇ APSIN VK ◇ ARCACIL ◇ ARCASIN ◇ BEROMYCIN ◇ BEROMYCIN 400 ◇ BEROMYCIN (penicillin) ◇ BETAPEN-VK ◇ CALCIOPEN K ◇ CLIACIL ◇ COMPOCILLIN-VK ◇ DISTAKAPS V-K ◇ DISTAQUAINE V-K ◇ DOWPEN V-K ◇ DQV-K ◇ FENOXYPEN ◇ ISOCILLIN ◇ ICIPEN ◇ ISPENORAL ◇ LEDERCILLIN VK ◇ MEGACILLIN ORAL ◇ ORACIL-VK ◇ ORAPEN ◇ OSPENEFF ◇ PEDIPEN ◇ PENAGEN ◇ PENCOMPREN ◇ PENICILLIN POTASSIUM PHENOXYMETHYL ◇ PENICILLIN V POTASSIUM ◇ PENICILLIN V POTASSIUM SALT ◇ PEN-VEE-K ◇ PEN-VEE-K POWDER ◇ PENVIKAL ◇ PEN-V-K POWDER ◇ PFIZERPEN VK ◇ PHENOXYMETHYLPENICILLIN POTASSIUM ◇ POTASSIUM PENICILLIN V SALT ◇ POTASSIUM PHENOXYMETHYLPENICILLIN ◇ PVK ◇ QIDPEN VK ◇ ROBICILLIN VK ◇ ROCILLIN-VK ◇ ROSCOPENIN ◇ SK-PENICILLIN VK ◇ STABILLIN VK SYRUP 125 ◇ STABILLIN VK SYRUP 62.5 ◇ SUMAPEN VK ◇ SUSPEN ◇ UTICILLIN VK ◇ V-CIL-K ◇ V-CILLIN K ◇ VEETIDS ◇ VEPEN

TOXICITY DATA with REFERENCE
orl-rat LD50:1040 mg/kg TXAPA9 18,185,71
ims-rat LD50:1124 mg/kg PJPPAA 29,39,77
ipr-mus LD50:1351 mg/kg NIIRDN 6,661,82
ivn-mus LD50:1000 mg/kg ARZNAD 12,751,62

CONSENSUS REPORTS: Reported in EPA TSCA Inventory.

SAFETY PROFILE: Moderately toxic by ingestion, intramuscular, intraperitoneal, and intravenous routes. When heated to decomposition it emits very toxic fumes of NO_x, K_2O, and SO_x. Used as a pharmaceutical and veterinary drug.

PDU250 CAS:17692-39-6 ***HR: 3***
4-(3-(p-PHENOXYMETHYLPHENYL)PROPYL) MORPHOLINE
mf: $C_{20}H_{25}NO_2$ mw: 311.46

SYNS: ERBOCAIN ◇ FOMOCAINE ◇ p-FOMOCAINE ◇ P 652 ◇ PANACAINE ◇ 4-(3-(4-(PHENOXYMETHYL)PHENYL)PROPYL)-MORPHOLINE, (9CI) ◇ N-(Γ-(5-PHENOXYMETHYL-PHENYL)-PROPYL)-MORPHOLIN (GERMAN) ◇ N-(4-(PHENOXY-METHYL)-Γ-PHENYL-PROPYL)-MORPHOLIN (GERMAN) ◇ 4-(3-(α-PHENOXY-p-TOLYL)PROPYL)-MORPHOLINE

TOXICITY DATA with REFERENCE
ivn-rat LD50:140 mg/kg ARZNAD 8,539,58
ipr-mus LD50:325 mg/kg ARZNAD 8,539,58
scu-mus LD50:1300 mg/kg ARZNAD 8,539,58
ivn-mus LD50:110 mg/kg ARZNAD 8,539,58

SAFETY PROFILE: Poison by intravenous and intraperitoneal routes. Moderately toxic by subcutaneous route. When heated to decomposition it emits toxic fumes of NO_x.

PDU500 CAS:69782-01-0 ***HR: 3***
((1-PHENOXYMETHYL)PROPYL)HYDRAZINE HYDROCHLORIDE
mf: $C_{10}H_{16}N_2O{\cdot}ClH$ mw: 216.74

TOXICITY DATA with REFERENCE
orl-mus LD50:200 mg/kg JMCMAR 6,63,63
ipr-mus LD50:200 mg/kg JMCMAR 6,63,63

SAFETY PROFILE: Poison by ingestion and intraperitoneal routes. When heated to decomposition it emits very toxic fumes of NO_x and HCl.

PDU750 CAS:39754-64-8 ***HR: 2***
PHENOXYMETHYL-6-TETRAHYDROXAZINE-1,3-THIONE-2
mf: $C_{11}H_{13}NO_2S$ mw: 223.31

SYNS: 6-PHENOXYMETHYL-3,4,5,6-TETRAHYDROXAZINE-1,3 THIONE-2 ◇ 3,4,5,6-TETRAHYDRO-6-PHENOXYMETHYL-2H-1,3-OXAZINE-2-THIONE ◇ TETRAHYDRO-6-(PHENOXYMETHYL)-2H-1,3-OXAZINE-2-THIONE ◇ TIFEMOXONE

TOXICITY DATA with REFERENCE
orl-rat LD50:2200 mg/kg ATSUDG 1,229,78
orl-mus LDLo:700 mg/kg EJMCA5 11,75,76

SAFETY PROFILE: Moderately toxic by ingestion. When heated to decomposition it emits very toxic fumes of NO_x and SO_x.

PDV000 CAS:69782-09-8 ***HR: 3***
(5-PHENOXYPENTYL)HYDRAZINE MALEATE
mf: $C_{11}H_{18}N_2O{\cdot}C_4H_4O_4$ mw: 310.39

TOXICITY DATA with REFERENCE
orl-mus LD50:500 mg/kg JMCMAR 6,63,63
ipr-mus LD50:125 mg/kg JMCMAR 6,63,63

SAFETY PROFILE: Poison by intraperitoneal route. Moderately toxic by ingestion. When heated to decomposition it emits toxic fumes of NO_x.

PDV250 CAS:713-68-8 ***HR: 2***
m-PHENOXYPHENOL
mf: $C_{12}H_{10}O_2$ mw: 186.22

TOXICITY DATA with REFERENCE
orl-rat LD50:1211 mg/kg HYSAAV 36,293,71
skn-rat LD50:2750 mg/kg HYSAAV 36,293,71
orl-mus LD50:493 mg/kg HYSAAV 36,293,71
orl-mam LD50:2307 mg/kg GISAAA 45(10),16,80
skn-mam LD50:1057 mg/kg GISAAA 45(10),16,80

CONSENSUS REPORTS: Reported in EPA TSCA Inventory.

SAFETY PROFILE: Moderately toxic by ingestion and skin contact. When heated to decomposition it emits acrid smoke and irritating fumes.

PDV700 CAS:3941-06-8 ***HR: 3***
PHENOXYPROPAZINE MALEATE
mf: $C_9H_{14}N_2O{\cdot}C_4H_4O_4$ mw: 282.33

SYNS: COMPOUND 1275 ◇ COMPOUND HP1275 ◇ DRAZINE ◇ HP 1275 ◇ HYDRAZINE, (1-METHYL-2-PHENOXYETHYL)-, (Z)-2-BUTENEDIOATE (1:1) (9CI) ◇ HYDRAZINE, (1-METHYL-2-PHENOXYETHYL)-, MALEATE ◇ HYDRAZINE, 1-(1-PHENOXY-2-PROPYL)-, MALEATE ◇ (1-METHYL-2-PHENOXYETHYL)HYDRAZINE MALEATE ◇ (1-METHYL-2-PHENOXYETHYL)HYDRAZINIUM

TOXICITY DATA with REFERENCE
orl-rat TDLo:108 mg/kg (female 9D pre):REP JRPFA4 2,362,61
orl-mus LD50:460 mg/kg 27ZQAG -,402,72
ipr-mus LD50:350 mg/kg JMCMAR 6,63,63

SAFETY PROFILE: Poison by ingestion and intraperitoneal routes. Experimental reproductive effects. When heated to decomposition it emits toxic fumes of NO_x.

PDV750 CAS:69781-96-0 ***HR: 3***
(3-PHENOXYPROPYL)HYDRAZINE MALEATE
mf: $C_9H_{14}N_2O \cdot C_4H_4O_4$ mw: 282.33

TOXICITY DATA with REFERENCE
orl-mus LD50:250 mg/kg JMCMAR 6,63,63
ipr-mus LD50:250 mg/kg JMCMAR 6,63,63

SAFETY PROFILE: Poison by ingestion and intraperitoneal routes. When heated to decomposition it emits toxic fumes of NO_x.

PDW250 CAS:55837-27-9 ***HR: 3***
4-PHENOXY-3-(PYRROLIDINYL)-5-SULFAMOYLBENZOIC ACID
mf: $C_{17}H_{18}N_2O_5S$ mw: 362.43

SYNS: 3-(AMINOSULFONYL)-4-PHENOXY-5-(1-PYRROLIDINYL)-BENZOIC ACID ◇ ARELIZ ◇ HOE 118 ◇ PIRETANIDE ◇ S 73 4118

TOXICITY DATA with REFERENCE
orl-rbt TDLo:13 mg/kg (female 6-18D post):REP KSRNAM 14,4367,80
orl-rbt TDLo:13 mg/kg (female 6-18D post):TER KSRNAM 14,4367,80
orl-rat LD50:5601 mg/kg DRFUD4 2,393,77
ipr-rat LD50:485 mg/kg KSRNAM 14,4254,80
scu-rat LD50:1100 mg/kg KSRNAM 14,4254,80
ivn-rat LD50:700 mg/kg KSRNAM 14,4254,80
orl-mus LD50:2 g/kg KSRNAM 14,4254,80
ipr-mus LD50:700 mg/kg KSRNAM 14,4254,80
scu-mus LD50:660 mg/kg KSRNAM 14,4254,80
ivn-mus LD50:618 mg/kg KSRNAM 14,4254,80
orl-rbt LD50:960 mg/kg KSRNAM 14,4254,80
ivn-rbt LD50:178 mg/kg KSRNAM 14,4254,80

SAFETY PROFILE: Poison by intravenous route. Moderately toxic by ingestion, intraperitoneal, and subcutaneous routes. An experimental teratogen. Experimental reproductive effects. Used as a diuretic agent. When heated to decomposition it emits very toxic fumes of SO_x and NO_x.

PDW400 CAS:50-60-2 ***HR: 3***
PHENTALAMINE
mf: $C_{17}H_{19}N_3O$ mw: 281.39

PROP: Crystals. Mp: 174-175°.

SYNS: C 7337 CIBA ◇ 2-((N-(m-HYDROXYPHENYL)-p-TOLUIDINO) METHYL)-2-IMIDAZOLINE ◇ 2-(m-HYDROXY-N-p-TOLYLANILINO-METHYL)-2-IMIDAZOLINE ◇ PHENOTOLAMINE ◇ PHENTOLAMINE ◇ REGITIN ◇ REGITINE ◇ REGITIPE ◇ ROGITINE ◇ 2-(N'-p-TOLYL-N'-m-HYDROXYPHENYLAMINOMETHYL)-2-IMIDAZOLINE

TOXICITY DATA with REFERENCE
ivn-rat TDLo:100 μg/kg (8D preg):REP DPTHDL 7(Suppl 1),72,84
orl-rat LD50:1250 mg/kg PSEBAA 71,70,49
scu-rat LD50:275 mg/kg PSEBAA 71,70,49
ivn-rat LD50:75 mg/kg PSEBAA 71,70,49
orl-mus LD50:1000 mg/kg AIPTAK 112,319,57
ipr-mus LD50:200 mg/kg ARZNAD 21,1727,71
ivn-mus LD50:35 mg/kg AIPTAK 105,317,56
orl-rbt LD50:2000 mg/kg CLDND* 81,352,51
scu-rbt LDLo:200 mg/kg SMWOAS 81,352,51
ivn-rbt LD50:28 mg/kg AIPTAK 174,243,68

SAFETY PROFILE: Poison by subcutaneous, intravenous, and intraperitoneal routes. Moderately toxic by ingestion. Experimental reproductive effects. When heated to decomposition it emits toxic fumes of NO_x. See also AMINES.

PDW500 CAS:437-38-7 ***HR: 3***
PHENTANYL
mf: $C_{22}H_{28}N_2O$ mw: 336.52

SYNS: FENTANEST ◇ FENTANIL ◇ FENTANYL ◇ N-PHENETHYL-4-(N-PROPIONYLANILINO)PIPERIDINE ◇ 1-PHENETHYL-4-N-PROPIONYLANILINOPIPERIDINE ◇ N-PHENYL-N-(1-(2-PHENYLETHYL)-4-PIPERIDINYL)-PROPANAMIDE (9CI) ◇ R 4263 ◇ SENTONIL

TOXICITY DATA with REFERENCE
ivn-rat LD50:3050 μg/kg DDREDK 1,83,81
ivn-hmn TDLo:2 μg/kg:PUL,CNS BMJOAE 1,278,78
ipr-mus LD50:76 mg/kg PCJOAU 10,1193,76
ivn-mus LD50:2910 μg/kg JMCMAR 17,1047,74

SAFETY PROFILE: Poison by intravenous and intraperitoneal routes. Human systemic effects by intravenous route: somnolence, respiratory depression. When heated to decomposition it emits toxic fumes of NO_x.

PDW750 CAS:990-73-8 ***HR: 3***
PHENTANYL CITRATE
mf: $C_{22}H_{28}N_2O \cdot C_6H_8O_7$ mw: 528.66

SYNS: FENTANEST ◇ FENTANYL CITRATE ◇ LEPTANAL ◇ McN-JR 4263 ◇ MCN-JR-4263-49 ◇ PENTANYL ◇ N-(1-PHENETHYL-4-PIPERIDI-NYL)PROPIONANILIDE DIHYDROGEN CITRATE ◇ N-(1-PHENETHYL-4-PIPERIDYL)PROPIONANILIDECITRATE ◇ N-(1-PHENETHYL-4-PIPERIDYL)PROPIONANILIDE DIHYDROGEN CITRATE ◇ R 4263 ◇ R 5240 ◇ SUBLIMAZE ◇ SUBLIMAZE CITRATE

TOXICITY DATA with REFERENCE
orl-rat LD50:18 mg/kg TXAPA9 18,185,71
scu-rat LD50:5460 μg/kg NIIRDN 6,666,82
ivn-rat LD50:990 μg/kg NIIRDN 6,666,82
ivn-mus LD50:10100 μg/kg NIIRDN 6,666,82

SAFETY PROFILE: Poison by ingestion, subcutaneous, and intravenous routes. When heated to decomposition it emits toxic fumes of NO_x. See also PHENTANYL.

PDW950 CAS:65-28-1 ***HR: 3***
PHENTOLAMINE MESYLATE
mf: $C_{17}H_{19}N_3O \cdot CH_4O_3S$ mw: 377.50

SYNS: PHENTOLAMINE MESILATE ◇ PHENTOLAMINE METHANESULFONATE ◇ PHENTOLAMINE METHANESULPHONATE ◇ REGITINE MESYLATE ◇ REGITINE METHANESULFONATE ◇ REGITIN METHANESULPHONATE

TOXICITY DATA with REFERENCE
ipr-rat TDLo:4 mg/kg (female 22D post):REP JRPFA4 76,415,86
scu-rat LDLo:275 mg/kg NIIRDN 6,667,82
ivn-rat LDLo:75 mg/kg NIIRDN 6,667,82
scu-rbt LDLo:200 mg/kg NIIRDN 6,667,82
ivn-rbt LDLo:35 mg/kg NIIRDN 6,667,82

SAFETY PROFILE: Poison by intravenous and subcutaneous routes. Experimental reproductive effects. When heated to decomposition it emits toxic fumes of SO_x and NO_x.

PDX000 CAS:101-48-4 ***HR: 2***
PHENYLACETALDEHYDE DIMETHYL ACETAL
mf: $C_{10}H_{14}O_2$ mw: 166.24

PROP: Colorless liquid; strong odor. D: 1.000-1.006, refr index: 1.493, flash p: 194°F. Sol in fixed oils, propylene glycol; insol in glycerin.

SYNS: (2,2-DIMETHOXYETHYL)-BENZENE (9CI) ◇ 1,1-DIMETHOXY-2-PHENYLETHANE ◇ FEMA No. 2876 ◇ HYSCYLENE P ◇ PHENACETALDEHYDE DIMETHYL ACETAL ◇ α-TOLYL ALDEHYDE DIMETHYL ACETAL ◇ VIRIDINE

TOXICITY DATA with REFERENCE
orl-rat LD50:3500 mg/kg FCTXAV 13,681,75

CONSENSUS REPORTS: Reported in EPA TSCA Inventory.

SAFETY PROFILE: Moderately toxic by ingestion. Combustible liquid. When heated to decomposition it emits acrid smoke and irritating fumes. See also ALDEHYDES.

PDX250 CAS:5694-72-4 ***HR: 2***
PHENYLACETALDEHYDE GLYCERYL ACETAL
mf: $C_{11}H_{14}O_3$ mw: 194.25

SYNS: 2-BENZYL-4-HYDROXYMETHYL-1,3-DIOXANE ◇ 2-BENZYL-4-HYDROXYMETHYL-1,3-DIOXOLANE ◇ 2-BENZYL-4-METHANOL-1,3-DIOXANE

TOXICITY DATA with REFERENCE
skn-rbt 500 mg/24H MOD FCTXAV 14,829,76
orl-rat LD50:1720 mg/kg FCTXAV 14,829,76
ipr-mus LD50:523 mg/kg AIPTAK 85,474,51

CONSENSUS REPORTS: Reported in EPA TSCA Inventory.

SAFETY PROFILE: Moderately toxic by ingestion and intraperitoneal routes. A skin irritant. When heated to decomposition it emits acrid smoke and irritating fumes. See also ALDEHYDES.

PDX500 CAS:519-87-9 ***HR: 2***
N-PHENYLACETAMIDE
mf: $C_{14}H_{13}NO$ mw: 211.28

SYNS: N-ACETYLDIPHENYLAMINE ◇ DIPHENYLACETAMIDE

TOXICITY DATA with REFERENCE
ipr-mus LD50:600 mg/kg PCJOAU 10,579,76

CONSENSUS REPORTS: Reported in EPA TSCA Inventory.

SAFETY PROFILE: Moderately toxic by intraperitoneal route. When heated to decomposition it emits toxic fumes of NO_x.

PDX750 CAS:103-81-1 ***HR: 2***
2-PHENYLACETAMIDE
mf: C_8H_9NO mw: 135.18

PROP: Plates or leaflets. Mp: 155°. Very sol in alc; sltly sol in water, benzene, ether.

SYNS: BENZENEACETAMIDE (9CI) ◇ α-PHENYLACETAMIDE ◇ PHENYLACETIC ACID AMIDE ◇ PHENYL-β-ACETYLAMINE ◇ α-TOLUAMIDE ◇ α-TOLUIMIDIC ACID

TOXICITY DATA with REFERENCE
ipr-mus LD50:430 mg/kg PCJOAU 10,579,76

CONSENSUS REPORTS: Reported in EPA TSCA Inventory.

SAFETY PROFILE: Moderately toxic by intraperitoneal route. When heated to decomposition it emits toxic fumes of NO_x.

PDY000 CAS:2113-47-5 ***HR: 3***
2'-PHENYLACETANILIDE
mf: $C_{14}H_{13}NO$ mw: 211.28

SYNS: ACETAMIDOBIPHENYL ◇ 2-ACETYLAMINOBIPHENYL ◇ N-(2-BIPHENYLYL)ACETAMIDE

TOXICITY DATA with REFERENCE
mma-sat 250 μg/plate MUREAV 118,49,83
orl-rat TDLo:4200 mg/kg/35W-C:ETA,REP CNREA8 16,525,56

SAFETY PROFILE: Experimental reproductive effects. Questionable carcinogen with experimental tumorigenic data. Mutation data reported. When heated to decomposition it emits toxic fumes of NO_x.

PDY250 CAS:2113-54-4 ***HR: 3***
3'-PHENYLACETANILIDE
mf: $C_{14}H_{13}NO$ mw: 211.28

SYNS: 3-ACETYLAMINOBIPHENYL ◇ N-(3-BIPHENYLYL)ACETAMIDE

TOXICITY DATA with REFERENCE
orl-rat TDLo:4200 mg/kg/35W-C:ETA CNREA8 16,525,56

SAFETY PROFILE: Questionable carcinogen with experimental tumorigenic data. When heated to decomposition it emits toxic fumes of NO_x.

PDY500 CAS:4075-79-0 ***HR: 3***
4'-PHENYLACETANILIDE
mf: $C_{14}H_{13}NO$ mw: 211.28

SYNS: 4-ACETAMIDOBIPHENYL ◇ 4-ACETYLAMINOBIPHENYL ◇ 4-BIPHENYLACETAMIDE ◇ N-4-BIPHENYLACETAMIDE ◇ N-(4-BIPHENYLYL)ACETAMIDE ◇ p-PHENYLACETANILIDE

TOXICITY DATA with REFERENCE
mma-sat 100 μg/plate BBRCA9 73,1025,76
dnr-ham:fbr 1 μmol/L JNCIAM 54,1287,75
orl-rat TDLo:2770 mg/kg/21W-C:CAR CNREA8 15,188,55
orl-dog TDLo:13 g/kg/3Y-I:ETA CNREA8 23,921,63
orl-rat TD:4070 mg/kg/34W-C:CAR CNREA8 16,525,56

CONSENSUS REPORTS: Reported in EPA TSCA Inventory.

SAFETY PROFILE: Questionable carcinogen with experimental carcinogenic and tumorigenic data. Mutation data reported. When heated to decomposition it emits toxic fumes of NO_x. Used in the manufacture of plastics, resins, rubber, synthetics, dyes, and pigments.

PDY750 CAS:122-79-2 ***HR: 2***
PHENYL ACETATE
mf: $C_8H_8O_2$ mw: 136.16

PROP: Water-white liquid. D: 1.073 @ 20°/40°, bp: 195-196°, vap d: 4.7, flash p: 176°F. Misc in alc, chloroform, and ether; very sltly sol in water; highly refractive.

SYNS: ACETYL PHENOL ◇ PHENOL ACETATE

TOXICITY DATA with REFERENCE
skn-rbt 535 mg open MLD UCDS** 11/4/64
orl-rat LD50:1630 mg/kg AIHAAP 30,470,69
skn-rbt LD50:8000 mg/kg UCDS** 11/4/69

CONSENSUS REPORTS: Reported in EPA TSCA Inventory.

SAFETY PROFILE: Moderately toxic by ingestion. Mildly toxic by skin contact. A skin irritant. Combustible when exposed to heat, flame or oxidizers. To fight fire, use alcohol foam. When heated to decomposition it emits acrid smoke and irritating fumes.

PDY850 CAS:103-82-2 ***HR: 2***
PHENYLACETIC ACID
mf: $C_8H_8O_2$ mw: 136.16

PROP: Leaflets on distillation in vac; plates, tablets from petr ether; disagreeable odor of geranium. Mp: 76.5°, bp: 265.5°,d (77/4) 1.091, flash p: +212°F. Sltly sol in cold water; freely sol in hot water; sol in alc and ether. Solubility @ 25° in chloroform (moles/L): 4.422; in carbon tetrachloride: 1.842; in acetylene tetrachloride: 4.513; in trichlorethylene: 3.299; in tetrachlorethylene: 1.558; in pentachloroethane: 3.252.

SYNS: BENZENACETIC ACID ◇ BENZENEACETIC ACID ◇ FEMA No. 2878 ◇ omega-PHENYLACETIC ACID ◇ α-TOLUIC ACID

TOXICITY DATA with REFERENCE
orl-rat TDLo:450 mg/kg (4D preg):TER VPITAR 32,50,73
orl-rat LD50:2250 mg/kg VPITAR 33(5),48,74
ipr-rat LD50:1600 mg/kg BCFAAI 112,53,73
orl-mus LD50:2250 mg/kg VPITAR 33(5),48,74
ipr-mus LD50:2270 mg/kg FRPSAX 13,286,58
scu-mus LD50:1500 mg/kg AIPTAK 116,154,58
orl-gpg LD50:2250 mg/kg VPITAR 33(5),48,74

CONSENSUS REPORTS: Reported in EPA TSCA Inventory.

SAFETY PROFILE: Moderately toxic by ingestion, subcutaneous, and intraperitoneal routes. An experimental teratogen. Combustible liquid. Used in production of drugs of abuse. When heated to decomposition it emits acrid smoke and irritating fumes.

PDY870 ***HR: 2***
PHENYLACETIC ACID 2-BENZYLHYDRAZIDE
mf: $C_{15}H_{16}N_2O$ mw: 240.33

SYNS: N-BENZYL-N'-PHENYLACETYLHYDRAZIDE ◇ P-5307

TOXICITY DATA with REFERENCE
orl-rat LD50:2500 mg/kg JJPAAZ 13,186,63
ipr-rat LD50:515 mg/kg JJPAAZ 13,186,63
orl-mus LD50:767 mg/kg JJPAAZ 13,186,63
ipr-mus LD50:525 mg/kg JJPAAZ 13,186,63

SAFETY PROFILE: Moderately toxic by ingestion and intraperitoneal routes. When heated to decomposition it emits toxic fumes of NO_x.

PEA500 CAS:4468-48-8 **HR: 3**
α-PHENYLACETOACETONITRILE
mf: $C_{10}H_9NO$ mw: 159.20

SYNS: PHENYLACETO-ACETONITRILE ◇ USAF PE-1

TOXICITY DATA with REFERENCE
ipr-mus LD50:200 mg/kg NTIS** AD277-689

CONSENSUS REPORTS: Cyanide and its compounds are on the Community Right-To-Know List. Reported in EPA TSCA Inventory.

SAFETY PROFILE: Poison by intraperitoneal route. When heated to decomposition it emits toxic fumes of NO_x and CN^-. See also NITRILES.

PEA750 CAS:140-29-4 **HR: 3**
PHENYLACETONITRILE
DOT: UN 2470
mf: C_8H_7N mw: 117.16

PROP: Oily liquid; aromatic odor. Mp: −23.8°, bp: 233.5°, d: 1.0214 @ 15°/15°, vap press: 1 mm @ 60.0°. Insol in water; misc in alc, ether.

SYNS: BENZENEACETONITRILE ◇ BENZYL CYANIDE ◇ BENZYL NITRILE ◇ (CYANOMETHYL)BENZENE ◇ α-CYANOTOLUENE ◇ ω-CYANOTOLUENE ◇ 2-PHENYLACETONITRILE ◇ PHENYLACETONITRILE, liquid (DOT) ◇ α-TOLUNITRILE ◇ USAF KF-21

TOXICITY DATA with REFERENCE
skn-rbt 500 mg/24H MLD FCTOD7 20(Suppl),803,82
orl-rat LD50:270 mg/kg GISAAA 32,20,67
ihl-rat LC50:430 mg/m^3/2H HYSAAV 32,176,67
skn-rat LD50:2 g/kg FCTOD7 20(Suppl),803,82
ipr-rat LDLo:74982 μg/kg TOLED5 10,265,82
orl-mus LD50:45500 μg/kg ARTODN 55,47,84
ihl-mus LCLo:100 mg/m^3 GISAAA 32,20,67
scu-mus LDLo:32 mg/kg AIPTAK 12,447,04
skn-rbt LD50:270 mg/kg FCTOD7 20(Suppl),803,82
ipr-mus LD50:10 mg/kg NTIS** AD277-689
scu-rbt LDLo:50 mg/kg AIPTAK 5,161,1899
scu-frg LDLo:1500 mg/kg AIPTAK 5,161,1899

CONSENSUS REPORTS: EPA Extremely Hazardous Substances List. Community Right-To-Know List. Reported in EPA TSCA Inventory.

DOT Classification: Poison B; Label: St. Andrews Cross.

SAFETY PROFILE: Poison by ingestion, inhalation, skin contact, subcutaneous, and intraperitoneal routes. A skin irritant. Explosive reaction with sodium hypochlorite. Used in production of drugs of abuse. When heated to decomposition it emits very toxic fumes of CN^- and NO_x. See also NITRILES.

PEB000 CAS:451-40-1 **HR: 3**
2-PHENYLACETOPHENONE
mf: $C_{14}H_{12}O$ mw: 196.26

SYNS: BENZYL PHENYL KETONE ◇ DEOXYBENZOIN ◇ DESOXYBENZOIN ◇ 1,2-DIPHENYLETHANONE ◇ PHENYL BENZYL KETONE

TOXICITY DATA with REFERENCE
ivn-mus LD50:320 mg/kg CSLNX* NX#03836

CONSENSUS REPORTS: Reported in EPA TSCA Inventory.

SAFETY PROFILE: Poison by intravenous route. When heated to decomposition it emits acrid smoke and irritating fumes. See also KETONES.

PEB250 CAS:63040-30-2 **HR: 3**
4'-PHENYL-o-ACETOTOLUIDE
mf: $C_{15}H_{15}NO$ mw: 225.31

SYN: 3-METHYL-4-ACETYLAMINOBIPHENYL

TOXICITY DATA with REFERENCE
orl-rat TDLo:2160 mg/kg/24W-C:CAR CBREA8 22,1002,62

SAFETY PROFILE: Questionable carcinogen with experimental carcinogenic data. When heated to decomposition it emits toxic fumes of NO_x.

PEB750 CAS:536-74-3 **HR: 3**
PHENYLACETYLENE
mf: C_8H_6 mw: 102.14

SYN: ETHYNYLBENZENE

TOXICITY DATA with REFERENCE
orl-rat LDLo:5000 mg/kg AMIHAB 19,403,59
ivn-mus LD50:100 mg/kg CSLNX* NX#04866

CONSENSUS REPORTS: Reported in EPA TSCA Inventory.

SAFETY PROFILE: Poison by intravenous route. Mildly toxic by ingestion. Explosive reaction with perchloric acid above −78°C. When heated to decomposition it emits acrid smoke and irritating fumes. See also ACETYLENE COMPOUNDS.

PEB775 CAS:25439-20-7 **HR: 2**
PHENYLACETYLGLYCINE DIMETHYLAMIDE
mf: $C_{12}H_{16}N_2O_2$ mw: 220.30

SYN: N,N-DIMETHYL-2-(2-PHENYLACETAMIDO)ACETAMIDE

TOXICITY DATA with REFERENCE
ipr-rat LD50:1180 mg/kg NIIRDN 6,658,82

orl-mus LD50:1600 mg/kg NIIRDN 6,658,82
ipr-mus LD50:1400 mg/kg NIIRDN 6,658,82
ivn-mus LD50:965 mg/kg NIIRDN 6,658,82

SAFETY PROFILE: Moderately toxic by ingestion, intravenous and intraperitoneal other routes. When heated to decomposition it emits toxic fumes of NO_x. See also AMIDES.

PEC250 CAS:63-98-9 ***HR: 2***
(PHENYLACETYL)UREA
mf: $C_9H_{10}N_2O_2$ mw: 178.21

SYNS: A-1348 ◇ ACETYLUREUM ◇ N-(AMINOCARBONYL) BENZENEACETAMIDE ◇ CARBAMIDE PHENYLACETATE ◇ CETYLUREUM ◇ COMITIADONE ◇ EFERON ◇ EPHERON ◇ EPICLASE ◇ FELUREA ◇ FENACEMID ◇ FENACEMIDE ◇ FENACETEAMIDE ◇ FENACETIL-KARBAMIDE ◇ FENILEP ◇ FENISED ◇ FENOSTENYL ◇ FENURAL ◇ FENUREA ◇ FENURONE ◇ FENYTAN ◇ NEOPHEDAN ◇ NEOPHENAL ◇ PHACETUR ◇ PHENACALUM ◇ PHENACEMIDE ◇ PHENACEREUM ◇ PHENACETUR ◇ PHENACETYLCARBAMIDE ◇ PHENACETYLUREA ◇ PHENARONE ◇ PHENICARB ◇ PHENURON ◇ PHENUTAL ◇ α-PHENYLACETYLUREA ◇ PHENYLACETYLUREE (FRENCH) ◇ PHENYRIT ◇ PHETYLUREUM ◇ TRIOXANONA

TOXICITY DATA with REFERENCE
ipr-mus TDLo:2781 mg/kg (female 8-10D post):TER TXAPA9 64,271,82
orl-rat LD50:1600 mg/kg ABMGAJ 2,335,59
orl-mus LD50:987 mg/kg NIIRDN 6,655,82
ipr-mus LD50:1550 mg/kg RPOBAR 2,309,70
orl-dog LDLo:3500 mg/kg 27ZIAQ -,188,73
orl-cat LDLo:2000 mg/kg 27ZIAQ -,-,65
orl-rbt LD50:2500 mg/kg 27ZQAG -,366,72
ipr-rbt LD50:2500 mg/kg 27ZQAG -,366,72

SAFETY PROFILE: Moderately toxic by ingestion and intraperitoneal routes. Experimental teratogenic effects. Used as an anticonvulsive agent. When heated to decomposition it emits toxic fumes of NO_x.

PEC500 CAS:673-06-3 ***HR: 1***
d-PHENYLALANINE
mf: $C_9H_{11}NO_2$ mw: 165.21

PROP: Needles from alc, white crystalline platlets. Mp: 104-105°; Sol in hot water; very sltly sol in alc; sltly sol petr ether.

SYNS: dl-α-AMINO-β-PHENYLPROPIONIC ACID ◇ NCI-C60195 ◇ d-β-PHENYLALANINE ◇ dl-PHENYLALANINE (FCC)

TOXICITY DATA with REFERENCE
orl-hmn TDLo:500 mg/kg/5W-I:GIT JACTDZ 1(3),124,82
ipr-rat LD50:5452 mg/kg ABBIA4 64,319,56

CONSENSUS REPORTS: Reported in EPA TSCA Inventory.

SAFETY PROFILE: Mildly toxic by intraperitoneal route. Human systemic effects by ingestion: nausea, hypermotility, diarrhea. When heated to decomposition it emits toxic fumes of NO_x.

PEC750 CAS:63-91-2 ***HR: 1***
l-PHENYLALANINE
mf: $C_9H_{11}NO_2$ mw: 165.21

PROP: White crystals or crystalline powder; slt odor and bitter taste. Mp: decomp @ 275-283°. Sol in water; very sltly sol in alc and ether.

SYNS: (S)-α-AMINOBENZENEPROPANOIC ACID ◇ α-AMINOHYDROCINNAMIC ACID ◇ α-AMINO-β-PHENYLPROPIONIC ACID ◇ ANTIBIOTIC FN 1636 ◇ PAL ◇ PHENYLALANINE ◇ PHENYL-α-ALANINE ◇ (S)-PHENYLALANINE ◇ β-PHENYLALANINE ◇ l-β-PHENYLALANINE ◇ 3-PHENYLALANINE

TOXICITY DATA with REFERENCE
mmo-esc 2 mmol/L MUREAV 161,113,86
orl-mky TDLo:33600 mg/kg (female 1-24W post):REP PEDIAU 42,27,68
orl-rat TDLo:21750 mg/kg (female 15-19D post):TER TOIZAG 30,518,83
ipr-rat LD50:5287 mg/kg ABBIA4 58,253,55

CONSENSUS REPORTS: Reported in EPA TSCA Inventory.

SAFETY PROFILE: Mildly toxic by intraperitoneal route. An experimental teratogen. Experimental reproductive effects. Mutation data reported. When heated to decomposition it emits toxic fumes of NO_x.

PED750 CAS:148-82-3 ***HR: 3***
l-PHENYLALANINE MUSTARD
mf: $C_{13}H_{18}Cl_2N_2O_2$ mw: 305.23

SYNS: ALANINE NITROGEN MUSTARD ◇ ALKERAN ◇ AT-290 ◇ l-3-(p-(BIS(2-CHLOROETHYL)AMINO)PHENYL)ALANINE◇ p-N-BIS(2-CHLOROETHYL)AMINO-l-PHENYLALANINE◇ 3-(p-(p-(BIS(2-CHLOROETHYL)AMINO)PHENYL)-l-ALANINE◇ 4-(BIS(2-CHLOROETHYL)AMINO)-l-PHENYLALANINE◇ CB 3025 ◇ p-N-DI(CHLOROETHYL)AMINOPHENYLALANINE◇ p-DI-(2-CHLOROETHYL)AMINO-l-PHENYLALANINE ◇ 3-p-(DI(2-CHLOROETHYL)AMINO)-PHENYL-l-ALANINE ◇ MELPHALAN ◇ NCI-C04853 ◇ NSC-8806 ◇ l-PAM ◇ PHENYLALANINE NITROGEN MUSTARD ◇ RCRA WASTE NUMBER U150 ◇ l-SARCOLYSIN ◇ p-l-SARCOLYSIN ◇ SK-15673

TOXICITY DATA with REFERENCE
skn-rbt 5 mg/24H rns TXCYAC 14,117,79
mmo-sat 100 μg/plate ONCODU 18,95,79
dnr-esc 100 μg/disc CNREA8 34,1658,74
orl-wmn TDLo:3750 μg/kg (25D pre):REP LANCAO 1,1174,77
orl-wmn TDLo:80 μg/kg/D-C:CAR BLOOAW 41,17,73
orl-man TDLo:57 μg/kg/D-C:CAR BLOOAW 41,17,73
ipr-rat TDLo:70 mg/kg/26W-I:NEO RRCRBU 52,1,75
skn-mus TDLo:58 mg/kg/9W-I:ETA BJCAAI 10,363,56
orl-hmn TDLo:700 mg/kg/7D:GIT 34ZIAG -,367,69
orl-hmn TDLo:1200 μg/kg/5D-I:GIT CCROBU 57,369,73

ivn-cld LDLo:4500 mg/kg NZMJAX 97,816,84
ivn-man LDLo:8140 mg/kg NZMJAX 97,816,84
ivn-inf TDLo:28 mg/kg:GIT,BLD LANCAO 2,1048,84
orl-rat LD50:11200 μg/kg IYKEDH 10,710,79
ivn-rat LD50:4100 μg/kg TYKEDH 10,710,79
ice-rat LD50:200 μg/kg JPPMAB 18,760,66
ivn-mus LD50:23 mg/kg NTIS** PB84-162486

CONSENSUS REPORTS: NTP Fifth Annual Report on Carcinogens. IARC Cancer Review: Group 1 IMEMDT 7,239,87; Animal Sufficient Evidence; Human Limited Evidence IMEMDT 9,167,75. NCI Carcinogenesis Studies (ipr); Clear Evidence: mouse, rat RRCRBU 52,1,75. EPA Genetic Toxicology Program.

SAFETY PROFILE: Confirmed human carcinogen producing leukemia and Hodgkin's disease. Poison by ingestion, intravenous, and intracerebral routes. Human systemic effects by ingestion: nausea, hypermotility, diarrhea, agranulocytosis, thrombocytopenia. Human reproductive effects by ingestion: menstrual changes. Mutation data reported. A skin irritant. Used as a poison gas. When heated to decomposition, it emits toxic fumes of Cl^- and NO_x.

PEE100 CAS:83997-16-4 ***HR: 3***
d-PHENYLALANYL-l-PROLYL-l-ARGININE ALDEHYDE SULFATE (1:1)
mf: $C_{20}H_{30}N_5O_3 \cdot H_2O_4S$ mw: 486.56

SYNS: N-(4-((AMINOIMINOMETHYL)AMINO)-1-FORMYLBUTYL)-3-PHENYLALANYL-l-PROLINAMIDE SULFATE, (S)- ◇ GYKI 14166 ◇ (S)-d-PHENYLALANYL-N-(4-((AMINOIMINOMETHYL)AMINO)-1-FORMYLBUTYL)-l-PROLINAMIDE SULFATE (1:1) ◇ RGH-2958

TOXICITY DATA with REFERENCE
orl-mus LD50:1250 mg/kg DRFUD4 10,829,85
ipr-mus LD50:225 mg/kg DRFUD4 10,829,85
ivn-mus LD50:45 mg/kg DRFUD4 10,829,85

SAFETY PROFILE: Poison by intravenous and intraperitoneal routes. Moderately toxic by ingestion. When heated to decomposition it emits toxic fumes of NO_x and SO_x.

PEE500 CAS:1078-21-3 ***HR: 2***
β-PHENYL-Γ-AMINOBUTYRIC ACID
mf: $C_{10}H_{13}NO_2$ mw: 179.24

SYNS: β-(AMINOMETHYL)BENZENEPROPANOIC ACID ◇ β-(AMINOMETHYL)HYDROCINNAMIC ACID ◇ 4-AMINO-3-PHENYLBUTYRIC ACID ◇ FENIBUT ◇ PGABA ◇ PHENIGAM ◇ PHENYL-Γ ◇ PhGABA

TOXICITY DATA with REFERENCE
orl-hmn TDLo:5 mg/kg:CNS BEXBAN 57,52,64
ipr-rat LD50:700 mg/kg BEXBAN 57,52,64
ipr-mus LD50:900 mg/kg BEXBAN 57,52,64

CONSENSUS REPORTS: Reported in EPA TSCA Inventory.

SAFETY PROFILE: Moderately toxic by intraperitoneal routes. Human systemic effects by ingestion: somnolence, hallucinations, distorted perception. Used as a mood elevator and tranquilizer. When heated to decomposition it emits toxic fumes of NO_x.

PEE600 CAS:43085-16-1 ***HR: 2***
17-β-PHENYLAMINOCARBONYLOXYOESTRA-1,3,5(10)-TRIENE-3-METHYL ETHER
mf: $C_{26}H_{31}NO_3$ mw: 405.58

SYN: STS 153

TOXICITY DATA with REFERENCE
orl-mus TDLo:250 mg/kg (female 12-16D post):REP VHAGAS 71,655,77
orl-mus TDLo:100 mg/kg (female 12-16D post):TER 40YJAX -,131,79
orl-mus TDLo:100 mg/kg (female 12-16D post):ETA,REP AMSHAR 28,209,80

SAFETY PROFILE: An experimental teratogen. Experimental reproductive effects. Questionable carcinogen with experimental tumorigenic data. When heated to decomposition it emits toxic fumes of NO_x. See also ETHERS.

PEE750 CAS:1698-60-8 ***HR: 2***
1-PHENYL-4-AMINO-5-CHLORPYRIDAZ-6-ONE
mf: $C_{10}H_8ClN_3O$ mw: 221.66

SYNS: 5-AMINO-4-CHLORO-2,3-DIHYDRO-3-OXO-2-PHENYL-PYRIDAZINE ◇ 5-AMINO-4-CHLORO-2-PHENYL-3(2H)-PYRIDAZINONE ◇ BUREX (CZECH) ◇ CHLORIDAZON ◇ 1-FENYL-4-AMINO-5-CHLOR-6-PYRIDAZINON (CZECH) ◇ HS-119-1 ◇ PCA ◇ PHENOSANE ◇ 1-PHENYL-4-AMINO-5-CHLOROPYRIDAZON-(6) (GERMAN) ◇ 1-PHENYL-4-AMINO-5-CHLORO-6-PYRIDAZONE ◇ 1-PHENYL-4-AMINO-5-CHLOROPYRIDAZONE-6 ◇ PYRAMINE ◇ PYRAMIN RB ◇ PYRAZON ◇ PYRAZONE ◇ PYRAZONL

TOXICITY DATA with REFERENCE
eye-rbt 500 mg/24H SEV 28ZPAK -,151,72
ivn-ham TDLo:175 mg/kg (female 8D post):REP ANREAK 175,503,73
orl-rat LD50:647 mg/kg FAATDF 7,299,86
ipr-rat LD50:600 mg/kg GTPZAB 23(8),48,79
orl-mus LD50:1000 mg/kg GTPZAB 23(8),48,79
ipr-mus LD50:410 mg/kg GTPZAB 23,8,48,79
orl-rbt LD50:2000 mg/kg GTPZAB 23(8),48,79
skn-rbt LD50:2500 mg/kg 85JFAN A080,85
orl-gpg LD50:760 mg/kg GTPZAB 23(8),48,79

CONSENSUS REPORTS: Reported in EPA TSCA Inventory. EPA Genetic Toxicology Program.

SAFETY PROFILE: Moderately toxic by ingestion and intraperitoneal routes. A severe eye irritant. Experimental reproductive effects. Used as a preemergence and early post-emergence herbicide. When heated to decomposition it emits very toxic fumes of Cl^- and NO_x.

PEF500 CAS:10535-87-2 ***HR: 3***
PHENYL-2-AMINO-1-PROPANE HYDROCHLORIDE
mf: $C_{22}H_{31}NO{\bullet}ClH$ mw: 362.00

SYNS: 1-(p-ETHOXYPHENYL)-1-DIETHYLAMINO-3-METHYL-3-PHENYLPROPANE HYDROCHLORIDE ◇ SP 725

TOXICITY DATA with REFERENCE
orl-rat LD50:1887 mg/kg ARZNAD 12,178,62
ipr-rat LD50:185 mg/kg ARZNAD 12,178,62
scu-rat LD50:2358 mg/kg ARZNAD 12,178,62
orl-mus LD50:270 mg/kg EJMCA5 12,459,77
ipr-mus LD50:90 mg/kg ARZNAD 12,178,62
scu-mus LD50:643 mg/kg ARZNAD 12,178,62

SAFETY PROFILE: Poison by ingestion and intraperitoneal route. Moderately toxic by subcutaneous route. When heated to decomposition it emits very toxic fumes of HCl and NO_x.

PEF750 CAS:24301-86-8 ***HR: 3***
PHENYL-1-AMINO-1-PROPANE HYDROCHLORIDE
mf: $C_9H_{13}N{\bullet}ClH$ mw: 171.69

SYNS: H 1672 ◇ 1-PHENYL-1-AMINO-PROPANE, HYDROCHLORIDE

TOXICITY DATA with REFERENCE
scu-rat LDLo:164 mg/kg JPETAB 71,62,41
scu-mus LD50:205 mg/kg ARZNAD 14(3),178,64
ivn-rbt LDLo:50 mg/kg JACSAT 53,1875,31

SAFETY PROFILE: Poison by subcutaneous and intravenous routes. When heated to decomposition it emits very toxic fumes of HCl and NO_x.

PEG000 CAS:86-26-0 ***HR: 2***
o-PHENYL ANISOLE
mf: $C_{13}H_{12}O$ mw: 184.25

PROP: Prisms from petr ether. Mp: 29-30°, bp: 274°.

SYNS: 2-METHOXY BIPHENYL ◇ METHYL DIPHENYL ETHER ◇ 2-PHENYLANISOLE

TOXICITY DATA with REFERENCE
skn-rbt 500 mg/24H MLD FCTXAV 13,681,75
orl-rat LD50:3600 mg/kg FCTXAV 13,681,75

CONSENSUS REPORTS: Reported in EPA TSCA Inventory.

SAFETY PROFILE: Moderately toxic by ingestion. A skin irritant. When heated to decomposition it emits acrid smoke and irritating fumes. See also ETHERS.

PEG250 CAS:613-37-6 ***HR: 3***
p-PHENYLANISOLE
mf: $C_{13}H_{12}O$ mw: 184.25

PROP: Leaves from alc. Mp: 90°. Sol in hot alc.

SYNS: p-METHOXYBIPHENYL ◇ 4-METHOXYBIPHENYL

TOXICITY DATA with REFERENCE
orl-rat TDLo:5450 mg/kg/52W-C:ETA ARZNAD 12,270,62

CONSENSUS REPORTS: Reported in EPA TSCA Inventory.

SAFETY PROFILE: Questionable carcinogen with experimental tumorigenic data. When heated to decomposition it emits acrid smoke and irritating fumes.

PEG500 CAS:91-40-7 ***HR: 3***
N-PHENYLANTHRANILIC ACID
mf: $C_{13}H_{11}NO_2$ mw: 213.25

PROP: Needles from alc. Mp: 185-187°, decomp 183-184°. Very sltly sol in hot water; sol in hot alc; very sltly sol in ether.

SYNS: o-ANILINOBENZOIC ACID ◇ 2-ANILINOBENZOIC ACID ◇ 2-CARBOXYDIPHENYLAMINE ◇ DIPHENYLAMINE-2-CARBOXYLIC ACID ◇ FENAMIC ACID ◇ PA ◇ 2-(PHENYLAMINO)BENZOIC ACID ◇ PHENYLANTHRANILIC ACID

TOXICITY DATA with REFERENCE
ipr-mus LD50:235 mg/kg RPTOAN 37,105,74
ivn-mus LD50:160 mg/kg YKKZAJ 89,1392,69

CONSENSUS REPORTS: Reported in EPA TSCA Inventory.

SAFETY PROFILE: Poison by intravenous and intraperitoneal routes. When heated to decomposition it emits toxic fumes of NO_x.

PEG750 CAS:637-03-6 ***HR: 3***
PHENYL ARSINE OXIDE
mf: C_6H_5AsO mw: 168.03

SYNS: ARSENOSOBENZENE ◇ ARZENE ◇ PHENYLARSENOXIDE

TOXICITY DATA with REFERENCE
ipr-mus LD50:1930 μg/kg JPETAB 74,210,42
ivn-mus LDLo:1570 μg/kg PHBUA9 2,19,54
ivn-rbt LD50:790 μg/kg JPETAB 69,342,40

CONSENSUS REPORTS: Arsenic and its compounds are on the Community Right-To-Know List. Reported in EPA TSCA Inventory.

OSHA PEL: TWA 0.5 mg(As)/m^3

SAFETY PROFILE: Poison by intravenous and intraperitoneal routes. When heated to decomposition it emits toxic fumes of As. See also ARSENIC COMPOUNDS.

PEH000 CAS:622-37-7 ***HR: 3***
PHENYL AZIDE
mf: $C_6H_5N_3$ mw: 119.13

SYN: AZIDOBENZENE

SAFETY PROFILE: Explodes when heated. Explosive or violent reactions with Lewis acids [e.g., aluminum chloride (explosive); sulfuric acid (violent)]. When heated to decomposition it emits toxic fumes of NO_x. See also AZIDES.

PEH250 CAS:13279-22-6 ***HR: 3***
N-PHENYL-1-AZIRIDINECARBOXAMIDE
mf: $C_9H_{10}N_2O$ mw: 162.21

SYNS: 1-AZIRIDINECARBOXANILIDE ◇ N-PHENYLAMINO-CARBONYL)AZIRIDINE ◇ 1-PHENYLCARBAMOYLAZIRIDINE ◇ PHENYL-N-CARBAMOYLAZIRIDINE

TOXICITY DATA with REFERENCE
ipr-mus TDLo:120 mg/kg/4W-I:NEO CNREA8 29,2184,69

SAFETY PROFILE: Questionable carcinogen with experimental neoplastigenic data. When heated to decomposition it emits toxic fumes of NO_x.

PEH500 CAS:17918-11-5 ***HR: 3***
α-PHENYL-1-AZIRIDINEETHANOL
mf: $C_{10}H_{13}NO$ mw: 163.24

SYNS: α-(1-AZIRIDINYLMETHYL)BENZYL ALCOHOL ◇ 2-(1-AZIRIDINYL)-1-PHENYLETHANOL

TOXICITY DATA with REFERENCE
orl-mus LD50:133 mg/kg EJMCA5 12,149,77
ipr-mus LD50:80 mg/kg EJMCA5 12,149,77
ivn-mus LD50:180 mg/kg CSLNX* NX#04518

CONSENSUS REPORTS: Reported in EPA TSCA Inventory.

SAFETY PROFILE: Poison by intravenous, intraperitoneal, and intravenous routes. When heated to decomposition it emits toxic fumes of NO_x. See also ALCOHOLS.

PEH750 CAS:4128-71-6 ***HR: 3***
4'-PHENYLAZOACETANILIDE
mf: $C_{14}H_{13}N_3O$ mw: 239.30

SYNS: p-ACETAMIDOAZOBENZENE ◇ 4-ACETYLAMINOAZOBENZENE ◇ p-PHENYLAZOACETANILIDE

TOXICITY DATA with REFERENCE
orl-rat TDLo:4700 mg/kg/52W-C:ETA JNCIAM 24,149,60

SAFETY PROFILE: Questionable carcinogen with experimental tumorigenic data. When heated to decomposition it emits toxic fumes of NO_x.

PEI000 CAS:60-09-3 ***HR: 3***
p-(PHENYLAZO)ANILINE
mf: $C_{12}H_{11}N_3$ mw: 197.26

PROP: Yellow crystals. Mp: 128°, bp: 360°. Sltly sol in hot water; sol in hot alc and ether.

SYNS: AAB ◇ AMINOAZOBENZENE ◇ p-AMINOAZOBENZENE ◇ 4-AMINOAZOBENZENE ◇ 4-AMINO-1,1'-AZOBENZENE ◇ p-AMINOAZOBENZOL ◇ 4-AMINOAZOBENZOL ◇ p-AMINODIPHENYLIMIDE ◇ ANILINE YELLOW ◇ 4-BENZENEAZOANILINE ◇ BRASILAZINA OIL YELLOW G ◇ CERES YELLOW R ◇ C.I. 11000 ◇ C.I. SOLVENT BLUE 7 ◇ C.I. SOLVENT YELLOW 1 ◇ FAST SPIRIT YELLOW AAB ◇ OIL SOLUBLE ANILINE YELLOW ◇ OIL YELLOW AAB ◇ ORGANOL YELLOW ◇ PARAPHENOLAZO ANILINE ◇ 4-(PHENYLAZO)ANILINE ◇ 4-(PHENYLAZO)BENZENAMINE ◇ p-PHENYLAZOPHENYLAMINE ◇ SOLVENT YELLOW 1 ◇ SUDAN YELLOW R ◇ USAF EK-1375

TOXICITY DATA with REFERENCE
mmo-sat 50 nmol/plate CRNGDP 3,113,82
otr-ham:kdy 80 μg/L BJCAAI 37,873,78
scu-mus TDLo:296 mg/kg (15-19D preg):NEO,TER CALEDQ 17,321,83
orl-rat TDLo:89 g/kg/57W-C:ETA JPBAA7 59,1,47
skn-rat TDLo:1965 mg/kg/2Y-I:NEO CNREA8 26,2406,66
ipr-mus LD50:200 mg/kg NTIS** AD277-689

CONSENSUS REPORTS: IARC Cancer Review: Group 2B IMEMDT 7,56,87; Animal Sufficient Evidence IMEMDT 8,53,75. Community Right-To-Know List. Reported in EPA TSCA Inventory.

SAFETY PROFILE: Suspected carcinogen with experimental neoplastigenic and tumorigenic data. Poison by intraperitoneal route. An experimental teratogen. Mutation data reported. Used as a dye for lacquer, varnish, wax products, oil stains, and styrene resins. When heated to decomposition it emits toxic fumes of NO_x. See also AMINES.

PEI250 CAS:3457-98-5 ***HR: 2***
p-(PHENYLAZO)ANILINE HYDROCHLORIDE
mf: $C_{12}H_{11}N_3 \cdot ClH$ mw: 233.72

PROP: Steel blue. Sol in water.

SYNS: p-AMINOAZOBENZENE HYDROCHLORIDE ◇ 4-AMINOAZOBENZENE HYDROCHLORIDE ◇ C.I. SOLVENT YELLOW 1, MONOHYDROCHLORIDE

TOXICITY DATA with REFERENCE
orl-rat LD50:1250 mg/kg MarJV# 29MAR77

CONSENSUS REPORTS: Reported in EPA TSCA Inventory.

SAFETY PROFILE: Moderately toxic by ingestion.

When heated to decomposition it emits very toxic fumes of NO_x and HCl.

PEI750 CAS:36368-30-6 ***HR: 3***
1-PHENYLAZO-2-ANTHROL
mf: $C_{20}H_{14}N_2O$ mw: 298.36

SYN: BENZENEAZO-2-ANTHROL

TOXICITY DATA with REFERENCE
scu-mus TDLo:25 g/kg/52W-I:CAR BJCAAI 10,653,56
imp-mus TDLo:80 mg/kg BJCAAI 17,127,63

SAFETY PROFILE: A poison by intramuscular route. Questionable carcinogen with experimental carcinogenic data. When heated to decomposition it emits toxic fumes of NO_x.

PEJ250 CAS:22670-79-7 ***HR: 3***
N-PHENYLAZO-N-METHYLTAURINE SODIUM SALT
mf: $C_9H_{12}N_3O_3S•Na$ mw: 265.29

SYNS: 3-METHYL-1-PHENYL-3-(2-SULFOETHYL)TRIAZENESODIUM SALT ◇ 1-PHENYL-3-METHYL-3-(2-SULFOAETHYL) NATRIUM SALZ (GERMAN) ◇ 1-PHENYL-3-METHYL-3-(2-SULFOETHYL) TRIAZENE, SODIUM SALT

TOXICITY DATA with REFERENCE
scu-rat TDLo:1170 mg/kg/78W-I:NEO,REP ZKKOBW 81,285,74
scu-rat LD50:150 mg/kg ZKKOBW 81,285,74

SAFETY PROFILE: Poison by subcutaneous route. Experimental reproductive effects. Questionable carcinogen with experimental neoplastigenic data. When heated to decomposition it emits very toxic fumes of NO_x, Na_2O, and SO_x.

PEJ500 CAS:842-07-9 ***HR: 3***
1-(PHENYLAZO)-2-NAPHTHOL
mf: $C_{16}H_{12}N_2O$ mw: 248.30

SYNS: ATUL ORANGE R ◇ BENZENEAZO-β-NAPHTHOL ◇ BENZENE-1-AZO-2-NAPHTHOL ◇ 1-BENZOAZO-2-NAPHTHOL ◇ BRILLIANT OIL ORANGE R ◇ CALCOGAS ORANGE NC ◇ CALCO OIL ORANGE 7078 ◇ CAMPBELLINE OIL ORANGE ◇ CARMINAPH ◇ CERES ORANGE R ◇ CEROTINORANGE G ◇ C.I. 12055 ◇ C.I. SOLVENT YELLOW 14 ◇ DISPERSOL YELLOW PP ◇ DUNKELGELB ◇ ENIAL ORANGE I ◇ FAST OIL ORANGE ◇ FAST ORANGE ◇ FETTORANGE R ◇ GRASAN ORANGE R ◇ HIDACO OIL ORANGE ◇ LACQUER ORANGE VG ◇ MOTIORANGE R ◇ NCI-C53929 ◇ OIL ORANGE ◇ OLEAL ORANGE R ◇ ORANGE A l'HUILE ◇ ORANGE INSOLUBLE OLG ◇ ORANGE PEL ◇ ORANGE RESENOLE No. 3 ◇ ORANGE SOLUBLE A l'HUILE ◇ ORGANOL ORANGE ◇ ORIENT OIL ORANGE PS ◇ PETROL ORANGE Y ◇ 1-(PHENYLAZO)-2-NAPHTHALENOL ◇ 1-PHENYLAZO-β-NAPHTHOL ◇ PLASTORESIN ORANGE F4A ◇ PYRONALORANGE ◇ RESINOL ORANGE R ◇ RESOFORM ORANGE G ◇ SANSEL ORANGE G ◇ SCHARLACH B ◇ SILOTRAS ORANGE TR ◇ SOLVENT YELLOW 14 ◇ SOMALIA ORANGE I ◇ SOUDAN I ◇ SPIRIT ORANGE ◇ SPIRIT YELLOW I ◇ STEARIX ORANGE ◇ SUDAN ORANGE R ◇ TERTROGRAS ORANGE SV ◇ TOYO OIL ORANGE ◇ WAXAKOL ORANGE GL ◇ WAXOLINE YELLOW I

TOXICITY DATA with REFERENCE
orl-rat TDLo:10815 mg/kg/2Y-C:NEO NTPTR* NTP-TR-226,82
scu-mus TDLo:6000 mg/kg/57W-I:ETA GMJOAZ 30,364,49
imp-mus TDLo:80 mg/kg:CAR BJCAAI 22,825,68
imp-mus TD:96 mg/kg:CAR GMCRDC 17,383,75

CONSENSUS REPORTS: IARC Cancer Review: Group 3 IMEMDT 7,56,87; Animal Sufficient Evidence IMEMDT 8,225,75. NTP Carcinogenesis Bioassay (feed); Clear Evidence: rat NTPTR* NTP-TR-226,82. Community Right-To-Know List. Reported in EPA TSCA Inventory. EPA Genetic Toxicology Program.

SAFETY PROFILE: Questionable carcinogen with experimental carcinogenic, neoplastigenic, and tumorigenic data. When heated to decomposition it emits toxic fumes of NO_x. Used for coloring hydrocarbon solvents, oils, fats, waxes, shoe and floor polishes, and gasoline.

PEK000 CAS:532-82-1 ***HR: 3***
4-PHENYLAZO-m-PHENYLENEDIAMINE
mf: $C_{12}H_{12}N_4•ClH$ mw: 248.74

SYNS: ASTRA CHRYSOIDINE R ◇ BRASILAZINA ORANGE Y ◇ BRILLIANT OIL ORANGE Y BASE ◇ CALCOZINE CHRYSOIDINE Y ◇ CALCOZINE ORANGE YS ◇ CHRYSOIDIN ◇ CHRYSOIDINE ◇ CHRYSOIDINE A ◇ CHRYSOIDINE B ◇ CHRYSOIDINE C CRYSTALS ◇ CHRYSOIDINE G ◇ CHRYSOIDINE GN ◇ CHRYSOIDINE HR ◇ CHRYSOIDINE(II) ◇ CHRYSOIDINE J ◇ CHRYSOIDINE M ◇ CHRYSOIDINE ORANGE ◇ CHRYSOIDINE PRL ◇ CHRYSOIDINE PRR ◇ CHRYSOIDINE SL ◇ CHRYSOIDINE SPECIAL (biological stain and indicator) ◇ CHRYSOIDINE SS ◇ CHRYSOIDINE Y ◇ CHRYSOIDINE Y BASE NEW ◇ CHRYSOIDINE Y CRYSTALS ◇ CHRYSOIDINE Y EX ◇ CHRYSOIDINE YGH ◇ CHRYSOIDINE YL ◇ CHRYSOIDINE YN ◇ CHRYSOIDINE Y SPECIAL ◇ CHRYSOIDIN FB ◇ CHRYSOIDIN Y ◇ CHRYSOIDIN YN ◇ CHRYZOIDYNA F.B. (POLISH) ◇ C.I. 11270 ◇ C.I. BASIC ORANGE 2 ◇ C.I. BASIC ORANGE 3 ◇ C.I. BASIC ORANGE 2, MONOHYDROCHLORIDE ◇ C.I. SOLVENT ORANGE 3 ◇ 2,4-DIAMINOAZOBENZENE HYDROCHLORIDE ◇ DIAZOCARD CHRYSOIDINE G ◇ ELCOZINE CHRYSOIDINE Y ◇ LEATHER ORANGE HR ◇ 4-(PHENYLAZO)-1,3-BENZENEDIAMINE MONOHYDROCHLORIDE ◇ 4-(PHENYLAZO)-m-PHENYLENEDIAMINE MONOHYDROCHLORIDE ◇ PURE CHRYSOIDINE YBH ◇ PURE CHRYSOIDINE YD ◇ PYRACRYL ORANGE Y ◇ SUGAI CHRYSOIDINE ◇ TERTROPHENE BROWN CG

TOXICITY DATA with REFERENCE
mma-sat 50 μg/plate MUREAV 44,9,77
orl-mus TDLo:93600 mg/kg/56W-C:ETA LANCAO 1,564,82
orl-mus LDLo:1670 mg/kg AEPPAE 195,295,40
scu-mus LDLo:1670 mg/kg AEPPAE 195,295,40

CONSENSUS REPORTS: IARC Cancer Review: Group 3 IMEMDT 7,169,87; Animal Sufficient Evidence IMEMDT 8,91,75. Reported in EPA TSCA Inventory. EPA Genetic Toxicology Program.

SAFETY PROFILE: Moderately toxic by ingestion and subcutaneous routes. Questionable carcinogen with experimental tumorigenic data. Mutation data reported. When heated to decomposition it emits very toxic fumes of NO_x and HCl. Used as a colorant in textiles, paper, leather, inks, wood, and biological stains.

PEK250 CAS:94-78-0 ***HR: 3***
3-(PHENYLAZO)-2,6-PYRIDINEDIAMINE
mf: $C_{11}H_{11}N_5$ mw: 213.27

SYNS: AP ◇ 2,6-DIAMINO-3-PHENYLAZOPYRIDINE ◇ DIRIDONE ◇ DPP ◇ GASTRACID ◇ GASTROTEST ◇ MALLOPHENE ◇ NC 150 ◇ PHENAZODINE ◇ PHENAZOPYRIDINE ◇ PHENYLAZO TABLET ◇ PIRID ◇ PYRAZOFEN ◇ PYRIDACIL ◇ PYRIDIUM ◇ PYRIPYRIDIUM ◇ SEDURAL ◇ URIDINAL ◇ URODINE ◇ W 1655

TOXICITY DATA with REFERENCE
imp-mus TDLo:80 mg/kg:NEO BJCAAI 11,212,57
ipr-rat LD50:560 mg/kg JPETAB 51,200,34

CONSENSUS REPORTS: IARC Cancer Review: Animal Inadequate Evidence IMEMDT 8,117,75.

SAFETY PROFILE: Moderately toxic by intraperitoneal route. Questionable carcinogen with experimental neoplastigenic data. Used as a local anesthetic. When heated to decomposition it emits toxic fumes of NO_x.

PEK675 CAS:33244-00-7 ***HR: 3***
N-PHENYLBENZAMIDRAZONE HYDROCHLORIDE
mf: $C_{13}H_{13}N_3 \cdot ClH$ mw: 247.75

SYNS: BENZAMIDEPHENYLHYDRAZONEHYDROCHLORIDE ◇ CBS-1114 ◇ 2-PHENYLHYDRAZIDEBENZENECARBOXIMIDIC ACID MONOHYDROCHLORIDE

TOXICITY DATA with REFERENCE
ipr-rat LD50:170 mg/kg DRFUD4 9,102,84
ivn-rat LD50:63 mg/kg DRFUD4 9,102,84
ipr-mus LD50:150 mg/kg DRFUD4 9,102,84
ivn-mus LD50:45 mg/kg DRFUD4 9,102,84

SAFETY PROFILE: Poison by intravenous and intraperitoneal routes. When heated to decomposition it emits toxic fumes of NO_x and HCl.

PEK750 CAS:19383-97-2 ***HR: 3***
5-PHENYL-1:2-BENZANTHRACENE
mf: $C_{24}H_{16}$ mw: 304.40

SYN: BENZ(a)ANTHRACENE, 8-PHENYL

TOXICITY DATA with REFERENCE
skn-mus TDLo:1650 mg/kg/69W-I:ETA PRLBA4 129,439,40

SAFETY PROFILE: Questionable carcinogen with experimental tumorigenic data. When heated to decomposition it emits acrid smoke and irritating fumes.

PEL000 CAS:29222-39-7 ***HR: 2***
PHENYL-1,2,4-BENZENETRIOL
mf: $C_{12}H_{10}O_3$ mw: 202.22

TOXICITY DATA with REFERENCE
ipr-mus LDLo:500 mg/kg CBCCT* 8,482,56

CONSENSUS REPORTS: Reported in EPA TSCA Inventory.

SAFETY PROFILE: Moderately toxic by intraperitoneal route. When heated to decomposition it emits acrid smoke and irritating fumes.

PEL250 CAS:716-79-0 ***HR: 3***
2-PHENYLBENZIMIDAZOLE
mf: $C_{13}H_{10}N_2$ mw: 194.25

PROP: Needles or plates. Mp: 291°. Sol in methanol; sltly sol in water, benzene, chloroform.

SYNS: PHENIZIDOLE ◇ PHENZIDOLE

TOXICITY DATA with REFERENCE
ipr-mus LD50:167 mg/kg FRPSAX 33,516,78

CONSENSUS REPORTS: Reported in EPA TSCA Inventory.

SAFETY PROFILE: Poison by intraperitoneal route. When heated to decomposition it emits toxic fumes of NO_x.

PEL500 CAS:93-99-2 ***HR: 2***
PHENYL BENZOATE
mf: $C_{17}H_{10}O_2$ mw: 246.27

PROP: Colorless crystals; geranium odor. Mp: 70°, bp: 314°, d: 1.235, vap press: 1 mm @ 106.8°. Insol in water; very sol in hot alc, ether; sltly sol in cold alc, ether.

TOXICITY DATA with REFERENCE
orl-mus LD50:1225 mg/kg YKKZAJ 89,1179,69

CONSENSUS REPORTS: Reported in EPA TSCA Inventory.

SAFETY PROFILE: Moderately toxic by ingestion. When heated it emits acrid smoke and irritating fumes. See also ESTERS, PHENOL and BENZOIC ACID.

PEL750 CAS:363-03-1 ***HR: 3***
PHENYL-p-BENZOQUINONE
mf: $C_{12}H_8O_2$ mw: 184.20

SYNS: PHENYLBENZOQUINONE ◇ PHENYL-1,4-BENZOQUINONE ◇ 2-PHENYLBENZOQUINONE ◇ PHENYLQUINONE

TOXICITY DATA with REFERENCE
dnd-rat-par 50 mg/L JJCREP 78,1027,87
scu-rat TDLo:5 mg/kg (female 1D pre):REP ENDOAO 57,466,55
ipr-mus LD50:19 mg/kg ARZNAD 20,1751,70

CONSENSUS REPORTS: Reported in EPA TSCA Inventory.

SAFETY PROFILE: Poison by intraperitoneal route. Experimental reproductive effects. Mutation data reported. When heated to decomposition it emits acrid smoke and irritating fumes.

PEM000 CAS:36774-74-0 ***HR: 2***
2-PHENYL-5-BENZOTHIAZOLEACETIC ACID
mf: $C_{15}H_{11}NO_2S$ mw: 269.33

TOXICITY DATA with REFERENCE
orl-rbt TDLo:450 mg/kg (8-16D preg):REP AMBNAS 24,173,77
orl-rbt TDLo:450 mg/kg (8-16D preg):TER AMBNAS 24,173,77
orl-mus LD50:1365 mg/kg JMCMAR 16,930,73
ipr-mus LD50:800 mg/kg JMCMAR 16,930,73

SAFETY PROFILE: Moderately toxic by ingestion and intraperitoneal routes. An experimental teratogen. Experimental reproductive effects. When heated to decomposition it emits very toxic fumes of SO_x and NO_x.

PEM750 CAS:511-55-7 ***HR: 3***
8-(p-PHENYLBENZYL)ATROPINIUM BROMIDE
mf: $C_{30}H_{34}NO_3{\cdot}Br$ mw: 536.56

PROP: Crystals. Decomp @ 220-222°.

SYNS: N-(p-BIPHENYLMETHYL)-ATROPINIUM BROMIDE ◇ N,4-BIPHENYL-METHYL-dl-TROPEYL-α-TROPINIUMBROMIDS (GERMAN) ◇ p-BIPHENYLMETHYL-(dl-TROPYL-α-TROPINIUM)BROMIDE ◇ DENDREPAR ◇ 4-DIPHENYLMETHYL-dl-TROPYLTROPINIUM BROMIDE ◇ 4-DIPHENYLMETHYLTROPYLTROPINIUM BROMIDE ◇ GASTRIPON ◇ 3-α-HYDROXY-8-(p-PHENYLBENZYL)-1-α-H,5-α-H-TROPANIUM BROMIDE, (±)-TROPATE ◇ N-399 ◇ XENYTROPIUM BROMIDE

TOXICITY DATA with REFERENCE
orl-rat TDLo:7500 mg/kg (30D pre):REP OYYAA2 8,533,74
orl-rat LD50:1880 mg/kg OYYAA2 8,533,74
ipr-rat LD50:66 mg/kg OYYAA2 8,533,74
scu-rat LD50:243 mg/kg OYYAA2 8,533,74
ivn-rat LD50:9400 μg/kg AIPTAK 113,1,57
orl-mus LD50:1650 mg/kg OYYAA2 8,533,74
ipr-mus LD50:37100 μg/kg OYYAA2 8,533,74
scu-mus LD50:173 mg/kg OYYAA2 8,533,74
ivn-mus LD50:9200 μg/kg ARZNAD 16,637,66
ivn-rbt LDLo:7 mg/kg AIPTAK 113,1,57
ipr-gpg LDLo:30 mg/kg AIPTAK 113,1,57

SAFETY PROFILE: Poison by intravenous, subcutaneous, and intraperitoneal routes. Moderately toxic by ingestion. Experimental reproductive effects. When heated to decomposition it emits very toxic fumes of NO_x and Br^-. See also BROMIDES.

PEN000 CAS:2045-52-5 ***HR: 3***
N-PHENYL-N-BENZYL-N′,N′-DIMETHYLETHYLENEDIAMINE HYDROCHLORIDE
mf: $C_{17}H_{22}N_2{\cdot}ClH$ mw: 290.87

SYNS: ANTERGAN HYDROCHLORIDE ◇ N-BENZYL-N-DIMETHYLAMINOETHYL, ANILINE, HYDROCHLORIDE ◇ COMPOUND 2339 RP ◇ CORPS 2339 R P (FRENCH) ◇ N-DIMETHYLAMINOETHYLBENZYLANILINE HYDROCHLORIDE ◇ PHENBENZAMINE HYDROCHLORIDE ◇ 2339 R.P. HYDROCHLORIDE

TOXICITY DATA with REFERENCE
orl-rat LD50:300 mg/kg AIPTAK 68,339,42
scu-rat LD50:15 mg/kg CRSBAW 144,887,50
scu-gpg LD50:110 mg/kg PHREA7 27,542,47
orl-mus LD50:325 mg/kg JPETAB 93,210,48
scu-mus LD50:125 mg/kg JPETAB 93,210,48
ivn-mus LD50:35 mg/kg JPETAB 93,210,48
ivn-rbt LD50:15 mg/kg JPETAB 93,210,47

SAFETY PROFILE: Poison by ingestion, intravenous, and subcutaneous routes. When heated to decomposition it emits very toxic fumes of HCl and NO_x. Used as an antihistaminic agent. See also AMINES.

PEO000 CAS:55-57-2 ***HR: 3***
PHENYLBIGUANIDE HYDROCHLORIDE
mf: $C_8H_{11}N_5{\cdot}ClH$ mw: 213.70

PROP: Prisms. Mp: 237°.

SYNS: N′-PHENYLBIGUANIDE MONOHYDROCHLORIDE ◇ 1-PHENYLBIGUANIDE MONOHYDROCHLORIDE ◇ N-PHENYL-IMIDODICARBONIMIDIC DIAMIDE MONOHYDROCHLORIDE

TOXICITY DATA with REFERENCE
ipr-mus LD50:290 mg/kg JMCMAR 10,521,67

CONSENSUS REPORTS: Reported in EPA TSCA Inventory.

SAFETY PROFILE: Poison by intraperitoneal route. When heated to decomposition it emits very toxic fumes of HCl and NO_x.

PEO500 CAS:108-86-1 ***HR: 2***
PHENYL BROMIDE
DOT: UN 2514
mf: C_6H_5Br mw: 157.02

PROP: Colorless, clear, mobile liquid. Mp: −30.7, bp: 156.2°, flash p: 124°F, d: 1.497, vap press: 10 mm @ 40°, vap d: 5.41, autoign temp: 1051°F.

SYNS: BROMOBENZENE (DOT) ◇ MONOBROMOBENZENE ◇ NCI-C55492

TOXICITY DATA with REFERENCE
mnt-mus-ipr:125 mg/kg/24H MUTAEX 2,111,87
ipr-mus LD50:817 mg/kg MUTAEX 2,111,87
orl-gpg LD50:1700 mg/kg 85GMAT -,28,82
orl-rbt LD50:3300 mg/kg 85GMAT -,28,82
orl-rat LD50:2699 mg/kg GTPZAB 19(9),36,75
ihl-rat LC50:20411 mg/m^3 GTPZAB 19(9),36,75
ipr-rat LD50:3882 mg/kg TXAPA9 83,108,86
orl-mus LD50:2700 mg/kg GTPZAB 13(9),56,69
ihl-mus LC50:21 g/m^3/2H GTPZAB 13(9),56,69
ipr-mus LD50:1000 mg/kg JJIND8 62,911,79
scu-mus LD50:2000 mg/kg JJPAAZ 3,99,54

CONSENSUS REPORTS: Reported in EPA TSCA Inventory. EPA Genetic Toxicology Program.

DOT Classification: Flammable or Combustible Liquid; Label: Flammable Liquid.

SAFETY PROFILE: Moderately toxic by ingestion, subcutaneous, and intraperitoneal routes. Mildly toxic by inhalation. An eye and mucous membrane irritant. Mutation data reported. Combustible when exposed to heat or flame. Can react with oxidizing materials. To fight fire, use water to blanket fire, foam, CO_2, water spray or mist, dry chemical. Violent reaction with bromobutane + sodium when heated above 30°C. When heated to decomposition it emits toxic fumes of Br^-. See also BROMIDES.

PEO750 CAS:70145-60-7 ***HR: 3***
PHENYLBUTAZONE CALCIUM SALT
mf: $C_{19}H_{20}N_2O_2$•Ca mw: 348.49

SYNS: DA-241 ◇ 1,2-DIPHENYL-4-BUTYL-3,5-DIKETOPYRAZOLIDINE CALCIUM SALT ◇ P 241

TOXICITY DATA with REFERENCE
orl-rat LD50:997 mg/kg FRPSAX 12,521,57
ivn-rat LD50:104 mg/kg FRPSAX 13,922,58

SAFETY PROFILE: Poison by intravenous route. Moderately toxic by ingestion. When heated to decomposition it emits toxic fumes of NO_x.

PEP250 CAS:90-27-7 ***HR: 3***
2-PHENYLBUTYRIC ACID
mf: $C_{10}H_{12}O_2$ mw: 164.22

SYN: α-PHENYL BUTYRIC ACID

TOXICITY DATA with REFERENCE
orl-mus LD50:1154 mg/kg BCFAAI 100,143,61
ipr-mus LD50:1320 mg/kg FRPSAX 13,286,58
scu-mus LD50:800 mg/kg AIPTAK 116,154,58
ivn-mus LD50:320 mg/kg CSLNX* NX#03832

CONSENSUS REPORTS: Reported in EPA TSCA Inventory.

SAFETY PROFILE: Poison by intravenous route. Moderately toxic by ingestion, intraperitoneal, and subcutaneous routes. When heated to decomposition it emits acrid smoke and irritating fumes.

PEQ000 CAS:63982-25-2 ***HR: 3***
N-PHENYLCARBAMIC ACID-3-(DIMETHYLAMINO)PHENYL ESTER HYDROCHLORIDE
mf: $C_{15}H_{16}N_2O_2$•ClH mw: 292.79

SYN: PHENYLCARBAMIC ESTER OF 3-OXYPHENYLDIMETHYLAMINE HYDROCHLORIDE

TOXICITY DATA with REFERENCE
orl-mus LDLo:500 mg/kg JPETAB 43,413,31
ivn-mus LDLo:20 mg/kg JPETAB 43,413,31

SAFETY PROFILE: Poison by intravenous route. Moderately toxic by ingestion. When heated to decomposition it emits very toxic fumes of HCl and NO_x. See also CARBAMATES and ESTERS.

PEQ250 CAS:100836-70-2 ***HR: 2***
N-PHENYLCARBAMIC ACID-2-(METHYL(1-PHENOXY-2-PROPYL)AMINO) ETHYL ESTER HYDROCHLORIDE
mf: $C_{19}H_{24}N_2O_3$•ClH mw: 364.91

SYN: C 6304

TOXICITY DATA with REFERENCE
eye-rbt 2% MLD ARZNAD 9,113,59
scu-mus LD50:437 mg/kg ARZNAD 9,113,59

SAFETY PROFILE: Moderately toxic by subcutaneous route. An eye irritant. When heated to decomposition it emits very toxic fumes of NO_x and HCl. See also CARBAMATES.

PEQ500 CAS:2828-42-4 ***HR: 2***
o-(N-PHENYLCARBAMOYL)PROPANONOXIME
mf: $C_{10}H_{12}N_2O_2$ mw: 192.24

SYNS: ACETONE-o-CARBANILOYLOXIME ◇ ACETONE-OXIME-N-PHENYLCARBAMATE ◇ ACETONE OXIME PHENYLURETHANE ◇ o-(N-PHENYLCARBAMOYL)-PROPANONOXIM (GERMAN) ◇ 2-PROPANONE, o-((PHENYLAMINO)CARBONYL)OXIME ◇ PROXIMPHAM (GERMAN)

TOXICITY DATA with REFERENCE
orl-rat LD50:1540 mg/kg FCTXAV 8,517,70
orl-mus LD50:1300 mg/kg 85DPAN -,-,71/76

SAFETY PROFILE: Moderately toxic by ingestion. When heated to decomposition it emits toxic fumes of NO_x. See also CARBAMATES.

PEQ750 CAS:104-68-7 ***HR: 2***
PHENYL CARBITOL
mf: $C_{10}H_{14}O_3$ mw: 182.24

PROP: Liquid. Bp: 207° @ 55 mm, fp: −50°, d: 1.1158 @ 20°/20°, vap press: <0.01 mm @ 20°, vap d: 6.28.

SYNS: DIETHYLENE GLYCOL MONOPHENYL ETHER ◇ DIETHYLENE GLYCOL PHENYL ETHER ◇ 2-(2-PHENOXYETHOXY) ETHANOL

TOXICITY DATA with REFERENCE
skn-rbt 505 mg open MLD UCDS** 8/16/61
eye-rbt 2 mg open SEV AMIHBC 10,61,54
orl-rat LD50:2140 mg/kg AMIHBC 10,61,54
skn-rbt LD50:2120 mg/kg UCDS** 8/16/61

CONSENSUS REPORTS: Glycol ether compounds are on the Community Right-To-Know List. Reported in EPA TSCA Inventory.

SAFETY PROFILE: Moderately toxic by ingestion and skin contact. A skin and severe eye irritant. Some glycol ethers have dangerous human reproductive effects. When heated to decomposition it emits acrid smoke and irritating fumes. See also GLYCOL ETHERS.

PER000 CAS:122-99-6 ***HR: 2***
PHENYL CELLOSOLVE
mf: $C_8H_{10}O_2$ mw: 138.18

PROP: Clear liquid. Mp: 14°, bp: 242°, flash p: 250°F.

SYNS: AROSOL ◇ DOWANOL EP ◇ DOWANOL EPH ◇ EMERESSENCE 1160 ◇ EMERY 6705 ◇ ETHYLENE GLYCOL MONOPHENYL ETHER ◇ ETHYLENE GLYCOL PHENYL ETHER ◇ 2-FENOXYETHANOL (CZECH) ◇ FENYL-CELLOSOLVE (CZECH) ◇ GLYCOL MONOPHENYL ETHER ◇ β-HYDROXYETHYL PHENYL ETHER ◇ 1-HYDROXY-2-PHENOXYETHANE ◇ PHENOXETHOL ◇ PHENOXETOL ◇ PHENOXYETHANOL ◇ 2-PHENOXYETHANOL ◇ PHENOXYETHYL ALCOHOL ◇ PHENOXYTOL ◇ PHENYLMONOGLYCOL ETHER ◇ ROSE ETHER

TOXICITY DATA with REFERENCE
skn-rbt 500 mg MLD UCDS** 6/24/58
skn-rbt 500 mg/24H MOD 28ZPAK -,99,72
eye-rbt 6 mg MOD UCDS** 6/24/58
eye-rbt 250 μg/24H SEV 28ZPAK -,99,72
dni-esc 2000 ppm MCBIA7 28,7,80
oms-esc 2000 ppm MCBIA7 28,7,80
orl-rat LD50:1260 mg/kg JIHTAB 23,259,41
skn-rbt LD50:5000 mg/kg UCDS** 6/24/58

CONSENSUS REPORTS: Glycol ether compounds are on the Community Right-To-Know List. Reported in EPA TSCA Inventory.

SAFETY PROFILE: Moderately toxic by ingestion and skin contact. A skin and severe eye irritant. Mutation data reported. Some glycol ethers have dangerous human reproductive effects. Combustible when exposed to heat or flame; can react vigorously with oxidizing materials. When heated to decomposition it emits acrid smoke and irritating fumes. To fight fire, use CO_2, dry chemical. Used as a solvent for ester type resins. See also GLYCOL ETHERS.

PER250 CAS:48145-04-6 ***HR: 2***
PHENYL CELLOSOLVE ACRYLATE
mf: $C_{11}H_{12}O_3$ mw: 192.23

SYN: 2-PHENOXY-ETHANOL, ACRYLATE

TOXICITY DATA with REFERENCE
skn-rbt 500 mg open MLD UCDS** 12/2/71
orl-rat LD50:5190 mg/kg UCDS** 12/2/71
skn-rbt LD50:1800 mg/kg UCDS** 12/2/71

CONSENSUS REPORTS: Reported in EPA TSCA Inventory.

SAFETY PROFILE: Moderately toxic by skin contact. Mildly toxic by ingestion. A skin irritant. When heated to decomposition it emits acrid smoke and irritating fumes.

PER500 CAS:4460-46-2 ***HR: 3***
PHENYLCHLORODIAZIRINE
mf: $C_7H_5ClN_2$ mw: 152.58

N=NC(Cl)C_6H_5

SAFETY PROFILE: The pure material is a much more shock-sensitive explosive than glycerylnitrate. May explode when heated above 80°C. Product of reaction with phenylacetylene is explosive. When heated to decomposition it emits very toxic fumes of NO_x and Cl^-.

PER600 CAS:64049-07-6 ***HR: 3***
PHENYL(β-CHLOROVINYL)CHLOROARSINE
mf: $C_8H_7AsCl_2$ mw: 248.97

SYNS: ARSINE, CHLORO(2-CHLOROVINYL)PHENYL- ◇ PHENYL(β-CHLOROVINYL)CHLORARSINE ◇ TL 59

TOXICITY DATA with REFERENCE
ihl-mus LC50:1 g/m³/10M NTIS** PB158-508

OSHA PEL: TWA 0.5 mg(As)/m³

SAFETY PROFILE: Poison by inhalation. When heated to decomposition it emits toxic fumes of As and Cl^-.

PER750 CAS:827-52-1 ***HR: 3***
PHENYLCYCLOHEXANE
mf: $C_{12}H_{16}$ mw: 160.28

PROP: Colorless, oily liquid. Mp: 7.5°, bp: 240.0°, flash p: 210°F (OC), d: 0.938 ± 0.02 @ 25°/15.6°, vap press: 1 mm @ 67.5°.

SYN: CYCLOHEXYLBENZENE

TOXICITY DATA with REFERENCE
orl-rat LDLo:5000 mg/kg 28ZRAQ -,55,60
ipr-mus LD50:248 mg/kg LIFSAK 31,803,82
ivn-mus LD50:67 mg/kg LIFSAK 31,803,82

CONSENSUS REPORTS: Reported in EPA TSCA Inventory.

SAFETY PROFILE: Poison by intraperitoneal and intravenous routes. Moderately toxic by ingestion. Combustible when exposed to heat or flame; can react with oxidizing materials. To fight fire, use alcohol foam, fog, mist, dry chemical. When heated to decomposition it emits acrid smoke and irritating fumes.

PES000 CAS:1444-64-0 ***HR: 2***
2-PHENYL CYCLOHEXANOL
mf: $C_{12}H_{16}O$ mw: 176.28

PROP: Colorless to pale straw-colored liquid. Bp: 276-281°, flash p: 280°F, d: 1.033 @ 25°/25°, vap d: 6.13.

SYN: INSECT REPELLENT 448

TOXICITY DATA with REFERENCE
orl-mus LD50:5400 mg/kg JPETAB 93,26,48
orl-rbt LD50:2700 mg/kg JPETAB 93,26,48
orl-gpg LD50:1600 mg/kg JPETAB 93,26,48

SAFETY PROFILE: Moderately toxic by ingestion. Can cause dermatitis. Combustible when exposed to heat or flame; can react with oxidizing materials. To fight fire, use foam, CO_2, dry chemical. When heated to decomposition it emits acrid smoke and irritating fumes. See also ALCOHOLS.

PES750 CAS:712-50-5 ***HR: 3***
PHENYL CYCLOHEXYL KETONE
mf: $C_{13}H_{16}O$ mw: 188.29

SYN: USAF KF-3

TOXICITY DATA with REFERENCE
ipr-mus LD50:200 mg/kg NTIS** AD277-689

CONSENSUS REPORTS: Reported in EPA TSCA Inventory.

SAFETY PROFILE: Poison by intraperitoneal route. When heated to decomposition it emits acrid smoke and irritating fumes. See also KETONES.

PET000 CAS:101-87-1 ***HR: 2***
N-PHENYL-N'-CYCLOHEXYL-p-PHENYLENEDIAMINE
mf: $C_{18}H_{22}N_2$ mw: 266.42

SYNS: N-FENYL-N'-CYKLOHEXYL-p-FENYLENDIAMIN(CZECH) ◇ VULKACIT 4010 (CZECH)

TOXICITY DATA with REFERENCE
eye-rbt 500 mg/24H MOD 28ZPAK -,73,72
orl-rat LD50:3150 mg/kg 28ZPAK -,73,72

CONSENSUS REPORTS: Reported in EPA TSCA Inventory.

SAFETY PROFILE: Moderately toxic by ingestion. An eye irritant. When heated to decomposition it emits toxic fumes of NO_x. See also AMINES.

PET250 CAS:125-85-9 ***HR: 3***
1-PHENYLCYCLOPENTANECARBOXYLIC ACID-2-DIETHYLAMINOETHYL ESTER HYDROCHLORIDE
mf: $C_{18}H_{27}NO_2 \cdot ClH$ mw: 325.92

SYNS: CARAMIPHEN ◇ CARAMIPHENE HYDROCHLORIDE ◇ DIETHYLAMINOETHYL-1-PHENYLCYCLOPENTANE-1-CARBOXYLATE HYDROCHLORIDE ◇ GEIGY 2747 ◇ HYDROCHLORIDE OF 1-PHENYLCYCLOPENTANECARBOXYLIC ACID DIETHYLAMINOETHYL ESTER ◇ PANPARNIT ◇ PARPANIT ◇ PENTAPHENE HYDROCHLORIDE

TOXICITY DATA with REFERENCE
ipr-mus LD50:209 mg/kg JPETAB 96,42,49
ivn-rat LD50:35 mg/kg SMWOAS 76,1282,46
orl-mus LD50:485 mg/kg AIPTAK 134,255,61
ipr-mus LD50:222 mg/kg JPETAB 96,42,49
scu-mus LDLo:500 mg/kg ARZNAD 6,695,56
ivn-mus LD50:36500 μg/kg EJPHAZ 9,304,70
orl-cat LD50:390 mg/kg JPETAB 96,42,49
ivn-rbt LD50:25 mg/kg JPETAB 96,42,49
scu-gpg LD50:130 mg/kg AIPTAK 137,375,62

SAFETY PROFILE: Poison by ingestion, intraperitoneal, subcutaneous, and intravenous routes. When heated to decomposition it emits very toxic fumes of HCl and NO_x. Used as an anticholinergic drug.

PET500 CAS:13492-01-8 ***HR: 3***
PHENYLCYCLOPROMINE SULFATE
mf: $C_{18}H_{20}N_2 \cdot O_4S$ mw: 360.46

SYNS: 1-AMINO-2-PHENYLCYCLOPROPANESULFATE ◇ CYCLOPROPANAMINE, 2-PHENYL-, trans-(±)-, SULFATE (2:1) ◇ PARNATE ◇ trans,D,L-2-PHENYLCYCLOPROPYLAMINE SULFATE ◇ TRANCYLPROMINE SULFATE ◇ TRANSAMINE SULFATE ◇ TRANYLCYPRAMINE SULFATE ◇ TRANYLCYPROMINE SULFATE ◇ TRANYLCYPROMINE SULPHATE

TOXICITY DATA with REFERENCE
orl-man TDLo:34 mg/kg/60D-I AJPSAO 140,1229,83
orl-wmn TDLo:1200 μg/kg/3D-I AJPSAO 144,119,87
ipr-rat LD50:45 mg/kg ABMGAJ 18,617,67
orl-mus LD50:38 mg/kg 27ZQAG -,359,72
ipr-mus LD50:41 mg/kg EJPHAZ 6,115,69
ivn-mus LD50:37 mg/kg 27ZQAG -,359,72

SAFETY PROFILE: Poison by ingestion, intravenous, and intraperitoneal routes. Human toxic effects by ingestion. When heated to decomposition it emits very toxic fumes of SO_x and NO_x.

PET750 CAS:3721-28-6 ***HR: 3***
trans-2-PHENYLCYCLOPROPYLAMINE
mf: $C_9H_{11}N$ mw: 133.21

SYNS: PARNATE ◇ trans-2-PHENYL-1-AMINOCYCLOPROPANE ◇ SKF 385 ◇ TRANILCYPROMINE ◇ TRANSAMINE ◇ TRANYLCYPRAMINE ◇ TRANYLCYPROMINE

TOXICITY DATA with REFERENCE
cyt-oin-urn 10 g/L JCLBA3 47,182a,70
ipr-rat TDLo:10 mg/kg (1D pre):REP FEPRA7 24,129,65
ipr-rat LD50:30 mg/kg FCTXAV 3,597,65
orl-mus LD50:20 mg/kg BCPCA6 17,369,68
ipr-mus LD50:11300 μg/kg BRXXAA #1460700
scu-mus LD50:30 mg/kg BCPCA6 17,369,68

SAFETY PROFILE: Poison by ingestion, intraperitoneal, and subcutaneous routes. Experimental reproductive effects. Mutation data reported. When heated to decomposition it emits toxic fumes of NO_x. See also AMINES.

PEU000 CAS:56974-46-0 ***HR: 3***
3-PHENYL-5-(DIBUTYLAMINOETHYLAMINO)-1,2,4-OXADIAZOLE HYDROCHLORIDE
mf: $C_{18}H_{28}N_4O \cdot ClH$ mw: 352.96

SYNS: ADREVIL ◇ BUTALAMINE HYDROCHLORIDE ◇ 5-((2-(DIBUTYLAMINO)-ETHYL)AMINO)-3-PHENYL-1,2,4-OXADIAZOLE HYDROCHLORIDE ◇ N,N-DIBUTYL-N'-(3-PHENYL-1,2,4-OXADIAZOL-5-YL)-1,2-ETHANEDIAMINE HYDROCHLORIDE ◇ HEMOTROPE ◇ LA 1221 ◇ SUREM ◇ SURHEME

TOXICITY DATA with REFERENCE
orl-rat LD50:1600 mg/kg THERAP 24,745,69
orl-mus LD50:625 mg/kg THERAP 24,745,69
scu-mus LD50:2500 mg/kg THERAP 24,745,69
ivn-mus LD50:43 mg/kg THERAP 24,745,69

SAFETY PROFILE: Poison by intravenous route. Moderately toxic by ingestion and subcutaneous routes. When heated to decomposition it emits very toxic fumes of NO_x and HCl. Used as a vasodilator. See also AMINES.

PEU500 CAS:13056-98-9 ***HR: 3***
1-PHENYL-3,3-DIETHYLTRIAZENE
mf: $C_{10}H_{15}N_3$ mw: 177.28

SYNS: 3,3-DIETHYL-1-PHENYLTRIAZENE ◇ 1-FENYL-3,3-DIETHYLTRIAZEN (CZECH) ◇ 1-PHENYL-3,3-DIAETHYLTRIAZEN (GERMAN)

TOXICITY DATA with REFERENCE
mnt-mus-ipr 71 mg/kg/24H MUREAV 56,319,78
cyt-ham:lng 10 mg/L MUREAV 88,197,81
scu-rat TDLo:220 mg/kg (female 15D post):TER IARCCD 4,45,73
scu-rat TDLo:300 mg/kg:CAR,REP ZKKOBW 81,285,74
scu-rat TDLo:110 mg/kg/(15D preg):CAR,TER IARCCD 4,45,73
scu-rat TD:1100 mg/kg/70W-I:CAR ZKKOBW 81,285,74
orl-rat LD50:520 mg/kg 28ZPAK -,77,72
scu-rat LD50:420 mg/kg ZKKOBW 81,285,74

CONSENSUS REPORTS: EPA Genetic Toxicology Program.

SAFETY PROFILE: Moderately toxic by ingestion and subcutaneous routes. Experimental teratogenic and reproductive effects. Questionable carcinogen with experimental carcinogenic data. Mutation data reported. When heated to decomposition it emits toxic fumes of NO_x.

PEU600 CAS:61001-31-8 ***HR: D***
2-PHENYL-5,6-DIHYDROPYRAZOLO(5,1-a)ISOQUINOLINE
mf: $C_{17}H_{14}N_2$ mw: 246.33

SYN: 5,6-DIHYDRO-2-PHENYLPYRAZOLO(5,1-a)ISOQUINOLINE

TOXICITY DATA with REFERENCE
scu-rat TDLo:100 mg/kg (female 6-10D post):REP EJMCA5 19,215,84

SAFETY PROFILE: Experimental reproductive effects. When heated to decomposition it emits toxic fumes of NO_x.

PEU650 CAS:55308-57-1 ***HR: D***
2-PHENYL-5,6-DIHYDRO-s-TRIAZOLO(5,1-a)ISOQUINOLINE
mf: $C_{16}H_{13}N_3$ mw: 247.32

SYNS: L 10499 ◇ 5,6-DIHYDRO-2-PHENYL-(1,2,4)TRIAZOLO(5,1-a)ISOQUINOLINE (9CI)

TOXICITY DATA with REFERENCE
scu-ham TDLo:5 mg/kg (female 4-8D post):REP ARZNAD 33,1222,83

SAFETY PROFILE: Experimental reproductive effects. When heated to decomposition it emits toxic fumes of NO_x.

PEV500 CAS:1754-58-1 ***HR: 3***
O-PHENYL-N,N'-DIMETHYL PHOSPHORODIAMIDATE
mf: $C_8H_{13}N_2O_2P$ mw: 200.20

SYNS: DIAMIDAFOS ◇ DIAMIDFOS ◇ DOWCO 169 ◇ NELLITE

TOXICITY DATA with REFERENCE
orl-rat LD50:140 mg/kg SPEADM 74-1,-,74
orl-rbt LD50:63 mg/kg SPEADM 74-1,-,74
skn-rbt LD50:100 mg/kg FMCHA2 -,C166,83
orl-bwd LD50:13 mg/kg TXAPA9 21,315,72

SAFETY PROFILE: Poison by ingestion and skin contact. When heated to decomposition it emits very toxic fumes of PO_x and NO_x. A pesticide used on tobacco to control rootknot nematodes.

PEV600 CAS:36981-93-8 ***HR: 3***
2-PHENYL-5,5-DIMETHYL-TETRAHYDRO-1,4-OXAZINE HYDROCHLORIDE
mf: $C_{12}H_{17}NO•ClH$ mw: 227.76

SYN: 2-PHENYL-5,5-DIMETHYL-TETRAHYDRO-1,4-OXAZIN HYDROCHLORID (GERMAN)

TOXICITY DATA with REFERENCE
orl-rbt TDLo:275 mg/kg (7-17D preg):TER ARZNAD 24,1627,74
orl-rbt TDLo:275 mg/kg (7-17D preg):REP ARZNAD 24,1627,74
orl-rat LD50:450 mg/kg ARZNAD 24,1627,74
orl-mus LD50:380 mg/kg ARZNAD 24,1627,74

SAFETY PROFILE: Poison by ingestion. An experimental teratogen. Experimental reproductive effects. When heated to decomposition it emits toxic fumes of NO_x and HCl.

PEV750 CAS:390-64-7 ***HR: 3***
1-PHENYL-2-(1',1'-DIPHENYLPROPYL-3'-AMINO)PROPANE
mf: $C_{24}H_{27}N$ mw: 329.52

SYNS: B-436 ◇ BISMETHIN ◇ CARDITIN ◇ CORPAX ◇ N-(3,3-DIPHENYLPROPYL)-α-METHYLPHENETHYLAMINE ◇ ELECOR ◇ HOSTAGINAN ◇ N-(1-METHYL-2-PHENYLETHYL)-Γ-PHENYLBENZENEPROPANAMINE ◇ PRENYLAMINE ◇ SEGONTIN ◇ SYNADRIN ◇ VALECOR

TOXICITY DATA with REFERENCE
orl-mus TDLo:180 mg/kg (female 7-12D post):TER GNRIDX 5,271,71
orl-mus TDLo:270 mg/kg (female 7-12D post):REP GNRIDX 5,271,71
orl-man TDLo:13 mg/kg ATXKA8 31,217,74
orl-man TDLo:2816 mg/kg/3Y-I BMJOAE 288,1048,84
orl-wmn TDLo:3600 mg/kg:CVS BMJOAE 2(6087),608,77
ivn-rat LD50:11 mg/kg ARZNAD 20,1362,70
ivn-mus LD50:250 mg/kg TXAPA9 6,676,64

SAFETY PROFILE: Poison by intravenous route. Human systemic effects by ingestion: arrythmia. Experimental teratogenic and reproductive effects. When heated to decomposition it emits toxic fumes of NO_x.

PEW000 CAS:30748-29-9 ***HR: 3***
4-PHENYL-1,2-DIPHENYL-3,5-PYRAZOLIDINEDIONE
mf: $C_{20}H_{20}N_2O_2$ mw: 320.42

SYNS: ANALUD ◇ DA 2370 ◇ FENILPRENAZONE ◇ FEPRAZONE ◇ 4-(β-ISOAMYLENYL)-1,2-DIPHENYL-3,5-PYRAZOLIDINEDIONE ◇ 4-(2-ISOPENTENYL)-1,2-DIPHENYL-3,5-PYRAZOLIDINEDIONE ◇ METHRAZONE ◇ 4-(3-METHYL-2-BUTENYL)-1,2-DIPHENYL-3,5-PYRAZOLIDINEDIONE ◇ PRENAZONE ◇ 4-PRENYL-1,2-DIPHENYL-3,5-PYRAZOLIDINEDIONE ◇ ZEPELIN

TOXICITY DATA with REFERENCE
orl-rat TDLo:1320 mg/kg (7-17D preg):REP IYKEDH 10,149,79
orl-rat TDLo:2640 mg/kg (7-17D preg):TER IYKEDH 10,149,79
orl-rat LD50:1343 mg/kg IYKEDH 10,133,79
ipr-rat LD50:362 mg/kg IYKEDH 10,133,79
orl-mus LD50:1067 mg/kg ARZNAD 22,196,72
ipr-mus LD50:270 mg/kg IYKEDH 10,133,79
scu-mus LDLo:374 mg/kg IYKEDH 10,133,79
ivn-mus LD50:180 mg/kg ARZNAD 22,196,72

SAFETY PROFILE: Poison by subcutaneous, intravenous, and intraperitoneal routes. Moderately toxic by ingestion. Experimental teratogenic and reproductive effects. When heated to decomposition it emits toxic fumes of NO_x.

PEW250 CAS:882-33-7 ***HR: 3***
PHENYL DISULFIDE
mf: $C_{12}H_{10}S_2$ mw: 218.34

PROP: White powder or needles from alc. Mp: 61°, bp: 310°. Insol in water; sol in alc; very sol in ether.

SYNS: DIPHENYL DISULFIDE ◇ DISULFIDE DIPHENYL ◇ USAF E-1

TOXICITY DATA with REFERENCE
ipr-mus LD50:100 mg/kg NTIS** AD277-689

CONSENSUS REPORTS: Reported in EPA TSCA Inventory.

SAFETY PROFILE: Poison by intraperitoneal route. When heated to decomposition it emits toxic fumes of SO_x. Flammable when exposed to heat or flame. See also SULFIDES.

PEW500 CAS:123-01-3 ***HR: 2***
1-PHENYLDODECANE
mf: $C_{18}H_{30}$ mw: 246.48

PROP: Liquid. Bp: 290-410°, flash p: 285°F, d: 0.9, vap d: 8.47.

SYNS: DETERGENT ALKYLATE ◇ DODECYLBENZENE ◇ PHENYLDODECAN (GERMAN)

CONSENSUS REPORTS: Reported in EPA TSCA Inventory.

SAFETY PROFILE: Probably moderately toxic. Combustible when exposed to heat or flame; can react with oxidizing materials. To fight fire, use foam, CO_2, dry chemical. When heated to decomposition it emits acrid smoke and irritating fumes.

PEW700 ***HR: 3***
PHENYLENE-p,p'-BIS(2-DIMETHYLBENZYL-AMMONIUMPROPYL) DICHLORIDE
mf: $C_{30}H_{42}N_2 \cdot 2Cl$ mw: 501.64

SYNS: p-PHENYLENEBIS(1-METHYLETHYLENE)BIS(BENZYL-DIMETHYLAMMONIUM CHLORIDE) ◇ (p-PHENYLENEBIS(1-METHYLETHYLENE))BIS(BENZYLDIMETHYLAMMONIUM)DICHLORIDE

TOXICITY DATA with REFERENCE
orl-mus LD50:96 mg/kg DIPHAH 13,131,61
scu-mus LD50:625 μg/kg AITEAT 14,104,66
ivn-mus LD50:525 μg/kg DIPHAH 13,131,61

SAFETY PROFILE: Poison by ingestion, subcutaneous, and intravenous routes. When heated to decomposition it emits toxic fumes of Cl^-, NO_x, and NH_3. See also CHLORIDES.

PEW750 CAS:64491-74-3 ***HR: 2***
(1,2-PHENYLENEBIS(IMINOCARBONOTHIOYL)) BIS-CARBAMIC ACID, DIMETHYL ESTER, mixed with 5-ETHOXY-3-(TRICHLOROMETHYL)-1,2,4-THIADIAZOLE
mf: $C_{12}H_{14}N_4O_4S_2 \cdot C_5H_5Cl_3N_2OS$ mw: 589.95

SYN: BANROT

TOXICITY DATA with REFERENCE
orl-rat LD50:5 g/kg FMCHA2 -,C23,83
skn-rbt LD50:2 g/kg FMCHA2 -,C23,83

SAFETY PROFILE: Moderately toxic by skin contact. Mildly toxic by ingestion. When heated to decomposition it emits very toxic fumes of NO_x, SO_x, and Cl^-. Used to control damping off, root and stem rot diseases. See also CARBAMATES.

PEX250 CAS:539-48-0 ***HR: 2***
p-PHENYLENEBIS(METHYLAMINE)
mf: $C_8H_{12}N_2$ mw: 136.22

SYNS: 1,4-BIS-AMINOMETHYLBENZEN (CZECH) ◇ p-XYLYLENDIAMINE (CZECH)

TOXICITY DATA with REFERENCE
skn-rbt 50 mg/24H SEV 28ZPAK -,64,72
eye-rbt 50 μg/24H SEV 28ZPAK -,64,72
orl-rat LD50:935 mg/kg 28ZPAK -,64,72

CONSENSUS REPORTS: Reported in EPA TSCA Inventory.

SAFETY PROFILE: Moderately toxic by ingestion. A severe skin and eye irritant. When heated to decomposition it emits toxic fumes of NO_x. See also AMINES.

PEX300 ***HR: 3***
m-PHENYLENEBIS(1-METHYLETHYLENE)BIS (BENZYLDIMETHYLAMMONIUM CHLORIDE)
mf: $C_{30}H_{42}N_2 \cdot 2Cl$ mw: 501.64

SYNS: (m-PHENYLENEBIS(1-METHYLETHYLENE)BIS(BENZYLDIMETHYLAMMONIUM) DICHLORIDE ◇ PHENYLENE-m,m'-DI-(2-DIMETHYLBENZYLAMMONIUMPROPYL)DICHLORIDE

TOXICITY DATA with REFERENCE
orl-mus LD50:87 mg/kg DIPHAH 13,131,61
scu-mus LD50:4500 μg/kg AITEAT 14,104,66
ivn-mus LD50:325 μg/kg DIPHAH 13,131,61

SAFETY PROFILE: Poison by ingestion, subcutaneous, and intravenous routes. When heated to decomposition it emits toxic fumes of Cl^-, NO_x, and NH_3. See also CHLORIDES.

PEX325 ***HR: 3***
m-PHENYLENEBIS(1-METHYLETHYLENE)BIS (DIMETHYLETHYLAMMONIUM BROMIDE)
mf: $C_{20}H_{38}N_2 \cdot 2Br$ mw: 466.42

SYNS: (m-PHENYLENEBIS(1-METHYLETHYLENE))BIS (DIMETHYLETHYLAMMONIUM) DIBROMIDE ◇ PHENYLENE-m,m'-DI(2-DIMETHYLETHYLAMMONIUMPROPYL)DIBROMIDE

TOXICITY DATA with REFERENCE
orl-mus LD50:35 mg/kg DIPHAH 13,131,61
scu-mus LD50:3 mg/kg AITEAT 14,104,66
ivn-mus LD50:1300 μg/kg DIPHAH 13,131,61

SAFETY PROFILE: Poison by ingestion, subcutaneous, and intravenous routes. When heated to decomposition it emits toxic fumes of Br^-, NO_x, and NH_3. See also BROMIDES.

PEX330 ***HR: 3***
p-PHENYLENEBIS(1-METHYLETHYLENE)BIS (DIMETHYLETHYLAMMONIUM BROMIDE
mf: $C_{20}H_{39}N_2 \cdot 2Br$ mw: 466.42

SYNS: (p-PHENYLENEBIS(1-METHYLETHYLENE)BIS(DIMETHYLETHYLAMMONIUM) DIBROMIDE ◇ PHENYLENE-p,p'-DI(2-DIMETHYLETHYLAMMONIUMPROPYL)DIBROMIDE

TOXICITY DATA with REFERENCE
orl-mus LD50:58 mg/kg DIPHAH 13,131,61
scu-mus LD50:4250 μg/kg AITEAT 14,104,66
ivn-mus LD50:1875 μg/kg DIPHAH 13,131,61

SAFETY PROFILE: Poison by ingestion, subcutaneous, and intravenous routes. When heated to decomposition it emits toxic fumes of Br^-, NO_x, and NH_3. See also BROMIDES.

PEX350 ***HR: 3***
m-PHENYLENEBIS(1-METHYLETHYLENE)BIS (TRIMETHYLAMMONIUM BROMIDE)
mf: $C_{18}H_{34}N_2 \cdot 2Br$ mw: 438.36

SYNS: (m-PHENYLENEBIS(1-METHYLETHYLENE)BIS(TRIMETHYL-AMMONIUM) DIBROMIDE ◇ PHENYLENE-m,m'-BIS(2-TRIMETHYLAMMONIUMPROPYL) DIBROMIDE

TOXICITY DATA with REFERENCE
orl-mus LD50:65 mg/kg DIPHAH 13,131,61
scu-mus LD50:6500 μg/kg AITEAT 14,104,66
ivn-mus LD50:2500 μg/kg DIPHAH 13,131,61

SAFETY PROFILE: Poison by ingestion, subcutaneous, and intravenous routes. When heated to decomposition it emits toxic fumes of Br^-, NO_x, and NH_3. See also BROMIDES.

PEX355 ***HR: 3***
p-PHENYLENEBIS(1-METHYLETHYLENE)BIS (TRIMETHYLAMMONIUM BROMIDE)
mf: $C_{18}H_{34}N_2•2Br$ mw: 438.36

SYNS: (p-PHENYLENEBIS(1-METHYLETHYLENE)BIS(TRIMETHYL-AMMONIUM) DIBROMIDE ◇ PHENYLENE-p,p'-BIS(2-TRIMETHYL-AMMONIUMPROPYL) DIBROMIDE

TOXICITY DATA with REFERENCE
orl-mus LD50:75 mg/kg DIPHAH 13,131,61
scu-mus LD50:10 mg/kg AITEAT 14,104,66
ivn-mus LD50:2500 μg/kg DIPHAH 13,131,61

SAFETY PROFILE: Poison by ingestion, subcutaneous, and intravenous routes. When heated to decomposition it emits toxic fumes of Br^-, NO_x, and NH_3. See also BROMIDES.

PEX500 CAS:23564-05-8 ***HR: 2***
4,4'-o-PHENYLENEBIS(3-THIOALLOPHANIC ACID)DIMETHYL ESTER
mf: $C_{12}H_{14}N_4O_4S_2$ mw: 342.42

SYNS: BAS 32500F ◇ o-BIS(3-METHOXYCARBONYL-2-THIOUREIDO)BENZENE ◇ 1,2-BIS(METHOXYCARBONYL- THIOUREIDO) BENZENE ◇ 1,2-BIS(3-(METHOXYCARBONYL)-2-THIOUREIDO)BENZENE ◇ CERCOBIN METHYL ◇ DIMETHYL-4,4'-PHENYLENE-BIS-(3-THIOALLOPHANATE) ◇ ENOVIT M ◇ FUNGITOX ◇ METHYL THIOPHANATE ◇ NEOTOPSIN ◇ NF 44 ◇ PELT-44 ◇ SIPCAVIT ◇ TD 1771 ◇ TOPSIN WP METHYL ◇ ZYBAN

TOXICITY DATA with REFERENCE
mmo-smc 5 ppm RSTUDV 6,161,76
sln-asn 14 μmol/L EVHPAZ 31,81,79
orl-mus TDLo:15 g/kg (female 1-15D post):REP TXAPA9 24,206,73
orl-rat LD50:6640 mg/kg TXAPA9 23,606,72
ipr-rat LD50:1140 mg/kg TXAPA9 23,606,72
orl-mus LD50:3400 mg/kg TXAPA9 23,606,72
ipr-mus LD50:790 mg/kg TXAPA9 23,606,72
orl-dog LDLo:4000 mg/kg TXAPA9 23,606,72
orl-rbt LD50:2270 mg/kg TXAPA9 23,606,72
orl-gpg LD50:3640 mg/kg TXAPA9 23,606,72

SAFETY PROFILE: Moderately toxic by ingestion and intraperitoneal routes. Experimental reproductive effects. Mutation data reported. When heated to decomposition it emits very toxic fumes of NO_x and SO_x. Used for plant disease control in vegetables, fruits, and turf. See also ESTERS.

PEX750 CAS:68772-49-6 ***HR: 3***
1,1'-(p-PHENYLENEBIS(VINYLENE-p-PHENYLENE))BIS(PYRIDINIUM) DI-p-TOLUENESULFONATE
mf: $C_{29}H_{26}N_2•2C_7H_7O_3S$ mw: 744.97

TOXICITY DATA with REFERENCE
dnd-mus:lym 760 nmol/L JMCMAR 22,134,79
ipr-mus LD10:7 mg/kg JMCMAR 22,134,79

SAFETY PROFILE: Poison by intraperitoneal route. Mutation data reported. When heated to decomposition it emits very toxic fumes of SO_x and NO_x.

PEY000 CAS:108-45-2 ***HR: 3***
m-PHENYLENEDIAMINE
DOT: UN 1673
mf: $C_6H_8N_2$ mw: 108.16

PROP: White crystals. Mp: 63°, bp: 286°, d: 1.139, vap press: 1 mm @ 99.8°. Sol in water, methanol, ethanol, chloroform, acetone; sltly sol in ether, carbon tetrachloride; very sltly sol in benzene, toluene.

SYNS: 3-AMINOANILINE ◇ m-AMINOANILINE ◇ APCO 2330 ◇ m-BENZENEDIAMINE ◇ 1,3-BENZENEDIAMINE ◇ C.I. 76025 ◇ DEVELOPER 11 ◇ m-DIAMINOBENZENE ◇ 1,3-DIAMINOBENZENE ◇ DIRECT BROWN BR ◇ m-FENYLENDIAMIN (CZECH) ◇ METAPHENYLENEDIAMINE ◇ 1,3-PHENYLENEDIAMINE ◇ m-PHENYLENEDIAMINE (DOT) ◇ PHENYLENEDIAMINE, META, solid (DOT)

TOXICITY DATA with REFERENCE
bfa-rat/sat 240 mg/kg MUREAV 138,137,84
cyt-ham:lng 12 mg/L GMCRDC 27,95,81
ipr-rat TDLo:375 mg/kg (30D male):TER SheCW# 25MAR77
scu-rat TDLo:1485 mg/kg/47W-I:ETA KJMSAH 13,175,62
ipr-rat LD50:283 mg/kg JTEHD6 2,657,77
scu-rat LDLo:30 mg/kg ZEPTAT 17,59,15
ipr-mus LDLo:400 mg/kg RBPMAZ 22,1,52
ivn-dog LDLo:17 mg/kg JIDHAN 4,386,23
orl-cat LDLo:300 mg/kg JIDHAN 4,386,23
orl-rbt LDLo:300 mg/kg JIDHAN 4,386,23
skn-rbt LDLo:5000 mg/kg JTEHD6 2,657,77
scu-rbt LDLo:200 mg/kg JIDHAN 4,386,23

CONSENSUS REPORTS: IARC Cancer Review: Group 3 IMEMDT 7,56,87; Animal Inadequate Evidence IMEMDT 16,111,78. EPA Genetic Toxicology Program. Reported in EPA TSCA Inventory.

ACGIH TLV: TWA 0.1 mg/m^3

DOT Classification: Poison B; Label: St. Andrews Cross.

SAFETY PROFILE: Poison by ingestion, intravenous, subcutaneous, and intraperitoneal routes. Mildly toxic by skin contact. Questionable carcinogen with experimental tumorigenic and teratogenic data. Mutation data reported. Combustible when exposed to heat or flame. A hair dye ingredient. When heated to decomposition it emits toxic fumes of NO_x. See also other phenylenediamine entries and AMINES.

PEY250 CAS:95-54-5 ***HR: 2***
o-PHENYLENEDIAMINE
DOT: UN 1673
mf: $C_6H_8N_2$ mw: 108.16

PROP: Tan crystals. Mp: 104°, bp: 257°. Sltly sol in water; very sol in alc, chloroform, ether.

SYNS: 2-AMINOANILINE ◇ o-BENZENEDIAMINE ◇ 1,2-BENZENEDIAMINE ◇ C.I. 76010 ◇ C.I. OXIDATION BASE 16 ◇ o-DIAMINOBENZENE ◇ 1,2-DIAMINOBENZENE ◇ EK 1700 ◇ NSC 5354 ◇ ORTHAMINE ◇ 1,2-PHENYLENEDIAMINE (DOT)

TOXICITY DATA with REFERENCE
mma-sat 1 μmol/plate JTSCDR 7,61,82
mnt-gpg-ipr 216 mg/kg/24H ARTODN 43,249,80
scu-rat LDLo:600 mg/kg AIPTAK 36,140,30
orl-rat LD50:1070 mg/kg JTEHD6 2,657,77
ipr-rat LD50:516 mg/kg JTEHD6 2,657,77
ipr-mus LD50:245 mg/kg NCISP* JAN86
scu-mus LD50:450 mg/kg KJMSAH 13,175,62
skn-rbt LDLo:5000 mg/kg JTEHD6 2,657,77
orl-bwd LD50:133 mg/kg AECTCV 12,355,83

CONSENSUS REPORTS: Reported in EPA TSCA Inventory. EPA Genetic Toxicology Program.

ACGIH TLV: TWA 0.1 mg/m^3; Suspected Human Carcinogen
DOT Classification: Poison B; Label: St. Andrews Cross.

SAFETY PROFILE: Poison by ingestion and intraperitoneal routes. Moderately toxic by subcutaneous route. Mildly toxic by skin contact. Mutation data reported. A pesticide and pharmaceutical. When heated to decomposition it emits toxic fumes of NO_x. See also other phenylenediamine entries and AMINES.

PEY500 CAS:106-50-3 ***HR: 3***
p-PHENYLENEDIAMINE
DOT: UN 1673
mf: $C_6H_8N_2$ mw: 108.16
PROP: White-sltly red crystals. Mp: 146°, flash p: 312°F, vap d: 3.72, bp: 267°. Sol in alc, chloroform, ether.

SYNS: p-AMINOANILINE ◇ 4-AMINOANILINE ◇ BASF URSOL D ◇ p-BENZENEDIAMINE ◇ 1,4-BENZENEDIAMINE ◇ BENZOFUR D ◇ C.I. 76060 ◇ C.I. DEVELOPER 13 ◇ C.I. OXIDATION BASE 10 ◇ DEVELOPER 13 ◇ DEVELOPER PF ◇ p-DIAMINOBENZENE ◇ 1,4-DIAMINOBENZENE ◇ DURAFUR BLACK R ◇ FENYLENODWUAMINA (POLISH) ◇ FOURAMINE D ◇ FOURRINE D ◇ FOURRINE 1 ◇ FUR BLACK 41867 ◇ FUR BROWN 41866 ◇ FURRO D ◇ FUR YELLOW ◇ FUTRAMINE D ◇ NAKO H ◇ ORSIN ◇ PARA ◇ PARAPHENYLEN-DIAMINE ◇ PELAGOL D ◇ PELAGOL DR ◇ PELAGOL GREY D ◇ PELTOL D ◇ 1,4-PHENYLENEDIAMINE ◇ PHENYLENEDIAMINE, PARA, solid (DOT) ◇ PPD ◇ RENAL PF ◇ SANTOFLEX IC ◇ TERTRAL D ◇ URSOL D ◇ USAF EK-394 ◇ VULKANOX 4020 ◇ ZOBA BLACK D

TOXICITY DATA with REFERENCE
skn-hmn 250 mg/24H MLD JSCCA5 23,371,72
skn-mus 250 mg/24H MLD JSCCA5 23,371,72
skn-dog 250 mg/24H MLD JSCCA5 23,371,72
skn-rbt 12500 μg/24H MLD FCTXAV 15,607,77
skn-rbt 250 mg/24H MOD JSCCA5 23,371,72
skn-pig 250 mg/24H MLD JSCCA5 23,371,72
skn-gpg 250 mg/24H MLD JSCCA5 23,371,72
mma-sat 10 μg/plate BCPCA6 26,729,77
otr-rat:emb 1850 ng/plate JJATDK 1,190,81
sln-dmg-orl 15500 μmol/L/3D MUREAV 48,181,77
scu-rat TDLo:2625 mg/kg/30W-C:ETA KJMSAH 9,94,58
orl-rat LD50:80 mg/kg JTEHD6 2,657,77
ivn-rat LDLo:50 mg/kg BSRSA6 41,302,72
scu-mus LDLo:140 mg/kg JSCCA5 12,500,61
ipr-rbt LDLo:150 mg/kg JIDIAQ 42,473,28
ivn-rbt LDLo:300 mg/kg AIMDAP 36,724,25
ipr-rat LD50:37 mg/kg JTEHD6 2,657,77
scu-rat LDLo:170 mg/kg JIDHAN 4,386,23
ipr-mus LD50:50 mg/kg NTIS** AD277-689
scu-dog LDLo:100 mg/kg XPHBAO 271,40,41
ivn-dog LDLo:17 mg/kg JIDHAN 4,386,23
orl-cat LDLo:100 mg/kg JIDHAN 4,386,23
orl-rbt LDLo:250 mg/kg JIDHAN 4,386,23
skn-rbt LDLo:5000 mg/kg JTEHD6 2,657,77
scu-rbt LDLo:200 mg/kg JIDHAN 4,386,23

CONSENSUS REPORTS: IARC Cancer Review: Group 3 IMEMDT 7,56,87; Animal Inadequate Evidence IMEMDT 16,125,78. Community Right-To-Know List. Reported in EPA TSCA Inventory. EPA Genetic Toxicology Program.

OSHA PEL: TWA 0.1 mg/m^3 (skin)
ACGIH TLV: TWA 0.1 mg/m^3 (skin)
DFG MAK: 0.1 mg/m^3
DOT Classification: ORM-A; Label: None: Poison B; Label: St. Andrews Cross.

SAFETY PROFILE: Poison by ingestion, subcutaneous, intravenous, and intraperitoneal routes. Mildly toxic by skin contact. A human skin irritant. Questionable carcinogen with experimental tumorigenic data. Mutation data reported. Implicated in aplastic anemia. Can cause fatal liver damage. The p-form is more toxic and a stronger irritant than the o- and m- isomers. When used as a hair dye it caused vertigo, anemia, gastritis, exfoliative dermatitis, and death. Has caused asthma and other respiratory symptoms in the fur dying industry. Combustible when exposed to heat or flame; can react vigorously with oxidizing materials. To fight fire, use water, CO_2, dry chemical. When heated to decomposition it emits acrid smoke and irritating fumes. See also other phenylenediamine entries and AMINES.

PEY600 CAS:615-28-1 ***HR: 3***
o-PHENYLENEDIAMINE, DIHYDROCHLORIDE
mf: $C_6H_8N_2$•2ClH mw: 181.08

PROP: Needles. Very sol in water.

SYN: USAF EK-678

TOXICITY DATA with REFERENCE
orl-rat TDLo:130 g/kg/78W-C:ETA JEPTDQ 2,325,78
orl-mus TDLo:260 g/kg/78W-C:CAR JEPTDQ 2,325,78
orl-mus TD:518 g/kg/78W-C:CAR JEPTDQ 2,325,78
ipr-rat LD50:290 mg/kg NCIBR* NIH-NCI-E-68-1311,10,73
ipr-mus LD50:200 mg/kg NTIS** AD414-344

CONSENSUS REPORTS: Reported in EPA TSCA Inventory.

SAFETY PROFILE: Poison by intraperitoneal route. Questionable carcinogen with experimental carcinogenic and tumorigenic data. When heated to decomposition it emits very toxic fumes of HCl and NO_x. See also other phenylenediamine entries and AMINES.

PEY650 CAS:624-18-0 ***HR: 3***
p-PHENYLENEDIAMINE DIHYDROCHLORIDE
mf: $C_6H_8N_2$•2ClH mw: 181.08

PROP: Colorless triclinic. Very sol in water; sltly sol in alc; insol in HCl.

SYNS: p-AMINOANILINE DIHYDROCHLORIDE ◇ 4-AMINOANILINE DIHYDROCHLORIDE ◇ p-BENZENEDIAMINE DIHYDROCHLORIDE ◇ 1,4-BENZENEDIAMINE DIHYDROCHLORIDE ◇ C.I. 76061 ◇ C.I. OXIDATION BASE 10A ◇ p-DIAMINOBENZENE DIHYDROCHLORIDE ◇ 1,4-DIAMINOBENZENE DIHYDROCHLORIDE ◇ DURAFUR BLACK RC ◇ FOURINE DS ◇ FOURRINE 64 ◇ NCI-C03930 ◇ OXIDATION BASE 10A ◇ p-PD HCl ◇ p-PDA HCl ◇ PELAGOL CD ◇ PELAGOL GREY CD ◇ 1,4-PHENYLENEDIAMINE DIHYDROCHLORIDE ◇ p-PHENYLENEDIAMINE HYDROCHLORIDE

TOXICITY DATA with REFERENCE
mma-sat 33300 ng/plate ENMUDM 7(Suppl 5),1,85
mma-esc 333 μg/plate ENMUDM 7(Suppl 5),1,85
orl-rat LD50:147 mg/kg NCILB* NIH-NCI-E-C-72-3252
orl-mus LD50:316 mg/kg NCILB* NIH-NCI-E-C-72-3252
scu-mus LDLo:50 mg/kg XPHBAO 271,42,41
scu-dog LDLo:10 mg/kg XPHBAO 271,42,41
scu-rbt LDLo:20 mg/kg XPHBAO 271,42,41
scu-frg LDLo:10 mg/kg XPHBAO 271,42,41

CONSENSUS REPORTS: IARC Cancer Review: Animal Inadequate Evidence IMEMDT 16,125,78. NCI Carcinogenesis Bioassay (feed); No Evidence: mouse, rat NCITR* NCI-CG-TR-174,79. Reported in EPA TSCA Inventory.

SAFETY PROFILE: Poison by ingestion and subcutaneous routes. Questionable carcinogen. Mutation data reported. When heated to decomposition it emits very toxic fumes of NO_x and HCl. Used as an analytical reagent. See also other phenylenediamine entries and AMINES.

PEY750 CAS:541-69-5 ***HR: 3***
m-PHENYLENEDIAMINE HYDROCHLORIDE
mf: $C_6H_8N_2$•2ClH mw: 181.08

PROP: Colorless needles. Very sol in water; sltly sol in alc, ether.

SYNS: m-AMINOANILINE DIHYDROCHLORIDE ◇ 3-AMINOANILINE DIHYDROCHLORIDE ◇ m-BENZENEDIAMINE DIHYDROCHLORIDE ◇ 1,3-BENZENEDIAMINE HYDROCHLORIDE ◇ m-DIAMINOBENZENE DIHYDROCHLORIDE ◇ 1,3-DIAMINOBENZENE DIHYDROCHLORIDE ◇ 1,3-PHENYLENEDIAMINE DIHYDROCHLORIDE ◇ USAF EK-206

TOXICITY DATA with REFERENCE
scu-rat TDLo:1800 mg/kg/21W-I:ETA KJMSAH 13,175,62
ipr-rat LD50:325 mg/kg NCIBR* NIH-NCI-E-68-1311,10,73
ipr-mus LD50:100 mg/kg NTIS** AD414-344

CONSENSUS REPORTS: IARC Cancer Review: Animal Inadequate Evidence IMEMDT 16,111,78. Reported in EPA TSCA Inventory.

SAFETY PROFILE: Poison by intraperitoneal route. Questionable carcinogen with experimental tumorigenic data. When heated to decomposition it emits very toxic fumes of HCl and NO_x. See also other phenylenediamine entries and AMINES.

PFA250 CAS:88-63-1 ***HR: 2***
1,3-PHENYLENEDIAMINE-4-SULFONIC ACID
mf: $C_6H_8N_2O_3S$ mw: 188.22

SYNS: o-AMINOSULFANILIC ACID ◇ 1,3-DIAMINOBENZENESULFONIC ACID ◇ 2,4-DIAMINOBENZENESULFONIC ACID ◇ 1,3-DIAMINOBENZENE-4-SULFONIC ACID ◇ 1,3-DIAMINOBENZENE-6-SULFONIC ACID ◇ KYSELINA-2,4-DIAMINOBENZENSULFONOVA (CZECH) ◇ KYSELINA-1,3-FENYLENDIAMIN-4-SULFONOVA (CZECH) ◇ m-PHENYLENEDIAMINESULFONIC ACID ◇ m-PHENYLENEDIAMINE-4-SULFONIC ACID

TOXICITY DATA with REFERENCE
skn-rbt 500 mg/24H MOD 28ZPAK -,180,72

eye-rbt 20 mg/24H SEV 28ZPAK -,180,72
orl-rat LD50:3480 mg/kg 28ZPAK -,180,72

CONSENSUS REPORTS: Reported in EPA TSCA Inventory.

SAFETY PROFILE: Moderately toxic by ingestion. A skin and severe eye irritant. When heated to decomposition it emits very toxic fumes of NO_x and SO_x.

PFA500 CAS:4044-65-9 ***HR: 3***
1,4-PHENYLENEDIISOTHIOCYANIC ACID
mf: $C_8H_4N_2S_2$ mw: 192.26

PROP: Tasteless, odorless, colorless crystals. Mp: 132°.

SYNS: BISCOMATE ◇ BITOSCANATE ◇ 1,4-DIISOTHIOCYANATOBENZENE ◇ ISOTHIOCYANIC ACID-p-PHENYLENE ESTER ◇ JONIT ◇ PHENYLENE-1,4-DIISOTHIOCYANATE ◇ PHENYLENE THIOCYANATE

TOXICITY DATA with REFERENCE
orl-hmn TDLo:3 mg/kg:CNS,GIT JTMHA9 72,252,69
orl-rat LD50:21 mg/kg JTMHA9 72,253,69
orl-mus LD50:230 mg/kg FAZMAE 17,108,73
ipr-mus LD50:21 mg/kg FAZMAE 17,108,73

CONSENSUS REPORTS: Cyanide and its compounds are on the Community Right-To-Know List. EPA Extremely Hazardous Substances List. Reported in EPA TSCA Inventory.

SAFETY PROFILE: Poison by ingestion and intraperitoneal routes. Human systemic effects by ingestion: hallucinations, nausea. When heated to decomposition it emits very toxic fumes of NO_x, CN^-, and SO_x. See also THIOCYANATES.

PFA600 CAS:1477-20-9 ***HR: D***
N,N'-(p-PHENYLENEDIMETHYLENE)BIS(2,2-DICHLORO-N-ETHYLACETAMIDE)
mf: $C_{16}H_{26}Cl_4N_2O_2$ mw: 420.24

SYNS: N,N'-BIS(DICHLOROACETYL)-N,N-DIETHYL-1,4-XYLYLENEDIAMINE ◇ N,N'-(1,4-PHENYLENEBIS(METHYLENE))BIS(2,2-DICHLORO-N-ETHYL-ACETAMIDE(9CI) ◇ WIN 13,099

TOXICITY DATA with REFERENCE
spm-hmn-orl 3800 mg/kg/19W TXAPA9 3,1,61
orl-man TDLo:600 mg/kg (42D male):REP TXAPA9 3,1,61

CONSENSUS REPORTS: EPA Genetic Toxicology Program.

SAFETY PROFILE: Human reproductive effects by ingestion: changes in testes, epididymis, sperm duct and spermatogenesis. Experimental reproductive effects. Human mutation data reported. When heated to decomposition it emits toxic fumes of Cl^- and NO_x.

PFA750 CAS:4561-43-7 ***HR: 3***
PHENYLETHANOL AMINE HYDROCHLORIDE
mf: $C_8H_{11}NO{\bullet}ClH$ mw: 173.66

SYNS: 1-PHENYL-2-AMINO-1-ETHANOL, HYDROCHLORIDE ◇ α-PHENYL-β-AMINOETHANOL HYDROCHLORIDE

TOXICITY DATA with REFERENCE
scu-rat LDLo:320 mg/kg JPETAB 71,62,41
orl-mus LDLo:1250 mg/kg QJPPAL 9,203,36
ipr-mus LD50:520 mg/kg AEPPAE 237,171,59
ivn-mus LDLo:275 mg/kg QJPPAL 9,203,36
ivn-rbt LDLo:38 mg/kg JPETAB 32,121,27
scu-gpg LDLo:1000 mg/kg JACSAT 52,3317,30

SAFETY PROFILE: Poison by intravenous and subcutaneous routes. Moderately toxic by ingestion and intraperitoneal routes. When heated to decomposition it emits very toxic fumes of HCl and NO_x. See also AMINES.

PFA850 CAS:101-84-8 ***HR: 2***
PHENYL ETHER
mf: $C_{12}H_{10}O$ mw: 170.22

PROP: Colorless crystals, geranium odor. Mp: 28°, bp: 257°, flash p: 239°F, d: 1.0728 @ 20°, vap d: 5.86, autoign temp: 1148°F, lel: 0.8%, uel: 1.5%.

SYNS: BIPHENYL OXIDE ◇ DIPHENYL ETHER ◇ DIPHENYL OXIDE ◇ GERANIUM CRYSTALS ◇ PHENOXYBENZENE

TOXICITY DATA with REFERENCE
eye-rat 10 ppm/140H TXAPA9 33,78,75
skn-rbt 500 mg/24H MLD FCTXAV 12,707,74
eye-rbt 10 ppm/140H TXAPA9 33,78,75
orl-rat LD50:3370 mg/kg FCTXAV 12,707,74

CONSENSUS REPORTS: Reported in EPA TSCA Inventory.

OSHA PEL: Vapor: TWA 1 ppm
ACGIH TLV: TWA 1 ppm; STEL 2 ppm (vapor)
DFG MAK: 1 ppm (7 mg/m^3)

SAFETY PROFILE: Moderately toxic by ingestion. Prolonged exposure damages liver, spleen, kidneys, and thyroids and upsets gastrointestinal tract. A skin and eye irritant. Combustible when exposed to heat or flame; can react with oxidizing materials. For explosion hazard, see ETHERS. To fight fire, use water, foam, CO_2, dry chemical. When heated to decomposition it emits acrid smoke and irritating fumes.

PFA860 CAS:8004-13-5 ***HR: 2***
PHENYL ETHER-BIPHENYL MIXTURE
mf: $C_{12}H_{10}{\bullet}C_{12}H_{10}O$ mw: 324.44

PROP: Eutectic mixture 73.5% phenylether and 26.5% biphenyl by weight (MELAAD 48,247,57).

SYNS: BIPHENYL, mixed with BIPHENYL OXIDE (3:7) ◇ 1,1'-BIPHENYL, mixed with 1,1'-OXYBIS(BENZENE) ◇ BIPHENYL-DIPHENYL ETHER mixture ◇ DINIL ◇ DINYL ◇ DIPHENYL mixed with DIPHENYL OXIDE ◇ DIPHYL ◇ DOWTHERM ◇ DOWTHERM A

TOXICITY DATA with REFERENCE
skn-rbt 500 mg/24H MLD 28ZPAK -,271,72
eye-rbt 500 mg/24H MLD 28ZPAK -,271,72
ihl-hmn TCLo:3 ppm:IRR 28ZRAQ -,204,60
orl-rat LD50:2460 mg/kg 28ZPAK -,271,72
orl-mus LD50:3210 mg/kg GTPZAB 1342,69
orl-rbt LD50:4200 mg/kg 85GMAT -,61,82
orl-gpg LD50:3 g/kg 85GMAT -,61,82

OSHA PEL: Vapor: TWA 1 ppm

SAFETY PROFILE: Poison by inhalation. Moderately toxic by ingestion. Human systemic effects by inhalation: unspecified effects on the sense of smell, conjunctiva irritation, and unspecified respiratory effects. A mild skin and eye irritant. When heated to decomposition it emits acrid smoke and irritating fumes.

PFB000 CAS:10402-90-1 ***HR: 3***
1-(2-PHENYL-2-ETHOXYETHYL)-4-(2-BENZYLOXYPROPYL)PIPERAZINE
mf: $C_{24}H_{32}N_2O_2$ mw: 380.58

SYNS: 3-(4-(β-ETHOXYPHENETHYL)-1-PIPERAZINYL)-2-METHYLPROPIOPHENONE ◇ 3-(4-(2-ETHOXY-2-PHENYLETHYL)-1-PIPERAZINYL)-2-METHYL-1-PHENYL-1-PROPANONE

TOXICITY DATA with REFERENCE
ipr-mus LD50:111 mg/kg OYYAA2 2,314,68
scu-mus LD50:246 mg/kg OYYAA2 2,314,68

SAFETY PROFILE: Poison by intraperitoneal and subcutaneous routes. When heated to decomposition it emits toxic fumes of NO_x.

PFB250 CAS:103-45-7 ***HR: 2***
2-PHENYLETHYL ACETATE
mf: $C_{10}H_{12}O_2$ mw: 164.22

PROP: Colorless liquid; sweet, rosy, honey odor. Mp: 164.2°, bp: 223.6°, fp: < −20°, flash p: 230°F, d: 1.032 @ 25°/25°, refr index: 1.497-1.501. Sol in alc, fixed oils, propylene glycol; insol in glycerin, water @ 232°.

SYNS: ACETIC ACID-2-PHENYLETHYL ESTER ◇ BENZYLCARBINYL ACETATE ◇ FEMA No. 2857 ◇ β-PHENETHYL ACETATE ◇ 2-PHENETHYL ACETATE ◇ β-PHENYLETHYL ACETATE

TOXICITY DATA with REFERENCE
orl-rat LD50:3670 mg/kg VPITAR 33(5),48,74
orl-mus LD50:3670 mg/kg VPITAR 33(5),48,74
orl-gpg LD50:3670 mg/kg VPITAR 33(5),48,74
skn-rbt LD50:6210 mg/kg FCTXAV 12,807,74

CONSENSUS REPORTS: Reported in EPA TSCA Inventory.

SAFETY PROFILE: Moderately toxic by ingestion. Mildly toxic by skin contact. Combustible when exposed to heat or flame; can react vigorously with oxidizing materials. To fight fire, use alcohol foam, CO_2 and dry chemical. When heated to decomposition it emits acrid smoke and irritating fumes. See also ESTERS.

PFB350 CAS:90-49-3 ***HR: 2***
PHENYLETHYLACETYLUREA
mf: $C_{11}H_{14}N_2O_2$ mw: 206.27

PROP: dl-Form: Needles from ethanol. Mp: 149-150°. d-Form: Needles from ethanol. Mp: 168-169°. l-Form: Crystals from 50% ethanol. Mp: 162-163°.

SYNS: BENURIDE ◇ ETHYLPHENACEMIDE ◇ 1-((ETHYL)PHENYLACETYL)UREA ◇ LIRCAPYL ◇ M 551 ◇ PHENETURIDE ◇ PHENURIDE ◇ N-(α-PHENYLBUTYRYL)UREA ◇ 2-PHENYLBUTYRYLUREA ◇ PHENYLETHYLACETYLUREE (FRENCH) ◇ S 46

TOXICITY DATA with REFERENCE
orl-rat LD50:1143 mg/kg ARZNAD 18,524,68
orl-mus LD50:910 mg/kg NIIRDN 6,114,82
orl-rbt LDLo:1000 mg/kg AIPTAK 91,437,52
orl-gpg LD50:641 mg/kg AIPTAK 91,437,52

SAFETY PROFILE: Moderately toxic by ingestion. When heated to decomposition it emits toxic fumes of NO_x.

PFB500 CAS:2842-37-7 ***HR: 2***
2-PHENYLETHYLAMINOETHANOL
mf: $C_{10}H_{15}NO$ mw: 165.26

TOXICITY DATA with REFERENCE
skn-rbt 10 mg/24H open MLD AMIHBC 10,61,54
eye-rbt 750 μg open SEV AMIHBC 10,61,54
orl-rat LD50:1870 mg/kg AMIHBC 10,61,54
skn-rbt LD50:3 g/kg AMIHBC 10,61,54

SAFETY PROFILE: Moderately toxic by ingestion and skin contact. A skin and severe eye irritant. When heated to decomposition it emits toxic fumes of NO_x.

PFB750 CAS:94-47-3 ***HR: 1***
2-PHENYLETHYL BENZOATE
mf: $C_{15}H_{14}O_2$ mw: 226.29

PROP: Yellow liquid. Bp: 204-206° @ 25 mm. Insol in water; sol in alc, ether.

SYNS: BENZYL CARBINYL BENZOATE ◇ PHENETHYL ALCOHOL, BENZOATE ◇ PHENETHYL BENZOATE ◇ PHENYLETHYL BENZOATE ◇ β-PHENYLETHYL BENZOATE

TOXICITY DATA with REFERENCE
orl-rat LD50:5000 mg/kg FCTXAV 13,681,75

CONSENSUS REPORTS: Reported in EPA TSCA Inventory.

SAFETY PROFILE: Mildly toxic by ingestion route. Combustible when exposed to flame. When heated to decomposition it emits acrid smoke and irritating fumes. An insecticide. See also BENZOIC ACID.

PFB800 CAS:103-52-6 ***HR: 2***
PHENYLETHYL BUTYRATE
mf: $C_{12}H_{16}O_2$ mw: 192.28

SYNS: BENZYLCARBINYL BUTYRATE ◇ 2-PHENETHYL BUTANOATE ◇ β-PHENETHYL N-BUTANOATE ◇ PHENETHYL BUTYRATE ◇ 2-PHENYLETHYL BUTYRATE

TOXICITY DATA with REFERENCE
skn-rbt 500 mg/24H MLD FCTXAV 17,859,79
orl-rat LD50:4600 mg/kg FCTXAV 17,889,79
orl-rbt LDLo:2 g/kg JEENAI 48,139,55

CONSENSUS REPORTS: Reported in EPA TSCA Inventory.

SAFETY PROFILE: Moderately toxic by ingestion. A skin irritant. When heated to decomposition it emits acrid smoke and irritating fumes. See also ESTERS.

PFC100 CAS:10138-63-3 ***HR: 1***
β-PHENYLETHYL ESTER HYDRACRYLIC ACID
mf: $C_{11}H_{14}O_3$ mw: 194.25

SYN: PHENETHYL ESTER HYDRACRYLIC ACID

TOXICITY DATA with REFERENCE
orl-rat LD50:7800 mg/kg JPETAB 93,26,48
skn-rat LD50:10 g/kg JPETAB 93,26,48
orl-mus LD50:4600 mg/kg JPETAB 93,26,48

SAFETY PROFILE: Mildly toxic by skin contact and ingestion. When heated to decomposition it emits acrid smoke and irritating fumes. See also ESTERS.

PFC250 CAS:104-62-1 ***HR: 2***
2-PHENYLETHYL FORMATE
mf: $C_9H_{10}O_2$ mw: 150.19

SYNS: BENZYLCARBINYL FORMATE ◇ PHENETHYL ALCOHOL, FORMATE ◇ PHENETHYL FORMATE ◇ β-PHENYLETHYL FORMATE

TOXICITY DATA with REFERENCE
orl-rat LD50:3220 mg/kg FCTXAV 12,807,74

CONSENSUS REPORTS: Reported in EPA TSCA Inventory.

SAFETY PROFILE: Moderately toxic by ingestion. When heated to decomposition it emits acrid smoke and irritating fumes.

PFC500 CAS:51-71-8 ***HR: 3***
2-PHENYLETHYLHYDRAZINE
mf: $C_8H_{12}N_2$ mw: 136.22

SYNS: 1-HYDRAZINO-2-PHENYLETHANE ◇ NARDIL ◇ PHENELZINE ◇ PHENETHYLHYDRAZINE ◇ β-PHENYLETHYLHYDRAZINE ◇ STINERVAL ◇ W 1544

TOXICITY DATA with REFERENCE
oms-bcs 10 mmol/L MUREAV 5,343,68
orl-rat TDLo:175 mg/kg (female 16-22D post):REP PENDAV 16,21,65
orl-rat TDLo:200 mg/kg (female 10-17D post):TER PENDAV 16,21,65
orl-chd TDLo:7500 μg/kg:CNS AJDCAI 130,507,76
orl-mus LD50:130 mg/kg BCPCA6 17,369,68
ipr-mus LD50:135 mg/kg FATOAO 32,526,69
scu-mus LD50:150 mg/kg BCPCA6 17,369,68

SAFETY PROFILE: Poison by ingestion, intraperitoneal, and subcutaneous routes. Human systemic effects by ingestion: ataxia, somnolence. An experimental teratogen. Experimental reproductive effects. Mutation data reported. Used as an antidepressant. When heated to decomposition it emits toxic fumes of NO_x.

PFC750 CAS:156-51-4 ***HR: 3***
β-PHENYLETHYLHYDRAZINE SULFATE
mf: $C_8H_{12}N_2{\cdot}H_2O_4S$ mw: 234.30

SYNS: ALACINE ◇ ALAZIN ◇ ALAZINE ◇ EP-411 ◇ ESTINERVAL ◇ FELAZINE ◇ FENELZIN ◇ 1-HYDRAZINO-2-PHENYLETHANE HYDROGEN SULPHATE ◇ KALGAN ◇ MAO-REM ◇ MONOPHEN ◇ MONOTEN ◇ N-1544A ◇ NARDELZINE ◇ NARDIL ◇ P 1531 ◇ PHENALZINE ◇ PHENALZINE DIHYDROGEN SULFATE ◇ PHENALZINE HYDROGEN SULPHATE ◇ PHENELZIN ◇ PHENELZINE ACID SULFATE ◇ PHENELZINE BISULPHATE ◇ PHENELZINE SULFATE ◇ PHENETHYLHYDRAZINE SULFATE (1:1) ◇ PHENLINE ◇ PHENODYNE ◇ PHENYLAETHYL-HYDRAZIN ◇ β-PHENYLETHYLHYDRAZINE DIHYDROGEN SULFATE ◇ 2-PHENYLETHYLHYDRAZINE DIHYDROGEN SULPHATE ◇ β-PHENYLETHYLHYDRAZINE HYDROGEN SULPHATE ◇ PHENYLETHYLHYDRAZINE SULPHATE ◇ S 1544 ◇ STINERVAL

TOXICITY DATA with REFERENCE
orl-man TDLo:2143 μg/kg (male 1D pre):REP NEJMAG 317,117,87
orl-mus TDLo:28 g/kg/77W-C:NEO CNREA8 36,917,76
orl-hmn TDLo:1070 μg/kg/D:CNS,GIT,CVS 34ZIAG -,461,69
orl-man TDLo:46 mg/kg/10W-I:LIV, AJMEAZ 80,689,86
orl-wmn TDLo:18 mg/kg ANASAB 41,53,86
orl-rat LD50:210 mg/kg ANYAA9 107,899,63
orl-mus LD50:156 mg/kg ANYAA9 107,899,63
ipr-mus LD50:162 mg/kg JMCMAR 18,20,75
scu-mus LD50:125 mg/kg JOENAK 30,205,64
ivn-mus LD50:157 mg/kg ANYAA9 107,899,63

CONSENSUS REPORTS: IARC Cancer Review: Group 3 IMEMDT 7,312,87; Human Inadequate Evidence IMEMDT 24,175,80; Animal Limited Evidence IMEMDT 24,175,80. EPA Genetic Toxicology Program.

SAFETY PROFILE: Poison by ingestion, intraperitoneal, intravenous, and subcutaneous routes. Human systemic effects by ingestion: wakefulness, blood pressure lowering, constipation, hepatitis, fibrous hepatitis. Experimental reproductive effects. Questionable carcinogen with experimental neoplastigenic data. Used as a drug for the treatment of depression. When heated to decomposition it emits very toxic fumes of SO_x and NO_x.

PFC775 CAS:27254-37-1 ***HR: 3***
2-PHENYLETHYL HYDROPEROXIDE
mf: $C_8H_{10}O_2$ mw: 138.17

SAFETY PROFILE: Decomposes violently at room temperature. When heated to decomposition it emits acrid smoke and irritating fumes. See also PEROXIDES.

PFD250 CAS:55719-85-2 ***HR: 1***
PHENYLETHYL-α-METHYLBUTENOATE
mf: $C_{13}H_{16}O_2$ mw: 204.29

SYNS: PHENETHYL TIGLATE ◇ PHENYLETHYL TIGLATE

TOXICITY DATA with REFERENCE
skn-rbt 500 mg/24H MLD FCTXAV 13,681,75

CONSENSUS REPORTS: Reported in EPA TSCA Inventory.

SAFETY PROFILE: A skin irritant. When heated to decomposition it emits acrid smoke and irritating fumes.

PFD325 CAS:3558-60-9 ***HR: 2***
PHENYLETHYL METHYL ETHER
mf: $C_9H_{12}O$ mw: 136.21

SYNS: METHYL PHENETHYL ETHER ◇ METHYL 2-PHENETHYL ETHER ◇ METHYL PHENYLETHYL ETHER ◇ β-PHENYLETHYL METHYL ETHER

TOXICITY DATA with REFERENCE
skn-rbt 500 mg/24H MLD FCTOD7 20(Suppl),807,82
orl-rat LD50:4100 mg/kg FCTOD7 20(Suppl),807,82
skn-rbt LD50:3970 mg/kg FCTOD7 20(Suppl),807,82

CONSENSUS REPORTS: Reported in EPA TSCA Inventory.

SAFETY PROFILE: Moderately toxic by skin contact. A skin irritant. When heated to decomposition it emits acrid smoke and irritating fumes. See also ETHERS.

PFD500 CAS:78219-38-2 ***HR: 3***
β-2-PHENYLETHYLPIPERIDINOETHYL BENZOATE HYDROCHLORIDE
mf: $C_{22}H_{27}NO_2{\cdot}ClH$ mw: 373.96

SYN: BENZOIC ACID, 2-(2-PHENYLETHYLPIPERIDINO) ETHYL ESTER, HYDROCHLORIDE

TOXICITY DATA with REFERENCE
scu-mus LD50:1250 mg/kg JACSAT 52,1633,30
ivn-mus LD50:28 mg/kg JACSAT 52,1633,30

SAFETY PROFILE: Poison by intravenous route. Moderately toxic by subcutaneous route. When heated to decomposition it emits very toxic fumes of HCl and NO_x.

PFD750 CAS:78219-45-1 ***HR: 3***
Γ-2-PHENYLETHYLPIPERIDINOPROPYL BENZOATE HYDROCHLORIDE
mf: $C_{23}H_{29}NO_2{\cdot}ClH$ mw: 387.99

SYN: BENZOIC ACID-3-(2-PHENYLETHYLPIPERIDINO) PROPYL ESTER, HYDROCHLORIDE

TOXICITY DATA with REFERENCE
ivn-rat LDLo:22 mg/kg JACSAT 55,4625,33
scu-mus LD50:550 mg/kg JACSAT 52,1633,30
ivn-mus LD50:15 mg/kg JACSAT 52,1633,30

SAFETY PROFILE: Poison by intravenous route. Moderately toxic by subcutaneous route. When heated to decomposition it emits very toxic fumes of HCl and NO_x.

PFE500 CAS:940-41-0 ***HR: 2***
(2-PHENYLETHYL)TRICHLOROSILANE
mf: $C_8H_9CL_3Si$ mw: 239.61

TOXICITY DATA with REFERENCE
skn-rbt 100 μg/24H open AIHAAP 23,95,62
orl-rat LDLo:2830 mg/kg AIHAAP 23,95,62
skn-rbt LD50:740 mg/kg AIHAAP 23,95,62

CONSENSUS REPORTS: Reported in EPA TSCA Inventory.

SAFETY PROFILE: Moderately toxic by ingestion and skin contact. A skin irritant. When heated to decomposition it emits toxic fumes of Cl^-. See also CHLOROSILANES.

PFE900 CAS:32228-97-0 ***HR: 2***
N-PHENYL-2-FLUORENAMINE
mf: $C_{19}H_{15}N$ mw: 257.35

SYN: N-PHENYL-9H-FLUORENAMINE

TOXICITY DATA with REFERENCE
ipr-rat TDLo:310 mg/kg/4W-I:ETA CNREA8 31,778,71

SAFETY PROFILE: Questionable carcinogen with experimental tumorigenic data. When heated to decomposition it emits toxic fumes of NO_x.

PFF000 CAS:31874-15-4 ***HR: 3***
N-PHENYL-2-FLUORENYLHYDROXYLAMINE
mf: $C_{19}H_{15}NO$ mw: 273.35

SYN: N-PHENYL-N-9H-FLUOREN-2-YLHYDROXYLAMINE

TOXICITY DATA with REFERENCE
ipr-rat TDLo:450 mg/kg/4W-I:ETA CNREA8 31,778,71

SAFETY PROFILE: Questionable carcinogen with experimental tumorigenic data. When heated to decomposition it emits toxic fumes of NO_x.

PFF300 CAS:7250-71-7 ***HR: 2***
PHENYLGLYCERYL ETHER DIACETATE
mf: $C_{13}H_{16}O_5$ mw: 252.29

SYNS: GLYCEROL PHENYL ETHER DIACETATE ◇ 3-PHENOXY-1,2-PROPANEDIOL DIACETATE ◇ 1-PHENYL ETHER-2,3-DIACETATE GLYCEROL

TOXICITY DATA with REFERENCE
orl-rat LD50:5250 mg/kg SCCUR* -,5,61
orl-mus LD50:9200 mg/kg SCCUR* -,5,61
ipr-mus LD50:1850 mg/kg JPETAB 97,414,49

SAFETY PROFILE: Moderately toxic by intraperitoneal route. Mildly toxic by ingestion. When heated to decomposition it emits acrid smoke and irritating fumes.

PFF500 CAS:78265-97-1 ***HR: 3***
dl-2-PHENYLGLYCINEDECYLESTERHYDROCHLORIDE
mf: $C_{18}H_{29}NO_2$•ClH mw: 327.94

SYN: (±)-α-AMINO-BENZENEACETIC ACID, DECYL ESTER, HYDROCHLORIDE

TOXICITY DATA with REFERENCE
ipr-mus LD50:350 mg/kg PHARAT 33,749,78
ivn-mus LD50:62 mg/kg PHARAT 33,749,78

SAFETY PROFILE: Poison by intraperitoneal and intravenous routes. When heated to decomposition it emits very toxic fumes of NO_x and Cl^-. See also ESTERS.

PFF750 CAS:25287-52-9 ***HR: 3***
dl-2-PHENYLGLYCINEHEPTYLESTERHYDROCHLORIDE
mf: $C_{15}H_{23}NO_2$•ClH mw: 285.85

SYN: (±)-α-AMINO-BENZENEACETIC ACID HEPTYL ESTER HYDROCHLORIDE

TOXICITY DATA with REFERENCE
orl-mus LD50:2500 mg/kg PHARAT 33,749,78
ipr-mus LD50:145 mg/kg PHARAT 33,749,78
ivn-mus LD50:39 mg/kg PHARAT 33,749,78

SAFETY PROFILE: Poison by intraperitoneal and intravenous routes. Moderately toxic by ingestion. When heated to decomposition it emits very toxic fumes of Cl^- and NO_x.

PFG000 CAS:69357-11-5 ***HR: 3***
dl-2-PHENYLGLYCINEHEXYLESTERHYDROCHLORIDE
mf: $C_{14}H_{21}NO_2$•ClH mw: 271.82

TOXICITY DATA with REFERENCE
ipr-mus LD50:182 mg/kg PHARAT 33,749,78
ivn-mus LD50:50 mg/kg PHARAT 33,749,78

SAFETY PROFILE: Poison by intraperitoneal and intravenous routes. When heated to decomposition it emits very toxic fumes of HCl and NO_x.

PFG250 CAS:69357-14-8 ***HR: 3***
dl-2-PHENYLGLYCINENONYLESTERHYDROCHLORIDE
mf: $C_{17}H_{27}NO_2$•ClH mw: 313.91

SYN: (±)-α-AMINO-BENZENEACETIC ACID NONYL ESTER HYDROCHLORIDE

TOXICITY DATA with REFERENCE
ipr-mus LD50:249 mg/kg PHARAT 33,749,78
ivn-mus LD50:44 mg/kg PHARAT 33,749,78

SAFETY PROFILE: Poison by intraperitoneal and intravenous routes. When heated to decomposition it emits very toxic fumes of Cl^-, HCl and NO_x.

PFG500 CAS:69357-13-7 ***HR: 3***
dl-2-PHENYLGLYCINEOCTYLESTERHYDROCHLORIDE
mf: $C_{16}H_{25}NO_2$•ClH mw: 299.88

SYN: (±)-α-AMINO-BENZENEACETIC ACID OCTYL ESTER HYDROCHLORIDE

TOXICITY DATA with REFERENCE
ipr-mus LD50:229 mg/kg PHARAT 33,749,78
ivn-mus LD50:42 mg/kg PHARAT 33,749,78

SAFETY PROFILE: Poison by intraperitoneal and intravenous routes. When heated to decomposition it emits very toxic fumes of HCl and NO_x.

PFG750 CAS:69357-10-4 ***HR: 3***
dl-2-PHENYLGLYCINEPENTYLESTERHYDROCHLORIDE
mf: $C_{13}H_{19}NO_2$•ClH mw: 257.79

SYN: (±)-α-AMINO-BENZENEACETIC ACID PENTYL ESTER HYDROCHLORIDE

TOXICITY DATA with REFERENCE
ipr-mus LD50:418 mg/kg PHARAT 33,749,78
ivn-mus LD50:69 mg/kg PHARAT 33,749,78

SAFETY PROFILE: Poison by intraperitoneal and intravenous routes. When heated to decomposition it emits very toxic fumes of NO_x and HCl.

PFH000 CAS:122-60-1 ***HR: 3***
PHENYL GLYCYDYL ETHER
mf: $C_9H_{10}O_2$ mw: 150.19

SYNS: 1,2-EPOXY-3-PHENOXYPROPANE ◇ 2,3-EPOXYPROPYL-PHENYL ETHER ◇ FENYL-GLYCIDYLETHER (CZECH) ◇ GLYCIDYL PHENYL ETHER ◇ PGE ◇ PHENOL-GLYCIDAETHER (GERMAN) ◇ PHENOL GLYCIDYL ETHER (MAK) ◇ 3-PHENOXY-1,2-EPOXY- PROPANE ◇ PHENOXYPROPENE OXIDE ◇ PHENOXYPROPYLENE OXIDE ◇ PHENYL-2,3-EPOXYPROPYL ETHER

TOXICITY DATA with REFERENCE
skn-rbt 10 mg/24H SEV AMIHBC 10,61,54
skn-rbt 500 mg/24H MOD 28ZPAK -,136,72
eye-rbt 250 μg/24H SEV 28ZPAK -,136,72
mmo-esc 20 μmol/L ARTODN 46,277,80
mmo-klp 100 μmol/L MUREAV 89,269,81
mmo-sat 50 μg/plate MUREAV 67,9,79
hma-mus/sat 2500 mg/kg MUREAV 67,9,79
ihl-rat TCLo:11 ppm/6H (19D male):REP TXAPA9 64,204,82
ihl-rat TCLo:12 ppm/6H/2Y-I:CAR AJPAA4 111,140,83
orl-mus LD50:1400 mg/kg AMIHAB 14,250,56
scu-mus LD50:760 mg/kg ARZNAD 15,1355,65
skn-rbt LD50:1500 mg/kg AMIHBC 10,61,54

CONSENSUS REPORTS: Reported in EPA TSCA Inventory. EPA Genetic Toxicology Program.

OSHA PEL: (Transitional: TWA 10 ppm) TWA 1 ppm
ACGIH TLV: TWA 1 ppm
DFG MAK: 1 ppm (6 mg/m^3), Suspected Carcinogen.
NIOSH REL: (Glycidyl Ethers) CL 5 mg/m^3/15M

SAFETY PROFILE: Suspected carcinogen with experimental carcinogenic data. Moderately toxic by ingestion, skin contact, and subcutaneous routes. A severe eye and skin irritant. Experimental reproductive effects. Mutation data reported. When heated to decomposition it emits acrid smoke and irritating fumes. Used as a chemical intermediate. See also ETHERS.

PFH250 ***HR: 2***
PHENYLGLYOXAL
mf: $C_8H_6O_2$ mw: 134.14

PROP: Flash p: <73.4°F.

SAFETY PROFILE: Dangerous fire hazard when exposed to heat or flame; can react vigorously with oxidizing materials. When heated to decomposition it emits acrid smoke and irritating fumes.

PFH275 ***HR: 3***
PHENYLGOLD
mf: C_6H_5Au mw: 274.07

SAFETY PROFILE: A touch-sensitive explosive, even at −70°C. When heated to decomposition it emits acrid smoke and irritating fumes.

PFI000 CAS:100-63-0 ***HR: 3***
PHENYLHYDRAZINE
DOT: UN 2572
mf: $C_6H_8N_2$ mw: 108.16

PROP: Yellow, monoclinic crystals or oil. Mp: 19.6°, bp: 243.5° (decomp), flash p: 192°F (CC), d: 1.0978 @ 20°/4°, vap press: 1 mm @ 71.8°, vap d: 3.7. Sltly sol in hot water; misc in alc, chloroform, ether, benzene.

SYNS: FENILIDRAZINA (ITALIAN) ◇ FENYLHYDRAZINE (DUTCH) ◇ HYDRAZINE-BENZENE ◇ HYDRAZINOBENEZENE ◇ PHENYL-HYDRAZIN (GERMAN)

TOXICITY DATA with REFERENCE
mmo-omi 150 μg/L MUREAV 173,233,86
dnd-mus-ipr 350 μmol/kg CNREA8 41,1469,81
ipr-rat TDLo:30 mg/kg (17-19D preg):REP TJADAB 8,97,73
scu-rat TDLo:5200 mg/kg/52W-I:CAR GTPZAB 19(6),28,75
orl-rat LD50:188 mg/kg HYSAAV 30,191,65
scu-rat LDLo:40 mg/kg AEPPAE 182,118,36
orl-mus LDLo:175 mg/kg HYSAAV 30,191,65
scu-mus LDLo:170 mg/kg JIDHAN 18,1,36
orl-dog LDLo:200 mg/kg XPHBAO 271,159,41
ivn-dog LDLo:120 mg/kg XPHBAO 271,159,41
orl-rbt LD50:80 mg/kg HYSAAV 30,191,65
scu-rbt LDLo:80 mg/kg AEPPAE 182,118,36
orl-gpg LD50:80 mg/kg HYSAAV 30,191,65

CONSENSUS REPORTS: Reported in EPA TSCA Inventory.

OSHA PEL: (Transitional: TWA 5 ppm (skin)) TWA 5 ppm (skin); STEL 10 ppm
ACGIH TLV: TWA 0.1 ppm (skin); Suspected Human Carcinogen
DFG MAK: 5 ppm (22 mg/m^3), Suspected Carcinogen.
NIOSH REL: CL 0.6 mg/m^3/2H
DOT Classification: Poison B; Label: Poison.

SAFETY PROFILE: Suspected carcinogen with experimental carcinogenic data. Poison by ingestion, subcutaneous, and intravenous routes. Experimental reproductive effects. Mutation data reported. Ingestion or subcutaneous injection can cause hemolysis of red blood cells. Other effects are damage to the spleen, liver, kidneys, and bone marrow. The most common effect of occupational exposure is the development of dermatitis which, in sensitized persons, may be quite severe. Systemic effects include anemia and general weakness, gastrointestinal disturbances and injury to the kidneys. Flammable when exposed to heat, flame, or oxidizers. To fight fire, use alcohol foam. Violent reaction with 2-phenylamino-3-phenyloxazirane. Reacts with perchloryl fluoride to form an explosive product. Vigorous reaction with lead(IV) oxide. Used as a chemical reagent, in organic synthesis, and in the manufacture of dyes and drugs. Dangerous; when heated to decomposition it emits highly toxic fumes of NO_x; can react with oxidizing materials.

PFI250 CAS:59-88-1 ***HR: 3***
PHENYLHYDRAZINE HYDROCHLORIDE
mf: $C_6H_8N_2•ClH$ mw: 144.62

PROP: Leaflet crystals from alc. Mp: 245°. Very sol in water; sol in alc; insol in ether.

SYNS: PHENYLHYDRAZINE MONOHYDROCHLORIDE ◇ PHENYLHYDRAZIN HYDROCHLORID (GERMAN) ◇ PHENYLHYDRAZINIUM CHLORIDE

TOXICITY DATA with REFERENCE
mmo-sat 800 μg/plate NEZAAQ 33,474,78
mma-sat 800 μg/plate NEZAAQ 33,474,78
ipr-rat TDLo:30 mg/kg (17-19D preg):REP SEIJBO 14,95,74
orl-mus TDLo:8000 mg/kg/42W-I:NEO 34ZRA9 -,869,65
ipr-mus TDLo:464 mg/kg/8W-I:ETA JNCIAM 42,337,69
ipr-rat LD50:161 mg/kg ABMGAJ 18,617,67
orl-mus LD50:2100 mg/kg QJDRAZ 9,1455,69
scu-mus LD50:89 mg/kg FRPSAX 9,274,54
orl-rbt LDLo:25 mg/kg HBAMAK 4,1289,35
scu-rbt LDLo:25 mg/kg HBAMAK 4,1289,35

CONSENSUS REPORTS: Reported in EPA TSCA Inventory. EPA Extremely Hazardous Substances List.

NIOSH REL: (Hydrazines) CL 0.6 mg/m³/2H

SAFETY PROFILE: Poison by ingestion, intraperitoneal, and subcutaneous routes. Experimental reproductive effects. Questionable carcinogen with experimental neoplastigenic and tumorigenic data. Mutation data reported. When heated to decomposition it emits very toxic fumes of NO_x and HCl.

PFI500 CAS:98-71-5 ***HR: 2***
PHENYLHYDRAZINE-p-SULFONIC ACID
mf: $C_6H_8N_2O_3S$ mw: 188.22

PROP: Crystals from alc. Mp: 286°. Sol in hot and cold water; sltly sol in alc.

TOXICITY DATA with REFERENCE
orl-rat LDLo:500 mg/kg JPETAB 90,260,47

CONSENSUS REPORTS: Reported in EPA TSCA Inventory.

SAFETY PROFILE: Moderately toxic by ingestion. When heated to decomposition it emits very toxic fumes of SO_x and NO_x.

PFI600 CAS:100482-34-6 ***HR: 3***
o-(PHENYLHYDROXYARSINO)BENZOIC ACID
mf: $C_{13}H_{11}AsO_3$ mw: 290.16

SYNS: BENZOIC ACID, o-(PHENYLHYDROXYARSINO)- ◇ 2-CARBOXYDIPHENYLARSINOUS ACID

TOXICITY DATA with REFERENCE
ivn-mus LDLo:10 mg/kg PHBUA9 2,19,54

OSHA PEL: TWA 0.5 mg(As)/m³

SAFETY PROFILE: Poison by intravenous route. When heated to decomposition it emits toxic fumes of As.

PFI750 CAS:63918-85-4 ***HR: 3***
1-PHENYL-2-(β-HYDROXYETHYL)AMINOPROPANE
mf: $C_{11}H_{17}NO$ mw: 179.29

SYNS: 1-PHENYL-2-β-OXY-AETHYL-AMINO-PROPAN(GERMAN) ◇ α-PHENYL-β-OXYETHYLAMINOPROPANE ◇ 2-(α-METHYLPHENETHYL)AMINOETHANOL

TOXICITY DATA with REFERENCE
orl-rat LDLo:250 mg/kg AEPPAE 195,647,40
ipr-mus LDLo:250 mg/kg AEPPAE 195,647,40
ipr-mus LDLo:750 mg/kg AEPPAE 195,647,40

SAFETY PROFILE: Poison by ingestion and intraperitoneal routes. When heated to decomposition it emits toxic fumes of NO_x.

PFJ000 CAS:94-35-9 ***HR: 2***
2-PHENYL-2-HYDROXYETHYL CARBAMATE
mf: $C_9H_{11}NO_3$ mw: 181.21

SYNS: AL 1076 ◇ CARBAMIC ACID-β-HYDROXYPHENETHYL ESTER ◇ β-HYDROXYPHENETHYL ALCOHOL-α-CARBAMATE ◇ β-HYDROXYPHENETHYL CARBAMATE ◇ 2-HYDROXY-2-PHENYLETHYL CARBAMATE ◇ LINAXAR ◇ MYOSPAZ ◇ SINAXAR ◇ STIRAMATO ◇ STYRAMATE

TOXICITY DATA with REFERENCE
orl-rat LDLo:1300 mg/kg JPETAB 126,318,59
orl-mus LD50:1240 mg/kg JPETAB 126,318,59
ipr-mus LD50:750 mg/kg 27ZQAG -,407,72
scu-cat LD50:1200 mg/kg JPETAB 126,318,59

SAFETY PROFILE: Moderately toxic by ingestion, intraperitoneal and subcutaneous routes. When heated to decomposition it emits toxic fumes of NO_x. Used as a skeletal muscle relaxant. See also CARBAMATES.

PFJ250 CAS:100-65-2 ***HR: 3***
β-PHENYLHYDROXYLAMINE
mf: C_6H_7NO mw: 109.14

PROP: Colorless needles. Mp: 81-82°. Sol in hot and cold water; very sol in alc and ether; very sltly sol in ligroin.

SYNS: NCI-C60093 ◇ N-PHENYLHYDROXYLAMINE

TOXICITY DATA with REFERENCE
orl-rat LD50:100 mg/kg GISAAA 44(3),68,79
orl-mus LD50:247 mg/kg GISAAA 44(3),68,79
orl-rbt LD50:125 mg/kg GISAAA 44(3),68,79
skn-hmn TDLo:500 μg/kg:SKN AEXPBL 35,401,1895
scu-rat LDLo:40 mg/kg AEPPAE 182,118,36
orl-dog LDLo:30 mg/kg HBAMAK 4,1289,35
scu-rbt LDLo:50 mg/kg AEXPBL 35,401,1895
scu-pgn LDLo:74 mg/kg AEXPBL 35,401,1895

CONSENSUS REPORTS: Reported in EPA TSCA Inventory.

SAFETY PROFILE: Poison by ingestion and subcutaneous routes. Human systemic effects by skin contact: primary irritation. Preparative hazard. When heated to decomposition it emits toxic fumes of NO_x.

PFJ275 ***HR: 3***
PHENYLHYDROXYLAMINIUM CHLORIDE
mf: C_6H_8ClNO mw: 145.59

SAFETY PROFILE: May explode in storage. When heated to decomposition it emits toxic fumes of Cl^- and NO_x.

PFJ750 CAS:83-12-5 ***HR: 3***
PHENYLINDIONE
mf: $C_{15}H_{10}O_2$ mw: 222.25

SYNS: ATHROMBON ◇ BINDAN ◇ DANILONE ◇ DIADILAN ◇ DINDEVAN ◇ DINEVAL ◇ EMANDIONE ◇ FENHYDREN ◇ FENILIN ◇ HEDULIN ◇ INDEMA ◇ INDION ◇ INDON ◇ PHENHYDREN ◇ PHENINDIONE ◇ 2-PHENYL-1,3-DIKETOHYDRINDENE ◇ PHENYLEN ◇ 2-PHENYLINDAN-1,3-DIONE ◇ 2-PHENYL-1,3-INDANDIONE ◇ PHENYLLIN ◇ PID ◇ PINDIONE

TOXICITY DATA with REFERENCE
orl-man TDLo:42500 μg/kg/17D-I:KID BMJOAE 1,1655,63
unr-man TDLo:22 mg/kg/17D-I:LVR,KID,SKN AIMEAS 52,706,60
unr-man TDLo:90 mg/kg/27D-I:LVR,SKN CMAJAX 77,1028,57
orl-man LDLo:48 mg/kg/27D-I:BLD,MET NZMJAX 57,283,58
unr-man LDLo:1683 mg/kg/31D-I:GIT,KID,SKN LANCAO 1,920,63
orl-rat LD50:163 mg/kg EJMCA5 9,519,74
ipr-rat LD50:530 mg/kg PHARAT 27,520,72
orl-mus LD50:175 mg/kg SMWOAS 90,213,60
ipr-mus LD50:150 mg/kg PHARAT 27,520,72

SAFETY PROFILE: A human poison by ingestion. Poison experimentally by ingestion and intraperitoneal routes. Moderately toxic to humans by an unspecified route. Human systemic effects by: blood effects; fever; hepatitis, hepatocellular necrosis; dermatitis, allergic effects, other skin effects; nausea or vomiting; tubule changes, urine volume decrease, interstitial nephritis. When heated to decomposition it emits acrid smoke and irritating fumes. Used to treat angina pectoris.

PFJ775 ***HR: 3***
PHENYLIODINE(III) CHROMATE
mf: $C_6H_5CrIO_4$ mw: 320.00

SYN: [(CHROMYLDIOXY)IODO]BENZENE

CONSENSUS REPORTS: Chromium and its compounds are on the Community Right-To-Know List.

SAFETY PROFILE: Explodes at 66°C. When heated to decomposition it emits toxic fumes of I^-. See also CHROMIUM COMPOUNDS and IODINE.

PFJ780 CAS:58776-08-2 ***HR: 3***
PHENYLIODINE(III) NITRATE
mf: $C_6H_5IN_2O_6$ mw: 328.02

SAFETY PROFILE: Explodes when heated above 100°C. When heated to decomposition it emits toxic fumes of I^- and NO_x. See also NITRATES and IODIDES.

PFK000 ***HR: 3***
9-PHENYL-9-IODOFLUORENE
mf: $C_{18}H_{13}I$ mw: 356.20

SAFETY PROFILE: Explodes. When heated to decomposition it emits toxic fumes of I^-. See also IODIDES.

PFK250 CAS:103-71-9 ***HR: 3***
PHENYL ISOCYANATE
DOT: UN 2487
mf: C_7H_5NO mw: 119.13

PROP: Liquid, acrid odor. Mp: −30° approx, bp: 158-168°, d: 1.1 @ 20°, vap press: 1 mm @ 10.6°, flash p: 132°. Decomp in water, alc; very sol in ether.

SYNS: CARBANIL ◇ ISOCYANIC ACID, PHENYL ESTER ◇ MONDUR P ◇ PHENYLCARBIMIDE ◇ PHENYL CARBONIMIDE

TOXICITY DATA with REFERENCE
mmo-sat 100 μg/plate ABCHA6 44,3017,80
orl-rat LD50:940 mg/kg MONS**
skn-rbt LD50:7130 mg/kg TXAPA9 42,417,77

CONSENSUS REPORTS: Reported in EPA TSCA Inventory.

DOT Classification: Poison B; Label: Flammable Liquid and Poison.

SAFETY PROFILE: A poison. Mutation data reported. An irritant. Flammable when exposed to heat or flame; can react vigorously with oxidizing materials. Has exploded when stirred with (cobalt pentammine triazoperchlorate + nitrosyl perchlorate). When heated to decomposition it emits toxic fumes of CN^- and NO_x. See also CYANATES.

PFL000 CAS:101-72-4 ***HR: 2***
N-PHENYL-N'-ISOPROPYL-p-PHENYLENEDIAMINE
mf: $C_{15}H_{18}N_2$ mw: 226.35

SYNS: CYZONE ◇ ELASTOZONE 34 ◇ FLEXZONE 3C ◇ 4-ISOPROPYLAMINODIPHENYLAMINE ◇ N-ISOPROPYL-N'-FENYL-p-FENYLENDIAMIN (CZECH) ◇ N-ISOPROPYL-N'-PHENYL-p-PHENYLENEDIAMINE ◇ NCI-C56304 ◇ NONOX ZA ◇ N-2-PROPYL-N'-PHENYL-p-PHENYLENEDIAMINE ◇ SANTOFLEX 36

TOXICITY DATA with REFERENCE
eye-rbt 100 mg/24H SEV 28ZPAK -,72,72
orl-rat LD50:555 mg/kg 28ZPAK -,72,72
orl-mus LD50:1122 mg/kg KOKABN 26,438,77

CONSENSUS REPORTS: Reported in EPA TSCA Inventory.

SAFETY PROFILE: Moderately toxic by ingestion. A severe eye irritant. When heated to decomposition it emits toxic fumes of NO_x. Used in the manufacture of plastics, resins, and rubber. See also AMINES.

PFL500 CAS:591-51-5 ***HR: 2***
PHENYLLITHIUM
mf: C_6H_5Li mw: 84.06

SAFETY PROFILE: Incompatible with air, water. Product of reaction with titanium tetraethoxide ignites spontaneously in air and reacts violently with water. When heated to decomposition it emits toxic fumes of Li_2O. See also LITHIUM COMPOUNDS and ORGANOMETALS.

PFL600 CAS:100-58-3 ***HR: 3***
PHENYLMAGNESIUM BROMIDE
mf: C_6H_5MgBr mw: 181.31

SAFETY PROFILE: Reacts with chlorine to form an explosive product. When heated to decomposition it emits toxic fumes of Br^-. See also MAGNESIUM COMPOUNDS and BROMIDES.

PFL750 CAS:941-69-5 ***HR: 3***
N-PHENYLMALEIMIDE
mf: $C_{10}H_7NO_2$ mw: 173.18

PROP: Yellow needles. Mp: 89-89.8°.

TOXICITY DATA with REFERENCE
orl-rat LD50:188 mg/kg SCCUR* -,8,61
ipr-rat LDLo:25 mg/kg NCNSA6 5,22,53
orl-mus LD50:78 mg/kg SCCUR* -,8,61
orl-rbt LDLo:100 mg/kg SCCUR* -,8,61

CONSENSUS REPORTS: Reported in EPA TSCA Inventory.

SAFETY PROFILE: Poison by ingestion and intraperitoneal routes. When heated to decomposition it emits toxic fumes of NO_x.

PFL850 CAS:108-98-5 ***HR: 3***
PHENYL MERCAPTAN
DOT: UN 2337
mf: C_6H_6S mw: 110.18

PROP: Liquid, repulsive odor. Bp: 168.3°, d: 1.0728 @ 25°/4°.

SYNS: BENZENETHIOL (DOT) ◇ RCRA WASTE NUMBER P014 ◇ THIOPHENOL (DOT) ◇ USAF XR-19

TOXICITY DATA with REFERENCE
eye-rbt 108 mg SEV AIHAAP 19,171,58
ipr-mus LD50:25 mg/kg NTIS** AD277-689
orl-rat LD50:46 mg/kg AIHAAP 19,171,58
ihl-rat LC50:33 ppm/4H AIHAAP 19,171,58
skn-rat LD50:300 mg/kg AIHAAP 19,171,58
ipr-rat LD50:10 mg/kg AIHAAP 19,171,58
ihl-mus LC50:28 ppm/4H AIHAAP 19,171,58
orl-bwd LD50:24 mg/kg TXAPA9 21,315,72

CONSENSUS REPORTS: Reported in EPA TSCA Inventory. EPA Extremely Hazardous Substances List.

OSHA PEL: TWA 0.5 ppm
ACGIH TLV: TWA 0.5 ppm
NIOSH REL: CL 0.5 mg/m^3/15M
DOT Classification: Poison B; Label: Flammable Liquid and Poison.

SAFETY PROFILE: Poison by ingestion, inhalation, skin contact, and intraperitoneal routes. A severe eye irritant. Can cause severe dermatitis. Exposure may cause headache and dizziness. When heated to decomposition or on contact with acids it emits toxic fumes of SO_x. See also MERCAPTANS.

PFM250 CAS:1192-89-8 ***HR: 3***
PHENYLMERCURIC BROMIDE
mf: C_6H_5BrHg mw: 357.61

SYNS: ARGONAL ◇ PHENYLMERCURY BROMIDE

TOXICITY DATA with REFERENCE
unr-rat LD50:55 mg/kg 30ZDA9 -,295,71

CONSENSUS REPORTS: Mercury and its compounds are on the Community Right-To-Know List.

OSHA PEL: (Transitional: CL 1 mg/10m³) CL 0.1 mg(Hg)/m³ (skin)
ACGIH TLV: TWA 0.1 mg(Hg)/m³ (skin)
NIOSH REL: TWA 0.05 mg(Hg)/m³

SAFETY PROFILE: Poison by an unspecified route. When heated to decomposition it emits very toxic fumes of Br^- and Hg. See also MERCURY COMPOUNDS and BROMIDES.

PFM500 CAS:100-56-1 ***HR: 3***
PHENYL MERCURIC CHLORIDE
mf: C_6H_5ClHg mw: 313.15

PROP: Colorless leaves from benzene. Mp: 251°, bp: sublimes. Insol in water; sltly sol in hot alc; sol in pyridine, ether, benzene.

SYNS: CHLORID FENYLRTUTNATY (CZECH) ◇ (CHLOROMERCURI)BENZENE ◇ FENYLMERCURICHLORID (CZECH) ◇ MERCURIPHENYL CHLORIDE ◇ MERFAZIN ◇ MERSOLITE 2 ◇ PHENYL CHLOROMERCURY ◇ PHENYLMERCURY CHLORIDE ◇ PHENYLQUECKSILBERCHLORID (GERMAN) ◇ PMC ◇ STOPSPOT

TOXICITY DATA with REFERENCE
cyt-hmn:hla 1 mg/L JJEMAG 39,47,69
orl-rat LD50:60 mg/kg PHJOAV 185,361,60
ipr-rat LDLo:50 mg/kg NCNSA6 5,30,53
scu-rat LD50:47 mg/kg JJEMAG 39,47,69

CONSENSUS REPORTS: Mercury and its compounds are on the Community Right-To-Know List.

OSHA PEL: (Transitional: CL 1 mg/10m³) CL 0.1 mg(Hg)/m³ (skin)
ACGIH TLV: TWA 0.1 mg(Hg)/m³ (skin)
NIOSH REL: TWA 0.05 mg(Hg)/m³

SAFETY PROFILE: Poison by ingestion, intraperitoneal, and subcutaneous routes. Human mutation data reported. When heated to decomposition it emits very toxic fumes of Cl^- and Hg. See also MERCURY COMPOUNDS and CHLORIDES.

PFN000 CAS:14235-86-0 ***HR: 3***
PHENYLMERCURIC DINAPHTHYLMETHANEDISULFONATE
mf: $C_{33}H_{24}Hg_2O_6S_2$ mw: 981.87

SYNS: BIS(PHENYLMERCURI)METHYLENEDINAPHTHALENESULFONATE ◇ CONOTRANE ◇ FIBROTAN ◇ HYDRAPHEN ◇ HYDRARGAPHEN ◇ METHYLENEDINAPHTHALENESULFONIC ACID BISPHENYLMERCURI SALT ◇ PENOTRANE ◇ PHENYL MERCURIC FIXTAN ◇ PHENYLMERCURIC 3,3'-METHYLENEBIS(2-NAPHTHALENESULFONATE) ◇ PHENYLMERCURY METHYLENEDINAPHTHALENESULFONATE ◇ P.M.F. ◇ SEPTOTAN ◇ VERSOTRANE

TOXICITY DATA with REFERENCE
eye-gpg 60 μg/2D SEV TOSUAH 89,863,69
orl-mus LD50:70 mg/kg JPPMAB 2,20,50
ipr-mus LD50:8 mg/kg JPPMAB 2,89,50

CONSENSUS REPORTS: Mercury and its compounds are on the Community Right-To-Know List.

OSHA PEL: (Transitional: CL 1 mg/10m³) CL 0.1 mg(Hg)/m³ (skin)
ACGIH TLV: TWA 0.1 mg(Hg)/m³ (skin)
NIOSH REL: (Mercury, Inorganic) TWA 0.05 mg (Hg)/m³

SAFETY PROFILE: Poison by intraperitoneal and ingestion routes. A severe eye irritant. When heated to decomposition it emits very toxic fumes of Hg and SO_x. See also MERCURY COMPOUNDS and SULFONATES.

PFN100 CAS:100-57-2 ***HR: 3***
PHENYLMERCURIC HYDROXIDE
DOT: UN 1894
mf: C_6H_6HgO mw: 294.71

PROP: Powder. Mp: 190°.

SYNS: HYDROXYPHENYLMERCURY ◇ MERCURY, HYDROXYPHENYL- ◇ MERSOLITE 1 ◇ PHENYL HYDROXYMERCURY ◇ PHENYLMERCURIC HYDROXIDE (DOT) ◇ PHENYLMERCURY HYDROXIDE

TOXICITY DATA with REFERENCE
sln-oin-dmg-orl 250 mg/L HEREAY 57,446,67
ivn-mus LD50:18 mg/kg CSLNX* NX#03648

CONSENSUS REPORTS: Reported in EPA TSCA Inventory.
ACGIH TLV: TWA 0.1 mg(Hg)/m³ (skin)
NIOSH REL: (Mercury, Inorganic): TWA 0.05 mg(Hg)/m³

DOT Classification: Poison B; Label: Poison

SAFETY PROFILE: Poison by intravenous route. Mutation data reported. When heated to decomposition it emits toxic fumes of Hg.

PFN250 CAS:14354-56-4 ***HR: 3***
PHENYL MERCURIC-8-HYDROXYQUINOLINATE
mf: $C_{15}H_{11}HgNO$ mw: 421.86

SYNS: 8-PHENYLMERCURIOXY-QUIN-OLINE ◇ PHENYL MERCURY OXYQUINOLATE ◇ QUINEX

TOXICITY DATA with REFERENCE
ipr-mus LDLo:4 mg/kg CBCCT* 3,308,51

CONSENSUS REPORTS: Mercury and its compounds are on the Community Right-To-Know List.

OSHA PEL: (Transitional: CL 1 mg/10m^3) CL 0.1 mg(Hg)/m^3 (skin)
ACGIH TLV: TWA 0.1 mg(Hg)/m^3 (skin)
NIOSH REL: TWA 0.05 mg(Hg)/m^3

SAFETY PROFILE: Poison by intraperitoneal route. When heated to decomposition it emits very toxic fumes of Hg and NO_x. See also MERCURY COMPOUNDS.

PFN500 CAS:5834-81-1 ***HR: 3***
N-(PHENYLMERCURI)-1,4,5,6,7,7-HEXA-CHLOROBICYCLO(2.2.1)HEPTENE-5-DICARBOXIMIDE
mf: $C_{15}H_7Cl_6HgNO_2$ mw: 646.52

SYN: PIMM

TOXICITY DATA with REFERENCE
unr-rat LD50:122 mg/kg 30ZDA9 -,295,71

CONSENSUS REPORTS: Mercury and its compounds are on the Community Right-To-Know List.

OSHA PEL: (Transitional: CL 1 mg/10m^3) CL 0.1 mg(Hg)/m^3 (skin)
ACGIH TLV: TWA 0.1 mg(Hg)/m^3 (skin)
NIOSH REL: TWA 0.05 mg(Hg)/m^3

SAFETY PROFILE: Poison by an unspecified route. When heated to decomposition it emits very toxic fumes of Cl^-, Hg, and NO_x. See also MERCURY COMPOUNDS.

PFN750 CAS:63869-08-9 ***HR: 2***
PHENYLMERCURILAURYL THIOETHER
mf: $C_{36}H_{58}Hg_2S$ mw: 924.18

SYN: BIS(PHENYLMERCURYLAURYL)SULFIDE

TOXICITY DATA with REFERENCE
ipr-mus LD50:500 mg/kg NTIS** AD691-490

CONSENSUS REPORTS: Mercury and its compounds are on the Community Right-To-Know List.

OSHA PEL: (Transitional: CL 1 mg/10m^3) CL 0.1 mg(Hg)/m^3 (skin)
ACGIH TLV: TWA 0.1 mg(Hg)/m^3 (skin)
NIOSH REL: TWA 0.05 mg(Hg)/m^3

SAFETY PROFILE: Moderately toxic by intraperitoneal route. When heated to decomposition it emits very toxic fumes of Hg and SO_x. See also MERCURY COMPOUNDS, ETHERS, and SULFIDES.

PFO000 CAS:103-27-5 ***HR: 3***
PHENYLMERCURI PROPIONATE
mf: $C_9H_{10}HgO_2$ mw: 350.78

SYNS: METASOL P-6 ◇ PHENYLMERCURY PROPIONATE ◇ PHENYL(PROPIONYLOXY)MERCURY ◇ PROPIONIC ACID, PHENYLMERCURY SALT

TOXICITY DATA with REFERENCE
ipr-mus LDLo:2 mg/kg CBCCT* 1,46,49

CONSENSUS REPORTS: Mercury and its compounds are on the Community Right-To-Know List. Reported in EPA TSCA Inventory.

OSHA PEL: (Transitional: CL 1 mg/10m^3) CL 0.1 mg (Hg)/m^3 (skin)
ACGIH TLV: TWA 0.1 mg(Hg)/m^3 (skin)
NIOSH REL: TWA 0.05 mg(Hg)/m^3

SAFETY PROFILE: Poison by intraperitoneal route. When heated to decomposition it emits toxic fumes of Hg. See also MERCURY COMPOUNDS.

PFO250 CAS:27360-58-3 ***HR: 3***
PHENYLMERCURIPYROCATECHIN
mf: $C_{12}H_{10}HgO_2$ mw: 386.81

SYNS: (DIHYDROXYPHENYL)PHENYL MERCURY ◇ GERMISAN

TOXICITY DATA with REFERENCE
unr-rat LD50:50 mg/kg 30ZDA9 -,295,71

CONSENSUS REPORTS: Mercury and its compounds are on the Community Right-To-Know List.

OSHA PEL: (Transitional: CL 1 mg/10m^3) CL 0.1 mg(Hg)/m^3 (skin)
ACGIH TLV: TWA 0.1 mg(Hg)/m^3 (skin)
NIOSH REL: TWA 0.05 mg(Hg)/m^3

SAFETY PROFILE: Poison by an unspecified route. When heated to decomposition it emits toxic fumes of Hg. See also MERCURY COMPOUNDS.

PFO500 ***HR: 3***
PHENYLMERCURY ACETATE 95% plus ETHYLMERCURY CHLORIDE 5%

SYN: (ACETATO)PHENYL-MERCURY mixed with CHLOROETHYL-MERCURY (19:1)

TOXICITY DATA with REFERENCE
orl-rat LD50:22 mg(Hg)/kg OCHRAI 15,5,63

CONSENSUS REPORTS: Mercury and its compounds are on the Community Right-To-Know List. Reported in EPA TSCA Inventory.

OSHA PEL: (Transitional: CL 1 mg/10m^3) CL 0.1 mg(Hg)/m^3 (skin)
ACGIH TLV: TWA 0.1 mg(Hg)/m^3 (skin)
NIOSH REL: (Mercury, Inorganic) TWA 0.05 mg(Hg)/m^3

SAFETY PROFILE: Poison by ingestion. When heated

to decomposition it emits very toxic fumes of Hg and Cl^-. See also MERCURY COMPOUNDS.

PFO550 CAS:34604-38-1 ***HR: 2***
PHENYL MERCURY AMMONIUM ACETATE
mf: $C_8H_8HgO_2 \cdot xH_3N$ mw: 456.03

SYNS: ACETATOPHENYLMERCURATE(1-) AMMONIUM SALT ◇ GALLOTOX ◇ MERCURATE(1-), ACETATOPHENYL-, AMMONIUM SALT ◇ MERCURY, (ACETATO-O)PHENYL-, AMMONIATE ◇ SETRETE

TOXICITY DATA with REFERENCE
orl-rat LD50:500 mg/kg FMCHA2-,C260,89
ACGIH TLV: TWA 0.1 mg(Hg)/m^3 (skin)

SAFETY PROFILE: Moderately toxic by ingestion. When heated to decomposition it emits toxic fumes of NO_x and Hg.

PFO750 CAS:3688-11-7 ***HR: 3***
PHENYL MERCURY CATECHOLATE
mf: $C_{12}H_{10}HgO_2$ mw: 386.81

SYNS: (1,2-BENZENEDIOLATO-O)PHENYLMERCURY ◇ (2-HYDROXYPHENOXY)PHENYLMERCURY (8CI) ◇ MERCUTAL ◇ PHENYLQUECKSILBERBRENZKATECHIN (GERMAN)

TOXICITY DATA with REFERENCE
orl-rat LD50:30 mg/kg 85GYAZ -,129,71
orl-mus LD50:70 mg/kg 85GYAZ -,129,71
ipr-mus LD50:50 mg/kg OCHRAI 15,5,63
orl-ckn LD50:116 mg/kg 85GYAZ -,129,71

CONSENSUS REPORTS: Mercury and its compounds are on the Community Right-To-Know List.

OSHA PEL: (Transitional: CL 1 mg/$10m^3$) CL 0.1 mg (Hg)/m^3 (skin)
ACGIH TLV: TWA 0.1 mg(Hg)/m^3 (skin)
NIOSH REL: TWA 0.05 mg(Hg)/m^3

SAFETY PROFILE: Poison by ingestion and intraperitoneal routes. When heated to decomposition it emits toxic fumes of Hg. See also MERCURY COMPOUNDS.

PFP000 CAS:31632-68-5 ***HR: 3***
PHENYLMERCURY NAPHTHENATE

PROP: Contains 10% mercury (AMIHAB 12,477,55).

SYN: NAPHTHENIC ACID, PHENYLMERCURY SALT

TOXICITY DATA with REFERENCE
orl-rat LD50:390 mg/kg AMIHAB 12,477,55
ipr-rat LD50:30 mg/kg AMIHAB 12,477,55

CONSENSUS REPORTS: Mercury and its compounds are on the Community Right-To-Know List.

OSHA PEL: (Transitional: CL 1 mg/$10m^3$) CL 0.1 mg (Hg)/m^3 (skin)
ACGIH TLV: TWA 0.1 mg(Hg)/m^3 (skin)
NIOSH REL: TWA 0.05 mg(Hg)/m^3

SAFETY PROFILE: Poison by ingestion and intraperitoneal routes. When heated to decomposition it emits toxic fumes of Hg. See also MERCURY COMPOUNDS.

PFP250 CAS:102-98-7 ***HR: 3***
PHENYL MERCURY SILVER BORATE
mf: $C_6H_6AgHgO_3$ mw: 434.58

SYNS: BORIC ACID, PHENYLMERCURY SILVER derivative ◇ MERFEN

TOXICITY DATA with REFERENCE
eye-rbt 150 mg MOD ARZNAD 9,349,59
eye-gpg 500 mg SEV ARZNAD 9,349,59
scu-mus LDLo:130 mg/kg MOLAAF 73,751,39

CONSENSUS REPORTS: Mercury and its compounds, as well as silver and its compounds, are on the Community Right-To-Know List.

OSHA PEL: TWA 0.01 mg(Ag)/m^3; (Transitional: CL 1 mg(Hg)/$10m^3$) CL 0.1 mg(Hg)/m^3 (skin)
ACGIH TLV: TWA 0.01 mg(Ag)/m^3; 0.1 mg(Hg)/m^3 (skin)
NIOSH REL: TWA 0.05 mg(Hg)/m^3

SAFETY PROFILE: Poison by subcutaneous route. A severe eye irritant. When heated to decomposition it emits toxic fumes of Hg. See also SILVER COMPOUNDS, MERCURY COMPOUNDS and BORON COMPOUNDS.

PFP500 CAS:2279-64-3 ***HR: 3***
PHENYL MERCURY UREA
mf: $C_7H_8HgN_2O$ mw: 336.76

SYNS: ABAVIT ◇ LEYTOSAN ◇ PHENYLMERCURIC UREA ◇ PHENYLMERCURIUREA

TOXICITY DATA with REFERENCE
unr-rat LD50:50 mg/kg 30ZDA9 -,295,71

CONSENSUS REPORTS: Mercury and its compounds are on the Community Right-To-Know List.

OSHA PEL: (Transitional: CL 1 mg/$10m^3$) CL 0.1 mg(Hg)/m^3 (skin)
ACGIH TLV: TWA 0.1 mg(Hg)/m^3 (skin)
NIOSH REL: TWA 0.05 mg(Hg)/m^3

SAFETY PROFILE: Poison by an unspecified route. When heated to decomposition it emits very toxic fumes of Hg and NO_x. See also MERCURY COMPOUNDS.

PFQ350 CAS:1131-18-6 ***HR: 2***
1-PHENYL-3-METHYL-5-AMINOPYRAZOLE
mf: $C_{10}H_{11}N_3$ mw: 173.24

SYN: 3-METHYL-1-PHENYL-1H-PYRAZOLE-5-AMINE

TOXICITY DATA with REFERENCE
orl-rat LD50:2500 mg/kg LONZA# 03FEB81
orl-mus LD50:1300 mg/kg LONZA# 03FEB81
orl-rbt LD50:1000 mg/kg LONZA# 03FEB81

CONSENSUS REPORTS: Reported in EPA TSCA Inventory.

SAFETY PROFILE: Moderately toxic by ingestion. When heated to decomposition it emits toxic fumes of NO_x.

PFR000 CAS:120-45-6 ***HR: 1***
PHENYLMETHYLCARBINYL PROPIONATE
mf: $C_{11}H_{14}O_2$ mw: 178.25

SYNS: α-METHYLBENZYL ALCOHOL, PROPIONATE ◇ α-METHYLBENZYL PROPANOATE ◇ METHYLPHENYLCARBINYL PROPIONATE ◇ 1-PHENYLETHYL PROPIONATE ◇ STYRALLYL PROPIONATE ◇ STYROLYL PROPIONATE

TOXICITY DATA with REFERENCE
skn-rbt 500 mg/24H MLD FCTXAV 14,613,76
orl-rat LD50:5200 mg/kg FCTXAV 14,613,76

CONSENSUS REPORTS: Reported in EPA TSCA Inventory.

SAFETY PROFILE: Mildly toxic by ingestion. A skin irritant. When heated to decomposition it emits acrid smoke and irritating fumes.

PFR100 ***HR: D***
PHENYLMETHYLCYCLOSILOXANE, mixed copolymer

TOXICITY DATA with REFERENCE
skn-rbt TDLo:650 g/kg (female 6-18D post):REP TXAPA9 21,15,72
orl-rat TDLo:1100 mg/kg (8-12D preg):TER TXAPA9 21,29,72

SAFETY PROFILE: An experimental teratogen. Experimental reproductive effects. When heated to decomposition it emits acrid smoke and irritating fumes.

PFR200 CAS:10415-87-9 ***HR: 2***
1-PHENYL-3-METHYL-3-PENTANOL
mf: $C_{12}H_{18}O$ mw: 178.30

SYNS: 3-PENTANOL, 3-METHYL-1-PHENYL- ◇ PHENETHYLMETHYLETHYLCARBINOL ◇ PHENYLETHYL METHYL ETHYL CARBINOL

TOXICITY DATA with REFERENCE
skn-rbt 500 mg/24H MOD FCTXAV 17,891,79
orl-rat LD50:2950 mg/kg FCTXAV 17,891,79

CONSENSUS REPORTS: Reported in EPA TSCA Inventory.

SAFETY PROFILE: Moderately toxic by ingestion. A skin irritant. When heated to decomposition it emits acrid smoke and irritating fumes.

PFR300 CAS:72007-81-9 ***HR: 1***
1-PHENYL-3-METHYL-3-PENTANYL ACETATE
mf: $C_{14}H_{20}O_2$ mw: 220.34

SYNS: METHYLETHYL PHENYLETHYL CARBINYL ACETATE ◇ PHENYLETHYL METHYLETHYLCARBINYL ACETATE

TOXICITY DATA with REFERENCE
skn-rbt 500 mg/24H MOD FCTXAV 14,833,76

CONSENSUS REPORTS: Reported in EPA TSCA Inventory.

SAFETY PROFILE: A skin irritant. When heated to decomposition it emits acrid smoke and irritating fumes.

PFR350 CAS:41363-50-2 ***HR: 3***
4-PHENYL-α-METHYLPHENYLACETATE-Γ-PROPYLSULFONATE SODIUM SALT
mf: $C_{18}H_{20}O_5S{\cdot}Na$ mw: 371.43

SYNS: 4-FENIL-α-METILFENILACETATO-Γ-PROPILSOLFONATO SALE SODICO (ITALIAN) ◇ α-METHYL-(1,1′-BIPHENYL)-4-ACETIC ACID 3-SULFOPROPYL ESTER SODIUM SALT

TOXICITY DATA with REFERENCE
orl-rat LD50:280 mg/kg FRPSAX 28,351,73
scu-rat LD50:270 mg/kg FRPSAX 28,351,73
orl-mus LD50:730 mg/kg FRPSAX 28,351,73
scu-mus LD50:590 mg/kg FRPSAX 28,351,73

SAFETY PROFILE: Poison by ingestion and subcutaneous routes. When heated to decomposition it emits toxic fumes of SO_x and Na_2O. See also ESTERS and SULFONATES.

PFS350 CAS:1943-79-9 ***HR: 3***
PHENYL MONOMETHYLCARBAMATE
mf: $C_8H_9NO_2$ mw: 151.18

SYN: METHYLCARBAMIC ACID PHENYL ESTER

TOXICITY DATA with REFERENCE
orl-rat LD50:540 mg/kg BWHOA6 44(1-3),241,71
ipr-rat LD50:357 mg/kg BWHOA6 44(1-3),241,71
ivn-rat LD50:13600 μg/kg BWHOA6 44(1-3),241,71

SAFETY PROFILE: Poison by intravenous and intraperitoneal routes. Moderately toxic by ingestion. When heated to decomposition it emits toxic fumes of NO_x. See also CARBAMATES and ESTERS.

PFS500 CAS:16033-21-9 ***HR: 3***
1-PHENYL-3-MONOMETHYLTRIAZENE
mf: $C_7H_9N_3$ mw: 135.19

SYN: PMT

TOXICITY DATA with REFERENCE
mmo-nsc 600 μmol/L MUREAV 13,276,71
skn-mus TDLo:284 mg/kg/8W-I:NEO CNREA8 34,1671,74
scu-mus LDLo:45 mg/kg CNREA8 34,1671,74

SAFETY PROFILE: Poison by subcutaneous route. Questionable carcinogen with experimental neoplastigenic data. Mutation data reported. When heated to decomposition it emits toxic fumes of NO_x.

PFS750 CAS:92-53-5 ***HR: 3***
PHENYL MORPHOLINE
mf: $C_{10}H_{13}NO$ mw: 163.24

PROP: Crystals from ethanol, ether. D: 1.058 @ 270°, mp: 57°, bp: 259.9°. Sol in water, alc, ether.

TOXICITY DATA with REFERENCE
eye-rbt 5 mg MLD UCDS** 1/13/72
orl-rat LD50:930 mg/kg JIHTAB 23,259,41
skn-rbt LD50:360 mg/kg UCDS** 1/13/72

CONSENSUS REPORTS: Reported in EPA TSCA Inventory.

SAFETY PROFILE: Poison by skin contact. Moderately toxic by ingestion. An eye irritant. When heated to decomposition it emits toxic fumes of NO_x.

PFT250 CAS:90-30-2 ***HR: 3***
N-PHENYL-1-NAPHTHYLAMINE
mf: $C_{16}H_{13}N$ mw: 219.30

PROP: Prisms from alc. Mp: 62°, bp: 335° @ 528 mm. Sol in water, benzene, alc, acetic acid, ether, chloroform.

SYNS: ACETO PAN ◇ ADDITIN 30 ◇ 1-ANILINONAPHTHALENE ◇ C.I. 44050 ◇ N-(1-NAPHTHYL)ANILINE ◇ NEOZONE A ◇ PANA ◇ PHENYLNAPHTHYLAMINE ◇ PHENYL-α-NAPHTHYLAMINE ◇ N-PHENYL-α-NAPHTHYLAMINE ◇ α-PHENYLNAPHTHYLAMINE ◇ VULKANOX PAN

TOXICITY DATA with REFERENCE
mmo-sat 10 μg/plate SYSWAE 12,41,79
mma-sat 500 nL/plate NTIS** AD-A041-973
otr-hmn:oth 27500 μg/L ITCSAF 17,719,81
dns-hmn:lng 50 mg/L NTIS** AD-A041-973
dlt-mus-ipr 830 mg/kg/5D-I NTIS** AD-A041-973
orl-mus TDLo:5400 mg/kg/9W-I:CAR CNREA8 44,3098,84
orl-mus TD:17280 mg/kg/9W-I:CAR CNREA8 44,3098,84
orl-rat LD50:1625 mg/kg AMRL** TR-74-78,74
orl-mus LD50:1231 mg/kg AMRL** TR-74-78,74

CONSENSUS REPORTS: Reported in EPA TSCA Inventory. EPA Genetic Toxicology Program.

SAFETY PROFILE: Moderately toxic by ingestion. Questionable carcinogen with experimental carcinogenic data. Human mutation data reported. When heated to decomposition it emits toxic fumes of NO_x. Used as a rubber antioxidant. See also AMINES.

PFT500 CAS:135-88-6 ***HR: 3***
N-PHENYL-β-NAPHTHYLAMINE
mf: $C_{16}H_{13}N$ mw: 219.30

PROP: Rhombic crystals from methanol. Mp: 107-108°, bp: 395.5°. Insol in water; sol in hot benzene; very sol in hot alc, ether.

SYNS: ACETO PBN ◇ AGERITE POWDER ◇ ANILINONAPHTHALENE ◇ 2-ANILINONAPHTHALENE ◇ ANTIOXIDANT 116 ◇ ANTIOXIDANT PBN ◇ N-(2-NAPHTHYL)ANILINE ◇ 2-NAPHTHYLPHENYLAMINE ◇ β-NAPHTHYLPHENYLAMINE ◇ NCI-C02915 ◇ NEOZONE D ◇ NILOX PBNA ◇ NONOX D ◇ PBNA ◇ PHENYL-β-NAPHTHYLAMINE ◇ PHENYL-2-NAPHTHYLAMINE ◇ N-PHENYL-2-NAPHTHYLAMINE ◇ STABILIZATOR AR

TOXICITY DATA with REFERENCE
otr-hmn:oth 23100 μg/L ITCSAF 17,719,81
cyt-mus-orl 360 mg/kg/18D GTPZAB 29(9),57,79
orl-mus TDLo:433 g/kg/2Y-C:ETA NTPTR* NTP-TR-333,88
orl-mus TDLo:208 g/kg/97W-C:CAR SYSWAE 14,129,81
scu-mus TDLo:464 mg/kg:NEO NTIS** PB223-159
orl-mus TD:17280 mg/kg/9W-I:CAR CNREA8 44,3098,84
orl-rat LD50:8730 mg/kg HYSAAV 31(2),183,66
orl-mus LD50:1450 mg/kg HYSAAV 31(2),183,66
orl-rbt LDLo:1000 mg/kg TPKVAL 10,91,68

CONSENSUS REPORTS: IARC Cancer Review: Group 3 IMEMDT 7,318,87; Human Inadequate Evidence IMEMDT 16,325,78; Animal Limited Evidence IMEMDT 16,325,78. Reported in EPA TSCA Inventory.

ACGIH TLV: Suspected Human Carcinogen.
DFG MAK: Suspected Carcinogen.

SAFETY PROFILE: Suspected carcinogen with experimental carcinogenic, neoplastigenic, and tumorigenic data. Moderately toxic by ingestion. Human mutation data reported. When heated to decomposition it emits toxic fumes of NO_x.

PFT600 CAS:6652-04-6 ***HR: 3***
4-PHENYLNITROSOPIPERIDINE
mf: $C_{11}H_{14}N_2O$ mw: 190.27

SYNS: NITROSO-4-PHENYLPIPERIDINE ◇ N-NITROSO-4-PHENYLPIPERIDINE ◇ 1-NITROSO-4-PHENYLPIPERIDINE

TOXICITY DATA with REFERENCE
mma-sat 25 μg/plate TCMUE9 1,13,84

pic-esc 2 mg/L TCMUE9 1,13,84
orl-rat TDLo:4175 mg/kg/2Y-I:CAR CRNGDP 2,1045,81

CONSENSUS REPORTS: EPA Genetic Toxicology Program.

SAFETY PROFILE: Questionable carcinogen with experimental carcinogenic data. Mutation data reported. When heated to decomposition it emits toxic fumes of NO_x. See also N-NITROSO COMPOUNDS.

PFU250 CAS:1841-19-6 ***HR: 3***
1-PHENYL-4-OXO-8-(4,4-BIS(4-FLUOROPHENYL)BUTYL-1,3,8-TRIAZASPIRO(4,5)DECANE
mf: $C_{29}H_{31}F_2N_3O$ mw: 475.63

SYN: 8-(4,4-BIS(p-FLUOROPHENYL)BUTYL)-1-PHENYL-1,3,8-TRIAZASPIRO(4,5)DECAN-4-ONE

TOXICITY DATA with REFERENCE
ivn-man TDLo:183 μg/kg:SKN BMJOAE 1,523,79
ims-mus LD50:125 mg/kg 27ZQAG -,231,72

SAFETY PROFILE: Poison by intramuscular route. Human systemic effects by intravenous route: corrosive to skin. When heated to decomposition it emits very toxic fumes of F^- and NO_x. See also FLUORIDES.

PFU500 CAS:101-54-2 ***HR: 2***
N-PHENYL-p-PHENYLENEDIAMINE
mf: $C_{12}H_{12}N_2$ mw: 184.26

SYNS: p-AMINODIFENYLAMIN (CZECH) ◇ p-AMINODIPHENYLAMINE ◇ C.I. 37240 ◇ C.I. 76085 ◇ N-PHENYL-p-AMINOANILINE

TOXICITY DATA with REFERENCE
eye-rbt 100 mg/24H SEV 28ZPAK -,70,72
orl-rat LD50:464 mg/kg NCILB* NIH-NCI-E-C-72-3252
orl-mus LD50:464 mg/kg NCILB* NIH-NCI-E-C-72-3252

CONSENSUS REPORTS: NCI Carcinogenesis Bioassay (feed); No Evidence: mouse, rat NCITR* NCI-CG-TR-82,78. Reported in EPA TSCA Inventory.

SAFETY PROFILE: Moderately toxic by ingestion. A severe eye irritant. When heated to decomposition it emits toxic fumes of NO_x. See also AMINES.

PFV000 CAS:63504-15-4 ***HR: 3***
N-PHENYL-N-(1-(PHENYLIMINO)ETHYL-N'-2,5-DICHLOROPHENYLUREA
mf: $C_{21}H_{22}Cl_2N_3O$ mw: 403.36

SYNS: N-CARBAMOYL-2-(2,6-DICHLOROPHENYL)ACETAMIDINE HYDROCHLORIDE ◇ 1-(2,6-DICHLOROBENZYLFORMIMIDOYL) UREA HYDROCHLORIDE ◇ LON-954

TOXICITY DATA with REFERENCE
orl-mus LD50:165 mg/kg ARZNAD 27,2326,77
ivn-mus LD50:180 mg/kg CSLNX* NX#01414

SAFETY PROFILE: Poison by ingestion and intravenous routes. When heated to decomposition it emits very toxic fumes of Cl^- and NO_x.

PFV250 CAS:638-21-1 ***HR: 3***
PHENYLPHOSPHINE
mf: C_6H_7P mw: 110.10

PROP: Needles from aq alc. Mp: 164-165°, bp: 305-308°. Insol in water; sol in alkali; very sol in alc and ether.

TOXICITY DATA with REFERENCE
ihl-rat LC50:38 ppm/4H AIHAAP 30,154,69

OSHA PEL: CL 0.05 ppm
ACGIH TLV: CL 0.05 ppm

SAFETY PROFILE: Poison by inhalation. Ignites spontaneously in air. When heated to decomposition it emits toxic fumes of PO_x. See also PHOSPHINE.

PFV500 CAS:1571-33-1 ***HR: 3***
PHENYLPHOSPHONIC ACID
mf: $C_6H_7O_3P$ mw: 158.10

PROP: Solid. Mp: 165°, d: 1.475.

SYN: BENZENEPHOSPHONIC ACID

TOXICITY DATA with REFERENCE
ipr-mus LD50:110 mg/kg CBCCT* 2,55,50

CONSENSUS REPORTS: Reported in EPA TSCA Inventory.

SAFETY PROFILE: Poison by intraperitoneal route. When heated to decomposition it emits toxic fumes of PO_x.

PFV750 CAS:1754-47-8 ***HR: 3***
PHENYLPHOSPHONIC ACID DIOCTYL ESTER
mf: $C_{22}H_{39}O_3P$ mw: 382.58

SYN: BENZENEPHOSPHONIC ACID, DIOCTYL ESTER

TOXICITY DATA with REFERENCE
ivn-mus LDLo:100 mg/kg CBCCT* 5,137,53

CONSENSUS REPORTS: Reported in EPA TSCA Inventory.

SAFETY PROFILE: Poison by intravenous route. When heated to decomposition it emits toxic fumes of PO_x. See also ESTERS.

PFV775 ***HR: 2***
PHENYLPHOSPHONIC AZIDE CHLORIDE
mf: $C_6H_5ClN_3OP$ mw: 201.55

SAFETY PROFILE: Potentially explosive during distillation at 80°C/0.5 mbar. When heated to decomposition it emits toxic fumes of Cl^-, PO_x and NO_x. See also AZIDES.

PFW000 ***HR: 3***
PHENYLPHOSPHONIC DIAZIDE
mf: $C_6H_5N_6OP$ mw: 208.12

SAFETY PROFILE: An impact-, heat-, and flame-sensitive explosive. When heated to decomposition it emits very toxic fumes of NO_x and PO_x. See also AZIDES.

PFW100 CAS:824-72-6 ***HR: 3***
PHENYL PHOSPHONYL DICHLORIDE
DOT: UN 2799
mf: $C_6H_5Cl_2OP$ mw: 178.99

DOT Classification: Corrosive Material; Label:Corrosive

SAFETY PROFILE: A storage hazard. May explode in a sealed bottle due to the release of hydrogen chloride gas. When heated to decomposition it emits toxic fumes of Cl^- and PO_x. See also CHLORIDES.

PFW500 CAS:14684-25-4 ***HR: 2***
PHENYLPHOSPHORODICHLORIDOTHIOUS ACID
DOT: UN 2799
mf: $C_6H_5Cl_2PS$ mw: 211.04

SYN: BENZENE PHOSPHORUS THIODICHLORIDE

DOT Classification: Corrosive Material; Label: Corrosive.

SAFETY PROFILE: A corrosive irritant to skin, eyes, and mucous membranes. When heated to decomposition it emits very toxic fumes of Cl^-, PO_x, and SO_x.

PFW750 CAS:5388-42-1 ***HR: 3***
PHENYLPHTHALIMIDINE
mf: $C_{14}H_{11}NO$ mw: 209.26

TOXICITY DATA with REFERENCE
orl-rat TDLo:450 mg/kg/(8-10D preg):TER TJADAB 19,341,79
orl-rat TDLo:450 mg/kg/(8-10D preg):REP TJADAB 19,341,79
ivn-mus LD50:56 mg/kg CSLNX* NX#04554

SAFETY PROFILE: Poison by intravenous route. Experimental teratogenic and reproductive effects. When heated to decomposition it emits toxic fumes of NO_x.

PFX000 CAS:92-54-6 ***HR: 3***
1-PHENYLPIPERAZINE
mf: $C_{10}H_{14}N_2$ mw: 162.26

PROP: Pale yellow oil. D: 1.0621 @ 20°/4°, bp: 286.5°, mp: 18.8°, flash p: 285°F. Insol in water; sol in alc, ether.

SYN: N-PHENYLPIPERAZINE

TOXICITY DATA with REFERENCE
skn-rbt 100 μg/24H open AIHAAP 23,95,62
eye-rbt 100 mg SEV 34ZIAG -,689,69
orl-rat LD50:210 mg/kg AIHAAP 23,95,62
skn-rbt LD50:140 mg/kg AIHAAP 23,95,62

CONSENSUS REPORTS: Reported in EPA TSCA Inventory.

SAFETY PROFILE: Poison by ingestion and skin contact. A skin and severe eye irritant. Combustible when exposed to heat or flame. It supports combustion and decomposes to yield toxic fumes of NO_x. To fight fire, use water, foam, dry chemical. See also PIPERAZINE.

PFX250 CAS:69103-95-3 ***HR: 3***
2-(3-(N-PHENYLPIPERAZINO)-PROPYL)-3-METHYLCHROMONE
mf: $C_{23}H_{26}N_2O_2$ mw: 362.51

TOXICITY DATA with REFERENCE
ipr-mus LD50:140 mg/kg EJMCA5 13,387,78
scu-mus LD50:250 mg/kg EJMCA5 13,387,78

SAFETY PROFILE: Poison by intraperitoneal and subcutaneous routes. When heated to decomposition it emits toxic fumes of NO_x.

PFX500 CAS:69103-97-5 ***HR: 3***
2-(3-(N-PHENYLPIPERAZINO)-PROPYL)-3-METHYL-7-FLUOROCHROMONE
mf: $C_{23}H_{25}FN_2O_2$ mw: 380.50

TOXICITY DATA with REFERENCE
ipr-mus LD50:90 mg/kg EJMCA5 13,387,78
scu-mus LD50:150 mg/kg EJMCA5 13,387,78

SAFETY PROFILE: Poison by intraperitoneal and subcutaneous routes. When heated to decomposition it emits very toxic fumes of F^- and NO_x.

PFY100 CAS:6952-94-9 ***HR: 3***
4-PHENYL-4-PIPERIDINECARBOXALDEHYDE
mf: $C_{12}H_{15}NO$ mw: 189.28

SYNS: 4-FENIL-4-FORMILPIPERIDINA (ITALIAN) ◇ 2723 I.S.

TOXICITY DATA with REFERENCE
ipr-mus LD50:40 mg/kg FRPSAX 19,14,64
ivn-mus LD50:13 mg/kg CSLNX* NX#07541
ivn-rbt LD50:3 mg/kg FRPSAX 19,14,64

SAFETY PROFILE: Poison by intravenous and intraperitoneal routes. When heated to decomposition it emits toxic fumes of NO_x. See also ALDEHYDES.

PFY105 ***HR: 3***
4-PHENYL-1-PIPERIDINECARBOXAMIDE
mf: $C_{11}H_{16}N_2O$ mw: 192.29

SYN: AH 1932

TOXICITY DATA with REFERENCE
orl-rat LD50:790 mg/kg JPPMAB 20,456,68
orl-mus LD50:1055 mg/kg JPPMAB 20,456,68
ipr-mus LD50:980 mg/kg APFRAD 70,339,82
orl-cat LDLo:400 mg/kg JPPMAB 20,456,68

SAFETY PROFILE: Poison by ingestion. Moderately toxic by intraperitoneal routes. When heated to decomposition it emits toxic fumes of NO_x. See also AMIDES.

PFY750 CAS:38940-46-4 ***HR: 3***
2-PHENYL-6-PIPERIDINOHEXYNOPHENONE
mf: $C_{23}H_{29}NO$ mw: 335.53

TOXICITY DATA with REFERENCE
orl-mus LD50:340 mg/kg CHTPBA 7,287,72
ivn-mus LD50:18500 μg/kg CHTPBA 7,287,72

SAFETY PROFILE: Poison by ingestion and intravenous routes. When heated to decomposition it emits toxic fumes of NO_x.

PGA250 CAS:74203-58-0 ***HR: 3***
1-PHENYL-4-(4-PIPERONYL-1-PIPERAZINYL-CARBONYL)-2-PYRROLIDINONE HYDROCHLORIDE
mf: $C_{17}H_{20}N_3O_4$•ClH mw: 366.86

SYN: CHLORHYDRATE de (PIPERONYL-4-PIPERAZINO)-1)-(PHENYL-1-PYRROLIDONE-2-CARBOXAMIDE-4)(FRENCH)

TOXICITY DATA with REFERENCE
ipr-mus LD50:65 mg/kg CHTPBA 7,398,72
ivn-mus LD50:51 mg/kg CHTPBA 7,398,72

SAFETY PROFILE: Poison by intraperitoneal and intravenous routes. When heated to decomposition it emits very toxic fumes of HCl and NO_x.

PGA500 CAS:579-07-7 ***HR: 3***
1-PHENYL-1,2-PROPANEDIONE
mf: $C_9H_8O_2$ mw: 148.16

PROP: Flash p: <69.8°F.

SAFETY PROFILE: A very dangerous fire hazard when exposed to heat or flame; can react vigorously with oxidizing materials. When heated to decomposition it emits acrid smoke and irritating fumes. See also KETONES.

PGA750 CAS:673-31-4 ***HR: 3***
3-PHENYL-1-PROPANOL CARBAMATE
mf: $C_{10}H_{13}NO_2$ mw: 179.24

SYNS: ACTOZINE ◇ ANSEPRON ◇ BENZENEPROPANOL CARBAMATE ◇ CARBAMIC ACID-3-PHENYLPROPYL ESTER ◇ 1-CARBAMOYLOXY-3-PHENYLPROPANE ◇ EIRENAL ◇ EXTACOL ◇ FENPROBAMATO ◇ GAMAQUIL ◇ Hg 532 ◇ MH-532 ◇ PALMITA ◇ Γ-PHENYLPROPYLCARBAMAT (GERMAN) ◇ Γ-PHENYLPROPYL CARBAMATE ◇ QUAMAQUIL ◇ SPANTOL ◇ TRANQUIL

TOXICITY DATA with REFERENCE
orl-rat LD50:1110 mg/kg ARZNAD 13,856,63
ipr-rat LD50:275 mg/kg AIPTAK 123,140,59
orl-mus LD50:840 mg/kg ARZNAD 12,340,62
ipr-mus LD50:150 mg/kg NTIS** AD691-490
ivn-mus LD50:320 mg/kg MEIEDD 10,1047,83
orl-rbt LD50:1125 mg/kg 27ZQAG -,402,72
ipr-rbt LD50:285 mg/kg 27ZQAG -,402,72

SAFETY PROFILE: Poison by intravenous and intraperitoneal routes. Moderately toxic by ingestion. Used as a tranquilizer and muscle relaxant. When heated to decomposition it emits toxic fumes of NO_x. See also CARBAMATES and ESTERS.

PGB500 ***HR: 3***
3-PHENYLPROPIONYL AZIDE
mf: $C_9H_9N_3O$ mw: 175.19

SAFETY PROFILE: A heat-sensitive explosive. When heated to decomposition it emits toxic fumes of NO_x. See also AZIDES.

PGB750 CAS:10402-52-5 ***HR: 1***
2-PHENYLPROPYL ACETATE
mf: $C_{11}H_{14}O_2$ mw: 178.25

SYNS: HYDRATROPIC ACETATE ◇ HYDRATROPYL ACETATE ◇ β-METHYLPHENYLETHYL ACETATE

TOXICITY DATA with REFERENCE
skn-rbt 500 mg/24H MLD FCTXAV 14,659,76
orl-rat LD50:4300 mg/kg FCTXAV 14,659,76

CONSENSUS REPORTS: Reported in EPA TSCA Inventory.

SAFETY PROFILE: Mildly toxic by ingestion. A skin irritant. When heated to decomposition it emits acrid smoke and irritating fumes.

PGC750 CAS:78128-85-5 ***HR: 2***
3-(3-PHENYLPROPYL)-4-HYDROXY-2(5H) FURANONE
mf: $C_{13}H_{14}O_3$ mw: 218.27

SYN: α-(β-PHENYL-PROPYL)-β-HYDROXY-Δ^{α},β-BUTENOLID (GERMAN)

TOXICITY DATA with REFERENCE
scu-mus LD50:1750 mg/kg ARZNAD 11,277,61
ivn-mus LD50:700 mg/kg ARZNAD 11,277,61

SAFETY PROFILE: Moderately toxic by subcutaneous

and intravenous routes. When heated to decomposition it emits acrid smoke and irritating fumes.

PGD000 CAS:3239-63-2 ***HR: 3***
O-PHENYL-S-PROPYL METHYL PHOSPHONODITHIOATE
mf: $C_{10}H_{15}OPS_2$ mw: 246.34

SYNS: ENT 27,186 ◇ V-C 3-764

TOXICITY DATA with REFERENCE
orl-rat LD50:208 mg/kg ARSIM* 20,26,66
orl-ckn LD50:8 mg/kg TXAPA9 11,49,67

SAFETY PROFILE: Poison by ingestion. When heated to decomposition it emits very toxic fumes of SO_x and PO_x.

PGE000 CAS:3567-38-2 ***HR: 3***
1-PHENYL-2-PROPYNYL CARBAMATE
mf: $C_{10}H_9NO_2$ mw: 175.20

SYNS: CARFIMAT ◇ CFC ◇ α-ETHYNYLBENZYL CARBAMATE ◇ EQUILIUM ◇ NIRVOTIN ◇ PHENYLETHYNLCARBINOL CARBAMATE

TOXICITY DATA with REFERENCE
ipr-rat LD50:150 mg/kg AIPTAK 99,245,54
orl-mus LD50:275 mg/kg THERAP 11,692,56
ipr-mus LD50:250 mg/kg AIPTAK 99,245,54

SAFETY PROFILE: Poison by ingestion and intraperitoneal routes. When heated to decomposition it emits toxic fumes of NO_x. See also CARBAMATES.

PGE250 CAS:1008-79-3 ***HR: 3***
1-PHENYL-4-PYRAZOLIN-3-ONE
mf: $C_9H_8N_2O$ mw: 160.19

SYN: 1-FENYL-3-PYRAZOLON(CZECH)

TOXICITY DATA with REFERENCE
skn-rbt 500 mg/24H MLD 28ZPAK -,144,72
eye-rbt 100 mg/24H SEV 28ZPAK -,144,72
orl-rat LD50:381 mg/kg 28ZPAK -,144,72
/THR **SAFETY PROFILE:** ingestion. A skin and severe eye irritant. When heated to decomposition it emits toxic fumes of NO_x. See also KETONES.

PGE300 CAS:61001-42-1 ***HR: D***
2-PHENYL-8H-PYRAZOLO(5,1-a)ISOINDOLE
mf: $C_{16}H_{12}N_2$ mw: 232.30

TOXICITY DATA with REFERENCE
scu-rat TDLo:375 mg/kg (6-10D preg):REP EJMCA5 17,223,82

SAFETY PROFILE: Experimental reproductive effects. When heated to decomposition it emits toxic fumes of NO_x.

PGE350 CAS:61001-36-3 ***HR: D***
2-PHENYL-PYRAZOLO(5,1-a)ISOQUINOLINE
mf: $C_{17}H_{12}N_2$ mw: 244.31

TOXICITY DATA with REFERENCE
scu-ham TDLo:5 mg/kg (4-8D preg):REP EJMCA5 19,215,84

SAFETY PROFILE: Experimental reproductive effects. When heated to decomposition it emits toxic fumes of NO_x.

PGE500 CAS:78109-90-7 ***HR: 2***
p-(3-PHENYL-1-PYRAZOLYL)BENZENESULFONIC ACID
mf: $C_{15}H_{12}N_2O_3S$ mw: 300.35

SYN: 1-(p-SULFOPHENYL)-3-PHENYL-PYRAZOL(GERMAN)

TOXICITY DATA with REFERENCE
scu-mus LD50:982 mg/kg ARZNAD 4,249,54
ivn-mus LD50:490 mg/kg ARZNAD 4,249,54

SAFETY PROFILE: Moderately toxic by intravenous and subcutaneous routes. When heated to decomposition it emits very toxic fumes of SO_x and NO_x.

PGE750 CAS:939-23-1 ***HR: 3***
4-PHENYLPYRIDINE
mf: $C_{11}H_9N$ mw: 155.21

PROP: Leaves from water. Mp: 77-78°, bp: 274-275°. Sol in hot water, alc, ether.

SYN: p-PHENYLPYRIDINE

TOXICITY DATA with REFERENCE
mma-sat 650 μmol/L/2H CNREA8 39,4152,79
ivn-mus LD50:89 mg/kg CSLNX* NX#12031

CONSENSUS REPORTS: Reported in EPA TSCA Inventory.

SAFETY PROFILE: Poison by intravenous route. Mutation data reported. When heated to decomposition it emits toxic fumes of NO_x.

PGE775 CAS:562-10-7 ***HR: 2***
PHENYL-2-PYRIDYLMETHYL-β-N,N-DIMETHYLAMINOETHYL ETHER SUCCINATE
mf: $C_{17}H_{22}N_2O•C_4H_6O_4$ mw: 388.51

SYNS: DECAPRYN ◇ DECAPRYN SUCCINATE ◇ DECARPYN SUCCINATE (1:1) ◇ DIMETHYLAMINOETHOXY-METHYL-BENZYL-PYRIDINE SUCCINATE ◇ 2-(α-(2-DIMETHYLAMINOETHOXY)-α-METHYLBENZYL)PYRIDINE SUCCINATE ◇ 2-DIMETHYLAMINOETHOXYPHENYLMETHYL-2-PICOLINESUCCINATE ◇ DOXYLAMINE SUCCINATE ◇ DOXYLAMINE SUCCINATE (1:1) ◇ HOGGAR N ◇ MEREPRINE ◇ PYRIDINE, 2-(α-(2-(DIMETHYLAMINO)ETHOXY)-α-METHYLBENZYL)-, SUCCINATE (1:1) ◇ UNISOM

TOXICITY DATA with REFERENCE
dns-rat:lvr 500 umol/L MUREAV 135,131,84
ims-mky TDLo:765 mg/kg (female 20-22D post):REP IMSCE2 13,225,85
ims-mky TDLo:1710 mg/kg (female 20-35D post):TER IMSCE2 13,225,85
orl-rat LDLo:600 mg/kg JAPMA8 37,311,48
orl-mus LD50:470 mg/kg JLCMAK 33,325,48

SAFETY PROFILE: Moderately toxic by ingestion. An experimental teratogen. Experimental reproductive effects. Mutation data reported. When heated to decomposition it emits toxic fumes of NO_x.

PGF000 CAS:25332-09-6 ***HR: 3***
PROPHENPYRIDAMINE HYDROCHLORIDE
mf: $C_{16}H_{20}N_2$•ClH mw: 276.84

SYNS: 1-PHENYL-1-(2-PYRIDYL)-3-DIMETHYLAMINOPROPANE HYDROCHLORIDE ◇ TRIMETON

TOXICITY DATA with REFERENCE
orl-mus LD50:185 mg/kg JPETAB 113,72,55
ipr-mus LD50:65 mg/kg JPETAB 113,72,55
scu-mus LD50:132 mg/kg JPETAB 113,72,55
ivn-mus LD50:60 mg/kg JPETAB 113,72,55
ivn-dog LDLo:95 mg/kg JPETAB 113,72,55
orl-gpg LD50:277 mg/kg JPETAB 113,72,55
scu-gpg LD50:109 mg/kg JPETAB 113,72,55

SAFETY PROFILE: Poison by ingestion, intravenous, intraperitoneal, and subcutaneous routes. When heated to decomposition it emits very toxic fumes of NO_x and HCl.

PGF100 CAS:3333-85-5 ***HR: D***
2-PHENYL-3-p-(β-PYRROLIDINOETHOXY)PHENYL-6-METHOXY-BENZOFURAN HYDROCHLORIDE
mf: $C_{27}H_{27}NO_3$•ClH mw: 450.01

SYNS: 1-(2-(p-(6-METHOXY-2-PHENYLBENZOFURAN-3-YL) PHENOXY)ETHYL)PYRROLIDINE HYDROCHLORIDE ◇ 1-(2-(p-(6-METHOXY-2-PHENYL-3-BENZOFURANYL)PHENOXY)ETHYL)PYRROLIDINE HYDROCHLORIDE ◇ 6-METHOXY-3-(p-(2-(N- PYRROLIDYL) ETHOXY)PHENYL)-2-PHENYLBENZO(b)FURANHYDROCHLORIDE

TOXICITY DATA with REFERENCE
orl-mus TDLo:90 mg/kg (female 4D pre):REP JMCMAR 14,1185,71

SAFETY PROFILE: Experimental reproductive effects. When heated to decomposition it emits toxic fumes of NO_x and HCl.

PGF125 CAS:24365-61-5 ***HR: 3***
2-PHENYL-3-p-(β-PYRROLIDINOETHOXY)PHENYL-(2:1-b)NAPHTHOFURAN
mf: $C_{30}H_{27}NO_2$ mw: 433.58

SYNS: 2-PHENYL-3-p-(β-PYRROLIDINOETHOXY)-PHENYL-NAPHTHO(2:1-b)FURAN ◇ 2-PHENYL-1-(p-(2-(1-PYRROLIDINYL)ETHOXY)PHENYL)NAPHTHO(2,1-b)FURAN

TOXICITY DATA with REFERENCE
ivg-rat TDLo:10 mg/kg (female 1D post):REP IJEBA6 12,370,74
ipr-mus LD50:400 mg/kg CCPTAY 1,29,70

SAFETY PROFILE: Poison by intraperitoneal route. Experimental reproductive effects. An oral antifertility agent. When heated to decomposition it emits toxic fumes of NO_x.

PGF800 CAS:1198-97-6 ***HR: 3***
PHENYLPYRROLIDONE
mf: $C_{10}H_{11}NO$ mw: 161.22

SYNS: 4-PHENYL-2-PYRROLIDINONE ◇ 4-PHENYLPYRROLIDONE-2

TOXICITY DATA with REFERENCE
ipr-mus LD50:325 mg/kg AITEAT 23,733,75

SAFETY PROFILE: Poison by intraperitoneal route. When heated to decomposition it emits toxic fumes of NO_x.

PGG000 CAS:132-60-5 ***HR: 3***
2-PHENYLQUINOLINE-4-CARBOXYLIC ACID
mf: $C_{16}H_{11}NO_2$ mw: 249.28

SYNS: ACIPHENOCHINOLINE ◇ ACIPHENOCHINOLINIUM ◇ AGOTAN ◇ ALUTYL ◇ ARTAM ◇ ATOCIN ◇ ATOPHAN ◇ CINCHOPHENE ◇ CINCHOPHENIC ACID ◇ CINCOPHEN ◇ IKTEROSAN ◇ MYLOFANOL ◇ PHENOQUIN ◇ 2-PHENYLCINCHONIC ACID ◇ 2-PHENYL-4-QUINOLINECARBOXYLIC ACID ◇ POLYPHLOGIN ◇ QUINOFEN ◇ QUINOPHEN ◇ RHEMATAN ◇ TOPHOL ◇ VANTYL

TOXICITY DATA with REFERENCE
orl-hmn LDLo:214 mg/kg 34ZIAG -,178,69
unr-man LDLo:74 mg/kg 85DCAI 2,73,70
orl-rat LDLo:500 mg/kg JPETAB 90,260,47
orl-mus LDLo:1500 mg/kg LDTU** -,-,31
ipr-mus LD50:200 mg/kg NTIS** AD691-490
scu-mus LD50:350 mg/kg AIPTAK 116,154,58

SAFETY PROFILE: A human poison by ingestion. An experimental poison by intraperitoneal and subcutaneous routes. Moderately toxic by an unspecified route. When heated to decomposition it emits toxic fumes of NO_x.

PGG350 CAS:553-69-5 ***HR: 3***
PHENYLRAMIDOL
mf: $C_{13}H_{14}N_2O$ mw: 214.29

PROP: Crystals from dil methanol. Mp: 82-85°.

SYNS: ABBOLEXIN ◇ ANALEXIN ◇ BONAPAR ◇ CABRAL ◇ ELAN ◇ EVASPIRINE ◇ EVASPRINE ◇ FENYRAMIDOL ◇ FENYRIPOL ◇ 2-(β-HYDROXYPHENETHYLAMINO)PYRIDINE ◇ IN 511 ◇ MIODAR

◇ MJ 505 ◇ PHENYRAMIDOL ◇ α-(2-PYRIDYLAMINOMETHYL)BENZYL ALCOHOL

TOXICITY DATA with REFERENCE
orl-rat LD50:756 mg/kg AIPTAK 130,280,61
orl-mus LD50:1850 mg/kg BCFAAI 103,245,64
ipr-mus LD50:355 mg/kg AIPTAK 141,83,63
scu-mus LD50:405 mg/kg APTOA6 19,247,62
ivn-mus LD50:124 mg/kg AIPTAK 141,83,63

SAFETY PROFILE: Poison by intravenous and intraperitoneal routes. Moderately toxic by ingestion and subcutaneous routes. When heated to decomposition it emits toxic fumes of NO_x.

PGG355 CAS:326-43-2 ***HR: 3***
PHENYLRAMIDOL HYDROCHLORIDE
mf: $C_{13}H_{14}N_2O•ClH$ mw: 250.75

SYNS: ANALEXIN ◇ FENYRAMIDOL HYDROCHLORIDE ◇ IN 511 ◇ MJ 505 ◇ NSC-17777 ◇ PHENYRAMIDOL HYDROCHLORIDE ◇ α-((2-PYRIDINYLAMINO)METHYL)-BENZENEMETHANOLMONOHYDROCHLORIDE ◇ α-((2-PYRIDYLAMINO)METHYL)-BENZYL ALCOHOL HYDROCHLORIDE

TOXICITY DATA with REFERENCE
orl-mus LD50:2425 mg/kg 29ZVAB -,95,69
ipr-mus LD50:450 mg/kg JPETAB 128,65,60
ivn-mus LD50:124 mg/kg JPETAB 128,65,60

SAFETY PROFILE: Poison by intravenous route. Moderately toxic by ingestion and intraperitoneal routes. When heated to decomposition it emits toxic fumes of NO_x and HCl.

PGG500 CAS:1457-46-1 ***HR: 3***
3-PHENYLRHODANINE
mf: $C_9H_7NOS_2$ mw: 209.29

SYN: 3-PHENYL-RHODANIN (GERMAN)

TOXICITY DATA with REFERENCE
unr-rat LDLo:300 mg/kg ARZNAD 16,1092,66

CONSENSUS REPORTS: Reported in EPA TSCA Inventory.

SAFETY PROFILE: Poison by an unspecified route. When heated to decomposition it emits very toxic fumes of NO_x and SO_x.

PGG750 CAS:118-55-8 ***HR: 2***
PHENYL SALICYLATE
mf: $C_{13}H_{10}O_3$ mw: 214.23

PROP: White, small crystals; pleasant odor and taste. D: 1.250 @ 20°/4°, mp: 41.4°, bp: 172-173° @ 12 mm. Sol in water, ether, benzene; very sol in hot alc.

SYN: SALOL

TOXICITY DATA with REFERENCE
orl-rat TDLo:600 mg/kg (7-12D preg):TER OCMJAJ 12,23,66
orl-rat TDLo:600 mg/kg (7-12D preg):REP OCMJAJ 12,23,66
orl-rat LD50:3 g/kg FCTXAV 14,659,76
orl-rbt LDLo:3000 mg/kg RMSRA6 15,561,1895

CONSENSUS REPORTS: Reported in EPA TSCA Inventory.

SAFETY PROFILE: Moderately toxic by ingestion. Experimental teratogenic and reproductive effects. When heated to decomposition it emits acrid smoke and irritating fumes. See also ESTERS.

PGH000 CAS:304-06-3 ***HR: 3***
3-PHENYLSALICYLIC ACID
mf: $C_{13}H_{10}O_3$ mw: 214.23

PROP: Rhombic crystals from alc. D: 1.250 @ 20°/4°, mp: 41.4°, bp: 172-173° @ 12 mm. Sol in water, ether, benzene; very sol in hot alc.

SYN: USAF DO-59

TOXICITY DATA with REFERENCE
ipr-mus LD50:125 mg/kg NTIS** AD277-689

CONSENSUS REPORTS: Reported in EPA TSCA Inventory.

SAFETY PROFILE: Poison by intraperitoneal route. When heated to decomposition it emits acrid smoke and irritating fumes.

PGH250 CAS:1132-39-4 ***HR: 3***
PHENYL SELENIDE
mf: $C_{12}H_{10}Se$ mw: 233.18

SYNS: BIPHENYL SELENIUM ◇ DIPHENYL SELENIDE

TOXICITY DATA with REFERENCE
orl-rat LD50:360 mg/kg TXAPA9 20,89,71

CONSENSUS REPORTS: Selenium and its compounds are on the Community Right-To-Know List. Reported in EPA TSCA Inventory.

OSHA PEL: TWA 0.2 mg(Se)/m^3
ACGIH TLV: TWA 0.2 mg(Se)/m^3
DFG MAK: 0.1 mg(Se)/m^3

SAFETY PROFILE: Poison by ingestion. When heated to decomposition it emits toxic fumes of Se. See also SELENIUM COMPOUNDS.

PGH500 CAS:39254-48-3 ***HR: 1***
PHENYLSELENONIC ACID
mf: $C_6H_6O_3Se$ mw: 205.08

SYN: BENZENESELENONIC ACID

CONSENSUS REPORTS: Selenium and its compounds are on the Community Right-To-Know List.

OSHA PEL: TWA 0.2 mg(Se)/m^3
ACGIH TLV: TWA 0.2 mg(Se)/m^3
DFG MAK: 0.1 mg(Se)/m^3

SAFETY PROFILE: Explodes feebly @ 180°. Silver salt is unstable. When heated to decomposition it emits toxic fumes of Se. See also SELENIUM COMPOUNDS.

PGH750 CAS:2097-19-0 ***HR: 3***
PHENYLSILATRANE
mf: $C_{12}H_{17}NO_3Si$ mw: 251.39

SYNS: FENYLSILATRAN (CZECH) ◇ PHENYL-2,8,9-TRIOXA-5-AZA-1-SILABICYCLO(3.3.3)UNDECANE

TOXICITY DATA with REFERENCE
orl-rat LDLo:1 mg/kg 28ZPAK -,221,72
ipr-mus LD50:330 μg/kg RCRVAB 38(12),975,69

CONSENSUS REPORTS: EPA Extremely Hazardous Substances List. Reported in EPA TSCA Inventory.

SAFETY PROFILE: Poison by ingestion and intraperitoneal routes. When heated to decomposition it emits toxic fumes of NO_x.

PGI000 CAS:5274-48-6 ***HR: 3***
PHENYLSILVER
mf: C_6H_5Ag mw: 184.98

CONSENSUS REPORTS: Silver and its compounds are on the Community Right-To-Know List.

SAFETY PROFILE: Dry solid can explode at room temperature. It is also sensitive to light friction or stirring. When heated to decomposition it emits acrid smoke and irritating fumes. See also SILVER COMPOUNDS.

PGI100 CAS:1623-99-0 ***HR: 3***
PHENYL SODIUM
mf: C_6H_5Na mw: 100.10

SAFETY PROFILE: Can ignite and burn violently in moist air. When heated to decomposition it emits toxic fumes of Na_2O. See also SODIUM COMPOUNDS and ORGANOMETALS.

PGI500 CAS:139-66-2 ***HR: 2***
PHENYL SULFIDE
mf: $C_{12}H_{10}S$ mw: 186.28

PROP: Colorless liquid; almost no odor. D: 1.118 @ 15°/15°, mp: approx −40°, bp: 295-297°. Insol in water; sol in hot alc; misc with benzene, ether, and carbon disulfide.

SYNS: DIPHENYL SULFIDE ◇ DIPHENYL THIOETHER ◇ 1,1'-THIOBIS(BENZENE)

TOXICITY DATA with REFERENCE
skn-rbt 10 mg/24H open SEV AIHAAP 23,95,62
orl-rat LD50:490 mg/kg AIHAAP 23,95,62
skn-rbt LD50:11300 mg/kg AIHAAP 23,95,62

CONSENSUS REPORTS: Reported in EPA TSCA Inventory.

SAFETY PROFILE: Moderately toxic by ingestion. Mildly toxic by skin contact. A severe skin irritant. Reacts violently with diazonium salts liberating SO_x. When heated to decomposition it emits highly toxic fumes of SO_x.

PGI750 CAS:127-63-9 ***HR: 3***
PHENYL SULFONE
mf: $C_{12}H_{10}O_2S$ mw: 218.28

SYNS: DIFENYLSULFON (CZECH) ◇ DIPHENYL SULFONE ◇ SULFOBENZIDE

TOXICITY DATA with REFERENCE
orl-rat LD50:1390 mg/kg MarJV# 29MAR77
ivn-mus LD50:320 mg/kg CSLNX* NX#01761

CONSENSUS REPORTS: Reported in EPA TSCA Inventory.

SAFETY PROFILE: Poison by intravenous route. Moderately toxic by ingestion. When heated to decomposition it emits toxic fumes of SO_x.

PGJ000 CAS:3368-13-6 ***HR: D***
2-(PHENYLSULFONYLAMINO)-1,3,4-THIADIAZOLE-5-SULFONAMIDE
mf: $C_8H_8N_2O_4S_3$ mw: 292.36

SYNS: BENZOLAMIDE ◇ CL 11366 ◇ W 1803

TOXICITY DATA with REFERENCE
scu-rat TDLo:1200 mg/kg (11-12D preg):REP JHMJAX 130,116,72
scu-rat TDLo:2400 mg/kg (11-12D preg):TER JHMJAX 130,116,72

SAFETY PROFILE: Experimental teratogenic and reproductive effects. When heated to decomposition it emits toxic fumes of NO_x and SO_x.

PGJ500 CAS:3999-10-8 ***HR: 3***
5-PHENYLTETRAZOLE
mf: $C_7H_6N_4$ mw: 146.15

HNN=NN=C(C_6H_5)

SAFETY PROFILE: Explodes on distillation. When heated to decomposition it emits toxic fumes of NO_x.

PGJ750 CAS:86-93-1 ***HR: 3***
1-PHENYLTETRAZOLE-5-THIOL
mf: $C_7H_6N_4S$ mw: 178.23

SYNS: 5-MERCAPTO-1-PHENYLTETRAZOLE ◇ 1-PHENYL-5-MERCAPTOTETRAZOLE ◇ 1-PHENYL-5-MERCAPTO-1,2,3,4-TETRAZOLE

TOXICITY DATA with REFERENCE
ipr-mus LD50:250 mg/kg NTIS** AD691-490

CONSENSUS REPORTS: Reported in EPA TSCA Inventory.

SAFETY PROFILE: Poison by intraperitoneal route. When heated to decomposition it emits very toxic fumes of NO_x and SO_x. See also MERCAPTANS.

PGK000 ***HR: 3***
3-PHENYL-1-TETRAZOLYL-1-TETRAZENE
mf: $C_7H_3N_8$ mw: 199.16

SAFETY PROFILE: Explodes on warming. When heated to decomposition it emits toxic fumes of NO_x.

PGK100 ***HR: 3***
PHENYLTHALLIUM DIAZIDE
mf: $C_6H_5N_6Tl$ mw: 365.53

CONSENSUS REPORTS: Thallium and its compounds are on the Community Right-To-Know List.

SAFETY PROFILE: Thallium compounds are poisons. Decomposes violently at 200°C. When heated to decomposition it emits toxic fumes of NO_x. See also THALLIUM COMPOUNDS and AZIDES.

PGL000 CAS:13354-35-3 ***HR: 2***
1-(PHENYLTHIO)ANTHRAQUINONE
mf: $C_{20}H_{12}O_2S$ mw: 316.38

SYN: 1-THIOPHENYLANTHRAQUINONE

TOXICITY DATA with REFERENCE
ipr-mus LD50:500 mg/kg NTIS** AD691-490

CONSENSUS REPORTS: Reported in EPA TSCA Inventory.

SAFETY PROFILE: Moderately toxic by intraperitoneal route. When heated to decomposition it emits toxic fumes of SO_x.

PGL250 CAS:636-04-4 ***HR: 3***
N-PHENYLTHIOBENZAMIDE
mf: $C_{13}H_{11}NS$ mw: 213.31

SYNS: THIOBENZANILIDE ◇ USAF EK-2010

TOXICITY DATA with REFERENCE
ipr-mus LD50:100 mg/kg NTIS** AD277-689

CONSENSUS REPORTS: Reported in EPA TSCA Inventory.

SAFETY PROFILE: Poison by intraperitoneal route. When heated to decomposition it emits very toxic fumes of NO_x and SO_x.

PGL500 CAS:69782-05-4 ***HR: 3***
(4-PHENYLTHIOBUTYL)HYDRAZINE MALEATE
mf: $C_{10}H_{16}N_2S{\cdot}C_4H_4O_4$ mw: 312.42

TOXICITY DATA with REFERENCE
orl-mus LD50:200 mg/kg JMCMAR 6,63,63
ipr-mus LD50:250 mg/kg JMCMAR 6,63,63

SAFETY PROFILE: Poison by ingestion and intraperitoneal routes. When heated to decomposition it emits very toxic fumes of SO_x and NO_x.

PGM250 CAS:69782-00-9 ***HR: 3***
((1-PHENYLTHIOMETHYL)ETHYL)HYDRAZINE MALEATE
mf: $C_9H_{14}N_2S{\cdot}C_4H_4O_4$ mw: 298.39

TOXICITY DATA with REFERENCE
orl-mus LD50:750 mg/kg JMCMAR 6,63,63
ipr-mus LD50:375 mg/kg JMCMAR 6,63,63

SAFETY PROFILE: Poison by intraperitoneal route. Moderately toxic by ingestion. When heated to decomposition it emits very toxic fumes of SO_x and NO_x.

PGM500 ***HR: 3***
PHENYLTHIOPHOSPHONIC DIAZIDE
mf: $C_6H_5N_6PS$ mw: 224.19

$C_6H_5P(:S)(N_3)_2$

SAFETY PROFILE: Explodes at 80°C/0.13 mbar. Violently explodes during distillation of crude material. When heated to decomposition it emits very toxic fumes of SO_x, NO_x and PO_x. See also AZIDES.

PGM750 CAS:645-48-7 ***HR: 3***
1-PHENYLTHIOSEMICARBAZIDE
mf: $C_7H_9N_3S$ mw: 167.25

PROP: Prisms from alc. Mp: 200-201°. Sltly sol in water, ether; sol in hot alc.

SYNS: USAF EK-5426 ◇ USAF EL-45

TOXICITY DATA with REFERENCE
dnd-hmn:hla 20 μmol/L BCPCA6 25,821,76
orl-rat LDLo:50 mg/kg JPETAB 93,287,48
ipr-mus LD50:15 mg/kg NTIS** AD277-689

CONSENSUS REPORTS: Reported in EPA TSCA Inventory.

SAFETY PROFILE: Poison by ingestion and intraperitoneal routes. Human mutation data reported. When heated to decomposition it emits very toxic fumes of NO_x and SO_x.

PGN000 CAS:5351-69-9 ***HR: 3***
4-PHENYL-3-THIOSEMICARBAZIDE
mf: $C_7H_9N_3S$ mw: 167.25

TOXICITY DATA with REFERENCE
orl-rat LDLo:50 mg/kg NCNSA6 5,44,53
ipr-mus LDLo:15 mg/kg NTIS** AD277-689
par-rat LD50:750 mg/kg ARZNAD 12,260,62

CONSENSUS REPORTS: Reported in EPA TSCA Inventory.

SAFETY PROFILE: Poison by ingestion and intraperitoneal routes. Moderately toxic by parenteral route. When heated to decomposition it emits very toxic fumes of NO_x and SO_x.

PGN250 CAS:103-85-5 ***HR: 3***
1-PHENYL-2-THIOUREA
mf: $C_7H_8N_2S$ mw: 152.23

PROP: Needle-like crystals; bitter taste. Mp: 154°, d: 1.3. Sol in water, alc, aq ether.

SYNS: NCI-C02017 ◇ PHENYLTHIOCARBAMIDE ◇ 1-PHENYLTHIOUREA ◇ N-PHENYLTHIOUREA ◇ α-PHNEYLTHIOUREA ◇ PTC ◇ PTU ◇ RCRA WASTE NUMBER P093 ◇ U 6324 ◇ USAF EK-1569

TOXICITY DATA with REFERENCE
orl-mus TDLo:168 mg/kg (1-21D preg):TER EXPEAM 34,1319,78
orl-rat LD50:3 mg/kg JMPCAS 4,109,61
ipr-rat LD50:5 mg/kg JPETAB 89,186,47
orl-mus LD50:10 mg/kg NCILB* NIH-NCI-E-C-72-3252
ipr-mus LD50:25 mg/kg NTIS** AD277-689
orl-rbt LD50:40 mg/kg JMPCAS 4,109,61

CONSENSUS REPORTS: NCI Carcinogenesis Bioassay (feed); No Evidence: mouse, rat NCITR* NCI-CG-TR-148,78. EPA Extremely Hazardous Substances List. Reported in EPA TSCA Inventory.

SAFETY PROFILE: Poison by ingestion and intraperitoneal routes. Experimental teratogenic effects. When heated to decomposition or on contact with acid or acid fumes it emits highly toxic fumes of SO_x and NO_x. Used in medical genetics and production of rodenticide.

PGN400 CAS:60510-57-8 ***HR: D***
5-PHENYL-3-(o-TOLYL)-s-TRIAZOLE
mf: $C_{15}H_{13}N_3$ mw: 235.31

TOXICITY DATA with REFERENCE
orl-rat TDLo:250 mg/kg (6-10D preg):REP JMCMAR 26,1187,83

SAFETY PROFILE: Experimental reproductive effects. When heated to decomposition it emits toxic fumes of NO_x.

PGN825 ***HR: D***
2-PHENYL-5H-1,2,4-TRIAZOLO(5,1-a)ISOINDOLE
mf: $C_{15}H_{11}N_3$ mw: 233.29

TOXICITY DATA with REFERENCE
scu-ham TDLo:17500 μg/kg (4-8D preg):REP ARZNAD 33,1222,83

SAFETY PROFILE: Experimental reproductive effects. When heated to decomposition it emits toxic fumes of NO_x.

PGN840 CAS:35257-18-2 ***HR: D***
2-PHENYL-s-TRIAZOLO(5,1-a)ISOQUINOLINE
mf: $C_{16}H_{11}N_3$ mw: 245.30

SYNS: L 11373 ◇ 2-PHENYL-(1,2,4)TRIAZOLO(5,1-a)ISOQUINOLINE (9CI)

TOXICITY DATA with REFERENCE
orl-ham TDLo:200 mg/kg (female 4-8D post):REP JOPHDQ 5,55,82

SAFETY PROFILE: Experimental reproductive effects. When heated to decomposition it emits toxic fumes of NO_x.

PGO000 CAS:780-69-8 ***HR: 2***
PHENYLTRIETHOXYSILANE
mf: $C_{12}H_{20}O_3Si$ mw: 240.41

SYN: TRIETHOXYPHENYLSILANE

TOXICITY DATA with REFERENCE
orl-rat LD50:2830 mg/kg TXAPA9 28,313,74
skn-rbt LD50:3180 mg/kg TXAPA9 28,313,74

CONSENSUS REPORTS: Reported in EPA TSCA Inventory.

SAFETY PROFILE: Moderately toxic by ingestion and skin contact. When heated to decomposition it emits acrid smoke and irritating fumes. See also SILANE.

PGO500 CAS:368-47-8 ***HR: 3***
PHENYLTRIFLUOROSILANE
mf: $C_6H_5F_3Si$ mw: 162.20

SYNS: FENYL-TRIFLUORSILAN ◇ SILANE, PHENYLTRIFLUORO-

TOXICITY DATA with REFERENCE
skn-rbt 10 mg/24H open SEV AIHAAP 23,95,62
skn-rbt 20 mg/24H MOD 85JCAE-,1228,86
eye-rbt 50 μg/24H SEV 85JCAE-,1228,86
orl-rat LD50:310 mg/kg AIHAAP 23,95,62
ihl-rat LCLo:1000 ppm/4H AIHAAP 23,95,62
skn-rbt LD50:640 mg/kg AIHAAP 23,95,62

SAFETY PROFILE: Poison by ingestion. Moderately toxic by skin contact. A severe skin and eye irritant. When heated to decomposition it emits toxic fumes of Si and F^-.

PGO750 CAS:2996-92-1 ***HR: 3***
PHENYLTRIMETHOXYSILANE
mf: $C_9H_{14}O_3Si$ mw: 198.32

TOXICITY DATA with REFERENCE
ivn-mus LD50:180 mg/kg CSLNX* NX#04066

CONSENSUS REPORTS: Reported in EPA TSCA Inventory.

SAFETY PROFILE: Poison by intravenous route. When heated to decomposition it emits acrid smoke and irritating fumes. See also SILANE.

PGP250 CAS:64-10-8 ***HR: 2***
1-PHENYLUREA
mf: $C_7H_8N_2O$ mw: 136.17

PROP: Monoclinic prisms. D: 1.302, mp: 147° (decomp), bp: 238°. Sol in hot water, alc; sltly sol in ether.

SYNS: MONOPHENYLUREA ◇ PHENYLCARBAMIDE

TOXICITY DATA with REFERENCE
orl-rat LD50:2000 mg/kg IIFBA4 12,195,69
orl-mus LD50:1580 mg/kg IIFBA4 12,195,69
ipr-mus LD50:1060 mg/kg IIFBA4 12,195,69

CONSENSUS REPORTS: Reported in EPA TSCA Inventory.

SAFETY PROFILE: Moderately toxic by ingestion and intraperitoneal routes. When heated to decomposition it emits toxic fumes of NO_x.

PGP500 CAS:132-45-6 ***HR: 3***
2-PHENYLVALERIC ACID-2-(DIETHYLAMINO)ETHYL ESTER HYDROCHLORIDE
mf: $C_{17}H_{27}NO_2{\cdot}ClH$ mw: 313.91

SYNS: 2-DIETHYLAMINOETHYL-α-PHENYLVALERATE HYDROCHLORIDE ◇ 2-DIETHYLAMINOETHYL-α-PROPYLTOLUATE HYDROCHLORIDE ◇ α-PHENYL-VALERATE du DIETHYLAMINO-ETHANOL CHLORHYDRATE (FRENCH) ◇ PROPIVANE ◇ PROSPASMIN ◇ PROSPASMINE ◇ PROSPASMINE HYDROCHLORIDE ◇ TROPISTON

TOXICITY DATA with REFERENCE
orl-mus LD50:600 mg/kg AIPTAK 59,149,38
ipr-mus LD50:220 mg/kg JPETAB 74,274,42
ivn-mus LD50:50 mg/kg AIPTAK 59,149,38
idr-mus LD50:1500 mg/kg AIPTAK 59,149,38
orl-rbt LD50:1500 mg/kg AIPTAK 59,149,38
ivn-rbt LD50:20 mg/kg AIPTAK 59,149,38
idr-rbt LD50:1300 mg/kg AIPTAK 59,149,38

SAFETY PROFILE: Poison by intraperitoneal and intravenous routes. Moderately toxic by ingestion and intradermal routes. When heated to decomposition it emits very toxic fumes of NO_x and HCl. See also ESTERS.

PGP600 CAS:28597-01-5 ***HR: 3***
PHENYLVANADIUM(V) DICHLORIDE OXIDE
mf: $C_6H_5Cl_2OV$ mw: 214.95

SAFETY PROFILE: May explode when heated. When heated to decomposition it emits toxic fumes of Cl^- and VO_x. See also VANADIUM COMPOUNDS and CHLORIDES.

PGP750 CAS:1322-78-7 ***HR: 1***
PHENYL XYLYL KETONE
mf: $C_{15}H_{14}O$ mw: 210.29

SYN: AR,AR-DIMETHYLBENZOPHENONE

TOXICITY DATA with REFERENCE
skn-rbt 500 mg/24H MLD 85JCAE-,294,86
eye-rbt 500 mg/24H MLD 85JCAE-,294,86
orl-rat LD50:4920 mg/kg AMIHBC 4,119,51
skn-rbt LD50:20 g/kg AMIHBC 4,119,51

SAFETY PROFILE: Mildly toxic by ingestion and skin contact. An eye irritant. Mild skin and eye irritant. When heated to decomposition it emits acrid smoke and irritating fumes. See also KETONES.

PGQ000 CAS:20921-50-0 ***HR: 3***
1-PHENYL-1-(3,4-XYLYL)-2-PROPYNYL N-CYCLOHEXYLCARBAMATE
mf: $C_{24}H_{27}NO_2$ mw: 361.52

SYN: N-CYCLOHEXYLCARBAMIC ACID 1-PHENYL-1-(3,4-XYLYL)-2-PROPYNYL ESTER

TOXICITY DATA with REFERENCE
mmo-sat 100 μg/plate PNASA6 72,979,75
dns-hmn:fbr 10 mmol/L/90M IJCNAW 16,284,75
orl-rat TDLo:20400 mg/kg/45W-C:ETA JJIND8 71,211,83

CONSENSUS REPORTS: EPA Genetic Toxicology Program.

SAFETY PROFILE: Questionable carcinogen with experimental tumorigenic data. Human mutation data reported. When heated to decomposition it emits toxic fumes of NO_x. See also ESTERS and CARBAMATES.

PGQ275 CAS:80611-44-5 ***HR: 3***
PHIC
mf: $C_{12}H_{18}N_2O_7Pt$ mw: 497.41

SYNS: (Z)-(CYCLOHEXANE-1,2-DIAMMINE)ISOCITRATO-PLATINUM(II) ◇ cis-ISOCITRATO-(1,2-DIAMINO-CYCLOHEXAN)-

PLATIN(II) (GERMAN) ◇ PLATINUM(II) (CYCLOHEXANE-1,2-DIAMMINE)ISOCITRATO-, (Z)-

TOXICITY DATA with REFERENCE
ipr-rat LD50:143 mg/kg GEXXA8 #158777
ivn-rat LD50:173 mg/kg GEXXA8 #158777
ipr-mus LD50:236 mg/kg GEXXA8 #158777
ivn-mus LD50:421 mg/kg GEXXA8 #158777

SAFETY PROFILE: Poison by intravenous and intraperitoneal routes. When heated to decomposition it emits toxic fumes of NO_x.

PGQ285 ***HR: 2***
PHILODENDRON (various species)

PROP: Climbing vines with aerial routes. The heart-shaped leaves may have smooth edges or be deeply notched, and may also be variegated. An extremely popular house plant in the United States. It grows wild and is cultivated outdoors in warm climates.

SYNS: BEJUCO de LOMBRIZ (CUBA) ◇ PAISAJE (PUERTO RICO)

SAFETY PROFILE: The leaves contain calcium oxalate crystals. Chewing these plant parts results in burning pain in the lips, mouth and throat, possibly followed by inflammation and blistering. Systemic effects are usually not seen because of the insolubility of calcium oxalate. See also OXALATES.

PGQ350 CAS:101997-27-7 ***HR: 2***
PHLEOCIDIN

PROP: An antibiotic produced by *S. kawachiensis* (85FZAT -,505,67).

TOXICITY DATA with REFERENCE
dnr-esc 20 μL/disc MUREAV 97,1,82
dnr-bcs 20 μL/disc MUREAV 97,1,82
ivn-mus LD50:800 mg/kg 85FZAT -,505,67

SAFETY PROFILE: Moderately toxic by intravenous route. Mutation data reported.

PGQ500 CAS:11006-33-0 ***HR: 3***
PHLEOMYCIN
mf: $C_{12}H_{28}N_3O_{12}$ mw: 418.44

SYN: NSC 616586

TOXICITY DATA with REFERENCE
cyt-mus-scu 2500 μg/kg MUREAV 6,289,68
cyt-mus:ovr 1 mg/L MUREAV 6,289,68
dnd-esc 500 μg/L/30M MUREAV 42,181,77
dnr-bcs 1 μg/disc AEMIDF 43,177,82
sln-dmg-orl 50 mg/L ENMUDM 5,633,83
cyt-hmn:leu 100 μg/L/24H MUREAV 7,251,69
orl-rat LD50:649 mg/kg TXAPA9 14,590,69
ivn-rat LD50:4 mg/kg TXAPA9 14,590,69
orl-mus LD50:371 mg/kg TXAPA9 14,590,69
ivn-mus LD50:21 mg/kg TXAPA9 14,590,69

CONSENSUS REPORTS: EPA Genetic Toxicology Program.

SAFETY PROFILE: Poison by ingestion and intravenous routes. Human mutation data reported. When heated to decomposition it emits toxic fumes of NO_x.

PGQ750 CAS:11006-33-0 ***HR: 3***
PHLEOMYCIN COMPLEX

TOXICITY DATA with REFERENCE
ivn-rat LDLo:25 mg/kg JAJAAA 19,260,66
orl-mus LDLo:500 mg/kg JAJAAA 19,260,66
ipr-mus LD50:13 mg/kg JAJAAA 19,260,66
scu-mus LD50:25 mg/kg JAJAAA 19,260,66
ivn-mus LD50:13 mg/kg 85ERAY 2,1394,78
ivn-gpg LDLo:13 mg/kg JAJAAA 19,260,66

SAFETY PROFILE: Poison by intravenous, intraperitoneal, and subcutaneous routes. Moderately toxic by ingestion. Used as an antibiotic.

PGR000 CAS:108-73-6 ***HR: 2***
PHLOROGLUCINOL
mf: $C_6H_6O_3$ mw: 126.12

PROP: White crystals; sweet taste. Mp: 218°, subl with decomp. Sol in ether; sltly water sol.

SYNS: BENZENE-s-TRIOL ◇ BENZENE-1,3,5-TRIOL ◇ 1,3,5-BENZENETRIOL ◇ 3,5-DIHYDROXYPHENOL ◇ DILOSPAN S ◇ 5-OXYRESORCINOL ◇ PHLOROGLUCIN ◇ s-TRIHYDROXYBENZENE ◇ sym-TRIHYDROXYBENZENE ◇ 1,3,5-TRIHYDROXYBENZENE ◇ 1,3,5-TRIHYDROXYCYCLOHEXATRIENE

TOXICITY DATA with REFERENCE
mrc-smc 3 g/L MUREAV 135,109,84
cyt-ham:ovr 3 g/L CALEDQ 14,251,81
scu-rat TDLo:5 mg/kg (female 1D pre):REP ENDOAO 57,466,55
orl-rat LD50:5200 mg/kg NIRRDN 6,740,82
ipr-rat LD50:3180 mg/kg NIIRDN 6,740,82
scu-rat LD50:4850 mg/kg NIIRDN 6,740,82
orl-mus LD50:4550 mg/kg OYYAA2 3,187,69
ipr-mus LD50:4050 mg/kg OYYAA2 3,187,69
scu-gpg LDLo:1000 mg/kg RMSRA6 15,561,1895

CONSENSUS REPORTS: Reported in EPA TSCA Inventory. EPA Genetic Toxicology Program.

SAFETY PROFILE: Moderately toxic by subcutaneous and intraperitoneal routes. Mildly toxic by ingestion. Experimental reproductive effects. Mutation data reported. When heated to decomposition it emits acrid smoke and irritating fumes. Used in diazo-type printing and textile dyeing, in microscopy as a bone specimen decalcifier.

PGR250 CAS:90-00-6 ***HR: 3***
PHLOROL
mf: $C_8H_{10}O$ mw: 122.18

PROP: Colorless liquid; phenol odor. Mp: −28° d: 1.037 @ 12°, bp: 204.52°, turns solid <18°. Insol in water; misc in alc, benzene, glacial acetic acid, ether.

SYNS: o-ETHYLPHENOL ◇ 2-ETHYLPHENOL

TOXICITY DATA with REFERENCE
skn-mus TDLo:3100 mg/kg/12W-I:NEO CNREA8 19,413,59
orl-mus LD50:600 mg/kg PHARAT 30,147,75
ipr-mus LD50:172 mg/kg JMCMAR 18,868,75

CONSENSUS REPORTS: Reported in EPA TSCA Inventory.

SAFETY PROFILE: Poison by intraperitoneal route. Moderately toxic by ingestion. Human toxic action similar to, but less severe than, that of phenol. Questionable carcinogen with experimental neoplastigenic data. When heated to decomposition it emits acrid smoke and irritating fumes. See also PHENOL.

PGR750 CAS:3520-42-1 ***HR: 3***
PHLOXINRHODAMINE

PROP: Fluorone conjugated with a rhodamine (TXAPA 9 44,225,78).

TOXICITY DATA with REFERENCE
ivn-mus LD50:310 mg/kg TXAPA9 44,225,78

CONSENSUS REPORTS: Reported in EPA TSCA Inventory.

SAFETY PROFILE: Poison by intravenous route.

PGR775 CAS:64925-80-0 ***HR: 3***
PHOMOPSIN

TOXICITY DATA with REFERENCE
dni-hmn:mmr 1 μmol/L JTEHD6 16,13,85
ipr-rat TDLo:400 μg/kg (6D preg):TER AJEBAK 61,105,83
ipr-rat TDLo:150 μg/kg (6-10D preg):REP AJEBAK 61,105,83
unr-rat LD50:5 mg/kg AJEBAK 61,105,83

SAFETY PROFILE: Poison by an unreported route. An experimental teratogen. Experimental reproductive effects. Human mutation data reported.

PGS000 CAS:298-02-2 ***HR: 3***
PHORATE
mf: $C_7H_{17}O_2PS_3$ mw: 260.39

PROP: Liquid. Bp: 118-120° @ 0.8 mm, d: 1.156. @ 25°/4°. Insol in water; misc with carbon tetrachloride, dioxane, xylene.

SYNS: O,O-DIAETHYL-S-(AETHYLTHIO-METHYL)-DITHIOPHOSPHAT (GERMAN) ◇ O,O-DIETHYL-S-ETHYLMERCAPTOMETHYL DITHIOPHOSPHONATE ◇ O,O-DIETHYL-S-ETHYLTHIOMETHYL DITHIOPHOSPHONATE ◇ O,O-DIETHYL-ETHYLTHIOMETHYL PHOSPHORODITHIOATE ◇ O,O-DIETHYL-S-ETHYLTHIOMETHYL THIOTHIONOPHOSPHATE ◇ O,O-DIETIL-S-(ETILTIO-METIL)-DITIOFOSFATO (ITALIAN) ◇ DITHIOPHOSPHATE de O,O-DIETHYLE et d'ETHYLTHIOMETHYLE (FRENCH) ◇ ENT 24,042 ◇ FORAAT (DUTCH) ◇ GRANUTOX ◇ PHORAT (GERMAN) ◇ PHORATE-10G ◇ RAMPART ◇ RCRA WASTE NUMBER P094 ◇ THIMET ◇ TIMET ◇ VEGFRU ◇ VERGFRU FORATOX

TOXICITY DATA with REFERENCE
mnt-mus-ipr 750 μg/kg MUREAV 155,131,85
ipr-grb TDLo:2500 μg/kg (male 1D pre):REP IJEBA6 18,1001,80
orl-rat LD50:1 mg/kg DOEAAH 35,25,79
ihl-rat LC50:11 mg/m^3/1H NTIS** PB277-077
skn-rat LD50:2500 μg/kg TXAPA9 2,88,60
orl-mus LD50:6590 μg/kg JPFCD2 B15,867,80
skn-rbt LD50:99 mg/kg FMCHA2 -,C226,89
skn-gpg LD50:20 mg/kg PCOC** -,895,66
orl-qal LD50:7 mg/kg EESADV 8,551,84
skn-dck LD50:203 mg/kg TXAPA9 47,451,79

CONSENSUS REPORTS: EPA Extremely Hazardous Substances List. EPA Genetic Toxicology Program.

OSHA PEL: TWA 0.05 mg/m^3 (skin)
ACGIH TLV: TWA 0.05 mg/m^3 (skin)

SAFETY PROFILE: Poison by ingestion and skin contact routes. Experimental reproductive effects. Mutation data reported. A cholinesterase inhibitor. When heated to decomposition it emits toxic fumes of PO_x and SO_x. See also PARATHION.

PGS250 CAS:17673-25-5 ***HR: 3***
PHORBOL
mf: $C_{20}H_{27}O_6$ mw: 363.47

PROP: Anhydrous crystals. Two forms: mp: 162-163° and 233-234°. decomp @ 250-251°.

TOXICITY DATA with REFERENCE
skn-mus 36 mg MLD PLMEAA 22,241,72
scu-mus TDLo:64 mg/kg (2D post):REP PSEBAA 165,394,80
ipr-mus TDLo:400 mg/kg/25W:CAR CNREA8 30,2744,70
ipr-mus TD:284 mg/kg/39W-I:ETA JCROD7 95,19,79

CONSENSUS REPORTS: EPA Genetic Toxicology Program.

SAFETY PROFILE: Experimental reproductive effects. A skin irritant. Questionable carcinogen with experimental carcinogenic and tumorigenic data. When heated to decomposition it emits acrid smoke and irritating fumes. See also other phorbol entries.

PGS500 CAS:20839-16-1 ***HR: 2***
PHORBOL ACETATE, LAURATE
mf: $C_{34}H_{52}O_8$ mw: 588.86

SYNS: 5H-CYCLOPROPA(3,4)BENZ(1,2-e)AZULEN-5-ONE, 1,1a,1b,4,4a,7a,7b,8,9,9a-DECAHYDRO-4a,7β,9,9a-TETRAHYDROXY-3-(HYDROXYMETHYL)-1,1,6,8-TETRAMETHYL-, 9-ACETATE 9a-LAURATE ◇ PHORBOL MONOACETATE MONOLAURATE

TOXICITY DATA with REFERENCE
skn-mus TDLo:20 mg/kg/25W-I:NEO NATWAY 54,282,67

SAFETY PROFILE: Questionable carcinogen with experimental neoplastigenic data. When heated to decomposition it emits acrid smoke and irritating fumes.

PGS750 CAS:24928-15-2 ***HR: 3***
PHORBOL-12,13-DIACETATE
mf: $C_{24}H_{32}O_8$ mw: 448.56

TOXICITY DATA with REFERENCE
dnd-esc 200 μg/tube CNREA8 33,3103,73
skn-mus TDLo:19 mg/kg/32W-I:ETA CNREA8 31,1074,71

SAFETY PROFILE: Questionable carcinogen with experimental tumorigenic data. Mutation data reported. When heated to decomposition it emits acrid smoke and irritating fumes. See also other phorbol entries.

PGT000 CAS:25405-85-0 ***HR: 3***
PHORBOL-12,13-DIBENZOATE
mf: $C_{34}H_{36}O_8$ mw: 572.70

TOXICITY DATA with REFERENCE
skn-mus TDLo:7760 μg/kg/10W-I:ETA CNREA8 31,1074,71

SAFETY PROFILE: Questionable carcinogen with experimental tumorigenic data. When heated to decomposition it emits acrid smoke and irritating fumes. See also other phorbol entries.

PGT250 CAS:24928-17-4 ***HR: 2***
PHORBOL-12,13-DIDECANOATE
mf: $C_{40}H_{55}O_8$ mw: 663.95

SYN: PDD

TOXICITY DATA with REFERENCE
skn-mus 7 ng MLD CCSUDL 2,11,78
dnd-esc 200 μg/tube CNREA8 33,3103,73
cyt-smc 1 mg/L NATUAS 294,263,81
dns-rat:lvr 1 μg/L CNREA8 40,4541,80
otr-ham:emb 100 μg/L CALEDQ 17,1,82
skn-mus TDLo:5472 μg/kg/6W-I:ETA CNREA8 31,1074,71

SAFETY PROFILE: A skin irritant. Questionable carcinogen with experimental tumorigenic data. Mutation data reported. When heated to decomposition it emits acrid smoke and irritating fumes. See also other phorbol entries.

PGT500 ***HR: 3***
PHORBOL-12,13-DIHEXA(Δ-2,4)-DIENOATE
mf: $C_{32}H_{57}O_8$ mw: 569.89

TOXICITY DATA with REFERENCE
skn-mus 6 μg MLD PLMEAA 22,241,72
skn-mus TDLo:55 g/kg/12W-I:ETA PLMEAA 22,241,72

SAFETY PROFILE: Questionable carcinogen with experimental tumorigenic data. A skin irritant. When heated to decomposition it emits acrid smoke and irritating fumes. See also other phorbol entries.

PGT750 CAS:37558-17-1 ***HR: 3***
PHORBOL-12,13-DIHEXANOATE
mf: $C_{32}H_{45}O_8$ mw: 557.77

TOXICITY DATA with REFERENCE
skn-mus 18 μg MLD PLMEAA 22,241,72
skn-mus TDLo:16 g/kg/12W-I:ETA PLMEAA 22,241,72

SAFETY PROFILE: Questionable carcinogen with experimental tumorigenic data. A skin irritant. When heated to decomposition it emits acrid smoke and irritating fumes. See also other phorbol entries.

PGU000 CAS:63040-44-8 ***HR: 3***
PHORBOL LAURATE, (+)-S-2-METHYLBUTYRATE
mf: $C_{37}H_{58}O_8$ mw: 630.95

SYN: PHORBOL MONOLAURATE MONO(S)-(+)-2-METHYLBUTYRATE

TOXICITY DATA with REFERENCE
skn-mus TDLo:24 mg/kg/30W-I:NEO NATWAY 54,282,67

SAFETY PROFILE: Questionable carcinogen with experimental neoplastigenic data. When heated to decomposition it emits acrid smoke and irritating fumes. See also other phorbol entries.

PGU250 CAS:16675-05-1 ***HR: 3***
PHORBOL MONOACETATE MONOLAURATE
mf: $C_{34}H_{52}O_8$ mw: 588.86

SYN: PHORBOL ACETATE LAURATE

TOXICITY DATA with REFERENCE
skn-mus TDLo:20 mg/kg/25W-I:NEO NATWAY 54,282,67

SAFETY PROFILE: Questionable carcinogen with ex-

perimental neoplastigenic data. When heated to decomposition it emits acrid smoke and irritating fumes. See also other phorbol entries.

PGU500 CAS:63040-43-7 ***HR: 3***
PHORBOL MONODECANOATE (S)-(+)-MONO(2-METHYLBUTYRATE)
mf: $C_{35}H_{54}O_8$ mw: 602.89

SYN: PHORBOL CAPRATE, (+)-(S)-2-METHYLBUTYRATE

TOXICITY DATA with REFERENCE
skn-mus TDLo:24 mg/kg/30W-I:NEO NATWAY 54,282,67

SAFETY PROFILE: Questionable carcinogen with experimental neoplastigenic data. When heated to decomposition it emits acrid smoke and irritating fumes. See also other phorbol entries.

PGU750 CAS:59086-92-9 ***HR: 3***
(E)-PHORBOL MONODECANOATE MONO(2-METHYLCROTONATE)
mf: $C_{35}H_{52}O_8$ mw: 600.87

SYN: PHORBOL CAPRATE, TIGLATE

TOXICITY DATA with REFERENCE
skn-mus TDLo:23 mg/kg/29W-I:NEO NATWAY 54,282,67

SAFETY PROFILE: Questionable carcinogen with experimental neoplastigenic data. When heated to decomposition it emits acrid smoke and irritating fumes. See also other phorbol entries.

PGV000 CAS:16561-29-8 ***HR: 3***
PHORBOL MYRISTATE ACETATE
mf: $C_{36}H_{56}O_8$ mw: 616.92

SYNS: PENTAHYDROXY-TIGLIADIENONE-MONOACETATE (C)MONOMYRISTATE(B) ◇ PHORBOL ACETATE, MYRISTATE ◇ PHORBOL MONOACETATE MONOMYRISTATE ◇ PMA ◇ 12-TETRADECANOYLPHORBOL-13-ACETATE ◇ 12-o-TETRADEKANOYLPHORBOL-13-ACETAT (GERMAN) ◇ TPA

TOXICITY DATA with REFERENCE
skn-mus 7 ng OPEN ARTODN 44,279,80
mmo-esc 1 mg/L JPPMAB 31,69P,79
dnd-hmn:fbr 1300 nmol/L CRNGDP 6,1667,85
dns-hmn:oth 10 μg/L CNREA8 44,4078,84
scu-mus TDLo:640 μg/kg (2D post):REP PSEBAA 265,394,80
ipr-mus TDLo:30 μg/kg (11-13D preg):TER RCOCB8 49,17,85
orl-mus TDLo:10 mg/kg:ETA CNREA8 39,1293,79
skn-mus TDLo:30204 μg/kg/72W-I:CAR ACPADQ 91,103,83
skn-mus TD:1234 μg/kg/25W-I:NEO VAAZA2 30,33,79
ivn-mus LD50:309 μg/kg AGACBH 16,535,85

CONSENSUS REPORTS: EPA Genetic Toxicology Program.

SAFETY PROFILE: Deadly poison by intravenous route. An experimental teratogen. Experimental reproductive effects. Human mutation data reported. A skin irritant. Questionable carcinogen with experimental carcinogenic, neoplastigenic, and tumorigenic data. When heated to decomposition it emits acrid smoke and irritating fumes.

PGV250 ***HR: 3***
PHORBOL-9-MYRISTATE-9a-ACETATE-3-ALDEHYDE
mf: $C_{36}H_{54}O_7$ mw: 598.90

SYN: PAMA

TOXICITY DATA with REFERENCE
skn-mus TDLo:32 mg/kg/27W-I:ETA CNREA8 39,2644,79

SAFETY PROFILE: Questionable carcinogen with experimental tumorigenic data. When heated to decomposition it emits acrid smoke and irritating fumes. See also other phorbol entries and ALDEHYDES.

PGV500 CAS:56937-68-9 ***HR: 1***
PHORBOLOL MYRISTATE ACETATE
mf: $C_{36}H_{58}O_8$ mw: 618.94

SYNS: PHORBOLOL ACETATE MYRISTATE ◇ TPA-3-β-OL

TOXICITY DATA with REFERENCE
skn-mus 29 ng MLD CCSUDL 2,11,78

SAFETY PROFILE: A skin irritant. When heated to decomposition it emits acrid smoke and irritating fumes. See also other phorbol entries.

PGV750 CAS:37415-55-7 ***HR: 3***
PHORBOL-12-o-TIGLYL-13-BUTYRATE
mf: $C_{28}H_{40}O_8$ mw: 504.68

SYN: 12-o-TIGLYL-PHORBOL-13-BUTYRATE

TOXICITY DATA with REFERENCE
skn-mus 520 ng MLD 85CVA2 5,213,70
skn-mus TDLo:29 mg/kg/12W-I:ETA 85CVA2 5,213,70

SAFETY PROFILE: A skin irritant. Questionable carcinogen with experimental tumorigenic data. When heated to decomposition it emits acrid smoke and irritating fumes. See also other phorbol entries.

PGW000 CAS:37394-32-4 ***HR: 3***
PHORBOL-12-o-TIGLYL-13-DODECANOATE
mf: $C_{37}H_{56}O_8$ mw: 628.93

SYN: 12-o-TIGLYL-PHORBOL-13-DODECANOATE

TOXICITY DATA with REFERENCE
skn-mus 1 ng MLD 85CVA2 5,213,70
skn-mus TDLo:12 mg/kg/12W-I:ETA 85CVA2 5,213,70

SAFETY PROFILE: Questionable carcinogen with experimental tumorigenic data. A skin irritant. When heated to decomposition it emits acrid smoke and irritating fumes. See also other phorbol entries.

PGW250 CAS:504-20-1 ***HR: 2***
PHORONE
mf: $C_9H_{14}O$ mw: 138.23

PROP: Solid or greenish liquid. Mp: 28°, flash p: 185°F (OC), d: 0.879, vap press: 1 mm @ 42.0°, vap d: 4.8, bp: 198-199°. Sol in water, alc, and ether.

SYNS: DIISOPROPYLIDENE ACETONE ◇ sym-DIISOPROPYLIDENE ACETONE ◇ 2,6-DIMETHYL-2,5-HEPTADIEN-4-ONE ◇ PHORON (GERMAN)

TOXICITY DATA with REFERENCE
scu-rbt LDLo:700 mg/kg AEXPBL 56,346,1906

CONSENSUS REPORTS: Reported in EPA TSCA Inventory.

SAFETY PROFILE: Moderately toxic by subcutaneous route. Combustible when exposed to heat or flame; can react with oxidizing materials. To fight fire, use foam, CO_2, dry chemical. When heated to decomposition it emits acrid smoke and irritating fumes. See also ISOACETOPHORONE.

PGW500 ***HR: 3***
PHOSAN-PLUS

PROP: A mixture of methoxychlor, dimethoate, and malathion (MUREAV 46,204,77).

TOXICITY DATA with REFERENCE
cyt-mus-ipr 30 mg/kg MUREAV 46,204,77

SAFETY PROFILE: A poison. Mutation data reported. A cholinesterase inhibitor. Some of the components are carcinogens. When heated to decomposition it emits toxic fumes of Cl^-. See also O,O-DIMETHYL METHYLCARBAMOYLMETHYL PHOSPHORODITHIOATE (dimethoate), CARBETHOXY MALATHION, and DIMETHOXY DDT (methoxychlor).

PGW600 CAS:2398-95-0 ***HR: 3***
PHOSCOLIC ACID
mf: $C_6H_{11}O_8P$ mw: 242.12

SYNS: FOSCOLIC ACID ◇ 2,2'-PHOSPHINICOBIS(2-HYDROXY-PROPANOIC) ACID (9CI) ◇ 2,2'-PHOSPHINICODILACTIC ACID

TOXICITY DATA with REFERENCE
orl-mus LD50:3900 mg/kg JPETAB 145,386,64
ipr-mus LD50:260 mg/kg JPETAB 145,386,64
ivn-mus LD50:152 mg/kg JPETAB 145,386,64

SAFETY PROFILE: Poison by intravenous and intraperitoneal routes. Moderately toxic by ingestion. When heated to decomposition it emits toxic fumes of PO_x.

PGW750 CAS:947-02-4 ***HR: 3***
PHOSFOLAN
mf: $C_7H_{14}NO_3PS_2$ mw: 255.31

SYNS: AC 47031 ◇ AMERICAN CYANAMID 47031 ◇ C.I. 47031 ◇ CYCLICETHYLENE(DIETHOXYPHOSPHINOTHIOYL)DITHIOIMIDOCARBONATE ◇ CYCLIC ETHYLENE P,P-DIETHYL PHOSPHONODITHIOIMIDOCARBONATE ◇ CYLAN ◇ CYOLANE ◇ CYOLANE INSECTICIDE ◇ (DIETHOXYPHOSPHINYL)DITHIOIMIDOCARBONIC ACID CYCLIC ETHYLENE ESTER ◇ 2-(DIETHOXYPHOSPHINYLIMINO)-1,3-DITHIOLANE ◇ P,P-DIETHYL CYCLIC ETHYLENE ESTER OF PHOSPHONODITHIOIMIDOCARBONIC ACID ◇ EI 47031 ◇ ENT 25,830 ◇ 1,2-ETHANEDITHIOL, CYCLIC ESTER with P,P-DIETHYL PHOSPHONODITHIOIMIDOCARBONATE ◇ 1,2-ETHANEDITHIOL, CYCLIC S,S-ESTER with PHOSPHONODITHIOIMIDOCARBONIC ACID P,P-DIETHYL ESTER

TOXICITY DATA with REFERENCE
orl-rat LD50:9 mg/kg ARSIM* 20,1,66
orl-mus LD50:12 mg/kg 28ZEAL 4,127,69
skn-rbt LD50:23 mg/kg GUCHAZ 6,200,73
skn-gpg LD50:54 mg/kg 31ZOAD 1,157,68
orl-pgn LD50:2370 μg/kg ASTTA8 (680),157,79
orl-qal LD50:23700 μg/kg ASTTA8 (680),157,79
orl-bwd LD50:2370 μg/kg ASTTA8 (680),157,79

CONSENSUS REPORTS: EPA Extremely Hazardous Substances List. Reported in EPA TSCA Inventory.

SAFETY PROFILE: Poison by ingestion and skin contact. An insecticide used against leaf-feeding larvae of cotton insect pests. When heated to decomposition it emits very toxic fumes of PO_x, SO_x, and NO_x. See also ESTERS.

PGX000 CAS:75-44-5 ***HR: 3***
PHOSGENE
DOT: UN 1076
mf: CCl_2O mw: 98.91

PROP: Colorless, poison gas or volatile liquid; odor of new mown hay or green corn. Mp: −118°, bp: 8.3°, d: 1.37 @ 20°, vap press: 1180 mm @ 20°, vap d: 3.4. Very sltly sol in water; very sol in benzene and acetic acid; decomp sltly in water.

SYNS: CARBONE (OXYCHLORURE de) (FRENCH) ◇ CARBONIO (OSSICLORURO di) (ITALIAN) ◇ CARBON OXYCHLORIDE ◇ CARBONYLCHLORID (GERMAN) ◇ CARBONYL CHLORIDE ◇ CHLOROFORMYL CHLORIDE ◇ DIPHOSGENE ◇ FOSGEEN (DUTCH) ◇ FOSGEN (POLISH) ◇ FOSGENE (ITALIAN) ◇ KOOLSTOFOXYCHLORIDE (DUTCH) ◇ NCI-C60219 ◇ PHOSGEN (GERMAN) ◇ RCRA WASTE NUMBER P095

TOXICITY DATA with REFERENCE
ihl-hmn LC50:3200 mg/m^3 SCJUAD 4,33,67
ihl-hmn TCLo:25 ppm/30M:PUL 29ZWAE -,207,68
ihl-man LCLo:360 mg/m^3/30M 85GMAT -,99,82
ihl-mam LCLo:50 ppm/5M AEPPAE 138,65,28
ihl-cat LCLo:190 mg/m^3/15M 85GMAT -,99,82

CONSENSUS REPORTS: EPA Extremely Hazardous Substances List. Community Right-To-Know List. Reported in EPA TSCA Inventory.

OSHA PEL: TWA 0.1 ppm
ACGIH TLV: TWA 0.1 ppm
DFG MAK: 0.1 ppm (0.4 mg/m^3)
NIOSH REL: (Phosgene) TWA 0.1 ppm; CL 0.2 ppm/15M
DOT Classification: Poison A; Label: Poison Gas.

SAFETY PROFILE: A human poison by inhalation. A severe eye, skin, and mucous membrane irritant. In the presence of moisture, phosgene decomposes to form hydrochloric acid and carbon monoxide. This occurs in the bronchioles and alveoli of the lungs resulting in pulmonary edema followed by bronchopneumonia and occasionally lung abscess. There is little immediate irritating effect upon the respiratory tract, and the warning properties of the gas are therefore very slight. There may be no immediate warning that dangerous concentrations are being inhaled. After a latent period of 2 to 24 hours, the patient complains of burning in the throat and chest, shortness of breath and increasing dyspnea. Where the exposure has been severe, the development of pulmonary edema may be so rapid that the patient dies within 36 hours after exposure. In cases where the exposure has been less, pneumonia may develop several days after the occurrence of the accident. In patients who recover, no permanent residual disability is thought to occur. A common air contaminant.

Under the appropriate conditions it undergoes hazardous reactions with Al; tert-butyl azido formate; 2,4-hexadiyn-1,6-diol; isopropyl alcohol; K; Na; sodium azide; hexafluoroisopropylideneamino lithium; lithium. When heated to decomposition or on contact with water or steam it will react to produce toxic and corrosive fumes of Cl^-. Caution: Arrangements should be made for monitoring its use.

PGX250 ***HR: 3***
PHOSPHAM
mf: HN_2P mw: 60.0

SAFETY PROFILE: Ignites in air at slightly elevated temperatures. Ignites in nitrogen dioxide vapors. Explodes when heated with chlorates or nitrates. Incandescent reaction with copper(II) oxide or mercury(II) oxide. Reaction with hydrogen sulfide + heat forms a pyrophoric solid. Incompatible with oxidants. When heated to decomposition it emits very toxic fumes of PO_x and NO_x. See also PHOSPHIDES.

PGX300 CAS:297-99-4 ***HR: 3***
trans-PHOSPHAMIDON
mf: $C_{10}H_{19}ClNO_5P$ mw: 299.69

SYNS: ENT 25,515 ◇ ML 97 ◇ OR 1191 ◇ (E)-PHOSPHAMIDON

TOXICITY DATA with REFERENCE
orl-mus LDLo:18 mg/kg AECTCV 14,111,85
orl-pgn LD50:4210 μg/kg ASTTA8 (680),157,79
orl-qal LD50:7500 μg/kg ASTTA8 (680),157,79
orl-bwd LD50:1780 μg/kg ASTTA8 (680),157,79

SAFETY PROFILE: Poison by ingestion. When heated to decomposition it emits toxic fumes of Cl^-, NO_x, and PO_x.

PGX500 ***HR: 2***
PHOSPHATES

SAFETY PROFILE: Alkali metal phosphates are strong caustics and therefore powerful irritants. Superphosphate is $Ca(H_2PO_4)_2/CaSO_4$. Triple superphosphate contains P_2O_5. Both are used as fertilizers. Organophosphates are often highly toxic pesticides. For an example of organic phosphates, see PARATHION. See also individual phosphates.

PGX750 ***HR: 3***
PHOSPHIDES

PROP: A combination of a cation + elemental phosphorus.

SAFETY PROFILE: Phosphides are particularly dangerous because they tend to decompose to the very toxic phosphine upon contact with moisture or acids. Dangerous fire hazard by chemical reaction, particularly with moisture. Moderate explosion hazard. They react with water, steam, acid, or acid fumes to produce toxic and flammable phosphine gas. Can react vigorously with oxidizing materials. Dangerous; when heated to decomposition they may emit highly toxic fumes of PO_x. See also PHOSPHINE.

PGY000 CAS:7803-51-2 ***HR: 3***
PHOSPHINE
DOT: UN 2199
mf: H_3P mw: 34.00

PROP: Colorless gas; foul odor of decaying fish. Mp: −132.5°, bp: −87.5°, d: 1.529 g/L @ 0°, autoign temp: 212°F, lel: 1%. Sltly sol in water.

SYNS: CELPHOS ◇ DELICIA ◇ DETIA GAS EX-B ◇ FOSFOROWODOR (POLISH) ◇ HYDROGEN PHOSPHIDE ◇ PHOS-

PHORUS TRIHYDRIDE ◇ PHOSPHORWASSERSTOFF (GERMAN) ◇ RCRA WASTE NUMBER P096

TOXICITY DATA with REFERENCE
ihl-hmn LCLo:1000 ppm GUCHAZ 6,412,73
ihl-rat LC50:11 ppm/4H AIHAAP 36,452,75
ihl-mus LCLo:380 mg/m^3/2H 85GMAT -,75,82
ihl-cat LCLo:70 mg/m^3/2H 85GMAT -,75,82
ihl-rbt LCLo:2500 ppm/20M AEXPBL 27,314,1890
ihl-gpg LCLo:140 mg/m^3/4H 85GMAT -,75,82
ihl-mam LCLo:1000 ppm/5M AEPPAE 138,65,28

CONSENSUS REPORTS: EPA Extremely Hazardous Substances List. Reported in EPA TSCA Inventory.

OSHA PEL: (Transitional: TWA 0.3 ppm) TWA 0.3 ppm; STEL 1 ppm
ACGIH TLV: TWA 0.3 ppm; STEL 1 ppm
DFG MAK: 0.1 ppm (0.15 mg/m^3)
DOT Classification: Poison A; Label: Flammable Gas and Poison Gas.

SAFETY PROFILE: A poison by inhalation. A very toxic gas whose effects are not completely understood. The chief effects are central nervous system depression and lung irritation. There may be pulmonary edema, dilation of the heart, and hyperemia of the visceral organs. Inhalation can cause coma and convulsions leading to death within 48 hours. However, most cases recover without after-effects. Chronic poisoning, characterized by anemia, bronchitis, gastrointestinal disturbances and visual, speech and motor disturbances, may result from continued exposure to very low concentrations.

Very dangerous fire hazard by spontaneous chemical reaction. Moderately explosive when exposed to flame. Explosive reaction with dichlorine oxide; silver nitrate; concentrated nitric acid; nitrogen trichloride; oxygen. Reacts with mercury(II) nitrate to form an explosive product. Ignition or violent reaction with air; boron trichloride; Br_2; Cl_2; aqueous halogen solutions; iodine; metal nitrates; NO; NCl_3; NO_3; N_2O; HNO_2; K + NH_3; oxidants. The organic derivatives of phosphine (phosphines) react vigorously with halogens. To fight fire, use CO_2, dry chemical, or water spray. Dangerous; when heated to decomposition it emits highly toxic fumes of PO_x. Used as a fumigant, doping agent for electronic components, and in chemical synthesis.

PGY250 CAS:6303-21-5 ***HR: 3***
PHOSPHINIC ACID
mf: H_3O_2P mw: 66.00

SYN: HYPOPHOSPHOROUS ACID

SAFETY PROFILE: Reaction with mercury(II) oxide is explosive. Violent reaction with mercury(II) nitrate. When heated to decomposition it emits toxic fumes of PO_x.

PGY600 CAS:6736-03-4 ***HR: 3***
PHOSPHOLINE
mf: $C_9H_{23}NO_3PS$ mw: 256.36

SYN: (2-MERCAPTOETHYL)TRIMETHYLAMMONIUM S-ESTER with O,O'-DIETHYLPHOSPHOROTHIOATE

TOXICITY DATA with REFERENCE
orl-rat LD50:88 mg/kg ARYPAY 32,507,81
scu-rat LD50:174 μg/kg ARYPAY 32,507,81
ipr-mus LD50:7079 μg/kg PHARAT 35,806,80

SAFETY PROFILE: Poison by ingestion, subcutaneous, and intraperitoneal routes. When heated to decomposition it emits toxic fumes of NH_3, PO_x, SO_x, and NO_x.

PGY750 CAS:51321-79-0 ***HR: 2***
PHOSPHONACETYL-l-ASPARTIC ACID
mf: $C_6H_{10}NO_8P$ mw: 255.14

SYNS: NSC-224131 ◇ PALA ◇ N-PHOSPHONACETYL-l-ASPARTATE

TOXICITY DATA with REFERENCE
dni-mus:leu 1 mmol/L CNREA8 42,4525,82
oms-dog-ivn 20 mg/kg CNREA8 43,2565,83
ipr-mus TDLo:5 mg/kg (8D preg):TER TJADAB 22,311,80
unr-hmn TDLo:862 mg/kg:GIT,MET DRFUD4 6,152,81

SAFETY PROFILE: Human systemic effects by an unspecified route: hypermotility, diarrhea, nausea or vomiting, weight loss or decreased weight gain. An experimental teratogen. Experimental reproductive effects. Mutation data reported. When heated to decomposition it emits very toxic fumes of NO_x and PO_x.

PGZ899 CAS:13598-36-2 ***HR: 2***
PHOSPHONIC ACID
DOT: UN 2834
mf: H_3O_3P mw: 82.00

CONSENSUS REPORTS: Reported in EPA TSCA Inventory.

DOT Classification: Corrosive Material; Label:Corrosive

SAFETY PROFILE: When heated to decomposition at 200°C it emits toxic fumes of PO_x and phosphine which may ignite. See also PHOSPHINE.

PGZ950 CAS:13590-71-1 ***HR: 2***
PHOSPHONIC ACID, MONOMETHYL
mf: CH_5O_3P mw: 96.03

SYNS: METHYLESTER KYSELINY FOSFORITE (CZECH) ◇ MONOMETHYLFOSFIT (CZECH)

TOXICITY DATA with REFERENCE
skn-rbt 500 mg/24H SEV 28ZPAK -,214,72

eye-rbt 50 μg/24H SEV 28ZPAK -,214,72
orl-rat LD50:1740 mg/kg 28ZPAK -,214,72

SAFETY PROFILE: Moderately toxic by ingestion. A severe eye and skin irritant. When heated to decomposition it emits toxic fumes of PO_x. See also ESTERS and PHOSPONIC ACID.

PHA000 CAS:12125-09-6 ***HR: 3***
PHOSPHONIUM IODIDE
mf: H_4IP mw: 161.91

PROP: Tetragonal, colorless, deliq crystals. Mp: subl @ 18.5°, d: 2.86, vap press: 40 mm @ 16.1°, 760 mm @ 62.5°.

SYNS: IODINE PHOSPHIDE ◇ JODPHOSPHONIUM (GERMAN)

TOXICITY DATA with REFERENCE
orl-rbt LDLo:5 mg/kg SAPHAO 15,420,04

SAFETY PROFILE: Poison by ingestion. Rapid heating causes detonation. Explosive reaction with antimony pentachloride. Violent reaction or ignition with bromates; $HClO_3$; chlorates; iodates; HIO_2, HNO_3; $AgNO_3$. Reaction with potassium hydroxide produces phosphine gas, reaction becomes explosive in the presence of air. When heated to decomposition it emits very toxic fumes of I^- and PO_x. See also IODIDES.

PHA250 ***HR: 3***
PHOSPHONIUM PERCHLORATE
mf: ClH_4O_4P mw: 134.46

$(PH_4)\cdot(ClO_4)$

SAFETY PROFILE: Poison by ingestion. A powerful irritant. A very explosive salt sensitive to moisture, heat, friction. Explodes on drying. See also PERCHLORATES.

PHA500 CAS:1071-83-6 ***HR: 3***
N-(PHOSPHONOMETHYL)GLYCINE
mf: $C_3H_8NO_5P$ mw: 169.09

SYNS: GLYPHOSATE ◇ MON 0573

TOXICITY DATA with REFERENCE
orl-mus LD50:1568 mg/kg TXAPA9 45,319,78
ipr-rat LD50:238 mg/kg TXAPA9 45,319,78
ipr-mus LD50:134 mg/kg TXAPA9 45,319,78
orl-rbt LD50:3800 mg/kg PESTC* 7,6,78
orl-rat LD50:470 mg/kg TOLED5 1000(Sp.1),148,80
ipr-rat LD50:235 mg/kg TOLED5 1000(Sp.1),148,80
orl-mus LD50:1581 mg/kg TOLED5 1000(Sp.1),148,80
ipr-mus LD50:130 mg/kg TOLED5 1000(Sp.1),148,80

SAFETY PROFILE: Poison by intraperitoneal route. Moderately toxic by ingestion. Used as an herbicide. When heated to decomposition it emits very toxic fumes of NO_x and PO_x.

PHA550 CAS:23155-02-4 ***HR: 2***
PHOSPHONOMYCIN
mf: $C_3H_7O_4P$ mw: 138.07

PROP: Crystals. Mp: approx. 94°. Sol in water.

SYNS: ANTIBIOTIC 833A ◇ FOSFOCINA ◇ FOSFOMYCIN ◇ FOSFONOMYCIN ◇ MK-955 ◇ PHOSPHOMYCIN

TOXICITY DATA with REFERENCE
pic-esc 10 μg/plate CNREA8 43,2819,83
mmo-omi 10 g/L MILEDM 8,27,78
ipr-mus LD50:4000 mg/kg 85GDA2 6,426,81

SAFETY PROFILE: Moderately toxic by intraperitoneal route. Mutation data reported. When heated to decomposition it emits toxic fumes of PO_x.

PHA575 CAS:6064-83-1 ***HR: 3***
2-PHOSPHONOXYBENZOIC ACID
mf: $C_7H_7O_6P$ mw: 218.11

SYNS: BENZOIC ACID, 2-(PHOSPHONOOXY)- (9CI) ◇ o-CARBOXYPHENYL PHOSPHATE ◇ DISDOLEN ◇ FOSFOSAL ◇ 2-PHOSPHONOXYBENZOESAEURE ◇ PHOSPHONOXYBENZOIC ACID ◇ SALICYLIC ACID, DIHYDROGEN PHOSPHATE ◇ SALICYL PHOSPHATE ◇ UR 1522

TOXICITY DATA with REFERENCE
orl-rat TDLo:65 g/kg (female 26W pre):REP ARZNAD 30,1098,80
orl-rat LD50:1104 mg/kg ARZNAD 30,1098,80
ipr-rat LD50:338 mg/kg ARZNAD 30,1098,80
ivn-rat LD50:153 mg/kg ARZNAD 30,1098,80
orl-mus LD50:1455 mg/kg ARZNAD 30,1098,80
ivn-mus LD50:94 mg/kg ARZNAD 30,1098,80

SAFETY PROFILE: Poison by intravenous and intraperitoneal routes. Moderately toxic by ingestion. Experimental reproductive effects. When heated to decomposition it emits toxic fumes of PO_x.

PHA750 CAS:5776-49-8 ***HR: 2***
PHOSPHORAMIDE MUSTARD CYCLOHEXYLAMINE SALT
mf: $C_4H_{11}Cl_2N_2O_2P\cdot C_6H_{13}N$ mw: 320.24

SYNS: N,N-BIS(2-CHLOROETHYL)PHOSPHORODIAMIDICACID, CYCLOHEXYL AMMONIUM SALT ◇ N-LOST-PHOSPHORSAURE-DIAMID (GERMAN) ◇ NLPD ◇ NSC-69945 ◇ OMF 59

TOXICITY DATA with REFERENCE
mma-sat 25 μg/plate CNREA8 42,3016,82
cyt-hmn:lym 339 nmol/L CNREA8 46,203,86
ipr-mus TDLo:61 mg/kg (11D preg):TER TJADAB 4,141,71

ipr-mus TDLo:61 mg/kg (11D preg):REP TJADAB 4,141,71
orl-mus LD50:624 mg/kg NCISP* JAN86

SAFETY PROFILE: Moderately toxic by ingestion. An experimental teratogen. Experimental reproductive effects. Human mutation data reported. When heated to decomposition it emits very toxic fumes of NH_3, Cl^-, PO_x, and NO_x.

PHB250 CAS:7664-38-2 ***HR: 3***
PHOSPHORIC ACID
DOT: UN 1805
mf: H_3O_4P mw: 98.00

PROP: Colorless liquid or rhombic crystals. Mp: 42.35°, loses $1/2H_2O$ @ 213°, fp: 42.4°, d: 1.864 @ 25°, vap press: 0.0285 mm @ 20°. Misc with water, alc.

SYNS: ACIDE PHOSPHORIQUE (FRENCH) ◇ ACIDO FOSFORICO (ITALIAN) ◇ FOSFORZUUROPLOSSINGEN (DUTCH) ◇ ORTHOPHOSPHORIC ACID ◇ PHOSPHORSAEURELOESUNGEN (GERMAN)

TOXICITY DATA with REFERENCE
skn-rbt 595 mg/24H SEV BIOFX* 17-4/70
eye-rbt 119 mg SEV BIOFX* 17-4/70
unr-man LDLo:220 mg/kg 85DCAI 2,73,70
orl-rat LD50:1530 mg/kg BIOFX* 17-4/70
skn-rbt LD50:2740 mg/kg BIOFX* 17-4/70

CONSENSUS REPORTS: Community Right-To-Know List. Reported in EPA TSCA Inventory. EPA Genetic Toxicology Program.

OSHA PEL: (Transitional: TWA 1 mg/m^3) TWA 1 mg/m^3; STEL 3 mg/m^3
ACGIH TLV: TWA 1 mg/m^3; STEL 3 mg/m^3

DOT Classification: Corrosive Material; Label: Corrosive.

SAFETY PROFILE: Human poison by an unspecified route. Moderately toxic by ingestion and skin contact. A corrosive irritant to eyes, skin, and mucous membranes, and a systemic irritant by inhalation. A common air contaminant. A strong acid. Mixtures with nitromethane are explosive. Reacts with chlorides + stainless steel to form explosive hydrogen gas. Potentially violent reaction with sodium tetrahydroborate. Dangerous; when heated to decomposition it emits toxic fumes of PO_x.

PHB500 CAS:7784-30-7 ***HR: 2***
PHOSPHORIC ACID, ALUMINUM SALT (1:1), (solution)
mf: $O_4P \cdot A_1$ mw:121.95
DOT: NA 1760

SYNS: ALUMINOPHOSPHORIC ACID ◇ ALUMINUM ACID PHOSPHATE ◇ ALUMINUM MONOPHOSPHATE ◇ ALUMINUM PHOSPHATE ◇ ALUMINUM PHOSPHATE (1:1) ◇ ALUMINUM PHOSPHATE, solution (DOT) ◇ ALUPHOS ◇ FFB 32 ◇ MONOALUMINUM PHOSPHATE ◇ PHOSPHALUGEL

CONSENSUS REPORTS: Reported in EPA TSCA Inventory.

ACGIH TLV: TWA 2 $mg(Al)/m^3$
DOT Classification: Corrosive Material; Label: Corrosive.

SAFETY PROFILE: Corrosive to the eyes, skin, and mucous membranes. When heated to decomposition it emits toxic fumes of PO_x. Used as an antacid and as a cement component, flux for ceramics, dental cement, glass and gels. See also ALUMINUM COMPOUNDS and PHOSPHATES.

PHC750 CAS:6465-92-5 ***HR: 3***
PHOSPHORIC ACID-2,2-DICHLOROVINYL METHYL ESTER, CALCIUM SALT mixed with 2,2-DICHLOROVINYL PHOSPHORIC ACID CALCIUM SALT
mf: $C_6H_8CaCl_4O_8P_2 \cdot C_8H_{14}Cl_4O_8P_2$ mw: 893.92

SYN: KRECALVIN

TOXICITY DATA with REFERENCE
orl-rat LD50:156 mg/kg HIRIA6 26,149,74
skn-rat LD50:3163 mg/kg HIRIA6 26,149,74
scu-rat LD50:34800 µg/kg HIRIA6 26,149,74
orl-mus LD50:250 mg/kg HIRIA6 26,149,74
skn-mus LD50:3040 mg/kg HIRIA6 26,149,74
scu-mus LD50:29300 µg/kg HIRIA6 26,149,74

SAFETY PROFILE: Poison by ingestion and subcutaneous routes. Moderately toxic by skin contact. When heated to decomposition it emits very toxic fumes of Cl^- and PO_x. See also ESTERS.

PHD250 CAS:3254-63-5 ***HR: 3***
PHOSPHORIC ACID DIMETHYL-p-(METHYLTHIO)PHENYL ESTER
mf: $C_9H_{13}O_4PS$ mw: 248.25

SYNS: O,O-DIMETHYLO-(4-METHYLMERCAPTOPHENYL)PHOSPHATE ◇ DIMETHYL-p-(METHYLTHIO)PHENYL PHOSPHATE ◇ ENT 25,734 ◇ 4-METHYLTHIOPHENYLDIMETHYL PHOSPHATE

TOXICITY DATA with REFERENCE
orl-rat LD50:7 mg/kg TXAPA9 21,315,72
orl-mus LD50:18 mg/kg ARTODN 34,103,75
skn-rbt LD50:48 mg/kg 28ZEAL 4,189,69
scu-rbt LD50:48 mg/kg FMCHA2 -,D156,80
orl-bwd LD50:560 µg/kg TXAPA9 21,315,72

CONSENSUS REPORTS: EPA Extremely Hazardous Substances List.

SAFETY PROFILE: Poison by ingestion, skin contact, and subcutaneous routes. When heated to decomposition it emits very toxic fumes of SO_x and PO_x. See also ESTERS.

PHD500 CAS:950-35-6 ***HR: 3***
PHOSPHORIC ACID, DIMETHYL-p-NITROPHENYL ESTER
mf: $C_8H_{10}NO_6P$ mw: 247.16

SYNS: DIMETHYL-p-NITROFENYLESTER KYSELINY FOSFORECNE (CZECH) ◇ DIMETHYL-p-NITROPHENYL PHOSPHATE ◇ DIMETHYL-4-NITROPHENYL PHOSPHATE ◇ DIMETHYL PARAOXON ◇ O,O-DIMETYL-O-p-NITROFENYLFOSFAT (CZECH) ◇ METHYL-E-600 ◇ METHYL PARAOXON ◇ PARAOXON-METHYL ◇ PHOSPHORIC ACID, DIMETHYL-4-NITROPHENYL ESTER (9CI)

TOXICITY DATA with REFERENCE
orl-rat LD50:3270 μg/kg 28ZPAK -,206,72
ivn-rat LD50:457 μg/kg BJIMAG 22,317,65
ims-rat LD50:1690 μg/kg BJIMAG 22,317,65
orl-mus LD50:21 mg/kg JAFCAU 17,243,69
ipr-mus LD50:708 μg/kg PHARAT 35,806,80
scu-mus LD50:1400 μg/kg AMIHAB 11,487,55
orl-gpg LD50:83 mg/kg ABCHA6 27,669,63
ivn-gpg LD50:2200 μg/kg ABCHA6 27,669,63

SAFETY PROFILE: Poison by ingestion, intravenous, subcutaneous, intraperitoneal, and intramuscular routes. When heated to decomposition it emits very toxic fumes of PO_x and NO_x. See also ESTERS.

PHD750 CAS:2255-17-6 ***HR: 3***
PHOSPHORIC ACID DIMETHYL-4-NITRO-m-TOLYL ESTER
mf: $C_9H_{12}NO_6P$ mw: 261.19

SYNS: O,O-DIMETHYL-O-(3-METHYL-4-NITROPHENYL)PHOSPHORATE ◇ FENITROXON ◇ 4-NITRO-m-CRESOL DIMETHYL PHOSPHATE ◇ OXOSUMITHION ◇ SUMIOXON

TOXICITY DATA with REFERENCE
orl-rat LD50:24 mg/kg ABCHA6 27,669,63
ivn-rat LD50:3300 μg/kg ABCHA6 27,669,63
orl-mus LD50:20 mg/kg ACPMAP 16,7,63
ipr-mus LD50:25 mg/kg NNGADV 3,35,78
orl-gpg LD50:221 mg/kg ABCHA6 27,669,63
ivn-gpg LD50:32 mg/kg ABCHA6 27,669,63

SAFETY PROFILE: Poison by ingestion, intravenous, and intraperitoneal routes. When heated to decomposition it emits very toxic fumes of NO_x and PO_x. See also ESTERS.

PHE250 CAS:5598-52-7 ***HR: 3***
PHOSPHORIC ACID, DIMETHYL-3,5,6-TRICHLORO- 2-PYRIDYL ESTER
mf: $C_7H_7Cl_3NO_4P$ mw: 306.47

SYNS: DIMETHYL-3,5,6-TRICHLOROPYRIDYLPHOSPHATE ◇ DIMETHYL-3,5,6-TRICHLORO-2-PYRIDYL PHOSPHATE ◇ ENT 27,521 ◇ FOSPIRATE ◇ FOSPIRATE METHYL ◇ NSC 195058 ◇ TORELLE

TOXICITY DATA with REFERENCE
orl-rat LD50:869 mg/kg SPEADM 78-1,45,78
orl-mus LD50:225 mg/kg ARTODN 34,103,75

SAFETY PROFILE: Poison by ingestion. When heated to decomposition it emits very toxic fumes of Cl^-, PO_x, and NO_x. See also ESTERS.

PHE500 CAS:1623-24-1 ***HR: 2***
PHOSPHORIC ACID, ISOPROPYL ESTER
DOT: UN 1793
mf: $C_3H_9O_4P$ mw: 140.09

SYNS: ISOPROPYL ACID PHOSPHATE solid ◇ ISOPROPYL PHOSPHORIC ACID

CONSENSUS REPORTS: Reported in EPA TSCA Inventory.

DOT Classification: Corrosive Material; Label: Corrosive.

SAFETY PROFILE: A highly corrosive material. Very irritating to the skin, eyes, and mucous membranes. When heated to decomposition it emits toxic fumes of PO_x. See also ESTERS.

PHE750 CAS:10497-05-9 ***HR: 2***
PHOSPHORIC ACID, TRIS(TRIMETHYLSILYL) ESTER
mf: $C_9H_{27}O_4PSi_3$ mw: 314.60

SYN: TRIS-(TRIMETHYLSILYL)FOSFAT(CZECH)

TOXICITY DATA with REFERENCE
orl-rat LD50:3440 mg/kg MarJV# 29MAR77

CONSENSUS REPORTS: Reported in EPA TSCA Inventory.

SAFETY PROFILE: Moderately toxic by ingestion. When heated to decomposition it emits toxic fumes of PO_x. See also ESTERS.

PHF250 CAS:13779-41-4 ***HR: 3***
PHOSPHORODIFLUORIDIC ACID
DOT: UN 1768
mf: F_2HO_2P mw: 101.98

CONSENSUS REPORTS: Reported in EPA TSCA Inventory.

OSHA PEL: TWA 2.5 mg(F)/m³
NIOSH REL: TWA 2.5 mg(F)/m³
DOT Classification: Corrosive Material; Label: Corrosive.
SAFETY PROFILE: Toxic and corrosive material. Very irritating to the skin, eyes, and mucous membranes.

When heated to decomposition it emits very toxic fumes of PO_x and F^-. Used as a catalyst. See also PHOSPHORODIFLUORIDIC ACID, anhydrous.

PHF500 CAS:13779-41-4 ***HR: 2***
PHOSPHORODIFLUORIDIC ACID (anhydrous)
mf: HPO_2F_2 mw: 102

PROP: Mp: −75°. Bp: 116°, d: 1.583 @ 25°/4°, vap d: 3.52.

SYN: DIFLUOROPHOSPHORIC ACID, anhydrous (DOT)

CONSENSUS REPORTS: Reported in EPA TSCA Inventory.

DOT Classification: Corrosive Material; Label: Corrosive.

SAFETY PROFILE: A corrosive irritant to the eyes, skin, and mucous membranes. When heated to decomposition it emits very toxic fumes of F^- and PO_x. See also FLUORIDES and PHOSPHORIC ACID.

PHF750 CAS:371-86-8 ***HR: 3***
PHOSPHORODI(ISOPROPYLAMIDIC) FLUORIDE
DOT: UN 2783
mf: $C_6H_{15}FN_2OP$ mw: 182.21

SYNS: BIS(ISOPROPYLAMIDO)FLUOROPHOSPHATE ◇ BIS(MONOISOPROPYLAMINO)FLUOROPHOSPHATE ◇ BIS(MONOISOPROPYLAMINO)FLUOROPHOSPHINE OXIDE ◇ N,N'-DIISOPROPIL-FOSFORODIAMMIDO-FLUORURO(ITALIAN) ◇ DI(ISOPROPYLAMIDO)PHOSPHORYLFLUORIDE ◇ N,N'-DIISOPROPYL-DIAMIDO-FOSFORZUUR-FLUORIDE (DUTCH) ◇ N,N'-DIISOPROPYL-DIAMIDO-PHOSPHORSAEURE-FLUORID(GERMAN) ◇ N,N'-DIISOPROPYLDIAMIDOPHOSPHORYL FLUORIDE ◇ N,N'-DIISOPROPYLPHOSPHORODIAMIDICFLUORIDE ◇ FLUOROBISISOPROPYLAMINO- PHOSPHINE OXIDE ◇ FLUORURE de N,N'-DIISOPROPYLE PHOSPHORODIAMIDE (FRENCH) ◇ ISOPESTOX ◇ MIPAFOX (DOT) ◇ PESTON XV ◇ PESTOX 15 ◇ PESTOX XV

TOXICITY DATA with REFERENCE
ipr-rat LD50:90 mg/kg PAREAQ 11,636,59
scu-rat LD50:75 mg/kg JEENAI 50(3),356,57
ipr-mus LD50:14 mg/kg BECTA6 2,163,67
orl-rbt LD50:100 mg/kg GUCHAZ 6,357,73
orl-gpg LD50:80 mg/kg PCOC** -,628,66

DOT Classification: ORM-A; Label: None.

SAFETY PROFILE: Poison by ingestion, subcutaneous, and intraperitoneal routes. When heated to decomposition it emits very toxic fumes of F^-, NO_x, and PO_x. See also PARATHION and FLUORIDES.

PHG500 CAS:298-06-6 ***HR: 1***
PHOSPHORODITHIOIC ACID, O,O-DIETHYL ESTER
mf: $C_4H_{11}O_2PS_2$ mw: 186.24

SYN: KYSELINA O,O-DIETHYLDITHIOFOSFORECNA (CZECH)

TOXICITY DATA with REFERENCE
skn-rbt 500 mg/24H MOD 28ZPAK -,212,72
eye-rbt 50 μg/24H SEV 28ZPAK -,212,72
orl-rat LD50:4510 mg/kg 28ZPAK -,212,72

CONSENSUS REPORTS: Reported in EPA TSCA Inventory.

SAFETY PROFILE: Mildly toxic by ingestion. A skin and severe eye irritant. When heated to decomposition it emits very toxic fumes of PO_x and SO_x. See also ESTERS.

PHG750 CAS:3338-24-7 ***HR: 1***
PHOSPHORODITHIOIC ACID, O,O-DIETHYL ESTER, SODIUM SALT
mf: $C_4H_{10}O_2PS_2$•Na mw: 208.22

SYN: O,O-DIETHYLDITHIOFOSFORECNAN SODNY (CZECH)

TOXICITY DATA with REFERENCE
eye-rbt 100 mg/24H SEV 28ZPAK -,212,72
orl-rat LD50:18 g/kg 28ZPAK -,212,72

CONSENSUS REPORTS: Reported in EPA TSCA Inventory.

SAFETY PROFILE: Mildly toxic by ingestion. An severe eye irritant. When heated to decomposition it emits very toxic fumes of PO_x, Na_2O, and SO_x. See also ESTERS.

PHH500 CAS:756-80-9 ***HR: 2***
PHOSPHORODITHIOIC ACID, O,O-DIMETHYL ESTER
mf: $C_2H_7O_2PS_2$ mw: 158.18

SYN: KYSELINA O,O-DIMETHYLDITHIOFOSFORCNA (CZECH)

TOXICITY DATA with REFERENCE
skn-rbt 500 mg/24H MOD 28ZPAK -,211,72
eye-rbt 250 μg/24H SEV 28ZPAK -,211,72
orl-rat LDLo:1000 mg/kg 28ZPAK -,211,72

CONSENSUS REPORTS: Reported in EPA TSCA Inventory.

SAFETY PROFILE: Moderately toxic by ingestion. A skin and severe eye irritant. When heated to decomposition it emits very toxic fumes of PO_x and SO_x. See also ESTERS.

PHI250 CAS:26377-29-7 ***HR: 2***
PHOSPHORODITHIOIC ACID, O,O-DIMETHYL ESTER, SODIUM SALT
mf: $C_2H_6O_2PS_2$•Na mw: 180.16

SYN: O,O-DIMETHYLDITHIOFOSFORECNAN SODNY (CZECH)

TOXICITY DATA with REFERENCE
skn-rbt 20 mg/24H MOD 85JCAE -,1175,86
eye-rbt 100 mg/24H SEV 28ZPAK -,211,72
ihl-rat TCLo:161 mg/m^3/6H (male 11W pre):REP EPASR* 8EHQ-0385-0547
orl-rat LD50:1 g/kg 85JCAE -,1175,86
orl-mus LD50:1550 mg/kg JAFCAU 28,599,80

CONSENSUS REPORTS: Reported in EPA TSCA Inventory.

SAFETY PROFILE: Mildly toxic by ingestion. A severe eye irritant. When heated to decomposition it emits very toxic fumes of PO_x, Na_2O, and SO_x. See also ESTERS.

PHI500 CAS:640-15-3 ***HR: 3***
PHOSPHORODITHIOIC ACID, O,O-DIMETHYL-S-(2-ETHYLTHIO)ETHYL ESTER
mf: $C_6H_{15}O_2PS_3$ mw: 246.36

PROP: Colorless liquid. Sol in acetone, dioxane, acetonitrile.

SYNS: BAY 23129 ◇ COMPOUND M-81 ◇ O,O-DIMETHYL-S-(2-AETHYLTHIO-AETHYL)-DITHIO PHOSPHAT (GERMAN) ◇ O,O-DIMETHYL-S-(CARBONYLMETHYLMORPHOLINO)PHOSPHORODITHIOATE ◇ O,O-DIMETHYL-S-(2-ETHYLMERCAPTOETHYL) DITHIOPHOSPHATE ◇ O,O-DIMETHYL-S-2-ETHYLMERKAPTOETHYLESTER KYSELINY DITHIOFOSFORECNE (CZECH) ◇ O,O-DIMETHYL-S-(2-ETHYLTHIO-ETHYL)-DITHIOFOSFAAT (DUTCH) ◇ O,O-DIMETHYL S-(2-(ETHYLTHIO)ETHYL) PHOSPHORODITHIOATE ◇ O,O-DIMETIL-S-(ETILTIO-ETIL)-DITIOFOSFATO (ITALIAN) ◇ DITHIOMETON (FRENCH) ◇ DITHIOPHOSPHATE de O,O-DIMETHYLE et de S-(2-ETHYLTHIOETHYLE) (FRENCH) ◇ EKATIN ◇ EKATIN AEROSOL ◇ EKATINE-25 ◇ EKATIN ULV ◇ 2-ETHYLTHIOETHYL O,O-DIMETHYL PHOSPHORODITHIOATE ◇ S-(2-(ETHYLTHIO)ETHYL) O,O-DIMETHYLPHOSPHORODITHIONATE ◇ S-(2-(ETHYLTHIO) ETHYL)DIMETHYL PHOSPHOROTHIOLOTHIONATE ◇ INTRATHION ◇ INTRATION ◇ LUXISTELM ◇ M 81 ◇ SAN 230 ◇ THIAMETON ◇ THIOMETON

TOXICITY DATA with REFERENCE
skn-rbt 500 mg/24H MOD 28ZPAK -,212,72
eye-rbt 750 μg/24H SEV 28ZPAK -,212,72
mma-sat 1 mg/plate MUREAV 116,185,83
mnt-mus-ipr 100 mg/kg MUREAV 155,131,85
orl-rat LD50:80500 μg/kg 28ZPAK -,212,72
skn-rat LD50:179 mg/kg 13ZGAF -,211,62
ivn-rat LD50:38500 μg/kg 13ZGAF -,206,62
orl-mus LD50:60 mg/kg PCOC** -,1132,66

SAFETY PROFILE: Poison by ingestion, skin contact, and intravenous routes. Mutation data reported. A skin and severe eye irritant. A cholinesterase inhibitor. When heated to decomposition it emits very toxic fumes of PO_x and SO_x. See also ESTERS and PARATHION.

PHJ250 CAS:13537-32-1 ***HR: 2***
PHOSPHOROFLUORIDIC ACID
DOT: UN 1776
mf: FH_2O_3P mw: 99.99

SYNS: FLUOROPHOSPHORIC ACID, anhydrous ◇ MONOFLUOROPHOSPHORIC ACID, anhydrous

NIOSH REL: TWA 2.5 mg(F)/m^3
DOT Classification: Corrosive Material; Label: Corrosive.

SAFETY PROFILE: A corrosive and irritating material to skin, eyes, and mucous membranes. When heated to decomposition it emits very toxic fumes of F^- and PO_x. See also FLUORIDES.

PHK000 CAS:42509-80-8 ***HR: 3***
PHOSPHOROTHIOIC ACID, O-(2-CHLORO-1-ISOPROPYLIMIDAZOL-4-YL) O,O-DIETHYL ESTER
mf: $C_{10}H_{18}ClN_2O_3PS$ mw: 312.78

SYNS: CGA-12223 ◇ O-(5-CHLORO-1-(METHYLETHYL)-1H 1,2,4-TRIAZOL-3-YL) O,O-DIETHYL PHOSPHOROTHIOATE ◇ ISAZOPHOS

TOXICITY DATA with REFERENCE
orl-rat LD50:40 mg/kg 85ARAE 1,131,77
skn-rat LD50:290 mg/kg FMCHA2 -,D205,80

SAFETY PROFILE: Poison by ingestion and skin contact. Used as a pesticide against pathogenic root nematodes on bananas, citrus, and various annual crops. When heated to decomposition it emits very toxic fumes of Cl^-, NO_x, PO_x, and SO_x. See also ESTERS.

PHK250 CAS:3734-95-0 ***HR: 3***
PHOSPHOROTHIOIC ACID-S-(((1-CYANO-1-METHYL-ETHYL)CARBAMOYL)METHYL)-O,O-DIETHYL ESTER
mf: $C_{10}H_{19}N_2O_4PS$ mw: 294.34

SYNS: α-CYANOISOPROPYLAMIDE OF THE O,O-DIETHYLTHIOPHOSPHORYL ACETIC ACID ◇ S-(((1-CYANO-1-METHYL-ETHYL)CARBAMOYL)METHYL) O,O-DIETHYL PHOSPHOROTHIOATE ◇ S-N-(1-CYANO-1-METHYLETHYL)CARBAMOYLMETHYL DIETHYL PHOSPHOROTHIOLATE ◇ O,O-DIAETHYL-S-1-METHYL)AETHYL)-CARBAMOYL-METHYL-MONOTHIOPHOSPHAT (GERMAN) ◇ O,O-DIETHYL-S-((2-CYAAN-2-METHYL-ETHYL)-CARBAMOYL)-METHYL-MONOTHIOFOSFAAT (DUTCH) ◇ O,O-DIETHYL-S-N-(A-CYANOISOPROPYL)CARBOMOYLMETHYLPHOSPHOROTHIOATE ◇ O,O-DIETIL-S-((2-CIAN-2-METIL-ETIL)-CARBAMOIL)METIL-MONOTIOFOSFATO (ITALIAN) ◇ TARTAN ◇ THIOPHOSPHATE de S-N-(1-CYANO-1-METHYLETHYL)CARBAMOYLMETHYLE et de O,O-DIETHYLE (FRENCH)

TOXICITY DATA with REFERENCE
orl-rat LD50:3500 μg/kg FMCHA2 -,D299,80
skn-rat LD50:105 mg/kg BESAAT 15,111,69
orl-mus LD50:12 mg/kg BESAAT 15,111,69
orl-dog LD50:20 mg/kg BESAAT 15,111,69

orl-rbt LD50:8 mg/kg BESAAT 15,111,69
orl-gpg LD50:13 mg/kg 28ZEAL 5,63,76

CONSENSUS REPORTS: Cyanide and its compounds are on the Community Right To Know List.

SAFETY PROFILE: Poison by ingestion and skin contact. When heated to decomposition it emits very toxic fumes of CN^-, PO_x, SO_x, and NO_x. See also ESTERS.

PHK750 CAS:3070-15-3 ***HR: 3***
PHOSPHOROTHIOIC ACID, O,O-DIETHYL O-(p-METHYLTHIO)PHENYL ESTER
mf: $C_{11}H_{17}O_3PS_2$ mw: 292.37

TOXICITY DATA with REFERENCE
ipr-rat LDLo:2500 μg/kg TXAPA9 6,78,64

CONSENSUS REPORTS: Reported in EPA TSCA Inventory.

SAFETY PROFILE: Poison by intraperitoneal route. When heated to decomposition it emits very toxic fumes of PO_x and SO_x. See also ESTERS.

PHL750 CAS:3735-33-9 ***HR: 3***
PHOSPHOROTHIOIC ACID, S-((1,3-DIHYDRO-1,3-DIOXO-2H-ISOINDOL-2-YL)METHYL) O,O-DIMETHYL ESTER
mf: $C_{11}H_{12}NO_5PS$ mw: 301.27

SYNS: ENT 25,707 ◇ STAUFFER R-1571

TOXICITY DATA with REFERENCE
orl-rat LD50:50 mg/kg ARSIM* 20,23,66
orl-ckn LD50:46 mg/kg TXAPA9 6,147,64

SAFETY PROFILE: Poison by ingestion. A pesticide. When heated to decomposition it emits very toxic fumes of NO_x, PO_x, and SO_x. See also ESTERS.

PHM000 CAS:74-60-2 ***HR: 3***
PHOSPHOROTHIOIC ACID, O,O-DIISOPROPYL O-(p-(METHYLSULFINYL)PHENYL) ESTER
mf: $C_{13}H_{21}O_4PS_2$ mw: 336.43

SYNS: BAYER 25142 ◇ ENT 25,507

TOXICITY DATA with REFERENCE
orl-rat LDLo:25 mg/kg ARSIM* 20,3,66
orl-ckn LD50:32600 μg/kg TXAPA9 6,147,64

SAFETY PROFILE: Poison by ingestion. A pesticide. When heated to decomposition it emits very toxic fumes of PO_x and SO_x. See also ESTERS.

PHM750 CAS:63980-89-2 ***HR: 3***
PHOSPHOROTHIOIC ACID, O-ETHYL S-(p-TOLYL) ESTER
mf: $C_9H_{13}O_3PS$ mw: 232.25

SYN: BAY 38156

TOXICITY DATA with REFERENCE
orl-rat LD50:50 mg/kg TXAPA9 21,315,72
orl-bwd LD50:1600 μg/kg TXAPA9 21,315,72

SAFETY PROFILE: Poison by ingestion. When heated to decomposition it emits very toxic fumes of PO_x and SO_x. See also ESTERS.

PHN000 CAS:13955-12-9 ***HR: 3***
PHOSPHOROTHIOIC ACID, O-ISOPROPYL O-METHYL O-(p-NITROPHENYL) ESTER
mf: $C_{10}H_{14}NO_5PS$ mw: 291.28

SYNS: BAYER 52553 ◇ ENT 27,248

TOXICITY DATA with REFERENCE
orl-rat LD50:5 mg/kg ARSIM* 20,5,66
orl-mus LD50:21 mg/kg JAFCAU 17,243,69

SAFETY PROFILE: Poison by ingestion. A pesticide. When heated to decomposition it emits very toxic fumes of NO_x, PO_x, and SO_x. See also ESTERS.

PHN250 CAS:1174-83-0 ***HR: 2***
PHOSPHOROTHIOIC ACID, O,O'-(SULFONYLDI-p-PHENYLENE) O,O,O',O'-TETRAMETHYL ESTER
mf: $C_{16}H_{20}O_8P_2S_3$ mw: 498.48

SYNS: AMERICAN CYANAMID AC 43,913 ◇ AMERICAN CYANAMID CL-43913 ◇ AMERICAN CYANAMID E.I. 43,913 ◇ ENT 27,161 ◇ O,O'-(SULFONYLDI-1,4-PHENYLENE) O,O,O',O'-TETRAMETHYL DIPHOSPHOROTHIOATE

TOXICITY DATA with REFERENCE
orl-rat LD50:2 g/kg TXAPA9 21,315,72
orl-mus LD50:580 mg/kg ARSIM* 20,1,66

SAFETY PROFILE: Moderately toxic by ingestion. A pesticide. When heated to decomposition it emits very toxic fumes of PO_x and SO_x. See ESTERS.

PHN500 CAS:10235-09-3 ***HR: 2***
PHOSPHOROTHIOIC ACID, O,O,O-TRIS(2-CHLOROETHYL) ESTER
mf: $C_6H_{12}Cl_3O_3PS$ mw: 301.56

SYNS: 2-CHLORO-ETHANOL, PHOSPHOROTHIOATE (3:1) ◇ PHOSPHOROTHIOIC ACID, O,O,O-TRI(2-CHLOROETHYL) ESTER

TOXICITY DATA with REFERENCE
skn-rbt 10 mg/24H open MLD AIHAAP 23,95,62
orl-rat LD50:820 mg/kg AIHAAP 23,95,62
skn-rbt LD50:1800 mg/kg AIHAAP 23,95,62

SAFETY PROFILE: Moderately toxic by ingestion and skin contact. A skin irritant. When heated to decomposition it emits very toxic fumes of Cl^-, PO_x, and SO_x. See also ESTERS.

PHO000 CAS:140-08-9 ***HR: 3***
PHOSPHOROUS ACID, TRIS(2-CHLOROETHYL) ESTER
mf: $C_6H_{12}Cl_3O_3P$ mw: 269.50

SYNS: 2-CHLOROETHANOL PHOSPHITE (3:1) ◇ TRIS(2-CHLOROETHYL)ESTER of PHOSPHORUS ACID ◇ TRIS(2-CHLOROETHYL) PHOSPHITE

TOXICITY DATA with REFERENCE
eye-rbt 100 mg SEV 34ZIAG -,612,69
orl-rat LD50:100 mg/kg ALBRW* #OPB-3,84
ipr-mus LDLo:250 mg/kg CBCCT* 2,185,50
skn-rbt LD50:810 mg/kg ALBRW* #OPB-3,84

CONSENSUS REPORTS: Reported in EPA TSCA Inventory.

SAFETY PROFILE: Poison by ingestion and intraperitoneal routes. Moderately toxic by skin contact. A severe eye irritant. When heated to decomposition it emits very toxic fumes of Cl^- and PO_x.

PHO250 CAS:63980-61-0 ***HR: 3***
PHOSPHOROUS ACID TRIS(2-FLUOROETHYL-ESTER)
mf: $C_6H_{12}F_3O_3P$ mw: 220.15

SYNS: 2-FLUOROETHANOL, PHOSPHITE (3:1) ◇ TL 833

TOXICITY DATA with REFERENCE
ihl-rat LCLo:500 mg/m^3/10M NDRC** No.9-4-1-9,43
ihl-mus LCLo:1000 mg/m^3/10M NDRC** No.9-4-1-9,43
ihl-dog LCLo:100 mg/m^3/10M NDRC** No.9-4-1-9,43
ihl-cat LCLo:100 mg/m^3/10M NDRC** No.9-4-1-9,43
ihl-rbt LCLo:150 mg/m^3/10M NDRC** No.9-4-1-9,43
ihl-gpg LCLo:100 mg/m^3/10M NDRC** No.9-4-1-9,43

SAFETY PROFILE: Poison by inhalation. When heated to decomposition it emits very toxic fumes of F^- and PO_x.

PHO500 CAS:7723-14-0 ***HR: 3***
PHOSPHORUS (red)
DOT: UN 1338/UN 1381/UN 2447
af: P aw: 30.97

PROP: Reddish-brown powder. Bp: 280° (with ignition), mp: 590° @ 43 atm, d: 2.34, autoign temp: 500°F in air, vap d: 4.77.

SYN: PHOSPHORUS, AMORPHOUS, RED (DOT)

TOXICITY DATA with REFERENCE
unr-man LDLo:4412 µg/kg 85DCAI 2,73,70

CONSENSUS REPORTS: EPA Extremely Hazardous Substances List.

DFG MAK: 0.1 mg/m^3
DOT Classification: Flammable Solid and Poison; Label: Flammable Solid and Poison.

SAFETY PROFILE: A human poison by an unspecified route. May have white phosphorus as an impurity. Generally less reactive than white phosphorus. Dangerous fire hazard when exposed to heat or by chemical reaction with oxidizers. Can also react with reducing materials. Moderate explosion hazard by chemical reaction or on contact with organic materials. May explode on impact. To fight fire, use water. Explosive reaction with chlorosulfuric acid; hydroiodic acid; magnesium perchlorate; chromyl chloride. Forms sensitive explosive mixtures with metal halogenates (e.g., chlorates, bromates, or iodates of barium, calcium, magnesium, potassium, sodium, zinc); ammonium nitrate; mercury(I) nitrate; silver nitrate; sodium nitrate; potassium permanganate. Violent reaction or ignition with alkalies + heat; fluorine; chlorine; liquid bromine; antimony pentachloride. Reacts with hot alkalies or hydroiodic acid to form phosphine gas which then ignites. Incompatible with cyanogen iodide; halogen azides; halogen oxides (e.g., chlorine dioxide; dichlorine oxide; oxygen difluoride; trioxygen difluoride); interhalogens (e.g., bromine trifluoride; bromine pentafluoride; chlorine trifluoride; iodine trichloride; iodine pentafluoride); hexalithium disilicide; hydrogen peroxide; metal acetylides (e.g., rubidium acetylide; cesium acetylide; lithium acetylide; sodium acetylide; potassium acetylide); antimony pentachloride; metal oxides (e.g., copper oxide; manganese dioxide; lead oxide; mercury oxide; silver oxide; chromium trioxide); metal peroxides (e.g., lead peroxide; potassium peroxide; sodium peroxide); metals (e.g., beryllium; copper; manganese; thorium; zirconium; cerium; lanthanum; neodymium; praseodymium; osmium; platinum); metal sulfates (e.g., barium sulfate; calcium sulfate); nitric acid; nitrogen halides; nitrosyl fluoride; nitryl fluoride; non-metal halides (e.g., boron triiodide; seleninyl chloride; sulfuryl chloride; disulfuryl chloride; disulfur dibromide); non-metal oxides (e.g., nitrogen oxide; dinitogen tetraoxide; dinitrogen pentaoxide; sulfur trioxide; oxygen; peroxyformic acid; potassium nitride; selenium; sodium chlorite; sulfur; sulfuric acid; peroxides; oxidizing materials. When heated to decomposition it emits toxic fumes of PO_x. See also PHOSPHORUS (white).

PHO740 CAS:7723-14-0 ***HR: 3***
PHOSPHORUS (yellow)
DOT: UN 1381/UN 2447
mf: P_4 mw: 123.88

PROP: Cubic crystals; colorless to yellow, wax-like solid. Mp: 44.1°, bp: 280°, flash p: spontaneously flammable in air, d: 1.82, autoign temp: 86°F, vap press: 1 mm @ 76.6°, vap d: 4.42.

SYNS: BONIDE BLUE DEATH RAT KILLER ◇ FOSFORO BIANCO (ITALIAN) ◇ GELBER PHOSPHOR (GERMAN) ◇ PHOSPHORE BLANC (FRENCH) ◇ PHOSPHORUS (white) ◇ RAT-NIP ◇ TETRAFOSFOR (DUTCH) ◇ TETRAPHOSPHOR (GERMAN) ◇ WEISS PHOSPHOR (GERMAN) ◇ WHITE PHOSPHORUS

TOXICITY DATA with REFERENCE
orl-rat TDLo:11 μg/kg (1-22D preg):REP ZDKAA8 36(5),87,76
orl-wmn LDLo:22 mg/kg:CVS AHJOA2 84,139,72
orl-wmn TDLo:11 mg/kg AJMSA9 209,223,44
orl-hmn LDLo:1400 μg/kg PCOC** -,901,66
orl-wmn LDLo:4600 μg/kg:PUL,GIT,SKN AIMDAP 83,164,49
orl-wmn TDLo:2600 μg/kg NEJMAG 232,247,45
orl-rat LD50:3030 μg/kg NTIS** AD-B011-150
orl-mus LD50:4820 μg/kg NTIS** AD-B011-150
orl-dog LDLo:10 mg/kg YKYUA6 28,329,77
scu-dog LDLo:2 mg/kg AEXPBL 52,173,1905
scu-rbt LDLo:10 mg/kg AEXPBL 64,274,11

CONSENSUS REPORTS: EPA Extremely Hazardous Substances List. Reported in EPA TSCA Inventory.

OSHA PEL: TWA 0.1 mg/m^3
ACGIH TLV: TWA 0.1 mg/m^3

DOT Classification: Flammable Solid and Poison; Label: Flammable Solid and Poison.

SAFETY PROFILE: Human poison by ingestion. Experimental poison by ingestion and subcutaneous routes. Experimental reproductive effects. Human systemic effects by ingestion: cardiomyopathy, cyanosis, nausea or vomiting, sweating. Toxic quantities have an acute effect on the liver and can cause severe eye damage. Inhalation can cause photophobia with myosis, dilation of the pupils, retinal hemorrhage, congestion of the blood vessels, and rarely an optic neuritis. Chronic exposure by inhalation or ingestion can cause anemia, gastrointestinal effects, and brittleness of the long bones leading to spontaneous fractures. The most common symptom, however, of chronic phosphorous poisoning is necrosis of the jaw (phossy-jaw).

More reactive than red phosphorus. Dangerous fire hazard when exposed to heat, flame or by chemical reaction with oxidizers. Ignites spontaneously in air. Very reactive. If combustion occurs in a confined space, it will remove the oxygen and cause asphyxiation. Dangerous explosion hazard by chemical reaction with: alkaline hydroxides; NH_4NO_3; SbF_5; $Ba(BrO_3)_2$; Be; BI_3; $Ca(BrO_3)_2$; $Mg(BrO_3)_2$; $K(BrO_3)$; $NaBrO_3$; $Zn(BrO_3)_2$; Br_2; halogens; BrF_3; BrN_3; (chlorates of Ba, Ca, Mg, K, Na, Zn); (iodates of Ba, Ca, Mg, K, Na, Zn); Ce; Cs; $CsHC_2$; Cs_3N; (charcoal + air); ClO_2; (Cl_2 + heptane); ClO; ClF_3; ClO_3; chlorosulfonic acid; CrO_3; $Cr(OCl)_2$; Cu; NCl; IBr; ICl; IF_5; Fe; La; PbO_2; Li; Li_2C_2; Li_6CS; $Mg(ClO_4)_2$; Mn; HgO; $HgNO_3$; Nd; Ni; nitrates; NBr; NO_2; NBr_3; NCl_3; NOF; FNO_2; O_2; performic acid; Pt; K; KOH; K_3N; $KMnO_4$; K_2O_2; Rb; $RbHC_2$; Se_2Cl_2; $SeOCl_2$; $SeOF_2$; SeF_4; $AgNO_3$; Ag_2O; Na; Na_2C_2; $NaClO_2$; NaOH; Na_2O_2; S; SO_3; H_2SO_4; Th; $VOCl_2$; Zr; peroxyformic acid; chloro sulfuric acid; halogen azides; hexalithium disilicide. Can react vigorously with oxidizing materials. To fight fire, use water. Used in fertilizers, tracer bullets, incendiaries manufacturing, rat poison, and gas analysis. When heated to decomposition it emits highly toxic fumes of PO_x. See also PHOSPHORUS (RED).

PHP000 CAS:7723-14-0 ***HR: 3***
PHOSPHORUS (white in water)

CONSENSUS REPORTS: EPA Extremely Hazardous Substances List. Community Right-To-Know List. Reported in EPA TSCA Inventory.

DOT Classification: Flammable Solid and Poison; Label: Flammable Solid and Poison.

SAFETY PROFILE: A poison. Very flammable solid which must be kept under water. When heated to decomposition it emits toxic fumes of PO_x. See also PHOSPHORUS (WHITE).

PHP250 CAS:37388-50-4 ***HR: 3***
PHOSPHORUS AZIDE DIFLUORIDE
mf: F_2N_3P mw: 110.99

SAFETY PROFILE: A very toxic material. May explode at 25°C. Unstable to light or heat. Ignites spontaneously in air. When heated to decomposition it emits very toxic fumes of PO_x, NO_x, and F^-. See also FLUORIDES, AZIDES, and PHOSPHIDES.

PHP500 CAS:38115-19-4 ***HR: 3***
PHOSPHORUS AZIDE DIFLUORIDE-BORANE
mf: $F_2N_3P \cdot BH_3$ mw: 124.82

SAFETY PROFILE: Probably very toxic. A very unstable explosive. When heated to decomposition it emits very toxic fumes of F^-, NO_x, PO_x, and BO_x. See also BORANE, FLUORIDES, AZIDES, and PHOSPHIDES.

PHQ000 ***HR: D***
PHOSPHORUS COMPOUNDS, INORGANIC

SAFETY PROFILE: Variable toxicity. Most inorganic phosphates (except phosphine) have low toxicity, but in large doses they may cause serious disturbances, particularly in calcium metabolism. Red phosphorus and phosphates are relatively harmless. White (yellow) phosphorus is highly toxic by several routes. The phosphorus halides decompose violently with water to form the halide acid and are thus severe irritants. Phosphorus sulfides behave similarly. Metaphosphates may be highly toxic, causing irritation and hemorrhages in the stomach, as well as liver and kidney damage. Phosphorus trichloride is the most used of the phosphorus halide compounds. Phosphoryl chloride is used to synthesize phosphate esters. Common air contaminants. When heated to decomposition it emits highly toxic fumes of PO_x. See also specific compounds.

PHQ250 CAS:1116-01-4 ***HR: 3***
PHOSPHORUS CYANIDE
mf: C_3N_3P mw: 110

$$P(C\equiv N)_3$$

PROP: White needles. Mp: Sublimes @ 130°.

SYN: PHOSPHORUS TRICYANIDE

CONSENSUS REPORTS: Cyanide and its compounds are on the Community Right-To-Know List.

SAFETY PROFILE: Ignites spontaneously in air. Explodes when heated to 100°C. Violent reaction with water. Upon decomposition it emits toxic fumes of PO_x and CN^-. See also CYANIDES.

PHQ500 CAS:7783-55-3 ***HR: 3***
PHOSPHORUS FLUORIDE
mf: F_3P mw: 87.97

PROP: Colorless gas. Mp: −152°, bp: −102°, d: 3.907 g/L.

SYN: PHOSPHOROUS TRIFLUORIDE

TOXICITY DATA with REFERENCE
ihl-mus LCLo:1900 mg/m^3/10M NDRC** NDCrc-132,June,42

CONSENSUS REPORTS: Reported in EPA TSCA Inventory.

OSHA PEL: TWA 2.5 mg(F)/m^3
ACGIH TLV: TWA 2.5 mg(F)/m^3
NIOSH REL: (Fluorides, Inorganic) TWA 2.5 mg(F)/m^3

SAFETY PROFILE: Moderately toxic by inhalation. A severe eye, skin, and mucous membrane irritant. Explodes on contact with dioxygen difluoride. Violent reaction or ignition with borane; diborane; F_2; hexafluoroisopropylideneamino lithium; O_2. Will react with water or steam to produce toxic and corrosive fumes. Dangerous; when heated to decomposition it emits highly toxic fumes of F^- and PO_x. See also HYDROFLUORIC ACID, FLUORIDES, and PHOSPHORUS PENTAFLUORIDE.

PHQ750 CAS:12037-82-0 ***HR: 3***
PHOSPHORUS HEPTASULFIDE
DOT: UN 1339

PROP: Light yellow crystals; light gray powder or fused solid. Mp: 310°, bp: 523°, d: 2.19 @ 17°.

DOT Classification: Flammable Solid; Label: Flammable Solid.

SAFETY PROFILE: A poison by ingestion. Flammable when exposed to heat or flame; can react vigorously with oxidizing materials. When heated to decomposition it emits very toxic fumes of PO_x and SO_x. See also SULFIDES and PHOSPHORUS.

PHQ800 CAS:10025-87-3 ***HR: 3***
PHOSPHORUS OXYCHLORIDE
DOT: UN 1810
mf: Cl_3OP mw: 153.32

PROP: Colorless to sltly yellow, fuming liquid. Mp: 1.2°, bp: 105.1°, d: 1.685 @ 15.5°, vap press: 40 mm @ 27.3°, vap d: 5.3.

SYNS: PHOSPHORUS OXYTRICHLORIDE ◇ PHOSPHORYL CHLORIDE

TOXICITY DATA with REFERENCE
orl-rat LD50:380 mg/kg TNICS* 13,104,73
ihl-rat LC50:48 ppm/4H AIHAAP 25,470,64
ihl-gpg LC50:53 ppm/4H AIHAAP 25,470,64

CONSENSUS REPORTS: Reported in EPA TSCA Inventory.

OSHA PEL: TWA 0.1 ppm
ACGIH TLV: TWA 0.1 ppm
DFG MAK: 0.2 ppm (1 mg/m^3)
DOT Classification: Corrosive; Label: Corrosive.

SAFETY PROFILE: Poison by inhalation and ingestion. A corrosive eye, skin, and mucous membrane irritant. Potentially explosive reaction with water evolves hydrogen chloride and phosphine which then ignites. Explosive reaction with 2,6-dimethylpyridine N-oxide; dimethyl sulfoxide; ferrocene-1,1′-dicarboxylic acid; pyridine N-oxide (above 60°C); sodium + heat. Violent reaction or ignition with BI_3; carbon disulfide; 2,5-dimethyl pyrrole + dimethyl formamide; organic matter; zinc powder. Reacts with water or steam to produce heat and toxic and corrosive fumes. Incompatible with carbon disulfide; N,N-dimethyl-formamide; 2,5-dimethylpyrrole; 2,6-dimethylpyridine N-oxide; dimethylsulfoxide; ferrocene-1,l-dicarboxylic acid; water; zinc. When heated to decomposition it emits highly toxic fumes of Cl^- and PO_x.

PHR000 ***HR: 2***
PHOSPHORUS OXYFLUORIDE
mf: POF_3 mw: 103.98

PROP: Colorless gas. Mp: −68°, bp: −39.8°, d: 4.69 g/L.

SYN: PHOSPHORYL FLUORIDE

SAFETY PROFILE: Hydrolysis of this material and halofluorides, i.e., (phosphoryl chlorodifluoride, phosphoryl dichlorofluoride and the bromo analogs) is vigorous to violent. When heated to decomposition it emits toxic fumes of F^- and PO_x.

PHR250 CAS:7789-69-7 ***HR: 3***
PHOSPHORUS PENTABROMIDE
DOT: UN 2691
mf: PBr_5 mw: 430.56

PROP: Yellow, crystalline mass. Mp: decomp, bp: decomp @ 106°. Sol in carbon disulfide.

SYNS: PENTABROMO PHOSPHORANE ◇ PENTABROMO PHOSPHORUS ◇ PHOSPHORIC BROMIDE

CONSENSUS REPORTS: Reported in EPA TSCA Inventory.

DOT Classification: Corrosive Material; Label: Corrosive.

SAFETY PROFILE: A poison. Corrosive to the eyes, skin and mucous membranes. Flammable by chemical reaction. Contact with moisture can cause a violent reaction and evolution of heat. Incompatible with water or steam to produce heat and toxic and corrosive fumes. When heated to decomposition it emits highly toxic fumes of Br^- and PO_x. See also BROMIDES.

PHR500 CAS:10026-13-8 ***HR: 3***
PHOSPHORUS PENTACHLORIDE
DOT: UN 1806
mf: Cl_5P mw: 208.22

PROP: Yellowish-white, fuming, crystalline mass; pungent odor. Mp: (under press) 148° decomp, bp: subl @ 160°, d: 4.65 g/L @ 296°, vap press: 1 mm @ 55.5°.

SYNS: FOSFORO(PENTACHLORURO di) (ITALIAN) ◇ FOSFORPENTACHLORIDE (DUTCH) ◇ PHOSPHORE(PENTACHLORURE de) (FRENCH) ◇ PHOSPHORIC CHLORIDE ◇ PHOSPHORPENTACHLORID (GERMAN) ◇ PHOSPHORUS PERCHLORIDE ◇ PIECIOCHLOREK FOSFORU (POLISH)

TOXICITY DATA with REFERENCE
orl-rat LD50:660 mg/kg TNICS* 13,104,73
ihl-rat LC50:205 mg/m³ TNICS* 13,105,73

CONSENSUS REPORTS: EPA Extremely Hazardous Substances List. Reported in EPA TSCA Inventory.

OSHA PEL: TWA 1 mg/m³
ACGIH TLV: TWA 0.85 mg/m³
DFG MAK: 1 mg/m³
DOT Classification: Corrosive; Label: Corrosive.

SAFETY PROFILE: Poison by inhalation. Moderately toxic by ingestion. A severe eye, skin, and mucous membrane irritant. Corrosive to body tissues. Flammable by chemical reaction. Explosive reaction with chlorine dioxide + chlorine; sodium; urea + heat. Reacts to form explosive products with carbamates; 3'-methyl-2-nitrobenzanilide (product explodes on contact with air). Ignites on contact with fluorine. Reacts violently with moisture; ClO_3; hydroxylamine; magnesium oxide; nitrobenzene; phosphorus(III) oxide; K. To fight fire, use CO_2, dry chemical. Incompatible with aluminum; chlorine dioxide; chlorine; diphosphorus trioxide; fluorine; hydroxylamine; magnesium oxide; 3'-methyl-2-nitrobenzanilide; nitrobenzene; sodium; urea; water. Will react with water or steam to produce heat and toxic and corrosive fumes. Used as a catalyst, chlorinating and dehydrating agent. When heated to decomposition it emits highly toxic fumes of Cl^- and PO_x.

PHR750 CAS:7647-19-0 ***HR: 3***
PHOSPHORUS PENTAFLUORIDE
DOT: UN 2198
mf: PF_5 mw: 125.98

PROP: Colorless gas, fumes strongly in air. Mp: −93.8°, bp: −84.6°, d: (gas) 5.805 g/L.

CONSENSUS REPORTS: EPA Extremely Hazardous Substances List. Reported in EPA TSCA Inventory.

OSHA PEL: TWA 2.5 mg(F)/m³
ACGIH TLV: TWA 2.5 mg(F)/m³

DOT Classification: Poison A; Label: Poison Gas.

SAFETY PROFILE: A poisonous gas. Violently irritating to skin, eyes, and mucous membranes. Inhalation may cause pulmonary edema. Reacts with water or steam to produce toxic and corrosive fumes. When heated to decomposition it emits highly toxic fumes of F^- and PO_x. See also FLUORIDES.

PHS000 CAS:1314-80-3 ***HR: 3***
PHOSPHORUS PENTASULFIDE
DOT: UN 1340
mf: P_2S_5 mw: 222.24

PROP: Gray to yellow-green, crystalline, deliquescent mass. Bp: 514°, d: 2.09, autoign temp: 287°F. Mp: 286-290°.

SYNS: PENTASULFURE de PHOSPHORE (FRENCH) ◇ PHOSPHORIC SULFIDE ◇ PHOSPHORUS PERSULFIDE ◇ RCRA WASTE NUM-

BER U189 ◇ SIRNIK FOSFORECNY (CZECH) ◇ SULFUR PHOSPHIDE ◇ THIOPHOSPHORIC ANHYDRIDE

TOXICITY DATA with REFERENCE
skn-rbt 500 mg/24H MOD 28ZPAK -,16,72
eye-rbt 20 mg/24H SEV 28ZPAK -,16,72
orl-rat LD50:389 mg/kg 28ZPAK -,16,72

CONSENSUS REPORTS: Reported in EPA TSCA Inventory.

OSHA PEL: (Transitional: TWA 1 mg/m^3) TWA 1 mg/m^3; STEL 3 mg/m^3
ACGIH TLV: TWA 1 mg/m^3; STEL 3 mg/m^3
DFG MAK: 1 mg/m^3

DOT Classification: Flammable Solid; Label: Flammable Solid and Dangerous When Wet.

SAFETY PROFILE: A poison by ingestion. A severe eye and skin irritant. Readily liberates toxic hydrogen sulfide and phosphorus pentoxide and evolves heat on contact with moisture. Dangerous fire hazard in the form of dust when exposed to heat or flame. Spontaneous heating in the presence of moisture. Moderate explosion hazard in solid form by spontaneous chemical reaction. Reacts with water, steam, or acids to produce toxic and flammable vapors; can react vigorously with oxidizing materials. Incompatible with air; alcohols; water. To fight fire use CO_2 snow, dry chemical or sand. Used as an intermediate in manufacturing lubricant additives, insecticides and fertilizer agents. When heated to decomposition it emits highly toxic fumes of SO_x and PO_x. See also HYDROGEN SULFIDE.

PHS250 CAS:1314-56-3 ***HR: 3***
PHOSPHORUS PENTOXIDE
DOT: NA 1807
mf: O_5P_2 mw: 141.94

PROP: Deliq crystals. D: 2.30; mp: 340°, subl @ 360°.

SYNS: DIPHOSPHORUS PENTOXIDE ◇ PHOSPHORIC ANHYDRIDE ◇ PHOSPHORUS(V) OXIDE ◇ POX ◇ PO_x

TOXICITY DATA with REFERENCE
ihl-rat LC50:1217 mg/m^3/1H TOXID9 1,140,81
ihl-mus LC50:271 mg/m^3/1H TOXID9 1,140,81
ihl-rbt LC50:1689 mg/m^3/1H TOXID9 1,140,81
ihl-gpg LC50:61 mg/m^3/1H TOXID9 1,140,81

CONSENSUS REPORTS: EPA Extremely Hazardous Substances List. Reported in EPA TSCA Inventory.

DFG MAK: 1 mg/m^3
DOT Classification: Corrosive Material; Label: Corrosive.

SAFETY PROFILE: Poison by inhalation. A corrosive irritant to the eyes, skin, and mucous membranes. With the appropriate conditions it undergoes hazardous reactions with formic acid; hydrogen fluoride; inorganic bases; iodides; metals; methyl hydroperoxide; oxidants (e.g., bromine; pentafluoride; chlorine trifluoride; perchloric acid; oxygen difluoride; hydrogen peroxide); 3-propynol; water. When heated to decomposition it emits toxic fumes of PO_x.

PHS500 CAS:1314-85-8 ***HR: 3***
PHOSPHORUS SESQUISULFIDE
DOT: UN 1341
mf: P_4S_3 mw: 220.06

PROP: Yellow, crystalline mass. Mp: 172.5°, bp: 407°, d: 2.03, autoign temp: 212°F.

SYNS: PHOSPHORUS(III) SULFIDE(IV) ◇ SESQUISULFURE de PHOSPHORE (FRENCH) ◇ TETRAPHOSPHORUS TRISULFIDE ◇ TRISULFURATED PHOSPHORUS

TOXICITY DATA with REFERENCE
orl-rbt LDLo:100 mg/kg SAPHAO 15,259,1904

DOT Classification: Flammable Solid; Label: Flammable Solid and Dangerous When Wet.

SAFETY PROFILE: Poison by ingestion. Flammable by spontaneous ignition. When heated to decomposition it emits very toxic fumes of PO_x and SO_x. See also SULFIDES and PHOSPHORUS.

PHS750 ***HR: 2***
PHOSPHORUS THIOCYANATE
mf: $P(SCN)_3$ mw: 205.23

PROP: Mp: approx −4°, bp: 265°, d: 1.625 @ 18°.

CONSENSUS REPORTS: Cyanide and its compounds are on the Community Right-To-Know List.

SAFETY PROFILE: Probably toxic. Combustible when exposed to heat or flame; can react vigorously with oxidizing materials. When heated to decomposition, or on contact with acids it emits highly toxic fumes of CN^- and PO_x. See also THIOCYANATES.

PHT000 CAS:56280-76-3 ***HR: 3***
PHOSPHORUS TRIAZIDE
mf: N_9P mw: 157.04

SAFETY PROFILE: Highly explosive. May explode at room temperature. When heated to decomposition it emits very toxic fumes of NO_x, PO_x, and PH_3. See also AZIDES and PHOSPHIDES.

PHT250 CAS:7789-60-8 ***HR: 3***
PHOSPHORUS TRIBROMIDE
DOT: UN 1808

SYNS: PHOSPHOROUS BROMIDE (DOT) ◇ TRIBROMOPHOSPHINE
mf: Br_3P mw: 270.70

PROP: Mp: −40°, bp: 175.3°, d: 2.852 @ 15°, vap press: 10 mm @ 47.8°.

CONSENSUS REPORTS: Reported in EPA TSCA Inventory.

DOT Classification: Corrosive Material; Label: Corrosive.

SAFETY PROFILE: Probably highly toxic. A corrosive irritant to the eyes, skin, and mucous membranes. Will react with water, steam, or acids to produce heat, toxic, and corrosive fumes. Violent reaction or ignition with calcium hydroxide + sodium carbonate, phenylpropanol, sulfuric acid, oleum, fluorosulfuric acid, chlorosulfuric acid, 1,1,1-tris(hydroxymethyl)methane, water, potassium, sodium, RuO_4. When heated to decomposition it emits very toxic fumes of Br^- and PO_x. See also PHOSPHIDES and BROMIDES.

PHT275 CAS:7719-12-2 ***HR: 3***
PHOSPHORUS TRICHLORIDE
DOT: UN 1809
mf: Cl_3P mw: 137.32

PROP: Clear, colorless, fuming liquid. Mp: −111.8°, bp: 76°, d: 1.574 @ 21°, vap press: 100 mm @ 21°, vap d: 4.75. Decomp by water and alc; sol in benzene, chloroform, and ether.

SYNS: CHLORIDE of PHOSPHORUS ◇ FOSFORO(TRICLORURO di) (ITALIAN) ◇ FOSFORTRICHLORIDE (DUTCH) ◇ PHOSPHORE(TRICHLORURE de) (FRENCH) ◇ PHOSPHORTRICHLORID (GERMAN) ◇ PHOSPHORUS CHLORIDE ◇ TROJCHLOREK FOSFORU (POLISH)

TOXICITY DATA with REFERENCE
orl-rat LD50:550 mg/kg TNICS* 13,104,73
ihl-rat LC50:104 ppm/4H AIHAAP 25,470,64
ihl-gpg LC50:50 ppm/4H AIHAAP 25,470,64

CONSENSUS REPORTS: EPA Extremely Hazardous Substances List. Reported in EPA TSCA Inventory.

OSHA PEL: (Transitional: TWA 0.5 ppm) TWA 0.2 ppm; STEL 0.5 ppm
ACGIH TLV: TWA 0.2 ppm; STEL 0.5 ppm
DFG MAK: 0.5 ppm (3 mg/m^3)
DOT Classification: Corrosive Material; Label: Corrosive.

SAFETY PROFILE: Poison by inhalation. Moderately toxic by ingestion. A corrosive irritant to skin, eyes (at 2 ppm), and mucous membranes. Potentially explosive reaction with chlorobenzene + sodium; dimethyl sulfoxide; molten sodium; chromyl chloride; nitric acid; sodium peroxide; oxygen (above 100°C); tetravinyl lead. Reacts with carboxylic acids (e.g., acetic acid) to form violently unstable products. Violent reaction or ignition with Al; chromium pentafluoride; diallyl phosphite + allyl alcohol; F_2; hexafluoroisopropylideneaminolithium; hydroxylamine; iodine chloride; PbO_2; HNO_2; organic matter; potassium; selenium dioxide; sulfur acids (e.g., sulfuric acid; fluorosulfuric acid; oleum). Violent reaction with water evolves hydrogen chloride and diphosphane gas which then ignite. Incompatible with metals or oxidants. Will react with water, steam, or acids to produce heat and toxic and corrosive fumes; can react with oxidizing materials. To fight fire, use CO_2, dry chemical. Used as a chlorinating agent, catalyst, and chemical intermediate. Dangerous; when heated to decomposition it emits highly toxic fumes of Cl^- and PO_x.

PHT500 CAS:1314-24-5 ***HR: 3***
PHOSPHORUS TRIOXIDE
DOT: UN 2578
mf: O_3P_2 mw: 110.0

PROP: Transparent, monoclinic crystals or colorless liquid. D: 2.135 @ 21°/4°, mp: 23.8°, bp: 173.1°. Sol in benzene.

SYN: DIPHOSPHORUS TRIOXIDE

DOT Classification: Corrosive Material; Label: Corrosive.

SAFETY PROFILE: A poison. A corrosive irritant to the eyes, skin, and mucous membranes. Melted material readily ignites in air. Incompatible with ammonia; disulfur dichloride; halogens; oxygen; phosphorus pentachloride; sulfur; sulfuric acid; water. When heated to decomposition it emits toxic fumes of PO_x. See also PHOSPHORUS PENTOXIDE.

PHT750 CAS:12165-69-4 ***HR: 3***
PHOSPHORUS TRISULFIDE
DOT: UN 1343
mf: P_2S_3 mw: 158.12

PROP: Gray-yellow crystals. Mp: 290°, bp: 490°.

DOT Classification: Flammable Solid; Label: Flammable Solid.

SAFETY PROFILE: Flammable solid which can react with oxidizers, water, or steam to emit toxic fumes of H_2S. When heated to decomposition it emits very toxic fumes of PO_x and SO_x. See also SULFIDES and PHOSPHORUS COMPOUNDS.

PHU000 CAS:7789-59-5 ***HR: 3***
PHOSPHORYL BROMIDE
DOT: UN 1939/UN 2576
mf: Br_3OP mw: 286.70

PROP: Colorless plates. Mp: 56°, bp: 190°, d: 2.882.

SYNS: PHOSPHOROUS OXYBROMIDE ◇ PHOSPHORYL TRIBROMIDE

CONSENSUS REPORTS: Reported in EPA TSCA Inventory.

DOT Classification: Corrosive Material; Label: Corrosive.

SAFETY PROFILE: Poison by ingestion, inhalation, and skin contact. A corrosive irritant to skin, eyes, and mucous membranes. A corrosive material. Reacts with steam, water to produce much heat with toxic fumes. When heated to decomposition it emits very toxic fumes of Br^- and PO_x. See also BROMIDES and PHOSPHORUS COMPOUNDS.

PHU500 CAS:520-52-5 ***HR: 3***
O-PHOSPHORYL-4-HYDROXY-N,N-DIMETHYLTRYPTAMINE
mf: $C_{12}H_{17}N_2O_4P$ mw: 284.28

SYNS: CY-39 ◇ 3-(2-(DIMETHYLAMINO)ETHYL)-1H-INDOL-4-OL DIHYDROGEN PHOSPHATE ESTER ◇ 3-2'-DIMETHYLAMINOETHYLINDOL-4-PHOSPHATE ◇ 3-(2-DIMETHYLAMINOETHYL) INDOL-4-YL DIHYDROGEN PHOSPHATE ◇ INDOCYBIN ◇ PSILOCIN PHOSPHATE ESTER ◇ PSILOCIPIN ◇ PSILOTSIBIN ◇ TEONANACATL

TOXICITY DATA with REFERENCE
ims-hmn TDLo:75 μg/kg:CNS PSYPAG 3,219,62
orl-hmn TDLo:60 μg/kg:CNS,GIT,EYE JNMDAN 131,428,60
ipr-hmn TDLo:37 μg/kg:CNS,GIT JNMDAN 131,428,60
ims-hmn TDLo:130 μg/kg:CNS PSDTAP 8,59,67
ivn-rat LD50:280 mg/kg 27ZQAG -,138,72
ipr-mus LD50:420 mg/kg MEIEDD 10,1143,83
ivn-mus LD50:285 mg/kg 27ZQAG -,138,72
ivn-rbt LD50:13 mg/kg 27ZQAG -,138,72

CONSENSUS REPORTS: EPA Genetic Toxicology Program.

SAFETY PROFILE: Poison by intravenous route. Moderately toxic by intraperitoneal route. Human systemic effects by ingestion and intraperitoneal routes: euphoria, hallucinations, toxic psychosis, muscle weakness; nausea or vomiting; visual field changes. When heated to decomposition it emits very toxic fumes of NO_x and PO_x.

PHU750 CAS:12067-99-1 ***HR: 2***
PHOSPHOTUNGSTIC ACID

PROP: White to sltly yellow crystals or powder. Sol in water, alc, ether.

SYN: TUNGSTOPHOSPHORIC ACID (8CI)

TOXICITY DATA with REFERENCE
orl-rat LD50:3300 mg/kg TXAPA9 42,417,77

CONSENSUS REPORTS: Reported in EPA TSCA Inventory.

NIOSH REL: (Tungsten) TWA 1 mg(W)/m^3

SAFETY PROFILE: Moderately toxic by ingestion. When heated to decomposition it emits toxic fumes of PO_x.

PHV250 CAS:13366-73-9 ***HR: D***
PHOTODIELDRIN
mf: $C_{12}H_8Cl_6O$ mw: 380.90

SYNS: 1,1,2,3,3A,7A-HEXACHLORO-5,6-EPOXYDECAHYDRO-2,4,7-METHENO-1H-CYCLOPENTA(a)PENTALENE, stereoisomer ◇ NCI-C00599

CONSENSUS REPORTS: NCI Carcinogenesis Bioassay (feed); No Evidence: mouse, rat NCITR* NCI-CG-TR-17,77.

SAFETY PROFILE: When heated to decomposition it emits toxic fumes of Cl^-. See also DIELDRIN.

PHV275 CAS:14459-29-1 ***HR: 3***
PHOTODYN
mf: $C_{34}H_{38}N_4O_6$ mw: 598.76

PROP: Deep red crystals. Insol in water; sol in alc; sparingly sol in ether, chloroform.

SYNS: HEMATOPORPHYRIN ◇ HP ◇ 1,3,5,8-TETRAMETHYL-2,4-BIS(α-HYDROXYETHYL)PROPHINE-6,7-DIPROPIONICACID

TOXICITY DATA with REFERENCE
dnd-nml 100 μmol/L CNREA8 41,3543,81
dni-mus:leu 1 mg/L TUMOAB 67,183,81
ivn-mus LD50:307 mg/kg NYKZAU 57,219,61

SAFETY PROFILE: Poison by intravenous route. Mutation data reported. When heated to decomposition it emits toxic fumes of NO_x.

PHV500 CAS:643-79-8 ***HR: 3***
PHTHALALDEHYDE
mf: $C_8H_6O_2$ mw: 134.14

SYN: PHTALALDEHYDES (FRENCH)

TOXICITY DATA with REFERENCE
unr-mus LDLo:7 mg/kg COREAF 246,851,58

CONSENSUS REPORTS: Reported in EPA TSCA Inventory.

SAFETY PROFILE: Poison by an unspecified route. When heated to decomposition it emits acrid smoke and irritating fumes. See also ALDEHYDES.

PHV725 CAS:56611-65-5 ***HR: 2***
PHTHALAZINOL
mf: $C_{14}H_{16}N_2O_4$ mw: 276.32

SYNS: 7-ETHOXYCARBONYL-4-HYDROXYMETHYL-6,8-DIMETHYL-1(2H)-PHTHALAZINONE ◇ PHTHALAZINOL (PHOSPHODIESTERASE INHIBITOR)

TOXICITY DATA with REFERENCE
orl-rat TDLo:8800 mg/kg (female 7-17D post):REP KSRNAM 16,6365,82
orl-rat TDLo:8800 mg/kg (female 7-17D post):TER KSRNAM 16,6365,82
orl-rat LD50:1750 mg/kg KSRNAM 16,6247,82
ipr-rat LD50:570 mg/kg KSRNAM 16,6247,82
orl-mus LD50:1850 mg/kg KSRNAM 16,6247,82
ipr-mus LD50:780 mg/kg KSRNAM 16,6247,82
orl-rbt LD50:1020 mg/kg KSRNAM 16,6247,82
ipr-rbt LD50:700 mg/kg KSRNAM 16,6247,82

SAFETY PROFILE: Moderately toxic by ingestion and intraperitoneal routes. An experimental teratogen. Experimental reproductive effects. When heated to decomposition it emits toxic fumes of NO_x.

PHV750 CAS:119-39-1 ***HR: 3***
1(2H)PHTHALAZINONE
mf: $C_8H_6N_2O$ mw: 146.16

SYNS: PHTHALAZINONE ◇ PHTHALAZONE

TOXICITY DATA with REFERENCE
orl-rat LD50:370 mg/kg IHFCAY 6,1,67
ivn-mus LD50:180 mg/kg CSLNX* NX#00139

CONSENSUS REPORTS: Reported in EPA TSCA Inventory.

SAFETY PROFILE: Poison by ingestion and intravenous routes. When heated to decomposition it emits toxic fumes of NO_x.

PHW250 CAS:88-99-3 ***HR: 2***
PHTHALIC ACID
mf: $C_8H_6O_4$ mw: 166.14

PROP: Crystals. Mp: >230°, d: 1.59, bp: 155° decomp. Sol in water, alc; sltly sol in ether; insol in chloroform.

SYNS: ACIDE PHTALIQUE (FRENCH) ◇ BENZENE-1,2-DICARBOXYLIC ACID ◇ o-BENZENEDICARBOXYLIC ACID ◇ 1,2-BENZENEDICARBOXYLIC ACID ◇ o-DICARBOXYBENZENE

TOXICITY DATA with REFERENCE
orl-rat LD50:7900 mg/kg JIHTAB 27,130,45
ipr-mus LD50:550 mg/kg PSEBAA 49,471,42

CONSENSUS REPORTS: Reported in EPA TSCA Inventory.

SAFETY PROFILE: Moderately toxic by intraperitoneal route. Slightly toxic by ingestion. Skin and mucous membrane irritant. Combustible when heated. In the form of dust (anhydride) it can explode. Mixtures with sodium nitrite explode when heated. Violent reaction with HNO_3. When heated to decomposition it emits acrid smoke and irritating fumes. Used in synthesis of dyes and dyestuffs, in medicines and perfumes.

PHW500 CAS:25724-58-7 ***HR: 1***
PHTHALIC ACID, DECYL HEXYL ESTER
mf: $C_{22}H_{38}O_4$ mw: 366.60

TOXICITY DATA with REFERENCE
orl-rat LD50:49 g/kg AIHAAP 30,470,69

CONSENSUS REPORTS: Reported in EPA TSCA Inventory.

SAFETY PROFILE: Mildly toxic by ingestion. When heated to decomposition it emits acrid smoke and irritating fumes. See also ESTERS.

PHW750 CAS:85-44-9 ***HR: 3***
PHTHALIC ANHYDRIDE
DOT: UN 2214
mf: $C_8H_4O_3$ mw: 148.12

$C_6H_4CO{\cdot}OCO$ (ring-closing bond drawn between first and last atoms)

PROP: White, crystalline needles. Mp: 131.2°, lel: 1.7%, uel: 10.4%, bp: 295° (sublimes), flash p: 305°F (CC), d: 1.527 @ 4°, autoign temp: 1058°F, vap press: 1 mm @ 96.5°, vap d: 5.10. Very sltly sol in water; sol in alc; sltly sol in ether.

SYNS: ANHYDRIDE PHTALIQUE (FRENCH) ◇ ANIDRIDE FTALICA (ITALIAN) ◇ 1,2-BENZENEDICARBOXYLIC ACID ANHYDRIDE ◇ 1,3-DIOXOPHTHALAN ◇ ESEN ◇ FTAALZUURANHYDRIDE (DUTCH) ◇ FTALOWY BEZWODNIK (POLISH) ◇ 1,3-ISOBENZOFURANDIONE ◇ NCI-C03601 ◇ 1,3-PHTHALANDIONE ◇ PHTHALIC ACID ANHYDRIDE ◇ PHTHALSAEUREANHYDRID (GERMAN) ◇ RCRA WASTE NUMBER U190 ◇ RETARDER AK ◇ RETARDER ESEN ◇ RETARDER PD

TOXICITY DATA with REFERENCE
skn-rbt 500 mg/24H MLD BIOFX* 13-4/70
eye-rbt 100 mg SEV BIOFX* 13-4/70
ipr-mus TDLo:203 mg/kg (8-10D preg):TER TCMUD8 2,61,82
orl-rat LD50:4020 mg/kg BIOFX* 13-4/70
orl-mus LD50:1500 mg/kg 85GMAT -,100,82
orl-cat LD50:800 mg/kg 85JCAE -,322,86
orl-gpg LDLo:100 mg/kg 29ZWAE -,410,68

CONSENSUS REPORTS: NCI Carcinogenesis Bioassay (feed); No Evidence: mouse, rat NCITR* NCI-CG-TR-159,79. Community Right-To-Know List. Reported in EPA TSCA Inventory.

OSHA PEL: (Transitional: TWA 2 ppm) TWA 1 ppm
ACGIH TLV: TWA 1 ppm
DFG MAK: 5 mg/m^3
DOT Classification: Corrosive; Label: Corrosive.

SAFETY PROFILE: Poison by ingestion. Experimental teratogenic effects. A corrosive eye, skin, and mucous membrane irritant. A common air contaminant. Combustible when exposed to heat or flame; can react with oxidizing materials. Moderate explosion hazard in the form of dust when exposed to flame. The production of this material has caused many industrial explosions. Mixtures with copper oxide or sodium nitrite explode when heated. Violent reaction with nitric acid + sulfuric acid above 80°C. To fight fire, use CO_2, dry chemical. Used in plasticizers, polyester resins, and alkyd resins, dyes and drugs. See also ANHYDRIDES.

PHX000 CAS:85-41-6 ***HR: 2***
PHTHALIMIDE
mf: $C_8H_5NO_2$ mw: 147.14

PROP: White to light-tan powder. Mp: 238°, bp: subl. Sol in water, alc, alkali, hot ether; insol in benzene.

SYNS: ISOINDOLE-1,3-DIONE ◇ 1,3-ISOINDOLEDIONE ◇ 1,3-ISOINDOLINEDIONE ◇ o-PHTHALIC IMIDE

TOXICITY DATA with REFERENCE
ipr-mus TDLo:100 mg/kg (9D preg):REP PHARAT 31,172,76
ipr-mus TDLo:6200 μg/kg (9D preg):TER PHARAT 31,172,76
orl-mus LD50:5000 mg/kg FRPSAX 20,3,65
ipr-mus LD50:1175 mg/kg MEXPAG 11,149,64

CONSENSUS REPORTS: Reported in EPA TSCA Inventory.

SAFETY PROFILE: Moderately toxic by intraperitoneal route. Mildly toxic by ingestion. An experimental teratogen. Other experimental reproductive effects. When heated to decomposition it emits toxic fumes of NO_x.

PHX100 CAS:3343-28-0 ***HR: D***
2-PHTHALIMIDOGLUTARIC ACID ANHYDRIDE
mf: $C_{13}H_9NO_5$ mw: 259.23

SYN: 2H-PYRAN-2,6(3H)-DIONE,DIHYDRO-2-PHTHALIMIDO-

TOXICITY DATA with REFERENCE
orl-rbt TDLo:900 mg/kg (female 7-12D post):REP LIFSAK 3,987,64

CONSENSUS REPORTS: Reported in EPA TSCA Inventory.

SAFETY PROFILE: Experimental reproductive effects. When heated to decomposition it emits toxic fumes of NO_x.

PHX250 CAS:732-11-6 ***HR: 3***
PHTHALIMIDOMETHYL-O,O-DIMETHYL PHOSPHORODITHIOATE
mf: $C_{11}H_{12}NO_4PS_2$ mw: 317.33

SYNS: APPA ◇ DECEMTHION P-6 ◇ (O,O-DIMETHYL-PHTHALIMIDIOMETHYL-DITHIOPHOSPHATE) ◇ O,O-DIMETHYL S-(N-PHTHALIMIDOMETHYL)DITHIOPHOSPHATE/SI O,O-DIMETHYL S-PHTHALIMIDOMETHYL PHOSPHORODITHIOATE ◇ ENT 25,705 ◇ FTALOPHOS ◇ IMIDAN ◇ KEMOLATE ◇ N-(MERCAPTOMETHYL) PHTHALIMIDE S-(O,O-DIMETHYL PHOSPHORODITHIOATE) ◇ PERCOLATE ◇ PHOSMET ◇ PHOSPHORODITHIOIC ACID, S-((1,3-DIHYDRO-1,3-DIOXO-ISOINDOL-2-YL)METHYL)O,O-DIMETHYL ESTER ◇ PHTHALIMIDO-O,O-DIMETHYL PHOSPHORODITHIOATE ◇ PHTHALOPHOS ◇ PMP ◇ PROLATE ◇ R 1504 ◇ SMIDAN ◇ STAUFFER R 1504

TOXICITY DATA with REFERENCE
mmo-sat 5 mg/plate MUREAV 116,185,83
mma-sat 5 mg/plate MUREAV 116,185,83
orl-rat TDLo:100 mg/kg (female 1-20D post):REP VETNAL 56(1),62,80
orl-rat TDLo:100 mg/kg (female 1-20D post):TER VETNAL 56(1),62,80
orl-hmn LDLo:50 mg/kg SPEADM 74-1,-,74
ihl-hmn TCLo:2 mg/m^3/8H:NOSE,EYE,CNS HYSAAV 34,192,69
orl-rat LD50:92500 μg/kg GTPZAB 14(12),46,70
ihl-rat LC50:54 mg/m^3/4H 85GMAT -,59,82
skn-rat LD50:1326 mg/kg 85GMAT -,59,82
orl-mus LD50:26 mg/kg HYSAAV 34,192,69
ihl-cat LCLo:65 mg/m^3/4H HYSAAV 34,192,69

CONSENSUS REPORTS: EPA Extremely Hazardous Substances List.

SAFETY PROFILE: A human poison by ingestion. Poison experimentally by inhalation and ingestion routes. Moderately toxic by skin contact. Human systemic effects by inhalation: lacrimation, somnolence, and olfaction effects. Experimental teratogenic and reproductive effects. Mutation data reported. When heated to decomposition it emits very toxic fumes of NO_x, PO_x, and SO_x. See also ESTERS.

PHX550 CAS:626-17-5 ***HR: 3***
m-PHTHALODINITRILE
mf: $C_8H_4N_2$ mw: 128.14

PROP: Colorless crystals; water insol; sol in benzene, acetone; vap d: 4.42; mp: 138°; bp: subl.

SYNS: 1,3-BENZENEDICARBONITRILE ◇ m-DICYANOBENZENE ◇ 1,3-DICYANOBENZENE ◇ DINITRILE of ISOPHTHALIC ACID ◇ IPN ◇ ISOFTALODINITRIL (CZECH) ◇ ISOPHTHALODINITRILE ◇ ISOPHTHALONITRILE ◇ NITRIL KYSELINY ISOFTALOVE (CZECH) ◇ m-PDN

TOXICITY DATA with REFERENCE
eye-rbt 500 mg/24H MLD 28ZPAK -,159,72
orl-rat LD50:1860 mg/kg 28ZPAK -,159,72
orl-mus LD50:178 mg/kg JMCMAR 21,906,78
ipr-mus LD50:481 mg/kg INHEAO 4,11,66
orl-rbt LD50:350 mg/kg 85GMAT -,77,82
orl-gpg LD50:370 mg/kg 85GMAT -,77,82

CONSENSUS REPORTS: Reported in EPA TSCA Inventory. Cyanide and its compounds are on the Community Right-To-Know List.

OSHA PEL: TWA 5 mg/m^3
ACGIH TLV: TWA 5 mg/m^3

SAFETY PROFILE: Poison by ingestion. An eye irritant. When heated to decomposition it emits toxic fumes of NO_x and CN^-. See also NITRILES.

PHY000 CAS:91-15-6 ***HR: 3***
PHTHALONITRILE
mf: $C_8H_4N_2$ mw: 128.14

SYNS: o-DICYANOBENZENE ◇ 1,2-DICYANOBENZENE ◇ PHTHALIC ACID DINITRILE ◇ PHTHALODINITRILE ◇ o-PHTHALODINITRILE ◇ USAF ND-09

TOXICITY DATA with REFERENCE
orl-rat TDLo:7425 mg/kg/66W-I:ETA VOONAW 18(1),81,72
ipr-rat LD50:62 mg/kg KJMDA6 11,63,65
orl-mus LD50:65 mg/kg INHEAO 4,11,66
ipr-mus LD50:25 mg/kg NTIS** AD277-689
scu-mus LD50:46 mg/kg INHEAO 4,11,66

CONSENSUS REPORTS: Cyanide and its compounds are on the Community Right-To-Know List. Reported in EPA TSCA Inventory.

SAFETY PROFILE: Poison by ingestion, subcutaneous, and intraperitoneal routes. Questionable carcinogen with experimental tumorigenic data. When heated to decomposition it emits toxic fumes of CN^- and NO_x. See also NITRILES.

PHY250 CAS:66968-12-5 ***HR: D***
N-PHTHALOYL-l-ASPARTIC ACID
mf: $C_{12}H_9NO_6$ mw: 263.22

TOXICITY DATA with REFERENCE
ipr-mus TDLo:200 mg/kg (9D preg):REP ARZNAD 27,126,77
ipr-mus TDLo:200 mg/kg (9D preg):TER ARZNAD 27,126,77

SAFETY PROFILE: An experimental teratogen. Experimental reproductive effects. When heated to decomposition it emits toxic fumes of NO_x.

PHY275 CAS:50906-29-1 ***HR: 3***
PHTHALOYL DIAZIDE
mf: $C_8H_4N_6O_2$ mw: 216.16

SAFETY PROFILE: Explodes when heated. When heated to decomposition it emits toxic fumes of NO_x. See also AZIDES.

PHY500 ***HR: 3***
PHTHALOYL PEROXIDE
mf: $C_8H_4O_4$ mw: 164.12

SYN: 2,3-BENZODIOXIN-1,4-DIONE

SAFETY PROFILE: An irritant to skin, eyes, mucous membranes. A powerful oxidizer. An explosive sensitive to impact or heating to 123°C. May be polymeric. When heated to decomposition it emits acrid smoke and irritating fumes. See also PEROXIDES.

PHY750 CAS:85-73-4 ***HR: 2***
PHTHALOYLSULFATHIAZOLE
mf: $C_{17}H_{13}N_3O_5S_2$ mw: 403.45

PROP: Crystals. Mp: 272-277° (decomp). Sltly sol in alc; very sltly sol in ether; very sol in NaOH; insol in chloroform, water.

SYNS: (o-CARBOXYBENZOYL)-p-AMINOPHENYLSULFONAMIDOTHIAZOLE ◇ N^4-(o-CARBOXYBENZOYL)-N'-2-THIAZOLYL-SULFANILAMIDE SULFAPHIHALAZOLE ◇ PHTHALAZOL ◇ 2-(N^4)-PHTHALYAMINOBENZENESULFONAMIDE)THIAZOLE ◇ 2-(N^4-PHTHALYAMINOBENZENESULFONAMIDO)THIAZOLE ◇ 2-(p-PHTHALYLAMINOBENZENESULFAMIDO)THIAZOLE ◇ PHTHALYLNORSULFAZOLE ◇ 2-(N^4-PHTHALYLSULFANILAMIDO)THIAZOLE ◇ 2-(p-N-PHTHALYLSULFANILYL) AMINOTHIAZOLE ◇ PHTHALYLSULFATHIAZOLE ◇ PHTHALYLSULFONAZOLE ◇ PHTHALYLSULPHATHIAZOLE ◇ SULFATHALIDINE ◇ 4'-(2-THIAZOLYLSULFAMOYL)PHTHALANILIC ACID ◇ 4'-(2-THIAZOLYLSULFAMYL)PHTHALANILIC ACID

TOXICITY DATA with REFERENCE
ipr-mus LD50:920 mg/kg JPETAB 81,116,44

CONSENSUS REPORTS: Reported in EPA TSCA Inventory.

SAFETY PROFILE: Moderately toxic by intraperitoneal route. When heated to decomposition it emits very toxic fumes of NO_x and SO_x.

PIA000 CAS:3982-20-5 ***HR: 2***
N-PHTHALYL-dl-ASPARTIMIDE
mf: $C_{12}H_8N_2O_4$ mw: 244.22

TOXICITY DATA with REFERENCE
orl-rat TDLo:2 g/kg (3-6D preg):REP JMCMAR 6,464,63
orl-rat LD50:5000 mg/kg JMCMAR 6,464,63
orl-mus LD50:3500 mg/kg JMCMAR 6,464,63

SAFETY PROFILE: Moderately toxic by ingestion. Experimental reproductive effects. When heated to decomposition it emits toxic fumes of NO_x.

PIA250 CAS:69352-40-5 ***HR: D***
N-PHTHALYLISOGLUTAMINE
mf: $C_{13}H_{12}N_2O_5$ mw: 276.27

SYNS: 4-CARBAMOYL-4-PHTHALYLBUTYRIC ACID ◇ 4-PHTHALYLGLUTARAMIC ACID

TOXICITY DATA with REFERENCE
ipr-mus TDLo:100 mg/kg (9D preg):REP ARTODN 35,63,76
ipr-mus TDLo:200 mg/kg (9D preg):TER ARTODN 35,63,76

SAFETY PROFILE: An experimental teratogen. Other experimental reproductive effects. When heated to decomposition it emits toxic fumes of NO_x.

PIA375 ***HR: 3***
PHYSALIA PHYSALIS TOXIN

SYNS: PORTUGUESE MAN-OF-WAR TOXIN ◇ P. PHYSALIS TOXIN ◇ TOXIN, NEMATOCYST, PHYSALIA PHYSALIS

TOXICITY DATA with REFERENCE
ivn-rat LD50:100 μg/kg TOXIA6 4,199,66
par-frg LDLo:15 g/kg 23EIAT 3,395,71
ipr-mus LDLo:37 mg/kg BIBUBX 115,219,58
scu-mus LDLo:2100 mg/kg BIBUBX 115,219,58

SAFETY PROFILE: Poison by intravenous and intraperitoneal routes. Moderately toxic by subcutaneous route.

PIA400 CAS:72497-31-5 ***HR: 3***
PHYSALIN-X

TOXICITY DATA with REFERENCE
orl-rat TDLo:700 mg/kg (8-14D preg):REP IJEBA6 17,690,79
orl-mus LD50:2 mg/kg IJEBA6 17,690,79
ipr-mus LD50:1 g/kg IJEBA6 17,690,79

SAFETY PROFILE: Poison by ingestion. Moderately toxic by intraperitoneal route. Experimental reproductive effects.

PIA500 CAS:57-47-6 ***HR: 3***
PHYSOSTIGMINE
mf: $C_{15}H_{21}N_3O_2$ mw: 275.39

SYNS: ERSERINE ◇ ESERINE ◇ ESEROLEIN, METHYLCARBAMATE (ESTER) ◇ METHYL-CARBAMIC ACID, ESTER with ESEROLINE ◇ PHYSOSTOL

TOXICITY DATA with REFERENCE
ipr-mus TDLo:50 μg/kg (13D preg):REP FEPRA7 31,596,72
orl-man TDLo:12 mg/kg/60D-I NEURAI 37,345,87
unr-man LDLo:882 μg/kg 85DCAI 2,73,70
orl-wmn TDLo:40 μg/kg/5D-I:BPR AJPSAO 143,910,85
orl-wmn TDLo:3920 μg/kg/2W-I NEURAI 37,345,87
orl-hmn TDLo:20 mg/kg:CNS,GIT,PUL 34ZIAG -,475,69
ipr-mus LD50:644 μg/kg FATOAO 43,717,80
scu-mus LD50:740 μg/kg:PNS JPETAB 123,121,58
ivn-mus LD50:400 μg/kg JPETAB 58,337,36
scu-dog LDLo:1138 μg/kg FEPRA7 5,184,46
orl-rbt LD50:11200 μg/kg DCTODJ 3,319,80
ims-rbt LD50:2200 μg/kg DCTODJ 3,319,80
ipr-rat LD50:2 mg/kg ARZNAD 22,1926,72
ivn-rbt LDLo:500 μg/kg JPETAB 43,413,31
scu-pgn LDLo:450 μg/kg HBAMAK 4,1289,35

CONSENSUS REPORTS: EPA Extremely Hazardous Substances List. Reported in EPA TSCA Inventory.

SAFETY PROFILE: A human poison by an unspecified route. Poison experimentally by ingestion, subcutaneous, intramuscular, intravenous, and intraperitoneal routes. Human systemic effects by ingestion: nausea, dyspnea, coma, blood pressure elevation, flaccid paralysis without anesthesia, muscle weakness. Normally administered by injection. Poisoning can occur as a result of a mistake in dosage or due to hypersensitivity of the patient within 5 to 25 minutes after administration. Death usually results from respiratory paralysis. Experimental reproductive effects. Combustible when exposed to heat or flame. When heated to decomposition it emits toxic fumes of NO_x. See also CARBAMATES.

PIA750 CAS:57-64-7 ***HR: 3***
PHYSOSTIGMINE SALICYLATE (1:1)
mf: $C_{15}H_{21}N_3O_2 \cdot C_7H_6O_3$ mw: 413.52

SYNS: ESERINE SALICYLATE ◇ PHYSOSTOL SALICYLATE ◇ SALICYLIC ACID with PHYSOSTIGMINE (1:1) ◇ TL-1380

TOXICITY DATA with REFERENCE
ipr-rat TDLo:200 μg/kg (1D male):REP PHBHA4 4,677,69
ivn-man TDLo:14 μg/kg:CVS,GIT JCLPDE 46,446,85
scu-rat LD50:2 mg/kg TXAPA9 25,569,73
ims-rat LD50:1280 μg/kg DCTODJ 7,507,84
orl-mus LD50:2500 μg/kg AIPTAK 81,726,50
ipr-mus LD50:640 μg/kg TXAPA9 28,227,74
scu-mus LD50:800 μg/kg JPPMAB 34,603,82
ivn-mus LD50:470 μg/kg AIPTAK 81,726,50
scu-dog LDLo:1200 μg/kg NTIS** PB158-508
scu-cat LDLo:800 μg/kg NTIS** PB158-508
scu-rbt LDLo:2 mg/kg AEXPBL 53,313,05
ims-rbt LD50:1570 μg/kg AIPTAK 81,726,50

CONSENSUS REPORTS: EPA Extremely Hazardous Substances List. Reported in EPA TSCA Inventory.

SAFETY PROFILE: Poison by ingestion, subcutaneous, intramuscular, intravenous, and intraperitoneal routes. Human systemic effects: arrhythmias, nausea or vomiting. Experimental reproductive effects. When heated to decomposition it emits toxic fumes of NO_x. See also PHYSOSTIGMINE.

PIB000 CAS:64-47-1 ***HR: 3***
PHYSOSTIGMINE SULFATE
mf: $C_{30}H_{42}N_6O_4S$ mw: 646.84

SYNS: ESERINE SULFATE ◇ ESERINE SULPHATE ◇ 1,2,3,3a,8,8a-HEXAHYDRO-5-HYDROXY-1,3a,8-TRIMETHYL-PYRROLO(2,3-b)INCOLE METHYLCARBAMATE (ester), (3aS-cis)-, SULFATE (2:1) ◇ PHYSOSTIGMINE SO4 ◇ PHYSOSTIGMINE SULFATE (2:1)

TOXICITY DATA with REFERENCE
par-mus TDLo:200 μg/kg (8D preg):REP JPMSAE 62,1626,73
par-mus TDLo:200 μg/kg (8D preg):TER JPMSAE 62,1626,73
scu-man TDLo:191 ng/kg:CNS,GIT AEXPBL 53,313,05
ipr-rat LD50:1621 μg/kg FAATDF 4(2,Pt 2),S195,84
orl-mus LD50:1680 mg/kg 27ZQAG -,137,72
ipr-mus LD50:510 mg/kg ATXKA8 29,39,72
scu-mus LD50:860 μg/kg 27ZQAG -,137,72
ivn-mus LD50:178 μg/kg CSLNX* NX#03167

CONSENSUS REPORTS: EPA Genetic Toxicology Program.

SAFETY PROFILE: Poison by subcutaneous, intravenous, and intraperitoneal routes. Moderately toxic by ingestion. Human systemic effects by subcutaneous route: hallucinations or distorted perceptions; nausea or vomiting, and other gastrointestinal changes. An experimental teratogen. Experimental reproductive effects. When heated to decomposition it emits toxic fumes of NO_x and SO_x. See also PHYSOSTIGMINE.

PIB250 CAS:83-86-3 ***HR: 3***
PHYTIC ACID
mf: $C_6H_{18}O_{24}P_6$ mw: 660.06

SYNS: ALKOVERT ◇ FYTIC ACID ◇ HEXAKIS(DIHYDROGEN PHOSPHATE) MYO-INOSITOL ◇ INOSITHEXAPHOSPHORSAURE (GERMAN) ◇ INOSITOL HEXAPHOSPHATE ◇ MYO-INOSISTOL HEXAKISPHOSPHATE ◇ MYO-INOSITOL HEXAPHOSPHATE ◇ SAURE DES PHYTINS (GERMAN)

TOXICITY DATA with REFERENCE
ivn-mus LD50:500 mg/kg TXCYAC 22,279,82
ivn-rbt LDLo:45 mg/kg HBAMAK 4,1289,35

CONSENSUS REPORTS: Reported in EPA TSCA Inventory.

SAFETY PROFILE: Poison by intravenous route. When heated to decomposition it emits toxic fumes of PO_x.

PIB575 CAS:9008-97-3 ***HR: D***
PHYTOHEMAGGLUTININ

SYNS: FITOHEMAGLUTYNINA (POLISH) ◇ PHA ◇ PHYTOHAEMAGGLUTININ ◇ PHYTOHEMAGLUTININS

TOXICITY DATA with REFERENCE
dns-mus:lym 10 mg/L PLMEAA 51,91,85
cyt-mus:oth 40 μL/plate FOMOAJ 33, 191,74
scu-mus TDLo:120 mg/kg (female 4-5D post):TER FOBGA8 24,213,76
scu-mus TDLo:10 mg/kg (4-5D preg):REP FOBGA8 24,213,76

SAFETY PROFILE: An experimental teratogen. Experimental reproductive effects. Mutation data reported.

PIB600 CAS:150-86-7 ***HR: 1***
PHYTOL
mf: $C_{20}H_{40}O$ mw: 296.60

PROP: Oily liquid. Bp: 203-204°, d: 0.859. Insol in water; sol in organic solvents.

SYNS: 2-HEXADECEN-1-OL, 3,7,11,15-TETRAMETHYL-, (R-(R*,R*-(E)))-(9CI) ◇ trans-PHYTOL ◇ 3,7,11,15-TETRAMETHYL-2-HEXADECEN-1-OL

TOXICITY DATA with REFERENCE
skn-rbt 500 mg/24H MOD FCTOD7 20,811,82

CONSENSUS REPORTS: Reported in EPA TSCA Inventory.

SAFETY PROFILE: A skin irritant. When heated to decomposition it emits acrid smoke and irritating fumes.

PIB700 CAS:79201-85-7 ***HR: 3***
PICENADOL
mf: $C_{16}H_{25}NO•ClH$ mw: 283.88

SYNS: (±)-3-(1,3-DIMETHYL-4-PROPYL-4-PIPERIDINYL)PHENOL HYDROCHLORIDE ◇ trans-(±)-3-(1,3-α-DIMETHYL-4-α-PROPYL-4-β-PIPERIDINYL)PHENOL HYDROCHLORIDE ◇ LY150720

TOXICITY DATA with REFERENCE
orl-rat LD50:887 mg/kg TOLED5 18(Suppl 1),143,83
scu-rat LD50:333 mg/kg TOLED5 18(Suppl 1),143,83
ivn-rat LD50:67 mg/kg TOLED5 18(Suppl 1),143,83
orl-mus LD50:305 mg/kg TOLED5 18(Suppl 1),143,83
scu-mus LD50:172 mg/kg TOLED5 18(Suppl 1),143,83
ivn-mus LD50:65 mg/kg TOLED5 18(Suppl 1),143,83
ims-dog LD50:80 mg/kg TOLED5 18(Suppl 1),143,83
ims-cat LD50:40 mg/kg TOLED5 18(Suppl 1),143,83

SAFETY PROFILE: Poison by ingestion, subcutaneous, intramuscular, and intravenous routes. A centrally acting analgesic. When heated to decomposition it emits toxic fumes of NO_x and HCl.

PIB750 CAS:213-46-7 ***HR: 3***
PICENE
mf: $C_{22}H_{14}$ mw: 278.36

PROP: Leaflets. Mp: 364°, bp: 520°. Insol in water; very sltly sol in alc, ether.

SYNS: 3,4-BENZCHRYSENE ◇ BENZO(a)CHRYSENE ◇ β,β-BINAPHTHYLENEETHENE ◇ DIBENZO(a,i)PHENANTHRENE ◇ 1,2:7,8-DIBENZOPHENANTHRENE

TOXICITY DATA with REFERENCE
skn-mus TDLo:111 mg/kg:NEO JNCIAM 50,1717,73

SAFETY PROFILE: Questionable carcinogen with experimental neoplastigenic data. When heated to decomposition it emits acrid smoke and irritating fumes.

PIB900 CAS:1918-02-1 ***HR: 3***
PICLORAM
mf: $C_6H_3Cl_3N_2O_2$ mw: 241.46

PROP: Crystals. Mp: 218°.

SYNS: AMDON GRAZON ◇ 4-AMINO-3,5,6-TRICHLOROPICOLINIC ACID ◇ 4-AMINO-3,5,6-TRICHLORO-2-PICOLINIC ACID ◇ 4-AMINO-3,5,6-TRICHLORPICOLINSAEURE (GERMAN) ◇ ATCP ◇ BOROLIN ◇ CHLORAMP (RUSSIAN) ◇ K-PIN ◇ NCI-C00237 ◇ TORDON ◇ TORDON 10K ◇ TORDON 22K ◇ TORDON 101 MIXTURE ◇ 3,5,6-TRICHLORO-4-AMINOPICOLINIC ACID

TOXICITY DATA with REFERENCE
mmo-smc 100 mg/L TGANAK 18,455,84
orl-rat TDLo:5 g/kg (6-15D preg):TER FCTXAV 10,797,72
orl-rat TDLo:209 mg/kg/80W-C:CAR JTEHD6 7,207,81
orl-mus TDLo:340 g/kg/80W-C:NEO JTEHD6 7,207,81
orl-rat TD:417 g/kg/80W-C:CAR JTEHD6 7,207,81
orl-rat TD:208 g/kg/80W-C:ETA NCITR* NCI-CG-TR-23,78
orl-rat LD50:2898 mg/kg GNAMAP 15,38,76
orl-mus LD50:1061 mg/kg GNAMAP 15,38,76
orl-rbt LD50:2000 mg/kg GUCHAZ 6,414,73
orl-gpg LD50:1922 mg/kg GNAMAP 15,38,76
orl-ckn LD50:4000 mg/kg PCOC** -,1144,66

CONSENSUS REPORTS: NCI Carcinogenesis Bioassay (feed); No Evidence: mouse NCITR* NCI-CG-TR-23,78; Clear Evidence: rat NCITR* NCI-CG-TR-23,78

OSHA PEL: (Transitional: Total Dust: 15 mg/m^3; Respirable Fraction: 5 mg/m^3) TWA Total Dust: 10 mg/m^3; Respirable Fraction: 5 mg/m^3
ACGIH TLV: TWA 10 mg/m^3

SAFETY PROFILE: Moderately toxic by ingestion. Questionable carcinogen with experimental carcinogenic, neoplastigenic, tumorigenic, and teratogenic data. An experimental teratogen. Mutation data reported. When heated to decomposition it emits very toxic fumes of Cl^- and NO_x.

PIB920 CAS:108-99-6 ***HR: 3***
3-PICOLINE
DOT: UN 2313
mf: C_6H_7N mw: 93.14

PROP: Colorless liquid; sweetish, not unpleasant odor. D: (15°/4°) 0.9613, bp: 143-144°, n (24/D) 1.5043. Misc with water, alc, ether.

SYNS: 3-METHYLPYRIDINE ◇ β-PICOLINE ◇ m-PICOLINE (DOT)

TOXICITY DATA with REFERENCE
ipr-rat LD50:150 mg/kg FAATDF 5,920,85
ipr-mus LD50:596 mg/kg JPETAB 88,82,46
ivn-mus LD50:298 mg/kg JPETAB 88,82,46
orl-qal LD50:1 g/kg AECTCV 12,355,83
orl-bwd LD50:1 g/kg AECTCV 12,355,83

CONSENSUS REPORTS: Reported in EPA TSCA Inventory.

DOT Classification: Flammable or Combustible Liquid; Label: Flammable Liquid.

SAFETY PROFILE: Poison by intravenous and intraperitoneal routes. Moderately toxic by ingestion. Flammable when exposed to heat or flame; can react vigorously with oxidizing materials. When heated to decomposition it emits toxic fumes of NO_x.

PIB925 CAS:3608-75-1 ***HR: 3***
PICOLINE-2-ALDEHYDE THIOSEMICARBAZONE
mf: $C_7H_8N_4S$ mw: 180.25

SYNS: 2-FORMYLPYRIDINE THIOSEMICARBAZONE ◇ NSC 729 ◇ 2-PYRIDINECARBOXALDEHYDE THIOSEMICARBAZONE ◇ PYRIDINE-2-CARBOXALDEHYDETHIOSEMICARBAZONE

TOXICITY DATA with REFERENCE
oms-esc 2500 μmol/L BCPCA6 21,3213,72
orl-mus LD50:70 mg/kg JMCMAR 8,676,65
ipr-mus LD50:30910 μg/kg NCISP* JAN86

SAFETY PROFILE: Poison by ingestion and intraperitoneal routes. Mutation data reported. When heated to decomposition it emits toxic fumes of NO_x and SO_x. See also ALDEHYDES.

PIB930 CAS:98-98-6 ***HR: 3***
PICOLINIC ACID
mf: $C_6H_5NO_2$ mw: 123.12

PROP: Needles from water, alc or benzene. Mp: 134-136°, sublimes. Very sol in glacial acetic acid; practically insol in ether, chloroform, carbon disulfide.

SYNS: ACIDE PICOLIQUE (FRENCH) ◇ 2-CARBOSYPYRIDINE ◇ o-PYRIDINECARBOXYLIC ACID ◇ α-PYRIDINECARBOXYLIC ACID ◇ 2-PYRIDINECARBOXYLIC ACID (9CI) ◇ PYRIDINE-CARBOXYLIQUE-2 (FRENCH)

TOXICITY DATA with REFERENCE
ipr-mus LD50:360 mg/kg CPBTAL 17,2377,69
ivn-mus LD50:487 mg/kg THERAP 23,1343,68
orl-qal LD50:562 mg/kg AECTCV 12,355,83
orl-bwd LD50:178 mg/kg AECTCV 12,355,83

CONSENSUS REPORTS: Reported in EPA TSCA Inventory.

SAFETY PROFILE: Poison by ingestion and intraperitoneal routes. Moderately toxic by intravenous route. When heated to decomposition it emits toxic fumes of NO_x.

PIC000 CAS:6959-47-3 ***HR: 3***
2-PICOLYL CHLORIDE HYDROCHLORIDE
mf: $C_6H_6ClN•ClH$ mw: 164.04

SYNS: 2-(CHLOROMETHYL) PYRIDINE HYDROCHLORIDE ◇ NCI-C03907 ◇ 2-PYRIDYLMETHYLCHLORIDE HYDROCHLORIDE

TOXICITY DATA with REFERENCE
mmo-sat 1 mg/plate NTPTB* APR 82
mma-sat 1 mg/plate NTPTB* APR 82
orl-rat LD50:316 mg/kg NCILB* NIH-NCI-E-C-72-3252
orl-mus LD50:316 mg/kg NCILB* NIH-NCI-E-C-72-3252

CONSENSUS REPORTS: NCI Carcinogenesis Bioassay (gavage); No Evidence: mouse, rat NCITR* NCI-CG-TR-178,79.

SAFETY PROFILE: Poison by ingestion. Mutation data reported. When heated to decomposition it emits very toxic fumes of NO_x and Cl^-. Used as a chemical intermediate.

PIC100 CAS:24656-22-2 ***HR: 3***
PICOPERINE TRIPALMITATE
mf: $C_{19}H_{25}N_3•C_{48}H_{96}O_6$ mw: 1067.91

SYNS: COBEN P ◇ N-(2-PICOLYL)-N-PHENYL-N-(2-PIPERIDINOETHYL)AMINE TRIPALMITATE ◇ N-(2-PIPERIDINOETHYL)-N-(2-PYRIDYLMETHYL)ANILINE TRIPALMITATE ◇ 1-(2-N-(2-PYRIDYLMETHYL)ANILINO)ETHYL)PIPERIDINE TRIPALMITATE ◇ N-(2-PYRIDYLMETHYL)-N-PHENYL-N-2-(PIPERIDINOETHYL)AMINE TRIPALMITATE ◇ TAT-3 TRIPALMITATE

TOXICITY DATA with REFERENCE
orl-rat LD50:3100 mg/kg KSRNAM 4,447,70
ipr-rat LD50:299 mg/kg KSRNAM 4,447,70
orl-mus LD50:374 mg/kg OYYAA2 4,171,70
ipr-mus LD50:275 mg/kg KSRNAM 4,447,70
orl-dog LDLo:500 mg/kg ARZNAD 19,1916,69

SAFETY PROFILE: Poison by ingestion and intraperitoneal routes. When heated to decomposition it emits toxic fumes of NO_x.

PIC250 CAS:466-24-0 ***HR: 3***
PICRACONITINE
mf: $C_{32}H_{45}NO_{10}$ mw: 603.78

SYNS: BENZACONINE ◇ BENZOYLACONINE ◇ ISACONITINE

TOXICITY DATA with REFERENCE
ipr-mus LD50:27 mg/kg ARZNAD 5,324,55
ivn-mus LD50:10100 μg/kg YHHPAL 19,641,84
ivn-cat LD50:24 mg/kg ARZNAD 5,324,55
ivn-gpg LD50:24 mg/kg ARZNAD 5,324,55

SAFETY PROFILE: Poison by intravenous and intraperitoneal routes. When heated to decomposition it emits toxic fumes of NO_x.

PIC500 CAS:831-52-7 ***HR: 3***
PICRAMIC ACID, SODIUM SALT
DOT: UN 0235/UN 1349
mf: $C_6H_4N_3O_5•Na$ mw: 221.12

PROP: Yellow, water-sol crystals.

SYN: SODIUM PICRAMATE, WET (DOT)

CONSENSUS REPORTS: Reported in EPA TSCA Inventory.

DOT Classification: Flammable Solid; Label: Flammable Solid (UN1349); Class B Explosive; Label: Explosive B (UN0235)

SAFETY PROFILE: A flammable solid. When heated to decomposition it emits toxic fumes of NO_x and Na_2O. See also PICRAMIC ACID.

PIC750 CAS:63868-82-6 ***HR: 3***
PICRAMIC ACID, ZIRCONIUM SALT (WET)
DOT: UN 0236/UN 1517
mf: $C_6H_5N_3O_5•1/4Zr$ mw: 221.95

SYN: ZIRCONIUM PICRAMATE, WET (DOT)

OSHA PEL: (Transitional: TWA 5 mg(Zr)/m^3) TWA 5 mg(Zr)/m^3; STEL 10 mg(Zr)/m^3
ACGIH TLV: TWA 5 mg(Zr)/m^3; STEL 10 mg(Zr)/m^3
DFG MAK: 5 mg(Zr)/m^3
DOT Classification: Flammable Solid; Label: Flammable Solid (UN1517); Oxidizer; Label: Oxidizer (UN1517); Class B Explosive; Label: Explosive B (UN0236).

SAFETY PROFILE: Flammable when exposed to heat or flame; can react vigorously with oxidizing materials. When heated to decomposition it emits toxic fumes of NO_x. See also ZIRCONIUM COMPOUNDS and PICRAMIC ACID.

PIC899 *HR: 3*
PICRATES

PROP: Salts of picric acid. The general chemical structure is $(O_2N)_3C_6H_2O^-$

SAFETY PROFILE: Most are poisons. Extremely explosive; sensitive to heat, friction, and impact. Picrates should not be heated above 210°C. The nickel salt is particularly dangerous. When heated to decomposition it emits toxic fumes of NO_x. See also PICRIC ACID, NITRO COMPOUNDS OF AROMATIC HYDROCARBONS and specific compounds.

PID000 CAS:88-89-1 *HR: 3*
PICRIC ACID
DOT: UN 0154/UN 1344
mf: $C_6H_3N_3O_7$ mw: 229.12

PROP: Yellow crystals or yellow liquid; very bitter. Mp: 121.8°, bp: explodes > 300°, flash p: 302° F, d: 1.763, autoign temp: 572°F, vap d: 7.90.

SYNS: ACIDE PICRIQUE (FRENCH) ◇ ACIDO PICRICO (ITALIAN) ◇ CARBAZOTIC ACID ◇ C.I. 10305 ◇ 2-HYDROXY-1,3,5-TRINITROBENZENE ◇ MELINITE ◇ NITROXANTHIC ACID ◇ PHENOL TRINITRATE ◇ PICRONITRIC ACID ◇ PIKRINEZUUR (DUTCH) ◇ PIKRINSAEURE (GERMAN) ◇ PIKRYNOWY KWAS (POLISH) ◇ 2,4,6-TRINITROFENOL (DUTCH) ◇ 2,4,6-TRINITROFENOLO (ITALIAN) ◇ 1,3,5-TRINITROPHENOL ◇ 2,4,6-TRINITROPHENOL

TOXICITY DATA with REFERENCE
mma-sat 5 μmol/plate MUREAV 90,91,81
sln-dmg-par 1500 ppm ENMUDM 7,677,85
scu-dog LDLo:60 mg/kg HBAMAK 4,1289,35
orl-cat LDLo:250 mg/kg HBAMAK 4,1289,35
orl-rbt LDLo:120 mg/kg XPHBAO 271,151,41
orl-gpg LDLo:100 mg/kg HBAMAK 4,1289,35
scu-pgn LDLo:200 mg/kg HBAMAK 4,1289,35

CONSENSUS REPORTS: Community Right-To-Know List. Reported in EPA TSCA Inventory. EPA Genetic Toxicology Program.

OSHA PEL: TWA 0.1 mg/m^3 (skin)
ACGIH TLV: TWA 0.1 mg/m^3
DFG MAK: 0.1 mg/m^3
DOT Classification: Class A Explosive; Label: Explosive A (UN0154); Flammable Solid; Label: Flammable Solid (NA1344, UN1344)

SAFETY PROFILE: Poison by ingestion and subcutaneous routes. Mutation data reported. An irritant and an allergen. Skin contact can cause local and systemic allergic reactions. Flammable solid when exposed to heat or flame; can react vigorously with oxidizing materials. Very unstable. A severe explosion hazard when shocked or exposed to heat. It forms salts easily, and many of its salts, known as picrates, are more sensitive explosives than picric acid. It forms unstable salts with concrete; NH_3; bases; and metals (e.g., copper; lead; mercury; and zinc). Many of these are heat-, friction-, or impact-sensitive. Mixtures with uronium perchlorate are extremely powerful explosives. Mixtures with aluminum and water ignite after a delay period. Can react vigorously with reducing materials. Used in synthesis of dyes, as a drug, to manufacture explosives and matches, to etch copper and make colored glass. See also NITRO COMPOUNDS OF AROMATIC HYDROCARBONS, and EXPLOSIVES, HIGH.

PID250 CAS:88-89-1 *HR: 3*
PICRIC ACID (dry)

CONSENSUS REPORTS: Community Right-To-Know List. Reported in EPA TSCA Inventory.

DOT Classification: Class A Explosive; Label: Explosive A.

SAFETY PROFILE: An explosive. Upon decomposition it emits highly toxic fumes of NO_x. See also EXPLOSIVES, HIGH; and PICRIC ACID.

PID500 CAS:14798-26-6 *HR: 3*
PICRIC ACID, ion(1$^-$) (dry)
mf: $C_6H_2N_3O_7$ mw: 228.11

SYNS: PICRATE, DRY (DOT) ◇ PICRATE ION

CONSENSUS REPORTS: Community Right-To-Know List.

DOT Classification: Class A Explosive; Label: Explosive A.

SAFETY PROFILE: A sensitive explosive. When heated to decomposition it emits toxic fumes of NO_x. See also EXPLOSIVES, HIGH; and PICRIC ACID.

PID750 CAS:88-89-1 *HR: 3*
PICRIC ACID (wet)

SYN: PICRIC ACID, WET (DOT)

CONSENSUS REPORTS: Community Right-To-Know List. Reported in EPA TSCA Inventory.

DOT Classification: Class A Explosive; Label: Explosive A.

SAFETY PROFILE: An explosive. See also PICRIC ACID and EXPLOSIVES, HIGH.

PIE000 CAS:550-74-3 *HR: D*
PICROLONIC ACID
mf: $C_{10}H_8N_4O_5$ mw: 264.22

PROP: Yellow needles from alc. Mp: 121-122°, bp: decomp @ 125°. Sol in cold and hot water, alc, ether, methyl alcohol.

SYN: 3-METHYL-4-NITRO-1-(p-NITROPHENYL)-2-PYRAZOLIN-5-ONE

TOXICITY DATA with REFERENCE
mmo-sat 1 mg/plate MUREAV 54,101,78
dnr-esc 1 mg/plate MUREAV 54,101,78

CONSENSUS REPORTS: Reported in EPA TSCA Inventory.

SAFETY PROFILE: Mutation data reported. When heated to decomposition it emits toxic fumes of NO_x.

PIE500 CAS:124-87-8 ***HR: 3***
PICROTOXIN
DOT: UN 1584
mf: $C_{15}H_{18}O_7 \cdot C_{15}H_{16}O_6$ mw: 602.64

PROP: Dried fruit of *Anamerta cocculus (L.)* containing meni-spermine, paramenispermine, 1% picrotoxin, pictrotoxic acid, cocculine alkaloid, and 5% fat.

SYNS: COCCULIN ◇ COCCULUS ◇ COCCULUS solid (DOT) ◇ COQUES DU LEVANT (FRENCH) ◇ FISH BERRY ◇ INDIAN BERRY ◇ ORIENTAL BERRY ◇ PICROTIN, compounded with PICROTOXININ (1:1) ◇ PICROTOXINE

TOXICITY DATA with REFERENCE
orl-hmn LDLo:357 μg/kg:CNS,GIT 34ZIAG -,476,69
ipr-rat LD50:1990 μg/kg ARZNAD 14,996,64
scu-rat LD50:2880 μg/kg JAPMA8 38,604,49
ivn-rat LD50:1600 μg/kg ARZNAD 14,996,64
orl-mus LD50:15 mg/kg JPETAB 128,176,60
ipr-mus LD50:11 mg/kg TXAPA9 13,307,68
scu-mus LD50:4100 μg/kg RPOBAR 1,423,64
ivn-mus LD50:2440 μg/kg AIPTAK 135,9,62
scu-dog LDLo:1500 μg/kg HBAMAK 4,1385,35
ims-dog LDLo:1 mg/kg HBAMAK 4,1385,35
orl-cat LDLo:1750 μg/kg HBAMAK 4,1385,35
scu-cat LDLo:2 mg/kg HBAMAK 4,1385,35
ice-cat LDLo:25 μg/kg AIPTAK 135,9,62

CONSENSUS REPORTS: EPA Extremely Hazardous Substances List.

DOT Classification: Poison B; Label: Poison.

SAFETY PROFILE: A human poison by ingestion. Poison experimentally by most routes. Human systemic effects by ingestion: somnolence, gastrointestinal effects. An alkaloid convulsant poison. When heated to decomposition it emits acrid smoke and irritating fumes.

PIE510 CAS:17617-45-7 ***HR: 3***
PICROTOXININ
mf: $C_{15}H_{16}O_6$ mw: 292.31

PROP: Very bitter large prisms or small crystals containing water. Mp: 209.5°. Sol in hot common organic solvents and in cold alc and chloroform.

SYN: PICROTOXININE

TOXICITY DATA with REFERENCE
scu-mus LDLo:1600 μg/kg JPHAA3 23,98,34
ice-mus LDLo:100 μg/kg TXAPA9 66,290,82
scu-dog LDLo:1100 μg/kg FDWU** -,-,31
scu-rbt LDLo:1350 μg/kg FDWU** -,-,31
scu-pgn LDLo:1600 μg/kg FDWU** -,-,31
ims-pgn LDLo:1600 μg/kg JPHAA3 23,98,34
scu-frg LD50:1100 μg/kg FDWU** -,-,31

SAFETY PROFILE: Poison by subcutaneous, intracerebral, and intramuscular routes. When heated to decomposition it emits acrid smoke and irritating fumes.

PIE525 CAS:1600-31-3 ***HR: 2***
PICRYL AZIDE
mf: $C_6H_2N_6O_6$ mw: 254.12

$(O_2N)_3C_6H_2N_3$

SYN: 2,4,6-TRINITROPHENYL AZIDE

SAFETY PROFILE: A weak explosive sensitive to impact. When heated to decomposition it emits toxic fumes of NO_x. See also AZIDES.

PIE550 CAS:82177-75-1 ***HR: 3***
2-PICRYL-5-NITROTETRAZOLE
mf: $C_7H_2N_8O_8$ mw: 326.14

$(O_2N)_3C_6H_2NN{=}NC(NO_2){=}N$ (ring closure between first N and terminal N)

SAFETY PROFILE: A powerful explosive which is relatively insensitive to impact. When heated to decomposition it emits toxic fumes of NO_x.

PIE750 CAS:102517-11-3 ***HR: 3***
PIFARNINE METHANESULFONATE
mf: $C_{27}H_{40}N_2O_2 \cdot xCH_4O_3S$ mw: 1097.46

SYNS: PIFAZINE METHANESULFONATE ◇ 1-PIPERONYL-4-(3,7,11-TRIMETHYL-2,6,10-DODECANTRIENYL)-PIPERAZINEMETHANESULFONATE ◇ U 27 METHANESULFONATE

TOXICITY DATA with REFERENCE
ivn-rat LD50:33300 μg/kg ARZNAD 25,580,75
orl-dog LD50:1 g/kg ARZNAD 25,580,75
orl-mus LD50:2175 mg/kg DRFUD4 1,354,76
orl-rat LD50:2610 mg/kg ARZNAD 25,580,75
orl-rbt LD50:1 g/kg ARZNAD 25,580,75
ivn-mus LD50:40600 μg/kg DRFUD4 1,354,76

SAFETY PROFILE: Poison by intravenous route. Moderately toxic by ingestion. When heated to decomposition it emits very toxic fumes of SO_x and NO_x.

PIF000 CAS:92-13-7 ***HR: 3***
PILOCARPINE
mf: $C_{11}H_{16}N_2O_2$ mw: 208.29

PROP: Colorless or yellow, hygroscopic, needle-like crystals. Mp: 34°, bp: 260° @ 5 mm.

SYNS: ALMOCARPINE ◇ (3S-cis)-3-ETHYLDIHYDRO-4-((1-METHYL-1H-IMIDAZOL-5-YL)METHYL)-2(3H)-FURANONE ◇ α-ETHYL-β-(HYDROXYMETHYL)-1-METHYL-IMIDAZOLE-5-BUTYRIC ACID, Γ-LACTONE ◇ PILOCARPOL

TOXICITY DATA with REFERENCE
scu-hmn LDLo:143 μg/kg CONEAT 10,8,49
orl-rat LD50:402 mg/kg IYKEDH 12,1204,81
ipr-rat LD50:166 mg/kg NIIRDN 6,APP-16,82
scu-rat LD50:366 mg/kg NIIRDN 6,APP-16,82
ivn-rat LD50:88500 μg/kg NIIRDN 6,APP-16,82
orl-mus LD50:119 mg/kg NIIRDN 6,APP-16,82
scu-mus LD50:90900 μg/kg NIIRDN 6,APP-16,82
ivn-mus LD50:61900 μg/kg NIIRDN 6,APP-16,82
ivn-rbt LDLo:120 mg/kg HBAMAK 4,1386,35

CONSENSUS REPORTS: Reported in EPA TSCA Inventory.

SAFETY PROFILE: A human poison by subcutaneous route. Poison experimentally by ingestion, intravenous, intraperitoneal, and subcutaneous routes. A very poisonous alkaloid which is used to remove excess fluid accumulations from the body. Its action on the sweat glands makes it a powerful sudorific. It very rarely causes death, but when it does, it is by paralysis of the heart or edema of the lungs. Dangerous; on heating to decomposition it emits toxic fumes of NO_x.

PIF250 CAS:54-71-7 ***HR: 3***
PILOCARPINE MONOHYDROCHLORIDE
mf: $C_{11}H_{16}N_2O_2{\cdot}ClH$ mw: 244.75

SYNS: ALMOCARPINE ◇ AMI-PILO ◇ AMISTURA P ◇ ISOPTO-CARPINE ◇ MI-PILO OPHTH SOL ◇ PILOCARPINE HYDROCHLORIDE ◇ PILOCARPINE MURIATE ◇ PILOCEL ◇ PILOMIOTIN ◇ PILOVISC

TOXICITY DATA with REFERENCE
scu-rat TDLo:130 mg/kg (7-19D preg):REP NYKZAU 85,79,85
scu-rbt TDLo:20 mg/kg (24-27D preg):TER AJANA2 154,163,79
ocu-man TDLo:200 μg/kg/7H-I:CVS AIMDAP 147,586,87
orl-mus LD50:200 mg/kg NIIRDN 6,646,82
ipr-mus LD50:155 mg/kg ATXKA8 29,39,72
ivn-pgn LDLo:353 mg/kg HBAMAK 4,1289,35

CONSENSUS REPORTS: Reported in EPA TSCA Inventory.

SAFETY PROFILE: Poison by ingestion, intraperitoneal, and intravenous routes. Experimental teratogenic and reproductive effects. Human systemic effects: cardiac changes. When heated to decomposition it emits very toxic fumes of HCl and NO_x. See also PILOCARPINE.

PIF500 CAS:148-72-1 ***HR: 2***
PILOCARPINE MONONITRATE
mf: $C_{11}H_{16}N_2O_2{\cdot}NO_3$ mw: 270.30

SYN: PILOCARPINE NITRATE

TOXICITY DATA with REFERENCE
orl-rat LD50:911 mg/kg PHMCAA 4,176,62

CONSENSUS REPORTS: Reported in EPA TSCA Inventory.

SAFETY PROFILE: Moderately toxic by ingestion. When heated to decomposition it emits toxic fumes of NO_x. See also PILOCARPINE and NITRATES.

PIF750 CAS:7681-93-8 ***HR: 3***
PIMARICIN
mf: $C_{33}H_{47}NO_{13}$ mw: 665.81

PROP: An antibiotic produced by a strain of *Streptomyces chattanoogensis* (85ERAY 2,956,78).

SYNS: ANTIBIOTIC A-5283 ◇ CL 12,625 ◇ MYCOPHYT ◇ MYPROZINE ◇ NATACYN ◇ NATAMYCIN ◇ PIMAFUCIN ◇ TENNECETIN

TOXICITY DATA with REFERENCE
orl-rat LD50:2730 mg/kg TXAPA9 8,97,66
ipr-rat LD50:85 mg/kg ANTCAO 9,406,59
scu-rat LD50:190 mg/kg ANTCAO 9,406,59
ivn-rat LD50:36 mg/kg ANTCAO 9,406,59
ims-rat LD50:128 mg/kg ANTCAO 9,406,59
orl-mus LD50:1500 mg/kg 85FZAT -,517,67
ipr-mus LD50:96 mg/kg NIIRDN 6,639,82
ivn-dog LD50:18 mg/kg ANTCAO 9,406,59
orl-rbt LD50:1420 mg/kg TXAPA9 8,97,66

SAFETY PROFILE: Poison by intravenous, intramuscular, subcutaneous, and intraperitoneal routes. Moderately toxic by ingestion. When heated to decomposition it emits toxic fumes of NO_x. Used as an antibacterial agent.

PIG000 CAS:111-16-0 ***HR: 1***
PIMELIC ACID
mf: $C_7H_{12}O_4$ mw: 160.19

PROP: Minerals from water. D: 1.291 @ 25°/4°, mp: 103-105°, bp: 272° @ 100 mm. Sol in water; very sol in alc, ether.

SYNS: HEPTANDIOIC ACID ◇ HEPTANEDIOIC ACID ◇ HEPTANE-1,7-DIOIC ACID ◇ 1,7-HEPTANEDIOIC ACID ◇ 1,5-PENTANEDICARBOXYLIC ACID

TOXICITY DATA with REFERENCE
orl-rat LD50:7000 mg/kg 34ZIAG -,477,69
orl-mus LD50:4800 mg/kg BIJOAK 34,1196,40

CONSENSUS REPORTS: Reported in EPA TSCA Inventory.

SAFETY PROFILE: Mildly toxic by ingestion. When heated to decomposition it emits acrid smoke and irritating fumes.

PIG250 CAS:60172-03-4 ***HR: 3***
4,4′-(PIMELOYLBIS(IMINO-p-PHENYLENEIMINO))BIS(1-ETHYLPYRIDINIUM) DIPERCHLORATE
mf: $C_{33}H_{40}N_6O_2 \cdot 2ClO_4$ mw: 751.69

TOXICITY DATA with REFERENCE
dnd-mus:lym 42 μmol/L JMCMAR 22,134,79
ipr-mus LD10:20 mg/kg JMCMAR 22,134,79

SAFETY PROFILE: Poison by intraperitoneal route. Mutation data reported. When heated to decomposition it emits very toxic fumes of NO_x and Cl^-. See also PERCHLORATES.

PIG500 CAS:60172-01-2 ***HR: 3***
4,4′-(PIMELOYLBIS(IMINO-p-PHENYLENEIMINO))BIS(1-METHYLPYRIDINIUM) DIBROMIDE
mf: $C_{31}H_{36}N_6O_2 \cdot 2Br$ mw: 684.55

TOXICITY DATA with REFERENCE
dnd-mus:lym 42 μmol/L JMCMAR 22,134,79
ipr-mus LD10:30 mg/kg JMCMAR 22,134,79

SAFETY PROFILE: Poison by intraperitoneal route. Mutation data reported. When heated to decomposition it emits very toxic fumes of Br^- and NO_x.

PIG730 CAS:8016-45-3 ***HR: 2***
PIMENTA LEAF OIL

PROP: Main constituent is eugenol. From steam distillation of the shrub *Pimenta officinalis* Lindl. (Fam. *Myrtaceae).* (FCTXAV 12,807,74). Pale yellow to brown liquid; spicy odor. D: 1.037-1.050, refr index: 1.531 @ 20°. Sol in propylene glycol, fixed oils; insol in glycerin, mineral oil.

SYN: OIL of PIMENTA LEAF

TOXICITY DATA with REFERENCE
skn-rbt 500 mg/24H SEV FCTXAV 12,807,74
orl-rat LD50:3600 mg/kg FCTXAV 12,807,74

CONSENSUS REPORTS: Reported in EPA TSCA Inventory.

SAFETY PROFILE: Moderately toxic by ingestion. A severe skin irritant. When heated to decomposition it emits acrid smoke and irritating fumes. See also EUGENOL.

PIG740 ***HR: 2***
PIMENTA OIL

PROP: Contains eugenol. Distilled from the fruit of *Pimenta officinalis* Lindley (Fam. *Myrtaceae*). Yellow to red-yellow liquid; odor and taste of allspice. D: 1.018-1.048, refr index: 1.527-1.540 @ 20°.

SYNS: ALLSPICE ◇ PIMENTA BERRIES OIL ◇ PIMENTO OIL

SAFETY PROFILE: A weak sensitizer which may cause dermatitis on local contact. Eugenol is moderately toxic. Combustible. See also EUGENOL.

PIH000 CAS:2062-78-4 ***HR: 3***
PIMOZIDE
mf: $C_{28}H_{29}F_2N_3O$ mw: 461.60

SYNS: 1-(4,4-BIS(p-FLUOROPHENYL)BUTYL)-4-(2-OXO-1-BENZIMIDAZOLINYL)PIPERIDINE ◇ 1-(1-(4,4-BIS(p-FLUOROPHENYL) BUTYL)-4-PIPERIDYL)-2-BENZIMIDAZOLINONE ◇ McN-JR-6238 ◇ ORAP ◇ R 6238

TOXICITY DATA with REFERENCE
orl-rat TDLo:49 mg/kg (male 2W pre):REP KSRNAM 14,2163,80
orl-mus TDLo:24 mg/kg (female 7-12D post):TER KSRNAM 6,629,72
orl-rat LD50:1100 mg/kg DRUGAY 6,639,82
ipr-rat LD50:350 mg/kg DRUGAY 6,639,82
ivn-rat LD50:90 mg/kg IYKEDH 5,106,74
orl-mus LD50:228 mg/kg CCCCAK 42,1179,77
ipr-mus LD50:1070 mg/kg IYKEDH 5,106,74
ivn-mus LD50:14 mg/kg 27ZQAG -,288,72
orl-dog LD50:40 mg/kg ARZNAD 18,261,68

SAFETY PROFILE: Poison by ingestion, intravenous, and intraperitoneal routes. An experimental teratogen. Experimental reproductive effects. When heated to decomposition it emits toxic fumes of NO_x and F^-.

PIH100 CAS:52463-83-9 ***HR: 3***
PINAZEPAM
mf: $C_{18}H_{13}ClN_2O$ mw: 308.78

PROP: Crystals from methanol/water. Mp: 140-142°.

SYNS: 7-CHLORO-1,3-DIHYDRO-5-PHENYL-1-(2-PROPYNYL)-2H-1,4-BENZODIAZEPIN-2-ONE ◇ 7-CHLORO-1-PROPARGYL-5-PHENYL-2H-1,4-BENZODIAZEPIN-2-ONE ◇ DOMAR ◇ Z-905 ◇ ZAMI 905

TOXICITY DATA with REFERENCE
orl-rat LD50:5819 mg/kg ARZNAD 25,934,75
ipr-rat LD50:622 mg/kg ARZNAD 25,934,75
orl-mus LD50:1355 mg/kg ARZNAD 25,934,75
ipr-mus LD50:266 mg/kg ARZNAD 25,934,75
orl-rbt LD50:494 mg/kg ARZNAD 25,934,75

SAFETY PROFILE: Poison by intraperitoneal route. Moderately toxic by ingestion. Used as an antidepressant. Note: This is a controlled substance (depressant) listed in the U.S. Code of Federal Regulations, Title 21 Part 1308.14 (1985). When heated to decomposition it emits toxic fumes of Cl^- and NO_x.

PIH175 CAS:83-26-1 ***HR: 3***
PINDONE
DOT: UN 2472
mf: $C_{14}H_{14}O_3$ mw: 230.28

PROP: Yellow crystals. Mp: 108°.

SYNS: CHEMRAT ◇ 2-(2,2-DIMETHYL-1-OXOPROPYL)-1H-INDENE-1,3(2H)-DIONE ◇ PINDON (DUTCH) ◇ PIVACIN ◇ PIVAL ◇ PIVALDION (ITALIAN) ◇ PIVALDIONE (FRENCH) ◇ 2-PIVALOYL-INDAAN-1,3-DION (DUTCH) ◇ 2-PIVALOYL-INDAN-1,3-DION (GERMAN) ◇ 2-PIVALOYL-1,3-INDANDIONE ◇ 2-PIVALOYLINDANE-1,3-DIONE ◇ 2-PIVALYL-1,3-INDANDIONE ◇ PIVALYL VALONE ◇ PIVALYN ◇ TRI-BAN ◇ 2-(TRIMETIL-ACETIL)-INDAN-1,3-DIONE (ITALIAN)

TOXICITY DATA with REFERENCE
orl-rat LD50:280 mg/kg TXAPA9 2,88,60
ivn-rat LD50:50 mg/kg YKYUA6 31,1385,80
par-rat LD50:50 mg/kg GUCHAZ 6,415,73
orl-dog LDLo:5 mg/kg APTOA6 42,81,78
orl-rbt LD50:150 mg/kg 85DPAN -,-,71/76
orl-dom LDLo:75 mg/kg AWLRAO 5,135,78

CONSENSUS REPORTS: Reported in EPA TSCA Inventory.

OSHA PEL: TWA 0.1 mg/m^3
ACGIH TLV: TWA 0.1 mg/m^3
DOT Classification: Poison B; Label: Poison; Poison B; Label: St. Andrews Cross.

SAFETY PROFILE: Poison by ingestion, intravenous, and parenteral routes. Causes reduced blood clotting which leads to hemorrhaging. Used as an anticoagulant and rodenticide. When heated to decomposition it emits acrid smoke and irritating fumes. See also WARFARIN.

PIH250 CAS:80-56-8 ***HR: 3***
2-PINENE
DOT: UN 2368
mf: $C_{10}H_{16}$ mw: 136.26

PROP: Liquid; odor of turpentine. Mp: −55°, bp: 155°, flash p: 91°F, d: 0.8592 @ 20°/4°, refr index: 1.464-1.468, vap press: 10 mm @ 37.3°, vap d: 4.7, autoign temp: 491°F. Insol in water; sol in alc, chloroform, ether, glacial acetic acid, fixed oils.

SYNS: ACINTENE A ◇ FEMA No. 2902 ◇ α-PINENE (FCC) ◇ 2,6,6-TRIMETHYLBICYCLO(3.1.1)-2-HEPT-2-ENE

TOXICITY DATA with REFERENCE
skn-man 100% SEV FCTXAV 16,637,78
skn-rbt 500 mg/24H MOD FCTXAV 16,637,78
orl-rat LD50:3700 mg/kg FCTXAV 16,637,78
ihl-rat LCLo:625 μg/m^3 FCTXAV 16,637,78
ihl-mus LCLo:364 μg/m^3 FCTXAV 16,637,78
ihl-gpg LCLo:572 μg/m^3 FCTXAV 16,637,78

CONSENSUS REPORTS: Reported in EPA TSCA Inventory.

DOT Classification: Flammable or Combustible Liquid; Label: Flammable Liquid.

SAFETY PROFILE: A deadly poison by inhalation. Moderately toxic by ingestion. An eye, mucous membrane, and severe human skin irritant. Flammable liquid. A dangerous fire hazard when exposed to heat, flame, or oxidizing materials. To fight fire, use foam, CO_2, dry chemical. Explodes on contact with nitrosyl perchlorate.

PIH400 CAS:8000-26-8 ***HR: 1***
PINE NEEDLE OIL, DWARF

PROP: From steam distillation of needles of *Pinus mugo* turra var. *pumilio* (Haenke) Zenari (Fam. *Pinaceae*) (FCTXAV 14,659,76). Colorless to yellow liquid; pleasant odor and a bitter, pungent taste. D: 0.853-0.871, refr index: 1.475 @ 20°.

SYNS: DWARF PINE NEEDLE OIL ◇ KNEE PINE OIL ◇ LATSCHENKIEFEROL ◇ OIL of MOUNTAIN PINE ◇ PINUS MONTANA OIL ◇ PINUS PUMILIO OIL

TOXICITY DATA with REFERENCE
skn-hmn 12% FCTXAV 14,843,76
orl-rat LD50:6880 mg/kg PHARAT 14,435,59

CONSENSUS REPORTS: Reported in EPA TSCA Inventory.

SAFETY PROFILE: Mildly toxic by ingestion. A human skin irritant. When heated to decomposition it emits acrid smoke and irritating fumes.

PIH500 CAS:8000-26-8 ***HR: 1***
PINE NEEDLE OIL, SCOTCH

PROP: Volatile oil from steam distillation of *Pinus sylvestris* L. (Fam. *Pinaceae*) constituted of dipentene, pinene, sylvestrene, cadinene and bornyl acetate. Yellow liquid; penetrating odor. Bp: 200-220°, flash p: 172°F (CC), d: 0.86, refr index: 1.473 @ 20°. Sol in fixed oils, mineral oil; sltly sol in propylene glycol; insol in glycerin.

SYNS: KIEFERNADEL OEL (GERMAN) ◇ SCOTCH PINE NEEDLE OIL

TOXICITY DATA with REFERENCE
orl-rat LD50:6880 mg/kg PHARAT 14,435,59

CONSENSUS REPORTS: Reported in EPA TSCA Inventory.

SAFETY PROFILE: Mildly toxic by ingestion. A weak allergen and a mild irritant. Flammable when exposed to heat or flame; can react vigorously with oxidizing materials. To fight fire, use foam, CO_2, dry chemical. When heated to decomposition it emits acrid smoke and irritating fumes. See also individual components.

PIH750 CAS:8002-09-3 ***HR: 2***
PINE OIL
DOT: UN 1272

PROP: Pale yellow liquid; penetrating odor. Bp: 200-220°, flash p: 172°F (CC), d: 0.86, flash p: (steam distilled): 138°F. Insol in water; sol in organic solvents.

SYNS: OIL of PINE ◇ OLEUM ABIETIS ◇ TERPENTIN OEL (GERMAN) ◇ UNIPINE ◇ YARMOR ◇ YARMOR PINE OIL

TOXICITY DATA with REFERENCE
skn-rbt 500 mg/24H SEV FCTOD7 21,875,83
orl-man TDLo:4700 mg/kg:CNS ARTODN 49,73,81
orl-rat LD50:3200 mg/kg FCTOD7 21,875,83
skn-rbt LD50:5 g/kg FCTOD7 21,875,83

CONSENSUS REPORTS: Reported in EPA TSCA Inventory.

DOT Classification: Combustible Liquid; Label: None.

SAFETY PROFILE: Moderately toxic by ingestion. Mildly toxic by skin contact. A weak allergen and a severe irritant to skin and mucous membranes. Human systemic effects by ingestion: excitement, ataxia, headache. Combustible when exposed to heat or flame; can react with oxidizing materials. Moderate spontaneous heating. To fight fire, use foam, CO_2, dry chemical. Used as an odorant, disinfectant, solvent, wetting agent, and frothing agent.

PIH800 ***HR: 3***
PINKROOT

PROP: *S. anthelmia* is an annual herb that grows to about 1.5 feet with oval leaves 6 inches long and 3 inches wide. The flowers have a white tube with magenta stripes and 5 pink petals. It grows in southern Florida, the Bahamas and the West Indies. *S. marilandica* is a perennial herb that grows to about 2 feet. The oval leaves are 4 inches long. The flowers are 2 inches long, red on the outside and yellow inside. It grows in the region bounded by Florida, Texas, southern Indiana, and South Carolina.

SYNS: CAROLINA PINK ◇ ESPIGELIA (CUBA) ◇ HERBE-A-BRINVILLIERS (HAITI) ◇ INDIAN PINK ◇ LOGGERHEAD WEED (BARBADOS) ◇ LOMBRICERA (PUERTO RICO) ◇ PINK WEED ◇ SPIGELIA ANTHELMIA ◇ SPIGELIA MARILANDICA ◇ WATER-WEED ◇ WEST INDIAN PINKROOT ◇ WORM GRASS

SAFETY PROFILE: The whole plant contains the poison spigeline. The liquid from the boiled plant is used as a vermifuge. Poisonings have resulted from eating the leaf or drinking this liquid. Ingestion of any part of the plant may result in vomiting, vertigo, muscle spasms, pupil dilation, and strychnine-like convulsions. See also STRYCHNINE.

PII100 CAS:54-47-7 ***HR: 2***
PIODEL
mf: $C_8H_{10}NO_PP$ mw: 247.16

PROP: Colorless in acid soln, bright-yellow in alkaline soln.

SYNS: APOLON B_6 ◇ BIOSECHS ◇ CODECARBOXYLASE ◇ HAIROXAL ◇ HEXERMIN P ◇ HIADELON ◇ HI-PYRIDOXIN ◇ 3-HYDROXY-2-METHYL-5-((PHOSPHONOOXY)METHYL)-4-PYRIDINECARBOXALDEHYDE ◇ PAL-P ◇ PHOSPHOPYRIDOXAL ◇ PHOSPHORIDOXAL COENZYME ◇ PLP ◇ PYDOXAL ◇ PYRIDOXALDEHYDE ◇ PYRIDOXALDEHYDE PHOSPHATE ◇ PYRIDOXAL MONOPHOSPHATE ◇ PYRIDOXAL PHOSPHATE ◇ PYRIDOXAL 5-PHOSPHATE ◇ PYRIDOXAL-5'-PHOSPHATE ◇ PYRIDOXYL PHOSPHATE ◇ PYROMIJIN ◇ SECHVITAN ◇ VITAHEXIN P ◇ VITAZECHS

TOXICITY DATA with REFERENCE
dnd-esc 50 μmol/L FEPRA7 34,530,75
orl-rat LD50:5900 mg/kg NIIRDN 6,643,82
scu-rat LD50:850 mg/kg NIIRDN 6,643,82
orl-mus LD50:4640 mg/kg NIIRDN 6,643,82
scu-mus LD50:870 mg/kg NIIRDN 6,643,82
ivn-mus LD50:530 mg/kg PCJOAU 15,303,81
ims-mus LD50:1150 mg/kg

CONSENSUS REPORTS: Reported in EPA TSCA Inventory.

SAFETY PROFILE: Moderately toxic subcutaneous, intravenous, and intramuscular routes. Mildly toxic by ingestion. Mutation data reported. When heated to decomposition it emits toxic fumes of NO_x and PO_x. See also ALDEHYDES.

PII150 CAS:11121-57-6 ***HR: 3***
PIOMY

SYNS: PIO ◇ PIOMYCIN

TOXICITY DATA with REFERENCE
orl-rat LD50:2150 mg/kg 85ARAE 4,41,76/77
ipr-mus LD50:400 mg/kg 85GDA2 5,254,81
scu-mus LD50:500 mg/kg 85GDA2 5,254,81

SAFETY PROFILE: Poison by intraperitoneal route. Moderately toxic by ingestion and subcutaneous routes. When heated to decomposition it emits acrid smoke and irritating fumes.

PII200 CAS:2448-68-2 ***HR: 3***
PIPAMPERONE DIHYDROCHLORIDE
mf: $C_{21}H_{30}FN_3O_2 \cdot 2ClH$ mw: 448.46

SYNS: PIPAMPERONE DICHLORHYDRATE ◇ PIPAMPERONE HYDROCHLORIDE ◇ PROPITAN

TOXICITY DATA with REFERENCE
orl-rat LD50:1120 mg/kg NIIRDN 6,631,82
scu-rat LD50:400 mg/kg NIIRDN 6,631,82
orl-mus LD50:910 mg/kg NIIRDN 6,631,82

scu-mus LD50:312 mg/kg NIIRDN 6,631,82
ivn-mus LD50:71 mg/kg NIIRDN 6,631,82

SAFETY PROFILE: Poison by subcutaneous and intravenous routes. Moderately toxic by ingestion. When heated to decomposition it emits toxic fumes of F^-, NO_x, and HCl.

PII250 CAS:52212-02-9 ***HR: 3***
PIPECURIUM BROMIDE
mf: $C_{35}H_{62}N_4O_4 \cdot 2Br$ mw: 762.83

SYNS: ARDUAN ◇ 2-β,16-β-(4'-DIMETHYL-1'-PIPERAZINO)-3-α,17-β-DIACETOXY-5-α-ANDROSTANE 2BR ◇ PIPECURONIUM BROMIDE ◇ RGH-1106

TOXICITY DATA with REFERENCE
ipr-rat LD50:450 μg/kg ARZNAD 30,346,80
scu-rat LD50:456 μg/kg ARZNAD 30,346,80
ivn-rat LD50:173 μg/kg ARZNAD 30,346,80
orl-mus LD50:22 μg/kg ARZNAD 30,346,80
ipr-mus LD50:71 μg/kg ARZNAD 30,346,80
scu-mus LD50:61 μg/kg ARZNAD 30,346,80
ivn-mus LD50:30 μg/kg ARZNAD 30,346,80
ims-mus LD50:44 μg/kg ARZNAD 30,346,80
ivn-rbt LD50:10 μg/kg ARZNAD 30,346,80

SAFETY PROFILE: A deadly poison by ingestion, intramuscular, intravenous, subcutaneous, and intraperitoneal routes. Used as a skeletal muscle relaxant. When heated to decomposition it emits very toxic fumes of Br^- and NO_x.

PII350 CAS:72571-82-5 ***HR: 2***
PIPEMIDIC ACID TRIHYDRATE
mf: $C_{14}H_{17}N_5O_3 \cdot 3H_2O$ mw: 357.42

SYNS: 8-ETHYL-5,8-DIHYDRO-5-OXO-2-(1-PIPERAZINYL) PYRIDO(2,3-d)PYRAMIDINE-6-CARBOXYLIC ACID 3H_2O ◇ 5,8-DIHYDRO-8-ETHYL-5-OXO-2-(1-PIPERAZINYL)-PYRIDO(2,3-d)PYRAMIDINE-6-CARBOXYLIC ACID TRIHYDRATE

TOXICITY DATA with REFERENCE
scu-rat LD50:1635 mg/kg NIIRDN 6,634,82
ivn-rat LD50:575 mg/kg NIIRDN 6,634,82
scu-mus LD50:1274 mg/kg NIIRDN 6,634,82
ivn-mus LD50:610 mg/kg NIIRDN 6,634,82
scu-mam LD50:1213 mg/kg IYKEDH 10,232,79
ivn-mam LD50:529 mg/kg IYKEDH 10,232,79

SAFETY PROFILE: Moderately toxic by subcutaneous and intravenous routes. When heated to decomposition it emits toxic fumes of NO_x.

PII500 CAS:3819-00-9 ***HR: 3***
PIPERACETAZINE
mf: $C_{24}H_{30}N_2O_2S$ mw: 410.62

SYNS: 2-ACETYL-10-(3-(4-(β-HYDROXYETHYL)PIPERIDINO)PROPYL)PHENOTHIAZINE ◇ ETHAN ◇ 10-(3-(4-(2-HYDROXYETHYL)PIPERIDINO)PROPYL)PHENOTHIAZIN-2-YL METHYL KETONE ◇ PC-1421 ◇ QUIDE ◇ SC 9794

TOXICITY DATA with REFERENCE
orl-rat LD50:390 mg/kg TXAPA9 5,49,63
ipr-rat LD50:93 mg/kg TXAPA9 5,49,63
orl-mus LD50:575 mg/kg AIPTAK 135,152,62
ipr-mus LD50:120 mg/kg TXAPA9 5,49,63
scu-mus LD50:365 mg/kg AIPTAK 149,374,64
ipr-dog LD50:100 mg/kg 27ZQAG -,39,72

SAFETY PROFILE: Poison by ingestion, subcutaneous, and intraperitoneal routes. When heated to decomposition it emits very toxic fumes of SO_x and NO_x.

PII750 CAS:71-78-3 ***HR: 3***
PIPERADROL HYDROCHLORIDE
mf: $C_{18}H_{21}NO \cdot ClH$ mw: 303.86

SYNS: α,α-DIPHENYL-2-PIPERIDINEMETHANOL HYDROCHLORIDE ◇ α-(2-PIPERIDYL)BENZHYDROL HYDROCHLORIDE ◇ PIPRADOL HYDROCHLORIDE ◇ PIPRADROL HYDROCHLORIDE ◇ PIRIDROL HYDROCHLORIDE ◇ PYRIDROL

TOXICITY DATA with REFERENCE
orl-rat LD50:180 mg/kg JPETAB 110,180,54
scu-rat LD50:240 mg/kg JPETAB 110,180,54
ivn-rat LD50:30 mg/kg JPETAB 110,180,54
orl-mus LD50:120 mg/kg JPETAB 118,153,56
ipr-mus LD50:94 mg/kg JPETAB 118,153,56
scu-mus LD50:147 mg/kg JPETAB 118,153,56
ivn-mus LD50:20 mg/kg CSLNX* NX#00031
orl-rbt LD50:180 mg/kg CLDND*
ivn-rbt LD50:15 mg/kg JPETAB 110,180,54

SAFETY PROFILE: Poison by ingestion, subcutaneous, intravenous, and intraperitoneal routes. When heated to decomposition it emits very toxic fumes of HCl and NO_x.

PIJ000 CAS:110-85-0 ***HR: 2***
PIPERAZINE
DOT: UN 2579
mf: $C_4H_{10}N_2$ mw: 86.16

$$\overline{HNC_2H_4NHCH_2CH_2}$$

PROP: Colorless, rhombic crystals. Mp: 106°, bp: 146°, flash p: 190°F (OC), d: 1.1, vap d: 3.0. Very sol in water, glycerol, glycols; insol in ether.

SYNS: ANTIREN ◇ 1,4-DIETHYLENEDIAMINE ◇ N,N-DIETHYLENE DIAMINE (DOT) ◇ DISPERMINE ◇ HEXAHYDRO-1,4-DIAZINE ◇ HEXAHYDROPYRAZINE ◇ LUMBRICAL ◇ PIPERAZIDINE ◇ PIPERAZIN (GERMAN) ◇ PIPERAZINE, anhydrous ◇ PYRAZINE HEXAHYDRIDE

TOXICITY DATA with REFERENCE
skn-rbt 500 mg open MLD UCDS** 7/16/67
eye-rbt 750 μg SEV AJOPAA 29,1363,46

eye-rbt 250 μg/24H SEV 85JCAE-,862,86
orl-cld TDLo:75 mg/kg 34ZIAG-,478,69
orl-rat LD50:1900 mg/kg TPKVAL 15,116,79
scu-rat LD50:3700 mg/kg DRUGAY 6,635,82
ivn-rat LD50:1340 mg/kg DRUGAY 6,635,82
orl-mus LD50:600 mg/kg BCFAAI 103,414,64
ihl-mus LC50:5400 mg/m^3/2H TPKVAL 15,116,79
orl-rat LD50:1900 mg/kg TPKVAL 15,116,79
scu-rat LD50:3700 mg/kg NIIRDN 6,635,82
skn-rbt LD50:4000 mg/kg UCDS** 7/16/65

CONSENSUS REPORTS: Reported in EPA TSCA Inventory.

DOT Classification: Corrosive Material; Label: Corrosive

SAFETY PROFILE: Moderately toxic by ingestion, skin contact, intravenous, and subcutaneous routes. Mildly toxic by inhalation. A skin and severe eye irritant. Excessive absorption can cause urticaria, vomiting, diarrhea, blurred vision, and weakness. Combustible when exposed to heat or flame; can react vigorously with oxidizing materials. Explodes on contact with dicyanofurazan. To fight fire, use alcohol foam, mist, dry chemical, water spray. When heated to decomposition it emits highly toxic fumes of NO_x.

PIJ250 ***HR: 3***
PIPERAZINE and SODIUM NITRITE (4 : 1)

SYN: SODIUM NITRITE and PIPERAZINE (1:4)

TOXICITY DATA with REFERENCE
orl-rat TDLo:90 g/kg/47W-C:ETA IGSBDO 5,321,79
orl-mus TDLo:183 g/kg/28W-C:CAR JNCIAM 46,1029,71

SAFETY PROFILE: Questionable carcinogen with experimental carcinogenic and tumorigenic data. When heated to decomposition it emits toxic fumes of NO_x and Na_2O. See also individual components and NITRITES.

PIJ500 CAS:144-29-6 ***HR: 3***
PIPERAZINE CITRATE (3 : 2)
mf: $C_{12}H_{30}N_6{\cdot}Cl_2H_{16}O_{14}$ mw: 642.76

PROP: Crystals. Decomp @ 182-187°. Very sol in water; insol in alc, ether, chloroform.

SYNS: ANTEPAR ◇ ANTHECOLE ◇ ANTOBAN ◇ ARPEZINE ◇ ASCAREX SYRUP ◇ EXELMIN ◇ MULTIFUGE CITRATE ◇ OXUCIDE ◇ OXYZINE ◇ PARAZINE ◇ PIN-TEGA ◇ PIPERAZINE CITRATE TELRA ◇ PIPIZAN CITRATE SYRUP ◇ RHOMEX ◇ TA-VERM ◇ TRIPIPERAZINE DICITRATE ◇ VERMAGO

TOXICITY DATA with REFERENCE
orl-man TDLo:150 mg/kg/5D:CNS,EYE,GIT JAMMAP 161,515,56
orl-chd LDLo:260 mg/kg/3D-I LANCAO 1,895,67
unr-inf TDLo:500 mg/kg/2D:CNS INPDAR 18,71,81
orl-rat LD50:11200 mg/kg JMCMAR 6,336,63
orl-mus LD50:8500 mg/kg JMCMAR 6,336,63
ipr-mus LD50:3548 mg/kg AIPTAK 274,253,85
ivn-mus LDLo:100 mg/kg CLDND*
ivn-rbt LDLo:175 mg/kg CLDND*

CONSENSUS REPORTS: Reported in EPA TSCA Inventory.

SAFETY PROFILE: A human poison by ingestion. A poison by intravenous route. Human systemic effects: nausea or vomiting, diplopia, somnolence, tremors, diarrhea, convulsions, coma. When heated to decomposition it emits very toxic fumes of Cl^- and NO_x. See also PIPERAZINE.

PIJ600 CAS:52195-07-0 ***HR: 3***
PIPERAZINE DIANTIMONY TARTRATE
$C_8H_4O_{12}Sb_2{\cdot}C_4H_{10}N_2{\cdot}2H$ mw: 623.80

SYN: BILHARCID

TOXICITY DATA with REFERENCE
dni-esc 4 μmol/L BCPCA6 23,1451,74
oms-esc 4 μmol/L BCPCA6 23,1451,74
cyt-hmn:fbr 28 μmol/L JDGRAX 7(3),27,75
cyt-rat-ipr 1 mg/kg ENMUDM 4,83,82

CONSENSUS REPORTS: Antimony and its compounds are on the Community Right-To-Know List.

OSHA PEL: 0.5 mg(Sb)/m^3
ACGIH TLV: TWA 0.5 mg(Sb)/m^3
NIOSH REL: TWA 0.5 mg(Sb)/m^3

SAFETY PROFILE: Antimony compounds are poisons. Human mutation data reported. When heated to decomposition it emits toxic fumes of NO_x and Sb. See also ANTIMONY COMPOUNDS.

PIJ750 CAS:122-96-3 ***HR: 2***
1,4-PIPERAZINEDIETHANOL
mf: $C_8H_{18}N_2O_2$ mw: 174.28

SYNS: N,N'-BIS(β-HYDROXYETHYL)PIPERAZINE ◇ 1,4-BIS(2-HYDROXYETHYL)PIPERAZINE ◇ N,N'-DI(2-HYDROXYETHYL)PIPERAZINE ◇ 1,4-DI(2-HYDROXYETHYL)PIPERAZINE

TOXICITY DATA with REFERENCE
skn-rbt 10 mg/24H open MLD AIHAAP 23,95,62
orl-rat LD50:3730 mg/kg AIHAAP 23,95,62

CONSENSUS REPORTS: Reported in EPA TSCA Inventory.

SAFETY PROFILE: Moderately toxic by ingestion. A skin irritant. When heated to decomposition it emits toxic fumes of NO_x.

PIK000 CAS:142-64-3 ***HR: 2***
PIPERAZINE DIHYDROCHLORIDE
mf: $C_4H_{10}N_2{\cdot}2ClH$ mw: 159.08

SYNS: DIHYDROCHLORIDE SALT OF DIETHYLENEDIAMINE ◇ DOWZENE DHC ◇ PIPERAZINE HYDROCHLORIDE

TOXICITY DATA with REFERENCE
orl-rat LD50:4900 mg/kg GUCHAZ 6,416,73
ipr-mus LD50:1970 mg/kg JJPAAZ 17,475,67

CONSENSUS REPORTS: Reported in EPA TSCA Inventory. EPA Genetic Toxicology Program.

OSHA PEL: TWA 5 mg/m^3
ACGIH TLV: TWA 5 mg/m^3

SAFETY PROFILE: Moderately toxic by intraperitoneal route. Mildly toxic by ingestion. When heated to decomposition it emits very toxic fumes of NO_x and HCl. Used in making fiber, pharmaceuticals, and insecticides. See also PIPERAZINE.

PIK075 CAS:41109-80-2 ***HR: 3***
PIPERAZINEDIONE
mf: $C_{14}H_{22}Cl_2H_4O_2 \cdot 2ClH$ mw: 422.22

SYNS: 593-A ◇ 3,6-BIS(5-CHLORO-2-PIPERIDYL)-2,5-PIPERAZINEDIONE DIHYDROCHLORIDE ◇ COMPOUND 593A ◇ NSC-135758 ◇ PIPERAZINEDIONE 593A

TOXICITY DATA with REFERENCE
pic-esc 2500 ng/plate CNREA8 43,2819,83
orl-mus LD50:11020 μg/kg NCISP* JAN86
ipr-mus LD50:5 mg/kg 85GDA2 4(1),153,80
scu-mus LD50:8160 μg/kg NCISP* JAN86
ivn-mus LD50:19970 μg/kg NTIS** PB82-165200
ivn-dog LD50:1500 μg/kg DRFUD4 3,610,78

SAFETY PROFILE: Poison by ingestion, subcutaneous, intravenous, and intraperitoneal routes. Mutation data reported. When heated to decomposition it emits toxic fumes of NO_x and HCl.

PIK250 CAS:21416-87-5 ***HR: 3***
2,6-PIPERAZINEDIONE-4,4'-PROPYLENE DIOXOPIPERAZINE
mf: $C_{11}H_{16}N_4O_4$ mw: 268.31

SYNS: (±)-1,2-BIS(3,5-DIOXOPIPERAZINE-1-YL)PROPANE◇ (±)-1,2-BIS(3,5-DIOXOPIPERAZINYL)PROPANE◇ ICRF-159 ◇ 4,4'-(1-METHYL-1,2-ETHANEDIYL)BIS-2,6-PIPERAZINEDIONE◇ NCI-C01627 ◇ NSC-129943 ◇ RAZOXIN ◇ (±)-(3,5,3',5'-TETRAOXO)-1,2-DIPIPERAZINOPROPANE

TOXICITY DATA with REFERENCE
dni-mus:emb 20 mg/L IJCNAW 5,47,70
msc-mus:lng 200 μg/L MUREAV 157,199,85
orl-rat TDLo:37500 μg/kg (6-8D preg):TER TJADAB 11,119,75
orl-rbt TDLo:105 mg/kg (female 6-8D post):REP TJADAB 11,119,75
ipr-rat TDLo:7488 mg/kg/52W-I:CAR,TER NCITR* NCI-CG-TR-78,78
ipr-rat TD:15 g/kg/Y-I:CAR,TER NCITR* NCI-CG-TR-78,78
ipr-mus TD:12 g/kg/Y-I:CAR NCITR* NCI-CG-TR-78,78
ipr-mus LD:1 g/kg/8W-I:ETA CNREA8 33,3069,73
orl-hmn TDLo:81 mg/kg/3D-I:GIT,BLD CCROBU 57,185,73
orl-hmn TDLo:500 mg/kg:GIT,BLD CTRRDO 62,465,78
ipr-mus LD50:861 mg/kg NCISP* JAN86

CONSENSUS REPORTS: NCI Carcinogenesis Bioassay (ipr); Clear Evidence: mouse, rat NCITR* NCI-CG-TR-78,78. EPA Genetic Toxicology Program.

SAFETY PROFILE: Suspected carcinogen with experimental carcinogenic and tumorigenic data. Moderately toxic by intraperitoneal route. Experimental teratogenic and reproductive effects. Human systemic effects by ingestion: nausea, thrombocytopenia, leukopenia. Mutation data reported. When heated to decomposition it emits toxic fumes of NO_x.

PIK375 CAS:4727-62-2 ***HR: 3***
1,1'-(1,4-PIPERAZINEDIYLDIETHYLENE)BIS(1-ETHYLPIPERIDINIUM IODIDE)
mf: $C_{22}H_{46}N_4 \cdot 2I$ mw: 620.52

SYN: 336 HC

TOXICITY DATA with REFERENCE
scu-mus LD50:50 mg/kg THERAP 9,314,54
ivn-mus LD50:10 mg/kg AIPTAK 94,1,53
par-mus LD50:10 mg/kg APFRAD 7,368,49

SAFETY PROFILE: Poison by subcutaneous, parenteral, and intravenous routes. When heated to decomposition it emits toxic fumes of NO_x and I^-. See also IODIDES.

PIK400 ***HR: 3***
(1,4-PIPERAZINEDIYLDIETHYLENE)BIS(TRIETHYL AMMONIUM IODIDE)
mf: $C_{20}H_{46}N_4 \cdot 2I$ mw: 596.50

SYN: 292 HC

TOXICITY DATA with REFERENCE
scu-mus LD50:80 mg/kg THERAP 9,314,54
ivn-mus LD50:16 mg/kg AIPTAK 94,1,53
par-mus LD50:16 mg/kg APFRAD 7,368,49

SAFETY PROFILE: Poison by subcutaneous, intravenous and parenteral routes. When heated to decomposition it emits toxic fumes of I^-, NO_x and NH_3. See also IODIDES.

PIK500 CAS:142-63-2 ***HR: 3***
PIPERAZINE HEXAHYDRATE
mf: $C_4H_{10}N_2 \cdot 6H_2O$ mw: 194.28

PROP: Crystals. Mp: 44°, bp: 125-130°. Very sol in water; sol in alc; insol in ether.

SYNS: USAF A-3803 ◇ VERMISOL

TOXICITY DATA with REFERENCE
orl-rat LDLo:500 mg/kg JPETAB 90,260,47
orl-mus LD50:11200 mg/kg JMCMAR 6,336,63
ipr-mus LD50:300 mg/kg NTIS** AD277-689
scu-mus LD50:2620 mg/kg ARZNAD 15,852,65

SAFETY PROFILE: Poison by intraperitoneal route. Moderately toxic by ingestion and subcutaneous routes. When heated to decomposition it emits toxic fumes of NO_x. See also PIPERAZINE.

PIK625 CAS:57775-27-6 ***HR: 2***
PIPERAZINE SULTOSILATE
mf: $C_{13}H_{12}O_7S_2 \cdot C_4H_{10}N_2$ mw: 430.53

PROP: Crystals from ethanol. Mp: 171-174°.

SYNS: A 585 ◇ MIMEDRAN ◇ SULTOSILATO de PIPERACINA (SPANISH) ◇ SULTOSILIC ACID, PIPERAZINE SALT ◇ 5-TOSILOXI 2-HYDROXIBENCENO SULFONATO de PIPERACINA (SPANISH)

TOXICITY DATA with REFERENCE
ipr-rat LD50:1272 mg/kg AFTOD7 5,281,79
ipr-mus LD50:834 mg/kg AFTOD7 5,281,79
ipr-dog LD50:605 mg/kg AFTOD7 5,281,79

SAFETY PROFILE: Moderately toxic by intraperitoneal route. When heated to decomposition it emits toxic fumes of SO_x and NO_x. See also PIPERAZINE and SULFONATES.

PIL500 CAS:110-89-4 ***HR: 3***
PIPERIDINE
DOT: UN 2401
mf: $C_5H_{11}N$ mw: 85.17

PROP: Clear, colorless liquid; amine-like odor. Mp: −7°, bp: 106°, flash p: 37.4°F, d: 0.8622 @ 20°/4°, vap press: 40 mm @ 29.2°, vap d: 3.0. Misc with water; sol in alc, benzene, chloroform.

SYNS: AZACYCLOHEXANE ◇ CYCLOPENTIMINE ◇ CYPENTIL ◇ HEXAHYDROPYRIDINE ◇ HEXAZANE ◇ PENTAMETHYLENEIMINE ◇ PIPERIDIN (GERMAN)

TOXICITY DATA with REFERENCE
skn-rbt 100 μg/24H open AIHAAP 23,95,62
mmo-sat 32 μg/plate JEPTDQ 4,345,80
ihl-rat TCLo:3 mg/m^3/24H (4D preg):TER TPKVAL 14,40,75
ihl-rat TCLo:100 mg/m^3/24H (4D preg):REP TPKVAL 14,40,75
orl-rat LD50:400 mg/kg GTPZAB 18(2),29,74
ihl-rat LCLo:4000 ppm/4H AIHAAP 23,95,62
orl-mus LD50:30 mg/kg TPKVAL 15,116,79
ihl-mus LC50:6000 mg/m^3/2H TPKVAL 15,116,79
ipr-mus LD50:50 mg/kg JEPTDQ 4,345,80
orl-mam LD50:22400 μg/kg TPKVAL 14,90,75
ihl-mam LD50:6500 mg/m^3 TPKVAL 14,90,75
scu-mus LDLo:460 mg/kg AEXPBL 50,199,1903
orl-rbt LD50:145 mg/kg 85GMAT -,100,82
skn-rbt LD50:320 mg/kg AIHAAP 23,95,62

CONSENSUS REPORTS: EPA Extremely Hazardous Substances List. Reported in EPA TSCA Inventory. EPA Genetic Toxicology Program.

DOT Classification: Flammable Liquid; Label: Flammable Liquid

SAFETY PROFILE: Poison by ingestion, skin contact, and intraperitoneal routes. Moderately toxic by subcutaneous route. Mildly toxic by inhalation. An experimental teratogen. Experimental reproductive effects by inhalation. A skin irritant. Mutation data reported. A very dangerous fire hazard when exposed to heat, flame, or oxidizers. Can react vigorously with oxidizing materials. To fight fire, use alcohol foam, CO_2, dry chemical. Explodes on contact with 1-perchloryl-piperidine; dicyanofurazan; N-nitrosoacetanilide. When heated to decomposition it emits highly toxic fumes of NO_x. Used in agriculture and pharmaceuticals, and as an intermediate for rubber accelerators. Used in production of drugs of abuse.

PIL550 ***HR: 3***
3-PIPERIDINE-1,1-DIPHENYL-PROPANOL-(1) METHANESULPHONATE
mf: $C_{20}H_{25}N \cdot CH_4O_3S$ mw: 375.57

SYNS: MYOLYSEEN ◇ PROPANOL, 1,1-DIPHENYL-3-PIPERIDINO-, METHANESULFONATE

TOXICITY DATA with REFERENCE
scu-rat TDLo:30 mg/kg (female 9-14D post):TER TOIZAG 14,178,67
orl-mus TDLo:600 mg/kg (female 7-12D post):REP TOIZAG 14,178,67
orl-rat LD50:305 mg/kg TOIZAG 14,178,67
scu-rat LD50:355 mg/kg TOIZAG 14,178,67
ivn-rat LD50:36800 μg/kg TOIZAG 14,178,67
orl-mus LD50:423 mg/kg TOIZAG 14,178,67
scu-mus LD50:324 mg/kg TOIZAG 14,178,67
ivn-mus LD50:36800 μg/kg TOIZAG 14,178,67

SAFETY PROFILE: Poison by ingestion, subcutaneous, and intravenous routes. An experimental teratogen. Experimental reproductive effects. When heated to decomposition it emits toxic fumes of NO_x and SO_x.

PIM000 CAS:4544-15-4 ***HR: 3***
1-PIPERIDINEETHANOL BENZILATE HYDROCHLORIDE
mf: $C_{21}H_{25}NO_3 \cdot ClH$ mw: 375.93

SYNS: 2-(1-PIPERIDINO)ETHYL BENZILATE HYDROCHLORIDE ◇ PIPERILATE HYDROCHLORIDE ◇ PIPETHANATE HYDROCHLORIDE ◇ SYCOTROL

TOXICITY DATA with REFERENCE
orl-rat TDLo:36400 mg/kg (26W pre):REP KSRNAM 9,659,75
orl-rat LD50:1210 mg/kg NIIRDN 6,633,82
ipr-rat LD50:89200 μg/kg NIIRDN 6,633,82
scu-rat LD50:235 mg/kg NIIRDN 6,633,82
orl-mus LD50:440 mg/kg BJPCAL 1,90,46
ipr-mus LD50:104 mg/kg ARTODN 36,139,76
ivn-mus LD50:40 mg/kg BJPCAL 1,90,46
orl-mam LD50:44 mg/kg 27ZQAG -,290,72
ivn-mam LD50:14 mg/kg 27ZQAG -,290,72

SAFETY PROFILE: Poison by ingestion, intraperitoneal, subcutaneous, and intravenous routes. Experimental reproductive effects. Mutation data reported. When heated to decomposition it emits very toxic fumes of HCl and NO_x.

PIM250 CAS:3088-41-3 ***HR: 3***
1-PIPERIDINEPROPIONITRILE
mf: $C_8H_{14}N_2$ mw: 138.24

SYNS: 1-PIPERIDINEPROPANENITRILE ◇ 3-(1-PIPERIDINE)PROPIONITRILE ◇ β-PIPERIDINOPROPIONITRILE ◇ 3-PIPERIDINOPROPIONITRILE

TOXICITY DATA with REFERENCE
ivn-mus LD50:10 mg/kg CSLNX* NX#07829

CONSENSUS REPORTS: Cyanide and its compounds are on the Community Right-To-Know List. Reported in EPA TSCA Inventory.

SAFETY PROFILE: Poison by intravenous route. When heated to decomposition it emits toxic fumes of NO_x and CN^-. See also NITRILES.

PIM500 CAS:56-12-2 ***HR: 3***
PIPERIDINIC ACID
mf: $C_4H_9NO_2$ mw: 103.14

PROP: Leaflets. Mp: decomp @ 285°, bp: subl @ >300°. Sol in cold water, hot alc; insol in ether.

SYNS: 4-AMINOBUTANOIC ACID ◇ Γ-AMINO-N-BUTYRIC ACID ◇ 4-AMINOBUTYRIC ACID ◇ DF 468 ◇ GABA ◇ GAMAREX ◇ GAMMALON

TOXICITY DATA with REFERENCE
ipr-rat LD50:5400 mg/kg AITEAT 13,70,65
ice-rat LDLo:18 mg/kg BCPCA6 14,1901,65
orl-mus LD50:12680 mg/kg YKKZAJ 85,463,65
ipr-mus LD50:4950 mg/kg AITEAT 13,70,65
scu-mus LD50:9210 mg/kg YKKZAJ 85,463,65
ivn-mus LD50:2748 mg/kg AIPTAK 145,233,63
ivn-rbt LDLo:2400 mg/kg AITEAT 13,70,65

CONSENSUS REPORTS: Reported in EPA TSCA Inventory.

SAFETY PROFILE: Poison by intracerebral route. Moderately toxic by intravenous route. Mildly toxic by ingestion. Used as an antihypertensive agent. When heated to decomposition it emits toxic fumes of NO_x. See also AMINES.

PIM750 CAS:73693-97-7 ***HR: 3***
2-PIPERIDINO-p-ACETOPHENETIDIDE HYDROCHLORIDE
mf: $C_{15}H_{22}N_2O_2•ClH$ mw: 298.85

SYN: C 3085

TOXICITY DATA with REFERENCE
eye-rbt 2% MLD ARZNAD 8,407,58
ipr-rat LD50:230 mg/kg ARZNAD 8,407,58
scu-mus LD50:680 mg/kg ARZNAD 8,407,58

SAFETY PROFILE: Poison by intraperitoneal route. Moderately toxic by subcutaneous route. An eye irritant. When heated to decomposition it emits very toxic fumes of NO_x and HCl.

PIN000 CAS:77966-90-6 ***HR: 3***
2-PIPERIDINO-2',6'-ACETOXYLIDIDE HYDROCHLORIDE
mf: $C_{15}H_{22}N_2O•ClH$ mw: 282.85

SYN: C 3062

TOXICITY DATA with REFERENCE
ipr-rat LD50:130 mg/kg ARZNAD 8,407,58
ipr-mus LD50:140 mg/kg ARZNAD 8,407,58
scu-mus LD50:240 mg/kg ARZNAD 8,407,58

SAFETY PROFILE: Poison by intraperitoneal and subcutaneous, routes. When heated to decomposition it emits very toxic fumes of HCl and NO_x.

PIN100 CAS:49830-98-0 ***HR: 3***
4-PIPERIDINOACETYL-3,4-DIHYDRO-2H-1,4-BENZOXAZINE HYDROCHLORIDE
mf: $C_{15}H_{20}N_2O_2•ClH$ mw: 296.83

SYN: BU 533

TOXICITY DATA with REFERENCE
orl-rat LD50:200 mg/kg OYYAA2 8,481,74
ipr-rat LD50:90 mg/kg OYYAA2 8,481,74
scu-rat LD50:300 mg/kg OYYAA2 8,481,74
orl-mus LD50:230 mg/kg OYYAA2 8,481,74
ipr-mus LD50:110 mg/kg OYYAA2 8,481,74
scu-mus LD50:185 mg/kg OYYAA2 8,481,74

SAFETY PROFILE: Poison by ingestion, subcutaneous and intraperitoneal routes. When heated to decomposition it emits toxic fumes of NO_x and HCl.

PIN225 CAS:3867-15-0 ***HR: 3***
1-PIPERIDINOCYCLOHEXANECARBONITRILE
mf: $C_{12}H_{20}N_2$ mw: 192.34

SYNS: 1-(1-CYANOCYCLOHEXYL)PIPERIDINE ◇ PCC ◇ PIPERIDINOCYCLOHEXANECARBONITRILE ◇ 1-(1-PIPERIDINYL)-CYCLOHEXANECARBONITRILE (9CI)

TOXICITY DATA with REFERENCE
orl-mus LD50:133 mg/kg JPPMAB 28,713,76
ipr-mus LD50:30 mg/kg JATOD3 4,119,80
ivn-mus LD50:18 mg/kg CSLNX* NX#03387

CONSENSUS REPORTS: Cyanide and its compounds are on the Community Right-To-Know List.

SAFETY PROFILE: Poison by ingestion, intravenous, and intraperitoneal routes. When heated to decomposition it emits toxic fumes of NO_x and CN^-. See also NITRILES.

PIN275 CAS:33265-79-1 ***HR: 3***
2-PIPERIDINOETHANOL HYDROCHLORIDE
mf: $C_7H_{15}NO•ClH$ mw: 165.69

SYNS: N-(HYDROXYETHYL)PIPERIDINE HYDROCHLORIDE ◇ N-(β-HYDROXYETHYL)PIPERIDINE HYDROCHLORIDE ◇ N-(2-HYDROXYETHYL)PIPERIDINE HYDROCHLORIDE ◇ 1-(2-HYDROXYETHYL)PIPERIDINE HYDROCHLORIDE ◇ 2-(1-PIPERIDINYL)ETHANOL HYDROCHLORIDE ◇ β-PIPERIDYLETHANOL HYDROCHLORIDE

TOXICITY DATA with REFERENCE
ipr-mus LD50:451 mg/kg JPETAB 94,249,48
scu-mus LD50:650 mg/kg AIPTAK 112,36,57
ivn-mus LD50:170 mg/kg AIPTAK 112,36,57

SAFETY PROFILE: Poison by intravenous route. Moderately toxic by subcutaneous and intraperitoneal routes. When heated to decomposition it emits toxic fumes of NO_x and HCl.

PIO000 CAS:77985-27-4 ***HR: 3***
N-(2-PIPERIDINOETHYL)CARBAMIC ACID, 6-CHLORO-o-TOLYL ESTER, HYDROCHLORIDE
mf: $C_{15}H_{21}ClN_2O_2•ClH$ mw: 333.29

SYN: C 5309

TOXICITY DATA with REFERENCE
eye-rbt 2% MLD ARZNAD 8,708,58
ipr-rat LD50:50 mg/kg ARZNAD 8,708,58
scu-mus LD50:62 mg/kg ARZNAD 8,708,58

SAFETY PROFILE: Poison by intraperitoneal and subcutaneous routes. An eye irritant. When heated to decomposition it emits very toxic fumes of Cl^- and NO_x. See also CARBAMATES.

PIO250 CAS:77985-28-5 ***HR: 3***
N-(2-PIPERIDINOETHYL)CARBAMIC ACID, MESITYL ESTER, HYDROCHLORIDE
mf: $C_{17}H_{26}N_2O_2•ClH$ mw: 326.91

SYN: C 5311

TOXICITY DATA with REFERENCE
eye-rbt 2% MLD ARZNAD 8,708,58
ipr-rat LD50:52 mg/kg ARZNAD 8,708,58
scu-mus LD50:115 mg/kg ARZNAD 8,708,58

SAFETY PROFILE: Poison by intraperitoneal and subcutaneous routes. An eye irritant. When heated to decomposition it emits very toxic fumes of HCl and NO_x. See also ESTERS and CARBAMATES.

PIO500 CAS:77985-29-6 ***HR: 3***
N-(2-(PIPERIDINO)ETHYL)CARBAMIC ACID, 2,6-XYLYL ESTER, HYDROCHLORIDE
mf: $C_{16}H_{24}N_2O_2•ClH$ mw: 312.88

SYN: C 5310

TOXICITY DATA with REFERENCE
ipr-rat LD50:26 mg/kg ARZNAD 8,708,58
scu-mus LD50:73 mg/kg ARZNAD 8,708,58

SAFETY PROFILE: Poison by intraperitoneal and subcutaneous routes. When heated to decomposition it emits very toxic fumes of HCl and NO_x. See also CARBAMATES.

PIO750 CAS:55792-21-7 ***HR: 3***
PIPERIDINOETHYL-2-HEPTOXYPHENYLCARBAMOATE HYDROCHLORIDE
mf: $C_{21}H_{34}N_2O_3•ClH$ mw: 399.03

SYNS: HEPTACAINE ◇ 2-HEPTYLOXYCARBANILIC ACID-2-(1-PIPERIDINYL)ETHYL ESTER HYDROCHLORIDE ◇ (2-(HEPTYLOXY) PHENYL)CARBAMIC ACID-2-(1-PIPERIDINYL)ETHYL ESTER HYDROCHLORIDE ◇ N-(2-(HEPTYLOXYPHENYLCARBAMOYLOXY)ETHYL) PIPERIDINIUM CHLORIDE

TOXICITY DATA with REFERENCE
ipr-mus LD50:150 mg/kg DRFUD4 4,489,79
scu-mus LD50:500 mg/kg AFPCAG 29,53,76
ivn-mus LD50:17600 μg/kg AFPCAG 29,81,76

SAFETY PROFILE: Poison by intravenous and intraperitoneal routes. Moderately toxic by subcutaneous route. When heated to decomposition it emits very toxic fumes of NO_x and HCl. See also CARBAMATES.

PIQ750 CAS:63916-54-1 ***HR: 3***
PIPERIDINOMETHYLCYCLOHEXANE CAMPHOSULFATE
mf: $C_{12}H_{23}N•C_{10}H_{16}O_4S$ mw: 413.68

SYNS: CYCLOHEXANE, PIPERIDINOMETHYL-, CAMPHOSULFATE ◇ SD 210-37

TOXICITY DATA with REFERENCE
orl-mus LD50:370 mg/kg AIPTAK 167,273,67
ipr-mus LD50:370 mg/kg AIPTAK 167,273,67
ivn-mus LD50:41 mg/kg AIPTAK 167,273,67

SAFETY PROFILE: Poison by ingestion, intraperitoneal, and intravenous routes. When heated to decomposition it emits very toxic fumes of SO_x and NO_x.

PIR000 CAS:5005-71-0 ***HR: 3***
PIPERIDINOMETHYLCYCLOHEXANE CHLORHYDRATE SALT
mf: $C_{12}H_{23}N{\cdot}ClH$ mw: 217.82

SYNS: CHLORHYDRATE de PIPERIDINOMETHYLCYCLOHEXANE (FRENCH) ◇ CYCLOHEXANE, PIPERIDINOMETHYL-, HYDROCHLORIDE ◇ 1-(CYCLOHEXYLMETHYL)PIPERIDINE HYDROCHLORIDE ◇ SD 210-32

TOXICITY DATA with REFERENCE
orl-mus LD50:205 mg/kg AIPTAK 167,273,67
ipr-mus LD50:65 mg/kg APFRAD 24,785,66
ivn-mus LD50:25 mg/kg AIPTAK 167,273,67

SAFETY PROFILE: Poison by ingestion, intravenous, and intraperitoneal routes. When heated to decomposition it emits toxic fumes of NO_x and HCl.

PIR100 CAS:69928-47-8 ***HR: 3***
2-PIPERIDINOMETHYL-4-METHYL-1-TETRALONE HYDROCHLORIDE
mf: $C_{17}H_{23}NO{\cdot}ClH$ mw: 293.87

SYNS: 3,4-DIHYDRO-4-METHYL-2-PIPERIDINYL-1(2H)-NAPTHALENONE HYDROCHLORIDE ◇ N 642

TOXICITY DATA with REFERENCE
ipr-mus LDLo:50 mg/kg YKKZAJ 84,395,64
scu-mus LD50:62500 μg/kg AIPTAK 130,155,61

SAFETY PROFILE: Poison by subcutaneous and intraperitoneal routes. When heated to decomposition it emits toxic fumes of NO_x and HCl.

PIS000 CAS:69766-15-0 ***HR: 3***
γ-PIPERIDINOPROPYL-p-AMINOBENZOATE HYDROCHLORIDE
mf: $C_{15}H_{22}N_2O_2{\cdot}ClH$ mw: 298.85

SYN: p-AMINOBENZOIC ACID-3-PIPERIDINOPROPYL ESTER HYDROCHLORIDE

TOXICITY DATA with REFERENCE
ivn-rat LDLo:10 mg/kg JACSAT 49,2835,27
scu-mus LDLo:100 mg/kg JACSAT 49,2835,27

SAFETY PROFILE: Poison by intravenous and subcutaneous routes. When heated to decomposition it emits very toxic fumes of NO_x and HCl. See also ESTERS.

PIT250 CAS:63918-29-6 ***HR: 3***
2-(1-PIPERIDINO)-2-(2-THENYL)ETHYLAMINE MALEATE
mf: $C_{11}H_{18}N_2S{\cdot}C_4H_4O_4$ mw: 326.45

SYN: CIBA CO. 2825

TOXICITY DATA with REFERENCE
orl-rat LD50:1483 mg/kg JETOAS 3,110,70
ivn-rat LD50:93 mg/kg JETOAS 3,110,70
orl-mus LD50:647 mg/kg JETOAS 3,110,70
ipr-mus LD50:116 mg/kg JETOAS 3,110,70
scu-mus LD50:243 mg/kg JETOAS 3,110,70
ivn-mus LD50:74 mg/kg JETOAS 3,110,70
ivn-dog LDLo:100 mg/kg JETOAS 3,110,70
ivn-mky LDLo:50 mg/kg JETOAS 3,110,70
orl-ham LD50:690 mg/kg JETOAS 3,110,70

SAFETY PROFILE: Poison by intravenous, intraperitoneal, and subcutaneous routes. Moderately toxic by ingestion. When heated to decomposition it emits very toxic fumes of NO_x and SO_x.

PIT600 ***HR: 3***
PIPERIDINYLETHYLMORPHINE
mf: $C_{24}H_{32}N_2O_3$ mw: 396.58

SYN: 7,8-DIDEHYDRO-4,5-α-EPOXY-17-METHYL-3-(2-PIPERIDINOETHOXY)MORPHINAN-6-α-OL

TOXICITY DATA with REFERENCE
ipr-mus LD50:30 mg/kg APFRAD 8,261,50
scu-mus LD50:27500 μg/kg THERAP 7,21,52
ivn-mus LD50:7750 μg/kg THERAP 7,21,52

SAFETY PROFILE: Poison by subcutaneous, intravenous, and intraperitoneal routes. When heated to decomposition it emits toxic fumes of NO_x.

PIT650 CAS:53912-89-3 ***HR: 3***
(S)-3-(2-PIPERIDINYL)PYRIDINE HYDROCHLORIDE
mf: $C_{10}H_{14}N_2{\cdot}ClH$ mw: 198.72

SYNS: ANABASIDE HYDROCHLORIDE ◇ ANABASIN CHLORIDE ◇ ANABASINE MONOHYDROCHLORIDE ◇ ANABASIN HYDROCHLORIDE ◇ PYRIDINE, 3-(2-PIPERIDINYL)-, MONOHYDROCHLORIDE, (S)-

TOXICITY DATA with REFERENCE
orl-rat TDLo:15 mg/kg (female 4D post):REP FATOAO 45(1),87,82
unr-mus LD50:39 mg/kg KHFZAN 11,30,77

SAFETY PROFILE: Poison by an unspecified route. Experimental reproductive effects. When heated to decomposition it emits toxic fumes of NO_x and HCl.

PIU000 CAS:675-20-7 ***HR: 2***
2-PIPERIDONE
mf: C_5H_9NO mw: 99.15

SYN: PIPERIDON (GERMAN)

TOXICITY DATA with REFERENCE
ivn-mus LD50:600 mg/kg AIPTAK 93,143,53
scu-frg LDLo:3000 mg/kg AEXPBL 50,199,1903

CONSENSUS REPORTS: Reported in EPA TSCA Inventory.

SAFETY PROFILE: Moderately toxic by subcutaneous and intravenous routes. When heated to decomposition it emits toxic fumes of NO_x.

PIU800 CAS:67196-02-5 ***HR: 3***
β-2-PIPERIDYLETHYLPHENYLURETHANE HYDROCHLORIDE
mf: $C_{14}H_{20}N_2O_2 \cdot ClH$ mw: 284.82

SYN: 2-PIPERIDINEETHANOL CARBANILATE (ester) HYDROCHLORIDE

TOXICITY DATA with REFERENCE
scu-mus LDLo:40 mg/kg JACSAT 61,1713,39
ivn-rbt LDLo:20 mg/kg ANESAV 1,305,40
isp-rbt LDLo:23450 μg/kg ANESAV 1,305,40
scu-gpg LDLo:456 mg/kg ANESAV 1,305,40

SAFETY PROFILE: Poison by subcutaneous, intravenous, and intraspinal routes. When heated to decomposition it emits toxic fumes of NO_x and HCl.

PIV500 CAS:886-06-6 ***HR: 3***
β-(1-PIPERIDYL)PROPIOPHENONE HYDROCHLORIDE
mf: $C_{14}H_{19}NO \cdot ClH$ mw: 253.80

SYNS: N-(β-BENZOYLETHYL)PIPERIDINEHYDROCHLORIDE ◇ NA 65 HYDROCHLORIDE ◇ 1-PHENYL-3-PIPERIDINOPROPAN-1-ONE HYDROCHLORIDE ◇ β-PIPERIDINOPROPIOPHENONE HYDROCHLORIDE

TOXICITY DATA with REFERENCE
ipr-mus LD50:61 mg/kg YKKZAJ 99,1155,79
scu-mus LD50:157 mg/kg ARZNAD 5,559,55
ivn-mus LD50:21 mg/kg JPETAB 115,419,55

SAFETY PROFILE: Poison by intraperitoneal, subcutaneous, and intravenous routes. When heated to decomposition it emits very toxic fumes of HCl and NO_x.

PIV600 CAS:94-62-2 ***HR: 3***
PIPERIN
mf: $C_{17}H_{19}NO_3$ mw: 285.37

PROP: Monoclinic prisms from alcohol; tasteless at first, but burning aftertaste. Mp: 130°. Insol in water (40 mg/liter at 18°) and in petr ether. One gram dissolves in 15 mL alc, 1.7 mL chloroform, 36 mL ether. Sol in benzene, acetic acid.

SYNS: 1-(5-(1,3-BENZODIOXOL-5-YL)-1-OXO-2,4-PENTADIENYL)PIPERIDINE (E,E)- (9CI) ◇ 1,3-BENZODIOXOL-5-YL-OXO-2,4-PENTADIENYL-PIPERINE ◇ PIPERINE ◇ 1-PIPEROYLPIPERIDINE

TOXICITY DATA with REFERENCE
orl-mus TDLo:150 mg/kg (15-20D preg):TER CCPTAY 26,625,82
orl-mus TDLo:125 mg/kg (female 8-12D post):REP CCPTAY 26,625,82
orl-rat LD50:514 mg/kg TOLED5 16,351,83
ipr-rat LD50:34 mg/kg TOLED5 16,351,83
orl-mus LD50:1637 mg/kg APHRDQ 2,141,79
ipr-mus LD50:43 mg/kg TOLED5 16,351,83
ipr-ham LD50:105 mg/kg TOLED5 16,351,83

CONSENSUS REPORTS: Reported in EPA TSCA Inventory.

SAFETY PROFILE: Poison by intraperitoneal route. Moderately toxic by ingestion. An experimental teratogen. Experimental reproductive effects. When heated to decomposition it emits toxic fumes of NO_x.

PIV650 ***HR: 1***
PIPER LONGUM L., fruit extract

PROP: Chinese and Indian plant belonging to the family *Piperaceae* JTCMSC 3,17,83

SYNS: FPL ◇ FRUCTUS PIPERIS LONGI

TOXICITY DATA with REFERENCE
orl-rat TDLo:700 mg/kg (female 1-7D post):REP IJSIDW 40,113,78
orl-mus LD50:87400 mg/kg JTCMSC 3,17,83

SAFETY PROFILE: Mildly toxic by ingestion. Experimental reproductive effects. When heated to decomposition it emits acrid smoke and irritating fumes.

PIV750 CAS:32248-37-6 ***HR: 3***
PIPEROCAINE
mf: $C_{16}H_{23}NO_2$ mw: 261.40

SYNS: 3-BENZOXY-1-(2-METHYLPIPERIDINO)PROPANE ◇ BENZOYL-Γ-(2-METHYLPIPERIDINO)PROPANOL ◇ ISOCAINE BASE ◇ 2-METHYL-1-PIPERIDINOPROPANOL, BENZOATE ◇ (2-METHYLPIPERIDINO)PROPYL BENZOATE ◇ Γ-(2-METHYLPIPERIDYL)PROPYL BENZOATE ◇ METYCAINE ◇ NEOTHESIN

TOXICITY DATA with REFERENCE
ipr-rat LD50:129 mg/kg ARZNAD 8,708,58
scu-mus LDLo:590 mg/kg JAPMA8 39,4,50
ivn-mus LDLo:26 mg/kg JAPMA8 39,4,50

SAFETY PROFILE: Poison by intravenous and intraperitoneal routes. Moderately toxic by subcutaneous route. When heated to decomposition it emits toxic fumes of NO_x.

PIW000 CAS:2622-26-6 **HR: 3**
PIPEROCYANOMAZINE
mf: $C_{21}H_{23}N_3OS$ mw: 365.53

SYNS: 2-CYANO-10-(3-(4-HYDROXYPIPERIDINO)PROPYL)PHENOTHIAZINE ◇ 2-CYANO-10-(3-(4-HYDROXY-1-PIPERIDYL)PROPYL)PHENOTHIAZINE ◇ CYANO-3-((HYDROXY-4 PIPERIDYL-1)-3 PROPYL)-10-PHENOTHIAZINE (FRENCH) ◇ F.I. 6145 ◇ 10-(3-(4-HYDROXYPIPERIDINO)PROPYL)PHENOTHIAZINE-2-CARBONITRILE ◇ IC 6002 ◇ NEMACTIL ◇ NEULACTIL ◇ NEULEPTIL ◇ PERICIAZINE ◇ PERICYAZINE ◇ PROPERICIAZINE ◇ 6909 RP ◇ RP 8908 ◇ SKF 20,716 ◇ WH 7508

TOXICITY DATA with REFERENCE
orl-rat LD50:395 mg/kg TXAPA9 21,315,72
ipr-rat LD50:85 mg/kg 27ZQAG -,43,72
scu-rat LD50:1200 mg/kg 27ZQAG -,43,72
ivn-rat LD50:35 mg/kg 27ZQAG -,43,72
orl-mus LD50:530 mg/kg CRSBAW 157,1242,63
ipr-mus LD50:115 mg/kg CRSBAW 157,1242,63
scu-mus LD50:375 mg/kg CRSBAW 157,1242,63
orl-bwd LD50:100 mg/kg TXAPA9 21,315,72

CONSENSUS REPORTS: Cyanide and its compounds are on the Community Right-To-Know List.

SAFETY PROFILE: Poison by ingestion, intraperitoneal, intravenous, and subcutaneous routes. Used as an antipsychotic agent. When heated to decomposition it emits very toxic fumes of CN^-, NO_x, and SO_x. See also NITRILES.

PIW250 CAS:120-57-0 **HR: 2**
PIPERONAL
mf: $C_8H_6O_3$ mw: 150.14

PROP: Colorless, lustrous crystals; floral odor. Mp: 37°, bp: 263°, vap press: 1 mm @ 87.0°. Very sol in alc, ether; sol in propylene glycol, fixed oils; insol water, glycerin.

SYNS: 3,4-BENZODIOXOLE-5-CARBOXALDEHYDE ◇ 3,4-DIHYDROXYBENZALDEHYDE METHYLENE KETAL ◇ DIOXYMETHYLENE-PROTOCATECHUIC ALDEHYDE ◇ FEMA No. 2911 ◇ HELIOTROPIN ◇ 3,4-METHYLENE-DIHYDROXYBENZALDEHYDE ◇ 3,4-METHYLENEDIOXYBENZALDEHYDE ◇ PIPERONALDEHYDE ◇ PIPERONYL ALDEHYDE ◇ PROTOCATECHUIC ALDEHYDE METHYLENE ETHER

TOXICITY DATA with REFERENCE
orl-rat LD50:2700 mg/kg TXAPA9 6,378,64
ipr-rat LDLo:1500 mg/kg RMSRA6 16,449,1896

CONSENSUS REPORTS: Reported in EPA TSCA Inventory.

SAFETY PROFILE: Moderately toxic by ingestion and intraperitoneal routes. Can cause central nervous system depression. A skin irritant. Combustible when exposed to heat or flame; can react with oxidizing materials. See also ALDEHYDES.

PIW500 CAS:40527-42-2 **HR: 2**
PIPERONAL DIETHYL ACETAL
mf: $C_{12}H_{16}O_4$ mw: 224.28

TOXICITY DATA with REFERENCE
ipr-mus LDLo:500 mg/kg CBCCT* 7,396,55

CONSENSUS REPORTS: Reported in EPA TSCA Inventory.

SAFETY PROFILE: Moderately toxic by intraperitoneal route. When heated to decomposition it emits acrid smoke and irritating fumes.

PIX000 CAS:326-61-4 **HR: 2**
PIPERONYL ACETATE
mf: $C_{10}H_{10}O_4$ mw: 194.20

PROP: Colorless to light yellow liquid; heliotrope odor.

SYNS: HELIOTROPYL ACETATE ◇ 3,4-METHYLENEDIOXYBENZYL ACETATE

TOXICITY DATA with REFERENCE
skn-rbt 500 mg/24H MLD FCTXAV 12,807,74
orl-rat LD50:2100 mg/kg FCTXAV 12,807,74

CONSENSUS REPORTS: Reported in EPA TSCA Inventory.

SAFETY PROFILE: Moderately toxic by ingestion. A skin irritant. When heated to decomposition it emits acrid smoke and irritating fumes.

PIX250 CAS:51-03-6 **HR: 3**
PIPERONYL BUTOXIDE
mf: $C_{19}H_{30}O_5$ mw: 338.49

PROP: Light brown liquid; mild odor. Bp: 180° @ 1 mm, flash p: 340°F, d: 1.04-1.07 @ 20°/20°. Misc with methanol, ethanol, benzene.

SYNS: BUTACIDE ◇ BUTOCIDE ◇ BUTOXIDE ◇ α-(2-(2-BUTOXYETHOXY)ETHOXY)-4,5-METHYLENEDIOXY-2-PROPYLTOLUENE ◇ α-(2-(2-n-BUTOXYETHOXY)-ETHOXY)-4,5-METHYLENEDIOXY-2-PROPYLTOLUENE ◇ 5-((2-(2-BUTOXYETHOXY)ETHOXY)METHYL)-6-PROPYL-1,3-BENZODIOXOLE ◇ BUTYL CARBITOL 6-PROPYLPIPERONYL ETHER ◇ BUTYL-CARBITYL (6-PROPYLPIPERONYL) ETHER ◇ ENT 14,250 ◇ FAC 5273 ◇ FMC 5273 ◇ 3,4-METHYLENDIOXY-6-PROPYLBENZYL-n-BUTYL-DIAETHYLENGLYKOLAETHER (GERMAN) ◇ (3,4-METHYLENEDIOXY-6-PROPYLBENZYL)(BUTYL)DIETHYLENE GLICOL ETHER ◇ 3,4-METHYLENEDIOXY-6-PROPYLBENZYL-n-BUTYL DIETHYLENEGLY- COL ETHER ◇ NCI-C02813 ◇ NIA 5273 ◇ NUSYN-NOXFISH ◇ PB ◇ PRENTOX ◇ 6-(PROPYLPIPERONYL)-BUTYL CARBITYL ETHER ◇ 6-PROPYLPIPERONYL BUTYL DIETHYLENE GLYCOL ETHER ◇ 5-PROPYL-4-(2,5,8-TRIOXA-DODECYL)-1,3-BENZODIOXOL (GERMAN) ◇ PYBUTHRIN ◇ PYRENONE 606 ◇ SYNPREN-FISH

TOXICITY DATA with REFERENCE
otr-ham:emb 500 μg/L CRNGDP 4,291,83
scu-mus TDLo:9 g/kg (female 6-14D post):TER NTIS** PB223-160

scu-mus TDLo:9 g/kg (female 6-14D post):REP NTIS** PB223-160
scu-mus TDLo:1000 mg/kg:ETA NTIS** PB223-159
orl-rat LD50:6150 mg/kg TXAPA9 14,515,69
orl-mus LD50:2600 mg/kg SKEZAP 24,268,83
ipr-mus LDLo:1000 mg/kg TXAPA9 23,288,72
orl-rbt LD50:2650 mg/kg YKYUA6 32,605,81
skn-rbt LD50:200 mg/kg PCOC** -,907,66

CONSENSUS REPORTS: IARC Cancer Review: Group 3 IMEMDT 7,56,87; Animal No Evidence IMEMDT 30,183,83. NCI Carcinogenesis Bioassay (feed); No Evidence: mouse, rat NCITR* NCI-CG-TR-120,79. Glycol ether compounds are on the Community Right-To-Know List. Reported in EPA TSCA Inventory.

SAFETY PROFILE: Poison by skin contact. Moderately toxic by ingestion and intraperitoneal routes. An experimental teratogen. Experimental reproductive effects. Many glycol ether compounds have dangerous human reproductive effects. Questionable carcinogen with experimental tumorigenic data. Mutation data reported. Combustible when exposed to heat or flame; can react with oxidizing materials. To fight fire, use foam, CO_2, dry chemical. When heated to decomposition it emits acrid smoke and irritating fumes. See also GLYCOL ETHERS.

PIX750 CAS:24951-05-1 ***HR: 3***
2-(4-PIPERONYL-1-PIPERAZINYL)-9H-PURINE-9-ETHANOL DIHYDROCHLORIDE
mf: $C_{19}H_{22}N_6O_3•2ClH$ mw: 455.39

SYN: 9-(2-HYDROXYETHYL)-2-(4-PIPERONYL-1-PIPERAZINYL)-9H-PURINE DIHYDROCHLORIDE

TOXICITY DATA with REFERENCE
orl-mus LDLo:1500 mg/kg CHTPBA 7,192,72
ipr-mus LDLo:400 mg/kg CHTPBA 7,192,72

SAFETY PROFILE: Poison by intraperitoneal route. Moderately toxic by ingestion. When heated to decomposition it emits very toxic fumes of HCl and NO_x.

PIX800 CAS:23182-46-9 ***HR: 3***
PIPETHANATE ETHYLBROMIDE
mf: $C_{23}H_{30}NO_3•Br$ mw: 448.45

SYNS: PB-106 ◇ 2-(1-PIPERIDINO)-ETHYL BENZILATE ETHYLBROMIDE

TOXICITY DATA with REFERENCE
scu-mus TDLo:900 mg/kg (female 7-12D post):TER OYYAA2 4,59,70
orl-rat TDLo:48 g/kg (female 30D pre):REP GNRIDX 3,601,69
orl-rat LD50:4600 mg/kg NIIRDN 6,633,82
ipr-rat LD50:121 mg/kg NIIRDN 6,633,82
scu-rat LD50:1350 mg/kg NIIRDN 6,633,82
ivn-rat LD50:21 mg/kg NIIRDN 6,633,82
ims-rat LD50:1570 mg/kg NIIRDN 6,633,82
orl-mus LD50:1001 mg/kg 85IPAE-,96,72
ipr-mus LD50:73 mg/kg NIIRDN 6,633,82
scu-mus LD50:288 mg/kg NIIRDN 6,633,82
ivn-mus LD50:18 mg/kg NIIRDN 6,633,82
ivn-rbt LD50:32 mg/kg NIIRDN 6,633,82
ims-rbt LD50:382 mg/kg NIIRDN 6,633,82

SAFETY PROFILE: Poison by intramuscular, subcutaneous, intravenous, and intraperitoneal routes. Moderately toxic by ingestion. An experimental teratogen. Experimental reproductive effects. When heated to decomposition it emits toxic fumes of NO_x and Br^-.

PIY000 CAS:18787-40-1 ***HR: 3***
PIPOCTANONE HYDROCHLORIDE
mf: $C_{22}H_{35}NO•ClH$ mw: 366.04

SYNS: 4′-OCTYL-3-PIPERIDINOPROPIOPHENONEHYDROCHLORIDE ◇ 1-PIPERIDINO-3-(p-OCTYLPHENYL)-3-PROPANONE HYDROCHLORIDE ◇ 1-PIPERIDINO-3-(4′-OCTYLPHENYL)-PROPAN-3-ON-HYDROCHLORID (GERMAN)

TOXICITY DATA with REFERENCE
orl-rat LD50:417 mg/kg ARZNAD 19,1011,69
ipr-rat LDLo:79 mg/kg ARZNAD 19,1011,69
ivn-rat LD50:18 mg/kg ARZNAD 19,1011,69
orl-mus LD50:410 mg/kg ARZNAD 19,1011,69
ipr-mus LD50:98 mg/kg ARZNAD 19,1011,69
ivn-mus LD50:21 mg/kg ARZNAD 19,1011,69
ipr-rbt LD50:165 mg/kg ARZNAD 19,1011,69
ivn-rbt LD50:11 mg/kg ARZNAD 19,1011,69

SAFETY PROFILE: Poison by intravenous and intraperitoneal routes. Moderately toxic by ingestion. When heated to decomposition it emits very toxic fumes of HCl and NO_x.

PIY500 CAS:98-77-1 ***HR: 3***
PIP-PIP
mf: $C_{11}H_{22}N_2S_2$ mw: 246.47

SYNS: PENTAMETHYLENEDITHIOCARBAMATE ◇ PIPERIDINIUM ◇ "522" RUBBER ACCELERATOR

TOXICITY DATA with REFERENCE
skn-hmn 500 mg/48H MLD AMIHBC 5,311,52
orl-rat LD50:250 mg/kg AMIHBC 5,311,52

CONSENSUS REPORTS: Reported in EPA TSCA Inventory.

SAFETY PROFILE: Poison by ingestion. A human skin irritant. An allergen. When heated to decomposition it emits very toxic fumes of NO_x and SO_x. See also CARBAMATES.

PIY750 CAS:1798-50-1 ***HR: 3***
γ-PIPRADOL
mf: $C_{18}H_{21}NO•ClH$ mw: 303.86

SYNS: AZACYCLONOL HYDROCHLORIDE ◇ α,α-DIPHENYL-4-PIPERIDINEMETHANOL HYDROCHLORIDE ◇ FRENQUEL HYDROCHLORIDE ◇ α-(4-PIPERIDYL)BENZHYDROL HYDROCHLORIDE

TOXICITY DATA with REFERENCE
orl-mus LD50:650 mg/kg JPETAB 118,153,56
ipr-mus LD50:220 mg/kg JPETAB 118,153,56
scu-mus LD50:355 mg/kg JPETAB 118,153,56
ivn-mus LD50:121 mg/kg KHFZAN 13(7),32,79

SAFETY PROFILE: Poison by intraperitoneal, subcutaneous, and intravenous routes. Moderately toxic by ingestion. Used as a tranquilizer. When heated to decomposition it emits very toxic fumes of HCl and NO_x.

PIZ000 CAS:51940-44-4 ***HR: 3***
PIPRAM
mf: $C_{14}H_{17}N_5O_3$ mw: 303.36

SYNS: DOLCOL ◇ PIPEDAC ◇ PIPEMIDIC ACID ◇ PIPERAMIC ACID ◇ RB 1489

TOXICITY DATA with REFERENCE
dnd-rat-orl 152 mg/kg MUTAEX 3,397,88
orl-rat TDLo:22400 mg/kg (8-14D preg):REP IYKEDH 7,321,76
orl-rat TDLo:22400 mg/kg (8-14D preg):TER IYKEDH 7,321,76
orl-rat LD50:16 g/kg PHINDQ 1,108,80
scu-rat LD50:1438 mg/kg YKYUA6 29,1231,78
ivn-rat LD50:529 mg/kg YKYUA6 29,1231,78
scu-mus LD50:2200 mg/kg 37ASAA 2,782,78
ivn-mus LD50:300 mg/kg 37ASAA 2,782,78
skn-rbt LD50:4000 mg/kg 37ASAA 2,782,78

SAFETY PROFILE: Poison by intravenous route. Moderately toxic by skin contact and subcutaneous routes. An experimental teratogen. Experimental reproductive effects. Mutation data reported. When heated to decomposition it emits toxic fumes of NO_x. Used as an antibacterial agent.

PIZ250 CAS:606-90-6 ***HR: 3***
PIPRINHYDRINATE
mf: $C_{19}H_{23}NO•C_7H_7ClN_4O_2$ mw: 496.06

PROP: Minute crystals. Mp: 151°. Sparingly sol in water; freely sol in alc.

SYNS: 8-CHLORO-THEOPHYLLINE compounded with 4-(DIPHENYLMETHOXY)-1-METHYLPIPERIDINE (1:1) ◇ DIPHENYLPYRALIN-8-CHLOR-THEOPHYLLINAT (GERMAN) ◇ DIPHENYLPYRALINE TEOCLATE ◇ KOLTON ◇ KOLTONAL ◇ MEPEDYL

TOXICITY DATA with REFERENCE
orl-mus LD50:275 mg/kg NIIRDN 6,334,82
ipr-mus LD50:86 mg/kg NIIRDN 6,334,82
ivn-mus LD50:75 mg/kg NIIRDN 6,334,82

SAFETY PROFILE: Poison by ingestion, intravenous, and intraperitoneal routes. When heated to decomposition it emits toxic fumes of Cl^- and NO_x.

PIZ499 CAS:5281-13-0 ***HR: 3***
PIPROTAL
mf: $C_{24}H_{40}O_8$ mw: 456.64

SYNS: ENT 28,344 ◇ HELIOTROPIN ACETAL ◇ PIPERONAL BIS(2-(2-BUTOXYETHOXY)ETHYL)ACETAL ◇ TROPITAL

TOXICITY DATA with REFERENCE
orl-rat LD50:4400 μg/kg GUCHAZ 6,419,73

CONSENSUS REPORTS: EPA Extremely Hazardous Substances List.

SAFETY PROFILE: Poison by ingestion. When heated to decomposition it emits acrid smoke and irritating fumes.

PJA000 CAS:125-51-9 ***HR: 3***
PIPTAL
mf: $C_{22}H_{28}NO_3•Br$ mw: 434.42

SYNS: BENZILIC ACID ESTER with 1-ETHYL-3-HYDROXY-1-METHYLPIPERIDINIUM BROMIDE ◇ 1-ETHYL-3-HYDROXY-1-METHYL-PIPERIDINIUM BROMIDE BENZILATE ◇ N-ETHYL-3-PIPERIDYLBENZILATE METHOBROMIDE ◇ 1-ETHYL-3-PIPERIDYL BENZILATE METHYLBROMIDE ◇ JB-323 ◇ PIPENZOLATE BROMIDE ◇ PIPENZOLATE METHYLBROMIDE ◇ QPB

TOXICITY DATA with REFERENCE
orl-rat LD50:916 mg/kg 29ZVAB -,97,69
scu-rat LD50:904 mg/kg CLDND*-,97,69
orl-mus LD50:1140 mg/kg CLDND* 112,64,54
ivn-gpg LD50:22 mg/kg CLDND* 112,64,54

SAFETY PROFILE: Poison by intravenous route. Moderately toxic by ingestion and subcutaneous routes. When heated to decomposition it emits very toxic fumes of Br^- and NO_x. See also ESTERS.

PJA120 CAS:135-14-8 ***HR: 3***
PIREVAN
mf: $C_{23}H_{24}N_4O•2HO_4S$ mw: 566.65

SYNS: ACAPRIN ◇ ATRAL ◇ BABURAN ◇ DIMETHYLQUINOLYL METHYLSULFATE UREA ◇ 1,3-DIQUINOLIN-6-YLUREA BISMETHOSULFATE ◇ PYROPLASMIN ◇ QUINURONIUM SULFATE ◇ 6,6'-UREYLENEBIS(1,1'-DIMETHYLQUINOLINIUM) SULFATE ◇ 6,6'-UREYLENEBIS(1-METHYLQUINOLINIUM)BIS(METHOSULFATE) ◇ ZOTHELONE

TOXICITY DATA with REFERENCE
ipr-rat LD50:6300 μg/kg TOLED5 20,69,84
scu-rat LD50:6500 μg/kg TOLED5 20,69,84
ipr-mus LD50:4800 μg/kg TOLED5 20,69,84
scu-mus LD50:5400 μg/kg TOLED5 20,69,84

SAFETY PROFILE: Poison by subcutaneous and intraperitoneal routes. When heated to decomposition it emits toxic fumes of NO_x and SO_x.

PJA130 CAS:3563-76-6 ***HR: 3***
PIREXYL PHOSPHATE
mf: $C_{21}H_{27}NO{\bullet}xH_3O_4P$ mw: 995.49

SYNS: ASA 158-5 ◇ BENPROPERINE PHOSPHATE ◇ 1-(2-(2-BENZIL-FENOSSI)-1-METILETIL)-PIPERIDINA FOSFATO (ITALIAN) ◇ 1-(2-BENZYLPHENOXY)-2-PIPERIDINOPROPANEPHOSPHATE ◇ BLASCORID ◇ 1-(1-METHYL-2-(2-(PHENYLMETHYL)PHENOXY) ETHYL)PIPERIDINE PHOSPHATE (9CI)

TOXICITY DATA with REFERENCE
orl-mus LD50:1100 mg/kg NIIRDN 6,782,82
ipr-mus LD50:139 mg/kg NIIRDN 6,782,82
scu-mus LD50:710 mg/kg NIIRDN 6,782,82
ivn-mus LD50:32 mg/kg NIIRDN 6,782,82

SAFETY PROFILE: Poison by intravenous and intraperitoneal routes. Moderately toxic by ingestion and subcutaneous routes. When heated to decomposition it emits toxic fumes of NO_x and PO_x.

PJA140 CAS:302-41-0 ***HR: 3***
PIRINITRAMIDE
mf: $C_{27}H_{34}N_4O$ mw: 430.65

PROP: Crystals from acetone. Mp: 149-150°.

SYNS: A65 ◇ 1'-(3-CYANO-3,3-DIPHENYLPROPYL)(1,4'-BIPIPERIDINE)-4'-CARBOXAMIDE ◇ 2,2-DIPHENYL-4-(4-PIPERIDINO-4-CARBAMOYLPIPERIDINO)BUTYRONITRILE ◇ DIPIDOLOR ◇ DIPIRITRAMIDE ◇ PIRIDOLAN ◇ PIRITRAMIDE ◇ R 3365

TOXICITY DATA with REFERENCE
orl-rat LD50:320 mg/kg JPPMAB 13,513,61
ivn-rat LD50:13 mg/kg JPPMAB 13,513,61
scu-mus LD50:280 mg/kg JPPMAB 13,513,61
ivn-mus LD50:34 mg/kg JPPMAB 13,513,61

CONSENSUS REPORTS: Cyanide and its compounds are on the Community Right-To-Know List.

SAFETY PROFILE: Poison by ingestion, subcutaneous, and intravenous routes. Caution: May be habit forming. This is a controlled substance (opiate) listed in the U.S. Code of Federal Regulations, Title 21 Part 1308.11 (1985). When heated to decomposition it emits toxic fumes of NO_x and CN^-. See also NITRILES.

PJA170 CAS:61477-94-9 ***HR: 3***
PIRMENOL HYDROCHLORIDE
mf: $C_{22}H_{30}N_2O{\bullet}ClH$ mw: 375.00

SYNS: CL-845 ◇ Z-(±)-2,6-DIMETHYL-α-PHENYL-α-(2-PYRIDYL)-1-PIPERIDINEBUTANOL HYDROCHLORIDE ◇ (±)-cis-2,6-DIMETHYL-α-PHENYL-α-2-PYRIDYL-1-PIPERIDINEBUTANOLMONOHYDROCHLORIDE

TOXICITY DATA with REFERENCE
orl-rat LD50:251 mg/kg TXAPA9 56,294,80
ivn-rat LD50:7900 μg/kg TXAPA9 56,294,80
orl-mus LD50:159 mg/kg TXAPA9 56,294,80
ivn-mus LD50:16 mg/kg TXAPA9 56,294,80

SAFETY PROFILE: Poison by ingestion and intravenous routes. When heated to decomposition it emits toxic fumes of NO_x and HCl.

PJA190 CAS:16378-21-5 ***HR: 3***
PIROHEPTINE
mf: $C_{22}H_{25}N$ mw: 303.48

PROP: Liquid. Bp: 167°.

SYN: 3-(10,11-DIHYDRO-5H-DIBENZO(a,d)CYCLOHEPTEN-5-YLIDENE)-1-ETHYL-2-METHYLPYRROLIDINE

TOXICITY DATA with REFERENCE
orl-rat LD50:600 mg/kg ARZNAD 22,961,72
ipr-rat LD50:100 mg/kg ARZNAD 22,961,72
scu-rat LD50:330 mg/kg ARZNAD 22,961,72
ivn-rat LD50:16 mg/kg ARZNAD 22,961,72
orl-mus LD50:127 mg/kg ARZNAD 22,961,72
ipr-mus LD50:78 mg/kg ARZNAD 22,961,72
scu-mus LD50:91 mg/kg ARZNAD 22,961,72
ivn-mus LD50:19 mg/kg ARZNAD 22,961,72
orl-dog LD50:195 mg/kg ARZNAD 22,961,72
ivn-dog LD50:13 mg/kg ARZNAD 22,961,72
orl-rbt LD50:383 mg/kg ARZNAD 22,961,72
ivn-rbt LD50:6 mg/kg ARZNAD 22,961,72

SAFETY PROFILE: Poison by ingestion, subcutaneous, intravenous, and intraperitoneal routes. When heated to decomposition it emits toxic fumes of NO_x.

PJA220 CAS:31793-07-4 ***HR: 3***
PIRPROFEN
mf: $C_{13}H_{14}ClNO_2$ mw: 251.73

PROP: Crystals from benzene-hexane. Mp: 98-100°.

SYNS: 3-CHLORO-4-(3-PYRROLIN-1-YL)HYDRATROPIC ACID ◇ RENGASIL ◇ SU 21524

TOXICITY DATA with REFERENCE
orl-rat LD50:351 mg/kg CTCEA9 30(Suppl 1),76,81
ivn-rat LD50:167 mg/kg CTCEA9 30(Suppl 1),76,81
orl-mus LD50:1350 mg/kg ATSUDG 7,365,84
orl-gpg LD50:193 mg/kg ATSUDG 7,365,84
orl-ham LD50:1190 mg/kg ATSUDG 7,365,84

SAFETY PROFILE: Poison by ingestion and intravenous routes. When heated to decomposition it emits toxic fumes of Cl^- and NO_x.

PJA250 CAS:9002-72-6 ***HR: 3***
PITUITARY GROWTH HORMONE

SYNS: ADENOHYPOPHYSEAL GROWTH HORMONE ◇ ANTERIOR PITUITARY GROWTH HORMONE ◇ HYPOPHYSEAL GROWTH HORMONE ◇ PHYOL ◇ PHYONE ◇ SOMACTON ◇ SOMATOTROPIC HORMONE ◇ SOMATOTROPIN

TOXICITY DATA with REFERENCE
cyt-mus-ipr 31 mg/kg/5D RRENAR 10,311,73
scu-rat TDLo:10 mg/kg (female 3-22D post):REP NATUAS 205,1136,65
unr-rat TDLo:700 mg/kg (female 3-16D post):TER COREAF 240,455,55
ipr-rat TDLo:3600 mg/kg/69W-I:CAR CNREA8 10,297,50

SAFETY PROFILE: Questionable carcinogen with experimental carcinogenic data. Experimental teratogenic and reproductive effects. Mutation data reported.

PJA500 CAS:75-98-9 ***HR: 3***
PIVALIC ACID
mf: $C_5H_{10}O_2$ mw: 102.15

PROP: Crystals. Mp: 35.5°, bp: 164°, d: 0.91. Very sol in alc, ether; somewhat sol in water.

SYNS: 2,2-DIMETHYLPROPANOIC ACID ◇ α,α-DIMETHYLPROPIONIC ACID ◇ 2,2-DIMETHYLPROPIONIC ACID ◇ NEOPENTANOIC ACID ◇ tert-PENTANOIC ACID ◇ PROPANOIC ACID ◇ TRIMETHYLACETIC ACID

TOXICITY DATA with REFERENCE
skn-mus TDLo:188 mg/kg/47W-I:ETA CALEDQ 17,61,82
orl-rat LD50:900 mg/kg 37ASAA 4,863,78
skn-rat LD50:1900 mg/kg 37ASAA 4,863,78

CONSENSUS REPORTS: Reported in EPA TSCA Inventory.

SAFETY PROFILE: Moderately toxic by ingestion and skin contact. Questionable carcinogen with experimental tumorigenic data. When heated to decomposition it emits acrid smoke and irritating fumes.

PJA750 CAS:630-18-2 ***HR: 3***
PIVALONITRILE
mf: C_5H_9N mw: 83.13

$(CH_3)_3CC{\equiv}N$

PROP: Flash p: 69.8°.

SYN: TRIMETHYLACETONITRILE

CONSENSUS REPORTS: Cyanide and its compounds are on the Community Right-To-Know List.

SAFETY PROFILE: Many nitriles are poisons. Dangerous fire hazard when exposed to heat or flame, can react vigorously with oxidizing materials. When heated to decomposition it emits very toxic fumes of NO_x and CN^-. See also NITRILES.

PJB000 CAS:4981-48-0 ***HR: 3***
PIVALOYL AZIDE
mf: $C_5H_9N_3O$ mw: 127.15

$(CH_3)_3CCO{\bullet}N_3$

SYN: TRIMETHYLACETYL AZIDE

SAFETY PROFILE: Explodes violently. May explode at room temperature. When heated to decomposition it emits toxic fumes of NO_x. See also AZIDES.

PJB500 CAS:63394-05-8 ***HR: 3***
PLAFIBRIDE
mf: $C_{16}H_{22}ClN_3O_4$ mw: 355.86

SYNS: N-2(p-CHLOROPHENOXY)ISOBUTYRYL-N'-MORPHOLINOMETHYLUREA ◇ N-2(p-CHLOROPHENOXY)-2-METHYLPROPIONYL-N'-MORPHOLINOMETHYLUREA ◇ IDONOR ◇ ITA-104 ◇ PERIFUNAL ◇ PLAFIBRIDA (SPANISH)

TOXICITY DATA with REFERENCE
orl-rbt TDLo:1300 mg/kg (6-18D preg):REP ARZNAD 31,1831,81
orl-rat TDLo:2500 mg/kg (6-15D preg):TER ARZNAD 31,1831,81
ivn-rat LD50:95 mg/kg ARZNAD 31,1816,81
orl-mus LD50:3365 mg/kg ARZNAD 31,1816,81
ipr-mus LD50:710 mg/kg ARZNAD 31,1816,81
orl-gpg LD50:2168 mg/kg DRFUD4 4,42,79

SAFETY PROFILE: Poison by intravenous route. Moderately toxic by ingestion and intraperitoneal routes. Experimental teratogenic and reproductive effects. Used as an antithrombotic agent. When heated to decomposition it emits very toxic fumes of Cl^- and NO_x.

PJB750 CAS:118-42-3 ***HR: 2***
PLAQUENIL
mf: $C_{18}H_{26}ClN_3O$ mw: 335.92

SYNS: 7-CHLORO-4-(4-(N-ETHYL-N-β-HYDROXYETHYLAMINO)-1-METHYLBUTYLAMINO)QUINOLINE ◇ 7-CHLORO-4-(4-(ETHYL(2-HYDROXYETHYL)AMINO-1-METHYLBUTYLAMINO)QUINOLINE ◇ 7-CHLORO-4-(5-(N-ETHYL-N-2-HYDROXYETHYLAMINO)-2-PENTYL)AMINOQUINOLINE ◇ 2-((4-(7-CHLORO-4-QUINOLYL)AMINO)PENTYL)-ETHYLAMINO)ETHANOL ◇ HYDROXYCHLOROQUINE ◇ WIN 1258

TOXICITY DATA with REFERENCE
orl-mus LD50:1240 mg/kg JMCMAR 12,184,69
orl-hmn TDLo:429 mg/kg/25D:SKN,BLD LANCAO 1,1275,65
orl-man TDLo:600 mg/kg/25D:GIT 34ZIAG -,321,69

SAFETY PROFILE: Moderately toxic by ingestion. Human systemic effects by ingestion: dermatitis, gastritis, angranulocytosis. When heated to decomposition it emits very toxic fumes of Cl^- and NO_x.

PJC000 CAS:102338-56-7 ***HR: 1***
PLASTICIZER G-316

TOXICITY DATA with REFERENCE
skn-rbt 500 mg open MLD UCDS** 8/30/62
orl-rat LD50:5660 mg/kg UCDS** 8/30/62

SAFETY PROFILE: Mildly toxic by ingestion. A skin irritant. When heated to decomposition it emits acrid smoke and irritating fumes.

PJC250 CAS:18268-70-7 ***HR: 3***
PLASTICIZER 4GO
mf: $C_{24}H_{46}O_7$ mw: 446.70

TOXICITY DATA with REFERENCE
skn-rbt 500 mg open MLD UCDS** 7/21/71
orl-rat LD50:18 mg/kg UCDS** 7/21/71

CONSENSUS REPORTS: Reported in EPA TSCA Inventory.

SAFETY PROFILE: Poison by ingestion. A skin irritant. When heated to decomposition it emits acrid smoke and irritating fumes.

PJC500 CAS:39306-82-6 ***HR: 1***
PLASTICIZER GPE

TOXICITY DATA with REFERENCE
skn-rbt 500 mg open MLD UCDS** 11/20/62
orl-rat LD50:45 g/kg UCDS** 11/20/62
skn-rbt LD50:16 g/kg UCDS** 11/20/62

SAFETY PROFILE: Mildly toxic by skin contact and ingestion. A skin irritant. When heated to decomposition it emits acrid smoke and irritating fumes.

PJC750 CAS:102338-57-8 ***HR: 1***
PLASTICIZER Z-88

TOXICITY DATA with REFERENCE
skn-rbt 500 mg open MLD UCDS** 1/20/72
orl-rat LD50:8720 mg/kg UCDS** 1/20/72
skn-rbt LD50:20 g/kg UCDS** 1/20/72

SAFETY PROFILE: Mildly toxic by ingestion and skin contact. A skin irritant. When heated to decomposition it emits acrid smoke and irritating fumes.

PJD000 CAS:15663-27-1 ***HR: 3***
cis-PLATINOUS DIAMMINE DICHLORIDE
mf: $Cl_2H_6N_2Pt$ mw: 300.07

SYNS: CACP ◇ CDDP ◇ CISPLATINO (SPANISH) ◇ CISPLATYL ◇ CPDC ◇ CPDD ◇ DDP ◇ cis-DDP ◇ cis-DIAMINEDICHLOROPLATINUM ◇ cis-DICHLORODIAMMINE PLATINUM(II) ◇ NCI-C55776 ◇ NEOPLATIN ◇ NSC-119875 ◇ PEYRONE'S CHLORIDE ◇ PLATIBLASTIN ◇ cis-PLATIN ◇ PLATINEX ◇ PLATINOL ◇ cis-PLATINUM(II) DIAMINEDICHLORIDE

TOXICITY DATA with REFERENCE
mmo-sat 250 ng/plate TAKHAA 44,96,85
sce-hmn:lym 250 ng/L/96H ARTODN 46,61,80
ivn-rat TDLo:660 μg/kg (female 7-17D post):REP KSRNAM 15,5782,81
ipr-mus TDLo:20 mg/kg (female 13D post):TER TJADAB 28,189,83
ipr-mus TDLo:16 mg/kg/19W-I:CAR CNREA8 39,913,79
ipr-mus TD:16204 μg/kg/10W-I:NEO CNREA8 41,4368,81
ivn-man TDLo:2140 μg/kg/5D-I:KID JJMDAT 23,283,84
ivn-hmn TDLo:1500 μg/kg/6D-I:EAR,KID,BLD CCROBU 57,191,73
ivn-hmn TDLo:500 μg/kg/13D-I:KID,BLD CTRRDO 62(5),693,78
ivn-hmn TDLo:2500 μg/kg:CNS:GIT:KID CCROBU 59,647,75
ivn-hmn TDLo:72 mg/kg/25D-I:GIT CTRRDO 62,1591,78
idr-hmn TDLo:40 ng/kg:SKN CNREA8 35,2766,75
par-man TDLo:2140 μg/kg/5D-I:KID NNGAAS 72,1426,83
unr-chd TDLo:19200 mg/kg/12W-I:EAR JOPDAB 103,1006,83
orl-rat LD50:25800 μg/kg YACHDS 10,723,82
scu-rat LD50:8100 μg/kg KSRNAM 15,5669,81
ivn-rat LD50:8 mg/kg JJIND8 67,201,81
ims-rat LD50:9200 μg/kg YACHDS 10,723,82
orl-mus LD50:32700 μg/kg KSRNAM 15,5669,81
ipr-mus LD50:8600 μg/kg RAEHDT 4(2),81,80
scu-mus LD50:16900 μg/kg YACHDS 10,723,82
ivn-mus LD50:11 mg/kg ARTODN 7,90,84
ims-mus LD50:17900 μg/kg YACHDS 10,723,82
par-mus LD50:22 mg/kg IOBPD3 5,1417,79
ivn-dog LDLo:2500 μg/kg TXAPA9 25,230,73

CONSENSUS REPORTS: IARC Cancer Review: Group 2A IMEMDT 7,170,87; Animal Limited Evidence IMEMDT 26,151,81. Reported in EPA TSCA Inventory. EPA Genetic Toxicology Program.

OSHA PEL: TWA 0.002 mg(Pt)/m^3
ACGIH TLV: TWA 0.002 mg(Pt)/m^3

SAFETY PROFILE: Suspected carcinogen with experimental carcinogenic and tumorigenic data. Poison by ingestion, intramuscular, subcutaneous, intravenous, and intraperitoneal routes. Human systemic effects: change in auditory acuity, change in kidney tubules, changes in bone marrow, corrosive to skin, depressed renal function tests, hallucinations, nausea or vomiting. Experimental teratogenic and reproductive effects. Human mutation data reported. When heated to decomposition it emits very toxic fumes of Cl^- and NO_x. See also PLATINUM COMPOUNDS.

PJD250 CAS:10025-99-7 ***HR: 3***
PLATINOUS POTASSIUM CHLORIDE
mf: $Cl_4Pt•K_2$ mw: 415.09

PROP: Ruby red. Mp: decomp @ 250°, d: 3.499 @ 24°. Sol in water.

SYNS: POTASSIUM CHLOROPLATINITE ◇ POTASSIUM PLATINOCHLORIDE ◇ POTASSIUM TETRACHLOROPLATINATE(II)

TOXICITY DATA with REFERENCE
mma-sat 100 ng/plate PCJOAU 16,721,82
sln-smc 50 μmol/L MUTAEX 1,21,86
idr-hmn TDLo:40 mg/kg:SKN CNREA8 35,2766,75
orl-chd LDLo:400 mg/kg BJIMAG 2,92,45
ipr-mus LD50:45 mg/kg VOONAW 25(11),47,79

CONSENSUS REPORTS: Reported in EPA TSCA Inventory.

OSHA PEL: TWA 0.002 mg(Pt)/m^3
ACGIH TLV: TWA 0.002 mg(Pt)/m^3

SAFETY PROFILE: Human poison by ingestion. Poison experimentally by intraperitoneal route. Corrosive to human skin by intradermal route. Mutation data reported. When heated to decomposition it emits toxic fumes of Cl^- and K_2O. Used as a catalyst for hydroformulations, photocatalysts, and dissociation of water. See also PLATINUM COMPOUNDS.

PJD500 CAS:7440-06-4 ***HR: 3***
PLATINUM
af: Pt aw: 195.09

PROP: Silvery-white, malleable, ductile metal; stable in air. Mp: 1772°, bp: 3827°, d: 21.45 @ 20°.

SYNS: C.I. 77795 ◇ LIQUID BRIGHT PLATINUM ◇ PLATIN (GERMAN) ◇ PLATINUM BLACK ◇ PLATINUM SPONGE

TOXICITY DATA with REFERENCE
imp-rat TDLo:5250 mg/kg:ETA NATWAY 42,75,55

CONSENSUS REPORTS: Reported in EPA TSCA Inventory.

OSHA PEL: TWA (metal) 1 mg/m^3; (soluble salts as Pt) 0.002 mg/m^3
ACGIH TLV: TWA (metal) 1 mg/m^3; (soluble salts as Pt) 0.002 mg/m^3
DFG MAK: 0.002 mg/m^3

SAFETY PROFILE: Questionable carcinogen with experimental tumorigenic data by implant route. Finely divided platinum is a powerful catalyst and can be dangerous to handle. Used catalysts are especially dangerous and may be explosive. May undergo hazardous reactions with aluminum; acetone; arsenic; carbon + methanol; nitrosyl chloride; dioxygen difluoride; ethanol; hydrazine; hydrogen + air; hydrogen peroxide; lithium; methyl hydroperoxide; ozonides; peroxymonosulphuric acid; phosphorus; selenium; tellurium; vanadium dichloride + water. See also PLATINUM COMPOUNDS.

PJD750 CAS:13820-91-2 ***HR: D***
PLATINUM(II) AMMINE TRICHLOROPOTASSIUM
mf: $Cl_3H_3NPt{\cdot}K$ mw: 341.459

SYNS: AMMINETRICHLORO-PLATINATE(1$^-$) POTASSIUM (SP-4-2) ◇ POTASSIUM, AMMINETRICHLOROPLATINATE (1$^-$) ◇ POTASSIUM, TRICHLOROAMMINEPLATINATE (2)

TOXICITY DATA with REFERENCE
msc-ham:ovr 50 μmol/L CNREA8 40,1463,80

SAFETY PROFILE: Mutation data reported. When heated to decomposition it emits very toxic fumes of K_2O, NO_x, and Cl^-. See also PLATINUM COMPOUNDS and AMINES.

PJE000 CAS:10025-65-7 ***HR: 3***
PLATINUM CHLORIDE
mf: Cl_2Pt mw: 265.99

PROP: Grayish-green powder. D: 5.87. Insol in water, alc, ether, benzene, chloroform.

SYNS: MURIATE of PLATINUM ◇ PLATINOUS CHLORIDE

TOXICITY DATA with REFERENCE
skn-rbt 100 mg/24H MLD AEHLAU 30,168,75
mrc-bcs 1 mmol/L MUREAV 77,109,80
dni-hmn:lym 300 μmol/L IAAAAM 79,83,86
orl-rat LD50:17547 μg/kg GTPZAB 21(7),55,77

CONSENSUS REPORTS: Reported in EPA TSCA Inventory.

SAFETY PROFILE: Poison by ingestion. A skin irritant. Human mutation data reported. When heated to decomposition it emits toxic fumes of Cl^-. See also PLATINUM COMPOUNDS.

PJE250 CAS:13454-96-1 ***HR: 3***
PLATINUM(IV) CHLORIDE
mf: Cl_4Pt mw: 336.89

SYN: PLATINUM TETRACHLORIDE

TOXICITY DATA with REFERENCE
skn-rbt 100 mg/24H SEV AEHLAU 30,168,75
sln-dmg-orl 300 μmol/L/72H ENMUDM 2,133,80
msc-ham:ovr 21 μmol/L MUREAV 151,293,85
itt-rat TDLo:26951 μg/kg (1D male):REP JRPFA4 7,21,64
orl-rat LD50:276 mg/kg GTPZAB 21(7),55,77
ivn-rat LDLo:26 mg/kg EVHPAZ 10,63,75

CONSENSUS REPORTS: Reported in EPA TSCA Inventory. EPA Genetic Toxicology Program.

OSHA PEL: TWA 0.002 mg(Pt)/m^3
ACGIH TLV: TWA 0.002 mg(Pt)/m^3

SAFETY PROFILE: Poison by ingestion and intravenous routes. Experimental reproductive effects. Mutation data reported. A severe skin irritant. When heated to decomposition it emits toxic fumes of Cl^-. See also PLATINUM COMPOUNDS.

PJE500 ***HR: 2***
PLATINUM COMPOUNDS

SAFETY PROFILE: cis-[$Pt(NH_3)_2Cl_2$] is an experimental carcinogen. Exposure to complex platinum salts has been shown to cause symptoms of intoxication such as wheezing, coughing, running of the nose, chest tightness, shortness of breath and cyanosis. Furthermore, many people working with platinum salts are troubled with dermatitis. They may become sensitized after years of exposure. Symptoms of platinum allergy include rhinitis, conjunctivitis, asthma, urticaria and contact dermatitis. Mainly the ionic platinum chloro compounds [e.g., $(NHN_4)_2(PtCl_6)$, $(NH_4)_2(PtCl_4)$, $H_2(PtCl_6)$] are responsible for this sensitivity. The bromide and iodide compounds are less effective. These platinum compounds form a platinum-protein conjugate which is the true allergen. Tetrachloroplatinates are mutagens. This seems only to be true of complex platinum salts. It does not include the complex salts of the other precious metals. Platinum amine nitrates and perchlorates either detonate when heated or are impact-sensitive.

PJE750 CAS:12044-52-9 ***HR: 3***
PLATINUM DIARSENIDE
mf: As_2Pt mw: 244.93

CONSENSUS REPORTS: Arsenic and its compounds are on the Community Right-To-Know List.

SAFETY PROFILE: Arsenic compounds are poisons. Unstable in preparation. When heated to decomposition it emits fumes of As. See also ARSENIC COMPOUNDS and PLATINUM COMPOUNDS.

PJF000 ***HR: 3***
PLATINUM FULMINATE
mf: $Pt(C_2N_2O_2)_2$ mw: 363.3

SAFETY PROFILE: Severe explosion hazard when shocked or exposed to heat. Dangerous; shock will explode it; when heated to decomposition it emits highly toxic fumes of NO_x. See also PLATINUM COMPOUNDS, EXPLOSIVES, HIGH; and FULMINATES.

PJF500 CAS:53231-79-1 ***HR: 3***
PLATINUM(II) SULFATE
mf: O_8PtS_2 mw: 387.21

SYN: PLATINUM SULFATE

TOXICITY DATA with REFERENCE
msc-ham:ovr 140 μmol/L MUREAV 67,65,79
orl-mus TDLo:397 mg/kg (8D preg):REP JTEHD6 13,879,84
orl-mus LD50:281 mg/kg EVHPAZ 34,203,80

OSHA PEL: TWA 0.002 mg(Pt)/m^3
ACGIH TLV: TWA 0.002 mg(Pt)/m^3

SAFETY PROFILE: Poison by ingestion. Experimental reproductive effects. Mutation data reported. When heated to decomposition it emits toxic fumes of SO_x. See also PLATINUM COMPOUNDS and SULFATES.

PJF750 CAS:69102-79-0 ***HR: 3***
PLATINUM SULFATE TETRAHYDRATE
mf: $O_8PtS_2•4H_2O$ mw: 459.29

SYN: PLATINUM(II) SULFATE TETRAHYDRATE

TOXICITY DATA with REFERENCE
orl-rat LD50:1010 mg/kg EVHPAZ 10,95,75
ipr-rat LD50:312 mg/kg EVHPAZ 10,95,75

OSHA PEL: TWA 0.002 mg(Pt)/m^3
ACGIH TLV: TWA 0.002 mg(Pt)/m^3

SAFETY PROFILE: Poison by intraperitoneal route. Moderately toxic by ingestion. When heated to decomposition it emits toxic fumes of SO_x. See also PLATINUM SULFATE.

PJG000 CAS:63748-54-9 ***HR: D***
PLATINUM THYMINE BLUE

SYNS: cis-PLATINUM-2-THYMINE ◇ PTB

TOXICITY DATA with REFERENCE
ipr-rat TDLo:50 mg/kg (6D preg):TER LIFSAK 31,757,82
ipr-rat TDLo:20 mg/kg (7D preg):REP LIFSAK 31,757,82

OSHA PEL: TWA 0.002 mg(Pt)/m^3
ACGIH TLV: TWA 0.002 mg(Pt)/m^3

SAFETY PROFILE: Experimental teratogenic and reproductive effects. When heated to decomposition it emits toxic fumes of NO_x. Used as an antitumor agent. See also PLATINUM COMPOUNDS.

PJG150 ***HR: 3***
PLATIPHILLIN HYDROCHLORIDE
mf: $C_{18}H_{27}NO_5•ClH$ mw: 373.92

SYNS: (1-α)-1,2-DIHYDRO-12-HYDROXYSENECIONAN-11,16-DIONE HYDROCHLORIDE ◇ PLATIPHYLLIN HYDROCHLORIDE

TOXICITY DATA with REFERENCE
sln-dmg-par 20 μmol/L ZEVBAS 91,74,60
ipr-rat LD50:160 mg/kg FRBGAT 9,142,68
ivn-rat LDLo:60 mg/kg JPETAB 68,130,40
ivn-mus LDLo:10 mg/kg JPETAB 68,130,40
ivn-mky LDLo:20 mg/kg JPETAB 68,130,40
ivn-gpg LDLo:55 mg/kg JPETAB 68,130,40

SAFETY PROFILE: Poison by intravenous and intraperitoneal routes. Mutation data reported. When heated to decomposition it emits toxic fumes of Cl^- and NO_x.

PJH500 CAS:9006-00-2 ***HR: 3***
PLIOFILM
mf: $(C_3H_5Cl)_n$

SYNS: PERMASEAL ◇ RUBBER HYDROCHLORIDE ◇ RUBBER HYDROCHLORIDE POLYMER

TOXICITY DATA with REFERENCE
imp-rat TDLo:18 mg/kg:ETA CNREA8 15,333,55

SAFETY PROFILE: Questionable carcinogen with experimental tumorigenic data. When heated to decomposition it emits toxic fumes of Cl^-.

PJH550 ***HR: 2***
PLUCHEA LANCEOLATA (DC.) Cl., extract excluding roots

PROP: Indian plant belonging to the family *Asteraceae* IJEBA6 22,312,84

TOXICITY DATA with REFERENCE
orl-ham TDLo:500 mg/kg (female 1-5D post):REP IJEBA6 22,312,84
ipr-mus LD50:681 mg/kg IJEBA6 22,312,84

SAFETY PROFILE: Moderately toxic by intraperitoneal route. Experimental reproductive effects. When heated to decomposition it emits acrid smoke and irritating fumes.

PJH610 CAS:481-42-5 ***HR: 3***
PLUMBAGIN
mf: $C_{11}H_8O_3$ mw: 188.19

PROP: Yellow needles from dil alc. Mp: 78-79°. Subl Sltly sol in hot water; sol in alc, acetone, chloroform, benzene, acetic acid.

SYNS: 5-HYDROXY-2-METHYL-1,4-NAPHTHALENEDIONE ◇ 5-HYDROXY-2-METHYL-1,4-NAPHTHOQUINONE ◇ 2-METHYL-5-HYDROXY-1,4-NAPHTHOQUINONE

TOXICITY DATA with REFERENCE
mma-sat 100 μg/plate MUREAV 124,25,83
mmo-esc 500 μmol/L JOBAAY 164,1309,85
orl-rat TDLo:35 mg/kg (female 5-11D post):REP IJMRAQ 65,829,77
orl-rat LD50:65 mg/kg IJMRAQ 65,829,77
orl-mus LD50:16 mg/kg IJEBA6 18,876,80
ipr-mus LD50:5 mg/kg 85GDA2 8(1),73,82

SAFETY PROFILE: Poison by ingestion and intraperitoneal routes. Experimental reproductive effects. Mutation data reported. When heated to decomposition it emits acrid smoke and irritating fumes.

PJH615 CAS:92202-07-8 ***HR: 2***
PLUMBAGO ZEYLANICA Linn., root extract

SYNS: CHITA ROOT EXTRACT ◇ CHITRAKA ROOT EXTRACT

TOXICITY DATA with REFERENCE
orl-rat TDLo:100 g/kg (female 10D pre):REP IJMRAQ 76(Suppl),99,82
ipr-mus LD50:500 mg/kg IJEBA6 7,250,69

SAFETY PROFILE: Moderately toxic by intraperitoneal route. Experimental reproductive effects. When heated to decomposition it emits acrid smoke and irritating fumes.

PJH630 CAS:9003-11-6 ***HR: 2***
PLURONIC L-81
mf: $(C_3H_6O{\bullet}C_2H_4O)_x$

SYN: POLYETHYLENE-POLYPROPYLENE GLYCOLS PLURONIC L-81

TOXICITY DATA with REFERENCE
orl-rat LD50:2300 mg/kg JPPMAB 34(Suppl),533,82
ipr-rat LD50:1140 mg/kg JPPMAB 34(Suppl),533,82
orl-mus LD50:1830 mg/kg JPPMAB 34(Suppl),533,82
ipr-mus LD50:420 mg/kg JPPMAB 34(Suppl),533,82

SAFETY PROFILE: Moderately toxic by ingestion and intraperitoneal routes. When heated to decomposition it emits acrid smoke and irritating fumes.

PJH750 ***HR: 3***
PLUTONIUM
af: Pu aw: 242

PROP: A silvery, radioactive metal; chemically reactive. Mp: 641°, bp: 3232°, d: 19.816 @ 20°/4°.

SAFETY PROFILE: An extremely poisonous radioactive material. The permissible levels for plutonium are the lowest for any of the radioactive elements. This is occasioned by the concentration of plutonium directly on bone surfaces, rather than the more uniform bone distribution shown by other heavy elements. This increases the possibility of damage from equivalent activities of plutonium and has led to the adoption of the extremely low permissible levels given. Radiation Hazard: Artificial isotope ^{238}Pu, $T_{0.5}$ = 86 Y, decays to radioactive ^{234}U by alphas of 5.5 MeV. Artificial isotope ^{239}Pu, $T_{0.5}$ = 24,000 Y decays to radioactive ^{235}U by alphas of 5.1 MeV. Artificial isotope ^{240}Up, $T_{0.5}$ = 6600 Y decays to radioactive ^{236}Pu (Neptunium Series), $T_{0.5}$ = 13 Y decays to radioactive ^{241}Am by betas of 0.02 MeV. Artificial isotope ^{242}Pu, $T_{0.5}$ = 3.8×10^5 Y decays to radioactive ^{238}U by alphas of 4.9 MeV. Ignites in air as low as 135°C. Explosive reaction with carbon tetrachloride. Particles exposed to air and moisture may ignite spontaneously. Corrosion products are usually pyrophoric. When heated to decomposition it emits toxic and radioactive fumes of Pu. See also PLUTONIUM COMPOUNDS.

PJH775 CAS:12010-53-6 *HR: 3*
PLUTONIUM BISMUTHIDE
mf: BiPu mw: 252.98

SAFETY PROFILE: Plutonium compounds are extremely dangerous due to the radioactivity of plutonium. Extremely pyrophoric. An extremely dangerous disaster hazard. Upon decompositon it emits toxic fumes of Bi and radioactive fumes of Pu. See also PLUTONIUM and BISMUTH COMPOUNDS.

PJI000 *HR: 3*
PLUTONIUM COMPOUNDS

SAFETY PROFILE: The toxicity of plutonium compounds is based first upon the very high radiotoxicity of the plutonium atom and secondly upon whatever atoms or combinations of atoms they might contain. Very dangerous!! Any disaster which could cause quantities of plutonium or plutonium compounds to be scattered about the environment can cause great ecological stress and render areas of the land unfit for public occupancy. Long-term storage in plastic containers is not recommended as the alpha particles can cause stress cracks and the potential for leakage. See also PLUTONIUM.

PJI250 CAS:15457-77-9 *HR: 3*
PLUTONIUM(III) HYDRIDE
mf: H_3Pu mw: 245.02

SAFETY PROFILE: Very toxic by radiotoxicity. The hydride is spontaneously flammable in air. Ignites on contact with water. Contamination dangerous due to radiation and toxic hazards. When heated to decomposition it emits extremely toxic and radioactive fumes of Pu. See also PLUTONIUM.

PJI500 CAS:14913-29-2 *HR: 3*
PLUTONIUM NITRATE (solution)
DOT: NA 9185

DOT Classification: Label: Radioactive.

SAFETY PROFILE: All plutonium compounds are extremely dangerous. When heated to decomposition it emits toxic fumes of NO_x and radioactive fumes of Pu. See also PLUTONIUM COMPOUNDS.

PJI575 CAS:2001-91-4 *HR: 3*
PMCG HYDROCHLORIDE
mf: $C_{20}H_{29}NO_3 \cdot ClH$ mw: 367.96

SYNS: α-CYCLOPENTYLMANDELIC ACID (1-ETHYL-2-PYRROLIDINYL)METHYL ESTER HYDROCHLORIDE ◇ PMCG

TOXICITY DATA with REFERENCE
ims-rat LD50:614 μg/kg BJPCBM 39,822,70
ims-mus LD50:896 μg/kg BJPCBM 39,822,70
ims-gpg LD50:175 μg/kg BJPCBM 39,822,70

SAFETY PROFILE: A deadly poison by intramuscular route. When heated to decomposition it emits toxic fumes of NO_x and HCl. See also ESTERS.

PJI600 CAS:53305-31-0 *HR: D*
(±)-PMHI MALEATE
mf: $C_{16}H_{23}NO \cdot C_4H_4O_4$ mw: 361.48

SYNS: (±)-(Z)-2-BUTENEDIOATE-1H-INDEN-5-OL,2,3-DIHYDRO-6-((2-METHYL-1-PIPERIDINYL)METHYL)- ◇ (±)-6-((2-METHYLPIPERIDINO)METHYL)-5-INDANOL MALEATE ◇ dl-6-(N-α-PIPECOLINOMETHYL)-5-HYDROXY-INDANEMALEATE

TOXICITY DATA with REFERENCE
orl-rat TDLo:150 mg/kg (1D male):REP JRPFA4 37,441,74

SAFETY PROFILE: Experimental reproductive effects. When heated to decomposition it emits toxic fumes of NO_x.

PJJ000 CAS:9000-55-9 *HR: 3*
PODOPHYLLIN

PROP: Light yellow powder or small yellow fragile lumps; bitter, acrid taste.

SYNS: PODOPHYLLUM ◇ PODOPHYLLUM RESIN

TOXICITY DATA with REFERENCE
ipr-mus TDLo:7500 μg/kg (female 9D post):TER CNJGA8 15,491,73
ipr-rat TDLo:10 mg/kg (female 11-12D post):REP PSEBAA 113,124,63
orl-mus TDLo:92 g/kg/60W-C:NEO CNREA8 28,2272,68
orl-wmn LDLo:6 mg/kg 34ZIAG -,482,69
orl-man LDLo:157 mg/kg SMJOAV 75,1269,82
ipr-rat LD50:15 mg/kg PSEBAA 113,124,63
scu-rat LDLo:18 mg/kg PSEBAA 77,269,51
orl-mus LD50:68 mg/kg PSEBAA 77,269,51
scu-mus LD50:58 mg/kg PSEBAA 77,269,51

SAFETY PROFILE: Poison by ingestion, subcutaneous, and intraperitoneal routes. An irritant to skin, eyes, and mucous membranes. Questionable carcinogen with experimental neoplastigenic data. An experimental teratogen. Other experimental reproductive effects. Combustible when exposed to heat or flames. When heated to decomposition it emits acrid smoke and irritating fumes.

PJJ225 CAS:4354-76-1 ***HR: 3***
PODOPHYLLOTOXIN
mf: $C_{22}H_{22}O_8$ mw: 414.44

SYNS: NSC 24818 ◇ PODOPHYLLINIC ACID LACTONE

TOXICITY DATA with REFERENCE
oms-nml:oth 50 nmol/L CNREA8 37,3071,77
dni-hmn:hla 10 μmol/L BICHAW 15,5435,76
ipr-rat TDLo:10 mg/kg (female 15-16D post):TER PSEBAA 113,124,63
orl-rat TDLo:500 mg/kg (1-5D preg):REP JRPFA4 14,534,67
skn-rat LD50:500 mg/kg PIXXD2 #86-04062
ipr-rat LD50:15 mg/kg FEPRA7 7,249,48
scu-rat LD50:8 mg/kg PSEBAA 77,269,51
ivn-rat LD50:8700 μg/kg FEPRA7 7,249,48
ims-rat LD50:3 mg/kg FEPRA7 7,249,48
orl-mus LD50:100 mg/kg PIXXD2 #86-04062
ipr-mus LD50:30 mg/kg ARZNAD 11,327,61
scu-dog LDLo:1 mg/kg AEXPBL 28,32,1891

SAFETY PROFILE: Poison by intraperitoneal, subcutaneous, intravenous, and intramuscular routes. Moderately toxic by skin route. An experimental teratogen. Experimental reproductive effects. Human mutation data reported. When heated to decomposition it emits acrid smoke and irritating fumes. Used an an antineoplastic agent.

PJJ300 ***HR: 3***
POISON HEMLOCK

PROP: It resembles a carrot plant but with a white root. The leaves may grow to 4 feet. The stem and leaves may have purple spots. It produces small white flowers. The leaves have an obnoxious smell when crushed. It grows wild in Canada and in the northern part of the United States.

SYNS: BUNK ◇ CALIFORNIA FERN ◇ CASHES ◇ CIGUE (CANADA) ◇ CONIUM MACULATUM ◇ HERB BONNETT ◇ KILL COW ◇ NEBRASKA FERN ◇ POISON PARSLEY ◇ POISON ROOT ◇ SNAKE WEED ◇ SPOTTED HEMLOCK ◇ SPOTTED PARSLEY ◇ ST. BENNET'S HERB ◇ WINTER FERN ◇ WODE WHISTLE

SAFETY PROFILE: The whole plant and especially the root and seeds contain the poison coniine and some related alkaloids. Ingestion may cause irritation of the mouth and throat, nausea, vomiting, headache, extreme dilation of the pupil, dizziness, convulsions, coma, and rarely death. The effects are similar to those produced by nicotine. See also NICOTINE.

PJJ315 ***HR: 2***
POKEWEED

PROP: An unpleasant smelling plant which grows to 12 feet from thick (to 6-inch diameter) roots. The sturdy stems have a purple tint and are heavily branched. The leaves are between 4 and 12 inches long. The small flowers range in color from green-white to purplish and grow on a vertical stalk. The berries are a dark purple or black. It grows in damp areas in the region bounded by Maine, southern Ontario, Texas, and Florida. It is also found in Hawaii and California.

SYNS: AMERICAN NIGHTSHADE ◇ BLEDO CARBONERO (CUBA) ◇ CANCER JALAP ◇ CHONGRASS ◇ COAKUM ◇ COCUM ◇ COKAN ◇ CROW BERRY ◇ GARGET ◇ INDIAN POLK ◇ INK BERRY ◇ PHYTOLACCA AMERICANA ◇ PIGEON-BERRY ◇ POCAN BUSH ◇ POKE ◇ POKEBERRY ◇ POLKWEED ◇ RED INK PLANT ◇ RED WEED ◇ SCOKE

SAFETY PROFILE: The leaves and roots contain the poisonous phytolaccatoxin and related triterpenes. Ingestion of these plant parts can cause after a 2 to 3 hour delay: nausea, abdominal cramps, profuse sweating and persistent vomiting and diarrhea. The young sprouts and stems are edible if boiled and the water discarded. These cooked plant parts are sold in cans. The ripe berries are relatively nontoxic. Poisonings generally result from using the uncooked leaves in salads or mistaking the roots for parsnips or horseradish.

PJJ325 CAS:2438-32-6 ***HR: 3***
POLARAMINE MALEATE
mf: $C_{16}H_{19}ClN_2 \cdot C_4H_4O_4$ mw: 390.90

SYNS: (+)-2-(p-CHLORO-α-(2-(DIMETHYLAMINO)ETHYL)BENZYL)PYRIDINE MALEATE ◇ (+)-CHLORPHENIRAMINE MALEATE ◇ d-CHLORPHENIRAMINE MALEATE ◇ S-(+)-CHLORPHENIRAMINE MALEATE ◇ DEXCHLOROPHENIRAMINE MALEATE ◇ DEXCHLORPHENIRAMINE MALEATE ◇ DEXTROCHLORPHENIRAMINE MALEATE ◇ POLARAMIN ◇ PORAMINE MALEATE

TOXICITY DATA with REFERENCE
orl-rat LD50:267 mg/kg JPETAB 124,347,58
ipr-rat LD50:119 mg/kg JPETAB 124,347,58
orl-mus LD50:189 mg/kg
ipr-mus LD50:117 mg/kg JPETAB 124,347,58
ivn-mus LD50:28 mg/kg JPETAB 124,347,58
orl-gpg LD50:240 mg/kg JPETAB 124,347,58 JPETAB 124,347,58

SAFETY PROFILE: Poison by ingestion, intravenous, and intraperitoneal routes. When heated to decomposition it emits toxic fumes of Cl^- and NO_x.

PJJ350 CAS:11016-29-8 ***HR: 3***
POLCILLIN

TOXICITY DATA with REFERENCE
orl-mus LD50:1200 mg/kg 85GDA2 4(1),376,80
ipr-mus LD50:49 mg/kg 85GDA2 4(1),376,80
ivn-mus LD50:38 mg/kg 85GDA2 4(1),376,80

SAFETY PROFILE: Poison by intravenous and intraperitoneal routes. Moderately toxic by ingestion.

PJJ500 CAS:37221-23-1 ***HR: 3***
POLIFUNGIN

PROP: A tetraene antibiotic produced by *Streptomyces noursei var. Polifungini* ATCC 21581 (85ERAY 2,979,78).

SYN: POLYFUNGIN

TOXICITY DATA with REFERENCE
ipr-rat LD50:470 mg/kg 85ERAY 2,979,78
ipr-mus LD50:210 mg/kg 85ERAY 2,979,78

SAFETY PROFILE: Poison by intraperitoneal route.

PJJ750 ***HR: 3***
POLONIUM
af: Po aw: 210

PROP: A low melting, volatile, radioactive, naturally occurring metallic element. Mp: 254°, bp: 962°, d: 9.4.

SYN: RADIUM F

SAFETY PROFILE: Suspected carcinogen. Severe radiotoxicity. Very dangerous to handle. Radiation Hazard: Natural isotope ^{210}Po (radium-F, Uranium Series), $T_{0.5}$ = 138 D. Decays to stable ^{206}Pb by alphas of 5.3 MeV. When heated to decomposition it emits toxic and radioactive fumes of Po. See also PLUTONIUM.

PJK000 ***HR: 3***
POLONIUM CARBONYL
mf: PoCO mw: 237.01

SAFETY PROFILE: Suspected carcinogen. Poison by ingestion, inhalation, intravenous, and subcutaneous routes. When heated to decomposition it emits toxic and radioactive fumes of Po. See also CARBONYLS and POLONIUM.

PJK250 CAS:25014-41-9 ***HR: 3***
POLYACRYLONITRILE, COMBUSTION PRODUCTS

PROP: Products of combustion of polyacrylonitrile in furnace maintained at 800°C (APFRAD 35, 461,77).

TOXICITY DATA with REFERENCE
ihl-mus LC50:160 mg/m³/10M APFRAD 35,461,77

CONSENSUS REPORTS: Reported in EPA TSCA Inventory.

SAFETY PROFILE: Poison by inhalation.

PJK750 CAS:25038-54-4 ***HR: 3***
POLYAMIDE-6 (combustion products)

PROP: Products of combustion of polyamide-6 in furnace maintained at 800°C (APFRAD 35,461,77).

TOXICITY DATA with REFERENCE
ihl-mus LC50:23 mg/m³/10M APFRAD 35,461,77

CONSENSUS REPORTS: Reported in EPA TSCA Inventory.

SAFETY PROFILE: Poison by inhalation.

PJL000 CAS:68822-50-4 ***HR: 2***
POLYAMINE D

TOXICITY DATA with REFERENCE
skn-rbt 500 mg open MLD UCDS** 4/1/64
eye-rbt 50 mg SEV UCDS** 4/1/64
orl-rat LD50:2590 mg/kg UCDS** 4/1/64
skn-rbt LD50:880 mg/kg UCDS** 4/1/64

SAFETY PROFILE: Moderately toxic by ingestion and skin contact. Skin and severe eye irritant. When heated to decomposition it emits toxic fumes of NO_x.

PJL100 CAS:37268-68-1 ***HR: 2***
POLYAMINE H SPECIAL

TOXICITY DATA with REFERENCE
skn-rbt 500 mg MLD 34ZIAG -,692,69
eye-rbt 100 mg MOD 34ZIAG -,692,69
orl-rat LD50:2500 mg/kg 34ZIAG -,692,69
skn-rbt LD50:620 mg/kg 34ZIAG -,692,69

SAFETY PROFILE: Moderately toxic by skin contact and ingestion. A skin and eye irritant. When heated to decomposition it emits toxic fumes of NO_x.

PJL325 CAS:13766-26-2 ***HR: 3***
POLY[BORANE(1)]
mf: $(BH)_n$

SAFETY PROFILE: Ignites spontaneously on contact with air. When heated to decomposition it emits toxic fumes of B. See also BORANES and BORON COMPOUNDS.

PJL335 ***HR: 3***
POLYBROMINATED BIPHENYLS

PROP: A class of aromatic compounds, related to polychlorinated biphenyls, containing two benzene nuclei with two or more substituent bromine atoms. Typically they are inert solids and thus have been used in industry as flame retardants. They do not occur as natural products, but are persistent in the environment and are concentrated in body fat.

SYN: PBB

CONSENSUS REPORTS: Community Right-To-Know List.

SAFETY PROFILE: The major isomer in production,

hexabromobiphenyl, was involved in a large-scale poisoning of dairy cattle in Michigan in 1973.

Generally acute LD_{50} doses are very high but in experimental animals subchronic poisoning may cause bodyweight decrease, liver hypertrophy, chloracne-like lesions, suppression of immune response, neuromuscular dysfunction, teratogenic and embryotoxic effects. In humans they cross the placental barrier and are concentrated and secreted in mothers' milk. When heated to decomposition it emits toxic fumes of Br^-. See also HEXABROMOBIPHENYL and OCTABROMODIPHENYL.

PJL350 CAS:9003-17-2 ***HR: 3***
cis-POLY(BUTADIENE)
mf: $(C_4H_6)_n$

$$(CH_2CH{=}CHCH_2)_n$$

SAFETY PROFILE: May explode when heated above 337°C. When heated to decomposition it emits acrid smoke and irritating fumes. See also BUTADIENE.

PJL375 ***HR: 3***
POLY(1,3-BUTADIENE PEROXIDE)
mf: $(C_4H_6O_2)_n$

SAFETY PROFILE: A powerful explosive very sensitive to shock. Formed by the reaction of butadiene with air. When heated to decomposition it emits acrid smoke and irritating fumes. See also PEROXIDES.

PJL600 CAS:25136-85-0 ***HR: 3***
POLY(CARBON MONOFLUORIDE)
mf: $(CF)_n$

SAFETY PROFILE: Explodes when heated to 500°C in inert atmospheres. Ignites when heated to 400°C in hydrogen atmospheres. When heated to decomposition it emits toxic fumes of F^-. See also FLUORIDES.

PJL750 CAS:1336-36-3 ***HR: 3***
POLYCHLORINATED BIPHENYL
DOT: UN 2315

PROP: Bp: 340-375°, flash p: 383°F (COC), d: 1.44 @ 30°. A series of technical mixtures consisting of many isomers and compounds that vary from mobile oily liquids to white crystalline solids and hard noncrystalline resins. Technical products vary in composition, in the degree of chlorination and possibly according to batch (IARC** 7,262,74).

SYNS: AROCLOR ◇ CHLOPHEN ◇ CHLOREXTOL ◇ CHLORINATED BIPHENYL ◇ CHLORINATED DIPHENYL ◇ CHLORINATED DIPHENYLENE ◇ CHLORO BIPHENYL ◇ CHLORO-1,1-BIPHENYL ◇ CLOPHEN ◇ DYKANOL ◇ FENCLOR ◇ INERTEEN ◇ KANECHLOR ◇ MONTAR ◇ NOFLAMOL ◇ PCB (DOT, USDA) ◇ PHENOCHLOR ◇ POLYCHLOROBIPHENYL ◇ PYRALENE ◇ PYRANOL ◇ SANTOTHERM ◇ SOVOL ◇ THERMINOL FR-1

TOXICITY DATA with REFERENCE
orl-mam TDLo:325 mg/kg (30D pre/1-36D preg):REP AMBOCX 6,239,77
orl-rat TDLo:16800 mg/kg/2Y-C:ETA TOERD9 1,159,78
orl-mus TDLo:1250 mg/kg/25W-I:CAR FCTOD7 21,688,83
orl-rat TD:1250 mg/kg/25W-I:CAR FCTOD7 21,688,83
orl-mus LD50:1900 mg/kg FKIZA4 60,544,69

CONSENSUS REPORTS: NTP Fifth Annual Report on Carcinogens. IARC Cancer Review: Group 2A IMEMDT 7,322,87; Human Limited Evidence IMEMDT 18,43,78. EPA Extremely Hazardous Substances List. Reported in EPA TSCA Inventory.

DFG MAK: Suspected Carcinogen.
NIOSH REL: TWA (Polychlorinated Biphenyls) 0.001 mg/m^3
DOT Classification: ORM-E; Label: None.

SAFETY PROFILE: Confirmed carcinogen with carcinogenic and tumorigenic data. Moderately toxic by ingestion. Some are poisons by other routes. Experimental reproductive effects.

Like the chlorinated naphthalenes, the chlorinated diphenyls have two distinct actions on the body, namely, a skin effect and a toxic action on the liver. This hepatotoxic action of the chlorinated diphenyls appears to be increased if there is exposure to carbon tetrachloride at the same time. The higher the chlorine content of the diphenyl compound, the more toxic it is liable to be. Oxides of chlorinated diphenyls are more toxic than the unoxidized materials. In persons who have suffered systemic intoxication, the usual signs and symptoms are nausea, vomiting, loss of weight, jaundice, edema, and abdominal pain. Where the liver damage has been severe the patient may pass into a coma and die.

Combustible when exposed to heat or flame. When heated to decomposition they emit highly toxic fumes of Cl^-. See also specific compounds.

PJL800 CAS:12674-11-2 ***HR: 2***
POLYCHLORINATED BIPHENYL (AROCLOR 1016)

SYNS: AROCLOR 1016 ◇ CHLORODIPHENYL (41% Cl)

TOXICITY DATA with REFERENCE
orl-uns TDLo:750 mg/kg (female 17W pre/28D post):REP AECTCV 9,627,80
orl-rat LD50:2300 mg/kg NTIS** PB85-143766

SAFETY PROFILE: Moderately toxic by ingestion. Experimental reproductive effects. When heated to decomposition it emits toxic fumes of PCB and Cl^-.

PJM000 CAS:11104-28-2 ***HR: 2***
POLYCHLORINATED BIPHENYL (AROCLOR 1221)

SYNS: AROCHLOR 1221 ◇ CHLORODIPHENYL (21% Cl)

TOXICITY DATA with REFERENCE
orl-rbt TDLo:28 mg/kg (female 1-28D post):REP BECTA6 6,120,71
orl-rat LD50:3980 mg/kg ARVPAX 14,139,74
skn-rbt LDLo:3169 mg/kg ARVPAX 14,139,74

CONSENSUS REPORTS: IARC Cancer Review: Human Limited Evidence IMEMDT 18,43,78.

NIOSH REL: TWA (Polychlorinated Biphenyls) 0.001 mg/m^3

SAFETY PROFILE: Suspected human carcinogen. Moderately toxic by ingestion and skin contact. Experimental reproductive effects. When heated to decomposition it emits toxic fumes of Cl^-. Used in heat transfer, hydraulic fluids, lubricants, and insecticides. See also POLYCHLORINATED BIPHENYLS.

PJM250 CAS:11141-16-5 ***HR: 2***
POLYCHLORINATED BIPHENYL (AROCLOR 1232)

SYNS: AROCLOR 1232 ◇ CHLORODIPHENYL (32% Cl)

TOXICITY DATA with REFERENCE
orl-rat LD50:4470 mg/kg ARVPAX 14,139,74
skn-rbt LDLo:2000 mg/kg ARVPAX 14,139,74

CONSENSUS REPORTS: IARC Cancer Review: Human Limited Evidence IMEMDT 18,43,78.

NIOSH REL: TWA (Polychlorinated Biphenyls) 0.001 mg/m^3

SAFETY PROFILE: Suspected human carcinogen. Moderately toxic by skin contact. Mildly toxic by ingestion. When heated to decomposition it emits toxic fumes of Cl^-. Used in heat transfer, hydraulic fluids, lubricants, and insecticides. See also POLYCHLORINATED BIPHENYLS.

PJM500 CAS:53469-21-9 ***HR: 3***
POLYCHLORINATED BIPHENYL (AROCLOR 1242)

SYNS: AROCHLOR 1242 ◇ AROCLOR 1242 ◇ CHLORIERTE BIPHENYLE, CHLORGEHALT 42% (GERMAN) ◇ CHLORODIPHENYL (42% Cl) (OSHA) ◇ CLORODIFENILI, CLORO 42% (ITALIAN) ◇ DIPHENYLE CHLORE, 42% de CHLORE (FRENCH) ◇ GECHLOREERDEDIFENYL (DUTCH) ◇ PCB's

TOXICITY DATA with REFERENCE
oms-mus:oth 25 ppm/4H EESADV 3,10,79
orl-pig TDLo:93 mg/kg (female 1-16W post):REP AJVRAH 36,23,75
ihl-hmn TCLo:10 mg/m^3:PUL,LIV 85CYAB 2,153,59
orl-rat LD50:4250 mg/kg TXAPA9 24,434,73
scu-gpg LDLo:345 mg/kg PHRPA6 59,1085,44

CONSENSUS REPORTS: IARC Cancer Review: Human Limited Evidence IMEMDT 18,43,78. EPA Genetic Toxicology Program.

OSHA PEL: TWA 1 mg/m^3 (skin)
ACGIH TLV: TWA 1 mg/m^3 (skin)
DFG MAK: 0.1 ppm (1 mg/m^3)
NIOSH REL: TWA (Polychlorinated Biphenyls) 0.001 mg/m^3

SAFETY PROFILE: Suspected human carcinogen. Poison by subcutaneous route. Mildly toxic by ingestion. Human systemic effects by inhalation: pulmonary and liver effects. Moderately toxic by ingestion. Experimental reproductive effects. Mutation data reported. When heated to decomposition it emits toxic fumes of Cl^-. Used in heat transfer, hydraulic fluids, lubricants, and insecticides. See also POLYCHLORINATED BIPHENYLS.

PJM750 CAS:12672-29-6 ***HR: 3***
POLYCHLORINATED BIPHENYL (AROCLOR 1248)

SYNS: AROCLOR 1248 ◇ CHLORODIPHENYL (48% Cl)

TOXICITY DATA with REFERENCE
orl-mky TDLo:32 mg/kg/(1-23W preg/91D post):REP NTOTDY 3,15,81
orl-rbt TDLo:165 mg/kg (female 1-31D post):TER DABBBA 40,2061,79
orl-rat LD50:11 g/kg ARVPAX 14,139,74
skn-rbt LDLo:1269 mg/kg ARVPAX 14,139,74

CONSENSUS REPORTS: IARC Cancer Review: Human Limited Evidence IMEMDT 18,43,78.

NIOSH REL: TWA (Polychlorinated Biphenyls) 0.001 mg/m^3

SAFETY PROFILE: Suspected human carcinogen. Moderately toxic by skin contact. Experimental teratogenic and reproductive effects. When heated to decomposition it emits toxic fumes of Cl^-. Used in heat transfer, hydraulic fluids, lubricants, and insecticides. See also POLYCHLORINATED BIPHENYLS.

PJN000 CAS:11097-69-1 ***HR: 3***
POLYCHLORINATED BIPHENYL (AROCLOR 1254)

PROP: Composed of 11% tetra-, 49% penta-, 34% hexa- and 6% heptachlorobiphenyls (FCTXAV 12,63,74).

SYNS: AROCHLOR 1254 ◇ AROCLOR 1254 ◇ CHLORIERTE

BIPHENYLE, CHLORGEHALT 54% (GERMAN) ◇ CHLORODIPHENYL (54% Cl) (OSHA) ◇ CLORODIFENILI, CLORO 54% (ITALIAN) ◇ DIPHENYLE CHLORE, 54% de CHLORE (FRENCH) ◇ NCI-C02664 ◇ PCB's

TOXICITY DATA with REFERENCE
cyt-ofs-ipr 50 mg/kg CBPCBB 82,489,85
otr-rat-orl 25 ppm/2Y-C EVHPAZ 60,89,85
dnd-rat-orl 1295 mg/kg BSIBAC 57,407,81
dnd-rat:lvr 300 μmol/L SinJF# 26OCT82
orl-mus TDLo:59400 μg/kg (female 3D pre-21D post):REP NETOD7 3,5,81
orl-rbt TDLo:350 mg/kg (female 1-28D post):TER EVPHBI 1,67,71
orl-rat TDLo:73500 mg/kg/2Y-C:CAR EVHPAZ 60,89,85
orl-mus TDLo:17 g/kg/48W-C:NEO JNCIAM 53,547,74
skn-mus TDLo:4 mg/kg:ETA BECTA6 18,552,77
orl-rat LD50:1010 mg/kg TXAPA9 60,33,81
ivn-rat LD50:358 mg/kg FCTXAV 12,63,74
ipr-mus LD50:2840 mg/kg BECTA6 8,245,72

CONSENSUS REPORTS: NTP Fifth Annual Report on Carcinogens. IARC Cancer Review: Group 2A IMEMDT 7,322,87; Animal Sufficient Evidence IMEMDT 7,261,74; Animal Limited Evidence IMEMDT 18,43,78; Human Limited Evidence IMEMDT 18,43,78. NCI Carcinogenesis Bioassay (feed); Some Evidence: rat NCITR* NCI-CG-TR-38,78. EPA Genetic Toxicology Program.

OSHA PEL: TWA 0.5 mg/m³ (skin)
ACGIH TLV: TWA 0.5 mg/m³ (skin)
NIOSH REL: TWA (Polychlorinated Biphenyls) 0.001 mg/m³

SAFETY PROFILE: Confirmed carcinogen with experimental carcinogenic and neoplastigenic data. Poison by intravenous route. Moderately toxic by ingestion and intraperitoneal routes. Experimental teratogenic and reproductive effects. Mutation data reported. When heated to decomposition it emits toxic fumes of Cl^-. Used in heat transfer, hydraulic fluids, lubricants, and insecticides. See also POLYCHLORINATED BIPHENYLS.

PJN250 CAS:11096-82-5 ***HR: 3***
POLYCHLORINATED BIPHENYL (AROCLOR 1260)

PROP: Composed of 12% penta-, 38% hexa-, 41% hepta-, 8% octa- and 1% nonachlorobiphenyls (FCTXAV 12,63,74).

SYNS: AROCHLOR 1260 ◇ AROCLOR 1260 ◇ CHLORODIPHENYL (60% Cl) ◇ CLOPHEN A60 ◇ PHENOCLOR DP6

TOXICITY DATA with REFERENCE
cyt-rat-orl 1080 mg/kg/26W-C APTOD9 19,A16,80
scu-mus TDLo:143 mg/kg (lactating female 21D post):REP ENPBBC 5,54,75
orl-rat TDLo:4380 mg/kg/83W-C:CAR JNCIAM 55,1453,75
orl-rat TD:4992 mg/kg/2Y-C:CAR TXAPA9 75,278,84
orl-rat TD:360 mg/kg/17W-C:NEO CALEDQ 39,59,88
orl-rat LD50:1315 mg/kg FCTXAV 12,63,74
skn-rbt LDLo:2000 mg/kg ARVPAX 14,139,74

CONSENSUS REPORTS: NTP Fifth Annual Report on Carcinogens. IARC Cancer Review: Animal Limited Evidence IMEMDT 18,43,78; Human Limited Evidence IMEMDT 18,43,78.

NIOSH REL: TWA (Polychlorinated Biphenyls) 0.001 mg/m³

SAFETY PROFILE: Confirmed carcinogen with carcinogenic and neoplastigenic data. Moderately toxic by ingestion and skin contact. Experimental reproductive effects. Mutation data reported. When heated to decomposition it emits highly toxic fumes of Cl^-. Used in heat transfer, hydraulic fluids, lubricants, and insecticides. See also POLYCHLORINATED BIPHENYLS.

PJN500 CAS:37324-23-5 ***HR: 3***
POLYCHLORINATED BIPHENYL (AROCLOR 1262)

SYNS: AROCLOR 1262 ◇ CHLORODIPHENYL (62% Cl)

TOXICITY DATA with REFERENCE
orl-rat LD50:11300 mg/kg ARVPAX 14,139,74
skn-rbt LDLo:3160 mg/kg ARVPAX 14,139,74

CONSENSUS REPORTS: IARC Cancer Review: Human Limited Evidence IMEMDT 18,43,78.

DFG MAK: 0.1 ppm (1 mg/m³)
NIOSH REL: (Polychlorinated Biphenyls) TWA 0.001 mg/m³

SAFETY PROFILE: Suspected human carcinogen. Moderately toxic by skin contact. When heated to decomposition it emits toxic fumes of Cl^-. Used in heat transfer, hydraulic fluids, lubricants, and insecticides. See also POLYCHLORINATED BIPHENYLS.

PJN750 CAS:11100-14-4 ***HR: 3***
POLYCHLORINATED BIPHENYL (AROCLOR 1268)

SYNS: AROCLOR 1268 ◇ CHLORODIPHENYL (68% Cl)

TOXICITY DATA with REFERENCE
orl-rat LD50:10900 mg/kg ARVPAX 14,139,74
skn-rbt LDLo:2500 mg/kg ARVPAX 14,139,74

CONSENSUS REPORTS: IARC Cancer Review: Human Limited Evidence IMEMDT 18,43,78.

NIOSH REL: (Polychlorinated Biphenyls) TWA 0.001 mg/m^3

SAFETY PROFILE: Suspected human carcinogen. Moderately toxic by skin contact. Used in heat transfer, hydraulic fluids, lubricants, and insecticides. When heated to decomposition it emits toxic fumes of Cl^-. See also POLYCHLORINATED BIPHENYLS.

PJO000 CAS:37324-24-6 ***HR: 3***
POLYCHLORINATED BIPHENYL (AROCLOR 2565)

SYN: AROCLOR 2565

TOXICITY DATA with REFERENCE
orl-rat LD50:6310 mg/kg ARVPAX 14,139,74
skn-rbt LDLo:3160 mg/kg ARVPAX 14,139,74

CONSENSUS REPORTS: IARC Cancer Review: Human Limited Evidence IMEMDT 18,43,78.

NIOSH REL: (Polychlorinated Biphenyls) TWA 0.001 mg/m^3

SAFETY PROFILE: Suspected human carcinogen. Moderately toxic by skin contact. Mildly toxic by ingestion. When heated to decomposition it emits toxic fumes of Cl^-. Used in heat transfer, hydraulic fluids, lubricants, and insecticides. See also POLYCHLORINATED BIPHENYLS.

PJO250 CAS:11120-29-9 ***HR: 3***
POLYCHLORINATED BIPHENYL (AROCLOR 4465)

SYN: AROCLOR 4465

TOXICITY DATA with REFERENCE
orl-rat LD50:16 g/kg ARVPAX 14,139,74
skn-rbt LDLo:3160 mg/kg ARVPAX 14,139,74

CONSENSUS REPORTS: IARC Cancer Review: Human Limited Evidence IMEMDT 18,43,78.

NIOSH REL: TWA (Polychlorinated Biphenyls) 0.001 mg/m^3

SAFETY PROFILE: Suspected human carcinogen. Moderately toxic by skin contact. Mildly toxic by ingestion. When heated to decomposition it emits toxic fumes of Cl^-. Used in heat transfer, hydraulic fluids, lubricants, and insecticides. See also POLYCHLORINATED BIPHENYLS.

PJO500 CAS:37353-63-2 ***HR: 3***
POLYCHLORINATED BIPHENYL (KANECHLOR 300)

PROP: Average content: 60% trichlorobiphenyl, 23% tetrachlorobiphenyl, 17% dichlorobiphenyl, 1% pentachlorobiphenyl (IARC** 7,262,74).

SYN: KANECHLOR 300

TOXICITY DATA with REFERENCE
orl-rat TDLo:22 mg/kg (female 1-22D post):TER OFAJAE 53,93,76
orl-rat LD50:1100 mg/kg SKEZAP 13,359,72
orl-rbt LD50:600 mg/kg SKEZAP 13,359,72

CONSENSUS REPORTS: IARC Cancer Review: Animal Limited Evidence IMEMDT 7,261,74, IMEMDT 18,43,78; Human Limited Evidence IMEMDT 18,43,78.

NIOSH REL: TWA (Polychlorinated Biphenyls) 0.001 mg/m^3

SAFETY PROFILE: Moderately toxic by ingestion. Suspected human carcinogen. An experimental teratogen. Used in heat transfer, hydraulic fluids, lubricants, and insecticides. When heated to decomposition it emits toxic fumes of Cl^-. See also POLYCHLORINATED BIPHENYLS.

PJO750 CAS:12737-87-0 ***HR: 3***
POLYCHLORINATED BIPHENYL (KANECHLOR 400)

PROP: Average content: 44% tetrachlorbiphenyl, 33% trichlorobiphenyl, 16% pentachlorobiphenyl, 5% hexachlorobiphenyl, 3% dichlorobiphenyl (IARC** 7,262,74).

SYNS: KANECHLOR 400 ◇ KC-400

TOXICITY DATA with REFERENCE
orl-rat TDLo:52500 μg/kg (1-21D preg):REP SKEZAP 15,252,74
orl-rat TDLo:10500 μg/kg (1-21D preg):TER SKEZAP 15,252,74
orl-rat TDLo:6750 mg/kg/69W-I:NEO GANNA2 64,105,73
orl-hmn TDLo:28 mg/kg:SKN FKIZA4 62,104,71
orl-rat LD50:1100 mg/kg SKEZAP 13,359,72
orl-mus LD50:1600 mg/kg SKEZAP 13,359,72

CONSENSUS REPORTS: IARC Cancer Review: Animal Limited Evidence IMEMDT 7,261,74, IMEMDT 18,43,78; Human Limited Evidence IMEMDT 18,43,78.

NIOSH REL: TWA (Polychlorinated Biphenyls) 0.001 mg/m^3

SAFETY PROFILE: Suspected carcinogen with experimental neoplastigenic data. Moderately toxic by ingestion. Experimental teratogenic and reproductive effects. Human systemic effects by ingestion: dermatitis, sweating. When heated to decomposition it emits toxic fumes of Cl^-. See also POLYCHLORINATED BIPHENYLS.

PJP000 CAS:37317-41-2 ***HR: 3***
POLYCHLORINATED BIPHENYL (KANECHLOR 500)

PROP: Average content, 55% pentachlorobiphenyl, 26.5% tetrachlorobiphenyl, 12.8% hexachloro biphenyl and 5% trichlorobiphenyl (JNCIAM 51,1637,73).

SYNS: KANECHLOR 500 ◇ KC-500

TOXICITY DATA with REFERENCE
orl-rat TDLo:140 mg/kg (15-21D preg):REP OFAJAE 53,105,76
scu-mus TDLo:4 g/kg (female 6-10D post):TER TXCYAC 19,49,81
orl-mus TDLo:13 g/kg/32W-C:CAR NAIZAM 25,635,74
orl-mus TD:23 g/kg/32W-C:CAR JNCIAM 51,1637,73
orl-mus TD:13440 mg/kg/32W-C:ETA SAIGBL 17,54,75

CONSENSUS REPORTS: IARC Cancer Review: Human Limited Evidence IMEMDT 18,43,78; Animal Limited Evidence IMEMDT 18,43,78; Animal Sufficient Evidence IMEMDT 7,261,74.

NIOSH REL: TWA (Polychlorinated Biphenyls) 0.001 mg/m^3.

SAFETY PROFILE: Suspected carcinogen with experimental carcinogenic and tumorigenic data. Experimental teratogenic and reproductive effects. When heated to decomposition it emits toxic fumes of Cl^-. Used in heat transfer, hydraulic fluids, lubricants, and insecticides. See also POLYCHLORINATED BIPHENYLS.

PJP100 ***HR: 3***
POLYCHLORINATED DIBENZOFURANS

PROP: Impurities in polychlorinated biphenyls-PCB. (TXAPA9 45,209,78).

SYN: PCDF

TOXICITY DATA with REFERENCE
orl-mus LD50:184 mg/kg TXAPA9 45,209.78
ipr-mus LDLo:100 mg/kg TXAPA9 45,209.78
scu-mus LDLo:200 mg/kg TXAPA9 45,209.78

SAFETY PROFILE: Poison by ingestion, subcutaneous, and intraperitoneal routes. When heated to decomposition they emit toxic fumes of Cl^- and NO_x.

PJP250 CAS:61788-33-8 ***HR: 3***
POLYCHLORINATED TERPHENYL

PROP: Kanechlor carbon consists of 95% polychlorinated terphenyl and 5% PCB (CALEDQ 4,271,78).

SYN: KANECHLOR 500

TOXICITY DATA with REFERENCE
orl-mus TD:10 g/kg/24W-C:ETA JTSCDR 3,259,78
orl-mus TDLo:11 g/kg/24W-C:CAR CALEDQ 4,271,78

CONSENSUS REPORTS: Reported in EPA TSCA Inventory.

SAFETY PROFILE: Questionable carcinogen with experimental carcinogenic and tumorigenic data. When heated to decomposition it emits toxic fumes of Cl^-. See also POLYCHLORINATED BIPHENYLS.

PJP750 CAS:12642-23-8 ***HR: 2***
POLYCHLORINATED TRIPHENYL (AROCLOR 5442)

SYN: AROCLOR 5442

TOXICITY DATA with REFERENCE
orl-rat LD50:10600 mg/kg ARVPAX 14,139,74
skn-rbt LD50:3160 mg/kg ARVPAX 14,139,74

SAFETY PROFILE: Moderately toxic by skin contact. Mildly toxic by ingestion. When heated to decomposition it emits toxic fumes of Cl^-. Used in heat transfer, hydraulic fluids, lubricants, and insecticides. See also POLYCHLORINATED BIPHENYLS.

PJQ000 CAS:1338-32-5 ***HR: 2***
POLYCHLOROBENZOIC ACID, DIMETHYLAMINE SALTS

PROP: Mixture of dimethylamine salt of 2,3,5,6-tetrachlorobenzoic acid, 2,3,6-trichlrobenzoic acid and other chlorinated benzoic acids (GUCHAZ 6,423,73).

SYNS: BENZAC ◇ DIMETHYLAMINE SALTS of mixed POLYCHLOROBENZOIC ACIDS ◇ PBA, DIMETHYLAMINE SALT ◇ ZOBAR

TOXICITY DATA with REFERENCE
orl-rat LD50:1140 mg/kg FMCHA2 -,D233,80
orl-mus LD50:1200 mg/kg GUCHAZ 6,423,73
orl-rbt LD50:480 mg/kg PCOC** -,919,66

SAFETY PROFILE: Moderately toxic by ingestion. When heated to decomposition it emits very toxic fumes of NO_x and Cl^-. See also CHLORINATED HYDROCARBONS, AROMATIC and AMINES.

PJQ250 CAS:25267-15-6 ***HR: 3***
POLYCHLOROPINENE
mf: $C_{10}H_{10}Cl_8$ mw: 413.80

SYN: STROBANE

TOXICITY DATA with REFERENCE
orl-rat LD50:165 mg/kg GISAAA 44(4),51,79

SAFETY PROFILE: Poison by by ingestion. When heated to decomposition it emits toxic fumes of Cl^-. See also CHLORINATED HYDROCARBONS, ALIPHATIC.

PJQ275 CAS:35398-20-0 ***HR: 3***
POLYCYCLOPENTADIENYLTITANIUM DICHLORIDE
mf: $(C_5H_5Cl_2Ti)_n$

SAFETY PROFILE: Ignites spontaneously in air. When heated to decomposition it emits toxic fumes of Cl^-. See also TITANIUM COMPOUNDS.

PJQ350 ***HR: 2***
POLYDAZOL

TOXICITY DATA with REFERENCE
orl-rat LD50:2432 mg/kg GTPZAB 26(9),54,82
orl-mus LD50:6760 mg/kg GTPZAB 26(9),54,82
orl-gpg LD50:700 mg/kg GTPZAB 26(9),54,82

SAFETY PROFILE: Moderately toxic by ingestion.

PJQ500 ***HR: 3***
POLYDIBROMOSILANE
mf: $(Br_2Si)_n$ mw: $(187.90)_n$

SYN: POLYDIBROMOSILYLENE

SAFETY PROFILE: An explosive. Explosive reaction with oxidants (e.g., nitric acid). Ignites in air when heated to 120°C. When heated to decomposition it emits toxic fumes of Br^-. See also SILANE.

PJQ750 CAS:26780-96-1 ***HR: 3***
POLY(1,2-DIHYDRO-2,2,4-TRIMETHYL-QUINOLINE)
mf: $(C_{11}H_{16}N)_n$

SYNS: TRIMETHYLDIHYDROQUINOLINE POLYMER ◇ 2,2,4-TRIMETHYL-1,2-DIHYDROQUINOLINEPOLYMER

TOXICITY DATA with REFERENCE
orl-rat TDLo:548 g/kg/2Y-C:NEO TXAPA9 9,583,66

CONSENSUS REPORTS: Reported in EPA TSCA Inventory.

SAFETY PROFILE: Questionable carcinogen with experimental neoplastigenic data. When heated to decomposition it emits toxic fumes of NO_x.

PJQ775 ***HR: 3***
POLY(DIMERCURYIMMONIUM ACETYLIDE)
mf: $(C_2HHg_2N)_n$

CONSENSUS REPORTS: Mercury and its compounds are on the Community Right-To-Know List.

SAFETY PROFILE: Highly explosive. When heated to decomposition it emits toxic fumes of NO_x and Hg. See also MERCURY COMPOUNDS and ACETYLIDES.

PJQ780 ***HR: 3***
POLY(DIMERCURYIMMONIUM BROMATE)
mf: $(BrHg_2NO_3)_n$

CONSENSUS REPORTS: Mercury and its compounds are on the Community Right-To-Know List.

SAFETY PROFILE: Highly explosive. When heated to decomposition it emits toxic fumes of Br^- and Hg. See also MERCURY COMPOUNDS and BROMATES.

PJR000 CAS:9016-00-6 ***HR: 3***
POLYDIMETHYL SILOXANE

PROP: A water-insoluble polymer of high viscosity (AMPLAO 67,589,59).

SYNS: DIMETHICONE 350 ◇ DOW CORNING 346 ◇ GEON ◇ GOOD-RITE ◇ GUM ◇ HYCAR ◇ LATEX ◇ METHYL SILICONE ◇ POLY(OXY(DIMETHYLSILYLENE))

TOXICITY DATA with REFERENCE
scu-rat TDLo:5 g/kg (6-15D preg):REP TJADAB 31,50A,85
imp-rat TDLo:1500 mg/kg:NEO AMPLAO 67,589,59

SAFETY PROFILE: Experimental reproductive effects. Questionable carcinogen with experimental neoplastigenic data. When heated to decomposition it emits acrid smoke and irritating fumes. Used as a release material, foam preventative, and surface active agent. See also POLYMERS, WATER INSOLUBLE.

PJR250 CAS:63394-02-5 ***HR: 3***
POLYDIMETHYLSILOXANE RUBBER

SYNS: POLYSILICONE ◇ SILASTIC ◇ SILICONE RUBBER

TOXICITY DATA with REFERENCE
imp-rat TDLo:1500 mg/kg:CAR AMPLAO 67,589,59
imp-rat TD:900 mg/kg:ETA JNCIAM 33,1005,64

SAFETY PROFILE: Questionable carcinogen with experimental carcinogenic and tumorigenic data. When heated to decomposition it emits acrid smoke and irritating fumes.

PJR500 CAS:9003-34-3 ***HR: 2***
POLY-p-DINITROSOBENZENE
mf: $(C_6H_4N_2O_2)_x$

SYNS: 1,4-DINITROSOBENZENE HOMOPOLYMER ◇ p-DINITROSOBENZENE POLYMERS ◇ POLYAC

TOXICITY DATA with REFERENCE
mmo-sat 150 μg/plate IAPUDO 59,289,84
mma-sat 150 μg/plate IAPUDO 59,289,84
orl-rat LDLo:1500 mg/kg RCTEA4 44,512,71

CONSENSUS REPORTS: Reported in EPA TSCA Inventory.

SAFETY PROFILE: Moderately toxic by ingestion.

Mutation data reported. When heated to decomposition it emits toxic fumes of NO_x. See also NITROSO COMPOUNDS.

PJR750 CAS:34828-67-6 ***HR: 3***
POLYESTRADIOL PHOSPHATE

SYNS: ESTRADURIN ◇ PEP

TOXICITY DATA with REFERENCE
ims-man TDLo:41 mg/kg (35D male):REP ASUPAZ 73,199,68
scu-mus TDLo:70 mg/kg/60D-I:ETA ATHBA3 12,209,73

SAFETY PROFILE: Human reproductive effects by intramuscular route: testes, epididymis, sperm duct effects. Questionable carcinogen with experimental tumorigenic data. When heated to decomposition it emits very toxic fumes of PO_x.

PJS000 ***HR: 3***
POLYETHER DIAMINE L-1000

TOXICITY DATA with REFERENCE
skn-rbt 500 mg open SEV UCDS** 1/20/72
eye-rbt 1 mg SEV UCDS** 1/20/72
orl-rat LD50:102 mg/kg UCDS** 1/20/72
skn-rbt LD50:50 mg/kg UCDS** 1/20/72

SAFETY PROFILE: Poison by ingestion and skin contact. A severe skin and eye irritant. When heated to decomposition it emits toxic fumes of NO_x. See also ETHERS and AMINES.

PJS750 CAS:9002-88-4 ***HR: 3***
POLYETHYLENE
mf: $(C_2H_4)_n$

PROP: Odorless. The high molecular weight compounds are tough, white leathery, resinous. D: 0.92 @ 20°/4°, mp: 85-110°. Sol in hot benzene; insol in water.

SYNS: AGILENE ◇ ALKATHENE ◇ BAKELITE DYNH ◇ DIOTHENE ◇ ETHENE POLYMER ◇ ETHYLENE HOMOPOLYMER ◇ ETHYLENE POLYMERS ◇ HOECHST PA 190 ◇ MICROTHENE ◇ POLYETHYLENE AS ◇ POLYWAX 1000 ◇ TENITE 800

TOXICITY DATA with REFERENCE
imp-rat TDLo:33 mg/kg:ETA CNREA8 15,333,55

CONSENSUS REPORTS: IARC Cancer Review: Group 3 IMEMDT 7,56,87; Animal Sufficient Evidence IMEMDT 19,157,79; Human Inadequate Evidence IMEMDT 19,157,79. Reported in EPA TSCA Inventory.

SAFETY PROFILE: Questionable carcinogen with experimental tumorigenic data by implant. Reacts violently with F_2. When heated to decomposition it emits acrid smoke and irritating fumes.

PJT000 CAS:25322-68-3 ***HR: 2***
POLYETHYLENE GLYCOL
mf: $H(OC_2H_4)_nOH$

PROP: Clear liquid or white solid. D: 1.110-1.140 @ 20°, mp: 4-10°, flash p: 471°F. Sol in organic solvents, aromatic hydrocarbons.

SYNS: ALKAPOL PEG-200 ◇ CARBOWAX ◇ α-HYDROXY-omega-HYDROXY-POLY(OXY-1,2-ETHANEDIYL) ◇ JEFFOX ◇ JORCHEM 400 ML ◇ LUTROL ◇ PEG ◇ PLURACOL P-410 ◇ POLY(ETHYLENE OXIDE) ◇ POLY-G SERIES ◇ POLYOX

TOXICITY DATA with REFERENCE
eye-rbt 500 mg/24H MLD 85JCAE-,1413,86
ivg-mus TDLo:416 mg/kg/Y-I:ETA BJCAAI 15,252,61
orl-rat LD50:32 g/kg DOWCC* MSD-937
ipr-rat LD50:15570 mg/kg ARZNAD 3,451,53
ipr-mus LD50:2000 mg/kg JPETAB 103,293,51
ivn-dog LDLo:3000 mg/kg JPETAB 103,293,51
orl-gpg LD50:22500 mg/kg JAPMA8 39,349,50

CONSENSUS REPORTS: Reported in EPA TSCA Inventory. EPA Genetic Toxicology Program.

SAFETY PROFILE: Moderately toxic by intraperitoneal and intravenous routes. Slightly toxic by ingestion. An eye irritant. Questionable carcinogen with experimental tumorigenic data. Combustible liquid when exposed to heat or flame. To fight fire, use water, foam, dry chemical. When heated to decomposition it emits acrid smoke and irritating fumes. See also other polyethylene glycol entries.

PJT200 CAS:25322-68-3 ***HR: 1***
POLYETHYLENE GLYCOL 200
mf: $H(OC_2H_4)_nOH$

PROP: Viscous, hydroscopic liquid with *n* about 4; slt characteristic odor. D (25°/25°) 1.127.

SYNS: CARBOWAX ◇ JEFFOX ◇ NYCOLINE ◇ PEG 200 ◇ PLURACOL E ◇ POLYAETHYLENGLYCOLE 200 (GERMAN) ◇ POLY-G ◇ POLYGLYCOL E ◇ SOLBASE

TOXICITY DATA with REFERENCE
orl-rat LD50:28900 mg/kg ARZNAD 3,451,53
orl-mus LD50:38300 mg/kg ARZNAD 3,451,53
ipr-mus LD50:7500 mg/kg NTIS** AD628-313
orl-rbt LD50:19900 mg/kg ARZNAD 3,451,53

CONSENSUS REPORTS: EPA Genetic Toxicology Program. Reported in EPA TSCA Inventory.

SAFETY PROFILE: Mildly toxic by ingestion. Caution: Solvent action on some plastics. When heated to decomposition it emits acrid smoke and irritating fumes. See also other polyethylene glycol entries.

PJT225 CAS:25322-68-3 *HR: 1*
POLYETHYLENE GLYCOL 300
mf: $(C_6H_{11}NO)_n$

SYNS: POLYAETHYLENGLYKOLE 300 (GERMAN) ◇ PEG 300

TOXICITY DATA with REFERENCE
orl-rat LD50:27500 mg/kg ARZNAD 3,451,53
ipr-rat LD50:170000 mg/kg ARZNAD 3,451,53
orl-rbt LD50:17300 mg/kg ARZNAD 3,451,53
orl-gpg LD50:19600 mg/kg ARZNAD 3,451,53

CONSENSUS REPORTS: EPA Genetic Toxicology Program. Reported in EPA TSCA Inventory.

SAFETY PROFILE: Mildly toxic by ingestion. When heated to decomposition it emits acrid smoke and irritating fumes. See also other polyethylene glycol entries.

PJT230 CAS:25322-68-3 *HR: 1*
POLYETHYLENE GLYCOL 400
mf: $H(OC_2H_4)_nOH$

PROP: Liquid with *n* about 8.2 to 9.1. Mw: 380-420, d: 1.128, mp: 4-8°.

SYNS: PEG 400 ◇ POLYAETHYLENGLYKOLE 400 (GERMAN) ◇ POLY G 400

TOXICITY DATA with REFERENCE
ipr-rat LD50:9708 mg/kg PESTD5 17,351,76
ivn-rat LD50:7312 mg/kg ARZNAD 26,1581,76
orl-mus LD50:28915 mg/kg PESTD5 17,351,76
ipr-mus LD50:9953 mg/kg PESTD5 17,351,76
ivn-mus LD50:8550 mg/kg ARZNAD 26,1581,76
orl-rbt LD50:26800 mg/kg ARZNAD 3,451,53
orl-gpg LD50:15700 mg/kg ARZNAD 3,451,53

CONSENSUS REPORTS: EPA Genetic Toxicology Program. Reported in EPA TSCA Inventory.

SAFETY PROFILE: Low toxicity by ingestion, intravenous, and intraperitoneal routes. When heated to decomposition it emits acrid smoke and irritating fumes. See also other polyethylene glycol entries.

PJT240 CAS:25322-68-3 *HR: 1*
POLYETHYLENE GLYCOL 600
mf: $H(OC_2H_4)_nOH$

PROP: Liquid with *n* about 12.5 to 13.9.mw: 570-630, d: 1.128, mp: 20-25°.

SYNS: PEG 600 ◇ POLYAETHYLENGLYKOLE 600 (GERMAN)

TOXICITY DATA with REFERENCE
eye-rbt 100 mg MLD 34ZIAG -,747,69
orl-rat LD50:38100 mg/kg 34ZIAG -,747,69
orl-mus LD50:47 g/kg ARZNAD 3,451,53

CONSENSUS REPORTS: EPA Genetic Toxicology Program. Reported in EPA TSCA Inventory.

SAFETY PROFILE: Low toxicity by ingestion. An eye irritant. When heated to decomposition it emits acrid smoke and irritating fumes. See also other polyethylene glycol entries.

PJT250 CAS:25322-68-3 *HR: 3*
POLYETHYLENE GLYCOL 1000
mf: $H(OC_2H_4)_nOH$

SYNS: CARBOWAX 1000 ◇ MACROGOL 1000 ◇ PEG 1000 ◇ POLYAETHYLENGLYKOLE 1000 (GERMAN) ◇ POLYGLYCOL 1000 ◇ POLYGLYCOL E1000

TOXICITY DATA with REFERENCE
eye-rbt 500 mg/24H MLD 85JCAE -,1413,86
ivg-mus TDLo:416 mg/kg/Y-I:ETA,REP BJCAAI 15,252,61
orl-rat LD50:32 g/kg DOWCC* MSD-937
ipr-rat LD50:15570 mg/kg ARZNAD 3,451,53
ipr-mus LD50:2000 mg/kg JPETAB 103,293,51
ivn-dog LDLo:3000 mg/kg JPETAB 103,293,51
orl-pig LD50:22500 mg/kg ARZNAD 3,451,53

CONSENSUS REPORTS: Reported in EPA TSCA Inventory. EPA Genetic Toxicology Program.

SAFETY PROFILE: Moderately toxic by intraperitoneal and intravenous routes. Mildly toxic by ingestion. Experimental reproductive effects. An eye irritant. Questionable carcinogen with experimental tumorigenic data. When heated to decomposition it emits acrid smoke and irritating fumes. See also other polyethylene glycol entries.

PJT500 CAS:25322-68-3 *HR: 1*
POLYETHYLENE GLYCOL 1500
mf: $H(OC_2H_4)_nOH$

PROP: White, free-flowing powder. D: 1.15-1.21 @ 25°/25°, fp: 44-48°.

SYNS: CARBOWAX 1500 ◇ α-HYDRO-omega-HYDROXY-POLY(OXY-1,2-ETHANEDIYL) ◇ PEG 1500 ◇ POLYAETHYLENGLYKOLE 1500 (GERMAN) ◇ POLYOXYETHYLENE 1500

TOXICITY DATA with REFERENCE
skn-hmn 500 mg/48H JIDEAE 19,423,52
orl-rat LD50:44200 mg/kg ARZNAD 3,451,53
ipr-rat LD50:17700 mg/kg ARZNAD 3,451,53
orl-rbt LD50:28900 mg/kg ARZNAD 3,451,53
ivn-rbt LD50:8 g/kg KRKRDT 4,71,78
orl-gpg LD50:28900 mg/kg ARZNAD 3,451,53

CONSENSUS REPORTS: Reported in EPA TSCA Inventory. EPA Genetic Toxicology Program.

SAFETY PROFILE: Mildly toxic by ingestion. A human skin irritant. When heated to decomposition it emits acrid smoke and irritating fumes. See also other polyethylene glycol entries.

PJT750 CAS:25322-68-3 ***HR: 1***
POLYETHYLENE GLYCOL 4000
mf: $H(OC_2H_4)_nOH$

PROP: White, free-flowing powder or white flakes. D: 1.20-1.21 @ 25°/25°Fp: 54-58°.

SYNS: CARBOWAX 4000 ◇ CARSONON PEG-4000 ◇ MACROGOL 4000 ◇ PEG 4000 ◇ POLYAETHYLENGLYKOLE 4000 (GERMAN) ◇ POLYGLYCOL 4000 ◇ POLYGLYCOL E-4000 ◇ POLYGLYCOL E-4000 USP ◇ POLYOXYETHYLENE (75)

TOXICITY DATA with REFERENCE
skn-rbt 500 mg open MLD UCDS** 4/13/65
orl-rat LD50:50 g/kg 34ZIAG -,747,69
ipr-rat LD50:11550 mg/kg ARZNAD 3,451,53
scu-mus LD50:18 g/kg ARZNAD 3,451,53
ivn-mus LD50:16 g/kg ARZNAD 3,451,53
orl-rbt LD50:76 g/kg ARZNAD 3,451,53
orl-gpg LD50:50900 mg/kg ARZNAD 3,451,53

CONSENSUS REPORTS: Reported in EPA TSCA Inventory. EPA Genetic Toxicology Program.

SAFETY PROFILE: Mildly toxic by ingestion. A skin irritant. When heated to decomposition it emits acrid smoke and irritating fumes. See also other polyethylene glycol entries.

PJU000 CAS:25322-68-3 ***HR: 1***
POLYETHYLENE GLYCOL 6000
mf: $H(OC_2H_4)_nOH$

PROP: White, waxy solid. Mp: 58-62°, flash p: >887°F. Water-sol.

SYNS: CARBOWAX 6000 ◇ PEG 6000 ◇ POLYAETHYLENGLYKOLE 6000 (GERMAN)

TOXICITY DATA with REFERENCE
skn-rbt 500 mg open MLD UCDS** 4/9/65
dnd-omi 100 g/L PNASA6 72,4288,75
cyt-ham:oth 50 pph DKBSAS 240,228,78
orl-rat LDLo:50 g/kg 34ZIAG -,747,69
ipr-rat LD50:6790 mg/kg ARZNAD 3,451,53
orl-gpg LD50:50 g/kg ARZNAD 3,451,53

CONSENSUS REPORTS: Reported in EPA TSCA Inventory. EPA Genetic Toxicology Program.

SAFETY PROFILE: Mildly toxic by ingestion. Mutation data reported. A skin irritant. Combustible when exposed to heat or flame. When heated to decomposition it emits acrid smoke and irritating fumes. See also other polyethylene glycol entries.

PJU250 CAS:52137-03-8 ***HR: 2***
POLYETHYLENE GLYCOL CHLORIDE 210
mf: $(C_2H_4O)_n$-C_2H_5ClO

PROP: Liquid. Mp: −90°, bp: 198.9°, flash p: 225°F (OC), d: 1.1753 @ 20°/20°, vap d: 4.31.

SYNS: α-(2-CHLOROETHYL)-omega-HDYROXY-POLY(OXY-1,2-ETHANEDIYL)POLYETHYLENE GLYCOL CHLORIDE 210 ◇ MONOCHLOROPOLYOXYETHYLENE

TOXICITY DATA with REFERENCE
skn-rbt 500 mg open MLD UCDS** 7/23/70
orl-rat LD50:1070 mg/kg UCDS** 1/20/72
skn-rbt LD50:3180 mg/kg UCDS** 1/20/72

SAFETY PROFILE: Moderately toxic by ingestion and skin contact. A skin irritant. Combustible when exposed to heat or flame; can react vigorously with oxidizing materials. To fight fire, use water, foam, CO_2, dry chemical. When heated to decomposition it emits highly toxic fumes of Cl^-.

PJU500 CAS:9005-08-7 ***HR: 3***
POLYETHYLENE GLYCOL DISTEARATE

PROP: Polyethylene glycol distearate, low molecular weight (JAPMA8 38,428,49).

SYNS: POLYETHYLENE GLYCOL 300 DISTEARATE ◇ POLYETHYLENE GLYCOL 400 (DI) STEARATE ◇ POLYETHYLENE GLYCOL 600 (DI) STEARATE ◇ POLYGLYCOL DISTEARATE

TOXICITY DATA with REFERENCE
ivn-mus LD50:365 mg/kg JAPMA8 38,428,49

CONSENSUS REPORTS: Reported in EPA TSCA Inventory.

SAFETY PROFILE: Poison by intravenous route. When heated to decomposition it emits acrid smoke and irritating fumes. See also POLYETHYLENE GLYCOL.

PJU750 CAS:9005-08-7 ***HR: 3***
POLYETHYLENE GLYCOL DISTEARATE 1000

PROP: Polyethylene glycol distearate, high molecular weight (JAPMA8 38,428,49).

SYN: CARBOWAX 1000 DISTEARATE

TOXICITY DATA with REFERENCE
ivn-mus LD50:220 mg/kg JAPMA8 38,428,49

CONSENSUS REPORTS: Reported in EPA TSCA Inventory.

SAFETY PROFILE: Poison by intravenous route. When heated to decomposition it emits acrid smoke and irritating fumes. See also POLYETHYLENE GLYCOL.

PJV000 CAS:25322-68-3 ***HR: 3***
POLYETHYLENE GLYCOL E 600

TOXICITY DATA with REFERENCE
ivn-mus LD50:7900 μg/kg RPOBAR 2,316,70

CONSENSUS REPORTS: Reported in EPA TSCA Inventory.

SAFETY PROFILE: Poison by intravenous route. When heated to decomposition it emits acrid smoke and irritating fumes. See also POLYETHYLENE GLYCOL.

PJV250 CAS:9004-99-3 ***HR: 3***
POLYETHYLENE GLYCOL MONOSTEARATE
mf: $(C_2H_4O)_n \cdot C_{18}H_{36}O_2$

SYNS: POLYOXYETHYLENE-8-MONOSTEARATE ◇ POLYOXYETHYLENE(8)STEARATE ◇ TRYDET SA SERIES

TOXICITY DATA with REFERENCE
orl-rat TDLo:635 g/kg (multi) :REP JONUAI 60,489,56
orl-rat TDLo:4015 g/kg/2Y-C:ETA AEHLAU 6,484,63
orl-rat LD50:64 g/kg FOREAE 21,348,56
orl-ham LD50:27 g/kg FOREAE 21,348,56

CONSENSUS REPORTS: Reported in EPA TSCA Inventory.

SAFETY PROFILE: Very slightly toxic by ingestion. Questionable carcinogen with experimental tumorigenic data. Experimental reproductive effects. When heated to decomposition it emits acrid smoke and irritating fumes. See also other polyethylene glycol monostearate entries and POLYETHYLENE GLYCOL.

PJV500 CAS:9004-99-3 ***HR: 3***
POLYETHYLENE GLYCOL MONOSTEARATE 200

SYNS: GLYCOL POLYETHYLENE MONOSTEARATE 200 ◇ USAF KE-12

TOXICITY DATA with REFERENCE
ipr-mus LD50:200 mg/kg NTIS** AD277-689

CONSENSUS REPORTS: Reported in EPA TSCA Inventory.

SAFETY PROFILE: Poison by intraperitoneal route. When heated to decomposition it emits acrid smoke and irritating fumes. See also other polyethylene glycol monostearate entries and POLYETHYLENE GLYCOL.

PJV750 CAS:9004-99-3 ***HR: 3***
POLYETHYLENE GLYCOL MONOSTEARATE 400

PROP: Polyethylene glycol monostearate, low molecular weight (JAPMA8 38,428,49).

TOXICITY DATA with REFERENCE
ivn-mus LD50:250 mg/kg JAPMA8 38,428,49

CONSENSUS REPORTS: Reported in EPA TSCA Inventory.

SAFETY PROFILE: Poison by intravenous route. When heated to decomposition it emits acrid smoke and irritating fumes. See also other polyethylene glycol monostearate entries and POLYETHYLENE GLYCOL.

PJW000 CAS:9004-99-3 ***HR: 3***
POLYETHYLENE GLYCOL MONOSTEARATE 1000

PROP: Polyethylene glycol monostearate, high molecular weight (JAPMA8 38,428,49).

SYNS: CARBOWAX 1000 MONOSTEARATE ◇ USAF KE-9

TOXICITY DATA with REFERENCE
ipr-mus LD50:200 mg/kg NTIS** AD277-689
ivn-mus LD50:870 mg/kg JAPMA8 38,428,49

CONSENSUS REPORTS: Reported in EPA TSCA Inventory.

SAFETY PROFILE: Poison by intraperitoneal route. Moderately toxic by intravenous route. When heated to decomposition it emits acrid smoke and irritating fumes. See also other polyethylene glycol monostearate entries and POLYETHYLENE GLYCOL.

PJW250 CAS:9004-99-3 ***HR: 3***
POLYETHYLENE GLYCOL MONOSTEARATE 6000

SYN: USAF KE-14

TOXICITY DATA with REFERENCE
ipr-mus LD50:200 mg/kg NTIS** AD277-689

CONSENSUS REPORTS: Reported in EPA TSCA Inventory.

SAFETY PROFILE: Poison by intraperitoneal route. When heated to decomposition it emits acrid smoke and irritating fumes. See also other polyethylene glycol monostearate entries and POLYETHYLENE GLYCOL.

PJW500 CAS:9004-98-2 ***HR: 2***
POLYETHYLENE GLYCOL 1000 OLEYL ETHER
mf: $(C_2H_4O)nC_{18}H_{36}O$

PROP: A polyoxyethylene alkyl ether of fatty alcohols (FCTXAV 2,509,64).

SYNS: AMEROX OE-20 ◇ BRIJ 98 ◇ EMERY 6802 ◇ EMULPHOR ON-870 ◇ ETHOXOL 20 ◇ LIPAL 20-OA ◇ LIPOCOL O-n20 ◇ NOVOL POE 20 ◇ OLEYL ALCOHOL EO (20) ◇ OLEYL ALCOHOL condensed with 20 MOLES ETHYLENE OXIDE ◇ α-9-OCTADECENYL-omega-HYDROXY-POLY(OXY-1,2-ETHANEDIYL, (Z) ◇ PEG-20 OLEYL ETHER ◇ POLYOXYETHYLENE (20) OLEYL ETHER ◇ PROCOL OA-20 ◇ SIPONIC Y-501 ◇ STANDAMUL O20 ◇ TRYCOL HCS ◇ VOLPO 20

TOXICITY DATA with REFERENCE
eye-rbt 5 mg/48H JANCA2 56,905,73
orl-rat LD50:2770 mg/kg SPCOAH 38,47,65

CONSENSUS REPORTS: Glycol ether compounds are on the Community Right-To-Know List. Reported in EPA TSCA Inventory.

SAFETY PROFILE: Moderately toxic by ingestion. An eye irritant. Many glycol ethers cause dangerous human reproductive effects. When heated to decomposition it emits acrid smoke and irritating fumes. See also GLYCOL ETHERS.

PJW750 CAS:9004-99-3 ***HR: 1***
POLYETHYLENEGLYCOLS MONOSTEARATE
mf: $(C_2H_4O)nC_{18}H_{36}O$

SYNS: α-1-(OXOOCTADECYL)-omega-HYDROXYPOLY(OXY-1,2-ETHANEDIYL) ◇ PMS No. 1 ◇ POLYOXYETHYLENE MONOSTEARATE

TOXICITY DATA with REFERENCE
orl-rat LD50:53 g/kg FOREAE 21,348,56
orl-ham LD50:20 g/kg FOREAE 21,348,56

CONSENSUS REPORTS: Reported in EPA TSCA Inventory.

SAFETY PROFILE: Mildly toxic by ingestion. When heated to decomposition it emits acrid smoke and irritating fumes.

PJX000 CAS:26913-06-4 ***HR: 3***
POLYETHYLENE IMINE
mf: $(C_2H_8N_2)_n$

SYNS: CORCAT ◇ POLYETHYLENEIMIN (CZECH)

TOXICITY DATA with REFERENCE
eye-rbt 500 mg/24H MOD 28ZPAK -,256,72
orl-rat LD50:3300 mg/kg 28ZPAK -,256,72
unr-mam LD50:30 mg/kg PCJOAU 17,349,84

SAFETY PROFILE: Poison by an unspecified route. Moderately toxic by ingestion. An eye irritant. When heated to decomposition it emits toxic fumes of NO_x.

PJX750 ***HR: 3***
POLYETHYLENE Y-141-A

TOXICITY DATA with REFERENCE
imp-rat TDLo:6750 mg/kg:ETA CNREA8 35,1591,75

SAFETY PROFILE: Questionable carcinogen with experimental tumorigenic data by implant route. When heated to decomposition it emits acrid smoke and irritating fumes.

PJX800 CAS:9002-98-6 ***HR: 2***
POLYETHYLENIMINE (10,000)

PROP: Molecular weight of 10,000 (GISAAA 41(7),19,76).

TOXICITY DATA with REFERENCE
orl-rat LD50:1350 mg/kg GISAAA 41(7),19,76
orl-mus LD50:1150 mg/kg GISAAA 41(7),19,76
orl-gpg LD50:940 mg/kg GISAAA 41(7),19,76

SAFETY PROFILE: Moderately toxic by ingestion. When heated to decomposition it emits toxic fumes of NO_x

PJX825 CAS:9002-98-6 ***HR: 2***
POLYETHYLENIMINE (20,000)

PROP: Molecular weight of 20,000 (GISAAA 41(7),19,76).

TOXICITY DATA with REFERENCE
orl-rat LD50:2200 mg/kg GISAAA 41(7),19,76
orl-mus LD50:1400 mg/kg GISAAA 41(7),19,76
orl-gpg LD50:1400 mg/kg GISAAA 41(7),19,76

CONSENSUS REPORTS: Reported in EPA TSCA Inventory.

SAFETY PROFILE: Moderately toxic by ingestion. When heated to decomposition it emits toxic fumes of NO_x

PJX835 CAS:9002-98-6 ***HR: 2***
POLYETHYLENIMINE (35,000)

PROP: Molecular weight of 35,000 (GISAAA 41(7), 19,76).

TOXICITY DATA with REFERENCE
orl-rat LD50:2300 mg/kg GISAAA 41(7),19,76
orl-mus LD50:1400 mg/kg GISAAA 41(7),19,76
orl-gpg LD50:1400 mg/kg GISAAA 41(7),19,76

CONSENSUS REPORTS: Reported in EPA TSCA Inventory.

SAFETY PROFILE: Moderately toxic by ingestion. When heated to decomposition it emits toxic fumes of NO_x

PJX845 CAS:9002-98-6 ***HR: 2***
POLYETHYLENIMINE (40,000)

PROP: Molecular weight of 40,000 (GISAAA 41(7), 19,76).

TOXICITY DATA with REFERENCE
orl-rat LD50:2200 mg/kg GISAAA 41(7),19,76
orl-mus LD50:1600 mg/kg GISAAA 41(7),19,76
orl-gpg LD50:1400 mg/kg GISAAA 41(7),19,76

CONSENSUS REPORTS: Reported in EPA TSCA Inventory.

SAFETY PROFILE: Moderately toxic by ingestion.

When heated to decomposition it emits toxic fumes of NO_x

PJX850 ***HR: 3***
POLY(ETHYLIDENE PEROXIDE)
mf: $(C_2H_4O_2)_n$

SAFETY PROFILE: Highly explosive. Formed by the peroxidation of diethyl ether. When heated to decomposition it emits acrid smoke and irritating fumes. See also PEROXIDES and ETHERS.

PJY000 CAS:9004-81-3 ***HR: 2***
POLYGLYCOL LAURATE

SYN: NOPALCOL 6-L

TOXICITY DATA with REFERENCE
ivn-mus LD50:500 mg/kg JAPMA8 38,428,49

CONSENSUS REPORTS: Reported in EPA TSCA Inventory.

SAFETY PROFILE: Moderately toxic by intravenous route. When heated to decomposition it emits acrid smoke and irritating fumes.

PJY100 CAS:9004-96-0 ***HR: 2***
POLYGLYCOL MONOOLEATE
mf: $(C_2H_4O)_n \cdot C_{18}H_{34}O_2$

SYNS: AKYPOROX O 50 ◇ ATLAS G-2142 ◇ ATLAS G-2144 ◇ CEMULSOL 1050 ◇ CEMULSOL A ◇ CEMULSOL C 105 ◇ CEMULSOL D-8 ◇ CHEMESTER 300-OC ◇ CITHROL PO ◇ CRODET O 6 ◇ E2 ◇ EMANON 4115 ◇ EMCOL H-2A ◇ EMCOL H 31A ◇ EMEREST 2646 ◇ EMEREST 2660 ◇ EMPILAN BP 100 ◇ EMPILAN BQ 100 ◇ EMULPHOR A ◇ EMULPHOR UN-430 ◇ EMULPHOR VN 430 ◇ ETHOFAT O ◇ ETHOFAT O 15 ◇ ETHYLAN A3 ◇ ETHYLAN A6 ◇ EXTREX P 60 ◇ IONET MO-400 ◇ LANNAGOL LF ◇ LIPAL 400-DL ◇ LIPAL 30W ◇ MACROGOL OLEATE 600 ◇ NIKKOL MYO 2 ◇ NIKKOL MYO 10 ◇ NIOGEN ES 160 ◇ NONEX 25 ◇ NONEX 30 ◇ NONEX 52 ◇ NONEX 64 ◇ NONION O2 ◇ NONION O4 ◇ NONION 06 ◇ NONISOL 200 ◇ NOPALCOL 1-0 ◇ NOPALCOL 6-0 ◇ OK 7 ◇ OLEIC ACID POLY(OXYETHYLENE) ESTER ◇ OLEOX 5 ◇ OLEPAL I ◇ OLEPAL III ◇ PEG 200MO ◇ PEG 600MO ◇ PEG 1000MO ◇ PEG-6 OLEATE ◇ PEG-20 OLEATE ◇ PEG-32 OLEATE ◇ PEGOSPERSE 400MO ◇ POLYETHYLENE GLYCOL MONOOLEATE ◇ POLYETHYLENE GLYCOL OLEATE ◇ POLYETHYLENE OXIDE MONOOLEATE ◇ POLY(ETHYLENE OXIDE) OLEATE ◇ POLY(OXYETHYLENE) MONOOLEATE ◇ POLY(OXYETHYLENE) OLEATE ◇ POLY(OXYETHYLENE) OLEIC ACID ESTER ◇ POOA ◇ PRODHYPHORE B ◇ ROKACET ◇ ROKACET O 7 ◇ S 1006 ◇ S 1132 ◇ SLOVASOL A ◇ TRYDET OS SERIES ◇ UNISOL 4-O ◇ WITCO 31 ◇ X-539-R

TOXICITY DATA with REFERENCE
skn-rbt 500 mg/24H MLD 28ZPAK -,299,72
eye-rbt 500 mg/24H MLD 28ZPAK -,299,72
ivn-mus LD50:500 mg/kg JAPMA8 38,428,49

CONSENSUS REPORTS: Reported in EPA TSCA Inventory.

SAFETY PROFILE: Moderately toxic by intravenous route. A skin and eye irritant. When heated to decomposition it emits acrid smoke and irritating fumes. A nonionic surfactant. See also ESTERS.

PJY250 CAS:9004-96-0 ***HR: 2***
POLYGLYCOL OLEATE
mf: $(C_2H_4O)_nC_{12}H_{24}O_2$

SYN: NOPALCOL 4-0

TOXICITY DATA with REFERENCE
eye-rbt 1% MLD JAPMA8 38,428,49
ivn-mus LD50:1080 mg/kg JAPMA8 38,428,49

CONSENSUS REPORTS: Reported in EPA TSCA Inventory.

SAFETY PROFILE: Moderately toxic by intravenous routes. An eye irritant. When heated to decomposition it emits acrid smoke and irritating fumes.

PJY500 CAS:25038-54-4 ***HR: 3***
POLY(IMINOCARBONYLPENTAMETHYLENE)
mf: $(C_6H_{11}NO)_n$

SYNS: AKULON ◇ ALKAMID ◇ AMILAN CM 1001 ◇ 6-AMINOHEXANOIC ACID HOMOPOLYMER ◇ BONAMID ◇ CAPRAN 80 ◇ CAPROAMIDE POLYMER ◇ CAPROLACTAM OLIGOMER ◇ epsilon-CAPROLACTAM POLYMERE (GERMAN) ◇ CAPRON ◇ CHEMLON ◇ DANAMID ◇ DULL 704 ◇ DURETHAN BK ◇ ERTALON 6SA ◇ GRILON ◇ HEXAHYDRO-2H-AZEPIN-2-ONE HOMOPOLYMER ◇ ITAMID ◇ KAPROLIT ◇ KAPROLON ◇ KAPROMIN ◇ KAPRON ◇ MARANYL F 114 ◇ METAMID ◇ MIRAMID WM 55 ◇ NYLON-6 ◇ ORGAMIDE ◇ PA 6 (polymer) ◇ PLASKON 201 ◇ POLICAPRAN ◇ POLYAMIDE 6 ◇ POLY(epsilon-AMINOCAPROIC ACID) ◇ POLYCAPROAMIDE ◇ POLY(epsilon-CAPROAMIDE) ◇ POLYCAPROLACTAM ◇ POLY(epsilon-CAPROLACTAM) ◇ POLY(IMINO(1-OXO-1,6-HEXANEDIYL)) ◇ RELON P ◇ SPENCER 401 ◇ STILON ◇ TARLON XB ◇ TARNAMID T ◇ ULTRAMID BMK ◇ VIDLON ◇ WIDLON ◇ ZYTEL 211

TOXICITY DATA with REFERENCE
imp-rat TDLo:5 film disc/rat:NEO ZENBAX 7B,353,52
orl-rat LD50:3200 mg/kg GISAAA 42(3),99,77
orl-mus LD50:1900 mg/kg GISAAA 42(3),99,77
ihl-mus LC50:11 g/m^3/30M PWPSA8 21,167,78

CONSENSUS REPORTS: IARC Cancer Review: Group 3 IMEMDT 7,56,87; Animal Inadequate Evidence IMEMDT 19,115,75. Reported in EPA TSCA Inventory.

SAFETY PROFILE: Moderately toxic by ingestion. Mildly toxic by inhalation. Questionable carcinogen with experimental neoplastigenic data by implant route. When heated to decomposition it emits toxic fumes of NO_x.

PJY750 CAS:24939-03-5 ***HR: 3***
POLYINOSINIC: POLYCYTIDYLIC ACID COPOLYMER

SYNS: 5'-INOSINIC ACID, HOMOPOLYMER complex with 5'-CYTIDYLIC ACID HOMOPOLYMER (1:1) ◇ NSC-120949 ◇ POLY C POLY I ◇ POLYCTYIDYLIC-POLYINOSINIC ACID ◇ POLY I:C ◇ POLYINOSINATE:POLYCYTIDYLATE

TOXICITY DATA with REFERENCE
eye-rbt 950 mg/6H ANOPB5 3,371,71
dni-hmn:hla 500 μg/L PNASA6 70,3904,73
dni-mus-ipr 4 mg/kg JNCIAM 54,219,75
cyt-mus-ipr 250 μg/kg VVIRAT (5),540,82
dlt-mus-ipr 250 μg/kg VVIRAT (5),540,82
ivn-rbt TDLo:50 μg/kg (1D pre):REP JRPFA4 63,81,81
ipr-rat LD50:365 mg/kg TXAPA9 18,220,71
ivn-rat LD50:185 mg/kg TXAPA9 18,220,71
ipr-mus LD50:35 mg/kg TXAPA9 23,579,72
ivn-mus LD50:9500 μg/kg TXAPA9 23,579,72
ivn-mky LDLo:15 mg/kg TXAPA9 23,579,72
ivn-rbt LD50:220 μg/kg VVIRAT 23(2),201,78

SAFETY PROFILE: Poison by intraperitoneal and intravenous routes. Experimental reproductive effects. Human mutation data reported. An eye irritant. When heated to decomposition it emits acrid smoke and irritating fumes. Used as an inducer of interferon and an antiviral agent.

PKA000 ***HR: 2***
POLYMERIC DIALDEHYDE
mf: $(C_6H_8O_5)_n$

TOXICITY DATA with REFERENCE
scu-mus TDLo:268 mg/kg/67W-I:ETA JNCIAM 46,143,71

SAFETY PROFILE: Questionable carcinogen with experimental tumorigenic data. When heated to decomposition it emits acrid smoke and irritating fumes. See also ALDEHYDES.

PKA850 ***HR: 2***
POLYMERS, WATER INSOLUBLE

SAFETY PROFILE: Many produce local tumors of the soft tissues surrounding the site of implantation. See also specific compounds.

PKA860 ***HR: 2***
POLYMERS, WATER SOLUBLE

SAFETY PROFILE: Many produce local tumors of the soft tissues surrounding the site of implantation and in the lungs, mucosal contact areas, organs, and tissues of retention and deposition. See also specific compounds.

PKB000 ***HR: 3***
POLY(METHYLENEMAGNESIUM)
mf: $(CH_2Mg)_n$ mw: $(38.34)_n$

SAFETY PROFILE: Ignites spontaneously in air. Very unstable. When heated to decomposition it emits acrid smoke and irritating fumes. See also MAGNESIUM COMPOUNDS.

PKB500 CAS:9011-14-7 ***HR: 3***
POLYMETHYLMETHACRYLATE
mf: $(C_5H_8O_2)_n$

SYNS: ACRYLITE ◇ ACRYPET ◇ ALUTOR M 70 ◇ CMW BONE CEMENT ◇ CRINOTHENE ◇ DEGALAN S 85 ◇ DELPET 50M ◇ DIAKON ◇ DISPASOL M ◇ DV 400 ◇ ELVACITE ◇ KALLOCRYL K ◇ KALLODENT CLEAR ◇ KORAD ◇ LPT ◇ LUCITE ◇ METAPLEX NO ◇ METHACRYLIC ACID METHYL ESTER POLYMERS ◇ METHYL METHACRYLATE HOMOPOLYMER ◇ METHYL METHACRYLATE POLYMER ◇ METHYL METHACRYLATE RESIN ◇ 2-METHYL-2-PROPENOIC ACID METHYL ESTER HOMOPOLYMER ◇ ORGANIC GLASS E 2 ◇ TEOBOND SURGICAL BONE CEMENT ◇ PALACOS ◇ PARAGLAS ◇ PARAPLEX P 543 ◇ PERSPEX ◇ PLEXIGLAS ◇ PLEXIGUM M 920 ◇ PMMA ◇ PONTALITE ◇ REPAIRSIN ◇ RESARIT 4000 ◇ RHOPLEX B 85 ◇ ROMACRYL ◇ SHINKOLITE ◇ SOL ◇ STELLON PINK ◇ SUMIPLEX LG ◇ SUPERACRYL AE ◇ SURGICAL SIMPLEX ◇ TENSOL 7 ◇ VEDRIL

TOXICITY DATA with REFERENCE
imp-rat TDLo:127 mg/kg:ETA CNREA8 15,333,55

CONSENSUS REPORTS: IARC Cancer Review: Group 3 IMEMDT 7,56,87; Human Inadequate Evidence IMEMDT 19,187,79; Animal Sufficient Evidence IMEMDT 19,187,79. Reported in EPA TSCA Inventory.

SAFETY PROFILE: Questionable carcinogen with experimental tumorigenic data by implant route. When heated to decomposition it emits acrid smoke and irritating fumes. Used as the main constituent of acrylic sheet, molding, and extrusion powers.

PKB775 CAS:11081-39-3 ***HR: 3***
POLYMYCIN

TOXICITY DATA with REFERENCE
orl-mus LD50:120 mg/kg 85GDA2 1,277,80
scu-mus LD50:4 mg/kg 85GDA2 1,277,80
ivn-mus LD50:50 μg/kg 85FZAT -,828,67

SAFETY PROFILE: Poison by ingestion, subcutaneous, and intravenous routes.

PKC000 CAS:1406-11-7 ***HR: 3***
POLYMYXIN

PROP: A series of antibiotic substances, polypeptide (basic), sol in water. Colorless powder. Decomp @ 228-230°.

SYN: B-71

TOXICITY DATA with REFERENCE
ipr-mus LD50:77 mg/kg ANYAA9 51,879,49
scu-mus LD50:250 mg/kg ANYAA9 51,935,49
ivn-mus LD50:18 mg/kg ANYAA9 51,879,49
ivn-dog LDLo:1300 μg/kg ANYAA9 51,935,49

SAFETY PROFILE: Poison by intraperitoneal, subcutaneous, and intravenous routes. An additive permitted in food for human consumption.

PKC250 CAS:1404-24-6 ***HR: 3***
POLYMYXIN A

TOXICITY DATA with REFERENCE
ipr-mus LDLo:13 mg/kg LANCAO 254,127,48
scu-mus LD50:88 mg/kg 85ERAY 3,1542,78
ivn-mus LDLo:6 mg/kg LANCAO 254,127,48
ice-rbt LDLo:600 μg/kg ANYAA9 51,952,49

SAFETY PROFILE: Poison by intraperitoneal, subcutaneous, intravenous, and intracerebral routes. When heated to decomposition it emits acrid smoke and irritating fumes.

PKC500 CAS:1404-26-8 ***HR: 3***
POLYMYXIN B
mf: $C_{43}H_{82}N_{16}O_{12}$ mw: 1015.41

SYN: AEROSPORIN

TOXICITY DATA with REFERENCE
orl-hmn TDLo:8570 mg/kg:GIT 34ZIAG -,488,69
ims-hmn TDLo:3 mg/kg/D:PNS,CNS 34ZIAG -,488,69
ipr-mus LDLo:13 mg/kg LANCAO 254,127,48
ivn-mus LD50:7940 μg/kg CSLNX* NX#12597

SAFETY PROFILE: Poison by intraperitoneal and intravenous routes. Human systemic effects by ingestion: diarrhea, nausea; by intramuscular route: paresthesia, ataxia. When heated to decomposition it emits toxic fumes of NO_x.

PKC550 CAS:4135-11-9 ***HR: 3***
POLYMYXIN B1
mf: $C_{56}H_{98}N_{16}O_{13}$ mw: 1203.70

TOXICITY DATA with REFERENCE
ipr-mus LD50:19 mg/kg 85GDA2 4(1),334,80
scu-mus LD50:80 mg/kg 85GDA2 4(1),334,80
ivn-mus LD50:1500 μg/kg 85GDA2 4(1),334,80

SAFETY PROFILE: Poison by subcutaneous, intravenous, and intraperitoneal routes. When heated to decomposition it emits toxic fumes of NO_x.

PKC750 CAS:1405-20-5 ***HR: 3***
POLYMYXIN B SULFATE
mf: $C_{43}H_{82}N_{16}O_{12}$•xH_2O_4S mw: 1701.97

TOXICITY DATA with REFERENCE
dnd-esc 50 mg/L MUREAV 89,95,81
orl-mus LD50:790 mg/kg NIIRDN 6,794,82
ipr-mus LD50:20500 μg/kg NIIRDN 6,794,82
scu-mus LD50:59500 μg/kg NIIRDN 6,794,82
ivn-mus LD50:5400 μg/kg NIIRDN 6,794,82
ivn-dog LDLo:8 mg/kg INURAQ 6,505,69
ice-dog LDLo:320 μg/kg BJPCAL 23,552,64

SAFETY PROFILE: Poison by intravenous, subcutaneous, intraperitoneal and intracerebral routes. Moderately toxic by ingestion. Mutation data reported. When heated to decomposition it emits very toxic fumes of SO_x and NO_x. See also POLYMYXIN B.

PKD050 CAS:10072-50-1 ***HR: 3***
POLYMYXIN D1
mf: $C_{50}H_{93}N_{15}O_{15}$ mw: 1144.58

TOXICITY DATA with REFERENCE
ipr-mus LD50:27 mg/kg 85GDA2 4(1),336,80
scu-mus LD50:35 mg/kg 85GDA2 4(1),336,80
ivn-mus LD50:3 mg/kg 85GDA2 4(1),336,80

SAFETY PROFILE: Poison by subcutaneous, intravenous, and intraperitoneal routes. When heated to decomposition it emits toxic fumes of NO_x.

PKD250 CAS:1066-17-7 ***HR: 3***
POLYMYXIN E
mf: $C_{42}H_{85}N_{13}O_{10}$ mw: 968.43

SYNS: COLIMYCIN ◇ COLISTICINA ◇ COLISTIN ◇ COLY-MYCIN ◇ COLYMYSIN S ◇ TOTAZINA

TOXICITY DATA with REFERENCE
oms-hmn:lym 142 units/ml TCMUD8 3,515,83
cyt-hmn:lym 142 unit/ml TCMUD8 3,515,83
ipr-mus LD50:236 mg/kg ANTBAL 5(4),10,60
scu-mus LD50:115 mg/kg JANTAJ 36,625,83
ivn-mus LD50:8800 μg/kg JANTAJ 36,625,83

CONSENSUS REPORTS: EPA Genetic Toxicology Program.

SAFETY PROFILE: Poison by subcutaneous, intravenous and intraperitoneal routes. Human mutation data reported. When heated to decomposition it emits toxic fumes of NO_x.

PKE100 CAS:19396-06-6 ***HR: 3***
POLYOXIN AL
mf: $C_{17}H_{25}N_5O_{13}$ mw: 507.47

SYNS: β-d-5-(2-AMINO-2-DEOXY-l-XYLONAMIDO)-1,5-DIDEOXY-1-(3,4-DIHYDRO-5-HYDROXYMETHYL)-2,4-DIOXO-1(2H)-PYRIMIDINYL ALLOFURANURONIC ACID, MONOCARBAMATE (ester) ◇ POLYOXIN B

TOXICITY DATA with REFERENCE
orl-rat LD50:14665 mg/kg FMCHA2 -,C192,83
orl-mus LD50:15638 mg/kg FMCHA2 -,C192,83
ivn-mus LD50:200 mg/kg 85DGA2 5,245,81

SAFETY PROFILE: Poison by intravenous route. Mildly toxic by ingestion. When heated to decomposition it emits toxic fumes of NO_x. See also CARBAMATES.

PKE250 CAS:25655-41-8 ***HR: 2***
POLY(1-(2-OXO-1-PYRROLIDINYL)ETHYLENE)IODINE COMPLEX
mf: $(C_6H_9NO)_n \cdot xI$

PROP: Yellowish-brown, amorphous powder with slt characteristic odor. Sol in alc, water; practically insol in chloroform, carbon tetrachloride, ether, solvent hexane, acetone.

SYNS: BETADINE ◇ BETAISODONA ◇ BRAUNOSAN H ◇ DISADINE ◇ DISPHEX ◇ EFO-DINE ◇ 1-ETHENYL-2-PYRROLIDINONE HOMOPOLYMER compounded with IODINE ◇ ISODINE ◇ POVIDONE-IODINE ◇ PVP-IODINE ◇ TRAUMASEPT ◇ ULTRADINE ◇ 1-VINYL-2-PYRROLIDINONE POLYMER, compounded with IODINE

TOXICITY DATA with REFERENCE
dnd-hmn:oth 200 ppm JTEHD6 1,977,76
otr-mus:fbr 5 g/L JEPTDQ 4(2-3),327,80
skn-hmn TDLo:3400 mg/kg/24H:BLD,SKN JAMAAP 240,249,78
scu-rat LD50:3450 mg/kg NIIRDN 6,788,82
ivn-rat LD50:640 mg/kg NIIRDN 6,788,82
orl-mus LD50:8100 mg/kg NIIRDN 6,788,82
scu-mus LD50:4100 mg/kg NIIRDN 6,788,82
ivn-mus LD50:480 mg/kg NIIRDN 6,788,82

CONSENSUS REPORTS: Reported in EPA TSCA Inventory. EPA Genetic Toxicology Program.

SAFETY PROFILE: Moderately toxic by subcutaneous and intravenous routes. Mildly toxic by ingestion. Human systemic effects by skin contact: hemorrhage and dermatitis. Human mutation data reported. When heated to decomposition it emits toxic fumes of NO_x and I^-.

PKE500 CAS:9004-98-2 ***HR: 1***
POLYOXYETHYLATED VEGETABLE OIL
mf: $(C_2H_4O)n\text{-}C_{18}\text{-}H_{36}\text{-}O$

SYNS: EL-620 ◇ EL-719 ◇ EMULPHOR ◇ EMULPHOR SURFACTANTS

TOXICITY DATA with REFERENCE
orl-rat LD50:70 g/kg FMCHA2 -,D124,80

CONSENSUS REPORTS: Reported in EPA TSCA Inventory.

SAFETY PROFILE: Mildly toxic by ingestion. When heated to decomposition it emits acrid smoke and irritating fumes.

PKE550 ***HR: 2***
POLYOXYETHYLENE ALKYLAMINE MONO-FATTY ACID ESTER

SYNS: GLYCOLS, POLYETHYLENE, (ALKYLIMINO)DIETHYLENE ETHER, MONOFATTY ACID ESTER ◇ PAFE

TOXICITY DATA with REFERENCE
orl-rat TDLo:5625 mg/kg (female 90D pre):REP OYYAA2 12,179,76
orl-rat LD50:2624 mg/kg OYYAA2 12,179,76
orl-mus LD50:4840 mg/kg OYYAA2 12,179,76

SAFETY PROFILE: Moderately toxic by ingestion. Experimental reproductive effects. When heated to decomposition it emits acrid smoke and irritating fumes.

PKE600 ***HR: 1***
POLYOXYETHYLENE-ALKYL CITRIC DIESTER-TRIETHANOLAMINE

SYNS: PAT ◇ POLYOXYETHYLENE-sec-ALKYL ETHER CITRIC DIESTER TRIETHANOLAMINE

TOXICITY DATA with REFERENCE
orl-rat LD50:14 g/kg YAHOA3 23,1,79
orl-mus LD50:8400 mg/kg YAHOA3 23,1,79
orl-ckn LD50:7100 mg/kg YAHOA3 23,1,79

SAFETY PROFILE: Mildly toxic by ingestion. When heated to decomposition it emits toxic fumes of NO_x. See also AMINES and ESTERS.

PKE750 CAS:9004-86-8 ***HR: 1***
POLYOXYETHYLENE DIBENZOATE
mf: $(C_2H_4O)_n\ C_{14}H_{10}O_3$

SYNS: BENZOFLEX P 200 ◇ BENZOFLEX P-600 ◇ BENZOIC ACID DIESTER with POLYETHYLENE GLYCOL 600 ◇ α-BENZOYL-omega-(BENZOYLOXY)POLY(OXY-1,2-ETHANEDIYL) ◇ POLYETHYLENE 600 DIBENZOATE ◇ POLYETHYLENE GLYCOL DIBENZOATE ◇ POLYETHYLENE GLYCOL 220 DIBENZOATE

TOXICITY DATA with REFERENCE
orl-rat LD50:5340 mg/kg NPIRI* 2,88,75

CONSENSUS REPORTS: Reported in EPA TSCA Inventory.

SAFETY PROFILE: Mildly toxic by ingestion. When heated to decomposition it emits acrid smoke and irritating fumes.

PKF000 CAS:9016-45-9 ***HR: 2***
POLYOXYETHYLENE (9) NONYL PHENYL ETHER

SYNS: ARKOPAL N-090 ◇ CARSONON N-9 ◇ CONCO NI-90 ◇ IGEPAL CO-630 ◇ NEUTRONYX 600 ◇ PEG-9 NONYL PHENYL ETHER

◇ POLYETHYLENE GLYCOL 450 NONYL PHENYL ETHER ◇ PROTACHEM 630 ◇ REWOPOL HV-9 ◇ TERGITOL TP-9 (NONIONIC)

TOXICITY DATA with REFERENCE
skn-hmn 15 mg/3D-I MLD 85DKA8 -,127,77
eye-mus 20 mg SEV FCTXAV 15,131,77
skn-rbt 500 mg open MLD UCDS** 4/11/63
eye-rbt 5 mg SEV UCDS** 4/11/63
orl-rat LD50:2590 mg/kg UCDS** 4/11/63
skn-rbt LD50:2830 mg/kg UCDS** 4/11/63

CONSENSUS REPORTS: Reported in EPA TSCA Inventory. Glycol ethers are on the Community Right-To-Know List.

SAFETY PROFILE: Moderately toxic by ingestion and skin contact. A severe eye and mild skin irritant in humans. Many glycol ethers cause dangerous human reproductive effects. When heated to decomposition it emits acrid smoke and irritating fumes. See also GLYCOL ETHERS.

PKF500 CAS:9002-93-1 ***HR: 2***
POLY(OXYETHYLENE)-p-tert-OCTYLPHENYL ETHER
mf: $(C_2H_4O)_n\ C_{14}H_{22}O$

PROP: Mixture in which *n* varies from 5 to 15. Pale yellow, viscous liquid. D: 1.0595. Miscible with water, alc, acetone; sol in benzene, toluene; insol in petr ether.

SYNS: ALFENOL 3 ◇ ALFENOL 9 ◇ ANTAROX A-200 ◇ CONCO NIX-100 ◇ HYDROL SW ◇ HYONIC PE-250 ◇ IGEPAL CA-63 ◇ MARLOPHEN 820 ◇ NEUTRONYX 605 ◇ OCTOXINOL ◇ OCTOXYNOL ◇ OCTOXYNOL 3 ◇ OCTOXYNOL 9 ◇ OCTYL PHENOL CONDENSED with 12-13 MOLES ETHYLENE OXIDE ◇ p-tert-OCTYLPHENOXYPOLYETHOXYETHANOL ◇ OPE 30 ◇ PEG-9 OCTYL PHENYL ETHER ◇ POLYETHYLENE GLYCOL MONOETHER with p-tert-OCTYLPHENYL ◇ POLYETHYLENE GLYCOL MONO(4-OCTYLPHENYL) ETHER ◇ POLYETHYLENE GLYCOL MONO(4-tert-OCTYLPHENYL) ETHER ◇ POLYETHYLENE GLYCOL MONO(p-tert-OCTYLPHENYL) ETHER ◇ POLYETHYLENE GLYCOL MONO(p-(1,1,3,3-TETRAMETHYLBUTYL)PHENYL) ETHER ◇ POLYETHYLENE GLYCOL OCTYLPHENOL ETHER ◇ POLYETHYLENE GLYCOL 450 OCTYL PHENYL ETHER ◇ POLYETHYLENE GLYCOL p-OCTYLPHENYL ETHER ◇ POLYETHYLENE GLYCOL p-tert-OCTYLPHENYL ETHER ◇ POLYETHYLENE GLYCOL p-1,1,3,3,-TETRAMETHYLBUTYLPHENYL ETHER ◇ POLYOXYETHYLENE MONO(OCTYLPHENYL) ETHER ◇ POLYOXYETHYLENE (9) OCTYLPHENYL ETHER ◇ POLYOXYETHYLENE (13) OCTYLPHENYL ETHER ◇ PRECEPTIN ◇ TRITON X 35 ◇ TRITON X 45 ◇ TRITON X 100 ◇ TRITON X 102 ◇ TRITON X 165 ◇ TRITON X 305 ◇ TRITON X 405 ◇ TRITON X 705 ◇ TX 100

TOXICITY DATA with REFERENCE
skn-hmn 2 mg/3D-I MLD 85DKAB -,127,77
eye-rbt 1 mg MOD PSTGAW 20,16,53
dni-hmn:hla 21 mg/L WATRAG 19,677,85
oms-hmn:hla 14 mg/L WATRAG 19,677,85
dns-mus:ast 200 ppm AMOKAG 32,1,78
orl-rat TDLo:65500 mg/kg (26W pre):REP JPPMAB 22,668,70
orl-rat LD50:1800 mg/kg PSTGAW 20,16,53
ivn-mus LD50:1200 mg/kg BCFAAI 101,173,62

CONSENSUS REPORTS: Glycol ether compounds are on the Community Right-To-Know List. Reported in EPA TSCA Inventory.

SAFETY PROFILE: Moderately toxic by ingestion and intravenous routes. Experimental reproductive effects. Human mutation data reported. An eye and human skin irritant. Many glycol ethers cause dangerous human reproductive effects. When heated to decomposition it emits toxic fumes of NO_x. A surfactant. See also GLYCOL ETHERS.

PKF750 CAS:25038-59-9 ***HR: 1***
POLY(OXYETHYLENEOXYTEREPHTHALOYL)
mf: $(C_{10}H_8O_4)_n$

SYNS: ALATHON ◇ AMILAR ◇ ARNITE A ◇ CASSAPPRET SR ◇ CELANAR ◇ CLEARTUF ◇ CRASTIN S 330 ◇ DAIYA FOIL ◇ DOWLEX ◇ ESTAR ◇ ESTROFOL ◇ ETHYLENE TEREPHTHALATE POLYMER ◇ FIBER V ◇ HOSTADUR ◇ HOSTAPHAN ◇ IAMBOLEN ◇ KLT 40 ◇ LAVSAN ◇ LAWSONITE ◇ LUMILAR 100 ◇ LUMIRROR ◇ MELIFORM ◇ MELINEX ◇ MYLAR ◇ NITRON LAVSAN ◇ NITRON (POLYESTER) ◇ PEGOTERATE ◇ POLYETHYLENE TEREPHTHALATE ◇ POLYETHYLENE TEREPHTHALATE FILM ◇ POLY(OXY-1,2-ETHANEDIYLOXYCARBONYL-1,4-PHENYLENECARBONYL) ◇ SCOTCH PAR ◇ SUPERFLOC ◇ TEREPHTAHLIC ACID-ETHYLENE GLYCOL POLYESTER ◇ TERFAN ◇ TERGAL ◇ TEROM ◇ TERPHAN ◇ VFR 3801 ◇ VITUF

TOXICITY DATA with REFERENCE
imp-rat TDLo:116 mg/kg:ETA CNREA8 15,333,55

CONSENSUS REPORTS: Reported in EPA TSCA Inventory.

SAFETY PROFILE: Questionable carcinogen with experimental tumorigenic data by implant route. When heated to decomposition it emits acrid smoke and irritating fumes.

PKG000 CAS:9005-64-5 ***HR: 2***
POLYOXYETHYLENE SORBITAN MONOLAURATE

SYNS: SORBITAN, MONOLAURATE POLYOXYETHYLENE derivative ◇ TWEEN 20

TOXICITY DATA with REFERENCE
ipr-mus TDLo:2500 mg/kg (9D preg):REP ZNCBDA 36C,904,81
ipr-mus TDLo:1 g/kg (9D preg):TER ZNCBDA 36C,904,81
ipr-rat LD50:3850 mg/kg ARZNAD 26,1581,76
ivn-rat LD50:770 mg/kg ARZNAD 26,1581,76
ipr-mus LD50:2640 mg/kg ARZNAD 26,1581,76
ivn-mus LD50:2970 mg/kg ARZNAD 26,1581,76

CONSENSUS REPORTS: Reported in EPA TSCA Inventory.

SAFETY PROFILE: Moderately toxic by intraperitoneal and intravenous routes. Experimental teratogenic and reproductive effects. When heated to decomposition it emits acrid smoke and irritating fumes. Used as a nonionic surfactant.

PKG500 CAS:9005-66-7 ***HR: 2***
POLYOXYETHYLENE SORBITAN MONOPALMITATE

SYN: TWEEN 40

TOXICITY DATA with REFERENCE
ivn-rat LD50:1580 mg/kg FAONAU 53A,257,74

CONSENSUS REPORTS: Reported in EPA TSCA Inventory.

SAFETY PROFILE: Moderately toxic by intravenous route. When heated to decomposition it emits acrid smoke and irritating fumes. See also SURFACTANTS.

PKH260 ***HR: 3***
POLY(PEROXYISOBUTYROLACTONE)
mf: $(C_4H_6O_3)_n$

$$(-C(CH_3)_2CO\bullet OO-)_n$$

SAFETY PROFILE: An unpredictable, violent explosive. The auto-oxidation product of dimethylketene. When heated to decomposition it emits acrid smoke and irritating fumes. See also PEROXIDES.

PKH850 ***HR: 2***
POLY p-PHENYLENE TEREPTHALAMIDE ARAMID FIBER

TOXICITY DATA with REFERENCE
ihl-rat TCLo:100 fibrils/cc/6H/2Y-I:ETA EPASR* 8EHQ-0485-0550

SAFETY PROFILE: Questionable carcinogen with experimental tumorigenic data. When heated to decomposition it emits acrid smoke and irritating fumes.

PKI000 CAS:63148-65-2 ***HR: 2***
POLY(2-PROPYL-m-DIOXANE-4,6-DIYLENE)
mf: $H_2\bullet(C_8H_{14}O_2)_n$

SYN: POLYVINYLBUTYRAL (CZECH)

TOXICITY DATA with REFERENCE
eye-rbt 100 mg/24H SEV 28ZPAK -,256,72

CONSENSUS REPORTS: Reported in EPA TSCA Inventory.

SAFETY PROFILE: A severe eye irritant. When heated to decomposition it emits acrid smoke and irritating fumes.

PKI250 CAS:9003-07-0 ***HR: 3***
POLYPROPYLENE, combustion products

PROP: Products of combustion of polypropylene in furnace maintained at 800° (APFRAD 35,461,77).

SYNS: PROPENE POLYMER ◇ PROPYLENE POLYMER

TOXICITY DATA with REFERENCE
ihl-mus LC50:30 mg/m^3/10M APFRAD 35,461,77

CONSENSUS REPORTS: Reported in EPA TSCA Inventory.

SAFETY PROFILE: Poison by inhalation.

PKI500 CAS:25322-69-4 ***HR: 1***
POLYPROPYLENE GLYCOL
mf: $(C_3H_8O_2)_n$

PROP: Clear, colorless liquid. Mw: 400-2000, mp: does not crystallize, flash p: +390°F, d: 1.002-1.007. Sol in water, aliphatic ketones and alcs; insol in ether, aliphatic hydrocarbons.

SYNS: ALKAPOL PPG-1200 ◇ JEFFOX ◇ POLYPROPYLENGLYKOL (CZECH)

TOXICITY DATA with REFERENCE
skn-rbt 500 mg/24H MLD 28ZPAK -,255,72
eye-rbt 500 mg AJOPAA29,1363,46
eye-rbt 500 mg/24H MLD 28ZPAK -,255,72
orl-rat LD50:4190 mg/kg 28ZPAK -,255,72

CONSENSUS REPORTS: Reported in EPA TSCA Inventory.

SAFETY PROFILE: Mildly toxic by ingestion. A skin and eye irritant. Combustible liquid when exposed to heat or flame; can react with oxidizing materials. To fight fire, use foam, CO_2, dry chemical. When heated to decomposition it emits acrid smoke and irritating fumes. See also GLYCOLS.

PKI550 CAS:25322-69-4 ***HR: 3***
POLYPROPYLENE GLYCOL 425
mf: $(C_3H_6O)_n\bullet H_2O$

TOXICITY DATA with REFERENCE
skn-rbt 500 mg MLD 34ZIAG -,731,69
eye-rbt 100 mg MLD 34ZIAG -,731,69
orl-rat LD50:2410 mg/kg UCDS** 10/16/73
ipr-rat LD50:460 mg/kg UCDS** 10/16/73
ivn-rat LD50:200 mg/kg UCDS** 10/16/73

CONSENSUS REPORTS: EPA Genetic Toxicology Program.

SAFETY PROFILE: Poison by intravenous route. Moderately toxic by ingestion and intraperitoneal routes. A skin and eye irritant. When heated to decomposition it emits toxic fumes of NO_x.

PKI750 CAS:25322-69-4 ***HR: 3***
POLYPROPYLENE GLYCOL 750

TOXICITY DATA with REFERENCE
orl-rat LD50:300 mg/kg 14CYAT 2,1524,63
ipr-mus LD50:195 mg/kg JPETAB 103,293,51
ivn-dog LDLo:20 mg/kg JPETAB 103,293,51

CONSENSUS REPORTS: Reported in EPA TSCA Inventory.

SAFETY PROFILE: Poison by ingestion, intraperitoneal, and intravenous routes. When heated to decomposition it emits acrid smoke and irritating fumes.

PKJ250 CAS:25322-69-4 ***HR: 3***
POLYPROPYLENE GLYCOL 1200

TOXICITY DATA with REFERENCE
orl-rat LD50:600 mg/kg 14CYAT 2,1525,63
ipr-mus LD50:113 mg/kg JPETAB 103,293,51
ivn-dog LDLo:20 mg/kg JPETAB 103,293,51

CONSENSUS REPORTS: Reported in EPA TSCA Inventory.

SAFETY PROFILE: Poison by intravenous and intraperitoneal routes. Moderately toxic by ingestion. When heated to decomposition it emits acrid smoke and irritating fumes. Used in cosmetic formulations, brake fluids, lubricating oils and greases, and rubber processing.

PKJ500 CAS:25322-69-4 ***HR: 2***
POLYPROPYLENE GLYCOL 2025

TOXICITY DATA with REFERENCE
skn-rbt 500 mg open MLD UCDS** 5/19/60
orl-rat LD50:9760 mg/kg UCDS** 5/19/60
ipr-rat LD50:4470 mg/kg AMIHBC 3,448,51
ivn-rat LD50:710 mg/kg AMIHBC 3,448,51

CONSENSUS REPORTS: Reported in EPA TSCA Inventory.

SAFETY PROFILE: Moderately toxic by intravenous route. Mildly toxic by ingestion and intraperitoneal routes. A skin irritant. When heated to decomposition it emits acrid smoke and irritating fumes.

PKK000 CAS:25322-69-4 ***HR: 1***
POLYPROPYLENE GLYCOL 4025

TOXICITY DATA with REFERENCE
skn-rbt 500 mg open MLD UCDS** 3/28/69
orl-rat LD50:57 g/kg UCDS** 3/28/69
skn-rbt LD50:20 g/kg UCDS** 3/28/69

CONSENSUS REPORTS: Reported in EPA TSCA Inventory.

SAFETY PROFILE: Mildly toxic by ingestion and skin contact. A skin irritant. When heated to decomposition it emits acrid smoke and irritating fumes.

PKK500 ***HR: 1***
POLYPROPYLENE GLYCOL 400, MONOBUTYL ETHER

SYN: BPG 400

TOXICITY DATA with REFERENCE
skn-rbt 80 mg/4H AMIHBC 4,261,51
eye-rbt 500 mg AMIHBC 4,261,51

CONSENSUS REPORTS: Glycol ether compounds are on the Community Right-To-Know List.

SAFETY PROFILE: A skin and eye irritant. Many glycol ethers cause dangerous human reproductive effects. When heated to decomposition it emits acrid smoke and irritating fumes. See also GLYCOL ETHERS.

PKK750 ***HR: 1***
POLYPROPYLENEGLYCOL 800, MONOBUTYL ETHER

SYN: BPG 800

TOXICITY DATA with REFERENCE
skn-rbt 80 mg/4H MLD AMIHBC 4,261,51
eye-rbt 500 mg AMIHBC 4,261,51

CONSENSUS REPORTS: Glycol ether compounds are on the Community Right-To-Know List.

SAFETY PROFILE: A skin and eye irritant. Many glycol ethers cause dangerous human reproductive effects. When heated to decomposition it emits acrid smoke and irritating fumes. See also GLYCOL ETHERS.

PKK775 CAS:32078-95-8 ***HR: 3***
POLYSILYLENE
mf: $(H_2Si)_n$

$$(-SiH_2-)_n$$

SAFETY PROFILE: Ignites spontaneously in air. Explodes on contact with sulfuric acid. Ignites on contact with concentrated nitric acid. See also SILANE.

PKL000 CAS:9005-64-5 ***HR: 2***
POLYSORBATE 20

PROP: Lemon to amber colored liquid; characteristic odor, bitter taste. Sol in water, alc, ethyl acetate, methanol, dioxane; insol in mineral oil, mineral spirits.

SYNS: GLYCOSPERSE L-20X ◇ POLYOXYETHYLENE (20) SORBITAN MONOLAURATE

TOXICITY DATA with REFERENCE
skn-hmn 15 mg/3D-I MLD 85DKA8 -,127,77
orl-rat LD50:37 g/kg FOREAE 21,348,56

ivn-mus LD50:1420 mg/kg RPOBAR 2,316,70
orl-ham LD50:18 g/kg FOREAE 21,348,56

CONSENSUS REPORTS: Reported in EPA TSCA Inventory.

SAFETY PROFILE: Moderately toxic by intravenous route. Mildly toxic by ingestion. A human skin irritant. When heated to decomposition it emits acrid smoke and irritating fumes.

PKL030 CAS:9005-67-8 ***HR: 3***
POLYSORBATE 60
mf: $C_{64}H_{126}O_{26}$ mw: 1311.90

PROP: Lemon to orange colored oily liquid; faint odor and bitter taste. Sol in water, aniline, ethyl acetate, toluene; insol in mineral oil, vegetable oil.

SYNS: CAPMUL ◇ LGYCOSPERSE S-20 ◇ LIPOSORB S-20 ◇ POLYOXYETHYLENE SORBITAN MONOSTEARATE ◇ POLYOXYETHYLENE 20 SORBITAN MONOSTEARATE ◇ SORBITAN, MONOOCTADECANOATE, POLY(OXY-1,2-ETHANEDIYL) DERIVATIVES ◇ TWEEN 60

TOXICITY DATA with REFERENCE
orl-mus TDLo:41600 mg/kg (female 7-14D post):REP NTIS** PB86-197605
scu-rat TDLo:2100 mg/kg/7W-I:ETA 13BYAH -,83,62
ivn-rat LD50:1220 mg/kg FAONAU 53A,256,74
ivn-mus LDLo:1 g/kg JAPMA8 45,685,56

CONSENSUS REPORTS: Reported in EPA TSCA Inventory.

SAFETY PROFILE: Moderately toxic by intravenous route. Experimental reproductive effects. Questionable carcinogen with experimental tumorigenic data. When heated to decomposition it emits acrid smoke and irritating fumes. See also SURFACTANTS.

PKL100 CAS:9005-65-6 ***HR: 3***
POLYSORBATE 80

PROP: Yellow to orange oily liquid; faint odor, bitter taste. Sol in water, alc, fixed oils, ethyl acetate, toluene; insol in mineral oil.

SYNS: ARMOTAN PMO-20 ◇ ATLOX 1087 ◇ CAPMUL POE-O ◇ CRILL 10 ◇ DREWMULSE POE-SMO ◇ DURFAX 80 ◇ EMSORB 6900 ◇ ETHOXYLATED SORBITAN MONOOLEATE ◇ GLYCOSPERSE O-20 ◇ HODAG SVO 9 ◇ LIPOSORB O-20 ◇ MONITAN ◇ MONTANOX 80 ◇ NCI-C60286 ◇ NIKKOL TO ◇ OLOTHORB ◇ POLYOXYETHYLENE SORBITAN MONOOLEATE ◇ POLYOXYETHYLENE SORBITAN OLEATE ◇ POLYSORBAN 80 ◇ POLYSORBATE 80, U.S.P. ◇ PROTASORB O-20 ◇ ROMULGIN O ◇ SORBIMACROGOL OLEATE ◇ SORBITAL O 20 ◇ SORETHYTAN (20) MONOOLEATE ◇ SORLATE ◇ SVO 9 ◇ TWEEN 80

TOXICITY DATA with REFERENCE
eye-rbt 150 mg MLD AROPAW 40,668,48
dni-hmn:lym 20 ppm BBRCA9 45,630,71
dni-mus:oth 20 ppm ENPBBC 5,84,75
orl-rat TDLo:635 g/kg (MGN):REP JONUAI 60,489,56
scu-rat TDLo:10 g/kg/27W-I:ETA FCTXAV 9,463,71
ipr-rat LD50:6804 mg/kg ARZNAD 35,804,85
ivn-rat LD50:1790 mg/kg FAONAU 53A,257,74
orl-mus LD50:25 g/kg BCFAAI 101,173,82
ipr-mus LD50:7600 mg/kg PHTHDT 5,467,79
ivn-mus LD50:4500 mg/kg ARZNAD 18,666,68
ivn-dog LDLo:500 mg/kg ARZNAD 28,1586,78
ivn-cat LDLo:500 mg/kg ARZNAD 28,1586,78

CONSENSUS REPORTS: Reported in EPA TSCA Inventory.

SAFETY PROFILE: Moderately toxic by intravenous route. Mildly toxic by ingestion. Experimental reproductive effects. Questionable carcinogen with experimental tumorigenic data. Human mutation data reported. An eye irritant. When heated to decomposition it emits acrid smoke and irritating fumes. See also SURFACTANTS.

PKL250 CAS:346-18-9 ***HR: 3***
POLYTHIAZIDE
mf: $C_{11}H_{13}ClF_3N_3O_4S_3$ mw: 439.90

PROP: Crystals. Mp: 202.5°. Insol in water; sol in aq sols.

SYN: 6-CHLORO-3,4-DIHYDRO-2-METHYL-3-(((2,2,2-TRIFLUOROETHYL)THIO)METHYL)2H-1,2,4-BENZOTHIADIAZINE-7-SULFONAMIDE, 1,1-DIOXIDE

TOXICITY DATA with REFERENCE
ipr-rat LD50:400 mg/kg 29ZVAB -,97,69
orl-dog LD50:450 mg/kg 29ZVAB -,97,69

SAFETY PROFILE: Poison by intraperitoneal route. Moderately toxic by ingestion. When heated to decomposition it emits very toxic fumes of Cl^-, SO_x, F^-, and NO_x.

PKL500 CAS:9009-54-5 ***HR: 3***
POLYURETHANE FOAM

SYNS: ETHERON SPONGE ◇ NCI-C56451 ◇ POLYFOAM PLASTIC SPONGE ◇ POLYFOAM SPONGE ◇ POLYURETHANE ESTER FOAM ◇ POLYURETHANE ETHER FOAM ◇ POLYURETHANE SPONGE

TOXICITY DATA with REFERENCE
itr-rat TDLo:225 mg/kg:ETA EVHPAZ 11,109,75

CONSENSUS REPORTS: IARC Cancer Review: Group 3 IMEMDT 7,56,87; Animal Sufficient Evidence IMEMDT 19,303,79.

SAFETY PROFILE: Questionable carcinogen with experimental tumorigenic data. When heated to decomposition it emits acrid toxic fumes of CN^- and NO_x.

PKL750 CAS:25931-01-5 ***HR: 3***
POLYURETHANE Y-195
mf: $(C_{15}H_{10}N_2O_2 \cdot C_6H_{10}O_4 \cdot C_2H_6O_2)_x$

SYNS: ADIPIC ACID, POLYMER with ETHYLENE GLYCOL and METHYLENEDI-p-PHENYLENE ISOCYANATE ◇ AMCHEM R 14 ◇ HEXANEDIOIC ACID, POLYMER with 1,3-ETHANEDIOL and 1,1'-METHYLENEBIS(4-ISOCYANATOBENZENE) ◇ MUL F 66 ◇ R 14 ◇ Y 195

TOXICITY DATA with REFERENCE
imp-rat TDLo:6750 mg/kg:ETA CNREA8 35,1591,75

CONSENSUS REPORTS: IARC Cancer Review: Animal Sufficient Evidence IMEMDT 19,303,79. Reported in EPA TSCA Inventory.

SAFETY PROFILE: Confirmed carcinogen with experimental tumorigenic data. When heated to decomposition it emits toxic fumes of NO_x.

PKM000 ***HR: 3***
POLYURETHANE Y-217

TOXICITY DATA with REFERENCE
imp-rat TDLo:6750 mg/kg:ETA CNREA8 35,1591,75

CONSENSUS REPORTS: IARC Cancer Review: Animal Sufficient Evidence IMEMDT 19,303,79.

SAFETY PROFILE: Confirmed carcinogen with experimental tumorigenic data. When heated to decomposition it emits very toxic fumes of NO_x and CN^-.

PKM250 CAS:26375-23-5 ***HR: 3***
POLYURETHANE Y-218
mf: $(C_{15}H_{10}N_2O_2 \cdot C_6H_{10}O_4 \cdot C_4H_{10}O_2)_x$

SYNS: ADIPIC ACID, POLYMER with 1,4-BUTANEDIOL and METHYLENEDI-p-PHENYLENE ISOCYANATE ◇ HEXANEDIOIC ACID, POLYMER with 1,4-BUTANEDIOL and 1,1'-METHYLENEBIS(4-ISOCYANATOBENZENE) ◇ PANDEX ◇ TEXIN 445D ◇ TPU 10M ◇ Y 218

TOXICITY DATA with REFERENCE
imp-rat TDLo:6750 mg/kg:ETA CNREA8 35,1591,75

CONSENSUS REPORTS: IARC Cancer Review: Animal Sufficient Evidence IMEMDT 19,303,79.

SAFETY PROFILE: Confirmed carcinogen with experimental tumorigenic data. When heated to decomposition it emits very toxic fumes of CN^- and NO_x.

PKM500 CAS:32238-28-1 ***HR: 3***
POLYURETHANE Y-221
mf: $(C_{15}H_{10}N_2O_2 \cdot C_{10}H_{14}O_4 \cdot C_6H_{10}O_4 \cdot C_4H_{10}O_2)_x$

SYNS: ADIPIC ACID, POLYMER with 1,4-BUTANEDIOL, METHYLENEDI-p-PHENYLENE ISOCYANATE and 2,2'-(p-PHENYLENEDIOXY)DIETHANOL ◇ Y 221

TOXICITY DATA with REFERENCE
imp-rat TDLo:6750 mg/kg:ETA CNREA8 35,1591,75

CONSENSUS REPORTS: IARC Cancer Review: Animal Sufficient Evidence IMEMDT 19,303,79. Reported in EPA TSCA Inventory.

SAFETY PROFILE: Confirmed carcinogen with experimental tumorigenic data. When heated to decomposition it emits very toxic fumes of CN^- and NO_x.

PKM750 ***HR: 3***
POLYURETHANE Y-222

TOXICITY DATA with REFERENCE
imp-rat TDLo:6750 mg/kg:ETA CNREA8 35,1591,75

CONSENSUS REPORTS: IARC Cancer Review: Animal Sufficient Evidence IMEMDT 19,303,79.

SAFETY PROFILE: Confirmed carcinogen with experimental tumorigenic data. When heated to decomposition it emits very toxic fumes of CN^- and NO_x.

PKN000 CAS:52292-20-3 ***HR: 3***
POLYURETHANE Y-223

SYNS: TECOFLEX HR ◇ Y-223

TOXICITY DATA with REFERENCE
imp-rat TDLo:6750 mg/kg:ETA CNREA8 35,1591,75

CONSENSUS REPORTS: IARC Cancer Review: Animal Sufficient Evidence IMEMDT 19,303,79. Reported in EPA TSCA Inventory.

SAFETY PROFILE: Confirmed carcinogen with experimental tumorigenic data. When heated to decomposition it emits very toxic fumes of CN^- and NO_x.

PKN250 ***HR: 3***
POLYURETHANE Y-224

TOXICITY DATA with REFERENCE
imp-rat TDLo:6750 mg/kg:ETA CNREA8 35,1591,75

CONSENSUS REPORTS: IARC Cancer Review: Animal Sufficient Evidence IMEMDT 19,303,79.

SAFETY PROFILE: Confirmed carcinogen with experimental tumorigenic data. When heated to decomposition it emits very toxic fumes of CN^- and NO_x.

PKN500 CAS:56779-19-2 ***HR: 3***
POLYURETHANE Y-225

SYN: 1,4-BUTANEDIAMINE, 2-METHYL-, POLYMER with α-HYDRO-omega-HYDROXYPOLY(OXY-1,4-BUTANEDIYL) and 1,1'-METHYLENEBIS(4-ISOCYANATOCYCLOHEXANE)

TOXICITY DATA with REFERENCE
imp-rat TDLo:6750 mg/kg:ETA CNREA8 35,1591,75

CONSENSUS REPORTS: IARC Cancer Review: Animal Sufficient Evidence IMEMDT 19,303,79.

SAFETY PROFILE: Confirmed carcinogen with experimental tumorigenic data. When heated to decomposition it emits very toxic fumes of CN^- and NO_x.

PKN750 CAS:56386-98-2 ***HR: 3***
POLYURETHANE Y-226

TOXICITY DATA with REFERENCE
imp-rat TDLo:6750 mg/kg:ETA CNREA8 35,1591,75

CONSENSUS REPORTS: IARC Cancer Review: Animal Sufficient Evidence IMEMDT 19,303,79.

SAFETY PROFILE: Confirmed carcinogen with experimental tumorigenic data. When heated to decomposition it emits very toxic fumes of CN^- and NO_x.

PKO000 CAS:56631-46-0 ***HR: 3***
POLYURETHANE Y-227

TOXICITY DATA with REFERENCE
imp-rat TDLo:6750 mg/kg:ETA CNREA8 35,1591,75

CONSENSUS REPORTS: IARC Cancer Review: Animal Sufficient Evidence IMEMDT 19,303,79.

SAFETY PROFILE: Confirmed carcinogen with experimental tumorigenic data. When heated to decomposition it emits very toxic fumes of CN^- and NO_x.

PKO500 CAS:27083-55-2 ***HR: 3***
POLYURETHANE Y-290
mf: $(C_{15}H_{10}N_2O_2{\cdot}C_6H_{10}O_4{\cdot}C_4H_{10}O_2{\cdot}C_2H_6O_2)_x$

SYNS: E6 ◇ PPE201 ◇ P07 ◇ TEXIN 192A ◇ TPU 2T

TOXICITY DATA with REFERENCE
imp-rat TDLo:6750 mg/kg:ETA CNREA8 35,1591,75

CONSENSUS REPORTS: IARC Cancer Review: Animal Sufficient Evidence IMEMDT 19,303,79.

SAFETY PROFILE: Confirmed carcinogen with experimental tumorigenic data. When heated to decomposition it emits very toxic fumes of CN^- and NO_x.

PKO750 CAS:25748-74-7 ***HR: 3***
POLYURETHANE Y-299

SYNS: 1,4-BUTANEDIOL, POLYMER with 1,6-DIISOCYANATOHEXANE ◇ DURANATE EXP-D 101 ◇ ISOCYANIC ACID, HEXAMETHYLENE ESTER, POLYMER with 1,4-BUTANEDIOL ◇ Y 299

TOXICITY DATA with REFERENCE
imp-rat TDLo:6000 mg/kg:ETA CNREA8 35,1591,75

SAFETY PROFILE: Questionable carcinogen with experimental tumorigenic data by implant route. When heated to decomposition it emits very toxic fumes of CN^- and NO_x.

PKP000 CAS:25805-16-7 ***HR: 3***
POLYURETHANE Y-302
mf: $(C_{15}H_{10}N_2O_2{\cdot}C_4H_{10}O_2)_x$

SYNS: 1,4-BUTANEDIOL POLYMER with 1,1'-METHYLENEBIS(4-ISOCYANATOBENZENE) ◇ ISOCYANIC ACID, METHYLENEDI-p-PHENYLENE ESTER, POLYMER with 1,4-BUTANEDIOL ◇ SANPRENE LQX 31 ◇ Y 302

TOXICITY DATA with REFERENCE
imp-rat TDLo:6750 mg/kg:ETA CNREA8 35,1591,75

CONSENSUS REPORTS: IARC Cancer Review: Animal Sufficient Evidence IMEMDT 19,303,79.

SAFETY PROFILE: Confirmed carcinogen with experimental tumorigenic data by implant route. When heated to decomposition it emits very toxic fumes of CN^- and NO_x.

PKP250 CAS:25036-33-3 ***HR: 3***
POLYURETHANE Y-304

TOXICITY DATA with REFERENCE
imp-rat TDLo:6000 mg/kg:ETA CNREA8 35,1591,75

CONSENSUS REPORTS: IARC Cancer Review: Animal Sufficient Evidence IMEMDT 19,303,79.

SAFETY PROFILE: Confirmed carcinogen with experimental tumorigenic data. When heated to decomposition it emits very toxic fumes of CN^- and NO_x.

PKP500 CAS:34149-92-3 ***HR: 3***
POLYVINYL ACETATE CHLORIDE

SYNS: ACETIC ACID, VINYL ESTER, CHLOROETHYLENE COPOLYMER ◇ POLYVINYLCHLORIDE ACETATE ◇ VINYL CHLORIDE ACETATE COPOLYMER ◇ VINYL CHLORIDE VINYL ACETATE COPOLYMER

TOXICITY DATA with REFERENCE
imp-mus TDLo:240 mg/kg:CAR JNCIAM 58,1443,77
imp-mus TD:1656 g/kg:ETA CNREA8 37,4367,77

SAFETY PROFILE: Questionable carcinogen with experimental carcinogenic and tumorigenic data by implant route. When heated to decomposition it emits toxic fumes of Cl^-.

PKP750 CAS:9002-89-5 ***HR: 3***
POLYVINYL ALCOHOL

PROP: Colorless, amorph powder. Mp: decomp over 200°, flash p: 175°F (OC), d: 1.329. Polymer of average molecular weight 120,000 (AMPLAO 67,589,59).

SYNS: ELVANOL ◇ ETHENOL HOMOPOLYMER (9CI)

◇ GELVATOLS ◇ GOHSENOLS ◇ POLY(VINYL ALCOHOL) ◇ VINYL ALCOHOL POLYMER

TOXICITY DATA with REFERENCE
scu-rat TDLo:2500 mg/kg:CAR AMPLAO 67,589,59
imp-rat TDLo:10 g/kg:ETA BJSUAM 52,49,65

CONSENSUS REPORTS: IARC Cancer Review: Group 3 IMEMDT 7,56,87; Animal Limited Evidence IMEMDT 19,341,79; Human Inadequate Evidence IMEMDT 19,341,79.

SAFETY PROFILE: Questionable carcinogen with experimental carcinogenic and tumorigenic data by implant route. Flammable when exposed to heat or flame; can react with oxidizing materials. Slight explosion hazard in the form of dust when exposed to flame. To fight fire, use alcohol foam, CO_2, dry chemical. When heated to decomposition it emits acrid smoke and irritating fumes.

PKQ000 CAS:25951-54-6 ***HR: 3***
POLYVINYLBROMIDE
mf: $(C_2H_3Br)_x$

PROP: Commercial PVBR is a 40% aqueous suspension in which PVBR constitutes about 90% of the solids (CNREA8 38,3236,78).

SYNS: BROMOETHYLENE POLYMER ◇ POLYBROMOETHYLENE ◇ PVBR

TOXICITY DATA with REFERENCE
scu-mus TDLo:44 g/kg/48W-I:CAR CNREA8 38,3236,78

CONSENSUS REPORTS: IARC Cancer Review: Animal Inadequate Evidence IMEMDT 19,367,79.

SAFETY PROFILE: Questionable carcinogen with experimental carcinogenic data. When heated to decomposition it emits toxic fumes of Br^-.

PKQ059 CAS:9002-86-2 ***HR: 2***
POLYVINYL CHLORIDE
mf: $(C_2H_3Cl)_n$

PROP: Polymers with molecular weights ranging from 60,000-150,000 (CNREA8 15,333,55). White powder, d: 1.406.

SYNS: ARMODOUR ◇ ARON COMPOUND HW ◇ ASTRALON ◇ ATACTIC POLY(VINYL CHLORIDE) ◇ BLACAR 1716 ◇ BOLATRON ◇ BONLOID ◇ BREON ◇ CARINA ◇ CHLOROETHENE HOMOPOLYMER ◇ CHLOROETHYLENE POLYMER ◇ CHLOROSTOP ◇ COBEX (polymer) ◇ CONTIZELL ◇ CORVIC 55/9 ◇ DACOVIN ◇ DANUVIL 70 ◇ DARVIC 110 ◇ DARVIS CLEAR 025 ◇ DECELITH H ◇ DENKA VINYL SS 80 ◇ DIAMOND SHAMROCK 40 ◇ DORLYL ◇ DUROFOL P ◇ DYNADUR ◇ E 62 ◇ E 66P ◇ EKAVYL SD 2 ◇ E-PVC ◇ ESCAMBIA 2160 ◇ EUROPHAN ◇ EXON 605 ◇ FC 4648 ◇ FLOCOR ◇ GAFCOTE ◇ GENOTHERM ◇ GEON ◇ GEON LATEX 151 ◇ GUTTAGENA ◇ HALVIC 223 ◇ HISHIREX 502 ◇ HISPAVIC 229 ◇ HOSTALIT ◇ IGELITE F ◇ IMPROVED WILT PRUF ◇ KAYLITE ◇ KLEGECELL ◇ KOROSEAL ◇ LONZA G ◇ LUCOFLEX ◇ LUCOVYL PE ◇ LUTOFAN ◇ MARVINAL ◇ MIRREX MCFD 1025 ◇ MOVINYL 100 ◇ MYRAFORM ◇ NCI-C60797 ◇ NIKA-TEMP ◇ NIKAVINYL SG 700 ◇ NIPEON A 21 ◇ NIPOL 576 ◇ NORVINYL ◇ NOVON 712 ◇ ONGROVIL S 165 ◇ OPALON ◇ ORTUDUR ◇ PANTASOTE R 873 ◇ PARCLOID ◇ PATTINA V 82 ◇ PEVIKON D 61 ◇ PLIOVIC ◇ POLIVINIT ◇ POLY(CHLOROETHYLENE) ◇ POLYTHERM ◇ POLYVINYLCHLORID (GERMAN) ◇ PROTOTYPE III SOFT ◇ PVC (MAK) ◇ QSAH 7 ◇ QUIRVIL ◇ QYSA ◇ RAVINYL ◇ RUCON B 20 ◇ S 65 (polymer) ◇ SCON 5300 ◇ SICRON ◇ S-LON ◇ SOLVIC ◇ SP 60 (CHLOROCARBON) ◇ SUMILIT EXA 13 ◇ SUMITOMO PX 11 ◇ TAKILON ◇ TECHNOPOR ◇ TENNECO 1742 ◇ TK 1000 ◇ TROVIDUR ◇ TROVITHERN HTL ◇ U 1 (polymer) ◇ ULTRON ◇ UNICHEM ◇ VERON P 130/1 ◇ VESTOLIT B 7021 ◇ VINIKA KR 600 ◇ VINIKULON ◇ VINIPLAST ◇ VINIPLEN P 73 ◇ VINNOL E 75 ◇ VINOFLEX ◇ VINYLCHLON 4000LL ◇ VINYL CHLORIDE HOMOPOLYMER ◇ VINYL CHLORIDE POLYMER ◇ VYGEN 85 ◇ WELVIC G 2/5 ◇ WILT PRUF ◇ WINIDUR ◇ X-AB ◇ YUGOVINYL

TOXICITY DATA with REFERENCE
orl-rat TDLo:210 g/kg/30W-C:ETA PATHAB 73,59,81

CONSENSUS REPORTS: IARC Cancer Review: Group 3 IMEMDT 7,56,87; Human Inadequate Evidence IMEMDT 19,377,79; IARC Cancer Review: Animal Inadequate Evidence IMEMDT 19,377,79. Reported in EPA TSCA Inventory.

DFG MAK: 6 mg/m^3 (dust)

SAFETY PROFILE: Chronic inhalation of dusts can cause pulmonary damage, blood effects, abnormal liver function. "Meat wrappers asthma" has resulted from the cutting of PVC films with a hot knife. Can cause allergic dermatitis. Questionable carcinogen with experimental tumorigenic data. Reacts violently with F_2. When heated to decomposition it emits toxic fumes of Cl^- and phosgene.

PKQ100 CAS:9045-81-2 ***HR: 1***
POLY(VINYLPYRIDINE N-OXIDE)
mf: $(C_7H_7NO)_x$

SYNS: ETHENYLPYRIDINE 1-OXIDE HOMOPOLYMER ◇ POLY(VINYLPYRIDINE 1-OXIDE) ◇ PVPO

TOXICITY DATA with REFERENCE
ivn-rat TDLo:5 g/kg (2W male):REP GTPZAB 23(12),38,79
ipr-mus LD50:16500 mg/kg GTPZAB 21(4),50,77
scu-mus LD50:14000 mg/kg GTPZAB 21(4),50,77

SAFETY PROFILE: Slightly toxic by intraperitoneal and subcutaneous routes. Experimental reproductive effects. When heated to decomposition it emits toxic fumes of NO_x.

PKQ250 CAS:9003-39-8 ***HR: 1***
POLY(1-VINYL-2-PYRROLIDINONE) HOMOPOLYMER
mf: $(C_6H_9ON)_n$

PROP: A free-flowing, white, amorphous powder. D:

1.23-1.29. Sol in water, chlorinated hydrocarbons, alc, amines, nitroparaffins, and lower molecular weight fatty acids.

SYNS: AGENT AT 717 ◇ ALBIGEN A ◇ ALDACOL Q ◇ AT 717 ◇ BOLINAN ◇ 1-ETHENYL-2-PYRROLIDINONE HOMOPOLYMER ◇ 1-ETHENYL-2-PYRROLIDINONE POLYMERS ◇ GANEX P 804 ◇ HEMODESIS ◇ HEMODEZ ◇ K25 (polymer) ◇ KOLLIDON ◇ LUVISKOL ◇ MPK 90 ◇ NCI C60582 ◇ NEOCOMPENSAN ◇ PERAGAL ST ◇ PERISTON ◇ PLASDONE ◇ POLYCLAR L ◇ POLY(1-(2-OXO-1-PYRROLIDINYL)ETHYLENE) ◇ POLYVIDONE ◇ POLY(n-VINYLBUTYROLACTAM) ◇ POLYVINYLPYRROLIDONE ◇ POVIDONE (USP XIX) ◇ PROTAGENT ◇ PVP (FCC) ◇ SUBTOSAN ◇ VINISIL ◇ N-VINYLBUTYROLACTAM POLYMER ◇ N-VINYLPYRROLIDONE POLYMER

TOXICITY DATA with REFERENCE
ipr-mus LD50:12 g/kg FAONAU 53A,487,74
ivn-mky LDLo:5300 mg/kg NCIHL* NIH-69-2067,70

CONSENSUS REPORTS: IARC Cancer Review: Group 3 IMEMDT 7,56,87. Reported in EPA TSCA Inventory.

SAFETY PROFILE: Mildly toxic by intraperitoneal and intravenous routes. Questionable carcinogen. When heated to decomposition it emits toxic fumes of NO_x.

PKQ500 CAS:9003-39-8 ***HR: 3***
POLY(1-VINYL-2-PYRROLIDINONE)
Hueper's polymer No. 1

PROP: Polymer of average molecular weight 20,000 (AMPLAO 67,589,59).

SYNS: NCI-C60582 ◇ PVP 1

TOXICITY DATA with REFERENCE
ipr-rat TDLo:2500 mg/kg:CAR,REP AMPLAO 67,589,59
scu-rat TDLo:2500 mg/kg:CAR AMPLAO 67,589,59
ivn-rat TDLo:750 mg/kg/I:CAR,REP AMPLAO 67,589,59

CONSENSUS REPORTS: IARC Cancer Review: Animal Limited Evidence IMEMDT 19,461,79. Reported in EPA TSCA Inventory.

SAFETY PROFILE: Suspected carcinogen with experimental carcinogenic data. Experimental reproductive effects. When heated to decomposition it emits toxic fumes of NO_x.

PKQ750 CAS:9003-39-8 ***HR: 3***
POLY(1-VINYL-2-PYRROLIDINONE)
Hueper's polymer No. 2
mf: $(C_6H_9NO)x$

PROP: Polymer of average molecular weight 20,000 (AMPLAO 67,589,59).

SYNS: NCI-C60582 ◇ PVP 2

TOXICITY DATA with REFERENCE
ipr-rat TDLo:2500 mg/kg:NEO,REP AMPLAO 67,589,59
scu-rat TDLo:2500 mg/kg:NEO,REP AMPLAO 67,589,59
ivn-rat TDLo:750 mg/kg/I:NEO,REP AMPLAO 67,589,59
scu-mus TDLo:8000 mg/kg:ETA AMPLAO 67,589,59
orl-mus LDLo:3 g/kg BIMADU 12,1,84

CONSENSUS REPORTS: IARC Cancer Review: Animal Limited Evidence IMEMDT 19,461,79. Reported in EPA TSCA Inventory.

SAFETY PROFILE: Suspected carcinogen with experimental neoplastigenic and tumorigenic data. Experimental reproductive effects. When heated to decomposition it emits toxic fumes of NO_x.

PKR000 CAS:9003-39-8 ***HR: 3***
POLY(1-VINYL-2-PYRROLIDINONE)
Hueper's polymer No. 3
mf: $(C_6H_9NO)_x$

PROP: Polymer of average molecular weight 50,000 (AMPLAO 67,589,59).

SYNS: NCI-C60582 ◇ PVP 3

TOXICITY DATA with REFERENCE
ipr-rat TDLo:2500 mg/kg:CAR,REP AMPLAO 67,589,59
scu-rat TDLo:2500 mg/kg:CAR AMPLAO 67,589,59
ivn-rat TDLo:750 mg/kg/I:CAR,REP AMPLAO 67,589,59
ipr-mus TDLo:8000 mg/kg:ETA AMPLAO 67,589,59

CONSENSUS REPORTS: IARC Cancer Review: Animal Limited Evidence IMEMDT 19,461,79. Reported in EPA TSCA Inventory.

SAFETY PROFILE: Suspected carcinogen with experimental carcinogenic and tumorigenic data. Experimental reproductive effects. When heated to decomposition it emits toxic fumes of NO_x.

PKR250 CAS:9003-39-8 ***HR: 3***
POLY(1-VINYL-2-PYRROLIDINONE) Hueper's polymer No. 4
mf: $(C_6H_9NO)_x$

PROP: Polymer of average molecular weight 300,000 (AMPLAO 67,589,59).

SYNS: NCI-C60582 ◇ PVP 4

TOXICITY DATA with REFERENCE
ipr-rat TDLo:2500 mg/kg:CAR,REP AMPLAO 67,589,59
scu-rat TDLo:2500 mg/kg:CAR AMPLAO 67,589,59
ivn-rat TDLo:750 mg/kg/I:CAR,REP AMPLAO 67,589,59

CONSENSUS REPORTS: IARC Cancer Review: Animal Limited Evidence IMEMDT 19,461,79. Reported in EPA TSCA Inventory.

SAFETY PROFILE: Suspected carcinogen with experimental carcinogenic data. Experimental reproductive effects. When heated to decomposition it emits toxic fumes of NO_x.

PKR500 CAS:9003-39-8 ***HR: 3***
POLY(1-VINYL-2-PYRROLIDINONE) Hueper's polymer No. 5
mf: $(C_6H_9NO)_x$

PROP: Polymer of average molecular weight 10,000 (AMPLAO 67,589,59).

SYN: PVP 5

TOXICITY DATA with REFERENCE
ipr-rat TDLo:2500 mg/kg:CAR AMPLAO 67,589,59
scu-rat TDLo:2500 mg/kg:CAR,REP AMPLAO 67,589,59
ipr-mus TDLo:8000 mg/kg:ETA AMPLAO 67,589,59

CONSENSUS REPORTS: IARC Cancer Review: Animal Limited Evidence IMEMDT 19,461,79. Reported in EPA TSCA Inventory.

SAFETY PROFILE: Suspected carcinogen with experimental carcinogenic and tumorigenic data. Experimental reproductive effects. When heated to decomposition it emits toxic fumes of NO_x.

PKR750 CAS:9003-39-8 ***HR: 3***
POLY(1-VINYL-2-PYRROLIDINONE) Hueper's polymer No. 6
mf: $(C_6H_9NO)_x$

PROP: Polymer of average molecular weight 50,000 (AMPLAO 67,589,59).

SYNS: NCI-C60582 ◇ PVP 6

TOXICITY DATA with REFERENCE
scu-rat TDLo:1000 mg/kg:CAR,REP AMPLAO 67,589,59
orl-mus LDLo:5 g/kg BIMADU 12,1,84

CONSENSUS REPORTS: IARC Cancer Review: Animal Limited Evidence IMEMDT 19,461,79. Reported in EPA TSCA Inventory.

SAFETY PROFILE: Suspected carcinogen with experimental carcinogenic data. Experimental reproductive effects. When heated to decomposition it emits toxic fumes of NO_x.

PKS000 CAS:9003-39-8 ***HR: 3***
POLY(1-VINYL-2-PYRROLIDINONE) Hueper's polymer No. 7

SYNS: NCI-C60582 ◇ PVP 7

TOXICITY DATA with REFERENCE
scu-rat TDLo:3000 mg/kg/I:NEO AMPLAO 67,589,59

CONSENSUS REPORTS: IARC Cancer Review: Animal Limited Evidence IMEMDT 19,461,79.

SAFETY PROFILE: Suspected carcinogen with experimental neoplastigenic data. When heated to decomposition it emits toxic fumes of NO_x.

PKS250 CAS:26837-42-3 ***HR: 3***
POLYVINYL SULFATE, POTASSIUM SALT

SYNS: POTASSIUM SALT OF POLYVINYL SULFATE ◇ PVSK

TOXICITY DATA with REFERENCE
ipr-mus LD50:225 mg/kg CRSBAW 166,121,72
scu-mus LD50:78 mg/kg OSDIAF 5,128,56

CONSENSUS REPORTS: Reported in EPA TSCA Inventory.

SAFETY PROFILE: Poison by intraperitoneal and subcutaneous routes. When heated to decomposition it emits toxic fumes of SO_x and K_2O.

PKS500 CAS:5586-87-8 ***HR: 3***
PONDINIL
mf: $C_{12}H_{18}ClN{\cdot}ClH$ mw: 248.22

SYNS: N-(3-CHLOROPROPYL)-α-METHYLPHENETHYLAMINE HYDROCHLORIDE ◇ N-(3-CHLORPROPYL)-1-METHYL-2-PHENYLAETHYLAMIN-HYDROCHLORID(GERMAN)

TOXICITY DATA with REFERENCE
orl-rat LD50:410 mg/kg ARZNAD 19,748,69
ivn-rat LD50:35 mg/kg ARZNAD 19,748,69
orl-mus LD50:230 mg/kg ARZNAD 19,748,69
ipr-mus LD50:144 mg/kg 27ZQAG -,356,72
scu-mus LD50:180 mg/kg ARZNAD 19,748,69
ivn-mus LD50:49 mg/kg 27ZQAG -,356,72
orl-rbt LD50:236 mg/kg 27ZQAG -,356,72

SAFETY PROFILE: Poison by ingestion, intravenous, intraperitoneal, and subcutaneous routes. When heated to decomposition it emits very toxic fumes of NO_x and Cl^-.

PKS600 ***HR: 2***
PORK TRYPSIN

SYN: PORCINE-TRYPSIN

TOXICITY DATA with REFERENCE
ipr-rat LD50:51 mg/kg KSRNAM 4,1875,70
scu-rat LD50:410 mg/kg KSRNAM 4,1875,70
ims-rat LD50:200 mg/kg KSRNAM 4,1875,70
orl-mus LD50:1450 mg/kg KSRNAM 4,1875,70
ipr-mus LD50:105 mg/kg KSRNAM 4,1875,70
scu-mus LD50:280 mg/kg KSRNAM 4,1875,70
ims-mus LD50:68 mg/kg KSRNAM 4,1875,70

SAFETY PROFILE: Poison by intramuscular, subcutaneous, and intraperitoneal routes. Moderately toxic by ingestion.

PKS750 CAS:65997-15-1 ***HR: 1***
PORTLAND CEMENT

PROP: Fine gray powder composed of compounds of lime, aluminum, silica and iron oxide as $(4CaO{\bullet}Al_2O_3{\bullet}Fe_2)_3$, $(3CaOAl_2O_3)$, $(3CaO{\bullet}SiO_2)$, and $(2CaOSiO_2)$. Small amounts of magnesia, sodium, potassium, chromium and sulfur are also present in combined form. Containing less than 1% crystalline silica (FEREAC 39,23540,74).

SYNS: CEMENT, PORTLAND ◇ PORTLAND CEMENT SILICATE

CONSENSUS REPORTS: Reported in EPA TSCA Inventory.

OSHA PEL: (Transitional: TWA 50 mppcf) TWA Total Dust: 10 mg/m^3; Respirable Fraction: 5 mg/m^3
ACGIH TLV: TWA (nuisance particulate) 10 mg/m^3 of total dust (when toxic impurities are not present, e.g., quartz < 1%).

SAFETY PROFILE: A nuisance dust. A skin irritant. See also NUISANCE DUSTS and AEROSOLS.

PKT000 CAS:299-45-6 ***HR: 3***
POTASAN
mf: $C_{14}H_{17}O_5PS$ mw: 328.34

PROP: Crystals; weak aromatic odor. Mp: 38°; bp: 210° @ 1 mm; d: 1.260 @ 38°/4°.

SYNS: O,O-DIAETHYL-O-(4-METHYL-COUMARIN-7-YL)-MONOTHIOPHOSPHAT (GERMAN) ◇ DIETHOXY THIOPHOSPHORIC ACID ESTER OF 7-HYDROXY-4-METHYL COUMARIN ◇ O,O-DIETHYL-O-(2-KETO-4-METHYL-7-α′,β′-BENZO-α′-PYRANYL) THIOPHOSPHATE ◇ O,O-DIETHYL-O-(4-METHYLCOUMARIN-7-YL)-MONOTHIOFOSFAAT (DUTCH) ◇ O,O-DIETHYL-O-(4-METHYL-7-COUMARINYL) PHOSPHOROTHIOATE ◇ O,O-DIETHYL-O-(4-METHYL-7-COUMARINYL) THIONOPHOSPHATE ◇ O,O-DIETHYL-O-(4-METHYLCOUMARINYL-7) THIOPHOSPHATE ◇ O,O-DIETHYL-O-(4-METHYL-7-KUMARINYL) ESTER KYSELINY THIOFOSFORESCNE (CZECH) ◇ O,O,DIETHYL O-(4-METHYL-2-OXO-2H-1-PHOSPHOROTHIOIC ACID BENZOPYRAN-7-YL)ESTER (9CI) ◇ O,O-DIETHYL-O-(4-METHYLUMBELLIFERONE) ESTER OF THIOPHOSPHORIC ACID ◇ O,O-DIETHYL-O-(4-METHYLUMBELLIFERONE) PHOSPHOROTHIOATE ◇ DIETHYL (4-METHYLUMBELLIFERYL) THIONOPHOSPHATE ◇ O,O-DIETIL-O-(4-METILCUMARIN-7-IL)-MONOTIOFOSFATO (ITALIAN) ◇ O,O-DIETYL-O-4-METHYL-KUMARINYL(7)TIOFOSFAT (CZECH) ◇ 7-HYDROXY-4-METHYL COUMARIN, O-ESTER with O,O,DIETHYL PHOSPHOROTHIOATE ◇ 4-METHYL-7-HYDROXY COUMARIN DIETHOXYTHIOPHOSPHATE ◇ 4-METHYLUMBELLIFERONE-O,O-DIETHYL THIOPHOSPHATE ◇ THIOPHOSPHATE de O,O-DIETHYLE et de O-(4-METHYL-7-COUMARINYLE) (FRENCH)

TOXICITY DATA with REFERENCE
orl-rat LD50:19 mg/kg JPETAB 105,156,52
ipr-rat LD50:15 mg/kg AMIHBC 6,9,52
orl-mus LD50:99 mg/kg JPETA8 105,156,52
scu-mus LD50:25 mg/kg PAREAQ 11,636,59
skn-rbt LD50:300 mg/kg WRPCA2 9,119,70
orl-gpg LD50:25 mg/kg JPETAB 105,156,52

SAFETY PROFILE: Poison by ingestion, skin contact, intraperitoneal, and subcutaneous routes. When heated to decomposition it emits toxic fumes of PO_x and SO_x. See also PARATHION.

PKT250 CAS:7440-09-7 ***HR: 3***
POTASSIUM
DOT: UN 1420/UN 2257
af: K aw: 39.10

PROP: Soft ductile, silvery-white, very reactive metal. Mp: 63.65°, bp: 774°, d: 0.862 @ 20°.

SYN: POTASSIUM, METAL (DOT)

CONSENSUS REPORTS: Reported in EPA TSCA Inventory.

DOT Classification: Label: Flammable Solid and Dangerous When Wet.

SAFETY PROFILE: The toxicity of potassium compounds is almost always that of the anion, not of potassium. A dangerous fire hazard. Metallic potassium reacts with moisture to form potassium hydroxide and hydrogen. The reaction evolves much heat, causing the potassium to melt and spatter. The reaction also ignites the hydrogen, which burns, or if there is any confinement, may explode. It can ignite spontaneously in moist air. Store under mineral oil. Potassium metal will form the peroxide (K_2O_2) and the superoxide (KO_3 or K_2O_4) at room temperature even when stored under mineral oil. These oxides can explode on contact with organic materials. Metal which has oxidized on storage under oil may explode violently when handled or cut. Oxide-coated potassium should be destroyed by burning.

Danger: burning potassium is difficult to extinguish; dry powdered soda ash or graphite or special mixtures of dry chemical are recommended.

A violent explosion hazard with the following materials under required conditions of temperature, pressure, and state of division: acetylene, air, moist air, alcohols (e.g., n-propanol through n-octanol, benzyl alcohol, cyclohexanol), $AlBr_3$, ammonium nitrate + ammonium sulfate, ammonium chlorocuprate, NH_4Br, NH_4I, antimony halides, arsenic halides, AsH_3 + NH_3, Bi_2O_3, boric acid, BBr_3, carbon disulfide (impact-sensitive), solid carbon dioxide, carbon monoxide, chlorinated hydrocarbons (e.g., chloroethane, dichloroethane, dichloromethane, trichloroethane, chloroform, pentachloroethane, carbontetrachloride, tetrachloroethane), halocarbons (e.g., bromoform, dibromomethane, diiodomethane), iodine (impact-sensitive), interhalogens (e.g., chlorine trifluoride, iodine bromide, iodine chloride, iodine pentafluoride, iodine trichloride), ClO, CrO_3, Cu_2OCl_2, CuO, ethylene oxide, fluorine, graphite, graphite + air, graphite + K_2O_2, hydrogen iodide,

H_2O_2, hydrogen chloride, hydrazine, Pb_2OCl_2, PbO_2, $PbSO_4$, maleic anhydride, metal halides (e.g., calcium bromide, iron(III) bromide, iron(III) chloride, iron(II) chloride, iron(II) bromide, iron(II) iodide, cobalt(II) chloride, chromium tetrachloride, silver fluoride, mercury(II) bromide, mercury(II) chloride, mercury(II) fluoride, mercury(II) iodide, copper(I) chloride, copper(I) iodide, copper(II) bromide, copper(II) chloride, ammonium tetrachlorocuprate, zinc chlorides, bromides or iodides, cadmium chlorides, bromides or iodides, aluminum fluorides, chlorides or bromides, thallium(I) bromide, tin chlorides, tin iodide, arsenic trichloride, arsenic triiodide, antimony tribromides, trichlorides or triiodides, bismuth tribromides, trichlorides, or triiodides, vanadium(V) chloride, manganese(II) chloride, nickel bromide, chloride or iodide), metal oxides (e.g., lead peroxide, mercury(I) oxide, MoO_3), nitric acid, nitrogen-containing explosives (e.g., ammonium nitrate, picric acid, nitrobenzene), non-metal halides (e.g., diselenium dichloride, seleninyl chloride, seleninyl bromide, sulfur dichloride, sulfur dibromide, phosphorus tribromide, phosphorus trichloride, phosgene, disulfur dichloride), non-metal oxides (e.g., dichlorine oxide, dinitrogen tetraoxide, dinitrogen pentaoxide, NO_2, P_2O_5), oxalyl dibromide, oxalyl dichloride, P_2NF, peroxides, $COCl_2$, $PH_3 + NH_3$, phosphorus, PCl_5, PBr_3, potassium chlorocuprate, potassium oxides (e.g., KO_3, K_2O_2, KO_2), selenium, $SeOCl_2$, $SiCl_4$, $AgIO_3$, $NaIO_3$, $NH_3 + NaNO_2$, Na_2O_2, $SnI_4 + S$, SnO_2, S, sulfuric acid, tellurium, thiophosphoryl fluoride, $VOCl_2$, water.

Other hazardous reactions may occur with carbon (e.g., soot, graphite, activated charcoal, dimethyl sulfoxide, ethylene oxide, chlorine, bromine vapor, hydrogen bromide, potassium iodide + magnesium bromide, chloride or iodide, maleic anhydride, mercury, copper(II) oxide, mercury(II) oxide, tin(IV) oxide, molybdenum(III) oxide, bismuth trioxide, phosphorus trichloride, sulfur dioxide, chromium trioxide.

When heated to decomposition it emits toxic fumes of K_2O.

PKT500 CAS:7440-09-7 ***HR: 3***
POTASSIUM (liquid alloy)
DOT: UN 1420

SYN: POTASSIUM, metal liquid alloy (DOT)

CONSENSUS REPORTS: Reported in EPA TSCA Inventory.

DOT Classification: Label: Flammable Solid and Dangerous When Wet.

SAFETY PROFILE: A very dangerous fire hazard. When heated to decomposition in air it emits toxic fumes of K_2O. See also POTASSIUM.

PKT750 CAS:127-08-2 ***HR: 2***
POTASSIUM ACETATE
mf: $C_2H_3O_2$•K mw: 98.15

PROP: White powder. Mp: 292°, d: 1.8 @ 20°/20°.

SYN: DIURETIC SALT

TOXICITY DATA with REFERENCE
orl-rat LD50:3250 mg/kg AIHAAP 30,470,69

CONSENSUS REPORTS: Reported in EPA TSCA Inventory.

SAFETY PROFILE: Moderately toxic by ingestion. When heated to decomposition it emits toxic fumes of K_2O.

PKT775 ***HR: 3***
POTASSIUM ACETYLENE-1,2-DIOXIDE
mf: $C_2K_2O_2$ mw: 134.22

SAFETY PROFILE: The yellow powder burns explosively on contact with air, halocarbons, halogens, alcohols, water, and any material with acidic hydrogen. When heated to decomposition it emits toxic fumes of K_2O. See also ACETYLENE COMPOUNDS.

PKU000 ***HR: 3***
POTASSIUM ACETYLIDE
mf: C_2HK mw: 64.13

SAFETY PROFILE: Will hydrolyze to KOH which is very caustic and an irritant. Reaction on contact with limited amounts of water evolves acetylene which may ignite and explode. Ignites on contact with chlorine. Incandescent reaction with sulfur dioxide; carbon dioxide + heat. When heated to decomposition it emits toxic fumes of K_2O. See also ACETYLIDES and POTASSIUM HYDROXIDE.

PKU250 CAS:7789-29-9 ***HR: 3***
POTASSIUM ACID FLUORIDE
DOT: NA 1811
mf: FK•FH mw: 78.11

PROP: Colorless crystals. Mp: decomp.

SYNS: BIFLUORURE de POTASSIUM (FRENCH) ◇ POTASSIUM BIFLUORIDE ◇ POTASSIUM HYDROGEN FLUORIDE

CONSENSUS REPORTS: Reported in EPA TSCA Inventory.

OSHA PEL: TWA 2.5 mg(F)/m^3
ACGIH TLV: TWA 2.5 mg(F)/m^3
NIOSH REL: TWA 2.5 mg(F)/m^3

SAFETY PROFILE: A poison by all routes. Corrosive to the eyes, skin, and mucous membranes. A very reactive, dangerous material. When heated to decomposition it emits toxic fumes of F^- and K_2O. See also FLUORIDES.

PKU500 CAS:7789-29-9 ***HR: 3***
POTASSIUM ACID FLUORIDE (solution)

SYNS: POTASSIUM BIFLUORIDE, solution (DOT) ◇ POTASSIUM HYDROGEN FLUORIDE, solution (DOT)

CONSENSUS REPORTS: Reported in EPA TSCA Inventory.

NIOSH REL: TWA 2.5 mg(F)/m^3
DOT Classification: Corrosive Material; Label: Corrosive.

SAFETY PROFILE: A poison. Very corrosive and reactive. A corrosive irritant to the eyes, skin, and mucous membranes. When heated to decomposition it emits toxic fumes of F^- and K_2O. See also FLUORIDES and HYDROFLUORIC ACID.

PKU600 CAS:868-14-4 ***HR: 1***
POTASSIUM ACID TARTRATE
mf: $C_4H_5KO_6$ mw: 188.18

KOCO•CHOHCHOHCO•OH

PROP: Colorless crystals or white crystalline powder; acid taste. Sol in water; sltly sol in alc.

SYNS: CREAM of TARTER ◇ POTASSIUM BITARTRATE

SAFETY PROFILE: Mixtures with carbon + nitrogen oxide ignite below 400°C. When heated to decomposition it emits toxic fumes of K_2O.

PKU750 ***HR: 3***
POTASSIUM AMALGAM

PROP: Silvery liquid or solid.

CONSENSUS REPORTS: Mercury and its compounds are on the Community Right-To-Know List.

SAFETY PROFILE: A poison. Flammable by spontaneous chemical reaction. Can react vigorously with oxidizing materials. Moderately explosive; liberates hydrogen upon contact with moisture, steam, acids, etc. When heated to decomposition it emits highly toxic fumes of K_2O, Hg, and PO_x. See also POTASSIUM.

PKV000 CAS:17242-52-3 ***HR: 3***
POTASSIUM AMIDE
mf: H_2KN mw: 16.02

SAFETY PROFILE: Violent or explosive reaction with water or potassium nitrite + heat. Reacts to form explosive products with tetraphenyl lead; ammonia + copper(II) nitrate. When heated to decomposition it emits very toxic fumes of NO_x, NH_3 and K_2O. See also POTASSIUM HYDROXIDE and AMMONIA which are hydrolysis products of KNH_2.

PKV500 CAS:10124-50-2 ***HR: 3***
POTASSIUM ARSENITE
DOT: UN 1678
mf: AsH_3O_3•xK mw: 399.65

PROP: White, hygroscopic powder. Sol in water.

SYNS: ARSENENOUS ACID, POTASSIUM SALT ◇ ARSENITE de POTASSIUM (FRENCH) ◇ ARSONIC ACID, POTASSIUM SALT ◇ KALIUMARSENIT (GERMAN) ◇ NSC 3060 ◇ POTASSIUM METAARSENITE

TOXICITY DATA with REFERENCE
cyt-hmn:leu 1 μmol/L/48H CNREA8 25,980,65
orl-man TDLo:214 mg/kg/15Y-C:CAR,LIV GASTAB 68,1582,75
skn-mus TDLo:576 mg/kg/12W-I:ETA BMJOAE 2,1107,22
orl-man TD:7560 mg/kg/26W-C:CAR,SKN ANSUA5 99,348,34
orl-man TD:441 mg/kg/3W-C:CAR,SKN ANSUA5 99,348,34
orl-chd TD:390 mg/kg/3Y-C:CAR,SKN AIMEAS 61,296,64
orl-hmn TDLo:74 mg/kg:LIV,SKN LANCAO 1,269,53
orl-rat LD50:14 mg/kg AFDOAQ 15,122,51
skn-rat LD50:150 mg/kg PHJOAV 185,361,60
scu-mus LDLo:16 mg/kg HBAMAK 4,1307,35
orl-dog LDLo:3 mg/kg HBAMAK 4,1307,35
scu-dog LDLo:700 μg/kg HBAMAK 4,1307,35
ivn-dog LDLo:2 mg/kg HBAMAK 4,1307,35
scu-cat LDLo:5 mg/kg HBAMAK 4,1307,35
scu-rbt LDLo:8 mg/kg HBAMAK 4,1307,35
ivn-rbt LDLo:6 mg/kg HBAMAK 4,1307,35
scu-gpg LDLo:9 mg/kg HBAMAK 4,1307,35
scu-pgn LDLo:12 mg/kg FDWU** -,-,31

CONSENSUS REPORTS: IARC Cancer Review: Human Sufficient Evidence IMEMDT 23,39,80; Animal Inadequate Evidence IMEMDT 23,39,80, IMEMDT 2,48,73. EPA Extremely Hazardous Substances List. Arsenic and its compounds are on the Community Right-To-Know List.

OSHA PEL: TWA 0.01 mg(As)/m^3; Cancer Hazard
ACGIH TLV: TWA 0.2 mg(As)/m^3
NIOSH REL: CL (Inorganic Arsenic) 0.002 mg(As)/m^3/15M
DOT Classification: Poison B; Label: Poison.

SAFETY PROFILE: Confirmed human carcinogen producing skin and liver tumors. Poison by ingestion, skin contact, subcutaneous, and intravenous routes. Human mutation data reported. Human systemic effects: dermatitis, liver changes. When heated to decomposition it emits toxic fumes of As and K_2O. Used in veterinary medicine and for chronic dermatitis in humans. See also ARSENIC COMPOUNDS.

PKV600 CAS:14007-45-5 ***HR: 2***
POTASSIUM ASPARTATE
mf: $C_4H_7NO_4 \cdot 7K$ mw: 406.82

SYNS: ASPARA K ◇ l-ASPARTIC ACID, POTASSIUM SALT (9CI) ◇ K-FLEBO ◇ POTASSIUM-l-ASPARTATE

TOXICITY DATA with REFERENCE
orl-rat LD50:7937 mg/kg NIIRDN 6,11,82
scu-rat LD50:4061 mg/kg NIIRDN 6,11,82
ivn-rat LD50:667 mg/kg NIIRDN 6,11,82

CONSENSUS REPORTS: Reported in EPA TSCA Inventory.

SAFETY PROFILE: Moderately toxic by intravenous route. Mildly toxic by ingestion. When heated to decomposition it emits toxic fumes of NO_x and K_2O.

PKW000 CAS:20762-60-1 ***HR: 2***
POTASSIUM AZIDE
mf: KN_3 mw: 81.12

SAFETY PROFILE: Hydrolysis yields KOH which is toxic and irritating. When heated it explodes weakly releasing nitrogen. Explosive reaction with sulfur dioxide at 120°C. Reacts with carbon disulfide to form an explosive product. Reacts violently with manganese dioxide when warmed. When heated to decomposition it emits very toxic fumes of NO_x and K_2O. See also AZIDES.

PKW250 CAS:67880-14-2 ***HR: 3***
POTASSIUM AZIDODISULFATE
mf: $KN_3O_6S_2$ mw: 241.25

$KOSO_2OSO_2N_3$

SAFETY PROFILE: A very unstable explosive. Explosive reaction with water. When heated to decomposition it emits very toxic fumes of SO_x, NO_x, and K_2O. See also AZIDES, SULFATES, and POTASSIUM HYDROXIDE.

PKW500 ***HR: 3***
POTASSIUM AZIDOSULFATE
mf: KN_3O_3S mw: 161.21

$KOSO_2N_3$

SAFETY PROFILE: A heat-sensitive explosive. When heated to decomposition it emits very toxic fumes of SO_x, NO_x, and K_2O. See also AZIDES, SULFATES, POTASSIUM HYDROXIDE.

PKW550 ***HR: 3***
POTASSIUM BENZENEHEXOXIDE
mf: $C_6K_6O_6$ mw: 402.65

SYN: POTASSIUM CARBONYL

SAFETY PROFILE: Explodes when heated in air or on contact with water. Reacts violently with oxygen. Reaction with moisture. Forms a very explosive product. When heated to decomposition it emits toxic fumes of K_2O. See also CARBONYLS.

PKW750 ***HR: 3***
POTASSIUM BENZENESULFONYLPEROXYSULFATE
mf: $C_6H_5KO_7S_2$ mw: 292.31

SAFETY PROFILE: A heat- and friction-sensitive explosive. When heated to decomposition it emits very toxic fumes of SO_x and K_2O. See also PEROXIDES.

PKW760 CAS:582-25-2 ***HR: 2***
POTASSIUM BENZOATE
mf: $C_7H_5O_2 \cdot K$ mw: 160.22

SAFETY PROFILE: Combustible when exposed to heat or flame. When heated to decomposition it emits acrid smoke and irritating fumes.

PKW775 ***HR: 3***
POTASSIUM-O-O-BENZOYLMONOPEROXOSULFATE
mf: $C_7H_5KO_5S$ mw: 240.27

SAFETY PROFILE: The anhyd salt is a friction-sensitive explosive. The monohydrate explodes when heated above 70°C or on contact with sulfuric acid. When heated to decomposition it emits toxic fumes of SO_x and K_2O. See also PEROXIDES and SULFATES.

PKX000 ***HR: 3***
POTASSIUM BENZOYLPEROXYSULFATE
mf: $C_7H_5KO_6S$ mw: 256.28

SAFETY PROFILE: A heat- and friction-sensitive explosive. When heated to decomposition it emits very toxic fumes of SO_x and K_2O.

PKX100 CAS:298-14-6 ***HR: 1***
POTASSIUM BICARBONATE
mf: $KHCO_3$ mw: 100.12

PROP: Colorless, transparent, monoclinic prisms or white granular powder; odorless. Sol in water; insol in alc.

SAFETY PROFILE: A nuisance dust.

PKX250 CAS:7778-50-9 ***HR: 3***
POTASSIUM BICHROMATE
DOT: UN 1479
mf: $Cr_2K_2O_7$ mw: 294.20

$K_2(OCrO_2OCrO_2O)$

PROP: Bright, yellowish-red, transparent crystals; bitter, metallic taste. Mp: 398°, bp: decomp @ 500°, d: 2.69.

SYNS: BICHROMATE OF POTASH ◇ CHROMIC ACID, DIPOTASSIUM SALT ◇ DIPOTASSIUM DICHROMATE ◇ IOPEZITE ◇ KALIUMDICHROMAT (GERMAN) ◇ POTASSIUM DICHROMATE(VI)

TOXICITY DATA with REFERENCE
dnr-bcs 1050 μg/L WATRAG 14,1613,80
dns-hmn:fbr 100 μmol/L MUREAV 117,279,83
ipr-mus TDLo:20 mg/kg (1D male):TER MUREAV 103,345,82
unr-mus TDLo:700 mg/kg (35W male):REP MUREAV 97,180,82
orl-chd LDLo:26 mg/kg ZEKIA5 81,417,58
orl-mus LD50:190 mg/kg SAIGBL 20,590,78
ipr-mus LD50:37 mg/kg CRNGDP 4,1535,83
scu-mus LDLo:100 mg/kg EQSSDX 1,1,75
orl-dog LDLo:2829 mg/kg EQSSDX 1,1,75
scu-mky LDLo:40 mg/kg AJPAA4 9,133,33
scu-rbt LDLo:10 mg/kg PSEBAA 9,13,11
ivn-rbt LDLo:27900 μg/kg EQSSDX 1,1,75
orl-gpg LDLo:163 mg/kg ZEKIA5 81,417,58
scu-gpg LDLo:29400 μg/kg EQSSDX 1,1,75

CONSENSUS REPORTS: NTP Fifth Annual Report on Carcinogens. IARC Cancer Review: Human Inadequate Evidence IMEMDT 23,205,80; Animal Inadequate Evidence IMEMDT 23,205,80. Chromium and its compounds are on the Community Right-To-Know List. Reported in EPA TSCA Inventory. EPA Genetic Toxicology Program.

OSHA PEL: CL 0.1 $mg(CrO_3)/m^3$
ACGIH TLV: TWA 0.05 $mg(CrO_3)/m^3$
NIOSH REL: TWA (Chromium(VI)) 0.025 $mg(Cr(VI))/m^3$; CL 0.05/15M
DOT Classification: ORM-A; Label: None.

SAFETY PROFILE: Confirmed carcinogen. Human poison by ingestion. An experimental poison by ingestion, intraperitoneal, intravenous, and subcutaneous routes. Human mutation data reported. An experimental teratogen. Other experimental reproductive effects. Flammable by chemical reaction. A powerful oxidizer. Explosive reaction with hydrazine. Reacts violently or ignites with H_2SO_4 + acetone; hydroxylamine; ethylene glycol (above 100°C). Forms pyrotechnic mixtures with boron + silicon; iron (ignites 1090°C); tungsten (ignites at 1700°C). Reacts with sulfuric acid to form the strong oxidant chromic acid. Used in photomechanical processing, chrome pigment production and wool preservation methods. When heated to decomposition it emits toxic fumes of K_2O. See also CHROMIUM COMPOUNDS.

PKX500 CAS:23746-34-1 ***HR: 3***
POTASSIUM BIS(2-HYDROXYETHYL) DITHIOCARBAMATE
mf: $C_5H_{10}NO_2S_2 \cdot K$ mw: 219.38

SYNS: BIS(2-HYDROXYETHYL)CARBAMODITHIOIC ACID, MONOPOTASSIUM SALT ◇ BIS(2-HYDROXYETHYL)DITHIOCARBAMIC ACID, MONOPOTASSIUM SALT ◇ BIS(2-HYDROXYETHYL) DITHOCARBAMIC ACID, POTASSIUM SALT

TOXICITY DATA with REFERENCE
orl-rat TDLo:82 g/kg/78W-C:ETA JJIND8 67,75,81
orl-mus TDLo:129 g/kg/79W-C:CAR JNCIAM 42,1101,69

CONSENSUS REPORTS: IARC Cancer Review: Group 3 IMEMDT 7,56,87; Animal Sufficient Evidence IMEMDT 12,183,76. Reported in EPA TSCA Inventory.

SAFETY PROFILE: Questionable carcinogen with experimental carcinogenic and tumorigenic data. When heated to decomposition it emits very toxic fumes of K_2O, SO_x, and NO_x. Used as an analytical reagent for quantitative determination of mercury, gold, and copper. See also CARBAMATES.

PKX639 ***HR: 3***
POTASSIUM BIS(PROPYNYL)PALLADATE
mf: $C_6H_6K_2Pd$ mw: 262.78

$K_2[Pd(C{\equiv}CCH_3)_2]$

SAFETY PROFILE: Ignites spontaneously in air. Explodes on contact with water. When heated to decomp it emits toxic fumes of K_2O. See also ACETYLENE COMPOUNDS and PALLADIUM.

PKX700 ***HR: 3***
POTASSIUM BIS(PROPYNYL)PLATINATE
mf: $C_6H_6K_2Pt$ mw: 351.39

$K_2[Pt(C{\equiv}CCH_3)_2]$

SAFETY PROFILE: Ignites spontaneously in air. Explodes on contact with water. When heated to decomposition it emits toxic fumes of K_2O. See also ACETYLENE COMPOUNDS and PLATINUM COMPOUNDS.

PKX750 CAS:7646-93-7 ***HR: 2***
POTASSIUM BISULFATE
DOT: UN 2509
mf: $HO_4S \cdot K$ mw: 136.17

PROP: White, deliquescent crystals. D: 2.24; mp: 197°. Sol in water.

SYNS: ACID POTASSIUM SULFATE ◇ MONOPOTASSIUM SULFATE ◇ POTASSIUM ACID SULFATE ◇ POTASSIUM BISULPHATE ◇ POTASSIUM HYDROGEN SULFATE, solid (DOT) ◇ SAL ENIXUM ◇ SULFURIC ACID, MONOPOTASSIUM SALT

TOXICITY DATA with REFERENCE
orl-rat LD50:2340 mg/kg MarJV# 29MAR77

CONSENSUS REPORTS: Reported in EPA TSCA Inventory.

DOT Classification: ORM-B; Label: None; Corrosive Material; Label: Corrosive.

SAFETY PROFILE: Moderately toxic by ingestion. A corrosive irritant to the skin, eyes, and mucous membranes. When heated to decomposition it emits toxic fumes of SO_x and K_2O. Can form an explosive mixture. See also SULFATES.

PKY000 CAS:14075-53-7 ***HR: 3***
POTASSIUM BOROFLUORIDE
mf: BF_4•K mw: 125.91

PROP: Rhombic or cubic, colorless crystals. Mp: 530°, d: 2.498.

SYNS: AVOGODRITE ◇ POTASSIUM FLUOBORATE ◇ POTASSIUM FLUOROBORATE ◇ TETRAFLUOROBORATE-(1−) POTASSIUM

TOXICITY DATA with REFERENCE
ipr-rat LD50:240 mg/kg 14KTAK -,693,64
ipr-mus LD50:590 mg/kg 14KTAK -,693,64
ipr-rbt LD50:380 mg/kg 14KTAK -,693,64

CONSENSUS REPORTS: Reported in EPA TSCA Inventory.

OSHA PEL: TWA 2.5 mg(F)/m^3
NIOSH REL: TWA (Inorganic Fluorides) 2.5 mg(F)/m^3

SAFETY PROFILE: Poison by intraperitoneal route. When heated to decomposition it emits very toxic fumes of F^-, K_2O, and BO_x. Used in sand casting of aluminum and magnesium, grinding, and in resinoid grinding wheels. See also FLUORIDES and BORON COMPOUNDS.

PKY250 CAS:13762-51-1 ***HR: 3***
POTASSIUM BOROHYDRATE
DOT: UN 1870
mf: BH_3•K mw: 52.94

PROP: White, water-sol crystals. D: 1.177, mp: >400° (decomp).

SYNS: BOROHYDRURE de POTASSIUM (FRENCH) ◇ POTASSIUM BOROHYDRIDE (DOT) ◇ TETRAHYDROBORATE(1−) POTASSIUM

TOXICITY DATA with REFERENCE
orl-rat LDLo:160 mg/kg 14KTAK -,693,64

CONSENSUS REPORTS: Reported in EPA TSCA Inventory.

DOT Classification: Flammable Solid; Label: Dangerous When Wet.

SAFETY PROFILE: Poison by ingestion. Burns quietly in air. When heated to decomposition it emits toxic fumes of K_2O. See also BORON COMPOUNDS and HYDRIDES.

PKY300 CAS:7758-01-2 ***HR: 3***
POTASSIUM BROMATE
DOT: UN 1484
mf: BrO_3•K mw: 167.01

PROP: White crystals or crystalline powder. Mp: 350° (approx), decomp @ 370°, d: 3.27 @ 17.5°. Sol in water; sltly sol in alc.

SYN: BROMIC ACID, POTASSIUM SALT

TOXICITY DATA with REFERENCE
mma-sat 1 mg/plate AMONDS 3,253,80
cyt-rat-ipr 500 μmol/kg MUREAV 147,274,85
cyt-rat-orl 3 mmol/kg MUREAV 147,274,85
cyt-ham:lng 85 mg/L GMCRDC 27,95,81
orl-rat TDLo:4200 mg/kg/13W-C:CAR JJCREP 78,358,87
orl-rat TDLo:38500 mg/kg/2Y-C:CAR GANNA2 73,335,82
orl-rat LD50:321 mg/kg ESKHA5 100,93,82

CONSENSUS REPORTS: IARC Cancer Review: Group 2B IMEMDT 7,56,87; Animal Sufficient Evidence IMEMDT 40,207,86. Reported in EPA TSCA Inventory.

DOT Classification: Oxidizer; Label: Oxidizer.

SAFETY PROFILE: Suspected carcinogen with experimental carcinogenic data. A poison by ingestion. A powerful oxidizer. An irritant to skin, eyes, and mucous membranes. Mutation data reported. Mixtures with sulfur may ignite. Violent reaction with Al; Al + dinitrotoluene @ 290°; As; C; Cu; $Pb(C_2H_3O_2)_2$; metal sulfides; organic matter; P; S. Aqueous solutions react violently with selenium. When heated to decomposition it emits very toxic fumes of Br^- and K_2O. See also BROMIDES.

PKY500 CAS:7758-02-3 ***HR: 1***
POTASSIUM BROMIDE
mf: BrK mw: 119.01

PROP: Colorless, cubic, sltly hygroscopic crystals. Mp: 730°, bp: 1380°, d: 2.75 @ 25°, vap press: 1 mm @ 795°.

SYN: BROMIDE SALT OF POTASSIUM

TOXICITY DATA with REFERENCE
hma-rat/ast 200 mg/kg GANNA2 54,155,63

CONSENSUS REPORTS: Reported in EPA TSCA Inventory.

SAFETY PROFILE: Large doses can cause central nervous system depression. Prolonged inhalation may cause skin eruptions. Mutation data reported. Violent reaction with BrF_3. When heated to decomposition it emits toxic fumes of K_2O and Br^-. See also BROMIDES.

PKY750 CAS:865-47-4 ***HR: 3***
POTASSIUM-tert-BUTOXIDE
mf: C_4H_9KO mw: 112.20

SAFETY PROFILE: Probably very toxic and irritating to skin, eyes, and mucous membranes. A powerful very reactive base. Ignites on contact with acids or reactive solvents (e.g., acetone; butanone; butyl acetate; acetic acid; ethanol; propanol; isopropanol; methanol; $CHCl_3$; carbon tetrachloride; 1-chloro-2,3-epoxypropane; chloroform; 1,2-dichloromethane; diethyl sulfate; dimethyl carbonate; epichlorohydrin; ethyl acetate; ethyl methyl ketone; sulfuric acid; isopropanol; 4-methyl-2-butanone; methyl isobutyl ketone; n-butyl acetate; n-propyl formate; propanol; propyl formate). Ignites when heated in air. When heated to decomposition it emits toxic fumes of K_2O.

PKY850 CAS:871-58-9 ***HR: 3***
POTASSIUM BUTYLXANTHATE
mf: $C_5H_9OS_2 \cdot K$ mw: 188.36

SYNS: BUTYL POTASSIUM XANTHATE ◇ BUTYL-XANTHIC ACID POTASSIUM SALT ◇ DITHIOCARBONIC ACID-o-BUTYL ESTER POTASSIUM SALT ◇ POTASSIUM-o-BUTYL XANTHATE ◇ POTASSIUM BUTYLXANTHOGENATE ◇ POTASSIUM XANTHOGENATE BUTYL ETHER

TOXICITY DATA with REFERENCE
orl-rat LD50:456 mg/kg 85GMAT -,102,82
ihl-rat LC50:7690 mg/m^3/2H 85GMAT -,102,82
ivn-mus LD50:158 mg/kg AIPTAK 135,330,62

SAFETY PROFILE: Poison by intravenous route. Moderately toxic by ingestion. Mildly toxic by inhalation. When heated to decomposition it emits toxic fumes of SO_x and K_2O. See also ESTERS.

PKZ000 CAS:2181-04-6 ***HR: 3***
POTASSIUM CANRENOATE
mf: $C_{22}H_{30}O_4 \cdot K$ mw: 397.62

SYNS: POTASSIUM-3-(17-β-HYDROXY-3-OXOANDROSTA-4,6-DIEN-17-YL)PROPIONATE ◇ POTASSIUM-17-HYDROXY-3-OXO-17-α-PREGNA-4,6-DIENE-21-CARBOXYLATE ◇ POTASSIUM-3-(3-OXO-17-β-HYDROXY-4,6-ANDROSTADIEN-17-α-YL)PROPANOATE

TOXICITY DATA with REFERENCE
ipr-mus TDLo:560 mg/kg (female 7-13D post):TER NICHAS 36,261,77
ipr-mus TDLo:560 mg/kg (female 7-13D post):REP NICHAS 36,261,77
orl-rat LD50:650 mg/kg NICHAS 36,7,77
ipr-rat LD50:183 mg/kg IYKEDH 11,811,80
scu-rat LD50:160 mg/kg IYKEDH 11,811,80
ivn-rat LD50:112 mg/kg IYKEDH 11,811,80
orl-mus LD50:740 mg/kg NICHAS 36,7,77
ipr-mus LD50:140 mg/kg AIPTAK 149,8,64
scu-mus LD50:165 mg/kg IYKEDH 11,811,80
ivn-mus LD50:125 mg/kg IYKEDH 11,811,80

SAFETY PROFILE: Poison by intravenous, subcutaneous, and intraperitoneal routes. Moderately toxic by ingestion. An experimental teratogen. Experimental reproductive effects. A steroid. When heated to decomposition it emits acrid smoke and irritating fumes of K_2O.

PLA000 CAS:584-08-7 ***HR: 3***
POTASSIUM CARBONATE (2:1)
mf: $CO_3 \cdot 2K$ mw: 138.21

PROP: White, deliquescent, granular, translucent powder; odorless with alkaline taste. D: 2.428 @ 19°, mp: 891°, bp: decomposes. Sol in water; insol in alc.

SYNS: CARBONIC ACID, DIPOTASSIUM SALT ◇ KALIUM-CARBONAT (GERMAN) ◇ K-GRAN ◇ PEARL ASH ◇ POTASH

TOXICITY DATA with REFERENCE
orl-rat LD50:1870 mg/kg AIHAAP 30,470,69
orl-bwd LD50:100 mg/kg AECTCV 12,355,83

CONSENSUS REPORTS: Reported in EPA TSCA Inventory.

SAFETY PROFILE: Poison by ingestion. A strong caustic. Incompatible with KCO; chlorine trifluoride; magnesium. When heated to decomposition it emits toxic fumes of K_2O.

PLA250 CAS:3811-04-9 ***HR: 3***
POTASSIUM CHLORATE
DOT: UN 1485/UN 2427
mf: $ClO_3 \cdot K$ mw: 122.55

PROP: Transparent, colorless crystals or white powder; cooling, saline taste. Mp: 368.4°, bp: decomp @ 400°, d: 2.32.

SYNS: BERTHOLLET SALT ◇ CHLORATE de POTASSIUM (FRENCH) ◇ CHLORATE OF POTASH (DOT) ◇ FEKABIT ◇ KALIUMCHLORAAT (DUTCH) ◇ KALIUMCHLORAT (GERMAN) ◇ OXYMURIATE OF POTASH ◇ PEARL ASH ◇ POTASH CHLORATE (DOT) ◇ POTASSIO (CHLORATO di) (ITALIAN) ◇ POTASSIUM CHLORATE (DOT) ◇ POTASSIUM (CHLORATE de) (FRENCH) ◇ POTASSIUM OXYMURIATE ◇ POTCRATE ◇ SALT OF TARTER

TOXICITY DATA with REFERENCE
unr-hmn LDLo:429 mg/kg AEXPBL 21,169,1886
orl-rat LDLo:1870 mg/kg AIHAAP 30,470,69
ipr-rat LDLo:1500 mg/kg JPETAB 35,1,29
ipr-gpg LDLo:1800 mg/kg JPETAB 35,1,29
orl-dog LDLo:1200 mg/kg HBTXAC 1,242,56
orl-rbt LDLo:2000 mg/kg AEXPBL 21,169,1886

CONSENSUS REPORTS: Reported in EPA TSCA Inventory.

DOT Classification: Oxidizer; Label: Oxidizer.

SAFETY PROFILE: Moderately toxic to humans by an unspecified route. Moderately toxic experimentally by ingestion and intraperitoneal routes. A gastrointestinal tract and kidney irritant. Can cause hemolysis of red blood cells and methemoglobinemia. Toxic dose to a human is about 5 grams.

A powerful oxidizer and very reactive material. It has been the cause of many industrial explosions. May explode on heating. Explosive reactions with ammonium chloride, aqua regia + rutheniun, sulfur dioxide solutions in ether or ethanol. Reacts with fluorine to form the explosive gas fluorine perchlorate.

Forms sensitive explosive mixtures with agricultural materials (e.g., peat, powdered sulfur, sawdust, thiuram), aluminum + antimony trisulfide powders, arsenic trisulfide, carbon, charcoal + potassium nitrate + sulfur, charcoal + sulfur, cyanides, cyanoguanidine, hydrocarbons, manganese dioxide + traces of organic matter, manganese dioxide + potassium hydroxide, metal + wood, metal phosphides (e.g., tricopper diphosphide, trimercury tetraphosphide), metal phosphinates (e.g., barium phosphinate), finely divided metals (e.g., aluminum, copper, magnesium, zinc, germanium, titanium, zirconium, steel, chromium), metal phosphides (e.g., tricopper diphosphide, trimercury tetraphosphide), metal sulfides (e.g., antimony trisulfide, silver sulfide), metal thiocyanates (e.g., ammonium thiocyanate, barium thiocyanate), nitric acid + organic materials, powdered non-metals (e.g., arsenic, carbon, phosphorus, sulfur, boron), reducing agents (e.g., calcium hydride, strontium hydride, sodium phosphinate, calcium phosphinate, barium phosphinate), sugars (e.g., glucose), sulfur, sulfur + metal derivatives (e.g., cobalt, cobalt oxide, copper nitride, copper sulfate, copper chlorate), sulfuric acid, sodium amide, tannic acid.

Violent reaction or ignition with NH_3, NH_4Cl, NH_4^- salts, ammonium sulfate, Sb_2S_3, As, barium hypophosphite, BaS, calcium hypophosphite, CaS, charcoal, Cu_3P_2, fabrics, gallic acid, hydrogen iodide, lactose, (Mg + $CuSO_4$ (anhydrous) + NH_4NO_3 + H_2O), MnO_2, dinickel trioxide, dibasic organic acids, organic matter, $NaNH_2$, sugar + sulfuric acid, sucrose, SO_2, H_2SO_4, thiocyanates, thorium dicarbide, sodium amide, fabrics, KOH, metal hypophosphites.

When heated to decomposition it emits very toxic fumes of Cl^- and K_2O. Used in the manufacture of soap, glass, and pottery. See also CHLORATES.

PLA500 CAS:7447-40-7 ***HR: 3***
POTASSIUM CHLORIDE
mf: ClK mw: 74.55

PROP: Colorless or white crystals or powder; odorless with salty taste. D: 1.987, mp: 773° (subl @ 1500°). Sol in water; sltly sol in alc; insol in abs alc.

SYNS: CHLORID DRASELNY (CZECH) ◇ CHLOROPOTASSURIL ◇ DIPOTASSIUM DICHLORIDE ◇ EMPLETS POTASSIUM CHLORIDE ◇ ENSEAL ◇ KALITABS ◇ KAOCHLOR ◇ KAON-Cl ◇ KAY CIEL ◇ K-LOR ◇ KLOTRIX ◇ K-PRENDE-DOME ◇ PFIKLOR ◇ POTASSIUM MONOCHLORIDE ◇ POTAVESCENT ◇ REKAWAN ◇ SLOW-K ◇ TRIPOTASSIUM TRICHLORIDE

TOXICITY DATA with REFERENCE
eye-rbt 500 mg/24H MLD 28ZPAK -,8,72
mrc-smc 400 mmol/L MUTAEX 1,21,86
cyt-ham:lng 12 g/L FCTOD7 22,501,84
orl-wmn TDLo:60 mg/kg/D:GIT,BLD LANCAO 2,919,80
orl-inf LDLo:938 mg/kg/2D JAMAAP 240,1339,78
orl-man LDLo:20 mg/kg:CVS,GIT,BLD LANCAO 2,919,80
orl-rat LD50:2600 mg/kg 28ZPAK -,8,72
ipr-rat LD50:660 mg/kg FCTXAV 3,597,65
ivn-rat LD50:39 mg/kg ARZNAD 14,1128,64
orl-mus LD50:383 mg/kg ARZNAD 14,1128,64
ipr-mus LD50:1181 mg/kg COREAF 256,1043,63
ivn-mus LD50:117 mg/kg EJTXAZ 8,188,75
ipr-dog LDLo:85 mg/kg AVERAG 44,555,37
orl-gpg LD50:2500 mg/kg JPETAB 35,1,29
ipr-gpg LDLo:900 mg/kg JPETAB 35,1,29

CONSENSUS REPORTS: Reported in EPA TSCA Inventory

SAFETY PROFILE: A human poison by ingestion. Poison experimentally by ingestion, intravenous, and intraperitoneal routes. Human systemic effects by ingestion: nausea, blood clotting changes, cardiac arrhythmias. An eye irritant. Mutation data reported. Explosive reaction with BrF_3; sulfuric acid + potassium permanganate. When heated to decomposition it emits toxic fumes of K_2O and Cl^-.

PLA525 CAS:14314-27-3 ***HR: 3***
POTASSIUM CHLORITE
mf: $ClKO_2$ mw: 106.55

SAFETY PROFILE: Violent reaction with sulfur. When heated to decomposition it emits toxic fumes of Cl^- and K_2O. See also CHLORITES.

PLA750 CAS:16919-73-6 ***HR: 1***
POTASSIUM CHLOROPALLADATE
mf: Cl_7K_2Pd mw: 397.30

PROP: Cubic, red crystals. Mp: (decomp), d: 2.738.

TOXICITY DATA with REFERENCE
skn-rbt 100 mg/24H MLD AEHLAU 30,168,75

CONSENSUS REPORTS: Reported in EPA TSCA Inventory.

SAFETY PROFILE: A skin irritant. When heated to decomposition it emits toxic fumes of K_2O and Cl^-. See also PALLADIUM.

PLB250 CAS:7789-00-6 ***HR: 3***
POTASSIUM CHROMATE(VI)
mf: CrO_4•2K mw: 194.20

PROP: Rhombic, yellow crystals. Mp: 975°, d: 2.73 @ 18°. Sol in water; insol in alc.

SYNS: BIPOTASSIUM CHROMATE ◇ CHROMATE OF POTASSIUM ◇ DIPOTASSIUM CHROMATE ◇ DIPOTASSIUM MONOCHROMATE ◇ NEUTRAL POTASSIUM CHROMATE ◇ TARAPACAITE

TOXICITY DATA with REFERENCE
dnr-ssp 60 nmol/L CNJGA8 24,771,82
dnd-hmn:lng 25 μmol/L CBINA8 36,345,81
ipr-mus TDLo:30 mg/kg (8-10D preg):TER APTOA6 47,66,80
ipr-mus TDLo:60 mg/kg (8-10D preg):REP APTOA6 47,66,80
orl-mus TDLo:1600 mg/kg/62W-C:ETA JONUAI 101,1431,71
scu-dog LDLo:19 mg/kg SMSJAR 26,131,1826
ivn-dog LDLo:2900 μg/kg EQSSDX 1,1,75
scu-rbt LDLo:12 mg/kg EQSSDX 1,1,75
ims-rbt LD50:11 mg/kg JPETAB 87,119,46
scu-gpg LDLo:60 mg/kg EQSSDX 1,1,75
ivn-mam LDLo:259 mg/kg SMSJAR 26,131,1826

CONSENSUS REPORTS: NTP Fifth Annual Report on Carcinogens. IARC Cancer Review: Human Inadequate Evidence IMEMDT 23,205,80; Animal Inadequate Evidence IMEMDT 23,205,80. Reported in EPA TSCA Inventory. EPA Genetic Toxicology Program. Chromium and its compounds are on the Community Right-To-Know List.

OSHA PEL: CL 0.1 mg(CrO_3)/m^3
ACGIH TLV: TWA 0.05 mg(Cr)/m^3
DOT Classification: ORM-E; Label: None.

SAFETY PROFILE: Confirmed carcinogen with experimental tumorigenic data. Poison by intravenous, subcutaneous, and intramuscular routes. An experimental teratogen. Other experimental reproductive effects. Human mutation data reported. A powerful oxidizer. When heated to decomposition it emits toxic fumes of K_2O. Used as a mordant for wool, in the oxidizing and treatment of dyes on materials. See also CHROMIUM COMPOUNDS.

PLB500 CAS:10141-00-1 ***HR: 3***
POTASSIUM CHROMIC SULFATE
mf: Cr•$2H_2O_4S$•K mw: 287.26

SYNS: CHROME ALUM ◇ CHROME POTASH ALUM ◇ CHROMIC POTASSIUM SULFATE ◇ CHROMIC POTASSIUM SULPHATE ◇ CHROMIUM POTASSIUM SULFATE (1:1:2) ◇ CHROMIUM POTASSIUM SULPHATE ◇ CRYSTAL CHROME ALUM ◇ POTASSIUM CHROMIC SULPHATE ◇ POTASSIUM CHROMIUM ALUM ◇ POTASSIUM DISULPHATOCHROMATE(III) ◇ SULFURIC ACID, CHROMIUM (3+) POTASSIUM SALT (2:1:1)

TOXICITY DATA with REFERENCE
cyt-ham:ovr 1 mg/L CRNGDP 3,1331,82
sce-ham:ovr 1 mg/L CRNGDP 3,1331,82

CONSENSUS REPORTS: Chromium and its compounds are on the Community Right-To-Know List. Reported in EPA TSCA Inventory. EPA Genetic Toxicology Program.

OSHA PEL: TWA 0.5 mg(Cr)/m^3
ACGIH TLV: TWA 0.5 mg(Cr)/m^3

SAFETY PROFILE: Many chromates are carcinogens. Mutation data reported. When heated to decomposition it emits toxic fumes of K_2O. See also CHROMIUM COMPOUNDS and SULFATES.

PLB750 CAS:866-84-2 ***HR: 3***
POTASSIUM CITRATE
mf: $C_6H_5O_7$•3K mw: 306.41

PROP: Colorless transparent crystals or white powder; odorless with salty taste. D: 1.98, decomp when heated to 230°. Deliq; sol in water and glycerol; almost insol in alc.

SYNS: CITRIC ACID, TRIPOTASSIUM SALT ◇ TRIPOTASSIUM CITRATE MONOHYDRATE

TOXICITY DATA with REFERENCE
ivn-dog LD50:167 mg/kg AVERAG 44,555,37

CONSENSUS REPORTS: Reported in EPA TSCA Inventory.

SAFETY PROFILE: Poison by intravenous route. When heated to decomposition it emits toxic fumes of K_2O.

PLB759 ***HR: 3***
POTASSIUM CITRATE TRI(HYDROGEN PEROXIDATE)
mf: $C_6H_5O_7$•3K•$3H_2O_2$ mw: 408.45

KOCO•COH(CH_2CO•OK$)_2$•$3H_2O_2$

SAFETY PROFILE: A touch-sensitive explosive. When heated to decomposition it emits toxic fumes of K_2O. See also POTASSIUM CITRATE and PEROXIDES.

PLB775 CAS:61177-45-5 ***HR: 2***
POTASSIUM CLAVULANATE
mf: $C_8H_9NO_5$•K mw: 238.28

SYN: BRL 14151K

TOXICITY DATA with REFERENCE
orl-rat TDLo:100 mg/kg (female 6-15D post):REP NKRZAZ 31(Suppl 2),238,83
ivn-rat TDLo:1650 mg/kg (female 7-17D post):TER NKRZAZ 34(Suppl 4),69,86
orl-rat LD50:7936 mg/kg NKRZAZ 31(Suppl 2),113,83
ipr-rat LD50:1399 mg/kg NKRZAZ 31(Suppl 2),113,83
scu-rat LD50:1398 mg/kg NKRZAZ 31(Suppl 2),113,83
orl-mus LD50:4526 mg/kg NKRZAZ 31(Suppl 2),113,83
ipr-mus LD50:1531 mg/kg NKRZAZ 31(Suppl 2),113,83
scu-mus LD50:2185 mg/kg NKRZAZ 31(Suppl 2),113,83

SAFETY PROFILE: Moderately toxic by intraperitoneal and subcutaneous routes. Mildly toxic by ingestion. An experimental teratogen. Experimental reproductive effects. When heated to decomposition it emits toxic fumes of Na_2O and K_2O.

PLC100 ***HR: D***
POTASSIUM COMPOUNDS

SAFETY PROFILE: Potassium is not toxic. Any toxicity of the salts is due to the anion. The following salts are known to be hazardous: arsenate, arsenite, bichromate, chromate, cyanide, hydroxide, permanganate. When heated to decomposition they emit toxic fumes of K_2O.

PLC175 CAS:13682-73-0 ***HR: 3***
POTASSIUM CUPROCYANIDE
DOT: UN 1679
mf: C_2CuN_2•K mw: 154.68

PROP: Monoclinic crystals. D: 2.38. Insol in water; sol in DMSO.

SYNS: COPPER(I) POTASSIUM CYANIDE ◇ CUPRATE(1-), DICYANO-, POTASSIUM ◇ CUPROUS POTASSIUM CYANIDE ◇ POTASSIUM COPPER(I) CYANIDE ◇ POTASSIUM DICYANOCUPRATE(1-)

CONSENSUS REPORTS: Reported in EPA TSCA Inventory.

ACGIH TLV: TWA 1 mg(Cu)/m^3
DOT Classification: Poison B; Label: Poison

SAFETY PROFILE: A poison. When heated to decomposition it emits toxic fumes of CN^-.

PLC250 CAS:590-28-3 ***HR: 3***
POTASSIUM CYANATE
mf: CNO•K mw: 81.12

PROP: Colorless crystals. Mp: 700-900° (decomp), d: 2.056 @ 20°. Sol in water; very sltly sol in alc.

SYNS: AERO CYANATE ◇ ALICYANATE ◇ BONIDE KRAB CRABGRASS KILLER ◇ BULPUR ◇ CYANIC ACID, POTASSIUM SALT ◇ DED-WEED CRABGRASS KILLER ◇ D & P DOUBLE O CRABGRASS KILLER ◇ DUPONT PC CRABGRASS KILLER ◇ GREEN CROSS CRABGRASS KILLER ◇ KALIUMCYANAT (GERMAN) ◇ MILLER P.C. WEEDKILLER ◇ P.C. 80 CRABGRASS KILLER ◇ POTASSIUM ISOCYANATE ◇ WEEDANOL CYANOL ◇ WEEDONE CRAB GRASS KILLER

TOXICITY DATA with REFERENCE
orl-mus LD50:841 mg/kg PCOC** -,925,66
ipr-mus LD50:320 mg/kg MEIEDD 10,1100,83

CONSENSUS REPORTS: Reported in EPA TSCA Inventory. EPA Genetic Toxicology Program. Cyanide and its compounds are on the Community Right-To-Know List.

SAFETY PROFILE: Poison by intraperitoneal route. Moderately toxic by ingestion. Causes irritation of the gastrointestinal tract. An herbicide. It is said to be slowly metabolized in the body to cyanide but does not have high toxicity of cyanides. When heated to decomposition it emits very toxic fumes of CN^- and K_2O.

PLC500 CAS:151-50-8 ***HR: 3***
POTASSIUM CYANIDE
DOT: UN 1680
mf: CN•K mw: 65.12

PROP: Colorless water soln. Slt odor of bitter almonds.

SYNS: CYANIDE of POTASSIUM ◇ CYANURE de POTASSIUM (FRENCH) ◇ HYDROCYANIC ACID, POTASSIUM SALT ◇ KALIUMCYANID (GERMAN) ◇ POTASSIUM CYANIDE, solution (DOT) ◇ RCRA WASTE NUMBER P098

TOXICITY DATA with REFERENCE
dni-mus:lym 1 mmol/L NEOLA4 28,423,81
cyt-mus:mmr 1 mmol/L/48H MUREAV 67,221,79
orl-dom TDLo:1767 mg/kg (8-20W preg/44D post):REP RJARAV 16,109,78
ipr-rat TDLo:45 mg/kg (1-15D preg):TER TJADAB 25(2),84A,82
orl-hmn LDLo:2857 μg/kg 34ZIAG -,191,69
orl-man TDLo:14 mg/kg NEURAI 35,921,85
orl-rat LD50:5 mg/kg ARTODN 54,275,83
ipr-rat LD50:4 mg/kg JAPYAA 32,315,72
scu-rat LD50:9 mg/kg AEPPAE 243,254,62
ims-rat LDLo:8 mg/kg 27ZIAQ -,209,73
orl-mus LD50:8500 μg/kg JPETAB 161,163,68
ipr-mus LD50:5991 μg/kg PCBPBS 2,95,72
scu-mus LD50:6500 μg/kg NYKZAU 54,1057,58
ivn-mus LD50:2600 μg/kg JJPAAZ 3,99,54

scu-dog LD50:6 mg/kg ATXKA8 21,89,65
ivn-dog LDLo:5 mg/kg 27ZIAQ -,209,73
orl-rbt LD50:5 mg/kg 27ZIAQ -,209,73
scu-rbt LD50:4 mg/kg JJPAAZ 3,99,54
ims-rbt LD50:3256 μg/kg JACTDZ 1(3),120,82
ocu-rbt LD50:7870 μg/kg JTOTDO 2,119,83
ipr-gpg LDLo:8 mg/kg CRSBAW 96,202,27
ims-pgn LD50:4 mg/kg JJPAAZ 3,99,54

CONSENSUS REPORTS: Cyanide and its compounds are on the Community Right-To-Know List. Reported in EPA TSCA Inventory. EPA Genetic Toxicology Program.

OSHA PEL: TWA 5 mg(CN)/m^3
ACGIH TLV: TWA 5 mg(CN)/m^3 (skin)
DFG MAK: 5 mg(CN)/m^3
NIOSH REL: CL (Cyanide) 5 mg(CN)/m^3/10M
DOT Classification: Poison B; Label: Poison, solid and solution.

SAFETY PROFILE: A deadly human poison by ingestion. A experimental poison by ocular, subcutaneous, intravenous, intramuscular, and intraperitoneal routes. Experimental teratogenic and reproductive effects. Mutation data reported. Reacts with acids or acid fumes to liberate deadly HCN. When heated to decomposition it emits very toxic fumes of K_2O, CN^-, and NO_x. See also CYANIDE.

PLC750 CAS:151-50-8 *HR: 3*
POTASSIUM CYANIDE (solid)
mf: CN•K mw: 65.12

PROP: White, deliq crystals; faint odor of bitter almonds. Mp: 622.5°, bp: 1625°, d: 1.52 @ 16°. Sol in water, glycerol; sltly sol in alc.

TOXICITY DATA with REFERENCE
unr-man LDLo:2941 μg/kg 85DCAI 2,73,70

CONSENSUS REPORTS: Cyanide and its compounds are on the Community Right-To-Know List. EPA Extremely Hazardous Substances List. Reported in EPA TSCA Inventory.

DOT Classification: Poison B; Label: Poison.

SAFETY PROFILE: A deadly human poison by an unspecified route. Ingestion, inhalation, or absorption through injured skin may cause poisoning. Strong solutions are corrosive to skin, eyes, and mucous membranes. Reacts with acids and acid fumes to liberate deadly HCN. Explosive reaction with nitrogen trichloride; sodium nitrite + heat; perchoryl fluoride (at 100-300°C); mercury(II) nitrate (if heated in a closed container). Incompatible with iodine. When heated to decomposition it emits very toxic fumes of K_2O, CN^-, and NO_x.

PLC775 *HR: 3*
POTASSIUM CYANIDE-POTASSIUM NITRITE
mf: $CKN \cdot KNO_2$ mw: 150.22

CONSENSUS REPORTS: Cyanide and its compounds are on the Community Right-To-Know List.

SAFETY PROFILE: Most cyanide compounds are poisons. Nitrites may be carcinogens. An explosive salt. When heated to decomposition it emits toxic fumes of K_2O, NO_x, and CN^-. See also CYANIDE and NITRITES.

PLC780 *HR: 3*
POTASSIUM CYCLOHEXANEHEXONE-1,3,5-TRIOXIMATE
mf: $C_6K_3N_3O_6$ mw: 327.38

SAFETY PROFILE: Explodes when heated above 130°C or on contact with sulfuric or nitric acids. The lead salt is a heat-sensitive explosive. When heated to decomposition it emits toxic fumes of NO_x and K_2O.

PLC800 CAS:30994-24-2 *HR: 3*
POTASSIUM CYCLOPENTADIENIDE
mf: C_5H_5K mw: 104.19

CH=CHCH=CHCHK (ring)

SAFETY PROFILE: The dry powder ignites spontaneously in air. When heated to decomposition it emits toxic fumes of K_2O.

PLD000 CAS:2244-21-5 *HR: 2*
POTASSIUM DICHLOROISOCYANURATE
DOT: NA 2465
mf: $C_3HCl_2N_3O_3 \cdot K$ mw: 237.07

PROP: White, sltly hygroscopic, crystalline powder or granules; chlorine odor. Mp: 250° (decomp).

SYNS: DICHLOROISOCYANURIC ACID POTASSIUM SALT ◇ DICHLORO-s-TRIAZINE-2,4,6(1H,3H,5H)-TRIONE POTASSIUM DERIV ◇ DICHLOR-s-TRIAZIN-2,4,6(1H,3H,5H)TRIONE POTASSIUM ◇ POTASSIUM DICHLORO-s-TRIAZINETRIONE

TOXICITY DATA with REFERENCE
skn-rbt 500 mg/24H MLD MONS** -,-,72
eye-rbt 10 mg/24H rns SEV MONS** -,-,72
orl-rat LD50:1215 mg/kg MONS**

CONSENSUS REPORTS: Reported in EPA TSCA Inventory.

DOT Classification: Label: Oxidizer.

SAFETY PROFILE: Moderately toxic by ingestion. A skin and severe eye irritant. Causes emaciation, weakness, lethargy, diarrhea, weight loss. Autopsy indicates gastrointestinal tract irritation, tissue edema, liver and kidney congestion. A powerful oxidizer. When heated to decomposition it emits very toxic fumes of K_2O, Cl^-, and NO_x.

PLD100 ***HR: 3***
POTASSIUM DIETHYNYLPALLADATE(2⁻)
mf: $C_4H_2K_2Pd$ mw: 234.68

$K_2[Pd(C{\equiv}CH)_2]$

SAFETY PROFILE: Ignites spontaneously in air. Explodes on contact with water. When heated to decomposition it emits toxic fumes of K_2O. See also ACETYLENE COMPOUNDS and PALLADIUM.

PLD150 ***HR: 3***
POTASSIUM DIETHYNYLPLATINATE(2⁻)
mf: $C_4H_2K_2Pt$ mw: 323.34

$K_2[Pt(C{\equiv}CH)_2]$

SAFETY PROFILE: Ignites spontaneously in air. Explodes on contact with water. When heated to decomposition it emits toxic fumes of K_2O. See also ACETYLENE COMPOUNDS and PLATINUM COMPOUNDS.

PLD250 CAS:13767-90-3 ***HR: 3***
POTASSIUM DIFLUOROPHOSPHATE
mf: $F_2O_2P{\bullet}K$ mw: 140.07

TOXICITY DATA with REFERENCE
ivn-mus LD50:56 mg/kg CSLNX* NX#00129

OSHA PEL: TWA 2.5 mg(F)/m^3
NIOSH REL: (Fluorides, Inorganic): 10H TWA 2.5 mg(F)/m^3

SAFETY PROFILE: Poison by intravenous route. When heated to decomposition it emits very toxic fumes of K_2O, F^-, and PO_x. See also FLUORIDES.

PLD500 CAS:128-03-0 ***HR: 3***
POTASSIUM DIMETHYL DITHIOCARBAMATE
mf: $C_3H_6NS_2{\bullet}K{\bullet}H_2O$ mw: 177.34

TOXICITY DATA with REFERENCE
ipr-mus LD50:350 mg/kg APTOA6 8,329,52

CONSENSUS REPORTS: Reported in EPA TSCA Inventory.

SAFETY PROFILE: Poison by intraperitoneal route. When heated to decomposition it emits very toxic fumes of K_2O, NO_x, and SO_x. See also CARBAMATES.

PLD550 CAS:26717-79-3 ***HR: 3***
POTASSIUM-2,5-DINITROCYCLOPENTANONIDE
mf: $C_5H_5KN_2O_5$ mw: 212.20

O:CCH(NO$_2$)(CH$_2$)$_2$C=N(O)OK (ring bond between the first and last ring carbons)

SAFETY PROFILE: Explodes when heated to 154°C. When heated to decomposition it emits toxic fumes of NO_x and K_2O. See also NITRO COMPOUNDS.

PLD575 CAS:32617-22-4 ***HR: 3***
POTASSIUM DINITROMETHANIDE
mf: $CHKN_2O_4$ mw: 144.13

SAFETY PROFILE: An impact-sensitive explosive. When heated to decomposition it emits toxic fumes of NO_x and K_2O. See also NITRO COMPOUNDS.

PLD600 CAS:15213-49-7 ***HR: 3***
POTASSIUM DINITROOXALATOPLATINATE(2⁻)
mf: $C_2K_2N_2O_8Pt$ mw: 453.30

SAFETY PROFILE: Decomposes violently when heated to 240°C. Upon decomposition it emits toxic fumes of NO_x and K_2O. See also OXALATES, NITRO COMPOUNDS, and PLATINUM COMPOUNDS.

PLD700 CAS:30533-63-2 ***HR: 3***
POTASSIUM-1,1-DINITROPROPANIDE
mf: $C_3H_5KN_2O_4$ mw: 172.18

$KC(NO_2)_2CH_2CH_3$

SYN: POTASSIUM aci-1,1-DINITROPROPANE

SAFETY PROFILE: A dangerous explosive. When heated to decomposition it emits toxic fumes of NO_x and K_2O. See also NITRO COMPOUNDS.

PLD710 CAS:26241-10-1 ***HR: 3***
POTASSIUM DINITROSOSULFITE
mf: $K_2N_2O_5S$ mw: 218.29

$KOSO_2N(NO)OK$

SYN: POTASSIUM N-NITROSOHYDROXYLAMINE-N-SULFONATE

SAFETY PROFILE: A heat-sensitive explosive. Many N-nitroso compounds are carcinogens. When heated to decomposition it emits very toxic fumes of SO_x, NO_x, and K_2O. See also N-NITROSO COMPOUNDS.

PLD730 CAS:70324-35-5 ***HR: 3***
POTASSIUM-3,5-DINITRO-2(1-TETRAZENYL) PHENOLATE
mf: $C_6H_5KN_6O_5$ mw: 280.25

SAFETY PROFILE: An explosive. When heated to decomposition it emits very toxic fumes of NO_x and K_2O. See also NITRO COMPOUNDS.

PLE260 CAS:12030-88-5 ***HR: 3***
POTASSIUM DIOXIDE
DOT: UN 2466

mf: KO_2 mw: 71.10

CONSENSUS REPORTS: Reported in EPA TSCA Inventory.

DOT Classification: Oxidizer; Label:Oxidizer

SAFETY PROFILE: Explosive reaction when heated with carbon; 2-aminophenol + tetrahydrofuran (at 65°C). Forms a friction-sensitive explosive mixture with hydrocarbons. Violent reaction with diselenium dichloride, ethanol, potassium-sodium alloy. May ignite on contact with organic compounds. Incandescent reaction with metals (e.g., arsenic, antimony, copper, potassium, tin, and zinc). When heated to decomposition it emits toxic fumes of K_2O. See also PEROXIDES.

PLE500 ***HR: 3***
POTASSIUM DIPEROXY ORTHOVANADATE
mf: K_2O_6V mw: 225.15

$K_2[(O_2)_2VO_2]$

SAFETY PROFILE: Explodes on heating. Upon decomposition it emits toxic fumes of K_2O. See also VANADIUM COMPOUNDS and PEROXIDES.

PLE575 CAS:917-58-8 ***HR: 2***
POTASSIUM ETHOXIDE
mf: C_2H_5OK mw: 84.16

SAFETY PROFILE: May ignite in moist air. When heated to decomposition it emits toxic fumes of K_2O.

PLE750 CAS:64048-06-2 ***HR: 3***
POTASSIUM ETHYLMERCURIC THIOGLYCOLLATE
mf: $C_4H_7HgO_2S{\bullet}K$ mw: 358.86

SYN: ETHYL(MERCAPTOACETATO(2−)-O,S)-MERCURATE(1−)-POTASSIUM

TOXICITY DATA with REFERENCE
ipr-rat LDLo:30 mg/kg JPETAB 35,343,29
ivn-rbt LDLo:20 mg/kg JPETAB 35,343,29

CONSENSUS REPORTS: Mercury and its compounds are on the Community Right-To-Know List.

OSHA PEL: (Transitional: CL 1 mg/10m^3) TWA 0.01 mg(Hg)/m^3; STEL 0.03 mg/m^3 (skin)
ACGIH TLV: TWA 0.01 mg(Hg)/m^3; STEL 0.03 mg(Hg)/m^3

SAFETY PROFILE: Poison by intraperitoneal and intravenous routes. When heated to decomposition it emits very toxic fumes of K_2O, Hg, and SO_x. See also MERCURY COMPOUNDS.

PLF000 CAS:140-89-6 ***HR: 3***
POTASSIUM ETHYL XANTHOGENATE
mf: $C_3H_6OS_2{\bullet}K$ mw: 161.31

PROP: Liquid. Mp: 200° (decomp), uel: 9.5%, flash p: 205°F (CC), d: 1.558.

SYNS: CARBONODITHIOIC ACID, O-ETHYL ESTER, POTASSIUM SALT ◇ (O-ETHYL DITHIOCARBONATO)POTASSIUM ◇ ETHYL POTASSIUM XANTHATE ◇ ETHYL POTASSIUM XANTHOGENATE ◇ ETHYLXANTHIC ACID POTASSIUM SALT ◇ POTASSIUM O-ETHYL DITHIOCARBONATE ◇ POTASSIUM ETHYLXANTHATE ◇ POTASSIUM XANTHATE ◇ POTASSIUM XANTHOGENATE ◇ Z 3 (PESTICIDE)

TOXICITY DATA with REFERENCE
orl-rat LD50:1700 mg/kg GISAAA 41(6),95,76
orl-mus LD50:683 mg/kg GISAAA 41(6),95,76
scu-mus LDLo:400 mg/kg AIPTAK 12,447,04
ivn-mus LD50:199 mg/kg AIPTAK 135,330,62

CONSENSUS REPORTS: Reported in EPA TSCA Inventory.

SAFETY PROFILE: Poison by subcutaneous and intravenous routes. Moderately toxic by ingestion. Combustible when exposed to heat or flame. Moderately explosive in the form of vapor when exposed to flame. Can react with oxidizing materials. To fight fire, use water, CO_2, dry chemical. Incompatible with diazonium salts. When heated to decomposition it emits highly toxic fumes of K_2O and SO_x.

PLF250 CAS:13746-66-2 ***HR: 2***
POTASSIUM FERRICYANATE
mf: $C_6FeN_6{\bullet}3K$ mw: 329.27

$K_3[Fe(CN)_6]$

PROP: Lemon yellow crystals. Mp: $-3H_2O$ @ 70°, bp: decomp, d: 1.85 @ 17°.

SYNS: HEXACYANOFERRATE(3−) TRIPOTASSIUM ◇ POTASSIUM FERRICYANIDE ◇ POTASSIUM HEXACYANOFERRATE(III) ◇ TRIPOTASSIUM HEXACYANOFERRATE

TOXICITY DATA with REFERENCE
mmo-smc 100 mmol/L MUREAV 117,149,83
mrc-smc 100 mmol/L MUREAV 117,149,83
orl-rat LDLo:1600 mg/kg KODAK* 21MAY71

CONSENSUS REPORTS: Cyanide and its compounds are on the Community Right-To-Know List. Reported in EPA TSCA Inventory. EPA Genetic Toxicology Program.

OSHA PEL: TWA 5 mg(CN)/m^3
ACGIH TLV: TWA 5 mg(CN)/m^3 (skin)
DFG MAK: 5 mg/m^3
NIOSH REL: (Cyanide) CL 5 mg(CN)/m^3/10M

SAFETY PROFILE: Moderately toxic by ingestion. Not as toxic as the simple cyanides. Mutation data reported.

Explosive reaction with ammonia, chromium trioxide (above 196°C); sodium nitrite + heat. Violent reaction with $Cu(NO_3)_2$. Mixtures with chromium trioxide + silver grains ignite with friction. When heated to decomposition or on contact with acid or acid fumes it emits highly toxic fumes of K_2O and CN^-. Used as a fixative in photography, as a metal cleaner, and for glass coatings.

PLF500 CAS:7789-23-3 ***HR: 3***
POTASSIUM FLUORIDE
DOT: UN 1812
mf: FK mw: 58.10

PROP: White, crystalline, deliq powder; sharp saline taste. Bp: 1500°, d: 2.48, vap press: 1 mm @ 885°, mp: 859.9°. Very sol in boiling water.

SYNS: FLUORURE de POTASSIUM (FRENCH) ◇ POTASSIUM FLUORIDE, solution (DOT) ◇ POTASSIUM FLUORURE (FRENCH)

TOXICITY DATA with REFERENCE
mma-mus:lyms 500 mg/L MUREAV 187,165,87
msc-mus:lyms 400 mg/L MUREAV 187,165,87
ipr-mus TDLo:1050 mg/kg (1-21D preg):TER DZZEA7 34,484,79
orl-rat LD50:245 mg/kg XEURAQ UR-154,1951
ipr-rat LD50:64 mg/kg XEURAQ UR-154,1951
ipr-mus LD50:40030 μg/kg DZZEA4 34,484,79
orl-gpg LDLo:250 mg/kg MEIEDD 10,1101,83
scu-frg LDLo:420 mg/kg CRSBAW 124,133,37

CONSENSUS REPORTS: Reported in EPA TSCA Inventory. EPA Genetic Toxicology Program.

OSHA PEL: TWA 2.5 mg(F)/m^3
ACGIH TLV: TWA 2.5 mg(F)/m^3
NIOSH REL: TWA (Inorganic Fluorides) 2.5 mg(F)/m^3
DOT Classification: Corrosive Material; Label: Corrosive, solution; ORM-B; Label: None; Poison B; Label: St. Andrews Cross.

SAFETY PROFILE: Poison by ingestion and intraperitoneal routes. Moderately toxic by subcutaneous route. Experimental teratogenic effects. A corrosive irritant to the eyes, skin, and mucous membranes. Mutation data reported. A very reactive material. When heated to decomposition it emits toxic fumes of K_2O and F^-. Used in etching glass, as a preservative, insecticide, and in organic synthesis. See also FLUORIDES and HYDROFLUORIC ACID.

PLF750 ***HR: 3***
POTASSIUM FLUORIDE, DIHYDRATE
mf: $FK \cdot 2H_2O$ mw: 94.14

TOXICITY DATA with REFERENCE
dnr-bcs 20 mg/plate SHKKAN 27,372,85

SAFETY PROFILE: A highly toxic, irritating and reactive solution. Mutation data reported. When heated to decomposition it emits toxic fumes of F^- and K_2O. See also POTASSIUM FLUORIDE.

PLG000 CAS:23745-86-0 ***HR: 3***
POTASSIUM FLUOROACETATE
DOT: UN 2628
mf: $C_2H_3FO_2 \cdot K$ mw:117.15

PROP: The potassium salt of monofluoroacetic acid was once designated as potassium cymonate.

SYNS: DICHAPETULUM CYMOSUM (HOOK) ENGL ◇ GIFBLAAR

TOXICITY DATA with REFERENCE
scu-mus LDLo:4 mg/kg OJVSA4 22,77,47
orl-rbt LDLo:500 μg/kg 11FYAN 3,74,63
scu-rbt LDLo:500 μg/kg 11FYAN 3,74,63
ivn-rbt LDLo:500 μg/kg 11FYAN 3,74,63
orl-ckn LDLo:50 mg/kg OJVSA4 22,77,47
scu-ckn LDLo:10 mg/kg OJVSA4 22,77,47
par-dck LDLo:20 mg/kg 11FYAN 3,73,63
scu-dom LDLo:2 mg/kg OJVSA4 22,77,47

DOT Classification: Poison B; Label: Poison.

SAFETY PROFILE: Poison by ingestion, intravenous, parenteral, and subcutaneous routes. When heated to decomposition it emits toxic fumes of F^- and K_2O. See also FLUORIDES.

PLG500 CAS:16923-95-8 ***HR: 3***
POTASSIUM FLUOZIRCONATE
mf: K_2ZrF_6 mw: 283.4

PROP: Monoclinic, colorless crystals. D: 3.48.

SYN: ZIRCONIUM POTASSIUM FLUORIDE

TOXICITY DATA with REFERENCE
orl-mus LD50:98 mg/kg HYSAAV 32,343,67

CONSENSUS REPORTS: Reported in EPA TSCA Inventory.

OSHA PEL: TWA 2.5 mg(F)/m^3
ACGIH TLV: TWA 2.5 mg(F)/m^3
NIOSH REL: TWA 2.5 mg(F)/m^3

SAFETY PROFILE: Poison by ingestion. When heated to decomposition it emits toxic fumes of F^- and K_2O. See also FLUORIDES and ZIRCONIUM COMPOUNDS.

PLG750 CAS:590-29-4 ***HR: 3***
POTASSIUM FORMATE
mf: $CHO_2 \cdot K$ mw: 84.12

PROP: Colorless, deliquescent crystals. Mp: 168°, bp: decomp, d: 1.91. Very sol in water.

TOXICITY DATA with REFERENCE
ivn-mus LD50:95 mg/kg ZERNAL 9,332,69

CONSENSUS REPORTS: Reported in EPA TSCA Inventory.

SAFETY PROFILE: Poison by intravenous route. Combustible when exposed to heat or flame. When heated to decomposition it emits toxic fumes of K_2O.

PLG800 CAS:299-27-4 ***HR: 2***
POTASSIUM GLUCONATE
mf: $C_6H_{12}O_7•K$ mw: 235.28

PROP: Yellowish-white crystals or powder; mild, sltly salty taste. Decomp at 180°. Freely sol in water, glycerin; practically insol in abs alc, ether, benzene, chloroform.

SYNS: d-GLUCONIC ACID, MONOPOTASSIUM SALT (9CI) ◇ GLUCONIC ACID POTASSIUM SALT ◇ GLUCONSAN K ◇ KALIUM-BETA ◇ KAON ◇ KAON ELIXIR ◇ KATORIN ◇ K-IAO ◇ POTALIUM ◇ POTASORAL ◇ POTASSIUM d-GLUCONATE ◇ POTASSURIL ◇ SIROKAL

TOXICITY DATA with REFERENCE
orl-rat LD50:10380 mg/kg NIIRDN 6,226,82
ipr-rat LD50:2664 mg/kg KSRNAM 6,810,72
scu-rat LD50:9650 mg/kg KSRNAM 6,810,72
orl-mus LD50:9100 mg/kg NIIRDN 6,226,82
ipr-mus LD50:2333 mg/kg KSRNAM 6,810,72
scu-mus LD50:8190 mg/kg KSRNAM 6,810,72

CONSENSUS REPORTS: Reported in EPA TSCA Inventory.

SAFETY PROFILE: Moderately toxic by intraperitoneal route. Mildly toxic by ingestion. When heated to decomposition it emits toxic fumes of K_2O.

PLG825 CAS:12081-88-8 ***HR: 3***
POTASSIUM GRAPHITE
mf: C_8K mw: 135.19

SAFETY PROFILE: May explode on contact with water. When heated to decomposition it emits toxic fumes of K_2O.

PLH000 CAS:16924-00-8 ***HR: 3***
POTASSIUM HEPTAFLUOROTANTALATE
mf: $F_7Ta•2K$ mw: 392.15

SYNS: POTASSIUM FLUOTANTALATE ◇ TANTALUM POTASSIUM FLUORIDE

TOXICITY DATA with REFERENCE
orl-rat LD50:2500 mg/kg AIHOAX 1,637,50
ipr-rat LD50:375 mg/kg AIHOAX 1,637,50
orl-mus LD50:110 mg/kg 20PKA3 -,115,67

CONSENSUS REPORTS: Reported in EPA TSCA Inventory.

OSHA PEL: TWA 2.5 mg(F)/m^3
NIOSH REL: TWA 2.5 mg(F)/m^3

SAFETY PROFILE: Poison by ingestion and intraperitoneal routes. When heated to decomposition it emits toxic fumes of F^- and K_2O.

PLH100 CAS:14459-95-1 ***HR: 3***
POTASSIUM HEXACYANOFERATE(II)
mf: $C_6FeN_6•4K$ mw: 368.35

$K_4[Fe(CN)_6]$

SYN: POTASSIUM FERROCYANIDE

CONSENSUS REPORTS: Cyanide and its compounds are on the Community Right-To-Know List.

OSHA PEL: TWA 5 mg(CN)/m^3
ACGIH TLV: TWA 5 mg(CN)/m^3 (skin)
DFG MAK: 5 mg/m^3
NIOSH REL: (Cyanide) CL 5 mg(CN)/m^3/10M

SAFETY PROFILE: Explodes when heated with sodium nitrite; copper(II) nitrate (at 220°C). When heated to decomposition it emits toxic fumes of CN^- and K_2O. See also CYANIDE.

PLH250 ***HR: 3***
POTASSIUM HEXAETHYNYLCOBALTATE
mf: $C_{12}H_6CoK_4$ mw: 365.52

$K_4[(HC{\equiv}C)_6Co]$

CONSENSUS REPORTS: Cobalt and its compounds are on the Community Right-To-Know List.

SAFETY PROFILE: A very shock- and friction-sensitive explosive. Explodes violently on contact with water. Product or reaction with ammonia explodes on contact with air. When heated to decomposition it emits toxic fumes of K_2O. See also COBALT COMPOUNDS and ACETYLENE COMPOUNDS.

PLH500 CAS:17029-22-0 ***HR: 3***
POTASSIUM HEXAFLUOROARSENATE
mf: AsF_6K mw: 228.02

SYNS: HEXAFLURATE ◇ NOPALMATE

TOXICITY DATA with REFERENCE
orl-rat LD50:1200 mg/kg 28ZEAL 5,129,76
ivn-mus LD50:56 mg/kg CSLNX* NX#00126

CONSENSUS REPORTS: Arsenic and its compounds are on the Community Right-To-Know List. Reported in EPA TSCA Inventory.

OSHA PEL: 8H TWA 2.5 mg(F)/m^3; Cancer Hazard
NIOSH REL: CL 2 μg/m^3/15M

SAFETY PROFILE: Confirmed human carcinogen.

Poison by intravenous route. Moderately toxic by ingestion. When heated to decomposition it emits very toxic fumes of K_2O, F^-, and As. See also FLUORIDES and ARSENIC COMPOUNDS.

PLH750 CAS:16871-90-2 ***HR: 3***
POTASSIUM HEXAFLUOROSILICATE
DOT: UN 2655
mf: $F_6Si•2K$ mw: 220.29

PROP: White, fine powder or crystals. D: 2.27, mp: decomp. Sltly sol in cold water; insol in alc.

SYNS: POTASSIUM FLUOSILICATE ◇ POTASSIUM SILICOFLUORIDE (DOT)

TOXICITY DATA with REFERENCE
orl-gpg LD50:500 mg/kg 28ZEAL 4,327,69
scu-gpg LDLo:500 mg/kg CRSBAW 124,133,37
scu-frg LDLo:448 mg/kg CRSBAW 124,133,37

CONSENSUS REPORTS: Reported in EPA TSCA Inventory.

OSHA PEL: TWA 2.5 mg(F)/m^3
NIOSH REL: TWA (Inorganic Fluorides) 2.5 mg(F)/m^3
DOT Classification: Poison B; Label: St. Andrews Cross.

SAFETY PROFILE: A poison. Moderately toxic by ingestion and subcutaneous routes. Ingestion can cause vomiting and diarrhea. A strong irritant. Incompatible with hydrofluoric acid. When heated to decomposition it emits toxic fumes of K_2O and F^-. Used as a porcelain enamel frit, a ceramic and glass ingredient, flux, and sand inhibitor.

PLI000 CAS:16919-27-0 ***HR: 3***
POTASSIUM HEXAFLUOROTITANATE
mf: $F_6Ti•2K$ mw: 240.10

SYNS: FLUOTITANATE de POTASSIUM (FRENCH) ◇ TITANIUM POTASSIUM FLUORIDE

TOXICITY DATA with REFERENCE
scu-frg LDLo:360 mg/kg CRSBAW 124,133,37

CONSENSUS REPORTS: Reported in EPA TSCA Inventory.

OSHA PEL: TWA 2.5 mg(F)/m^3
NIOSH REL: TWA 2.5 mg(F)/m^3

SAFETY PROFILE: Poison by subcutaneous route. When heated to decomposition it emits toxic fumes of K_2O and F^-. See also FLUORIDES.

PLI250 CAS:16893-93-9 ***HR: 3***
POTASSIUM HEXAFLUORSTANNATE
mf: $F_6Sn•2K$ mw: 310.89

TOXICITY DATA with REFERENCE
ivn-mus LD50:56 mg/kg CSLNX* NX#00133

OSHA PEL: TWA 2 mg(Sn)/m^3; TWA 2.5 mg(F)/m^3
NIOSH REL: (Fluorides, Inorganic):10H TWA 2.5 mg(F)/m^3

SAFETY PROFILE: Poison by intravenous route. When heated to decomposition it emits toxic fumes of K_2O and F^-. See also TIN COMPOUNDS and FLUORIDES.

PLI500 CAS:17083-63-5 ***HR: 3***
POTASSIUM HEXAHYDRATE ALUMINATE
mf: AlH_6K_3 mw: 150.33

SAFETY PROFILE: After storage, dry samples have exploded violently when disturbed. When heated to decomposition it emits toxic fumes of K_2O. See also ALUMINUM COMPOUNDS.

PLI750 CAS:13782-01-9 ***HR: 3***
POTASSIUM HEXANITROCOBALTATE
mf: $CoK_3N_6O_{12}$ mw: 452.28

SYNS: POTASSIUMTRIAZIDOCOBALTATE

CONSENSUS REPORTS: Cobalt and its compounds are on the Community Right-To-Know List.

SAFETY PROFILE: An explosive. When heated to decomposition it emits very toxic fumes of NO_x and K_2O. See also COBALT COMPOUNDS.

PLJ000 CAS:12273-50-6 ***HR: 3***
POTASSIUM HEXAOXYXENONATE(4−) XENON TRIOXIDE
mf: $K_4O_6Xe•2O_3Xe$ mw: 611.02

SAFETY PROFILE: A powerful, shock-sensitive explosive. When heated to decomposition it emits toxic fumes of K_2O.

PLJ250 CAS:7693-26-7 ***HR: 3***
POTASSIUM HYDRIDE
mf: HK mw: 40.11

PROP: White needles. Mp: decomp, d: 1.43-1.47.

SAFETY PROFILE: Dangerous fire hazard by chemical reaction. Ignites spontaneously in air. Moderate explosion hazard when exposed to heat or by chemical reaction. Will react with water, steam, or acids to produce H_2 which then ignites. Can react vigorously with oxidizing materials. To fight fire, use CO_2, dry chemical. Potentially explosive reactions with o-2,4-dinitrophenylhydroxylamine; fluoroalkenes. Ignites on contact with air; oxygen + moisture; fluorine. Incompatible with Cl_2;

acetic acid; acrolein; acrylonitrile; ($CaC + Cl_2$); ClO_2; ($H_2O_2 + Cl_2$); ($CHFl_3 + CH_3OH$); 1,2-dichloroethylene; maleic anhydride; (n-methyl-n-nitrosourea + CH_2Cl_2); nitroethane; NCl_3; nitromethane; nitroparaffins; o-nitrophenol; nitropropane; n-nitrosomethylurea; (nitrosomethylurea + CH_2Cl_2); H_2O; trichloroethylene; tetrahydrofuran; tetrachlorethane. When heated to decomposition it emits highly toxic fumes of K_2O. See also POTASSIUM and HYDRIDES.

PLJ500 CAS:1310-58-3 ***HR: 3***
POTASSIUM HYDROXIDE
DOT: UN 1813/UN 1814
mf: HKO mw: 56.11

PROP: White, deliquescent pieces, lumps or sticks having crystalline fracture. Mp: 360° ± 7°, bp: 1320°, d: 2.044. Sol in water, alc.

SYNS: CAUSTIC POTASH ◇ CAUSTIC POTASH, dry, solid, flake, bead, or granular (DOT) ◇ CAUSTIC POTASH, liquid or solution (DOT) ◇ HYDROXYDE de POTASSIUM (FRENCH) ◇ KALIUMHYDROXID (GERMAN) ◇ KALIUMHYDROXYDE (DUTCH) ◇ LYE ◇ POTASSA ◇ POTASSE CAUSTIQUE (FRENCH) ◇ POTASSIO (IDROSSIDO di) (ITALIAN) ◇ POTASSIUM HYDRATE (DOT) ◇ POTASSIUM HYDROXIDE, dry, solid, flake, bead, or granular (DOT) ◇ POTASSIUM HYDROXIDE, liquid or solution (DOT) ◇ POTASSIUM (HYDROXYDE de) (FRENCH)

TOXICITY DATA with REFERENCE
skn-hmn 50 mg/24H SEV TXAPA9 31,481,75
skn-rbt 50 mg/24H SEV TXAPA9 31,481,75
eye-rbt 1 mg/24H rns MOD TXAPA9 32,239,75
cyt-rat/ast 1800 mg/kg GANNA2 54,155,63
orl-rat LD50:365 mg/kg TXAPA9 32,239,75

CONSENSUS REPORTS: Reported in EPA TSCA Inventory.

OSHA PEL: CL 2 mg/m^3
ACGIH TLV: CL 2 mg/m^3
DOT Classification: Corrosive Material; Label: Corrosive; Label: Corrosive, solution.

SAFETY PROFILE: Poison by ingestion. An eye irritant and severe human skin irritant. Very corrosive to the eyes, skin, and mucous membranes. Mutation data reported. Ingestion may cause violent pain in throat and epigastrium, hematemesis, collapse. Stricture of esophagus may result if not immediately fatal. Above 84° it reacts with reducing sugars to form the poisonous carbon monoxide gas. Violent, exothermic reaction with water. Potentially explosive reaction with bromoform + crown ethers; chlorine dioxide; nitrobenzene; nitromethane; nitrogen trichloride; peroxidized tetrahydrofuran; 2,4,6-trinitrotoluene. Reaction with ammonium hexachloroplatinate(2−) + heat forms a heat-sensitive explosive product. Violent reaction or ignition under the appropriate conditions with acids; alcohols; p-bis(1,3-dibromoethyl)benzene; cyclopentadiene; germanium; hyponitrous acid; maleic anhydride; nitroalkanes; 2-nitrophenol; potassium peroxodisulphate; sugars; 2,2,3,3-tetrafluoropropanol; thorium dicarbide. When heated to decomposition it emits toxic fumes of K_2O. See also SODIUM HYDROXIDE.

PLJ750 CAS:1310-58-3 ***HR: 3***
POTASSIUM HYDROXIDE (solution)
DOT: UN 1813/UN 1814
mf: HKO mw: 56.11

PROP: Clear liquid.

SYN: POTASSIUM HYDRATE (solution)

TOXICITY DATA with REFERENCE
skn-rbt 5 mg/24H MOD TXAPA9 32,239,75
eye-rbt 1 mg/24H rns MOD TXAPA9 32,239,75

CONSENSUS REPORTS: Reported in EPA TSCA Inventory.

DOT Classification: Corrosive Material; Label: Corrosive.

SAFETY PROFILE: Very corrosive to the eyes, skin, and mucous membranes. When heated to decomposition it emits toxic fumes of K_2O. See also POTASSIUM HYDROXIDE and SODIUM HYDROXIDE.

PLJ775 CAS:86341-95-9 ***HR: 3***
POTASSIUM-4-HYDROXYAMINE-5,7-DINITRO-4,5-DIHYDROBENZO-FURAZANIDE-3-OXIDE
mf: $C_6H_4KN_5O_7$ mw: 297.23

SAFETY PROFILE: A primary explosive sensitive to mechanical, electrostatic, and thermal shock. When heated to decomposition it emits toxic fumes of NO_x and K_2O. See also NITRO COMPOUNDS and EXPLOSIVES, HIGH.

PLJ780 CAS:57891-85-7 ***HR: 3***
POTASSIUM-4-HYDROXY-5,7-DINITRO-4,5-DIHYDROBENZOFURAZANIDE
mf: $C_6H_3KN_4O_7$ mw: 282.21

SAFETY PROFILE: An explosive. When heated to decomposition it emits toxic fumes of NO_x and K_2O. See also EXPLOSIVES, HIGH.

PLJ790 ***HR: 3***
POTASSIUM HYPOBORATE
mf: $B_2H_6K_2O_2$ mw: 137.86

SAFETY PROFILE: A strong reductant which can react vigorously with oxidants. When heated to decomposition it emits toxic fumes of K_2O. See also BORON COMPOUNDS.

PLK000 CAS:7778-66-7 ***HR: 3***
POTASSIUM HYPOCHLORITE (solution)
DOT: UN 1791
mf: KOCl mw: 90.55

SYN: HYPOCHLORITE, solution (DOT)

CONSENSUS REPORTS: Reported in EPA TSCA Inventory.

DOT Classification: Corrosive Material; Label: Corrosive.

SAFETY PROFILE: A poison by all routes. Powerful irritant and corrosive to skin, eyes, and mucous membranes. When heated to decomposition it emits toxic fumes of K_2O and Cl^-. See also HYPOCHLORITES.

PLK250 CAS:7758-05-6 ***HR: 3***
POTASSIUM IODATE
mf: IO_3•K mw: 214.00

PROP: Colorless crystals or white crystalline powder. Mp: 560°, d: 3.89. Sol in water; insol in alc.

SYN: IODIC ACIODIC ACID, POTASSIUM SALT

TOXICITY DATA with REFERENCE
orl-mus LDLo:531 mg/kg JPETAB 120,171,57
ipr-mus LD50:136 mg/kg JPETAB 120,171,57
orl-dog LDLo:200 mg/kg 34ZIAG -,492,69
orl-gpg LDLo:400 mg/kg FAONAU 40,113,67

CONSENSUS REPORTS: Reported in EPA TSCA Inventory.

SAFETY PROFILE: Poison by ingestion and intraperitoneal routes. A trace mineral added to animal feeds. Potentially explosive reaction with charcoal + ozone; metals (e.g., powdered aluminum; copper); arsenic carbon; phosphorus; sulfur; alkali metal hydrides; alkaline earth metal hydrides; antimony sulfide; arsenic sulfide; copper sulfide; tin sulfide; metal cyanides; metal thiocyanates; manganese dioxide; phosphorus. Violent reaction with organic matter. When heated to decomposition it emits very toxic fumes of I^- and K_2O. See also IODATES.

PLK500 CAS:7681-11-0 ***HR: 2***
POTASSIUM IODIDE
mf: IK mw: 166.00

PROP: Colorless or white granules. Mp: 723°, bp: 1420°, d: 3.13, vap press: 1 mm @ 745°. Sltly hygroscopic. Sol in water, glycerin, alc.

TOXICITY DATA with REFERENCE
cyt-rat/ast 500 mg/kg GANNA2 54,155,63
orl-wmn TDLo:3240 mg/kg (1-39W preg):REP ADCHAK 43,702,68
orl-wmn TDLo:2700 mg/kg (1-39W preg):TER JOPDAB 67,353,65
ivn-rat LDLo:120 mg/kg AEXPBL 96,292,23
orl-mus LDLo:1862 mg/kg JPETAB 120,171,57
ipr-mus LDLo:1117 mg/kg JPETAB 120,171,57

CONSENSUS REPORTS: Reported in EPA TSCA Inventory.

SAFETY PROFILE: Poison by intravenous route. Moderately toxic by ingestion and intraperitoneal routes. Human teratogenic effects by ingestion: developmental abnormalities of the endocrine system. Experimental teratogenic and reproductive effects. Mutation data reported. Explosive reaction with charcoal + ozone; trifluoroacetyl hypofluorite; fluorine perchlorate. Violent reaction or ignition on contact with diazonium salts; diisopropyl peroxydicarbonate; bromine pentafluoride; chlorine trifluoride. Incompatible with oxidants; BrF_3; FClO; metallic salts; calomel. When heated to decomposition it emits very toxic fumes of K_2O and I^-. See also IODIDES.

PLK580 CAS:928-70-1 ***HR: 3***
POTASSIUM ISOAMYL XANTHATE
mf: $C_{23}H_{30}N_4O_5$ mw: 442.57

SYNS: CARBONIC ACID, DIESTER with 1-(2,3-DIMETHYLPHENYL)-3-(2-HYDROXYETHYL)UREA ◇ ISOAMYL POTASSIUM XANTHATE ◇ POTASSIUM ISOAMYL XANTHOGENATE ◇ POTASSIUM ISOPENTYL XANTHATE

TOXICITY DATA with REFERENCE
orl-rat LD50:765 mg/kg GISAAA 41(6),95,76
orl-mus LD50:480 mg/kg GISAAA 41(6),95,76
ivn-mus LD50:158 mg/kg AIPTAK 135,330,62

SAFETY PROFILE: Poison by intravenous route. Moderately toxic by ingestion. When heated to decomposition it emits toxic fumes of NO_x and K_2O. See also ESTERS.

PLK600 CAS:13001-46-2 ***HR: 3***
POTASSIUM ISOBUTYL XANTHATE
mf: $C_5H_9OS_2$•K mw: 188.36

SYNS: ISOBUTYL POTASSIUM XANTHATE ◇ POTASSIUM-o-ISOBUTYL XANTHATE ◇ POTASSIUM ISOBUTYL XANTHOGENATE

TOXICITY DATA with REFERENCE
orl-rat LD50:1290 mg/kg GISAAA 41(6),95,76
orl-mus LD50:480 mg/kg GISAAA 41(6),95,76
ivn-mus LD50:158 mg/kg AIPTAK 135,330,62

SAFETY PROFILE: Poison by intravenous route. Moderately toxic by ingestion. When heated to decomposition it emits toxic fumes of SO_x and K_2O. See also ESTERS.

PLK750 CAS:10124-65-9 ***HR: 1***
POTASSIUM LAURATE
mf: $C_{12}H_{23}O_2•K$ mw: 238.45

SYN: POTASSIUM DODECANOATE

TOXICITY DATA with REFERENCE
skn-hmn 9560 mg BJDEAZ 75,113,63

CONSENSUS REPORTS: Reported in EPA TSCA Inventory.

SAFETY PROFILE: A human skin irritant. When heated to decomposition it emits toxic fumes of K_2O.

PLK800 CAS:13355-00-5 ***HR: 3***
POTASSIUM MELARSONYL
mf: $C_{13}H_{11}AsN_6O_4S_2•ClH$ mw: 532.54

SYNS: MELARSONYL POTASSIUM SALT ◇ MEL W ◇ POTASSIUM PENTYL THIARSAPHENYLMELAMINE ◇ TRIMELARSEN

TOXICITY DATA with REFERENCE
scu-mus TDLo:10 mg/kg (2D male):REP JTSCDR 3,262,78
ivn-mus LD50:105 mg/kg BWHOA6 42,115,70

CONSENSUS REPORTS: Arsenic and its compounds are on the Community Right-To-Know List.

SAFETY PROFILE: Poison by intravenous route. Experimental reproductive effects. When heated to decomposition it emits toxic fumes of SO_x, NO_x, K_2O, and As. See also ARSENIC COMPOUNDS.

PLK825 CAS:19416-93-4 ***HR: 3***
POTASSIUM METHANEDIZOATE
mf: CH_3KN_2O mw: 98.15

SYN: POTASSIUM METHYLDIAZENE OXIDE

SAFETY PROFILE: Explosive reaction on contact with water. When heated to decomposition it emits toxic fumes of NO_x and K_2O.

PLK850 CAS:865-33-8 ***HR: 3***
POTASSIUM METHOXIDE
mf: CH_3KO mw: 70.13

SAFETY PROFILE: Ignites in moist air. When heated to decomposition it emits toxic fumes of K_2O.

PLK860 CAS:1270-21-9 ***HR: 3***
POTASSIUM-4-METHOXY-1-aci-NITRO-3,5-DINITRO-2,5-CYCLOHEXADIENONE
mf: $C_7H_6KN_3O_7$ mw: 283.24

SAFETY PROFILE: An explosive. When heated to decomposition it emits toxic fumes of NO_x and K_2O. See also NITRO COMPOUNDS.

PLK900 CAS:13769-43-2 ***HR: 3***
POTASSIUM METAVANADATE
DOT: UN 2864
mf: $O_3V•K$ mw: 138.04

SYNS: POTASSIUM METAVANADATE ◇ POTASSIUM VANADIUM TRIOXIDE ◇ VANADIC ACID, POTASSIUM SALT

ACGIH TLV: TWA 0.05 mg(V_2O_5)/m^3
DOT Classification: Poison B; Label: Poison

CONSENSUS REPORTS: Reported in EPA TSCA Inventory.

SAFETY PROFILE: A poison. When heated to decomposition it emits toxic fumes of V_2O_5

PLL000 CAS:54448-39-4 ***HR: 3***
POTASSIUM METHYLAMIDE
mf: CH_4KN mw: 69.15

SAFETY PROFILE: Highly irritating. Ignites spontaneously in air. Extremely hygroscopic. May explode on contact with air. When heated to decomposition it emits very toxic fumes of K_2O and NO_x.

PLL100 ***HR: 3***
POTASSIUM-4-METHYLFURAZAN-5-CARBOXYLATE-2-OXIDE
mf: $C_4H_3KN_2O_4$ mw: 182.18

ON=C(CO•OK)C(CH_3)=N:O

SAFETY PROFILE: The dry salt is a powerful explosive sensitive to heating, impact, or friction. When heated to decomposition it emits toxic fumes of NO_x and K_2O.

PLL250 CAS:12030-85-2 ***HR: 3***
POTASSIUM NIOBATE
mf: $8K•Nb_6O_{19}$ mw: 1174.26

SYNS: NIOBATE, OCTAPOTASIUM ◇ NIOBIUM POTASSIUM OXIDE ◇ POTASSIUM COLUMBATE

TOXICITY DATA with REFERENCE
orl-rat LD50:3000 mg/kg NTIS** AEC-TR-6710
ipr-rat LD50:125 mg/kg 34ZIAG -,422,69
orl-mus LD50:940 mg/kg 34ZIAG -,422,69
ipr-mus LD50:18 mg/kg 34ZIAG -,422,69

CONSENSUS REPORTS: Reported in EPA TSCA Inventory.

SAFETY PROFILE: Poison by intraperitoneal route. Moderately toxic by ingestion. When heated to decomposition it emits toxic fumes of K_2O. See also NIOBIUM.

PLL500 CAS:7757-79-1 ***HR: 3***
POTASSIUM NITRATE
DOT: UN 1486
mf: KNO_3 mw: 140.21

PROP: Transparent, colorless or white crystalline powder or crystals; odorless with a cooling, pungent, salty taste. Mp: 334°, bp: decomp @ 400°, d: 2.109 @ 16°. Sol in glycerol, water; moderately sol in alc.

SYNS: KALIUMNITRAT (GERMAN) ◇ NITER ◇ NITRE ◇ NITRIC ACID, POTASSIUM SALT ◇ SALTPETER ◇ VICKNITE

TOXICITY DATA with REFERENCE mrc-esc 5 pph JGMIAN 8,45,53
orl-gpg TDLo:15 g/kg (female 24W pre):REP TXAPA9 12,179,68
orl-gpg TDLo:1670 g/kg (female 29W pre):TER TXAPA9 12,179,68
orl-rat LD50:3750 mg/kg NYKZAU 81,469,83
ivn-cat LDLo:100 mg/kg HBAMAK 4,1289,35
orl-rbt LD50:1901 mg/kg SOVEA7 27,246,74

CONSENSUS REPORTS: Reported in EPA TSCA Inventory.

DOT Classification: Oxidizer; Label: Oxidizer.

SAFETY PROFILE: Poison by intravenous route. Moderately toxic by ingestion. An experimental teratogen. Experimental reproductive effects. Mutation data reported. Ingestion of large quantities may cause gastroenteritis. Chronic exposure can cause anemia, nephritis, and methemoglobinemia. Heated reaction with calcium hydroxide + polychlorinated phenols forms extremely toxic chlorinated benzodioxins.

A powerful oxidizer. Gunpowder is a mixture of potassium nitrate + sulfur + charcoal. Explosive reaction with aluminum + barium nitrate + potassium perchlorate + water (in storage); boron + laminac + trichloroethylene. Forms explosive mixtures with lactose; powdered metals (e.g., titanium; antimony; germanium); metal sulfides (e.g., antimony trisulfide; barium sulfide; calcium sulfide; germanium monosulfide; titanium disulfide; arsenic disulfide; molybdenum disulfide); nonmetals (e.g., boron; carbon; white phosphorus; arsenic); organic materials; phosphides (e.g., copper(II) phosphide; copper monophosphide); reducing agents (e.g., sodium phosphinate; sodium thiosulfate); sodium acetate. Can react violently under the appropriate conditions with 1,3-bis(trichloromethyl)benzene; boron phosphide; F_2; calcium silicide; charcoal; chromium nitride; Na hypophosphite; (Na_2O_2 + dextrose); red phosphorus; (S + As_2S_3); thorium dicarbide; trichloroethylene; zinc; zirconium. When heated to decomposition it emits very toxic fumes of NO_x and K_2O. See also NITRATES.

PLL750 ***HR: 3***
POTASSIUM NITRATE mixed with CHARCOAL and SULFUR (15:3:2)
DOT: UN 0027/UN 0028

SYNS: BLACK POWDER (DOT) ◇ BLACK POWDER, compressed (DOT) ◇ BLACK POWDER, granular or as meal (DOT) ◇ BLASTING POWDER (DOT) ◇ GUNPOWDER ◇ RIFLE POWDER (DOT)

DOT Classification: Class A Explosive; Label: Explosive A.

SAFETY PROFILE: Heat and/or shock can cause an explosion. When heated to decomposition it emits toxic fumes of SO_x, K_2O, and NO_x. See also POTASSIUM NITRATE and EXPLOSIVES, LOW.

PLM000 ***HR: 3***
POTASSIUM NITRATE mixed with SODIUM NITRITE
DOT: UN 1487

SYN: POTASSIUM NITRATE mixed (fused) with SODIUM NITRITE (DOT)

DOT Classification: Oxidizer; Label: Oxidizer.

SAFETY PROFILE: A powerful oxidizer. Nitrites have been implicated as possible carcinogens. When heated to decomposition it emits toxic fumes of NO_x, K_2O, and Na_2O. See also POTASSIUM NITRATE and SODIUM NITRITE.

PLM250 CAS:29285-24-3 ***HR: 3***
POTASSIUM NITRIDE
mf: K_3N mw: 131.32

PROP: Greenish-black crystals. Mp: decomp.

SAFETY PROFILE: Ignites in air and forms a very flammable mixture with phosphorous or sulfur. Mixture with sulfur ignites on contact with water and evolves ammonia and hydrogen sulfide. When heated to decomposition it emits very toxic fumes of K_2O, NO_x, and NH_3. See also AMMONIA and POTASSIUM HYDROXIDE.

PLM500 CAS:7758-09-0 ***HR: 3***
POTASSIUM NITRITE (1:1)
DOT: UN 1488
mf: $NO_2{\cdot}K$ mw: 85.11

PROP: White or sltly yellowish, deliquescent prisms or sticks. Mp: 387°, bp: decomp, d: 1.915. Very sol in water; sltly sol in alc.

SYNS: NITROUS ACID, POTASSIUM SALT ◇ POTASSIUM NITRITE (DOT)

TOXICITY DATA with REFERENCE
mmo-omi 500 mmol/L JMOBAK 9,352,64
dni-omi 20 mmol/L JMOBAK 5,442,62

orl-gpg TDLo:137 g/kg (18W pre):REP TXAPA9 12,179,68
orl-gpg TDLo:201 g/kg (19W pre):TER TXAPA9 12,179,68
orl-hmn TDLo:1428 μg/kg:EAR,BLD GISAAA 48(1),62,83
ihl-mus LC50:85 g/m^3/2H GISAAA 48(1),62,83
orl-rbt LD50:200 mg/kg SOVEA7 27,246,74

CONSENSUS REPORTS: Reported in EPA TSCA Inventory.

DOT Classification: Oxidizer; Label: Oxidizer.

SAFETY PROFILE: Poison by ingestion. Human systemic effects: tinnitus, pulse rate increase, blood pressure lowering. Experimental teratogenic and reproductive effects. Nitrites have been implicated in an increased incidence of cancer. Mutation data reported. Flammable when exposed to heat or flame. A powerful oxidizing material. Slight explosion hazard when exposed to heat. It will explode at 1000°F. Explosive reaction with potassium amide + heat; potassium cyanide or other cyanide salts + heat. Violent reaction or ignition with ammonium salts (e.g., ammonium sulfate); boron. Upon decomposition it emits toxic fumes of K_2O. See also NITRITES.

PLM550 ***HR: 3***
POTASSIUM-4-NITROBENZENEAZOSULFONATE
mf: $C_6H_4KN_3O_5S$ mw: 269.27

$O_2NC_6H_4N{=}NSO_2OK$

SAFETY PROFILE: The crystalline solid is an unstable explosive. When heated to decomposition it emits toxic fumes of SO_x, NO_x, and K_2O. See also NITRO COMPOUNDS OF AROMATIC HYDROCARBONS and SULFONATES.

PLM575 CAS:12244-59-6 ***HR: 3***
POTASSIUM-6-aci-NITRO-2,4-DINITRO-2,4-CYCLOHEXADIENIMINIDE
mf: $C_6H_3KN_4O_6$ mw: 266.21

SAFETY PROFILE: Explodes violently when heated to 110°C. When heated to decomposition it emits toxic fumes of NO_x and K_2O. See also NITRO COMPOUNDS.

PLM650 ***HR: 3***
POTASSIUM-1-NITROETHOXIDE
mf: $C_2H_4KNO_3$ mw: 90.06

SAFETY PROFILE: The solid explodes when heated. When heated to decomposition it emits toxic fumes of NO_x and K_2O. See also NITRO COMPOUNDS.

PLM700 CAS:1124-31-8 ***HR: 3***
POTASSIUM-4-NITROPHENOXIDE
mf: $C_6H_4KNO_3$ mw: 177.20

SAFETY PROFILE: It may explode spontaneously in storage. When heated to decomposition it emits toxic fumes of NO_x and K_2O. See also NITRO COMPOUNDS.

PLM750 CAS:14293-70-0 ***HR: 3***
POTASSIUM NITROSODISULFATE
mf: $K_2NO_7S_2$ mw: 268.34

SAFETY PROFILE: Slow decomposition can lead to violent explosion during storage. When heated to decomposition it emits very toxic fumes of SO_x, NO_x, and K_2O. See also NITROSO COMPOUNDS.

PLN000 ***HR: 3***
POTASSIUM NITROSOOSMATE(1$^-$)
mf: KNO_3Os mw: 149.17

SAFETY PROFILE: Will explode. When heated to decomposition it emits very toxic fumes of K_2O and NO_x. See also OSMIUM and NITROSO COMPOUNDS.

PLN050 CAS:15611-84-4 ***HR: 3***
POTASSIUM NITROTRICHLOROPLATINATE
mf: $Cl_3NO_2Pt{\cdot}2K$ mw: 425.65

SYNS: DIPOTASSIUM TRICHLORONITROPLATINATE ◇ PLATINATE(2-), NITROTRICHLORO-, DIPOTASSIUM ◇ PLATINATE(2-), TRICHLORO(NITRITO-N-), DIPOTASSIUM (SP-4-2)- ◇ PLATINUM (2-), NITROTRICHLORO-, DIPOTASSIUM

TOXICITY DATA with REFERENCE
ipr-mus LDLo:75 mg/kg BICHBX 2,187,73

OSHA PEL: TWA 0.002 mg(Pt)/m^3
ACGIH TLV: TWA 0.002 mg(Pt)/m^3

SAFETY PROFILE: Poison by intravenous route. When heated to decomposition it emits toxic fumes of NO_x and Pt.

PLN100 CAS:23705-25-1 ***HR: 3***
POTASSIUM OCTACYANODICOBALTATE
mf: $C_8Co_2K_8N_8$ mw: 638.79

$K_8[Co_2(CN)_8]$

CONSENSUS REPORTS: Cyanide and its compounds, as well as cobalt and its compounds, are on the Community Right-To-Know List.

OSHA PEL: TWA 5 mg(CN)/m^3
ACGIH TLV: TWA 5 mg(CN)/m^3 (skin)
DFG MAK: 5 mg/m^3
NIOSH REL: (Cyanide) CL 5 mg(CN)/m^3/10M

SAFETY PROFILE: Many cyanide compounds are poisons. Ignites spontaneously in air. A very unstable material. When heated to decomposition it emits toxic fumes of CN^-, NO_x, and K_2O. See also COBALT COMPOUNDS and CYANIDE.

PLN250 CAS:764-71-6 ***HR: 1***
POTASSIUM OCTANOATE
mf: $C_8H_{16}O_2$•K mw: 183.34

SYNS: OCTANOIC ACID, POTASSIUM SALT ◇ POTASSIUM CAPRYLATE

TOXICITY DATA with REFERENCE
skn-hmn 7320 mg BJDEAZ 75,113,63

CONSENSUS REPORTS: Reported in EPA TSCA Inventory.

SAFETY PROFILE: A human skin irritant. When heated to decomposition it emits toxic fumes of K_2O.

PLN500 CAS:17523-77-2 ***HR: 3***
POTASSIUM OXOTETRAFLUORONIOBATE(V)
mf: F_5NbO•2K mw: 282.11

SYNS: COLUMBIUM POTASSIUM FLUORIDE ◇ NIOBIUM POTASSIUM FLUORIDE ◇ POTASSIUM HYDROXYFLUORONIOBATE

TOXICITY DATA with REFERENCE
orl-mus LD50:130 mg/kg HYSAAV 32,343,67

OSHA PEL: TWA 2.5 mg(F)/m^3
NIOSH REL: TWA 2.5 mg(F)/m^3

SAFETY PROFILE: Poison by ingestion. When heated to decomposition it emits toxic fumes of F^- and K_2O. See also FLUORIDES and NIOBIUM.

PLN750 CAS:10025-98-6 ***HR: 3***
POTASSIUM PALLADIUM CHLORIDE
mf: Cl_4K_2Pd mw: 326.40

SYNS: DIPOTASSIUM TETRACHLOPALLADATE ◇ POTASSIUM PALLADOUS CHLORIDE ◇ POTASSIUM TETRACHLOROPALLADATE ◇ TETRACHLOROPALLADATE(2-) DIPOTASSIUM

TOXICITY DATA with REFERENCE
skn-rbt 100 mg/24H MLD AEHLAU 30,168,75
dnr-esc 100 μg/L PCJOAU 16,721,82
oms-omi 1 μmol/L SOGEBZ 11,911,75
dnd-omi 1 μmol/L SOGEBZ 11,911,75
ivn-rat LD50:6400 μg/kg EVHPAZ 10,63,75

CONSENSUS REPORTS: Reported in EPA TSCA Inventory.

SAFETY PROFILE: Poison by intravenous route. Mutation data reported. A skin irritant. When heated to decomposition it emits toxic fumes of K_2O and Cl^-. See also PALLADIUM.

PLO000 CAS:2624-31-9 ***HR: 1***
POTASSIUM PALMITATE
mf: $C_{16}H_{31}O_2$•K mw: 294.57

TOXICITY DATA with REFERENCE
skn-hmn 11800 mg BJDEAZ 75,113,63

CONSENSUS REPORTS: Reported in EPA TSCA Inventory.

SAFETY PROFILE: A human skin irritant. When heated to decomposition it emits toxic fumes of K_2O.

PLO100 CAS:78937-14-1 ***HR: 3***
ASSIUM PENTACARBONYL VANADATE(3⁻)
mf: $C_5K_3O_5V$ mw: 308.29

$K_3[V(CO)_5]$

SAFETY PROFILE: Ignites spontaneously in air. An extremely shock-sensitive explosive. Ignites on contact with polychlorotrifluoroethylene. When heated to decomposition it emits toxic fumes of K_2O and V_2O. See also VANADIUM COMPOUNDS and CARBONYLS.

PLO150 ***HR: 3***
POTASSIUM PENTACYANODIPEROXOCHROMATE(5⁻)
mf: $C_5CrK_5N_5O_4$ mw: 441.57

$K_5[(CN)_5Cr(O_2)_2]$

CONSENSUS REPORTS: Chromium and its compounds, as well as cyanide and its compounds, are on the Community Right-To-Know List.

OSHA PEL: TWA 5 mg(CN)/m^3
ACGIH TLV: TWA 5 mg(CN)/m^3 (skin)
DFG MAK: 5 mg/m^3
NIOSH REL: (Cyanide) CL 5 mg(CN)/m^3/10M

SAFETY PROFILE: Many cyanide compounds are poisons. Many chromates are carcinogens. Highly explosive. When heated to decomposition it emits toxic fumes of NO_x and K_2O. See also CYANIDE, PEROXIDES, and CHROMIUM COMPOUNDS.

PLO250 ***HR: 3***
POTASSIUM PENTAPEROXYDICHROMATE
mf: $Cr_2K_2O_{12}$ mw: 274.20

SYN: POTASSIUM PENTAPEROXODICHROMATE

CONSENSUS REPORTS: Chromium and its compounds are on the Community Right-To-Know List.

SAFETY PROFILE: Many cyanide compounds are poisons. An oxidizer. Very unstable. It explodes above 0°C. Upon decomposition it emits toxic fumes of K_2O. See also CHROMIUM COMPOUNDS.

PLO500 CAS:7778-74-7 ***HR: 3***
POTASSIUM PERCHLORATE
DOT: UN 1489
mf: $ClO_4 \cdot K$ mw: 138.55

PROP: Colorless crystals or white powder. Decomp @ 400° and with organic matter. D: 2.52, mp: 610° ± 10°. Insol in alc.

SYNS: PERIODIN ◇ POTASSIUM HYPERCHLORIDE

TOXICITY DATA with REFERENCE
orl-rat TDLo:27675 mg/kg (female 1-9D post):TER JRPFA4 27,265,71

CONSENSUS REPORTS: Reported in EPA TSCA Inventory.
DOT Classification: Oxidizer; Label: Oxidizer.

SAFETY PROFILE: An experimental teratogen. A powerful oxidizer. Severe irritant to skin, eyes, and mucous membranes. Has been implicated in aplastic anemia. Absorption can cause methemoglobinemia and kidney injury.

It has been involved in many industrial explosions. Explodes on contact with aluminum + barium nitrate + potassium nitrate + water. Forms explosive mixtures with aluminum powder + titanium dioxide; ethylene glycol (240°C); cotton lint (245°C); furfural (270°C); lactose; metal powders (e.g., aluminum; iron; magnesium; molybdenum; nickel; tantalum; titanium); sulfur; titanium hydride. Reaction with ethanol + heat forms the explosive ethyl perchlorate. Violent reaction or ignition under the proper conditions with aluminum + aluminum fluoride; barium chromate + tungsten or titanium; boron + magnesium + silicone rubber; ferrocenium diamminetetrakis(thiocyanato-N) chromate(1−); potassium hexacyanocobaltate(3−); Al + Mg; charcoal; F_2; Ni + Ti; reducing agents. When heated to decomposition it emits very toxic fumes of K_2O and Cl^-. See also PERCHLORATES.

PLO750 CAS:7790-21-8 ***HR: 3***
POTASSIUM PERIODATE
mf: IKO_4 mw: 230.01

PROP: Colorless crystals. Mp: 582° (decomp), d: 3.618 @ 15°, loses O_2 @ 300°.

SAFETY PROFILE: Powerful oxidizer and irritating to skin, eyes, and mucous membranes. Flammable when exposed to heat or flame. Traces of the periodate increase the impact sensitivity with ammonium perchlorate. When heated to decomposition it emits very toxic fumes of K_2O and I^-. See also IODATES.

PLP000 CAS:7722-64-7 ***HR: 3***
POTASSIUM PERMANGANATE
DOT: UN 1490
mf: $MnO_4 \cdot K$ mw: 158.04

PROP: Dark purple crystals with a blue metallic sheen; sweetish astringent taste. Mp: decomp @ <240°, d: 2.703.

SYNS: CAIROX ◇ CHAMELEON MINERAL ◇ C.I. 77755 ◇ CONDY'S CRYSTALS ◇ KALIUMPERMANGANAAT (DUTCH) ◇ KALIUMPERMANGANAT (GERMAN) ◇ PERMANGANATE de POTASSIUM (FRENCH) ◇ PERMANGANATE of POTASH (DOT) ◇ POTASSIO (PERMANGANATO di) (ITALIAN) ◇ POTASSIUM (PERMANGANATE de) (FRENCH)

TOXICITY DATA with REFERENCE
dnd-esc 200 μmol/L MUREAV 89,95,81
dnr-bcs 17 mg/L WATRAG 14,1613,80
mmo-omi 10 ppm JOBAAY 54,767,47
cyt-mus:mmr 1 mmol/L/48H MUREAV 67,221,79
itt-rat TDLo:400 mg/kg (1D male):REP FESTAS 24,884,73
orl-wmn TDLo:2400 μg/kg/D:GIT AIPTAK 44,446,33
orl-hmn LDLo:143 mg/kg:PUL,GIT 34ZIAG -,493,69
orl-rat LD50:1090 mg/kg AIHAAP 30,470,69
scu-mus LD50:500 mg/kg 27ZWAY 3.2,1346,-
orl-dog LDLo:400 mg/kg YKYUA6 31,855,80
orl-rbt LDLo:600 mg/kg YKYUA6 31,855,80
ivn-rbt LDLo:70 mg/kg EQSSDX 1,1,75

CONSENSUS REPORTS: Manganese and its compounds are on the Community Right-To-Know List. Reported in EPA TSCA Inventory. EPA Genetic Toxicology Program.

OSHA PEL: CL 5 $mg(Mn)/m^3$
ACGIH TLV: TWA 5 $mg(Mn)/m^3$
DOT Classification: Oxidizer; Label: Oxidizer.

SAFETY PROFILE: A human poison by ingestion. Poison experimentally by ingestion and intravenous routes. Moderately toxic by subcutaneous route. Human systemic effects by ingestion: dyspnea, nausea, other gastrointestinal effects. Experimental reproductive effects. Mutation data reported. A strong irritant due to its oxidizing properties. Used in production of drugs of abuse, as a topical antibacterial agent, and a chemical reagent.

Flammable by chemical reaction. A powerful oxidizer. A dangerous explosion hazard; handle with care. Explosions may occur in contact with organic or readily oxidizable materials, either when dry or in solution. Dangerous; keep away from combustible materials.

Explodes on contact with acetic acid; acetic anhydride; ammonium nitrate; dimethylformamide; formaldehyde; concentrated hydrochloric acid; potassium chloride + sulfuric acid; sulfuric acid + water. Forms sensitive explosive mixtures with aluminum powder + ammonium nitrate + glyceryl nitrate + nitrocellulose;

ammonium perchlorate; arsenic; phosphorus; sulfur; slag wool; titanium.

Ignites on contact with Al_4C_3; dimethyl sulfoxide; ethylene glycol; H_2S_3; HCl; H_2SO_4; (H_2SO_4 + organic matter); (H_2SO_4 + KCl); NH_4ClO_4; NH_3; NH_4; NO_3; NH_2OH; organic matter; wood; oxygenated organic compounds (e.g., ethylene glycol; propane-1,2-diol; erythritol; mannitol; triethanolamine; 3-chloropropane-1,2-diol; acetaldehyde; isobutyraldehyde; benzaldehyde; acetylacetone; esters of ethylene glycol; lactic acid; acetic acid; oxalic acid).

Violent reaction or ignition under the proper conditions with acetone + tert-butylamine; alcohols + nitric acid; aluminum carbide; ammonia + sulfuric acid; antimony; coal + peroxomonosulfuric acid; dichloromethylsilane; dimethyl sulfoxide; ethanol + sulfuric acid; glycerol; concentrated hydrofluoric acid; hydrogen peroxide; hydrogen trisulfide; hydroxylamine; carbon; organic nitro compounds; polypropylene; 3,4,4′-trimethyldiphenyl sulfone.

When heated to decomposition it emits toxic fumes of K_2O. See also PERMANGANATES.

PLP250 CAS:17014-71-0 ***HR: 3***
POTASSIUM PEROXIDE
DOT: UN 1491
mf: KO_2 mw: 71.1

PROP: Yellow, amorph mass (white crystals). Mp: 490°.

CONSENSUS REPORTS: Reported in EPA TSCA Inventory.

DOT Classification: Oxidizer; Label: Oxidizer.

SAFETY PROFILE: Dangerous fire hazard by spontaneous chemical reaction. It is a very powerful oxidizer. Fires of this material should be handled like sodium peroxide fires. Moderate explosion hazard by spontaneous chemical reaction. Explodes on contact with water. Violent reactions with air; Sb; As; O_2; K. Vigorous reaction on contact with reducing materials. On contact with acid or acid fumes, it can emit toxic fumes. Incompatible with carbon; diselenium dichloride; ethanol; hydrocarbons; metals. When heated to decomposition it emits toxic fumes of K_2O. See also PEROXIDES, INORGANIC.

PLP500 ***HR: 3***
POTASSIUM PEROXYFERRATE
mf: FeK_2O_5 mw: 214.06

SYN: POTASSIUM PEROXOFERRATE(2−)

SAFETY PROFILE: A heat- and impact-sensitive explosive. Explodes on contact with charcoal; phosphorus; sulfur; sulfuric acid. When heated to decomposition it emits toxic fumes of K_2O.

PLP750 CAS:10361-76-9 ***HR: 2***
POTASSIUM PEROXYSULFATE
mf: $HO_5S{\bullet}K$ mw: 152.17

SYNS: PEROXYSULFURIC ACID, POTASSIUM SALT ◇ POTASSIUM MONOPERSULFATE

TOXICITY DATA with REFERENCE
skn-gpg 25% SEV 27ZTAP 3,118,69

SAFETY PROFILE: Very toxic and irritating to skin, eyes and mucous membranes. When heated to decomposition it emits toxic fumes of SO_x and K_2O. Used in laundry bleaches, scouring powders, denture cleaners, in general oxidizing reactions.

PLQ000 CAS:10466-65-6 ***HR: 2***
POTASSIUM PERRHENATE
mf: KO_4Re mw: 289.30

PROP: White crystals. Mp: 350°, d: 4.887.

TOXICITY DATA with REFERENCE
ipr-mus LD50:692 mg/kg 20PKA3 -,45,67

CONSENSUS REPORTS: Reported in EPA TSCA Inventory.

SAFETY PROFILE: Moderately toxic by intraperitoneal route. A powerful oxidizer. May undergo hazardous reactions with flammable materials. When heated to decomposition it can emit K_2O. See also RHENIUM.

PLQ275 ***HR: 3***
POTASSIUM PHENYL DINITROMETHANIDE
mf: $C_7H_5KN_2O_4$ mw: 220.22

SAFETY PROFILE: An explosive. When heated to decomposition it emits toxic fumes of NO_x and K_2O. See also NITRO COMPOUNDS.

PLQ400 ***HR: 1***
POTASSIUM PHOSPHATE, DIBASIC
mf: K_2HPO_4 mw: 174.18

PROP: Colorless or white granular solid. Deliq; sol in water; insol in alc.

SYNS: DIPOTASSIUM MONOPHOSPHATE ◇ DIPOTASSIUM PHOSPHATE

SAFETY PROFILE: A nuisance dust.

PLQ405 ***HR: 1***
POTASSIUM PHOSPHATE, MONOBASIC
mf: KH_2PO_4 mw: 136.09

PROP: Colorless crystals or white crystalline powder; odorless. Sol in water; insol in alc.

SYNS: POTASSIUM BIPHOSPHATE ◇ POTASSIUM DIHYDROGEN PHOSPHATE ◇ MONOPOTASSIUM PHOSPHATE

SAFETY PROFILE: A nuisance dust.

PLQ410 ***HR: 1***
POTASSIUM PHOSPHATE, TRIBASIC
mf: K_3PO_4 mw: 212.27

PROP: White crystals. Hygroscopic, sol in water; insol in alc.

SYN: TRIPOTASSIUM PHOSPHATE

SAFETY PROFILE: A nuisance dust.

PLQ500 CAS:20770-41-6 ***HR: 3***
POTASSIUM PHOSPHIDE
DOT: UN 2012
mf: K_3P mw: 148.27

PROP: A solid.

SYN: PHOSPHURE de POTASSIUM (FRENCH)

TOXICITY DATA with REFERENCE
ihl-rat LCLo:580 ppm/1H ZGSHAM 25,279,33
ihl-cat LCLo:173 ppm/2H ZGSHAM 25,279,33
ihl-gpg LCLo:288 ppm/2H ZGSHAM 25,279,33

CONSENSUS REPORTS: Reported in EPA TSCA Inventory.

DOT Classification: Flammable Solid; Label: Dangerous When Wet, Poison

SAFETY PROFILE: Paison by ingestion. Moderately toxic by inhalation. Flammable solid. When heated to decomposition it emits very toxic fumes of PO_x, K_2O, and PH_3. See also PHOSPHIDES and PHOSPHINE which is released upon contact of phosphide with water, steam.

PLQ750 CAS:7782-87-8 ***HR: 3***
POTASSIUM PHOSPHINATE
mf: H_2KO_2P mw: 104.09

PROP: White, opaque, very deliq crystals or powder; pungent saline taste. Mp: decomp.

SYN: POTASSIUM HYPOPHOSPHITE

SAFETY PROFILE: When heated it evolves phosphine which then ignites in air. Explodes on evaporation with HNO_3. When heated to decomposition it emits very toxic fumes of PO_x and K_2O.

PLQ775 CAS:573-83-1 ***HR: 3***
POTASSIUM PICRATE
mf: $C_6H_2KNO_7$ mw: 239.18

SYN: POTASSIUM-2,4,6-TRINITROPHENOXIDE

SAFETY PROFILE: An explosive. Reaction with 2,2-dinitro-2-fluoroethoxycarbonyl chloride forms the explosive ester 2,2-dinitro-2-fluoroethyl-2,4,6-trinitrophenyl carbonate. When heated to decomposition it emits toxic fumes of K_2O. See also PICRATES.

PLR000 CAS:16921-30-5 ***HR: 1***
POTASSIUM PLATINIC CHLORIDE
mf: $Cl_6Pt•K_2$ mw: 485.99

PROP: Yellow crystals. Mp: decomp @ 250°, d: 3.499 @ 24°.

SYNS: HEXACHLOROPLATINATE(2−) DIPOTASSIUM ◇ PLATINIC POTASSIUM CHLORIDE ◇ POTASSIUM CHLOROPLATINATE ◇ POTASSIUM HEXACHLOROPLATINATE(IV)

TOXICITY DATA with REFERENCE
mma-sat 100 ng/plate PCJOAU 16,721,82
dnr-esc 100 μg/L PCJOAU 16,721,82
msc-ham:ovr 20 μmol/L/20H MUREAV 67,65,79
idr-hmn TDLo:40 mg/kg:SKN CNREA8 35,2766,75

CONSENSUS REPORTS: Reported in EPA TSCA Inventory.

OSHA PEL: TWA 0.002 mg(Pt)/m^3
ACGIH TLV: TWA 0.002 mg(Pt)/m^3

SAFETY PROFILE: Mutation data reported. Human systemic effects by intradermal route: dermatitis. When heated to decomposition it emits toxic fumes of K_2O and Cl^-. Used as a catalyst for carbonylation of alkynes. See also PLATINUM COMPOUNDS.

PLR125 ***HR: 1***
POTASSIUM POLYMETAPHOSPHATE
mf: $(KPO_3)_x$

PROP: White powder; odorless. Insol in water; sol in dilute solutions of sodium salts.

SYNS: POTASSIUM KURROL'S SALT ◇ POTASSIUM METAPHOSPHATE

SAFETY PROFILE: A nuisance dust.

PLR175 CAS:71939-10-1 ***HR: 3***
POTASSIUM-O-PROPIONOHYDROXAMATE
mf: $C_3H_6KNO_2$ mw: 127.18

SAFETY PROFILE: The solid explodes when dried. When heated to decomposition it emits toxic fumes of NO_x and K_2O.

PLR200 ***HR: 1***
POTASSIUM PYROPHOSPHATE
mf: $K_4P_2O_7$ mw: 330.34

PROP: Colorless crystals or white granular solid. Hygroscopic, sol in water; insol in alc.

SYN: TETRAPOTASSIUM PYROPHOSPHATE

SAFETY PROFILE: A nuisance dust.

PLR250 CAS:16731-55-8 ***HR: 3***
POTASSIUM PYROSULFITE
DOT: NA 2693
mf: O_5S_2•K mw: 183.22

PROP: Monoclinic plates or white crystalline powder; sulfur dioxide odor. Mp: decomp; d: 2.3. Sol in water; insol in alc.

SYNS: POTASSIUM METABISULFITE (DOT, FCC) ◇ PYROSULFUROUS ACID, DIPOTASSIUM SALT

TOXICITY DATA with REFERENCE
orl-rat TDLo:35 g/kg (49D pre/1-21D preg):REP CRSBAW 172,470,78
orl-mus TDLo:1440 g/kg/2Y-C:ETA EESADV 3,451,79

CONSENSUS REPORTS: Reported in EPA TSCA Inventory. EPA Genetic Toxicology Program.

DOT Classification: ORM-B; Label: None.

SAFETY PROFILE: Experimental reproductive effects. A very irritating material. Questionable carcinogen with experimental tumorigenic data. When heated to decomposition it emits toxic fumes of SO_x and K_2O. See also SULFITES.

PLR500 CAS:54328-07-3 ***HR: 3***
POTASSIUM RHENATE
mf: KO_4Re mw: 289.30

SYN: RHENIC ACID, POTASSIUM SALT

TOXICITY DATA with REFERENCE
ipr-mus LD50:2800 mg/kg EQSSDX 1,1,75
ivn-cat LDLo:70 mg/kg EQSSDX 1,1,75

SAFETY PROFILE: Poison by intravenous route. Moderately toxic by intraperitoneal route. When heated to decomposition it emits toxic fumes of K_2O. See also RHENIUM.

PLR750 CAS:7790-59-2 ***HR: 3***
POTASSIUM SELENATE
mf: O_4Se•2K mw: 221.16

PROP: Colorless crystals. D: 3.07. Sol in water.

SYN: SELENIC ACID, DIPOTASSIUM SALT

TOXICITY DATA with REFERENCE
orl-rat TDLo:126 mg/kg (15-22D preg/21D post):REP PSEBAA 87,295,54
ivn-rat LDLo:4300 μg/kg JPETAB 33,270,28
orl-rbt LDLo:1800 mg/kg JPETAB 33,270,28
ivn-rbt LDLo:1100 mg/kg JPETAB 33,270,28

CONSENSUS REPORTS: Selenium and its compounds are on the Community Right-To-Know List. Reported in EPA TSCA Inventory. EPA Genetic Toxicology Program.

OSHA PEL: TWA 0.2 mg(Se)/m^3
ACGIH TLV: TWA 0.2 mg(Se)/m^3
DFG MAK: 0.1 mg(Se)/m^3

SAFETY PROFILE: Poison by intravenous route. Moderately toxic by ingestion. Experimental reproductive effects. When heated to decomposition it emits toxic fumes of Se and K_2O. See also SELENIUM COMPOUNDS.

PLS000 CAS:3425-46-5 ***HR: 3***
POTASSIUM SELENOCYANATE
mf: CHNSe•K mw: 145.09

SYN: SELENOCYANIC ACID, POTASSIUM SALT

TOXICITY DATA with REFERENCE
ipr-rat LDLo:200 mg/kg NCNSA6 5,28,53

CONSENSUS REPORTS: Cyanide and its compounds, as well as selenium and its compounds, are on the Community Right-To-Know List. Reported in EPA TSCA Inventory.

OSHA PEL: TWA 5 mg(CN)/m^3; TWA 0.2 mg(Se)/m^3
ACGIH TLV: TWA 5 mg/m^3 (skin); TWA 0.2 mg(Se)/m^3
DFG MAK: 5 mg/m^3
NIOSH REL: TWA CL 5 mg/m^3/10M

SAFETY PROFILE: Poison by intraperitoneal route. When heated to decomposition it emits very toxic fumes of NO_x, CN^-, K_2O, and Se. See also SELENIUM COMPOUNDS and CYANATES.

PLS250 CAS:506-61-6 ***HR: 3***
POTASSIUM SILVER CYANIDE
mf: C_2AgN_2•K mw: 199.01

PROP: White crystals, light-sensitive. Sol in water, acids.

SYNS: KYANOSTRIBRNAN DRASELNY (CZECH) ◇ RCRA WASTE NUMBER P099 ◇ SILVER POTASSIUM CYANIDE

TOXICITY DATA with REFERENCE
skn-rbt 50 mg/24H SEV 28ZPAK -,14,72
eye-rbt 250 μg/24H SEV 28ZPAK -,14,72
orl-rat LD50:20900 μg/kg 28ZPAK -,14,72

CONSENSUS REPORTS: EPA Extremely Hazardous Substances List. Silver and its compounds, as well as cyanide and its compounds, are on the Community Right-To-Know List. Reported in EPA TSCA Inventory.

SAFETY PROFILE: Poison by ingestion. A severe skin and eye irritant. When heated to decomposition it emits very toxic fumes of CN^-, K_2O, and NO_x. See also CYANIDES and SILVER COMPOUNDS.

PLS500 CAS:11135-81-2 ***HR: 3***
POTASSIUM SODIUM ALLOY
DOT: UN 1422

PROP: Low-melting alloy of sodium and potassium metals.

SYN: SODIUM POTASSIUM ALLOY, liquid and solid (DOT)

DOT Classification: Label: Flammable Solid; Label: Flammable Solid and Dangerous When Wet.

SAFETY PROFILE: A low melting alloy of Na and K. Its toxicity is due to either Na or K alone. Corrosive to the eyes, skin, and mucous membranes. Upon contact with moisture it reacts violently to evolve H_2; much heat; and a highly caustic residue of NaOH or KOH. Oxidation forms Na_2O and K_2O which are powerful caustics.

A dangerous fire and explosion hazard. Violent or explosive reaction with O_2; water; moisture; steam; halogens; oxidizers; acids or acid fumes; giving off much heat, hydrogen; toxic and corrosive fumes; often spattering either red-hot particles or actually flaming particles. A severe explosion hazard; will react explosively under the appropriate conditions with moisture; acids; acid fumes; solid CO_2; carbon disulfide; halocarbons (e.g., CH_3Cl; carbon tetrachloride; chloroform; bromoform; 1,1,1-trichloroethane; 1,1,2-trichlorotrifluoroethane; tetrachloroethane; CH_2Cl_2; CH_2I_2); ammonium sulfate + NH_4 + NO_3; HgO; metal halides (e.g., silver halides; zinc chloride; iron(III) chloride); metal oxides (e.g., silver oxide; mercury oxide); nitrogen containing explosives (e.g., ammonium nitrate; ammonium sulfate; picric acid; nitrobenzene); oxalyl bromide; oxalyl chloride; pentachloroethane; K oxides; KO_2; Si; $NaHCO_3$; polytetrafluoroethylene. Reacts vigorously with oxidizing materials.

To fight fire, use G-1 powder, dry sodium chloride, dry sodium carbonate, dry calcium carbonate, dry sand, resin-coated sodium chloride, or dry soda ash. Never use water, graphite, carbon dioxide, halocarbons or foam.

Dangerous; when heated it emits highly toxic fumes of Na_2O and K_2O. Used as a liquid coolant for nuclear reactor cores. See also SODIUM and POTASSIUM.

PLS750 CAS:590-00-1 ***HR: 2***
POTASSIUM SORBATE
mf: $C_6H_7O_2$•K mw: 150.23

PROP: White crystals, crystalline powder, or pellets. Mp: 270° (decomp): d: 1.363 @ 25°/20°. Sol in alc, water.

SYNS: 2,4-HEXADIENOIC ACID POTASSIUM SALT ◇ SORBIC ACID, POTASSIUM SALT ◇ SORBISTAT-K ◇ SORBISTAT-POTASSIUM

TOXICITY DATA with REFERENCE
cyt-ham:lng 10 g/L ATSUDG (4),41,80
cyt-ham:fbr 4 g/L/48H MUREAV 48,337,77
sce-ham:lng 10 g/L FCTOD7 22,501,84
orl-rat LD50:4920 mg/kg FAONAU 40,61,67
ipr-mus LD50:1300 mg/kg FAONAU 53A,121,74

CONSENSUS REPORTS: Reported in EPA TSCA Inventory. EPA Genetic Toxicology Program.

SAFETY PROFILE: Moderately toxic by intraperitoneal route. Mildly toxic by ingestion. Mutation data reported. When heated to decomposition it emits toxic fumes of K_2O.

PLS760 CAS:12125-03-0 ***HR: 3***
POTASSIUM STANNATE TRIHYDRATE
mf: K_2OSn•$3H_2O$ mw: 266.95

TOXICITY DATA with REFERENCE
ivn-mus LD50:178 mg/kg CSLNX* NX#02352

OSHA PEL: TWA 2 mg(Sn)/m^3
ACGIH TLV: TWA 2 mg(Sn)/m^3

SAFETY PROFILE: Poison by intravenous route. When heated to decomposition it emits toxic fumes of Sn.

PLT000 CAS:7778-80-5 ***HR: 2***
POTASSIUM SULFATE (2:1)
mf: O_4S•2K mw: 174.26

PROP: Colorless to white, odorless crystals; bitter salty taste. D: 2.66, mp: 1067°. Sol in water; insol in alc.

SYN: SULFURIC ACID, DIPOTASSIUM SALT

TOXICITY DATA with REFERENCE
orl-wmn LDLo:800 mg/kg AEXPBL 21,169,1886
orl-rat LD50:6600 mg/kg GISAAA 50(7),24,85
scu-gpg LDLo:3000 mg/kg HBAMAK 4,1360,35

CONSENSUS REPORTS: Reported in EPA TSCA Inventory.

SAFETY PROFILE: Moderately toxic to humans by ingestion. Moderately toxic experimentally by subcutaneous route. Swallowing large doses causes severe gastrointestinal tract effects. When heated to decomposition it emits toxic fumes of K_2O and SO_x. See also SULFATES.

PLT250 CAS:1312-73-8 ***HR: 3***
POTASSIUM SULFIDE (2:1)
DOT: UN 1382/UN 1847
mf: K_2S mw: 110.26

PROP: Red, crystalline mass; deliq in air. Mp: 912°, d: 1.805 @ 14°.

SYNS: HEPAR SULFUROUS ◇ POTASSIUM MONOSULFIDE

CONSENSUS REPORTS: Reported in EPA TSCA Inventory.

DOT Classification: Flammable Solid; Label: Flammable Solid (UN1382); Flammable Solid; Label: Spontaneously Combustible (UN1382); Corrosive Material; Label: Corrosive (UN1847).

SAFETY PROFILE: Poison by ingestion and inhalation. Emits H_2S in contact with acids; steam. A flammable solid. Unstable; may explode on percussion or rapid heating. Ignites on contact with nitrogen oxide. When heated to decomposition it emits very toxic fumes of K_2O and SO_x. See also SULFIDES.

PLT500 CAS:10117-38-1 ***HR: D***
POTASSIUM SULFITE
mf: O_3S•2K mw: 158.26

PROP: White crystals or granular powder; odorless. Sol in water; sltly sol in alc.

SYN: SULFUROUS ACID, DIPOTASSIUM SALT

CONSENSUS REPORTS: Reported in EPA TSCA Inventory.

SAFETY PROFILE: When heated to decomposition it emits toxic fumes of SO_x and K_2O. See also SULFUROUS ACID and SULFITES.

PLT750 CAS:79796-14-8 ***HR: 3***
POTASSIUM SULFUR DIIMIDE
mf: K_2N_2S mw: 138.28

KN:S:NK

SYN: POTASSIUMSULFURDIIMIDATE

SAFETY PROFILE: Ignites in air. Explodes on contact with water, methanol, chloromethane, methylene chloride, carbon tetrachloride. When heated to decomposition it emits very toxic fumes of SO_x, NO_x, and K_2O. See also SULFIDES.

PLU000 CAS:15571-91-2 ***HR: 3***
POTASSIUM TELLURITE
mf: O_3Te•2K mw: 253.80

TOXICITY DATA with REFERENCE
orl-man TDLo:40 mg/kg/7D:CNS,GIT NRTTA8 1283,1,67
unr-mam LDLo:20 mg/kg NRTTA8 1283,1,67

CONSENSUS REPORTS: Reported in EPA TSCA Inventory. EPA Genetic Toxicology Program.

OSHA PEL: TWA 0.1 mg(Te)/m^3
ACGIH TLV: TWA 0.1 mg(Te)/m^3

SAFETY PROFILE: A poison by an unspecified route. Human systemic effects by ingestion: sleep disturbance, anorexia, nausea. When heated to decomposition it emits toxic fumes of K_2O and Te. See also TELLURIUM compounds.

PLU250 CAS:13826-94-3 ***HR: 3***
POTASSIUM TETRABROMOPLATINATE
mf: Br_4Pt•2K mw: 592.93

SYN: TETRABROMO-PLATINUM(2-), DIPOTASSIUM

TOXICITY DATA with REFERENCE
dnr-esc 100 μg/L PCJOAU 16,721,82
ipr-mus LDLo:45 mg/kg BICHBX 2,187,73

CONSENSUS REPORTS: Reported in EPA TSCA Inventory.

OSHA PEL: TWA 0.002 mg(Pt)/m^3
ACGIH TLV: TWA 0.002 mg(Pt)/m^3

SAFETY PROFILE: Poison by intraperitoneal route. Mutation data reported. When heated to decomposition it emits toxic fumes of K_2O and Br^-. See also PLATINUM COMPOUNDS and BROMIDES.

PLU500 CAS:591-89-9 ***HR: 3***
POTASSIUM TETRACYANOMERCURATE(II)
DOT: UN 1626
mf: C_2HgN_2•2CKN mw: 382.87

SYNS: MERCURIC POTASSIUM CYANIDE (DOT) ◇ MERCURIC POTASSIUM CYANIDE, solid (DOT)

CONSENSUS REPORTS: Cyanide and its compounds, as well as mercury and its compounds, are on the Community Right-To-Know List.

OSHA PEL: (Transitional: CL 1 mg/10m^3) CL 0.1 mg(Hg)/m^3 (skin)
ACGIH TLV: TWA 0.1 mg(Hg)/m^3 (skin)
NIOSH REL: TWA (Inorganic Mercury) 0.05 mg(Hg)/m^3
DOT Classification: Poison B; Label: Poison.

SAFETY PROFILE: A poison. May explode on contact with ammonia. When heated to decomposition it emits very toxic fumes of NO_x, Hg, K_2O, and CN^-. See also MERCURY COMPOUNDS and CYANIDE.

PLU550 CAS:75038-71-0 ***HR: 3***
POTASSIUM TETRACYANOTITANATE(IV)
mf: $C_4K_4N_4Ti$ mw: 464.85

CONSENSUS REPORTS: Cyanide and its compounds are on the Community Right-To-Know List.

SAFETY PROFILE: Many cyanide compounds are poisons. Violent reaction with water. When heated to decomposition it emits toxic fumes of NO_x, CN^- and K_2O. See also CYANIDE and TITANIUM COMPOUNDS.

PLU575 CAS:65664-23-5 ***HR: 3***
POTASSIUM TETRAETHYNYL NICKELATE(2⁻)
mf: $C_8H_4K_2Ni$ mw: 237.01

$K_2[(HC\equiv C)_4Ni]$

CONSENSUS REPORTS: Nickel and its compounds are on the Community Right-To-Know List.

SAFETY PROFILE: Ignites spontaneously in air. Explodes on contact with water. When heated to decomposition it emits toxic fumes of K_2O. See also NICKEL COMPOUNDS and ACETYLENE COMPOUNDS.

PLU580 ***HR: 3***
POTASSIUM TETRAETHYNYL NICKELATE(4⁻)
mf: $C_8H_4K_4Ni$ mw: 315.20

$K_4[(HC\equiv C)_4Ni]$

CONSENSUS REPORTS: Nickel and its compounds are on the Community Right-To-Know List.

SAFETY PROFILE: An explosive sensitive to impact, friction or exposure to to flame. When heated to decomposition it emits toxic fumes of K_2O. See also NICKEL COMPOUNDS and ACETYLENE COMPOUNDS.

PLU590 CAS:14244-61-2 ***HR: 3***
POTASSIUM TETRAKISTHIOCYANATOPLATINATE
mf: $C_4K_2N_4PtS_4$ mw: 505.61

SYN: PLATINATE(2-), TETRAKIS(THIOCYANATO)-, DIPOTASSIUM

TOXICITY DATA with REFERENCE
ipr-mus LD50:50 mg/kg VOONAW 25(11),47,79

OSHA PEL: TWA 0.002 mg(Pt)/m^3
ACGIH TLV: TWA 0.002 mg(Pt)/m^3

SAFETY PROFILE: Poison by intraperitoneal route. When heated to decomposition it emits toxic fumes of SO_x, CN^-, and Pt.

PLU600 CAS:32607-31-1 ***HR: 3***
POTASSIUM-1,1,2,2-TETRANITROETHANDIIDE
mf: $C_2K_2N_4O_8$ mw: 286.24

SAFETY PROFILE: A very impact-sensitive explosive. When heated to decomposition it emits toxic fumes of NO_x and K_2O. See also NITRO COMPOUNDS.

PLU750 CAS:12331-76-9 ***HR: 3***
POTASSIUM TETRAPEROXYCHROMATE
mf: CrK_3O_8 mw: 297.31

CONSENSUS REPORTS: Chromium and its compounds are on the Community Right-To-Know List.

SAFETY PROFILE: Many cyanide compounds are poisons. A poison. An oxidizer. Explodes @ 178°. Explodes on contact with sulfuric acid. The impure salt is explosive. When heated to decomposition it emits toxic fumes of K_2O. See also CHROMIUM COMPOUNDS and PEROXIDES.

PLV000 CAS:42489-15-6 ***HR: 3***
POTASSIUM TETRAPEROXYMOLYBDATE
mf: K_2MoO_8 mw: 302.15

SAFETY PROFILE: Explosive. When heated to decomposition it emits toxic fumes of K_2O. See also PEROXIDES and MOLYBDENUM COMPOUNDS.

PLV250 CAS:37346-96-6 ***HR: 3***
POTASSIUM TETRAPEROXYTUNGSTATE
mf: K_2O_8W mw: 390.06

SAFETY PROFILE: Explodes on friction or rapid heating to 80°C. An oxidizer. When heated to decomposition it emits toxic fumes of K_2O. See also PEROXIDES and TUNGSTEN COMPOUNDS.

PLV275 CAS:51286-83-0 ***HR: 3***
POTASSIUM-1-TETRAZOLACETATE
mf: $C_3H_3KN_4O_2$ mw: 166.18

SAFETY PROFILE: An explosive sensitive to spark, flame, or heating over 200°C. When heated to decomposition it emits toxic fumes of NO_x and K_2O.

PLV500 ***HR: 3***
POTASSIUM THALLIUM(I)AMIDE AMMONIATE
mf: $K_2NTl\cdot NH_3$ mw: 313.61

CONSENSUS REPORTS: Thallium and its compounds are on the Community Right-To-Know List.

SAFETY PROFILE: Thallium compounds are considered to be poisons. Self-explodes. Incompatible with acids or water. When heated to decomposition it emits very toxic fumes of K_2O, NH_3, and NO_x. See also THALLIUM COMPOUNDS and AMMONIA.

PLV750 CAS:333-20-0 ***HR: 3***
POTASSIUM THIOCYANATE
mf: $CNS\cdot K$ mw: 97.18

$KSC\equiv N$

PROP: Colorless, deliq crystals. D: 1.89, mp: about 173°.

SYNS: ARTEROCYN ◇ ATEROCYN ◇ KYONATE ◇ POTASSIUM ISOTHIOCYANATE ◇ POTASSIUM RHODANATE ◇ POTASSIUM RHODANIDE ◇ POTASSIUM SULFOCYANATE ◇ POTASSIUM THIOCYANIDE ◇ RODANCA ◇ RHODANIDE

TOXICITY DATA with REFERENCE
orl-dom TDLo:1779 mg/kg (female 1-20W post):TER RJARAV 14,67,76
orl-hmn LDLo:80 mg/kg JAMAAP 119,1177,42
orl-man TDLo:428 mg/kg ATXKA8 23,66,67
orl-rat LD50:854 mg/kg JAPMA8 29,152,40
orl-mus LD50:594 mg/kg JAPMA8 29,152,40
scu-rbt LDLo:550 mg/kg HBAMAK 4,1391,35
ivn-rbt LDLo:150 mg/kg HBAMAK 4,1391,35

CONSENSUS REPORTS: Reported in EPA TSCA Inventory.

SAFETY PROFILE: A human poison by ingestion. Poison experimentally by intravenous route. An experimental teratogen. Moderately toxic by subcutaneous and ingestion routes. Large doses can cause skin eruptions, psychoses, and collapse. Incompatible with calcium chlorite and perchloryl fluoride. When heated to decomposition it emits very toxic fumes of CN^-, K_2O, SO_x, and NO_x. See also THIOCYANATES.

PLW000 CAS:10294-66-3 ***HR: D***
POTASSIUM THIOSULFATE
mf: $O_3S_2 \cdot 2K$ mw: 190.32

PROP: Colorless, hygroscopic crystals. Sol in water; insol in alc.

SYNS: POTASSIUM HYPOSULFITE ◇ THIOSULFURIC ACID, DIPOTASSIUM SALT

CONSENSUS REPORTS: Reported in EPA TSCA Inventory.

SAFETY PROFILE: When heated to decomposition it emits toxic fumes of SO_x and K_2O. See also THIOSULFATES.

PLW150 CAS:12056-53-0 ***HR: 2***
POTASSIUM TITANIUM OXIDE
mf: KO_8Ti_4 mw: 358.70

SYN: POTASSIUM OCTATITANATE

TOXICITY DATA with REFERENCE
imp-rat TDLo:200 mg/kg:NEO JJIND8 67,965,81

SAFETY PROFILE: Questionable carcinogen with experimental neoplastigenic data.

PLW200 CAS:12012-50-9 ***HR: 3***
POTASSIUM TRICHLOROETHYLENEPLATINATE
mf: $C_2H_4Cl_3Pt \cdot K$ mw: 368.60

SYNS: PLATINATE(1-), TRICHLOROETHYLENE-, DIPOTASSIUM ◇ Pt-93

TOXICITY DATA with REFERENCE
ipr-mus LD50:85 mg/kg VOONAW 25(11),47,79

CONSENSUS REPORTS: Reported in EPA TSCA Inventory.

OSHA PEL: TWA 0.002 mg(Pt)/m^3
ACGIH TLV: TWA 0.002 mg(Pt)/m^3

SAFETY PROFILE: Poison by intraperitoneal route. When heated to decomposition it emits toxic fumes of Pt and Cl^-.

PLW275 CAS:65521-60-0 ***HR: 3***
POTASSIUM TRICYANODIPEROXOCHROMATE (3⁻)
mf: $C_3CrK_3N_3O_4$ mw: 217.15

CONSENSUS REPORTS: Cyanide and its compounds, as well as chromium and its compounds, are on the Community Right-To-Know List.

SAFETY PROFILE: Many cyanide compounds are poisons. Many cyanide compounds are poisons. Very explosive. When heated to decomposition it emits toxic fumes of NO_x, CN^-, and K_2O. See also CYANIDE, CHROMIUM COMPOUNDS, and PEROXIDES.

PLW285 CAS:12298-68-9 ***HR: D***
POTASSIUM TRIIODIDE
mf: I_3K mw: 419.80

SYNS: IODINE-POTASSIUM IODIDE ◇ LUGOL'S IODINE ◇ LUGOL'S SOLUTION ◇ MICROIODIDE

TOXICITY DATA with REFERENCE
iut-rat TDLo:500 mg/kg (female 1D pre):REP AJOGAH 116,167,73

CONSENSUS REPORTS: Reported in EPA TSCA Inventory.

SAFETY PROFILE: Experimental reproductive effects. When heated to decomposition it emits toxic fumes of I^-.

PLW300 CAS:14268-23-6 ***HR: 3***
POTASSIUM TRINITROMETHANIDE
mf: CKN_3O_6 mw: 189.13

SAFETY PROFILE: Potentially explosive. Keep damp and do not store. When heated to decomposition it emits toxic fumes of NO_x and K_2O. See also NITRO COMPOUNDS.

PLW400 ***HR: 1***
POTASSIUM TRIPOLYPHOSPHATE
mf: $K_5P_3O_{10}$ mw: 448.41

PROP: White granules or powder. Hygroscopic, sol in water.

SYNS: PENTAPOTASSIUM TRIPHOSPHATE ◇ POTASSIUM TRIPHOSPHATE

SAFETY PROFILE: A nuisance dust.

PLW500 CAS:11103-86-9 ***HR: 3***
POTASSIUM ZINC CHROMATE HYDROXIDE
mf: $Cr_2HO_9Zn_2$•K mw: 418.85

SYNS: BUTTERCUP YELLOW ◇ CHROMIC ACID, POTASSIUM ZINC SALT (2:2:1) ◇ CITRON YELLOW ◇ POTASSIUM ZINC CHROMATE ◇ ZINC CHROME ◇ ZINC YELLOW

CONSENSUS REPORTS: IARC Cancer Review: Animal Inadequate Evidence IMEMDT 23,205,80. Chromium and its compounds, as well as zinc and its compounds, are on the Community Right-To-Know List. Reported in EPA TSCA Inventory.

OSHA PEL: (Transitional: 1 mg(CrO_3)/10m^3) CL 0.1 mg(CrO_3)/m^3
ACGIH TLV: TWA 0.01 mg(Cr)/M^3; Confirmed Human Carcinogen
DFG MAK: Human Carcinogen.
NIOSH REL: (Chromium (VI)) TWA 0.001 mg (Cr(VI))/m^3

SAFETY PROFILE: Confirmed carcinogen. When heated to decomposition it emits toxic fumes of ZnO and K_2O. Used as a corrosion inhibiting pigment and in steel priming. See also CHROMIUM COMPOUNDS and ZINC COMPOUNDS.

PLW550 ***HR: 3***
POTATO BLOSSOMS, GLYCOALKALOID EXTRACT

PROP: Total glycoalkaloids comprised of 20% α-solanine, α-chaconine and β-chaconine and 80% inorganic substances which are isolated from the blossoms of the potato *Solanum tuberosum L.* TXAPA9 36,227,76

SYNS: GLYCOALKALOID EXTRACT from POTATO BLOSSOMS ◇ TGA-EXTRACT

TOXICITY DATA with REFERENCE
ipr-rat TDLo:40 mg/kg (female 5-12D post):TER TXAPA9 36,227,76
ipr-rat TDLo:40 mg/kg (female 5-12D post):REP TXAPA9 36,227,76
ipr-rat LD50:60 mg/kg TXAPA9 36,227,76

SAFETY PROFILE: Poison by intraperitoneal route. An experimental teratogen. Experimental reproductive effects. When heated to decomposition it emits acrid smoke and irritating fumes.

PLW750 ***HR: 3***
POTATO, GREEN PARTS

SYN: SOLANUM TUBEROSUM L

TOXICITY DATA with REFERENCE
orl-ham TDLo:6250 mg/kg (female 8D post):REP TJADAB 30,371,84
orl-ham TDLo:2700 mg/kg (8D preg):TER TJADAB 17,327,78
ipr-rat TDLo:136 g/kg/2Y-I:CAR EXPEAM 34,645,78
orl-ham LDLo:2700 mg/kg TJADAB 17,327,78

SAFETY PROFILE: Questionable carcinogen with experimental carcinogenic data. Moderately toxic by ingestion. Experimental teratogenic effects. Other experimental reproductive effects.

PLW800 ***HR: 1***
POTHOS

PROP: A climbing vine with very large heart-shaped leaves. The leaves sometimes have yellow streaks. It is a common indoor plant and in small pots the leaves are only a few inches long. It is cultivated outdoors in southern Florida, Hawaii, the West Indies, and Guam.

SYNS: AMAPALO AMRILLO (PUERTO RICO) ◇ DEVIL'S IVY ◇ EPIPREMNUM AUREUM ◇ GOLDEN CEYLON CREEPER ◇ GOLDEN HUNTER'S ROBE ◇ GOLDEN POTHOS ◇ HUNTER'S ROBE ◇ IVY ARUM ◇ MALANGA TREPADORA (CUBA) ◇ SOLOMON ISLAND IVY ◇ TARO VINE ◇ VARIEGATED PHILODENDRON

SAFETY PROFILE: The whole plant contains poisonous crystals of calcium oxalate. Chewing any part of the plant results in burning pain in the lips, mouth, and throat, possibly followed by inflammation and blistering. Cut or crushed pieces may also cause contact dermatitis or conjunctivitis. Systemic effects are usually not seen because of the insolubility of calcium oxalate. See also OXALATES.

PLX000 ***HR: 3***
POWDERED METALS

SAFETY PROFILE: Dangerous fire hazard in dispersed form when exposed to flame or sparks or by chemical reaction with oxidizers. Many powdered metals can ignite spontaneously and explode when suspended in air. In general the more finely divided the metal the greater its reactivity. To fight fire, use no water, use powdered graphite, dolomite, sodium chloride, etc. Get instructions from the supplier of the powdered metal. See also specific metals.

PLX100 ***HR: 3***
PRAIRIE RATTLESNAKE VENOM

SYNS: CROTALUS VIRIDIS VIRIDIS VENOM ◇ C. VIRIDIS VIRIDIS VENOM ◇ VENOM, SNAKE, CROTALUS VIRIDIS VIRIDIS

TOXICITY DATA with REFERENCE
ipr-mus LD50:2 mg/kg TOXIA6 9,131,71
scu-mus LD50:5500 μg/kg TOXIA6 10,81,72
ivn-mus LD50:1010 μg/kg BCPCA6 20,1549,71
ims-mus LD50:3 mg/kg TOXIA6 21,35,83
ipr-mam LD50:2250 μg/kg CLPTAT 8,849,67
ivn-mam LD50:1610 μg/kg CLPTAT 8,849,67

SAFETY PROFILE: Poison by subcutaneous, intramuscular, intravenous, and intraperitoneal routes.

PLX250 CAS:154-97-2 ***HR: 3***
PRALIDOXIME MESYLATE
mf: $C_7H_{10}N_2O{\cdot}CH_3O_3S$ mw: 232.28

SYNS: 2-ALDOXIME PYRIDINIUM-N-METHYL METHANESULPHONATE ◇ CONTRATHION ◇ 2-FORMYL-N-METHYLPYRIDINIUM OXIME METHANESULFONATE ◇ 2-FORMYL-1-METHYLPYRIDINIUM METHANESULFONATE OXIME ◇ 2-((HYDROXYIMINO)METHYL)-1-METHYLPYRIDINIUM EMTHANESULFONATE ◇ 2-HYDROXYIMINOMETHYL-N-METHYLPYRIDINIUM METHANESULPHONATE ◇ 1-METHYLPYRIDINIUM-2-ALDOXIME METHANESULFONATE ◇ N-METHYLPYRIDINIUM-2-ALDOXIMEMETHANESULPHONATE ◇ N-METHYLPYRIDINIUM METHANE SULFONATE-2-ALDOXIME ◇ 2-PAM METHANESULFONATE ◇ PRALIDOXIME METHANESULFONATE ◇ PRALIDOXIME METHOSULFATE

TOXICITY DATA with REFERENCE
orl-hmn TDLo:52500 mg/kg/4W-I:GIT AEHLAU 15,599,67
ivn-hmn TDLo:45 mg/kg:CVS AEHLAU 15,599,67
orl-rat LD50:7000 mg/kg CJBPAZ 39,351,61
ipr-rat LD50:262 mg/kg BJPCAL 13,202,58
scu-rat LD50:332 mg/kg BJPCAL 13,202,58
ivn-rat LD50:109 mg/kg BJPCAL 13,202,58
ims-rat LD50:188 μg/kg BJPCBM 39,822,70
orl-mus LD50:3700 mg/kg CJBPAZ 39,351,61
scu-mus LD50:165 mg/kg BJPCAL 13,202,58
ivn-mus LD50:122 mg/kg BJPCAL 13,202,58
ims-mus LD50:418 μg/kg BJPCBM 39,882,70
ims-mky LD50:356 mg/kg BJPCAL 13,202,58

SAFETY PROFILE: Poison by intraperitoneal, subcutaneous, intravenous, and intramuscular routes. Moderately toxic by ingestion. Human systemic effects by ingestion: gastrointestinal changes; by intravenous route: blood pressure elevation. When heated to decomposition it emits very toxic fumes of NO_x and SO_x.

PLX400 CAS:52549-17-4 ***HR: 3***
PRANOPROFEN
mf: $C_{15}H_{13}NO_3$ mw: 255.29

PROP: Crystals from aq dioxane. Mp: 182-183°.

SYNS: 2-(5H-(1)BENZOPYRANO(2,3-b)PYRIDIN-7-YL)PROPIONIC ACID ◇ α-METHYL-5H-(1)-BENZOPYRANO(2,3-b)PYRIDINE-7-ACETIC ACID ◇ NIFLAN ◇ Y-8004

TOXICITY DATA with REFERENCE
orl-rat TDLo:67500 μg/kg (female 17-22D post):REP IYKEDH 9,194,78
orl-mus TDLo:30 mg/kg (female 7-12D post):TER IYKEDH 7,301,76
orl-rat LD50:59500 μg/kg NIIRDN 6,APP-18,82
ipr-rat LD50:51200 μg/kg NIIRDN 6,APP-18,82
scu-rat LD50:51500 μg/kg NIIRDN 6,APP-18,82
orl-mus LD50:447 mg/kg IYKEDH 7,211,76
ipr-mus LD50:354 mg/kg IYKEDH 7,211,76
scu-mus LD50:503 mg/kg IYKEDH 12,1204,81

SAFETY PROFILE: Poison by ingestion, subcutaneous, and intraperitoneal routes. An experimental teratogen. Experimental reproductive effects. When heated to decomposition it emits toxic fumes of NO_x.

PLX500 CAS:7440-10-0 ***HR: 3***
PRASEODYMIUM
af: Pr aw: 140.9077

PROP: Yellowish metal. Mp: 935°, bp: 3290°, d: (a) 6.772, (b) 6.64.

SAFETY PROFILE: As a lanthanon, it may depress coagulation of the blood. Limited data suggest moderate toxicity. Flammable in the form of dust when exposed to heat or flame or by chemical reaction. Fine dust ignites readily. Incompatible with air or halogens. See also LANTHANUM, POWDERED METALS, and RARE EARTHS.

PLX750 CAS:10361-79-2 ***HR: 3***
PRASEODYMIUM CHLORIDE
mf: Cl_3Pr mw: 247.26

TOXICITY DATA with REFERENCE
skn-rbt 500 mg/24H MLD TXAPA9 6,614,64
eye-rbt 50 mg TXAPA9 6,614,64
ipr-rat LD50:1318 mg/kg EQSSDZ 1,1,75
scu-rat LDLo:5000 mg/kg AIPTAK 37,199,30
ivn-rat LD50:3500 μg/kg AMIHAB 16,475,57
orl-mus LD50:2987 mg/kg EQSSDX 1,1,75
ipr-mus LD50:359 mg/kg AEHLAU 5,437,62
scu-mus LD50:1659 mg/kg EQSSDX 1,1,75
scu-rbt LD50:351 mg/kg EQSSDX 1,1,75
ivn-rbt LDLo:200 mg/kg AEPPAE 188,465,38
ipr-gpg LD50:125 mg/kg AMIHAB 15,9,57
ivn-gpg LDLo:440 mg/kg AIPTAK 37,199,30
scu-frg LD50:879 mg/kg EQSSDX 1,1,75

CONSENSUS REPORTS: Reported in EPA TSCA Inventory.

SAFETY PROFILE: Poison by intraperitoneal, subcutaneous, and intravenous routes. Moderately toxic by ingestion. A skin and eye irritant. When heated to decomposition it emits toxic fumes of Cl^-. See also PRASEODYMIUM.

PLY250 CAS:10361-80-5 ***HR: 3***
PRASEODYMIUM(III) NITRATE (1:3)
mf: $N_3O_9 \cdot Pr$ mw: 326.94

SYN: NITRIC ACID, PRASEODYMIUM(3+) SALT

TOXICITY DATA with REFERENCE
orl-rat LD50:1859 mg/kg EQSSDX 1,1,75
ipr-rat LD50:184 mg/kg EQSSDX 1,1,75
ivn-rat LD50:5576 µg/kg EQSSDX 1,1,75
ipr-mus LD50:218 mg/kg EQSSDX 1,1,75

CONSENSUS REPORTS: Reported in EPA TSCA Inventory.

SAFETY PROFILE: Poison by intraperitoneal and intravenous routes. Moderately toxic by ingestion. When heated to decomposition it emits toxic fumes of NO_x. See also PRASEODYMIUM and NITRATES.

PLY300 CAS:2920-86-7 ***HR: D***
PREDNISOLONE SUCCINATE
mf: $C_{25}H_{32}O_8$ mw: 460.57

SYNS: PREDNISOLONBISUCCINAT ◇ PREDNISOLONE BISUCCINATE ◇ PREDNISOLONE HEMISUCCINATE ◇ PREDNISOLONE 21-HEMISUCCINATE ◇ PREDNISOLONE 21-SUCCINATE ◇ PREDNISOLUT ◇ PREGNA-1,4-DIENE-3,20-DIONE, 21-(3-CARBOXY-1-OXOPROPOXY)-11,17-DIHYDROXY-,(11-β)- (9CI) ◇ PREGNA-1,4-DIENE-3,20-DIONE, 11-β,17,21-TRIHYDROXY-, 21-(HYDROGEN SUCCINATE)

TOXICITY DATA with REFERENCE
par-rat TDLo:50 mg/kg (female 13D post):TER ARZNAD 36,216,86

SAFETY PROFILE: An experimental teratogen. When heated to decomposition it emits acrid smoke and irritating fumes.

PLZ000 CAS:53-03-2 ***HR: 3***
PREDNISONE
mf: $C_{21}H_{26}O_5$ mw: 358.47

PROP: White, odorless, crystalline powder. Mp: 235° (with some decomp). Very sltly sol in water; sltly sol in alc, chloroform, methanol, and dioxane.

SYNS: ANCORTONE ◇ BICORTONE ◇ COLISONE ◇ CORTAN ◇ CORTANCYL ◇ Δ-CORTELAN ◇ CORTIDELT ◇ Δ-CORTISONE ◇ Δ^1-CORTISONE ◇ Δ-CORTONE ◇ COTONE ◇ DACORTIN ◇ DECORTANCYL ◇ DECORTIN ◇ DECORTISYL ◇ Δ-1-DEHYDROCORTISONE ◇ 1-DEHYDROCORTISONE ◇ DEKORTIN ◇ DELTACORTELAN ◇ DELTACORTISONE ◇ DELTACORTONE ◇ DELTA-DOME ◇ DELTISONE ◇ 17,21-DIHYDROXYPREGNA-1,4-DIENE-3,11,20-TRIONE ◇ ENCORTON ◇ HOSTACORTIN ◇ IN-SONE ◇ JUVASON ◇ LISACORT ◇ METACORTANDRACIN ◇ NCI-C04897 ◇ NSC 10023 ◇ ORASONE ◇ PARACORT ◇ PRECORT ◇ PREDNICEN-M ◇ PREDNILONGA ◇ PREDNISON ◇ PREDNIZON ◇ 1,4-PREGNADIENE-17-α,21-DIOL-3,11,20-TRIONE ◇ RECTODELT ◇ SERVISONE ◇ SK-PREDNISONE ◇ SUPERCORTIL ◇ U 6020 ◇ ULTRACORTEN ◇ WOJTAB ◇ ZENADRID (VETERINARY)

TOXICITY DATA with REFERENCE
mmo-sat 3333 µg/plate NTPTB*J JAN82
mma-sat 333 µg/plate ENMUDM 5(Suppl 1),3,83
scu-mus TDLo:24 mg/kg (13-18D preg):REP PBBHAU 12,213,80
ipr-rat TDLo:860 mg/kg/26W-I:ETA CANCAR 40S,1935,77
orl-wmn TDLo:2400 µg/kg/2D-I:PNS NEURAI 36,729,86
orl-man TDLo:857 µg/kg:PNS NEURAI 36,729,86
orl-man LDLo:400 µg/kg:SKN IJMSAT 155,234,86
unr-wmn TDLo:113 mg/kg:PNS JAMAAP 243,1260,80
ipr-mus LD50:135 mg/kg NCISP* JAN86
scu-mus LD50:101 mg/kg NCISP* JAN86
ims-mus LD50:600 mg/kg CNREA8 42,122,82

CONSENSUS REPORTS: IARC Cancer Review: Group 3 IMEMDT 7,326,87; Human Inadequate Evidence IMEMDT 26,293,81; Animal Inadequate Evidence IMEMDT 26,293,81. NCI Carcinogenesis Studies (ipr); No Evidence: mouse CANCAR 40,1935,77; (ipr); Equivocal Evidence: rat CANCAR 40,1935,77. Reported in EPA TSCA Inventory.

SAFETY PROFILE: Poison by intraperitoneal and subcutaneous routes. Moderately toxic by intramuscular route. Human systemic effects: sensory change involving peripheral nerves, dermatitis. Experimental reproductive effects. Questionable carcinogen with experimental tumorigenic data. Mutation data reported. Has been implicated in aplastic anemia.

PLZ100 CAS:125-10-0 ***HR: D***
PREDNISONE 21-ACETATE
mf: $C_{23}H_{28}O_6$ mw: 400.51

SYNS: CORTANCYL ◇ Δ^1-DEHYDROCORTISONE ACETATE ◇ DELCORTIN ◇ Δ-CORLIN ◇ DELTALONE ◇ FERROSAN ◇ NISONE ◇ PREDNISONE ACETATE ◇ 1,4-PREGNADIEN-17-α-21-DIOL-3,11,20-TRIONE-21-ACETATE ◇ PREGNA-1,4-DIENE-3,11,20-TRIONE, 17,21-DIHYDROXY-, 21-ACETATE ◇ PREGNA-1,4-DIENE-3,11,20-TRIONE, 21-(ACETYLOXY)-17-HYDROXY-(9CI)

TOXICITY DATA with REFERENCE
ims-mus TDLo:330 mg/kg (female 10D post):TER TXAPA9 56,23,80
ims-mus TDLo:135 mg/kg (female 10D post):REP TXAPA9 56,23,80

SAFETY PROFILE: An experimental teratogen. Experimental reproductive effects. When heated to decomposition it emits acrid smoke and irritating fumes.

PMA000 CAS:50-24-8 ***HR: 3***
PREDONIN
mf: $C_{21}H_{28}O_5$ mw: 360.49

SYNS: CODELCORTONE ◇ CO-HYDELTRA ◇ Δ^1-CORTISOL ◇ DECORTIN H ◇ Δ^1-DEHYDROCORTISOL ◇ Δ^1-DEHYDROHYDROCORTISONE ◇ 1-DEHYDROHYDROCORTISONE ◇ DELCORTOL ◇ DELTA-CORTEF ◇ DELTACORTENOL ◇ DELTACORTRIL ◇ DELTA F ◇ DELTA-STAB ◇ DEXA-CORTIDELT HOSTACORTIN H ◇ DI-ADRESON F ◇ DICORTOL ◇ DYDELTRONE ◇ FERNISOLONE ◇ HOSTACORTIN ◇ HYDELTRA ◇ HYDELTRONE ◇ Δ^1-HYDROCORTISONE ◇ HYDRODELTALONE ◇ HYDRODELTISONE ◇ HYDRORETROCORTIN ◇ METACORTANDRALONE ◇ METICORTELONE ◇ METI-DERM ◇ PARACORTOL ◇ PARACOTOL ◇ PRECORTANCYL ◇ PRECORTISYL ◇ PREDNE-DOME ◇ PREDNELAN ◇ PREDNIS ◇ PREDNISOLONE ◇ PREDONINE ◇ 1,4-PREGNADIENE-3,20-DIONE-11-β,17-α,21-TRIOL ◇ 1,4-PREGNADIENE-11-β,17-α,21-TRIOL-3,20-DIONE ◇ SCHERISOLON ◇ STERANE ◇ STEROLONE ◇ 11-β,17,21-TRIHYDROXYPREGNA-1,4-DIENE-3,20-DIONE ◇ 11-β,17-α,21-TRIHYDROXYPREGNA-1,4-DIENE-3,20-DIONE ◇ 11-β,17-α,21-TRIHYDROXY-1,4-PREGNADIENE-3,20-DIONE ◇ ULACORT ◇ ULTRACORTENE-H

TOXICITY DATA with REFERENCE
dni-hmn-unr 6300 μg/kg/8W STBIBN 50,172,75
sce-hmn:lym 4560 μg/L EJCODS 18,533,82
unr-wmn TDLo:56 mg/kg (1-40W preg):REP LANCAO 1,117,68
unr-wmn TDLo:23 mg/kg (1-33W preg):TER LANCAO 1,117,68
orl-man TDLo:9 mg/kg/2W-I DICPBB 18,603,84
orl-wmn TDLo:14 mg/kg/13D-I DICPBB 18,603,84
ipr-rat LD50:2000 mg/kg ADTEAS 3,181,68
scu-rat LD50:147 mg/kg TXAPA9 8,250,66
ivn-rat LD50:120 mg/kg PCJOAU 16,63,82
orl-mus LD50:1680 mg/kg ARZNAD 20,111,70
ivn-mus LD50:180 mg/kg PCJOAU 16,63,82

CONSENSUS REPORTS: Reported in EPA TSCA Inventory. EPA Genetic Toxicology Program.

SAFETY PROFILE: A poison by intravenous and subcutaneous routes. Moderately toxic by ingestion and intraperitoneal routes. Human teratogenic effects by an unspecified route: developmental abnormalities of the central nervous system; effects on embryo or fetus: fetal death, extra embryonic structures. Human reproductive effects by an unspecified route: stillbirth. An experimental teratogen. Experimental reproductive effects. Human mutation data reported. When heated to decomposition it emits acrid smoke and irritating fumes.

PMA100 CAS:1715-33-9 ***HR: 2***
PREDONINE SOLUBLE
mf: $C_{25}H_{31}O_8 \cdot Na$ mw: 482.55

SYNS: DI-ADRESON-F-AQUOSUM ◇ METICORTELANE ◇ METICORTELONE SOLUBLE ◇ PREDNISOLONE SODIUM HEMISUCCINATE ◇ PREDNISOLONE SODIUM SUCCINATE ◇ PREDNISOLONE 21-SODIUM SUCCINATE ◇ PREDNISOLONE 21-SUCCINATE SODIUM ◇ PREGNA-1,4-DIENE-3,20-DIONE, 11-β,17,21-TRIHYDROXY-, 21-(HYDROGEN SUCCINATE), MONOSODIUM SALT ◇ SOLU-DECORTIN

TOXICITY DATA with REFERENCE
scu-mus TDLo:96 mg/kg (11D preg):TER SEIJBO 13,245,73
scu-mus TDLo:96 mg/kg (11D preg):REP SEIJBO 13,245,73
ivn-rat LD50:770 mg/kg PCJOAU 16,63,82
ivn-mus LD50:1125 mg/kg PCJOAU 16,63,82

SAFETY PROFILE: Moderately toxic by intravenous route. An experimental teratogen. Experimental reproductive effects. When heated to decomposition it emits toxic fumes of Na_2O. See also PREDONIN.

PMA250 ***HR: 3***
9-β,10-α-PREGNA-4,6-DIENE-3,20-DIONE and 17-α-HYDROXYPREGN-4-ENE-3,20-DIONE (9:10)
mf: $C_{21}H_{28}O_2 \cdot C_{27}H_{40}O_4$ mw: 741.16

SYN: DYDROGESTERONE and HYDROXYPROGESTERONE (9:10)

TOXICITY DATA with REFERENCE
mul-hmn TDLo:124 mg/kg/(8-36W preg):TER LANCAO 2,982,77

SAFETY PROFILE: Human teratogenic effects by multiple routes: developmental abnormalities of the urogenital system. A steroid. When heated to decomposition it emits acrid smoke and irritating fumes.

PMA450 CAS:124-85-6 ***HR: D***
5-α-17-α-PREGNA-2-EN-20-YN-17-OL, ACETATE
mf: $C_{23}H_{32}O_2$ mw: 340.55

SYNS: 17-β-ACETOXY-5-α-17-α-PREGN-2-ENE-20-YNE ◇ PREGN-2-EN-20-YN-17-OL, ACETATE, (5-α-17α— ◇ TX 380

TOXICITY DATA with REFERENCE
orl-rat TDLo:15600 μg/kg (female 13D pre):REP ARZNAD 36,1069,86
orl-rat TDLo:15500 μg/kg (female 16-20D post):TER ARZNAD 36,1069,86

SAFETY PROFILE: An experimental teratogen. Other experimental reproductive effects. When heated to decomposition it emits acrid smoke and irritating fumes.

PMA750 CAS:134-49-6 ***HR: 3***
PRELUDIN
mf: $C_{11}H_{15}NO$ mw: 177.27

SYNS: A 66 ◇ OXAZIMEDRINE ◇ PHENMETRAZIN ◇ PHENMETRAZINE ◇ dl-2-PHENYL-3-METHYLTETRAHYDRO-1,4-OXAZINE ◇ PROBESE-P ◇ PSYCHAMINE A 66

TOXICITY DATA with REFERENCE
orl-man TDLo:2857 μg/kg:ANS THERAP 34,205,79
orl-rat LD50:370 mg/kg ARZNAD 23,810,73
orl-mus LD50:125 mg/kg ARZNAD 23,810,73

ipr-mus LD50:125 mg/kg
ARZNAD 21,1727,71
scu-mus LD50:195 mg/kg AEPPAE 222,540,54

SAFETY PROFILE: Poison by ingestion, subcutaneous, and intraperitoneal routes. Human systemic effects by ingestion: stimulates the adrenergic neurons of the sympathetic nervous system. When heated to decomposition it emits toxic fumes of NO_x.

PMB000 CAS:12126-59-9 ***HR: 3***
PREMARIN

SYNS: CEE ◇ CONJUGATED EQUINE ESTROGEN ◇ ESTROGENS, CONJUGATES

TOXICITY DATA with REFERENCE
orl-wmn TDLo:3 mg/kg (2-6D preg):REP CCPTAY 17,443,78
scu-rat TDLo:7 mg/kg (15-17D preg):TER BBACAQ 385,257,75
orl-rat TDLo:51 mg/kg/2Y-C:ETA TXAPA9 11,489,67
ipr-rat LD50:325 mg/kg TXAPA9 18,185,71

CONSENSUS REPORTS: NTP Fifth Annual Report on Carcinogens. IARC Cancer Review: Human Limited Evidence IMEMDT 21,147,79; Animal Inadequate Evidence IMEMDT 21,147,79.

SAFETY PROFILE: Confirmed carcinogen with experimental tumorigenic data. Poison by intraperitoneal route. Human reproductive effects by ingestion: changes in female fertility. Experimental teratogenic effects. A steroid. When heated to decomposition it emits acrid smoke and irritating fumes.

PMB250 CAS:115-76-4 ***HR: 2***
PRENDIOL
mf: $C_7H_{16}O_2$ mw: 132.23

PROP: Crystals. Bp: 125° @ 10 mm, fp: 61.3°, flash p: 215°F (OC), d: 1.052 @ 20°.

SYNS: 3,3-BIS(HYDROXYMETHYL)PENTANE ◇ DEP ◇ DI-AETHYL-PROPANEDIOL (GERMAN) ◇ 2,2-DIETHYLPROPANEDIOL-1,3 ◇ 2,2-DIETHYLPROPANE-1,3-DIOL ◇ 2,2-DIETHYL-1,3-PROPANEDIOL ◇ DIETILPROPANDIOLO ◇ MC 1415 ◇ PENDEROL ◇ PRENDEROL

TOXICITY DATA with REFERENCE
skn-rbt 500 mg open MLD UCDS** 2/21/58
eye-rbt 20 mg open SEV AMIHBC 10,61,54
orl-rat LD50:850 mg/kg UCDS** 2/21/58
ipr-rat LD50:700 mg/kg JPETAB 100,27,50
ivn-rat LD50:635 mg/kg JPETAB 100,27,50
orl-mus LD50:1550 mg/kg JPETAB 100,27,50
ipr-mus LD50:1220 mg/kg JPETAB 100,27,50
ivn-mus LD50:1170 mg/kg JPETAB 100,27,50
skn-rbt LD50:4240 mg/kg AMIHBC 10,61,54

SAFETY PROFILE: Moderately toxic by ingestion, intraperitoneal, and intravenous routes. Mildly toxic by skin contact. A skin and severe eye irritant. Large doses can cause drowsiness, vertigo, nausea, and vomiting. Combustible when exposed to heat or flame; can react with oxidizing materials. To fight fire, use alcohol foam, foam, CO_2, dry chemical. Used as a skeletal muscle relaxant. When heated to decomposition it emits acrid smoke and irritating fumes.

PMB500 CAS:50765-87-2 ***HR: 3***
α-PRENYL-α-(2-DIMETHYLAMINOETHYL)-1-NAPHTHYLACETAMIDE
mf: $C_{21}H_{28}N_2O$ mw: 324.51

SYNS: α-(2-DIMETHYLAMINOETHYL)-α-(3-METHYL-2-BUTENYL)-1-NAPHTHALENEACETAMIDE ◇ 2-(2-DIMETHYLAMINOETHYL)-2-(3-METHYL-2-BUTENYL)-2-(1-NAPHTHYL)ACETAMIDE

TOXICITY DATA with REFERENCE
orl-rat LD50:143 mg/kg JMCMAR 16,720,73
orl-mus LD50:78 mg/kg JMCMAR 16,720,73

SAFETY PROFILE: Poison by ingestion. When heated to decomposition it emits toxic fumes of NO_x.

PMB600 CAS:68555-58-8 ***HR: 2***
PRENYL SALICYLATE
mf: $C_{12}H_{14}O_3$ mw: 206.26

SYNS: BENZOIC ACID, 2-HYDROXY-, 3-METHYL-2-BUTENYL ESTER ◇ 2-BUTEN-1-OL, 3-METHYL-, SALICYLATE ◇ 3-METHYL-2-BUTENYL SALICYLATE

TOXICITY DATA with REFERENCE
skn-rbt 500 mg/24H MOD FCTOD7 20,821,82
orl-rat LD50:3200 mg/kg FCTOD7 20,821,82

CONSENSUS REPORTS: Reported in EPA TSCA Inventory.

SAFETY PROFILE: Moderately toxic by ingestion. A skin irritant. When heated to decomposition it emits acrid smoke and irritating fumes.

PMB800 ***HR: 3***
PREPARATION 48-80
mf: $(C_{11}H_{15}NO)_n$

SYNS: No. 48-80 ◇ POLY(2-METHOXY-5(2-(METHYLAMINO)ETHYL)-m-PHENYLENEMETHYLENE)

TOXICITY DATA with REFERENCE
ipr-rat LD50:5 mg/kg JPETAB 107,1,53
ipr-mus LD50:7250 μg/kg JPETAB 107,1,53
ivn-mus LD50:3100 μg/kg JPETAB 107,1,53

SAFETY PROFILE: Poison by intravenous and intraperitoneal routes. When heated to decomposition it emits toxic fumes of NO_x.

PMB850 CAS:51218-49-6 ***HR: 2***
PRETILACHLOR
mf: $C_{17}H_{26}ClNO_2$ mw: 311.89

PROP: Colorless liquid. Bp: 135°. Insol in water; sol in most organic solvents.

SYNS: ACETAMIDE,2-CHLORO-N-(2,6-DIETHYLPHENYL)-N-(2-PROPOXYETHYL)- ◇ CG 113 ◇ CGA 26423 ◇ 2-CHLORO-N-(2,6-DIETHYLPHENYL)-N-(2-PROPOXYETHYL)ACETAMIDE ◇ 2-CHLORO-2′,6′-DIETHYL-N-(2-PROPOXYETHYL)ACETANILIDE ◇ PRETILACHLORE ◇ RIFIT ◇ SOLNET

TOXICITY DATA with REFERENCE
skn-rbt 500 mg MOD NNGADV 11,139,86
eye-rbt 100 mg MLD NNGADV 11,139,86
orl-rat LD50:2200 mg/kg NNGADV 11,139,86
ipr-rat LD50:1120 mg/kg NNGADV 11,139,86
orl-mus LD50:1800 mg/kg NNGADV 11,139,86
ipr-mus LD50:1120 mg/kg NNGADV 11,139,86

SAFETY PROFILE: Moderately toxic by ingestion and intraperitoneal routes. A skin irritant. When heated to decomposition it emits toxic fumes of NO_x and Cl^-.

PMC100 ***HR: 2***
PREVOCEL #12

TOXICITY DATA with REFERENCE
orl-rat LD50:1800 mg/kg GISAAA 48(1),84,83
orl-mus LD50:1800 mg/kg GISAAA 48(1),84,83
orl-rbt LD50:1800 mg/kg GISAAA 48(1),84,83
orl-gpg LD50:1800 mg/kg GISAAA 48(1),84,83

SAFETY PROFILE: Moderately toxic by ingestion.

PMC250 CAS:968-58-1 ***HR: 3***
PRIDINOL HYDROCHLORIDE
mf: $C_{20}H_{25}NO \cdot ClH$ mw: 331.92

PROP: Crystals. Decomp @ 238°. Sol in alc.

SYNS: α,α-DIPHENYL-1-PIPERIDINEPROPANOLHYDROCHLORIDE ◇ 1,1-DIPHENYL-3-PIPERIDINO-1-PROPANOL HYDROCHLORIDE ◇ 1,1-DIPHENYL-3-(1-PIPERIDYL)-1-PROPANOL HYDROCHLORIDE ◇ MITANOLINE ◇ PAR KS-12 ◇ α-(2-PIPERIDYLETHYL)BENZHYDROL HYDROCHLORIDE ◇ PRIDINOL

TOXICITY DATA with REFERENCE
ipr-rat LD50:91 mg/kg JPETAB 96,151,49
ivn-rat LD50:33 mg/kg JPETAB 96,151,49
ipr-mus LD50:131 mg/kg JPETAB 96,151,49
ivn-mus LD50:25 mg/kg JPETAB 116,177,56

SAFETY PROFILE: Poison by intraperitoneal and intravenous routes. When heated to decomposition it emits very toxic fumes of NO_x and HCl.

PMC275 CAS:53639-82-0 ***HR: 3***
PRIDINOL MESILATE
mf: $C_{21}H_{27}NO_3S$ mw: 373.55

SYN: α,α-DIPHENYL-1-PIPERIDINEPROPANOLMETHANESULFONATE (ester)

TOXICITY DATA with REFERENCE
orl-rat LD50:1240 mg/kg NIIRDN 6,690,82
ivn-rat LD50:40 mg/kg NIIRDN 6,690,82
ivn-mus LD50:37 mg/kg NIIRDN 6,690,82
ims-rat LD50:1000 mg/kg NIIRDN 6,690,82
orl-mus LD50:790 mg/kg NIIRDN 6,690,82
ims-mus LD50:490 mg/kg NIIRDN 6,690,82

SAFETY PROFILE: Poison by intravenous route. Moderately toxic by ingestion and intramuscular routes. When heated to decomposition it emits toxic fumes of NO_x and SO_x.

PMC300 CAS:90-34-6 ***HR: 3***
PRIMAQUINE
mf: $C_{15}H_{21}N_3O$ mw: 259.39

PROP: Viscous liquid. Bp: 175-179°. Sol in ether.

SYNS: 8-(4-AMINO-1-METHYLBUTYLAMINO)-6-METHOXYQUINOLINE ◇ 6-METHOXY-8-(4-AMINO-1-METHYLBUTYLAMINO)QUINOLINE ◇ SN 13,272

TOXICITY DATA with REFERENCE
sce-rat-orl 8 mg/kg SYXUD2 23,26,85
orl-mus LD50:100 mg/kg AMACCQ 12,51,77
ivn-mus LD50:15900 μg/kg ANTCAO 12,103,62

SAFETY PROFILE: Poison by ingestion and intravenous routes. Mutation data reported. When heated to decomposition it emits toxic fumes of NO_x.

PMC310 CAS:63-45-6 ***HR: 3***
PRIMAQUINE PHOSPHATE
mf: $C_{15}H_{21}N_3O \cdot 2H_3O_4P$ mw: 455.39

SYNS: N^4-(6-METHOXY-8-QUINOLINYL)-1,4-PENTANEDIAMINE PHOSPHATE (1:2) (9CI) ◇ PRIMAQUINE DIPHOSPHATE

TOXICITY DATA with REFERENCE
dnd-esc 50 μmol/L MUREAV 89,95,81
orl-rat TDLo:540 mg/kg (8-16D preg):TER TRSTAZ 74,43,80
orl-mus LD50:68 mg/kg TRSTAZ 74,43,80

SAFETY PROFILE: Poison by ingestion. An experimental teratogen. Mutation data reported. When heated to decomposition it emits toxic fumes of NO_x and PO_x.

PMC400 CAS:68955-54-4 ***HR: 3***
PRIMENE JM-T

TOXICITY DATA with REFERENCE
skn-rbt 500 mg SEV RHPC** PC-81-MAY82
orl-rat LD50:400 mg/kg RHPC** PC-81-MAY82

CONSENSUS REPORTS: Reported in EPA TSCA Inventory.

SAFETY PROFILE: Poison by ingestion. A severe skin irritant.

PMC600 CAS:40778-40-3 ***HR: 3***
PRIMIDOLOL HYDROCHLORIDE
mf: $C_{17}H_{23}N_3O_4 \cdot ClH$ mw: 369.89

TOXICITY DATA with REFERENCE
orl-mus TDLo:27000 mg/kg/77W-C:CAR TXCYAC 21,279,81
orl-mus TD:6750 mg/kg/77W-C:CAR TXCYAC 21,279,81
orl-mus TD:13500 mg/kg/77W-C:ETA,REP TXCYAC 21,279,81

SAFETY PROFILE: Questionable carcinogen with experimental carcinogenic and tumorigenic data. Experimental reproductive effects. When heated to decomposition it emits toxic fumes of NO_x and HCl.

PMC700 CAS:434-05-9 ***HR: 2***
PRIMOBOLAN
mf: $C_{22}H_{32}O_3$ mw: 344.54

SYNS: 17-β-HYDROXY-1-METHYL-5-α-ANDROST-1-EN-3-ONE ACETATE ◇ METENOLONE ACETATE ◇ METHENOLONE ACETATE ◇ NIBAL ◇ PREMOBOLAN ◇ PRIMOBOLONE ◇ PRIMONABOL ◇ SH 567 ◇ SH 567a ◇ SQ 16496

TOXICITY DATA with REFERENCE
orl-rat TDLo:200 mg/kg (male 10D pre):REP ARZNAD 14,330,64
scu-rat TDLo:60 mg/kg (female 17-20D post):TER AVBIB9 13,71,74
orl-rat LD50:4000 mg/kg NIIRDN 6,835,82
orl-mus LD50:4000 mg/kg NIIRDN 6,835,82
ipr-mus LD50:2000 mg/kg NIIRDN 6,835,82

SAFETY PROFILE: Moderately toxic by ingestion and intraperitoneal routes. An experimental teratogen. Other experimental reproductive effects. When heated to decomposition it emits acrid smoke and irritating fumes.

PMC750 CAS:3750-26-3 ***HR: 3***
PRIMOCARCIN
mf: $C_8H_{12}N_2O_3$ mw: 184.22

PROP: Needles. Mp: 130-131°. Sol in water, acetone; sltly sol in organic solvents.

SYNS: 4-ACETAMIDO-4-PENTEN-3-ONE-1-CARBOXAMIDE ◇ 5-(ACETYLAMINO)-4-OXO-5-HEXENAMIDE

TOXICITY DATA with REFERENCE
ipr-mus LD50:55500 μg/kg JAJAAA 15,250,62
ivn-mus LD50:300 mg/kg JAJAAA 15,77,62

SAFETY PROFILE: Poison by intraperitoneal and intravenous routes. When heated to decomposition it emits toxic fumes of NO_x.

PMD325 CAS:1239-04-9 ***HR: 3***
PRINADOL HYDROBROMIDE
mf: $C_{22}H_{27}NO \cdot BrH$ mw: 402.42

SYNS: 2'-HYDROXY-5,9-DIMETHYL-2-PHENETHYL-6,7-BENZOMORPHAN HYDROBROMIDE ◇ NARPHEN ◇ NIH 7519 HYDROBROMIDE ◇ PHENAZOCINE HYDROBROMIDE

TOXICITY DATA with REFERENCE
scu-ham TDLo:185 mg/kg (8D preg):TER AJOGAH 123,705,75
scu-mus LD50:22500 μg/kg JPETAB 130,431,60
ivn-mus LD50:11 mg/kg TXAPA9 6,334,64

SAFETY PROFILE: Poison by subcutaneous and intravenous routes. An experimental teratogen. An analgesic. When heated to decomposition it emits toxic fumes of NO_x and HBr.

PMD350 CAS:650-42-0 ***HR: 3***
PRISMANE
mf: C_6H_6 mw: 78.11

SAFETY PROFILE: An explosive liquid. When heated to decomposition it emits acrid smoke and irritating fumes.

PMD500 CAS:1921-70-6 ***HR: 3***
PRISTANE
mf: $C_{19}H_{40}$ mw: 268.59

PROP: Mobile, transparent, stable liquid. D: 0.782 @ 20°/4°, fp: −100°, bp: 296°. Sol in ether, petr ether, benzene, chloroform.

SYN: 2,6,10,14-TETRAMETHYLPENTADECANE

TOXICITY DATA with REFERENCE
ipr-mus TDLo:47 mg/kg:ETA CNREA8 40,579,80
ipr-mus TD:60 mg/kg/26W-I:NEO NATUAS 222,994,69

CONSENSUS REPORTS: Reported in EPA TSCA Inventory.

SAFETY PROFILE: Questionable carcinogen with experimental neoplastigenic and tumorigenic data. When heated to decomposition it emits acrid smoke and irritating fumes.

PMD525 CAS:1258-84-0 ***HR: 3***
PRISTIMERIN
mf: $C_{30}H_{40}O_4$ mw: 464.70

SYN: CELASTROL-METHYLETHER

TOXICITY DATA with REFERENCE
orl-mus LD50:8000 mg/kg 85GDA2 8(1),90,82
ipr-mus LD50:200 mg/kg 85GDA2 8(1),90,82
scu-mus LD50:400 mg/kg 85GDA2 8(1),90,82

SAFETY PROFILE: Poison by subcutaneous and intraperitoneal routes. Mildly toxic by ingestion. When heated to decomposition it emits acrid smoke and irritating fumes. See also ETHERS.

PMD550 ***HR: 3***
PRIVET

PROP: A deciduous shrub with opposite, smooth-edged, elliptical leaves about 2 inches long, dark green on top and lighter green on the bottom. It blooms in clusters of small white flowers and later produces large numbers of blue or black waxy berries. The berries stay on the plant through the winter. It is native to the Mediterranean countries and grows wild in eastern North America. It is cultivated as a hedge plant in most of the United States, Canada, and Hawaii.

SYNS: HEDGE PLANT ◇ LIGUSTRUM VULGARE ◇ LOVAGE ◇ PRIM

SAFETY PROFILE: The whole plant contains the glucosides syringin (ligustrin), a direct irritant, and nuzhenids. Ingestion of large amounts of the berries can cause colic, vomiting, diarrhea and death, especially in children.

PMD800 CAS:27605-76-1 ***HR: 2***
PROBENAZOLE
mf: $C_{10}H_9NO_3S$ mw: 223.26

SYNS: ORYZAEMATE ◇ PO-20 ◇ 3-(2-PROPENYLOXY)-1,2-BENZISOTHIAZOLE 1,1-DIOXIDE (9CI)

TOXICITY DATA with REFERENCE
orl-rat LD50:2030 mg/kg NNGADV 7,307,82
ipr-rat LD50:850 mg/kg OYYAA2 13,219,77
orl-mus LD50:2750 mg/kg OYYAA2 13,219,77
ipr-mus LD50:745 mg/kg OYYAA2 13,219,77

SAFETY PROFILE: Moderately toxic by ingestion and intraperitoneal routes. When heated to decomposition it emits toxic fumes of SO_x and NO_x.

PMD825 CAS:28610-84-6 ***HR: 3***
PROBONAL
mf: $C_{13}H_{19}N_2O_3•CH_3O_4S$ mw: 362.44

SYNS: PROBON ◇ RIMAZOLIUM METHYL SULFATE ◇ 6,7,8,9-TETRAHYDRO-3-CARBOXY-1,6-DIMETHYL-4-OXO-4H-PYRIDO(1,2-a)PYRIMIDINIUM ETHYL ESTER, METHYL SULFATE

TOXICITY DATA with REFERENCE
orl-rat LD50:1500 mg/kg ARZNAD 21,719,71
ipr-rat LD50:720 mg/kg ARZNAD 21,719,71
scu-rat LD50:750 mg/kg ARZNAD 21,727,71
ivn-rat LD50:217 mg/kg ARZNAD 21,717,71
orl-mus LD50:1100 mg/kg ARZNAD 21,719,71
scu-mus LD50:500 mg/kg ARZNAD 21,719,71
ivn-mus LD50:200 mg/kg ARZNAD 21,719,71
orl-dog LD50:500 mg/kg ARZNAD 21,719,71
ipr-gpg LD50:500 mg/kg ARZNAD 21,719,71
scu-gpg LD50:420 mg/kg ARZNAD 21,719,71

SAFETY PROFILE: Poison by intravenous route. Moderately toxic by ingestion, intraperitoneal, and subcutaneous routes. When heated to decomposition it emits toxic fumes of NO_x and SO_x.

PME000 CAS:614-39-1 ***HR: 3***
PROCAINAMIDE HYDROCHLORIDE
mf: $C_{13}H_{21}N_3O•ClH$ mw: 271.83

PROP: Crystals. Mp: 165-169°. Very sol in water; sol in alc; sltly sol in chloroform; very sltly sol in benzene, ether.

SYNS: p-AMINO-N-(2-(DIETHYLAMINO)ETHYL))BENZAMIDEHYDROCHLORIDE ◇ PROCAINE AMIDE HYDROCHLORIDE ◇ PRONESTYL HYDROCHLORIDE

TOXICITY DATA with REFERENCE
orl-man TDLo:200 mg/kg/4W-I:BLD AIMDAP 145,700,85
orl-man TDLo:2786 mg/kg/56W-I:BLD,MSK AJMEAZ 80,999,86
ivn-rat LD50:95 mg/kg ARZNAD 18,1127,68
orl-mus LD50:1110 mg/kg ARZNAD 18,1127,68
ipr-mus LD50:325 mg/kg APTOA6 28,17,70
scu-mus LD50:800 mg/kg PSEBAA 73,236,50
ivn-mus LD50:94640 μg/kg NIIRDN 6,719,82
ims-mus LD50:860 mg/kg THERAP 9,332,54
ivn-mus LD50:150 mg/kg ARZNAD 18,1127,68

SAFETY PROFILE: Poison by intravenous and intraperitoneal routes. Moderately toxic by ingestion and subcutaneous routes. Human systemic effects by ingestion: thrombocytopenia, leukopenia, joint effects. Used as a local anesthetic. When heated to decomposition it emits very toxic fumes of HCl and NO_x.

PME250 CAS:671-16-9 ***HR: 3***
PROCARBAZINE
mf: $C_{12}H_{19}N_3O$ mw: 221.34

SYNS: IBENZMETHYZINE ◇ 2-(p-ISOPROPYL CARBAMOYL BENZYL)-1-METHYLHYDRAZINE ◇ N-ISOPROPYL-α-(2-METHYLHYDRAZINO)-p-TOLUAMIDE ◇ MATULANE ◇ 4-((2-METHYLHYDRAZINO)METHYL)-N-ISOPROPYLBENZAMIDE ◇ 1-METHYL-2-(-

ISOPROPYLCARBAMOYL)BENZYL)HYDRAZINE ◇ MIH ◇ NATULAN ◇ NSC-77213 ◇ PCB ◇ RO 4-6467

TOXICITY DATA with REFERENCE
mmo-smc 12 g/L BSIBAC 56,1322,80
spm-mus-ipr 120 mg/kg CNREA8 42,122,82
scu-mus TDLo:20 mg/kg (female 12D post):REP FEPRA7 32,745,73
ipr-mus TDLo:10 mg/kg (female 16D post):TER HUGEDQ 51,183,79
ivn-rat TDLo:1250 mg/kg/1Y-I:CAR ARZNAD 20,1461,70
ivn-rat TDLo:125 mg/kg (22D preg):ETA IARCCD 4,92,73
ipr-mus TDLo:1200 mg/kg/4W-I:NEO JNCIAM 59,423,77
ivn-rat LD50:350 mg/kg ARZNAD 20,1467,70
ipr-mus LD50:614 mg/kg JJIND8 62,911,79

SAFETY PROFILE: Questional carcinogen with experimental carcinogenic, neoplastigenic, tumorigenic data. Poison by intravenous route. Moderately toxic by intraperitoneal route. Has been implicated as a brain carcinogen. An experimental teratogen. Experimental reproductive effects. Mutation data reported. When heated to decomposition it emits toxic fumes of NO_x.

PME500 CAS:366-70-1 ***HR: 3***
PROCARBAZINE HYDROCHLORIDE
mf: $C_{12}H_{19}N_3O•ClH$ mw: 257.80

PROP: Crystals. Mp: 223-236°.

SYNS: IBENZMETHYZINE HYDROCHLORIDE ◇ IBENZMETHYZIN HYDROCHLORIDE ◇ IBZ ◇ 1-(p-ISOPROPYLCARBAMOYLBENZYL)-2-METHYLHYDRAZINE HYDROCHLORIDE ◇ 2-(p-(ISOPROPYLCARBAMOYL)BENZYL)-1-METHYLHYDRAZINEHYDROCHLORIDE ◇ N-ISOPROPYL-p-(2-METHYLHYDRAZINOMETHYL) BENZAMIDE-HYDROCHLORIDE ◇ N-ISOPROPYL-α-(2-METHYLHYDRAZINO)-p-TOLUAMIDE HYDROCHLORIDE ◇ MATULANE ◇ MBH ◇ N-(1-METHYLETHYL)-4-((2-METHYLHYDRAZINO)METHYL)BENZAMIDE MONOHYDROCHLORIDE ◇ p-(N'-METHYLHYDRAZINOMETHYL)-N-ISOPROPYL)BENZAMIDE ◇ p-(N'-METHYLHYDRAZINOMETHYL)-N-ISOPROPYLBENZAMIDE HYDROCHLORIDE ◇ 1-METHYL-2-p-(ISOPROPYLCARBAMOYL)BENZOHYDRAZINEHYDROCHLORIDE ◇ 1-METHYL-2-(p-ISOPROPYLCARBAMOYLBENZYL)HYDRAZINE HYDROCHLORIDE ◇ MIH HYDROCHLORIDE ◇ NATHULANE ◇ NATULAN ◇ NATULANAR ◇ NATULAN HYDROCHLORIDE ◇ NCI-C01810 ◇ NSC-77213 ◇ PCB HYDROCHLORIDE ◇ PROCARBAZIN (GERMAN) ◇ RO 4-6467

TOXICITY DATA with REFERENCE
dnd-rat-ipr 25 mg/kg ENMUDM 7,563,85
hma-mus/esc 100 mg/kg MUREAV 164,19,86
orl-rat TDLo:30 mg/kg (female 12-15D post):REP TJADAB 32,203,85
orl-rat TDLo:120 mg/kg (9-14D preg):TER OYYAA2 6,381,72
orl-rat TDLo:500 mg/kg:ETA JNCIAM 40,1027,68
ipr-rat TDLo:1170 mg/kg/26W-I:CAR NCITR* NCI-CG-TR-19,79
scu-rat TDLo:300 mg/kg:CAR CALEDQ 21,155,83
ivn-rat TDLo:125 mg/kg (22D preg):ETA,REP ARZNAD 22,905,72
ipr-mus TD:1950 mg/kg/26W-I:NEO RRCRBU 52,1,75
orl-rat LD50:570 mg/kg OYYAA2 6,355,72
scu-rat LD50:490 mg/kg NIIRDN 6,720,82
ivn-rat LD50:350 mg/kg ARZNAD 20,1467,70
orl-rat LD50:785 mg/kg EVSRBT 24,977,81
orl-mus LD50:560 mg/kg IYKEDH 4,467,73
ipr-mus LD50:699 mg/kg ARTODN 41,287,79
scu-mus LD50:710 mg/kg IYKEDH 4,467,73
ivn-mus LD50:540 mg/kg IYKEDH 4,467,73

CONSENSUS REPORTS: NTP Fifth Annual Report on Carcinogens. IARC Cancer Review: Group 2A IMEMDT 7,327,87; Human Limited Evidence IMEMDT 26,311,81; Animal Sufficient Evidence IMEMDT 26,311,81. NCI Carcinogenesis Bioassay (ipr); Clear Evidence: mouse, rat NCITR* NCI-CG-TR-19,79; (ipr); Clear Evidence: mouse, rat RRCRBU 52,1,75. EPA Genetic Toxicology Program.

SAFETY PROFILE: Confirmed carcinogen with experimental carcinogenic, neoplastigenic, tumorigenic, and teratogenic data. Poison by an unspecified route. Moderately toxic by ingestion, subcutaneous, intravenous, and intraperitoneal routes. Experimental reproductive effects. Mutation data reported. When heated to decomposition it emits very toxic fumes of NO_x and HCl. Used as a chemotherapeutic agent.

PME600 CAS:62929-91-3 ***HR: 3***
PROCATEROL HYDROCHLORIDE
mf: $C_{16}H_{22}N_2O_3•ClH$ mw: 326.86

SYNS: 8-HYDROXY-5-(1-HYDROXY-2-((1-METHYLETHYL)AMINO) BUTYL)-2(1H)-QUINOLINONE ◇ 5-(1-HYDROXY-2-ISOPROPYL-0AMINO)BUTYL-8-HYDROXYCARBOSTYRILHYDROCHLORIDE ◇ MEPTIN ◇ OPC 2009

TOXICITY DATA with REFERENCE
orl-rat TDLo:6750 mg/kg (female 17-22D post):REP IYKEDH 10,112,79
orl-rat TDLo:55 mg/kg (female 7-17D post):TER IYKEDH 10,80,79
orl-rat LD50:2600 mg/kg USXXAM #4026897
ipr-rat LD50:487 mg/kg OYYAA2 22,191,81
scu-rat LD50:900 mg/kg NIIRDN 6,719,82
ivn-rat LD50:80 mg/kg USXXAM #4026897
orl-mus LD50:3200 mg/kg NIIRDN 6,719,82
ipr-mus LD50:330 mg/kg NIIRDN 6,719,82
scu-mus LD50:370 mg/kg NIIRDN 6,719,82
ivn-mus LD50:83 mg/kg NIIRDN 6,719,82

SAFETY PROFILE: Poison by subcutaneous, intravenous, and intraperitoneal routes. Moderately toxic by ingestion. An experimental teratogen. Other experimental reproductive effects. When heated to decomposition it emits toxic fumes of NO_x and HCl.

PME700 CAS:1257-78-9 *HR: 3*
PROCHLORPERAZINE ETHANE DISULFONATE
mf: $C_{20}H_{24}ClN_3S \cdot C_2H_4O_6S_2$ mw: 562.16

SYNS: 2-CHLORO-10-(3-(1-METHYL-4-PIPERAZINYL)PROPYL)PHENOTHIAZINE EDISYLATE ◇ 2-CHLORO-10-(3-(4-METHYL-1-PIPERAZINYL)PROPYL)PHENOTHIAZINE1,2-ETHANEDISULFONATE (1:1) ◇ PHENOTHIAZINE, 2-CHLORO-10-(3-(1-METHYL-4-PIPERAZINYL)PROPYL)-, ETHANEDISULFONATE ◇ PROCHLORPERAZINE EDISYLATE

TOXICITY DATA with REFERENCE
orl-rat TDLo:350 mg/kg (female 7-20D post):REP SCIEAS 205,1220,79
orl-mus LD50:400 mg/kg 27ZQAG -,40,72
scu-mus LD50:320 mg/kg 27ZQAG -,40,72

SAFETY PROFILE: Poison by ingestion and subcutaneous routes. Experimental reproductive effects. When heated to decomposition it emits toxic fumes of NO_x, SO_x, and Cl^-.

PMF250 CAS:84-02-6 *HR: 3*
PROCHLORPERAZINE HYDROGEN MALEATE
mf: $C_{20}H_{24}ClN_3S \cdot 2C_4H_4O_4$ mw: 606.14

SYNS: 2-CHLORO-10-(3-(4-METHYL-1-PIPERAZINYL)-PROPYL-10H-PHENOTHIAZINE-(Z)-2-BUTENEDIOATE(1:2) ◇ 2-CHLORO-10-(3-(1-METHYL-4-PIPERAZINYL)PROPYL)PHENOTHIAZINE,DIMALEATE ◇ 2-CHLORO-10-(3-(4-METHYL-1-PIPERAZINYL)PROPYL)PHENOTHIAZINE DIMALEATE ◇ 2-CHLORO-10-(3-(4-METHYL-1-PIPERAZINYL)PROPYL)PHENOTHIAZINE MALEATE ◇ COMPAZINE ◇ EMETIRAL ◇ METERAZIN MALEATE ◇ PASOTOMIN ◇ PROCHLOROPROAZINE HYDROGEN MALEATE ◇ PROCHLORPERAZINE BIMALEATE ◇ PROCHLORPERAZINE DIMALEATE ◇ PROCHLORPERAZINE MALEATE ◇ PROCHLORPERIZINE MALEATE ◇ STEMETIL DIMALEATE

TOXICITY DATA with REFERENCE
orl-rat LD50:750 mg/kg AIPTAK 123,78,59
orl-mus LD50:400 mg/kg CRSBAW 152,1371,58
scu-mus LD50:320 mg/kg 27ZQAG -,40,72
ivn-mus LD50:142 mg/kg AIPTAK 123,78,59

SAFETY PROFILE: Poison by ingestion, intravenous, and subcutaneous routes. When heated to decomposition it emits very toxic fumes of Cl^-, NO_x, and SO_x.

PMF500 CAS:58-38-8 *HR: 3*
PROCHLORPROMAZINE
mf: $C_{20}H_{24}ClN_3S$ mw: 373.98

SYNS: CHLORO-3 (N-METHYLPIPERAZINYL-3 PROPYL)-10 PHENOTHIAZINE (FRENCH) ◇ 2-CHLORO-10-(3-(1-METHYL-4-PIPERAZINYL)-PROPYL)-PHENOTHIAZINE ◇ 2-CHLORO-10-(3-(4-METHYL-1-PIPERAZINYL)PROPYL)PHENOTHIAZINE ◇ 3-CHLORO-10-(3-(1-METHYL-4-PIPERAZINYL)PROPYL)PHENOTHIAZINE ◇ CHLORPERAZINE ◇ COMPAZINE ◇ N-(Γ-(4'-METHYLPIPERAZINYL-1')PROPYL)-3-CHLOROPHENOTHIAZINE ◇ NIPODAL ◇ NOVAMIN ◇ PROCHLOROPERAZINE ◇ PROCHLORPEMAZINE ◇ PROCHLORPERAZINE ◇ 6140 RP ◇ STEMETIL ◇ TEMENTIL

TOXICITY DATA with REFERENCE
scu-rat TDLo:20400 μg/kg (51D pre):REP PSYPAG 16,5,69
ipr-mus TDLo:100 mg/kg (10D preg):TER CAJPBD 3,2,63
orl-hmn TDLo:560 μg/kg/10H-I:CNS,CVS NEJMAG 312,1125,85
orl-rat LD50:1800 mg/kg 27ZIAQ -,213,73
ivn-rat LDLo:20 mg/kg AIPTAK 120,450,59
orl-mus LD50:400 mg/kg NIIRDN 6,722,82
ipr-mus LD50:120 mg/kg CRSBAW 152,1371,58
scu-mus LD50:400 mg/kg NIIRDN 6,722,82
orl-dog LDLo:102 mg/kg 27ZIAQ -,213,73
ivn-dog LDLo:100 mg/kg 27ZIAQ -,213,73
ivn-rbt LDLo:5 mg/kg AIPTAK 120,450,59

SAFETY PROFILE: Poison by ingestion, subcutaneous, intravenous, and intraperitoneal routes. Experimental teratogenic and reproductive effects. Human systemic effects by ingestion: headache, blood pressure elevation. Implicated in aplastic anemia. When heated to decomposition it emits very toxic fumes of SO_x, NO_x, and Cl^-.

PMF525 CAS:27325-18-4 *HR: 3*
PROCINOLOL HYDROCHLORIDE
mf: $C_{15}H_{23}NO_2 \cdot ClH$ mw: 285.85

SYNS: 1-(o-CYCLOPROPYLPHENOXY)-3-(ISOPROPYLAMINO)-2-PROPANOL HYDROCHLORIDE ◇ SD 2124-01 HYDROCHLORIDE

TOXICITY DATA with REFERENCE
orl-mus LD50:382 mg/kg JMCMAR 13,971,70
ipr-mus LD50:131 mg/kg JMCMAR 13,971,70
ivn-mus LD50:32 mg/kg JMCMAR 13,971,70

SAFETY PROFILE: Poison by ingestion, intravenous, and intraperitoneal routes. When heated to decomposition it emits toxic fumes of NO_x and HCl.

PMF535 CAS:32752-13-9 *HR: 3*
(±)-PROCINOLOL HYDROCHLORIDE
mf: $C_{15}H_{23}NO_2 \cdot ClH$ mw: 285.85

SYNS: dl-1-(o-CYCLOPROPYLPHENOXY)-3-ISOPROPYLAMINO-2-PROPANOL HYDROCHLORIDE ◇ (±)-SD 2124-01 HYDROCHLORIDE

TOXICITY DATA with REFERENCE
orl-mus LD50:382 mg/kg EJPHAZ 15,151,71
ipr-mus LD50:131 mg/kg EJPHAZ 15,151,71
ivn-mus LD50:32 mg/kg EJPHAZ 15,151,71

SAFETY PROFILE: Poison by ingestion, intravenous, and intraperitoneal routes. When heated to decomposition it emits toxic fumes of NO_x and HCl.

PMF540 CAS:17804-49-8 *HR: 2*
PROCION RED MX 5B
mf: $C_{19}H_{10}Cl_2N_6O_7S_2 \cdot 2Na$ mw: 615.33

SYNS: BRILLIANT RED 5SKH ◇ CHEMICTIVE BRILLIANT RED 5B ◇ C.I. REACTIVE RED 2 ◇ MIKACION BRILLIANT RED 5BS ◇ OSTAZIN BRILLIANT RED S 5B ◇ PROCION BRILLIANT RED 5BS ◇ PROCION BRILLIANT RED M 5B ◇ PROCION BRILLIANT RED MX 5B ◇ REACTIVE BRILLIANT RED 5SKH

TOXICITY DATA with REFERENCE
orl-rat LD50:8317 mg/kg GISAAA 50(2),88,85
orl-mus LD50:5700 mg/kg GISAAA 50(2),88,85
orl-gpg LD50:2100 mg/kg GISAAA 50(2),88,85

SAFETY PROFILE: Moderately toxic by ingestion. When heated to decomposition it emits toxic fumes of Cl^-, SO_x, SO_x, and Na_2O. See also SULFONATES.

PMF550 CAS:14088-71-2 ***HR: D***
PROCLONOL
mf: $C_{15}H_{14}Cl_2O$ mw: 293.20

SYNS: 4-CHLORO-α-(4-CHLOROPHENYL)-α-CYCLOPROPYL-BENZENEMETHANOL (9CI) ◇ CL 69049 ◇ DPX 3654 ◇ DUPONT DPX 3654 ◇ KILACAR ◇ R 8284

TOXICITY DATA with REFERENCE
mma-sat 2500 ng/plate MUREAV 157,13,85
cyt-ham-orl 80 mg/kg MUREAV 157,13,85

SAFETY PROFILE: Mutation data reported. When heated to decomposition it emits toxic fumes of Cl^-.

PMF600 CAS:32889-48-8 ***HR: 3***
PROCYAZINE
mf: $C_{10}H_{13}ClN_6$ mw: 252.74

SYNS: 2-((4-CHLORO-6-(CYCLOPROPYLAMINO)-1,3,5-TRIAZIN-2-YL)AMINO)-2-METHYLPROPANENITRILE ◇ 2-(4-CHLORO-6-(CYCLOPROPYLAMINO)-s-TRIAZIN-2-YL)AMINO-2-METHYLPROPIONITRILE ◇ CGA-18762 ◇ CYCLE

TOXICITY DATA with REFERENCE
mma-sat 10 μg/plate MUREAV 136,233,84
mrc-smc 131 μmol/L MUREAV 136,233,84
orl-rat LD50:290 mg/kg 85ARAE 2,136,77

CONSENSUS REPORTS: Cyanide and its compounds are on the Community Right-To-Know List. EPA Genetic Toxicology Program.

SAFETY PROFILE: Poison by ingestion. Mutation data reported. When heated to decomposition it emits toxic fumes of Cl^-, CN^-, and NO_x. See also NITRILES.

PMF750 CAS:32809-16-8 ***HR: 1***
PROCYMIDONE
mf: $C_{13}H_{11}Cl_2NO_2$ mw: 284.15

SYNS: 3-(3,5-DICHLOROPHENYL)-1,5-DIMETHYL-3-AZABICYCLO(3.1.0)HEXANE-2,4-DIONE ◇ N-(3′,5′-DICHLOROPHENYL)-1,2-DIMETHYLCYCLOPROPANE-1,2-DICARBOXIMIDE ◇ S 7131 ◇ SUMILEX ◇ SUMISCLEX

TOXICITY DATA with REFERENCE
orl-rat LD50:6800 mg/kg FMCHA2 -,C223,83
skn-mus LD50:7800 mg/kg FMCHA2 -,C223,83

SAFETY PROFILE: Mildly toxic by ingestion and skin contact. When heated to decomposition it emits very toxic fumes of Cl^- and NO_x.

PMG000 CAS:25046-79-1 ***HR: 3***
PRO-DIABAN
mf: $C_{20}H_{27}N_5O_5S$ mw: 449.58

SYNS: BS 4231 ◇ GLISOXEPID ◇ GLISOXEPIDE ◇ RP 22410

TOXICITY DATA with REFERENCE
ivn-rat LD50:196 mg/kg ARZNAD 24,409,74
ivn-mus LD50:283 mg/kg ARZNAD 24,409,74

SAFETY PROFILE: Poison by intravenous route. When heated to decomposition it emits very toxic fumes of NO_x and SO_x. Used as an antidiabetic agent.

PMG600 CAS:302-23-8 ***HR: D***
PRODOX ACETATE
mf: $C_{23}H_{32}O_4$ mw: 372.55

SYNS: ACETOXYPROGESTERONE ◇ 17-ACETOXYPROGESTERONE ◇ 17-α-ACETOXYPROGESTERONE ◇ 17-AP ◇ 17-HYDROXYPREGN-4-ENE-3,20-DIONE ACETATE ◇ 17-HYDROXYPROGESTERONE ACETATE ◇ U 5533

TOXICITY DATA with REFERENCE
orl-rbt TDLo:2500 μg/kg (female 1D pre):REP ACEDAB 73,17,63

SAFETY PROFILE: Experimental reproductive effects. A steroid. When heated to decomposition it emits acrid smoke and irritating fumes.

PMG750 ***HR: 3***
PRODUCER GAS

PROP: Composed of carbon monoxide, hydrogen, air and steam. Lel: 20-30%, uel: 70-80%.

SAFETY PROFILE: Poison. Dangerous fire hazard when exposed to flame. Explosive in the form of vapor by spark or flame when mixed with air in the range of 20.7-73.7%. Dangerous; can react vigorously with oxidizing materials. To fight fire, use CO_2, dry chemical, water spray. See also CARBON MONOXIDE, METHANE, and HYDROGEN.

PMH100 CAS:1811-28-5 ***HR: 3***
PROFLAVINE HEMISULPHATE
mf: $C_{13}H_{11}N_3 \cdot 1/2H_2O_4S$ mw: 258.29

SYNS: 3,6-DIAMINOACRIDINE HEMISULFATE ◇ PROFLAVINE

TOXICITY DATA with REFERENCE
dnd-omi 100 pph ZNCBDA 29,128,74

sce-ham:lng 10 mg/L CNREA8 44,3270,84
scu-mus LD50:200 mg/kg QJPPAL 10,649,37

CONSENSUS REPORTS: IARC Cancer Review: Group 3 IMEMDT 7,56,87; Animal Inadequate Evidence IMEMDT 24,195,80

SAFETY PROFILE: Poison by subcutaneous route. Questionable carcinogen. Mutation data reported. When heated to decomposition it emits toxic fumes of SO_x and NO_x.

PMH250 CAS:952-23-8 ***HR: 3***
PROFLAVINE MONOHYDROCHLORIDE
mf: $C_{13}H_{11}N_3$•ClH mf: 245.73

SYNS: 3,6-ACRIDINEDIAMINE, MONOHYDROCHLORIDE (9CI) ◇ 3,6-DIAMINOACRIDINE MONOHYDROCHLORIDE ◇ 3,6-DIAMINOACRIDINIUM CHLORIDE ◇ 3,6-DIAMINOACRIDINIUM CHLORIDE HYDROCHLORIDE ◇ 2,8-DIAMINOACRIDINIUM CHLORIDE MONOHYDROCHLORIDE ◇ PROFLAVINE HYDROCHLORIDE

TOXICITY DATA with REFERENCE
mmo-sat 300 ng/plate IAPUDO 27,283,80
mma-sat 3 μg/plate IAPUDO 27,283,80
mma-esc 10 μg/plate ENMUDM 7(Suppl 5),1,85
scu-mus LD50:190 mg/kg QJPPAL 10,649,37

CONSENSUS REPORTS: IARC Cancer Review: Animal Indefinite Evidence IMEMDT 24,195,80.

SAFETY PROFILE: Poison by subcutaneous route. Questionable carcinogen. Mutation data reported. When heated to decomposition it emits very toxic fumes of NO_x and HCl. Used as a drug, as a disinfectant, and as a topical antiseptic.

PMH500 CAS:57-83-0 ***HR: 3***
PROGESTERONE
mf: $C_{31}H_{30}O_2$ mw: 314.51

PROP: A female sex hormone. White, crystalline powder; odorless. D: 1.166 @ 23°, mp: 127-131°. Practically insol in water; sol in alc, acetone, and dioxane; sparingly sol in oils.

SYNS: CORLUTIN ◇ CORLUVITE ◇ CORPORIN ◇ CORPUS LUTEUM HORMONE ◇ CYCLOGEST ◇ Δ^4-PREGNENE-3,20-DIONE ◇ GLANDUCORPIN ◇ HORMOFLAVEINE ◇ HORMOLUTON ◇ LINGUSORBS ◇ LIPO-LUTIN ◇ LUCORTEUM SOL ◇ LUTEAL HORMONE ◇ LUTEOHORMONE ◇ LUTEOSAN ◇ LUTEX ◇ LUTOCYCLIN ◇ LUTROMONE ◇ NALUTRON ◇ NSC-9704 ◇ PERCUTACRINE ◇ PIAPONON ◇ 3,20-PREGNENE-4 ◇ PREGNENEDIONE ◇ PREGNENE-3,20-DIONE ◇ PREGN-4-ENE-3,20-DIONE ◇ 4-PREGNENE-3,20-DIONE ◇ PROGEKAN ◇ PROGESTEROL ◇ β-PROGESTERONE ◇ PROGESTERONUM ◇ PROGESTIN ◇ PROGESTONE ◇ PROLIDON ◇ SYNGESTERONE ◇ SYNOVEX S ◇ SYNTOLUTAN

TOXICITY DATA with REFERENCE
dni-hmn:lym 5 μmol/L PSEBAA 146,401,74
dni-hmn:kdy 100 μg/L CNJGA8 10,299,68
orl-wmn TDLo:120 mg/kg (20D pre):REP ACEDAB 28,18,56
orl-wmn TDLo:113 mg/kg (6-32W preg):TER JCEMAZ 19,1369,59
scu-mus TDLo:40 mg/kg:NEO BJCAAI 19,824,65
ims-dog TDLo:26643 mg/kg/4Y-I:ETA FESTAS 31,340,79
ipr-rat LDLo:327 mg/kg MEIEDD 10,1120,83
ivn-mus LDLo:100 mg/kg JMCMAR 11,117,68

CONSENSUS REPORTS: NTP Fifth Annual Report on Carcinogens. IARC Cancer Review: Animal Limited Evidence IMEMDT 21,491,79; Animal Sufficient Evidence IMEMDT 6,135,74. EPA Genetic Toxicology Program. Reported in EPA TSCA Inventory.

SAFETY PROFILE: Confirmed carcinogen with experimental carcinogenic, neoplastigenic, tumorigenic, and teratogen data. Poison by intravenous and intraperitoneal routes. Human teratogenic effects by ingestion and parenteral routes: developmental abnormalities of the urogenital system. Human male reproductive effects by intramuscular route: changes in spermatogenesis, the prostate, seminal vesicle, Cowper's gland, and accessory glands; impotence, and breast development. Human female reproductive effects by ingestion, parenteral, and intravaginal routes: fertility changes; menstrual cycle changes and disorders; uterus, cervix, and vagina changes. Experimental reproductive effects. Human mutation data reported. When heated to decomposition it emits acrid smoke and irritating fumes.

PMH575 ***HR: 2***
PROGLUMIDE SODIUM
mf: $C_{18}H_{25}N_2O_4$•Na mw: 356.44

SYN: dl-4-BENZAMIDO-N,N-DIPROPYLGLUTARAMIC ACID SODIUM SALT

TOXICITY DATA with REFERENCE
orl-rat LD50:8200 mg/kg OYYAA2 5,203,71
ipr-rat LD50:2360 mg/kg OYYAA2 5,203,71
scu-rat LD50:6110 mg/kg OYYAA2 5,203,71
orl-mus LD50:8900 mg/kg OYYAA2 5,203,71
ipr-mus LD50:2811 mg/kg OYYAA2 5,203,71
scu-mus LD50:5886 mg/kg OYYAA2 5,203,71
ivn-mus LD50:2968 mg/kg OYYAA2 5,203,71

SAFETY PROFILE: Moderately toxic by intraperitoneal and intravenous routes. Mildly toxic by ingestion. When heated to decomposition it emits toxic fumes of NO_x and Na_2O.

PMH600 CAS:303-53-7 ***HR: 3***
PROHEPTATRIENE
mf: $C_{20}H_{21}N$ mw: 275.38

PROP: Bp: 175-180°.

SYNS: CYCLOBENZAPRINE ◇ N,N-DIMETHYL-5H-

DIBENZO(a,d)CYCLOHEPTENE-Δ^5,Γ-PROPYLAMINE ◇ 5-(3-DIMETILAMINOPROPILIDEN)-5H-DIBENZO-(a,d)-CICLOPENTENE (ITALIAN) ◇ PROEPTATRIENE (ITALIAN)

TOXICITY DATA with REFERENCE
orl-mus LD50:250 mg/kg FRPPAO 25,519,70
ipr-mus LD50:90 mg/kg JMCMAR 6,338,63
ivn-mus LD50:36 mg/kg FRPPAO 25,519,70

CONSENSUS REPORTS: EPA Genetic Toxicology Program.

SAFETY PROFILE: Poison by ingestion, intravenous, and intraperitoneal routes. When heated to decomposition it emits toxic fumes of NO_x.

PMH625 CAS:9002-62-4 ***HR: D***
PROLACTIN

PROP: Polypeptide hormone of mw about 23,000. Crystals. Insol in water; sol in abs methanol or ethanol with addition of small amount of acid.

SYNS: ADENOHYPOPHYSEAL LUTEOTROPIN ◇ ANTERIOR PITUITARY LUTEOTROPIN ◇ BOVINE LACTOGENIC HORMONE ◇ BOVINE PROLACTIN ◇ GALCTIN ◇ LACTOGEN ◇ LACTOGENIC HORMONE ◇ LACTOSOMATOTROPIC HORMONE ◇ LUTEOTROPHIN ◇ LUTEOTROPIC HORMONE ◇ LUTEOTROPIC HORMONE LTH ◇ LUTEOTROPIN ◇ MAMMOTROPIN ◇ PARALCTIN ◇ PITUITARY LACTOGENIC HORMONE

TOXICITY DATA with REFERENCE
dns-sat-scu 40 mg/kg DNREA8 36,2223,76
dns-rat:mmr 5 mg/L DNREA8 36,2162,76
dns-rbt-par 62 μg/kg JNCSAI 14,105,74
scu-rat TDLo:12500 μg/kg (16-20D preg):REP PSEBAA 94,746,57

SAFETY PROFILE: Experimental reproductive effects. Mutation data reported.

PMI000 CAS:37106-99-3 ***HR: 3***
PROLINOMETHYLTETRACYCLINE
mf: $C_{28}H_{33}N_3O_{10}$ mw: 571.64

SYNS: 1-((((4-DIMETHYLAMINO)-1,4,4A,5,5A,6,11,12A-OCTAHYDRO-3,6,10,12,12A-PENTAHYDROXY-6-METHYL-1,11-DIOXO-2-NAPHTHACENYL)CARBONYL)AMINO)METHYL)-l-PROLINE◇ PM-TC

TOXICITY DATA with REFERENCE
ipr-rat TDLo:300 mg/kg (female 10-15D post):REP IYKEDH 3,69,72
ipr-rat TDLo:600 mg/kg (10-15D preg):TER IYKEDH 3,69,72
ipr-rat LD50:115 mg/kg JJANAX 25,95,72
scu-rat LD50:1733 mg/kg KSRNAM 6,51,72
ivn-rat LD50:144 mg/kg KSRNAM 6,51,72
ims-rat LD50:788 mg/kg KSRNAM 6,51,72
ipr-mus LD50:353 mg/kg KSRNAM 6,51,72
scu-mus LD50:1465 mg/kg KSRNAM 6,51,72
ivn-mus LD50:85100 μg/kg KSRNAM 6,51,72
ims-mus LD50:283 mg/kg JJANAX 24,156,71

SAFETY PROFILE: Poison by intramuscular, intraperitoneal, and intravenous routes. Moderately toxic by subcutaneous routes. Experimental teratogenic and reproductive effects. When heated to decomposition it emits toxic fumes of NO_x.

PMI250 CAS:2746-81-8 ***HR: 3***
PROLIXIN ENANTHATE
mf: $C_{29}H_{38}F_3N_3O_2S$ mw: 549.76

SYNS: EUTIMOX ◇ FLUPHENAZINE ENANTHATE ◇ MODITEN ENANTHATE ◇ MODITEN-RETARD ◇ OF ◇ SQ 16,114 ◇ 4-(3-(2-TRIFLUOROMETHYL-10-PHENOTHIAZINYL)PROPYL)-1-PIPERAZINEETHANOL ENANTHATE ◇ 2-((4-(3-(2-(TRIFLUOROMETHYL) PHENOTHIAZIN-10-YL)PROPYL)-1-PIPERAZINYL))ETHYLHEPTANOATE

TOXICITY DATA with REFERENCE
scu-hmn TDLo:357 μg/kg:CNS AJPSAO 119,779,63
ims-hmn TDLo:1633 μg/kg:MSK BMJOAE 2,1071,66
ipr-mus LD50:300 mg/kg NIIRDN 6,701,82
scu-mus LD50:50 mg/kg NIIRDN 6,701,82
ivn-mus LD50:56 mg/kg CSLNX* NX#03337
ims-mus LD50:17200 μg/kg NIIRDN 6,701,82

SAFETY PROFILE: Poison by intravenous, intramuscular, intraperitoneal, and subcutaneous routes. Human systemic effects by subcutaneous route: muscle weakness; by intramuscular route: musculo-skeletal changes. When heated to decomposition it emits very toxic fumes of F^-, SO_x, and NO_x.

PMI500 CAS:53-60-1 ***HR: 3***
PROMAZINE HYDROCHLORIDE
mf: $C_{17}H_{20}N_2S{\cdot}ClH$ mw: 320.91

PROP: White to sltly yellow, practically odorless, crystalline powder. Decomp @ 181°. Sol in water, methanol, ethanol, chloroform; insol in ether, benzene.

SYNS: 10-(Γ-DIMETHYLAMINO-N-PROPYL)PHENOTHIAZINEHYDROCHLORIDE ◇ 10-(3-(DIMETHYLAMINO)PROPYL)PHENOTHIAZINE HYDROCHLORIDE ◇ SPARINE HYDROCHLORIDE

TOXICITY DATA with REFERENCE
orl-hmn TDLo:20 mg/kg/D-I:CNS JNMDAN 123,553,56
orl-rat LD50:400 mg/kg FAZMAE 5,269,63
scu-rat LD50:300 mg/kg 27ZMA4 2,-,67
ivn-rat LD50:29 mg/kg 27ZQAG -,42,72
ipr-mus LDLo:225 mg/kg ARZNAD 4,171,54
scu-mus LD50:300 mg/kg 27ZQAG -,42,72
ivn-mus LD50:38 mg/kg 27ZQAG -,42,72
ims-dog LD50:4400 μg/kg 27ZIAQ -,-,65
ivn-rbt LD50:21 mg/kg 27ZQAG -,42,72

CONSENSUS REPORTS: Reported in EPA TSCA Inventory.

SAFETY PROFILE: Poison by ingestion, subcutaneous, intravenous, intraperitoneal, and intramuscular routes. Human systemic effects by ingestion: general anesthesia, tremors, antipsychotic effects. An additive permitted in food for human consumption; also permitted in the feed and drinking water of animals and/or for the treatment of food-producing animals. When heated to decomposition it emits toxic fumes of NO_x, SO_x, and HCl.

PMI750 CAS:58-33-3 ***HR: 3***
PROMETHAZINE HYDROCHLORIDE
mf: $C_{17}H_{20}N_2S{\bullet}ClH$ mw: 320.91

PROP: Crystals. Mp: 230-232° (decomp). Very sol in water; sol in alc, chloroform; insol in acetone, ether, ethyl acetate.

SYNS: 10-(3-DIMETHYLAMINOISOPROPYL)PHENOTHIAZINEHYDROCHLORIDE ◇ N-(2'-DIMETHYLAMINO-2'-METHYL) ETHYLPHENOTHIAZINE HYDROCHLORIDE ◇ 10-(2-(DIMETHYLAMINO) PROPYL)PHENOTHIAZINE MONOHYDROCHLORIDE ◇ N-(2-DIMETHYLAMINOPROPYL-1)PHENOTHIAZINEHYDROCHLORIDE ◇ 10-(2-DIMETHYLAMINOPROPYL)PHENOTHIAZINE HYDROCHLORIDE ◇ DIPHERGAN ◇ DORME ◇ FARGAN ◇ FELLOZINE ◇ FENAZIL ◇ FENERGAN ◇ GANPHEN ◇ HL 8700 ◇ LERGIGAN ◇ PHENCEN ◇ PHENERGAN HYDROCHLORIDE ◇ PLLETIA ◇ PRIMINE ◇ PROMANTINE ◇ PROMETHAZINE N-(2'-DIMETHYLAMINO-2'-METHYLETHYL)PHENOTHIAZINE HYDROCHLORIDE ◇ PROMETHIAZIN (GERMAN) ◇ PROREX ◇ PROTAZINE ◇ REMSED ◇ 3277 R.P. ◇ N,N,α-TRIMETHYL-10H-PHENOTHIAZINE-10-ETHANAMINE MONOHYDROCHLORIDE

TOXICITY DATA with REFERENCE
orl-rat TDLo:60 mg/kg (5-16D preg):TER RCOCB8 7,701,74
orl-rat TDLo:240 mg/kg (5-16D preg):REP RCOCB8 7,701,74
orl-hmn TDLo:3500 μg/kg/D:CNS AJPSAO 113,654,57
orl-chd TDLo:20 mg/kg:CNS LANCAO 1,368,80
orl-mus LD50:255 mg/kg AIPTAK 135,364,62
ipr-mus LD50:160 mg/kg CKFRAY 15,526,66
scu-mus LD50:240 mg/kg YAKUD5 22,375,80
ivn-mus LD50:50 mg/kg JPETAB 108,340,53
ipr-gpg LD50:35 mg/kg PHARAT 38,749,83
ivn-gpg LD50:42500 μg/kg AIPTAK 113,313,58

CONSENSUS REPORTS: Reported in EPA TSCA Inventory.

SAFETY PROFILE: Poison by ingestion, subcutaneous, intraperitoneal, and intravenous routes. Human systemic effects by ingestion: excitement, sleep, convulsions, rigidity. Experimental teratogenic effects. Other experimental reproductive effects. When heated to decomposition it emits very toxic fumes of HCl, SO_x, and NO_x. Used as an antihistamine.

PMJ000 ***HR: 3***
PROMETHIUM
af: Pm aw: 147.

PROP: ^{147}Pm: Metallic solid. Mp: 1080°, bp: 2460°, d: 7.22. A rare earth. The 145 isotope has a half life of 18 years. The 147 isotope has a half life of 2.64 years. The 147 isotope is the only one available.

SAFETY PROFILE: A poison. Radiotoxic metal. See also RARE EARTHS.

PMJ100 CAS:9036-06-0 ***HR: 3***
PRONASE

SYNS: PROTELINE ◇ PROTENAZA 1 ◇ S. GRISEUS PROTEASE ◇ S. GRISEUS PROTEINASE ◇ STREPTOMYCES GRISEUS PROTEASE ◇ STREPTOMYCES GRISEUS PROTEINASE

TOXICITY DATA with REFERENCE
orl-rat LD50:3290 mg/kg NIIRDN 6,730,82
ipr-rat LD50:16200 μg/kg NIIRDN 6,730,82
scu-rat LD50:100 mg/kg NIIRDN 6,730,82
ivn-rat LD50:15800 μg/kg NIIRDN 6,730,82
orl-mus LD50:4010 mg/kg NIIRDN 6,730,82
ipr-mus LD50:14100 μg/kg NIIRDN 6,730,82
scu-mus LD50:26800 μg/kg NIIRDN 6,730,82
ivn-mus LD50:20500 μg/kg NIIRDN 6,730,82

SAFETY PROFILE: Poison by subcutaneous, intravenous, and intraperitoneal routes. Moderately toxic by ingestion.

PMJ500 CAS:154-41-6 ***HR: 3***
PROPADRINE HYDROCHLORIDE
mf: $C_9H_{13}NO{\bullet}ClH$ mw: 187.69

SYNS: α-(1-AMINOETHYL)BENZENEMETHANOLHYDROCHLORIDE ◇ α-(1-AMINOETHYL)BENZYL ALCOHOL HYDROCHLORIDE ◇ (±)-2-AMINO-1-PHENYL-1-PROPANOL HYDROCHLORIDE ◇ α-HYDROXY-β-AMINOPROPYLBENZENE HYDROCHLORIDE ◇ MONHYDRIN ◇ MUCORAMA ◇ MYDRIATINE ◇ dl-NOREPHEDRINE HYDROCHLORIDE ◇ dl-1-PHENYL-2-AMINO-1-PROPANOL MONOHYDROCHLORIDE ◇ PHENYLPROPANOLAMINE HYDROCHLORIDE

TOXICITY DATA with REFERENCE
scu-mus TDLo:6400 μg/kg (16D preg):TER PRGLBA 2,295,72
orl-rat LD50:1490 mg/kg TXAPA9 18,185,71
ipr-rat LD50:160 mg/kg JPETAB 86,284,46
scu-rat LDLo:80 mg/kg JPETAB 71,62,41
orl-mus LD50:150 mg/kg THERAP 20,297,65
ipr-mus LD50:428 mg/kg JPETAB 92,207,48
scu-mus LD50:600 mg/kg JPETAB 86,280,46
ivn-mus LDLo:275 mg/kg QJPPAL 9,203,36
scu-rbt LD50:255 mg/kg JPETAB 86,284,46
ivn-rbt LD50:50 mg/kg JPETAB 86,284,46
ims-rbt LD50:320 mg/kg JPETAB 86,284,46

CONSENSUS REPORTS: Reported in EPA TSCA Inventory.

SAFETY PROFILE: Poison by ingestion, subcutaneous, intravenous, intraperitoneal and intramuscular routes. An experimental teratogen. When heated to decomposition it emits very toxic fumes of HCl and NO_x. Used as a raw material in cold and diet tablets.

PMJ525 CAS:34183-22-7 ***HR: 3***
PROPAFENONE HYDROCHLORIDE
mf: $C_{21}H_{27}N_2O_3$•ClH mw: 391.96

SYNS: BAXARYTMON ◇ FENOPRAIN ◇ 2'-(2-HYDROXY-3-(PROPYLAMINO)PROPOXY)-3-PHENYLPROPIOPHENONEHYDROCHLORIDE ◇ PROPAFENON HYDROCHLORID ◇ 1-PROPANONE, 1-(2-(2-HYDROXY-3-(PROPYLAMINO)PROPOXY)PHENYL)-3-PHENYL-, HYDROCHLORIDE (9CI) ◇ RHYTHMONORM ◇ RYTMONORM ◇ SA 79 ◇ WZ 884

TOXICITY DATA with REFERENCE
orl-rat TDLo:1600 mg/kg (female 17-20D post):REP KSRNAM 21,3239,87
orl-wmn TDLo:12 mg/kg/1D-I:PNS JCLPDE 46,104,85
orl-cld TDLo:133 mg/kg AEMED3 16,437,87
orl-man LDLo:4286 μg/kg/15H-I:CVS PGMJAO 60,155,84
orl-rat LD50:700 mg/kg ARZNAD 26,1849,76
ivn-rat LD50:18800 μg/kg ARZNAD 26,1849,76
ivn-dog LD50:10 mg/kg ARZNAD 26,1849,76

SAFETY PROFILE: Poison by intravenous route. Moderately toxic by ingestion, subcutaneous, and intraperitoneal routes. Human systemic effects: paresthesia, wakefulness, hallucinations, distorted perceptions, pulse rate increase. Experimental reproductive effects. When heated to decomposition it emits toxic fumes of NO_x and HCl.

PMJ550 CAS:57619-29-1 ***HR: 3***
PRO-PAM
mf: $C_7H_{10}N_2O$•ClH mw: 174.65

SYNS: 1-METHYL-1,6-DIHYDROPICOLINALDEHYDE OXIME HYDROCHLORIDE ◇ N-METHYL-1,6-DIHYDROPYRIDINE-2-CARBALDOXIME HYDROCHLORIDE ◇ 1-METHYL-1,6-DIHYDRO-2-PYRIDINECARBOXYALDEHYDE OXIME HYDROCHLORIDE

TOXICITY DATA with REFERENCE
ipr-mus LD50:173 mg/kg TXAPA9 47,305,79
ivn-mus LD50:168 mg/kg TXAPA9 47,305,79
ims-mus LD50:125 mg/kg TXAPA9 47,305,79

SAFETY PROFILE: Poison by intramuscular, intravenous, and intraperitoneal routes. When heated to decomposition it emits toxic fumes of NO_x and HCl.

PMJ750 CAS:74-98-6 ***HR: 3***
PROPANE
DOT: UN 1075/UN 1978
mf: C_3H_8 mw: 44.11

PROP: Colorless gas. Bp: −42.1°, lel: 2.3%, uel: 9.5%, fp: −187.1°, flash p: −156°F, d: 0.5852 @ −44.5°/4°, autoign temp: 842°F, vap d: 1.56. Sol in water, alc, ether.

SYNS: DIMETHYLMETHANE ◇ PROPYL HYDRIDE

CONSENSUS REPORTS: Reported in EPA TSCA Inventory.

OSHA PEL: TWA 1000 ppm
ACGIH TLV: Asphyxiant
DFG MAK: 1000 ppm (1800 mg/m^3)
DOT Classification: Flammable Gas; Label: Flammable Gas.

SAFETY PROFILE: Central nervous system effects at high concentrations. An asphyxiant. Flammable gas. Highly dangerous fire hazard when exposed to heat or flame; can react vigorously with oxidizers. Explosive in the form of vapor when exposed to heat or flame. Explosive reaction with ClO_2. Violent exothermic reaction with barium peroxide + heat. To fight fire, stop flow of gas. When heated to decomposition it emits acrid smoke and irritating fumes.

PMK000 CAS:542-78-9 ***HR: 3***
PROPANEDIAL
mf: $C_3H_4O_2$ mw: 72.07

SYNS: MALONALDEHYDE ◇ MALONDIALDEHYDE ◇ MALONIC ALDEHYDE ◇ MALONIC DIALDEHYDE ◇ MALONODIALDEHYDE ◇ MALONYLDIALDEHYDE ◇ NCI-C54842 ◇ 1,3-PROPANEDIAL ◇ 1,3-PROPANEDIALDEHYDE ◇ 1,3-PROPANEDIONE

TOXICITY DATA with REFERENCE
mmo-sat 13850 nmol/plate BTERDG 2,81,80
mmo-esc 2 mmol/L MUREAV 88,23,81
dnd-hmn:leu 1 mmol/L CLREAS 23(5),595A,75
mnt-rat:fbr 100 μmol/L MUREAV 101,237,82
skn-mus TDLo:7488 mg/kg/2Y-I:CAR AUODDK 55,3,80
skn-mus TD:30 g/kg/9W-I:CAR JNCIAM 53,1771,74
orl-rat LD50:632 mg/kg TXAPA9 7,826,65
orl-mus LD50:606 mg/kg AUODDK 55,3,80

CONSENSUS REPORTS: IARC Cancer Review: Group 3 IMEMDT 7,56,87; Animal Inadequate Evidence IMEMDT 36,163,85. EPA Genetic Toxicology Program.

SAFETY PROFILE: Moderately toxic by ingestion. Questionable carcinogen with experimental carcinogenic data. Human mutation data reported. When heated to decomposition it emits acrid smoke and irritating fumes. See also ALDEHYDES.

PMK250 CAS:78-90-0 ***HR: 3***
1,2-PROPANEDIAMINE
DOT: UN 2258
mf: $C_3H_{10}N_2$ mw: 74.15

PROP: Flash p: 92°F (OC), d: 0.9, vap d: 2.6, bp: 118.9°.

SYNS: 1,2-DIAMINOPROPANE ◇ PROPYLENEDIAMINE ◇ PROPYLENE DIAMINE (DOT)

TOXICITY DATA with REFERENCE
skn-rbt 10 mg/24H open AMIHBC 10,61,54
skn-rbt 435 mg open SEV UCDS** 3/12/69
eye-rbt 87 mg SEV UCDS** 3/12/69
orl-rat LD50:2230 mg/kg UCDS** 3/12/69
scu-rat LDLo:2250 mg/kg ZEPTAT 17,59,15
skn-rbt LD50:500 mg/kg AMIHBC 10,61,54

CONSENSUS REPORTS: Reported in EPA TSCA Inventory.

DOT Classification: Flammable Liquid; Label: Flammable Liquid, Corrosive; Flammable or Combustible Liquid; Label: Flammable Liquid, Corrosive.

SAFETY PROFILE: Moderately toxic by ingestion, skin contact, and subcutaneous routes. A corrosive irritant to eyes, skin, and mucous membranes. Dangerous fire hazard when exposed to heat, flames, oxidizers. To fight fire, use alcohol foam. When heated to decomposition it emits toxic fumes of NO_x. Used as an intermediate in production of petroleum and polymer additives, and surfactants. See also AMINES.

PMK500 CAS:109-76-2 ***HR: 3***
1,3-PROPANEDIAMINE
mf: $C_3H_{10}N_2$ mw: 74.15

PROP: Water-white liquid, amine odor. D: 0.8881 @ 20°/20°, bp: 139.7°, fp: −12°, flash p: 120°F (TOC). Completely sol in water, methanol, and ether.

SYNS: 1,3-DIAMINOPROPANE ◇ 1,3-PROPYLENEDIAMINE ◇ TRIMETHYLENEDIAMINE

TOXICITY DATA with REFERENCE
skn-rbt 50 mg open SEV UCDS** 1/28/63
eye-rbt 1 mg SEV UCDS** 1/28/63
ipr-mus TDLo:264 mg/kg (12D preg):TER TJADAB 28,237,83
orl-rat LD50:350 mg/kg AIHAAP 23,95,62
skn-rbt LD50:200 mg/kg AIHAAP 23,95,62

CONSENSUS REPORTS: Reported in EPA TSCA Inventory.

SAFETY PROFILE: Poison by ingestion and skin contact. Experimental teratogenic effects. A severe skin and eye irritant. Flammable when exposed to heat or flame. When heated to decomposition it emits toxic fumes of NO_x. See also AMINES.

PMK750 CAS:5442-32-0 ***HR: 3***
PROPANE DIISOTHIOUREA DIHYDROBROMIDE
mf: $C_5H_{12}N_4S_2$•2BrH mw: 354.17

SYN: 2,2′-TRIMETHYLENE-BIS-(2-THIOPSEUDOUREA), DIHYDROBROMIDE

TOXICITY DATA with REFERENCE
ipr-mus LD50:250 mg/kg NTIS** AD691-490

CONSENSUS REPORTS: Reported in EPA TSCA Inventory.

SAFETY PROFILE: Poison by intraperitoneal route. When heated to decomposition it emits very toxic fumes of HBr, SO_x and NO_x.

PMK800 CAS:63843-89-0 ***HR: 2***
PROPANEDIOIC ACID, ((3,5-BIS(1,1-DIMETHYLETHYL)-4-HYDROXYPHENYL)METHYL)BUTYL-, BIS(1,2,2,6,6- PENTAMETHYL-4-PIPERIDINYL) ESTER
mf: $C_{42}H_{72}N_2O_5$ mw: 685.16

SYN: TINUVIN 144

TOXICITY DATA with REFERENCE
eye-rbt 100 mg MLD EPASR* 8EHQ-0186-0585S
orl-rat TDLo:8400 mg/kg (male 28D pre):REP EPASR* 8EHQ-0186-0585S
orl-rat LD50:1500 mg/kg EPASR* 8EHQ-0186-0585S

CONSENSUS REPORTS: Reported in EPA TSCA Inventory.

SAFETY PROFILE: Moderately toxic by ingestion. Experimental reproductive effects. An eye irritant. When heated to decomposition it emits toxic fumes of NO_x.

PML000 CAS:57-55-6 ***HR: 2***
1,2-PROPANEDIOL
mf: $C_3H_8O_2$ mw: 76.11

$CH_3CHOHCH_2OH$

PROP: Colorless viscous liquid; practically odorless. Bp: 188.2°, flash p: 210°F (OC), lel: 2.6%, uel: 12.6%, d: 1.0362 @ 25°/25°, autoign temp: 700°F, vap press: 0.08 mm @ 20°, vap d: 2.62, fp: −59°. Hygroscopic; misc with water, acetone, chloroform; sol in essential oils; immisc with fixed oils.

SYNS: 1,2-DIHYDROXYPROPANE ◇ DOWFROST ◇ METHYLETHYLENE GLYCOL ◇ METHYL GLYCOL ◇ MONOPROPYLENE GLYCOL ◇ PG 12 ◇ PROPANE-1,2-DIOL ◇ PROPYLENE GLYCOL (FCC) ◇ PROPYLENE GLYCOL USP ◇ α-PROPYLENEGLYCOL ◇ 1,2-PRO-

PYLENE GLYCOL ◇ SIRLENE ◇ SOLAR WINTER BAN ◇ TRIMETHYL GLYCOL

TOXICITY DATA with REFERENCE
skn-hmn 500 mg/7D MLD JIDEAE 55,190,70
skn-hmn 104 mg/3D-I MOD 85DKA8 -,127,77
skn-man 10%/2D JIDEAE 19,423,52
eye-rbt 100 mg MLD FCTOD7 20,573,82
eye-rbt 500 mg/24H MLD 28ZPAK -,37,72
dni-mus-scu 8000 mg/kg APMUAN S274,304,81
cyt-mus-scu 8000 mg/kg APMUAN S274,304,81
cyt-ham:fbr 32 g/L FCTOD7 23,623,84
ipr-mus TDLo:100 mg/kg (15D preg):TER KAIZAN 37,239,62
ipr-mus TDLo:100 mg/kg (11D preg):REP KAIZAN 37,239,62
orl-chd TDLo:79 g/kg/56W-I:CNS,BRN JOPDAB 93,515,78
par-inf TDLo:10 g/kg/3D-C:SYS PEDIAU 72,353,83
orl-rat LD50:20 g/kg TXAPA9 45,362,78
ipr-rat LD50:6660 mg/kg KRKRDT 9,36,81
scu-rat LD50:22500 mg/kg IAEC** 17JUN74
ivn-rat LD50:6423 mg/kg ARZNAD 26,1581,76
ims-rat LD50:14 g/kg IAEC** 17JUN74
orl-mus LD50:22 g/kg JPETAB 65,89,39
ipr-mus LD50:9718 mg/kg FEPRA7 6,342,47
scu-mus LD50:17370 mg/kg KRKRDT 8,46,81
ivn-mus LD50:6630 mg/kg ARZNAD 26,1581,76

CONSENSUS REPORTS: Reported in EPA TSCA Inventory. EPA Genetic Toxicology Program.

SAFETY PROFILE: Slightly toxic by ingestion, skin contact, intraperitoneal, intravenous, subcutaneous, and intramuscular routes. Human systemic effects by ingestion: general anesthesia, convulsions, changes in surface EEG. Experimental teratogenic and reproductive effects. An eye and human skin irritant. Mutation data reported. Combustible liquid when exposed to heat or flame; can react with oxidizing materials. Explosive in the form of vapor when exposed to heat or flame. May react with hydrofluoric acid + nitric acid + silver nitrate to form the explosive silver fulminate. To fight fire, use alcohol foam. When heated to decomposition it emits acrid smoke and irritating fumes.

PML250 CAS:504-63-2 ***HR: 1***
1,3-PROPANEDIOL
mf: $C_3H_8O_2$ mw: 76.11

PROP: Colorless, odorless liquid. D: 1.0536 @ 25°, bp: 210-211°, vap d: 2.6, autoign temp: 752°F. Sol in water, alc, and ether.

SYNS: 2-DEOXYGLYCEROL ◇ 1,3-DIHYDROXYPROPANE ◇ 2-(HYDROXYMETHYL)ETHANOL ◇ NSC 65426 ◇ PG ◇ PROPANE-1,3-DIOL ◇ β-PROPYLENE GLYCOL ◇ 1,3-PROPYLENE GLYCOL ◇ TRIMETHYLENE GLYCOL

TOXICITY DATA with REFERENCE
dnd-rat-orl 2100 mg/kg/10W-C CBINA8 50,87,84
orl-mus LD50:4773 mg/kg TXAPA9 49,385,79

CONSENSUS REPORTS: Reported in EPA TSCA Inventory.

SAFETY PROFILE: Mildly toxic by ingestion. Mutation data reported. When heated to decomposition it emits acrid smoke and irritating fumes. See also 1,2-PROPANEDIOL.

PML400 CAS:1120-71-4 ***HR: 3***
PROPANE SULTONE
mf: $C_3H_6O_3S$ mw: 122.15

SYNS: 3-HYDROXY-1-PROPANESULFONIC ACID Γ-SULTONE ◇ 3-HYDROXY-1-PROPANESULPHONIC ACID SULFONE ◇ 3-HYDROXY-1-PROPANESULPHONIC ACID SULTONE ◇ 1,2-OXATHIOLANE-2,2-DIOXIDE ◇ 1-PROPANESULFONIC ACID-3-HYDROXY-Γ-SULTONE ◇ 1,3-PROPANE SULTONE (MAK) ◇ RCRA WASTE NUMBER U193

TOXICITY DATA with REFERENCE
skn-rbt 500 mg MLD 34ZIAG -,498,69
dnd-esc 10 μmol/L MUREAV 89,95,81
otr-hmn:oth 7500 μg/L CNREA8 41,5096,81
cyt-hmn:lym 1 mmol/L TOLED5 28,139,85
ivn-rat TDLo:80 mg/kg (15D preg):TER IARCCD 4,45,73
orl-rat TDLo:7840 mg/kg/60W-I:CAR NATUAS 230,460,71
scu-rat TDLo:10 mg/kg:ETA ZEKBAI 75,69,70
ivn-rat TDLo:20 mg/kg/(15D preg):CAR,TER ZEKBAI 75,69,70
skn-mus TDLo:1000 mg/kg:CAR TXCYAC 6,139,76
scu-mus TDLo:756 mg/kg/63W-I:NEO JNCIAM 46,143,71
scu-rat LD50:135 mg/kg ZEKBAI 75,69,70
skn-mus LDLo:1000 mg/kg TXCYAC 6,139,76
ipr-mus LD50:467 mg/kg JJIND8 62,911,79

CONSENSUS REPORTS: NTP Fifth Annual Report on Carcinogens. IARC Cancer Review: Group 2B IMEMDT 7,56,87; Animal Sufficient Evidence IMEMDT 4,253,74. Community Right-To-Know List. Reported in EPA TSCA Inventory. EPA Genetic Toxicology Program.

ACGIH TLV: Suspected Human Carcinogen.
DFG MAK: Animal Carcinogen, Suspected Human Carcinogen.

SAFETY PROFILE: Confirmed carcinogen with experimental carcinogenic, neoplastigenic, tumorigenic, and teratogenic data. Poison by subcutaneous route. Moderately toxic by skin contact and intraperitoneal routes. Human mutation data reported. Implicated as a human brain carcinogen. A skin irritant. When heated to decomposition it emits toxic fumes of SO_x.

PML500 CAS:107-03-9 ***HR: 3***
PROPANETHIOL
DOT: UN 2402
mf: C_3H_8S mw: 76.17

PROP: Flash p: $-4°F$.

SYNS: 3-MERCAPTOPROPANOL ◇ PROPANE-1-THIOL ◇ PROPYL MERCAPTAN ◇ N-PROPYL MERCAPTAN

TOXICITY DATA with REFERENCE
eye-rbt 83 mg SEV AIHAAP 19,171,58
orl-rat LD50:1790 mg/kg AIHAAP 19,171,58
ihl-rat LC50:7300 ppm/4H AIHAAP 19,171,58
ipr-rat LD50:515 mg/kg AIHAAP 19,171,58
ihl-mus LC50:4010 ppm/4H AIHAAP 19,171,58

CONSENSUS REPORTS: Reported in EPA TSCA Inventory.
NIOSH REL: (Thiols (n-Alkane Mono)) CL 0.5 ppm/15M

DOT Classification: Label: Flammable Liquid.

SAFETY PROFILE: Moderately toxic by ingestion and intraperitoneal routes. Mildly toxic by inhalation. A severe eye irritant. Explodes on contact with calcium hypochlorite. Very dangerous fire hazard when exposed to heat or flame. When heated to decomposition it emits toxic fumes of SO_x. See also MERCAPTANS.

PMM000 CAS:1421-14-3 ***HR: 3***
PROPANIDIDE
mf: $C_{18}H_{27}NO_5$ mw: 337.46

SYNS: BAYER 1420 ◇ 4-(2-(DIETHYLAMINO)-2-OXOETHOXY)-3-METHOXYBENZENEACETIC ACID, PROPYL ESTER ◇ (p-((DIETHYLCARBAMOYL)METHOXY)-3-METHOXYPHENYL)ACETIC ACID PROPYL ESTER ◇ (4-((DIETHYLCARBAMOYL)METHOXY)-3-METHOXYPHENYL)ACETIC ACID PROPYL ESTER ◇ EPONTHOL ◇ FABANTOL ◇ FBA 1420 ◇ (3-METHOXY-4-((N,N-DIETHYLCARBAMIDO)METHOXY)PHENYL)ACETIC ACID n-PROPYL ESTER ◇ PROPANIDID ◇ PROPANTAN ◇ PROPYL(4-((DIETHYLCARBAMOYL)METHOXY)-3-METHOXYPHENYL)ACETATE◇ 13245 R. P. ◇ SOMBREVIN ◇ 2180 TH ◇ WH 5668

TOXICITY DATA with REFERENCE
ivn-dog TDLo:1200 mg/kg (1-60D preg):REP NYSJAM 64,2177,64
ivn-dog TDLo:1200 mg/kg (1-60D preg):TER NYSJAM 64,2177,64
ivn-hmn TDLo:5 mg/kg:CNS BJANAD 45,1097,73
orl-rat LD50:700 mg/kg TXAPA9 18,185,71
ivn-mus LD50:113 mg/kg OYYAA2 19,845,80

SAFETY PROFILE: Poison by intravenous route. Moderately toxic by ingestion. An experimental teratogen. Experimental reproductive effects. Human systemic effects by intravenous route: general anesthesia. When heated to decomposition it emits toxic fumes of NO_x. Used as an anesthetic.

PMM250 CAS:156-87-6 ***HR: 2***
3-PROPANOLAMINE
mf: C_3H_9NO mw: 75.13

PROP: Colorless liquid; fishy odor. Bp: 168° @ 500 mm, flash p: $>175°F$ (TOC), fp: 12.4°, d: 0.9786 @ 30°, vap press: 2.1 mm @ 60°, vap d: 2.59. Very sol in water; sol in alc; insol in ether.

SYNS: β-ALANINOL ◇ Γ-AMINOPROPANOL ◇ 3-AMINOPROPANOL ◇ 3-AMINO-1-PROPANOL ◇ 3-AMINOPROPYL ALCOHOL ◇ 3-HYDROXYPROPYLAMINE ◇ PROPANOLAMINE ◇ 1,3-PROPANOLAMINE

TOXICITY DATA with REFERENCE
skn-rbt 10 mg/24H open SEV AIHAAP 23,95,62
ipr-mus TDLo:5700 μg/kg (11D preg):TER TJADAB 4,141,71
orl-rat LDLo:2830 mg/kg AIHAAP 23,95,62
skn-rbt LD50:1250 mg/kg AIHAAP 23,95,62

CONSENSUS REPORTS: Reported in EPA TSCA Inventory.

SAFETY PROFILE: Moderately toxic by ingestion and skin contact. An experimental teratogen. A severe skin irritant. Combustible when exposed to heat or flame; can react with oxidizing materials. To fight fire, use foam, CO_2, dry chemical. When heated to decomposition it emits toxic fumes of NO_x. See also AMINES.

PMN250 CAS:3692-90-8 ***HR: 3***
3-PROPARGLOXYPHENYL-N-METHYL-CARBAMATE
mf: $C_{11}H_{11}NO_3$ mw: 205.23

SYNS: ENT 25,732 ◇ H 8717 ◇ HERCULES 8717 ◇ m-(2-PROPYNYLOXY)PHENYL ESTER METHYLCARBAMIC ACID ◇ 3-(2-PROPYNYLOXY)PHENYL-N-METHYLCARBAMATE

TOXICITY DATA with REFERENCE
orl-rat LD50:150 mg/kg TXAPA9 21,315,72
ivn-mus LD50:31600 μg/kg CSLNX* NX#02076
orl-bwd LD50:15 mg/kg TXAPA9 21,315,72

SAFETY PROFILE: Poison by ingestion and intravenous routes. When heated to decomposition it emits toxic fumes of NO_x. See also CARBAMATES.

PMN450 CAS:107-19-7 ***HR: 3***
PROPARGYL ALCOHOL
DOT: NA 1986
mf: C_3H_4O mw: 56.07

$$HC\equiv CCH_2OH$$

PROP: Moderately volatile liquid; geranium odor. D: 0.9715 @ 20°/4°, mp: $-48°$ to $-52°$, bp: 114-115°, flash p: 33°C (97°F) (OC), vap press: 11.6 mm @ 20°, vap d: 1.93.

SYNS: ETHYNYLCARBINOL ◇ ETHYNYLMETHANOL ◇ 1-PROPYNE-3-OL ◇ 2-PROPYN-1-OL ◇ 3-PROPYNOL ◇ 2-PROPYNYL ALCOHOL ◇ RCRA WASTE NUMBER P102

TOXICITY DATA with REFERENCE
orl-rat LD50:55 mg/kg JPFCD2 20,593,85
ihl-rat LC50:2000 mg/m^3/2H TPKVAL 8,97,66
orl-mus LD50:50 mg/kg TPKVAL 8,97,66
ihl-mus LC50:2000 mg/m^3/2H TPKVAL 8,97,66
skn-rbt LD50:88 mg/kg TXAPA9 42,417,77
scu-rbt LDLo:25 mg/kg BDBU** -,-,30
orl-gpg LD50:60 mg/kg DTLVS* 4,346,80

CONSENSUS REPORTS: Reported in EPA TSCA Inventory.

OSHA PEL: TWA 1 ppm (skin)
ACGIH TLV: TWA 1 ppm (skin)
DFG MAK: 2 ppm (5 mg/m^3)
DOT Classification: Flammable Liquid; Label: Flammable Liquid and Poison.

SAFETY PROFILE: Poison by ingestion, skin contact and subcutaneous routes. Moderately toxic by inhalation. A central nervous system depressant. A skin and mucous membrane irritant. Dangerous fire hazard when exposed to heat or flame; can ignite. To fight fire, use foam, CO_2, dry chemical. Potentially explosive reactions with alkalies (when dried); sulfuric acid. Ignites on contact with phosphorus pentaoxide. Violent reaction with mercury(II) sulfate + sulfuric acid + water (at 70°C). Incompatible with oxidizing materials. When heated to decomposition it emits acrid smoke and irritating fumes. Used as a corrosion inhibitor, solvent stabilizer, soil fumigant, and chemical intermediate. See also ACETYLENE COMPOUNDS.

PMN500 CAS:106-96-7 ***HR: 3***
PROPARGYL BROMIDE
DOT: UN 2345
mf: C_3H_3Br mw: 118.97

PROP: An almost colorless liquid; sharp odor. Bp: 88-90°, fp: −61.07°, flash p: 65°F (COC), d: 1.564-1.570, vap d: 6.87.

SYNS: Γ-BROMOALLYLENE ◇ 3-BROMOPROPYNE (DOT) ◇ 3-BROMO-1-PROPYNE

TOXICITY DATA with REFERENCE
orl-rat LDLo:53 mg/kg BESAAT 12,161,66
orl-gpg LD50:29 μg/kg 28ZEAL 4,76,69

CONSENSUS REPORTS: EPA Extremely Hazardous Substances List. Reported in EPA TSCA Inventory.

DOT Classification: Flammable Liquid; Label: Flammable Liquid.

SAFETY PROFILE: A deadly poison by ingestion. A dangerous fire hazard when exposed to heat or flame. The aerated liquid may be ignited by pressure. A dangerous, extremely shock-sensitive explosive. It can detonate when heated to 220°C, by impact (especially when mixed with chloropicrin) or when heated while confined. May explode on contact with copper; high copper alloys; mercury; or silver. Mixtures with trichloronitromethane are shock- and heat-sensitive explosives. Can react vigorously with oxidizing materials. To fight fire, use water, foam, CO_2, dry chemical. When heated to decomposition it emits highly toxic fumes of Br^-. See also ACETYLENE COMPOUNDS and BROMIDES.

PMN700 CAS:27223-49-0 ***HR: 1***
PROPARTHRIN
mf: $C_{19}H_{24}O_3$ mw: 300.43

SYNS: CYCLOPROPANECARBOXYLIC ACID, 2,2-DIMETHYL-3-(2-METHYLPROPENYL)-,(2-METHYL-5-(2-PROPYNYL)-3-FURYL) METHYL ESTER ◇ KIKUTHRIN ◇ 2-METHYL-5-(2-PROPYNYL)-3-FURYLMETHYL-cis-trans-CHRYSANTHEMATE

TOXICITY DATA with REFERENCE
scu-mus TDLo:30 g/kg (female 7-12D post):REP BOCKAE 35,113,70
scu-mus TDLo:6 g/kg (female 7-12D post):TER BOCKAE 35,113,70
orl-rat LD50:14 g/kg YKYUA6 36,533,85
scu-rat LD50:20 g/kg BWHOA6 44,325,71
orl-mus LD50:8200 mg/kg BOCKAE 35,103,70

SAFETY PROFILE: Slightly toxic by ingestion. An experimental teratogen. Experimental reproductive effects. When heated to decomposition it emits acrid smoke and irritating fumes.

PMN850 CAS:139-40-2 ***HR: 2***
PROPAZINE
mf: $C_9H_{16}ClN_7O_2$ mw: 229.75

SYNS: 2,4-BIS(ISOPROPYLAMINO)-6-CHLORO-s-TRIAZINE ◇ 2,4-BIS(PROPYLAMINO)-6-CHLOR-1,3,5-TRIAZIN (GERMAN) ◇ GESAMIL ◇ MILOGARD ◇ PLANTULIN ◇ PRIMATOL P ◇ PROPASIN ◇ PROZINEX

TOXICITY DATA with REFERENCE
eye-rbt 400 mg open CIGET* -,-,77
orl-gpg TDLo:11 g/kg/78W-I:ETA NTIS** PB223-159
orl-gpg LD50:1200 mg/kg 85GMAT -,35,82
orl-mus LD50:3180 mg/kg 85GMAT -,35,82

SAFETY PROFILE: Moderately toxic by ingestion. Moderate eye irritation. Questionable carcinogen with experimental tumorigenic data. When heated to decomposition it emits toxic fumes of NO_x and Cl^-.

PMO250 CAS:695-53-4 ***HR: 2***
PROPAZONE
mf: $C_5H_7NO_3$ mw: 129.13

SYNS: AC 1198 ◇ BAX 1400Z ◇ DIMETHADIONE ◇ DIMETHYLOXAZOLIDINEDIONE ◇ 5,5-DIMETHYLOXAZOLIDINE-2,4-DIONE ◇ 5,5-DIMETHYL-2,4-OXAZOLIDINEDIONE ◇ DMO ◇ EU-PRACTONE ◇ NSC-30152

TOXICITY DATA with REFERENCE
ipr-mus TDLo:1800 mg/kg (female 11-12D post):REP TXAPA9 97,406,89
orl-rat TDLo:1134 mg/kg (1-21D preg):TER TXCYAC 9,155,78
ipr-mus LD50:850 mg/kg DIPHAH 18,337,66
ivn-mus LD50:450 mg/kg MEIEDD 10,468,83

CONSENSUS REPORTS: Reported in EPA TSCA Inventory.

SAFETY PROFILE: Moderately toxic by intraperitoneal and intravenous routes. Experimental teratogenic and reproductive effects. When heated to decomposition it emits toxic fumes of NO_x. Used as an anticonvulsant.

PMO500 CAS:115-07-1 ***HR: 3***
PROPENE
DOT: UN 1075/UN 1077
mf: C_3H_6 mw: 42.09

$H_2C{=}CHCH_3$

PROP: A gas. D: (gas) 1.49 (air = 1.0), d: (liquid) 0.581 @ 0°. Mp: −185°, bp: −47.7°, autoign temp: 860°F, vap press: 10 atm @ 19.8°, lel: 2.4%, uel: 10.1%, vap d: 1.5, flash p: −162°F.

SYNS: METHYLETHENE ◇ METHYLETHYLENE ◇ NCI-C50077 ◇ 1-PROPENE ◇ PROPYLENE (DOT)

CONSENSUS REPORTS: IARC Cancer Review: Group 3 IMEMDT 7,56,87. NTP Carcinogenesis Studies (inhalation); No Evidence: mouse, rat NTPTR* NTP-TR-272,85. EPA Extremely Hazardous Substances List. Reported in EPA TSCA Inventory.

DOT Classification: Flammable Gas; Label: Flammable Gas.

SAFETY PROFILE: A simple asphyxiant. No irritant effects from high concentrations in gaseous form. When compressed to liquid form, can cause skin burns from freezing effects on tissue of rapid evaporation. Questionable carcinogen. Very dangerous fire hazard when exposed to heat, flame, or oxidizers. Explosive in the form of vapor when exposed to heat or flame. Under unusual conditions, i.e., 955 atm. pressure and 327°C, it has been known to explode. Explodes on contact with trifluoromethyl hypofluorite. Explosive polymerization is initiated by lithium nitrate + sulfur dioxide. Reacts with oxides of nitrogen to form an explosive product. Dangerous; can react vigorously with oxidizing materials. To fight fire, stop flow of gas. Used in production of fabricated polymers, fibers, and solvents, in production of plastic products and resins. For effects of simple asphyxiants, see ARGON.

PMO800 CAS:1945-91-1 ***HR: 3***
PROPENE-1,3-DIOL DIACETATE
mf: $C_7H_{10}O_4$ mw: 158.17

SYNS: 1,3-DIACETOXYPROPENE ◇ 1,3-DIACETYLOXYPROPENE ◇ 1-PROPENE-1,3-DIOL, DIACETATE

TOXICITY DATA with REFERENCE
skn-rbt 5 mg/24H SEV 85JCAE-,358,86
eye-rbt 50 µg/24H SEV 85JCAE-,358,86
orl-rat LD50:150 mg/kg JIDHAN 31,60,49
ihl-rat LCLo:8 ppm/4H JIDHAN 31,343,49
skn-rbt LD50:670 mg/kg JIDHAN 31,60,49

SAFETY PROFILE: A poison by ingestion and inhalation. Moderately toxic by skin contact. A severe skin and eye irritant. When heated to decomposition it emits acrid smoke and irritating fumes.

PMP250 CAS:38787-96-1 ***HR: 3***
PROPENE OZONIDE
mf: $C_3H_6O_3$ mw: 90.08

$OCH_2OOCHCH_3$ (ring)

SYN: 3-METHYL-1,2,4-TRIOXOLANE

SAFETY PROFILE: A powerful oxidant. Explosively decomposes at ambient temperatures. When heated to decomposition it emits acrid smoke and irritating fumes. See also OZONIDES.

PMP500 CAS:9003-07-0 ***HR: 2***
PROPENE POLYMERS
mf: $(C_3H_6)_n$

PROP: Solid material. Mp: about 165°, d: 0.90-0.92. Insol in organic materials.

SYNS: ADMER PB 02 ◇ AMCO ◇ AMERFIL ◇ AMOCO 1010 ◇ ATACTIC POLYPROPYLENE ◇ AVISUN ◇ AZDEL ◇ BEAMETTE ◇ BICOLENE P ◇ CARLONA P ◇ CELGARD 2500 ◇ CHISSO 507B ◇ CLYSAR ◇ COATHYLENE PF 0548 ◇ DAPLEN AD ◇ DEXON E 117 ◇ EASTBOND M 5 ◇ ELPON ◇ ENJAY CD 460 ◇ EPOLENE M 5K ◇ GERFIL ◇ HERCOFLAT 135 ◇ HERCULON ◇ HOSTALEN PP ◇ HULS P 6500 ◇ ICI 543 ◇ ISOTACTIC POLYPROPYLENE ◇ J 400 ◇ LAMBETH ◇ LUPAREEN ◇ MARLEX 9400 ◇ MAURYLENE ◇ MERAKLON ◇ MOPLEN ◇ MOSTEN ◇ NOBLEN ◇ NOVAMONT 2030 ◇ NOVOLEN ◇ OLETAC 100 ◇ PAISLEY POLYMER ◇ PELLON 2506 ◇ POLYPRO 1014 ◇ POLYPROPENE ◇ POLYPROPYLENE ◇ POLYTAC ◇ POPROLIN ◇ PROFAX ◇ PROPATHENE ◇ 1-PROPENE HOMOPOLYMER (9CI) ◇ PROPOLIN ◇ PROPOPHANE ◇ PROPYLENE POLYMER ◇ REXALL 413S ◇ REXENE ◇ SHELL 5520 ◇ SHOALLOMER ◇ SYNDIOTACTIC POLYPROPYLENE ◇ TENITE 423 ◇ TRESPAPHAN ◇ TUFF-LITE ◇ ULSTRON ◇ VISCOL 350P ◇ W 101 ◇ WEX 1242

CONSENSUS REPORTS: IARC Cancer Review: Group 3 IMEMDT 7,56,87; Animal Limited Evidence

IMEMDT 19,213,79; Human Inadequate Evidence IMEMDT 19,213,79. Reported in EPA TSCA Inventory.

SAFETY PROFILE: Questionable carcinogen. When heated to decomposition it emits acrid smoke and irritating fumes. Used in injection molding for auto parts, in bottle caps, and container closures.

PMP750 CAS:6842-15-5 ***HR: 2***
PROPENE TETRAMER
DOT: UN 2850
mf: $C_{12}H_{24}$ mw: 168.36

PROP: Colorless liquid. Mp: −31.5°, bp: 213°, d: 0.76 @ 20°/4°, vap press: 1 mm @ 47.2°, vap d: 5.81, flash p: <212°F, autoign temp: 491°F.

SYNS: AMSCO TETRAMER ◇ DODECENE ◇ DODECYLENE ◇ PROPYLENE TETRAMER (DOT) ◇ TETRAPROPYLENE

CONSENSUS REPORTS: Reported in EPA TSCA Inventory.

DOT Classification: Flammable or Combustible Liquid; Label: Flammable Liquid.

SAFETY PROFILE: Probably irritating and narcotic in high concentration. Flammable or combustible when exposed to heat or flame; can react with oxidizing materials. To fight fire, use foam, CO_2, dry chemical. When heated to decomposition it emits acrid smoke and irritating fumes.

PMQ000 CAS:5411-08-5 ***HR: 3***
4-p-PROPENONE-OXY-PHENYLARSONIC ACID
mf: $C_9H_{13}AsO_5$ mw: 276.14

SYN: 4-(1-(2-HYDROXY)PROPYOXY)BENZENEARSONICACID

TOXICITY DATA with REFERENCE
orl-rat LD50:3500 mg/kg UCPHAQ 1,291,40
ivn-rat LDLo:750 mg/kg JPETAB 63,122,38
ims-rat LDLo:400 mg/kg JPETAB 63,122,38

CONSENSUS REPORTS: Arsenic and its compounds are on the Community Right-To-Know List.

OSHA PEL: TWA 0.5 mg(As)/m^3

SAFETY PROFILE: Poison by intramuscular route. Moderately toxic by ingestion and intravenous routes. When heated to decomposition it emits toxic fumes of As.

PMQ250 CAS:768-03-6 ***HR: 3***
2-PROPENOPHENONE
mf: C_9H_8O mw: 132.17

SYNS: ACRYLOPHENONE ◇ PHENYLVINYL KETONE

TOXICITY DATA with REFERENCE
scu-rat TDLo:2520 mg/kg/63W-I:ETA BJCAAI 19,392,65

SAFETY PROFILE: Questionable carcinogen with experimental tumorigenic data. When heated to decomposition it emits acrid smoke and irritating fumes. See also KETONES.

PMQ750 CAS:104-46-1 ***HR: 3***
p-PROPENYLANISOLE
mf: $C_{10}H_{12}O$ mw: 148.22

PROP: Leaves from alc or light yellow liquid above 23°; sweet taste with anise odor. D: 0.991 @ 20°/20°, refr index: 1.557-1.561, mp: 22.5°, bp: 235.3°, flash p: 198°F. Very sltly sol in water; misc in abs alc, ether, chloroform.

SYNS: ACINTENE O ◇ ANETHOLE (FCC) ◇ ANISE CAMPHOR ◇ ARIZOLE ◇ FEMA No. 2086 ◇ ISOESTRAGOLE ◇ p-METHOXY-β-METHYLSTYRENE ◇ 1-(p-METHOXYPHENYL)PROPENE ◇ 1-METHOXY-4-PROPENYLBENZENE ◇ 4-METHOXYPROPENYLBENZENE ◇ MONASIRUP ◇ NAULI "GUM" ◇ OIL of ANISEED ◇ p-1-PROPENYLANISOLE ◇ 4-PROPENYLANISOLE ◇ p-PROPENYLPHENYL METHYL ETHER

TOXICITY DATA with REFERENCE
ipr-mus TDLo:2400 mg/kg/8W-I:ETA CNREA8 33,3069,73
orl-rat LD50:2090 mg/kg FCTXAV 2,327,64
orl-mus LD50:3050 mg/kg FCTXAV 2,327,64
orl-gpg LD50:2160 mg/kg FCTXAV 2,327,64
orl-bwd LD50:316 mg/kg AECTCV 12,355,83

CONSENSUS REPORTS: Reported in EPA TSCA Inventory.

SAFETY PROFILE: Poison by ingestion. Questionable carcinogen with experimental tumorigenic data. Combustible liquid. When heated to decomposition it emits acrid smoke and irritating fumes. See also ETHERS.

PMR000 CAS:25679-28-1 ***HR: 3***
cis-p-PROPENYLANISOLE
mf: $C_{10}H_{12}O$ mw: 148.22

PROP: Prisms from alc. Mp: −22.5°, bp: = 79° @ 2.3 mm, d: 0.9878 @ 20°/4°. Insol in water; sltly sol in alc; sol in ether and benzene.

SYN: cis-ANETHOLE

TOXICITY DATA with REFERENCE
skn-rbt 100 μg/24H open AIHAAP 23,95,62
orl-rat LD50:150 mg/kg FAONAU 44A,7,67
ipr-rat LD50:93 mg/kg THERAP 22,309,67
ipr-mus LD50:95 mg/kg COREAF 246,1465,58

SAFETY PROFILE: Poison by ingestion and intraperitoneal routes. A skin irritant. When heated to decomposition it emits acrid smoke and irritating fumes.

PMR250 CAS:4180-23-8 ***HR: 2***
trans-p-PROPENYLANISOLE
mf: $C_{10}H_{12}O$ mw: 148.22

PROP: Crystals. Mp: 21.4°, d: 0.9883 @ 20°/4°, bp: 81° @ 2.3 mm. Nearly insol in water; misc ether, chloroform; sol in benzene, ethylacetate, acetone, CS_2, petro ether.

SYNS: trans-ANETHOL ◇ trans-ANETHOLE ◇ (E)-1-METHOXY-4-(1-PROPENYL)BENZENE ◇ (E)-p-PROPENYLANISOLE

TOXICITY DATA with REFERENCE
mmo-sat 200 ng/plate NTIS** AD-A116,715
mma-sat 1 μmol/plate MUREAV 60,143,79
orl-rat LD50:2090 mg/kg FCTXAV 2,327,64
ipr-rat LD50:900 mg/kg THERAP 22,309,67
orl-mus LD50:3050 mg/kg FCTXAV 2,327,64
ipr-mus LD50:650 mg/kg THERAP 22,309,67
orl-gpg LD50:2167 mg/kg FCTXAV 2,327,64

CONSENSUS REPORTS: Reported in EPA TSCA Inventory.

SAFETY PROFILE: Moderately toxic by ingestion and intraperitoneal routes. Mutation data reported. When heated to decomposition it emits acrid smoke and irritating fumes. Used as a flavoring agent.

PMR750 CAS:590-21-6 ***HR: 3***
PROPENYL CHLORIDE
mf: C_3H_5Cl mw: 76.53

PROP: Liquid. Mp: −137.4°, bp: 22.65°, flash p: <21°F, d: 0.9189°, lel: 4.5%, uel: 16%. Insol in water.

SYNS: 1-CHLOROPROPENE ◇ 1-CHLORO-1-PROPENE

TOXICITY DATA with REFERENCE
eye-rbt 500 mg open AMIHBC 10,61,54
mmo-sat 250 μL/plate DTESD7 2,249,77
mma-sat 100 μmol/plate BCPCA6 29,2611,80
orl-mus TDLo:3560 mg/kg/89W-I:NEO JJIND8 63,1433,79
orl-rat LD50:1950 mg/kg AMIHBC 10,61,54
ihl-mus LCLo:230 g/m^3/10M UCPHAQ 1,119,38
skn-rbt LD50:20 g/kg AMIHBC 10,61,54

CONSENSUS REPORTS: Reported in EPA TSCA Inventory.

SAFETY PROFILE: Moderately toxic by ingestion. Very mildly toxic by skin contact and inhalation. An eye irritant. Mutation data reported. Questionable carcinogen with experimental neoplastigenic data. Very dangerous fire hazard when exposed to heat, flames (sparks), or oxidizers. Explosive in the form of vapor when exposed to heat or flame. To fight fire, use alcohol foam, dry chemical, mist spray, fog. When heated to decomposition it emits toxic fumes of Cl^-. See also CHLORINATED HYDROCARBONS, ALIPHATIC.

PMR800 ***HR: 3***
(8-β)-6-(2-PROPENYL)ERGOLINE-8-ACETAMIDE TARTRATE
mf: $C_{19}H_{23}N_3O \cdot C_4H_6O_6$ mw: 459.55

SYN: ERGOLINE-8-ACETAMIDE, 6-(2-PROPENYL)-, (8-β)-, (R-(R*,R*))-2,3-DIHYDROXYBUTANEDIOATE

TOXICITY DATA with REFERENCE
orl-rat TDLo:50 μg/kg (female 5D post):REP CCCCAK 44,3385,79
orl-rat LD50:60 mg/kg CCCCAK 44,3385,79

SAFETY PROFILE: Poison by ingestion. Experimental reproductive effects. When heated to decomposition it emits toxic fumes of NO_x.

PMS250 CAS:6380-21-8 ***HR: 3***
2-PROPENYLPHENOL
mf: $C_9H_{10}O$ mw: 134.19

SYNS: o-PROPENYLPHENOL ◇ 2-(1-PROPENYL)PHENOL

TOXICITY DATA with REFERENCE
sce-hmn:lym 250 μmol/L MUREAV 169,129,86
skn-mus TDLo:3400 mg/kg/12W-I:NEO CNREA8 19,413,59

SAFETY PROFILE: Questionable carcinogen with experimental neoplastigenic data. Human mutation data reported. When heated to decomposition it emits acrid smoke and irritating fumes.

PMS500 CAS:1797-74-6 ***HR: 2***
2-PROPENYL PHENYLACETATE
mf: $C_{11}H_{12}O_2$ mw: 176.23

PROP: Colorless to light yellow liquid; fruity odor of banana and honey.

SYNS: ALLYL PHENYLACETATE ◇ BENZENEACETIC ACID, 2-PROPENYL ESTER ◇ PHENYLACETIC ACID ALLYL ESTER

TOXICITY DATA with REFERENCE
skn-hmn 30 mg/48H FCTXAV 15,621,77
skn-rbt 310 mg/kg/24H MOD FCTXAV 15,621,77
orl-rat LD50:650 mg/kg FCTXAV 15,621,77

CONSENSUS REPORTS: Reported in EPA TSCA Inventory.

SAFETY PROFILE: Moderately toxic by ingestion. A human skin irritant. When heated to decomposition it emits acrid smoke and irritating fumes. See also ESTERS and ALLYL COMPOUNDS.

PMS775 CAS:41029-45-2 ***HR: 3***
PROP-2-ENYL TRIFLUOROMETHANE SULFONATE
mf: $C_4H_5F_3O_3S$ mw: 190.14

$H_2C{=}CHCH_2OSO_2CF_3$

SAFETY PROFILE: Explodes above 20°C. When heated to decomposition it emits toxic fumes of F^- and SO_x. See also SULFONATES and FLUORIDES.

PMS800 CAS:1420-03-7 ***HR: 3***
PROPENZOLATE HYDROCHLORIDE
mf: $C_{20}H_{29}NO_3{\cdot}ClH$ mw: 367.96

SYNS: (+)-α-CYCLOHEXYL-α-HYDROXY-BENZENEACETIC ACID-1-METHYL-3-PIPERIDINYL ESTER, HCl ◇ d-1-METHYL-3-PIPERIDYL-dl-α-PHENYLCYCLOHEXANEGLYCOLATE HYDROCHLORIDE

TOXICITY DATA with REFERENCE
orl-rat LD50:700 mg/kg AIPTAK 153,139,65
ivn-rat LD50:25 mg/kg AIPTAK 153,139,65
ipr-mus LD50:225 mg/kg AIPTAK 153,139,65
orl-dog LD50:330 mg/kg AIPTAK 153,139,65
ivn-dog LD50:30 mg/kg AIPTAK 153,139,65

SAFETY PROFILE: Poison by ingestion, intravenous, and intraperitoneal routes. When heated to decomposition it emits toxic fumes of NO_x and HCl.

PMS825 CAS:60364-26-3 ***HR: 2***
PROPHAM
mf: $C_8H_{15}N_2O_2$ mw: 170.24

SYNS: CENESTIL ◇ INPC ◇ ISO PPC ◇ (2-ISOPROPILCROTONIL) UREA (ITALIAN) ◇ 2-ISOPROPYLCROTONYLUREA ◇ ISOPROPYL-N-PHENYLCARBAMAT (GERMAN)

TOXICITY DATA with REFERENCE
orl-rat LD50:9 g/kg 85GYAZ -,69,71
ipr-rat LD50:600 mg/kg 85GYAZ -,69,71
orl-mus LD50:940 mg/kg 27ZQAG -,426,72
ipr-mus LD50:1 g/kg 85GYAZ -,69,71

SAFETY PROFILE: Moderately toxic by ingestion and intraperitoneal routes. When heated to decomposition it emits toxic fumes of NO_x. See also CARBAMATES.

PMT000 CAS:25535-16-4 ***HR: 3***
PROPIDIUM DIIODIDE
mf: $C_{27}H_{34}N_4{\cdot}2I$ mw: 668.45

SYNS: 3,8-DIAMINO-5-(3-(DIETHYLMETHYLAMMONIO)PROPYL)-6-PHENYLPHENANTHRIDINIUM DIIODIDE ◇ PROPIDIUM IODIDE

TOXICITY DATA with REFERENCE
mma-sat 200 μg/plate MUREAV 127,31,84
mmo-smc 25 μmol/L MUREAV 82,87,81
dnd-ham:ovr 5 mg/L/4H JULRA7 57,43,76
scu-mus LD50:16 mg/kg JULRA7 54,43,76

CONSENSUS REPORTS: EPA Genetic Toxicology Program.

SAFETY PROFILE: Poison by subcutaneous route. Mutation data reported. When heated to decomposition it emits very toxic fumes of NO_x, NH_3, and I^-.

PMT100 CAS:57-57-8 ***HR: 3***
β-PROPIOLACTONE
mf: $C_3H_4O_2$ mw: 72.07

SYNS: BETAPRONE ◇ BPL ◇ HYDRACRYLIC ACID β-LACTONE ◇ 3-HYDROXYPROPIONIC ACID LACTONE ◇ PROPANOLIDE ◇ PROPIOLACTONE ◇ 1,3-PROPIOLACTONE ◇ 3-PROPIOLACTONE ◇ β-PROPIONOLACTONE ◇ β-PROPRIOLACTONE (OSHA) ◇ β-PROPROLACTONE

TOXICITY DATA with REFERENCE
slt-dmg-orl 200 mmol/L ENMUDM 6,153,84
otr-hmn:fbr 28 μmol/L PNASA6 80,7219,83
orl-rat TDLo:2868 mg/kg/1Y-I:CAR BJCAAI 46,924,82
ihl-rat TCLo:5 ppm/6H/30D-I:CAR JJIND8 79,285,87
itr-rat TDLo:72 mg/kg/30W-I:ETA BJCAAI 20,134,66
scu-mus TDLo:69 mg/kg/43W-I:NEO BJCAAI 19,392,65 DTLVS* 4,347,80
ipr-mus LD50:405 mg/kg JJIND8 62,911,79

CONSENSUS REPORTS: NTP Fifth Annual Report on Carcinogens. IARC Cancer Review: Group 2B IMEMDT 7,56,87; Animal Sufficient Evidence IMEMDT 4,259,74. EPA Genetic Toxicology Program. Community Right-To-Know List. EPA Extremely Hazardous Substances List. Reported in EPA TSCA Inventory.

OSHA: Carcinogen.
ACGIH TLV: TWA 0.5 ppm; Suspected Human Carcinogen.
DFG MAK: Animal Carcinogen, Suspected Human Carcinogen.

SAFETY PROFILE: Confirmed carcinogen with experimental carcinogenic, neoplastigenic, and tumorigenic data. Poison by inhalation. Moderately toxic by intraperitoneal route. An initiator. Human mutation data reported. When heated to decomposition it emits acrid smoke and irritating fumes.

PMT250 CAS:624-67-9 ***HR: 3***
PROPIOLALDEHYDE
mf: C_3H_2O mw: 54.05

SYNS: PROPARGYL ALDEHYDE ◇ 2-PROPYNAL (9CI)

TOXICITY DATA with REFERENCE
mmo-sat 1 μmol/plate PAACA3 24,91,83

CONSENSUS REPORTS: Reported in EPA TSCA Inventory.

SAFETY PROFILE: Mutation data reported. A storage hazard especially in glass. Explosive reaction with pyridine. Violent polymerization reaction on contact with alkalies. When heated to decomposition it emits acrid smoke and irritating fumes. See also ALDEHYDES.

PMT275 CAS:471-25-0 ***HR: 2***
PROPIOLIC ACID
mf: $C_3H_2O_2$ mw: 70.05

SYN: PROPYNOIC ACID

SAFETY PROFILE: Reacts with solutions of copper(I) or silver salts in ammonia to form an explosive precipitate. When heated to decomposition it emits acrid smoke and irritating fumes. See also ACETYLENE COMPOUNDS.

PMT300 CAS:50277-65-1 ***HR: 2***
PROPIOLOYL CHLORIDE
mf: C_3HClO mw: 88.49

SAFETY PROFILE: The presence of chloroacetylene will cause the chloride to ignite in air. When heated to decomposition it emits toxic fumes of Cl^-. See also ACETYLENE COMPOUNDS.

PMT500 CAS:64-89-1 ***HR: 3***
PROPIOMAZINE HYDROCHLORIDE
mf: $C_{20}H_{24}N_2OS•ClH$ mw: 376.98

PROP: Yellow, odorless powder. Very sol in alc, water; insol in benzene.

SYNS: 1-(10-(2-(DIMETHYLAMINO)-1-METHYLETHYL)PHENOTHIAZIN-2-YL)-1-PROPANONEHYDROCHLORIDE ◇ LARGON HYDROCHLORIDE

TOXICITY DATA with REFERENCE
ivn-mus LD50:48 mg/kg 27ZQAG -,43,72
ims-mus LD50:210 mg/kg 27ZQAG -,43,72
skn-rbt LDLo:2980 mg/kg 31ZOAD 1,250,68

SAFETY PROFILE: Poison by intravenous and intramuscular routes. Moderately toxic by skin contact. When heated to decomposition it emits very toxic fumes of SO_x, NO_x, and HCl.

PMT750 CAS:123-38-6 ***HR: 3***
PROPIONALDEHYDE
DOT: UN 1275
mf: C_3H_6O mw: 58.09

$CH_3CH_2CO•H$

PROP: Colorless, mobile liquid; suffocating odor. Mp: −81°, bp: 48°, flash p: 15-19°F (OC), d: 0.807 @ 20°/4°, lel: 2.9%, uel: 17%, vap d: 2.0, autoign temp: 405°F. Misc with alc, ether, water @ 49°.

SYNS: ALDEHYDE PROPIONIQUE (FRENCH) ◇ FEMA No. 2923 ◇ METHYLACETALDEHYDE ◇ NCI-C61029 ◇ PROPALDEHYDE ◇ PROPANAL ◇ PROPIONIC ALDEHYDE ◇ PROPYL ALDEHYDE ◇ PROPYLIC ALDEHYDE

TOXICITY DATA with REFERENCE
skn-rbt 500 mg open MLD UCDS** 4/25/58
eye-rbt 41 mg SEV UCDS** 4/25/58
orl-rat LD50:1410 mg/kg AMIHBC 4,119,51
ihl-rat LCLo:8000 ppm/4H AMIHBC 4,119,51
scu-rat LD50:820 mg/kg APTOA6 6,299,50
orl-mus LDLo:800 mg/kg KODAK* 21MAY71
scu-mus LD50:680 mg/kg APTOA6 6,299,50
skn-rbt LD50:5040 mg/kg AMIHBC 4,119,51
ihl-mam LC50:21800 mg/m^3 GTPZAB 12(7),16,68

CONSENSUS REPORTS: Community Right-To-Know List. Reported in EPA TSCA Inventory.

DOT Classification: Flammable Liquid; Label: Flammable Liquid.

SAFETY PROFILE: Moderately toxic by skin contact, ingestion and subcutaneous routes. Mildly toxic by inhalation. A skin and severe eye irritant. Flammable liquid. Dangerous fire hazard when exposed to heat or flame; reacts vigorously with oxidizers. Explosive in the form of vapor when exposed to heat or flame. Vigorous polymerization reaction with methyl methacrylate. To fight fire, use alcohol foam, CO_2, dry chemical. When heated to decomposition it emits acrid smoke and irritating fumes. See also ALDEHYDES.

PMU250 CAS:79-05-0 ***HR: 3***
PROPIONAMIDE
mf: C_3H_7NO mw: 73.11

PROP: Orthorhombic platelets. D: 1.0335 @ 20°/4°, mp: 79°, bp: 222.2°. Very sol in water, alc, ether, chloroform.

SYNS: PROPIONIC ACID AMIDE ◇ PROPIONIC AMIDE

TOXICITY DATA with REFERENCE
ihl-rat LCLo:8000 ppm MEIEDD 11,1243,89
ivn-rbt LDLo:230 mg/kg COREAF 153,895,11

CONSENSUS REPORTS: Reported in EPA TSCA Inventory.

SAFETY PROFILE: Poison by intravenous route. Slightly toxic by inhalation. When heated to decomposition it emits toxic fumes of NO_x. See also AMIDES.

PMU500 CAS:620-71-3 ***HR: 2***
PROPIONANILIDE
mf: $C_9H_{11}NO$ mw: 149.21

TOXICITY DATA with REFERENCE
orl-mus LD50:1100 mg/kg TXAPA9 19,20,71

CONSENSUS REPORTS: Reported in EPA TSCA Inventory.

SAFETY PROFILE: Moderately toxic by ingestion. When heated to decomposition it emits toxic fumes of NO_x.

PMU750 CAS:79-09-4 ***HR: 3***
PROPIONIC ACID
DOT: UN 1848
mf: $C_3H_6O_2$ mw: 74.09

PROP: Oily liquid; pungent, disagreeable, rancid odor. D: 0.998 @ 15°/4°, mp: −21.5°, bp: 141.1°, vap press: 10 mm @ 39.7°, vap d: 2.56, autoign temp: 955°F. Misc in water, alc, ether, chloroform.

SYNS: ACIDE PROPIONIQUE (FRENCH) ◇ CARBOXYETHANE ◇ ETHANECARBOXYLIC ACID ◇ ETHYLFORMIC ACID ◇ METACETONIC ACID ◇ METHYL ACETIC ACID ◇ PROPANOIC ACID ◇ PROPIONIC ACID, solution containing not less than 80% acid (DOT) ◇ PROPIONIC ACID GRAIN PRESERVER ◇ PROZOIN ◇ PSEUDOACETIC ACID ◇ SENTRY GRAIN PRESERVER ◇ TENOX P GRAIN PRESERVATIVE

TOXICITY DATA with REFERENCE
skn-rbt 495 mg open SEV UCDS** 3/24/70
eye-rbt 990 μg SEV UCDS** 3/24/70
orl-rat LD50:3500 mg/kg FMCHA2 -,C198,83
ipr-rat LD50:200 mg/kg FMCHA2 -,C230,83
ivn-mus LD50:625 mg/kg APTOA6 18,141,61
skn-rbt LD50:500 mg/kg AIHAAP 23,95,62

CONSENSUS REPORTS: Reported in EPA TSCA Inventory.

OSHA PEL: TWA 10 ppm
ACGIH TLV: TWA 10 ppm
DFG MAK: 10 ppm (30 mg/m^3)
DOT Classification: Corrosive Material; Label: Corrosive; Label: Corrosive, solution; Label: Corrosive, Flammable Liquid.

SAFETY PROFILE: Poison by intraperitoneal route. Moderately toxic by ingestion, skin contact, and intravenous routes. A corrosive irritant to eyes, skin, and mucous membranes. Flammable liquid. Highly flammable when exposed to heat, flame, or oxidizers. To fight fire, use alcohol foam. When heated to decomposition it emits acrid smoke and irritating fumes.

PMV000 CAS:79-09-4 ***HR: 2***
PROPIONIC ACID (solution)

CONSENSUS REPORTS: Reported in EPA TSCA Inventory.

DOT Classification: Corrosive Material; Label: Corrosive.

SAFETY PROFILE: A corrosive material that attacks skin, eyes, and mucous membranes. See also PROPIONIC ACID.

PMV250 CAS:540-42-1 ***HR: 3***
PROPIONIC ACID, ISOBUTYL ESTER
DOT: UN 2394
mf: $C_7H_{14}O_2$ mw: 130.21

SYNS: ISOBUTY PROPIONATE (DOT) ◇ 2-METHYLPROPYL PROPIONATE ◇ PROPANOIC ACID, 2-METHYLPROPYL ESTER

TOXICITY DATA with REFERENCE
orl-rbt LD50:5599 mg/kg IMSUAI 41,31,72

CONSENSUS REPORTS: Reported in EPA TSCA Inventory.

DOT Classification: Flammable Liquid; Label: Flammable Liquid.

SAFETY PROFILE: Mildly toxic by ingestion. Flammable when exposed to heat or flame, can react vigorously with oxidizing materials. When heated to decomposition it emits acrid smoke and irritating fumes. See also ESTERS, PROPIONIC ACID, and ISOBUTYL ALCOHOL.

PMV500 CAS:123-62-6 ***HR: 2***
PROPIONIC ANHYDRIDE
DOT: UN 2496
mf: $C_6H_{10}O_3$ mw: 130.16

PROP: Liquid; very rancid odor. Mp: −45°, bp: 167.0°, flash p: 165°F (OC), d: 1.012, vap press: 1 mm @ 20.6°, vap d: 4.49. Decomp in water, alc; sol in methanol, ethanol, ether, chloroform.

SYNS: METHYLACETIC ANHYDRIDE ◇ PROPANOIC ANHYDRIDE ◇ PROPIONIC ACID ANHYDRIDE ◇ PROPIONYL OXIDE

TOXICITY DATA with REFERENCE
skn-rbt 10 mg/24H MLD AMIHBC 10,61,54
skn-rbt 510 mg open MOD UCDS** 12/15/70
eye-rbt 750 μg SEV AMIHBC 10,61,54
orl-rat LD50:2360 mg/kg AMIHBC 10,61,54
skn-rbt LD50:10 g/kg AMIHBC 10,61,54

CONSENSUS REPORTS: Reported in EPA TSCA Inventory.

DOT Classification: Corrosive Material; Label: Corrosive.

SAFETY PROFILE: Moderately toxic by ingestion. Mildly toxic by skin contact. A corrosive irritant to skin, eyes, and mucous membranes. Combustible when exposed to heat or flame; can react with oxidizing materials. To fight fire, use CO_2, dry chemical. When heated to decomposition it emits acrid smoke and irritating fumes. Used as an esterifying agent and dehydrating agent. See also ANHYDRIDES.

PMV750 CAS:107-12-0 ***HR: 3***
PROPIONONITRILE
DOT: UN 2404
mf: C_3H_5N mw: 55.09

$CH_3CH_2C{\equiv}N$

PROP: Colorless liquid; ethereal odor. Bp: 97.1°, d: 0.783 @ 21°/4°, vap d: 1.9, flash p: 36°F, lel: 3.1%; mp: −91.8°. Misc with alc and ether.

SYNS: CYANOETHANE ◇ ETHER CYANATUS ◇ ETHYL CYANIDE ◇ HYDROCYANIC ETHER ◇ PROPANENITRILE ◇ PROPIONIC NITRILE ◇ RCRA WASTE NUMBER P101

TOXICITY DATA with REFERENCE
skn-rbt 500 mg/24H MLD 85JCAE -,898,86
eye-rbt 20 mg open AMIHBC 4,119,51
eye-rbt 100 mg/24H MOD 85JCAE -,898,86
ipr-ham TDLo:238 mg/kg (female 8D post):TER TJADAB 23,325,81
orl-rat TDLo:1120 mg/kg (female 6-19D post):REP FAATDF 7,33,86
orl-rat LD50:39 mg/kg AMIHBC 4,119,51
ihl-rat LCLo:500 ppm/4H AMIHBC 4,119,51
orl-mus LD50:35797 μg/kg NEZAAQ 39,423,84
ihl-mus LC50:163 ppm/1H CTOXAO 18,991,81
ipr-mus LD50:28 mg/kg TXAPA9 59,589,81
skn-rbt LD50:210 mg/kg AMIHBC 4,119,51
ivn-rbt LD50:50 mg/kg CRSBAW 1,251,1889
scu-frg LDLo:8000 mg/kg AIPTAK 5,161,1899

CONSENSUS REPORTS: Cyanide and its compounds are on the Community Right-To-Know List. EPA Extremely Hazardous Substances List. Reported in EPA TSCA Inventory.

NIOSH REL: TWA (Nitriles) 14 mg/m^3

DOT Classification: Flammable Liquid; Label: Flammable Liquid and Poison.

SAFETY PROFILE: Poison by ingestion, skin contact, intravenous, and intraperitoneal routes. Moderately toxic by inhalation. Experimental teratogenic effects. Other experimental reproductive effects. A skin and eye irritant. Dangerous fire hazard when exposed to heat, flame (sparks), oxidizers. Mixture with N-bromosuccinimide may explode when heated. To fight fire, use water spray, foam, mist, CO_2, dry chemical. When heated to decomposition it emits toxic fumes of NO_x and CN^-. Used as a solvent in petroleum refining, and as a raw material for drug manufacture. See also NITRILES.

PMW500 CAS:79-03-8 ***HR: 3***
PROPIONYL CHLORIDE
DOT: UN 1815
mf: C_3H_5ClO mw: 92.53

$CH_3CH_2CO{\bullet}Cl$

PROP: Mp: −94°, bp: 80°, flash p: 53.6°, d: 1.065, vap d: 3.2.

SYNS: PROPANOYL CHLORIDE ◇ PROPIONIC ACID CHLORIDE ◇ PROPIONIC CHLORIDE

CONSENSUS REPORTS: Reported in EPA TSCA Inventory.

DOT Classification: Flammable liquid; Label: Corrosive, Flammable Liquid.

SAFETY PROFILE: A corrosive irritant to skin, eyes, and mucous membranes. Dangerous fire hazard when exposed to heat or flame; can react vigorously with oxidizing materials. Reacts with water or steam to produce toxic and corrosive fumes. Exothermic reaction with diisopropyl ether produces much gas. The reaction may be dangerous if confined. To fight fire, use CO_2, dry chemical; do not use water. When heated to decomposition it emits highly toxic fumes of Cl^-. See also HYDROCHLORIC ACID.

PMW750 CAS:2494-55-5 ***HR: 3***
PROPIONYLCHOLINE IODIDE
mf: $C_8H_{18}NO_2{\bullet}I$ mw: 287.17

SYN: CHOLINE, IODIDE, PROPIONATE

TOXICITY DATA with REFERENCE
ivn-mus LD50:56 mg/kg CSLNX* NX#03798

CONSENSUS REPORTS: Reported in EPA TSCA Inventory.

SAFETY PROFILE: Poison by intravenous route. When heated to decomposition it emits very toxic fumes of I^- and NO_x. See also IODIDES.

PMW760 CAS:75464-10-7 ***HR: 3***
10-PROPIONYL DITHRANOL
mf: $C_{17}H_{14}O_4$ mw: 282.31

SYNS: 9(10H)-ANTHRACENONE, 1,8-DIHYDROXY-10-(1-OXOPROPYL)- ◇ DITHRANOL, 10-PROPIONYL-

TOXICITY DATA with REFERENCE
mmo-sat 250 μg/plate ARTODN 59,180,86
mma-sat 25 μg/plate ARTODN 59,180,86
cyt-hmn:lyms 30 μg/L ARTODN 59,180,86
skn-mus TDLo:126 mg/kg/50W-I:ETA JPETAB 229,255,84
orl-rat LD50:138 mg/kg ARTODN 59,180,86
orl-mus LD50:46 mg/kg ARTODN 59,180,86

SAFETY PROFILE: Questionable carcinogen with experimental tumorigenic data. Poison by ingestion. Human mutation data reported. When heated to decomposition it emits acrid smoke and irritating fumes.

PMW770 CAS:82198-80-9 ***HR: 3***
PROPIONYL HYPOBROMITE
mf: $C_3H_5BrO_2$ mw: 152.98

$CH_3CH_2CO•OBr$

SAFETY PROFILE: An unpredictable explosive solid. When heated to decomposition it emits toxic fumes of Br^-.

PMX250 CAS:13717-04-9 ***HR: 3***
N-PROPIONYL-2-(1-PIPERIDINOISOPROPYL)AMINOPYRIDINE FUMARATE
mf: $C_{16}H_{25}N_3O•C_4H_4O_4$ mw: 391.52

SYNS: ALGERIL ◇ BAY 4503 ◇ DIRAME ◇ FBA 4503 ◇ N-(1-METHYL-2-PIPERIDINOETHYL)-N-2-PYRIDYLPROPIONAMIDE FUMARATE ◇ N-(1-METHYL-2-(1-PIPERIDINYL)ETHYL)-N-2-PYRIDINYLPROPANAMIDE-(E)-2-BUTENEDIOATE(1:1) ◇ N-PROPIONYL-N-(2-PYRIDYL)-1-PIPERIDINO-2-AMINOPROPANEFUMARATE ◇ PROPIRAM FUMARATE

TOXICITY DATA with REFERENCE
orl-rat TDLo:4950 mg/kg (6-16D preg):REP ARZNAD 24,624,74
orl-rat LD50:1289 mg/kg ARZNAD 24,600,74
scu-rat LD50:366 mg/kg ARZNAD 24,600,74
ivn-rat LD50:63800 μg/kg ARZNAD 24,624,74
orl-mus LD50:874 mg/kg ARZNAD 24,624,74
scu-mus LD50:290 mg/kg ARZNAD 24,600,74
ivn-mus LD50:48200 μg/kg ARZNAD 24,600,74

SAFETY PROFILE: Poison by subcutaneous and intravenous routes. Moderately toxic by ingestion. Experimental reproductive effects. When heated to decomposition it emits toxic fumes of NO_x. Used as an analgesic.

PMX500 CAS:25333-83-9 ***HR: 3***
PROPIOPROMAZINE MALEATE
mf: $C_{20}H_{24}N_2OS•C_4H_4O_4$ mw: 456.60

SYNS: 1678 CB ◇ 1-(10-(3-DIMETHYLAMINO)PROPYL)PHENOTHIAZIN-2-YL)-1-PROPANONEMALEATE

TOXICITY DATA with REFERENCE
orl-mus LD50:650 mg/kg 27ZQAG -,44,72
ipr-mus LD50:170 mg/kg 27ZQAG -,44,72
orl-bwd LD50:316 mg/kg AECTCV 12,355,83

SAFETY PROFILE: Poison by ingestion and intraperitoneal routes. When heated to decomposition it emits very toxic fumes of NO_x and SO_x.

PMY000 CAS:10350-81-9 ***HR: 3***
PROPOQUIN DIHYDROCHLORIDE
mf: $C_{20}H_{20}ClN_3O•2ClH$ mw: 426.80

SYNS: AMOPYROQUIN DIHYDROCHLORIDE ◇ 4-((7-CHLORO-4-QUINOLINYL)AMINO)-2-(1-PYRROLIDINYLMETHYL)PHENOLDIHYDROCHLORIDE ◇ 4-((7-CHLORO-4-QUINOLYL)AMINO)-α-1-PYRROLIDINYL-o-CRESOL DIHYDROCHLORIDE

TOXICITY DATA with REFERENCE
orl-rat LD50:410 mg/kg ANTCAO 8,450,58
ivn-rat LD50:34 mg/kg ANTCAO 8,450,58
orl-mus LD50:580 mg/kg ANTCAO 8,450,58
ivn-mus LD50:35 mg/kg ANTCAO 8,450,58
ivn-mky LDLo:30 mg/kg AJTHAB 6,987,57

SAFETY PROFILE: Poison by intravenous route. Moderately toxic by ingestion. When heated to decomposition it emits very toxic fumes of Cl^- and NO_x.

PMY300 CAS:114-26-1 ***HR: 3***
PROPOXUR
mf: $C_{11}H_{15}NO_3$ mw: 209.27

PROP: A white to tan, crystalline solid; sltly sol in water; sol in all polar organic solvents.

SYNS: APROCARB ◇ BAY 9010 ◇ BAYER 39007 ◇ BAYGON ◇ BIFEX ◇ BLATTANEX ◇ BOYGON ◇ ENT 25,671 ◇ o-IMPC ◇ INVISI-GARD ◇ ISOCARB ◇ o-ISOPROPOXYPHENYL METHYLCARBAMATE ◇ o-ISOPROPOXYPHENYL-N-METHYLCARBAMATE ◇ 2-ISOPROPOXYPHENYL-N-METHYLCARBAMATE ◇ 2-(1-METHYLETHOXY)PHENOL METHYLCARBAMATE ◇ N-METHYL-2-ISOPROPOXYPHENYLCARBAMATE ◇ OMS-33 ◇ PHC ◇ PROPOKSURU (POLISH) ◇ PROPYON ◇ SENDRAN ◇ SUNCIDE ◇ TUGON FLIEGENKUGEL ◇ UNDEN

TOXICITY DATA with REFERENCE
mmo-sat 250 ng/plate RPZHAW 30,81,79
cyt-ham-ipr 250 mg/kg ARTODN 58,152,85
orl-rat TDLo:1600 mg/kg (6-22D preg/15D post):REP BECTA6 20,624,78
orl-mus TDLo:600 mg/kg (female 7-16D post):TER JESEDU 20,373,85
orl-wmn LDLo:24 mg/kg PGMJAO 63,311,87
orl-rat LD50:70 mg/kg IYKEDH 10,884,79
ihl-rat LC50:1440 mg/m^3/1H 85GYAZ -,70,71
skn-rat LD50:800 mg/kg RPZHAW 22,579,71
ipr-rat LD50:30 mg/kg BWHOA6 44,241,71
scu-rat LD50:56 mg/kg YKYUA6 32,471,81
ivn-rat LD50:11 mg/kg BJIMAG 22,317,65
ims-rat LD50:53 mg/kg BJIMAG 22,317,65
orl-mus LD50:23500 μg/kg TOIZAG 17,60,70

CONSENSUS REPORTS: EPA Genetic Toxicology Program. Community Right-To-Know List.

OSHA PEL: TWA 0.5 mg/m^3
ACGIH TLV: TWA 0.5 mg/m^3
DFG MAK: 2 mg/m^3

SAFETY PROFILE: A human poison by ingestion. An experimental poison by ingestion, subcutaneous, intraperitoneal, intravenous, and intramuscular routes. Moderately toxic by inhalation and skin contact. An experimental teratogen. Experimental reproductive effects. Mutation data reported. Moderately irritating to skin. When heated to decomposition it emits toxic fumes of NO_x. See also CARBAMATES and ESTERS.

PMY310 CAS:38777-13-8 ***HR: 3***
PROPOXUR NITROSO
mf: $C_{11}H_{14}N_2O_4$ mw: 238.27

SYNS: BAYGON, NITROSO derivative ◇ CARBAMIC ACID, METHYLNITROSO-, o-ISOPROPOXYPHENYL ESTER ◇ CARBAMIC ACID, METHYLNITROSO-, 2-(1-METHYLETHOXY)PHENYL ESTER (9CI) ◇ 2-ISOPROPOXYPHENYL N-METHYLCARBAMATE, nitrosated ◇ o-ISOPROPOXYPHENYL N-METHYL-N-NITROSOCARBAMATE ◇ o-ISOPROPOXYPHENYLMETHYLNITROSOCARBAMATE ◇ METHYLNITROSOCARBAMIC ACID o-ISOPROPOXYPHENYL ESTER ◇ NITROSO-BAYGON ◇ NITROSOPROPOXUR ◇ SUNCIDE, nitrosated

TOXICITY DATA with REFERENCE
mmo-sat 92 ng/plate MUREAV 56,1,77
dnd-hmn:fbr 10 μmol/L MUREAV 44,1,77
orl-rat TDLo:625 mg/kg/31W-I:CAR EESADV 2,413,78

SAFETY PROFILE: Questionable carcinogen with experimental carcinogenic data. Many N-nitroso compounds are carcinogens. Human mutation data reported. When heated to decomposition it emits toxic fumes of NO_x. See also N-NITROSO COMPOUNDS and CARBAMATES.

PMY500 CAS:5736-85-6 ***HR: 2***
4-PROPOXYBENZALDEHYDE
mf: $C_{10}H_{12}O_2$ mw: 164.22

SYNS: p-PROPOXYBENZALDEHYDE ◇ p-(N-PROPOXY)BENZALDEHYDE

TOXICITY DATA with REFERENCE
orl-rat LD50:1600 mg/kg JPETAB 93,26,48
orl-mus LD50:1800 mg/kg JPETAB 93,26,48

CONSENSUS REPORTS: Reported in EPA TSCA Inventory.

SAFETY PROFILE: Moderately toxic by ingestion. When heated to decomposition it emits acrid smoke and irritating fumes. See also ALDEHYDES.

PMY750 CAS:59643-84-4 ***HR: 2***
2-N-PROPOXYBENZAMIDE
mf: $C_{10}H_{13}NO_2$ mw: 179.24

SYNS: H.P. 206 ◇ o-PROPOXYBENZAMIDE

TOXICITY DATA with REFERENCE
orl-mus LD50:1260 mg/kg JPPMAB 4,872,52
ipr-mus LD50:510 mg/kg JPPMAB 4,872,52

SAFETY PROFILE: Moderately toxic by ingestion and intraperitoneal routes. When heated to decomposition it emits toxic fumes of NO_x.

PNA200 CAS:3686-69-9 ***HR: 3***
p-PROPOXYBENZOIC ACID 2-(1-PYRROLIDINYL)ETHYL ESTER HYDROCHLORIDE
mf: $C_{16}H_{23}NO_3•ClH$ mw: 313.86

SYN: 1-RBO-12-A

TOXICITY DATA with REFERENCE
skn-rbt 1 pph MLD AIPTAK 137,410,62
eye-rbt 1 pph MLD AIPTAK 137,410,62
ipr-mus LD50:158 mg/kg AIPTAK 137,410,62

SAFETY PROFILE: Poison by intraperitoneal route. A skin and eye irritant. When heated to decomposition it emits toxic fumes of NO_x and HCl.

PNA225 CAS:20706-25-6 ***HR: 1***
2-PROPOXYETHYL ACETATE
mf: $C_7H_{14}O_3$ mw: 146.21

SYNS: EGPEA ◇ ETHYLENE GLYCOL MONOPROPYL ETHER ACETATE ◇ 2-PROPOXYETHANOL ACETATE

TOXICITY DATA with REFERENCE
eye-rbt 100 mg MLD EVHPAZ 57,165,84
skn-gpg 500 mg MLD EVHPAZ 57,165,84
ihl-rat TCLo:800 ppm/6H (6-15D preg):REP EVHPAZ 57,25,84
ihl-rat TCLo:200 ppm/6H (6-15D preg):TER EVHPAZ 57,25,84
orl-rat LD50:9456 mg/kg EVHPAZ 57,165,84

CONSENSUS REPORTS: Glycol ether compounds are on the Community Right-To-Know List.

SAFETY PROFILE: Low toxicity by ingestion. An experimental teratogen. Experimental reproductive effects by inhalation. A skin and eye irritant. Many glycol ethers have dangerous human reproductive effects. When heated to decomposition it emits acrid smoke and irritating fumes. See also GLYCOL ETHERS.

PNA250 CAS:77-50-9 ***HR: 3***
PROPOXYPHENE
mf: $C_{22}H_{29}NO_2$ mw: 339.52

PROP: Crystals. Mp: 75-76°.

SYNS: α-4-DIMETHYLAMINO-1,2-DIPHENYL-3-METHYL-2-BUTANOL PROPIONATE ◇ α-4-DIMETHYLAMINO-3-METHYL-1,2-DIPHENYL-2-BUTANOLPROPIONATE ◇ 4-(DIMETHYLAMINO)-3-METHYL-1,2-DIPHENYL-2-BUTANOL PROPIONATE ◇ 4-DIMETHYLAMINO- 3-METHYL-1,2-DIPHENYL-2-PROPOXYBUTANE ◇ α-N,N-DIMETHYL-3,4-DIPHENYL-2-METHYL-3-PROPIONOXY-1-BUTYLAMINE ◇ 1,2-DIPHENYL-2-PROPIONOXY-3-METHYL-4-DIMETHYLAMINOBUTANE ◇ LETUSIN

TOXICITY DATA with REFERENCE
orl-rat TDLo:350 mg/kg (7-20D preg):REP TJADAB 17,39A,78

orl-hmn LDLo:20 mg/kg CTOXAO 10,327,77
scu-mus LD50:35 mg/kg CBINA8 53,77,85

SAFETY PROFILE: A human poison by ingestion. An experimental poison by subcutaneous route. Experimental reproductive effects. When heated to decomposition it emits toxic fumes of NO_x. See also d-PROPOXYPHENE HYDROCHLORIDE.

PNA500 CAS:1639-60-7 ***HR: 3***
d-PROPOXYPHENE HYDROCHLORIDE
mf: $C_{22}H_{29}NO_2 \cdot ClH$ mw: 375.98

PROP: Bitter crystals. Mp: 163-168.5°. Sol in water, alc, chloroform, acetone; insol in benzene, ether.

SYNS: ALGAFAN ◇ ANTALVIC ◇ DARVON HYDROCHLORIDE ◇ DEPRANCOL ◇ DEPROMIC ◇ DEVELIN ◇ DEXTROPROPOXYPHENE HYDROCHLORIDE ◇ DEXTROPROXYPHEN HYDROCHLORIDE ◇ d-4-DIMETHYLAMINO-3-METHYL-1,2-DIPHENYL-2-BUTANOL PROPIONATE HYDROCHLORIDE ◇ s-α-(2-(DIMETHYLAMINO)-1-METHYLETHYL)-α-PHENYLBENZENEETHANOL PROPIOATE HYDROCHLORIDE ◇ (+)-1,2-DIPHENYL-2-PROPIONOXY-3-METHYL-4-DIMETHYLAMINOBUTANE HYDROCHLORIDE ◇ DOLENE ◇ DOLOCAP ◇ DOLOXENE ◇ DORAPHEN ◇ ERANTIN ◇ FEMADOL ◇ HARMAR ◇ PROPOX ◇ PROPOXYCHEL ◇ PROPOXYPHENE HYDROCHLORIDE ◇ (+)-PROPOXYPHENE HYDROCHLORIDE ◇ α-PROPOXYPHENE HYDROCHLORIDE ◇ α-d-PROPOXYPHENE HYDROCHLORIDE ◇ d-PROPOXYPHENE MONOHYDROCHLORIDE ◇ PROXAGESIC

TOXICITY DATA with REFERENCE
orl-rat TDLo:1050 mg/kg (7-20D preg):REP SCIEAS 205,1220,79
scu-ham TDLo:225 mg/kg (8D preg):TER AJOGAH 123,705,75
orl-man TDLo:11 mg/kg:CNS,PUL AIMDAP 129,62,72
orl-man LDLo:15 mg/kg:PUL JFSCAS 19,72,74
orl-wmn TDLo:650 μg/kg:GIT,LIV,SYS GUTTAK 27,444,86
orl-wmn LDLo:15 mg/kg:CNS,PUL JFSCAS 19,72,74
orl-wmn LDLo:39 mg/kg:CNS,PUL AIMDAP 129,62,72
orl-chd TDLo:27 mg/kg:PUL,CNS JOPDAB 63,158,63
orl-chd LDLo:70 mg/kg JOPDAB 63,158,63
orl-rat LD50:84 mg/kg JPETAB 134,332,61
ipr-rat LD50:58 mg/kg TXAPA9 19,445,71
scu-rat LD50:134 mg/kg TXAPA9 19,445,71
ivn-rat LD50:15 mg/kg TXAPA9 19,445,71
ims-rat LD50:157 mg/kg TXAPA9 19,445,71
orl-mus LD50:84 mg/kg JPETAB 134,332,61
ipr-mus LD50:111 mg/kg TXAPA9 19,445,71
ivn-mus LD50:25 mg/kg TXAPA9 6,334,64

CONSENSUS REPORTS: Reported in EPA TSCA Inventory.

SAFETY PROFILE: A human poison by ingestion. An experimental poison by ingestion, intraperitoneal, subcutaneous, intravenous, and intramuscular routes. Human systemic effects by ingestion: anorexia, ataxia, chronic pulmonary edema, cyanosis, increased body temperature, jaundice, nausea or vomiting, sleep disturbance. Experimental teratogenic and reproductive effects. When heated to decomposition it emits very toxic fumes of HCl and NO_x. See also PROPOXYPHENE.

PNB250 CAS:1155-49-3 ***HR: 3***
4-PROPOXY-β-(1-PIPERIDYL)PROPIOPHENONE HYDROCHLORIDE
mf: $C_{17}H_{25}NO_2 \cdot ClH$ mw: 311.89

SYNS: FALICAIN ◇ FALICAINE HYDROCHLORIDE ◇ FELICAIN (GERMAN) ◇ 3-PIPERIDINO-4'-PROPOXYPROPIOPHENONE HYDROCHLORIDE ◇ 3-(1-PIPERIDINYL)-1-(4-PROPOXYPHENYL)-1-PROPANONE HYDROCHLORIDE ◇ 4'-PROPOXY-3-PIPERIDINO PROPIOPHENONE HYDROCHLORIDE ◇ PROPOXYPIPEROCAINE ◇ PROPOXYPIPEROCAINE HYDROCHLORIDE ◇ S 142

TOXICITY DATA with REFERENCE
orl-mus LD50:125 mg/kg PHARAT 31,21,76
scu-mus LD50:57 mg/kg ARZNAD 5,559,55
ivn-mus LD50:15 mg/kg JPETAB 115,419,55

SAFETY PROFILE: Poison by ingestion, subcutaneous, and intravenous routes. When heated to decomposition it emits very toxic fumes of HCl and NO_x.

PNB500 CAS:1569-01-3 ***HR: 2***
1-PROPOXY-2-PROPANOL
mf: $C_6H_{14}O_2$ mw: 118.20

SYN: PROPASOL SOLVENT P

TOXICITY DATA with REFERENCE
orl-rat LD50:3250 mg/kg NPIRI* 1,107,74
skn-rbt LD50:3560 mg/kg NPIRI* 1,107,74

CONSENSUS REPORTS: Reported in EPA TSCA Inventory.

SAFETY PROFILE: Moderately toxic by ingestion and skin contact. When heated to decomposition it emits acrid smoke and irritating fumes.

PNB750 CAS:30136-13-1 ***HR: 2***
n-PROPOXYPROPANOL (mixed isomers)
mf: $C_6H_{14}O_2$ mw: 118.20

PROP: Liquid. D: 0.8865 (20/20°C), bp 149.8°C, fp −80°C, (sets to glass below this), flash p 128°F (53.3°C). Sol in water.

TOXICITY DATA with REFERENCE
skn-rbt 500 mg open MLD UCDS** 4/25/67
eye-rbt 15 mg SEV UCDS** 12/15/70
orl-rat LD50:3250 mg/kg AIHAAP 30,470,69
skn-rbt LD50:4000 mg/kg UCDS** 4/25/67

SAFETY PROFILE: Moderately toxic by ingestion and skin contact. A skin and severe eye irritant. When heated to decomposition it emits acrid smoke and irritating fumes.

PNB790 CAS:13013-17-7 *HR: 3*
dl-PROPRANOLOL
mf: $C_{16}H_{21}NO_2$ mw: 259.38

SYNS: (±)-1-((1-METHYLETHYL)AMINO)-3-(1-NAPHTHALENYLOXY)-2-PROPANOL (9CI) ◇ (±)-PROPRANOLOL ◇ racemic-PROPRANOLOL

TOXICITY DATA with REFERENCE
dnd-rat:lvr 10 mmol/L JJIND8 70,747,83
dns-rat-ipr 75 mg/kg TOPADD 13,18,85
oms-rat-ipr 75 mg/kg TOPADD 13,18,85
orl-rat TDLo:229 mg/kg (female 8-22D post):REP JPETAB 208,118,79
ivn-rat LD50:2750 μg/kg PJPPAA 32,823,80

SAFETY PROFILE: Poison by intravenous route. Experimental reproductive effects. Mutation data reported. When heated to decomposition it emits toxic fumes of NO_x.

PNB800 CAS:3506-09-0 *HR: 3*
dl-PROPRANOLOL HYDROCHLORIDE
mf: $C_{16}H_{21}NO_2$•ClH mw: 295.84

SYN: (±)-1-(ISOPROPYLAMINO)-3-(1-NAPHTHYLOXY)-2-PROPANOL HYDROCHLORIDE

TOXICITY DATA with REFERENCE
orl-mus LD50:380 mg/kg EJPHAZ 15,151,71
ipr-mus LD50:87600 μg/kg ARZNAD 24,1751,74
ivn-mus LD50:33300 μg/kg ARZNAD 24,1751,74

SAFETY PROFILE: Poison by ingestion, intravenous, and intraperitoneal routes. When heated to decomposition it emits toxic fumes of HCl and NO_x.

PNC250 CAS:109-60-4 *HR: 3*
n-PROPYL ACETATE
DOT: UN 1276
mf: $C_5H_{10}O_2$ mw: 102.15

PROP: Clear, colorless liquid; pleasant odor. Mp: −92.5°, bp: 101.6°, flash p: 58°F, lel: 2.0%, uel: 8.0%, d: 0.887, autoign temp: 842°F, vap press: 40 mm @ 28.8°, vap d: 3.52. Misc with alc, ether; sol in water.

SYNS: ACETATE de PROPYLE NORMAL (FRENCH) ◇ ACETIC ACID, n-PROPYL ESTER ◇ 1-ACETOXYPROPANE ◇ OCTAN PROPYLU (POLISH) ◇ PROPYL ACETATE ◇ 1-PROPYL ACETATE

TOXICITY DATA with REFERENCE
skn-rbt 500 mg open MLD UCDS** 1/25/65
ihl-hmn TCLo:1000 mg/m^3:EYE,PUL AGGHAR 5,1,33
orl-rat LD50:9370 mg/kg FCTXAV 2,327,64
ihl-rat LCLo:8000 ppm/4H AIHAAP 30,470,69
orl-mus LD50:8300 mg/kg FCTXAV 2,327,64
ipr-mus LD50:1420 mg/kg SCCUR* -,8,61
ihl-cat LCLo:38 g/m^3/5H AGGHAR 5,1,33
scu-cat LDLo:3000 mg/kg AGGHAR 5,1,33
orl-rbt LD50:6640 mg/kg IMSUAI 41,31,72

CONSENSUS REPORTS: Reported in EPA TSCA Inventory.

OSHA PEL: (Transitional: TWA 200 ppm) TWA 200 ppm; STEL 250 ppm
ACGIH TLV: TWA 200 ppm; STEL 250 ppm
DFG MAK: 200 ppm (840 mg/m^3)
DOT Classification: Flammable Liquid; Label: Flammable Liquid.

SAFETY PROFILE: Moderately toxic by intraperitoneal and subcutaneous routes. Mildly toxic by ingestion and inhalation. Human systemic effects by inhalation: lachrimation, cough. A skin irritant. A narcotic at high concentrations. Isopropyl acetate is slightly less narcotic than normal propyl acetate. Dangerous fire hazard when exposed to heat, flame, or oxidizers. Explosive in the form of vapor when exposed to heat or flame. Can react vigorously with oxidizing materials. To fight fire, use alcohol foam, CO_2, dry chemical. When heated to decomposition it emits acrid smoke and irritating fumes.

PNC500 CAS:6728-26-3 *HR: 3*
β-PROPYL ACROLEIN
mf: $C_6H_{10}O$ mw: 98.16

$CH_3CH_2CH_2CH{=}CHCO•H$

SYNS: 2-HEXENAL, (E)- ◇ trans-2-HEXENAL

TOXICITY DATA with REFERENCE
skn-rbt 500 mg/24H MOD FCTXAV 13,453,75
orl-rat LD50:780 mg/kg FCTXAV 9,775,71
ipr-rat LD50:180 mg/kg FCTXAV 9,775,71
orl-mus LD50:1550 mg/kg FCTXAV 9,775,71
ipr-mus LD50:100 mg/kg FCTXAV 9,775,71
skn-rbt LD50:600 mg/kg FCTXAV 13,449,75

CONSENSUS REPORTS: Reported in EPA TSCA Inventory.

SAFETY PROFILE: Poison by intraperitoneal route. Moderately toxic by ingestion and skin contact. A skin irritant. Reacts with nitric acid to form a heat-sensitive explosive product. Can form peroxides in storage. When heated to decomposition it emits acrid smoke and irritating fumes.

PNC875 CAS:35080-11-6 *HR: 3*
N-PROPYLAJMALINE
mf: $C_{23}H_{33}N_2O_2$ mw: 369.58

SYNS: (17R,21-α)-17,21-DIHYDROXY-4-PROPYLAJMALANIUM ◇ PRAJMALINE ◇ PRAJMALIUM ◇ N-PROPYLAJMALINIUM ◇ N^4-PROPYLAJMALINIUM

TOXICITY DATA with REFERENCE
ivn-rat LD50:1400 μg/kg PHARAT 31,36,76
orl-mus LD50:65500 μg/kg CPBTAL 22,2329,74
ivn-mus LD50:3400 μg/kg CPBTAL 22,2329,74

SAFETY PROFILE: Poison by ingestion and intravenous routes. When heated to decomposition it emits toxic fumes of NO_x. See also other ajmalines.

PNC925 CAS:14046-99-2 ***HR: 3***
1-PROPYLAJMALINIUM BROMIDE
mf: $C_{23}H_{36}N_2O{\cdot}Br$ mw: 449.49

SYNS: (17R,21-α)-17,21-DIHYDROXY-4-PROPYLAJMALINIUMBROMIDE ◇ NPAB ◇ N-PROPYLAJMALINE BROMIDE

TOXICITY DATA with REFERENCE
ipr-rat LD50:12 mg/kg RPTOAN 33,147,70
ipr-mus LD50:19 mg/kg RPTOAN 33,147,70
ivn-cat LDLo:5 mg/kg RPTOAN 33,147,70
ivn-gpg LDLo:7 mg/kg RPTOAN 33,147,70

SAFETY PROFILE: Poison by intraperitoneal and intravenous routes. When heated to decomposition it emits very toxic fumes of NO_x and Br^-. See also BROMIDES.

PND000 CAS:71-23-8 ***HR: 3***
n-PROPYL ALCOHOL
DOT: UN 1274
mf: C_3H_8O mw: 60.11

PROP: Clear liquid; alc-like odor. Mp: −127°, bp: 97.19°, flash p: 59°F (CC), ULC: 55-60, d: 0.8044 @ 20°/4°, lel: 2.1%, uel: 13.5%, autoign temp: 824°F, vap press: 10 mm @ 14.7°, vap d: 2.07. Misc in water, alc, and ether.

SYNS: ALCOOL PROPILICO (ITALIAN) ◇ ALCOOL PROPYLIQUE (FRENCH) ◇ ETHYL CARBINOL ◇ 1-HYDROXYPROPANE ◇ OPTAL ◇ OSMOSOL EXTRA ◇ n-PROPANOL ◇ PROPANOL-1 ◇ 1-PROPANOL ◇ PROPANOLE (GERMAN) ◇ PROPANOLEN (DUTCH) ◇ PROPANOLI (ITALIAN) ◇ PROPYL ALCOHOL ◇ 1-PROPYL ALCOHOL ◇ n-PROPYL ALKOHOL (GERMAN) ◇ PROPYLIC ALCOHOL ◇ PROPYLOWY ALKOHOL (POLISH)

TOXICITY DATA with REFERENCE
skn-rbt 500 mg open MLD UCDS** 6/28/72
eye-rbt 4 mg open SEV AMIHBC 10,61,54
eye-rbt 20 mg/24H MOD 28ZPAK -,34,72
mmo-esc 4 pph ABMGAJ 23,843,69
cyt-smc 100 mmol/tube HEREAY 33,457,47
mmo-esc 4 pph
orl-rat TDLo:50 g/kg/81W-I:CAR ARGEAR 45,19,75
scu-rat TDLo:6 g/kg/95W-I:CAR ARGEAR 45,19,75
orl-wmn LDLo:5700 mg/kg ATXKA8 16,84,56
orl-rat LD50:1870 mg/kg AMIHBC 10,61,54
ipr-rat LD50:2164 mg/kg EVHPAZ 61,321,85
ivn-rat LD50:590 mg/kg EVHPAZ 61,321,85
orl-mus LD50:6800 mg/kg PSDTAP 9,276,68
ihl-mus LC50:48 g/m³ GTPZAB 18(3),48,74
ipr-mus LD50:3125 mg/kg EVHPAZ 61,321,85
scu-mus LD50:4700 mg/kg TXAPA9 18,185,71

CONSENSUS REPORTS: Reported in EPA TSCA Inventory. EPA Genetic Toxicology Program.

OSHA PEL: (Transitional: TWA 200 ppm) TWA 200 ppm; STEL 250 ppm
ACGIH TLV: TWA 200 ppm; STEL 250 ppm (skin)
DOT Classification: Flammable Liquid; Label: Flammable Liquid.

SAFETY PROFILE: Poison by subcutaneous route. Moderately toxic by inhalation, ingestion, intraperitoneal, and intravenous routes. A skin and severe eye irritant. Questionable carcinogen with experimental carcinogenic data. Mutation data reported. Dangerous fire hazard when exposed to heat, flame, or oxidizers. Explosive in the form of vapor when exposed to heat or flame. Ignites on contact with potassium-tert-butoxide. Dangerous upon exposure to heat or flame; can react vigorously with oxidizing materials. To fight fire, use alcohol foam, CO_2, dry chemical. When heated to decomposition it emits acrid smoke and irritating fumes.

PND250 CAS:107-10-8 ***HR: 3***
PROPYLAMINE
DOT: UN 1277
mf: C_3H_9N mw: 59.13

PROP: Colorless, alkaline liquid; strong ammonia odor. D: 0.7191 @ 20°/20°, mp: −83°, bp: 48-49°, vap press: 248 mm @ 20°, flash p: −35°F, autoign temp: 604°F, lel: 2.0%, uel: 10.4%. Misc water, alc, ether.

SYNS: 1-AMINOPROPANE ◇ MONO-N-PROPYLAMINE ◇ PROPANAMINE ◇ N-PROPYLAMINE ◇ RCRA WASTE NUMBER U194

TOXICITY DATA with REFERENCE
skn-rbt 100 μg/24H open AIHAAP 23,95,62
eye-rbt 720 μg SEV UCDS** 8/12/68
orl-rat LDLo:570 mg/kg AIHAAP 23,95,62
ihl-rat LC50:2310 ppm/4H AEHLAU 1,343,60
ihl-mus LC50:2500 mg/m³/2H 85GMAT -,102,82
skn-rbt LD50:560 mg/kg AIHAAP 23,95,62
ihl-mam LD50:2500 mg/m³ TPKVAL 14,80,75

CONSENSUS REPORTS: Reported in EPA TSCA Inventory.

DOT Classification: Flammable Liquid; Label: Flammable Liquid.

SAFETY PROFILE: Moderately toxic by inhalation, ingestion, and skin contact routes. A skin and severe eye irritant. Possibly a skin sensitizer. Very dangerous fire hazard when exposed to heat, flame, or oxidizers. Explosive in the form of vapor when exposed to heat or flame. To fight fire, use alcohol foam. When heated to decomposition it emits toxic fumes of NO_x. Incompatible with triethynyl aluminum. See also AMINES.

PNE000 CAS:2696-84-6 ***HR: 3***
4-PROPYLANILINE
mf: $C_9H_{13}N$ mw: 135.23

PROP: Liquid. D: 0.949 @ 18°, bp: 222°. Insol in water; very sol in alc and ether.

SYNS: 1-AMINO-4-PROPYLBENZENE ◇ p-PROPYLANILINE ◇ p-N-PROPYLANILINE ◇ 4-N-PROPYLANILINE

TOXICITY DATA with REFERENCE
ipr-mus LD50:201 mg/kg JMCMAR 17,900,74

CONSENSUS REPORTS: Reported in EPA TSCA Inventory.

SAFETY PROFILE: Poison by intraperitoneal route. When heated to decomposition it emits toxic fumes of NO_x.

PNE250 CAS:104-45-0 ***HR: 1***
p-n-PROPYL ANISOLE
mf: $C_{10}H_{14}O$ mw: 150.24

PROP: Colorless to pale yellow liquid; anise odor. D: 0.940, refr index: 1.502-1.506, flash p: 185°F. Sol in fixed oils; insol in glycerin, propylene glycol.

SYNS: DIHYDROANETHOLE ◇ FEMA No. 2930 ◇ 1-METHOXY-4-PROPYLBENZENE ◇ 4-PROPYLANISOLE ◇ 4-n-PROPYLANISOLE

TOXICITY DATA with REFERENCE
sln-dmg-orl 5 mmol/L FCTOD7 21,707,83
orl-rat LD50:4400 mg/kg TXAPA9 6,378,64
orl-mus LD50:7300 mg/kg FCTXAV 2,327,64

CONSENSUS REPORTS: Reported in EPA TSCA Inventory.

SAFETY PROFILE: Mildly toxic by ingestion. Mutation data reported. Combustible liquid. When heated to decomposition it emits acrid smoke and irritating fumes.

PNE500 CAS:28785-06-0 ***HR: 2***
4-PROPYLBENZALDEHYDE
mf: $C_{10}H_{12}O$ mw: 148.22

PROP: Liquid. D: 0.978 @ 20°/4°, bp: 236-237°. Insol in water; sol in alc and ether.

SYN: p-PROPYLBENZALDEHYDE

TOXICITY DATA with REFERENCE
orl-rat LD50:1600 mg/kg JPMSAE 63,1068,74

CONSENSUS REPORTS: Reported in EPA TSCA Inventory.

SAFETY PROFILE: Moderately toxic by ingestion. When heated to decomposition it emits acrid smoke and irritating fumes. See also ALDEHYDES.

PNE750 CAS:54889-82-6 ***HR: 3***
5-n-PROPYL-1,2-BENZANTHRACENE
mf: $C_{21}H_{18}$ mw: 270.39

SYN: 8-PROPYLBENZ(a)ANTHRACENE

TOXICITY DATA with REFERENCE
skn-mus TDLo:240 mg/kg/10W-I:ETA PRLBA4 123,343,37

SAFETY PROFILE: Questionable carcinogen with experimental tumorigenic data. When heated to decomposition it emits acrid smoke and irritating fumes.

PNF000 CAS:63020-32-6 ***HR: 3***
5-PROPYLBENZO(c)PHENANTHRENE
mf: $C_{21}H_{18}$ mw: 270.39

SYN: 2-n-PROPYL-3:4-BENZPHENANTHRENE

TOXICITY DATA with REFERENCE
skn-mus TDLo:1100 mg/kg/46W-I:ETA PRLBA4 131,170,42

SAFETY PROFILE: Questionable carcinogen with experimental tumorigenic data. When heated to decomposition it emits acrid smoke and irritating fumes.

PNF500 CAS:1114-71-2 ***HR: 3***
S-PROPYL BUTYLETHYLTHIOCARBAMATE
mf: $C_{10}H_{21}NOS$ mw: 203.38

PROP: Liquid. Bp: 142° @ 20 mm.

SYNS: BUTYLETHYLTHIOCARBAMIC ACID S-PROPYL ESTER ◇ PEBC ◇ PEBULATE ◇ S-PROPYL-N-AETHYL-N-BUTYL-THIOCARBAMAT (GERMAN) ◇ PROPYL-ETHYLBUTYLTHIOCARBAMATE ◇ N-PROPYL-N-ETHYL-N-(N-BUTYL)THIOCARBAMATE ◇ PROPYLETHYL-N-BUTYLTHIOCARBAMATE ◇ PROPYL N-ETHYL-N-BUTYLTHIOCARBAMATE ◇ S-(N-PROPYL)-N-ETHYL-N-N-BUTYLTHIOCARBAMATE ◇ N-PROPYL-N-ETHYL-N-(N-BUTYL)THIOLCARBAMATE ◇ PROPYL ETHYLBUTYLTHIOL-CARBAMATE ◇ R-2061 ◇ STAUFFER R-2061 ◇ TILLAM (RUSSIAN) ◇ TILLAM-6-E

TOXICITY DATA with REFERENCE
scu-mus TDLo:10 mg/kg:ETA NTIS** PB223-159
orl-rat LD50:921 mg/kg JMCHA2 -,C236,83
orl-mus LD50:1652 mg/kg PCOC** -,1138,66

SAFETY PROFILE: Moderately toxic by ingestion. Causes violent vomiting when accompanied by alcohol ingestion. Questionable carcinogen with experimental tumorigenic data. When heated to decomposition it emits highly toxic fumes of SO_x and NO_x. See also CARBAMATES.

PNF750 CAS:25413-64-3 ***HR: 3***
N-PROPYL-N-BUTYLNITROSAMINE
mf: $C_7H_{16}N_2O$ mw: 144.25

SYNS: N-NITROSO-N-PROPYL-1-BUTANAMINE ◇ N-NITROSO-N-PROPYLBUTYLAMINE

TOXICITY DATA with REFERENCE
mma-sat 5 μmol/plate MUREAV 48,121,77
orl-rat TDLo:46 g/kg/16W-C:ETA GANNA2 67,825,76

CONSENSUS REPORTS: EPA Genetic Toxicology Program.

SAFETY PROFILE: Questionable carcinogen with experimental tumorigenic data. Mutation data reported. Many N-nitroso compounds are carcinogens. When heated to decomposition it emits toxic fumes of NO_x. See also N-NITROSO COMPOUNDS.

PNG250 CAS:627-12-3 ***HR: 3***
PROPYL CARBAMATE
mf: $C_4H_9NO_2$ mw: 103.14

PROP: Crystals. Bp: 196°, mp: 60°, vap press: 1 mm @ 52.4°. Very sol in water, alc, ether.

SYNS: CARBAMIC ACID, PROPYL ESTER ◇ N-PROPYL CARBAMATE ◇ PROPYL URETHANE

TOXICITY DATA with REFERENCE
mmo-esc 2 pph/3H AMNTA4 85,119,51
ipr-ham TDLo:248 mg/kg (8D preg):TER CNREA8 27,1696,67
ipr-mus TDLo:650 mg/kg/13W-I:NEO JNCIAM 8,99,47
ipr-mus TD:5700 mg/kg/10W-I:ETA IJCNAW 4,318,69
scu-mus LD50:1300 mg/kg AJEBAK 45,507,67

CONSENSUS REPORTS: IARC Cancer Review: Group 3 IMEMDT 7,56,87; Animal Sufficient Evidence IMEMDT 12,201,76. Reported in EPA TSCA Inventory.

SAFETY PROFILE: Moderately toxic by subcutaneous route. An experimental teratogen. Questionable carcinogen with experimental neoplastigenic and tumorigenic data. Mutation data reported. When heated to decomposition it emits toxic fumes of NO_x. See also CARBAMATES.

PNG500 CAS:56316-37-1 ***HR: D***
N-PROPYL-N-(3-CARBOXYPROPYL)NITROSAMINE
mf: $C_7H_{14}N_2O_3$ mw: 174.23

SYN: 4-(N-NITROSOPROPYLAMINO)BUTYRICACID

TOXICITY DATA with REFERENCE
mma-sat 10 μmol/plate CNREA8 37,399,77

SAFETY PROFILE: Mutation data reported. Many N-nitroso compounds are carcinogens. When heated to decomposition it emits toxic fumes of NO_x. See also N-NITROSO COMPOUNDS.

PNG750 CAS:2807-30-9 ***HR: 3***
PROPYL CELLOSOLVE
mf: $C_5H_{12}O_2$ mw: 104.17

SYNS: EKTASOLVE EP ◇ ETHYLENE GLYCOL-MONO-PROPYL ETHER ◇ ETHYLENE GLYCOL-MONO-n-PROPYL ETHER ◇ MONOPROPYL ETHER of ETHYLENE GLYCOL ◇ 2-PROPOXYETHANOL

TOXICITY DATA with REFERENCE
eye-rbt 100 mg SEV EVHPAZ 57,165,84
skn-gpg 500 mg MLD EVHPAZ 57,165,84
orl-mus TDLo:16 g/kg (female 7-14D post):REP NTIS** PB86-197605
ihl-rat TCLo:100 ppm/6H (6-15D preg):TER TJADAB 32,93,85
orl-rat LD50:3089 mg/kg EVHPAZ 57,165,84
ihl-rat LCLo:2000 ppm/4H JIHTAB 31,343,49
ihl-mus LC50:1530 ppm/7H JIHTAB 25,157,43
skn-rbt LD50:960 mg/kg AIHAAP 30,470,69
skn-gpg LD50:1 g/kg EVHPAZ 57,165,84

CONSENSUS REPORTS: Glycol ether compounds are on the Community Right-To-Know List. Reported in EPA TSCA Inventory.

SAFETY PROFILE: Moderately toxic by ingestion and skin contact. Mildly toxic by inhalation. An experimental teratogen. Experimental reproductive effects. Some glycol ethers have dangerous human reproductive effects. A skin and severe eye irritant. Flammable; can react with oxidizing materials. When heated to decomposition it emits acrid smoke and irritating fumes. See also GLYCOL ETHERS.

PNH000 CAS:109-61-5 ***HR: 3***
PROPYL CHLOROCARBONATE
DOT: UN 2740
mf: $C_4H_7ClO_2$ mw: 122.56

PROP: Colorless liquid. D: 1.090 @ 20°/4°, bp: 114-115° @ 768 mm. Insol and sltly decomp in water, alc; misc in ether, benzene.

SYNS: CARBONOCHLORIDIC ACID, PROPYL ESTER ◇ CHLOROFORMIC ACID PROPYL ESTER ◇ PROPYL CHLOROFORMATE ◇ n-PROPYL CHLOROFORMATE (DOT)

TOXICITY DATA with REFERENCE
orl-mus LD50:650 mg/kg 37ASAA 4,758,78
ihl-mus LD50:319 ppm/1H 37ASAA 4,758,78
skn-mus LD50:10 mg/kg 37ASAA 4,758,78

CONSENSUS REPORTS: EPA Extremely Hazardous Substances List. Reported in EPA TSCA Inventory.

DOT Classification: Flammable or Combustible Liquid; Label: Flammable, Poison, Corrosive.

SAFETY PROFILE: Poison by skin contact. Moderately toxic by ingestion and inhalation. A corrosive irritant to skin, eyes, and mucous membranes. Flammable when exposed to heat or flame, can react vigorously with oxidizing materials. When heated to decomposition it emits toxic fumes of Cl^-. Used as a reactive intermediate to polymerization initiators.

PNH150 CAS:13889-92-4 ***HR: 2***
S-PROPYL CHLOROTHIOFORMATE
mf: C_4H_7ClOS mw: 138.62

TOXICITY DATA with REFERENCE
orl-rat LD50:1129 mg/kg STCC**
ihl-rat LCLo:1300 mg/m^3 STCC**
skn-rbt LD50:4500 mg/kg STCC**

CONSENSUS REPORTS: Reported in EPA TSCA Inventory.

SAFETY PROFILE: Moderately toxic by ingestion and inhalation. Mildly toxic by skin contact. When heated to decomposition it emits toxic fumes of Cl^- and SO_x. See also ESTERS.

PNH250 CAS:7778-83-8 ***HR: 2***
n-PROPYL CINNAMATE
mf: $C_{12}H_{14}O_2$ mw: 190.26

SYN: CINNAMIC ACID, PROPYL ESTER

TOXICITY DATA with REFERENCE
orl-gpg LD50:3000 mg/kg JPETAB 93,26,48

CONSENSUS REPORTS: Reported in EPA TSCA Inventory.

SAFETY PROFILE: Moderately toxic by ingestion. When heated to decomposition it emits acrid smoke and irritating fumes. See also ESTERS.

PNH500 CAS:18365-12-3 ***HR: 3***
PROPYLCOPPER(I)
mf: C_3H_7Cu mw: 106.63

CONSENSUS REPORTS: Copper and its compounds are on the Community Right-To-Know List.

SAFETY PROFILE: A very unstable explosive. When heated to decomposition it emits acrid smoke and irritating fumes. See also COPPER COMPOUNDS.

PNH550 CAS:70348-66-2 ***HR: 3***
3-PROPYLDIAZIRINE
mf: $C_4H_8N_2$ mw: 84.12

$CH_3CH_2CH_2CHN{=}N$ (ring)

SAFETY PROFILE: May explode at 75°C. When heated to decomposition it emits toxic fumes of NO_x.

PNH650 CAS:926-53-4 ***HR: 2***
PROPYLDICHLORARSINE
mf: $C_3H_7AsCl_2$ mw: 188.92

SYNS: ARSINE, DICHLOROPROPYL- ◇ TL 295

TOXICITY DATA with REFERENCE
ihl-mus LC50:1400 mg/m^3/10M NTIS** PB158-508

OSHA PEL: TWA 0.5 mg(As)/m^3

SAFETY PROFILE: Moderately toxic by inhalation. When heated to decomposition it emits toxic fumes of As and Cl^-.

PNH750 CAS:5836-10-2 ***HR: 2***
PROPYL-p,p'-DICHLOROBENZILATE
mf: $C_{17}H_{16}Cl_2O_3$ mw: 339.23

SYNS: ACARALATE ◇ CHLORMITE ◇ CHLOROPROPYLATE ◇ ENT 26,999 ◇ G 24,163 ◇ GESAKUR ◇ ISOPROPYL-4,4'-DICHLOROBENZILATE ◇ 1-METHYLETHYL-4-CHLORO-α-(4-CHLOROPHENYL)-α-HYDROXYBENZENEACETATE ◇ ROSPAN ◇ ROSPIN

TOXICITY DATA with REFERENCE
skn-rbt 125 mg open MLD CIGET* -,-,77
eye-rbt 25 mg SEV CIGET* -,-,77
orl-rat LD50:5 g/kg 28ZEAL 4,256,69
orl-mus LD50:5 g/kg 28ZEAL 4,256,69
skn-rbt LD50:10200 mg/kg CIGET* -,-,77
orl-ckn LD50:2500 mg/kg 31ZOAD 1,94,68
orl-bwd LD50:2500 mg/kg 28ZEAL 4,256,69

SAFETY PROFILE: Moderately toxic by ingestion. Mildly toxic by skin contact. A skin and severe eye irritant. When heated to decomposition it emits toxic fumes of Cl^-. Used as a miticide for apples and pears.

PNI250 CAS:25606-41-1 ***HR: 3***
PROPYL(3-(DIMETHYLAMINO)PROPYL)CARBAMATE MONOHYDROCHLORIDE
mf: $C_9H_{20}N_2O_2{\cdot}ClH$ mw: 224.77

SYNS: BANOL TURF FUNGICIDE ◇ PREVEX ◇ PREVICUR-N ◇ PROPAMOCARB HYDROCHLORIDE ◇ SH 66752

TOXICITY DATA with REFERENCE
orl-rat LD50:7860 μg/kg FMCHA2 -,C196,83
orl-mus LD50:1440 mg/kg FMCHA2 -,C196,83

SAFETY PROFILE: Poison by ingestion. When heated to decomposition it emits very toxic fumes of NO_x and HCl. See also CARBAMATES.

PNI500 CAS:63020-33-7 ***HR: 3***
5-n-PROPYL-9,10-DIMETHYL-1,2-BENZANTHRACENE
mf: $C_{23}H_{22}$ mw: 298.45

SYN: 7,12-DIMETHYL-8-PROPYL-BENZ(a)ANTHRACENE

TOXICITY DATA with REFERENCE
scu-mus TDLo:200 mg/kg:ETA CNREA8 1,685,41

SAFETY PROFILE: Questionable carcinogen with experimental tumorigenic data. When heated to decomposition it emits acrid smoke and irritating fumes.

PNI750 CAS:1929-77-7 ***HR: 2***
PROPYL-N,N-DIPROPYLTHIOLCARBAMATE
mf: $C_{10}H_{21}NOS$ mw: 203.38

SYNS: DIPROPYLTHIOCARBAMIC ACID-S-PROPYL ESTER ◇ PPTC ◇ S-PROPYL DIPROPYLTHIOCARBAMATE ◇ N-PROPYL-DI-N-PROPYLTHIOLCARBAMATE ◇ R-1607 ◇ VERNAM ◇ VERNOLATE

TOXICITY DATA with REFERENCE
cyt-mus-unr 500 mg/kg TGANAK 14(6),41,80
cyt-mus-orl 500 mg/kg CYGEDX 14(6),38,80
orl-rat LD50:1200 mg/kg FMCHA2 -,C252,83

SAFETY PROFILE: Moderately toxic by ingestion. Mutation data reported. When heated to decomposition it emits very toxic fumes of NO_x and SO_x. See also CARBAMATES.

PNI850 CAS:7361-89-9 ***HR: 3***
PROPYL DISELENIDE
mf: $C_6H_{14}Se_2$ mw: 244.12

SYNS: DI-n-PROPYL DISELENIDE ◇ DISELENIDE, DIPROPYL-(9CI)

TOXICITY DATA with REFERENCE
ivn-mus LD50:56 mg/kg CSLNX* NX#09238

OSHA PEL: TWA 0.2 mg(Se)/m^3
ACGIH TLV: TWA 0.2 mg(Se)/m^3

SAFETY PROFILE: Poison by intravenous route. When heated to decomposition it emits toxic fumes of Se.

PNJ000 CAS:60784-42-1 ***HR: 3***
1,1'-PROPYLENEBIS(3-(2-CHLOROETHYL)-3-NITROSOUREA)
mf: $C_9H_{16}Cl_2N_6O_4$ mw: 343.21

SYN: 1,1'-PROPYLENEBIS-CNU

TOXICITY DATA with REFERENCE
sln-dmg-orl 100 μmol/L/24H MUREAV 57,297,78
mrc-smc 500 μmol/L/16H MUREAV 42,45,77
ipr-rat LDLo:23 mg/kg JNCIAM 60,345,78

SAFETY PROFILE: Poison by intraperitoneal route. Mutation data reported. Many N-nitroso compounds are carcinogens. When heated to decomposition it emits very toxic fumes of Cl^- and NO_x. See also N-NITROSO COMPOUNDS.

PNJ400 CAS:78-87-5 ***HR: 3***
PROPYLENE DICHLORIDE
mf: $C_3H_6Cl_2$ mw: 112.99

PROP: Colorless liquid. Bp: 96.8°, flash p: 60°F, d: 1.1593 @ 20°/20°, vap press: 40 mm @ 19.4°, vap d: 3.9, autoign temp: 1035°F, lel: 3.4%, uel: 14.5%.

SYNS: BICHLORURE de PROPYLENE (FRENCH) ◇ 1,2-DICHLOROPROPANE ◇ α,β-DICHLOROPROPANE ◇ DWUCHLOROPROPAN (POLISH) ◇ ENT 15,406 ◇ NCI-C55141 ◇ PROPYLENE CHLORIDE ◇ α,β-PROPYLENE DICHLORIDE ◇ RCRA WASTE NUMBER U083

TOXICITY DATA with REFERENCE
eye-rbt 500 mg MLD AIHAAP 30,470,69
mmo-sat 100 μg/plate ENNUDM 5(Suppl 1),3,83
mma-sat 333 μg/plate ENMUDM 5(Suppl 1),3,83
orl-mus TDLo:130 g/kg/2Y-I:CAR NTPTR* NTP-TR-263,86
orl-rat LD50:1947 mg/kg AIHAAP 20,364,59
ihl-rat LC50:14 g/m^3/8H AIHAAP 20,364,59
orl-rat LD50:2196 mg/kg AIHAAP 30,470,69
ihl-mus LCLo:1000 ppm/2H JIHTAB 28,1,46
orl-dog LDLo:5000 mg/kg AJHYA2 16,325,32
skn-rbt LD50:8750 mg/kg UCDS** 7/28/66
orl-gpg LD50:2000 mg/kg FMCHA2 -,C198,83

CONSENSUS REPORTS: IARC Cancer Review: Group 3 IMEMDT 7,56,87; Animal Limited Evidence IMEMDT 41,131,86. NTP Carcinogenesis Studies (gavage); Equivocal Evidence: rat NTPTR* NTP-TR-263,86; Some Evidence: mouse NTPTR* NTP-TR-263,86. Reported in EPA TSCA Inventory. EPA Genetic Toxicology Program. Community Right-To-Know List.

OSHA PEL: (Transitional: TWA 75 ppm) TWA 75 ppm; STEL 110 ppm
ACGIH TLV: TWA 75 ppm; STEL 110 ppm
DFG MAK: 75 ppm (350 mg/m^3)

DOT Classification: Flammable Liquid; Label: Flammable Liquid.

SAFETY PROFILE: Moderately toxic by inhalation and ingestion. Mildly toxic by skin contact. An eye irritant. Questionable carcinogen with experimental carcinogenic data. Mutation data reported. Can cause liver, kidney, and heart damage. Can cause dermatitis. One of the more toxic chlorinated hydrocarbons. A suggested order of increasing toxicity is dichloromethane, trichloroethylene, carbon tetrachloride, dichloropropane, dichloroethane. Animals exposed to high concentrations often showed marked visceral congestion, fatty degeneration of the liver, kidney, and, less frequently, of the heart. They also showed areas of coagulation and necrosis of the liver. There was found to be a heavy mortality among mice exposed to 400 ppm concentrations. A very dangerous fire hazard when exposed to heat or flame. Reacts with aluminum to form aluminum chloride. This reaction, when confined, can lead to explosion. Can react vigorously with oxidizing materials. To fight fire, use water, foam, CO_2, dry chemical. When heated to decomposition it emits toxic fumes of Cl^-. See also CHLORINATED HYDROCARBONS, ALIPHATIC.

PNJ500 CAS:21962-24-3 ***HR: 2***
PROPYLENE FORMAL
mf: $C_6H_{12}O_2$ mw: 116.18

SYN: 1,1-DIMETHOXY-2-BUTENE

TOXICITY DATA with REFERENCE
ihl-rat LC50:400 ppm/4H AMIHAB 12,623,55

CONSENSUS REPORTS: Reported in EPA TSCA Inventory.

SAFETY PROFILE: Moderately toxic inhalation. When heated to decomposition it emits acrid smoke and irritating fumes.

PNJ750 CAS:9005-37-2 ***HR: 1***
PROPYLENE GLYCOL ALGINATE
mf: $(C_9H_{14}O_7)_8$ mw: 1873.6

PROP: White fibrous or granular powder; odorless and tasteless. Sol in water and dilute organic acids.

SYNS: HYDROXY PROPYL ALGINATE ◇ KELCOLOID

TOXICITY DATA with REFERENCE
orl-rat LD50:7200 mg/kg FDRLI* 124,-,76
orl-mus LD50:7800 mg/kg FDRLI* 124,-,76
orl-rbt LD50:7600 mg/kg FDRLI* 124,-,76
orl-ham LD50:7000 mg/kg FDRLI* 124,-,76

CONSENSUS REPORTS: Reported in EPA TSCA Inventory.

SAFETY PROFILE: Mildly toxic by ingestion. When heated to decomposition it emits acrid smoke and irritating fumes.

PNK000 CAS:1331-17-5 ***HR: 3***
PROPYLENE GLYCOL, ALLYL ETHER
mf: $C_6H_{12}O_2$ mw: 116.18

SYNS: ALLYL ETHER OF PROPYLENE GLYCOL ◇ 1,2-PROPANEDIOL, ALLYL ETHER

TOXICITY DATA with REFERENCE
skn-rbt 10 mg/24H MLD JIHTAB 31,60,49
eye-rbt 2 mg SEV JIHTAB 31,60,49
orl-rat LD50:510 mg/kg JIHTAB 31,60,49
skn-rbt LD50:1100 mg/kg JIHTAB 31,60,49

CONSENSUS REPORTS: Glycol ether compounds are on the Community Right-To-Know List. Reported in EPA TSCA Inventory.

SAFETY PROFILE: Moderately toxic by ingestion and skin contact. A skin and severe eye irritant. Some glycol ethers have dangerous human reproductive effects. When heated to decomposition it emits acrid smoke and irritating fumes. See also GLYCOL ETHERS and ALLYL COMPOUNDS.

PNK500 ***HR: D***
PROPYLENE GLYCOL-sec-BUTYL PHENYL ETHER
mf: $C_{13}H_{16}O_2$ mw: 208.3

PROP: Sltly yellow liquid. Mp: −20°, bp: 276.8°, flash p: 270°F, d: 0.992 @ 25°/25°, vap d: 7.2.

SYN: 1-(o-SEC-BUTYLPHENOXY)-2-PROPANOL

CONSENSUS REPORTS: Glycol ether compounds are on the Community Right-To-Know List.

SAFETY PROFILE: Many glycol ethers have dangerous human reproductive effects. Combustible when exposed to heat or flame. Incompatible with oxidizing materials. To fight fire, use CO_2, dry chemical. When heated to decomposition it emits acrid smoke and irritating fumes. See also GLYCOL ETHERS.

PNK750 CAS:623-84-7 ***HR: 2***
PROPYLENE GLYCOL DIACETATE
mf: $C_7H_{13}O_4$ mw: 160.19

PROP: Colorless liquid. Mp: −31°, bp: 190.2°. Sol in water.

SYN: α-PROPYLENE GLYCOL DIACETATE

TOXICITY DATA with REFERENCE
skn-hmn 107 mg/3D-I MLD 85DKA8 -,127,77
eye-rbt 530 mg AJOPAA 29,1363,46
orl-rat LD50:13530 mg/kg JIDHAN 23,259,41
orl-gpg LD50:3420 mg/kg JIHTAB 23,259,41

CONSENSUS REPORTS: Reported in EPA TSCA Inventory.

SAFETY PROFILE: Moderately toxic by ingestion. Experimental teratogenic effects. An eye and human skin irritant. When heated to decomposition it emits acrid smoke and irritating fumes.

PNL000 CAS:6423-43-4 ***HR: 3***
PROPYLENE GLYCOL DINITRATE
mf: $C_3H_6N_2O_6$ mw: 166.11

SYNS: PGDN ◇ 1,2-PROPYLENE GLYCOL DINITRATE ◇ PROPYLENEGLYCOL-1,2-DINITRATE

TOXICITY DATA with REFERENCE
eye-rbt 100 mg MLD TXAPA9 22,128,72
ihl-hmn TCLo:200 ppb/6H:CNS TXAPA9 30,377,74
ihl-hmn TCLo:1500 ppb/5M:EYE TXAPA9 30,377,74
orl-rat LD50:250 mg/kg AIHAAP 34,526,73
scu-rat LD50:463 mg/kg TXAPA9 15,175,69
ipr-rat LD50:479 mg/kg AIHAAP 34,526,73
scu-mus LD50:1208 mg/kg TXAPA9 15,175,69
ipr-mus LD50:1047 mg/kg AIHAAP 34,526,73
ivn-mky LDLo:410 mg/kg TXAPA9 22,128,72
scu-cat LDLo:200 mg/kg TXAPA9 15,175,69
ipr-gpg LD50:402 mg/kg AIHAAP 34,526,73

CONSENSUS REPORTS: Reported in EPA TSCA Inventory.

OSHA PEL: TWA 0.05 ppm
ACGIH TLV: TWA 0.05 ppm (skin)
DFG MAK: 0.05 ppm (0.3 mg/m^3)

SAFETY PROFILE: Poison by ingestion and subcutaneous routes. Moderately toxic by intraperitoneal and intravenous routes. Human systemic effects by inhalation: conjunctiva irritation, headache. An eye irritant. When heated to decomposition it emits toxic fumes of NO_x. See also NITRATES.

PNL250 CAS:107-98-2 ***HR: 2***
PROPYLENE GLYCOL MONOMETHYL ETHER
mf: $C_4H_{10}O_2$ mw: 90.14

PROP: Colorless liquid. Mp: −96.7°, bp: 120°, flash p: 100°F, d: 0.919 @ 25°/25°.

SYNS: DOWANOL 33B ◇ DOWTHERM 209 ◇ METHOXY ETHER of PROPYLENE GLYCOL ◇ 1-METHOXY-2-PROPANOL ◇ POLY-SOLVE MPM ◇ PROPYLENE GLYCOL METHYL ETHER ◇ α-PROPYLENE GLYCOL MONOMETHYL ETHER ◇ PROPYLENGLYKOL-MONOMETHYLAETHER (GERMAN)

TOXICITY DATA with REFERENCE
skn-rbt 500 mg open MLD UCDS** 11/15/71
eye-rbt 230 mg MLD AMIHBC 9,509,54
ihl-rat TCLo:3000 ppm/6H (6-15D preg):TER FAATDF 4,784,84
ihl-hmn TCLo:3000 ppm:NOSE,CNS,GIT NPIRI* 1,105,74
orl-rat LD50:5660 mg/kg AIHAAP 23,95,62
ihl-rat LCLo:7000 ppm/4H AMIHBC 9,509,54
ipr-rat LD50:3720 mg/kg 38MKAJ 2C,3977,82
scu-rat LD50:7800 mg/kg ARZNAD 22,569,72
ivn-rat LD50:4200 mg/kg ARZNAD 22,569,72
orl-mus LD50:11700 mg/kg ARZNAD 22,569,72
ivn-mus LD50:5300 mg/kg ARZNAD 22,569,72

CONSENSUS REPORTS: Glycol ether compounds are on the Community Right-To-Know List. Reported in EPA TSCA Inventory.

OSHA PEL: TWA 100 ppm; STEL: 150 ppm
ACGIH TLV: TWA 100 ppm; STEL: 150 ppm
DFG MAK: 100 ppm (375 mg/m^3)

SAFETY PROFILE: Moderately toxic by intravenous route. Mildly toxic by ingestion, inhalation, and skin contact. Human systemic effects by inhalation: general anesthesia, nausea. A skin and eye irritant. An experimental teratogen. Many glycol ethers have dangerous human reproductive effects. Very dangerous fire hazard when exposed to heat or flame; can react with oxidizing materials. To fight fire, use foam, CO_2, dry chemical. When heated to decomposition it emits acrid smoke and irritating fumes. Used as a solvent and in solvent-sealing of cellophane. See also GLYCOL ETHERS and ETHYLENE GLYCOL MONOMETHYL ETHER.

PNL400 CAS:75-55-8 ***HR: 3***
PROPYLENE IMINE
DOT: UN 1921
mf: C_3H_7N mw: 57.11

$CH_3CHNHCH_2$ (ring: CH–NH–CH$_2$)

PROP: Liquid. Vap d: 2.0, flash p: 14°F.

SYNS: 2-METHYLAZACYCLOPROPANE ◇ 2-METHYLAZIRIDINE ◇ METHYLETHYLENIMINE ◇ 2-METHYLETHYLENIMINE ◇ 1,2-PROPYLENEIMINE ◇ PROPYLENE IMINE, INHIBITED (DOT) ◇ RCRA WASTE NUMBER P067

TOXICITY DATA with REFERENCE
eye-rbt 250 μg open SEV JIHTAB 30,63,48
mmo-sat 3300 ng/plate ENMUDM 6(Suppl 2),1,84
mmo-esc 1 μg/plate ENMUDM 6(Suppl 2),1,84
orl-rat TDLo:1120 mg/kg/28W-I:CAR NATUAS 230,460,71
orl-rat TD:3920 mg/kg/27W-C:CAR JJIND8 67,75,81
orl-rat LD50:19 mg/kg JIHTAB 30,63,48
ihl-rat LCLo:500 ppm/4H JIHTAB 31,343,49
ihl-gpg LCLo:500 ppm/1H JIHTAB 30,2,48
skn-gpg LD50:43 mg/kg JIHTAB 30,63,48

CONSENSUS REPORTS: NTP Fifth Annual Report on Carcinogens. IARC Cancer Review: Group 2B IMEMDT 7,56,87; Animal Limited Evidence IMEMDT 9,61,75. EPA Genetic Toxicology Program. Reported in EPA TSCA Inventory. EPA Extremely Hazardous Substances List. Community Right-To-Know List.

OSHA PEL: TWA 2 ppm (skin)
ACGIH TLV: TWA 2 ppm (skin), Suspected Human Carcinogen.
DFG MAK: Animal Carcinogen, Suspected Human Carcinogen.

DOT Classification: Flammable Liquid; Label: Flammable Liquid.

SAFETY PROFILE: Confirmed carcinogen with experimental carcinogenic data. Poison by ingestion and skin contact. Moderately toxic by inhalation. Mutation data reported. Severe eye irritant. Implicated as a brain carcinogen. A very dangerous fire hazard when exposed to heat or flame; can react vigorously with oxidizing materials. Polymerizes explosively on exposure to acids or acid fumes. A storage hazard. When heated to decomposition it emits toxic fumes of NO_x.

PNL500 CAS:62641-66-1 ***HR: D***
PROPYLENENITROSOUREA
mf: $C_4H_7N_3O_2$ mw: 129.14

SYN: 1-NITROSO-TETRAHYDRO-2(1H)-PYRIMIDINONE

TOXICITY DATA with REFERENCE
mmo-sat 3300 nmol/L MUREAV 48,131,77

SAFETY PROFILE: Mutation data reported. Many N-nitroso compounds are carcinogens. When heated to decomposition it emits toxic fumes of NO_x. See also N-NITROSO COMPOUNDS.

PNL600 CAS:75-56-9 ***HR: 3***
PROPYLENE OXIDE
DOT: UN 1280
mf: C_3H_6O mw: 58.09

PROP: Colorless liquid; ethereal odor. Bp: 33.9°, lel: 2.8%, uel: 37%, fp: −104.4°, flash p: −35°F (TOC), d: 0.8304 @ 20°/20°, vap press: 400 mm @ 17.8°, vap d: 2.0. Sol in water, alc, and ether.

SYNS: EPOXYPROPANE ◇ 1,2-EPOXYPROPANE ◇ 2,3-EPOXYPROPANE ◇ METHYL ETHYLENE OXIDE ◇ METHYL OXIRANE ◇ NCI-C50099 ◇ OXYDE de PROPYLENE (FRENCH) ◇ PROPENE OXIDE ◇ PROPYLENE EPOXIDE ◇ 1,2-PROPYLENE OXIDE

TOXICITY DATA with REFERENCE
skn-rbt 415 mg open MOD UCDS** 12/13/63
skn-rbt 50 mg/6M SEV AMIHAB 13,228,56
eye-rbt 5 mg SEV AJOPAA 29,1363,46
mmo-sat 350 μg/plate ABCHA6 47,2461,83
sce-hmn:lym 25000 ppm ENMUDM 7(Suppl 3),48,85
ihl-rat TCLo:500 ppm/7H (7-16D preg):TER NTIS** PB83-258038
ihl-rat TCLo:500 ppm/7H (15D pre/1-16D preg):REP SWEHDO 9,94,83
orl-rat TDLo:10798 mg/kg/2Y-I:CAR BJCAAI 46,924,82
ihl-mus TCLo:400 ppm/6H/2Y-I:CAR NTPTR* NTP-TR-267,85
ihl-rat TCLo:100 ppm/7H/2Y-I:NEO TXAPA9 76,69,84
scu-rat TDLo:1500 mg/kg/46W-I:ETA ANYAA9 68,750,58
orl-rat LD50:380 mg/kg GTPZAB 14(11),55,70
ihl-rat LCLo:4000 ppm/4H AIHAAP 30,470,69
ipr-rat LD50:150 mg/kg 85GMAT -,103,82
orl-mus LD50:440 mg/kg GTPZAB 14(11),55,70
ihl-mus LC50:1740 ppm/4H AMIHAB 13,237,56
ihl-dog LCLo:2005 ppm/4H AMIHAB 13,237,56
skn-rbt LD50:1245 mg/kg AIHAAP 30,470,69
orl-gpg LD50:660 mg/kg GISAAA 46(7),76,81
ihl-gpg LCLo:4000 ppm/4H AMIHAB 13,228,56

CONSENSUS REPORTS: NTP Fifth Annual Report on Carcinogens. IARC Cancer Review: Group 2A IMEMDT 7,328,87; Human Inadequate Evidence IMEMDT 36,227,85; Animal Sufficient Evidence IMEMDT 36,227,85; Animal Limited Evidence IMEMDT 11,191,76. Carcinogenesis Studies (inhalation); Some Evidence: rat NTPTR* NTP-TR-267,85; Clear Evidence: mouse NTPTR* NTP-TR-267,85. Reported in EPA TSCA Inventory. EPA Genetic Toxicology Program. Community Right-To-Know List. EPA Extremely Hazardous Substances List.

OSHA PEL: (Transitional: TWA 100 ppm) TWA 20 ppm
ACGIH TLV: TWA 20 ppm
DFG MAK: Animal Carcinogen, Suspected Human Carcinogen.
DOT Classification: Flammable Liquid; Label: Flammable Liquid.

SAFETY PROFILE: Confirmed carcinogen with experimental carcinogenic, neoplastigenic, and tumorigenic data. Poison by intraperitoneal route. Moderately toxic by ingestion, inhalation, and skin contact. An experimental teratogen. Experimental reproductive effects. Human mutation data reported. A severe skin and eye irritant. Flammable liquid. A very dangerous fire and explosion hazard when exposed to heat or flame. Explosive reaction with epoxy resin and sodium hydroxide. Forms explosive mixtures with oxygen. Reacts with ethylene oxide + polyhydric alcohol to form the thermally unstable polyether alcohol. Incompatible with NH_4OH; chlorosulfonic acid; HCl; HF; HNO_3; oleum; H_2SO_4. Dangerous; can react vigorously with oxidizing materials. Keep away from heat and open flame. To fight fire, use alcohol foam, CO_2, dry chemical. When heated to decomposition it emits acrid smoke and fumes.

PNL750 CAS:1072-43-1 ***HR: 3***
PROPYLENE SULFIDE
mf: C_3H_6S mw: 74.15

SYN: 2-METHYLTHIIRANE

TOXICITY DATA with REFERENCE
orl-rat LD50:254 mg/kg AIHAAP 25,560,64
ihl-rat LC50:660 ppm/6H BECTA6 6,509,71
ipr-rat LD50:44 mg/kg AIHAAP 25,560,64

CONSENSUS REPORTS: Reported in EPA TSCA Inventory.

SAFETY PROFILE: Poison by ingestion and intraperitoneal routes. Moderately toxic by inhalation. When heated to decomposition it emits toxic fumes of SO_x.

PNL800 CAS:67658-46-2 ***HR: 3***
(8-β)-6-PROPYLERGOLINE-8-ACETAMIDE TARTRATE (2:1)
mf: $C_{38}H_{50}N_6O_2 \cdot C_4H_6O_6$ mw: 773.04

SYN: ERGOLINE-8-ACETAMIDE, 6-PROPYL-, (8-β)-, (R-(R*,R*))-2,3-DIHYDROXYBUTANEDIOATE(2:1)

TOXICITY DATA with REFERENCE
orl-rat TDLo:5 μg/kg (female 5D post):REP CCCCAK 44,3385,79
orl-rat LD50:9100 μg/kg CCCCAK 44,3385,79

SAFETY PROFILE: Poison by ingestion. Experimental reproductive effects. When heated to decomposition it emits toxic fumes of NO_x.

PNM000 CAS:111-43-3 ***HR: 3***
PROPYL ETHER
DOT: UN 2384
mf: $C_6H_{14}O$ mw: 102.20

PROP: Colorless liquid. Mp: −122°, bp: 90°, d: 0.736 @ 20°/4°, flash p: 70°F, autoign temp: 419°F. Sltly sol in water; sol in alc, ether; very volatile.

SYNS: DIPROPYL ETHER ◇ DI-n-PROPYLETHER ◇ DIPROPYL OXIDE ◇ 1,1'-OXYBISPROPANE

TOXICITY DATA with REFERENCE
ivn-mus LD50:204 mg/kg JPMSAE 67,566,78

CONSENSUS REPORTS: Reported in EPA TSCA Inventory.

DOT Classification: Flammable Liquid; Label: Flammable Liquid.

SAFETY PROFILE: Poison by intravenous route. Possibly narcotic. Dangerous fire hazard when exposed to heat, flame, or oxidizers. Forms explosive peroxides. Dangerous upon exposure to heat or flame; can react vigorously with oxidizing materials. When heated to decomposition it emits acrid smoke and irritating fumes. See also ETHERS.

PNM500 CAS:110-74-7 ***HR: 3***
n-PROPYL FORMATE
DOT: UN 1281
mf: $C_4H_8O_2$ mw: 88.12

PROP: Colorless liquid, pleasant odor. Mp: −93°, bp: 82°, flash p: 27°F (CC), d: 0.901 @ 20°, vap press: 100 mm @ 29.5°, vap d: 3.03, autoign temp: 851°F, lel: 2.3%; misc alc, ether; sltly sol in water.

SYNS: FORMIATE de PROPYLE (FRENCH) ◇ PROPYL FORMATE (DOT) ◇ PROPYL METHANOATE

TOXICITY DATA with REFERENCE
orl-rat LD50:3980 mg/kg FCTXAV 2,327,64
orl-mus LD50:3400 mg/kg FCTXAV 2,327,64

CONSENSUS REPORTS: Reported in EPA TSCA Inventory.

DOT Classification: Flammable Liquid; Label: Flammable Liquid.

SAFETY PROFILE: Moderately toxic by ingestion. An irritant to skin, eyes, and mucous membranes. Narcotic in high concentration. Dangerous fire hazard when exposed to heat, flame, or oxidizers. Ignites on contact with potassium-tert-butoxide. Explosive in the form of vapor when exposed to heat or flame. To fight fire, use alcohol foam. When heated to decomposition it emits acrid smoke and irritating fumes. See also ESTERS.

PNM650 CAS:77337-54-3 ***HR: 2***
N-n-PROPYL-N-FORMYLHYDRAZINE
mf: $C_4H_{10}N_2O$ mw: 102.16

SYNS: FORMIC ACID, 1-PROPYLHYDRAZIDE ◇ PFH

TOXICITY DATA with REFERENCE
orl-mus TDLo:34944 mg/kg/60W-C:CAR BJCAAI 42,922,80
orl-mus TD:34944 mg/kg/60W-C:CAR BJCAAI 42,922,80

SAFETY PROFILE: Questionable carcinogen with experimental carcinogenic data. When heated to decomposition it emits toxic fumes of NO_x.

PNM750 CAS:121-79-9 ***HR: 3***
n-PROPYL GALLATE
mf: $C_{10}H_{12}O_5$ mw: 212.22

PROP: Odorless, fine, ivory powder or crystals; sltly bitter taste. Mp: 147-149°. Sltly sol in water; sol in alc and ether.

SYNS: GALLIC ACID, PROPYL ESTER ◇ NIPA 49 ◇ NIPAGALLIN P ◇ PROGALLIN P ◇ n-PROPYL ESTER of 3,4,5-TRIHYDROXYBENZOIC ACID ◇ PROPYL GALLATE ◇ n-PROPYL-3,4,5-TRIHYDROXYBENZOATE ◇ TENOX PG ◇ 3,4,5-TRIHYDROXYBENZENE-1-PROPYLCARBOXYLATE ◇ 3,4,5-TRIHYDROXYBENZOIC ACID, n-PROPYL ESTER

TOXICITY DATA with REFERENCE
mmo-sat 200 μg/plate SYSWAE 12,41,79
cyt-ham:fbr 40 mg/L ESKHA5 96,55,78
orl-rat TDLo:19 g/kg (1-22D preg):REP SKEZAP 20,378,79
orl-rat TDLo:45 g/kg (1-22D preg):TER SKEZAP 20,378,79
orl-mus TDLo:168 g/kg/2Y-C:ETA NKEZA4 29,25,82
orl-rat LD50:2100 mg/kg NTIS** PB245-441
ipr-rat LD50:380 mg/kg FOTEAO 2,308,48
orl-mus LD50:1700 mg/kg JAOCA7 54,239,77
orl-cat LD50:400 mg/kg 14CYAT 2,1897,63

CONSENSUS REPORTS: NTP Carcinogenesis Bioassay (feed); No Evidence: mouse, rat NTPTR* NTP-TR-240,82. Reported in EPA TSCA Inventory.

SAFETY PROFILE: Poison by ingestion and intraperitoneal routes. Experimental teratogenic and reproductive effects. Questionable carcinogen with experimental tumorigenic data. Mutation data reported. Combustible when exposed to heat or flame; can react with oxidizing materials. When heated to decomposition it emits acrid smoke and irritating fumes.

PNN250 CAS:10042-59-8 ***HR: 1***
2-PROPYLHEPTANOL
mf: $C_{10}H_{22}O$ mw: 158.32

TOXICITY DATA with REFERENCE
skn-rbt 10 mg/24H open MLD AIHAAP 23,95,62
orl-rat LD50:6730 mg/kg AIHAAP 23,95,62

CONSENSUS REPORTS: Reported in EPA TSCA Inventory.

SAFETY PROFILE: Mildly toxic by ingestion. A skin irritant. When heated to decomposition it emits acrid smoke and irritating fumes. See also ALCOHOLS.

PNN300 CAS:1007-33-6 ***HR: 3***
PROPYLHEXADRINE HYDROCHLORIDE
mf: $C_{10}H_{21}N{\bullet}ClH$ mw: 191.78

SYNS: BENZEDREX (SKF) ◇ CHP ◇ CYCLOHEXYLISOPROPYLMETHYLAMINE HYDROCHLORIDE ◇ 1-CYCLOHEXYL-2-METHYLAMINOPROPANEHYDROCHLORIDE ◇ N,α-DIMETHYLCYCLOHEXANEETHYLAMINE HYDROCHLORIDE ◇ E-111 ◇ EGGOBESIN ◇ EVENTIN ◇ EVENTIN HYDROCHLORIDE ◇ OBESIN ◇ PERNSATOR-WIRKSTOFF ◇ PROPYLHEXADRINE

TOXICITY DATA with REFERENCE
ipr-rat LD50:65 mg/kg JPETAB 100,267,50
ipr-mus LD50:85 mg/kg AEPPAE 237,171,59
ipr-rbt LDLo:80 mg/kg JPETAB 100,267,50
ipr-gpg LD50:85 mg/kg JPETAB 100,267,50

SAFETY PROFILE: Poison by intraperitoneal route. When heated to decomposition it emits toxic fumes of NO_x and HCl.

PNN400 CAS:101-40-6 ***HR: 3***
PROPYLHEXEDRINE
mf: $C_{10}H_{21}N$ mw: 155.32

SYNS: 1-CYCLOHEXYL-2-METHYLAMINOPROPAN (GERMAN) ◇ 1-CYCLOHEXYL-N-METHYL-2-PROPANAMINE ◇ (±)N,α-DIMETHYL-CYCLOHEXANEETHANAMINE ◇ α,N-DIMETHYLCYCLOHEXANEETHYLAMINE ◇ N,α-DIMETHYLCYCLOHEXANEETHYLAMINE ◇ DRISTAN INHALER

TOXICITY DATA with REFERENCE
orl-rat LD50:83 mg/kg ARZNAD 13,613,63
orl-mus LD50:190 mg/kg ARZNAD 13,613,63
ivn-mus LD50:32 mg/kg CSLNX* NX#05178

SAFETY PROFILE: Poison by ingestion and intravenous routes. When heated to decomposition it emits toxic fumes of NO_x.

PNO000 CAS:56795-66-5 ***HR: 3***
N-PROPYLHYDRAZINE HYDROCHLORIDE
mf: $C_3H_{10}N_2{\bullet}ClH$ mw: 110.61

TOXICITY DATA with REFERENCE
orl-mus TDLo:20 g/kg/58W-C:NEO EJCAAH 11,473,75
orl-mus TD:50 mg/kg/W-C:ETA PAACA3 16,61,75

SAFETY PROFILE: Questionable carcinogen with experimental neoplastigenic and tumorigenic data. When heated to decomposition it emits very toxic fumes of HCl and NO_x.

PNO250 CAS:35285-69-9 ***HR: 3***
PROPYL-p-HYDROXYBENZOATE, SODIUM SALT
mf: $C_{10}H_{11}O_3{\bullet}Na$ mw: 202.20

SYN: p-HYDROXYBENZOIC ACID, PROPYL ESTER, SODIUM DERIVATIVE

TOXICITY DATA with REFERENCE
orl-mus LD50:3700 mg/kg JAPMA8 45,260,56
ipr-mus LD50:490 mg/kg JAPMA8 45,260,56
ivn-mus LD50:180 mg/kg JAPMA8 45,260,56
ivn-dog LDLo:85 mg/kg DRSTAT 20,89,52

SAFETY PROFILE: Poison by intravenous route. Moderately toxic by ingestion and intraperitoneal routes. When heated to decomposition it emits toxic fumes of Na_2O.

PNO500 CAS:17369-59-4 ***HR: 2***
PROPYLIDENE PHTHALIDE
mf: $C_{11}H_{10}O_2$ mw: 174.21

SYN: 3-PROPYLIDENE-1(3H)-ISOBENZOFURANONE

TOXICITY DATA with REFERENCE
skn-rbt 500 mg/24H MOD FCTXAV 16,865,78
orl-rat LD50:1650 mg/kg FCTXAV 16,865,78

CONSENSUS REPORTS: Reported in EPA TSCA Inventory.

SAFETY PROFILE: Moderately toxic by ingestion. A skin irritant. When heated to decomposition it emits acrid smoke and irritating fumes.

PNO750 CAS:107-08-4 ***HR: 3***
n-PROPYL IODIDE
mf: C_3H_7I mw: 170.00

PROP: Colorless liquid. D: 1.743 @ 20°/4°, mp: about −98.7°, bp: 102.5°. Sltly sol in water; misc in alc, ether.

SYN: 1-IODOPROPANE

TOXICITY DATA with REFERENCE
ipr-mus TDLo:3000 mg/kg/8W-I:NEO CNREA8 35,1411,75
ihl-rat LC50:73000 mg/m^3/30M FAVUAI 7,35,75
ipr-rat LD50:650 mg/kg 85GMAT -,103,82
ipr-mus LD50:297 mg/kg 85GMAT -,103,82
ipr-gpg LD50:595 mg/kg 85GMAT -,103,82

CONSENSUS REPORTS: Reported in EPA TSCA Inventory.

SAFETY PROFILE: Poison by intraperitoneal route. Questionable carcinogen with experimental neoplastigenic data. Very mildly toxic by inhalation. When heated to decomposition it emits toxic fumes of I^-. See also IODIDES.

PNP000 CAS:110-78-1 ***HR: 3***
PROPYL ISOCYANATE
DOT: UN 2482
mf: C_4H_7NO mw: 85.12

SYNS: 1-ISOCYANATOPROPANE ◇ ISOCYANIC ACID, PROPYL ESTER ◇ m-PROPYL ISOCYANATE ◇ 1-PROPYL ISOCYANATE

TOXICITY DATA with REFERENCE
ivn-mus LD50:56 mg/kg CSLNX* NX#02916

CONSENSUS REPORTS: Reported in EPA TSCA Inventory.

DOT Classification: Flammable Liquid; Label: Flammable Liquid and Poison.

SAFETY PROFILE: Poison by intravenous route. Flammable when exposed to heat or flame, can react vigorously with oxidizing materials. When heated to decomposition it emits toxic fumes of NO_x. See also ISOCYANATES.

PNP250 CAS:83-59-0 ***HR: 3***
n-PROPYL ISOMER
mf: $C_{20}H_{26}O_6$ mw: 362.46

SYNS: DI-n-PROPYL MALEATE-ISOSAFROLE CONDENSATE ◇ DI-n-PROPYL6,7-METHYLENEDIOXY-3-METHYL-1,2,3,4-TETRAHYDRONAPHTHALENE ◇ DI-n-PROPYL-3-METHYL-6,7-METHYLENEDIOXY-1,2,3,4-TETRAHYDRONAPHTHALENE-1,2-DICARBOXYLATE ◇ DIPROPYL-5,6,7,8-TETRAHYDRO-7-METHYLNAPHTHO(2,3-d)-1,3-DIOXOLE-5,6-DICARBOXYLATE ◇ ENT 15,266 ◇ PROPYL ISOMER

TOXICITY DATA with REFERENCE
orl-mus TDLo:655 g/kg/78W-I:ETA NTIS** PB223-159
scu-mus TDLo:1000 mg/kg:CAR NTIS** PB223-159
orl-rat LD50:1500 mg/kg GUCHAZ 6,435,73
skn-rbt LD50:375 mg/kg YKYUA6 30,1365,79

SAFETY PROFILE: Poison by skin contact. Moderately toxic by ingestion. Questionable carcinogen with experimental carcinogenic and tumorigenic data. When heated to decomposition it emits acrid smoke and irritating fumes.

PNP275 CAS:2417-93-8 ***HR: 3***
PROPYL LITHIUM
mf: C_3H_7Li mw: 50.03

SAFETY PROFILE: Ignites spontaneously in air. See also LITHIUM COMPOUNDS.

PNP750 CAS:2210-28-8 ***HR: 2***
n-PROPYL METHACRYLATE
mf: $C_7H_{12}O_2$ mw: 128.19

SYN: PROPYL METHACRYLATE

TOXICITY DATA with REFERENCE
ipr-mus LD50:1121 mg/kg JPMSAE 62,778,73

CONSENSUS REPORTS: Reported in EPA TSCA Inventory.

SAFETY PROFILE: Moderately toxic by intraperitoneal route. When heated to decomposition it emits acrid smoke and irritating fumes. See also ESTERS.

PNQ000 CAS:1912-31-8 ***HR: D***
n-PROPYL METHANESULFONATE
mf: $C_4H_{10}O_3S$ mw: 138.20

SYNS: METHANESULFONIC ACID, PROPYL ESTER ◇ PMS

TOXICITY DATA with REFERENCE
msc-hmn:lym 1 mmol/L CNREA8 38,1595,78
dnd-rat-ipr 400 mg/kg ENMUDM 7,563,85
ipr-mus TDLo:600 mg/kg (1D pre):REP MUREAV 13,171,71

CONSENSUS REPORTS: EPA Genetic Toxicology Program.

SAFETY PROFILE: Experimental reproductive effects. Human mutation data reported. When heated to decomposition it emits toxic fumes of SO_x. See also SULFONATES.

PNQ250 CAS:2532-49-2 *HR: 3*
2-n-PROPYL-4-METHYLPYRIMIDYL-(6)-N,N-DIMETHYL CARBAMATE
mf: $C_{11}H_{17}N_3O_2$ mw: 223.31

SYNS: ENT 19,059 ◇ 6-METHYL-2-PROPYL-4-PYRIMIDINYL DIMETHYL CARBAMATE

TOXICITY DATA with REFERENCE
orl-rat LD50:200 mg/kg ARSIM* 20,17,66
orl-mus LD50:225 mg/kg 28ZEAL 5,192,76
ipr-mus LDLo:96 mg/kg TXAPA9 6,402,64
orl-ckn LD50:60 mg/kg TXAPA9 11,49,67

SAFETY PROFILE: Poison by ingestion and intraperitoneal routes. When heated to decomposition it emits toxic fumes of NO_x. See also CARBAMATES.

PNQ500 CAS:627-13-4 *HR: 3*
n-PROPYL NITRATE
DOT: UN 1865
mf: $C_3H_7NO_3$ mw: 105.11

PROP: Pale yellow liquid; sickly odor. Bp: 110.5°, d: 1.054 @ 20°/4°, flash p: 68°F, autoign temp: 347°F (in air), lel: 2%, uel: 100%. Very sltly sol in water; sol in alc, ether.

SYNS: NITRATE de PROPYLE NORMAL (FRENCH) ◇ NITRIC ACID, PROPYL ESTER ◇ PROPYL NITRATE

TOXICITY DATA with REFERENCE
ivn-dog LDLo:100 mg/kg JPETAB 118,77,56
ivn-cat LDLo:100 mg/kg JPETAB 118,77,56
ivn-rbt LD50:200 mg/kg JPETAB 118,77,56

CONSENSUS REPORTS: Reported in EPA TSCA Inventory.

OSHA PEL: (Transitional: TWA 25 ppm) TWA 25 ppm; STEL 40 ppm
ACGIH TLV: TWA 25 ppm; STEL 40 ppm
DFG MAK: 25 ppm (110 mg/m^3)

DOT Classification: Flammable Liquid; Label: Flammable Liquid.

SAFETY PROFILE: Poison by intravenous route. Inhalation can cause a hypotension and methemoglobinemia. Dangerous fire hazard when exposed to heat, flame, or oxidizers. Explosive in the form of vapor when exposed to heat or flame. A shock-sensitive explosive. It can be desensitized by the addition of 1-2% propane, butane, chloroform, dimethyl ether, or diethyl ether. When heated to decomposition it emits toxic fumes of NO_x. Used as a fuel ignition promoter, chemical intermediate, and in the manufacture of rocket fuels. See also NITRATES and ESTERS.

PNQ750 CAS:543-67-9 *HR: 1*
PROPYL NITRITE
mf: $C_3H_7NO_2$ mw: 89.11

PROP: Liquid. D: 0.8864 @ 20°/4°, bp: 46-48°. Sol in alc, ether.

SYNS: NITROUS ACID, PROPYL ESTER ◇ PROPANOL NITRITE

TOXICITY DATA with REFERENCE
mmo-sat 1 mg/plate BSIBAC 56,816,80

SAFETY PROFILE: Mutation data reported. An irritant by inhalation and ingestion which causes vasodilation, smooth muscle relaxation, hypotension. Nitrites have been implicated in increased incidences of cancer. When heated to decomposition it emits toxic fumes of NO_x. See also NITRITES.

PNR250 CAS:66017-91-2 *HR: 3*
PROPYLNITROSAMINOMETHYL ACETATE
mf: $C_6H_{12}N_2O_3$ mw: 160.20

SYNS: ACETOXYMETHYLPROPYLNITROSAMINE ◇ N-(ACETOXY)METHYL-N-n-PROPYLNITROSAMINE ◇ N-NITROSO-N-(1-ACETOXYMETHYL)PROPYL AMINE ◇ PAMN ◇ PROPYL ACETOXYMETHYLNITROSAMINE ◇ N-PROPYL-N-(ACETOXYMETHYL)NITROSAMINE

TOXICITY DATA with REFERENCE
mmo-esc 1 μmol/plate GANNA2 71,124,80
dns-rat:oth 10 μmol/L CBINA8 53,99,85
orl-rat TDLo:450 mg/kg/90D-I:ETA ZKKOBW 91,317,78
scu-rat TDLo:50 mg/kg/10W-I:CAR JCROD7 104,13,82
scu-rat TD:61 mg/kg/10W-I:CAR IAPUDO 41,619,82
orl-rat LD50:1000 mg/kg ZKKOBW 91,317,78

SAFETY PROFILE: Moderately toxic by ingestion. Questionable carcinogen with experimental carcinogenic and tumorigenic data. Mutation data reported. When heated to decomposition it emits toxic fumes of NO_x. See also N-NITROSO COMPOUNDS.

PNR500 CAS:19935-86-5 *HR: 3*
N-PROPYL-N-NITROSOURETHANE
mf: $C_6H_{12}N_2O_3$ mw: 160.20

SYN: N-NITROSO-N-PROPYLCARBAMIC ACID ETHYL ESTER

TOXICITY DATA with REFERENCE
cyt-ham:fbr 30 mg/L/20H MUREAV 48,337,77
cyt-ham:lng 19 mg/L GMCRDC 27,95,81
orl-rat TDLo:1680 mg/kg/6W-C:CAR BJCAAI 46,423,82
orl-rat TDLo:1370 mg/kg/29W-C:ETA GANNA2 67,549,76

SAFETY PROFILE: Questionable carcinogen with experimental carcinogenic and tumorigenic data. Mutation data reported. When heated to decomposition it emits toxic fumes of NO_x. See also N-NITROSO COMPOUNDS and CARBAMATES.

PNR750 CAS:99-66-1 ***HR: 2***
2-PROPYLPENTANOIC ACID
mf: $C_8H_{16}O_2$ mw: 144.24

PROP: Colorless liquid. D: 0.922 @ 0°/4°, bp: 221-222°. Very sltly sol in water.

SYNS: ABBOTT 44090 ◇ DEPAKENE ◇ DEPAKINE ◇ DIPROPYLACETIC ACID ◇ N-DIPROPYLACETIC ACID ◇ DI-n-PROPYLESSIGSAURE (GERMAN) ◇ n-DPA ◇ EPILIM ◇ 2-PROPYLVALERIC ACID ◇ VALPROATE ◇ VALPROIC ACID

TOXICITY DATA with REFERENCE
unr-wmn TDLo:2394 mg/kg (female 1-36W post):REP JPETAB 219,768,81
unr-wmn TDLo:5320 mg/kg (female 2-39W post):TER JOPDAB 97,332,80 CPEDAM 22,336,83
orl-cld TDLo:10500 mg/kg/30W-I:GIT AJDCAI 138,912,84
orl-man TDLo:21 mg/kg/2D-I NEURAI 37,886,87
unr-chd TDLo:1800 mg/kg/60D:GIT LANCAO 1,1196,80
orl-hmn TDLo:13333 μg/kg/D-I:CNS NEJMAG 301,435,79
orl-rat LD50:670 mg/kg FCTXAV 2,327,64
ipr-rat LD50:704 mg/kg ARZAND 33,1155,83
ivn-rat LD50:509 mg/kg ARZNAD 33,1155,83
orl-mus LD50:1098 mg/kg EPXXDW #78785
ipr-mus LD50:470 mg/kg CHTPBA 3,430,68
scu-mus LD50:860 mg/kg 85GNAW -,38,68
orl-gpg LD50:824 mg/kg 85GNAW -,39,68

SAFETY PROFILE: Moderately toxic by ingestion, subcutaneous, intravenous, and intraperitoneal routes. Human systemic effects: changes in exocrine pancreas, nausea, diarrhea, sleep disturbance, changes in structure or function of exocrine pancreas. Human teratogenic and reproductive effects by ingestion: developmental abnormalities of the central nervous system, eyes, ears, craniofacial features, skin and skin appendages, musculoskeletal system and cardiovascular system; effects on newborn: weaning or lactation index, Apgar score, physical effects. An experimental teratogen. Other experimental reproductive effects. Used as an anticonvulsant for the control of epileptic seizures. When heated to decomposition it emits acrid smoke and irritating fumes.

PNR800 CAS:74512-62-2 ***HR: 2***
N-(4-PROPYLPHENAZOL-5-YL)-2-ACETOXYBENZAMIDE
mf: $C_{23}H_{25}N_3O_4$ mw: 407.51

SYNS: AIA ◇ ASPIRIN-ISOPROPYLANTIPYRINE ◇ 1-PHENYL-2-METHYL-3-(((2-ACETOXYBENZOYL)AMINO)METHYL)PYRAZOLONE ◇ N-3′-a-PROPYPHENAZONYL-2-ACETOXYBENZAMIDE

TOXICITY DATA with REFERENCE
orl-mus LD50:5 g/kg DRFUD4 9,91,84
ipr-mus LD50:3680 mg/kg CPBTAL 28,1237,80
scu-mus LD50:5 g/kg DRFUD4 9,91,84

SAFETY PROFILE: Moderately toxic by intraperitoneal route. Mildly toxic by ingestion. When heated to decomposition it emits toxic fumes of NO_x.

PNS000 CAS:1211-28-5 ***HR: 3***
1-(α-PROPYLPHENETHYL)PYRROLIDINE HYDROCHLORIDE
mf: $C_{15}H_{23}N{\cdot}ClH$ mw: 253.85

SYNS: 1-(1-BENZYLBUTYL)PYRROLIDINEHYDROCHLORIDE ◇ CATOVITAN ◇ KATOVIT HYDROCHLORIDE ◇ 1-(1-(PHENYLMETHYL)BUTYL)PYRROLIDINE HYDROCHLORIDE ◇ PHENYL-2-PYRROLIDINOPENTANE HYDROCHLORIDE ◇ 1-PHENYL-2-N-PYRROLIDINOPENTANE HYDROCHLORIDE ◇ PROLINTANE HYDROCHLORIDE ◇ PROMOTIL ◇ VILESCON ◇ VILLESCON

TOXICITY DATA with REFERENCE
orl-rat LD50:278 mg/kg OYYAA2 9,601,75
ipr-rat LD50:78 mg/kg OYYAA2 9,601,75
scu-rat LD50:142 mg/kg OYYAA2 9,601,75
ivn-rat LD50:40 mg/kg OYYAA2 9,601,75
orl-mus LD50:230 mg/kg OYYAA2 9,601,75
ipr-mus LD50:66 mg/kg OYYAA2 9,601,75
scu-mus LD50:99 mg/kg ARZNAD 7,344,57
ivn-mus LD50:25 mg/kg OYYAA2 9,601,75

SAFETY PROFILE: Poison by ingestion, subcutaneous, intraperitoneal, and intravenous routes. When heated to decomposition it emits very toxic fumes of HCl and NO_x.

PNS250 CAS:644-35-9 ***HR: 3***
o-PROPYLPHENOL
mf: $C_9H_{12}O$ mw: 136.21

PROP: Liquid. D: 1.015 @ 0°, bp: 221-226°. Very sltly sol in water; sol in alc, ether.

TOXICITY DATA with REFERENCE
orl-rat LD50:500 mg/kg ETATAW 24,281,75
orl-mus LD50:356 mg/kg GISAAA 39(7),94,74
orl-gpg LD50:450 mg/kg GISAAA 39(7),94,74

CONSENSUS REPORTS: Reported in EPA TSCA Inventory.

SAFETY PROFILE: Poison by ingestion. When heated to decomposition it emits acrid smoke and irritating fumes. See also p-PROPYLPHENOL.

PNS500 CAS:645-56-7 ***HR: 3***
p-PROPYLPHENOL
mf: $C_9H_{12}O$ mw: 136.21

PROP: Crystals. D: 1.009 @ 0°, mp: 21-22°, bp: 230-232°. Very sltly sol in water; sol in alc.

TOXICITY DATA with REFERENCE
orl-rat LD50:500 mg/kg ETATAW 24,281,75
orl-mus LD50:348 mg/kg GISAAA 39(7),94,74
ipr-mus LD50:81 mg/kg JMCMAR 18,868,75
orl-gpg LD50:678 mg/kg GISAAA 39(7),94,74
orl-mam LD50:540 mg/kg GISAAA 45(10),16,80
skn-mam LD50:2150 mg/kg GISAAA 45(10),16,80

CONSENSUS REPORTS: Reported in EPA TSCA Inventory.

SAFETY PROFILE: Poison by ingestion and intraperitoneal routes. Moderately toxic by skin contact. When heated to decomposition it emits acrid smoke and irritating fumes.

PNS750 CAS:622-85-5 ***HR: 2***
PROPYL PHENYL ETHER
mf: $C_9H_{12}O$ mw: 136.21

PROP: Colorless liquid. D: 0.953 @ 15°/15°, bp: 189-190°.

SYN: PROPOXYPHENYL

TOXICITY DATA with REFERENCE
orl-mus LD50:3400 mg/kg JPETAB 88,400,46

CONSENSUS REPORTS: Reported in EPA TSCA Inventory.

SAFETY PROFILE: Moderately toxic by ingestion. When heated to decomposition it emits acrid smoke and irritating fumes. See also ETHERS.

PNT000 CAS:458-88-8 ***HR: 3***
2-PROPYLPIPERIDINE
mf: $C_8H_{17}N$ mw: 127.26

PROP: Colorless liquid; mousy odor. Mp: -2°, bp: 166-166.5°. Sol in alc, ether, acetone, benzene, and amyl alcohol; sltly sol in chloroform.

SYNS: CICUTIN ◇ CICUTINE ◇ d-CONICINE ◇ CONIIN ◇ CONIINE ◇ (+)-CONIINE ◇ α-CONINE ◇ CONINE ◇ PIPERIDINE, 2-PROPYL-, (S)- ◇ β-PROPYLPIPERIDINE

TOXICITY DATA with REFERENCE
orl-ctl TDLo:69300 μg/kg (55-75D preg):TER COVEAZ 70,19,80
orl-mus LD50:100 mg/kg JPPMAB 15,1,63
scu-mus LD50:80 mg/kg JPPMAB 15,1,63
ivn-mus LD50:19 mg/kg JPPMAB 15,1,63
ivn-dog LDLo:25 mg/kg 85IXA4 -,588,48
scu-rbt LDLo:80 mg/kg JPMRAB 7,128,33

SAFETY PROFILE: Poison by ingestion, subcutaneous, and intravenous routes. Experimental teratogenic effects. When heated to decomposition it emits toxic fumes of NO_x.

PNT500 CAS:78219-52-0 ***HR: 3***
1-PROPYL-4-PIPERIDYL-p-AMINOBENZOATE HYDROCHLORIDE
mf: $C_{15}H_{22}N_2O_2$•ClH mw: 298.85

SYN: p-AMINO-1-PROPYL-4-PIPERIDYL ESTER HYDROCHLORIDE BENZOIC ACID

TOXICITY DATA with REFERENCE
ivn-rat LDLo:20 mg/kg JACSAT 51,922,29
scu-mus LDLo:25 mg/kg JACSAT 51,922,29

SAFETY PROFILE: Poison by intravenous and subcutaneous routes. When heated to decomposition it emits very toxic fumes of HCl and NO_x.

PNT750 CAS:78219-53-1 ***HR: 3***
1-PROPYL-4-PIPERIDYL BENZOATE HYDROCHLORIDE
mf: $C_{15}H_{21}NO_2$•ClH mw: 283.83

SYN: BENZOIC ACID, 1-PROPYL-1-PIPERIDYL ESTER HYDROCHLORIDE

TOXICITY DATA with REFERENCE
ivn-rat LDLo:13 mg/kg JACSAT 51,922,29
scu-mus LDLo:250 mg/kg JACSAT 51,922,29

SAFETY PROFILE: Poison by intravenous and subcutaneous routes. When heated to decomposition it emits very toxic fumes of NO_x and HCl. See also ESTERS.

PNU000 CAS:106-36-5 ***HR: 2***
PROPYL PROPIONATE
mf: $C_6H_{12}O_2$ mw: 116.18

PROP: Clear liquid. Mp: −76°, bp: 122.4°, flash p: 175°F (OC), d: 0.885, vap press: 10 mm @ 19.4°, vap d: 4.0. Sltly sol in water; misc with alc, ether.

SYNS: PROPIONIC ACID, PROPYL ESTER ◇ PROPYL PROPANOATE ◇ n-PROPYL PROPIONATE

TOXICITY DATA with REFERENCE
skn-rbt 500 mg/24H MOD FCTXAV 17,899,79
ihl-mus LC50:24 g/m³/2H 85GMAT -,104,82
orl-rbt LD50:3950 mg/kg IMSUAI 41,31,72

CONSENSUS REPORTS: Reported in EPA TSCA Inventory.

SAFETY PROFILE: Moderately toxic by ingestion. Very mildly toxic by inhalation. A skin irritant. Combustible when exposed to heat or flame; can react with oxidizing materials. To fight fire, use foam, CO_2, dry chemical. When heated to decomposition it emits acrid smoke and irritating fumes. See also ESTERS.

PNV250 CAS:19056-03-2 ***HR: 3***
1-PROPYL-6-((p-(p-((1-PROPYLQUINOLINIUM-6-YL)CARBAMOYL)BENZAMIDO) BENZAMIDO)QUINLINIUM), DI-p-TOLUENESULFONATE
mf: $C_{39}H_{37}N_5O_3 \cdot 2C_7H_7O_3S$ mw: 966.21

TOXICITY DATA with REFERENCE
dnd-mus:lym 370 nmol/L JMCMAR 22,134,79
ipr-mus LD10:95 mg/kg JMCMAR 22,134,79

SAFETY PROFILE: Poison by intraperitoneal route. Mutation data reported. When heated to decomposition it emits very toxic fumes of SO_x and NO_x. See also SULFONATES.

PNV750 CAS:18705-22-1 ***HR: 3***
PROPYL-2-PROPYNYLPHENYLPHOSPHONATE
mf: $C_{12}H_{15}O_3P$ mw: 238.24

SYNS: FMC-16388 ◇ NIA 16,388 ◇ O-n-PROPYL O-(2-PROPYNYL) PHENYLPHOSPHONATE

TOXICITY DATA with REFERENCE
orl-rat LD50:362 mg/kg 28ZEAL 4,332,69
skn-rbt LD50:2650 μg/kg 28ZEAL 4,332,69
orl-ckn LD50:341 mg/kg JTEHD6 10,907,82

SAFETY PROFILE: Poison by ingestion and skin contact. When heated to decomposition it emits toxic fumes of PO_x.

PNV760 CAS:67465-26-3 ***HR: 3***
n-PROPYLSELENINIC ACID
mf: $C_3H_8O_2Se$ mw: 155.07

SYN: SELENINIC ACID, PROPYL-

TOXICITY DATA with REFERENCE
ipr-rat LDLo:20 mg(Se)/kg JPETAB 63,357,38

OSHA PEL: TWA 0.2 mg(Se)/m^3
ACGIH TLV: TWA 0.2 mg(Se)/m^3

SAFETY PROFILE: Poison by intraperitoneal route. When heated to decomposition it emits toxic fumes of Se.

PNV775 CAS:13154-66-0 ***HR: 3***
PROPYL SILANE
mf: $C_3H_{10}Si$ mw: 74.20

SAFETY PROFILE: Ignites spontaneously in air. See also SILANE.

PNV800 CAS:15790-54-2 ***HR: 3***
PROPYL SODIUM
mf: C_3H_7Na mw: 66.08

SAFETY PROFILE: The powder ignites spontaneously in air. When heated to decomposition it emits toxic fumes of Na_2O.

PNW250 CAS:63906-63-8 ***HR: 3***
1-PROPYL THEOBROMINE
mf: $C_{10}H_{14}N_4O_2$ mw: 222.28

SYN: 1-n-PROPYL-3,7-DIMETHYLXANTHINE

TOXICITY DATA with REFERENCE
dni-hmn:oth 4 mmol/L BIOJAU 35,665,81
dns-hmn:oth 1 mmol/L BIOJAU 35,665,81
orl-hmn TDLo:26 mg/kg:CNS,GIT,SKN JPETAB 86,113,46
ivn-mus LD50:126 mg/kg JPETAB 86,113,46

SAFETY PROFILE: Poison by intravenous route. Human systemic effects by ingestion: change in motor activity, sweating, nausea. When heated to decomposition it emits toxic fumes of NO_x.

PNW750 CAS:14222-60-7 ***HR: 2***
2-PROPYL-THIOISONICOTINAMIDE
mf: $C_9H_{12}N_2S$ mw: 180.29

SYNS: EKTEBIN ◇ PETEHA ◇ 2-PROPYLISONICOTINYLTHIOAMIDE ◇ 2-PROPYL-4-PYRIDINECARBOTHIOAMIDE ◇ 2-PROPYL-4-THIOCARBAMOYLPYRIDINE ◇ PROTHIONAMIDE ◇ PROTION ◇ PROTIONAMID ◇ PROTIONAMIDE ◇ PROTIONIZINA ◇ 9778 R.P. ◇ TEBEFORM ◇ 1321 TH ◇ TREVINTIX ◇ TUBEREX

TOXICITY DATA with REFERENCE
orl-rat LD50:1930 mg/kg NIIRDN 6,726,82
ipr-rat LD50:1320 mg/kg MEIEDD 10,1137,83
scu-rat LD50:760 mg/kg NIIRDN 6,726,82
orl-mus LD50:1450 mg/kg NIIRDN 6,726,82
ipr-mus LD50:1 g/kg MEIEDD 10,1137,83
scu-mus LD50:660 mg/kg NIIRDN 6,726,82

SAFETY PROFILE: Moderately toxic by ingestion, intraperitoneal, and subcutaneous routes. When heated to decomposition it emits very toxic fumes of NO_x and SO_x. Used as an antibacterial agent.

PNW800 CAS:6288-93-3 ***HR: 3***
6-(PROPYLTHIO)PURINE
mf: $C_8H_{10}N_4S$ mw: 194.28

SYNS: NSC 11595 ◇ 6-PROPYLMERCAPTOPURINE ◇ 6-PROPYL-MP ◇ QA 71

TOXICITY DATA with REFERENCE
orl-rat TDLo:100 mg/kg (7D preg):REP JRPFA4 4,291,62
ipr-mus LD50:92880 μg/kg NCISP* JAN86
par-mus LD50:120 mg/kg JPMSAE 71,618,82

SAFETY PROFILE: Poison by parenteral and intraperitoneal routes. Experimental reproductive effects. When heated to decomposition it emits toxic fumes of NO_x and SO_x. See also MERCAPTANS.

PNX000 CAS:51-52-5 ***HR: 3***
6-PROPYL-2-THIOURACIL
mf: $C_7H_{10}N_2OS$ mw: 170.25

PROP: White, bitter, crystalline powder. Mp: 219-221°. Insol in ether, chloroform, benzene; very sol in aq solns of ammonia; very sltly sol in water.

SYNS: 2,3-DIHYDRO-6-PROPYL-2-THIOXO-4(1H)-PYRIMIDINONE ◇ 2-MERCAPTO-4-HYDROXY-6-N-PROPYLPYRIMIDINE ◇ 2-MERCAPTO-6-PROPYL-4-PYRIMIDONE ◇ 2-MERCAPTO-6-PROPYLPYRIMID-4-ONE ◇ PROCASIL ◇ PROPACIL ◇ PROPILTHIOURACIL ◇ 6-PROPIL-TIOURACILE (ITALIAN) ◇ PROPYCIL ◇ 6-PROPYL-2-THIO-2,4(1H,3H)PYRIMIDINEDIONE ◇ PROPYLTHIORIST ◇ PROPYLTHIOURACIL ◇ 4-PROPYL-2-THIOURACIL ◇ 6-N-PROPYLTHIOURACIL ◇ 6-N-PROPYL-2-THIOURACIL ◇ PROPYLTHYRACIL ◇ PROPYTHIOURACIL ◇ PROTHIUCIL ◇ PROTHIURONE ◇ PROTHYCIL ◇ PROTHYRAN ◇ PROTIURAL ◇ PTU (THYREOSTATIC) ◇ 2-THIO-4-OXO-6-PROPYL-1,3-PYRIMIDINE ◇ 2-THIO-6-PROPYL-1,3-PYRIMIDIN-4-ONE ◇ 6-THIO-4-PROPYLURACIL ◇ THIURAGYL ◇ THYREOSTAT II ◇ T 72

TOXICITY DATA with REFERENCE
mmo-smc 1 g/L CHINAG (21),847,80
dni-hmn:lym 100 mg/L JCEMAZ 43,1046,76
orl-wmn TDLo:1818 mg/kg (33D pre/1-39W preg):REP JAMAAP 159,848,55
orl-wmn TDLo:1818 mg/kg (33D pre/1-39W preg):TER JAMAAP 159,848,55
orl-mus TDLo:600 g/kg/73W-C:ETA PSEBAA 112,365,63
orl-gpg TDLo:37 g/kg/2Y-C:NEO GROWAH 27,305,63
orl-ham TDLo:653 g/kg/70W-C:CAR CANCAR 21,952,68
orl-wmn LDLo:480 mg/kg/9W-I:CLD SMJOAV 75,1297,82
orl-cld TDLo:900 mg/kg/60D-I:LIV AIMEAS 82,228,75
orl-wmn TDLo:900 mg/kg/21W-I:LIV AIMDAP 140,1184,80
orl-cld TDLo:165 mg/kg/2W-I:LIV DICPBB 19,669,85
orl-wmn LDLo:84 mg/kg/2W-I:LIV AIMDAP 142,838,82
orl-man TDLo:116 mg/kg/11W-I:LIV AIMDAP 135,319,75
orl-rat LD50:1980 mg/kg FSTEAI 7,313,52
ipr-rat LDLo:400 mg/kg JPETAB 97,478,49

CONSENSUS REPORTS: NTP Fifth Annual Report on Carcinogens. IARC Cancer Review: Group 2B IMEMDT 7,329,87; Animal Sufficient Evidence IMEMDT 7,67,74. Reported in EPA TSCA Inventory.

SAFETY PROFILE: Confirmed carcinogen with experimental carcinogenic, neoplastigenic, and tumorigenic data. Poison by intraperitoneal route. Moderately toxic by ingestion. Human systemic effects: agranulocytosis, hepatitis, jaundice. Human teratogenic effects by ingestion: developmental abnormalities of the endocrine system and changes in newborn viability. Human and experimental teratogenic and reproductive effects. Human mutation data reported. When heated to decomposition it emits very toxic fumes of SO_x and NO_x. See also MERCAPTANS.

PNX100 ***HR: D***
PROPYLTHIOURACIL and IODINE
mf: $C_7H_{10}N_2OS{\cdot}I_2$ mw: 424.05

SYNS: IODINE and PROPYLTHIOURACIL ◇ URACIL, 6-PROPYL-2-THIO-, and IODINE

TOXICITY DATA with REFERENCE
mul-wmn TDLo:10260 mg/kg (female 1-39W post):REP JAMAAP 149,1399,52
mul-wmn TDLo:10260 mg/kg (female 1-39W post):TER JAMAAP 149,1399,52

SAFETY PROFILE: A human teratogen. Human reproductive effects. When heated to decomposition it emits toxic fumes of NO_x, SO_x, and I^-.

PNX250 CAS:141-57-1 ***HR: 3***
n-PROPYLTRICHLOROSILANE
DOT: UN 1816
mf: $C_3H_7Cl_3Si$ mw: 177.54

PROP: Vap d: 6.15, flash p: 100°F.

SYNS: PROPYLTRICHLOROSILANE (DOT) ◇ TRICHLOROPROPYLSILANE

CONSENSUS REPORTS: Reported in EPA TSCA Inventory.

DOT Classification: Corrosive Material; Label: Corrosive; Label: Corrosive, Flammable Liquid.

SAFETY PROFILE: A corrosive and irritating material to skin, eyes, and mucous membranes. A dangerous fire hazard when exposed to heat or flame. Will react with water or steam to produce toxic and corrosive fumes; can react with oxidizing materials. To fight fire, use foam, CO_2, dry chemical. When heated to decomposition it emits toxic fumes of Cl^-. See also CHLOROSILANES.

PNX500 CAS:60062-60-4 ***HR: 3***
4-PROPYL-2,6,7-TRIOXA-1-STIBABICYCLO(2.2.2)OCTANE
mf: $C_7H_{13}O_3Sb$ mw: 266.95

SYN: 1,3-PROPANEDIOL, 2-(HYDROXYMETHYL)-2-PROPYL-, CYCLIC ESTER with ANTIMONIC ACID

TOXICITY DATA with REFERENCE
ipr-mus LD50:250 mg/kg TXAPA9 36,261,76

OSHA PEL: TWA 0.5 mg(Sb)/m^3
ACGIH TLV: TWA 0.5 mg(Sb)/m^3
NIOSH REL: (Antimony):10H TWA 0.5 mg(Sb)/m^3

SAFETY PROFILE: Poison by intraperitoneal route.

When heated to decomposition it emits toxic fumes of Sb.

PNX600 CAS:2430-27-5 ***HR: 2***
2-PROPYLVALERAMIDE
mf: $C_8H_{17}NO$ mw: 143.26

PROP: White, odorless, bitter, crystalline powder. Mp: 125-126°. Practically insol in water.

SYNS: DEPAMID ◇ DEPAMIDE ◇ DIPROPYLACETAMIDE ◇ 2-PROPYLPENTAMIDE ◇ α-PROPYLVALERAMIDE ◇ PROPYL-2-VALERAMIDE ◇ VALPROMIDE

TOXICITY DATA with REFERENCE
scu-mus TDLo:600 mg/kg (female 8D post):TER FAATDF 6,669,86
orl-rat LD50:890 mg/kg 85GNAW -,72,68
orl-mus LD50:950 mg/kg 85GNAW -,71,68
ipr-mus LD50:430 mg/kg NEPHBW 24,427,85

SAFETY PROFILE: Moderately toxic by ingestion and intraperitoneal routes. An experimental teratogen. When heated to decomposition it emits toxic fumes of NO_x.

PNX750 CAS:1069-66-5 ***HR: 2***
2-PROPYLVALERIC ACID SODIUM SALT
mf: $C_8H_{15}O_2 \cdot Na$ mw: 166.22

SYNS: CONVULEX ◇ DEPAKENE ◇ DEPAKINE ◇ DIPROPYLACETATE SODIUM ◇ DPA SODIUM ◇ EPILIN ◇ ERGENYL ◇ EUREKENE ◇ KW-066 ◇ LABAZENE ◇ SODIUM BISPROPYLACETATE ◇ SODIUM DIPROPYLACETATE ◇ SODIUM-α,α-DIPROPYLACETATE ◇ SODIUM-2-PROPYLPENTANOATE ◇ SODIUM-2-PROPYLVALERATE ◇ SODIUM VALPROATE ◇ VALPROATE SODIUM ◇ VALPROIC ACID SODIUM SALT

TOXICITY DATA with REFERENCE
orl-wmn TDLo:2700 mg/kg (female 1-39W post):REP LANCAO 2,1518,87
orl-wmn TDLo:2700 mg/kg (female 1-39W post):TER LANCAO 2,1518,87
orl-inf LDLo:250 mg/kg/10D-I:GIT ADCHAK 58,543,83
orl-cld LDLo:750 mg/kg LANCAO 1,221,84
orl-wmn TDLo:500 mg/kg PGMJAO 62,409,86
orl-chd TDLo:1820 mg/kg/12W-I:CNS,GIT,MET LANCAO 2,1110,80
orl-rat LD50:670 mg/kg BJPCBM 74,957P,81
ipr-rat LD50:970 mg/kg PBPSDY 2,58,79
scu-rat LD50:1029 mg/kg KSRNAM 5,41,71
ivn-rat LD50:509 mg/kg ARZNAD 33,1155,83
orl-mus LD50:1020 mg/kg PBPSDY 2,58,79
ipr-mus LD50:470 mg/kg PBPSDY 2,58,79
scu-mus LD50:860 mg/kg PBPSDY 2,58,79
ivn-mus LD50:750 mg/kg ARZNAD 26,299,76
ims-mus LD50:832 mg/kg BCPCA6 22,1701,73

SAFETY PROFILE: Moderately toxic by ingestion, intramuscular, intraperitoneal, intravenous, and subcutaneous routes. Human systemic effects by ingestion: coma, sleep, nausea, dehydration, changes in structure or function of exocrine pancreas, hemorrhage. Human and experimental teratogenic and reproductive effects. When heated to decomposition it emits toxic fumes of Na_2O.

PNY275 CAS:27846-30-6 ***HR: 3***
3-PROPYNETHIOL
mf: C_3H_4S mw: 72.12

SAFETY PROFILE: Polymerizes explosively when heated. The polymer formed by exposure to air explodes when heated. When heated to decomposition it emits toxic fumes of SO_x. See also ACETYLENE COMPOUNDS.

PNY750 ***HR: 3***
2-PROPYN-1-THIOL
mf: C_3H_4S mw: 72.13

SAFETY PROFILE: Can polymerize explosively. When heated to decomposition it emits toxic fumes of SO_x. See also ACETYLENE COMPOUNDS and MERCAPTANS.

POA000 CAS:2450-71-7 ***HR: 3***
2-PROPYNYLAMINE
mf: C_3H_5N mw: 55.09

SYN: 2-PROPYN-1-AMINE

TOXICITY DATA with REFERENCE
orl-rat LD50:780 mg/kg TXAPA9 28,313,74
ihl-rat LCLo:4500 ppm/4H TXAPA9 28,313,74
skn-rbt LD50:77 mg/kg TXAPA9 28,313,74

CONSENSUS REPORTS: Reported in EPA TSCA Inventory.

SAFETY PROFILE: Poison by skin contact. Moderately toxic by ingestion. Mildly toxic inhalation. When heated to decomposition it emits toxic fumes of NO_x. See also ACETYLENE COMPOUNDS and AMINES.

POA100 CAS:30645-13-7 ***HR: 3***
1-PROPYNYL COPPER(I)
mf: C_3H_3Cu mw: 102.60

CONSENSUS REPORTS: Copper and its compounds are on the Community Right-To-Know List.

SAFETY PROFILE: Very explosive. See also COPPER COMPOUNDS and ACETYLENE COMPOUNDS.

POA250 CAS:358-52-1 ***HR: 2***
1-(2-PROPYNYL)CYCLOHEXYL CARBAMATE
mf: $C_{10}H_{15}NO_2$ mw: 181.26

SYNS: CARBAMATE du PROPINYLCYCLOHEXANOL (FRENCH) ◇ 1-CARBAMOYLOXY-1-(2-PROPYNYL)CYCLOHEXANE ◇ ESAPROPIMATO ◇ HEXAPROPYMATE ◇ HEXAPROPYNATE ◇ L-2103 ◇ LF 62 ◇ LUNAMIN ◇ MERINAX ◇ MODIRAX ◇ PROPYNYLCYCLOHEXANOL CARBAMATE ◇ 1-(2-PROPYNYL)CYCLOHEXANOL CARBAMATE

TOXICITY DATA with REFERENCE
orl-hmn TDLo:228 mg/kg:CNS BMJOAE 1,1593,78
orl-mus LD50:900 mg/kg 27ZQAG -,414,72
ipr-mus LD50:450 mg/kg 27ZQAG -,414,72

SAFETY PROFILE: Moderately toxic by ingestion and intraperitoneal routes. Human systemic effects by ingestion: general anesthesia, blood pressure lowering. When heated to decomposition it emits toxic fumes of NO_x. See also CARBAMATES.

POA500 CAS:6921-27-3 ***HR: 3***
2-PROPYNYL ETHER
mf: C_6H_6O mw: 94.12

$(HC{\equiv}CCH_2)_2O$

SYNS: DIPROPARGYL ETHER ◇ DI(2-PROPYNYL) ETHER ◇ PROPARGYL ETHER

TOXICITY DATA with REFERENCE
orl-rat LD50:539 mg/kg FEPRA7 19,389,60
orl-dog LD50:913 mg/kg FEPRA7 19,389,60
orl-rbt LD50:168 mg/kg FEPRA7 19,389,60

SAFETY PROFILE: Poison by ingestion. Forms thermally unstable explosive peroxides in air. When heated to decomposition it emits acrid smoke and irritating fumes. See also ACETYLENE COMPOUNDS and ETHERS.

POB000 CAS:37882-31-8 ***HR: 2***
2-PROPYNYL(2E,4E)-3,7,11-TRIMETHYL-2,4-DODECADIENOATE
mf: $C_{18}H_{28}O_2$ mw: 276.46

SYNS: ALTODEL ◇ ENSTAR ◇ ENT 70,531 ◇ KINOPRENE ◇ PROP-2-YNYL-3,7,11-TRIMETHYL-2,4-DODECADIENOATE ◇ 3,7,11-TRIMETHYL-2,4-DODECADIENOIC ACID 2-PROPYNYL ESTER ◇ ZR-777

TOXICITY DATA with REFERENCE
orl-rat LD50:4900 mg/kg FMCHA2 -,C136,83
orl-dog LD50:3160 mg/kg SPEADM 78-19-,78
skn-rbt LD50:9000 mg/kg FMCHA2 -,C136,83

SAFETY PROFILE: Moderately toxic by ingestion. Mildly toxic by skin contact. When heated to decomposition it emits acrid smoke and irritating fumes. Used for control of scales, mealybugs, whiteflies. See also ACETYLENE COMPOUNDS.

POB250 CAS:21916-66-5 ***HR: 3***
2-PROPYNYL VINYL SULFIDE
mf: C_5H_6S mw: 98.17

$HC{\equiv}CCH_2SCH{=}CH_2$

SAFETY PROFILE: Explodes when heated above 85°C. When heated to decomposition it emits toxic fumes of SO_x. See also ACETYLENE COMPOUNDS and SULFIDES.

POB300 CAS:22760-18-5 ***HR: 3***
PROQUAZONE
mf: $C_{18}H_{18}N_2O_2$ mw: 294.38

PROP: Yellow crystals from ethyl acetate. Mp: 137-138°. Sol in chloroform; insol in water.

SYNS: BIARISON ◇ BIARSAN ◇ ISOPROPYL-7-METHYL-4-PHENYL-2(1H)-QUINAZOLINONE ◇ RU 43715 ◇ SaH 43-715 ◇ SANDOZ 43-715

TOXICITY DATA with REFERENCE
ipr-rat LD50:375 mg/kg DCTODJ 3,361,80
orl-mus LD50:240 mg/kg DCTODJ 3,361,80
ipr-mus LD50:339 mg/kg ARZNAD 34,879,84
orl-rbt LD50:260 mg/kg DCTODJ 3,361,80

SAFETY PROFILE: Poison by ingestion and intraperitoneal routes. When heated to decomposition it emits toxic fumes of NO_x.

POB500 CAS:466-06-8 ***HR: 3***
PROSCILLARIDIN
mf: $C_{30}H_{42}O_8$ mw: 530.72

SYNS: A-32686 ◇ CARADRIN ◇ CARDION ◇ CARDIOVITE ◇ CARMAZON ◇ CORATOL ◇ (3-β)-2-((6-DEOXY-α-l-MANNOPYRANOSYL)OXY)-14-HYDROXYBUFA-4,20,22-TRIENOLIDE ◇ 3-β-((6-DEOXY-α-l-MANNOSYL)OXY)-14-HYDROXYBUFA-4,20,22-TRIENOLIDE ◇ DESGLUCO-TRANSVAALIN ◇ 3-β,14-β-DIHYDROXYBUFA-4,20,22-TRIENOLIDE 3-RHAMNOSIDE ◇ HERZO ◇ HERZO PROSCILLAN ◇ 14-HYDROXY-3-β-(RHAMNOSYLOXY)BUFA-4,20,22-TRIENOLIDE ◇ PROCARDIN ◇ PROCILAN ◇ PROSCILLAN ◇ PROSTOSIN ◇ PROSZIN ◇ PROTASIN ◇ PSC-801 ◇ PUROSIN-TC ◇ 3-β-RHAMNOSIDO-14-β-HYDROXY-$\Delta^{4,20,22}$-BUFATRIENOLIDE ◇ SANDOSCILL ◇ SCILLACRIST ◇ SCILLA "DIDIER" ◇ α-l-SCILLARENIN-3-RHAMNOPYRANOSIDE ◇ SCILLARENIN-RHAMNOSE (GERMAN) ◇ SIMEON ◇ SOLESTRIL ◇ STELLARID ◇ TALUCARD ◇ TALUSIN ◇ TRADENAL ◇ URGILAN ◇ WIRNESIN

TOXICITY DATA with REFERENCE
orl-rat LD50:56 mg/kg TXAPA9 20,599,71
ivn-rat LD50:9 mg/kg TXAPA9 20,599,71
orl-mus LD50:30500 μg/kg NIIRDN 6,723,82
ivn-mus LD50:23200 μg/kg NIIRDN 6,723,82
ivn-cat LD50:111 μg/kg ARZNAD 20,1,70
idu-cat LDLo:268 μg/kg ARZNAD 14,716,64
orl-gpg LD50:6810 μg/kg USXXAM #4230702
ivn-gpg LD50:280 μg/kg ARZNAD 20,1,70

SAFETY PROFILE: Poison by ingestion, intravenous

and intraduodenal routes. When heated to decomposition it emits acrid smoke and irritating fumes.

POC000 CAS:23476-83-7 ***HR: 2***
PROSPIDIN
mf: $C_{18}H_{36}Cl_2N_4O_2$ mw: 411.48

SYNS: 3,12-BIS(3-CHLORO-2-HYDROXYPROPYL)-3,12-DIAZA-6,9-DIAZONIADISPIRO(5.2.5.2)HEXADECANE, DICHLORIDE ◇ NSC 166100 ◇ PROSPIDINE ◇ PROSPIDIUM CHLORIDE

TOXICITY DATA with REFERENCE
mma-sat 1 g/plate PCJOAU 12,35,78
oms-bcs 500 μg/L CHTHBK 26,309,80
ipr-rat LD50:1200 mg/kg PCJOAU 16,880,82
ipr-mus LD50:994 mg/kg NCISP* JAN86

SAFETY PROFILE: Moderately toxic by intraperitoneal route. Mutation data reported. When heated to decomposition it emits very toxic fumes of Cl^- and NO_x.

POC250 CAS:14152-28-4 ***HR: D***
PROSTAGLANDIN A1
mf: $C_{20}H_{32}O_4$ mw: 336.52

SYNS: 2-(3-HYDROXY-1-OCTENYL)-5-OXO-3-CYCLOPENTENE-1-HEPTANOIC ACID ◇ (13E,15S)-15-HYDROXY-9-OXO-PROSTA-10,13-DIEN-1-OIC ACID (9CI) ◇ PGA^1 ◇ PROSTAGLANDIN A^1 ◇ PROSTAGLANDIN E^1-217

TOXICITY DATA with REFERENCE
oms-hmn:oth 2500 μg/L CNREA8 43,513,83
scu-mus TDLo:10 mg/kg (female 16D post):TER PRGLBA 1,191,72
scu-mus TDLo:40 mg/kg (female 16D post):REP PRGLBA 1,191,72

SAFETY PROFILE: Experimental teratogenic and reproductive effects. Human mutation data reported. When heated to decomposition it emits acrid smoke and irritating fumes. See also other prostaglandins.

POC275 CAS:41598-07-6 ***HR: D***
PROSTAGLANDIN D2
mf: $C_{20}H_{32}O_5$ mw: 352.52

SYNS: 11-DEHYDROPROSTAGLANDIN F2-α ◇ 9,15-DIHYDROXY-11-OXO-PROSTA-5,13-DIEN-1-OIC ACID, (5Z,9-α,13E,15S)- ◇ PGD2

TOXICITY DATA with REFERENCE
dni-mus:oth 5 mg/L/24H CNREA8 46,1688,86
oms-mus:oth 5 mg/L/24H CNREA8 46,1688,86
scu-ham TDLo:400 μg/kg (4D preg):REP JMCMAR 26,790,83

SAFETY PROFILE: Experimental reproductive effects. Mutation data reported. When heated to decomposition it emits acrid smoke and irritating fumes.

POC350 CAS:745-65-3 ***HR: 3***
PROSTAGLANDIN E1
mf: $C_{20}H_{34}O_5$ mw: 354.54

PROP: A primary prostaglandin; easily crystallized from purified biological extracts. Crystals from ethyl acetate + heptane. Mp: 115-116°.

SYNS: ALPROSTADIL ◇ (11-α,13E,15S)-11,15-DIHYDROXY-9-OXO-PROST-13-EN-1-OIC ACID (9CI) ◇ 3-HYDROXY-2-(3-HYDROXY-1-OCTENYL)-5-OXOCYCLOPENTANEHEPTANOIC ACID ◇ 1-3-HYDROXY-2-(3-HYDROXY-1-OCTENYL)-5-OXOCYCLOPENTANEHEPTANOIC ACID ◇ MINPROG ◇ PGE-1 ◇ PROSTANDIN ◇ PROSTIN VR ◇ (−)-PROTAGLANDIN E1

TOXICITY DATA with REFERENCE
oms-rat-scu 200 μg/kg JDGRAX 13(1-2),1,81
dni-mus:oth 1 μmol/L CNREA8 43,3514,83
oms-mus:oth 100 nmol/L JIDEAE 66,313,76
cyt-mus:oth 19 mg/L JRPFA4 47,1,76
ivn-wmn TDLo:64 μg/kg (20W preg):REP BMJOAE 2,258,70
scu-mus TDLo:5 mg/kg (female 16D post):TER PRGLBA 1,191,72
orl-rat LD50:228 mg/kg NIIRDN 6,47,82
ipr-rat LD50:24900 μg/kg NIIRDN 6,47,82
scu-rat LD50:18600 μg/kg NIIRDN 6,47,82
ivn-rat LD50:19200 μg/kg NIIRDN 6,47,82
orl-mus LD50:186 mg/kg NIIRDN 6,47,82
ipr-mus LD50:19800 μg/kg NIIRDN 6,47,82
scu-mus LD50:26400 μg/kg NIIRDN 6,47,82
ivn-mus LD50:21 mg/kg NIIRDN 6,47,82

SAFETY PROFILE: Poison by ingestion, subcutaneous, intravenous, and intraperitoneal routes. Human reproductive effects by intravenous route: terminates pregnancy; changes in the uterus, cervix, vagina. Experimental teratogenic and reproductive effects. Mutation data reported. When heated to decomposition it emits acrid smoke and irritating fumes. See also other prostaglandin entries.

POC360 CAS:53697-17-9 ***HR: D***
PROSTAGLANDIN E2 SODIUM SALT
mf: $C_{20}H_{32}O_5$•Na mw: 375.51

SYNS: (5Z,11-α,13E,15S)-11,15-DIHYDROXY-9-OXOPROSTA-5,13-DIEN-1-OIC ACID MONOSODIUM SALT ◇ PGE2 SODIUM SALT ◇ U 12062A

TOXICITY DATA with REFERENCE
orl-rat TDLo:2 mg/kg (7-12D preg):REP IJEBA6 14,322,76
ivg-ham TDLo:2 mg/kg (female 8D post):TER CCPTAY 11,479,75

SAFETY PROFILE: An experimental teratogen. Experimental reproductive effects. When heated to decomposition it emits toxic fumes of Na_2O. See also other prostaglandin entries.

POC400 CAS:745-62-0 ***HR: D***
PROSTAGLANDIN F1 $-\alpha$
mf: $C_{20}H_{36}O_5$ mw: 356.56

SYNS: PGF 1-α ◇ PROTAGLANDIN F1 ◇ (9-α,11-α,13E,15S)-11,15-TRIHYDROXY-PROST-13-EN-1-OIC ACID (9CI)

TOXICITY DATA with REFERENCE
spm-mus-ipr 3 mg/kg INJFA3 21,82,76
scu-mus TDLo:10 mg/kg (female 16D post):TER PRGLBA 1,191,72
scu-rat TDLo:150 mg/kg (4-6D preg):REP BIREBV 1,367,69

SAFETY PROFILE: An experimental teratogen. Experimental reproductive effects. Mutation data reported. When heated to decomposition it emits acrid smoke and irritating fumes. See also other prostaglandin entries.

POC500 CAS:551-11-1 ***HR: 3***
PROSTAGLANDIN F2 $-\alpha$
mf: $C_{20}H_{34}O_5$ mw: 354.54

SYNS: AMOGLANDIN ◇ 7-(3,5-DIHYDROXY-2-(3-HYDROXY-1-OCTENYL)CYCLOPENTYL)-5-HEPTENOIC ACID ◇ DINOPROST ◇ ENZAPROST ◇ ENZAPROST F ◇ PANACELAN ◇ PGF2-α ◇ PROSTALMON F ◇ PROSTARMON F ◇ PROSTIN F2-α ◇ (5Z,9,α,11,α,13E,15S)-9,11,15-TRIHYDROXYPROSTA-5,13-DIEN-1-OIC ACID ◇ 9,11,15-TRIHYDROXYPROSTA-5,13-DIEN-1-OIC ACID ◇ U-14583

TOXICITY DATA with REFERENCE
dns-hmn-idr 28 ng/kg ADREDL 271,143,81
oms-mus:oth 1 μmol/L JIDEAE 66,313,76
cyt-mus:oth 50 mg/L JRPFA4 47,1,76
ipc-wmn TDLo:488 mg/kg (female 16W post):TER OBGNAS 48,216,76
ims-wmn TDLo:200 μg/kg (female 14W post):REP JOGBAS 79,737,72
ivn-wmn TDLo:20 μg/kg:GIT JMGZAI 10(7),9,73
orl-rat LD50:1170 mg/kg JMGZAI 10(7),9,73
scu-rat LD50:95 mg/kg IYKEDH 4,193,73
ivn-rat LD50:106 mg/kg JMGZAI 10(7),9,73
ims-rat LD50:112 mg/kg JMGZAI 10(7),9,73
orl-mus LD50:1300 mg/kg JMGZAI 10(7),9,73
scu-mus LD50:212 mg/kg JMGZAI 10(7),9,73
ivn-mus LD50:56 mg/kg NIIRDN 6,325,82
ims-mus LD50:152 mg/kg JMGZAI 10(7),9,73
ivn-rbt LD50:2500 μg/kg IYKEDH 9,261,78
ims-rbt LD50:2500 μg/kg IYKEDH 9,261,78

CONSENSUS REPORTS: EPA Genetic Toxicology Program.

SAFETY PROFILE: Poison by subcutaneous, intravenous, and intramuscular routes. Moderately toxic by ingestion. Human and experimental teratogenic and experimental reproductive effects. Human reproductive effects by subcutaneous, intravenous, intramuscular, intraperitoneal, intravaginal, and intraplacental routes: postpartum depression and other maternal effects, abortion and changes in measures of fertility. Human teratogenic effects by intraplacental route: extra embryonic structures. Human systemic effects by intravenous route: hypermotility, diarrhea, nausea or vomiting. Human mutation data reported. When heated to decomposition it emits acrid smoke and fumes.

POC525 CAS:23518-25-4 ***HR: D***
dl-PROSTAGLANDIN F2-α
mf: $C_{20}H_{34}O_5$ mw: 354.54

SYNS: dl-7-(3,5-DIHYDROXY-2-(3-HYDROXY-1-OCTENYL)CYCLOPENTYL)-5-HEPTENOIC ACID ◇ PGF2-α racemic mixture ◇ (±)-PGF2-α ◇ dl-PGF2-α ◇ PROSTAGLANDIN F2-α, racemic mixture ◇ (5Z,9-α,11-α,13E,15S)-(±)-8,11,15-TRIHYDROXY-PROSTA-5,13-DIEN-1-OIC ACID (9CI)

TOXICITY DATA with REFERENCE
orl-ham TDLo:36 mg/kg (female 4-6D post):REP JOENAK 53,201,72

SAFETY PROFILE: Experimental reproductive effects. When heated to decomposition it emits acrid smoke and irritating fumes. See also other prostaglandin entries.

POC750 CAS:38562-01-5 ***HR: 3***
PROSTAGLANDIN F2-α-THAM
mf: $C_{20}H_{34}NO_5 \cdot C_4H_{11}NO_3$ mw: 475.70

SYNS: 7-(3,5-DIHYDROXY-2-(3-HYDROXY-1-OCTENYL)CYCLOPENTYL)-5-HEPTENOIC ACID, THAM ◇ 7-(3,5-DIHYDROXY-2-(3-HYDROXY-1-OCTENYL)CYCLOPENTYL)-5-HEPTENOIC ACID, TRIMETHAMINE SALT ◇ DINOPROST TROMETHAMINE (USDA) ◇ 583E ◇ LUTALYSE ◇ PGF2-α THAM ◇ PGF2-α TRIS SALT ◇ PGF2-α TROMETHAMINE ◇ PROSTAGLANDIN F2-α THAM SALT ◇ PROSTAGLANDIN F2a TROMETHAMINE ◇ THAM ◇ TROMETHAMINE PROSTAGLANDIN F2-α ◇ U-14

TOXICITY DATA with REFERENCE
ivg-wmn TDLo:12 mg/kg (5W preg):REP AJOGAH 117,346,73
ipr-mus TDLo:1500 μg/kg (female 7-12D post):TER KSRNAM 7,640,73
orl-rat LD50:665 mg/kg YKYUA6 32,1129,81
ipr-rat LD50:101 mg/kg TOIZAG 22,517,75
scu-rat LD50:66 mg/kg YKYUA6 32,1129,81
ivn-rat LD50:101 mg/kg NIIRDN 6,327,82
ims-rat LD50:59 mg/kg NIIRDN 6,327,82
orl-mus LD50:711 mg/kg TOIZAG 21,623,74
ipr-mus LD50:101 mg/kg TOIZAG 21,623,74
scu-mus LD50:214 mg/kg TOIZAG 21,623,74
ims-mus LD50:111 mg/kg TOIZAG 21,623,74

SAFETY PROFILE: Poison by intraperitoneal, subcutaneous, intravenous, and intramuscular routes. Moderately toxic by ingestion. Human reproductive effects by intervaginal route: terminates pregnancy, effects on fertility. Experimental teratogenic and reproductive effects.

When heated to decomposition it emits toxic fumes of NO_x. See also other prostaglandin entries.

POD000 CAS:114-80-7 ***HR: 3***
PROSTIGMINE BROMIDE
mf: $C_{12}H_{19}N_2O_2 \cdot Br$ mw: 303.24

SYNS: BENZENAMINIUM,3-(((DIMETHYLAMINO)CARBONYL)OXY)-N,N,N-TRIMETHYL-, BROMIDE (9CI) ◇ CARBAMIC ACID, DIMETHYL-, ester with (m-HYDROXYPHENYL) TRIMETHYLAMMONIUM BROMIDE ◇ EUSTIGMIN BROMIDE ◇ (m-HYDROXYPHENYL)TRIMETHYLAMMONIUMBROMIDE DIMETHYLCARBAMATE ◇ 3-HYDROXYPHENYLTRIMETHYLAMMONIUM BROMIDE DIMETHYLCARBAMIC ESTER ◇ KIRKSTIGMINE BROMIDE ◇ LEOSTIGMINE BROMIDE ◇ NEOESERINE BROMIDE ◇ NEOSERINE BROMIDE ◇ NEOSTIGMINE BROMIDE ◇ NEOSTIGMINE METHYL BROMIDE ◇ PHILOSTIGMIN BROMIDE ◇ PROSERINE ◇ PROSERINE BROMIDE ◇ PROSTIGMIN BROMIDE ◇ RCRA WASTE NUMBER U053 ◇ STIGMANOL BROMIDE ◇ STIGMOSAN BROMIDE ◇ SYNSTIGMIN BROMIDE ◇ SYNTHOSTIGMINE BROMIDE ◇ SYNTOSTIGMIN (tablet) ◇ SYNTOSTIGMIN BROMIDE ◇ SYNTOSTIGMINE BROMIDE ◇ VAGOSTIGMINE BROMIDE

TOXICITY DATA with REFERENCE
orl-rat LD50:51 mg/kg PSEBAA 120,511,65
ipr-rat LDLo:835 μg/kg JPETAB 136,20,62
scu-rat LD50:370 μg/kg JAPMA8 39,701,50
ivn-rat LD50:165 μg/kg JAPMA8 39,701,50
orl-mus LD50:7 mg/kg PSEBAA 120,511,65
ipr-mus LD50:610 μg/kg JPMSAE 57,172,68
scu-mus LD50:420 μg/kg JPETAB 99,16,50
ivn-mus LD50:130 μg/kg PCJOAU 16,164,82

SAFETY PROFILE: Poison by ingestion, subcutaneous, intravenous, and intraperitoneal routes. When heated to decomposition it emits very toxic fumes of Br^-, NH_3, and NO_x. See also CARBAMATES and BROMIDES.

POD500 ***HR: 3***
PROTACTINIUM
af: Pa aw: 231.036

PROP: A bright, lustrous metal. Mp: 1600°, d: 15.37, vap press: 5×10^{-5} mm @ 1927°. Natural isotope ^{231}Pa (Actinium series), $T_{0.5} = 3 \times 10^4$ Y., decays to radioactive ^{227}Ac by alphas of 5.0 MeV. Artificial isotope ^{233}Pa (Neptunium Series), $T_{0.5}$ = 27D, decays to radioactive ^{233}U by betas of 0.15 (37%), 0.26 (58%), 0.57 (5%) MeV; emits gammas of 0.02-0.42 MeV. Natural isotope ^{234}Pa (Uranium Series), $T_{0.5}$ = 6.7H, decays to radioactive 234U by betas of 0.23-1.36 MeV, emits gammas of 0.04-0.8 MeV.

SAFETY PROFILE: Confirmed carcinogen. A highly radiotoxic metallic element. An alpha emitter. It is a general hazard if absorbed systemically. The dust and fumes are hazardous if inhaled. A severe radiation hazard.

POD750 CAS:9009-65-8 ***HR: 3***
PROTAMINE SULFATE

SYNS: ANGIOCICLAN ◇ DILANGIO ◇ FLUXEMA ◇ HALIDOR ◇ USAF D-4 ◇ VASORELAX

TOXICITY DATA with REFERENCE
ipr-rat LD50:120 mg/kg NIIRDN 6,726,82
ivn-rat LD50:75 mg/kg AIPTAK 165,374,67
ipr-mus LD50:94 mg/kg NIIRDN 6,726,82
scu-mus LD50:200 mg/kg NIIRDN 6,726,82
ivn-mus LD50:44 mg/kg TXAPA9 1,185,59
ipr-gpg LD50:120 mg/kg NIIRDN 6,726,82
scu-gpg LD50:200 mg/kg NIIRDN 6,726,82
ivn-gpg LD50:120 mg/kg NIIRDN 6,726,82

CONSENSUS REPORTS: Reported in EPA TSCA Inventory.

SAFETY PROFILE: Poison by intraperitoneal, subcutaneous, and intravenous routes. When heated to decomposition it emits very toxic fumes of NO_x and SO_x.

POD800 CAS:25627-41-2 ***HR: 3***
cis-PROTHIADENE-S-OXIDE HYDROGEN MALEATE
mf: $C_{19}H_{21}NOS \cdot C_4H_4O_4$ mw: 427.55

SYNS: DIBENZO(b,e)THIEPIN-μ!MN11(6H),Γ-PROPYLAMINE, N,N-DIMETHYL-, 5-OXIDE, MALEATE (1:1), (Z)- ◇ cis-11-(3-DIMETHYLAMINOPROPYLIDENE)-6,11-DIHYDRODIBENZO(b,e)THIEPIN 5-OXIDE HYDROGEN MALEATE

TOXICITY DATA with REFERENCE
ipr-rat TDLo:36 mg/kg (female 14-19D post):REP JOAND3 6,386,85
ivn-mus LD50:50 mg/kg CCCCAK 34,1963,69

SAFETY PROFILE: Poison by intravenous route. Experimental reproductive effects. When heated to decomposition it emits toxic fumes of NO_x and SO_x.

POD875 CAS:23031-38-1 ***HR: 2***
PROTHRIN
mf: $C_{18}H_{22}O_3$ mw: 286.40

SYNS: FURAMETHRIN ◇ 5-PROPARGYLFURFURYL CHRYSANTHEMATE ◇ 5-PROPARGYL-2-FURYLMETHYL, dl-cis,trans-CHRYSANTHEMATE

TOXICITY DATA with REFERENCE
orl-rat TDLo:32760 mg/kg (91D pre):REP OYYAA2 5,605,71
orl-rat LD50:10 g/kg YKYUA6 36,533,85
orl-mus LD50:3130 mg/kg OYYAA2 8,1067,84
ipr-mus LD50:600 mg/kg OYYAA2 8,1067,84

SAFETY PROFILE: Moderately toxic by ingestion and intraperitoneal routes. Experimental reproductive effects. When heated to decomposition it emits acrid smoke and irritating fumes. An insecticide. See also ESTERS.

POE050 CAS:53935-32-3 ***HR: 2***
PROTIRELIN TARTRATE
mf: $C_{16}H_{22}N_6O_4{\cdot}C_4H_6O_4$ mw: 480.54

SYNS: 5-OXO-l-PROLYL-l-HISTIDYL-l-PROLINAMIDETARTRATE ◇ l-PYROGLUTAMYL-l-HISTIDYL-l-PROLINEAMIDE-l-TARTRATE ◇ TRH-T

TOXICITY DATA with REFERENCE
ipr-rat TDLo:4380 mg/kg (30D pre):REP YACHDS 4,1085,76
orl-rat LD50:18 g/kg TAKHAA 36,22A,77
ipr-rat LD50:2700 mg/kg TAKHAA 36,22A,77
scu-rat LD50:6600 mg/kg NIIRDN 6,727,82
ivn-rat TDLo:580 mg/kg NIIRDN 6,727,82
ims-rat TDLo:2900 mg/kg NIIRDN 6,727,82
orl-mus LD50:16 g/kg TAKHAA 36,22A,77
scu-mus LD50:5400 mg/kg NIIRDN 6,727,82
ivn-mus LD50:520 mg/kg NIIRDN 6,727,82
ims-mus LD50:3300 mg/kg NIIRDN 6,727,82

SAFETY PROFILE: Moderately toxic by intravenous and intramuscular routes. Experimental reproductive effects. When heated to decomposition it emits toxic fumes of NO_x.

POE100 CAS:54323-85-2 ***HR: 3***
PROTIZINIC ACID
mf: $C_{17}H_{17}NO_3S$ mw: 315.41

PROP: Crystals from diisopropyl ether or acetonitrile. Mp: 124-125°.

SYNS: 7-METHOXY-α,10-DIMETHYLPHENOTHIAZINE-2-ACETIC ACID ◇ PIROCRID ◇ PROTHIZINIC ACID ◇ PRT ◇ 17190RP

TOXICITY DATA with REFERENCE
orl-mus TDLo:350 mg/kg (female 7-13D post):REP IYKEDH 6,77,75
orl-rat TDLo:700 mg/kg (8-14D preg):TER IYKEDH 6,77,75
orl-rat LD50:738 mg/kg IYKEDH 6,60,75
ipr-rat LD50:350 mg/kg IYKEDH 11,181,80
scu-rat LD50:375 mg/kg IYKEDH 6,60,75
ivn-rat TDLo:326 mg/kg IYKEDH 6,60,75
orl-mus LD50:1250 mg/kg IYKEDH 11,181,80
ipr-mus LD50:400 mg/kg IYKEDH 11,181,80
scu-mus LD50:470 mg/kg IYKEDH 6,60,75
ivn-mus LD50:232 mg/kg IYKEDH 6,408,75

SAFETY PROFILE: Poison by subcutaneous, intravenous, and intraperitoneal routes. Moderately toxic by ingestion. An experimental teratogen. Experimental reproductive effects. When heated to decomposition it emits toxic fumes of SO_x and NO_x. An anti-inflammatory agent.

POF000 CAS:143-57-7 ***HR: 3***
PROTOVERATRINE A
mf: $C_{41}H_{63}NO_{14}$ mw: 794.05

PROP: Sltly bitter crystals. Decomp @ 266-267°. Sol in alc; sltly sol in ether; insol in water, petr ether.

SYNS: PROTALBA ◇ PROTOVERATRIN ◇ PROTOVERATRINE ◇ PUROVERINE

TOXICITY DATA with REFERENCE
orl-rat LD50:5 mg/kg AEPPAE 189,397,38
scu-rat LD50:600 μg/kg AEPPAE 189,397,38
ipr-mus LD50:370 μg/kg PSEBAA 76,847,51
scu-mus LD50:290 μg/kg JPETAB 113,89,55
ivn-mus LD50:45 μg/kg APFRAD 14,783,56
scu-dog LDLo:500 μg/kg AEXPBL 29,440,1892
scu-cat LDLo:500 μg/kg AEXPBL 29,440,1892
scu-rbt LDLo:250 μg/kg APFRAD 14,783,56
ivn-rbt LDLo:50 μg/kg AEXPBL 29,440,1892
scu-frg LD50:4500 μg/kg AEPPAE 189,397,38

SAFETY PROFILE: Poison by ingestion, subcutaneous, intraperitoneal, and intravenous routes. When heated to decomposition it emits toxic fumes of NO_x.

POF250 CAS:1225-55-4 ***HR: 3***
PROTRIPTYLINE HYDROCHLORIDE
mf: $C_{19}H_{21}N{\cdot}ClH$ mw: 299.87

SYNS: 5-(3-METHYLAMINOPROPYL)-5H-DIBENZO(a,d)CYCLOHEPTENE HYDROCHLORIDE ◇ N-METHYL-5H-DIBENZO(A,D)CYCLOHEPTENE-5-PROPYLAMINEHYDROCHLORIDE ◇ NORMETHYL EX4442 ◇ TRIPTIL HYDROCHLORIDE ◇ VIVACTIL

TOXICITY DATA with REFERENCE
orl-rat LD50:299 mg/kg TXAPA9 18,185,71
ipr-rat LD50:44 mg/kg 27ZQAG -,88,72
orl-mus LD50:211 mg/kg TXAPA9 18,185,71
scu-mus LD50:192 mg/kg 27ZQAG -,88,72
ivn-mus LD50:49 mg/kg APTOA6 24,121,66
orl-dog LDLo:400 mg/kg 27ZQAG -,88,72
orl-rbt LD50:310 mg/kg 27ZQAG -,88,72
ivn-rbt LD50:8200 μg/kg 27ZQAG -,88,72
orl-gpg LD50:180 mg/kg 27ZQAG -,88,72
ipr-gpg LD50:21 mg/kg 27ZQAG -,88,72

SAFETY PROFILE: Poison by ingestion, intraperitoneal, intravenous, and subcutaneous routes. When heated to decomposition it emits very toxic fumes of NO_x and HCl.

POF275 CAS:8003-08-5 ***HR: 1***
PROVEST
mf: $C_{24}H_{34}O_4{\cdot}C_{20}H_{24}O_2$ mw: 683.02

SYNS: 17-(ACETYLOXY)-6-METHYLPREGN-4-ENE-3,20-DIONE,(6-α)-mixed with (17-α)-19-NORPREGNA-1,3,5(10)-TRIEN-20-YNE-3,17-DIOL ◇ ETHINYLESTRADIOL and MEDROXYPROGESTERONE ACETATE

◇ MEDROXYPROGESTERONE ACETATE and ETHINYLESTRADIOL ◇ NOGEST

TOXICITY DATA with REFERENCE
orl-wmn TDLo:4020 μg/kg (20D pre):REP INJFA3 8,725,63
orl-mus LD50:4216 mg/kg INJFA3 8,589,63

SAFETY PROFILE: Mildly toxic by ingestion. Human reproductive effects by ingestion: effects on fertility; effects on uterus, vagina, cervix; menstrual changes or disorders. Experimental reproductive effects. When heated to decomposition it emits acrid smoke and irritating fumes.

POF500 CAS:132-35-4 ***HR: 3***
PROXAZOLE CITRATE
mf: $C_{17}H_{25}N_3O \cdot C_6H_8O_7$ mw: 479.59

SYNS: AERBRON ◇ 5-(2-(DIETHYLAMINO)ETHYL)-3-(α-ETHYLBENZYL)-1,2,4-OXADIAZOLECITRATE◇ 5-(2-(DIETHYLAMINO)ETHYL)-3-(α-PHENYLPROPYL)-1,2,4-OXADIAZOLE CITRATE ◇ N,N-DIETHYL-3-(1-PHENYLPROPYL)-1,2,4-OXADIAZOLE-5-ETHANAMINE CITRATE ◇ FLOU ◇ PIRECIN ◇ PROPAXOLIN CITRATE ◇ PROPAXOLINE CITRATE ◇ SOLACIL

TOXICITY DATA with REFERENCE
orl-rat LD50:1400 mg/kg ARZNAD 13,798,63
ipr-rat LD50:390 mg/kg ARZNAD 13,798,63
orl-mus LD50:1270 mg/kg ARZNAD 13,798,63
ipr-mus LD50:390 mg/kg ARZNAD 13,798,63
scu-mus LD50:720 mg/kg ARZNAD 13,798,63
ivn-mus LD50:68 mg/kg ARZNAD 13,798,63

SAFETY PROFILE: Poison by intraperitoneal and intravenous routes. Moderately toxic by ingestion and subcutaneous routes. When heated to decomposition it emits toxic fumes of NO_x. A smooth muscle relaxant, analgesic, and anti-inflammatory agent.

POF550 CAS:59209-40-4 ***HR: 3***
PROXIL
mf: $C_{46}H_{58}ClN_5O_8 \cdot 2C_4H_4O_4$ mw: 1076.70

SYNS: AFLOXAN ◇ CR-604 ◇ PROGLUMETACINA (SPANISH) ◇ PROGLUMETACIN MALEATE ◇ PROTACINE

TOXICITY DATA with REFERENCE
orl-rat TDLo:132 mg/kg (female 7-17D post):REP OYYAA2 32,1017,86
orl-rat TDLo:132 mg/kg (female 7-17D post):TER OYYAA2 32,1017,86
orl-rat LD50:170 mg/kg OYYAA2 32,1173,86
ipr-rat LD50:247 mg/kg OYYAA2 32,1173,86
scu-rat LD50:332 mg/kg OYYAA2 32,1173,86
orl-mus LD50:67 mg/kg OYYAA2 32,1173,86
ipr-mus LD50:243 mg/kg OYYAA2 32,1173,86

SAFETY PROFILE: Poison by ingestion, intraperitoneal, and subcutaneous routes. An experimental teratogen. Other experimental reproductive effects. When heated to decomposition it emits toxic fumes of Cl^- and NO_x. An anti-inflammatory agent.

POF800 CAS:56299-00-4 ***HR: 3***
PR TOXIN
mf: $C_{17}H_{20}O_6$ mw: 320.37

SYNS: PRT ◇ PR TOXINE ◇ SPIRO(NAPHTHALENE-2(1H),2'-OXIRANE)-3'-CARBOXALDEHYDE,3,5,6,7,8,8a-HEXAHYDRO-7-ACETOXY-5,6-EPOXY-3',8,8a-TRIMETHYL-3-OXO- ◇ TOXIN (PENICILLIUM ROQUEFORTII) ◇ TOXIN PR

TOXICITY DATA with REFERENCE
mmo-sat 625 ng/plate MUREAV 130,79,84
dnr-bcs 100 μg/disc CNREA8 36,445,76
mmo-nsc 2 mg/L ENMUDM 1,45,79
orl-rat TDLo:100 mg/kg/52D-C:ETA MYCPAH 78,125,82
orl-rat TD:139 mg/kg/52D-C:ETA,REP MYCPAH 78,125,82
orl-rat LDLo:115 mg/kg APMBAY 25,111,73
ipr-rat LD50:11600 μg/kg TOXIA6 20,433,82
ivn-rat LD50:8200 μg/kg TOXIA6 20,433,82
orl-mus LD50:72 mg/kg FCTXAV 16,369,78
ipr-mus LD50:2 mg/kg FCTXAV 16,369,78
ivn-mus LD50:2 mg/kg 85GDA2 6,116,81

SAFETY PROFILE: Poison by ingestion, intravenous, and intraperitoneal routes. Questionable carcinogen with experimental tumorigenic data. Experimental reproductive effects. Mutation data reported. When heated to decomposition it emits acrid smoke and irritating fumes.

POG000 CAS:102648-46-4 ***HR: 3***
PRUNACETIN A

PROP: Produced by the strain *Streptomyces griseus var. Purpureus* No. CD-270 (85ERAY 2,1294,78).

TOXICITY DATA with REFERENCE
ipr-mus LD50:93 mg/kg 85ERAY 2,1294,78
ivn-mus LD50:220 mg/kg 85ERAY 2,1294,78

SAFETY PROFILE: Poison by intraperitoneal and intravenous routes.

POG250 ***HR: 3***
PSEUDOACONITINE
mf: $C_{36}H_{51}NO_{12}$ mw: 689.78

PROP: White crystals or syrupy mass. Mp: 214° (decomp). Insol in water; sol in alc, ether.

SAFETY PROFILE: Poison by ingestion, inhalation, and skin contact. When heated to decomposition it emits highly toxic fumes of NO_x. See also ACONITINE.

POG300 CAS:700-13-0 ***HR: D***
PSEUDOCUMOHYDROQUINONE
mf: $C_9H_{12}O_2$ mw: 152.21

SYNS: 1,4-BENZENEDIOL, 2,3,5-TRIMETHYL- (9CI) ◇ psi-CUMOHYDROQUINONE ◇ HYDROQUINONE, TRIMETHYL- ◇ TRIMETHYLHYDROQUINONE ◇ 2,3,5-TRIMETHYLHYDROQUINONE

TOXICITY DATA with REFERENCE
scu-rat TDLo:20 mg/kg (female 4D pre):REP JSICAZ 19,264,60

CONSENSUS REPORTS: Reported in EPA TSCA Inventory.

SAFETY PROFILE: Experimental reproductive effects. When heated to decomposition it emits acrid smoke and irritating fumes.

POH000 CAS:90-82-4 ***HR: 3***
l-(+)-PSEUDOEPHEDRINE
mf: $C_{10}H_{15}NO$ mw: 165.26

SYNS: BESAN ◇ d-psi-EPHEDRINE ◇ psi-EPHEDRINE ◇ d-ISOEPHEDRINE ◇ α-(1-(METHYLAMINO)ETHYL)BENZYL ALCOHOL ◇ d-psi-2-METHYLAMINO-1-PHENYL-1-PROPANOL ◇ PSEUDO-EPHEDRINE ◇ NOVAFED ◇ SUDAFED

TOXICITY DATA with REFERENCE
orl-rat LD50:660 mg/kg JPMSAE 60,1523,71
scu-rat LDLo:650 mg/kg 27ZIAQ -,225,73
orl-mus LD50:500 mg/kg JPMSAE 60,1523,71
ivn-dog LDLo:125 mg/kg 27ZIAQ -,225,73
scu-rbt LDLo:400 mg/kg 27ZIAQ -,225,73
ivn-rbt LDLo:85 mg/kg 27ZIAQ -,225,73

CONSENSUS REPORTS: Reported in EPA TSCA Inventory.

SAFETY PROFILE: Poison by intravenous and subcutaneous routes. Moderately toxic by ingestion. Used in production of drugs of abuse. When heated to decomposition it emits toxic fumes of NO_x. See also l(+)-PSEUDOEPHEDRINE HYDROCHLORIDE.

POH250 CAS:345-78-8 ***HR: 3***
l(+)-PSEUDOEPHEDRINE HYDROCHLORIDE
mf: $C_{10}H_{15}NO•ClH$ mw: 201.72

SYNS: S-(R,R)-d-(α-(1-METHYLAMINO)ETHYL)BENZENEMETHANOLHYDROCHLORIDE ◇ d-(α-(1-METHYLAMINO)ETHYL)BENZYL ALCOHOL HYDROCHLORIDE ◇ NOVAFED ◇ d-PSEUDOEPHEDRINE HYDROCHLORIDE ◇ SINUFED ◇ SUDAFED ◇ SYMPTOM 2 ◇ TUSSAPHED

TOXICITY DATA with REFERENCE
orl-hmn TDLo:714 μg/kg:CVS AIMDAP 39,404,27
scu-rat LDLo:80 mg/kg JPETAB 71,62,41
orl-mus LD50:371 mg/kg JPMSAE 60,1523,71
ipr-mus LD50:202 mg/kg JPETAB 158,135,67
par-mus LDLo:400 mg/kg MAIZAB 3,1,25
ivn-rbt LDLo:75 mg/kg JPETAB 36,363,29
par-frg LDLo:800 mg/kg 85DHAX Mn,128,73

CONSENSUS REPORTS: Reported in EPA TSCA Inventory.

SAFETY PROFILE: Poison by ingestion, subcutaneous, intraperitoneal, parenteral, and intravenous routes. Human systemic effects by ingestion: blood pressure elevation. When heated to decomposition it emits very toxic fumes of NO_x and HCl.

POH500 CAS:670-40-6 ***HR: 3***
1-PSEUDOEPHEDRINE HYDROCHLORIDE
mf: $C_{10}H_{15}NO•ClH$ mw: 201.72

SYN: (−)-PSEUDOEPHEDRINE HYDROCHLORIDE

TOXICITY DATA with REFERENCE
scu-rat LDLo:160 mg/kg JPETAB 71,62,41
ipr-mus LD50:284 mg/kg JPETAB 158,135,67
ivn-rbt LDLo:80 mg/kg JPETAB 36,363,29

SAFETY PROFILE: Poison by subcutaneous, intraperitoneal, and intravenous routes. When heated to decomposition it emits very toxic fumes of HCl and NO_x. See also l(+)-PSEUDOEPHEDRINE HYDROCHLORIDE.

POH550 CAS:82508-32-5 ***HR: 3***
PSEUDOLARIC ACID A
mf: $C_{22}H_{28}O_6$ mw: 388.50

SYN: 2,4-PENTADIENOIC ACID, 5-(4a-(ACETYLOXYL)-3,4,4a,5,6,9-HEXAHYDRO-3,7-DIMETHYL-1-OXO-1H-4,9a-ETHANO-CYCLOHEPTA(c)PYRAN-3-YL)-2-METHYL-,(3-α(2E,4E),4-α-9a-α)-(-)-(9CI)

TOXICITY DATA with REFERENCE
orl-ham TDLo:200 mg/kg (female 7-11D post):TER CYLPDN 9,445,88
orl-rat TDLo:24600 μg/kg (female 7-9D post):REP CYLPDN 9,445,88
orl-rat LD50:219 mg/kg CYLPDN 9,445,88
ipr-mus LD50:397 mg/kg CYLPDN 9,445,88
ivn-mus TDLo:486 mg/kg CYLPDN 9,445,88

SAFETY PROFILE: Poison by ingestion and intraperitoneal routes. Moderately toxic by intravenous route. An experimental teratogen. Experimental reproductive effects. When heated to decomposition it emits acrid smoke and irritating fumes.

POH600 CAS:82508-31-4 ***HR: 3***
PSEUDOLARIC ACID B
mf: $C_{23}H_{28}O_8$ mw: 432.51

SYN: 1H-4,9a-ETHANOCYCLOHEPTA(c)PYRAN-7-CARBOXYLIC ACID,4a-(ACETYLOXY)-3-(4-CARBOXY-1,3-PENTADIENYL) 3,4,4a,5,6,9-HEXAHYDRO-3-METHYL-1-OXO-, 7-METHYL ESTER, (3-α(1E,3E),4-α-4a-α-9a-α)-(-)- (9CI)

TOXICITY DATA with REFERENCE
orl-ham TDLo:100 mg/kg (female 7-11 post):TER CYLPDN 9,445,88
orl-rat TDLo:13500 μg/kg (female 7-9D post):REP CYLPDN 9,445,88
orl-rat LD50:130 mg/kg CYLPDN 9,445,88
ipr-mus LD50:316 mg/kg CYLPDN 9,445,88
ivn-mus LD50:423 mg/kg CYLPDN 9,445,88

SAFETY PROFILE: Poison by ingestion and intraperitoneal routes. Moderately toxic by intravenous route. An experimental teratogen. Experimental reproductive effects. When heated to decomposition it emits acrid smoke and irritating fumes.

POH750 CAS:127-91-3 ***HR: 1***
PSEUDOPINENE
mf: $C_{10}H_{16}$ mw: 136.26

PROP: Colorless liquid; pine odor. D: 0.864, refr index: 1.477, flash p: 88°F. Sol in fixed oils; insol in water, propylene glycol, glycerin

SYNS: 6,6-DIMETHYL-2-METHYLENEBICYCLO(3.1.1)HEPTANE ◇ FEMA No. 2903 ◇ NOPINEN ◇ NOPINENE ◇ β-PINENE (FCC) ◇ 2(10)-PINENE ◇ PSEUDOPINEN

TOXICITY DATA with REFERENCE
skn-rbt 500 mg/24H MOD FCTXAV 16,859,78
orl-rat LD50:4700 mg/kg FCTXAV 16,859,78

CONSENSUS REPORTS: Reported in EPA TSCA Inventory.

SAFETY PROFILE: Mildly toxic by ingestion. A skin irritant. Flammable liquid. When heated to decomposition it emits acrid smoke and irritating fumes.

POH800 ***HR: 3***
PSIDIUM GUAJAVA Linn., extract excluding roots

PROP: Indian plant belonging to the family *Myrtaceae* IJEBA6 15,208,77

SYN: AMROOD, extract

TOXICITY DATA with REFERENCE
scu-rat TDLo:1111 mg/kg (male 10D pre):REP TJEMAO 62,287,55
ipr-mus LD50:188 mg/kg IJEBA6 15,208,77

SAFETY PROFILE: Poison by intraperitoneal route. Experimental reproductive effects. When heated to decomposition it emits acrid smoke and irritating fumes.

POI100 CAS:87625-62-5 ***HR: 3***
PTAQUILOSIDE
mf: $C_{19}H_{28}O_8$ mw: 384.47

SYN: 1',3'-α,4',7'-α-TETRAHYDRO-7'-α-β-3-GLUCOPYRANOSYLOXY)-4'-HYDROXY-2',4',6'-TRIMETHYL-SPIRO(CYCLOPROPANE-1,5'-(5H)INDEN)-3',(2'H)-ONE

TOXICITY DATA with REFERENCE
dns-rat:lvr 1 μmol/L MUREAV 143,75,85
orl-rat TDLo:1580 mg/kg/9W-I:CAR JJCREP 75,833,84
orl-rat TD:1930 mg/kg/10W-I:CAR CALEDQ 21,239,84

CONSENSUS REPORTS: IARC Cancer Review: Group 3 IMEMDT 7,56,87; Animal Limited Evidence IMEMDT 40,47,86

SAFETY PROFILE: Questionable carcinogen with experimental carcinogenic data. Mutation data reported. When heated to decomposition it emits acrid smoke and irritating fumes.

POI550 ***HR: 3***
PTT 119
mf: $C_{29}H_{39}Cl_2FN_4O_4S \cdot ClH$ mw: 666.14

SYN: l-METHIONINE,N-(N-(3-(p-FLUOROPHENYL)-l-ANANYL)-3-(m-(BIS(2-CHLOROETHYL)AMINO)PHENYL)-l-ALANYL)-, ETHYL ESTER, HYDROCHLORIDE

TOXICITY DATA with REFERENCE
ipr-rat LD50:9200 μg/kg FRPSAX 38,205,83
ivn-rat LD50:9300 μg/kg FRPSAX 38,205,83
ipr-mus LD50:21 mg/kg FRPSAX 38,205,83
ivn-mus LD50:17 mg/kg FRPSAX 38,205,83

SAFETY PROFILE: Poison by intravenous and intraperitoneal routes. When heated to decomposition it emits toxic fumes of F^-, SO_x, NO_x, and Cl^-. See also ESTERS.

POI575 ***HR: 3***
PUERARIA PHASEOLOIDES (Roxb.) Benth., extract excluding roots

PROP: Indian plant belonging to the family *Fabaceae* IJEBA6 22,487,84

TOXICITY DATA with REFERENCE
orl-rat TDLo:150 mg/kg (female 12-14D post):REP IJEBA6 22,487,84
ipr-mus LD50:825 mg/kg IJEBA6 22,487,84

SAFETY PROFILE: Moderately toxic by intraperitoneal route. Experimental reproductive effects. When heated to decomposition it emits acrid smoke and irritating fumes.

POI600 ***HR: 3***
PUFFER POISON, HYDROCHLORIDE

PROP: Extracted from liver, gonads, stomach and intestines of *Sphoeroides pardalis, Sphoeroides vermicularis porphyreus* and *Sphoeroides vermicularis radiatus* (JPETAB 122,247,57).

TOXICITY DATA with REFERENCE
ipr-mus LD50:20 μg/kg JPETAB 122,247,57
ipr-cat LDLo:10 μg/kg JPETAB 122,247,57
ivn-cat LDLo:10 μg/kg JPETAB 122,247,57

SAFETY PROFILE: Poison by intravenous and intraperitoneal routes.

POI650 ***HR: 2***
PULICARIA ANGUSTIFOLIA DC., extract

PROP: Indian plant belonging to the family *Asteraceae* IJEBA6 22,312,84

TOXICITY DATA with REFERENCE
orl-rat TDLo:150 mg/kg (female 12-14D post):REP IJEBA6 22,312,84
ipr-mus LD50:825 mg/kg IJEBA6 22,312,84

SAFETY PROFILE: Moderately toxic by intraperitoneal route. Experimental reproductive effects. When heated to decomposition it emits acrid smoke and irritating fumes.

POI800 ***HR: D***
PUNICA GRANATUM Linn., fruit skin

TOXICITY DATA with REFERENCE
orl-gpg TDLo:504 g/kg (28D male):REP IJMRAQ 48,46,60

SAFETY PROFILE: Experimental reproductive effects.

POJ000 CAS:126-17-0 ***HR: 3***
PURAPURIDINE
mf: $C_{27}H_{43}NO_2$ mw: 413.71

SYNS: SOLANCARPIDINE ◇ SOLANIDINE-S ◇ SOLASOD-5-EN-3-β-OL ◇ SOLASODINE

TOXICITY DATA with REFERENCE
orl-rat TDLo:120 mg/kg (6-9D preg):REP LANCAO 1,11877,73
orl-ham TDLo:1500 mg/kg (female 7D post):TER BECTA6 15,522,76
orl-rat LD50:4978 mg/kg PCJOAU 11,1095,77
ipr-rat LD50:396 mg/kg PCJOAU 11,1095,77
orl-mus LD50:27500 μg/kg FATOAO 24,469,61
ipr-mus LD50:899 mg/kg PCJOAU 11,1095,77
orl-ham LDLo:1200 mg/kg TJADAB 17,327,78

SAFETY PROFILE: Poison by intraperitoneal route. Moderately toxic by ingestion. Experimental teratogenic and reproductive effects. When heated to decomposition it emits toxic fumes of NO_x.

POJ100 CAS:52175-10-7 ***HR: 3***
1H-PURIN-6-AMINE PHOSPHATE
mf: $C_5H_5N_4•7H_3O_4P$ mw: 807.14

TOXICITY DATA with REFERENCE
orl-mus LD50:750 mg/kg NIIRDN 6,20,82
ipr-mus LD50:417 mg/kg NIIRDN 6,20,82
ivn-mus LD50:400 mg/kg NIIRDN 6,20,82

SAFETY PROFILE: Poison by intravenous route. Moderately toxic by ingestion and intraperitoneal routes. When heated to decomposition it emits toxic fumes of PO_x and NO_x.

POJ250 CAS:120-73-0 ***HR: 3***
PURINE
mf: $C_5H_4N_4$ mw: 120.13

PROP: Needles. Mp: 216-217°. Very sol in hot alc; sltly sol in hot ethyl acetate, acetone; insol in ether, chloroform.

SYNS: 7H-IMIDAZO(4,5-D)PYRIMIDINE ◇ ISOPURINE ◇ NSC 753 ◇ 7H-PURINE ◇ 9H-PURINE ◇ 3,5,7-TRIAZAINDOLE

TOXICITY DATA with REFERENCE
pic-esc 1 g/L ZAPOAK 12,583,72
dnd-mam:lym 50 mmol/L PNASA6 48,686,62
scu-rat TDLo:156 mg/kg/8W-I:ETA CNREA8 38,2229,78
ipr-rat LDLo:800 mg/kg CPCHAO 18,307,62
ipr-mus LD50:150 mg/kg JMCMAR 6,480,63

CONSENSUS REPORTS: EPA Genetic Toxicology Program.

SAFETY PROFILE: Poison by intraperitoneal route. Questionable carcinogen with experimental tumorigenic data. Mutation data reported. When heated to decomposition it emits toxic fumes of NO_x.

POJ500 CAS:1904-98-9 ***HR: 3***
1H-PURINE-2,6-DIAMINE
mf: $C_5H_6N_6$ mw: 150.17

SYNS: 2-AMINOADENINE ◇ DAP ◇ 2,6-DIAMINOPURINE ◇ NSC 743 ◇ SQ 21065 ◇ X 79

TOXICITY DATA with REFERENCE
mmo-sat 10 μL/plate ANYAA9 76,475,58
mmo-esc 2500 mg/L/1H CRSUBM 3,69,55
mmo-omi 500 mg/L SOGEBZ 6,1509,70
ipr-mus LD50:202 mg/kg NCISP* JAN86

CONSENSUS REPORTS: Reported in EPA TSCA Inventory. EPA Genetic Toxicology Program.

SAFETY PROFILE: Poison by intraperitoneal route. Mutation data reported. When heated to decomposition it emits toxic fumes of NO_x.

POJ750 CAS:28199-55-5 ***HR: 3***
PURINE-3-OXIDE
mf: $C_5H_4N_4O$ mw: 136.13

TOXICITY DATA with REFERENCE
scu-rat TDLo:1800 mg/kg/8W-I:CAR CNREA8 38,2229,78
scu-rat TD:18 mg/kg/8W-I:ETA PAACA3 17,147,76

SAFETY PROFILE: Questionable carcinogen with experimental carcinogenic and tumorigenic data. When heated to decomposition it emits toxic fumes of NO_x.

POK000 CAS:50-44-2 ***HR: 3***
PURINE-6-THIOL
mf: $C_5H_4N_4S$ mw: 152.19

SYNS: 1,7-DIHYDRO-6H-PURINE-6-THIONE ◇ ISMIPUR ◇ LEUKERAN ◇ LEUPURIN ◇ MERCALEUKIN ◇ MERCAPTOPURIN (GERMAN) ◇ 6-MERCAPTOPURIN ◇ 6-MERCAPTOPURINE ◇ 7-MERCAPTO-1,3,4,6-TETRAZAINDENE ◇ MERCAPURIN ◇ MERN ◇ MP ◇ NCI-C04886 ◇ NSC 755 ◇ PURIMETHOL ◇ 3H-PURINE-6-THIOL ◇ 6-PURINETHIOL ◇ PURINETHOL ◇ THIOHYPOXANTHINE ◇ 6-THIOXOPURINE ◇ U-4748

TOXICITY DATA with REFERENCE
mma-sat 150 μmol/L EXPEAM 40,370,84
oms-bcs 10 μmol/L CNREA8 40,4381,80
dni-hmn:hla 10 μmol/L BBACAQ 366,333,74
ipr-rat TDLo:30 mg/kg (female 9D post):REP BEXBAN 76,1289,73
scu-mus TDLo:25 mg/kg (female 11D post):TER TJADAB 32,31A,85
orl-wmn TDLo:360 mg/kg/34W-C:NEO,BLD AJGAAR 78,316,83
ipr-rat TDLo:500 mg/kg/7W-I:ETA CANCAR 40,1935,77
orl-chd TD:675 mg/kg/39W-C:CAR,BLD JOPDAB 72,409,68
orl-man TDLo:40 mg/kg/1W-I:SKN ARDEAC 122,1413,86
ipr-rat LD50:159 mg/kg APPHAX 41,571,84
ivn-rat LD50:250 mg/kg ARZNAD 20,1467,70
par-rat LD50:250 mg/kg RRCRBU 52,76,75
orl-mus LD50:260 mg/kg CKFRAY 14,389,65
ipr-mus LD50:80 mg/kg CNREA8 42,122,82
scu-mus LD50:100 mg/kg JMCMAR 18,320,75
ivn-mus LD50:80 mg/kg CNREA8 42,122,82
ipr-ham LD50:364 mg/kg ARTODN 32,1,74

CONSENSUS REPORTS: IARC Cancer Review: Group 3 IMEMDT 7,240,87; Animal Inadequate Evidence IMEMDT 26,249,81; Human Inadequate Evidence IMEMDT 26,249,81. NCI Carcinogenesis Studies (ipr); Equivocal Evidence: rat CANCAR 40,1935,77; (ipr); Clear Evidence: mouse CANCAR 40,1935,77. EPA Genetic Toxicology Program.

SAFETY PROFILE: Poison by ingestion, intraperitoneal, subcutaneous, parenteral, and intravenous routes. Human systemic effects by ingestion: dermatitis. Experimental teratogenic and reproductive effects. Questionable human carcinogen producing Hodgkin's disease and leukemia. Human mutation data reported. When heated to decomposition it emits very toxic fumes of SO_x and NO_x. See also MERCAPTANS.

POK250 CAS:3506-23-8 ***HR: 3***
PUROMYCIN CHLOROHYDRATE
mf: $C_{22}H_{29}N_7O_5$•ClH mw: 508.04

SYNS: 3′-(α-AMINO-p-METHOXYHYDROCINNAMAMIDO)-3′-DEOXY-N,N-DIMETHYLADENOSINE MONOHYDROCHLORIDE ◇ PUROMYCIN HYDROCHLORIDE ◇ STYLOMYCIN

TOXICITY DATA with REFERENCE
dni-mus-ipr 333 mg/kg JCLBA3 30,13,66
oms-mus-ipr 333 mg/kg JCLBA3 30,13,66
orl-mus LD50:720 mg/kg ABANAE 2,757,54/55
ipr-mus LD50:580 mg/kg ABANAE 2,757,54/55
ivn-mus LD50:335 mg/kg ABANAE 2,757,54/55
orl-gpg LD50:600 mg/kg ABANAE 2,757,54/55
ipr-gpg LD50:287 mg/kg ABANAE 2,757,54/55
ims-gpg LD50:202 mg/kg ABANAE 2,757,54/55

SAFETY PROFILE: Poison by intravenous, intraperitoneal, and intramuscular routes. Moderately toxic by ingestion. Mutation data reported. When heated to decomposition it emits very toxic fumes of NO_x and HCl.

POK300 CAS:58-58-2 ***HR: D***
PUROMYCIN DIHYDROCHLORIDE
mf: $C_{22}H_{29}N_7O_5$•2ClH mw: 544.50

SYNS: 3′-(α-AMINO-p-METHOXYHYDROCINNAMAMIDO)-3′-DEOXY-N,N-DIMETHYLADENOSINE DIHYDROCHLORIDE ◇ CL 16,536 ◇ PDH ◇ PUROMYCIN HYDROCHLORIDE

TOXICITY DATA with REFERENCE
dns-mus:mmr 5 mg/L CNREA8 29,391,69
oms-mus:oth 5 mg/L CNREA8 29,391,69

SAFETY PROFILE: Mutation data reported. When heated to decomposition it emits toxic fumes of NO_x and HCl.

POK325 CAS:333-93-7 ***HR: 2***
PUTRESCINE DIHYDROCHLORIDE
mf: $C_4H_{12}N_2$•2ClH mw: 161.10

SYNS: 1,4-DIAMINOBUTANE DIHYDROCHLORIDE ◇ PUTRESCINE HYDROCHLORIDE

TOXICITY DATA with REFERENCE
scu-rat LD50:1625 mg/kg OYYAA2 25,489,83
ivn-rat LD50:760 mg/kg OYYAA2 25,489,83
scu-mus LD50:1880 mg/kg OYYAA2 25,489,83
ivn-mus LD50:510 mg/kg OYYAA2 25,489,83

CONSENSUS REPORTS: Reported in EPA TSCA Inventory.

SAFETY PROFILE: Moderately toxic by subcutaneous and intravenous routes. When heated to decomposition it emits toxic fumes of NO_x and HCl. See also AMINES.

POK575 CAS:22204-24-6 ***HR: 2***
PYRANTEL PAMOATE
mf: $C_{23}H_{16}O_6 \cdot C_{11}H_{14}N_2S$ mw: 594.72

SYNS: COMBANTRIN ◇ 2-NAPHTHOIC ACID, 4,4'-METHYLENEBIS(3-HYDROXY-, compounded with (E)-1,4,5,6-TETRAHYDRO-1-METHYL-2-(2-(2-THIENYL)VINYL)PYRIMIDINE(1:1)

TOXICITY DATA with REFERENCE
orl-rat TDLo:1800 mg/kg (9-14D preg):TER OYYAA2 5,41,71
orl-rat TDLo:91 g/kg (female 91D pre):REP OYYAA2 5,335,71
ipr-rat LD50:535 mg/kg NIIRDN 6,642,82
scu-rat LDLo:6600 mg/kg OYYAA2 5,305,71
ipr-mus LD50:620 mg/kg NIIRDN 6,642,82
scu-mus LDLo:4 g/kg OYYAA2 5,305,71

SAFETY PROFILE: Moderately toxic by intraperitoneal and subcutaneous routes. An experimental teratogen. Experimental reproductive effects. When heated to decomposition it emits toxic fumes of SO_x and NO_x.

POL000 CAS:522-25-8 ***HR: 3***
PYRATHIAZINE HYDROCHLORIDE
mf: $C_{18}H_{20}N_2S \cdot ClH$ mw: 332.92

PROP: Cubes. Mp: 200-201°. Sol in water, alc.

SYNS: 10-(2-(2-PYROLIDYL)ETHYL)PHENOTHIAZINEHYDROCHLORIDE ◇ PYRROLAZOTE ABERGIC ◇ PYRROLAZOTE HYDROCHLORIDE ◇ N-(2-PYRROLIDINOETHYL)PHENOTHIAZINE HYDROCHLORIDE ◇ 10-(2-(1-PYRROLIDYL)ETHYL)PHENOTHIAZINE HYDROCHLORIDE ◇ ROLAZOTE

TOXICITY DATA with REFERENCE
ipr-rat TDLo:100 mg/kg (4D preg):REP JRPFA4 8,225,64
ivn-rat LD50:26 mg/kg JPETAB 94,197,48
ipr-mus LD50:177 mg/kg 27ZQAG -,44,72
scu-mus LD50:1340 mg/kg JPETAB 94,197,48
ivn-mus LD50:37 mg/kg JPETAB 94,197,48
ivn-rbt LD50:36 mg/kg JPETAB 94,197,48

SAFETY PROFILE: Poison by intravenous and intraperitoneal routes. Moderately toxic by subcutaneous route. Experimental reproductive effects. An antihistamine. When heated to decomposition it emits very toxic fumes of SO_x, NO_x, and HCl.

POL475 CAS:26308-28-1 ***HR: 2***
PYRAZAPON
mf: $C_{15}H_{16}N_4O$ mw: 268.35

SYNS: CI-683 ◇ 1-ETHYL-4,6-DIHYDRO-3-METHYL-8-PHENYLPYRAZOLO(4,3-e)(1,4)DIAZEPINE ◇ RIPAZEPAM

TOXICITY DATA with REFERENCE
orl-mus TDLo:81900 mg/kg/78W-C:CAR FAATDF 4,178,84

SAFETY PROFILE: Questionable carcinogen with experimental carcinogenic data. When heated to decomposition it emits toxic fumes of NO_x.

POL500 CAS:98-96-4 ***HR: 3***
PYRAZINECARBOXAMIDE
mf: $C_5H_5N_3O$ mw: 123.13

SYNS: ALDINAMID ◇ 2-CARBAMYL PYRAZINE ◇ D-50 ◇ EPRAZIN ◇ MK 56 ◇ NCI-C01785 ◇ PYRAZINAMIDE ◇ PYRAZINEAMIDE ◇ PYRAZINE CARBOXYLAMIDE ◇ PYRAZINOIC ACID AMIDE ◇ TEBRAZID

TOXICITY DATA with REFERENCE
cyt-hmn:lym 120 mg/L MUREAV 48,215,77
orl-mus TDLo:328 g/kg/78W-C:ETA NCITR* NCI-CG-TR-48,78
orl-rat LDLo:3 g/kg ARTUA4 70,423,54
orl-mus LDLo:3 g/kg ARTUA4 70,423,54
ipr-mus LD50:1680 mg/kg CMTRAG 5,39,62
scu-mus LD50:2793 mg/kg NIIRDN 6,641,82

CONSENSUS REPORTS: NCI Carcinogenesis Bioassay (feed); No Evidence: rat NCITR* NCI-CG-TR-48,78; (feed); Inadequate Studies: mouse NCITR* NCI-CG-TR-48,78. Reported in EPA TSCA Inventory.

SAFETY PROFILE: Moderately toxic by ingestion, subcutaneous, and intraperitoneal routes. Questionable carcinogen with experimental tumorigenic data. Human mutation data reported. When heated to decomposition it emits toxic fumes of NO_x.

POL750 CAS:4933-19-1 ***HR: D***
PYRAZINE-2,3-DICARBOXYLIC ACID IMIDE
mf: $C_6H_3N_3O_2$ mw: 149.12

SYN: 3,6-DIAZAPHTHALIMID(GERMAN)

TOXICITY DATA with REFERENCE
ipr-mus TDLo:50 mg/kg (9D preg):TER PHARAT 31,172,76

SAFETY PROFILE: An experimental teratogen. When heated to decomposition it emits toxic fumes of NO_x.

POM000 CAS:35250-53-4 ***HR: 3***
2-PYRAZINYLETHANETHIOL
mf: $C_6H_8N_2S$ mw: 140.22

SYN: FEMA 3230

TOXICITY DATA with REFERENCE
orl-rat LD50:158 mg/kg FCTXAV 13,487,75

CONSENSUS REPORTS: Reported in EPA TSCA Inventory.

SAFETY PROFILE: Poison by ingestion. When heated to decomposition it emits very toxic fumes of NO_x and SO_x.

POM275 CAS:58011-68-0 ***HR: 2***
PYRAZOLATE
mf: $C_{19}H_{16}Cl_2N_2O_4S$ mw: 403.86

SYNS: 4-(2,4-DICHLOROBENZOYL)-1,3-DIMETHYL-5-PYRAZOLYL-p-TOLUENE SULFONATE ◇ 4-(2,4-DICHLOROBENZOYL)-1,3-DIMETHYL-p-TOLUENE SULFONATE-1H-PYRAZOL-4-OL ◇ SW-751

TOXICITY DATA with REFERENCE
orl-rat LD50:9550 mg/kg SKKNAJ 36,44,84
ipr-rat LD50:1950 mg/kg SKKNAJ 36,44,84
orl-mus LD50:10070 mg/kg SKKNAJ 36,44,84
ipr-mus LD50:3681 mg/kg SKKNAJ 36,44,84

SAFETY PROFILE: Moderately toxic by intraperitoneal route. Mildly toxic by ingestion. When heated to decomposition it emits toxic fumes of Cl^-, SO_x, and NO_x. See also SULFONATES.

POM500 CAS:288-13-1 ***HR: 2***
PYRAZOLE
mf: $C_3H_4N_2$ mw: 68.09

PROP: Needles from ether. Mp: 70°, bp: 186-188°. Sol in water, alc, ether, benzene.

SYN: 1,2-DIAZOLE

TOXICITY DATA with REFERENCE
ipr-mus TDLo:24 mg/kg (8-14D preg):TER TJADAB 25(2),43A,82
orl-rat TDLo:100 mg/kg (9D preg):REP RCOCB8 26,65,79
orl-rat LD50:1010 mg/kg EXPEAM 25,816,69
orl-mus LD50:1456 mg/kg JAPMA8 47,70,58
ipr-mus LD50:538 mg/kg JPETAB 178,199,71

CONSENSUS REPORTS: Reported in EPA TSCA Inventory.

SAFETY PROFILE: Moderately toxic by ingestion and intraperitoneal routes. Experimental teratogenic and reproductive effects. When heated to decomposition it emits toxic fumes of NO_x.

POM600 CAS:2380-63-4 ***HR: 3***
PYRAZOLOADENINE
mf: $C_5H_5N_5$ mw: 135.15

SYNS: 4-AMINOPYRAZOLOPYRIMIDINE ◇ 4-AMINOPYRAZOLO(3,4-d)PYRIMIDINE ◇ 4-APP ◇ NSC 1393 ◇ 1H-PYRAZOLO(3,4-d)PYRIMIDIN-4-AMINE(9CI)

TOXICITY DATA with REFERENCE
dni-mus:ast 5 μmol/L ECREAL 48,319,67
ipr-mus LD50:181 mg/kg NCISP* JAN86

CONSENSUS REPORTS: Reported in EPA TSCA Inventory.

SAFETY PROFILE: Poison by intraperitoneal route. Mutation data reported. When heated to decomposition it emits toxic fumes of NO_x.

POM700 CAS:76263-73-5 ***HR: 2***
2H-PYRAZOLO(4,3-c)QUINOLINE, 3,3a,4,5-TETRAHYDRO-2-ACETYL-8-METHOXY-5-((4-METHYLPHENYL) SULFONYL)-3-PHENYL-, cis-
mf: $C_{26}H_{25}N_3O_4S$ mw: 475.60

TOXICITY DATA with REFERENCE
orl-rat TDLo:100 mg/kg (female 1-5D post):REP IJOCAP 19,297,80
ipr-mus LD50:681 mg/kg IJOCAP 19,297,80

SAFETY PROFILE: Moderately toxic by intraperitoneal route. Experimental reproductive effects. When heated to decomposition it emits toxic fumes of NO_x and SO_x.

POM710 CAS:78431-47-7 ***HR: 2***
2H-PYRAZOLO(4,3-c)QUINOLINE, 3,3a,4,5-TETRAHYDRO-2-ACETYL-8-METHOXY-5-((4-METHYLPHENYL) SULFONYL)-3-(4-PYRIDINYL)-, trans-
mf: $C_{25}H_{24}N_4O_4S$ mw: 476.59

TOXICITY DATA with REFERENCE
orl-rat TDLo:100 mg/kg (female 1-5D post):REP IJOCAP 20,135,81
uns-mus LD50:681 mg/kg IJOCAP 20,135,81

SAFETY PROFILE: Moderately toxic by an unspecified route. Experimental reproductive effects. When heated to decomposition it emits toxic fumes of NO_x and SO_x.

POM800 CAS:38029-10-6 ***HR: 3***
PYRBUTEROL HYDROCHLORIDE
mf: $C_{12}H_{20}N_2O_3 \cdot 2ClH$ mw: 313.26

SYNS: 6-(2-(tert-BUTYLAMINO)-1-HYDROXYETHYL)-3-HYDROXY-2-PYRIDINEMETHANOL DIHYDROCHLORIDE ◇ α^6-(((1,1-DIMETHYLETHYL)AMINO)METHYL)-3-HYDROXY-2,6-PYRIDINEDIMETHANOL DIHYDROCHLORIDE ◇ PIRBUTEROL DIHYDROCHLORIDE

TOXICITY DATA with REFERENCE
orl-rat TDLo:1 g/kg (female 17-22D post):REP YACHDS 8,731,80
orl-rbt TDLo:390 mg/kg (female 6-18D post):TER YACHDS 8,731,80

orl-rat LD50:2 g/kg MDACAP 19,382,83
scu-rat LD50:645 mg/kg YACHDS 8,689,80
ivn-rat LD50:53 mg/kg IYKEDH 13,637,82
orl-mus LD50:2 g/kg MDACAP 19,382,83
scu-mus LD50:280 mg/kg IYKEDH 13,637,82
ivn-mus LD50:42 mg/kg IYKEDH 13,637,82

SAFETY PROFILE: Poison by subcutaneous and intravenous routes. Moderately toxic by ingestion. An experimental teratogen. Experimental reproductive effects. When heated to decomposition it emits NO_x and HCl.

PON000 CAS:1606-67-3 ***HR: 3***
1-PYRENAMINE
mf: $C_{16}H_{11}N$ mw: 217.28

SYNS: 1-AMINOPYRENE ◇ 3-AMINOPYRENE

TOXICITY DATA with REFERENCE
mmo-sat 500 ng/plate MUREAV 138,113,84
dnd-esc 10 μmol/L PNCCA2 2,39,65
orl-rat LD50:1070 mg/kg AIHAAP 30,470,69
ipr-mus LD50:250 mg/kg AIPTAK 135,376,62

CONSENSUS REPORTS: Reported in EPA TSCA Inventory. EPA Genetic Toxicology Program.

SAFETY PROFILE: Poison by intraperitoneal route. Moderately toxic by ingestion. Mutation data reported. When heated to decomposition it emits toxic fumes of NO_x.

PON250 CAS:129-00-0 ***HR: 3***
PYRENE
mf: $C_{16}H_{10}$ mw: 202.26

PROP: Colorless solid, solutions have a slight blue color. Mp: 156°, d: 1.271 @ 23°, bp: 404°. Insol in water; fairly sol in organic solvents. (A condensed ring hydrocarbon).

SYNS: BENZO(def)PHENANTHRENE ◇ PYREN (GERMAN) ◇ β-PYRINE

TOXICITY DATA with REFERENCE
skn-rbt 500 mg/24H MLD 28ZPAK -,26,72
mma-sat 300 ng/plate ENMUDM 6(Suppl 2),1,84
dns-hmn:fbr 100 mg/L TXCYAC 21,151,81
skn-mus TDLo:10 g/kg/3W-I:ETA BJCAAI 10,363,56
orl-rat LD50:2700 mg/kg GTPZAB 15(2),59,71
ihl-rat LC50:170 mg/m^3 GTPZAB 15(2),59,71
orl-mus LD50:800 mg/kg GTPZAB 15(2),59,71
ipr-mus LD50:514 mg/kg PMRSDJ 1,682,81

CONSENSUS REPORTS: IARC Cancer Review: Group 3 IMEMDT 7,56,87; Animal No Evidence IMEMDT 32,431,83. EPA Extremely Hazardous Substances List. Reported in EPA TSCA Inventory. EPA Genetic Toxicology Program.

OSHA PEL: TWA 0.2 mg/m^3

SAFETY PROFILE: Poison by inhalation. Moderately toxic by ingestion and intraperitoneal routes. A skin irritant. Questionable carcinogen with experimental tumorigenic data. Human mutation data reported. When heated to decomposition it emits acrid smoke and irritating fumes.

PON500 CAS:1732-14-5 ***HR: 3***
N-PYREN-2-YLACETAMIDE
mf: $C_{18}H_{13}NO$ mw: 259.32

SYN: 2-ACETYLAMINOPYRENE

TOXICITY DATA with REFERENCE
orl-rat TDLo:5508 mg/kg/32W-C:NEO CNREA8 15,188,55

SAFETY PROFILE: Questionable carcinogen with experimental neoplastigenic data. When heated to decomposition it emits toxic fumes of NO_x.

PON750 CAS:73529-24-5 ***HR: D***
4-PYRENYLOXIRANE
mf: $C_{18}H_{12}O$ mw: 244.30

SYN: 4-EPOXYETHYLPYRENE

TOXICITY DATA with REFERENCE
mmo-sat 100 pmol/plate CNREA8 40,642,80
msc-ham:lng 1 μmol/L CNREA8 40,642,80

SAFETY PROFILE: Mutation data reported. When heated to decomposition it emits acrid smoke and irritating fumes.

POO000 CAS:97-11-0 ***HR: 2***
PYRETHRIN
mf: $C_{21}H_{28}O_3$ mw: 328.49

SYNS: 2-CYCLOPENTENYL-4-HYDROXY-3-METHYL-2-CYCLOPENTEN-1-ONE CHRYSANTHEMATE ◇ 3-(2-CYCLOPENTEN-1-YL)-2-METHYL-4-OXO-2-CYCLOPENTEN-1-YLCHRYSANTHEMUMATE ◇ 3-(2-CYCLOPENTENYL)-2-METHYL-4-OXO-2-CYCLOPENTENYL CHRYSANTHEMUMMONOCARBOXYLATE ◇ CYCLOPENTENYLRETHONYL CHRYSANTHEMATE ◇ ENT 22,952

TOXICITY DATA with REFERENCE
orl-rat LD50:1410 mg/kg ARSIM* 20,7,66

SAFETY PROFILE: Moderately toxic by ingestion. When heated to decomposition it emits acrid smoke and irritating fumes. See also other pyrethrin entries.

POO050 CAS:121-21-1 ***HR: 3***
PYRETHRIN I
mf: $C_{21}H_{28}O_3$ mw: 328.49

SYNS: CHRYSANTHEMUM MONOCARBOXYLIC ACID PYRETHROLONE ESTER ◇ PIRETRINA I (PORTUGUESE) ◇ PYRETHROLONE, CHRYSANTHEMUM MONOCARBOXYLIC ACID

ESTER ◇ (+)-PYRETHRONYL (+)-trans-CHRYSANTHEMATE ◇ RCRA WASTE NUMBER P008

TOXICITY DATA with REFERENCE
orl-rat LD50:260 mg/kg PCBPBS 2,308,72
ivn-rat LDLo:5 mg/kg BIOGAL 41(10),283,75

SAFETY PROFILE: Poison by ingestion and intravenous routes. When heated to decomposition it emits acrid smoke and irritating fumes. An insecticide. See also other pyrethrin entries.

POO100 CAS:121-29-9 ***HR: 3***
PYRETHRIN II
mf: $C_{22}H_{28}O_5$ mw: 372.50

PROP: Viscous liquid. Bp: 200° @ 0.1 mm (decomp).

SYNS: CHRYSANTHEMUMDICARBOXYLIC ACID MONOMETHYL ESTER PYRETHROLONE ESTER ◇ ENT 7,543 ◇ PYRETHRIN ◇ PYRETHROLONE CHRYSANTHEMUM DICARBOXLIC ACIDMETHYL ESTER ESTER ◇ PYRETHROLONE ESTER of CHRYSANTHEMUMDICARBOXYLIC ACID MONOMETHYL ESTER ◇ (+)-PYRETHRONYL (+)-PYRETHRATE ◇ PYRETRIN II

TOXICITY DATA with REFERENCE
unr-man LDLo:1029 mg/kg 85DCAI 2,73,70
orl-rat LD50:200 mg/kg ARSIM* 20,17,66
ivn-rat LD50:1 mg/kg PCBPBS 2,308,72

CONSENSUS REPORTS: Reported in EPA TSCA Inventory.

SAFETY PROFILE: Poison experimentally by ingestion and intravenous routes. Moderately toxic to humans by unspecified route. An allergen. When heated to decomposition it emits acrid smoke and irritating fumes. An insecticide. See also other pyrethrin entries.

POO250 CAS:8003-34-7 ***HR: 3***
PYRETHRINS

PROP: Viscous liquid. Bp: 170° @ 0.1 mm (decomp).

SYNS: BUHACH ◇ CHRYSANTHEMUM CINERAREAEFOLIUM ◇ CINERIN I or II ◇ DALMATION INSECT FLOWERS ◇ FIRMOTOX ◇ INSECT POWDER ◇ JASMOLIN I or II ◇ PYRETHRIN I or II ◇ PYRETHRUM (ACGIH) ◇ PYRETHRUM (insecticide) ◇ TRIESTE FLOWERS

TOXICITY DATA with REFERENCE
orl-rat TDLo:500 mg/kg (6-15D preg):REP JTEHD6 10,111,82
orl-hmn LDLo:1 g/kg 85GYAZ -,142,71
orl-chd LDLo:750 mg/kg CMEP** -,1,56
orl-rat LD50:200 mg/kg GUCHAZ 6,442,73
ipr-rat LD50:200 mg/kg WRPCA2 3,28,64
ivn-rat LDLo:5 mg/kg PCBPBS 2,301,72
orl-mus LD50:370 mg/kg EVHPAZ 14,15,76
ipr-mus LDLo:25 mg/kg PSEBAA 34,135,36

OSHA PEL: TWA 5 mg/m^3
ACGIH TLV: TWA 5 mg/m^3
DFG MAK: 5 mg/m^3

SAFETY PROFILE: Moderately toxic to humans by ingestion. Poison experimentally by ingestion, intraperitoneal, and intravenous route. Experimental reproductive effects. An allergen. It is rapidly detoxified in the gastrointestinal tract, but can cause gastrointestinal, respiratory, and central nervous system effects. A dose of 15 grams has caused the death of a child. Chronic exposures can cause liver damage. Combustible when exposed to heat or flame. When heated to decomposition it emits acrid smoke and irritating fumes. An insecticide. See also other pyrethrin entries.

POO750 CAS:154-69-8 ***HR: 3***
PYRIBENZAMINE HYDROCHLORIDE
mf: $C_{16}H_{21}N_3{\cdot}ClH$ mw: 291.86

PROP: Insol in benzene, ether, ethylacetate.

SYNS: 2-(BENZYL(2-(DIMETHYLAMINO)ETHYL)AMINO)PYRIDINE HYDROCHLORIDE ◇ 2-(BENZYL(2-(DIMETHYLAMINO)ETHYL) AMINO)PYRIDINE MONOHYDROCHLORIDE ◇ N-BENZYL-N-DIMETHYLAMINOETHYL α-AMINOPYRIDINE MONOHYDROCHLORIDE ◇ N-BENZYL-N',N'-DIMETHYL-N-2-PYRIDYL-ETHYLENEDIAMINE HYDROCHLORIDE ◇ N-BENZYL-N-α-PYRIDYL-N',N'-DIMETHYL-AETHYLENDIAMIN-HYDROCHLORID (GERMAN) ◇ DEHISTIN HYDROCHLORIDE ◇ N,N-DIMETHYL-N'-(2-PYRIDYL)-N'-BENZYL-ETHYLENEDIAMINE HYDROCHLORIDE ◇ PIRISTIN ◇ PYRABENZAMINE ◇ N^1-α-PYRIDYL-N^1-BENZYL-N,N-DIMETHYL ETHYLENEDIAMINE MONOHYDROCHLORIDE ◇ PYRINAMINE ◇ RESISTAMINE ◇ STANZAMINE ◇ TRIPELENNAMINE HYDROCHLORIDE ◇ TRIPELENNAMINE MONOHYDROCHLORIDE

TOXICITY DATA with REFERENCE
dns-rat:lvr 100 μmol/L ENMUDM 3,11,81
unr-man LDLo:15 mg/kg 85DCAI 2,73,70
orl-rat LD50:515 mg/kg JLCMAK 31,749,46
scu-rat LD50:225 mg/kg JLCMAK 31,749,46
ivn-rat LD50:12 mg/kg JLCMAK 31,749,46
orl-mus LD50:97 mg/kg JPETAB 99,488,50
ipr-mus LD50:47 mg/kg CTOXAO 16,17,80
scu-mus LD50:41 mg/kg JPETAB 113,72,55
ivn-mus LD50:9 mg/kg JPETAB 92,249,48
ivn-dog LDLo:49 mg/kg JPETAB 113,72,55
scu-rbt LD50:33 mg/kg JLCMAK 31,749,46
ivn-rbt LD50:9 mg/kg JLCMAK 31,749,46
orl-gpg LD50:155 mg/kg JPETAB 113,72,55
scu-gpg LD50:30 mg/kg JPETAB 113,72,55
ivn-ham LD50:13 mg/kg JPETAB 97,371,49

SAFETY PROFILE: A human poison by an unspecified route. An experimental poison by ingestion, intravenous, intraperitoneal, and subcutaneous routes. Mutation data reported. When heated to decomposition it emits very toxic fumes of NO_x and HCl.

POP000 CAS:119-12-0 **HR: 3**
PYRIDAPHENTHION
mf: $C_{14}H_{17}N_2O_4PS$ mw: 340.36

SYNS: AMERICAN CYANAMID 12,503 ◇ CL 12503 ◇ O,O-DIETHYL O-(2,3-DIHYDRO-3-OXO-2-PHENYL-6-PYRIDAZINYL)PHOSPHOROTHIOATE ◇ O,O-DIETHYLPHOSPHOROTHIOATE, O-ESTER with 6-HYDROXY-2-PHENYL-3(2H)-PYRIDAZINONE ◇ O-(1,6)-DIHYDRO-6-OXO-1-PHENYLPYRIDAZIN-3-LY), O,O-DIETHYL PHOSPHOROTHIOATE ◇ ENT 23,968 ◇ OFNACK ◇ OFUNACK ◇ PYRIDAFENTHION

TOXICITY DATA with REFERENCE
orl-rat LD50:850 mg/kg FMCHA2 -,C172,83
skn-rat LD50:2100 mg/kg FMCHA2 -,C172,83
orl-mus LD50:459 mg/kg NEZAAQ 27,111,72
ipr-mus LD50:64 mg/kg 28ZEAL 4,170,69

SAFETY PROFILE: Poison by intraperitoneal route. Moderately toxic by ingestion and skin contact. When heated to decomposition it emits very toxic fumes of SO_x, PO_x, and NO_x. Used to control chewing and sucking insects on rice, fruits, vegetables, and cereals.

POP100 CAS:26445-05-6 **HR: 2**
PYRIDINAMINE (9CI)
mf: $C_5H_6N_2$ mw: 94.13

SYNS: AMINOPYRIDINE ◇ AMINOPYRINE ◇ PYRIDINE, AMINO-

TOXICITY DATA with REFERENCE
scu-mus TDLo:600 mg/kg (female 7-9D post):TER RCSADO 1,263,80
orl-mus LD50:658 mg/kg NYKZAU 62(1),24S,66
orl-rbt LD50:741 mg/kg NYKZAU 62(1),24S,66

SAFETY PROFILE: Moderately toxic by ingestion. An experimental teratogen. When heated to decomposition it emits toxic fumes of NO_x.

POP250 CAS:110-86-1 **HR: 3**
PYRIDINE
DOT: UN 1282
mf: C_5H_5N mw: 79.11

N=CHCH=CHCH=CH (ring)

PROP: Colorless liquid; sharp, penetrating, empyreumatic odor; burning taste. Bp: 115.3°, lel: 1.8%, uel: 12.4%, fp: −42°, flash p: 68°F (CC), d: 0.982, autoign temp: 900°F, vap press: 10 mm @ 13.2°, vap d: 2.73. Volatile with steam; misc with water, alc, ether.

SYNS: AZABENZENE ◇ AZINE ◇ NCI-C55301 ◇ PIRIDINA (ITALIAN) ◇ PIRYDYNA (POLISH) ◇ PYRIDIN (GERMAN) ◇ RCRA WASTE NUMBER U196

TOXICITY DATA with REFERENCE
skn-rbt 10 mg/24H open MLD AMIHBC 4,119,51
eye-rbt 2 mg open SEV AMIHBC 4,119,51
mma-sat 6 mmol/L/2H CNREA8 39,4152,79
orl-rat LD50:891 mg/kg BIOFX* 14-4/70
ihl-rat LC50:4000 ppm/4H AMIHBC 4,119,51
ipr-rat LD50:866 mg/kg NTIS** PB195-158
scu-rat LD50:1000 mg/kg PSEBAA 62,19,46
ipr-mus LD50:1108 mg/kg JPETAB 88,82,46
scu-mus LDLo:1000 mg/kg CNREA8 22,483,62
ivn-mus LD50:538 mg/kg JPETAB 88,82,46
ivn-dog LD50:880 mg/kg TXCYAC 4,165,75
skn-rbt LD50:1121 mg/kg BIOFX* 14-4/70
ipr-rbt LDLo:15 mg/kg JIDIAQ 42,473,28
orl-gpg LDLo:4000 mg/kg JPHYA7 17,272,1894
ipr-gpg LDLo:870 mg/kg JPHYA7 17,272,1894

CONSENSUS REPORTS: Community Right-To-Know List. Reported in EPA TSCA Inventory. EPA Genetic Toxicology Program.

OSHA PEL: TWA 5 ppm
ACGIH TLV: TWA 5 ppm
DFG MAK: 5 ppm (15 mg/m^3)
DOT Classification: Flammable Liquid; Label: Flammable Liquid; Flammable Liquid; Label: Flammable Liquid, Poison.

SAFETY PROFILE: Poison by intraperitoneal route. Moderately toxic by ingestion, skin contact, intravenous, and subcutaneous routes. Mildly toxic by inhalation. A skin and severe eye irritant. Mutation data reported. Can cause central nervous system depression, gastrointestinal upset, and liver and kidney damage. Dangerous fire hazard when exposed to heat, flame, or oxidizers. Severe explosion hazard in the form of vapor when exposed to flame or spark. Reacts violently with chlorosulfonic acid; chromium trioxide; dinitrogen tetraoxide; HNO_3; oleum; perchromates; β-propiolactone; $AgClO_4$; H_2SO_4. Incandescent reaction with fluorine. Reacts to form pyrophoric or explosive products with bromine trifluoride; trifluoromethyl hypofluorite. Mixtures with formamide + iodine + sulfur trioxide are storage hazards which release carbon dioxide and sulfuric acid. Incompatible with oxidizing materials. Reacts with maleic anhydride (above 150°C) evolving carbon dioxide. To fight fire, use alcohol foam. When heated to decomposition it emits highly toxic fumes of NO_x.

POP750 CAS:156-25-2 **HR: 3**
PYRIDINE-3-AZO-p-DIMETHYLANILINE
mf: $C_{13}H_{14}N_4$ mw: 226.31

SYNS: 3′-(4-DIMETHYLAMINOPHENYL)AZOPYRIDINE ◇ N,N-DIMETHYL-p-(3-PYRIDYLAZO)ANILINE ◇ N,N-DIMETHYL-4-(3′-PYRIDYLAZO)ANILINE

TOXICITY DATA with REFERENCE
orl-rat TDLo:3530 mg/kg/28W-C:ETA JNCIAM 15,67,54

SAFETY PROFILE: Questionable carcinogen with ex-

perimental tumorigenic data. When heated to decomposition it emits toxic fumes of NO_x.

POQ000 CAS:63019-82-9 ***HR: 3***
PYRIDINE-4-AZO-p-DIMETHYLANILINE
mf: $C_{13}H_{14}N_4$ mw: 226.31

TOXICITY DATA with REFERENCE
orl-rat TDLo:6480 mg/kg/26W-C:ETA CNREA8 14,22,54

SAFETY PROFILE: Questionable carcinogen with experimental tumorigenic data. When heated to decomposition it emits toxic fumes of NO_x.

POQ250 CAS:110-51-0 ***HR: 3***
PYRIDINE BORANE
mf: C_5H_8BN mw: 92.95

SYNS: BORANE-PYRIDINE ◇ PB ◇ PYRIDINE compounded with BORON HYDRIDE

TOXICITY DATA with REFERENCE
orl-rat LDLo:94500 μg/kg 14KTAK -,693,64
ipr-rat LDLo:64800 μg/kg 14KTAK -,693,64
skn-gpg LD50:200 mg/kg 14KTAK -,693,64
ipr-gpg LD50:54 mg/kg 14KTAK -,693,64

SAFETY PROFILE: Poison by ingestion, skin contact, and intraperitoneal routes. Potentially hazardous rapid decomposition at 120°C/7.5 mbar. When heated to decomposition it emits toxic fumes of NO_x. See also BORON COMPOUNDS and PYRIDINE.

POQ500 CAS:586-92-5 ***HR: 3***
3-PYRIDINEDIAZONIUM TETRAFLUOROBORATE
mf: $C_5H_4BF_4N_3$ mw: 192.91

SAFETY PROFILE: The dry material is spontaneously explosive above 15°C. When heated to decomposition it emits very toxic fumes of NO_x, F^-, and BO_x. See also BORON COMPOUNDS and FLUORIDES.

POQ750 CAS:4664-00-0 ***HR: D***
2,3-PYRIDINEDICARBOXIMIDE
mf: $C_7H_4N_2O_2$ mw: 148.13

SYNS: 3-AZAPHTHALIMID (GERMAN) ◇ QUINOLINIC ACID IMIDE

TOXICITY DATA with REFERENCE
ipr-mus TDLo:25 mg/kg (9D preg):REP PHARAT 31,172,76
ipr-mus TDLo:25 mg/kg (9D preg):TER PHARAT 31,172,76

SAFETY PROFILE: An experimental teratogen. Experimental reproductive effects. When heated to decomposition it emits toxic fumes of NO_x.

POR000 CAS:4664-01-1 ***HR: D***
3,4-PYRIDINEDICARBOXIMIDE
mf: $C_7H_4N_2O_2$ mw: 148.13

SYNS: 4-AZAPHTHALIMID (GERMAN) ◇ CINCHOMERONIC ACID IMIDE

TOXICITY DATA with REFERENCE
ipr-mus TDLo:25 mg/kg (9D preg):REP PHARAT 31,172,76
ipr-mus TDLo:12500 μg/kg (9D preg):TER PHARAT 31,172,76

SAFETY PROFILE: An experimental teratogen. Experimental reproductive effects. When heated to decomposition it emits toxic fumes of NO_x.

POR250 ***HR: D***
2,6-PYRIDINEDIMETHANOL, METHYL CARBAMATE NITROSOCARBAMATE
mf: $C_{11}H_{13}N_4O_5$ mw: 281.28

SYN: N-MONONITROSOPYRIDINOL CARBAMATE

TOXICITY DATA with REFERENCE
mmo-sat 10 nmol/plate CALEDQ 18,143,83
mmo-esc 10 nmol/plate CALEDQ 18,143,83

SAFETY PROFILE: Mutation data reported. Many N-nitroso compounds are carcinogens. When heated to decomposition it emits toxic fumes of NO_x. See also N-NITROSO COMPOUNDS, CARBAMATES, and ESTERS.

POR500 CAS:5344-27-4 ***HR: 3***
4-PYRIDINEETHANOL
mf: C_7H_9NO mw: 123.17

SYN: 4-ETHANOLPYRIDINE

TOXICITY DATA with REFERENCE
ivn-mus LD50:320 mg/kg CSLNX* NX#01586

CONSENSUS REPORTS: Reported in EPA TSCA Inventory.

SAFETY PROFILE: Poison by intravenous route. When heated to decomposition it emits toxic fumes of NO_x.

POR750 CAS:628-13-7 ***HR: 2***
PYRIDINE HYDROCHLORIDE
mf: $C_5H_5N•ClH$ mw: 115.57

PROP: Hygroscopic plates from alc. Mp: 82°, bp: 218-219°. Sol in water, alc, benzene; insol in ether.

TOXICITY DATA with REFERENCE
orl-rat LDLo:1600 mg/kg KODAK* -,-,71
skn-rat LDLo:2500 mg/kg KODAK* -,-,71
ipr-rat LDLo:800 mg/kg KODAK* -,-,71

CONSENSUS REPORTS: Reported in EPA TSCA Inventory.

SAFETY PROFILE: Moderately toxic by ingestion, skin contact and intraperitoneal routes. When heated to decomposition it emits very toxic fumes of NO_x and HCl.

POR800 CAS:586-98-1 ***HR: 2***
2-PYRIDINEMETHANOL
mf: C_6H_7NO mw: 109.14

SYNS: 2-(HYDROXYMETHYL)PYRIDINE ◇ α-PICOLYL ALCOHOL ◇ PYRIDINE-2-CARBINOL ◇ W-PYRIDINYLMETHANOL ◇ 2-PYRIDYLCARBINOL ◇ 2-PYRIDYLMETHANOL

TOXICITY DATA with REFERENCE
ivn-mus LD50:1000 mg/kg PHARAT 11,242,56
orl-qal LD50:1 g/kg AECTCV 12,355,83
orl-bwd LD50:750 mg/kg AECTCV 12,355,83

SAFETY PROFILE: Moderately toxic by ingestion and intravenous routes. When heated to decomposition it emits toxic fumes of NO_x.

POS000 CAS:694-59-7 ***HR: 3***
PYRIDINE-1-OXIDE
mf: C_5H_5NO mw: 95.11

PROP: Crystals, water-sol. Fp: 67°, bp: 102° @ 1 mm.

SYN: PYRIDINE N-OXIDE

TOXICITY DATA with REFERENCE
ipr-mus LD50:1425 mg/kg JPPMAB 16,472,64
ivn-mus LD50:180 mg/kg CSLNX* NX#04070
orl-qal LD50:1000 mg/kg JRPFA4 48,371,76
orl-bwd LD50:1 g/kg AECTCV 12,355,83

CONSENSUS REPORTS: Reported in EPA TSCA Inventory. EPA Genetic Toxicology Program.

SAFETY PROFILE: Poison by intravenous route. Moderately toxic by ingestion and intraperitoneal routes. Flammable. Potentially explosive reaction with hexamethyldisilane + tetrabutylammonium fluoride. Incompatible with oxidizing materials. When heated to decomposition it emits toxic fumes of NO_x.

POS250 CAS:59405-47-9 ***HR: 3***
PYRIDINE-1-OXIDE-3-AZO-p-DIMETHYLANILINE
mf: $C_{13}H_{14}N_4O$ mw: 242.31

TOXICITY DATA with REFERENCE
orl-rat TDLo:4300 mg/kg/17W-C:ETA CNREA8 14,715,54

SAFETY PROFILE: Questionable carcinogen with experimental tumorigenic data. When heated to decomposition it emits toxic fumes of NO_x.

POS500 CAS:13520-96-2 ***HR: 3***
PYRIDINE-1-OXIDE-4-AZO-p-DIMETHYLANILINE
mf: $C_{13}H_{14}N_4O$ mw: 242.31

TOXICITY DATA with REFERENCE
orl-rat TDLo:1134 mg/kg/9W-C:NEO JNCIAM 37,365,66
orl-rat TD:4320 mg/kg/17W-C:ETA CNREA8 14,22,54

SAFETY PROFILE: Questionable carcinogen with experimental neoplastigenic and tumorigenic data. When heated to decomposition it emits toxic fumes of NO_x.

POS750 CAS:94-63-3 ***HR: 3***
PYRIDINIUM-2-ALDOXIME-N-METHYLIODIDE
mf: $C_7H_9N_2O \cdot I$ mw: 264.08

PROP: Water-sol crystals. Mp: 214°.

SYNS: 2-FORMYL-1-METHYLPYRIDINIUM IODIDE OXIME ◇ 2-FORMYL-N-METHYLPYRIDINIUM OXIME IODIDE ◇ 2-HYDROXYIMINOMETHYL-1-METHYLPYRIDINIUM IODIDE ◇ 1-METHYL-2-ALDOXIMINOPYRIDINIUM IODIDE ◇ 1-METHYL-2-HYDROXYIMINOMETHYLPYRIDINIUM IODIDE ◇ N-METHYLPYRIDINE-2-ALDOXIME IODIDE ◇ N-METHYLPYRIDINIUM-2-ALDOXIME IODIDE ◇ NSC-7760 ◇ PAM (CZECH) ◇ 2-PAM IODIDE ◇ PRALIDOXIME IODIDE ◇ PRALIDOXIME METHIODIDE ◇ PROTOPAM IODIDE ◇ 2-PYRIDINALDOXIM METHOJODID (GERMAN) ◇ PYRIDIN-2-ALDOXIN (CZECH) ◇ 2-PYRIDINE ALDOXIME IODOMETHYLATE ◇ PYRIDINE-2-ALDOXIME METHIODIDE ◇ PYRIDINE-2-ALDOXIME METHYL IODIDE

TOXICITY DATA with REFERENCE
ipr-rat LD50:305 mg/kg BJPCAL 13,202,58
ivn-rat LD50:178 mg/kg JPETAB 129,31,60
ims-rat LD50:249 mg/kg JPETAB 129,31,60
orl-mus LD50:1500 mg/kg JPMSAE 53,1143,64
ipr-mus LD50:136 mg/kg JPMSAE 53,1143,64
scu-mus LD50:140 mg/kg AEHLAU 5,21,62
ivn-mus LD50:134 mg/kg 28ZPAK -,148,72
ims-mus LD50:240 mg/kg ARZNAD 14,5,64

SAFETY PROFILE: Poison by subcutaneous, intravenous, intramuscular, and intraperitoneal routes. Moderately toxic by ingestion. Used as an antidote to the cholinesterase inhibitors of the parathion group. When heated to decomposition it emits highly toxic fumes of NO_x and I^-.

POT000 CAS:19142-70-2 ***HR: 3***
PYRIDINIUM, 3,3'-(2-AMINOTEREPHTHALOYLBIS(IMINO-p-PHENYLENECARBONYLIMINO))BIS(1-PROPYL-, DI-p-TOLUENE SULFONATE
mf: $C_{38}H_{39}N_7O_4 \cdot 2C_7H_7O_3S$ mw: 1000.24

TOXICITY DATA with REFERENCE
dnd-mus:lym 530 nmol/L JMCMAR 22,134,79
ipr-mus LD10:47 mg/kg JMCMAR 22,134,79

SAFETY PROFILE: Poison by intraperitoneal route. Mutation data reported. When heated to decomposition it emits very toxic fumes of SO_x and NO_x.

POT500 CAS:60172-15-8 ***HR: 3***
PYRIDINIUM, 3,3'-(BICYCLO(2.2.2)OCTANE-1,4-DIYLBIS(CARBONYLIMINO-4,1-PHENYLENE-CARBONYLIMINO))BIS-1-PROPYL-, salt with 4-METHYLBENZENESULFONIC ACID (1:2)
mf: $C_{40}H_{46}N_6O_4 \cdot 2C_7H_7O_3S$ mw: 1017.32

TOXICITY DATA with REFERENCE
dnd-mus:lym 6800 nmol/L JMCMAR 22,134,79
ipr-mus LD10:30 mg/kg JMCMAR 22,134,79

SAFETY PROFILE: Poison by intraperitoneal route. Mutation data reported. When heated to decomposition it emits very toxic fumes of SO_x and NO_x.

POT750 CAS:19060-45-8 ***HR: 3***
PYRIDINIUM, 1-BUTYL-3-(p-(p-((p-((1-BUTYL-PYRIDINIUM-3-YL)CARBAMOYL)PHENYL)CARBAMOYL)CINNAMAMIDO) BENZAMIDO)-, DI-p-TOLUENE SULFONATE
mf: $C_{42}H_{44}N_6O_4 \cdot 2C_7H_7O_3S$ mw: 1039.32

TOXICITY DATA with REFERENCE
dnd-mus:lym 300 nmol/L JMCMAR 22,134,79
ipr-mus LD10:24 mg/kg JMCMAR 22,134,79

SAFETY PROFILE: Poison by intraperitoneal route. Mutation data reported. When heated to decomposition it emits very toxic fumes of SO_x and NO_x.

POU000 CAS:16802-50-9 ***HR: 3***
PYRIDINIUM, 3,3'-(2-CHLOROTEREPHTHALOYLBIS (IMINO-p-PHENYLENECARBONYLIMINO))BIS(1-ETHYL-, DI-p-TOLUENE SULFONATE
mf: $C_{36}H_{33}ClN_6O_4 \cdot 2C_7H_7O_3S$ mw: 991.60

TOXICITY DATA with REFERENCE
dnd-mus:lym 1 μmol/L JMCMAR 22,134,79
ipr-mus LD10:29 mg/kg JMCMAR 22,134,79

SAFETY PROFILE: Poison by intraperitoneal route. Mutation data reported. When heated to decomposition it emits very toxic fumes of Cl^-, NO_x, and SO_x.

POU250 CAS:16760-11-5 ***HR: 3***
PYRIDINIUM, 3,3'-(2-CHLOROTEREPHTHALOYLBIS(IMINO-p-PHENYLENECARBONYLIMINO))BIS(1-METHYL-, DI-p-TOLUENE SULFONATE)
mf: $C_{34}H_{29}ClN_6O_4 \cdot 2C_7H_7O_3S$ mw: 963.54

TOXICITY DATA with REFERENCE
dnd-mus:lym 1 μmol/L JMCMAR 22,134,79
ipr-mus LD10:30 mg/kg JMCMAR 22,134,79

SAFETY PROFILE: Poison by intraperitoneal route. Mutation data reported. When heated to decomposition it emits very toxic fumes of SO_x, Cl^-, and NO_x.

POV250 CAS:19060-43-6 ***HR: 3***
PYRIDINIUM, 1-ETHYL-3- p-(-((p-((1-ETHYL-PYRIDINIUM-3-YL)CARBAMYL)PHENYLOCARBAMOYL)CINNAMAMIDO)BENZAMIDO)-, DI-p-TOLUENE SULFONATE
mf: $C_{38}H_{36}N_6O_4 \cdot 2C_7H_7O_3S$ mw: 983.20

TOXICITY DATA with REFERENCE
dnd-mus:lym 300 nmol/L JMCMAR 22,134,79
ipr-mus LD10:55 mg/kg JMCMAR 22,134,79

SAFETY PROFILE: Poison by intraperitoneal route. Mutation data reported. When heated to decomposition it emits very toxic fumes of NO_x and SO_x.

POV500 CAS:20302-25-4 ***HR: 3***
PYRIDINIUM, 1-ETHYL-3-(p-(p-((p-(1-ETHYLPYRIDINIUM-3-YL)PHENYL)CARBAMOYL)CINNAMAMIDO)PHENYL)-DI-p-TOLUENE SULFONATE
mf: $C_{36}H_{34}N_4O_2 \cdot 2C_7H_7O_3S$ mw: 897.14

TOXICITY DATA with REFERENCE
dnd-mus:lym 300 nmol/L JMCMAR 22,134,79
ipr-mus LD10:60 mg/kg JMCMAR 22,134,79

SAFETY PROFILE: Poison by intraperitoneal route. Mutation data reported. When heated to decomposition it emits very toxic fumes of NO_x and SO_x.

POV750 CAS:74203-42-2 ***HR: 3***
PYRIDINIUM, 2-FORMYL-1-METHYL-, CHLORIDE, OXIME mixed with 1,1'-TRIMETHYLENEBIS(4-FORMYLPYRIDINIUM), DICHLORIDE, DIOXIME (1:1)

TOXICITY DATA with REFERENCE
ivn-mus LD50:61 mg/kg JPETAB 132,50,61
ivn-rbt LD50:61 mg/kg JPETAB 132,50,61

SAFETY PROFILE: Poison by intravenous route. When heated to decomposition it emits very toxic fumes of Cl^- and NO_x.

POW000 CAS:19060-75-4 ***HR: 3***
PYRIDINIUM, 3,3'-(2-METHOXYTEREPHTHALOYLBIS(IMINO-p-PHENYLENECARBONYLIMINO))BIS(1-ETHYL)-, DI-p-TOLUENE SULFONATE
mf: $C_{37}H_{36}N_6O_5 \cdot 2C_7H_7O_3S$ mw: 987.19

TOXICITY DATA with REFERENCE
dnd-mus:lym 640 nmol/L JMCMAR 22,134,79
ipr-mus LD10:20 mg/kg JMCMAR 22,134,79

SAFETY PROFILE: Poison by intraperitoneal route. Mutation data reported. When heated to decomposition it emits very toxic fumes of NO_x and SO_x.

POW250 CAS:19060-74-3 ***HR: 3***
PYRIDINIUM, 3,3'-(2-METHOXY-TEREPHTHALOYLBIS(IMINO-p-PHENYLENECARBONYLIMINO))BIS(1-METHYL)-DI-p-TOLUENE SULFONATE
mf: $C_{35}H_{32}N_6O_5{\bullet}2C_7H_7O_3S$ mw: 959.13

TOXICITY DATA with REFERENCE
dnd-mus:lym 640 nmol/L JMCMAR 22,134,79
ipr-mus LD10:18 mg/kg JMCMAR 22,134,79

SAFETY PROFILE: Poison by intraperitoneal route. Mutation data reported. When heated to decomposition it emits very toxic fumes of SO_x and NO_x.

POW500 CAS:19060-76-5 ***HR: 3***
PYRIDINIUM, 3,3'-(2-METHOXYTEREPHTHALOYLBIS(IMINO-p-PHENYLENECARBONYLIMINO))BIS(1-PROPYL)-DI-p-TOLUENE SULFONATE
mf: $C_{39}H_{40}N_6O_5{\bullet}2C_7H_7O_3S$ mw: 1015.25

TOXICITY DATA with REFERENCE
dnd-mam:lym 640 nmol/L JMCMAR 22,134,79
ipr-mus LD10:22 mg/kg JMCMAR 22,134,79

SAFETY PROFILE: Poison by intraperitoneal route. Mutation data reported. When heated to decomposition it emits very toxic fumes of SO_x and NO_x.

POX250 CAS:102584-11-2 ***HR: 3***
PYRIDINIUM, 1-METHYL-3-(p-(p-((p-((1-METHYLPYRIDINIUM-3-YL)AMINO)PHENYL)CARBAMOYL)BENZAMIDO) BENZAMIDO)-, DIIODIDE
mf: $C_{33}H_{30}N_6O_3{\bullet}2I$ mw: 812.49

TOXICITY DATA with REFERENCE
dnd-mus:lym 630 nmol/L JMCMAR 22,134,79
ipr-mus LD10:4500 μg/kg JMCMAR 22,134,79

SAFETY PROFILE: Poison by intraperitoneal route. Mutation data reported. When heated to decomposition it emits very toxic fumes of NO_x and I^-.

POX500 CAS:68797-91-1 ***HR: 3***
PYRIDINIUM, 1-METHYL-3-(p-(p-((p-((1-METHYLPYRIDINIUM-4-YL)AMINO)PHENYL)CARBAMOYL)BENZAMIDO) BENZAMIDO)-, DI-p-TOLUENE SULFONATE
mf: $C_{33}H_{30}N_6O_3{\bullet}2C_7H_7O_3S$ mw: 901.09

TOXICITY DATA with REFERENCE
dnd-mus:lym 840 nmol/L JMCMAR 22,134,79
ipr-mus LD10:9 mg/kg JMCMAR 22,134,79

SAFETY PROFILE: Poison by intraperitoneal route. Mutation data reported. When heated to decomposition it emits very toxic fumes of SO_x and NO_x.

POX750 CAS:20719-22-6 ***HR: 3***
PYRIDINIUM, 1-METHYL-4-(p-(p-((p-((1-METHYLPYRIDINIUM-4-YL)AMINO)PHENYL)CARBAMOYL)CINNAMAMIDO)ANILINO)-, DI-p-TOLUENE SULFONATE
mf: $C_{34}H_{32}N_6O_2{\bullet}2C_7H_7O_3S$ mw: 899.12

TOXICITY DATA with REFERENCE
dnd-mus:lym 430 nmol/L JMCMAR 22,134,79
ipr-mus LD10:11 mg/kg JMCMAR 22,134,79

SAFETY PROFILE: Poison by intraperitoneal route. Mutation data reported. When heated to decomposition it emits very toxic fumes of SO_x and NO_x.

POY250 CAS:68771-76-6 ***HR: 3***
PYRIDINIUM, 1-METHYL-3-(p-(p-((((1-METHYLPYRIDINIUM-3-YL)CARBAMOYL)1,4-NAPHTHYL)CARBAMOYL)BENZAMIDO)BENZAMIDO)-, DI-p-TOLUENE SULFONATE
mf: $C_{38}H_{32}N_6O_4{\bullet}2C_7H_7O_3S$ mw: 979.16

TOXICITY DATA with REFERENCE
dnd-mus:lym 2200 nmol/L JMCMAR 22,134,79
ipr-mus LD10:75 mg/kg JMCMAR 22,134,79

SAFETY PROFILE: Poison by intraperitoneal route. Mutation data reported. When heated to decomposition it emits very toxic fumes of SO_x and NO_x.

POY750 CAS:102584-14-5 ***HR: 3***
PYRIDINIUM, 1-METHYL-3-(m-(p-((p-((1-METHYLPYRIDINIUM-3-YL)CARBAMOYL)PHENYL)CARBAMOYL)BENZAMIDO) BENZAMIDO)-, DIIODIDE
mf: $C_{34}H_{30}N_6O_3{\bullet}2I$ mw: 824.50

TOXICITY DATA with REFERENCE
dnd-mus:lym 830 nmol/L JMCMAR 22,134,79
ipr-mus LD10:17 mg/kg JMCMAR 22,134,79

SAFETY PROFILE: Poison by intraperitoneal route. Mutation data reported. When heated to decomposition it emits very toxic fumes of I^- and NO_x.

PPA000 CAS:19060-37-8 ***HR: 3***
PYRIDINIUM, 1-METHYL-3-(p-(p-((p-(1-METHYLPYRIDINIUM-3-YL)PHENYL)CARBAMOYL)BENZAMIDO)BENZAMIDO)-, DI-p-TOLUENE SULFONATE
mf: $C_{33}H_{29}N_5O_3{\bullet}2C_7H_7O_3S$ mw: 886.07

TOXICITY DATA with REFERENCE
dnd-mus:lym 630 nmol/L JMCMAR 22,134,79
ipr-mus LD10:14 mg/kg JMCMAR 22,134,79

SAFETY PROFILE: Poison by intraperitoneal route.

Mutation data reported. When heated to decomposition it emits very toxic fumes of NO_x and SO_x.

PPA250 CAS:19060-39-0 ***HR: 3***
PYRIDINIUM, 1-METHYL-3-(p-(p-((p-(1-METHYLPYRIDINIUM-3-YL)PHENYL)CARBAMOYL)CINNAMAMIDO)PHENYL)-, DI-p-TOLUENE SULFONATE
mf: $C_{34}H_{30}N_4O_2 \cdot 2C_7H_7O_3S$ mw: 869.08

TOXICITY DATA with REFERENCE
dnd-mus:lym 300 nmol/L JMCMAR 22,134,79
ipr-mus LD10:85 mg/kg JMCMAR 22,134,79

SAFETY PROFILE: Poison by intraperitoneal route. Mutation data reported. When heated to decomposition it emits very toxic fumes of NO_x and SO_x.

PPA750 CAS:16760-22-8 ***HR: 3***
PYRIDINIUM, 3,3'-(2-METHYLTEREPHTHALOYLBIS(IMINO-p-PHENYLENECARBONYLIMINO))BIS(1-ETHYL-DI-p-TOLUENE SULFONATE
mf: $C_{37}H_{36}N_6O_4 \cdot 2C_7H_7O_3S$ mw: 971.19

TOXICITY DATA with REFERENCE
dnd-mus:lym 1200 nmol/L JMCMAR 22,134,79
ipr-mus LD10:26 mg/kg JMCMAR 22,134,79

SAFETY PROFILE: Poison by intraperitoneal route. Mutation data reported. When heated to decomposition it emits very toxic fumes of NO_x and SO_x.

PPB000 CAS:16760-12-6 ***HR: 3***
PYRIDINIUM, 3,3'-(2-METHYL-TEREPHTHALOYLBIS(IMINO-p-PHENYLENECARBONYLIMINO))BIS(1-METHYL-DI-p-TOLUENE SULFONATE
mf: $C_{35}H_{32}N_6O_4 \cdot 2C_7H_7O_3S$ mw: 943.13

TOXICITY DATA with REFERENCE
dnd-mus:lym 1200 nmol/L JMCMAR 22,134,79
ipr-mus LD10:14 mg/kg JMCMAR 22,134,79

SAFETY PROFILE: Poison by intraperitoneal route. Mutation data reported. When heated to decomposition it emits very toxic fumes of NO_x and SO_x.

PPB250 CAS:16760-23-9 ***HR: 3***
PYRIDINIUM, 3,3'-(2-METHYLTEREPHTHALOYLBIS(IMINO-p-PHENYLENECARBONYLIMINO))BIS(1-PROPYL-, DI-p-TOLUENE SULFONATE
mf: $C_{39}H_{40}N_6O_4 \cdot 2C_7H_7O_3S$ mw: 999.25

TOXICITY DATA with REFERENCE
dnd-mus:lym 1200 nmol/L JMCMAR 22,134,79
ipr-mus LD10:30 mg/kg JMCMAR 22,134,79

SAFETY PROFILE: Poison by intraperitoneal route. Mutation data reported. When heated to decomposition it emits very toxic fumes of NO_x and SO_x.

PPB500 CAS:68771-70-0 ***HR: 3***
PYRIDINIUM, 3,3'-(1,4-NAPHTHYLENEBIS(CARBANYLIMINO-p-PHENYLENECARBONYLIMINO))BIS(1-METHYL-DI-p-TOLUENE SULFONATE
mf: $C_{38}H_{32}N_6O_4 \cdot 2C_7H_7O_3S$ mw: 979.16

TOXICITY DATA with REFERENCE
dnd-mus:lym 1500 nmol/L JMCMAR 22,134,79
ipr-mus LD10:140 mg/kg JMCMAR 22,134,79

SAFETY PROFILE: Poison by intraperitoneal route. Mutation data reported. When heated to decomposition it emits very toxic fumes of NO_x and SO_x.

PPB550 CAS:543-53-3 ***HR: 3***
PYRIDINIUM NITRATE
mf: $C_5H_6N_2O_3$ mw: 142.11

$$C_5H_5N^-H \cdot NO_3^-$$

SAFETY PROFILE: Explodes when heated, but is not shock- or friction-sensitive. When heated to decomposition it emits toxic fumes of NO_x. See also NITRATES.

PPC000 CAS:34211-26-2 ***HR: 3***
PYRIDINIUM, 1,1'-(OXYDIMETHYLENE)BIS(2-FORMYL-, DICHLORIDE DIOXIME
mf: $C_{14}H_{16}N_4O_3 \cdot 2Cl$ mw: 359.24

SYNS: BIS((HYDROXYIMINO-METHYL)-PYRIDINIUM-(1)-METHYL-AETHER-DICHLORID (GERMAN) ◇ BIS(2-HYDROXYIMINOMETHYL-PYRIDINIUM-1-METHYL)ETHER DICHLORIDE ◇ HS 4 ◇ PYRIDINIUM, 1,1'-(OXYBIS(METHYLENE)BIS(2-((HYDROXYIMINO)METHYL), DICHLORIDE (9CI)

TOXICITY DATA with REFERENCE
ipr-rat LD50:168 mg/kg ARTODN 41,301,79
ivn-rat LD50:67 mg/kg ARTODN 41,301,79

SAFETY PROFILE: Poison by intraperitoneal and intravenous routes. When heated to decomposition it emits very toxic fumes of Cl^- and NO_x.

PPC100 CAS:15598-34-2 ***HR: 3***
PYRIDINIUM PERCHLORATE
mf: $C_5H_6ClNO_4$ mw: 179.56

$$C_5H_5N^-H \cdot ClO_4^-$$

DOT Classification: Forbidden

SAFETY PROFILE: An explosive sensitive to impact or heating above 335°C. Addition of ammonium perchlorate decreases the temperature required for initiation.

When heated to decomposition it emits toxic fumes of Cl^- and NO_x. See also PERCHLORATES.

PPC250 CAS:21595-62-0 ***HR: 3***
PYRIDINIUM, 4,4'-(p-PHENYLENEBIS (ACRYLOYLIMINO-p-PHENYLENEIMINO)) BIS(1-ETHYL-, DI-p-TOLUENE SULFONATE)
mf: $C_{38}H_{38}N_6O_2•2C_7H_7O_3S$ mw: 953.22

TOXICITY DATA with REFERENCE
dnd-mus:lym 310 nmol/L JMCMAR 22,134,79
ipr-mus LD10:32 mg/kg JMCMAR 22,134,79

SAFETY PROFILE: Poison by intraperitoneal route. Mutation data reported. When heated to decomposition it emits very toxic fumes of NO_x and SO_x.

PPD000 CAS:19060-44-7 ***HR: 3***
PYRIDINIUM, 1-PROPYL-3-(p-(p-((p-((1-PROPYLPYRIDINIUM-3-YL)CARBAMOYL)PHENYL)CARBAMOYL)CINNAMAMIDO) BENZAMIDO)-, DI-p-TOLUENE SULFONATE
mf: $C_{40}H_{40}N_6O_4•2C_7H_7O_3S$ mw: 1011.26

TOXICITY DATA with REFERENCE
dnd-mus:lym 300 nmol/L JMCMAR 22,134,79
ipr-mus LD10:46 mg/kg JMCMAR 22,134,79

SAFETY PROFILE: Poison by intraperitoneal route. Mutation data reported. When heated to decomposition it emits very toxic fumes of NO_x and SO_x.

PPD250 CAS:19060-40-3 ***HR: 3***
PYRIDINIUM, 1-PROPYL-3-(p-(p-((p-(1-PROPYLPYRIDINIUM-3-YL)PHENYL)CARBAMOYL)CINNAMAMIDO)PHENYL)-,DI-p-TOLUENE SULFONATE
mf: $C_{32}H_{34}N_4O_2•2C_7H_7O_3S$ mw: 849.10

TOXICITY DATA with REFERENCE
dnd-mus:lym 300 nmol/L JMCMAR 22,134,79
ipr-mus LD10:20 mg/kg JMCMAR 22,134,79

SAFETY PROFILE: Poison by intraperitoneal route. Mutation data reported. When heated to decomposition it emits very toxic fumes of NO_x and SO_x.

PPD500 CAS:60172-05-6 ***HR: 3***
PYRIDINIUM, 4,4'-(SUBEROYLBIS(IMINO-p-PHENYLENEIMINO))BIS(1-METHYL-, DIBROMIDE
mf: $C_{32}H_{38}N_6O_2•2Br$ mw: 698.58

SYN: 4,4'-(SUBEROYLBIS(IMINO-p-PHENYLENEIMINO))BIS(1-METHYLPYRIDINIUM)DIBROMIDE

TOXICITY DATA with REFERENCE
dnd-mus:lym 40 μmol/L JMCMAR 22,134,79
ipr-mus LD10:30 mg/kg JMCMAR 22,134,79

SAFETY PROFILE: Poison by intraperitoneal route. Mutation data reported. When heated to decomposition it emits very toxic fumes of NO_x and Br^-.

PPE500 CAS:19142-71-3 ***HR: 3***
PYRIDINIUM, 3,3'-(TEREPHTHALOYLBIS (IMINO(3-CHLORO-p-PHENYLENE)CARBONYLIMINO))BIS(1-METHYL-DI-p-TOLUENE SULFONATE)
mf: $C_{34}H_{28}Cl_2N_6O_4•2C_7H_7O_3S$ mw: 997.98

TOXICITY DATA with REFERENCE
dnd-mus:lym 1400 nmol/L JMCMAR 22,134,79
ipr-mus LD10:26 mg/kg JMCMAR 22,134,79

SAFETY PROFILE: Poison by intraperitoneal route. Mutation data reported. When heated to decomposition it emits very toxic fumes of Cl^-, NO_x, and SO_x.

PPF250 CAS:19083-81-9 ***HR: 3***
PYRIDINIUM, 3,3'-(TEREPHTHALOYLBIS (IMINO(3-METHOXY-p-PHENYLENE)CARBONYLIMINO))BIS(1-ETHYL-, DI-p-TOLUENE SULFONATE)
mf: $C_{38}H_{38}N_6O_6•2C_7H_7O_3S$ mw: 1017.22

TOXICITY DATA with REFERENCE
dnd-mus:lym 400 nmol/L JMCMAR 22,134,79
ipr-mus LD10:40 mg/kg JMCMAR 22,134,79

SAFETY PROFILE: Poison by intraperitoneal route. Mutation data reported. When heated to decomposition it emits very toxic fumes of SO_x and NO_x.

PPG250 CAS:19083-82-0 ***HR: 3***
PYRIDINIUM, 3,3'-(TEREPHTHALOYLBIS (IMINO(3-METHOXY-p-PHENYLENE)CARBONYLIMINO))BIS(1-PROPYL-, DI-p-TOLUENE SULFONATE
mf: $C_{40}H_{42}N_6O_6•2C_7H_7O_3S$ mw: 1045.28

TOXICITY DATA with REFERENCE
dnd-mus:lym 400 nmol/L JMCMAR 22,134,79
ipr-mus LD10:37 mg/kg JMCMAR 22,134,79

SAFETY PROFILE: Poison by intraperitoneal route. Mutation data reported. When heated to decomposition it emits very toxic fumes of NO_x and SO_x.

PPH000 CAS:19146-99-7 ***HR: 3***
PYRIDINIUM, 3,3'-(TEREPHTHALOYLBIS (IMINO-p-PHENYLENECARBONYLIMINO))BIS(1-BUTYL, DI-p-TOLUENESULFONATE)
mf: $C_{40}H_{42}N_6O_4•2C_7H_7O_3S$ mw: 1013.28

TOXICITY DATA with REFERENCE
dnd-mus:lym 460 nmol/L JMCMAR 22,134,79
ipr-mus LD10:58 mg/kg JMCMAR 22,134,79

SAFETY PROFILE: Poison by intraperitoneal route. Mutation data reported. When heated to decomposition it emits very toxic fumes of NO_x and SO_x.

PPH050 CAS:1882-26-4 ***HR: 2***
PYRIDINOL CARBAMATE
mf: $C_{11}H_{15}N_3O_4$ mw: 253.29

PROP: Needles from methanol or acetone. Mp: 136-137°. Sparingly sol in cold water; freely sol in hot water.

SYNS: ANGININ ◇ ANGININE ◇ ANGIOXINE ◇ ATEROSAN ◇ ATOVER ◇ COLESTERINEX ◇ DUVALINE ◇ METHYLCARBAMIC ACID-2,6-PYRIDINEDIYLDIMETHYLENE ESTER ◇ PARMIDIN ◇ PARMIDINE ◇ PARMIDINE R ◇ PIRIDINOL CARBAMATO (SPANISH) ◇ PRODECTINE ◇ SOSPITAN ◇ VASAPRIL ◇ VASOCIL ◇ VASOVERIN ◇ VERANTEROL

TOXICITY DATA with REFERENCE
orl-rat LD50:1230 mg/kg AFTOD7 6,255,80
orl-mus LD50:3100 mg/kg PCJOAU 11,1573,77
orl-rbt LD50:4250 mg/kg NIIRDN 6,642,82

SAFETY PROFILE: Moderately toxic by ingestion. When heated to decomposition it emits toxic fumes of NO_x. See also CARBAMATES and ESTERS.

PPH100 CAS:71799-98-9 ***HR: D***
PYRIDINOL NITROSOCARBAMATE
mf: $C_{11}H_{13}N_5O_6$ mw: 311.29

SYNS: DINITROSOPRODECTIN ◇ N,N'-DINITROSOPYRIDINOL CARBAMATE ◇ N-DINITROSOPYRIDINOLCARBAMATE ◇ N-DINITROZO-PIRIDINOLKARBAMAT (HUNGARIAN)

TOXICITY DATA with REFERENCE
mmo-sat 1 μg/plate MGONAD 24,1,80
slt-dmg-orl 640 μmol/kg MUREAV 144,177,85

SAFETY PROFILE: Mutation data reported. Many N-nitroso compounds are carcinogens. When heated to decomposition it emits toxic fumes of NO_x. See also N-NITROSO COMPOUNDS, CARBAMATES, and ESTERS.

PPI750 CAS:101651-44-9 ***HR: D***
N-(5H-PYRIDO(4,3-b)INDOL-3-YL)ACETAMIDE
mf: $C_{13}H_{11}N_3O$ mw: 225.27

TOXICITY DATA with REFERENCE
mma-sat 100 μmol/L PNASA6 77,1427,80
dnd-mam:lym 100 μmol/L PNASA6 77,1427,80

SAFETY PROFILE: Mutation data reported. When heated to decomposition it emits toxic fumes of NO_x.

PPI775 CAS:18791-21-4 ***HR: 3***
PYRIDOMYCIN
mf: $C_{27}H_{32}N_4O_8$ mw: 540.63

PROP: Crystals from ethanol. Mp: 222°. Sol in lower alcs, ethyl or butyl acetate, benzene, acetone, dioxane, tetrahydrofuran; practically insol in water.

SYNS: ERIZOMYCIN ◇ U-24544

TOXICITY DATA with REFERENCE
orl-mus LD50:1000 mg/kg 85GDA2 4(2),144,80
ipr-mus LD50:1000 mg/kg 85GDA2 4(2),144,80
ivn-mus LD50:300 mg/kg 85FZAT -,541,67

SAFETY PROFILE: Poison by intravenous route. Moderately toxic by ingestion and intraperitoneal routes. When heated to decomposition it emits toxic fumes of NO_x.

PPI800 CAS:155-97-5 ***HR: 3***
PYRIDOSTIGMINE
mf: $C_9H_{13}N_2O_2$ mw: 181.24

SYNS: 3-(((DIMETHYLAMINO)CARBONYL)OXY)-1-METHYL-PYRIDINIUM (9CI) ◇ 3-HYDROXY-1-METHYL-PYRIDINIUM DIMETHYLCARBAMATE (ester)

TOXICITY DATA with REFERENCE
ims-rat LD50:2790 μg/kg DCTODJ 1,355,78
ipr-mus LD50:2700 μg/kg FAATDF 1,217,81
orl-rbt LD50:53900 μg/kg DCTODJ 3,319,80
ims-rbt LD50:2800 μg/kg DCTODJ 3,319,80

SAFETY PROFILE: Poison by ingestion, intramuscular, and intraperitoneal routes. When heated to decomposition it emits toxic fumes of NO_x. See also CARBAMATES and ESTERS.

PPI815 CAS:240-39-1 ***HR: 3***
12H-PYRIDO(2,3-a)THIENO(2,3-i)CARBAZOLE
mf: $N_2SC_{17}H_{10}$ mw: 274.35

TOXICITY DATA with REFERENCE
scu-mus TDLo:72 mg/kg/9W-I:NEO CHDDAT 271,1474,70

SAFETY PROFILE: Questionable carcinogen with experimental neoplastigenic data. When heated to decomposition it emits very toxic fumes of NO_x and SO_x.

PPJ900 CAS:447-05-2 ***HR: 1***
PYRIDOXINE PHOSPHATE
mf: $C_8H_{12}NO_6P$ mw: 249.18

SYNS: PYRIDOXINE-5-PHOSPHATE ◇ PYRIDOXINE-5'-PHOSPHATE ◇ PYRIDOXOL-5-(DIHYDROGEN PHOSPHATE) ◇ PYRIDOXOL-5-PHOSPHATE ◇ PYRIDOXOL-5'-PHOSPHATE

TOXICITY DATA with REFERENCE
ipr-rat LD50:7400 mg/kg GNRIDX 1,493,67
orl-mus LD50:7800 mg/kg GNRIDX 1,493,67
ipr-mus LD50:5600 mg/kg GNRIDX 1,493,67
scu-mus LD50:7100 mg/kg GNRIDX 1,493,67

SAFETY PROFILE: Mildly toxic by ingestion and other

routes. When heated to decomposition it emits toxic fumes of NO_x and PO_x.

PPK250 CAS:65-23-6 ***HR: 2***
PYRIDOXOL
mf: $C_8H_{11}NO_3$ mw: 169.20

SYNS: ADERMINE ◇ BEESIX ◇ GRAVIDOX ◇ HYDOXIN ◇ 3-HYDROXY-4,5-DIMETHYLOL-α-PICOLINE ◇ 5-HYDROXY-6-METHYL-3,4-PYRIDINEDIMETHANOL ◇ 3-HYDROXY-2-PICOLINE-4,5-DIMETHANOL ◇ 2-METHYL-4,5-BIS(HYDROXYMETHYL)-3-HYDROXYPYRIDINE ◇ 2-METHYL-3-HYDROXY-4,5-BIS(HYDROXYMETHYL)PYRIDINE ◇ 2-METHYL-3-HYDROXY-4,5-DIHYDROXYMETHYL-PYRIDIN (GERMAN) ◇ 2-METHYL-3-HYDROXY-4,5-DI(HYDROXYMETHYL) PYRIDINE ◇ PYRODOXIN ◇ PYRIDOXINE ◇ VITAMIN B6

TOXICITY DATA with REFERENCE
orl-rat LD50:4000 mg/kg PSEBAA 43,116,40
ipr-rat LD50:1500 mg/kg PSDTAP 4,179,64
scu-rat LD50:3100 mg/kg PSEBAA 43,116,40
ivn-rat LD50:657 mg/kg PSEBAA 44,147,40
ipr-mus LD50:966 mg/kg FATOAO 40(3),371,77
ivn-mus LD50:1150 mg/kg PSDTAP 4,179,64

CONSENSUS REPORTS: Reported in EPA TSCA Inventory.

SAFETY PROFILE: Moderately toxic by ingestion, subcutaneous, intravenous, and intraperitoneal routes. When heated to decomposition it emits toxic fumes of NO_x.

PPK500 CAS:58-56-0 ***HR: 3***
PYRIDOXOL HYDROCHLORIDE
mf: $C_8H_{11}NO_3$•ClH mw: 205.66

PROP: Commercial form of pyridoxine (Vitamin B_6). Colorless to white platelets or crystalline powder; odorless. Mp: 204-206° (decomp). Sol in water, alc, acetone; sltly sol in other organic solvents; insol in ether.

SYNS: ADERMINE HYDROCHLORIDE ◇ BECILAN ◇ BENADON ◇ CAMPOVITON 6 ◇ HEBABIONE HYDROCHLORIDE ◇ HEXABETALIN ◇ HEXAVIBEX ◇ HEXERMIN ◇ HEXOBION ◇ 3-HYDROXY-4,5-DIMETHYLOL-α-PICOLINE HYDROCHLORIDE ◇ 5-HYDROXY-6-METHYL-3,4-PYRIDINEDICARBINOLHYDROCHLORIDE ◇ 5-HYDROXY-6-METHYL-3,4-PYRIDINEDIMETHANOL HYDROCHLORIDE ◇ 2-METHYL-3-HYDROXY-4,5-BIS(HYDROXYMETHYL)PYRIDINE HYDROCHLORIDE ◇ PYRIDIPCA ◇ PYRIDOXINE HYDROCHLORIDE (FCC) ◇ PYRIDOXINIUM CHLORIDE ◇ PYRIDOXINUM HYDROCHLORICUM (HUNGARIAN) ◇ VITAMIN B6-HYDROCHLORIDE

TOXICITY DATA with REFERENCE
sce-hmn:lym 2 mg/L MUREAV 124,175,83
orl-wmn TDLo:2 mg/kg (38W preg):REP APSVAM 72,525,83
orl-rat LD50:4000 mg/kg ARZNAD 11,922,61
scu-rat LD50:3 g/kg ARZNAD 11,922,61
ivn-rat LD50:530 mg/kg ARZNAD 11,922,61
orl-mus LD50:5500 mg/kg ARZNAD 11,922,61
scu-mus LD50:2450 mg/kg ARZNAD 11,922,61
ivn-mus LD50:660 mg/kg ARZNAD 11,922,61
orl-cat LDLo:1000 mg/kg ARZNAD 11,922,61
ivn-cat LD50:560 mg/kg ARZNAD 11,922,61
ims-cat LD50:500 mg/kg ARZNAD 11,922,61
ivn-rbt LD50:464 mg/kg ARZNAD 11,922,61
ivn-pgn LD50:145 mg/kg ARZNAD 11,922,61

CONSENSUS REPORTS: Reported in EPA TSCA Inventory.

SAFETY PROFILE: Poison by intravenous route. Moderately toxic by ingestion, intramuscular, and subcutaneous routes. Human reproductive effects by ingestion and intramuscular routes: postpartum changes. Experimental reproductive effects. Human mutation data reported. When heated to decomposition it emits very toxic fumes of NO_x and HCl.

PPL500 CAS:19992-69-9 ***HR: 3***
1-(PYRIDYL-3)-3,3-DIMETHYL TRIAZENE
mf: $C_7H_{10}N_4$ mw: 150.21

SYNS: 1-(PYRIDYL-3)-3,3-DIMETHYL-TRIAZEN (GERMAN) ◇ 1-(m-PYRIDYL)-3,3-DIMETHYL-TRIAZENE

TOXICITY DATA with REFERENCE
hma-mus/lng 6250 μg/kg PSEBAA 158,269,78
mnt-mus-ipr 1250 μg/kg/24H MUREAV 56,319,78
scu-rat TDLo:10 mg/kg (female 10D post):TER IARCCD 4,45,73
orl-rat TDLo:160 mg/kg:NEO ZKKOBW 81,285,74
orl-rat LD50:160 mg/kg ZKKOBW 81,285,74
scu-rat LD50:130 mg/kg ZKKOBW 81,285,74

SAFETY PROFILE: Poison by ingestion and subcutaneous routes. Questionable carcinogen with experimental neoplastigenic data. An experimental teratogen. Mutation data reported. When heated to decomposition it emits toxic fumes of NO_x.

PPM550 ***HR: 3***
N-(2-(2-PYRIDYL)ETHYL)-PHTHALIMIDE HYDROCHLORIDE
mf: $C_{15}H_{12}N_2O_2$•ClH mw: 288.75

SYN: WIN 3074

TOXICITY DATA with REFERENCE
orl-rat LD50:860 mg/kg AIPTAK 88,142,51
ivn-rat LD50:72 mg/kg AIPTAK 88,142,51
orl-mus LD50:360 mg/kg AIPTAK 88,142,51
ivn-mus LD50:60 mg/kg AIPTAK 88,142,51
ivn-gpg LD50:100 mg/kg AIPTAK 88,142,51

SAFETY PROFILE: Poison by ingestion and intravenous routes. When heated to decomposition it emits toxic fumes of NO_x and HCl.

PPM725 ***HR: 3***
N-2-PYRIDYLMANDELAMIDE HYDROCHLORIDE
mf: $C_{13}H_{12}N_2O_2 \cdot ClH$ mw: 264.73

SYN: MANDELAMIDE, N-2-PYRIDYL-, HYDROCHLORIDE

TOXICITY DATA with REFERENCE
orl-mus LD50:2125 mg/kg JPETAB 128,65,60
ipr-mus LD50:345 mg/kg JPETAB 128,65,60
ivn-mus LD50:133 mg/kg JPETAB 128,65,60

SAFETY PROFILE: Poison by intravenous and intraperitoneal routes. Moderately toxic by ingestion. When heated to decomposition it emits toxic fumes of NO_x and HCl.

PPN000 CAS:10058-07-8 ***HR: 2***
7-(2-((3-PYRIDYLMETHYL)AMINO)ETHYL) THEOPHYLLINE NICOTINATE
mf: $C_{21}H_{23}N_7O_4$ mw: 437.51

SYNS: ES 902 ◇ 7-(2-(PIRIDIL)-METILAMMINO-ETIL)-TEOFILLINO NICOTINATO (ITALIAN) ◇ TEONICON

TOXICITY DATA with REFERENCE
orl-rat LD50:3700 mg/kg GMITAB 132,36,73
scu-rat LD50:2300 mg/kg GMITAB 132,36,73
orl-mus LD50:2530 mg/kg PAHEAA 48,133,73
ivn-mus LD50:470 mg/kg PAHEAA 48,133,73

SAFETY PROFILE: Moderately toxic by ingestion, subcutaneous, and intravenous routes. When heated to decomposition it emits toxic fumes of NO_x.

PPN250 CAS:64059-53-6 ***HR: D***
1-PYRIDYL-3-METHYL-3-ETHYLTRIAZENE
mf: $C_8H_{12}N_4$ mw: 164.24

SYN: 3-ETHYL-3-METHYL-1-PYRIDYL-TRIAZENE

TOXICITY DATA with REFERENCE
unr-rat TDLo:15 mg/kg (10D preg):TER XENOBH 3,271,73

SAFETY PROFILE: An experimental teratogen. When heated to decomposition it emits toxic fumes of NO_x.

PPN500 CAS:2845-82-1 ***HR: 3***
5-(4-PYRIDYL)-1,3,4-OXADIAZOL-2-OL
mf: $C_8H_5N_3O_2$ mw: 175.16

SYN: 2-PYRIDYL-(4)-1,3,4-OXADIAZOLON-(5)(GERMAN)

TOXICITY DATA with REFERENCE
orl-rbt LD50:1000 mg/kg ARZNAD 12,22,62
scu-rbt LD50:400 mg/kg ARZNAD 12,22,62
orl-gpg LD50:700 mg/kg ARZNAD 12,22,62
scu-gpg LD50:750 mg/kg ARZNAD 12,22,62

SAFETY PROFILE: Poison by subcutaneous route. Moderately toxic by ingestion. When heated to decomposition it emits toxic fumes of NO_x.

PPN750 CAS:599-79-1 ***HR: 2***
5-(p-(2-PYRIDYLSULFAMOYL)PHENYLAZO)SALICYLIC ACID
mf: $C_{18}H_{14}N_4O_5S$ mw: 398.42

SYNS: ACCUCOL ◇ ASULFIDINE ◇ AZOPYRIN ◇ AZULFIDINE ◇ 4-(PYRIDYL-2-AMIDOSULFONYL)-3'-CARBOXY-4'-HYDROXYAZOBENZENE ◇ 5-(4-(2-PYRIDYLSULFAMOYL)PHENYLAZO)-2-HYDROXYBENZOIC ACID ◇ RORASUL ◇ SALAZOPYRIN ◇ SALAZOSULFAPYRIDINE ◇ SALICYLAZOSULFAPYRIDINE ◇ S.A.S.-500 ◇ SASP ◇ SULCOLON ◇ SULFASALAZINE ◇ SULPHASALAZINE ◇ W-T SASP ORAL

TOXICITY DATA with REFERENCE
orl-wmn TDLo:8100 mg/kg (female 1-40W post):TER NEJMAG 318,1128,88
orl-rat TDLo:8400 mg/kg (male 28D pre):REP CCPTAY 28,273,83
orl-cld TDLo:3800 mg/kg/19D-I:LIV AJDDAL 23,956,78
orl-man TDLo:7143 μg/kg:GIT NEJMAG 306,409,82
orl-man TDLo:143 mg/kg/10D-I:BLD AIMEAS 102,277,85
orl-man LDLo:3200 mg/kg/31D-I:BLD,SKN JRSMD9 73,587,80
unr-man TDLo:1300 mg/kg/4W-I:CNS,GIT PRACAK 204,850,70
rec-man TDLo:7143 μg/kg:GIT NEJMAG 306,409,82
orl-wmn TDLo:140 mg/kg/2W-I LANCAO 1,917,84
unr-wmn TDLo:2700 mg/kg/13W-I:SKN,ALR BMJOAE 286,1547,83
orl-hmn TDLo:429 mg/kg/10D-I:BLD NEJMAG 289,491,73

CONSENSUS REPORTS: EPA Genetic Toxicology Program.

SAFETY PROFILE: Human systemic effects by ingestion: allergic dermatitis, allergic reaction with multiple organ involvement, angranulocytosis, ataxia, dermatitis, diarrhea, dyspnea, hallucinations, headache, hemolysis, hepatitis, hypermotility, nausea or vomiting, thrombocytopenia. Human teratogenic effects. Experimental reproductive effects. When heated to decomposition it emits very toxic fumes of SO_x and NO_x.

PPO000 CAS:144-83-2 ***HR: 2***
N-2-PYRIDYLSULFANILAMIDE
mf: $C_{11}H_{11}N_3O_2S$ mw: 249.31

PROP: Colorless prisms in water. Mp: 191-192°. Sol in water and alc; sltly sol in ether; very sol in alkali and aqueous HCl.

SYNS: ADIPLON ◇ COCCOCLASE ◇ DAGENAN ◇ EUBASIN ◇ EUBASINUM ◇ HAPTOCIL ◇ M+B 695 ◇ N^1-2-PYRIDYLSULFANILAMIDE ◇ RELBAPIRIDINA ◇ RONIN ◇ SEPTIPULMON ◇ 2-SULFANILYL

AMINOPYRIDINE ◇ SULFAPYRIDINE ◇ 2-SULFAPYRIDINE ◇ SULFIDINE ◇ THIOSEPTAL ◇ TRIANON

TOXICITY DATA with REFERENCE
sce-hmn:lyms 200 mg/L MUREAV 222,27,89
orl-rat TDLo:13482 mg/kg (male 6W pre):REP JRPFA4 81,259,87
scu-rat TDLo:135 mg/kg/9W-I:ETA PSEBAA 68,330,48
orl-rat LD50:15800 mg/kg QJPPAL 11,217,38
ivn-rat LD50:800 mg/kg AEPPAE 211,367,50
orl-mus LD50:16600 mg/kg QJPPAL 11,217,38
ipr-mus LD50:1150 mg/kg AITEAT 16,804,68
ivn-dog LDLo:500 mg/kg JPETAB 67,454,39

CONSENSUS REPORTS: Reported in EPA TSCA Inventory.

SAFETY PROFILE: Moderately toxic by intraperitoneal and intravenous routes. Slightly toxic by ingestion. Experimental reproductive effects. Human mutation data reported. Questionable carcinogen with experimental tumorigenic data. When heated to decomposition it emits very toxic fumes of NO_x and SO_x.

PPO250 CAS:127-57-1 ***HR: 2***
N^1-2-PYRIDYLSULFANILAMIDE SODIUM SALT
mf: $C_{11}H_{10}N_3O_2S•Na$ mw: 271.29

SYNS: SODIUM SULFAPYMONOHYDRATE ◇ SODIUM SULFAPYRIDINE ◇ SOLUBLE SULFAPYRIDINE

TOXICITY DATA with REFERENCE
ipr-rat LDLo:1120 mg/kg PSEBAA 45,15,40
orl-mus LD50:2000 mg/kg JPETAB 68,259,40
scu-mus LD50:872 mg/kg JPETAB 71,138,41
ivn-mus LD50:847 mg/kg JPETAB 79,127,43
par-mus LD50:1000 mg/kg PSEBAA 43,328,40

CONSENSUS REPORTS: Reported in EPA TSCA Inventory.

SAFETY PROFILE: Moderately toxic by ingestion, intraperitoneal, subcutaneous, intravenous, and parenteral routes. When heated to decomposition it emits very toxic fumes of NO_x, Na_2O, and SO_x.

PPO750 CAS:289-95-2 ***HR: D***
PYRIMIDINE
mf: $C_4H_4N_2$ mw: 80.10

PROP: Crystals; penetrating odor. Mp: 20-22°, bp: 123-124°. Misc in water; sol in alc and ether.

SYNS: 1,3-DIAZABENZENE ◇ m-DIAZINE ◇ METADIAZINE

TOXICITY DATA with REFERENCE
dnd-mam:lym 150 mmol/L PNASA6 48,686,62

CONSENSUS REPORTS: Reported in EPA TSCA Inventory.

SAFETY PROFILE: Mutation data reported. When heated to decomposition it emits toxic fumes of NO_x.

PPP000 CAS:56606-38-3 ***HR: D***
PYRIMIDINE-4,5-DICARBOXYLIC ACID IMIDE
mf: $C_6H_3N_3O_2$ mw: 149.12

SYN: 3,5-DIAZAPHTHALIMID(GERMAN)

TOXICITY DATA with REFERENCE
ipr-mus TDLo:200 mg/kg (9D preg):REP PHARAT 31,172,76
ipr-mus TDLo:200 mg/kg (9D preg):TER PHARAT 31,172,76

SAFETY PROFILE: An experimental teratogen. Experimental reproductive effects. When heated to decomposition it emits toxic fumes of NO_x.

PPP250 CAS:36505-84-7 ***HR: 3***
8-(4-(4-(2-PYRIMIDINYL)-1-PIPERIZINYL)BUTYL)-8-AZASPIRO(4,5) DECANE-7,9-DIONE
mf: $C_{21}H_{31}N_5O_2$ mw: 385.57

SYN: BUSPIRONE

TOXICITY DATA with REFERENCE
unr-hmn TDLo:399 mg/kg/19D-I:CNS DRFUD4 1,409,76
ipr-rat LD50:136 mg/kg DRFUD4 1,409,76

SAFETY PROFILE: Poison by intraperitoneal route. Human systemic effects by an unspecified route: change in motor activity. When heated to decomposition it emits toxic fumes of NO_x. Used as an anxiolytic agent.

PPP500 CAS:68-35-9 ***HR: 3***
N^1-2-PYRIMIDINYL-SULFANILAMIDE
mf: $C_{10}H_{10}N_4O_2S$ mw: 250.30

SYNS: ADIAZINE ◇ 4-AMINO-N-2-PYRIMIDINYLBENZENESULFONAMIDE ◇ COCO-DIAZINE ◇ CODIAZINE ◇ CREMODIAZINE ◇ DEBENAL ◇ DELTAZINA ◇ DIAZOLONE ◇ DIAZYL ◇ ESKADIAZINE ◇ HONEY DIAZINE ◇ LIPO-DIAZINE ◇ LIPO-LEVAZINE ◇ LIQUADIAZINE ◇ MIRCOSULFON ◇ NEAZINE ◇ PECTA-DIAZINE, suspension ◇ PIRIDISIR ◇ PYRIMAL ◇ RP 2616 ◇ SANODIAZINE ◇ SILVADENE ◇ S.N. 112 ◇ SPOFADRIZINE ◇ STERAZINE ◇ SULFADIAZINE ◇ SULFANILAMIDOPYRIMIDINE ◇ 2-SULFANILYLAMINOPYRIMIDINE ◇ SULFAPYRIMIDIN (GERMAN) ◇ 2-SULFAPYRIMIDINE ◇ SULPHADIAZINE ◇ THERADIAZINE

TOXICITY DATA with REFERENCE
orl-mus TDLo:6 g/kg (female 7-12D post):TER SEIJBO 13,7,73
orl-rat TDLo:6 g/kg (9-14D preg):REP SEIJBO 13,7,73
orl-man TDLo:229 mg/kg/8D-I:GIT,KID IJMDAI 6,561,70
unr-chd TDLo:138 mg/kg:CNS,KID PGMJAO 53,103,77
ipr-rat LDLo:446 mg/kg KLWOAZ 27,449,49
orl-mus LD50:1500 mg/kg ARZNAD 21,571,71
ipr-mus LD50:750 mg/kg 29ZVAB -,109,69

scu-mus LD50:1600 mg/kg NIIRDN 6,386,82
ivn-mus LD50:180 mg/kg CSLNX* NX#03347

CONSENSUS REPORTS: Reported in EPA TSCA Inventory.

SAFETY PROFILE: Poison by intravenous route. Moderately toxic by ingestion and intraperitoneal routes. Human systemic effects by ingestion: hematuria, anuria, general anesthesia, gastrointestinal effects. Experimental teratogenic and reproductive effects. When heated to decomposition it emits very toxic fumes of NO_x and SO_x.

PPP750 CAS:53558-25-1 ***HR: 3***
PYRIMINYL
mf: $C_{13}H_{12}N_4O_3$ mw: 272.29

SYNS: N-(4-NITROPHENYL)-N'-(3-PYRIDINYLMETHYL)UREA ◇ N-3-PYRIDYLMETHYL-N'-p-NITROPHENYLUREA

TOXICITY DATA with REFERENCE
orl-man TDLo:5571 μg/kg:CNS NEJMAG 302,73,80
orl-man LDLo:11142 μg/kg NEJMAG 302,73,80
orl-man TDLo:5571 μg/kg:GIT NEJMAG 302,73,80
orl-rat LD50:6700 μg/kg JOHYAY 80,401,78

CONSENSUS REPORTS: EPA Extremely Hazardous Substances List. Reported in EPA TSCA Inventory.

SAFETY PROFILE: Human poison by ingestion. Human systemic effects by ingestion: hallucinations, distorted perceptions, muscle weakness, nausea. When heated to decomposition it emits toxic fumes of NO_x.

PPQ000 CAS:14222-46-9 ***HR: D***
PYRITIDIUM BROMIDE
mf: $C_{26}H_{27}N_7$•2Br mw: 597.42

SYN: PROTHIDIUM

TOXICITY DATA with REFERENCE
mma-sat 220 nmol/plate MUREAV 48,103,77
oms-mus:leu 500 μmol/L BCPCA6 20,2921,71

CONSENSUS REPORTS: EPA Genetic Toxicology Program.

SAFETY PROFILE: Mutation data reported. When heated to decomposition it emits toxic fumes of Br^- and NO_x.

PPQ500 CAS:87-66-1 ***HR: 3***
PYROGALLOL
mf: $C_6H_6O_3$ mw: 126.12

PROP: White, lustrous crystals. Bp: 309°, d: 1.453 @ 4°/4°, vap press: 10 mm @ 167.7°, mp: 131-133°. Sltly sol in benzene, chloroform.

SYNS: 1,2,3-BENZENETRIOL ◇ C.I. 76515 ◇ C.I. OXIDATION BASE 32 ◇ FOURAMINE BROWN AP ◇ FOURRINE PG ◇ PYROGALLIC ACID ◇ 1,2,3-TRIHYDROXYBENZEN (CZECH) ◇ 1,2,3-TRIHYDROXYBENZENE

TOXICITY DATA with REFERENCE
skn-rbt 500 mg/24H SEV 28ZPAK -,59,72
eye-rbt 20 mg/24H MOD 28ZPAK -,59,72
mmo-sat 1 μmol/plate NEZAAQ 35,533,83
mrc-smc 300 mg/L MUREAV 135,109,84
orl-rat TDLo:3 g/kg (6-15D preg):TER JACTDZ 2(4),325,83
orl-rat TDLo:3 g/kg (6-15D preg):REP JACTDZ 2(4),325,83
scu-rat TDLo:3950 mg/kg/58W-I:ETA PAACA3 20,117,79
orl-hmn LDLo:28 mg/kg:CNS,PUL,GIT 34ZIAG -,507,69
scu-man LDLo:120 mg/kg ZMWIAJ 19,545,1881
scu-rat LDLo:650 mg/kg JAFCAU 17,497,69
orl-mus LD50:300 mg/kg GISAAA 38(8),6,73
ipr-mus LD50:400 mg/kg NTIS** AD691-490
scu-mus LD50:566 mg/kg ZGIMAL 2,333,47
orl-dog LDLo:25 mg/kg HBTXAC 1,254,56
scu-dog LDLo:300 mg/kg HBTXAC 1,254,56
ivn-dog LDLo:80 mg/kg HBTXAC 1,254,56
orl-rbt LD50:1600 mg/kg AJVRAH 23,1264,62

CONSENSUS REPORTS: Reported in EPA TSCA Inventory.

SAFETY PROFILE: Human poison by ingestion and subcutaneous routes. An experimental poison by ingestion, subcutaneous, intravenous, and intraperitoneal routes. Experimental teratogenic and reproductive effects. Questionable carcinogen with experimental tumorigenic data. Mutation data reported. Readily absorbed through the skin. Human systemic effects by ingestion: convulsions, dyspnea, gastrointestinal effects. A severe skin and eye irritant. Incompatible with alkalies; NH_3; antipyrine; phenol; iron and lead salts; iodine; $KMnO_4$. When heated to decomposition it emits acrid smoke and irritating fumes. Used as a topical antibacterial agent, as an intermediate, hair dye component, and analytical reagent.

PPQ600 ***HR: 3***
PYROGEN

SYNS: LIPOPOLYSACCHARIDE ◇ LPS

TOXICITY DATA with REFERENCE
dns-mus:leu 50 mg/L CNREA8 43,2536,83
par-rat TDLo:2 mg/kg (6D preg):REP OYYAA2 9,579,75
ivn-rat LD50:4900 μg/kg NYKZAU 79,357,82
ipr-mus LD50:20 mg/kg NYKZAU 85,99P,85

SAFETY PROFILE: Poison by intravenous and intraperitoneal routes. Experimental reproductive effects. Mutation data reported. When heated to decomposition it emits acrid smoke and irritating fumes.

PPQ625 CAS:87-47-8 ***HR: 3***
PYROLAN
mf: $C_{13}H_{15}N_3O_2$ mw: 245.31

PROP: Crystals. Mp: 50°, bp: 160-162° @ 0.2 mm.

SYNS: DIMETHYL CARBAMIC ACID-3-METHYL-1-PHENYL PYRAZOL-5-YL ESTER ◇ DIMETHYL-5-(3-METHYL-1-PHENYLPYRAZOLYL) CARBAMATE ◇ ENT 17,588 ◇ 3-METHYL-1-PHENYL PYRAZOL-5-YL DIMETHYL CARBAMATE ◇ 3-METHYL-1-PHENYL-5-PYRAZOLYL DIMETHYL CARBAMATE ◇ 1-PHENYL-3-METHYL-5-PYRAZOLYL- N,N-DIMETHYL CARBAMATE

TOXICITY DATA with REFERENCE
orl-mus LD50:90 mg/kg PCOC** -,955,66
orl-ckn LD50:11 mg/kg TXAPA9 3,521,61
orl-bwd LD50:39 mg/kg TXAPA9 21,315,72

SAFETY PROFILE: Poison by ingestion. A cholinesterase inhibitor type insecticide. When heated to decomposition it emits toxic fumes of NO_x. See also CARBAMATES.

PPQ650 CAS:74847-35-1 ***HR: 2***
PYRONARIDINE
mf: $C_{29}H_{32}ClN_5O_2$ mw: 518.11
SYN: MALARIDINE

TOXICITY DATA with REFERENCE
mmo-sat 100 μg/plate CYLPDN 3,51,82
orl-rat TDLo:1200 mg/kg (male 60D pre):REP CYLPDN 6,131,85
orl-rat TDLo:1100 mg/kg (7D preg):TER YHHPAL 17,401,82
orl-mus LD50:1100 mg/kg YHHPAL 17,401,82

SAFETY PROFILE: Moderately toxic by ingestion. Experimental reproductive effects. Mutation data reported. When heated to decomposition it emits toxic fumes of Cl^- and NO_x. An antimalarial drug.

PPQ750 CAS:92-32-0 ***HR: D***
PYRONIN YELLOW
mf: $C_{17}H_{19}N_2O \cdot Cl$ mw: 302.83

PROP: Green powder. Sol in water; sltly sol in alc; insol in ether.

SYNS: C.I. 45005 ◇ N-(6-(DIMETHYLAMINO-3H-XANTHEN-3-YLIDENE)-N-METHYLMETHANAMINIUM CHLORIDE ◇ METHYL PYRONIN ◇ PYRONINE ◇ PYRONIN G ◇ SCHULTZ No. 853 ◇ TETRAMETHYL PYRONIN

TOXICITY DATA with REFERENCE
mma-sat 80 μg/plate TRENAF 27,153,76
dnd-esc 20 μmol/L MUREAV 89,95,81

CONSENSUS REPORTS: Reported in EPA TSCA Inventory. EPA Genetic Toxicology Program.

SAFETY PROFILE: Mutation data reported. When heated to decomposition it emits very toxic fumes of NO_x and Cl^-. Used as a stain.

PPR500 CAS:7791-27-7 ***HR: 3***
PYROSULFURYL CHLORIDE
DOT: UN 1817
mf: $Cl_2O_5S_2$ mw: 215.02

$(ClSO_2)_2O$

PROP: Colorless, mobile, fuming liquid. Mp: −37°, bp: 151°, d: 1.83, (gas): 9.6 g/L.

SYNS: CHLOROSULFONIC ANHYDRIDE ◇ DISULFUR PENTOXYDICHLORIDE ◇ DISULFURYL CHLORIDE ◇ DISULFURYL DICHLORIDE ◇ PYRO SULFURYL CHLORIDE (DOT) ◇ PYROSULPHURYL CHLORIDE (DOT)

DOT Classification: Corrosive Material; Label: Corrosive.

SAFETY PROFILE: A very poisonous material which is also corrosive to the eyes, skin, and mucous membranes. Violent reaction with water. Vigorous reaction with phosphorus. When heated to decomposition it emits very toxic fumes of Cl^- and SO_x. See also CHLOROSULFONIC ACID.

PPS250 CAS:109-97-7 ***HR: 3***
PYRROLE
mf: C_4H_5N mw: 67.10

CH=CHNHCH=CH (ring)

PROP: Colorless liquid, darkens on standing; mild nutty odor. Fp: −24°, flash p: 102°F (TCC), d: 0.968 @ 20°/4°, refr index: 1.507, vap d: 2.31, bp: 130-131° @ 761 mm. Sltly sol in water; very sol in alc, fixed oils, benzene, ether; insol in alkali.

SYNS: 1-AZA-2,4-CYCLOPENTADIENE ◇ AZOLE ◇ DIVINYLENIMINE ◇ FEMA No. 3386 ◇ IMIDOLE ◇ MONOPYRROLE

TOXICITY DATA with REFERENCE
scu-mus LD50:61 mg/kg 28ZEAL 4,335,69
ipr-rbt LDLo:150 mg/kg AIPTAK 36,387,30
scu-rbt LDLo:250 mg/kg AIPTAK 36,387,30

CONSENSUS REPORTS: Reported in EPA TSCA Inventory.

SAFETY PROFILE: Poison by subcutaneous and intraperitoneal routes. Flammable when exposed to heat or flame; can react with oxidizing materials. To fight fire, use foam, CO_2, dry chemical. Violent reaction with 2-nitrobenzaldehyde. When heated to decomposition it emits highly toxic fumes of NO_x.

PPS500 CAS:123-75-1 ***HR: 3***
PYRROLIDINE
DOT: UN 1922
mf: C_4H_9N mw: 71.14

PROP: Colorless, mobile liquid; penetrating, amine-like odor. Fp: −63°, flash p: 37°F (TCC), d: 0.8618 @ 20°/4°, vap press: 128 mm @ 39°, vap d: 2.45; bp: 88.5-89°. Fumes in air. Misc with water; sol in alc, ether, chloroform.

SYNS: AZACYCLOPENTANE ◇ TETRAHYDROPYRROLE ◇ TETRAMETHYLENIMINE

TOXICITY DATA with REFERENCE
orl-rat LD50:300 mg/kg GTPZAB 18,29,74
ihl-mus LC50:1300 mg/m^3/2H GTPZAB 18,29,74
ivn-mus LD50:56 mg/kg CSLNX* NX#06769

CONSENSUS REPORTS: Reported in EPA TSCA Inventory.

DOT Classification: Flammable Liquid; Label: Flammable Liquid.

SAFETY PROFILE: Poison by ingestion and intravenous routes. Moderately toxic by inhalation. Dangerous fire hazard when exposed to heat or flame; can react vigorously with oxidizing materials. To fight fire, use alcohol foam, CO_2, dry chemical. When heated to decomposition it emits highly toxic fumes of NO_x.

PPT325 CAS:13758-99-1 ***HR: 3***
2-(2-PYRROLIDINOETHYL)-3a,4,7,7a-TETRAHYDRO-4,7-ETHANEISOINDOLINE DIMETHIODIDE
mf: $C_{20}H_{32}N_2•2I$ mw: 554.34

TOXICITY DATA with REFERENCE
orl-rat LD50:1806 mg/kg SKNEA7 10,15,60
ivn-rat LD50:176 mg/kg SKNEA7 10,15,60
orl-mus LD50:1664 mg/kg SKNEA7 10,15,60
ivn-mus LD50:155 mg/kg SKNEA7 10,15,60

SAFETY PROFILE: Poison by intravenous route. Moderately toxic by ingestion. When heated to decomposition it emits toxic fumes of I^- and NO_x. See also IODIDES.

PPT400 CAS:66839-97-2 ***HR: 3***
3-PYRROLIDINOMETHYL-4-HYDROXYBIPHENYL
mf: $C_{17}H_{19}NO$ mw: 253.37

SYNS: (1,1'-BIPHENYL)-4-OL, 3-(1-PYRROLIDINYLMETHYL)- ◇ 3-(1-PYRROLIDINYLMETHYL)(1,1'-BIPHENYL)-4-OL

TOXICITY DATA with REFERENCE
orl-rat TDLo:10 mg/kg (female 4-5D post):REP IJMRAQ 67,392,78
orl-rat LD50:30 mg/kg IJMRAQ 67,392,78

SAFETY PROFILE: Poison by ingestion. Experimental reproductive effects. When heated to decomposition it emits toxic fumes of NO_x.

PPT500 CAS:616-45-5 ***HR: 1***
2-PYRROLIDINONE
mf: C_4H_7NO mw: 85.12

SYNS: PYRROLIDON (GERMAN) ◇ 2-PYRROLIDONE

TOXICITY DATA with REFERENCE
orl-rat LD50:6500 mg/kg 34ZIAG -,508,69
scu-mus LDLo:10 g/kg AEXPBL 50,199,1903
orl-gpg LD50:6500 mg/kg 34ZIAG -,508,69

CONSENSUS REPORTS: Reported in EPA TSCA Inventory.

SAFETY PROFILE: Mildly toxic by ingestion and subcutaneous routes. When heated to decomposition it emits toxic fumes of NO_x.

PPU750 CAS:77966-28-0 ***HR: 3***
2-PYRROLIDINYL-p-ACETOPHENETIDIDE HYDROCHLORIDE
mf: $C_{14}H_{20}N_2O_2•ClH$ mw: 284.82

SYNS: C 3087 ◇ 4'-ETHOXY-2-PYRROLIDINYLACETANILIDE HYDROCHLORIDE

TOXICITY DATA with REFERENCE
eye-rbt 2% MLD ARZNAD 8,270,58
ipr-rat LD50:350 mg/kg ARZNAD 8,270,58
scu-mus LD50:705 mg/kg ARZNAD 8,270,58

SAFETY PROFILE: Poison by intraperitoneal route. Moderately toxic by subcutaneous route. An eye irritant. When heated to decomposition it emits very toxic fumes of HCl and NO_x.

PPV750 CAS:75348-49-1 ***HR: 2***
4-PYRROLIDINYLCARBONYLMETHYL-PIPERAZINYL-3,4,5-TRIMETHOXYCINNAMYL KETONE MALEATE
mf: $C_{22}H_{31}N_3O_5•C_4H_4O_4$ mw: 533.64

TOXICITY DATA with REFERENCE
orl-mus LD50:1000 mg/kg CHTPBA 4,293,69
ivn-mus LD50:630 mg/kg CHTPBA 4,293,69

SAFETY PROFILE: Moderately toxic by ingestion and intravenous routes. When heated to decomposition it emits toxic fumes of NO_x.

PPW000 CAS:38198-35-5 ***HR: 3***
trans-2-(1-PYRROLIDINYL)CYCLOHEXYL-3-PENTYLOXYCARBANILATE HYDROCHLORIDE
mf: $C_{22}H_{34}N_2O_3•ClH$ mw: 411.04

SYNS: K-1902 ◇ PENTACAINE ◇ PENTACAINE HYDROCHLORIDE

◇ PENTAKAIN ◇ m-PENTYLOXYCARBANILIC ACID, trans-2-(1-PYRROLIDINYL)CYCLOHEXYL ESTER HYDROCHLORIDE ◇ (3-(PENTYLOXY)PHENYL)CARBAMIC ACID, 2-(1-PYRROLIDINYL) CYCLOHEXYL ESTER, HCl, (E)-

TOXICITY DATA with REFERENCE
scu-mus LD50:120 mg/kg ARPMAS 305,648,72
ivn-mus LD50:15 mg/kg ARPMAS 306,648,72

SAFETY PROFILE: Poison by subcutaneous and intravenous routes. When heated to decomposition it emits very toxic fumes of HCl and NO_x. Used as a local anesthetic. See also CARBAMATES.

PPW750 CAS:77985-30-9 ***HR: 3***
N-(2-PYRROLIDINYLETHYL)CARBAMIC ACID, 6-CHLORO-o-TOLYL ESTER, HYDROCHLORIDE
mf: $C_{14}H_{19}ClN_2O_2$•ClH mw: 319.26

SYN: C 5312

TOXICITY DATA with REFERENCE
eye-rbt 2% MLD ARZNAD 8,708,58
ipr-rat LD50:55 mg/kg ARZNAD 8,708,58
scu-mus LD50:92 mg/kg ARZNAD 8,708,58

SAFETY PROFILE: Poison by intraperitoneal and subcutaneous routes. An eye irritant. When heated to decomposition it emits very toxic fumes of NO_x and Cl^-. See also CARBAMATES and ESTERS.

PPX000 CAS:77985-31-0 ***HR: 3***
N-(2-PYRROLIDINYLETHYL)CARBAMIC ACID, MESITYL ESTER, HYDROCHLORIDE
mf: $C_{16}H_{24}N_2O_2$•ClH mw: 312.88

SYN: C 5319

TOXICITY DATA with REFERENCE
eye-rbt 2% MLD ARZNAD 8,708,58
ipr-rat LD50:74 mg/kg ARZNAD 8,708,58
scu-mus LD50:225 mg/kg ARZNAD 8,708,58

SAFETY PROFILE: Poison by intraperitoneal and subcutaneous routes. An eye irritant. When heated to decomposition it emits very toxic fumes of NO_x and HCl. See also CARBAMATES and ESTERS.

PPX250 CAS:77985-32-1 ***HR: 3***
N-(2-(PYRROLIDINYL)ETHYL)CARBAMIC ACID, 2,6-XYLYL ESTER HYDROCHLORIDE
mf: $C_{15}H_{22}N_2O_2$•ClH mw: 298.85

SYN: C 5318

TOXICITY DATA with REFERENCE
ipr-rat LD50:45 mg/kg ARZNAD 8,708,58
scu-mus LD50:175 mg/kg ARZNAD 8,708,58

SAFETY PROFILE: Poison by intraperitoneal and subcutaneous routes. When heated to decomposition it emits very toxic fumes of HCl and NO_x. See also CARBAMATES and ESTERS.

PPY250 CAS:751-97-3 ***HR: 3***
N-(1-PYRROLIDINYLMETHYL)-TETRACYCLINE
mf: $C_{27}H_{33}N_3O_8$ mw: 527.63

SYNS: BRISTACIN ◇ PIRROLIDINOMETIL-TETRACICLINA (ITALIAN) ◇ PRM-TC ◇ PYRROLIDINO-METHYL-TETRACYCLINE ◇ N-(PYRROLIDINOMETHYL)TETRACYCLINE ◇ REVERIN ◇ ROLITETRACYCLINE ◇ SQ 15,659 ◇ SUPERCICLIN ◇ SYNOTODECIN ◇ SYNTETREX ◇ SYNTETRIN ◇ TRANSCYCLINE ◇ VELACICLINE ◇ VELACYCLINE

TOXICITY DATA with REFERENCE
ipr-mus TDLo:450 mg/kg (female 8-13D post):TER IYKEDH 3,75,72
ipr-mus TDLo:450 mg/kg (female 8-13D post):REP IYKEDH 3,75,72
orl-mus LD50:1320 mg/kg JPPMAB 16,33,64
ivn-mus LD50:75 mg/kg ARZNAD 9,63,59
ivn-dog LD50:93 mg/kg NIIRDN 6,914,82
ivn-rbt LD50:47 mg/kg NIIRDN 6,914,82

SAFETY PROFILE: Poison by intravenous route. Experimental teratogenic effects. Other experimental reproductive effects. When heated to decomposition it emits toxic fumes of NO_x.

PQB275 CAS:67230-67-5 ***HR: D***
7-(((2-PYRROLYL)METHYL)AMINO)-ACTINOMYCIN D
mf: $C_{67}H_{92}N_{14}O_{16}$ mw: 1349.73

SYN: 7-((1H-PYRROL-2-YLMETHYL)AMINO)ACTINOMYCIND

TOXICITY DATA with REFERENCE
dni-mus:lym 370 nmol/L JMCMAR 24,1052,81
oms-mus:lym 184 nmol/L JMCMAR 24,1052,81

SAFETY PROFILE: Mutation data reported. When heated to decomposition it emits toxic fumes of NO_x.

PQB500 CAS:494-98-4 ***HR: 3***
3-PYRROL-2-YLPYRIDINE
mf: $C_9H_8N_2$ mw: 144.19

SYN: NORNICOTYRINE

TOXICITY DATA with REFERENCE
itr-rat TDLo:23 mg/kg:ETA BJCAAI 16,453,62

SAFETY PROFILE: Questionable carcinogen with experimental tumorigenic data. When heated to decomposition it emits toxic fumes of NO_x.

PQB750 CAS:19089-24-8 ***HR: 3***
PYRROMECAINE HYDROCHLORIDE
mf: $C_{18}H_{28}N_2O$•ClH mw: 324.94

SYNS: N-BUTYL-α-PYRROLIDINE-CARBOXY-MESIDIDE HYDROCHLORIDE ◇ 1-BUTYL-N-(2,4,6-TRIMETHYLPHENYL)-2-PYRROLIDINECARBOXAMIDE MONOHYDROCHLORIDE ◇ PYROMECAINE

TOXICITY DATA with REFERENCE
orl-rat LD50:13958 mg/kg STOAAT 59,27,80
ipr-rat LD50:13512 mg/kg STOAAT 59,27,80
orl-mus LD50:290 mg/kg RPTOAN 46,202,83
ipr-mus LD50:128 mg/kg RPTOAN 46,202,83
scu-mus LD50:285 mg/kg RPTOAN 46,202,83
ivn-mus LD50:42 mg/kg RPTOAN 35,114,72
orl-gpg LD50:6283 mg/kg STOAAT 59,27,80
ipr-gpg LD50:4712 mg/kg STOAAT 59,27,80

SAFETY PROFILE: Poison by ingestion, intraperitoneal, subcutaneous, and intravenous routes. When heated to decomposition it emits very toxic fumes of HCl and NO_x.

PQB800 CAS:4435-60-3 ***HR: 3***
PYRROMELITIC ACID DIANHYDRIDE
mf: $C_{10}H_2O_6$ mw: 218.12

TOXICITY DATA with REFERENCE
orl-rat LD50:2200 mg/kg 85GMAT -,105,82
ihl-rat LCLo:150 mg/m^3/4H 85GMAT -,105,82
orl-mus LD50:2400 mg/kg 85GMAT -,105,82

SAFETY PROFILE: Poison by inhalation. Moderately toxic by ingestion. When heated to decomposition it emits acrid smoke and irritating fumes. See also ANHYDRIDES.

PQC000 CAS:78-98-8 ***HR: 3***
PYRUVALDEHYDE
mf: $C_3H_4O_2$ mw: 72.07

PROP: Mobile, yellow liquid; pungent odor. Bp: 72°, d: 1.06 @ 20°/20°. Sol in alc, ether, benzene.

SYNS: ACETYLFORMALDEHYDE ◇ ACETYLFORMYL ◇ α-KETOPROPIONALDEHYDE ◇ 1-KETOPROPIONALDEHYDE ◇ 2-KETOPROPIONALDEHYDE ◇ METHYLGLYOXAL ◇ NSC 79019 ◇ 2-OXOPROPANAL ◇ PROPANEDIONE ◇ PROPANOLONE ◇ PYRORACEMIC ALDEHYDE ◇ PYRUVIC ALDEHYDE

TOXICITY DATA with REFERENCE
eye-rbt 403 mg AJOPAA 29,1363,46
mmo-sat 10 μg/plate ENMUDM 8,9,86
dns-rat-orl 300 mg/kg CRNGDP 6,91,85
dnd-ham:ovr 1500 μmol/L CRNGDP 6,683,85
sce-ham:ovr 100 μmol/L MUREAV 144,189,85
orl-rat LD50:1165 mg/kg EESADV 2,369,78
ipr-mus LD50:179 mg/kg NCISP* JAN86

CONSENSUS REPORTS: Reported in EPA TSCA Inventory.

SAFETY PROFILE: Poison by intraperitoneal route. Moderately toxic by ingestion. An eye irritant. Mutation data reported. Combustible when exposed to heat or flame. When heated to decomposition it emits acrid smoke and irritating fumes. See also ALDEHYDES.

PQC100 CAS:127-17-3 ***HR: 2***
PYRUVIC ACID
mf: $C_3H_4O_3$ mw: 88.06

$CH_3CO•CO•OH$

SYN: 2-OXOPROPANOIC ACID

SAFETY PROFILE: Unstable in storage. When heated to decomposition it emits acrid smoke and irritating fumes.

PQC500 CAS:3546-41-6 ***HR: 3***
PYRVINIUM-4,4'-METHYLENEBIS(3-HYDROXYNAPHTHALENE-2-CARBOXYLATE)
mf: $C_{75}H_{70}N_6O_6$ mw: 1151.51

SYNS: ALNOXIN ◇ ALTOTAL ◇ MOLEVAC ◇ NEO-OXYPATE ◇ PAMOVIN ◇ PIRVINIUM PAMOATE ◇ POVAN ◇ POVANYL ◇ PYRCON ◇ PYRVINIUM EMBONATE ◇ PYRVINIUM PAMOATE ◇ SN-4395 ◇ TOLAPIN ◇ TRU ◇ VANQUIN ◇ VIPRYNIUM EMBONATE

TOXICITY DATA with REFERENCE
mmo-sat 2500 ng/plate MUREAV 117,79,83
mma-sat 25 μg/plate JACTDZ 3(4),285,84
orl-rat TDLo:90 g/kg (female 30D pre):REP OYYAA2 5,313,71
orl-mus LDLo:50 mg/kg BSIBAC 44,1032,68
ipr-mus LDLo:5 mg/kg OYYAA2 5,305,71
scu-mus LD50:200 mg/kg ARZNAD 15,1349,65

CONSENSUS REPORTS: EPA Genetic Toxicology Program.

SAFETY PROFILE: Poison by ingestion, subcutaneous, and intraperitoneal routes. Experimental reproductive effects. Mutation data reported. When heated to decomposition it emits toxic fumes of NO_x.

PQC525 CAS:38082-89-2 ***HR: 1***
PYX EXPLOSIVE
mf: $C_{17}H_7N_{11}O_{16}$ mw: 621.35

SYNS: 2,6-BIS(PICRYLAMINO)-3,5-DINITROPYRIDINE◇ 3,5-DINITRO-N,N'-BIS(2,4,6-TRINITROPHENYL)-2,6-PYRIDINEDIAMINE ◇ 2,6-PYRIDINEDIAMINE, 3,5-DINITRO-N,N'-BIS(2,4,6-TRINITROPHENYL)- ◇ PYX

TOXICITY DATA with REFERENCE
eye-rbt 100 mg/1H MLD NTIS** LA-8695-MS

CONSENSUS REPORTS: Reported in EPA TSCA Inventory.

SAFETY PROFILE: An eye irritant. When heated to decomposition it emits toxic fumes of NO_x.

Q

QAK000 CAS:72-44-6 ***HR: 3***
QUAALUDE
mf: $C_{16}H_{14}N_2O$ mw: 250.32

PROP: Crystals. Mp: 120°. Sol in ethanol and chloroform; practically insol in water.

SYNS: CATEUDYL ◇ CITEXAL ◇ CI-705 ◇ CN 38703 ◇ 3,4-DIHYDRO-2-METHYL-4-OXO-3-o-TOLYLQUINAZOLINE◇ DORMIGOA ◇ DORMOGEN ◇ DORMUTIL ◇ DORSEDIN ◇ FADORMIR ◇ HOLODORM ◇ HYMINAL ◇ HYPCOL ◇ HYPTOR BASE ◇ IPNOFIL ◇ MAOA ◇ MEQUIN ◇ MELSEDIN BASE ◇ MELSOMIN ◇ METAQUALON ◇ METHAQUALONE ◇ METHAQUALONEINONE ◇ 2-METHYL-3-(2-METHYLPHENYL)-4-QUINAZOLINONE◇ 2-METHYL-3-(2-METHYLPHENYL)-4(3H)-QUINAZOLINONE ◇ 2-METHYL-3-o-TOLYL-4 (3H)-CHINAZOLINON (GERMAN) ◇ 2-METHYL-3-o-TOLYL-4(3H)-CHINAZOLONE ◇ (2-METHYL-3-(o-TOLYL)-3,4-DIHYDRO-4-(QUINAZOLINONE) ◇ 2-METHYL-3-(o-TOLYL)-3,4-DIHYDRO-4-QUINAZOLINONE ◇ 2-METHYL-3-TOLYL-4-OXYBENSDIAZINE ◇ 2-METHYL-3-o-TOLYL-4 (3H)-QUINAZOLINONE ◇ 2-METHYL-3-o-TOLYL-4-QUINAZOLONE ◇ 2-METHYL-3-(2-TOLYL)QUINAZOL-4-ONE ◇ METOLQUIZOLONE ◇ MOLLINOX ◇ MOTOLON ◇ MOZAMBIN ◇ MTQ ◇ NOBEDORM ◇ NOCTILENE ◇ NORMI-NOX ◇ OMNYL ◇ OPTINOXAN ◇ ORTHONAL ◇ ORTONAL ◇ PAREST ◇ PARMINAL ◇ PRO-DORM ◇ QZ 2 ◇ REVONAL ◇ RORER 148 ◇ ROUQUALONE ◇ SINDESVEL ◇ SOMBEROL ◇ SOMNAFAC ◇ SOMNOMED ◇ SONAL ◇ SOVERIN ◇ TORINAL ◇ TUAZOLE ◇ TUAZOLONE

TOXICITY DATA with REFERENCE
orl-rat TDLo:11 g/kg (3D male/3D pre-22D preg):REP EXPEAM 19,183,63
orl-rbt TDLo:900 mg/kg (female 8-16D post):TER TXAPA9 10,244,67
orl-hmn TDLo:57 mg/kg:CNS,PUL,GIT ATXKA8 20,31,63
orl-man LDLo:114 mg/kg ATXKA8 20,31,63
orl-rat LD50:230 mg/kg ARZNAD 17,229,67
ipr-rat LD50:125 mg/kg ARZNAD 17,242,67
orl-mus LD50:420 mg/kg TXAPA9 1,42,59
ipr-mus LD50:180 mg/kg IJMRAQ 69,1008,79
par-mus LD50:500 mg/kg PCJOAU 7,626,73
ivn-rbt LD50:100 mg/kg ATXKA8 20,31,63

CONSENSUS REPORTS: Reported in EPA TSCA Inventory.

SAFETY PROFILE: Human poison by ingestion. Experimental poison by ingestion, intravenous, and intraperitoneal routes. Moderately toxic by parenteral route. Human systemic effects by ingestion: convulsions or effect on seizure threshold, nausea or vomiting, and pulmonary changes. An experimental teratogen. Experimental reproductive effects. A controlled drug under 21CFR 1308.11 which is often abused. When heated to decomposition it emits toxic fumes of NO_x.

QAT000 CAS:102-60-3 ***HR: 2***
QUADROL
mf: $C_{14}H_{32}N_2O_4$ mw: 292.48

PROP: A viscous liquid. Bp: 190°. Misc with water; sol in ethanol, methanol, toluene, ethylene glycol, and perchloroethylene.

SYNS: ENTPROL ◇ 1,1′,1″,1‴-(ETHYLENEDINITRIOLO)TETRA-2-PROPANOL ◇ N,N,N′,N′-TETRAKIS(2-HYDROXY-PROPYL)ETHYLENEDIAMINE

TOXICITY DATA with REFERENCE
unr-mam LD50:3900 mg/kg FMCHA2 -,D261,80

CONSENSUS REPORTS: Reported in EPA TSCA Inventory.

SAFETY PROFILE: Moderately toxic by an unspecified route. When heated to decomposition it emits toxic fumes of NO_x.

QBJ000 CAS:1401-55-4 ***HR: 3***
QUEBRACHO TANNIN

SYNS: SCHINOPSIS LORENTZII TANNIN ◇ TANNIN from QUEBRACHO

TOXICITY DATA with REFERENCE
scu-rat TDLo:350 mg/kg/12W-I:ETA BJCAAI 14,147,60
ipr-mus LD50:360 mg/kg JPPMAB 9,98,57
ivn-mus LD50:130 mg/kg JPPMAB 9,98,57

SAFETY PROFILE: Poison by intraperitoneal and intravenous routes. Questionable carcinogen with experimental tumorigenic data. When heated to decomposition it emits acrid smoke and irritating fumes. See also TANNIN.

QBS000 CAS:64719-39-7 ***HR: 3***
QUELAMYCIN
mf: $C_{27}H_{27}O_{11}•2Fe(2+)•Fe(3+)$ mw: 709.07

SYNS: NSC-267703 ◇ TRIFERRIC ADRIAMYCIN ◇ TRIFERRIC DOXORUBICIN

TOXICITY DATA with REFERENCE
dni-mus:leu 15 μmol/L EJCAAH 14,1185,78
ivn-hmn TDLo:60 mg/kg:BLD CTRRDO 62,1527,78
ipr-rat LD50:45 mg/kg DRFUD4 4,356,79
ivn-rat LD50:45 mg/kg DRFUD4 4,356,79

ipr-mus LD50:45 mg/kg EJCAAH 14,1185,78
ivn-mus LD50:45 mg/kg DRFUD4 4,356,79

SAFETY PROFILE: Poison by intravenous and intraperitoneal routes. Human systemic effects by intravenous route: blood effects. Mutation data reported. When heated to decomposition it emits acrid smoke and irritating fumes. See also IRON DUST.

QCA000 CAS:117-39-5 ***HR: 3***
QUERCETIN
mf: $C_{15}H_{10}O_7$ mw: 302.25

SYNS: C.I. 75670 ◇ C.I. NATURAL RED 1 ◇ C.I. NATURAL YELLOW 10 ◇ CYANIDELONON 1522 ◇ 2-(3,4-DIHYDROXYPHENYL)-3,5,7-TRIHYDROXY-4H-1-BENZOPYRAN-4-ONE ◇ MELETIN ◇ NCI-C60106 ◇ 3,5,7,3',4'-PENTAHYDROXYFLAVONE ◇ QUERCETINE ◇ QUERCETOL ◇ QUERCITIN ◇ QUERTINE ◇ SOPHORETIN ◇ 3',4',5,7-TETRAHYDROXYFLAVAN-3-OL ◇ T-GELB BZW, GRUN 1 ◇ XANTHAURINE

TOXICITY DATA with REFERENCE
dni-hmn:fbr 50 mg/L BCPCA6 33,3823,84
sce-ham:ovr 15 mg/L MUREAV 113,45,83
orl-rat TDLo:20 mg/kg (6-15D preg):TER FCTOD7 20,75,82
ipr-mus TDLo:16 mg/kg (male 5D pre):REP ENMUDM 9,79,87
orl-rat TDLo:33610 mg/kg/58W-C:CAR CNREA8 40,3468,80
orl-mus TDLo:966 g/kg/23W-C:ETA GANNA2 72,327,81
orl-rat TD:38235 mg/kg/58W-C:CAR CNREA8 40,3468,80
orl-rat TD:243 g/kg/3Y-C:NEO PAACA3 25,95,84
orl-rat LD50:161 mg/kg PSEBAA 77,269,51
orl-mus LD50:159 mg/kg PSEBAA 77,269,51
scu-mus LD50:97 mg/kg PSEBAA 77,269,51
ivn-mus LD50:18 mg/kg CSLNX* NX#02589

CONSENSUS REPORTS: IARC Cancer Review: Group 3 IMEMDT 7,56,87; Animal Limited Evidence IMEMDT 31,213,83. Reported in EPA TSCA Inventory. EPA Genetic Toxicology Program.

SAFETY PROFILE: Poison by ingestion, subcutaneous, and intravenous routes. Experimental teratogenic and reproductive effects. Questionable carcinogen with experimental carcinogenic, neoplastigenic, and tumorigenic data. Human mutation data reported. Used as a pharmaceutical and veterinary drug. When heated to decomposition it emits acrid smoke and irritating fumes.

QCA175 CAS:6151-25-3 ***HR: 3***
QUERCETIN DIHYDRATE
mf: $C_{15}H_{10}O_7 \cdot 2H_2O$ mw: 338.29

PROP: Mp: >300°

SYNS: 3,3',4',5,7-PENTAHYDROXYFLAVONE

TOXICITY DATA with REFERENCE
mmo-sat 15 μg/plate MUREAV 206,201,88
mma-sat 7500 ng/plate MUREAV 206,201,88
orl-rat TDLo:1350 g/kg/77W-C:ETA,REP CALEDQ 13,15,81
orl-rat TD:4250 g/kg/121W-C:ETA,REP CALEDQ 13,15,81
orl-mus LD50:159 mg/kg PSEBAA 77,269,51

SAFETY PROFILE: Poison by ingestion. Experimental reproductive effects. Questionable carcinogen with experimental tumorigenic data. Mutation data reported. When heated to decomposition it emits acrid smoke and irritating fumes. See also QUERCETIN.

QCJ000 CAS:522-12-3 ***HR: 3***
QUERCITRIN
mf: $C_{21}H_{20}O_{11}$ mw: 448.41

PROP: Yellow crystals from dil methanol or ethanol. Mp: 176-179°. Crystals from water, mp: 167°. Insol in cold water, ether; sol in alc and in aqueous alkaline solns; sltly sol in hot water.

SYNS: C.I. 75720 ◇ NCI-C60102 ◇ 3,3',4',5,7-PENTAHYDROXYFLAVONE-3-l-RHAMNOSIDE ◇ QUERCETIN, 3-(6-DEOXY-α-l-MANNOPYRANOSIDE) ◇ QUERCETIN-3-l-RHAMNOSIDE ◇ USAF CF-2

TOXICITY DATA with REFERENCE
mma-sat 166 nmol/plate MUREAV 54,297,78
ipr-mus LD50:200 mg/kg NTIS** AD277-689

SAFETY PROFILE: Poison by intraperitoneal route. Mutation data reported. When heated to decomposition it emits acrid smoke and irritating fumes.

QCJ275 CAS:1916-59-2 ***HR: 3***
QUESTIOMYCIN A
mf: $C_{12}H_8N_2O_2$ mw: 212.22

SYNS: 2-AMINO-3H-PHENOXAZIN-3-ONE ◇ 2-AMINOPHENOXAZON ◇ 2-AMINOPHENOXAZONE

TOXICITY DATA with REFERENCE
dni-mus:ast 20 μmol/L CPBTAL 17,105,69
dnd-mam lym 100 μmol/L CPBTAL 17,105,69
ipr-mus LD50:200 mg/kg 85GDA2 5,174,81

SAFETY PROFILE: Poison by intraperitoneal route. Mutation data reported. When heated to decomposition it emits toxic fumes of NO_x.

QCS000 CAS:545-93-7 ***HR: 3***
QUIETALUM
mf: $C_{10}H_{13}BrN_2O_3$ mw: 289.16

PROP: Crystals; sltly bitter taste. Mp: 177-179. Sltly sol in water; freely sol in alc, glacial acetic acid, acetone, and alkalies; sparingly sol in ether, chloroform, and benzene.

SYNS: 5-(2'-BROMALLYL)-5-ISOPROPYLBARBITURIC ACID

◇ BROMOAPROBARBITAL ◇ 5-(2-BROMO-2-PROPENYL)-5-(1-METHYLETHYL)-2,4,6(1H,3H,5H)-PYRIMIDINETRIONE ◇ IBOMAL ◇ 5-ISOPROPYL-5-BROMALLYLBARBITURIC ACID ◇ 5-ISOPROPYL-5-(2-BROMOALLYL)BARBITUATE ◇ KWIETAL ◇ NOCTAL ◇ NOCTENAL ◇ NOSTAL ◇ NOSTRAL ◇ PROPALDON ◇ PROPALLYLONAL ◇ QUIETAL

TOXICITY DATA with REFERENCE
ipr-rbt LDLo:60 mg/kg JPETAB 44,325,32
scu-rat LD50:90 mg/kg AEPPAE 152,341,30
scu-mus LDLo:100 mg/kg HBAMAK 4,1289,35
orl-rbt LDLo:225 mg/kg JPETAB 44,325,32
ipr-rbt LDLo:120 mg/kg JPETAB 44,325,32
scu-gpg LDLo:80 mg/kg HBAMAK 4,1289,35
scu-frg LDLo:300 mg/kg HBAMAK 4,1289,35
orl-mam LDLo:300 mg/kg JPETAB 42,253,31

SAFETY PROFILE: Poison by ingestion, intraperitoneal, and subcutaneous routes. Used as a sedative and hypnotic agent. When heated to decomposition it emits very toxic fumes of NO_x and Br^-. A controlled drug under 21 CFR 1308.11, which may be habit forming. See also BARBITURATES.

QCS875 CAS:10072-24-9 ***HR: 2***
QUINACRINE ETHYL M/2
mf: $C_{18}H_{19}Cl_2N_3O{\cdot}2ClH{\cdot}H_2O$ mw: 455.24

SYNS: ACRIDINE,9-(2-((2-CHLOROETHYL)AMINO)ETHYLAMINO)-6-CHLORO-2-METHOXY-, DIHYDROCHLORIDE, HYDRATE ◇ 9-(2-((2-CHLOROETHYL)AMINO)ETHYLAMINO)-6-CHLORO-2-METHOXYACRIDINE, DIHYDROCHLORIDE ◇ ICR-125

TOXICITY DATA with REFERENCE
ipr-mus TDLo:570 mg/kg/4W-I:NEO JNCIAM 36,915,66

SAFETY PROFILE: Questionable carcinogen with experimental neoplastigenic data. When heated to decomposition it emits toxic fumes of NO_x and HCl.

QDJ000 CAS:64046-79-3 ***HR: 3***
QUINACRINE MUSTARD
mf: $C_{23}H_{28}Cl_3N_3O$ mw: 468.89

SYNS: 9-(4-(BIS-β-CHLOROETHYLAMINO)-1-METHYLBUTYLAMINO)-6-CHLORO-2-METHOXYACRIDINE ◇ NSC-3424

TOXICITY DATA with REFERENCE
cyt-hmn-lym 1 μg/L/72 H ARTODN 46,61,80
msc-ham-lng 1 mg/L CNREA8 44,3270,84
ipr-rat LD10:970 μg/kg CNCRA6 17,1,62
ivn-dog LDLo:910 μg/kg CCSUBJ 2,202,65
ivn-mus LDLo:910 μg/kg CCSUBJ 2,202,65

SAFETY PROFILE: A deadly poison by intravenous and intraperitoneal routes. Human mutagenic data reported. When heated to decomposition it emits very toxic fumes of Cl^- and NO_x.

QDS000 CAS:4213-45-0 ***HR: 3***
QUINACRINE MUSTARD DIHYDROCHLORIDE
mf: $C_{23}H_{28}Cl_3N_3O{\cdot}2ClH$ mw: 541.81

SYNS: 9-(4-BIS(2-CHLOROETHYL)AMINO-1-METHYLBUTYLAMINO)-6-CHLORO-2-METHOXYACRIDINEDIHYDROCHLORIDE ◇ ICR 10 ◇ 2-METHOXY-6-CHLORO-9-(4-BIS(2-CHLOROETHYL)AMINO-1-METHYLBUTYLAMINO)ACRIDINEDIHYDROCHLORIDE ◇ 2-METHOXY-6-CHLORO-9-(3-(ETHYL-2-CHLOROETHYL)AMINOPROPYLAMINO)ACRIDINE DIHYDROCHLORIDE ◇ QUINACRINE MUSTARD

TOXICITY DATA with REFERENCE
dnr-esc 20 μL/disc MUREAV 97,1,82
dnr-bcs 20 μL/disc MUREAV 97,1,82
slt-dmg-par 2 mmol/L MUREAV 1,437,64
sce-ham:ovr 468 μg/L/2H-C ENMUDM 4,647,82
cyt-mam:lng 50 mg/L HEREAY 69,217,71
sce-hmn-lym 1 mg/L MUREAV 30,273,75
ivn-mus TDLo:2700 μg/kg:NEO CNREA8 36,2423,76

CONSENSUS REPORTS: EPA Genetic Toxicology Program. Reported in EPA TSCA Inventory.

SAFETY PROFILE: Questionable carcinogen with experimental neoplastigenic data. Human mutation data reported. Corrosive. When heated to decomposition it emits very toxic fumes of Cl^- and NO_x. See also QUINACRINE MUSTARD.

QDS225 CAS:7054-25-3 ***HR: 3***
QUINAGLUTE
mf: $C_{20}H_{24}N_2O_2{\cdot}C_6H_{12}O_7$ mw: 520.64

PROP: Triboluminescent. Mp: 174-175° after drying of solvated crystals.

SYNS: QUINIDINE GLUCONATE ◇ QUINIDINE-d-GLUCONATE (salt) ◇ QUINIDINE MONO-d-GLUCONATE (salt)

TOXICITY DATA with REFERENCE
orl-man TDLo:7609 mg/kg/78W-I:SKN AIMDAP 145,446,85
orl-wmn TDLo:1773 mg/kg/13W-I:SKN AIMDAP 145,446,85
ipr-mus LDLo:150 mg/kg TXAPA9 23,288,72

SAFETY PROFILE: Poison by intraperitoneal route. Human systemic effects by ingestion: allergic dermatitis. When heated to decomposition it emits toxic fumes of NO_x. See also QUINIDINE.

QEA000 CAS:93-10-7 ***HR: 3***
QUINALDIC ACID
mf: $C_{10}H_7NO_2$ mw: 173.18

PROP: Dihydrate: crystals. Mp: 155-157°.

SYNS: QUINALDINIC ACID ◇ QUINOLINE-2-CARBOXYLIC ACID ◇ 2-QUINOLINECARBOXYLIC ACID

TOXICITY DATA with REFERENCE
orl-bwd LD50:100 mg/kg TXAPA9 21,315,72

CONSENSUS REPORTS: Reported in EPA TSCA Inventory.

SAFETY PROFILE: Poison by ingestion. When heated to decomposition it emits toxic fumes of NO_x.

QEJ000 CAS:91-63-4 ***HR: 2***
QUINALDINE
mf: $C_{10}H_9N$ mw: 143.20

PROP: Colorless, oily liquid; quinoline odor. D: 1.06, bp: 246-247°. Insol in water; sol in chloroform and ether.

SYNS: CHINALDINE ◇ 2-METHYLQUINOLINE

TOXICITY DATA with REFERENCE
skn-rbt 10 mg/24H MLD AMIHBC 4,119,51
eye-rbt 750 μg SEV AMIHBC 4,119,51
orl-rat LD50:1230 mg/kg AMIHBC 4,119,51
skn-rbt LD50:1870 mg/kg AMIHBC 4,119,51

CONSENSUS REPORTS: Reported in EPA TSCA Inventory.

SAFETY PROFILE: Moderately toxic by ingestion and skin contact. A skin and severe eye irritant. When heated to decomposition it emits toxic fumes of NO_x.

QEJ800 CAS:86-96-4 ***HR: 2***
2,4(1H,3H)-QUINAZOLINEDIONE
mf: $C_8H_6N_2O_2$ mw: 162.16

SYNS: BENZOURACIL ◇ BENZOYLENEUREA ◇ 2,4-DIHYDROXYQUINAZOLINE ◇ 2,4-DIOXOTETRAHYDROQUINAZOLINE ◇ 2-KETO-4-QUINAZOLINONE ◇ QUINAZOLINEDIONE ◇ QUINAZOLINE-2,4-DIONE ◇ 2,4-QUINAZOLINEDIONE ◇ (1H,3H)QUINAZOLINE DIONE-2,4

TOXICITY DATA with REFERENCE
orl-mus TDLo:17400 mg/kg (male 8D pre):REP MPHEAE 15,7,66
ipr-rat LD50:1200 mg/kg CHTPBA 3,100,68
ipr-mus LD50:1447 mg/kg ARZNAD 12,1204,62

SAFETY PROFILE: Moderately toxic by intraperitoneal route. Experimental reproductive effects. When heated to decomposition it emits toxic fumes of NO_x.

QFA000 CAS:491-36-1 ***HR: 2***
4(3H)-QUINAZOLINONE
mf: $C_8H_6N_2O$ mw: 146.16

SYN: 4-QUINAZOLINONE

TOXICITY DATA with REFERENCE
orl-mus LD50:609 mg/kg ARZNAD 12,1204,62
ipr-mus LD50:450 mg/kg ARZNAD 12,1204,62

SAFETY PROFILE: Moderately toxic by ingestion and intraperitoneal routes. When heated to decomposition it emits toxic fumes of NO_x.

QFA250 CAS:152-43-2 ***HR: 2***
QUINESTROL
mf: $C_{25}H_{32}O_2$ mw: 364.57

PROP: Crystals. Mp: 107-108°.

SYNS: 3-(CYCLOPENTYLOXY)-19-NOR-17-α-PREGNA-1,3,5(10)-TRIEN-20-YN-17-OL ◇ EECPE ◇ ESTON ◇ ESTRADIOL-17-β 3-CYCLOPENTYL ETHER ◇ ESTROVIS ◇ ESTROVIS 4000 ◇ ESTROVISTER ◇ 17-α-ETHINYLESTRADIOL 3-CYCLOPENTYL ETHER ◇ PLESTROVIS ◇ QUI-LEA ◇ W 3566

TOXICITY DATA with REFERENCE
orl-wmn TDLo:80 μg/kg (2D pre):REP INJFA3 14,295,69

SAFETY PROFILE: Human female reproductive effects by ingestion of extremely small amounts: postpartum disorders and changes in fertility. Experimental reproductive effects. A steroid. When heated to decomposition it emits acrid smoke and irritating fumes.

QFA275 CAS:3000-39-3 ***HR: 2***
QUINGESTANOL ACETATE
mf: $C_{27}H_{36}O_3$ mw: 408.63

SYNS: 3-CYCLOPENTYL ENOL ETHER of NORETHINDRONE ACETATE ◇ 3-(CYCLOPENTYLOXY)-19-NOR-17-α-PREGNA-3,5-DIEN-20-YN-17-OL ACETATE (ester) ◇ NORETHINDRONE ACETATE 3-CYCLOPENTYL ENOL ETHER

TOXICITY DATA with REFERENCE
orl-wmn TDLo:1090 μg/kg (26W pre):REP CCPTAY 9,213,74

SAFETY PROFILE: Human female reproductive effects by ingestion of very small amounts: changes in menstrual cycle, ovaries, fallopian tubes, and fertility index. When heated to decomposition it emits acrid smoke and irritating fumes.

QFJ000 CAS:106-34-3 ***HR: 3***
QUINHYDRONE
mf: $C_6H_6O_2 \cdot C_6H_4O_2$ mw: 220.24

PROP: Dark green crystals. D: 1.40, mp: 171°. Sltly sol in cold water; sol in alc, ether, hot water, ammonia; insol in petr ether. Subl with partial decomp.

SYNS: p-BENZOQUINONE, compounded with HYDROQUINONE ◇ GREEN HYDROQUINONE ◇ HYDROQUINONE, compounded with p-BENZOQUINONE

TOXICITY DATA with REFERENCE
orl-rat LDLo:225 mg/kg FEPRA7 8,348,49
ivn-rat LD50:35 mg/kg FEPRA7 8,348,49

CONSENSUS REPORTS: Reported in EPA TSCA Inventory.

SAFETY PROFILE: Poison by ingestion and intrave-

nous routes. Small doses caused lowered metabolism. When heated to decomposition it emits acrid smoke and irritating fumes.

QFJ300 ***HR: 3***
QUINICINE OXALATE
mf: $C_{20}H_{24}N_2O_2 \cdot C_2H_2O_4$ mw: 414.50

SYN: 1-(6-METHOXY-4-QUINOLYL)-3-(3-VINYL-4-PIPERIDYL)-1-PROPANONE OXALATE

TOXICITY DATA with REFERENCE
orl-mus LD50:420 mg/kg APFRAD 24,39,66
scu-mus LD50:155 mg/kg APFRAD 24,39,66
ivn-mus LD50:48 mg/kg APFRAD 24,39,66
orl-gpg LDLo:700 mg/kg APFRAD 24,39,66
scu-gpg LDLo:250 mg/kg APFRAD 24,39,66
ivn-gpg LDLo:60 mg/kg APFRAD 24,39,66

SAFETY PROFILE: Poison by subcutaneous and intravenous routes. Moderately toxic by ingestion. When heated to decomposition it emits toxic fumes of NO_x. See also OXALATES.

QFS000 CAS:56-54-2 ***HR: 3***
QUINIDINE
mf: $C_{20}H_{24}N_2O_2$ mw: 324.46

SYNS: CHINIDIN (GERMAN) ◇ CIN-QUIN ◇ CONCHININ ◇ CONQUININE ◇ 6'-METHOXYCINCHONAN-9-OL ◇ α-(6-METHOXY-4-QUINOLYL)-5-VINYL-2-QUINUCLIDINEMETHANOL ◇ 6-METHOXY-α-(5-VINYL-2-QUINUCLIDINYL)-4-QUINOLINEMETHANOL ◇ NCI-C56246 ◇ PITAYINE ◇ QUINICARDINE ◇ QUINIDEX ◇ (+)-QUINIDINE ◇ β-QUININE

TOXICITY DATA with REFERENCE
eye-rbt 3% MLD AIPTAK 137,410,62
orl-rat LD50:263 mg/kg ARZNAD 27,589,77
ipr-rat LDLo:174 mg/kg AEPPAE 205,129,48
ivn-rat LD50:23 mg/kg JPETAB 128,22,60
orl-mus LD50:535 mg/kg JPETAB 105,291,52
ipr-mus LD50:135 mg/kg AIPTAK 137,410,62
scu-mus LDLo:400 mg/kg AEPPAE 205,129,48
ivn-mus LD50:53600 μg/kg JMCMAR 27,1142,84
ims-mus LD50:200 mg/kg 27ZIAQ -,232,73

CONSENSUS REPORTS: Reported in EPA TSCA Inventory.

SAFETY PROFILE: Poison by ingestion, subcutaneous, intravenous, intramuscular, and intraperitoneal routes. An eye irritant. Implicated in aplastic anemia. When heated to decomposition it emits toxic fumes of NO_x.

QHA000 CAS:50-54-4 ***HR: 3***
QUINIDINE SULFATE (2:1) (salt)
mf: $C_{40}H_{48}N_4O_4 \cdot H_2O_4S$ mw: 747.00

SYNS: CIN-QUIN ◇ QUINICARDINE ◇ QUINIDATE ◇ QUINIDEX ◇ QUINITEX ◇ QUINORA ◇ SYSTODIN

TOXICITY DATA with REFERENCE
orl-wmn TDLo:272 mg/kg/17D-I:CVS AIMDAP 145,2051,85
orl-man TDLo:40 mg/kg/2W-I:CVS AIMDAP 145,2051,85
orl-hmn TDLo:9600 mg/kg/8W:CNS,PUL JAMAAP 238,884,77
orl-wmn TDLo:409 g/kg/14Y:CNS JAMAAP 237,2093,77
orl-man TDLo:120 mg/kg/7D:LIV,MET JAMAAP 234,310,75
mul-man LDLo:30 mg/kg/2D-I:CVS,PUL AIMEAS 16,571,42
orl-rat LD50:456 mg/kg ARZNAD 18,1127,68
ipr-rat LDLo:140 mg/kg TXAPA9 1,156,59
scu-rat LD50:610 mg/kg ARZNAD 18,1127,68
ivn-rat LD50:56 mg/kg ARZNAD 18,1127,68
orl-mus LD50:540 mg/kg JPETAB 136,114,62
ipr-mus LD50:165 mg/kg CYLPDN 6,213,85
ivn-mus LD50:54 mg/kg AIPTAK 105,221,56
ivn-dog LDLo:19 mg/kg DECRDP 10,197,84
ivn-cat LD50:22 mg/kg AEPPAE 192,639,39
ivn-rbt LDLo:26 mg/kg DECRDP 10,197,84

CONSENSUS REPORTS: Reported in EPA TSCA Inventory.

SAFETY PROFILE: Human poison by multiple routes. Poison by intraperitoneal and intravenous routes. Moderately toxic by ingestion and subcutaneous routes. Human systemic effects by ingestion and other routes: somnolence, hallucinations and distorted perceptions, tremors, arrythmias, dyspnea, cyanosis, cholestatic jaundice, liver function impairment, fever, other vascular and pulmonary changes, death. When heated to decomposition it emits very toxic fumes of NO_x and SO_x. See also QUINIDINE.

QHJ000 CAS:130-95-0 ***HR: 3***
QUININE
mf: $C_{20}H_{24}N_2O_2$ mw: 324.46

PROP: Bulky, white, amorph powder or crystals; bitter taste. Mp: 174.9°.

SYNS: CHININ (GERMAN) ◇ (8-α,9R)-6'-METHOXYCINCHONAN-9-OL ◇ 6-METHOXYCINCHONINE ◇ α-(6-METHOXY-4-QUINOYL)-5-VINYL-2-QUINCLIDINEMETHANOL ◇ (−)-QUININE

TOXICITY DATA with REFERENCE
dnd-esc 30 μmol/L MUREAV 89,95,81
dnd-mam:lym 100 μmol/L PMSBA4 2,134,71
orl-wmn TDLo:20 mg/kg/(4-5W preg):TER MMWOAU 108,2293,66
ims-rat TDLo:10 mg/kg (5D preg):REP PSEBAA 100,555,59

orl-wmn TDLo:74 mg/kg:EYE,EAR,GIT AJOPAA 90,403,80
unr-man LDLo:294 mg/kg 85DCAI 2,73,70
orl-rat LDLo:800 mg/kg JPETAB 100,408,50
scu-rat LDLo:200 mg/kg AEPPAE 205,129,48
ims-rat LDLo:300 mg/kg JPETAB 63,122,38
ipr-mus LD50:115 mg/kg ARZNAD 35,1760,85
scu-mus LDLo:200 mg/kg AEPPAE 205,129,48
scu-dog LDLo:180 mg/kg HBAMAK 4,1320,35
scu-cat LDLo:100 mg/kg AEPPAE 205,129,48
ivn-cat LDLo:100 mg/kg RIMAAX 11,3,32
orl-rbt LDLo:500 mg/kg RIMAAX 11,3,32
scu-rbt LDLo:231 mg/kg HBAMAK 4,1320,35
ivn-rbt LDLo:70 mg/kg HBAMAK 4,1320,35
orl-gpg LD50:1800 mg/kg SMWOAS 84,351,54

CONSENSUS REPORTS: Reported in EPA TSCA Inventory.

SAFETY PROFILE: Human poison by unspecified route. Experimental poison by subcutaneous, intravenous, intramuscular, and intraperitoneal routes. Moderately toxic experimentally by ingestion. An experimental teratogen. Human systemic effects by ingestion: visual field changes, tinnitus and nausea or vomiting. Human teratogenic effects by ingestion: developmental abnormalities of the central nervous system; body wall; musculoskeletal, cardiovascular and hepatobiliary systems. Experimental reproductive effects. Mutation data reported. Can cause temporary loss of vision. Quinine dermatitis is an occupational hazard to barbers particularly, and generally to people who work with quinine tonics, medicaments, or cosmetics. An irritant to mucous membranes. Combustible when exposed to heat or flame. Decomposes on exposure to light. When heated to decomposition it emits toxic fumes of NO_x. Used to treat malaria.

QIJ000 CAS:60-93-5 ***HR: 3***
QUININE DIHYDROCHLORIDE
mf: $C_{20}H_{24}N_2O_2{\cdot}2ClH$ mw: 397.38

PROP: White needles or crystalline powder; odorless with very bitter taste. Sol in water, alc, glycerin; sltly sol in chloroform; very sltly sol in ether.

SYNS: ACID QUININE HYDROCHLORIDE ◇ CHININDIHYDROCHLORID (GERMAN) ◇ 6'-METHOXYCINCHONAN-9-OL DIHYDROCHLORIDE ◇ QUININE BIMURIATE ◇ (−)-QUININE DIHYDROCHLORIDE

TOXICITY DATA with REFERENCE
mma-sat 2800 nmol/plate MUREAV 66,33,79
orl-rat LD50:1392 mg/kg JPETAB 91,157,47
ivn-rat LD50:78 mg/kg JPETAB 91,157,47
orl-mus LD50:660 mg/kg JPETAB 91,157,47
ivn-mus LD50:96 mg/kg JPETAB 91,157,47
orl-rbt LD50:640 mg/kg JPETAB 91,157,47
ivn-rbt LD50:35 mg/kg JPETAB 91,157,47
scu-gpg LDLo:199 mg/kg PSEBAA 32,595,35
ivn-gpg LD50:57 mg/kg JPETAB 91,157,47

CONSENSUS REPORTS: Reported in EPA TSCA Inventory.

SAFETY PROFILE: Poison by intravenous and subcutaneous routes. Moderately toxic by ingestion. Mutation data reported. When heated to decomposition it emits very toxic fumes of NO_x and HCl. See also QUININE.

QIS000 CAS:73771-81-0 ***HR: 3***
QUININE ETHIODIDE
mf: $C_{22}H_{29}N_2O_2{\cdot}I$ mw: 480.43

SYNS: (8-α,9R)-1-ETHYL-9-HYDROXY-6'-METHOXYCINCHONAN-1-IUM IODIDE ◇ 6-(1-HYDROXY-1-(6-METHOXY-4-QUINOLINYL) METHYL-1-ETHYL-3-VINYLQUINUCLIDINIUM, IODIDE ◇ 6-(HYDROXY(6-METHOXY-4-QUINOLINYL)METHYL)-1-ETHYL-3-VINYLQUINUCLIDINIUM, IODIDE

TOXICITY DATA with REFERENCE
ivn-rat LDLo:23 mg/kg JPETAB 91,127,47
ivn-rbt LDLo:9 mg/kg JPETAB 91,127,47

SAFETY PROFILE: Poison by intravenous route. When heated to decomposition it emits very toxic fumes of I^- and NO_x. See also QUININE and IODIDES.

QIS300 CAS:130-90-5 ***HR: 3***
QUININE FORMATE
mf: $C_{20}H_{24}N_2O_2{\cdot}CH_2O_2$ mw: 370.49

PROP: White, cryst, powder. Mp: 113°. Sol in 30 parts water, in alc, and chloroform; sltly in ether.

SYNS: FORMIC ACID, compounded with QUININE (1:1) ◇ QUININE, FORMATE (SALT) ◇ QUINOFORM

TOXICITY DATA with REFERENCE
ims-dog TDLo:450 mg/kg (female 18-47D post):REP THERAP 26,563,71
ims-dog LD50:290 mg/kg THERAP 26,563,71

SAFETY PROFILE: Poison by intramuscular route. Experimental reproductive effects. When heated to decomposition it emits toxic fumes of NO_x.

QJJ100 CAS:549-49-5 ***HR: 3***
QUININE HYDROBROMIDE
mf: $C_{20}H_{24}N_2O_2{\cdot}BrH$ mw: 405.38

PROP: Monohydrate: white, odorless, bitter silky needles. Darkens in light. Sol in water, alc, chloroform, and glycerol; sltly sol in ether.

SYNS: BROMOQUIN[INE ◇ CHININ HYDROBROMID (GERMAN)

TOXICITY DATA with REFERENCE
orl-rbt LDLo:750 mg/kg FDWU** -,-,31
scu-rbt LDLo:250 mg/kg FDWU** -,-,31

orl-pgn LDLo:6 g/kg FDWU** -,-,31
scu-pgn LDLo:1 g/kg FDWU** -,-,31
orl-frg LDLo:30 g/kg FDWU** -,-,31
scu-frg LDLo:18 g/kg FDWU** -,-,31

SAFETY PROFILE: Poison by subcutaneous route. Moderately toxic by ingestion. When heated to decomposition it emits toxic fumes of NO_x and HBr. See also QUININE.

QJS000 CAS:130-89-2 ***HR: 3***
QUININE HYDROCHLORIDE
mf: $C_{20}H_{24}N_2O_2•ClH$ mw: 360.92

SYNS: QUININE CHLORIDE ◇ QUININE MONOHYDROCHLORIDE ◇ QUININE MURIATE

TOXICITY DATA with REFERENCE
mnt-mus-orl 110 mg/kg TXCYAC 26,173,83
sce-mus-orl 75 mg/kg TXCYAC 26,173,83
ivn-hmn LDLo:230 μg/kg:CNS,GIT AEXPBL 17,363,1883
orl-rat LDLo:500 mg/kg JPETAB 63,122,38
ipr-rat LD50:170 mg/kg TXAPA9 24,37,73
scu-rat LDLo:790 mg/kg ZGEMAZ 11,257,20
ivn-rat LDLo:75 mg/kg JPETAB 63,122,38
ims-rat LDLo:300 mg/kg JPETAB 63,122,38
orl-mus LD50:1160 mg/kg BJPCAL 6,185,51
ipr-mus LD50:240 mg/kg TXAPA9 24,37,73
scu-mus LDLo:700 mg/kg JPETAB 8,53,16
ivn-mus LD50:68300 μg/kg TXAPA9 1,454,56

CONSENSUS REPORTS: Reported in EPA TSCA Inventory.

SAFETY PROFILE: Poison by ingestion, subcutaneous, intravenous, intramuscular, and intraperitoneal routes. Human systemic effects by intravenous route: convulsions or effect on seizure threshold, muscle contraction or spasticity, and nausea or vomiting. Mutation data reported. Used as a local anesthetic. When heated to decomposition it emits very toxic fumes of NO_x and HCl. See also QUININE.

QMA000 CAS:804-63-7 ***HR: 3***
QUININE SULFATE
mf: $C_{20}H_{24}N_2O_2•O_4S$ mw: 420.52

PROP: Fine white needlelike crystals; odorless with a very bitter taste. Sol in water, alc; sltly sol in chloroform.

SYNS: QUININE BISULFATE ◇ QUININE HYDROGEN SULFATE

TOXICITY DATA with REFERENCE
pic-esc 100 μg/plate CNREA8 43,2819,83
orl-rat TDLo:1425 mg/kg (14D pre-21D post):REP BNEOBV 36,273,79
orl-man TDLo:129 mg/kg:EYE HUTODJ 3,399,84
orl-man TDLo:27 mg/kg:EYE,EAR BMJOAE 287,1700,83
orl-wmn TDLo:12 mg/kg/1D-I:LIV BMJOAE 286,264,83
orl-wmn LDLo:220 mg/kg CTOXAO 7,129,74
orl-hmn TDLo:4300 μg/kg:BLD,PNS BMJOAE 1,605,77
orl-wmn TDLo:80 mg/kg:EYE,EAR,GIT BMJOAE 287,1700,83
orl-mus LDLo:800 mg/kg JPETAB 78,159,43

CONSENSUS REPORTS: Reported in EPA TSCA Inventory.

SAFETY PROFILE: Human poison by ingestion. Human systemic effects by ingestion: acuity changes, blood angranulocytosis, fibrous hepatitis, flaccid paralysis without anesthesia, motor activity changes, mydriasis (pupillary dilation), nausea or vomiting, tinnitus, visual field changes. Experimental reproductive effects. Mutation data reported. When heated to decomposition it emits very toxic fumes of SO_x and NO_x. See also QUININE.

QMJ000 CAS:91-22-5 ***HR: 3***
QUINOLINE
DOT: UN 2656
mf: C_9H_7N mw: 129.17

$C_6H_4N{=}CHCH{=}CH$

PROP: Refractive, colorless liquid; peculiar odor. Mp: −14.5°, bp: 237.7°, d: 1.0900 @ 25°/4°, autoign temp: 896°F, vap press: 1 mm @ 59.7°, vap d: 4.45. Sol in water, CS_2; misc in alc, ether.

SYNS: 1-AZANAPHTHALENE ◇ 1-BENZAZINE ◇ 1-BENZINE ◇ BENZO(b)PYRIDINE ◇ CHINOLEINE ◇ CHINOLIN (CZECH) ◇ CHINOLINE ◇ LEUCOL ◇ LEUCOLINE ◇ LEUKOL ◇ USAF EK-218

TOXICITY DATA with REFERENCE
skn-rbt 10 mg/24H open MLD AMIHBC 4,119,51
eye-rbt 250 μg open SEV AMIHBC 4,119,51
mma-sat 1 μmol/plate ABCHA6 42,861,78
dnd-esc 30 μmol/L MUREAV 89,95,81
mma-ham:ovr 80 μmol/L ENMUDM 4,395,82
sce-ham:ovr 110 μg/L ENMUDM 7,1,85
orl-rat TDLo:7770 mg/kg/37W-C:NEO CNREA8 36,329,76
orl-mus TDLo:50 g/kg/30W-C:ETA GANNA2 68,785,77
ipr-mus TDLo:9042 μg/kg/15D-I:CAR JJCREP 78,139,87
orl-rat LD50:331 mg/kg MarJV# 29MAR77
ipr-mus LDLo:64 mg/kg CBCCT* 2,190,50
skn-rbt LD50:540 mg/kg AMIHBC 4,119,51
scu-rbt LDLo:200 mg/kg HBAMAK 4,1289,35
scu-frg LDLo:150 mg/kg HBAMAK 4,1289,35

CONSENSUS REPORTS: Reported in EPA TSCA Inventory. EPA Genetic Toxicology Program. Community Right-To-Know List.

DOT Classification: Poison B; Label: St. Andrews Cross, Flammable Liquid; ORM-E; Label: None.

SAFETY PROFILE: Poison by ingestion, subcutaneous, and intraperitoneal routes. Moderately toxic by skin contact. A skin and severe eye irritant. Mutation data reported. Questionable carcinogen with experimental neoplastigenic and tumorigenic data. It can cause retinitis similar to that caused by naphthalene but without causing opacity of the lens. Combustible when exposed to heat or flame. Its preparation has caused many industrial explosions. Potentially explosive reaction with hydrogen peroxide. Violent reaction with dinitrogen tetraoxide; perchromates. Incompatible with linseed oil + thionyl chloride; maleic anhydride. Unpredictably violent. When heated to decomposition it emits toxic fumes of NO_x.

QMS000 ***HR: 3***
6-QUINOLINE CARBONYL AZIDE
mf: $C_{10}H_6N_4O$ mw: 198.19

PROP: Mp 88°C.

SAFETY PROFILE: Explodes when heated above 88°C. When heated to decomposition it emits toxic fumes of NO_x. See also AZIDES.

QNA000 CAS:86-95-3 ***HR: D***
2,4-QUINOLINEDIOL
mf: $C_9H_7NO_2$ mw: 161.17

TOXICITY DATA with REFERENCE
mma-sat 1 mg/plate JNCIAM 60,405,78

CONSENSUS REPORTS: Reported in EPA TSCA Inventory.

SAFETY PROFILE: Mutation data reported. When heated to decomposition it emits toxic fumes of NO_x.

QNJ000 CAS:1011-50-3 ***HR: 2***
2-QUINOLINEETHANOL
mf: $C_{11}H_{11}NO$ mw: 173.23

SYN: 2-(2-HYDROXYETHYL)QUINOLINE

TOXICITY DATA with REFERENCE
ipr-mus LDLo:500 mg/kg CBCCT* 5,288,53

CONSENSUS REPORTS: Reported in EPA TSCA Inventory.

SAFETY PROFILE: Moderately toxic by intraperitoneal route. When heated to decomposition it emits toxic fumes of NO_x.

QOJ000 CAS:69365-68-0 ***HR: D***
2-QUINOLINE THIOACETAMIDE HYDROCHLORIDE
mf: $C_{11}H_{10}N_2S•ClH$ mw: 238.75

TOXICITY DATA with REFERENCE
orl-rat TDLo:200 mg/kg (17D preg):TER TJADAB 14,254,76

SAFETY PROFILE: An experimental teratogen. When heated to decomposition it emits very toxic fumes of NO_x, SO_x and HCl.

QOJ250 CAS:50308-94-6 ***HR: 3***
QUINOLINIUM DIBROMIDE
mf: $C_{29}H_{28}N_6O•2Br$ mw: 636.45

SYNS: CAIN'S QUINOLINIUM ◇ NSC 176319 ◇ QUINOLINIUM, 6-AMINO-1-METHYL-4-((4-((((1-METHYLPYRIDINIUM-4-YL)AMINO) PHENYL)CARBAMOYL)PHENYL)AMINO)-, DIBROMIDE

TOXICITY DATA with REFERENCE
dnd-mus:lym 660 nmol/L JMCMAR 22,134,79
dnd-mam:lym 67 mg/L PHMGBN 17,61,78
ipr-mus LD50:7399 μg/kg NCISP* JAN86
scu-mus LD50:35910 μg/kg NCISP* JAN86

SAFETY PROFILE: Poison by subcutaneous and intraperitoneal routes. Mutation data reported. When heated to decomposition it emits toxic fumes of Br^- and NO_x.

QPA000 CAS:148-24-3 ***HR: 3***
8-QUINOLINOL
mf: C_9H_7NO mw: 145.17

PROP: White crystals or powder. Mp: 76°, bp: 267°. Very sltly sol in cold water; sltly sol in ether; sol in alc, dilute alkali.

SYNS: BIOQUIN ◇ FENNOSAN ◇ HYDROXYBENZOPYRIDINE ◇ 8-HYDROXY-CHINOLIN (GERMAN) ◇ 8-HYDROXYQUINOLINE ◇ NCI-C55298 ◇ 8-OQ ◇ OXINE ◇ OXYBENZOPYRIDINE ◇ OXYCHINOLIN ◇ o-OXYCHINOLIN (GERMAN) ◇ OXYQUINOLINE ◇ 8-OXYQUINOLINE ◇ PHENOPYRIDINE ◇ 8-QUINOL ◇ QUINOPHENOL ◇ TUMEX ◇ USAF EK-794

TOXICITY DATA with REFERENCE
dni-hmn:hla 25 μmol/L MUREAV 92,427,82
bfa-rat/sat 600 mg/kg TXCYAC 34,231,85
ivg-rat TDLo:33 g/kg/82W-I:ETA,REP ARPAAQ 79,245,65
scu-mus TDLo:900 mg/kg/21W-I:ETA,REP VOONAW 16(8),67,70
skn-mus TDLo:7200 mg/kg/50W-I:ETA,REP VOONAW 16(8),67,70
orl-rat TDLo:29 g/kg/48W-I:ETA,REP JNCIAM 41,985,68
imp-mus TDLo:50 mg/kg:CAR BJCAAI 11,212,57
ivg-mus TDLo:5600 mg/kg/35W-I:ETA VOONAW 16(8),67,70
imp-mus TD:80 mg/kg:NEO BJCAAI 11,212,57
imp-mus TD:100 mg/kg/:CAR BMBUAQ 14,1475,68
orl-rat LD50:1200 mg/kg PCOC** -,602,66
orl-mus LD50:20 g/kg NIIRDN 6,271,82
ipr-mus LD50:43 mg/kg FATOAO 42(4),396,79

scu-mus LD50:83600 μg/kg PHARAT 1,150,46
orl-gpg LD50:1205 mg/kg TJADAB 20,413,79

CONSENSUS REPORTS: IARC Cancer Review: Group 3 IMEMDT 7,56,87; Animal Inadequate Evidence IMEMDT 13,101,77. NTP Carcinogenesis Studies (feed); No Evidence: mouse, rat NTPTR* NTP-TR-276,85. Reported in EPA TSCA Inventory. EPA Genetic Toxicology Program.

SAFETY PROFILE: Poison by intraperitoneal and subcutaneous routes. Moderately toxic by ingestion. Questionable carcinogen with experimental carcinogenic, neoplastigenic, tumorigenic data. Experimental reproductive effects. A central nervous system stimulant. Human mutation data reported. Combustible when exposed to heat or flame. When heated to decomposition it emits highly toxic fumes of NO_x.

QPJ000 CAS:63716-63-2 ***HR: 3***
8-QUINOLINOLIUM-4',7'-DIBROMO-3'-HYDROXY-2'-NAPHTHOATE
mf: $C_{20}H_{13}Br_2NO_3$ mw: 475.16

TOXICITY DATA with REFERENCE
sce-hmn:fbr 5 mg/L MUREAV 58,317,78
ipr-mus LD50:85 mg/kg TXAPA9 5,599,63

SAFETY PROFILE: Poison by intraperitoneal route. Human mutation data reported. When heated to decomposition it emits very toxic fumes of Br^- and NO_x.

QPS000 CAS:134-31-6 ***HR: 3***
8-QUINOLINOL SULFATE (2:1) (SALT)
mf: $C_{18}H_{14}N_2O_2 \cdot H_2O_4S$ mw: 388.42

SYNS: 8-HYDROXY-CHINOLIN-SULFAT (GERMAN) ◇ 8-HYDROXYQUINOLINE SULFATE ◇ OXINE SULFATE ◇ OXYQUINOLINE SULFATE ◇ 8-QUINOLINOL HYDROGEN SULFATE (2:1) ◇ 8-QUINOLINOL SULFATE

TOXICITY DATA with REFERENCE
mma-sat 20 μg/plate MUREAV 39,285,77
sce-hmn:lym 10 μmol/L ENMUDM 2,191,80
orl-rat LD50:2038 mg/kg 85DPAN -,-,71/76
orl-mus LD50:280 mg/kg 85DPAN -,-,71/76
orl-mam LD50:1200 mg/kg FMCHA2 -,D56,77

CONSENSUS REPORTS: Reported in EPA TSCA Inventory.

SAFETY PROFILE: Poison by ingestion. Human mutation data reported. When heated to decomposition it emits very toxic fumes of SO_x and NO_x. See also 8-QUINOLINOL.

QQA000 CAS:63040-20-0 ***HR: 3***
N-(4-QUINOLYL)ACETOHYDROXAMIC ACID
mf: $C_{11}H_{10}N_2O_2$ mw: 202.23

SYN: MONOACETYL 4-HYDROXYAMINOQUINOLINE

TOXICITY DATA with REFERENCE
scu-mus TDLo:96 mg/kg/20W-I:NEO JJEMAG 40,475,70

SAFETY PROFILE: Questionable carcinogen with experimental neoplastigenic data. When heated to decomposition it emits toxic fumes of NO_x.

QQS075 CAS:11001-74-4 ***HR: 3***
QUINOMYCIN C
mf: $C_{55}H_{72}N_{12}O_{12}S_2$ mw: 1157.51

SYNS: ANTIBIOTIC U 48160 ◇ 4-(N,4-DIMETHYL-l-ALLOISOLEUCINE)-8-(N,4-DIMETHYL-l-ALLOISOLEUCINE)-QUINOMYCINA ◇ U 48160

TOXICITY DATA with REFERENCE
orl-mus LD50:5 mg/kg 85GDA2 4(2),83,80
ipr-mus LD50:25 μg/kg 85FZA5 -,550,67
scu-mus LD50:790 μg/kg ITGDA2 4(2),83,80
ivn-mus LD50:480 μg/kg ITGDA2 4(2),83,80

SAFETY PROFILE: Poison by ingestion, subcutaneous, intravenous, and intraperitoneal routes. When heated to decomposition it emits toxic fumes of NO_x and SO_x.

QQS200 CAS:106-51-4 ***HR: 3***
QUINONE
DOT: UN 2587
mf: $C_6H_4O_2$ mw: 108.10

O:CCH=CHCO•CH=CH (ring closure bracket between first C and last CH)

PROP: Yellow crystals; characteristic irritating odor. Mp: 115.7°, bp: sublimes, d: 1.318 @ 20°/4°.

SYNS: BENZO-CHINON (GERMAN) ◇ 1,4-BENZOQUINE ◇ 1,4-BENZOQUINONE ◇ BENZOQUINONE (DOT) ◇ p-BENZOQUINONE ◇ CHINON (DUTCH, GERMAN) ◇ p-CHINON (GERMAN) ◇ CHINONE ◇ CYCLOHEXADEINEDIONE ◇ 1,4-CYCLOHEXADIENEDIONE ◇ 2,5-CYCLOHEXADIENE-1,4-DIONE ◇ 1,4-CYCLOHEXADIENE DIOXIDE ◇ 1,4-DIOSSIBENZENE (ITALIAN) ◇ 1,4-DIOXYBENZENE ◇ 1,4-DIOXY-BENZOL (GERMAN) ◇ NCI-C55845 ◇ p-QUINONE ◇ RCRA WASTE NUMBER U197 ◇ USAF P-220

TOXICITY DATA with REFERENCE
oms-hmn:lym 5 μmol/L CNREA8 45,2471,85
sce-hmn:lum 5 μmol/L CNREA8 45,2471,85
skn-mus TDLo:800 mg/kg/29W-C:ETA PIATA8 16,309,40
orl-rat LD50:130 mg/kg FEPRA7 8,348,49
ivn-rat LD50:25 mg/kg FEPRA7 8,348,49
ipr-mus LD50:8500 μg/kg BCPCA6 12,885,63
scu-mus LD50:93800 μg/kg ZGIMAL 2,333,47

CONSENSUS REPORTS: IARC Cancer Review: Group 3 IMEMDT 7,56,87; Animal Inadequate Evidence IMEMDT 15,255,77. Reported in EPA TSCA Inventory. Community Right-To-Know List. EPA Genetic Toxicology Program.

OSHA PEL: TWA 0.1 ppm
ACGIH TLV: TWA 0.1 ppm
DFG MAK: 0.1 ppm (0.4 mg/m^3)

DOT Classification: Poison B; Label: Poison.

SAFETY PROFILE: Poison by ingestion, subcutaneous, intraperitoneal, and intravenous routes. Questionable carcinogen with experimental tumorigenic data by skin contact. Human mutation data reported. Quinone has a characteristic, irritating odor. Causes severe damage to the skin and mucous membranes by contact with it in the solid state, in solution, or in the form of condensed vapors. Locally, it causes discoloration, severe irritation, erythema, swelling, and the formation of papules and vesicles, whereas prolonged contact may lead to necrosis. When the eyes become involved, it causes dangerous disturbances of vision. The moist material self heats and decomposes exothermically above 60°C. When heated to decomposition it emits acrid smoke and fumes.

QRJ000 CAS:91-19-0 ***HR: 3***
QUINOXALINE
mf: $C_8H_6N_2$ mw: 130.16

PROP: Crystals. D: 1.133 @ 45°/4°, mp: 29-30°, bp: 225-226°. Sol in water; misc in alc, ether, benzene.

SYNS: 1,4-BENZODIAZINE ◇ BENZOPARADIAZINE ◇ BENZO(A) PYRAZINE ◇ 1,4-DIAZANAPHTHALENE ◇ QUINAZINE ◇ USAF EK-7094

TOXICITY DATA with REFERENCE
ipr-mus LD50:250 mg/kg NTIS** AD607-952

CONSENSUS REPORTS: Reported in EPA TSCA Inventory.

SAFETY PROFILE: Poison by intraperitoneal route. When heated to decomposition it emits toxic fumes of NO_x.

QRS000 CAS:15804-19-0 ***HR: 3***
2,3-QUINOXALINEDIOL
mf: $C_8H_6N_2O_2$ mw: 162.16

SYNS: 2,3-DIHYDROXYQUINOXALINE ◇ USAF EK-6232

TOXICITY DATA with REFERENCE
ipr-rat LDLo:500 mg/kg NCNSA6 5,26,53
ipr-mus LD50:200 mg/kg NTIS** AD607-952

CONSENSUS REPORTS: Reported in EPA TSCA Inventory.

SAFETY PROFILE: Poison by intraperitoneal route. When heated to decomposition it emits toxic fumes of NO_x.

QSA000 CAS:2423-66-7 ***HR: 3***
QUINOXALINE-1,4-DI-N-OXIDE
mf: $C_8H_6N_2O_2$ mw: 162.16

SYNS: GROFAS ◇ QUINDOXIN ◇ QUINOXALINE DIOXIDE ◇ QUINOXALINE DI-N-OXIDE ◇ QUINOXALINE 1,4-DIOXIDE ◇ USAF H-1

TOXICITY DATA with REFERENCE
mmo-sat 5 μg/plate CPBTAL 27,1954,79
mma-sat 5 μg/plate CPBTAL 27,1954,79
dnr-sat 100 μg/plate AMACCQ 20,151,81
sce-ham:lng 500 mg/L MUREAV 139,199,84
orl-rat TDLo:5400 mg/kg/77W-C:CAR JNCIAM 55,137,75
ipr-mus LD50:500 mg/kg NTIS** AD277-689

CONSENSUS REPORTS: EPA Genetic Toxicology Program.

SAFETY PROFILE: Moderately toxic by intraperitoneal route. Questionable carcinogen with experimental carcinogenic data. Mutation data reported. When heated to decomposition it emits toxic fumes of NO_x.

QSJ000 CAS:1199-03-7 ***HR: 3***
2,3-QUINOXALINEDITHIOL
mf: $C_8H_6N_2S_2$ mw: 194.28

SYNS: 2,3-QUINOXALINETHIOL ◇ USAF EK-7317

TOXICITY DATA with REFERENCE
ipr-mus LD50:100 mg/kg NTIS** AD607-952

CONSENSUS REPORTS: Reported in EPA TSCA Inventory.

SAFETY PROFILE: Poison by intraperitoneal route. When heated to decomposition it emits very toxic fumes fumes of NO_x and SO_x.

QSJ800 CAS:73927-90-9 ***HR: 3***
(2,3-QUINOXALINYLDITHIO)DIMETHYLTIN
mf: $C_{10}H_{10}N_2S_2Sn$ mw: 341.03

SYN: STANNANE,DIMETHYL(2,3-QUINOXALINYLDITHIO)-

TOXICITY DATA with REFERENCE
ivn-mus LD50:180 mg/kg CSLNX* NX#01824

OSHA PEL: TWA 0.1 mg(Sn)/m^3 (skin)
ACGIH TLV: TWA 0.1 mg(Sn)/m^3; STEL 0.2 mg/m^3 (skin)
NIOSH REL: (Organotin Compound):10H TWA 0.1 mg(Sn)/m^3

SAFETY PROFILE: Poison by intravenous route. When heated to decomposition it emits toxic fumes of NO_x, SO_x, and Sn.

QTJ000 CAS:73927-96-5 ***HR: 3***
(2,3-QUINOXALINYLDITHIO)DIPHENYLTIN
mf: $C_{20}H_{14}N_2S_2Sn$ mw: 465.17

SYN: (2,3-QUINOXALINYLDITHIO)DIPHENYLSTANNANE

TOXICITY DATA with REFERENCE
ivn-mus LD50:100 mg/kg CSLNX* NX#01825

OSHA PEL: TWA 0.1 mg(Sn)/m^3 (skin)
ACGIH TLV: TWA 0.1 mg(Sn)/m^3 (skin) (Proposed: TWA 0.1 mg(Sn)/m^3; STEL 0.2 mg(Sn)/m^3 (skin))
NIOSH REL: (Organotin Compounds) TWA 0.1 mg(Sn)/m^3

SAFETY PROFILE: Poison by intravenous route. When heated to decomposition it emits very toxic fumes of NO_x and SO_x. See also TIN COMPOUNDS.

QTS000 CAS:59-40-5 ***HR: 2***
N-(2-QUINOXALINYL)SULFANILAMIDE
mf: $C_{14}H_{12}N_4O_2S$ mw: 300.36

SYNS: 2-p-AMINOBENZENESULFONAMIDOQUINOXALINE ◇ 2-p-AMINOBENZENESULPHONAMIDOQUINOXALINE ◇ N^1-2-QUINOXALINYLSULFANILAMIDE ◇ N'-2-QUINOXALYLSULFANILAMIDE ◇ SULFABENZPYRAZINE ◇ 2-SULFANILAMIDOQUINOXALINE ◇ SULFAQUINOXALINE

TOXICITY DATA with REFERENCE
orl-rat TD50:1370 mg/kg MahWM# 16NOV82

CONSENSUS REPORTS: Reported in EPA TSCA Inventory.

SAFETY PROFILE: Moderately toxic by ingestion. When heated to decomposition it emits very toxic fumes of NO_x and SO_x.

QUJ300 CAS:32226-69-0 ***HR: 3***
QUINPYRROLIDINE
mf: $C_{15}H_{18}N_2O$ mw: 242.35

SYN: 4-(2,1-PYRROLIDINOETHOXY)QUINOLINEHYDROCHLORIDE

TOXICITY DATA with REFERENCE
orl-rat LD50:280 mg/kg IJMRAQ 60,604,72
ipr-rat LD50:135 mg/kg IJMRAQ 60,604,72
orl-mus LD50:315 mg/kg IJMRAQ 59,614,71
ipr-mus LD50:123 mg/kg IJMRAQ 60,604,72

SAFETY PROFILE: Poison by ingestion and intraperitoneal routes. When heated to decomposition it emits toxic fumes of NO_x.

QUS000 CAS:6109-70-2 ***HR: 3***
3-QUINUCLIDINOL ACETATE (ESTER) HYDROCHLORIDE
mf: $C_9H_{15}NO_2\bullet ClH$ mw: 205.71

SYN: ACECLIDIN-HCL

TOXICITY DATA with REFERENCE
ipr-rat LD50:105 mg/kg ARZNAD 18,322,68
ipr-mus LD50:116 mg/kg ARZNAD 18,322,68
ivn-mus LD50:27 mg/kg ARZNAD 18,322,68

SAFETY PROFILE: Poison by intraperitoneal and intravenous routes. When heated to decomposition it emits very toxic fumes of NO_x and HCl. See also ESTERS.

QVA000 CAS:6581-06-2 ***HR: 3***
3-QUINUCLIDINOL BENZILATE
mf: $C_{21}H_{23}NO_3$ mw: 337.45

PROP: Crystals from acetone-ether. Mp: 164-165°

SYNS: 1-AZABICYCLO(2.2.2)OCTAN-3-OL BENZILATE (9CI) ◇ 3-CHINUCLIDYLBENZILATE

TOXICITY DATA with REFERENCE
mmo-smc 5 mmol/L ACNSAX 17,252,75
cyt-mus-ipr 30 mg/kg ACNSAX 17,252,75
dlt-mus-ipr 100 mg/kg ACNSAX 17,252,75
cyt-ham-ipr 40 mg/kg/6H-I ACNSAX 17,252,75
ivn-mus LD50:25 mg/kg CSLNX* NX#11998

SAFETY PROFILE: Poison by intravenous route. Mutation data reported. When heated to decomposition it emits toxic fumes of NO_x.

QVJ000 CAS:63716-96-1 ***HR: 3***
3-QUINUCLIDINOL, DIPHENYLACETATE (ESTER), HYDROGEN SULFATE (2:1) DIHYDRATE
mf: $C_{42}H_{46}N_2O_4\bullet H_2O_4S\bullet 2H_2O$ mw: 777.02

SYNS: AZABICYCLO(2.2.1)OCTAN-3-OL,DIPHENYLACETATE (ester), HYDROGEN SULFATE (2:1) DIHYDRATE ◇ DIPHENYLACETIC ACID, ESTER with 3-QUINUCLIDINOL, HYDROGEN SULFATE (2:1) DIHYDRATE

TOXICITY DATA with REFERENCE
ipr-mus LD50:159 mg/kg JPETAB 104,284,52
ivn-mus LD50:28 mg/kg JPETAB 104,284,52
ivn-dog LD50:20 mg/kg JPETAB 104,284,52

SAFETY PROFILE: Poison by intravenous and intraperitoneal routes. When heated to decomposition it emits very toxic fumes of NO_x and SO_x.

QWJ000 CAS:13004-56-3 ***HR: 3***
QUINUCLIDYL BENZYLATE
mf: $C_{21}H_{23}NO_3\bullet ClH$ mw: 373.91

SYNS: 1-AZABICYCLO(2.2.2)OCTAN-3-OL, BENZYLATE (ESTER), HYDROCHLORIDE ◇ BENZILIC ACID, 3-QUINUCLIDINYL ESTER, HYDROCHLORIDE ◇ RO 2-3308

TOXICITY DATA with REFERENCE
scu-hmn TDLo:3 μg/kg:CNS JPETAB 104,284,52
ipr-mus LD50:110 mg/kg JPETAB 104,284,52
ivn-mus LD50:18 mg/kg CSLNX* NX#03527
ivn-dog LD50:15 mg/kg JPETAB 104,291,52

SAFETY PROFILE: Poison by intravenous and intraperitoneal routes. Human systemic effects by subcutaneous route: somnolence and distorted perceptions. When heated to decomposition it emits very toxic fumes of NO_x and HCl. See also ESTERS.

QWJ500 CAS:5786-68-5 ***HR: 3***
QUIPAZINE MALEATE
mf: $C_{13}H_{15}N_3 \cdot C_4H_4O_4$ mw: 329.39

SYNS: MA 1291 ◇ 2-(1-PIPERAZINYL)-QUINOLINE (Z)-2-BUTENEDIOATE (1:1) (9CI) ◇ 2-(1-PIPERAZINYL)-QUINOLINE MALEATE (1:1)

TOXICITY DATA with REFERENCE
orl-mus LD50:225 mg/kg FATOAO 43(5),530,80
ipr-mus LD50:102 mg/kg FATOAO 43(5),530,80
ivn-mus LD50:75 mg/kg FATOAO 43(5),530,80

SAFETY PROFILE: Poison by ingestion, intravenous, and intraperitoneal routes. When heated to decomposition it emits toxic fumes of NO_x.

R

RAF100 CAS:22248-79-9 ***HR: 3***
RABOND
mf: $C_{10}H_9Cl_4O_4P$ mw: 365.96

PROP: Mp 97-98°. Solubility in water: 11 ppm, in xylene: < 15%, 40-50% in chloroform at room temp.

SYNS: APPEX ◇ (Z)-2-CHLORO-1-(2,4,5-TRICHLOROPHENYL) VINYL DIMETHYL PHOSPHATE ◇ CVMP ◇ DEBANTIC ◇ DIETREEN ◇ DUST M ◇ ENT 25,841 ◇ GARDCIDE ◇ GARDONA ◇ GORDONA ◇ (Z)-PHOSPHORIC ACID-2-CHLORO-1-(2,4,5-TRICHLOROPHENYL) ETHENYL DIMETHYL ESTER ◇ RABON ◇ ROL ◇ SD 8447 ◇ STIROFOS ◇ STIROPHOS ◇ 2,4,5-TRICHLORO-α-(CHLOROMETHYLENE) BENZYL PHOSPHATE ESTER

TOXICITY DATA with REFERENCE
mnt-mus-ipr 200 mg/kg/7D-I MUREAV 117,329,83
mnt-mus-orl 720 mg/kg/24H-C MUREAV 117,329,83
orl-rat LD50:1100 mg/kg TXAPA9 11,49,67
orl-ctl LD50:360 mg/kg VETNAL 57(5),55,81
orl-bwd LD50:100 mg/kg TXAPA9 21,315,72

CONSENSUS REPORTS: IARC Cancer Review: Group 3 IMEMDT 7,56,87; Animal Limited Evidence IMEMDT 30,197,83

SAFETY PROFILE: Poison by ingestion. Questionable carcinogen. Mutation data reported. Used as an insecticide. A cholinesterase inhibitor. When heated to decomposition it emits toxic fumes of Cl^- and PO_x. See also PARATHION.

RAF300 ***HR: 3***
RACEMETHORPHAN HYDROBROMIDE
mf: $C_{18}H_{25}NO \cdot BrH$ mw: 352.61

PROP: Crystals. Mp: 124-126°.

SYNS: (±)-3-METHOXY-17-METHYLMORPHINANHYDROBROMIDE ◇ RO 1-5470

TOXICITY DATA with REFERENCE
orl-rat LD50:235 mg/kg JPETAB 109,189,53
scu-rat LD50:165 mg/kg JPETAB 109,189,53
orl-mus LD50:175 mg/kg JPETAB 109,189,53
scu-mus LD50:160 mg/kg JPETAB 109,189,53
ivn-mus LD50:31 mg/kg JPETAB 109,189,53
ivn-rbt LD50:16700 μg/kg JPETAB 109,189,53

SAFETY PROFILE: Poison by ingestion, subcutaneous, and intravenous routes. When heated to decomposition it emits toxic fumes of NO_x and HBr.

RAG300 CAS:3808-42-2 ***HR: 3***
RACEMOMYCIN A
mf: $C_{19}H_{34}N_8O_8$ mw: 502.61

SYNS: ANTIBIOTIC S 15-1A ◇ 2-((2-DEOXY-2-(3,6-DIAMINOHEXANAMIDO)-α-d-GLUOPYRANOSYL)AMINO)-3,3a,5,6,7,7a-HEXAHYDRO-7-HYDROXY-4H-IMIDAZO(4,5-C)PYRIDIN-4-ONE-6-CARBAMATE ◇ STREPTOTHRICIN F ◇ STREPTOTHRICIN VI ◇ YAZUMYCIN A

TOXICITY DATA with REFERENCE
ipr-mus LD50:250 mg/kg 85GDA2 1,250,80
scu-mus LD50:1370 μg/kg ANTBAL 14,48,69
ivn-mus LD50:150 mg/kg 85GDA2 1,285,80

SAFETY PROFILE: Poison by subcutaneous, intravenous, and intraperitoneal routes. When heated to decomposition it emits toxic fumes of NO_x. See also CARBAMATES.

RAQ000 ***HR: D***
RADIATION

PROP: Electromagnetic radiation (also called *radiant energy*) is emitted from matter in the form of photons (quanta), each having an associated electromagnetic wave having frequency (ν) and wavelength (λ). The various forms of radiant energy are characterized by their wavelength, and together they comprise the electromagnetic spectrum, the components of which are as follows: (1) cosmic gamma rays, (2) gamma rays from radioactive disintegration of atomic nuclei, (3) x-rays, (4) ultraviolet rays, (5) visible light rays, (6) infrared, (7) microwave, and (8) radio (Hertzian) and electric rays. Radiation having the shortest wavelength is the most penetrating. Quanta are not electrically charged and have no mass, their velocity of propagation is the same, and all display the properties characteristic of light having a dual nature (wave-like and corpuscular). Infrared radiation is that part of the electromagnetic spectrum between visible light and the microwave region, i.e., 7000 Å $- 2.2 \times 10^6$ Å. All objects at a temperature greater than 0°K emit IR radiation to cooler surfaces, and the hotter the emitter the shorter the emitted IR wavelength. When the emitter is hot enough, visible (4000 Å – 7000 Å), and even UV (100 Å – 4000 Å), radiation is also emitted.

SAFETY PROFILE: The main physical effect of exposure to infrared radiation is heating. This is also true for biological tissue. In the case of the eye, there is very sensitive tissue available for exposure to IR radiation. "Near IR," (7800 Å – 14000 Å) is blamed for many eye cataracts. The eyes may be easily protected by wearing goggles. Ultraviolet (UV) radiation is that part of the EM spectrum between 100Å and 4000Å. The UV-A band of UV extends from 3150Å – 4000Å and is called "Black light" or "near UV." This band can cause thermal skin burns, skin pigmentation and photoreactions. It does not, in general, cause eye injury. From 2800Å to 3150Å is "mid-UV," or erythemal region. This band produces photokeratitis and possibly skin cancer. The UV band from 1000Å – 2800Å is the UV-C band. It is known as "far UV" or "short UV." This band of UV is germicidal and viricidal, and destroys molds and yeasts as well. There is a sub-region of UV-C from 1700Å – 2200Å which produces ozone. The whole UV region can damage human skin and eyes. In eyes, it can cause blepharitis, conjunctivitis, keratitis, and keratoconjunctivitis. Skin exposure to solar UV can cause erythema, tanning; chronic skin exposure to solar UV leads to tanning, elastosis (dry, leathery, deeply wrinkled skin) and an incidence of non-melanoma skin cancer.

Type of radiation	Wavelength Å
cosmic	0.0005-0.005
gamma	0.005 -1.4
X	0.1 -100
UV	100 -4000
visible	4000 -7000
infrared	7000 -2,000,000

RAQ010 ***HR: D***

RADIATION, IONIZING

PROP: Extremely short-wavelength, highly energetic, penetrating rays of the following types: (a) gamma rays emitted by radioactive elements and radioisotopes (decay of atomic nucleus); (b) x-rays generated by sudden stoppage of fast-moving electrons; (c) subatomic charged particles (electrons, protons, deuterons) when accelerated in a cyclotron or betatron. The term is restricted to electromagnetic radiation at least as energetic as x-rays, and to charged particles of similar energies. Neutrons also may induce ionization. Such radiation is strong enough to remove electrons from any atoms in its path, leading to the formation of free radicals.

SAFETY PROFILE: These short-lived but highly reactive particles initiate decomposition of many organic compounds. Thus, ionizing radiation can cause mutations in DNA and in cell nuclei; adversely affect protein and amino acid mechanisms; impair or destroy body tissue; and attack bone marrow, the source of red blood cells. Exposure to ionizing radiation for even a short period is highly dangerous, and for an extended period may be lethal. The study of the chemical effects of such radiation is called radiation chemistry or (in the case of body reactions) radiation biochemistry.

RAV000 ***HR: 3***

RADIUM

af: Ra aw: 226.025

PROP: A radioactive earth metal. Brilliant white, tarnishes in air. Decomp in water. Mp: 700°, bp: 1737°, d: 5.5.

SAFETY PROFILE: A highly radiotoxic element. 1 gram = 3.7 10^{10} disintegrations per second. Inhalation, ingestion, or bodily exposure can lead to lung cancer, bone cancer, osteitis, skin damage, and blood dyscrasias. A common air contaminant. Radium replaces calcium in the bone structure and is a source of irradiation to the blood-forming organs. The ingestion of luminous dial paint prepared from radium caused death in many of the early dial painters before the hazard was fully understood. The data on these workers have been the source of many of the radiation precautions and the maximum permissible levels for internal emitters which are now accepted. ^{226}Ra is the parent of radon and the precautions described under ^{222}Rn should be followed. ^{228}Ra is a member of the thorium series. It was a common constituent of luminous paints, and while its low beta energy was not a hazard, its daughters in the series may have been a causative agent in the deaths of the radium dial painters following World War I. It is metabolized the same as any other radium isotope and it is a source of thoron. The precautions recommended under ^{220}Rn should be followed. Highly dangerous; must be kept heavily shielded and stored away from possible dissemination by explosion, flood, etc.

Radiation Hazard: Natural isotope ^{223}Ra (Actinium-X, Actinium Series), $T\frac{1}{2}$= 11.4 D, decays to radioactive ^{219}Rn by alphas of 5.5-5.7 MeV. Natural isotope ^{224}Ra (Thorium-X, Thorium Series), $T\frac{1}{2}$ = 3.6 D, decays to radioactive ^{220}Rn by alphas of 5.7 MeV. Natural isotope 226Ra (Uranium Series), $T\frac{1}{2}$ = 1600 Y, decays to radioactive ^{222}Rn by alphas of 4.8 MeV. Natural isotope ^{228}Ra (Mesothorium = 1, Thorium Series), $T\frac{1}{2}$ = 6.7 Y, decays to radioactive ^{228}Ac by betas of 0.05 MeV.

RBA000 ***HR: 3***

RADON

af: Rn aw: 222

PROP: Colorless, odorless, inert gas; very dense. Bp: −62°; d (gas @ 1 atm and 0°): 9.73 g/L, (liquid @ bp): 4.4.

SAFETY PROFILE: A common air contaminant. Radon is a noble gas and thus is relatively unreactive. Radiation Hazard: Natural isotope ^{220}Rn (Thoron, Thorium Series), $T\frac{1}{2}$ = 55s, decays to radioactive $^{216}P_o$ by alphas of 6.3 MeV. Natural isotope $^{222}R_n$ (Uranium Series), $T\frac{1}{2}$ = 3.8 D, decays to radioactive ^{218}Po by alphas of 5.5 MeV. The permissible levels are given for ^{222}Rn in equilibrium with its daughters. The chief hazard from this isotope is inhalation of the gaseous element and its solid daughters, which are collected on the normal dust of the air. This material is deposited in the lungs and has been considered to be a major causative agent in the high incidence of lung cancer found in uranium miners. Radon and its daughters build up to an equilibrium value in about a month from radium compounds, while the build-up from uranium compounds is negligible. Good ventilation of areas where radium is handled or stored is recommended to prevent accumulation of hazardous concentration of Rn and its daughters. Accumulation of radon in homes has been implicated in increased incidence of lung cancers. This accumulation is found in well insulated buildings located over land which has concentrations of uranium.

RBA400 ***HR: 2***
RAGWORT

PROP: *S. jacobaea* is a biennial or perennial herb up to 4 feet tall and produces clusters of yellow flowers. It is native to Europe and grows wild in the region from Massachusetts to Newfoundland, Quebec, and Ontario, and on the west coast of North America from British Columbia to Washington and Oregon. *S. longilobus* is a perennial shrub with white flowers. It grows wild in the area from Utah to northern Mexico. *S. vulgaris* is an annual which grows to 1 foot high with golden yellow flowers. It is native to Europe and is a weed in Alaska, all of Canada, the region bounded by New England, South Carolina and Wisconsin; California, New Mexico, and Texas.

SYNS: COMMON GROUNDSEL ◇ GORDOLOBO YERBA (MEXICO) ◇ GROUNDSEL ◇ HIERBA de SANTIAGO (MEXICO) ◇ SENECIA JACOBAEA ◇ SENECIA LONGILOBUS ◇ SENECIO VULGARIS ◇ STINKING WILLIE ◇ TANSY RAGWORT ◇ THREADLEAF GROUNDSEL

SAFETY PROFILE: The whole plant contains poisonous pyrrolizidine alkaloids. The milk from cows which have eaten the plant and honey from the nectar are toxic. Chronic consumption of teas made from this plant may cause anorexia with nausea, vomiting, diarrhea, and liver and kidney damage due to blockage of the veins in these organs.

RBA500 ***HR: 1***
RAIN LILY

PROP: Bulb-producing plants with 1-foot long grassy leaves that grow directly from the bulb. It produces white flowers (sometimes tinged with purple) which grow at the end of a leafless stalk. It grows in marshy areas in the region bounded by Virginia, Florida, and Alabama.

SYNS: ATAMASCO LILY ◇ FAIRY LILY ◇ ZEPHYR LILY ◇ ZEPHYRANTHES ATAMASCO

SAFETY PROFILE: The bulbs contain the poison lycorine. Ingestion of large amounts of the bulbs can cause nausea, persistent vomiting, and diarrhea.

RBF100 CAS:26538-44-3 ***HR: D***
RALGRO
mf: $C_{18}H_{26}O_5$ mw: 322.44

SYNS: 6-(6,10-DIHYDROXYUNDECYL)-β-RESORCYLIC ACID-mu-LACTONE ◇ FRIDERON ◇ MK-188 ◇ P1496 ◇ RALABOL ◇ RALONE ◇ ZEARALANOL ◇ ZEARANOL ◇ ZERANOL (USDA)

TOXICITY DATA with REFERENCE
orl-rat TDLo:52 mg/kg (6-18D preg):REP TJADAB 25,37,82
orl-rat TDLo:52 mg/kg (female 6-18D post):TER TJADAB 26,229,82

CONSENSUS REPORTS: Reported in EPA TSCA Inventory.

SAFETY PROFILE: An experimental teratogen. Experimental reproductive effects. When heated to decomposition it emits acrid smoke and irritating fumes.

RBF400 CAS:66357-35-5 ***HR: 3***
RANIDIL
mf: $C_{13}H_{24}N_4O_3S$•ClH mw: 352.93

SYNS: AH 19065 ◇ RANITIDINE HYDROCHLORIDE ◇ ZANTAC

TOXICITY DATA with REFERENCE
orl-wmn TDLo:48 mg/kg/8D-I:LVR AIMEAS 101,207,84
orl-man TDLo:90 mg/kg/3W-I:LVR,MET AIMEAS 101,208,84
orl-rat LD50:4190 mg/kg OYYAA2 26,147,83
ipr-rat LD50:441 mg/kg YKYUA6 36,521,85
scu-rat LD50:1700 mg/kg YKYUA6 36,521,85
ivn-rat LD50:136 mg/kg OYYAA2 26,147,83
ims-rat LD50:1530 mg/kg YKYUA6 36,521,85
orl-mus LD50:1440 mg/kg OYYAA2 26,147,83
ipr-mus LD50:300 mg/kg YKYUA6 36,521,85
scu-mus LD50:630 mg/kg YKYUA6 36,521,85
ivn-mus LD50:83 mg/kg OYYAA2 26,147,83
ims-mus LD50:400 mg/kg YKYUA6 36,521,85
orl-rbt LD50:2500 mg/kg YKYUA6 36,521,85
ivn-rbt LD50:109 mg/kg YKYUA6 36,521,85

SAFETY PROFILE: Poison by intravenous, intramuscular, and intraperitoneal routes. Moderately toxic by ingestion and subcutaneous routes. Human systemic effects by ingestion: diffuse hepatitis, fibrous hepatitis and fever. When heated to decomposition it emits toxic fumes of SO_x, NO_x, and HCl.

RBK000 CAS:53123-88-9 ***HR: 3***
RAPAMYCIN
mf: $C_{56}H_{89}NO_{14}$ mw: 1000.46

SYNS: ANTIBIOTIC AY 22989 ◇ AY 22989 ◇ NSC 226080

TOXICITY DATA with REFERENCE
dnd-mus:leu 60 μmol/L PAACA3 24,321,83
dni-mus:leu 10 μmol/L PAACA3 34,321,83
oms-mus:leu 1 mmol/L PAACA3 24,321,83
ipr-rat LD50:18220 μg/kg NTIS** PB83-228577
orl-mus LD50:2500 mg/kg 85ERAY 2,947,78
ipr-mus LD50:597 mg/kg JANTAJ 31,539,78

SAFETY PROFILE: Poison by intraperitoneal route. Moderately toxic by ingestion. Mutation data reported. When heated to decomposition it emits toxic fumes of NO_x.

RBP000 ***HR: 3***
RARE EARTHS

SAFETY PROFILE: Modern ion exchange techniques have eased the separation of rare earths from their ores and from one another. The elements to be considered are listed in Table 1 below.

Table 1
RARE EARTHS
Rare Earth Elements

Name	Symbol	At. wt:	At. no.	Abundance (g/metric ton)
Scandium	Sc	44.956	21	5.0
Yttrium	Y	88.905	39	28.10
Lanthanum	La	138.910	57	18.30
Cerium	Ce	140.120	58	46.10
Praseodymium	Pr	140.907	59	5.53
Neodymium	Nd	144.240	60	23.90
Promethium	Pm	147.000	61	ca.0
Samarium	Sm	150.350	62	6.47
Europium	Eu	151.960	63	1.06
Gadolinium	Gd	157.250	64	6.36
Terbium	Tb	158.924	65	0.91
Dysprosium	Dy	162.500	66	4.47
Holmium	Ho	164.930	67	1.15
Erbium	Er	167.260	68	2.47
Thulium	Tm	168.934	69	0.20
Ytterbium	Yb	173.040	70	2.66
Lutetium	Lu	174.970	71	0.75

SAFETY PROFILE: The rare earths are moderately to highly toxic. The acute lethal dose by various routes of administration are given below. The symptoms of toxicity of the rare earth elements including writhing, ataxia, labored respiration, walking on the toes with arched back, and sedation. There is a delayed lethality with the death rate peaking between 48 and 96 hours. A sex difference is apparent with the males being less susceptible than the females. If the animals survive for 30 days, there is generalized peritonitis, adhesions, and hemorrhagic ascitic fluid (Bruse et al., 1963) and also true granulomatous peritonitis and focal hepatic necrosis (Steffee 1959). Chelating agents, citrate or EDTA act to obscure the lethal effects of the rare earths by either decreasing their rate of release or by increasing their lethality by exchanging with other elements such as calcium. The effect of atomic weight on lethality is difficult to assess, but the transition elements (terbium group) appear to have a lesser toxicity than those above or below them in the periodic table (Haley, 1965). The low toxicity is probably related to poor intestinal absorption. Czech investigators reported high mortality at 7 and 16 days following intraperitoneal injection of $ScCl_3$, but their LD_{50} value was only 32 (28.6-36.8) mg/kg which does not agree with value in table 40.2 (Donozal et al., 1966). Intravenous administration of any of the rare earth chlorides to rats followed by local injection of epinephrine causes topical hemorrhagic lesions at the site or in the kidneys. This effect can be blocked by pretreatment with α-adrenergic blocking drugs in the case of $ScCl_3$ (Gabbiani et al., 1966). When given intravenously to mice, the toxicity of rare earth elements increased in the following order: La < Nd < Y. The toxicity of Nd salts increased as follows: chloride < propionate < acetate < 3-sulfoisonicotinate < sulfate < nitrate (Zimakov, 1973). Rats pretreated with polymyxin, then given $CdCl_3$, can be prevented by reticuloendothelial system activators such as zymosan, triolein or BCG, while $CeCl_3$ induced fatty degeneration of the liver can be prevented by reticuloendothelial inhibition with methyl palmitate (Lazar, 1973a). Glucocorticoids plus $GdCl_3$ produce hepatic peliosis in the rat liver (Selye et al., 1972). Oral or intraperitoneal doses of 5 to 10 g/kg of Dy_2O_3 or Gd_2O_3 had no pathological effects in mice and daily doses of 2 g/kg were harmless. Rare earth oxides are much less toxic than chlorides or citrates (Mogilevskaya and Roshchira, 1976). Reticuloendothelial system activity in rats was depressed by doses of 0.2 mg/100 g of $LaCl_3$, $CeCl_3$, $NdCl_3$, $HoCl_3$ or $YbCl_3$ (Lazar, 1973b).

The rare earth elements exhibit low toxicity by ingestion exposure. However, the intraperitoneal route is highly toxic while the subcutaneous route is poison to moderately toxic. The production of skin and lung granulomas after exposure to them requires extensive protection to prevent such exposure. Toxicity from exposure to rare earth radionuclides is related to absorbed radiation dose. The rare earth radionuclides have proven useful clinically in radiohypophysectomy, treatment of mammary and prostatic carcinoma and Cushing's syndrome, diabetic retinopathy and carcinoma in other body tissues and organs. Rare earth chelates have proven useful diagnostic agents in brain, lung and renal scanning, and in determining regional blood flow and renal function. (Haley T. J., 1965 *J. Pharm Sci* 54, 663.) They were first used for cigarette lighter flints, in Welsbach mantles for

Table 2
Rare Earth Toxicity Data Table

Chemical	Species and sex	LD_{50} (mg/kg)	Adm. rt.[b]
	Rat ♀	4200 (3684-4788)	P.O.
	Rat ♀	4.3 (3.4-5.6)	i.v.
	Rat ♂	49.6 (32.8-74.4)	i.v.
Praseodymium chloride	Frog	about 1000-1500[a]	s.c.
	Mouse	2500[a]	s.c.
	Mouse	358.9 (297.2-433.5)	i.p.
	Mouse	900-1500[a]	s.c.
	Mouse ♂	600 (552-652)	i.p.
	Mouse ♂	4500 (4054-4995)	P.O.
	Rat	<2000	i.p.
	Guinea pig	125 (78.2-200)	i.p.
	Rabbit	200-250	s.c.
citrate chloride	Mouse	140.6 (126.2-156.7)	i.p.
	Guinea pig	53 (29.9-70.3)	i.p.
nitrate	Mouse ♀	200 (259-325)	i.p.
	Rat ♀	245 (209-287)	i.p.
	Rat ♀	3500 (3017-4060)	P.O.
	Rat ♀	7.4 (5.1-10.8)	i.v.
	Rat ♂	77.2 (49.7-119.8)	i.v.
	Rat	10.8-13.9[a]	i.v.
Neodymium chloride	Frog	250[a]	s.c.
	Mouse	4000[a]	s.c.
	Mouse ♂	600 (562.0-640)	i.p.
	Mouse ♂	5250 (4730-5830)	P.O.
	Mouse	348.3 (297.2-408.3)	i.p.
	Rat	150-250[a]	i.p.
	Guinea pig	70[a]	i.v
	Guinea pig	139.6 (99.3-196.3)	i.p.
	Rabbit	200-250[a]	i.v.
citrate chloride	Mouse	138 (94.4-201.8)	i.p.
	Guinea pig	40.5 (4.7-348)	i.p.
nitrate	Mouse ♀	270 (221-329)	i.p.
	Rat ♀	270 (231-316)	i.p.
	Rat ♀	2750 (1896-3988)	P.O.
	Rat ♀	6.4 (5.5-7.3)	i.v.
	Rat ♂	66.8 (53.5-83.6)	i.v.
Samarium chloride	Frog	about 150[a]	s.c.
	Mouse ♂	585 (508.7-672.7)	i.p.
	Mouse ♂	>2000	P.O.
	Rat	>2000[a]	s.c.
	Guinea pig	750-1000[a]	s.c.
nitrate	Frog	1600[a]	s.c.
	Mouse ♀	315 (258-384)	i.p.
	Rat ♀	285 (254-319)	i.p.
	Rat ♀	2900 (2660-3161)	P.O.
	Rat ♀	8.9 (6.8-11.8)	i.v.
	Rat ♂	59.1 (40.5-86.3)	i.v.
	Guinea pig	about 500[a]	s.c.
Europium chloride	Mouse ♂	550 (515.5-586.9)	i.p.
	Mouse ♂	5000 (4505-5500)	i.p.
nitrate	Mouse ♀	320 (294-349)	i.p.
	Rat ♀	210 (172-256)	i.p.
	Rat ♀	>5000	P.O.
Gadolinium chloride	Mouse ♂	550 (495.5-610.5)	i.p.
	Mouse ♂	>2000	P.O.
nitrate	Mouse ♀	300 (261-345)	i.p.
	Rat ♀	230 (204-260)	i.p.
	Rat ♀	>5000	P.O.
Terbium chloride	Mouse ♂	550 (521.3-580.3)	i.p.
	Mouse ♂	5100 (5049.5-5151)	P.O.
nitrate	Mouse ♀	480 (444-518)	i.p.
	Rat ♀	260 (232-291)	i.p.
	Rat ♀	>5000	P.O.
Dysprosium chloride	Mouse ♂	585 (552-620)	i.p.
	Mouse ♂	7650 (7150-8186)	P.O.
nitrate	Mouse ♀	310 (261-369)	i.p.
	Rat ♀	295 (236-369)	i.p.
	Rat ♀	3100 (2870-3348)	P.O.
Holmium chloride	Mouse ♂	560 (541-580)	i.p.
	Mouse ♂	7200 (6667-7776)	P.O.

Continues on following page

Table 2 *(Continued)*

Chemical	Species and sex	L.D.$_{50}$ (mg/kg)	Adm. rt.[b]
nitrate	Mouse ♀	320 (302-339)	i.p.
	Rat ♀	270 (237-308)	i.p.
	Rat ♀	3000 (2804-3210)	P.O.
Erbium chloride	Frog	300-400[a]	s.c.
	Mouse ♂	535 (509-562)	i.p.
	Mouse ♂	6200 (5390-7140)	P.O.
nitrate	Mouse ♀	225 (194-261)	i.p.
	Rat	82.8-96.6[a]	i.v.
	Rat ♀	230 (195-271)	i.p.
	Rat ♀	35.8 (27.8-49.9)	i.v.
	Rat ♂	52.4 (37.0-74.5)	i.v.
Thulium chloride	Mouse ♂	48.5 (466.3-504.4)	i.p.
	Mouse ♂	6250 (5430-7190)	P.O.
nitrate	Mouse ♀	255 (226-288)	i.p.
	Rat ♀	285 (252-322)	i.p.
Scandium chloride	Mouse ♂	755 (741.7-768.6)	i.p.
	Mouse ♂	4000 (3960-4040)	P.O.
Yttrium chloride	Mouse	88 (67.7-114.4)	i.p.
	Rat	450	i.p.
	Rat	45 (41.1-49.3)	i.p.
nitrate	Frog	350[a]	s.c.
	Mouse	1660	s.c.
	Rat	20-30[a]	i.v.
	Rat	350	i.p.
	Rabbit	500	i.v.
oxide	Rat	500	i.p.
Lanthanum acetate	Rat	10,000	P.O.
	Rat	475	i.p.
ammonium nitrate	Rat	3400	P.O.
	Rat	625	i.p.
citrate chloride	Mouse	78.2 (70.0-100.3)	i.p.
	Guinea pig	60.7 (17.7-207.0)	i.p.
chloride	Frog	about 1000[a]	s.c.
	Mouse	3500[a]	s.c.
	Mouse	3500[a]	s.c.
	Mouse	372.4 (323.6-428.5)	i.p.
	Mouse	>500[a]	s.c.
	Mouse	>160[a]	i.p.
	Rat	106 (91.4-123)	i.p.
	Rat	350	i.p.
	Guinea pig	129.7 (105.4-159.6)	i.p.
	Rat	4200	P.O.
	Rabbit	200-250[a]	i.v.
nitrate	Rat	4500	P.O.
	Rat	450	i.p.
	Mouse ♀	410 (353-475)	i.p.
oxide	Rat	>10,000	P.O.
sulfate	Rat	>5000	P.O.
	Rat	275	i.p.
Cerium chloride	Frog	about 300[a]	s.c.
	Mouse	5000-10 000[a]	s.c.
	Mouse	353.2 (296.5-420.7)	i.p.
	Rat	2000-4000[a]	s.c.
	Rat	50-60[a]	i.v.
	Guinea pig	55.7 (40.1-77.4)	i.p.
citrate chloride	Rat	146.6 (128.6-167.1)	i.p.
	Guinea pig	103.5 (73.1-146.5)	i.p.
nitrate	Mouse ♀	470 (435-508)	i.p.
	Rat ♀	290 (238-354)	i.p.
Ytterbium chloride	Mouse ♂	395 (375-416.7)	i.p.
	Mouse ♂	6700 (6374.9-7041.7)	P.O.
nitrate	Mouse ♀	250 (185-338)	i.p.
	Rat ♀	255 (220-296)	i.p.
	Rat ♀	3100 (2924-3286)	P.O.
Lutetium chloride	Mouse ♂	315 (267-372)	i.p.
	Mouse ♂	7100 (6630-7590)	P.O.
	Mouse ♀	290 (259-325)	i.p.
	Rat ♀	325 (294-382)	i.p.

[a] Minimum lethal dose; [b] s.c., subcutaneous; P.O., oral; i.p., intraperitoneal; i.v., intravenous. Source: Haley, T. J., 1965. *J. Pharm. Sci.* **54**, 663.

increasing the brightness of gas lights and in Coleman Lanterns. Additional uses include control rods for atomic reactors utilizing their large cross-section capture values for neutrons, the addition of cerium to increase the life of nickel-chrome resistant wire, radiothulium in portable roentgenographic equipment, new types of alloys, lasers, masers, microwave devices, phosphors, insulators, capacitors, semiconductors, ferroelectrics and color television.

References

Bruce, D. W., B. E. Heitbrink and K. P. DuBois, *Toxicol, Appl. Pharmacol.* **5,** 750 (1963).
Donozal, V., B. Hajek, K. Kacl. I. Manova and F. Petru, *Z. Chem.* **6,** 154 (1966).
Gabbiani, G., B. Solymoss and R. M. Richard, *Arzneimittel-Frosch,* **17,** 505 (1967).
Haley, T. J., *J. Pharm. Sci.,* **54,** 663 (1965).
Lazar, G., *Experientia* **29,** 818 (1973a).
Z. I., ed. Gosudarstyennoe Izdatelstvo Meditsinskoi Literatury, Moscow, pp. 195-208.
Savitskiy, Y. M., *Vestn. Akad. Nauk SSSR* **6,** 81 (1960).
Selye, H., S. Szabo, B. Tuchweber and F. LeFebvre, *Proc. Soc. Exp. Biol. Med.* **139,** 887 (1972).
Spedding, F. H. and K. A. Gschneider, Jr., *Ind. Res.* **6,** 20 (1964).
Spiller, R., *Atom Industry* **9,** 13 (1960).
Steffee, C. H., *AMA Arch. Ind. Health* **20** 414 (1959).
Zimakov, Yu A., *Mater, Povolzh, Konf. Fisol. Uchastiem Biokhim. Farmakol. Morfol. 6th,* **2,** 28 (1973).

RBU000 CAS:5471-51-2 ***HR: 3***
RASPBERRY KETONE
mf: $C_{10}H_{12}O_2$ mw: 164.22

PROP: White solid; raspberry odor. Mp: 81-86°, flash p: +212°F.

SYNS: FEMA No. 2588 ◇ FRAMBINONE ◇ 4-(4-HYDROXPHENYL)-2-BUTANONE ◇ p-HYDROXYBENZYL ACETONE ◇ 1-(p-HYDROXYPHENYL)-3-BUTANONE ◇ 4-(p-HYDROXYPHENYL)-2-BUTANONE (FCC) ◇ OXYPHENALON ◇ RHEOSMIN

TOXICITY DATA with REFERENCE
orl-rat LD50:1320 mg/kg FCTXAV 8,349,70
ipr-rat LD50:350 mg/kg FCTXAV 8,349,70

CONSENSUS REPORTS: Reported in EPA TSCA Inventory.

SAFETY PROFILE: Poison by intraperitoneal route. Moderately toxic by ingestion. Combustible liquid. When heated to decomposition it emits acrid smoke and irritating fumes. See also KETONES.

RBZ000 ***HR: 3***
RATON

PROP: Aqueous extract from the dried leaves of the plant (JNCIAM 46,1131,71).

SYN: GLIRICIDIAL SEPIUM

TOXICITY DATA with REFERENCE
scu-rat TDLo:300 g/kg/1Y-I:ETA JNCIAM 46,1131,71

SAFETY PROFILE: Questionable carcinogen with experimental tumorigenic data. When heated to decomposition it emits acrid smoke and irritating fumes.

RBZ400 ***HR: 3***
RATTLE BOX

PROP: Stout yellow herbs with flowers that are normally yellow. When dry, the seeds rattle inside the seed pods. They are common weeds in the United States and the West Indies, but not in Canada and Alaska. *C. retusa* is cultivated in Florida.

SYNS: ALA de PICO (MEXICO) ◇ CASCABELILLO (PUERTO RICO) ◇ CROTALARIA BERTEROANA ◇ CROTALARIA INCANA ◇ CROTALARIA JUNCEA ◇ CROTALARIA RETUSA ◇ CROTALARIA SPECTABILIS ◇ MAROMERA (CUBA) ◇ PETE-PETE (HAITI) ◇ RABBIT-BELLS ◇ RATTLEWEED (JAMAICA) ◇ SHAKE-SHAKE

SAFETY PROFILE: The whole plant contains poisonous pyrrolizidine alkaloids. Poisonings have been reported from contamination of grain with Crotalaria seeds and from herbal teas made with the plant. Ingestion of any part of the plant may cause abdominal pains and accumulation of fluid, nausea, vomiting, diarrhea, and cirrhosis of and blood clots in the liver. Chronic ingestion may cause hypertension.

RCA200 CAS:84-36-6 ***HR: 3***
RAUNOVA
mf: $C_{35}H_{42}N_2O_{11}$ mw: 666.79

PROP: Crystals from acetone. Mp: 175-179°.

SYNS: CARBETHOXYSYRINGOYL METHYLRESERPATE ◇ ETROPRES ◇ 4-HYDROXY-3,5-DIMETHOXY-BENZOIC ACID ETHYL CARBONATE, ester with METHYL RESERPATE ◇ IPORES ◇ ISOTENSE ◇ LONDOMIN ◇ MENATENSINA ◇ METHYL CARBETHOXYSYRINGOYL RESERPATE ◇ METHYL RESERPATE-4-ETHOXYCARBONYL-3,5-DIMETHOXYBENZOIC ACID ESTER ◇ METHYL RESERPAT ESTER of SYRINGIG ACID ETHYL CARBONATE ◇ NEORESERPAN ◇ NOVOSERPINA ◇ RESERPIC ACID-METHYL ESTER, ESTER with 4-HYDROXY-3,5-DIMETHOXYBENZOIC ACID ETHYL CARBONATE ◇ REVELOX-WIRKSTOFF ◇ SENIRAMIN ◇ SERINGINA ◇ SERPAGON ◇ SINGOSERP ◇ SIRINGAL ◇ SIRINGINA ◇ SIRINGONE ◇ SIRISERPIN ◇ SU 3118 ◇ SYRINGIC ACID ETHYL CARBONATE ESTER with METHYL RESERPATE ◇ SYRINGOPINE ◇ SYROSINGOPIN ◇ SYROSINGOPINE ◇ VEGASERPINA

TOXICITY DATA with REFERENCE
scu-rat TDLo:10 mg/kg (1D pre):REP ENDOAO 78,225,66
ipr-rat LD50:286 mg/kg NIIRDN 6,365,82
ivn-rat LD50:50 mg/kg PSEBAA 76,847,51
orl-mus LD50:1293 mg/kg NIIRDN 6,365,82
ipr-mus LD50:101 mg/kg NIIRDN 6,365,82
scu-mus LD50:281 mg/kg NIIRDN 6,365,82

SAFETY PROFILE: Poison by subcutaneous, intravenous, and intraperitoneal routes. Moderately toxic by ingestion. Experimental reproductive effects. When heated to decomposition it emits toxic fumes of NO_x.

RCA275 CAS:8059-82-3 ***HR: 2***
RAUTRAX

TOXICITY DATA with REFERENCE
orl-wmn TDLo:1825 mg/kg/5Y-C:NEO LANCAO 2,672,74

SAFETY PROFILE: Questionable human carcinogen producing lung and skin tumors. When heated to decomposition it emits acrid smoke and irritating fumes.

RCA375 CAS:21416-67-1 ***HR: 3***
RAZOXANE
mf: $C_{11}H_{16}N_4O_4$ mw: 268.31

SYNS: ICI 59118 ◇ ICRF 159 ◇ 4,4′-PROPYLENEDI-2,6-PIPERAZINE-DIONE ◇ RAZOXIN

TOXICITY DATA with REFERENCE
dni-hmn:lym 20 mg/L INNDDK 1,283,83
oms-hmn:lym 20 mg/L INNDDK 1,283,83
mnt-mus-ipr 200 mg/kg BJCAAI 52,725,85
cyt-ham-orl 100 mg/kg BJCAAI 52,725,85
orl-man TDLo:693 mg/kg/77W-I:CAR LANCAO 2,1343,81
orl-wmn TD:4650 mg/kg/2Y-C:CAR LANCAO 2,1343,81
orl-wmn TDLo:3650 mg/kg/2Y-I:BLD LANCAO 2,1085,87
unr-wmn TDLo:600 mg/kg/34W-I:BLD BJDEAZ 113,131,85
unr-man TDLo:6467 mg/kg/6Y-I:BLD BJDEAZ 113,131,85
ipr-mus LD50:500 mg/kg CPBTAL 29,1594,81

SAFETY PROFILE: Suspected human carcinogen producing leukemia and skin tumors. Moderately toxic by intraperitoneal route. Human effects: normocytic anemia and thrombocytopenia. Human mutation data reported. When heated to decomposition it emits toxic fumes of NO_x.

RCA435 CAS:35133-58-5 ***HR: 3***
RC 72-01
mf: $C_{22}H_{35}NO_2$ mw: 345.58

SYNS: 4-(CYCLOHEXYLMETHYL)-α-(4-METHOXYPHENYL)-β-METHYL-1-PIPERIDINEETHANOL ◇ 2-(4-CYCLOHEXYLMETHYL-PIPERIDINO)-1-(4′-METHOXYPHENYL)-1-PROPANOL

TOXICITY DATA with REFERENCE
orl-mus LD50:115 mg/kg ARZNAD 21,1992,71
ipr-mus LD50:37 mg/kg ARZNAD 21,1992,71
ivn-mus LD50:8 mg/kg ARZNAD 21,1992,71

SAFETY PROFILE: Poison by ingestion, intravenous, and intraperitoneal routes. When heated to decomposition it emits toxic fumes of NO_x.

RCA450 CAS:35133-59-6 ***HR: 3***
RC 72-02
mf: $C_{22}H_{27}NO_2$ mw: 337.50

SYNS: 2-(4-BENZYL-1,2,3,6-TETRAHYDROPYRIDINO)-1-(4-METHOXYPHENYL)-1-PROPANOL ◇ 3,6-DIHYDRO-α-(4-METHOXYPHENYL)-β-METHYL-4-(PHENYLMETHYL)-1(2H)-PYRIDINEETHANOL

TOXICITY DATA with REFERENCE
orl-mus LD50:760 mg/kg ARZNAD 21,1992,71
ipr-mus LD50:75 mg/kg ARZNAD 21,1992,71
ivn-mus LD50:7500 μg/kg ARZNAD 21,1992,71

SAFETY PROFILE: Poison by intravenous and intraperitoneal routes. Moderately toxic by ingestion. When heated to decomposition it emits toxic fumes of NO_x.

RCF000 ***HR: 3***
RED SQUILL

PROP: Sea onion bulbs contain a potent concentration of *Scilliroside,* a glycoside bearing a close chemical resemblance to the *Scillarens* (JCPTA9 62,23,52).

SYNS: BONIDE TOPZOL RAT BAITS and KILLING SYRUP ◇ RAT-O-CIDE RAT BAIT ◇ RAT'S END ◇ RODINE ◇ ROUGH & READY RAT BAIT & RAT PASTE ◇ SCILLIROSIDE GLYCOSIDE ◇ SILMURIN ◇ SQUILL ◇ TOPZOL ◇ URGENEA MARITIMA

TOXICITY DATA with REFERENCE
orl-man TDLo:1414 mg/kg:GIT,CVS JAMAAP 75,971,20
orl-rat LD50:430 mg/kg 31ZOAD 1,371,68
ipr-rat LD50:150 mg/kg JAPMA8 37,307,48
orl-mus LD50:50 mg/kg APTOA6 8,391,52
orl-cat LDLo:100 mg/kg JCPTA9 62,23,52
orl-rbt LDLo:300 mg/kg JCPTA9 62,23,52
orl-pig LDLo:300 mg/kg JCPTA9 62,23,52
orl-dom LDLo:250 mg/kg JCPTA9 62,23,52
orl-rat LD50:125 mg/kg JAFCAV 34,973,86

SAFETY PROFILE: Poison by ingestion and intraperitoneal routes. Human systemic effects by ingestion: nausea or vomiting, decreased pulse rate and fall in blood pressure. When heated to decomposition it emits acrid smoke and irritating fumes.

RCK000 CAS:63521-15-3 ***HR: 2***
REFOSPOREN
mf: $C_{18}H_{14}Cl_2N_5O_5S_3 \cdot Na$ mw: 570.44

PROP: Crystals from acetonitrile.

SYN: CEFAZEDONE SODIUM SALT

TOXICITY DATA with REFERENCE
ims-rat TDLo:29 g/kg (female 15-22D post):REP ARZNAD 29,419,79

ims-rat TDLo:5 g/kg (female 6-15D post):TER ARZNAD 29,419,79
ivn-rat LD50:4225 mg/kg ARZNAD 29,424,79
ivn-mus LD50:6800 mg/kg ARZNAD 29,424,79
ivn-dog LD50:3000 mg/kg ARZNAD 29,424,79
ivn-rbt LD50:3200 mg/kg ARZNAD 29,424,79

SAFETY PROFILE: Moderately toxic by intravenous route. An experimental teratogen. Experimental reproductive effects. Used as an antibacterial agent. When heated to decomposition, it emits very toxic fumes of Cl^-, NO_x, Na_2O, and SO_x.

RCK725 ***HR: 2***
REFRACTORY CERAMIC FIBERS

PROP: A mixture of ALUMINA and SILICA (1:1).

SYN: FIBERS, REFRACTORY CERAMIC

TOXICITY DATA with REFERENCE
ipr-rat TDLo:125 mg/kg:ETA EPASR* 8EHQ-0485-0553

SAFETY PROFILE: Questionable carcinogen with experimental tumorigenic data.

RCU000 CAS:17095-24-8 ***HR: 3***
REMAZOL BLACK B
mf: $C_{26}H_{25}N_5O_{19}S_6$•4Na mw: 995.88

PROP: Mp: >300°.

TOXICITY DATA with REFERENCE
orl-rat TDLo:72 g/kg/86W-I:ETA TKORAS 3,53,67

CONSENSUS REPORTS: Reported in EPA TSCA Inventory.

SAFETY PROFILE: Questionable carcinogen with experimental tumorigenic data. When heated to decomposition it emits very toxic fumes of Na_2O, NO_x, and SO_x.

RCZ000 CAS:18976-74-4 ***HR: 3***
REMAZOL YELLOW G
mf: $C_{20}H_{19}ClN_4O_{11}S_3$•2Na mw: 669.04

SYNS: C.I. REACTIVE YELLOW 14 ◇ p-((4-(5-((2-HYDROXYETHYL) SULFONYL)-2-METHOXYPHENYL)AZO)-5-HYDROXY-3-METHYL-PYRAZOL-1-YL)-3-CHLORO-5-METHYL-BENZENESULFONICACID, HYDROGEN SULFATE (ESTER), DISODIUM SALT ◇ PROCION YELLOW MX 4R

TOXICITY DATA with REFERENCE
orl-rat TDLo:52 g/kg/58W-I:ETA TKORAS 3,53,67

SAFETY PROFILE: Questionable carcinogen with experimental tumorigenic data. When heated to decomposition it emits very toxic fumes of SO_x, Cl^-, Na_2O, and NO_x.

RDA350 CAS:9039-61-6 ***HR: 3***
REPTILASE

PROP: Rectangular crystals. Sol in physiological saline; practically insol in distilled water. Forms complexes with phenol and phenol derivatives that are practically insol in water.

SYNS: BATROXOBIN ◇ BOTHROPS VENOM PROTEINASE ◇ BOTROPASE ◇ DEFIBRASE ◇ DEFIBRASE R ◇ DF-521 ◇ REPTILASE R

TOXICITY DATA with REFERENCE
ivn-rat TDLo:1320 μg/kg (female 7-12D post):REP OYYAA2 25,549,83
ivn-rat TDLo:6 mg/kg (75D pre):TER OYYAA2 25,537,83
ivn-rat LD50:210 μg/kg OYYAA2 25,339,83
ivn-mus LD50:384 μg/kg OYYAA2 25,339,83
ivn-dog LD50:380 μg/kg OYYAA2 25,339,83

SAFETY PROFILE: A deadly poison by intravenous route. An experimental teratogen. Experimental reproductive effects.

RDA375 CAS:125-60-0 ***HR: 3***
RESANTIN
mf: $C_{22}H_{29}N_2O$•Br mw: 417.44

PROP: Crystals. Mp: 177.5-178.5° (dimorphic crystals from isopropanol + ethyl acetate, mp: 216-216.5°). Freely sol in water (water soln is neutral).

SYNS: 1-(4-AMINO-4-OXO-3,3-DIPHENYLBUTYL)-1-METHYLPIPERIDINIUM BROMIDE (9CI) ◇ DIPHENYL-PIPERIDINO-AETHYL-ACETAMID-BROMMETHYLAT (GERMAN) ◇ FENPIVERIMIUM BROMIDE ◇ 12494 HOECHST

TOXICITY DATA with REFERENCE
scu-rat LD50:350 mg/kg AEPPAE 224,357,55
orl-mus LD50:800 mg/kg AEPPAE 224,357,55
scu-mus LD50:75 mg/kg AEPPAE 224,357,55
ivn-mus LD50:13500 μg/kg AEPPAE 224,357,55
ivn-rbt LD50:22500 μg/kg AEPPAE 224,357,55

SAFETY PROFILE: Poison by subcutaneous and intravenous routes. Moderately toxic by ingestion. When heated to decomposition it emits toxic fumes of NO_x and Br^-.

RDF000 CAS:131-01-1 ***HR: 3***
RESERPIDINE
mf: $C_{32}H_{38}N_2O_8$ mw: 578.72

PROP: Three crystal forms from methanol; α form, Mp: 228-232°, β form, Mp: 230-232°, and Γ form, 138° with resolidification at 175°.

SYNS: A-11025 ◇ CANESCINE ◇ 11-DEMETHOXYRESERPINE ◇ DERESPERINE ◇ DESERPINE ◇ DESMETHOXYRESERPINE ◇ 11-DESMETHOXYRESERPINE ◇ ENDURONYL ◇ HARMONYL ◇ LILLY 22641 ◇ RAUNORINE ◇ RECANESCIN ◇ TRANQUINIL

TOXICITY DATA with REFERENCE
unr-rat TDLo:4950 μg/kg (6-16D preg):REP CRSBAW 155,2291,61
orl-wmn TDLo:16 mg/kg/9Y-C:NEO LANCAO 2,672,74
orl-rat TDLo:54 mg/kg/77W-C:CAR COREAF 254,1535,62
ivn-rat LD50:15 mg/kg 27ZQAG -,104,72
orl-mus LD50:500 mg/kg JAPMA8 44,688,55
ipr-mus LD50:60 mg/kg JAPMA8 44,688,55

SAFETY PROFILE: Poison by intravenous and intraperitoneal routes. Moderately toxic by ingestion. Experimental reproductive effects. Questionable human carcinogen producing skin tumors. When heated to decomposition it emits toxic fumes of NO_x.

RDK000 CAS:50-55-5 ***HR: 3***
RESERPINE
mf: $C_{33}H_{40}N_2O_9$ mw: 608.75

PROP: White or pale buff to sltly yellow powder, odorless. Mp: 264-265° (decomp). Insol in water; very sltly sol in alc; sol in chloroform and acetic acid.

SYNS: ENT 50,146 ◇ METHYL RESERPATE 3,4,5-TRIMETHOXYBENZOIC ACID ◇ METHYL RESERPATE 3,4,5-TRIMETHOXYBENZOIC ACID ESTER ◇ NCI-C50157 ◇ RAUSERPIN ◇ RAUWOLEAF ◇ SERPASIL ◇ SERPASIL APRESOLINE ◇ 3,4,5-TRIMETHOXYBENZOYL METHYL RESERPATE ◇ USAF CB-27 ◇ YOHIMBAN-16-CARBOXYLIC ACID DERIVATIVE of BENZ(G)INDOLO(2,3-A)QUINOLIZINE

TOXICITY DATA with REFERENCE
cyt-mus-ipr 30 mg/kg/48H PCJOAU 12,298,78
dlt-mus-ipr 300 μg/kg PCJOAU 12,298,78
ivn-wmn TDLo:200 μg/kg (female 39W post):REP AJDCAI 90,286,55
ipr-rat TDLo:2500 μg/kg (female 19D post):TER NISFAY 30,161,78
orl-wmn TDLo:456 μg/kg/Y-C:CAR LANCAO 2,672,74
orl-rat TDLo:54 mg/kg/77W-C:CAR COREAF 254,1535,62
scu-rat TDLo:132 mg/kg/2.5Y-I:NEO VOONAW 32(7),76,86
orl-rat LD50:420 mg/kg PSSCBG 11,555,80
ipr-rat LD50:44 mg/kg AIPTAK 110,20,57
scu-rat LD50:25 mg/kg NYKZAU 56,377,60
ivn-rat LD50:15 mg/kg 27ZQAG-,111,72
orl-mus LD50:200 mg/kg PSSCBG 11,555,80
ipr-mus LD50:5 mg/kg EJMCA5 10,519,75
scu-mus LD50:52 mg/kg RPTOAN 31,53,68
ivn-mus LD50:21 mg/kg ARZNAD 14,1040,64
ivn-dog LD50:500 μg/kg 27ZIAQ-,234,73
ipr-rbt LD50:7 mg/kg JDGRAX 6(3),19,74
ivn-rbt LD50:15 mg/kg 27ZQAG-,111,72

CONSENSUS REPORTS: NTP Fifth Annual Report on Carcinogens. IARC Cancer Review: Group 3 IMEMDT 7,330,87; Animal Inadequate Evidence IMEMDT 10,217,76; Human Limited Evidence IMEMDT 24,211,80; Animal Limited Evidence IMEMDT 24,211,80. NCI Carcinogenesis Bioassay (feed); Clear Evidence: mouse, rat NCITR* NCI-CG-TR-193,80. Reported in EPA TSCA Inventory.

SAFETY PROFILE: Confirmed human carcinogen producing tumors of the skin and brain. Poison by ingestion, intravenous, subcutaneous, and intraperitoneal routes. Mutation data reported. An experimental teratogen. Human and experimental reproductive effects by ingestion: stillbirth, reduced viability, and other neonatal measures or effects. In humans, 0.014 mg/kg causes psychotropic effects. Experimental reproductive effects. A medicine with side effects. Used as an additive permitted in the feed and drinking water of animals and/or for the treatment of food-producing animals. Also permitted in food for human consumption. A sedative. When heated to decomposition it emits toxic fumes of NO_x.

RDK050 CAS:1263-94-1 ***HR: 3***
RESERPINE PHOSPHATE
mf: $C_{33}H_{40}N_2O_9 \cdot 7H_3O_4P$ mw: 1294.75

TOXICITY DATA with REFERENCE
scu-rat TDLo:400 μg/kg (11-14D preg):REP JNEUAY 4,87,62
ivn-rat LD50:28 mg/kg JPETAB 138,78,62

SAFETY PROFILE: Poison by intravenous route. Experimental reproductive effects. When heated to decomposition it emits toxic fumes of PO_x and NO_x. See also RESERPINE.

RDP000 ***HR: 3***
RESIN (solution)
DOT: UN 1866

SYNS: RESIN, solution, in flammable liquid (DOT) ◇ RESIN, solution (resin compound, liquid) (DOT) ◇ SOLUBOND 0-869 ◇ SOLUBOND 3520

DOT Classification: Flammable Liquid; Label: Flammable Liquid; Flammable or Combustible Liquid; Label: Flammable Liquid.

SAFETY PROFILE: Flammable when exposed to heat or flame, can react vigorously with oxidizing materials. When heated to decomposition it emits acrid smoke and irritating fumes.

RDP300 CAS:99-30-9 ***HR: 3***
RESISAN
mf: $C_6H_4Cl_2N_2O_2$ mw: 207.02

SYNS: AL-50 ◇ ALLISAN ◇ BORTRAN ◇ BOTRAN ◇ CDNA ◇ CNA ◇ DCNA ◇ DCNA (fungicide) ◇ DICHLORAN ◇ DICHLORAN (amine fungicide) ◇ 2,6-DICHLOR-4-NITROANILIN (CZECH) ◇ 2,6-DICHLORO-4-NITROANILINE ◇ 2,6-DICHLORO-4-NITROBENZENAMINE (9CI)

◇ DICLORAN ◇ DITRANIL ◇ 4-NITROANILINE, 2,6-DICHLORO- ◇ RD-6584 ◇ U-2069

TOXICITY DATA with REFERENCE
mmo-asn 14 μmol/L PHYTAJ 66,217,76
sln-asn 38 μmol/L EVHPAZ 31,81,79
orl-mus TDLo:80 g/kg/78W-I:ETA NTIS** PB223-159
orl-rat LDLo:1500 mg/kg WRPCA2 9,119,70
orl-mus LD50:1500 mg/kg PCOC** -,343,66
ivn-mus LD50:56 mg/kg CSLNX* NX#03022
orl-gpg LD50:1450 mg/kg PCOC** -,343,66

CONSENSUS REPORTS: EPA Genetic Toxicology Program. Reported in EPA TSCA Inventory.

SAFETY PROFILE: Poison by intravenous route. Moderately toxic by ingestion. Questionable carcinogen with experimental tumorigenic data. Mutation data reported. Used as a fungicide. When heated to decomposition it emits toxic fumes of Cl^- and NO_x.

RDU000 CAS:63-56-9 ***HR: 3***
RESISTAB
mf: $C_{16}H_{22}N_4O{\bullet}ClH$ mw: 322.88

SYNS: ANAHIST ◇ ANDHIST ◇ 2-((2-(DIMETHYLAMINO)ETHYL) (p-METHOXY-BENZYL)AMINO)-PYRIMIDINEHYDROCHLORIDE ◇ N,N-DIMETHYL-N'-(4-METHOXYBENZYL)-N'-(2-PYRIMIDYL) ETHYLENEDIAMINE HYDROCHLORIDE ◇ N-p-METHOXYBENZYL-N',N'-DIMETHYL-N-2-PYRIMIDINYLETHYLENE DIAMINE HYDROCHLORIDE ◇ NEOHETRAMINE HYDROCHLORIDE ◇ NH 188 ◇ NOVOHETRAMIN ◇ THONZYLAMINE HYDROCHLORIDE ◇ THONZYLAMINIUM CHLORIDE

TOXICITY DATA with REFERENCE
orl-mus LD50:245 mg/kg NIIRDN 6,536,82
ipr-mus LD50:110 mg/kg AIPTAK 123,419,60
orl-gpg LD50:493 mg/kg NIIRDN 6,536,82
ivn-gpg LD50:38200 μg/kg AIPTAK 113,313,58

SAFETY PROFILE: Poison by ingestion, intravenous, and intraperitoneal routes. Used as an antihistamine. When heated to decomposition it emits very toxic fumes of HCl and NO_x.

RDZ875 CAS:35764-59-1 ***HR: 3***
(+)-cis-RESMETHRIN
mf: $C_{22}H_{26}O_3$ mw: 338.48

SYNS: 5-BENZYL-3-FURYLMETHYL(+)-cis-CHRYSANTHEMATE ◇ CISMETHRIN ◇ NRDC 119

TOXICITY DATA with REFERENCE
orl-rat LD50:63 mg/kg PCBPBS 2,308,72
ivn-rat LDLo:6 mg/kg ARTODN 45,325,80
orl-mus LD50:152 mg/kg EVHPAZ 14,15,76

SAFETY PROFILE: Poison by ingestion and intravenous routes. When heated to decomposition it emits acrid smoke and irritating fumes. See also ESTERS.

RDZ900 CAS:102-29-4 ***HR: 3***
RESORCIN MONOACETATE

PROP: Bp: 283°, d: 1.226, flash p: >230° F.
mf: $C_8H_8O_3$ mw: 152.16

SYNS: 3-ACETOXYPHENOL ◇ ACETYLRESORCINOL ◇ 1,3-BENZENEDIOL, MONOACETATE ◇ EURESOL ◇ m-HYDROXYPHENYL ACETATE ◇ REMONOL ◇ RESORCIN ACETATE ◇ RESORCINOL, MONOACETATE ◇ RESORCITATE

TOXICITY DATA with REFERENCE
eye-rbt 5% SEV JAPMA8 46,185,57
ipr-mus LD50:400 mg/kg JAPMA8 46,185,57

CONSENSUS REPORTS: Reported in EPA TSCA Inventory.

SAFETY PROFILE: Poison by intraperitoneal route. A severe eye irritant. Combustible liquid. When heated to decomposition it emits acrid smoke and irritating fumes.

REA000 CAS:108-46-3 ***HR: 3***
RESORCINOL
DOT: UN 2876
mf: $C_6H_6O_2$ mw: 110.12

$C_6H_4(OH)_2$

PROP: Very white crystals, become pink on exposure to light when not perfectly pure; unpleasant sweet taste. Mp: 110°, bp: 280.5°, flash p: 261°F (CC), d: 1.285 @ 15°, autoign temp: 1126°F, vap press: 1 mm @ 108.4°, vap d: 3.79. Very sol in alc, ether, glycerol; sltly sol in chloroform; sol in water.

SYNS: m-BENZENEDIOL ◇ 1,3-BENZENEDIOL ◇ C.I. 76505 ◇ C.I. DEVELOPER 4 ◇ C.I. OXIDATION BASE 31 ◇ DEVELOPER R ◇ m-DIHYDROXYBENZENE ◇ 1,3-DIHYDROXYBENZENE ◇ m-DIOXYBENZENE ◇ DURAFUR DEVELOPER G ◇ FOURAMINE RS ◇ FOURRINE 79 ◇ m-HYDROQUINONE ◇ 3-HYDROXYCYCLOHEXADIEN-1-ONE ◇ m-HYDROXYPHENOL ◇ 3-HYDROXYPHENOL ◇ NAKO TGG ◇ NCI-C05970 ◇ PELAGOL GREY RS ◇ RCRA WASTE NUMBER U201 ◇ RESORCIN ◇ RESORCINE

TOXICITY DATA with REFERENCE
skn-rbt 500 mg IHFCAY 6,1,67
eye-rbt 100 mg SEV BIOFX* 11-4/70
mma-sat 20 μmol/plate MUREAV 90,91,81
mrc-smc 1 g/L MUREAV 135,109,84
cyt-hmn:lym 80 mg/L ARZNAD 32,533,82
cyt-hmn:oth 40 mg/L ARZNAD 32,533,82
skn-mus TDLo:4800 mg/kg/12W-I:ETA CNREA8 19,413,59
orl-hmn LDLo:29 mg/kg 34ZIAG -,519,69
orl-rat LD50:301 mg/kg BIOFX* 11-4/70
scu-rat LDLo:400 mg/kg RMSRA6 15,561,1895
ipr-mus LD50:240 mg/kg JPAMA8 46,185,57
scu-mus LD50:213 mg/kg ZGIMAL 2,333,47
skn-rbt LD50:3360 mg/kg AIHAAP 37,596,76

scu-gpg LDLo:400 mg/kg RMSRA6 15,561,1895
par-frg LDLo:270 mg/kg AEPPAE 166,437,32

CONSENSUS REPORTS: IARC Cancer Review: Group 3 IMEMDT 7,56,87; Animal Inadequate Evidence IMEMDT 15,155,77. Reported in EPA TSCA Inventory. EPA Genetic Toxicology Program.

OSHA PEL: TWA 10 ppm; STEL 20 ppm
ACGIH TLV: TWA 10 ppm; STEL 20 ppm

DOT Classification: ORM-E; Label: None; Poison B; Label: St. Andrews Cross.

SAFETY PROFILE: Human poison by ingestion. Experimental poison by ingestion, intraperitoneal, parenteral and subcutaneous routes. Moderately toxic experimentally by skin contact and intravenous routes. Questionable carcinogen with experimental tumorigenic data. Human mutation data reported. A skin and severe eye irritant. It can cause systemic poisoning by acting both as a blood and nerve poison. In a suitable solvent, this material can readily be absorbed through human skin and can cause local hyperemia, itching, dermatitis, edema, and corrosion associated with enlargement of regional lymph glands as well as serious systemic disorders such as restlessness, methemoglobinemia, cyanosis, convulsions, tachycardia, dyspnea and death. These same symptoms can be induced by ingestion of the material. For poisoning, treat symptomatically. Get medical advice. Used as a topical antiseptic and keratolytic agent.

Combustible when exposed to heat or flame; can react with oxidizing materials. To fight fire, use water, CO_2, dry chemical. Potentially explosive reaction with concentrated nitric acid. Incompatible with acetanilide, alkalies, ferric salts, spirit nitrous ether, urethan. When heated to decomposition it emits acrid smoke and irritating fumes.

REA100 CAS:108-58-7 ***HR: 2***
RESORCINOL, DIACETATE
mf: $C_{10}H_{10}O_4$ mw: 194.20

PROP: Bp: 146°/12 mm Hg, d: 1.178

SYNS: 1,3-BENZENEDIOL, DIACETATE ◇ 1,3-DIACETOXYBENZENE ◇ 1,3-DIHYDROXYBENZENE DIACETATE ◇ m-PHENYLENEDIACETATE

TOXICITY DATA with REFERENCE
eye-rbt 5% SEV JAPMA8 46,185,57
ipr-mus LD50:660 mg/kg JAPMA8 46,185,57

CONSENSUS REPORTS: Reported in EPA TSCA Inventory.

SAFETY PROFILE: Moderately toxic by intraperitoneal route. A severe eye irritant. When heated to decomposition it emits acrid smoke and irritating fumes.

REF000 CAS:101-90-6 ***HR: 3***
RESORCINOL DIGLYCIDYL ETHER
mf: $C_{12}H_{14}O_4$ mw: 222.26

SYNS: ARALDITE ERE 1359 ◇ m-BIS(2,3-EPOXYPROPOXY)BENZENE ◇ 1,3-BIS(2,3-EPOXYPROPOXY)BENZENE ◇ m-BIS(GLYCIDYLOXY)BENZENE ◇ 1,3-DIGLYCIDYLOXYBENZENE ◇ DIGLYCIDYL RESORCINOL ETHER ◇ ERE 1359 ◇ NCI-C54966 ◇ 2,2'-(1,3-PHENYLENEBIS(OXYMETHYLENE))BISOXIRANE ◇ RDGE ◇ RESORCINOL BIS(2,3-EPOXYPROPYL)ETHER ◇ RESORCINYL DIGLYCIDYL ETHER

TOXICITY DATA with REFERENCE
skn-rbt 500 mg/24H MOD AMIHAB 17,129,58
mmo-sat 50 μg/plate MUREAV 135,159,84
sln-dmg-orl 5 pph ENMUDM 7,325,85
trn-dmg-orl 5 pph ENMUDM 7,325,85
cyt-ham:ovr 8 mg/L MUREAV 135,159,84
orl-rat TDLo:6180 mg/kg/2Y-I:CAR NTPTR* NTP-TR-257,86
orl-mus TDLo:25750 mg/kg/2Y-I:CAR NTPTR* NTP-TR-257,86
unr-mus TDLo:6700 mg/kg:ETA RARSAM 3,193,63
orl-rat LD50:2570 mg/kg AMIHAB 17,129,58
ipr-rat LD50:178 mg/kg AMIHAB 17,129,58
orl-mus LD50:980 mg/kg AMIHAB 17,129,58
ipr-mus LD50:243 mg/kg AMIHAB 17,129,58
orl-rbt LD50:1240 mg/kg AMIHAB 17,129,58

CONSENSUS REPORTS: NTP Fifth Annual Report on Carcinogens. IARC Cancer Review: Group 2B IMEMDT 7,56,87; Animal Sufficient Evidence IMEMDT 36,181,85; Animal Inadequate Evidence IMEMDT 11,125,76. NTP Carcinogenesis Studies (gavage); Clear Evidence: mouse, rat NTPTR* NTP-TR-257,86. Reported in EPA TSCA Inventory.

SAFETY PROFILE: Confirmed carcinogen with experimental carcinogenic and tumorigenic data. Poison by intraperitoneal route. Moderately toxic by ingestion. Mutation data reported. A skin irritant. When heated to decomposition it emits acrid smoke and irritating fumes. See also ETHERS.

REF050 CAS:150-19-6 ***HR: 3***
RESORCINOL MONOMETHYL ETHER
mf: $C_7H_8O_2$ mw: 124.15

SYNS: m-GUAIACOL ◇ m-HYDROXYANISOLE ◇ 3-HYDROXYANISOLE ◇ m-METHOXYPHENOL ◇ 3-METHOXYPHENOL ◇ RESORCINOL METHYL ETHER

TOXICITY DATA with REFERENCE
eye-rbt 5% SEV JAPMA8 46,185,57
ihl-wmn TCLo:230 μg/m³ (22W pre):REP GISAAA 37(3),108,72
ihl-hmn TCLo:230 μg/m³/22:CNS GISAAA 37(3),108,72
orl-rat LD50:597 mg/kg GISAAA 37(3),108,72
ihl-rat LC50:11500 mg/m³/4H 85GMAT -,105,82
skn-rat LD50:682 mg/kg GISAAA 37(3),108,72

orl-mus LD50:312 mg/kg GISAAA 37(3),108,72
ihl-mus LC50:11500 mg/m^3/4H 85GMAT -,105,82
ipr-mus LD50:320 mg/kg JAPMA8 46,185,57

CONSENSUS REPORTS: Reported in EPA TSCA Inventory.

SAFETY PROFILE: Poison by ingestion and intraperitoneal routes. Moderately toxic by skin contact. Human systemic effects by inhalation: muscle weakness, headache and irritability. Human female reproductive effects by inhalation: menstrual cycle changes and disorders. A severe eye irritant. When heated to decomposition it emits acrid smoke and irritating fumes. See also ETHERS.

REK325 ***HR: 1***
RESURRECTION LILY

PROP: Bulb-producing plant with long-thin leaves that grow directly from the bulb. The leaves die before flowers are borne on a tall, leafless stalk. A seed capsule contains a few black seeds.

SYNS: GOLDEN HURRICANE LILY ◇ GOLDEN SPIDER LILY ◇ LYCORIS AFRICANA ◇ LYCORIS RADIATA ◇ LYCORIS SQUAMIGERA ◇ MAGIC LILY ◇ RED SPIDER LILY ◇ SPIDER LILY

SAFETY PROFILE: The bulbs contain low concentrations of the poison lycorine. Ingestion of large amounts of the bulbs may cause nausea, persistent vomiting, and diarrhea.

REP400 CAS:55079-83-9 ***HR: D***
RETINOID ETRETIN
mf: $C_{21}H_{26}O_3$ mw: 326.47

SYNS: all-trans-3,7-DIMETHYL-9-(4-METHOXY-2,3,6-TRIMETHYLPHENYL)-2,4,6,8-NONATETRAENOIC ACID ◇ ETRETIN ◇ (all-E)-9-(4-METHOXY-2,3,6-TRIMETHYLPHENYL)-3,7-DIMETHYL-2,4,6,8-NONATETRAENOIC ACID ◇ RO 10-1670

TOXICITY DATA with REFERENCE
orl-rat TDLo:150 mg/kg (female 7-16D post):REP ARTODN 58,50,85
orl-rbt TDLo:2600 μg/kg (female 7-19D post):TER ARTODN 58,50,85

SAFETY PROFILE: An experimental teratogen. Experimental reproductive effects. When heated to decomposition it emits acrid smoke and irritating fumes.

REZ200 ***HR: D***
all-trans-RETINYLIDENE METHYL NITRONE
mf: $C_{21}H_{31}NO$ mw: 313.53

TOXICITY DATA with REFERENCE
orl-ham TDLo:75 mg/kg (8D preg):TER TXCYAC 33,331,84
orl-ham TDLo:12500 μg/kg (female 8D post):REP TXCYAC 33,331,84

SAFETY PROFILE: An experimental teratogen. Experimental reproductive effects. When heated to decomposition it emits toxic fumes of NO_x.

RFK000 CAS:875-22-9 ***HR: 3***
RETRONECINE HYDROCHLORIDE
mf: $C_8H_{13}NO_2 \cdot ClH$ mw: 191.68

TOXICITY DATA with REFERENCE
scu-rat TDLo:600 mg/kg:ETA JNCIAM 49,665,72
ipr-rat LDLo:600 mg/kg NATUAS 185,842,60
scu-rat LDLo:1000 mg/kg JNCIAM 49,665,72

SAFETY PROFILE: Moderately toxic by intraperitoneal and subcutaneous routes. Questionable carcinogen with experimental tumorigenic data. When heated to decomposition it emits very toxic fumes of NO_x and HCl.

RFP000 CAS:480-54-6 ***HR: 3***
RETRORSINE
mf: $C_{18}H_{25}NO_6$ mw: 351.44

PROP: Crystals from ethyl acetate. Mp: 212°. Readily sol in alc and chloroform; sltly sol in water, acetone, and ethyl acetate; practically insol in ether.

SYNS: 12,18-DIHYDROXY-SENECIONAN-11,16-DIONE ◇ β-LONGILOBINE ◇ cis-RETRONECIC ACID ESTER of RETRONECINE

TOXICITY DATA with REFERENCE
ctr-ham:kdy 25 μg/L CRNGDP 1,161,80
sln-dmg-par 10 mmol/L JOGNAU 59,273,66
orl-rat TDLo:560 mg/kg/62W-I:NEO BJCAAI 8,458,54
ipr-rat TDLo:150 mg/kg/22W-I:ETA BJCAAI 11,535,57
orl-rat LDLo:30 mg/kg JPBAA7 78,471,59
ipr-rat LD50:34 mg/kg CBINA8 5,227,72
ivn-rat LD50:38 mg/kg RETOAE 5,53,49
ivn-mus LD50:59 mg/kg RETOAE 5,53,49
orl-mam LDLo:33 mg/kg PAREAQ 22,429,70

CONSENSUS REPORTS: IARC Cancer Review: Group 3 IMEMDT 7,56,87; Animal Limited Evidence IMEMDT 10,303,76.

SAFETY PROFILE: Poison by ingestion, intraperitoneal, and intravenous routes. Questionable carcinogen with experimental neoplastigenic and tumorigenic data. Mutation data reported. When heated to decomposition it emits toxic fumes of NO_x.

RFU000 CAS:15503-86-3 ***HR: 3***
RETRORSINE-N-OXIDE
mf: $C_{18}H_{25}NO_7$ mw: 367.44

PROP: Crystals from ethanol. Mp: 140.5-141.5°.

SYNS: ISATIDINE ◇ cis-RETRONECIC ACID ESTER of RETRONECINE-N-OXIDE

TOXICITY DATA with REFERENCE
sln-dmg-par 10 mmol/L JOGNAU 59,273,66
orl-rat TDLo:540 mg/kg/60W-I:NEO BJCAAI 8,458,54
mul-rat TDLo:440 mg/kg/64W-I:ETA BJCAAI 8,458,54
orl-rat LD50:48 mg/kg CBINA8 5,227,72
ipr-rat LD50:250 mg/kg CBINA8 5,227,72
ivn-mus LD50:835 mg/kg JPETAB 75,83,42

CONSENSUS REPORTS: IARC Cancer Review: Group 3 IMEMDT 7,56,87; Animal Sufficient Evidence IMEMDT 10,269,76.

SAFETY PROFILE: Poison by ingestion and intraperitoneal routes. Moderately toxic by intravenous route. Questionable carcinogen with experimental neoplastigenic and tumorigenic data. Mutation data reported. When heated to decomposition it emits toxic fumes of NO_x. See also ESTERS.

RFU600 CAS:55102-44-8 ***HR: 3***
RFCNU
mf: $C_{18}H_{21}ClN_4O_9$ mw: 472.88

SYNS: (CHLORO-2-ETHYL)-1-(RIBOFURANOSYLISOPROPYLIDENE-2'-3'-PARANITROBENZOATE-5')-3-NITROSOUREA ◇ (CHLORO-2-ETIL)-1-(RIBOFURANOSILISOPROPILIDENE-2,3'-PARANITROBENZOATO)-3-NITROSOUREA ◇ I.C.I.G. 1105

TOXICITY DATA with REFERENCE
mmo-sat 200 μg/plate INSSDM 19,165,81
dns-hmn:lym 10 mg/L FRPSAX 36,947,81
dni-hmn:lym 10 mg/L FRPSAX 36,947,81
oms-mus:oth 2 mg/L INSSDM 19,229,81
ipr-mus LD50:90 mg/kg INSSDM 19,123,81

SAFETY PROFILE: Poison by intraperitoneal route. Human mutation data reported. Many N-nitroso compounds are carcinogens. When heated to decomposition it emits toxic fumes of Cl^- and NO_x. See also N-NITROSO COMPOUNDS.

RFU800 CAS:69579-13-1 ***HR: 3***
RGH-5526
mf: $C_{16}H_{25}N_5O_3$ mw: 335.46

SYNS: GYKI 11679 ◇ 1-(6-MORPHOLINO-3-PYRIDAZINYL)-2-(1-(tert-BUTOXYCARBONYL)-2-PROPYLIDENE)HYDRAZINE◇ 3-((6-(4-MORPHOLINYL)-3-PYRIDAZINYL)HYDRAZONO)BUTANOICACID 1,1- DIMETHYL ETHYL ESTER

TOXICITY DATA with REFERENCE
orl-rat LD50:970 mg/kg DRFUD4 10,32,85
ivn-rat LD50:670 mg/kg DRFUD4 10,32,85
orl-dog LD50:300 mg/kg DRFUD4 10,32,85

SAFETY PROFILE: Poison by ingestion. Moderately toxic by intravenous route. When heated to decomposition it emits toxic fumes of NO_x. See also ESTERS.

RFU875 ***HR: 3***
RHABDOPHIS TIGRINUS TIGRINUS VENOM

SYN: VENOM, SNAKE, RHABDOPHIS TIGRINUS TIGRINUS

TOXICITY DATA with REFERENCE
scu-mus LD50:9200 μg/kg NJGKBV 15,7,83
ivn-mus LD50:265 μg/kg NJGKBV 15,7,83
ims-mus LD50:7350 μg/kg NJGKBV 15,7,83

SAFETY PROFILE: Poison by subcutaneous, intramuscular, and intravenous routes.

RFZ000 CAS:6869-51-8 ***HR: 3***
3-β(α-l-RHAMNOPYRANOSIDE)-5,11-α,14-β-TRIHYDROXY-5-β-CARD(20,22)ENOLIDE
mf: $C_{29}H_{44}O_{10}$ mw: 552.73

SYNS: BIPINDOGENIN-l-RHAMNOSID (GERMAN) ◇ LOCUNDIEZIDE ◇ LOCUNDIOSIDE ◇ LOKUNDJOSID (GERMAN) ◇ LOKUNDJOSIDE

TOXICITY DATA with REFERENCE
orl-cat LD50:790 μg/kg RPTOAN 32,185,69
ivn-cat LD50:97 μg/kg RPTOAN 32,185,69
orl-pgn LD50:1860 μg/kg RPTOAN 32,185,69
ivn-pgn LD50:154 μg/kg RPTOAN 32,185,69

SAFETY PROFILE: A deadly poison by ingestion and intravenous routes. When heated to decomposition it emits acrid smoke and irritating fumes.

RGA000 CAS:84775-95-1 ***HR: 3***
RHATHANI

PROP: Aqueous extract from the root of the plant (JNCIAM 52,1579,74).

SYN: KRAMERIA TRIANDRA

TOXICITY DATA with REFERENCE
scu-rat TDLo:1058 mg/kg/49W-I:NEO JNCIAM 52,1579,74

SAFETY PROFILE: Questionable carcinogen with experimental neoplastigenic data. When heated to decomposition it emits acrid smoke and irritating fumes.

RGF000 CAS:7440-15-5 ***HR: 3***
RHENIUM
af: Re aw: 186.20

PROP: Hexagonal, close-packed crystals; black to silver gray. Mp: 3180°, bp: approx 5900°, d: 21.02.

SAFETY PROFILE: No reported cases of human toxicity. Experimentally, the Re^{3+} cation is more toxic than the ReO_4^- anion. Symptoms of Re^{3+} poisoning in rats are sedation, abdominal irritation, and death from cardiovascular collapse. Symptoms of ReO_4^- toxicity in rats include severe sedation and ataxia, tonic convulsions, and cardiovascular collapse. In cats, rhenium causes transient hypertension with tachycardia and transient auricular and ventricular fibrillations. In experimental animals, inhalation of rhenium dust causes pulmonary fibrosis.

Radiation Hazard: Natural (63%) isotope ^{187}Re, $T\frac{1}{2}$ = 4 × 10^{10} Y, decays to stable ^{187}Os by betas of less than 0.10 MeV. Flammable in the form of dust when exposed to heat or flame. Violent reaction with F_2 @ 125°. Ignites in oxygen at 300°C. See also various rhenium compounds and RARE EARTHS.

RGK000 CAS:12038-67-4 ***HR: 3***
RHENIUM(VII) SULFIDE
mf: Re_2S_7 mw: 596.85

SAFETY PROFILE: Will ignite spontaneously in air. When heated to decomposition it emits toxic fumes of SO_x. See also SULFIDES and RHENIUM.

RGP000 CAS:13569-63-6 ***HR: 3***
RHENIUM TRICHLORIDE
mf: Cl_3Re mw: 292.55

TOXICITY DATA with REFERENCE
ipr-mus LD50:280 mg/kg JPMSAE 57,321,68

CONSENSUS REPORTS: Reported in EPA TSCA Inventory.

SAFETY PROFILE: Poison by intraperitoneal route. When heated to decomposition it emits toxic fumes of Cl^-. See also RHENIUM and CHLORIDES.

RGP450 CAS:74195-73-6 ***HR: 3***
RHINASPRAY
mf: $C_{13}H_{17}N_3$•ClH•H_2O mw: 269.81

SYNS: BICIRON ◇ 4,5-DIHYDRO-N-(4,6,7,8-TETRAHYDRO-1-NAPHTHALENYL)-1H-IMIDAZOL-2-AMINE HYDROCHLORIDE HYDRATE ◇ KB 227 ◇ RHINOGUTT ◇ RHINOSPRAY ◇ 2-((5,6,7,8-TETRAHYDRO-1-NAPHTHYL)AMINO)-2-IMIDAZOLINE HYDROCHLORIDE HYDRATE ◇ TOWK ◇ TRAMAZOLINE HYDROCHLORIDE MONOHYDRATE

TOXICITY DATA with REFERENCE
orl-mus LD50:195 mg/kg ARZNAD 12,971,62
ipr-mus LD50:57 mg/kg ARZNAD 12,971,62
scu-mus LD50:77 mg/kg ARZNAD 12,971,62

SAFETY PROFILE: Poison by ingestion, subcutaneous, and intraperitoneal routes. An adrenergic agent. When heated to decomposition it emits toxic fumes of NO_x and HCl.

RGW000 CAS:989-38-8 ***HR: 3***
RHODAMINE 6G EXTRA BASE
mf: $C_{28}H_{30}N_2O_3$•ClH mw: 479.06

SYNS: C.I. 45160 ◇ C.I. BASIC RED 1, MONOHYDROCHLORIDE ◇ NCI-C56122 ◇ RHODAMINE 6G (biological stain) ◇ RHODAMINE 6GEX ETHYL ESTER

TOXICITY DATA with REFERENCE
mma-sat 20 nmol/plate CNREA8 39,4412,79
dnd-ham:ovr 90 μmol/L CNREA8 39,4412,79
ipr-mus TDLo:4 mg/kg (female 7-10D post):TER TJADAB 33,67C,86
scu-rat TDLo:100 mg/kg/1Y-I:ETA GANNA2 47,51,56
orl-mus LDLo:50 mg/kg GTPZAB 7(2),34,63

CONSENSUS REPORTS: IARC Cancer Review: Group 3 IMEMDT 7,56,87; Animal Sufficient Evidence IMEMDT 16,233,78. Reported in EPA TSCA Inventory. Community Right-To-Know List.

SAFETY PROFILE: Poison by ingestion and intraperitoneal routes. An experimental teratogen. Mutation data reported. Questionable carcinogen with experimental tumorigenic data. When heated to decomposition it emits very toxic fumes of Cl^- and NO_x.

RGZ100 CAS:37299-86-8 ***HR: 2***
RHODAMINE WT
mf: $C_{29}H_{29}N_2O_5$•Cl•2Na mw: 567.03

SYN: ACID RED 388

TOXICITY DATA with REFERENCE
dnd-ham:ovr 80 nmol/L MUREAV 118,117,83
sce-ham:ovr 12 nmol/L MUREAV 118,117,83
ivn-mus LD50:430 mg/kg TXAPA9 44,225,78

SAFETY PROFILE: Moderately toxic by intravenous route. Mutation data reported. When heated to decomposition it emits toxic fumes of Cl^-, NO_x, and Na_2O. See also CHLORIDES.

RGZ550 CAS:141-84-4 ***HR: 3***
RHODANINE
mf: $C_3H_3NOS_2$ mw: 133.19

SCS•NHCO•CH_2 (ring closure bracket from first S to CH_2)

PROP: Yellow crystals. Mp: 168° (can explode), d: 0.868. Sol in water, alkali, NH_4OH; very sol in alc, ether.

SYNS: 4-OXO-2-THIONOTHIAZOLIDINE ◇ RHODANIC ACID ◇ RHODANIN (CZECH) ◇ RHODANINIC ACID ◇ 2-THIO-4-KETOTHIAZOLIDINE ◇ 2-THIOXO-4-THIAZOLIDINONE ◇ USAF HA-2

TOXICITY DATA with REFERENCE
skn-rbt 500 mg/24H MOD 28ZPAK -,202,72
eye-rbt 750 μg/24H SEV 28ZPAK -,202,72
orl-rat LD50:326 mg/kg 28ZPAK -,202,72
ipr-mus LD50:200 mg/kg NTIS** AD277-689
scu-mus LDLo:200 mg/kg AIPTAK 12,447,04
ivn-mus LD50:32 mg/kg CSLNX* NX#04798

CONSENSUS REPORTS: Reported in EPA TSCA Inventory.

SAFETY PROFILE: Poison by ingestion, intraperitoneal, subcutaneous, and intravenous routes. A skin and severe eye irritant. May explode on rapid heating. When heated to decomposition it emits very toxic fumes of NO_x and SO_x.

RHA000 CAS:141-11-7 ***HR: 1***
RHODINYL ACETATE
mf: $C_{12}H_{22}O_2$ mw: 198.34

PROP: Mixture of acetates of geraniol and l-citronellol, found in geranium oil (FCTXAV 12,807,74). Colorless to sltly yellow liquid; fresh rose odor. D: 0.895-0.908, refr index: 1.450-1.458. Sol in alc and fixed oils; insol in glycerin, propylene glycol, and water @ 237°.

SYNS: α-CITRONELLYL ACETATE ◇ 3,7-DIMETHYL-7-OCTEN-1-OL ACETATE ◇ FEMA No. 2981 ◇ RHODINOL ACETATE

TOXICITY DATA with REFERENCE
skn-rbt 500 mg/24H MLD FCTXAV 12,975,74

CONSENSUS REPORTS: Reported in EPA TSCA Inventory.

SAFETY PROFILE: A skin irritant. When heated to decomposition it emits acrid smoke and irritating fumes.

RHA125 CAS:64253-73-2 ***HR: 3***
RHODIRUBIN A
mf: $C_{42}H_{55}NO_{16}$ mw: 829.98

SYNS: 1-HYDROXY-MA-144-N1 ◇ MA144 N2

TOXICITY DATA with REFERENCE
dni-mus:leu 350 nmol/L JANTAJ 34,1596,81
oms-mus:leu 43 nmol/L JANTAJ 34,1596,81
ipr-mus LD50:7500 μg/kg JANTAJ 30,616,77

SAFETY PROFILE: Poison by intraperitoneal route. Mutation data reported. When heated to decomposition it emits toxic fumes of NO_x.

RHA150 CAS:64502-82-5 ***HR: 3***
RHODIRUBIN B
mf: $C_{43}H_{55}NO_{15}$ mw: 813.98

TOXICITY DATA with REFERENCE
dni-mus:leu 370 nmol/L JANTAJ 34,1596,81
oms-mus:leu 55 nmol/L JANTAJ 34,1596,81
ipr-mus LD50:10 mg/kg JANTAJ 30,616,77

SAFETY PROFILE: Poison by intraperitoneal route. Mutation data reported. When heated to decomposition it emits toxic fumes of NO_x.

RHF000 CAS:7440-16-6 ***HR: 3***
RHODIUM
af: Rh aw: 102.91

PROP: A silvery-white, metallic element. Mp: 1966°, bp: 3727°, d: 2.41 @ 20°.

CONSENSUS REPORTS: Reported in EPA TSCA Inventory.

OSHA PEL: TWA Metal, Fume, Insol Compounds: 0.1 mg(Rh)/m^3; Sol Compounds: 0.001 mg(Rh)/m^3
ACGIH TLV: TWA (Metal) 1 mg/m^3, (insoluble compounds as Rh) 1 mg/m^3, (soluble compounds as Rh) 0.01 mg/m^3

SAFETY PROFILE: Handle carefully. It may be a sensitizer but not to the same extent as platinum. Most rhodium compounds have only moderate toxicity by ingestion. Flammable when exposed to heat or flame. Violent reaction with chlorine bromine pentafluoride, trifluoride and, OF_2. A catalytic metal.

RHF150 CAS:5503-41-3 ***HR: 3***
RHODIUM(II) ACETATE
mf: $C_4H_8O_4$•Rh mw: 223.03

SYNS: Rh-110 ◇ RHODIUM DIACETATE

TOXICITY DATA with REFERENCE
dni-mus:ast 10 μmol/L PSEBAA 145,1278,74
dnd-mam:lym 100 μmol/L CCROBU 59,611,75
ipr-mus LD50:27 mg/kg PSEBAA 145,1278,74

OSHA PEL: TWA 0.001 mg(Rh)/m^3
ACGIH TLV: TWA 1 mg(Rh)/m^3

SAFETY PROFILE: Poison by intraperitoneal route. Mutation data reported. When heated to decomposition it emits acrid smoke and irritating fumes. See also RHODIUM.

RHK000 CAS:10049-07-7 ***HR: 3***
RHODIUM(III) CHLORIDE (1 : 3)
mf: Cl_3Rh mw: 209.26

SYNS: RHODIUM CHLORIDE ◇ RHODIUM TRICHLORIDE

TOXICITY DATA with REFERENCE
mrc-bcs 5 mmol/L MUREAV 77,109,80
itt-rat TDLo:16741 μg/kg (1D male):REP JRPFA4 7,21,64
orl-mus TDLo:940 mg/kg/66W-C:CAR JONUAI 101,1431,71
orl-rat LD50:1302 mg/kg GTPZAB 21(7),55,77
ipr-rat LD50:280 mg/kg EQSSDX 1,1,75
ivn-rat LD50:198 mg/kg TXAPA9 21,589,72
ivn-rbt LD50:215 mg/kg TXAPA9 21,589,72

CONSENSUS REPORTS: Reported in EPA TSCA Inventory. EPA Genetic Toxicology Program.

OSHA PEL: TWA 0.001 mg(Rh)/m^3
ACGIH TLV: TWA 1 mg(Rh)/m^3

SAFETY PROFILE: Poison by ingestion, intraperitoneal, and intravenous routes. Experimental reproductive effects. Questionable carcinogen with experimental carcinogenic data. Mutation data reported. Incompatible with penta carbonyl iron + zinc. When heated to decomposition it emits toxic fumes of Cl^-. See also RHODIUM and CHLORIDES.

RHK250 CAS:56047-14-4 ***HR: 3***
RHODIUM DIBUTYRATE
mf: $C_8H_{16}O_4$•Rh mw: 279.15

SYN: RHODIUM(II) BUTYRATE

TOXICITY DATA with REFERENCE
dni-mus/ast 600 μg/kg CCROBU 59,611,75
oms-mus/ast 600 μg/kg CCROBU 59,611,75
ipr-mus LD10:700 μg/kg CCROBU 59,611,75

OSHA PEL: TWA 0.001 mg(Rh)/m^3
ACGIH TLV: TWA 1 mg(Rh)/m^3

SAFETY PROFILE: Poison by intraperitoneal route. Mutation data reported. When heated to decomposition it emits acrid smoke and irritating fumes. See also RHODIUM.

RHK850 CAS:31126-81-5 ***HR: 3***
RHODIUM(II) PROPIONATE
mf: $C_{12}H_{20}O_8Rh_2$ mw: 498.14

SYNS: TETRAKIS(mu-(PROPANOATO-O:O'))DI-RHODIUM (Rh-Rh) ◇ TETRAKIS(mu-(PROPIONATO)DI-RHODIUM (Rh-Rh)

TOXICITY DATA with REFERENCE
dni-mus/ast 2 mg/kg CCROBU 59,611,75
dni-mus:ast 10 μmol/L PSEBAA 145,1278M74
ipr-mus LD50:4500 μg/kg PSEBAA 145,1278,74

OSHA PEL: TWA 0.1 mg(Rh)/m^3
ACGIH TLV: TWA 1 mg(Rh)/m^3

SAFETY PROFILE: Poison by intraperitoneal route. Mutation data reported. When heated to decomposition it emits acrid smoke and irritating fumes. See also RHODIUM.

RHP000 CAS:13569-65-8 ***HR: D***
RHODIUM TRICHLORIDE TRIHYDRATE
mf: $Cl_3Rh•3H_2O$ mw: 263.32

SYNS: RHODIUM CHLORIDE ◇ RHODIUM CHLORIDE, TRIHYDRATE

TOXICITY DATA with REFERENCE
mmo-sat 1 μmol/plate MUREAV 88,165,81
mmo-esc 10 mg/L MUREAV 77,109,80

OSHA PEL: TWA 0.1 mg(Rh)/m^3
ACGIH TLV: TWA 1 mg(Rh)/m^3

SAFETY PROFILE: Mutation data reported. When heated to decomposition it emits toxic fumes of Cl^-. See also RHODIUM and CHLORIDES.

RHU500 ***HR: 3***
RHODODENDRON

PROP: An extremely large genus of shrubs which grow wild and are cultivated outdoors in Canada and most of the United States, except the north central states and southern Florida.

SYNS: AZALEA ◇ RHODORA (CANADA) ◇ ROSA LAUREL (MEXICO) ◇ ROSEBAY

SAFETY PROFILE: The leaves and nectar contain poisonous grayanotoxins (andromedotoxins). Ingestion of the leaves or honey made from the nectar produces an immediate burning pain in the mouth and may cause after a delay period of several hours: vomiting, diarrhea, headache, muscle weakness, impaired vision, slowed heart rate, severe low blood pressure, convulsions, coma, and death.

RHZ000 CAS:491-92-9 ***HR: 3***
RHODOQUINE
mf: $C_{19}H_{29}N_3O$ mw: 315.51

SYNS: AMINOQUIN ◇ BEPROCHINE ◇ 8-((4-(DIETHYLAMINO)-1-METHYLBUTYL)AMINO)-6-METHOXYQUINOLINE ◇ GAMEFAR ◇ PAMAQUIN ◇ PLASMOCHIN ◇ PLASMOCIDE ◇ PLASMOQUINE ◇ PRAEQUINE ◇ PREQUINE ◇ QUIPENYL

TOXICITY DATA with REFERENCE
unr-man LDLo:7353 μg/kg 85DCAI 2,73,70
orl-mus LD50:68 mg/kg BJPCAL 6,185,51
scu-mus LDLo:12500 μg/kg AEPPAE 144,341,29
scu-dog LDLo:20 mg/kg AEPPAE 144,341,29
orl-cat LDLo:7500 μg/kg AEPPAE 144,341,29
scu-cat LDLo:5 mg/kg AEPPAE 144,341,29
ivn-cat LDLo:5 mg/kg AEPPAE 144,341,29
scu-rbt LDLo:20 mg/kg AEPPAE 144,341,29
ivn-rbt LDLo:3500 μg/kg AEPPAE 144,341,29
ims-brd LDLo:33 mg/kg AEPPAE 144,341,29

SAFETY PROFILE: Human poison by an unspecified route. Experimental poison by ingestion, subcutaneous, intravenous, and intramuscular routes. When heated to decomposition it emits toxic fumes of NO_x.

RHZ600 ***HR: 2***
RHUBARB

PROP: A perennial with long stalks that become red when mature. The large, oval leaves are wrinkled and have wavy edges. It is commonly grown for the edible stalk.

SYNS: PIE PLANT ◇ RHEUM RHABARBARUM ◇ RHUBARBE (CANADA) ◇ WINE PLANT

SAFETY PROFILE: The raw or canned leaves contain poisonous anthraquinone glycosides and soluble oxalates. The most common symptom from ingestion of the leaves is diarrhea. Oxalate poisoning from ingestion of large amounts of the leaves may have caused kidney damage. See also OXALATES.

RIF000 CAS:488-81-3 ***HR: 1***
RIBITOL
mf: $C_5H_{12}O_5$ mw: 152.17

SYNS: ADONITOL ◇ 1,2,3,4,5-PENTANEPENTOL

TOXICITY DATA with REFERENCE
ipr-mus LDLo:10 g/kg PSEBAA 35,98,36

CONSENSUS REPORTS: Reported in EPA TSCA Inventory.

SAFETY PROFILE: Mildly toxic by intraperitoneal route. When heated to decomposition it emits acrid smoke and irritating fumes.

RIK000 CAS:83-88-5 ***HR: 3***
RIBOFLAVINE
mf: $C_{17}H_{20}N_4O_6$ mw: 376.37

PROP: Orange to yellow crystals; slt odor. Mp: 282° (decomp). Sltly sol in water, alc; insol in ether, chloroform.

SYNS: BEFLAVINE ◇ 6,7-DIMETHYL-9-d-RIBITYLISOALLOXAZINE ◇ 7,8-DIMETHYL-10-d-RIBITYLISOALLOXAZINE ◇ 7,8-DIMETHYL-10-(d-RIBO-2,3,4,5-TETRAHYDROXYPENTYL)ISOALLOXAZINE◇ FLAVAXIN ◇ HYFLAVIN ◇ HYRE ◇ LACTOFLAVIN ◇ LACTOFLAVINE ◇ RIBIPCA ◇ RIBODERM ◇ RIBOFLAVIN ◇ RIBOFLAVINEQUINONE ◇ VITAMIN B2 ◇ VITAMIN G

TOXICITY DATA with REFERENCE
cyt-ham:lng 300 mg/L GMCRDC 27,95,81
ipr-rat LD50:560 mg/kg JPETAB 76,75,42
scu-rat LD50:5000 mg/kg JPETAB 76,75,42
ivn-mus LDLo:365 mg/kg TXAPA9 44,225,78

CONSENSUS REPORTS: Reported in EPA TSCA Inventory.

SAFETY PROFILE: Poison by intravenous route. Moderately toxic by intraperitoneal and subcutaneous routes. Mutation data reported. When heated to decomposition it emits toxic fumes of NO_x.

RIP000 CAS:53797-35-6 ***HR: 3***
RIBOFLAVINE SULFATE
mf: $C_{17}H_{34}N_4O_{10}\cdot xH_2O_4S$ mw: 1141.11

SYNS: LANDAMYCINE ◇ RIBOMYCINE ◇ RIBOSTAMIN ◇ RIBOSTAMYCIN SULFATE ◇ SF 733 ANTIBIOTIC SULFATE ◇ VISTAMYCIN

TOXICITY DATA with REFERENCE
ipr-rat LD50:3080 mg/kg JMGZAI 10(2),5,73
scu-rat LD50:5600 mg/kg IYKEDH 4,90,73
ivn-rat LD50:375 mg/kg JMGZAI 10(2),5,73
ims-rat LD50:2030 mg/kg NIIRDN 6,885,82
ipr-mus LD50:1981 mg/kg JMGZAI 10(2),5,73
scu-mus LD50:2345 mg/kg IYKEDH 4,90,73
ivn-mus LD50:210 mg/kg IYKEDH 4,90,73
ims-mus LD50:1561 mg/kg IYKEDH 4,90,73

SAFETY PROFILE: Poison by intravenous route. Moderately toxic by intraperitoneal, intramuscular, and subcutaneous routes. Used as an antibiotic. When heated to decomposition it emits very toxic fumes of NO_x and SO_x.

RIU000 CAS:16755-07-0 ***HR: 3***
2-β-d-RIBOFURANOSYLMALEIMIDE
mf: $C_9H_{11}NO_6$ mw: 229.21

SYNS: 3-β-d-RIBOFURANOSYL-1H-PYRROLE-2,5-DIONE ◇ SHOWDOMYCIN

TOXICITY DATA with REFERENCE
ipr-mus LD50:25 mg/kg JAJAAA 17,148,64
scu-mus LD50:18 mg/kg JAJAAA 17,148,64
ivn-mus LD50:110 mg/kg 85ERAY 2,1443,78

SAFETY PROFILE: Poison by intraperitoneal, subcutaneous, and intravenous routes. When heated to decomposition it emits toxic fumes of NO_x.

RJA000 CAS:54-25-1 ***HR: 3***
2-β-d-RIBOFURANOSYL-as-TRIAZINE-3,5(2H,4H)-DIONE
mf: $C_8H_{11}N_3O_6$ mw: 245.22

SYNS: 6-AZAURACILRIBOSIDE ◇ 6-AZAURACIL-β-d-RIBOSIDE ◇ AZAURIDINE ◇ 6-AZAURIDINE ◇ AZUR ◇ 6-AZUR ◇ 6-AZURIDINE ◇ 3,5-DIOXO-2,3,4,5-TETRAHYDRO-1,2,4-TRIAZINE RIBOSIDE ◇ NSC 32074 ◇ RIBO-AZAURACIL ◇ RIBO-AZURACIL ◇ 2-β-d-RIBOFURANOSYL-1,2,4-TRIAZINE-3,5(2H,4H)-DIONE

TOXICITY DATA with REFERENCE
sln-dmg-par 20 mmol/L BCPCA6 15,299,66
spm-rat-ipr 2 g/kg CEFYAD 21,278,72
dni-mus-ipr 3 g/kg NEOLA4 20,243,73
spm-mus-ipr 5 g/kg CEFYAD 21,278,72
ivn-wmn TDLo:150 mg/kg (47D preg):TER CLPTAT 7,162,66
ivn-rat TDLo:900 mg/kg (female 7-9D post):REP IJMRAQ 62,1888,74
ipr-rat LD50:9400 mg/kg TJADAB 4,287,71
ipr-mus LD50:11250 mg/kg BCPCA6 14,1517,65
ipr-dog LD50:3400 mg/kg BCPCA6 14,1537,65
ipr-cat LD50:2400 mg/kg BCPCA6 14,1537,65

CONSENSUS REPORTS: EPA Genetic Toxicology Program.

SAFETY PROFILE: Moderately toxic by intraperitoneal route. Human teratogenic effects by intravenous route: unspecified developmental abnormalities or effects on the embryo or fetus. An experimental teratogen. Experimental reproductive effects. Mutation data reported. When heated to decomposition it emits toxic fumes of NO_x. Used as an antineoplastic agent.

RJA500 CAS:36791-04-5 ***HR: 2***
1-β-d-RIBOFURANOSYL-1,2,4-TRIAZOLE-3-CARBOXAMIDE
mf: $C_8H_{12}N_4O_5$ mw: 244.24

SYNS: ICN-1229 ◇ RIBAVIRIN ◇ RCTA ◇ VIRAMID ◇ VIRAZOLE

TOXICITY DATA with REFERENCE
oms-mus:lym 20 μmol/L CNREA8 45,5512,85
dni-mus:lym 10 μmol/L CNREA8 45,5512,85
ipr-mus TDLo:25 mg/kg (female 10D post):REP TXAPA9 52,99,80
ipr-rat TDLo:37500 μg/kg (female 9D post):TER TJADAB 17,93,78
orl-rat LD50:2700 mg/kg PCJOAU 18,667,84
orl-mus LDLo:4 g/kg PCJOAU 18,667,84
ipr-mus LD50:1300 mg/kg JMCMAR 15,1150,72

SAFETY PROFILE: Moderately toxic by intraperitoneal route. Mildly toxic by ingestion. An experimental teratogen. Experimental reproductive effects. Mutation data reported. Used as an antiviral agent. When heated to decomposition it emits toxic fumes of NO_x.

RJF000 CAS:550-33-4 ***HR: 3***
RIBOSYLPURINE
mf: $C_{10}H_{12}N_4O_4$ mw: 252.26

PROP: Small crystals from ethyl methyl ketone and methanol. Mp: 181-182°. Nneedles from methanol. Mp: 182-183°. Sol in water; sltly sol in cold ethanol; very sltly sol in acetone, ether, and chloroform.

SYNS: NEBULARIN(E) ◇ NEBULARINE ◇ PURINE RIBONUCLEOSIDE ◇ 9-PURINE RIBONUCLEOSIDE ◇ PURINE RIBOSIDE ◇ 9-(β-d-RIBOFURANOSYL)PURINE ◇ 9-(β-d-RIBOFURANOSYL)-9H-PURINE

TOXICITY DATA with REFERENCE
scu-rat LD50:220 mg/kg FEPRA7 14,391,55
scu-mus LD50:200 mg/kg 85GDA2 5,289,81
ivn-mus LD50:50 mg/kg 85GDA2 5,289,81
scu-gpg LD50:15 mg/kg FEPRA7 14,391,55

CONSENSUS REPORTS: EPA Genetic Toxicology Program.

SAFETY PROFILE: Poison by intravenous and subcutaneous routes. When heated to decomposition it emits toxic fumes of NO_x.

RJF400 CAS:80702-47-2 ***HR: 2***
RIBOTIDE
mf: $C_{10}H_{12}N_4O_8P \cdot 2Na \cdot C_{10}H_{11}N_4O_8P \cdot 2Na$ mw: 1100.66

SYNS: DISODIUM-5′-GUANYLATE mixed with DISODIUM 5′-INOSINATE (1:1) ◇ DISODIUM-5′-INOSINATE mixed with DISODIUM 5′-GUANYLATE (1:1) ◇ DISODIUM-5′-RIBONUCLEOTIDE ◇ 5′-INOSINIC ACID, DISODIUM SALT, mixed with DISODIUM-5′-GUANYLATE (1:1) ◇ RB

TOXICITY DATA with REFERENCE
orl-rat TDLo:8750 mg/kg (MGN):REP TXCYAC 3,333,75
ivn-mus LD50:3800 mg/kg TAKHAA 33,24,74

SAFETY PROFILE: Moderately toxic by intravenous route. Experimental reproductive effects. When heated to decomposition it emits toxic fumes of NO_x, PO_x, and Na_2O.

RJF500 CAS:60084-10-8 ***HR: 2***
RIBOXAMIDE
mf: $C_9H_{12}N_2O_5S$ mw: 260.29

SYNS: NSC-286193 ◇ 2-β-d-RIBOFURANOSYLTHIAZOLE-4-CARBOXAMIDE ◇ 2-β-d-RIBOFURANOSYL-4-THIAZOLECARBOXAMIDE ◇ TIAZOFURIN

TOXICITY DATA with REFERENCE
dni-mus:lym 10 μmol/L CNREA8 45,5512,85
dni-mus:leu 10 μmol/L BCPCA6 34,1109,85
ivn-mus TDLo:3425 mg/kg (1D male):REP NTIS** PB83-156901
ivn-hmn TDLo:1135 mg/kg/15D-I:CNS,MSK CNREA8 45,5169,85
ivn-hmn TDLo:892 mg/kg/10D-I:CNS CNREA8 45,5169,85
ipr-mus LD50:1684 mg/kg NCISP* JAN86
ivn-mus LD50:3400 mg/kg NTIS** PB83-156901
ivn-dog LDLo:1090 mg/kg NTIS** PB83-156901

SAFETY PROFILE: Moderately toxic by intravenous and intraperitoneal routes. Human systemic effects by intravenous route: headache, convulsions, and musculoskeletal changes. Experimental reproductive effects. Mutation data reported. When heated to decomposition it emits toxic fumes of SO_x and NO_x.

RJK000 CAS:9009-86-3 ***HR: 3***
RICIN

PROP: White powder. Found in castor beans (BIPAA8 74,85,76).

TOXICITY DATA with REFERENCE
eye-rbt 1500 ng NTIS** PB158-508
dni-rat:lvr 300 μg/L ARTODN 44,175,80
dni-mus/ast 25 μg/kg TOXIA6 11,379,73
ipr-mus TDLo:1350 ng/kg (3D preg):TER JTEHD6 12,193,83

ipr-mus TDLo:1350 ng/kg (3D preg):REP JTEHD6 12,193,83
orl-hmn LDLo:2 mg/kg PCOC** -,963,66
orl-man TDLo:900 μg/kg:PUL,GIT HUTODJ 2,239,83
orl-man LDLo:300 μg/kg:CNS,GIT NTIS** PB158-508
unr-chd LDLo:500 μg/kg:CNS,GIT 34ZIAG -,158,69
orl-rat LDLo:100 mg/kg NCNSA6 5,47,53
ihl-rat LC50:50 mg/m^3 NTIS** PB158-508
ipr-rat LD50:1500 ng/kg TOXIA6 18,649,80
par-rat LD50:336 ng/kg NEREDZ 4,259,79
itr-rat LD50:5 μg/kg NTIS** PB158-508
ihl-mus LC50:9 mg/m^3 NTIS** PB158-508
ipr-mus LD50:2 μg/kg TOXIA6 18,649,80
scu-mus LD50:22100 ng/kg NTIS** PB158-508
ivn-mus LD50:2200 ng/kg NTIS** PB158-508
ihl-dog LC50:24 mg/m^3 NTIS** PB158-508
itr-dog LD50:5 μg/kg NTIS** PB158-508
ihl-mky LC50:100 mg/m^3 NTIS** PB158-508

SAFETY PROFILE: A deadly poison to humans by ingestion. A deadly experimental poison by ingestion, inhalation, intratracheal, subcutaneous, parenteral, intravenous, and intraperitoneal routes. An experimental teratogen. Human systemic effects by ingestion: convulsions or effect on seizure threshold, cyanosis, death, dehydration, diarrhea, headache, hypermotility, nausea or vomiting, other gastrointestinal effects, somnolence. Experimental reproductive effects. Mutation data reported. An eye irritant. Inhalation or ingestion of minute amounts causes violent purging which may lead to collapse and death. Small particles in the eyes, nose, or any skin abrasion may prove fatal. May cause destruction of red blood cells. A very active military poison. Has been used in bizarre assassinations involving introduction of a tiny amount into the victim's body by subcutaneous route. A naturally occuring toxin. When heated to decomposition it emits toxic fumes of NO_x. See also ABRIN.

RJP000 CAS:141-22-0 ***HR: 3***
RICINOLEIC ACID
mf: $C_{18}H_{34}O_3$ mw: 298.52

PROP: Liquid. D: 0.940 @ 27.4°/4°, mp: 5.5° bp: 245° @ 10 mm. Insol in water; misc in alc, chloroform, and ether.

SYNS: l'ACIDE RICINOLEIQUE (FRENCH) ◇ 12-HYDROXY-cis-9-OCTADECENOIC ACID ◇ RICINIC ACID ◇ RICINOLIC ACID

TOXICITY DATA with REFERENCE
scu-rbt TDLo:390 mg/kg/17W-I:ETA CRSBAW 137,760,43

CONSENSUS REPORTS: Reported in EPA TSCA Inventory.

SAFETY PROFILE: Questionable carcinogen with experimental tumorigenic data. When heated to decomposition it emits acrid smoke and irritating fumes.

RJU000 CAS:4722-99-0 ***HR: 3***
RICINOLEIC ACID, BARIUM SALT
mf: $C_{36}H_{66}O_6$•Ba mw: 732.36

TOXICITY DATA with REFERENCE
ipr-mus LDLo:125 mg/kg CBCCT* 4,111,52

CONSENSUS REPORTS: Barium and its compounds are on the Community Right-To-Know List. Reported in EPA TSCA Inventory.

SAFETY PROFILE: Poison by intraperitoneal route. When heated to decomposition it emits acrid smoke and irritating fumes. See also RICINOLEIC ACID and BARIUM COMPOUNDS.

RJZ000 CAS:23246-96-0 ***HR: 3***
RIDDELLINE
mf: $C_{18}H_{23}NO_6$ mw: 349.42

PROP: An alkaloid isolated from *S. riddellii* (RETOAE 5,55,49).

TOXICITY DATA with REFERENCE
dns-rat-orl 125 mg/kg ENMUDM 7(Suppl 3),73,85
dns-rat:lvr 3200 μg/L ENMUDM 5,482,83
mul-rat TDLo:210 mg/kg/52W-I:ETA BJCAAI 11,535,57
orl-rat LDLo:25 mg/kg JPBAA7 78,471,59
ivn-mus LD50:105 mg/kg JPETAB 78,372,43

CONSENSUS REPORTS: IARC Cancer Review: Group 3 IMEMDT 7,56,87; Animal Inadequate Evidence IMEMDT 10,313,76.

SAFETY PROFILE: Poison by ingestion and intravenous routes. Questionable carcinogen with experimental tumorigenic data. Mutation data reported. When heated to decomposition it emits toxic fumes of NO_x.

RKA000 CAS:2750-76-7 ***HR: 3***
RIFAMIDE
mf: $C_{43}H_{58}N_2O_{13}$ mw: 811.03

PROP: Yellow-orange solid. Mp: 170° (decomp).

SYNS: 4-o-(2-(DIETHYLAMINO)-2-OXOETHYL)RIFAMYCIN ◇ N,N-DIETHYLRIFAMYCIN B AMIDE ◇ NCI 143-418 ◇ NSC-133099 ◇ RIFAMPICIN M/14 ◇ RIFAMYCIN DIETHYLAMIDE ◇ RIFAMYCIN B N,N-DIETHYLAMIDE ◇ RIFAMYCIN M14 ◇ RIFOCINA M ◇ RIFOMIDE ◇ RIFOMYCIN B DIETHYLAMIDE

TOXICITY DATA with REFERENCE
scu-rbt TDLo:1 g/kg (2-12D preg):REP TXAPA9 8,126,66
ipr-rat LD50:535 mg/kg TXAPA9 8,126,66
scu-rat LD50:2500 mg/kg TXAPA9 8,126,66
ivn-rat LD50:380 mg/kg TXAPA9 8,126,66
orl-mus LD50:2450 mg/kg TXAPA9 8,126,66

ipr-mus LD50:320 mg/kg TXAPA9 8,126,66
scu-mus LD50:640 mg/kg TXAPA9 8,126,66
ivn-mus LD50:315 mg/kg TXAPA9 8,126,66
ivn-dog LDLo:425 mg/kg TXAPA9 8,126,66

SAFETY PROFILE: Poison by intraperitoneal and intravenous routes. Moderately toxic by ingestion and subcutaneous routes. Experimental reproductive effects. When heated to decomposition it emits toxic fumes of NO_x. Used as an antibacterial agent.

RKK000 CAS:13929-35-6 ***HR: 2***
RIFAMYCIN
mf: $C_{39}H_{49}NO_{14}$ mw: 755.89

PROP: Yellow, prismatic needles. Mp: 300°, decomp @ 160-164°.

SYNS: 4-o-(CARBOXYMETHYL)RIFAMYCIN ◇ NACIMYCIN ◇ NCI 145-604 ◇ RIFAMYCIN B ◇ RIFOMYCIN B

TOXICITY DATA with REFERENCE
ivn-rat LD50:1680 mg/kg 85ERAY 1,865,78
ivn-mus LD50:2040 mg/kg MEIEDD 10,1187,83
ivn-dog LD50:1200 mg/kg 85ERAY 1,865,78
ipr-gpg LD50:3000 mg/kg 85ERAY 1,865,78

SAFETY PROFILE: Moderately toxic by intravenous and intraperitoneal routes. When heated to decomposition it emits toxic fumes of NO_x. Used as an antibiotic.

RKP000 CAS:13292-46-1 ***HR: 3***
RIFAMYCIN AMP
mf: $C_{43}H_{58}N_4O_{12}$ mw: 823.05

SYNS: ARCHIDYN ◇ ARFICIN ◇ DIONE 21-ACETATE ◇ L-5103 ◇ 3-(4-METHYLPIPERAZINYLIMINOMETHYL)-RIFAMYCIN SV ◇ 8-(4-METHYLPIPERAZINYLIMINOMETHYL) RIFAMYCIN SV ◇ 8-(((4-METHYL-1-PIPERAZINYL)IMINO)METHYL)RIFAMYCIN SV ◇ NSC 113926 ◇ R/AMP ◇ RIFA ◇ RIFADINE ◇ RIFAGEN ◇ RIFALDAZINE ◇ RIFALDIN ◇ RIFAMATE ◇ RIFAMPICIN ◇ RIFAMPICINE (FRENCH) ◇ RIFAMPICINUM ◇ RIFAMPIN ◇ RIFAPRODIN ◇ RIFINAH ◇ RIFOBAC ◇ RIFOLDIN ◇ RIFORAL ◇ RIMACTAN ◇ RIMACTAZID ◇ TUBOCIN

TOXICITY DATA with REFERENCE
dnr-esc 20 μL/disc MUREAV 97,1,82
dni-hmn:hla 250 mg/L IJEBA6 22,350,84
ihl-rat TCLo:6100 μg/m³/24H (female 1-22D post):REP ANTBAL 25,280,80
orl-rat TDLo:1400 mg/kg/(4-10D preg):TER CHDDAT 269,2147,69
orl-mus TDLo:8400 mg/kg/60W-C:NEO TXAPA9 43,293,78
orl-hmn TDLo:180 mg/kg:SKN,EYE BMJOAE 2,1189,77
orl-man TDLo:13 mg/kg/2D:EYE BMJOAE 1,199,76
orl-man LDLo:857 mg/kg JAMAAP 240,2283,78
orl-wmn TDLo:315 mg/kg/5W-I:SKN JAADDB 17,303,87
orl-rat LD50:1570 mg/kg JJANAX 23,257,70
ipr-rat LD50:390 mg/kg TXAPA9 24,37,73
scu-rat LD50:534 mg/kg JJANAX 23,242,70
orl-mus LD50:500 mg/kg CHDDAT 269,2147,69
ipr-mus LD50:480 mg/kg TXAPA9 24,37,73
scu-mus LD50:621 mg/kg JJANAX 23,242,70
ivn-mus LD50:260 mg/kg ANBCB3 16,316,70

CONSENSUS REPORTS: IARC Cancer Review: Animal Limited Evidence IMEMDT 24,243,80; Human Inadequate Evidence IMEMDT 24,243,80.

SAFETY PROFILE: Suspected carcinogen with experimental neoplastigenic and teratogenic data. Poison by intraperitoneal and intravenous routes. Moderately toxic to humans by ingestion. Moderately experimentally toxic by ingestion and subcutaneous routes. Human systemic effects by ingestion: conjunctiva irritation, iritis (inflammation of the iris), other eye effects, dermatitis. Experimental reproductive effects. Human mutation data reported. When heated to decomposition it emits toxic fumes of NO_x.

RKP400 CAS:14487-05-9 ***HR: 2***
RIFAMYCIN O
mf: $C_{39}H_{49}NO_{14}$ mw: 755.89

PROP: Pale yellow crystals from methanol. Mp: 300°. Practically insol in dil acids; sol in alkaline solns and acetone tetrahydrofuran; sltly sol in methanol, ethanol, and ethyl acetate; practically insol in ether and petr ether.

SYNS: 4-o-(CARBOXYMETHYL)-1-DEOXY-1,4-DIHYDRO-4-HYDROXY-1-OXO-RIFAMYCIN Γ-LACTONE ◇ RIFOMYCIN O

TOXICITY DATA with REFERENCE
ipr-mus LD50:1000 mg/kg 85GDA2 2,462,80
scu-mus LD50:3000 mg/kg 85GDA2 2,462,80
ivn-mus LD50:1800 mg/kg 85FZAT -,566,67

SAFETY PROFILE: Moderately toxic by intraperitoneal, subcutaneous, and intravenous routes. When heated to decomposition it emits toxic fumes of NO_x.

RKU000 CAS:13553-79-2 ***HR: 3***
RIFAMYCIN S
mf: $C_{37}H_{45}NO_{12}$ mw: 695.83

PROP: Yellow-orange crystals. Decomp @ 179-180°.

SYNS: 1,4-DIDEOXY-1,4-DIHYDRO-1,4-DIOXORIFAMYCIN ◇ RIFOMYCIN S

TOXICITY DATA with REFERENCE
orl-mus LD50:3 g/kg MEIEDD 10,1188,83
ipr-mus LD50:258 mg/kg 85ERAY 1,865,78
ivn-mus LD50:122 mg/kg 85ERAY 1,865,78

SAFETY PROFILE: Poison by intraperitoneal and intravenous routes. Moderately toxic by ingestion. When heated to decomposition it emits toxic fumes of NO_x.

RKZ000 CAS:6998-60-3 ***HR: 3***
RIFAMYCIN SV
mf: $C_{37}H_{47}NO_{12}$ mw: 697.85

PROP: Yellow-orange crystals. Decomp @ 140°. Sltly sol in water, petroleum ether; sol in ether, bicarbonate solutions; very sol in methanol, ethanol, acetone, ethyl acetate.

SYNS: M-14 ◇ RIFAMICINE SV ◇ RIFAMYCIN ◇ RIFOCIN ◇ RIFOCYN ◇ RIFOMYCIN SV

TOXICITY DATA with REFERENCE
pic-esc 100 ng/plate CNREA8 43,2819,83
cyt-mus-mmr 6300 mmol/L/24H-C JTSCDR 5,141,80
orl-rbt TDLo:900 mg/kg (female 8-16D post):REP GNRIDX 4,328,70
orl-mus TDLo:500 mg/kg (female 2D post):TER ARZNAD 36,219,86
orl-mus LD50:2120 mg/kg JJANAX 23,242,70
ipr-mus LD50:625 mg/kg MEIEDD 10,1188,83
ivn-mus LD50:550 mg/kg JMCMAR 7,596,64
ice-mus LD50:5800 μg/kg ARZNAD 15,951,65

SAFETY PROFILE: Poison by intracerebral route. Moderately toxic by ingestion, intraperitoneal, and intravenous routes. An experimental teratogen. Experimental reproductive effects. Mutation data reported. When heated to decomposition it emits toxic fumes of NO_x. Used as an antibacterial agent.

RLF350 CAS:52315-07-8 ***HR: 3***
RIPCORD
mf: $C_{22}H_{19}Cl_2NO_3$ mw: 416.30

SYNS: AMMO ◇ ARDAP ◇ AVICADE ◇ BARRICADE ◇ CCN52 ◇ (±)-α-CYANO-3-PHENOXYBENZYL 2,2-DIMETHYL-3-(2,2-DICHLOROVINYL)CYCLOPROPANE CARBOXYLATE ◇ CYMBUSH ◇ CYPERKILL ◇ CYPERMETHRIN ◇ FMC 30980 ◇ FMC 45497 ◇ FMC 45806 ◇ IMPERATOR ◇ JF 5705F ◇ KAFIL SUPER ◇ NRDC 149 ◇ NRDC 160 ◇ NRDC 166 ◇ PP383 ◇ SIPERIN ◇ STOCKADE ◇ WL 43467

TOXICITY DATA with REFERENCE
mnt-mus-orl 756 mg/kg/7D-C MUREAV 155,135,85
cyt-mus-unr 10 mg/kg TGANAK 18,455,84
unr-rat TDLo:400 mg/kg (6-15D preg):TER GISAAA 49(10),70,84
ipr-mus TDLo:30 mg/kg (male 1-4D pre):REP TOLED5 41,223,88
orl-rat LD50:70 mg/kg 85JFAN A649,86
skn-rat LD50:1600 mg/kg FMCHA2 -,C87,89
ivn-rat LDLo:6 mg/kg ARTODN 45,325,80

CONSENSUS REPORTS: EPA Genetic Toxicology Program. Cyanide and its compounds are on the Community Right-To-Know List.

SAFETY PROFILE: Poison by ingestion and intravenous routes. Moderately toxic by skin contact. An experimental teratogen. Experimental reproductive effects. Mutation data reported. When heated to decomposition it emits toxic fumes of CN^-, NO_x, and Cl^-.

RLK000 CAS:298-59-9 ***HR: 3***
RITALIN HYDROCHLORIDE
mf: $C_{14}H_{19}NO_2{\cdot}ClH$ mw: 269.80

SYNS: CENTEDRIN ◇ METHYLPHENIDATE HYDROCHLORIDE ◇ METHYLPHENIDYLACETATE HYDROCHLORIDE ◇ METHYL α-PHENYL-2-PIPERIDINEACETATE HYDROCHLORIDE ◇ RITALIN

TOXICITY DATA with REFERENCE
otr-ham:emb 500 mg/L ENMUDM 8(Suppl 6),4,86
orl-cld TDLo:32 mg/kg/6W-I:PSY BIPCBF 20,1332,85
orl-rat LD50:367 mg/kg 27ZQAG -,264,72
scu-rat LD50:170 mg/kg CLDND*-,264,72
orl-mus LD50:60 mg/kg JMCMAR 18,71,75
ipr-mus LD50:167 mg/kg JMCMAR 14,1106,71
scu-mus LD50:150 mg/kg JMCMAR 14,1106,71
ivn-mus LDLo:40 mg/kg KLWOAZ 32,445,54

SAFETY PROFILE: Poison by ingestion, subcutaneous, intravenous, and intraperitoneal routes. Human systemic effects by ingestion: toxic psychosis. Mutation data reported. Used as a central nervous system stimulant. When heated to decomposition it emits very toxic fumes of HCl and NO_x.

RLK700 CAS:23239-51-2 ***HR: 3***
RITODRINE HYDROCHLORIDE
mf: $C_{17}H_{21}NO_3{\cdot}ClH$ mw: 323.85

SYNS: DU 21220 ◇ erythro-p-HYDROXY-α-(1-((p-HYDROXYPHENETHYL)AMINO)ETHYL)BENZYL ALCOHOL HYDROCHLORIDE ◇ PRE-PAR ◇ UTOPAR ◇ YUTOPAR

TOXICITY DATA with REFERENCE
ivn-wmn TDLo:450 μg/kg (female 39W post):REP GOBIDS 23,160,87
ivn-wmn TDLo:4 mg/kg (female 39W post):TER LANCAO 1,1468,82
ivn-wmn TDLo:14508 μg/kg/15H-C:SYS BMJOAE 2,1194,78
orl-mus LD50:687 mg/kg YIGODN 16,122,85
ipr-mus LD50:265 mg/kg YIGODN 16,122,85
ivn-mus LD50:69 mg/kg YIGODN 16,122,85

SAFETY PROFILE: Poison by intravenous and intraperitoneal routes. Moderately toxic by ingestion. Human and experimental teratogen. Human and experimental reproductive effects. Human systemic effects: metabolic acidosis. When heated to decomposition it emits toxic fumes of NO_x and HCl.

RLK800 CAS:53902-12-8 ***HR: 3***
RIZABEN
mf: $C_{18}H_{17}NO_5$ mw: 327.36

PROP: Crystals from chloroform. Mp: 211-213°.

SYNS: N-(3,4-DIMETHOXYCINNAMOYL)ANTHRANILICACID ◇ N-(3',4'-DIMETHOXYCINNAMOYL)ANTHRANILIC ACID ◇ TRANILAST

TOXICITY DATA with REFERENCE
orl-rat TDLo:1100 mg/kg (7-17D preg):REP IYKEDH 9,161,78
orl-rat TDLo:8500 mg/kg (male 9W pre):TER IYKEDH 9,148,78
orl-rat LD50:1100 mg/kg IYKEDH 13,1128,82
ipr-rat LD50:395 mg/kg IYKEDH 13,1128,82
scu-rat LD50:3060 mg/kg IYKEDH 13,1128,82
orl-mus LD50:680 mg/kg OYYAA2 19,503,80
ipr-mus LD50:385 mg/kg OYYAA2 19,503,80
scu-mus LD50:2630 mg/kg IYKEDH 13,1128,82

SAFETY PROFILE: Poison by intraperitoneal route. Moderately toxic by ingestion and subcutaneous routes. An experimental teratogen. Experimental reproductive effects. When heated to decomposition it emits toxic fumes of NO_x.

RLK875 ***HR: 1***
ROBAVERON

SYN: SWINE PROSTATE EXTRACT

TOXICITY DATA with REFERENCE
orl-mus TDLo:500 mg/kg (6-15D preg):REP KSRNAM 19,371,85
orl-mus TDLo:500 mg/kg (6-15D preg):TER KSRNAM 19,371,85
orl-rat LD50:184 g/kg NIIRDN 6,912,82
scu-rat LD50:146 g/kg NIIRDN 6,912,82
ivn-rat LD50:51 g/kg NIIRDN 6,912,82
orl-mus LD50:128 g/kg NIIRDN 6,912,82
scu-mus LD50:118 g/kg NIIRDN 6,912,82
ivn-mus LD50:107 g/kg NIIRDN 6,912,82

SAFETY PROFILE: Very mildly toxic by ingestion and other routes. An experimental teratogen. Experimental reproductive effects.

RLK890 CAS:25875-51-8 ***HR: 2***
ROBENIDINE
mf: $C_{15}H_{13}Cl_2N_5$ mw: 334.23

PROP: Crystals from ethanol. Mp: 289-290°.

SYNS: 1,3-BIS((p-CHLOROBENZYLIDENE)AMINO)GUANIDINE ◇ CARBONIMIDIC DIHYDRAZIDE, BIS((4-CHLOROPHENYL)-METHYLENE)- ◇ CHEMCOCCIDE ◇ CHEMOCCIDE ◇ CHIMCOCCIDE ◇ KHIMCOCCID ◇ KHIMCOECID ◇ KHIMKOKTSID ◇ KHIMKOKTSIDE

TOXICITY DATA with REFERENCE
orl-rat LD50:1350 mg/kg VETNAL 57(9),53,81
orl-mus LD50:1212 mg/kg VETNAL 53(11),69,76
orl-rbt LD50:1245 mg/kg VETNAL 57(9),53,81
orl-ckn LD50:500 mg/kg VETNAL 57(9),53,81
orl-dck LD50:1017 mg/kg VETNAL 57(9),53,81

SAFETY PROFILE: Moderately toxic by ingestion. When heated to decomposition it emits toxic fumes of Cl^- and NO_x.

RLP000 CAS:490-31-3 ***HR: D***
ROBINETIN
mf: $C_{15}H_{10}O_7$ mw: 302.25

SYN: 3,3',4',5',7-PENTAHYDROXYFLAVANONE

TOXICITY DATA with REFERENCE
mmo-sat 100 μg/plate BCSTB5 5,1489,77
mma-sat 100 μg/plate BCSTB5 5,1489,77

SAFETY PROFILE: Mutation data reported. When heated to decomposition it emits acrid smoke and irritating fumes.

RLU000 CAS:93-14-1 ***HR: 2***
ROBITUSSIN
mf: $C_{10}H_{14}O_4$ mw: 198.24

PROP: Colorless crystals or powder. Mp: 78-79°. Sol in water, alc, ether, chloroform.

SYNS: ARESOL ◇ CORTUSSIN ◇ CRESON ◇ 1,2-DIHYDROXY-3-(2-METHOXYPHENOXY)PROPANE ◇ DILYN ◇ GAIMAR ◇ GLYCERIN GUAIACOLATE ◇ GLYCERINMONOGUAIACOL ETHER ◇ GLYCEROL GUAIACOLATE ◇ GLYCEROL-α-(o-METHOXYPHENYL)ETHER ◇ GLYCEROL-α-MONOGUAIACOL ETHER ◇ GLYCEROL MONO(2-METHOXYPHENYL)ETHER ◇ GLYCERYL GUAIACOLATE ◇ α-GLYCERYL GUAIACOLATE ETHER ◇ GLYCERYL GUAIACYL ETHER ◇ GLYCOTUSS ◇ GUAJACOL-GLYCERINAETHER (GERMAN) ◇ GUAIACOLGLICERINETERE ◇ GUAIACOL GLYCERYL ETHER ◇ GUAIACURANE ◇ GUAIACYL GLYCERYL ETHER ◇ GUAIAMAR ◇ GUAIANESIN ◇ GUAIFENESIN ◇ GUAIPHENESINE ◇ GUAJACOL-α-GLYCERINETHER ◇ HUSTODIL ◇ HUSTOSIL ◇ METFENOSSIDIOLO ◇ METHOXYPROPANEDIOL ◇ METHOXYPHENOXYDIOL ◇ 3-o-METHOXYPHENOXYPROPANE 1:2-DIOL ◇ 3-(o-METHOXYPHENOXY)-1,2-PROPANEDIOL ◇ o-METHOXYPHENYL GLYCERYL ETHER ◇ METOSSIPROPANDIOLO ◇ MIOCURIN ◇ MIORELAX ◇ MYOCAINE ◇ MYORELAX ◇ MYOSCAINE ◇ NEUROTONE ◇ ORESOL ◇ ORESON ◇ PROPANOSEDYL ◇ REDUTON ◇ RELAXYL-G ◇ REORGANIN ◇ RESIL ◇ RESPENYL ◇ RITUSSIN ◇ SIROTOL ◇ SL-90 ◇ TOLSERON ◇ TOLYN ◇ TULYL ◇ XL-90

TOXICITY DATA with REFERENCE
orl-rat LD50:1510 mg/kg PLRCAT 1,413,69
ipr-rat LDLo:1000 mg/kg AIPTAK 191,147,71
scu-rat LD50:2550 mg/kg NIIRDN 6,210,82
orl-mus LD50:690 mg/kg 27ZQAG -,398,72
ipr-mus LD50:495 mg/kg 27ZQAG -,398,72
scu-mus LD50:800 mg/kg ARZNAD 15,1355,65

CONSENSUS REPORTS: Reported in EPA TSCA Inventory.

SAFETY PROFILE: Moderately toxic by ingestion, intraperitoneal and subcutaneous routes. Used as a cough medicine. When heated to decomposition it emits acrid smoke and irritating fumes.

RLU550 ***HR: D***
ROC-101

PROP: An herbal preparation composed of a mixture of three different plants indigenous to India (IJMRAQ 60,1054,72).

TOXICITY DATA with REFERENCE
orl-mus TDLo:336 g/kg (male 28D pre):REP IJMRAQ 60,1054,72

SAFETY PROFILE: Experimental reproductive effects. When heated to decomposition it emits acrid smoke and irritating fumes.

RLZ000 CAS:1866-43-9 ***HR: 3***
ROLODINE
mf: $C_{14}H_{14}N_4$ mw: 238.32

SYNS: 4-(BENZYLAMINO)-2-METHYL-7H-PYRROLO(2,3-d)PYRIMIDINE ◇ BW 58-271 ◇ 2-METHYL-4-BENZYLAMINO-PYRROLO-(2,3d)PYRIMIDINE ◇ ROLODIN

TOXICITY DATA with REFERENCE
orl-mus LD50:88 mg/kg 27ZQAG -,297,72
ipr-mus LD50:24 mg/kg 27ZQAG -,297,72

SAFETY PROFILE: Poison by ingestion and intraperitoneal routes. When heated to decomposition it emits toxic fumes of NO_x.

RMA000 CAS:50471-44-8 ***HR: 1***
RONILAN
mf: $C_{12}H_9Cl_2NO_3$ mw: 286.12

SYNS: BAS 352 F ◇ 3-(3,5-DICHLOROPHENYL)-5-ETHENYL-5-METHYL-2,4-OXAZOLIDINEDIONE ◇ 3-(3,5-DICHLOROPHENYL)-5-METHYL-5-VINYL-2,4-OXAZOLIDINEDIONE ◇ VINCLOZOLIN (GERMAN)

TOXICITY DATA with REFERENCE
mma-sat 100 mg/L ATSUDG (5),345,82
mma-ssp 100 mg/L ATSUDG (5),345,82
orl-gpg LD50:8000 mg/kg 85DPAN -,-,71/76

SAFETY PROFILE: Mildly toxic by ingestion. Mutation data reported. When heated to decomposition it emits very toxic fumes of Cl^- and NO_x.

RMA500 CAS:299-84-3 ***HR: 3***
RONNEL
mf: $C_8H_8Cl_3O_3PS$ mw: 321.54

PROP: White powder. Mp: 41°, vap press: 8×10^{-4} mm.

SYNS: DERMAFOSU (POLISH) ◇ DERMAPHOS ◇ O,O-DIMETHYL-O-2,4,5-TRICHLOROPHENYL PHOSPHOROTHIOATE ◇ DIMETHYL TRICHLOROPHENYL THIOPHOSPHATE ◇ O,O-DIMETHYL-O-(2,4,5-TRICHLOROPHENYL)THIOPHOSPHATE ◇ O,O-DIMETHYL-O-(2,4,5-TRICHLORPHENYL)-THIONOPHOSPHAT(GERMAN) ◇ DOW ET 14 ◇ DOW ET 57 ◇ ECTORAL ◇ ENT 23,284 ◇ ET 14 ◇ ET 57 ◇ ETROLENE ◇ FENCHLOORFOS (DUTCH) ◇ FENCHLORFOS ◇ FENCHLORFOSU (POLISH) ◇ FENCHLOROPHOS ◇ FENCHLORPHOS ◇ KARLAN ◇ KORLAN ◇ KORLANE ◇ NANCHOR ◇ NANKER ◇ NANKOR ◇ THIOPHOSPHATE de O,O-DIMETHYLE et de O-(2,4,5-TRICHLOROPHENYLE) (FRENCH) ◇ O-(2,4,5-TRICHLOOR-FENYL)-O,O-DIMETHYL-MONOTHIOFOSFAAT (DUTCH) ◇ TRICHLOROMETAFOS ◇ 2,4,5-TRICHLOROPHENOL, O-ESTER with O,O-DIMETHYL PHOSPHOROTHIOATE ◇ O-(2,4,5-TRICHLOR-PHENYL)-O,O-DIMETHYL-MONOTHIOPHOSPHAT (GERMAN) ◇ O-(2,4,5-TRICLORO-FENIL)-O,O-DIMETIL-MONOTIOFOSFATO (ITALIAN) ◇ TROLEN ◇ TROLENE ◇ VIOZENE

TOXICITY DATA with REFERENCE
sce-hmn:lyms 2 mg/L MUREAV 102,89,82
orl-rat TDLo:368 mg/kg (16-22D preg):REP NUPOBT 17,577,79
orl-rbt TDLo:650 mg/kg (female 6-18D post):TER ACVTA8 24,295,83
orl-rat LD50:625 mg/kg GISAAA 45(6),14,80
skn-rat LD50:2000 mg/kg WRPCA2 9,119,70
ipr-rat LD50:2823 mg/kg PSEBAA 129,699,68
orl-mus LD50:2000 mg/kg PCOC** -,965,66
ipr-mus LD50:118 mg/kg PSEBAA 129,699,68
orl-dog LD50:500 mg/kg 85GYAZ -,26,71
orl-rbt LD50:420 mg/kg PCOC** -,965,66
skn-rbt LD50:1 g/kg FMCHA2 -,C252,89
skn-gpg LD50:2000 mg/kg 85DPAN -,-,71/76

CONSENSUS REPORTS: Chlorophenol compounds are on the Community Right-To-Know List.

OSHA PEL: (Transitional: TWA 15 mg/m^3) TWA 10 mg/m^3
ACGIH TLV: TWA 10 mg/m^3

SAFETY PROFILE: Poison by ingestion and intraperitoneal routes. Moderately toxic by skin contact. A cholinesterase inhibitor. An experimental teratogen. Experimental reproductive effects. Human mutation data reported. When heated to decomposition it emits very toxic fumes of Cl^-, PO_x, and SO_x. See also PARATHION and CHLOROPHENOLS.

RMF000 CAS:35834-26-5 ***HR: 3***
ROSAMICIN
mf: $C_{31}H_{51}NO_9$ mw: 581.83

PROP: Crystals. Mp: 119-122°. Very sol in methanol, acetone, chloroform, benzene; sltly sol in ether, water.

SYNS: ANTIBIOTIC 67-694 ◇ ANTIBIOTIC M 4365A2 ◇ 4'-DEOXY-CIRRAMYCIN A[1] ◇ JUVENIMICIN A3 ◇ M-4365A2 ◇ ROSARAMICIN ◇ Sch 14947

TOXICITY DATA with REFERENCE
orl-mus LD50:1000 mg/kg 85ESA3 9,1071,76
ipr-mus LD50:350 mg/kg 85ERAY 1,62,78
scu-mus LD50:625 mg/kg 85ERAY 1,62,78

SAFETY PROFILE: Poison by intraperitoneal route. Moderately toxic by ingestion and subcutaneous routes.

When heated to decomposition it emits toxic fumes of NO_x. Used as an antibacterial agent.

RMK000 ***HR: D***
p-ROSANILINE
mf: $C_{20}H_{19}N_3$ mw: 301.42

PROP: Brownish-red crystals. Decomp @ 186°. Sltly sol in water; sol in alc, acids.

SYNS: 4-((4-AMINOPHENYL)(4-IMINO-2,5-CYCLOHEXADIEN-1-YLIDENE)METHYL)-2-METHYLBENZENAMINE ◇ 4-((p-AMINO-PHENYL)(4-IMINO-2,5-CYCLOHEXADIEN-1-YLIDENE)METHYL)-o-TOLUIDINE

TOXICITY DATA with REFERENCE
mmo-esc 5 g/L MUREAV 130,97,84
otr-ham:emb 2 mg/L NCIMAV 58,243,81

SAFETY PROFILE: Mutation data reported. When heated to decomposition it emits toxic fumes of NO_x.

RMK020 CAS:569-61-9 ***HR: 3***
p-ROSANILINE HYDROCHLORIDE
mf: $C_{19}H_{17}N_3$•ClH mw: 323.85

SYNS: 4-((4-AMINOPHENYL)(4-IMINO-2,5-CYCLOHEXADIEN-1-YLIDENE)METHYL), MONOCHLORIDE ◇ BASIC PARAFUCHSINE ◇ CALCOZINE MAGENTA N ◇ C.I. 42500 ◇ C.I. BASIC RED 9, MONO-HYDROCHLORIDE ◇ p-FUCHSIN ◇ FUCHSINE DR-001 ◇ FUCHSINE SPC ◇ 4,4'-((4-IMINO-2,5-CYCLOHEXADIEN-1-YLIDENE)METHYLENE)DIANILINE MONOHYDROCHLORIDE-o-TOLUIDINE ◇ NCI-C54739 ◇ PARAFUCHSIN (GERMAN) ◇ PARA-MAGENTA ◇ PARAROSANILINE ◇ PARAROSANILINE CHLORIDE ◇ PARAROSANILINE HYDROCHLORIDE ◇ p-ROSANILINE HCL ◇ SCHULTZ-TAB No. 779 (GERMAN) ◇ 4,4'4''-TRIAMINOTRIPHENYLMETHAN-HYDROCHLORID (GERMAN)

TOXICITY DATA with REFERENCE
mma-sat 100 μg/plate ENMUDM 6(Suppl 2),1,84
otr-rat:emb 1400 μg/L JJIND8 67,1303,81
orl-rat TDLo:728 mg/kg/2Y-C:CAR NTPTR* NTP-TR-285,86
scu-rat TDLo:1714 mg/kg/43W-I:ETA NATWAY 43,543,56
orl-mus TDLo:364 mg/kg/2Y-C:CAR NTPTR* NTP-TR-285,86
orl-mus LD50:5000 mg/kg FAZMAE 17,108,73

CONSENSUS REPORTS: NTP Fifth Annual Report on Carcinogens. IARC Cancer Review: Group 3 IMEMDT 7,238,87; Animal Limited Evidence IMEMDT 4,57,74; Human Inadequate Evidence IMEMDT 4,57,74. EPA Genetic Toxicology Program. Reported in EPA TSCA Inventory.

SAFETY PROFILE: Confirmed carcinogen with experimental carcinogenic and tumorigenic data. Mildly toxic by ingestion. Mutation data reported. When heated to decomposition it emits very toxic fumes of HCl and NO_x.

RMK200 CAS:52934-83-5 ***HR: 2***
ROSANOMYCIN A
mf: $C_{16}H_{14}O_6$ mw: 302.30

SYNS: ANTIBIOTIC OS 3966A ◇ NANAOMYCIN A ◇ OS-3966-A

TOXICITY DATA with REFERENCE
scu-rat LD50:52 mg/kg JJANAX 33,728,80
skn-mus LD50:52 mg/kg JJANAX 33,728,80
ipr-mus LD50:28 mg/kg 85ERAY 1,551,78
scu-mus LD50:56 mg/kg JJANAX 33,728,80
ivn-mus LD50:52 mg/kg JJANAX 33,728,80

SAFETY PROFILE: Poison by skin contact, subcutaneous, intravenous, and intraperitoneal routes. When heated to decomposition it emits acrid smoke and irritating fumes.

RMK250 ***HR: 2***
ROSARY PEA

PROP: A slender climbing vine considered a common weed in Hawaii, Guam, and the Caribbean, including Florida. A pea-shaped pod about 1.5 inches long contains 3 to 5 bright red pea-size seeds with a black spot. They are used in jewelry and charms.

SYNS: ABRUS PRECATORIUS L. (SEED) ◇ BLACK EYED SUSAN ◇ CORAL BEAD PLANT ◇ CRAB'S EYES ◇ GRAINES D'EGLISE (GUADELOUPE) ◇ LICORICE VINE ◇ KOLALES HALOMTANO (GUAM) ◇ JUMBEE BEADS (VIRGIN ISLANDS) ◇ LOVE BEAD ◇ OJO de CANGREJO (PUERTO RICO) ◇ PEONIA (CUBA) ◇ PERONIAS (PUERTO RICO) ◇ PRAYER BEADS ◇ PUKIAWE-LEI (HAWAII) ◇ RED BEAD VINE ◇ REGLISSE (GUADELOUPE and HAITI) ◇ SEMILLA de CULEBRA (MEXICO) ◇ SEMINOLE BEAD ◇ WEATHER PLANT ◇ WILD LICORICE

SAFETY PROFILE: The seeds contain the poison abrin, a plant lectin which inhibits cell growth. The seed must be chewed or broken open to be poisonous. However, chewing and ingesting one seed could be fatal. Human systemic effects by ingestion: nausea, vomiting, diarrhea, loss of intestinal function, and lesions of the gastrointestinal tract. Symptoms may develop over a period from several hours to three days, depending on the dose. See also ABRIN.

RMP000 CAS:84604-12-6 ***HR: 1***
ROSE ABSOLUTE FRENCH

PROP: Isolated from the flowers of *Rosa centifolia L. (Fam. Rosaceae)* (FCTXAV 13,68175).

SYNS: ABSOLUTE FRENCH ROSE ◇ FRENCH ROSE ABSOLUTE ◇ ROSE de MAI ABSOLUTE

TOXICITY DATA with REFERENCE
skn-rbt 500 mg/24H MOD FCTXAV 13,681,75

SAFETY PROFILE: A skin irritant. When heated to decomposition it emits acrid smoke and irritating fumes.

RMP175 CAS:632-69-9 ***HR: D***
ROSE BENGAL SODIUM
mf: $C_{20}H_2Cl_4I_4O_5\cdot 2Na$ mw: 1017.60

SYNS: FOOD RED COLOR No. 105, SODIUM SALT ◇ FOOD RED No. 105, SODIUM SALT ◇ R105 SODIUM ◇ 9-(3',4',5',6'-TETRACHLORO-o-CARBOXYPHENYL)-6-HYDROXY-2,4,5,7-TETRAIODO-3-ISOXANTHONE•2Na ◇ 4,5,6,7-TETRACHLORO-2',4',5',7'-TETRAIODOFLUORESCEIN DISODIUM SALT

TOXICITY DATA with REFERENCE
orl-rat TDLo:400 mg/kg (female 8-15D post):REP OYYAA2 24,391,82
orl-rat TDLo:650 mg/kg (8-20D preg):TER OYYAA2 24,391,82
orl-mus TDLo:707 g/kg/2Y-C:NEO JJIND8 77,277,86

CONSENSUS REPORTS: Reported in EPA TSCA Inventory.

SAFETY PROFILE: An experimental teratogen. Experimental reproductive effects. Questionable carcinogen with experimental neoplastigenic data. When heated to decomposition it emits toxic fumes of Cl^-, I^-, and Na_2O.

RMU000 CAS:8000-25-7 ***HR: 1***
ROSEMARY OIL

PROP: Constituents are α-pinene, camphene, and cineole. From steam distillation of flowering tops of *Rosmarinus officinalis* L. (Fam. *Labiatae*) (FCTXAV 12,807,74). Colorless to pale yellow liquid; odor of rosemary. D: 0.894-0.912, refr index: 1.464 @ 20°.

SYNS: ROSEMARIE OIL ◇ ROSMARIN OIL (GERMAN)

TOXICITY DATA with REFERENCE
skn-rbt 500 mg/24H MOD FCTXAV 12,977,74
orl-rat LD50:5000 mg/kg FCTXAV 12,977,74

CONSENSUS REPORTS: Reported in EPA TSCA Inventory.

SAFETY PROFILE: Mildly toxic by ingestion. A skin irritant. When heated to decomposition it emits acrid smoke and irritating fumes. See also individual components.

RNA000 CAS:8007-01-0 ***HR: 1***
ROSE OIL

PROP: Volatile oil from steam distillation of fresh flowers of *Rosa gallica* L. and *Rosa Damascena* Mill. and varieties of these species (Fam. *Rosaceae*). Colorless to yellow liquid; odor and taste of rose. D: 0.848-0.863 @ 30°/15°, refr index: 1.457 @ 30°.

SYN: ROSEN OEL (GERMAN)

TOXICITY DATA with REFERENCE
orl-rat LD50:12560 mg/kg PHARAT 14,435,59

CONSENSUS REPORTS: Reported in EPA TSCA Inventory.

SAFETY PROFILE: Mildly toxic by ingestion. When heated to decomposition it emits acrid smoke and irritating fumes.

RNF000 CAS:8007-01-0 ***HR: 2***
ROSE OIL BULGARIAN

PROP: Main constituents are 1-citronellol and geraniol, found in plant *Rosa damascena Mill. Var. Alba.* (FCTXAV 12,807,74).

SYNS: BULGARIAN ◇ OIL of ROSE BULGARIAN ◇ OTTO of ROSE

TOXICITY DATA with REFERENCE
skn-rbt 500 mg/24H MLD FCTXAV 12,807,74
skn-rbt LD50:2500 mg/kg FCTXAV 12,807,74

CONSENSUS REPORTS: Reported in EPA TSCA Inventory.

SAFETY PROFILE: Moderately toxic by skin contact. A skin irritant. When heated to decomposition it emits acrid smoke and irritating fumes. See also individual components.

RNK000 CAS:84604-12-6 ***HR: 1***
ROSE OIL MOROCCAN

PROP: By steam distillation of the flowers of *Rosa centifolia L.*, mainly 1-citronellol and geraniol (FCTXAV 12,807,74).

SYN: OIL of ROSE MOROCCAN

TOXICITY DATA with REFERENCE
skn-rbt 500 mg/24H MOD FCTXAV 12,807,74

SAFETY PROFILE: A skin irritant. When heated to decomposition it emits acrid smoke and irritating fumes. See also individual components.

RNU000 CAS:16409-43-1 ***HR: 1***
ROSE OXIDE LEVO
mf: $C_{10}H_{18}O$ mw: 154.28

SYN: TETRAHYDRO-2-(2-METHYL-1-PROPENYL)-4-METHYLPYRAN

TOXICITY DATA with REFERENCE
skn-rbt 500 mg/24H MOD FCTXAV 14,659,76
orl-rat LD50:4300 mg/kg FCTXAV 14,659,76

CONSENSUS REPORTS: Reported in EPA TSCA Inventory.

SAFETY PROFILE: Mildly toxic by ingestion. A skin irritant. When heated to decomposition it emits acrid smoke and irritating fumes.

RNZ000 CAS:83-79-4 ***HR: 3***
ROTENONE
mf: $C_{23}H_{22}O_6$ mw: 394.45

PROP: Orthorhombic plates. Mp: 165-166° (dimorphic form mp: 185-186°). D: 1.27 @ 20°. Almost insol in water; sol in alc, acetone, carbon tetrachloride, chloroform, ether, and other organic solvents. Decomp on exposure to light and air.

SYNS: BARBASCO ◇ CENOL GARDEN DUST ◇ CHEM FISH ◇ CHEM-MITE ◇ CUBE ◇ CUBE EXTRACT ◇ CUBE-PULVER ◇ CUBE ROOT ◇ CUBOR ◇ CUREX FLEA DUSTER ◇ DACTINOL ◇ DERIL ◇ DERRIN ◇ DERRIS ◇ DRI-KIL ◇ ENT 133 ◇ EXTRAX ◇ FISH-TOX ◇ GREEN CROSS WARBLE POWDER ◇ HAIARI ◇ LIQUID DERRIS ◇ MEXIDE ◇ NCI-C55210 ◇ NICOULINE ◇ NOXFISH ◇ PARADERIL ◇ POWDER and ROOT ◇ PRENTOX ◇ RO-KO ◇ RONONE ◇ ROTEFIVE ◇ ROTEFOUR ◇ ROTENONA (SPANISH) ◇ ROTESSENOL ◇ ROTOCIDE ◇ TUBATOXIN

TOXICITY DATA with REFERENCE
eye-rbt 1% MLD PSEBAA 34,135,36
mnt-mus:oth 1 mg/L JNCIAM 56,357,76
orl-rat TDLo:98 mg/kg (female 6-15D post):REP BECTA6 28,360,82
orl-rat TDLo:50 mg/kg (6-15D preg):TER JTEHD6 10,111,82
orl-rat TDLo:3245 mg/kg/2Y-C:ETA NTPTR* NTP-TR-320,88
ipr-rat TDLo:71 mg/kg/42D-I:NEO CNREA8 33,3047,73
orl-hmn LDLo:143 mg/kg 34ZIAG -,521,79
unr-man LDLo:294 mg/kg 85DCAI 2,73,70
orl-rat LD50:60 mg/kg DOEAAH 36,25,79
ipr-rat LD50:5 mg/kg JAFCAU 17,497,69
orl-mus LD50:350 mg/kg 31ZOAD 1,372,68
ipr-mus LD50:2650 μg/kg RAREAE 91,186,82
orl-dog LDLo:300 mg/kg JPETAB 43,193,31
orl-gpg LDLo:100 mg/kg PSEBAA 34,135,36
ipr-gpg LDLo:10 mg/kg PSEBAA 34,135,36

OSHA PEL: TWA 5 mg/m^3
ACGIH TLV: TWA 5 mg/m^3
DFG MAK: 5 mg/m^3

SAFETY PROFILE: Human poison by ingestion. Experimental poison by ingestion and intraperitoneal routes. Experimental reproductive effects. Mutation data reported. A skin and eye irritant. Acute poisoning causes numbness, nausea, vomiting, and tremors. Questionable carcinogen with experimental neoplastigenic, tumorigenic, and teratogenic data. Chronic exposure injures liver and kidneys. It is toxic to animals and very toxic to fish, but leaves no harmful residue on vegetable crops. When heated to decomposition it emits acrid smoke and irritating fumes. Used as an insecticide and as a fish poison.

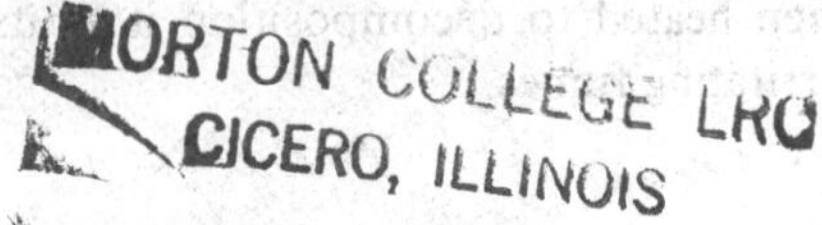

ROA300 ***HR: 1***
ROUGE PLANT

PROP: A small shrub which grows to 3 feet. The pink-white flowers hang in clusters. The berries are brilliant orange or red. They grow wild in the southern United States from New Mexico to Florida, Hawaii and the West Indies. They are a popular house plant.

SYNS: BABY PEPPER ◇ BLOODBERRY ◇ CAIMONICILLO (DOMINICAN REPUBLIC) ◇ CARMIN (PUERTO RICO) ◇ CAT'S BLOOD ◇ CORAL BERRY ◇ CORALITOS (CUBA) ◇ PIGEON BERRY ◇ RIVINA HUMILIS

SAFETY PROFILE: The leaves and roots contain the poisons phytolaccatoxin and related triterpenes. The berries apparently contain little or no toxin. Ingestion may cause nausea, vomiting, and diarrhea.

ROA400 CAS:65546-74-9 ***HR: 2***
ROWACHOL

TOXICITY DATA with REFERENCE
orl-rat TDLo:9600 mg/kg (9-14D preg):REP OYYAA2 15,1109,78
orl-rat TDLo:9600 mg/kg (9-14D preg):TER OYYAA2 15,1109,78
orl-rat LD50:6500 mg/kg NIIRDN 6,916,82
ipr-rat LD50:2250 mg/kg YACHDS 5,943,77
orl-mus LD50:4400 mg/kg YACHDS 5,943,77
ipr-mus LD50:2500 mg/kg YACHDS 5,943,77

SAFETY PROFILE: Moderately toxic by intraperitoneal route. Mildly toxic by ingestion. An experimental teratogen. Experimental reproductive effects.

ROA425 ***HR: 2***
ROWATIN

PROP: Contains 31% pinene, 1% camphene, 10% borneol, 4% anethol, 4% fenchon, 3% eucalyptol, in olive oil (NIIRDN 6,916,82).

TOXICITY DATA with REFERENCE
orl-rat TDLo:2400 mg/kg (-14D preg):REP YACHDS 5,2439,77
orl-rat TDLo:2400 mg/kg (-14D preg):TER YACHDS 5,2439,77
orl-rat LD50:5 g/kg NIIRDN 6,916,82
ipr-rat LD50:2600 mg/kg YACHDS 5,943,77
orl-mus LD50:5400 mg/kg YACHDS 5,943,77
ipr-mus LD50:4400 mg/kg YACHDS 5,943,77

SAFETY PROFILE: Moderately toxic by intraperitoneal route. Mildly toxic by ingestion. An experimental teratogen. Experimental reproductive effects. When heated to decomposition it emits acrid smoke and irritating fumes. See also individual components.

ROF200 CAS:55102-43-7 ***HR: 3***
RPCNU
mf: $C_{15}H_{22}ClN_3O_{10}$ mw: 439.85

SYNS: 1-(2-CHLOROETHYL)-1-NITROSO-3-RIBOPYRANOSYLUREA-2',3',4'-TRIACETATE ◇ (CHLORO-2-ETHYL)-1-(RIBOPYRANOSYLTRIACETATE-2',3',4')-3-NITROSOUREA ◇ (CLORO-2-ETIL)-1-(RIBOPIRANOSILTRIACETATO-2',3',4')-3-NITROSOUREA (ITALIAN) ◇ I.C.I.G. 1163

TOXICITY DATA with REFERENCE
mmo-sat 200 μg/plate INSSDM 19,165,81
dns-hmn:lym 10 mg/L FRPSAX 36,947,81
dni-hmn:lym 10 mg/L FRPSAX 36,947,81
oms-mus:oth 50 mg/L INSSDM 19,229,81
ipr-mus LD50:50 mg/kg INSSDM 19,123,81

SAFETY PROFILE: Poison by intraperitoneal route. Human mutation data reported. Many N-nitroso compounds are carcinogens. When heated to decomposition it emits toxic fumes of Cl^- and NO_x. See also N-NITROSO COMPOUNDS.

ROF300 CAS:135-51-3 ***HR: 3***
R SALT
mf: $C_{10}H_6O_7S_2 \cdot 2Na$ mw: 348.26

SYNS: FERRICON ◇ 2-NAPHTHOL-3,6-DISULFONIC ACID SODIUM SALT

TOXICITY DATA with REFERENCE
ivn-mus TDLo:600 mg/kg (female 7-12D post):REP KSRNAM 8,3476,74
orl-rat TDLo:300 mg/kg (female 1-20D post):TER FCTXAV 11,355,73
ivn-rat LD50:385 mg/kg KSRNAM 8,3454,74
ivn-mus LD50:403 mg/kg KSRNAM 8,3454,74
ivn-rbt LD50:280 mg/kg KSRNAM 8,3454,74

CONSENSUS REPORTS: Reported in EPA TSCA Inventory.

SAFETY PROFILE: Poison by intravenous route. An experimental teratogen. Experimental reproductive effects. When heated to decomposition it emits toxic fumes of SO_x and Na_2O. See also SULFONATES.

ROH900 CAS:9006-04-6
RUBBER, NATURAL
DOT: UN 1287/UN 1345

SYNS: CAOUTCHOUC ◇ GUM NAFKACRYSTAL ◇ INDIA RUBBER ◇ NAFKA ◇ NAFKA CRYSTAL GUM ◇ NAFKA KRISTALGOM ◇ NATURAL RUBBER ◇ RUBBER ◇ RUBBER (DOT) ◇ THIOKOL NVT

DOT Classification: Flammable Liquid; Label: Flammable Liquid (UN1287); Flammable or Combustible Liquid; Label: Flammable Liquid (UN1287); Flammable Solid, Label: Flammable Solid (UN1345)

SAFETY PROFILE: A dangerous fire hazard when exposed to heat or flame. When heated to decomposition it emits toxic fumes of SO_x.

ROK000 ***HR: 3***
RUBBER SCRAP
DOT: UN 1345

SYN: RUBBER BUFFINGS (DOT) ◇ RUBBER SHODDY (DOT)

DOT Classification: Label: Flammable Solid.

SAFETY PROFILE: A very dangerous fire hazard when exposed to heat or flame. Moderately explosive in the form of dust when exposed to flame. When heated to decomposition it emits toxic fumes of SO_x.

ROU000 ***HR: 3***
RUBBER SOLVENT

PROP: A petroleum cut distilling between 38 to 149°C consisting chiefly of C_5 to C_9 aliphatic hydrocarbons used in making rubber cements and in tire manufacturer. Flash p: −40°F (varies with manufacturer), autoign temp: 450°F (varies with manufacturer), lel: 1.0%, uel: 7.0%, d: <1, bp: 100-280°F. Insol in water. (27ZTAP 3,124,69).

SYNS: LACQUER DILUENT ◇ NAPHTHA ◇ SKELLY-SOLVE-L

TOXICITY DATA with REFERENCE
ihl-rat LC50:61 g/m^3/4H TXAPA9 33,526,75
ihl-cat LCLo:49 g/m^3/4H TXAPA9 33,526,75
ACGIH TLV: TWA 400 ppm
NIOSH REL: TWA (Petroleum Solvent) 350 mg/m^3; CL 1800 mg/m^3/15M

SAFETY PROFILE: Mildly toxic by inhalation. A very dangerous fire hazard when exposed to heat or flame. Explosive in the form of vapor when exposed to heat or flame. To fight fire, use foam, alcohol foam. When heated to decomposition it emits acrid smoke and irritating fumes.

ROU450 ***HR: 3***
RUBBER VINE

PROP: Woody vines with shiny green leaves 3 to 4 inches long and flowers ranging in color from lilac to purple to red depending on the species. The seeds are contained in milkweed-like pods. They are cultivated in southern Florida and grow wild in the West Indies and Guam.

SYNS: ALAMANDA MORADA FALSA (PUERTO RICO) ◇ CAOUTCHOUC (HAITI) ◇ CRYPTOSTEGIA GRANDIFLORA ◇ CRYPTOSTEGIA MADAGASCARIENSIS ◇ ESTRELLA DEL NORTE (CUBA) ◇ INDIA RUBBER VINE ◇ PICHUCO (MEXICO) ◇ PURPLE ALLAMANDA

SAFETY PROFILE: The whole plant contains poisonous cardiac glycosides similar to digitalis. Human systemic effects by ingestion include: mouth pain, nausea, vomiting, abdominal pain, cramps and diarrhea. Cardiac glycosides may cause death by their effect on heart function. See also DIGITALIS.

ROU800 CAS:54083-22-6 ***HR: D***
RUBIDAZONE
mf: $C_{34}H_{35}N_3O_{10}$ mw: 645.72

SYNS: BENZOYLHYDRAZONE DAUNORUBICIN ◇ DAUNOMYCIN BENZOYLHYDRAZONE ◇ RUBIDAZON ◇ ZORUBICIN

TOXICITY DATA with REFERENCE
mmo-sat 10 μg/plate EJCODS 19,641,83
dni-hmn:oth 500 μg/L CCPHDZ 2,31,79

SAFETY PROFILE: Human mutation data reported. When heated to decomposition it emits toxic fumes of NO_x. See also RUBIDAZONE MONOHYDROCHLORIDE.

ROZ000 CAS:36508-71-1 ***HR: 3***
RUBIDAZONE MONOHYDROCHLORIDE

PROP: A semisynthetic daunorubicin derivative (CTRRDO 62,1053,78).

SYNS: 22050 R.P. ◇ BENZOIC ACID HYDRAZIDE, 3-HYDRAZONE with DAUNORUBICIN, MONOHYDROCHLORIDE ◇ DAUNORUBICIN, BENZOYLHYDRAZONE, MONOHYDROCHLORIDE ◇ NSC-164011 ◇ RP 22,050 HYDROCHLORIDE ◇ ZORUBICIN HYDROCHLORIDE

TOXICITY DATA with REFERENCE
mmo-sat 20 μg/plate EJCAAH 19,641,83
mma-sat 4 μg/plate EJCAAH 19,641,83
pic-esc 700 μg/L MUREAV 77,197,80
oms-hmn:lym 1 mg/L EJCAAH 14,741,78
ivn-hmn TDLo:48 mg/kg/21D-I:CVS,GIT,BLD CTRRDO 62,1053,78
ipr-mus LD50:28710 μg/kg NCISP* JAN86

SAFETY PROFILE: Poison by intraperitoneal route. Human systemic effects by intravenous route: cardiovascular, blood, and gastrointestinal system effects. Human mutation data reported. When heated to decomposition it emits very toxic fumes of NO_x and HCl. Used as an antineoplastic agent. See also RUBIDAZONE.

RPA000 CAS:7440-17-7 ***HR: 3***
RUBIDIUM
DOT: UN 1423
af: Rb aw: 85.47

PROP: Soft, silvery-white metal. Mp: 38.89°, bp: 688°, d (solid): 1.532 @ 20°, d (liquid): 1.475 @ 39°.

SYNS: RUBIDIUM METAL (DOT) ◇ RUBIDIUM METAL, IN CARTRIDGES (DOT)

CONSENSUS REPORTS: Reported in EPA TSCA Inventory.

DOT Classification: Flammable Solid; Label: Flammable Solid & Dangerous When Wet.

SAFETY PROFILE: A very reactive alkali metal (more reactive than potassium or cesium). In the body, rubidium substitutes for potassium as an intracellular ion. The ratio of Rb/K intake is important in the toxicology of rubidium. A ratio above 40% is dangerous. In rats, a failure to gain weight is the first symptom, followed by ataxia and hyperirritability. Symptoms include: skin ulcers, poor hair coat, sensitivity, and extreme nervousness leading to convulsions and death.

A very dangerous fire and explosion hazard when exposed to heat or flame or by chemical reaction with oxidizers. Ignites on contact with air; oxygen, and halogens. Reaction with water, moisture or steam forms explosive hydrogen gas which then ignites. Explodes in contact with liquid bromine. Can react explosively with air; halogens; mercury; non-metals; vanadium chloride oxide; moisture; acids; oxidizers. Violent reaction with vanadium trichloride oxide (at 60°C); Cl_2O_2; P. Molten rubidium ignites in sulfur vapor and reacts vigorously with carbon. RbOH is more basic than KOH. Storage and handling: Keep under benzene, petroleum, or other liquids not containing O_2. When heated to decomposition it emits toxic fumes of Rb_2O. See also SODIUM and SODIUM POTASSIUM ALLOY.

RPB100 CAS:22754-97-8 ***HR: 3***
RUBIDIUM ACETYLIDE
mf: C_2Rb_2 mw: 194.96

SAFETY PROFILE: Explodes on contact with concentrated nitric acid, lead oxide. Ignites on contact with fluorine, chlorine, bromine, iodine, concentrated hydrochloric acid, arsenic, sulfur, selenium vapors. Ignites when warmed with carbon dioxide, nitrogen oxide, sulfur dioxide. Incandescent reaction when heated to 350°C with copper oxide; manganese dioxide. Vigorous exothermic reaction with iron(III) chloride, chromium(III) oxide, boron + heat, silicon + heat. When heated to decomposition it emits toxic fumes of Rb_2O. See also ACETYLIDES and RUBIDIUIM.

RPF000 CAS:7791-11-9 ***HR: 2***
RUBIDIUM CHLORIDE
mf: ClRb mw: 120.92

PROP: White, crystalline powder. Mp: 715°, bp: 1390°, d: 2.76.

TOXICITY DATA with REFERENCE
sln-smc 50 mmol/L MUTAEX 1,21,86
orl-mus LD50:3800 mg/kg 20PKA3 -,56,67
ipr-mus LD50:1625 mg/kg COREAF 256,1043,63

CONSENSUS REPORTS: Reported in EPA TSCA Inventory.

SAFETY PROFILE: Moderately toxic by ingestion and intraperitoneal routes. Mutation data reported. Reacts violently with BrF_3. When heated to decomposition it emits toxic fumes of Cl^-, RbCl, and Rb_2O. See also RUBIDIUM and CHLORIDES.

RPK000 CAS:13446-73-6 ***HR: 3***
RUBIDIUM DICHROMATE
mf: $Cr_2O_7Rb_2$ mw: 386.94

PROP: Crystals. D: 3.02-3.13

CONSENSUS REPORTS: Chromium and its compounds are on the Community Right-To-Know List.

OSHA PEL: CL 0.1 $mg(CrO_3)/m^3$
ACGIH TLV: TWA 0.05 $mg(Cr)/m^3$
NIOSH REL: TWA 0.025 $mg(Cr(VI))/m^3$; CL 0.05/15M

SAFETY PROFILE: Suspected carcinogen. A poison. A powerful oxidizer. When heated to decomposition it emits toxic fumes of Rb_2O. See also CHROMATES and RUBIDIUM.

RPP000 CAS:13446-74-7 ***HR: 3***
RUBIDIUM FLUORIDE
mf: FRb mw: 104.47

PROP: Colorless crystals. Mp: 775°, bp: 1410°, d: 3.557, vap press: 1 mm @ 921°.

CONSENSUS REPORTS: Reported in EPA TSCA Inventory.

OSHA PEL: TWA 2.5 $mg(F)/m^3$
NIOSH REL: TWA 2.5 $mg(F)/m^3$

SAFETY PROFILE: Poison as a soluble fluoride. When heated to decomposition it emits toxic fumes of Rb_2O and F^-. See also RUBIDIUM and FLUORIDES.

RPU000 CAS:13446-75-8 ***HR: 3***
RUBIDIUM HYDRIDE
mf: HRb mw: 86.48

SAFETY PROFILE: Violent reaction with water. Ignites on contact with moist air or oxygen. Vigorous reaction with acetylene. When heated to decomposition it emits toxic fumes of Rb_2O. See also RUBIDIUM HYDROXIDE, RUBIDIUM, and HYDRIDES.

RPZ000 CAS:1310-82-3 ***HR: 2***
RUBIDIUM HYDROXIDE
DOT: UN 2677/UN 2678
mf: HORb mw: 102.48

PROP: Grayish-white, deliqu mass; strong base. Mp: 300°, d: 3.203 @ 11°.

SYN: RUBIDIUM HYDROXIDE, solid and solution (DOT)

TOXICITY DATA with REFERENCE
orl-rat LD50:586 mg/kg TXAPA9 32,239,75
orl-mus LD50:900 mg/kg 20PKA3 -,56,67

CONSENSUS REPORTS: Reported in EPA TSCA Inventory.

OSHA PEL: TWA 2.5 $mg(F)/m^3$

DOT Classification: Corrosive Material; Label: Corrosive.

SAFETY PROFILE: Moderately toxic by ingestion. A powerful, corrosive irritant to skin, eyes, and mucous membranes. When heated to decomposition it emits toxic fumes of Rb_2O. See also POTASSIUM HYDROXIDE and RUBIDIUM.

RQA000 CAS:7790-29-6 ***HR: 1***
RUBIDIUM IODIDE
mf: IRb mw: 212.37

PROP: Colorless crystals. Mp: 642°, bp: 1300°, d: 3.55, d (liquid): 2.87 @ 825°, vap press: 1 mm @ 748°. Sol in alc; very sol in water.

TOXICITY DATA with REFERENCE
orl-rat LD50:4708 mg/kg NIOSH* TR-74,1,72

CONSENSUS REPORTS: Reported in EPA TSCA Inventory.

SAFETY PROFILE: Mildly toxic by ingestion. When heated to decomposition it emits toxic fumes of Rb_2O and I^-. See also IODIDES and RUBIDIUM.

RQF000 CAS:12136-85-5 ***HR: 3***
RUBIDIUM NITRIDE
mf: NRb mw: 99.47

SAFETY PROFILE: Alkali nitrides burn in air. When heated to decomposition it emits very toxic fumes of Rb_2O and NO_x. See also RUBIDIUM and NITRIDES.

RQF350 CAS:11016-71-0 ***HR: 3***
RUBIFLAVIN

SYNS: B 17476 ◇ NSC 105023

TOXICITY DATA with REFERENCE
dnd-esc 10 mg/L MUREAV 89,95,81
pic-esc 1 ng/plate CNREA8 43,2819,83
ipr-mus LD50:15 mg/kg 85ERAY 2,1438,78
scu-mus LD50:14310 μg/kg NCISP* JAN86

SAFETY PROFILE: Poison by subcutaneous and intraperitoneal routes. Mutation data reported.

RQK000 CAS:11016-72-1 ***HR: 3***
RUBOMYCIN

TOXICITY DATA with REFERENCE
pic-esc 60 mg/L ZAPOAK 8,139,68
cyt-hmn:lym 40 μg/L CYGEDX 15(2),74,81
cyt-mus:oth 9 nmol/L IPPABX 20,1,84
ivn-rat LD50:18 mg/kg ANTBAL 28,298,83
orl-mus LD50:24400 μg/kg ANTBAL 11,126,66
ipr-mus LD50:5400 μg/kg ANTBAL 20,897,75
scu-mus LD50:5400 μg/kg ANTBAL 11,126,66
ivn-mus LD50:935 μg/kg ANTBAL 11,126,66
ivn-rbt LDLo:100 μg/kg ANTBAL 11,126,66

SAFETY PROFILE: Poison by ingestion, intraperitoneal, subcutaneous, and intravenous routes. Human mutation data reported.

RQP000 CAS:21794-01-4 ***HR: 3***
RUBRATOXIN B
mf: $C_{26}H_{30}O_{11}$ mw: 518.56

PROP: Is produced by isolates of *Penicillium rubrum* and *P. purpurogenum* (BECTA6 10,200,73).

SYN: 10-((3,6-DIHYDRO-6-OXO-2H-PYRAN-2-YL)HYDROXYMETHYL)-5,9,10,11-tert-TRAHYDRO-4-HYDROXY-5-(1-HYDROXYHEPTYL)-1H-CYCLONONA(1,2-C :5,6-C')DIFURAN-1,3,6,8(4H)-TETRONE

TOXICITY DATA with REFERENCE
dni-hmn:hla 32 mg/L JJEMAG 40,409,70
oms-hmn:hla 32 mg/L JJEMAG 40,409,70
orl-mus TDLo:100 mg/kg (female 7D post):TER 85EGD4 -,739,78
ipr-mus TDLo:400 μg/kg (female 8D post):REP 85EGD4 -,739,78
orl-rat LD50:400 mg/kg TXAPA9 19,712,71
ipr-rat LD50:350 μg/kg TXAPA9 19,712,71
ipr-mus LD50:270 μg/kg TXAPA9 19,712,71
scu-mus LD50:6800 mg/kg 85GDA2 6,135,81
ipr-dog LDLo:500 μg/kg JEPTDQ 1(1),59,77
ipr-cat LD50:200 mg/kg TXAPA9 19,712,71
ipr-gpg LD50:480 mg/kg TXAPA9 19,712,71

CONSENSUS REPORTS: EPA Genetic Toxicology Program.

SAFETY PROFILE: A deadly poison by ingestion and intraperitoneal routes. An experimental teratogen. Experimental reproductive effects. Human mutation data reported. When heated to decomposition it emits acrid smoke and irritating fumes.

RQU300 ***HR: 2***
RUBUS ELLIPTICUS Smith, extract excluding roots

PROP: Indian plant belonging to the family *Rosaceae* IJEBA6 9,91,71

SYNS: HINSALU ◇ ZARDANCHU

TOXICITY DATA with REFERENCE
orl-rat TDLo:350 mg/kg (female 1-7D post):REP IJCREE 21,183,83
unr-rat LD50:1 g/kg IJCREE 21,183,83
ipr-mus LD50:500 mg/kg IJEBA6 9,91,71

SAFETY PROFILE: Moderately toxic by intraperitoneal route. Experimental reproductive effects. When heated to decomposition it emits acrid smoke and irritating fumes.

RQU650 ***HR: 2***
RUDBECKIA BICOLOR nutt., extract

PROP: Indian plant belonging to the family *Compositae* (IJEBA6 18,594,80).

TOXICITY DATA with REFERENCE
orl-rat TDLo:150 mg/kg (12-14D preg):REP IJEBA6 18,594,80
ipr-mus LD50:1 g/kg IJEBA6 18,594,80

SAFETY PROFILE: Moderately toxic by intraperitoneal route. Experimental reproductive effects. When heated to decomposition it emits acrid smoke and irritating fumes.

RRA000 CAS:23537-16-8 ***HR: 3***
RUGULOSIN
mf: $C_{30}H_{20}O_{10}$ mw: 540.50

PROP: Anthraquinoid hepatotoxin of *Penicillium rugulosum Thom* (JJEMAG 41,177,71).

SYNS: RADICALISIN ◇ (+)-RUGULOSIN

TOXICITY DATA with REFERENCE
mmo-sat 10 μg/7H JEPTDQ 2(2),313,78
mrc-bcs 20 μg/disc CNREA8 36,445,76
orl-mus TDLo:4400 mg/kg/1Y-C:ETA ARMIAZ 26,279,72
ipr-rat LD50:44 mg/kg JJEMAG 41,177,71
ipr-mus LD50:55 mg/kg JJEMAG 41,177,71

CONSENSUS REPORTS: IARC Cancer Review: Group 3 IMEMDT 7,56,87; Animal Inadequate Evidence IMEMDT 40,99,86. EPA Genetic Toxicology Program.

SAFETY PROFILE: Poison by intraperitoneal route. Questionable carcinogen with experimental tumorigenic data. Mutation data reported. When heated to decomposition it emits acrid smoke and irritating fumes.

RRK000 ***HR: 3***
RUSSIAN COMFREY LEAVES

PROP: Fresh leaves dried, milled, and mixed with diet (JNCIAM 61,865,78).

SYNS: COMFREY, RUSSIAN ◇ SYMPHYTUM OFFICINALE L

TOXICITY DATA with REFERENCE
orl-rat TDLo:4800 g/kg/86W-C:CAR JJIND8 61(3),865,78
orl-rat TD:9900 g/kg/86W-C:CAR JJIND8 61(3),865,78

CONSENSUS REPORTS: IARC Cancer Review: Animal Limited Evidence IMEMDT 31,239,83

SAFETY PROFILE: Questionable carcinogen with experimental carcinogenic data. When heated to decomposition it emits acrid smoke and irritating fumes.

RRP000 ***HR: 3***
RUSSIAN COMFREY ROOTS

PROP: Fresh roots dried, milled and, mixed with diet (JNCIAM 61,86578).

SYNS: COMFREY, RUSSIAN ◇ SYMPHYTUM OFFICINALE L

TOXICITY DATA with REFERENCE
orl-rat TDLo:140 g/kg/43W-I:CAR JJIND8 61(3),865,78
orl-rat TD:91 g/kg/52W-C:CAR JJIND8 61(3),865,78

CONSENSUS REPORTS: IARC Cancer Review: Animal Limited Evidence IMEMDT 31,239,83

SAFETY PROFILE: Questionable carcinogen with experimental carcinogenic data. When heated to decomposition it emits acrid smoke and irritating fumes.

RRP675 ***HR: D***
RUTA GRAVEOLENS, extract

TOXICITY DATA with REFERENCE
mmo-sat 10 uL/plate MUTAEX 2,271,87
mma-sat 20 uL/plate MUTAEX 2,271,87
orl-rat TDLo:8 g/kg (female 1-10D post):REP PLMEAA 55,176,89

SAFETY PROFILE: Experimental reproductive effects. Mutation data reported.

RRU000 CAS:7440-18-8 ***HR: 3***
RUTHENIUM
af: Ru aw: 101.07

PROP: Lustrous, hard metal, hexagonal crystals. D: 12.45 @ 20°/4°, mp: approx 2450°, bp: approx 4150°. Stable in air.

SAFETY PROFILE: Most ruthenium compounds are poisons. Ruthenium is retained in the bones for a long time. Flammable in the form of dust when exposed to heat or flame. Violent reaction with ruthenium oxide. Explosive reaction with aqua regia + potassium chlorate. When heated to decomposition it emits very toxic fumes of RuO_x and Ru which are highly injurious to the eyes and lung and can produce nasal ulcerations. See also RUTHENIUM COMPOUNDS.

RRZ000 CAS:10049-08-8 ***HR: 3***
RUTHENIUM CHLORIDE
mf: Cl_3Ru mw: 207.42

PROP: α Form: Black lustrous crystals. Insol in alc, water. β Form: Dark brown, fluffy, hexagonal crystals. Sol in alc.

SYN: RUTHENIUM TRICHLORIDE

TOXICITY DATA with REFERENCE
ipr-rat LD50:360 mg/kg EQSSDX 1,1,75
ipr-mus LD50:108 mg/kg TXAPA9 48,A112,79

CONSENSUS REPORTS: EPA Genetic Toxicology Program. Reported in EPA TSCA Inventory.

SAFETY PROFILE: Poison by intraperitoneal route. Incompatible with penta carbonyl iron, zinc. When heated to decomposition it emits toxic fumes of RuO_x and Cl^-. See also RUTHENIUM COMPOUNDS.

RSA000 CAS:16845-29-7 ***HR: 3***
RUTHENIUM CHLORIDE HYDROXIDE
mf: Cl_3HORu mw: 224.43

SYNS: RUTHENIUMHYDROXIDE TRICHLORIDE ◇ RUTHENIUM HYDROXYCHLORIDE ◇ RUTHENIUM TRICHLORIDE HYDROXIDE

TOXICITY DATA with REFERENCE
orl-rat LD50:1250 mg/kg GTPZAB 23(6),54,79
ihl-rat LCLo:9500 μg/m^3 41HTAH -,36,78
orl-mus LD50:463 mg/kg GTPZAB 23(6),54,79
ipr-mus LD50:225 mg/kg GTPZAB 23(6),54,79
orl-gpg LD50:210 mg/kg GTPZAB 23(6),54,79

SAFETY PROFILE: Poison by ingestion, inhalation, and intraperitoneal routes. When heated to decomposition it emits toxic fumes of RuO_x and Cl^-.

RSF000 ***HR: 3***
RUTHENIUM COMPOUNDS

SAFETY PROFILE: Most ruthenium compounds are poisons or moderately toxic. Ruthenium red is an antagonist of Ca^{2-}, inhibits Ca^{2-} transport and binding in mitochondrial membranes, and inhibits Ca^{2-}-ATPase activity. They resemble osmium compounds in that when heated in air, they evolve fumes which are injurious to the eyes and lungs and can produce nasal ulcerations. When heated to decomposition they emit toxic fumes of RuO_x and Ru. See also RUTHENIUM and specific compounds.

RSF875 CAS:12036-10-1 ***HR: 3***
RUTHENIUM OXIDE
mf: O_2Ru mw: 133.07

TOXICITY DATA with REFERENCE
orl-rat LD50:4580 mg/kg GTPZAB 25(1),46,81

ihl-rat LCLo:34 mg/m^3 41HTAH -,36,78
orl-mus LD50:5570 mg/kg GTPZAB 25(1),46,81
ipr-mus LD50:3050 mg/kg GTPZAB 25(1),46,81

CONSENSUS REPORTS: EPA Genetic Toxicology Program.

SAFETY PROFILE: Poison by inhalation. Moderately toxic by intraperitoneal route. Mildly toxic by ingestion. When heated to decomposition it emits toxic fumes of RuO_x. See also RUTHENIUM and RUTHENIUM COMPOUNDS.

RSK000 CAS:20427-56-9 ***HR: 3***
RUTHENIUM(VIII) OXIDE
mf: O_4Ru mw: 165.07

PROP: Golden yellow, monoclinic prisms. D: 3.29 @ 20°, mp: 25.4°, bp: 40°. Very volatile, subl at room temp. Sltly sol in water; very sol in carbon tetrachloride, chlorinated solvents; sol in bromine, liquid SO_2.

SYN: RUTHENIUM TETRAOXIDE

SAFETY PROFILE: A poison. Fumes are highly injurious to the eyes and mucous membranes. Handle in hood only. Flammable by chemical reaction with reducing agents. A powerful oxidizing agent. The liquid explodes above 106°C. Explosive reaction with hydriodic acid, charcoal, ethanol, cellulose fibers, other organic materials, sulfur. Violent reaction or ignition with ammonia. Vigorous reaction with phosphorus tribromide. When heated to decomposition it emits toxic fumes of RuO_x. See also RUTHENIUM COMPOUNDS.

RSP000 ***HR: 3***
RUTHENIUM SALT of TETRAMETHYLPHENANTHRENE

TOXICITY DATA with REFERENCE
scu-mus TDLo:1000 mg/kg/I:ETA BECCAN 40,30,62

SAFETY PROFILE: Questionable carcinogen with experimental tumorigenic data. When heated to decomposition it emits very toxic fumes of NO_x and RuO_x. See also RUTHENIUM COMPOUNDS.

RSU000 CAS:153-18-4 ***HR: 3***
RUTIN
mf: $C_{27}H_{30}O_{16}$ mw: 610.57

PROP: Pale yellow needles. Sol in pyridine, formamide, and alkaline solns; sltly sol in alc, acetone, ethyl acetate; insol in chloroform, carbon bisulfide, ether, benzene.

SYNS: BIOFLAVONOID ◇ BIRUTAN ◇ C.I. 75730 ◇ ELDRIN ◇ GLOBULARIACITRIN ◇ ILIXATHIN ◇ MELIN ◇ MYRITICALORIN ◇ MYRITICOALORIN ◇ OSYRITRIN ◇ OXYRITIN ◇ PALIUROSIDE ◇ 3,3',4',5,7-PENTAHYDROXYFLAVONE-3-(o-RHAMNOSYLGLUCOSIDE) ◇ 3,3',4',5,7-PENTAHYDROXYFLAVONE-3-RUTINOSIDE ◇ PHYTOMELIN ◇ QUERCETIN-3-(6-o-(6-DEOXY-α-l-MANNOPYRANOSYL-β-d-GLUCOPYRANOSIDE) ◇ QUERCETIN RHAMNOGLUCOSIDE ◇ QUERCETIN-3-RHAMNOGLUCOSINE ◇ QUERCETIN-3-(6-o-α-l-RHAMNOPYRANOSYL-β-d-GLUCOPYRANOSIDE) ◇ QUERCETIN-3-RUTINOSIDE ◇ RUTINIC ACID ◇ RUTOSIDE ◇ SOPHORIN ◇ TANRUTIN ◇ USAF CF-5 ◇ VIOLAQUERCITRIN ◇ VITAMIN P

TOXICITY DATA with REFERENCE
mma-sat 2 mg/plate MUREAV 66,223,79
mma-sat 80 μg/plate FCTOD7 23,669,85
dnr-esc 100 mg/L FCTXAV 18,223,83
slt-dmg-unr 71300 ppm/48H MUREAV 120,233,83
orl-rat TDLo:973 g/kg/3Y-C:NEO PAACA3 25,95,84
ipr-rat LD50:2000 mg/kg EKMMA8 19,207,80
ipr-mus LD50:200 mg/kg NTIS** AD277-689
ivn-mus LD50:950 mg/kg JAPMA8 39,556,50
ipr-gpg LD50:2000 mg/kg EKMMA8 19,207,80

CONSENSUS REPORTS: Reported in EPA TSCA Inventory.

SAFETY PROFILE: Poison by intraperitoneal route. Moderately toxic by intravenous route. Questionable carcinogen with experimental neoplastigenic data. Mutation data reported. Used as a pharmaceutical and veterinary drug. When heated to decomposition it emits acrid smoke and irritating fumes.

RSU450 CAS:20228-27-7 ***HR: 3***
RUVAZONE
mf: $C_{12}H_{14}N_2O_4$ mw: 250.28

SYNS: o-ETHOXYBENZOIC ACID (1-CARBOXYETHYLIDENE) HYDRAZIDE ◇ o-ETHOXY-BENZOYL-HYDRAZONE of PYRUVIC ACID ◇ o-ETHOXY-BENZOIL-IDRAZONE DELL'ACIDO PIRUVICO (ITALIAN) ◇ M 6/42

TOXICITY DATA with REFERENCE
orl-rat LD50:507 mg/kg BCFAAI 107,769,68
ipr-rat LD50:157 mg/kg BCFAAI 107,769,68
scu-rat LD50:600 mg/kg BCFAAI 107,769,68
ims-rat LD50:206 mg/kg BCFAAI 107,769,68
orl-mus LD50:324 mg/kg BCFAAI 107,769,68
ipr-mus LD50:166 mg/kg BCFAAI 107,769,68
scu-mus LD50:207 mg/kg BCFAAI 107,769,68
ims-mus LD50:185 mg/kg BCFAAI 107,769,68

SAFETY PROFILE: Poison by ingestion, subcutaneous, intramuscular, and intraperitoneal routes. When heated to decomposition it emits toxic fumes of NO_x.

RSZ000 CAS:15662-33-6 ***HR: 3***
RYANIA
mf: $C_{25}H_{35}NO_9$ mw: 493.61

PROP: The powdered stem of *Ryania speciosa,* of proven insecticidal activity (JPETAB 93,407,48).

SYNS: BONIDE RYATOX ◇ GROUND RYANIA SPECISA(VAHL) STEMWOOD (ALKOLOID RYANODINE) ◇ RYANEXEL ◇ RYANIA POWDER ◇ RYANIA SPECIOSA ◇ RYANICIDE ◇ RYANODINE

TOXICITY DATA with REFERENCE
orl-hmn LDLo:143 mg/kg:CNS,GIT,PUL 34ZIAG -,522,69
orl-rat LD50:750 mg/kg WRPCA2 9,119,70
skn-rat LD50:750 mg/kg SPEADM 78-1,6,78
orl-mus LD50:650 mg/kg JPETAB 93,407,48
orl-dog LD50:150 mg/kg JPETAB 93,407,48
orl-rbt LD50:650 mg/kg GUCHAZ 6,450,73
orl-gpg LD50:2500 mg/kg JPETAB 93,407,48
orl-pgn LD50:2310 μg/kg ASTTA8 (680),157,79
orl-qal LD50:13300 μg/kg ASTTA8 (680),157,79
orl-bwd LD50:1780 μg/kg ASTTA8 (680),157,79

SAFETY PROFILE: Human poison by ingestion. Experimental poison by ingestion. Moderately toxic experimentally by skin contact. Human systemic effects by ingestion: weakness, respiratory changes, diarrhea, gastrointestinal disturbances, tremors, convulsions, coma and death. Used as an insecticide. Flammable when exposed to heat or flame. To fight fire, use CO_2, mist, spray, foam. When heated to decomposition it emits toxic fumes of NO_x.

RSZ375 CAS:71653-63-9 ***HR: 3***
RYODIPINE
mf: $C_{18}H_{19}F_2NO_5$ mw: 367.38

SYNS: 1,4-DIHYDRO-4-(2-(DIFLUOROMETHOXY)PHENYL)-2,6-DIMETHYL-3,5-PYRIDINEDICARBOXYLIC ACID DIMETHYL ESTER ◇ 2,6-DIMETHYL-3,5-DIMETHOXYCARBONYL-4-(o-DIFLUOROMETHOXYPHENYL)-1,4-DIHYDROPYRIDINE ◇ PP-1466

TOXICITY DATA with REFERENCE
orl-rat LD50:11 g/kg ARZNAD 35,672,85
orl-mus LD50:721 mg/kg ARZNAD 35,672,85
ipr-mus LD50:395 mg/kg PCJOAU 16,817,82
orl-dog LD50:1198 mg/kg ARZNAD 35,915,85
orl-rbt LD50:209 mg/kg ARZNAD 35,915,85

SAFETY PROFILE: Poison by ingestion and intraperitoneal routes. When heated to decomposition it emits toxic fumes of F^- and NO_x.

RSZ600 CAS:22059-60-5 ***HR: 3***
RYTHMODAN
mf: $C_{21}H_{29}N_3O•H_3O_4P$ mw: 437.53

SYNS: α-(2-BIS(1-METHYLETHYL)AMINO)ETHYL)-α-PHENYL-2-PYRIDINEACETAMIDE PHOSPHATE (9CI) ◇ α-(2-DIISOPROPYLAMINOETHYL)-α-PHENYL-2-PYRIDINEACETAMIDEPHOSPHATE ◇ DIISOPYRAMIDE PHOSPHATE ◇ NORPACE ◇ SC 13957 ◇ SC 7031 PHOSPHATE

TOXICITY DATA with REFERENCE
ivn-rat TDLo:330 mg/kg (female 7-17D post):REP JZKEDZ 7,157,81
ipr-rat TDLo:1820 mg/kg (male 10W pre):TER JZKEDZ 7,145,81
orl-wmn LDLo:60 mg/kg/10D-I:SYS NYSJAM 83,1057,83
orl-man TDLo:43 mg/kg/5D-I:SYS NYSJAM 83,1057,83
orl-wmn TDLo:8 mg/kg/1D-I:LIV,BLD SMJOAV 75,496,82
orl-man TDLo:1429 μg/kg:SYS NNGAAS 72,1177,83
orl-chd LDLo:30 mg/kg MJAUAJ 2,335,78
orl-man TDLo:45 mg/kg/1W-I SMJOAV 76,1453,83
orl-wmn TDLo:20 mg/kg/3D-I:CVS,BLD AHJOA2 105,870,83
orl-rat LD50:880 mg/kg YACHDS 9(Suppl 1),5,81
ipr-rat LD50:255 mg/kg YACHDS 9(Suppl 1),5,81
scu-rat LD50:1000 mg/kg YACHDS 9(Suppl 1),5,81
ivn-rat LD50:88 mg/kg YACHDS 9(Suppl 1),5,81
orl-mus LD50:820 mg/kg YACHDS 9(Suppl 1),5,81
ipr-mus LD50:190 mg/kg YACHDS 9(Suppl 1),5,81
scu-mus LD50:680 mg/kg YACHDS 9(Suppl 1),5,81
ivn-mus LD50:81 mg/kg YACHDS 9(Suppl 1),5,81

SAFETY PROFILE: Human poison by ingestion. Experimental poison by intravenous and intraperitoneal routes. Moderately toxic experimentally by ingestion and subcutaneous routes. Human systemic effects: arrhythmias, blood clotting factor change, hypoglycemia, liver function tests impaired, pulse rate increase, thrombocytopenia. An experimental teratogen. Experimental reproductive effects. When heated to decomposition it emits toxic fumes of NO_x and PO_x.

RSZ675 ***HR: D***
RYUTAN (JAPANESE)

PROP: Crude drug extract from *Gentiana Scabra* (MUREAV 97,81,82).

SYN: GENTIANAE SCABRAE RADIX (LATIN)

TOXICITY DATA with REFERENCE
mmo-sat 10 mg/plate MUREAV 97,81,82
dnr-bcs 100 g/L MUREAV 97,81,82

SAFETY PROFILE: Mutation data reported.

S

SAA000 ***HR: 3***
S 151
mf: $C_{20}H_{18}NO_2 \cdot Cl$ mw: 339.82

SYNS: 1-BENZYL-3-CARBOXYPYRIDINIUM CHLORIDE BENZYL ESTER ◇ CHLORURE de 1-BENZYL-3-BENZYL-CARBOXY-PYRIDINIUM (FRENCH)

TOXICITY DATA with REFERENCE
ivn-mus LD50:229 mg/kg THERAP 13,508,58
ivn-dog LDLo:4 mg/kg THERAP 13,508,58
ivn-cat LDLo:10 mg/kg THERAP 13,508,58

SAFETY PROFILE: Poison by intravenous route. When heated to decomposition it emits toxic fumes of NO_x and Cl^-.

SAA040 CAS:72150-17-5 ***HR: 3***
1-ST-2121
mf: $C_{15}H_{21}NO \cdot BrH$ mw: 312.29

SYNS: (−)-1,4-DIMETHYL-10-HYDROXY-2,3,4,5,6,7-HEXAHYDRO-1,6-METHANO-1H-4-BENZAZONINE HYDROBROMIDE ◇ 1-1,4-DIMETHYL-10-HYDROXY-2,3,4,5,6,7-HEXAHYDRO-1,6-METHANO-1H-4-BENZAZONINE HYDROBROMIDE ◇ (−)-2,3,4,5,6,7-HEXAHYDRO-1,4-DIMETHYL-1,6-METHANO-1H-4-BENZAZONIN-10-OLHYDROBROMIDE

TOXICITY DATA with REFERENCE
scu-mus TDLo:150 mg/kg (female 6-15D post):REP OYYAA2 20,511,80
scu-mus TDLo:750 mg/kg (female 6-15D post):TER OYYAA2 20,511,80
orl-rat LD50:320 mg/kg OYYAA2 20,299,80
ipr-rat LD50:96 mg/kg OYYAA2 20,299,80
scu-rat LD50:540 mg/kg OYYAA2 20,299,80
ivn-rat LD50:64 mg/kg OYYAA2 20,299,80
orl-mus LD50:310 mg/kg OYYAA2 20,299,80
ipr-mus LD50:102 mg/kg OYYAA2 20,299,80
scu-mus LD50:124 mg/kg OYYAA2 20,299,80
ims-mus LD50:135 mg/kg OYYAA2 20,299,80

SAFETY PROFILE: Poison by ingestion, subcutaneous, intramuscular, intravenous, and intraperitoneal routes. An experimental teratogen. Experimental reproductive effects. Used as an analgesic. When heated to decomposition it emits toxic fumes of NO_x and Br^-.

SAB800 CAS:40225-02-3 ***HR: 3***
SAD-128
mf: $C_{20}H_{30}N_2O \cdot 2Cl$ mw: 385.42

SYNS: BIS-(4-tert-BUTYLPYRIDINE)-1-METHYLETHERDICHLORIDE ◇ 1,1'-(OXYBIS(METHYLENE))BIS(4-(1,1-DIMETHYLETHYL)-PYRIDINIUM, DICHLORIDE (9CI) ◇ 1,1'-OXYDIMETHYLENE BIS(4-tert-BUTYLPYRIDINIUM CHLORIDE)

TOXICITY DATA with REFERENCE
ipr-mus LD50:49100 μg/kg FAATDF 1,193,81
ims-mus LD50:106 mg/kg ATXKA8 26,293,70
ivn-rbt LD50:12 mg/kg BCPCA6 27,757,78

SAFETY PROFILE: Poison by intramuscular, intravenous, and intraperitoneal routes. When heated to decomposition it emits toxic fumes of Cl^- and NO_x.

SAC000 CAS:8001-23-8 ***HR: 1***
SAFFLOWER OIL

PROP: From *Carthanus tinctorius*, consists of triglycerides of linoleic acid (85DIA2 2,287,77). Light yellow oil. D: 0.9211 @ 25°/25°. Sol in oil and fat solvents.

SYN: SAFFLOWER OIL (UNHYDROGENATED) (FCC)

TOXICITY DATA with REFERENCE
skn-hmn 300 mg/3D-I MLD 85DKA8 -,127,77

CONSENSUS REPORTS: Reported in EPA TSCA Inventory.

SAFETY PROFILE: A human skin irritant. Ingestion of large doses can cause vomiting. When heated to decomposition it emits acrid smoke and irritating fumes.

SAC875 CAS:33419-68-0 ***HR: D***
SAFRASIN
mf: $C_{11}H_{16}N_2O_2$ mw: 208.29

SYN: SAPHRAZINE

TOXICITY DATA with REFERENCE
sln-dmg-orl 2 mg/8D SOGEBZ 11,718,75

SAFETY PROFILE: Mutation data reported. When heated to decomposition it emits toxic fumes of NO_x.

SAD000 CAS:94-59-7 ***HR: 3***
SAFROL
mf: $C_{10}H_{10}O_2$ mw: 162.20

PROP: Colorless liquid or crystals; sassafras odor. Mp: 11°, bp: 234.5°, d: 1.0960 @ 20°, vap press: 1 mm @ 63.8°. Insol in water; very sol in alc; misc with chloroform, ether.

SYNS: 5-ALLYL-1,3-BENZODIOXOLE ◇ ALLYLCATECHOL METHYLENE ETHER ◇ ALLYLDIOXYBENZENE METHYLENE ETHER ◇ 1-

ALLYL-3,4-METHYLENEDIOXYBENZENE ◇ 4-ALLYL-1,2-METHYLENEDIOXYBENZENE ◇ m-ALLYLPYROCATECHIN METHYLENE ETHER ◇ 4-ALLYLPYROCATECHOL FORMALDEHYDE ACETAL ◇ ALLYLPYROCATECHOL METHYLENE ETHER ◇ 1,2-METHYLENEDIOXY-4-ALLYLBENZENE ◇ 3,4-METHYLENEDIOXY-ALLYBENZENE ◇ 5-(2-PROPENYL)-1,3-BENZODIOXOLE ◇ RCA WASTE NUMBER U203 ◇ RHYUNO OIL ◇ SAFROLE ◇ SAFROLE MF ◇ SHIKIMOLE ◇ SHIKOMOL

TOXICITY DATA with REFERENCE
skn-rbt 500 mg/24H MOD FCTXAV 12,983,74
dns:hmn:hla 10 μL/L PMRSDJ 5,347,85
otr-mus:emb 100 mg/L PMRSDJ 5,659,85
ipr-mus TDLo:1 g/kg (5D male):REP PMRSDJ 1,712,81
orl-rat TDLo:200 g/kg/94W-C:CAR CNREA8 37,1883,77
orl-mus LDLo:22 g/kg/90W-I:CAR CNREA8 39,4378,79
orl-mus TDLo:480 mg/kg (12-18D post):NEO CNREA8 39,4378,79
orl-rat LD50:1950 mg/kg TXAPA9 7,18,65
orl-mus LD50:2350 mg/kg FCTXAV 2,327,64
scu-mus LD50:1020 mg/kg SIZSAR 3,73,52
scu-cat LDLo:500 mg/kg AEXPBL 35,342,1895
orl-rbt LDLo:1000 mg/kg AEXPBL 35,342,1895
scu-rbt LDLo:1000 mg/kg AEXPBL 35,342,1895
ivn-rbt LDLo:200 mg/kg AEXPBL 35,342,1895

CONSENSUS REPORTS: NTP Fifth Annual Report on Carcinogens. IARC Cancer Review: Group 3 IMEMDT 7,56,87; Animal Sufficient Evidence IMEMDT 10,231,76, IMEMDT 1,169,72. Community Right-To-Know List. EPA Genetic Toxicology Program. Reported in EPA TSCA Inventory.

SAFETY PROFILE: Confirmed carcinogen with experimental carcinogenic and neoplastigenic data. Poison by intravenous routed. Moderately toxic by ingestion and subcutaneous routes. Experimental reproductive effects. Human mutation data reported. A skin irritant. Combustible when exposed to heat or flame. When heated to decomposition it emits acrid smoke and irritating fumes. See also ALDEHYDES and ALLYL COMPOUNDS.

SAD100 CAS:77491-30-6 ***HR: D***
SAFROTIN
mf: $C_{10}H_{20}NO_4PS \cdot C_4H_7Cl_2O_4P$ mw: 502.32

SYN: 2-BUTENOIC ACID, 3-(((ETHYLAMINO)METHOXYPHOSPHINOTHIOYL)OXY)-, 1-METHYLETHYL ESTER, (E)-, mixt. with 2,2-DICHLOROETHENYL DIMETHYL PHOSPHATE

TOXICITY DATA with REFERENCE
mnt-mus-orl 80 mg/kg ENVRAL 41,44,86
spm-mus-orl 80 mg/kg ENVRAL 41,44,86
orl-mus TDLo:80 mg/kg (male 5D pre):REP ENVRAL 41,44,86

SAFETY PROFILE: Experimental reproductive effects. Mutation data reported. When heated to decomposition it emits toxic fumes of NO_x, PO_x, SO_x, and Cl^-.

SAE500 CAS:8022-56-8 ***HR: 2***
SAGE OIL, DALMATIAN TYPE

PROP: Main constituent is thujone. From steam distillation of leaves from *Salvia officinalis* l. (FCTXAV 12,807,74). Yellow liquid; thujone odor and taste. D: 0.903-0.925, refr index: 1.457 @ 20°. Sol in fixed oils, mineral oil; sltly sol in propylene glycol; insol in glycerin.

SYNS: DALMATIAN SAGE OIL ◇ SAGE OIL ◇ SALBEI OEL (GERMAN)

TOXICITY DATA with REFERENCE
skn-hmn 100% FCTXAV 12,987,74
skn-rbt 500 mg/24H MOD FCTXAV 12,987,74
dnr-bcs 10 mg/disc TOFOD5 8,91,85
orl-rat LD50:2600 mg/kg FCTXAV 12,987,74

CONSENSUS REPORTS: Reported in EPA TSCA Inventory.

SAFETY PROFILE: Moderately toxic by ingestion. Mutation data reported. A human skin irritant. When heated to decomposition it emits acrid smoke and irritating fumes. See also THUJONE.

SAE550 CAS:8022-56-8 ***HR: 2***
SAGE OIL, SPANISH TYPE

PROP: From steam distillation of plants from *Salvia lavandulaefolia* Vahl. or *Salvia hispanorium* Lag. (Fam. *Labiatae*) (FCTXAV 12,807,74). Colorless to yellow oil. D: 0.909-0.932, refr index: 1.468 @ 20°. Sol in fixed oils, glycerin, mineral oil, propylene glycol.

SYNS: SAGE OIL ◇ SALBEI OEL (GERMAN)

TOXICITY DATA with REFERENCE
orl-rat LD50:2600 mg/kg PHARAT 14,435,59

CONSENSUS REPORTS: Reported in EPA TSCA Inventory.

SAFETY PROFILE: Moderately toxic by ingestion. When heated to decomposition it emits acrid smoke and irritating fumes.

SAF000 CAS:89997-47-7 ***HR: 3***
SAGRADO

SYNS: CHENOPODIUM AMBROSIOIDES ◇ JERUSALEM OAK ◇ WORMWOOD PLANT

TOXICITY DATA with REFERENCE
scu-rat TDLo:2320 mg/kg/58W-I:NEO JNCIAM 60,683,78

SAFETY PROFILE: Questionable carcinogen with experimental neoplastigenic data. When heated to decomposition it emits acrid smoke and irritating fumes.

SAF300 CAS:89958-12-3 ***HR: 3***
SAIKOSIDE, CRUDE

SYNS: BUPLEURUM FALCATUM L ◇ CRUDE SAIKOSIDE

TOXICITY DATA with REFERENCE
orl-mus LD50:4700 mg/kg YKKZAJ 89,712,69
ipr-mus LD50:112 mg/kg YKKZAJ 89,712,69
scu-mus LD50:175 mg/kg YKKZAJ 89,712,69
ivn-mus LD50:70 mg/kg YKKZAJ 89,712,69
ipr-gpg LD50:58 mg/kg YKKZAJ 89,712,69

SAFETY PROFILE: Poison by subcutaneous, intravenous, and intraperitoneal routes. Mildly toxic by ingestion.

SAF400 CAS:51022-70-9 ***HR: 3***
SALBUTAMOL SULFATE
mf: $C_{13}H_{21}NO_3 \cdot 1/2H_2O_4S$ mw: 288.35

SYN: SALBUTAMOL HEMISULFATE

TOXICITY DATA with REFERENCE
orl-rat TDLo:1 g/kg (1D pre):REP OYYAA2 22,191,81
ipr-rat LD50:295 mg/kg IYKEDH 2,128,71
ivn-rat LD50:59100 μg/kg IYKEDH 2,128,71
orl-mus LD50:3800 mg/kg JTSCDR 6,301,81
ipr-mus LD50:239 mg/kg IYKEDH 2,128,71
scu-mus LD50:737 mg/kg IYKEDH 2,128,71
ivn-mus LD50:48700 μg/kg IYKEDH 2,128,71

SAFETY PROFILE: Poison by intravenous and intraperitoneal routes. Moderately toxic by ingestion and subcutaneous routes. Experimental reproductive effects. When heated to decomposition it emits toxic fumes of SO_x and NO_x.

SAF500 CAS:81295-38-7 ***HR: 3***
SALI

PROP: Aq extract from the dried leaves of the plant *Heliotropium ternatum* (JNCIAM 46,1131,71).

SYNS: HELIOTROPIUM TERNATUM ◇ H. TERNATUM

TOXICITY DATA with REFERENCE
scu-rat TDLo:300 g/kg/1Y-I:ETA JNCIAM 46,1131,71

SAFETY PROFILE: Questionable carcinogen with experimental tumorigenic data. When heated to decomposition it emits acrid smoke and irritating fumes.

SAG000 CAS:90-02-8 ***HR: 2***
SALICYLALDEHYDE
mf: $C_7H_6O_2$ mw: 122.13

PROP: Colorless, oily liquid; bitter, almond-like odor; burning taste. Mp: −7°, bp: 196-197°, d: 1.167 @ 25°/4°. Sltly sol in water; sol in alc and ether.

SYNS: o-FORMYLPHENOL ◇ 2-FORMYLPHENOL ◇ o-HYDROXYBENZALDEHYDE ◇ 2-HYDROXYBENZALDEHYDE ◇ SAH ◇ SALICYLADEHYDE ◇ SALICYLAL ◇ SALICYLIC ALDEHYDE

TOXICITY DATA with REFERENCE
skn-rbt 500 mg/24H MOD FCTXAV 17,903,79
scu-rat TDLo:400 mg/kg (11D preg):TER RCOCB8 38,209,82
scu-rat TDLo:400 mg/kg (11D preg):REP RCOCB8 38,209,82
orl-rat LD50:520 mg/kg FCTXAV 17,903,79
skn-rat LD50:600 mg/kg FCTXAV 17,903,79
scu-rat LD50:900 mg/kg FCTXAV 17,903,79
skn-rbt LD50:3000 mg/kg FCTXAV 17,903,79

CONSENSUS REPORTS: Reported in EPA TSCA Inventory.

SAFETY PROFILE: Moderately toxic by ingestion, skin contact and subcutaneous routes. An experimental teratogen. Experimental reproductive effects. A skin irritant. A fumigant. Used in perfumery. Flammable when exposed to heat or flame; can react with oxidizing materials. To fight fire, use alcohol foam, spray, mist, dry chemical. When heated to decomposition it emits acrid smoke and irritating fumes. See also ALDEHYDES.

SAG500 CAS:94-67-7 ***HR: 3***
SALICYLALDEHYDE OXIME
mf: $C_7H_7NO_2$ mw: 137.15

PROP: Colorless prisms in benzene or petro ether. Mp: 57-59°, bp: decomp. Very sltly sol in cold water; sol in alc, dil HCl, ether, benzene; insol in ligroin.

SYNS: o-HYDROXYBENZALDEHYDEOXIME ◇ SALICYLALDOXIME

TOXICITY DATA with REFERENCE
orl-rat LDLo:400 mg/kg NCNSA6 5,36,53

CONSENSUS REPORTS: Reported in EPA TSCA Inventory.

SAFETY PROFILE: Poison by ingestion. When heated to decomposition it emits toxic fumes of NO_x. See also ALDEHYDES.

SAH000 CAS:65-45-2 ***HR: 3***
SALICYLAMIDE
mf: $C_7H_7NO_2$ mw: 137.15

PROP: White to sltly pink crystals or powder; somewhat bitter taste. Mp: 140°. Sol in hot water, alc, chloroform, and ether.

SYNS: ACKET ◇ AFKO-SAL ◇ ALGAMON ◇ ALGIAMIDA ◇ AMIDOSAL ◇ AMID-SAL ◇ ANAMID ◇ BENESAL ◇ CIDAL ◇ DOLOMIDE ◇ DROPSPRIN ◇ H.P. 34 ◇ o-HYDROXYBENZAMIDE ◇ 2-HYDROXYBENZAMIDE ◇ LIQUIPRIN ◇ NOVECYL ◇ OHB ◇ ORAMID ◇ PANITHAL ◇ RASPBERIN ◇ SALAMID ◇ SALAMIDE ◇ SALICILAMIDE (ITALIAN) ◇ SALICIM ◇ SALICYLAMID

◇ SALIPUR ◇ SALIZELL ◇ SALRIN ◇ SALYMID ◇ SAM ◇ SAMID ◇ URTOSAL

TOXICITY DATA with REFERENCE
eye-rbt 100 mg/24H MOD 85JCAE -,658,86
orl-rat TDLo:7 g/kg (female 12-18D post):TER TJADAB 18,17,78
orl-rat TDLo:7 g/kg (5-11D preg):REP TJADAB 18,17,78
orl-rat LD50:980 mg/kg JAPMA8 47,479,58
ipr-rat LD50:600 mg/kg JPETAB 108,450,53
orl-mus LD50:300 mg/kg ARZNAD 5,572,55
ipr-mus LD50:180 mg/kg BCFAAI 111,293,72
ivn-mus LD50:313 mg/kg JPETAB 101,119,51
ipr-dog LDLo:1000 mg/kg JPETAB 101,119,51
ipr-cat LDLo:1000 mg/kg JPETAB 101,119,51

CONSENSUS REPORTS: Reported in EPA TSCA Inventory.

SAFETY PROFILE: Poison by ingestion, intravenous, and intraperitoneal routes. An experimental teratogen. Experimental reproductive effects. Can cause dizziness, drowsiness, nausea, vomiting, epigastric distress, allergic reactions and blood dyscrasias in average to large doses. An eye irritant. Used as an analgesic, antipyretic, and anti-inflammatory agent. When heated to decomposition it emits toxic fumes of NO_x. See also AMIDES.

SAH500 CAS:87-17-2 ***HR: 1***
SALICYLANILIDE
mf: $C_{13}H_{11}NO_2$ mw: 213.25

PROP: White, odorless crystals. Mp: 135°, bp: decomp. Sltly sol in water; very sol in alc, ether, chloroform, benzene.

SYNS: ANSADOL ◇ SHIRLAN EXTRA

CONSENSUS REPORTS: Reported in EPA TSCA Inventory.

SAFETY PROFILE: In concentrated form, it may cause irritation to skin and mucous membranes. When heated to decomposition it emits toxic fumes of NO_x. See also SALICYLIC ACID, ANILINE, and AMIDES.

SAI000 CAS:69-72-7 ***HR: 3***
SALICYLIC ACID
mf: $C_7H_6O_3$ mw: 138.13

PROP: Powder. D: 1.443 @ 20°/4°, mp: 158.3°, bp: 211° @ 20 mm ±. Sol in water, alc, ether.

SYNS: ACIDO SALICILICO (ITALIAN) ◇ o-HYDROXYBENZOIC ACID ◇ 2-HYDROXYBENZOIC ACID ◇ KERALYT ◇ ORTHO-HYDROXYBENZOIC ACID ◇ RETARDER W ◇ SA ◇ SAX

TOXICITY DATA with REFERENCE
skn-rbt 500 mg/24H MLD BIOFX* 21-3/71
eye-rbt 100 mg SEV BIOFX* 21-3/71
mmo-smc 1 mmol/L/3H MUREAV 60,291,79
dni-mus-orl 100 mg/kg MUREAV 46,305,77
orl-rat TDLo:350 mg/kg (female 8-14D post):TER SKEZAP 14,549,73
orl-rat TDLo:1050 mg/kg (female 8-14D post):REP SEIJBO 13,73,73
skn-man TDLo:57 mg/kg:EAR JAMAAP 244,660,80
orl-rat LD50:891 mg/kg BIOFX* 21-3/71
orl-mus LD50:480 mg/kg HBTXAC 5,148,59
ipr-mus LD50:300 mg/kg GNRIDX 3,675,69
scu-mus LD60:520 mg/kg AIPTAK 38,9,30
ivn-mus LD50:184 mg/kg YKKZAJ 91,550,71
orl-cat LD50:400 mg/kg HBTXAC 5,148,59
orl-rbt LDLo:1300 mg/kg NIIRDN 6,291,82
scu-rbt LDLo:6 g/kg HBAMAK 4,1392,35

CONSENSUS REPORTS: Reported in EPA TSCA Inventory. EPA Genetic Toxicology Program.

SAFETY PROFILE: Poison by ingestion, intravenous, and intraperitoneal routes. Moderately toxic by subcutaneous route. An experimental teratogen. Human systemic effects by skin contact: ear tinnitus. Mutation data reported. A skin and severe eye irritant. Experimental reproductive effects. Incompatible with iron salts; spirit nitrous ether; lead acetate; iodine. Used in the manufacture of aspirin. When heated to decomposition it emits acrid smoke and irritating fumes.

SAI100 CAS:147-90-0 ***HR: 2***
SALICYLIC ACID, compounded with MORPHOLINE (1:1)
mf: $C_{11}H_{15}NO_4$ mw: 225.27

SYNS: BENZOIC ACID, 2-HYDROXY-, compounded with MORPHOLINE (1:1) ◇ MORPHOLINE SALICYLATE ◇ MORPHOLINE, componded with SALICYLIC ACID (1:1) ◇ MORPHOLIN SALICYLAT ◇ RETARCYL

TOXICITY DATA with REFERENCE
orl-rat TDLo:2400 mg/kg (female 9-14D post):TER GNRIDX 1,503,67
orl-rat TDLo:2400 mg/kg (female 9-14D post):REP GNRIDX 1,503,67
uns-mus LD50:1250 mg/kg ZBPMAL 124,733,85

SAFETY PROFILE: Moderately toxic by unspecified route. An experimental teratogen. Experimental reproductive effects. When heated to decomposition it emits toxic fumes of NO_x.

SAI500 CAS:99-96-7 ***HR: 3***
p-SALICYLIC ACID
mf: $C_7H_6O_3$ mw: 138.13

SYNS: 4-CARBOXYPHENOL ◇ p-HYDROXYBENZOIC ACID ◇ 4-HYDROXYBENZOIC ACID ◇ p-OXYBENZOESAURE (GERMAN)

TOXICITY DATA with REFERENCE
orl-mus LD50:2200 mg/kg DRSTAT 20,89,52

ipr-mus LD50:210 mg/kg JAPMA8 45,260,56
scu-mus LD50:1050 mg/kg AIPTAK 128,135,60

CONSENSUS REPORTS: Reported in EPA TSCA Inventory.

SAFETY PROFILE: Poison by intraperitoneal route. Moderately toxic by ingestion and subcutaneous routes. When heated to decomposition it emits acrid smoke and irritating fumes. See also SALICYLIC ACID.

SAJ000 CAS:65405-77-8 ***HR: 1***
SALICYLIC ACID-3-HEXEN-1-YL ESTER
mf: $C_{13}H_{16}O_3$ mw: 220.29

SYNS: β,Γ-cis-HEXENYL SALICYLATE ◇ cis-3-HEXENYL SALICYLATE

TOXICITY DATA with REFERENCE
skn-rbt 500 mg/24H MOD FCTXAV 17,373,79
orl-rat LD50:5 g/kg FCTXAV 17,373,79

CONSENSUS REPORTS: Reported in EPA TSCA Inventory.

SAFETY PROFILE: Mildly toxic by ingestion. A skin irritant. When heated to decomposition it emits acrid smoke and irritating fumes. See also ESTERS.

SAJ500 CAS:6969-49-9 ***HR: 1***
SALICYLIC ACID OCTYL ESTER
mf: $C_{15}H_{22}O_3$ mw: 250.37

SYNS: N-OCTYL-o-HYDROXYBENZOATE ◇ OCTYL SALICYLATE

TOXICITY DATA with REFERENCE
skn-rbt 500 mg/24H MOD FCTXAV 16,637,78

CONSENSUS REPORTS: Reported in EPA TSCA Inventory.

SAFETY PROFILE: A skin irritant. When heated to decomposition it emits acrid smoke and irritating fumes. See also ESTERS.

SAK000 CAS:2050-08-0 ***HR: 2***
SALICYLIC ACID PENTYL ESTER
mf: $C_{12}H_{16}O_3$ mw: 208.28

PROP: Liquid. D: 1.065 @ 15°, bp: 265°. Insol in water; misc in alc and ether.

SYN: AMYL SALICYLATE

TOXICITY DATA with REFERENCE
ivn-dog LD50:500 mg/kg 14CYAT 2,1897,63

CONSENSUS REPORTS: Reported in EPA TSCA Inventory.

SAFETY PROFILE: Moderately toxic by intravenous route. When heated to decomposition it emits acrid smoke and irritating fumes. See also ESTERS.

SAK500 CAS:24781-13-3 ***HR: 2***
SALICYLIC ACID-3-PHENYLPROPYL ESTER
mf: $C_{16}H_{16}O_3$ mw: 256.32

SYN: S 22

TOXICITY DATA with REFERENCE
par-rbt LD50:4000 mg/kg CHTPBA 4,453,69

CONSENSUS REPORTS: Reported in EPA TSCA Inventory.

SAFETY PROFILE: Moderately toxic by parenteral route. When heated to decomposition it emits acrid smoke and irritating fumes. See also ESTERS.

SAL000 CAS:118-61-6 ***HR: 2***
SALICYLIC ETHYL ESTER
mf: $HO{\bullet}C_6H_4{\bullet}CO_2{\bullet}C_2H_5$ mw: 166.18

PROP: Colorless liquid; wintergreen odor. D: 1.127, refr index: 1.520, mp: 1.3°, bp: 233-234°. Sol in alc, ether, acetic acid, fixed oils; sltly sol in water, glycerin.

SYNS: ETHYL-o-HYDROXYBENZOATE ◇ ETHYL SALICYLATE (FCC) ◇ FEMA No. 2458 ◇ SALICYLIC ETHER

TOXICITY DATA with REFERENCE
skn-rbt 500 mg/24H MOD FCTXAV 16,637,78
orl-rat LD50:1320 mg/kg FCTXAV 16,637,78
orl-gpg LDLo:1400 mg/kg FCTXAV 16,637,78
scu-gpg LDLo:1500 mg/kg AJPHAP 13,331,1905

CONSENSUS REPORTS: Reported in EPA TSCA Inventory.

SAFETY PROFILE: Moderately toxic by ingestion and subcutaneous routes. A skin irritant. When heated to decomposition it emits acrid smoke and irritating fumes. See also ESTERS.

SAL500 CAS:89-73-6 ***HR: 2***
SALICYLOHYDROXAMIC ACID
mf: $C_7H_7NO_3$ mw: 153.15

SYNS: N,2-DIHYDROXYBENZAMIDE ◇ 2-HYDROXYBENZHYDROXAMIC ACID ◇ o-HYDROXYBENZOHYDROXAMIC ACID ◇ 2-HYDROXYBENZOHYDROXAMIC ACID ◇ NSC 5088 ◇ SALICYLHYDROXAMIC ACID ◇ SALICYLOHYDROXIMIC ACID ◇ SHA

TOXICITY DATA with REFERENCE
mmo-sat 1 mg/plate AMACCQ 11,753,77
ipr-mus LD50:860 mg/kg NCISP* JAN86

CONSENSUS REPORTS: EPA Genetic Toxicology Program.

SAFETY PROFILE: Moderately toxic by intraperitoneal route. Mutation data reported. Used as a pharmaceutical and veterinary drug. When heated to decomposition it emits toxic fumes of NO_x.

SAM000 CAS:4342-30-7 ***HR: 3***
SALICYLOYLOXYTRIBUTYLSTANNANE
mf: $C_{19}H_{32}O_3Sn$ mw: 427.20

SYNS: STANNANE, TRIBUTYLSALICYLOYLOXY ◇ TRIBUTYLTIN SALICYLATE ◇ TRI-N-BUTYLTIN SALICYLATE ◇ TRI-N-BUTYLZINN SALICYLAT (GERMAN)

TOXICITY DATA with REFERENCE
orl-rat LD50:137 mg/kg ARZNAD 19,934,69
ivn-mus LD50:8910 μg/kg CSLNX* NX#02333

CONSENSUS REPORTS: Reported in EPA TSCA Inventory.
OSHA PEL: TWA 0.1 mg(Sn)/m^3 (skin)
ACGIH TLV: TWA 0.1 mg(Sn)/m^3 (skin) (Proposed: TWA 0.1 mg(Sn)/m^3; STEL 0.2 mg(Sn)/m^3 (skin))
NIOSH REL: (Organotin Compounds) TWA 0.1 mg(Sn)/m^3

SAFETY PROFILE: Poison by ingestion and intravenous routes. Tributyl tin compounds are extremely toxic to marine life. When heated to decomposition it emits acrid smoke and irritating fumes. See also TIN COMPOUNDS.

SAN000 CAS:552-94-3 ***HR: 2***
o-SALICYLSALICYLIC ACID
mf: $C_{14}H_{10}O_5$ mw: 258.24

PROP: Colorless crystals in benzene or chloroform. Mp: 148-149°, bp: decomp. Insol in water; sol in benzene, alc, acetyl alc, ether; very sol in CCl_4.

SYNS: DISALICYLIC ACID ◇ DISALYL ◇ SALICYLOYLSALICYLIC ACID ◇ SASAPYRIN

TOXICITY DATA with REFERENCE
orl-mus TDLo:665 mg/kg (female 17D post):TER APTOA6 29,250,71
orl-man TDLo:2229 mg/kg/2Y-I:KID AIMEAS 107,116,87
scu-mus LD50:1020 mg/kg YAKUD5 22,1101,80

CONSENSUS REPORTS: Reported in EPA TSCA Inventory.

SAFETY PROFILE: Moderately toxic by subcutaneous route. An experimental teratogen. Human systemic effects: changes kidney in tubules. When heated to decomposition it emits acrid smoke and irritating fumes.

SAN200 CAS:487-54-7 ***HR: 2***
SALICYLURIC ACID
mf: $C_9H_9NO_4$ mw: 195.19

SYNS: N-(2-HYDROXYBENZOYL)-GLYCINE (9CI) ◇ N-o-HYDROXYBENZOYLGLYCINE ◇ o-HYDROXYHIPPURIC ACID ◇ SALICYLOYLGLYCINE

TOXICITY DATA with REFERENCE
scu-rat LD50:3 g/kg AIPTAK 38,9,30
ivn-mus LDLo:600 mg/kg AIPTAK 38,9,30
scu-pgn LDLo:5730 mg/kg AIPTAK 38,9,30
ivn-pgn LDLo:2080 mg/kg AIPTAK 38,9,30

SAFETY PROFILE: Moderately toxic by subcutaneous and intravenous routes. When heated to decomposition it emits toxic fumes of NO_x.

SAN600 CAS:5003-48-5 ***HR: 2***
SALIPRAN
mf: $C_{17}H_{15}NO_5$ mw: 313.33

PROP: Crystals from methanol or ethanol. Mp: 175-176°.

SYNS: 4'-(ACETAMIDO)PHENYL-2-ACETOXYBENZOATE ◇ p-ACETAMIDOPHENYL ACETYLSALICYLATE ◇ 2-ACETOXY-4'-ACETAMINO)PHENYLBENZOATE ◇ p-N-ACETYLAMINOPHENYL-ACETYLSALICYLATE ◇ 2-(ACETYLOXY)BENZOIC ACID 4-(ACETYLAMINO)PHENYL ESTER ◇ ASPIRIN ACETAMINOPHEN ESTER ◇ BENORAL ◇ BENORILATE ◇ BENORTAN ◇ BENORYLATE ◇ FENASPARATE ◇ QUINEXIN ◇ TO 125 ◇ WIN 11450

TOXICITY DATA with REFERENCE
orl-rat TDLo:3960 mg/kg (female 7-17D post):TER SEIJBO 20,143,80
unr-wmn TDLo:1280 mg/kg/8D-I:CNS,PUL BMJOAE 288,1344,84
orl-rat LD50:10 g/kg USXXAM #3431293
ipr-rat LD50:1830 mg/kg USXXAM #3431293
orl-mus LD50:1551 mg/kg ARZNAD 28,1692,78
ipr-mus LD50:1255 mg/kg USXXAM #3431293

SAFETY PROFILE: Moderately toxic by ingestion and intraperitoneal routes. Human systemic effects by an unspecified route: respiratory stimulation, dehydration and distorted perceptions. An experimental teratogen. When heated to decomposition it emits toxic fumes of NO_x.

SAO200 CAS:53597-25-4 ***HR: 3***
SALMINE SULFATE

PROP: Salmine base is a powder, sol in disodium phosphate buffer. The sulfate separates from cold water as a clear, immisc liquid.

SYN: SALMINE SULFATE (1:1)

TOXICITY DATA with REFERENCE
par-rat LDLo:120 mg/kg PSEBAA 50,300,42
ipr-gpg LDLo:120 mg/kg PSEBAA 50,300,42
ivn-gpg LDLo:120 mg/kg PSEBAA 50,300,42
par-gpg LDLo:60 mg/kg PSEBAA 50,300,42

SAFETY PROFILE: Poison by intravenous, parenteral, and intraperitoneal routes. When heated to decomposition it emits toxic fumes of SO_x and Na_2O.

SAO250 ***HR: 3***
SALMONELLA ENTERITIDIS ENDOTOXIN

PROP: Extracted and purified from *Salmonella enteriditis ser typhimurium strain* SR-11 (ADSRDV 2,113,79).

TOXICITY DATA with REFERENCE
ipr-mus TDLo:16 μg/kg (female 16-19D post):TER AIPAAV 102,77,62
ipr-mus TDLo:16 μg/kg (female 16-19D post):REP AIPAAV 102,77,62
ipr-mus LD50:10 mg/kg AIPAAV 102,77,62

SAFETY PROFILE: Poison by intraperitoneal route. An experimental teratogen. Experimental reproductive effects.

SAO475 CAS:85886-25-5 ***HR: 2***
SALSOCAIN

TOXICITY DATA with REFERENCE
ipr-rat TDLo:1310 mg/kg (91D pre):REP KSRNAM 12,1868,78
ipr-rat LD50:646 mg/kg KSRNAM 12,1868,78
ipr-mus LD50:946 mg/kg KSRNAM 12,1868,78
ivn-mus LD50:813 mg/kg KSRNAM 12,1868,78

SAFETY PROFILE: Moderately toxic by intraperitoneal and intravenous routes. Experimental reproductive effects. When heated to decomposition it emits toxic fumes of Cl^-, SO_x, NO_x, and Na_2O.

SAO500 ***HR: 3***
SALT BATHS (NITRATE OR NITRITE)

SAFETY PROFILE: A very dangerous fire hazard by spontaneous chemical reaction. These baths are oxidizing in nature. Moderately explosive by chemical reaction due to contamination by cyanides or easily oxidizable materials or when heated to over 1000°F. Highly dangerous; in molten form will react with water, steam, or acids to produce heat, hydrogen, and toxic and corrosive fumes; can react vigorously with reducing materials. See also NITRATES and NITRITES.

SAP500 CAS:139-93-5 ***HR: 3***
SALVARSAN
mf: $C_{12}H_{12}As_2N_2O_2 \cdot 2ClH$ mw: 439.02

SYNS: ARSAMINOL ◇ ARSENPHENOLAMINE HYDROCHLORIDE ◇ ARSPHENAMINE ◇ 3,3'-DIAMINO-4,4'-DIHYDROXYARSENOBENZENE DIHYDROCHLORIDE ◇ EHRLICH 606 ◇ PHENARSENAMINE ◇ SIX HUNDRED SIX

TOXICITY DATA with REFERENCE
ipr-rat LDLo:112 mg/kg JPETAB 9,354,17
scu-rat LDLo:175 mg/kg JPETAB 9,354,17
ivn-rat LD50:37 mg/kg JPETAB 73,12,41
orl-mus LDLo:2 g/kg JPETAB 76,358,42
ipr-mus LDLo:162 mg/kg JPETAB 9,354,17
scu-mus LDLo:125 mg/kg JPETAB 9,354,17
ivn-mus LDLo:91 mg/kg HBAMAK 4,1289,35

CONSENSUS REPORTS: Arsenic and its compounds are on the Community Right-To-Know List.

OSHA PEL: TWA 0.5 $mg(As)/m^3$
ACGIH TLV: TWA 0.2 $mg(As)/m^3$

SAFETY PROFILE: Poison by intravenous, intraperitoneal, and intravenous routes. Implicated in aplastic anemia. When heated to decomposition it emits very toxic fumes of As, NO_x, and HCl. See also ARSENIC COMPOUNDS.

SAQ000 CAS:7681-34-7 ***HR: 2***
SALYRGAN THEOPHYLLINE
mf: $(C_{13}H_{17}HgNNaO_6)_n \cdot C_7H_8N_4O_2$

SYNS: (3-(o-(CARBOXYMETHOXY)BENZAMIDO)-2-METHOXYPROPYL)HYDROXY MERCURY, MONOSODIUM SALT compounded with THEOPHYLLINE ◇ MERSALYL with THEOPHYLLINE

TOXICITY DATA with REFERENCE
ivn-mus LD50:950 mg/kg JPETAB 99,149,50

CONSENSUS REPORTS: Mercury and its compounds are on the Community Right-To-Know List.

OSHA PEL: (Transitional: CL 1 $mg/10m^3$) CL 0.1 mg $(Hg)/m^3$ (skin)
ACGIH TLV TWA: 0.1 $mg(Hg)/m^3$ (skin)
NIOSH REL: (Inorganic Mercury) TWA 0.05 $mg(Hg)/m^3$

SAFETY PROFILE: Moderately toxic by intravenous route. When heated to decomposition it emits very toxic fumes of Hg, Na_2O, and NO_x. See also MERCURY COMPOUNDS and THEOPHYLLINE.

SAQ500 ***HR: 3***
SAMARIUM
af: Sm aw: 150.36

PROP: Bright, yellow, lustrous, stable metal. Mp: 1072°, bp: 1778°, d (α): 7.536, d (β): 7.40.

SAFETY PROFILE: As a lanthanon, it may cause impairment of blood clotting. Flammable in the form of dust when exposed to flame or by spontaneous chemical reaction with oxidizers. Ignites at 150° in air. Reacts with water to form explosive hydrogen gas. Can react violently with halogens. Potentially explosive reaction with 1,1,2-trichlorotrifluoroethane. See also LANTHANUM, RARE EARTHS, and POWDERED METALS.

SAR000 CAS:10465-27-7 ***HR: 3***
SAMARIUM ACETATE
mf: $C_6H_9O_6 \cdot Sm$ mw: 327.50

SYNS: ACETIC ACID, SAMARIUM SALT ◇ SAMARIUMACETAT (GERMAN)

TOXICITY DATA with REFERENCE
scu-mus LD50:10 mg/kg ZGEMAZ 113,536,44
ivn-cat LDLo:50 mg/kg ZGEMAZ 113,536,44

CONSENSUS REPORTS: Reported in EPA TSCA Inventory.

SAFETY PROFILE: Poison by intravenous and subcutaneous routes. When heated to decomposition it emits acrid smoke and irritating fumes. See also SAMARIUM and RARE EARTHS.

SAR500 CAS:10361-82-7 ***HR: 3***
SAMARIUM(III) CHLORIDE
mf: Cl_3Sm mw: 256.70

PROP: White-yellowish powder. D: 4.465, mp: 686°.

TOXICITY DATA with REFERENCE
skn-rbt 500 mg MOD BJPCAL 17,526,61
eye-rbt 1 mg/1H MLD BJPCAL 17,526,61
scu-rat LDLo:2000 mg/kg AEPPAE 145,19,29
ipr-mus LD50:365 mg/kg AEHLAU 5,437,62
scu-gpg LD50:703 mg/kg EQSSDX 1,1,75
scu-frg LD50:256 mg/kg EQSSDX 1,1,75

CONSENSUS REPORTS: Reported in EPA TSCA Inventory.

SAFETY PROFILE: Poison by intraperitoneal and subcutaneous routes. A skin and eye irritant. When heated to decomposition it emits toxic fumes of Cl^-. See also SAMARIUM and RARE EARTHS.

SAS000 CAS:13074-85-6 ***HR: 3***
SAMARIUM CITRATE

SYN: CITRIC ACID, SAMARIUM SALT

TOXICITY DATA with REFERENCE
ipr-mus LD50:164 mg/kg AEHLAU 5,437,62
ipr-gpg LD50:75 mg/kg AEHLAU 5,437,62

SAFETY PROFILE: Poison by intraperitoneal route. When heated to decomposition it emits acrid smoke and irritating fumes. See also SAMARIUM.

SAT000 CAS:13759-83-6 ***HR: 3***
SAMARIUM(III) NITRATE, HEXAHYDRATE (1:3:6)
mf: N_3O_9•Sm•$6H_2O$ mw: 444.50

PROP: Pale yellow crystals. Mp: 78-79°, d: 2.375.

SYNS: NITRIC ACID, SAMARIUM(3+) SALT, HEXAHYDRATE ◇ SAMARIUM NITRAT (GERMAN)

TOXICITY DATA with REFERENCE
orl-rat LD50:2900 mg/kg TXAPA9 5,750,63
ipr-rat LD50:285 mg/kg TXAPA9 5,750,63
ivn-rat LD50:9 mg/kg TXAPA9 5,750,63
ipr-mus LD50:315 mg/kg TXAPA9 5,750,63
scu-frg LDLo:1600 mg/kg AEPPAE 145,19,29

SAFETY PROFILE: Poison by intraperitoneal and intravenous routes. Moderately toxic by ingestion and subcutaneous routes. When heated to decomposition it emits toxic fumes of NO_x. See also SAMARIUM and NITRATES.

SAT200 CAS:10361-83-8 ***HR: 3***
SAMARIUM TRINITRATE
mf: SmN_3O_9 mw: 336.38

SYN: SAMARIUM NITRATE

TOXICITY DATA with REFERENCE
orl-rat LD50:2160 mg/kg EQSSDX 1,1,75
ipr-rat LD50:217 mg/kg EQSSDX 1,1,75
ivn-rat LD50:6771 μg/kg EQSSDX 1,1,75
ipr-mus LD50:239 mg/kg EQSSDX 1,1,75
scu-gpg LD50:1128 mg/kg EQSSDX 1,1,75
scu-frg LD50:1489 mg/kg EQSSDX 1,1,75

SAFETY PROFILE: Poison by intraperitoneal and intravenous routes. Moderately toxic by ingestion and subcutaneous routes. When heated to decomposition it emits toxic fumes of NO_x. See also NITRATES and SAMARIUM.

SAT875 ***HR: 3***
SANDBOX TREE

PROP: A large tree which may grow higher than 60 feet with a trunk over 3 feet thick. Its trunk is covered with short thorns. The oval leaves are about 8 inches long. The male flowers are green, about 2 inches long, and the female flowers are red. The seed pod pops open with a loud noise when dry. It grows commonly in the West Indies and is more rare in Florida and Hawaii.

SYNS: HURA CREPITANS ◇ JAVILLO (DOMINICAN REPUBLIC, PUERTO RICO) ◇ MOLINILLO (PUERTO RICO) ◇ MONKEY PISTOL ◇ MONKEY'S DINNER BELL ◇ POSSUM WOOD ◇ SABLIER (HAITI) ◇ SALVADERA (CUBA)

SAFETY PROFILE: The seeds contain the poison hurin (crepitin) a toxalbumin which inhibits protein synthesis in the intestinal wall, and huratoxin a direct irritant. Ingestion of the seeds may cause, after several hours, nausea, vomiting, diarrhea, and extensive fluid and electrolyte loss. Eating more than 2 or 3 seeds may cause death. The sap may be an eye and skin irritant. See also ABRIN.

SAU000 CAS:18417-89-5 ***HR: 3***
SANGIVAMYCIN
mf: $C_{12}H_{15}N_5O_5$ mw: 309.32

SYNS: 4-AMINO-7-β-d-RIBOFURANOSYL-7H-PYRROLO(2,3-d)PYRIMIDINE-5-CARBOXAMIDE ◇ ANTIBIOTIC B-14437 ◇ B-14437 ◇ B 90912 ◇ 7-DEAZAADENOSINE-7-CARBOXAMIDE ◇ NSC-65346 ◇ SKI 27013

TOXICITY DATA with REFERENCE
dni-hmn:hla 20 mg/L JANTAJ 35,119,82
oms-hmn:hla 20 mg/L JANTAJ 35,119,82
oms-mus:leu 10 μmol/L BCPCA6 29,305,80
ipr-rat LDLo:1200 μg/kg NCINS* -,269,67
orl-mus LD50:11 mg/kg NCISP* JAN86
ipr-mus LDLo:4 mg/kg NCINS* -,269,67
ivn-mus LD50:5 mg/kg 85GDA2 5,313,81

SAFETY PROFILE: Poison by ingestion, intravenous and intraperitoneal routes. Human mutation data reported. When heated to decomposition it emits toxic fumes of NO_x.

SAU350 CAS:39404-28-9 ***HR: 3***
SANGUIRITRIN
mf: $C_{21}H_{18}NO_4$•$C_{20}H_{14}NO_4$•HO_4S•HO_4S mw: 874.89

SYNS: SANGUIRITRINE ◇ SANGUIRYTHRINE

TOXICITY DATA with REFERENCE
orl-rat LD50:500 mg/kg PCJOAU 16,925,82
ipr-rat LD50:12 mg/kg PCJOAU 16,925,82
orl-mus LD50:470 mg/kg PCJOAU 16,925,82
ipr-mus LD50:14200 μg/kg PCJOAU 16,925,82

SAFETY PROFILE: Poison by intraperitoneal route. Moderately toxic by ingestion. When heated to decomposition it emits toxic fumes of NO_x and SO_x. See also SULFATES.

SAU400 ***HR: 1***
SANTALYL ACETATE

PROP: Mixture of α- and β-isomers from acetylation of santalol. Colorless to sltly yellow liquid; sandalwood odor. D: 0.980, refr index: 1.488-1.491, flash p: +212°F. Sol in alc; insol in water.

SYN: FEMA No. 3007

SAFETY PROFILE: Combustible liquid. When heated to decomposition it emits acrid smoke and irritating fumes.

SAU475 CAS:39456-78-5 ***HR: 1***
SANTOFLEX 134

TOXICITY DATA with REFERENCE
skn-rbt 500 mg/24H MOD 28ZPAK -,295,72
eye-rbt 500 mg/24H MLD 28ZPAK -,295,72
orl-rat LD50:4110 mg/kg 28ZPAK -,295,72

SAFETY PROFILE: Mildly toxic by ingestion. A skin and eye irritant.

SAU500 CAS:481-06-1 ***HR: 3***
SANTONIN
mf: $C_{15}H_{18}O_3$ mw: 246.33

PROP: Glossy, colorless crystals or white powder turning yellow on exposure to light; odorless, tasteless at first, then bitter. Mp: 170°, bp: subl, d: 1.187. Sol in water, alc and ether.

SYNS: 6α-HYDROXY-3-OXO-11-EPIISOEUSANTONA-1,4-DIENIC ACID, Γ-LACTONE ◇ 6α-HYDROXY-3-OXO-EUDESMA-1,4-DIEN-12-OIC ACID, Γ-LACTONE (11S)-(−)- ◇ α-SANTONIN ◇ l-α-SANTONIN ◇ SANTONINIC ANHYDRIDE ◇ 3a,5,5a,9b-TETRAHYDRO-3,5a,9-TRIMETHYL-NAPHTHO(1,2-b)PURAN-2,8(3H,4H)-DIONE

TOXICITY DATA with REFERENCE
unr-man LDLo:15 mg/kg 85DCAI 2,73,70
orl-mus LD50:900 mg/kg NIIRDN 6,301,82
scu-mus LDLo:250 mg/kg 27ZXA3 4.7b,1394,-
ivn-mus LD50:180 mg/kg CXLNX* NX#02213

CONSENSUS REPORTS: Reported in EPA TSCA Inventory.

SAFETY PROFILE: Human poison by an unspecified route. Experimental poison by intravenous and subcutaneous routes. Moderately toxic experimentally by ingestion. It can cause disturbance of color vision. Objects first show bluish tinge, then yellow which is most prominent. Complete blindness may occur lasting perhaps for nearly a week. Dizziness, drowsiness, and nausea may also occur. Recovery is spontaneous. Combustible when exposed to heat or flame. When heated to decomposition it emits acrid smoke and irritating fumes.

SAV000 CAS:91-53-2 ***HR: 3***
SANTOQUINE
mf: $C_{14}H_{19}NO$ mw: 217.34

PROP: Clear, light yellow liquid. Mp: <0°, bp: 125° @ 2 mm, vap d: 7.48, d: 1.030 @ 25°, refr index: 1.57.

SYNS: 1,2-DIHYDRO-6-ETHOXY-2,2,4-TRIMETHYLQUINOLINE ◇ 1,2-DIHYDRO-2,2,4-TRIMETHYL-6-ETHOXYQUINOLINE ◇ EMQ ◇ EQ ◇ 6-ETHOXY-1,2-DIHYDRO-2,2,4-TRIMETHYLQUINOLINE ◇ ETHOXYQUIN (FCC) ◇ ETHOXYQUINE ◇ 6-ETHOXY-2,2,4-TRIMETHYL-1,2-DIHYDROQUINOLINE ◇ NIFLEX ◇ NIX-SCALD ◇ SANTOFLEX A ◇ SANTOFLEX AW ◇ SANTOQUIN ◇ STOP-SCALD ◇ 2,2,4-TRIMETHYL-6-ETHOXY-1,2-DIHYDROQUINOLINE ◇ USAF B-24

TOXICITY DATA with REFERENCE
mma-sat 200 μg/plate PCBRD2 141,407,84
orl-rat LD50:800 mg/kg RCTEA4 45,627,72
orl-mus LD50:1730 mg/kg MEIEDD 10,545,83
ipr-mus LD50:200 mg/kg NTIS** AD277-689

CONSENSUS REPORTS: EPA Genetic Toxicology Program. Reported in EPA TSCA Inventory.

SAFETY PROFILE: Poison by intraperitoneal route. Moderately toxic by ingestion. Mutation data reported. Combustible when exposed to heat or flame; can react with oxidizing materials. When heated to decomposition it emits toxic fumes of NO_x.

SAV500 CAS:8047-15-2 ***HR: 3***
SAPONIN

PROP: Bitter taste.

TOXICITY DATA with REFERENCE
ipr-rat TDLo:350 mg/kg (9-22D preg):REP TRBMAV 30,319,72
orl-mus LDLo:3000 mg/kg HBAMAK 4,1395,35
scu-mus LDLo:900 mg/kg HBAMAK 4,1395,35
ivn-mus LDLo:1000 mg/kg HBAMAK 4,1395,35
ivn-cat LDLo:46 mg/kg HBAMAK 4,1394,35
ivn-rbt LDLo:40 mg/kg HBAMAK 4,1394,35

SAFETY PROFILE: Poison by intravenous route. Moderately toxic by ingestion and subcutaneous routes. Experimental reproductive effects. Injection can cause rapid and severe destruction of red blood cells. Saponins are natural plant toxins. When heated to decomposition it emits acrid smoke and irritating fumes.

SAW300 ***HR: 2***
SAPWOOD of LINDEN

SYNS: l'AUBIER de TILIA SYLVESTRIS (FRENCH) ◇ l'AUBIER de TILLEUL (FRENCH)

TOXICITY DATA with REFERENCE
ipr-rat LD50:1150 mg/kg AIPTAK 129,319,60
orl-mus LD50:20 g/kg AIPTAK 129,319,60
ipr-mus LD50:1350 mg/kg AIPTAK 129,319,60

SAFETY PROFILE: Moderately toxic by intraperitoneal route. Mildly toxic by ingestion.

SAX000 CAS:8013-77-2 ***HR: 3***
SARAN
mf: $(C_4H_5Cl_3)_n$

TOXICITY DATA with REFERENCE
imp-rat TDLo:36 mg/kg:ETA CNREA8 15,333,55

SAFETY PROFILE: Questionable carcinogen with experimental tumorigenic by implant data. When heated to decomposition it emits toxic fumes of Cl^-. See also POLYMERS, INSOLUBLE.

SAX200 CAS:13045-94-8 ***HR: 3***
d-SARCOLYSINE
mf: $C_{13}H_{18}Cl_2N_2O_2$ mw: 305.23

SYNS: 4-(BIS(2-CHLOROETHYL)AMINO)-d-PHENYLALANINE ◇ (+)-3-(p-(BIS(2-CHLOROETHYL)AMINO)PHENYL)ALANINE ◇ d-3-(p-(BIS(2-CHLOROETHYL)AMINO)PHENYL)ALANINE ◇ 3026 C.B. ◇ CB-3026 ◇ p-DI-(2-CHLOROETHYL)-AMINO-d-PHENYLALANINE ◇ p-DI(2-CHLOROETHYL)AMINO-d-PHENYLALANINE ◇ MEDFALAN ◇ MEDPHALAN ◇ NSC-35051 ◇ d-PHENYLALANINE MUSTARD

TOXICITY DATA with REFERENCE
cyt-dmg-par 10 mmol/L JOGNAU 54,146,56
sln-dmg-par 8 mmol/L GENRA8 1,173,60
skn-mus TDLo:120 mg/kg/9W-I:ETA BJCAAI 10,363,56
ice-rat LD50:113 μg/kg JPPMAB 18,760,66

CONSENSUS REPORTS: EPA Genetic Toxicology Program.

SAFETY PROFILE: A deadly poison by intracerebral route. Questionable carcinogen with experimental tumorigenic data. Mutation data reported. When heated to decomposition it emits toxic fumes of Cl^- and NO_x.

SAX210 CAS:1088-80-8 ***HR: D***
m-l-SARCOLYSINE
mf: $C_{13}H_{18}Cl_2N_2O_2$ mw: 305.23

SYN: l-3-(m-(BIS(2-CHLOROETHYL)AMINO)PHENYL)ALANINE

TOXICITY DATA with REFERENCE
dns-hmn:leu 10 μmol/L EJCODS 17,991,81
oms-hmn:leu 10 μmol/L EJCODS 17,991,81
sce-hmn:fbr 1 μmol/L EJCODS 17,991,81

SAFETY PROFILE: Human mutation data reported. When heated to decomposition it emits toxic fumes of Cl^- and NO_x.

SAX275 ***HR: 3***
SARIFA, seed extract

PROP: Indian plant belonging to the family *Annonaceae* (IJEBA6 22,312,84).

TOXICITY DATA with REFERENCE
orl-ham TDLo:500 mg/kg (1-5D preg):REP IJEBA6 22,312,84
ipr-mus LDLo:50 mg/kg PLMEAA 28,97,75

SAFETY PROFILE: Poison by intraperitoneal route. Experimental reproductive effects.

SAX500 CAS:11031-48-4 ***HR: 3***
SARKOMYCIN
mf: $C_7H_8O_3$ mw: 140.15

PROP: Oily liquid. Sol in water, methanol, ethanol, butanol, ethyl acetate; sltly sol in ether. Isolated from *Streptomyces sp.* (ANTCAO 4,514,54).

SYNS: 2-METHYLENE-3-OXO-CYCLOPENTANECARBOXYLICACID ◇ SARCOMYCIN

TOXICITY DATA with REFERENCE
mmo-eug 10 μg/L NEOLA4 19,579,72
dni-eug 10 μg/L NEOLA4 19,579,72
scu-rat TDLo:25 mg/kg (6-10D preg):TER OSDIAF 14,107,65
scu-rat TDLo:25 mg/kg (6-10D preg):REP OSDIAF 14,107,65

scu-rat TDLo:8220 μg/kg/21W-I:ETA JAJAAA 8,168,55
orl-mus LD50:5600 mg/kg 85ERAY 2,1261,78
ipr-mus LDLo:700 mg/kg TDKNAF 14,60,55
scu-mus LD50:600 mg/kg 85ERAY 2,1261,78
ivn-mus LD50:1200 mg/kg 85ERAY 2,1261,78

SAFETY PROFILE: Moderately toxic by intravenous, subcutaneous, and intraperitoneal routes. Mildly toxic by ingestion. Experimental teratogenic and reproductive effects. Questionable carcinogen with experimental tumorigenic data. Mutation data reported. Used as an antibiotic. When heated to decomposition it emits acrid smoke and irritating fumes.

SAY000 CAS:101952-95-8 ***HR: 3***
SARKOMYCIN compounded with NICOTINIC HYDRAZIDE
mf: $C_7H_8O_3 \cdot C_6H_{13}N_3O$ mw: 283.37

SYNS: COMPOUND SN ◇ 2-METHYLENE-3-OXO-1-CYCLOPENTANECARBOXYLIC ACID compounded with NICOTINIC HYDRAZIDE

TOXICITY DATA with REFERENCE
ipr-mus LD50:4 mg/kg 85ERAY 2,1261,78
ivn-mus LD50:700 mg/kg 85ERAY 2,1261,78

SAFETY PROFILE: Poison by intraperitoneal route. Moderately toxic by intravenous route. When heated to decomposition it emits toxic fumes of NO_x.

SAY875 ***HR: 2***
SA 96 SODIUM SALT
mf: $C_7H_{12}NO_3S_2 \cdot Na$ mw: 245.31

SYNS: N-(2-MERCAPTO-2-METHYL-1-OXOPROPYL)-l-CYSTEINESODIUM SALT ◇ N-(2-MERCAPTO-2-METHYLPROPANOYL)-l-CYSTEINE SODIUM SALT

TOXICITY DATA with REFERENCE
orl-rat LD50:5220 mg/kg IYKEDH 14,346,83
ipr-rat LD50:1130 mg/kg IYKEDH 14,346,83
scu-rat LD50:1172 mg/kg IYKEDH 14,346,83
ivn-rat LD50:1088 mg/kg IYKEDH 14,346,83
orl-mus LD50:6600 mg/kg IYKEDH 14,346,83
ipr-mus LD50:1600 mg/kg IYKEDH 14,346,83
scu-mus LD50:1350 mg/kg IYKEDH 14,346,83
ivn-mus LD50:1090 mg/kg IYKEDH 14,346,83

SAFETY PROFILE: Moderately toxic by intraperitoneal, intravenous, and subcutaneous routes. Mildly toxic by ingestion. An anti-rheumatic agent. When heated to decomposition it emits toxic fumes of SO_x, NO_x, and Na_2O. See also l-CYSTEINE and MERCAPTANS.

SAY900 ***HR: 3***
SASSAFRAS

PROP: A yellowish-reddish, volatile oil; pungent, aromatic odor and taste. D: 1.065-1.077 @ 25°/25°. Sol in alc, ether, chloroform, glacial acetic acid, CS_2. Safrole-free ethanol extract of *Sassafras albidum* root bark (JNCIAM 60,683,78).

SYN: SASSAFRAS ALBIDUM

TOXICITY DATA with REFERENCE
skn-rbt 500 mg/24H MOD FCTOD7 20(Suppl),825,82
scu-rat TDLo:3540 mg/kg/59W-I:NEO JNCIAM 60,683,78

SAFETY PROFILE: A skin irritant. Questionable carcinogen with experimental neoplastigenic data. When heated to decomposition it emits acrid smoke and irritating fumes.

SAY950 CAS:56302-13-7 ***HR: 1***
SATRANIDAZOLE
mf: $C_8H_{11}N_5O_5S$ mw: 289.30

SYNS: CG 10213 GO ◇ GO 10213 ◇ 2-IMIDAZOLIDINONE, 1-(1-METHYL-5-NITRO-1H-IMIDAZOL-2-YL)-3-(METHYLSULFONYL)-◇ 1-(1-METHYL-5-NITRO-1H-IMIDAZOL-2-YL)-3-(METHYLSULFONYL)-2-IMIDAZOLIDINO NE ◇ 1-METHYLSULFONYL-3-(1-METHYL-5-NITRO-2-IMIDAZOLYL)-2-IMIDAZOLIDINONE

TOXICITY DATA with REFERENCE
orl-rat TDLo:16800 mg/kg (male 4W pre):REP ARZNAD 35,1692,85
orl-dog LDLo:5 g/kg ARZNAD 35,1692,85

SAFETY PROFILE: Slightly toxic by ingestion. Experimental reproductive effects. When heated to decomposition it emits toxic fumes of NO_x and SO_x.

SAZ000 CAS:28249-77-6 ***HR: 2***
SATURN
mf: $C_{12}H_{16}ClNOS$ mw: 257.80

PROP: Bp: 127° @ 0.008 mm.

SYNS: B-3015 ◇ BENTHIOCARB ◇ BOLERO ◇ S-(4-CHLOROBENZYL)-N,N-DIETHYLTHIOCARBAMATE ◇ S-((4-CHLOROPHENYL) METHYL)DIETHYLCARBAMOTHIOATE ◇ DIETHYLCARBAMOTHIOIC ACID S-((4-CHLOROPHENYL)METHYL) ESTER ◇ IMC 3950 ◇ SATURNO ◇ SIACARB ◇ TAMARIZ ◇ THIOBENCARB

TOXICITY DATA with REFERENCE
orl-rat LD50:1903 mg/kg FMCHA2 -,C233,83
orl-mus LD50:560 mg/kg GUCHAZ 6,37,73

SAFETY PROFILE: Moderately toxic by ingestion. Used as an herbicide. When heated to decomposition it emits very toxic fumes of Cl^-, NO_x, and SO_x. See also CARBAMATES.

SBA000 CAS:8016-68-0 ***HR: 3***
SAVORY OIL (summer variety)

PROP: From steam distillation of *Saturiea hortensis* L. (Fam. *Labiatae*) (FCTXAV 14,659,76). Light yellow to dark brown liquid; spicy odor. D: 0.875-0.954, refr index: 1.486-1.505 @ 20°. Sol in fixed oils, mineral oil; insol in glycerin, propylene glycol.

TOXICITY DATA with REFERENCE
skn-mus 50 mg FCTXAV 14,659,76
skn-rbt 500 mg/24H SEV FCTXAV 14,659,76
skn-gpg 500 mg/24H SEV FCTXAV 14,659,76
orl-rat LD50:1370 mg/kg FCTXAV 14,659,76
skn-gpg LD50:340 mg/kg FCTXAV 14,659,76

CONSENSUS REPORTS: Reported in EPA TSCA Inventory.

SAFETY PROFILE: Poison by skin contact. Moderately toxic by ingestion. A severe skin irritant. When heated to decomposition it emits acrid smoke and irritating fumes.

SBA500 CAS:35554-08-6 ***HR: 3***
SAXITOXIN DIHYDROCHLORIDE
mf: $C_{10}H_{17}N_7O_4 \cdot 2ClH$ mw: 372.26

PROP: White, hyroscopic solid. Very sol in water, methanol; sltly sol in ethanol, glacial acetic acid; insol in alkalies.

SYNS: CLAM POISON DIHYDROCHLORIDE ◇ GONYAULAX TOXIC DIHYDROCHLORIDE ◇ MUSSEL POISON DIHYDROCHLORIDE ◇ PARALYTIC SHELLFISH POISON DIHYDROCHLORIDE ◇ SAXITOXIN HYDROCHLORIDE ◇ STX DIHYDROCHLORIDE

TOXICITY DATA with REFERENCE
orl-mus LD50:263 μg/kg MEIEDD 10,1206,83
ipr-mus LD50:10 μg/kg MEIEDD 10,1206,83
ivn-mus LD50:3400 ng/kg MEIEDD 10,1206,83

SAFETY PROFILE: A very deadly poison by ingestion, intravenous, and intraperitoneal routes. Used as a neuromuscular blocking agent. When heated to decomposition it emits very toxic fumes of NO_x and HCl.

SBA600 CAS:35523-89-8 ***HR: 3***
SAXITOXIN HYDRATE
mf: $C_{10}H_{17}N_7O_4$ mw: 299.34

SYNS: 2,6-DIAMINO-4-(((AMINO-CARBONYL)OXY)METHYL)-3a,4,8,9-TETRAHYDRO-1H,10H-PYRROLO(1,2-c)PURINE-10,10-DIOL (3aS-(3a-α-4-α,10aR*)) ◇ SAXITOXIN (8CI)

TOXICITY DATA with REFERENCE
ipr-mus LD50:5 μg/kg BIBUDZ 7,151,80
scu-mus LDLo:16500 ng/kg TOXID9 4,13,84
ivn-mus LD50:8 μg/kg TOXIA6 7,315,69

SAFETY PROFILE: A very deadly poison by subcutaneous, intravenous, and intraperitoneal routes. When heated to decomposition it emits toxic fumes of NO_x.

SBA625 ***HR: D***
SC-11800M
mf: $C_{24}H_{32}O_4 \cdot C_{21}H_{26}O_2$ mw: 695.03

TOXICITY DATA with REFERENCE
orl-rbt TDLo:6 mg/kg (female 1D pre):REP YACHDS 6,1321,78

SAFETY PROFILE: Experimental reproductive effects. When heated to decomposition it emits acrid smoke and irritating fumes.

SBA875 CAS:37519-65-6 ***HR: 3***
SC/3123
mf: $C_{21}H_{27}N_3O_4 \cdot BrH$ mw: 466.43

SYNS: 3,4-DIHYDRO-6,7-DIMETHOXY-N-(2-(2-METHOXYPHENOXY)ETHYL)-2(1H)-ISOQUINOLINECARBOXIMIDAMIDE ◇ N-(2-o-METHOXYPHENOXYETHYL)-6,7-DIMETHOXY-3,4-DIHYDRO-2-(1H)-ISOQUINOLINE CARBOXAMIDINE HBr

TOXICITY DATA with REFERENCE
orl-mus LD50:3300 mg/kg BCFAAI 111,353,72
ipr-mus LD50:160 mg/kg BCFAAI 111,353,72
ivn-mus LD50:12 mg/kg BCFAAI 111,353,72

SAFETY PROFILE: Poison by intravenous and intraperitoneal routes. Moderately toxic by ingestion. When heated to decomposition it emits toxic fumes of NO_x and Br^-.

SBB000 CAS:96-88-8 ***HR: 3***
SCANDICAINE
mf: $C_{15}H_{22}N_2O$ mw: 246.39

SYNS: CARBOCAINE ◇ MEPIVACAINE ◇ dl-MEPIVACAINE ◇ N-METHYLHEXAHYDRO-2-PICOLINIC ACID, 2,6-DIMETHYLANILIDE ◇ N-METHYL-2-PIPECOLIC ACID, 2,6-DIMETHYLANILIDE ◇ N-METHYL-2-PIPECOLIC ACID, 2,6-XYLIDIDE ◇ (±)-1-METHYL-2',6'-PIPECOLOXYLIDIDE ◇ SCANDICAIN ◇ SCANDICANE

TOXICITY DATA with REFERENCE
ims-rat TDLo:6 mg/kg (female 11D post):REP NETOD7 8,61,86
scu-rat LD50:500 mg/kg APTOA6 31,273,72
ivn-rat LD50:30 mg/kg ARVPAX 9,503,69
ivn-mus LD50:35 mg/kg APTOA6 31,273,72
ivn-rbt LDLo:53 mg/kg APTOA6 31,273,72

CONSENSUS REPORTS: Reported in EPA TSCA Inventory.

SAFETY PROFILE: Poison by intravenous route. Moderately toxic by subcutaneous route. Experimental reproductive effects. When heated to decomposition it emits toxic fumes of NO_x.

SBB500 ***HR: 2***
SCANDIUM
af: Sc aw: 44.9559

PROP: A naturally occurring isotope.

SAFETY PROFILE: Should be handled carefully. Flammable in the form of dust when exposed to heat or flame or by chemical reaction with oxidizers. Can react violently with halogens; air. See also POWDERED METALS and RARE EARTHS.

SBC000 CAS:10361-84-9 ***HR: 3***
SCANDIUM CHLORIDE
mf: Cl_3Sc mw: 151.31

PROP: White, deliq solid. Mp: 960°. Sol in water; insol in alc.

SYN: SCANDIUM (3+) CHLORIDE

TOXICITY DATA with REFERENCE
orl-mus LD50:3980 mg/kg EQSSDX 1,1,75
ipr-mus LD50:314 mg/kg COREAF 256,1043,63

CONSENSUS REPORTS: Reported in EPA TSCA Inventory.

SAFETY PROFILE: Poison by intraperitoneal route. Moderately toxic by ingestion. When heated to decomposition it emits toxic fumes of Cl^-. See also SCANDIUM.

SBC500 CAS:85-83-6 ***HR: 3***
SCARLET RED
mf: $C_{24}H_{20}N_4O$ mw: 380.48

SYNS: BRASILAZINA OIL RED B ◇ CALCO OIL RED D ◇ C.I. 258 ◇ C.I. SOLVENT RED 24 ◇ 2',3-DIMETHYL-4-(2-HYDROXYNAPHTHYLAZO)AZOBENZENE ◇ FAST OIL RED B ◇ FAT RED B ◇ 1-((2-METHYL-4-((2-METHYLPHENYL)AZO)PHENYL)AZO)-2-NAPHTHALENOL ◇ PHENOPLASTE ORGANOL RED B ◇ RUBRUM SCARLATINUM ◇ o-TOLUENEAZO-o-TOLUENEAZO-β-NAPHTHOL ◇ o-TOLUENEAZO-o-TOLUENE-β-NAPHTHOL ◇ o-TOLYLAZO-o-TOLYLAZO-β-NAPHTHOL ◇ o-TOLYLAZO-o-TOLYLAZO-2-NAPHTHOL ◇ 1-((4-(o-TOLYLAZO)-o-TOLYL)AZO)-2-NAPHTHOL)

TOXICITY DATA with REFERENCE
mma-sat 100 μg/plate MUREAV 56,249,78
scu-rat TDLo:512 mg/kg/58W-C:ETA GANNA2 49,27,58

CONSENSUS REPORTS: IARC Cancer Review: Group 3 IMEMDT 7,56,87. Reported in EPA TSCA Inventory.

SAFETY PROFILE: Questionable carcinogen with experimental tumorigenic data. Mutation data reported. When heated to decomposition it emits toxic fumes of NO_x.

SBC550 ***HR: 2***
SCARLET WISTERIA TREE

PROP: Woody annuals with green stems that grow 3 to 8 feet tall. The compound leaves have many thin leaflets. The small flowers are yellow with purple dots.

SYNS: BACULO (PUERTO RICO) ◇ COLORADO RIVER HEMP ◇ EGYPTIAN RATTLEPOD ◇ GALLITO (CUBA) ◇ 'OHAI (HAWAII) ◇ 'OHAI-KE'OKE'O (HAWAII) ◇ 'OHAI-'ULA'ULA (HAWAII) ◇ POIS VALLIERE (HAITI) ◇ SESBANIA (VARIOUS SPECIES) ◇ VEGETABLE-HUMMING-BIRD

SAFETY PROFILE: All parts of the plant contain poisonous pyrrolizidine alkaloids. Chronic consumption of teas made from this plant may cause anorexia with nausea, vomiting, diarrhea and liver and kidney damage due to blockage of the veins in these organs.

SBC700 ***HR: 3***
SCH 5802 B
mf: $C_{20}H_{31}NO_4$ mw: 349.52

SYNS: 2-ETHYL-2-PHENYLSUCCINICACID-2-(DIETHYLAMINO) ETHYL ETHYL ESTER ◇ PHENYLAETHYLBERNSTEINSAEURE-AETHYL-DIAETHYL-AMINOAETHYL-DI-ESTER(GERMAN)

TOXICITY DATA with REFERENCE
orl-rat LD50:160 mg/kg AEPPAE 237,264,59
ipr-rat LD50:79 mg/kg AEPPAE 237,264,59
orl-mus LD50:215 mg/kg AEPPAE 237,264,59
ipr-mus LD50:151 mg/kg AEPPAE 237,264,59

SAFETY PROFILE: Poison by ingestion and intraperitoneal routes. When heated to decomposition it emits toxic fumes of NO_x.

SBC800 CAS:28241-06-7 ***HR: 3***
SCH 12650
mf: $C_{15}H_{14}ClN_3O \cdot 7C_4H_4O_4$ mw: 1100.33

SYNS: α-(p-CHLOROPHENYL)-α-2-IMIDAZOLIN-2-YL-2-PYRIDINEMETHANOL MALEATE ◇ MALEATE de 2-((p-CHLOROPHENYL)-2-(PYRIDYL)HYDROXY METHYL)IMIDAZOLINE (FRENCH)

TOXICITY DATA with REFERENCE
orl-mus LD50:285 mg/kg THERAP 26,805,71
ipr-mus LD50:149 mg/kg THERAP 26,805,71
ivn-mus LD50:45 mg/kg THERAP 26,805,71

SAFETY PROFILE: Poison by ingestion, intravenous, and intraperitoneal routes. When heated to decomposition it emits toxic fumes of Cl^- and NO_x.

SBD000 CAS:56391-56-1 ***HR: 3***
SCH 20569
mf: $C_{21}H_{41}N_5O_7$ mw: 475.67

SYN: 1-N-AETHYLSISOMICIN ◇ 1-N-ETHYLSISOMICIN ◇ NETILMICIN ◇ NTL

TOXICITY DATA with REFERENCE
ims-rat TDLo:3040 mg/kg (female 6-22D post):REP APTOA6 45,145,79
ims-rat TDLo:2720 mg/kg (14D pre/1-20D preg):TER APTOA6 45,145,79
ivn-hmn TDLo:52 mg/kg/7D-C:KID CMAJAX 120,161,79

ipr-rat LD50:231 mg/kg JJANAX 35,461,82
scu-rat LD50:169 mg/kg JJANAX 35,461,82
ivn-rat LD50:25200 μg/kg ARZNAD 31,816,81
ivn-mus LD50:22 mg/kg JJANAX 35,461,82
ims-mus LD50:63100 μg/kg JJANAX 35,461,82
ims-mus LD50:63100 μg/kg JJANAX 35,461,82
scu-dog LDLo:500 mg/kg ARZNAD 31,816,81
ims-dog LD50:125 mg/kg ARZNAD 31,816,81
scu-mky LDLo:75 mg/kg JIDIAQ 137,476,78
scu-gpg LD50:207 mg/kg ARZNAD 31,816,81
ims-gpg LD50:181 mg/kg ARZNAD 31,816,81

SAFETY PROFILE: Poison by intravenous, intramuscular, and subcutaneous routes. An experimental teratogen. Human systemic effects by intravenous route: changes to kidney, ureter and bladder. Experimental reproductive effects. When heated to decomposition it emits acrid smoke and irritating fumes.

SBE000 CAS:68917-52-2 ***HR: 1***
SCHINUS MOLLE OIL

PROP: Distilled from the berries of *Schinus molle* (FCTXAV 14,659,76).

TOXICITY DATA with REFERENCE
skn-rbt 500 mg/24H MOD FCTXAV 14,659,76

CONSENSUS REPORTS: Reported in EPA TSCA Inventory.

SAFETY PROFILE: A skin irritant. When heated to decomposition it emits acrid smoke and irritating fumes.

SBE400 CAS:7432-28-2 ***HR: 2***
SCHISANDRIN
mf: $C_{24}H_{32}O_7$ mw: 432.56

SYNS: SCHIZANDRIN ◇ SCHIZANDROL B ◇ WUWEIZI ALCOHOL A ◇ WUWEIZICHUN A

TOXICITY DATA with REFERENCE
orl-mus LDLo:250 mg/kg CMJODS 93,41,80
ipr-mus LDLo:250 mg/kg CMJODS 93,41,80
scu-mus LD50:1861 mg/kg YKKZAJ 101,1030,81

SAFETY PROFILE: Poison by ingestion and intraperitoneal routes. Moderately toxic by subcutaneous route. When heated to decomposition it emits acrid smoke and irritating fumes.

SBE450 CAS:58546-54-6 ***HR: 3***
SCHISANDROL B
mf: $C_{23}H_{28}O_7$ mw: 416.51

SYNS: GOMISIN A ◇ SCHIZANDROL B ◇ WUWEIZI ALCOHOL B ◇ WUWEIZICHUN B

TOXICITY DATA with REFERENCE
orl-mus LD50:777 mg/kg YKKZAJ 101,1030,81
ipr-mus LD50:390 mg/kg YKKZAJ 101,1030,81
scu-mus LD50:500 mg/kg YKKZAJ 101,1030,81

SAFETY PROFILE: Poison by intraperitoneal route. Moderately toxic by ingestion and subcutaneous routes. When heated to decomposition it emits acrid smoke and irritating fumes.

SBE500 CAS:548-57-2 ***HR: 3***
SCHISTOSOMICIDE
mf: $C_{20}H_{24}N_2OS{\cdot}ClH$ mw: 376.98

SYNS: B.W. 57-233 ◇ CBC 900139 ◇ 1-DIAETHYLAMINOAETHYLAMINO-4-METHYL-THIOXANTHONHYDROCHLORID(GERMAN) ◇ 1-(β-DIETHYLAMINOETHYLAMINO)-4-METHYLTHIAXANTHONE HYDROCHLORIDE ◇ 1-(2-DIETHYLAMINOETHYLAMINO)-4-METHYLTHIAXANTHONE HYDROCHLORIDE ◇ 1-((2-DIETHYLAMINOETHYL)AMINO)-4-METHYLTHIOXANTHEN-9-ONE HYDROCHLORIDE ◇ 1-((2-DIETHYLAMINO)ETHYL)AMINO)-4-METHYL-9H-THIOXANTHEN-9-ONE, MONOHYDROCHLORIDE ◇ DR-15771 ◇ LUCANTHONE HYDROCHLORIDE ◇ LUCANTHONE MONOHYDROCHLORIDE ◇ MIRACIL D ◇ MIRACIL D HYDROCHLORIDE ◇ MIRACOL ◇ MS. 752 ◇ NIH 3127 ◇ NILODIN ◇ NSC 14574 ◇ 3735 R.P. ◇ SCAPUREN ◇ 79T61 ◇ TIXANTONE

TOXICITY DATA with REFERENCE
dnd-esc 10 μmol/L MUREAV 89,95,81
cyt-hmn:oth 10 μmol/L ACIEAY 10,302,71
cyt-hmn:leu 5 μmol/L ZENBAX 25B,115,70
ipr-mus LD50:181 mg/kg NCISP* JAN86
ivn-mus LD50:45 mg/kg QJPPAL 20,31,47

CONSENSUS REPORTS: EPA Genetic Toxicology Program.

SAFETY PROFILE: Poison by intravenous and intraperitoneal routes. Human mutation data reported. Used as an antischisosomal agent. When heated to decomposition it emits very toxic fumes of HCl, SO_x, and NO_x.

SBE800 CAS:67814-76-0 ***HR: 2***
SCH 21420 SULFATE
mf: $C_{22}H_{43}N_5O_{12}{\cdot}H_2O_4S$ mw: 667.78

SYN: 1-N-(S-3-AMINO-2-HYDROXYPROPIONYL)-GENTAMICINB SULFATE ◇ HAPA-B SULFATE ◇ ISEPAMICIN SULFATE

TOXICITY DATA with REFERENCE
ims-rat TDLo:5400 mg/kg (female 17-22D post):REP JJANAX 39,3311,86
ipr-rat LD50:1591 mg/kg JJANAX 39,3164,86
scu-rat LD50:3392 mg/kg JJANAX 39,3164,86
ims-rat LD50:2088 mg/kg JJANAX 39,3164,86
ipr-mus LD50:2244 mg/kg JJANAX 39,3164,86
scu-mus LD50:3320 mg/kg JJANAX 39,3164,86

SAFETY PROFILE: Moderately toxic by intraperitoneal and intramuscular route. Experimental reproductive effects. When heated to decomposition it emits toxic fumes of SO_x and NO_x. See also SULFATES.

SBF500 CAS:507-60-8 ***HR: 3***
(3-β,6-β)SCILLIROSIDE
mf: $C_{32}H_{44}O_{12}$ mw: 620.76

PROP: Long prisms. Very sol in alc, ethylene glycol, dioxane; sltly sol in water, acetone, chloroform; insol in ether, petr ether. From the bulb of *Urginea maritma* (JPETAB 103,420,51).

SYNS: 3-β,6-β-6-ACETYLOXY-3-(β-d-GLUCOPYRANOSYLOXY)-8,14-DIHYDROXYBUFA-4,20,22-TRIENOLIDE ◇ 6-β-(ACETYLOXY)-3-β-(β-d-GLUCOPYRANOSYLOXY)-8,14-DIHYDROXYBUFA-4,20,22-TRIENOLIDE ◇ SCILLIROSID ◇ SCILLIROSIDE ◇ SCILLIROSIDIN + GLUCOSE (GERMAN) ◇ SILMURIN

TOXICITY DATA with REFERENCE
orl-rat LD50:430 μg/kg GUCHAZ 6,466,73
orl-mus LD50:440 μg/kg APTOA6 8,391,52
scu-mus LD50:471 μg/kg APTOA6 8,391,52
ivn-cat LD50:130 μg/kg JPETAB 103,420,51
orl-ckn LD50:400 mg/kg 85DPAN -,-,71/76

SAFETY PROFILE: A deadly poison by ingestion, subcutaneous, and intravenous routes. Used as a rodenticide. When heated to decomposition it emits acrid smoke and irritating fumes.

SBF550 ***HR: 3***
SCIRPUS ARTICULATUS Linn., extract

PROP: Indian plant belonging to the family *Cyperaceae* IJEBA6 22,312,84

SYN: SCHOENOPLECTUS ARTICULATUS (Linn.) Palla, extract

TOXICITY DATA with REFERENCE
orl-ham TDLo:500 mg/kg (female 1-5D post):REP IJEBA6 22,312,84
ipr-mus LD50:21500 μg/kg IJEBA6 22,312,84

SAFETY PROFILE: Poison by intraperitoneal route. Experimental reproductive effects. When heated to decomposition it emits acrid smoke and irritating fumes.

SBG000 CAS:51-34-3 ***HR: 3***
SCOPOLAMINE
mf: $C_{17}H_{21}NO_4$ mw: 303.39

PROP: Thick, colorless, syrupy liquid alkaloid. Mp: 55°. Very sol in hot water, alc, ether, chloroform, acetone; sltly sol in benzene, petr ether. Decomp on standing.

SYNS: ATROCHIN ◇ ATROQUIN ◇ 6-β,7-β-EPOXY-3-α-TROPANYL S-(−)-TROPATE ◇ EPOXYTROPINE TROPATE ◇ HYOSCINE ◇ (−)-HYOSCINE ◇ HYOSOL ◇ ISOPTO HYOSCINE ◇ 9-METHYL-3-OXA-9-AZATRICYCLO(3.3.1.$0^{2,4}$)NONAN-7-OL,TROPATE (ESTER) ◇ OSCINE ◇ SCOPINE TROPATE ◇ (−)-SCOPOLAMINE ◇ TROPIC ACID, ESTER with SCOPINE ◇ TROPIC ACID, 9-METHYL-3-OXA-9-AZATRICYCLO(3.3.1.$0^{2,4}$)NON-7-YL ESTER

TOXICITY DATA with REFERENCE
cyt-hmn:leu 1 pph/22H 22XWAN -,367,70
ipr-mus TDLo:1 mg/kg (13D preg):REP FEPRA7 31,596,72
par-mus TDLo:50600 μg/kg (female 6-16D post):TER CMJODS 101,339,88
scu-hmn TDLo:2 μg/kg:BRN,CNS JPETAB 137,133,62
scu-man TDLo:14 μg/kg AJPSAO 141,1010,84
scu-wmn TDLo:13 μg/kg AJPSAO 141,1010,84
ims-man TDLo:24 μg/kg:CNS FEPRA7 32,250,73
ims-hmn TDLo:4 μg/kg:EYE,CNS 85IVAW 1,L1,82
orl-mus LD50:1275 mg/kg AIPTAK 134,255,61
ipr-mus LD50:400 mg/kg ARZNAD 21,1727,71
scu-mus LD50:1700 mg/kg JPETAB 56,85,36
ivn-mus LD50:100 mg/kg JMCMAR 26,1772,83
scu-rbt LDLo:75 mg/kg HBAMAK 4,1289,35
ivn-rbt LDLo:50 mg/kg HBAMAK 4,1289,35

CONSENSUS REPORTS: EPA Genetic Toxicology Program.

SAFETY PROFILE: Poison by intravenous, intraperitoneal, and subcutaneous routes. Moderately toxic by ingestion. Human systemic effects from very small amounts by subcutaneous and intramuscular routes: changes in surface EEG, distorted perceptions, excitement, hallucinations, and mydriasis. It can cause the individual who is affected to lose a certain amount of his normal inhibitory control. It is for that reason that it has been called "truth serum". An experimental teratogen. Experimental reproductive effects. Human mutation data reported. In many cases of poisoning from this material, and even to a certain extent following its medical application, there is retention of the urine caused by paralysis of the bladder, and catheterization is necessary. The fatal dose is variable. Death has occurred from as little as 0.6 mg, while recovery has occurred from doses of 7-15 mg. An anticholinergic drug. When heated to decomposition it emits highly toxic fumes of NO_x. See also ESTERS.

SBG500 CAS:149-64-4 ***HR: 3***
SCOPOLAMINE-N-BUTYL BROMIDE
mf: $C_{21}H_{30}NO_4 \cdot Br$ mw: 440.43

SYNS: AMISEPAN ◇ BUSCAPINA ◇ BUSCAPINE ◇ BUSCOL ◇ BUSCOLAMIN ◇ BUSCOLYSINE ◇ BUSCOPAN ◇ BUTYLHYOSCINE ◇ N-BUTYLHYOSCINE BROMIDE ◇ N-BUTYLHYOSCINIUM BROMIDE ◇ BUTYLMIN ◇ BUTYLSCOPOLAMINE BROMIDE ◇ N-BUTYLSCOPOLAMINE BROMIDE ◇ N-BUTYLSCOPOLAMINIUM BROMIDE ◇ BUTYLSCOPOLAMMONIUM BROMIDE ◇ N-BUTYLSCOPOLAMMONIUM BROMIDE ◇ DONOPON ◇ HYOSCIN-N-BUTYLBROMID (GERMAN) ◇ HYOSCIN-N-BUTYL BROMIDE ◇ HYOSCINE BUTOBROMIDE ◇ HYOSCINE BUTYL BROMIDE ◇ HYOSCINE-N-BUTYL BROMIDE ◇ JOSCINE ◇ MONOSPAN ◇ SCOBRO ◇ SCOBRON ◇ SCOBUTIL ◇ SCOBUTYL ◇ SCOPOLAMINE BROMOBUTYLATE ◇ SCOPOLAMINE BUTOBROMIDE ◇ SCOPOLAMINE BUTYL BRO-

MIDE ◇ SCOPOLAN ◇ SPARICON ◇ SPORAMIN ◇ STILBRON ◇ TIRANTIL

TOXICITY DATA with REFERENCE
ims-rat TDLo:2010 mg/kg (30D pre):REP OYYAA2 9,615,75
orl-rat LD50:1040 mg/kg AIPTAK 180,155,69
scu-rat LD50:510 mg/kg AIPTAK 180,155,69
ivn-rat LD50:24 mg/kg OYYAA2 23,461,82
idu-rat LD50:180 mg/kg AIPTAK 180,155,69
orl-mus LD50:1170 mg/kg THERAP 14,1096,59
ipr-mus LD50:73 mg/kg JMCMAR 16,1063,73
scu-mus LD50:304 mg/kg NIIRDN 6,354,82
ivn-mus LD50:10300 μg/kg OYYAA2 5,518,71
orl-gpg LD50:750 mg/kg FRPSAX 39,3,84
scu-gpg LD50:65 mg/kg FRPSAX 39,3,84
scu-gpg LD50:155 mg/kg AIPTAK 137,375,62

SAFETY PROFILE: Poison by intravenous, subcutaneous, intraperitoneal, and intraduodenal routes. Moderately toxic by ingestion. Experimental reproductive effects. Used as an antispasmodic agent. When heated to decomposition it emits very toxic fumes of NO_x, NH_3, and Br^-. See also BROMIDES and SCOPOLAMINE.

SBH000 CAS:6106-46-3 ***HR: 3***
SCOPOLAMINE compounded with METHYL NITRATE (1:1)
mf: $C_{17}H_{21}NO_4 \cdot CH_3NO_3$ mw: 380.44

SYNS: EPOXYTROPINE TROPATE METHYLNITRATE ◇ METHYL SCOPOLAMINE NITRATE ◇ SCOPOLAMINE METHYL NITRATE

TOXICITY DATA with REFERENCE
orl-rat LD50:1560 mg/kg TXAPA9 1,42,59
ipr-rat LD50:126 mg/kg CLDND*

CONSENSUS REPORTS: Reported in EPA TSCA Inventory.

SAFETY PROFILE: Poison by intraperitoneal route. Moderately toxic by ingestion. When heated to decomposition it emits toxic fumes of NO_x. See also SCOPOLAMINE and METHYL NITRATE.

SBH500 CAS:155-41-9 ***HR: 3***
SCOPOLAMINE METHYLBROMIDE
mf: $C_{18}H_{24}NO_4 \cdot Br$ mw: 398.34

SYNS: AMPYROX ◇ BLOCAN ◇ DIOPAL ◇ EPOXYMETHAMINE BROMIDE ◇ EPOXYTROPINE TROPATE METHYLBROMIDE ◇ HOLOPAN ◇ HYOSCINE METHYL BROMIDE ◇ LESCOPINE BROMIDE ◇ MESCOPIL ◇ METHOSCOPYLAMINE BROMIDE ◇ METHSCOPOLAMINE BROMIDE ◇ N-METHYLHYOSCINE BROMIDE ◇ METHYLSCOPOLAMINE BROMIDE ◇ METHYLSCOPOLAMINE HYDROBROMIDE ◇ N-METHYLSCOPOLAMMONIUM BROMIDE ◇ NEO-AVAGAL ◇ NUTROP ◇ PAMINE ◇ PAMINE BROMIDE ◇ PARASPAN ◇ PROSCOMIDE ◇ RESTROPIN ◇ SCOPOLAMINE METHOBROMIDE ◇ (−)-SCOPOLAMINE METHYL BROMIDE

TOXICITY DATA with REFERENCE
orl-rat LD50:3400 mg/kg AIPTAK 180,155,69
scu-rat LD50:2060 mg/kg AIPTAK 180,155,69
idu-rat LD50:870 mg/kg AIPTAK 180,155,69
orl-mus LD50:1214 mg/kg AIPTAK 156,467,65
scu-mus LD50:880 mg/kg ARZNAD 18,1132,68
ivn-mus LD50:26806 μg/kg AIPTAK 103,100,55

CONSENSUS REPORTS: Reported in EPA TSCA Inventory.

SAFETY PROFILE: Poison by intravenous route. Moderately toxic by ingestion, subcutaneous, and intraduodenal routes. When heated to decomposition it emits very toxic fumes of NO_x, NH_3, and Br^-. See also BROMIDES and SCOPOLAMINE.

SBI800 ***HR: 3***
SEA ANEMONE VENOM

SYNS: AIPTASIA PALLIDA VENOM ◇ VENOM, SEA ANEMONE, AIPTASIA PALLIDA

TOXICITY DATA with REFERENCE
ipr-mus LD50:56 μg/kg TOXIA6 22,308,84
ivn-mus LD50:136 μg/kg TOXIA6 17,109,79
ice-mus LD50:1700 ng/kg TOXIA6 22,308,84

SAFETY PROFILE: Deadly poison by intravenous, intraperitoneal, and intracerebral routes.

SBI860 CAS:9074-07-1 ***HR: 3***
SEAPROSE S

SYNS: ASPERFILLUS ALKALINE PROTEINASE ◇ ASPERGILLOPEPTIDASE B ◇ E.C. 3.4.21.15 ◇ PROTEINASE, ASPERFILLUS ALKALINE ◇ PROZIME 10

TOXICITY DATA with REFERENCE
orl-rat LD50:8750 mg/kg NIIRDN 6,396,82
scu-rat LD50:510 mg/kg NIIRDN 6,396,82
orl-mus LD50:7400 mg/kg NIIRDN 6,396,82
scu-mus LD50:48 mg/kg NIIRDN 6,396,82
ivn-mus LD50:21 mg/kg NIIRDN 6,396,82

SAFETY PROFILE: Poison by subcutaneous and intravenous routes. Mildly toxic by ingestion.

SBI875 ***HR: 2***
SEARS HEAVY DUTY LAUNDRY DETERGENT

PROP: Sodium carbonate, Sears Roebuck and Co. (FCTXAV 14,78,76).

TOXICITY DATA with REFERENCE
orl-dog LDLo:2500 mg/kg FCTXAV 14,78,76
orl-pig LDLo:2500 mg/kg FCTXAV 14,78,76

SAFETY PROFILE: Moderately toxic by ingestion. When heated to decomposition it emits toxic fumes of Na_2O.

SBI880 ***HR: 3***
SEA SNAKE VENOM, AIPYSURUS LAEVIS

SYNS: AIPYSURUS LAEVIS VENOM ◇ VENOM, SEA SNAKE, AIPYSURUS LAEVIS

TOXICITY DATA with REFERENCE
scu-mus LD50:262 μg/kg TOXIA6 14,347,76
ivn-mus LD50:155 μg/kg TOXIA6 14,347,76
ims-mus LD50:90 μg/kg 35FVAR 2,161,76

SAFETY PROFILE: Deadly poison by subcutaneous, intramuscular, and intravenous routes.

SBI890 ***HR: 3***
SEA SNAKE VENOM, ASTROTIA STOKESII

SYNS: ASTROTIA STOKESII VENOM ◇ VENOM, SEA SNAKE, ASTROTIA STOKESII

TOXICITY DATA with REFERENCE
scu-mus LD50:260 μg/kg TOXIA6 14,347,76
ivn-mus LD50:164 μg/kg TOXIA6 14,347,76
ims-mus LD50:250 μg/kg 35FUAR 2,161,76

SAFETY PROFILE: Deadly poison by subcutaneous, intramuscular, and intravenous routes.

SBI900 ***HR: 3***
SEA SNAKE VENOM, HYDROPHIS CYANOCINCTUS

SYNS: H. CYANOCINCTUS VENOM ◇ HYDROPHIS CYANOCINCTUS VENOM ◇ VENOM, SEA SNAKE, HYDROPHIS CYANOCINCTUS

TOXICITY DATA with REFERENCE
ipr-mus LD50:200 μg/kg 85EGD4 -,415,78
scu-mus LD50:465 μg/kg TOXIA6 14,347,76
ivn-mus LD50:275 μg/kg TOXIA6 14,347,76

SAFETY PROFILE: Deadly poison by subcutaneous, intravenous, and intraperitoneal routes.

SBI910 ***HR: 3***
SEA SNAKE VENOM, LAPEMIS HARDWICKII

SYNS: LAPEMIS HARDWICKII VENOM ◇ VENOM, SEA SNAKE, LAPEMIS HARDWICKII

TOXICITY DATA with REFERENCE
ipr-mus LD50:260 μg/kg 85EGD4 -,341,78
scu-mus LD50:541 μg/kg TOXIA6 14,347,76
ivn-mus LD50:303 μg/kg TOXIA6 14,347,76

SAFETY PROFILE: Deadly poison by subcutaneous, intravenous, and intraperitoneal routes.

SBI929 ***HR: 3***
SEA SNAKE VENOM, MICROCEPHALOPHIS GRACILIS

SYN: VENOM, SEA SNAKE, MICROCEPHALOPHIS GRACILIS

TOXICITY DATA with REFERENCE
scu-mus LD50:480 μg/kg TOXIA6 14,347,76
ivn-mus LD50:125 μg/kg 14FHAR -,373,63
par-mus LDLo:125 μg/kg SCNEBK 110,355,76

SAFETY PROFILE: Deadly poison by subcutaneous, parenteral, and intravenous routes.

SBJ500 CAS:111-20-6 ***HR: 2***
SEBACIC ACID
mf: $C_{10}H_{18}O_4$ mw: 202.28

PROP: Monoclinic, prismatic tablets. Commercial product is a white, free-flowing powder with a mild fatty acid odor. D: 1.207 @ 20°/4°; mp: 134.5°; bp: 294.5° @ 100 mm. vap press: 1 mm @ 183.0°. Very sol in alc, esters, ketones; sltly sol in hydrocarbons, chlorinated hydrocarbons.

SYNS: DECANEDIOIC ACID ◇ 1,8-OCTANEDICARBOXYLIC ACID ◇ USAF HC-1

TOXICITY DATA with REFERENCE
orl-rat LD50:3400 mg/kg GISAAA 48(9),72,83
orl-mus LD50:6000 mg/kg BIJOAK 34,1196,40
ipr-mus LD50:500 mg/kg NTIS** AD277-689
ipr-mam LD50:1175 mg/kg GTPZAB 24(5),48,80

CONSENSUS REPORTS: Reported in EPA TSCA Inventory.

SAFETY PROFILE: Moderately toxic by ingestion and intraperitoneal routes. When heated to decomposition it emits acrid smoke and irritating fumes.

SBK000 CAS:925-83-7 ***HR: 3***
SEBACIC ACID, DIHYDRAZIDE
mf: $C_{10}H_{22}N_4O_2$ mw: 230.36

TOXICITY DATA with REFERENCE
orl-rat LD50:2500 mg/kg OMCDS*
orl-mus LD50:400 mg/kg BIJOAK 34,1196,40
ivn-mus LD50:56 mg/kg CSLNX* NX#04470

CONSENSUS REPORTS: Reported in EPA TSCA Inventory.

SAFETY PROFILE: Poison by ingestion and intravenous routes. When heated to decomposition it emits toxic fumes of NO_x.

SBK500 CAS:1871-96-1 ***HR: 3***
SEBACONITRILE
mf: $C_{10}H_{16}N_2$ mw: 164.28

TOXICITY DATA with REFERENCE
orl-rat LDLo:500 mg/kg JPETAB 90,260,47
orl-mus LD50:396 mg/kg ARTODN 57,88,85

CONSENSUS REPORTS: Cyanide and its compounds

are on the Community Right-To-Know List. Reported in EPA TSCA Inventory.

SAFETY PROFILE: Poison by ingestion. When heated to decomposition it emits toxic fumes of NO_x and CN^-. See also NITRILES.

SBL500 CAS:51165-36-7 ***HR: 3***
R(+)-SECOBARBITAL SODIUM
mf: $C_{12}H_{17}N_2O_3$•Na mw: 260.30

PROP: White, hygroscopic powder; bitter taste. Very sol in water; sol in alc; insol in ether. The free acid melts @ 100°.

SYN: (R+)-5-ALLYL-5-(1-METHYLBUTYL)BARBITURIC ACID SODIUM SALT

TOXICITY DATA with REFERENCE
ipr-mus LD50:137 mg/kg TXAPA9 26,495,73
ivn-mus LD50:130 mg/kg TXAPA9 26,495,73

SAFETY PROFILE: Poison by intraperitoneal and intravenous routes. Abuse may lead to addiction. When heated to decomposition it emits toxic fumes of NO_x and Na_2O. See also BARBITURATES.

SBM500 CAS:76-73-3 ***HR: 3***
SECONAL
mf: $C_{12}H_{18}N_2O_3$ mw: 238.32

SYNS: 5-ALLYL-5-(1-METHYLBUTYL)BARBITURIC ACID ◇ 5-ALLYL-5-(1-METHYLBUTYL)MALONYLUREA ◇ BARBOSEC ◇ EVRONAL ◇ HYPOTROL ◇ IMESONAL ◇ IMMENOCTAL ◇ IMMENOX ◇ MEBALLYMAL ◇ 5-(1-METHYLBUTYL)-5-(2-PROPENYL)-2,4,6(1H,3H,5H)-PYRIMIDINITRIONE ◇ QUINALBARBITAL ◇ QUINALBARBITONE ◇ SECOBARBITAL ◇ SECOBARBITONE ◇ TRISOMNIN

TOXICITY DATA with REFERENCE
orl-wmn TDLo:32 mg/kg:CNS,GIT BMJOAE 1,1238,55
orl-hmn LDLo:33 mg/kg CTOXAO 10,327,77
orl-rat LDLo:125 mg/kg JPHAA3 26,1248,37
ipr-rat LDLo:110 mg/kg JPHAA3 26,1248,37
scu-rat LDLo:140 mg/kg JPHAA3 26,1248,37
orl-mus LD50:267 mg/kg OYYAA2 7,1349,73
ipr-mus LD50:116 mg/kg TXAPA9 1,65,59
scu-mus LDLo:160 mg/kg JPHAA3 26,1248,37
ivn-mus LDLo:80 mg/kg JPHAA3 26,1248,37
orl-dog LDLo:90 mg/kg JPHAA3 26,1248,37

SAFETY PROFILE: Human poison by ingestion. Experimental poison by ingestion, intraperitoneal, subcutaneous, and intravenous routes. Human systemic effects by ingestion: changes in motor activity, coma and nausea or vomiting. When heated to decomposition it emits toxic fumes of NO_x. See also BARBITURATES.

SBN000 CAS:309-43-3 ***HR: 3***
SECONAL SODIUM
mf: $C_{12}H_{17}N_2O_3$•Na mw: 260.30

SYNS: 5-ALLYL-5(1-METHYLBUTYL)BARBITURIC ACID SODIUM DERIVATIVE ◇ 5-ALLYL-5-(1-METHYLBUTYL)BARBITURIC ACID SODIUM SALT ◇ 5-ALLYL-5-(1-METHYLBUTYL)MALONYLUREA SODIUM SALT ◇ BARBOSEC ◇ BIPANAL ◇ BIPINAL SODIUM ◇ EVRONAL SODIUM ◇ EVRRONAL ◇ HYPOTROL ◇ IMESONAL ◇ IMMENOCTAL ◇ MEBALLYMAL SODIUM ◇ 5-(1-METHYLBUTYL)-5-(2-PROPENYL)-2,4,6(1H,3H,5H)-PYRIMIDINETRIONEMONOSODIUM SALT ◇ PRAMIL ◇ PROPYLMETHYLCARBINYLALLYL BARBITURIC ACID SODIUM SALT ◇ QUINALBARBITONE SODIUM ◇ QUINALSPAN ◇ SEBAR ◇ SECO 8 ◇ SECOBARBITAL SODIUM ◇ SECOBARBITONE ◇ SECOBARBITONE SODIUM ◇ SEDUTAIN ◇ SEOTAL ◇ SODIUM-5-ALLYL-5-(1-METHYLBUTYL)BARBITURATE ◇ SODIUM QUINALBARBITONE ◇ SODIUM SECOBARBITAL ◇ SODIUM SECONAL ◇ SYNATE ◇ TRISOMNIN

TOXICITY DATA with REFERENCE
orl-rat LD50:125 mg/kg TXAPA9 21,315,72
ipr-rat LDLo:110 mg/kg PSEBAA 32,1563,35
scu-rat LDLo:140 mg/kg PSEBAA 32,1563,35
ivn-rat LD50:65 mg/kg JAPMA8 44,152,55
ipr-mus LD50:112 mg/kg TXAPA9 26,495,73
ivn-mus LD50:100 mg/kg TXAPA9 26,495,73
orl-dog LD50:85 mg/kg JAPMA8 44,152,55

SAFETY PROFILE: Poison by ingestion, intraperitoneal, subcutaneous, and intravenous routes. A much-used sedative and hypnotic drug. It is also used as a veterinary pre-anesthetic and sedative. Often overused or abused. When heated to decomposition it emits toxic fumes of NO_x and Na_2O. See also BARBITURATES.

SBN300 CAS:57558-46-0 ***HR: 3***
SECOVERINE HYDROCHLORIDE
mf: $C_{22}H_{35}NO_2$•ClH mw: 382.04

SYNS: 1-CYCLOHEXYL-4-(ETHYL-p-METHOXY-α-METHYLPHENETHYL)AMINO)-1-BUTANONE HYDROCHLORIDE ◇ SECOVERINE

TOXICITY DATA with REFERENCE
orl-rat LD50:369 mg/kg ARZNAD 30,1526,80
ivn-rat LD50:17 mg/kg ARZNAD 30,1526,80
orl-mus LD50:369 mg/kg ARZNAD 30,1526,80
ivn-mus LD50:32 mg/kg ARZNAD 30,1526,80

SAFETY PROFILE: Poison by ingestion and intravenous routes. When heated to decomposition it emits toxic fumes of NO_x and HCl.

SBN350 CAS:5610-40-2 ***HR: 3***
(-)-SECURININE
mf: $C_{13}H_{15}NO_2$ mw: 217.29

SYNS: SECURINAN-11-ONE (9CI) ◇ SECURININ ◇ SECURININE

TOXICITY DATA with REFERENCE
scu-rat TDLo:50 mg/kg (male 5D pre):REP NATWAY 52,483,65

ipr-rat LD50:41 mg/kg CHHTAT 54,234,74
ivn-rat LD50:15100 μg/kg CHHTAT 54,234,74
ims-rat LD50:17400 μg/kg CHHTAT 54,234,74
orl-mus LD50:270 mg/kg CHHTAT 54,234,74
ipr-mus LD50:31800 μg/kg CHHTAT 54,234,74
scu-mus LD50:20420 μg/kg CHHTAT 54,234,74

SAFETY PROFILE: Poison by ingestion, intravenous, and intraperitoneal routes. Experimental reproductive effects. When heated to decomposition it emits toxic fumes of NO_x.

SBN440 CAS:38821-53-3 ***HR: 2***
SEFRIL
mf: $C_{16}H_{19}N_3O_4S$ mw: 349.44

PROP: Monohydrate: Small, colorless crystals. Mp: 140-142° (decomp). Sol in propylene glycol; sltly sol in acetone, ethanol; insol in ether, chloroform, benzene, hexane.

SYNS: CEFRADINE ◇ CEPHRADIN ◇ CEPHRADINE ◇ VELOSEF

TOXICITY DATA with REFERENCE
orl-mus TDLo:600 mg/kg (7-12D preg):REP NKRZAZ 23,37,75
orl-rat TDLo:3 g/kg (9-14D preg):TER NKRZAZ 23,37,75
orl-wmn TDLo:5 mg/kg:SKN CUTIBC 38,58,86
orl-mus LD50:3549 mg/kg NKRZAZ 23,37,75
ipr-mus LD50:597 mg/kg IYKEDH 8,680,77
scu-mus LD50:2890 mg/kg IYKEDH 8,680,77
ivn-mus LD50:3539 mg/kg IYKEDH 8,680,77

SAFETY PROFILE: Moderately toxic by ingestion, intraperitoneal, subcutaneous, and intravenous routes. Human systemic effects: dermatitis. An experimental teratogen. Experimental reproductive effects. When heated to decomposition it emits toxic fumes of SO_x and NO_x.

SBN450 CAS:31828-50-9 ***HR: 2***
SEFRIL HYDRATE
mf: $C_{16}H_{19}N_3O_4S \cdot xH_20$ mw: 475.58

SYNS: d-7-(2-AMINO-2-(1,4-CYCLOHEXADIEN-1-YL)ACETAMIDO)-3-METHYL-8-OXO-5-THIA-1-AZABICYCLO(4.2.0)OCT-2-ENE-2-CARBOXYLIC ACID HYDRATE ◇ DEFRADIN HYDRATE ◇ DEPHRADINE HYDRATE ◇ VELOSEF HYDRATE

TOXICITY DATA with REFERENCE
ipr-rat LD50:4000 mg/kg AMACCQ 3,682,73
orl-mus LD50:5000 mg/kg AMACCQ 3,682,73
ipr-mus LD50:700 mg/kg AMACCQ 3,682,73
ivn-mus LD50:3000 mg/kg AMACCQ 3,682,73

SAFETY PROFILE: Moderately toxic by intraperitoneal and intravenous routes. Mildly toxic by ingestion. When heated to decomposition it emits toxic fumes of SO_x and NO_x.

SBN475 ***HR: 3***
SELECTROL
mf: $C_{20}H_{33}N_3O_4 \cdot ClH$ mw: 416.02

SYNS: 3-(3-ACETIL-4-(3-tert-BUTILAMINO)-2-HIDROXIPROPOXI) FENIL)-1,1-DIETILUREA HCl (SPANISH) ◇ 3-(3-ACETYL-4-(3-tert-BUTYLAMINO-2-HYDROXYPROPOXY)PHENYL)-1,1-DIETHYL-HARNSTOFF HCl (GERMAN) ◇ 3-(3-ACETYL-4-(3-tert-BUTYLAMINO-2-HYDROXYPROPOXY)PHENYL)-1,1-DIETHYLUREAHYDROCHLORIDE ◇ CELIPROLOL HYDROCHLORID (GERMAN) ◇ CELIPROLOL HYDROCHLORIDE ◇ CLORHIDRATO de CELIPROLOL (SPANISH) ◇ ST 1396 CLORHIDRATO (SPANISH) ◇ ST 1396 HYDROCHLORID (GERMAN) ◇ ST 1396 HYDROCHLORIDE

TOXICITY DATA with REFERENCE
orl-rat LD50:3826 mg/kg ARZNAD 33,41,83
ivn-rat LD50:68300 μg/kg ARZNAD 33,41,83
orl-mus LD50:1834 mg/kg ARZNAD 33,41,83
ivn-mus LD50:56200 μg/kg ARZNAD 33,41,83
orl-dog LDLo:1250 mg/kg ARZNAD 33,41,83
ivn-dog LDLo:100 mg/kg ARZNAD 33,41,83

SAFETY PROFILE: Poison by intravenous route. Moderately toxic by ingestion. When heated to decomposition it emits toxic fumes of NO_x and HCl.

SBN500 CAS:7783-08-6 ***HR: 3***
SELENIC ACID
DOT: UN 1905
mf: H_2O_4Se mw: 144.98

PROP: Colorless liquid or colorless, hexagonal prisms. Mp: 58°, bp: 260° (decomp), d (solid): 2.951 @ 15°, (liquid): 2.609 @ 15°. Very sol in water; sol in sulfuric acid; insol in ammonia; decomp in alc. Very deliq.

SYN: SELENIC ACID, liquid (DOT)

CONSENSUS REPORTS: Selenium and its compounds are on the Community Right-To-Know List. EPA Genetic Toxicology Program. Reported in EPA TSCA Inventory.

OSHA PEL: TWA 0.2 mg(Se)/m^3
ACGIH TLV: TWA 0.2 mg(Se)/m^3
DFG MAK: 0.1 mg(Se)/m^3
DOT Classification: Corrosive Material; Label: Corrosive.

SAFETY PROFILE: Selenium compounds are poisons. A corrosive irritant to skin, eyes, and mucous membranes. When heated to decomposition it emits toxic fumes of Se. See also SELENIUM COMPOUNDS.

SBN505 CAS:10102-23-5 ***HR: 3***
SELENIC ACID, DISODIUM SALT, DECAHYDRATE
mf: $O_4Se \cdot 2Na \cdot 10H_2O$ mw: 369.14

TOXICITY DATA with REFERENCE
ipr-mus LD50:18 mg/kg COREAF 257,791,63

OSHA PEL: TWA 0.2 mg(Se)/m^3
ACGIH TLV: TWA 0.2 mg(Se)/m^3

SAFETY PROFILE: Poison by intraperitoneal route. When heated to decomposition it emits toxic fumes of Se.

SBN510 CAS:55509-78-9 ***HR: 3***
β-SELENINOPROPIONIC ACID
mf: $C_3H_6O_4Se$ mw: 185.05

SYN: PROPIONIC ACID, 3-SELENINO-

TOXICITY DATA with REFERENCE
ipr-rat LDLo:25 mg(Se)/kg JPETAB 63,357,38

OSHA PEL: TWA 0.2 mg(Se)/m^3
ACGIH TLV: TWA 0.2 mg(Se)/m^3

SAFETY PROFILE: Poison by intravenous route. When heated to decomposition it emits toxic fumes of Se.

SBN525 CAS:2424-09-1 ***HR: 3***
SELENINYL BIS(DIMETHYLAMIDE)
mf: $C_4H_{12}N_2OSe$ mw: 183.11

$SeO[N(CH_3)_2]_2$

CONSENSUS REPORTS: Selenium and its compounds are on the Community Right-To-Know List.

OSHA PEL: TWA 0.2 mg(Se)/m^3
ACGIH TLV: TWA 0.2 mg(Se)/m^3
DFG MAK: 0.1 mg(Se)/m^3

SAFETY PROFILE: Selenium compounds are poisons. May decompose explosively when heated above 50°C. When heated to decomposition it emits toxic fumes of NO_x and Se. See also SELENIUM COMPOUNDS.

SBN550 CAS:7789-51-7 ***HR: 3***
SELENINYL BROMIDE
mf: Br_2OSe mw: 254.77

CONSENSUS REPORTS: Selenium and its compounds are on The Community Right-To-Know List.

OSHA PEL: TWA 0.2 mg(Se)/m^3
ACGIH TLV: TWA 0.2 mg(Se)/m^3
DFG MAK: 0.1 mg(Se)/m^3

SAFETY PROFILE: Selenium compounds are poisons. Explosive reaction with white phosphorous. Violent reaction with sodium, potassium, zinc, red phosphorous. When heated to decomposition it emits toxic fumes of Br^- and Se. See also SELENIUM COMPOUNDS.

SBO000 CAS:7783-00-8 ***HR: 3***
SELENIOUS ACID
mf: H_2O_3Se mw: 128.98

PROP: Transparent, colorless crystals. Mp: decomp, d: 3.004 @ 15°/4°, vap press: 2 mm @ 15°. Very sol in alc; insol in ammonia.

SYNS: RCRA WASTE NUMBER U204 ◇ SELENIUM DIOXIDE

TOXICITY DATA with REFERENCE
cyt-hmn:lym 10 μmol/L ESKGA2 26,99,80
orl-rat LDLo:25 mg/kg NCNSA6 5,28,53
ipr-rat LDLo:10 mg/kg NCNSA6 5,28,53
ivn-mus LD50:11 mg/kg CSLNX* NX#05656

CONSENSUS REPORTS: EPA Extremely Hazardous Substances List. Selenium and its compounds are on the Community Right-To-Know List. Reported in EPA TSCA Inventory.

OSHA PEL: TWA 0.2 mg(Se)/m^3
ACGIH TLV: TWA 0.2 mg(Se)/m^3
DFG MAK: 0.1 mg(Se)/m^3

SAFETY PROFILE: Poison by ingestion, intraperitoneal, and intravenous routes. Human mutation data reported. Used as an oxidizing agent. When heated to decomposition it emits toxic fumes of Se. See also SELENIUM COMPOUNDS.

SBO500 CAS:7782-49-2 ***HR: 3***
SELENIUM
DOT: UN 2658
af: Se aw: 78.96

PROP: Steel gray, nonmetallic element. Mp: 170-217°, bp: 690°, d: 4.81-4.26, vap press: 1 mm @ 356°. Insol in water and alc; very sltly sol in ether.

SYNS: C.I. 77805 ◇ COLLOIDAL SELENIUM ◇ ELEMENTAL SELENIUM ◇ SELEN (POLISH) ◇ SELENIUM ALLOY ◇ SELENIUM BASE ◇ SELENIUM DUST ◇ SELENIUM ELEMENTAL ◇ SELENIUM HOMOPOLYMER ◇ SELENIUM METAL POWDER, NON-PYROPHORIC (DOT) ◇ VANDEX

TOXICITY DATA with REFERENCE
orl-mus TDLo:134 mg/kg (MGN):TER AEHLAU 23,102,71
orl-mus TDLo:480 mg/kg/60D-C:ETA YMBUA7 11,368,60
orl-rat LD50:6700 mg/kg TXAPA9 20,89,71
ihl-rat LDLo:33 mg/kg/8H AMIHBC 4,458,51
ivn-rat LD50:6 mg/kg AMIHBC 4,458,51
ivn-rbt LDLo:2500 μg/kg JOGBAS 35,693,28

CONSENSUS REPORTS: IARC Cancer Review: Group 3 IMEMDT 7,56,87. Selenium and its compounds are on the Community Right-To-Know List. Reported in EPA TSCA Inventory.

OSHA PEL: TWA 0.2 mg(Se)/m^3
ACGIH TLV: TWA 0.2 mg(Se)/m^3
DFG MAK: 0.1 mg(Se)/m^3
DOT Classification: Poison B; Label: St. Andrews Cross.

SAFETY PROFILE: Poison by inhalation and intravenous routes. Questionable carcinogen with experimental tumorigenic and teratogenic data. Occupational exposure has caused pallor, nervousness, depression, garlic odor of breath and sweat, gastrointestinal disturbances, and dermatitis. Liver damage in experimental animals. Chronic ingestion of 5 mg of selenium per day resulted in 49% morbidity in 5 Chinese villages. The main symptoms were brittle hair with intact follicles, new hair with no pigment, brittle nails with spots and streaks, skin lesions, peripheral anesthesia, acroparaesthesia, pain, and hyperreflexia. Similar effects have been seen in populations with selenium blood levels of 800 μg/L. In cattle, ''alkali disease'' is associated with consumption of grain or plants containing 5-25 mg/kg of selenium. The symptoms are lack of vitality, loss of appetite, emaciation, deformation and shedding of hoofs, loss of hair, and erosion of joints. Consumption of plants grown in seleniferous areas can cause effects in humans and animals. Selenosis in humans has occurred from ingestion of 3.2 mg selenium per day. Selenium is an essential trace element for many species.

Reacts to form explosive products with metal amides. Can react violently with barium carbide, bromine pentafluoride, calcium carbide, chlorates, chlorine trifluoride, chromic oxide (CrO_3), fluorine, lithium carbide, lithium silicon (Li_6 Si_2), metals, nickel, nitric acid, sodium, nitrogen trichloride, oxygen, potassium, potassium bromate, rubidium carbide, zinc, silver bromate, strontium carbide, thorium carbide, uranium. When heated to decomposition it emits toxic fumes of Se. See also SELENIUM COMPOUNDS.

SBP000 CAS:7782-49-2 ***HR: 3***
SELENIUM (colloidal)
af: Se aw: 78.96

TOXICITY DATA with REFERENCE
ivn-rat LDLo:6 mg/kg JPETAB 33,270,28

CONSENSUS REPORTS: Selenium and its compounds are on the Community Right-To-Know List. Reported in EPA TSCA Inventory.

OSHA PEL: TWA 0.2 mg(Se)/m^3
ACGIH TLV: TWA 0.2 mg(Se)/m^3
DFG MAK: 0.1 mg(Se)/m^3

SAFETY PROFILE: Poison by intravenous route. When heated to decomposition it emits toxic fumes of Se. See also SELENIUM and SELENIUM COMPOUNDS.

SBP500 ***HR: 3***
SELENIUM COMPOUNDS

CONSENSUS REPORTS: Selenium and its compounds are on the Community Right-To-Know List.

OSHA PEL: TWA 0.2 mg(Se)/m^3
ACGIH TLV: TWA 0.2 mg(Se)/m^3
DFG MAK: 0.1 mg(Se)/m^3

SAFETY PROFILE: Poison by inhalation and intravenous routes. Some selenium compounds are experimental carcinogens. Selenium in small amounts is essential for normal growth of some animals. Deficiency or excess is associated with serious disease in livestock. Long-term exposure may be a cause of amyotrophic lateral sclerosis in humans, just as it may cause ''blind staggers'' in cattle. Elemental selenium has low acute systemic toxicity, but dust or fumes can cause serious irritation of the respiratory tract. Hydrogen selenide resembles other hydrides in being highly toxic, and selenium oxychloride is a vesicant. Some organoselenium compounds have the high toxicity of other organometals. Inorganic selenium compounds can cause dermatitis. Garlic odor of breath is a common symptom. Pallor, nervousness, depression, digestive disturbances, and death have been reported in cases of chronic exposure. Selenium compounds are common air contaminants. When heated to decomposition they emit toxic fumes of Se. See also SELENIUM and specific compounds.

SBP600 CAS:1464-43-3 ***HR: 3***
SELENIUM CYSTINE
mf: $C_6H_{12}N_2O_4Se_2$ mw: 334.12

SYNS: ALANINE, 3,3'-DISELENOBIS-(9CI) ◇ 3,3'-DISELENODIALANINE ◇ SELENOCYSTINE

TOXICITY DATA with REFERENCE
slt-dmg-orl 10 umol/L CNJGA8 17,55,75
ipr-rat LDLo:4 mg/kg CTOXAO 17,171,80

OSHA PEL: TWA 0.2 mg(Se)/m^3
ACGIH TLV: TWA 0.2 mg(Se)/m^3

SAFETY PROFILE: Poison by intraperitoneal route. Mutation data reported. When heated to decomposition it emits toxic fumes of NO_x and Se.

SBP900 CAS:5456-28-0 ***HR: 2***
SELENIUM DIETHYLDITHIOCARBAMATE
mf: $C_{20}H_{40}N_4S_8$•Se mw: 672.08

SYNS: ETHYL SELENAC ◇ ETHYL SELERAM ◇ TETRAKIS (DIETHYLCARBAMODITHIOATO-S,S')SELENIUM ◇ TETRAKIS (DIETHYLDITHIOCARBAMATO)SELENIUM

TOXICITY DATA with REFERENCE
orl-mus TDLo:3060 mg/kg/81W-I:CAR JNCIAM 42,1101,69

CONSENSUS REPORTS: IARC Cancer Review: Group 3 IMEMDT 7,56,87; Animal Inadequate Evidence IMEMDT 9,245,75; IMEMDT 12,107,76

OSHA PEL: TWA 0.2 mg(Se)/m^3
ACGIH TLV: TWA 0.2 mg(Se)/m^3
DFG MAK: 0.1 mg(Se)/m^3

SAFETY PROFILE: Questionable carcinogen with experimental carcinogenic data. When heated to decomposition it emits toxic fumes of NO_x ans Se.

SBQ000 CAS:144-34-3 ***HR: 3***
SELENIUM DIMETHYLDITHIOCARBAMATE
mf: $C_{12}H_{24}N_4S_8$•Se mw: 559.84

PROP: Yellow powder, crystals. D: 1.58, melting range: 140-172°.

SYNS: METHYL SELENAC ◇ TETRAKIS(DIMETHYLCARBAMODITHIOATO-S,S')SELENIUM

CONSENSUS REPORTS: IARC Cancer Review: Group 2B IMEMDT 7,56,87; Animal Inadequate Evidence IMEMDT 12,161,76. Selenium and its compounds are on the Community Right-To-Know List. Reported in EPA TSCA Inventory.

OSHA PEL: TWA 0.2 mg(Se)/m^3
ACGIH TLV: TWA 0.2 mg(Se)/m^3
DFG MAK: 0.1 mg(Se)/m^3

SAFETY PROFILE: Suspected carcinogen. Selenium compounds are poisons. When heated to decomposition it emits very toxic fumes of Se, SO_x, and NO_x. See also SELENIUM COMPOUNDS and CARBAMATES.

SBQ500 CAS:7446-08-4 ***HR: 3***
SELENIUM(IV) DIOXIDE (1:2)
mf: O_2Se mw: 110.96

PROP: White to sltly reddish, lustrous, crystalline powder or needles. Mp: 340-350° (subl), d: 3.95 @ 15°/15°, vap press: 1 mm @ 157.0°.

SYNS: RCRA WASTE NUMBER U204 ◇ SELENIOUS ANHYDRIDE ◇ SELENIUM DIOXIDE ◇ SELENIUM OXIDE

TOXICITY DATA with REFERENCE
mrc-bcs 10 mmol/L MUREAV 77,109,80
scu-grb TDLo:5600 μg/kg (1D male):REP JRMSAS 85,297,65
ihl-rbt LCLo:5890 mg/m^3/20M NTIS** PB158-508
scu-rbt LD50:4 mg/kg CTOXAO 17,171,80
scu-grb LDLo:7500 μg/kg JRMSAS 85,297,65
ihl-dom LCLo:6590 mg/m^3/10M NTIS** PB158-508

CONSENSUS REPORTS: Selenium and its compounds are on the Community Right-To-Know List. Reported in EPA TSCA Inventory. EPA Genetic Toxicology Program.

OSHA PEL: TWA 0.2 mg(Se)/m^3
ACGIH TLV: TWA 0.2 mg(Se)/m^3
DFG MAK: 0.1 mg(Se)/m^3

SAFETY PROFILE: Poison by subcutaneous route. Mildly toxic by inhalation. Mutation data reported. Experimental reproductive effects. Incompatible with PCl_3. Used as an oxidizing agent. When heated to decomposition it emits toxic fumes of Se. See also SELENIUM COMPOUNDS.

SBR000 CAS:7488-56-4 ***HR: 3***
SELENIUM(IV) DISULFIDE (1:2)
DOT: UN 2657
mf: S_2Se mw: 143.08

PROP: Red-yellow crystals. Mp: <100°, bp: decomp.

SYNS: EXSEL ◇ RCA WASTE NUMBER U205 ◇ SELENIUM DISULPHIDE (DOT) ◇ SELENIUM SULFIDE ◇ SELSUN BLUE

TOXICITY DATA with REFERENCE
orl-rat LD50:138 mg/kg TXAPA9 20,89,71

CONSENSUS REPORTS: Selenium and its compounds are on the Community Right-To-Know List. Reported in EPA TSCA Inventory.

OSHA PEL: TWA 0.2 mg(Se)/m^3
ACGIH TLV: TWA 0.2 mg(Se)/m^3
DFG MAK: 0.1 mg(Se)/m^3
DOT Classification: Poison B; Label: Poison.

SAFETY PROFILE: Poison by ingestion. Used in shampoos. When heated to decomposition it emits very toxic fumes of SO_x and Se. See also SELENIUM COMPOUNDS and SULFIDES.

SBR500 ***HR: 2***
SELENIUM(IV) DISULFIDE SHAMPOO (2.5%)

SYN: SELENIUM DISULFIDE (2.5%) SHAMPOO

TOXICITY DATA with REFERENCE
eye-rbt 2500 μg SEV SMJOAV 55,318,62

CONSENSUS REPORTS: Selenium and its compounds are on the Community Right-To-Know List.

OSHA PEL: TWA 0.2 mg(Se)/m^3
ACGIH TLV: TWA 0.2 mg(Se)/m^3
DFG MAK: 0.1 mg(Se)/m^3

SAFETY PROFILE: A severe eye irritant. When heated to decomposition it emits very toxic fumes of Se and SO_x. See also SELENIUM(IV) DISULFIDE (1:2), SELENIUM COMPOUNDS, and SULFIDES.

SBS000 CAS:7783-79-1 ***HR: 3***
SELENIUM HEXAFLUORIDE
DOT: UN 2194
mf: F_6Se mw: 192.96

PROP: Colorless gas. Mp: −39° (subl @ −40.6°), bp: −34.5°, d: 3.25 @ −25°.

SYN: SELENIUM FLUORIDE

TOXICITY DATA with REFERENCE
ihl-rat LCLo:10 ppm/H ATXKA8 18,140,60
ihl-mus LCLo:10 ppm/3H ATXKA8 18,140,60
ihl-rbt LCLo:10 ppm DTLVS* 3,226,71
ihl-gpg LCLo:10 ppm/1H ATXKA8 18,140,60

CONSENSUS REPORTS: Selenium and its compounds are on the Community Right-To-Know List.

OSHA PEL: TWA 0.05 ppm (Se)
ACGIH TLV: TWA 0.05 ppm (Se)
DFG MAK: 0.1 mg(Se)/m^3
DOT Classification: Poison A; Label: Poison Gas.

SAFETY PROFILE: Poison by inhalation. When heated to decomposition it emits very toxic fumes of F^- and Se. See also SELENIUM COMPOUNDS and FLUORIDES.

SBS500 CAS:10025-68-0 ***HR: 3***
SELENIUM MONOCHLORIDE
mf: Cl_2Se_2 mw: 228.83

PROP: Deep red, oily liquid. Bp: 127° @ 733° (decomp), mp: −85°. Sol in chloroform, benzene, carbon tetrachloride, carbon disulfide, fuming sulfuric acid; decomp in water.

SYNS: DISELENIUM DICHLORIDE ◇ SELENIUM CHLORIDE

CONSENSUS REPORTS: Selenium and its compounds are on the Community Right-To-Know List.

OSHA PEL: TWA 0.2 mg(Se)/m^3
ACGIH TLV: TWA 0.2 mg(Se)/m^3
DFG MAK: 0.1 mg(Se)/m^3

SAFETY PROFILE: Selenium compounds are poisons. Explosive reaction with potassium. Reacts to form an explosive product with trimethylsilyl azide. Violent reaction with potassium dioxide; sodium peroxide. Vigorous or Incandescent reaction with aluminum (above 80°C); sodium + heat. When heated to decomposition it emits toxic fumes of Se and Cl^-. See also SELENIUM COMPOUNDS and CHLORIDES.

SBT000 CAS:7446-34-6 ***HR: 3***
SELENIUM MONOSULFIDE
mf: SSe mw: 111.02

PROP: Orange-yellow tablets or powder. Mp: 111.03°, bp: decomp @ 118-119°, d: 3.056 @ 0°.

SYNS: NCI-C50033 ◇ SELENIUM SULFIDE ◇ SELENIUM SULPHIDE ◇ SELENSULFID (GERMAN) ◇ SULFUR SELENIDE

TOXICITY DATA with REFERENCE
mmo-sat 1 mg/plate SCIEAS 236,933,87
cyt-ham:ovr 5 mg/L SCIEAS 236,933,87
orl-rat TDLo:11 g/kg/2Y-C:CAR NCITR* NCI-CG-TR-194,80
orl-mus TDLo:72 g/kg/2Y-C:CAR NCITR* NCI-CG-TR-194,80
orl-rat LD50:38 mg/kg CTOXAO 17,171,80
orl-mus LD50:370 mg/kg ATXKA8 24,341,69
orl-rbt LDLo:55 mg/kg CLDND* NCI-TR-194,80

CONSENSUS REPORTS: NTP Fifth Annual Report on Carcinogens. NCI Carcinogenesis Bioassay (dermal); Inadequate Studies: mouse NCITR* NCI-CG-TR-197,80; (gavage); Clear Evidence: mouse, rat NCITR* NCI-CG-TR-194,80. Selenium and its compounds are on the Community Right-To-Know List.

OSHA PEL: TWA 0.2 mg(Se)/m^3
ACGIH TLV: TWA 0.2 mg(Se)/m^3
DFG MAK: 0.1 mg(Se)/m^3

SAFETY PROFILE: Confirmed carcinogen with experimental carcinogenic data. Poison by ingestion. Mutation data reported. When heated to decomposition it emits very toxic fumes of SO_x and Se. See also SELENIUM COMPOUNDS and SULFIDES.

SBT500 CAS:7791-23-3 ***HR: 3***
SELENIUM OXYCHLORIDE
DOT: UN 2879
mf: Cl_2OSe mw: 165.86

PROP: Colorless-yellowish liquid. Mp: 8.5°, bp: 176.4°, d: 2.42 @ 22°, vap press: 1 mm @ 34.8°.

SYNS: SELENINYL CHLORIDE ◇ SELENIUM CHLORIDE OXIDE

TOXICITY DATA with REFERENCE
skn-man TDLo:710 μg/kg PHRPA6 53,94,38
skn-rbt LDLo:2 mg/kg CTOXAO 17,171,80
scu-rbt LD50:7 mg/kg CTOXAO 17,171,80

CONSENSUS REPORTS: Selenium and its compounds are on the Community Right-To-Know List. EPA Extremely Hazardous Substances List. Reported in EPA TSCA Inventory.

OSHA PEL: TWA 0.2 mg(Se)/m^3
ACGIH TLV: TWA 0.2 mg(Se)/m^3
DFG MAK: 0.1 mg(Se)/m^3

DOT Classification: Corrosive Material; Label:Corrosive and Poison

SAFETY PROFILE: Poison by skin contact and subcutaneous routes. Human systemic effects by skin contact with very small amounts: primary irritant, corrosive. Explodes on contact with potassium, white phosphorus. Ignites on contact with antimony. Vigorous reaction with metal oxides (e.g., silver oxide, lead(II) oxide, lead(IV) oxide, Lead(II)(IV) oxide). When heated to decomposition it emits very toxic fumes of Cl^- and Se. See also SELENIUM COMPOUNDS and CHLORIDES.

SBU000 CAS:10026-03-6 ***HR: 3***
SELENIUM TETRACHLORIDE
mf: Cl_4Se mw: 220.76

PROP: Cubic, white-yellow; deliq crystals. Mp: 300° (subl @ 170-196°), bp: decomp @ 288°, vap press: 1 mm @ 74°. D: 2.6. Decomp in water and moist air. Insol in liquid bromine; decomp by dry ammonia.

SYN: SELENIUM(IV) CHLORIDE (1:4)

TOXICITY DATA with REFERENCE
scu-gpg LD50:19 mg/kg NDRC** -,126,43

CONSENSUS REPORTS: Selenium and its compounds are on the Community Right-To-Know List. Reported in EPA TSCA Inventory.

OSHA PEL: TWA 0.2 mg(Se)/m^3
ACGIH TLV: TWA 0.2 mg(Se)/m^3
DFG MAK: 0.1 mg(Se)/m^3

SAFETY PROFILE: Poison by subcutaneous route. When heated to decomposition it emits very toxic fumes of Se and Cl^-. See also SELENIUM COMPOUNDS.

SBU100 CAS:66472-85-3 ***HR: 3***
SELENOASPIRINE
mf: $C_9H_8O_3Se$ mw: 243.13

SYNS: 2-(ACETYLSELENO)BENZOIC ACID ◇ ACETYLSELENO-2 BENZOIC ACID ◇ ACIDE ACETYL SELENO-2 BENZOIQUE ◇ BENZOIC ACID, 2-(ACETYLSELENO)-

TOXICITY DATA with REFERENCE
ipr-rat LDLo:100 mg/kg CRSBAW 172,383,78
ivn-rat LDLo:75 mg/kg CRSBAW 172,383,78

OSHA PEL: TWA 0.2 mg(Se)/m^3
ACGIH TLV: TWA 0.2 mg(Se)/m^3

SAFETY PROFILE: Poison by intraperitoneal and intravenous routes. When heated to decomposition it emits toxic fumes of Se.

SBU150 CAS:6512-83-0 ***HR: 3***
2,2′-SELENOBIS(BENZOIC ACID)
mf: $C_{14}H_{10}O_4Se_2$ mw: 400.16

SYNS: ACIDE DISELINO SALICYLIQUE ◇ BENZOIC ACID, 2,2′-DISELENOBIS- ◇ DISELENO SALICYLIC ACID ◇ o,o′-SELENOBIS(BENZOIC ACID)

TOXICITY DATA with REFERENCE
ipr-rat LD50:75 mg/kg CRSBAW 172,383,78
ivn-rat LD50:50 mg/kg CRSBAW 172,383,78

OSHA PEL: TWA 0.2 mg(Se)/m^3
ACGIH TLV: TWA 0.2 mg(Se)/m^3

SAFETY PROFILE: Poison by intraperitoneal and intravenous routes. When heated to decomposition it emits toxic fumes of Se.

SBU200 CAS:2897-21-4 ***HR: 3***
SELENOCYSTINE
mf: $C_6H_{12}N_2O_4Se_2$ mw: 334.12

SYNS: ALANINE, 3,3′-DISELENODI-, dl- ◇ d,l-SELENOCYSTINE ◇ SELENO-dl-CYSTINE

TOXICITY DATA with REFERENCE
ipr-rat LD50:8463 μg/kg JPETAB 108,437,53
scu-rat LDLo:13 mg/kg ARTODN 45,207,80

OSHA PEL: TWA 0.2 mg(Se)/m^3
ACGIH TLV: TWA 0.2 mg(Se)/m^3

SAFETY PROFILE: Poison by subcutaneous and intraperitoneal routes. When heated to decomposition it emits toxic fumes of NO_x and Se.

SBU500 CAS:33944-90-0 ***HR: D***
SELENODIGLUTATHIONE
mf: $C_{20}H_{32}N_6O_{12}S_2Se$ mw: 691.66

SYNS: GLUTAMINE, N,N′-((SELENODITHIO)BIS(1-((CARBOXYMETHYL)CARBAMOYL)ETHYLENE))DI-, L- ◇ l-GLUTAMINE,N,N′-(SELENOBIS(THIO(1-(((CARBOXYMETHYL)AMINO)CARBONYL)-2,1-ETHANEDIYL)))BIS-

TOXICITY DATA with REFERENCE
ivn-mus TDLo:11342 μg/kg (female 12D post):TER TOLED5 21,35,84

OSHA PEL: TWA 0.2 mg(Se)/m^3
ACGIH TLV: TWA 0.2 mg(Se)/m^3

SAFETY PROFILE: An experimental teratogen. When heated to decomposition it emits toxic fumes of NO_x, SO_x, and Se.

SBU600 CAS:57897-99-1 ***HR: 3***
2-SELENOETHYLGUANIDINE
mf: $C_3H_9N_3Se$ mw:166.11

SYNS: SELENOMERCAPTOAETHYLGUANIDIN (GERMAN) ◇ SE-MEG

TOXICITY DATA with REFERENCE
ipr-mus LD50:55 mg/kg STRAAA 151,78,76
scu-mus LD50:120 mg/kg STRAAA 151,78,76
ivn-mus LD50:48 mg/kg STRAAA 151,78,76

CONSENSUS REPORTS: Selenium and its compounds are on the Community Right-To-Know List.

OSHA PEL: TWA 0.2 mg(Se)/m^3
ACGIH TLV: TWA 0.2 mg(Se)/m^3
DFG MAK: 0.1 mg(Se)/m^3

SAFETY PROFILE: Poison by intraperitoneal, subcutaneous, and intravenous routes. When heated to decomposition it emits toxic fumes of Se, SO_x, and NO_x. See also SELENIUM COMPOUNDS and MERCAPTANS.

SBU700 CAS:29411-74-3 ***HR: 3***
6-SELENOGUANOSINE
mf: $C_{10}H_{13}N_5O_4Se$ mw: 346.24

SYNS: 2-AMINO-9-β-d-RIBOFURANOSYL-6-SELENO-OH-PURIN-6(1H)-ONE ◇ NSC 137679 ◇ 6-SELENO-GUANOSINE (9CI)

TOXICITY DATA with REFERENCE
orl-mus LD50:137 mg/kg NCISP* JAN86
ipr-mus LD50:126 mg/kg NCISP* JAN86
scu-mus LD50:58840 μg/kg NCISP* JAN86

CONSENSUS REPORTS: Selenium and its compounds are on the Community Right-To-Know List.

OSHA PEL: TWA 0.2 mg(Se)/m^3
ACGIH TLV: TWA 0.2 mg(Se)/m^3
DFG MAK: 0.1 mg(Se)/m^3

SAFETY PROFILE: Poison by ingestion, subcutaneous, and intraperitoneal routes. When heated to decomposition it emits toxic fumes of NO_x and Se. See also SELENIUM COMPOUNDS.

SBU710 CAS:7776-33-2 ***HR: 3***
SELENOHOMOCYSTINE
mf: $C_8H_{16}N_2O_4Se_2$ mw: 362.18

SYNS: BUTANOIC ACID, 4,4′-DISELENOBIS(2-AMINO-(9CI) ◇ BUTYRIC ACID, 4,4′-DISELENOBIS(2-AMINO- ◇ 4,4′-DISELENOBIS(2-AMINOBUTYRIC ACID)

TOXICITY DATA with REFERENCE
ipr-rat LD50:8 mg/kg JPETAB 108,437,53

ACGIH TLV: TWA 0.2 mg(Se)/m^3

SAFETY PROFILE: Poison by intraperitoneal route. When heated to decomposition it emits toxic fumes of NO_x and Se.

SBU725 CAS:1464-42-2 ***HR: 3***
SELENOMETHIONINE
mf: $C_5H_{11}NO_2Se$ mw: 196.13

PROP: dl-Form: Transparent, hexagonal sheets or plates from methanol and water; metallic luster. Mp: 265° (decomp). l-Form: Crystals from aq acetone. Mp: 266-268° (decomp).

SYN: 2-AMINO-4-(METHYLSELENYL)BUTYRIC ACID

TOXICITY DATA with REFERENCE
dns-rat:lvr 100 μmol/L CALEDQ 10,75,80
ipr-rat LDLo:4250 μg/kg PSDAA2 28,117,49
ivn-mus LD50:22 mg/kg NRTXDN 2,383,81
icv-mus LD50:13 mg/kg NRTXDN 2,838,81
ivn-pig LDLo:7450 μg/kg AJVRAH 34,1227,73

CONSENSUS REPORTS: Selenium and its compounds are on the Community Right-To-Know List.

OSHA PEL: TWA 0.2 mg(Se)/m^3
ACGIH TLV: TWA 0.2 mg(Se)/m^3
DFG MAK: 0.1 mg(Se)/m^3

SAFETY PROFILE: Poison by intravenous, intraperitoneal and intracervical routes. Mutation data reported. When heated to decomposition it emits toxic fumes of NO_x and Se. See also AMINES and SELENIUM COMPOUNDS.

SBU900 CAS:10161-84-9 ***HR: 3***
SELENOPHOS
mf: $C_{10}H_{24}NO_2PSe$ mw: 300.28

SYNS: O-ETHYLSe-(2-DIETHYLAMINOETHYL)PHOSPHONOSELENOATE ◇ PHOSPHONOSELENOIC ACID, ETHYL-, Se-(2-(DIETHYLAMINO)ETHYL) O-ETHYL ESTER

TOXICITY DATA with REFERENCE
scu-mus LD50:21 μg/kg JMCMAR 10,115,67

OSHA PEL: TWA 0.2 mg(Se)/m^3
ACGIH TLV: TWA 0.2 mg(Se)/m^3

SAFETY PROFILE: Poison by subcutaneous route. When heated to decomposition it emits toxic fumes of NO_x, PO_x, and Se.

SBU950 CAS:53184-19-3 ***HR: 3***
SELENO-TOLUIDINE BLUE
mf: $C_{15}H_{16}N_3Se{\cdot}Cl$ mw: 352.75

SYNS: 3-AMINO-7-(DIMETHYLAMINO)-2-METHYL-PHENOSELENAZIN-5-IUM CHLORIDE ◇ PHENOSELENAZIN-5-IUM, 3-AMINO-7-(DIMETHYLAMINO)-2-METHYL-,CHLORIDE

TOXICITY DATA with REFERENCE
unr-mus LD50:33 mg/kg JMCMAR 17,902,74

OSHA PEL: TWA 0.2 mg(Se)/m^3
ACGIH TLV: TWA 0.2 mg(Se)/m^3

SAFETY PROFILE: Poison by unreported route. When heated to decomposition it emits toxic fumes of NO_x and Se.

SBV000 CAS:630-10-4 ***HR: 3***
SELENOUREA
mf: CH_4N_2Se mw: 123.03

PROP: Prisms in water. Mp: decomp @ 200° +. Sol in water, alc, ether.

SYN: RCRA WASTE NUMBER P103

TOXICITY DATA with REFERENCE
orl-rat LD50:50 mg/kg TXAPA9 20,89,71
ivn-mus LD50:56 mg/kg CSLNX* NX#05665

CONSENSUS REPORTS: Selenium and its compounds are on the Community Right-To-Know List. Reported in EPA TSCA Inventory.

OSHA PEL: TWA 0.2 mg(Se)/m^3
ACGIH TLV: TWA 0.2 mg(Se)/m^3
DFG MAK: 0.1 mg(Se)/m^3

SAFETY PROFILE: Poison by ingestion and intravenous routes. When heated to decomposition it emits very toxic fumes of NO_x and Se. See also SELENIUM COMPOUNDS.

SBV500 CAS:74220-04-5 ***HR: 3***
SELOKEN
mf: $C_{15}H_{25}NO_3$•$C_4H_6O_6$ mw: 417.51

SYNS: BELOC ◇ BETALOC ◇ 1-ISOPROPYLAMINO-3-(p-(2-METHOXYETHYL)PHENOXY)-2-PROPANOL TARTRATE ◇ LOPRESOR ◇ LOPRESSOR ◇ METOPROLOL TARTRATE ◇ PRELIS

TOXICITY DATA with REFERENCE
orl-rat TDLo:12 g/kg (9W male/2W pre-3W post):REP APTSAI 36,96,75
orl-hmn TDLo:160 mg/kg:CVS,PUL BMJOAE 1,222,76
orl-rat LD50:2000 mg/kg BMJOAE 1,222,76
ivn-rat LD50:72 mg/kg APTSAI 36,96,75
orl-mus LD50:2090 mg/kg DRUGAY 14,321,77
ivn-mus LD50:69 mg/kg APTSAI 36,96,75

SAFETY PROFILE: Poison by intravenous route. Moderately toxic by ingestion. Human systemic effects by ingestion: decreased blood pressure along with possible decrease in pulse rate, and cyanosis. Experimental reproductive effects. When heated to decomposition it emits toxic fumes of NO_x.

SBW000 ***HR: 2***
SELSUN

PROP: 2.4% w/v selenium sulfide in aq suspension, also contains bentonite, sodium alkyl aryl sulfonate, sodium phosphate, glycerol monoricinoleate, citric acid, captan, and perfume (FEREAC 41,7218,76).

SYN: NCI-C54546

TOXICITY DATA with REFERENCE
eye-rbt 50%/1W SEV AJOPAA 38,560,54

CONSENSUS REPORTS: NCI Carcinogenesis Bioassay (dermal); No Evidence: mouse NCITR* NCI-CG-TR-199,80. Selenium and its compounds are on the Community Right-To-Know List.

OSHA PEL: TWA 0.2 mg(Se)/m^3
ACGIH TLV: TWA 0.2 mg(Se)/m^3
DFG MAK: 0.1 mg(Se)/m^3

SAFETY PROFILE: A severe eye irritant. Used as a pharmaceutical and veterinary drug. When heated to decomposition it emits very toxic fumes of Se, SO_x, PO_x, Na_2O, and NO_x. See also SELENIUM COMPOUNDS, BENTONITE, TRISODIUM o-PHOSPHATE, CITRIC ACID, and CAPTAN.

SBW500 CAS:563-41-7 ***HR: 3***
SEMICARBAZIDE HYDROCHLORIDE
mf: CH_5N_3O•ClH mw: 111.55

PROP: Prisms from dilute alc. Decomp @ 175-185°, mp: 176° (decomp). Very sol in water; very sltly sol in hot alc; insol in anhyd ether.

SYNS: AMIDOUREA HYDROCHLORIDE ◇ AMINOUREA HYDROCHLORIDE ◇ CARBAMYLHYDRAZINE HYDROCHLORIDE ◇ CH ◇ HYDRAZINECARBOXAMIDE MONOHYDROCHLORIDE

TOXICITY DATA with REFERENCE
cyt-grh-par 100 mmol/L MUREAV 40,237,76
sln-oin-grh-par 100 mmol/L MUREAV 40,237,76
ipr-rat TDLo:74 mg/kg (female 10D post):TER BNEOBV 49,150,86
ipr-rat TDLo:74 mg/kg (female 10D post):REP BNEOBV 49,150,86
orl-mus TDLo:67 g/kg/76W-C:NEO EJCAAH 11,17,75
orl-rat LDLo:10 mg/kg NCNSA6 5,44,53
orl-mus LD50:225 mg/kg ARZNAD 10,686,60
ipr-mus LD50:145 mg/kg FATOAO 24,623,61
scu-mus LD50:167 mg/kg ABMGAJ 21,635,68

CONSENSUS REPORTS: IARC Cancer Review: Group 3 IMEMDT 7,56,87; Animal Sufficient Evidence IMEMDT 12,209,76. Reported in EPA TSCA Inventory. EPA Extremely Hazardous Substances List.

SAFETY PROFILE: Poison by ingestion and intraperitoneal routes. Experimental reproductive effects. Questionable carcinogen with experimental neoplastigenic and teratogenic data. Mutation data reported. When heated to decomposition it emits very toxic fumes of NO_x and HCl.

SBW950 ***HR: 2***
SENECIO CANNABIFOLIUS, leaves and stalks

PROP: Herb of the family SENECIONEAE CALEDQ 20,191,83

TOXICITY DATA with REFERENCE
orl-rat TD:240 g/kg/69W-C:CAR CALEDQ 20,191,83

SAFETY PROFILE: Questionable carcinogen with experimental carcinogenic data. When heated to decomposition it emits acrid smoke and irritating fumes.

SBX000 ***HR: 3***
SENECIO LONGILOBUS
PROP: Contains pyrrolizidine alkaloids (CNREA8 28,2237,68).

TOXICITY DATA with REFERENCE
orl-rat TDLo:33 g/kg/31W-I:ETA CNREA8 30,2881,70

SAFETY PROFILE: Questionable carcinogen with experimental tumorigenic data. When heated to decomposition it emits toxic fumes of NO_x.

SBX200 ***HR: 3***
SENECIO NEMORENSIS FUCHSII, alkaloidal extract

PROP: Alkaloidal extract contains the two pyrrolizidine alkaloids fuchsisenecionine and senecionine (ARZNAD 32,144,82).

TOXICITY DATA with REFERENCE
msc-ham:lng 156 mg/L ARZNAD 32,144,82
orl-rat TDLo:4160 mg/kg/2Y-I:CAR ARZNAD 32,144,82
orl-rat TD:20800 mg/kg/2Y-I:CAR ARZNAD 32,144,82

SAFETY PROFILE: Questionable carcinogen with experimental carcinogenic data. Mutation data reported. See also AUREINE.

SBX500 CAS:480-81-9 ***HR: D***
SENECIPHYLLINE
mf: $C_{18}H_{23}NO_5$ mw: 333.42

PROP: Small, rhombic platelets from hot alc or acetone. Mp: 217-218°. Easily sol in chloroform, ethylene chloride; less sol in alc, acetone; difficultly sol in ether, ligroin. An alkaloid isolated from *S. stenocephalus* (RETOAE 5,55,49).

SYNS: 13,19-DIDEHYDRO-12-HYDROXY-SENECIONAN-11,16-DIONE ◇ JACOBINE ◇ SENECIPHYLLIN

TOXICITY DATA with REFERENCE
dns-rat:lvr 1 μmol/L CNREA8 45,3125,85
dns-ham:lvr 1 μmol/L CNREA8 45,3125,85

CONSENSUS REPORTS: IARC Cancer Review: Group 3 IMEMDT 7,56,87; Animal Inadequate Evidence IMEMDT 10,319,76

SAFETY PROFILE: Questionable carcinogen. Mutation data reported. When heated to decomposition it emits toxic fumes of NO_x.

SBX525 ***HR: 3***
SENECIPHYLLIN HYDROCHLORIDE
mf: $C_{18}H_{23}NO_5 \cdot ClH$ mw: 369.88

SYNS: 13,19-DIDEHYDRO-12-HYDROXY-SENECIONAN-11,16-DIONE HYDROCHLORIDE ◇ JACODINE HYDROCHLORIDE ◇ NCI-C61165 ◇ NSC 30622

TOXICITY DATA with REFERENCE
orl-rat LDLo:25 mg/kg NATUAS 179,361,57
ipr-rat LD50:77 mg/kg FRBGAT 9,142,68
ivn-rat LDLo:60 mg/kg JPETAB 68,130,40
ivn-mus LDLo:50 mg/kg JPETAB 68,130,40
ivn-gpg LDLo:50 mg/kg JPETAB 68,130,40

SAFETY PROFILE: Poison by ingestion, intravenous, and intraperitoneal routes. When heated to decomposition it emits toxic fumes of NO_x and HCl.

SBY000 CAS:2469-34-3 ***HR: 3***
SENEGENIN
mf: $C_{30}H_{45}ClO_6$ mw: 537.20

SYNS: 12-(CHLOROMETHYL)-2-β,3-β-DIHYDROXY-27-NORO-13-ENE-23,28-DIOIC ACID ◇ SENEGIN

TOXICITY DATA with REFERENCE
orl-mus LDLo:1 g/kg HBAMAK 4,1289,35
ipr-mus LD50:3 mg/kg 85GDA2 8(2),259,82g
scu-mus LDLo:30 mg/kg HBAMAK 4,1289,35
ivn-mus LDLo:45 mg/kg HBAMAK 4,1289,35

SAFETY PROFILE: Poison by intraperitoneal, subcutaneous, and intravenous routes. Moderately toxic by ingestion. When heated to decomposition it emits toxic fumes of Cl^-.

SBZ000 CAS:62362-59-8 ***HR: 3***
SEPTACIDIN
mf: $C_{30}H_{51}N_7O_7$ mw: 621.88

PROP: An antibiotic produced by a strain of *Streptomyces fimbriatus* (85ERAY 2,1090,78).

SYN: NSC-65104

TOXICITY DATA with REFERENCE
ipr-mus LD50:500 μg/kg 85GDA2 5,303,81
scu-mus LD02:3 mg/kg 85ERAY 2,1090,78

SAFETY PROFILE: Poison by intraperitoneal and subcutaneous routes. When heated to decomposition it emits toxic fumes of NO_x.

SCA000 CAS:54927-63-8 ***HR: 3***
SEPTAMYCIN
mf: $C_{48}H_{83}O_{16}$ mw: 916.31

SYN: ANTIBIOTIC A 28695 A

TOXICITY DATA with REFERENCE
ivn-rat LD50:10 mg/kg 85ERAY 1,800,78

orl-mus LD50:41 mg/kg 85GDA2 5,486,81
ivn-dog LD50:5 mg/kg 85ERAY 1,800,78

SAFETY PROFILE: Poison by ingestion and intravenous routes. When heated to decomposition it emits acrid smoke and irritating fumes.

SCA400 CAS:14919-77-8 ***HR: 2***
dl-SERINE 2-(2,3,4-TRIHYDROXYBENZYL) HYDRAZINE HYDROCHLORIDE
mf: $C_{10}H_{15}N_3O_5 \cdot ClH$ mw: 293.74

SYNS: BENSERAZIDE HYDROCHLORIDE ◇ RO 4-4602 ◇ dl-SERINE 2-((2,3,4-TRIHYDROXYPHENYL)METHYL)HYDRAZIDE MONOHYDROCHLORIDE (9CI) ◇ 2-(2,3,4-TRIHYDROXYBENZYL)HYDRAZIDE SERINE MONOHYDROCHLORIDE, dl-

TOXICITY DATA with REFERENCE
mmo-sat 1 mg/plate RCOCB8 49,415,85
mma-sat 1 mg/plate RCOCB8 49,415,85
orl-rat LD50:5800 mg/kg IYKEDH 11,181,80
ipr-rat LD50:2150 mg/kg IYKEDH 11,181,80
scu-rat LD50:2910 mg/kg IYKEDH 11,181,80
orl-mus LD50:5 g/kg KSRNAM 11,62,77
ipr-mus LD50:1700 mg/kg KSRNAM 11,62,77
scu-mus LD50:3300 mg/kg IYKEDH 11,181,80

SAFETY PROFILE: Moderately toxic by intraperitoneal and subcutaneous routes. Mildly toxic by ingestion. Mutation data reported. When heated to decomposition it emits toxic fumes of NO_x and HCl.

SCA475 CAS:131-07-7 ***HR: 3***
SERPENTINE
mf: $C_{21}H_{29}N_2O_3$ mw: 348.43

SYNS: METHYL ESTER of SERPENTINIC ACID ◇ SERPENTINE HYDROXIDE, inner salt ◇ SERPENTINE (alkaloid) HYDROXIDE, inner salt

TOXICITY DATA with REFERENCE
ipr-rat LD50:42 mg/kg AIPTAK 123,168,59
ipr-mus LD50:144 mg/kg AIPTAK 123,168,59
scu-mus LD50:166 mg/kg AIPTAK 123,168,59

SAFETY PROFILE: Poison by subcutaneous and intraperitoneal routes. When heated to decomposition it emits toxic fumes of NO_x.

SCA525 ***HR: 3***
SERPENTINIC ACID ISOBUTYL ESTER
mf: $C_{24}H_{27}N_2O_3$ mw: 391.53

SYN: Ph. 458

TOXICITY DATA with REFERENCE
ipr-rat LD50:29 mg/kg AIPTAK 123,168,59
ipr-mus LD50:46500 μg/kg AIPTAK 123,168,59
scu-mus LD50:137 mg/kg AIPTAK 123,168,59

SAFETY PROFILE: Poison by subcutaneous and intraperitoneal routes. When heated to decomposition it emits toxic fumes of NO_x. See also ESTERS.

SCA550 CAS:78329-75-6 ***HR: 3***
SERPENTINTARTRAT (GERMAN)
mf: $C_{33}H_{40}N_2O_9 \cdot C_4H_6O_6$ mw: 758.75

SYN: 3-β,20-α-YOHIMBAN-16-β-CARBOXYLIC ACID, 18-β-HYDROXY-11,17-α-DIMETHOXY-METHYL ESTER, 3,4,5-TRIMETHOXYBENZOATE (ESTER), TARTRATE

TOXICITY DATA with REFERENCE
scu-mus LD50:164 mg/kg ARZNAD 5,715,55
ivn-mus LD50:20 mg/kg ARZNAD 5,715,55

SAFETY PROFILE: Poison by subcutaneous and intravenous routes. When heated to decomposition it emits toxic fumes of NO_x. See also ESTERS.

SCA600 ***HR: 3***
SERRATIA MARCESCENS ENDOTOXIN

TOXICITY DATA with REFERENCE
ipr-rat TDLo:1 g/kg (1D male):REP JRPFA4 22,161,70
ipr-rat LD50:18182 μg/kg INFIBR 3,444,71
ivn-mus LD50:17 mg/kg EXPEAM 37,174,81

SAFETY PROFILE: Poison by intraperitoneal and intravenous routes. Moderately toxic by intraperitoneal route. Experimental reproductive effects.

SCA610 ***HR: 3***
SERRATIA PISCVATORUM POLYSACCHARIDE

TOXICITY DATA with REFERENCE
scu-mus LD50:575 mg/kg JJPAAZ 24,109,74
ivn-mus LD50:44 mg/kg JJPAAZ 24,109,74
ims-mus LD50:185 mg/kg JJPAAZ 24,109,74

SAFETY PROFILE: Poison by intramuscular and intravenous routes. Moderately toxic by subcutaneous route.

SCA625 CAS:37312-62-2 ***HR: 3***
SERRATIO PEPTIDASE

SYNS: DASEN ◇ SERRATIA EXTRACELLULAR PROTEINASE

TOXICITY DATA with REFERENCE
ipr-rat LD50:34620 μg/kg NIIRDN 6,415,82
scu-rat LD50:89 mg/kg NIIRDN 6,415,82
ivn-rat LD50:6950 μg/kg NIIRDN 6,415,82
ipr-mus LD50:15200 μg/kg NIIRDN 6,415,82
scu-mus LD50:44 mg/kg NIIRDN 6,415,82
ivn-mus LD50:14230 μg/kg NIIRDN 6,415,82

SAFETY PROFILE: Poison by subcutaneous, intravenous, and intraperitoneal routes.

SCA750 CAS:9002-70-4 ***HR: D***
SERUM GONADOTROPIN

SYNS: ANTERON ◇ ANTEX-490 ◇ ANTOSTAB ◇ ELEAGOL ◇ EQUINE CYONIN ◇ EQUINE GONADOTROPHIN ◇ EQUINE GONADOTROPIN ◇ GESTYL ◇ GONADOTRAPHON FSH ◇ GONADYL ◇ GORMAN ◇ LOBULANTINA ◇ PMS ◇ PMSG ◇ PREDALON-S ◇ PREGNANT MARE SERUM GONADOTROPIN ◇ PRIATIN ◇ PRIMANTRON ◇ SERAGON ◇ SERAGONIN ◇ SEROGAN ◇ SEROTROPIN ◇ SERUM GONADOTROPHIN ◇ SERUM GONADOTROPIC HORMONE

TOXICITY DATA with REFERENCE
cyt-mus-ipr 750 μg/kg JRPFA4 50,275,77 JRPFA4 50,275,77
ipr-ham TDLo:37500 μg/kg (female 4-6D post):TER ENDOAO 83,217,68
scu-mus TDLo:50 mg/kg (female 8-11D post):REP IRLCDZ 8,117,80

SAFETY PROFILE: Mutation data reported. An experimental teratogen. Experimental reproductive effects.

SCB000 CAS:8008-74-0 ***HR: 3***
SESAME OIL

PROP: Flash p: 491°F, d: 0.9. From seed of *Sesamum indicum* (85DIA2 2,290,77).

SYNS: GINGILLI OIL ◇ SEXTRA

TOXICITY DATA with REFERENCE
skn-hmn 300 mg/3D-I MLD 85DKA8 -,127,77
scu-mus TDLo:2000 mg/kg/W-I:CAR AVBIB9 22/23,359,79
scu-mus TD:2000 mg/kg/W-I:ETA FEPRA7 38,1450,79
ivn-rbt LD50:678 μg/kg APTOA6 45,352,79

CONSENSUS REPORTS: Reported in EPA TSCA Inventory.

SAFETY PROFILE: Poison by intravenous route. Questionable carcinogen with experimental carcinogenic and tumorigenic data. A human skin irritant. Combustible when exposed to heat or flame. To fight fire, use CO_2, dry chemical. Used in cosmetics, lotions, injectables, and flavorants. When heated to decomposition it emits acrid smoke and irritating fumes.

SCB100 ***HR: 3***
SESBANIA SESBAN (L.) Merr. var. BICOLOR W. & A., extract excluding roots

PROP: Indian plant belonging to the family *Leguminosae* IJEBA6 18,594,80

SYN: JAYANTI, extract

TOXICITY DATA with REFERENCE
orl-rat TDLo:500 mg/kg (female 1-5D post):REP IJEBA6 18,594,80
ipr-mus LD50:45 mg/kg IJEBA6 18,594,80

SAFETY PROFILE: Poison by intraperitoneal route. Experimental reproductive effects. When heated to decomposition it emits acrid smoke and irritating fumes.

SCB500 CAS:3563-36-8 ***HR: 3***
SESQUIMUSTARD
mf: $C_6H_{12}Cl_2S_2$ mw: 219.20

SYNS: 1,2-BIS(2-CHLOROETHYLMERCAPTO)ETHANE ◇ BIS(2-CHLOROETHYLTHIO)ETHANE ◇ 1,2-BIS(β-CHLOROETHYLTHIO) ETHANE ◇ SESQUIMUSTARD Q ◇ TL 86

TOXICITY DATA with REFERENCE
ihl-hmn LCLo:300 mg/m³ SCJUAD 4,33,67
ihl-rat LC50:11 mg/m³/10M TXAPA9 5,677,63
ihl-mus LC50:6 mg/m³/10M TXAPA9 5,677,63
scu-mus LDLo:10 mg/kg NTIS** PB158-507
ihl-dog LC50:90 mg/m³/2M TXAPA9 5,677,63
ihl-cat LC50:900 mg/m³/10M NTIS** PB158-508
ihl-rbt LC50:2 g/m³/10M NTIS** PB158-508
ihl-gpg LC50:8 mg/m³/10M TXAPA9 5,677,63
scu-gpg LDLo:40 mg/kg NTIS** PB158-507
ihl-ham LC50:22 mg/m³/10M TXAPA9 5,677,63
ihl-pgn LC50:61 mg/m³/10M TXAPA9 5,677,63

SAFETY PROFILE: Human poison by inhalation. Experimental poison by inhalation and subcutaneous routes. When heated to decomposition it emits very toxic fumes of SO_x and Cl^-. See also MERCAPTANS.

SCC000 CAS:3792-59-4 ***HR: 3***
S-SEVEN
mf: $C_{14}H_{13}Cl_2O_2PS$ mw: 347.20

SYNS: O-ETHYL-O-2,4-DICHLOROPHENYLTHIONOBENZENEPHOSPHONATE ◇ PHENYLPHOSPHONOTHIOIC ACID, O-(2,4-DICHLOROPHENYL), O-ETHYL ESTER

TOXICITY DATA with REFERENCE
orl-mus LD50:274 mg/kg FMCHA2 -,D287,80
scu-mus LD50:784 mg/kg 28ZEAL 5,105,76

SAFETY PROFILE: Poison by ingestion. Moderately toxic by subcutaneous route. When heated to decomposition it emits very toxic fumes of Cl^-, SO_x, and PO_x. See also ESTERS.

SCC550 CAS:63551-77-9 ***HR: 3***
SFERICASE

SYN: BACILLUS SPHAERICUS ALKALINE PROTEINASE

TOXICITY DATA with REFERENCE
orl-mus TDLo:500 mg/kg (female 6-15D post):REP OYYAA2 16,941,78
orl-mus TDLo:2500 mg/kg (female 6-15D post):TER OYYAA2 16,941,78
orl-rat LD50:6100 mg/kg OYYAA2 16,885,78
scu-rat LD50:52500 μg/kg OYYAA2 16,885,78
ivn-rat LD50:7400 μg/kg OYYAA2 16,885,78

orl-mus LD50:3800 mg/kg OYYAA2 16,885,78
scu-mus LD50:12800 μg/kg OYYAA2 16,885,78
ivn-mus LD50:8 mg/kg OYYAA2 16,885,78

SAFETY PROFILE: Poison by subcutaneous and intravenous routes. Moderately toxic by ingestion. An experimental teratogen. Experimental reproductive effects.

SCD500 CAS:2589-15-3 ***HR: 2***
SHELL SD-10576
mf: $C_{10}H_7Cl_7$ mw: 375.32

SYNS: ENT 27,313 ◇ 1,4,5,6,7,8,8-HEPTACHLORO-3a,4,5,6,7,7a-HEXAHYDRO-4,7-METHANO-1H-INDENE ◇ SD-10576

TOXICITY DATA with REFERENCE
orl-rat LD50:2000 mg/kg ARSIM* 20,21,66
orl-mus LD50:2000 mg/kg ARSIM* 20,21,66

SAFETY PROFILE: Moderately toxic by ingestion. When heated to decomposition it emits toxic fumes of Cl^-.

SCD750 ***HR: 3***
SHIGA TOXIN

SYNS: S. DYSENTERIAE TOXIN ◇ SHIGELLA DYSENTERIAE TOXIN ◇ TOXIN, SHIGELLA DYSENTERIAE

TOXICITY DATA with REFERENCE
ipr-mus LD50:250 ng/kg JOIMA3 130,380,83
ivn-mus LD50:450 ng/kg JOIMA3 130,380,83
ivn-rbt LD50:2200 ng/kg JOIMA3 130,380,83

SAFETY PROFILE: An extremely deadly poison by intravenous and intraperitoneal routes.

SCE000 CAS:138-59-0 ***HR: 3***
SHIKIMIC ACID
mf: $C_7H_{10}O_5$ mw: 174.17

PROP: Isolated from Bracken (NATUAS 250,348,74).

SYNS: BRACKEN FERN TOXIC COMPONENT ◇ SHIKIMATE ◇ 3,4,5-TRIHYDROXY-1-CYCLOHEXENE-1-CARBOXYLIC ACID

TOXICITY DATA with REFERENCE
dlt-mus-ipr 1000 mg/kg BJLSAF 73,105,76
dlt-mus-orl 3200 mg/kg NATUAS 250,348,74
otr-ham:kdy 250 mg/L TOLED5 19,43,83
orl-mus TDLo:4000 mg/kg:ETA NATUAS 250,348,74
ipr-mus LD50:1000 mg/kg NATUAS 250,348,74

CONSENSUS REPORTS: IARC Cancer Review: Group 3 IMEMDT 7,56,87; Animal Inadequate Evidence IMEMDT 40,47,86. EPA Genetic Toxicology Program.

SAFETY PROFILE: Moderately toxic by intraperitoneal route. Questionable carcinogen with experimental tumorigenic data. Mutation data reported. When heated to decomposition it emits acrid smoke and irritating fumes.

SCF000 ***HR: 3***
SHINING SUMAC

PROP: Hot water extract of shining sumac root (JNCIAM 60,683,78).

SYN: RHUS COPALLINA

TOXICITY DATA with REFERENCE
scu-rat TDLo:3900 mg/kg/65W-I:NEO JNCIAM 60,683,78

SAFETY PROFILE: Questionable carcinogen with experimental neoplastigenic data. When heated to decomposition it emits acrid smoke and irritating fumes.

SCF025 ***HR: 3***
cis-SHP
mf: $C_6H_{17}N_2O_5Pt$ mw: 392.34

SYN: cis-SULFATO-1,2-DIAMINOCYCLOHEXANEPLATINUM(II)

TOXICITY DATA with REFERENCE
mmo-sat 20 nmol/plate CNREA8 41,4368,81
dnd-sat 10 mg/L/20H-C CNREA8 41,4368,81
oms-bcs 1600 nmol/L/3H-C CNREA8 41,4368,81
ipr-mus TDLo:10590 μg/kg/10W-I:NEO CNREA8 41,4368,81

SAFETY PROFILE: Questionable carcinogen with experimental neoplastigenic data. Mutation data reported. When heated to decomposition it emits toxic fumes of NO_x. See also PLATINUM COMPOUNDS.

SCF050 ***HR: 3***
trans(−)-SHP
mf: $C_6H_{17}N_2O_5Pt$ mw: 392.34

SYN: trans(−)-SULFATO-1,2-DIAMINOCYCLOHEXANE-PLATINUM(II)

TOXICITY DATA with REFERENCE
mmo-sat 20 nmol/plate CNREA8 41,4368,81
dnd-sat 10 mg/L/20H-C CNREA8 41,4368,81
oms-bcs 1100 nmol/L/3H-C CNREA8 41,4368,81
ipr-mus TDLo:10590 μg/kg/10W-I:NEO CNREA8 41,4368,81
ipr-mus TD:42368 μg/kg/10W-I:ETA CNREA8 41,4368,81

SAFETY PROFILE: Questionable carcinogen with experimental neoplastigenic data. Mutation data reported. When heated to decomposition it emits toxic fumes of NO_x. See also PLATINUM COMPOUNDS.

SCF075 ***HR: 3***
trans(+)-SHP
mf: $C_6H_{17}N_2O_5Pt$ mw: 392.34

SYN: trans(+)-SULFATO-1,2-DIAMINOCYCLOHEXANE-PLATINUM(II)

TOXICITY DATA with REFERENCE
mmo-sat 20 nmol/plate CNREA8 41,4368,81
dnd-sat 10 mg/L/20H-C CNREA8 41,4368,81
oms-bcs 1600 nmol/L/3H-C CNREA8 41,4368,81
ipr-mus TDLo:10590 μg/kg/10W-I:NEO CNREA8 41,4368,81

SAFETY PROFILE: Questionable carcinogen with experimental neoplastigenic data. Mutation data reported. When heated to decomposition it emits toxic fumes of NO_x. See also PLATINUM COMPOUNDS.

SCF500 CAS:12684-33-2 ***HR: 3***
SIBIROMYCIN
mf: $C_{24}H_{31}N_3O_7$ mw: 473.53

SYNS: 7-((4,6-DIDEOXY-3-(METHYL-4-(METHYLAMINO)β-ALTROPYRANOSYL)OXY)-10,11-DIHYDRO-9,11-DIHYDROXY-8-METHYL-2-(1-PROPENYL)-1-PYRROLO(2,1-C)(1,4)BENZODIAZEPIN-5-ONE ◇ SYBIROMYCIN

TOXICITY DATA with REFERENCE
dni-mus:ast 1 μg/L NEOLA4 22,105,75
oms-mus:ast 1 μg/L NEOLA4 22,105,75
dnd-mam:lym 400 μmol/L BBACAQ 475,521,77
orl-mus LD50:459 mg/kg 85GDA2 5,185,81
ipr-mus LD50:50 μg/kg JANTAJ 29,981,76
scu-mus LD50:84 mg/kg 85GDA2 5,185,81
ivn-mus LD50:58 mg/kg 85GDA2 5,185,81

SAFETY PROFILE: Poison by intraperitoneal, intravenous, and subcutaneous routes. Moderately toxic by ingestion. Mutation data reported. When heated to decomposition it emits toxic fumes of NO_x.

SCH000 CAS:112945-52-5 ***HR: 1***
SILICA, AMORPHOUS FUMED
mf: O_2Si mw: 60.09

PROP: A finely powdered microcellular silica foam with minimum SiO_2 content of 89.5%. Insol in water; sol in hydrofluoric acid.

SYNS: ACTICEL ◇ AEROSIL ◇ AMORPHOUS SILICA DUST ◇ AQUAFIL ◇ CAB-O-GRIP II ◇ CAB-O-SIL ◇ CAB-O-SPERSE ◇ CATALOID ◇ COLLOIDAL SILICA ◇ COLLOIDAL SILICON DIOXIDE ◇ DAVISON SG-67 ◇ DICALITE ◇ DRI-DIE INSECTICIDE 67 ◇ ENT 25,550 ◇ FLO-GARD ◇ FOSSIL FLOUR ◇ FUMED SILICA ◇ FUMED SILICON DIOXIDE ◇ HI-SEL ◇ LO-VEL ◇ LUDOX ◇ NALCOAG ◇ NYACOL ◇ NYACOL 830 ◇ NYACOL 1430 ◇ SANTOCEL ◇ SG-67 ◇ SILICA AEROGEL ◇ SILICA, AMORPHOUS ◇ SILICIC ANHYDRIDE ◇ SILICON DIOXIDE (FCC) ◇ SILIKILL ◇ SYNTHETIC AMORPHOUS SILICA ◇ VULKASIL

TOXICITY DATA with REFERENCE
dns-rat-itr 120 mg/kg ENVRAL 41,61,86
bfa-rat:lng 120 mg/kg ENVRAL 41,61,86
ihl-rat TCLo:50 mg/m³/6H/2Y-I:CAR CNREA8 2,255,86
orl-rat LD50:3160 mg/kg ARSIM* 20,9,66
ipr-rat LDLo:50 mg/kg AHBAAM 136,1,52
ivn-rat LD50:15 mg/kg BSIBAC 44,1685,68
itr-rat LDLo:10 mg/kg AHBAAM 136,1,52
ipr-gpg LDLo:120 mg/kg BJEPA5 3,75,22

CONSENSUS REPORTS: IARC Cancer Review: Group 3 IMEMDT 7,341,87; Animal Inadequate Evidence IMEMDT 42,209,88; Human Inadequate Evidence IMEMDT 42,209,88. Reported in EPA TSCA Inventory.

OSHA PEL: (Transitional: TWA 80 mg/m³/%SiO_2) TWA 6 mg/m³
ACGIH TLV: (Proposed TWA 2 mg/m³ (Respirable Dust))

SAFETY PROFILE: Poison by intraperitoneal, intravenous, and intratracheal routes. Moderately toxic by ingestion. Much less toxic than crystalline forms. Questionable carcinogen with experimental carcinogenic data. Mutation data reported. Does not cause silicosis. See also other silica entries.

SCI000 CAS:7631-86-9 ***HR: 1***
SILICA, AMORPHOUS HYDRATED
mf: O_2Si mw: 60.09

SYNS: SILICA AEROGEL ◇ SILICA GEL ◇ SILICA XEROGEL ◇ SILICIC ACID

CONSENSUS REPORTS: IARC Cancer Review: Animal Inadequate Evidence IMEMDT 42,209,88; Human Inadequate Evidence IMEMDT 42,209,88.

OSHA PEL: (Transitional: TWA 80 mg/m³/%SiO_2) TWA 6 mg/m³
ACGIH TLV: TWA (nuisance particulate) 10 mg/m³ of total dust (when toxic impurities are not present, e.g., quartz < 1%).

SAFETY PROFILE: The pure unaltered form is considered nontoxic. Some deposits contain small amounts of crystalline quartz which is therefore fibrogenic. When diatomaceous earth is calcined (with or without fluxing agents) some silica is converted to cristobalite and is therefore fibrogenic. Tridymite has never been detected in calcined diatomaceous earth. See also other silica entries.

SCI500 ***HR: 3***
SILICA (crystalline)
mf: SiO_2 mw: 60.09

PROP: Transparent, tasteless crystals or amorph powder. Mp: 1710°, bp: 2230°, d (amorph): 2.2, d (crystalline): 2.6, vap press: 10 mm @ 1732°. Practically insol in water or acids. Dissolves readily in HF, forming silicon tetrafluoride.

SYNS: AGATE ◇ AMETHYST ◇ CHALCEDONY ◇ CHERTS ◇ CRISTOBALITE ◇ FLINT ◇ ONYX ◇ PURE QUARTZ ◇ ROSE QUARTZ ◇ SAND ◇ SILICA FLOUR ◇ SILICON DIOXIDE ◇ TRIDYMITE ◇ TRIPOLI

CONSENSUS REPORTS: IARC Cancer Review: Group 2A IMEMDT 7,341,87; Animal Sufficient Evidence IMEMDT 42,209,88; Human Limited Evidence IMEMDT 42,209,88.

OSHA PEL: (Transitional: TWA Respirable Fraction: 10 mg/m^3/2(%SiO_2+2); Total Dust: 30 mg/m^3/2 (%SiO_2 +2)) Respirable Fraction: TWA 0.05 mg/m^3
ACGIH TLV: TWA Respirable Fraction: 0.05 mg/m^3
DFG MAK: 0.15 mg/m^3
NIOSH REL: (Silica, Crystalline) TWA 50 μg/m^3

SAFETY PROFILE: Moderately toxic as an acute irritating dust. From the point of view of numbers of workers exposed and cases of disability produced, silica is the chief cause of pulmonary dust disease. The prolonged inhalation of dusts containing free silica may result in the development of a disabling pulmonary fibrosis known as silicosis. The Committee on Pneumoconiosis of the American Public Health Association defines silicosis as "a disease due to the breathing of air containing silica (SiO_2) characterized by generalized fibrotic changes and the development of miliary nodules in both lungs, and clinically by shortness of breath, decreased chest expansion, lessened capacity for work, absence of fever, increased susceptibility to tuberculosis (some or all of which symptoms may be present), and characteristic x-ray findings."

Silica occurs in the pure state in nature as highly fibrogenic quartz. It is the main constituent of relatively much less toxic sand, sandstone, tripoli and diatomaceous earth. It is present in crystalline form in high amounts (up to 35%) in granite. Exposure to silica occurs in hard rock mining, in foundries, in manufacture of porcelain and pottery, in the spraying of vitreous enamels, in sandblasting, in granite-cutting and tombstone-making, in the manufacture of silica firebrick and other refractories, in grinding and polishing operations where natural abrasive wheels are used, and other occupations.

The duration of exposure which is associated with the development of silicosis varies widely for different occupations. Thus, the average duration of exposure required for the development of silicosis in sand-blasters is 2-10 years, in moulders and granite cutters, about 30 years, and in hard rock miners, 10-15 years. There is also much variation in individual susceptibility; certain workers show radiological evidence of the disease years before their fellow workmen who are similarly exposed. Such susceptible individuals are, fortunately, rather rare.

The action of crystalline silica on the lungs results in the production of a diffuse, nodular fibrosis in which the parenchyma and the lymphatic systems are involved. This fibrosis is, to a certain extent, progressive, and may continue to increase for several years after exposure is terminated. Where the pulmonary reserve is sufficiently reduced, the worker complains of shortness of breath on exertion. This is the first and most common symptom in cases of uncomplicated silicosis. If severe, it may incapacitate the worker for heavy, or even light, physical exertion, and in extreme cases there may be shortness of breath even while at rest. The most common physical sign of silicosis is a limitation of expansion of the chest. There may be a dry cough, sometimes very troublesome. The characteristic radiographic appearance is one of diffuse, discrete nodulation, scattered throughout both lung fields. Where the disease advances, the shortness of breath becomes worse, and the cough more productive and troublesome. There is no fever or other evidence of systemic reaction. Further progress of the disease results in marked fatigue, extreme dyspnea and cyanosis, loss of appetite, pleuritic pain and total incapacity to work. If tuberculosis does not supervene, the condition may eventually cause death either from cardiac failure or from destruction of lung tissue, with resultant anoxemia. In the later stages, the x-ray may show large conglomerate shadows, due to the coalescence of the silicotic nodules, with areas of emphysema between them.

Silica in some forms is used as an additive permitted in the feed and drinking water of animals and/or for the treatment of food-producing animals. It is also permitted in food for human consumption. It is a common air contaminant. Reacts violently with ClF_3; MnF_3; OF_2. See also other silica entries.

SCJ000 CAS:14464-46-1 ***HR: 3***
SILICA, CRYSTALLINE-CRISTOBALITE
mf: O_2Si mw: 60.09

PROP: White, cubic-system crystals formed from quartz at temperatures above 1000°C (NTIS** PB246-697).

SYNS: CALCINED DIATOMITE ◇ CRISTOBALITE

TOXICITY DATA with REFERENCE
ipl-rat TDLo:90 mg/kg:CAR JNCIAM 57,509,76
ipl-rat TD:100 mg/kg:ETA BJCAAI 41,908,80
ihl-hmn TCLo:400 particles/cc/4Y-I:PUL BJIMAG 5,148,48
ihl-hmn TCLo:16 mppcf/8H/17.9Y-I:PUL NTIS** PB246-697
itr-rat LDLo:200 mg/kg BJIMAG 10,9,53

CONSENSUS REPORTS: IARC Cancer Review: Group 2A IMEMDT 7,341,87; Animal Sufficient Evidence IMEMDT 42,209,88; Human Limited Evidence IMEMDT 42,209,88. Reported in EPA TSCA Inventory.

OSHA PEL: (Transitional: TWA Respirable Fraction: (10 mg/m^3/2(%SiO_2+2); Total Dust: 30 mg/m^3/2(%SiO_2+2)) TWA Respirable Fraction: 0.05 mg/m^3
ACGIH TLV: TWA Respirable Fraction: 0.05 mg/m^3
DFG MAK: 0.15 mg/m^3
NIOSH REL: (Silica, Crystalline) TWA 50 μg/m^3

SAFETY PROFILE: Confirmed carcinogen with experimental carcinogenic and tumorigenic data. Poison by intratracheal route. Human systemic effects by inhalation: cough, dyspnea, fibrosis. About twice as toxic as silica in causing silicosis. See also other silica entries.

SCJ500 CAS:14808-60-7 ***HR: 3***
SILICA, CRYSTALLINE-QUARTZ
mf: O_2Si mw: 60.09

PROP: Mp: 1710°, bp: 2230°, d: 2.6.

SYNS: AGATE ◇ AMETHYST ◇ CHALCEDONY ◇ CHERTS ◇ FLINT ◇ ONYX ◇ PURE QUARTZ ◇ QUARTZ ◇ QUAZO PURO (ITALIAN) ◇ ROSE QUARTZ ◇ SAND ◇ SILICA FLOUR (powdered crystalline silica) ◇ SILICIC ANHYDRIDE

TOXICITY DATA with REFERENCE
ihl-rat TCLo:50 mg/m^3/6H/71W-I:CAR ENVRAL 40,499,86
itr-rat TDLo:100 mg/kg/19W-I:ETA EVHPAZ 34,47,80
ipr-rat TDLo:45 mg/kg:CAR ZHPMAT 162,467,76
imp-rat TDLo:900 mg/kg:NEO AICCA6 10,119,54
ihl-hmn TCLo:16 mppcf/8H/17.9Y-I:PUL NTIS** PB246-697
ihl-hmn LCLo:300 μg/m^3/10Y-I:LVR ANYAA9 271,324,76
ivn-rat LDLo:90 mg/kg JNCIAM 57,509,76
itr-rat LDLo:200 mg/kg BJIMAG 10,9,53
ivn-mus LDLo:40 mg/kg JNCIAM 1,241,40
ivn-dog LDLo:20 mg/kg BIJOAK 27,1007,33

CONSENSUS REPORTS: IARC Cancer Review: Animal Sufficient Evidence IMEMDT 42,209,88; Human Limited Evidence IMEMDT 42,209,88; Group 2A IMEMDT 7,341,87. Reported in EPA TSCA Inventory.

OSHA PEL: (Transitional: TWA Respirable Fraction: 10 mg/m^3/2(%SiO_2+2); Total Dust: 30 mg/m^3/2 (%SiO_2+2)) TWA Respirable Fraction: 0.1 mg/m^3
ACGIH TLV: TWA Respirable Fraction: 0.1 mg/m^3
DFG MAK: 0.15 mg/m^3
NIOSH REL: TWA 50 μg/m^3; 3000000 fibers/m^3

SAFETY PROFILE: Confirmed carcinogen with experimental carcinogenic, tumorigenic, and neoplastigenic data. Experimental poison by intratracheal and intravenous routes. Human systemic effects by inhalation: cough, dyspnea, liver effects. Incompatible with OF_2, vinylacetate. See also other silica entries.

SCK000 CAS:15468-32-3 ***HR: 3***
SILICA, CRYSTALLINE-TRIDYMITE
mf: O_2Si mw: 60.09

PROP: White or colorless platelets or orthorhombic (crystals) formed from quartz @ temperatures >870° (NTIS** PB246-697).

SYNS: TRIDIMITE (FRENCH) ◇ TRIDYMITE

TOXICITY DATA with REFERENCE
ipl-rat TDLo:100 mg/kg:ETA BJCAAI 41,908,80
ihl-hmn TCLo:16 mppcf/8H/17.9Y-I:PUL NTIS** PB246-697
itr-rat LDLo:200 mg/kg BJIMAG 10,9,53

CONSENSUS REPORTS: IARC Cancer Review: Group 2A IMEMDT 7,341,87; Animal Sufficient Evidence IMEMDT 42,209,88; Human Limited Evidence IMEMDT 42,209,88.

OSHA PEL: (Transitional: TWA Respirable: 10 mg/m^3/2(%SiO_2+2); Total Dust: TWA 30 mg/m^3/2(%SiO_2+2)) TWA 0.05 mg/m^3
ACGIH TLV: (Silica, Crystalline) TWA Respirable Fraction: 0.05 mg/m^3
DFG MAK: 0.15 mg/m^3
NIOSH REL: TWA 50 μg/m^3

SAFETY PROFILE: Confirmed carcinogen with experimental tumorigenic data. Poison by intratracheal route. Human systemic effects by inhalation: cough, dyspnea. About twice as toxic as silica in causing silicosis. See also other silica entries.

SCK500 ***HR: 3***
SILICA FLOUR

PROP: A finely ground *crystalline* silica sometimes marketed as "Amorphous." It is *not* amorphous.

SAFETY PROFILE: Toxic by inhalation. It has shown a very high incidence of silicosis among "silica flour" workers. See also other silica entries.

SCK600 CAS:60676-86-0 ***HR: 3***
SILICA, FUSED
mf: O_2Si mw: 60.09

PROP: Made up of spherical submicroscopic particles under 0.1 micron in size (AMIHBC 9,389,54).

SYNS: AMORPHOUS FUSED SILICA ◇ FUSED QUARTZ ◇ FUSED SILICA (ACGIH) ◇ QUARTZ GLASS ◇ SILICA, AMORPHOUS FUSED ◇ SILICA, VITREOUS ◇ SILICON DIOXIDE ◇ VITREOUS QUARTZ

TOXICITY DATA with REFERENCE
imp-rat TDLo:400 mg/kg:ETA NATWAY 41,534,54
ipr-rat LDLo:400 mg/kg AMIHBC 9,389,54
itr-rat LDLo:120 mg/kg AMIHBC 9,389,54
ipr-mus LDLo:40 mg/kg BJEPA5 3,75,22

ivn-cat LDLo:15 mg/kg JLCMAK 26,774,41
ivn-rbt LDLo:35 mg/kg BJEPAS 3,75,22

CONSENSUS REPORTS: IARC Cancer Review: Group 3 IMEMDT 7,341,87; Animal Inadequate Evidence IMEMDT 42,39,87; Human Inadequate Evidence IMEMDT 42,39,87. Reported in EPA TSCA Inventory.

OSHA PEL: (Transitional: TWA Respirable: 10 mg/m^3/2(%SiO_2+2); Total Dust: TWA 30 mg/m^3/2 (%SiO_2+2)) TWA 0.1 mg/m^3
ACGIH TLV: TWA Respirable Fraction: 0.1 mg/m^3
NIOSH REL: (Silica, Crystalline): TWA 0.05 mg/m^3

SAFETY PROFILE: Questionable carcinogen with experimental tumorigenic data. Poison by intraperitoneal, intravenous, and intratracheal routes. See also other silica entries.

SCL000 CAS:7699-41-4 ***HR: 1***
SILICA, GEL and AMORPHOUS-PRECIPITATED
mf: H_2O_3Si mw: 78.11

SYNS: KIESELSAURE (GERMAN) ◇ METASILICIC ACID ◇ PRECIPITATED SILICA ◇ SILICA GEL ◇ SILICIC ACID

TOXICITY DATA with REFERENCE
eye-rbt 8300 μg/48H JANCA2 56,905,73
ivn-mus LDLo:234 mg/kg AMIHAB 17,204,58

CONSENSUS REPORTS: IARC Cancer Review: Group 3 IMEMDT 7,341,87; Animal Inadequate Evidence IMEMDT 42,39,87; Human Inadequate Evidence IMEMDT 42,39,87. Reported in EPA TSCA Inventory.

OSHA PEL: (Transitional: TWA 80 mg/m^3/%SiO_2) TWA 6 mg/m^3
ACGIH TLV: TWA (nuisance particulate) 10 mg/m^3 of total dust (when toxic impurities are not present, e.g., quartz $< 1\%$).

SAFETY PROFILE: Poison by intravenous route. An eye irritant and nuisance dust. Questionable carcinogen. See also other silica entries and SILICATES.

SCM500 ***HR: D***
SILICATES

PROP: Widely occurring compounds containing silicon, oxygen, and one or more metals with or without hydrogen.

SAFETY PROFILE: Soluble alkaline silicates act locally like mild alkalies. The dust of certain silicates, such as asbestos (hydrated magnesium silicate) and talc, can produce fibrotic changes in the lungs and are implicated as experimental carcinogens. React violently with Li. See also mica, soapstone, talc.

SCN000 ***HR: 2***
SILICATE SOAPSTONE
mf: $3MgO{\bullet}4SiO_2{\bullet}H_2O$ mw: 379.31

PROP: Containing less than 1% crystalline silica (FEREAC 39,23540,74).

SYN: SOAPSTONE

OSHA PEL: (Transitional: TWA 20 mppcf) TWA Total Dust: 6 mg/m^3; Respirable Fraction: 3 mg/m^3
ACGIH TLV: TWA Respirable Fraction: 3 mg/m^3; 6 mg/m^3 of total dust (when toxic impurities are not present, e.g., quartz $< 1\%$).

SAFETY PROFILE: Less toxic than quartz. See SILICATES.

SCN500 CAS:15191-85-2 ***HR: 3***
SILICIC ACID, BERYLLIUM SALT
mf: $O_4Si{\bullet}2Be$ mw: 110.11

PROP: Colorless crystals. D: 3.0.

SYNS: BERYLLIUM ORTHOSILICATE ◇ BERYLLIUM SILICATE ◇ BERYLLIUM SILICIC ACID ◇ ORTHOSILICATE ◇ PHENACITE ◇ PHENAKITE ◇ PHENAZITE

TOXICITY DATA with REFERENCE
ivn-rbt TDLo:500 mg/kg:ETA AICCA6 7,171,50

CONSENSUS REPORTS: Beryllium and its compounds are on the Community Right-To-Know List. IARC Cancer Review: Group 2A IMEMDT 7,127,87; Animal Sufficient Evidence IMEMDT 23,143,80.

OSHA PEL: (Transitional: TWA 0.002 mg(Be)/m^3; CL 0.005; Pk 0.025/30M/8H) TWA 0.002 mg(Be)/m^3; STEL 0.005 mg(Be)/m^3/30M; CL 0.025 mg(Be)/m^3
ACGIH TLV: TWA 0.002 mg(Be)/m^3; Suspected Carcinogen
NIOSH REL: CL not to exceed 0.0005 mg/(Be)/m^3

SAFETY PROFILE: Suspected carcinogen with experimental carcinogenic and tumorigenic data. When heated to decomposition it emits toxic fumes of BeO. See also BERYLLIUM COMPOUNDS and SILICATES.

SCO500 CAS:16961-83-4 ***HR: 3***
SILICOFLUORIC ACID
DOT: NA 1778
mf: $F_6Si{\bullet}2H$ mw: 144.11

PROP: Transparent, colorless, fuming liquid. Bp: decomp.

SYNS: ACIDE FLUOROSILICIQUE (FRENCH) ◇ ACIDE FLUOSILICIQUE (FRENCH) ◇ ACIDO FLUOSILICICO (ITALIAN) ◇ FLUOROSILICIC ACID ◇ FLUOSILICIC ACID ◇ HEXAFLUOROKIESELSAIURE (GERMAN) ◇ HEXAFLUOROKIEZELZUUR (DUTCH) ◇ HEXAFLUOROSILICATE(2-) DIHYDROGEN ◇ HEXAFLUOSILICIC ACID

◇ HYDROFLUOSILICIC ACID ◇ HYDROGEN HEXAFLUOROSILICATE ◇ HYDROSILICOFLUORIC ACID ◇ KIEZELFLUORWATERSTOFZUUR (DUTCH) ◇ SAND ACID

TOXICITY DATA with REFERENCE
scu-frg LDLo:140 mg/kg CRSBAW 124,133,37

CONSENSUS REPORTS: Reported in EPA TSCA Inventory.

NIOSH REL: TWA 2.5 mg(F)/m^3
DOT Classification: Corrosive Material; Label: Corrosive.

SAFETY PROFILE: Poison by subcutaneous route. A corrosive irritant to skin, eyes, and mucous membranes. Will react with water or steam to produce toxic and corrosive fumes. When heated to decomposition it emits toxic fumes of F^-. See also FLUORIDES.

SCP000 CAS:7440-21-3 ***HR: 3***
SILICON
DOT: UN 1346
af: Si aw: 28.09

PROP: Cubic, steel-gray crystals or dark brown powder. Mp: 1420°, bp: 2600°, d: 2.42 or 2.3 @ 20°vap press: 1 mm @ 1724°. Almost insol in water; sol in molten alkali oxides.

CONSENSUS REPORTS: Reported in EPA TSCA Inventory.

OSHA PEL: (Transitional: TWA Total Dust: 15 mg/m^3; Respirable Fraction: 5 mg/m^3) TWA Total Dust: 10 mg/m^3 of total; Respirable Fraction: 5 mg/m^3
ACGIH TLV: TWA (nuisance particulate) 10 mg/m^3 of total dust (when toxic impurities are not present, e.g., quartz < 1%).
DOT Classification: Flammable Solid; Label: Flammable Solid

SAFETY PROFILE: Does not occur freely in nature, but is found as silicon dioxide (silica) and as various silicates. Elemental Si is flammable when exposed to flame or by chemical reaction with oxidizers. Violent reactions with alkali carbonates; oxidants; (Al + PbO); Ca; Cs_2C_2; Cl_2; CoF_2; F_2; IF_5; MnF_3; Rb_2C_2; FNO; AgF; NaK alloy. When heated it will react with water or steam to produce H_2; can react with oxidizing materials. See also various silica entries, SILICATES, and POWDERED METALS.

SCP500 ***HR: 2***
SILICON BROMIDE
mf: $SiBr_4$ mw: 347.72

PROP: Colorless, fuming liquid; disagreeable odor. Mp: 5°, bp: 153°, d: 2.814, vap d: 2.82, fp: −12°.

SYNS: TETRABROMOSILANE ◇ TETRABROMOSILICANE

SAFETY PROFILE: Probably very irritating to skin, eyes and mucous membranes. Will react with water or steam to produce heat and toxic and corrosive fumes. When heated to decomposition it emits highly toxic fumes of hydrobromic acid. See also BROMIDES.

SCQ000 CAS:409-21-2 ***HR: 3***
SILICON CARBIDE
mf: CSi mw: 40.10

PROP: Bluish-black, iridescent crystals. Mp: 2600°, bp: subl > 2000°, decomp @ 2210°, d: 3.17.

SYNS: CARBOLON ◇ CARBON SILICIDE ◇ CARBORUNDEUM ◇ CARBORUNDUM ◇ KZ 3M ◇ KZ 5M ◇ KZ 7M ◇ SILICON MONOCARBIDE ◇ SILUNDUM

TOXICITY DATA with REFERENCE
imp-rat TDLo:200 mg/kg:NEO JJIND8 67,965,81

CONSENSUS REPORTS: Reported in EPA TSCA Inventory.

OSHA PEL: (Transitional: TWA Total Dust: 15 mg/m^3; Respirable Fraction: 5 mg/m^3) TWA Total Dust: 10 mg/m^3; Respirable Fraction: 5 mg/m^3
ACGIH TLV: TWA (nuisance particulate) 10 mg/m^3 of total dust (when toxic impurities are not present, e.g., quartz < 1%).
DFG MAK: 4 mg/$m^3$0

SAFETY PROFILE: Questionable carcinogen with experimental neoplastigenic data.

SCQ500 CAS:10026-04-7 ***HR: 3***
SILICON CHLORIDE
DOT: UN 1818
mf: Cl_4Si mw: 169.89

PROP: Colorless, fuming liquid; suffocating odor. Mp: −70°, bp: 57.57°, d: 1.482. Misc with benzene, ether, chloroform, petr ether.

SYNS: CHLORID KREMICITY (CZECH) ◇ EXTREMA ◇ SILICIO (TETRACLORURO di) ◇ SILICIUMTETRACHLORID (GERMAN) ◇ SILICIUMTETRACHLORIDE (DUTCH) ◇ SILICIUM(TETRACHLORURE de) (FRENCH) ◇ SILICON TETRACHLORIDE (DOT) ◇ TETRACHLOROSILANE ◇ TETRACHLORURE de SILICIUM (FRENCH)

TOXICITY DATA with REFERENCE
skn-rbt 500 mg/24H SEV 28ZPAK -,14,72
eye-rbt 20 mg/24H MOD 28ZPAK -,14,72
ihl-rat LC50:8000 ppm/4H JIHTAB 31,343,49

CONSENSUS REPORTS: Reported in EPA TSCA Inventory.

DOT Classification: Corrosive Material; Label: Corrosive.

SAFETY PROFILE: Mildly toxic by inhalation. A corrosive irritant to eyes, skin, and mucous membranes. Reacts with water to form HCl. Violent reaction with Na; K. When heated to decomposition it emits toxic fumes of Cl^-. See also CHLOROSILANES.

SCR000 CAS:11133-78-1 ***HR: D***
SILICONCHROME (exothermic)

SYN: EXOTHERMIC SILICON CHROME (DOT)

CONSENSUS REPORTS: Chromium and its compounds are on the Community Right-To-Know List.

DOT Classification: ORM-C; Label: None.

SAFETY PROFILE: See SILICON and CHROMIUM COMPOUNDS.

SCR100 CAS:13520-74-6 ***HR: 3***
SILICON DIBROMIDE SULFIDE
mf: Br_2SSi mw: 219.95

SAFETY PROFILE: The hydrolysis reaction is explosive. When heated to decomposition it emits toxic fumes of Br^- and SO_x. See also SULFIDES and BROMIDES.

SCR400 CAS:63148-62-9 ***HR: D***
SILICONE 360

SYNS: ANTIFOAM FD 62 ◇ DC 360 ◇ KO 08 ◇ PMS 1.5 ◇ PMS 300 ◇ PMS 154A ◇ PMS 200A ◇ PNS 25 ◇ S DC 200 ◇ SILAK M 10 ◇ SILICONE DC 200 ◇ SILICONE DC 360 ◇ SILICONE DC 360 FLUID ◇ SILICONE RELEASE L 45 ◇ SILIKON ANTIFOAM FD 62 ◇ SILOXANES and SILICONES, DI Me ◇ UC LIQUID G ◇ UNION CARBIDE LIQUID G ◇ XF-13-563

TOXICITY DATA with REFERENCE
scu-rat TDLo:8 g/kg (15-22D preg):REP JTEHD6 1,909,76
scu-rbt TDLo:260 mg/kg (6-18D preg):TER JTEHD6 1,909,76

CONSENSUS REPORTS: EPA Genetic Toxicology Program. Reported in EPA TSCA Inventory.

SAFETY PROFILE: An experimental teratogen. Experimental reproductive effects. See also SILICONES.

SDC000 ***HR: D***
SILICONES

SYN: SILOXANES

CONSENSUS REPORTS: Organosilicon oxide polymers such as $-R_2Si-O$, where R is a monovalent organic radical.

SAFETY PROFILE: Most of the silicones that have been studied are only slightly toxic and mildly irritating, however some may be severe irritants. May be spontaneously flammable in air. There can be toxicity due to contamination of silicones by components of manufacture.

SDF000 CAS:67762-92-9 ***HR: 1***
SILICONE Y-6607

SYNS: SILICONES ◇ SILOXANES ◇ DIMETHYL ◇ (DIMETHYL-AMINO)-TERMINATED

TOXICITY DATA with REFERENCE
ihl-rat LC50:774 ppm/4H EPASR* 8EHQ-0680-0349

CONSENSUS REPORTS: Reported in EPA TSCA Inventory.

SAFETY PROFILE: Mildly toxic by inhalation. See also SILICONES.

SDF650 CAS:7783-61-1 ***HR: 3***
SILICON FLUORIDE
DOT: UN 1859
mf: F_4Si mw: 104.09

PROP: Colorless gas, very pungent odor. Mp: −77°, bp: −65° @ 181 mm, d: 4.67.

SYNS: SILICON TETRAFLUORIDE (DOT) ◇ TETRAFLUOROSILANE

CONSENSUS REPORTS: Reported in EPA TSCA Inventory.

OSHA PEL: TWA 2.5 mg(F)/m^3
ACGIH TLV: TWA 2.5 mg(F)/m^3
NIOSH REL: (Inorganic Fluorides) TWA 2.5 mg(F)/m^3
DOT Classification: Nonflammable Gas; Label: Nonflammable Gas; Poison A; Label: Poison Gas, Corrosive.

SAFETY PROFILE: A poison. A corrosive irritant to skin, eyes, and mucous membranes. When heated to decomposition it emits toxic fumes of F^-. See also FLUORIDES and HYDROFLUORIC ACID.

SDH000 CAS:10097-28-6 ***HR: 3***
SILICON OXIDE
mf: OSi mw: 44.09

SAFETY PROFILE: Ignites spontaneously in air.

SDH500 CAS:27890-58-0 ***HR: 3***
SILICON TETRAAZIDE
mf: $N_{12}Si$ mw: 196.17

SAFETY PROFILE: Has exploded spontaneously. When heated to decomposition it emits toxic fumes of NO_x. See also AZIDES.

SDH575 CAS:7803-62-5 ***HR: 3***
SILICON TETRAHYDRIDE
DOT: UN 2203
mf: H_4Si mw: 32.13

PROP: Gas with repulsive odor; slowly decomp by

water. D: 0.68 @ −185°, mp: −185°, bp: 112°, fp: −200°.

SYNS: MONOSILANE ◇ SILANE ◇ SILICANE

TOXICITY DATA with REFERENCE
ihl-rat LC50:9600 ppm/4H TXAPA9 42,417,77
ihl-mus LCLo:9600 ppm/4H AMRL** TR-72-62,72

CONSENSUS REPORTS: Reported in EPA TSCA Inventory.

OSHA PEL: TWA 5 ppm
ACGIH TLV: TWA 5 ppm
DOT Classification: Flammable Gas; Label: Flammable Gas.

SAFETY PROFILE: Mildly toxic by inhalation. Silanes are irritating to skin, eyes, and mucous membranes. Easily ignited in air. Explosive reaction or ignition on contact with halogens or covalent halides (e.g., bromine, chlorine, carbonyl chloride, antimony pentachloride, tin(IV) chloride). Ignites in oxygen. Can react with oxidizers. It may self-explode. When heated to decomposition it burns or explodes.

SDH670 CAS:42959-18-2 ***HR: 2***
SILICON TRIETHANOLAMIN
mf: $C_8H_{19}NO_3Si$ mw: 205.37

SYNS: 2,2-DIMETHYL-1,3-DIOXA-6-AZA-2-SILACYCLOOCTANE-6-ETHANOL ◇ DIMETHYL-1,1-DIOXA-2,8-HYDROXYETHYL-5 SILA-1 AZA-5 CYCLOOCTANE (FRENCH) ◇ T.E.A.S. ◇ TRIETHANOLAMINE SILICIEE (FRENCH)

TOXICITY DATA with REFERENCE
orl-rat LD50:3400 mg/kg THERAP 32,517,77
ipr-rat LD50:1920 mg/kg THERAP 32,517,77
orl-mus LD50:3900 mg/kg EJMCA5 16,425,81
ipr-mus LD50:2230 mg/kg EJMCA5 16,425,81

SAFETY PROFILE: Moderately toxic by ingestion and intraperitoneal routes. When heated to decomposition it emits toxic fumes of NO_x.

SDI000 ***HR: 2***
SILK

TOXICITY DATA with REFERENCE
imp-rat TDLo:36 mg/kg:ETA CNREA8 15,333,55

SAFETY PROFILE: In the form of dust it is an allergen and a nuisance dust. Questionable carcinogen with experimental tumorigenic data by implant. Flammable when exposed to heat or flame. A moderate explosion hazard. When heated to decomposition it emits acrid smoke and irritating fumes. See also POLYMERS, INSOLUBLE.

SDI500 CAS:7440-22-4 ***HR: 2***
SILVER
af: Ag aw: 107.868

PROP: Soft, ductile, malleable, lustrous, white metal. Mp: 961.93°, bp: 2212°, d: 10.50 @ 20°.

SYNS: ARGENTUM ◇ C.I. 77820 ◇ SHELL SILVER ◇ SILBER (GERMAN) ◇ SILVER ATOM

TOXICITY DATA with REFERENCE
mul-rat TDLo:330 mg/kg/43W-I:ETA ZEKBAI 63,586,60
ihl-hmn TCLo:1 mg/m^3:SKN DTLVS* 3,231,71

CONSENSUS REPORTS: Silver and its compounds are on the Community Right-To-Know List. Reported in EPA TSCA Inventory.

OSHA PEL: Metal, Dust, and Fume: TWA 0.01 mg/m^3
ACGIH TLV: TWA (metal) 0.1 mg/m^3, (soluble compounds as Ag) 0.01 mg/m^3
DFG MAK: 0.01 mg/m^3

SAFETY PROFILE: Human systemic effects by inhalation: skin effects. Inhalation of dusts can cause argyrosis. Questionable carcinogen with experimental tumorigenic data. Flammable in the form of dust when exposed to flame or by chemical reaction with C_2H_2; NH_3; bromoazide; ClF_3; ethylene imine; H_2O_2; oxalic acid; H_2SO_4; tartaric acid. Incompatible with acetylene; acetylene compounds; aziridine; bromine azide; 3-bromopropyne; carboxylic acids; copper + ethylene glycol; electrolytes + zinc; ethanol + nitric acid; ethylene oxide; ethyl hydroperoxide; ethyleneimine; iodoform; nitric acid; ozonides; peroxomonosulfuric acid; peroxyformic acid. See also POWDERED METALS and SILVER COMPOUNDS.

SDI750 CAS:7440-22-4 ***HR: 3***
SILVER (colloidal)
af: Ag aw: 107.868

SYNS: ARGENTIUM CREDE ◇ COLLARGOL

TOXICITY DATA with REFERENCE
orl-mus LD50:100 mg/kg JPPMAB 2,20,50
ivn-rbt LDLo:49 mg/kg ZGEMAZ 52,33,26

CONSENSUS REPORTS: Silver and its compounds are on the Community Right-To-Know List. Reported in EPA TSCA Inventory.

OSHA PEL: Metal, Dust, and Fume: TWA 0.01 mg/m^3
ACGIH TLV: TWA (metal) 0.1 mg/m^3, (soluble compounds as Ag) 0.01 mg/m^3
DFG MAK: 0.01 mg/m^3

SAFETY PROFILE: Poison by ingestion and intravenous routes. See also SILVER and SILVER COMPOUNDS.

SDJ000 CAS:13092-75-6 ***HR: 3***
SILVER ACETYLIDE
mf: C_2Ag_2 mw: 239.76

AgC≡CAg

CONSENSUS REPORTS: Silver and its compounds are on the Community Right-To-Know List.

DOT Classification: Forbidden (dry).

SAFETY PROFILE: Severe explosion hazard. A more powerful detonator than copper acetylide. Explodes when heated to 120-140°C. Formed when silver-containing solutions contact acetylene. Upon decomposition it emits acrid smoke and irritating fumes. See also SILVER COMPOUNDS and ACETYLIDES.

SDJ025 CAS:15336-58-0 ***HR: 3***
SILVER ACETYLIDE-SILVER NITRATE
mf: $C_2Ag_2 \cdot AgNO_3$ mw: 409.63

AgC≡CAg•$AgNO_3$

SYN: DISILVER ACETYLIDE SILVER NITRATE

CONSENSUS REPORTS: Silver and its compounds are on the Community Right-To-Know List.

SAFETY PROFILE: The dry material may explode when exposed to high intensity light, heat, or sparks. When heated to decomposition it emits toxic fumes of NO_x. See also SILVER COMPOUNDS, ACETYLIDES, and NITRATES.

SDJ500 CAS:65235-79-2 ***HR: 3***
SILVER AMIDE
mf: AgH_2N mw: 123.89

CONSENSUS REPORTS: Silver and its compounds are on the Community Right-To-Know List.

SAFETY PROFILE: Very explosive when dry. When heated to decomposition it emits toxic fumes of NO_x. See also SILVER COMPOUNDS.

SDK000 CAS:50577-64-5 ***HR: 3***
SILVER 5-AMINOTETRAZOLIDE
mf: CH_2AgN_5 mw: 191.93

CONSENSUS REPORTS: Silver and its compounds are on the Community Right-To-Know List.

SAFETY PROFILE: When heated it explodes. When heated to decomposition it emits toxic fumes of NO_x. See also SILVER COMPOUNDS.

SDK500 ***HR: 3***
SILVER AMMONIUM COMPOUNDS

CONSENSUS REPORTS: Silver and its compounds are on the Community Right-To-Know List.

SAFETY PROFILE: Severe explosion hazard when shocked, exposed to heat, or by chemical reaction. Upon decomposition they can yield toxic fumes of NH_3 and NO_x. See also SILVER COMPOUNDS.

SDL000 CAS:102492-24-0 ***HR: 2***
SILVER AMMONIUM LACTATE
mf: $C_6H_{14}AgNO_6$ mw: 304.08

TOXICITY DATA with REFERENCE
eye-rbt 1500 μg SEV AROPAW 25,839,41

CONSENSUS REPORTS: Silver and its compounds are on the Community Right-To-Know List. Reported in EPA TSCA Inventory.

OSHA PEL: TWA 0.01 mg(Ag)/m^3
ACGIH TLV: TWA 0.01 mg(Ag)/m^3

SAFETY PROFILE: A severe eye irritant. When heated to decomposition it emits toxic fumes of NH_3 and NO_x. See also SILVER AMMONIUM COMPOUNDS and SILVER COMPOUNDS.

SDL500 CAS:23606-32-8 ***HR: 2***
SILVER AMMONIUM NITRATE
mf: $AgH_4N_3O_6$ mw: 249.94

TOXICITY DATA with REFERENCE
eye-rbt 1 mg SEV AROPAW 25,839,41

CONSENSUS REPORTS: Silver and its compounds are on the Community Right-To-Know List. Reported in EPA TSCA Inventory.

OSHA PEL: TWA 0.01 mg(Ag)/m^3
ACGIH TLV: TWA 0.01 mg(Ag)/m^3

SAFETY PROFILE: A severe eye irritant. When heated to decomposition it emits very toxic fumes of NH_3 and NO_x. See also SILVER AMMONIUM COMPOUNDS, SILVER COMPOUNDS, and NITRATES.

SDM000 CAS:102262-34-0 ***HR: 2***
SILVER AMMONIUM SULFATE
mf: AgH_4NO_4S mw: 221.98

TOXICITY DATA with REFERENCE
eye-rbt 1 mg SEV AROPAW 25,839,41

CONSENSUS REPORTS: Silver and its compounds are on the Community Right-To-Know List.

OSHA PEL: TWA 0.01 mg(Ag)/m^3
ACGIH TLV: TWA 0.01 mg(Ag)/m^3

SAFETY PROFILE: A severe eye irritant. When heated to decomposition it emits very toxic fumes of NO_x, NH_3, and SO_x. See also SILVER AMMONIUM COMPOUNDS, SILVER COMPOUNDS, and SULFATES.

SDM100 CAS:7784-08-9 ***HR: 3***
SILVER ARSENITE
DOT: UN 1683
mf: AsO_3•3Ag mw: 446.53

SYNS: ARSENIOUS ACID, TRISILVER(1+) SALT ◇ ARSENOUS ACID, TRISILVER(1+) SALT (9CI)

OSHA: Cancer Hazard
DOT Classification: Poison B; Label: Poison

SAFETY PROFILE: Confirmed carcinogen. When heated to decomposition it emits toxic fumes of As.

SDM500 CAS:13863-88-2 ***HR: 3***
SILVER AZIDE
mf: AgN_3 mw: 149.87

CONSENSUS REPORTS: Silver and its compounds are on the Community Right-To-Know List.

OSHA PEL: TWA 0.01 mg(Ag)/m^3
ACGIH TLV: TWA 0.01 mg(Ag)/m^3

DOT Classification: Forbidden.

SAFETY PROFILE: Explodes when heated above 270°C or on impact. Pure silver azide, explodes @ 340°. An electric field or irradiation by electron pulses can explode the crystals. Shock-sensitive when dry and has detonated @ 250 C. Solutions in aqueous ammonia explode above 100°C. Reacts to form more explosive products with iodine (forms iodine azide); bromine and other halogens. The presence of metal oxides or metal sulfides increases the azides sensitivity to explosion. Mixtures with sulfur dioxide are explosive. When heated to decomposition it emits toxic fumes of NO_x. See also AZIDES and SILVER COMPOUNDS.

SDM525 CAS:82177-80-8 ***HR: 3***
SILVER 2-AZIDO-4,6-DINITROPHENOXIDE
mf: $C_6H_2AgN_5O_5$ mw: 331.98

CONSENSUS REPORTS: Silver and its compounds are on the Community Right-To-Know List.

SAFETY PROFILE: Ignites at 122°C. When heated to decomposition it emits toxic fumes of NO_x. See also SILVER COMPOUNDS and AZIDES.

SDM550 CAS:74093-43-9 ***HR: 3***
SILVER AZIDODITHIOFORMATE
mf: $CAgN_3S_2$ mw: 226.02

$AgSC(S)N_3$

CONSENSUS REPORTS: Silver and its compounds are on the Community Right-To-Know List.

SAFETY PROFILE: A friction-sensitive explosive. When heated to decomposition it emits toxic fumes of SO_x and NO_x. See also SILVER COMPOUNDS and AZIDES.

SDM575 ***HR: 3***
SILVER BENZO-1,2,3-TRIAZOLE-1-OXIDE
mf: $C_6H_4AgN_3O$ mw: 241.99

$C_6H_4N(OAg)N{=}N$ (ring)

CONSENSUS REPORTS: Silver and its compounds are on the Community Right-To-Know List.

SAFETY PROFILE: Explodes when heated. When heated to decomposition it emits toxic fumes of NO_x. See also SILVER COMPOUNDS.

SDN000 CAS:14104-20-2 ***HR: 3***
SILVER BOROFLUORIDE
mf: $AgBF_4$ mw: 194.68

SYNS: SILVER FLUOROBORATE ◇ SILVER TETRAFLUOROBORATE

TOXICITY DATA with REFERENCE
ivn-mus LD50:56 mg/kg CSLNX* NX#06756

CONSENSUS REPORTS: Silver and its compounds are on the Community Right-To-Know List. Reported in EPA TSCA Inventory.

OSHA PEL: TWA 0.01 mg(Ag)/m^3; 2.5 mg(F)/m^3
ACGIH TLV: TWA 0.01 mg(Ag)/m^3; 2.5 mg(F)/m^3
NIOSH REL: (Inorganic Fluorides) TWA 2.5 mg(F)/m^3

SAFETY PROFILE: Poison by intravenous route. When heated to decomposition it emits toxic fumes of F^-. Used for plating baths. See also SILVER COMPOUNDS, BORON COMPOUNDS, and FLUORIDES.

SDN100 CAS:15383-68-3 ***HR: 3***
SILVER BUTEN-3-YNIDE
mf: C_4H_3Ag mw: 158.94

$AgC{\equiv}CCH{=}CH_2$

CONSENSUS REPORTS: Silver and its compounds are on the Community Right-To-Know List.

SAFETY PROFILE: Deflagrates when heated. Explodes on contact with ammonia; nitric acid. See also

SILVER COMPOUNDS and ACETYLENE COMPOUNDS.

SDN399 CAS:7783-92-8 ***HR: 3***
SILVER CHLORATE
mf: $AgClO_3$ mw: 185.34

CONSENSUS REPORTS: Silver and its compounds are on the Community Right-To-Know List.

OSHA PEL: TWA 0.01 mg(Ag)/m^3
ACGIH TLV: TWA 0.01 mg(Ag)/m^3

SAFETY PROFILE: An explosive. A powerful oxidant. Upon decomposition it emits toxic fumes of Cl^-. See also SILVER COMPOUNDS and CHLORATES.

SDN500 CAS:7783-91-7 ***HR: 3***
SILVER CHLORITE, dry
mf: $AgClHO_2$ mw: 176.33

SYNS: CHLOROUS ACID ◇ SILVER(II) SALT

DOT Classification: Forbidden.

SAFETY PROFILE: Impact-sensitive, cannot be ground. Explodes @ 105°. Explosive with ethyl or methyl iodide. Explodes on contact with HCl or rubbing with S. When heated to decomposition it emits toxic fumes of Cl^-. See also SILVER COMPOUNDS, an oxidant.

SDN525 ***HR: 3***
SILVER CHLOROACETYLIDE
mf: C_2AgCl mw: 167.34

$$AgC{\equiv}CCl$$

CONSENSUS REPORTS: Silver and its compounds are on the Community Right-To-Know List.

SAFETY PROFILE: A sensitive explosive when wet. When heated to decomposition it emits toxic fumes of Cl^-. See also SILVER COMPOUNDS and ACETYLIDES.

SDO500 ***HR: 3***
SILVER COMPOUNDS

CONSENSUS REPORTS: Silver and its compounds are on the Commnuity Right-To-Know List.

SAFETY PROFILE: The water-soluble silver compounds are irritating to the skin and mucous membranes and may cause death if ingested. 50 mg of silver collargol is lethal after intravenous injection. Autopsy shows pulmonary edema, hemorrhage, and necrosis of the bone marrow, liver and kidney. The absorption of silver compounds into the circulation and the subsequent deposition of the reduced silver in various tissues of the body may result in the production of a generalized greyish pigmentation of the skin and mucous membranes, a condition known as argyria. Ingestion of 1-30 grams of soluble silver salts or long-term inhalation of a total 1-8 grams of silver can cause argyrosis. The introduction of fine particles of silver through breaks in the skin produces a local pigmentation at the site of the injury. 1 mg/m^3 of silver dust causes skin effects. The condition develops slowly, usually after 2-25 years of exposure. Pigmentation is noticeable first in conjunctivae, and later in the mucous membranes of the mouth and gums and in the skin. There are no constitutional symptoms or physical disability. Persons exhibiting the condition, and who subsequently died from unrelated disease, showed, on autopsy, a deposition of silver in the blood vessel walls, kidneys, testes, pituitary, choroid plexus, and mucous membranes of the nose, maxillary antra, trachea and bronchi. Once deposited, there is no known method by which the silver can be eliminated; the pigmentation is permanent. See also SILVER.

SDO525 CAS:3315-16-0 ***HR: 3***
SILVER CYANATE
mf: CAgNO mw: 149.89

CONSENSUS REPORTS: Silver and its compounds, as well as cyanide and its compounds, are on the Community Right-To-Know List.

SAFETY PROFILE: Explodes when heated. When heated to decomposition it emits toxic fumes of NO_x and CN^-. See also SILVER COMPOUNDS and CYANATES.

SDP000 CAS:506-64-9 ***HR: 3***
SILVER CYANIDE
DOT: UN 1684
mf: CAgN mw: 133.89

PROP: White, odorless, tasteless powder which darkens upon exposure to light. Mp: 320° (decomp), d: 3.95.

SYNS: CYANURE d'ARGENT (FRENCH) ◇ KYANID STRIBRNY (CZECH) ◇ RCRA WASTE NUMBER P104

TOXICITY DATA with REFERENCE
skn-rbt 500 mg/24H MLD 28ZPAK -,13,72
eye-rbt 5 mg/24H SEV 28ZPAK -,13,72
orl-rat LD50:123 mg/kg 28ZPAK -,13,72

CONSENSUS REPORTS: Silver and its compounds, as well as cyanide and its compounds, are on the Community Right-To-Know List. Reported in EPA TSCA Inventory.

DOT Classification: Poison B; Label: Poison.

SAFETY PROFILE: Deadly poison by ingestion. A skin and severe eye irritant. When heated to decomposition it emits very toxic fumes of CN^- and NO_x. Incompatible with phosphorus tricyanide; fluorine. Used in silver plating. See also SILVER COMPOUNDS and CYANIDES.

SDP025 CAS:70324-20-8 ***HR: 3***
SILVER 3-CYANO-1-PHENYLTRIAZEN-3-IDE
mf: $C_7H_5AgN_4$ mw: 253.01

$C_6H_5N{=}NN(CN)Ag$

CONSENSUS REPORTS: Silver and its compounds, as well as cyanide and its compounds, are on The Community Right-To-Know List.

SAFETY PROFILE: An explosive. When heated to decomposition it emits toxic fumes of CN^- and NO_x. See also SILVER COMPOUNDS and CYANIDES.

SDP100 ***HR: 3***
SILVER CYCLOPROPYLACETYLIDE
mf: C_5H_5Ag mw: 172.96

$AgC \equiv CCHCH_2CH_2$

CONSENSUS REPORTS: Silver and its compounds are on the Community Right-To-Know List.

SAFETY PROFILE: Explodes when heated. See also SILVER COMPOUNDS and ACETYLIDES.

SDP500 ***HR: 3***
SILVER DINITRITODIOXYSULFATE
mf: $Ag_4N_2O_2S$ mw: 523.55

SYN: TETRASILVER DIIMIDODIOXOSULFATE

CONSENSUS REPORTS: Silver and its compounds are on the Community Right-To-Know List.

SAFETY PROFILE: The dry salt is a friction- and impact-sensitive explosive. When heated to decomposition it emits very toxic fumes of SO_x and NO_x. See also SILVER COMPOUNDS and SULFATES.

SDP550 CAS:26163-27-9 ***HR: 3***
SILVER DINITROACETAMIDE
mf: $C_2H_2AgN_3O_5$ mw: 255.92

CONSENSUS REPORTS: Silver and its compounds are on the Community Right-To-Know List.

SAFETY PROFILE: Explodes when heated to 130°C. When heated to decomposition it emits toxic fumes of NO_x. See also SILVER COMPOUNDS and NITRO COMPOUNDS.

SDP600 CAS:58302-42-4 ***HR: 3***
SILVER 3,5-DINITROANTHRANILATE
mf: $C_7H_4AgN_3O_6$ mw: 333.99

$(O_2N)_2(H_2N)C_6H_2CO{\bullet}OAg$

CONSENSUS REPORTS: Silver and its compounds are on the Community Right-To-Know List.

SAFETY PROFILE: Explodes when heated to 394°C. It is not impact-sensitive. When heated to decomposition it emits toxic fumes of NO_x. See also SILVER COMPOUNDS and NITRO COMPOUNDS.

SDQ000 CAS:7775-41-9 ***HR: 3***
SILVER(I) FLUORIDE
mf: AgF mw: 126.88

PROP: Yellow, crystalline mass. Mp: 435°, bp: 1150°, d: 5.852 @ 15.5°. Sol in HF, NH_3, CH_3, and CN.

CONSENSUS REPORTS: Silver and its compounds are on the Community Right-To-Know List.

SAFETY PROFILE: Prolonged absorption may cause mottling of teeth and skeletal changes. Explosive reaction when ground with boron, silicon, and other nonmetals. Reacts violently with calcium hydride, dimethyl sulfoxide, potassium, sodium. Incandescent reaction in air at 320°C and with titanium at 320°C. Very hygroscopic. When heated to decomposition it emits toxic fumes of F^-. See also FLUORIDES and SILVER COMPOUNDS.

SDQ500 CAS:7783-95-1 ***HR: 3***
SILVER(II) FLUORIDE
mf: AgF_2 mw: 145.87

PROP: White when pure; usually a grey-black or brownish solid. D: 4.7, mp: 690°.

SYNS: ARGENT FLUORURE (FRENCH) ◇ ARGENTIC FLUORIDE ◇ SILVER DIFLUORIDE

TOXICITY DATA with REFERENCE
scu-frg LDLo:224 mg/kg CRSBAW 124,133,37

CONSENSUS REPORTS: Silver and its compounds are on the Community Right-To-Know List. Reported in EPA TSCA Inventory.

OSHA PEL: TWA 0.01 mg(Ag)/m^3; 2.5 mg(F)/m^3
ACGIH TLV: TWA 0.01 mg(Ag)/m^3; 2.5 mg(F)/m^3
NIOSH REL: (Inorganic Fluorides) TWA 2.5 mg(F)/m^3

SAFETY PROFILE: Poison by subcutaneous route. Powerful oxidizing agent. Mixtures with boron + water are explosive. When heated to decomposition it emits toxic fumes of F^-. See also FLUORIDES and SILVER COMPOUNDS.

SDR000 CAS:5610-59-3 ***HR: 3***
SILVER FULMINATE, dry
mf: AgCNO mw: 149.89

CONSENSUS REPORTS: Silver and its compounds are on the Community Right-To-Know List.

OSHA PEL: TWA 0.01 mg(Ag)/m^3
ACGIH TLV: TWA 0.01 mg(Ag)/m^3
DOT Classification: Forbidden.

SAFETY PROFILE: A powerful explosive more sensitive than mercury fulminate. Explosive reaction with hydrogen sulfide. Detonation temperature is 175°C. When heated to decomposition it emits toxic fumes of NO_x. See also SILVER COMPOUNDS and FULMINATES.

SDR150 ***HR: 3***
SILVER 1,3,5-HEXATRIENIDE
mf: C_6Ag_2 mw: 287.80

$$Ag(C\equiv C)_3Ag$$

CONSENSUS REPORTS: Silver and its compounds are on the Community Right-To-Know List.

SAFETY PROFILE: The dry material explodes when touched. See also SILVER COMPOUNDS and ACETYLENE COMPOUNDS.

SDR175 ***HR: 3***
SILVER 3-HYDROXYPROPYNIDE
mf: C_3H_3AgO mw: 162.92

$$AgC\equiv CCH_2OH$$

CONSENSUS REPORTS: Silver and its compounds are on the Community Right-To-Know List.

SAFETY PROFILE: An explosive. See also SILVER COMPOUNDS and ACETYLIDES.

SDR350 CAS:57421-56-4 ***HR: 3***
SILVER MALONATE
mf: $C_3H_2Ag_2O_4$ mw: 317.78

CONSENSUS REPORTS: Silver and its compounds are on the Community Right-To-Know List.

SAFETY PROFILE: Explodes when heated. See also SILVER COMPOUNDS.

SDR400 CAS:70247-51-7 ***HR: 3***
SILVER 3-METHYLISOXAZOLIN-4,5-DIONE-4-OXIMATE
mf: $C_4H_3AgN_2O_3$ mw: 234.95

AgON:CCO•ON=CCH$_3$ (ring)

CONSENSUS REPORTS: Silver and its compounds are on the Community Right-To-Know List.

SAFETY PROFILE: Explodes when heated rapidly. When heated slowly it reacts to form the dangerous silver fulminate. When heated to decomposition it emits toxic fumes of NO_x. See also SILVER COMPOUNDS.

SDR500 CAS:2386-52-9 ***HR: 3***
SILVER METHYLSULFONATE
mf: $CH_3O_3S•Ag$ mw: 202.97

SYNS: METHANESULFONIC ACID, SILVER SALT ◇ METHANESULFONIC ACID, SILVER(1+) SALT ◇ SILVER METHANESULFONATE

TOXICITY DATA with REFERENCE
ivn-mus LD50:18 mg/kg CSLNX* NX#00622

CONSENSUS REPORTS: Silver and its compounds are on the Community Right-To-Know List. Reported in EPA TSCA Inventory.

OSHA PEL: TWA 0.01 mg(Ag)/m^3
ACGIH TLV: TWA 0.01 mg(Ag)/m^3

SAFETY PROFILE: Poison by intravenous route. When heated to decomposition it emits toxic fumes of SO_x. See also SULFONATES and SILVER COMPOUNDS.

SDR759 CAS:13092-75-6 ***HR: 3***
SILVER MONOACETYLIDE
mf: C_2HAg mw: 132.90

CONSENSUS REPORTS: Silver and its compounds are on the Community Right-To-Know List.

DOT Classification: Forbidden (dry)

SAFETY PROFILE: Severe explosion hazard. When heated to decomposition it emits acrid smoke and irritating fumes. See also SILVER COMPOUNDS and ACETYLIDES.

SDS000 CAS:7761-88-8 ***HR: 3***
SILVER(I) NITRATE (1:1)
DOT: UN 1493
mf: $NO_3•Ag$ mw: 169.88

PROP: Mp: 212°, bp: 444° (decomp), d: 4.352 @ 19°. Very sol in ammonia, water; sltly sol in ether.

SYNS: LUNAR CAUSTIC ◇ NITRATE d'ARGENT (FRENCH) ◇ NITRIC ACID, SILVER(1+) SALT ◇ SILBERNITRAT ◇ SILVER(1+) NITRATE ◇ SILVER NITRATE (DOT)

TOXICITY DATA with REFERENCE
eye-rbt 1 mg SEV AROPAW 25,839,41
eye-rbt 10 mg MOD TXAPA9 55,501,80
sln-smc 140 ppb ANYAA9 407,186,83
dni-hmn:lym 76 μmol/L IAAAAM 79,83,86
dni-mus-ipr 20 g/kg ARGEAR 51,605,81
itt-rat TDLo:400 mg/kg (male 1D pre):REP FESTAS 24,884,73
skn-mus TDLo:15 g/kg/19W-I:ETA NCIMAV 10,489,63
unr-man LDLo:29 mg/kg 85DCAI 2,73,70

orl-mus LD50:50 mg/kg JPPMAB 2,20,50
ipr-mus LD50:34500 μg/kg COREAF 256,1043,63
orl-dog LDLo:20 mg/kg HBAMAK 4,1289,35
orl-rbt LDLo:800 mg/kg HBAMAK 4,1289,35
ivn-rbt LDLo:8800 μg/kg EQSSDX 1,1,75
ipr-gpg LDLo:216 mg/kg AEHLAU 11,201,65
scu-gpg LDLo:62 mg/kg BMJOAE 2,217,13

CONSENSUS REPORTS: Silver and its compounds are on the Community Right-To-Know List. EPA Genetic Toxicology Program. Reported in EPA TSCA Inventory.

OSHA PEL: TWA 0.01 mg(Ag)/m^3
ACGIH TLV: TWA 0.01 mg(Ag)/m^3
DOT Classification: Oxidizer; Label: Oxidizer.

SAFETY PROFILE: Human poison by an unspecified route. Experiemental poison by ingestion, intravenous, subcutaneous, and intraperitoneal routes. Experimental reproductive effects. Human mutation data reported. A severe eye irritant. A powerful caustic and irritant to skin, eyes, and mucous membranes. Swallowing can cause severe gastroenteritis that may be fatal. Questionable carcinogen with experimental tumorigenic data. A powerful oxidizer. Incompatible with acetylene; acetylides; alkalies; aluminum; antimony salts; arsenic; arsenites; bromides; carbon; carbonates; chlorides; ClF_3; chlorosulfuric acid; copper; creosote; ethanol; ferrous salts; hypophosphites; iodides; Mg powder with H_2O; morphine salts; NH_3 with KOH to yield black Ag_3N; oils; PH_3; phosphates; phosphonium iodide; phosphorous; plastics; sulfur; tannic acid; tartrates; thiocyanates; vegetable decoctions and extracts; zinc with NH_3 with KOH. When heated to decomposition it emits toxic fumes of NO_x. See also SILVER COMPOUNDS and NITRATES.

SDS500 CAS:20737-02-4 ***HR: 3***
SILVER NITRIDE
mf: Ag_3N mw: 337.7

PROP: Colorless solid.

CONSENSUS REPORTS: Silver and its compounds are on the Community Right-To-Know List.

SAFETY PROFILE: Severe explosion hazard when shocked or heated to above 100°C. Impact-sensitive when wet. Upon decomposition it emits toxic fumes of NO_x. See also SILVER COMPOUNDS, NITRIDES, and EXPLOSIVES, HIGH.

SDT000 ***HR: 3***
SILVER NITRIDOOSMITE
mf: $OsAgNO_3$ mw: 360.1

CONSENSUS REPORTS: Silver and its compounds are on the Community Right-To-Know List.

SAFETY PROFILE: Violently explodes @ 80° or from shock. Upon decomposition it emits toxic fumes of NO_x.

SDT300 CAS:86255-25-6 ***HR: 3***
SILVER 4-NITROPHENOXIDE
mf: $C_6H_4AgNO_3$ mw: 245.97

$AgOC_6H_4NO_2$

CONSENSUS REPORTS: Silver and its compounds are on the Community Right-To-Know List.

SAFETY PROFILE: Explodes when heated to 110°C. When heated to decomposition it emits toxic fumes of NO_x. See also SILVER COMPOUNDS and NITRO COMPOUNDS.

SDT500 ***HR: 3***
SILVER NITROPRUSSIDE
mf: $Ag_2[FeNO(CN)_5]$ mw: 431.71

CONSENSUS REPORTS: Silver and its compounds are on the Community Right-To-Know List.

SAFETY PROFILE: A poison. When heated to decomposition it emits highly toxic fumes of NO_x and CN^-. See also SILVER COMPOUNDS and HYDROCYANIC ACID.

SDT750 ***HR: 3***
SILVER OSMATE
mf: Ag_2O_4Os mw: 469.93

CONSENSUS REPORTS: Silver and its compounds are on The Community Right-To-Know List.

SAFETY PROFILE: Explodes on impact or heating. See also SILVER COMPOUNDS and OSMIUM.

SDU000 CAS:533-51-7 ***HR: 3***
SILVER OXALATE, dry
mf: $C_2H_2O_4{\cdot}2Ag$ mw: 305.78

PROP: White, crystalline powder. D: 5.03. Sol in moderate concentrated nitric acid, ammonia.

CONSENSUS REPORTS: Silver and its compounds are on the Community Right-To-Know List.

OSHA PEL: TWA 0.01 mg(Ag)/m^3
ACGIH TLV: TWA 0.01 mg(Ag)/m^3
DOT Classification: Forbidden.

SAFETY PROFILE: Explodes when heated above 140°C and produces Ag + CO_2. The dry material may explode when ground. Exothermic decomposition to metal and carbon dioxide. See also SILVER COMPOUNDS and OXALATES.

SDU500 CAS:20667-12-3 ***HR: 3***
SILVER (1+) OXIDE
mf: Ag_2O mw: 231.74

PROP: Brownish-black, heavy, odorless powder. D: 7.22 @ 25°/4°, Decomp at approx 200°. Very sol in dilute nitric acid, ammonia; less sol in NaOH solns; insol in alc.

SYNS: ARGENTOUS OXIDE ◇ DISILVER OXIDE

TOXICITY DATA with REFERENCE
orl-rat LDLo:2820 mg/kg AIHAAP 30,470,69
orl-rat LD50:2820 mg/kg AIHAAP 30,470,69

CONSENSUS REPORTS: Silver and its compounds are on the Community Right-To-Know List. Reported in EPA TSCA Inventory.

OSHA PEL: TWA 0.01 mg(Ag)/m^3
ACGIH TLV: TWA 0.01 mg(Ag)/m^3

SAFETY PROFILE: Moderately toxic by ingestion. Flammable by chemical reaction; an oxidizing agent. Explodes in contact with ammonia. Incompatible with CuO; (NH_3 + ethanol); (hydrazine + ethanol); CO; H_2S; Mg; auric sulfide; Sb sulfide; Hg sulfide; nitroalkanes; Se; S; P; K; Na; NaK; seleninyl chloride. See also SILVER COMPOUNDS.

SDV000 CAS:7783-93-9 ***HR: 3***
SILVER PERCHLORATE
mf: $AgClO_4$ mw: 207.32

$AgOClO_3$

CONSENSUS REPORTS: Silver and its compounds are on the Community Right-To-Know List.

SAFETY PROFILE: Explosive reaction with 1,2-diaminoethane. Concentrated solutions in alkynes (e.g., 2-pentyne or 3-hexyne) explode on contact with mercury. It forms solid, explosive complexes with acetic acid; aniline; carbon tetrachloride + hydrochloric acid; diethyl ether; dimethyl sulfoxide; dioxane; ethanol; 1,4-oxathiane; pyridine; toluene; benzene and many other aromatic hydrocarbons. Incompatible with tetrasulfurtetraimide. Self-reactive. Upon crystallization from diethyl ether it explodes on crushing. When heated to decomposition it emits toxic fumes of Cl^-. See also PERCHLORATES and SILVER COMPOUNDS.

SDV500 CAS:25870-02-4 ***HR: 3***
SILVER PERCHLORYL AMIDE
mf: $AgClHNO_3$ mw: 206.34

CONSENSUS REPORTS: Silver and its compounds are on the Community Right-To-Know List.

SAFETY PROFILE: Shock-sensitive when dry and may detonate. When heated to decomposition it emits very toxic fumes of Cl^- and NO_x. See also SILVER COMPOUNDS and PERCHLORATES.

SDV600 ***HR: 3***
SILVER N-PERCHLORYL BENZYLAMIDE
mf: $C_7H_7AgClNO_3$ mw: 296.45

CONSENSUS REPORTS: Silver and its compounds are on The Community Right-To-Know List.

SAFETY PROFILE: An explosive sensitive to impact or heating above 105°C. When heated to decomposition it emits toxic fumes of Cl^- and NO_x. See also SILVER COMPOUNDS and PERCHLORATES.

SDV700 CAS:25455-73-6 ***HR: 3***
SILVER PEROXIDE
mf: Ag_2O_2 mw: 247.73

CONSENSUS REPORTS: Silver and its compounds are on The Community Right-To-Know List.

SAFETY PROFILE: Mixtures with 1% polyisobutene are explosive. See also SILVER COMPOUNDS and PEROXIDES.

SDW000 ***HR: 3***
SILVER PEROXYCHROMATE
mf: $AgCrO_5$ mw: 239.87

CONSENSUS REPORTS: Silver and its compounds, as well as chromium and its compounds, are on the Community Right-To-Know List.

SAFETY PROFILE: Confirmed carcinogen. An oxidant. When mixed with H_2SO_4 @ −80° it explodes on slow warming to −30°. See also CHROMIUM and SILVER COMPOUNDS.

SDW100 CAS:61514-68-9 ***HR: 3***
SILVER PHENOXIDE
mf: C_6H_5AgO mw: 200.97

CONSENSUS REPORTS: Silver and its compounds are on The Community Right-To-Know List.

SAFETY PROFILE: The dry solid explodes when warmed. See also SILVER COMPOUNDS.

SDW500 CAS:13086-63-0 ***HR: 3***
SILVER TETRAZOLIDE
mf: $CHAgN_4$ mw: 176.87

CONSENSUS REPORTS: Silver and its compounds are on the Community Right-To-Know List.

SAFETY PROFILE: Unstable. Explodes on heating. When heated to decomposition it emits toxic fumes of NO_x. See also SILVER COMPOUNDS.

SDW600 ***HR: 3***
SILVER TRICHLOROMETHANEPHOSPHONATE
mf: $CAg_2Cl_3O_3P$ mw: 413.08

CONSENSUS REPORTS: Silver and its compounds are on The Community Right-To-Know List.

SAFETY PROFILE: Explodes when heated. When heated to decomposition it emits toxic fumes of Cl^- and PO_x. See also SILVER COMPOUNDS.

SDX000 ***HR: 3***
SILVER TRIFLUORO METHYL ACETYLIDE
mf: C_3AgF_3 mw: 200.00

$AgC \equiv CCF_3$

SYN: SILVER TRIFLUOROPROPYNIDE

CONSENSUS REPORTS: Silver and its compounds are on the Community Right-To-Know List.

SAFETY PROFILE: Decomposes explosively on heating. When heated to decomposition it emits toxic fumes of F^-. See also FLUORIDES and SILVER COMPOUNDS.

SDX200 CAS:25987-94-4 ***HR: 3***
SILVER TRINITROMETHANIDE
mf: $CAgN_3O_6$ mw: 257.90

$AgC(NO_2)_3$

CONSENSUS REPORTS: Silver and its compounds are on The Community Right-To-Know List.

SAFETY PROFILE: An explosive. When heated to decomposition it emits toxic fumes of NO_x. See also SILVER COMPOUNDS and NITRO COMPOUNDS.

SDX300 CAS:5356-88-7 ***HR: 3***
SILYL PEROXIDE Y-5712
mf: $C_{14}H_{30}O_3Si$ mw: 274.53

SYN: Y-5712 SILYL PEROXIDE

TOXICITY DATA with REFERENCE
skn-rbt 500 mg open MLD UCDS** 10/12/71
eye-rbt 100 mg MLD UCDS** 10/12/71
orl-rat LD50:2830 mg/kg UCDS** 10/12/71
skn-rbt LD50:318 mg/kg UCDS** 10/12/71

SAFETY PROFILE: Poison by skin contact. Moderately toxic by ingestion. A skin and eye irritant. When heated to decomposition it emits acrid smoke and irritating fumes.

SDX625 CAS:65666-07-1 ***HR: 3***
SILYMARIN

SYN: CARSIL

TOXICITY DATA with REFERENCE
ivn-rat LD50:370 mg/kg ARZNAD 25,82,75
ivn-mus LD50:391 mg/kg ARZNAD 25,82,75
ivn-rbt LDLo:142 mg/kg ARZNAD 25,82,75

SAFETY PROFILE: Poison by intravenous route.

SDX630 CAS:65496-97-1 ***HR: 3***
SILYMARIN SODIUM HEMISUCCINATE

SYN: SILYMARIN HYDROGEN BUTANEDIOATE SODIUM SALT

TOXICITY DATA with REFERENCE
ivn-rat LD50:825 mg/kg ARZNAD 25,89,75
ivn-mus LD50:970 mg/kg ARZNAD 25,89,75
ivn-rat LDLo:300 mg/kg ARZNAD 25,89,75

SAFETY PROFILE: Poison by intravenous route. When heated to decomposition it emits toxic fumes of Na_2O.

SDY500 CAS:14929-11-4 ***HR: 2***
SINIFIBRATE
mf: $C_{23}H_{26}Cl_2O_6$ mw: 469.39

SYNS: 2-(4-CHLOROPHENOXY)-2-METHYLPROPANOICACID-1,3-PROPANEDIYL ESTER ◇ 2-(p-CHLOROPHENOXY)-2-METHYLPROPIONIC ACID TRIMETHYLENE ESTER ◇ CHOLESOLVIN ◇ CLY-503 ◇ 1,3-PROPANEDIOL BIS(α-(p-CHLOROPHENOXY)ISOBUTYRATE) ◇ 1,3-PROPANEDIOL BIS(2-(4-CHLOROPHENOXY)-2-METHYLPROPIONATE) ◇ SIMFIBRATE ◇ SINFIBRATE

TOXICITY DATA with REFERENCE
orl-rat LD50:7300 mg/kg MEIEDD 10,1225,83
ipr-rat LD50:4450 mg/kg NIIRDN 6,369,82
orl-mus LD50:3300 mg/kg NIIRDN 6,369,82
ipr-mus LD50:3300 mg/kg NIIRDN 6,369,82

SAFETY PROFILE: Moderately toxic by ingestion and intraperitoneal routes. Used as an anticholesteremic. When heated to decomposition it emits toxic fumes of Cl^-. See also ESTERS.

SDY600 ***HR: D***
SINOMENIUM ACUTUM, crude extract

PROP: Crude drug extract from *Sinomenium acutum* (MUREAV 97,81,82).

SYNS: BIO (JAPANESE) ◇ SINOMENI CAULIS et RHIZOMA (LATIN)

TOXICITY DATA with REFERENCE
mmo-sat 5 mg/plate MUREAV 97,81,82
dnr-bcs 100 g/L MUREAV 97,81,82

SAFETY PROFILE: Mutation data reported.

SDY675 CAS:99814-12-7 ***HR: D***
SIPPR-113
mf: $C_{30}H_{38}O_8$ mw: 526.68

SYN: 2-α,17-α-DIETHYNYL-A-NOR-5-α-ANDROSTANE-2-β,17-β-DIOL DIHEMISUCCINATE

TOXICITY DATA with REFERENCE
orl-rat TDLo:1 mg/kg (1D preg):REP CCPTAY 32,301,85

SAFETY PROFILE: Experimental reproductive effects. When heated to decomposition it emits acrid smoke and irritating fumes.

SDY750 CAS:32385-11-8 ***HR: 3***
SISOMICIN
mf: $C_{19}H_{37}N_5O_7$ mw: 447.55

SYNS: ANTIBIOTIC 66-40 ◇ o-3-DEOXY-4-C-METHYL-3-(METHYLAMINO)-β-l-ARABINOPYRANOSYL-(1-6)-o-(2,6-DIAMINO-2,3,4,6-TETRADEOXY-α-d-glycero-HEX-4-ENOPYRANOSYL-(1-4)-2-DEOXY-d-STREPTAMINE ◇ RICKAMICIN ◇ SCH 13475 ◇ SISEPTIN ◇ SISOLLINE

TOXICITY DATA with REFERENCE
ivn-rat TDLo:66 mg/kg (female 7-17D post):REP YACHDS 12,2727,84
ivn-rat TDLo:66 mg/kg (female 7-17D post):TER YACHDS 12,2727,84
ivn-rat LD50:32 mg/kg ARTODN 54,275,83
ipr-mus LD50:221 mg/kg 85GDA2 1,178,80
scu-mus LD50:288 mg/kg 85GDA2 1,178,80
ivn-mus LD50:34 mg/kg 38KLAC -,239,77

SAFETY PROFILE: Poison by subcutaneous, intravenous, and intraperitoneal routes. An experimental teratogen. Experimental reproductive effects. When heated to decomposition it emits toxic fumes of NO_x.

SDY755 ***HR: 3***
SISOMICIN HYDROCHLORIDE
mf: $C_{19}H_{37}N_5O_7$•ClH mw: 484.07

SYN: o-3-DEOXY-4-C-METHYL-3-(METHYLAMINO)-β-l-ARABINOPYRANOSYL-(1-6)-o-(2,6-DIAMINO-2,3,4,6-TETRADEOXY-α-d-glycero-HEX-4-ENDOPYRANOSYL-(1-4)-2-DEOXY-d-STREPTAMINE HYDROCHLORIDE

TOXICITY DATA with REFERENCE
scu-rat LD50:560 mg/kg IYKEDH 12,933,81
ivn-rat LD50:58 mg/kg IYKEDH 12,933,81
ims-rat LD50:450 mg/kg IYKEDH 12,933,81
scu-mus LD50:280 mg/kg IYKEDH 12,933,81
ivn-mus LD50:37 mg/kg IYKEDH 12,933,81
ims-mus LD50:280 mg/kg IYKEDH 12,933,81

SAFETY PROFILE: Poison by subcutaneous, intramuscular, and intravenous routes. When heated to decomposition it emits toxic fumes of NO_x and HCl.

SDZ000 CAS:14334-41-9 ***HR: 3***
SISTALGIN
mf: $C_{21}H_{27}N$•ClH mw: 329.95

SYNS: 4,4-DIPHENYL-N-ISOPROPYLCYCLOHEXYLAMINEHYDROCHLORIDE ◇ EMD 9806 ◇ HSP 2986 ◇ MONOVERIN ◇ PRAMIVERINE HYDROCHLORIDE ◇ PRAMIVERIN HYDROCHLORIDE

TOXICITY DATA with REFERENCE
orl-mus TDLo:500 mg/kg (6-15D preg):REP ARZNAD 26,703,76
orl-rat LD50:623 mg/kg ARZNAD 26,703,76
ivn-rat LD50:26 mg/kg ARZNAD 26,703,76
orl-mus LD50:346 mg/kg ARZNAD 26,703,76
ivn-mus LD50:25 mg/kg ARZNAD 26,703,76
orl-dog LD50:140 mg/kg ARZNAD 26,703,76
ivn-dog LD50:20 mg/kg ARZNAD 26,703,76

SAFETY PROFILE: Poison by ingestion and intravenous routes. Experimental reproductive effects. When heated to decomposition it emits very toxic fumes of NO_x and HCl.

SDZ300 ***HR: 3***
SISTRURUS MILARIUS BARBOURI VENOM

SYNS: S. MILARIUS BARBOURI VENOM ◇ VENOM, SNAKE, SISTRURUS MILARIUS BARBOURI

TOXICITY DATA with REFERENCE
ipr-mus LD50:6844 μg/kg TOXIA6 9,131,71
scu-mus LD50:24250 μg/kg PAASAH 44,145,56
ivn-mus LD50:2800 μg/kg TOXIA6 9,131,71

SAFETY PROFILE: Poison by subcutaneous, intravenous, and intraperitoneal routes.

SDZ450 ***HR: 2***
SKUNK CABBAGE

PROP: The large, single flower can be green, purple, or brown with stripes or spots. It emerges before the leaves which can grow to 3 feet long by 1 foot wide. The whole plant smells badly. It grows in the region bounded by Quebec, Nova Scotia, North Carolina, and Iowa.

SYNS: CHOU PUANT (CANADA) ◇ POLECAT WEED ◇ SYMPLOCARPUS FOETIDUS ◇ TABAC du DIABLE (CANADA)

SAFETY PROFILE: The leaves contain poisonous crystals of calcium oxalate and possibly other toxins. Chewing the leaves results in burning pain in the lips, mouth, and throat, possibly followed by inflammation and blistering. Systemic effects are usually not seen because of the insolubility of calcium oxalate. See also OXALATES.

SDZ475 ***HR: 2***
SLIPPER FLOWER

PROP: A succulent shrub with many zigzag stems. The pointed, oval leaves are about 3 inches long and alternate along the stem. They are dropped soon after developing. Red shoe-shaped flowers grow at the ends of the stems.

A common house plant, it grows wild in Florida, California, Hawaii, Guam, and the West Indies.

SYNS: CANDELILLA (MEXICO) ◇ CHRISTMAS CANDLE ◇ DEVIL'S BACKBONE ◇ FIDDLE FLOWER ◇ ITAMO REAL (PUERTO RICO) ◇ JAPANESE POINTSETTIA ◇ JEW BUSH ◇ PEDILANTHUS TITHYMALOIDES ◇ REDBIRD FLOWER OR CACTUS ◇ RIBBON CACTUS ◇ SLIPPER PLANT

SAFETY PROFILE: The sap contains the poisonous euphorbol and other terpenes. Ingestion of plant parts may cause protracted gastritis including vomiting and diarrhea.

SEA000 ***HR: 2***
SLUDGE ACID
DOT: UN 1906

SYNS: ACID, SLUDGE (DOT)

DOT Classification: Corrosive Material; Label: Corrosive.

SAFETY PROFILE: A corrosive irritant to skin, eyes, and mucous membranes. When heated to decomposition it emits very toxic fumes of SO_x and NO_x. See also NITRIC ACID and SULFURIC ACID.

SEA400 ***HR: 3***
S. MARCESCENS LIPOPOLYSACCHARIDE

TOXICITY DATA with REFERENCE
ipr-rat TDLo:2 mg/kg (11-12D preg):TER PSEBAA 109,429,62
ipr-rat TDLo:2 mg/kg (11-12D preg):REP PSEBAA 109,429,62
ipr-rat LD50:25 mg/kg PSEBAA 109,429,62

SAFETY PROFILE: Poison by intraperitoneal route. An experimental teratogen. Experimental reproductive effects.

SEA500 CAS:9011-13-6 ***HR: 1***
SMA 1440-H RESIN

PROP: The 35% (W/V) ammoniacal solution of SMA 1440-H which are copolymers of styrene and maleic anhydride highly esterified with water-sol alc (CTOXAO 10,255,77).

SYN: RESIN, SMA 1440-H

TOXICITY DATA with REFERENCE
eye-rbt 100 mg MLD CTOXAO 10,255,77
orl-rat LD50:21 g/kg CTOXAO 10,255,77

CONSENSUS REPORTS: Reported in EPA TSCA Inventory.

SAFETY PROFILE: Mildly toxic by ingestion. An eye irritant. When heated to decomposition it emits acrid smoke and irritating fumes.

SEB000 ***HR: 2***
SMOG

PROP: An atmospheric combination of smoke, fog, and industrial gases. Composition: Contents vary, but sulfur dioxide, oxides of nitrogen, and ozone are common components; others are sulfides, fluorides, chlorides, carbon particles and various hydrocarbons.

SAFETY PROFILE: Moderately irritating to eyes and mucous membranes. Numerous chronic effects have been reported in susceptible populations. A common air contaminant. Possibly carcinogenic.

SEC000 ***HR: 3***
SMOKE CONDENSATE, cigarette

SYNS: CIGARETTE SMOKE CONDENSATE ◇ CSC ◇ TOBACCO SMOKE CONDENSATE ◇ TOBACCO TAR

TOXICITY DATA with REFERENCE
mma-sat 300 μg/plate ENMUDM 7,471,85
mmo-esc 10 g/L MUREAV 130,97,84
dni-hmn:fbr 10 mg/L LIFSAK 17,767,75
sce-hmn-ihl 10 cigarettes/D HEREAY 88,147,78
mnt-mus-ipr 320 mg/kg IJEBA6 23,145,85
ihl-rat TCLo:360 g/kg/2Y-I:NEO JJIND8 64,383,80
scu-rat TDLo:160 mg/kg/64W-I:ETA ARZNAD 18,814,68
skn-mus TDLo:374 g/kg/2Y-I:CAR TXCYAC 23,177,82
ipr-ham TDLo:150 mg/kg (10-14D preg):NEO,TER JCROD7 94,249,79
skn-mus TD:499 g/kg/2Y-I:CAR TXCYAC 23,177,82

CONSENSUS REPORTS: IARC Cancer Review: Group 1 IMEMDT 7,359,87; Human Sufficient Evidence IMEMDT 38,309,86, Animal Sufficient Evidence IMEMDT 38,309,86.

SAFETY PROFILE: Confirmed carcinogen with experimental carcinogenic, neoplastigenic, and tumorigenic data. An experimental teratogen. Human mutation data reported. See also NICOTINE, SMOKELESS TOBACCO, and various tobacco entries.

SED000 ***HR: 3***
SMOKELESS POWDER

PROP: Nitrocellulose containing about 13.1% nitrogen produced by blending material of somewhat lower (12.6%) and sltly higher (13.2%) nitrogen content, converting to a dough with alc-ether mixture, extruding, cutting and drying to a hard horny product. Small amounts of stabilizers (amines) and plasticizers are usually present, as well as various modifying agents (nitrotoluene, nitroglycerin, salts).

SAFETY PROFILE: Explosive. See also CELLULOSE TETRANITRATE and EXPLOSIVES, HIGH.

SED400 *HR: 3*
SMOKELESS TOBACCO

PROP: A variety of habituating substances containing tobacco as the major ingredient and used without burning. Tobacco is a product of the leaves and stems of two species of Nicotiana, *N. Tabacum* (grown in North America and Western Europe) and *N. Rustica* (grown in the USSR and India). There is considerable evidence that many if not all of the forms of smokeless tobacco are human carcinogens.

The smokeless tobaccos are introduced into the body through the mouth (chewing tobacco, snuff, misshri, gudakhu, shammah, khaini, nass, naswar or in combination with betel quid) or nose (snuff).

The various smokeless tobacco products are:

Chewing tobacco is placed between the cheek and gum and chewed slowly. There are three main types: plug, twist/roll, and loose-leaf.

Fine-Cut tobacco was formerly classified in the United States as chewing tobacco and is now placed in the category of moist fine-cut snuff.

Gudakhu is a paste of powdered tobacco, molasses and other ingredients used in parts of India to clean teeth.

Khaini is a mixture of tobacco and lime formed into a ball and placed in the mouth.

Kiwam is made from processed tobacco leaves. After the stalks and stems are removed, the leaves are soaked and boiled in water with flavorings and spices, crushed, then strained, leaving a paste which is chewed.

Loose-Leaf tobacco is prepared from fermented cigar leaves, sweetened with sugars, syrups, liquors, and other flavoring materials. It is packaged as batches of loose pieces or cut strips.

Mainpuri tobacco is a chewed mixture of tobacco with slaked lime, areca nut, camphor, and cloves. It is used in India.

Mishri is prepared from roasted or half-burnt tobacco which has been baked till black on a hot metal plate and then powdered. It is used primarily to clean teeth but is also used as chewing tobacco. Synonyms are masheri and misheri.

Nass is a mixture of tobacco, lime, wood-ash, and cottonseed oil, chewed in Iran and the central Asian region of the USSR.

Naswar is a mixture of powdered tobacco, slaked lime, and indigo placed on the bottom of the mouth or behind the lower lip. It is used in Afghanistan and Pakistan.

Pattiwala tobacco is a sun-cured tobacco leaf chewed with or without lime. It is used in India.

Pill is dried and pelleted Kiwam paste.

Plug tobacco is made from enriched tobacco leaves or leaf fragments wrapped in fine tobacco and pressed into flat bars or rolls. It is chewed.

Shammah is a mixture of powdered tobacco leaves with calcium or sodium carbonate and other materials, including ash, placed in the cheek or behind the lower lip. It is used in southern Saudi Arabia.

Snuff is taken through the mouth or the nose. Moist snuff is finely cut tobacco plus flavorings with a moisture content of up to 50%. It is placed in the cheek. Dry snuff has a moisture content of less than 10% and may have flavorings. It may be sniffed through the nose, placed behind the lower lip or in the cheek. Oriental snuff is about 50% heated calcium carbonate and calcium phosphate with some powdered cuttle-fish bone. In southern Africa, snuff is made from powdered tobacco leaves, plant ash, and sometimes oils, lemon juice, and herbs. In the United States, "dipping", refers to the ingestion use of snuff.

Twist/Roll tobacco is stripped tobacco leaves rolled or twisted like a length of rope.

Zarda is tobacco leaf broken into small pieces and boiled in water with lime and spices to dryness and then colored with vegetable dyes. It is usually chewed mixed with areca nut and spices.

SYNS: CHEWING TOBACCO ◇ GUDAKHU (INDIA) ◇ KHAINI (INDIA) ◇ KIWAM (INDIA) ◇ MASHERI (INDIA) ◇ MISHERI (INDIA) ◇ MISHRI (INDIA) ◇ NASS (IRAN) ◇ NASWAR (PAKISTAN and AFGHANISTAN) ◇ PILLS (INDIA) ◇ SHAMMAH (SAUDI ARABIA) ◇ SNUFF ◇ ZARDA (INDIA)

SAFETY PROFILE: Tobaccos contain from 0.5-5% alkaloids predominantly as l-nicotine (>85%). Nicotine is strongly addictive and is the chief cause of tobacco dependence. It is a mild stimulant. It readily forms salts with most acids. These salts are poorly absorbed through the mucous membranes whereas the base is easily absorbed. This explains the practice of combining lime or other alkali in conjunction with ingestion tobacco use. Nicotine and some of the other tobacco alkaloids are experimental teratogens and mutagens.

There are several known classes of carcinogens present in the smokeless tobaccos: N-nitrosamines, polynuclear aromatic hydrocarbons (PAH's), heavy metals (arsenic trioxide, lead, cadmium and nickel compounds), and radionuclides (^{226}Ra, ^{210}Pb and ^{210}Po). Of these, nitrosamines are present in the highest concentration (in the range of mg/kg). The concentrations of the nitrosamines are 100 times higher in tobacco than in other consumer products. Nitrosamine concentrations are higher in chewing tobacco than in cigarette smoke. The major nitrosamines in tobacco are N′-nitrosonornicotine (NNN), 4-(methylnitrosamine)-1-(3-pyridyl)-1-butanone (NNK) and N′-nitrosoanatabine (NAT). They are probably generated during curing, fermentation, and aging of the tobacco leaf from the tobacco alkaloids: nicotine, nornicotine, anatabine, anabasine, continine, myosmine, 2,3′-dipyridyl and N′-formyl-nornicotine. They may also form in the mouth.

There is sufficient evidence that the ingestion use of snuff, chewing tobacco, and tobacco mixed with lime is carcinogenic to humans. Evidence suggests that the ingestion use of other smokeless tobacco preparations and the nasal use of snuff is carcinogenic to humans. Oral precancerous lesions are commonly observed in smokeless tobacco users.

See also NICOTINE, ARECA NUT, BETEL QUID, N-NITROSO COMPOUNDS, NITROSAMINES, and individual compounds.

SED500 CAS:8016-69-1 ***HR: 1***
SNAKEROOT OIL, CANADIAN

PROP: Consists of linalool, geraniol, 1-α-terrineol, eugenol and methyl eugenol (FCTXAV 16,637,78).

SYN: WILD GINGER OIL

TOXICITY DATA with REFERENCE
orl-rat LD50:4480 mg/kg FCTXAV 16,637,78

CONSENSUS REPORTS: Reported in EPA TSCA Inventory.

SAFETY PROFILE: Mildly toxic by ingestion. When heated to decomposition it emits acrid smoke and irritating fumes. See also constituents as listed above.

SED550 ***HR: 1***
SNOWBERRY

PROP: A deciduous shrub about 3 feet tall with oval leaves. It produces clusters of small, bell-shaped pink flowers and white berries. It grows in region bounded by southeastern Alaska, California, Virginia, and Quebec.

SYNS: BELLUAINE (CANADA) ◇ BUCK BRUSH ◇ SYMPHORICARPOS (various species) ◇ WAXBERRY

SAFETY PROFILE: The berries contain an unidentified toxin. Ingestion of large quantities of the berries may cause vomiting and diarrhea.

SED575 ***HR: 1***
SNOWDROP

PROP: A bulb- and berry-producing plant which blooms in early spring with a single white flower with green markings. It is native to Europe and is cultivated in the northern United States.

SYN: GALANTHUS NIVALIS

SAFETY PROFILE: The bulb contains the poison lycorine. Ingestion of large quantities of the bulbs may cause nausea, vomiting, and diarrhea.

SED700 CAS:77680-87-6 ***HR: 3***
SOAz

mf: $C_{10}H_{20}N_8OP_2S$ mw: 362.38

SYN: 1,3,3,5,5-PENTAAZIRIDINO-1-THIA-2,4,6-TRIAZA-3,5-DIPHOSPHORINE-1-OXIDE

TOXICITY DATA with REFERENCE
ivn-rat TDLo:75 mg/kg (1D male):REP JJATDK 4,1,84
ivn-rat LD50:82 mg/kg JJATDK 4,1,84
ivn-mus LD50:325 mg/kg JJATDK 4,1,84
ivn-dog LD50:12 mg/kg JJATDK 4,1,84

SAFETY PROFILE: Poison by intravenous route. Experimental reproductive effects. When heated to decomposition it emits toxic fumes of PO_x, SO_x, and NO_x.

SEE000 CAS:8006-28-8 ***HR: 2***
SODA LIME (solid)

DOT: UN 1907

PROP: White to gray granules. Rapidly deteriorates on exposure to air.

DOT Classification: Corrosive Material; Label: Corrosive

SAFETY PROFILE: A corrosive irritant to skin, eyes, and mucous membranes. See also SODIUM HYDROXIDE and CALCIUM OXIDE.

SEE250 CAS:54-87-5 ***HR: 3***
3-SODIO-5-(5′-NITRO-2′-FURFURYLIDENAMINO) IMIDAZOLIDIN-2,4-DIONE

mf: $C_8H_5N_4NaO_5$ mw: 260.14

SAFETY PROFILE: An explosive. When heated to decomposition it emits toxic fumes of NO_x and Na_2O. See also NITRO COMPOUNDS.

SEE500 CAS:7440-23-5 ***HR: 3***
SODIUM

DOT: UN 1428/UN 1429
af: Na aw: 22.9898

PROP: Light, soft, ductile, malleable, silver-white metal. Mp: 97.81°, bp: 881.4°, d: 0.9710 @ 20°, autoign temp: > 115° in dry air, vap press: 1.2 mm @ 400°.

SYNS: NATRIUM ◇ SODIUM METAL (DOT)

CONSENSUS REPORTS: Reported in EPA TSCA Inventory.

DOT Classification: Label: Flammable Solid and Dangerous When Wet

SAFETY PROFILE: Metallic sodium reacts exothermally with the moisture of body or tissue surfaces, causing thermal and chemical burns. Sodium in elemental form is highly reactive. Sodium reacts violently with water to form sodium hydroxide. A very dangerous fire hazard when exposed to heat and moisture. Under the appropriate conditions, it can react violently with moisture; air; $AlBr_3$; $AlCl_3$; AlF_3; NH_4 chlorocuprate; NH_4NO_3; $SbBr_3$; $SbCl_3$; SbI_3; $AsCl_3$; AsI_3; $BiBr_3$; $BiCl_3$; BiI_3; Bi_2O_3; BBr_3; bromoazide; CO_2; CO + NH_3; CCl_4; Cl_2; ClF_3; $CrCl_4$; CrO_3; CoBr; CoCl; $CuCl_2$; CuO; $FeBr_3$; $FeCl_3$; $FeBr_2$; $FeCl_2$; FeI_2; hydrazine hydrate; H_2O_2; H_2S; HCl; HF; F_2; 1,2-dichloroethylene; dichloromethane; Br_2; hydroxylamine; iodine; iodine monochloride; iodine pentafluoride; lead oxide; maleic anhydride; manganous chloride; mercuric bromide; mercuric chloride; mercuric fluoride; mercuric iodide; mercurous chloride; mercurous oxide; methyl chloride; molybdenum trioxide; monoammonium phosphate; nitric acid; nitrogen peroxide; nitrosyl fluoride; nitrous oxide; phosgene; phosphorus; phosphorous pentafluoride; phosphorus pentoxide; phosphorus tribromide; phosphorus trichloride; phosphoryl chloride; potassium oxides; potassium ozonide; potassium superoxide; selenium; silicon tetrachloride; silver bromide; silver chloride; silver fluoride; silver iodide; sodium peroxide; stannic chloride; stannic iodide with sulfur; stannic oxide; stannous chloride; sulfur; sulfur dibromide; sulfur dichloride; sulfur dioxide; sulfuric acid; tellurium; tetrachloroethane; thallous bromide; thiophosphoryl bromide; trichlorethylene; vanadium pentachloride; vanadyl chloride; zinc bromide; any oxidizing material. Decomposes moisture to evolve hydrogen and heat. Reacts exothermally with halogens, acids, and halogenated hydrocarbons.

Heated sodium is spontaneously flammable in air. Can be safely stored under liquid hydrocarbons. Dangerous explosion hazard when exposed to moisture in any form!! Keep away from water at all times!! When heated in air it emits toxic fumes of sodium oxide. Reacts with water or steam to produce heat, hydrogen, and flammable vapors. Can react vigorously to explosively with oxidizing materials. To fight fire, use soda ash, dry sodium chloride or graphite, in order of preference. When heated to decomposition it emits toxic fumes of Na_2O. See also SODIUM HYDROXIDE and HYDROGEN.

SEF500 CAS:7440-23-5 ***HR: 3***
SODIUM (dispersions)

PROP: Finely divided metallic sodium suspended in toluene, xylene, naphtha, kerosene, etc.

SYN: SODIUM, METAL DISPERSION IN ORGANIC SOLVENT

CONSENSUS REPORTS: Reported in EPA TSCA Inventory.

DOT Classification: Label: Flammable solid and Dangerous When Wet

SAFETY PROFILE: A very dangerous fire hazard when exposed to heat or flame or by chemical reaction. These are very reactive forms of sodium which, if carelessly handled, may catch fire. After sodium has been extinguished, the burning organic vapor can be dealt with by very cautious use of a carbon dioxide extinguisher. To extinguish, see SODIUM. Do not use carbon tetrachloride. Moderate explosion hazard by chemical reaction; will react with water or steam to produce heat and hydrogen; on contact with oxidizing materials it can react vigorously, and on contact with acid or acid fumes it can emit toxic fumes. When heated it loses the solvent and emits highly toxic fumes of Na_2O. See also SODIUM and individual dispersant.

SEF600 CAS:7440-23-5 ***HR: 3***
SODIUM (liquid alloy)
DOT: NA 1421
mf: Na mw: 22.99

SYN: SODIUM, metal liquid alloy (DOT)

DOT Classification: Label: Flammable Solid and Dangerous When Wet

SAFETY PROFILE: Flammable when exposed to heat or flame; can react vigorously with oxidizing materials. When heated to decomposition it emits toxic fumes of Na_2O. See also SODIUM.

SEG000 CAS:5892-48-8 ***HR: 3***
SODIUM ACETARSONE
mf: $C_8H_9AsNO_5 \cdot Na$ mw: 297.09

SYN: N-ACETYL-4-HYDROXY-m-ARSANILIC ACID SODIUM SALT

TOXICITY DATA with REFERENCE
ivn-rat LDLo:750 mg/kg ADSYAF 25,799,32
orl-cat LDLo:125 mg/kg PSEBAA 27,267,30
orl-rbt LDLo:150 mg/kg PSEBAA 27,267,30
ivn-rbt LDLo:600 mg/kg ADSYAF 25,799,32
orl-gpg LDLo:100 mg/kg PSEBAA 29,125,31

CONSENSUS REPORTS: Arsenic and its compounds are on the Community Right To Know List.

OSHA PEL: TWA 0.5 mg(As)/m^3
ACGIH TLV: TWA 0.2 mg(As)/m^3

SAFETY PROFILE: Poison by ingestion. Moderately toxic by intravenous route. When heated to decomposition it emits very toxic fumes of As, Na_2O, and NO_x. See also ARSENIC COMPOUNDS.

SEG500 CAS:127-09-3 ***HR: 3***
SODIUM ACETATE
mf: $C_2H_3O_2 \cdot Na$ mw: 82.04

PROP: White granular powder. Autoign temp: 1125°F,

d: 1.45, mp: 58°. Decomp @ higher temp. Sol in water, alc.

SYNS: ACETIC ACID, SODIUM SALT ◇ NATRIUMACETAT (GERMAN) ◇ SODIUM ACETATE, anhydrous (FCC)

TOXICITY DATA with REFERENCE
skn-rbt 500 mg/24H MLD BIOFX* 19-3/71
eye-rbt 10 mg MLD BIOFX* 19-3/71
orl-rat LD50:3530 mg/kg FAONAU 40,127,67
orl-mus LD50:6891 mg/kg JIHTAB 23,78,41
scu-mus LD50:8000 mg/kg ZGEMAZ 113,536,44
ivn-mus LD50:335 mg/kg JLCMAK 29,809,44
ivn-rbt LDLo:1300 mg/kg AEXPBL 21,119,1886

CONSENSUS REPORTS: Reported in EPA TSCA Inventory. EPA Genetic Toxicology Program.

SAFETY PROFILE: Poison by intravenous route. Moderately toxic by ingestion. A skin and eye irritant. Migrates to food from packaging materials. Violent reaction with F_2, KNO_3, diketene. When heated to decomposition it emits toxic fumes of Na_2O.

SEG650 CAS:31304-44-6 ***HR: D***
SODIUM ACETATE MONOHYDRATE
mf: $C_2H_3O_2 \cdot Na \cdot H_2O$ mw: 100.06

TOXICITY DATA with REFERENCE
orl-mus TDLo:40 mg/kg (female 1D post):REP IJMRAQ 60,48,72

SAFETY PROFILE: Experimental reproductive effects. When heated to decomposition it emits toxic fumes of Na_2O.

SEG700 CAS:2881-62-1 ***HR: 3***
SODIUM ACETYLIDE
mf: C_2Na_2 mw: 70.00

SYN: SODIUM ETHYNIDE

SAFETY PROFILE: Burns in fluorine or chlorine. Ignites in contact with bromine or iodine when warmed. Violent reaction when ground with metals (e.g., aluminum, iron, lead, mercury); metal salts (e.g., chlorides, iodides, sulfates, nitrates). Incandescent reaction with carbon dioxide, sulfur dioxide, oxygen + heat, dinitrogen pentaoxide (at 150°C). When heated to decomposition it emits toxic fumes of Na_2O. See also ACETYLIDES.

SEG800 CAS:7681-38-1 ***HR: 2***
SODIUM ACID SULFATE (solid)
DOT: UN 1821/UN 2837
mf: $HO_4S \cdot Na$ mw: 120.06

PROP: White crystals or granules. Mp: >315° (decomp), d: 2.435 @ 13°. Sol in water.

SYNS: GBS ◇ NITRE CAKE ◇ SODIUM ACID SULFATE ◇ SODIUM ACID SULFATE, solution (DOT) ◇ SODIUM BISULFATE, fused ◇ SODIUM BISULFATE, solid (DOT, FCC) ◇ SODIUM BISULFATE, solution (DOT) ◇ SODIUM HYDROGEN SULFATE, solid (DOT) ◇ SODIUM HYDROGEN SULFATE, solution (DOT) ◇ SODIUM PYROSULFATE ◇ SULFURIC ACID, MONOSODIUM SALT

TOXICITY DATA with REFERENCE
mmo-omi 1000 ppm POASAD 34,114,53

CONSENSUS REPORTS: Reported in EPA TSCA Inventory.

DOT Classification: ORM-B; Label: None (UN1821); Corrosive Material; Label: Corrosive (UN2837); Corrosive Material; Label: Corrosive (UN1821, UN2837).

SAFETY PROFILE: A corrosive irritant to skin, eyes, and mucous membranes. Mutation data reported. Reacts with moisture to form sulfuric acid. Mixtures with calcium hypochlorite + starch + sodium carbonate explode when compressed. Violent reaction with acetic anhydride + ethanol may lead to ignition and a vapor explosion. Incompatible with calcium hypochlorite. When heated to decomposition it emits toxic fumes of SO_x and Na_2O. See also SULFATES.

SEH000 CAS:9005-38-3 ***HR: 3***
SODIUM ALGINATE
mf: $(C_6H_7O_6Na)_n$ mw: 198.11

PROP: Colorless to slt yellow filamentous or granular solid or powder; odorless and tasteless. In water it forms a viscous colloidal soln; insol in ether, alc, chloroform.

SYNS: ALGIN ◇ ALGINATE KMF ◇ ALGIN (polysaccharide) ◇ ALGIPON L-1168 ◇ AMNUCOL ◇ ANTIMIGRANT C 45 ◇ CECALGINE TBV ◇ COHASAL-IH ◇ DARID QH ◇ DARILOID QH ◇ DUCKALGIN ◇ HALLTEX ◇ K'-ALGILINE ◇ KELCO GEL LV ◇ KELCOSOL ◇ KELGIN ◇ KELGUM ◇ KELSET ◇ KELSIZE ◇ KELTEX ◇ KELTONE ◇ LAMITEX ◇ MANUCOL ◇ MANUCOL DM ◇ MANUTEX ◇ MEYPRALGIN R/LV ◇ MINUS ◇ MOSANON ◇ NOURALGINE ◇ OG 1 ◇ PECTALGINE ◇ PROCTIN ◇ PROTACELL 8 ◇ PROTANAL ◇ PROTATEK ◇ SNOW ALGIN H ◇ SODIUM POLYMANNURONATE ◇ STIPINE ◇ TAGAT ◇ TRAGAYA

TOXICITY DATA with REFERENCE
ivn-rat LD50:1000 mg/kg FAONAU 53A,382,74
ivn-mus LD50:200 mg/kg FAONAU 53A,382,74
ipr-cat LD50:250 mg/kg FAONAU 53A,382,74
ivn-rbt LD50:100 mg/kg FAONAU 53A,382,74

CONSENSUS REPORTS: Reported in EPA TSCA Inventory.

SAFETY PROFILE: Poison by intravenous and intraperitoneal routes. When heated to decomposition it emits toxic fumes of Na_2O.

SEH450 CAS:9010-06-4 ***HR: 2***
SODIUM ALGINATE SULFATE

SYNS: ALGINIC ACID HYDROGEN SULFATE SODIUM SALT (9CI) ◇ ATEROID ◇ HEPARINOID ◇ HEPINOID ◇ HEPTARINOID ◇ PARITOL ◇ SODIUM POLYANHYDROMANNURONIC ACID SULFATE ◇ THROMBOCID

TOXICITY DATA with REFERENCE
orl-rat LD50:18100 mg/kg KSRNAM 17,2539,83
scu-rat LD50:500 mg/kg KSRNAM 17,2539,83
orl-mus LD50:17500 mg/kg KSRNAM 17,2539,83
ivn-mus LD50:1898 mg/kg AJMSA9 216,234,48

SAFETY PROFILE: Moderately toxic by intravenous and subcutaneous routes. Mildly toxic by ingestion. When heated to decomposition it emits toxic fumes of SO_x and Na_2O. See also SODIUM ALGINATE.

SEM000 CAS:1344-00-9 ***HR: 1***
SODIUM ALUMINOSILICATE

PROP: Fine, white, amorphous powder or beads; odorless and tasteless. Insol in water, alc, and other organic solvents.

SYNS: NCI-C55505 ◇ SODIUM SILICOALUMINATE

CONSENSUS REPORTS: Reported in EPA TSCA Inventory.

SAFETY PROFILE: An irritant to skin, eyes, and mucous membranes. When heated to decomposition it emits toxic fumes of Na_2O. See also SILICATES and ALUMINUM COMPOUNDS.

SEM300 ***HR: 1***
SODIUM ALUMINUM PHOSPHATE, ACIDIC
mf: $NaAl_3H_{14}(PO_4)_8 \cdot 4H_2O$ mw: 949.88

PROP: White powder; odorless. Insol in water; sol in hydrochloric acid.

SYN: SALP

SAFETY PROFILE: A nuisance dust.

SEM305 ***HR: 1***
SODIUM ALUMINUM PHOSPHATE, BASIC

PROP: White powder; odorless. Insol in water; sol in hydrochloric acid.

SYN: KASAL

SAFETY PROFILE: A nuisance dust.

SEM500 CAS:13770-96-2 ***HR: 3***
SODIUM ALUMINUM TETRAHYDRIDE
DOT: UN 2835
mf: $AlH_4 \cdot Na$ mw: 54.01

PROP: White, crystalline material; stable in dry air but sensitive to moisture. Mp: 183°; d: 1.24. Sol in tetrahydrofuran.

SYNS: ALUMINUM SODIUM HYDRIDE ◇ SAH 22 ◇ SODIUM ALUMINUM HYDRIDE (DOT) ◇ SODIUM TETRAHYDROALUMINATE(1−) ◇ (T-4) SODIUM, TETRAHYDROALUMINATE(1−) (9CI)

ACGIH TLV: TWA 2 mg(Al)/m^3
DOT Classification: Flammable Solid; Label: Flammable Solid and Dangerous When Wet

SAFETY PROFILE: Flammable when exposed to heat or flame. May ignite and explode on contact with water. Reacts violently with tetrahydrofuran when heated. When heated to decomposition it emits toxic fumes of Na_2O. See also ALUMINUM COMPOUNDS and HYDRIDES.

SEN000 CAS:7782-92-5 ***HR: 3***
SODIUM AMIDE
DOT: UN 1425
mf: H_2NNa mw: 39.02

PROP: White, crystalline powder. Mp: 210°, bp: 400°.

SYN: SODAMIDE

CONSENSUS REPORTS: Reported in EPA TSCA Inventory.

DOT Classification: Label: Flammable Solid and Dangerous When Wet

SAFETY PROFILE: An intense irritant to tissue, skin, and eyes. Flammable by chemical reaction. Ignites or explodes with heat or grinding. Explosive reaction with moisture; chromium trioxide; potassium chlorate; halocarbons (e.g., 1,1-diethoxy-2-chloroethane); oxidants; sodium nitrite; air. Can become explosive in storage. Violent reaction with dinitrogen tetraoxide. Will react with water or steam to produce heat and toxic and corrosive fumes of sodium hydroxide and ammonia. When heated to decomposition it emits highly toxic fumes of NH_3 and Na_2O. See also AMIDES.

SEN500 CAS:737-31-5 ***HR: 2***
SODIUM AMIDOTRIZOATE
mf: $C_{11}H_8I_3N_2O_4 \cdot Na$ mw: 635.90

SYNS: 3,5-DIACETAMIDO-2,4,6-TRIIODOBENZOIC ACID, SODIUM SALT ◇ 3,5-DIACETYLAMINO-2,4,6-TRIJODBENZOSAEURE NATRIUM (GERMAN) ◇ DIATRIZOATE SODIUM SALT ◇ HYPAQUE ◇ HYPAQUE SODIUM ◇ SODIUM-3,5-DIACETAMIDO-2,4,6-TRIIODOBENZOATE ◇ SODIUM DIACETYLDIAMINETRIIODOBENZOATE ◇ SODIUM DIATRIZOATE ◇ TRIOMBRIN ◇ UROVISON ◇ WIN 8308-3

TOXICITY DATA with REFERENCE
mnt-hmn:lym 40 mg/L RADLAX 129,199,78
cyt-hmn:lym 2000 ppm RADLAX 129,199,78
ivn-rat LD50:14700 mg/kg AEPPAE 222,584,54
ivn-mus LD50:1400 mg/kg JPETAB 116,394,56

ims-mus LD50:20349 mg/kg JPETAB 117,307,56
ivn-dog LD50:13200 mg/kg JPETAB 116,394,56
ivn-cat LD50:11300 mg/kg CLDND* 116,394,56

SAFETY PROFILE: Moderately toxic by intravenous route. Human mutation data reported. When heated to decomposition it emits very toxic fumes of NO_x, Na_2O, and I^-.

SEO500 CAS:555-06-6 ***HR: 2***
SODIUM p-AMINOBENZOATE
mf: $C_7H_6NO_2 \cdot Na$ mw: 159.13

SYNS: 4-AMINOBENZOIC ACID, MODOSODIUM SALT ◇ p-AMINO-BENZOIC ACID SODIUM SALT ◇ ANTERGYL ◇ PAB ◇ PABAVJT ◇ SODIUM-4-AMINOBENZOATE

TOXICITY DATA with REFERENCE
ivn-rat LD50:2760 mg/kg PSEBAA 49,184,42
scu-mus LD50:3 g/kg CRSBAW 146,466,52
ivn-mus LD50:2840 mg/kg MECHAN 6,343,63

CONSENSUS REPORTS: Reported in EPA TSCA Inventory.

SAFETY PROFILE: Moderately toxic by intravenous and subcutaneous routes. When heated to decomposition it emits toxic fumes of NO_x and Na_2O.

SEP000 CAS:133-10-8 ***HR: 2***
SODIUM p-AMINOSALICYLATE
mf: $C_7H_6NO_3 \cdot Na$ mw: 175.13

SYN: p-AMINOSALICYLATE SODIUM ◇ p-AMINOSALICYLIC ACID SODIUM SALT ◇ BACTYLAN ◇ LEPASEN ◇ NATRI-PAS ◇ NIPPAS ◇ PAMISYL SODIUM ◇ PASADE ◇ PASALON-RAKEET ◇ PASNAL ◇ PASSODICO ◇ SALVIS ◇ SANIPIROL ◇ SODIOPAS ◇ SODIUM AMINOSALICYLATE ◇ SODIUM p-AMINOSALICYLIC ACID ◇ TUBERSAN

TOXICITY DATA with REFERENCE
cyt-hmn:lym 600 mg/L MUREAV 48,215,77
dnd-rat:ast 6 pph BBACAQ 335,69,74
orl-rbt TDLo:40 mg/kg (7-14D preg):REP DPHFAK 23,383,71
scu-rat TDLo:69300 μg/kg (6-14D preg):TER DPHFAK 23,383,71
orl-rat LD50:8 g/kg ZENBAX 6B,183,51
ipr-rat LD50:4950 mg/kg ZENBAX 6B,183,51
orl-mus LD50:6900 mg/kg ZENBAX 6B,183,51
ipr-mus LD50:3850 mg/kg DPHFAK 23,383,71
scu-mus LD50:2800 mg/kg CRSBAW 146,446,52
ivn-mus LD50:3380 mg/kg YHTPAD 15(12),17,80

CONSENSUS REPORTS: Reported in EPA TSCA Inventory.

SAFETY PROFILE: Moderately toxic by intraperitoneal, subcutaneous, and intravenous routes. Mildly toxic by ingestion. An experimental teratogen. Experimental reproductive effects. Human mutation data reported. When heated to decomposition it emits toxic fumes of NO_x and Na_2O. See also 4-AMINO SALICYLIC ACID.

SEP500 CAS:10042-84-9 ***HR: 2***
SODIUM AMINOTRIACETATE
mf: $C_6H_9NO_6 \cdot xNa$ mw: 352.09

SYNS: NITRILOTRIACETIC ACID SODIUM SALT ◇ SODIUM NITRILOACETATE ◇ SODIUM NITRILOTRIACETATE

TOXICITY DATA with REFERENCE
ipr-mus LD50:460 mg/kg REPMBN 10,391,62

CONSENSUS REPORTS: Reported in EPA TSCA Inventory.

SAFETY PROFILE: Moderately toxic by intraperitoneal route. When heated to decomposition it emits toxic fumes of NO_x and Na_2O.

SEQ000 CAS:69-52-3 ***HR: 2***
SODIUM AMPICILLIN
mf: $C_{16}H_{18}N_3O_4S \cdot Na$ mw: 371.42

SYNS: ALPEN-N ◇ AMCILL-S ◇ d-(−)-α-AMINOBENZYL-PENICILLIN SODIUM SALT ◇ AMPICILLIN SODIUM ◇ AMPICILLIN SODIUM SALT ◇ BINOTAL SODIUM ◇ CITTERAL ◇ DOMICILLIN ◇ OMNIPEN-N ◇ PEN A/N ◇ PENBRITIN-S ◇ PENIALMEN ◇ POLY-CILLIN-N ◇ PRINCIPEN/N ◇ SODIUM d-(−)-α-AMINOBENZYLPENICILLIN ◇ SODIUM BINOTAL ◇ SODIUM P-50

TOXICITY DATA with REFERENCE
orl-wmn TDLo:100 mg/kg/5D:GIT LANCAO 2,707,78
ipr-mus LD50:3480 mg/kg TXAPA9 18,185,71

SAFETY PROFILE: Moderately toxic by intraperitoneal route. Human systemic effects by ingestion: hypermotility, diarrhea, and other gastrointestinal changes. When heated to decomposition it emits very toxic fumes of NO_x, Na_2O, and SO_x.

SER000 CAS:131-08-8 ***HR: 1***
SODIUM-9,10-ANTHRAQUINONE-2-SULFONATE
mf: $C_{14}H_7O_5S \cdot Na$ mw: 310.26

SYNS: ANTHRACHINON-1-SULFONAN SODNY (CZECH) ◇ ANTHRAQUINONE-2-SULFONATE SODIUM SALT ◇ 9,10-ANTHRAQUINONE-2-SODIUM SULFONATE ◇ 2-ANTHRAQUINONESULFONATE SODIUM ◇ 2-ANTHRAQUINONESULFONIC ACID SODIUM SALT ◇ SODIUM-2-ANTHRACHINONESULPHONATE ◇ SODIUM-β-ANTHRAQUINONESULFONATE ◇ SODIUM-2-ANTHRAQUINONE-SULFONATE ◇ 2-SULFOANTHRAQUINONE SODIUM SALT

TOXICITY DATA with REFERENCE
eye-rbt 500 mg/24H MLD 28ZPAK -,192,72

CONSENSUS REPORTS: Reported in EPA TSCA Inventory.

SAFETY PROFILE: An eye irritant. When heated to decomposition it emits very toxic fumes of Na_2O and SO_x. See also SULFONATES.

SER500 ***HR: 3***
SODIUM ANTIMONY ERYTHRITOL

TOXICITY DATA with REFERENCE
ipr-mus LD50:34 mg(Sb)/kg AJTMAQ 25,263,45

CONSENSUS REPORTS: Antimony and its compounds are on the Community Right To Know List.

OSHA PEL: TWA 0.5 mg(Sb)/m^3
ACGIH TLV: TWA 0.5 mg(Sb)/m^3
NIOSH REL: TWA (Antimony) 0.5 mg(Sb)/m^3

SAFETY PROFILE: Poison by intraperitoneal route. When heated to decomposition it emits toxic fumes of Sb and Na_2O. See also ANTIMONY COMPOUNDS.

SES000 ***HR: 3***
SODIUM ANTIMONYL ADONITOL

TOXICITY DATA with REFERENCE
ipr-mus LD50:21 mg(Sb)/kg AJTMAQ 25,263,45

CONSENSUS REPORTS: Antimony and its compounds are on the Community Right To Know List.

OSHA PEL: TWA 0.5 mg(Sb)/m^3
ACGIH TLV: TWA 0.5 mg(Sb)/m^3
NIOSH REL: (Antimony) TWA 0.5 mg(Sb)/m^3

SAFETY PROFILE: Poison by intraperitoneal route. When heated to decomposition it emits toxic fumes of Sb and Na_2O. See also ANTIMONY COMPOUNDS.

SES500 ***HR: 3***
SODIUM ANTIMONYL-d-ARABITOL

TOXICITY DATA with REFERENCE
ipr-mus LD50:20 mg(Sb)/kg AJTMAQ 25,263,45

CONSENSUS REPORTS: Antimony and its compounds are on the Community Right To Know List.

OSHA PEL: TWA 0.5 mg(Sb)/m^3
ACGIH TLV: TWA 0.5 mg(Sb)/m^3
NIOSH REL: (Antimony) TWA 0.5 mg(Sb)/m^3

SAFETY PROFILE: Poison by intraperitoneal route. When heated to decomposition it emits toxic fumes of Sb and Na_2O. See also ANTIMONY COMPOUNDS.

SET000 ***HR: 3***
SODIUM ANTIMONYL BISCATECHOL

TOXICITY DATA with REFERENCE
ipr-mus LD50:19 mg(Sb)/kg AJTMAQ 25,263,45

CONSENSUS REPORTS: Antimony and its compounds are on the Community Right To Know List.

OSHA PEL: TWA 0.5 mg(Sb)/m^3
ACGIH TLV: TWA 0.5 mg(Sb)/m^3
NIOSH REL: (Antimony) TWA 0.5 mg(Sb)/m^3

SAFETY PROFILE: Poison by intraperitoneal route. When heated to decomposition it emits toxic fumes of Sb and Na_2O. See also ANTIMONY COMPOUNDS.

SET500 ***HR: 3***
SODIUM ANTIMONYL tert-BUTYL CATECHOL

TOXICITY DATA with REFERENCE
ipr-mus LD50:25 mg(Sb)/kg AJTMAQ 25,263,45

CONSENSUS REPORTS: Antimony and its compounds are on the Community Right To Know List.

OSHA PEL: TWA 0.5 mg(Sb)/m^3
ACGIH TLV: TWA 0.5 mg(Sb)/m^3
NIOSH REL: (Antimony) TWA 0.5 mg(Sb)/m^3

SAFETY PROFILE: Poison by intraperitoneal route. When heated to decomposition it emits toxic fumes of Sb and Na_2O. See also ANTIMONY COMPOUNDS.

SEU000 ***HR: 3***
SODIUM ANTIMONYL CATECHOL THIOSALICYLATE

SYNS: SODIUM ANTIMONY(III)-3-CATECHOL THIOSALICYLATE ◇ STIBSOL

TOXICITY DATA with REFERENCE
ipr-mus LD50:24 mg(Sb)/kg AJTMAQ 25,263,45
ivn-mus LD50:56 mg/kg JPETAB 81,224,44

CONSENSUS REPORTS: Antimony and its compounds are on the Community Right To Know List.

OSHA PEL: TWA 0.5 mg(Sb)/m^3
ACGIH TLV: TWA 0.5 mg(Sb)/m^3
NIOSH REL: (Antimony) TWA 0.5 mg(Sb)/m^3

SAFETY PROFILE: Poison by intraperitoneal and intravenous routes. When heated to decomposition it emits very toxic fumes of Sb, Na_2O, and SO_x. See also ANTIMONY COMPOUNDS.

SEU500 ***HR: 3***
SODIUM ANTIMONYL CITRATE

TOXICITY DATA with REFERENCE
ipr-mus LD50:18 mg(Sb)/kg AJTMAQ 25,263,45

CONSENSUS REPORTS: Antimony and its compounds are on the Community Right To Know List.

OSHA PEL: TWA 0.5 mg(Sb)/m^3
ACGIH TLV: TWA 0.5 mg(Sb)/m^3
NIOSH REL: (Antimony) TWA 0.5 mg(Sb)/m^3

SAFETY PROFILE: Poison by intraperitoneal route.

When heated to decomposition it emits toxic fumes of Sb and Na_2O. See also ANTIMONY COMPOUNDS.

SEV000 ***HR: 3***
SODIUM ANTIMONYL-d-FUNCITOL

TOXICITY DATA with REFERENCE
ipr-mus LD50:46 mg(Sb)/kg AJTMAQ 25,263,45

CONSENSUS REPORTS: Antimony and its compounds are on the Community Right To Know List.

OSHA PEL: TWA 0.5 mg(Sb)/m^3
ACGIH TLV: TWA 0.5 mg(Sb)/m^3
NIOSH REL: (Antimony) TWA 0.5 mg(Sb)/m^3

SAFETY PROFILE: Poison by intraperitoneal route. When heated to decomposition it emits toxic fumes of Sb and Na_2O. See also ANTIMONY COMPOUNDS.

SEV500 ***HR: 3***
SODIUM ANTIMONYL GLUCO-GULOHEPTITOL

TOXICITY DATA with REFERENCE
ipr-mus LD50:35 mg(Sb)/kg AJTMAQ 25,263,45

CONSENSUS REPORTS: Antimony and its compounds are on the Community Right To Know List.

OSHA PEL: TWA 0.5 mg(Sb)/m^3
ACGIH TLV: TWA 0.5 mg(Sb)/m^3
NIOSH REL: (Antimony) TWA 0.5 mg(Sb)/m^3

SAFETY PROFILE: Poison by intraperitoneal route. When heated to decomposition it emits toxic fumes of Sb and Na_2O. See also ANTIMONY COMPOUNDS.

SEW000 ***HR: 3***
SODIUM ANTIMONYL GLYCEROL

TOXICITY DATA with REFERENCE
ipr-mus LD50:20 mg(Sb)/kg AJTMAQ 25,263,45

CONSENSUS REPORTS: Antimony and its compounds are on the Community Right To Know List.

OSHA PEL: TWA 0.5 mg(Sb)/m^3
ACGIH TLV: TWA 0.5 mg(Sb)/m^3
NIOSH REL: (Antimony) TWA 0.5 mg(Sb)/m^3

SAFETY PROFILE: Poison by intraperitoneal route. When heated to decomposition it emits toxic fumes of Sb and Na_2O. See also ANTIMONY COMPOUNDS.

SEW500 ***HR: 3***
SODIUM ANTIMONYL-d-MANNITOL

TOXICITY DATA with REFERENCE
ipr-mus LD50:17 mg(Sb)/kg AJTMAQ 25,263,45

CONSENSUS REPORTS: Antimony and its compounds are on the Community Right To Know List.

OSHA PEL: TWA 0.5 mg(Sb)/m^3
ACGIH TLV: TWA 0.5 mg(Sb)/m^3
NIOSH REL: (Antimony) TWA 0.5 mg(Sb)/m^3

SAFETY PROFILE: Poison by intraperitoneal route. When heated to decomposition it emits toxic fumes of Sb and Na_2O. See also ANTIMONY COMPOUNDS.

SEX000 ***HR: 3***
SODIUM ANTIMONYL-2,5-METHYLENE-d-MANNITOL

TOXICITY DATA with REFERENCE
ipr-mus LD50:44 mg(Sb)/kg AJTMAQ 25,263,45

CONSENSUS REPORTS: Antimony and its compounds are on the Community Right To Know List.

OSHA PEL: TWA 0.5 mg(Sb)/m^3
ACGIH TLV: TWA 0.5 mg(Sb)/m^3
NIOSH REL: (Antimony) TWA 0.5 mg(Sb)/m^3

SAFETY PROFILE: Poison by intraperitoneal route. When heated to decomposition it emits toxic fumes of Sb and Na_2O. See also ANTIMONY COMPOUNDS.

SEX500 ***HR: 3***
SODIUM ANTIMONYL-2,4-METHYLENE-d-SORBITOL

TOXICITY DATA with REFERENCE
ipr-mus LD50:32 mg(Sb)/kg AJTMAQ 25,263,45

CONSENSUS REPORTS: Antimony and its compounds are on the Community Right To Know List.

OSHA PEL: TWA 0.5 mg(Sb)/m^3
ACGIH TLV: TWA 0.5 mg(Sb)/m^3
NIOSH REL: (Antimony) TWA 0.5 mg(Sb)/m^3

SAFETY PROFILE: Poison by intraperitoneal route. When heated to decomposition it emits toxic fumes of Sb and Na_2O. See also ANTIMONY COMPOUNDS.

SEY000 ***HR: 3***
SODIUM ANTIMONYL XYLITOL

TOXICITY DATA with REFERENCE
ipr-mus LD50:33 mg(Sb)/kg AJTMAQ 25,263,45

CONSENSUS REPORTS: Antimony and its compounds are on the Community Right To Know List.

OSHA PEL: TWA 0.5 mg(Sb)/m^3
ACGIH TLV: TWA 0.5 mg(Sb)/m^3
NIOSH REL: (Antimony) TWA 0.5 mg(Sb)/m^3

SAFETY PROFILE: Poison by intraperitoneal route. When heated to decomposition it emits toxic fumes of Sb and Na_2O. See also ANTIMONY COMPOUNDS.

SEY500 CAS:7784-46-5 ***HR: 3***
SODIUM ARSENITE
mf: $AsO_2 \cdot Na$ mw: 129.91
DOT: UN 1686/UN 2027

PROP: White or grayish white powder. Commercially: 95-98% pure. Very sol in water; sltly sol in alc.

SYNS: ARSENENOUS ACID, SODIUM SALT (9CI) ◇ ARSENIOUS ACID, SODIUM SALT ◇ ARSENITE de SODIUM (FRENCH) ◇ ATLAS "A" ◇ CHEM PELS C ◇ CHEM-SEN 56 ◇ KILL-ALL ◇ PENITE ◇ PRODALUMNOL ◇ PRODALUMNOL DOUBLE ◇ SODANIT ◇ SODIUM ARSENITE, liquid (solution) (DOT) ◇ SODIUM ARSENITE, solid (DOT) ◇ SODIUM METAARSENITE

TOXICITY DATA with REFERENCE
dnr-esc 63 μg/well ENMUDM 3,429,81
mmo-smc 100 mmol/L MUREAV 117,149,83
oms-hmn:lym 700 nmol/L SWEHDO 7,277,81
cyt-hmn:lym 1 mg/L ENMUDM 3,597,81
orl-ham TDLo:25 mg/kg (female 12D post):TER BECTA6 29,671,82
orl-ham TDLo:5 mg/kg (female 9D post):REP TJADAB 23,40A,81
orl-chd TDLo:1 mg/kg CTOXAO 10,477,77
orl-chd LDLo:2 mg/kg CTOXAO 10,477,77
orl-rat LD50:41 mg/kg AIHAAP 30,470,69
skn-rat LD50:150 mg/kg PHJOAV 185,361,60
ivn-rat LDLo:6 mg/kg JPETAB 33,270,28
ipr-mus LD50:1170 μg/kg COREAF 257,791,63
ims-mus LD50:14 mg/kg EXMDA4 (440),312,78
orl-rbt LDLo:12 mg/kg JPETAB 33,270,28
ivn-rbt LD50:7600 μg/kg TXCYAC 51,213,88

CONSENSUS REPORTS: NTP Fifth Annual Report on Carcinogens. IARC Cancer Review: Group 1 IMEMDT 7,100,87; Animal Inadequate Evidence IMEMDT 23,39,80; Human Sufficient Evidence IMEMDT 23,39,80; Animal No Evidence IMEMDT 2,48,73. Arsenic and its compounds are on the Community Right To Know List. Reported in EPA TSCA Inventory. EPA Genetic Toxicology Program. EPA Extremely Hazardous Substances List.

OSHA PEL: TWA 0.01 mg(As)/m^3; Cancer Hazard
ACGIH TLV: TWA 0.2 mg(As)/m^3
NIOSH REL: CL (Inorganic Arsenic) 0.002 mg(As)/m^3/15M
DOT Classification: Poison B; Label: Poison

SAFETY PROFILE: Confirmed human carcinogen. Human poison by ingestion. Experimental poison by ingestion, skin contact, intravenous, intramuscular, and intraperitoneal routes. An experimental teratogen. Experimental reproductive effects. Human mutation data reported. Used as a herbicide and pesticide. When heated to decomposition it emits toxic fumes of As and Na_2O. See also ARSENIC COMPOUNDS.

SEZ000 CAS:7784-46-5 ***HR: 3***
SODIUM ARSENITE (liquid)

CONSENSUS REPORTS: Arsenic and its compounds are on the Community Right To Know List. Reported in EPA TSCA Inventory.

OSHA PEL: TWA 0.01 mg(As)/m^3; Cancer Hazard
ACGIH TLV: TWA 0.2 mg(As)/m^3
NIOSH REL: CL (Inorganic Arsenic) 0.002 mg(As)/m^3/15M
DOT Classification: Poison B; Label: Poison

SAFETY PROFILE: Confirmed human carcinogen. A deadly poison. When heated to decomposition it emits toxic fumes of As and Na_2O. See also SODIUM ARSENITE.

SEZ350 CAS:5598-53-8 ***HR: D***
SODIUM ASPARTATE
mf: $C_4H_5NO_4 \cdot 2Na$ mw: 177.08

SYNS: ASPARTIC ACID DISODIUM SALT ◇ l-ASPARTIC ACID, DISODIUM SALT (9CI) ◇ SODIUM l-ASPARTATE

TOXICITY DATA with REFERENCE
scu-mus TDLo:500 mg/kg (1D male):REP BECTA6 32,410,84

SAFETY PROFILE: Experimental reproductive effects. When heated to decomposition it emits toxic fumes of NO_x and Na_2O.

SEZ355 CAS:17090-93-6 ***HR: D***
SODIUM l-ASPARTATE
mf: $C_4H_7NO_4 \cdot 7Na$ mw: 293.96

SYN: l-ASPARTIC ACID, SODIUM SALT (9CI)

TOXICITY DATA with REFERENCE
scu-mus TDLo:34100 mg/kg (10D male):REP PBBHAU 9,481,78

SAFETY PROFILE: Experimental reproductive effects. When heated to decomposition it emits toxic fumes of NO_x and Na_2O.

SFA000 CAS:26628-22-8 ***HR: 3***
SODIUM AZIDE
DOT: UN 1687
mf: N_3Na mw: 65.02

PROP: Colorless, hexagonal crystals. Mp: decomp, d: 1.846. Insol in ether; sol in liquid ammonia.

SYNS: AZIDE ◇ AZIUM ◇ AZOTURE de SODIUM (FRENCH) ◇ KAZOE ◇ NATRIUMAZID (GERMAN) ◇ NATRIUMMAZIDE (DUTCH) ◇ NCI-C06462 ◇ NSC 3072 ◇ RCA WASTE NUMBER P105 ◇ SODIUM, AZOTURE de (FRENCH) ◇ SODIUM, AZOTURO di (ITALIAN) ◇ U-3886

TOXICITY DATA with REFERENCE
mmo-sat 10 μg/plate MUREAV 144,231,85
dni-hmn:fbr 50 mg/L STBIBN 78,165,80
orl-rat TDLo:2730 mg/kg/78W-C:ETA JJIND8 67,75,81
orl-hmn TDLo:710 μg/kg:CNS,KID JCPAAK 28,350,75
orl-rat LD50:27 mg/kg FMCHA2 -,C21,83
ipr-rat LDLo:30 mg/kg PHRPA6 58,607,43
scu-rat LDLo:35 mg/kg PHRPA6 58,607,43
orl-mus LD50:27 mg/kg CLDND* 30,98,48
scu-mus LDLo:17 mg/kg ATXKA8 22,160,66
ivn-mus LD50:19 mg/kg CLDND* 6,160,52
ivn-mky LDLo:12 mg/kg BRAIAK 95,505,72
skn-rbt LD50:20 mg/kg FMCHA2 -,C21,83

CONSENSUS REPORTS: Reported in EPA TSCA Inventory. EPA Genetic Toxicology Program. EPA Extremely Hazardous Substances List.

OSHA PEL: As NH_3: CL 0.1 ppm; As NaN_3: Cl 0.3 mg/m^3 (skin)
ACGIH TLV: CL 0.3 mg/m^3
DFG MAK: 0.07 ppm (0.2 mg/m^3)
DOT Classification: Poison B; Label: Poison

SAFETY PROFILE: Poison by ingestion, skin contact, intraperitoneal, intravenous, and subcutaneous routes. Human systemic effects by ingestion: general anesthesia, somnolence, and kidney changes. Questionable carcinogen with experimental tumorigenic data. Human mutation data reported.

Violent reaction with benzoyl chloride combined with KOH; Br_2; barium carbonate; CS_2; $Cr(OCl)_2$; Cu; Pb; HNO_3; $BaCO_3$; H_2SO_4; hot water; $(CH_3)_2SO_4$; dibromomalononitrile; sulfuric acid. Incompatible with acids; ammonium chloride + trichloroacetonitrile; phosgene; cyanuric chloride; 2,5-dinitro-3-methylbenzoic acid + oleum; trifluroracryloyl chloride. Reacts with heavy metals (e.g., brass; copper; lead) to form dangerously explosive heavy metal azides, a particular problem in laboratory equipment and drain traps. When heated to decomposition it emits very toxic fumes of NO_x and Na_2O. See also AZIDES.

SFA100 CAS:35038-45-0 ***HR: 3***
SODIUM-5-AZIDOTETRAZOLIDE
mf: CN_7Na mw: 133.05

SAFETY PROFILE: An explosive which is extremely sensitive to friction, heat, or pressure. When heated to decomposition it emits toxic fumes of NO_x and Na_2O. See also AZIDES.

SFA600 ***HR: 3***
SODIUM BENZENE HEXOIDE
mf: $C_6Na_6O_6$ mw: 306.00

SAFETY PROFILE: An explosive sensitive to shock or heating to 90°C. Ignites or explodes on contact with water. Upon decomposition it emits toxic fumes of Na_2O. See also SODIUM COMPOUNDS.

SFB000 CAS:532-32-1 ***HR: 3***
SODIUM BENZOATE
mf: $C_7H_5O_2$•Na mw: 144.11

PROP: White crystalline solid; odorless. Sol in water, alc.

SYNS: ANTIMOL ◇ BENZOATE of SODA ◇ BENZOATE SODIUM ◇ BENZOESAEURE (NA-SALZ) (GERMAN) ◇ BENZOIC ACID, SODIUM SALT ◇ SOBENATE ◇ SODIUM BENZOIC ACID

TOXICITY DATA with REFERENCE
cyt-ham:lng 1 g/L ATSUDG (4)41,80
cyt-ham:fbr 2 g/L/48H MUREAV 48,337,77
orl-rat TDLo:44 g/kg (1-22D preg):REP ESKHA5 96,47,78
ipr-rat TDLo:300 mg/kg (female 12-14D post):TER TXAPA9 19,373,71
orl-rat LD50:4070 mg/kg JIHTAB 30,63,48
ivn-rat LD50:1714 μg/kg JAPMA8 32,44,43
orl-mus LD50:1600 mg/kg GISAAA 51(1),75,86
ivn-mus LD50:1440 mg/kg MECHAN 6,290,63
ims-mus LD50:2306 mg/kg JPETAB 117,307,56
orl-rbt LD50:2 g/kg FAONAU 53A,34,74
ipr-gpg LDLo:1400 mg/kg HBAMAK 4,1313,35
scu-frg LDLo:100 mg/kg HBAMAK 4,1313,35

CONSENSUS REPORTS: Reported in EPA TSCA Inventory. EPA Genetic Toxicology Program.

SAFETY PROFILE: Poison by subcutaneous and intravenous routes. Moderately toxic by ingestion, intramuscular, and intraperitoneal routes. An experimental teratogen. Experimental reproductive effects. Mutation data reported. Larger doses of 8-10 grams by mouth may cause nausea and vomiting. Small doses have little or no effect. Combustible when exposed to heat or flame. When heated to decomposition it emits toxic fumes of Na_2O. See also BENZOIC ACID.

SFB100 CAS:26249-01-4 ***HR: 3***
SODIUM-2-BENZOTHIAZOLYLSULFIDE
mf: $C_7H_4NNaS_2$ mw: 189.23

$C_6H_4SC(SNa){=}N$

SAFETY PROFILE: Explosive reaction with maleic anhydride. When heated to decomposition it emits toxic fumes of SO_x, NO_x, and Na_2O. See also SULFIDES.

SFB500 CAS:63915-76-4 ***HR: 3***
SODIUM BERYLLIUM MALATE
mf: $C_8H_6Be_4Na_2O_{12}$•$7H_2O$ mw: 502.30

TOXICITY DATA with REFERENCE
ivn-mus LD50:36 μg/kg BJEPA5 30,375,49

CONSENSUS REPORTS: Beryllium and its compounds are on the Community Right To Know List.

OSHA PEL: (Transitional: TWA 0.002 mg(Be)/m^3; CL 0.005; Pk 0.025/30M/8H) TWA 0.002 mg(Be)/m^3; STEL 0.005 mg(Be)/m^3/30M; CL 0.025 mg(Be)/m^3
ACGIH TLV: TWA 0.002 mg(Be)/m^3, Suspected Carcinogen
NIOSH REL: CL (Beryllium) not to exceed 0.0005 mg(Be)/m^3

SAFETY PROFILE: Confirmed carcinogen. Poison by intravenous route. When heated to decomposition it emits toxic fumes of BeO and Na_2O. See also BERYLLIUM COMPOUNDS.

SFC000 CAS:63915-77-5 ***HR: 3***
SODIUM BERYLLIUM TARTRATE
mf: $C_8H_4Be_4Na_2O_{13}{\cdot}10H_2O$ mw: 570.34

TOXICITY DATA with REFERENCE
scu-mus LDLo:5 mg/kg BJEPA5 30,375,49
ivn-mus LDLo:32 μg/kg BJEPA5 30,375,49

CONSENSUS REPORTS: Beryllium and its compounds are on the Community Right To Know List.

OSHA PEL: (Transitional: TWA 0.002 mg(Be)/m^3; CL 0.005; Pk 0.025/30M/8H) TWA 0.002 mg(Be)/m^3; STEL 0.005 mg(Be)/m^3/30M; CL 0.025 mg(Be)/m^3
ACGIH TLV: TWA 0.002 mg(Be)/m^3, Suspected Carcinogen
NIOSH REL: CL (Beryllium) not to exceed 0.0005 mg(Be)/m^3

SAFETY PROFILE: Confirmed carcinogen. Poison by subcutaneous and intravenous routes. When heated to decomposition it emits toxic fumes of BeO and Na_2O. See also BERYLLIUM COMPOUNDS.

SFC500 CAS:144-55-8 ***HR: 1***
SODIUM BICARBONATE
mf: $NaHCO_3$ mw: 84.01

PROP: White crystalline powder. Sol in water; insol in alc.

SYNS: BAKING SODA ◇ BICARBONATE of SODA ◇ CARBONIC ACID MONOSODIUM SALT ◇ COL-EVAC ◇ JUSONIN ◇ MONOSODIUM CARBONATE ◇ NEUT ◇ SODA MINT ◇ SODIUM ACID CARBONATE ◇ SODIUM HYDROGEN CARBONATE

TOXICITY DATA with REFERENCE
skn-hmn 30 mg/3D-I MLD 85DKA8 -,127,77
eye-rbt 100 mg/30S MLD TXCYAC 23,281,82
dns-rat-orl 50400 mg/kg/4W-C CRNGDP 9,1203,88
ipr-mus TDLo:40 mg/kg (female 7D post):TER POASAD 56,10,76
orl-inf TDLo:1260 mg/kg:PUL,KID AJDCAI 135,965,81
orl-rat LD50:4220 mg/kg TXAPA9 6,726,64

CONSENSUS REPORTS: Reported in EPA TSCA Inventory.

SAFETY PROFILE: An experimental teratogen. A nuisance dust. Human systemic effects: respiratory changes, increased urine volume, sodium level changes. Mutation data reported.

SFD000 CAS:12232-99-4 ***HR: 3***
SODIUM BISMUTHATE
mf: $BiNaO_3$ mw: 279.97

PROP: Yellow to yellowish-brown, hygroscopic. Insol in cold water.

TOXICITY DATA with REFERENCE
mrc-smc 5 mmol/L MUTAEX 1,21,86
sln-smc 5 mmol/L MUTAEX 1,21,86
orl-rat LDLo:720 mg/kg EQSSDX 1,1,75
ivn-rat LDLo:25 mg/kg EQSSDX 1,1,75
ims-rat LDLo:250 mg/kg EQSSDX 1,1,75
orl-rbt LDLo:510 mg/kg EQSSDX 1,1,75
ivn-rbt LDLo:9 mg/kg EQSSDX 1,1,75
ims-rbt LDLo:110 mg/kg EQSSDX 1,1,75

CONSENSUS REPORTS: Reported in EPA TSCA Inventory.

SAFETY PROFILE: Poison by intravenous and intramuscular routes. Moderately toxic by ingestion. Mutation data reported. When heated to decomposition it emits toxic fumes of Bi and Na_2O. See also BISMUTH COMPOUNDS.

SFD300 ***HR: 3***
SODIUM-3,5-BIS(aci-NITRO)CYCLOHEXENE-4,6-DIIMINIDE
mf: $C_6H_4N_4Na_2O_2$ mw: 210.10

SAFETY PROFILE: Decomposes violently when exposed to heat or moisture. When heated to decomposition it emits toxic fumes of NO_x and Na_2O.

SFE000 CAS:7631-90-5 ***HR: 3***
SODIUM BISULFITE
DOT: NA 2693
mf: $HO_3S{\cdot}Na$ mw: 104.06

PROP: White, crystalline powder; odor of sulfur dioxide, disagreeable taste. D: 1.48, mp: decomp. Very sol in hot or cold water; sltly sol in alc.

SYNS: BISULFITE de SODIUM (FRENCH) ◇ HYDROGEN SULFITE SODIUM ◇ SODIUM ACID SULFITE ◇ SODIUM BISULFITE ◇ SODIUM BISULFITE (1 : 1) ◇ SODIUM BISULFITE, solid (DOT) ◇ SODIUM BISULFITE, solution (DOT) ◇ SODIUM HYDROGEN SULFITE ◇ SODIUM HYDROGEN SULFITE, solid (DOT) ◇ SODIUM HYDROGEN SULFITE, solution (DOT) ◇ SODIUM SULHYDRATE ◇ SULFUROUS ACID, MONOSODIUM SALT

TOXICITY DATA with REFERENCE
mmo-sat 1 mmol/L TCMUD8 5,195,85
mmo-hmi 200 mmol/L ABMPAC 17,115,80
orl-rat LD50:2000 mg/kg DTLVS* 4,369,80
ipr-rat LD50:650 mg/kg CLPTAT 9,328,68
ivn-rat LD50:115 mg/kg JPETAB 101,101,51
ipr-mus LD50:675 mg/kg CLPTAT 9,328,68
ivn-mus LD50:130 mg/kg JPETAB 101,101,51
ipr-dog LD50:244 mg/kg CLPTAT 9,328,68
ipr-rbt LD50:300 mg/kg CLPTAT 9,328,68
ivn-rbt LD50:65 mg/kg JPETAB 101,101,51

CONSENSUS REPORTS: Reported in EPA TSCA Inventory. EPA Genetic Toxicology Program.

OSHA PEL: TWA 5 mg/m^3
ACGIH TLV: TWA 5 mg/m^3
DOT Classification: ORM-B; Label: None; Corrosive Material; Label: Corrosive

SAFETY PROFILE: Poison by intravenous and intraperitoneal routes. Moderately toxic by ingestion. A corrosive irritant to skin, eyes, and mucous membranes. Mutation data reported. An allergen. When heated to decomposition it emits toxic fumes of SO_x and Na_2O. See also SULFUROUS ACID and SULFITES.

SFE500 CAS:1303-96-4 ***HR: D***
SODIUM BORATE
mf: B_4O_7•2Na mw: 201.22

PROP: White crystals. Mp: 741°, bp: 1575° (decomp), d: 2.367. Slowly soluble in water.

SYNS: BORATES, TETRA, SODIUM SALT, anhydrous (OSHA, ACGIH) ◇ SODIUM BORATE anhydrous

TOXICITY DATA with REFERENCE
orl-rat TDLo:16750 μg/kg (30D male):REP EVHPAZ 13,59,76

OSHA PEL: 10 mg/m^3 (anhydrous, decahydrate, pentahydrate)
ACGIH TLV: TWA 1 mg/m^3

SAFETY PROFILE: Experimental reproductive effects. When heated to decomposition it emits toxic fumes of Na_2O, boron. See also BORON COMPOUNDS.

SFF000 CAS:1303-96-4 ***HR: 3***
SODIUM BORATE DECAHYDRATE
mf: B_4O_7•2Na•10H_2O mw: 381.42

PROP: Hard, odorless crystals, granules or crystalline powder. D: 1.73, mp: 75° (when rapidly heated).

SYNS: ANTIPYONIN ◇ BORACSU ◇ BORATES, TETRA, SODIUM SALT, anhydrous (OSHA, ACGIH) ◇ BORAX (8CI) ◇ BORAX DECAHYDRATE ◇ BORICIN ◇ GERTLEY BORATE ◇ JAIKIN ◇ NEOBOR ◇ POLYBOR ◇ SODIUM BIBORATE ◇ SODIUM BIBORATE DECAHYDRATE ◇ SODIUM PYROBORATE ◇ SODIUM PYROBORATE DECAHYDRATE ◇ SODIUM TETRABORATE ◇ SODIUM TETRABORATE DECAHYDRATE

TOXICITY DATA with REFERENCE
slt-dmg-orl 714 ppm SOGEBZ 10,601,74
cyt-dmg-orl 714 ppm SOGEBZ 10,601,74
orl-rat TDLo:37 g/kg (multi):REP TXAPA9 23,351,72
orl-inf LDLo:1000 mg/kg PCOC** -,144,66
orl-man LDLo:709 mg/kg JAMAAP 76,378,21
orl-rat LD50:2660 mg/kg FMCHA2 -,C46,89
orl-mus LD50:2000 mg/kg 14KTAK -,693,64
ipr-mus LD50:2711 mg/kg AIPTAK 143,144,63
ivn-mus LD50:1320 mg/kg 14KTAK -,693,64
scu-rbt LDLo:150 mg/kg HBAMAK 4,1289,35
orl-gpg LD50:5330 mg/kg 14KTAK -,693,64

CONSENSUS REPORTS: Reported in EPA TSCA Inventory.

OSHA PEL: TWA 10 mg/m^3
ACGIH TLV: TWA 5 mg/m^3

SAFETY PROFILE: Experimental poison by subcutaneous route. Moderately toxic to humans by ingestion. Moderately toxic experimentally by ingestion, intravenous, and intraperitoneal routes. Experimental reproductive effects. Mutation data reported. Ingestion of 5-10 grams of borax by children can cause severe vomiting, diarrhea, shock, death. Incompatible with acids; metallic salts. When heated to decomposition it emits toxic fumes of Na_2O, boron. See also BORON COMPOUNDS. Used in ant poisons, for fly control around refuse and manure piles, as a larvicide, in manufacture of glazes, enamels, cleaning compounds, and in soldering metals.

SFF500 CAS:16940-66-2 ***HR: 3***
SODIUM BOROHYDRIDE
mf: BH_4•Na mw: 37.84
DOT: UN 1426

PROP: White to gray-white, microcrystalline powder or lumps. Hygroscopic. Mp: >400°C (vacuum) (decomp), d: 1.07. Reacts with hot water; sol in liquid ammonia and "Cellosolve" ether.

SYNS: BOROHYDRURE de SODIUM (FRENCH) ◇ SODIUM TETRAHYDROBORATE(1-)

TOXICITY DATA with REFERENCE
orl-rat LDLo:160 mg/kg 14KTAK -,693,64
ipr-rat LD50:18 mg/kg AMIHAB 17,124,58
ipr-mus LDLo:18 mg/kg 14KTAK -,693,64
ipr-rbt LDLo:60 mg/kg 14KTAK -,693,64

CONSENSUS REPORTS: Reported in EPA TSCA Inventory.

DOT Classification: Flammable Solid; Label: Dangerous When Wet

SAFETY PROFILE: Poison by ingestion and intraperitoneal routes. A strong alkali. A severe eye, skin, and mucous membrane irritant.

Ignites in air above 288°C when exposed to spark. Potentially explosive reaction with aluminum chloride + bis(2-methoxyethyl) ether. Reacts with ruthenium salts to form a solid product which explodes when touched or on contact with water. Reacts to form dangerously explosive hydrogen gas on contact with alkali; water and other protic solvents (e.g., methanol; ethanol; ethylene glycol; phenol); aluminum chloride + bis(2-methoxyethyl)ether. Reacts violently with anhydrous acids (e.g., sulfuric; phosphoric; fluorophosphoric) to form diborane. Violent exothermic reaction with dimethyl formamide has caused industrial explosions. Mixtures with sulfuric acid may ignite. Incompatible with palladium; diborane + bis(2-methoxyethyl) ether; polyglycols; dimethylacetamide; oxidizers; metal salts; finely divided metallic precipitates of cobalt; nickel; copper; iron and possibly other metals. Emits flammable vapors on contact with acid fumes. Materials sensitive to polymerization under alkaline conditions, such as acrylonitrile, may polymerize upon contact with sodium borohydride. Avoid storage in glass containers. When heated to decomposition it emits toxic fumes of Na_2O. See also HYDRIDES, BORON COMPOUNDS, and SODIUM COMPOUNDS.

SFG000 CAS:7789-38-0 ***HR: 3***
SODIUM BROMATE
DOT: UN 1494
mf: $BrO_3 \cdot Na$ mw: 150.90

PROP: White crystals or crystalline powder. Odorless. Mp: 381°, d: 3.339 @ 17.5°.

SYNS: BROMATE de SODIUM (FRENCH) ◇ BROMIC ACID, SODIUM SALT ◇ DYETONE

TOXICITY DATA with REFERENCE
ipr-mus LD50:140 mg/kg COREAF 257,791,63
scu-dog LDLo:120 mg/kg SAPHAO 30,337,13
orl-rbt LDLo:250 mg/kg SAPHAO 30,337,13
ivn-rbt LDLo:360 mg/kg SAPHAO 30,337,13
scu-gpg LDLo:100 mg/kg SAPHAO 30,337,13

CONSENSUS REPORTS: Reported in EPA TSCA Inventory.

DOT Classification: Oxidizer; Label: Oxidizer

SAFETY PROFILE: Poison by ingestion, intravenous, subcutaneous, and intraperitoneal routes. A powerful oxidizer. Violent reactions with Al, As, C, Cu, oil, F_2, metal sulfides, organic matter, P, S. Mixtures with grease are shock-sensitive explosives at 120°C. When heated to decomposition it emits toxic fumes of Na_2O and Br^-. See also BROMATES.

SFG500 CAS:7647-15-6 ***HR: 2***
SODIUM BROMIDE
mf: BrNa mw: 102.90

PROP: White crystals, granules or powder; saline bitter taste. D: 3.21, mp: 755°, volatilizes at higher temp.

SYNS: BROMIDE SALT of SODIUM ◇ BROMNATRIUM (GERMAN) ◇ SEDONEURAL

TOXICITY DATA with REFERENCE
orl-rat TDLo:720 mg/kg (female 3-20D post):REP JOPSAM 19,17,45
orl-rat LD50:3500 mg/kg JPETAB 55,200,35
orl-mus LD50:7000 mg/kg SMWOAS 85,305,55
scu-mus LD50:5020 mg/kg JPMSAE 50,858,61
orl-rbt LDLo:580 mg/kg 27ZIAQ -,243,73

CONSENSUS REPORTS: Reported in EPA TSCA Inventory.

SAFETY PROFILE: Moderately toxic by ingestion. Experimental reproductive effects. Incompatible with acids, alkaloidal and heavy metal salts. When heated to decomposition it emits toxic fumes of Br^- and Na_2O. See also BROMIDES.

SFG600 ***HR: 3***
SODIUM BROMOACETYLIDE
mf: C_2BrNa mw: 126.92

$$NaC \equiv CBr$$

SAFETY PROFILE: An extremely shock-sensitive explosive. When heated to decomposition it emits toxic fumes of Br^- and Na_2O. See also ACETYLIDES.

SFJ500 CAS:64048-05-1 ***HR: 3***
SODIUM BUTYLMERCURIC THIOGLYCOLLATE
mf: $C_6H_{11}HgO_2S \cdot Na$ mw: 370.81

SYN: S-(BUTYLMERCURIC)-THIOGLYCOLIC ACID, SODIUM SALT

TOXICITY DATA with REFERENCE
ipr-rat LDLo:30 mg/kg JPETAB 35,343,29
ivn-rbt LDLo:20 mg/kg JPETAB 35,343,29

CONSENSUS REPORTS: Mercury and its compounds are on the Community Right To Know List.

OSHA PEL: (Transitional: CL 1 mg/10m^3) CL 0.1 mg(Hg)/m^3 (skin)
ACGIH TLV: TWA 0.1 mg(Hg)/m^3 (skin)
NIOSH REL: (Inorganic Mercury) TWA 0.05 mg(Hg)/m^3

SAFETY PROFILE: Poison by intraperitoneal and intravenous routes. When heated to decomposition it emits very toxic fumes of Hg, SO_x, and Na_2O. See also MERCURY COMPOUNDS.

SFJ875 CAS:67050-64-0 ***HR: 3***
SODIUM-5-(1-(BUTYLTHIO)ETHYL)-5-ETHYLBARBITURATE
mf: $C_{12}H_{19}N_2O_3S{\cdot}Na$ mw: 294.38

SYN: 5-(1-(BUTYLTHIO)ETHYL)-5-ETHYLBARBITURIC ACID SODIUM SALT

TOXICITY DATA with REFERENCE
orl-rat LD50:582 mg/kg JAPMA8 35,231,46
ivn-rat LD50:90 mg/kg JPETAB 88,343,46
ivn-rbt LD50:55 mg/kg JAPMA8 35,244,46

SAFETY PROFILE: Poison by intravenous route. Moderately toxic by ingestion. When heated to decomposition it emits toxic fumes of SO_x, NO_x, and Na_2O. See also BARBITURATES.

SFN600 CAS:156-54-7 ***HR: D***
SODIUM BUTYRATE
mf: $C_4H_8O_2{\cdot}Na$ mw: 111.11

SYNS: BUTYRATE SODIUM ◇ SODIUM BUTANOATE ◇ SODIUM n-BUTYRATE

TOXICITY DATA with REFERENCE
dni-hmn:fbr 5 mmol/L CNREA8 46,713,86
oms-hmn:fbr 100 mmol/L CNREA8 46,713,86

CONSENSUS REPORTS: Reported in EPA TSCA Inventory.

SAFETY PROFILE: Human mutation data reported. When heated to decomposition it emits toxic fumes of Na_2O.

SFN700 ***HR: 1***
SODIUM CALCIUM ALUMINOSILICATE, HYDRATED

SAFETY PROFILE: A nuisance dust.

SFO000 CAS:497-19-8 ***HR: 3***
SODIUM CARBONATE (2:1)
mf: $CO_3{\cdot}2Na$ mw: 105.99

PROP: White, odorless, small crystals or crystalline powder; alkali taste. Mp: 851°, bp: decomp, d: 2.509 @ 0°. Hygroscopic; sol in water.

SYNS: CARBONIC ACID, DISODIUM SALT ◇ CRYSTOL CARBONATE ◇ DISODIUM CARBONATE ◇ SODA ASH ◇ TRONA

TOXICITY DATA with REFERENCE
skn-rbt 500 mg/24H MLD 28ZPAK -,7,72
eye-rbt 100 mg/24H MOD 28ZPAK -,8,72
eye-rbt 100 mg rns MLD TXCYAC 23,281,82
iut-mus TDLo:84800 ng/kg (4D preg):REP JRPFA4 63,365,81
orl-rat LD50:4090 mg/kg 28ZPAK -,8,72
ihl-rat LC50:2300 mg/m^3/2H ENVRAL 31,138,83
ihl-mus LC50:1200 mg/m^3/2H ENVRAL 31,138,83
ipr-mus LD50:117 mg/kg COREAF 257,791,63
scu-mus LD50:2210 mg/kg RPTOAN 33,266,70
ihl-gpg LC50:800 mg/m^3/2H ENVRAL 31,138,83

CONSENSUS REPORTS: Reported in EPA TSCA Inventory. EPA Genetic Toxicology Program.

SAFETY PROFILE: Poison by intraperitoneal route. Moderately toxic by inhalation and subcutaneous routes. Mildly toxic by ingestion. Experimental reproductive effects. A skin and eye irritant. It migrates to food from packaging materials. Can react violently with Al; P_2O_5; H_2SO_4; F_2; Li; 2,4,6-trinitro-toluene. When heated to decomposition it emits toxic fumes of Na_2O.

SFO500 CAS:9004-32-4 ***HR: 3***
SODIUM CARBOXYMETHYL CELLULOSE

PROP: A synthetic cellulose gum (the sodium salt of carboxy methyl cellulose not less than 99.5% on a dry weight basis, with maximum substitution of 0.95 carboxymethyl groups per anhydroglucose unit, and with a minimum viscosity of 25 centipoises for 2% weight aq solutions at 25°). Colorless, odorless, hygroscopic powder or granules. Insol in most organic solvents.

SYNS: AC-DI-SOL NF ◇ AQUAPLAST ◇ B10 ◇ BLANOSE BWM ◇ B 10 (polysaccharide) ◇ CARBOXYMETHYL CELLULOSE ◇ CARBOXYMETHYL CELLULOSE, SODIUM ◇ CARBOXYMETHYL CELLULOSE, SODIUM SALT ◇ CARMETHOSE ◇ CELLOFAS ◇ CELLOGEL C ◇ CELLPRO ◇ CELLUFIX FF 100 ◇ CELLUGEL ◇ CELLULOSE GLYCOLIC ACID, SODIUM SALT ◇ CELLULOSE GUM ◇ CELLULOSE SODIUM GLYCOLATE ◇ CMC ◇ CM-CELLULOSE Na SALT ◇ CMC 7H ◇ CMC SODIUM SALT ◇ COLLOWELL ◇ COPAGEL PB 25 ◇ COURLOSE A 590 ◇ DAICEL 1150 ◇ FINE GUM HES ◇ GLIKOCEL TA ◇ KMTS 212 ◇ LOVOSA ◇ LUCEL (polysaccharide) ◇ MAJOL PLX ◇ MODOCOLL 1200 ◇ NACM-CELLULOSE SALT ◇ NYMCEL S ◇ POLYFIBRON 120 ◇ SANLOSE SN 20A ◇ SARCELL TEL ◇ S 75M ◇ SODIUM CELLULOSE GLYCOLATE ◇ SODIUM CMC ◇ SODIUM CM-CELLULOSE ◇ SODIUM SALT of CARBOXYMETHYLCELLULOSE ◇ TYLOSE 666 ◇ UNISOL RH

TOXICITY DATA with REFERENCE
orl-rat TDLo:140 mg/kg (14D male):REP OYYAA2 14,623,77
scu-rat TDLo:1900 mg/kg/19W-I:NEO 13BYAH -,83,62
orl-rat LD50:27000 mg/kg FOREAE 13,29,48
orl-gpg LD50:16000 mg/kg FOREAE 13,29,48

CONSENSUS REPORTS: Reported in EPA TSCA Inventory.

SAFETY PROFILE: Mildly toxic by ingestion. Experimental reproductive effects. Questionable carcinogen with experimental neoplastigenic data. It migrates to food from packaging materials. When heated to decomposition it emits toxic fumes of Na_2O. See also POLYMERS, SOLUBLE.

SFP000 CAS:9061-82-9 *HR: 1*
SODIUM CARRAGHEENATE

PROP: A mixture of highly sulfated polygalactosides extracted from seaweeds (FAONAU 53A,598,74).

SYN: CARRAGEENAN, SODIUM SALT

TOXICITY DATA with REFERENCE
orl-rat LD50:5650 mg/kg FAONAU 53A,398,74
orl-mus LD50:9200 mg/kg FDRLI* 124,-,76
orl-rbt LD50:4670 mg/kg FAONAU 53A,398,74
orl-ham LD50:7530 mg/kg FAONAU 53A,398,74

CONSENSUS REPORTS: Reported in EPA TSCA Inventory.

SAFETY PROFILE: Mildly toxic by ingestion. When heated to decomposition it emits toxic fumes of SO_x and Na_2O.

SFP500 CAS:66007-89-4 *HR: 3*
SODIUM CARRIOMYCIN
mf: $C_{47}H_{80}O_{15}$•Na mw: 908.26

SYNS: CARRIOMYCIN, SODIUM SALT ◇ T-42082

TOXICITY DATA with REFERENCE
orl-mus LD50:2000 mg/kg JANTAJ 31,7,78
ipr-mus LD50:125 mg/kg JANTAJ 31,7,78

SAFETY PROFILE: Poison by intraperitoneal route. Moderately toxic by ingestion. When heated to decomposition it emits toxic fumes of Na_2O.

SFQ000 CAS:9005-46-3 *HR: 3*
SODIUM CASEINATE

PROP: Coarse, white powder; odorless. Insol in water, alc.

SYNS: CASEIN and CASEINATE SALTS (FCC) ◇ CASEIN-SODIUM ◇ CASEIN, SODIUM COMPLEX ◇ CASEINS, SODIUM COMPLEXES ◇ NUTROSE

TOXICITY DATA with REFERENCE
scu-mus TDLo:45 g/kg/15D-I:ETA JNCIAM 57,1367,76

CONSENSUS REPORTS: Reported in EPA TSCA Inventory.

SAFETY PROFILE: Questionable carcinogen with experimental tumorigenic data. When heated to decomposition it emits toxic fumes of Na_2O.

SFQ300 CAS:56238-63-2 *HR: 2*
SODIUM CEFUROXIME
mf: $C_{16}H_{16}N_4O_8S$•Na mw: 447.41

SYNS: (6R-(6-α,7-β(Z)))-3-(((AMINOCARBONYL)OXY)METHYL)-7-((2-FURANYL(METHYOXYIMINO)ACETYL)AMINO)-8-OXO-5-THIA-1-AZABICYCLO(4.2.0)OCT-2-ENE-2-CARBOXYLIC ACID MONOSODIUM SALT ◇ CEFUROXIME SODIUM ◇ CEFUROXIME SODIUM SALT ◇ CXM

TOXICITY DATA with REFERENCE
scu-mus TDLo:43200 mg/kg (16-21D preg/21D post):REP TXCYAC 13,1,79
ipr-rat LD50:10 g/kg NKRZAZ 27(Suppl 6),124,79
ivn-rat LD50:4000 mg/kg TXCYAC 13,1,79
ivn-mus LD50:10400 mg/kg TXCYAC 13,1,79

SAFETY PROFILE: Moderately toxic by intravenous route. Experimental reproductive effects. When heated to decomposition it emits toxic fumes of SO_x, NO_x, and Na_2O.

SFQ500 CAS:58-71-9 *HR: 2*
SODIUM CEPHALOTHIN
mf: $C_{16}H_{15}N_2O_6S_2$•Na mw: 418.44

SYNS: CEFALOTHINE SODIUM ◇ CEFALOTINA SODICA (SPANISH) ◇ CEPHALOTHIN SODIUM ◇ CEPOVENIN ◇ KEFLIN ◇ LILLY 38253 ◇ MICROTIN ◇ SODIUM CEPHALOTIN ◇ 5-THIA-1-AZABICYCLO(4.2.0)OCT-2-ENE-2-CARBOXYLIC ACID, 3-(HYDROXYMETHYL)-8-OXO-7-(2-(2-IHIENYL)ACETAMIDO)-, ACETATE, MONOSODIUM SALT ◇ 7-(THIOPHENE-2-ACETAMIDO)CEPHALOSPORANIC ACID SODIUM SALT

TOXICITY DATA with REFERENCE
ivn-wmn TDLo:20 mg/kg:BLD AIMEAS 77,401,72
ims-chd TDLo:89 mg/kg/19D-I:BLD AACHAX -,272,63
ims-man TDLo:955 mg/kg/22D-I:BLD AACHAX -,272,63
ipr-rat LD50:7000 mg/kg JJANAX 28,188,75
scu-rat LD50:7500 mg/kg JIDIAQ 137,S51,78
ipr-mus LD50:5670 mg/kg JJANAX 27,746,74
scu-mus LD50:7500 mg/kg JACHDX 6(Suppl A),79,80
ivn-mus LD50:4800 mg/kg JJANAX 27,746,74

CONSENSUS REPORTS: EPA Genetic Toxicology Program.

SAFETY PROFILE: Human systemic effects by intravenous and intramuscular routes: blood agranulocytosis and thrombocytopenia. When heated to decomposition it emits very toxic fumes of NO_x, SO_x, and Na_2O.

SFR000 CAS:63937-26-8 *HR: 3*
SODIUM CHAULMOOGRATE
mf: $C_{18}H_{32}O_2$•Na mw: 303.49

SYNS: CHAULMOOGRIC ACID, SODIUM SALT ◇ 2-CYCLOPENTENE-1-TRIDECANOIC ACID, SODIUM SALT

TOXICITY DATA with REFERENCE
scu-rat LDLo:2000 mg/kg IJLEAG 2,39,34
ivn-rat LDLo:200 mg/kg IJLEAG 2,39,34

SAFETY PROFILE: Poison by intravenous route. Moderately toxic by subcutaneous route. When heated to decomposition it emits toxic fumes of Na_2O.

SFS000 CAS:7775-09-9 ***HR: 3***
SODIUM CHLORATE
DOT: UN 1495/UN 2428
mf: $ClO_3{\cdot}Na$ mw: 106.44

PROP: Colorless, odorless crystals; cooling, saline taste. Mp: 248-261°, bp: decomp, d: 2.490 @ 15°.

SYNS: ASEX ◇ ATLACIDE ◇ ATRATOL ◇ B-HERBATOX ◇ CHLORATE of SODA (DOT) ◇ CHLORATE SALT of SODIUM ◇ CHLORAX ◇ CHLORSAURE (GERMAN) ◇ DE-FOL-ATE ◇ DESOLET ◇ DREXEL DEFOL ◇ DROP LEAF ◇ EVAU-SUPER ◇ FALL ◇ GRAIN SORGHUM HARVEST-AID ◇ GRANEX O ◇ HARVEST-AID ◇ KLOREX ◇ KUSATOHRU ◇ KUSATOL ◇ NATRIUMCHLORAAT (DUTCH) ◇ NATRIUMCHLORAT (GERMAN) ◇ ORTHO C-1 DEFOLIANT & WEED KILLER ◇ OXYCIL ◇ RASIKAL ◇ SHED-A-LEAF ◇ SHED-A-LEAF "L" ◇ SODA CHLORATE (DOT) ◇ SODIO (CLORATO di) (ITALIAN) ◇ SODIUM (CHLORATE de) (FRENCH) ◇ SODIUM CHLORATE, aqueous solution (DOT) ◇ TRAVEX ◇ TUMBLEAF ◇ UNITED CHEMICAL DEFOLIANT No. 1 ◇ VAL-DROP

TOXICITY DATA with REFERENCE
skn-rbt 500 mg/24H MLD BIOFX* 24,3/71
eye-rbt 10 mg MLD BIOFX* 21-3/71
mma-sat 40 μmol/plate MUREAV 90,91,81
sln-dmg-orl 250 mmol/L MUREAV 90,91,81
orl-wmn TDLo:800 mg/kg:PUL,BLD 34ZIAG -,539,69
unr-hmn LDLo:214 mg/kg GUCHAZ 6,461,73
unr-chd LD50:185 mg/kg GUCHAZ 6,461,73
orl-rat LD50:1200 mg/kg PHJOAV 185,361,60
ipr-mus LD50:596 mg/kg COREAF 257,791,63
orl-dog LDLo:700 mg/kg HBAMAK 4,1289,35
orl-cat LDLo:1350 mg/kg PCOC** -,1013,66
orl-rbt LDLo:8000 mg/kg AEXPBL 21,169,1886

CONSENSUS REPORTS: Reported in EPA TSCA Inventory.

DOT Classification: Oxidizer; Label: Oxidizer

SAFETY PROFILE: Human poison by unspecified routes. Moderately toxic experimentally by ingestion and intraperitoneal routes. Human systemic effects by ingestion: blood hemolysis with or without anemia, methemoglobinemia-carboxhemoglobinemia and pulmonary changes. Mutation data reported. A skin, mucous membrane, and eye irritant. Damages the red blood cells of humans when ingested.

A powerful oxidizer. It can explode on contact with flame or sparks (static discharge) and has caused many industrial explosions. May react explosively with agricultural materials (e.g., peat; powdered sulfur; sawdust; urotropine; thiuram); alkenes + potassium osmate; aluminum + rubber; ammonium salts; grease; leather; powdered metals; non-metals; sulfides; cyanides; cyanoborane oligomer; nitrobenzene; organic matter; paint + polyethylene; phosphorus; sodium phosphinate. Violent reaction or ignition with aluminum; ammonium sulfate; Sb_2S_3; arsenic; arsenic trioxide; 1,3-bis(trichloromethylbenzene) + heat; carbon; charcoal; MnO_2; phosphorus; potassium cyanide; osmium + heat; paper; sulfuric acid; thiocyanates; triethylene glycol + wood; wood; zinc. Can also react violently with nitrobenzene, paper, metal sulfides, dibasic organic acids, organic matter. When heated to decomposition it emits toxic fumes of Cl^- and Na_2O. See also CHLORATES.

SFS500 CAS:52623-84-4 ***HR: 1***
SODIUM CHLORATE BORATE
mf: $BNaO_2{\cdot}ClNaO_3$ mw: 172.24

PROP: A complex of a sodium borate with sodium chlorate (FMCHA2 -,D281,80).

SYNS: CHEM-FROST ◇ LEAF DROP ◇ MONOBOR-CHLORATE ◇ OXY MBC ◇ POLYBOR-CHLORATE

TOXICITY DATA with REFERENCE
mmo-sat 100 μg/plate MUREAV 51,151,78
orl-rat LD50:4330 mg/kg FMCHA2 -,C216,83
skn-rbt LD50:20 g/kg FMCHA2 -,C161,83

SAFETY PROFILE: Mildly toxic by ingestion and skin contact. Mutation data reported. A powerful oxidizer. When heated to decomposition it emits toxic fumes of Cl^-, Na_2O, and BO_x. See also CHLORATES and BORON COMPOUNDS.

SFT000 CAS:7647-14-5 ***HR: 2***
SODIUM CHLORIDE
mf: ClNa mw: 58.44

PROP: Colorless, transparent crystals or white, crystalline powder. Mp: 801°, bp: 1413°, d: 2.165, vap press: 1 mm @ 865°. Sol in water, glycerin.

SYNS: COMMON SALT ◇ DENDRITIS ◇ EXTRA FINE 200 SALT ◇ EXTRA FINE 325 SALT ◇ HALITE ◇ H.G. BLENDING ◇ NATRIUMCHLORID (GERMAN) ◇ PUREX ◇ ROCK SALT ◇ SALINE ◇ SALT ◇ SEA SALT ◇ STERLING ◇ TABLE SALT ◇ TOP FLAKE ◇ USP SODIUM CHLORIDE ◇ WHITE CRYSTAL

TOXICITY DATA with REFERENCE
skn-rbt 50 mg/24H MLD BIOFX* 20-3/71
skn-rbt 500 mg/24H MLD 28ZPAK -,7,72
eye-rbt 100 mg MLD BIOFX* 20-3/71
eye-rbt 100 mg/24H MOD 28ZPAK -,7,72
eye-rbt 10 mg MOD TXAPA9 55,501,80
dni-hmn:fbr 125 mmol/L CNREA8 46,713,86
ipc-wmn TDLo:27 mg/kg (15W preg):REP AJOGAH 118,218,74
ipr-rat TDLo:1710 mg/kg (female 13D post):TER SEIJBO 8,197,68
orl-hmn TDLo:12357 mg/kg/23D-C:CVS AJDDAL 21,180,54
orl-rat LD50:3000 mg/kg TXAPA9 20,57,71
scu-rat LDLo:3500 mg/kg ENDOAO 24,523,39
orl-mus LD50:4000 mg/kg FRPPAO 27,19,72
ipr-mus LD50:6614 mg/kg COREAF 256,1043,63

scu-mus LD50:3 g/kg ARZNAD 7,445,57
ivn-mus LD50:645 mg/kg ARZNAD 7,445,57
icv-mus LD50:131 mg/kg TYKNAQ 27,131,80
ipr-dog LDLo:364 mg/kg AVERAG 44,555,37
ivn-dog LDLo:2 g/kg AEXPBL 21,119,1886
orl-rbt LDLo:8 g/kg HBAMAK 4,1289,35
ivn-rbt LDLo:1100 mg/kg AEXPBL 21,119,1886
scu-gpg LDLo:2160 mg/kg HBAMAK 4,1289,35
ivn-gpg LDLo:2910 mg/kg JPETAB 6,595,15

CONSENSUS REPORTS: Reported in EPA TSCA Inventory. EPA Genetic Toxicology Program.

SAFETY PROFILE: Poison by intraperitoneal and intracervical routes. Moderately toxic by ingestion, intravenous, and subcutaneous routes. An experimental teratogen. Human systemic effects by ingestion: blood pressure increase. Human reproductive effects by intraplacental route: terminates pregnancy. Experimental reproductive effects. Human mutation data reported. A skin and eye irritant. When bulk sodium chloride is heated to high temperature, a vapor is emitted which is irritating, particularly to the eyes. Ingestion of large amounts of sodium chloride can cause irritation of the stomach. Improper use of salt tablets may produce this effect. Potentially explosive reaction with dichloromaleic anhydride + urea. Electrolysis of mixtures with nitrogen compounds may form the explosive nitrogen trichloride. Reaction with burning lithium forms the dangerously reactive sodium. The molten salt at 1100° reacts explosively with water. Violent reaction with BrF_3. When heated to decomposition it emits toxic fumes of Cl^- and Na_2O.

SFT500 CAS:7758-19-2 ***HR: 3***
SODIUM CHLORITE
DOT: UN 1496/UN 1908
mf: $ClNaO_2$ mw: 90.44

PROP: White crystals or crystalline powder. Bp: decomp @ 180-200°.

SYN: TEXTILE

TOXICITY DATA with REFERENCE
mma-sat 300 μg/plate FCTOD7 22,623,84
cyt-ham:fbr 20 mg/L FCTOD7 22,623,84
orl-rat TDLo:1130 mg/kg (male 8W pre):REP TJADAB 35,43A,87
orl-rat TDLo:800 mg/kg (8-15D preg):TER EVHPAZ 46,25,82
orl-mus TDLo:29750 mg/kg/85W-C:CAR EVHPAZ 69,221,86
orl-rat LD50:165 mg/kg YKYUA6 31,959,80
orl-mus LD50:350 mg/kg GISAAA 45(4),6,80
orl-gpg LD50:300 mg/kg GISAAA 45(4),6,80

CONSENSUS REPORTS: Reported in EPA TSCA Inventory.

DOT Classification: Oxidizer; Label: Oxidizer (UN1496); Corrosive Material; Label: Corrosive (UN1908)

SAFETY PROFILE: Poison by ingestion. An experimental teratogen. Experimental reproductive effects. Questionable carcinogen with experimental carcinogenic data. Mutation data reported. May act as an irritant due to its oxidizing power. A powerful oxidizing agent; ignited by friction, heat, or shock. An explosive sensitive to impact or heating to 200°. Potentially explosive reaction with acids, oils, organic matter, oxalic acid + water, zinc. Violent reaction or ignition with carbon (above 60°); ethylene glycol (at 100°); phosphorus (above 50°); sodium dithionate; sulfur containing materials. Can react vigorously on contact with reducing materials. When heated to decomposition it emits highly toxic fumes of Cl^- and Na_2O. Used as a bleaching agent. See also CHLORITES.

SFU000 CAS:7758-19-2 ***HR: 3***
SODIUM CHLORITE (solution)
DOT: UN 1908
mf: $ClNaO_2$ mw: 90.44

PROP: Solution contains 42% or less sodium chlorite (FEREAC 41,15972,76).

TOXICITY DATA with REFERENCE
orl-rat LD50:350 mg/kg GISAAA 45(4),6,80
orl-mus LD50:350 mg/kg GISAAA 45(4),6,80
orl-gpg LD50:300 mg/kg GISAAA 45(4),6,80

CONSENSUS REPORTS: Reported in EPA TSCA Inventory.

DOT Classification: Corrosive Material; Label: Corrosive

SAFETY PROFILE: Poison by ingestion. A corrosive irritant to skin, eyes, and mucous membranes. When heated to decomposition it emits toxic fumes of Cl^- and Na_2O. See also SODIUM CHLORITE.

SFU500 CAS:3926-62-3 ***HR: 3***
SODIUM CHLOROACETATE
DOT: UN 2659
mf: $C_2H_2ClO_2 \cdot Na$ mw: 116.48

PROP: White, free-flowing, odorless powder. Mp: decomp @ 200°.

SYNS: CHLOROACETIC ACID SODIUM SALT ◇ CHLOROCTAN SODNY (CZECH) ◇ DOW DEFOLIANT ◇ MONOXONE ◇ SMA ◇ SMCA ◇ SODIUM MONOCHLORACETATE

TOXICITY DATA with REFERENCE
orl-rat LD50:95 mg/kg JIHTAB 23,78,41

orl-mus LD50:318 mg/kg JIHTAB 23,78,41
ivn-mus LD50:109 mg/kg 28ZPAK -,90,72
orl-rbt LD50:156 mg/kg PCOC** -,1047,66
orl-gpg LD50:99 mg/kg JIHTAB 23,78,41

CONSENSUS REPORTS: Reported in EPA TSCA Inventory.

DOT Classification: Poison B; Label: St. Andrews Cross

SAFETY PROFILE: Poison by ingestion and intravenous routes. When heated to decomposition it emits toxic fumes of Cl^- and Na_2O. Used as an herbicide.

SFU600 ***HR: 3***
SODIUM-4-CHLOROACETOPHENONE OXIMATE
mf: $C_8H_4ClNNaO$ mw: 191.59

$ClC_6H_4C(:NONa)CH_3$

SAFETY PROFILE: Explodes on contact with air. When heated to decomposition it emits toxic fumes of Cl^-, NO_x, and Na_2O. See also SODIUM COMPOUNDS.

SFV000 ***HR: 3***
SODIUM CHLOROACETYLIDE
mf: C_2ClNa mw: 82.47

SYN: SODIUM CHLOROETHYNIDE

SAFETY PROFILE: It is dangerously explosive in the solid state. When heated to decomposition it emits very toxic fumes of Na_2O and Cl^-. See also ACETYLIDES.

SFV250 CAS:4312-97-4 ***HR: 3***
SODIUM cis-β-CHLOROACRYLATE
mf: $C_3H_2ClO_2$•Na mw: 128.49

SYNS: ACROFOL ◇ AKROFOL ◇ cis-β-CHLOROACRYLIC ACID SODIUM SALT ◇ PREP ◇ PREP-DEFOLIANT ◇ SODIUM cis-3-CHLOROACRYLATE ◇ UC 20,299 ◇ UNION CARBIDE 20299

TOXICITY DATA with REFERENCE
orl-rat LD50:320 mg/kg SCIEAS 141,1192,63
orl-mus LD50:750 mg/kg 31ZOAD 1,386,68
skn-rbt LD50:400 mg/kg PCOC** -,1017,66
orl-ckn LD50:600 mg/kg 31ZOAD 1,386,68

SAFETY PROFILE: Poison by ingestion and skin contact. When heated to decomposition it emits toxic fumes of Cl^- and Na_2O.

SFV275 CAS:127-52-6 ***HR: 3***
SODIUM N-CHLOROBENZENESULFONAMIDE
mf: $C_6H_5ClNNaO_2S$ mw: 213.61

$NaN(Cl)SO_2C_6H_5$

SYN: CHLORAMINE B

SAFETY PROFILE: Explodes when heated to 185°C. When heated to decomposition it emits toxic fumes of Cl^-, SO_x, NO_x, and Na_2O.

SFV300 CAS:54976-93-1 ***HR: 2***
SODIUM-4-CHLORO-2-METHYL PHENOXIDE
mf: C_7H_6ClNaO mw: 164.57

$NaOC_6H_3(CH_3)Cl$

SAFETY PROFILE: Undergoes rapid exothermic decomposition when heated to 200°C. When heated to decomposition it emits toxic fumes of Cl^- and Na_2O.

SFV500 CAS:63992-41-6 ***HR: 2***
SODIUM-2-CHLORO-6-PHENYL PHENATE
mf: $C_{12}H_8ClO$•Na mw: 226.64

SYNS: 6-CHLORO-2-PHENYLPHENOL, SODIUM SALT ◇ DOWICIDE 31 ◇ STOP-MOLD

TOXICITY DATA with REFERENCE
orl-rat LD50:3500 mg/kg 28ZEAL 4,110,69
orl-gpg LDLo:1800 mg/kg PCOC** -,1015,66

CONSENSUS REPORTS: Chlorophenol compounds are on the Community Right To Know List.

SAFETY PROFILE: Moderately toxic by ingestion. A fungicide. When heated to decomposition it emits toxic fumes of Cl^- and Na_2O. See also CHLOROPHENOLS.

SFV550 CAS:127-65-1 ***HR: 3***
SODIUM N-CHLORO-4-TOLUENE SULFONAMIDE
mf: $C_7H_7ClNNaO_2S$ mw: 227.64

$CH_3C_6H_4SO_2N(Cl)Na$

SYN: CHLORAMINE-T

SAFETY PROFILE: The anhydrous material explodes when heated to 175°C. Mixtures with calcium carbonate + isonitriles explode when warmed. When heated to decomposition it emits toxic fumes of Cl^-, NO_x, SO_x and Na_2O.

SFW000 CAS:361-09-1 ***HR: 3***
SODIUM CHOLATE
mf: $C_{24}H_{39}O_5$•Na mw: 430.62

SYNS: CHOLIC ACID, MONOSODIUM SALT ◇ CHOLIC ACID, SODIUM SALT ◇ DS-Na ◇ OX BILE EXTRACT ◇ PURIFIED OXGALL ◇ SODIUM CHOLIC ACID ◇ TRIHYDROXY-3,7,12-CHOLANATE de Na (FRENCH) ◇ (3-α,5-β,7-α,12-α)3,7,12-TRIHYDROXY-CHOLAN-24-OIC ACID, MONOSODIUM SALT

TOXICITY DATA with REFERENCE
ivn-mus LD50:200 mg/kg AIPTAK 90,18,52
ivn-rbt LDLo:25 mg/kg ZGEMAZ 52,779,26

CONSENSUS REPORTS: Reported in EPA TSCA Inventory.

SAFETY PROFILE: Poison by intravenous route. When heated to decomposition it emits toxic fumes of Na_2O. See also CHOLIC ACID.

SFW300 CAS:9082-07-9 ***HR: 2***
SODIUM CHONDROITIN POLYSULFATE

SYNS: CHONDROITIN POLYSULFATE SODIUM ◇ CHONDRON ◇ SODIUM CHONDROITIN SULFATE

TOXICITY DATA with REFERENCE
orl-mus TDLo:1500 mg/kg (7-12D preg):REP OYYAA2 6,589,72
orl-mus TDLo:30 g/kg (female 7-12D post):TER OYYAA2 6,589,72
ipr-rat LD50:2900 mg/kg OYYAA2 6,573,72
scu-rat LD50:3700 mg/kg OYYAA2 6,573,72
ipr-mus LD50:9800 mg/kg OYYAA2 6,573,72
ivn-mus LD50:4980 mg/kg KSRNAM 6,433,72

SAFETY PROFILE: Moderately toxic by intraperitoneal and subcutaneous routes. An experimental teratogen. Experimental reproductive effects. When heated to decomposition it emits toxic fumes of SO_x and Na_2O.

SFW500 CAS:13517-17-4 ***HR: 3***
SODIUM CHROMATE DECAHYDRATE
mf: $CrO_42Na \cdot 10H_2O$ mw: 342.18

SYN: CHROMIC ACID, DISODIUM SALT, DECAHYDRATE

CONSENSUS REPORTS: Chromium and its compounds are on the Community Right To Know List.

OSHA PEL: CL 0.1 mg(CrO_3)/m^3
ACGIH TLV: TWA 0.05 mg(Cr)/m^3
NIOSH REL: TWA 0.025 mg(Cr(VI))/m^3; CL 0.05 mg/m^3/15M

SAFETY PROFILE: Confirmed human carcinogen. When heated to decomposition it emits toxic fumes of Na_2O. See also CHROMIUM COMPOUNDS and SODIUM CHROMATE.

SFX000 CAS:538-42-1 ***HR: 2***
SODIUM CINNAMATE
mf: $C_9H_7O_2 \cdot Na$ mw: 170.15

PROP: White crystals or powder. Mp: decomp @ >115°; sol in hot or cold water, alc, glycerol.

SYN: CINNAMIC ACID, SODIUM SALT

TOXICITY DATA with REFERENCE
ipr-mus LD50:2000 mg/kg PHMCAA 3,62,61

CONSENSUS REPORTS: Reported in EPA TSCA Inventory.

SAFETY PROFILE: Moderately toxic by intraperitoneal route. Combustible when exposed to heat or flame. When heated to decomposition it emits toxic fumes of Na_2O.

SFX725 CAS:55049-48-4 ***HR: 3***
SODIUM CITRATE, POTASSIUM CITRATE, CITRIC ACID (2:2:1)

SYNS: CG-120 ◇ URALYT-U

TOXICITY DATA with REFERENCE
orl-rat LD50:7200 mg/kg OYYAA2 21,715,81
scu-rat LD50:2900 mg/kg OYYAA2 21,715,81
ivn-rat LD50:143 mg/kg OYYAA2 21,715,81
orl-mus LD50:9100 mg/kg OYYAA2 21,715,81
scu-mus LD50:2160 mg/kg OYYAA2 21,715,81
ivn-mus LD50:231 mg/kg OYYAA2 21,715,81

SAFETY PROFILE: Poison by intravenous route. Moderately toxic by subcutaneous route. Mildly toxic by ingestion. When heated to decomposition it emits toxic fumes of K_2O and Na_2O.

SFX750 CAS:13600-98-1 ***HR: D***
SODIUM COBALTINITRITE
mf: $CoN_6O_{12} \cdot 3Na$ mw: 403.96

SYNS: COBALTATE(3-), HEXAKIS(NITRITO-N)-, TRISODIUM (OC-6-11)- (9CI) ◇ COBALTATE(3-), HEXANITRO-, TRISODIUM ◇ SODIUM HEXANITROCOBALTATE ◇ SODIUM NITROCOBALTATE (III) ◇ TRISODIUM HEXANITRITOCOBALTATE ◇ TRISODIUM HEXANITRITOCOBALTATE(3-) ◇ TRISODIUM HEXANITROCOBALTATE ◇ TRISODIUM HEXANITROCOBALTATE(3-)

TOXICITY DATA with REFERENCE
ipr-mus TDLo:50 mg/kg (female 10D post):TER JPMSAE 58,766,69

SAFETY PROFILE: Experimental reproductive effects. When heated to decomposition it emits toxic fumes of NO_x and Co.

SFY000 CAS:61791-42-2 ***HR: 1***
SODIUM COCOMETHYLAMINOETHYL-2-SULFONATE

SYN: SODIUM COCO METHYL TAURIDE

TOXICITY DATA with REFERENCE
skn-rbt 10 mg/24H DCTODJ 1,305,78
eye-rbt 2 mg DCTODJ 1,305,78

CONSENSUS REPORTS: Reported in EPA TSCA Inventory.

SAFETY PROFILE: A skin and eye irritant. When heated to decomposition it emits toxic fumes of Na_2O and SO_x.

SFY500 CAS:8068-28-8 ***HR: 3***
SODIUM COLISTINEMETHANESULFONATE
mf: $C_{49}H_{93}N_{13}O_{22}S_4 \cdot 4Na$ mw: 1436.75

SYNS: CLM ◇ COLIMYCIN M ◇ COLISTIMETHATE SODIUM ◇ COLISTIN SODIUM METHANESULFONATE ◇ COLISTIN SULFOMETHAT ◇ COLISTIN SULFOMETHATE SODIUM ◇ COLISTRIMETHATE SODIUM ◇ COLY-MYCIN INJECTABLE ◇ COLYSTINMETHANSULFONAT (GERMAN) ◇ PENTASODIUM COLISTINMETHANESULFONATE ◇ SODIUM COLISTIMETHATE ◇ SODIUM COLISTIN METHANESULFONATE ◇ W 1929

TOXICITY DATA with REFERENCE
ivn-mus TDLo:1250 mg/kg (female 6-15D post):REP NKRZAZ 29,1051,81
ivn-rat TDLo:275 mg/kg (7-17D preg):TER NKRZAZ 29,149,81
ims-hmn TDLo:56700 μg/kg/9D-I:KID AIMEAS 66,1052,67
ims-chd LDLo:90 mg/kg/18H-I:CNS,KID JAMAAP 207,2099,69
orl-rat LD50:5450 mg/kg OYYAA2 7,541,73
ipr-rat LD50:86 mg/kg OYYAA2 7,541,73
scu-rat LD50:87 mg/kg OYYAA2 7,541,73
ims-rat LD50:44 mg/kg TXAPA9 18,185,71
ipr-mus LD50:126 mg/kg NIIRDN 6,277,82
scu-mus LD50:138 mg/kg NIIRDN 6,277,82
ivn-mus LD50:222 mg/kg NIIRDN 6,277,82
ims-mus LD50:748 mg/kg OYYAA2 21,737,81
ims-rbt LD50:465 mg/kg OYYAA2 21,737,81
ims-gpg LD50:444 mg/kg OYYAA2 21,737,81

CONSENSUS REPORTS: Reported in EPA TSCA Inventory.

SAFETY PROFILE: Human poison by intramuscular route. Experimental poison by intramuscular, intraperitoneal, subcutaneous, and intravenous routes. Mildly toxic by ingestion. An experimental teratogen. Human systemic effects by intramuscular route: convulsions or effect on seizure threshold, change in motor activity, change in kidney tubules and urine volume decrease or anuria. Experimental reproductive effects. Used as an antibiotic. When heated to decomposition it emits very toxic fumes of NO_x, SO_x, and Na_2O.

SFZ000 ***HR: D***
SODIUM COMPOUNDS

SAFETY PROFILE: Variable toxicity. Sodium ion as such is practically nontoxic. The toxicity of sodium compounds is frequently, though not always, due to the anion involved. The hydroxide is very corrosive, being strongly basic. Even here it is the concentration of hydroxyl ion which is responsible for the caustic action of this material. When heated to decomposition it emits toxic fumes of Na_2O.

SFZ100 CAS:14264-31-4 ***HR: 3***
SODIUM CUPROCYANIDE (DOT)
DOT: UN 2316/UN 2317
mf: C_3CuN_3•2Na mw: 187.58

SYNS: COPPER SODIUM CYANIDE ◇ CUPRATE(2-), TRIS(CYANO-C)-, DISODIUM ◇ SODIUM CUPROCYANIDE, solution (DOT)

CONSENSUS REPORTS: Reported in EPA TSCA Inventory.
ACGIH TLV: TWA 1 mg(Cu)/m^3

DOT Classification: Poison B; Label: Poison

SAFETY PROFILE: A poison. When heated to decomposition it emits toxic fumes of CN^-.

SGA500 CAS:143-33-9 ***HR: 3***
SODIUM CYANIDE
DOT: UN 1689
mf: CNNa mw: 49.01

PROP: White, deliquescent, crystalline powder. Mp: 563.7°, bp: 1496°, vap press: 1 mm @ 817°.

SYNS: CIANURO di SODIO (ITALIAN) ◇ CYANIDE of SODIUM ◇ CYANOBRIK ◇ CYANOGRAN ◇ CYANURE de SODIUM (FRENCH) ◇ CYMAG ◇ HYDROCYANIC ACID, SODIUM SALT ◇ KYANID SODNY (CZECH) ◇ RCRA WASTE NUMBER P106 ◇ SODIUM CYANIDE, solid and solution (DOT)

TOXICITY DATA with REFERENCE
imp-ham TDLo:5999 mg/kg (6-9D preg):TER TXAPA9 64,456,82
imp-ham TDLo:5928 mg/kg (6-9D preg):REP TXAPA9 64,456,82
orl-man TDLo:714 μg/kg:CNS APTOA6 20,291,63
orl-man LDLo:6557 μg/kg:CNS,GIT APTOA6 1,18,45
orl-hmn LDLo:2857 μg/kg 34ZIAG -,191,69
unr-man LDLo:2206 μg/kg 85DCAI 2,73,70
orl-rat LD50:6440 μg/kg 28ZPAK -,13,72
ipr-rat LD50:4300 μg/kg GISAAA 43(12),90,78
ipr-mus LD50:5881 μg/kg TXAPA9 15,505,69
scu-mus LD50:3660 μg/kg AJPHAP 179,60,54
scu-dog LDLo:6 mg/kg JAMAAP 100,1920,33
ivn-dog LDLo:1300 μg/kg FEPRA7 6,349,47
scu-rbt LDLo:2200 μg/kg JAMAAP 100,1920,33
ims-rbt LD50:1666 μg/kg JACTDZ 1(3),120,82
ocu-rbt LD50:5048 μg/kg JACTDZ 1(3),120,82
par-frg LDLo:60 mg/kg AEPPAE 166,437,32

CONSENSUS REPORTS: Cyanide and its compounds are on the Community Right To Know List. Reported in EPA TSCA Inventory.

OSHA PEL: TWA 5 mg(CN)/m^3 (skin)
ACGIH TLV: TWA 5 mg(CN)/m^3 (skin)
DFG MAK: 5 mg(CN)/m^3
NIOSH REL: CL 5 mg(CN)/m^3/10M
DOT Classification: Poison B; Label: Poison, solid and solution

SAFETY PROFILE: A deadly human poison by ingestion. A deadly experimental poison by ingestion, intraperitoneal, subcutaneous, intravenous, parenteral, intramuscular, and ocular routes. An experimental teratogen. Human systemic effects by ingestion: hallucinations, distorted perceptions, muscle weakness and gastritis. Experimental reproductive effects.

The volatile cyanides resemble hydrocyanic acid physiologically, inhibiting tissue oxidation and causing death through asphyxia. Cyanogen is probably as toxic as hydrocyanic acid; the nitriles are generally considered somewhat less toxic, probably because of their lower volatility. The nonvolatile cyanide salts appear to be relatively nonhazardous systemically, so long as they are not ingested and care is taken to prevent the formation of hydrocyanic acid. Workers, such as electroplaters and picklers, who are daily exposed to cyanide solutions may develop a "cyanide" rash, characterized by itching and by macular, papular, and vesicular eruptions. Frequently there is secondary infection. Exposure to small amounts of cyanide compounds over long periods of time is reported to cause loss of appetite, headache, weakness, nausea, dizziness, and symptoms of irritation of the upper respiratory tract and eyes.

Flammable by chemical reaction with heat, moisture, acid. Many cyanides evolve hydrocyanic acid rather easily. This is a flammable gas and is highly toxic. Carbon dioxide from the air is sufficiently acidic to liberate hydrocyanic acid from cyanide solutions. Explodes if melted with nitrite or chlorate @ about 450°. Violent reaction with F_2; Mg; nitrates; HNO_3; nitrites. Upon contact with acid, acid fumes, water or steam, they will produce toxic and flammable vapors of CN^- and Na_2O. Used in the extraction of gold and silver ores, in electroplating, and in insecticides. See also CYANIDE and HYDROCYANIC ACID.

SGB000 CAS:143-33-9 ***HR: 3***
SODIUM CYANIDE (solution)
DOT: UN 1689
mf: CNNa mw: 49.01

CONSENSUS REPORTS: Cyanide and its compounds are on the Community Right To Know List. EPA Extremely Hazardous Substances List. Reported in EPA TSCA Inventory.

DOT Classification: Poison B; Label: Poison

SAFETY PROFILE: A deadly poison. When heated to decomposition it emits toxic fumes of CN^- and Na_2O. See also SODIUM CYANIDE.

SGB500 CAS:23239-41-0 ***HR: 3***
SODIUM-7-(2-CYANOACETAMIDO)CEPHALOSPORANIC ACID
mf: $C_{13}H_{12}N_3O_6S$•Na mw: 361.33

SYNS: Ba 36278 ◇ CEFACETRILE SODIUM ◇ CELOSPOR ◇ CEPHACETRILE SODIUM ◇ CIBA 36278-BA ◇ SODIUM CEFACETRIL ◇ SODIUM CEPHACETRILE

TOXICITY DATA with REFERENCE
scu-rat TDLo:364 g/kg (female 26W pre):REP JZKEDZ 2,1,76
orl-rat LD50:15100 mg/kg IYKEDH 9,222,78
scu-rat LD50:11600 mg/kg AACHAX -,150,70
ivn-rat LD50:3100 mg/kg JZKEDZ 2,1,76
ims-rat LD50:7700 mg/kg JZKEDZ 2,1,76
scu-mus LD50:7500 mg/kg ARZNAD 24,1459,74
ivn-mus LD50:3700 mg/kg ARZNAD 24,1459,74
orl-dog LDLo:500 mg/kg AACHAX -,150,70
ims-dog LDLo:500 mg/kg AACHAX -,150,70
ivn-rbt LDLo:300 mg/kg AACHAX -,150,70
scu-gpg LD50:240 mg/kg AACHAX -,150,70

CONSENSUS REPORTS: Cyanide and its compounds are on the Community Right To Know List.

SAFETY PROFILE: Poison by intravenous and subcutaneous routes. Moderately toxic by ingestion and intramuscular routes. Experimental reproductive effects. When heated to decomposition it emits very toxic fumes of NO_x, SO_x, Na_2O, and CN^-. See also NITRILES.

SGC000 CAS:139-05-9 ***HR: 2***
SODIUM CYCLAMATE
mf: $C_6H_{12}NO_3S$•Na mw: 201.24

PROP: White, crystalline powder; practically odorless. Sol in water; almost insol in alc, benzene, chloroform, and ether.

SYNS: ASSUGRIN ◇ ASSUGRIN FEINUSS ◇ ASSUGRIN VOLLSUSS ◇ ASUGRYN ◇ CYCLAMATE ◇ CYCLAMATE SODIUM ◇ CYCLAMIC ACID SODIUM SALT ◇ CYCLOHEXANESULFAMIC ACID, MONOSODIUM SALT ◇ CYCLOHEXANESULPHAMIC ACID, MONOSODIUM SALT ◇ CYCLOHEXYL SULPHAMATE SODIUM ◇ DULZOR-ETAS ◇ HACHI-SUGAR ◇ IBIOSUC ◇ NATREEN ◇ NATRIUMZYKLAMATE (GERMAN) ◇ SODIUM CYCLOHEXANESULFAMATE ◇ SODIUM CYCLOHEXANESULPHAMATE ◇ SODIUM CYCLOHEXYL AMIDOSULPHATE ◇ SODIUM CYCLOHEXYL SULFAMATE ◇ SODIUM CYCLOHEXYL SULFAMIDATE ◇ SODIUM CYCLOHEXYL SULPHAMATE ◇ SODIUM SUCARYL ◇ SUCARYL SODIUM ◇ SUCCARIL ◇ SUCROSA ◇ SUESSETTE ◇ SUESTAMIN ◇ SUGARIN ◇ SUGARON

TOXICITY DATA with REFERENCE
cyt-hmn:fbr 500 mg/L ACYTAN 16,41,72
cyt-ham:fbr 10 mg/L MUREAV 39,1,76
orl-mus TDLo:420 g/kg (MGN):REP TXCYAC 8,285,77
orl-mus TDLo:180 mg/kg (7D preg):TER IIZAAX 16,330,64
orl-rat TDLo:63 g/kg/9W-C:NEO CNREA8 37,2943,77
orl-rat TD:610 g/kg/87W-C:ETA CBINA8 11,255,75
orl-rat LD50:15250 mg/kg FCTXAV 6,313,68
ipr-rat LD50:1350 mg/kg NCIBR* NIH-NCI-E-68-1311,10,73
ivn-rat LD50:3500 mg/kg FCTXAV 6,313,68
orl-mus LD50:17 g/kg FCTXAV 6,313,68

ipr-mus LD50:1154 mg/kg NCIBR* NIH-NCI-E-68-1311,10,73
ivn-mus LD50:4800 mg/kg FCTXAV 6,313,68

CONSENSUS REPORTS: IARC Cancer Review: Group 3 IMEMDT 7,178,87; Animal Limited Evidence IMEMDT 22,55,80. Reported in EPA TSCA Inventory. EPA Genetic Toxicology Program.

SAFETY PROFILE: Moderately toxic by intravenous and intraperitoneal routes. Mildly toxic by ingestion. Experimental reproductive effects. Questionable carcinogen with experimental neoplastigenic, tumorigenic, and teratogenic data. Human mutation data reported. When heated to decomposition it emits very toxic fumes of Na_2O, SO_x, and NO_x.

SGD000 CAS:4418-26-2 ***HR: 3***
SODIUM DEHYDROACETIC ACID
mf: $C_8H_7O_4$•Na mw: 190.14

PROP: White powder; odorless with slt characteristic taste. Mp: 109-111°. Sol in water, propylene glycol, glycerin.

SYNS: DEHYDROACETIC ACID, SODIUM SALT ◇ DHA-SODIUM ◇ HARVEN ◇ 4-HEXENOIC ACID, 2-ACETYL-5-HYDROXY-3-OXO, Δ-LACTONE, SODIUM derivative ◇ 3-(1-HYDROXYETHYLIDENE)-6-METHYL-2H-PYRAN-2,4(3H)-DIONE, SODIUM SALT ◇ SODIUM DEHYDROACETATE (FCC)

TOXICITY DATA with REFERENCE
eye-rbt 100 mg MLD JACTDZ 4(3),123,85
cyt-ham:fbr 30 g/L FCTOD7 22,623,84
cyt-ham:lng 2500 mg/L GMCRDC 27,95,81
orl-mus TDLo:2 g/kg (6-15D preg):TER NKEZA4 27,91,80
orl-mus TDLo:2 g/kg (6-15D preg):REP NKEZA4 27,91,80
orl-rat LD50:500 mg/kg JACTDZ 4(3),123,85
orl-mus LD50:1050 mg/kg JACTDZ 4(3),123,85
ivn-dog LD50:400 mg/kg JPETAB 99,69,50
orl-rat LD50:500 mg/kg LONZA# 02JUN80

CONSENSUS REPORTS: Reported in EPA TSCA Inventory. EPA Genetic Toxicology Program.

SAFETY PROFILE: Poison by intravenous route. Moderately toxic by ingestion. An experimental teratogen. Experimental reproductive effects. An eye irritant. Mutation data reported. When heated to decomposition it emits toxic fumes of Na_2O.

SGD500 CAS:145-41-5 ***HR: 2***
SODIUM DEHYDROCHOLATE
mf: $C_{24}H_{34}O_5$•Na mw: 425.57

SYNS: BILIRON ◇ CARACHOL ◇ DECHOLIN SODIUM SALT ◇ DEHYDROCHOLATE SODIUM ◇ DEHYDROCHOLIC ACID, SODIUM SALT ◇ DILABIL SODIUM ◇ DYCHOLIUM ◇ NATRIUM-DEHYDROCHOLAT (GERMAN) ◇ SODIUM-3,7,12-TRIKETOCHOLANATE ◇ SODIUM-3,7,12-TRIOXO-5-β-CHOLANATE ◇ SUPRACHOL ◇ TRICETO 3-7-12 CHOLANATE de Na (FRENCH) ◇ 3,7,12-TRIOXO-5-β-CHOLAN-24-OIC ACID MONOSODIUM SALT

TOXICITY DATA with REFERENCE
orl-rat LD50:3755 mg/kg CWXXBX #2630169
ipr-rat LD50:1100 mg/kg CHTPBA 5,188,70
scu-rat LD50:2300 mg/kg YOIZA3 25(1),70,74
ivn-rat LD50:887 mg/kg RPOBAR 2,280,70
orl-mus LD50:2150 mg/kg YOIZA3 25(1),70,74
ipr-mus LD50:1480 mg/kg RPOBAR 2,280,70
scu-mus LD50:1350 mg/kg YOIZA3 25(1),70,74
ivn-mus LD50:1100 mg/kg YOIZA3 25(1),70,74
ivn-rbt LDLo:550 mg/kg ZGEMAZ 52,779,26

CONSENSUS REPORTS: Reported in EPA TSCA Inventory.

SAFETY PROFILE: Moderately toxic by ingestion, intraperitoneal, subcutaneous, and intravenous routes. When heated to decomposition it emits toxic fumes of Na_2O.

SGE000 CAS:302-95-4 ***HR: 3***
SODIUM DESOXYCHOLATE
mf: $C_{24}H_{39}O_4$•Na mw: 414.62

SYNS: DEOXYCHOLATE SODIUM ◇ DEOXYCHOLIC ACID SODIUM SALT ◇ DIHYDROXY 3-12 CHOLANATE de Na (FRENCH) ◇ (3-α,5-β,12-α)-3,12-DIHYDROXY-CHOLAN-24-OIC ACID MONOSODIUM SALT ◇ 3-α,12-α-DIHYDROXY-5-β-CHOLAN-24-OIC ACID SODIUM SALT ◇ NA-DESOXYCHOLAT (GERMAN) ◇ NATRIUM-3-α,12-α-DIHYDROXYCHOLANAT (GERMAN) ◇ SODIUM DEOXYCHOLATE ◇ SODIUM DEOXYCHOLIC ACID

TOXICITY DATA with REFERENCE
mmo-esc 5 pph/3H CSHSAZ 12,256,47
cyt-asn 640 mg/L MUREAV 93,101,82
dns-rat-rec 10 μmol/kg GANNA2 75,29,84
ipr-rat TDLo:1400 mg/kg (28D male):REP OYYAA2 3,45,69
orl-rat LD50:1370 mg/kg OYYAA2 3,45,69
ipr-rat LD50:123 mg/kg OYYAA2 3,45,69
scu-rat LD50:2430 mg/kg OYYAA2 3,45,69
ivn-rat LD50:150 mg/kg ARZNAD 20,323,70
orl-mus LD50:1050 mg/kg ESKHA5 (103),29,85
ipr-mus LD50:36 mg/kg ESKHA5 (103),29,85
scu-mus LD50:815 mg/kg OYYAA2 3,45,69
ivn-mus LD50:107 mg/kg AIPTAK 90,18,52
ivn-rbt LDLo:7500 μg/kg ZGEMAZ 52,779,26

CONSENSUS REPORTS: Reported in EPA TSCA Inventory.

SAFETY PROFILE: Poison by intraperitoneal and intravenous routes. Moderately toxic by ingestion and subcutaneous routes. Experimental reproductive effects. Mutation data reported. When heated to decomposition it emits toxic fumes of Na_2O.

SGF000 CAS:63041-43-0 ***HR: 3***
SODIUM-1,2:5,6-DIBENZANTHRACENE-9,10-endo-α,β-SUCCINATE
mf: $C_{26}H_{16}O_4$•2Na mw: 438.40

SYN: 1,2,5,6-DIBENZANTHRACENE-9,10-endo-α,β-SUCCINIC ACID, SODIUM SALT

TOXICITY DATA with REFERENCE
mmo-esc 500 ppm CRSBAW 142,453,48
scu-mus TDLo:2304 mg/kg/16W-I:CAR JPBAA7 54,321,42
ipr-mus TDLo:900 mg/kg/8W-I:ETA AJCAA7 27,267,36

SAFETY PROFILE: Questionable carcinogen with experimental carcinogenic and tumorigenic data. Mutation data reported. When heated to decomposition it emits toxic fumes of Na_2O.

SGF500 CAS:136-30-1 ***HR: 3***
SODIUM DIBUTYLDITHIOCARBAMATE
mf: $C_9H_{18}NS_2$•Na mw: 227.39

SYNS: BUTYL NAMATE ◇ DIBUTYLDITHIOCARBAMIC ACID SODIUM SALT ◇ PENNAC ◇ SODIUM DBDT ◇ TEPIDONE ◇ TEPIDONE RUBBER ACCELERATOR ◇ USAF B-35 ◇ VULCACURE

TOXICITY DATA with REFERENCE
ipr-mus LD50:300 mg/kg NTIS** AD277-689

CONSENSUS REPORTS: Reported in EPA TSCA Inventory.

SAFETY PROFILE: Poison by intraperitoneal route. When heated to decomposition it emits very toxic fumes of NO_x, SO_x, and Na_2O. See also CARBAMATES.

SGG000 CAS:2156-56-1 ***HR: 1***
SODIUM DICHLOROACETATE
mf: $C_2HCl_2O_2$•Na mw: 150.92

SYNS: DICHLOROACETATE SODIUM SALT ◇ DICHLOROACETIC ACID SODIUM SALT ◇ DICHLOROCTAN SODNY (CZECH)

TOXICITY DATA with REFERENCE
mmo-sat 5 μg/plate AJCNAC 33,1179,80
mma-sat 5 μg/plate AJCNAC 33,1179,80
orl-rat TDLo:182 g/kg (13W male):REP TXAPA9 57,273,81
orl-rat LD50:5281 mg/kg JIHTAB 23,78,41
orl-mus LD50:4845 mg/kg JPETAB 222,501,82

SAFETY PROFILE: Mildly toxic by ingestion. Experimental reproductive effects. Mutation data reported. When heated to decomposition it emits toxic fumes of Cl^- and Na_2O.

SGG500 CAS:2893-78-9 ***HR: 2***
SODIUM DICHLOROCYANURATE
DOT: UN 2465
mf: $C_3HCl_2N_3O_3$•Na mw: 220.96

PROP: White crystals; chlorine odor. Mp: 230-250°, water-sol.

SYNS: ACL 60 ◇ CDB 63 ◇ DICHLOROISOCYANURIC ACID SODIUM SALT (DOT) ◇ DIKONIT ◇ DIMANIN C ◇ FI CLOR 60S ◇ OCI 56 ◇ SDIC ◇ SIMPLA ◇ SODIUM DICHLORISOCYANURATE ◇ SODIUM DICHLOROISOCYANURATE ◇ SODIUM-1,3-DICHLORO-1,3,5-TRIAZINE-2,4-DIONE-6-OXIDE ◇ 1-SODIUM-3,5-DICHLORO-s-TRIAZINE-2,4,6-TRIONE ◇ 1-SODIUM-3,5-DICHLORO-1,3,5-TRIAZINE-2,4,6-TRIONE ◇ SODIUM DICHLORO-s-TRIAZINETRIONE, dry, containing more than 39% available chlorine (DOT) ◇ SODIUM SALT of DICHLORO-s-TRIAZINETRIONE

TOXICITY DATA with REFERENCE
skn-rbt 500 mg/34H MLD MONS** -,-,72
skn-rbt 500 mg SEV 34ZIAG -,167,69
eye-rbt 10 mg/34H rns SEV MONS** -,-,72
orl-mus TDLo:4 g/kg (6-15D preg):REP YKRYAH 13,353,80
orl-mus TDLo:4 g/kg (6-15D preg):TER YKRYAH 13,353,80
orl-hmn LDLo:3570 mg/kg:GIT 34ZIAG -,167,69

CONSENSUS REPORTS: Reported in EPA TSCA Inventory.

DOT Classification: Oxidizer; Label: Oxidizer

SAFETY PROFILE: Moderately toxic to humans and animals by ingestion. An experimental teratogen. Experimental reproductive effects. A severe skin and eye irritant. Human systemic effects by ingestion: ulceration or bleeding from stomach. The other main toxic effects were gastrointestinal irritation, salivation, lacrimation, dyspnea, weakness, emaciation, lethargy, diarrhea, coma and (following very high dosage) deaths after 1-8 days, with autopsy showing irritation of stomach and gastrointestinal tract, liver dysfunction, and lung congestion. The concentrated material may be a little more toxic, due to greater gastrointestinal irritation. In the dry form, it is not appreciably irritating to dry skin. However, when moist, the concentrated material is irritating to skin, and also may cause severe eye irritation.

A powerful oxidizer. Incompatible with combustible materials; ammonium salts; nitrogenous materials. Used to chlorinate swimming pools and in cleaning, bleaching, disinfecting, sanitizing. When heated to decomposition it emits very toxic fumes of Cl^-, NO_x, and Na_2O.

SGH500 CAS:2702-72-9 ***HR: 3***
SODIUM-2,4-DICHLOROPHENOXYACETATE
mf: $C_8H_5Cl_2O_3$•Na mw: 243.02

SYNS: AGRION ◇ 2,4-DICHLOROPHENOXYACETIC ACID, SODIUM SALT ◇ DICONIRT D ◇ 2,4-D SODIUM SALT ◇ FERNOXENE ◇ HORMIT ◇ PIELIK E ◇ SODIUM-2,4-D ◇ SPRAY-HORMITE ◇ SPRITZ-HORMIT

TOXICITY DATA with REFERENCE
dns-ckn-oth 2500 μmol/L ARTODN 33,91,75

mrc-sce 300 mg/L MUREAV 42,3,77
orl-rat TDLo:100 μg/kg (10D preg):TER GISAAA 44(4),70,79
ihl-man TCLo:23 mg/m³/3-5Y-I:CNS,GIT,LIV HYSAAV 31,383,66
orl-rat LD50:555 mg/kg GTPZAB 17(3),35,73
skn-rat LD50:1500 mg/kg 85JCAE -,544,86
ipr-rat LD50:424 mg/kg BCTKAG 14,17,81
orl-mus LD50:375 mg/kg JIDHAN 29,85,47
ipr-mus LD50:375 mg/kg JIDHAN 29,85,47
scu-mus LD50:280 mg/kg AEHLAU 7,202,63
orl-rbt LD50:800 mg/kg JIDHAN 29,85,47
ipr-rbt LD50:400 mg/kg JIDHAN 29,85,47
ivn-rbt LD50:400 mg/kg JIDHAN 29,85,47

CONSENSUS REPORTS: Reported in EPA TSCA Inventory. EPA Genetic Toxicology Program.

SAFETY PROFILE: Poison by ingestion, intraperitoneal, subcutaneous, and intravenous routes. Moderately toxic by skin contact. An experimental teratogen. Human systemic effects by inhalation: anorexia, gastrointestinal, and liver changes. Mutation data reported. When heated to decomposition it emits toxic fumes of Cl^- and Na_2O.

SGI000 CAS:10588-01-9 ***HR: 3***
SODIUM DICHROMATE
DOT: UN 1479
mf: Cr_2O_7•2Na mw: 261.98

$Na_2(OCrO_2OCrO_2O)$

PROP: Anhydrous. Mp: 356.7°, decomp @ about 400°, d: 2.35 @ 13°. Very sol in water.

SYNS: BICHROMATE de SODIUM (FRENCH) ◇ BICHROMATE of SODA ◇ CHROMIC ACID, DISODIUM SALT ◇ CHROMIUM SODIUM OXIDE ◇ DISODIUM DICHROMATE ◇ NATRIUMBICHROMAAT (DUTCH) ◇ NATRIUMDICHROMAAT (DUTCH) ◇ NATRIUMDICHROMAT (GERMAN) ◇ SODIO (DICROMATO di) (ITALIAN) ◇ SODIUM BICHROMATE ◇ SODIUM CHROMATE ◇ SODIUM DICHROMATE(VI) ◇ SODIUM DICHROMATE de (FRENCH)

TOXICITY DATA with REFERENCE
slt-dmg-orl 2340 μmol/L MUREAV 157,157,85
cyt-hmn:lym 2 μmol/L CARYAB 33,239,80
ihl-rat TCLo:252 μg/m³/78W-I:ETA TXCYAC 42,219,86
orl-cld LDLo:50 mg/kg YAKUD5 22,291,80
orl-cld TDLo:250 mg/kg:LNG,GIT,SKN AUPJB7 21,65,85
orl-rat LD50:50 mg/kg GTPZAB 22(8),38,78
scu-rat LDLo:80 mg/kg TOXID9 4,75,84
ivn-mus LDLo:26200 μg/kg EQSSDX 1,1,75
ivn-rbt LDLo:18400 μg/kg EQSSDX 1,1,75
skn-gpg LDLo:335 mg/kg AEHLAU 11,201,65
ipr-gpg LDLo:335 mg/kg AEHLAU 11,201,65
scu-gpg LDLo:23 mg/kg HBAMAK 4,1330,35

CONSENSUS REPORTS: NTP Fifth Annual Report on Carcinogens. IARC Cancer Review: Group 1 IMEMDT 7,165,87; Animal Inadequate Evidence IMEMDT 2,100,73; IMEMDT 23,205,80; Human Inadequate Evidence IMEMDT 23,205,80. Chromium and its compounds are on the Community Right To Know List. Reported in EPA TSCA Inventory. EPA Genetic Toxicology Program.

OSHA PEL: CL 0.1 mg/(CrO_3)/m³
ACGIH TLV: TWA 0.05 mg(Cr)/m³
NIOSH REL: TWA 0.025 mg(Cr(VI))/M³; CL 0.05 mg/M³/15M
DOT Classification: ORM-A; Label: None

SAFETY PROFILE: Confirmed carcinogen with experimental tumorigenic data. Poison by ingestion, skin contact, intravenous, intraperitoneal, and subcutaneous routes. Human systemic effects by ingestion: cough, nausea or vomiting, and sweating. Human mutation data reported. A caustic and irritant. A powerful oxidizer. Potentially explosive reaction with acetic anhydride; ethanol + sulfuric acid + heat; hydrazine. Violent reaction or ignition with boron + silicon (pyrotechnic); organic residues + sulfuric acid; 2-propanol + sulfuric acid; sulfuric acid + trinitrotoluene. Incompatible with hydroxylamine. When heated to decomposition it emits toxic fumes of Na_2O. See also CHROMIUM COMPOUNDS.

SGI500 CAS:7789-12-0 ***HR: 3***
SODIUM DICHROMATE DIHYDRATE
mf: $Cr_2Na_2O_7$•$2H_2O$ mw: 298.02

PROP: Red crystals. Mp: loses $2H_2O$ @ 100°, mp (anhydrous): 356.7°, bp: decomp @ 400°, d: 2.35 @ 13°.

SYN: CHROMIS ACID ($H_2Cr_2O_7$), DISODIUM SALT, DIHYDRATE (9CI)

TOXICITY DATA with REFERENCE
slt-dmg-orl 2340 μm/L MUREAV 157,157,85
itr-rat TDLo:163 mg/kg/130W-I:CAR EXPADD 30,129,86

CONSENSUS REPORTS: Chromium and its compounds are on the Community Right To Know List.

OSHA PEL: CL 0.1 mg(CrO_3)/m³
ACGIH TLV: TWA 0.05 mg(Cr)/m³
NIOSH REL: TWA 0.025 mg(Cr(VI))/m³; CL 0.05/15M

SAFETY PROFILE: Probably a poison. Questionable carcinogen with experimental carcinogenic data. A caustic irritant. Mutation data reported. It can react violently with hydrazine. When heated to decomposition it emits toxic fumes of Na_2O. See also SODIUM DICHROMATE and CHROMIUM COMPOUNDS.

SGJ000 CAS:148-18-5 ***HR: 3***
SODIUM DIETHYLDITHIOCARBAMATE
mf: $C_5H_{10}NS_2 \cdot Na$ mw: 171.27

PROP: Crystals. Mp: 95°, d: 1.1 @ 20°/20°, vap d: 5.9.

SYNS: CUPRAL ◇ DDC ◇ DEDC ◇ DEDK ◇ DIETHYLCARBAMODITHIOIC ACID, SODIUM SALT ◇ DIETHYLDITHIOCARBAMATE SODIUM ◇ DIETHYLDITHIOCARBAMIC ACID SODIUM ◇ DIETHYLDITHIOCARBAMIC ACID, SODIUM SALT ◇ DIETHYL SODIUM DITHIOCARBAMATE ◇ DITHIOCARB ◇ DITHIOCARBAMATE ◇ NCI-C02835 ◇ SODIUM DEDT ◇ SODIUM N,N-DIETHYLDITHIOCARBAMATE ◇ SODIUM SALT of N,N-DIETHYLDITHIOCARBAMIC ACID ◇ THIOCARB ◇ USAF EK-2596

TOXICITY DATA with REFERENCE
oms-omi 100 μmol/L BBACAQ 519,65,78
dnd-hmn:hla 100 μmol/L BBACAQ 519,65,78
cyt-rat-orl 5200 mg/kg MUREAV 53,212,78
dni-ckn:emb 4 μmol/L BBACAQ 519,65,78
oms-ckn:emb 40 μmol/L BBACAQ 519,65,78
scu-mus TDLo:1935 mg/kg (female 6-14D post):TER NTIS** PB223-160
scu-mus TDLo:1935 mg/kg (female 6-14D post):REP NTIS** PB223-160
orl-mus TDLo:76 g/kg/78W-I:NEO NTIS** PB223-159
orl-rat LD50:1500 mg/kg DRFUD4 6,225,81
ipr-rat LD50:1250 mg/kg DRFUD4 6,225,81
orl-mus LD50:1500 mg/kg DRFUD4 6,225,81
ipr-mus LD50:1302 mg/kg DRFUD4 6,225,81
scu-rbt LD50:500 mg/kg APTOA6 8,329,52

CONSENSUS REPORTS: IARC Cancer Review: Group 3 IMEMDT 7,56,87; Animal Inadequate Evidence IMEMDT 12,217,76. NCI Carcinogenesis Bioassay (feed); No Evidence: mouse, rat NCITR* NCI-CG-TR-172,79. Reported in EPA TSCA Inventory.

SAFETY PROFILE: Moderately toxic by ingestion, intraperitoneal, and subcutaneous routes. Experimental reproductive effects. Questionable carcinogen with experimental neoplastigenic and teratogenic data. Human mutation data reported. When heated to decomposition it emits very toxic fumes of NO_x, SO_x, and Na_2O. Used as a pesticide. See also CARBAMATES.

SGJ500 CAS:20624-25-3 ***HR: 2***
SODIUM DIETHYLDITHIOCARBAMATE TRIHYDRATE
mf: $C_5H_{10}NS_2 \cdot Na \cdot 3H_2O$ mw: 225.33

SYNS: DIETHYLDITHIOCARBAMIC ACID SODIUM SALT TRIHYDRATE ◇ DIETHYLDITHIOCARBAMIC SODIUM TRIHYDRATE ◇ DITHIOCARB

TOXICITY DATA with REFERENCE
dni-rat:lym 1 mg/L ARZNAD 35,1052,85
orl-rat LD50:1500 mg/kg AJMSA9 235,26,58
ipr-rat LD50:1500 mg/kg AJMSA9 235,26,58
orl-mus LD50:1500 mg/kg AJMSA9 235,26,58
ipr-mus LD50:1500 mg/kg AJMSA9 235,26,58
ipr-mus LD50:952 mg/kg EJPHAZ 9,183,70

SAFETY PROFILE: Moderately toxic by ingestion and intraperitoneal routes. Mutation data reported. When heated to decomposition it emits very toxic fumes of NO_x, SO_x, and Na_2O. See also CARBAMATES.

SGK600 CAS:34461-00-2 ***HR: 3***
SODIUM DIFORMYLNITROMETHANIDE HYDRATE
mf: $C_3H_2NNaO_4 \cdot H_2O$ mw: 157.06

SAFETY PROFILE: An impact-sensitive explosive solid. When heated to decomposition it emits toxic fumes of NO_x and Na_2O.

SGK800 CAS:22722-98-1 ***HR: 2***
SODIUM DIHYDROBIS(2-METHOXYETHOXY)ALUMINATE
mf: $C_6H_{16}AlNaO_4$ mw: 202.16

$Na[H_2Al(OC_2H_4OCH_3)_2]$

SAFETY PROFILE: Concentrated solutions (above 70%) may ignite when large surface areas are exposed to air. When heated to decomposition it emits toxic fumes of Na_2O. See also HYDRIDES and ALUMINUM COMPOUNDS.

SGM000 CAS:24167-76-8 ***HR: 3***
SODIUM DIHYDROGENPHOSPHIDE
mf: H_2NaP mw: 55.98

SAFETY PROFILE: Ignites spontaneously in air. When heated to decomposition it emits very toxic fumes of Na_2O and PO_x. See also PHOSPHINE.

SGM100 CAS:12275-58-0 ***HR: 3***
SODIUM-4,4-DIMETHOXY-1-aci-NITRO-3,5-DINITRO-2,5-CYCLOHEXADIENE
mf: $C_8H_8N_3NaO_8$ mw: 297.16

SAFETY PROFILE: Explodes when exposed to heat or flame. When heated to decomposition it emits toxic fumes of NO_x and Na_2O. See also NITRO COMPOUNDS of AROMATIC HYDROCARBONS.

SGM500 CAS:128-04-1 ***HR: 2***
SODIUM N,N-DIMETHYLDITHIOCARBAMATE
mf: $C_3H_6NS_2 \cdot Na$ mw: 143.21

PROP: Crystals.

SYNS: ACETO SDD 40 ◇ ALCOBAM NM ◇ BROGDEX 555 ◇ CARBON S ◇ DIBAM ◇ DIMETHYLDITHIOCARBAMIC ACID, SODIUM SALT ◇ DMDK ◇ METHYL NAMATE ◇ SDDC ◇ SHARSTOP 204 ◇ STA-FRESH 615 ◇ STERISEAL LIQUID #40 ◇ THIOSTOP N ◇ VINSTOP ◇ VULNOPOL NM ◇ WING STOP B

TOXICITY DATA with REFERENCE
mmo-sat 50 μg/plate MUREAV 116,185,83
mma-sat 5 μg/plate PCBRD2 141,407,84
orl-rat LD50:1000 mg/kg PCOC** -,1029,66
ipr-rat LD50:1000 mg/kg PHARAT 27,139,72
orl-mus LD50:1500 mg/kg FATOAO 28,230,65
ipr-mus LD50:573 mg/kg YKKZAJ 94,1419,74

CONSENSUS REPORTS: Reported in EPA TSCA Inventory.

SAFETY PROFILE: Moderately toxic by ingestion and intraperitoneal routes. Mutation data reported. When heated to decomposition it emits very toxic fumes of NO_x, SO_x, and Na_2O. See also CARBAMATES.

SGP500 CAS:6373-74-6 ***HR: 2***
SODIUM-4-(2,4-DINITROANILINO)DIPHENYLAMINE-2-SULFONATE
mf: $C_{18}H_{14}N_4O_7S{\cdot}Na$ mw: 453.41

SYNS: ACID FAST YELLOW AG ◇ ACID LEATHER LIGHT BROWN G ◇ ACID ORANGE No. 3 ◇ ACID YELLOW E ◇ AIREDALE YELLOW E ◇ AMIDO YELLOW EA-CF ◇ ANTHRALAN YELLOW RRT ◇ C.I. 10385 ◇ C.I. ACID ORANGE 3 ◇ DERMA YELLOW P ◇ ERIO FAST YELLOW AEN ◇ FAST LIGHT YELLOW E ◇ FENALAN YELLOW E ◇ KITON FAST YELLOW A ◇ LIGHT FAST YELLOW ES ◇ LISSAMINE FAST YELLOW AE ◇ NCI-C54911 ◇ NYLOMINE ACID YELLOW B-RD ◇ SUPERIAN YELLOW R ◇ TECTILON ORANGE 3GT ◇ TERTRACID LIGHT YELLOW 2R ◇ VONDACID FAST YELLOW AE ◇ XYLENE FAST YELLOW ES

TOXICITY DATA with REFERENCE
mmo-sat 100 μg/plate EMMUEG 11(Suppl 12),1,88
mma-sat 667 μg/plate EMMUEG 11(Suppl 12),1,88
orl-rat TDLo:386 g/kg/2Y-I:CAR NTPTR* NTP-TR-335,88

CONSENSUS REPORTS: NTP Carcinogenesis Studies (gavage): Clear Evidence: rat, NTPTR* NTP-TR-335,88; No Evidence: mouse NTPTR* NTP-TR-335,88. Reported in EPA TSCA Inventory.

SAFETY PROFILE: Questionable carcinogen with experimental carcinogenic data. Mutation data reported. When heated to decomposition it emits very toxic fumes of SO_x, NO_x, and Na_2O.

SGP600 CAS:25854-41-5 ***HR: 3***
SODIUM DINITROMETHANIDE
mf: $CHNNaO_4$ mw: 114.01

SAFETY PROFILE: An impact-sensitive explosive. When heated to decomposition it emits toxic fumes of NO_x and Na_2O. See also NITRO COMPOUNDS.

SGQ000 CAS:2783-96-2 ***HR: 3***
SODIUM-5-DINITROMETHYLTETRAZOLIDE
mf: $C_2HN_6NaO_4$ mw: 196.06

SAFETY PROFILE: Very explosive at its melting point (160°C). When heated to decomposition it emits very toxic fumes of NO_x and Na_2O.

SGQ100 CAS:38892-09-0 ***HR: 3***
SODIUM-2,4-DINITROPHENOXIDE
mf: $C_6H_3N_2NaO_5$ mw: 206.09

$NaOC_6H_3(NO_2)_2$

SAFETY PROFILE: Explosive. When heated to decomposition it emits toxic fumes of NO_x and Na_2O. See also NITRO COMPOUNDS.

SGQ500 CAS:573-58-0 ***HR: 3***
SODIUM DIPHENYLDIAZO-BIS(α-NAPHTHYLAMINESULFONATE)
mf: $C_{32}H_{24}N_6O_6S_2{\cdot}2Na$ mw: 698.72

SYNS: ATLANTIC CONGO RED ◇ ATUL CONGO RED ◇ AZOCARD RED CONGO ◇ BENZO CONGO RED ◇ BRASILAMINA CONGO 4B ◇ C.I. 22120 ◇ C.I. DIRECT RED 28 ◇ C.I. DIRECT RED 28, DISODIUM SALT ◇ CONGO RED ◇ COTTON RED L ◇ DIACOTTON CONGO RED ◇ DIRECT RED 28 ◇ ERIE CONGO 4B ◇ HISPAMIN CONGO 4B ◇ KAYAKU CONGO RED ◇ MITSUI CONGO RED ◇ PEERAMINE CONGO RED ◇ SUGAI CONGO RED ◇ TERTRODIRECT RED C ◇ TRISULFON CONGO RED ◇ VONDACEL RED CL

TOXICITY DATA with REFERENCE
mma-sat 33 μg/plate CRNGDP 3,21,82
dnd-esc 10 μmol/L MUREAV 89,95,81
orl-mus TDLo:5 g/kg (female 8-12D post):REP TOXID9 4,166,84
orl-mus TDLo:5 g/kg (female 8-12D post):TER TOXID9 4,166,84
orl-man LDLo:143 mg/kg:CVS 34ZIAG -,185,69
ivn-rat LDLo:160 mg/kg AJMSA9 198,73,39
ivn-mus LDLo:250 mg/kg JOIMA3 68,53,52
ivn-cat LDLo:100 mg/kg AJMSA9 198,73,39
ivn-rbt LDLo:230 mg/kg AJMSA9 198,73,39
ivn-gpn LDLo:120 mg/kg AJMSA9 198,73,39

CONSENSUS REPORTS: Reported in EPA TSCA Inventory.

SAFETY PROFILE: Human poison by ingestion with cardiovascular effects. Experimental poison by intravenous route. An experimental teratogen. Experimental reproductive effects. Mutation data reported. When heated to decomposition it emits very toxic fumes of NO_x, SO_x and Na_2O.

SGR500 CAS:22868-13-9 ***HR: 3***
SODIUM DISULFIDE
mf: Na_2S_2 mw: 110.11

SAFETY PROFILE: Probably highly toxic. Incompatible with diazonium salts. When heated to decomposition it emits very toxic fumes of Na_2O and SO_x. See also SULFIDES.

SGR800 CAS:141-52-6 *HR: 3*
SODIUM ETHOXIDE
mf: C_2H_5ONa mw: 68.05

SAFETY PROFILE: May ignite in moist air. When heated to decomposition it emits toxic fumes of Na_2O.

SGS000 CAS:73506-39-5 *HR: 3*
SODIUM ETHOXYACETYLIDE
mf: C_4H_5NaO mw: 92.07

$$NaC{\equiv}COCH_2CH_3$$

SAFETY PROFILE: Ignites spontaneously in air above −70°C. May explode spontaneously in air. When heated to decomposition it emits toxic fumes of Na_2O. See also ACETYLIDES.

SGS500 CAS:985-16-0 *HR: 2*
SODIUM-6-(2-ETHOXY-1-NAPHTHAMIDO)PENICILLANATE
mf: $C_{21}H_{22}N_2O_5S{\bullet}Na$ mw: 437.50

SYNS: BRL 1383 ◇ 6-(2-ETHOXY-1-NAPHTHAMIDO)PENICILLIN SODIUM ◇ NAFCILLIN SODIUM SALT ◇ NAFTOPEN ◇ SODIUM NAFCILLIN ◇ UNIPEN ◇ WY 3277

TOXICITY DATA with REFERENCE
ivn-wmn TDLo:4560 mg/kg/19D-I:BLD JAMAAP 232,1150,75
ivn-chd TDLo:450 mg/kg/3D:LIV,BLD AJDCAI 135,52,81
ivn-man TDLo:3100 mg/kg/18D-I:KID JAMAAP 244,2609,80
ivn-man TDLo:429 mg/kg/23D-I:BLD JAMAAP 232,1152,75
ipr-rat LD50:920 mg/kg TXAPA9 24,37,73
ipr-mus LD50:1600 mg/kg TXAPA9 24,37,73
ivn-mus LD50:1000 mg/kg RPTOAN 37,69,74

SAFETY PROFILE: Moderately toxic by intraperitoneal and intravenous routes. Human systemic effects by intravenous route: interstitial nephritis (kidney damage), proteinuria (protein in the urine), hematuria (blood in the urine), leukopenia (reduced number of leukocytes in the blood), rosinophilia, angranulocytosis (reduced levels of granulocytes) and liver changes. When heated to decomposition it emits very toxic fumes of NO_x, SO_x, and Na_2O.

SHE500 CAS:140-90-9 *HR: 3*
SODIUM ETHYLXANTHATE
mf: $C_3H_5OS_2{\bullet}Na$ mw: 144.19

PROP: Yellow powder; sol in alc, water.

SYNS: ETHYLXANTHIC ACID SODIUM SALT ◇ SODIUM ETHYLXANTHOGENATE ◇ SODIUM XANTHATE ◇ SODIUM XANTHOGENATE

CONSENSUS REPORTS: Reported in EPA TSCA Inventory.

SAFETY PROFILE: Flammable when exposed to heat or flame. When heated to decomposition it emits toxic fumes of Na_2O and SO_x. See also SULFIDES.

SHE700 *HR: 1*
SODIUM FERRIC PYROPHOSPHATE
mf: $Na_8Fe_4(P_2O_7)_5{\bullet}xH_2O$ mw: 1277.02

PROP: White to tan powder; odorless. Insol in water; sol in hydrochloric acid.

SYNS: FERRIC SODIUM PYROPHOSPHATE ◇ SODIUM IRON PYROPHOSPHATE

SAFETY PROFILE: A nuisance dust.

SHF000 CAS:15096-52-3 *HR: 3*
SODIUM FLUOALUMINATE
mf: $AlF_6{\bullet}3Na$ mw: 209.95

PROP: Very white, vitreous masses. Mp: 1000°, d: 2.95. Sol in concentrated H_2SO_4.

SYNS: ALUMINUM SODIUM FLUORIDE ◇ CRYOLITE ◇ ENT 24,984 ◇ KRYOLITH (GERMAN) ◇ NATRIUMALUMINUMFLUORID (GERMAN) ◇ NATRIUMHEXAFLUOROALUMINATE (GERMAN) ◇ SODIUM ALUMINOFLUORIDE ◇ SODIUM ALUMINUM FLUORIDE ◇ SODIUM HEXAFLUOROALUMINATE ◇ VILLIAUMITE

TOXICITY DATA with REFERENCE
orl-rat LD50:200 mg/kg AFDOAQ 15,122,51
orl-rbt LDLo:9 g/kg JIHTAB 30,92,48

CONSENSUS REPORTS: Reported in EPA TSCA Inventory.

OSHA PEL: TWA 2.5 $mg(F)/m^3$
ACGIH TLV: TWA 2 $mg(Al)/m^3$
NIOSH REL: TWA 2.5 $mg(F)/m^3$

SAFETY PROFILE: Poison by ingestion. Used as a pesticide. When heated to decomposition it emits toxic fumes of F^- and Na_2O. See also FLUORIDES.

SHF500 CAS:7681-49-4 *HR: 3*
SODIUM FLUORIDE
DOT: UN 1690
mf: FNa mw: 41.99

PROP: Clear, lustrous crystals or white powder or balls. Mp: 993°, bp: 1700°, d: 2 @ 41°, vap press: 1 mm @ 1077°.

SYNS: ALCOA SODIUM FLUORIDE ◇ ANTIBULIT ◇ CAVI-TROL ◇ CHEMIFLUOR ◇ CREDO ◇ DISODIUM DIFLUORIDE ◇ FDA 0101 ◇ F1-TABS ◇ FLORIDINE ◇ FLOROCID ◇ FLOZENGES ◇ FLUORAL ◇ FLUORIDENT ◇ FLUORID SODNY (CZECH) ◇ FLUORIGARD ◇ FLUORINEED ◇ FLUORINSE ◇ FLUORITAB ◇ FLUOR-O-KOTE ◇ FLUORURE de SODIUM (FRENCH) ◇ FLURA-GEL ◇ FLURCARE ◇ FUNGOL B ◇ GEL II ◇ GELUTION ◇ GLEEM ◇ IRADICAV

◇ KARIDIUM ◇ KARIGEL ◇ KARI-RINSE ◇ LEA-COV ◇ LEMOFLUR ◇ LURIDE ◇ NAFEEN ◇ NaFPAK ◇ Na FRINSE ◇ NATRIUM FLUORIDE ◇ NCI-C55221 ◇ NUFLUOR ◇ OSSALIN ◇ OSSIN ◇ PEDIAFLOR ◇ PEDIDENT ◇ PENNWHITE ◇ PERGANTENE ◇ PHOS-FLUR ◇ POINT TWO ◇ PREDENT ◇ RAFLUOR ◇ RESCUE SQUAD ◇ ROACH SALT ◇ SODIUM FLUORIDE, solid and solution (DOT) ◇ SODIUM FLUORURE (FRENCH) ◇ SODIUM HYDROFLUORIDE ◇ SODIUM MONOFLUORIDE ◇ SO-FLO ◇ STAY-FLO ◇ STUDAFLUOR ◇ SUPER-DENT ◇ T-FLUORIDE ◇ THERA-FLUR-N ◇ TRISODIUM TRIFLUORIDE ◇ VILLIAUMITE

TOXICITY DATA with REFERENCE
eye-rbt 20 mg/24H MOD 28ZPAK -,20,72
dnr-bcs 86 mg/L WATRAG 14,1613,80
dns-hmn:fbr 100 mg/L MUREAV 139,193,84
orl-mus TDLo:1240 mg/kg (male 8W pre):REP JONUAI 103,1319,73
ipr-rat TDLo:9 mg/kg (female 10-18D post):TER AOBIAR 17,371,72
orl-mus TDLo:14 mg/kg/43W-C:ETA IARC** 27,237,82
orl-wmn LDLo:90 mg/kg JAMAAP 81,811,23
orl-wmn LDLo:360 mg/kg:PUL,GIT JAMAAP 100,97,33
orl-wmn TDLo:7 mg/kg:EYE,PUL BMJOAE 1,886,36
orl-man TDLo:1662 mg/kg :CVS,PUL,GIT JAMAAP 222,816,72
orl-hmn TDLo:214 μg/kg:CNS,GIT RMLIAC 12,408,57
idr-hmn TDLo:14 μg/kg:PNS,IMM RMLIAC 12,408,57
orl-hmn LDLo:71 mg/kg:CNS,MSK MEIEDD 10,1235,83
unr-man LDLo:75 mg/kg FMCHA2 -,C264,89
orl-rat LD50:52 mg/kg NTIS** UR-3490-95
ipr-rat LD50:22 mg/kg XEURAQ UR-154,1951
scu-rat LD50:175 mg/kg OYYAA2 2,411,68
ivn-rat LD50:26 mg/kg IARC** 27,237,82
orl-mus LD50:57 mg/kg SHGKA3 80,1519,80
skn-mus LDLo:300 mg/kg AEPPAE 183,427,36
ipr-mus LD50:38 mg/kg IARC** 27,237,82
scu-mus LD50:70 mg/kg OYYAA2 2,411,68
ivn-mus LD50:50830 μg/kg IARC** 27,237,82
orl-dog LDLo:75 mg/kg 27ZWAY 7,-,38
ipr-dog LDLo:50 mg/kg XEURAQ UR-154,1951
scu-dog LDLo:155 mg/kg HBAMAK 4,1349,35
ivn-dog LDLo:80 mg/kg 27ZWAY 7,-,38
ims-dog LDLo:40 mg/kg 27ZWAY 7,-,38
ivn-mky LD50:26600 μg/kg APTOA6 22,99,65
scu-cat LDLo:14 mg/kg HBAMAK 4,1349,35
orl-rbt LD50:200 mg/kg JEENAI 61,751,68
ipr-rbt LDLo:250 mg/kg 14CYAT 2,835,63

CONSENSUS REPORTS: Reported in EPA TSCA Inventory. EPA Genetic Toxicology Program.

OSHA PEL: TWA 2.5 mg(F)/m^3
ACGIH TLV: TWA 2.5 mg(F)/m^3
NIOSH REL: TWA (Inorganic Fluorides) 2.5 mg(F)/m^3
DOT Classification: ORM-B; Label: None; Corrosive Material; Label: Corrosive, solution; Poison B; Label: St. Andrews Cross.

SAFETY PROFILE: Human poison by ingestion. Experimental poison by ingestion, skin contact, intravenous, intraperitoneal, subcutaneous, and intramuscular routes. Human systemic effects: changes in teeth and supporting structures, cyanosis, diarrhea, EKG changes, fluid intake, headache, hypermotility, increased immune response, muscle weakness, musculo-skeletal changes, nausea or vomiting, paresthesia, ptosis (drooping of the eyelid from sympathetic innervation), respiratory depression, salivary gland changes, tremors. Experimental teratogenic and reproductive effects. Human mutation data reported. A corrosive irritant to skin, eyes, and mucous membranes. Experimental reproductive effects. Questionable carcinogen with experimental tumorigenic data. It is very phytotoxic. When heated to decomposition it emits toxic fumes of F^- and Na_2O. Used in chemical cleaning, for fluoridation of drinking water, as a fungicide and insecticide. See also FLUORIDES.

SHG000 CAS:7681-49-4 ***HR: 2***
SODIUM FLUORIDE (solution)
DOT: UN 1690
mf: FNa mw: 41.99

CONSENSUS REPORTS: IARC Cancer Review: Group 3 IMEMDT 7,208,87; Animal Inadequate Evidence IMEMDT 27,237,82. Reported in EPA TSCA Inventory.

NIOSH REL: TWA 2.5 mg(F)/m^3
DOT Classification: Corrosive Material; Label: Corrosive

SAFETY PROFILE: A corrosive irritant to skin, eyes and mucous membranes. Questionable carcinogen. When heated to decomposition it emits very toxic fumes of F^- and Na_2O. See also SODIUM FLUORIDE.

SHG500 CAS:62-74-8 ***HR: 3***
SODIUM FLUOROACETATE
DOT: UN 2629
mf: $C_2H_2FO_2$•Na mw: 100.03

PROP: Fine, white powder. Sol in water.

SYNS: 1080 ◇ COMPOUND No. 1080 ◇ FLUORACETATO di (ITALIAN) ◇ FLUOROACETIC ACID, SODIUM SALT ◇ FLUORESSIGAEURE (GERMAN) ◇ FRATOL ◇ FURATOL ◇ MONOFLUORESSIGSAURES NATRIUM (GERMAN) ◇ NATRIUMFLUORACETAAT (DUTCH) ◇ NATRIUMFLUORACETAT (GERMAN) ◇ RATBANE 1080 ◇ RCRA WASTE NUMBER P058 ◇ SODIO, FLUORACETATO di (ITALIAN) ◇ SODIUM FLUOACETATE ◇ SODIUM FLUOACETIC ACID ◇ SODIUM FLUORACETATE de (FRENCH) ◇ SODIUM MONOFLUOROACETATE ◇ TL 869 ◇ YASOKNOCK

TOXICITY DATA with REFERENCE
orl-rat TDLo:210 μg/kg (3D male):REP JRPFA4 56,201,79
orl-hmn LDLo:714 μg/kg 34ZIAG -,542,69
unr-man LDLo:5 mg/kg AJPEAG 36,1427,46

orl-rat LD50:100 μg/kg AJPEAG 36,1427,46
ipr-rat LDLo:3 mg/kg JAPMA8 36,59,47
orl-mus LD50:500 μg/kg JAPMA8 37,307,48
ipr-mus LD50:7 mg/kg NEZAAQ 34,193,79
scu-mus LDLo:7200 μg/kg NDRC** 30101,2,45
orl-dog LD50:66 μg/kg JPETAB 101,82,50
ivn-dog LD50:60 μg/kg YKYUA6 31,1385,80
ivn-mky LD50:5 mg/kg AJPEAG 36,1427,46

CONSENSUS REPORTS: Reported in EPA TSCA Inventory. EPA Extremely Hazardous Substances List.

OSHA PEL: (Transitional: TWA 0.05 mg/m^3 (skin)) TWA 0.05 mg/m^3 (skin); STEL 0.15 mg/m^3 (skin)
ACGIH TLV: TWA 0.05 mg/m^3 (skin); STEL 0.15 mg/m^3 (skin)
DFG MAK: 0.05 mg/m^3
DOT Classification: Poison B: Label: Poison.

SAFETY PROFILE: A deadly human poison by ingestion. Experimental poison by ingestion, intraperitoneal, subcutaneous, and intravenous, routes. A very highly toxic water-soluble salt used mainly as an immediate action rodenticide. It is rapidly absorbed by the gastrointestinal tract but slowly by the skin unless the skin is abraided or cut. It operates by blocking the Krebs cycle by formation of fluorocitric acid, which inhibits aconitase. It has an effect on either or both the cardiovascular and nervous systems in all species and, in some species, the skeletal muscles. Humans have mixed responses with the cardiac feature predominating. By a direct action on the heart, contractile power is lost which leads to declining blood pressure. Ventricular premature contractions and arrhythmias are seen in all species, including humans. The central nervous system is directly attacked by sodium fluoroacetate. In humans, the action on the central nervous system produces epileptiform convulsive seizures followed by severe depression. The dangerous dose for humans is 0.5-2 mg/kg. Other species vary considerably in their response to this material with primates and birds being the most resistant and carnivora and rodents being the most susceptible. Most domestic animals show a susceptibility falling between the two extremes indicated above. Experimental reproductive effects. When heated to decomposition it emits highly toxic fumes of Na_2O and F^-.

SHI000 CAS:69780-81-0 ***HR: 3***
SODIUM γ-FLUORO-β-HYDROXYBUTYRATE
mf: $C_4H_6FO_3$•Na mw: 144.09

SYN: 4-FLUORO-2-HYDROXYBUTYRIC ACID SODIUM SALT

TOXICITY DATA with REFERENCE
orl-rat LDLo:1 mg/kg NDRC** 30101,3,45
scu-mus LDLo:500 μg/kg NDRC** 30101,3,45

SAFETY PROFILE: Poison by ingestion and subcutaneous routes. When heated to decomposition it emits toxic fumes of F^- and Na_2O.

SHI500 CAS:870-72-4 ***HR: 2***
SODIUM FORMALDEHYDE BISULFITE
mf: CH_3O_4S•Na mw: 134.09

TOXICITY DATA with REFERENCE
orl-rat LDLo:3200 mg/kg KODAK* -,-,71
ipr-rat LDLo:3200 mg/kg KODAK* -,-,71

CONSENSUS REPORTS: Reported in EPA TSCA Inventory.

SAFETY PROFILE: Moderately toxic by ingestion and intraperitoneal routes. When heated to decomposition it emits toxic fumes of SO_x and Na_2O. See also FORMALDEHYDE and SULFITES.

SHI625 ***HR: 2***
SODIUM FORMALDEHYDE SULFOXYLATE of 3-AMINO-4-HYDROXYPHENYLARSONIC ACID
mf: $C_7H_9AsNO_6$•Na mw: 301.08

TOXICITY DATA with REFERENCE
ivn-rat LDLo:1500 mg/kg JCHAAE 11,34,34
orl-rbt LDLo:500 mg/kg JCHAAE 11,34,34
ivn-rbt LDLo:500 mg/kg JCHAAE 11,34,34

CONSENSUS REPORTS: Arsenic and its compounds are on the Community Right To Know List.

SAFETY PROFILE: Moderately toxic by ingestion and intravenous routes. When heated to decomposition it emits toxic fumes of NO_x, Na_2O, and As. See also ARSENIC COMPOUNDS.

SHJ000 CAS:141-53-7 ***HR: 2***
SODIUM FORMATE
mf: CHO_2•Na mw: 68.01

PROP: White, deliq crystals. Mp: 253°, d: 1.92 @ 20°.

SYN: SALACHLOR

TOXICITY DATA with REFERENCE
orl-mus LD50:11200 mg/kg ZERNAL 9,332,69
ivn-mus LD50:807 mg/kg ZERNAL 9,332,69
orl-dog LDLo:4000 mg/kg JPETAB 16,463,21
ivn-dog LDLo:3000 mg/kg JPETAB 16,463,21
scu-cat LDLo:1140 mg/kg AEXPBL 21,119,1886
ivn-rbt LDLo:1250 mg/kg AEXPBL 21,119,1886

CONSENSUS REPORTS: Reported in EPA TSCA Inventory.

SAFETY PROFILE: Moderately toxic by ingestion, intravenous, and subcutaneous routes. Combustible when exposed to heat or flame. When heated to decomposition it emits toxic fumes of Na_2O. See also FORMIC ACID.

SHJ500 CAS:15736-98-8 ***HR: 2***
SODIUM FULMINATE
mf: CNNaO mw: 65.01

SAFETY PROFILE: An extremely touch-sensitive explosive. When heated to decomposition it emits very toxic fumes of NO_x and Na_2O. See also FULMINATES.

SHK000 CAS:751-94-0 ***HR: 3***
SODIUM FUSIDATE
mf: $C_{31}H_{47}O_6$•Na mw: 538.77

SYNS: FUCIDINA ◇ FUCIDINE ◇ FUSIDATE SODIUM ◇ FUSIDIN ◇ FUSIN ◇ INTERTULLE FUCIDIN ◇ SODIUM FUSIDIN ◇ SQ 16360 ◇ ZN 6 ◇ ZN 6-NA

TOXICITY DATA with REFERENCE
ipr-mam LD50:170 mg/kg 85ERAY 3,1958,78
scu-mam LD50:313 mg/kg 85ERAY 3,1958,78
ivn-mam LD50:205 mg/kg 85ERAY 3,1958,78

SAFETY PROFILE: Poison by intraperitoneal, subcutaneous, and intravenous routes. Used as an antibacterial agent. When heated to decomposition it emits toxic fumes of Na_2O.

SHK500 CAS:12265-93-9 ***HR: 3***
SODIUM GERMANIDE
mf: GeNa mw: 95.58

SAFETY PROFILE: Ignites spontaneously in air. May ignite on contact with water. When heated to decomposition it emits toxic fumes of Na_2O. See also GERMANIUM COMPOUNDS.

SHK800 CAS:527-07-1 ***HR: 1***
SODIUM GLUCONATE
mf: $C_6H_{12}O_7$•Na mw: 219.17

PROP: White to tan granular or crystalline powder. Very sol in water; sltly sol in alc; insol in ether.

SYNS: GLONSEN ◇ GLUCONATO di SODIO (ITALIAN) ◇ GLUCONIC ACID SODIUM SALT ◇ MONOSODIUM GLUCONATE ◇ PASEXON 100T ◇ PMP SODIUM GLUCONATE ◇ SODIUM d-GLUCONATE

TOXICITY DATA with REFERENCE
ivn-rbt LDLo:7630 mg/kg AFSPA2 68,1,39

CONSENSUS REPORTS: Reported in EPA TSCA Inventory.

SAFETY PROFILE: Low toxicity by intravenous route. When heated to decomposition it emits acrid smoke and irritating fumes.

SHL500 CAS:7009-49-6 ***HR: 3***
SODIUM HEXACYCLONATE
mf: $C_9H_{15}O_3$•Na mw: 194.23

SYNS: CYCLOHEXANEACETIC ACID, 1-(HYDROXYMETHYL)-, MONOSODIUM SALT (9CI) ◇ ESACICLONATO ◇ GEVILON ◇ GO 186 ◇ HEXACYCLONAS ◇ HEXACYCLONATE SODIUM ◇ 1-(HYDROXYMETHYL)CYCLOHEXANEACETIC ACID, SODIUM SALT ◇ NEURYL ◇ REPRISCAL ◇ SODIUM-1-(HYDROXYMETHYL) CYCLOHEXANEACETATE ◇ SODIUM-β,β-PENTAMETHYLENE-Γ-HYDROXYBUTYRATE

TOXICITY DATA with REFERENCE
orl-rat LD50:20 mg/kg 27ZQAG -,421,72
ipr-rat LD50:6 mg/kg 27ZQAG -,421,72
ivn-rat LD50:3 mg/kg 27ZQAG -,421,72
orl-mus LD50:36 mg/kg 27ZQAG -,421,72
ivn-mus LD50:7 mg/kg 27ZQAG -,421,72
ivn-cat LD50:4 mg/kg 27ZQAG -,421,72
ivn-gpg LD50:3 mg/kg 27ZQAG -,421,72

SAFETY PROFILE: Poison by ingestion, intraperitoneal, and intravenous routes. When heated to decomposition it emits toxic fumes of Na_2O.

SHM000 CAS:12005-86-6 ***HR: 3***
SODIUM HEXAFLUOROARSENATE
mf: AsF_6•Na mw: 211.91

TOXICITY DATA with REFERENCE
ivn-mus LD50:56 mg/kg CSLNX* NX#00127

CONSENSUS REPORTS: Arsenic and its compounds are on the Community Right To Know List.

OSHA PEL: Cancer Hazard
ACGIH TLV: TWA 0.2 mg(As)/m^3
NIOSH REL: (Arsenic, Inorganic) CL 2 $\mu g/m^3$/15M

SAFETY PROFILE: Confirmed human cacinogen. Poison by intravenous route. When heated to decomposition it emits very toxic fumes of As, F^-, and Na_2O. See also FLUORIDES and ARSENIC COMPOUNDS.

SHM500 CAS:10124-56-8 ***HR: 3***
SODIUM HEXAMETAPHOSPHATE
mf: $O_{18}P_6$•6Na mw: 611.76

PROP: White powder or flakes. Sol in water.

SYNS: CALGON ◇ CHEMI-CHARL ◇ HEXAMETAPHOSPHATE, SODIUM SALT ◇ HMP ◇ MEDI-CALGON ◇ PHOSPHATE, SODIUM HEXAMETA ◇ POLYPHOS ◇ SHMP

TOXICITY DATA with REFERENCE
orl-mus LD50:7250 mg/kg ARZNAD 7,445,57
ipr-mus LD50:870 mg/kg REPMBN 10,391,62
scu-mus LD50:1300 mg/kg ARZNAD 7,445,57
ivn-mus LD50:62 mg/kg ARZNAD 7,445,57
ivn-rbt LDLo:140 mg/kg AEPPAE 169,238,33

CONSENSUS REPORTS: Reported in EPA TSCA Inventory.

SAFETY PROFILE: Poison by intravenous route. Moderately toxic by intraperitoneal and subcutaneous routes. Mildly toxic by ingestion. When heated to decomposition it emits toxic fumes of PO_x and Na_2O. See also PHOSPHATES.

SHN000 CAS:12436-28-1 ***HR: 3***
SODIUM HEXAVANADATE
mf: $O_{15}V_6{\cdot}4Na$ mw: 637.60

TOXICITY DATA with REFERENCE
scu-rat LDLo:140 mg/kg AJSNAO 1,347,17
scu-mus LDLo:350 mg/kg AJSNAO 1,347,17
ivn-rbt LDLo:105 mg/kg AJSNAO 1,347,17
scu-gpg LDLo:140 mg/kg AJSNAO 1,347,17

ACGIH TLV: TWA 0.05 mg(V_2O_5)/m^3
NIOSH REL: CL 0.05 mg(V)/m^3/15M

SAFETY PROFILE: Poison by subcutaneous and intravenous routes. When heated to decomposition it emits toxic fumes of Na_2O and VO_x. See also VANADIUM COMPOUNDS.

SHN150 CAS:101952-86-7 ***HR: 3***
SODIUM HEXESTROL DIPHOSPHATE
mf: $C_{18}H_{24}O_8P_2{\cdot}Na$ mw: 453.35

SYN: HEXESTROL DIPHOSPHATE SODIUM

TOXICITY DATA with REFERENCE
orl-rat LD50:2000 mg/kg NIIRDN 6,744,82
ipr-rat LD50:200 mg/kg NIIRDN 6,744,82
scu-rat LD50:1000 mg/kg NIIRDN 6,744,82
orl-mus LD50:1000 mg/kg NIIRDN 6,744,82
ipr-mus LD50:200 mg/kg NIIRDN 6,744,82

SAFETY PROFILE: Poison by intraperitoneal route. Moderately toxic by ingestion and subcutaneous routes. When heated to decomposition it emits toxic fumes of PO_x and Na_2O.

SHN275 CAS:64038-10-4 ***HR: 3***
SODIUM HEXYLETHYL THIOBARBITURATE
mf: $C_{12}H_{19}N_2O_2S{\cdot}Na$ mw: 278.38

SYN: 5-ETHYL-5-HEXYL-2-THIOBARBITURIC ACID SODIUM SALT

TOXICITY DATA with REFERENCE
orl-rat LDLo:600 mg/kg JPETAB 60,125,37
ipr-rat LDLo:200 mg/kg JPETAB 60,125,37
ivn-mus LDLo:220 mg/kg JPETAB 60,125,37
ivn-dog LDLo:100 mg/kg JPETAB 60,125,37
ivn-rbt LDLo:70 mg/kg JPETAB 60,125,37

SAFETY PROFILE: Poison by intraperitoneal and intravenous routes. Moderately toxic by ingestion. When heated to decomposition it emits toxic fumes of SO_x, NO_x, and Na_2O. See also BARBITURATES.

SHN500 CAS:532-94-5 ***HR: 1***
SODIUM HIPPURATE
mf: $C_9H_8NO_3{\cdot}Na$ mw: 201.17

PROP: White powder. Sol in water and alc.

SYNS: N-BENZOYLGLYCINE, MONOSODIUM SALT ◇ HIPPURIC ACID SODIUM SALT

TOXICITY DATA with REFERENCE
ivn-mus LD50:4150 mg/kg JPETAB 86,147,45

CONSENSUS REPORTS: Reported in EPA TSCA Inventory. EPA Genetic Toxicology Program.

SAFETY PROFILE: Mildly toxic by intravenous route. When heated to decomposition it emits toxic fumes of NO_x and Na_2O.

SHO000 CAS:13598-47-5 ***HR: 3***
SODIUM HYDRAZIDE
mf: H_3N_2Na mw: 54.03

$NaNHNH_2$

SAFETY PROFILE: Explodes on contact with air, ethanol, or water. Explodes when heated to 100°C. When heated to decomposition it emits very toxic fumes of Na_2O and NO_x.

SHO500 CAS:7646-69-7 ***HR: 3***
SODIUM HYDRIDE
DOT: UN 1427
mf: HNa mw: 24.00

PROP: Microcrystalline, white to brownish-gray powder; reacts with water. Mp: 800° (decomp), d: 0.9.

SYN: NAH 80

CONSENSUS REPORTS: Reported in EPA TSCA Inventory.

DOT Classification: Flammable Solid; Label: Flammable Solid and Dangerous When Wet

SAFETY PROFILE: The powder ignites spontaneously in air. Flammable when exposed to heat or flame. Potentially explosive reaction with water; diethyl succinate + ethyltrifluoroacetate (above 60°C); dimethyl sulfoxide + heat; sulfur dioxide. Ignition or violent reaction with dimethylformamide (above 50°C); ethyl 2,2,3-trifluoropropionate; oxygen (at 230°C). Incompatible with acetylene + moisture; glycerol; halogens; sulphur. Normal fire extinguishers are unsuitable; use sand, ashes, sodium chloride. The commercial material may contain traces of sodium. When heated to decomposition it emits toxic fumes of Na_2O. See also HYDRIDES.

SHQ000 CAS:67293-75-8 ***HR: 3***
SODIUM HYDROGEN S-((N-CYCLOOCTYL-METHYLAMIDINO)METHYL) PHOSPHOROTHIOATE HYDRATE (4:4:5)
mf: $C_{11}H_{22}N_2O_3PS•Na•5/4H_2O$ mw: 338.89

TOXICITY DATA with REFERENCE
orl-mus LD50:80 mg/kg JMCMAR 15,1313,72
ipr-mus LD50:38 mg/kg JMCMAR 15,1313,72

SAFETY PROFILE: Poison by ingestion and intraperitoneal routes. When heated to decomposition it emits very toxic fumes of NO_x, PO_x, SO_x, and Na_2O.

SHQ500 CAS:1333-83-1 ***HR: 3***
SODIUM HYDROGEN FLUORIDE
DOT: UN 2439
mf: F_2HNa mw: 62.00

PROP: White powder. D: 2.08. Sol in water to 42,000 ppm @ 20°C.

SYNS: HYDROFLUORIC ACID, SODIUM SALT (2:1) ◇ SODIUM ACID FLUORIDE ◇ SODIUM BIFLUORIDE (VAN) ◇ SODIUM FLUORIDE(Na(HF_2)) ◇ SODIUM HYDROGEN DIFLUORIDE

CONSENSUS REPORTS: Reported in EPA TSCA Inventory.

OSHA PEL: TWA 2.5 mg(F)/m^3
ACGIH TLV: TWA 2.5 mg(F)/m^3
NIOSH REL: TWA 2.5 mg(F)/m^3
DOT Classification: Corrosive Material; Label:Corrosive

SAFETY PROFILE: This material is very toxic to humans by ingestion; between 1 teaspoonful and 1 ounce may be fatal. Inhalation of dust may cause irritation to respiratory tract. Skin contact may result in irritation and ulceration; eye contact may cause burns. To fight fire, use water, foam, CO_2, dry chemicals. When heated to decomposition it emits toxic fumes of F^- and Na_2O. See also FLUORIDES and HYDROFLUORIC ACID.

SHR000 CAS:16721-80-5 ***HR: 3***
SODIUM HYDROSULFIDE
DOT: UN 2318/NA 2949
mf: HNaS mw: 56.06

SYNS: SODIUM BISULFIDE ◇ SODIUM HYDROGEN SULFIDE ◇ SODIUM HYDROSULFIDE, solution (DOT) ◇ SODIUM HYDROSULPHIDE, with less than 25% water of crystallization (DOT) ◇ SODIUM HYDROSULPHIDE, solid (DOT) ◇ SODIUM MERCAPTAN ◇ SODIUM MERCAPTIDE ◇ SODIUM SULFHYDRATE

TOXICITY DATA with REFERENCE
mmo-sat 2 μmol/plate MUREAV 90,91,81
sln-dmg-orl 50 mmol/L MUREAV 90,91,81
ipr-rat LD50:30 mg/kg TXAPA9 55,198,80
ipr-mus LD50:18 mg/kg DCTODJ 1,327,78
scu-mus LD50:200 mg/kg JJPAAZ 3,99,54

CONSENSUS REPORTS: Reported in EPA TSCA Inventory.

DOT Classification: Corrosive Material; Label: Corrosive (NA2922, NA2923, NA2949); Flammable Solid; Label: Flammable Solid (UN2318); Flammable Solid; Label: Spontaneously Combustible (UN2318); Corrosive Material; Label: Corrosive.

SAFETY PROFILE: Poison by intraperitoneal and subcutaneous routes. Mutation data reported. A corrosive irritant to skin, eyes, and mucous membranes. Flammable when exposed to heat or flame. Spontaneous combustion. Reacts violently with diazonium salts. Readily yields H_2S. When heated to decomposition it emits toxic fumes of SO_x and Na_2O. See also SULFIDES and MERCAPTANS.

SHR500 CAS:7775-14-6 ***HR: 3***
SODIUM HYDROSULPHITE
DOT: UN 1384
mf: $O_4S_2•2Na$ mw: 174.10

PROP: White or yellow-white crystals. Mp: decomp @ 52°. Decomp in water (hot); sltly sol in cold water; insol in acids.

SYNS: D-OX ◇ HYDROLIN ◇ K-BRITE ◇ REDUCTONE ◇ SODIUM DITHIONITE (DOT) ◇ SODIUM HYDROSULFITE (DOT) ◇ SODIUM SULFOXYLATE ◇ VATROLITE ◇ V-BRITE ◇ VIRCHEM ◇ VIRTEX CC ◇ VIRTEX D ◇ VIRTEX L ◇ VIRTEX RD

CONSENSUS REPORTS: Reported in EPA TSCA Inventory.

DOT Classification: Flammable Solid; Label: Spontaneously Combustible; Flammable Solid; Label: Flammable Solid

SAFETY PROFILE: Toxic and an irritant. An allergen. Flammable when exposed to heat or flame. Ignites on contact with water or sodium chlorite. Decomposes violently when heated to 190°C and emits toxic fumes of SO_x and Na_2O.

SHS000 CAS:1310-73-2 ***HR: 3***
SODIUM HYDROXIDE
DOT: UN 1823/UN 1824
mf: HNaO mw: 40.00

PROP: White, pieces, lumps or sticks. Mp: 318.4°, bp: 1390°, d: 2.120 @ 20°/4°, vap press: 1 mm @ 739°. Deliq; sol in water and alc.

SYNS: CAUSTIC SODA ◇ CAUSTIC SODA, bead (DOT) ◇ CAUSTIC SODA, dry (DOT) ◇ CAUSTIC SODA, flake (DOT) ◇ CAUSTIC SODA, granular (DOT) ◇ CAUSTIC SODA, liquid (DOT) ◇ CAUSTIC SODA, solid (DOT) ◇ CAUSTIC SODA, solution (DOT) ◇ HYDROXYDE de SODIUM

(FRENCH) ◇ LEWIS-RED DEVIL LYE ◇ LYE (DOT) ◇ NATRIUMHYDROXID (GERMAN) ◇ NATRIUMHYDROXYDE (DUTCH) ◇ SODA LYE ◇ SODIO(IDROSSIDO di) (ITALIAN) ◇ SODIUM HYDRATE (DOT) ◇ SODIUM HYDROXIDE, bead (DOT) ◇ SODIUM HYDROXIDE, dry (DOT) ◇ SODIUM HYDROXIDE, flake (DOT) ◇ SODIUM HYDROXIDE, granular (DOT) ◇ SODIUM HYDROXIDE, solid (DOT) ◇ SODIUM(HYDROXYDE de) (FRENCH) ◇ WHITE CAUSTIC

TOXICITY DATA with REFERENCE
eye-mky 1%/24H SEV TXAPA9 6,701,64
skn-rbt 500 mg/24H SEV 28ZPAK -,7,72
eye-rbt 4 g MLD OYYAA2 26,627,83
eye-rbt 1% SEV AJOPAA 29,1363,46
eye-rbt 50 μg/24H SEV 28ZPAK -,7,72
eye-rbt 1 mg/24H SEV TXAPA9 6,701,64
eye-rbt 100 mg rns SEV TXCYAC 23,281,82
cyt-grh-par 20 mg NULSAK 9,119,66
ipr-mus LD50:40 mg/kg COREAF 257,791,63
orl-rbt LDLo:500 mg/kg AEPPAE 184,587,37

CONSENSUS REPORTS: Reported in EPA TSCA Inventory. EPA Genetic Toxicology Program.

OSHA PEL: (Transitional: TLV 2 mg/m^3) CL 2 mg/m^3
ACGIH TLV: Cl 2 mg/m^3
DFG MAK: 2 mg/m^3
NIOSH REL: (Sodium Hydroxide) CL 2 mg/m^3/15M
DOT Classification: Corrosive Material; Label: Corrosive

SAFETY PROFILE: Poison by intraperitoneal route. Moderately toxic by ingestion. Mutation data reported. A corrosive irritant to skin, eyes, and mucous membranes. This material, both solid and in solution, has a markedly corrosive action upon all body tissue causing burns and frequently deep ulceration, with ultimate scarring. Mists, vapors, and dusts of this compound cause small burns, and contact with the eyes rapidly causes severe damage to the delicate tissue. Ingestion causes very serious damage to the mucous membranes or other tissues with which contact is made. It can cause perforation and scarring. Inhalation of the dust or concentrated mist can cause damage to the upper respiratory tract and to lung tissue, depending upon the severity of the exposure. Thus, effects of inhalation may vary from mild irritation of the mucous membranes to a severe pneumonitis.

A strong base. Vigorous reaction with 1,2,4,5-tetrachlorobenzene has caused many industrial explosions and forms the extremely toxic 2,3,7,8-tetrachlorodibenzodioxin. Mixtures with aluminum + arsenic compounds form the poisonous gas arsine. Potentially explosive reaction with bromine; 4-chlorobutyronitrile; 4-chloro-2-methylphenol (in storage); nitrobenzene + heat; sodium tetrahydroborate; 2,2,2-trichloroethanol; zirconium + heat. Reacts to form explosive products with ammonia + silver nitrate (forms silver nitride); N,N′-bis(trinitroethyl)urea (in storage); cyanogen azide; glycols above 230° (e.g., ethylene glycol; diethylene glycol); 3-methyl-2-penten-4-yn-1-ol; trichloroethylene (forms dichloroacetylene). Caution: Under the proper conditions of temperature, pressure, and state of division, it can ignite or react violently with acetic acid, acetaldehyde; acetic anhydride; acrolein; acrylonitrile; allyl alcohol; allyl chloride; Al; benzene-1,4-diol; chlorine trifluoride; chloroform + methanol; chlorohydrin; chloronitro-toluenes; chlorosulfonic acid; 1,2-dichloroethylene; ethylene cyanhydrin; glyoxal; HCl; HF; hydroquinone; maleic anhydride; HNO_3; nitroethane; nitromethane; nitroparaffins; nitropropane; pentol; oleum; P; P_2O_5; β-propiolactone; H_2SO_4; (CH_3OH + tetrachloro-benzene); tetrahydrofuran; water; cinnamaldehyde; diborane + octanol oxime; 2,2-dichloro-3,3-dimethylbutane; 4-methyl-2-nitrophenol; 1,1,1-trichloroethanol; trichloronitromethane; zinc.

Dangerous material to handle. When heated to decomposition it emits toxic fumes of Na_2O.

SHS500 CAS:1310-73-2 ***HR: 3***
SODIUM HYDROXIDE (liquid)
DOT: UN 1823/UN 1824
mf: HNaO mw: 40.00

PROP: Clear to slightly turbid, colorless liquid.

SYNS: CAUSTIC SODA, solution ◇ LYE, solution ◇ SODA LYE ◇ SODIUM HYDRATE, solution ◇ SODIUM HYDROXIDE, solution (FCC) ◇ WHITE CAUSTIC, solution

TOXICITY DATA with REFERENCE
cyt-grh-par 20 μL NULSAK 9,119,66
ipr-mus LD50:40 mg/kg COREAF 257,791,63
orl-rbt LDLo:500 mg/kg AEPPAE 184,587,37

CONSENSUS REPORTS: Reported in EPA TSCA Inventory. Community Right-To-Know List.

DOT Classification: Corrosive Material; Label: Corrosive

SAFETY PROFILE: Poison by intraperitoneal route. Moderately toxic by ingestion. Mutation data reported. A corrosive irritant to skin, eyes, and mucous membranes. When heated to decomposition it emits toxic fumes of Na_2O.

SHT000 CAS:2836-32-0 ***HR: 1***
SODIUM HYDROXYACETATE
mf: $C_2H_3O_3 \cdot Na$ mw: 98.04

SYNS: GLYKOKOLAN SODNY (CZECH) ◇ HYDROXYACETIC ACID, MONOSODIUM SALT ◇ MONOSODIUM GLYCOLATE ◇ SODIUM α-HYDROXYACETATE

TOXICITY DATA with REFERENCE
orl-rat LD50:7110 mg/kg 28ZPAK -,104,72

CONSENSUS REPORTS: Reported in EPA TSCA Inventory.

SAFETY PROFILE: Mildly toxic by ingestion. When heated to decomposition it emits toxic fumes of Na_2O.

SHT100 CAS:7388-28-5 ***HR: 3***
SODIUM-2-HYDROXYETHOXIDE
mf: $C_2H_5NaO_2$ mw: 84.05

SAFETY PROFILE: Explosive reaction when heated to 230° in the presence of polychlorobenzenes, e.g., 1,2,4,5-tetrachlorobenzene; trichlorobenzenes. When heated to decomposition it emits toxic fumes of Na_2O. See also SODIUM COMPOUNDS.

SHT500 CAS:55939-60-1 ***HR: 3***
SODIUM o-HYDROXYMERCURIBENZOATE
mf: $C_7H_5HgO_3{\cdot}Na$ mw: 360.70

SYN: (o-CARBOXYPHENYL)HYDROXY-MERCURY SODIUM SALT

TOXICITY DATA with REFERENCE
ipr-rat LD50:15500 μg/kg JAPMA8 36,30,47

CONSENSUS REPORTS: Mercury and its compounds are on the Community Right To Know List.

OSHA PEL: (Transitional: CL 1 mg/10m^3) CL 0.1 mg(Hg)/m^3 (skin)
ACGIH TLV: TWA 0.1 mg(Hg)/m^3 (skin)
NIOSH REL: (Mercury, Inorganic) TWA 0.05 mg(Hg)/m^3

SAFETY PROFILE: Poison by intraperitoneal route. When heated to decomposition it emits toxic fumes of Hg and Na_2O. See also MERCURY COMPOUNDS.

SHU000 CAS:138-85-2 ***HR: 3***
SODIUM p-HYDROXYMERCURIBENZOATE
mf: $C_7H_5HgO_3{\cdot}Na$ mw: 360.70

TOXICITY DATA with REFERENCE
ipr-rat LD50:4960 μg/kg JAPMA8 36,30,47

CONSENSUS REPORTS: Mercury and its compounds are on the Community Right To Know List. Reported in EPA TSCA Inventory.

OSHA PEL: (Transitional: CL 1 mg/10m^3) CL 0.1 mg(Hg)/m^3 (skin)
ACGIH TLV: TWA 0.1 mg(Hg)/m^3 (skin)
NIOSH REL: (Mercury, Inorganic) TWA 0.05 mg(Hg)/m^3

SAFETY PROFILE: Poison by intraperitoneal route. When heated to decomposition it emits toxic fumes of Hg and Na_2O. See also MERCURY COMPOUNDS.

SHU175 ***HR: 3***
SODIUM-3-HYDROXYMERCURIO-2,6-DINITRO-4-aci-NITRO-2,5-CYCLOHEXADIENONIDE
mf: $C_6H_2HgN_3NaO_5$ mw: 419.68

CONSENSUS REPORTS: Mercury and its compounds are on The Community Right To Know List.

SAFETY PROFILE: Mercury compounds are generally poisons. Explodes when heated rapidly. When heated to decomposition it emits toxic fumes of Hg, NO_x, and Na_2O. See also MERCURY COMPOUNDS and NITRO COMPOUNDS.

SHU250 ***HR: 3***
SODIUM-2-HYDROXYMERCURIO-4-aci-NITRO-2,5-CYCLOHEXADIENONIDE
mf: $C_6H_4HgNNaO_4$ mw: 377.68

CONSENSUS REPORTS: Mercury and its compounds are on The Community Right To Know List.

SAFETY PROFILE: Explodes when heated. When heated to decomposition it emits toxic fumes of Hg, NO_x, and Na_2O. See also MERCURY COMPOUNDS.

SHU275 ***HR: 3***
SODIUM-2-HYDROXYMERCURIO-6-NITRO-4-aci-NITRO-2,5-CYCLOHEXADIENONIDE
mf: $C_6H_3HgN_2NaO_6$ mw: 422.68

CONSENSUS REPORTS: Mercury and its compounds are on The Community Right To Know List.

SAFETY PROFILE: Explodes violently when heated. When heated to decomposition it emits toxic fumes of Hg, NO_x, and Na_2O. See also MERCURY COMPOUNDS.

SHU300 ***HR: 3***
SODIUM-5-(5'-HYDROXYTETRAZOL-3'-YLAZO) TETRAZOLIDE
mf: $C_2HN_{10}NaO$ mw: 204.09

SAFETY PROFILE: Explodes violently when heated. Upon decomposition it emits toxic fumes of NO_x and Na_2O.

SHU475 ***HR: 3***
SODIUM HYPOBORATE
mf: $B_2H_6Na_2O_2$ mw: 105.65

SAFETY PROFILE: A powerful reducing agent which can react violently with oxidants. Probably a strong irritant to the eyes, skin, and mucous membranes. When heated to decomposition it emits toxic fumes of Na_2O. See also SODIUM COMPOUNDS and BORON COMPOUNDS.

SHU500 CAS:7681-52-9 ***HR: 3***
SODIUM HYPOCHLORITE
DOT: UN 1791
mf: $ClHO{\cdot}Na$ mw: 75.45

PROP: Mp: decomp.

SYNS: ANTIFORMIN ◇ B-K LIQUID ◇ CARREL-DAKIN SOLUTION ◇ CHLOROS ◇ CHLOROX ◇ CLOROX ◇ DAKINS SOLUTION ◇ HYCLORITE ◇ MILTON ◇ SURCHLOR

TOXICITY DATA with REFERENCE
eye-rbt 10 mg MOD TXAPA9 55,501,80
mma-sat 1 mg/plate AMONDS 3,253,80
cyt-hmn:lym 100 ppm/24H ARMCAH 21,409,70

CONSENSUS REPORTS: Reported in EPA TSCA Inventory. EPA Genetic Toxicology Program.

DOT Classification: ORM-B; Label: None

SAFETY PROFILE: Human mutation data reported. An eye irritant. Corrosive and irritating by ingestion and inhalation. The anhydrous salt is highly explosive and sensitive to heat or friction. Explosive reaction with formic acid (at 55°); phenylacetonitrile. Reacts to form explosive products with amines; ammonium salts [e.g., ammonium acetate; $(NH_4)_2CO_3$; ammonium nitrate; ammonium oxalate; $(NH_4)_3PO_4$]; aziridine; methanol. Violent reaction with phenyl acetonitrile; cellulose; ethylene imine. Solutions in water are storage hazards due to oxygen evolution. When heated to decomposition it emits toxic fumes of Na_2O and Cl^-. Used as a bleach.

SHU525 ***HR: 1***
SODIUM HYPOCHLORITE PENTAHYDRATE
mf: $ClHO{\cdot}Na{\cdot}5H_2O$ mw: 165.55

SYN: HYDROCHLOROUS ACID, SODIUM SALT, PENTAHYDRATE

TOXICITY DATA with REFERENCE
skn-rbt 500 mg/24H MOD BIOFX* 18-4/70
eye-rbt 100 mg MOD BIOFX* 18-4/70
orl-rat LD50:8910 mg/kg BIOFX* 18-4/70

SAFETY PROFILE: Mildly toxic by ingestion. An eye and skin irritant. When heated to decomposition it emits toxic fumes of Cl^- and Na_2O.

SHV000 CAS:7681-53-0 ***HR: 3***
SODIUM HYPOPHOSPHITE
mf: $H_2O_2P{\cdot}Na$ mw: 87.98

PROP: Colorless, pearly, crystalline plates or white granular powder; bittersweet, saline taste. Deliq; sol in water; sltly sol in alc.

SYNS: NATRIUMHYPOPHOSPHIT (GERMAN) ◇ SODIUM PHOSPHINATE

TOXICITY DATA with REFERENCE
ipr-mus LD50:1584 mg/kg COREAF 257,791,63
scu-rbt LDLo:50 mg/kg HBAMAK 4,1289,35

CONSENSUS REPORTS: Reported in EPA TSCA Inventory.

SAFETY PROFILE: Poison by subcutaneous route. Moderately toxic by intraperitoneal route. Flammable when exposed to heat or flame. Aqueous solutions may explode on evaporation. Potentially explosive reaction with oxidants (e.g., chlorates; nitrates). Heat causes it to evolve phosphine. It can explode. When heated to decomposition it emits toxic fumes of PO_x and Na_2O. See also PHOSPHITES, PHOSPHINE, and HYPOPHOSPHITES.

SHV100 ***HR: 2***
SODIUM (E)-3-(p-(1H-IMIDAZOL-1-YLMETHYL) PHENYL)-2-PROPENOATE
mf: $C_{13}H_{10}N_2O_2{\cdot}Na$ mw: 249.24

SYNS: CINNAMIC ACID, p-(1H-IMIDAZOL-1-YLMETHYL)-, SODIUM SALT, (E)- ◇ OKY-046 SODIUM ◇ OZAGREL

TOXICITY DATA with REFERENCE
ivn-rat TDLo:27 g/kg (female 17-22D post):REP KSRNAM 20,2891,86
ivn-rat TDLo:3300 mg/kg (female 7-17D post):TER KSRNAM 20,2873,86
orl-rat LD50:5700 mg/kg KSRNAM 20,2671,86
scu-rat LD50:2250 mg/kg KSRNAM 20,2671,86
ivn-rat LD50:1150 mg/kg KSRNAM 20,2671,86
orl-mus LD50:3600 mg/kg KSRNAM 20,2671,86
scu-mus LD50:2100 mg/kg KSRNAM 20,2671,86
ivn-dog LD50:733 mg/kg KSRNAM 20,2671,86

SAFETY PROFILE: Moderately toxic by ingestion, intravenous, and intraperitoneal route. Experimental reproductive effects. When heated to decomposition it emits toxic fumes of NO_x.

SHV500 CAS:7681-55-2 ***HR: 3***
SODIUM IODATE
mf: $IO_3{\cdot}Na$ mw: 197.89

PROP: Rhombic, white crystals. Mp: decomp, d: 4.277 @ 17.5°.

SYNS: IODIC ACID, SODIUM SALT ◇ NATRIUMJODAT (GERMAN)

TOXICITY DATA with REFERENCE
ipr-mus LD50:119 mg/kg JPETAB 120,171,57
ivn-mus LD50:108 mg/kg JPETAB 120,171,57
ivn-dog LDLo:200 mg/kg HBTXAC 1,274,55
ivn-rbt LDLo:75 mg/kg JPETAB 40,451,30

CONSENSUS REPORTS: Reported in EPA TSCA Inventory.

SAFETY PROFILE: Poison by intraperitoneal and intravenous routes. A trace mineral added to animal feeds. May react violently with Al; As; C; Cu; H_2O_2; metal sulfides; organic matter; P; K; S. Used as a mucous membrane antiseptic. When heated to decomposition it emits very toxic fumes of I^- and Na_2O.

SHW000 CAS:7681-82-5 ***HR: 2***
SODIUM IODIDE
mf: INa mw: 149.89

PROP: Cubic, colorless crystals. Mp: 651°, bp: 1300°, d: 3.667, vap press: 1 mm @ 767°.

SYNS: ANAYODIN ◇ IODURIL ◇ JODID SODNY ◇ NATRIUMJODID (GERMAN) ◇ SODIUM IODINE ◇ SODIUM MONOIODIDE

TOXICITY DATA with REFERENCE
skn-rbt 500 mg/24H MOD 28ZPAK -,21,72
eye-rbt 100 mg/24H MOD 28ZPAK -,21,72
orl-wmn TDLo:9240 mg/kg (1-43W preg):TER ADCHAK 38,526,63
orl-wmn TDLo:9240 mg/kg (1-43W preg):REP ADCHAK 38,526,63
orl-rat LD50:4340 mg/kg 28ZPAK -,21,72
ivn-rat LD50:1060 mg/kg AEPPAE 222,584,54
orl-mus LD50:1 g/kg 85GMAT -,105,82
ipr-mus LD50:430 mg/kg PSEBAA 115,551,64
ivn-dog LDLo:760 mg/kg HBAMAK 4,1289,35

CONSENSUS REPORTS: Reported in EPA TSCA Inventory.

SAFETY PROFILE: Moderately toxic by ingestion, intravenous, and intraperitoneal routes. Human teratogenic effects by ingestion: developmental abnormalities of the endocrine system. Human reproductive effects by ingestion: effects on newborn including postnatal measurements. A skin and eye irritant. Reacts violently with BrF_3, $HClO_4$, oxidants. When heated to decomposition it emits toxic fumes of I^- and Na_2O. See also IODIDES.

SHX000 CAS:126-31-8 ***HR: 2***
SODIUM IODOMETHANESULFONATE
mf: $CH_3IO_3S•Na$ mw: 244.99

SYNS: ABRODEN ◇ ABRODIL ◇ CONTUREX ◇ DIAGNORENOL ◇ IODOMETHANESULFONIC ACID SODIUM SALT ◇ KONTRAST-U ◇ METHIODAL SODIUM ◇ METHOIDAL SODIUM ◇ MONOIODOMETHANESULFONIC ACID, SODIUM SALT ◇ MYELOTRAST ◇ NCI-C03849 ◇ RADIOGRAPHOL ◇ SERGOSIN ◇ SKIODAN ◇ SODIUM METHIODAL ◇ SODIUM MONOIODOMETHANESULFONATE

TOXICITY DATA with REFERENCE
ivn-rat LD50:4800 mg/kg JPETAB 116,394,56
ivn-mus LD50:3900 mg/kg JPETAB 116,394,56

SAFETY PROFILE: Moderately toxic by intravenous route. When heated to decomposition it emits very toxic fumes of I^-, SO_x, and Na_2O. See also SULFONATES and IODIDES.

SHX500 CAS:3565-15-9 ***HR: 3***
SODIUM-5-IODO-2-THIOURACIL
mf: $C_4H_3IN_2OS•Na$ mw: 277.04

SYN: 5-IODO-2-THIOURACIL, SODIUM SALT

TOXICITY DATA with REFERENCE
orl-rat TDLo:30 g/kg/43W-C:ETA CANCAR 10,690,57

SAFETY PROFILE: Questionable carcinogen with experimental tumorigenic data. When heated to decomposition it emits very toxic fumes of I^-, NO_x, SO_x, and Na_2O.

SHY000 CAS:67114-26-5 ***HR: 3***
SODIUM-5-ISOBUTYL-5-(METHYLTHIOMETHYL) BARBITURATE
mf: $C_{10}H_{15}N_2O_3S•Na$ mw: 266.32

SYN: 5-ISOBUTYL-5-(METHYLTHIOMETHYL)BARBITURICACID SODIUM SALT

TOXICITY DATA with REFERENCE
orl-rat LD50:505 mg/kg JAPMA8 35,231,46
ivn-rat LD50:170 mg/kg JPETAB 88,343,46
ipr-mus LD50:354 mg/kg JAPMA8 35,231,46

SAFETY PROFILE: Poison by intravenous and intraperitoneal routes. Moderately toxic by ingestion. When heated to decomposition it emits toxic fumes of SO_x, NO_x, and Na_2O. See also BARBITURATES.

SHY300 CAS:683-60-3 ***HR: 3***
SODIUM ISOPROPOXIDE
mf: C_3H_7NaO mw: 82.08

$NaOCH(CH_3)_2$

SAFETY PROFILE: Ignites on contact with dimethyl sulfoxide. When heated to decomposition it emits toxic fumes of Na_2O. See also SODIUM COMPOUNDS.

SHY500 CAS:36993-63-2 ***HR: 3***
SODIUM ISOPROPYLETHYL THIOBARBITURATE
mf: $C_9H_{13}N_2O_2S•Na$ mw: 236.29

SYN: 5-ETHYL-5-ISOPROPYL-2-THIOBARBITURIC ACID SODIUM SALT

TOXICITY DATA with REFERENCE
orl-rat LDLo:150 mg/kg JPETAB 60,125,37
ipr-rat LDLo:20 mg/kg JPETAB 60,125,37
ivn-rat LDLo:120 mg/kg JPETAB 60,125,37

SAFETY PROFILE: Poison by ingestion, intraperitoneal, and intravenous routes. When heated to decomposition it emits very toxic fumes of NO_x, SO_x, and Na_2O. See also BARBITURATES.

SIA000 CAS:140-93-2 ***HR: 2***
SODIUM ISOPROPYLXANTHATE
mf: $C_4H_7OS_2•Na$ mw: 158.22

PROP: White, deliq powder.

SYNS: ISOPROPYLXANTHIC ACID, SODIUM SALT ◇ PROXANSODIUM

TOXICITY DATA with REFERENCE
par-mus LDLo:600 mg/kg CBCCT* 7,696,55

CONSENSUS REPORTS: Reported in EPA TSCA Inventory.

SAFETY PROFILE: Moderately toxic by parenteral route. A combustible material. When heated to decomposition it emits toxic fumes of SO_x and Na_2O. See also SULFIDES.

SIA500 CAS:540-72-7 ***HR: 3***
SODIUM ISOTHIOCYANATE
mf: CNS•Na mw: 81.07

PROP: Colorless, deliq crystals or white powder. Mp: 287°.

SYNS: HAIMASED ◇ NATRIUMRHODANID (GERMAN) ◇ SCYAN ◇ SODIUM RHODANATE ◇ SODIUM RHODANIDE ◇ SODIUM SULFOCYANATE ◇ SODIUM SULFOCYANIDE ◇ SODIUM THIOCYANATE ◇ SODIUM THIOCYANIDE ◇ THIOCYANATE SODIUM ◇ USAF EK-T-434

TOXICITY DATA with REFERENCE
orl-rat LD50:764 mg/kg JAPMA8 29,152,40
ipr-rat LD50:540 mg/kg JPETAB 96,416,49
orl-mus LD50:362 mg/kg 85GMAT -,106,82
ipr-mus LD50:500 mg/kg JAPMA8 29,152,40
scu-mus LDLo:400 mg/kg JAPMA8 29,152,40
ivn-mus LD50:484 mg/kg JAPMA8 29,152,40
ivn-dog LDLo:360 mg/kg AEPPAE 169,429,33
orl-rbt LDLo:750 mg/kg AEPPAE 169,429,33
scu-rbt LDLo:300 mg/kg AEPPAE 152,250,30
ivn-rbt LDLo:250 mg/kg AEPPAE 169,429,33
orl-gpg LDLo:600 mg/kg JAPMA8 29,152,40
ipr-gpg LDLo:500 mg/kg AEPPAE 169,429,33
scu-gpg LD50:500 mg/kg JAPMA8 29,152,40

CONSENSUS REPORTS: Reported in EPA TSCA Inventory.

SAFETY PROFILE: Poison by ingestion, intravenous, and subcutaneous routes. Moderately toxic by intraperitoneal route. Large doses taken internally cause vomiting, convulsions. Chronic poisoning is manifested by weakness, confusion, diarrhea, and skin rashes. When heated to decomposition it emits very toxic fumes of NO_x, SO_x, and Na_2O. See also THIOCYANATES.

SIB500 CAS:9004-82-4 ***HR: 2***
SODIUM LAURYL ETHER SULFATE
mf: $(C_2H_4O)n•C_{12}H_{26}O_4S•Na$

TOXICITY DATA with REFERENCE
skn-rbt 25 mg/24H MOD 33NFA8 -,12,75
skn-gpg 5%/9H-I imm 33NFA8 -,12,75
orl-rat LD50:1600 mg/kg JACTDZ 2(5),1,83

CONSENSUS REPORTS: Reported in EPA TSCA Inventory.

SAFETY PROFILE: Moderately toxic by ingestion. A skin and eye irritant. A surfactant. When heated to decomposition it emits toxic fumes of SO_x and Na_2O. See also SULFATES.

SIB600 CAS:151-21-3 ***HR: 3***
SODIUM LAURYL SULFATE
mf: $C_{12}H_{26}O_4S•Na$ mw: 289.43

PROP: White to cream-colored crystals, flakes or powder; slt odor. Sol in water.

SYNS: AQUAREX METHYL ◇ AVIROL 118 CONC ◇ CARSONOL SLS ◇ CONCO SULFATE WA ◇ CYCLORYL 21 ◇ DETERGENT 66 ◇ DODECYL ALCOHOL, HYDROGEN SULFATE, SODIUM SALT ◇ DODECYL SODIUM SULFATE ◇ DODECYL SULFATE, SODIUM SALT ◇ DREFT ◇ DUPONOL ◇ EMERSAL 6400 ◇ EMULSIFIER No. 104 ◇ HEXAMOL SLS ◇ IRIUM ◇ LANETTE WAX-S ◇ LAURYL SODIUM SULFATE ◇ LAURYL SULFATE, SODIUM SALT ◇ MAPROFIX 563 ◇ MAPROFIX WAC-LA ◇ NCI-C50191 ◇ NEUTRAZYME ◇ ORVUS WA PASTE ◇ PRODUCT No. 161 ◇ QUOLAC EX-UB ◇ REWOPOL NLS 30 ◇ RICHONOL C ◇ SIPEX OP ◇ SIPON WD ◇ SLS ◇ SODIUM DODECYL SULFATE ◇ SODIUM MONODODECYL SULFATE ◇ SOLSOL NEEDLES ◇ STANDAPOL 112 CONC ◇ STEPANOL WAQ ◇ STERLING WAQ-COSMETIC ◇ SULFOPON WA 1 ◇ SULFOTEX WALA ◇ SULFURIC ACID, MONODODECYL ESTER, SODIUM SALT ◇ TARAPON K 12 ◇ TEXAPON ZHC ◇ TREPENOL WA ◇ ULTRA SULFATE SL-1

TOXICITY DATA with REFERENCE
skn-hmn 250 mg/24H MLD TXAPA9 31,481,75
skn-hmn 25 mg/24H MLD JSCCA5 23,371,72
skn-mus 25 mg/24H MOD JSCCA5 23,371,72
skn-rbt 50 mg/24H SEV BIOFX* 23-3/71
skn-rbt 25 mg/24H MOD JSCCA5 23,371,72
skn-rbt 250 mg/24H MOD TXAPA9 31,481,75
skn-rbt 50 mg/24H MLD TXAPA9 21,369,72
eye-rbt 100 mg/24H MOD 28ZPAK -,305,72
eye-rbt 250 μg MLD AROPAW 34,99,45
eye-rbt 2 mg DCTODJ 1,305,78
eye-rbt 10 mg MOD TXAPA9 55,501,80
mmo-omi 200 mg/L JDREAF 55,266,76
sln-asn 900 ppm MUREAV 142,179,85
skn-mus TDLo:480 mg/kg (6-13D preg):TER TRENAF 27(2),113,76
orl-rat LD50:1288 mg/kg FCTXAV 5,763,67
ipr-rat LD50:210 mg/kg PSTGAW 3,1,45
ivn-rat LD50:118 mg/kg JPMSAE 52,803,63
ipr-mus LD50:250 mg/kg JAPMA8 42,283,53
ivn-mus LD50:118 mg/kg JPMSAE 52,803,63

CONSENSUS REPORTS: Reported in EPA TSCA Inventory.

SAFETY PROFILE: Poison by intravenous and intraperitoneal routes. Moderately toxic by ingestion. An experimental teratogen. A human skin irritant. An experimental eye and severe skin irritant. A mild allergen. Mutation data reported. When heated to decomposition it emits toxic fumes of SO_x and Na_2O. See also ESTERS and SULFATES.

SIC000 CAS:13150-00-0 ***HR: 2***
SODIUM LAURYL TRIOXYETHYLENE SULFATE
mf: $C_{18}H_{37}O_7S{\cdot}Na$ mw: 420.60

SYNS: 2-(2-(2-(DODECYLOXY)ETHOXY)ETHOXY)ETHANESULFONIC ACID, SODIUM SALT ◇ 2-(2-(2-(DODECYLOXY)ETHOXY)ETHOXY)ETHANOL HYDROGEN SULFATE SODIUM SALT ◇ SULFURIC ACID, MONO(2-(2-(2-(DODECYLOXY)ETHOXY)ETHOXY)ETHYL) ESTER, SODIUM SALT

TOXICITY DATA with REFERENCE
orl-rat LD50:1820 mg/kg TXAPA9 4,402,62

CONSENSUS REPORTS: Reported in EPA TSCA Inventory.

SAFETY PROFILE: Moderately toxic by ingestion. When heated to decomposition it emits very toxic fumes of SO_x and Na_2O.

SIC250 ***HR: 2***
SODIUM LINOLEIC ACID HYDROPEROXIDE
mf: $C_{18}H_{32}O_3{\cdot}Na$ mw: 319.49

SYNS: PEROXYLINOLEIC ACID, SODIUM SALT ◇ SODIUM LINOLEATE HYDROPEROXIDE

TOXICITY DATA with REFERENCE
scu-rat TDLo:1 g/kg/45W-I:CAR FCTXAV 12,451,74
scu-rat TD:1765 mg/kg/95W-I:NEO FCTXAV 12,451,74

SAFETY PROFILE: Questionable carcinogen with experimental carcinogenic and neoplastigenic data. When heated to decomposition it emits acrid smoke and irritating fumes.

SIC500 CAS:13284-86-1 ***HR: 3***
SODIUM LITHOCHOLATE
mf: $C_{24}H_{40}O_3{\cdot}Na$ mw: 399.63

SYNS: 3-α-HYDROXY-5-β-CHOLAN-24-OIC ACID, MONOSODIUM SALT ◇ (3-α,5-β)-3-HYDROXYCHOLAN-24-OIC ACID, MONOSODIUM SALT

TOXICITY DATA with REFERENCE
rec-mus TDLo:115 g/kg/48W-I:NEO CNREA8 39,1521,79

SAFETY PROFILE: Questionable carcinogen with experimental neoplastigenic data. When heated to decomposition it emits toxic fumes of Na_2O.

SID000 CAS:57-30-7 ***HR: 3***
SODIUM LUMINAL
mf: $C_{12}H_{12}N_2O_3{\cdot}Na$ mw: 255.25

PROP: White crystals.

SYNS: 5-ETHYL-5-PHENYLBARBITURIC ACID SODIUM ◇ 5-ETHYL-5-PHENYLBARBITURIC ACID SODIUM SALT ◇ 5-ETHYL-5-PHENYL-2,4,6-(1H,3H,5H)PYRIMIDINETRIONE MONOSODIUM SALT ◇ GARDENAL SODIUM ◇ LUMINAL SODIUM ◇ PBS ◇ PHENEMALUM ◇ PHENOBAL SODIUM ◇ PHENOBARBITAL ELIXIR ◇ PHENOBARBITAL Na ◇ PHENOBARBITAL SODIUM ◇ PHENOBARBITAL SODIUM SALT ◇ PHENOBARBITONE SODIUM ◇ PHENOBARBITONE SODIUM SALT ◇ PHENYLETHYLBARBITURIC ACID, SODIUM SALT ◇ SODIUM-5-ETHYL-5-PHENYLBARBITURATE ◇ SODIUM PHENOBARBITAL ◇ SODIUM PHENOBARBITONE ◇ SODIUM PHENYLETHYLBARBITURATE ◇ SODIUM PHENYLETHYLMALONYLUREA ◇ SOL PHENOBARBITAL ◇ SOL PHENOBARBITONE ◇ SOLUBLE PHENOBARBITAL ◇ SOLUBLE PHENOBARBITONE

TOXICITY DATA with REFERENCE
dnr-esc 400 μg/well ENMUDM 3,429,81
msc-mus:lym 5 mmol/L MUREAV 110,147,83
scu-mus TDLo:120 mg/kg (female 16-21D post):REP PBBHAU 3,1137,75
orl-rat TDLo:3623 mg/kg (1-21D preg):TER TJADAB 12,291,75
orl-rat TDLo:25 g/kg/2Y-C:NEO IJCNAW 19,179,77
orl-rat TD:11650 mg/kg/33W-C:ETA JJIND8 71,815,83
orl-man TDLo:36 mg/kg:CNS,GIT BMJOAE 1,1238,55
orl-rat LD50:150 mg/kg JPETAB 138,224,62
ipr-rat LD50:152 mg/kg ARZNAD 33,1155,83
scu-rat LD50:195 mg/kg NYKZAU 56,377,60
ivn-rat LD50:83 mg/kg JMCMAR 8,220,65
idu-rat LDLo:100 mg/kg ARZNAD 25,1037,75
orl-mus LD50:200 mg/kg JPETAB 127,318,59
ipr-mus LD50:123 mg/kg JPETAB 87,265,46
scu-mus LD50:180 mg/kg ARZNAD 8,42,58
ivn-mus LD50:226 mg/kg ARZNAD 33,1555,83
ivn-rbt LD50:40 mg/kg JPETAB 87,265,46

CONSENSUS REPORTS: IARC Cancer Review: Animal Sufficient Evidence IMEMDT 13,157,77. EPA Genetic Toxicology Program.

SAFETY PROFILE: Confirmed carcinogen with experimental carcinogenic, neoplastigenic, tumorigenic, and teratogenic data. Poison by ingestion, intravenous, intraperitoneal, intraduodenal, and subcutaneous routes. Human systemic effects by ingestion: nausea or vomiting and coma. Experimental reproductive effects. Mutation data reported. Used to treat epilepsy, as an hypnotic and sedative. When heated to decomposition it emits toxic fumes of NO_x and Na_2O. See also BARBITURATES.

SIE000 CAS:371-47-1 ***HR: 2***
SODIUM MALEATE
mf: $C_4H_4O_4{\cdot}xNa$ mw: 277.01

SYN: MALEINAN SODNY (CZECH)

TOXICITY DATA with REFERENCE
eye-rbt 500 mg/24H MLD 28ZPAK -,50,72
orl-rat LD50:3380 mg/kg 28ZPAK -,50,72
ipr-rat LD50:600 mg/kg JPPMAB 17,814,65

SAFETY PROFILE: Moderately toxic by ingestion and intraperitoneal routes. An eye irritant. When heated to decomposition it emits toxic fumes of Na_2O.

SIE500 CAS:141-95-7 ***HR: 2***
SODIUM MALONATE
mf: $C_3H_3O_4{\cdot}Na$ mw: 126.05

TOXICITY DATA with REFERENCE
ipr-rat LD50:1100 mg/kg JPPMAB 17,814,65
scu-rbt LDLo:1584 mg/kg JBCHA3 161,55,45

CONSENSUS REPORTS: Reported in EPA TSCA Inventory.

SAFETY PROFILE: Moderately toxic by intraperitoneal and subcutaneous routes. When heated to decomposition it emits toxic fumes of Na_2O.

SIF000 CAS:6963-70-8 ***HR: 1***
SODIUM MANNITOL ANTIMONATE

TOXICITY DATA with REFERENCE
ivn-mus LD50:5100 mg/kg CLDND* 37(S-7),5,88

CONSENSUS REPORTS: Antimony and its compounds are on the Community Right To Know List.

OSHA PEL: TWA 0.5 mg(Sb)/m^3
ACGIH TLV: TWA 0.5 mg(Sb)/m^3
NIOSH REL: (Antimony) TWA 0.5 mg(Sb)/m^3

SAFETY PROFILE: Most antimony compounds are poisons by ingestion, inhalation, and intraperitoneal routes. When heated to decomposition it emits toxic fumes of Sb and Na_2O. See also ANTIMONY COMPOUNDS.

SIF425 CAS:6385-02-0 ***HR: 2***
SODIUM MECLOFENAMATE
mf: $C_{14}H_{10}Cl_2NO_2{\cdot}Na$ mw: 318.14

SYNS: CI-583 NA ◇ 2-((2,6-DICHLORO-3-METHYLPHENYL)AMINOBENZOIC ACID MONOSODIUM SALT ◇ MECLOFENAMATE SODIUM ◇ MECLOMEN ◇ SODIUM-2-((2,6-DICHLORO-3-METHYLPHENYL) AMINO)BENZOATE ◇ SODIUM MECLOPHENATE

TOXICITY DATA with REFERENCE
orl-rat TDLo:112 mg/kg (14D pre/1-22D preg):REP TXAPA9 15,46,69
orl-hmn TDLo:120 mg/kg/6W:GIT,SKN ARZNAD 33,631,83
orl-wmn TDLo:3 mg/kg:SKN,MET JRHUA9 10,169,83

SAFETY PROFILE: Human systemic effects by ingestion: diarrhea, nausea or vomiting, allergic dermatitis, and fever. Experimental reproductive effects. When heated to decomposition it emits toxic fumes of Cl^-, NO_x, and Na_2O.

SIF500 ***HR: 2***
SODIUM MERALEIN
mf: $C_{19}H_8HgI_2O_7S{\cdot}2Na$ mw: 880.70

SYNS: HYDROXY(6-HYDROXY-2,7-DIIODO-3-OXO-9-(o-SULFOPHENYL)-3H-XANTHEN-5-YL)MERCURY DISODIUM SALT ◇ MERALEIN DISODIUM ◇ MERODICEIN ◇ MONOHYDROXY-MERCURI-DIIODORESORCIN-SULPHONPHTHALEINDISODIUM

CONSENSUS REPORTS: Mercury and its compounds are on the Community Right To Know List.
NIOSH REL: TWA 0.05 mg(Hg)/m^3

SAFETY PROFILE: When heated to decomposition it emits very toxic fumes of Hg, I^-, SO_x, and Na_2O. See also MERCURY COMPOUNDS and IODIDES.

SIG000 CAS:129-99-7 ***HR: 3***
SODIUM MERALLURIDE
mf: $C_{16}H_{21}HgN_6O_7{\cdot}Na$ mw: 633.01

SYNS: MERCARDAN ◇ MERCURHYDRIN SODIUM

TOXICITY DATA with REFERENCE
ims-rat LD50:3160 μg/kg FEPRA7 8,316,49
ivn-mus LD50:1070 mg/kg JPETAB 99,149,50

CONSENSUS REPORTS: Mercury and its compounds are on the Community Right To Know List.

ACGIH TLV: TWA 0.1 mg(Hg)/m^3 (skin)
NIOSH REL: (Mercury, Inorganic) TWA 0.05 mg(Hg)/m^3

SAFETY PROFILE: Poison by intramuscular route. Moderately toxic by intravenous route. When heated to decomposition it emits very toxic fumes of Hg, NO_x, and Na_2O. See also MERCURY COMPOUNDS.

SIG500 CAS:2492-26-4 ***HR: 2***
SODIUM-2-MERCAPTOBENZOTHIAZOLE
mf: $C_7H_4NS_2{\cdot}Na$ mw: 189.23

SYNS: 2-MERCAPTOBENZOTHIAZOLE SODIUM DERIVATIVE ◇ 2-MERCAPTOBENZOTHIAZOLE SODIUM SALT

TOXICITY DATA with REFERENCE
orl-rat LD50:3120 mg/kg FMCHA2 -,D327,80

CONSENSUS REPORTS: Reported in EPA TSCA Inventory.

SAFETY PROFILE: Moderately toxic by ingestion. When heated to decomposition it emits very toxic fumes of NO_x, SO_x, and Na_2O. See also MERCAPTANS.

SIH500 CAS:492-18-2 ***HR: 3***
SODIUM MERSALYL
mf: $C_{13}H_{16}HgNO_6{\cdot}Na$ mw: 505.88

SYNS: 3-(α-CARBOXY-o-ANISAMIDO)-2-METHOXYPROPYL HYDROXYMERCURY, MONOSODIUM SALT ◇ o-((3-HYDROXYMERCURI-2-METHOXYPROPYL)CARBAMOYL)PHENOXYACETIC

ACID MONOSODIUM SALT ◇ N-(Γ-HYDROXYMERCURI-β-METHOXY-PROPYL)SALICYLAMIDE-o-ACETIC ACID SODIUM SALT ◇ IGROSIN ◇ MERCURAMIDE ◇ MERCURITAL ◇ MERCUSAL ◇ MERSALIN ◇ MERSALYL ◇ SALURIN ◇ SALYRGAN ◇ SODIUM o-((3-(HYDROXY-MERCURI)-2-METHOXYPROPYL)CARBAMOYL)PHENOXYACETATE ◇ SODIUM SALICYL-(Γ-HYDROXYMERCURI-β-METHOXY-PROPYL)AMIDE-o-ACETATE ◇ URAGAN

TOXICITY DATA with REFERENCE
ivn-rat LD50:17 mg/kg JAPMA8 40,249,51
ims-rat LD50:24 mg/kg JAPMA8 40,249,51
ipr-mus LD50:118 mg/kg JOCEAH 16,1879,51
ivn-mus LD50:73 mg/kg JAPMA8 40,249,51
ivn-rbt LDLo:7 mg/kg AEPPAE 152,341,30
ims-rbt LD50:24 mg/kg JAPMA8 40,249,51

CONSENSUS REPORTS: Mercury and its compounds are on the Community Right To Know List.

OSHA PEL: (Transitional: CL 1 mg/10m^3) CL 0.1 mg (Hg)/m^3 (skin)
ACGIH TLV: TWA 0.1 mg(Hg)m^3 (skin)
NIOSH REL: (Inorganic Mercury) TWA 0.05 mg(Hg)/m^3

SAFETY PROFILE: Poison by intravenous, intramuscular, and intraperitoneal routes. Used as a diuretic agent. When heated to decomposition it emits very toxic fumes of Hg, NO_x, and Na_2O. See also MERCURY COMPOUNDS.

SII000 CAS:7681-57-4 ***HR: 3***
SODIUM METABISULFITE
DOT: NA 2693
mf: $O_5S_2 \cdot 2Na$ mw: 190.10

PROP: Colorless crystals or white to yellowish powder; odor of sulfur dioxide. Sol in water; sltly sol in alc.

SYNS: DISODIUM PYROSULFITE ◇ SODIUM METABOSULPHITE ◇ SODIUM PYROSULFITE

TOXICITY DATA with REFERENCE
cyt-ham:ovr 180 μg/L ENMUDM 7,1,85
sce-ham:ovr 200 μg/L ENMUDM 7,1,85
orl-rat TDLo:20 g/kg (MGN):REP FCTXAV 10,291,72
ivn-rat LD50:115 mg/kg DTLVS* 4,371,80
par-mus LD50:910 mg/kg RPTOAN 31,120,68
ivn-rbt LDLo:192 mg/kg TXAPA9 24,266,73

CONSENSUS REPORTS: Reported in EPA TSCA Inventory. EPA Genetic Toxicology Program.

OSHA PEL: TWA 5 mg/m^3
ACGIH TLV: TWA 5 mg/m^3
DOT Classification: ORM-B; Label: None

SAFETY PROFILE: Poison by intravenous route. Moderately toxic by parenteral route. Experimental reproductive effects. Mutation data reported. When heated to decomposition it emits toxic fumes of SO_x and Na_2O.

SII500 CAS:10361-03-2 ***HR: 2***
SODIUM METAPHOSPHATE
mf: $O_3P \cdot Na$ mw: 101.96

PROP: Sodium metaphosphate exists as a number of different molecular species, some of which exhibit various crystalline forms. The vitreous sodium phosphates having a Na_2O/P_2O_5 mole ratio near unity are classified as sodium metaphosphates. The term also extends to short-chain vitreous compositions, the compounds of which exhibit the polyphosphate formula $Na_{n+2}P_nO_{3n+1}$ with n as low as 4-5. In such as $(NaPO_3)$, n may be a small integer <3 (cyclic molecules) or a large number (polymers). Amorphous white solids. Very sol in water.

SYNS: GRAHAM'S SALT ◇ METAFOS ◇ SODIUM HEXAMETA-PHOSPHATE ◇ SODIUM POLYPHOSPHATES, GLASSY ◇ SODIUM TETRAPOLYPHOSPHATE

TOXICITY DATA with REFERENCE
ipr-mus LD50:830 mg/kg REPMBN 10,391,62

CONSENSUS REPORTS: Reported in EPA TSCA Inventory.

SAFETY PROFILE: Moderately toxic by intraperitoneal route. When heated to decomposition it emits toxic fumes of Na_2O and PO_x. See also PHOSPHATES.

SIJ000 CAS:3804-89-5 ***HR: 3***
SODIUM METHANESULFONATE
mf: $C_7H_8N_3O_4S \cdot Na$ mw: 253.23

SYNS: ISONIAZID SODIUM METHANESULFONATE ◇ NEOISCOTIN ◇ NEOTIZIDE ◇ NEO-TIZIDE SODIUM SALT ◇ SODIUM ISONICO-TINYL HYDRAZINE METHANSULFONATE

TOXICITY DATA with REFERENCE
orl-mus TDLo:101 g/kg/30W-C:CAR GANNA2 51,83,60
orl-mus LD50:1100 mg/kg NIIRDN 6,74,82
scu-mus LD50:1500 mg/kg TXAPA9 8,325,66
orl-dog LDLo:250 mg/kg TXAPA9 10,183,67

SAFETY PROFILE: Poison by ingestion. Moderately toxic by subcutaneous route. Questionable carcinogen with experimental carcinogenic data. When heated to decomposition it emits very toxic fumes of Na_2O, SO_x, and NO_x. See also SULFONATES.

SIJ600 ***HR: 3***
SODIUM METHOXYACETYLIDE
mf: C_3H_3NaO mw: 78.05

$$NaC \equiv COCH_3$$

SAFETY PROFILE: May ignite spontaneously in air. Reaction with brine is weakly explosive. When heated to decomposition it emits toxic fumes of Na_2O. See also SODIUM COMPOUNDS and ACETYLIDES.

SIK000 CAS:2126-70-7 ***HR: 3***
SODIUM β-4-METHOXYBENZOYL-β-BROMOACRYLATE
mf: $C_{11}H_8BrO_4{\bullet}Na$ mw: 307.09

SYNS: 3-p-ANISOYL-3-BROMOACRYLIC ACID, SODIUM SALT ◇ 3-BROMO-3-(4-METHOXYBENZOYL)ACRYLIC ACID SODIUM SALT ◇ (Z)-3-BROMO-4-(4-METHOXYPHENYL)-4-OXO-2-BUTENOIC ACID, SODIUM SALT ◇ CYTEMBENA ◇ NSC-104801 ◇ SODIUM BROMEBRATE ◇ SODIUM SALT of β-4-METHOXYBENZOYL-β-BROMOACRYLIC ACID

TOXICITY DATA with REFERENCE
cyt-rat-ipr 400 mg/kg 40YJAX -,277,76
mnt-mus-ipr 3200 mg/kg 40YJAX -,277,76
dni-mus:lym 160 μmol/L NEOLA4 22,259,75
dlt-mus-ipr 350 mg/kg/14D-I NEOLA4 25,523,78
cyt-ham-ipr 640 mg/kg 40YJAX -,277,76
ivn-hmn TDLo:60 mg/kg/5D:CNS JNCIAM 59,1619,77
ipr-mus LD50:49720 μg/kg NCISP* JAN86
ivn-dog LDLo:1200 mg/kg CCROBU 59,1071,75

CONSENSUS REPORTS: EPA Genetic Toxicology Program.

SAFETY PROFILE: Poison by intraperitoneal route. Moderately toxic by intravenous route. Human systemic effects by intravenous route: hallucinations, distorted perceptions, motor activity changes and nausea or vomiting. Mutation data reported. When heated to decomposition it emits toxic fumes of Br^- and Na_2O.

SIK450 CAS:124-41-4 ***HR: 3***
SODIUM METHYLATE
DOT: NA 1289/UN 1431
mf: $CH_3O{\bullet}Na$ mw: 54.03

PROP: White, amorphous, free-flowing powder. Decomp in air above 127°; decomp by water. Sol in methyl and ethyl alc, fats, esters.

SYNS: METHANOL, SODIUM SALT ◇ SODIUM METHOXIDE ◇ SODIUM METHYLATE, DRY (DOT)

CONSENSUS REPORTS: Reported in EPA TSCA Inventory.

DOT Classification: Flammabel Solid; Label: Flammable Solid & Danger When Wet (UN1431); Combustible Liquid; Label: None (NA1289); Flammable Liquid; Label: Flammable Liquid (NA1289, UN1289); Corrosive Material; Label: Corrosive (NA1289); Flammable or Combustible Liquid; Label: Flammable Liquid (UN1289).

SAFETY PROFILE: A corrosive and irritating material. It hydrolyzes into methanol and sodium hydroxide. May ignite spontaneously in moist air. Flammable when exposed to heat or flame. Ignites on contact with water. Violent reaction with ($CHCl_3$ + CH_3OH); (methyl azide + dimethylmalonate); $FClO_3$. When heated to decomposition it emits toxic fumes of Na_2O.

SIK500 CAS:124-41-4 ***HR: 2***
SODIUM METHYLATE (alcohol mixture)
DOT: NA 1289

CONSENSUS REPORTS: Reported in EPA TSCA Inventory.

DOT Classification: Combustible Liquid; Label: None; Corrosive Material; Label: Corrosive, Flammable Liquid.

SAFETY PROFILE: A corrosive irritant to skin, eyes and mucous membranes. Combustible when exposed to heat or flame; can react vigorously with oxidizing materials. When heated to decomposition it emits toxic fumes of Na_2O. See also ALCOHOLS.

SIL500 CAS:3653-48-3 ***HR: 3***
SODIUM (2-METHYL-4-CHLOROPHENOXY) ACETATE
mf: $C_9H_8ClO_3{\bullet}Na$ mw: 222.61

SYNS: AGROXONE 3 ◇ 4-CHLORO-2-METHYLPHENOXYACETIC ACID SODIUM SALT ◇ (p-CHLORO-o-TOLYLOXY)ACETIC ACID SODIUM SALT ◇ CHWASTOKS ◇ CHWASTOX ◇ DIAMET ◇ DICOTEX 80 ◇ DIKOTEKS ◇ DIKOTEX 30 ◇ MCPA SODIUM SALT ◇ METAXONE ◇ METHOXONE ◇ (2-METHYL-4-CHLOROPHENOXY)ACETIC ACID, SODIUM SALT ◇ 2M-4KH SODIUM SALT ◇ 2M-4X ◇ Na MCPA ◇ PHENOXYLENE ◇ SODIUM (4-CHLORO-2-METHYLPHENOXY)ACETATE ◇ SODIUM MCPA ◇ SYS 67ME

TOXICITY DATA with REFERENCE
orl-rat TDLo:542 mg/kg (female 5D post):REP GTPZAB 31(2),31,87
orl-rat LD50:800 mg/kg WRPCA2 9,119,70
ihl-rat LDLo:520 mg/m³/4H GTPZAB 31(2),31,87
ipr-rat LD50:400 mg/kg FCTXAV 3,883,65
scu-rat LD50:500 mg/kg BECTA6 18,152,77
orl-mus LD50:450 mg/kg GTPZAB 31(2),31,87
ipr-mus LD50:500 mg/kg FCTXAV 3,883,65

CONSENSUS REPORTS: Reported in EPA TSCA Inventory.

SAFETY PROFILE: Poison by intraperitoneal route. Moderately toxic by ingestion and subcutaneous routes. Experimental reproductive effects. When heated to decomposition it emits toxic fumes of Cl^- and Na_2O.

SIL600 CAS:70247-50-6 ***HR: 3***
SODIUM-3-METHYLISOXAZOLIN-4,5-DIONE-4-OXIMATE
mf: $C_4H_3N_2NaO_3$ mw: 150.07

NaON:CCO•ON=CCH_3

SAFETY PROFILE: Explodes when heated rapidly. When heated to decomposition it emits toxic fumes of NO_x and Na_2O.

SIM000 CAS:64048-07-3 ***HR: 3***
SODIUM METHYLMERCURIC THIOGLY COLLATE
mf: $C_3H_5HgO_2S$•Na mw: 328.72

SYNS: (MERCAPTOACETATO(2-)-O,S)METHYL-MERCURATE(1-), SODIUM ◇ S-(METHYLMERCURIC)THIOGLYCOLIC ACID SODIUM SALT

TOXICITY DATA with REFERENCE
ipr-rat LDLo:40 mg/kg JPETAB 35,343,29
ivn-rbt LDLo:20 mg/kg JPETAB 35,343,29

CONSENSUS REPORTS: Mercury and its compounds are on the Community Right To Know List.

OSHA PEL: (Transitional: CL 1 mg/$10m^3$) TWA 0.01 mg(Hg)/m^3; STEL 0.03 mg/m^3 (skin)
ACGIH TLV: TWA 0.01 mg(Hg)/m^3; STEL 0.03 mg(Hg)/m^3
NIOSH REL: TWA 0.05 mg(Hg)/m^3

SAFETY PROFILE: Poison by intraperitoneal and intravenous routes. When heated to decomposition it emits very toxic fumes of Hg, SO_x, and Na_2O. See also MERCURY COMPOUNDS and MERCAPTANS.

SIM100 CAS:22113-51-5 ***HR: 3***
SODIUM-4-METHYLPHENOXIDE
mf: C_7H_7NaO mw: 130.12

$NaOC_6H_4CH_3$

SYNS: SODIUM-4-CRESOLATE ◇ SODIUM p-CRESYLATE

SAFETY PROFILE: Ignites when heated in air even when moist. Oxidizes vigorously when heated in air. When heated to decomposition it emits toxic fumes of Na_2O. See also SODIUM COMPOUNDS.

SIN000 CAS:305-53-3 ***HR: 3***
SODIUM MONOIODOACETATE
mf: $C_2H_2IO_2$•Na mw: 207.93

SYNS: IODOACETATE SODIUM SALT ◇ IODOACETIC ACID SODIUM SALT ◇ USAF EK-6279

TOXICITY DATA with REFERENCE
dni-hmn:lyms 10 mmol/L STBIBN 50,97,75
orl-mus TDLo:500 mg/kg (female 8-12D post):REP TCMUD8 6,361,86
ipr-rat LDLo:30 mg/kg JPPMAB 17,814,65
orl-mus LD50:63 mg/kg JPETAB 86,336,46
ipr-mus LD50:75 mg/kg NTIS** AD277-689

CONSENSUS REPORTS: Reported in EPA TSCA Inventory.

SAFETY PROFILE: Poison by ingestion and intraperitoneal routes. Experimental reproductive effects. Human mutation data reported. When heated to decomposition it emits toxic fumes of I^- and Na_2O.

SIN500 CAS:12401-86-4 ***HR: 3***
SODIUM MONOXIDE
DOT: UN 1825
mf: Na_2O mw: 61.98

PROP: White-gray, deliq crystals. Bp: 1275° (subl), d. 2.27.

SYNS: CALCINED SODA ◇ DISODIUM MONOXIDE ◇ DISODIUM OXIDE ◇ SODIUM MONOXIDE, solid (DOT) ◇ SODIUM OXIDE

DOT Classification: Corrosive Material; Label: Corrosive

SAFETY PROFILE: Very corrosive and irritating to skin, eyes, and mucous membranes. Can react violently with water, nitric oxide (above 100°C). Ignites when mixed with 2,4-dinitrotoluene. Mixtures with phosphorus(V) oxide react violently when warmed or on contact with moisture. When heated to decomposition it emits toxic fumes of Na_2O. See also SODIUM HYDROXIDE.

SIN650 CAS:873-58-5 ***HR: 2***
SODIUM MORPHOLINECARBODITHIOATE
mf: $C_5H_8NOS_2$•Na mw: 185.25

SYNS: SODIU-4-MORPHOLINECARBODITHIOATE ◇ SODIUM MORPHOLINEDITHIOCARBAMATE ◇ SODIUM MORPHOLINODITHIOCARBAMATE ◇ SODIUM SALT of 4-MORPHOLINECARBODITHIOIC ACID

TOXICITY DATA with REFERENCE
ipr-rat LD50:520 mg/kg PHARAT 27,139,72
orl-mus LD50:1830 mg/kg AIPTAK 112,36,57
ipr-mus LD50:600 mg/kg PHARAT 27,139,72

CONSENSUS REPORTS: Reported in EPA TSCA Inventory.

SAFETY PROFILE: Moderately toxic by ingestion and intraperitoneal routes. When heated to decomposition it emits toxic fumes of SO_x, NO_x, and Na_2O. See also CARBAMATES.

SIN675 ***HR: 3***
SODIUM MORPHOLINE and NITRITE (1:1)

SYN: MORPHOLINE and SODIUM NITRITE (1:1)

TOXICITY DATA with REFERENCE
hma-mus/sat 1450 μmol/kg CNREA8 37,4572,77
otr-ham-orl 1 g/kg BBRCA9 81,310,78
mse-ham-orl 1 g/kg BBRCA9 81,310,78

orl-rat TDLo:18800 mg/kg/27W-C:ETA FCTXAV 10,887,72
orl-mus TDLo:17200 mg/kg/96W-I:CAR EKSODD 8(1),41,86

SAFETY PROFILE: Questionable carcinogen with experimental carcinogenic and tumorigenic data. Mutation data reported. When heated to decomposition it emits toxic fumes of NO_x and Na_2O. See also SODIUM NITRITE and MORPHOLINE.

SIN850 CAS:37415-62-6 ***HR: 3***
SODIUM MYCOPHENOLATE
mf: $C_{17}H_{19}O_6 \cdot Na$ mw: 342.35

SYN: 6-(4-HYDROXY-6-METHOXY-7-METHYL-3-OXO-5-PHTHALANYL)- 4-METHYL-4-HEXENOIC ACID, SODIUM SALT

TOXICITY DATA with REFERENCE
orl-mus LD50:1176 mg/kg AACHAX -,229,68
ipr-mus LD50:568 mg/kg AACHAX -,229,68
ivn-mus LDLo:400 mg/kg LANCAO 250,46,46

SAFETY PROFILE: Poison by intravenous route. Moderately toxic by ingestion and intraperitoneal routes. When heated to decomposition it emits toxic fumes of Na_2O.

SIO000 CAS:1191-50-0 ***HR: 3***
SODIUM MYRISTYL SULFATE
mf: $C_{14}H_{29}O_4S \cdot Na$ mw: 316.48

SYNS: 7-ETHYL-2-METHYL-4-HENDECANOL SULFATE SODIUM SALT ◇ MYRISTYL SULFATE, SODIUM SALT ◇ NIAPROOF 4 ◇ SODIUM SOTRADECOL ◇ SODIUM TETRADECYL SULFATE ◇ STS ◇ SULFURIC ACID, MONOTETRADECYL ESTER, SODIUM SALT ◇ SULFURIC ACID, MYRISTYL ESTER, SODIUM SALT ◇ TERGITOL 4 ◇ TETRADECYL SODIUM SULFATE ◇ TETRADECYL SULFATE, SODIUM SALT ◇ TROMBAVAR ◇ TROMBOVAR

TOXICITY DATA with REFERENCE
ivg-rbt TDLo:2500 μg/kg (female 1D pre):REP JRPFA4 68,257,83
ipr-mus LD50:342 mg/kg JAPMA8 42,283,53
ivn-mus LD50:56 mg/kg CSLNX* NX#00154

CONSENSUS REPORTS: Reported in EPA TSCA Inventory.

SAFETY PROFILE: Poison by intraperitoneal and intravenous route. Experimental reproductive effects. When heated to decomposition it emits toxic fumes of SO_x and Na_2O. See also SULFATES.

SIO500 CAS:132-67-2 ***HR: 2***
SODIUM N-1-NAPHTHYLPHTHALAMIC ACID
mf: $C_{18}H_{12}NO_3 \cdot Na$ mw: 313.30

SYNS: α-NAPHTHYLPHTHALAMIC ACID SODIUM SALT ◇ N-1-NAPHTHYLPHTHALAMIC ACID SODIUM SALT ◇ SODIUM N-1-NAPHTHYLPHTHALAMATE

TOXICITY DATA with REFERENCE
orl-rat LD50:1770 mg/kg GUCHAZ 6,372,73

CONSENSUS REPORTS: Reported in EPA TSCA Inventory.

SAFETY PROFILE: Moderately toxic by ingestion. When heated to decomposition it emits toxic fumes of NO_x and Na_2O.

SIO788 CAS:28643-80-3 ***HR: 3***
SODIUM NIGERICIN
mf: $C_{40}H_{68}O_{11} \cdot Na$ mw: 748.07

SYN: NIGERICIN SODIUM SALT

TOXICITY DATA with REFERENCE
ipr-mus TDLo:5 mg/kg (9D preg):TER TJADAB 27,81A,83
ipr-mus TDLo:5 mg/kg (female 10D post):REP TJADAB 33,47,86
ipr-mus LD50:48 mg/kg JANTAJ 30,77-5,77

SAFETY PROFILE: Poison by intraperitoneal route. An experimental teratogen. Experimental reproductive effects. When heated to decomposition it emits toxic fumes of Na_2O.

SIO900 CAS:7631-99-4 ***HR: 3***
SODIUM(I) NITRATE (1:1)
DOT: UN 1498
mf: $NO_3 \cdot Na$ mw: 85.00

PROP: Colorless, transparent, odorless crystals; saline, sltly bitter taste. Mp: 306.8°, bp: decomp @ 380°, d: 2.261. Deliq in moist air; sol in water, sltly sol in alc.

SYNS: CHILE SALTPETER ◇ CUBIC NITER ◇ NITRATE de SODIUM (FRENCH) ◇ NITRATINE ◇ NITRIC ACID, SODIUM SALT ◇ SODA NITER ◇ SODIUM NITRATE (DOT)

TOXICITY DATA with REFERENCE
mnt-ham-orl 250 mg/kg MUREAV 66,149,79
cyt-ham:lng 125 mg/L GMCRDC 27,95,81
orl-mus TDLo:16800 mg/kg (male 14D pre):REP MUREAV 204,689,88
orl-rat TDLo:100 g/kg/2Y-C:ETA FCTOD7 22,715,84
orl-man LDLo:114 mg/kg FAONAU 38A,31,65
orl-rat LD50:3236 mg/kg FAONAU 38A,31,65
ivn-mus LD50:175 mg/kg ATXKA8 21,89,65
orl-rbt LD50:2680 mg/kg SOVEA7 27,246,74

CONSENSUS REPORTS: Reported in EPA TSCA Inventory. EPA Genetic Toxicology Program.

DOT Classification: Oxidizer; Label: Oxidizer

SAFETY PROFILE: Human poison by ingestion. Poison by intravenous route. Questionable carcinogen with experimental tumorigenic data. Human mutation data reported. A powerful oxidizer. It will ignite with heat or friction. Explodes when heated to over 1000°F, or when mixed with cyanides, sodium hypophosphite, boron phosphide. Forms explosive mixtures with aluminum powder; antimony powder; barium thiocyanate; metal amidosulfates; sodium; sodium phosphinate; sodium thiosulfate; sulfur + charcoal (gunpowder). Potentially violent reaction or ignition when mixed with bitumen; organic matter, calcium-silicon alloy; jute + magnesium chloride; magnesium; metal cyanides; non-metals; perosyformic acid; phenol + trifluoroacetic acid. Incompatible with acetic anhydride; barium thiocyanate; wood. A dangerous disaster hazard. Experimental reproductive effects. When heated to decomposition it emits toxic fumes of NO_x and Na_2O. See also NITRATES.

SIP000 CAS:12136-83-3 ***HR: 3***
SODIUM NITRIDE
mf: Na_3N mw: 82.98

SAFETY PROFILE: Decomposes explosively on warming. A very heat-sensitive explosive. When heated to decomposition it emits very toxic fumes of Na_2O and NO_x. See also AMMONIA and SODIUM HYDROXIDE.

SIP500 CAS:5064-31-3 ***HR: 3***
SODIUM NITRILOTRIACETATE
mf: $C_6H_6NO_6 \cdot 3Na$ mw: 257.10

SYNS: HAMPSHIRE NTA ◇ NITRILOTRIACETIC ACID, TRISODIUM SALT ◇ NTA ◇ TRISODIUM NITRILOTRIACETATE ◇ TRISODIUM NITRILOTRIACETIC ACID

TOXICITY DATA with REFERENCE
or-rat:emb 495 μg/plate JJATDK 1,190,81
orl-rat TDLo:39 g/kg (8W male/8W pre-3W post):REP FCTXAV 9,509,71
orl-rat TDLo:70300 mg/kg/2Y-C:NEO JJIND8 66,869,81
orl-rat LD50:1100 mg/kg TXAPA9 18,398,71
ipr-rat LD50:254 mg/kg TXAPA9 18,398,71
orl-mus LD50:681 mg/kg NCILB* NIH-NCI-E-C-72-3252
orl-mky LD50:750 mg/kg TXAPA9 18,398,71

CONSENSUS REPORTS: Reported in EPA TSCA Inventory.

SAFETY PROFILE: Poison by intraperitoneal route. Moderately toxic by ingestion. Experimental reproductive effects. Questionable carcinogen with experimental neoplastigenic data. Mutation data reported. When heated to decomposition it emits toxic fumes of NO_x and Na_2O.

SIQ500 CAS:7632-00-0 ***HR: 3***
SODIUM NITRITE
DOT: UN 1500
mf: $NO_2 \cdot Na$ mw: 69.00

PROP: Sltly yellowish or white crystals, sticks or powder; slt salty taste. Mp: 271°, bp: decomp @ 320°, d: 2.168. Deliq in air; sol in water, sltly sol in alc.

SYNS: ANTI-RUST ◇ DIAZOTIZING SALTS ◇ DUSITAN SODNY (CZECH) ◇ ERINITRIT ◇ FILMERINE ◇ NATRIUM NITRIT (GERMAN) ◇ NCI-C02084 ◇ NITRITE de SODIUM (FRENCH) ◇ NITROUS ACID, SODIUM SALT

TOXICITY DATA with REFERENCE
eye-rbt 500 mg/24H MLD 28ZPAK -,15,72
mmo-omi 50 mmol/L JGMIAN 128,1401,82
dns-hmn:hla 6 mmol/L FCTOD7 21,551,83
orl-rat TDLo:11 g/kg (1-22D preg/21D post):REP TOXID9 4,89,84
orl-rat TDLo:660 mg/kg (1-22D preg):TER AJHEAA 62,1045,72
orl-rat TDLo:2190 g/kg/2Y-C:CAR HKXUDL 6,246,86
orl-rat TD:63 g/kg/95W-C:ETA JJIND8 64,1435,80
orl-rat TD:91 g/kg/2Y-C:ETA,REP FCTOD7 20,25,82
orl-rat TD:183 g/kg/2Y-C:ETA,REP FCTOD7 20,25,82
orl-rat TD:100 g/kg/2Y-I:NEO CRNGDP 4,1189,83
orl-mus TDLo:2149 mg/(pre-post-birth):CAR,TER CNREA8 45,3561,85
orl-man TDLo:1714 μg/kg/70M:CVS JCINAO 16,73,37
orl-hmn LDLo:71 mg/kg:CNS,GIT,BLD 34ZIAG -,543,69
orl-chd LDLo:22 mg/kg LANCAO 2,162,17
orl-rat LD50:85 mg/kg AIHAAP 30,470,69
ihl-rat LC50:5500 $\mu g/m^3$ GTPZAB 16(10),36,72
scu-rat LDLo:10 mg/kg AEPPAE 201,197,43
ivn-rat LD50:65 mg/kg ARZNAD 13,320,63
orl-mus LD50:175 mg/kg FAONAU 53A,97,74
ipr-mus LD50:158 mg/kg JPETAB 165,30,69
scu-mus LDLo:150 mg/kg AIPTAK 35,480,29
orl-dog LDLo:330 mg/kg 27ZIAQ -,245,73
scu-dog LDLo:60 mg/kg AIPTAK 21,425,11
ivn-dog LDLo:15 mg/kg 27ZIAQ -,245,73
orl-cat LDLo:1500 mg/kg HBAMAK 4,1372,35
scu-cat LDLo:35 mg/kg AEPPAE 126,209,27
orl-rbt LD50:186 mg/kg SOVEA7 27,246,74

CONSENSUS REPORTS: Reported in EPA TSCA Inventory. EPA Genetic Toxicology Program.

DOT Classification: Oxidizer; Label: Oxidizer

SAFETY PROFILE: Human poison by ingestion. Experimental poison by ingestion, inhalation, subcutaneous, intravenous, and intraperitoneal routes. Human systemic effects by ingestion: motor activity changes, coma, decreased blood pressure with possible pulse rate increase without fall in blood pressure, arteriolar or venous dilation, nausea or vomiting, and blood methemoglobinemia-carboxhemoglobinemia. Experimental teratogenic and reproductive effects. An eye irritant. Questionable carcinogen with experimental neoplastigenic and tumorigenic data. Human mutation data reported. They may react with organic amines in the body to form carcinogenic nitrosamines.

Flammable; a strong oxidizing agent. In contact with organic matter, will ignite by friction. Explodes when heated to over 1000°F or on contact with cyanides; NH_4^- salts; cellulose; Li; (K + NH_3); $Na_2S_2O_3$. Incompatible with aminoguanidine salts; butadiene; phthalic acid; phthalic anhydride; reducants; sodium amide; sodium disulphite; sodium thiocyanate; urea; wood. When heated to decomposition it emits toxic fumes of NO_x and Na_2O. See also NITRITES.

SIQ675 ***HR: 2***
SODIUM NITRITE mixed with 1-(p-BROMOPHENYL)-3-METHYLUREA

SYNS: 1-(p-BROMOPHENYL)-3-METHYLUREA mixed with SODIUM NITRITE ◇ 1-METHYL-3-(p-BROMOPHENYL)UREA mixed with SODIUM NITRITE ◇ SODIUM NITRITE mixed with 1-METHYL-3-(p-BROMOPHENYL)UREA

TOXICITY DATA with REFERENCE
orl-rat TDLo:1186 mg/kg/82W-I:CAR ARGEAR 53,329,83

SAFETY PROFILE: Questionable carcinogen with experimental carcinogenic data. When heated to decomposition it emits toxic fumes of NO_x.

SIQ700 ***HR: 2***
SODIUM NITRITE and CARBENDAZIME (1:1)
mf: $C_9H_7N_5O_4$ mw: 249.21

SYNS: CARBENDAZIME and SODIUM NITRITE (1:1) ◇ 2-METHYL-N-NITROSO-BENZIMIDAZOLE CARBAMATE and SODIUM NITRITE (1:1) ◇ SODIUM NITRITE and 2-METHYL-N-NITROSO-BENZIMIDAZOLE CARBAMATE (1:1)

TOXICITY DATA with REFERENCE
orl-mus TDLo:4200 mg/kg (8-14D post):TER MGONAD 20,163,76

SAFETY PROFILE: An experimental teratogen. When heated to decomposition it emits toxic fumes of NO_x.

SIS000 CAS:104639-49-8 ***HR: 3***
SODIUM NITRITE mixed with CHLORDIAZEPOXIDE (1:1)

SYNS: CHLORDIAZEPOXIDE, NITROSATED ◇ CHLORDIAZEPOXIDE mixed with SODIUM NITRITE (1:1) ◇ 7-CHLORO-2-METHYLAMINO-5-PHENYL-3H-1,4-BENZODIAZEPINE-4-OXIDE, mixed with SODIUM NITRITE (1:1)

TOXICITY DATA with REFERENCE
mmo-sat 800 nmol/plate CALEDQ 12,81,81
mma-sat 800 nmol/plate CALEDQ 12,81,81
orl-rat TDLo:90 g/kg/50W-I:CAR FCTXAV 15,269,77

SAFETY PROFILE: Questionable carcinogen with experimental carcinogenic data. Mutation data reported. When heated to decomposition it emits very toxic fumes of NO_x, Cl^-, and Na_2O. See also SODIUM NITRITE.

SIS100 ***HR: 2***
SODIUM NITRITE and l-CITRULLINE (1:2)
mf: $C_6H_{13}N_3O_3 \cdot 1/2NNaO_2$ mw: 239.68

SYN: l-CITRULLINE and SODIUM NITRITE (2:1)

TOXICITY DATA with REFERENCE
orl-rat TDLo:1650 mg/kg (13-23D post):NEO,TER NCIMAV 51,103,79

SAFETY PROFILE: An experimental teratogen. Questionable carcinogen with experimental tumorigenic data. When heated to decomposition it emits toxic fumes of NO_x.

SIS150 ***HR: 2***
SODIUM NITRITE mixed with DIMETHYLDODECYLAMINE (8:7)

SYN: DIMETHYLDODECYLAMINE HYDROCHLORIDE mixed with SODIUM NITRITE (7:8)

TOXICITY DATA with REFERENCE
orl-rat TDLo:135 g/kg/80W-I:CAR FCTXAV 15,269,77

SAFETY PROFILE: Questionable carcinogen with experimental carcinogenic data. When heated to decomposition it emits toxic fumes of NO_x.

SIS200 ***HR: 2***
SODIUM NITRITE mixed with DISULFIRAM
mf: $C_{10}H_{20}N_2S_4 \cdot NNaO_3$ mw: 381.56

SYNS: BIS(DIETHYLTHIOCARBAMOYL)DISULFIDE mixed with SODIUM NITRITE ◇ DISULFIRAM mixed with SODIUM NITRITE ◇ SODIUM NITRITE mixed with BIS(DIETHYLTHIOCARBAMOYL)DISULFIDE

TOXICITY DATA with REFERENCE
orl-rat TDLo:164 g/kg/78W-C:ETA FCTXAV 18,85,80

SAFETY PROFILE: Questionable carcinogen with experimental tumorigenic data. When heated to decomposition it emits toxic fumes of NO_x.

SIS500 ***HR: 3***
SODIUM NITRITE mixed with ETHAMBUTOL (1:1)

SYNS: d-N,N'-DI(1-HYDROXYMETHYLPROPYL)ETHYLENEDIAMINE DIHYDROCHLORIDE mixed with SODIUM NITRITE ◇ ETHAMBUTOL mixed with SODIUM NITRITE (1:1)

TOXICITY DATA with REFERENCE
orl-mus TDLo:8640 mg/kg/36W-C:CAR LAPPA5 35,45,75

SAFETY PROFILE: Questionable carcinogen with experimental carcinogenic data. When heated to decomposition it emits toxic fumes of NO_x, HCl, and Na_2O. See also SODIUM NITRITE.

SIS650 ***HR: 2***
SODIUM NITRITE mixed with N-METHYL-ADENOSINE (4:1)
mf: $C_{11}H_{15}N_5O_4 \cdot 4NNaO_2$ mw: 557.31

SYNS: ADENOSINE, N-METHYL-, mixed with SODIUM NITRITE (1:4) ◇ N[6]-METHYLADENOSINE mixed with SODIUM NITRITE (1:4) ◇ N-METHYLADENOSINE mixed with SODIUM NITRITE (1:4)

TOXICITY DATA with REFERENCE
orl-mus TDLo:33918 mg/kg/34W-I:CAR IJCNAW 24,319,79
orl-mus TD:34 g/kg/34W-I:NEO IAPUDO 31,787,80

SAFETY PROFILE: Questionable carcinogen with experimental carcinogenic and neoplastigenic data. When heated to decomposition it emits toxic fumes of NO_x.

SIS675 ***HR: 2***
SODIUM NITRITE and 1-(METHYLETHYL)UREA
mf: $C_8H_{20}N_4O_2 \cdot NNaO_2$ mw: 273.32

SYNS: ISOPROPYLUREA and SODIUM NITRITE ◇ 1-(METHYLETHYL)UREA and SODIUM NITRITE ◇ SODIUM NITRITE and ISOPROPYLUREA

TOXICITY DATA with REFERENCE
orl-rat TDLo:180 mg/kg (21D post):NEO,TER ZAPPAN 121,61,77
orl-rat TDLo:25850 mg/kg/47D-C:NEO ZAPPAN 121,61,77

SAFETY PROFILE: An experimental teratogen. Questionable carcinogen with experimental neoplastigenic data. When heated to decomposition it emits toxic fumes of NO_x.

SIS700 ***HR: 3***
SODIUM NITRITE and 1-METHYL-1-NITROSO-3-PHENYLUREA

TOXICITY DATA with REFERENCE
orl-rat TDLo:8250 mg/kg/32W-I:CAR CALEDQ 4,299,78
orl-rat TD:200 mg/kg:ETA MVMZA8 33,128,78

SAFETY PROFILE: Questionable carcinogen with experimental carcinogenic and tumorigenic data. When heated to decomposition it emits toxic fumes of NO_x and Na_2O. See also SODIUM NITRITE and N-METHYL-N-NITROSOUREA.

SIT000 ***HR: 3***
SODIUM NITRITE mixed with OCTAHYDROAZOCINE HYDROCHLORIDE (1:1)

SYNS: HEPTAMETHYLENEIMINE mixed with SODIUM NITRITE ◇ OCTAHYDROAZOCINE HYDROCHLORIDE mixed with SODIUM NITRITE (1:1) ◇ SODIUM NITRITE mixed with HEPTAMETHYLENEIMINE

TOXICITY DATA with REFERENCE
orl-rat TDLo:15 g/kg/24W-I:CAR NATUAS 244,176,73

SAFETY PROFILE: Questionable carcinogen with experimental carcinogenic data. When heated to decomposition it emits very toxic fumes of HCl, NO_x, and Na_2O. See also SODIUM NITRITE.

SIT500 ***HR: 2***
SODIUM NITRITE and 1-PROPYLUREA
mf: $C_4H_{10}N_2O \cdot 1/3NNaO_2$ mw: 125.13

SYNS: n-PROPYLUREA and SODIUM NITRITE ◇ 1-PROPYLUREA and SODIUM NITRITE ◇ SODIUM NITRITE and n-PROPYLUREA

TOXICITY DATA with REFERENCE
orl-rat TDLo:31350 mg/kg/57D-C:NEO ZAPPAN 121,61,77

SAFETY PROFILE: Questionable carcinogen with experimental neoplastigenic data. When heated to decomposition it emits toxic fumes of NO_x.

SIT750 ***HR: 3***
SODIUM NITRITE, mixed with SODIUM NITRATE and POTASSIUM NITRATE
DOT: NA 1487

DOT Classification: Oxidizer; Label: Oxidizer

SAFETY PROFILE: Both components are poisons. A powerful oxidizer. When heated to decomposition it emits toxic fumes of NO_x, K_2O, and Na_2O. See also SODIUM NITRITE, POTASSIUM NITRITE, and NITRITES.

SIT800 ***HR: 3***
SODIUM NITRITE and TRIFORINE

SYN: TRIFORINE and SODIUM NITRITE

TOXICITY DATA with REFERENCE
orl-mus TDLo:19 g/kg/15W-I:CAR IARCCD 19,477,78

SAFETY PROFILE: Questionable carcinogen with experimental carcinogenic data. When heated to decomposition it emits toxic fumes of NO_x and Na_2O. See also SODIUM NITRITE.

SIU000 CAS:1718-34-9 ***HR: D***
SODIUM p-NITROBENZENEAZOSALICYLATE
mf: $C_{13}H_8N_3O_5 \cdot Na$ mw: 309.23

SYNS: C.I. MORDANT ORANGE 1, MONOSODIUM SALT ◇ MORDANT YELLOW 3R

TOXICITY DATA with REFERENCE
mma-sat 500 μg/plate MUREAV 56,249,78

CONSENSUS REPORTS: Reported in EPA TSCA Inventory.

SAFETY PROFILE: Mutation data reported. When heated to decomposition it emits toxic fumes of NO_x and Na_2O.

SIU500 CAS:14402-89-2 ***HR: 3***
SODIUM NITROFERRICYANIDE
mf: $C_5FeN_6O•2Na$ mw: 261.94

$Na_2[(NC)_5FeNO]$

SYNS: DISODIUM NITROSYLPENTACYANOFERRATE ◇ NIPRIDE ◇ NITROPRUSSIDNATRIUM (GERMAN) ◇ SODIUM NITROPRUSSATE ◇ SODIUM NITROPRUSSIDE ◇ SODIUM NITROSYLPENTACYANOFERRATE ◇ SODIUM NITROSYLPENTACYANOFERRATE(III)

TOXICITY DATA with REFERENCE
ivn-dom TDLo:909 ng/kg (20W preg):TER AJOGAH 139,708,81
ivn-wmn TDLo:5 μg/kg/5M-C:BRN JAMAAP 246,2679,81
unr-wmn TDLo:23 mg/kg:CNS,CVS,MET ANESAV 49,428,78
ihl-hmn LDLo:11 mg/kg/5H BJANAD 46,324,74
ivn-hmn TDLo:3360 μg/kg/14H-C AIMEAS 104,895,86
ivn-man LDLo:29 mg/kg/1W-I SMJOAV 77,1035,84
ivn-hmn LDLo:10 mg/kg CANJAE 22,547,75
orl-rat LD50:99 mg/kg ARZNAD 24,308,74
ipr-rat LD50:9400 μg/kg ARZNAD 24,308,74
ivn-rat LD50:6286 μg/kg TOXID9 1,92,81
orl-mus LD50:61 mg/kg ARZNAD 24,308,74
ipr-mus LD50:9400 μg/kg ARZNAD 24,308,74
ivn-dog LDLo:20800 mg/kg ARZNAD 24,308,74
orl-rbt LD50:34 mg/kg TOXID9 1,92,81
ivn-rbt LD50:4191 μg/kg TOXID9 1,92,81

CONSENSUS REPORTS: Cyanide and its compounds are on the Community Right To Know List. Reported in EPA TSCA Inventory.

SAFETY PROFILE: Human poison by inhalation and intravenous routes. Experimental poison by ingestion, intraperitoneal, and intravenous routes. Human systemic effects: increased intracranial pressure, general anesthesia, change in heart rate and metabolic acidosis. An experimental teratogen. Used as a vasodilator for short-term treatment of severe hypertension. Mixtures with sodium nitrite explode when heated. When heated to decomposition it emits toxic fumes of NO_x, CN^-, and Na_2O.

SIV000 ***HR: 3***
SODIUM NITROMALONALDEHYDE
mf: $C_3H_2NNaO_4$ mw: 139.06

SAFETY PROFILE: An impact-sensitive solid. When heated to decomposition it emits very toxic fumes of Na_2O and NO_x. See also ALDEHYDES.

SIV500 ***HR: 3***
SODIUM aci-NITROMETHANE
mf: CH_2NNaO_2 mw: 83.01

SYN: SODIUM aci-NITROMETHANIDE

SAFETY PROFILE: A very unstable powerful explosive. Explodes on contact with water. Reacts with mercury(II) chloride + acids to form mercury fulminate. Incompatible with 1,1,3,3-tetramethyl-2,4-cyclobutanedione. When heated to decomposition it emits very toxic fumes of NO_x and Na_2O.

SIV600 CAS:824-78-2 ***HR: 3***
SODIUM-4-NITROPHENOXIDE
mf: $C_6H_4NNaO_3$ mw: 161.09

SAFETY PROFILE: Decomposes violently when heated. When heated to decomposition it emits toxic fumes of NO_x and Na_2O. See also NITRO COMPOUNDS of AROMATIC HYDROCARBONS and SODIUM COMPOUNDS.

SIW000 CAS:63919-22-2 ***HR: 3***
SODIUM NITROPRUSSIATE
mf: $C_5FeN_6O•4Na•x(H_2O)$ mw: 434.06

SYN: SODIUM NITROSOPENTACYANOFERRATE (3)

TOXICITY DATA with REFERENCE
orl-rat LD50:99 mg/kg AIPTAK 172,487,68
ipr-rat LD50:9400 μg/kg AIPTAK 172,487,68
orl-mus LD50:61 mg/kg AIPTAK 172,487,68
ipr-mus LD50:9400 μg/kg AIPTAK 172,487,68
ivn-rbt LDLo:3040 μg/kg AIPTAK 172,487,68

SAFETY PROFILE: Poison by ingestion, intravenous, and intraperitoneal routes. When heated to decomposition it emits toxic fumes of Na_2O and NO_x. See also CYANIDE and NITROSO COMPOUNDS.

SIW500 CAS:13755-38-9 ***HR: 3***
SODIUM NITROPRUSSIDE DIHYDRATE
mf: $C_5FeN_6O•2Na•2H_2O$ mw: 297.98

PROP: Rhombic, red crystals.

SYNS: DISODIUM NITROPRUSSIDE DIHYDRATE ◇ FERRATE(2-), PENTAKIS(CYANO-C)NITROSYL-, DISODIUM, DIHYDRATE (OC-6-22)-, (9CI) ◇ NIPRIDE DIHYDRATE ◇ NITROPRUSSIATE de SODIUM ◇ NITROPRUSSIDNATRIUM ◇ SODIUM NITROSYLPENTACYANOFERRATE(III) DIHYDRATE

TOXICITY DATA with REFERENCE
orl-rat LDLo:20 mg/kg YAKUD5 22,883,80
ipr-rat LD50:7 mg/kg AANFAE 18(2),153,77
ivn-rat LD50:9300 μg/kg ARZNAD 29,1092,79
ipr-mus LD50:9 mg/kg AANFAE 18(2),153,77
scu-mus LDLo:9500 μg/kg AIPTAK 35,480,29

SAFETY PROFILE: Poison by ingestion, intraperitoneal, subcutaneous, and intravenous routes. The effects of this material are similar to that of nitrites, causing fall in blood pressure but no formation of methemoglobin. Large amounts, when taken internally, may form cyanide upon being metabolized. When heated to decomposition it emits toxic fumes of Na_2O and NO_x. See also CYANIDES.

SIW550 CAS:823-87-0 ***HR: 3***
SODIUM-4-NITROSOPHENOXIDE
mf: $C_6H_4NNaO_2$ mw: 145.09

$NaOC_6H_4N:O$

SAFETY PROFILE: Ignites spontaneously when dry. When heated to decomposition it emits toxic fumes of NO_x and Na_2O. See also SODIUM COMPOUNDS and NITROSO COMPOUNDS.

SIW600 CAS:67312-43-0 ***HR: 3***
SODIUM-5-NITROTETRAZOLIDE
mf: CN_5NaO_2 mw: 137.03

SAFETY PROFILE: A primary explosive extremely sensitive to pressure and friction. It may explode spontaneously. When heated to decomposition it emits toxic fumes of NO_x and Na_2O. See also NITRO COMPOUNDS and SODIUM COMPOUNDS.

SIW625 CAS:22755-25-5 ***HR: 3***
SODIUM-2-NITROTHIOPHENOXIDE
mf: $C_6H_4NNaO_2S$ mw: 177.15

SAFETY PROFILE: May explode when heated. When heated to decomposition it emits toxic fumes of SO_x, NO_x, and Na_2O. See also SODIUM COMPOUNDS and NITRO COMPOUNDS of AROMATIC HYDROCARBONS.

SIX000 CAS:13968-14-4 ***HR: 3***
SODIUM NITROXYLATE
mf: NNa_2O_2 mw: 92.0

Na(O:NONa)

SAFETY PROFILE: A very sensitive explosive. Explodes on contact with air, moisture, carbon dioxide. When heated to decomposition it emits very toxic fumes of NO_x and Na_2O.

SIX500 CAS:1072-15-7 ***HR: 3***
SODIUM NONYL SULFATE
mf: $C_9H_{19}O_4S{\cdot}Na$ mw: 246.33

SYN: SULFURIC ACID, MONONONYL ESTER, SODIUM SALT

TOXICITY DATA with REFERENCE
ivn-mus LD50:180 mg/kg CSLNX* NX#00169

CONSENSUS REPORTS: Reported in EPA TSCA Inventory.

SAFETY PROFILE: Poison by intravenous route. When heated to decomposition it emits toxic fumes of SO_x and Na_2O. See also SULFATES.

SIX550 CAS:12007-46-4 ***HR: 3***
SODIUM OCTAHYDROTRIBORATE
mf: B_3H_8Na mw: 63.48

SAFETY PROFILE: Ignites on contact with air, ether, dioxane. When heated to decomposition it emits toxic fumes of Na_2O. See also BORON COMPOUNDS and SODIUM COMPOUNDS.

SIY000 CAS:137-20-2 ***HR: 3***
SODIUM OLEYLMETHYLTAURIDE
mf: $C_{21}H_{40}NO_4S{\cdot}Na$ mw: 425.67

SYNS: ADINOL T ◇ CONCOGEL 2 CONCENTRATE ◇ HOSTAPON T ◇ IGEPON T-33 ◇ IGEPON T-43 ◇ IGEPON T 51 ◇ IGEPON T-71 ◇ IGEPON T-73 ◇ IGEPON T-77 ◇ IGEPON TE ◇ METAUPON PASTE ◇ N-METHYL-N-OLEOYLTAURINE SODIUM SALT ◇ Z-2-(METHYL(1-OXO-9-OCTADECENYL)AMINO)-ETHANESULFONIC ACID, SODIUM SALT ◇ NISSAN DIAPION S ◇ NISSAN DIAPON T ◇ OLEOYLMETHYLTAURINE SODIUM SALT ◇ OMT ◇ SODIUM-2-(N-METHYLOLEAMIDO)ETHANE-1-SULFONATE ◇ SODIUM METHYL OLEOYL TAURATE ◇ SODIUM N-METHYL-N-OLEOYLTAURATE ◇ SODIUM N-OLEOYL-N-METHYLATAURINE ◇ SODIUM N-OLEOYL-N-METHYLTAURATE

TOXICITY DATA with REFERENCE
eye-rbt 1% SEV JAPMA8 38,428,49
ivn-mus LD50:350 mg/kg JAPMA8 38,428,49
orl-mam LD50:5190 mg/kg FMCHA2 -,C130,83

CONSENSUS REPORTS: Reported in EPA TSCA Inventory.

SAFETY PROFILE: Poison by intravenous route. Mildly toxic by ingestion. A severe eye irritant. When heated to decomposition it emits very toxic fumes of NO_x, SO_x, and Na_2O.

SIY250 CAS:13721-39-6 ***HR: 3***
SODIUM ORTHOVANADATE
mf: $O_4V{\cdot}3Na$ mw: 183.91

PROP: Colorless, hexagonal prisms. Mp: 850-866°.

SYNS: SODIUM VANADATE ◇ SODIUM VANADIUM OXIDE ◇ TRISODIUM ORTHOVANADATE ◇ VANADIC(II) ACID, TRISODIUM SALT

TOXICITY DATA with REFERENCE
oms-mam:kdy 10 μmol/L JCLBA3 79,573,78
scu-rat LDLo:50 mg/kg 30ZIAO -,140,64
ipr-mus LD50:36300 μg/kg JTEHD6 12,749,83
scu-mus LDLo:50 mg/kg AJSNAO 1,347,17
orl-rbt LDLo:100 mg/kg EQSSDX 1,1,75
scu-rbt LDLo:9280 μg/kg PHRPA6 53,765,38

ivn-rbt LDLo:2 mg/kg 30ZIAO -,140,64
scu-gpg LDLo:1 mg/kg AJSNAO 1,347,17

CONSENSUS REPORTS: Reported in EPA TSCA Inventory.

ACGIH TLV: TWA 0.05 $mg(V_2O_5)/m^3$
NIOSH REL: (Vanadium Compounds) CL 0.05 $mg(V)/m^3/15M$

SAFETY PROFILE: Poison by ingestion, intraperitoneal, intravenous, and subcutaneous routes. Mutation data reported. When heated to decomposition it emits toxic fumes of VO_x and Na_2O. See also VANADIUM COMPOUNDS.

SIY500 CAS:62-76-0 ***HR: 3***
SODIUM OXALATE
mf: C_2O_4•2Na mw: 134.00

PROP: White, crystalline powder. D: 2.34.

SYNS: ETHANEDIOIC ACID, DISODIUM SALT ◇ NATRIUMOXALAT (GERMAN) ◇ OXALIC ACID, DISODIUM SALT

TOXICITY DATA with REFERENCE
ivn-hmn LDLo:17 mg/kg:CNS,CVS,KID MLDCAS 4,178,71
scu-mus LDLo:100 mg/kg AIPTAK 8,255,01
scu-cat LDLo:100 mg/kg HDWU** -,-,33

CONSENSUS REPORTS: Reported in EPA TSCA Inventory.

SAFETY PROFILE: Poison by subcutaneous route. Human systemic effects by intravenous route: coma, arrythmias, and decreased urine volume or anuria. When heated to decomposition it emits toxic fumes of Na_2O. See also OXALATES.

SIZ000 CAS:35249-69-5 ***HR: 2***
SODIUM OXYDIACETATE
mf: $C_4H_4O_5$•2Na mw: 178.06

SYN: OXYDIACETIC ACID, DISODIUM SALT

TOXICITY DATA with REFERENCE
orl-rat LD50:2440 mg/kg TXAPA9 28,313,74
skn-rbt LD50:1270 mg/kg TXAPA9 28,313,74

CONSENSUS REPORTS: Reported in EPA TSCA Inventory.

SAFETY PROFILE: Moderately toxic by ingestion and skin contact. When heated to decomposition it emits toxic fumes of Na_2O.

SIZ050 ***HR: 1***
SODIUM PANTOTHENATE
SAFETY PROFILE: A nuisance dust.

SIZ100 CAS:33634-75-2 ***HR: 3***
SODIUM PENTACARBONYL RHENATE
mf: C_5NaO_5Re mw: 349.25
$Na[Re(CO)_5]$

SAFETY PROFILE: Ignites spontaneously in air. When heated to decomposition it emits toxic fumes of Na_2O and Re. See also SODIUM COMPOUNDS, CARBONYLS, and RHENIUM COMPOUNDS.

SJA000 CAS:131-52-2 ***HR: 3***
SODIUM PENTACHLOROPHENATE
DOT: UN 2567
mf: C_6Cl_5O•Na mw: 288.30

PROP: Tan powder.

SYNS: DOW DORMANT FUNGICIDE ◇ DOWICIDE G-ST ◇ NAPCLOR-G ◇ PENTACHLOROPHENATE SODIUM ◇ PENTACHLOROPHENOL, SODIUM SALT ◇ PENTACHLOROPHENOXY SODIUM ◇ PENTAPHENATE ◇ SANTOBRITE ◇ SODIUM PCP ◇ SODIUM PENTACHLOROPHENATE (DOT) ◇ SODIUM PENTACHLOROPHENOL ◇ SODIUM PENTACHLOROPHENOLATE ◇ SODIUM PENTACHLOROPHENOXIDE ◇ WEEDBEADS

TOXICITY DATA with REFERENCE
mrc-bcs 5 ng/disc/24H MUREAV 40,19,76
orl-rat TDLo:360 mg/kg (8-19D preg):TER CHYCDW 13,8,79
orl-rat TDLo:360 mg/kg (8-19D preg):REP CHYCDW 13,8,79
orl-rat LD50:126 mg/kg CHYCDW 13,8,79
ihl-rat LD50:11700 μg/kg BECTA6 15,463,76
scu-rat LD50:66 mg/kg JPETAB 76,104,42
itr-rat LDLo:146 mg/kg MZUZA8 (9),29,59
orl-mus LD50:197 mg/kg 85GMAT -,106,82
ihl-mus LC50:240 $mg/m^3/2H$ 85GMAT -,106,82
skn-mus LD50:124 mg/kg 85GMAT -,106,82
scu-mus LDLo:56 mg/kg HBTXAC 5,125,59
itr-mus LDLo:164 mg/kg MZUZA8 (9),29,59
scu-dog LDLo:135 mg/kg HBTXAC 5,125,59
orl-rbt LD50:328 mg/kg MZUZA8 (9),29,59
skn-rbt LDLo:250 mg/kg JPETAB 76,104,42
ipr-rbt LDLo:50 mg/kg JIHTAB 23,239,41
scu-rbt LDLo:108 mg/kg JPETAB 76,104,42
ivn-rbt LDLo:22 mg/kg JPETAB 76,104,42
orl-gpg LDLo:250 mg/kg MZUZA8 (9),29,59
ihl-gpg LC50:341 $mg/m^3/2H$ 85GMAT -,106,82
skn-gpg LDLo:266 mg/kg HBTXAC 5,125,59
itr-gpg LDLo:120 mg/kg MZUZA8 (9),29,59

CONSENSUS REPORTS: EPA Extremely Hazardous Substances List. Chlorophenol compounds are on the Community Right-To-Know List. Reported in EPA TSCA Inventory. EPA Genetic Toxicology Program.

DOT Classification: ORM-A; Label: None: Poison B; Label: Poison

SAFETY PROFILE: Poison by ingestion, inhalation, skin contact, intravenous, intraperitoneal, subcutaneous, and intratracheal routes. An experimental teratogen. Experimental reproductive effects. Mutation data reported. When heated to decomposition it emits toxic fumes of Cl^- and Na_2O. See also CHLOROPHENOLS.

SJA500 CAS:22578-17-2 ***HR: 3***
SODIUM PENTAFLUOROSTANNITE
mf: $F_5Sn_2 \cdot Na$ mw: 355.37

SYN: mu-FLUOROTETRAFLUORODISTANNATE(1-),SODIUM

TOXICITY DATA with REFERENCE
orl-rat TDLo:447 mg/kg (1-20D preg):REP JONUAI 101,525,71
orl-rat LD50:221 mg/kg TXAPA9 22,304,72
ipr-rat LD50:70 mg/kg TXAPA9 22,304,72
ivn-rat LD50:12 mg/kg TXAPA9 22,304,72
orl-mus LD50:595 mg/kg TXAPA9 22,304,72
ivn-mus LD50:19 mg/kg TXAPA9 22,304,72

OSHA PEL: TWA 2 mg(Sn)/m^3; 2.5 mg(F)/m^3
ACGIH TLV: TWA 2 mg(Sn)/m^3
NIOSH REL: (Inorganic Fluorides) TWA 2.5 mg(F)/m^3

SAFETY PROFILE: Poison by ingestion, intraperitoneal and intravenous routes. Experimental reproductive effects. When heated to decomposition it emits toxic fumes of F^- and Na_2O. See also TIN COMPOUNDS and FLUORIDES.

SJB000 CAS:64057-57-4 ***HR: 3***
SODIUM PERACETATE
mf: $C_2H_3O_3 \cdot Na$ mw: 98.04

SYNS: SODIUM ETHANEPEROXOATE ◇ SODIUM PEROXYACETATE

TOXICITY DATA with REFERENCE
unr-mus TDLo:314 mg/kg:ETA RARSAM 3,193,63

SAFETY PROFILE: Questionable carcinogen with experimental tumorigenic data. An oxidizer. The dry salt explodes spontaneously at room temperature. When heated to decomposition it emits toxic fumes of Na_2O. See also PEROXIDES.

SJB350 CAS:10486-00-7 ***HR: 3***
SODIUM PERBORATE TETRAHYDRATE
mf: $BO_3 \cdot Na \cdot 4H_2O$ mw: 153.88

TOXICITY DATA with REFERENCE
eye-rbt 50 mg MOD GTPZAB 28(6),44,84
orl-hmn LDLo:214 mg/kg YKYUA6 31,855,80
orl-chd LDLo:250 mg/kg YKYUA6 31,855,80
orl-inf LDLo:400 mg/kg YKYUA6 31,855,80
orl-rat LD50:1200 mg/kg GTPZAB 28(6),44,84
orl-mus LD50:1060 mg/kg GTPZAB 28(6),44,84
ipr-mus LD50:538 mg/kg COREAF 257,791,63

SAFETY PROFILE: Human poison by ingestion. Moderately toxic experimentally by ingestion and intraperitoneal routes. An eye irritant. When heated to decomposition it emits toxic fumes of Na_2O. See also BORON COMPOUNDS.

SJB500 CAS:7790-28-5 ***HR: 3***
SODIUM PERIODATE
mf: $IO_4 \cdot Na$ mw: 213.89

SYN: SODIUM METAPERIODATE

TOXICITY DATA with REFERENCE
ipr-mus LD50:58 mg/kg COREAF 257,791,63

CONSENSUS REPORTS: Reported in EPA TSCA Inventory.

SAFETY PROFILE: Poison by intraperitoneal route. A powerful oxidizer. When heated to decomposition it emits toxic fumes of I^- and Na_2O. See also IODINE COMPOUNDS.

SJC000 CAS:10101-50-5 ***HR: 3***
SODIUM PERMANGANATE
DOT: UN 1503
mf: $MnO_4 \cdot Na$ mw: 141.93

PROP: Purple to red-black crystals. Mp: decomp.

SYNS: PERMANGANATE de SODIUM (FRENCH) ◇ PERMANGANIC ACID, SODIUM SALT

CONSENSUS REPORTS: Manganese and its compounds are on the Community Right To Know List. Reported in EPA TSCA Inventory.

OSHA PEL: CL 5 mg(Mn)/m^3
ACGIH TLV: TWA 5 mg(Mn)/m^3
DOT Classification: Oxidizer; Label: Oxidizer

SAFETY PROFILE: Probably a severe irritant to the skin, eyes, and mucous membranes. A powerful oxidizer and fire hazard. Explosive reaction with acetic acid; acetic anhydride. Reacts vigorously with combustibles. When heated to decomposition it emits toxic fumes of Na_2O. See also MANGANESE COMPOUNDS, PERMANGANATES, and POTASSIUM PERMANGANATE.

SJC500 CAS:1313-60-6 ***HR: 3***
SODIUM PEROXIDE
DOT: UN 1504
mf: Na_2O_2 mw: 77.98

PROP: White powder turning yellow when heated. Mp: decomp @ 460°, bp: decomp, d: 2.805.

SYNS: DISODIUM DIOXIDE ◇ DISODIUM PEROXIDE ◇ FLOCOOL 180 ◇ SODIUM DIOXIDE ◇ SODIUM OXIDE (Na2-O2) ◇ SOLOZONE

CONSENSUS REPORTS: Reported in EPA TSCA Inventory.

DOT Classification: Oxidizer; Label: Oxidizer

SAFETY PROFILE: A severe irritant to skin, eyes, and mucous membranes. Dangerous fire hazard by chemical reaction; a powerful oxidizing agent. Reacts explosively or violently under the appropriate conditions with water; acids; powdered metals; acetic acid; acetic anhydride; Al; (Al + CO_2); aluminum + aluminum chloride; almond oil; $(NH_4)_2S_2O_8$; aniline; Sb; As; benzene; boron nitride; calcium acetylide; charcoal; Cu; cotton wool; (KNO_3 + dextrose); diethyl ether; fibrous materials + water; glucose + potassium nitrate; hexamethylenetetramine; hydrogen sulfide; hydroxy compounds (e.g., ethanol, ethylene glycol, glycerol, sugar); magnesium; (Mg + CO_2); MnO_2; metals; metals + carbon dioxide + water; non-metals (e.g., carbon; phosphorus; antimony; arsenic; boron; sulfur; selenium); non-metal halides (e.g., diselenium dichloride; disulfur dichloride; phosphorus trichloride); organic matter; paraffin; K; silver chloride + charcoal; soap; Na; sodium dioxide; SCl; Sn; Zn; wood; peroxyformic acid; reducing materials. Will react with water or steam to produce heat and toxic fumes. To fight fire, use carbon dioxide or dry chemical. Combustible materials ignited by contact with sodium peroxide should be smothered with soda ash, salt or dolomite mixtures. Chemical fire extinguishers should not be used. If the fire cannot be smothered, it should be flooded with large quantities of water from a hose. When heated to decomposition it emits toxic fumes of Na_2O. See also SODIUM HYDROXIDE and PEROXIDES, INORGANIC.

SJD000 ***HR: 2***
SODIUM PEROXYBORATE
mf: BO_3•Na mw: 81.80

SYNS: BORIC ACID, SODIUM SALT ◇ SODIUM PERBORATE ◇ SODIUM PEROXOBORATE

TOXICITY DATA with REFERENCE
dnr-esc 3300 pmol/plate MUREAV 21,171,73

CONSENSUS REPORTS: Reported in EPA TSCA Inventory.

SAFETY PROFILE: Mutation data reported. A powerful oxidizer. Can be detonated by light friction. Catalytically decomposed by heavy metals and their salts. When heated to decomposition it emits toxic fumes of Na_2O. See also BORON COMPOUNDS and PEROXIDES.

SJD500 CAS:13472-33-8 ***HR: 2***
SODIUM PERRHENATE
mf: NaO_4Re mw: 273.19

SYN: RHENIUM(VII) SODIUM OXIDE

TOXICITY DATA with REFERENCE
ipr-rat LDLo:1320 mg/kg PSEBAA 45,576,40

CONSENSUS REPORTS: Reported in EPA TSCA Inventory.

SAFETY PROFILE: Moderately toxic by intraperitoneal route. When heated to decomposition it emits toxic fumes of Na_2O and Re. See also RHENIUM COMPOUNDS and SODIUM MONOXIDE.

SJE000 CAS:7775-27-1 ***HR: 3***
SODIUM PERSULFATE
DOT: UN 1505
mf: O_8S_2•2Na mw: 238.10

PROP: White, crystalline powder. Sol in water; decomp by alc.

SYNS: PERSULFATE de SODIUM (FRENCH) ◇ SODIUM PEROXYDISULFATE

TOXICITY DATA with REFERENCE
ipr-mus LD50:226 mg/kg COREAF 257,791,63
ivn-rbt LDLo:178 mg/kg MEIEDD 10,1239,83

CONSENSUS REPORTS: Reported in EPA TSCA Inventory.

DOT Classification: Oxidizer; Label: Oxidizer

SAFETY PROFILE: Poison by intraperitoneal and intravenous routes. A powerful oxidizer; can cause fires. When heated to decomposition it emits toxic fumes of SO_x and Na_2O. See also SULFATES.

SJF000 CAS:139-02-6 ***HR: 3***
SODIUM PHENOXIDE
DOT: UN 2497
mf: C_6H_5O•Na mw: 116.10

PROP: White, deliq crystals.

SYNS: PHENOL SODIUM SALT ◇ SODIUM CARBOLATE ◇ SODIUM PHENATE ◇ SODIUM PHENOLATE, solid (DOT)

TOXICITY DATA with REFERENCE
scu-mus LDLo:350 mg/kg AEXPBL 52,220,05
scu-frg LDLo:100 mg/kg AEXPBL 52,220,05

CONSENSUS REPORTS: Reported in EPA TSCA Inventory.

DOT Classification: Corrosive Material; Label: Corrosive

SAFETY PROFILE: Poison by subcutaneous route. A

corrosive irritant to skin, eyes, and mucous membranes. When heated to decomposition it emits toxic fumes of Na_2O. See also PHENOL and SODIUM HYDROXIDE.

SJF500 CAS:1004-22-4 ***HR: 2***
SODIUM PHENYLACETYLIDE
mf: C_8H_5Na mw: 124.12

$$NaC \equiv CC_6H_5$$

SAFETY PROFILE: A dangerous fire hazard when exposed to heat or flame. The powder moistened with ether ignites in air. When heated to decomposition it emits toxic fumes of Na_2O. See also ACETYLIDES.

SJG000 CAS:64046-97-5 ***HR: 3***
SODIUM PHENYL-β-AMINOPROPIONAMIDE-p-ARSONATE
mf: $C_9H_{12}AsN_2O_4$•Na mw: 310.14

SYNS: N-(2-CARBAMOYLETHYL)ARSANILIC ACID SODIUM SALT ◇ PROPARSAMIDE

TOXICITY DATA with REFERENCE
orl-cat LDLo:200 mg/kg PSEBAA 29,125,31
orl-rbt LDLo:200 mg/kg PSEBAA 29,125,31
orl-gpg LDLo:100 mg/kg PSEBAA 29,125,31

CONSENSUS REPORTS: Arsenic and its compounds are on the Community Right To Know List.

OSHA PEL: TWA 0.5 mg(As)/m^3
ACGIH TLV: TWA 0.2 mg(As)/m^3

SAFETY PROFILE: Poison by ingestion. When heated to decomposition it emits very toxic fumes of As, NO_x, and Na_2O. See also ARSENIC COMPOUNDS.

SJH000 CAS:5949-18-8 ***HR: 2***
SODIUM-2-PHENYLCINCHONINATE
mf: $C_{16}H_{11}NO_2$•Na mw: 272.27

SYNS: ATOPHAN-NATRIUM (GERMAN) ◇ ATOPHAN SODIUM ◇ CINCHOPHEN SODIUM ◇ CINCHOPHEN, SODIUM SALT ◇ 2-PHENYL-4-QUINOLONECARBOXYLIC AICD SODIUM SALT ◇ SODIUM CINCHOPHEN

TOXICITY DATA with REFERENCE
scu-mus LDLo:1000 mg/kg HBAMAK 4,1310,35
orl-dog LDLo:3600 mg/kg HBAMAK 4,1310,35
scu-dog LDLo:3600 mg/kg HBAMAK 4,1310,35
orl-cat LDLo:3600 mg/kg HBAMAK 4,1310,35
scu-cat LDLo:3600 mg/kg HBAMAK 4,1310,35
orl-rbt LDLo:1000 mg/kg HBAMAK 4,1310,35
scu-rbt LDLo:500 mg/kg HBAMAK 4,1310,35

SAFETY PROFILE: Moderately toxic by ingestion and subcutaneous routes. When heated to decomposition it emits toxic fumes of NO_x and Na_2O.

SJH090 CAS:7558-79-4 ***HR: 3***
SODIUM PHOSPHATE, DIBASIC
mf: HO_4P•2Na mw: 141.96

PROP: Colorless, translucent crystals or white powder. Sol in water; very sltly sol in alc.

SYNS: DIBASIC SODIUM PHOSPHATE ◇ DISODIUM HYDROGEN PHOSPHATE ◇ DISODIUM MONOHYDROGEN PHOSPHATE ◇ DISODIUM ORTHOPHOSPHATE ◇ DISODIUM PHOSPHATE ◇ DISODIUM PHOSPHORIC ACID ◇ DSP ◇ EXSICCATED SODIUM PHOSPHATE ◇ NATRIUMPHOSPHAT (GERMAN) ◇ PHOSPHORIC ACID, DISODIUM SALT ◇ SODA PHOSPHATE ◇ SODIUM HYDROGEN PHOSPHATE ◇ SODIUM MONOHYDROGEN PHOSPHATE (2 : 1 : 1)

TOXICITY DATA with REFERENCE
skn-rbt 500 mg/24H MLD 28ZPAK -,16,72
eye-rbt 500 mg/24H MLD 28ZPAK -,16,72
orl-rat LD50:17 g/kg 28ZPAK -,16,72
ipr-rat LDLo:1000 mg/kg BIZEA2 163,226,25
scu-rat LDLo:1000 mg/kg BIZEA2 163,226,25
ims-rat LDLo:1000 mg/kg BIZEA2 163,226,25

CONSENSUS REPORTS: Reported in EPA TSCA Inventory.

SAFETY PROFILE: Poison by intravenous route. Moderately toxic by intraperitoneal, subcutaneous, and intramuscular routes. Mildly toxic by ingestion. A skin and eye irritant. When heated to decomposition it emits toxic fumes of PO_x and Na_2O. See also PHOSPHATES.

SJH100 CAS:7558-80-7 ***HR: 3***
SODIUM PHOSPHATE, MONOBASIC
mf: H_2O_4P•Na mw: 119.98

PROP: White crystalline powder or granules; odorless. Hygroscopic; sol in water; insol in alc.

SYNS: MONOSODIUM DIHYDROGEN PHOSPHATE ◇ MONOSODIUM PHOSPHATE ◇ MONOSORB XP-4 ◇ PRIMARY SODIUM PHOSPHATE ◇ SODIUM ACID PHOSPHATE ◇ SODIUM BIPHOSPHATE ◇ SODIUM BIPHOSPHATE anhydrous ◇ SODIUM DIHYDROGEN PHOSPHATE (1 : 2 : 1)

TOXICITY DATA with REFERENCE
eye-hmn 50 mg MLD ARZNAD 9,349,59
eye-rbt 150 mg MLD ARZNAD 9,349,59
orl-rat LD50:8290 mg/kg 28ZPAK -,16,72
ims-rat LD50:250 mg/kg EMSUA8 4,223,46

CONSENSUS REPORTS: Reported in EPA TSCA Inventory.

SAFETY PROFILE: Poison by intramuscular route. Mildly toxic by ingestion. A human and experimental eye irritant. When heated to decomposition it emits toxic fumes of PO_x and Na_2O. See also PHOSPHATES.

SJH200 CAS:7601-54-9 ***HR: 2***
SODIUM PHOSPHATE, TRIBASIC
mf: O_4P•3Na mw: 163.94

PROP: White crystals or crystalline powder; odorless. Sol in water; insol in alc.

SYNS: DRI-TRI ◇ EMULSIPHOS 440/660 ◇ NUTRIFOS STP ◇ PHOSPHORIC ACID, TRISODIUM SALT ◇ SODIUM PHOSPHATE ◇ SODIUM PHOSPHATE, anhydrous ◇ TRIBASIC SODIUM PHOSPHATE ◇ TRINATRIUMPHOSPHAT (GERMAN) ◇ TRISODIUM ORTHOPHOSPHATE ◇ TRISODIUM PHOSPHATE ◇ TROMETE ◇ TSP

TOXICITY DATA with REFERENCE
sln-dmg-orl 11 pph DRISAA 20,87,46
ivn-rbt LDLo:1580 mg/kg HBAMAK 4,1289,35

CONSENSUS REPORTS: Reported in EPA TSCA Inventory.

SAFETY PROFILE: Moderately toxic by intravenous route. Mutation data reported. A strong, caustic material. When heated to decomposition it emits toxic fumes of Na_2O and PO_x. See also PHOSPHATES.

SJI500 CAS:12058-85-4 ***HR: 2***
SODIUM PHOSPHIDE
DOT: UN 1432
mf: PNa_3 mw: 99.94

PROP: Red crystals. Mp: decomp.

SYN: PHOSPHURE de SODIUM (FRENCH)

TOXICITY DATA with REFERENCE
ihl-rat LCLo:580 ppm/1H ZGSHAM 25,279,33
ihl-cat LCLo:173 ppm/2H ZGSHAM 25,279,33
ihl-gpg LCLo:288 ppm/2H ZGSHAM 25,279,33

CONSENSUS REPORTS: Reported in EPA TSCA Inventory.

DOT Classification: Flammable Solid; Label: Flammable Solid and Dangerous When Wet

SAFETY PROFILE: Moderately toxic by inhalation. Flammable when exposed to heat or flame. Reacts violently with water to yield phosphine. When heated to decomposition it emits toxic fumes of PO_x and Na_2O. See also PHOSPHIDES.

SJJ000 CAS:51312-42-6 ***HR: 2***
SODIUM PHOSPHOTUNGSTATE
mf: $Na_4O_2{\cdot}O_5P_2{\cdot}O_{36}W_{12}{\cdot}18H_2O$ mw: 3372.46

SYNS: SODIUM TUNGSTOPHOSPHATE ◇ TUNGSTOPHOSPHORIC ACID, SODIUM SALT

TOXICITY DATA with REFERENCE
orl-rat LD50:1600 mg/kg HYSAAV 31,197,66
orl-mus LD50:700 mg/kg HYSAAV 31,197,66

CONSENSUS REPORTS: Reported in EPA TSCA Inventory.

ACGIH TLV: TWA 1 mg(W)/m^3; STEL 3 mg(W)/m^3
NIOSH REL: TWA 1 mg(W)/m^3

SAFETY PROFILE: Moderately toxic by ingestion. When heated to decomposition it emits toxic fumes of PO_x and Na_2O. See also TUNGSTEN COMPOUNDS.

SJJ175 CAS:10040-45-6 ***HR: 2***
SODIUM PICOSULFATE
mf: $C_{18}H_{13}NO_8S_2{\cdot}2Na$ mw: 481.42

PROP: White, crystalline solid from ethanol or methanol. Mp: 272-275° (decomp). Readily sol in water; sltly sol in alc; practically insol in most organic solvents.

SYNS: DA-1773 ◇ DISODIUM-4,4'-DISULFOXYDIPHENYL-(2-PYRIDYL)METHANE ◇ EVANOL ◇ GUTTALAX ◇ LAXIDOGOL ◇ LAXOBERAL ◇ LAXOBERON ◇ NEOPAX ◇ 2-PICOLYIDENEBIS(p-PHENYL SODIUM SULFATE) ◇ 4,4'-(2-PICOLYLIDENE)BIS (PHENYLSULFURIC ACID) DISODIUM SALT ◇ PICOSULFATE SODIUM ◇ PICOSULFOL ◇ 4,4'-(2-PYRIDINYLMETHYLENE)BISPHENOL BIS(HYDROGEN SULFATE) (ESTER) DISODIUM SALT ◇ 4,4'-(2-PYRIDYLMETHYLENE)DIPHENOLBIS(HYDROGENSULFATE) (ESTER) DISODIUM SALT

TOXICITY DATA with REFERENCE
orl-rat TDLo:270 mg/kg (female 17-22D post):REP IYKEDH 8,366,77
orl-rat TDLo:6 g/kg (female 9-14D post):TER IYKEDH 8,366,77
orl-rat LD50:5 g/kg MEIEDD 10,1067,83
ipr-rat LD50:3 g/kg IYKEDH 8,341,77
scu-rat LD50:6980 mg/kg NAIZAM 28(2),258,77
orl-mus LD50:14500 mg/kg IYKEDH 8,341,77
ipr-mus LD50:2900 mg/kg IYKEDH 8,341,77
scu-mus LD50:6420 mg/kg NAIZAM 28(2),258,77

SAFETY PROFILE: Moderately toxic by intraperitoneal route. Mildly toxic by ingestion. An experimental teratogen. Other experimental reproductive effects. A cathartic. When heated to decomposition it emits toxic fumes of SO_x, NO_x, and Na_2O. See also SULFATES.

SJJ190 CAS:73771-13-8 ***HR: 3***
SODIUM PICRATE
mf: $C_6H_2N_3NaO_7$ mw: 251.09

$NaOC_6H_2(NO_2)_3$

SYN: SODIUM-2,4,6-TRINITROPHENOXIDE

SAFETY PROFILE: Explodes on impact. When heated to decomposition it emits toxic fumes of NO_x and Na_2O. See also PICRIC ACID and SODIUM COMPOUNDS.

SJJ200 CAS:59703-84-3 ***HR: 2***
SODIUM PIPERACILLIN
mf: $C_{23}H_{26}N_5O_7S{\cdot}Na$ mw: 539.59

SYNS: CL 227193 ◇ PIPERACILLIN SODIUM ◇ PIPRACIL ◇ PIPRIL ◇ T 1220

TOXICITY DATA with REFERENCE
ivn-mus TDLo:10 g/kg (6-15D preg):TER NKRZAZ 25,915,77
ivn-wmn TDLo:2180 mg/kg/6D-I:BLD SMJOAV 79,255,86
ims-wmn TDLo:960 mg/kg/3D-I:BLD SMJOAV 78,363,85
ipr-rat LD50:7600 mg/kg NKRZAZ 25,816,77
scu-rat LD50:8800 mg/kg IYKEDH 10,884,79
ivn-rat LD50:2260 mg/kg IYKEDH 10,884,79
ipr-mus LD50:9770 mg/kg NKRZAZ 25,816,77
ivn-mus LD50:4900 mg/kg NKRZAZ 25,816,77

SAFETY PROFILE: Moderately toxic by intravenous route. An experimental teratogen. Human systemic effects: normocytic anemia, leukopenia, hemorrhage. When heated to decomposition it emits toxic fumes of SO_x, NO_x and Na_2O.

SJJ500 CAS:1307-82-0 ***HR: 3***
SODIUM PLATINIC CHLORIDE
mf: $Cl_6Pt•2Na•4H_2O$ mw: 571.83

SYNS: PLATINIC SODIUM CHLORIDE ◇ SODIUM CHLOROPLATINATE

TOXICITY DATA with REFERENCE
ihl-hmn TCLo:0.9 $\mu g/m^3$:PUL BJIMAG 2,92,45
scu-rbt LDLo:180 mg/kg BSIBAC 8,1152,33

OSHA PEL: TWA 0.002 mg(Pt)/m^3
ACGIH TLV: TWA 0.002 mg(Pt)/m^3

SAFETY PROFILE: Poison by subcutaneous route. Human systemic effects by inhalation of very small amounts: pulmonary changes. When heated to decomposition it emits toxic fumes of Cl^- and Na_2O. See also PLATINUM COMPOUNDS.

SJK000 CAS:9003-04-7 ***HR: 1***
SODIUM POLYACRYLATE

TOXICITY DATA with REFERENCE
eye-rbt 2 mg MOD PSTGAW 20,16,53

CONSENSUS REPORTS: Reported in EPA TSCA Inventory.

SAFETY PROFILE: An eye irritant. When heated to decomposition it emits toxic fumes of Na_2O.

SJK200 ***HR: 3***
SODIUM POLYOXYETHYLENE ALKYL ETHER SULFATE

TOXICITY DATA with REFERENCE
orl-rat LD50:3870 mg/kg TOIZAG 25,876,78
scu-rat LD50:5050 mg/kg TOIZAG 25,876,78
ivn-rat LD50:146 mg/kg TOIZAG 25,876,78
orl-mus LD50:5100 mg/kg TOIZAG 25,876,78
scu-mus LD50:4340 mg/kg TOIZAG 25,876,78
ivn-mus LD50:223 mg/kg TOIZAG 25,876,78

SAFETY PROFILE: Poison by intravenous route. Moderately toxic by ingestion. When heated to decomposition it emits toxic fumes of SO_x and Na_2O. See also ETHERS and SULFATES.

SJK375 CAS:25704-18-1 ***HR: 1***
SODIUM POLYSTYRENE SULFONATE
mf: $(C_8H_8O_3S•Na)_n$

SYNS: 4-ETHENYL-BENZENESULFONIC ACID SODIUM SALT, HOMOPOLYMER (9CI) ◇ KAYEXALATE ◇ POLY(SODIUM p-STYRENESULFONATE)

TOXICITY DATA with REFERENCE
orl-rat TDLo:28 g/kg (8-14D preg):REP OYYAA2 4,79,70
orl-rat LD50:16 g/kg OYYAA2 4,79,70
orl-mus LD50:10 g/kg OYYAA2 4,79,70

CONSENSUS REPORTS: Reported in EPA TSCA Inventory.

SAFETY PROFILE: Mildly toxic by ingestion. Experimental reproductive effects. When heated to decomposition it emits toxic fumes of SO_x and Na_2O. See also SULFONATES.

SJK400 CAS:1099-87-2 ***HR: 3***
SODIUM PRASTERONE SULFATE
mf: $C_{19}H_{26}O_5S•Na$ mw: 389.50

SYN: (3-β)-3-(SULFOOXY)-ANDROST-5-EN-17-ONE SODIUM SALT (9CI)

TOXICITY DATA with REFERENCE
ipr-rat LD50:523 mg/kg NIIRDN 6,687,82
scu-rat LD50:1005 mg/kg NIIRDN 6,687,82
ivn-rat LD50:468 mg/kg NIIRDN 6,687,82
ipr-mus LD50:460 mg/kg NIIRDN 6,687,82
scu-mus LD50:899 mg/kg NIIRDN 6,687,82
ivn-mus LD50:274 mg/kg NIIRDN 6,687,82

SAFETY PROFILE: Poison by intravenous route. Moderately toxic by intraperitoneal and subcutaneous routes. A steroid. When heated to decomposition it emits toxic fumes of SO_x and Na_2O.

SJK410 CAS:1099-87-2 ***HR: 3***
SODIUM PRASTERONE SULFATE DIHYDRATE
mf: $C_{19}H_{26}O_5S•Na$ mw: 426.55

SYN: 3-β-HYDROXY-5-ANDROSTEN-17-ONE SODIUM SULFATE DIHYDRATE

TOXICITY DATA with REFERENCE
ipr-rat LD50:559 mg/kg IYKEDH 12,668,81
scu-rat LD50:1005 mg/kg IYKEDH 12,668,81

ivn-rat LD50:468 mg/kg IYKEDH 12,668,81
ipr-mus LD50:460 mg/kg IYKEDH 12,668,81
scu-mus LD50:899 mg/kg IYKEDH 12,668,81
ivn-mus LD50:274 mg/kg IYKEDH 12,668,81

SAFETY PROFILE: Poison by intravenous route. Moderately toxic by intraperitoneal and subcutaneous routes. A steroid. When heated to decomposition it emits toxic fumes of SO_x and Na_2O.

SJK475 CAS:74203-61-5 ***HR: 2***
SODIUM p-2-PROPANOL-OXY-PHENYLARSONATE
mf: $C_9H_{10}AsO_5 \cdot Na$ mw: 296.10

SYNS: p-ACETONYLOXYBENZENEARSONIC ACID SODIUM SALT ◇ BENZENEARSONIC ACID, p-ACETONYLOXY-, SODIUM SALT

TOXICITY DATA with REFERENCE
orl-rat LDLo:4000 mg/kg UCPHAQ 1,291,40

OSHA PEL: TWA 0.5 mg(As)/m^3

SAFETY PROFILE: Moderately toxic by ingestion. When heated to decomposition it emits toxic fumes of As.

SJL500 CAS:137-40-6 ***HR: 2***
SODIUM PROPIONATE
mf: $C_3H_5O_2 \cdot Na$ mw: 96.07

PROP: Transparent crystals or granules; nearly odorless. Very sol in water; sltly sol in alc.

SYNS: NATRIUMPROPIONAT (GERMAN) ◇ PROPANOIC ACID, SODIUM SALT

TOXICITY DATA with REFERENCE
scu-mus LD50:2100 mg/kg ZGEMAZ 113,536,44
skn-rbt LD50:1640 mg/kg JIHTAB 31,60,49
unr-rbt LDLo:6750 mg/kg HYSAAV 35,433,70

CONSENSUS REPORTS: Reported in EPA TSCA Inventory.

SAFETY PROFILE: Moderately toxic by skin contact and subcutaneous routes. Mildly toxic by unspecified routes. An allergen. When heated to decomposition it emits toxic fumes of Na_2O.

SJM500 CAS:13517-26-5 ***HR: 3***
SODIUM PYROVANADATE
mf: $O_7V_2 \cdot 4Na$ mw: 305.84

PROP: Colorless, hexagonal plates. Mp: 632-654°.

TOXICITY DATA with REFERENCE
scu-rat LDLo:67 mg/kg AJSNAO 1,347,17
scu-mus LDLo:94 mg/kg AJSNAO 1,347,17
ivn-rbt LDLo:5 mg/kg AJSNAO 1,347,17
scu-gpg LDLo:2 mg/kg AJSNAO 1,347,17

CONSENSUS REPORTS: Reported in EPA TSCA Inventory.

ACGIH TLV: TWA 0.05 mg(V_2O_5)/m^3
NIOSH REL: CL 0.05 mg(V)/m^3/15M

SAFETY PROFILE: Poison by subcutaneous and intravenous routes. When heated to decomposition it emits toxic fumes of Na_2O and VO_x. See also VANADIUM COMPOUNDS.

SJN000 CAS:13497-05-7 ***HR: 3***
SODIUM RETINOATE
mf: $C_{20}H_{27}O_2 \cdot Na$ mw: 322.46

SYN: RETINOIC ACID, SODIUM SALT

TOXICITY DATA with REFERENCE
orl-ham TDLo:23 mg/kg (female 7D post):TER TJADAB 5,103,72
ivn-ham LDLo:34 mg/kg TJADAB 5,103,72

SAFETY PROFILE: Poison by intravenous route. An experimental teratogen. When heated to decomposition it emits toxic fumes of Na_2O.

SJN500 CAS:5323-95-5 ***HR: 3***
SODIUM RICINOLEATE
mf: $C_{18}H_{33}O_3 \cdot Na$ mw: 320.50

PROP: White-yellow powder. Sol in water and alc.

SYN: RICINOLEIC ACID, SODIUM SALT

TOXICITY DATA with REFERENCE
ipr-mus LDLo:250 mg/kg CBCCT* 8,491,56

CONSENSUS REPORTS: Reported in EPA TSCA Inventory.

SAFETY PROFILE: Poison by intraperitoneal route. When heated to decomposition it emits toxic fumes of Na_2O.

SJN650 CAS:15105-92-7 ***HR: 3***
SODIUM RIFOMYCIN SV
mf: $C_{37}H_{47}NO_{12} \cdot Na$ mw: 720.84

TOXICITY DATA with REFERENCE
orl-rat LD50:2680 mg/kg FRPSAX 16,235,61
ipr-rat LD50:480 mg/kg FRPSAX 16,235,61
scu-rat LD50:1120 mg/kg FRPSAX 16,235,61
orl-mus LD50:2120 mg/kg FRPSAX 16,235,61
ipr-mus LD50:625 mg/kg FRPSAX 16,235,61
scu-mus LD50:1080 mg/kg FRPSAX 16,235,61
ivn-mus LD50:550 mg/kg FRPSAX 16,235,61
ivn-dog LD50:350 mg/kg FRPSAX 16,235,61

SAFETY PROFILE: Poison by intravenous route. Moderately toxic by ingestion, subcutaneous, and intraperitoneal routes. When heated to decomposition it emits toxic fumes of Na_2O and NO_x.

SJN700 CAS:128-44-9 ***HR: 3***
SODIUM SACCHARIN
mf: $C_7H_4NO_3S \cdot Na$ mw: 205.17

PROP: White crystals or crystalline powder; odorless, very sweet taste. Sol in water, alc.

SYNS: ARTIFICIAL SWEETENING SUBSTANZ GENDORF 450 ◇ CRISTALLOSE ◇ CRYSTALLOSE ◇ DAGUTAN ◇ KRISTALLOSE ◇ MADHURIN ◇ ODA ◇ SACCHARIN ◇ SACCHARIN SOLUBLE ◇ SACCHARIN, SODIUM ◇ SACCHARIN, SODIUM SALT ◇ SACCHARINE SOLUBLE ◇ SACCHARINNATRIUM ◇ SACCHAROIDUM NATRICUM ◇ SAXIN ◇ SODIUM-1,2 BENZISOTHIAZOLIN-3-ONE-1,1-DIOXIDE ◇ SODIUM o-BENZOSULFIMIDE ◇ SODIUM BENZOSULPHIMIDE ◇ SODIUM-o-BENZOSULPHIMIDE ◇ SODIUM-2-BENZOSULPHIMIDE ◇ SODIUM SACCHARIDE ◇ SODIUM SACCHARINATE ◇ SODIUM SACCHARINE ◇ SOLUBLE GLUSIDE ◇ SOLUBLE SACCHARIN ◇ SUCCARIL ◇ SUCRA ◇ o-SULFONBENZOIC ACID IMIDE SODIUM SALT ◇ SULPHOBENZOIC IMIDE, SODIUM SALT ◇ SWEETA ◇ SYKOSE ◇ WILLOSETTEN

TOXICITY DATA with REFERENCE
mrc-smc 2 g/L MUREAV 67,215,79
cyt-hmn:leu 500 mg/L MUREAV 32,81,75
sce-hmn:leu 20 μmol/L ENMUDM 1,177,79
orl-mus TDLo:103 g/kg (30D male):TER IJMRAQ 60,599,72
orl-mus TDLo:2 g/kg (MGN):REP TOLED5 19,267,83
orl-rat TDLo:1092 g/kg/1Y-C:CAR GANNA2 74,8,83
imp-mus TDLo:176 mg/kg:NEO SCIEAS 168,1238,70
orl-rat TD:1330 g/kg/95W-C:ETA CBINA8 11,225,75
orl-rat LD50:14200 mg/kg FCTXAV 6,313,68
ipr-rat LD50:7100 mg/kg FCTXAV 6,313,68
orl-mus LD50:17500 mg/kg FCTXAV 6,313,68
ipr-mus LDLo:512 mg/kg CBCCT* 2,302,50
scu-mus LDLo:7 g/kg YKYUA6 32,1367,81
orl-rbt LDLo:4 g/kg YKYUA6 32,1367,81

CONSENSUS REPORTS: IARC Cancer Review: Group 2B IMEMDT 7,334,87; Animal Sufficient Evidence IMEMDT 22,111,80. EPA Genetic Toxicology Program. Reported in EPA TSCA Inventory.

SAFETY PROFILE: Suspected carcinogen with experimental carcinogenic, neoplastigenic, tumorigenic, and teratogenic data. Moderately toxic by ingestion and intraperitoneal routes. A promoter. Experimental reproductive effects. Human mutation data reported. When heated to decomposition it emits very toxic fumes of SO_x, Na_2O, and NO_x.

SJO000 CAS:54-21-7 ***HR: 3***
SODIUM SALICYLATE
mf: $C_7H_5O_3 \cdot Na$ mw: 160.11

PROP: White, odorless crystals, scales or powder.

SYNS: ALYSINE ◇ ARDALL ◇ AROALL ◇ CLIN ◇ DIURETIN ◇ ENTEROSALICYL ◇ ENTEROSALIL ◇ 2-HYDROXYBENZOIC ACID MONOSODIUM SALT ◇ o-HYDROXYBENZOIC SODIUM SALT ◇ IDOCYL NOVUM ◇ KERASALICYL ◇ KEROSAL ◇ MAGSALYL ◇ NADISAL ◇ NEO-SALICYL ◇ PARBOCYL-REV ◇ SALICYLIC ACID, SODIUM SALT ◇ SALISOD ◇ SALSONIN ◇ SODIUM-o-HYDROXYBENZOATE ◇ SODIUM SALICYLIC ACID

TOXICITY DATA with REFERENCE
dnd-rat:lvr 240 mg/L BIORAK 34,934,69
dnd-rat:emb 30 mg/L BIORAK 34,934,69
orl-rat TDLo:375 mg/kg (female 8-10D post):REP NETOD7 6,171,84
orl-mus TDLo:665 mg/kg (female 17D post):TER APTOA6 29,250,71
unr-wmn TDLo:1400 mg/kg:CNS,PUL,SKN JAMAAP 126,806,44
mul-chd TDLo:2970 mg/kg/13D:CNS,GIT AJDCAI 69,37,45
orl-hmn LDLo:700 mg/kg NEJMAG 232,617,45
orl-rat LD50:1200 mg/kg ARZNAD 5,572,55
ipr-rat LD50:542 mg/kg KSRNAM 13,791,79
scu-rat LDLo:800 mg/kg RMSRA6 15,561,1895
orl-mus LD50:540 mg/kg ARZNAD 5,572,55
ipr-mus LD50:560 mg/kg PJPPAA 25,127,73
ivn-mus LD50:560 mg/kg KSRNAM 13,791,79
ims-mus LD50:760 mg/kg APSVAM 59,517,70
orl-dog LDLo:450 mg/kg HBAMAK 4,1392,35
scu-dog LDLo:200 mg/kg RMSRA6 15,561,1895
ivn-rbt LD50:415 mg/kg KSRNAM 13,791,79

CONSENSUS REPORTS: Reported in EPA TSCA Inventory.

SAFETY PROFILE: Experimental poison by subcutaneous route. Moderately toxic to humans by ingestion. Moderately toxic experimentally by ingestion, intraperitoneal, and intravenous routes. An experimental teratogen. Human systemic effects by multiple and unspecified routes: toxic psychosis, excitement, respiratory stimulation, nausea or vomiting and sweating. Experimental reproductive effects. Mutation data reported. A powerful irritant which affects the central nervous system. Incompatible with ferric salts; mineral acids; iodine; lead acetate; silver nitrate; sodium phosphate powder. When heated to decomposition it emits toxic fumes of Na_2O.

SJP000 CAS:3019-89-4 ***HR: 2***
SODIUM SALT-m-CRESOL
mf: $C_7H_7O \cdot Na$ mw: 130.13

TOXICITY DATA with REFERENCE
scu-mus LDLo:450 mg/kg AEXPBL 52,220,05
scu-frg LDLo:250 mg/kg AEXPBL 52,220,05

CONSENSUS REPORTS: Reported in EPA TSCA Inventory.

SAFETY PROFILE: Poison by subcutaneous route. When heated to decomposition it emits toxic fumes of Na_2O.

SJP500 CAS:73940-79-1 ***HR: 3***
SODIUM SALT of HYDROXYMERCURIACETYLAMINOBENZOIC ACID

SYNS: 4-ACETYLAMINO-2-HYDROXYMERCURIBENZOIC ACID, SODIUM SALT ◇ HYDROXY(o-ACETAMINOBENZOATO)MERCURY SODIUM SALT ◇ TOXYNON

TOXICITY DATA with REFERENCE
ivn-rbt LDLo:20 mg/kg JPETAB 41,21,31

CONSENSUS REPORTS: Mercury and its compounds are on the Community Right To Know List.

OSHA PEL: (Transitional: CL 1 mg/10m^3) CL 0.1 mg (Hg)/m^3 (skin)
ACGIH TLV: TWA 0.1 mg(Hg)/m^3 (skin)
NIOSH REL: (Mercury, Inorganic) TWA 0.05 mg(Hg)/m^3

SAFETY PROFILE: Poison by intravenous route. When heated to decomposition it emits very toxic fumes of Hg, NO_x, and Na_2O. See also MERCURY COMPOUNDS.

SJQ000 CAS:64025-05-4 ***HR: 3***
SODIUM SALT of HYDROXYMERCURIPROPANOL PHENYLACETIC ACID
mf: $C_{11}H_{13}HgO_4 \cdot Na$ mw: 432.82

SYN: 4-HYDROXY-5-(HYDROXYMERCURI)-2-PHENYLPENTANOIC ACID, SODIUM SALT

TOXICITY DATA with REFERENCE
ivn-rbt LDLo:6800 μg/kg JPETAB 41,21,31

CONSENSUS REPORTS: Mercury and its compounds are on the Community Right To Know List.

OSHA PEL: (Transitional: CL 1 mg/10m^3) CL 0.1 mg (Hg)/m^3 (skin)
ACGIH TLV: TWA 0.1 mg(Hg)/m^3 (skin)
NIOSH REL: (Inorganic Mercury) TWA 0.05 mg(Hg)/m^3

SAFETY PROFILE: Poison by intravenous route. When heated to decomposition it emits toxic fumes of Hg and Na_2O. See also MERCURY COMPOUNDS.

SJQ500 CAS:63919-18-6 ***HR: 3***
SODIUM SALT of HYDROXYMERCURIPROPANOL PHENYL MALONIC ACID
mf: $C_{12}H_{12}HgO_6 \cdot 2Na$ mw: 498.81

SYN: (2-HYDROXY-3-HYDROXYMERCURI)PROPYL(PHENYL)MALONIC ACID SODIUM SALT

TOXICITY DATA with REFERENCE
ivn-rbt LDLo:10 mg/kg JPETAB 41,21,31

CONSENSUS REPORTS: Mercury and its compounds are on the Community Right To Know List.

OSHA PEL: (Transitional: CL 1 mg/10m^3) CL 0.1 mg (Hg)/m^3 (skin)
ACGIH TLV: TWA 0.1 mg(Hg)/m^3 (skin)
NIOSH REL: (Inorganic Mercury) TWA 0.05 mg(Hg)/m^3

SAFETY PROFILE: Poison by intravenous route. When heated to decomposition it emits toxic fumes of Hg and Na_2O. See also MERCURY COMPOUNDS.

SJR000 CAS:63992-02-9 ***HR: 3***
SODIUM SALT of MERCURI-BIS SALICYLIC ACID
mf: $C_{14}H_8HgO_6 \cdot 2Na$ mw: 518.79

SYN: MERCURIDISALICYLIC ACID, DISODIUM SALT

TOXICITY DATA with REFERENCE
ivn-rbt LDLo:10 mg/kg JPETAB 41,21,31

CONSENSUS REPORTS: Mercury and its compounds are on the Community Right To Know List.

ACGIH TLV: TWA 0.1 mg(Hg)/m^3 (skin)
NIOSH REL: TWA 0.05 mg(Hg)/m^3

SAFETY PROFILE: Poison by intravenous route. When heated to decomposition it emits toxic fumes of Hg and Na_2O. See also MERCURY COMPOUNDS.

SJR500 CAS:64046-01-1 ***HR: 3***
SODIUM SALT of MERCURY DITHIOCARBAMATE PIPERAZINE ACETIC ACID
mf: $C_{14}H_{20}HgN_4O_4S_4 \cdot 2Na$ mw: 683.19

SYN: BIS(4-(DITHIOCARBOXY)-1-PIPERAZINEACETATO(2−))-MERCURY(2−), DISODIUM

TOXICITY DATA with REFERENCE
ivn-rbt LDLo:7500 μg/kg JPETAB 41,21,31

CONSENSUS REPORTS: Mercury and its compounds are on the Community Right To Know List.

OSHA PEL: (Transitional: CL 1 mg/10m^3) CL 0.1 mg (Hg)/m^3 (skin)
ACGIH TLV: TWA 0.1 mg(Hg)/m^3 (skin)
NIOSH REL: (Inorganic Mercury) TWA 0.05 mg(Hg)/m^3

SAFETY PROFILE: Poison by intravenous route. When heated to decomposition it emits very toxic fumes of SO_x, NO_x, Hg, and Na_2O. See also MERCURY COMPOUNDS and CARBAMATES.

SJS500 CAS:874-21-5 ***HR: 3***
SODIUM SARKOMYCIN
mf: $C_7H_7O_3$•Na mw: 162.13

PROP: A metabolite of *Streptomyces erythrochromogehes* (BJCAAI 19,392,65).

SYNS: 2-METHYLENE-3-OXOCYCLOPENTANECARBOXYLICACID, SODIUM SALT ◇ SARKOMYCIN B*, SODIUM SALT ◇ SARKOMYCIN, SODIUM SALT

TOXICITY DATA with REFERENCE
scu-rat TDLo:1680 mg/kg/42W-I:ETA BJCAAI 19,392,65

SAFETY PROFILE: Questionable carcinogen with experimental tumorigenic data. When heated to decomposition it emits toxic fumes of Na_2O.

SJT000 CAS:1313-85-5 ***HR: 3***
SODIUM SELENIDE
mf: Na_2Se mw: 124.94

PROP: White to red deliq crystals. Mp: >875°, d: 2.625 @ 10°.

TOXICITY DATA with REFERENCE
sce-ham:lng 2 mg/L CALEDQ 18,109,83
ipr-mus LD50:4 mg/kg COREAF 257,791,63

CONSENSUS REPORTS: Selenium and its compounds are on the Community Right To Know List. Reported in EPA TSCA Inventory.

OSHA PEL: TWA 0.2 mg(Se)/m^3
ACGIH TLV: TWA 0.2 mg(Se)/m^3
DFG MAK: 0.1 mg(Se)/m^3

SAFETY PROFILE: Poison by intraperitoneal route. Mutation data reported. When heated to decomposition it emits toxic fumes of Se and Na_2O. See also SELENIUM COMPOUNDS.

SJT500 CAS:10102-18-8 ***HR: 3***
SODIUM SELENITE
DOT: UN 2630
mf: O_3Se•2Na mw: 172.94

PROP: White crystals.

SYNS: DISODIUM SELENITE ◇ NATRIUMSELENIT (GERMAN) ◇ SELENIOUS ACID, DISODIUM SALT

TOXICITY DATA with REFERENCE
dnr-sat 10 μg/plate CALEDQ 10,75,80
dni-hmn:hla 25 μmol/L TOLED5 25,219,85
cyt-hmn:fbr 80 μmol/L BTERDG 2,81,80
orl-mus TDLo:112 mg/kg (female 6-13D post):REP TCMUD8 7,29,87
scu-mus TDLo:1729 μg/kg (female 16D post):TER INHEAO 23,95,85
orl-rat LD50:7 mg/kg TXAPA9 20,89,71
ipr-rat LDLo:5476 μg/kg JPETAB 58,454,36
ipr-rat LD50:3 mg/kg JPETAB 58,454,36
par-rat LD50:6570 μg/kg CTOXAO 17,171,80
orl-mus LD50:7 mg/kg HYSAAV 35,176,70
ivn-mus LD50:5 mg/kg NRXTDN 2,383,81
icv-mus LD50:300 μg/kg NRXTDN 2,383,81
scu-dog LDLo:4 mg/kg EQSSDX 1,1,75
ivn-dog LD50:1916 μg/kg PSDAA2 36,173,57
ims-rbt LD50:2530 μg/kg AXVMAW 30,627,76

CONSENSUS REPORTS: IARC Cancer Review: Group 3 IMEMDT 7,56,87; Animal Inadequate Evidence IMEMDT 9,245,75. Reported in EPA TSCA Inventory. EPA Genetic Toxicology Program. EPA Extremely Hazardous Substances List. Selenium and its compounds are on the Community Right To Know List.

OSHA PEL: TWA 0.2 mg(Se)/m^3
ACGIH TLV: TWA 0.2 mg(Se)/m^3
DFG MAK: 0.1 mg(Se)/m^3
DOT Classification: Poison B; Label: Poison

SAFETY PROFILE: Poison by ingestion, intraperitoneal, intravenous, subcutaneous, intracervical, parenteral, and intramuscular routes. Experimental teratogenic and reproductive effects. Questionable carcinogen. Human mutation data reported. When heated to decomposition it emits toxic fumes of Se and Na_2O. See also SELENIUM COMPOUNDS.

SJT600 CAS:26970-82-1 ***HR: 3***
SODIUM SELENITE PENTAHYDRATE
mf: O_3Se•2Na•$5H_2O$ mw: 332.01

SYN: SELENIOUS ACID DISODIUM SALT PENTAHYDRATE

TOXICITY DATA with REFERENCE
scu-mus TDLo:10522 μg/kg (12D preg):REP TJADAB 28,333,83
orl-mus TDLo:38 mg/kg (30D pre/1-18D preg):TER TXAPA9 47,79,79
ipr-mus LD50:9 mg/kg COREAF 257,791,63
ims-mus LD50:10070 μg/kg EXMDA4 (440),312,78

CONSENSUS REPORTS: Selenium and its compounds are on the Community Right To Know List.

OSHA PEL: TWA 0.2 mg(Se)/m^3
ACGIH TLV: TWA 0.2 mg(Se)/m^3
DFG MAK: 0.1 mg(Se)/m^3

SAFETY PROFILE: Poison by intramuscular and intraperitoneal routes. An experimental teratogen. Experimental reproductive effects. When heated to decomposition it emits toxic fumes of Se and Na_2O. See also SELENIUM COMPOUNDS.

SJT750 CAS:533-96-0 ***HR: 1***
SODIUM SESQUICARBONATE
mf: $Na_2CO_3NaHCO_3 \cdot 2H_2O$ mw: 226.03

PROP: White crystals, flakes, or crystalline powder. Sol in water.

SAFETY PROFILE: A nuisance dust.

SJU000 CAS:6834-92-0 ***HR: 3***
SODIUM SILICATE
mf: $O_3Si \cdot 2Na$ mw: 122.07

SYNS: B-W ◇ CRYSTAMET ◇ DISODIUM METASILICATE ◇ DISODIUM MONOSILICATE ◇ METSO 20 ◇ METSO BEADS 2048 ◇ METSO BEADS, DRYMET ◇ METSO PENTABEAD 20 ◇ ORTHOSIL ◇ SODIUM METASILICATE ◇ SODIUM METASILICATE, anhydrous ◇ WATER GLASS

TOXICITY DATA with REFERENCE
skn-hmn 250 mg/24H SEV TXAPA9 31,481,75
skn-rbt 250 mg/24H SEV TXAPA9 31,481,75
skn-gpg 250 mg/24H MOD TXAPA9 31,481,75
orl-rat TDLo:15 g/kg (14W male/14W pre-3W post):REP JANSAG 36,271,73
orl-rat LD50:1153 mg/kg TOLED5 31(Suppl),44,86
orl-mus LD50:770 mg/kg TOLED5 31(Suppl),44,86
orl-dog LDLo:250 mg/kg FCTXAV 14,78,76
orl-pig LDLo:250 mg/kg FCTXAV 14,78,76
ipr-gpg LDLo:200 mg/kg JAMAAP 111,1925,38

CONSENSUS REPORTS: Reported in EPA TSCA Inventory.

SAFETY PROFILE: Poison by ingestion and intraperitoneal routes. A caustic material which is a severe eye, skin, and mucous membrane irritant. Experimental reproductive effects. Ingestion causes gastrointestinal tract upset. Violent reaction with F_2. When heated to decomposition it emits toxic fumes of Na_2O. Used in cosmetics. See also SILICATES.

SJU500 CAS:12164-12-4 ***HR: 3***
SODIUM SILICIDE
mf: NaSi mw: 51.08

SAFETY PROFILE: Ignites spontaneously in air. Explodes on contact with water or dilute acids. When heated to decomposition it emits toxic fumes of Na_2O.

SJV000 CAS:7757-81-5 ***HR: 2***
SODIUM SORBATE
mf: $C_6H_7O_2 \cdot Na$ mw: 134.12

SYN: SORBIC ACID, SODIUM SALT

TOXICITY DATA with REFERENCE
cyt-ham:lng 400 mg/L FCTOD7 22,501,84
sce-ham:lng 200 mg/L FCTOD7 22,501,84
msc-ham:lng 1 g/L CNREA8 44,3270,84
orl-rat LD50:7160 mg/kg JIHTAB 30,63,48
ipr-mus LD50:2500 mg/kg FAONAU 53A,121,74

SAFETY PROFILE: Moderately toxic by intraperitoneal route. Mildly toxic by ingestion. Mutation data reported. Migrates to food from packaging material. When heated to decomposition it emits toxic fumes of Na_2O.

SJV500 CAS:822-16-2 ***HR: 3***
SODIUM STEARATE
mf: $C_{18}H_{36}O_2 \cdot Na$ mw: 306.52

SYNS: OCTADECANOIC ACID, SODIUM SALT ◇ SODIUM OCTADECANOATE ◇ STEARIC ACID, SODIUM SALT

TOXICITY DATA with REFERENCE
ivn-dog LDLo:10 mg/kg FCTXAV 17,357,79

CONSENSUS REPORTS: Reported in EPA TSCA Inventory.

SAFETY PROFILE: Poison by intravenous route. When heated to decomposition it emits toxic fumes of Na_2O.

SJW000 ***HR: 3***
SODIUM STIBINIVANADATE
mf: $Na_{28}O_{14} \cdot O_5Sb_2 \cdot 22O_5V_2$ mw: 5192.42

TOXICITY DATA with REFERENCE
ivn-rbt LDLo:142 mg/kg AJSNAO 1,347,17

CONSENSUS REPORTS: Antimony and its compounds are on the Community Right To Know List.

OSHA PEL: TWA 0.2 mg(Sb)/m^3
ACGIH TLV: TWA 0.5 mg(Sb)/m^3; 0.05 mg(V_2O_5)/m^3
NIOSH REL: (Antimony) TWA 0.5 mg(Sb)/m^3; (Vanadium Compounds) CL 0.05 mg(V)/m^3/15M

SAFETY PROFILE: Poison by intravenous route. When heated to decomposition it emits toxic fumes of VO_x, Sb, and Na_2O. See also ANTIMONY COMPOUNDS and VANADIUM COMPOUNDS.

SJW300 CAS:63979-86-2 ***HR: 2***
SODIUM SULFAETHYLTHIADIAZOLE
mf: $C_{10}H_{12}N_4O_2S_2 \cdot Na$ mw: 307.37

SYNS: BENZENESULFONAMIDE,4-AMINO-N-(5-ETHYL-1,3,4-THIADIAZOL-2-YL)-, MONOSODIUM SALT (9CI) ◇ SULFAETHYLTHIADIAZOLE SODIUM ◇ SULFANILAMIDE, N^1-(5-ETHYL-1,3,4-THIADIAZOL-2-YL)-, MONOSODIUM SALT

TOXICITY DATA with REFERENCE
scu-rat TDLo:1200 mg/kg (female 10-11D post):REP JHMJAX 130,95,72
scu-mus LD50:1243 mg/kg AMSSAQ 142,64,43

SAFETY PROFILE: Moderately toxic by subcutaneous

route. Experimental reproductive effects. When heated to decomposition it emits toxic fumes of NO_x and SO_x.

SJW475 CAS:127-58-2 ***HR: 2***
SODIUM SULFAMERAZINE
mf: $C_{11}H_{12}N_4O_2S \cdot Na$ mw: 287.32

PROP: Crystals; bitter, caustic taste. Hygroscopic. On prolonged exposure to humid air, it absorbs CO_2 with the liberation of sulfamerazine and becomes incompletely sol in water. Its solns are alkaline to phenolphthalein (pH 10 or more). One gram dissolves in 3.6 mL water. Sltly sol in alc; insol in ether, chloroform.

SYNS: 4-AMINO-N-(4-METHYL-2-PYRIMIDINYL)-BENZENESULFONAMIDE MONOSODIUM SALT ◇ N[1]-(4-METHYL-2-PYRIMIDINYL) SULFANILAMIDE SODIUM SALT ◇ SODIUM SULPHAMERAZINE ◇ SOLUBLE SULFAMERAZINE ◇ SOLUMEDINE ◇ SULFAMERAZINE SODIUM

TOXICITY DATA with REFERENCE
orl-mus LD50:2800 mg/kg AIPTAK 94,338,53
ipr-mus LD50:1522 mg/kg JPETAB 81,17,44
scu-mus LD50:1739 mg/kg JPETAB 81,17,44
ivn-mus LD50:900 mg/kg AIPTAK 94,338,53

CONSENSUS REPORTS: Reported in EPA TSCA Inventory.

SAFETY PROFILE: Moderately toxic by ingestion, subcutaneous, intraperitoneal, and intravenous routes. When heated to decomposition it emits toxic fumes of SO_x, NO_x, and Na_2O.

SJW500 CAS:1981-58-4 ***HR: 2***
SODIUM SULFAMETHIAZINE
mf: $C_{12}H_{14}N_4O_2S \cdot Na$ mw: 301.35

SYNS: 4-AMINO-N-(4,6-DIMETHYL-2-PYRIMIDINYL)BENZENESULFONAMIDE, MONOSODIUM SALT ◇ N[1]-(4,6-DIMETHYL-2-PYRIDINYL)SULFANILAMIDE, MONOSODIUM SALT ◇ (N[1]-(4,6-DIMETHYL-2-PYRIMIDINYL)SULFANILAMIDO) SODIUM ◇ SODIUM SULFAMETAZINE ◇ SODIUM SULFAMETHAZINE ◇ SODIUM SULFAMEZATHINE ◇ SULFADIMIDINE SODIUM ◇ SULFAMETHAZINE SODIUM ◇ SULMET ◇ VESADIN

TOXICITY DATA with REFERENCE
orl-mus LD50:2057 mg/kg JPETAB 81,17,44
ipr-mus LD50:974 mg/kg JPETAB 81,17,44
scu-mus LD50:1191 mg/kg JPETAB 81,17,44
ivn-mus LD50:728 mg/kg RPOBAR 2,329,70

CONSENSUS REPORTS: Reported in EPA TSCA Inventory.

SAFETY PROFILE: Moderately toxic by ingestion, intravenous, intraperitoneal, and subcutaneous routes. When heated to decomposition it emits very toxic fumes of NO_x, SO_x, and Na_2O.

SJY000 CAS:7757-82-6 ***HR: 2***
SODIUM SULFATE (2:1)
mf: $O_4S \cdot 2Na$ mw: 142.04

PROP: White crystals or powder; odorless. Mp: 888°, d: 2.671. Sol in water, glycerin; insol alc.

SYNS: DISODIUM SULFATE ◇ NATRIUMSUFAT (GERMAN) ◇ SALT CAKE ◇ SODIUM SULFATE anhydrous ◇ SODIUM SULPHATE ◇ SULFURIC ACID, DISODIUM SALT ◇ THENARDITE ◇ TRONA

TOXICITY DATA with REFERENCE
orl-mus TDLo:14 g/kg (female 8-12D post):REP TCMUD8 6,361,86
par-mus TDLo:60 mg/kg (female 8D post):TER JPMSAE 62,1626,73
scu-mus TDLo:806 mg/kg/26W-I:ETA IVIVE4 1,39,87
orl-mus LD50:5989 mg/kg SKEZAP 4,15,63
ivn-mus LDLo:1220 mg/kg CLDND*

CONSENSUS REPORTS: Reported in EPA TSCA Inventory. EPA Genetic Toxicology Program.

SAFETY PROFILE: Moderately toxic by intravenous route. Mildly toxic by ingestion. An experimental teratogen. Experimental reproductive effects. Questionable carcinogen with experimental tumorigenic effects. Violent reaction with Al. When heated to decomposition it emits toxic fumes of SO_x and Na_2O. See also SULFATES.

SJY500 CAS:1313-82-2 ***HR: 3***
SODIUM SULFIDE (anhydrous)
DOT: UN 1385
mf: Na_2S mw: 78.04

PROP: Amorph, yellow-pink or white, deliq crystals. Mp: 1180°, d: 1.856 @ 14°.

SYNS: SODIUM MONOSULFIDE ◇ SODIUM SULPHIDE

CONSENSUS REPORTS: Reported in EPA TSCA Inventory.

DOT Classification: Label: Flammable Solid

SAFETY PROFILE: Flammable when exposed to heat or flame. Unstable and can explode on rapid heating or percussion. Reacts violently with carbon; diazonium salts; n,n-dichloromethylamine; o-nitroaniline diazonium salt; water. When heated to decomposition it emits toxic fumes of SO_x and Na_2O. See also SULFIDES.

SJZ000 CAS:7757-83-7 ***HR: 3***
SODIUM SULFITE (2:1)
mf: $O_3S \cdot 2Na$ mw: 126.04

PROP: Hexagonal prisms or white powder; odorless with salty, sulfurous taste. Bp: decomp, d: 2.633 @ 15.4°. Sol in water; sltly sol in alc.

SYNS: DISODIUM SULFITE ◇ EXSICATED SODIUM SULFITE ◇ NATRIUMSULFID (GERMAN) ◇ SODIUM SULFITE, anhydrous ◇ SODIUM SULPHITE ◇ SULFTECH ◇ SULFUROUS ACID, SODIUM SALT (1:2)

TOXICITY DATA with REFERENCE
dni-hmn:lym 10 mmol/L CHRTBC 4,211,73
cyt-mus:oth 25 mg/L ENVRAL 9,84,75
ivn-rat LD50:115 mg/kg JPETAB 101,101,51
ipr-mus LD50:950 mg/kg ARZNAD 31,1713,81
ivn-mus LD50:130 mg/kg JPETAB 101,101,51
scu-dog LDLo:1300 mg/kg HBAMAK 4,1289,35
scu-cat LDLo:1300 mg/kg HBAMAK 4,1289,35
ivn-cat LDLo:200 mg/kg AHYGAJ 57,87,06
orl-rbt LDLo:2825 mg/kg AHYGAJ 57,87,06
scu-rbt LDLo:300 mg/kg AHYGAJ 57,87,06
ivn-rbt LD50:65 mg/kg JPETAB 101,101,51

CONSENSUS REPORTS: Reported in EPA TSCA Inventory. EPA Genetic Toxicology Program.

SAFETY PROFILE: Poison by intravenous and subcutaneous routes. Moderately toxic by ingestion and intraperitoneal routes. Human mutation data reported. When heated to decomposition it emits very toxic fumes of Na_2O and SO_x. A reducing agent. See also SULFITES.

SKB000 ***HR: 3***
SODIUM TARTRATE
mf: $C_4H_4O_6 \cdot Na$ mw: 171.07

SYNS: NATRIUMTARTRAT (GERMAN) ◇ TARTARIC ACID, MONOSODIUM SALT

TOXICITY DATA with REFERENCE
scu-cat LDLo:3000 mg/kg JPETAB 25,467,25
scu-mus LDLo:1000 mg/kg JPETAB 25,467,25
scu-rbt LDLo:1 g/kg HBAMAK 4,1289,35
ivn-rbt LDLo:200 mg/kg JAPMA8 31,1,42

SAFETY PROFILE: Poison by intravenous route. Moderately toxic by subcutaneous route. When heated to decomposition it emits toxic fumes of Na_2O.

SKB500 CAS:145-42-6 ***HR: 3***
SODIUM TAUROCHOLATE
mf: $C_{26}H_{44}NO_7S \cdot Na$ mw: 537.76

TOXICITY DATA with REFERENCE
ivn-rbt LDLo:55 mg/kg ZGEMAZ 52,779,26

CONSENSUS REPORTS: Reported in EPA TSCA Inventory.

SAFETY PROFILE: Poison by intravenous route. When heated to decomposition it emits very toxic fumes of NO_x, SO_x, and Na_2O.

SKC000 CAS:10101-83-4 ***HR: 3***
SODIUM TELLURATE
mf: $O_4Te \cdot 2Na$ mw: 237.58

PROP: White powder.

SYN: SODIUM TELLURATE VI

TOXICITY DATA with REFERENCE
orl-rat LD50:385 mg/kg HYSAAV 32,15,67
ipr-rat LD50:56780 μg/kg EQSSDX 1,1,75
ivn-rat LD50:55850 μg/kg EQSSDX 1,1,75
orl-mus LD50:165 mg/kg HYSAAV 32,15,67
orl-rbt LD50:104 mg/kg EQSSDX 1,1,75

CONSENSUS REPORTS: Reported in EPA TSCA Inventory.

OSHA PEL: TWA 0.1 mg(Te)/m^3
ACGIH TLV: TWA 0.1 mg(Te)/m^3

SAFETY PROFILE: Poison by ingestion, intraperitoneal and intravenous routes. When heated to decomposition it emits toxic fumes of Te and Na_2O. See also TELLURIUM COMPOUNDS.

SKC100 CAS:26006-71-3 ***HR: 3***
SODIUM TELLURATE, DIHYDRATE
mf: $O_4Te \cdot 2Na \cdot 2H_2O$ mw: 273.62

PROP: Crystals. Mp: decomp.

TOXICITY DATA with REFERENCE
ipr-mus LD50:74 mg/kg COREAF 257,791,63

OSHA PEL: TWA 0.1 mg(Te)/m^3
ACGIH TLV: TWA 0.1 mg(Te)/m^3

SAFETY PROFILE: Poison by intraperitoneal route. When heated to decomposition it emits toxic fumes of Te.

SKC500 CAS:10102-20-2 ***HR: 3***
SODIUM TELLURITE
mf: $O_3Te \cdot 2Na$ mw: 221.58

SYNS: SODIUM TELLURATE(IV) ◇ TELLUROUS ACID, DISODIUM SALT

TOXICITY DATA with REFERENCE
mmo-sat 10 mmol/L MUREAV 77,109,80
cyt-hmn:leu 1 nmol/L AEMBAP 91,117,78
orl-hmn LDLo:30 mg/kg BJIMAG 3,175,46
par-man LDLo:29 mg/kg BJIMAG 3,175,46
orl-rat LD50:83 mg/kg HYSAAV 32,15,67
ipr-rat LD50:2400 μg/kg EQSSDX 1,1,75
ivn-rat LDLo:2431 μg/kg JPETAB 33,270,28
orl-mus LD50:20 mg/kg HYSAAV 32,15,67
orl-rbt LD50:53600 μg/kg EQSSDX 1,1,75

CONSENSUS REPORTS: EPA Extremely Hazardous Substances List. Reported in EPA TSCA Inventory.

OSHA PEL: TWA 0.1 mg(Te)/m^3
ACGIH TLV: TWA 0.1 mg(Te)/m^3

SAFETY PROFILE: Human poison by ingestion and parenteral routes. Experimental poison by ingestion, intravenous, and intraperitoneal routes. Human mutation data reported. When heated to decomposition it emits toxic fumes of Te and Na_2O. See also TELLURIUM.

SKD600 ***HR: 3***
SODIUM TETRACYANATOPALLADATE(II)
mf: $C_4N_4Na_2O_4Pd$ mw: 320.47

$Na_2[Pd(OCN)_4]$

CONSENSUS REPORTS: Cyanide and its compounds are on The Community Right To Know List.

SAFETY PROFILE: An explosive, sensitive to heat or impact. When heated to decomposition it emits toxic fumes of NO_x, CN^-, and Na_2O. See also CYANATES, SODIUM COMPOUNDS, and PALLADIUM.

SKE000 CAS:13755-29-8 ***HR: 3***
SODIUM TETRAFLUOROBORATE
mf: $BF_4 \cdot Na$ mw: 109.80

SYNS: STB ◇ TETRAFLUOROBORATE(1-) SODIUM

TOXICITY DATA with REFERENCE
scu-rat LD50:550 mg/kg ZKKOBW 80,17,73

CONSENSUS REPORTS: Reported in EPA TSCA Inventory.
OSHA PEL: TWA 2.5 mg(F)/m^3
NIOSH REL: (Inorganic Fluorides) TWA 2.5 mg(F)/m^3

SAFETY PROFILE: Moderately toxic by subcutaneous route. When heated to decomposition it emits toxic fumes of F^- and Na_2O.

SKE500 CAS:32106-51-7 ***HR: 3***
SODIUM TETRAHYDROGALLATE
mf: GaH_4Na mw: 96.74

SAFETY PROFILE: Explodes on contact with moisture. When heated to decomposition it emits toxic fumes of Na_2O. See also GALLIUM COMPOUNDS.

SKF000 CAS:12206-14-3 ***HR: 3***
SODIUM TETRAPEROXYCHROMATE
mf: $CrNa_3O_8$ mw: 248.97

CONSENSUS REPORTS: Chromium and its compounds are on the Community Right To Know List.
OSHA PEL: CL 0.1 mg(CrO_3)/m^3
ACGIH TLV: TWA 0.05 mg(Cr)/m^3
NIOSH REL: TWA 0.025 mg(Cr(VI))/m^3; CL 0.05 mg/m^3/15M

SAFETY PROFILE: Confirmed human carcinogen. Explodes when heated to 115°C. When heated to decomposition it emits toxic fumes of Na_2O. See also CHROMIUM COMPOUNDS and PEROXIDES.

SKF500 CAS:42489-15-6 ***HR: 3***
SODIUM TETRAPEROXYMOLYBDATE
mf: $MoNa_2O_8$ mw: 269.92

SAFETY PROFILE: Explodes in vacuum. When heated to decomposition it emits toxic fumes of Na_2O. See also MOLYBDENUM COMPOUNDS and PEROXIDES, INORGANIC.

SKG500 CAS:12058-74-1 ***HR: 3***
SODIUM TETRAVANADATE
mf: $O_{11}V_4 \cdot 2Na$ mw: 425.74

TOXICITY DATA with REFERENCE
ivn-hmn LDLo:1 mg/kg AJSNAO 1,347,17
scu-rat LDLo:70 mg/kg AJSNAO 1,347,17
scu-mus LDLo:58 mg/kg AJSNAO 1,347,17
ivn-rbt LDLo:14 mg/kg AJSNAO 1,347,17
scu-gpg LDLo:42 mg/kg AJSNAO 1,347,17
ims-pgn LDLo:21 mg/kg AJSNAO 1,347,17
ims-ckn LDLo:35 mg/kg AJSNAO 1,347,17
par-frg LDLo:117 mg/kg AJSNAO 1,347,17
scu-dom LDLo:10 mg/kg AJSNAO 1,347,17
ACGIH TLV: TWA 0.05 mg(V_2O_5)/m^3
NIOSH REL: CL 0.05 mg(V)/m^3/15M

SAFETY PROFILE: Human poison by intravenous route. Experimental poison by intravenous, subcutaneous, intramuscular, and parenteral routes. When heated to decomposition it emits toxic fumes of Na_2O and VO_x. See also VANADIUM COMPOUNDS.

SKH000 CAS:3485-82-3 ***HR: 3***
SODIUM THEOPHYLLINE
mf: $C_7H_8N_4O_2 \cdot Na$ mw: 203.18

TOXICITY DATA with REFERENCE
scu-mus LD50:189 mg/kg ARZNAD 4,649,54
ivn-mus LD50:171 mg/kg RPOBAR 2,335,70
scu-gpg LD50:192 mg/kg ARZNAD 4,649,54

CONSENSUS REPORTS: Reported in EPA TSCA Inventory.

SAFETY PROFILE: Poison by subcutaneous and intravenous routes. When heated to decomposition it emits toxic fumes of NO_x and Na_2O. See also THEOPHYLLINE.

SKH500 CAS:367-51-1 ***HR: 3***
SODIUM THIOGLYCOLATE
mf: $C_2H_3O_2S{\cdot}Na$ mw: 114.10

PROP: Hygroscopic crystals.

SYNS: MERCAPTOACETIC ACID SODIUM SALT ◇ SODIUM MERCAPTOACETATE ◇ SODIUM THIOGLYCOLLATE ◇ THIOGLYCOLATESODIUM ◇ THIOGLYCOLLIC ACID, SODIUM SALT ◇ USAF EK-5199

TOXICITY DATA with REFERENCE
skn-hmn 7%/24H INMEAF 15,669,46
ipr-rat LD50:126 mg/kg JPETAB 118,296,56
orl-mus LD50:504 mg/kg JPETAB 118,296,56
ipr-mus LD50:200 mg/kg NTIS** AD277-689
ivn-mus LD50:422 mg/kg JJANAX 38,137,85
ivn-dog LDLo:500 mg/kg JPETAB 118,296,56
ivn-mky LDLo:300 mg/kg JPETAB 118,296,56
ivn-rbt LDLo:100 mg/kg JPETAB 35,343,29

CONSENSUS REPORTS: Reported in EPA TSCA Inventory.

SAFETY PROFILE: Poison by intravenous and intraperitoneal routes. Moderately toxic by ingestion. A human skin irritant. This material yields hydrogen sulfide on decomposition. A death has been attributed to the absorption of toxic decomposition products from the use of this material in a hair permanent waving solution. When heated to decomposition it emits toxic fumes of SO_x and Na_2O. See also SULFIDES and MERCAPTANS.

SKI000 CAS:7772-98-7 ***HR: 2***
SODIUM THIOSULFATE
mf: $O_3S_2{\cdot}2Na$ mw: 158.10

PROP: Colorless crystals or crystalline powder. Sol in water; insol in alc.

SYNS: HYPO ◇ SODIUM HYPOSULFITE ◇ SODIUM THIOSULFATE, anhydrous

TOXICITY DATA with REFERENCE
scu-rbt LDLo:4000 mg/kg AIPTAK 5,161,1899
scu-frg LDLo:6 g/kg AIPTAK 5,161,1899

CONSENSUS REPORTS: Reported in EPA TSCA Inventory.

SAFETY PROFILE: Moderately toxic by subcutaneous route. Incompatible with metal nitrates, sodium nitrite. When heated to decomposition it emits very toxic fumes of Na_2O and SO_x. See also SODIUM THIOSULFATE and PENTAHYDRATE.

SKI500 CAS:10102-17-7 ***HR: 2***
SODIUM THIOSULFATE, PENTAHYDRATE
mf: $O_3S_2{\cdot}2Na{\cdot}5H_2O$ mw: 248.20

PROP: Monoclinic, colorless, odorless crystals. Mp: 48° (rapid heating), d: 1.69.

SYNS: AMETOX ◇ ANTICHLOR ◇ HYPO ◇ NSC-45624 ◇ SODIUM HYPOSULFITE ◇ SODOTHIOL ◇ SULFOTHIORINE ◇ THIOSULFURIC ACID, DISODIUM SALT, PENTAHYDRATE

TOXICITY DATA with REFERENCE
orl-hmn TDLo:300 mg/kg/7D:PUL TJEMAO 63,383,56
ipr-mus LD50:5600 mg/kg ARZNAD 31,1713,81
ivn-mus LD50:2350 mg/kg KSRNAM 13,89,79
ivn-dog LDLo:3000 mg/kg CNCRA6 50,255,66

SAFETY PROFILE: Moderately toxic by intravenous route. Human systemic effects by ingestion: cyanosis. Large doses internally have a cathartic action. Violent reaction with $NaNO_2$. When heated to decomposition it emits toxic fumes of SO_x and Na_2O. See also THIOSULFATES and SODIUM MONOXIDE.

SKJ300 CAS:137-53-1 ***HR: 1***
SODIUM d-THYROXINE
mf: $C_{15}H_{10}I_4NO_4{\cdot}Na$ mw: 798.85

SYNS: BIOTIRMONE ◇ CHOLOXIN ◇ DEBETROL ◇ DETHYRONA ◇ DETYROXIN ◇ DEXTROID ◇ DEXTROTHYROXINE SODIUM ◇ DEXTROXIN ◇ DYNOTHEL ◇ o-(4-HYDROXY-3,5-DIIODOPHENYL)-3,5-DIIODO-d-TYROSINE MONOSODIUM SALT ◇ LISOLIPIN ◇ NADROTHYRON D ◇ SODIUM DEXTROTHYROXINE ◇ SODIUM d-T4 ◇ 3:5:3′:5′-TETRAIODO-d-THYRONINE SODIUM ◇ d-T4 SODIUM ◇ d-THYROXINE SODIUM ◇ d-THYROXINE SODIUM SALT ◇ TRAVENON

TOXICITY DATA with REFERENCE
orl-rbt TDLo:15 mg/kg (female 15-29D post):REP DABBBA 32,1182,71
orl-rbt TDLo:29 mg/kg (female 1-29D post):TER DABBBA 32,1182,71
orl-wmn TDLo:12 μg/kg/12D-I:LIV,SYS JJMDAT 23,282,84

SAFETY PROFILE: An experimental teratogen. Experimental reproductive effects. Human systemic effects: hepatitis, liver function tests, increased body temperature. When heated to decomposition it emits toxic fumes of I^-, NO_x, and Na_2O.

SKJ340 CAS:35711-34-3 ***HR: 2***
SODIUM TOLMETIN
mf: $C_{14}H_{12}NO_3{\cdot}Na$ mw: 265.26

SYNS: 1-METHYL-5-(4-METHYLBENZOYL)-1H-PYRROLE-2-ACETIC ACID SODIUM SALT ◇ 1-METHYL-5-(p-TOLUOYL)-2-PYRROLEACETIC ACID ◇ TOLECTIN ◇ TOLMETIN SODIUM

TOXICITY DATA with REFERENCE
orl-rbt TDLo:100 mg/kg (1D pre):REP FESTAS 38,238,82
orl-man TDLo:538 mg/kg/61D-I:SKN JAMAAP 243,1263,80
orl-hmn TDLo:8 mg/kg:CNS,CVS JAMAAP 240,246,78

orl-wmn TDLo:8 mg/kg:SPN,EYE,LIV JAMAAP 245,67,81
orl-cld TDLo:2520 mg/kg/8W-I:BLD ARHEAW 25,1144,82
orl-rat LD50:914 mg/kg PBPSDY 1,233,77
ipr-rat LD50:770 mg/kg NIIRDN 6,532,82
scu-rat LD50:1050 mg/kg NIIRDN 6,532,82
orl-mus LD50:899 mg/kg PBPSDY 1,233,77
ipr-mus LD50:690 mg/kg NIIRDN 6,532,82
scu-mus LD50:670 mg/kg NIIRDN 6,532,82

SAFETY PROFILE: Moderately toxic by ingestion, intraperitoneal and subcutaneous routes. Human systemic effects by ingestion: allergic dermatitis, blood pressure depression, eye irritation, hallucinations or distorted perceptions, liver function impairment, shock, spinal cord effects, thrombocytopenia. Experimental reproductive effects. When heated to decomposition it emits toxic fumes of NO_x and Na_2O.

SKJ350 CAS:64490-92-2 ***HR: 2***
SODIUM TOLMETIN DIHYDRATE
mf: $C_{15}H_{14}NO_3 \cdot Na \cdot 2H_2O$ mw: 315.33

SYNS: SODIUM-1-METHYL-5-p-TOLUOYLPYRROLE-2-ACETATEDIHYDRATE ◇ TOLMETIN SODIUM DIHYDRATE ◇ TOLUMETIN SODIUM DIHYDRATE

TOXICITY DATA with REFERENCE
orl-rbt TDLo:1300 mg/kg (6-18D preg):REP IYKEDH 8,158,77
orl-rbt TDLo:163 mg/kg (female 6-18D post):TER IYKEDH 8,158,77
orl-rat LD50:858 mg/kg IYKEDH 10,232,79
ipr-rat LD50:688 mg/kg IYKEDH 10,232,79
scu-rat LD50:887 mg/kg IYKEDH 10,232,79
orl-mus LD50:1260 mg/kg IYKEDH 10,232,79
ipr-mus LD50:590 mg/kg IYKEDH 10,232,79
scu-mus LD50:603 mg/kg IYKEDH 10,232,79

SAFETY PROFILE: Moderately toxic by ingestion, intraperitoneal, and subcutaneous routes. An experimental teratogen. Experimental reproductive effects. When heated to decomposition it emits toxic fumes of NO_x and Na_2O. See also SODIUM TOLMETIN.

SKJ500 CAS:824-79-3 ***HR: 2***
SODIUM-p-TOLUENE SULFINIC ACID
mf: $C_7H_8O_2S \cdot Na$ mw: 178.19

SYNS: SODIUM-4-METHYLBENZENESULFINATE ◇ SODIUM-p-TOLUENESULFINATE ◇ SODIUM-4-TOLUENESULFINATE

TOXICITY DATA with REFERENCE
orl-rat LDLo:3200 mg/kg KODAK* -,-,71
ipr-rat LDLo:1600 mg/kg KODAK* -,-,71

CONSENSUS REPORTS: Reported in EPA TSCA Inventory.

SAFETY PROFILE: Moderately toxic by ingestion and intraperitoneal routes. When heated to decomposition it emits toxic fumes of SO_x and Na_2O.

SKK000 CAS:657-84-1 ***HR: 2***
SODIUM-p-TOLYL SULFONATE
mf: $C_7H_8O_3S \cdot Na$ mw: 194.19

SYNS: 4-METHYL-BENZENESULFONIC ACID, SODIUM SALT ◇ SODIUM-p-METHYLBENZENESULFONATE ◇ SODIUM PARATOLUENE SULPHONATE ◇ SODIUM-p-TOLUENESULFONATE ◇ SODIUM TOSYLATE ◇ 4-TOLUENESULFINIC ACID SODIUM SALT

TOXICITY DATA with REFERENCE
ivn-mus LD50:1700 mg/kg BJPCAL 14,536,59

CONSENSUS REPORTS: Reported in EPA TSCA Inventory.

SAFETY PROFILE: Moderately toxic by intravenous route. When heated to decomposition it emits toxic fumes of SO_x and Na_2O. See also SULFONATES.

SKK100 ***HR: 3***
SODIUM TRIAZIDOAURATE
mf: AuN_9Na mw: 346.02

SAFETY PROFILE: Explodes when heated to 130°C. When heated to decomposition it emits toxic fumes of NO_x and Na_2O. See also AZIDES and GOLD COMPOUNDS.

SKK500 CAS:136-32-3 ***HR: 2***
SODIUM-2,4,5-TRICHLOROPHENATE
mf: $C_6H_2Cl_3O \cdot Na$ mw: 219.42

SYNS: DOWICIDE B ◇ PREVENTOL 1 ◇ SODIUM SALT of 2,4,5-TRICHLOROPHENOL ◇ 2,4,5-TRICHLOROPHENOL, SODIUM SALT ◇ (2,4,5-TRICHLOROPHENOXY)SODIUM

TOXICITY DATA with REFERENCE
orl-rat LD50:1620 mg/kg DOWCC* -,-,69

CONSENSUS REPORTS: Chlorophenol compounds are on the Community Right To Know List. Reported in EPA TSCA Inventory.

SAFETY PROFILE: Moderately toxic by ingestion. When heated to decomposition it emits toxic fumes of Cl^- and Na_2O. Used as fungicide for pulp and papermill wet-end systems. See also CHLOROPHENOLS.

SKL000 CAS:3570-61-4 ***HR: 3***
SODIUM-2,4,5-TRICHLOROPHENOXYETHYL SULFATE
mf: $C_8H_6Cl_3O_5S \cdot Na$ mw: 343.54

SYNS: NATRIN HERBICIDE ◇ 2,4,5-TES

TOXICITY DATA with REFERENCE
eye-rbt 1 mg MOD UCDS** 3/23/73

orl-rat LD50:720 mg/kg RREVAH 10,97,65
skn-rbt LD50:400 mg/kg UCDS** 3/23/73

SAFETY PROFILE: Poison by skin contact. Moderately toxic by ingestion. An eye irritant. When heated to decomposition it emits very toxic fumes of SO_x, Cl^-, and Na_2O. See also SULFATES.

SKL500 CAS:13782-22-4 ***HR: 3***
SODIUM TRIFLUOROSTANNITE
mf: $F_3Sn•Na$ mw: 198.68

TOXICITY DATA with REFERENCE
ivn-mus LD50:18 mg/kg CSLNX* NX#00131

OSHA PEL: TWA 2 mg(Sn)/m^3
ACGIH TLV: TWA 2 mg(Sn)/m^3
NIOSH REL: TWA 2.5 mg(F)/m^3

SAFETY PROFILE: Poison by intravenous route. When heated to decomposition it emits toxic fumes of F^- and Na_2O. See also TIN COMPOUNDS and FLUORIDES.

SKM000 CAS:1221-56-3 ***HR: 3***
SODIUM TRIIODOHYDROCINNAMATE
mf: $C_{12}H_{12}I_3N_2•Na$ mw: 619.95

SYNS: BILIMIN ◇ BILOPTIN ◇ BILOPTINON ◇ 3-((DIMETHYLAMINOMETHYLENE)AMINO)-2,4,6-TRIIODOHYDROCINNAMICACID SODIUM SALT ◇ IOPODATE SODIUM ◇ IPODATE SODIUM ◇ NSC-106962 ◇ ORAGRAFIN-SODIUM ◇ SH 514 ◇ SODIUM-3-(DIMETHYLAMINOMETHYLENEAMINO)-2,4,6-TRIIODOHYDROCINNAMATE ◇ SODIUM IOPODATE ◇ SODIUM IPODATE ◇ SODIUM SALT of 3-(3-DIMETHYLAMINOMETHYLENEAMINO-2,4,6-TRIIODOPHENYL)PROPIONIC ACID ◇ SQ-15761 ◇ β-(2,4,6-TRIJOD-3-DIMETHYLAMINOMETHYLENAMINO-PHENYL)-PROPIONSAEURE(GERMAN)

TOXICITY DATA with REFERENCE
orl-rat LD50:2500 mg/kg FRPPAO 31,397,76
ivn-rat LD50:303 mg/kg FRPPAO 31,397,76
orl-mus LD50:1100 mg/kg KHFZAN 14(12),62,80
ivn-mus LD50:290 mg/kg JMCMAR 13,997,70
ivn-dog LDLo:300 mg/kg FRPPAO 31,397,76

SAFETY PROFILE: Poison by intravenous route. Moderately toxic by ingestion. When heated to decomposition it emits very toxic fumes of I^-, NO_x, and Na_2O.

SKM500 CAS:7785-84-4 ***HR: 3***
SODIUM TRIMETAPHOSPHATE
mf: $O_9P_3•3Na$ mw: 305.88

PROP: White crystals or white crystalline powder. Sol in water.

SYN: TRIMETAPHOSPHATE SODIUM

TOXICITY DATA with REFERENCE
ipr-rat LD50:3650 mg/kg JPETAB 108,117,53
ivn-rbt LDLo:240 mg/kg AEPPAE 169,238,33

CONSENSUS REPORTS: Reported in EPA TSCA Inventory.

SAFETY PROFILE: Poison by intravenous route. Moderately toxic by intraperitoneal route. When heated to decomposition it emits toxic fumes of PO_x and Na_2O. See also PHOSPHATES.

SKN000 CAS:13573-18-7 ***HR: 3***
SODIUM TRIPOLYPHOSPHATE
mf: $O_{10}P_3•5Na$ mw: 367.86

PROP: White granules or powder. Sltly hygroscopic; sol in water.

SYNS: ARMOFOS ◇ NATRIUMTRIPOLYPHOSPHAT (GERMAN) ◇ PENTASODIUM TRIPHOSPHATE ◇ POLY ◇ POLYGON ◇ SODIUM TRIPHOSPHATE ◇ STPP ◇ TRIPHOSPHORIC ACID, SODIUM SALT ◇ TRIPOLY ◇ TRIPOLYPHOSPHATE

TOXICITY DATA with REFERENCE
orl-rat LD50:6500 mg/kg AIHAAP 30,470,69
ipr-rat LD50:525 mg/kg JPETAB 108,117,53
orl-mus LD50:3210 mg/kg ARZNAD 7,445,57
ipr-mus LD50:700 mg/kg REPMBN 10,391,62
scu-mus LD50:900 mg/kg ARZNAD 7,445,57
ivn-mus LD50:74 mg/kg ARZNAD 7,445,57

SAFETY PROFILE: Poison by intravenous route. Moderately toxic by ingestion, subcutaneous, and intraperitoneal routes. Ingestion of large doses of sodium phosphates causes catharsis. Sodium meta and pyrophosphates can cause hemorrhages from the intestine if taken internally in large doses. When heated to decomposition it emits toxic fumes of PO_x and Na_2O.

SKN500 CAS:13472-45-2 ***HR: 3***
SODIUM TUNGSTATE
mf: $O_4W•2Na$ mw: 293.83

PROP: White, rhombic crystals. Mp: 698°, d: 4.179.

SYN: TUNGSTIC ACID, DISODIUM SALT

TOXICITY DATA with REFERENCE
pic-esc 5 mmol/L ENMUDM 6,59,84
sln-smc 100 mmol/L MUTAEX 1,21,86
orl-rat LD50:1190 mg/kg HYSAAV 31,197,66
scu-rat LD50:240 mg/kg EQSSDX 1,1,75
orl-mus LD50:240 mg/kg HYSAAV 31,197,66
scu-dog LDLo:140 mg/kg EQSSDX 1,1,75
ivn-cat LD50:139 mg/kg AGSOA6 8,51,67
orl-rbt LD50:875 mg/kg HYSAAV 31,197,66
scu-rbt LDLo:78 mg/kg EQSSDX 1,1,75
ims-rbt LDLo:105 mg/kg EQSSDX 1,1,75
orl-gpg LD50:1152 mg/kg HYSAAV 31,197,66
scu-gpg LDLo:810 mg/kg EQSSDX 1,1,75

CONSENSUS REPORTS: Reported in EPA TSCA Inventory.

ACGIH TLV: TWA 5 mg(W)/m^3
NIOSH REL: (Tungsten) TWA 1 mg(W)/m^3

SAFETY PROFILE: Poison by ingestion, intravenous, intramuscular, and subcutaneous routes. Mutation data reported. When heated to decomposition it emits toxic fumes of Na_2O. See also TUNGSTEN COMPOUNDS.

SKO000 CAS:53125-86-3 ***HR: 3***
SODIUM TUNGSTATE, DIHYDRATE
mf: $O_4W•2Na•2H_2O$ mw: 329.87

SYN: TUNGSTIC ACID, SODIUM SALT, DIHYDRATE

TOXICITY DATA with REFERENCE
cyt-hmn:lym 10 mg/L CYGEDX 12(3),46,78
ivn-mus TDLo:330 g/kg (8D preg):REP ENVRAL 33,47,84
ipr-rat LD50:204 mg/kg AIPTAK 154,243,65
ipr-mus LD50:145 mg/kg AIPTAK 154,243,65

ACGIH TLV: TWA 5 mg(W)/m^3; STEL 10 mg(W)/m^3
NIOSH REL: (Tungsten) TWA 1 mg(W)/m^3

SAFETY PROFILE: Poison by intraperitoneal route. Experimental reproductive effects. Human mutation data reported. When heated to decomposition it emits toxic fumes of Na_2O. See also SODIUM TUNGSTATE.

SKO500 CAS:7246-21-1 ***HR: 2***
SODIUM TYROPANOATE
mf: $C_{15}H_{17}I_3NO_3•Na$ mw: 663.02

SYNS: BILOPAC ◇ BILOPAQUE ◇ 3-BUTYRAMIDO-α-ETHYL-2,4,6-TRIIODOHYDROCINNAMIC ACID SODIUM SALT ◇ α-ETHYL-β-(2,4,6-TRIIODO-3-BUTYRAMIDOPHENYL)PROPIONIC ACID SODIUM SALT ◇ NSC-107434 ◇ SODIUM-3-BUTYRAMIDO-α-ETHYL-2,4,6-TRIIODOHYDROCINNAMATE ◇ TRYOPANOATE SODIUM ◇ WIN 8851-2

TOXICITY DATA with REFERENCE
orl-rat LD50:6500 mg/kg NIIRDN 6,467,82
scu-rat LD50:2850 mg/kg NIIRDN 6,467,82
ivn-rat LD50:860 mg/kg NIIRDN 6,467,82
orl-mus LD50:4200 mg/kg NIIRDN 6,467,82
scu-mus LD50:2300 mg/kg NIIRDN 6,467,82
ivn-mus LD50:720 mg/kg JMCMAR 13,997,70

SAFETY PROFILE: Moderately toxic by intravenous and subcutaneous route. Mildly toxic by ingestion. When heated to decomposition it emits very toxic fumes of I^-, NO_x, and Na_2O.

SKO575 CAS:1198-77-2 ***HR: D***
SODIUM URATE
mf: $C_5H_3N_4O_3•Na$ mw: 190.11

SYNS: MONOSODIUM URATE ◇ 1H-PURINE-2,6,8(3H)-TRIONE, 7,9-DIHYDRO-, MONOSODIUM SALT (9CI) ◇ URIC ACID, MONOSODIUM SALT

TOXICITY DATA with REFERENCE
ivn-mus TDLo:300 mg/kg (female 8-10D post):REP TXCYAC 6,289,76

CONSENSUS REPORTS: Reported in EPA TSCA Inventory.

SAFETY PROFILE: Experimental reproductive effects. When heated to decomposition it emits toxic fumes of NO_x.

SKP000 CAS:13718-26-8 ***HR: 3***
SODIUM VANADATE
mf: $O_3V•Na$ mw: 121.93

SYNS: SODIUM METAVANADATE ◇ VANADIC ACID, MONOSODIUM SALT

TOXICITY DATA with REFERENCE
orl-rat LDLo:98 mg/kg TOLED5 23,227,84
orl-mus LD50:74600 μg/kg TOLED5 23,227,84
ipr-mus LD50:36 mg/kg TOLED5 23,227,84
ivn-cat LD50:7180 μg/kg AGSOA6 8,51,67
orl-rbt LDLo:200 mg/kg EQSSDX 1,1,75
ivn-rbt LDLo:17 mg/kg EQSSDX 1,1,75
scu-dog LDLo:17 mg/kg EQSSDX 1,1,75
ivn-dog LDLo:11 mg/kg EQSSDX 1,1,75

CONSENSUS REPORTS: Reported in EPA TSCA Inventory.

ACGIH TLV: TWA 0.05 mg(V_2O_5)/m^3
NIOSH REL: (Vanadium Compounds) CL 0.05 mg(V)/m^3/15M

SAFETY PROFILE: Poison by ingestion, intraperitoneal, subcutaneous, and intravenous routes. When heated to decomposition it emits toxic fumes of Na_2O and VO_x. See also VANADIUM COMPOUNDS.

SKQ000 CAS:64082-34-4 ***HR: 3***
SODIUM VANADITE
mf: $O_9V_4•2Na$ mw: 393.74

SYN: VANADIOUS(4+) ACID, DISODIUM SALT

TOXICITY DATA with REFERENCE
scu-rat LDLo:21 mg/kg AJSNAO 1,347,17
ipr-rat LDLo:11 mg/kg EQSSDX 1,1,75
scu-mus LDLo:216 mg/kg AJSNAO 1,347,17
scu-gpg LDLo:65 mg/kg AJSNAO 1,347,17

ACGIH TLV: TWA 0.05 mg(V_2O_5)/m^3
NIOSH REL: CL 0.05 mg(V)/m^3/15M

SAFETY PROFILE: Poison by subcutaneous and intraperitoneal routes. When heated to decomposition it emits toxic fumes of Na_2O and VO_x. See also VANADIUM COMPOUNDS.

SKQ400 CAS:2307-68-8 ***HR: 2***
SOLAN
mf: $C_{13}H_{18}ClNO$ mw: 239.77

PROP: Crystals. Mp: 79-80°. Sol in pine oil, diisobutylketone, isophorone, xylene; practically insol in water.

SYNS: N-(3-CHLOR-METHYLPHENYL)-2-METHYLPENTANAMID (GERMAN) ◇ N-(3-CHLORO-4-METHYLPHENYL)-2-METHYLPENTANAMIDE ◇ 3'-CHLORO-2-METHYL-p-VALEROTOLUIDIDE ◇ CMMP ◇ DAKURON ◇ DUTOM ◇ NIAGARA 4512 ◇ PENTANOCHLOR

TOXICITY DATA with REFERENCE
orl-rat LD50:5620 mg/kg EKMMA8 13,123,74
orl-mus LD50:1800 mg/kg GTPZAB 21(12),30,77

SAFETY PROFILE: Moderately toxic by ingestion. When heated to decomposition it emits toxic fumes of Cl^- and NO_x.

SKQ500 CAS:11004-30-1 ***HR: 3***
(22s,25r)-5-α-SOLANIDAN-3-β-OL
mf: $C_{27}H_{46}NO$ mw: 400.74

TOXICITY DATA with REFERENCE
orl-ham TDLo:81 mg/kg (8D preg):REP JAFCAU 26(3),566,78
orl-ham TDLo:81 mg/kg (8D preg):TER JAFCAU 26(3),566,78
orl-ham LDLo:177 mg/kg JAFCAU 26(3),566,78

SAFETY PROFILE: Poison by ingestion. An experimental teratogen. Experimental reproductive effects. When heated to decomposition it emits toxic fumes of NO_x.

SKQ510 ***HR: 3***
(22r,25s)-5-α-SOLANIDAN-3-β-OL
mf: $C_{27}H_{46}NO$ mw: 400.74

SYN: (22r,25s)-5-α-SOLANIDANINE-3-β-OL

TOXICITY DATA with REFERENCE
orl-ham TDLo:343 mg/kg (7D preg):REP 41CIAR -,409,78
orl-ham LDLo:343 mg/kg JAFCAU 26,566,78

SAFETY PROFILE: Poison by ingestion. Experimental reproductive effects. When heated to decomposition it emits toxic fumes of NO_x.

SKR000 CAS:79-58-3 ***HR: 3***
SOLANID-5-ENE-3-β, 12-α-DIOL
mf: $C_{27}H_{43}NO_2$ mw: 413.71

PROP: An alkaloid isolated from *Veratrum album*. (JPETAB 82,167,44).

SYN: RUBIJERVINE

TOXICITY DATA with REFERENCE
orl-rbt TDLo:66.7 mg/kg (7D preg):TER PSEBAA 136,1174,71
ivn-mus LD50:70 mg/kg JPETAB 82,167,44

SAFETY PROFILE: Poison by intravenous route. An experimental teratogen. When heated to decomposition it emits toxic fumes of NO_x.

SKR500 CAS:566-09-6 ***HR: D***
(22s,25r)-SOLANID-5-EN-3-β-OL
mf: $C_{27}H_{44}NO$ mw: 398.72

TOXICITY DATA with REFERENCE
orl-ham TDLo:88 mg/kg (8D preg):REP JAFCAU 26,566,78
orl-ham TDLo:88 mg/kg (8D preg):TER JAFCAU 26,566,78

SAFETY PROFILE: An experimental teratogen. Experimental reproductive effects. When heated to decomposition it emits toxic fumes of NO_x.

SKR875 CAS:51938-42-2 ***HR: 3***
SOLANIN

TOXICITY DATA with REFERENCE
ivn-dog LDLo:40 mg/kg AEXPBL 26,88,1890
ivn-cat LDLo:27 mg/kg AEXPBL 26,88,1890
orl-rbt LDLo:300 mg/kg AEXPBL 26,88,1890
ivn-rbt LDLo:25 mg/kg AEXPBL 26,88,1890
par-frg LDLo:303 mg/kg AEXPBL 26,88,1890

SAFETY PROFILE: Poison by ingestion, intravenous, and parenteral routes.

SKS000 CAS:20562-02-1 ***HR: 3***
SOLANINE
mf: $C_{45}H_{73}NO_{15}$ mw: 868.19

SYNS: α-SOLANIN ◇ α-SOLANINE

TOXICITY DATA with REFERENCE
oth-hmn:fbr 66600 μg/L HUMAA7 26,105,75
ipr-rat TDLo:40 mg/kg (5-12D preg):TER TXAPA9 36,227,76
ipr-mus TDLo:100 mg/kg (female 7-11D post):REP JRPFA4 46,257,76
orl-hmn TDLo:2800 μg/kg VHTODE 25,13,83
unr-man LDLo:2941 μg/kg 85DCAI 2,73,70
orl-rat LD50:590 mg/kg TJADAB 8,349,73
ipr-rat LD50:67 mg/kg TXAPA9 36,227,76
ipr-mus LD50:32 mg/kg FCTXAV 10,395,72
ipr-rbt LDLo:20 mg/kg TXAPA9 19,81,71
scu-rbt LDLo:5 mg/kg HBAMAK 4,1398,35
ivn-rbt LDLo:10 mg/kg TXPA9 19,81,71

SAFETY PROFILE: Human poison by unspecified routes. Experimental poison by intraperitoneal, subcutaneous, and intravenous routes. Moderately toxic by ingestion. An experimental teratogen. Experimental reproductive effects. Human mutation data reported. When heated to decomposition it emits toxic fumes of NO_x.

SKS100 CAS:40816-40-8 ***HR: 3***
SOLANINE HYDROCHLORIDE
mf: $C_{45}H_{73}NO_{15}$•ClH mw: 904.65

SYNS: β-d-GALACTOPYRANOSIDE, (3-β)-SOLANID-5-EN-3-YL O-6-DEOXY-α-l-MANNOPYRANOSYL-(1-2)-O-(β-d-GLUCOPYRANOSYL-(1-3))-, HYDROCHLORIDE ◇ SOLANIN HYDROCHLORIDE

TOXICITY DATA with REFERENCE
unr-rat TDLo:2500 mg/kg (female 28D pre):REP TVLAA5 14,97,71
ipr-mus LD50:42 mg/kg MEIEDD 10,1244,83

SAFETY PROFILE: Poison by intraperitoneal route. Experimental reproductive effects. When heated to decomposition it emits toxic fumes of NO_x and HCl.

SKS299 CAS:68070-90-6 ***HR: 3***
SOLCOSERYL

SYN: SS-094

TOXICITY DATA with REFERENCE
orl-rat TDLo:7200 mg/kg (9-14D preg):REP KSRNAM 7,2239,73
orl-mus TDLo:240 mg/kg (female 7-12D post):TER KSRNAM 7,2239,73
orl-rat LD50:1200 mg/kg KSRNAM 7,2298,73
scu-rat LD50:800 mg/kg KSRNAM 7,2298,73
ivn-rat LD50:400 mg/kg KSRNAM 7,2298,73
orl-mus LD50:4 g/kg KSRNAM 7,2298,73
scu-mus LD50:2 g/kg KSRNAM 7,2298,73
ivn-mus LD50:2 g/kg KSRNAM 7,2298,73

SAFETY PROFILE: Poison by intravenous route. Moderately toxic by ingestion and subcutaneous routes. An experimental teratogen. Experimental reproductive effects.

SKS500 CAS:96-64-0 ***HR: 3***
SOMAN
mf: $C_7H_{16}FO_2P$ mw: 182.20

SYNS: 3,3-DIMETHYL-2-BUTANOLMETHYLPHOSPHONOFLUORIDATE ◇ 3,3-DIMETHYL-n-BUT-2-YL METHYLPHOSPHONOFLUORIDATE ◇ 3,3-DIMETHYL-2-BUTYL METHYLPHOSPHONOFLUORIDATE ◇ FLUOROMETHYL(1,2,2-TRIMETHYLPROPOXY) PHOSPHINE OXIDE ◇ GD ◇ METHYLFLUORPHOSPHORSAEUREPINAKOLYLESTER (GERMAN) ◇ METHYLPHOSPHONOFLUORIDIC ACID, 3,3-DIMETHYL-2-BUTYL ESTER ◇ METHYLPHOSPHONOFLUORIDIC ACID 1,2,2-TRIMETHYLPROPYL ESTER ◇ METHYL PINACOLYLOXY PHOSPHORYLFLUORIDE ◇ METHYL PINACOLYL PHOSPHONOFLUORIDATE ◇ PINACOLOXYMETHYLPHOSPHORYL FLUORIDE ◇ PINACOLYL METHYLFLUOROPHOSPHONATE ◇ PINACOLYL METHYLPHOSPHONOFLUORIDATE ◇ PINACOLYL METHYLPHOSPHONOFLUORIDE ◇ PINACOLYLOXY METHYLPHOSPHORYL FLUORIDE ◇ PMFP ◇ PYNACOLYL METHYLFLUOROPHOSPHONATE ◇ 1,2,2-TRIMETHYLPROPYL METHYLPHOSPHONOFLUORIDATE

TOXICITY DATA with REFERENCE
ihl-hmn LCLo:70 mg/m^3 SCJUAD 4,33,67
skn-hmn LDLo:18 mg/kg SCJUAD 4,33,67
ipr-rat LD50:177 μg/kg FAATDF 5,S84,85
scu-rat LD50:75 μg/kg FAATDF 4(2, Pt 2),S106,84
ivn-rat LD50:53 μg/kg FAATDF 5,S84,85
ims-rat LD50:62 μg/kg FAATDF 5,S84,85
ihl-mus LC50:1 mg/m^3/30M DEGEA3 15,2179,60
skn-mus LD50:7800 μg/kg TXAPA9 5,685,63
ipr-mus LD50:393 μg/kg FAATDF 4(2, Pt 2),S96,84
sku-mky LD50:13 μg/kg FAATDF 1,419,81
ims-mky LD50:9500 ng/kg ECNEAZ 32,557,72

SAFETY PROFILE: A deadly human poison by inhalation and skin contact. A deadly experimental poison by inhalation, skin contact, subcutaneous, intravenous, intramuscular, and intraperitoneal routes. An extremely toxic military nerve gas. When heated to decomposition it emits very toxic fumes of F^- and PO_x.

SKS600 CAS:51110-01-1 ***HR: 3***
SOMATOSTATIN 14

SYNS: PANHIBIN ◇ SOMATOSTATIN ◇ SRIF

TOXICITY DATA with REFERENCE
ipr-mus TDLo:197 μg/kg (35D male):REP KSRNAM 14,3419,80
ivn-rat LD50:21 mg/kg TOLED5 17,145,83
ivn-mus LD50:33 mg/kg TOLED5 17,145,83

SAFETY PROFILE: Poison by intravenous route. Experimental reproductive effects.

SKS700 CAS:480-30-8 ***HR: 2***
SOMINAT
mf: $C_{11}H_{12}N_2O$•$2C_2H_3Cl_3O_2$ mw: 519.05

PROP: Small, prismatic needles from water. Mp: 68°.

SYNS: DICHLORALANTIPYRIN ◇ DICHLORALANTIPYRINE ◇ DICHLORALPHENAZONE ◇ DORMUPHAR ◇ DORMWELL ◇ IBIODORM ◇ KLORALFENAZON ◇ SEDOR ◇ WELLDORM

TOXICITY DATA with REFERENCE
orl-rat LD50:1338 mg/kg JAPMA8 45,137,56
orl-mus LD50:1437 mg/kg JAPMA8 45,137,56
ipr-mus LD50:982 mg/kg JAPMA8 45,137,56

SAFETY PROFILE: Moderately toxic by ingestion and intraperitoneal routes. When heated to decomposition it emits toxic fumes of Cl^- and NO_x. See also ANTIPYRINE and CHLORAL HYDRATE.

SKS750 ***HR: 3***
SOOT

PROP: Soot is defined as a brown-to-black substance incidentally produced during the incomplete and uncontrolled combustion of any carbonaceous material. It is a mixture of colloidal carbon, organic tars, and refractory inorganics whose composition depends on combustion conditions. It is not unusual for the tarry component to account for more than 50 wt% of the soot, particularly, when produced by inefficient combustion of coal or wood. Can be distinguished from carbon black on the basis of differences in physical and chemical properties.

SAFETY PROFILE: Confirmed human carcinogen producing skin, scrotum, or lung tumors. The tarry component and, to a lesser extent, trace inorganic impurities, are believed responsible for the known health hazards attributed to soot, i.e., chronic contact or long-term inhalation can lead to cancer. See also CARBON BLACK.

SKS800 CAS:78003-71-1 ***HR: 3***
SOPHOCARPINE HYDROBROMIDE
mf: $C_{15}H_{22}N_2O \cdot BrH$ mw: 327.31

SYNS: 13,14-DIDEHYDROMATRIDIN-15-ONEHYDROBROMIDE ◇ 13,14-DIDEHYDRO-MATRIDIN-15-ONE MONOHYDROBROMIDE

TOXICITY DATA with REFERENCE
orl-mus LD50:298 mg/kg CTYAD8 11,555,80
ivn-mus LD50:73640 μg/kg CTYAD8 11,555,80
ims-mus LD50:101 mg/kg CTYAD8 11,555,80

SAFETY PROFILE: Poison by ingestion, intravenous, and intramuscular routes. When heated to decomposition it emits toxic fumes of NO_x and Br^-.

SKT500 CAS:142-83-6 ***HR: 3***
SORBALDEHYDE
mf: C_6H_8O mw: 96.14

SYNS: HEXA-2,4-DIENAL ◇ 2,4-HEXADIENAL ◇ 1,3-PENTADIENE-1-CARBOXALDEHYDE ◇ SOBIC ALDEHYDE

TOXICITY DATA with REFERENCE
skn-rbt 10 mg/24H SEV AMIHBC 10,61,54
eye-rbt 750 μg SEV AMIHBC 10,61,54
mmo-sat 500 nmol/L MUREAV 148,25,85
orl-rat LD50:730 mg/kg AMIHBC 10,61,54
ihl-rat LCLo:2000 ppm/4H AMIHBC 10,61,54
skn-rbt LD50:270 mg/kg AMIHBC 10,61,54

CONSENSUS REPORTS: Reported in EPA TSCA Inventory.

SAFETY PROFILE: Poison by skin contact. Moderately toxic by ingestion. Mildly toxic by inhalation. Mutation data reported. A severe eye and skin irritant. The presence of 1-amino-2-propanol as an impurity may cause explosive polymerization when heated. When heated to decomposition it emits acrid smoke and irritating fumes. See also ALDEHYDES.

SKU000 CAS:110-44-1 ***HR: 3***
SORBIC ACID
mf: $C_6H_8O_2$ mw: 112.14

PROP: Colorless needles or white powder; characteristic odor. Bp: 228° (decomp), mp: 134.5°, flash p: 260°F (COC), vap press: 0.01 mm @ 20°, vap d: 3.87. Sol in hot water; very sol in alc, ether.

SYNS: (2-BUTENYLIDENE)ACETIC ACID ◇ CROTYLIDENE ACETIC ACID ◇ HEXADIENIC ACID ◇ HEXADIENOIC ACID ◇ 2,4-HEXADIENOIC ACID ◇ trans-trans-2,4-HEXADIENOIC ACID ◇ 1,3-PENTADIENE-1-CARBOXYLIC ACID ◇ 2-PROPENYLACRYLIC ACID ◇ SORBISTAT

TOXICITY DATA with REFERENCE
skn-man 150 mg/1H SEV JPPMAB 10,719,58
skn-rbt 1 mg SEV UCDS** 7/14/65
cyt-ham:lng 1050 mg/L FCTOD7 22,501,84
sce-ham:lng 1050 mg/L FCTOD7 22,501,84
orl-rat TDLo:4154 g/kg (2Y pre):REP FCTXAV 13,31,75
scu-rat TDLo:1040 mg/kg/65W-I:ETA BJCAAI 20,134,66
orl-rat LD50:7360 mg/kg JIHTAB 30,63,48
orl-mus LD50:3200 mg/kg 38MKAJ 2C,4953,82
ipr-mus LD50:2820 mg/kg JPPMAB 21,85,69
scu-mus LD50:2820 mg/kg JPPMAB 21,85,69

CONSENSUS REPORTS: Reported in EPA TSCA Inventory.

SAFETY PROFILE: Moderately toxic by intraperitoneal and subcutaneous routes. Mildly toxic by ingestion. Experimental reproductive effects. A severe human and experimental skin irritant. Questionable carcinogen with experimental tumorigenic data. Mutation data reported. Combustible when exposed to heat or flame; can react with oxidizing materials. To fight fire, use water. When heated to decomposition it emits acrid smoke and irritating fumes.

SKV000 CAS:1338-39-2 ***HR: 1***
SORBITAN MONOLAURATE
mf: $C_{18}H_{34}O_6$ mw: 346.52

SYNS: EMSORB 2515 ◇ RADIASURF 7125 ◇ SORBITAN MONODODECANOATE ◇ SPAN 20

TOXICITY DATA with REFERENCE
orl-rat TDLo:86240 mg/kg (14D male):REP FCTXAV 16,519,78
skn-mus TDLo:1350 mg/kg/24W-I:NEO SCIEAS 120,1075,54
orl-rat LD50:33600 mg/kg JACTDZ 4(3),65,85

CONSENSUS REPORTS: Reported in EPA TSCA Inventory.

SAFETY PROFILE: Slightly toxic by ingestion. Experimental reproductive effects. Questionable carcinogen with experimental neoplastigenic data. When heated to decomposition it emits acrid smoke and irritating fumes. See also SURFACTANTS.

SKV100 CAS:1338-43-8 ***HR: 1***
SORBITAN MONOOLEATE
mf: $C_{24}H_{44}O_6$ mw: 428.68

SYNS: ARLACEL 80 ◇ ARMOTAN MO ◇ EMSORB 2500 ◇ GLYCOMUL O ◇ IONET S-80 ◇ LIPOSORB O ◇ LIPOSORB O-20 ◇ ML 33F ◇ ML 55F ◇ MONODEHYDROSORBITOL MONOOLEATE ◇ MONTAN 80 ◇ NIKKOL SO 10 ◇ NIKKOL SO-15 ◇ NIKKOL SO-30 ◇ NONION OP80R ◇ O 250 ◇ RADIASURF 7155 ◇ SORBESTER P 17 ◇ SORBITAN MONOOLEIC ACID ESTER ◇ SORBITAN O ◇ SORBITAN OLEATE ◇ SORGEN 40 ◇ SPAN 80

TOXICITY DATA with REFERENCE
skn-rbt 250 ug MLD JACTDZ 4(3),65,85
dni-hmn:lyms 100 ppm BBRCA9 45,630,71
orl-rat TDLo:254 g/kg (male 42D pre):REP FCTXAV 16,535,78

CONSENSUS REPORTS: Reported in EPA TSCA Inventory.

SAFETY PROFILE: Experimental reproductive effects. Skin irritant. Mutation data reported. Human mutation data reported. When heated to decomposition it emits acrid smoke and irritating fumes.

SKV150 CAS:1338-41-6 ***HR: 1***
SORBITAN MONOSTEARATE
mf: $C_{24}H_{46}O_6$ mw: 430.70

PROP: Cream to tan-colored waxy solid; bland odor and taste. Insol in cold water, mineral spirits, acetone; dispersible in warm water; sol above 50° in mineral oil, ethyl acetate.

SYNS: ANHYDRO-d-GLUCITOL MONOOCTADECANOATE ◇ ANHYDROSORBITOL STEARATE ◇ ARLACEL 60 ◇ ARMOTAN MS ◇ CRILL 3 ◇ CRILL K 3 ◇ DREWSORB 60 ◇ DURTAN 60 ◇ EMSORB 2505 ◇ GLYCOMUL S ◇ HODAG SMS ◇ IONET S 60 ◇ LIPOSORB S ◇ LIPOSORB S-20 ◇ MONTANE 60 ◇ MS 33 ◇ MS 33F ◇ NEWCOL 60 ◇ NIKKOL SS 30 ◇ NISSAN NONION SP 60 ◇ NONION SP 60 ◇ NONION SP 60R ◇ RIKEMAL S 250 ◇ SORBITAN C ◇ SORBITAN MONOOCTADECANOATE ◇ SORBITAN STEARATE ◇ SORBON S 60 ◇ SORGEN 50 ◇ SPAN 55 ◇ SPAN 60

TOXICITY DATA with REFERENCE
skn-rbt 800 ug MLD JACTDZ 4(3),65,85
orl-rat TDLo:635 g/kg (MGN):REP JONUAI 60,489,56
orl-rat LD50:31 g/kg FOREAE 21,348,56

CONSENSUS REPORTS: EPA Genetic Toxicology Program.

SAFETY PROFILE: Very mildly toxic by ingestion. Experimental reproductive effects. A skin irritant. When heated to decomposition it emits acrid smoke and irritating fumes.

SKV195 CAS:9005-71-4 ***HR: D***
SORBITAN, TRISTEARATE, POLYOXYETHYLENE derivatives.

SYNS: EMSORB 6907 ◇ GLYCOSPERSE TS 20 ◇ POLYSORBATE 65 ◇ SORBIMACROGOL TRISTEARATE 300 ◇ SORBITAN, TRIOCTADECANOATE, POLY(OXY-1,2-ETHANEDIYL) derivs. (9CI) ◇ TWEEN 65

TOXICITY DATA with REFERENCE
orl-rat TDLo:635 g/kg (multi) :REP JONUAI 60,489,56

CONSENSUS REPORTS: Reported in EPA TSCA Inventory.

SAFETY PROFILE: Experimental reproductive effects. When heated to decomposition it emits acrid smoke and irritating fumes.

SKV200 CAS:50-70-4 ***HR: 1***
SORBITOL
mf: $C_6H_{14}O_6$ mw: 182.20

PROP: White crystalline powder; odorless with sweet taste. D: 1.47 @ −5°, mp: 93° (metastable form); 97.5°, (stable form), bp: 105°. Sol in water; sltly sol in methanol, ethanol, acetic acid, phenol, and acetamide; almost insol in other organic solvents.

SYNS: CHOLAXINE ◇ DIAKARMON ◇ GLUCITOL ◇ d-GLUCITOL ◇ GULITOL ◇ l-GULITOL ◇ KARION ◇ NIVITIN ◇ SIONIT ◇ SIONON ◇ SORBICOLAN ◇ SORBITE ◇ d-SORBITOL ◇ SORBO ◇ SORBOL ◇ SORBOSTYL ◇ SORVILANDE

TOXICITY DATA with REFERENCE
orl-wmn TDLo:1700 mg/kg/D:GIT AJDDAL 23,568,78
orl-rat LD50:15900 mg/kg FAONAU 53A,500,74
ivn-rat LD50:7100 mg/kg FAONAU 53A,500,74
orl-mus LD50:17800 mg/kg YKYUA6 32,1367,81
ipr-mus LDLo:15 g/kg PSEBAA 35,98,36
ivn-mus LD50:9480 mg/kg YKYUA6 32,1367,81

CONSENSUS REPORTS: Reported in EPA TSCA Inventory. EPA Genetic Toxicology Program.

SAFETY PROFILE: Mildly toxic by ingestion. Human systemic effects by ingestion: hypermotility and diarrhea. When heated to decomposition it emits acrid smoke and irritating fumes.

SKV500 ***HR: 3***
SORSAKA

PROP: Aq extract from the dried leaves of the plant *Annona muricata* (JNCIAM 46,1131,71).

SYNS: ANNONA MURICATA ◇ l-SORSAKA, LEAF and STEM EXTRACT

TOXICITY DATA with REFERENCE
scu-rat TDLo:300 g/kg/1Y-I:ETA JNCIAM 46,1131,71
ipr-mus LDLo:40 g/kg JPPMAB 14,556,62

SAFETY PROFILE: Questionable carcinogen with experimental tumorigenic data. When heated to decomposition it emits acrid smoke and irritating fumes.

SKW000 CAS:13642-52-9 ***HR: 3***
SOTERENOL
mf: $C_{12}H_{20}N_2O_4S$ mw: 288.40

SYN: MJ 1992

TOXICITY DATA with REFERENCE
orl-mus LD50:1140 mg/kg JPETAB 164,290,68
ipr-mus LD50:335 mg/kg JPETAB 164,290,68
ivn-mus LD50:47 mg/kg JPETAB 164,290,68
ivn-dog LD50:1 mg/kg JPETAB 164,290,68
orl-gpg LD50:41 mg/kg JPETAB 164,290,68
scu-gpg LD50:14 mg/kg JPETAB 164,290,68

SAFETY PROFILE: Poison by ingestion, subcutaneous, intravenous and intraperitoneal routes. When heated to decomposition it emits very toxic fumes of SO_x and NO_x. See also SOTERENOL HYDROCHLORIDE.

SKW500 CAS:14816-67-2 ***HR: 3***
SOTERENOL HYDROCHLORIDE
mf: $C_{12}H_{20}N_2O_4S•ClH$ mw: 324.86

SYNS: 2'-HYDROXY-5'-(1-HYDROXY-2-(ISOPROPYLAMINO)ETHYL)-METHANESULFONANILIDE MONOHYDROCHLORIDE ◇ MJ 1992

TOXICITY DATA with REFERENCE
orl-rat TDLo:2512 mg/kg/78W-C:ETA TXAPA9 22,279,72
orl-mus LD50:660 mg/kg MEIEDD 10,1248,83
ipr-mus LD50:315 mg/kg MEIEDD 10,1248,83
ivn-mus LD50:41 mg/kg MEIEDD 10,1248,83

SAFETY PROFILE: Poison by intravenous and intraperitoneal routes. Moderately toxic by ingestion. Questionable carcinogen with experimental tumorigenic data. When heated to decomposition it emits very toxic fumes of NO_x, SO_x, and HCl. A bronchodilator.

SKW775 ***HR: 3***
SOUTHERN COPPERHEAD VENOM

SYNS: A. CONTORTRIX CONTORTRIX VENOM ◇ AGKISTRODON CONTORTRIX CONTROTRIX VENOM ◇ VENOM, SNAKE, AGKISTRODON CONTORTRIX CONTORTRIX

TOXICITY DATA with REFERENCE
ipr-mus LD50:268 μg/kg CBPBB8 64,201,79
scu-mus LD50:735 μg/kg CBPBB8 64,201,78
ivn-mus LDLo:9520 μg/kg 19DDA6 1,269,67
ivn-dog LDLo:750 μg/kg 19DDA6 1,269,67

SAFETY PROFILE: Poison by subcutaneous, intravenous, and intraperitoneal routes.

SKX000 CAS:1404-64-4 ***HR: 3***
SPARSOMYCIN
mf: $C_{13}H_{19}N_3O_5S_2$ mw: 361.47

SYNS: NSC 59729 ◇ U-19183

TOXICITY DATA with REFERENCE
dnd-esc 20 μmol/L MUREAV 89,95,81
orl-mus LD50:22 mg/kg UPHOH* 2(6),-,71
ipr-mus LD50:2400 μg/kg CNCRA6 30,9,63
ivn-mus LD50:4 mg/kg UPJOH* 2(6),-,71
ivn-dog LD50:1 mg/kg UPJOH* 2(6),-,71

SAFETY PROFILE: Poison by ingestion, intraperitoneal, and intravenous routes. Mutation data reported. When heated to decomposition it emits very toxic fumes of NO_x and SO_x. Used as an antifungal agent.

SKX500 CAS:90-39-1 ***HR: 3***
SPARTEINE
mf: $C_{15}H_{26}N_2$ mw: 234.43

PROP: A colorless oil. Bp: 173° @ 8 mm, 181° @ 20 mm in air, d: 1.020 @ 20°/4°.

SYNS: DODECAHYDRO-7,14-METHANO-2H,6H-DIPYRIDO(1,2-a:1',2'-e)(1,5)DIAZOCINE ◇ LUPINIDINE ◇ 6-β,7-α,9-α,11-α-PACHYCARPINE ◇ (−)-SPARTEINE ◇ 1-SPARTEINE

TOXICITY DATA with REFERENCE
orl-rat LD50:960 mg/kg AIPTAK 210,27,74
ipr-rat LD50:42 mg/kg CLDND*
orl-mus LD50:220 mg/kg PLMEAA 50,420,84
ipr-mus LD50:36 mg/kg PLMEAA 50,420,84
ivn-mus LD50:17400 μg/kg ARZNAD 30,1497,80
scu-rbt LDLo:100 mg/kg HBAMAK 4,1398,35
ivn-rbt LDLo:30 mg/kg HBAMAK 4,1398,35
ipr-gpg LDlo:23 mg/kg JAGRAC 32,51,26
ivn-gpg LDLo:27 mg/kg PLMEAA 50,420,84
scu-pgn LDLo:86 mg/kg HBAMAK 4,1398,35
scu-frg LDLo:3330 mg/kg HBAMAK 4,1398,35
ims-frg LDLo:1665 mg/kg HBAMAK 4,1398,35

SAFETY PROFILE: Poison by ingestion, subcutaneous, intravenous, and intraperitoneal routes. Moderately toxic by intramuscular route. When heated to decomposition it emits toxic fumes of NO_x.

SKX750 CAS:299-39-8 ***HR: 3***
SPARTOCIN
mf: $C_{15}H_{26}N_2•H_2O_4S$ mw: 332.51

SYN: SPARTEINE SULFATE

TOXICITY DATA with REFERENCE
ims-wmn TDLo:6 mg/kg (39W preg):REP AJOGAH 89,263,64

ivn-rat LD50:30400 μg/kg JJPAAZ 16,353,66
orl-mus LD50:350 mg/kg ARZNAD 30,1497,80
ipr-mus LD50:43 mg/kg JMCMAR 20,1668,77
scu-mus LD50:117 mg/kg JJPAAZ 16,353,66
ivn-mus LD50:15 mg/kg JJPAAZ 16,353,66
ipr-gpg LDLo:42 mg/kg JAGRAC 32,51,26

SAFETY PROFILE: Poison by ingestion, subcutaneous, intravenous, and intraperitoneal routes. Human reproductive effects by intramuscular route: stillbirth; changes in uterus, cervix and vagina. When heated to decomposition it emits toxic fumes of NO_x and SO_x.

SKX775 ***HR: 1***
SPATHE FLOWER

PROP: Tropical perennial herbs which grow to 2 feet with clusters of oval leaves. The flower is a white or greenish spathe with a short, white spadix that looks like a small ear of corn. They are native to South America and are cultivated outdoors in subtropical climates. They are common indoor plants.

SYNS: SPATHIPHYLLUM (VARIOUS SPECIES) ◇ WHITE ANTHURIUM

SAFETY PROFILE: All parts of the plant contain poisonous crystals of calcium oxalate. Chewing parts of the plant results in burning pain in the lips, mouth and throat, possibly followed by inflammation and blistering. Systemic effects are usually not seen because of the insolubility of calcium oxalate. See also OXALATES.

SKY000 CAS:8008-79-5 ***HR: 1***
SPEARMINT OIL

PROP: From steam distillation of the plant *Mentha spicata* L. (Common Spearmint), or of *Mentha cardiaca* Gerard ex Baker (Scotch Spearmint) (Fam. *Labiatae*). Contains principally carvone, phellandrene, limonene, and either dihydrocarveol acetate or dihydrocuminic acetate (FCTXAV 16, 637,78). Colorless or greenish-yellow liquid; odor and taste of spearmint.

SYN: OIL of SPEARMINT

TOXICITY DATA with REFERENCE
skn-rbt 500 mg/24H MOD FCTXAV 16,637,78
skn-gpg 100% MLD FCTXAV 16,871,78
dnr-bcs 10 mg/disc TOFOD5 8,91,85
orl-rat LD50:5 g/kg FCTXAV 16,637,78

CONSENSUS REPORTS: Reported in EPA TSCA Inventory.

SAFETY PROFILE: Mildly toxic by ingestion. Mutation data reported. A skin irritant and an allergen. When heated to decomposition it emits acrid smoke and irritating fumes. Used as a flavoring agent.

SKY500 CAS:22189-32-8 ***HR: 2***
SPECTOGARD
mf: $C_{14}H_{24}N_2O_7 \cdot 2ClH \cdot 5H_2O$ mw: 494.41 &QH 5

SYNS: ACTINOSPECTACIN DIHYDROCHLORIDE PENTAHYDRATE ◇ ESPECTINOMICINA DIHYDROCHLORIDE PENTAHYDRATE ◇ M 141 DIHYDROCHLORIDE PENTAHYDRATE ◇ SPECTINOMYCIN, DIHYDROCHLORIDE, PENTAHYDRATE ◇ STANILO ◇ TOGAMYCIN ◇ TROBICIN

TOXICITY DATA with REFERENCE
ipr-rat TDLo:2400 mg/kg (9-14D preg):TER KSRNAM 8,3018,74
ipr-rat LD50:2020 mg/kg YKYUA6 30,421,79
ipr-mus LD50:2350 mg/kg YKYUA6 30,421,79
scu-mus LD50:8400 mg/kg YKYUA6 30,421,79

SAFETY PROFILE: Moderately toxic by intraperitoneal route. An experimental teratogen. When heated to decomposition it emits very toxic fumes of NO_x and HCl. Used as an antibacterial agent.

SKZ000 ***HR: 2***
SPENT OXIDE

CONSENSUS REPORTS: Composition: Iron sponge + 5%-10% nitrogen.

SYN: GASHOUSE TANKAGE

SAFETY PROFILE: Can contain toxic materials. Flammable when exposed to heat or flame; can react with oxidizing materials.

SLA000 CAS:124-20-9 ***HR: 3***
SPERMIDINE
mf: $C_7H_{19}N_3$ mw: 145.29

SYNS: N-(3-AMINOPROPYL)-1,4-BUTANEDIAMINE ◇ N-(3-AMINOPROPYL)-1,4-DIAMINOBUTANE

TOXICITY DATA with REFERENCE
oms-bcs 50 mg/L MGGEAE 178,21,80
dnd-ham:ovr 300 μmol/L CNREA8 46,175,86
ivn-mus LD50:78 mg/kg AIPTAK 198,36,72

CONSENSUS REPORTS: Reported in EPA TSCA Inventory. EPA Genetic Toxicology Program.

SAFETY PROFILE: Poison by intravenous route. Mutation data reported. When heated to decomposition it emits toxic fumes of NO_x.

SLB000 CAS:27007-85-8 ***HR: 2***
SPICLOMAZINE HYDROCHLORIDE
mf: $C_{22}H_{24}ClN_3OS_2 \cdot ClH$ mw: 482.52

SYN: 8-(3-(2-CHLOROPHENOTHIAZIN-10-YL)PROPYL-1-THIA-4,8-DIAZASPIRO(4.5)DECAN-3-ONEHYDROCHLORIDE

TOXICITY DATA with REFERENCE
orl-rat TDLo:60 mg/kg (9-14D preg):REP OYYAA2 4,497,70
orl-rat TDLo:300 mg/kg (female 9-14D post):TER OYYAA2 4,497,70
orl-rat LD50:3800 mg/kg OYYAA2 4,487,70
ipr-rat LD50:2950 mg/kg OYYAA2 4,487,70
orl-mus LD50:3800 mg/kg OYYAA2 4,487,70
ipr-mus LD50:3200 mg/kg OYYAA2 4,487,70

SAFETY PROFILE: Moderately toxic by ingestion and intraperitoneal routes. An experimental teratogen. Experimental reproductive effects. When heated to decomposition it emits very toxic fumes of SO_x, NO_x, and Cl^-.

SLB250 ***HR: 1***
SPIDER LILY

PROP: Bulb-producing plants similar to onions with long thin leaves arranged in a spiral and white, pink or red lily-like flowers. Various species are native to the southern United States, West Indies, southern Africa and tropical Asia and many are cultivated.

SYNS: CRINUM (VARIOUS SPECIES) ◇ LIRIO (SPANISH) ◇ LYS (HAITI) ◇ POISON BULB

SAFETY PROFILE: All parts of the plant contain the poison lycorine. Ingestion may cause nausea, vomiting, and diarrhea.

SLB500 CAS:84837-04-7 ***HR: 2***
SPIKE LAVENDER OIL

PROP: From steam distillation of the plant *Lavandula latifolia* Vill. (*Lavandula spica*, D.C.)(Fam. *Labiatae*). The main constituents are linalool and cineole. (FCTXAV 14,443,76). Yellow liquid; lavender odor. D: 0.893-0.909, refr index: 1.463 @ 20°. Sol in fixed oils, propylene glycol; sltly sol in glycerin, mineral oil.

SYNS: LAVENDER OIL, SPIKE ◇ OIL of SPIKE LAVENDER

TOXICITY DATA with REFERENCE
skn-rbt 500 mg/24H MOD FCTXAV 14,443,76
orl-rat LD50:3800 mg/kg FCTXAV 14(5),443,76

SAFETY PROFILE: Moderately toxic by ingestion. A skin irritant. When heated to decomposition it emits acrid smoke and irritating fumes. See also LINALOOL and CAJEPUTOL.

SLC000 CAS:8025-81-8 ***HR: 3***
SPIRAMYCIN

PROP: An antibiotic substance produced by *Streptomyces ambofaciens* from soil of northern France.

SYNS: FOROMACIDIN ◇ KITASAMYCIN ◇ LEUCOMYCIN ◇ NSC-64393 ◇ PROVAMYCIN ◇ ROVAMICINA ◇ 5337 R.P. ◇ SELECTOMYCIN ◇ SEQUAMYCIN ◇ SPIRAMYCINS

TOXICITY DATA with REFERENCE
spm-rat-unr 200 mg/kg/8D JOURAA 112,348,74
orl-rat TDLo:6 g/kg (10-15D preg):TER JJANAX 25,187,72
orl-man TDLo:133 mg/kg/5D:GIT LANCAO 2,993,78
orl-rat LD50:3550 mg/kg JJANAX 23,429,70
ipr-rat LD50:575 mg/kg JJANAX 23,429,70
scu-rat LD50:1 g/kg ARZNAD 8,386,58
ivn-rat LD50:170 mg/kg ARZNAD 8,386,58
orl-mus LD50:2900 mg/kg THERAP 23,161,68
ipr-mus LD50:322 mg/kg JJANAX 23,429,70
scu-mus LD50:1470 mg/kg JJANAX 23,429,70
ivn-mus LD50:150 mg/kg 85ERAY 1,71,78
ipr-rbt LD50:1130 mg/kg JJANAX 23,429,70
ivn-rbt LD50:182 mg/kg JJANAX 23,429,70

SAFETY PROFILE: Poison by intravenous and intraperitoneal routes. Moderately toxic by ingestion and subcutaneous routes. An experimental teratogen. Human systemic effects by ingestion: hypermotility, diarrhea, nausea or vomiting. Experimental reproductive effects. Mutation data reported. When heated to decomposition it emits acrid smoke and irritating fumes.

SLC300 CAS:68880-55-7 ***HR: 3***
SPIRAMYCIN ADIPATE

SYNS: MAMIVERT ◇ SPIRAMYCIN, HEXANEDIOATE ◇ SUANOVIL

TOXICITY DATA with REFERENCE
orl-rat LD50:4650 mg/kg ANTCAO 10,273,60
orl-mus LD50:3130 mg/kg ANTCAO 10,273,60
ivn-mus LD50:188 mg/kg RPOBAR 2,327,70
scu-cat LD50:950 mg/kg ANTCAO 9,353,59
orl-rbt LD50:4330 mg/kg ANTCAO 10,273,60
orl-gpg LD50:250 mg/kg ANTCAO 10,273,60

SAFETY PROFILE: Poison by ingestion and intravenous routes. Moderately toxic by subcutaneous route.

SLD000 CAS:55-63-0 ***HR: 3***
SPIRIT of GLYCERYL TRINITRATE
DOT: UN 0143/UN 1204

PROP: Clear, colorless liquid. Composition: 1.0-1.1% glycerol trinitrate in alcoholic solution. D: 0.814-0.820 @ 25°. Misc with alc, chloroform, ether; very sltly sol in water.

SYNS: GLYCERYL TRINITRATE, solution ◇ RCRA WASTE NUMBER P081 ◇ SPIRIT of GLONOIN ◇ SPIRITS of NITROGLYCERIN (DOT) ◇ SPIRIT of TRINITROGLYCERIN

CONSENSUS REPORTS: Reported in EPA TSCA Inventory.

DOT Classification: Flammable Liquid; Label: Flammable Liquid; Class A Explosive; Label: Explosive A

SAFETY PROFILE: Likely to produce violent headache

when tasted or applied to the skin. Flammable when exposed to heat or flame. If the alcohol evaporates, the residue is nitroglycerin; a shock-sensitive explosive. Upon contact with oxidizing materials the mixture can react vigorously. If spilled, immediately pour NaOH solution over the dry residue; friction or shock will explode it. Incompatible with alkalies; carbonates; HCl; HI. When heated to decomposition it emits highly toxic fumes. See also NITROGLYCERIN and ETHYL ALCOHOL.

SLD500 ***HR: 3***
SPIRIT of NITER (sweet)
mf: $C_2H_5NO_2$ mw: 75.067

PROP: Composition: Alc solution: 3.5-4.5%. Pale straw-colored liquid; fragrant pungent odor, burning taste. D: not over 0.823 @ 25°.

SAFETY PROFILE: An irritant. Flammable when exposed to heat or flame. Explosive in the form of vapor when exposed to heat or flame. Can react with oxidizing materials. When heated to decomposition it emits highly toxic fumes of NO_x. See also NITRITES.

SLD600 CAS:86641-76-1 ***HR: 2***
SPIROBROMIN
mf: $C_{18}H_{32}Br_2N_4O_2•2Cl$ mw: 567.26

SYNS: 3,12-BIS(3-BROMO-1-OXOPROPYL)-3,12-DIAZA-6,9-DIAZONIADISPIRO(5.2.5.2)HEXADECANE DICHLORIDE ◇ N,N^3-DI(β-BROMOPROPIONYL)-N^1,N^2-DISPIROTRIPIPERAZINIUM DICHLORIDE

TOXICITY DATA with REFERENCE
ipr-rat LD50:800 mg/kg DRFUD4 9,121,84
ivn-rat LD50:825 mg/kg DRFUD4 9,121,84
ipr-mus LD50:1924 mg/kg DRFUD4 9,121,84

SAFETY PROFILE: Moderately toxic by intravenous and intraperitoneal routes. When heated to decomposition it emits toxic fumes of Cl^-, Br^-, and NO_x.

SLD800 CAS:41992-22-7 ***HR: 3***
SPIROGERMANIUM DIHYDROCHLORIDE
mf: $C_{17}H_{36}GeN_2•2ClH$ mw: 414.06

SYNS: 8,8-DIETHYL-N,N-DIMETHYL-2-AZA-8-GERMASPIRO(4.5)DECANE-2-PROPANAMINE DIHYDROCHLORIDE ◇ NSC 192965 ◇ S 99 A ◇ SPIRO-32 ◇ SPIROGERMANIUM HYDROCHLORIDE

TOXICITY DATA with REFERENCE
ipr-rat LD50:107 mg/kg NCISP* JAN86
scu-rat LD50:210 mg/kg NCISP* JAN86
ivn-mus LD50:41500 μg/kg NTIS** PB264-117
ims-mus LD50:119 mg/kg CTRRDO 64,1207,80
ivn-dog LDLo:40 mg/kg CTRRDO 64,1207,80
ims-dog LDLo:40 mg/kg CTRRDO 64,1207,80

SAFETY PROFILE: Poison by subcutaneous, intramuscular, intravenous, and intraperitoneal routes. When heated to decomposition it emits toxic fumes of NO_x and HCl. See also GERMANIUM COMPOUNDS.

SLD900 CAS:56605-16-4 ***HR: 3***
SPIROMUSTINE
mf: $C_{14}H_{23}Cl_2N_3O_2$ mw: 336.30

SYNS: 3-(2-(BIS(2-(CHLOROETHYL)AMINO)ETHYL)-1,3-DIAZASPIRO(4.5)DECANE-2,4-DIONE ◇ 3-(2-(BIS(2-(CHLOROETHYL)AMINO)ETHYL)-5,5-PENTAMETHYLENEHYDANTOIN ◇ NSC-172112 ◇ SHM ◇ SPIROHYDANTOIN MUSTARD

TOXICITY DATA with REFERENCE
mma-sat 50 μg/plate BCPCA6 32,523,83
bfa-mus/sat 60 mg/kg BCPCA6 32,523,83
ivn-rat LD50:9100 μg/kg NTIS** PB83-149518
ipr-mus LD50:33244 μg/kg NCISP* JAN86
ivn-mus LD50:48280 μg/kg NTIS** PB83-148395
ivn-dog LDLo:3750 μg/kg NTIS** PB83-149518

SAFETY PROFILE: Poison by intravenous and intraperitoneal routes. Mutation data reported. When heated to decomposition it emits toxic fumes of Cl^- and NO_x.

SLE500 CAS:749-02-0 ***HR: 3***
SPIROPERIDOL
mf: $C_{23}H_{26}FN_3O_2$ mw: 395.52

SYNS: 8-(3-p-FLUOROBENZOYL-1-PROPYL)-4-OXO-1-PHENYL-1,3,8-TRIAZASPIRO(4,5)DECANE ◇ 8-(3-(p-FLUOROBENZOYL)PROPYL)-1-PHENYL-1,3,8-TRIAZASPIRO(4.5)DECAN-4-ONE ◇ 4'-FLUORO-4-(4-OXO-1-PHENYL-1,3,8-TRIAZASPIRO(4,5)DECAN-8-YL)-BUTYROPHENONE ◇ 8-(4-(4-FLUOROPHENYL)-4-OXOBUTYL)-1-PHENYL-1,3,8-TRIAZASPIRO(4.5)DECAN-4-ONE ◇ SPIPERONE ◇ SPIROPITAN

TOXICITY DATA with REFERENCE
orl-mus TDLo:750 mg/kg (10-12D preg):TER TOIZAG 28,621,81
ivn-rat LD50:14 mg/kg 27ZQAG -,193,72
ims-rat LD50:168 mg/kg NIIRDN 6,380,82
orl-mus LD50:600 mg/kg CHTPBA 7,493,72
ivn-mus LD50:25500 μg/kg NIIRDN 6,380,82
ims-mus LD50:125 mg/kg NIIRDN 6,380,82

SAFETY PROFILE: Poison by intravenous and intramuscular routes. Moderately toxic by ingestion. An experimental teratogen. When heated to decomposition it emits very toxic fumes of F^- and NO_x. See also FLUORIDES.

SLE875 CAS:74790-08-2 ***HR: 3***
SPIROPLATIN
mf: $C_8H_{18}N_2O_4PtS$ mw: 433.43

SYNS: NSC 311056 ◇ TNO 6

TOXICITY DATA with REFERENCE
ivn-rat LD50:4 mg/kg EJCODS 20,1087,84
ipr-mus LD50:13 mg/kg EJCODS 20,1087,84
ivn-mus LD50:8900 μg/kg EJCODS 20,1087,84

SAFETY PROFILE: Poison by intravenous and intraperitoneal routes. When heated to decomposition it emits toxic fumes of NO_x and SO_x.

SLE880 ***HR: 2***
SPIZOFURONE
mf: $C_{12}H_{10}O_3$ mw: 202.21

SYNS: 5-ACETYLSPIRO(BENZOFURAN-2(3H),1'-CYCLOPROPAN)-3-ONE ◇ AG-629

TOXICITY DATA with REFERENCE
orl-rat LD50:5440 mg/kg KSRNAM 20,541,86
ipr-rat LD50:609 mg/kg KSRNAM 20,541,86
scu-rat LD50:1610 mg/kg KSRNAM 20,541,86
orl-mus LD50:1740 mg/kg KSRNAM 20,541,86
ipr-mus LD50:840 mg/kg KSRNAM 20,541,86
scu-mus LD50:1770 mg/kg KSRNAM 20,541,86

SAFETY PROFILE: Moderately toxic by ingestion, intraperitoneal and subcutaneous routes. When heated to decomposition it emits acrid smoke and irritating fumes.

SLE890 ***HR: 1***
SPLIT LEAF PHILODENDRON

PROP: A climbing plant with large, thick, irregularly perforated leaves. It is native to Mexico; cultivated outdoors in Hawaii, Guam, and the West Indies, and is a common greenhouse plant.

SYNS: BREADFRUIT VINE ◇ CASIMAN (MEXICO, PUERTO RICO) ◇ CERIMAN ◇ CERIMAN de MEJICO (CUBA) ◇ CUT LEAF PHILODENDRON ◇ FRUIT SALAD PLANT ◇ HURRICANE PLANT ◇ MEXICAN BREADFRUIT ◇ MONSTERA DELICIOSA ◇ PINANONA (MEXICO, PUERTO RICO) ◇ SHINGLE PLANT ◇ SWISS CHEESE PLANT ◇ WINDOWLEAF ◇ WINDOW PLANT

SAFETY PROFILE: The leaves have a direct irritant effect. Ingestion of the leaves causes an immediate burning pain in the lips, mouth, and throat, followed by an acute inflammatory reaction which may result in blistering and edema.

SLF500 CAS:68743-79-3 ***HR: 3***
SPORARICIN A
mf: $C_{17}H_{35}N_5O_5$ mw: 389.57

PROP: Produced by *Saccharopohyspora hirsuta* subsp. *Kobensis,* strain KC-6606 (JANTAJ 32,173,79).

SYN: ANTIBIOTIC KA 66061

TOXICITY DATA with REFERENCE
scu-mus LD50:310 mg/kg JANTAJ 32,173,79
ivn-mus LD50:73 mg/kg JANTAJ 32,173,79
ipr-mus LD50:100 mg/kg 85GDA2 1,239,80

SAFETY PROFILE: Poison by intraperitoneal, subcutaneous, and intravenous routes. When heated to decomposition it emits toxic fumes of NO_x.

SLG500 ***HR: 3***
SPRENGEL EXPLOSIVES

SAFETY PROFILE: This type of explosive is a mixture of nitrobenzene and fuming nitric acid. It is a powerful and cheap explosive and would have many uses except that it is limited by practical disadvantages. The components have to be mixed in glass shortly before the explosive is used. This requires preparation and equipment not always available at the site of the explosion. This material can be destroyed by burning small quantities at a time. See also EXPLOSIVES, HIGH.

SLH200 ***HR: 3***
SQUILL

PROP: A hyacinth-like plant which produces blue, purple, or white flowers. Various species are native to Europe and Asia, and grow as perennials in Canada from Newfoundland to British Columbia. They are also commonly cultivated.

SYNS: CUBAN LILLY ◇ HYACINTH-OF-PERU ◇ JACINTO de PERU (CUBA) ◇ PERUVIAN JACINTH ◇ SCILLA (VARIOUS SPECIES) ◇ SEA ONION ◇ STAR HYACINTH

SAFETY PROFILE: The whole plant contains cardiac glycosides similar to digitalis. Human systemic effects by ingestion include: mouth pain, nausea, vomiting, abdominal pain and cramps, diarrhea. Cardiac glycosides may cause death by their effect on heart function. See also DIGITALIS.

SLH300 ***HR: 2***
SRC-II, HEAVY DISTILLATE

PROP: High-boiling (550-850°F) liquid material from solvent refined coal processes (TJADAB 27,471,83).

SYNS: SOLVENT REFINED COAL MATERIALS, HEAVY DISTILLATE ◇ SOLVENT REFINED COAL-II, HEAVY DISTILLATE

TOXICITY DATA with REFERENCE
otr-ham:emb 5 mg/L JTEHD6 11,591,83
orl-rat TDLo:3300 mg/kg (12-16D preg):REP TJADAB 25(2),26A,82
orl-rat TDLo:420 mg/kg (12-14D preg):TER TOLED5 18(Suppl 1),77,83
orl-rat LD50:1840 mg/kg JACTDZ 2(2),248,83
ihl-rat LC50:3340 mg/m^3/4H JACTDZ 2(2),248,83

SAFETY PROFILE: Moderately toxic by inhalation and ingestion. An experimental teratogen. Experimental reproductive effects. Mutation data reported. Combustible when exposed to heat or flame. When heated to decomposition it emits acrid smoke and irritating fumes.

SLI300 CAS:636-47-5 ***HR: 3***
STALLIMYCIN
mf: $C_{22}H_{27}N_9O_4$ mw: 481.58

PROP: Mp: 154-156°.

SYNS: N″-(2-AMIDINOETHYL)-4-FORMAMIDO-1,1′,1″-TRIMETHYL-(N,4′:N′,4″-TERPYRROLE)-2-CARBOXAMIDE ◇ DISTAMYCIN A ◇ DISTAMICINA A (ITALIAN)

TOXICITY DATA with REFERENCE
dni-omi 25 mg/L JMCMAR 23,1144,80
oms-hmn:fbr 100 mg/L GICTAL 30,39,83
ipr-mus LD50:500 mg/kg 85FZAT -,258,67
orl-bwd LD50:72 mg/kg AECTCV 12,355,83

SAFETY PROFILE: Poison by ingestion. Moderately toxic by intraperitoneal route. Human mutation data reported. When heated to decomposition it emits toxic fumes of NO_x.

SLI325 CAS:21736-83-4 ***HR: 2***
STANILO
mf: $C_{14}H_{24}N_2O_7$•2ClH mw: 405.32

SYNS: DECAHYDRO-4a,7,9-TRIHYDROXY-2-METHYL-6,8-BIS(METHYLAMINO)-4H-PYRANO(2,3-b)(1,4)BENZODIOXIN-4-ONEDIHYDROCHLORIDE, (2R-(2-α,4a-β,5a-β,6-β,7-β,8-β,9-α,9a-α,10a-β))- ◇ SPECTINOMYCIN DIHYDROCHLORIDE ◇ SPECTINOMYCIN HYDROCHLORIDE

TOXICITY DATA with REFERENCE
ipr-rat LD50:2020 mg/kg NIIRDN 6,382,82
ipr-mus LD50:2350 mg/kg NIIRDN 6,382,82
scu-mus LD50:8400 mg/kg NIIRDN 6,382,82

SAFETY PROFILE: Moderately toxic by intraperitoneal route. When heated to decomposition it emits toxic fumes of NO_x and HCl.

SLI335 CAS:6493-69-2 ***HR: 2***
STANNIC CITRIC ACID

TOXICITY DATA with REFERENCE
ivn-rbt LDLo:500 mg/kg JLCMAK 28,1344,43

OSHA PEL: TWA 2 mg(Sn)/m^3
ACGIH TLV: TWA 2 mg(Sn)/m^3

SAFETY PROFILE: Moderately toxic by intravenous route. When heated to decomposition it emits toxic fumes of Sn.

SLJ000 CAS:7081-44-9 ***HR: 2***
STAPHYBIOTIC
mf: $C_{19}H_{17}ClN_3O_5S$•Na•H_2O mw: 475.91

SYNS: BACTOPEN ◇ BRL-1621 ◇ 6-(3-(o-CHLOROPHENYL)-5-METHYL-4-ISOXAZOLECARBOXAMIDEO)-3,3-DIMETHYL-7-OXO-4-THIA-1-AZABICYCLO(3.2.0)HEPTANE-2-CARBOXYLIC ACID, SODIUM SALT, MONOHYDRATE ◇ CLOXACILLIN SODIUM MONOHYDRATE ◇ CLOXAPEN ◇ CLOXYPEN ◇ EKVACILLIN ◇ GELSTAPH ◇ METHOCILLIN-S ◇ ORBENIN SODIUM HYDRATE ◇ P-25 ◇ PROSTAPHLIN-A ◇ SODIUM CLOXACILLIN MONOHYDRATE ◇ STAPHOBRISTOL-250 ◇ TEGOPEN ◇ TEPOGEN

TOXICITY DATA with REFERENCE
orl-rat LD50:5 g/kg NIIRDN 6,232,82
ipr-rat LD50:1350 mg/kg ARZNAD 15,322,65
scu-rat LD50:4200 mg/kg NIIRDN 6,232,82
ivn-rat LD50:1660 mg/kg NIIRDN 6,232,82
ims-rat LD50:2600 mg/kg NIIRDN 6,232,82
orl-mus LD50:5 g/kg NIIRDN 6,232,82
ipr-mus LD50:1170 mg/kg NIIRDN 6,232,82
ivn-mus LD50:1100 mg/kg NIIRDN 6,232,82
ims-mus LD50:1220 mg/kg NIIRDN 6,232,82

SAFETY PROFILE: Moderately toxic by intraperitoneal, intramuscular, subcutaneous, and intravenous routes. Mildly toxic by ingestion. When heated to decomposition it emits very toxic fumes of Cl^-, NO_x, Na_2O, and SO_x.

SLJ050 CAS:642-78-4 ***HR: 2***
STAPHYBIOTIC I
mf: $C_{19}H_{18}ClN_3O_5S$•Na mw: 435.46

SYNS: ANKERBIN ◇ AUSTRASTAPH ◇ BRL-1621 SODIUM SALT ◇ 6-(3-(o-CHLOROPHENYL)-5-METHYL-4-ISOXAZOLECARBOXAMIDEO)-3,3-DIMETHYL-7-OXO-4-THIA-1-AZABICYCLO(3.2.0)HEPTANE-2-CARBOXYLIC ACID, MONOSODIUM SALT ◇ 3-o-CHLOROPHENYL-5-METHYL-4-ISOXAZOLYLPENICILLIN SODIUM ◇ CLOXACILLIN SODIUM SALT ◇ CLOXAPEN ◇ EKVACILLIN ◇ GELSTAPH ◇ ORBENIN SODIUM ◇ PREVENCILINA P ◇ PROSTAPHILIN A ◇ SODIUM CLOXACILLIN ◇ SODIUM ORBENIN ◇ SODIUM SYNTARPEN ◇ TEGOPEN

TOXICITY DATA with REFERENCE
orl-rat LD50:5 g/kg NIIRDN 6,232,82
ipr-rat LD50:1600 mg/kg NIIRDN 6,232,82
scu-rat LD50:4200 mg/kg NIIRDN 6,232,82
ivn-rat LD50:1660 mg/kg NIIRDN 6,232,82
orl-mus LD50:5 g/kg NIIRDN 6,232,82
ipr-mus LD50:1170 mg/kg NIIRDN 6,232,82
scu-mus LD50:1500 mg/kg NIIRDN 6,232,82
ivn-mus LD50:916 mg/kg FATOAO 31,232,68
ims-mus LD50:1220 mg/kg NIIRDN 6,232,82

SAFETY PROFILE: Moderately toxic by intraperitoneal, subcutaneous, intramuscular, and intravenous routes. Mildly toxic by ingestion. When heated to decomposition it emits very toxic fumes of Cl^-, NO_x, SO_x, Na_2O.

SLJ100 ***HR: D***
STAPHYLOCOCCAL PHAGE LYSATE

TOXICITY DATA with REFERENCE
scu-rat TDLo:220 mg/kg (female 6-16D post):REP OYYAA2 20,575,80

SAFETY PROFILE: Experimental reproductive effects.

SLJ500 CAS:9005-25-8 ***HR: 1***
STARCH DUST

SYNS: AMAIZO W 13 ◇ AMYLOMAIZE VII ◇ AMYLUM

◇ AQUAPEL (POLYSACCHARIDE) ◇ ARGO BRAND CORN STARCH ◇ ARROWROOT STARCH ◇ CLARO 5591 ◇ CLEARJEL ◇ CLEARJREL ◇ CORN PRODUCTS ◇ CPC 3005 ◇ CPC 6448 ◇ FARINEX 100 ◇ GALACTASOL A ◇ GENVIS ◇ HRW 13 ◇ KEESTAR ◇ MAIZENA ◇ MARANTA ◇ MELOGEL ◇ MELUNA ◇ OK PRE-GEL ◇ PENFORD GUM 380 ◇ REMYLINE Ac ◇ RICE STARCH ◇ SORGHUM GUM ◇ STARAMIC 747 ◇ STARCH ◇ α-STARCH ◇ STARCH, CORN ◇ STARCH (OSHA) ◇ STA-RX 1500 ◇ TAPIOCA STARCH ◇ TAPON ◇ TROGUM ◇ W-GUM ◇ W-13 STABILIZER

CONSENSUS REPORTS: Reported in EPA TSCA Inventory.

OSHA PEL: Total Dust: 15 mg/m^3; Respirable Fraction: 5 mg/m^3
ACGIH TLV: TWA (nuisance particulate) 10 mg/m^3 of total dust (when toxic impurities are not present, e.g., quartz < 1%).

SAFETY PROFILE: A nuisance dust. An allergen. Flammable when exposed to flame, can react with oxidizing materials. Moderately explosive when exposed to flame.

SLJ650 ***HR: 2***
STAR-OF-BETHLEHEM

PROP: A perennial herb with few branches, 8-inch long leaves, and white flowers shaped like a narrow funnel 2 to 3 inches long. It grows as a weed in the tropics including Hawaii, Guam, and the West Indies.

SYNS: CIPRIL (PUERTO RICO) ◇ FEUILLES CRABE (HAITI) ◇ GINBEY (DOMINICAN REPUBLIC) ◇ HIPPOBROMA LONGIFLORA ◇ HORSE POISON (JAMAICA) ◇ MADAME FATE (JAMAICA) ◇ PUAHOKU (HAWAII) ◇ REVIENTA CABALLOS (CUBA) ◇ TIBEY (PUERTO RICO)

SAFETY PROFILE: The whole plant contains the poison diphenyl lobelidiol (similar to lobeline). Ingestion may cause convulsions. See also LOBELINE.

SLK000 CAS:57-11-4 ***HR: 3***
STEARIC ACID
mf: $C_{18}H_{36}O_2$ mw: 284.54

PROP: White, amorph solid; slt odor and taste of tallow. Mp: 69.3°, bp: 383°, flash p: 385°F (CC), d: 0.847, autoign temp: 743°F, vap press: 1 mm @ 173.7°, vap d: 9.80. Sol in alc, ether, acetone, chloroform; insol in water.

SYNS: CENTURY 1240 ◇ DAR-CHEM 14 ◇ EMERSOL 120 ◇ GLYCON DP ◇ GLYCON S-70 ◇ GLYCON TP ◇ GROCO 54 ◇ 1-HEPTADECANECARBOXYLIC ACID ◇ HYDROFOL ACID 1655 ◇ HY-PHI 1199 ◇ HYSTRENE 80 ◇ INDUSTRENE 5016 ◇ KAM 1000 ◇ KAM 2000 ◇ KAM 3000 ◇ NEO-FAT 18-61 ◇ NEO-FAT 18-S ◇ OCTADECANOIC ACID ◇ PEARL STEARIC ◇ STEAREX BEADS ◇ STEAROPHANIC ACID ◇ TEGOSTEARIC 254

TOXICITY DATA with REFERENCE
skn-hmn 75 mg/3D-I MLD 85DKA8 -,127,77
skn-rbt 500 mg/24H MOD FCTXAV 17,357,79
imp-mus TDLo:400 mg/kg:ETA BJCAAI 17,127,63
ivn-rat LD50:21500 μg/kg FCTXAV 17,357,79
ivn-mus LD50:23 mg/kg FCTXAV 17,357,79

CONSENSUS REPORTS: Reported in EPA TSCA Inventory. EPA Genetic Toxicology Program.

SAFETY PROFILE: Poison by intravenous route. A human skin irritant. Questionable carcinogen with experimental tumorigenic data by implantation route. Combustible when exposed to heat or flame. Heats spontaneously. To fight fire, use CO_2, dry chemical. When heated to decomposition it emits acrid smoke and irritating fumes.

SLK500 CAS:7460-84-6 ***HR: 3***
STEARIC ACID-2,3-EPOXYPROPYL ESTER
mf: $C_{21}H_{40}O_3$ mw: 340.61

SYNS: 2,3-EPOXY-1-PROPANOL STEARATE ◇ 2,3-EPOXYPROPYL ESTER of STEARIC ACID ◇ 2,3-EPOXYPROPYL STEARATE ◇ GLYCIDOL STEARATE ◇ GLYCIDYL OCTADECANOATE ◇ GLYCIDYL STEARATE ◇ OXIRANYLMETHYL ESTER of OCTADECANOIC ACID

TOXICITY DATA with REFERENCE
cyt-rat-ipr 5 mg/kg BJPCAL 6,235,51
scu-rat TDLo:2500 mg/kg/5W-I:ETA ANYAA9 68,750,58

CONSENSUS REPORTS: IARC Cancer Review: Group 3 IMEMDT 7,56,87; Animal Inadequate Evidence IMEMDT 11,187,76.

SAFETY PROFILE: Questionable carcinogen with experimental tumorigenic data. Mutation data reported. When heated to decomposition it emits acrid smoke and irritating fumes. See also ESTERS.

SLL000 CAS:1323-39-3 ***HR: 3***
STEARIC ACID, MONOESTER with 1,2-PROPANEDIOL
mf: $C_{21}H_{42}O_3$ mw: 342.63

SYNS: ATLAS G 924 ◇ CERASYNT PA ◇ CERASYNT PN ◇ CRILL 26 ◇ DRAGIL-P ◇ EMCOL PS-50 RHP ◇ EMEREST 2381 ◇ MONOSTEOL ◇ NOCA ◇ OCTADECANOIC ACID, MONOESTER with 1,2-PROPANEDIOL ◇ 1,2-PROPANEDIOL, MONOSTEARATE ◇ PROPYLENE GLYCOL MONOSTEARATE ◇ PROSTEARIN ◇ TEGIN P ◇ USAF KE-13

TOXICITY DATA with REFERENCE
ipr-mus LD50:200 mg/kg NTIS** AD277-689

CONSENSUS REPORTS: Reported in EPA TSCA Inventory.

SAFETY PROFILE: Poison by intraperitoneal route. When heated to decomposition it emits acrid smoke and irritating fumes. Used in hand care products and other cosmetics. See also ESTERS.

SLL400 CAS:502-26-1 ***HR: 2***
Γ-STEAROLACTONE
mf: $C_{18}H_{34}O_2$ mw: 282.52

SYN: DIHYDRO-5-TETRADECYL-2(3H)-FURANONE

TOXICITY DATA with REFERENCE
scu-mus TDLo:160 mg/kg/40W-I:NEO CNREA8 30,1037,70
scu-mus TD:52 mg/kg/26W-I:ETA CNREA8 32,880,72

SAFETY PROFILE: Questionable carcinogen with experimental neoplastigenic and tumorigenic data. When heated to decomposition it emits acrid smoke and irritating fumes.

SLL500 CAS:638-65-3 ***HR: 1***
STEARONITRILE
mf: $C_{18}H_{35}N$ mw: 265.54

PROP: Colorless crystals. D: 0.818 @ 41°/4° mp: 41°, bp: 274.5° @ 100 mm. Insol in water.

SYNS: HEPTADECYL CYANIDE ◇ NITRIL KYSELINY STEAROVE (CZECH) ◇ OCTADECANENITRILE ◇ OCTADECANONITRILE

TOXICITY DATA with REFERENCE
eye-rbt 500 mg/24H MLD 28ZPAK -,160,72

CONSENSUS REPORTS: Cyanide and its compounds are on the Community Right To Know List. Reported in EPA TSCA Inventory.

SAFETY PROFILE: An eye irritant. When heated to decomposition it emits toxic fumes of NO_x and CN^-. See also NITRILES.

SLM000 CAS:3891-30-3 ***HR: 3***
1-STEAROYLAZIRIDINE
mf: $C_{20}H_{39}NO$ mw: 309.60

SYN: STEAROYL ETHYLENEIMINE

TOXICITY DATA with REFERENCE
cyt-rat-ipr 200 mg/kg BJPCAL 9,306,54
scu-rat TDLo:1000 mg/kg/5W-I:NEO BJPCAL 6,357,51

SAFETY PROFILE: Questionable carcinogen with experimental neoplastigenic data. Mutation data reported. When heated to decomposition it emits toxic fumes of NO_x.

SLM500 CAS:9005-00-9 ***HR: 2***
STEARYL ALCOHOL condensed with 10 moles ETHYLENE OXIDE
mf: $(C_2H_4O)_nC_{18}H_{38}O$

SYN: STEARYL ALCOHOL EO (10)

TOXICITY DATA with REFERENCE
orl-rat LD50:2900 mg/kg 34ZIAG -,690,69

CONSENSUS REPORTS: Reported in EPA TSCA Inventory.

SAFETY PROFILE: Moderately toxic by ingestion. When heated to decomposition it emits acrid smoke and irritating fumes. See also STEARYL ALCOHOL and ETHYLENE OXIDE.

SLN000 CAS:9005-00-9 ***HR: 2***
STEARYL ALCOHOL condensed with 20 moles ETHYLENE OXIDE
mf: $(C_2H_4O)_nC_{18}H_{38}O$

SYN: STEARYL ALCOHOL EO (20)

TOXICITY DATA with REFERENCE
orl-rat LD50:1900 mg/kg 34ZIAG -,690,69

CONSENSUS REPORTS: Reported in EPA TSCA Inventory.

SAFETY PROFILE: Moderately toxic by ingestion. When heated to decomposition it emits acrid smoke and irritating fumes. See also STEARYL ALCOHOL and ETHYLENE OXIDE.

SLN509 CAS:102583-65-3 ***HR: 3***
STENDOMYCIN A
mf: $C_{95}H_{172}N_{20}O_{31}$ mw: 2090.87

PROP: An antibiotic produced by *S. Species res. S. Sndus* or *S. Antimycoticus* (85FZAT -,610,67).

TOXICITY DATA with REFERENCE
orl-mus LD50:500 g/kg 85FZAT -,610,67
scu-mus LD50:62 mg/kg 85GDA2 4(2),113,80
ivn-mus LD50:14 mg/kg 85FZAT -,610,67

SAFETY PROFILE: Poison by subcutaneous and intravenous routes. Moderately toxic by ingestion. When heated to decomposition it emits toxic fumes of NO_x.

SLO000 CAS:62536-49-6 ***HR: 3***
STENDOMYCIN SALICLATE

PROP: Produced by an *Actinomycete* strain resembling *Streptomyces endus* and *S. Antimycoticus* (85ERAY 1,399,78).

TOXICITY DATA with REFERENCE
orl-rat LD50:400 mg/kg 85ERAY 1,399,78
ipr-rat LD50:9 mg/kg 85ERAY 1,399,78
orl-mus LD50:626 mg/kg 85ERAY 1,399,78
ipr-mus LD50:12 mg/kg 85ERAY 1,399,78

SAFETY PROFILE: Poison by ingestion and intraperitoneal routes. When heated to decomposition it emits acrid smoke and irritating fumes.

SLO450 ***HR: D***
STERCULIA FOETIDA OIL

PROP: Contains 50% cyclopropene fatty acids (JONUAI 107,574,77).

TOXICITY DATA with REFERENCE
orl-rat TDLo:43 g/kg (female 1-22D post):REP JONUAI 107,574,77
orl-rat TDLo:41 g/kg (female 60D pre):TER PSEBAA 131,61,69

SAFETY PROFILE: An experimental teratogen. Experimental reproductive effects.

SLO500 CAS:9000-36-6 ***HR: 1***
STERCULIA GUM

SYN: GUM STERCULIA

TOXICITY DATA with REFERENCE
orl-rat LD50:9100 mg/kg FDRLI* 124,-,76
orl-mus LD50:8200 mg/kg FDRLI* 124,-,76
orl-rbt LD50:6400 mg/kg FDRLI* 124,-,76
orl-ham LD50:6900 mg/kg FDRLI* 124,-,76

CONSENSUS REPORTS: Reported in EPA TSCA Inventory.

SAFETY PROFILE: Mildly toxic by ingestion.

SLP000 CAS:10048-13-2 ***HR: 3***
STERIGMATOCYSTIN
mf: $C_{18}H_{12}O_6$ mw: 328.34

PROP: A metabolite of *Aspergillus versicolor* (BJCAAI 20,134,66).

SYN: 3a,12c-DIHYDRO-8-HYDROXY-6-METHOXY-7H-FURO(3',2':4,5)FURO(2,3-C)XANTHEN-7-ONE

TOXICITY DATA with REFERENCE
dnr-bcs 10 µg/disc MUREAV 97,339,82
dns-hmn:hla 1 µmol/L CNREA8 38,2621,78
msc-hmn:emb 100 µg/L KIKNAJ (29),38,78
orl-rat TDLo:175 mg/kg/52W-C:CAR FCTXAV 8,289,70
skn-rat TDLo:560 mg/kg/70W-I:CAR TXAPA9 26,274,73
scu-rat TDLo:96 mg/kg/24W-I:ETA BJCAAI 20,134,66
orl-rat LD50:120 mg/kg FCTXAV 7,135,69
ipr-rat LD50:65 mg/kg FCTXAV 7,135,69
orl-mus LD50:800 mg/kg MMAPAP 35,373,68
orl-mky LD50:32 mg/kg BJEPA5 51,183,70

CONSENSUS REPORTS: IARC Cancer Review: Group 2B IMEMDT 7,56,87; Animal Sufficient Evidence IMEMDT 10,245,76; Animal Limited Evidence IMEMDT 1,175,72. EPA Genetic Toxicology Program.

SAFETY PROFILE: Suspected carcinogen with experimental carcinogenic and tumorigenic data. Poison by ingestion and intraperitoneal routes. Human mutation data reported. When heated to decomposition it emits acrid smoke and irritating fumes.

SLP199 ***HR: 3***
STEROIDS

PROP: Lipids which contain a hydrogenated cyclopentophenanthrene-ring system. Closely related to terpenes. Examples are progesterone, adrenocortical hormones, the gonadal hormones, cardiac aglycones, digitalis derivatives, bile acids, cholesterol, other sterols, toad poisons, saponins, and some carcinogenic hydrocarbons. Most steroids have 2 methyl groups and an aliphatic side-chain attached to the nucleus. The length of the side-chain varies and generally contains 8,9, or 10 carbon atoms in the sterols, 5 carbon atoms in the bile acids, 2 in the adrenal cortical steroids, and none in the estrogens and androgens.

SAFETY PROFILE: Some sex hormones (e.g., estrogens, progestins, and androgens) have carcinogenic potential. They may also have teratogenic as well as reproductive effects. Anabolic steroids are synthetic derivatives of testosterone. They have pronounced anabolic properties and relatively weak androgenic properties. They are used clinically to promote growth and repair of body tissues in senility, debilitating illness, and convalescence. They are much abused by athletes and can cause serious systemic effects. See also specific entries.

SLP350 ***HR: 3***
STEVIA REBAUDIANA Bertoni, extract

TOXICITY DATA with REFERENCE
orl-rat TDLo:45 g/kg (18D pre):REP SCIEAS 162,1007,68
orl-rat TDLo:41475 mg/kg/79W-C:ETA SKEZAP 26,169,85

SAFETY PROFILE: Questionable carcinogen with experimental tumorigenic data. Experimental reproductive effects.

SLP500 CAS:138-31-8 ***HR: 3***
STIBACETIN
mf: $C_8H_9NO_4Sb \cdot Na$ mw: 327.92

PROP: Light yellow powder. Antimony content 35%.

SYNS: p-ACETAMIDOBENZENESTIBONIC ACID SODIUM SALT ◇ ACETYL-p-AMINOPHENYLSTIBINSAURES NATRIUM (GERMAN) ◇ SODIUM p-ACETYLAMINOPHENYLANTIMONATE ◇ STIBENYL

TOXICITY DATA with REFERENCE
ipr-mus LDLo:133 mg/kg CLDND* 81,224,44
scu-mus LDLo:1 g/kg HBAMAK 4,1289,35

CONSENSUS REPORTS: Antimony and its compounds are on the Community Right To Know List.

OSHA PEL: TWA 0.5 mg(Sb)/m^3
ACGIH TLV: TWA 0.5 mg(Sb)/m^3
NIOSH REL: (Antimony) TWA 0.5 mg(Sb)/m^3

SAFETY PROFILE: Poison by intraperitoneal route. When heated to decomposition it emits very toxic fumes of Sb, Na_2O, and NO_x. See also ANTIMONY COMPOUNDS.

SLP600 CAS:554-76-7 ***HR: 2***
STIBANILIC ACID
mf: $C_6H_8NO_3Sb$ mw: 263.90

SYNS: p-AMINOBENZENESTIBONIC ACID ◇ 4-AMINOBENZENESTIBONIC ACID ◇ ASTARIL ◇ 693B ◇ BAYER 693 ◇ BENZENAMINE, 4-STIBONO-(9CI) ◇ ETHYLSTIBAMINE ◇ NEOSTIBOSAN ◇ STIBOSAMIN

TOXICITY DATA with REFERENCE
ivn-mus LD50:472 mg/kg JPETAB 81,224,44

OSHA PEL: TWA 0.5 mg(Sb)/m^3
ACGIH TLV: TWA 0.5 mg(Sb)/m^3

SAFETY PROFILE: Moderately toxic by intravenous route. When heated to decomposition it emits toxic fumes of NO_x and Sb.

SLQ000 CAS:7803-52-3 ***HR: 3***
STIBINE
DOT: UN 2676
mf: H_3Sb mw: 124.78

PROP: Colorless gas, disagreeable odor. Mp: −88°, bp: −18.4°, d: 2.204 g/mL @ bp. Gas is sltly sol in water; very sol in alc, carbon disulfide, and organic solvents.

SYNS: ANTIMONWASSERSTOFFES (GERMAN) ◇ ANTIMONY HYDRIDE ◇ ANTIMONY TRIHYDRIDE ◇ ANTYMONOWODOR (POLISH) ◇ HYDROGEN ANTIMONIDE

TOXICITY DATA with REFERENCE
ihl-mus LCL0:10 ppm 34ZIAG -,553,69
ihl-gpg LCLo:92 ppm/1H 28ZLA8 -,30,61

CONSENSUS REPORTS: Antimony and its compounds are on the Community Right To Know List.

OSHA PEL: TWA 0.1 ppm
ACGIH TLV: TWA 0.1 ppm
DFG MAK: 0.1 ppm (0.5 mg/m^3)
DOT Classification: Poison A; Label: Poison Gas and Flammable Gas

SAFETY PROFILE: Poison by inhalation. Potentially explosive decomposition at 200°C. Flammable when exposed to heat or flame. Explosive reaction with ammonia + heat; chlorine; concentrated nitric acid; ozone. Incompatible with oxidants. The decomposition products are hydrogen and metallic antimony. When heated to decomposition it emits toxic fumes of Sb. Used as a fumigating agent. See also ANTIMONY COMPOUNDS and HYDRIDES.

SLQ500 CAS:102585-62-6 ***HR: 3***
((2-STIBONOPHENYL)THIO)ACETIC ACID DIETHANOLAMINE SALT
mf: $C_8H_8O_5SSb{\cdot}C_4H_{12}NO_2$ mw: 444.14

TOXICITY DATA with REFERENCE
ivn-mus LD50:430 mg/kg AIPTAK 85,100,51
ivn-rbt LD50:50 mg/kg AIPTAK 85,100,51
ivn-ham LD50:200 mg/kg AIPTAK 85,100,51

CONSENSUS REPORTS: Antimony and its compounds are on the Community Right To Know List.

OSHA PEL: TWA 0.5 mg(Sb)/m^3
ACGIH TLV: TWA 0.5 mg(Sb)/m^3
NIOSH REL: (Antimony) TWA 0.5 mg(Sb)/m^3

SAFETY PROFILE: Poison by intravenous route. When heated to decomposition it emits very toxic fumes of SO_x, Sb, and NO_x. See also ANTIMONY COMPOUNDS.

SLQ625 CAS:59297-18-6 ***HR: D***
5-α-STIGMASTANE-3-β,5,6-β-TRIOL 3-MONOBENZOATE
mf: $C_{36}H_{56}O_4$ mw: 552.92

TOXICITY DATA with REFERENCE
orl-mus TDLo:60 mg/kg (female 6-7D post):REP JRPFA4 46,461,76

SAFETY PROFILE: Experimental reproductive effects. When heated to decomposition it emits acrid smoke and irritating fumes.

SLQ650 CAS:91682-96-1 ***HR: 3***
STIGMATELLIN
mf: $C_{30}H_{42}O_7$ mw: 514.72

TOXICITY DATA with REFERENCE
oms-smc 3300 μg/L JANTAJ 37,454,84
orl-mus LD50:30 mg/kg JANTAJ 34,454,84
scu-mus LD50:2 mg/kg JANTAJ 34,454,84

SAFETY PROFILE: Poison by ingestion and subcutaneous routes. Mutation data reported. When heated to decomposition it emits acrid smoke and irritating fumes.

SLQ900 CAS:834-24-2 ***HR: 3***
4-STILBENAMINE
mf: $C_{14}H_{13}N$ mw: 195.28

SYNS: p-AMINOSTILBENE ◇ 4-AMINOSTILBENE ◇ 4-(2-PHENYLETHENYL)BENZENAMINE ◇ 4-N-STILBENAMINE ◇ p-STYRYLANILINE

TOXICITY DATA with REFERENCE
imp-mus TDLo:80 mg/kg:ETA BJCAAI 17,127,63

SAFETY PROFILE: Questionable carcinogen with experimental tumorigenic data. When heated to decomposition it emits toxic fumes of NO_x.

SLR000 CAS:588-59-0 ***HR: 3***
STILBENE
mf: $C_{14}H_{12}$ mw: 180.26

PROP: Colorless or sltly yellow crystals. Mp: 124-125°, bp: 306-307°, d: 0.9707. Insol in water; sol in 90 parts cold alc and 13 parts boiling alc; freely sol in benzene and ether.

SYNS: DIPHENYLETHYLENE ◇ STILBEN (GERMAN)

TOXICITY DATA with REFERENCE
ipr-mus LD50:1150 mg/kg ARZNAD 19,617,69
ivn-mus LD50:34 mg/kg ARZNAD 19,617,69

CONSENSUS REPORTS: Reported in EPA TSCA Inventory.

SAFETY PROFILE: Poison by intravenous route. Moderately toxic by intraperitoneal route. Violent reaction with O_2. When heated to decomposition it emits acrid smoke and irritating fumes.

SLR500 CAS:621-96-5 ***HR: 3***
4,4'-STILBENEDIAMINE
mf: $C_{14}H_{14}N_2$ mw: 210.30

SYNS: 4:4'-DIAMINOSTILBENE ◇ 4,4'-VINYLENEDIANILINE

TOXICITY DATA with REFERENCE
scu-rat TDLo:200 mg/kg/W-I:ETA BMBUAQ 14,141,58

SAFETY PROFILE: Questionable carcinogen with experimental tumorigenic data. When heated to decomposition it emits toxic fumes of NO_x. See also AMINES.

SLS000 CAS:122-06-5 ***HR: 3***
4,4'-STILBENEDICARBOXAMIDINE
mf: $C_{16}H_{16}N_4$ mw: 264.36

SYNS: DIAMIDINO STILBENE ◇ 4,4'-DIAMIDINOSTILBENE

TOXICITY DATA with REFERENCE
ivn-man TDLo:39 mg/kg/29D:CNS LANCAO 246,531,44
ipr-rat LD50:43 mg/kg FEPRA7 1,167,42
ipr-mus LD50:317 mg/kg JPETAB 119,444,57

SAFETY PROFILE: Poison by intraperitoneal route. Human systemic effects by intravenous route: central nervous system effects. When heated to decomposition it emits toxic fumes of NO_x.

SLS500 CAS:6935-63-3 ***HR: 3***
4,4'-STILBENEDICARBOXAMIDINE, DIHYDROCHLORIDE
mf: $C_{16}H_{16}N_4 \cdot 2ClH$ mw: 337.28

SYNS: 4,4'-DIAMIDINOSTILBENE DIHYDROCHLORIDE ◇ STILBAMIDINE DIHYDROCHLORIDE ◇ 4,4'-VINYLENEDIBENZAMIDINE, DIHYDROCHLORIDE

TOXICITY DATA with REFERENCE
ipr-mus LD50:91 mg/kg ANTCAO 2,581,52
scu-mus LD50:180 mg/kg ATMPA2 37,1,43
ivn-mus LD50:18900 μg/kg ANTCAO 2,581,52

SAFETY PROFILE: Poison by intraperitoneal, subcutaneous, and intravenous routes. When heated to decomposition it emits very toxic fumes of HCl and NO_x.

SLU000 CAS:7081-53-0 ***HR: 3***
STIMULEXIN
mf: $C_{24}H_{30}N_2O_2 \cdot H_2O \cdot ClH$ mw: 433.04

SYNS: AHR-619 ◇ DOPRAM ◇ DOXAPRAM HYDROCHLORIDE HYDRATE ◇ 1-ETHYL-4-(2-MORPHOLINOETHYL)-3,3-DIPHENYL-2-PYRROLIDINONE HYDROCHLORIDE HYDRATE

TOXICITY DATA with REFERENCE
ipr-mus TDLo:864 mg/kg (7-12D preg):TER OYYAA2 8,229,74
ipr-mus TDLo:864 mg/kg (7-12D preg):REP OYYAA2 8,229,74
ivn-man TDLo:41 mg/kg/24H-C:LIV PGMJAO 61,833,85
orl-rat LD50:261 mg/kg TXAPA9 18,185,71
ipr-rat LD50:174 mg/kg TXAPA9 13,242,68
scu-rat LD50:312 mg/kg TXAPA9 13,242,68
ivn-rat LD50:72 mg/kg TXAPA9 13,242,68
orl-mus LD50:270 mg/kg TXAPA9 13,242,68
ipr-mus LD50:153 mg/kg TXAPA9 13,242,68
scu-mus LD50:312 mg/kg NIIRDN 6,504,82
ivn-mus LD50:85 mg/kg TXAPA9 13,242,68
orl-dog LD50:150 mg/kg 27ZQAG -,225,72
ivn-dog LDLo:40 mg/kg TXAPA9 18,185,71

SAFETY PROFILE: Poison by ingestion, intraperitoneal, subcutaneous and intravenous routes. An experimental teratogen. Experimental reproductive effects. Human systemic effects: hepatitis. Used as a respiratory stimulant. When heated to decomposition it emits very toxic fumes of HCl and NO_x.

SLU500 CAS:8052-41-3 ***HR: 3***
STODDARD SOLVENT

PROP: Clear, colorless liquid. Composed of 85% nonane and 15% trimethyl benzene. Bp: 220-300°, flash p: 100-110°F, lel: 1.1%, uel: 6%, autoign temp: 450°F, d: 1.0. Insol in water; misc with abs alc, benzene, ether, chloroform, carbon tetrachloride, carbon disulfide, and some oils (not castor oil). Stoddard solvent to a first approximation contains 85% nonane and 15% trimethylbenzene.

SYNS: NAPHTHA SAFETY SOLVENT ◇ VARNOLINE ◇ WHITE SPIRITS

TOXICITY DATA with REFERENCE
eye-hmn 470 ppm/15M TXAPA9 32,282,75
ihl-cat LCLo:10 g/m³/2.5H TXAPA9 32,282,75

CONSENSUS REPORTS: Reported in EPA TSCA Inventory.

OSHA PEL: (Transitional: TWA 500 ppm) TWA 100 ppm
ACGIH TLV: TWA 100 ppm
NIOSH REL: (Refined Petroleum Solvents) TWA 350 mg/m³; CL 1800 mg/m³/15M

SAFETY PROFILE: Mildly toxic by inhalation. A human eye irritant. Flammable when exposed to heat or flame. Explosive in the form of vapor when exposed to heat or flame. When heated to decomposition it emits acrid fumes and may explode; can react with oxidizing materials. To fight fire, use foam, CO_2, dry chemical. See also N-NONANE and TRIMETHYL BENZENE (MIXED ISOMERS).

SLV500 CAS:8063-18-1 ***HR: 3***
STRAMONIUM

PROP: *Datura stramonium* have 0.25-0.45% alkaloids consisting of atropine, hyoscyamine, and scopolamine.

SYNS: ANGEL TULIP ◇ DATURA STRAMONIUM ◇ DEVIL'S APPLE ◇ DHUTRA ◇ JAMESTOWN WEED ◇ JIMSON WEED ◇ POMME EPINEUSE (FRENCH) ◇ STECKAPFUL (GERMAN) ◇ STRAMONA (ITALIAN) ◇ THORN APPLE

TOXICITY DATA with REFERENCE
orl-hmn LDLo:57 mg/kg POMDAS 28,364,60
orl-bwd LD50:202 mg/kg POMDAS 28,364,60

SAFETY PROFILE: Human and experimental poison by ingestion. When heated to decomposition it emits acrid smoke and irritating fumes. See also individual components.

SLW450 CAS:9002-01-1 ***HR: 2***
STREPTOKINASE

SYNS: AWELYSIN ◇ KABIKINASE ◇ STREPTASE ◇ STREPTOCOCCAL FIBRINOLYSIN ◇ STREPTOKINASE (ENZYME-ACTIVATING)

TOXICITY DATA with REFERENCE
scu-hmn TDLo:600 iu/kg:IMM JAMAAP 252,1314,84
ivn-man TDLo:562500 iu/kg:NOSE,PUL JAMAAP 252,1314,84
ivn-rat LD50:2 g/kg FATOAO 43,671,80
ipr-mus LD50:4200 mg/kg FOHEAW 103,445,76
ivn-mus LD50:3700 mg/kg FATOAO 43,671,80

SAFETY PROFILE: Moderately toxic by intravenous route. Human systemic effects by subcutaneous or intravenous routes: effects on the nose, dyspnea and unspecified allergic affects. A thrombolytic agent used to dissolve blood clots in vivo.

SLW475 CAS:7229-50-7 ***HR: 3***
STREPTOLYDIGIN
mf: $C_{33}H_{48}N_2O_9$ mw: 616.83

PROP: Crystals from acetone + water. Mp: 147-148°. Sol in ethanol, ethyl acetate, ether, chloroform. Practically insol in water, hydrocarbons.

SYNS: ANTIBIOTIC D-45 ◇ FREE ACID ◇ PORTAMYCIN

TOXICITY DATA with REFERENCE
oms-bcs 288 mg/L FOMIAZ 22,329,77
ipr-mus LD50:533 mg/kg UPJOH* 2(6),-,71
scu-ckn LDLo:10 mg/kg CBPCBB 66,83,80

SAFETY PROFILE: Poison by subcutaneous route. Moderately toxic by intraperitoneal route. Mutation data reported. When heated to decomposition it emits toxic fumes of NO_x.

SLW500 CAS:57-92-1 ***HR: 3***
STREPTOMYCIN
mf: $C_{21}H_{39}N_7O_{12}$ mw: 581.67

PROP: An antibiotic. It is a base and readily forms salts with anions.

SYNS: AGRIMYCIN 17 ◇ CHEMFORM ◇ GEROX ◇ HOKKO-MYCIN ◇ NSC 14083 ◇ STREPCEN ◇ STREPTOMICINA (ITALIAN) ◇ STREPTOMYCIN A ◇ STREPTOMYCINE ◇ STREPTOMYCINUM ◇ STREPTOMYZIN (GERMAN)

TOXICITY DATA with REFERENCE
mmo-esc 100 mg/L NATUAS 271,385,78
mmo-omi 40 mg/L ANTBAL 25,822,80
cyt-hmn:leu 9200 mg/L MGGEAE 107,361,70
unr-wmn TDLo:320 mg/kg (6-14W preg):REP NEJMAG 271,949,64
unr-wmn TDLo:620 mg/kg (23-39W preg):TER SJRDAH 50,61,69
orl-hmn TDLo:400 mg/kg/28D-I:EAR RMSRA6 73,820,53
ipr-hmn TDLo:143 mg/kg:CVS,PUL BMJOAE 1,556,61
par-hmn TDLo:28 mg/kg/D:BLD RMSRA6 73,820,53
orl-rat LD50:9 g/kg 85ARAE 4,35,76/77
scu-rat LDLo:600 mg/kg 85ERAY 1,570,78
ivn-rat LDLo:175 mg/kg 85ERAY 1,570,78
orl-mus LD50:9000 mg/kg 31ZOAD 1,390,68
ipr-mus LD50:1250 mg/kg DPHFAK 23,383,71
scu-mus LD50:520 mg/kg BJPCAL 15,496,60
ivn-mus LD50:90200 μg/kg ANTBAL 18,249,73
scu-dog LDLo:300 mg/kg 85ERAY 1,570,78
scu-mky LDLo:400 mg/kg 85ERAY 1,570,78

orl-cat LDLo:2000 mg/kg 85ERAY 1,570,78
scu-cat LDLo:600 mg/kg 85ERAY 1,570,78

CONSENSUS REPORTS: EPA Genetic Toxicology Program.

SAFETY PROFILE: Poison by intravenous and subcutaneous routes. Moderately toxic by ingestion and intraperitoneal routes. An experimental teratogen. Human systemic effects by ingestion and intraperitoneal routes: change in vestibular functions, blood pressure decrease, eosinophiis, respiratory depression, and other pulmonary changes. Human reproductive and teratogenic effects by unspecified routes: developmental abnormalities of the eye and ear and effects on newborn including post-natal measures or effects. Toxic to kidneys and central nervous system. Has been implicated in aplastic anemia. Experimental reproductive effects. Human mutation data reported. When heated to decomposition it emits toxic fumes of NO_x.

SLX500 CAS:485-19-8 ***HR: 3***
STREPTOMYCIN C
mf: $C_{21}H_{39}N_7O_{13}$ mw: 597.67

PROP: Produced by *S. Reticuli, S. Griseocarneus* NRRL B-1068, *Streptomyces sp.* MA-232-MI, and also by *S. Subrutilus* (85ERAY 1,570,78).

SYNS: HYDROXYSTREPTOMYCIN ◇ RETICULIN

TOXICITY DATA with REFERENCE
scu-mus LD50:865 mg/kg 85ERAY 1,570,78
ivn-mus LD50:154 mg/kg 85ERAY 1,570,78

SAFETY PROFILE: Poison by intravenous route. Moderately toxic by subcutaneous route. When heated to decomposition it emits toxic fumes of NO_x.

SLY000 CAS:15493-35-3 ***HR: 3***
STREPTOMYCIN CALCIUM CHLORIDE
mf: $C_{21}H_{39}N_7O_{12}{\bullet}CaCl_2$ mw: 692.65

SYNS: CALCIUM CHLORIDE with STREPTOMYCIN (1:1) ◇ STREPTOMYCIN compounded with CALCIUM CHLORIDE (1:1)

TOXICITY DATA with REFERENCE
orl-mus LD50:224 mg/kg ANTBAL 27,535,82
scu-mus LD50:758 mg/kg ANTBAL 18,444,73
ivn-mus LD50:165 mg/kg ANTBAL 18,444,73

SAFETY PROFILE: Poison by ingestion and intravenous routes. Moderately toxic by subcutaneous route. When heated to decomposition it emits very toxic fumes of Cl^- and NO_x. See also STREPTOMYCIN.

SLY200 CAS:8017-59-2 ***HR: 3***
STREPTOMYCIN and DIHYDROSTREPTOMYCIN
mf: $C_{21}H_{41}N_7O_{12}{\bullet}C_{21}H_{39}N_7O_{12}$ mw: 1165.36

SYNS: AMBISTRIN ◇ DIHYDROSTREPTOMYCIN and STREPTOMYCIN ◇ DUO-STREPTOMYCIN

TOXICITY DATA with REFERENCE
unr-wmn TDLo:540 mg/kg (27-39W preg):REP SJRDAH 50,101,69
unr-wmn TDLo:540 mg/kg (27-39W preg):TER SJRDAH 50,101,69

SAFETY PROFILE: Human reproductive effects by unspecified route: developmental abnormalities of the eye and ear. A human teratogen. When heated to decomposition it emits toxic fumes of NO_x. See also STREPTOMYCIN.

SLY500 CAS:3810-74-0 ***HR: 3***
STREPTOMYCIN SESQUISULFATE
mf: $C_{42}H_{78}N_{14}O_{24}{\bullet}H_6O_{12}S_3$ mw: 1457.58

SYNS: AGRI-MYCIN ◇ AGRISTREP ◇ AS-15 ◇ PHYTOMYCIN ◇ PLANTOMYCIN ◇ STREPCIN ◇ STREP-GRAN ◇ STREPSULFAT ◇ STREPTOMYCIN SULFATE ◇ STREPTOMYCIN SULPHATE B.P. ◇ STREPTOREX ◇ STREPVET ◇ VETSTREP

TOXICITY DATA with REFERENCE
mmo-sat 10 μg/plate ENMUDM 5(Suppl 1),3,83
mma-sat 1 mg/plate NTPTB* JAN 82
orl-rat TDLo:37 g/kg (MGN):REP OYYAA2 17,109,79
unr-mus TDLo:1200 mg/kg (9-11D preg):TER TJADAB 14,250,76
ipr-inf TDLo:170 mg/kg:PNS,CNS,PUL BMJOAE 1,557,61
scu-mus LD50:500 mg/kg ARZNAD 6,379,56
ivn-mus LD50:90200 μg/kg ANTBAL 18,444,73
orl-ham LD50:400 mg/kg TXAPA9 14,510,69
orl-mam LD50:9000 mg/kg FMCHA2 -,C221,83

CONSENSUS REPORTS: Reported in EPA TSCA Inventory. EPA Genetic Toxicology Program.

SAFETY PROFILE: Poison by ingestion and intravenous routes. Moderately toxic by subcutaneous route. An experimental teratogen. Human systemic effects by intraperitoneal route: flaccid paralysis without anesthesia, motor activity changes and other pulmonary changes. Experimental reproductive effects. Mutation data reported. When heated to decomposition it emits very toxic fumes of NO_x and SO_x. See also STREPTOMYCIN and SULFATES.

SLZ000 CAS:298-39-5 ***HR: 3***
STREPTOMYCIN SULPHATE
mf: $C_{21}H_{39}N_7O_{12}{\bullet}3H_2O_4S$ mw: 875.91

SYNS: STREPTOMYCIN, SULFATE (1:3) SALT ◇ STRYCIN

TOXICITY DATA with REFERENCE
ims-wmn TDLo:1700 mg/kg (195-280D preg):TER APLAAQ 11,1,58
orl-rat LD50:430 mg/kg 85GMAT -,106,82

scu-rat LD50:800 mg/kg TXAPA9 9,445,66
orl-mus LD50:430 mg/kg 85GMAT -,106,82
scu-mus LD50:970 mg/kg PSEBAA 76,466,51

SAFETY PROFILE: Moderately toxic by ingestion and subcutaneous routes. Human teratogenic effects by intramuscular route: developmental abnormalities of the eye and ear. When heated to decomposition it emits very toxic fumes of SO_x and NO_x. See also STREPTOMYCIN and SULFATES.

SMA000 CAS:3930-19-6 ***HR: 3***
STREPTONIGRAN
mf: $C_{25}H_{22}N_4O_8$ mw: 506.51

SYNS: BRUNEOMYCIN ◇ NIGRIN ◇ NSC 45383 ◇ 5278 R.P. ◇ RUFOCHROMOMYCIN ◇ RUFOCROMOMYCINE ◇ SN ◇ STREPTONIGRIN ◇ VALACIDIN

TOXICITY DATA with REFERENCE
mmo-sat 100 ng/plate PNASA6 79,7445,82
mmo-bcs 600 μg/L CMMUAO 9,165,84
par-rat TDLo:250 μg/kg (female 11D post):TER JEEMAF 18,215,67
ivn-rbt TDLo:90 μg/kg (female 1D pre):REP MUREAV 127,73,84
orl-mus LD50:2330 μg/kg ANTBAL 12,132,67
ipr-mus LD50:750 μg/kg 85GDA2 3,332,80
scu-mus LD50:1000 μg/kg COREAF 261,4911,65
ivn-mus LD50:500 μg/kg 85ERAY 2,1359,78
orl-dog LD50:500 μg/kg ANTBAL 11,1090,66
ivn-rat LD50:600 μg/kg ANTBAL 12,132,67

CONSENSUS REPORTS: EPA Genetic Toxicology Program.

SAFETY PROFILE: A deadly poison by ingestion, intraperitoneal, subcutaneous, and intravenous routes. An experimental teratogen. Experimental reproductive effects. Human mutation data reported. When heated to decomposition it emits toxic fumes of NO_x.

SMA500 CAS:3398-48-9 ***HR: 3***
STREPTONIGRIN METHYL ESTER
mf: $C_{26}H_{24}N_4O_8$ mw: 520.54

SYNS: METHYL ESTER STREPONIGRIN ◇ METHYL STREPONIGRIN ◇ NSC-45384

TOXICITY DATA with REFERENCE
ipr-rat TDLo:10 mg/kg (9D preg):TER CCROBU 53,23,69
ipr-mus LD50:88960 μg/kg NCISP* JAN 86

SAFETY PROFILE: Poison by intraperitoneal route. An experimental teratogen. When heated to decomposition it emits toxic fumes of NO_x. See also STREPTONIGRAN.

SMB000 CAS:303-81-1 ***HR: 3***
STREPTONIVICIN
mf: $C_{31}H_{36}N_2O_{11}$ mw: 612.69

PROP: Light yellow to white antibiotic crystals. D: 1.345, mp: decomp @ 152°.

SYNS: ALBAMIX ◇ ALBAMYCIN ◇ ANTIBIOTIC PA-93 ◇ CARDELMYCIN ◇ CATHOCIN ◇ CATHOMYCIN ◇ CRYSTALLINIC ACID ◇ INAMYCIN ◇ NOVOBIOCIN ◇ NOVO-R ◇ PA 93 ◇ ROBIOCINA ◇ SIRBIOCINA ◇ SPHEROMYCIN ◇ STILBIOCINA

TOXICITY DATA with REFERENCE
mmo-esc 200 mg/L NATUAS 271,385,78
pic-esc 10 μg/plate CNREA8 43,2819,83
dni-hmn:emb 600 mmol/L CRNGDP 3,1171,82
cyt-mus-par 50 mg/kg NULSAK 2,161,71
dnr-ham:oth 200 mg/L CRNGDP 5,187,84
unr-hmn TDLo:171 mg/kg/12D-C:EYE,SKN 85ERAY 1,137,78
orl-mus LD50:1500 mg/kg 85GDA2 6,338,81
ipr-mus LD50:262 mg/kg 85FZAT -,471,67
ivn-mus LD50:407 mg/kg 85FZAT -,471,67

SAFETY PROFILE: Poison by intraperitoneal route. Moderately toxic by ingestion and intravenous routes. Human systemic effects by unspecified route: conjuctiva irritation and skin dermatitis. Human mutation data reported. An antibiotic with serious side effects which include liver and blood disease. It may also cause the development of resistant strains of staphylococcus. Used as a food additive permitted in the feed and drinking water of animals and/or for the treatment of food-producing animals; also permitted in food for human consumption. When heated to decomposition it emits toxic fumes of NO_x.

SMB850 CAS:23344-17-4 ***HR: 3***
STREPTOVARICIN C
mf: $C_{40}H_{51}NO_{14}$ mw: 769.92

TOXICITY DATA with REFERENCE
pic-esc 100 pg/plate CNREA8 43,2819,83
orl-mus LD50:2000 mg/kg 85GDA2,477,80
ipr-mus LD50:233 mg/kg UPJOH* 2(6),-,71

SAFETY PROFILE: Poison by intraperitoneal route. Moderately toxic by ingestion. Mutation data reported. When heated to decomposition it emits toxic fumes of NO_x.

SMC375 CAS:56833-74-0 ***HR: 3***
STREPTOVIRUDIN

PROP: White, amorph solid. Sol in methanol, pyridine; sltly sol in water, ethanol; insol in chloroform, acetone, benzene, ethyl acetate.

TOXICITY DATA with REFERENCE
orl-mus LD50:150 mg/kg JANTAJ 28,514,75
scu-mus LD50:15 mg/kg JANTAJ 28,514,75
ivn-mus LD50:17500 μg/kg JANTAJ 28,514,75

SAFETY PROFILE: Poison by ingestion, subcutaneous, and intravenous routes.

SMC500 CAS:523-86-4 ***HR: 3***
STREPTOVITACIN A
mf: $C_{15}H_{23}NO_5$ mw: 297.39

SYNS: 3-(2-HYDROXY-2-(5-HYDROXY-3,5-DIMETHYL-2-OXOCYCLOHEXYL)ETHYL)GLUTARIMIDE ◇ 4-(2-HYDROXY-2-(5-HYDROXY-3,5-DIMETHYL-2-OXOCYCLOHEXYL)ETHYL)-2,6-PIPERIDINEDIONE ◇ NSC 39147 ◇ RESACTIN A ◇ U-9361

TOXICITY DATA with REFERENCE
orl-rat LD50:3 mg/kg UPJOH* 2(6),-,71
ipr-rat LD50:920 μg/kg JPETAB 136,400,62
orl-mus LD50:3952 μg/kg NCISP* JAN86
ipr-mus LD50:6500 μg/kg CNCRA6 30,9,63
ivn-mus LD50:11 mg/kg UPJOH* 2(6),-,71

SAFETY PROFILE: Poison by ingestion, intravenous, and intraperitoneal routes. When heated to decomposition it emits toxic fumes of NO_x.

SMD000 CAS:18883-66-4 ***HR: 3***
STREPTOZOTICIN
mf: $C_8H_{15}N_3O_7$ mw: 265.26

PROP: Platelets. Mp: 115° (decomp).

SYNS: 2-DEOXY-2-(((METHYLNITROSOAMINO)CARBONYL)AMINO)-d-GLUCOPYRANOSE ◇ 2-DEOXY-2-(3-METHYL-3-NITROSOUREIDO)-d-GLUCOPYRANOSE ◇ 2-DEOXY-2-(3-METHYL-3-NITROSOUREIDO)-α(and β)-d-GLUCOPYRANOSE ◇ N-d-GLUCOSYL (2)-N'-NITROSOMETHYLHARNSTOFF (GERMAN) ◇ N-d-GLUCOSYL-(2)-N'-NITROSOMETHYLUREA ◇ NCI-C03167 ◇ NSC 85598 ◇ NSC-85998 ◇ RCRA WASTE NUMBER U206 ◇ STR ◇ STREPTOZOCIN ◇ STRZ ◇ STZ ◇ U-9889 ◇ ZANOSAR

TOXICITY DATA with REFERENCE
mmo-sat 20 ng/plate TAKHAA 44,96,85
dns-rat-ipr 250 mg/kg ENMUDM 7,889,85
ivn-rat TDLo:40 mg/kg (female 5D post):REP DIAEAZ 19,610,70
ipr-rat TDLo:50 mg/kg (1D pre):TER DBTGAJ 10,89,74
ipr-rat TDLo:470 mg/kg/26W-I:NEO RRCRBU 52,1,75
ivn-mus TDLo:250 mg/kg:CAR CNREA8 45,703,85
ipr-ham TDLo:100 mg/kg/I:ETA JNCIAM 51,1287,73
ivn-wmn TDLo:13513 μg/kg NEJMAG 311,798,84
ivn-wmn LDLo:440 mg/kg/65W-I GUTTAK 12,717,71
ivn-hmn TDLo:1044 mg/kg/5D:GIT,LIV,KID CANCAR 34,993,74
ivn-rat LD50:138 mg/kg CNCRA6 29,91,63
orl-mus LD50:264 mg/kg CNCRA6 48,1,65
ipr-mus LD50:360 mg/kg JMCMAR 19,918,76
scu-mus LD50:335 mg/kg NCISP* JAN86
ivn-mus LD50:275 mg/kg JMCMAR 19,918,76
par-mus LD50:264 mg/kg CNCRA6 48,1,65
ivn-dog LDLo:50 mg/kg DCTODJ 3,201,80
ivn-mky LDLo:80 mg/kg DCTODJ 3,201,80

CONSENSUS REPORTS: NTP Fifth Annual Report on Carcinogens. IARC Cancer Review: Group 2B IMEMDT 7,56,87; Human Limited Evidence IMEMDT 17,337,78; Animal Sufficient Evidence IMEMDT 17,337,78. NCI Carcinogenesis Studies (ipr); Clear Evidence: mouse, rat RRCRBU 52,1,75. EPA Genetic Toxicology Program.

SAFETY PROFILE: Confirmed carcinogen with experimental carcinogenic, neoplastigenic, tumorigenic, and teratogenic data. Experimental poison by ingestion, intravenous, parenteral, subcutaneous, and intraperitoneal routes. Moderately toxic to humans by intravenous route. Human systemic effects: nausea or vomiting, impaired liver function, kidney changes. Human mutation data reported. Experimental reproductive effects. When heated to decomposition it emits toxic fumes of NO_x. See also NITROSAMINES.

SMD500 ***HR: 3***
STRONTIUM
af: Sr aw: 87.62

PROP: Silvery-white metal. D: 2.6, mp: 757° ± 1°, bp: 1366°; vap press: 10 mm @ 898°.

SAFETY PROFILE: It resembles calcium in its metabolism and behavior. The stable form has low toxicity. Ignites spontaneously in air. Moderately explosive in the form of dust when exposed to flame or by spontaneous chemical reaction. Reacts vigorously with water or steam to evolve hydrogen. Reaction with halogens may lead to ignition. Can be stored under liquid hydrocarbons. Vigorous reaction on contact with oxidizing materials. Highly dangerous in the form of the radioactive isotopes ^{90}Sr. See also STRONTIUM COMPOUNDS and POWDERED METALS.

SME000 CAS:543-94-2 ***HR: 3***
STRONTIUM ACETATE
mf: $C_4H_6O_4 \cdot Sr$ mw: 205.72

TOXICITY DATA with REFERENCE
ivn-rat LDLo:123 mg/kg JPETAB 71,1,41
ivn-mus LDLo:239 mg/kg JPETAB 71,1,41

CONSENSUS REPORTS: Reported in EPA TSCA Inventory.

SAFETY PROFILE: Poison by intravenous route. When heated to decomposition it emits acrid smoke and irritating fumes. See also STRONTIUM COMPOUNDS.

SME100 CAS:12071-29-3 ***HR: 2***
STRONTIUM ACETYLIDE
mf: C_2Sr mw: 111.64

SAFETY PROFILE: Incandescent reaction with chlorine (at 197°C); bromine (at 174°C); iodine (at 182°C). See also STRONTIUM COMPOUNDS and ACETYLIDES.

SME500 CAS:91724-16-2 ***HR: 3***
STRONTIUM ARSENITE
DOT: UN 1691
mf: As_2O_4Sr mw: 301.46

PROP: White powder.

SYNS: ARSENIOUS ACID, STRONTIUM SALT ◇ STRONTIUM ARSENITE, solid (DOT)

CONSENSUS REPORTS: Arsenic and its compounds are on the Community Right To Know List.

OSHA PEL: Cancer Hazard
NIOSH REL: (Arsenic, Inorganic) CL 0.002 mg(As)/m^3/15M
DOT Classification: Poison B; Label: Poison

SAFETY PROFILE: Confirmed human carcinogen. A deadly poison. When heated to decomposition it emits toxic fumes of As. See also ARSENIC COMPOUNDS and STRONTIUM COMPOUNDS.

SMF000 CAS:10476-81-0 ***HR: 2***
STRONTIUM BROMIDE
mf: Br_2Sr mw: 247.44

PROP: White, hygroscopic needles. Mp: 643°, bp: decomp, d: 4.216 @ 24°.

SYN: SR E2

TOXICITY DATA with REFERENCE
ipr-rat LD50:1 g/kg MEIEDD 10,1267,83
ivn-rat LDLo:500 mg/kg EQSSDX 1,1,75

CONSENSUS REPORTS: Reported in EPA TSCA Inventory.

SAFETY PROFILE: Moderately toxic by intraperitoneal and intravenous routes. When heated to decomposition it emits toxic fumes of Br^-. See also STRONTIUM COMPOUNDS and BROMIDES.

SMF500 CAS:7791-10-8 ***HR: 2***
STRONTIUM CHLORATE
DOT: UN 1506
mf: Cl_2O_6Sr mw: 254.52

PROP: White, crystalline powder. Mp: (decomp) 120°, d: 3.152.

SYNS: CHLORIC ACID, STRONTIUM SALT ◇ STRONTIUM CHLORATE, WET (DOT)

CONSENSUS REPORTS: Reported in EPA TSCA Inventory.

DOT Classification: Oxidizer; Label: Oxidizer and Oxidizer, wet

SAFETY PROFILE: A powerful oxidizer. When heated to decomposition it emits toxic fumes of Cl^-. See also CHLORATES.

SMG000 CAS:7791-10-8 ***HR: 2***
STRONTIUM CHLORATE (wet)
mf: Cl_2O_6Sr mw: 285.59

CONSENSUS REPORTS: Reported in EPA TSCA Inventory.

DOT Classification: Oxidizer; Label: Oxidizer

SAFETY PROFILE: A powerful oxidizer. When heated to decomposition it emits toxic fumes of Cl^-. See also STRONTIUM CHLORATE.

SMG500 CAS:10476-85-4 ***HR: 3***
STRONTIUM CHLORIDE
mf: Cl_2Sr mw: 158.52

TOXICITY DATA with REFERENCE
sln-smc 80 mmol/L MUTAEX 1,21,86
orl-rat LD50:2250 mg/kg FOREAE 7,313,42
ipr-rat LD50:405 mg/kg EQSSDX 1,1,75
ivn-rat LD50:222 mg/kg EQSSDX 1,1,75
orl-mus LD50:3100 mg/kg FOREAE 7,313,42
ipr-mus LD50:1643 mg/kg COREAF 256,1043,63
ivn-mus LD50:148 mg/kg TXAPA9 22,150,72
orl-rbt LDLo:7500 mg/kg HBAMAK 4,1289,35

CONSENSUS REPORTS: Reported in EPA TSCA Inventory. EPA Genetic Toxicology Program.

SAFETY PROFILE: Poison by intravenous route. Moderately toxic by ingestion and intraperitoneal routes. Mutation data reported. When heated to decomposition it emits toxic fumes of Cl^-. See also CHLORIDES and STRONTIUM COMPOUNDS.

SMH000 CAS:7789-06-2 ***HR: 3***
STRONTIUM CHROMATE (1:1)
mf: $CrO_4•Sr$ mw: 203.62

PROP: Monoclinic, yellow crystals. D: 3.895 @ 15°.

SYNS: CHROMIC ACID, STRONTIUM SALT (1:1) ◇ C.I. PIGMENT YELLOW 32 ◇ DEEP LEMON YELLOW ◇ STRONTIUM CHROMATE (VI) ◇ STRONTIUM CHROMATE 12170 ◇ STRONTIUM YELLOW

TOXICITY DATA with REFERENCE
mmo-sat 800 ng/plate MUREAV 156,219,85

sce-ham:ovr 100 μg/L MUREAV 156,219,85
itr-rat TDLo:40 mg/kg/34W-I:ETA AEHLAU 5,445,62
orl-rat LD50:3118 mg/kg GISAAA 45(10),76,80
itr-rat LD50:16600 mg/kg GISAAA 45(10),76,80

CONSENSUS REPORTS: NTP Fifth Annual Report on Carcinogens. IARC Cancer Review: Group 1 IMEMDT 7,165,87; Animal Sufficient Evidence IMEMDT 2,100,73; IMEMDT 23,205,80; Human Sufficient Evidence IMEMDT 23,205,80. Chromium and its compounds are on the Community Right To Know List. Reported in EPA TSCA Inventory.

OSHA PEL: CL 0.1 mg(CrO_3)/m^3
ACGIH TLV: (Proposed: TWA 0.0005 ppm; Suspected Human Carcinogen)
DFG TRK: 0.1 mg/m^3; Animal Carcinogen, Suspected Human Carcinogen.
NIOSH REL: TWA 0.0001 mg(Cr(VI))/m^3

SAFETY PROFILE: Confirmed human carcinogen with experimental carcinogenic and tumorigenic data. Moderately toxic by ingestion. Mutation data reported. See also CHROMIUM COMPOUNDS and STRONTIUM COMPOUNDS.

SMH500 ***HR: 1***
STRONTIUM COMPOUNDS

SAFETY PROFILE: The strontium ion has a low order of toxicity. It is chemically and biologically similar to calcium. Strontium salicylate is the most toxic compound. The oxides and hydroxides are moderately caustic materials. Symptoms of acute toxicity are excessive salivation, vomiting, colic and diarrhea, and possibly respiratory failure. The gastrointestinal absorption of soluble strontium ranges from 5 to 25%. Workers in strontium salt plants have reduced activity of choline esterase and acetylcholine. Drinking water with 13 mg Sr/L caused impaired tooth development in 1 year old children. As with other compounds, the toxicity of a given compound may be a function of the anion. Compounds are highly dangerous if they contain the radioactive isotope ^{90}Sr.

SMH525 CAS:10025-70-4 ***HR: 2***
STRONTIUM DICHLORIDE HEXAHYDRATE
mf: $Cl_2Sr{\cdot}6H_2O$ mw: 266.64

SYN: STRONTIUM CHLORIDE, HEXAHYDRATE

TOXICITY DATA with REFERENCE
orl-rat TDLo:405 mg/kg (male 90D pre):REP TXCYAC 7,11,77
ivn-rat LDLo:400 mg/kg MEIEDD 10,1267,83
ipr-mus LD50:1253 mg/kg TXAPA9 63,461,82

SAFETY PROFILE: Moderately toxic by intravenous and intraperitoneal route. Experimental reproductive effects. When heated to decomposition it emits toxic fumes of Sr and Cl^-.

SMI000 CAS:13814-98-7 ***HR: 2***
STRONTIUM FLUOBORATE
mf: $BF_4{\cdot}1/2Sr$ mw: 130.62

SYNS: TETRAFLUOROBORATE(1-)STRONTIUM (2:1) ◇ TL 1139

TOXICITY DATA with REFERENCE
orl-rat LDLo:500 mg/kg NCNSA6 5,28,53
ihl-mus LCLo:650 mg/m^3/10M NDRC** No.9-4-1-19,44

CONSENSUS REPORTS: Reported in EPA TSCA Inventory.

OSHA PEL: TWA 2.5 mg(F)/m^3
NIOSH REL: (Inorganic Fluorides) TWA 2.5 mg(F)/m^3

SAFETY PROFILE: Moderately toxic by ingestion and inhalation. When heated to decomposition it emits toxic fumes of F^-. See also STRONTIUM COMPOUNDS and FLUORIDES.

SMI500 CAS:7783-48-4 ***HR: 2***
STRONTIUM FLUORIDE
mf: F_2Sr mw: 125.62

PROP: Cubic, odorless; crystals or white powder. Mp: 1190°, d: 4.24, bp: 2460. Decomp by strong acids.

TOXICITY DATA with REFERENCE
orl-rat LD50:10600 mg/kg VAMNAQ 32,28,77
ivn-rat LDLo:625 mg/kg EQSFAP 1,1,75
ipr-mus LD50:4400 mg/kg VAMNAQ 32,28,77

CONSENSUS REPORTS: Reported in EPA TSCA Inventory.

OSHA PEL: TWA 2.5 mg(F)/m^3
ACGIH TLV: TWA 2.5 mg(F)/m^3
NIOSH REL: TWA 2.5 mg(F)/m^3

SAFETY PROFILE: Moderately toxic by intravenous route. Mildly toxic by ingestion. When heated to decomposition it emits toxic fumes of F^-. See also FLUORIDES and STRONTIUM COMPOUNDS.

SMJ000 CAS:18943-30-1 ***HR: 3***
STRONTIUM FLUOSILICATE
mf: $F_6Si{\cdot}Sr$ mw: 229.71

PROP: Monoclinic crystals. Mp: decomp, d: 2.99 @ 17.5°.

SYN: HEXAFLUOROSILICATE(2-)STRONTIUM

TOXICITY DATA with REFERENCE
orl-rat LDLo:250 mg/kg NCNSA6 5,28,53

OSHA PEL: TWA 2.5 mg(F)/m^3
NIOSH REL: TWA 2.5 mg(F)/m^3

SAFETY PROFILE: Poison by ingestion. When heated to decomposition it emits toxic fumes of F^-. See also FLUOSILICATES and STRONTIUM COMPOUNDS.

SMJ500 CAS:10476-86-5 ***HR: 2***
STRONTIUM IODIDE
mf: I_2Sr mw: 341.42

PROP: Colorless plates. Bp: decomp, d: 4.549 @ 25°.

TOXICITY DATA with REFERENCE
ipr-rat LD50:800 mg/kg AIHOAX 1,637,50

CONSENSUS REPORTS: Reported in EPA TSCA Inventory.

SAFETY PROFILE: Moderately toxic by intraperitoneal route. Reacts violently with K. When heated to decomposition it emits toxic fumes of I^-. See also IODIDES and STRONTIUM COMPOUNDS.

SMK000 CAS:10042-76-9 ***HR: 2***
STRONTIUM(II) NITRATE (1:2)
DOT: UN 1507
mf: N_2O_6•Sr mw: 211.64

PROP: White powder. Mp: 570°, d: 2.986.

SYNS: NITRATE de STRONTIUM (FRENCH) ◇ NITRIC ACID, STRONTIUM SALT

TOXICITY DATA with REFERENCE
orl-rat LD50:2750 mg/kg GISAAA 41(5),28,76
ipr-rat LD50:540 mg/kg AIHOAX 1,637,50
orl-mus LD50:1826 mg/kg GISAAA 41(5),28,76

CONSENSUS REPORTS: Reported in EPA TSCA Inventory.

DOT Classification: Oxidizer; Label: Oxidizer

SAFETY PROFILE: Moderately toxic by ingestion and intraperitoneal routes. A powerful oxidizer. When heated to decomposition it emits toxic fumes of NO_x. See also NITRATES and STRONTIUM COMPOUNDS.

SMK500 CAS:1314-18-7 ***HR: 2***
STRONTIUM PEROXIDE
DOT: UN 1509
mf: O_2Sr mw: 119.62

PROP: White powder. Mp: decomp, d: 4.56.

CONSENSUS REPORTS: Reported in EPA TSCA Inventory.

DOT Classification: Oxidizer; Label: Oxidizer

SAFETY PROFILE: A powerful oxidizer. A skin, eye, and mucous membrane irritant. Mixtures with organic materials readily ignite with friction or on contact with moisture. See also PEROXIDES and STRONTIUM COMPOUNDS.

SML000 CAS:12504-13-1 ***HR: 2***
STRONTIUM PHOSPHIDE
mf: PSr mw: 118.59
DOT: UN 2013

SYN: PHOSPHURE de STRONTIUM (FRENCH)

TOXICITY DATA with REFERENCE
ihl-rat LCLo:580 ppm/1H ZGSHAM 25,279,33
ihl-cat LCLo:173 ppm/2H ZGSHAM 25,279,33
ihl-gpg LCLo:288 ppm/2H ZGSHAM 25,279,33

SAFETY PROFILE: Moderately toxic by inhalation. When heated to decomposition it emits toxic fumes of PO_x. See also PHOSPHIDES and STRONTIUM COMPOUNDS.

SML500 CAS:526-26-1 ***HR: 3***
STRONTIUM SALICYLATE
mf: $C_7H_5O_3$•1/2Sr mw: 180.93

SYNS: 2-HYDROXYBENZOIC ACID STRONTIUM SALT (2:1) ◇ STRONCYLATE

TOXICITY DATA with REFERENCE
ipr-rat LD50:400 mg/kg AIHOAX 1,637,50

CONSENSUS REPORTS: Reported in EPA TSCA Inventory.

SAFETY PROFILE: Poison by intraperitoneal route. When heated to decomposition it emits acrid smoke and irritating fumes.

SMM000 CAS:1314-96-1 ***HR: 3***
STRONTIUM SULFIDE
mf: SSr mw: 119.68

PROP: Cubic, light gray crystals. D: 3.70 @ 15°.

SYNS: C.I. 77847 ◇ STRONTIUM MONOSULFIDE ◇ STRONTIUM SULPHIDE

CONSENSUS REPORTS: Reported in EPA TSCA Inventory.

SAFETY PROFILE: Poison by inhalation and ingestion. Readily decomposes to yield H_2S. Incompatible with lead(IV) oxide. When heated to decomposition it emits toxic fumes of SO_x. See also SULFIDES and STRONTIUM COMPOUNDS.

SMM500 CAS:66-28-4 ***HR: 3***
k-STROPHANTHIDINE
mf: $C_{23}H_{32}O_6$ mw: 404.55

PROP: From the seeds of *Corchorus capsularis* (JPETAB 103,420,51).

SYNS: ANTIBIOTIC XS-89 ◇ APOCYNAMARIN ◇ CONVALLATOXIGENIN ◇ CORCHORGENIN ◇ CORCHORIN ◇ CORCHOSIDE A AGLYCON ◇ CORCHSULARIN ◇ CYANOTOXIN ◇ CYMARIGENIN ◇ CYNOTOXIN ◇ ERYSIMUPICRONE ◇ ERYSIMUPIKRON ◇ 5-β-HYDROXY-19-OXODIGITOXIGENIN ◇ NSC-86078 ◇ 19-OXO-CARDOGENEN-(20:22)-TRIOL(3-β,5,14) (GERMAN) ◇ STROPHANTHIDIN ◇ K-STROPHANTHIDIN ◇ STROPHANTHIDINE ◇ 3,5,14-TRIHYDROXY-19-OXOCARD-20(22)-ENOLIDE ◇ 3-β,5,14-TRIHYDROXY-19-OXO-5-β-CARD-20(22)-ENOLIDE ◇ 3-β,5,14-TRIOXY-19-OXO-CARDEN-(20:22)-OLID (GERMAN) ◇ 3-β,5,14-TRIOXY-19-OXO-DIGEN-(20:22)-OLID (GERMAN) ◇ XS-89

TOXICITY DATA with REFERENCE
ivn-mus LD50:330 μg/kg 85GDA2 8(2),224,82
ivn-cat LD50:224 μg/kg JPETAB 103,420,51
ivn-rbt LDLo:110 μg/kg AEPPAE 185,329,37
ivn-gpg LDLo:6465 μg/kg BRCAB7 76,62,81

SAFETY PROFILE: A deadly poison by intravenous. When heated to decomposition it emits acrid smoke and irritating fumes.

SMN000 CAS:11005-63-3 ***HR: 3***
k-STROPHANTHIN
mf: $C_{36}H_{54}O_{14}$ mw: 710.90

PROP: White or yellowish powder; very bitter taste.

SYNS: COMBETIN ◇ EUSTROPHINUM ◇ KOMBÉ-STROPHANTHIN ◇ KOMBETIN ◇ STROPHANTHIN

TOXICITY DATA with REFERENCE
scu-rat LDLo:50 mg/kg HBAMAK 4,1289,35
ivn-rat LD50:7 mg/kg ARTODN 54,275,83
ipr-mus LD50:3500 μg/kg AIPTAK 155,165,65
ivn-mus LD50:2500 μg/kg CLDND* 184,181,37
orl-rbt LDLo:20 mg/kg FDWU** -,-,31
scu-rbt LDLo:250 μg/kg FDWU** -,-,31
ivn-rbt LDLo:200 μg/kg AEPPAE 185,329,37

SAFETY PROFILE: Poison by ingestion, intraperitoneal, subcutaneous, and intravenous routes. When heated to decomposition it emits acrid smoke and irritating fumes.

SMN002 CAS:560-53-2 ***HR: 3***
STROPHANTHIN K
mf: $C_{36}H_{54}O_{14}$ mw: 710.90

SYNS: NSC 4320 ◇ STROPHANTHIDIN-GLUCOCYMAROSID (GERMAN) ◇ β-k-STROPHANTHIN ◇ K-STROPHANTHIN-β ◇ STROPHANTHIN K (crystalline) ◇ β-STROPHANTHOBIOSIDE, STROPHANTHIDIN-3

TOXICITY DATA with REFERENCE
scu-rat LDLo:50 mg/kg 27ZWAY E,1,78,-
ipr-mus LD50:1071 μg/kg NCISP* JAN86
scu-mus LD50:213 μg/kg NCISP* JAN86

SAFETY PROFILE: Poison by subcutaneous and intraperitoneal routes. When heated to decomposition it emits acrid smoke and irritating fumes.

SMN100 ***HR: 3***
STROPHANTHUS GRATUS Franch., leaf and stem bark extract

PROP: Indian plant belonging to the family *Apocynaceae* IJEBA6 18,594,80

TOXICITY DATA with REFERENCE
orl-rat TDLo:150 mg/kg (female 12-14D post):REP IJEBA6 18,594,80
ipr-mus LD50:178 mg/kg IJEBA6 18,594,80

SAFETY PROFILE: Poison by intraperitoneal route. Experimental reproductive effects. When heated to decomposition it emits acrid smoke and irritating fumes.

SMN275 CAS:595-21-1 ***HR: 3***
STROSPESIDE
mf: $C_{30}H_{46}O_9$ mw: 550.76

SYNS: DESGLUCO-DIGITALINUM VERUM (GERMAN) ◇ GITOXIGENIN-3-o-MONODIGITALOSIDE ◇ GITOXOGENIN-d-DIGITALOSID (GERMAN) ◇ STROSPESID (GERMAN)

TOXICITY DATA with REFERENCE
orl-mus LD50:41200 μg/kg AIPTAK 153,436,65
scu-mus LD50:40 mg/kg AIPTAK 153,436,65
ivn-cat LD50:410 μg/kg 85ELDJ -,194,63
ivn-gpg LDLo:1990 μg/kg AIPTAK 153,436,65

SAFETY PROFILE: Poison by ingestion, subcutaneous, and intravenous routes. When heated to decomposition it emits acrid smoke and irritating fumes.

SMN500 CAS:57-24-9 ***HR: 3***
STRYCHNINE
DOT: UN 1692
mf: $C_{21}H_{22}N_2O_2$ mw: 334.45

PROP: Hard, white, crystalline alkaloid; very bitter taste. Mp: 268°, bp: 270°, d: 1.359 @ 18°.

SYNS: CERTOX ◇ DOLCO MOUSE CEREAL ◇ KWIK-KIL ◇ MOLE DEATH ◇ MOUSE-NOTS ◇ MOUSE-RID ◇ MOUSE-TOX ◇ PIED PIPER MOUSE SEED ◇ RCRA WASTE NUMBER P108 ◇ RO-DEX ◇ SANASEED ◇ STRICNINA (ITALIAN) ◇ STRYCHNIDIN-10-ONE ◇ STRYCHNIN (GERMAN) ◇ STRYCHNINE, solid and liquid (DOT) ◇ STRYCHNOS

TOXICITY DATA with REFERENCE
ims-rat TDLo:500 μg/kg (5D preg):REP PSEBAA 100,555,59
orl-hmn LDLo:30 mg/kg PCOC** -,1073,66
orl-man LDLo:30 mg/kg PEMNDP 8,753,87
orl-rat LD50:2350 μg/kg JAPMA8 31,113,42
ipr-rat LD50:1100 μg/kg JPETAB 58,239,36
scu-rat LD50:1200 μg/kg RPTOAN 31,260,68
ivn-rat LD50:582 μg/kg CHHTAT 54,234,74

orl-mus LD50:2 mg/kg IJEBA6 19,1075,81
ipr-mus LD50:980 μg/kg JPETAB 128,176,60
scu-mus LD50:474 μg/kg APSXAS 7,329,70
ivn-mus LD50:410 μg/kg AIPTAK 128,204,60
orl-dog LD50:500 μg/kg RMVEAG 154,137,78
scu-dog LDLo:350 μg/kg HBAMAK 4,1403,35
ivn-dog LD50:800 μg/kg JPETAB 126,41,59

CONSENSUS REPORTS: Reported in EPA TSCA Inventory. EPA Extremely Hazardous Substances List.

OSHA PEL: TWA 0.15 mg/m^3
ACGIH TLV: TWA 0.15 mg/m^3
DFG MAK: 0.15 mg/m^3
DOT Classification: Poison B; Label: Poison; Poison B; Label: St. Andrews Cross

SAFETY PROFILE: Human poison by ingestion. Experimental poison by ingestion, intravenous, subcutaneous, and intraperitoneal routes. Experimental reproductive effects. An allergen. Lethal dose to man: 30-60 mg/kg. If ingested, the time of action depends upon the condition of the stomach, that is, whether empty or full, and the nature of the food present. If taken by subcutaneous injection, the place of administration of the injection will affect the time of action. The first symptoms are a feeling of uneasiness with a heightened reflex of irritability, followed by muscular twitching in some parts of the body. With larger doses, this is followed by a sense of impending suffocation. Convulsive movements begin which have the effect of mechanically causing the patient to cry out or to shriek; then follow the characteristic spasms which set in with violence. These are at first clonic and then tonic. There are successive attacks of spasms. With each successive attack, the symptoms become more violent, eventually resulting in death. A rodenticide. When heated to decomposition it emits toxic fumes of NO_x.

SMO000 CAS:66-32-0 ***HR: 3***
STRYCHNINE MONONITRATE
mf: $C_{21}H_{22}N_2O_2{\cdot}HNO_3$ mw: 397.47

PROP: Colorless, odorless needles or white, crystalline powder. Insol in ether.

SYN: STRYCHNINE NITRATE

TOXICITY DATA with REFERENCE
orl-rat LD50:16200 μg/kg NIIRDN 6,377,82
scu-rat LDLo:3 mg/kg AEPPAE 187,607,37
ivn-rat LDLo:600 μg/kg ARZNAD 20,1778,70
orl-mus LD50:930 μg/kg JTCTDW 22,425,84
ivn-mus LDLo:600 μg/kg ARZNAD 20,1778,70

SAFETY PROFILE: Poison by ingestion, intravenous, and subcutaneous routes. When heated to decomposition it emits very toxic fumes of NO_x. See also STRYCHNINE.

SMO500 ***HR: 3***
STRYCHNINE SALT (solid)
DOT: UN 1692

DOT Classification: Poison B; Label: St. Andrews Cross.

SAFETY PROFILE: A deadly poison. See also STRYCHNINE.

SMP000 CAS:60-41-3 ***HR: 3***
STRYCHNINE SULFATE (2:1)
mf: $C_{21}H_{22}N_2O_2{\cdot}1/2H_2O_4S$ mw: 383.49

SYNS: STRYCHININE SULFATE ◇ STRYCHNIDIN-10-ONE, SULFATE (2:1)

TOXICITY DATA with REFERENCE
orl-rat LD50:5 mg/kg TXAPA9 21,315,72
ipr-rat LD50:1 mg/kg ARZNAD 23,1125,73
scu-rat LD50:1700 μg/kg TXAPA9 18,185,71
ivn-rat LD50:570 μg/kg AIPTAK 144,416,63
orl-mus LDLo:8 mg/kg AECTCV 14,111,85
ipr-mus LD50:900 μg/kg IVEBBN 16,31,77
scu-mus LD50:1250 μg/kg JPETAB 67,153,39
ivn-mus LD50:475 μg/kg JPETAB 79,127,43
ipr-gpg LD50:10900 μg/kg AIPTAK 144,416,63
scu-gpg LD50:4800 μg/kg AIPTAK 144,416,63
ivn-gpg LD50:390 μg/kg AIPTAK 144,416,63
orl-bwd LD50:5600 μg/kg TXAPA9 21,315,72

CONSENSUS REPORTS: EPA Extremely Hazardous Substances List.

SAFETY PROFILE: Poison by ingestion, intraperitoneal, intravenous, and subcutaneous routes. When heated to decomposition it emits very toxic fumes of SO_x and NO_x. See also STRYCHNINE and SULFATES.

SMP400 CAS:65928-58-7 ***HR: 2***
STS 557
mf: $C_{20}H_{25}NO_2$ mw: 311.46

SYN: 17-α-CYANOMETHYL-17-β-HYDROXY-ESTRA-4,9(10)-DIEN-3-ONE

TOXICITY DATA with REFERENCE
dlt-mus-scu 700 mg/kg/35D ENDKAC 76,13,80
orl-mus TDLo:500 mg/kg (female 13-17D post):TER EXCEDS 81,197,83
orl-mus TDLo:500 mg/kg (female 8-12D post):REP EXCEDS 81,197,83
orl-mus LD50:4000 mg/kg PHARAT 34,319,79
ipr-mus LDLo:1 g/kg EXCEDS 81,175,83
scu-mus LDLo:5000 mg/kg PHARAT 34,319,79
orl-rbt LDLo:1 g/kg EXCEDS 81,175,83
ipr-rbt LDLo:1500 mg/kg EXCEDS 81,175,83

CONSENSUS REPORTS: Cyanide and its compounds are on the Community Right To Know List. EPA Genetic Toxicology Program.

SAFETY PROFILE: Moderately toxic by ingestion, subcutaneous, and intraperitoneal routes. An experimental teratogen. Experimental reproductive effects. Mutation data reported. When heated to decomposition it emits toxic fumes of NO_x and CN^-. See also NITRILES.

SMP500 CAS:82-71-3 ***HR: 3***
STYPHNIC ACID
DOT: UN 0219/UN 0394
mf: $C_6H_3N_3O_8$ mw: 245.12

PROP: Hexagonal, yellow crystals; astringent taste. Mp: (dry) 175.5°. Very sol in alc, ether.

SYNS: 2,4-DIHYDROXY-1,3,5-TRINITROBENZENE ◇ 1,3-DIHYDROXY-2,4,6-TRINITROBENZENE ◇ 3-HYDROXY-2,4,6-TRINITROPHENOL ◇ 2,4,6-TRINITROBENZENE-1,3-DIOL ◇ 2,4,6-TRINITRO-1,3-BENZENEDIOL ◇ 2,4,6-TRINITRORESORCINOL ◇ TRINITRORESORCINOL (DOT) ◇ TRINITRORESORCINOL, DRY (DOT) ◇ TRINITRORESORCINOL, wetted with less than 20% water (DOT)

CONSENSUS REPORTS: Reported in EPA TSCA Inventory.

DOT Classification: Class A Explosive; Label: Explosive A

SAFETY PROFILE: Very explosive. Upon decomposition it emits toxic fumes of NO_x. See also NITRO COMPOUNDS of AROMATIC HYDROCARBONS and EXPLOSIVES, HIGH.

SMQ000 CAS:100-42-5 ***HR: 3***
STYRENE
DOT: UN 2055
mf: C_8H_8 mw: 104.16

$$C_6H_5CH{=}CH_2$$

PROP: Colorless, refractive, oily liquid. Mp: −31°, bp: 146°, lel: 1.1%, uel: 6.1%, flash p: 88°F, d: 0.9074 @ 20°/4°, autoign temp: 914°F, vap d: 3.6, fp: −33°, ULC: 40-50. Very sltly sol in water; misc in alc and ether.

SYNS: CINNAMENE ◇ CINNAMENOL ◇ DIAREX HF 77 ◇ ETHENYLBENZENE ◇ NCI-C02200 ◇ PHENETHYLENE ◇ PHENYLETHENE ◇ PHENYLETHYLENE ◇ STIROLO (ITALIAN) ◇ STYREEN (DUTCH) ◇ STYREN (CZECH) ◇ STYRENE MONOMER (ACGIH) ◇ STYRENE MONOMER, inhibited (DOT) ◇ STYROL (GERMAN) ◇ STYROLE ◇ STYROLENE ◇ STYRON ◇ STYROPOR ◇ VINYLBENZEN (CZECH) ◇ VINYLBENZENE ◇ VINYLBENZOL

TOXICITY DATA with REFERENCE
skn-hmn 500 mg nse INMEAF 17,199,48
skn-rbt 500 mg open MLD UCDS** 12/13/63
skn-rbt 100% MOD AMIHAB 14,387,56
eye-rbt 18 mg AJOPAA 29,1363,46
orl-rat TDLo:8600 mg/kg (1-22D preg/21D post):REP NTOTDY 7,23,85
ihl-rat TCLo:1500 μg/m³/24H (female 1-22D post):TER GISAAA 39(11),65,74
ihl-rat TCLo:100 ppm/4H/5D/1Y-I:CAR Blood - leukemia ANYAA9 534,203,88
ihl-hmn LCLo:10000 ppm/30M 29ZWAE -,77,68
ihl-hmn TCLo:600 ppm:NOSE,EYE AMIHAB 14,387,56
ihl-hmn TCLo:20 μg/m³:EYE GISAAA 26(8),11,61
orl-rat LD50:5000 mg/kg AMIHAB 14,387,56
ihl-rat LC50:24 g/m³/4H GTPZAB 26(8),53,82
ipr-rat LD50:898 mg/kg ENVRAL 40,411,86
orl-mus LD50:316 mg/kg NCILB* NCI-E-C-72-3252,73
ihl-mus LC50:9500 mg/m³/4H 85GMAT -,106,82
ipr-mus LD50:660 mg/kg ARZNAD 19,617,69
ivn-mus LD50:90 mg/kg ARZNAD 19,617,69
ihl-gpg LCLo:12 mg/m³/14H JIHTAB 24,295,42

CONSENSUS REPORTS: IARC Cancer Review: Group 2B IMEMDT 7,345,87; Animal Sufficient Evidence IMEMDT 19,231,79; Human Inadequate Evidence IMEMDT 19,231,79. NCI Carcinogenesis Bioassay (gavage); Inadequate Studies: mouse, rat NCITR* NCI-CG-TR-170,79; (gavage). Reported in EPA TSCA Inventory. EPA Genetic Toxicology Program. Community Right To Know List.

OSHA PEL: (Transitional: TWA 100 ppm; CL 200; Pk 600/5M/3H) TWA 50 ppm; STEL: 100 ppm
ACGIH TLV: TWA 50 ppm; STEL: 100 ppm (skin); BEI: 1 g(mandelic acid)/L in urine at end of shift; 40 ppb styrene in mixed-exhaled air prior to shift; 18 ppm styrene in mixed-exhaled air during shift; 0.55 mg/L styrene in blood end of shift; 0.02 mg/L styrene in blood prior to shift.
DFG MAK: 20 ppm (85 mg/m³); BAT: 2g/L of mandelic acid in urine at end of shift.
NIOSH REL: (Styrene) TWA 50 ppm; CL 100 ppm
DOT Classification: Flammable Liquid; Label: Flammable Liquid; Flammable or Combustible Liquid; Label: Flammable Liquid

SAFETY PROFILE: Suspected carcinogen. Experimental poison by ingestion, inhalation, and intravenous routes. Moderately toxic experimentally by intraperitoneal route. Mildly toxic to humans by inhalation. An experimental teratogen. Human systemic effects by inhalation: eye and olfactory changes. It can cause irritation and violent itching of the eyes @ 200 ppm, lacrimation, and severe human eye injuries. Its toxic effects are usually transient and result in irritation and possible narcosis. Experimental reproductive effects. Human mutation data reported. A human skin irritant. An experimental skin and eye irritant.

The monomer has been involved in several industrial explosions. It is a storage hazard above 32°C. A very dangerous fire hazard when exposed to flame, heat, or oxidants. Explosive in the form of vapor when exposed to heat or flame. Reacts with oxygen above 40°C to form a heat-sensitive explosive peroxide. Violent or explosive polymerization may be initiated by alkali metal-graphite composites; butyllithium; dibenzoyl peroxide; other initiators (e.g., azoisobutyronitrile; di-tert-butyl peroxide). Reacts violently with chlorosulfonic acid; oleum; sulfuric acid; chlorine + iron(III) chloride (above 50°C). May ignite when heated with air + polymerizing polystyrene. Can react vigorously with oxidizing materials. To fight fire, use foam, CO_2, dry chemical. When heated to decomposition it emits acrid smoke and irritating fumes.

SMQ500 CAS:9003-53-6 ***HR: 2***
STYRENE POLYMER
mf: $(C_8H_8)_n$
DOT: UN 2211

SYNS: A 3-80 ◇ AFCOLENE ◇ ATACTIC POLYSTYRENE ◇ BACTOLATEX ◇ BAKELITE SMD 3500 ◇ BASF III ◇ BEXTRENE XL 750 ◇ BICOLASTIC A 75 ◇ BUSTREN ◇ CADCO 0115 ◇ CARINEX GP ◇ COPAL Z ◇ COSDEN 550 ◇ DENKA QP3 ◇ DIAREX 43G ◇ DORVON ◇ DOW 860 ◇ DYLENE ◇ DYLITE F 40 ◇ ESBRITE ◇ ESCOREZ 7404 ◇ ESTYRENE G 20 ◇ ETHENYLBENZENE HOMOPOLYMER ◇ FOSTER GRANT 834 ◇ GEDEX ◇ HI-STYROL ◇ HOSTYREN S ◇ HT-F 76 ◇ IT 40 ◇ KB (POLYMER) ◇ KRASTEN 1.4 ◇ LACQREN 550 ◇ LUSTREX ◇ MX 5517-02 ◇ NBS 706 ◇ OWISPOL GF ◇ PICCOLASTIC ◇ POLIGOSTYRENE ◇ POLYSTROL D ◇ POLYSTYRENE ◇ POLYSTYRENE BEADS (DOT) ◇ POLYSTYRENE LATEX ◇ POLYSTYROL ◇ PRINTEL'S ◇ REXOLITE 1422 ◇ RHODOLNE ◇ SHELL 300 ◇ STYRAFOIL ◇ STYRAGEL ◇ STYRENE POLYMERS ◇ STYROFOAM ◇ STYROLUX ◇ STYRON ◇ TOPOREX 855-51 ◇ TROLITUL ◇ UBATOL U 2001 ◇ VESTYRON ◇ VINYLBENZENE POLYMER ◇ VINYL PRODUCTS R 3612

TOXICITY DATA with REFERENCE
imp-rat TDLo:19 mg/kg:ETA CNREA8 15,333,55
ihl-mus LC50:120 mg/m^3/10M APFRAD 35,461,77

CONSENSUS REPORTS: IARC Cancer Review: Group 3 IMEMDT 7,56,87; Animal Limited Evidence IMEMDT 19,231,79. Reported in EPA TSCA Inventory.

DOT Classification: ORM; Label: None.

SAFETY PROFILE: Questionable carcinogen with experimental tumorigenic data by implant. When heated to decomposition it emits acrid smoke and irritating fumes. See also POLYMERS, INSOLUBLE.

SMR000 CAS:9003-55-8 ***HR: 1***
STYRENE POLYMER with 1,3-BUTADIENE

SYNS: AFCOLAC B 101 ◇ ANDREZ ◇ BASE 661 ◇ 1,3-BUTADIENE-STYRENE COPOLYMER ◇ BUTADIENE-STYRENE POLYMER ◇ 1,3-BUTADIENE-STYRENE POLYMER ◇ BUTADIENE-STYRENE RESIN ◇ BUTADIENE-STYRENE RUBBER (FCC) ◇ BUTAKON 85-71 ◇ DIAREX 600 ◇ DIENOL S ◇ DOW 209 ◇ DOW LATEX 612 ◇ DST 50 ◇ DURANIT ◇ EDISTIR RB 268 ◇ ETHENYLBENZENE POLYMER with 1,3-BUTADIENE ◇ GOODRITE 1800X73 ◇ HISTYRENE S 6F ◇ HYCAR LX 407 ◇ K 55E ◇ KOPOLYMER BUTADIEN STYRENOVY (CZECH) ◇ KRO 1 ◇ LITEX CA ◇ LYTRON 5202 ◇ MARBON 9200 ◇ NIPOL 407 ◇ PHAROS 100.1 ◇ PLIOFLEX ◇ PLIOLITE S5 ◇ POLYBUTADIENE-POLYSTYRENE COPOLYMER ◇ POLYCO 2410 ◇ RICON 100 ◇ SBS ◇ SD 354 ◇ S6F HISTYRENE RESIN ◇ SKS 85 ◇ SOIL STABILIZER 661 ◇ SOLPRENE 300 ◇ STYRENE-BUTADIENE COPOLYMER ◇ STYRENE-1,3-BUTADIENE COPOLYMER ◇ STYRENE-BUTADIENE POLYMER ◇ SYNPOL 1500 ◇ THERMOPLASTIC 125 ◇ TR 201 ◇ UP 1E ◇ VESTYRON HI

TOXICITY DATA with REFERENCE
eye-rbt 500 mg/24H MLD 28ZPAK -,257,72

CONSENSUS REPORTS: IARC Cancer Review: Group 3 IMEMDT 7,56,87; Human Inadequate Evidence IMEMDT 19,231,79. Reported in EPA TSCA Inventory.

SAFETY PROFILE: An eye irritant. Questionable carcinogen. When heated to decomposition it emits acrid smoke and irritating fumes. See also POLYMERS, INSOLUBLE.

SMR500 CAS:841-18-9 ***HR: 3***
trans-4'-STYRYLACETANILIDE
mf: $C_{16}H_{15}NO$ mw: 237.32

SYNS: 4-ACETAMIDOSTILBENE ◇ trans-4-ACETAMIDOSTILBENE ◇ trans-4-ACETAMINOSTILBENE ◇ 4-ACETYLAMINOSTILBENE ◇ trans-4-ACETYLAMINOSTILBENE ◇ N-(4-(2-PHENYLETHENYL)PHENYL)ACETAMIDE

TOXICITY DATA with REFERENCE
mma-sat 5 nmol/plate JMCMAR 25,593,82
dnr-ham:fbr 1 μmol/L JNCIAM 54,1287,75
dnd-rat-orl 25 μmol/kg CBINA8 24,355,79
oms-rat-orl 25 μmol/kg CBINA8 24,355,79
orl-rat TDLo:48 mg/kg/3W-I:CAR CRNGDP 4,1519,83
scu-rat TDLo:31 mg/kg/26W-I:CAR CNREA8 24,128,64
scu-rat TD:27 mg/kg/13W-I:ETA PTRMAD 241,147,48

CONSENSUS REPORTS: EPA Genetic Toxicology Program.

SAFETY PROFILE: Questionable carcinogen with experimental carcinogenic and tumorigenic data. Mutation data reported. When heated to decomposition it emits toxic fumes of NO_x.

SMS000 CAS:63019-73-8 ***HR: 3***
5-STYRYL-3,4-BENZOPYRENE
mf: $C_{28}H_{18}$ mw: 354.46

SYN: 6-STYRYL-BENZO(a)PYRENE

TOXICITY DATA with REFERENCE
scu-mus TDLo:200 mg/kg:ETA COREAF 245,876,57

SAFETY PROFILE: Questionable carcinogen with ex-

perimental tumorigenic data. When heated to decomposition it emits acrid smoke and irritating fumes.

SMS500 CAS:122-57-6 ***HR: 3***
STYRYL METHYL KETONE
mf: $C_{10}H_{10}O$ mw: 146.20

PROP: Plates. D: 1.035 @ 20°/2°, mp: 40-42°, bp: 260-262°. Sol in H_2SO_4, alc, chloroform, ether, benzene

SYNS: BENZALACETON (GERMAN) ◇ BENZALACETONE ◇ BENZYLIDENE ACETONE ◇ 4-PHENYL-3-BUTEN-2-ONE

TOXICITY DATA with REFERENCE
skn-rbt 500 mg/24H MLD FCTXAV 11,1021,73
skn-gpg 1%/48H MOD JSCCA5 28,357,77
ipr-mus LD50:1210 mg/kg ARZNAD 19,617,69
ivn-mus LD50:112 mg/kg ARZNAD 19,617,69

CONSENSUS REPORTS: Reported in EPA TSCA Inventory.

SAFETY PROFILE: Poison by intravenous route. Moderately toxic by intraperitoneal route. A skin irritant. When heated to decomposition it emits acrid smoke and irritating fumes. See also KETONES.

SMT000 CAS:18559-95-0 ***HR: 3***
N-(p-STYRYLPHENYL)ACETOHYDROXAMIC ACID
mf: $C_{16}H_{15}NO_2$ mw: 253.32

SYNS: ACETIC ACID, ESTER with N-(p-STYRYLPHENYL) ACETOHYDROXAMIC ACID ◇ N-ACETOXY-4-ACETAMIDO-STILBENE ◇ 4-(N-HYDROXYACETAMIDO)STILBENE ◇ N-HYDROXY-4-ACETYLAMINOSTILBENE ◇ N-HYDROXY-N-(4-(2-PHENYL-ETHENYL)PHENYL)ACETAMIDE

TOXICITY DATA with REFERENCE
dns-hmn:fbr 10 μmol/L/5H IJCNAW 16,284,75
ivn-mus LD50:100 mg/kg CSLNX* NX#02989

CONSENSUS REPORTS: EPA Genetic Toxicology Program.

SAFETY PROFILE: Poison by intravenous route. Human mutation data reported. When heated to decomposition it emits toxic fumes of NO_x.

SMT500 CAS:843-23-2 ***HR: 3***
trans-N-(p-STYRYLPHENYL)ACETOHYDROXAMIC ACID
mf: $C_{16}H_{15}NO_2$ mw: 253.32

SYNS: HAAS ◇ trans-N-HYDROXY-4-ACETAMIDOSTILBENE ◇ trans-N-HYDROXY-4-ACETYLAMINOSTILBENE ◇ trans-N-HYDROXY-ASS

TOXICITY DATA with REFERENCE
mma-sat 2500 pmol/plate JMCMAR 25,593,82
mrc-smc 10 ppm ZEKBAI 74,412,70
dnd-rat-ipr 40 mg/kg CRNGDP 5,231,84
orl-rat TDLo:140 mg/kg/20W-I:CAR ZEKBAI 74,200,70
scu-rat TDLo:34 mg/kg/4W-I:CAR CNREA8 24,128,64

SAFETY PROFILE: Questionable carcinogen with experimental carcinogenic data. Mutation data reported. When heated to decomposition it emits toxic fumes of NO_x.

SMU000 CAS:63021-62-5 ***HR: 3***
trans-N-(p-STYRYLPHENYL)ACETOHYDROXAMIC ACID, COPPER(2+) COMPLEX
mf: $C_{32}H_{28}N_2O_4{\cdot}Cu$ mw: 568.16

SYNS: COPPER complex with trans-N-(p-STYRYLPHENYL)ACETOHYDROXAMIC ACID ◇ trans-N-HYDROXY-4-ACETYLAMINOSTILBENE CUPRIC CHELATE

TOXICITY DATA with REFERENCE
scu-rat TDLo:30 mg/kg/1W-I:CAR CNREA8 24,128,64
scu-rat TD:75 mg/kg/4W-I:CAR CNREA8 24,128,64

CONSENSUS REPORTS: Copper and its compounds are on the Community Right To Know List.

SAFETY PROFILE: Questionable carcinogen with experimental carcinogenic data. When heated to decomposition it emits toxic fumes of NO_x. See also COPPER COMPOUNDS.

SMV000 CAS:502-42-1 ***HR: 2***
SUBERONE
mf: $C_7H_{12}O$ mw: 112.2

PROP: Liquid, nearly water-insol. Bp: 181°, d: 0.9490 @ 20°/4°. Sol in alc and ether.

SYN: CYCLOHEPTANONE

TOXICITY DATA with REFERENCE
ipr-mus LD50:750 mg/kg COREAF 254,2245,62
scu-mus LDLo:930 mg/kg AEXPBL 50,199,1903

SAFETY PROFILE: Moderately toxic by subcutaneous and intraperitoneal routes. Can cause central nervous system depression. When heated to decomposition it emits acrid smoke and irritating fumes.

SMV500 CAS:1121-92-2 ***HR: 3***
SUBERONE OXIME
mf: $C_7H_{13}NO$ mw: 127.21

SYNS: OCTAHYDROAZOCINE ◇ SUBERONISOXIM (GERMAN)

TOXICITY DATA with REFERENCE
scu-frg LDLo:250 mg/kg AEXPBL 50,199,1903

CONSENSUS REPORTS: Reported in EPA TSCA Inventory.

SAFETY PROFILE: Poison by subcutaneous route. When heated to decomposition it emits toxic fumes of NO_x.

SMW000 CAS:60172-05-6 ***HR: 3***
4,4'-(SUBEROYLBIS(IMINO-p-PHENYLENEIMINO))BIS(1-ETHYLPYRIDINIUM) DIBROMIDE
mf: $C_{34}H_{42}N_6O_2$•2Br mw: 726.64

TOXICITY DATA with REFERENCE
dnd-mus:lym 40 μmol/L JMCMAR 22,134,79
ipr-mus LD10:10 mg/kg JMCMAR 22,134,79

SAFETY PROFILE: Poison by intraperitoneal route. Mutation data reported. When heated to decomposition it emits very toxic fumes of Br^- and NO_x.

SMW400 ***HR: 1***
SUBTIGEN

SYN: SV-052

TOXICITY DATA with REFERENCE
scu-mus TDLo:60 g/kg (female 7-12D post):TER TOIZAG 20,369,73
orl-rat LD50:17 g/kg TOIZAG 16,258,69
ipr-rat LD50:4790 mg/kg TOIZAG 16,258,69
scu-rat LD50:13 g/kg TOIZAG 16,258,69
orl-mus LD50:21 g/kg TOIZAG 16,258,69
ipr-mus LD50:5730 mg/kg TOIZAG 16,258,69
scu-mus LD50:11 g/kg TOIZAG 16,258,69

SAFETY PROFILE: Mildly toxic by intraperitoneal route. An experimental teratogen. When heated to decomposition it emits acrid smoke and irritating fumes.

SMW500 CAS:1393-38-0 ***HR: 3***
SUBTILIN
mf: $C_{148}H_{227}N_{39}O_{38}S_5$ mw: 3321.44

TOXICITY DATA with REFERENCE
orl-mus LDLo:5000 mg/kg SCIEAS 103,419,46
scu-mus LD50:670 mg/kg SCIEAS 103,419,46
ivn-mus LD50:60 mg/kg SCIEAS 103,419,46

SAFETY PROFILE: Poison by intravenous route. Moderately toxic by subcutaneous route. Mildly toxic by ingestion. When heated to decomposition it emits very toxic fumes of SO_x and NO_x.

SMX000 CAS:5450-96-4 ***HR: 2***
SUCCINALDEHYDE DISODIUM BISULFITE
mf: $C_4H_8O_8S_2$•2Na mw: 294.22

SYN: 1,4-DIHYDROXY-1,4-BUTANEDISULFONIC ACID, DISODIUM SALT

TOXICITY DATA with REFERENCE
orl-rat LDLo:3200 mg/kg KODAK* -,-,71

CONSENSUS REPORTS: Reported in EPA TSCA Inventory.

SAFETY PROFILE: Moderately toxic by ingestion. When heated to decomposition it emits toxic fumes of SO_x and Na_2O. See also ALDEHYDES.

SMY000 CAS:110-15-6 ***HR: 2***
SUCCINIC ACID
mf: $C_4H_6O_4$ mw: 118.10

PROP: Colorless or white crystals; odorless with acid taste. Mp: 185°, bp: 235° (decomp), d: 1.564 @ 15°/4°. Sol in water; very sol in alc, ether, acetone, glycerin.

SYNS: AMBER ACID ◇ BERNSTEINSAURE (GERMAN) ◇ BUTANEDIOIC ACID ◇ 1,2-ETHANEDICARBOXYLIC ACID ◇ ETHYLENESUCCINIC ACID

TOXICITY DATA with REFERENCE
eye-rbt 1179 μg SEV AJOPAA 29,1363,46
orl-rat LD50:2260 mg/kg KODAK* #82-0158

CONSENSUS REPORTS: Reported in EPA TSCA Inventory.

SAFETY PROFILE: Moderately toxic by subcutaneous route. A severe eye irritant. When heated to decomposition it emits acrid smoke and irritating fumes.

SMZ000 CAS:10058-20-5 ***HR: 1***
SUCCINIC ACID, BIS(2-(HEXYLOXY)ETHYL) ESTER
mf: $C_{20}H_{38}O_6$ mw: 374.58

SYN: SUCCINIC ACID, DI-2-HEXYLOXYETHYL ESTER

TOXICITY DATA with REFERENCE
skn-rbt 10 mg/24H open MLD AMIHBC 10,61,54
eye-rbt 500 mg open AMIHBC 10,61,54
orl-rat LD50:4280 mg/kg AMIHBC 10,61,54
skn-rbt LD50:12310 mg/kg AMIHBC 10,61,54

SAFETY PROFILE: Mildly toxic by ingestion and skin contact. A skin and eye irritant. When heated to decomposition it emits acrid smoke and irritating fumes. See also ESTERS.

SNA500 CAS:141-03-7 ***HR: 1***
SUCCINIC ACID, DIBUTYL ESTER
mf: $C_{12}H_{22}O_4$ mw: 230.34

SYNS: BUTYL BUTANEDIOATE ◇ DIBUTYL SUCCINATE ◇ DI-N-BUTYLSUCCINATE ◇ SUCCINIC ACID DI-N-BUTYL ESTER

TOXICITY DATA with REFERENCE
orl-rat LD50:8000 mg/kg FMCHA2 -,D296,80

CONSENSUS REPORTS: Reported in EPA TSCA Inventory.

SAFETY PROFILE: Mildly toxic by ingestion. When heated to decomposition it emits acrid smoke and irritating fumes. See also ESTERS.

SNB000 CAS:123-25-1 ***HR: 1***
SUCCINIC ACID, DIETHYL ESTER
mf: $C_8H_{14}O_4$ mw: 174.22

$(-CH_2CO{\bullet}OCH_2CH_3)_2$

PROP: Colorless, mobile liquid; pleasant odor. Flash p: 230°F. Sol in alc, ether, fixed oils, water.

SYNS: BUTANEDIOIC ACID, DIETHYL ESTER ◇ DIETHYL SUCCINATE (FCC) ◇ ETHYL SUCCINATE ◇ FEMA No. 2377

TOXICITY DATA with REFERENCE
skn-rat 500 mg/24H MLD FCTXAV 16,637,78
eye-rat 500 mg/24H MLD FCTXAV 16,637,78
skn-rbt 500 mg/24H MLD AMIHBC 4,119,51
eye-rbt 500 mg/24H MLD AMIHBC 4,119,51
orl-rat LD50:8530 mg/kg AMIHBC 4,119,51

CONSENSUS REPORTS: Reported in EPA TSCA Inventory.

SAFETY PROFILE: Mildly toxic by ingestion. A skin and eye irritant. Combustible liquid. Reaction with ethyl trifluoroacetate + sodium hydride may cause a fire or explosion. When heated to decomposition it emits acrid smoke and irritating fumes. See also ESTERS.

SNB500 CAS:60634-59-5 ***HR: 3***
SUCCINIC ACID, (4-ETHOXY-1-NAPTHYLCARBONYLMETHYL) ESTER
mf: $C_{18}H_{18}O_6$ mw: 330.36

SYNS: 1-AETHOXY-4-(1-CAETO-2-HYDROXYAETHYL)-NAPHTALAENE SUCCINATE ◇ 1-ETHOXY-4-(1-KETO-2-HYDROXYETHYL)-NAPHTHALENE SUCCINATE

TOXICITY DATA with REFERENCE
orl-rat LD50:3500 mg/kg NEPSBV 1,286,59
ipr-rat LD50:290 mg/kg NEPSBV 1,286,59

SAFETY PROFILE: Poison by intraperitoneal route. Moderately toxic by ingestion. When heated to decomposition it emits acrid smoke and irritating fumes. See also ESTERS.

SNC000 CAS:108-30-5 ***HR: 3***
SUCCINIC ANHYDRIDE
mf: $C_4H_4O_3$ mw: 100.08

PROP: Colorless needles. Mp: 119.6°, bp: 261°, d: 1.104, vap press: 1 mm @ 92.0°. Very sltly sol in water, petr ether; sltly sol in ether.

SYNS: BERNSTEINSAURE-ANHYDRID(GERMAN) ◇ BUTANEDIOIC ANHYDRIDE ◇ DIHYDRO-2,5-FURANDIONE ◇ 2,5-DIKETOTETRAHYDROFURAN ◇ NCI-C55696 ◇ SUCCINIC ACID ANHYDRIDE ◇ SUCCINYL OXIDE ◇ TETRAHYDRO-2,5-DIOXOFURAN

TOXICITY DATA with REFERENCE
eye-rbt 750 µg SEV AJOPAA 29,1363,46
mmo-esc 2 g/L MUREAV 130,97,84
otr-ham:emb 100 µg/L IJCNAW 19,642,77
ipr-mus TDLo:79 mg/kg (8-10D preg):TER TCMUD8 2,61,82
scu-rat TDLo:2600 mg/kg/65W-I:NEO BJCAAI 19,392,65
orl-rat LD50:1510 mg/kg EPASR* 8EHQ-0282-0433

CONSENSUS REPORTS: IARC Cancer Review: Group 3 IMEMDT 7,56,87; Animal Inadequate Evidence IMEMDT 15,265,77. Reported in EPA TSCA Inventory. EPA Genetic Toxicology Program.

SAFETY PROFILE: Experimental teratogenic effects. Moderately toxic by ingestion. A severe eye irritant. Mutation data reported. Questionable carcinogen with experimental neoplastigenic data. When heated to decomposition it emits acrid smoke and irritating fumes. See also ANHYDRIDES.

SNC500 CAS:123-23-9 ***HR: 3***
SUCCINIC PEROXIDE
DOT: UN 2135/UN 2962
mf: $C_8H_{10}O_8$ mw: 234.18

$(HOCO{\bullet}C_2H_4CO{\bullet}O-)_2$

PROP: Fine white powder, odorless with tart taste. Mp: 125° (decomp). Mod sol in water.

SYNS: BIS(3-CARBOXYPROPIONYL) PEROXIDE ◇ 3,3′-(DIOXYDICARBONYL)DIPROPIONIC ACID ◇ SUCCINIC ACID PEROXIDE (DOT) ◇ SUCCINYL PEROXIDE

TOXICITY DATA with REFERENCE
ipr-mus LD50:16 mg/kg CBCCT* 2,302,50

CONSENSUS REPORTS: Reported in EPA TSCA Inventory.

DOT Classification: Organic Peroxide; Label: Organic Peroxide

SAFETY PROFILE: Poison by intraperitoneal route. An irritant. Explodes on contact with flame. When heated to decomposition it emits acrid smoke and irritating fumes. See also PEROXIDES, ORGANIC.

SND000 CAS:123-56-8 ***HR: 3***
SUCCINIMIDE
mf: $C_4H_5NO_2$ mw: 99.10

PROP: Crystals in acetone. D: 1.412 @ 16°, mp: 125-126°, bp: 287-288°. Very sol in water; sol in alc; insol in chloroform and ether.

SYNS: BUTANIMIDE ◇ 3,4-DIHYDROPYRROLIDINE ◇ 2,5-DIKETOPYRROLIDINE ◇ 2,5-PYRROLIDINEDIONE ◇ SUCCINIC IMIDE

TOXICITY DATA with REFERENCE
orl-rat LD50:14 g/kg MDACAP 10,224,74
ivn-mus LD50:320 mg/kg CSLNX* NX#03763

CONSENSUS REPORTS: Reported in EPA TSCA Inventory.

SAFETY PROFILE: Poison by intravenous route. Mildly toxic by ingestion. When heated to decomposition it emits toxic fumes of NO_x.

SND500 CAS:128-09-6 ***HR: 3***
SUCCINOCHLORIMIDE
mf: $C_4H_4ClNO_2$ mw: 133.54

PROP: Rhombic crystals from benzene; odor of chlorine. D: 1.65, mp: 148°. Acid to litmus (1:50 aq soln). One gram dissolves in about 70 mL water, 150 mL alc, 50 mL benzene. Sparingly sol in ether, chloroform, carbon tetrachloride.

SYNS: 1-CHLORO-2,5-PYRROLIDINEDIONE ◇ N-CHLORSUCCINIMIDE ◇ SUCCINCHLORIMIDE

TOXICITY DATA with REFERENCE
scu-mus TDLo:840 mg/kg/70W-I:ETA JNCIAM 53,695,74
orl-rat LDLo:1000 mg/kg PHRPA6 59,541,44
ivn-rat LDLo:200 mg/kg PHRPA6 59,541,44

CONSENSUS REPORTS: EPA Genetic Toxicology Program.

SAFETY PROFILE: Poison by intravenous route. Moderately toxic by ingestion. Questionable carcinogen with experimental tumorigenic data. Stored as a dust it heats spontaneously. Explosive reaction with aliphatic alcohols, benzylamine or hydrazine hydrate. When heated to decomposition it emits toxic fumes of Cl^- and NO_x.

SNE000 CAS:110-61-2 ***HR: 3***
SUCCINONITRILE
mf: $C_4H_4N_2$ mw: 80.10

$(-CH_2C\equiv N)_2$

PROP: Colorless, odorless, waxy material. Mp: 58.1°, bp: 267°, flash p: 270°F, d: 1.022 @ 25°, vap press: 2 mm @ 100°, vap d: 2.1. Sltly sol in ether, water, alc; sol in acetone.

SYNS: 1,4-BUTANEDINITRILE ◇ DEPRELIN ◇ s-DICYANOETHANE ◇ 1,2-DICYANOETHANE ◇ DINILE ◇ ETHYLENE CYANIDE ◇ ETHYLENE DICYANIDE ◇ SUCCINIC ACID DINITRILE ◇ SUCCINIC DINITRILE ◇ SUCCINODINITRILE ◇ SUXIL ◇ USAF A-9442

TOXICITY DATA with REFERENCE
ipr-ham TDLo:365 mg/kg (8D preg):TER FAATDF 3,41,83
orl-rat LD50:450 mg/kg 38MKAJ 2C,4879,82
orl-mus LD50:129 mg/kg ARTODN 57,88,85
ipr-mus LD50:63110 μg/kg BCPCA6 27,1135,78
scu-dog LDLo:150 mg/kg AIPTAK 3,77,1897
scu-rbt LDLo:36 mg/kg AIPTAK 3,77,1897
scu-pgn LDLo:2200 mg/kg AIPTAK 3,77,1897
scu-frg LDLo:1000 mg/kg AIPTAK 3,77,1897

CONSENSUS REPORTS: Cyanide and its compounds are on the Community Right To Know List. Reported in EPA TSCA Inventory.
NIOSH REL: (Nitriles) TWA 20 mg/m^3

SAFETY PROFILE: Poison by ingestion, intraperitoneal, and subcutaneous routes. An experimental teratogen. Combustible when exposed to heat or flame. Decomposes exothermically above 195°C. Can react with oxidizing materials. To fight fire, use alcohol foam, CO_2, dry chemical. When heated to decomposition, or on contact with acid or acid fumes, it emits highly toxic fumes of NO_x and CN^-. See also NITRILES.

SNF000 CAS:40428-75-9 ***HR: 3***
SUCCINOYL DIAZIDE
mf: $C_4H_4N_6O_2$ mw: 168.12

$(-CH_2CO{\bullet}N_3)_2$

SAFETY PROFILE: An sensitive explosive. When heated to decomposition it emits toxic fumes of NO_x. An intermediate in the preparation of ethylene diisocyanate. See also AZIDES.

SNF500 CAS:63042-13-7 ***HR: 3***
4'-SUCCINYLAMINO-2,3'-DIMETHYLAZOBENZOL
mf: $C_{18}H_{19}N_3O_3$ mw: 325.40

SYNS: 4'-SUCCINOYLAMINO-2,3'-DIMETHYLAZOBENZENE ◇ N-(4'-o-TOLYL-o-TOLYLAZOSUCCINAMIC ACID

TOXICITY DATA with REFERENCE
orl-rat TDLo:94 g/kg/54W-C:ETA GANNA2 33,196,39

SAFETY PROFILE: Questionable carcinogen with experimental tumorigenic data. When heated to decomposition it emits toxic fumes of NO_x.

SNG000 CAS:543-20-4 ***HR: 3***
SUCCINYL CHLORIDE
mf: $C_4H_4Cl_2O_2$ mw: 154.98

PROP: Colorless crystals. D: 1.377 @ 20°/4°, mp: 16.7°, bp: 192-193°. Decomp in water, alc; sol in benzene; insol in petr ether.

SYNS: SUCCINIC ACID DICHLORIDE ◇ SUCCINIC CHLORIDE ◇ SUCCINOYL CHLORIDE ◇ SUCCINYL DICHLORIDE

TOXICITY DATA with REFERENCE
ipr-mus LD50:62 mg/kg CBCCT* 4,111,52

CONSENSUS REPORTS: Reported in EPA TSCA Inventory.

SAFETY PROFILE: Poison by intraperitoneal route.

When heated to decomposition it emits toxic fumes of Cl^-. See also CHLORIDES.

SNG500 CAS:63979-84-0 ***HR: 3***
SUCCINYLNITRILE
mf: $C_6H_4N_2O_2$ mw: 136.12

TOXICITY DATA with REFERENCE
scu-rbt LDLo:36 mg/kg AIPTAK 3,77,1897
scu-frg LDLo:1000 mg/kg AIPTAK 3,77,1897

CONSENSUS REPORTS: Cyanide and its compounds are on the Community Right To Know List.

SAFETY PROFILE: Poison by subcutaneous route. When heated to decomposition it emits toxic fumes of NO_x and CN^-. See also NITRILES.

SNH000 CAS:57-50-1 ***HR: 1***
SUCROSE
mf: $C_{12}H_{22}O_{11}$ mw: 342.34

PROP: White crystals; sweet taste. D: 1.587 @ 25°/4°, mp: 170-186° (decomp). Sol in water, alc; insol in ether.

SYNS: BEET SUGAR ◇ CANE SUGAR ◇ CONFECTIONER'S SUGAR ◇ α-d-GLUCOPYRANOSYL β-d-FRUCTOFURANOSIDE ◇ (α-d-GLUCOSIDO)-β-d-FRUCTOFURANOSIDE ◇ GRANULATED SUGAR ◇ NCI-C56597 ◇ ROCK CANDY ◇ SACCHAROSE ◇ SACCHARUM ◇ SUGAR

TOXICITY DATA with REFERENCE
mma-sat 600 μg/plate PMRSDJ 1,343,81
dnr-smc 300 mg/L PMRSDJ 1,502,81
cyt-ham:lng 10 g/L ATSUDG (4),41,80
orl-mam TDLo:54810 mg/kg (15-35D preg):TER TJADAB 30,203,84
orl-rat LD50:29700 mg/kg TXAPA9 7,609,65
ipr-mus LD50:14000 mg/kg PCJOAU 15,139,81
orl-dom LDLo:40 g/kg NAREA4 30,503,60

CONSENSUS REPORTS: Reported in EPA TSCA Inventory. EPA Genetic Toxicology Program.

OSHA PEL: TWA Total Dust: 15 mg/m^3; Respirable Fraction: 5 mg/m^3
ACGIH TLV: TWA (nuisance particulate) 10 mg/m^3 of total dust (when toxic impurities are not present, e.g., quartz < 1%).

SAFETY PROFILE: Mildly toxic by ingestion. An experimental teratogen. Mutation data reported. Vigorous reaction with nitric acid or sulfuric acid (forms carbon monoxide and carbon dioxide). When heated to decomposition it emits acrid smoke and irritating fumes.

SNH150 CAS:60561-17-3 ***HR: 3***
SUFENTANIL CITRATE
mf: $C_{22}H_{30}N_2O_2S \cdot C_6H_8O_7$ mw: 578.74

SYNS: N-(4-(METHOXYMETHYL)-1-(2-(2-THIENYL)ETHYL)-4-PIPERIDINYL)-N-PHENYLPROPIONAMIDE CITRATE ◇ N-(4-(METHOXYMETHYL)-1-(2-(2-THIENYL)ETHYL)-4-PIPERIDYL)PROPIONANILIDE CITRATE ◇ R 30 730 CITRATE SALT ◇ SUFENTA ◇ SUFENTANYL ◇ SUFENTANYL CITRATE ◇ SULFENTANIL

TOXICITY DATA with REFERENCE
ivn-rat LD50:17900 μg/kg ARZNAD 26,1548,76
ivn-mus LD50:18700 μg/kg ARZNAD 26,1548,76
ivn-dog LD50:14100 μg/kg ARZNAD 26,1548,76

SAFETY PROFILE: Poison by intravenous route. When heated to decomposition it emits toxic fumes of NO_x and SO_x.

SNH450 ***HR: D***
SUKHTEH

PROP: Opium pipe residues (LANCAO 2,494,78).

TOXICITY DATA with REFERENCE
sce-hmn:lym 10 mg/L CRNGDP 4,227,83
otr-ham:emb 70 mg/L MUREAV 150,177,85

SAFETY PROFILE: Human mutation data reported. See also OPIUM.

SNH480 CAS:61318-91-0 ***HR: 2***
SULCONAZOLE NITRATE
mf: $C_{18}H_{15}Cl_3N_2S \cdot HNO_3$ mw: 460.78

SYNS: (±)-1-(2-(((4-CHLOROPHENYL)METHYL)THIO)-2-(2,4-DICHLOROPHENYL)ETHYL)-1H-IMIDAZOLE MONONITRATE ◇ (±)-1-(2,4-DICHLORO-β-((4-CHLOROBENZYL)THIO)PHENETHYL)IMIDAZOLE NITRATE ◇ RS 44872

TOXICITY DATA with REFERENCE
scu-rat TDLo:33 mg/kg (female 7-17D post):REP OYYAA2 30,451,85
orl-rat LD50:1741 mg/kg IYKEDH 17,365,86
ipr-rat LD50:735 mg/kg IYKEDH 17,365,86
orl-mus LD50:2475 mg/kg IYKEDH 17,365,86
ipr-mus LD50:810 mg/kg IYKEDH 17,365,86

SAFETY PROFILE: Moderately toxic by ingestion and intraperitoneal routes. Experimental reproductive effects. An antimycotic agent. When heated to decomposition it emits toxic fumes of Cl^-, SO_x, and NO_x. See also NITRATES.

SNH800 CAS:127-71-9 ***HR: 3***
SULFABENZAMIDE
mf: $C_{13}H_{12}N_2O_3S$ mw: 276.33

SYNS: N^1-BENZOYLSULFANILAMIDE ◇ SULFABENZID ◇ SULFABENZIDE ◇ SULFABENZOYLAMIDE ◇ N-SULFAMYLBENZAMIDE ◇ N-SULFANILYLBENZAMIDE

TOXICITY DATA with REFERENCE
ivn-mus LD50:320 mg/kg CSLNX* NX#03348

CONSENSUS REPORTS: Reported in EPA TSCA Inventory.

SAFETY PROFILE: Poison by intravenous route. When heated to decomposition it emits very toxic fumes of NO_x and SO_x.

SNI000 CAS:58098-08-1 ***HR: 3***
SULFACOMBIN

PROP: Mixture of sulfamerazine (26%), sulfadiazine (37%), and sulfathiazole (37%).

TOXICITY DATA with REFERENCE
par-rat TDLo:400 mg/kg:ETA ACRAAX 37,258,52

SAFETY PROFILE: Questionable carcinogen with experimental tumorigenic data. When heated to decomposition it emits very toxic fumes of NO_x and SO_x. See also components as listed above.

SNI425 CAS:22199-08-2 ***HR: 3***
SULFADIAZINE SILVER SALT
mf: $C_{10}H_{10}N_4O_2S{\cdot}Ag$ mw: 358.17

SYNS: 4-AMINO-N-2-PYRIMIDINYL-BENZENESULFONAMIDE MONOSILVER(1+) SALT ◇ 4-AMINO-N-(2-PYRIMIDINYL) BENZENESULFONAMIDE SILVER SALT ◇ FLAMAZINE ◇ SILVADENE ◇ SILVER SULFADIAZINE ◇ SILVER SULPHADIAZINE ◇ SULFADIAZINE SILVER

TOXICITY DATA with REFERENCE
dnd-omi 5 mg/L AMACCQ 2,367,72
dns-omi 4 mg/L PAMIAD 39,434,73
dni-omi 2500 μg/L AMACCQ 2,367,72
scu-rat TDLo:11 g/kg (female 7-17D post):REP YACHDS 8,3637,80
scu-rat TDLo:11 g/kg (female 7-17D post):TER YACHDS 8,3637,80
ipr-rat LD50:95 mg/kg NIIRDN 6,APP-7,82
orl-mus LDLo:5000 mg/kg NIIRDN 6,APP-7,82
ipr-mus LD50:100 mg/kg IYKEDH 13,637,82

CONSENSUS REPORTS: Silver and its compounds are on the Community Right To Know List. EPA Genetic Toxicology Program.

OSHA PEL: TWA 0.01 mg(Ag)/m^3
ACGIH TLV: TWA 0.01 mg(Ag)/m^3

SAFETY PROFILE: Poison by intraperitoneal route. Mildly toxic by ingestion. An experimental teratogen. Experimental reproductive effects. Mutation data reported. When heated to decomposition it emits toxic fumes of SO_x and NO_x. See also SILVER COMPOUNDS.

SNI500 CAS:155-91-9 ***HR: D***
SULFADIMETHOXYPYRIMIDINE
mf: $C_{12}H_{14}N_4O_4S$ mw: 310.36

SYNS: 2-p-AMINOBENZENESULPHONAMIDO-4,6-DIMETHOXYPYRIMIDINE ◇ 4-AMINO-N-(4,6-DIMETHOXY-2-PYRIMIDINYL)BENZENESUL-FONAMIDE ◇ N^1-(4,6-DIMETHOXY-PYRIMIDIN-2-YL)SULFANILAMIDE ◇ ICI 3435 ◇ SDmP ◇ SULPHAMOPRINE

TOXICITY DATA with REFERENCE
orl-rat TDLo:538 mg/kg (1-22D preg/21D post):REP BJPCAL 23,305,64
orl-rat TDLo:538 mg/kg (1-22D preg/21D post):TER BJPCAL 23,305,64

SAFETY PROFILE: An experimental teratogen. Experimental reproductive effects. When heated to decomposition it emits very toxic fumes of NO_x and SO_x.

SNJ000 CAS:57-68-1 ***HR: 3***
SULFADIMETHYLDIAZINE
mf: $C_{12}H_{14}N_4O_2S$ mw: 278.36

PROP: Crystals; odorless. Mp: 176° (also a range reported of from 178-179°, 198-199°, and 205-207°). Sol in acetone, water, ether; sltly sol in alc.

SYNS: A-502 ◇ 2-(p-AMINOBENZENESULFONAMIDO)-4,6-DIMETHYLPYRIMIDINE ◇ 6-(4'-AMINOBENZOL-SULFONAMIDO)-2,4-DIMETHYLPYRIMIDIN (GERMAN) ◇ (p-AMINOBENZOLSULFONYL)-2-AMINO-4,6-DIMETHYLPYRIMIDIN (GERMAN) ◇ AZOLMETAZIN ◇ CREMOMETHAZINE ◇ DIAZYL ◇ N^1-(4,6-DIMETHYL-2-PYRIMIDINYL)SULFANILAMIDE ◇ N-(4,6-DIMETHYL-2-PYRIMIDYL)SULFANILAMIDE ◇ 4,6-DIMETHYL-2-SULFANILAMIDOPYRIMIDINE ◇ DIMEZATHINE ◇ MERMETH ◇ METAZIN ◇ NCI-C56600 ◇ NEASINA ◇ PIRMAZIN ◇ PRIMAZIN ◇ SA 111 ◇ SEAZINA ◇ SPANBOLET ◇ SULFADIMERAZINE ◇ SULFADIMETHYLPYRIMIDINE ◇ SULFADIMETINE ◇ SULFADIMEZINE ◇ SULFADIMIDINE ◇ SULFADINE ◇ SULFADSIMESINE ◇ SULFA-ISODIMERAZINE ◇ SULFAISODIMIDINE ◇ SULFAMETHIAZINE ◇ SULFAMETHIN ◇ SULFAMEZATHINE ◇ 2-SULFANILAMIDO-4,6-DIMETHYLPYRIMIDINE ◇ SULFISOMIDIN ◇ SULFISOMIDINE ◇ SULFODIMESIN ◇ SULFODIMEZINE ◇ SULMET ◇ SULPHADIMETHYL-PYRIMIDINE ◇ SULPHADIMIDINE ◇ SUPERSEPTIL ◇ VERTOLAN

TOXICITY DATA with REFERENCE
orl-rbt TDLo:25200 mg/kg (female 6-19D post):TER NTIS** PB85-105690
orl-rat TDLo:46200 mg/kg (male 6W pre):REP JRPFA4 81,259,87
imp-rat TDLo:5000 mg/kg:ETA ACRAAX 37,258,52
orl-mus LD50:50 g/kg NIIRDN 6,392,82
ipr-mus LD50:1060 mg/kg AITEAT 16,804,68
ivn-mus LD50:1776 mg/kg HBTXAC 5,163,59

CONSENSUS REPORTS: Reported in EPA TSCA Inventory.

SAFETY PROFILE: Moderately toxic by intravenous and intraperitoneal routes. Mildly toxic by ingestion. Experimental teratogenic and reproductive effects. Questionable carcinogen with experimental tumorigenic data. When heated to decomposition it emits very toxic fumes of SO_x and NO_x.

SNJ350 CAS:515-64-0 ***HR: 3***
SULFAISODIMERAZINE
mf: $C_{12}H_{14}N_4O_2S$ mw: 278.36

PROP: Needles from ethanol. Mp: 243°. Solubility in water at 15°: 0.12 g/100 mL, at 30°: 0.30 g/100 mL. More sol in hot water (about 1:60). Aq solns are neutral to litmus. Sltly sol in alc and acetone; practically insol in benzene, ether, chloroform; freely sol in dil HCl and NaOH. Do not confuse with sulfamethazine.

SYNS: 6-(p-AMINOBENZENESULFONAMIDO)-2,4-DIMETHYLPYRIMIDINE ◇ 6-(4-AMINOBENZENESULFONAMIDO)-2,4-DIMETHYLPYRIMIDINE ◇ 6-(p-AMINOBENZENESULFONYL)AMINO-2,4-DIMETHYLPYRIMIDINE ◇ (p-AMINOBENZOLSULFONYL)-4-AMINO-2,6-DIMETHYLPYRIMIDIN (GERMAN) ◇ 4-AMINO-N-(2,6-DIMETHYL-4-PYRIMIDINYL)BENZENESULFONAMIDE(9CI) ◇ ARISTAMID ◇ ARISTAMIDE ◇ ARISTOGYN ◇ N[1]-(2,6-DIMETHYL-4-PYRIMIDINYL)SULFANILAMIDE ◇ 2,4-DIMETHYL-6-SULFANILAMIDOPYRIMIDINE ◇ 2,6-DIMETHYL-4-SULFANILAMIDOPYRIMIDINE ◇ DOMAIN ◇ ELCOSINE ◇ ELKOSIL ◇ ELKOSIN ◇ ELKOSINE ◇ MEFENAL ◇ 4-SULFA-2,6-DIMETHYLPYRIMIDINE ◇ SULFADIMETINE ◇ SULFAISODIMIDINE ◇ SULFAMETHIN ◇ 4-SULFANILAMIDO-2,6-DIMETHYLPYRIMIDINE ◇ 6-SULFANILAMIDO-2,4-DIMETHYLPYRIMIDINE ◇ SULFASOMIDINE ◇ SULFISOMIDIN ◇ SULFISOMIDINE ◇ SULPHASOMIDINE

TOXICITY DATA with REFERENCE
orl-rat TDLo:6 g/kg (9-14D preg):REP SEIJBO 13,7,73
orl-mus LD50:50 g/kg AIPTAK 94,338,53
ivn-mus LD50:36 mg/kg ARZNAD 3,66,53

SAFETY PROFILE: Poison by intravenous route. Mildly toxic by ingestion. Experimental reproductive effects. When heated to decomposition it emits toxic fumes of SO_x and NO_x.

SNJ400 CAS:62907-78-2 ***HR: 2***
SULFALENE N-METHYLGLUCAMINE SALT
mf: $C_7H_{17}NO_5 \cdot C_{11}H_{12}N_4O_3S$ mw: 475.58

SYN: N-METHYLGLUCAMINE and SULFALENE

TOXICITY DATA with REFERENCE
ivn-rat LD50:1600 mg/kg RPTOAN 46,30,83
ivn-mus LD50:870 mg/kg RPTOAN 46,30,83
ims-mus LD50:1450 mg/kg RPTOAN 46,30,83

SAFETY PROFILE: Moderately toxic by intravenous and intramuscular routes. When heated to decomposition it emits toxic fumes of SO_x and NO_x.

SNK000 CAS:723-46-6 ***HR: 3***
SULFAMETHOXAZOL
mf: $C_{10}H_{11}N_3O_3S$ mw: 253.30

SYNS: 4-AMINO-N-(5-METHYL-3-ISOXAZOLYL)BENZENESULFONAMIDE ◇ 3-(p-AMINOPHENYLSULPHONAMIDO)-5-METHYLISOXAZOLE ◇ AZO-GANTANOL ◇ BACTRIM ◇ CO-TRIMOXAZOLE ◇ EUSAPRIM ◇ FECTRIM ◇ GANTANOL ◇ N'-(5-METHYL-3-ISOXAZOLE)SULFANILAMIDE ◇ N'-(5-METHYL-3-ISOXAZOLYL)SULFANILAMIDE ◇ N'-(5-METHYLISOXAZOL-3-YL)SULPHANILAMIDE ◇ N[1]-(5-METHYL-3-ISOXAZOLYL)SULPHANILAMIDE ◇ 5-METHYL-3-SULFANILAMIDOISOXAZOLE ◇ 5-METHYL-3-SULPHANIL-AMIDOISOXAZOLE ◇ METOXAL ◇ MS 53 ◇ RADONIL ◇ RO 4-2130 ◇ SIM ◇ SINOMIN ◇ SEPTRA ◇ SEPTRAN ◇ SULFAMETHALAZOLE ◇ SULFAMETHOXAZOLE ◇ SULFAMETHYLISOXAZOLE ◇ 3-SULFANILAMIDO-5-METHYLISOXAZOLE ◇ SULFISOMEZOLE ◇ SULPHAMETHALAZOLE ◇ SULPHAMETHOXAZOL ◇ SULPHAMETHOXAZOLE ◇ SULPHAMETHYLISOXAZOLE ◇ 3-SULPHANILAMIDO-5-METHYLISOXAZOLE ◇ SULPHISOMEZOLE ◇ TRIB ◇ TRIMETOPRIM-SULFA

TOXICITY DATA with REFERENCE
orl-rat TDLo:21 g/kg/60W-C:ETA TXAPA9 24,351,73
orl-rat LD50:6370 mg/kg KSRNAM 12,1861,78
ipr-rat LD50:2690 mg/kg NIIRDN 6,388,82
orl-mus LD50:2650 mg/kg ARZNAD 15,1441,65
ipr-mus LD50:2300 mg/kg NIIRDN 6,388,82

CONSENSUS REPORTS: IARC Cancer Review: Group 3 IMEMDT 7,348,87; Human Inadequate Evidence IMEMDT 24,285,80; Animal Limited Evidence IMEMDT 24,285,80. Reported in EPA TSCA Inventory.

SAFETY PROFILE: Moderately toxic by ingestion and intraperitoneal routes. Questionable carcinogen with experimental tumorigenic data. When heated to decomposition it emits very toxic fumes of NO_x and SO_x.

SNK500 CAS:5329-14-6 ***HR: 3***
SULFAMIC ACID
mf: H_3NO_3S mw: 97.10
DOT: UN 2967

PROP: White crystals. Mp: 200° (decomp), bp: decomp, d: 203 @ 12°.

SYNS: AMIDOSULFONIC ACID ◇ AMIDOSULFURIC ACID ◇ AMINOSULFONIC ACID ◇ KYSELINA AMIDOSULFONOVA (CZECH) ◇ KYSELINA SULFAMINOVA (CZECH) ◇ SULFAMIDIC ACID ◇ SULPHAMIC ACID (DOT)

TOXICITY DATA with REFERENCE
skn-hmn 4%/5D-I MLD JIHTAB 25,26,43
skn-rbt 500 mg/24H SEV 28ZPAK -,18,72
eye-rbt 20 mg MOD JIHTAB 25,26,43
eye-rbt 250 μg/24H SEV 28ZPAK -,18,72
orl-rat LD50:3160 mg/kg 28ZPAK -,18,72
ipr-rat LDLo:100 mg/kg JIHTAB 25,26,43

CONSENSUS REPORTS: Reported in EPA TSCA Inventory.

DOT Classification: Corrosive Material; Label: Corrosive

SAFETY PROFILE: Poison by intraperitoneal route. Moderately toxic by ingestion. A human skin irritant. A corrosive irritant to skin, eyes, and mucous membranes. A substance which migrates to food from packaging materials. Violent or explosive reactions with chlorine;

metal nitrates + heat; metal nitrites + heat; fuming HNO_3. When heated to decomposition it emits very toxic fumes of SO_x and NO_x. See also SULFONATES.

SNL800 CAS:1220-83-3 ***HR: 2***
SULFAMONOMETHOXIN
mf: $C_{11}H_{12}N_4O_3S$ mw: 280.33

SYNS: 4-AMINO-N-(6-METHOXY-4-PYRIMIDINYL)-BENZENESULFONAMIDE (9CI) ◇ DAIMETON ◇ DJ-1550 ◇ DS-36 ◇ ICI 32525 ◇ RO 4-3476 ◇ SULFAMONOMETHOXINE

TOXICITY DATA with REFERENCE
orl-rat TDLo:3 g/kg (9-14D preg):TER SEIJBO 13,7,73
orl-rat TDLo:3 g/kg (9-14D preg):REP SEIJBO 13,7,73
orl-mus LD50:4680 mg/kg NIIRDN 6,391,82
ivn-mus LD50:1010 mg/kg NIIRDN 6,391,82

SAFETY PROFILE: Moderately toxic by intravenous route. Mildly toxic by ingestion. An experimental teratogen. Experimental reproductive effects. When heated to decomposition it emits toxic fumes of SO_x and NO_x.

SNL830 CAS:53607-04-8 ***HR: 2***
SULFAMONOMETOXINE N-METHYLGLUCAMINE SALT
mf: $C_7H_{17}NO_5 \cdot C_{11}H_{12}N_4O_3S$ mw: 475.58

SYN: N-METHYLGLUCAMINE and SULFAMONOMETOXINE

TOXICITY DATA with REFERENCE
ivn-rat LD50:1555 mg/kg RPTOAN 46,30,83
ivn-mus LD50:940 mg/kg RPTOAN 46,30,83
ims-mus LD50:1090 mg/kg RPTOAN 46,30,83

SAFETY PROFILE: Moderately toxic by intravenous and intramuscular routes. When heated to decomposition it emits toxic fumes of SO_x and NO_x.

SNL850 CAS:57197-43-0 ***HR: 2***
SULFAMOXOLE-TRIMETHOPRIM mixture
mf: $C_{14}H_{18}N_4O_3 \cdot C_{11}H_{13}N_3O_3S$ mw: 557.69

SYNS: CN 3123 ◇ CO-FRAM ◇ CO-TRIFAMOLE ◇ NEVIN ◇ NEVIN (MIXTURE) ◇ SULFAMOXOLE and TRIMETHOPRIM ◇ SUPRISTOL ◇ TRIMETHOPRIM and SULFAMOXOLE ◇ TRIMETHOPRIM-SULFAMOXOLE mixture

TOXICITY DATA with REFERENCE
orl-rat TDLo:21 g/kg (16-22D preg/28D post):REP ARZNAD 26,643,76
orl-rat TDLo:3360 mg/kg (8-15D preg):TER ARZNAD 26,643,76
orl-rat LD50:14 g/kg ARZNAD 26,643,76
ipr-rat LD50:2000 mg/kg ARZNAD 26,634,76
ipr-mus LD50:1870 mg/kg ARZNAD 26,634,76

SAFETY PROFILE: Moderately toxic by intraperitoneal route. Mildly toxic by ingestion. An experimental teratogen. Experimental reproductive effects. When heated to decomposition it emits toxic fumes of NO_x and SO_x. See also AROMATIC AMINES and AMIDES.

SNM500 CAS:63-74-1 ***HR: 3***
SULFANILAMIDE
mf: $C_6H_8N_2O_2S$ mw: 172.22

PROP: Crystals. Mp: 164.5-166.5°. Sol in glycerol, propylene glycol, HCl; almost insol in chloroform, ether, benzene, petr ether.

SYNS: ALBEXAN ◇ ALBOSAL ◇ AMBESIDE ◇ p-AMINOBENZENESULFAMIDE ◇ p-AMINOBENZENESULFONAMIDE ◇ 4-AMINOBENZENESULFONAMIDE ◇ p-AMINOPHENYLSULFONAMIDE ◇ 4-AMINOPHENYLSULFONAMIDE ◇ p-ANILINESULFONAMIDE ◇ ANILINE-p-SULFONIC AMIDE ◇ ANTISTREPT ◇ BACTERAMID ◇ COLLOMIDE ◇ COLSULANYDE ◇ COPTICIDE ◇ DIPRON ◇ ESTREPTOCIDA ◇ F 1162 ◇ FOURNEAU 1162 ◇ GERISON ◇ GOMBARDOL ◇ LUSIL ◇ LYSOCOCCINE ◇ NEOCOCCYL ◇ ORGASEPTINE ◇ PABS ◇ PRONTALBIN ◇ PRONTOSIL I ◇ PROSEPTINE ◇ PROSEPTOL ◇ PYSOCOCCINE ◇ RUBIAZOL A ◇ SEPTAMIDE ALBUM ◇ SEPTINAL ◇ SEPTOPLEX ◇ STOPTON ALBUM ◇ STREPAMIDE ◇ STREPTAGOL ◇ STREPTOCLASE ◇ STREPTOL ◇ STREPTOSIL ◇ STREPTOZONE ◇ STREPTROCIDE ◇ p-SULFAMIDOANILINE ◇ SULFAMIDYL ◇ SULFANA ◇ SULFANALONE ◇ SULFANIL ◇ SULFOCIDINE ◇ SULFONAMIDE ◇ SULFONAMIDE P ◇ SULPHANILAMIDE ◇ THERAPOL ◇ WHITE STREPTOCIDE

TOXICITY DATA with REFERENCE
mmo-sat 10 μL/plate ANYAA9 76,475,58
dnd-esc 50 μmol/L MUREAV 89,95,81
unr-wmn TDLo:974 mg/kg (34-37W preg):TER JAMAAP 117,1314,41
orl-rat TDLo:22 mg/kg (1-22D preg):REP JHHBAI 66,139,40
scu-rat TDLo:135 mg/kg/9W-I:CAR PSEBAA 68,330,48
orl-rat LD50:3900 mg/kg 85JCAE -,1072,86
orl-mus LD50:3 g/kg NYKZAU 48,336,52
ipr-mus LD50:5 mg/kg 85ERAY 2,1364,78
scu-mus LD50:2900 mg/kg ARZNAD 5,213,55
ivn-mus LD50:587 mg/kg JIDXA3 33,503,40
orl-dog LD50:2000 mg/kg PCOC** -,1079,66

CONSENSUS REPORTS: Reported in EPA TSCA Inventory. EPA Genetic Toxicology Program.

SAFETY PROFILE: Poison by intraperitoneal routes. Moderately toxic by ingestion, subcutaneous, and intravenous routes. Human teratogenic effects by unspecified route: developmental abnormalities of the blood and lymphatic systems (including the spleen and bone marrow). Experimental reproductive effects. Questionable carcinogen with experimental carcinogenic data. Mutation data reported. Implicated in aplastic anemia. When heated to decomposition it emits very toxic fumes of NO_x and SO_x.

SNN300 CAS:122-11-2 ***HR: 2***
6-SULFANILAMIDO-2,4-DIMETHOXYPYRIMIDINE
mf: $C_{12}H_{14}N_4O_4S$ mw: 310.36

PROP: Crystals from dil alc. Mp: 201-203°. Sol in dil HCl and in aq solns of sodium carbonate. Solubility in water at 37° (mg/100 mL): 4.6 at pH 4.10; 29.5 at pH 6.7; 58.0 at pH 7.06.

SYNS: ABCID ◇ AGRIBON ◇ ALBON ◇ 4-AMINO-N-(2,6-DIMETHOXY-4-PYRIMIDINYL)BENZENESULFONAMIDE ◇ ARNOSULFAN ◇ BACTROVET ◇ DEPOSUL ◇ DIASULFA ◇ DIASULFYL ◇ DIMETAZINA ◇ 2,6-DIMETHOXY-4-(p-AMINOBENZENESULFONAMIDO)PYRIMIDINE ◇ N[1]-(2,6-DIMETHOXY-4-PYRIMIDINYL)SULFANILAMIDE ◇ DIMETHOXYSULFADIAZINE ◇ 2,4-DIMETHOXY-6-SULFANILAMIDO-1,3-DIAZINE ◇ 2,6-DIMETHOXY-4-SULFANILAMIDOPYRIMIDINE ◇ DINOSOL ◇ DORISUL ◇ FUXAL ◇ MADRIBON ◇ MADRIGID ◇ MADRIQID ◇ MADROXIN ◇ MADROXINE ◇ MAXULVET ◇ MEMCOZINE ◇ METOXIDON ◇ NEOSTREPAL ◇ OMNIBON ◇ PERSULFEN ◇ RADONIN ◇ REDIFAL ◇ ROSCOSULF ◇ SCANDISIL ◇ SDM ◇ SDMO ◇ SUDINE ◇ SULDIXINE ◇ SULFADIMETHOXIN ◇ SULFADIMETHOXINE ◇ SULFADIMETHOXYDIAZINE ◇ SULFADIMETOSSINA (ITALIAN) ◇ SULFADIMETOXIN ◇ SULFASOL ◇ SULFASTOP ◇ SULFOPLAN ◇ SULPHADIMETHOXINE ◇ SULXIN ◇ SYMBIO ◇ THERACANZAN

TOXICITY DATA with REFERENCE
orl-mus TDLo:12 g/kg (female 7-12D post):TER OYYAA2 7,1005,73
orl-rat TDLo:11970 mg/kg (male 6W pre):REP JRPFA4 81,259,87
ipr-mus LD50:866 mg/kg NIIRDN 6,386,82
scu-mus LD50:791 mg/kg NIIRDN 6,386,82
ivn-mus LD50:844 mg/kg NIIRDN 6,386,82
ivn-rbt LD50:1000 mg/kg NIIRDN 6,386,82

SAFETY PROFILE: Moderately toxic by intraperitoneal, intravenous, and subcutaneous routes. An experimental teratogen. Experimental reproductive effects. When heated to decomposition it emits toxic fumes of SO_x and NO_x.

SNN500 CAS:127-69-5 ***HR: 1***
5-SULFANILAMIDO-3,4-DIMETHYL-ISOXAZOLE
mf: $C_{11}H_{13}N_3O_3S$ mw: 267.33

SYNS: ACCUZOLE ◇ AMIDOXAL ◇ 5-(p-AMINOBENZENESULFONAMIDO)-3,4-DIMETHYLISOOXALE ◇ 5-(p-AMINOBENZENESULFONAMIDO)-3,4-DIMETHYLISOXAZOLE ◇ 5-(p-AMINOBENZENESULPHONAMIDO)-3,4-DIMETHYLISOXAZOLE ◇ 5-(p-AMINOBENZENESULPHONAMIDE)-3,4-DIMETHYLISOXAZOLE ◇ 4-AMINO-N-(3,4-DIMETHYL-5-ISOXAZOLYL)BENZENESULPHONAMIDE ◇ 5-(4-AMINOPHENYLSULFONAMIDO)-3,4-DIMETHYLISOXAZOLE ◇ 5-(4-AMINOPHENYLSULPHONAMIDO)-3,4-DIMETHYLISOXAZOLE ◇ ASTRAZOLO ◇ AZO GANTRISIN ◇ AZOSULFIZIN ◇ BACTESULF ◇ BARAZAE ◇ CHEMOUAG ◇ 3,4-DIMETHYLISOXALE-5-SULFANILAMIDE ◇ 3,4-DIMETHYLISOXALE-5-SULPHANILAMIDE ◇ N(1)-(3,4-DIMETHYL-5-ISOXAZOLYL)SULFANILAMIDE ◇ N'-(3,4)DIMETHYLISOXAZOL-5-YL-SULPHANILAMIDE ◇ N(1)-(3,4-DIMETHYL-5-ISOXAZOLYL)SULPHANILAMIDE ◇ 3,4-DIMETHYL-5-SULFANILAMIDOISOXAZOLE ◇ 3,4-DIMETHYL-5-SULPHANILAMIDOISOXAZOLE ◇ 3,4-DIMETHYL-5-SULPHONAMIDOISOXAZOLE ◇ DORSULFAN ◇ ENTUSIL ◇ GANTRISINE ◇ ISOXAMIN ◇ J-SUL ◇ KORO-SULF ◇ NEOXAZOL ◇ NORILGAN-S ◇ NOVOSAXAZOLE ◇ PANCID ◇ RENOSULFAN ◇ RESOXOL ◇ ROXOSUL TABLETS ◇ SAXOSOZINE ◇ SODIZOLE ◇ SOXISOL ◇ STANSIN ◇ SULFADIMETHYLISOXAZOLE ◇ SULFAFURAZOL ◇ SULFAGAN ◇ SULFASOXAZOLE ◇ SULFAZOLE ◇ SULFISIN ◇ SULFISOXAZOLE ◇ SULFIZOLE ◇ SULPHADIMETHYLISOXAZOLE ◇ SULPHAFURAZ ◇ 5-SULPHANILAMIDO-3,4-DIMETHYLISOXAZOLE ◇ SULPHISOXAZOL ◇ SULPHOFURAZOLE ◇ THIASIN ◇ UNISULF ◇ URISOXIN ◇ URITRISIN ◇ UROGAN ◇ VAGILIA

TOXICITY DATA with REFERENCE
msc-mus:lyms 750 mg/L SCIEAS 236,933,87
sce-ham:ovr 100 mg/L SCIEAS 236,933,87
orl-rat TDLo:6 g/kg (9-14D preg):TER SEIJBO 13,7,73
orl-rat LD50:10 g/kg NIIRDN 6,391,82
orl-mus LD50:6800 mg/kg ARZNAD 15,1441,65
orl-rbt LD50:20 g/kg NIIRDN 6,391,82

CONSENSUS REPORTS: IARC Cancer Review: Group 3 IMEMDT 7,347,87; Human Inadequate Evidence IMEMDT 24,275,80; Animal Inadequate Evidence IMEMDT 24,275,80. NCI Carcinogenesis Bioassay (gavage); No Evidence: mouse, rat NCITR* NCI-CG-TR-138,79. Reported in EPA TSCA Inventory.

SAFETY PROFILE: Mildly toxic by ingestion. An experimental teratogen. Questionable carcinogen. Mutation data reported. When heated to decomposition it emits very toxic fumes of SO_x and NO_x. See also SULFONAMIDES.

SNO000 CAS:121-47-1 ***HR: 1***
m-SULFANILIC ACID
mf: $C_6H_7NO_3S$ mw: 173.20

PROP: Monohydrate: orthorhombic plates. Decomp without melting at approx 288°. Practically insol in ethanol, benzene, ether; sltly sol in hot methanol.

SYNS: m-AMINOBENZENESULFONIC ACID ◇ 1-AMINOBENZENE-3-SULFONIC ACID ◇ 3-AMINO-BENZENESULFONIC ACID ◇ m-ANILINESULFONIC ACID ◇ KYSELINA ANILIN-3-SULFONOVA (CZECH) ◇ KYSELINA METANILOVA (CZECH) ◇ METANILIC ACID

TOXICITY DATA with REFERENCE
eye-rbt 500 mg/24H MLD 28ZPAK -,179,72
mmo-omi 1 pph POASAD 34,114,53
orl-rat LD50:12 g/kg 28ZPAK -,179,72

CONSENSUS REPORTS: Reported in EPA TSCA Inventory.

SAFETY PROFILE: Mildly toxic by ingestion. Mutation data reported. An eye irritant. When heated to decomposition it emits very toxic fumes of NO_x and SO_x. See also SULFONATES.

SNP500 CAS:144-80-9 ***HR: 1***
N-SULFANILYLACETAMIDE
mf: $C_8H_{10}N_2O_3S$ mw: 214.26

SYNS: ACETOCID ◇ ACETOSULFAMIN ◇ N-ACETYL-4-AMINOBENZENESULFONAMIDE ◇ N(1)-ACETYL-4-AMINOPHENYLSULFONAMIDE ◇ N-ACETYLSULFANILAMIDE ◇ N'-ACETYLSULFANILAMIDE ◇ N1-ACETYLSULFANILAMIDE ◇ N-ACETYLSULFANILAMINE ◇ ALBAMINE ◇ ALBUCID ◇ ALESTEN ◇ p-AMINOBENZENESULFONACETAMIDE ◇ N-((4-AMINOPHENYL)SULFONYL)ACETAMIDE ◇ BLEPH-10 ◇ FORMOSULFACETAMIDE ◇ ISOPTO CETAMIDE ◇ OP-SULFA 30 ◇ OPTHEL-S ◇ REGION ◇ SEBIZON ◇ STERAMIDE ◇ SULAMYD ◇ SULF-10 ◇ SULFACET ◇ SULFACETAMIDE ◇ SULFACETIMIDE ◇ SULFACYL ◇ SULFANILACETAMIDE ◇ SULPHACETAMIDE ◇ SULPHASIL ◇ UROSULFON ◇ UROSULFONE

TOXICITY DATA with REFERENCE
sln-nsc 250 mg/L MUREAV 167,35,86
sln-smc 2 g/L EVHPAZ 31,97,79
orl-rat TDLo:10 g/kg (10D preg):TER PROEAS 14,89,68
ivn-rat LD50:6600 mg/kg AEPPAE 211,367,50
orl-mus LD50:16500 mg/kg JOURAA 47,183,42
orl-dog LDLo:8 g/kg JOURAA 47,183,42
orl-rbt LDLo:5 g/kg JOURAA 47,183,42

CONSENSUS REPORTS: Reported in EPA TSCA Inventory. EPA Genetic Toxicology Program.

SAFETY PROFILE: Mildly toxic by ingestion and intravenous routes. An experimental teratogen. Mutation data reported. When heated to decomposition it emits very toxic fumes of NO_x and SO_x.

SNQ000 CAS:127-56-0 ***HR: 1***
N-SULFANILYLACETAMIDE, SODIUM SALT
mf: $C_8H_9N_2O_3S•Na$ mw: 236.24

SYNS: N(1)-ACETYLSULFANILAMIDE SODIUM SALT ◇ SODIUM SULFACETAMIDE ◇ SODIUM SULFANILYLACETAMIDE ◇ SODIUM N-SULFANILYLACETAMIDE ◇ SOLUBLE SULFACETAMIDE ◇ SULFACYL SODIUM SALT ◇ N-SULFANILYLACETAMIDE SODIUM ◇ SUPRACID

TOXICITY DATA with REFERENCE
scu-mus LDLo:9700 mg/kg JPETAB 73,170,41

CONSENSUS REPORTS: Reported in EPA TSCA Inventory.

SAFETY PROFILE: When heated to decomposition it emits very toxic fumes of NO_x, Na_2O and SO_x.

SNQ500 CAS:3337-71-1 ***HR: 2***
SULFANILYLCARBAMIC ACID, METHYL ESTER
mf: $C_8H_{10}N_2O_4S$ mw: 230.26

SYNS: 4-AMINO-BENZOLSULFONYL-METHYLCARBAMAT(GERMAN) ◇ ASILAN ◇ ASULAM ◇ ASULFOX F ◇ ASULOX ◇ ASULOX 40 ◇ JONNIX ◇ MB 9057 ◇ METHYL-N-(4-AMINOBENZENESULFONYL)CARBAMATE ◇ METHYL ((4-AMINOPHENYL)SULFONYL)CARBAMATE ◇ METHYL SULFANILYL CARBAMATE

TOXICITY DATA with REFERENCE
orl-rat LD50:2000 mg/kg 85ARAE 2,90,77
unr-mus LD50:5000 mg/kg 30ZDA9 -,204,71

SAFETY PROFILE: Moderately toxic by ingestion. When heated to decomposition it emits very toxic fumes of NO_x and SO_x.

SNQ550 CAS:547-44-4 ***HR: 3***
SULFANILYLUREA
mf: $C_7H_9N_3O_3S$ mw: 215.25

SYNS: A 435 ◇ p-AMINOBENZENESULFONYLUREA ◇ 1-(4-AMINOBENZENESULFONYL)UREA ◇ BENZENESULFONAMIDE, 4-AMINO-N-(AMINOCARBONYL)- (9CI) ◇ EUVERNIL ◇ SULFACARBAMIDE ◇ 4-SULFACARBAMIDE ◇ SULFANILCARBAMID ◇ N-SULFANILCARBAMIDE ◇ SULFANYLHARNSTOFF ◇ SULFANYLUREE ◇ SULFAUREA ◇ SULPHAUREA ◇ URACTYL ◇ URAMID ◇ UREA, SULFANILYL- ◇ URENIL ◇ UROSULFAN ◇ UROSULFANE

TOXICITY DATA with REFERENCE
orl-rat TDLo:10 g/kg (female 10D post):TER PROEAS 14,89,68
ipr-mus LD50:405 mg/kg FRZKAP (6),26,87

SAFETY PROFILE: Poison by intraperitoneal route. An experimental teratogen. When heated to decomposition it emits toxic fumes of NO_x and SO_x.

SNQ700 CAS:12765-82-1 ***HR: 2***
SULFANOL NP 1

SYNS: SULFONAL NP 1 ◇ SULPHONAL NP 1

TOXICITY DATA with REFERENCE
orl-rat LD50:1575 mg/kg GISAAA 47(9),33,82

SAFETY PROFILE: Moderately toxic by ingestion.

SNR000 CAS:618-82-6 ***HR: 3***
SULFARSPHENAMINE
mf: $C_{14}H_{14}As_2N_2O_8S_2•2Na$ mw: 598.24

PROP: Yellow powder. Very sol in water; very sltly sol in alc.

SYNS: (1,2-DIARSENEDIYLBIS((6-HYDROXY-3,1-PHENYLENE)IMINO))BISMETHANESULFONIC ACID DISODIUM SALT ◇ DISODIUM-3,3'-DIAMINO-4,4'-DIHYDROXYARSENOBENZENE-N,N'-DIMETHYLENEBISULFITE ◇ DISODIUM-3,3'-DIAMINO-4,4'-DIHYDROXYARSENOBENZENE-N-DIMETHYLENESULFONATE ◇ KARSULPHAN ◇ METARSENOBILLON ◇ MYARSENOL ◇ MYOSALVARSAN ◇ SULFARSENOBENZENE ◇ SULFARSENOL ◇ THIOARSMINE

TOXICITY DATA with REFERENCE
scu-mus LDLo:40 g/kg SDUU** -,-,34
ivn-rbt LDLo:320 mg/kg PHRPA6 37,2783,22

CONSENSUS REPORTS: Arsenic and its compounds are on the Community Right To Know List.

OSHA PEL: TWA 0.5 mg(As)/m^3
ACGIH TLV: TWA 0.2 mg(As)/m^3

SAFETY PROFILE: Poison by intravenous route. When heated to decomposition it emits very toxic fumes of As, NO_x, Na_2O, and SO_x. See also ARSENIC COMPOUNDS and SULFONATES.

SNS000 ***HR: D***
SULFATES

SAFETY PROFILE: Variable toxicity. In general the toxic properties of substances containing the sulfate radical is that of the material (cation) with which the sulfate (anion) is combined. See specific compound. Violent reaction with Al; Mg. When heated to decomposition they emit toxic fumes of SO_x.

SNS100 CAS:73825-85-1 ***HR: 3***
SULFATOBIS(DIMETHYLSELENIDE) PLATINUM(II) HYDRATE

SYNS: PLATINUM (II), BIS(METHYL SELENIDE)SULFATO-, HYDRATE ◇ NSC 281278

TOXICITY DATA with REFERENCE
ivn-mus LD50:100 mg/kg CTRRDO 61,1519,77

OSHA PEL: TWA 0.2 mg(Se)/m^3
ACGIH TLV: TWA 0.2 mg(Se)/m^3

SAFETY PROFILE: Poison by intravenous route. When heated to decomposition it emits toxic fumes of SO_x and Se.

SNT000 ***HR: D***
SULFIDES

SAFETY PROFILE: Variable toxicity. The alkaline sulfides (potassium, calcium, ammonium, and sodium) are similar in action to alkalies. They cause softening and irritation of the skin. If ingested, they are corrosive and irritating through the liberation of hydrogen sulfide and free alkali. Hydrogen sulfide is especially toxic. Sulfides of the heavy metals are generally insoluble and hence have little toxic action except through the liberation of hydrogen sulfide. Sulfides are used as fungicides. Flammable when exposed to flame or by spontaneous chemical reaction. Many sulfides ignite easily in air at room temperature. Others require a higher temperature or the presence of an oxidizer. Upon contact with moisture or acids, hydrogen sulfide is evolved. Many powerful oxidizers on contact with sulfides ignite violently. Many sulfides react violently and explosively on contact with powerful oxidizers. Hydrogen sulfide evolved can form explosive mixtures with air. They react with water, steam, or acids to produce toxic and flammable vapors of hydrogen sulfide. When heated to decomposition they emit highly toxic fumes of SO_x. See also HYDROGEN SULFIDE.

SNT200 CAS:507-16-4 ***HR: 1***
SULFINYL BROMIDE
mf: Br_2OS mw: 207.87

SYN: THIONYL BROMIDE

SAFETY PROFILE: An unstable material which must be stored refrigerated. When heated to decomposition it emits toxic fumes of Br^- and SO_x. See also BROMIDES.

SNT300 ***HR: 3***
SULFINYL CYANAMIDE
mf: CN_2OS mw: 88.08

CONSENSUS REPORTS: Cyanide and its compounds are on the Community Right To Know List.

SAFETY PROFILE: Explodes on contact with traces of water or ethanol. Polymerizes on exposure to light at room temperature. When heated to decomposition it emits toxic fumes of SO_x, CN^-, and NO_x. See also CYANIDE.

SNT500 ***HR: 2***
SULFITES

SAFETY PROFILE: Fairly large doses of sulfites can be tolerated since they are rapidly oxidized to sulfates, although if swallowed they may cause irritation of the stomach by liberating sulfurous acid. Experimentally, large doses of sodium sulfite have been shown to cause retarded growth, nerve irritation, atrophy of bone marrow, depression, and paralysis. They will react with water, steam, or acids to produce a toxic and corrosive material. When heated to decomposition they emit highly toxic fumes of SO_x.

SNU000 CAS:123-43-3 ***HR: 2***
SULFOACETIC ACID
mf: $C_2H_4O_5S$ mw: 140.12

PROP: Monohydrate: hygroscopic crystals. Mp: 84-86° (anhyd), bp: 245° (decomp). Sol in water, alc; insol in ether, chloroform.

SYN: SULFOETHANOIC ACID

TOXICITY DATA with REFERENCE
orl-rat LD50:3160 mg/kg JIHTAB 31,60,49
skn-rbt LD50:1570 mg/kg JIHTAB 31,60,49

CONSENSUS REPORTS: Reported in EPA TSCA Inventory.

SAFETY PROFILE: Moderately toxic by ingestion and skin contact. When heated to decomposition it emits toxic fumes of SO_x.

SNU500 CAS:22326-31-4 ***HR: 2***
5-SULFOBENZEN-1,3-DICARBOXYLIC ACID
mf: $C_8H_6O_7S$ mw: 246.20

SYN: KYSELINA 3,5-DIKARBOXYBENZENSULFONOVA (CZECH)

TOXICITY DATA with REFERENCE
skn-rbt 500 mg/24H MLD 28ZPAK -,184,72
eye-rbt 250 μg/24H SEV 28ZPAK -,184,72
orl-rat LD50:2600 mg/kg 28ZPAK -,184,72

CONSENSUS REPORTS: Reported in EPA TSCA Inventory.

SAFETY PROFILE: Moderately toxic by ingestion. A skin and severe eye irritant. When heated to decomposition it emits toxic fumes of SO_x.

SNV000 CAS:28002-18-8 ***HR: 1***
α-SULFOBENZYLPENICILLIN DISODIUM
mf: $C_{16}H_{16}N_2O_7S_2$•2Na mw: 458.44

SYNS: DISODIUM SULBENICILLIN ◇ DISODIUM SULFOBENZYLPENICILLIN ◇ DISODIUM α-SULFOBENZYLPENICILLIN ◇ KEDACILLIN ◇ LILACILLIN ◇ SULBENICILLIN ◇ SULBENICILLIN DISODIUM ◇ SULFOBENZYLPENICILLIN ◇ SULFOCILLIN ◇ SULPELIN

TOXICITY DATA with REFERENCE
ims-rat TDLo:182 g/kg (male 91D pre):REP TAKHAA 30,262,71
ipr-rat TDLo:14 g/kg (9-15D preg):TER TAKHAA 30,332,71
ipr-rat LD50:7200 mg/kg JMGZAI 10(2),4,73
scu-rat LD50:11 g/kg JMGZAI 10(2),4,73
ivn-rat LD50:6 g/kg TAKHAA 30,262,71
ims-rat LD50:8300 mg/kg TAKHAA 30,262,71
ipr-mus LD50:9600 mg/kg JMGZAI 10(2),4,73
scu-mus LD50:11500 mg/kg JMGZAI 10(2),4,73
ivn-mus LD50:7900 mg/kg JMGZAI 10(2),4,73
ims-mus LD50:10500 mg/kg NIIRDN 6,393,82
ivn-mky LD50:15 g/kg TAKHAA 34,405,75

SAFETY PROFILE: An experimental teratogen. Experimental reproductive effects. When heated to decomposition it emits very toxic fumes of NO_x, SO_x, and Na_2O.

SNW500 CAS:126-33-0 ***HR: 2***
SULFOLANE
mf: $C_4H_8O_2S$ mw: 120.18

PROP: Liquid. D: 1.2606 @ 30°/4°, mp: 27.4-27.8°, bp: 285°, flash p: 330°F. Misc with water, acetone, toluene (at 30°); sltly misc with octanes, olefins, naphthenes.

SYNS: CYCLIC TETRAMETHYLENE SULFONE ◇ CYCLOTETRAMETHYLENE SULFONE ◇ DIHYDRO BUTADIENE SULFONE ◇ 1,1-DIOXIDE TETRA HYDROTHIOFURAN ◇ DIOXOTHIOLAN ◇ TETRAMETHYLENE SULFONE ◇ THIACYCLOPENTANE DIOXIDE ◇ THIOPHANE DIOXIDE

TOXICITY DATA with REFERENCE
eye-rbt 253 mg MLD BJIMAG 23,302,66
orl-rat LD50:1941 mg/kg AIHAAP 30,470,69
ihl-rat LCLo:4700 mg/m³/24H TXAPA9 40,463,77
ipr-rat LD50:1600 mg/kg TOXID9 4,22,84
orl-mus LDLo:1900 mg/kg BJIMAG 23,302,66
ipr-mus LD50:1250 mg/kg RPTOAN 42,97,79
ivn-mus LD50:1080 mg/kg AIPTAK 119,423,59
skn-rbt LD50:4009 mg/kg AIHAPP 30,470,69

CONSENSUS REPORTS: Reported in EPA TSCA Inventory.

SAFETY PROFILE: Moderately toxic by ingestion, intraperitoneal, and intravenous routes. Mildly toxic by inhalation and skin contact. An eye irritant. Combustible when exposed to heat or flame. When heated to decomposition it emits toxic fumes of SO_x.

SNW550 CAS:74222-97-2 ***HR: 1***
SULFOMETURON METHYL
mf: $C_{15}H_{16}N_4O_5S$ mw: 364.41

PROP: White solid. Mp: 198-202°. Slt sol in water.

SYNS: METHYL 2-(((((4,6-DIMETHYL-2-PYRIMIDINYL)AMINO)CARBONYL)AMINO)SULFONYL)BENZOATE ◇ BENZOIC ACID, o-((3-(4,6-DIMETHYL-2-PYRIMIDINYL)UREIDO)SULFONYL)-, METHYL ESTER

TOXICITY DATA with REFERENCE
eye-rbt 10 mg MLD DUPON* AG-361

SAFETY PROFILE: An eye irritant. When heated to decomposition it emits toxic fumes of NO_x and SO_x

SNW800 CAS:2435-64-5 ***HR: 2***
4-SULFONAMIDE-4′-DIMETHYLAMINOAZOBENZENE
mf: $C_{14}H_{16}N_4O_2S$ mw: 304.40

SYN: BENZENESULFONAMIDE,p-((p-(DIMETHYLAMINO)PHENYL)AZO)-

TOXICITY DATA with REFERENCE
orl-rat TDLo:15 g/kg/55W-C:ETA BJCAAI 10,129,56

SAFETY PROFILE: Questionable carcinogen with experimental tumorigenic data. When heated to decomposition it emits toxic fumes of NO_x and SO_x.

SNX000 CAS:63019-42-1 ***HR: 3***
4-SULFONAMIDO-3′-METHYL-4′-AMINOAZOBENZENE
mf: $C_{13}H_{14}N_4O_2S$ mw: 290.37

SYN: p-((4-AMINO-m-TOLYL)AZO)BENZENESULFONAMIDE

TOXICITY DATA with REFERENCE
orl-rat TDLo:20 g/kg/87W-C:ETA BJCAAI 10,129,56

SAFETY PROFILE: Questionable carcinogen with ex-

perimental tumorigenic data. When heated to decomposition it emits very toxic fumes of SO_x and NO_x.

SNX250 ***HR: 3***
SULFO-3-NAPHTHALENEFURANE
mf: $C_{30}H_{11}Na_3O_{14}S_3$ mw: 760.56

TOXICITY DATA with REFERENCE
orl-rat LD50:15633 mg/kg GTPZAB 26(8),53,82
ipr-rat LD50:105 mg/kg GTPZAB 26(8),53,82
scu-rat LD50:551 mg/kg GTPZAB 26(8),53,82
orl-mus LD50:10200 mg/kg GTPZAB 26(8),53,82
ipr-mus LD50:46500 μg/kg GTPZAB 26(8),53,82
scu-mus LD50:220 mg/kg GTPZAB 26(8),53,82

SAFETY PROFILE: Poison by subcutaneous and intraperitoneal routes. Mildly toxic by ingestion. When heated to decomposition it emits toxic fumes of SO_x and Na_2O.

SNY000 ***HR: 1***
SULFONATES

SAFETY PROFILE: Variable toxicity. See specific compounds. Usually irritating. When heated to decomposition or on contact with acid or acid fumes, they emit highly toxic fumes of SO_x.

SNY500 CAS:77-46-3 ***HR: 3***
4',4'''-SULFONYLBIS(ACETANILIDE)
mf: $C_{16}H_{16}N_2O_4S$ mw: 332.40

SYNS: ACEDAPSONE ◇ ACETAMIN ◇ ATILON ◇ BA 2650 ◇ BIS(p-ACETAMIDOPHENYL) SULFONE ◇ BIS(4-ACETAMIDOPHENYL)SULFONE ◇ CAMILAN ◇ C.I. 556 ◇ DADDS ◇ 4,4'-DIACETAMIDODIPHENYL SULFONE ◇ DI-(p-ACETYLAMINOPHENYL)SULFONE ◇ DIACETYLDAPSONE ◇ N,N'-DIACETYLDAPSONE ◇ 4,4'-DIACETYLDIAMINODIPHENYL SULFONE ◇ N,N'-DIACETYL-4,4'-DIAMINODIPHENYL SULFONE ◇ 1399 F ◇ HANSOLAR ◇ RODILONE ◇ SULFADIAMINE ◇ SULFODIAMINE ◇ p,p'-SULFONYLBISACETANILIDE ◇ 4,4'-SULFONYLBISACETANILIDE

TOXICITY DATA with REFERENCE
orl-rat TDLo:4200 mg/kg/43W-C:CAR JNCIAM 24,149,60

SAFETY PROFILE: Questionable carcinogen with experimental carcinogenic data. When heated to decomposition it emits very toxic fumes of NO_x and SO_x.

SNZ000 CAS:13080-89-2 ***HR: 3***
4,4'-SULFONYLBIS(4-PHENYLENEOXY) DIANILINE
mf: $C_{24}H_{20}N_2O_4S$ mw: 432.52

SYN: BIS(4-AMINOPHENOXYPHENYL)SULFONE

TOXICITY DATA with REFERENCE
orl-rat LD50:220 mg/kg TXAPA9 28,313,74
skn-rbt LD50:560 mg/kg TXAPA9 28,313,74

CONSENSUS REPORTS: Reported in EPA TSCA Inventory.

SAFETY PROFILE: Poison by ingestion. Moderately toxic by skin contact. When heated to decomposition it emits very toxic fumes of NO_x and SO_x. See also SULFONATES.

SOA000 CAS:599-61-1 ***HR: 3***
3,3'-SULFONYLDIANILINE
mf: $C_{12}H_{12}N_2O_2S$ mw: 248.32

SYNS: BIS(m-AMINOPHENYL) SULFONE ◇ 3,3'-DIAMINODIFENYLSULFON (CZECH) ◇ 3,3'-DIAMINODIPHENYL SULFONE ◇ 3,3'-DIAMINOPHENYL SULFONE ◇ 3,3'-SULFONYLBIS(ANILINE)

TOXICITY DATA with REFERENCE
orl-rat LD50:4920 mg/kg MarJV# 29MAR77
ivn-mus LD50:178 mg/kg CSLNX* NX#02275

CONSENSUS REPORTS: Reported in EPA TSCA Inventory.

SAFETY PROFILE: Poison by intravenous route. Mildly toxic by ingestion. When heated to decomposition it emits very toxic fumes of NO_x and SO_x. See also SULFONATES.

SOA500 CAS:80-08-0 ***HR: 3***
4,4'-SULFONYLDIANILINE
mf: $C_{12}H_{12}N_2O_2S$ mw: 248.32

PROP: Crystals. Mp: 176°, vap d: 8.3. Nearly insol in water; sol in acetone and alc.

SYNS: AVLOSULPHONE ◇ BIS(p-AMINOPHENYL) SULFONE ◇ BIS(4-AMINOPHENYL) SULFONE ◇ BIS(p-AMINOPHENYL)SULPHONE ◇ BIS(4-AMINOPHENYL)SULPHONE ◇ CROYSULFONE ◇ DADPS ◇ DAPSONE ◇ DDS ◇ DIAMINODIFENILSULFONA (SPANISH) ◇ DIAMINO-4,4'-DIPHENYL SULFONE ◇ p,p'-DIAMINODIPHENYL SULFONE ◇ 4,4'-DIAMINODIPHENYL SULFONE ◇ DIAMINO-4,4'-DIPHENYL SULPHONE ◇ p,p-DIAMINODIPHENYL SULPHONE ◇ DI(p-AMINOPHENYL) SULFONE ◇ DI(4-AMINOPHENYL)SULFONE ◇ DI(p-AMINOPHENYL)SULPHONE ◇ DI(4-AMINOPHENYL)SULPHONE ◇ DIAPHENYLSULFONE ◇ DIAPHENYLSULPHON ◇ DIAPHENYLSULPHONE ◇ DIPHONE ◇ DISULONE ◇ DSS ◇ DUBRONAX ◇ DUMITONE ◇ EPORAL ◇ 1358F ◇ F 1358 ◇ MALOPRIM ◇ METABOLITE C ◇ NCI-C01718 ◇ NOVOPHONE ◇ NSC-6091 ◇ SULFONA ◇ 1,1'-SULFONYLBIS(4-AMINOBENZENE) ◇ 4,4'-SULFONYLBISANILINE ◇ p,p-SULFONYLBISBENZAMINE ◇ 4,4'-SULFONYLBISBENZAMINE ◇ p,p-SULFONYLBISBENZENAMINE ◇ p,p'-SULFONYLDIANILINE ◇ SULPHADIONE ◇ SULPHON-MERE ◇ 1,1'-SULPHONYLBIS(4-AMINOBENZENE) ◇ p,p-SULPHONYLBISBENZAMINE ◇ 4,4'-SULPHONYLBISBENZAMINE ◇ p,p-SULPHONYLBISBENZENAMINE ◇ 4,4'-SULPHONYLBISBENZENAMINE ◇ SULPHONYLDIANILINE ◇ p,p-SULPHONYLDIANILINE ◇ TARIMYL ◇ UDOLAC ◇ WR 448

TOXICITY DATA with REFERENCE
cyt-hmn-leu 4 mg/L IJLEAG 43,41,75
orl-rat TDLo:2100 mg/kg(6 W pre):REP JRPFA4 81,259,87

orl-rat TDLo:20 g/kg/80W-C:CAR NCITR* NCI-CG-TR-20,77
ipr-mus TDLo:1312 mg/kg/8W-I:NEO CNREA8 33,3069,73
orl-rat TD:39 g/kg/78W-C:CAR NCITR* NCI-CG-TR-20,77
orl-hmn TDLo:18 g/kg/15Y:KID,MSK BMJOAE 1,78,78
orl-man TDLo:36 mg/kg:BLD HUTODJ 2,507,83
orl-wmn LDLo:32 mg/kg:BLD BMJOAE 286,1244,83
orl-wmn TDLo:28 mg/kg:BLD BMJOAE 286,1244,83
orl-wmn TDLo:18 mg/kg:PUL,BLD AMSVAZ 214,215,83
orl-cld TDLo:5 mg/kg:PUL JTCTDW 19,1061,82/83
orl-rat LD50:1 g/kg ARZNAD 5,213,55
ipr-rat LD50:196 mg/kg AIPTAK 183,36,70
orl-mus LD50:375 mg/kg NIIRDN 6,305,82
ipr-mus LD50:313 mg/kg NIIRDN 6,305,82
scu-mus LD50:329 mg/kg NIIRDN 6,305,82
orl-cat LDLo:357 mg/kg ANPSAI 54,319,45

CONSENSUS REPORTS: IARC Cancer Review: Group 3 IMEMDT 7,185,87; Animal Limited Evidence IMEMDT 24,59,80; Human Inadequate Evidence IMEMDT 24,59,80. NCI Carcinogenesis Bioassay (feed); No Evidence: mouse NCITR* NCI-CG-TR-20,77; (feed); Clear Evidence: rat NCITR* NCI-CG-TR-20,77. Reported in EPA TSCA Inventory.

SAFETY PROFILE: Poison by ingestion, intraperitoneal, and subcutaneous routes. Human systemic effects by ingestion: agranulocytosis, change in tubules and other kidney changes, cyanosis, effect on joints, hemolysis with or without anemia, jaundice, methemoglobinemia-carboxyhemoglobin, retinal changes, somnolence. Experimental reproductive effects. Can cause hepatitis, dermatitis, and neuritis. Questionable carcinogen with experimental carcinogenic and neoplastigenic data. Human mutation data reported. Used in leprosy treatment and veterinary medicine. When heated to decomposition it emits very toxic fumes of NO_x and SO_x. See also SULFONATES.

SOB000 CAS:80-09-1 ***HR: 3***
4,4'-SULFONYLDIPHENOL
mf: $C_{12}H_{10}O_4S$ mw: 250.28

TOXICITY DATA with REFERENCE
skn-rbt 500 mg/24H MLD BIOFX* 601-05501,74
eye-rbt 100 mg MOD BIOFX* 601-05501,74
ipr-mus LDLo:32 mg/kg CBCCT* 2,135,50
orl-rat LD50:4556 mg/kg BIOFX* 601-05501,74

CONSENSUS REPORTS: Reported in EPA TSCA Inventory.

SAFETY PROFILE: Poison by intraperitoneal route. Mildly toxic by ingestion. A skin and eye irritant. When heated to decomposition it emits toxic fumes of SO_x.

SOB500 CAS:34681-23-7 ***HR: 3***
3-(SULFONYL)-o-((METHYLAMINO)CARBONYL)OXIME-2-BUTANONE
mf: $C_7H_{14}N_2O_4S$ mw: 222.29

SYNS: BUTOXICARBOXIM (GERMAN) ◇ 2-METHYLSULFONYL-o-(N-METHYL-CARBAMOYL)-BUTANON-(3)-OXIM (GERMAN) ◇ PLANT PIN

TOXICITY DATA with REFERENCE
orl-rat LD50:458 mg/kg FMCHA2 -,C40,83
scu-rat LD50:288 mg/kg 85DPAN -,-,71/76
orl-rbt LD50:275 mg/kg 85DPAN -,-,71/76

SAFETY PROFILE: Poison by ingestion and subcutaneous routes. When heated to decomposition it emits very toxic fumes of SO_x and NO_x.

SOC500 CAS:97-05-2 ***HR: 2***
5-SULFOSALICYLIC ACID
mf: $C_7H_6O_6S$ mw: 218.19

PROP: Dihydrate, white crystals or powder. Mp: 120° (decomp @ higher temp). Very sol in water or alc; sol in ether.

SYNS: 3-CARBOXY-4-HYDROXYBENZENESULFONIC ACID ◇ 2-HYDROXYBENZOIC-5-SULFONIC ACID ◇ SALICYLSULFONIC ACID ◇ SULFOSALICYLIC ACID

TOXICITY DATA with REFERENCE
orl-rat LD50:2450 mg/kg 34ZIAG -,565,69
orl-rbt LDLo:1300 mg/kg 34ZIAG -,565,69

CONSENSUS REPORTS: Reported in EPA TSCA Inventory.

SAFETY PROFILE: Moderately toxic by ingestion. Irritating to skin and mucous membranes. When heated to decomposition it emits toxic fumes of SO_x. See also SULFONATES.

SOD000 CAS:37753-10-9 ***HR: 3***
SULFOSFAMIDE
mf: $C_8H_{18}ClN_2O_5PS$ mw: 320.76

SYNS: ASTA 5122 ◇ 3-(2-CHLOROETHYL)-2-(2-MESYLOXYETHYLAMINO)TETRAHYDRO-2H-1,3,2-OXAZAPHOSPHORIN-2-OXIDE◇ 2-((3-(2-CHLOROETHYL)TETRAHYDRO-2H-1,3,2-OXAZAPHOSPHORIN-2-YL)AMINO)-ETHANOL, METHANESULFONATE (ester), p-OXIDE

TOXICITY DATA with REFERENCE
mnt-mus-ipr 160 mg/kg/24H MUREAV 56,319,78
ipr-rat LD50:120 mg/kg ARZNAD 24,1149,74
ivn-rat LD50:102 mg/kg ARZNAD 24,1149,74
ipr-mus LD50:136 mg/kg ARZNAD 24,1149,74
ivn-mus LD50:91 mg/kg ARZNAD 24,1149,74

CONSENSUS REPORTS: EPA Genetic Toxicology Program.

SAFETY PROFILE: Poison by intravenous and in-

traperitoneal routes. Mutation data reported. When heated to decomposition it emits very toxic fumes of PO_x, SO_x, Cl^-, and NO_x.

SOD100 CAS:3689-24-5 ***HR: 3***
SULFOTEP
DOT: UN 1704
mf: $C_8H_{20}O_5P_2S_2$ mw: 322.34

PROP: A liquid almost insol in water.

SYNS: BAYER-E 393 ◇ BIS-O,O-DIETHYLPHOSPHOROTHIONIC ANHYDRIDE ◇ BLADAFUME ◇ BLADAFUN ◇ DITHIO ◇ DITHIODIPHOSPHORIC ACID, TETRAETHYL ESTER ◇ DITHIOFOS ◇ DITHIONE ◇ DI(THIOPHOSPHORIC) ACID, TETRAETHYL ESTER ◇ DITHIOPYROPHOSPHATE de TETRAETHYLE (FRENCH) ◇ DITHIOTEP ◇ E393 ◇ ENT 16,273 ◇ ETHYL THIOPYROPHOSPHATE ◇ LETHALAIRE G-57 ◇ PIROFOS ◇ PLANT DITHIO AEROSOL ◇ PLANTFUME 103 SMOKE GENERATOR ◇ PYROPHOSPHORODITHIOIC ACID, TETRAETHYL ESTER ◇ PYROPHOSPHORODITHIOIC ACID-O,O,O,O-TETRAETHYL ESTER ◇ RCRA WASTE NUMBER P109 ◇ SULFOTEPP ◇ TEDP (OSHA, MAK) ◇ O,O,O,O-TETRAAETHYL-DITHIONOPYROPHOSPHAT (GERMAN) ◇ O,O,O,O-TETRAETHYL-DITHIO-DIFOSFAAT (DUTCH) ◇ TETRAETHYL DITHIONOPYROPHOSPHATE ◇ TETRAETHYL DITHIOPYROPHOSPHATE ◇ O,O,O,O-TETRAETHYL DITHIOPYROPHOSPHATE ◇ TETRAETHYL DITHIO PYROPHOSPHATE, liquid (DOT) ◇ O,O,O,O-TETRAETIL-DITIO-PIROFOSFATO (ITALIAN) ◇ THIOTEPP

TOXICITY DATA with REFERENCE
orl-rat LD50:5 mg/kg ARSIM* 20,24,66
ihl-rat LC50:38 mg/m^3/4H ARTODN 33,1,74
skn-rat LD50:65 mg/kg ARTODN 33,1,74
ipr-rat LD50:6600 μg/kg ARTODN 33,1,74
ims-rat LD50:55 μg/kg CJCHAG 34,1819,56
orl-mus LD50:22 mg/kg ARTODN 33,1,74
ihl-mus LC50:40 mg/m^3/4H ARTODN 33,1,74
ipr-mus LD50:940 μg/kg CJCHAG 34,1819,56
scu-mus LD50:8 mg/kg PAREAQ 11,636,59
ivn-mus LD50:300 μg/kg CJCHAG 34,1819,56
ims-mus LD50:500 μg/kg CJCHAG 34,1819,56
orl-dog LD50:5 mg/kg ARTODN 33,1,74
orl-cat LD50:3 mg/kg ARTODN 33,1,74
orl-rbt LD50:25 mg/kg ARTODN 33,1,74
skn-rbt LD50:20 mg/kg WRPCA2 9,119,70

OSHA PEL: TWA 0.2 mg/m^3 (skin)
ACGIH TLV: TWA 0.2 mg/m^3 (skin)
DFG MAK: 0.015 ppm (0.2 mg/m^3)
DOT Classification: Poison B; Label: Poison; Poison B; Label: St. Andrews Cross.

SAFETY PROFILE: Poison by ingestion, skin contact, inhalation, intramuscular, intraperitoneal, subcutaneous, and intravenous routes. A cholinesterase inhibitor type of insecticide. When heated to decomposition it emits toxic fumes of PO_x and SO_x. See also PARATHION.

SOD200 ***HR: 3***
SULFOTRINAPHTHYLENOFURAN, SODIUM SALT
mf: $C_{30}H_{14}O_{14}S_3$•3Na mw: 763.59

TOXICITY DATA with REFERENCE
ipr-mus TDLo:367 mg/kg/43W-I:ETA VOONAW 29(2),67,83
orl-rat LD50:15633 mg/kg GISAAA 49(10),67,84
ipr-rat LD50:105 mg/kg GISAAA 49(10),67,84
scu-rat LD50:551 mg/kg VOONAW 29(2),67,83
orl-mus LD50:10200 mg/kg GISAAA 49(10),67,84
ipr-mus LD50:46500 μg/kg GISAAA 49(10),67,84
scu-mus LD50:220 mg/kg VOONAW 29(2),67,83

SAFETY PROFILE: Poison by subcutaneous and intraperitoneal routes. Mildly toxic by ingestion. Questionable carcinogen with experimental tumorigenic data. When heated to decomposition it emits toxic fumes of SO_x and Na_2O.

SOD500 CAS:7704-34-9 ***HR: 3***
SULFUR
DOT: UN 1350/UN 2448
af: S aw: 32.06

PROP: Rhombic yellow crystals or yellow powder. Mp: 119°, bp: 444.6°, flash p: 405°F (CC), d: 2.07, d (liquid): 1.803, autoign temp: 450°F, vap press: 1 mm @ 183.8°. Insol in water; sltly sol in alc, ether; sol in carbon disulfide, benzene, toluene.

SYNS: BENSULFOID ◇ BRIMSTONE ◇ COLLOIDAL SULFUR ◇ COLLOKIT ◇ COLSUL ◇ COROSUL D AND S ◇ COSAN ◇ CRYSTEX ◇ FLOWERS of SULPHUR (DOT) ◇ GROUND VOCLE SULPHUR ◇ HEXASUL ◇ KOCIDE ◇ KOLOFOG ◇ KOLOSPRAY ◇ KUMULUS ◇ MAGNETIC 70,90 and 95 ◇ MICROFLOTOX ◇ PRECIPITATED SULFUR ◇ SOFRIL ◇ SPERLOX-S ◇ SPERSUL ◇ SPERSUL THIOVIT ◇ SUBLIMED SULFUR ◇ SULFIDAL ◇ SULFORON ◇ SULFUR FLOWER (DOT) ◇ SULKOL ◇ SUPER COSAN ◇ SULPHUR (DOT) ◇ SULPHUR, lump or power (DOT) ◇ SULPHUR, molten (DOT) ◇ SULSOL ◇ TECHNETIUM TC 99M SULFUR COLLOID ◇ TESULOID ◇ THIOLUX ◇ THIOVIT

TOXICITY DATA with REFERENCE
eye-hmn 8 ppm ANCHAM 21,1411,49
ivn-rat LDLo:8 mg/kg JAPMA8 29,289,40
ivn-dog LDLo:10 mg/kg JAPMA8 29,289,40
orl-rbt LDLo:175 mg/kg JAPMA8 29,289,40
ivn-rbt LDLo:5 mg/kg JAPMA8 29,289,40
ipr-gpg LDLo:55 mg/kg JAPMA8 29,289,40

CONSENSUS REPORTS: Reported in EPA TSCA Inventory.

DOT Classification: ORM-C; Label: None; Flammable Solid; Label: Flammable Solid

SAFETY PROFILE: Poison by ingestion, intravenous and intraperitoneal routes. A human eye irritant. A fungicide. Chronic inhalation can cause irritation of mucous membranes. Combustible when exposed to heat or flame or by chemical reaction with oxidizers. Explosive in the form of dust when exposed to flame. Can react violently with halogens; carbides; halogenates; halogenites; zinc; uranium; tin; sodium; lithium; nickel; palladium; phosphorus; potassium; indium; calcium; boron; aluminum; (aluminum + niobium pentoxide); ammonia; ammonium nitrate; ammonium perchlorate; BrF_5; BrF_3; (Ca + VO + H_2O); $Ca(OCl)_2$; Ca_3P_2; Cs_3N; charcoal; (Cu + chlorates); ClO_2; ClO; ClF_3; CrO_3; $Cr(OCl)_2$; hydrocarbons; IF_5; IO_5; PbO_2; $Hg(NO_3)_2$; HgO; Hg_2O; NO_2; P_2O_3; (KNO_3 + As_2S_3); K_3N; $KMnO_4$; $AgNO_3$; Ag_2O; NaH; ($NaNO_3$ + charcoal); (Na + SnI_4); SCl_2; Tl_2O_3; F_2. Can react with oxidizing materials. To fight fire, use water or special mixtures of dry chemical. When heated it burns and emits highly toxic fumes of SO_x. See also NUISANCE DUSTS.

SOE000 ***HR: 3***
SULFUR BROMIDE
mf: S_2Br_2 mw: 223.96

PROP: Mp: −40°, bp: 54° @ 0.2 mm, d: 2.635.

SYN: SULFUR MONOBROMIDE

SAFETY PROFILE: Poison irritant to skin, eyes, and mucous membranes. Combustible when exposed to heat or flame. It will react with water or steam to produce toxic and corrosive fumes. When heated to decomposition it emits highly toxic fumes of SO_x and Br^-. See also BROMIDES.

SOF000 CAS:13780-57-9 ***HR: 3***
SULFUR CHLORIDE PENTAFLUORIDE
mf: ClF_5S mw: 162.51

SYNS: SULFUR CHLORIDE FLUORINE ◇ SULFUR CHLOROPENTAFLUORIDE

TOXICITY DATA with REFERENCE
ihl-rat LCLo:100 ppm/1H BJIMAG 27,1,70

OSHA PEL: TWA 2.5 mg(F)/m^3
ACGIH TLV: TWA 2.5 mg(F)/m^3
NIOSH REL: (Inorganic Fluorides) TWA 2.5 mg(F)/m^3

SAFETY PROFILE: Poison by inhalation. An irritant to skin, eyes, and mucous membranes. When heated to decomposition it emits toxic fumes of Cl^-, F^-, and SO_x. See also FLUORIDES.

SOF500 ***HR: D***
SULFUR COMPOUNDS

SAFETY PROFILE: Variable toxicity. See specific material as listed. Common air contaminants. When heated to decomposition these materials can evolve highly toxic fumes contining SO_x. See also SULFIDES.

SOG500 CAS:10545-99-0 ***HR: 3***
SULFUR DICHLORIDE
DOT: UN 1828
mf: Cl_2S mw: 102.96

PROP: Reddish-brown liquid; pungent odor. Mp: −78°, bp: 59°, d: 1.621 @ 15°/15°, vap d: 3.55.

SYNS: CHLORIDE of SULFUR (DOT) ◇ CHLORINE SULFIDE ◇ DICHLOROSULFANE ◇ MONOSULFUR DICHLORIDE ◇ SULFUR CHLORIDE ◇ SULFUR CHLORIDE (MONO) (DOT)

CONSENSUS REPORTS: Reported in EPA TSCA Inventory.

DOT Classification: Corrosive Material; Label: Corrosive

SAFETY PROFILE: Poison irritant and corrosive to skin, eyes, and mucous membranes. Flammable when exposed to heat or flame. Reacts violently with Al; NH_3; K; Na; acetone; dimethyl sulfoxide; water; oxidants; metals; hexafluoro isopropylidene amino lithium, and toluene. Reactive with water; steam. When heated to decomposition it emits very toxic fumes of SO_x and Cl^-.

SOH000 ***HR: 3***
SULFUR DINITRIDE
mf: S_4N_2 mw: 156.3

PROP: Red liquid or grey solid. D: 1.901 @ 18°, mp: 23°, bp: explodes with decomp @ 100°. Sol in ether; sltly sol in alc, CS_2.

SAFETY PROFILE: Unstable. Explodes @ 100°. When heated to decomposition it emits very toxic fumes of SO_x and NO_x. See also SULFUR NITRIDE.

SOH500 CAS:7446-09-5 ***HR: 3***
SULFUR DIOXIDE
DOT: UN 1079
mf: O_2S mw: 64.06

PROP: Colorless gas or liquid under pressure; pungent odor. Mp: −75.5°, bp: −10.0°, d (liquid): 1.434 @ 0°, vap d: 2.264 @ 0°, vap press: 2538 mm @ 21.1°. Sol in water.

SYNS: BISULFITE ◇ FERMENICIDE LIQUID ◇ FERMENICIDE POWDER ◇ SCHWEFELDIOXYD (GERMAN) ◇ SIARKI DWUTLENEK (POLISH) ◇ SULFUROUS ACID ANHYDRIDE ◇ SULFUROUS ANHYDRIDE ◇ SULFUROUS OXIDE ◇ SULFUR OXIDE ◇ SULPHUR DIOXIDE, LIQUEFIED (DOT)

TOXICITY DATA with REFERENCE
eye-rbt 6 ppm/4H/32D MLD JPCAAC 10,17,60
mmo-omi 10 mmol/L MUREAV 39,149,77
dnd-hmn:lym 5700 ppb MUREAV 39,149,77

ihl-mus TCLo:32 ppm/24H (female 7-18D post):REP TJADAB 31,9B,85
ihl-mus TCLo:25 ppm/7H (female 6-15D post):TER FCTXAV 18,743,80
ihl-mus TCLo:500 ppm/5M/30W-I:ETA BJCAAI 21,606,67
ihl-hmn LCLo:1000 ppm/10M:PUL CTOXAO 5,198,72
ihl-hmn TCLo:3 ppm/5D:PUL TXAPA9 22,319,72
ihl-hmn TCLo:12 ppm/1H:PUL SAIGBL 14,449,72
ihl-hmn LCLo:3000 ppm/5M TABIA2 3,231,33
ihl-rat LC50:2520 ppm/1H NTIS** AD-A148-952
ihl-mus LC50:3000 ppm/30M JCTODH 4,236,77
ihl-gpg LCLo:1039 ppm/24H CBTIAE 10,281,39
ihl-frg LCLo:1 pph/15M HBAMAK 4,1396,35

CONSENSUS REPORTS: EPA Extremely Hazardous Substances List. Reported in EPA TSCA Inventory. EPA Genetic Toxicology Program.

OSHA PEL: (Transitional: TWA 5 ppm) TWA 2 ppm; STEL 5 ppm
ACGIH TLV: TWA 2 ppm; STEL 5 ppm
DFG MAK: 2 ppm (5 mg/m^3)
NIOSH REL: (Sulfur Dioxide) TWA 0.5 ppm
DOT Classification: Nonflammable Gas; Label: Nonflammable Gas; Poison A; Label: Poison Gas

SAFETY PROFILE: A poison gas. Experimental reproductive effects. Human mutation data reported. Human systemic effects by inhalation: pulmonary vascular resistance, respiratory depression and other pulmonary changes. Questionable carcinogen with experimental tumorigenic and teratogenic data. It chiefly affects the upper respiratory tract and the bronchi. It may cause edema of the lungs or glottis, and can produce respiratory paralysis. A corrosive irritant to eyes, skin, and mucous membranes. This material is so irritating that it provides its own warning of toxic concentration. Levels of 400-500 ppm are immediately dangerous to life. Its toxicity is comparable to that of hydrogen chloride. However, less than fatal concentration can be borne for fair periods of time with no apparent permanent damage. It is a common air contaminant.

A nonflammable gas. It reacts violently with acrolein; Al; $CsHC_2$; Cs_2O; chlorates; ClF_3; Cr; FeO; F_2; Mn; KHC_2; $KClO_3$; Rb_2C_2; Na; Na_2C_2; SnO; lithium acetylene carbide diammino. Will react with water or steam to produce toxic and corrosive fumes. Incompatible with halogens, or interhalogens; lithium nitrate; metal acetylides; metal oxides; metals; polymeric tubing; potassium chlorate; sodium hydride. When heated to decomposition it emits toxic fumes of SO_x.

SOI000 CAS:2551-62-4 ***HR: 1***
SULFUR HEXAFLUORIDE
DOT: UN 1080
mf: F_6S mw: 146.06

PROP: Colorless gas. Mp: −51° (subl @ −64°), vap d: 6.602, d (liquid): 1.67 @ −100°.

SYNS: HEXAFLUORURE de SOUFRE (FRENCH) ◇ SULFUR FLUORIDE

TOXICITY DATA with REFERENCE
ivn-rbt LD50:5790 mg/kg TOLED5 1000(Sp 1),100,80

CONSENSUS REPORTS: Reported in EPA TSCA Inventory.

OSHA PEL: TWA 1000 ppm
ACGIH TLV: TWA 1000 ppm
DFG MAK: 1000 ppm (6000 mg/m^3)
DOT Classification: Nonflammable Gas; Label: Nonflammable Gas

SAFETY PROFILE: This material is chemically inert in the pure state and is considered to be physiologically inert as well. However, as it is ordinarily obtainable, it can contain variable quantities of the low sulfur fluorides. Some of these are toxic, very reactive chemically and corrosive in nature. These materials can hydrolyze on contact with water to yield hydrogen fluoride, which is highly toxic and very corrosive. In high concentrations and when pure it may act as a simple asphyxiant. Incompatible with disilane. Vigorous reaction with disilane. May explode. When heated to decomposition emits highly toxic fumes of F^- and SO_x.

SOI200 CAS:57670-85-6 ***HR: 3***
SULFUR THIOCYANATE
mf: $C_2N_2S_3$ mw: 148.22

SAFETY PROFILE: Decomposes explosively on storage at room temperature. When heated to decomposition it emits toxic fumes of SO_x and NO_x. See also THIOCYANATES.

SOI500 CAS:7664-93-9 ***HR: 3***
SULFURIC ACID
DOT: UN 1830/UN 1832/UN 2796
mf: H_2O_4S mw: 98.08

PROP: Colorless oily liquid; odorless. Mp: 10.49°, d: 1.834, vap press: 1 mm @ 145.8°, bp: 290°, decomp @ 340°. Misc with water and alc (liberating great heat).

SYNS: ACIDE SULFURIQUE (FRENCH) ◇ ACIDO SOLFORICO (ITALIAN) ◇ BOV ◇ DIPPING ACID ◇ HYDROOT ◇ MATTING ACID (DOT) ◇ NORDHAUSEN ACID (DOT) ◇ OIL of VITRIOL (DOT) ◇ SCHWEFELSAEURELOESUNGEN (GERMAN) ◇ SPENT SULFURIC ACID (DOT) ◇ SULPHURIC ACID ◇ VITRIOL BROWN OIL ◇ VITRIOL, OIL of (DOT) ◇ ZWAVELZUUROPLOSSINGEN (DUTCH)

TOXICITY DATA with REFERENCE
eye-rbt 1380 μg SEV AJOPAA 29,1363,46
eye-rbt 100 mg rns SEV TXCYAC 23,281,82

ihl-rbt TCLo:20 mg/m^3/7H (female 6-18D post):TER JESEDU 13,251,79
ihl-hmn TCLo:3 mg/m^3/24W BJIMAG 18,63,61
unr-man LDLo:135 mg/kg 85DCAI 2,73,70
orl-rat LD50:2140 mg/kg AIHAAP 30,470,69
ihl-rat LC50:510 mg/m^3/2H 85GMAT -,107,82
ihl-mus LC50:320 mg/m^3/2H 85GMAT -,107,82
ihl-gpg LC50:18 mg/m^3 MELAAD 45,590,54

CONSENSUS REPORTS: Reported in EPA TSCA Inventory.

OSHA PEL: TWA 1 mg/m^3
ACGIH TLV: TWA 1 mg/m^3; STEL 3 ppm
DFG MAK: 1 mg/m^3
NIOSH REL: (Sulfuric Acid) TWA 1 mg/m^3
DOT Classification: Corrosive Material; Label: Corrosive

SAFETY PROFILE: Human poison by unspecified route. Experimental poison by inhalation. Moderately toxic by ingestion. A severe eye irritant. Extremely irritating, corrosive, and toxic to tissue resulting in rapid destruction of tissue, causing severe burns. If much of the skin is involved, it is accompanied by shock, collapse and symptoms similar to those seen in severe burns. Repeated contact with dilute solutions can cause a dermatitis, and repeated or prolonged inhalation of a mist of sulfuric acid can cause inflammation of the upper respiratory tract leading to chronic bronchitis. Sensitivity to sulfuric acid or mists or vapors varies with individuals. Normally 0.125-0.50 ppm may be mildly annoying and 1.5-2.5 ppm can be definitely unpleasant, 10-20 ppm is unbearable. Workers exposed to low concentrations of the vapor gradually lose their sensitivity to its irritating action. Inhalation of concentrated vapor or mists from hot acid or oleum can cause rapid loss of consciousness with serious damage to lung tissue. Severe exposure may cause a chemical pneumonitis; erosion of the teeth due to exposure to strong acid fumes has been recognized in industry. An experimental teratogen.

This is a very powerful, acidic oxidizer which can ignite or explode on contact with many materials, i.e., acetic acid; acetone cyanhydrin; (acetone + HNO_3); (acetone + $K_2Cr_2O_7$); acetonitrile; acrolein; acrylonitrile; (acrylonitrile + H_2O); (alcohols + H_2O_2); allyl alcohol; allyl chloride; NH_4OH; 2-amino ethanol; NH_4; triperchromate; aniline; (bromates + metals); BrF_5; n-butyraldehyde; carbides; $CoHC_2$; chlorates; (metals + chlorates); ClF_3; chlorosulfonic acid; Cu_3N; diisobutylene; (dimethyl benzylcarbinol + H_2O_2); epichlorohydrin; ethylene cyanhydrin; ethylene diamine; ethylene glycol; ethylene imine; fulminates; HCl; H_2; IF_7; (indene + HNO_3); Fe; isoprene; Li_6Si_2; Hg_3N_2; mesityl oxide; metals; (HNO_3 + glycerides); p-nitrotoluene; perchlorates; $HClO_4$; (C_6H_6 + permanganates); pentasilver trihydroxydiamino phosphate; (1-phenyl-2-methyl propyl alcohol + H_2O_2); P; $P(OCN)_3$; picrates; potassium-tert-butoxide; $KClO_3$; $KMnO_4$; ($KMnO_4$ + KCl); ($KMnO_4$ + H_2O); β-propiolactone; $RbHC_2$; propylene oxide; pyridine; Na; Na_2CO_3; NaOH; steel; styrene monomer; water; vinyl acetate; (HNO_3 + toluene). When heated it emits highly toxic fumes; will react with water or steam to produce heat; can react with oxidizing or reducing materials. When heated to decomposition it emits toxic fumes of SO_x. See also SULFATES.

SOI510 ***HR: 3***
SULFURIC ACID, aromatic

PROP: Clear, reddish-brown liquid; peculiar aromatic odor; pleasant acid taste when diluted.

SYN: ELIXIR of VITRIOL

SAFETY PROFILE: Corrosive. Flammable when exposed to heat or flame. Explosive in the form of vapor (ethyl alcohol) when exposed to heat or flame. When heated to decomposition it emits toxic fumes of SO_x. See also ETHYL ALCOHOL and SULFURIC ACID.

SOI520 CAS:8014-95-7 ***HR: 3***
SULFURIC ACID, fuming
DOT: NA 1831
mf: $H_2O_4S{\cdot}O_3S$ mw: 178.14

PROP: Heavy, fuming, yellow liquid. H_2SO_4 + up to 80% SO_3. A solution of sulfuric anhydride (sulfur trioxide) in anhydrous sulfuric acid (NTIS** PB233-098).

SYNS: DISULPHURIC ACID ◇ DITHIONIC ACID ◇ FUMING SULFURIC ACID (DOT) ◇ OLEUM (DOT) ◇ PYROSULPHURIC ACID

TOXICITY DATA with REFERENCE
ihl-rat LC50:347 ppm/1H TXAPA9 42,417,77

NIOSH REL: TWA 1 mg/m^3
DOT Classification: Corrosive Material; Label: Corrosive and Corrosive, Poison

SAFETY PROFILE: A poison. Moderately toxic by inhalation. A corrosive irritant to skin, eyes, and mucous membranes. A very dangerous fire hazard by chemical reaction with reducing agents and carbohydrates. A severe explosion hazard by chemical reaction with acetic acid; acetic anhydride; acetonitrile; acrolein; acrylic acid; acrylonitrile; allyl alcohol; allyl chloride; 2-amino ethanol; NH_4OH; aniline; cresol; n-butyraldehyde; cumene; dichloroethyl ether; diethylene glycol monomethyl ether; diisobutylene; epichlorohydrin; ethyl acetate; ethylene cyanohydrin; ethylene diamine; ethylene glycol; ethylene glycol monoethyl ether acetate; ethylene imine; glyoxal; HCl; HF; isoprene; isopropyl alcohol; mesityl oxide; methyl ethyl ketone; HNO_3; 2-nitropropane; β-propiolacetone; propylene oxide; pyridine;

NaOH; styrene monomer; vinylidene chloride; sulfolane; vinyl acetate. Will react with water or steam to produce heat and toxic and corrosive fumes. Can react vigorously with reducing materials. When heated to decomposition it emits highly toxic fumes of SO_x. See also SULFUROUS ACID.

SOI530 CAS:7664-93-9 ***HR: 3***
SULFURIC ACID (mist)
mf: H_2O_4S mw: 98.08

PROP: The airborne form of sulfuric acid is an aerosol of droplets of varying diameter of aq sulfuric acid solution.

TOXICITY DATA with REFERENCE
ihl-hmn TCLo:3 mg/m^3/24W:MTH BJIMAG 18,63,61
ihl-rat LCLo:178 ppm/7H AMIHBC 2,716,50
ihl-mus LCLo:140 ppm/3.5H AMIHBC 2,716,50
ihl-gpg LCLo:48 ppm/H AMIHBC 2,716,50

CONSENSUS REPORTS: EPA Extremely Hazardous Substances List. Reported in EPA TSCA Inventory.

ACGIH TLV: TWA 1 mg/m^3
NIOSH REL: TWA 1 mg/m^3

SAFETY PROFILE: Poison by inhalation. Human systemic effects by inhalation: mouth effects. When heated to decomposition it emits toxic fumes of SO_x. See also SULFURIC ACID.

SOJ000 CAS:68130-43-8 ***HR: 1***
SULFURIC ACID, sec-ALKYL ESTER, SODIUM SALT

PROP: Alkyl containing C_{8-18}.

TOXICITY DATA with REFERENCE
skn-hmn 2500 μg/24H MLD AKEDAX 235,180,69
skn-rat 10 mg/16H MLD JSCCA5 22,411,71
skn-rbt 10 mg MLD JSCCA5 22,411,71

CONSENSUS REPORTS: Reported in EPA TSCA Inventory.

SAFETY PROFILE: A human and experimental skin irritant. When heated to decomposition it emits toxic fumes of SO_x and Na_2O. See also ESTERS and SULFATES.

SOK000 CAS:142-87-0 ***HR: 3***
SULFURIC ACID, DECYL ESTER, SODIUM SALT
mf: $C_{10}H_{21}O_4S{\cdot}Na$ mw: 260.36

SYNS: SODIUM DECYL SULFATE ◇ SULFURIC ACID, MONODECYL ESTER, SODIUM SALT

TOXICITY DATA with REFERENCE
orl-rat LD50:1950 mg/kg JAPMA8 42,283,53
ipr-mus LD50:284 mg/kg JAPMA8 42,283,53
ivn-mus LD50:56 mg/kg CSLNX* NX#00153

CONSENSUS REPORTS: Reported in EPA TSCA Inventory.

SAFETY PROFILE: Poison by intravenous and intraperitoneal routes. Moderately toxic by ingestion. When heated to decomposition it emits toxic fumes of SO_x and Na_2O. See also SULFATES and ESTERS.

SOM500 CAS:2235-54-3 ***HR: 1***
SULFURIC ACID, MONODODECYL ESTER, AMMONIUM SALT
mf: $C_{12}H_{26}O_4S{\cdot}H_3N$ mw: 283.48

SYNS: AMMONIUM DODECYL SULFATE ◇ AMMONIUM-N-DODECYL SULFATE ◇ AMMONIUM LAURYL SULFATE ◇ DODECYL AMMONIUM SULFATE ◇ LAURYL AMMONIUM SULFATE ◇ LAURYL SULFATE AMMONIUM SALT ◇ SULFURIC ACID, LAURYL ESTER, AMMONIUM SALT

TOXICITY DATA with REFERENCE
skn-rbt 10 mg/24H DCTODJ 1,305,78
eye-rbt 2 mg DCTODJ 1,305,78

CONSENSUS REPORTS: Reported in EPA TSCA Inventory.

SAFETY PROFILE: A skin and eye irritant. When heated to decomposition it emits very toxic fumes of NH_3, NO_x, and SO_x. See also ESTERS and SULFATES.

SON000 CAS:139-96-8 ***HR: 1***
SULFURIC ACID, MONODODECYL ESTER, compounded with 2,2',2''-NITRILOTRIS(ETHANOL)
mf: $C_{12}H_{26}O_4S{\cdot}C_6H_{15}NO_3$ mw: 415.66

SYNS: AKYPOSAL TLS ◇ CYCLORYL TAWF ◇ CYCLORYL WAT ◇ DRENE ◇ ELFAN 4240 T ◇ EMAL T ◇ EMERSAL 6434 ◇ MAPROFIX TLS 65 ◇ MELANOL LP20T ◇ PROPASTE T ◇ REWOPOL TLS 40 ◇ RICHONOL T ◇ SIPON LT ◇ STANDAPOL TLS 40 ◇ STEINAPOL TLS 40 ◇ STEPANOL WAT ◇ STERLING WAT ◇ SULFURIC ACID, DODECYL ESTER, TRIETHANOLAMINE SALT ◇ SULFURIC ACID, MONODODECYL ESTER, compounded with 2,2',2''-NITRILOTRIETHANOL (1 : 1) ◇ TEA LAURYL SULFATE ◇ TEXAPON T-42 ◇ TRIETHANOLAMINE DODECYL SULFATE ◇ TRIETHANOLAMINE LAURYL SULFATE ◇ TYLOROL LT 50

TOXICITY DATA with REFERENCE
skn-rbt 10 mg/34H DCTODJ 1,305,78
eye-rbt 46 mg TXAPA9 19,276,71
eye-rbt 2 mg DCTODJ 1,305,78

CONSENSUS REPORTS: Reported in EPA TSCA Inventory.

SAFETY PROFILE: A skin and eye irritant. When heated to decomposition it emits very toxic fumes of NO_x, SO_x, and CN^-. See also NITRILES and SULFATES.

SON510 CAS:10025-67-9 ***HR: 3***
SULFUR MONOCHLORIDE
DOT: UN 1828
mf: Cl_2S_2 mw: 135.02

PROP: Amber to yellowish-red, oily, fuming liquid; penetrating odor. mp: −80°, bp: 138.0°, flash p: 245°F (CC), d: 1.6885 @ 15.5°/15.5°, autoign temp: 453°F, vap press: 10 mm @ 27.5°, vap d: 4.66. Decomp in water.

SYNS: CHLORIDE of SULFUR (DOT) ◇ DISULFUR DICHLORIDE ◇ SIARKI CHLOREK (POLISH) ◇ SULFUR CHLORIDE ◇ SULFUR CHLORIDE(DI) (DOT) ◇ SULFUR SUBCHLORIDE ◇ THIOSULFUROUS DICHLORIDE

CONSENSUS REPORTS: Reported in EPA TSCA Inventory.

OSHA PEL: (Transitional: TWA 1 ppm) CL 1 ppm
ACGIH TLV: CL 1 ppm
DFG MAK: 1 ppm (6 mg/m^3)
DOT Classification: Corrosive Material; Label: Corrosive

SAFETY PROFILE: Poison by ingestion and inhalation. A fuming, corrosive liquid very irritating to skin, eyes, and mucous membranes. It decomposes on contact with water to form the highly irritating hydrogen chloride, thiosulfuric acid, and sulfur. Its toxic effects are irritating to the upper respiratory tract, although the results of intoxication are usually transitory in nature. However, if hydrolysis is not complete in the upper respiratory tract, injury to the bronchioles and alveoli can result. A fire hazard when in contact with organic matter; P_2O_3; Na_2O_2; water; $Cr(OCl)_2$. Combustible when exposed to heat or flame. Will react with water or steam to produce heat and toxic and corrosive fumes. Can react with oxidizing materials. To fight fire, use CO_2, dry chemical. When heated to decomposition it emits highly toxic fumes of Cl^- and SO_x.

SON520 CAS:78173-90-7 ***HR: 3***
SULFURMYCIN A
mf: $C_{43}H_{53}NO_{16}$ mw: 839.97

TOXICITY DATA with REFERENCE
dni-mus:leu 480 nmol/L JANTAJ 34,1596,81
oms-mus:leu 43 nmol/L JANTAJ 34,1596,81
ipr-mus LD50:100 mg/kg JANTAJ 35,82-67,82

SAFETY PROFILE: Poison by intraperitoneal route. Mutation data reported. When heated to decomposition it emits toxic fumes of NO_x.

SON525 CAS:78193-30-3 ***HR: 3***
SULFURMYCIN B
mf: $C_{43}H_{51}NO_{16}$ mw: 837.95

TOXICITY DATA with REFERENCE
dni-mus:leu 660 nmol/L JANTAJ 34,1596,81
oms-mus:leu 60 nmol/L JANTAJ 34,1596,81
ipr-mus LD50:25 mg/kg JANTAJ 35,82-68,82

SAFETY PROFILE: Poison by intraperitoneal route. Mutation data reported. When heated to decomposition it emits toxic fumes of NO_x.

SOO000 CAS:28950-34-7 ***HR: 3***
SULFUR NITRIDE
mf: S_4N_4 mw: 184.3

PROP: Orange-red crystals. D: 2.22 @ 15°, mp: subl @ 179°, bp: explodes @ 160°. Decomp by cold water; sol in CS_2, $CHCl_3$, benzene, NH_3.

SYN: TETRASULFUR TETRANITRIDE

SAFETY PROFILE: An explosive sensitive to friction, shock, or heating above 100°C or contact with $Ba(ClO_3)_2$. Sensitivity increases with purity. Endothermic nitride can explosively decompose on friction, shock, or heating. When heated to decomposition it emits very toxic fumes of SO_x and NO_x.

SOO500 CAS:7782-99-2 ***HR: 3***
SULFUROUS ACID
DOT: UN 1833
mf: H_2SO_3 mw: 82.08

PROP: Colorless liquid; suffocating sulfur odor (in solution only). D: approx 1.03.

SYNS: SULFUR DIOXIDE, solution ◇ SCHWEFLIGE SAURE (GERMAN)

TOXICITY DATA with REFERENCE
orl-hmn TDLo:500 μg/kg:GIT AEXPBL 54,421,06

CONSENSUS REPORTS: Reported in EPA TSCA Inventory.

DOT Classification: Corrosive Material; Label: Corrosive

SAFETY PROFILE: A poison by ingestion and inhalation. A corrosive irritant to skin, eyes, and mucous membranes. Human systemic effects by ingestion: nausea or vomiting, hypermotility, diarrhea, and other gastrointestinal effects. When heated to decomposition it emits highly toxic fumes of SO_x.

SOP000 CAS:2312-35-8 ***HR: 3***
SULFUROUS ACID, 2-(p-tert-BUTYLPHENOXY) CYCLOHEXYL-2-PROPYNYL ESTER
mf: $C_{19}H_{26}O_4S$ mw: 350.51

SYNS: BPPS ◇ 2-(p-tert-BUTYLPHENOXY)CYCLOHEXYL PROPARGYL SULFITE ◇ 2-(p-tert-BUTYLPHENOXY)CYCLOHEXYL 2-PROPYNYL SULFITE ◇ COMITE ◇ 2-(4-(1,1-DIMETHYLETHYL) PHENOXY)CYCLOHEXYL 2-PROPYNYL ESTER, SULFUROUS ACID ◇ 2-(4-(1,1-DIMETHYLETHYL)PHENOXY)CYCLOHEXYL 2-PROPYNYL SULFITE ◇ DO 14 ◇ ENT 27,226 ◇ NAUGATUCK D-014

◇ OMAIT ◇ OMITE ◇ PROPARGITE (DOT) ◇ UNIROYAL D014 ◇ U.S. RUBBER D-014

TOXICITY DATA with REFERENCE
orl-rat LD50:1480 mg/kg TXAPA9 14,515,69
skn-rat LD50:250 mg/kg WRPCA2 9,119,70

SAFETY PROFILE: Poison by skin contact. Moderately toxic by ingestion. When heated to decomposition it emits toxic fumes of SO_x. See also ESTERS and SULFUROUS ACID.

SOP500 CAS:140-57-8 ***HR: 3***
SULFUROUS ACID, 2-(p-tert-BUTYLPHENOXY)-1-METHYLETHYL-2-CHLOROETHYL ESTER
mf: $C_{15}H_{23}ClO_4S$ mw: 334.89

PROP: Liquid. D: 1.145-1.1620, mp: −31.7°, bp: 175° @ 0.1 mm, vap press: <10 mm @ 25°. Misc with many organic solvents; insol in water.

SYNS: ACARACIDE ◇ ARACIDE ◇ ARAMITE ◇ ARAMITEAR-ARAMITE-15W ◇ ARATRON ◇ BUTYLPHENOXYISOPROPYL CHLOROETHYL SULFITE ◇ 2-(p-BUTYLPHENOXY)ISOPROPYL 2-CHLOROETHYL SULFITE ◇ 2-(4-tert-BUTYLPHENOXY)ISOPROPYL-2-CHLOROETHYL SULFITE ◇ 2-(p-tert-BUTYLPHENOXY)ISOPROPYL 2′-CHLOROETHYL SULPHITE ◇ 2-(p-tert-BUTYLPHENOXY)-1-METHYLETHYL 2-CHLOROETHYL ESTER of SULPHUROUS ACID ◇ 2-(p-BUTYLPHENOXY)-1-METHYLETHYL 2-CHLOROETHYL SULFITE ◇ 2-(p-tert-BUTYLPHENOXY)-1-METHYLETHYL-2-CHLOROETHYL SULFITE ESTER ◇ 2-(p-tert-BUTYLPHENOXY)-1-METHYLETHYL 2′-CHLOROETHYL SULPHITE ◇ 2-(p-tert- BUTYLPHENOXY)-1-METHYLETHYL SULPHITE of 2-CHLOROETHANOL ◇ 1-(p-tert-BUTYLPHENOXY)-2-PROPANOL-2-CHLOROETHYLSULFITE ◇ CES ◇ 2-CHLOROETHANOL-2-(p-tert-BUTYLPHENOXY)-1-METHYLETHYL SULFITE ◇ 2-CHLORO- ETHANOL ESTER with 2-(p-tert-BUTYLPHENOXY)-1-METHYLETHYL SULFITE ◇ β-CHLOROETHYL-β′-(p-tert-BUTYLPHENOXY)-α′- METHYLETHYL SULFITE ◇ β-CHLOROETHYL-β-(p-tert-BUTYLPHENOXY)-α-METHYLETHYL SULPHITE ◇ 2-CHLOROETHYL1-METHYL-2-(p-tert-BUTYLPHENOXY)ETHYL SULPHATE ◇ 2-CHLOROETHYL SULFUROUS ACID-2-(4-(1,1-DIMETHYLETHYL) PHENOXY)-1-METHYLETHYL ESTER ◇ 2-CHLOROETHYL SULPHITE of 1-(p-tert-BUTYLPHENOXY)-2-PROPANOL ◇ COMPOUND 88R ◇ ENT 16,519 ◇ NIAGARAMITE ◇ ORTHO-MITE ◇ 88-R

TOXICITY DATA with REFERENCE
orl-rat TDLo:2520 mg/ (12 male W pre):REP TXAPA9 2,441,60
orl-rat TDLo:21900 mg/kg/2Y-C:CAR TXAPA9 2,441,60
orl-rat TDLo:15 g/kg/2Y-C:NEO CANCAR 13,1035,60
orl-mus TDLo:130 g/kg/80W-C:CAR JNCIAM 42,1101,69
orl-dog TDLo:20 g/kg/2Y-C:ETA CANCAR 13,780,60
orl-hmn LDLo:429 mg/kg 34ZIAG -,111,69
orl-rat LD50:3900 mg/kg PCOC** -,61,66
ipr-mus LDLo:200 mg/kg TXAPA9 23,288,72
orl-gpg LD50:3900 mg/kg PCOC** -,61,66

CONSENSUS REPORTS: NTP Fifth Annual Report on Carcinogens. IARC Cancer Review: Group 2B IMEMDT 7,56,87; Animal Sufficient Evidence IMEMDT 5,39,74.

SAFETY PROFILE: Confirmed carcinogen with experimental carcinogenic, neoplastigenic, and tumorigenic data. Experimental poison by intraperitoneal route. Moderately toxic to humans by ingestion. Moderately toxic experimentally by ingestion. Experimental reproductive effects. A pesticide. When heated to decomposition it emits toxic fumes of Cl^- and SO_x. See also ESTERS and SULFUROUS ACID.

SOQ000 ***HR: 3***
SULFUR OXIDE (N-FLUOROSULPHONYL)IMIDE
mf: FNO_3S_2 mw: 145.13

SAFETY PROFILE: Reacts explosively with water. When heated to decomposition it emits very toxic fumes of F^-, NO_x, and SO_x.

SOQ450 CAS:5714-22-7 ***HR: 3***
SULFUR PENTAFLUORIDE
mf: $F_{10}S_2$ mw: 254.12

SYNS: SULFUR DECAFLUORIDE ◇ TL 70

TOXICITY DATA with REFERENCE
ihl-rat LC50:2 g/m^3/10M NTIS** PB158-508
ihl-mus LC50:1 g/m^3/10M NTIS** PB158-508
ihl-dog LC50 4 g/m^3/10M NTIS** PB158-508
ivn-dog LDLo:1 mg/kg AMIHBC 8,436,53
ihl-mky LC50:9 g/m^3/10M NTIS** PB158-508
ihl-cat LC50:4500 mg/m^3/10M NTIS** PB158-508
ihl-rbt LC50:4 g/m^3/10M NTIS** PB158-508
ivn-rbt LDLo:5790 μg/kg AMIHBC 8,436,53

OSHA PEL: (Transitional: TWA 0.025 ppm) CL 0.01 ppm
ACGIH TLV: CL 0.01 ppm
DFG MAK: 0.025 ppm (0.25 mg/m^3)

SAFETY PROFILE: Poison by intravenous route. Moderately toxic by inhalation. When heated to decomposition it emits very toxic fumes of F^- and SO_x. See also FLUORIDES and SULFIDES.

SOQ500 ***HR: 3***
SULFUR TETRACHLORIDE
mf: SCl_4 mw: 173.89

PROP: Yellow-brown liquid or gas at ordinary temp. Mp: −30°, bp: −15° (decomp).

SAFETY PROFILE: Poison irritant to skin, eyes, and mucous membranes. Will react with water or steam to produce toxic and corrosive fumes. When heated to decomposition it emits highly toxic fumes of Cl^- and SO_x. See also HYDROCHLORIC ACID.

SOR000 CAS:7783-60-0 ***HR: 3***
SULFUR TETRAFLUORIDE
DOT: UN 2418
mf: F_4S mw: 108.06

PROP: Gas. Bp: −40°, mp: −124°.

SYN: TETRAFLUOROSULFURANE

TOXICITY DATA with REFERENCE
ihl-rat LCLo:19 ppm/4H JOCMA7 4,262,62

CONSENSUS REPORTS: Reported in EPA TSCA Inventory. EPA Extremely Hazardous Substances List.

OSHA PEL: CL 0.1 ppm
ACGIH TLV: CL 0.1 ppm
NIOSH REL: (Inorganic Fluorides) TWA 2.5 mg(F)/m^3
DOT Classification: Poison A; Label: Poison Gas

SAFETY PROFILE: Poison by inhalation. A powerful irritant. Will react with water, steam, or acids to yield toxic and corrosive fumes. Incompatible with dioxygen difluoride. When heated to decomposition it emits very toxic fumes of F^- and SO_x. See also FLUORIDES.

SOR500 CAS:7446-11-9 ***HR: 3***
SULFUR TRIOXIDE
DOT: UN 1829
mf: O_3S mw: 80.06

PROP: It exists in three forms; the most valuable commercially is the Γ form (mp: 16.8°, bp: 44.8°) which has a strong tendency to polymerize to the straight chain β form (mp β: 32.5°) and subsequently to the cross-linked α form (mp α: 62°). When the β or α forms are melted they tend to revert to the Γ form liquid or ice-like crystals. SO_3 (β) asbestos-like crystals; vap press: (β) 433 mm @ 250°, vap press (α): 344 mm, vap d: 2.76.

SYNS: SULFAN ◇ SULFURIC ANHYDRIDE (DOT) ◇ SULFURIC OXIDE ◇ SULFUR TRIOXIDE, STABILIZED (DOT)

TOXICITY DATA with REFERENCE
ihl-hmn TCLo:30 mg/m^3:NOSE,PUL JPBAA7 68,197,54
ihl-gpg LCLo:30 mg/m^3/6H JPBAA7 68,197,54

CONSENSUS REPORTS: EPA Extremely Hazardous Substances List. Reported in EPA TSCA Inventory.

DOT Classification: Corrosive Material; Label: Corrosive

SAFETY PROFILE: Poison by inhalation. Human systemic effects by inhalation: cough and other pulmonary and olfactory changes. A corrosive irritant to skin, eyes, and mucous membranes. Violent reaction with O_2F_2; PbO; $NClO_2$; $HClO_4$; P; tetrafluorethylene; acetonitrile; sulfuric acid; dimethyl sulfoxide; dioxan; water; diphenylmercury; formamide; iodine; pyridine; metal oxides. Reacts with steam to form corrosive, toxic fumes of sulfuric acid. When heated to decomposition it emits toxic fumes of SO_x. See also SULFURIC ACID.

SOS000 CAS:7446-11-9 ***HR: 3***
SULFUR TRIOXIDE (stabilized)
DOT: UN 1829

CONSENSUS REPORTS: Reported in EPA TSCA Inventory.

DOT Classification: Corrosive Material; Label: Corrosive and Poison

SAFETY PROFILE: A toxic and corrosive material to skin, eyes, and mucous membranes. When heated to decomposition it emits toxic fumes of SO_x. See also SULFUR TRIOXIDE.

SOS500 ***HR: 3***
SULFURYL AZIDE CHLORIDE
mf: ClN_3O_2S mw: 141.54

SAFETY PROFILE: Very toxic and irritating. Explosive. Very unstable compound. When heated to decomposition it emits very toxic fumes of Cl^-, NO_x, and SO_x. See also AZIDES.

SOT000 CAS:7791-25-5 ***HR: 3***
SULFURYL CHLORIDE
DOT: UN 1834
mf: Cl_2O_2S mw: 134.96

PROP: Colorless liquid; pungent odor. Mp: −54.1°, bp: 69.1°, d: 1.6674, vap press: 100 mm @ 17.8°, vap d: 4.65.

SYNS: SULFONYL CHLORIDE ◇ SULFURIC OXYCHLORIDE

TOXICITY DATA with REFERENCE
skn-mus TDLo:2850 mg/kg/1Y-I:ETA GISAAA 47(7),9,82

CONSENSUS REPORTS: Reported in EPA TSCA Inventory.

DOT Classification: Corrosive Material; Label: Corrosive

SAFETY PROFILE: A corrosive irritant to skin, eyes, and mucous membranes. Questionable carcinogen with experimental tumorigenic data. Can explode with PbO_2. Will react with water or steam to produce heat and toxic and corrosive fumes. Incompatible with alkalies; diethyl ether; dimethyl sulfoxide; dinitrogen pentaoxide; lead dioxide; phosphorus. When heated to decomposition it emits highly toxic fumes of Cl^- and SO_x. See also SULFURIC ACID and HYDROCHLORIC ACID, which are formed upon hydrolysis.

SOT500 CAS:13637-84-8 ***HR: 3***
SULFURYL CHLORIDE FLUORIDE
mf: $ClFO_2S$ mw: 118.51

PROP: Colorless gas. Mp: −124.7°, bp: 7.1°.

SYNS: CHLORO FLUORO SULFONE ◇ CHLOROSULFONYL FLUO-

RIDE ◇ FLUOROSULFONYL CHLORIDE ◇ SULFONYL CHLORIDE FLUORIDE ◇ SULFURYL CHLOROFLUORIDE ◇ SULFURYL FLUOROCHLORIDE ◇ TL 212

TOXICITY DATA with REFERENCE
ihl-mus LCLo:520 mg/m^3/10M NDRC** NDCrc-132,June,42

CONSENSUS REPORTS: Reported in EPA TSCA Inventory.

OSHA PEL: TWA 2.5 mg(F)/m^3
ACGIH TLV: TWA 2.5 mg(F)/m^3
NIOSH REL: (Inorganic Fluorides) TWA 2.5 mg(F)/m^3

SAFETY PROFILE: Poison by inhalation. When heated to decomposition it emits very toxic fumes of Cl^-, F^-, and SO_x. See also FLUORIDES and CHLORIDES and SULFIDES.

SOU000 ***HR: 3***
SULFURYL DIAZIDE
mf: N_6O_2S mw: 148.11

SAFETY PROFILE: Explodes violently on heating and spontaneously at ambient temperatures. When heated to decomposition it emits very toxic fumes of SO_x and NO_x. See also AZIDES.

SOU500 CAS:2699-79-8 ***HR: 3***
SULFURYL FLUORIDE
DOT: UN 2191
mf: F_2O_2S mw: 102.06

PROP: Colorless gas. Mp: −137°, bp: −55°, d: 3.72 g/L.

SYNS: FLUORURE de SULFURYLE (FRENCH) ◇ SULFURIC OXYFLUORIDE ◇ VIKANE ◇ VIKANE FUMIGANT

TOXICITY DATA with REFERENCE
orl-rat LD50:100 mg/kg 85ARAE 3,15,76/77
ihl-rat LC50:3020 ppm/1H TXAPA9 42,417,77
ihl-mus LCLo:1200 ppm/1H AMPMAR 34(10-11),581,73
ihl-rbt LCLo:5000 ppm/1H AMPMAR 34(10-11),581,73
orl-gpg LD50:100 mg/kg DOWCC*

CONSENSUS REPORTS: Reported in EPA TSCA Inventory.

OSHA PEL: (Transitional: TWA 5 ppm) TWA 5 ppm; STEL 10 ppm
ACGIH TLV: TWA 5 ppm; STEL 10 ppm
DOT Classification: Nonflammable Gas; Label: Nonflammable Gas

SAFETY PROFILE: Poison by ingestion. Mildly toxic by inhalation. Accidental human exposure caused nausea, vomiting, cramps, itching. May be narcotic in high concentration. Can react with water, steam. When heated to decomposition it emits very toxic fumes of F^- and SO_x. See also FLUORIDES.

SOU550 CAS:38194-50-2 ***HR: 3***
SULINDAC
mf: $C_{20}H_{17}FO_3S$ mw: 356.42

PROP: Yellow, odorless crystals from ethyl acetate. Mp: 182-185° (decomp). Sparingly sol in methanol, alc; sltly sol in ethyl acetate; practically insol in water.

SYNS: ARTHROBID ◇ ARTHROCINE ◇ ARTRIBID ◇ CITIREUMA ◇ CLINORIL ◇ cis-5-FLUORO-2-METHYL-1-((4-METHYLSULFINYL) PHENYL)METHYLENE)-1H-INDENE-3-ACETIC ACID ◇ IMBARAL ◇ MK 231 ◇ REUMOFIL ◇ SULINOL

TOXICITY DATA with REFERENCE
orl-man TDLo:43 mg/kg/10D-I:PUL,LIV,BLD JAMAAP 244,269,80
orl-wmn TDLo:56 mg/kg/1W-C:MET,CNS,GIT NEJMAG 300,735,79
orl-man TDLo:133 mg/kg/31D-I:MET,GIT AIMDAP 142,1292,82
orl-wmn LDLo:126 mg/kg/3W-I:MET AIMDAP 142,1292,82
orl-rat LD50:264 mg/kg ARZNAD 30,1398,80
ipr-rat LD50:289 mg/kg IYKEDH 13,637,82
scu-rat LD50:336 mg/kg IYKEDH 13,637,82
orl-mus LD50:507 mg/kg NIIRDN 6,APP-6,82
ipr-mus LD50:305 mg/kg IYKEDH 13,637,82
scu-mus LD50:398 mg/kg IYKEDH 13,637,82

SAFETY PROFILE: Human poison by ingestion. Experimental poison by ingestion, subcutaneous, and intraperitoneal routes. Human systemic effects by ingestion: anorexia, respiratory effects, salivary gland effects, nausea or vomiting, impaired liver function, luekopenia, fever, and metabolic acidosis. An anti-inflammatory agent. When heated to decomposition it emits toxic fumes of F^- and SO_x.

SOU600 CAS:54767-75-8 ***HR: 3***
SULOCTIDYL
mf: $C_{20}H_{35}NOS$ mw: 337.62

SYNS: CP 556S ◇ erythro-p-(ISOPROPYLTHIO)-α-(1-(OCTYLAMINO) ETHYL)BENZYL ALCOHOL ◇ erythro-1-(4-ISOPROPYLTHIOPHENYL)-2-n-OCTYLAMINOPROPANOL ◇ (R*,S*)-4-((1-METHYLETHYL)THIO)-α-(1-OCTYLAMINO)ETHYL)BENZENEMETHANOL ◇ SULOCTIDIL ◇ SULOCTON

TOXICITY DATA with REFERENCE
orl-rat TDLo:600 mg/kg (female 17-22D post):REP IYKEDH 14,761,83
orl-rbt TDLo:1820 mg/kg (female 6-18D post):TER IYKEDH 14,770,83
orl-rat LD50:1740 mg/kg IYKEDH 15,10,84
scu-rat LD50:2600 mg/kg IYKEDH 15,10,84
ivn-rat LD50:8500 μg/kg IYKEDH 15,10,84
orl-mus LD50:3700 mg/kg GWXXBX #2334404
scu-mus LD50:1720 mg/kg IYKEDH 15,10,84
ivn-mus LD50:8400 μg/kg IYKEDH 15,10,84
orl-rbt LD50:1195 mg/kg IYKEDH 15,10,84

SAFETY PROFILE: Poison by intravenous route. Moderately toxic by ingestion and subcutaneous route. An experimental teratogen. Experimental reproductive effects. When heated to decomposition it emits toxic fumes of SO_x and NO_x. A vasodilator.

SOU625 CAS:35400-43-2 ***HR: 3***
SULPROFOS
mf: $C_{12}H_{19}O_2PS_3$ mw: 322.46

SYNS: BAY-NTN-9306 ◇ BOLSTAR ◇ O-ETHYL-O-(4-METHYLMERCAPTO)PHENYL)-S-N-PROPYLPHOSPHOROTHIONOTHIOLATE ◇ O-ETHYL-O-(4-(METHYLTHIO)PHENYL)PHOSPHORODITHIOIC ACID-S-PROPYL ESTER ◇ O-ETHYL-O-(4-(METHYLTHIO)PHENYL) S-PROPYL PHOSPHORODITHIOATE ◇ HELOTHION

TOXICITY DATA with REFERENCE
orl-rat LD50:65 mg/kg 85ARAE 1,205,77
orl-mus LD50:1600 mg/kg DTLWS* 4,380,82
skn-rbt LD50:820 mg/kg FMCHA2 -,C34,83

OSHA PEL: TWA 1 mg/m^3
ACGIH TLV: TWA 1 mg/m^3

SAFETY PROFILE: Poison by ingestion. Moderately toxic by skin contact. When heated to decomposition it emits very toxic fumes of PO_x and SO_x.

SOU650 CAS:60325-46-4 ***HR: D***
SULPROSTONE
mf: $C_{23}H_{31}NO_7S$ mw: 465.61

PROP: Colorless oil.

SYNS: CP 34089 ◇ NALADOR ◇ 16-PHENOXY-omega-17,18,19,20-TETRANOR PROSTAGLANDIN E2 METHYLSULFONYLAMIDE ◇ 16-PHENOXY-17,18,19,20 TETRANOR PROSTAGLANDIN E_2 METHYL SULFONYLAMIDE ◇ SCH 286 ◇ SHB 286 ◇ ZK 57671

TOXICITY DATA with REFERENCE
icv-wmn TDLo:1 μg/kg (10D preg):REP CCPTAY 16,377,77

SAFETY PROFILE: Human reproductive effects by intravenous, intramuscular, intraplacental, and intracervical routes: pregnancy termination. Experimental reproductive effects. When heated to decomposition it emits toxic fumes of SO_x and NO_x.

SOU675 ***HR: 2***
SULTAMICILLIN TOSILATE
mf: $C_{25}H_{32}N_4O_9S_2{\cdot}C_7H_7O_3S$ mw: 767.93

SYNS: CP 49952 p-TOLUENESULFONATE ◇ SBTPC ◇ 4-THIA-1-AZABICYCLO(3.2.0)HEPTANE-2-CARBOXYLIC ACID, 6-((AMINOPHENYLACETYL)AMINO)-3,3-DIMETHYL-7-OXO-,(((3,3-DIMETHYL-7-OXO-4-THIA-1-AZABICYCLO(3.2.0)HEPT-2-YL)CARBOXYL)OXY) METHYL ESTER, S,S-DIOXIDE, p-TOLUENESULFONATE, (2S-(2-α(2R*,5S*),5-α-6-β(S*)))- ◇ VD 1827 p-TOLUENESULFONATE

TOXICITY DATA with REFERENCE
orl-rat TDLo:44400 mg/kg (male 63D pre):REP NKRZAZ 33(Suppl 2),112,85
ipr-rat LD50:1319 mg/kg NKRZAZ 33(Suppl 2),102,85
scu-rat LD50:4769 mg/kg NKRZAZ 33(Suppl 2),102,85
ipr-mus LD50:2356 mg/kg NKRZAZ 33(Suppl 2),102,85
scu-mus LD50:3181 mg/kg NKRZAZ 33(Suppl 2),102,85

SAFETY PROFILE: Moderately toxic by intraperitoneal route. Experimental reproductive effects. When heated to decomposition it emits toxic fumes of NO_x and SO_x.

SOU725 ***HR: 3***
SULTOPRIDE HYDROCHLORIDE
mf: $C_{17}H_{26}N_2O_4S{\cdot}ClH$ mw: 390.97

SYNS: N-((1-ETHYL-2-PYRROLIDINYL)METHYL)-2-METHOXY-5-ETHYLSULFONYLBENZAMIDE HYDROCHLORIDE ◇ N-(1-ETHYL-2-PYRROLIDYLMETHYL)-2-METHOXY-5-ETHYLSULFONYLBENZAMIDE CHLORHYDRATE ◇ SULTROPRIDE CHLORHYDRATE

TOXICITY DATA with REFERENCE
orl-rat TDLo:180 mg/kg (female 17-22D post):REP OYYAA2 28,663,84
scu-mus TDLo:3 g/kg (female 6-15D post):TER OYYAA2 28,663,84
orl-rat LD50:2100 mg/kg OYYAA2 25,1025,83
scu-rat LD50:570 mg/kg OYYAA2 25,1025,83
ivn-rat LD50:136 mg/kg OYYAA2 25,1025,83
ims-rat LD50:640 mg/kg OYYAA2 25,1025,83
orl-mus LD50:590 mg/kg OYYAA2 25,1025,83
ipr-mus LD50:242 mg/kg OYYAA2 25,1025,83
scu-mus LD50:480 mg/kg OYYAA2 25,1025,83
ivn-mus LD50:115 mg/kg OYYAA2 25,1025,83

SAFETY PROFILE: Poison by intravenous and intraperitoneal routes. Moderately toxic by ingestion, intramuscular, and subcutaneous routes. An experimental teratogen. Experimental reproductive effects. When heated to decomposition it emits toxic fumes of SO_x, NO_x, and HCl.

SOU750 ***HR: 3***
SUMAC TANNIN

SYN: TANNIN from SUMAC

TOXICITY DATA with REFERENCE
ipr-mus LD50:150 mg/kg JPPMAB 9,98,57
scu-mus LD50:150 mg/kg JPPMAB 9,98,57
ivn-mus LD50:150 mg/kg JPPMAB 9,98,57
ims-mus LD50:150 mg/kg JPPMAB 9,98,57

SAFETY PROFILE: Poison by subcutaneous, intramuscular, intravenous, and intraperitoneal routes.

SOV000 ***HR: 3***
SUNLIGHT

PROP: A source of actinic (i.e., ultraviolet) radiation in the region of 100A to 3900A.

SAFETY PROFILE: Exposure to eyes is dangerous. Over exposure of skin can result in severe skin burns and often skin cancer.

SOV100 CAS:52-21-1 ***HR: 2***
SUPERCORTYL
mf: $C_{23}H_{30}O_6$ mw: 402.53

SYNS: CORMALONE ◇ CORTIPRED ◇ DELTILEN ◇ METICORTELONE ACETATE ◇ NISOLONE ◇ PREDIACORTINE ◇ PREDICORT ◇ PREDNELAN-N ◇ PREDNIDOREN ◇ PREDNISOLONE ACETATE ◇ PREDNISOLONE 21-ACETATE ◇ PREDONINE INJECTION ◇ PREGNA-1,4-DIENE-3,20-DIONE,21-(ACETYLOXY)-11,17-DIHYDROXY-, (11-β)- ◇ PRENEMA ◇ 11-β,17,21-TRIHYDROXYPREGNA-1,4-DIENE-3,20-DIONE, 21-ACETATE

TOXICITY DATA with REFERENCE
scu-mus TDLo:96 mg/kg (11D preg):TER SEIJBO 13,245,73
orl-rat TDLo:1800 mg/kg (2-19D preg):REP FESTAS 22,735,71
scu-mus LD50:3500 mg/kg NIIRDN 6,713,82

SAFETY PROFILE: Moderately toxic by subcutaneous route. An experimental teratogen. Experimental reproductive effects. When heated to decomposition it emits acrid smoke and irritating fumes.

SOV500 CAS:8011-76-5 ***HR: 1***
SUPERPHOSPHATE

TOXICITY DATA with REFERENCE
orl-dom LD50:5 g/kg NEZTAF 30,165,82

CONSENSUS REPORTS: Reported in EPA TSCA Inventory.

SAFETY PROFILE: Mildly toxic by ingestion. When heated to decomposition it emits toxic fumes of PO_x. See also PHOSPHATES.

SOW000 CAS:50-02-2 ***HR: 3***
SUPERPREDNOL
mf: $C_{22}H_{29}FO_5$ mw: 392.51

SYNS: AEROSEB-DEX ◇ AZIUM ◇ CORSONE ◇ DECADERM ◇ DECADRON ◇ DECASONE ◇ DECASPRAY ◇ DECTANCYL ◇ 1-DEHYDRO-16-α-METHYL-9-α-FLUOROHYDROCORTISONE ◇ DELTAFLUORENE ◇ DERGRAMIN ◇ DERONIL ◇ DESADRENE ◇ DESAMETASONE ◇ DEXA ◇ DEXACORT ◇ DEXA-CORTIDELT ◇ DEXADELTONE ◇ DEXAMETH ◇ DEXAMETHASONE ALCOHOL ◇ DEXONE ◇ DEXTELAN ◇ DEZONE ◇ DXMS ◇ Δ^1-9-α-FLUORO-16-α-METHYLCORTISOL ◇ 9-α-FLUORO-16-α-METHYLPREDNISOLONE ◇ 9-α-FLUORO-16-α-METHYL-1,4-PREGNADIENE-11-β,17-α,21-TRIOL-3,20-DIONE ◇ 4-α-FLUORO-16-α-METHYL-11-β,17,21-TRIHYDROXY-PREGNA-1,4-DIENE-3,20-DIONE ◇ 9-FLUORO-11-β,17,21-TRIHYDROXY-16-α-METHYLPREGNA-1,4-DIENE-3,20-DIONE ◇ 9-α-FLUORO-11-β,17-α,21-TRIHYDROXY-16-α-METHYLPREGNA-1,4-DIENE-3,20-DIONE ◇ FORTECORTIN ◇ GAMMACORTEN ◇ HEXADECADROL ◇ HEXADROL ◇ MAXIDEX ◇ 16-α-METHYL-9-α-FLUORO-1-DEHYDROCORTISOL ◇ 16-α-METHYL-9-α-FLUORO-Δ^1-HYDROCORTISONE ◇ 16-α-METHYL-9-α-FLUOROPREDNISOLONE ◇ 16-α-METHYL-9-α-FLUORO-1,4-PREGNADIENE-11-β,17-α,21-TRIOL-3,20-DIONE ◇ 16-α-METHYL-9-α-FLUORO-11-β,17-α,21-TRIHYDROXYPREGNA-1,4-DIENE-3,20-DIONE ◇ MEXIDEX ◇ MILLICORTEN ◇ MK 125 ◇ ORADEXON ◇ SK-DEXAMETHASONE

TOXICITY DATA with REFERENCE
dns-mus:bmr 1 mmol/L NTIS** AD-A080-636
dni-mus-skn 3 mg/kg JNCIAM 54,931,75
ipr-rat TDLo:3 mg/kg (female 14-15D post):REP BBRCA9 75,125,77
orl-rat TDLo:2 mg/kg (female 6-15D post):TER 28QFAD -,85,73
ipr-rat LD50:54 mg/kg OYYAA2 28,687,84
scu-rat LD50:14 mg/kg TXAPA9 8,250,66
ipr-mus LD50:410 mg/kg RPOBAR 2,281,70
scu-rbt LD50:7200 μg/kg OYYAA2 28,687,84

CONSENSUS REPORTS: Reported in EPA TSCA Inventory.

SAFETY PROFILE: Poison by intraperitoneal and subcutaneous routes. An experimental teratogen. Experimental reproductive effects. Mutation data reported. When heated to decomposition it emits toxic fumes of F^-.

SOW500 CAS:551-58-6 ***HR: 3***
SUPININE
mf: $C_{15}H_{25}NO_4$ mw: 283.41

SYN: SUPININ

TOXICITY DATA with REFERENCE
sln-dmg-par 20 μmol/L ZEVBA5 91,74,60
ivn-rat LD50:212 mg/kg JPETAB 126,179,59
ivn-mus LD50:149 mg/kg JPETAB 126,179,59

CONSENSUS REPORTS: EPA Genetic Toxicology Program.

SAFETY PROFILE: Poison by intravenous. Mutation data reported. When heated to decomposition it emits toxic fumes of NO_x.

SOX400 CAS:33005-95-7 ***HR: 3***
SURGAM
mf: $C_{14}H_{12}O_3S$ mw: 260.32

PROP: Mp: 96° (isopropyl ether).

SYNS: 5-BENZOYL-α-METHYL-2-THIOPHENEACETIC ACID ◇ FC 3001 ◇ RU 15060 ◇ TIAPROFENIC ACID

TOXICITY DATA with REFERENCE
orl-mus TDLo:7200 mg/kg (male 10W pre):TER JZKEDZ 6,195,80

orl-rbt TDLo:975 mg/kg (female 6-18D post):REP YACHDS 8,1773,80
orl-wmn TDLo:672 mg/kg/7W-I LANCAO 1,803,86
orl-wmn TDLo:16 mg/kg/2D-I CMAJAX 137,1022,87
orl-rat LD50:181 mg/kg OYYAA2 9,715,75
ipr-rat LD50:225 mg/kg OYYAA2 9,715,75
scu-rat LD50:215 mg/kg IYKEDH 15,688,84
orl-mus LD50:690 mg/kg OYYAA2 9,715,75
ipr-mus LD50:430 mg/kg OYYAA2 9,715,75
scu-mus LD50:675 mg/kg IYKEDH 15,688,84

SAFETY PROFILE: Poison by ingestion, subcutaneous, and intraperitoneal routes. An experimental teratogen. Experimental reproductive effects. When heated to decomposition it emits toxic fumes of SO_x.

SOX500 CAS:337-47-3 ***HR: 3***
SURITAL SODIUM
mf: $C_{12}H_{18}N_2O_2S{\bullet}Na$ mw: 277.37

SYNS: 5-ALLYL-5-(1-METHYLBUTYL)-2-THIOBARBITURATESODIUM ◇ 5-ALLYL-5-(1-METHYLBUTYL)-2-THIO-BARBITURIC ACID SODIUM SALT ◇ BURITAL SODIUM ◇ SODIUM-5-ALLYL-5-(1-METHYLBUTYL)-2-THIOBARBITURATE ◇ SODIUM THIAMYLAL ◇ SURITAL ◇ SURITAL SODIUM (derivative) ◇ SURITAL SODIUM SALT ◇ THIAMYLAL SODIUM ◇ THIOMYLAL SODIUM

TOXICITY DATA with REFERENCE
ipr-mus TDLo:40 mg/kg (10D preg):TER KAIZAN 40,323,65
scu-rat LD50:52 mg/kg 29ZVAB -,115,69
ivn-rat LD50:51 mg/kg 29ZVAB -,115,69
orl-mus LD50:180 mg/kg 29ZVAB -,115,69
ipr-mus LD50:109 mg/kg TXAPA9 26,495,73
scu-mus LD50:112 mg/kg 29ZVAB -,115,69
ivn-mus LD50:85 mg/kg TXAPA9 26,495,73
orl-dog LD50:134 mg/kg 29ZVAB -,115,69
ivn-dog LD50:32 mg/kg 29ZVAB -,115,69
ivn-rbt LD50:24 mg/kg 29ZVAB -,115,69
scu-rbt LD50:225 mg/kg 29ZVAB -,115,69

SAFETY PROFILE: Poison by ingestion, subcutaneous, intravenous, and intraperitoneal routes. An experimental teratogen. When heated to decomposition it emits very toxic fumes of NO_x, SO_x, and Na_2O.

SOX550 CAS:521-78-8 ***HR: 3***
SURMONTIL MALEATE
mf: $C_{20}H_{26}N_2{\bullet}C_4H_4O_4$ mw: 410.56

SYNS: 10,11-DIHYDRO-5-(3-(DIMETHYLAMINO)-2-METHYLPROPYL)-5H-DIBENZ(b,f)AZEPINE MALEATE (1:1) ◇ TRIMEPRIMINE MALEATE ◇ TRIMEPRIMINE MONOMALEATE ◇ TRIMIPRAMINE ACID MALEATE ◇ TRIMIPRAMINE MALEATE

TOXICITY DATA with REFERENCE
scu-mus TDLo:85 mg/kg (9D preg):TER DGDFA5 22,61,80
orl-rat LD50:800 mg/kg FRPPAO 25,519,70
ipr-rat LD50:150 mg/kg FRPPAO 25,519,70
scu-rat LD50:308 mg/kg FRPPAO 25,519,70
ivn-rat LD50:38 mg/kg FRPPAO 25,519,70
orl-mus LD50:425 mg/kg NIIRDN 6,526,82
ipr-mus LD50:140 mg/kg FRPPAO 25,519,70
scu-mus LD50:240 mg/kg FRPPAO 25,519,70
ivn-mus LD50:40 mg/kg FRPPAO 25,519,70

SAFETY PROFILE: Poison by subcutaneous, intravenous, and intraperitoneal routes. Moderately toxic by ingestion. An experimental teratogen. When heated to decomposition it emits toxic fumes of NO_x.

SOX875 CAS:27470-51-5 ***HR: 3***
SUXIBUZONE
mf: $C_{24}H_{26}N_2O_6$ mw: 438.52

PROP: White, bitter powder. Mp: 126-127°. Sol in most organic solvents; insol in water.

SYNS: ALFIDE ◇ 4-BUTYL-4-(β-CARBOXYPROPIONYL-OXYMETHYL)-1,2-DIPHENYL-3,5-PYRAZOLIDINEDIONE ◇ 4-BUTYL-4-HYDROXYMETHYL-1,2-DIPHENYL-3,5-PYRAZOLIDINEDIONEHYDROGEN SUCCINATE ◇ CALIBENE ◇ DANILON ◇ FLOGOS ◇ MONOESTER with 4-BUTYL-4-(HYDROXYMETHYL)-1,2-DIPHENYL-SUCCINIC ACID 3,5-PYRAZOLIDINEDIONE ◇ SOLUROL

TOXICITY DATA with REFERENCE
orl-rat TDLo:900 mg/kg (female 17-20D post):REP OYYAA2 20,393,80
orl-rat TDLo:1562 mg/kg (7-17D preg):TER OYYAA2 20,377,80
orl-rat LD50:1700 mg/kg IYKEDH 12,1204,81
scu-rat LD50:410 mg/kg IYKEDH 12,1204,81
ivn-rat LD50:305 mg/kg IYKEDH 12,1204,81
orl-mus LD50:1200 mg/kg IYKEDH 12,1204,81
scu-mus LD50:520 mg/kg IYKEDH 12,1204,81
ivn-mus LD50:285 mg/kg IYKEDH 12,1204,81
orl-dog LD50:373 mg/kg OYYAA2 20,265,80
orl-rbt LD50:1266 mg/kg OYYAA2 20,265,80

SAFETY PROFILE: Poison by ingestion and intravenous routes. Moderately toxic by subcutaneous route. An experimental teratogen. Experimental reproductive effects. When heated to decomposition it emits toxic fumes of NO_x.

SOY000 CAS:122-10-1 ***HR: 3***
SWAT
mf: $C_9H_{15}O_8P$ mw: 282.21

SYNS: BOMYL ◇ DIMETHYL 1,3-BIS(CARBOMETHOXY)-1-PROPEN-2-YL PHOSPHATE ◇ DIMETHYL-1,3-DI(CARBOMETHOXY)-1-PROPEN-2-YL PHOSPHATE ◇ DIMETHYL 3-(DIMETHOXYPHOSPHINYLOXY)GLUTACONATE ◇ DIMETHYL 3-HYDROXYGLUTACONATE DIMETHYL PHOSPHATE ◇ ENT 24,833 ◇ FLY BAIT GRITS ◇ GC 3707 ◇ GENERAL CHEMICALS 3707 ◇ 3-HYDROXYGLUTACONIC ACID, DIMETHYL ESTER, DIMETHYL PHOSPHATE ◇ 3-HYDROXY-2-PENTANEDIOIC ACID, DIMETHYL ESTER, DIMETHYL PHOSPHATE

◇ PHOSPHORIC ACID, DIMETHYL ESTER, ESTER with DIMETHYL 3-HYDROXYGLUTACONATE

TOXICITY DATA with REFERENCE
orl-rat LD50:31 mg/kg ARSIM* 20,6,66
skn-rbt LDLo:20 mg/kg BESAAT 12,161,66
orl-ckn LD50:9 mg/kg TXAPA9 11,49,67
orl-bwd LD50:950 μg/kg TXAPA9 21,315,72

SAFETY PROFILE: Poison by ingestion and skin contact. Used as an insecticide. When heated to decomposition it emits very toxic fumes of PO_x.

SOY100 CAS:68917-50-0 ***HR: 2***
SWEET BIRCH OIL

SYNS: BLACK BIRCH OIL ◇ OILS, SWEET BIRCH

TOXICITY DATA with REFERENCE
skn-rbt 500 mg/24H MOD FCTXAV 17,907,79
orl-rat LD50:1700 mg/kg FCTXAV 17,907,79

CONSENSUS REPORTS: Reported in EPA TSCA Inventory.

SAFETY PROFILE: Moderately toxic by ingestion. A skin irritant. When heated to decomposition it emits acrid smoke and irritating fumes.

SOY500 CAS:1401-55-4 ***HR: 3***
SWEET GUM

PROP: Tannin containing fraction of leaf used.

SYNS: LIQUIDAMBAR STYRACIFLUA ◇ TANNIN from SWEET GUM

TOXICITY DATA with REFERENCE
scu-rat TDLo:1100 mg/kg/56W-I:NEO JNCIAM 57,207,76

CONSENSUS REPORTS: Reported in EPA TSCA Inventory.

SAFETY PROFILE: Questionable carcinogen with experimental neoplastigenic data.

SOZ000 ***HR: D***
SWEET PEA SEEDS

PROP: Ground seeds from *L. odoratus* (TJADAB 5,33,72).

SYN: LATHYRUS ODORATUS, SEEDS

TOXICITY DATA with REFERENCE
orl-rat TDLo:500 g/kg (1-20D preg):TER AOBIAR 13,823,68
orl-rat TDLo:175 g/kg/(10-16D preg):REP TJADAB 5,33,72

SAFETY PROFILE: An experimental teratogen. Experimental reproductive effects.

SPA000 CAS:3441-64-3 ***HR: 3***
SYDNOPHEN HYDROCHLORIDE
mf: $C_{11}H_{13}N_3O{\bullet}ClH$ mw: 239.73

SYNS: 3-(α-METHYLPHENETHYL)SYDONE IMINE MONOHYDROCHLORIDE ◇ 3-(β-PHENYLISOPROPYL)-SIDNONIMINE HYDROCHLORIDE ◇ SIDNOFEN ◇ SYDNONE IMINE, 3-(1-METHYL-2-PHENYLETHYL)-, MONOHYDROCHLORIDE ◇ SYDNOPHENE

TOXICITY DATA with REFERENCE
ipr-mus LD50:225 mg/kg RPTOAN 35(4),169,72
scu-mus LD50:123 mg/kg RPTOAN 35(4),169,72
ivn-mus LD50:51 mg/kg RPTOAN 35(4),169,72

SAFETY PROFILE: Poison by intraperitoneal, intravenous, and subcutaneous routes. When heated to decomposition it emits very toxic fumes of NO_x and HCl.

SPA500 CAS:64245-83-6 ***HR: 3***
SYEP
mf: $C_7H_8F_6N_6O_{10}$ mw: 450.21

SYNS: BIS(2,2-DIFLUOROAMINO)-1,3-BIS(DINITRO-FLUOROETHOXY)PROPANE ◇ 1,3-BIS(2,2,-DINITRO-2-FLUOROETHOXY)-N,N,N'N'-TETRAFLUORO-2,2-PROPANEDIAMINE

TOXICITY DATA with REFERENCE
orl-rat LD50:812 mg/kg NTIS** AD-A062-138
ipr-mus LD50:235 mg/kg NTIS** AD-A062-138

CONSENSUS REPORTS: Reported in EPA TSCA Inventory.

SAFETY PROFILE: Poison by intraperitoneal route. Moderately toxic by ingestion. When heated to decomposition it emits very toxic fumes of F^- and NO_x.

SPA650 CAS:13517-49-2 ***HR: D***
SYGETHIN
mf: $C_{18}H_{20}O_6S_2{\bullet}2K$ mw: 474.70

SYNS: meso-α,α'-DIETHYLBIBENZYL-4,4'-DISULFONIC ACID DIPOTASSIUM SALT ◇ (R*,S*)-4,4'-(1,2-DIETHYL-1,2-ETHANEDIYL) BIS-BENZENESULFONIC ACID DIPOTASSIUM SALT ◇ DIPOTASSIUM-meso-N,N-DISULFO-3,4-DIPHENYLHEXANE ◇ DISULFO-meso-4,4-DIPHENYLHEXANE DIPOTASSIUM ◇ SIGETIN ◇ SYGETIN

TOXICITY DATA with REFERENCE
scu-rat TDLo:30 mg/kg (19-21D preg):REP BEXBAN 90,1597,80
ivn-rat TDLo:100 mg/kg (female 13-16D post):TER PAFEAY 19,66,75

SAFETY PROFILE: An experimental teratogen. Experimental reproductive effects. When heated to decomposition it emits toxic fumes of SO_x and K_2O.

SPB000 CAS:63394-02-5 ***HR: 1***
SYLGARD 184 CURING AGENT

TOXICITY DATA with REFERENCE
skn-rbt 500 mg/24H MLD NTIS** LA-7368-MS
eye-rbt 500 mg/1H MLD NTIS** LA-7368-MS

SAFETY PROFILE: A skin and eye irritant.

SPB500 CAS:22571-95-5 ***HR: 3***
SYMPHYTINE
mf: $C_{20}H_{31}NO_6$ mw: 381.52

SYNS: 2-METHYLBUTENOICACID-7-((2,3-DIHYDROXY-2-(1-METHYLETHYL)-1-OXOBUTOXY)METHYL-2,3,5,7a-TETRAHYDRO-1H-PYRROLIZIN-1-YL ESTER ◇ 7-TIGLYLRETRONECINE VIRIDIFLORATE ◇ 7-TIGLYL-9-VIRIDIFLORYLRETRONECINE

TOXICITY DATA with REFERENCE
ipr-rat TDLo:780 mg/kg/56W-I:ETA JJIND8 63,469,79
ipr-rat LD50:130 mg/kg JJIND8 63,469,79
ipr-mus LD50:300 mg/kg EXPEAM 36,377,80

CONSENSUS REPORTS: IARC Cancer Review: Group 3 IMEMDT 7,56,87; Animal Inadequate Evidence IMEMDT 31,239,83.

SAFETY PROFILE: Poison by intraperitoneal route. Questionable carcinogen with experimental tumorigenic data. When heated to decomposition it emits toxic fumes of NO_x.

SPB800 CAS:15016-15-6 ***HR: 3***
SYMPOCAINE HYDROCHLORIDE
mf: $C_{17}H_{28}N_2O_3$•ClH mw: 344.93

SYNS: 4-AMINO-2-BUTOXY-BENZOIC ACID 2-(DIETHYLAMINO)ETHYL ESTER HYDROCHLORIDE ◇ 4-AMINO-2-BUTOXY-BENZOIC ACID 2-(DIETHYLAMINO)ETHYL ESTER, MONOHYDROCHLORIDE ◇ SUMPOCAINE ◇ WIN 3706

TOXICITY DATA with REFERENCE
scu-mus LD50:100 mg/kg APFRAD 40,133,82
ivn-mus LD50:4019 μg/kg JPETAB 104,40,52

SAFETY PROFILE: Poison by subcutaneous and intravenous routes. Used as a local anesthetic. When heated to decomposition it emits toxic fumes of NO_x and HCl.

SPB875 ***HR: 3***
SYNANCEJA HORRIDA Linn. VENOM

SYN: VENOM, STONEFISH, SYNANCEJA HORRIDA Linn.

TOXICITY DATA with REFERENCE
ipr-mus LD50:1800 μg/kg AIPTAK 123,195,59
ivn-mus LD50:180 μg/kg AIPTAK 123,195,59
ivn-rbt LD50:15 μg/kg AIPTAK 123,195,59

SAFETY PROFILE: Poison by intravenous and intraperitoneal routes.

SPC500 CAS:61-76-7 ***HR: 3***
m-SYNEPHRINE HYDROCHLORIDE
mf: $C_9H_{13}NO_2$•ClH mw: 203.6

SYNS: ADRIANOL ◇ ALMEFRIN ◇ BIOMYDRIN ◇ CONSDRIN ◇ CONSDRIN HYDROCHLORIDE ◇ DERIZENE ◇ EFRICEL ◇ EMAGRIN ◇ FENOX ◇ FURACIN ◇ HISTABID ◇ FENILFAR ◇ (−)-α-HYDROXY-β-(METHYLAMINO)ETHYL-α-(3-HYDROXYBENZENE) HYDROCHLORIDE ◇ (R)-3-HYDROXY-α-((METHYLAMINO) METHYL)BENZENEMETHANOL HYDROCHLORIDE ◇ 1-m-HYDROXY-α-(METHYLAMINOMETHYL)BENZYL ALCOHOL HYDROCHLORIDE ◇ l-1-(m-HYDROXYPHENYL)-2-METHYLAMINOETHANOL HYDROCHLORIDE ◇ IDRIANOL ◇ ISOPHRINE ◇ ISOPHRIN HYDROCHLORIDE ◇ LEXATOL ◇ METAOKSEDRIN ◇ METAOXEDRIN ◇ METAOXEDRINUM ◇ m-METHYLAMINOETHANOLPHENOL HYDROCHLORIDE ◇ METROXEDRINE ◇ MEZATON ◇ MYDFRIN ◇ NCI-C55641 ◇ NEOOXEDRINE ◇ NEOPHRYN ◇ NEO-SINEFRINA ◇ NEOSYMPATOL ◇ NEOSYNEPHRINE ◇ NEOSYNEPHRINE HYDROCHLORIDE ◇ NEOSYNESINE ◇ NEWPHRINE ◇ OCUSOL ◇ OFTALFRINE ◇ m-OXEDRINE ◇ PHENISTAN ◇ PHENYL-DRANE ◇ PHENYLEPHRINE HYDROCHLORIDE ◇ d-(−)-PHENYLEPHRINE HYDRO- CHLORIDE ◇ PREFRIN ◇ PYRISTAN ◇ RHINALL ◇ STANEPHRIN ◇ SUCRAPHEN ◇ m-SYMPATHOL ◇ m-SYMPATOL ◇ m-SYNEPHRINE ◇ SYNASAL ◇ SYNETHENATE ◇ URI ◇ VISADRON

TOXICITY DATA with REFERENCE
mmo-sat 100 μg/L FOMIAZ 25,388,80
dnr-esc 10 mg/L FOMIAZ 25,388,80
scu-rbt TDLo:15 mg/kg (female 22-31D post):TER SGOBA9 129,341,69
scu-rbt TDLo:15 mg/kg (female 22-31D post):REP SGOBA9 129,341,69
orl-rat LD50:350 mg/kg AIPTAK 180,155,69
ipr-rat LD50:17 mg/kg JPETAB 86,284,46
scu-rat LD50:27 mg/kg JPETAB 86,284,46
ivn-rat LD50:440 μg/kg JPETAB 113,341,55
orl-mus LD50:120 mg/kg IJNEAQ 4,219,65
ipr-mus LD50:89 mg/kg TXAPA9 28,227,74
scu-mus LD50:22 mg/kg JPETAB 86,284,46
ivn-mus LD50:1120 μg/kg EJPHAZ 9,289,70
ipr-dog LDLo:16 mg/kg IJNEAQ 4,219,65
scu-rbt LD50:22 mg/kg JPETAB 86,284,46
ivn-rbt LD50:500 μg/kg JPETAB 86,284,46
ims-rbt LD50:7200 μg/kg JPETAB 86,284,46

CONSENSUS REPORTS: Reported in EPA TSCA Inventory.

SAFETY PROFILE: Poison by ingestion, intraperitoneal, subcutaneous, intravenous, and intramuscular routes. Mutation data reported. When heated to decomposition it emits very toxic fumes of HCl and NO_x.

SPD000 CAS:67-04-9 ***HR: 2***
SYNEPHRINE TARTRATE
mf: $C_9H_{13}NO_2$•$C_4H_6O_6$ mw: 317.33

SYNS: CORVASYMTON ◇ (+ −)-1-(4-HYDROXYPHENYL)-2-METHYLAMINOETHANOL TARTRATE ◇ (+ −)-1-(4-IDROSSIFENIL)-2-METILAMINOETANOLO TARTRATO (ITALIAN) ◇ OXEDRINE TAR-

TRATE ◇ SIMPADREN ◇ SINEFRINA TARTRATO (ITALIAN) ◇ SYMPATHOL ◇ SYMPATOL ◇ SYMPATOL TARTRATE ◇ dl-SYNEPHRINE ◇ dl-p-SYNEPHRINE ◇ (+ −)-SYNEPHRINE TARTRATE

TOXICITY DATA with REFERENCE
ims-rat TDLo:2087 mg/kg (7-16D preg):TER AAMLAR 25,203,70
orl-rat LD50:4450 mg/kg AIPTAK 180,155,69
scu-rat LD50:2730 mg/kg AIPTAK 180,155,69
idu-rat LD50:3050 mg/kg AIPTAK 180,155,69
ipr-mus LD50:2 g/kg AEPPAE 237,171,59
scu-mus LDLo:3200 mg/kg AEPPAE 166,437,32

SAFETY PROFILE: Moderately toxic by intraperitoneal, subcutaneous, and intraduodenal routes. Mildly toxic by ingestion. An experimental teratogen. When heated to decomposition it emits toxic fumes of NO_x.

SPD100 CAS:88026-65-7 ***HR: 1***
SYNFEROL AH EXTRA

TOXICITY DATA with REFERENCE
skn-rbt 500 mg/24H MOD 28ZPAK -,303,72
eye-rbt 20 mg/24H MOD 28ZPAK -,303,72
orl-rat LD50:13 g/kg 28ZPAK -,303,72

SAFETY PROFILE: Mildly toxic by ingestion. An eye and skin irritant.

SPD500 CAS:67-73-2 ***HR: 3***
SYNSAC
mf: $C_{24}H_{30}F_2O_6$ mw: 452.54

SYNS: DERMALAR ◇ 6-α,9-α-DIFLUORO-16-α-HYDROXYPREDNISOLONE-16,17-ACETONIDE ◇ FLUCINAR ◇ FLUCORT ◇ FLUOCINOLONE ACETONIDE ◇ FLUOCINOLONE 16,17-ACETONIDE ◇ FLUOVITIF ◇ JELLIN ◇ LOCALYN ◇ PERCUTINA ◇ RADIOCIN ◇ RS-1401 AT ◇ SINALAR ◇ SYNALAR ◇ SYNAMOL ◇ SYNANDONE ◇ SYNANDRONE

TOXICITY DATA with REFERENCE
oms-hmn-skn 2000 ppm ARDEAC 103,39,71
dni-rat:emb 10 g/L CRNGDP 4,653,83
scu-rbt TDLo:1200 μg/kg (8-31D preg):REP ARZNAD 27,2102,77
scu-rat TDLo:1 mg/kg (1-20D preg):TER ARZNAD 27,2102,77
ipr-rat LD50:42 mg/kg NIIRDN 6,694,82
scu-rat LD50:108 mg/kg NIIRDN 6,694,82
ipr-mus LD50:103 mg/kg NIIRDN 6,694,82
scu-mus LD50:200 mg/kg NIIRDN 6,694,82

CONSENSUS REPORTS: EPA Genetic Toxicology Program.

SAFETY PROFILE: Poison by subcutaneous and intraperitoneal routes. An experimental teratogen. Experimental reproductive effects. Human mutation data reported. When heated to decomposition it emits toxic fumes of F^-.

SPE000 CAS:20685-78-3 ***HR: 3***
SYNTETRIN NITRATE
mf: $C_{27}H_{33}N_3O_8{\cdot}HNO_3$ mw: 590.65

SYNS: PYRROCYCLINE N ◇ ROLITETRACYCLINE

TOXICITY DATA with REFERENCE
orl-mus LD50:3000 mg/kg NIIRDN 6,915,82
ipr-mus LD50:431 mg/kg NIIRDN 6,915,82
ivn-mus LD50:274 mg/kg NIIRDN 6,915,82

SAFETY PROFILE: Poison by intravenous route. Moderately toxic by ingestion and intraperitoneal routes. When heated to decomposition it emits toxic fumes of NO_x. See also NITRATES.

SPE500 CAS:12679-83-3 ***HR: 1***
SYNTHAMIDE 5

PROP: A detergent (GISAAA 43(3),14,78).

TOXICITY DATA with REFERENCE
par-rat LD50:19 g/kg GISAAA 43(3),14,78
par-mus LD50:18 g/kg GISAAA 43(3),14,78

SAFETY PROFILE: Mildly toxic.

SPF000 CAS:4891-54-7 ***HR: 3***
SYSTOX SULFONE
mf: $C_8H_{19}O_5PS_2$ mw: 290.36

SYNS: O,O-DIETHYL-2-ETHYLMERCAPTOETHYL THIOPHOSPHATE, THIONO ISOMER ◇ O,O-DIETHYL-O-(2-ETHYLSULFONYLETHYL)PHOSPHOROTHIOATE ◇ DIETHYL-2-ETHYLSULFONYLETHYL THIONOPHOSPHATE ◇ THIONODEMETON SULFONE

TOXICITY DATA with REFERENCE
orl-rat LD50:90 mg/kg 28ZEAL 4,164,69

SAFETY PROFILE: Poison by ingestion. When heated to decomposition it emits very toxic fumes of PO_x and SO_x. See also MERCAPTANS.

SPF200 ***HR: 2***
SYZYGIUM JAMBOS (Linn.) Alston, extract excluding roots

PROP: Indian plant belonging to the family *Myrtaceae* IJEBA6 24,48,86

SYN: EUGENIA JAMBOS Linn., extract excluding roots

TOXICITY DATA with REFERENCE
orl-rat TDLo:700 mg/kg (female 1-7D post):REP IJEBA6 24,48,86
ipr-mus LD50:825 mg/kg IJEBA6 24,48,86

SAFETY PROFILE: Moderately toxic by intraperitoneal route. Experimental reproductive effects. When heated to decomposition it emits acrid smoke and irritating fumes.

T

TAA100 CAS:93-76-5 ***HR: 3***
2,4,5-T
DOT: UN 2765
mf: $C_8H_5Cl_3O_3$ mw: 255.48

PROP: Crystals; light tan solid. Mp: 151-153°. Usually has 2,3,7,8-TCDD as a minor component.

SYNS: ACIDE 2,4,5-TRICHLORO PHENOXYACETIQUE (FRENCH) ◇ ACIDO (2,4,5-TRICLORO-FENOSSI)-ACETICO (ITALIAN) ◇ AMINE 2,4,5-T FOR RICE ◇ BCF-BUSHKILLER ◇ BRUSH-OFF 445 MLD VOLATILE BRUSH KILLER ◇ BRUSH RHAP ◇ BRUSHTOX ◇ DACAMINE ◇ DEBROUSSAILLANT CONCENTRE ◇ DECAMINE 4T ◇ DED-WEED BRUSH KILLER ◇ DED-WEED LV-6 BRUSH KIL and T-5 BRUSH KIL ◇ DINOXOL ◇ ENVERT-T ◇ ESTERCIDE T-2 and T-245 ◇ ESTERON 245 BE ◇ ESTERON BRUSH KILLER ◇ FARMCO FENCE RIDER ◇ FORRON ◇ FORST U 46 ◇ FORTEX ◇ FRUITONE A ◇ INVERTON 245 ◇ LINE RIDER ◇ PHORTOX ◇ RCRA WASTE NUMBER U232 ◇ REDDON ◇ REDDOX ◇ SPONTOX ◇ SUPER D WEEDONE ◇ TIPPON ◇ TORMONA ◇ TRANSAMINE ◇ TRIBUTON ◇ (2,4,5-TRICHLOOR-FENOXY)-AZIJNZUUR (DUTCH) ◇ 2,4,5-TRICHLOROPHENOXYACETIC ACID ◇ (2,4,5-TRICHLOR-PHENOXY)-ESSIGSAEURE (GERMAN) ◇ TRINOXOL ◇ TRIOXON ◇ TRIOXONE ◇ U 46 ◇ VEON 245 ◇ VERTON 2T ◇ VISKO RHAP LOW VOLATILE ESTER ◇ WEEDAR ◇ WEEDONE

TOXICITY DATA with REFERENCE
mmo-bcs 1 nmol/plate MSERDS 5,93,81
cyt-grb-ipr 250 mg/kg MUREAV 65,83,79
orl-rat TDLo:6 mg/kg (female 8D post):REP EXPEAM 39,891,83
orl-rat TDLo:100 mg/kg (female 11D post):TER BEXBAN 76,831,73
orl-mus TDLo:3379 mg/kg/33W-C:NEO BJCAAI 33,626,76
scu-mus TDLo:215 mg/kg:ETA NTIS** PB223-159
orl-rat LD50:300 mg/kg RREVAH 10,97,65
skn-rat LD50:1535 mg/kg FAATDF 7,299,86
orl-mus LD50:242 mg/kg RPZHAW 31,373,80
orl-dog LD50:100 mg/kg PAREAQ 14,225,62
orl-gpg LD50:381 mg/kg PCOC** -,1178,66
orl-ham LD50:425 mg/kg MUREAV 65,83,79
orl-ckn LD50:310 mg/kg AJVRAH 15,622,54

CONSENSUS REPORTS: IARC Cancer Review: Group 2B IMEMDT 7,156,87; Animal Inadequate Evidence IMEMDT 15,273,77; Human Inadequate Evidence IMEMDT 15,273,77; Human Limited Evidence IMEMDT 41,357,86. Reported in EPA TSCA Inventory. EPA Genetic Toxicology Program.

OSHA PEL: TWA 10 mg/m^3
ACGIH TLV: TWA 10 mg/m^3
DFG MAK: 10 mg/m^3
DOT Classification: ORM-A; Label: None.

SAFETY PROFILE: Suspected carcinogen with experimental neoplastigenic, tumorigenic, and teratogenic data. Poison by ingestion. Moderately toxic by skin contact. Experimental reproductive effects. Mutation data reported. A highly toxic chlorinated phenoxy acid herbicide which is rapidly excreted after ingestion. Readily absorbed by inhalation and ingestion routes, slowly by skin contact. Signs of intoxication include weakness, lethargy, anorexia, diarrhea, ventricular fibrillation and/or cardiac arrest and death. The teratogenicity is due in part to 2,3,7,8-TCDD, which is present as a contaminant (ARENAA 17,123,72). When heated to decomposition it emits toxic fumes of Cl^-. See also TCDD and CHLOROPHENOLS.

TAA400 CAS:76648-01-6 ***HR: 1***
T-1982
mf: $C_{22}H_{27}N_9O_9S{\cdot}Na$ mw: 616.63

TOXICITY DATA with REFERENCE
scu-rat TDLo:13500 mg/kg (17-22D preg/21D post):REP NKRZAZ 30(Suppl 3),319,82
ivn-rat TDLo:11 g/kg (7-17D preg):TER NKRZAZ 30(Suppl 3),319,82
ivn-rat LD50:9550 mg/kg NKRZAZ 30(Suppl 3),232,82

SAFETY PROFILE: Experimental reproductive effects. An experimental teratogen. When heated to decomposition it emits toxic fumes of SO_x, NO_x and Na_2O.

TAA420 CAS:82547-81-7 ***HR: 1***
T-2588
mf: $C_{22}H_{27}N_9O_7S_2$ mw: 593.70

SYN: 5-THIA-1-AZABICYCLO(4.2.0)OCT-2-ENE-2-CARBOXYLIC ACID, 7-(((2-AMINO-4-THIAZOLYL)(METHOXYIMINO)ACETYL) AMINO)-3-((5-METHYL-2H-TETRAZOL-2-YL)METHYL)-8-OXO-,(2,2-DIMETHYL-1-OXOPROPOXY)METHYL ESTER, (6R-(6-α-7-β(Z)))-

TOXICITY DATA with REFERENCE
orl-rat TDLo:11 g/kg (female 7-17D post):REP NKRZAZ 34(Suppl 2),250,86
orl-rat TDLo:11 g/kg (female 7-17D post):TER NKRZAZ 34(Suppl 2),250,86
ipr-rat LD50:5090 mg/kg NKRZAZ 34(Suppl 2),166,86
ipr-mus LD50:5860 mg/kg NKRZAZ 34(Suppl 2),166,86

SAFETY PROFILE: Mildly toxic by intraperitoneal route. An experimental teratogen. Experimental reproductive effects. When heated to decomposition it emits toxic fumes of NO_x and SO_x.

TAA875 ***HR: 1***
TABASCO HOT PEPPER SAUCE

PROP: Produced by the McIlhenny Company, New Iberia, LA (DCTODJ 5,89,82).

TOXICITY DATA with REFERENCE
skn-rbt 500 mg MLD DCTODJ 5,89,82
eye-rbt 100 mg MOD DCTODJ 5,89,82
orl-rat LD50:20 g/kg DCTODJ 5,89,82

SAFETY PROFILE: Mildly toxic by ingestion. A skin and eye irritant. Causes burning sensation in mouth upon ingestion.

TAB250 CAS:51481-61-9 ***HR: 3***
TAGAMET
mf: $C_{10}H_{16}N_6S$ mw: 252.38

PROP: Mp: 141-143°.

SYNS: ACIBILIN ◇ CIMETIDINE ◇ N-CYANO-N'-METHYL-N''-(2-(((5-METHYL-1H-IMIDAZOL-4-YL)METHYL)THIO)ETHYL)GUANIDINE ◇ 1-CYANO-2-METHYL-3-(2-(((5-METHYL-4-IMIDAZOLYL)METHYL) THIO)ETHYL)GUANIDINE ◇ 2-CYANO-1-METHYL-3-(2-(((5-METHYL-IMIDAZOL-4-YL)METHYL)THIO) ETHYL)GUANIDINE ◇ EURECEPTOR ◇ FPF 1002 ◇ GASTROMET ◇ SKF 92334 ◇ TAMETIN ◇ TRATUL ◇ ULCEDINE ◇ ULCIMET ◇ ULCOMET ◇ VALMAGEN ◇ VENOPEX

TOXICITY DATA with REFERENCE
dnd-rat:lvr 3 mmol/L MUREAV 120,133,83
dns-rat:lvr 1 mmol/L MUREAV 120,133,83
unr-wmn TDLo:720 mg/kg (33-37W preg):REP AJDCAI 134,87,80
unr-wmn TDLo:720 mg/kg (33-37W preg):TER AJDCAI 134,87,80
orl-rat TDLo:15600 mg/kg/2Y-C:ETA TXAPA9 61,119,81
orl-man TDLo:2857 μg/kg:CVS LANCAO 1,225,87
orl-man TDLo:57 mg/kg/5D-I:LIV DDSCDJ 32,333,87
ivn-man TDLo:4286 μg/kg/30M-C:PUL,SKN AIMEAS 97,374,82
ivn-wmn LDLo:1008 mg/kg/6W-I:CVS DICPBB 17,126,83
ivn-wmn TDLo:4 mg/kg/2M-C:CVS LANCAO 1,99,87
orl-wmn TDLo:30 mg/kg/2D-I AEMED3 16,1162,87
ivn-cld TDLo:100 mg/kg:CVS AIMEAS 97,283,82
unr-man TDLo:600 mg/kg/6W:SKN LANCAO 1,1160,81
orl-hmn TDLo:80 mg/kg/8D:CVS,SYS LANCAO 1,444,78
orl-rat LD50:5 g/kg JIMRBV 3,86,75
ipr-rat LD50:328 mg/kg YACHDS 11,1685,83
scu-rat LD50:860 mg/kg YACHDS 11,1685,83
ivn-rat LD50:106 mg/kg JIMRBV 3,86,75
orl-mus LD50:2550 mg/kg YACHDS 11,1685,83
ipr-mus LD50:306 mg/kg ARZNAD 33,1655,83
scu-mus LD50:437 mg/kg YACHDS 11,1685,83
ivn-mus LD50:150 mg/kg MEIEDD 10,323,83
ivn-dog LD50:206 mg/kg KSRNAM 14,2806,80
ipr-rbt LD50:1063 mg/kg YACHDS 11,1685,83

CONSENSUS REPORTS: Cyanide and its compounds are on the Community Right-To-Know List. EPA Genetic Toxicology Program.

SAFETY PROFILE: Questionable carcinogen with experimental tumorigenic and teratogenic data. Experimental poison by intravenous and intraperitoneal routes. Moderately toxic to humans by ingestion. Moderately toxic experimentally by ingestion and subcutaneous routes. Human systemic effects: blood pressure lowering, cholestatic jaundice, dyspnea, hair changes, increased body temperature, leukopenia, pulse rate increase, sweating. Human reproductive and teratogenic effects by unspecified route: developmental abnormalities of the hepatobiliary system and effects on newborn including apgar score. Experimental reproductive effects. An antagonist to histamine H2 receptors used in the treatment of peptic ulcers and allergic dermatitis. Mutation data reported. When heated to decomposition it emits very toxic fumes of NO_x, CN^-, and SO_x.

TAB275 ***HR: 2***
TAGETES OIL

PROP: Found in various species of *Tagetes*, particularly *T. Glandulifera* (FCTOD7 20(Suppl),829,82).

TOXICITY DATA with REFERENCE
skn-rbt 500 mg/24H SEV FCTOD7 20(Suppl),829,82
orl-rat LD50:3700 mg/kg FCTOD7 20(Suppl),829,82
ipr-rat LD50:450 mg/kg FCTOD7 20(Suppl),829,82

SAFETY PROFILE: Moderately toxic by ingestion and intraperitoneal routes. A severe skin irritant.

TAB300 CAS:11006-31-8 ***HR: 3***
TAKACIDIN
mf: $C_{72}H_{133}NO_{22}$ mw: 1363.94

SYN: MONZAOMYCIN

TOXICITY DATA with REFERENCE
orl-mus LD50:100 mg/kg 85FZAT -,635,67
ipr-mus LD50:5 mg/kg 85FZAT -,635,67
ivn-mus LD50:5 mg/kg 85ERAY 1,218,78

SAFETY PROFILE: Poison by ingestion, intravenous, and intraperitoneal routes. When heated to decomposition it emits toxic fumes of NO_x.

TAB750 CAS:14807-96-6 ***HR: 3***
TALC
mf: $H_2O_3Si \cdot 3/4Mg$ mw: 96.33

PROP: White to grayish-white, fine powder; odorless and tasteless. Powdered native hydrous magnesium silicate. Insol in water, cold acids, or alkalies. Containing less than 1% crystalline silica.

SYNS: AGALITE ◇ AGI TALC, BC 1615 ◇ ALPINE TALC USP, BC 127 ◇ ALPINE TALC USP, BC 141 ◇ ALPINE TALC USP, BC 662 ◇ ASBESTINE ◇ C.I. 77718 ◇ DESERTALC 57 ◇ EMTAL 596 ◇ FIBRENE C 400 ◇ LO MICRON TALC 1 ◇ LO MICRON TALC, BC 1621 ◇ LO MICRON TALC USP, BC 2755 ◇ METRO TALC 4604 ◇ METRO TALC 4608 ◇ METRO TALC 4609 ◇ MISTRON FROST P ◇ MISTRON RCS ◇ MISTRON 2SC ◇ MISTRON STAR ◇ MISTRON SUPER FROST ◇ MISTRON VAPOR ◇ MP 12-50 ◇ MP 25-38 ◇ MP 45-26 ◇ NCI-C06008 ◇ No. 907 METRO TALC ◇ NYTAL ◇ OOS ◇ OXO ◇ PURTALC USP ◇ SIERRA C-400 ◇ SNOWGOOSE ◇ STEAWHITE ◇ SUPREME DENSE ◇ TALCUM

TOXICITY DATA with REFERENCE
skn-hmn 300 μg/3D-I MLD 85DKA8 -,127,77
ihl-rat TCLo:11 mg/m³/1Y-I:ETA 43GRAK -,389,79

CONSENSUS REPORTS: IARC Cancer Review: Animal Inadequate Evidence IMEMDT 42,185,87; Human Inadequate Evidence IMEMDT 42,185,87; Group 3 IMEMDT 7,349,87. Reported in EPA TSCA Inventory.

OSHA PEL: (Transitional: TWA 20 mppcf (containing no asbestos fibers)) TWA 2 mg/m³
ACGIH TLV: TWA 2 mg/m³, respirable dust (use asbestos TLV if asbestos fibers are present)
DFG MAK: 2 mg/m³

SAFETY PROFILE: The talc with less than 1 percent asbestos is regarded as a nuisance dust. A human skin irritant. Prolonged or repeated exposure can produce a form of pulmonary fibrosis (talc pneumoconiosis) which may be due to asbestos content. Questionable carcinogen with experimental tumorigenic data. A common air contaminant.

TAB775 CAS:14807-96-6 ***HR: 3***
TALC, containing asbestos fibers
mf: $H_2O_2Si{\cdot}3/4Mg$ mw: 96.33

TOXICITY DATA with REFERENCE
ipr-rat TDLo:200 mg/kg:ETA JJIND8 67,965,81

CONSENSUS REPORTS: IARC Cancer Review: Group 1 IMEMDT 7,349,87; Human Sufficient Evidence IMEMDT 42,185,87; Reported in EPA TSCA Inventory.

ACGIH TLV: Human Carcinogen; TWA > 2 mg/m³, Respirable Dust

SAFETY PROFILE: Confirmed human carcinogen with experimental tumorigenic data.

TAB785 CAS:76069-32-4 ***HR: 3***
TALISOMYCIN S[10b]
mf: $C_{59}H_{91}N_{19}O_{26}S_2$ mw: 1546.81

SYN: TALLYSOMYCIN S[10b]

TOXICITY DATA with REFERENCE
dnd-omi 107 nmol/L CNREA8 40,4173,80
dni-hmn:lym 50 mg/L MUREAV 112,119,83
ipr-mus LD50:123 mg/kg JANTAJ 34,665,81
ivn-mus LD50:104 mg/kg DCTODJ 7,259,84

SAFETY PROFILE: Poison by intravenous and intraperitoneal routes. Human mutation data reported. When heated to decomposition it emits toxic fumes of NO_x and SO_x.

TAC000 CAS:8002-26-4 ***HR: 1***
TALL OIL

PROP: Composition: Rosin acids, oleic and linoleic acids. Dark brown liquid; acrid odor. D: 0.95, flash p: 360°F.
SYNS: LIQUID ROSIN ◇ TALLOL

SAFETY PROFILE: A mild allergen. A substance which migrates to food from packaging materials. Combustible when exposed to heat or flame; can react with oxidizing materials. To fight fire, use dry chemical, CO_2. When heated to decomposition it emits acrid smoke and irritating fumes.

TAC200 CAS:70084-70-7 ***HR: 1***
N-TALLOWPYRROLIDINONE

SYN: 2-PYRROLIDINONE, 1-TALLOW ALKYL derivs.

TOXICITY DATA with REFERENCE
skn-rbt 500 mg/24H SEV FCTOD7 26,475,88
eye-rbt 100 mg MLD FCTOD7 26,475,88
orl-rat LD50:12600 mg/kg FCTOD7 26,475,88

CONSENSUS REPORTS: Reported in EPA TSCA Inventory.

SAFETY PROFILE: Mildly toxic by ingestion. A severe skin and eye irritant. When heated to decomposition it emits acrid smoke and irritating fumes.

TAC500 CAS:65057-90-1 ***HR: 3***
TALLYSOMYCIN A
mf: $C_{68}H_{11}N_{22}O_{27}S_2$ mw: 1732.12

PROP: White amorph solid. No definite mp, gradually decomp > 210°. Sol in water, methanol, and DMF; sltly sol in ethanol; practically insol in other organic solvents.

SYNS: ANTIBIOTIC BU 2231A ◇ BU 2231A ◇ TALISOMYCIN

TOXICITY DATA with REFERENCE
mmo-sat 50 ng/plate MUREAV 117,9,83
dni-hmn:lym 50 mg/L MUREAV 112,119,83
cy-hmn:lym 25 mg/L MUREAV 112,119,83
ipr-mus LD50:19 mg/kg JANTAJ 31,667,78
scu-mus LD50:28 mg/kg USXXAM #4051237
ivn-mus LD50:17 mg/kg JANTAJ 31,667,78

SAFETY PROFILE: Poison by intraperitoneal, subcutaneous and intravenous routes. Human mutation data reported. When heated to decomposition it emits very toxic fumes of SO_x and NO_x.

TAC750 CAS:65057-91-2 ***HR: 3***
TALLYSOMYCIN B
mf: $C_{62}H_{98}N_{20}O_{26}S_2$ mw: 1603.92

PROP: No definite mp, gradually decomp > 210°. Sol in water, methanol, and DMF; slightly sol in ethanol; practically insol in other organic solvents.

SYNS: ANTIBIOTIC BU 2231B ◇ TALISOMYCIN B

TOXICITY DATA with REFERENCE
mmo-sat 50 ng/plate MUREAV 117,9,83
dnd-nml:lvr 25 mg/L CNREA8 38,3322,78
ipr-mus LD50:46 mg/kg JANTAJ 30,779,77
ivn-mus LD50:30 mg/kg JANTAJ 30,779,77
itr-mus LDLo:16 μg/kg TXAPA9 56,326,80

SAFETY PROFILE: Poison by intraperitoneal, intratracheal, and intravenous routes. Mutation data reported. When heated to decomposition it emits very toxic fumes of SO_x and NO_x.

TAC800 CAS:56073-10-0 ***HR: 3***
TALON RODENTICIDE
mf: $C_{31}H_{23}BrO_3$ mw: 523.45

PROP: Off-white powder. Mp: 228-230°. Insol in water; sltly sol in alc, benzene; sol in acetone, chloroform.

SYNS: BRODIFACOUM ◇ 3-(3-(4'-BROMOBIPHENYL-4-YL)-1,2,3,4-TETRAHYDRONAPHTH-1-YL)-4-HYDROXYCOUMARIN ◇ 3-(3-(4'-BROMO-1,1'-BIPHENYL-4-YL)-1,2,3,4-TETRAHYDRO-1-NAPHTHYL)-4-HYDROXYCOUMARIN ◇ HAVOC ◇ KLERAT ◇ PP 581 ◇ RATAK ◇ RATAK PLUS ◇ TALON ◇ VOLID ◇ WBA 8119

TOXICITY DATA with REFERENCE
orl-man TDLo:120 μg/kg:BLD JAMAAP 252,3005,84
orl-rat LD50:160 μg/kg MAGJAL 52(4),1,79
skn-rat LD50:50 mg/kg PMCHA2 -,C227,83
orl-dog LD50:1090 μg/kg NZJEA3 9,147,81
orl-rbt LD50:200 μg/kg NZJEA3 9,23,81
orl-grb LD50:1 mg/kg JOHYAY 87,179,81

SAFETY PROFILE: Poison by ingestion and skin contact. Human systemic effects by ingestion: change in clotting factors. When heated to decomposition it emits toxic fumes of Br^-.

TAD175 CAS:54965-24-1 ***HR: 3***
TAMOXIFEN CITRATE
mf: $C_{26}H_{29}NO \cdot C_6H_8O_7$ mw: 563.70

PROP: Fine, white, crystalline powder. Mp: 140-142°. Sltly sol in water; sol in ethanol, methanol, and acetone. Hygroscoopic at high relative humidities.

SYNS: trans-1-(p-β-DIMETHYLAMINOETHOXYPHENYL)-1,2-DIPHENYLBUT-1-ENE CITRATE ◇ ICI 46,474 ◇ NOLVADEX

TOXICITY DATA with REFERENCE
orl-rbt TDLo:34 mg/kg (female 10-26D post):REP JRPFA4 48,367,76
orl-wmn TDLo:154 mg/kg/1Y-I:EYE AJOPAA 104,185,87
orl-rat LD50:1190 mg/kg JZKEDZ 6,1,80
ipr-rat LD50:575 mg/kg JZKEDZ 6,1,80
orl-mus LD50:3100 mg/kg JRPFA4 13,101,67
ipr-mus LD50:218 mg/kg NIIRDN 6,440,82

SAFETY PROFILE: Poison by intraperitoneal route. Moderately toxic by ingestion. Experimental reproductive effects. Human systemic effects: visual field changes, retinal changes. An anti-estrogenic drug. When heated to decomposition it emits toxic fumes of NO_x.

TAD250 CAS:72869-73-9 ***HR: 1***
TANGELO OIL

TOXICITY DATA with REFERENCE
skn-rbt 500 mg/24H MOD FCTOD7 20,831,82

CONSENSUS REPORTS: Reported in EPA TSCA Inventory.

SAFETY PROFILE: A skin irritant. When heated to decomposition it emits acrid smoke and irritating fumes.

TAD500 CAS:8008-31-9 ***HR: 1***
TANGERINE OIL

PROP: Expressed from the peels of Dancy and related varieties of *Citrus reticulata Blanco*. The components include d-limonene, n-octylaldehyde, n-decylaldehyde, citral, linalool, citronella, cadinene, terpenes, aldehydes, alcohols, and esters (FCTXAV 16, 637,78). Red orange to brown orange liquid; orange-like odor. Sol in fixed oils, mineral oil; sltly sol in propylene glycol; insol in glycerin.

SYNS: TANGERINE OIL, COLDPRESSED (FCC) ◇ TANGERINE OIL, EXPRESSESED (FCC)

TOXICITY DATA with REFERENCE
skn-mus 500 mg open FCTXAV 16,637,78
skn-rbt 500 mg/24H FCTXAV 16,637,78
skn-pig 500 mg open FCTXAV 16,637,78

CONSENSUS REPORTS: Reported in EPA TSCA Inventory.

SAFETY PROFILE: A skin irritant. When heated to decomposition it emits acrid smoke and irritating fumes. See also individual components, ALDEHYDES, ALCOHOLS, and ESTERS.

TAD750 CAS:1401-55-4 *HR: 3*
TANNIC ACID
mf: $C_{76}H_{52}O_{46}$ mw: 1701.28

PROP: From the nutgalls of *Quercus infectoria Oliver* or seed pods of *Caesalpinia spinosa* or the nutgalls of various sumac species. Yellowish-white or brown, bulky powder or flakes; odorless with astringent taste. Mp: 200-218°, flash p: 390°F (OC), autoign temp: 980°F. Very sol in water, alc, acetone; almost insol in benzene, chloroform, ether, petr ether, carbon disulfide.

SYNS: d'ACIDE TANNIQUE (FRENCH) ◇ GALLOTANNIC ACID ◇ GALLOTANNIN ◇ GLYCERITE ◇ TANNIN

TOXICITY DATA with REFERENCE
dns-rat-orl 25 g/kg JJIND8 74,1283,85
dni-mus-ipr 76 mg/kg IJEBA6 17,1141,79
orl-rat TDLo:112 g/kg (49D pre/1-21D preg):REP CRSBAW 172,470,78
scu-rat TDLo:4450 mg/kg/17W-I:CAR AMSHAR 3,353,53
scu-mus TDLo:750 mg/kg/12W-I:ETA BJCAAI 14,147,60
orl-rat LD50:2260 mg/kg 34ZIAG -,571,69
orl-rat LDLo:200 mg/kg WKWOAO 62,270,50
par-rat LDLo:1400 mg/kg FCTXAV 14,565,76
orl-mus LDLo:2000 mg/kg JPETAB 77,63,43
scu-mus LDLo:75 mg/kg JPETAB 77,63,43
ivn-mus LDLo:10 mg/kg JPETAB 77,63,43
ims-mus LD50:350 mg/kg JPPMAB 9,98,57
orl-rbt LD50:5000 mg/kg AJVRAH 23,1264,62

CONSENSUS REPORTS: IARC Cancer Review: Group 3 IMEMDT 7,56,87; Animal Sufficient Evidence IMEMDT 10,253,76. Reported in EPA TSCA Inventory. EPA Genetic Toxicology Program.

SAFETY PROFILE: Poison by ingestion, intramuscular, intravenous, and subcutaneous routes. Moderately toxic by parenteral route. Experimental reproductive effects. Questionable carcinogen with experimental carcinogenic and tumorigenic data. Mutation data reported. Combustible when exposed to heat or flame. To fight fire, use water. Incompatible with salts of heavy metals; oxidizing materials. When heated to decomposition it emits acrid smoke and irritating fumes.

TAE250 *HR: 3*
TANNIN-FREE FRACTION of BRACKEN FERN

SYN: BRACKEN FERN, TANNIN-FREE

TOXICITY DATA with REFERENCE
orl-rat TDLo:2000 g/kg/56W-C:CAR JJIND8 65,131,80

SAFETY PROFILE: Questionable carcinogen with experimental carcinogenic data. When heated to decomposition it emits toxic and irritating fumes.

TAE500 CAS:8016-87-3 *HR: 2*
TANSY OIL

PROP: Yellowish liquid, strong odor, D: 0.925-0.950 @ 15/15°. Sol in alc, ether chloroform; sltly sol in water. From *Tamanacetum vulgare,* contained thujone, comphor and borneol (FCTXAV 14,659,76).

TOXICITY DATA with REFERENCE
skn-rbt 500 mg/24H MLD FCTXAV 14,659,76
orl-rat LD50:1150 mg/kg FCTXAV 14,659,76

CONSENSUS REPORTS: Reported in EPA TSCA Inventory.

SAFETY PROFILE: Moderately toxic by ingestion. A skin irritant. When heated to decomposition it emits acrid smoke and irritating fumes. See also components as listed.

TAE750 CAS:7440-25-7 *HR: 3*
TANTALUM
af: Ta aw: 180.948

PROP: Gray, very hard, malleable, ductile metal. Mp: 2996°; bp: 5429°; d: 16.69. Insol in water.

SYN: TANTALUM-181

TOXICITY DATA with REFERENCE
imp-rat TDLo:3760 mg/kg:ETA CNREA8 16,439,56

CONSENSUS REPORTS: Reported in EPA TSCA Inventory.

OSHA PEL: TWA 5 mg/m^3
ACGIH TLV: TWA 5 mg/m^3
DFG MAK: 5 mg/m^3

SAFETY PROFILE: Some industrial skin injuries from tantalum have been reported. Systemic industrial poisoning however, is apparently unknown. Questionable carcinogen with experimental tumorigenic data. The dry powder ignites spontaneously in air. Incompatible with bromine trifluoride; fluorine; lead chromate. See also specific tantalum compounds.

TAF000 CAS:7721-01-9 *HR: 3*
TANTALUM CHLORIDE
mf: Cl_5Ta mw: 358.20

PROP: White or light yellow crystals or powder. Monoclinic, decomp in moisture. D: 3.68, mp: 216-220°, bp: 239.3°. Decomposed by water; sol in alc. Volatilizes at 144°.

SYN: TANTALUM PENTACHLORIDE

TOXICITY DATA with REFERENCE
orl-rat LD50:1900 mg/kg AIHOAX 1,637,50
ipr-rat LD50:75 mg/kg AIHOAX 1,637,50

CONSENSUS REPORTS: Reported in EPA TSCA Inventory.

SAFETY PROFILE: Poison by intraperitoneal route. Moderately toxic by ingestion. When heated to decomposition it emits toxic fumes of Cl^-.

TAF250 CAS:7783-71-3 ***HR: 3***
TANTALUM FLUORIDE
mf: F_5Ta mw: 275.95

PROP: Deliq, refractive prisms. D: 4.74 @ 20°; mp: 96.8°; bp: 229.5°; vap press: 100 mm @ 130°. Sol in water, concentrated nitric acid; very sol in fuming nitric acid; sltly sol in hot carbon disulfide, hot carbon tetrachloride.

SYN: TANTALIUM PENTAFLUORIDE

TOXICITY DATA with REFERENCE
ivn-mus LD50:110 mg/kg 19UQAS -,30,65

CONSENSUS REPORTS: Reported in EPA TSCA Inventory.

OSHA PEL: TWA 2.5 mg(F)/m^3
ACGIH TLV: TWA 2.5 mg(F)/m^3
NIOSH REL: TWA 2.5 mg(F)/m^3

SAFETY PROFILE: Poison by intravenous route. Corrosive. When heated to decomposition it emits toxic fumes of F^-. See also TANTALUM and FLUORIDES.

TAF500 CAS:1314-61-0 ***HR: 1***
TANTALUM OXIDE
mf: O_5Ta_2 mw: 441.90

PROP: White, microcrystalline, infusible powder. D: 8.200. Insol in water, alc, and mineral acids; sol in HF.

SYNS: TANTALIC ACID ANHYDRIDE ◇ TANTALUM(V) OXIDE ◇ TANTALUM PENTAOXIDE ◇ TANTALUM PENTOXIDE

TOXICITY DATA with REFERENCE
orl-rat LD50:8000 mg/kg EQSSDX 1,1,75

CONSENSUS REPORTS: Reported in EPA TSCA Inventory.

OSHA PEL: TWA 5 mg/m^3
ACGIH TLV: TWA 5 mg/m^3

SAFETY PROFILE: Mildly toxic by ingestion. Incompatible with ClF_3; BrF_3; Li. See also TANTALUM.

TAF675 CAS:113-59-7 ***HR: 3***
TARASAN
mf: $C_{18}H_{18}ClNS$ mw: 315.88

PROP: Pale yellow crystals. Mp: 97-98°. Practically insol in water; sol in alc, ether, chloroform.

SYNS: (α-2-CHLORO-9-omega-DIMETHYLAMINO-PROPYLAMINE)THIOXANTHENE ◇ 2-CHLORO-9-(omega-DI-METHYLAMINO-PROPYLIDENE)THIOXANTHENE ◇ 2-CHLORO-9-(3-(DIMETHYL-AMINO)PROPYLIDENE)-THIOXANTHENE ◇ 2-CHLORO-N,N-DIMETHYLTHIOXANTHENE-Δ^9-Γ-PROPYLAMINE ◇ CHLOROPROTHIXENE ◇ (Z)-3-(2-CHLORO-9H-THIOXANTHEN-9-YLIDENE)-N,N-DIMETHYL-1-PROPANAMINE ◇ CHLORPROTHIXEN ◇ CHLORPROTHIXENE ◇ α-CHLORPROTHIXENE ◇ cis-CHLORPROTHIXENE ◇ CHLORPROTIXEN ◇ CHLORPROTIXENE ◇ CHLORPROTIXINE ◇ CHLOTHIXEN ◇ CPT ◇ CPX ◇ IARACTAN ◇ MK 184 ◇ N 714 ◇ N 714C ◇ PAXYL ◇ RENTOVET ◇ RO 4-0403 ◇ TACTARAN ◇ TARACTAN ◇ TRANQUILAN ◇ TRAQUILAN ◇ TRICTAL ◇ TRUXAL ◇ TRUXALETTEN ◇ TRUXIL ◇ VETACALM

TOXICITY DATA with REFERENCE
orl-man TDLo:360 mg/kg (15W male):REP AJPSAO 120,1004,64
orl-rat LD50:200 mg/kg NIIRDN 6,252,82
orl-mus LD50:50100 μg/kg CNREA8 47,5944,87
ipr-mus LD50:56200 μg/kg CNREA8 47,5944,87
ivn-mus LD50:75 mg/kg APTOA6 41,369,77
orl-rbt LD50:182 mg/kg 29ZVAB -,33,69

CONSENSUS REPORTS: EPA Genetic Toxicology Program.

SAFETY PROFILE: Poison by ingestion, intravenous, and intraperitoneal routes. Human reproductive effects by ingestion: impotence. Incompatible with acids; alkalies. When heated to decomposition it emits toxic fumes of Cl^-, SO_x, and NO_x.

TAF700 CAS:8016-88-4 ***HR: 2***
TARRAGON OIL

PROP: From steam distillation of leaves, stems, and flowers from *Artemesia dracunculus* L. Pale yellow to amber liquid; spicy licorice and sweet basil odor. Sol in fixed oils, mineral oil; insol in propylene glycol, glycerin.

SYN: ESTRAGON OIL

TOXICITY DATA with REFERENCE
skn-mus 100 % FCTXAV 12,709,74
skn-rbt 500 mg/24H FCTXAV 12,709,74
orl-rat LD50:1900 mg/kg FCTXAV 12,709,74

CONSENSUS REPORTS: Reported in EPA TSCA Inventory.

SAFETY PROFILE: Moderately toxic by ingestion. A skin irritant. When heated to decomposition it emits acrid smoke and irritating fumes.

TAF750 CAS:87-69-4 ***HR: 2***
TARTARIC ACID
mf: $C_4H_6O_6$ mw: 150.10

$$(-CHOHCO \cdot OH)_2$$

PROP: Colorless to translucent crystals or white powder; odorless with an acid taste. Mp: 170-172°. Sol in water and alc.

SYNS: 2,3-DIHYDROSUCCINIC ACID ◇ 2,3-DIHYDROXY-BUTANEDIOC ACID

TOXICITY DATA with REFERENCE
ivn-mus LD50:485 mg/kg JAFCAU 5,759,57
orl-dog LDLo:5000 mg/kg JAPMA8 39,275,50
orl-rbt LDLo:5000 mg/kg IECHAD 15,628,23

CONSENSUS REPORTS: Reported in EPA TSCA Inventory.

SAFETY PROFILE: Moderately toxic by intravenous route. Mildly toxic by ingestion. Reaction with silver produces the unstable silver tartrate. When heated to decomposition it emits acrid smoke and irritating fumes.

TAF775 CAS:5976-95-4 ***HR: 2***
meso-TARTRATE
mf: $C_4H_4O_6$ mw: 148.08

SYNS: (R*,S*)-2,3-DIHYDROXY-BUTANEDIOIC ACID ION(2−) (9CI) ◇ meso-TARTARIC ACID, ION(2−)

TOXICITY DATA with REFERENCE
orl-rat LD50:4748 mg/kg OYYAA2 14,963,77
ipr-rat LD50:499 mg/kg OYYAA2 14,963,77
scu-rat LD50:2308 mg/kg OYYAA2 14,963,77
orl-mus LD50:2497 mg/kg OYYAA2 14,963,77
ipr-mus LD50:434 mg/kg OYYAA2 14,963,77
scu-mus LD50:2044 mg/kg OYYAA2 14,963,77

SAFETY PROFILE: Moderately toxic by ingestion, intraperitoneal, and subcutaneous routes.

TAG250 ***HR: 3***
TARWEED

PROP: The seeds containing pyrrolizidine alkaloids are from *Amsinckia intermedia fisch* and *Mey* (JNCIAM 47,1037,71).

SYNS: AMSINCKIA INTERMEDIA ◇ FIDDLE-NECK ◇ SEED of FIDDLENECK

TOXICITY DATA with REFERENCE
orl-rat TDLo:275 g/kg (1-22D preg):ETA,TER JNCIAM 47,1037,71
orl-rat LDLo:1500 mg/kg JAVMA4 150,1305,67

SAFETY PROFILE: Questionable carcinogen with experimental tumorigenic data. Moderately toxic by ingestion. An experimental teratogen. When heated to decomposition it emits acrid smoke and irritating fumes.

TAG750 CAS:107-35-7 ***HR: D***
TAURINE
mf: $C_2H_7NO_3S$ mw: 125.16

PROP: Colorless prisms from water. Mp: decomp @ 300°, bp: subl @ 110°. Very sol in water and alc; insol in anhydrous ether.

SYNS: 2-AMINOETHANESULFONIC ACID ◇ NCI-C60606

TOXICITY DATA with REFERENCE
otr-rat-orl 7560 mg/kg/6W CRNGDP 9,387,88
orl-mus TDLo:878 g/kg (MGN):REP OYYAA2 6,535,72
scu-mus LD50:6 g/kg QJPPAL 19,48,46

CONSENSUS REPORTS: Reported in EPA TSCA Inventory.

SAFETY PROFILE: Experimental reproductive effects. Mutation data reported. When heated to decomposition it emits very toxic fumes of SO_x and NO_x.

TAG875 CAS:22103-31-7 ***HR: 2***
TAURINOPHENETIDINE HYDROCHLORIDE
mf: $C_{10}H_{16}N_2O_3S \cdot ClH$ mw: 280.80

SYN: 2-AMINOETHANESULFONO-p-PHENETIDINEHYDROCHLORIDE

TOXICITY DATA with REFERENCE
orl-mus LD50:1550 mg/kg JPMSAE 60,249,71
ipr-mus LD50:570 mg/kg JMPSAE 60,249,71
scu-mus LD50:590 mg/kg JMPSAE 60,249,71

SAFETY PROFILE: Moderately toxic by ingestion, intraperitoneal, and subcutaneous routes. When heated to decomposition it emits toxic fumes of SO_x, NO_x, and HCl.

TAH250 CAS:81-24-3 ***HR: 2***
TAUROCHOLIC ACID
mf: $C_{26}H_{45}NO_7S$ mw: 515.78

PROP: Cluster of slender, four-sided prisms. Mp: 125° (decomp). Very sol in water; sol in alc; insol in ether, ethyl acetate. Occurs as a sodium salt in the bile.

SYNS: CHOLAIC ACID ◇ CHOLYLTAURINE ◇ TAUROCHOLATE ◇ 2-((3-α,7-α,12-α-TRIHYDROXY-24-OXO-5-β-CHOLAN-24-YL)AMINO) ETHANESULFONIC ACID

TOXICITY DATA with REFERENCE
ipr-rat LD50:450 mg/kg TXAPA9 24,37,73
ipr-mus LD50:620 mg/kg TXAPA9 24,37,73

SAFETY PROFILE: Moderately toxic by intraperitoneal route. An emulsifying agent used as a food additive. When heated to decomposition it emits very toxic fumes of NO_x and SO_x.

TAH500 CAS:102418-13-3 ***HR: 3***
TAUROMYCETIN

PROP: An antibiotic with a pronounced antitumor activity in vivo with respect to some transplantable tumors of mice (ANTBAL 12,360,67).

TOXICITY DATA with REFERENCE
orl-rat LD50:50 mg/kg ANTBAL 12,360,67
ipr-rat LD50:11 mg/kg ANTBAL 12,360,67

orl-mus LD50:66 mg/kg ANTBAL 12,360,67
ipr-mus LD50:24 mg/kg ANTBAL 12,360,67
scu-mus LD50:45 mg/kg ANTBAL 12,360,67
ivn-mus LD50:15 mg/kg ANTBAL 12,360,67
orl-dog LD50:20 mg/kg ANTBAL 12,360,67
ivn-rbt LDLo:3 mg/kg ANTBAL 12,360,67

SAFETY PROFILE: Poison by ingestion, intraperitoneal, subcutaneous, and intravenous routes.

TAH650 CAS:35906-51-5 ***HR: 3***
TAUROMYCETIN-III
mf: $C_{42}H_{51}NO_{16}$ mw: 825.94

SYNS: ANTIBIOTIC MA 144B2 ◇ CINERUBIN B ◇ CINERUBINE B ◇ MA144 B2 ◇ TAVROMYCETIN III

TOXICITY DATA with REFERENCE
pic-esc 24 mg/L ZAPOAK 8,139,68
dni-mus:leu 360 nmol/L JANTAJ 34,1596,81
oms-mus:leu 48 nmol/L JANTAJ 34,1596,81
orl-mus LD50:67 mg/kg 85GDA2 3,96,80
ipr-mus LD50:24 mg/kg 85GDA2 3,96,80
scu-mus LD50:45 mg/kg 85GDA2 3,96,80
ivn-mus LD50:15 mg/kg 85GDA2 3,96,80

SAFETY PROFILE: Poison by ingestion, subcutaneous, intravenous, and intraperitoneal routes. Mutation data reported. When heated to decomposition it emits toxic fumes of NO_x.

TAH675 CAS:34044-10-5 ***HR: 3***
TAUROMYCETIN-IV
mf: $C_{42}H_{53}NO_{16}$ mw: 827.96

SYNS: ANTIBIOTIC MA 144A2 ◇ CINERUBIN A ◇ CINERUBINE A ◇ MA-144A2 ◇ RHODIRUBIN C ◇ RYEMYCIN-B2 ◇ TAVROMYCETIN-IV

TOXICITY DATA with REFERENCE
dni-mus:leu 310 nmol/L JANTAJ 34,1596,81
oms-mus:leu 36 nmol/L JANTAJ 34,1596,81
ivn-mus LD50:2500 μg/kg 85GDA2 3,94,80

SAFETY PROFILE: Poison by intravenous route. Mutation data reported. When heated to decomposition it emits toxic fumes of NO_x.

TAH750 CAS:12607-93-1 ***HR: 3***
TAXINE
mf: $C_{37}H_{51}NO_{10}$ mw: 669.89

PROP: Amorphous, granular powder. Mp: 121-124°. Sol in ether, chloroform, and alc; practically insol in water.

SYN: TAXIN (GERMAN)

TOXICITY DATA with REFERENCE
ipr-mus LDLo:11 mg/kg QJPPAL 5,205,32
ipr-dog LDLo:9 mg/kg QJPPAL 5,205,32
ipr-cat LDLo:25 mg/kg QJPPAL 5,205,32
orl-rbt LDLo:22 mg/kg ZGEMAZ 29,310,22
ipr-rbt LDLo:6 mg/kg QJPPAL 5,205,32
ivn-rbt LDLo:3 mg/kg QJPPAL 5,205,32
ipr-gpg LDLo:14 mg/kg ZGEMAZ 29,310,32
scu-gpg LDLo:10 mg/kg QJPPAL 5,205,32
par-frg LDLo:17 mg/kg QJPPAL 5,205,32

SAFETY PROFILE: Poison by ingestion, intraperitoneal, intravenous, subcutaneous, and parenteral routes. When heated to decomposition it emits toxic fumes of NO_x.

TAH775 CAS:33069-62-4 ***HR: 3***
TAXOL
mf: $C_{47}H_{51}NO_{14}$ mw: 853.99

PROP: Needles from aq methanol. Mp: 213-216° (decomp).

SYNS: 5-β,20-EPOXY-1,2-α,4,7-β,10-β,13-α-HEXAHYDROXY-TAX-11-EN-9-ONE 4,10-DIACETATE 2-BENZOATE 13-ESTER with (2R,3S)-N-BENZOYL-3-PHENYLISOSERINE ◇ NSC-125973

TOXICITY DATA with REFERENCE
ipr-rat LD50:32530 μg/kg NTIS** PB83-17096
ipr-mus LD50:128 mg/kg NTIS** PB83-17096
ivn-dog LDLo:15 mg/kg NTIS** PB83-17096

SAFETY PROFILE: Poison by intravenous and intraperitoneal routes. When heated to decomposition it emits toxic fumes of NO_x.

TAH900 CAS:2545-59-7 ***HR: D***
2,4,5-T BUTOXYETHANOL ESTER
mf: $C_{14}H_{17}Cl_3O_4$ mw: 355.66

SYNS: BLADEX H ◇ BUTOXYETHYL 2,4,5-T ◇ 2,4,5-T BUTOXYETHYL ESTER ◇ HORMOSLYR 500T ◇ (2,4,5-TRICHLOROPHENOXY)ACETIC ACID-2-BUTOXYETHYL ESTER ◇ TRINOXOL

TOXICITY DATA with REFERENCE
cyt-mus-ipr 2400 μg/kg HEREAY 85,123,77

SAFETY PROFILE: Mutation data reported. See also ESTERS. When heated to decomposition it emits toxic fumes of Cl^-.

TAI000 CAS:1746-01-6 ***HR: 3***
TCDD
mf: $C_{12}H_4Cl_4O_2$ mw: 321.96

PROP: Colorless needles. Mp: 305°.

SYNS: 2,3,7,8-CZTEROCHLORODWUBENZO-p-DWUOKSYNY(POLISH) ◇ DIOKSYNY (POLISH) ◇ DIOXINE ◇ DIOXIN (herbicide contaminant) ◇ NCI-C03714 ◇ TCDBD ◇ 2,3,7,8-TCDD ◇ 2,3,7,8-TETRACHLORODIBENZO(b,e)(1,4)DIOXAN ◇ 2,3,6,7-TETRACHLORODIBENZO-p-DIOXIN ◇ 2,3,7,8-TETRACHLORODIBENZO-p-DIOXIN ◇ 2,3,7,8-TETRACHLORODIBENZO-1,4-DIOXIN ◇ TETRADIOXIN

TOXICITY DATA with REFERENCE
eye-rbt 2 mg MOD EVHPAZ 5,87,73
mmo-smc 10 mg/L CMSHAF 12,549,83
oms-mus:oth 1 nmol/L JBCHA3 259,12357,84
ipr-rat TDLo:6 μg/kg (female 17D post):REP BCPCA6 26,1383,77
orl-mus TDLo:12 μg/kg (female 10-13D post):TER TJADAB 33,29,86
orl-rat TDLo:52 μg/kg/2Y-I:CAR NTPTR* NTP-TR-209,82
orl-mus TDLo:52 μg/kg/2Y-I:CAR NTPTR* NTP-TR-209,82
orl-rat TD:328 μg/kg/78W-C:ETA PESTC* 5,12,77
orl-mus TD:36 μg/kg/52W-I:NEO NATUAS 278,548,79
skn-hmn TDLo:107 μg/kg:SKN SCAMAC 154,29,86
orl-rat LD50:20 μg/kg EXPEAM 38,879,82
ipr-rat LD50:60 μg/kg TOXID9 4,189,84
orl-mus LD50:114 μg/kg TXAPA9 29,229,74
skn-mus LDLo:80 μg/kg RCOCB8 20,101,78
ipr-mus LD50:120 μg/kg SCAMAC 254,29,86
orl-dog LD50:1 μg/kg EXPEAM 38,879,82
orl-mky LD50:2 μg/kg EXPEAM 38,879,82
skn-rbt LD50:275 μg/kg EVHPAZ 5,87,73

CONSENSUS REPORTS: NTP Fifth Annual Report on Carcinogens. IARC Cancer Review: Group 2B IMEMDT 7,350,87; Human Inadequate Evidence IMEMDT 15,41,77; Animal Inadequate Evidence IMEMDT 15,41,77. NTP Carcinogenesis Bioassay (gavage); Clear Evidence: mouse, rat NTPTR* NTP-TR-209,82; (dermal). EPA Genetic Toxicology Program.

DFG MAK: Animal Carcinogen, Suspected Human Carcinogen.
NIOSH REL: (Dioxin) Reduce to lowest feasible level.

SAFETY PROFILE: Confirmed carcinogen with experimental carcinogenic, neoplastigenic, tumorigenic, and teratogenic data. One of the most toxic synthetic chemicals. A deadly experimental poison by ingestion, skin contact, and intraperitoneal routes. Human systemic effects by skin contact: allergic dermatitis. Experimental reproductive effects. Human mutation data reported. An eye irritant.

TCDD is the most toxic member of the 75 dioxins. It causes death in rats by hepatic cell necrosis. Death can follow a lethal dose by weeks. Acute and subacute exposure result in wasting, hepatic necrosis, thymic atrophy, hemorrhage, lymphoid depletion, chloracne. A by-product of the manufacture of polychlorinated phenols. It is found at low levels in 2,4,5-T; 2,4,5-trichlorophenol and hexachlorophene. It is also formed during various combustion processes. Incineration of chemical wastes, including chlorophenols, chlorinated benzenes, and biphenyl ethers, may result in the presence of TCDD in flue gases, fly ash, and soot particles. It is immobile in contaminated soil and may be retained for years. TCDD has the potential for bio-accumulation in animals. An accident in Seveso, Italy and inadvertent soil contamination in Missouri have resulted in abandonment of the contaminated areas. When heated to decomposition it emits toxic fumes of Cl^-.

TAI050 CAS:22604-10-0 ***HR: 3***
TCNP SODIUM SALT
mf: $C_7HN_4 \cdot Na$ mw: 164.11

SYNS: SODIUM SALT of 1,1,3,3-TETRACYANOPROPENE ◇ (1,1,3,3-TETRACYANOALLYL)SODIUM

TOXICITY DATA with REFERENCE
orl-mus LD50:153 mg/kg AACHAX -,642,67
ipr-mus LD50:66 mg/kg AACHAX -,642,67
scu-mus LD50:75 mg/kg AACHAX -,642,67

SAFETY PROFILE: Poison by ingestion, subcutaneous, and intraperitoneal routes. When heated to decomposition it emits toxic fumes of NO_x and Na_2O.

TAI100 CAS:34521-14-7 ***HR: 3***
TEAEI
mf: $C_9H_{22}N_3S \cdot Br \cdot BrH$ mw: 365.23

SYNS: (2-((AMINOIMINOMETHYL)THIO)ETHYL)TRIETHYLAMMONIUMBROMIDE HYDROBROMIDE ◇ 2-((AMINOIMINOMETHYL)THIO)-N,N,N-TRIETHYL-ETHANAMINIUM BROMIDE, MONOHYDROBROMIDE (9CI) ◇ s-(2-TRIETHYLAMINOETHYL)ISOTHIURONIUM BROMIDE HYDROBROMIDE

TOXICITY DATA with REFERENCE
ipr-mus LD50:90 mg/kg CPBTAL 23,1639,75
scu-mus LD50:99 mg/kg CPBTAL 23,1639,75
ivn-mus LD50:48800 μg/kg CPBTAL 23,1639,75

SAFETY PROFILE: Poison by subcutaneous, intravenous, and intraperitoneal routes. When heated to decomposition it emits toxic fumes of SO_x, NO_x, NH_3, and Br^-.

TAI250 CAS:9002-84-0 ***HR: 1***
TEFLON
mf: $(C_2F_4)_n$

PROP: Grayish-white, tough plastic. Chemically very inert. Mp: 342°, d: 2.1-2.3.

SYNS: AFLON ◇ ALGLOFLON ◇ ALGOFLON SV ◇ ALKATHENE RXDG33 ◇ AMIP 15m ◇ BALFON 7000 ◇ BDH 29-801 ◇ CHROMOSORB T ◇ DIXON 164 ◇ DLX-6000 ◇ DUROID 5870 ◇ EK 1108GY-A ◇ ETHICON PTFE ◇ FLUO-KEM ◇ FLUON ◇ FLUOROFLEX ◇ FLUOROLON 4 ◇ FLUOROPAK 80 ◇ FLUORPLAST 4 ◇ FTORLON 4 ◇ FTOROPLAST 4 ◇ GORE-TEX ◇ HALON TFEG 180 ◇ HEYDEFLON ◇ HOSTAFLON ◇ MOLYKOTE 522 ◇ POLIFEN ◇ POLITEF ◇ POLY(ETHYLENE TETRAFLUORIDE) ◇ POLYFENE ◇ POLYFLON ◇ POLYTEF ◇ POLYTETRAFLUOROETHENE ◇ POLYTETRAFLUOROETHYLENE ◇ PTFE ◇ SOREFLON 604 ◇ TARFLEN ◇ TEFLON (various) ◇ TETRAFLUOROETHENE HOMOPOLYMER ◇ TETRAFLUOROETHENE POLYMER ◇ TETRAFLUOROETHYLENE HOMOPOLYMER

◇ TETRAFLUOROETHYLENE POLYMERS ◇ TETRAN PTFE ◇ UNON P ◇ VALFLON ◇ VELFLON ◇ ZITEX H 662-124

TOXICITY DATA with REFERENCE
imp-rat TDLo:80 mg/kg:ETA CNREA8 15,333,55

CONSENSUS REPORTS: IARC Cancer Review: Group 3 IMEMDT 7,56,87; Animal Sufficient Evidence IMEMDT 19,285,79; Human Inadequate Evidence IMEMDT 19,285,79. Reported in EPA TSCA Inventory.

SAFETY PROFILE: The finished polymerized compound is inert under ordinary conditions. There have been reports of "polymer fume fever" in humans exposed to pyrolysis products which also are irritants. Smoking should be prohibited in areas where this material is being fabricated or, in general, where there may be dust from it. Exposure to pyrolysis or decomposition products appear to be the chief health-related problem. Questionable carcinogen with experimental tumorigenic data by implant. Incompatible with fluorine; sodium potassium alloy. Under the proper conditions it undergoes hazardous reactions with boron, magnesium, or titanium. When heated to above 750°F it decomposes to yield highly toxic fumes of F^-.

TAI400 CAS:61036-64-4 ***HR: 3***
TEICHOMYCIN A2

TOXICITY DATA with REFERENCE
ipr-mus LD50:1000 mg/kg 85GDA2 1,323,80
scu-mus LD50:2860 mg/kg 48HGAR 12,342,82
ivn-mus LD50:275 mg/kg JANTAJ 31,276,78

SAFETY PROFILE: Poison by intravenous route. Moderately toxic by subcutaneous and intraperitoneal routes.

TAI450 CAS:50679-08-8 ***HR: 2***
TELDANE
mf: $C_{32}H_{41}NO_2$ mw: 471.74

PROP: Crystals from acetone. Mp: 146.5-148.5°.

SYNS: ALDABAN ◇ α-(4-(1,1-DIMETHYLETHYL)PHENYL)-4-(HYDROXYDIPHENYLMETHYL)-1-PIPERIDINEBUTANOL ◇ RMI 9918 ◇ TERFENADINE ◇ TRILUDAN

TOXICITY DATA with REFERENCE
orl-rat TDLo:8550 mg/kg (female 14D pre-21D post):REP ARZNAD 32,1179,82
orl-man TDLo:156 mg/kg/13W-I:SKN BMJOAE 291,940,85
orl-man TDLo:857 μg/kg:SKN,ALR BMJOAE 293,536,86
orl-wmn TDLo:1200 μg/kg:SKN,ALR BMJOAE 293,536,86
orl-wmn TDLo:190 mg/kg/23W-I AIMEAS 103,634,85
orl-cld TDLo:3 mg/kg:SKN,ALR BMJOAE 293,536,86
orl-rat LD50:5 g/kg PHINDQ 4,252,83
ipr-rat LD50:474 mg/kg ARZNAD 32,1179,82
orl-mus LD50:5 g/kg PHINDQ 4,252,83
ipr-mus LD50:510 mg/kg ARZNAD 32,1179,82

SAFETY PROFILE: Moderately toxic by intraperitoneal route. Human systemic effects: allergic reactions, allergic rhinitis, dermatitis, serum sickness, urticaria. Experimental reproductive effects. When heated to decomposition it emits toxic fumes of NO_x. Used as an antihistamine.

TAI500 CAS:113-92-8 ***HR: 3***
TELDRIN
mf: $C_{16}H_{19}ClN_2 \cdot C_4H_4O_4$ mw: 390.90

SYNS: ALLERCLOR ◇ ALLERGIN ◇ ALLERGISAN ◇ ALUNEX ◇ ANTAGONATE ◇ CARBINOXAMIDE MALEATE ◇ CHLORMENE ◇ dl-2(-p-CHLORO-α-2-(DIMETHYLAMINO)ETHYLBENZYL)PYRIDINE BIMALEATE ◇ 1-p-CHLOROPHENYL-1-(2-PYRIDYL)-3-DIMETHYLAMINOPROPANE MALEATE ◇ CHLOROPROPHENYPYRIDAMINE MALEATE ◇ CHLORPHENIRAMINE MALEATE ◇ CHLOR-TRIMETON ◇ CHLOR-TRIMETON MALEATE ◇ CHLOR-TRIPOLON ◇ CLOROPIRIL ◇ C-METON ◇ 1-(N,N-DIMETHYLAMINO)-3-(p-CHLOROPHENYL-3-α-PYRIDYL)PROPANE MALEATE ◇ HISTADUR ◇ HISTADUR DURA-TABS ◇ HISTALEN ◇ HISTAPAN ◇ IBIOTON ◇ LORPHEN ◇ M.P. CHLORCAPS T.D. ◇ NCI-C55265 ◇ NEORESTAMIN ◇ PIRIEX ◇ PIRITON ◇ POLARONIL (GERMAN) ◇ PYRIDAMAL-100 ◇ SYNISTAMIN

TOXICITY DATA with REFERENCE
cyt-ham:ovr 500 mg/L NTPTR* NTP-TR-317,86
sce-ham:ovr 125 mg/L NTPTR* NTP-TR-317,86
orl-mus TDLo:420 mg/kg (1-21D preg):REP ARZNAD 18,188,68
orl-rat LD50:306 mg/kg KSRNAM 12,1371,78
scu-rat LD50:365 mg/kg TXAPA9 18,185,71
orl-mus LD50:130 mg/kg ARZNAD 28,1644,78
ipr-mus LD50:76700 μg/kg JPETAB 113,72,55
ivn-mus LD50:26100 μg/kg TXAPA9 23,537,72
ivn-dog LD50:97600 μg/kg JPETAB 113,72,55
orl-gpg LD50:198 mg/kg JPETAB 113,72,55

CONSENSUS REPORTS: NTP Carcinogenesis Studies (gavage); No Evidence: mouse, rat NTPTR* NTP-TR-317,86. Reported in EPA TSCA Inventory.

SAFETY PROFILE: Poison by ingestion, intravenous, and subcutaneous routes. Experimental reproductive effects. Used as an antihistamine. Mutation data reported. When heated to decomposition it emits very toxic fumes of Cl^- and NO_x.

TAI600 CAS:37244-85-2 ***HR: 3***
TELEBRIX 38
mf: $C_{12}H_{11}I_3N_2O_5 \cdot C_{12}H_{11}I_3N_2O_5 \cdot C_7H_{17}NO_5 \cdot Na$ mw: 1506.14

SYN: AG 58107

TOXICITY DATA with REFERENCE
ivn-rat LD50:16800 mg/kg THERAP 26,595,71
ipr-mus LD50:21900 mg/kg THERAP 26,595,71
ivn-mus LD50:13700 mg/kg THERAP 26,595,71
ice-mus LD50:200 mg/kg THERAP 26,595,71

SAFETY PROFILE: Poison by intracerebral route. When heated to decomposition it emits toxic fumes of I^-, NO_x, and Na_2O.

TAI725 CAS:66813-55-6 ***HR: 2***
TELEMIN-SOFT

TOXICITY DATA with REFERENCE
orl-rat TDLo:45 g/kg (90D pre):REP OYYAA2 15,85,78
orl-rat LD50:4 g/kg OYYAA2 15,85,78
orl-mus LD50:4600 mg/kg OYYAA2 15,85,78

SAFETY PROFILE: Moderately toxic by ingestion. Experimental reproductive effects. When heated to decomposition it emits toxic fumes of SO_x, NO_x, and Na_2O.

TAI735 ***HR: 3***
TELLURANE-1,1-DIOXIDE
mf: $C_5H_{10}O_2Te$ mw: 229.73

$$(O:)_2\overline{Te(CH_2)_4CH_2}$$

SAFETY PROFILE: Explodes when heated rapidly. Violent reaction on contact with acids (e.g., sulfuric acid; nitric acid). When heated to decomposition it emits toxic fumes of Te. See also TELLURIUM COMPOUNDS.

TAI750 CAS:7803-68-1 ***HR: 3***
TELLURIC ACID
mf: H_6O_6Te mw: 229.66

PROP: White solid. D: (monoclinic) 3.068; (cubic) 3.163. Sltly sol in concentrated nitric acid; very sol in water.

SYNS: ORTHOTELLURIC ACID ◇ TELLURATE ◇ TELLURIC(VI) ACID ◇ TELLURIUM HYDROXIDE

TOXICITY DATA with REFERENCE
ivn-rat LDLo:31 mg/kg JPETAB 33,270,28
orl-rbt LDLo:56 mg/kg JPETAB 33,270,28
ivn-rbt LDLo:5600 μg/kg JPETAB 33,270,28

CONSENSUS REPORTS: Reported in EPA TSCA Inventory.

OSHA PEL: TWA 0.1 mg(Te)/m^3
ACGIH TLV: TWA 0.1 mg(Te)/m^3

SAFETY PROFILE: Poison by ingestion and intravenous routes. When heated to decomposition it emits toxic fumes of Te. See also TELLURIUM COMPOUNDS.

TAI800 CAS:22451-06-5 ***HR: 3***
TELLURIC ACID, DISODIUM SALT, PENTAHYDRATE
mf: $O_3Te•2Na•5H_2O$ mw: 380.65

TOXICITY DATA with REFERENCE
ipr-mus LD50:7 mg/kg COREAF 257,791,63

OSHA PEL: TWA 0.1 mg(Te)/m^3
ACGIH TLV: TWA 0.1 mg(Te)/m^3

SAFETY PROFILE: Poison by intraperitoneal route. When heated to decomposition it emits toxic fumes of Te.

TAJ000 CAS:13494-80-9 ***HR: 3***
TELLURIUM
af: Te aw: 127.60

PROP: Silvery-white, metallic, lustrous element; quite brittle. Mp: 449.5°; bp: 989.8°; d: 6.24 @ 20°; vap press: 1 mm @ 520°. Insol in water, benzene, carbon disulfide.

SYNS: NCI-C60117 ◇ TELLOY ◇ TELLUR (POLISH)

TOXICITY DATA with REFERENCE
ims-rat TDLo:13 mg/kg (10D preg):REP EXPEAM 28,1444,72
orl-rat TDLo:3300 mg/kg (1-22D preg):TER EXNEAC 31,1,71
orl-rat LD50:83 mg/kg 85GMAT -,107,82
itr-rat LDLo:200 mg/kg AIHAAP 39,100,78
orl-mus LD50:20 mg/kg 85GMAT -,107,82
orl-rbt LD50:67 mg/kg 85GMAT -,107,82
orl-gpg LD50:45 mg/kg 85GMAT -,107,82

CONSENSUS REPORTS: Reported in EPA TSCA Inventory.

OSHA PEL: TWA 0.1 mg(Te)/m^3
ACGIH TLV: TWA 0.1 mg(Te)/m^3
DFG MAK: 0.1 mg/m^3

SAFETY PROFILE: Poison by ingestion and intratracheal routes. An experimental teratogen. Exposure causes nausea, vomiting, tremors, convulsions, respiratory arrest, central nervous system depression, and garlic odor to breath. Aerosols of tellurium, tellurium dioxide, and hydrogen telluride cause irritation of the respiratory system and may lead to the development of bronchitis and pneumonia. Experimental reproductive effects. Under the proper conditions it undergoes hazardous reactions with halogens (e.g., chlorine; fluorine); interhalogens (e.g., bromine pentafluoride; chlorine fluorine; chlorine trifluoride); metals (e.g., cadmium; potassium; sodium; platinum; tin; zinc); hexalithium disilicide; silver bromate; silver iodate. When heated to decomposition it emits toxic fumes of Te. See also TELLURIUM COMPOUNDS.

TAJ010 CAS:13494-80-9 ***HR: 1***
TELLURIUM (dust or fume)

CONSENSUS REPORTS: Reported in EPA TSCA Inventory. EPA Extremely Hazardous Substances List.

ACGIH TLV: TWA 0.1 mg/m^3
DFG MAK: 0.1 mg/m^3

SAFETY PROFILE: May cause irritation of the respiratory system and lead to bronchitis and pneumonia. When heated to decomposition it emits toxic fumes of Te. See also TELLURIUM and TELLURIUM COMPOUNDS.

TAJ250 CAS:10026-07-0 ***HR: D***
TELLURIUM CHLORIDE
mf: Cl_4Te mw: 269.40

PROP: Mp: 224°, bp: 380°, d: 3.260. Hygroscopic.

SYNS: TELLURIC CHLORIDE ◇ TELLURIUM TETRACHLORIDE ◇ TETRACHLOROTELLURIUM

TOXICITY DATA with REFERENCE
mrc-bcs 1 mmol/L MUREAV 77,109,80
itt-rat TDLo:21552 μg/kg (1D male):REP JRPFA4 7,21,64

CONSENSUS REPORTS: Reported in EPA TSCA Inventory. EPA Genetic Toxicology Program.

OSHA PEL: TWA 0.1 $mg(Te)/m^3$
ACGIH TLV: TWA 0.1 $mg(Te)/m^3$

SAFETY PROFILE: Mutation data reported. Experimental reproductive effects. Incompatible with ammonia. Irritant. When heated to decomposition it emits very toxic fumes of Cl^- and Te. See also TELLURIUM COMPOUNDS.

TAJ500 ***HR: 2***
TELLURIUM COMPOUNDS

SAFETY PROFILE: Elemental tellurium has relatively low toxicity. It is converted in the body to dimethyl telluride which imparts a garlic-like odor to the breath and sweat. Heavy exposures may, in addition, result in headache, drowsiness, metallic taste, loss of appetite, nausea, tremors, convulsions, and respiratory arrest. Various tellurium salts may also produce similar symptoms. Large doses can be fatal, as was the case following accidental administration of 2 grams of sodium tellurite. Workers in an iron foundry exposed to less than 0.1 mg Te/m^3 developed a garlic-like odor in breath, sweat, and urine, as well as anorexia, nausea, depression, somnolence, itchy skin, and metallic taste. When heated or on contact with acid or acid fumes they emit highly toxic fumes. See also TELLURIUM and specific compounds.

TAJ750 CAS:7446-07-3 ***HR: 3***
TELLURIUM DIOXIDE
mf: O_2Te mw: 159.60

SYN: TELLURIUM OXIDE

TOXICITY DATA with REFERENCE
scu-rat TDLo:79800 μg/kg (female 15-19D post):REP TJADAB 31,66A,85
scu-rat TDLo:798 mg/kg (15-19D preg):TER TJADAB 29(2),50A,84
itr-rat LDLo:120 mg/kg AIHAAP 39(2),100,78

CONSENSUS REPORTS: Reported in EPA TSCA Inventory.

OSHA PEL: TWA 0.1 $mg(Te)/m^3$
ACGIH TLV: TWA 0.1 $mg(Te)/m^3$

SAFETY PROFILE: Poison by intratracheal route. An experimental teratogen. Experimental reproductive effects. When heated to decomposition it emits toxic fumes of Te. See also TELLURIUM COMPOUNDS.

TAK250 CAS:7783-80-4 ***HR: 3***
TELLURIUM HEXAFLUORIDE
DOT: UN 2195
mf: F_6Te mw: 241.60

PROP: Colorless gas; repulsive odor. Mp: −37.6°, bp: −38.9° (subl); d: (solid) 4.006 @ −191°; (liquid) 2.499 @ −10°.

TOXICITY DATA with REFERENCE
ihl-rat LCLo:5 ppm/24H ATXKA8 18,140,60
ihl-mus LCLo:5 ppm/24H ATXKA8 18,140,60
ihl-rbt LCLo:5 ppm/8H ATXKA8 18,140,60
ihl-gpg LCLo:5 ppm/6H ATXKA8 18,140,60

CONSENSUS REPORTS: EPA Extremely Hazardous Substances List. Reported in EPA TSCA Inventory.

OSHA PEL: TWA 0.02 ppm (Te)
ACGIH TLV: TWA 0.02 ppm (Te)
DOT Classification: Poison A; Label: Poison Gas.

SAFETY PROFILE: Poison by inhalation. Human skin (systemic) effects. When heated to decomposition it emits very toxic fumes of F^- and Te. See also FLUORIDES and TELLURIUM COMPOUNDS.

TAK500 ***HR: 3***
TELLURIUM NITRIDE

PROP: A solid.

SAFETY PROFILE: Poison. A severe explosion hazard when shocked or exposed to heat. Will react with water or steam to produce toxic fumes. When heated or on contact with acid or acid fumes it emits highly toxic fumes of Te and NO_x. See also TELLURIUM COMPOUNDS and NITRIDES.

TAK600 CAS:10031-27-3 ***HR: 3***
TELLURIUM TETRABROMIDE
mf: $TeBr_4$ mw: 447.22

SAFETY PROFILE: Reaction with ammonia forms a heat-sensitive explosive product. When heated to decomposition it emits toxic fumes of Br^- and Te. See also TELLURIUM COMPOUNDS and BROMIDES.

TAK750 CAS:11032-05-6 ***HR: 3***
TELOCIDIN B
mf: $C_{28}H_{41}N_3O_2$ mw: 451.72

TOXICITY DATA with REFERENCE
dns-hmn:lng 100 nmol/L CRNGDP 5,209,84
oms-hmn:lng 100 nmol/L CRNGDP 5,209,84
dni-hmn:lym 10 μg/L ONCOBS 40,273,83
dns-rat:lvr 100 pmol/L CRNGDP 5,1547,84
ipr-mus LDLo:300 μg/kg SCIEAS 204,193,79
ivn-mus LD50:220 μg/kg SCIEAS 204,193,79

SAFETY PROFILE: A deadly poison by intraperitoneal and intravenous routes. Human mutation data reported. When heated to decomposition it emits toxic fumes of NO_x.

TAK800 CAS:19246-24-3 ***HR: 2***
TELOMYCIN
mf: $C_{59}H_{77}N_{13}O_{19}$ mw: 1272.36

PROP: Amorph, gray solid. Minimum solubility in water at pH 3.0 to 3.3 = 4 mg/mL. Maximum solubility is above pH 8.5 = >150 mg/mL. Sol in 10% sodium chloride soln less than 1 mg/mL; mod sol in methanol, ethanol; very sltly sol in acetone, ethyl acetate; insol in ether, chloroform, hydrocarbons. Aq solns are stable to heat. aureus.

TOXICITY DATA with REFERENCE
scu-mus LD50:1000 mg/kg 85GDA2 4(2),161,80
ivn-mus LD50:750 mg/kg 85GDA2 4(2),161,80

SAFETY PROFILE: Moderately toxic by subcutaneous and intravenous routes. When heated to decomposition it emits toxic NO_x fumes.

TAL000 CAS:3589-21-7 ***HR: 3***
TEMARIL
mf: $C_{20}H_{26}N_2$•ClH mw: 330.94

SYNS: 5-(3-DIMETHYLAMINO-2-METHYLPROPYL)-10,11-DIHYDRO-5H-DIBENZ(b,f)AZEPINE HYDROCHLORIDE ◇ TRIMEPRAMINE HYDROCHLORIDE ◇ TRIMEPROPIMINE HYDROCHLORIDE ◇ TRIMEPROPRIMINE HYDROCHLORIDE ◇ TRIMIPROMINE HYDROCHLORIDE

TOXICITY DATA with REFERENCE
orl-rat LD50:800 mg/kg 27ZQAG -,91,72
ipr-rat LD50:150 mg/kg 27ZQAG -,91,72
ivn-rat LD50:38 mg/kg 27ZQAG -,91,72
orl-mus LD50:250 mg/kg 27ZQAG -,91,72
ipr-mus LD50:135 mg/kg 27ZQAG -,91,72
scu-mus LD50:200 mg/kg 27ZQAG -,91,72
ivn-mus LD50:58 mg/kg 27ZQAG -,91,72
ims-mus LD50:223 mg/kg SKIZAB 25,644,69

SAFETY PROFILE: Poison by ingestion, intraperitoneal, intravenous, and subcutaneous routes. When heated to decomposition it emits very toxic fumes of HCl and NO_x.

TAL250 CAS:3383-96-8 ***HR: 3***
TEMEPHOS
mf: $C_{16}H_{20}O_6P_2S_3$ mw: 466.48

PROP: White crystals. Mp: 30°.

SYNS: ABATE ◇ ABATHION ◇ AC 52160 ◇ AMERICAN CYANAMID AC 52,160 ◇ BIOTHION ◇ BITHION ◇ CL 52160 ◇ DIFENTHOS ◇ O,O-DIMETHYL PHOSPHOROTHIOATE-O,O-DIESTER with 4,4'-THIODIPHENOL ◇ ECOPRO ◇ EI 52160 ◇ ENT 27,165 ◇ EXPERIMENTAL INSECTICIDE 52160 ◇ NIMITEX ◇ NIMITOX ◇ SWEBATE ◇ TEMEFOS ◇ TEMOPHOS ◇ TETRAMETHYL-O,O'-THIODI-p-PHENYLENE PHOSPHOROTHIOATE ◇ O,O,O'O',-TETRAMETHYL-O,O'-THIODI-p-PHENYLENE PHOSPHOROTHIOATE ◇ O,O'-(THIODI-4,1-PHENYLENE)BIS(O,O-DIMETHYLPHOSPHOROTHIOATE) ◇ O,O'-(THIODI-p-PHENYLENE)-O,O,O',O'-TETRAMETHYL BIS(PHOSPHOROTHIOATE)

TOXICITY DATA with REFERENCE
skn-rbt 500 mg/24H MLD TXAPA9 21,369,72
skn-rbt TDLo:2132 mg/kg (6-18D preg):TER NTIS** AD-A134-545
orl-rat LD50:1000 mg/kg GUCHAZ 6,493,73
skn-rat LD50:1370 mg/kg WRPCA2 9,119,70
ipr-rat LD50:912 mg/kg TOIZAG 16,297,69
scu-rat LD50:2302 mg/kg TOIZAG 16,297,69
orl-mus LD50:223 mg/kg ARSIM* 20,1,66
orl-rbt LD50:313 mg/kg VETNAL 60(11),63,84
skn-rbt LD50:970 mg/kg AMCYC* 8/66

CONSENSUS REPORTS: Reported in EPA TSCA Inventory.

OSHA PEL: (Transitional: Total Dust: 15 mg/m^3; Respirable Fraction: 5 mg/m^3) TWA Total Dust: 10 mg/m^3; Respirable Fraction: 5 mg/m^3
ACGIH TLV: TWA 10 mg/m^3

SAFETY PROFILE: Poison by ingestion. Moderately toxic by intraperitoneal, subcutaneous, and skin contact routes. An experimental teratogen. A skin irritant. A cholinesterase inhibitor type of insecticide. When heated to decomposition it emits toxic fumes of PO_x and SO_x. See also PARATHION.

TAL275 CAS:30015-57-7 *HR: 3*
TEMEQUINE
mf: $C_{11}H_{21}N \cdot BrH$ mw: 248.25

SYNS: TEMECHINE ◇ 2,2,6,6-TETRAMETHYLQUINUCLIDINE HYDROBROMIDE

TOXICITY DATA with REFERENCE
ipr-mus LD50:345 mg/kg PCJOAU 4,588,70
scu-mus LD50:130 mg/kg RPTOAN 35,278,72
ivn-mus LD50:52 mg/kg PCJOAU 4,588,70

SAFETY PROFILE: Poison by subcutaneous, intravenous, and intraperitoneal routes. When heated to decomposition it emits toxic fumes of NO_x and Br^-.

TAL325 CAS:52485-79-7 *HR: 3*
TEMGESIC
mf: $C_{29}H_{41}NO_4$ mw: 467.71

PROP: Crystals. Mp: 209°.

SYNS: BUPRENORPHINE ◇ RX 6029M

TOXICITY DATA with REFERENCE
ipr-rat LD50:197 mg/kg DRUGAY 17,81,79
ivn-rat LD50:31 mg/kg DRUGAY 17,81,79
orl-mus LD50:260 mg/kg DRUGAY 17,81,79
ipr-mus LD50:90 mg/kg DRUGAY 17,81,79
ivn-mus LD50:24 mg/kg DRUGAY 17,81,79

SAFETY PROFILE: Poison by ingestion, intravenous, and intraperitoneal routes. Note: This is a controlled substance (narcotic) listed in the U.S. Code of Federal Regulations, Title 21 Part 1308.15 (1985). When heated to decomposition it emits toxic fumes of NO_x.

TAL350 CAS:62169-70-4 *HR: 3*
TENEMYCIN

SYNS: A-12253 ◇ TENEBRIMYCIN

TOXICITY DATA with REFERENCE
orl-mus LD50:8000 mg/kg 85GDA2 1,163,80
ipr-mus LD50:1250 mg/kg 85GDA2 1,163,80
scu-mus LD50:1500 mg/kg 85GDA2 1,163,80
ivn-mus LD50:300 mg/kg 85GDA2 1,163,80

SAFETY PROFILE: Poison by intravenous route. Moderately toxic by intraperitoneal and subcutaneous routes. Mildly toxic by ingestion.

TAL475 CAS:29122-68-7 *HR: 3*
TENORMIN
mf: $C_{14}H_{22}N_2O_3$ mw: 266.38

PROP: Crystals from ethyl acetate. Mp: 146-148°.

SYNS: ATENOLOL ◇ 1-p-CARBAMOYLMETHYLPHENOXY-3-ISOPROPYLAMINO-2-PROPANOL ◇ 2-(p-(2-HYDROXY-3-(ISOPROPYLAMINO)PROPOXY)PHENYL)ACETAMIDE ◇ 4-(2-HYDROXY-3-((1-METHYLETHYL)AMINO)PROPOXY)BENZENEACETAMIDE ◇ IBINOLO ◇ ICI 66,082 ◇ PRENORMINE ◇ MYOCORD ◇ SELES BETA

TOXICITY DATA with REFERENCE
orl-wmn TDLo:16 mg/kg (lactating female 5D post):REP JOPDAB 114,476,89
orl-rat TDLo:2200 mg/kg (female 7-17D post):TER JZKEDZ 6,247,80
orl-man TDLo:49 mg/kg/30D-I JCPYDR 6,390,86
orl-wmn TDLo:16 mg/kg:PUL AEMED3 14,161,85
orl-wmn TDLo:1080 mg/kg/78W-I:BLD,SKN JRHUA9 13,446,86
orl-wmn TDLo:42 mg/kg/3W-I:SYS BMJOAE 294,1324,87
orl-rat LD50:3000 mg/kg DRUGAY 17,425,79
ivn-rat LD50:59 mg/kg DRUGAY 17,425,79
orl-mus LD50:2000 mg/kg DRUGAY 17,425,79
ivn-mus LD50:57 mg/kg YACHDS 8,4579,80
ivn-rbt LD50:50 mg/kg DRUGAY 17,425,79

SAFETY PROFILE: Poison by intravenous route. Moderately toxic by ingestion. Human systemic effects: calcium level changes, dermatitis, leukopenia, respiratory obstruction. An experimental teratogen. Human and experimental reproductive effects. When heated to decomposition it emits toxic fumes of NO_x.

TAL485 CAS:59804-37-4 *HR: 3*
TENOXICAM
mf: $C_{13}H_{11}N_3O_4S_2$ mw: 337.37

PROP: Crystals from xylene. Mp: 209-213° (decomp).

SYNS: 4-HYDROXY-2-METHYL-N-2-PYRIDINYL-2H-THIENO(2,3-e)-1,2-THIAZINE-3-CARBOXAMIDE 1,1-DIOXIDE ◇ RO 12-0068 ◇ TILCOTIL

TOXICITY DATA with REFERENCE
orl-rat LD50:79 mg/kg YACHDS 12(Suppl 5),759,84
ipr-rat LD50:234 mg/kg YACHDS 12(Suppl 5),759,84
scu-rat LD50:186 mg/kg YACHDS 12(Suppl 5),759,84
orl-mus LD50:297 mg/kg YACHDS 12(Suppl 5),759,84
ipr-mus LD50:446 mg/kg YACHDS 12(Suppl 5),759,84
scu-mus LD50:449 mg/kg YACHDS 12(Suppl 5),759,84

SAFETY PROFILE: Poison by ingestion, subcutaneous, and intraperitoneal routes. When heated to decomposition it emits toxic fumes of SO_x and NO_x.

TAL490 CAS:302-83-0 *HR: 3*
TENSILON BROMIDE
mf: $C_{10}H_{16}NO \cdot Br$ mw: 246.18

PROP: Crystals from ethanol + ether; decomp @ 151-152°. Bitter taste. Sol in water (more than 10%); mod sol in alc; practically insol in ether. Solns are stable.

SYNS: EDROPHONIUM BROMIDE ◇ N-ETHYL-3-HYDROXY-N,N-

DIMETHYL-BENZENAMINIUM BROMIDE (9CI) ◇ ETHYL(m-HYDROXYPHENYL)DIMETHYLAMMONIUM BROMIDE (8CI) ◇ 3-HYDROXYPHENYLDIMETHYLETHYLAMMONIUM BROMIDE ◇ OEDROPHANIUM ◇ TENSILON

TOXICITY DATA with REFERENCE
orl-mus LD50:600 mg/kg JPETAB 100,83,50
ipr-mus LD50:37 mg/kg JPETAB 100,83,50
scu-mus LD50:130 mg/kg JPETAB 100,83,50
ivn-mus LD50:9 mg/kg JPETAB 100,83,50
ivn-dog LD50:15 mg/kg JPETAB 100,83,50
inv-rbt LD50:29 mg/kg JPETAB 100,83,50

SAFETY PROFILE: Poison by subcutaneous, intravenous, and intraperitoneal routes. Moderately toxic by ingestion. When heated to decomposition it emits toxic fumes of Br^-, NH_3, and NO_x.

TAL550 CAS:19202-92-7 ***HR: 3***
TENULIN
mf: $C_{17}H_{22}O_5$ mw: 306.39

TOXICITY DATA with REFERENCE
dni-mus-unr 30 mg/kg/3D JMCMAR 20,333,77
dni-mus:ast 2143 μmol/L JPMSAE 67,1235,78
ipr-mus LD50:185 mg/kg RCOCB8 28,189,80

CONSENSUS REPORTS: EPA Genetic Toxicology Program.

SAFETY PROFILE: Poison by intraperitoneal route. Mutation data reported.

TAL560 CAS:65184-10-3 ***HR: 3***
TEOPROLOL
mf: $C_{23}H_{30}N_6O_4$ mw: 454.59

SYNS: D-13,312 ◇ 3,7-DIHYDRO-7-(3-((2-HYDROXY-3-((2-METHYL-1H-INDOL-4-YL)OXY)PROPYL)AMINO)BUTYL)-1,3-DIMETHYL-1H-PURINE-2,6-DIONE ◇ 7-(3-((2-HYDROXY-3-((2-METHYLINDOL-4-YL)OXY)PROPYL)AMINO)BUTYL)THEOPHYLLINE

TOXICITY DATA with REFERENCE
ivn-rat LD50:110 mg/kg DRFUD4 7,824,82
orl-mus LD50:1870 mg/kg DRFUD4 7,824,82
ivn-mus LD50:39 mg/kg DRFUD4 7,824,82
ivn-dog LD50:38 mg/kg DRFUD4 7,824,82

SAFETY PROFILE: Poison by intravenous route. Moderately toxic by ingestion. When heated to decomposition it emits toxic fumes of NO_x.

TAL575 CAS:15589-31-8 ***HR: 3***
TERALLETHRIN
mf: $C_{17}H_{24}O_3$ mw: 276.41

SYNS: (±)-3-ALLYL-2-METHYL-4-OXO-2-CYCLOPENTENYL2,2,3,3-TETRAMETHYLCYCLOPROPANE CARBOXYLATE ◇ M-108

TOXICITY DATA with REFERENCE
orl-rat LD50:174 mgkg IYKEDH 13,1128,82
ipr-rat LD50:81 mg/kg IYKEDH 13,1128,82
scu-rat LD50:252 mg/kg IYKEDH 13,1128,82
orl-mus LD50:177 mg/kg IYKEDH 13,1128,82
ipr-mus LD50:81 mg/kg IYKEDH 13,1128,82
scu-mus LD50:293 mg/kg IYKEDH 13,1128,82

SAFETY PROFILE: Poison by ingestion, subcutaneous, and intraperitoneal routes. When heated to decomposition it emits acrid smoke and irritating fumes. See also ESTERS.

TAL750 ***HR: 2***
TERBIUM
af: Tb aw: 158.9254

PROP: A silvery-gray, soft ductile, malleable metallic element. Easily oxidized in air. Mp: 1356°, bp: 3041°, d: 8.234.

SAFETY PROFILE: As a lanthanon it may impair blood coagulation. Fire hazard in the form of dust in air or on contact with halogens. Incompatible with air; halogens. See also POWDERED METALS, LANTHANONS, and RARE EARTHS.

TAM000 CAS:10042-88-3 ***HR: 3***
TERBIUM CHLORIDE
mf: Cl_3Tb mw: 265.27

TOXICITY DATA with REFERENCE
skn-rbt 500 mg/24H SEV TXAPA9 5,427,63
eye-rbt 50 mg TXAPA9 5,427,63
orl-mus LD50:3631 mg/kg EQSSDX 1,1,75
ipr-mus LD50:332 mg/kg AEHLAU 5,437,62

CONSENSUS REPORTS: Reported in EPA TSCA Inventory.

SAFETY PROFILE: Poison by intraperitoneal route. Moderately toxic by ingestion. An eye and severe skin irritant. When heated to decomposition it emits very toxic fumes of Cl^-. See also TERBIUM, CHLORIDES, and RARE EARTHS.

TAM500 CAS:13482-49-0 ***HR: 3***
TERBIUM CITRATE
mf: $C_6H_8O_7 \cdot Tb$ mw: 351.06

SYN: 2-HYDROXY-1,2,3-PROPANECARBOXYLIC ACID TERBIUM (3+) SALT (1:1)

TOXICITY DATA with REFERENCE
ipr-mus LD50:121 mg/kg AEHLAU 5,437,62
ipr-gpg LD50:74 mg/kg AEHLAU 5,437,62

SAFETY PROFILE: Poison by intraperitoneal route. When heated to decomposition it emits acrid smoke and irritating fumes. See also TERBIUM and RARE EARTHS.

TAN000 CAS:12738-76-0 ***HR: 2***
TERBIUM OXIDE

TOXICITY DATA with REFERENCE
orl-rat LDLo:1000 mg/kg CURL** 35,25,60

CONSENSUS REPORTS: Reported in EPA TSCA Inventory.

SAFETY PROFILE: Moderately toxic by ingestion. See also TERBIUM and RARE EARTHS.

TAN100 CAS:23031-25-6 ***HR: 3***
TERBUTALINE
mf: $C_{12}H_{19}NO_3$ mw: 225.32

SYNS: 1,3-BENZENEDIOL,5-(2-((1,1-DIMETHYLETHYL)AMINO)-1-HYDROXYETHYL)- (9CI) ◇ BENZYL ALCOHOL, α-((t-BUTYLAMINO) METHYL)-3,5-DIHYDROXY- ◇ BRICAN ◇ BRICANYL ◇ BRICAR ◇ BRICARIL ◇ BRICYN ◇ 5-(2-((1,1-DIMETHYLETHYL)AMINO)-1-HYDROXYETHYL)-1,3-BENZENEDIOL ◇ KWD 2019 ◇ TERBUTALIN

TOXICITY DATA with REFERENCE
mul-wmn TDLo:5660 μg/kg (female 32-34W post):REP JOPDAB 94,449,79
mul-wmn TDLo:380 μg/kg:CVS JAMAAP 244,692,80
orl-man TDLo:714 μg/kg NEJMAG 300,143,79

SAFETY PROFILE: Poison by ingestion. Human reproductive effects. Human systemic effects by multiple routes: cardiac changes. When heated to decomposition it emits toxic fumes of NO_x.

TAN250 CAS:23031-32-5 ***HR: 3***
TERBUTALINE SULPHATE
mf: $C_{24}H_{36}N_2O_8S$ mw: 512.68

SYNS: BRETHINE ◇ BRICANYL ◇ α-(BUTYLAMINO)METHYL)-3,5-DIHYDROXYBENZYL ALCOHOL, SULFATE (2:1) ◇ 1-(3,5-DIHYDROXYPHENYL)-2-tert-BUTYLAMINOETHANOLSULPHATE ◇ 5-(2-((1,1-DIMETHYLETHYL)AMINO)-1-HYDROXYETHYL)-1,3-BENZENEDIOL, SULFATE (2:1) (SALT) ◇ KWD 2019 ◇ TERBUTALINE SULFATE

TOXICITY DATA with REFERENCE
scu-rat TDLo:24 mg/kg (9-14D preg):TER KSRNAM 6,757,72
scu-mus TDLo:192 mg/kg (female 7-12D post):REP KSRNAM 6,757,72
orl-wmn TDLo:10 mg/kg HUTODJ 6,525,87
scu-wmn TDLo:50 μg/kg:CVS NEJMAG 296,821,77
orl-rat LD50:8700 mg/kg TYKEDH 5,106,74
ipr-rat LD50:220 mg/kg IYKEDH 5,106,74
scu-rat LD50:130 mg/kg NIIRDN 6,500,82
ivn-rat LD50:69 mg/kg IYKEDH 5,106,74
orl-mus LD50:205 mg/kg APTOA6 31,49,72
ipr-mus LD50:130 mg/kg NIIRDN 6,500,82
scu-mus LD50:260 mg/kg IYKEDH 5,106,74
ivn-mus LD50:36 mg/kg IYKEDH 5,106,74
orl-dog LD50:1520 mg/kg KSRNAM 6,757,72
ivn-dog LD50:116 mg/kg KSRNAM 6,757,72
ivn-rbt LD50:110 mg/kg KSRNAM 6,757,72

SAFETY PROFILE: Poison by ingestion, intravenous, subcutaneous, and intraperitoneal routes. Human systemic effects: EKG changes. An experimental teratogen. Experimental reproductive effects. When heated to decomposition it emits very toxic fumes of SO_x and NO_x.

TAN500 CAS:623-27-8 ***HR: 2***
TEREPHTHALALDEHYDE
mf: $C_8H_6O_2$ mw: 134.14

SYNS: p-BENZENEDICARBOXALDEHYDE ◇ 1,4-BENZENEDICARBOXALDEHYDE (9CI) ◇ 1,4-DIFORMYLBENZENE ◇ p-FORMYLBENZALDEHYDE ◇ 4-FORMYLBENZALDEHYDE ◇ p-PHTHALALDEHYDE ◇ TEREPHTALDEHYDE ◇ TEREPHTALDEHYDES (FRENCH) ◇ TEREPHTHALIC ALDEHYDE

TOXICITY DATA with REFERENCE
unr-mus LDLo:805 mg/kg COREAF 246,851,58

CONSENSUS REPORTS: Reported in EPA TSCA Inventory.

SAFETY PROFILE: Moderately toxic by an unspecified route. When heated to decomposition it emits acrid smoke and irritating fumes. See also ALDEHYDES.

TAN750 CAS:100-21-0 ***HR: 2***
TEREPHTHALIC ACID
mf: $C_8H_6O_4$ mw: 166.14

PROP: White crystals or powder. D: 1.51, sublimes @ 300°. Insol in water, chloroform, ether, acetic acid; sltly sol in alc; sol in alkalies.

SYNS: ACIDE TEREPHTALIQUE (FRENCH) ◇ p-BENZENEDICARBOXYLIC ACID ◇ 1,4-BENZENEDICARBOXYLIC ACID ◇ KYSELINA TERFTALOVA (CZECH) ◇ TA 12 ◇ TA-33MP

TOXICITY DATA with REFERENCE
eye-rbt 500 mg/24H MOD 28ZPAK -,52,72
orl-rat LD50:18800 mg/kg 28ZPAK -,52,72
orl-mus LDLo:10 g/kg 85GMAT -,107,82
ipr-mus LD50:1430 mg/kg CPBTAL 16,1655,68
ivn-dog LDLo:767 mg/kg TXAPA9 18,469,71

CONSENSUS REPORTS: Community Right-To-Know List. Reported in EPA TSCA Inventory. EPA Genetic Toxicology Program.

SAFETY PROFILE: Moderately toxic by intravenous and intraperitoneal routes. Mildly toxic by ingestion. An eye irritant. Can explode during preparation. When heated to decomposition it emits acrid smoke and irritating fumes.

TAO750 CAS:17794-13-7 ***HR: 3***
5,5′-(TEREPHTHALOYLBIS(IMINO-p-PHENYLENE))BIS(2,4-DIAMINO-1-ETHYLPYRIMIDINIUM)-DI-p-TOLUENESULFONATE
mf: $C_{32}H_{34}N_{10}O_2 \cdot 2C_7H_7O_3S$ mw: 933.16

TOXICITY DATA with REFERENCE
dnd-mus:lym 840 nmol/L JMCMAR 22,134,79
ipr-mus LD10:15 mg/kg JMCMAR 22,134,79

SAFETY PROFILE: Poison by intraperitoneal route. Mutation data reported. When heated to decomposition it emits very toxic fumes of NO_x and SO_x.

TAP000 CAS:20719-34-0 ***HR: 3***
5,5′-(TEREPHTHALOYLBIS(IMINO-p-PHENYLENE))BIS(2,4-DIAMINO-1-METHYLPYRIMIDINIUM)-DI-p-TOLUENESULFONATE
mf: $C_{30}H_{30}N_{10}O_2 \cdot 2C_7H_7O_3S$ mw: 905.10

TOXICITY DATA with REFERENCE
dnd-mus:lym 840 nmol/L JMCMAR 22,134,79
ipr-mus LD10:12 mg/kg JMCMAR 22,134,79

SAFETY PROFILE: Poison by intraperitoneal route. Mutation data reported. When heated to decomposition it emits very toxic fumes of SO_x and NO_x.

TAP250 CAS:20719-36-2 ***HR: 3***
5,5′-(TEREPHTHALOYLBIS(IMINO-p-PHENYLENE))BIS(2,4-DIAMINO-1-PROPYLPYRIMIDINIUM)-DI-p-TOLUENESULFONATE
mf: $C_{34}H_{38}N_{10}O_2 \cdot 2C_7H_7O_3S$ mw: 961.22

TOXICITY DATA with REFERENCE
dnd-mus:lym 840 nmol/L JMCMAR 22,134,79
ipr-mus LD10:50 mg/kg JMCMAR 22,134,79

SAFETY PROFILE: Poison by intraperitoneal route. Mutation data reported. When heated to decomposition it emits very toxic fumes of SO_x and NO_x.

TAR000 CAS:18520-54-2 ***HR: 3***
3,3′-(TEREPHTHALOYLBIS(IMINO-p-PHENYLENE))BIS(1-PROPYLPYRIDINIUM)-DI-p-TOLUENESULFONATE
mf: $C_{36}H_{36}N_4O_2 \cdot 2C_7H_7O_3S$ mw: 899.16

SYN: 3,3′-(TEREPHTHALOYLDIIMINOBIS(p-PHENYLENE))BIS(1-PROPYLPYRIDINIUM)-DI-p-TOLUENESULFONATE

TOXICITY DATA with REFERENCE
dnd-mus:lym 290 nmol/L JMCMAR 22,134,79
ipr-mus LD10:10 mg/kg JMCMAR 22,134,79

SAFETY PROFILE: Poison by intraperitoneal route. Mutation data reported. When heated to decomposition it emits very toxic fumes of NO_x and SO_x.

TAR250 CAS:18520-57-5 ***HR: 3***
4,4′-(TEREPHTHALOYLBIS(IMINO-p-PHENYLENE))BIS(1-PROPYLPYRIDINIUM)-DI-p-TOLUENESULFONATE
mf: $C_{36}H_{36}N_4O_2 \cdot 2C_7H_7O_3S$ mw: 899.16

TOXICITY DATA with REFERENCE
dnd-mus:lym 550 nmol/L JMCMAR 22,134,79
ipr-mus LD10:10 mg/kg JMCMAR 22,134,79

SAFETY PROFILE: Poison by intraperitoneal route. Mutation data reported. When heated to decomposition it emits very toxic fumes of SO_x and NO_x.

TAV250 CAS:100-20-9 ***HR: 2***
TEREPHTHALOYL DICHLORIDE
mf: $C_8H_4Cl_2O_2$ mw: 203.02

PROP: White, crystalline material; musty odor. Bp: 266°, flash p: 356°F (COC), fp: 81.4°.

SYNS: 1,4-BENZENEDICARBONYL CHLORIDE ◇ 1,4-BENZENEDICARBONYL DICHLORIDE ◇ p-PHENYLENEDICARBONYL DICHLORIDE ◇ p-PHTHALOYL CHLORIDE ◇ p-PHTHALOYL DICHLORIDE ◇ TEREPHTHALIC ACID CHLORIDE ◇ TEREPHTHALIC ACID DICHLORIDE ◇ TEREPHTHALIC DICHLORIDE

TOXICITY DATA with REFERENCE
orl-rat TDLo:4122 mg/kg (female 26W pre):REP GISAAA 49(12),35,84
orl-rat LD50:2500 mg/kg 34ZIAG -,475,69
orl-mus LD50:2140 mg/kg GISAAA 49(12),35,84
orl-rbt LD50:950 mg/kg GISAAA 49(12),35,84

CONSENSUS REPORTS: Reported in EPA TSCA Inventory.

SAFETY PROFILE: Moderately toxic by ingestion. Corrosive. Combustible when exposed to heat or flame. When heated to decomposition it emits toxic fumes of Cl^-. See also CHLORIDES.

TAV750 CAS:126-92-1 ***HR: 3***
TERGITOL 08
mf: $C_8H_{18}O_4S \cdot Na$ mw: 233.31

SYNS: EMERSAL 6465 ◇ 2-ETHYL-1-HEXANOL HYDROGEN SULFATE, SODIUM SALT ◇ 2-ETHYL-1-HEXANOL SULFATE SODIUM SALT ◇ 2-ETHYLHEXYL SODIUM SULFATE ◇ MONO(2-ETHYLHEXYL)SULFATE SODIUM SALT ◇ NCI-C50204 ◇ NIA PROOF 08 ◇ PROPASTE 6708 ◇ SIPEX BOS ◇ SODIUM ETASULFATE ◇ SODIUM ETHASULFATE ◇ SODIUM(2-ETHYLHEXYL)ALCOHOL SULFATE ◇ SODIUM-2-ETHYLHEXYL SULFATE ◇ SULFURIC ACID, MONO(2-ETHYLHEXYL)ESTER, SODIUM SALT (8CI) ◇ TERGEMIST ◇ TERGIMIST ◇ TERGITOL ANIONIC 08

TOXICITY DATA with REFERENCE
skn-rbt 500 mg open MOD UCDS** 1/20/72
eye-rbt 250 μg MLD AROPAW 34,99,45
orl-mus LD50:1550 mg/kg GISAAA 43(7),70,78
orl-rat LD50:4 g/kg FMCHA2 -,D279,80

ipr-rat LD50:320 mg/kg TXAPA9 17,53,70
scu-rat LD50:4730 mg/kg TXAPA9 17,53,70
orl-rbt LD50:3580 mg/kg TXAPA9 17,53,70
orl-gpg LD50:1300 mg/kg TXAPA9 17,53,70
skn-gpg LDLo:1520 mg/kg JIHTAB 23,478,41

CONSENSUS REPORTS: Reported in EPA TSCA Inventory.

SAFETY PROFILE: Poison by intraperitoneal route. Moderately toxic by ingestion and skin contact. A skin and eye irritant. When heated to decomposition it emits very toxic fumes of SO_x and Na_2O.

TAW250 CAS:9016-45-9 ***HR: 2***
TERGITOL NP-14

TOXICITY DATA with REFERENCE
skn-rbt 500 mg open MLD UCDS** 4/25/58
eye-rbt 100 mg SEV UCDS** 4/25/58
orl-rat LD50:4290 mg/kg UCDS** 4/25/68
skn-rbt LD50:2520 mg/kg UCDS** 4/25/68

CONSENSUS REPORTS: Reported in EPA TSCA Inventory.

SAFETY PROFILE: Moderately toxic by skin contact. Mildly toxic by ingestion. A skin and severe eye irritant.

TAW500 CAS:9016-45-9 ***HR: 2***
TERGITOL NP-27

PROP: D: 1.06, refr index: 1.489.

TOXICITY DATA with REFERENCE
skn-rbt 500 mg open MLD UCDS** 7/14/65
eye-rbt 15 mg SEV UCDS** 7/14/65
orl-rat LD50:3670 mg/kg UCDS** 7/14/65
skn-rbt LD50:1780 mg/kg UCDS** 7/14/65

CONSENSUS REPORTS: Reported in EPA TSCA Inventory.

SAFETY PROFILE: Moderately toxic by ingestion and skin contact. A skin and severe eye irritant.

TAX000 CAS:9016-45-9 ***HR: 2***
TERGITOL NP-35 (nonionic)

TOXICITY DATA with REFERENCE
orl-rat LD50:4000 mg/kg UCDS** 9/22/64

CONSENSUS REPORTS: Reported in EPA TSCA Inventory.

SAFETY PROFILE: Moderately toxic by ingestion.

TAX250 CAS:9016-45-9 ***HR: 1***
TERGITOL NP-40 (nonionic)

TOXICITY DATA with REFERENCE
skn-rbt 500 mg open MLD UCDS** 11/8/71
orl-rat LD50:16 g/kg UCDS** 11/8/71
skn-rbt LD50:4490 mg/kg UCDS** 11/8/71

CONSENSUS REPORTS: Reported in EPA TSCA Inventory.

SAFETY PROFILE: Mildly toxic by ingestion and skin contact. A skin irritant.

TAX500 CAS:9014-92-0 ***HR: 2***
TERGITOL 12-P-9

TOXICITY DATA with REFERENCE
skn-rbt 500 mg open MLD UCDS** 5/15/68
eye-rbt 15 mg SEV UCDS** 5/15/68
orl-rat LD50:1870 mg/kg UCDS** 5/15/68
skn-rbt LD50:1110 mg/kg UCDS** 5/15/68

CONSENSUS REPORTS: Reported in EPA TSCA Inventory.

SAFETY PROFILE: Moderately toxic by ingestion and skin contact. A skin and severe eye irritant.

TAY250 CAS:68131-40-8 ***HR: 2***
TERGITOL 15-S-5

TOXICITY DATA with REFERENCE
skn-rbt 500 mg open MLD UCDS** 12/16/68
eye-rbt 100 mg MLD UCDS** 12/16/68
orl-rat LD50:8570 mg/kg UCDS** 12/16/68
skn-rbt LD50:4000 mg/kg UCDS** 12/16/68

CONSENSUS REPORTS: Reported in EPA TSCA Inventory.

SAFETY PROFILE: Moderately toxic by skin contact. Mildly toxic by ingestion. A skin and eye irritant. When heated to decomposition it emits toxic fumes of SO_x. See also SULFATES.

TAZ000 CAS:68131-40-8 ***HR: 2***
TERGITOL 15-S-9 (nonionic)

TOXICITY DATA with REFERENCE
skn-rbt 500 mg open MLD UCDS** 2/7/66
eye-rbt 1 mg SEV UCDS** 2/7/66
orl-rat LD50:2380 mg/kg UCDS** 2/7/66
skn-rbt LD50:2000 mg/kg UCDS** 2/7/66

CONSENSUS REPORTS: Reported in EPA TSCA Inventory.

SAFETY PROFILE: Moderately toxic by ingestion and skin contact. A skin and severe eye irritant. When heated to decomposition it emits toxic fumes of SO_x. See also SULFATES.

TBA500 CAS:68131-40-8 ***HR: 1***
TERGITOL 15-S-20

TOXICITY DATA with REFERENCE
skn-rbt 500 mg open MLD UCDS** 11/22/68
orl-rat LD50:5660 mg/kg UCDS** 11/22/68
skn-rbt LD50:11 g/kg UCDS** 11/22/68

CONSENSUS REPORTS: Reported in EPA TSCA Inventory.

SAFETY PROFILE: Mildly toxic by skin contact and ingestion. A skin irritant. When heated to decomposition it emits toxic fumes of SO_x. See also SULFATES.

TBA750 CAS:141-65-1 ***HR: 2***
TERGITOL ANIONIC P-28

TOXICITY DATA with REFERENCE
skn-rbt 500 mg open SEV UCDS** 4/29/64
eye-rbt 5 mg MLD UCDS** 4/29/64
orl-rat LD50:7720 mg/kg UCDS** 4/29/64
skn-rbt LD50:3050 mg/kg UCDS** 4/29/64

CONSENSUS REPORTS: Reported in EPA TSCA Inventory.

SAFETY PROFILE: Moderately toxic by skin contact. Mildly toxic by ingestion. A eye and severe skin irritant.

TBA800 CAS:103331-86-8 ***HR: 2***
TERGITOL MIN-FOAM 1X

SYN: ALCOHOLS, C12-14-SECONDARY, ETHOXYLATED PROPOXYLATED

TOXICITY DATA with REFERENCE
skn-rbt 500 mg MLD UCDS** 6/13/68
eye-rbt 100 mg MOD UCDS** 6/13/68
orl-rat LD50:3250 mg/kg UCDS** 6/13/68

SAFETY PROFILE: Moderately toxic by ingestion. A skin and eye irritant. When heated to decomposition it emits acrid smoke and irritating fumes.

TBB750 ***HR: 2***
TERGITOL PENETRANT 7

TOXICITY DATA with REFERENCE
eye-rbt 450 mg SEV AROPAW 40,668,48
orl-rat LD50:5660 mg/kg JIHTAB 30,63,48

CONSENSUS REPORTS: Reported in EPA TSCA Inventory.

SAFETY PROFILE: Mildly toxic by ingestion. A severe eye irritant.

TBB775 CAS:60828-78-6 ***HR: 1***
TERGITOL TMN-10

TOXICITY DATA with REFERENCE
skn-rbt 500 mg open MLD UCDS** 12/23/68
eye-rbt 5 mg SEV UCDS** 12/23/68
orl-rat LD50:5650 mg/kg UCDS** 12/23/68
skn-rbt LD50:4780 mg/kg UCDS** 12/23/68

CONSENSUS REPORTS: Reported in EPA TSCA Inventory.

SAFETY PROFILE: Mildly toxic by ingestion and skin contact. A skin and severe eye irritant. When heated to decomposition it emits acrid smoke and irritating fumes.

TBC200 CAS:7082-21-5 ***HR: 3***
TERODILINE HYDROCHLORIDE
mf: $C_{20}H_{27}N{\cdot}ClH$ mw: 317.94

SYNS: BENZENEPROPANAMINE,N-(1,1-DIMETHYLETHYL)-α-METHYL-Γ-PHENYL-, HYDROCHLORIDE (9CI) ◇ PROPYLAMINE, N-tert-BUTYL-1-METHYL-3,3-DIPHENYL-, HYDROCHLORIDE ◇ TD-758 ◇ TERODILINE CHLORIDE

TOXICITY DATA with REFERENCE
orl-rat TDLo:2700 mg/kg (female 17-22D post):REP KSRNAM 21,1196,87
orl-rat TDLo:1100 mg/kg (female 7-17D post):TER KSRNAM 21,1169,87
orl-rat LD50:465 mg/kg KSRNAM 21,979,87
scu-rat LD50:370 mg/kg KSRNAM 21,979,87
ivn-rat LD50:27 mg/kg KSRNAM 21,979,87
orl-mus LD50:330 mg/kg KSRNAM 21,979,87
scu-mus LD50:170 mg/kg KSRNAM 21,979,87
ivn-mus LD50:28 mg/kg KSRNAM 21,979,87
orl-dog LD50:63300 μg/kg KSRNAM 21,979,87

SAFETY PROFILE: Poison by ingestion, subcutaneous, and intravenous routes. An experimental teratogen. Experimental reproductive effects. When heated to decomposition it emits toxic fumes of NO_x and HCl.

TBC450 CAS:59653-73-5 ***HR: 3***
TEROXIRONE
mf: $C_{12}H_{15}N_3O_6$ mw: 297.30

SYNS: HENKEL'S COMPOUND ◇ NSC-296934 ◇ α-TRIGLYCIDYL ISOCYANURATE ◇ α-1,3,5-TRIS(2,3-EPOXYPROPYL)-s-TRIAZINE-2,4,6-(1H,3H,5H)-TRIONE ◇ (α)-1,3,5-TRIS(OXIRANYLMETHYL)-1,3,5-TRIAZINE-2,4,6-(1H,3H,5H)-TRIONE ◇ XB 2615

TOXICITY DATA with REFERENCE
ipr-mus LD50:111 mg/kg NCISP* JAN86
ivn-mus LD50:145 mg/kg NTIS** PB82-110537
ivn-dog LDLo:36200 μg/kg NTIS** PB82-193442

SAFETY PROFILE: Poison by intravenous and intraperitoneal routes. When heated to decomposition it emits toxic fumes of NO_x.

TBC500 CAS:8001-50-1 ***HR: 3***
TERPENE POLYCHLORINATES

PROP: Chlorinated mixed terpenes (IARC** 4,219,75).

SYNS: DICHLORICIDE MOTHPROOFER ◇ ENT 19,442 ◇ STROBANE

TOXICITY DATA with REFERENCE
orl-mus TDLo:1272 mg/kg/79W-I-C:CAR JNCIAM 42,1101,69
orl-rat LD50:200 mg/kg PCOC** -,1071,66
orl-dog LD50:200 mg/kg PCOC** -,1071,66

CONSENSUS REPORTS: IARC Cancer Review: Group 3 IMEMDT 7,56,87; Animal Sufficient Evidence IMEMDT 5,219,74.

SAFETY PROFILE: Poison by ingestion. Questionable carcinogen with experimental carcinogenic data. When heated to decomposition it emits toxic fumes of Cl^-. See also CHLORINATED HYDROCARBONS, AROMATIC.

TBC620 CAS:92-06-8 ***HR: 2***
m-TERPHENYL
mf: $C_{18}H_{14}$ mw: 230.32

PROP: Mp: 86-87°, bp: 379°.

SYNS: m-DIPHENYLBENZENE ◇ ISODIPHENYLBENZENE ◇ SANTOWAX M ◇ 1,3-TERPHENYL ◇ m-TRIPHENYL

TOXICITY DATA with REFERENCE
orl-rat LD50: 2400 mg/kg AIHAAP 23,372,62

CONSENSUS REPORTS: Reported in EPA TSCA Inventory.

OSHA PEL: (Transitional: TWA CL 1 ppm) CL 0.5 ppm
ACGIH TLV: TWA CL 0.5 ppm

SAFETY PROFILE: Moderately toxic by ingestion. Combustible when exposed to heat or flame. To fight fire, use water, CO_2, dry chemical. When heated to decomposition it emits acrid smoke and irritating fumes.

TBC640 CAS:84-15-1 ***HR: 2***
o-TERPHENYL
mf: $C_{18}H_{14}$ mw: 230.32

PROP: Mp: 58-59°, bp: 337°, flash p: >230°F.

SYN: 1,2-DIPHENYLBENZENE

TOXICITY DATA with REFERENCE
orl-rat LD50: 1900 mg/kg AIHAAP 23,372,62

CONSENSUS REPORTS: Reported in EPA TSCA Inventory.

OSHA PEL: (Transitional: TWA CL 1 ppm) CL 0.5 ppm
ACGIH TLV: TWA CL 0.5 ppm

SAFETY PROFILE: Moderately toxic by ingestion. Combustible when exposed to heat or flame. To fight fire, use water, CO_2, dry chemical. When heated to decomposition it emits acrid smoke and irritating fumes.

TBC750 CAS:92-94-4 ***HR: 2***
p-TERPHENYL
mf: $C_{18}H_{14}$ mw: 230.32

PROP: Leaves or needles. D: 1.234 @ 0/4°, mp: 212-213°, bp: 389°, flash p: 405°F (OC), vap d: 7.95. Sol in hot benzene; very sol in hot alc, sltly sol in ether.

SYNS: p-DIPHENYLBENZENE ◇ 1,4-DIPHENYLBENZENE ◇ 4-PHENYLBIPHENYL ◇ 4-PHENYLDIPHENYL ◇ SANTOWAX ◇ p-TRIPHENYL

TOXICITY DATA with REFERENCE
orl-rat LDLo:500 mg/kg NCNSA6 5,26,53

CONSENSUS REPORTS: Reported in EPA TSCA Inventory.

OSHA PEL: (Transitional: TWA CL 1 ppm) CL 0.5 ppm
ACGIH TLV: TWA CL 0.5 ppm

SAFETY PROFILE: Moderately toxic by ingestion. Combustible when exposed to heat or flame. To fight fire, use water, CO_2, dry chemical. When heated to decomposition it emits acrid smoke and irritating fumes.

TBD000 CAS:26140-60-3 ***HR: 2***
TERPHENYLS
mf: $C_{18}H_{14}$ mw: 230.32

SYNS: DELOWAS S ◇ DELOWAX OM ◇ DIPHENYLBENZENE ◇ GILOTHERM OM 2 ◇ TERBENZENE ◇ TRIPHENYL

TOXICITY DATA with REFERENCE
orl-mus LD50: 13200 mg/kg SHHUE8 15,305,86

CONSENSUS REPORTS: Reported in EPA TSCA Inventory.

OSHA PEL: (Transitional: TWA CL 1 ppm) CL 0.5 ppm
ACGIH TLV: TWA CL 0.5 ppm

SAFETY PROFILE: Moderately toxic by ingestion. Combustible when exposed to heat or flame. To fight fire, use water, CO_2, dry chemical. When heated to decomposition it emits acrid smoke and irritating fumes.

TBD250 CAS:64058-92-0 ***HR: 3***
p-TERPHENYL-4-YLACETAMIDE
mf: $C_{20}H_{17}NO$ mw: 287.38

SYNS: 4-ACETYLAMINO-p-DIPHENYLBENZENE ◇ 4-ACETYL-AMINO-p-TERPHENYL

TOXICITY DATA with REFERENCE
orl-rat TDLo:7000 mg/kg/43W-C:ETA CNREA8 16,525,56

SAFETY PROFILE: Questionable carcinogen with experimental tumorigenic data. When heated to decomposition it emits toxic fumes of NO_x.

TBD500 CAS:8006-39-1 ***HR: 1***
TERPINEOL
mf: $C_{10}H_{18}O$ mw: 154.28

PROP: A mixture of α, β, and Γ isomers (FCTXAV 12,807,74). Colorless, viscous liquid; lilac odor. D: 0.930-0.936, refr index: 1.482, flash p: 196°F. Sltly sol in water, glycerin.

SYNS: FEMA No. 3045 ◇ p-MENTH-1-EN-8-OL ◇ MIXTURE of p-METHENOLS ◇ α-TERPINEOL (FCC) ◇ TERPINEOLS

TOXICITY DATA with REFERENCE
skn-rbt 500 mg/24H MOD FCTXAV 12,807,74
orl-rat LD50:4300 mg/kg FCTXAV 12,807,74

SAFETY PROFILE: Mildly toxic by ingestion. A skin irritant. Combustible liquid. When heated to decomposition it emits acrid smoke and irritating fumes. See also α-TERPINEOL.

TBD750 CAS:98-55-5 ***HR: 2***
α-TERPINEOL
mf: $C_{10}H_{18}O$ mw: 154.28

PROP: Colorless solid. D: 0.935 @ 20/20°, mp: 35-40°, bp: 218-221°. Insol in water; very sol in alc, ether.

SYNS: p-MENTH-1-EN-8-OL (8CI) ◇ TERPINEOL SCHLECHTHIN ◇ α,α,4-TRIMETHYL-3-CYCLOHEXENE-1-METHANOL

TOXICITY DATA with REFERENCE
ims-mus LD50:2000 mg/kg JSICAZ 21,342,62

CONSENSUS REPORTS: Reported in EPA TSCA Inventory.

SAFETY PROFILE: Moderately toxic by intramuscular route. When heated to decomposition it emits acrid smoke and irritating fumes.

TBD825 CAS:562-74-3 ***HR: 2***
4-TERPINEOL
mf: $C_{10}H_{18}O$ mw: 154.28

SYNS: 4-CARVOMENTHENOL ◇ 1-p-MENTHEN-4-OL ◇ 4-METHYL-1-(1-METHYLETHYL)-3-CYCLOHEXEN-1-OL (9CI) ◇ TERPINENOL-4

TOXICITY DATA with REFERENCE
skn-rbt 500 mg/24H MOD FCTOD7 20(Suppl),833,82
orl-rat LD50:1300 mg/kg FCTOD7 20 (Suppl),833,82
orl-mus LD50:1016 mg/kg CCHCDE 4(4),203,81

CONSENSUS REPORTS: Reported in EPA TSCA Inventory.

SAFETY PROFILE: Moderately toxic by ingestion. A skin irritant. When heated to decomposition it emits acrid smoke and irritating fumes. See also other terpineol entries.

TBE000 CAS:586-62-9 ***HR: 3***
TERPINOLENE
DOT: UN 2541
mf: $C_{10}H_{16}$ mw: 136.26

PROP: Colorless liquid. Bp: 185°, d: 0.855, flash p: 100°F (CC). Insol in water; misc in alc, ether. Mixture of p-mentha-1,4(8)-diene and p-mentha-2,4(8)-diene (FCT XAV 14,659,76).

SYN: 1-METHYL-4-(1-METHYLETHYLIDENE)CYCLOHEXENE

TOXICITY DATA with REFERENCE
orl-rat LD50:4390 mg/kg FCTXAV 14,877,76

CONSENSUS REPORTS: Reported in EPA TSCA Inventory.

DOT Classification: Flammable or Combustible Liquid; Label: Flammable Liquid.

SAFETY PROFILE: Mildly toxic by ingestion. A very dangerous fire hazard when exposed to heat or flame. To fight fire, use foam, CO_2, dry chemical. Can react with oxidizing materials. When heated to decomposition it emits acrid smoke and irritating fumes.

TBE250 CAS:80-26-2 ***HR: 1***
TERPINYL ACETATE
mf: $C_{12}H_{20}O_2$ mw: 196.32

PROP: Colorless liquid; sweet, herbaceous odor. D: 0.966 @ 20/4°, refr index: 1.464, mp: $< -50°$, bp: 220° decomp, flash p: 212°F. Insol in water; sol in alc, fixed oils, mineral oil, propylene glycol.

SYNS: FEMA No. 3047 ◇ α-TERPINEOL ACETATE

TOXICITY DATA with REFERENCE
orl-rat LDLo:4160 mg/kg FCTXAV 2,327,64

CONSENSUS REPORTS: Reported in EPA TSCA Inventory.

SAFETY PROFILE: Mildly toxic by ingestion. Combustible liquid. When heated to decomposition it emits acrid smoke and irritating fumes.

TBE500 CAS:2153-26-6 ***HR: 1***
TERPINYL FORMATE
mf: $C_{11}H_{18}O_2$ mw: 182.29

SYN: p-MENTH-1-EN-8-OL, FORMATE (mixed isomers)

TOXICITY DATA with REFERENCE
skn-rbt 500 mg/24H MOD FCTXAV 14,659,76

CONSENSUS REPORTS: Reported in EPA TSCA Inventory.

SAFETY PROFILE: A skin irritant. When heated to decomposition it emits acrid smoke and irritating fumes. See also ESTERS.

TBE600 ***HR: 1***
TERPINYL PROPIONATE
mf: $C_{13}H_{22}O_2$ mw: 210.32

PROP: Colorless to sltly yellow liquid; sweet, floral, lavender-like odor. D: 0.944, refr index: 1.461, flash p: +212°F. Sol in glycerin; misc in alc, chloroform, ether, fixed oils; sltly sol in propylene glycol; insol in water @ 240°.

SYNS: FEMA No. 3053 ◇ MENTHEN-1-YL-8 PROPIONATE

SAFETY PROFILE: Combustible liquid. When heated to decomposition it emits acrid smoke and irritating fumes.

TBF000 CAS:69766-62-7 ***HR: 3***
TERRAMYCIN SODIUM
mf: $C_{22}H_{24}N_2O_9 \cdot xNa$

TOXICITY DATA with REFERENCE
orl-mus LD50:4101 mg/kg ANYAA9 53,238,50
scu-mus LD50:251 mg/kg ANYAA9 53,238,50
ivn-mus LD50:163 mg/kg ANYAA9 53,238,50
ivn-dog LD50:150 mg/kg ANYAA9 53,238,50
ivn-rbt LD50:112 mg/kg ANYAA9 53,238,50

SAFETY PROFILE: Poison by subcutaneous and intravenous routes. Mildly toxic by ingestion. When heated to decomposition it emits toxic fumes of NO_x and Na_2O.

TBF325 ***HR: 3***
TERREIC ACID
mf: $C_7H_6O_4$ mw: 154.13

SYNS: 5,6-EPOXY-3-HYDROXY-p-TOLUQUINONE ◇ 2-HYDROXY-3-METHYL-1,4-BENZOQUINONE 5,6-EPOXIDE ◇ 3-HYDROXY-4-METHYL-7-OXABICYCLO(4.1.0)HEPT-3-ENE-2,5-DIONE ◇ (1R-cis)-3-HYDROXY-4-METHYL-7-OXABICYCLO(4.1.0)HEPT-3-ENE-2,5-DIONE

TOXICITY DATA with REFERENCE
dnd-mus-ipr 250 mg/kg TOLED5 10,249,82
dnd-mus:lvr 200 mg/L TOLED5 10,249,82
ipr-mus LD50:75 mg/kg JJANAX 33,320,80

SAFETY PROFILE: Poison by intraperitoneal route. Mutation data reported. When heated to decomposition it emits acrid smoke and irritating fumes.

TBF350 CAS:37338-91-3 ***HR: 3***
TERRILYTIN

SYNS: PROTEINASE, ASPERGILLUS TERRICOLA NEUTRAL ◇ TERRILITIN

TOXICITY DATA with REFERENCE
ivn-rat LD50:161 mg/kg PCJOAU 10,1130,76
orl-mus LD50:7500 mg/kg PCJOAU 10,1130,76
ipr-mus LD50:206 mg/kg PCJOAU 10,1130,76
scu-mus LD50:438 mg/kg PCJOAU 10,1130,76
ivn-mus LD50:206 mg/kg PCJOAU 10,1130,76
ims-mus LD50:462 mg/kg PCJOAU 10,1130,76
ivn-rbt LD50:88 mg/kg PCJOAU 10,1130,76

SAFETY PROFILE: Poison by intravenous and intraperitoneal routes. Moderately toxic by subcutaneous and intramuscular routes. Mildly toxic by ingestion.

TBF500 CAS:58-22-0 ***HR: 3***
TESTOSTERONE
mf: $C_{19}H_{28}O_2$ mw: 288.47

PROP: Crystals. Mp: 155°. Insol in water; sol in alc and ether.

SYNS: ANDROLIN ◇ ANDRONAQ ◇ ANDROST-4-EN-17β-OL-3-ONE ◇ Δ^4-ANDROSTEN-17(β)-OL-3-ONE ◇ ANDRUSOL ◇ CRISTERONE T ◇ GENO-CRISTAUZ GREMY ◇ HOMOSTERONE ◇ 17-β-HYDROXY-Δ^4-ANDROSTEN-3-ONE ◇ 17-β-HYDROXYANDROST-4-EN-3-ONE ◇ 17-β-HYDROXY-4-ANDROSTEN-3-ONE ◇ 17-HYDROXY-(17-β)-ANDROST-4-EN-3-ONE ◇ 7-β-HYDROXYANDROST-4-EN-3-ONE ◇ MALESTRONE (AMPS) ◇ MERTESTATE ◇ NEO-TESTIS ◇ ORETON-F ◇ ORQUISTERONE ◇ PERANDREN ◇ PERCUTACRINE ANDROGENIQUE ◇ PRIMOTEST ◇ PROMOTESTON ◇ SUSTANONE ◇ SYNANDROL F ◇ TESLEN ◇ TESTANDRONE ◇ TESTICULOSTERONE ◇ TESTOBASE ◇ TESTOPROPON ◇ TESTOSTEROID ◇ trans-TESTOSTERONE ◇ TESTOSTERONE HYDRATE ◇ TESTOSTOSTERONE ◇ TESTOVIRON SCHERING ◇ TESTOVIRON T ◇ TESTRONE ◇ TESTRYL ◇ VIRORMONE ◇ VIROSTERONE

TOXICITY DATA with REFERENCE
dni-hmn:lym 50 μmol/L PSEBAA 146,401,74
cyt-hmn:kdy 100 μg/L CNJGA8 10,299,68
dnd-mam:lym 10 μmol/L ENZYAS 41,183,71
unr-wmn TDLo:34600 μg/kg (7-13W preg):TER AMDCA5 95,9,58
orl-mus TDLo:15 g/kg (female 8-12D post):REP TCMUD8 6,361,86
orl-mus TDLo:6240 mg/kg/52D-C:NEO CNREA8 48,2788,88
scu-mus TDLo:30 mg/kg/5D-I:NEO,TER JNCIAM 39,75,67
ipr-rat LDLo:326 mg/kg PSEBAA 46,116,41

CONSENSUS REPORTS: IARC Cancer Review: Animal Sufficient Evidence IMEMDT 6,209,74, IMEMDT 21,519,79; Human Limited Evidence IMEMDT

21,519,79. Reported in EPA TSCA Inventory. EPA Genetic Toxicology Program.

SAFETY PROFILE: Confirmed carcinogen with experimental neoplastigenic and teratogenic data. Poison by intraperitoneal route. Human teratogenic effects by unspecified route: developmental abnormalities of the urogenital system. Experimental reproductive effects. Human mutation data reported. Workers engaged in manufacture and packaging have shown effects from this hormone, i.e., enlargement of the breasts in male workers. A promoter. When heated to decomposition it emits acrid smoke and irritating fumes. Used as a drug for the treatment of hypogonadism and metastatic breast cancer.

TBF600 CAS:58-20-8 ***HR: D***
TESTOSTERONE CYCLOPENTYLPROPIONATE
mf: $C_{27}H_{40}O_3$ mw: 412.67

SYNS: ANDROST-4-EN-3-ONE, 17-(3-CYCLOPENTYL-1-OXOPROPOXY)-, (17-β)- (9CI) ◇ DEPO-TESTOSTERONE ◇ DEPOVIRIN ◇ PERTESTIS ◇ TESTODRIN PROLONGATUM ◇ TESTOSTERONE, CYCLOPENTANEPROPIONATE ◇ TESTOSTERONE CYPIONATE ◇ TESTOSTERONE 17-β-CYPIONATE

TOXICITY DATA with REFERENCE
ims-man TDLo:4286 μg/kg (male 2W pre):REP ANYAA9 55,725,52

SAFETY PROFILE: Experimental reproductive effects. When heated to decomposition it emits acrid smoke and irritating fumes.

TBF750 CAS:315-37-7 ***HR: 3***
TESTOSTERONE HEPTANOATE
mf: $C_{26}H_{40}O_3$ mw: 400.66

SYNS: ANDROTARDYL ◇ ATLATEST ◇ DELATESTRYL ◇ HEPTANOIC ACID, ester with TESTOSTERONE ◇ MALOGEN L.A.200 ◇ NSC-17591 ◇ (17-β)17-((1-OXOHEPTYL)OXY)-ANDROST-4-EN-3-ONE ◇ REPOSO-TMD ◇ TE ◇ TESTATE ◇ TESTOSTERONE ENANTHATE ◇ TESTOSTERONE ETHANATE ◇ TESTOSTERONE HEPTOATE ◇ TESTOSTERONE HEPTYLATE ◇ TESTOSTERONE OENANTHATE ◇ TESTOSTROVAL

TOXICITY DATA with REFERENCE
orl-man TDLo:34 mg/kg (52W male):REP CCPTAY 21,631,80
ivg-dom TDLo:35455 μg/kg (female 4-14W post):TER THGNBO 6,21,76
ims-rbt TDLo:409 mg/kg/2Y-I:ETA,TER CNREA8 26,474,66
ims-man TDLo:7519 mg/kg/5W-I:PUL NEJMAG 308,508,83
ipr-rat LD50:2000 mg/kg DRUGAY 6,486,82
ipr-mus LD50:4 mg/kg CLDND*

CONSENSUS REPORTS: EPA Genetic Toxicology Program.

SAFETY PROFILE: Poison by intraperitoneal route. Human systemic effects by ingestion: dyspnea. Human reproductive effects by ingestion, intramuscular, and parenteral routes: changes in spermatogenesis and paternal effects to testes, epididymis, and sperm duct. Experimental reproductive effects. Questionable carcinogen with experimental tumorigenic and teratogenic data. When heated to decomposition it emits acrid smoke and irritating fumes. A drug used to treat hypoganadism and metastatic breast cancer. See also TESTOSTERONE.

TBG000 CAS:57-85-2 ***HR: 3***
TESTOSTERONE PROPIONATE
mf: $C_{22}H_{32}O_3$ mw: 344.54

PROP: Stout prisms from alc and water. Mp: 118-122°. Insol in water; freely sol in alc, ether, pyridine, and other organic solvents. Sol in vegetable oils.

SYNS: AGOVIRIN ◇ ANDROGEN ◇ ANDROSAN ◇ Δ^4-ANDROSTENE-17-β-PROPIONATE-3-ONE ◇ ANDROTESTON ◇ ANDROTEST P ◇ ANDRUSOL-P ◇ ANERTAN ◇ AQUAVIRON ◇ BIO-TESTICULINA ◇ ENARMON ◇ HOMANDREN (amps) ◇ HORMOTESTON ◇ MASENATE ◇ NASDOL ◇ NEO-HOMBREOL ◇ NSC 9166 ◇ OKASA-MASCUL ◇ ORCHIOL ◇ ORCHISTIN ◇ ORETON ◇ ORETON PROPIONATE ◇ 17-(1-OXOPROPOXY)-(17-β)-ANDROST-4-EN-3-ONE ◇ PANESTIN ◇ PERANDREN ◇ PROPIOKAN ◇ RECTHORMONE TESTOSTERONE ◇ STERANDRYL ◇ SYNANDROL ◇ SYNERONE ◇ TELIPEX ◇ TESTAFORM ◇ TESTEX ◇ TESTODET ◇ TESTODRIN ◇ TESTOGEN ◇ TESTONIQUE ◇ TESTORMOL ◇ TESTOSTERON PROPIONATE ◇ TESTOSTERONE-17-PROPIONATE ◇ TESTOSTERONE-17-β-PROPIONATE ◇ TESTOVIRON ◇ TESTOXYL ◇ TESTREX ◇ TOSTRIN ◇ TP ◇ UNITESTON ◇ VULVAN

TOXICITY DATA with REFERENCE
cyt-ofs-unr 500 μg BEXBBO 15,329,80
spm-ofs-unr 500 μg BEXBBO 15,329,80
dnd-rat:lvr 300 μmol/L SinJF# 26OCT82
spm-rat-scu 10 mg/kg AMTUA3 8,68,71
par-wmn TDLo:18500 μg/kg (31D pre):REP PSEBAA 44,319,40
unr-rat TDLo:3500 mg/kg (female 10-16D post):TER SAKNAH 64,737,71
scu-rat TDLo:10 mg/kg:NEO JSONDX 5,396,84
imp-rat TDLo:432 mg/kg/48W-C:ETA CNREA8 37,1929,77
orl-rat LD50:1000 mg/kg NIIRDN 6,487,82
ipr-rat LD50:585 mg/kg NIIRDN 6,487,82
orl-mus LD50:1350 mg/kg NIIRDN 6,487,82
ipr-mus LD50:970 mg/kg NIIRDN 6,487,82

CONSENSUS REPORTS: IARC Cancer Review: Animal Sufficient Evidence IMEMDT 21,519,79. Reported in EPA TSCA Inventory.

SAFETY PROFILE: Confirmed carcinogen with experimental neoplastigenic, tumorigenic, and teratogenic data. Moderately toxic by ingestion and intraperitoneal routes. Human male reproductive effects by intramuscular and parenteral routes: changes in spermatogenesis, testes, epididymis, and sperm duct. Human female reproductive effects by intramuscular and parenteral routes: menstrual cycle changes or disorders and effects on fertility. Experimental reproductive effects. Mutation data reported. When heated to decomposition it emits acrid smoke and irritating fumes. See also TESTOSTERONE.

TBG250 ***HR: 3***
TETRAACRYLONITRILECOPPER(I) PERCHLORATE
mf: $C_{12}H_{12}ClCuN_4O_4$ mw: 375.25

CONSENSUS REPORTS: Cyanide and its compounds, as well as copper and its compounds, are on the Community Right-To-Know List.

SAFETY PROFILE: Explodes on heating. When heated to decomposition it emits very toxic fumes of NO_x, Cl^-, and CN^-. See also NITRILES, COPPER COMPOUNDS, and PERCHLORATES.

TBG500 ***HR: 3***
TETRAACRYLONITRILECOPPER(II) PERCHLORATE
mf: $C_{12}H_{12}Cl_2CuN_4O_8$ mw: 474.70

CONSENSUS REPORTS: Cyanide and its compounds, as well as copper and its compounds, are on the Community Right-To-Know List.

SAFETY PROFILE: Decomposes explosively on heating. When heated to decomposition it emits very toxic fumes of Cl^-, CN^-, and NO_x. See also NITRILES, COPPER COMPOUNDS, and PERCHLORATES.

TBG600 CAS:56370-81-1 ***HR: 3***
TETRAAMINE-2,3-BUTANEDIIMINE RUTHENIUM(III) PERCHLORATE
mf: $C_4H_{20}Cl_3N_6O_{12}Ru$ mw: 551.66

$[(H_3N)_4RuC_4H_8N_2][ClO_4]_3$

SAFETY PROFILE: An explosive. When heated to decomposition it emits toxic fumes of Cl^-, RuO_4, and NO_x. See also RUTHENIUM COMPOUNDS, AMINES, and PERCHLORATES.

TBG625 CAS:36294-69-6 ***HR: 3***
TETRAAMINEDITHIOCYANATO COBALT(III) PERCHLORATE
mf: $C_2H_{12}ClCoN_6O_4S_2$ mw: 342.66

CONSENSUS REPORTS: Cobalt and its compounds are on The Community Right-To-Know List.

SAFETY PROFILE: Explodes at 335°C. It is moderately impact-sensitive. When heated to decomposition it emits toxic fumes of Cl^-, NO_x, and SO_x. See also THIOCYANATES, COBALT COMPOUNDS, and PERCHLORATES.

TBG700 CAS:2475-45-8 ***HR: 2***
1,4,5,8-TETRAAMINO-9,10-ANTHRACENEDIONE
mf: $C_{14}H_{12}N_4O_2$ mw: 268.30

SYNS: ACETATE BLUE G ◇ ACETOQUINONE BLUE L ◇ ACETOQUINONE BLUE R ◇ ACETYLON FAST BLUE G ◇ AMACEL BLUE GG ◇ AMACEL PURE BLUE B ◇ ARTISIL BLUE SAP ◇ BRASILAZET BLUE GR ◇ CELANTHRENE PURE BLUE BRS ◇ CELLITON BLUE G ◇ C.I. 64500 ◇ CIBACET SAPPHIRE BLUE G ◇ C.I. DISPERSE BLUE 1 ◇ CILLA BLUE EXTRA ◇ C.I. SOLVENT BLUE 18 ◇ DIACELLITON FAST BLUE R ◇ DISPERSE BLUE NO 1 ◇ DURANOL BRILLIANT BLUE CB ◇ FENACET BLUE G ◇ GRASOL BLUE 2GS ◇ KAYALON FAST BLUE BR ◇ MICROSETILE BLUE EB ◇ MIKETON FAST BLUE ◇ NACELAN BLUE G ◇ NCI-C54900 ◇ NEOSETILE BLUE EB ◇ NYLOQUINONE BLUE 2J ◇ ORACET SAPPHIRE BLUE G ◇ PERLITON BLUE B ◇ SERINYL BLUE 2G ◇ SUPRACET BRILLIANT BLUE 2GN ◇ 1,4,5,8-TETRAAMINOANTHRAQUINONE ◇ 1,4,5,8-TETRAMINOANTHRAQUINONE

TOXICITY DATA with REFERENCE
mmo-sat 33 μg/plate NTPTR* NTP-TR-299,86
mmo-sat 100 μg/plate MUREAV 40,203,76
orl-rat TDLo:90125 mg/kg/2Y-C:CAR NTPTR* NTP-TR-299,86
orl-rat TD:180 g/kg/2Y-C:CAR,REP NTPTR* NTP-TR-299,86

CONSENSUS REPORTS: NTP Carcinogenesis Studies (feed); Equivocal Evidence: mouse NTPTR* NTP-TR-299-86; (feed); Clear Evidence: rat NTPTR* NTP-TR-299,86. Reported in EPA TSCA Inventory.

SAFETY PROFILE: Experimental reproductive effects. Questionable carcinogen with experimental carcinogenic data. Mutation data reported. When heated to decomposition it emits toxic fumes of NO_x.

TBG710 CAS:64049-03-2 ***HR: 3***
TETRA-(4-AMINOPHENYL)ARSONIUM CHLORIDE
mf: $C_{24}H_{24}AsN_4•Cl$ mw: 478.89

TOXICITY DATA with REFERENCE
ivn-rat LDLo:30 mg/kg JACSAT 63,1496,41

CONSENSUS REPORTS: Arsenic and its compounds are on the Community Right-To-Know List.

OSHA PEL: TWA 0.5 mg(As)/m^3

SAFETY PROFILE: Poison by intravenous route. When heated to decomposition it emits very toxic fumes of As, NO_x, and Cl^-. See also ARSENIC COMPOUNDS.

TBG750 CAS:70992-03-9 ***HR: 3***
TETRAAMMINECOPPER(II) AZIDE
mf: $CuH_{12}N_{10}$ mw: 215.71

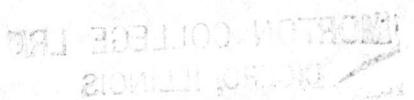

$[(H_3N)_4Cu](N_3)_2$

CONSENSUS REPORTS: Copper and its compounds are on the Community Right-To-Know List.

SAFETY PROFILE: Explodes on heating or impact. When heated to decomposition it emits toxic fumes of NO_x. See also AZIDES and COPPER COMPOUNDS.

TBH000 CAS:31058-64-7 ***HR: 3***
TETRAAMMINECOPPER(II) NITRATE
mf: $CuH_{12}N_6O_6$ mw: 255.68

$[(H_3N)_4Cu](NO_3)_2$

CONSENSUS REPORTS: Copper and its compounds are on the Community Right-To-Know List.

SAFETY PROFILE: Explodes at 330°C or on impact. Upon decomposition it emits toxic fumes of NO_x. See also NITRATES and COPPER COMPOUNDS.

TBH250 CAS:39729-81-2 ***HR: 3***
TETRAAMMINECOPPER(II) NITRITE
mf: $CuH_{12}N_6O_4$ mw: 223.68

$[(H_3N)_4Cu](NO_2)_2$

CONSENSUS REPORTS: Copper and its compounds are on the Community Right-To-Know List.

SAFETY PROFILE: A very shock-sensitive explosive. Traces of nitrate render it heat-sensitive. When heated to decomposition it emits toxic fumes of NO_x. See also NITRITES and COPPER COMPOUNDS.

TBH500 ***HR: 3***
TETRAAMMINEHYDROXYNITRATOPLATINUM (IV) NITRATE
mf: $H_{13}N_7O_{10}Pt$ mw: 466.24

SAFETY PROFILE: Explodes violently on heating. When heated to decomposition it emits toxic fumes of NO_x. See also PLATINUM COMPOUNDS and NITRATES.

TBH750 ***HR: 2***
TETRAAMMINELITHIUM DIHYDROGEN-PHOSPHIDE
mf: $H_{14}LiN_4P$ mw: 108.05

$[(H_3N)_4Li]PH_2$

SAFETY PROFILE: Reacts vigorously with water, evolving phosphine and ammonia gases which may ignite. When heated to decomposition it emits very toxic fumes of PO_x, NO_x, and Li_2O. See also PHOSPHIDES and LITHIUM COMPOUNDS.

TBI000 CAS:13601-08-6 ***HR: 2***
TETRAAMMINEPALLADIUM(II) NITRATE
mf: $H_{12}N_6O_6Pd$ mw: 298.54

$[(H_3N)_4Pd](NO_3)_2$

SAFETY PROFILE: Can ignite and burn violently. When heated to decomposition it emits toxic fumes of NO_x. See also PALLADIUM COMPOUNDS and NITRATES.

TBI100 CAS:13933-32-9 ***HR: D***
TETRAAMMINEPLATINUM DICHLORIDE
mf: $Cl_2H_{12}N_4Pt$ mw: 334.15

SYNS: TETRAAMMINEDICHLOROPLATINUM(II) ◇ TETRAAMMINEPLATINNIUM(II) DICHLORIDE (SP-4-1)- (9CI) ◇ TETRAAMMINEPLATINUM(2+) DICHLORIDE ◇ TETRAAMMINEPLATINUM(2+) DICHLORIDE (SP-4-1)- ◇ TETRAAMMINEPLATINUM(II) DICHLORIDE

TOXICITY DATA with REFERENCE
mma-sat 100 ng/plate PCJOAU 16,721,82
pic-esc μg/plate CNREA8 43,2819,83

SAFETY PROFILE: Mutation data reported. When heated to decomposition it emits toxic fumes of Cl^- and NO_x.

TBI600 CAS:3765-65-9 ***HR: 3***
TETRAAMYLSTANNANE
mf: $C_{20}H_{44}Sn$ mw: 403.33

SYN: TETRAPENTYLSTANNANE

TOXICITY DATA with REFERENCE
orl-mus LDLo:807 mg/kg APFRAD 14,88,56
ipr-mus LDLo:202 mg/kg APFRAD 14,88,56
ivn-mus LDLo:121 mg/kg APFRAD 14,88,56
ims-mus LDLo:1613 mg/kg APFRAD 14,88,56
par-mus LDLo:403 mg/kg COREAF 239,1091,54

OSHA PEL: TWA 0.1 mg(Sn)/m³ (skin)
ACGIH TLV: TWA 0.1 mg(Sn)/m³ (skin) (Proposed: TWA 0.1 mg(Sn)/m³; STEL 0.2 mg(Sn)/m³ (skin))
NIOSH REL: (Organotin Compounds) TWA 0.1 mg(Sn)/m³

SAFETY PROFILE: Poison by intravenous and intraperitoneal routes. Moderately toxic by ingestion and parenteral routes. When heated to decomposition it emits acrid smoke and irritating fumes. See also TIN COMPOUNDS.

TBI750 CAS:63040-21-1 ***HR: 3***
9,10,14c-15-TETRAAZANAPHTHO(1,2,-3-fg) NAPHTHACENE NITRATE
mf: $C_{21}H_{12}N_4 \cdot xHNO_3$

SYN: ISOTRICYCLOQUINAZOLINENITRATE

TOXICITY DATA with REFERENCE
skn-mus TDLo:1240 mg/kg/Y-I:ETA BJCAAI 17,266,63

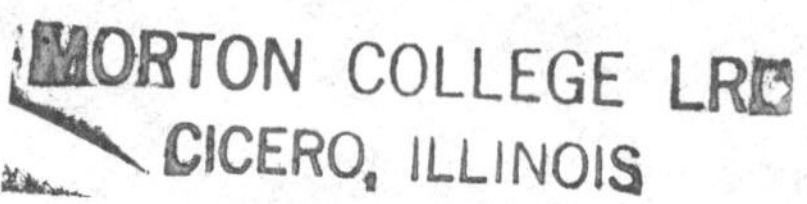

SAFETY PROFILE: Questionable carcinogen with experimental tumorigenic data. When heated to decomposition it emits very toxic fumes of NO_x.

TBI775 CAS:262-38-4 ***HR: 3***
4a,8a,9a,10a-TETRAAZA-2,3,6,7-TETRAOXAPERHYDROANTHRACENE
mf: $C_6H_{12}N_4O_4$ mw: 204.19

SAFETY PROFILE: Explodes when heated to 120°C. Upon decomposition it emits toxic fumes of NO_x. See also PEROXIDES.

TBJ000 CAS:51-46-7 ***HR: 3***
1,3,6,8-TETRAAZATRICYCLO(4.4.1.1^{3,8})DODECANE
mf: $C_8H_{16}N_4$ mw: 168.28

SYN: 1,4,6,9-TETRAAZA-TRICYCLO-(4.4.1.1^{9,4}) DODECAN (GERMAN)

TOXICITY DATA with REFERENCE
ivn-rat LD50:210 mg/kg ARZNAD 16,734,66
ipr-mus LD50:273 mg/kg ARZNAD 16,734,66

CONSENSUS REPORTS: Reported in EPA TSCA Inventory.

SAFETY PROFILE: Poison by intravenous and intraperitoneal routes. When heated to decomposition it emits toxic fumes of NO_x.

TBJ250 CAS:22826-61-5 ***HR: 3***
TETRAAZIDO-p-BENZOQUINONE
mf: $C_6N_{12}O_2$ mw: 272.15

$O:C_6(N_3)_4:O$

DOT Classification: Forbidden

SAFETY PROFILE: Extremely explosive, sensitive to heat (explodes at 120°C), impact or friction. When heated to decomposition it emits toxic fumes of NO_x. See also AZIDES.

TBJ275 CAS:58-46-8 ***HR: 3***
TETRABENZAINE
mf: $C_{19}H_{27}NO_3$ mw: 317.47

SYNS: 2H-BENZO(a)QUINOLIZIN-2-ONE,1,3,4,6,7,11b-HEXAHYDRO-3-ISOBUTYL-9,10-DIMETHOXY- ◇ 1,3,4,6,7,11b-HEXAHYDRO-3-ISOBUTYL-9,10-DIMETHOXY-2H-BENZO(a)QUINOLIZIN-2-ONE ◇ NITOMAN ◇ 2-OXO-3-ISOBUTYL-9,10-DIMETHOXY-1,3,4,6,7,11-β-HEXAHYDRO-2H-BENZOQUIN OLIZINE ◇ REGULIN ◇ RO 1-9569 ◇ RO 1-9569/12 ◇ RUBIGEN ◇ TETRABENAZINE ◇ TETRABENZINE

TOXICITY DATA with REFERENCE
scu-rat TDLo:15 mg/kg (female 1D pre):REP ENDOAO 78,225,66
orl-mus LD50:550 mg/kg 27ZQAG -,308,72
ipr-mus LD50:250 mg/kg NTIS** AD691-490
scu-mus LD50:400 mg/kg 27ZQAG -,308,72
ivn-mus LD50:150 mg/kg PSCBAY 2,17,63

SAFETY PROFILE: Poison by subcutaneous and intraperitoneal routes. Moderately toxic by ingestion. Experimental reproductive effects. When heated to decomposition it emits toxic fumes of NO_x.

TBJ300 CAS:18283-93-7 ***HR: 3***
TETRABORANE(10)
mf: B_4H_{10} mw: 53.32

PROP: Gas; disagreeable odor. Mp: 120°, bp: 18°. Decomp at room temp in a few hours; decomp rapidly at 100°.

SAFETY PROFILE: Ignites spontaneously in air or oxygen. Explodes on contact with concentrated nitric acid. See also BORANES and BORON COMPOUNDS.

TBJ400 CAS:68436-99-7 ***HR: 3***
TETRA(BORON NITRIDE)FLUOROSULFATE
mf: $B_4FN_4O_3S$ mw: 198.32

SAFETY PROFILE: Explodes when heated above 40°C. When heated to decomposition it emits toxic fumes of F^-, SO_x, and NO_x. See also BORON COMPOUNDS, NITRIDES, FLUORIDES, and SULFATES.

TBJ475 CAS:17156-85-3 ***HR: 3***
TETRABORON TETRACHLORIDE
mf: B_4Cl_4 mw: 185.05

SAFETY PROFILE: Ignites spontaneously in air. When heated to decomposition it emits toxic fumes of Cl^-. See also BORON COMPOUNDS and CHLORIDES.

TBJ500 CAS:576-55-6 ***HR: 1***
3,4,5,6-TETRABROMO-o-CRESOL
mf: $C_7H_4Br_4O$ mw: 423.75

PROP: White to buff crystals. Mp: 205-208° (decomp), d. 3.027 at 20°. Insol in water.

SYN: 2-METHYL-3,4,5,6-TETRABROMOPHENOL

CONSENSUS REPORTS: Reported in EPA TSCA Inventory.

SAFETY PROFILE: Animal experiments show corrosive action on skin but low systemic toxicity. When heated to decomposition it emits highly toxic fumes of Br^-. See also CRESOL and BROMIDES.

TBK000 CAS:36323-28-1 ***HR: 3***
α,α,α′,α′-TETRABROMO-m-XYLENE
mf: $C_8H_6Br_4$ mw: 421.78

PROP: Mp: 105-108°.

SYN: 1,3-BIS(DIBROMOMETHYL)BENZENE

TOXICITY DATA with REFERENCE
ivn-mus LD50:56 mg/kg CSLNX* NX#00642

CONSENSUS REPORTS: Reported in EPA TSCA Inventory.

SAFETY PROFILE: Poison by intravenous route. Corrosive. Lacrymator. When heated to decomposition it emits toxic fumes of Br^-. See also BROMIDES.

TBK250 CAS:23488-38-2 ***HR: 2***
TETRABROMO-p-XYLENE
mf: $C_8H_6Br_4$ mw: 421.78

PROP: Mp: 254-256°

SYNS: 1,4-BIS(DIBROMMETHYL)BENZEN (CZECH) ◇ TETRABROMO-p-XYLEN (CZECH)

TOXICITY DATA with REFERENCE
skn-rbt 500 mg/24H MOD 28ZPAK -,31,72
eye-rbt 500 mg/24H SEV 28ZPAK -,31,72

CONSENSUS REPORTS: Reported in EPA TSCA Inventory.

SAFETY PROFILE: A skin and severe eye irritant. When heated to decomposition it emits toxic fumes of Br^-. See also BROMIDES.

TBK500 CAS:1643-19-2 ***HR: D***
TETRABUTYLAMMONIUM BROMIDE
mf: $C_{16}H_{36}N{\cdot}Br$ mw: 322.44

PROP: Crystals. Mp: 102-104°.

SYN: TETRA-N-BUTYLAMMONIUM BROMIDE

TOXICITY DATA with REFERENCE
dnd-mam:lym 50 mmol/L CBINA8 19,197,77

CONSENSUS REPORTS: Reported in EPA TSCA Inventory.

SAFETY PROFILE: Mutation data reported. When heated to decomposition it emits very toxic fumes of NO_x, Br^-, and NH_3. See also BROMIDES.

TBK750 CAS:2052-49-5 ***HR: 3***
TETRABUTYLAMMONIUM HYDROXIDE
mf: $C_{16}H_{36}N{\cdot}HO$ mw: 259.54

PROP: D: 0.968, refr index: 1.435.

TOXICITY DATA with REFERENCE
scu-mus LDLo:19 mg/kg JPETAB 28,367,26

CONSENSUS REPORTS: Reported in EPA TSCA Inventory.

SAFETY PROFILE: Poison by subcutaneous route. When heated to decomposition it emits toxic fumes of NO_x and NH_3.

TBL000 CAS:311-28-4 ***HR: 2***
TETRABUTYLAMMONIUM IODIDE
mf: $C_{16}H_{36}N{\cdot}I$ mw: 369.43

PROP: Crystals or leaves from benzene. Mp: 114-148°. Sltly sol in water; sol in alc, ether.

SYN: TETRA-N-BUTYLAMMONIUMJODID (CZECH)

TOXICITY DATA with REFERENCE
orl-rat LD50:1990 mg/kg MarJV# 29MAR77

CONSENSUS REPORTS: Reported in EPA TSCA Inventory.

SAFETY PROFILE: Moderately toxic by ingestion. When heated to decomposition it emits very toxic fumes of NO_x, NH_3, and I^-. See also IODIDES.

TBL250 CAS:1941-27-1 ***HR: 3***
TETRABUTYLAMMONIUM NITRATE
mf: $C_{16}H_{36}N{\cdot}NO_3$ mw: 304.54

PROP: Hygroscopic solid. Mp: 118-120°.

TOXICITY DATA with REFERENCE
ivn-mus LD50:10 mg/kg CSLNX* NX#02073

CONSENSUS REPORTS: Reported in EPA TSCA Inventory.

SAFETY PROFILE: Poison by intravenous route. An irritant. When heated to decomposition it emits very toxic fumes of NO_x and NH_3. See also NITRATES.

TBL500 CAS:10428-19-0 ***HR: 3***
TETRABUTYL DICHLOROSTANNOXANE
mf: $C_{16}H_{36}Cl_2OSn_2$ mw: 552.80

TOXICITY DATA with REFERENCE
ivn-mus LD50:32 mg/kg CSLNX* NX#00183

CONSENSUS REPORTS: Reported in EPA TSCA Inventory.

OSHA PEL: TWA 0.1 mg(Sn)/m^3 (skin)
ACGIH TLV: TWA 0.1 mg(Sn)/m^3 (skin) (Proposed: TWA 0.1 mg(Sn)/m^3; STEL 0.2 mg(Sn)/m^3 (skin))
NIOSH REL: TWA (Organotin Compounds) 0.1 mg (Sn)/m^3

SAFETY PROFILE: Poison by intravenous route. When heated to decomposition it emits toxic fumes of Cl^-. See also TIN COMPOUNDS.

TBL750 CAS:3115-68-2 ***HR: 3***
TETRA-N-BUTYLPHOSPHONIUM BROMIDE
mf: $C_{16}H_{36}P{\cdot}Br$ mw: 339.40

PROP: Crystals. Mp: 102°.

TOXICITY DATA with REFERENCE
ivn-mus LD50:56 mg/kg CSLNX* NX#03131

CONSENSUS REPORTS: Reported in EPA TSCA Inventory.

SAFETY PROFILE: Poison by intravenous route. When heated to decomposition it emits very toxic fumes of PO_x and Br^-. See also BROMIDES.

TBM000 CAS:2304-30-5 ***HR: 3***
TETRA-N-BUTYLPHOSPHONIUM CHLORIDE
mf: $C_{16}H_{36}P•Cl$ mw: 294.94

PROP: Mp: 80-83°. Hygroscopic.

TOXICITY DATA with REFERENCE
ivn-mus LD50:32 mg/kg CSLNX* NX#03140

CONSENSUS REPORTS: Reported in EPA TSCA Inventory.

SAFETY PROFILE: Poison by intravenous route. Irritant. When heated to decomposition it emits very toxic fumes of PO_x and Cl^-.

TBM250 CAS:1461-25-2 ***HR: 3***
TETRABUTYLSTANNANE
mf: $C_{16}H_{36}Sn$ mw: 347.21

SYNS: TETRA-n-BUTYLCIN (CZECH) ◇ TETRABUTYLTIN

TOXICITY DATA with REFERENCE
eye-rbt 500 mg/24H MLD 28ZPAK -,233,72
orl-mus LDLo:1389 mg/kg APFRAD 14,88,56
ipr-mus LDLo:174 mg/kg APFRAD 14,88,56
ivn-mus LD50:56 mg/kg CSLNX* NX#02221
ims-mus LDLo:1389 mg/kg APFRAD 14,88,56
par-mus LDLo:347 mg/kg COREAF 239,1091,54
skn-rbt LDLo:2000 mg/kg SAIGBL 15,3,73

CONSENSUS REPORTS: Reported in EPA TSCA Inventory.

OSHA PEL: TWA 0.1 mg(Sn)/m³ (skin)
ACGIH TLV: TWA 0.1 mg(Sn)/m³ (skin) (Proposed: TWA 0.1 mg(Sn)/m³; STEL 0.2 mg(Sn)/m³ (skin))
NIOSH REL: (Organotin Compounds) TWA 0.1 mg(Sn)/m³

SAFETY PROFILE: Poison by intravenous, intraperitoneal, and parenteral routes. Moderately toxic by ingestion and skin contact. An eye irritant. When heated to decomposition it emits acrid smoke and irritating fumes. See also TIN COMPOUNDS.

TBM750 CAS:1634-02-2 ***HR: 2***
TETRABUTYLTHIURAM DISULPHIDE
mf: $C_{18}H_{36}N_2S_4$ mw: 408.80

SYN: BIS(DIBUTYLTHIOCARBAMOYL)DISULFIDE

TOXICITY DATA with REFERENCE
ipr-mus LD50:2350 mg/kg APTOA6 8,329,52

CONSENSUS REPORTS: Reported in EPA TSCA Inventory.

SAFETY PROFILE: Moderately toxic by intraperitoneal route. When heated to decomposition it emits very toxic fumes of NO_x and SO_x. See also SULFIDES.

TBM850 CAS:4559-86-8 ***HR: 2***
1,1,3,3-TETRABUTYLUREA

SYNS: TETRABUTYLUREA ◇ UREA, 1,1,3,3-TETRABUTYL- ◇ UREA, TETRABUTYL-

TOXICITY DATA with REFERENCE
skn-rat TDLo:5 g/kg (female 6-15D post):REP FCTOD7 25,173,87
skn-rat TDLo:2500 mg/kg (female 6-15D post):TER FCTOD7 25,173,87
orl-rat LD50:17 g/kg FCTOD7 25,173,87
ihl-rat LCLo:700 mg/m³/4H FCTOD7 25,173,87

SAFETY PROFILE: Moderately toxic by inhalation. An experimental teratogen. Experimental reproductive effects. When heated to decomposition it emits toxic fumes of NO_x.

TBN000 CAS:136-47-0 ***HR: 3***
TETRACAINE HYDROCHLORIDE
mf: $C_{15}H_{24}N_2O_2•ClH$ mw: 300.87

PROP: Bitter crystals causing numbness of mouth. Mp: 147-150°. Sol in water and alc; insol in ether and benzene.

SYNS: AMETHOCAINE HYDROCHLORIDE ◇ ANACEL ◇ ANETHAÏNE ◇ BUTETHANOL ◇ p-(BUTYLAMINO)BENZOIC ACID, 2-(DIMETHYLAMINO)ETHYL ESTER, HYDROCHLORIDE ◇ p-BUTYLAMINOBENZOYL-2-DIMETHYLAMINOETHANOLHYDROCHLORIDE ◇ BUTYLOCAINE ◇ CURTACAIN ◇ DECICAINE ◇ DIKAIN HYDROCHLORIDE ◇ 2-DIMETHYLAMINOETHANOL-4-N-BUTYLAMINOBENZOATE HYDROCHLORIDE ◇ DIMETHYLAMINOETHYL-p-N-BUTYLAMINOBENZOATE HYDROCHLORIDE ◇ 2-(DIMETHYLAMINO)ETHYL-p-(BUTYLAMINO)BENZOATEHYDROCHLORIDE ◇ GINGICAIN M ◇ MENONASAL ◇ PANTOCAINE HYDROCHLORIDE ◇ PONTOCAINE HYDROCHLORIDE ◇ STERILE TETRACAINE HYDROCHLORIDE ◇ TONEXOL

TOXICITY DATA with REFERENCE
eye-rbt 2% MOD ARZNAD 8,181,58
ipr-rat LD50:23500 μg/kg OYYAA2 9,413,75
scu-rat LD50:24 mg/kg OYYAA2 9,413,75
ivn-rat LD50:4500 μg/kg OYYAA2 9,413,75
orl-mus LD50:160 mg/kg OYYAA2 9,413,75
ipr-mus LD50:32 mg/kg OYYAA2 9,413,75
scu-mus LD50:28 mg/kg OYYAA2 9,413,75
ivn-mus LD50:6600 μg/kg OYYAA2 9,413,75

scu-rbt LDLo:15 mg/kg ODFU** -,-,31
ivn-rbt LD50:2150 μg/kg AIPTAK 113,313,58
isp-rbt LDLo:5 mg/kg JPETAB 57,221,36
ivn-gpg LD50:15600 μg/kg AIPTAK 113,313,58

CONSENSUS REPORTS: EPA Genetic Toxicology Program.

SAFETY PROFILE: Poison by ingestion, intraperitoneal, intravenous, intraspinal, and subcutaneous routes. An eye irritant. When heated to decomposition it emits very toxic fumes of HCl and NO_x. An FDA over-the-counter drug used as a local anesthetic.

TBN100 CAS:12774-81-1 ***HR: 3***
TETRACARBON MONOFLUORIDE
mf: C_4F mw: 67.04

SAFETY PROFILE: Deflagrates (nearly explosive combustion) when heated rapidly. When heated to decomposition it emits toxic fumes of F^-. See also FLUORIDES.

TBN150 CAS:15712-13-7 ***HR: 3***
TETRACARBONYLMOLYBDENUM DICHLORIDE
mf: $C_4Cl_2MoO_4$ mw: 278.89

$(OC)_4MoCl_2$

SAFETY PROFILE: A storage hazard. The complex decomposes at room temperature to form a dark pyrophoric powder. When heated to decomposition it emits toxic fumes of Cl^-. See also MOLYBDENUM COMPOUNDS and CARBONYLS.

TBN200 CAS:21239-57-6 ***HR: 3***
TETRACARBONYL(TRIFLUOROMETHYLTHIO) MANGANESE dimer
mf: $C_{10}F_6Mn_2O_8S_2$ mw: 536.10

SYN: MANGANESE, TETRACARBONYL(TRIFLUOROMETHYLTHIO)-,dimer

TOXICITY DATA with REFERENCE
ivn-mus LD50:180 mg/kg CSLNX* NX#07141

OSHA PEL: CL 5 mg(Mn)/m^3
ACGIH TLV: TWA 5 mg(Mn)/m^3

SAFETY PROFILE: Poison by intravenous route. When heated to decomposition it emits toxic fumes of SO_x, Mn, and F^-.

TBN250 ***HR: D***
TETRACHLOROACETONE
mf: $C_3H_2Cl_4O$ mw: 195.85

PROP: Liquid. Bp: 180-182°. Sltly decomp; very sol in benzene, alc, ether.

TOXICITY DATA with REFERENCE
mmo-sat 100 ng/plate BECTA6 24,590,80
orl-mus TDLo:150 mg/kg (6-15D preg):TER TJADAB 19,32A,79

SAFETY PROFILE: An experimental teratogen. Mutation data reported. When heated to decomposition it emits toxic fumes of Cl^-.

TBN300 CAS:632-21-3 ***HR: 3***
1,1,3,3-TETRACHLOROACETONE
mf: $C_3H_2Cl_4O$ mw: 195.85

SYNS: 1,1,3,3-TETRACHLOROPROPANONE ◇ 1,1,3,3-TETRACHLORO-2-PROPANONE

TOXICITY DATA with REFERENCE
mmo-sat 50 μL/plate ENMUDM 7,163,85
mma-smc 1 μL/L MUREAV 155,53,85
orl-mus TDLo:150 mg/kg (female 6-15D post):REP FAATDF 2,220,82
orl-mus TDLo:500 mg/kg (6-15D preg):TER FAATDF 2,220,82
orl-mus LD50:176 mg/kg FAATDF 2,220,82
skn-rbt LD50:80 mg/kg FAATDF 2,220,82

SAFETY PROFILE: Poison by ingestion and skin contact. An experimental teratogen. Experimental reproductive effects. Mutation data reported. When heated to decomposition it emits toxic fumes of Cl^-.

TBN400 ***HR: 3***
N,N,N',N'-TETRACHLOROADIPAMIDE
mf: $C_6H_8Cl_4N_2O_2$ mw: 281.95

$Cl_2NCO\cdot(CH_2)_4CO\cdot NCl_2$

SAFETY PROFILE: Reacts with water to form an explosive product. When heated to decomposition it emits toxic fumes of Cl^- and NO_x. See also AMIDES.

TBN500 CAS:14047-09-7 ***HR: D***
3,3',4,4'-TETRACHLOROAZOBENZENE
mf: $C_{12}H_6Cl_4N_2$ mw: 320.00

SYNS: BIS(3,4-DICHLOROPHENYL)-DIAZENE (9CI) ◇ TCAB ◇ 3,4,3',4'-TETRACHLOROAZOBENZENE

TOXICITY DATA with REFERENCE
mma-sat 100 μg/plate RCOCB8 17,225,77
dns-rat:lvr 1 μmol/L CALEDQ 6,207,79

CONSENSUS REPORTS: EPA Genetic Toxicology Program.

SAFETY PROFILE: Mutation data reported. When heated to decomposition it emits very toxic fumes of Cl^- and NO_x.

TBN550 CAS:21232-47-3 ***HR: D***
3,3',4,4'-TETRACHLOROAZOXYBENZENE
mf: $C_{12}H_6Cl_4N_2O$ mw: 336.00

SYNS: BIS(3,4-DICHLOROPHENYL)-DIAZENE 1-OXIDE (9CI) ◇ TCAOB ◇ 3,4,3',4'-TETRACHLOROAZOXYBENZENE

TOXICITY DATA with REFERENCE
cyt-mus-orl 134 mg/kg/4W-C AECTCV 14,677,85
orl-mus TDLo:7560 μg/kg (female 14D pre-28D post):REP AECTCV 14,677,85
ipr-mus TDLo:16 mg/kg (female 12D post):TER ARTODN 55,20,84

SAFETY PROFILE: An experimental teratogen. Experimental reproductive effects. Mutation data reported. When heated to decomposition it emits toxic fumes of Cl^- and NO_x.

TBN740 CAS:634-66-2 ***HR: 2***
1,2,3,4-TETRACHLOROBENZENE
mf: $C_6H_2Cl_4$ mw: 215.88

PROP: Mp: 46-47°, bp: 254°/ 761 mm, flash p: > 230°F.

TOXICITY DATA with REFERENCE
orl-rat TDLo:1500 mg/kg (9-13D preg):TER TXCYAC 26,243,83
orl-rat TDLo:2 g/kg (6-15D preg):REP TJADAB 29,21,84
orl-rat LD50:1167 mg/kg JTEHD6 11,663,83

CONSENSUS REPORTS: Reported in EPA TSCA Inventory.

SAFETY PROFILE: Moderately toxic by ingestion. An experimental teratogen. Experimental reproductive effects. Irritant. Combustible liquid. When heated to decomposition it emits toxic fumes of Cl^-. See also CHLORINATED HYDROCARBONS, AROMATIC.

TBN750 CAS:95-94-3 ***HR: 2***
1,2,4,5-TETRACHLOROBENZENE
mf: $C_6H_2Cl_4$ mw: 215.88

PROP: Crystals. Mp: 138°, bp: 245°, d: 1.734, vap press: < 0.1 mm @ 25°, vap d: 7.4, flash p: 311°F (CC).

SYN: RCRA WASTE NUMBER U207

TOXICITY DATA with REFERENCE
orl-rat LD50:1500 mg/kg HYSAAV 30,8,65
orl-mus LD50:1035 mg/kg HYSAAV 30,8,65

CONSENSUS REPORTS: Reported in EPA TSCA Inventory.

SAFETY PROFILE: Moderately toxic by ingestion. Combustible when exposed to heat or flame. Can react vigorously with oxidizing materials. A reaction mixture with sodium hydroxide + a solvent (e.g., methanol or ethylene glycol) is used commercially to prepare 2,4,5-trichlorophenol. Improperly controlled the reaction can be explosive. A side product is the extremely toxic 2,3,7,8-tetrachlorodibenzodioxin. All plant-scale accidents have distributed the dioxin widely. The Seveso, Italy accident in 1976 is the most recent example. To fight fire, use CO_2, dry chemical. When heated to decomposition it emits highly toxic fumes of Cl^-. See also CHLORINATED HYDROCARBONS, AROMATIC.

TBO000 CAS:15721-02-5 ***HR: 3***
TETRACHLOROBENZIDINE
mf: $C_{12}H_8Cl_4N_2$ mw: 322.02

SYNS: 2,2',5,5'-TETRACHLOROBENZIDINE ◇ 3,3',6,6'-TETRACHLOROBENZIDINE ◇ 2,2',5,5'-TETRACHLORO-(1,1'-BIPHENYL)-4,4'-DIAMINE, (9CI) ◇ 2,2',5,5'-TETRACHLORO-4,4'-DIAMINODIPHENYL

TOXICITY DATA with REFERENCE
orl-rat TDLo:54 mg/kg/43W-C:ETA JMEJAS 25,123,78
orl-mus TDLo:36 mg/kg/43W-C:CAR JMEJAS 25,123,78
orl-mus TD:300 g/kg/43W-C:CAR IARC** 27,141,82

CONSENSUS REPORTS: IARC Cancer Review: Group 3 IMEMDT 7,56,87; Animal Inadequate Evidence IMEMDT 27,141,82. Reported in EPA TSCA Inventory.

SAFETY PROFILE: Suspected carcinogen with experimental carcinogenic and tumorigenic data. When heated to decomposition it emits very toxic fumes of NO_x and Cl^-.

TBO500 CAS:118-75-2 ***HR: 3***
2,3,5,6-TETRACHLORO-1,4-BENZOQUINONE
mf: $C_6Cl_4O_2$ mw: 245.86

PROP: Yellow crystals. Mp: 290°. Insol in water.

SYNS: CHLORANIL ◇ DOW SEED DISINFECTANT No. 5 ◇ ENT 3,797 ◇ G-25804 ◇ GEIGY-444E ◇ RERANIL ◇ SPERGON I ◇ SPERGON TECHNICAL ◇ TETRACHLOROBENZOQUINONE ◇ TETRACHLORO-p-BENZOQUINONE ◇ TETRACHLORO-1,4-BENZOQUINONE ◇ 2,3,5,6-TETRACHLORO-p-BENZOQUINONE ◇ 2,3,5,6-TETRACHLORO-2,5-CYCLOHEXADIENE-1,4-DIONE ◇ TETRACHLOROQUINONE ◇ TETRACHLORO-p-QUINONE ◇ VULKLOR

TOXICITY DATA with REFERENCE
orl-mus TDLo:71 g/kg/78W-I:NEO NTIS** PB223-159
orl-rat LD50:4000 mg/kg PCOC** -,218,66
ipr-rat LDLo:500 mg/kg JIHTAB 25,98,43

CONSENSUS REPORTS: Reported in EPA TSCA Inventory.

SAFETY PROFILE: Moderately toxic by ingestion and intraperitoneal routes. Can cause central nervous system depression. Questionable carcinogen with experimental neoplastigenic data. May be irritating to skin and mucous membranes. Used as a fungicide. When heated to decomposition it emits highly toxic fumes of Cl^-.

TBO700 CAS:32598-13-3 ***HR: 3***
3,3',4,4'-TETRACHLOROBIPHENYL
mf: $C_{12}H_6Cl_4$ mw: 291.98

SYNS: 4-CB ◇ TCB ◇ 3,4,3',4'-TETRACHLOROBIPHENYL

TOXICITY DATA with REFERENCE
orl-mus TDLo:224 mg/kg (female 10-16D post):REP TOLED5 7,417,81
orl-rat TDLo:110 mg/kg (female 8-18D post):TER TJADAB 25,45,82
orl-gpg LD50:1 mg/kg EVHPAZ 60,57,85

SAFETY PROFILE: Poison by ingestion. An experimental teratogen. Experimental reproductive effects. When heated to decomposition it emits toxic fumes of Cl^-. See also POLYCHLORINATED BIPHENYLS.

TBO750 CAS:79-95-8 ***HR: 1***
2,2',6,6'-TETRACHLOROBISPHENOL A
mf: $C_{15}H_{12}Cl_4O_2$ mw: 366.07

SYNS: 2,2-BIS(3,5-DICHLORO-4-HYDROXYPHENYL)PROPANE ◇ 2,2-BIS(4-HYDROXY-3,5-DICHLOROPHENYL)PROPANE ◇ 4,4'-ISOPROPYLIDENEBIS(2,6-DICHLOROPHENOL)◇ 4,4'-(1-METHYLETHYLIDENE)BIS(2,6-DICHLOROPHENOL)◇ TETRACHLORDIAN (CZECH)

TOXICITY DATA with REFERENCE
skn-rbt 500 mg/24H MLD 28ZPAK -,83,72
orl-rat LD50:7432 mg/kg GISAAA 49(2),64,84
orl-mus LD50:5050 mg/kg GISAAA 49(2),64,84

CONSENSUS REPORTS: Chlorophenol compounds are on the Community Right-To-Know List. Reported in EPA TSCA Inventory.

SAFETY PROFILE: Mildly toxic by ingestion. A skin irritant. When heated to decomposition it emits toxic fumes of Cl^-. See also CHLOROPHENOLS.

TBO760 CAS:921-09-5 ***HR: 2***
1,1,2,3-TETRACHLORO-1,3-BUTADIENE
mf: $C_4Cl_4H_2$ mw: 191.86

TOXICITY DATA with REFERENCE
orl-rat LD50:421 mg/kg GTPZAB 26(2),53,82
ihl-rat LC50:1600 mg/m^3 GTPZAB 26(2),53,82
orl-mus LD50:486 mg/kg GTPZAB 26(2),53,82
ihl-mus LC50:670 mg/m^3 GTPZAB 26(2),53,82

SAFETY PROFILE: Moderately toxic by ingestion and inhalation. When heated to decomposition it emits toxic fumes of Cl^-. See also CHLORINATED HYDROCARBONS, ALIPHATIC.

TBO765 CAS:19792-18-8 ***HR: 3***
1,1,4,4-TETRACHLOROBUTATRIENE
mf: C_4Cl_4 mw: 189.86

$Cl_2C{=}C{=}C{=}CCl_2$

SAFETY PROFILE: Forms an explosive polymer when heated. When heated to decomposition it emits toxic fumes of Cl^-. See also CHLORINATED HYDROCARBONS, ALIPHATIC.

TBO776 CAS:16893-05-3 ***HR: 3***
cis-TETRACHLORODIAMMINE PLATINUM(IV)
mf: $Cl_4H_6N_2Pt$ mw: 370.97

SYNS: NSC 119876 ◇ DIAMMINETETRACHLOROPLATINUM (OC-6-22) (9CI) ◇ cis-DIAMMINOTETRACHLOROPLATINUM ◇ cis-PLATINUMDIAMMINE TETRACHLORIDE ◇ cis-PLATINUM(IV) DIAMMINOTETRACHLORIDE ◇ Pt-09

TOXICITY DATA with REFERENCE
mmo-sat 3 μg/plate MUREAV 95,79,82
dni-hmn:oth 25 μmol/L IJCNAW 6,207,70
ipr-mus LD50:22840 μg/kg NCISP* JAN86
ivn-mus LD50:26820 μg/kg NCISP* JAN86

SAFETY PROFILE: Poison by intravenous and intraperitoneal routes. Human mutation data reported. When heated to decomposition it emits toxic fumes of Cl^- and NO_x. See also PLATINUM COMPOUNDS and AMINES.

TBO777 CAS:16893-06-4 ***HR: D***
trans-TETRACHLORODIAMMINE PLATINUM(IV)
mf: $Cl_4H_6N_2Pt$ mw: 370.97

PROP: Crystals. Hygroscopic.

TOXICITY DATA with REFERENCE
mmo-sat 30 μg/plate MUREAV 95,79,82
dnd-mam:lym 50 μmol/L CBINA8 10,27,75

SAFETY PROFILE: Mutation data reported. Irritant. When heated to decomposition it emits toxic fumes of Cl^- and NO_x. See also PLATINUM COMPOUNDS and AMINES.

TBO700 CAS:634-90-2 ***HR: 2***
1,2,3,5-TETRACHLOROBENZENE
mf: $C_6H_2Cl_4$ mw: 215.88

SYN: BENZENE, 1,2,3,5-TETRACHLORO-

TOXICITY DATA with REFERENCE
orl-rat TDLo:2 g/kg (female 6-15D post):REP TJADAB 29,21,84
orl-rat LD50:1727 mg/kg JTEHD6 11,663,83

CONSENSUS REPORTS: Reported in EPA TSCA Inventory.

SAFETY PROFILE: Moderately toxic by ingestion. Experimental reproductive effects. When heated to decomposition it emits toxic fumes of Cl^-.

TBO778 CAS:21572-61-2 ***HR: 3***
TETRACHLORODIAZOCYCLOPENTADIENE
mf: $C_5Cl_4N_2$ mw: 229.88

$ClC{=}CClCCl{=}CClC{:}N_2$

SAFETY PROFILE: Explodes when heated to 150°C. When heated to decomposition it emits toxic fumes of Cl^- and NO_x.

TBO780 CAS:51207-31-9 ***HR: 3***
2,3,7,8-TETRACHLORODIBENZOFURAN
mf: $C_{12}H_4Cl_4O$ mw: 305.96

SYN: NCI-C56611

TOXICITY DATA with REFERENCE
orl-mus TDLo:250 μg/kg (10D preg):TER TOLED5 20,183,84
ipr-mus TDLo:600 μg/kg (female 12D post):REP TOLED5 23,37,84
orl-mky LDLo:1 mg/kg ANYAA9 320,151,79
orl-gpg LD50:5 μg/kg TXAPA9 55,342,80

CONSENSUS REPORTS: EPA Genetic Toxicology Program.

SAFETY PROFILE: Poison by ingestion. An experimental teratogen. Experimental reproductive effects. When heated to decomposition it emits toxic fumes of Cl^-.

TBP000 CAS:76-11-9 ***HR: 1***
1,1,1,2-TETRACHLORO-2,2-DIFLUOROETHANE
mf: $C_2Cl_4F_2$ mw: 203.82

SYNS: HALOCARBON 112a ◇ REFRIGERANT 112a

TOXICITY DATA with REFERENCE
ihl-rat LCLo:20000 ppm/30M TXAPA9 19,1,71

CONSENSUS REPORTS: Reported in EPA TSCA Inventory.

OSHA PEL: TWA 500 ppm
ACGIH TLV: TWA 500 ppm
DFG MAK: 1000 ppm (8340 mg/m^3)

SAFETY PROFILE: Mildly toxic by inhalation. When heated to decomposition it emits very toxic fumes of Cl^- and F^-. Used as a refrigerant. See also FLUORIDES.

TBP050 CAS:76-12-0 ***HR: 2***
1,1,2,2-TETRACHLORO-1,2-DIFLUOROETHANE
mf: $C_2Cl_4F_2$ mw: 203.82

PROP: Liquid. Bp: 92.8°, d: 1.6447 @ 25°, vap d: 7.03.

SYNS: 1,2-DIFLUORO-1,1,2,2-TETRACHLOROETHANE ◇ F-112 ◇ FREON 112 ◇ GENETRON 112 ◇ HALOCARBON 112 ◇ REFRIGERANT 112

TOXICITY DATA with REFERENCE
skn-gpg 100%/24H open MLD AIHAAP 27,332,66
eye-gpg 100 mg/20S rns MLD AIHAAP 27,332,66
ihl-rat LCLo:15000 ppm/4H FLCRAP 1,197,67
orl-mus LD50:800 mg/kg 85GMAT -,54,82
ihl-mus LC50:123 g/m^3/2H 85GMAT -,54,82

CONSENSUS REPORTS: Reported in EPA TSCA Inventory.

OSHA PEL: TWA 500 ppm
ACGIH TLV: TWA 500 ppm
DFG MAK: 500 ppm (4170 mg/m^3)

SAFETY PROFILE: Moderately toxic by ingestion. Mildly toxic by inhalation. A skin and eye irritant. When heated to decomposition it emits toxic fumes of F^- and Cl^-. See also CHLORINATED HYDROCARBONS, ALIPHATIC; and FLUORIDES.

TBP075 ***HR: 3***
3,3,4,5-TETRACHLORO-3,6-DIHYDRO-1,2-DIOXIN
mf: $C_4H_4Cl_4O_2$ mw: 225.89

$OOCCl_2CHClCHClCH_2$

SAFETY PROFILE: An unstable explosive. It is the autooxidation product of 1,1,2,3-tetrachloro-1,3-butadiene. When heated to decomposition it emits toxic fumes of Cl^-.

TBP250 CAS:31242-94-1 ***HR: 3***
TETRACHLORODIPHENYL OXIDE
mf: $C_{12}H_6Cl_4O$ mw: 307.98

SYNS: PHENYL ETHER TETRACHLORO ◇ TETRACHLOROPHENYL ETHER

TOXICITY DATA with REFERENCE
orl-gpg LDLo:50 mg/kg 14CYAT 2,1707,63

OSHA PEL: TWA 0.5 mg/m^3

SAFETY PROFILE: Poison by ingestion. When heated to decomposition it emits toxic fumes of Cl^-. See also ETHERS.

TBP500 CAS:13497-91-1 ***HR: 3***
TETRACHLORODIPHOSPHANE
mf: Cl_4P_2 mw: 203.75

SAFETY PROFILE: Oxidizes rapidly in air and may ignite. When heated to decomposition it emits very toxic fumes of Cl^- and PO_x.

TBP750 CAS:25322-20-7 ***HR: 3***
TETRACHLOROETHANE
DOT: UN 1702
mf: $C_2H_2Cl_4$ mw: 167.84

TOXICITY DATA with REFERENCE
dlt-mus-ipr 250 mg/kg BIRUAA 18,94,80
par-mus TDLo:3200 mg/kg (7-14D preg):REP BIRUAA 14,220,76
orl-rat LD50:317 mg/kg AIHAAP 30,470,69
ihl-rat LCLo:1000 ppm/4H AIHAAP 30,470,69
skn-rbt LD50:6330 mg/kg AIHAAP 30,470,69
ihl-gpg LCLo:11 mg/m^3/30M AISFAR 15,1,37
ipr-gpg LDLo:500 mg/kg AIHAAP 35,21,74

CONSENSUS REPORTS: Reported in EPA TSCA Inventory.

NIOSH REL: (Tetrachloroethane) Reduce to lowest feasible level
DOT Classification: ORM-A; Label: None.

SAFETY PROFILE: Poison by ingestion and inhalation. Moderately toxic by intraperitoneal route. Mildly toxic by skin contact. Experimental reproductive effects. Mutation data reported. When heated to decomposition it emits toxic fumes of Cl^-. See also CHLORINATED HYDROCARBONS, ALIPHATIC.

TBQ000 CAS:630-20-6 ***HR: 3***
1,1,1,2-TETRACHLOROETHANE
mf: $C_2H_2Cl_4$ mw: 167.84

PROP: Liquid. D: 1.588 @ 20/4°, bp: 129-130°. Sol in water; misc in alc, ether.

SYNS: NCI-C52459 ◇ RCRA WASTE NUMBER U208

TOXICITY DATA with REFERENCE
skn-rbt 500 mg/24H AMPMAR 35,593,74
eye-rbt 100 mg SEV AMPMAR 35,593,74
orl-mus TDLo:129 g/kg/2Y-I:CAR NTPTR* NTP-TR-237,82
orl-mus TD:258 g/kg/2Y-I:CAR NTPTR* NTP-TR-237,82

CONSENSUS REPORTS: IARC Cancer Review: Group 3 IMEMDT 7,56,87; Animal Limited Evidence IMEMDT 41,87,86. NTP Carcinogenesis Bioassay (gavage); Clear Evidence: mouse NTPTR* NTP-TR-237,82; (gavage); No Evidence: rat NTPTR* NTP-TR-237,82. Reported in EPA TSCA Inventory.

SAFETY PROFILE: A skin and severe eye irritant. Questionable carcinogen with experimental carcinogenic data. Incompatible with dinitrogen tetraoxide; 2,4-dinitrophenyl disulfide; potassium; potassium hydroxide; nitrogen tetroxide; sodium; sodium potassium alloy. When heated to decomposition it emits very toxic fumes of Cl^-.

TBQ100 CAS:79-34-5 ***HR: 3***
1,1,2,2-TETRACHLOROETHANE
DOT: UN 1702
mf: $C_2H_2Cl_4$ mw: 167.84

PROP: Heavy, colorless, mobile liquid; chloroform-like odor. Mp: −43.8°, bp: 146.4°, d: 1.600 @ 20°/4°.

SYNS: ACETYLENE TETRACHLORIDE ◇ BONOFORM ◇ CELLON ◇ 1,1,2,2-CZTEROCHLOROETAN (POLISH) ◇ 1,1-DICHLORO-2,2-DICHLOROETHANE ◇ NCI-C03554 ◇ RCRA WASTE NUMBER U209 ◇ TCE ◇ TETRACHLORETHANE ◇ 1,1,2,2-TETRACHLOORETHAAN (DUTCH) ◇ 1,1,2,2-TETRACHLORAETHAN (GERMAN) ◇ 1,1,2,2-TETRACHLORETHANE (FRENCH) ◇ sym-TETRACHLOROETHANE ◇ TETRACHLORURE d'ACETYLENE (FRENCH) ◇ 1,1,2,2-TETRACLOROETANO (ITALIAN) ◇ WESTRON

TOXICITY DATA with REFERENCE
mmo-sat 10 μg/plate TECSDY 15,101,87
mma-sat 10 μg/plate TECSDY 15,101,87
dnr-esc 15866 μg/plate CNREA8 34,2576,74
sce-ham:ovr 56 mg/L EMMUEG 10(Suppl 10),1,87
orl-rat TDLo:42 g/kg/78W-I:ETA NCITR* NCI-TR-27,78
orl-mus TDLo:55 g/kg/78W-I:CAR NCITR* NCI-TR-27,78
orl-mus TD:110 g/kg/78W-I:CAR NCITR* NCI-TR-27,78
orl-hmn TDLo:30 mg/kg:CNS PCOC**-,1110,66
ihl-hmn TCLo:1000 mg/m^3/30M:CNS AHBAAM 116,131,36
orl-rat LD50:800 mg/kg 85GMAT-,107,82
ihl-rat LCLo:1000 ppm/4H JIDHAN 31,343,49
ihl-mus LC50:4500 mg/m^3/2H 85GMAT-,107,82
ipr-mus LDLo:30 mg/kg CBCCT* 4,378,52
scu-mus LD50:1108 mg/kg JPETAB 123,224,58
orl-dog LDLo:300 mg/kg AJHYA2 16,325,32
ivn-dog LDLo:50 mg/kg QJPPAL 7,205,34
ihl-cat LCLo:19 g/m^3/45M AHBAAM 116,131,36
scu-rbt LDLo:500 mg/kg QJPPAL 7,205,34

CONSENSUS REPORTS: IARC Cancer Review: Group 3 IMEMDT 7,354,87; Animal Limited Evidence IMEMDT 20,477,79; NCI Carcinogenesis Bioassay (gavage); Clear Evidence: mouse NCITR* NCI-CG-TR-27,78; Some Evidence: rat NCITR* NCI-CG-TR-27,78. Reported in EPA TSCA Inventory. EPA Genetic Toxicology Program. Community Right-To-Know List.

OSHA PEL: (Transitional: TWA 5 ppm (skin)) TWA 1 ppm (skin)
ACGIH TLV: TWA 1 ppm (skin)
DFG MAK: 1 ppm (7 mg/m^3); Suspected Carcinogen.
NIOSH REL: (1,1,2,2-Tetrachlorethane) Reduce to lowest level.
DOT Classification: IMO: Poison B; Label: Poison.

SAFETY PROFILE: Suspected carcinogen with experimental carcinogenic and tumorigenic data. Poison by inhalation, ingestion, and intraperitoneal routes. Moderately toxic by several other routes. Mutation data reported. Human central nervous system effects by ingestion and inhalation: general anesthesia, somnolence, hallucinations, and distorted perceptions. Considered the most toxic of the common chlorinated hydrocarbons. Considered to be a very severe industrial hazard and its use has been restricted or even forbidden in certain countries. It is not an inert solvent. Reacts violently with N_2O_4, 2,4-dinitrophenyl disulfide and on contact with sodium or potassium. When heated in contact with solid potassium hydroxide, spontaneously flammable chloro- or dichloroacetylene gas is evolved. Any water can cause appreciable hydrolysis, even at room temperature, and both hydrolysis and oxidation become comparatively rapid above 110°. When heated to decomposition it emits toxic fumes of Cl^-.

A strong irritant of eyes and mucous membranes. A concentration of 3 ppm produces a detectable odor, thus an initial warning effect. Its narcotic action is stronger than that of chloroform, but because of its low volatility, narcosis is less severe and much less common in industrial poisoning than in the case of other chlorinated hydrocarbons. The toxic action of this material is chiefly on the liver where it produces acute yellow atrophy and cirrhosis. Fatty degeneration of the kidneys and heart, hemorrhage into the lungs and serous membranes, and edema of the brain have also been found in fatal cases. Some reports indicate a toxic action on the central nervous system with changes in the brain and in the peripheral nerves. The effect on the blood is one of hemolysis with appearance of young cells in the circulation and a monocytosis. Due to its solvent action on the natural skin oils, dermatitis is not uncommon.

The initial symptoms resulting from exposure to the vapor are lacrimation, salivation and irritation of the nose and throat. Continued exposure to high concentrations results in restlessness, dizziness, nausea, vomiting, and narcosis. The latter, however, is rare in industry. More commonly, exposure is less severe and most complaints are vague and related to the digestive and nervous systems. The patient's symptoms gradually progress to a more serious illness with development of toxic jaundice, liver tenderness, etc., and possibly albuminuria and edema. With serious liver damage, the jaundice increases and toxic symptoms appear with somnolence, delirium, convulsions and coma usually preceding death. See also ACETYLENE COMPOUNDS and CHLORIDES.

TBQ255 CAS:22432-68-4 ***HR: 2***
TETRACHLOROETHYLENE CARBONATE
mf: $C_3Cl_4O_3$ mw: 225.84

SAFETY PROFILE: Reacts with tributylamine to form the toxic phosgene gas. When heated to decomposition it emits toxic fumes of Cl^-.

TBQ275 CAS:16650-10-5 ***HR: 3***
TETRACHLOROETHYLENE OXIDE
mf: C_2Cl_4O mw: 181.82

SYNS: EPOXYPERCHLOROVINYL ◇ PCEO ◇ TETRACHLOROEPOXYETHANE

TOXICITY DATA with REFERENCE
otr-ham:emb 4300 μmol/L JJIND8 69,531,82
skn-mus TDLo:300 mg/kg/66W-I:CAR CNREA8 43,159,83
scu-mus TDLo:20 mg/kg/70W-I:ETA CNREA8 43,159,83

SAFETY PROFILE: Questionable carcinogen with experimental carcinogenic and tumorigenic data. Mutation data reported. When heated to decomposition it emits toxic fumes of Cl^-.

TBQ300 ***HR: 3***
2,3,4,5-TETRACHLOROHEXATRIENE
mf: $C_6H_4Cl_4$ mw: 217.90

TOXICITY DATA with REFERENCE
orl-rat LD50:370 mg/kg 85GMAT -,108,82
ihl-rat LCLo:670 mg/m³/2H 85GMAT -,108,82
orl-mus LD50:290 mg/kg 85GMAT -,108,82
ihl-mus LCLo:190 mg/m³/2H 85GMAT -,108,82

SAFETY PROFILE: Poison by inhalation and ingestion. When heated to decomposition it emits toxic fumes of Cl^-. See also CHLORINATED HYDROCARBONS, ALIPHATIC.

TBQ500 CAS:87-87-6 ***HR: 3***
TETRACHLOROHYDROQUINONE
mf: $C_6H_2Cl_4O_2$ mw: 247.88

SYN: USAF DO-62

TOXICITY DATA with REFERENCE
dnd-omi 100 μmol/L MUREAV 145,71,85
dnd-mam:lym 50 mmol/L MUREAV 145,71,85
orl-mus LD50:500 mg/kg ARTODN 40,63,78
ipr-mus LD50:25 mg/kg NTIS** AD277-689

CONSENSUS REPORTS: Reported in EPA TSCA Inventory.

SAFETY PROFILE: Poison by intraperitoneal route. Moderately toxic by ingestion. Mutation data reported. When heated to decomposition it emits toxic fumes of Cl^-.

TBQ750 CAS:1897-45-6 ***HR: 3***
TETRACHLOROISOPHTHALONITRILE
mf: $C_8Cl_4N_2$ mw: 265.90

SYNS: BRAVO ◇ BRAVO 6F ◇ BRAVO-W-75 ◇ CHLOROALONIL ◇ CHLOROTHALONIL ◇ CHLORTHALONIL (GERMAN) ◇ DAC 2797 ◇ DACONIL ◇ DACONIL 2787 FLOWABLE FUNGICIDE ◇ DACOSOIL ◇ 1,3-DICYANOTETRACHLOROBENZENE ◇ EXOTHERM ◇ EXOTHERM TERMIL ◇ FORTURF ◇ NCI-C00102 ◇ NOPCOCIDE ◇ SWEEP ◇ TCIN ◇ m-TCPN ◇ TERMIL ◇ 2,4,5,6-TETRACHLORO-3-CYANOBENZONITRILE ◇ m-TETRACHLOROPHTHALONITRILE ◇ TPN (pesticide)

TOXICITY DATA with REFERENCE
orl-rat TDLo:142 g/kg/80W-C:CAR NCITR* NCI-CG-TR-41,78
orl-rat LD50:10 mg/kg 85ARAE 4,75,76
orl-mus LD50:6 g/kg INHEAO 4,11,66
ipr-mus LD50:2500 mg/kg INHEAO 4,11,66

CONSENSUS REPORTS: IARC Cancer Review: Group 3 IMEMDT 7,56,87; Animal Limited Evidence IMEMDT 30,319,83. NCI Carcinogenesis Bioassay (feed); Clear Evidence: rat NCITR* NCI-CG-TR-41,78. Cyanide and its compounds are on the Community Right-To-Know List. Reported in EPA TSCA Inventory. EPA Genetic Toxicology Program.

SAFETY PROFILE: Moderately toxic by intraperitoneal route. Mildly toxic by ingestion. Questionable carcinogen with experimental carcinogenic data. When heated to decomposition it emits very toxic fumes of Cl^-, NO_x, and CN^-. See also NITRILES.

TBR000 CAS:1335-88-2 ***HR: 3***
TETRACHLORONAPHTHALENE
mf: $C_{10}H_4Cl_4$ mw: 265.94

PROP: Crystals. Mp: 182°.

SYN: HALOWAX

CONSENSUS REPORTS: Reported in EPA TSCA Inventory.

OSHA PEL: TWA 2 mg/m^3
ACGIH TLV: TWA 2 mg/m^3

SAFETY PROFILE: Probably a poison. When heated to decomposition it emits highly toxic fumes of Cl^-. See also CHLORINATED HYDROCARBONS, AROMATIC; and POLYCHLORINATED BIPHENYLS.

TBR250 CAS:2438-88-2 ***HR: 3***
TETRACHLORONITROANISOLE
mf: $C_7H_3Cl_4NO_3$ mw: 290.91

SYNS: ENT 22,335 ◇ NCI-C03032 ◇ 4-NITRO-2,3,5,6-TETRACHLORANISOLE ◇ TCNA ◇ 1,2,4,5-TETRACHLORO-3-METHOXY-6-NITROBENZENE (9CI) ◇ 2,3,5,6-TETRACHLORO-4-NITROANISOLE

TOXICITY DATA with REFERENCE
orl-rat LD50:260 mg/kg IHFCAY 6,1,67

CONSENSUS REPORTS: NCI Carcinogenesis Bioassay (feed); No Evidence: mouse, rat NCITR* NCI-CG-TR-114,78.

SAFETY PROFILE: Poison by ingestion. When heated to decomposition it emits very toxic fumes of Cl^- and NO_x.

TBR500 CAS:3714-62-3 ***HR: 3***
2,3,4,6-TETRACHLORONITROBENZENE
mf: $C_6HCl_4NO_2$ mw: 260.88

SYNS: 4-NITRO-1,2,3,5-TETRACHLORO-BENZENE ◇ 1,2,3,5-TETRACHLORO-4-NITROBENZENE

TOXICITY DATA with REFERENCE
skn-mus TDLo:576 mg/kg/12W-I:NEO CNREA8 26,12,66

SAFETY PROFILE: Questionable carcinogen with experimental neoplastigenic data. When heated to decomposition it emits very toxic fumes of HCl and NO_x. See also NITRO COMPOUNDS of AROMATIC HYDROCARBONS.

TBR750 CAS:117-18-0 ***HR: 3***
2,3,5,6-TETRACHLORONITROBENZENE
mf: $C_6HCl_4NO_2$ mw: 260.88

PROP: Mp: 98-101°, bp: 304°.

SYNS: CHIPMAN 3,142 ◇ FOLOSAN ◇ FUSAREX ◇ 3-NITRO-1,2,4,5-TETRACHLOROBENZENE ◇ TCNB ◇ TECNAZEN (GERMAN) ◇ TECNAZENE ◇ 2,3,5,6-TETRACHLOR-3-NITROBENZOL (GERMAN) ◇ 1,2,4,5-TETRACHLORO-3-NITROBENZENE

TOXICITY DATA with REFERENCE
mmo-sat 3 μg/plate ENMUDM 5(Suppl 1),3,83
mma-sat 75 μg/plate ENMUDM 5(Suppl 1),3,83
mmo-asn 12 μmol/L PHYTAJ 66,217,76
sln-asn 24 μmol/L EVHPAZ 31,81,79
skn-mus TDLo:576 mg/kg/12W-I:NEO CNREA8 26,12,66
orl-rat LD50:7500 mg/kg 85DPAN -,-,71/76
unr-rat LD50:250 mg/kg 30ZDA9 -,82,71

SAFETY PROFILE: Poison by unspecified routes. Mildly toxic by ingestion. An irritant. Questionable carcinogen with experimental neoplastigenic data. Mutation data reported. When heated to decomposition it emits very toxic fumes of HCl and NO_x. Used as a pesticide. See also NITRO COMPOUNDS of AROMATIC HYDROCARBONS.

TBS000 CAS:879-39-0 ***HR: 3***
1,2,3,4-TETRACHLORO-5-NITROBENZENE
mf: $C_6HCl_4NO_2$ mw: 260.88

SYNS: DB-905 ◇ FOLOSAN DB-905 FUMITE ◇ FOLSAN ◇ FUSAREX ◇ TCBN ◇ 2,3,4,5-TETRACHLORONITROBENZENE

TOXICITY DATA with REFERENCE
skn-mus TDLo:576 mg/kg/12W-I:NEO CNREA8 26,12,66

SAFETY PROFILE: Questionable carcinogen with experimental neoplastigenic data. When heated to decomposition it emits very toxic fumes of Cl^- and NO_x. See also NITRO COMPOUNDS of AROMATIC HYDROCARBONS.

TBS250 CAS:25167-83-3 ***HR: 3***
TETRACHLOROPHENOL
mf: $C_6H_2Cl_4O$ mw: 231.88

PROP: Brown solid; phenol odor. Mp: 50°, d: 1.65 @ 60°.

TOXICITY DATA with REFERENCE
orl-rat LD50:140 mg/kg FEPRA7 2,76,43
scu-rat LD50:210 mg/kg FEPRA7 2,76,43

CONSENSUS REPORTS: Chlorophenol compounds are on the Community Right-To-Know List. Reported in EPA TSCA Inventory.

SAFETY PROFILE: Poison by ingestion and subcutaneous routes. When heated to decomposition it emits toxic fumes of Cl^-. See also CHLOROPHENOLS.

TBS500 CAS:4901-51-3 ***HR: 3***
2,3,4,5-TETRACHLOROPHENOL
mf: $C_6H_2Cl_4O$ mw: 231.88

PROP: Needles from ligroin. D: 1.6 @ 60°/4°, mp: 69-70°, bp: 164° at 23 mm. Very sltly sol in water; very sol in alc, ether.

TOXICITY DATA with REFERENCE
orl-mus LD50:400 mg/kg ARTODN 40,63,78
ipr-mus LD50:97 mg/kg ARTODN 40,63,78

CONSENSUS REPORTS: Chlorophenol compounds are on the Community Right-To-Know List.

SAFETY PROFILE: Poison by ingestion and intraperitoneal routes. When heated to decomposition it emits toxic fumes of Cl^-. See also CHLOROPHENOLS.

TBS750 CAS:935-95-5 ***HR: 3***
2,3,5,6-TETRACHLOROPHENOL
mf: $C_6H_2Cl_4O$ mw: 231.88

TOXICITY DATA with REFERENCE
orl-mus LD50:109 mg/kg ARTODN 40,63,78
ipr-mus LDLo:500 mg/kg CBCCT* 7,788,55

CONSENSUS REPORTS: Chlorophenol compounds are on the Community Right-To-Know List.

SAFETY PROFILE: Poison by ingestion. Moderately toxic by intraperitoneal route. When heated to decomposition it emits toxic fumes of Cl^-. See also CHLOROPHENOLS.

TBT000 CAS:58-90-2 ***HR: 3***
2,4,5,6-TETRACHLOROPHENOL
mf: $C_6H_2Cl_4O$ mw: 231.88

SYNS: DOWICIDE 6 ◇ RCRA WASTE NUMBER U212 ◇ 2,3,4,6-TETRACHLOROPHENOL

TOXICITY DATA with REFERENCE
msc-ham:lng 3500 μg/L CMSHAF 14,1617,85
orl-rat TDLo:300 mg/kg (6-15D preg):TER TXAPA9 28,146,74
scu-mus TDLo:100 mg/kg:CAR NTIS** PB223-159
orl-rat LD50:140 mg/kg IMSUAI 39,56,70
skn-rat LD50:485 mg/kg BECTA6 31,680,83
ipr-rat LD50:130 mg/kg BJPCAL 13,20,58
scu-rat LDLo:210 mg/kg HBTXAC 5,129,59
ipr-mus LDLo:250 mg/kg CBCCT* 4,231,52
skn-rbt LD50:250 mg/kg HBTXAC 5,129,59
orl-gpg LD50:250 mg/kg FMCHA2 -,D200,75

CONSENSUS REPORTS: IARC Cancer Review: Human Limited Evidence IMEMDT 41,319,86. Chlorophenol compounds are on the Community Right-To-Know List. Reported in EPA TSCA Inventory.

SAFETY PROFILE: Suspected carcinogen with experimental carcinogenic and teratogenic data. Poison by ingestion, skin contact, intraperitoneal, and subcutaneous routes. Mutation data reported. When heated to decomposition it emits toxic fumes of Cl^-. Used as a disinfectant and a preservative for wood, latex, and leather. See also CHLOROPHENOLS.

TBT200 CAS:1953-99-7 ***HR: 3***
TETRACHLOROPHTHALONITRILE
mf: $C_8Cl_4N_2$ mw: 265.90

SYNS: o-TCPN ◇ o-TETRACHLOROPHTHALODINITRILE ◇ 3,4,5,6-TETRACHLOROPHTHALONITRILE

TOXICITY DATA with REFERENCE
orl-mus LD50:13720 mg/kg INHEAO 4,11,66
ipr-mus LD50:66 mg/kg INHEAO 4,11,66
scu-mus LD50:3500 mg/kg INHEAO 4,11,66

CONSENSUS REPORTS: Cyanide and its compounds are on the Community Right-To-Know List.

SAFETY PROFILE: Poison by intraperitoneal route. Moderately toxic by subcutaneous route. Mildly toxic by ingestion. When heated to decomposition it emits toxic fumes of NO_x, Cl^-, and CN^-. See also NITRILES.

TBT250 CAS:1070-78-6 ***HR: 2***
1,1,1,3-TETRACHLOROPROPANE
mf: $C_3H_4Cl_4$ mw: 181.87

TOXICITY DATA with REFERENCE
orl-rat LDLo:1600 mg/kg GISAAA 27,3,62

CONSENSUS REPORTS: Reported in EPA TSCA Inventory.

SAFETY PROFILE: Moderately toxic by ingestion. When heated to decomposition it emits toxic fumes of

Cl^-. See also CHLORINATED HYDROCARBONS, ALIPHATIC.

TBT500 CAS:10436-39-2 ***HR: 3***
1,1,2,3-TETRACHLOROPROPENE
mf: $C_3H_2Cl_4$ mw: 179.85

TOXICITY DATA with REFERENCE
mmo-sat 10 µg/plate 37KXA7 3,865,80
mmo-esc 500 µmol/L CNJGA8 23,17,81
mmo-smc 200 µg/L MUREAV 119,273,83
cyt-ham:ovr 100 µmol/L CNJGA8 23,17,81
orl-rat LD50:350 mg/kg AIHAAP 30,470,69
ihl-rat LCLo:250 ppm/4H AIHAAP 30,470,69
orl-mus LD50:800 mg/kg GISAAA 43(4),15,78
skn-rbt LD50:400 mg/kg AIHAAP 30,470,69

CONSENSUS REPORTS: Reported in EPA TSCA Inventory.

SAFETY PROFILE: Poison by ingestion and skin contact. Moderately toxic by inhalation. Mutation data reported. When heated to decomposition it emits toxic fumes of Cl^-. See also CHLORINATED HYDROCARBONS, ALIPHATIC.

TBT750 CAS:2808-86-8 ***HR: 2***
2,3,4,5-TETRACHLOROPYRIDINE
mf: C_5HCl_4N mw: 216.87

TOXICITY DATA with REFERENCE
ipr-mus LD50:425 mg/kg TXAPA9 11,361,67

CONSENSUS REPORTS: Reported in EPA TSCA Inventory.

SAFETY PROFILE: Moderately toxic by intraperitoneal route. When heated to decomposition it emits very toxic fumes of Cl^- and NO_x.

TBU000 CAS:2402-79-1 ***HR: 2***
2,3,5,6-TETRACHLOROPYRIDINE
mf: C_5HCl_4N mw: 216.87

TOXICITY DATA with REFERENCE
ipr-mus LD50:1150 mg/kg TXAPA9 11,361,67

CONSENSUS REPORTS: Reported in EPA TSCA Inventory.

SAFETY PROFILE: Moderately toxic by intraperitoneal route. When heated to decomposition it emits very toxic fumes of Cl^- and NO_x.

TBU250 CAS:1780-40-1 ***HR: 3***
2,4,5,6-TETRACHLOROPYRIMIDINE
mf: $C_4Cl_4N_2$ mw: 217.86

PROP: Mp: 68-70°.

SYNS: PERCHLOROPYRIMIDINE ◇ TETRACHLOROPYRIMIDINE

TOXICITY DATA with REFERENCE
ivn-mus LD50:32 mg/kg CSLNX* NX#07768

CONSENSUS REPORTS: Reported in EPA TSCA Inventory.

SAFETY PROFILE: Poison by intravenous route. A lachrymator and an irritant. When heated to decomposition it emits very toxic fumes of Cl^- and NO_x.

TBU500 CAS:1198-55-6 ***HR: 3***
TETRACHLOROPYROCATECHOL
mf: $C_6H_2Cl_4O_2$ mw: 247.88

SYNS: 3,4,5,6-TETRACHLORO-1,2-BENZENEDIOL ◇ TETRACHLOROCATECHOL

TOXICITY DATA with REFERENCE
orl-mus LD50:612 mg/kg ARTODN 40,63,78
ipr-mus LD50:136 mg/kg ARTODN 40,63,78

CONSENSUS REPORTS: Reported in EPA TSCA Inventory.

SAFETY PROFILE: Poison by intraperitoneal route. Moderately toxic by ingestion. When heated to decomposition it emits toxic fumes of Cl^-.

TBU750 ***HR: 2***
TETRACHLOROPYROCATECHOL HYDRATE
mf: $C_6H_2Cl_4O_2 \cdot H_2O$ mw: 265.90

SYN: TETRACHLORPYROKATECHIN (CZECH)

TOXICITY DATA with REFERENCE
skn-rbt 500 mg/24H MOD 28ZPAK -,81,72
eye-rbt 50 µg/24H SEV 28ZPAK -,81,72
orl-rat LD50:758 mg/kg 28ZPAK -,81,72

CONSENSUS REPORTS: Reported in EPA TSCA Inventory.

SAFETY PROFILE: Moderately toxic by ingestion. A skin and severe eye irritant. When heated to decomposition it emits toxic fumes of Cl^-.

TBV000 CAS:1154-59-2 ***HR: 3***
3,3',4',5-TETRACHLOROSALICYLANILIDE
mf: $C_{13}H_7Cl_4NO_2$ mw: 351.01

SYN: TCSA

TOXICITY DATA with REFERENCE
skn-rbt 150 µg open MLD JSCCA5 25,113,74
skn-gpg 150 µg open MLD JSCCA5 25,113,74
orl-rat LD50:243 mg/kg PCOC** -,618,66

CONSENSUS REPORTS: Reported in EPA TSCA Inventory.

SAFETY PROFILE: Poison by ingestion. A skin irritant. When heated to decomposition it emits very toxic fumes of Cl^- and NO_x.

TBV250 CAS:1861-32-1 ***HR: 3***
TETRACHLOROTEREPHTHALIC ACID DIMETHYL ESTER
mf: $C_{10}H_6Cl_4O_4$ mw: 331.96

SYNS: CHLORTHAL-DIMETHYL ◇ CHLORTHAL-METHYL ◇ DACTHAL ◇ DIMETHYL TETRACHLOROTEREPHTHALATE ◇ DIMETHYL 2,3,5,6-TETRACHLOROTEREPHTHALATE ◇ FATAL ◇ 2,3,5,6-TETRACHLORO-1,4-BENZENEDICARBOXYLIC ACID, DIMETHYL ESTER ◇ 2,3,5,6-TETRACHLORPHTHALSAURE-DIMETHYLESTER (GERMAN)

TOXICITY DATA with REFERENCE
ivn-mus LD50:320 mg/kg CSLNX* NX#02634
unr-mus LD50:3500 mg/kg TGANAK 16(1),45,82

SAFETY PROFILE: Poison by intravenous route. Moderately toxic by an unspecified route. When heated to decomposition it emits toxic fumes of Cl^-. Used as a pesticide. See also ESTERS.

TBV750 CAS:6012-97-1 ***HR: 3***
2,3,4,5-TETRACHLOROTHIOPHENE
mf: C_4Cl_4S mw: 221.90

PROP: Mp: 28-30°, bp: 75°/2 mm, refr index: 1.5910, d: 1.704, flash p: 230°F.

SYNS: 2,3,4,5-CHLOROTHIOPHENE ◇ ENT 25,764 ◇ IF (fumigant) ◇ PENN SALT TD-183 ◇ PENPHENE ◇ PERCHLOROTHIOPHENE ◇ TCTP ◇ TD-183 ◇ TETRACHLOROTHIOFENE ◇ TETRACHLOR-OTHIOPHENE ◇ TETRACHLOROTHIOPHENE

TOXICITY DATA with REFERENCE
orl-rat LD50:70 mg/kg ARSIM* 20,17,66
ihl-rat LC50:146 ppm 28ZEAL 5,219,76
ipr-mus LDLo:64 mg/kg CBCCI* 1,47,49
ivn-mus LD50:75 mg/kg CSLNX* ND#00063
skn-rbt LD50:256 mg/kg MRPCA2 9,119,70

SAFETY PROFILE: Poison by ingestion, skin contact, intravenous, and intraperitoneal routes. Moderately toxic by inhalation. Combustible when exposed to heat or flame. When heated to decomposition it emits very toxic fumes of Cl^- and SO_x.

TBW000 CAS:2338-29-6 ***HR: 3***
4,5,6,7-TETRACHLORO-2-(TRIFLUOROMETHYL)BENZIMIDAZOLE
mf: $C_8HCl_4F_3N_2$ mw: 323.91

SYN: TTFB

TOXICITY DATA with REFERENCE
orl-rat LD50:245 μg/kg PSSCBG 15,31,84
ipr-mus LD50:23 mg/kg BCPCA6 18,1389,69
ivn-mus LD50:13 mg/kg CSLNX* NX#07537

SAFETY PROFILE: Poison by ingestion, intraperitoneal, and intravenous routes. When heated to decomposition it emits very toxic fumes of F^-, Cl^-, and NO_x.

TBW025 CAS:1066-48-4 ***HR: 3***
TETRACHLOROTRIFLUOROMETHYLPHOSPHORANE
mf: CCl_4F_3P mw: 241.79

SAFETY PROFILE: Explosive reaction with tetramethyllead. When heated to decomposition it emits toxic fumes of Cl^-, F^-, and PO_x. See also CHLORIDES and FLUORIDES.

TBW100 CAS:961-11-5 ***HR: 3***
TETRACHLORVINPHOS
mf: $C_{10}H_9Cl_4O_4P$ mw: 365.96

SYNS: 2-CHLORO-1-(2,4,5-TRICHLOROPHENYL)VINYLDIMETHYL PHOSPHATE ◇ 2-CHLORO-1-(2,4,5-TRICHLOROPHENYL(VINYL PHOSPHORIC ACID DIMETHYL ESTER ◇ O,O-DIMETHYL-O-2-CHLOR-1-(2,4,5-TRICHLORPHENYL)-VINYL-PHOSPHAT(GERMAN) ◇ IPO 8 ◇ NCI C00168 ◇ PHOSPHORIC ACID, 2-CHLORO-1-(2,4,5-TRICHLOROPHENYL)ETHENYL DIMETHYL ESTER ◇ 2,4,5-TRICHLORO-α-(CHLOROMETHYLENE)BENZYL PHOSPHATE

TOXICITY DATA with REFERENCE
orl-mus TDLo:692 g/kg (2Y male):REP FAATDF 5,840,85
orl-rat TDLo:240 g/kg,80W-C:NEO NCITR* NCI-CG-TR-33,78
orl-mus TDLo:450 g/kg/67W-C:CAR NCITR* NCI-CG-TR-33,78
orl-mus TD:1057 g/kg/80W-C:CAR NCITR* NCI-CG-TR-33,78
orl-rat TD:120 g/kg/80W-C:ETA NCITR* NCI-CG-TR-33,78
orl-rat LD50:4 g/kg 85GYAZ -,34,71
orl-mus LD50:4200 mg/kg SAIGBL 16,523,74
ipr-mus LD50:1170 mg/kg SAIGBL 16,523,74
orl-bwd LD50: 100 mg/kg AECTCV 12,355,83

CONSENSUS REPORTS: NCI Carcinogenesis Bioassay (feed); Results Positive: Mouse, Rat NCITR* NCI-CG-TR-33,78. Community Right-To-Know List.

SAFETY PROFILE: Suspected carcinogen with experimental carcinogenic, neoplastigenic, and tumorigenic data. Poison by ingestion. Moderately toxic by ipr route. Experimental reproductive effects. When heated to decomposition it emits toxic fumes of Cl^- and PO_x.

TBW250 CAS:14323-41-2 ***HR: 3***
TETRACYANONICKELATE(2−) DIPOTASSIUM, HYDRATE
mf: $C_4N_4Ni•K•H_2O$ mw: 219.91

SYN: POTASSIUM CYANONICKELATE HYDRATE

TOXICITY DATA with REFERENCE
cyt-mus:mmr 1600 μmol/L MUREAV 68,337,79

CONSENSUS REPORTS: Nickel and its compounds, as well as cyanide and its compounds, are on the Community Right-To-Know List.

OSHA PEL: (Transitional: TWA 1 mg/m^3) TWA 0.1 mg (Ni)/m^3
ACGIH TLV: TWA 0.1 mg(Ni)/m^3; (Proposed: TWA 0.05 mg(Ni)/m^3; Human Carcinogen)

SAFETY PROFILE: Suspected human carcinogen. Mutation data reported. Many nickel compounds are poisons. When heated to decomposition it emits very toxic fumes of CN^-, K_2O, and NO_x. See also NICKEL COMPOUNDS and CYANIDES.

TBW500 ***HR: 3***
TETRACYANOOCTAETHYLTETRAGOLD
mf: $C_{20}H_{40}Au_4N_4$ mw: 1124.45

CONSENSUS REPORTS: Cyanide and its compounds are on the Community Right-To-Know List.

SAFETY PROFILE: An explosive sensitive to friction and heating to 80°C. The tetramer of cyanodiethylgold. When heated to decomposition it emits very toxic fumes of NO_x and CN^-. See also GOLD and CYANIDE.

TBW750 CAS:1518-16-7 ***HR: 3***
TETRACYANOQUINODIMETHAN
mf: $C_{12}H_4N_4$ mw: 204.20

TOXICITY DATA with REFERENCE
ivn-mus LD50:56 mg/kg CSLNX* NX#00796

CONSENSUS REPORTS: Cyanide and its compounds are on the Community Right-To-Know List. Reported in EPA TSCA Inventory.

SAFETY PROFILE: Poison by intravenous route. When heated to decomposition it emits toxic fumes of NO_x and CN^-. See also CYANIDE.

TBX000 CAS:60-54-8 ***HR: 3***
TETRACYCLINE
mf: $C_{22}H_{24}N_2O_8$ mw: 444.48

PROP: Produced by *Streptomyces albo-niger*. Trihydrate: crystals. Decomp @ 170-175°.

SYNS: ABRAMYCIN ◇ ABRICYCLINE ◇ ACHROMYCIN ◇ AGROMICINA ◇ AMBRAMICINA ◇ AMBRAMYCIN ◇ BIO-TETRA ◇ BRISTACICLIN α ◇ BRISTACYCLINE ◇ CEFRACYCLINE SUSPENSION ◇ CRISEOCICLINE ◇ CYCLOMYCIN ◇ DEMOCRACIN ◇ DESCHLOROBIOMYCIN ◇ HOSTACYCLIN ◇ LIQUAMYCIN ◇ 6-METHYL-1,11-DIOXY-2-NAPHTHACENECARBOXAMIDE ◇ NEOCYCLINE ◇ OLETETRIN ◇ ωMYCIN ◇ PANMYCIN ◇ POLYCYCLINE ◇ PUROCYCLINA ◇ ROBITET ◇ SANCLOMYCINE ◇ SIGMAMYCIN ◇ SK-TETRACYCLINE ◇ STECLIN ◇ T-125 ◇ TETRABON ◇ TETRACYCLINE I ◇ TETRACYN ◇ TETRADECIN ◇ TETRAVERINE ◇ TSIKLOMITSIN

TOXICITY DATA with REFERENCE
dnr-esc 20 μL/disc MUREAV 97,1,82
mrc-smc 3 g/L MUREAV 12,357,71
cyt-grh-orl 5 mg PISCAD 58(Pt. 3),623,71
unr-wmn TDLo:200 mg/kg (34-35W preg):REP BJDEAZ 76,374,64
ihl-rat TCLo:100 μg/m^3/24H (female 1-20D post):TER ANTBAL 20,839,75
orl-wmn TDLo:600 mg/kg/15D:CNS,GIT,KID SMJOAV 71,961,78
mul-wmn LDLo:310 mg/kg NEJMAG 270,157,64
orl-rat LD50:807 mg/kg TXAPA9 18,185,71
ipr-rat LD50:310 mg/kg NIIRDN 6,493,82
ivn-rat LD50:129 mg/kg ANTCAO 4,411,54
orl-mus LD50:678 mg/kg AJTHAB 2,254,53
ipr-mus LD50:125 mg/kg TDKNAF 14,60,55
scu-mus LD50:400 mg/kg 85GDA2 3,42,80
ivn-mus LD50:157 mg/kg FRPSAX 10,346,55
orl-gpg LD50:1875 mg/kg 85GMAT -,109,82

CONSENSUS REPORTS: EPA Genetic Toxicology Program.

SAFETY PROFILE: Human poison by multiple routes. Experimental poison by intraperitoneal, intravenous, and subcutaneous routes. Moderately toxic by ingestion. Human systemic effects by ingestion: somnolence, decreased motility or constipation and urine volume decrease or anuria. Human reproductive effects by unspecified routes: effects on newborn including postnatal measures or effects and delayed effects on newborn. An experimental teratogen. Experimental reproductive effects. Mutation data reported. When heated to decomposition it emits toxic fumes of NO_x.

TBX250 CAS:64-75-5 ***HR: 3***
TETRACYCLINE HYDROCHLORIDE
mf: $C_{22}H_{24}N_2O_8 \cdot ClH$ mw: 480.94

PROP: Very sol in water; sol in methanol, ethanol; insol in ether, hydrocarbon solvents.

SYNS: ACHROMYCIN ◇ ACHROMYCIN HYDROCHLORIDE ◇ AMBRACYN ◇ ARTOMYCIN ◇ BRISTACYCLINE ◇ CEFRACYCLINE TABLETS ◇ CHLORHYDRATE de TETRACYCLINE (FRENCH) ◇ CYCLOPAR ◇ DIACYCINE ◇ DUMOCYCIN ◇ MEDAMYCIN ◇ MEPHACYCLIN ◇ NCI-C55561 ◇ PALTET ◇ PANMYCIN HYDROCHLORIDE ◇ PARTREX ◇ PIRACAPS ◇ POLYCYCLINE HYDROCHLORIDE ◇ QIDTET ◇ QUADRACYCLINE ◇ QUATREX ◇ REMICYCLIN ◇ RICYCLINE ◇ RO-CYCLINE ◇ SK-TETRACYCLINE ◇ STECLIN HYDROCHLORIDE ◇ SUBAMYCIN ◇ SUPRAMYCIN ◇ T-250 CAPSULES ◇ TC HYDROCHLORIDE ◇ TEFILIN ◇ TELINE ◇ TELOTREX ◇ TETRABAKAT ◇ TETRABLET ◇ TETRACAPS ◇ TETRACICLINA CLORIDRATO (ITALIAN) ◇ TETRACOMPREN ◇ TETRACYCLINE CHLORIDE ◇ TETRA-D ◇ TETRALUTION ◇ TETRA-WEDEL ◇ TETROSOL ◇ TOPICYCLINE ◇ TOTOMYCIN

◇ TRIPHACYCLIN ◇ U-5965 ◇ UNICIN ◇ UNIMYCIN ◇ VETQUAMYCIN-324

TOXICITY DATA with REFERENCE
dnd-esc 50 μmol/L MUREAV 89,95,81
msc-mus:mmr 100 mg/L MUREAV 40,261,76
scu-mus TDLo:2250 mg/kg (female 10-18D post):REP CRSBAW 161,300,67
ipr-mus TDLo:900 mg/kg (female 8-13D post):TER IYKEDH 3,75,72
orl-man TDLo:200 mg/kg/7D-I:SYS CPHADV 4,455,85
orl-rat LD50:6443 mg/kg TXAPA9 18,185,71
ipr-rat LD50:318 mg/kg GNRIDX 2,26,68
scu-rat LD50:700 mg/kg TXAPA9 9,445,66
ivn-rat LD50:128 mg/kg FRPSAX 10,197,55
orl-mus LD50:2759 mg/kg JPPMAB 16,33,64
ipr-mus LD50:337 mg/kg GNRIDX 2,26,68
ivn-mus LD50:157 mg/kg FRPSAX 10,197,55

CONSENSUS REPORTS: Reported in EPA TSCA Inventory.

SAFETY PROFILE: Poison by intraperitoneal and intravenous routes. Moderately toxic by ingestion and subcutaneous routes. Human systemic effects: change in taste function. An experimental teratogen. Experimental reproductive effects. Mutation data reported. When heated to decomposition it emits very toxic fumes of HCl and NO_x. See also TETRACYCLINE.

TBX500 CAS:124-25-4 ***HR: 1***
1-TETRADECANAL
mf: $C_{14}H_{28}O$ mw: 212.42

PROP: Found in several essential oils (FCTXAV 11,477,73).

SYNS: MYRISTIC ALDEHYDE ◇ 1-TETRADECYL ALDEHYDE

TOXICITY DATA with REFERENCE
skn-rbt 500 mg MOD FCTXAV 11,1079,73

CONSENSUS REPORTS: Reported in EPA TSCA Inventory.

SAFETY PROFILE: A skin irritant. When heated to decomposition it emits acrid smoke and irritating fumes. See also ALDEHYDES.

TBX750 CAS:629-59-4 ***HR: 3***
TETRADECANE
mf: $C_{14}H_{30}$ mw: 198.44

PROP: Colorless liquid. D: 0.765 @ 20/4°, mp: 5.5°, bp: 252-255°, lel: 0.5%, flash p: 212°F, vap press: 1 mm @ 76.4°, vap d: 6.83, autoign temp: 396°F. Insol in water; very sol in alc and ether.

TOXICITY DATA with REFERENCE
skn-mus TDLo:9600 mg/kg/20W-I:ETA TXAPA9 9,70,66

SAFETY PROFILE: Probably irritating and narcotic in high concentrations. Questionable carcinogen with experimental tumorigenic data. Combustible when exposed to heat or flame. Moderate explosion hazard in the form of vapor when exposed to heat or flame. Can react with oxidizing materials. To fight fire, use foam, CO_2, dry chemical. When heated to decomposition it emits acrid smoke and irritating fumes.

TBY250 CAS:112-72-1 ***HR: 3***
1-TETRADECANOL
mf: $C_{14}H_{30}O$ mw: 214.44

PROP: Opaque leaflets. Mp: 37.62°, bp: 264.1°, flash p: 285°F (OC), d: (solid) 0.8355 @ 20/20°, d: (liquid) 0.8236 @ 38°/4°, vap press: 0.01 mm @ 20°, vap d: 7.39.

SYNS: MYRISIIC ALCOHOL ◇ N-TETRADECANOL-1 ◇ TETRADECYL ALCOHOL ◇ N-TETRADECYL ALCOHOL

TOXICITY DATA with REFERENCE
skn-hmn 75 mg/3D-I MOD 85DKA8 -,127,77
skn-mus TDLo:12 g/kg/24W-I:ETA TXAPA9 9,70,66

CONSENSUS REPORTS: Reported in EPA TSCA Inventory.

SAFETY PROFILE: Questionable carcinogen with experimental tumorigenic data. A human skin irritant. Combustible when exposed to heat or flame; can react with oxidizing materials. To fight fire, use CO_2, dry chemical. When heated to decomposition it emits acrid smoke and irritating fumes. See also ALCOHOLS and TETRADECANOL, mixed isomers.

TBY500 CAS:27196-00-5 ***HR: 1***
TETRADECANOL, mixed isomers
mf: $C_{14}H_{30}O$ mw: 214.44

SYNS: MYRISTYL ALCOHOL (mixed isomers) ◇ TETRADECYL ALCOHOL

TOXICITY DATA with REFERENCE
orl-rat LD50:33 g/kg AIHAAP 30,470,69
skn-rbt LD50:7130 mg/kg AIHAAP 30,470,69

CONSENSUS REPORTS: Reported in EPA TSCA Inventory.

SAFETY PROFILE: Mildly toxic by ingestion and skin contact. Combustible when exposed to heat or flame; can react with oxidizing materials. To fight fire, use CO_2, dry chemical. When heated to decomposition it emits acrid smoke and irritating fumes. See also 1-TETRADECANOL.

TBY750 CAS:27306-79-2 ***HR: 1***
TETRADECANOL condensed with 7 moles ETHYLENE OXIDE

SYNS: ALKYL(C-14) POLYETHOXYLATES (ETHOXY-7) ◇ POLYOXYETHYLENE (7) ALKYL (14) ETHER

TOXICITY DATA with REFERENCE
skn-rbt 800 mg/4W MLD FCTXAV 15,319,77

CONSENSUS REPORTS: Reported in EPA TSCA Inventory.

SAFETY PROFILE: A skin irritant. When heated to decomposition it emits acrid smoke and irritating fumes. See also ETHYLENE OXIDE and TETRADECANOL.

TCA250 CAS:64604-09-7 ***HR: 3***
12-o-TETRADECA-2-cis-4-trans,6,8-TETRAENOYLPHORBOL-13-ACETATE
mf: $C_{36}H_{48}O_8$ mw: 608.84

SYN: TI-8

TOXICITY DATA with REFERENCE
dns-mus-skn 400 nmol/kg RCOCB8 24,533,79
skn-mus TDLo:19 mg/kg/48W-I:ETA CNREA8 39,4183,79

SAFETY PROFILE: Questionable carcinogen with experimental tumorigenic by skin contact. Mutation data reported. When heated to decomposition it emits acrid smoke and irritating fumes.

TCA500 CAS:139-08-2 ***HR: 1***
TETRADECYL DIMETHYL BENZYLAMMONIUM CHLORIDE
mf: $C_{23}H_{42}N{\bullet}Cl$ mw: 368.11

SYNS: ARQUAD DM14B-90 ◇ N,N-DIMETHYL-N-TETRADECYLBENZENEMETHANAMINIUM, CHLORIDE (9CI) ◇ NISSAN CATION M2-100

TOXICITY DATA with REFERENCE
skn-rbt 1 mg/24H OYYAA2 6(2),329,72
eye-rbt 1 mg OYYAA2 6(2),329,72

CONSENSUS REPORTS: Reported in EPA TSCA Inventory.

SAFETY PROFILE: A skin and eye irritant. When heated to decomposition it emits very toxic fumes of NO_x, NH_3, and Cl^-.

TCA750 CAS:40036-79-1 ***HR: 3***
TETRADECYLHEPTAETHOYLATE

TOXICITY DATA with REFERENCE
skn-hmn 140 mg/12H-I MLD FCTXAV 15,309,77
eye-mky 100% FCTXAV 15,309,77
skn-rbt 50 mg/24H MOD FCTXAV 15,309,77
eye-rbt 10 mg MOD FCTXAV 15,309,77
orl-rat LD50:2600 mg/kg FCTXAV 15,309,77
ipr-rat LD50:209 mg/kg FCTXAV 15,309,77
ipr-mus LDLo:200 mg/kg FCTXAV 15,309,77
orl-mky LD50:10 g/kg FCTXAV 15,309,77
ipr-mky LDLo:300 mg/kg FCTXAV 15,309,77
skn-rbt LD50:2 g/kg FCTXAV 15,309,77
ipr-gpg LD50:100 mg/kg FCTXAV 15,309,77

SAFETY PROFILE: Poison by intraperitoneal route. Moderately toxic by ingestion and skin contact. A human skin irritant. An experimental skin and eye irritant. When heated to decomposition it emits acrid smoke and irritating fumes.

TCB000 CAS:4671-75-4 ***HR: 2***
TETRADECYL PHOSPHONIC ACID
mf: $C_{14}H_{31}O_3P$ mw: 278.42

PROP: Crystals.

TOXICITY DATA with REFERENCE
orl-mus LD50:2100 mg/kg AMIHAB 11,487,55

CONSENSUS REPORTS: Reported in EPA TSCA Inventory.

SAFETY PROFILE: Moderately toxic by ingestion. When heated to decomposition it emits toxic fumes of PO_x.

TCB100 CAS:17661-50-6 ***HR: 1***
TETRADECYL STEARATE
mf: $C_{32}H_{64}O_2$ mw: 480.96

SYNS: MYRISTYL STEARATE ◇ OCTADECANOIC ACID, TETRADECYL ESTER (9CI) ◇ STEARIC ACID, TETRADECYL ESTER ◇ TETRADECYL OCTADECANOATE

TOXICITY DATA with REFERENCE
eye-rbt 100 mg MLD JACTDZ 4(5),107,85

CONSENSUS REPORTS: Reported in EPA TSCA Inventory.

SAFETY PROFILE: An eye irritant. When heated to decomposition it emits acrid smoke and irritating fumes.

TCB500 CAS:631-41-4 ***HR: 2***
TETRAETHANOL AMMONIUM HYDROXIDE
mf: $C_8H_{20}NO_4{\bullet}HO$ mw: 211.30

PROP: Crystals. Mp: 123°, vap press: < 0.01 mm @ 20°, vap d: 7.28. Misc in water.

TOXICITY DATA with REFERENCE
eye-rbt 2 mg SEV AJOPAA 29,1363,46
orl-rat LD50:2250 mg/kg JIHTAB 23,259,41
orl-gpg LD50:3510 mg/kg JIHTAB 23,259,41

CONSENSUS REPORTS: Reported in EPA TSCA Inventory.

SAFETY PROFILE: Moderately toxic by ingestion. A

severe eye irritant. When heated to decomposition it emits toxic fumes of NO_x and NH_3.

TCB725 CAS:66-40-0 ***HR: 3***
TETRAETHYLAMMONIUM
mf: $C_8H_{20}N$ mw: 130.29

SYNS: TEA ◇ TETRAETILAMMONIO (ITALIAN) ◇ TETRAMON ◇ TETRYLAMMONIUM ◇ N,N,N-TRIETHYL-ETHANAMINIUM (9CI)

TOXICITY DATA with REFERENCE
ipr-rat LD50:115 mg/kg FRPSAX 10,1027,55
ivn-rat LD50:63 mg/kg FRPSAX 10,1027,55
ipr-mus LD50:60 mg/kg FRPSAX 10,1027,55
ivn-mus LD50:36 mg/kg JMPCAS 3,167,61
ivn-dog LD50:55 mg/kg FRPSAX 10,1027,55
ivn-rbt LD50:72 mg/kg FRPSAX 10,1027,55

SAFETY PROFILE: Poison by intravenous and intraperitoneal routes. When heated to decomposition it emits toxic fumes of NO_x and NH_3.

TCB750 CAS:17083-85-1 ***HR: 3***
TETRAETHYLAMMONIUM BOROHYDRIDE
mf: $C_8H_{20}N \cdot BH_4$ mw: 145.14

TOXICITY DATA with REFERENCE
ivn-mus LD50:32 mg/kg CSLNX* NX#03076

CONSENSUS REPORTS: Reported in EPA TSCA Inventory.

SAFETY PROFILE: Poison by intravenous route. When heated to decomposition it emits toxic fumes of NO_x and NH_3. See also HYDRIDES and BORON COMPOUNDS.

TCC000 CAS:71-91-0 ***HR: 3***
TETRAETHYL AMMONIUM BROMIDE
mf: $C_8H_{20}N \cdot Br$ mw: 210.20

PROP: Crystals in abs alc. Mp: 285-290° (decomp). Very sol in alc; sol in chloroform.

SYNS: BEPARON ◇ ETAMBRO ◇ ETYLON ◇ SYMPATEKTOMAN ◇ TEA ◇ TEAB ◇ TETRYLAMMONIUM BROMIDE ◇ USAF DO-32

TOXICITY DATA with REFERENCE
dnd-mam:lym 50 mmol/L CBINA8 19,197,77
ipr-rat LD50:108 mg/kg PHMGEN 21,256,80
ipr-mus LD50:50 mg/kg NTIS** AD277-689
ivn-mus LD50:31 mg/kg AIPTAK 105,221,56

CONSENSUS REPORTS: Reported in EPA TSCA Inventory.

SAFETY PROFILE: Poison by intraperitoneal and intravenous routes. Mutation data reported. When heated to decomposition it emits very toxic fumes of NO_x, NH_3, and Br^-. See also BROMIDES.

TCC250 CAS:56-34-8 ***HR: 3***
TETRAETHYLAMMONIUM CHLORIDE
mf: $C_8H_{20}N \cdot Cl$ mw: 165.74

SYNS: ETAMON CHLORIDE ◇ TEAC ◇ TEA CHLORIDE ◇ N,N,N-TRIETHYLETHANAMINIUM CHLORIDE

TOXICITY DATA with REFERENCE
ivn-hmn TDLo:5 mg/kg:PNS,EYE AJMSA9 213,572,47
orl-rat LD50:2630 mg/kg JPETAB 92,103,48
ivn-rat LD50:56 mg/kg JPETAB 92,103,48
ims-rat LD50:110 mg/kg JPETAB 92,103,48
orl-mus LD50:833 mg/kg JPETAB 117,169,56
ipr-mus LD50:65 mg/kg JPETAB 92,103,48
scu-mus LDLo:120 mg/kg JPETAB 25,315,25
ivn-mus LD50:37 mg/kg JPETAB 117,169,56
ivn-dog LD50:36 mg/kg JPETAB 92,103,48
ims-dog LD50:58 mg/kg JPETAB 92,103,48
scu-frg LDLo:3 mg/kg SAPHAO 10,201,1900

CONSENSUS REPORTS: Reported in EPA TSCA Inventory.

SAFETY PROFILE: Poison by intravenous, intramuscular, intraperitoneal, and subcutaneous routes. Moderately toxic by ingestion. Human systemic effects by intravenous route: paresthesia, pupillary dilation and ptosis. When heated to decomposition it emits very toxic fumes of NO_x, Cl^-, and NH_3.

TCC500 CAS:77-98-5 ***HR: 3***
TETRAETHYLAMMONIUM HYDROXIDE
mf: $C_8H_{20}N \cdot HO$ mw: 147.30

PROP: Only in solution (crystals + $4H_2O$). Mp: 40-50°, bp: (decomp), refr index: 1.4164, d: 1.023, flash p: none. Sol in water.

TOXICITY DATA with REFERENCE
scu-mus LDLo:107 mg/kg JPETAB 28,367,26

CONSENSUS REPORTS: Reported in EPA TSCA Inventory.

SAFETY PROFILE: Poison by subcutaneous route. Corrosive. When heated to decomposition it emits toxic fumes of NO_x and NH_3.

TCC750 CAS:68-05-3 ***HR: 3***
TETRAETHYLAMMONIUM IODIDE
mf: $C_8H_{20}N \cdot I$ mw: 257.19

PROP: Colorless crystals from water. D: 1.559 @ 4°, mp: >200°. Sol in water, ethyl, alc, chloroform; insol in ether.

SYN: TETAMON IODIDE

TOXICITY DATA with REFERENCE
ivn-mus LD50:56 mg/kg CSLNX* NX#00818

CONSENSUS REPORTS: Reported in EPA TSCA Inventory.

SAFETY PROFILE: Poison by intravenous route. When heated to decomposition it emits very toxic fumes of NO_x, NH_3, and I^-. See also IODIDES.

TCD000 CAS:2537-36-2 ***HR: 3***
TETRAETHYLAMMONIUM PERCHLORATE
mf: $C_8H_{20}N{\bullet}ClO_4$ mw: 229.74

TOXICITY DATA with REFERENCE
ivn-mus LD50:56 mg/kg CSLNX* NX#01268

CONSENSUS REPORTS: Reported in EPA TSCA Inventory.

SAFETY PROFILE: Poison by intravenous route. When heated to decomposition it emits very toxic fumes of NO_x, NH_3, and Cl^-. See also PERCHLORATES.

TCD250 CAS:78-13-7 ***HR: 1***
TETRA(2-ETHYLBUTYL) ORTHOSILICATE
mf: $C_{24}H_{52}O_4Si$ mw: 432.85

PROP: D: 0.8920-0.9018 @ 20/20°, mp: < −100°, bp: 238° @ 50 mm, flash p: 335°F (OC). Insol in water; sltly sol in methanol; misc with most organic solvents.

SYNS: 2-ETHYL-1-BUTANOL, SILICATE ◇ TETRA(2-ETHYLBUTOXY)SILANE

TOXICITY DATA with REFERENCE
skn-rbt 500 mg open MLD UCDS** 2/15/66
orl-rat LD50:20 g/kg UCDS** 2/15/66

CONSENSUS REPORTS: Reported in EPA TSCA Inventory.

SAFETY PROFILE: Mildly toxic by ingestion. A skin irritant. Combustible when exposed to heat or flame. To fight fire, use mist, spray, dry chemical. When heated to decomposition it emits acrid smoke and irritating fumes. See also SILICATES and SILANE.

TCD500 CAS:612-08-8 ***HR: 3***
TETRAETHYLDIARSANE
mf: $C_8H_{20}As_2$ mw: 266.09

CONSENSUS REPORTS: Arsenic and its compounds are on the Community Right-To-Know List.

SAFETY PROFILE: Ignites spontaneously in air. When heated to decomposition it emits toxic fumes of As. See also ARSENIC COMPOUNDS.

TCE250 CAS:112-60-7 ***HR: 1***
TETRAETHYLENE GLYCOL
mf: $C_8H_{18}O_5$ mw: 194.26

PROP: Colorless to pale straw-colored liquid. Bp: 327.3°, fp: −6°, flash p: 360°F (OC), d: 1.1248 @ 20/20°, vap press: 1 mm @ 153.9°. Misc in water.

SYNS: HI-DRY ◇ 2,2′-(OXYBIS(ETHYLENEOXY))DIETHANOL

TOXICITY DATA with REFERENCE
skn-rbt 550 mg open MLD UCDS** 3/3/69
eye-rbt 565 mg AJOPAA 29,1363,46
orl-rat LD50:29 g/kg UCDS** 3/3/69

CONSENSUS REPORTS: Reported in EPA TSCA Inventory.

SAFETY PROFILE: Mildly toxic by ingestion. A skin and eye irritant. Combustible when exposed to heat or flame; can react with oxidizing materials. To fight fire, use alcohol foam, water, CO_2, dry chemical. When heated to decomposition it emits acrid smoke and irritating fumes.

TCE350 CAS:112-98-1 ***HR: 3***
TETRAETHYLENE GLYCOL, DIBUTYL ETHER
mf: $C_{16}H_{34}O_5$ mw: 306.50

SYN: ETHYLENE GLYCOL, DIBUTOXYTETRA

TOXICITY DATA with REFERENCE
orl-rat LD50:6500 mg/kg AMIHBC 4,119,51
ivn-mus LD50:320 mg/kg CSLNX* NX#03906
skn-rbt LD50:10 g/kg AMIHBC 4,119,51

CONSENSUS REPORTS: Glycol ether compounds are on the Community Right-To-Know List.

SAFETY PROFILE: Poison by intravenous route. Mildly toxic by ingestion and skin contact. Some glycol ether compounds have dangerous human reproductive effects. When heated to decomposition it emits acrid smoke and irritating fumes. See also GLYCOL ETHERS.

TCE375 CAS:70729-68-9 ***HR: 1***
TETRAETHYLENE GLYCOL-DI-n-HEPTANOATE
mf: $C_{22}H_{42}O_7$ mw: 418.64

SYN: TEGDH

TOXICITY DATA with REFERENCE
skn-rbt 500 mg/24H MLD DCTODJ 8,409,85
orl-rat LD50:25 g/kg DCTODJ 8,409,85
ihl-rat LCLo:14200 mg/m³/4H DCTODJ 8,409,85

SAFETY PROFILE: Mildly toxic by ingestion and inhalation. A skin irritant. When heated to decomposition it emits acrid smoke and irritating fumes. See also ESTERS.

TCE500 CAS:112-57-2 ***HR: 3***
TETRAETHYLENEPENTAMINE
DOT: UN 2320
mf: $C_8H_{23}N_5$ mw: 189.36

PROP: Viscous, hygroscopic liquid. Bp: 333°, flash p: 325°F (OC), d: 0.9980 @ 20/20°, vap press: <0.01 mm @ 20°.

SYNS: D.E.H. 26 ◇ 1,4,7,10,13-PENTAAZATRIDECANE

TOXICITY DATA with REFERENCE
skn-rbt 495 mg open SEV UCDS** 3/20/73
eye-rbt 5 mg MOD UCDS** 3/20/73
mmo-sat 333 μg/plate ENMUDM 8(Suppl 7),1,86
mma-sat 333 μg/plate ENMUDM 8(Suppl 7),1,86
orl-rat LD50:205 mg/kg ICHAA3 91,L51,84
ivn-mus LD50:320 mg/kg CSLNX* NX#03522
skn-rbt LD50:660 mg/kg JIHTAB 31,60,49

CONSENSUS REPORTS: Reported in EPA TSCA Inventory.

DOT Classification: Corrosive Material; Label: Corrosive.

SAFETY PROFILE: Poison by ingestion and intravenous routes. Moderately toxic by skin contact. Mutation data reported. A corrosive irritant to skin, eyes, and mucous membranes. Combustible when exposed to heat or flame. Can react with oxidizing materials. To fight fire, use CO_2, dry chemical. When heated to decomposition it emits toxic fumes of NO_x.

TCE750 CAS:597-63-7 ***HR: 2***
TETRAETHYL GERMANE
mf: $C_8H_{20}Ge$ mw: 188.87

PROP: Colorless oil. Decomp by water. D: 1.198 @ 0°, mp: −90°, bp: 163°.

SYN: TETRAETHYL GERMANIUM

TOXICITY DATA with REFERENCE
orl-rat LDLo:700 mg/kg CHDDAT 262,1302,66
ipr-rat LDLo:590 mg/kg CHDDAT 262,1302,66
orl-mus LDLo:2870 mg/kg CHDDAT 262,1302,66

CONSENSUS REPORTS: Reported in EPA TSCA Inventory.

SAFETY PROFILE: Moderately toxic by ingestion and intraperitoneal routes. Animal experiments show stimulation of blood formation. When heated to decomposition it emits acrid smoke and irritating fumes. See also GERMANIUM COMPOUNDS.

TCF000 CAS:78-00-2 ***HR: 3***
TETRAETHYL LEAD
DOT: NA 1649
mf: $C_8H_{20}Pb$ mw: 323.47

PROP: Colorless, oily liquid; pleasant characteristic odor. Mp: 125-150°, bp: 198-202° with decomp, d: 1.659 @ 18°, vap press: 1 mm @ 38.4°, flash p: 200°F.

SYNS: CZTEROETHLEK OLOWIU (POLISH) ◇ NCI-C54988 ◇ RCRA WASTE NUMBER P110 ◇ TEL ◇ TETRAETHYLPLUMBANE

TOXICITY DATA with REFERENCE
otr-ham:emb 12500 μg/L AETODY 1,241,80
orl-rat TDLo:11 mg/kg (6-16D preg):TER FCTXAV 13,629,75
orl-rat TDLo:7500 μg/kg (female 12-14D post):REP TXAPA9 21,265,72
scu-mus TDLo:100 mg/kg/21D-I:CAR EXPEAM 24,580,68
ihl-hmn TCLo:1749 g/m^3/30M:NOSE,EYE,GIT SAIGBL 17,223,75
unr-man LDLo:1470 μg/kg 85DCAI 2,73,70
orl-rat LD50:12300 μg/kg EXPEAM 28,923,72
ihl-rat LC50:850 mg/m^3/60M BJIMAG 18,277,61
ipr-rat LD50:15 mg/kg MELAAD 54,486,63
ivn-rat LD50:14400 μg/kg MELAAD 54,486,63
par-rat LD50:15 mg/kg AOHYA3 3,226,61
ihl-mus LCLo:650 mg/m^3/7H SAIGBL 15,3,73
skn-dog LDLo:547 mg/kg SAIGBL 15,3,73
orl-rbt LDLo:30 mg/kg SAIGBL 15,3,73
skn-rbt LDLo:830 mg/kg SAIGBL 15,3,73
scu-rbt LDLo:32 mg/kg EQSSDX 1,1,75
ivn-rbt LDLo:22 mg/kg EQSSDX 1,1,75
skn-gpg LDLo:995 mg/kg SAIGBL 15,3,73

CONSENSUS REPORTS: IARC Cancer Review: Group 3 IMEMDT 7,230,87; Animal Inadequate Evidence IMEMDT 23,325,80, IMEMDT 2,150,73. EPA Extremely Hazardous Substances List. Reported in EPA TSCA Inventory. EPA Genetic Toxicology Program.

OSHA PEL: TWA 0.075 mg(Pb)/m^3 (skin)
ACGIH TLV: TWA 0.1 mg(Pb)/m^3 (skin)
DFG MAK: 0.01 ppm (0.075 mg/m^3)
DOT Classification: Poison B; Label: Poison and Poison, Flammable Liquid.

SAFETY PROFILE: Human poison by an unspecified route. Experimental poison by ingestion, intraperitoneal, intravenous, subcutaneous, and parenteral routes. Moderately toxic by inhalation and skin contact. Experimental teratogenic and reproductive effects. Questionable carcinogen with experimental carcinogenic data. Mutation data reported. Lead compounds are particularly toxic to the central nervous system. It is a solvent for fatty materials and has some solvent action on rubber as well. The fact that it is a lipoid solvent makes it an industrial hazard because it can cause intoxication not only by inhalation but also by absorption through the skin. Decomposes when exposed to sunlight or allowed to evaporate; forms triethyl lead, which is also a poisonous compound, as one of its decomposition products. May cause lead exposure intoxication by coming in contact with the skin. A common air contaminant.

Flammable when exposed to heat, flame, or oxidizers.

Can react vigorously with oxidizing materials. Exposure to air for several days may cause explosive decomposition. To fight fire, use dry chemical, CO_2, mist, foam. When heated to decomposition it emits toxic fumes of Pb. See also LEAD COMPOUNDS.

TCF250 CAS:107-49-3 ***HR: 3***
TETRAETHYL PYROPHOSPHATE
DOT: NA 2783
mf: $C_8H_{20}O_7P_2$ mw: 290.22

PROP: Water-white to amber hygroscopic liquid. D: 1.20.

SYNS: BIS-O,O-DIETHYLPHOSPHORIC ANHYDRIDE ◇ BLADAN ◇ DIPHOSPHORIC ACID TETRAETHYL ESTER ◇ ENT 18,771 ◇ FOSVEX ◇ GRISOL ◇ HEPT ◇ HEXAMITE ◇ KILLAX ◇ KILMITE 40 ◇ LETHALAIRE G-52 ◇ LIROHEX ◇ MORTOPAL ◇ NIFOS T ◇ PYROPHOSPHATE de TETRAETHYLE (FRENCH) ◇ RCRA WASTE NUMBER P111 ◇ TEPP ◇ O,O,O,O-TETRAAETHYL-DIPHOSPHAT, BIS(O,O-DIAETHYLPHOSPHORSAEURE-ANHYDRID (GERMAN) ◇ O,O,O,O-TETRAETHYL-DIFOSFAAT (DUTCH) ◇ TETRAETHYL PYROFOSFAAT (BELGIAN) ◇ TETRAETHYL PYROPHOSPHATE, liquid (DOT) ◇ O,O,O,O-TETRAETIL-PIROFOSFATO (ITALIAN) ◇ TETRASTIGMINE ◇ TETRON ◇ TETRON-100 ◇ VAPOTONE

TOXICITY DATA with REFERENCE
par-hmn TDLo:71 μg/kg:CNS,GIT CMEP** -,1,56
orl-hmn LDLo:1429 μg/kg:PNS,CNS,GIT CMEP** -,1,56
orl-hmn TDLo:309 μg/kg:CNS,GIT CMEP** -,1,56
ims-hmn LDLo:286 μg/kg:PNS,CNS,GIT CMEP** -,1,56
orl-rat LD50:500 μg/kg PHJOAV 185,361,60
skn-rat LD50:2400 μg/kg TXAPA9 14,515,69
ipr-rat LD50:650 μg/kg FEPRA7 6,353,47
scu-rat LD50:279 μg/kg FAATDF 4(2,Pt 2),S195,84
ivn-rat LD50:300 μg/kg CJBPAZ 34,197,56
ims-rat LD50:1800 μg/kg JCINAO 37,350,58
orl-mus LD50:7 mg/kg TXAPA9 21,153,72
ipr-mus LD50:830 μg/kg AIPTAK 121,104,59
scu-mus LD50:520 μg/kg AIPTAK 124,212,60
ivn-mus LD50:200 μg/kg BJPCAL 9,299,54
ivn-rbt LDLo:300 μg/kg BJPCAL 8,466,53
skn-dck LD50:64 mg/kg TXAPA9 47,451,79
par-frg LD50:34 mg/kg AIPTAK 124,212,60

CONSENSUS REPORTS: EPA Extremely Hazardous Substances List.

OSHA PEL: TWA 0.05 mg/m^3 (skin)
ACGIH TLV: TWA 0.004 mg/m^3 (skin)
DFG MAK: 0.0005 ppm (0.05 mg/m^3)
DOT Classification: Poison B; Label: Poison, liquid.

SAFETY PROFILE: Human poison by ingestion and intramuscular routes. Experimental poison by ingestion, skin contact, intraperitoneal, intramuscular, subcutaneous, parenteral, and intravenous routes. Human systemic effects by ingestion, intramuscular, and parenteral routes: paresthesia, wakefulness, excitement, muscle contraction or spasticity, nausea or vomiting and other gastrointestinal changes. The action is similar to that of parathion; causing an irreversible inhibition of the cholinesterase molecules and the consequent accumulation of large amounts of acetylcholine. Small doses at frequent intervals are largely additive. When heated to decomposition it emits toxic fumes of PO_x. See also PARATHION.

TCF260 ***HR: 3***
TETRAETHYLPYROPHOSPHATE and compressed gas mixtures
DOT: UN 1705
mf: $C_8H_{20}O_7P_2$ mw: 290.22

DOT Classification: Poison A; Label: Poison Gas.

SAFETY PROFILE: A poison gas. When heated to decomposition it emits toxic fumes of PO_x. See also TETRAETHYLPYROPHOSPHATE.

TCF270 CAS:107-49-3 ***HR: 3***
TETRAETHYLPYROPHOSPHATE MIXTURE (dry)
DOT: NA 2783
mf: $C_8H_{20}O_7P_2$ mw: 290.22

PROP: Mobile liquid; pleasant odor. Hygroscopic, d: 1.185.

SYN: PYROPHOSPHORIC ACID, TETRAETHYL ESTER (dry mixture)

DOT Classification: Poison B; Label: Poison.

SAFETY PROFILE: A poison. A cholinesterase inhibitor. When heated to decomposition it emits toxic fumes of PO_x. See also TETRAETHYLPYROPHOSPHATE.

TCF280 CAS:107-49-3 ***HR: 3***
TETRAETHYL PYROPHOSPHATE MIXTURE (liquid)
DOT: NA 2783
mf: $C_8H_{20}O_7P_2$ mw: 290.22

PROP: Water white to amber, hygroscopic liquid. D: 1.20, decomp @ 170-213°, bp: 138° @ 2.3 mm. Misc with water, acetone, methanol, ethanol, benzene, chloroform.

SYNS: FOSVEX ◇ PYROPHOSPHORIC ACID, TETRAETHYL ESTER (liquid mixture) ◇ TEP ◇ TETRON ◇ VAPTONE

TOXICITY DATA with REFERENCE
orl-rat LD50:1050 μg/kg TXAPA9 14,515,69
skn-rat LD50:2400 μg/kg TXAPA9 14,515,69
orl-mus LD50:7 mg/kg JPETAB 105,156,52
orl-gpg LD50:2300 μg/kg JPETAB 105,156,52

DOT Classification: Poison B; Label: Poison.

SAFETY PROFILE: Poison by ingestion and skin contact. A cholinesterase inhibitor type of insecticide. The effects of chronic exposure to small doses is additive. When heated to decomposition it emits toxic fumes of PO_x. See also TETRAETHYLPYROPHOSPHATE.

TCF750 CAS:597-64-8 ***HR: 3***
TETRAETHYLSTANNANE
mf: $C_8H_{20}Sn$ mw: 234.97

PROP: Colorless liquid. D: 1.187 @ 23°, mp: −112°, bp: 181°. Insol in water; sol in organic solvents.

SYN: TETRAETHYL TIN

TOXICITY DATA with REFERENCE
orl-rat LD50:16 mg/kg HYSAAV 32,11,67
ivn-rat LDLo:25 mg/kg BJPCAL 10,16,55
ipr-mus LDLo:32 mg/kg CBCCT* 4,234,52
orl-rbt LD50:7 mg/kg HYSAAV 32,11,67
orl-gpg LD50:37 mg/kg HYSAAV 32,11,67

CONSENSUS REPORTS: Reported in EPA TSCA Inventory.

OSHA PEL: TWA 0.1 mg(Sn)/m^3 (skin)
ACGIH TLV: TWA 0.1 mg(Sn)/m^3 (skin) (Proposed: TWA 0.1 mg(Sn)/m^3; STEL 0.2 mg(Sn)/m^3 (skin))
NIOSH REL: (Organotin Compounds) TWA 0.1 mg (Sn)/m^3

SAFETY PROFILE: Poison by ingestion, intravenous and intraperitoneal routes. When heated to decomposition it emits acrid smoke and irritating fumes. See also TIN COMPOUNDS.

TCG000 CAS:4531-35-5 ***HR: 3***
TETRAETHYNYLGERMANIUM
mf: C_8H_4Ge mw: 172.67

$$(HC{\equiv}C)_4Ge$$

SAFETY PROFILE: Explodes on rapid heating or friction. When heated to decomposition it emits acrid smoke and irritating fumes. See also GERMANIUM COMPOUNDS and ACETYLENE COMPOUNDS.

TCG250 CAS:16413-88-0 ***HR: 3***
TETRAETHYNYLTIN
mf: C_8H_4Sn mw: 218.77

$$(HC{\equiv}C)_4Sn$$

SAFETY PROFILE: Explodes on rapid heating. When heated to decomposition it emits acrid smoke and irritating fumes. See also TIN COMPOUNDS and ACETYLENE COMPOUNDS.

TCG450 CAS:25483-10-7 ***HR: 3***
TETRAFLUOROAMMONIUM PERBROMATE
mf: BrF_4NO_4 mw: 233.90

SAFETY PROFILE: Explodes on contact with hydrogen fluoride. When heated to decomposition it emits toxic fumes of F^-, Br^-, NO_x, and NH_3. See also FLUORIDES and BROMATES.

TCG500 CAS:15640-93-4 ***HR: 3***
TETRAFLUORO AMMONIUM TETRAFLUOROBORATE
mf: BF_8N mw: 176.81

SAFETY PROFILE: Mixtures with dioxygenyl tetrafluoroborate are powerful oxidizers and impact-sensitive explosives. The mixture causes 2-propanol to ignite on contact. When heated to decomposition it emits very toxic fumes of NO_x, NH_3, and F^-. See also FLUORIDES and BORON COMPOUNDS.

TCG750 CAS:42794-87-6 ***HR: D***
3,3′,5,5′-TETRAFLUOROBENZIDINE
mf: $C_{12}H_8F_4N_2$ mw: 256.22

SYN: 3,3′,5,5′-TETRAFLUORO(1,1′-BIPHENYL)-4,4′-DIAMINE

TOXICITY DATA with REFERENCE
mmo-sat 50 μg/plate CALEDQ 1,39,75
dns-hmn:hla 100 nmol/L CNREA8 38,2621,78

CONSENSUS REPORTS: EPA Genetic Toxicology Program.

SAFETY PROFILE: Human mutation data reported. When heated to decomposition it emits very toxic fumes of F^- and NO_x.

TCH000 ***HR: 3***
TETRAFLUOROBORATE(1−) compound with p-AMINOBENZOIC ACID 2-(DIETHYLAMINO) ETHYL ESTER
mf: $BF_4{\cdot}C_{13}H_{20}N_2O_2$

PROP: Colorless liquid. Bp: decomp @ 130°.

SYN: PROCAINE FLUOBORATE

TOXICITY DATA with REFERENCE
ivn-mus LD50:65 mg/kg THERAP 10,473,55

SAFETY PROFILE: Poison by intravenous route. A corrosive irritant to skin, eyes, and mucous membranes. When heated to decomposition it emits toxic fumes of F^- and NO_x. See also FLUORIDES and BORON COMPOUNDS.

TCH100 CAS:2252-95-1 ***HR: 3***
1,1,4,4-TETRAFLUOROBUTATRIENE
mf: C_4F_4 mw: 124.04

$$F_2C{=}C{=}C{=}CF_2$$

SAFETY PROFILE: The liquid explodes above −5°C.

Upon decomposition it emits toxic fumes of F^-. See also FLUORIDES.

TCH250 CAS:17224-09-8 ***HR: 3***
TETRAFLUORODIAZIRIDINE
mf: CF_4N_2 mw: 116.02

SAFETY PROFILE: Powerful explosive very sensitive to shock or phase changes. A powerful oxidizer. When heated to decomposition it emits very toxic fumes of NO_x and F^-. See also FLUORIDES.

TCH325 CAS:71292-84-7 ***HR: 3***
3,5,3',5'-TETRAFLUORODIETHYLSTILBESTROL
mf: $C_{18}H_{16}F_4O_2$ mw: 340.34

SYNS: trans-α,α'-DIETHYL-3,3',5,5'-TETRAFLUORO-4,4'-STILBENEDIOL ◇ α,α'-DIETHYL-3,3',5,5'-TETRAFLUORO-4,4'-STILBENEDIOL (E)-

TOXICITY DATA with REFERENCE
otr-ham:emb 10 μg/L CNREA8 42,3040,82
dnd-ham:emb 7 mg/L CRNGDP 7,1329,86
imp-ham TDLo:360 mg/kg/12W-I:CAR CNREA8 43,2678,83

SAFETY PROFILE: Questionable carcinogen with experimental carcinogenic data. Mutation data reported. When heated to decomposition it emits toxic fumes of F^-. See also DIETHYLSTILBESTEROL.

TCH500 CAS:116-14-3 ***HR: 3***
TETRAFLUOROETHYLENE
DOT: UN 1081
mf: C_2F_4 mw: 100.02

PROP: Colorless gas. Mp: −142.5°, bp: −78.4°. lel: 11%; uel: 60%.

SYNS: FLUOROPLAST 4 ◇ PERFLUOROETHENE ◇ PERFLUOROETHYLENE ◇ TETRAFLUORETHYLENE ◇ TETRAFLUOROETHENE ◇ TETRAFLUOROETHYLENE, inhibited (DOT)

TOXICITY DATA with REFERENCE
ihl-rat LC50:40000 ppm/4H JOCMA7 4,262,62
ihl-mus LC50:143 g/m³/kg GTPZAB 21(5),36,77
ihl-gpg LC50:116 g/m³/kg GTPZAB 21(5),36,77

CONSENSUS REPORTS: IARC Cancer Review: Group 3 IMEMDT 7,56,87. Reported in EPA TSCA Inventory.

DOT Classification: Flammable Gas; Label: Flammable Gas.

SAFETY PROFILE: Mildly toxic by inhalation. Can act as an asphyxiant and may have other toxic properties. Questionable carcinogen. The gas is flammable when exposed to heat or flame. The inhibited monomer will explode if ignited. Explosive in the form of vapor when exposed to heat or flame. Will explode at pressures above 2.7 bar if terpene inhibitor is not added. Iodine pentafluoride depletes the limonene inhibitor and then causes explosive polymerization of the monomer. Mixtures with hexafluoropropene and air form an explosive peroxide. Reacts violently with SO_3; air; difluoromethylene dihypofluorite; dioxygen difluoride; iodine pentafluoride; oxygen. When heated to decomposition it emits highly toxic fumes of F^-. See also FLUORIDES.

TCH750 CAS:116-14-3 ***HR: 3***
TETRAFLUOROETHYLENE (inhibited)
DOT: UN 1081

CONSENSUS REPORTS: Reported in EPA TSCA Inventory.

DOT Classification: Label: Flammable Gas.

SAFETY PROFILE: Flammable gas. When heated to decomposition it emits very toxic fumes of F^-. See also TETRAFLUOROETHYLENE.

TCI000 CAS:10036-47-2 ***HR: 3***
TETRAFLUORO HYDRAZINE
DOT: UN 1955
mf: N_2F_4 mw: 104.0

F_2NNF_2

PROP: Colorless gas or liquid; white solid when pure. Mp: −163°, bp: −73°, d (liquid): 1.5 @ −100°.

SYNS: DINITROGEN TETRAFLUORIDE ◇ PERFLUORO HYDRAZINE

TOXICITY DATA with REFERENCE
ihl-gpg LC50:900 ppg/1H TXAPA9 6,447,64
ihl-rat LCLo:50 ppm/4H 34ZIAG -,580,69

CONSENSUS REPORTS: Reported in EPA TSCA Inventory.

OSHA PEL: TWA 2.5 mg(F)/m³
ACGIH TLV: TWA 2.5 mg(F)/m³
NIOSH REL: TWA (Inorganic Fluorides) 2.5 mg(F)/m³
DOT Classification: Poison A; Label: Poison Gas.

SAFETY PROFILE: A poison. An unstable explosive gas sensitive to light, heat, or contact with air or steel. At high pressures it can explode due to shock or blast. Flammable when exposed to heat or flame. Potentially explosive reaction with hydrocarbons; hydrogen; organic materials; reducing agents; oxygen. Forms explosive mixtures with alkenyl nitrates; nitrogen trifluoride. When heated to decomposition it emits highly toxic fumes of F^- and NO_x. See also FLUORIDES, HYDROFLUORIC ACID, and HYDRAZINE.

TCI100 CAS:102489-70-3 ***HR: 3***
2,2,3,3-TETRAFLUORO-4,7-METHANO-2,3,5,6,8,9-HEXAHYDROBENZOSELENOPHENE
mf: $C_9H_{10}F_4Se$ mw: 273.15

SYN: 4,7-METHANOSELENOPHENE,OCTAHYDRO-2,2,3,3-TETRAFLUORO-

TOXICITY DATA with REFERENCE
ivn-mus LD50;1 mg/kg CSLNX* NX#10021

OSHA PEL: TWA 0.2 mg(Se)/m^3
ACGIH TLV: TWA 0.2 mg(Se)/m^3

SAFETY PROFILE: Poison by intravenous route. When heated to decomposition it emits toxic fumes of Se and F^-.

TCI250 CAS:63886-77-1 ***HR: 3***
TETRAFLUORO-m-PHENYLENE DIAMINE DIHYDROCHLORIDE
mf: $C_6H_4F_4N_2$•2ClH mw: 253.04

TOXICITY DATA with REFERENCE
orl-mus TDLo:65 g/kg/78W-C:CAR JEPTDQ 2,325,78
orl-mus TD:130 g/kg/78W-C:CAR JEPTDQ 2,325,78
ipr-rat LD50:270 mg/kg NCIBR* NIH-NCI-E-68-1311,10,73

SAFETY PROFILE: Poison by intraperitoneal route. Questionable carcinogen with experimental carcinogenic data. When heated to decomposition it emits very toxic fumes of F^-, NO_x, and HCl.

TCI500 CAS:56973-16-1 ***HR: 2***
TETRAFLUORO-m-PHENYLENE DIMALEIMIDE
mf: $C_{14}H_4F_4N_2O_4$ mw: 340.20

SYNS: 1,1'-(TETRAFLUORO-m-PHENYLENE)DIMALEIMIDE ◇ TFPHM

TOXICITY DATA with REFERENCE
orl-rat LD50:2100 mg/kg GISAAA 40(11),109,75
orl-mus LD50:1650 mg/kg GISAAA 40(11),109,75

SAFETY PROFILE: Moderately toxic by ingestion. When heated to decomposition it emits very toxic fumes of F^- and NO_x.

TCI750 CAS:76-37-9 ***HR: 2***
2,2,3,3-TETRAFLUORO-1-PROPANOL
mf: $C_3H_4F_4O$ mw: 132.07

$F_2CHCF_2CH_2OH$

PROP: Bp: 109°, refr index: 1.3210, d: 1.471, flash p: 110° F.

TOXICITY DATA with REFERENCE
orl-rat LDLo:3400 mg/kg JOCMA7 4,262,62
ihl-rat LCLo:2000 ppm/4H JOCMA7 4,262,62
ihl-mus LC50:8600 mg/m^3/2H 85GMAT -,109,82

CONSENSUS REPORTS: Reported in EPA TSCA Inventory.

SAFETY PROFILE: Moderately toxic by ingestion. Mildly toxic by inhalation. An irritant. Combustible liquid. Violent reaction and ignition with sodium hydroxide or potassium hydroxide. When heated to decomposition it emits toxic fumes of F^-. See also FLUORIDES.

TCI800 ***HR: 3***
TETRAFLUOROSUCCINAMIDE
mf: $C_4H_4F_4N_2O_2$ mw: 188.08

$(-CF_2CO•NH_2)_2$

SAFETY PROFILE: Forms an explosive complex with lithium tetrahydroaluminate. When heated to decomposition it emits toxic fumes of F^- and NO_x. See also FLUORIDES and AMIDES.

TCJ000 CAS:377-38-8 ***HR: 3***
TETRAFLUOROSUCCINIC ACID
mf: $C_4H_2F_4O_4$ mw: 190.06

$(HOCO•CF_2-)_2$

SYNS: PERFLUOROSUCCINIC ACID ◇ TETRAFLUOROBUTANEDIOIC ACID

TOXICITY DATA with REFERENCE
unr-mus LDLo:200 mg/kg 11FYAN 3,85,63

CONSENSUS REPORTS: Reported in EPA TSCA Inventory.

SAFETY PROFILE: Poison by an unspecified route. Reacts to form an explosive product with cesium fluoride + fluorine. When heated to decomposition it emits toxic fumes of F^-. See also FLUORIDES.

TCJ025 CAS:10256-92-5 ***HR: 3***
TETRAFLUOROUREA
mf: CF_4N_2O mw: 132.02

SAFETY PROFILE: Solutions in acetonitrile form the explosive difluorodiazine at room temperature. When heated to decomposition it emits toxic fumes of F^- and NO_x.

TCJ050 ***HR: 3***
TETRAFORMYL-DIGITOXIN
mf: $C_{45}H_{64}O_{17}$ mw: 877.09

SYN: TETRAFORMATE DIGITOXIN

TOXICITY DATA with REFERENCE
orl-mus LD50:5920 μg/kg AIPTAK 153,436,65
scu-mus LD50:4460 μg/kg AIPTAK 153,436,65
ivn-gpg LDLo:2900 μg/kg AIPTAK 153,436,65

SAFETY PROFILE: Poison by ingestion, subcutaneous, and intravenous routes. When heated to decomposition it emits acrid smoke and irritating fumes. See also DIGITOXIN.

TCJ075 CAS:321-64-2 ***HR: 3***
TETRAHYDROAMINOCRINE
mf: $C_{13}H_{14}N_2$ mw: 198.29

SYNS: 5-AMINO-6,7,8,9-TETRAHYDROACRIDINE (EUROPEAN) ◇ 9-AMINO-1,2,3,4-TETRAHYDROACRIDINE ◇ TACRINE ◇ 1,2,3,4-TETRAHYDRO-9-ACRIDINAMINE (9CI) ◇ TETRAHYDROAMINACRINE ◇ TETRAHYDROAMINOCRIN

TOXICITY DATA with REFERENCE
cyt-mus-ipr 75 mg/kg ACNSAX 26,84,84
cyt-mus-orl 100 mg/kg ACNSAX 26,84,84
scu-mus LD50:25 mg/kg BJEPA5 28,1,47

SAFETY PROFILE: Poison by subcutaneous. Mutation data reported. When heated to decomposition it emits toxic fumes of NO_x.

TCJ100 CAS:1321-16-0 ***HR: 2***
TETRAHYDROBENZALDEHYDE
mf: C_7H_9N mw: 110.17

SYN: CYCLOHEXENECARBOXALDEHYDE

TOXICITY DATA with REFERENCE
skn-rbt 500 mg open MLD UCDS** 6/28/65
eye-rbt 5 mg SEV UCDS** 6/28/65
orl-rat LD50:1050 mg/kg GTPZAB 18(11),41,74
orl-mus LD50:1000 mg/kg GTPZAB 18(11),41,74
ihl-mus LC50:556 mg/m³/4H GTPZAB 18(11),41,74
orl-rbt LD50:1600 mg/kg GTPZAB 18(11),41,74
skn-rbt LD50:1770 mg/kg UCDS** 6/28/65
orl-gpg LD50:1750 mg/kg GTPZAB 18(11),41,74

SAFETY PROFILE: Moderately toxic by ingestion, skin contact, and inhalation. A skin and severe eye irritant. The liquid is flammable when exposed to heat or flame. When heated to decomposition it emits toxic fumes of NO_x. See also ALDEHYDES.

TCJ500 CAS:3570-54-5 ***HR: 3***
1,2,5,6-TETRAHYDROBENZO(j)CYCLOPENT(fg)ACEANTHRYLENE
mf: $C_{22}H_{16}$ mw: 280.38

SYNS: NORSTEARANTHRENE ◇ F-NORSTEARANTHRENE

TOXICITY DATA with REFERENCE
scu-mus TDLo:72 mg/kg/13W-I:ETA NATWAY 53,583,66

SAFETY PROFILE: Questionable carcinogen with experimental tumorigenic data. When heated to decomposition it emits acrid smoke and irritating fumes.

TCJ775 CAS:17750-93-5 ***HR: 2***
7,8,9,10-TETRAHYDROBENZO(a)PYRENE
mf: $C_{20}H_{16}$ mw: 256.36

SYN: 1′,2′,3′,4′-TETRAHYDRO-3,4-BENZOPYRENE

TOXICITY DATA with REFERENCE
scu-mus TDLo:72 mg/kg/9W-I:ETA COREAF 251,1322,60

SAFETY PROFILE: Questionable carcinogen with experimental neoplastigenic data. When heated to decomposition it emits acrid smoke and irritating fumes.

TCJ800 CAS:5096-57-1 ***HR: 3***
TETRAHYDROBERBERINE
mf: $C_{20}H_{21}NO_4$ mw: 339.42

SYNS: CANADINE ◇ (−)-CANADINE ◇ α-CANADINE ◇ 9,10-DIMETHOXY-2,3-(METHYLENEDIOXY)BERBINE ◇ (−)-TETRAHYDROBERBERINE ◇ (s)-5,8,13,13a-TETRAHYDRO-9,10-DIMETHOXY-6H-BENZO(g)-1,3-BENZODIOXOLO(5,6-a)QUINOLIZINE ◇ 5,6,13,13a-TETRAHYDRO-9,10-DIMETHOXY-2,3-(METHYLENEDIOXY)-8H-DIBENZO(a,g)QUINOLIZINE ◇ XANTHOPUCCINE

TOXICITY DATA with REFERENCE
orl-mus LD50:940 mg/kg MEIEDD 10,239,83
scu-mus LD50:790 mg/kg MEIEDD 10,239,83
ivn-mus LD50:100 mg/kg MEIEDD 10,239,83

SAFETY PROFILE: Poison by intravenous route. Moderately toxic by ingestion and subcutaneous routes. When heated to decomposition it emits toxic fumes of NO_x.

TCM000 CAS:5957-75-5 ***HR: 3***
1-trans-Δ⁸-TETRAHYDROCANNABINOL
mf: $C_{21}H_{30}O_2$ mw: 314.51

SYNS: (−)-Δ^6-3,4-trans-TETRAHYDROCANNABINOL ◇ (−)-Δ^8-trans-TETRAHYDROCANNABINOL ◇ Δ^6-THC ◇ Δ^8-THC

TOXICITY DATA with REFERENCE
dni-hmn:hla 10 μmol/L ANTRD4 3,211,83
oms-mus-ipr 200 mg/kg RCOCB8 17,703,77
orl-rat LD50:860 mg/kg TXAPA9 25,363,73
ipr-rat LD50:560 mg/kg AIPTAK 196,133,72
ivn-rat LD50:97 mg/kg AIPTAK 196,133,72
ipr-mus LD50:210 mg/kg AIPTAK 196,133,72
ivn-mus LD50:27500 μg/kg JMCMAR 21,1079,78

CONSENSUS REPORTS: EPA Genetic Toxicology Program.

SAFETY PROFILE: Poison by intravenous and intraperitoneal routes. Moderately toxic by ingestion. Human mutation data reported. An hallucinatory drug. When heated to decomposition it emits acrid smoke and irritating fumes. See also CANNABIS.

TCM250 CAS:1972-08-3 ***HR: 3***
1-trans-Δ⁹-TETRAHYDROCANNABINOL
mf: $C_{21}H_{30}O_2$ mw: 314.51

SYNS: ABBOTT 40566 ◇ 3-PENTYL-6,6,9-TRIMETHYL-6a,7,8,10a-TETRAHYDRO-6H-DIBENZO(b,d)PYRAN-1-OL ◇ SP 104 ◇ (l)-Δ^1-TETRAHYDROCANNABINOL ◇ Δ^1-TETRAHYDROCANNABINOL ◇ (−)-Δ^1-3,4-trans-TETRAHYDROCANNABINOL ◇ (−)-Δ^9-trans-TETRAHYDROCANNABINOL ◇ trans-Δ^9-TETRAHYDROCANNABINOL ◇ Δ^9-TETRAHYDROCANNABINON ◇ THC ◇ Δ^1-THC ◇ Δ^9-THC ◇ 6,6,9-TRIMETHYL-3-PENTYL-7,8,9,10-TETRAHYDRO-6H-DIBENZO(B,D)PYRAN-1-OL

TOXICITY DATA with REFERENCE
dni-hmn:hla 40 μmol/L ANTRD4 3,211,83
oms-hmn:fbr 5 μmol/L ANTRD4 3,211,83
orl-mky TDLo:648 mg/kg (female 1-23W post):REP NETOD7 4,469,82
orl-mky TDLo:396 mg/kg (female 1-23W post):TER AVBIB9 22/23,501,79
scu-mus TDLo:800 μg/kg/W-I:ETA FEPRA7 38,1450,79
orl-rat LD50:666 mg/kg PSEBAA 136,260,71
ipr-rat LD50:373 mg/kg PSEBAA 136,260,71
ivn-rat LD50:29 mg/kg ANYAA9 191,74,71
orl-mus LD50:482 mg/kg PSEBAA 136,260,71
ipr-mus LD50:168 mg/kg AIPTAK 196,133,72
ivn-mus LD50:42 mg/kg ANYAA9 191,74,71
orl-dog LDLo:525 mg/kg TXAPA9 25,363,73
ivn-mky LD50:128 mg/kg TXAPA9 27,648,74

CONSENSUS REPORTS: EPA Genetic Toxicology Program.

SAFETY PROFILE: Poison by intraperitoneal and intravenous routes. Moderately toxic by ingestion. Experimental reproductive effects. Questionable carcinogen with experimental tumorigenic and teratogenic data. Human mutation data reported. An hallucinatory drug. When heated to decomposition it emits acrid smoke and irritating fumes. See also CANNABIS.

TCM400 CAS:21001-46-7 ***HR: 3***
5,6,7,8-TETRAHYDRO-3-CARBAZOLECARBOXYLIC ACID 2-(DIETHYLAMINO)ETHYL ESTER HYDROCHLORIDE
mf: $C_{19}H_{26}N_2O_2{\cdot}ClH$ mw: 350.93

TOXICITY DATA with REFERENCE
skn-rbt 1000 ppm MOD JAPMA8 40,373,51
eye-rbt 1000 ppm MLD JAPMA8 40,373,51
scu-mus LD50:178 mg/kg JAPMA8 40,373,51
ivn-mus LD50:33 mg/kg JAPMA8 40,373,51

SAFETY PROFILE: Poison by subcutaneous and intravenous routes. A skin and eye irritant. When heated to decomposition it emits toxic fumes of NO_x and HCl.

TCN250 CAS:13073-86-4 ***HR: D***
2-(p-(1,2,3,4-TETRAHYDRO-2-(p-CHLOROPHENYL)NAPHTHYL)PHENOXY)TRIETHYLAMINE
mf: $C_{28}H_{32}ClNO$ mw: 434.06

SYN: 2-(p-(2-(p-CHLOROPHENYL)-1,2,3,4-TETRAHYDRONAPHTHYL)PHENOXY)TRIETHYLAMINE ◇ SU-13320

TOXICITY DATA with REFERENCE
orl-ham TDLo:225 mg/kg (female 1-9D post):REP CCPTAY 3,347,71
orl-rat TDLo:90 mg/kg (6-14D preg):TER TXAPA9 10,565,67

SAFETY PROFILE: An experimental teratogen. Experimental reproductive effects. When heated to decomposition it emits very toxic fumes of Cl^- and NO_x.

TCN750 CAS:153-39-9 ***HR: 3***
1,2,3,4-TETRAHYDRODIBENZ(a,h)ANTHRACENE
mf: $C_{22}H_{18}$ mw: 282.40

TOXICITY DATA with REFERENCE
skn-mus TDLo:74 mg/kg/29W-I:NEO JNCIAM 34,1,65

SAFETY PROFILE: Questionable carcinogen with experimental neoplastigenic data. When heated to decomposition it emits acrid smoke and irritating fumes.

TCO000 CAS:16310-68-2 ***HR: 3***
1,2,3,4-TETRAHYDRODIBENZ(a, j)ANTHRACENE
mf: $C_{22}H_{18}$ mw: 282.40

TOXICITY DATA with REFERENCE
skn-mus TDLo:104 mg/kg/35W-I:ETA JNCIAM 44,641,70

SAFETY PROFILE: Questionable carcinogen with experimental tumorigenic data. When heated to decomposition it emits acrid smoke and irritating fumes.

TCP000 ***HR: 3***
1,2,3,11a-TETRAHYDRO-3,8-DIHYDROXY-7-METHOXY-5H-PYRROLO(2,1-c)(1,4)BENZODIAZEPIN-5-ONE-(3S-cis)-
mf: $C_{13}H_{14}N_2O_4$ mw: 262.29

SYN: NEOTHRAMYCIN A

TOXICITY DATA with REFERENCE
ipr-mus LD50:6 mg/kg 85GDA2 5,189,81
ivn-mus LD50:5 mg/kg 85GDA2 5,189,81

SAFETY PROFILE: Poison by intraperitoneal and intravenous routes. When heated to decomposition it emits toxic fumes of NO_x.

TCP600 CAS:67242-54-0 ***HR: 3***
1,2,3,4-TETRAHYDRO-7,12-DIMETHYLBENZ(a)ANTHRACENE
mf: $C_{20}H_{20}$ mw: 260.40

SYNS: 7,12-DIMETHYL-1,2,3,4-TETRAHYDROBENZ(a)ANTHRACENE ◇ 1,2,3,4-TETRAHYDRO-DMBA ◇ TH-DMBA

TOXICITY DATA with REFERENCE
mmo-sat 50 μg/plate JMCMAR 23,278,80
mma-sat 2500 ng/plate CRNGDP 4,1221,83
otr-hmn:fbr 500 μg/L CALEDQ 13,119,81
dnd-ham:lng 100 nmol/L CRNGDP 3,651,82
skn-mus TDLo:200 mg/kg/25W-I:CAR CRNGDP 4,1221,83
skn-mus TD:104 μg/kg:ETA CRNGDP 3,651,82

SAFETY PROFILE: Questionable carcinogen with experimental carcinogenic and tumorigenic data. Human mutation data reported. When heated to decomposition it emits acrid smoke and irritating fumes.

TCQ250 CAS:1320-94-1 ***HR: 2***
TETRAHYDRODIMETHYLFURAN
mf: $C_6H_{12}O$ mw: 100.18

PROP: A liquid.
SYN: TETRAHYDRODIMETHYL FURANE

TOXICITY DATA with REFERENCE
skn-rbt 10 mg/24H open JIHTAB 26,269,44
eye-rbt 20 mg AJOPAA 29,1363,46
orl-rat LD50:4000 mg/kg JIHTAB 26,269,44
skn-gpg LD50:4000 mg/kg JIHTAB 26,269,44

SAFETY PROFILE: Moderately toxic by ingestion and skin contact. A skin and eye irritant. Flammable when exposed to heat or flame; can react with oxidizing materials. To fight fire, use spray, foam, dry chemical, CO_2. When heated to decomposition it emits acrid smoke and irritating fumes. See also TETRAHYDROFURAN.

TCQ260 CAS:3613-73-8 ***HR: 3***
2,3,4,5-TETRAHYDRO-2,8-DIMETHYL-5-(2-(6-METHYL-3-PYRIDYL)ETHYL)-1H-PYRID O(4,3-b) INDOLE
mf: $C_{21}H_{25}N_3$ mw: 319.49

SYNS: DIMEBOLIN ◇ DIMEBOLINE ◇ DIMEBON ◇ DIMEBONE ◇ PREPARATION 84 ◇ 1H-PYRIDO(4,3-b)INDOLE, 2,3,4,5-TETRAHYDRO-2,8-DIMETHYL-5-(2-(6-METHYL-3-PYRIDYL)ETHYL)-

TOXICITY DATA with REFERENCE
orl-rat TDLo:1500 mg/kg (female 8-13D post):TER FATOAO 48(6),89,85
orl-mus LDLo:256 mg/kg FATOAO 46(5),90,83
ipr-mus LD50:178 mg/kg FATOAO 31,105,68

SAFETY PROFILE: Poison by ingestion and intraperitoneal routes. An experimental teratogen. When heated to decomposition it emits toxic fumes of NO_x.

TCQ275 CAS:25952-35-6 ***HR: 2***
TETRAHYDRO-3,5-DIMETHYL-4H,1,3,5-OXADIAZINE-4-THIONE
mf: $C_5H_{10}N_2OS$ mw: 146.23

SYNS: 4H-1,3,5-OXADIAZINE-4-THIONE,TETRAHYDRO-3,5-DIMETHYL- ◇ TDOT

TOXICITY DATA with REFERENCE
skn-rat TDLo:1 g/kg (female 10-11D post):TER TXAPA9 41,35,77
orl-mus LDLo:1600 mg/kg AECTCV 14,111,85

CONSENSUS REPORTS: Reported in EPA TSCA Inventory.

SAFETY PROFILE: Moderately toxic by ingestion. An experimental teratogen. When heated to decomposition it emits toxic fumes of NO_x and SO_x.

TCQ350 CAS:1073-79-6 ***HR: 2***
TETRAHYDRO-2,6-DIMETHYL-4H-PYRAN-4-ONE
mf: $C_7H_{12}O_2$ mw: 128.19

SYNS: DIMETHYLTETRAHYDROPYRONE ◇ 2,6-DIMETHYL TETRAHYDRO-1,4-PYRONE ◇ 4H-PYRAN-4-ONE, TETRAHYDRO-2,6-DIMETHYL-

TOXICITY DATA with REFERENCE
skn-rbt 500 mg/24H MLD JIDHAN 31,60,49
eye-rbt 750 μg/24H SEV JIDHAN 31,60,49
orl-rat LD50:3400 mg/kg JIDHAN 31,60,49
ihl-rat LC50:4000 ppm/4H JIDHAN 31,60,49

SAFETY PROFILE: Moderately toxic by ingestion. A skin and severe eye irritant. When heated to decomposition it emits acrid smoke and irritating fumes.

TCQ500 CAS:703-95-7 ***HR: 3***
1,2,3,6-TETRAHYDRO-2,6-DIOXO-5-FLUORO-4-PYRIMIDINECARBOXYLIC ACID
mf: $C_5H_3FN_2O_4$ mw: 174.10

SYNS: ENT 26,398 ◇ 5-FLUOROOROTATE ◇ 5-FLUOROOROTIC ACID ◇ 5-FLUORO-1,2,3,6-TETRAHYDRO-2,6-DIOXO-4-PYRIMIDINECARBOXYLIC ACID ◇ FO ◇ FOA ◇ NSC 31712 ◇ RO 2-9945

TOXICITY DATA with REFERENCE
ipr-rat LDLo:300 mg/kg CPCHAO 18,307,62
orl-mus LD50:981 mg/kg NCISP* JAN86

SAFETY PROFILE: Poison by intraperitoneal route. Moderately toxic by ingestion. When heated to decomposition it emits very toxic fumes of F^- and NO_x. See also FLUORIDES.

TCR250 CAS:80-81-9 ***HR: 3***
1,2,3,4-TETRAHYDRO-6-ETHYL-1,1,4,4-TETRAMETHYLNAPHTHALENE
mf: $C_{16}H_{24}$ mw: 216.40

SYN: 1,1,4,4-TETRAMETHYL-1,2,3,4-TETRAHYDRO-6-ETHYL-NAPHTHALENE

TOXICITY DATA with REFERENCE
ivn-mus LD50:56 mg/kg CSLNX* NX#00104

CONSENSUS REPORTS: Reported in EPA TSCA Inventory.

SAFETY PROFILE: Poison by intravenous route. When heated to decomposition it emits acrid smoke and irritating fumes.

TCR400 CAS:135-16-0 ***HR: D***
5,6,7,8-TETRAHYDROFOLIC ACID
mf: $C_{19}H_{23}N_7O_6$ mw: 445.49

SYNS: GLUTAMIC ACID, N-(p-(((2-AMINO-3,4,5,6,7,8-HEXAHYDRO-4-OXO-6-PTERIDINYL)METHYL)AMINO) BENZOYL)-, L- ◇ TETRAHYDROFOLIC ACID ◇ TETRAHYDROPTEROYLGLUTAMIC ACID ◇ THFA

TOXICITY DATA with REFERENCE
scu-rat TDLo:70 mg/kg (female 8-14D post):REP KSRNAM 7,1132,73
scu-mus TDLo:70 mg/kg (female 7-13D post):TER KSRNAM 7,1132,73

SAFETY PROFILE: An experimental teratogen. Other experimental reproductive effects. When heated to decomposition it emits toxic fumes of NO_x.

TCR750 CAS:109-99-9 ***HR: 3***
TETRAHYDROFURAN
DOT: UN 2056
mf: C_4H_8O mw: 72.12

$O(CH_2)_3CH_2$ (ring)

PROP: Colorless, mobile liquid; ether-like odor. Bp: 65.4°, flash p: 1.4°F (TCC), lel: 1.8%, uel: 11.8%, fp: −108.5°, d: 0.888 @ 20/4°, vap press: 114 mm @ 15°, vap d: 2.5, autoign temp: 610°F. Misc with water, alc, ketones, esters, ethers, and hydrocarbons.

SYNS: BUTYLENE OXIDE ◇ CYCLOTETRAMETHYLENE OXIDE ◇ DIETHYLENE OXIDE ◇ 1,4-EPOXYBUTANE ◇ FURANIDINE ◇ HYDROFURAN ◇ NCI-C60560 ◇ OXACYCLOPENTANE ◇ OXOLANE ◇ RCRA WASTE NUMBER U213 ◇ TETRAHYDROFURAAN (DUTCH) ◇ TETRAHYDROFURANNE (FRENCH) ◇ TETRAIDROFURANO (ITALIAN) ◇ TETRAMETHYLENE OXIDE ◇ THF

TOXICITY DATA with REFERENCE
mmo-esc 1 μmol/L GTPZAB 26(1),43,82
ihl-hmn TCLo:25000 ppm:CNS 34ZIAG -,580,69
orl-rat LDLo:3000 mg/kg TPKVAL 5,21,63
ihl-rat LC50:21000 ppm/3H SSEIBV 20,141,84
ipr-rat LD50:2900 mg/kg SAIGBL 24,373,82
ihl-mus LCLo:24000 mg/m³/2H TPKVAL 5,21,63
ipr-mus LD50:1900 mg/kg SAIGBL 24,373,82
ipr-gpg LDLo:500 mg/kg AIHAAP 35,21,74

CONSENSUS REPORTS: Reported in EPA TSCA Inventory.

OSHA PEL: (Transitional: TWA 200 ppm) TWA 200 ppm; STEL 250 ppm
ACGIH TLV: TWA 200 ppm; STEL 250 ppm
DFG MAK: 200 ppm (590 mg/m³)
DOT Classification: Flammable Liquid; Label: Flammable Liquid.

SAFETY PROFILE: Moderately toxic by ingestion and intraperitoneal routes. Mildly toxic by inhalation. Human systemic effects by inhalation: general anesthesia. Mutation data reported. Irritant to eyes and mucous membranes. Narcotic in high concentrations. Reported as causing injury to liver and kidneys.

Flammable liquid. A very dangerous fire hazard when exposed to heat, flames, oxidizers. Explosive in the form of vapor when exposed to heat or flame. In common with ethers, unstabilized tetrahydrofuran forms thermally explosive peroxides on exposure to air. Stored THF must always be tested for peroxide prior to distillation. Peroxides can be removed by treatment with strong ferrous sulfate solution made slightly acidic with sodium bisulfate. Caustic alkalies deplete the inhibitor in THF and may subsequently cause an explosive reaction. Explosive reaction with KOH; $NaAlH_2$; NaOH; sodium tetrahydroaluminate. Reacts with 2-aminophenol + potassium dioxide to form an explosive product. Reacts with lithium tetrahydroaluminate or borane to form explosive hydrogen gas. Violent reaction with metal halides (e.g., hafnium tetrachloride; titanium tetrachloride; zirconium tetrachloride). Vigorous reaction with bromine; calcium hydride + heat. Can react with oxidizing materials. To fight fire, use foam, dry chemical, CO_2. When heated to decomposition it emits acrid smoke and irritating fumes. See also 2-TETRAHYDROFURYL HYDROPEROXIDE.

TCS000 CAS:767-08-8 ***HR: 2***
TETRAHYDRO-2-FURANPROPANOL
mf: $C_7H_{14}O_2$ mw: 130.21

TOXICITY DATA with REFERENCE
ipr-mus LDLo:500 mg/kg CBCCT* 6,220,54
par-mus LDLo:4000 mg/kg CBCCT* 7,688,55

CONSENSUS REPORTS: Reported in EPA TSCA Inventory.

SAFETY PROFILE: Moderately toxic by intraperi-

toneal and parenteral routes. When heated to decomposition it emits acrid smoke and irritating fumes.

TCS500 CAS:4795-29-3 ***HR: 3***
TETRAHYDROFURFURYLAMINE
DOT: UN 2943
mf: $C_5H_{11}NO$ mw: 101.17

PROP: Bp: 153-155°, d: 0.977, refr index: 1.454.

SYN: USAF Q-2

TOXICITY DATA with REFERENCE
ipr-mus LD50:200 mg/kg NTIS** AD277-689

CONSENSUS REPORTS: Reported in EPA TSCA Inventory.

DOT Classification: Flammable or Combustible Liquid; Label: Flammable Liquid.

SAFETY PROFILE: Poison by intraperitoneal route. Flammable liquid when exposed to heat or flame. When heated to decomposition it emits toxic fumes of NO_x. See also AMINES.

TCS750 CAS:4676-82-8 ***HR: 3***
2-TETRAHYDROFURYL HYDROPEROXIDE
mf: $C_4H_8O_3$ mw: 104.11

$O(CH_2)_3CHOOH$ (ring)

SAFETY PROFILE: A powerful oxidizer. The first autooxidation product of tetrahydrofuran. Readily changes to a highly explosive polyalkylidene peroxide. When heated to decomposition it emits acrid smoke and irritating fumes.

TCT000 CAS:97-99-4 ***HR: 2***
TETRAHYDRO-2-FURYLMETHANOL
mf: $C_5H_{10}O_2$ mw: 102.15

PROP: A hygroscopic liquid. Water-sol. Mp: $< -80°$, lel: 1.5%, uel: 9.7% @ 72 to 122°F, bp: 178°, d: 1.0485 @ 20/4°, autoign temp: 540°F, vap d: 3.5, flash p: 183°F. Misc with water, alc, ether, acetone, benzene.

SYNS: QO THFA ◇ TETRAHYDRO-2-FURANCARBINOL ◇ TETRAHYDRO-2-FURANMETHANOL ◇ TETRAHYDROFURFURYL ALCOHOL ◇ TETRAHYDROFURYLALKOHOL (CZECH) ◇ THFA

TOXICITY DATA with REFERENCE
eye-rbt 20 mg/24H MOD 28ZPAK -,138,72
eye-rbt 20 mg/24H SEV 28ZPAK -,138,72
orl-rat LD50:2500 mg/kg HYSAAV 32,273,67
ipr-rat LDLo:1000 mg/kg JPPMAB 11,150,59
orl-mus LD50:2300 mg/kg HYSAAV 32,273,67
ivn-rbt LD50:725 mg/kg FEPRA7 8,294,49
orl-gpg LD50:3000 mg/kg HYSAAV 32,273,67

CONSENSUS REPORTS: Reported in EPA TSCA Inventory.

SAFETY PROFILE: Moderately toxic by ingestion, intravenous, and intraperitoneal routes. A severe eye irritant. Irritating to skin and mucous membranes. Violent or explosive reaction with 3-nitro-N-bromophthalimide. Combustible when exposed to heat or flame; can react with oxidizing materials. Explosive in the form of vapor when exposed to heat or flame. To fight fire, use alcohol foam, water, CO_2, dry chemical. When heated to decomposition it emits acrid smoke and irritating fumes.

TCT250 CAS:40959-16-8 ***HR: 3***
TETRAHYDROHARMINE
mf: $C_{13}H_{16}N_2O{\cdot}ClH$ mw: 252.77

SYN: 1,2,3,4-TETRAHYDRO-7-METHOXY-1-METHYL-9H-PYRIDO(3,4-b)INDOLEHYDROCHLORIDE

TOXICITY DATA with REFERENCE
ipr-mus LDLo:240 mg/kg QJPPAL 3,1,30
scu-mus LDLo:540 mg/kg QJPPAL 3,1,30
scu-rbt LDLo:320 mg/kg QJPPAL 3,1,30
scu-gpg LDLo:120 mg/kg QJPPAL 3,1,30
scu-frg LDLo:270 mg/kg QJPPAL 3,1,30

SAFETY PROFILE: Poison by intraperitoneal and subcutaneous routes. When heated to decomposition it emits very toxic fumes of NO_x and HCl.

TCU000 CAS:65860-38-0 ***HR: 3***
1,2,3,4-TETRAHYDRO-5-(HYDROXYMETHYL)-2-(ISOPROPYLAMINO)-1,6-NAPHTHALENEDIOL SESQUIHYDRATE (6-E), HYDROCHLORIDE
mf: $C_{14}H_{21}NO_3{\cdot}ClH{\cdot}3/2H_2O$ mw: 314.85

SYN: AA-497

TOXICITY DATA with REFERENCE
orl-mus LD50:4984 mg/kg ARZNAD 30,276,80
ivn-mus LD50:132 mg/kg ARZNAD 30,276,80
ivn-pig LD50:4564 mg/kg ARZNAD 30,276,80

SAFETY PROFILE: Poison by intravenous route. Mildly toxic by ingestion. When heated to decomposition it emits very toxic fumes of NO_x and HCl.

TCU250 CAS:3048-65-5 ***HR: 2***
3a,4,7,7a-TETRAHYDRO-1H-INDENE
mf: C_9H_{12} mw: 120.21

SYNS: BICYCLO(4,3,0)NONA-3,7-DIENE ◇ 3a,4,7,7a-TETRAHYDROINDENE ◇ 7,7,8,9-TETRAHYDROINDENE ◇ TETRAHYDROINDENE (RUSSIAN)

TOXICITY DATA with REFERENCE
orl-rat LD50:3730 mg/kg AIHAAP 30,470,69
orl-mus LD50:3500 mg/kg GTPZAB 18(10)52,74
skn-rbt LD50:14100 mg/kg AIHAAP 30,470,69

CONSENSUS REPORTS: Reported in EPA TSCA Inventory.

SAFETY PROFILE: Moderately toxic by ingestion. Mildly toxic by skin contact. When heated to decomposition it emits acrid smoke and irritating fumes.

TCU500 CAS:91-21-4 ***HR: 3***
TETRAHYDROISOQUINOLINE
mf: $C_9H_{11}N$ mw: 133.21

SYN: 1,2,3,4-TETRAHYDROISOQUINOLINE

TOXICITY DATA with REFERENCE
ipr-mus LDLo:128 mg/kg CBCCT* 3,52,51

CONSENSUS REPORTS: Reported in EPA TSCA Inventory.

SAFETY PROFILE: Poison by intraperitoneal route. When heated to decomposition it emits toxic fumes of NO_x.

TCU600 ***HR: 2***
TETRAHYDROLINALOOL
mf: $C_{10}H_{22}O_2$ mw: 158.29

PROP: Colorless liquid; floral odor. D: 0.923, refr index: 1.431, flash p: 183°F. Sol in alc, fixed oils; insol in water.

SYNS: 3,7-DIMETHYL-3-OCTANOL ◇ FEMA No. 3060

SAFETY PROFILE: Combustible liquid. When heated to decomposition it emits acrid smoke and irritating fumes.

TCU700 CAS:120-15-0 ***HR: 3***
1,2,3,4-TETRAHYDRO-6-METHOXYQUINOLINE
mf: $C_{10}H_{13}NO$ mw: 163.24

SYN: THALLINE

TOXICITY DATA with REFERENCE
ipr-mus LDLo:128 mg/kg CBCCT* 2,62,50

CONSENSUS REPORTS: Reported in EPA TSCA Inventory.

SAFETY PROFILE: Poison by intraperitoneal route. When heated to decomposition it emits toxic fumes of NO_x.

TCV375 CAS:69462-56-2 ***HR: 1***
1,2,3,6-TETRAHYDRO-1-((6-METHYL-3-CYCLOHEXEN-1-YL)CARBONYL)PYRIDINE
mf: $C_{13}H_{19}NO$ mw: 205.33

SYNS: AI3-36570 ◇ PYRIDINE, 1,2,3,6-TETRAHYDRO-1-((6-METHYL-3-CYCLOHEXEN-1-YL)CARBONYL)-

TOXICITY DATA with REFERENCE
skn-rbt 500 mg/24H MLD AEHA** 51-029-76
eye-rbt 100 mg/24H MLD AEHA** 51-029-76

SAFETY PROFILE: A skin and eye irritant. When heated to decomposition it emits toxic fumes of NO_x.

TCV400 CAS:69462-48-2 ***HR: 1***
1,2,3,6-TETRAHYDRO-1-((2-METHYLCYCLOHEXYL)CARBONYL)PYRIDINE
mf: $C_{13}H_{21}NO$ mw: 207.35

SYNS: AI3-36563 ◇ PYRIDINE, 1,2,3,6-TETRAHYDRO-1-((2-METHYLCYCLOHEXYL)CARBONYL)-

TOXICITY DATA with REFERENCE
skn-rbt 500 mg/24H MLD AEHA** 51-029-76
eye-rbt 100 mg/24H SEV AEHA** 51-029-76

SAFETY PROFILE: A skin and severe eye irritant. When heated to decomposition it emits toxic fumes of NO_x.

TCW500 CAS:26427-28-1 ***HR: 3***
TETRAHYDRO-2-METHYL-6-(TETRAHYDRO-2,5-DIOXO-3-FURYL)PYRAN-3,4-DICARBOXYLIC ANHYDRIDE POLYMER

SYN: NSC-46015

TOXICITY DATA with REFERENCE
dns-mus:lvr 2 g/L CNREA8 38,1610,78
dni-mus:lvr 3 g/L CNREA8 38,1610,78
orl-rat LDLo:1000 mg/kg NCIAL* -,38,64
ivn-rat LD50:121 mg/kg NCIAL* -,38,64
ivn-mus LD50:106 mg/kg NCIAL* -,38,64

SAFETY PROFILE: Poison by intravenous route. Moderately toxic by ingestion. Mutation data reported. When heated to decomposition it emits toxic fumes of NO_x. See also ANHYDRIDES.

TCW750 CAS:33401-94-4 ***HR: 3***
(E)-4,5,6-TETRAHYDRO-1-METHYL-2-(2-(2-THIENYL)ETHENYL)PYRIMIDINE
mf: $C_{11}H_{14}N_2S \cdot C_4H_6O_6$ mw: 356.43

SYNS: BANMINTH ◇ CP 10423-18 ◇ PYRANTEL TARTRATE ◇ PYREQUAN TARTRATE ◇ (E)-1,4,5,6-TETRAHYDRO-1-METHYL-2-(2-(2-THIENYL)VINYL)PYRIMIDINE TARTARATE (1:1)

TOXICITY DATA with REFERENCE
orl-rat LD50:170 mg/kg AUVJA2 46,297,70
orl-mus LD50:123 mg/kg OYYAA2 5,305,71
ivn-mus LD50:2220 µg/kg OYYAA2 5,305,71

SAFETY PROFILE: Poison by ingestion and intravenous routes. When heated to decomposition it emits very toxic fumes of NO_x and SO_x.

TCX500 CAS:119-64-2 ***HR: 2***
1,2,3,4-TETRAHYDRONAPHTHALENE
mf: $C_{10}H_{12}$ mw: 132.22

PROP: Colorless liquid; menthol odor. Bp: 207.2°, d: 0.981, vap press: 1 mm @ 38.0°, vap d: 4.55, autoign temp: 725°F, lel: 0.8% @ 212°F, uel: 5.0% @ 302°F, mp: −31.0°, flash p: 171°F (OC). Insol in water; misc with ethanol, butanol, acetone, benzene, ether.

SYNS: NAPHTHALENE-1,2,3,4-TETRAHYDRIDE ◇ $\Delta^{5,7,9}$-NAPHTHANTRIENE ◇ TETRAHYDRONAPHTHALENE ◇ TETRALIN ◇ TETRALINA (POLISH) ◇ TETRALINE ◇ TETRANAP

TOXICITY DATA with REFERENCE
skn-rbt 500 mg open SEV UCDS** 1/12/72
eye-rbt 500 mg open AMIHBC 4,119,51
orl-rat LD50:2860 mg/kg AMIHBC 4,119,51
skn-rbt LD50:17 g/kg UCDS** 1/12/72

CONSENSUS REPORTS: Reported in EPA TSCA Inventory.

SAFETY PROFILE: Moderately toxic by ingestion. Mildly toxic by skin contact. An eye and severe skin irritant. Narcotic in high concentration. Reported as causing cataracts and kidney injury in experimental animals. Irritating to mucous membranes. Combustible when exposed to heat or flame; can react with oxidizing materials. Explosive in the form of vapor when exposed to heat or flame. To fight fire, use water, foam, CO_2, dry chemical. Prolonged, close contact with air may cause an explosion. When heated to decomposition it emits acrid smoke and irritating fumes.

TCX750 CAS:530-91-6 ***HR: 3***
1,2,3,4-TETRAHYDRO-2-NAPHTHOL
mf: $C_{10}H_{12}O$ mw: 148.22

SYN: β-TETRALOL

TOXICITY DATA with REFERENCE
orl-rat LD50:1000 mg/kg JPETAB 93,26,48
orl-mus LD50:2000 mg/kg JPETAB 93,26,48
scu-dog LDLo:330 mg/kg ZMWIAJ 19,545,1881
orl-rbt LD50:2800 mg/kg JPETAB 93,26,48
orl-gpg LD50:1000 mg/kg JPETAB 93,26,48

CONSENSUS REPORTS: Reported in EPA TSCA Inventory.

SAFETY PROFILE: Poison by subcutaneous route. Moderately toxic by ingestion. When heated to decomposition it emits acrid smoke and irritating fumes.

TCY000 CAS:1125-78-6 ***HR: 3***
5,6,7,8-TETRAHYDRO-2-NAPHTHOL
mf: $C_{10}H_{12}O$ mw: 148.22

SYNS: 5,6,7,8-TETRAHYDRO-β-NAPHTHOL ◇ TETRALOL ◇ ac-β-TETRALOL

TOXICITY DATA with REFERENCE
skn-mus TDLo:9600 mg/kg/30W-I:CAR CNREA8 19,413,59
skn-mus TD:4144 mg/kg/14W-I:NEO CNREA8 19,413,59

SAFETY PROFILE: Questionable carcinogen with experimental carcinogenic and neoplastigenic data. When heated to decomposition it emits acrid smoke and irritating fumes.

TCY250 CAS:2954-50-9 ***HR: 3***
1,2,3,4-TETRAHYDRO-2-NAPHTHYLAMINE
mf: $C_{10}H_{13}N$ mw: 147.24

SYNS: 2-AMINO-1,2,3,4-TETRAHYDRONAFTALEN(CZECH) ◇ AMINOTETRALIN (CZECH) ◇ 2-AMINOTETRALIN ◇ TETRAHYDRO-β-NAPHTHYLAMINE ◇ β-TETRAHYDRONAPHTHYLAMINE ◇ β-1,2,3,4-TETRAHYDRONAPHTHYLAMINE ◇ THN

TOXICITY DATA with REFERENCE
skn-rbt 500 mg/24H MLD 28ZPAK -,67,72
eye-rbt 500 mg/24H MLD 28ZPAK -,67,72
ims-rat TDLo:20 mg/kg (5D preg):REP PSEBAA 100,555,59
scu-hmn TDLo:10 mg/kg:EYE,CNS APAVAY 115,14,1889
par-hmn TDLo:10 mg/kg:EYE,CNS APAVAY 115,14,1889
scu-mus LD50:64 mg/kg NYKZAU 61,168,65
ivn-dog LDLo:10 mg/kg AIPTAK 35,70,28
scu-frg LDLo:420 mg/kg APAVAY 115,14,1889

SAFETY PROFILE: Poison by intravenous and subcutaneous routes. Human systemic effects by subcutaneous and parenteral routes: pupillary dilation, sleep disorders and hallucinations or distorted perceptions. Experimental reproductive effects. A skin and eye irritant. When heated to decomposition it emits toxic fumes of NO_x.

TCY260 ***HR: 3***
2-(1,2,3,4-TETRAHYDRO-1-NAPHTHYLAMINO)-2-IMIDAZOLINE HYDROCHLORIDE
mf: $C_{13}H_{17}N_3 \cdot ClH$ mw: 251.79

SYN: KB-227

TOXICITY DATA with REFERENCE
orl-rat TDLo:200 mg/kg (7-14D preg):REP NYKZAU 61,490,66
orl-mus TDLo:100 mg/kg (7-14D preg):TER NYKZAU 61,490,66
orl-mus LD50:237 mg/kg NYKZAU 61,490,66
ipr-mus LD50:57 mg/kg ARZNAD 12,971,62
scu-mus LD50:77 mg/kg ARZNAD 12,971,62

SAFETY PROFILE: Poison by ingestion, subcutaneous, and intraperitoneal routes. An experimental teratogen. Experimental reproductive effects. When heated to decomposition it emits toxic fumes of NO_x and HCl.

TCY275 CAS:1136-84-1 ***HR: 3***
5,6,7,8-TETRAHYDRO-1-NAPHTHYL METHYLCARBAMATE
mf: $C_{12}H_{15}NO_2$ mw: 205.28

SYNS: ENT 27,253 ◇ MAM-TOX ◇ 5,6,7,8-TETRAHYDRO-1-NAFTYL-N-METHYLKARBAMAT (CZECH) ◇ UC 8,454 ◇ UNION CARBIDE UC-8454

TOXICITY DATA with REFERENCE
eye-rbt 500 mg/24H MLD 28ZPAK -,163,72
orl-rat LD50:199 mg/kg 28ZPAK -,163,72
skn-rbt LD50:1130 mg/kg BESAAT 12,161,66
orl-gpg LDLo:100 mg/kg JEENAI 61,1261,68
scu-gpg LDLo:50 mg/kg JEENAI 61,1261,68

SAFETY PROFILE: Poison by ingestion and subcutaneous routes. Moderately toxic by skin contact. An eye irritant. When heated to decomposition it emits toxic fumes of NO_x. See also CARBAMATES and ESTERS.

TCY500 CAS:58911-02-7 ***HR: 3***
TETRAHYDRONORHARMAN
mf: $C_{11}H_{12}N_2$•ClH mw: 208.71

SYN: 1,2,3,4-TETRAHYDRO-9H-PYRIDO(3,4-b)INDOLE,HYDROCHLORIDE

TOXICITY DATA with REFERENCE
ipr-mus LDLo:290 mg/kg QJPPAL 2,525,29
scu-mus LDLo:420 mg/kg QJPPAL 2,525,29
scu-rbt LDLo:300 mg/kg QJPPAL 2,525,29
scu-gpg LDLo:320 mg/kg QJPPAL 2,525,29
scu-frg LDLo:350 mg/kg QJPPAL 2,525,29

SAFETY PROFILE: Poison by intraperitoneal and subcutaneous routes. When heated to decomposition it emits very toxic fumes of NO_x and HCl.

TCY750 CAS:509-67-1 ***HR: 3***
TETRAHYDRO-1,4-OXAZINYLMETHYLCODEINE
mf: $C_{23}H_{30}N_2O_4$ mw: 398.55

SYNS: CODYLIN ◇ DIA-TUSS ◇ 7,8-DIDEHYDRO-4,5-α-EPOXY-17-METHYL-3-(2-MORPHOLINOETHOXY)MORPHINAN-6α-OL ◇ ETHNINE ◇ FOLCODINE ◇ GLYCODINE ◇ HIBERNYL ◇ HOMOCODEINE ◇ MEMINE ◇ β-MORPHOLINOETHYLMORPHINE ◇ O^3-(2-MORPHOLINOETHYL)MORPHINE ◇ 3-O-(2-MORPHOLINO-ETHYL)MORPHINE ◇ MORPHOLINYLETHYLMORPHINE ◇ 3-(2-(4-MORPHOLINYL)ETHYL)MORPHINE ◇ 3-MORPHOLYLAETHYLMORPHIN (GERMAN) ◇ PECTOLIN ◇ PHOLCODIN ◇ PHOLCODINE ◇ PRODROMINE ◇ WEIFACODINE

TOXICITY DATA with REFERENCE
ipr-mus LD50:400 mg/kg THERAP 7,21,52
scu-mus LD50:540 mg/kg ARZNAD 8,325,58
ivn-mus LD50:230 mg/kg THERAP 7,21,52
ivn-rbt LD50:300 mg/kg ARZNAD 8,325,58

SAFETY PROFILE: Poison by intraperitoneal and intravenous routes. Moderately toxic by subcutaneous route. When heated to decomposition it emits toxic fumes of NO_x. Used as an antitussive agent. See also CODEINE.

TCZ000 CAS:1195-16-0 ***HR: 2***
N-(TETRAHYDRO-2-OXO-3-THIENYL) ACETAMIDE
mf: $C_6H_9NO_2S$ mw: 159.22

SYNS: 2-ACETAMIDE-4-MERCAPTOBUTYRIC ACID Γ-THIOLACTONE ◇ 2-ACETAMIDO-4-MERCAPTOBUTYRIC ACID THIOLACTONE ◇ α-ACETAMIDO-Γ-THIOBUTYROLACTONE ◇ N-ACETYLHOMOCYSTEINE THIOLACTONE ◇ N-ACETYLHOMOCYSTEINTHIOLAKTON (GERMAN) ◇ ACHTL ◇ AHCTL ◇ BO 714 ◇ CITIOLASE ◇ CITIOLONE ◇ THIOXIDRENE

TOXICITY DATA with REFERENCE
ipr-rat LD50:1950 mg/kg MEIEDD 10,329,83
ivn-mus LD50:1200 mg/kg ARZNAD 8,72,58

CONSENSUS REPORTS: Reported in EPA TSCA Inventory.

SAFETY PROFILE: Moderately toxic by intraperitoneal and intravenous routes. When heated to decomposition it emits very toxic fumes of NO_x and SO_x. Used to treat hepatic disorders. Used as a photographic defogging agent and radiation protectant. See also MERCAPTANS.

TDA500 CAS:52-31-3 ***HR: 3***
TETRAHYDROPHENOBARBITAL
mf: $C_{12}H_{16}N_2O_3$ mw: 236.30

SYNS: ADORM ◇ AMNOSED ◇ CAVONLY ◇ CYCLOBARBITAL ◇ CYCLOBARBITOL ◇ CYCLOBARBITONE ◇ CYCLODORM ◇ CYCLOHEXENYL-ETHYL BARBITURIC ACID ◇ 5-(1-CYCLOHEXENYL)-5-ETHYLBARBITURIC ACID ◇ 5-(1-CYCLOHEXEN-1-YL)-5-ETHYLBARBITURIC ACID ◇ 5-(1-CYCLOHEXEN-1-YL)-5-ETHYL-2,4,6(1H,3H,5H)-PYRIMIDINETRIONE ◇ 5-ETHYL-5-CYCLOHEXENYL-BARBITURIC ACID ◇ ETHYLHEXABITAL ◇ FANODORMO ◇ HEXEMAL ◇ IRIFAN ◇ NAMURON ◇ PALINUM ◇ PHANODORM ◇ PHANODORN ◇ PHILODORM ◇ PRALUMIN ◇ PRO-SONIL ◇ SONAFORM

TOXICITY DATA with REFERENCE
orl-wmn TDLo:2 g/kg:PUL AIMEAS 37,290,52
orl-rat LDLo:300 mg/kg AEPPAE 182,348,36
ipr-rat LD50:290 mg/kg ARZNAD 17,242,67
scu-rat LD50:210 mg/kg AEPPAE 152,341,30
orl-mus LD50:840 mg/kg ARZNAD 17,242,67
ipr-mus LD50:350 mg/kg ARZNAD 17,242,67
scu-mus LDLo:300 mg/kg HDTU** -,-,33
orl-dog LDLo:200 mg/kg HBAMAK 4,1289,35
scu-dog LDLo:100 mg/kg HBAMAK 4,1289,35
orl-cat LDLo:120 mg/kg PHREA7 19,472,39
scu-cat LDLo:100 mg/kg HBAMAK 4,1289,35
orl-rbt LDLo:450 mg/kg JPETAB 44,325,32
ipr-rbt LDLo:130 mg/kg JPETAB 44,325,32

scu-rbt LDLo:300 mg/kg HBAMAK 4,1289,35
ivn-rbt LDLo:90 mg/kg JPPGAR 30,364,32

SAFETY PROFILE: Poison by ingestion, subcutaneous, intravenous, and intraperitoneal routes. Human systemic effects by ingestion: pulmonary consolidation. Used as a central nervous system depressant, hypnotic and sedative. When heated to decomposition it emits toxic fumes of NO_x. See also BARBITURATES.

TDB000 CAS:85-43-8 ***HR: 2***
TETRAHYDROPHTHALIC ACID ANHYDRIDE
DOT: UN 2698
mf: $C_8H_8O_3$ mw: 152.16

PROP: White powder. Mp: 101.9°, bp: 195° @ 50 mm, flash p: 315°F (OC), d: 1.375 @ 25/20°, vap press: 0.01 mm @ 20°, vap d: 5.25.

SYNS: ANHYDRID KYSELINY TETRAHYDROFTALOVE (CZECH) ◇ MALEIC ANHYDRIDE adduct of BUTADIENE ◇ TETRAHYDROFTALANHYDRID (CZECH) ◇ 3a,4,7,7a-TETRAHYDRO-1,3-ISOBENZOFURANDIOINE ◇ TETRAHYDROPHTHALIC ANHYDRIDE ◇ Δ^4-TETRAHYDROPHTHALIC ANHYDRIDE ◇ 1,2,3,6-TETRAHYDRO PHTHALIC ANHYDRIDE ◇ THPA

TOXICITY DATA with REFERENCE
skn-rbt 500 mg/24H MLD 28ZPAK -,140,72
eye-rbt 20 mg/24H MOD 28ZPAK -,140,72
orl-rat LD5O:5410 mg/kg 28ZPAK -,140,72
ipr-mus LDLo:500 mg/kg CBCCT* 7,780,55

CONSENSUS REPORTS: Reported in EPA TSCA Inventory.

DOT Classification: Corrosive Material; Label: None.

SAFETY PROFILE: Moderately toxic by intraperitoneal route. Mildly toxic by ingestion. A corrosive irritant to skin, eyes, and mucous membranes. Combustible when exposed to heat or flame. Will react with water or steam to produce heat; can react with oxidizing materials. To fight fire, use water, foam, CO_2, dry chemical. When heated to decomposition it emits acrid smoke and irritating fumes. See also ANHYDRIDES.

TDB200 CAS:69352-90-5 ***HR: 2***
α-(1,2,3,6-TETRAHYDROPHTHALIMIDO) GLUTARIMIDE
mf: $C_{13}H_{12}N_2O_4$ mw: 260.27

SYN: 4-CYCLOHEXENE-1,2-DICARBOXIMIDE,N-(2,6-DIOXO-3-PIPERIDYL)-

TOXICITY DATA with REFERENCE
ipr-mus TDLo:400 mg/kg (female 9D post):REP MOPMA3 13,133,77
ipr-mus LD50:700 mg/kg MOPMA3 13,133,77

SAFETY PROFILE: Moderately toxic by intraperitoneal route. Experimental reproductive effects. When heated to decomposition it emits toxic fumes of NO_x.

TDB600 ***HR: 3***
4'-O-TETRAHYDROPYRANYLADRIAMYCIN HYDROCHLORIDE
mf: $C_{32}H_{37}NO_{12}$•ClH mw: 664.16

SYN: (2''-R)-4'-O-TETRAHYDROPYRANYLADRIAMYCINMONO-HYDROCHLORIDE

TOXICITY DATA with REFERENCE
ivn-rat TDLo:3300 μg/kg (female 7-17D post):TER JJANAX 39,477,86
ivn-rat TDLo:8400 μg/kg (male 9W pre):REP JJANAX 39,463,86
unr-hmn TDLo:405 μg/kg:CNS,GIT,BLD INNDDK 1,169,83
orl-rat LDLo:300 mg/kg JJANAX 39,259,86
ipr-rat LD50:20300 μg/kg JJANAX 39,259,86
scu-rat LD50:21760 μg/kg JJANAX 39,259,86
ivn-rat LD50:18070 μg/kg JJANAX 39,259,86
orl-mus LD50:419 mg/kg JJANAX 39,250,86
ipr-mus LD50:13900 μg/kg JJANAX 39,250,86
scu-mus LD50:16900 μg/kg JJANAX 39,250,86
ivn-mus LD50:14 mg/kg JJANAX 39,250,86
ivn-dog LDLo:1 mg/kg JJANAX 39,265,86

SAFETY PROFILE: Poison by ingestion, subcutaneous, intravenous, and intraperitoneal routes. Human systemic effects by an unspecified route: anorexia, nausea or vomiting, and leukopenia. An experimental teratogen. Experimental reproductive effects. Used as an anticancer agent. When heated to decomposition it emits toxic fumes of NO_x, and HCl.

TDB750 CAS:102583-64-2 ***HR: 3***
TETRAHYDROSPIRAMYCIN

TOXICITY DATA with REFERENCE
scu-mus LD50:1350 mg/kg JJANAX 23,429,70
ivn-mus LD50:181 mg/kg JJANAX 23,429,70

SAFETY PROFILE: Poison by intravenous route. Moderately toxic by subcutaneous route.

TDC725 ***HR: 3***
o-((4,5,6,7-TETRAHYDROTHIENO(3,2-c)PYRIDIN-5-YL)METHYL)BENZONITRILE METHANE-SULFONATE
mf: $C_{15}H_{14}N_2S$•CH_4SO_3 mw: 350.48

SYN: 5-o-CYANOBENZYL-4,5,6,7-TETRAHYDROTHIENO(3,2-c)PYRIDINE METHANESULFONATE

TOXICITY DATA with REFERENCE
orl-rat LD50:228 mg/kg USXXAM #4400384
ipr-rat LD50:719 mg/kg USXXAM #4400384
ivn-rat LD50:80200 μg/kg USXXAM #4400384

orl-mus LD50:883 mg/kg USXXAM #4400384
ipr-mus LD50:898 mg/kg USXXAM #4400384
ivn-mus LD50:134 mg/kg USXXAM #4400384

CONSENSUS REPORTS: Cyanide and its compounds are on the Community Right-To-Know List.

SAFETY PROFILE: Poison by ingestion and intravenous routes. Moderately toxic by intraperitoneal route. When heated to decomposition it emits toxic fumes of SO_x, NO_x and CN^-. See also NITRILES.

TDC730 CAS:110-01-0 ***HR: 3***
TETRAHYDROTHIOPHENE
DOT: UN 2412
mf: C_4H_8S mw: 88.17

$S(CH_2)_3CH_2$ (ring)

PROP: Liquid. Mp: -96°, bp: 119°, flash p: 55° F, d: 1.01, refr index: 1.5030.

DOT Classification: Flammable Liquid; Label:Flammable Liquid

SAFETY PROFILE: Flammable liquid. Potentially explosive reaction with hydrogen peroxide. When heated to decomposition it emits toxic fumes of SO_x.

TDD000 CAS:81-61-8 ***HR: D***
1,2,5,8-TETRAHYDROXYANTHRAQUINONE
mf: $C_{14}H_8O_6$ mw: 272.22

PROP: Yellow-red. Mp: subl, decomp. Sltly sol in hot water; sol in alc, hot acetic acid, H_2SO_4; very sltly sol in ether.

SYN: QUINALIZARIN

TOXICITY DATA with REFERENCE
mmo-sat 50 μg/plate BCSTB5 5,1489,77
mma-sat 50 μg/plate BCSTB5 5,1489,77

CONSENSUS REPORTS: Reported in EPA TSCA Inventory.

SAFETY PROFILE: Mutation data reported. When heated to decomposition it emits acrid smoke and irritating fumes.

TDD250 CAS:81-60-7 ***HR: 2***
1,4,5,8-TETRAHYDROXYANTHRAQUINONE
mf: $C_{14}H_8O_6$ mw: 272.22

PROP: Red needles from phenyl nitrate. Mp: >275°, bp: subl. Sol in alkali, H_2SO_4.

SYNS: 1,4,5,8-LEUCOTETRAOXYANTHRAQUINONE ◇ 1,4,5,8-LTOA (RUSSIAN) ◇ 1,4,5,8-TETRAHYDROXY-9,10-ANTHRACENEDIONE ◇ 1,4,5,8-TETRAHYDROXYANTHRACHINON (CZECH)

TOXICITY DATA with REFERENCE
skn-rbt 500 mg/24H MLD 28ZPAK -,104,72
eye-rbt 500 mg/24H MLD 28ZPAK -,104,72
ipr-rat LD50:2200 mg/kg GTPZAB 21(12),27,77

SAFETY PROFILE: Moderately toxic by intraperitoneal route. A skin and eye irritant. When heated to decomposition it emits acrid smoke and irritating fumes.

TDD500 CAS:569-77-7 ***HR: D***
2,3,4,6-TETRAHYDROXY-5H-BENZOCYCLOHEPTENE-5-ONE
mf: $C_{11}H_8O_5$ mw: 220.19

SYN: PURPUROGALLIN

TOXICITY DATA with REFERENCE
mma-sat 50 μg/plate PMRSDJ 1,11,81
mma-sat 10 μg/plate PMRSDJ 1,11,81

CONSENSUS REPORTS: EPA Genetic Toxicology Program.

SAFETY PROFILE: Mutation data reported. When heated to decomposition it emits acrid smoke and irritating fumes.

TDD750 CAS:61490-68-4 ***HR: 3***
7-β,8-α-9-β,10-β-TETRAHYDROXY-7,8,9,10-TETRAHYDROBENZO(a)PYRENE
mf: $C_{20}H_{16}O_4$ mw: 320.36

SYN: BENZO(a)PYRENE,7,8,9,10-TETRAHYDRO-7-β,8-α-9-α-10-α-TETRAHYDROXY- ◇ BENZO(a)PYRENE-7-β,8-α-9-α-10-α-TETRAOL

TOXICITY DATA with REFERENCE
skn-mus TDLo:5126 μg/kg:NEO CRNGDP 3,371,78

SAFETY PROFILE: Questionable carcinogen with experimental neoplastigenic data. When heated to decomposition it emits acrid smoke and irritating fumes.

TDE000 CAS:64043-55-6 ***HR: 3***
TETRAIODO-α,α'-DIETHYL-4,4'-STILBENEDIOL
mf: $C_{18}H_{16}I_4O_2$ mw: 771.94

SYN: DIETHYLSTILBESTROL, IODINE DERIVATIVE

TOXICITY DATA with REFERENCE
scu-mus TDLo:85 mg/kg/3W-I:CAR AIPUAN 8,207,65

SAFETY PROFILE: Questionable carcinogen with experimental carcinogenic data. When heated to decomposition it emits toxic fumes of I^-. See also DIETHYLSTILBESTEROL and IODIDES.

TDE250 CAS:513-92-8 ***HR: 3***
TETRAIODOETHYLENE
mf: C_2I_4 mw: 531.62

PROP: Light yellow; heavy, small, practically odorless crystals; characteristic odor. D: 2.98, mp: 187°. Insol in water; sol in benzene, chloroform, toluene, CS_2; sltly sol in ether.

TOXICITY DATA with REFERENCE
ivn-mus LD50:56 mg/kg CSLNX* NX#04958

CONSENSUS REPORTS: Reported in EPA TSCA Inventory.

SAFETY PROFILE: Poison by intravenous route. Reacts violently with IF_5. When heated to decomposition it emits toxic fumes of I^-. See also IODIDES.

TDE750 CAS:632-73-5 ***HR: 3***
TETRAIODOTHALEIN SODIUM
mf: $C_{20}H_8I_4O_4•2Na$ mw: 865.86

SYNS: ANTINOSIN ◇ BILITRAST ◇ CHOLEPULVIS ◇ CHOLUMBRIN ◇ FORIOD ◇ IODEIKON ◇ IODOGNOST ◇ IODOPHENE SODIUM ◇ IODOPHTHALEIN SODIUM ◇ IODORAYORAL ◇ IODTETRAGNOST ◇ KERAPHEN ◇ NOSOPHENE SODIUM ◇ OPACIN ◇ PHOTOBILINE ◇ PILIOPHEN ◇ RADIOTETRANE ◇ SHADOCOL ◇ SOLUBLE IODOPHTHALEIN ◇ STIPOLAC ◇ TETIOTHALEIN SODIUM ◇ TETRAIODE ◇ TETRAIODOPHENOLPHTHALEIN SODIUM ◇ TIPPS ◇ VIDEOPHEL

TOXICITY DATA with REFERENCE
ivn-mus LD50:360 mg/kg JMCMAR 13,997,70

SAFETY PROFILE: Poison by intravenous route. When heated to decomposition it emits toxic fumes of I^- and Na_2O. See also IODIDES.

TDE765 CAS:26562-01-6 ***HR: 3***
TETRAISOAMYLSTANNANE
mf: $C_{20}H_{44}Sn$ mw: 403.33

SYN: TETRAISOPENTYLSTANNANE

TOXICITY DATA with REFERENCE
orl-mus LDLo:202 mg/kg APFRAD 14,88,56
ipr-mus LDLo:20 mg/kg APFRAD 14,88,56
ivn-mus LDLo:40 mg/kg APFRAD 14,88,56
ims-mus LDLo:403 mg/kg APFRAD 14,88,56

OSHA PEL: TWA 0.1 mg(Sn)/m^3 (skin)
ACGIH TLV: TWA 0.1 mg(Sn)/m^3 (skin) (Proposed: TWA 0.1 mg(Sn)/m^3; STEL 0.2 mg(Sn)/m^3 (skin))
NIOSH REL: (Organotin Compounds) TWA 0.1 mg(Sn)/m^3

SAFETY PROFILE: Poison by ingestion, intravenous, and intraperitoneal routes. Moderately toxic by intramuscular route. When heated to decomposition it emits acrid smoke and irritating fumes. See also TIN COMPOUNDS.

TDE775 CAS:3531-43-9 ***HR: 3***
TETRAISOBUTYLTIN
mf: $C_{16}H_{36}Sn$ mw: 347.21

SYN: TETRAISOBUTYLSTANNANE

TOXICITY DATA with REFERENCE
orl-mus LDLo:347 mg/kg APFRAD 14,88,56
ipr-mus LDLo:87 mg/kg APFRAD 14,88,56
ivn-mus LDLo:17 mg/kg APFRAD 14,88,56
ims-mus LDLo:174 mg/kg APFRAD 14,88,56

OSHA PEL: TWA 0.1 mg(Sn)/m^3 (skin)
ACGIH TLV: TWA 0.1 mg(Sn)/m^3 (skin) (Proposed: TWA 0.1 mg(Sn)/m^3; STEL 0.2 mg(Sn)/m^3 (skin))
NIOSH REL: (Organotin Compounds) TWA 0.1 mg(Sn)/m^3

SAFETY PROFILE: Poison by ingestion, intramuscular, intravenous, and intraperitoneal routes. When heated to decomposition it emits acrid smoke and irritating fumes. See also TIN COMPOUNDS.

TDF000 CAS:5424-26-0 ***HR: 3***
TETRAISOPENTYLAMMONIUM IODIDE
mf: $C_{20}H_{44}N•I$ mw: 425.55

TOXICITY DATA with REFERENCE
ivn-mus LD50:100 mg/kg CSLNX* NX#02831

CONSENSUS REPORTS: Reported in EPA TSCA Inventory.

SAFETY PROFILE: Poison by intravenous route. When heated to decomposition it emits very toxic fumes of NO_x, NH_3, and I^-. See also IODIDES.

TDF250 CAS:61614-71-9 ***HR: 3***
TETRAISOPROPYL DITHIONOPYROPHOSPHATE
mf: $C_{12}H_{28}O_5P_2S_2$ mw: 378.46

SYN: TETRAISOPROPYLTHIOPYROPHOSPHATE

TOXICITY DATA with REFERENCE
ipr-rat LD50:50 mg/kg CJCHAG 34,1819,56
ipr-mus LD50:25 mg/kg CJCHAG 34,1819,56

SAFETY PROFILE: Poison by intraperitoneal route. When heated to decomposition it emits very toxic fumes of PO_x and SO_x. See also PHOSPHATES.

TDF500 CAS:4593-82-2 ***HR: 2***
TETRAISOPROPYL GERMANE
mf: $C_{12}H_{28}Ge$ mw: 244.99

TOXICITY DATA with REFERENCE
orl-rat LDLo:2000 mg/kg CHDDAT 262,1302,66
ipr-rat LDLo:430 mg/kg CHDDAT 262,1302,66

orl-mus LDLo:2180 mg/kg CHDDAT 262,1302,66
ipr-mus LDLo:620 mg/kg CHDDAT 262,1302,66

SAFETY PROFILE: Moderately toxic by ingestion and intraperitoneal routes. When heated to decomposition it emits acrid smoke and irritating fumes. See also GERMANIUM COMPOUNDS.

TDF750 CAS:5836-28-2 ***HR: 3***
TETRAISOPROPYL PYROPHOSPHATE
mf: $C_{12}H_{28}O_7P_2$ mw: 346.34

SYN: ISOPROPYL PYROPHOSPHATE

TOXICITY DATA with REFERENCE
orl-rat LDLo:5 mg/kg NCNSA6 5,38,53
ipr-mus LD50:16 mg/kg PAREAQ 11,636,59

SAFETY PROFILE: Poison by ingestion and intraperitoneal routes. When heated to decomposition it emits toxic fumes of PO_x. See also PHOSPHATES and ESTERS.

TDG000 CAS:2949-42-0 ***HR: 3***
TETRAISOPROPYLTIN
mf: $C_{12}H_{28}Sn$ mw: 291.09

SYN: TETRAISOPROPYLSTANNANE

TOXICITY DATA with REFERENCE
ivn-mus LD50:56 mg/kg CSLNX* NX#00226

OSHA PEL: TWA 0.1 mg(Sn)/m^3 (skin)
ACGIH TLV: TWA 0.1 mg(Sn)/m^3 (skin) (Proposed: TWA 0.1 mg(Sn)/m^3; STEL 0.2 mg(Sn)/m^3 (skin))
NIOSH REL: (Organotin Compounds) TWA 0.1 mg (Sn)/m^3

SAFETY PROFILE: Poison by intravenous route. When heated to decomposition it emits acrid smoke and irritating fumes. See also TIN COMPOUNDS.

TDG250 CAS:16646-44-9 ***HR: 2***
1,1,2,2-TETRAKIS(ALLYLOXY)ETHANE
mf: $C_{14}H_{22}O_4$ mw: 254.36

TOXICITY DATA with REFERENCE
orl-rat LD50:1410 mg/kg TXAPA9 28,313,74
skn-rbt LD50:1410 mg/kg TXAPA9 28,313,74

SAFETY PROFILE: Moderately toxic by ingestion and skin contact. When heated to decomposition it emits acrid smoke and irritating fumes.

TDG500 CAS:4239-06-9 ***HR: D***
2,3,5,6-TETRAKIS(1-AZIRIDINYL)-1,4-BENZOQUINONE

SYNS: TEB ◇ 2,3,5,6-TETRA-ETHYLENEIMINO-1,4-BENZOQUINONE

TOXICITY DATA with REFERENCE
mmo-nsc 200 μg/plate MUREAV 53,297,78
cyt-hmn:fbr 1800 μmol/L MUREAV 34,299,76

CONSENSUS REPORTS: EPA Genetic Toxicology Program.

SAFETY PROFILE: Human mutation data reported. When heated to decomposition it emits toxic fumes of NO_x.

TDG750 CAS:63918-49-0 ***HR: 3***
N,N,N'N'-TETRAKIS(2-CHLOROETHYL)ETHYLENE DIAMINE DIHYDROCHLORIDE
mf: $C_{10}H_{20}Cl_4N_2 \cdot 2ClH$ mw: 383.04

TOXICITY DATA with REFERENCE
scu-rat LD50:19 mg/kg JPETAB 91,224,47
ivn-rat LD50:3800 μg/kg JPETAB 91,224,47
ipr-mus LD50:16 mg/kg CANCAR 2,1075,49
scu-mus LD50:26 mg/kg JPETAB 91,224,47
ivn-mus LD50:7500 μg/kg JPETAB 91,224,47
ivn-rbt LD50:2500 μg/kg JPETAB 91,224,47

SAFETY PROFILE: Poison by subcutaneous, intravenous, and intraperitoneal routes. When heated to decomposition it emits very toxic fumes of HCl and NO_x.

TDG775 ***HR: 3***
TETRAKIS(CHLOROETHYNYL)SILANE
mf: C_8Cl_4Si mw: 265.99

$$(ClC \equiv C)_4Si$$

SAFETY PROFILE: Explodes on grinding or impact. When heated to decomposition it emits toxic fumes of Cl^-. See also SILANE and ACETYLENE COMPOUNDS.

TDH000 ***HR: 3***
TETRAKIS-(N,N-DICHLOROAMINOMETHYL) METHANE
mf: $C_5H_8Cl_8N_4$ mw: 407.77

$$(Cl_2NCH_2)_4C$$

SAFETY PROFILE: Shock- and heat-sensitive explosive. More powerful than mercury fulminate. When heated to decomposition it emits very toxic fumes of NO_x and Cl^-.

TDH100 CAS:18928-76-2 ***HR: 1***
TETRAKIS-(2-HYDROXYETHYL)SILANE
mf: $C_8H_{20}O_4Si$ mw: 208.37

SYNS: ETHANOL, 2,2',2'',2'''-SILANETETRAYLTETRAKIS- ◇ 2,2',2'',2'''-SILANETETRAYLTETRAKISETHANOL ◇ TETRA-(2-HYDROXYETHYL)SILANE

TOXICITY DATA with REFERENCE
skn-rbt 500 mg/24H MLD 85JCAE-,1232,86

eye-rbt 500 mg/24H MLD 85JCAE-,1232,86
orl-rat LD50:7460 mg/kg TXAPA9 28,313,74

SAFETY PROFILE: Mildly toxic by ingestion. A skin and eye irritant. When heated to decomposition it emits acrid smoke and irritating fumes.

TDH250 CAS:7580-37-2 ***HR: 3***
TETRAKIS(HYDROXYMETHYL)PHOSPHONIUM ACETATE
mf: $C_4H_{12}O_4P{\bullet}C_2H_3O_2$ mw: 214.18

TOXICITY DATA with REFERENCE
ivn-mus LD50:100 mg/kg CSLNX* NX#03094

CONSENSUS REPORTS: Reported in EPA TSCA Inventory.

SAFETY PROFILE: Poison by intravenous route. When heated to decomposition it emits toxic fumes of PO_x.

TDH500 CAS:55818-96-7 ***HR: 3***
TETRAKIS(HYDROXYMETHYL)PHOSPHONIUM ACETATE mixed with TETRAKIS (HYDROXYMETHYL)PHOSPHONIUM DIHYDROGEN PHOSPHATE (76:24)
mf: $C_4H_{12}O_4P{\bullet}C_4H_{12}O_4P{\bullet}C_2H_3O_2{\bullet}1/3O_4P$ mw: 400.97

SYNS: PYROSET FLAME RETARDANT TKP ◇ PYROSET TKP

TOXICITY DATA with REFERENCE
skn-mus TDLo:14 g/kg/57W-I:ETA JTEHD6 2,539,77

SAFETY PROFILE: Questionable carcinogen with experimental tumorigenic data. When heated to decomposition it emits toxic fumes of PO_x.

TDH750 CAS:124-64-1 ***HR: 3***
TETRAKIS(HYDROXYMETHYL)PHOSPHONIUM CHLORIDE
mf: $C_4H_{12}O_4P{\bullet}Cl$ mw: 190.58

SYNS: NCI-C55061 ◇ THPC

TOXICITY DATA with REFERENCE
otr-mus:emb 5000 ppm JTEHD6 6,259,80
otr-ham:kdy 5000 ppm JTEHD6 6,259,80
cyt-ham:lng 87 mg/L GMCRDC 27,95,81
msc-ham:lng 5000 ppm JTEHD6 6,259,80
orl-mus LD50:400 mg/kg KOBUA3 24,788,75
ipr-mus LDLo:125 mg/kg CBCCT* 7,789,55

CONSENSUS REPORTS: Reported in EPA TSCA Inventory.

SAFETY PROFILE: Poison by ingestion and intraperitoneal routes. Mutation data reported. When heated to decomposition it emits very toxic fumes of PO_x and Cl^-.

TDH775 CAS:24748-25-2 ***HR: 3***
TETRAKIS(HYDROXYMETHYL)PHOSPHONIUM NITRATE
mf: $C_4H_{12}NO_7P$ mw: 217.12

$$(HOCH_2)_4P^-NO_3^-$$

SAFETY PROFILE: May explode when heated above 105°C. When heated to decomposition it emits toxic fumes of PO_x and NO_x. See also NITRATES.

TDI000 CAS:55566-30-8 ***HR: D***
TETRAKIS(HYDROXYMETHYL)PHOSPHONIUM SULFATE
mf: $C_8H_{24}O_8P_2{\bullet}O_4S$ mw: 406.32

SYNS: NCI-C55050 ◇ PYROSET TKO ◇ THPS

TOXICITY DATA with REFERENCE
otr-mus:emb 5000 ppm JTEHD6 6,259,80
msc-ham:lng 5000 ppm JTEHD6 6,259,80

CONSENSUS REPORTS: Reported in EPA TSCA Inventory.

SAFETY PROFILE: Mutation data reported. When heated to decomposition it emits very toxic fumes of PO_x and SO_x. See also SULFATES.

TDI100 CAS:83195-98-6 ***HR: 3***
1,3,4,6-TETRAKIS(2-METHYLTETRAZOL-5-YL)-HEXAAZA-1,5-DIENE
mf: $C_8H_{12}N_{22}$ mw: 416.33

SAFETY PROFILE: An explosive sensitive to shock, friction or rapid heating above 121°C. When heated to decomposition it emits toxic fumes of NO_x.

TDI250 CAS:6452-62-6 ***HR: 3***
TETRAKIS(p-PHENOXYPHENYL)TIN
mf: $C_{48}H_{36}O_4Sn$ mw: 795.53

SYN: TETRAKIS(p-PHENOXYPHENYL)STANNANE

TOXICITY DATA with REFERENCE
ivn-mus LD50:56 mg/kg CSLNX* NX#01768

OSHA PEL: TWA 0.1 mg(Sn)/m^3 (skin)
ACGIH TLV: TWA 0.1 mg(Sn)/m^3 (skin) (Proposed: TWA 0.1 mg(Sn)/m^3; STEL 0.2 mg(Sn)/m^3 (skin))
NIOSH REL: (Organotin Compounds) TWA 0.1 mg (Sn)/m^3

SAFETY PROFILE: Poison by intravenous route. When heated to decomposition it emits acrid smoke and irritating fumes. See also TIN COMPOUNDS.

TDI300 CAS:50831-29-3 ***HR: 3***
TETRAKIS(THIOUREA)MANGANESE(II) PERCHLORATE
mf: $C_4H_{16}Cl_2MnN_8O_8S_4$ mw: 558.30

$[(CH_4N_2S)_4Mn][ClO_4]_2$

CONSENSUS REPORTS: Manganese and its compounds are on The Community Right-To-Know List.

SAFETY PROFILE: Explodes when heated to 257°C. When heated to decomposition it emits toxic fumes of Cl^-, SO_x, and NO_x. See also MANGANESE COMPOUNDS and PERCHLORATES.

TDI350 CAS:529-33-9 ***HR: 2***
α-TETRALOL
mf: $C_{10}H_{12}O$ mw: 148.22

PROP: Liquid. Mp: 15.5°, bp: 140°.

SYNS: 1-NAPHTHOL, 1,2,3,4-TETRAHYDRO- ◇ 1,2,3,4-TETRAHYDRO-α-NAPHTHOL ◇ 1,2,3,4-TETRAHYDRO-1-NAPHTHOL

TOXICITY DATA with REFERENCE
skn-rbt 500 mg/24H MLD 85JCAE-,217,86
eye-rbt 500 mg/24H MLD 85JCAE-,217,86
orl-rat LD50:1620 mg/kg TXAPA9 28,313,74

SAFETY PROFILE: Moderately toxic by ingestion. A skin and eye irritant. When heated to decomposition it emits acrid smoke and irritating fumes.

TDI475 CAS:475-81-0 ***HR: 3***
1,2,9,10-TETRAMETHOXY-6a-α-APORPHINE
mf: $C_{21}H_{25}NO_4$ mw: 355.47

PROP: d-Form found in nature. d-Form: Orthorhombic plates, prisms from ethyl acetate or ether. Mp: 120°. Sol in acetone, alc, chloroform, ethyl acetate; mod sol in ether, petr ether; practically insol in water and benzene.

SYNS: BOLDINE DIMETHYL ETHER ◇ BROMCHOLITIN ◇ (s)-5,6,6a,7-TETRAHYDRO-1,2,9,10-TETRAMETHOXY-6-METHYL-4H-DIBENZO(de,g)QUINOLINE (9CI) ◇ GLAUCINE ◇ (+)-GLAUCINE ◇ d-GLAUCINE ◇ s-(+)-GLAUCINE ◇ GLAUVENT ◇ (s)-5,6,6a,7-TETRAHYDRO-1,2,9,10-TETRAMETHOXY-6-METHYL-4H-DIBENZO(de,g)QUINOLINE

TOXICITY DATA with REFERENCE
orl-rat LD50:545 mg/kg FATOAO 46(4),100,83
ipr-rat LD50:143 mg/kg FATOAO 46(4),100,83
orl-mus LD50:434 mg/kg FATOAO 46(4),100,83
ipr-mus LD50:167 mg/kg FATOAO 46(4),100,83
ipr-gpg LD50:140 mg/kg FATOAO 46(4),100,83

SAFETY PROFILE: Poison by intraperitoneal route. Moderately toxic by ingestion. When heated to decomposition it emits toxic fumes of NO_x.

TDI500 CAS:2269-06-9 ***HR: 3***
1,2,9,10-TETRAMETHOXY-6a-α-APORPHINE HYDROCHLORIDE
mf: $C_{21}H_{25}NO_4{\bullet}ClH$ mw: 391.93

SYN: GLAUCINE HYDROCHLORIDE

TOXICITY DATA with REFERENCE
orl-rat LD50:545 mg/kg FATOAO 46(4),100,83
ipr-rat LD50:143 mg/kg FATOAO 46(4),100,83
orl-mus LD50:430 mg/kg PCJOAU 10,134,76
ipr-mus LD50:120 mg/kg APPBDI 5,3,79
scu-mus LD50:420 mg/kg PCJOAU 10,134,76
ivn-mus LD50:54 mg/kg PCJOAU 10,134,76
ipr-gpg LD50:140 mg/kg FATOAO 46(4),100,83

SAFETY PROFILE: Poison by intraperitoneal and intravenous routes. Moderately toxic by ingestion and subcutaneous routes. When heated to decomposition it emits very toxic fumes of NO_x and HCl.

TDI600 ***HR: 3***
dl-1,2,9,10-TETRAMETHOXY-6a-α-APORPHINE PHOSPHATE
mf: $C_{21}H_{28}NO_4{\bullet}H_3O_4P$ mw: 456.50

SYNS: dl-832 ◇ dl-GLAUCINE PHOSPHATE ◇ dl-GLAUCINPHOSPHAT (GERMAN)

TOXICITY DATA with REFERENCE
orl-mus LD50:401 mg/kg ARZAND 33,936,83
ipr-mus LD50:272 mg/kg ARZNAD 33,936,83
scu-mus LD50:311 mg/kg ARZNAD 33,936,83
ivn-mus LD50:98 mg/kg ARZNAD 33,936,83

SAFETY PROFILE: Poison by subcutaneous, intravenous, and intraperitoneal routes. Moderately toxic by ingestion. When heated to decomposition it emits toxic fumes of NO_x and PO_x.

TDI750 CAS:478-15-9 ***HR: 3***
dl-2,3,9,10-TETRAMETHOXYBERBIN-1-OL
mf: $C_{21}H_{25}NO_5$ mw: 371.47

SYNS: CAPAURIDINE ◇ CAPAURINE, (±)- ◇ CAPAURINE, dl-

TOXICITY DATA with REFERENCE
ipr-rat LDLo:500 mg/kg NCNSA6 5,22,53
ivn-mus LD50:82 mg/kg FEPRA7 5,163,46

SAFETY PROFILE: Poison by intravenous route. Moderately toxic by intraperitoneal route. When heated to decomposition it emits toxic fumes of NO_x.

TDJ000 CAS:60223-52-1 ***HR: D***
3,6,11,14-TETRAMETHOXYDIBENZO(g,p)CHRYSENE
mf: $C_{30}H_{24}O_4$ mw: 448.54

TOXICITY DATA with REFERENCE
mma-sat 100 μg/plate MUREAV 40,225,76

CONSENSUS REPORTS: EPA Genetic Toxicology Program.

SAFETY PROFILE: Mutation data reported. When heated to decomposition it emits acrid smoke and irritating fumes.

TDJ250 CAS:73728-78-6 ***HR: 3***
3,2',4',6'-TETRAMETHYLAMINODIPHENYL
mf: $C_{16}H_{19}N$ mw: 225.36

SYN: 3,2',4',6'-TETRAMETHYLBIPHENYLAMINE

TOXICITY DATA with REFERENCE
scu-rat TDLo:1400 mg/kg/W-I:ETA BMBUAQ 14,141,58

SAFETY PROFILE: Questionable carcinogen with experimental tumorigenic data. When heated to decomposition it emits toxic fumes of NO_x.

TDJ300 CAS:13422-81-6 ***HR: 2***
TETRAMETHYLAMMONIUM AMIDE
mf: $C_4H_{14}N_2$ mw: 90.17

$(CH_3)_4N^-NH_2^-$

SAFETY PROFILE: Violent reaction with ammonia at room temperature. When heated to decomposition it emits toxic fumes of NO_x and NH_3. See also AMIDES.

TDJ325 CAS:68574-15-2 ***HR: 3***
TETRAMETHYLAMMONIUM AZIDOCYANATOIODATE(I)
mf: $C_5H_{12}IN_5O$ mw: 285.09

$(CH_3)_4N^-[(OCN)IN_3]^-$

CONSENSUS REPORTS: Cyanide and its compounds are on The Community Right-To-Know List.

SAFETY PROFILE: Explodes when exposed to intense light. When heated to decomposition it emits toxic fumes of I^-, CN^-, NH_3, and NO_x. See also CYANATES, AZIDES, and IODATES.

TDJ350 CAS:68574-17-4 ***HR: 3***
TETRAMETHYLAMMONIUM AZIDOCYANOIODATE(I)
mf: $C_5H_{12}IN_5$ mw: 269.09

$(CH_3)_4N^-[(NC)IN_3]^-$

CONSENSUS REPORTS: Cyanide and its compounds are on The Community Right-To-Know List.

SAFETY PROFILE: Explodes when exposed to intense light. When heated to decomposition it emits toxic fumes of I^-, CN^-, NH_3, and NO_x. See also CYANIDE, IODATES, and AZIDES.

TDJ375 ***HR: 3***
TETRAMETHYLAMMONIUM AZIDOSELENOCYANATOIODATE(I)
mf: $C_5H_{12}IN_5Se$ mw: 348.05

$(CH_3)_4N^-[(CNSe)IN_3]^-$

CONSENSUS REPORTS: Selenium and its compounds, as well as Cyanide and its compounds, are on The Community Right-To-Know List.

OSHA PEL: TWA 0.2 mg(Se)/m^3
ACGIH TLV: TWA 0.2 mg(Se)/m^3
DFG MAK: 0.1 mg(Se)/m^3

SAFETY PROFILE: Explodes when exposed to intense light. When heated to decomposition it emits toxic fumes of I^-, Se, CN^-, NH_3, and NO_x. See also SELENIUM COMPOUNDS, CYANATES, and IODATES.

TDJ500 CAS:16883-45-7 ***HR: 3***
TETRAMETHYLAMMONIUM BOROHYDRATE
mf: $C_4H_{12}N{\cdot}BH_4$ mw: 89.02

SYN: TMAB

TOXICITY DATA with REFERENCE
skn-rbt 10% SEV NTIS** ORNL-5576
skn-rbt 1% MLD NTIS** ORNL-5576
eye-rbt 24 mg MOD NTIS** ORNL-5576
orl-rat LD50:77 mg/kg NTIS** ORNL-5576
orl-mus LD50:67 mg/kg NTIS** ORNL-5576
itr-mus LD50:31 mg/kg NTIS** ORNL-5576
skn-rbt LDLo:100 mg/kg NTIS** ORNL-5576

CONSENSUS REPORTS: Reported in EPA TSCA Inventory.

SAFETY PROFILE: Poison by ingestion, skin contact and intratracheal routes. An eye and severe skin irritant. When heated to decomposition it emits toxic fumes of NO_x and NH_3. See also BORON COMPOUNDS.

TDJ750 CAS:64-20-0 ***HR: 3***
TETRAMETHYLAMMONIUM BROMIDE
mf: $C_4H_{12}N{\cdot}Br$ mw: 154.08

PROP: Crystals. D: 1.50, mp: decomp @ >230°, bp: subl >360°. Sol in water, SO_2; sltly sol in abs alc; insol in ether.

TOXICITY DATA with REFERENCE
dnd-mam:lym 50 mmol/L CBINA8 19,197,77
ipr-rat LD50:24600 μg/kg PHMGBN 21,256,80
ivn-mus LD50:1800 μg/kg CSLNX* NX#01652

CONSENSUS REPORTS: Reported in EPA TSCA Inventory.

SAFETY PROFILE: Poison by intraperitoneal and intravenous routes. Mutation data reported. When heated to decomposition it emits very toxic fumes of NO_x, NH_3, and Br^-. See also BROMIDES.

TDK000 CAS:75-57-0 ***HR: 3***
TETRAMETHYLAMMONIUM CHLORIDE
mf: $C_4H_{12}N{\cdot}Cl$ mw: 109.62

PROP: Crystals. Mp: > 300°. Hygroscopic.

SYN: USAF AN-8

TOXICITY DATA with REFERENCE
ipr-mus LD50:25 mg/kg NTIS** AD414-344
scu-mus LD50:40 mg/kg NTIS** ORNL-5576
scu-rbt LDLo:6 mg/kg AIPTAK 7,183,1900
scu-frg LDLo:2 mg/kg SAPHAO 10,201,1900

CONSENSUS REPORTS: Reported in EPA TSCA Inventory.

SAFETY PROFILE: Poison by intraperitoneal and subcutaneous routes. When heated to decomposition it emits very toxic fumes of NO_x, NH_3, and Cl^-. See also CHLORIDES.

TDK250 ***HR: 3***
TETRAMETHYLAMMONIUM CHLORITE
mf: $C_4H_{13}ClNO_2$ mw: 142.58

SAFETY PROFILE: The dry solid is an impact-sensitive explosive. Upon decomposition it emits very toxic fumes of NO_x, NH_3, and Cl^-. See also CHLORITES.

TDK300 CAS:68574-13-0 ***HR: 3***
TETRAMETHYLAMMONIUM DIAZIDOIODATE(I)
mf: $C_4H_{12}IN_7$ mw: 285.09

$(CH_3)_4N^-[I(N_3)_2]^-$

SAFETY PROFILE: A crystalline high explosive. When heated to decomposition it emits toxic fumes of NO_x, NH_3, and I^-. See also IODATES, AZIDES, and EXPLOSIVES, HIGH.

TDK500 CAS:75-59-2 ***HR: 3***
TETRAMETHYLAMMONIUM HYDROXIDE
DOT: UN 1835
mf: $C_4H_{12}N{\cdot}HO$ mw: 91.18

PROP: Liquid. D: 1.

SYNS: HYDROXYDE de TETRAMETHYLAMMONIUM (FRENCH) ◇ TM ◇ TETRAMETHYL AMMONIUM HYDROXIDE, liquid (DOT)

TOXICITY DATA with REFERENCE
scu-mus LDLo:20 mg/kg AIPTAK 7,183,1900

CONSENSUS REPORTS: Reported in EPA TSCA Inventory.

DOT Classification: Corrosive Material; Label: Corrosive.

SAFETY PROFILE: Poison by subcutaneous route. A powerful caustic. A corrosive irritant to skin, eyes, and mucous membranes. When heated to decomposition it emits toxic fumes of NO_x and NH_3.

TDK750 CAS:75-58-1 ***HR: 3***
TETRAMETHYLAMMONIUM IODIDE
mf: $C_4H_{12}N{\cdot}I$ mw: 201.07

PROP: Hygroscopic crystals. Mp: > 300°.

TOXICITY DATA with REFERENCE
ivn-rat LD50:3560 μg/kg BJPCAL 4,381,49
ipr-mus LD50:30 mg/kg NTIS** AD691-490
scu-mus LDLo:33 mg/kg JPETAB 25,315,25
ivn-mus LD50:1800 μg/kg CSLNX* NX#01073

CONSENSUS REPORTS: Reported in EPA TSCA Inventory.

SAFETY PROFILE: Poison by intraperitoneal, subcutaneous, and intravenous routes. An irritant. When heated to decomposition it emits very toxic fumes of NO_x, NH_3, and I^-. See also IODIDES.

TDK885 CAS:66552-77-0 ***HR: 3***
2,3,9,10-TETRAMETHYLANTHRACENE
mf: $C_{18}H_{18}$ mw: 234.34

TOXICITY DATA with REFERENCE
mmo-sat 50 μg/plate CRNGDP 6,1483,85
mma-sat 100 μg/plate CRNGDP 6,1483,85
skn-mus TDLo:40 mg/kg/20D-I:CAR CRNGDP 6,1483,85

SAFETY PROFILE: Questionable carcinogen with experimental carcinogenic data. Mutation data reported. When heated to decomposition it emits acrid smoke and irritating fumes.

TDL000 CAS:10563-70-9 ***HR: 3***
N,N,10,10-TETRAMETHYL-$\Delta^{9(10H),\gamma}$-ANTHRACENEPROPYLAMINE HYDROCHLORIDE
mf: $C_{21}H_{25}N{\cdot}ClH$ mw: 327.93

SYNS: 9-(3-DIMETHYLAMINOPROPYLIDENE)-10,10-DIMETHYL-9,10-DIHYDROANTHRACENE HYDROCHLORIDE ◇ 3-(10,10-DIMETHYL-9(10H)-ANTHRACENYLIDENE)-N,N-DIMETHYL-1-PROPANAMINE HYDROCHLORIDE ◇ DIXERAN ◇ MELITRACENE HYDROCHLORIDE ◇ MELITRACEN HYDROCHLORIDE ◇ TRAUSABUN ◇ U-24973A

TOXICITY DATA with REFERENCE
orl-rat LD50:170 mg/kg 27ZQAG -,80,72
ipr-rat LD50:96 mg/kg 27ZQAG -,80,72
scu-rat LD50:760 mg/kg OYYAA2 3,390,69

orl-mus LD50:315 mg/kg 27ZQAG -,80,72
ipr-mus LD50:131 mg/kg 27ZQAG -,80,72
scu-mus LD50:275 mg/kg OYYAA2 3,390,69
ivn-mus LD50:52 mg/kg APTOA6 24,121,66

SAFETY PROFILE: Poison by ingestion, intraperitoneal, subcutaneous, and intravenous routes. When heated to decomposition it emits very toxic fumes of NO_x and HCl.

TDL250 CAS:5814-20-0 ***HR: 3***
TETRAMETHYLARSONIUM IODIDE
mf: $C_4H_{12}As•I$ mw: 261.98

TOXICITY DATA with REFERENCE
ipr-mus LDLo:125 mg/kg CBCCT* 4,316,52
scu-mus LDLo:150 mg/kg JPETAB 25,315,25
ivn-mus LDLo:50 mg/kg JPETAB 25,315,25

CONSENSUS REPORTS: Arsenic and its compounds are on the Community Right-To-Know List.

OSHA PEL: TWA 0.5 mg(As)/m^3

SAFETY PROFILE: Poison by intraperitoneal, subcutaneous, and intravenous routes. When heated to decomposition it emits very toxic fumes of As and I^-. See also ARSENIC COMPOUNDS and IODIDES.

TDL500 CAS:51787-44-1 ***HR: 3***
7,8,9,11-TETRAMETHYLBENZ(c)ACRIDINE
mf: $C_{21}H_{19}N$ mw: 285.41

SYN: 1,3,4,10-TETRAMETHYL-7,8-BENZACRIDINE(FRENCH)

TOXICITY DATA with REFERENCE
skn-mus TDLo:350 mg/kg/29W-I:ETA ACRSAJ 4,315,56

SAFETY PROFILE: Questionable carcinogen with experimental tumorigenic data. When heated to decomposition it emits toxic fumes of NO_x.

TDL750 CAS:63020-39-3 ***HR: 3***
5,6,9,10-TETRAMETHYL-1,2-BENZANTHRACENE
mf: $C_{22}H_{20}$ mw: 284.42

SYN: 7,8,9,12-TETRAMETHYLBENZ(a)ANTHRACENE

TOXICITY DATA with REFERENCE
skn-mus TDLo:340 mg/kg/14W-I:ETA PRLBA4 129,439,40

SAFETY PROFILE: Questionable carcinogen with experimental tumorigenic data. When heated to decomposition it emits acrid smoke and irritating fumes.

TDM000 CAS:63019-70-5 ***HR: 3***
6,7,9,10-TETRAMETHYL-1,2-BENZANTHRACENE
mf: $C_{22}H_{20}$ mw: 284.42

SYN: 7,9,10,12-TETRAMETHYLBENZ(a)ANTHRACENE

TOXICITY DATA with REFERENCE
skn-mus TDLo:270 mg/kg/34W-I:ETA CRSBAW 148,812,54

SAFETY PROFILE: Questionable carcinogen with experimental tumorigenic data. When heated to decomposition it emits acrid smoke and irritating fumes.

TDM250 CAS:488-23-3 ***HR: 1***
1,2,3,4-TETRAMETHYLBENZENE
mf: $C_{10}H_{14}$ mw: 134.24

PROP: Liquid. D: 0.904 (20/4°), mp: −6.3°, bp: 205.0°.

TOXICITY DATA with REFERENCE
skn-rbt 100 mg/24H MLD DCTODJ 1,219,78
orl-rat LD50:6408 mg/kg DCTODJ 1,219,78

CONSENSUS REPORTS: Reported in EPA TSCA Inventory.

SAFETY PROFILE: Mildly toxic by ingestion. A skin irritant. A flammable material. When heated to decomposition it emits acrid smoke and irritating fumes.

TDM500 CAS:527-53-7 ***HR: 1***
1,2,3,5-TETRAMETHYLBENZENE
mf: $C_{10}H_{14}$ mw: 134.24

PROP: Liquid. D: 0.990 @ 20/4°, mp: −20°, bp: 197.9°. Insol in water, sol in alc.

SYN: ISODURENE

TOXICITY DATA with REFERENCE
skn-rbt 100 mg/24H MLD DCTODJ 1,219,78
orl-rat LD50:5157 mg/kg DCTODJ 1,219,78

CONSENSUS REPORTS: Reported in EPA TSCA Inventory.

SAFETY PROFILE: Mildly toxic by ingestion. A skin irritant. A flammable material. When heated to decomposition it emits acrid smoke and irritating fumes.

TDM750 CAS:95-93-2 ***HR: 3***
1,2,4,5-TETRAMETHYLBENZENE
mf: $C_{10}H_{14}$ mw: 134.2

PROP: D: 0.838 (81°/4°), mp: 79.3°, bp: 196°. Insol in water; sol in benzene, alc, ether.

TOXICITY DATA with REFERENCE
orl-rat LD50:6989 mg/kg DCTODJ 1,219,78
ivn-mus LD50:180 mg/kg CSLNX* NX#00212

CONSENSUS REPORTS: Reported in EPA TSCA Inventory.

SAFETY PROFILE: Poison by intravenous route.

Mildly toxic by ingestion. When heated to decomposition it emits acrid smoke and irritating fumes.

TDM800 CAS:54827-17-7 ***HR: 3***
3,3′,5,5′-TETRAMETHYLBENZIDINE
mf: $C_{16}H_{20}N_2$ mw: 240.38

SYNS: 3,5,3′,5′-TETRAMETHYLBENZIDINE ◇ TMB

TOXICITY DATA with REFERENCE
dnd-esc 20 µmol/L MUREAV 89,95,81
mnt-mus-ipr 112500 µg/kg PMRSDJ 1,698,81
ipr-mus LD50:135 mg/kg PMRSDJ 1,682,81

CONSENSUS REPORTS: EPA Genetic Toxicology Program. Reported in EPA TSCA Inventory.

SAFETY PROFILE: Poison by intraperitoneal route. Mutation data reported. When heated to decomposition it emits toxic fumes of NO_x.

TDN000 CAS:97-84-7 ***HR: 3***
N,N,N′,N′-TETRAMETHYL-1,3-BUTANEDIAMINE
mf: $C_8H_{20}N_2$ mw: 144.30

SYN: TETRAMETHYL BUTANEDIAMINE

TOXICITY DATA with REFERENCE
skn-rbt 395 mg open SEV UCDS** 3/12/69
skn-rbt 1 mg open MOD TXAPA9 4,522,62
eye-rbt 790 µg MOD UCDS** 3/12/69
eye-rbt 5 mg MOD TXAPA9 4,522,62
orl-rat LD50:750 mg/kg TXAPA9 4,522,62
ivn-mus LD50:180 mg/kg CSLNX* NX#00707
skn-rbt LD50:320 mg/kg TXAPA9 4,522,62

CONSENSUS REPORTS: Reported in EPA TSCA Inventory.

SAFETY PROFILE: Poison by skin contact and intravenous routes. Moderately toxic by ingestion. An eye and severe skin irritant. When heated to decomposition it emits toxic fumes of NO_x. See also AMINES.

TDN250 CAS:107-45-9 ***HR: 3***
1,1,3,3-TETRAMETHYLBUTYLAMINE
mf: $C_8H_{19}N$ mw: 129.28

SYN: BIS(1,3-DIMETHYL)BUTYLAMINE

TOXICITY DATA with REFERENCE
orl-rat LDLo:70 mg/kg BJIMAG 1,213,44
ihl-rat LCLo:1278 ppm SHELL* -,53,61
ihl-mus LCLo:400 ppm SHELL* 16,-,61

CONSENSUS REPORTS: Reported in EPA TSCA Inventory.

SAFETY PROFILE: Poison by ingestion. Moderately toxic by inhalation. When heated to decomposition it emits toxic fumes of NO_x. See also AMINES.

TDN500 CAS:140-66-9 ***HR: 3***
p-(1,1,3,3-TETRAMETHYLBUTYL)PHENOL
mf: $C_{14}H_{22}O$ mw: 206.36

SYNS: p-tert-OCTYLPHENOL ◇ p-terc.OKTYLFENOL (CZECH) ◇ p-(1′,1′,3′,3′-TETRAMETHYLBUTYL)FENOL

TOXICITY DATA with REFERENCE
skn-rbt 500 mg/24H MOD 28ZPAK -,56,72
eye-rbt 50 µg/24H SEV 28ZPAK -,56,72
skn-mus TDLo:5280 mg/kg/12W-I:ETA CNREA8 19,413,59
orl-rat LD50:2160 mg/kg 28ZPAK -,56,72
orl-mus LD50:3210 mg/kg KCRZAE 26(9),28,67

CONSENSUS REPORTS: Reported in EPA TSCA Inventory.

SAFETY PROFILE: Moderately toxic by ingestion. Questionable carcinogen with experimental tumorigenic data. A skin and severe eye irritant. When heated to decomposition it emits acrid smoke and irritating fumes. See also PHENOL.

TDN750 CAS:25301-02-4 ***HR: 2***
p-(1,1,3,3-TETRAMETHYLBUTYL)PHENOL, polymer with ETHYLENE OXIDE and FORMALDEHYDE
mf: $(C_{14}H_{22}O•C_2H_4O•CH_2O)_x$

SYNS: ALEVAIRE ◇ p-ISOOCTYLPOLYOXYETHYLENEPHENOL FORMALDEHYDE POLYMER ◇ MACROCYCLON ◇ OXYETHYLATEED TERTIARY OCTYL-PHENOL-FORMALDEHYDE POLYMER ◇ SUPERINONE ◇ SUPERIUONE (FRENCH) ◇ 4-(1,1,3,3-TETRAMETHYLBUTYL)PHENOL, polymer with FORMALDEHYDE and OXIRANE ◇ TRITON A-20 ◇ TRITON WR 1339 ◇ TYLOXAPOL ◇ TYLOXYPAL

TOXICITY DATA with REFERENCE
ivn-rat TDLo:1 g/kg (10D preg):TER PSEBAA 122,874,66
ipr-mus TDLo:600 mg/kg (6-8D preg):REP CHDDAT 266,2171,68
ivn-rat LD50:1800 mg/kg NIIRDN 6,467,82

SAFETY PROFILE: Moderately toxic by intravenous route. An experimental teratogen. Experimental reproductive effects. When heated to decomposition it emits acrid smoke and irritating fumes.

TDO000 CAS:73927-97-6 ***HR: 3***
(2-(2,2,3,3-TETRAMETHYLBUTYLTHIO)ACETOXY)TRIBUTYLSTANNANE
mf: $C_{22}H_{46}O_2SSn$ mw: 493.43

SYN: TRIBUTYLTIN ISOOCTYLTHIOACETATE

TOXICITY DATA with REFERENCE
ivn-mus LD50:178 mg/kg CSLNX* NX#02290

OSHA PEL: TWA 0.1 mg(Sn)/m^3 (skin)
ACGIH TLV: TWA 0.1 mg(Sn)/m^3 (skin) (Proposed: TWA 0.1 mg(Sn)/m^3; STEL 0.2 mg(Sn)/m^3 (skin))
NIOSH REL: (Organotin Compounds) TWA 0.1 mg(Sn)/m^3

SAFETY PROFILE: Poison by intravenous route. Tributyl tin compounds are extremely toxic to marine life. When heated to decomposition it emits toxic fumes of SO_x. See also TIN COMPOUNDS and SULFIDES.

TDO250 CAS:933-52-8 ***HR: 3***
TETRAMETHYL-1,3-CYCLOBUTANEDIONE
mf: $C_8H_{12}O_2$ mw: 140.20

PROP: White, crystalline solid. Mp: 116° (sublimes), bp: 159°, vap press: 6 mm @ 52°, d: 1.11. Insol in water; sol in alc, acetic acid.

SYNS: TETRAMETHYLCYCLOBUTA-1,3-DIONE ◇ 2,2,4,4-TETRAMETHYL-1,3-CYCLOBUTANEDIONE

TOXICITY DATA with REFERENCE
scu-mus TDLo:308 mg/kg/77W-I:NEO JNCIAM 46,143,71

CONSENSUS REPORTS: Reported in EPA TSCA Inventory.

SAFETY PROFILE: Questionable carcinogen with experimental neoplastigenic data. When heated to decomposition it emits acrid smoke and irritating fumes.

TDO260 CAS:35449-36-6 ***HR: 2***
2,2,9,9-TETRAMETHYL-1,10-DECANEDIOL
mf: $C_{14}H_{30}O_2$ mw: 230.44

SYNS: CI 720 ◇ 1,10-DECANEDIOL, 2,2,9,9-TETRAMETHYL- ◇ GEMCADIOL

TOXICITY DATA with REFERENCE
orl-rat TDLo:2 g/kg (female 6-15D post):TER FAATDF 6,520,86
ipr-rat LDLo:1023 mg/kg FAATDF 6,520,86
ipr-mus LDLo:1023 mg/kg FAATDF 6,520,86

SAFETY PROFILE: Moderately toxic by intraperitoneal route. An experimental teratogen. When heated to decomposition it emits acrid smoke and irritating fumes.

TDO275 ***HR: 3***
N,N,N',N'-TETRAMETHYLDEUTERO-FORMAMIDINIUM PERCHLORATE
mf: $C_5H_{12}ClDN_2O_4$ mw: 201.63

SAFETY PROFILE: Moist salts may explode during vacuum evaporation. When heated to decomposition it emits toxic fumes of Cl^- and NO_x. See also PERCHLORATES.

TDO500 CAS:33196-65-5 ***HR: 3***
TETRAMETHYLDIALUMINUM DIHYDRIDE
mf: $C_4H_{14}Al_2$ mw: 116.12

$(CH_3)_2Al{:}H_2{:}Al(CH_3)_2$

SAFETY PROFILE: Ignites spontaneously in air and burns explosively. When heated to decomposition it emits acrid smoke and irritating fumes. See also ALUMINUM COMPOUNDS.

TDO750 CAS:119-58-4 ***HR: 2***
TETRAMETHYLDIAMINOBENZHYDROL
mf: $C_{17}H_{22}N_2O$ mw: 270.41

SYNS: 4,4'-BIS(DIMETHYLAMINO)BENZOHYDROL ◇ α,α-BIS(p-DIMETHYLAMINOPHENYL)METHANOL ◇ p,p'-MICHLER'S HYDROL

TOXICITY DATA with REFERENCE
orl-rat LDLo:500 mg/kg JPETAB 90,260,47
ipr-mus LDLo:500 mg/kg CBCCT* 4,227,52

CONSENSUS REPORTS: Reported in EPA TSCA Inventory.

SAFETY PROFILE: Moderately toxic by ingestion and intraperitoneal routes. When heated to decomposition it emits toxic fumes of NO_x.

TDP250 CAS:471-35-2 ***HR: 3***
TETRAMETHYLDIARSANE
mf: $C_4H_{12}As_2$ mw: 209.96

$(CH_3)_2AsAs(CH_3)_2$

CONSENSUS REPORTS: Arsenic and its compounds are on the Community Right-To-Know List.

SAFETY PROFILE: Arsenic compounds are poisons. Ignites spontaneously in air. Incompatible with chlorine. When heated to decomposition it emits toxic fumes of As. See also ARSENIC COMPOUNDS.

TDP275 CAS:21482-59-7 ***HR: 3***
TETRAMETHYLDIBORANE
mf: $C_4H_{14}B_2$ mw: 83.77

$(CH_3)_2B{:}H_2{:}B(CH_3)_2$

SAFETY PROFILE: Ignites spontaneously in air. When heated to decomposition it emits acrid smoke and irritating fumes. See also BORON COMPOUNDS and BORANES.

TDP300 CAS:65313-37-3 ***HR: 3***
TETRAMETHYLDIGALLANE
mf: $C_4H_{12}Ga_2$ mw: 199.58

$(CH_3)_2GaGa(CH_3)_2$

SAFETY PROFILE: Ignites spontaneously in air. Upon

decomposition it emits acrid smoke and fumes. See also GALLIUM COMPOUNDS.

TDP325 CAS:35856-82-7 ***HR: 3***
TETRAMETHYL-1,2-DIOXETANE
mf: $C_6H_{12}O_2$ mw: 116.16

$(CH_3)_2COOC(CH_3)_2$

SAFETY PROFILE: May explode if stored at room temperature. When heated to decomposition it emits acrid smoke and irritating fumes.

TDP500 CAS:22653-19-6 ***HR: 3***
TETRAMETHYLDIGOLD DIAZIDE
mf: $C_4H_{12}Au_2N_6$ mw: 538.09

$(CH_3)_2Au:(N_3)_2:Au(CH_3)_2$

SAFETY PROFILE: An extremely sensitive explosive. Explodes under water if touched. When heated to decomposition it emits toxic fumes of NO_x. See also AZIDES and GOLD COMPOUNDS.

TDP525 CAS:3676-91-3 ***HR: 3***
TETRAMETHYLDIPHOSPHANE
mf: $C_4H_{12}P_2$ mw: 122.09

$(CH_3)_2PP(CH_3)_2$

SYN: TETRAMETHYLDIPHOSPHINE

SAFETY PROFILE: Ignites spontaneously in air. When heated to decomposition it emits toxic fumes of PO_x. See also PHOSPHINE.

TDP750 CAS:6711-48-4 ***HR: 3***
N,N,N',N'-TETRAMETHYL-DIPROPYLENETRIAMINE
mf: $H_2NCHCH_3CHCH_3CH_2NHCHCH_3$ mw: 189.

TOXICITY DATA with REFERENCE
skn-rbt 100 μg/24H open AIHAAP 23,95,62
orl-rat LD50:1620 mg/kg AIHAAP 23,95,62
skn-rbt LD50:310 mg/kg AIHAAP 23,95,62

CONSENSUS REPORTS: Reported in EPA TSCA Inventory.

SAFETY PROFILE: Poison by skin contact. Moderately toxic by ingestion. A skin irritant. When heated to decomposition it emits toxic fumes of NO_x. See also AMINES.

TDP775 CAS:3277-26-7 ***HR: 3***
TETRAMETHYLDISILOXANE
mf: $C_4H_{14}OSi_2$ mw: 134.32

$H(CH_3)_2SiOSi(CH_3)_2H$

PROP: Bp: 71-72°, d: 0.757, refr index: 1.370.

SYN: 1,1,3,3-TETRAMETHYLDISILOXANE

SAFETY PROFILE: Mixtures with oxygen + sodium hydroxide + water explode when ignited. When heated to decomposition it emits acrid smoke and irritating fumes.

TDQ000 CAS:41422-43-9 ***HR: 3***
TETRAMETHYLDISTIBINE
mf: $C_4H_{12}Sb_2$ mw: 303.61

CONSENSUS REPORTS: Antimony and its compounds are on the Community Right-To-Know List.

SAFETY PROFILE: Most antimony compounds are poisons. Ignites spontaneously in air. When heated to decomposition it emits toxic fumes of Sb. See also ANTIMONY COMPOUNDS.

TDQ225 CAS:6611-01-4 ***HR: 3***
N,N'-TETRAMETHYLENEBIS(1-AZIRIDINECARBOXAMIDE)
mf: $C_{10}H_{18}N_4O_2$ mw: 226.32

SYNS: N,N'-(1,4-BUTANEDIYL)BIS-1-AZIRIDIENCARBOXAMIDE ◇ ENT-50,838 ◇ TMAC

TOXICITY DATA with REFERENCE
dlt-oin-par 100 ng MUREAV 7,63,69
dlt-oin-unr 1000 ppm/5M-C AESAAI 62,790,69
dlt-oin-par 10 ppm AESAAI 63,71,70
orl-qal LD50:133 mg/kg JRPFA4 48,371,76

SAFETY PROFILE: Poison by ingestion. Mutation data reported. When heated to decomposition it emits toxic fumes of NO_x.

TDQ230 CAS:10397-75-8 ***HR: 3***
5,5'-(TETRAMETHYLENEBIS(CARBONYLIMINO))BIS(N-METHYL-2,4,6-TRIIODOISOPHTHALAMIC ACID)
mf: $C_{24}H_{20}I_6N_4O_8$ mw: 1253.85

PROP: Crystals from dimethylformamide. Mp: 302° (decomp).

SYNS: IOCARMIC ACID ◇ MYELOTRAST

TOXICITY DATA with REFERENCE
iat-rat LD50:6587 mg/kg USXXAM #4001323
ivn-mus LD50:9057 mg/kg USXXAM #4001323
ice-mus LD50:461 mg/kg USXXAM #4001323
par-rbt LD50:61 mg/kg USXXAM #4001323

SAFETY PROFILE: Poison by parenteral route. Moderately toxic by intracerebral route. When heated to decomposition it emits toxic fumes of I^- and NO_x.

TDQ250 CAS:60784-43-2 ***HR: 3***
1,1′-TETRAMETHYLENEBIS(3-(2-CHLOROETHYL)-3-NITROSOUREA)
mf: $C_{10}H_{18}Cl_2N_6O_4$ mw: 357.24

TOXICITY DATA with REFERENCE
sln-dmg-orl 100 μmol/L/24H MUREAV 57,297,78
mrc-smc 500 μmol/L/16H MUREAV 42,45,77
ipr-rat LD50:46 mg/kg EJCAAH 13,937,77

SAFETY PROFILE: Poison by intraperitoneal route. Mutation data reported. Many N-nitroso compounds are carcinogens. When heated to decomposition it emits very toxic fumes of Cl^- and NO_x. See also N-NITROSO COMPOUNDS.

TDQ255 ***HR: 3***
1,1′-TETRAMETHYLENEBIS(1-ETHYLPIPERIDINIUM)DIBROMIDE
mf: $C_{18}H_{38}N_2$•2Br mw: 442.40

SYN: 1,4-TETRAMETHYLEN-BIS-(N-AETHYLPIPERIDINIUM)-DIBROMID (GERMAN)

TOXICITY DATA with REFERENCE
orl-mus LD50:1 g/kg AEPPAE 232,219,57
scu-mus LD50:135 mg/kg AEPPAE 232,219,57
ivn-mus LD50:21 mg/kg AEPPAE 232,219,57

SAFETY PROFILE: Poison by subcutaneous and intravenous routes. Moderately toxic by ingestion. When heated to decomposition it emits toxic fumes of NO_x and Br^-.

TDQ260 ***HR: 3***
1,1′-TETRAMETHYLENEBIS(1-ETHYLPYRROLIDINIUM DIBROMIDE
mf: $C_{16}H_{34}N_2$•2Br mw: 414.34

SYN: 1,4-TETRAMETHYLEN-BIS-(N-AETHYLPYRROLIDINIUM)-DIBROMID (GERMAN)

TOXICITY DATA with REFERENCE
orl-mus LD50:775 mg/kg AEPPAE 232,219,57
scu-mus LD50:106 mg/kg AEPPAE 232,219,57
ivn-mus LD50:26 mg/kg AEPPAE 232,219,57

SAFETY PROFILE: Poison by subcutaneous and intravenous routes. Moderately toxic by ingestion. When heated to decomposition it emits toxic fumes of Br^- and NO_x.

TDQ263 ***HR: 3***
1,1′-TETRAMETHYLENEBIS(1-METHYLPIPERIDINIUM)DIBROMIDE
mf: $C_{16}H_{34}N_2$•2Br mw: 414.34

SYN: 1,4-TETRAMETHYLEN-BIS-(N-METHYLPIPERIDINIUM)-DIBROMID (GERMAN)

TOXICITY DATA with REFERENCE
orl-mus LD50:825 mg/kg AEPPAE 232,219,57
scu-mus LD50:74 mg/kg AEPPAE 232,219,57
ivn-mus LD50:33 mg/kg AEPPAE 232,219,57

SAFETY PROFILE: Poison by subcutaneous and intravenous routes. Moderately toxic by ingestion. When heated to decomposition it emits toxic fumes of NO_x and Br^-.

TDQ265 ***HR: 3***
1,1′-TETRAMETHYLENEBIS(1-METHYLPYRROLIDINIUM DIBROMIDE
mf: $C_{14}H_{30}N_2$•2Br mw: 386.28

SYN: 1,4-TETRAMETHYLEN-BIS-(N-METHYLPYRROLIDINIUM)-DIBROMID (GERMAN)

TOXICITY DATA with REFERENCE
orl-mus LD50:1450 mg/kg AEPPAE 232,219,57
scu-mus LD50:190 mg/kg AEPPAE 232,219,57
ivn-mus LD50:41 mg/kg AEPPAE 232,219,57

SAFETY PROFILE: Poison by subcutaneous and intravenous routes. Moderately toxic by ingestion. When heated to decomposition it emits toxic fumes of Br^- and NO_x.

TDQ325 CAS:14208-08-3 ***HR: 3***
1,1′-TETRAMETHYLENEBIS(PYRIDINIUM BROMIDE)
mf: $C_{14}H_{18}N_2$•2Br mw: 374.16

SYNS: 1,4-TETRAMETHYLEN-BIS-(PYRIDINIUM)-DIBROMID(GERMAN) ◇ 1,1′-TETRAMETHYLENBIS-PYRIDINIUM DIBROMIDE

TOXICITY DATA with REFERENCE
orl-mus LD50:2750 mg/kg AEPPAE 232,219,57
scu-mus LD50:212 mg/kg AEPPAE 232,219,57
ivn-mus LD50:55 mg/kg AIPTAK 90,271,52

SAFETY PROFILE: Poison by subcutaneous and intravenous routes. Moderately toxic by ingestion. When heated to decomposition it emits toxic fumes of Br^- and NO_x.

TDQ500 CAS:1600-44-8 ***HR: 2***
TETRAMETHYLENE SULPHOXIDE
mf: C_4H_8OS mw: 104.18

PROP: Refr index: 1.5200, d: 1.158, flash p: >230°F.

SYN: TETRAHYDRO-THIOPHENE-1-OXIDE

TOXICITY DATA with REFERENCE
ipr-mus LD50:3500 mg/kg IJRBA3 3,41,61

CONSENSUS REPORTS: Reported in EPA TSCA Inventory.

SAFETY PROFILE: Moderately toxic by intraperitoneal route. Combustible when exposed to heat, sparks, or flame. When heated to decomposition it emits toxic fumes of SO_x.

TDQ750 CAS:110-18-9 ***HR: 3***
TETRAMETHYL ETHYLENE DIAMINE
DOT: UN 2372
mf: $C_6H_{16}N_2$ mw: 116.24

PROP: Mp: -55°, bp: 120-122°, refr index: 1.4179, d: 0.770, flash p: 50°F.

SYNS: 1,2-BIS-(DIMETHYLAMINO-ETHANE (DOT) ◇ 1,2-DI-(DIMETHYLAMINO)ETHANE (DOT) ◇ PROPAMINE D ◇ TEMED ◇ TETRAMEEN ◇ N,N,N',N'-TETRAMETHYL-1,2-DIAMINOETHANE ◇ N,N,N',N'-TETRAMETHYLETHYLENEDIAMINE ◇ TMEDA

TOXICITY DATA with REFERENCE
skn-rbt 10 mg/24H open JIHTAB 30,63,48
eye-rbt 750 μg SEV AJOPAA 29,1363,46
orl-rat LD50:1580 mg/kg JIHTAB 30,63,48
skn-rbt LD50:5390 mg/kg AIHAAP 30,470,69

CONSENSUS REPORTS: Reported in EPA TSCA Inventory.

DOT Classification: Flammable Liquid; Label: Flammable Liquid.

SAFETY PROFILE: Moderately toxic by ingestion. Mildly toxic by skin contact. A skin and severe eye irritant. Flammable when exposed to heat or flame; can react with oxidizing materials. When heated to decomposition it emits toxic fumes of NO_x. See also AMINES.

TDR000 CAS:76-09-5 ***HR: 2***
TETRAMETHYLETHYLENE GLYCOL
mf: $C_6H_{14}O_2$ mw: 118.20

SYNS: 2,3-DIMETHYL-2,3-BUTANEDIOL ◇ PINACOL

TOXICITY DATA with REFERENCE
skn-rbt 500 mg/24H MOD FCTXAV 14,659,76
orl-mus LD50:3380 mg/kg FCTXAV 14,841,76

CONSENSUS REPORTS: Reported in EPA TSCA Inventory.

SAFETY PROFILE: Moderately toxic by ingestion. A skin irritant. When heated to decomposition it emits acrid smoke and irritating fumes.

TDR250 CAS:61556-82-9 ***HR: 3***
TETRAMETHYLHYDRAZINE HYDROCHLORIDE
mf: $C_4H_{12}N_2$•ClH mw: 124.64

TOXICITY DATA with REFERENCE
orl-mus TDLo:49 g/kg/34W-C:CAR JNCIAM 57,1179,76

SAFETY PROFILE: Questionable carcinogen with experimental carcinogenic data. When heated to decomposition it emits very toxic fumes of HCl and NO_x.

TDR500 CAS:75-74-1 ***HR: 3***
TETRAMETHYL LEAD
mf: $C_4H_{12}Pb$ mw: 267.35

PROP: Colorless liquid. Mp: −18°F, lel: 1.8%, bp: 110°, d: 1.99, vap d: 9.2, flash p: 100°F.

SYNS: TETRAMETHYLPLUMBANE ◇ TML

TOXICITY DATA with REFERENCE
orl-rat TDLo:112 mg/kg (9-11D preg):TER TXAPA9 21,265,72
orl-rat TDLo:112 mg/kg (9-11D preg):REP TXAPA9 21,265,72
orl-rat LD50:105 mg/kg MELAAD 54,486,63
ipr-rat LD50:90 mg/kg MELAAD 54,486,63
ivn-rat LD50:88 mg/kg MELAAD 54,486,63
par-rat LD50:105 mg/kg AOHYA3 3,226,61
ihl-mus LC50:8500 mg/m³/30M 85JCAE -,1258,86
orl-rbt LDLo:24 mg/kg SAIGBL 15,3,73
skn-rbt LDLo:3391 mg/kg SAIGBL 15,3,73
ivn-rbt LDLo:90 mg/kg JPETAB 38,161,30

CONSENSUS REPORTS: EPA Extremely Hazardous Substances List. Lead and its compounds are on the Community Right-To-Know List. Reported in EPA TSCA Inventory.

OSHA PEL: TWA 0.075 mg(Pb)/m³ (skin)
ACGIH TLV: TWA 0.15 mg(Pb)/m³ (skin)
DFG MAK: 0.01 ppm (0.075 mg/m³)

SAFETY PROFILE: Poison by ingestion, intraperitoneal, parenteral, and intravenous routes. Moderately toxic by skin contact. An experimental teratogen. Experimental reproductive effects. Lead and its compounds have dangerous central nervous system effects. A very dangerous fire hazard when exposed to heat, flame, or oxidizers. Moderate explosion hazard in the form of vapor when exposed to flame. May explode when heated above 90°C. Explosive reaction with tetrachlorotrifluoromethyl phosphorane. Can react vigorously with oxidizing materials. To fight fire, use water, foam, CO_2, dry chemical. When heated to decomposition it emits toxic fumes of Pb. Used as an octane enhancer for gasoline. See also LEAD COMPOUNDS.

TDR750 CAS:51-80-9 ***HR: 3***
N,N,N'N'-TETRAMETHYLMETHANEDIAMINE
DOT: NA 9069
mf: $C_5H_{14}N_2$ mw: 102.21

PROP: Liquid. Bp: 85°, d: 0.749, refr index: 1.4005, flash p: 35° F.

SYNS: N,N,N',N'-TETRAMETHYLDIAMINOMETHAN(GERMAN) ◇ TETRAMETHYL METHYLENE DIAMINE (DOT)

TOXICITY DATA with REFERENCE
ipr-mus LD50:220 mg/kg ARZNAD 16,734,66

CONSENSUS REPORTS: Reported in EPA TSCA Inventory.

DOT Classification: ORM-A; Label: None.

SAFETY PROFILE: Poison by intraperitoneal route. Flammable liquid and very dangerous fire hazard when exposed to powerful oxidizers, heat or open flame. When heated to decomposition it emits toxic fumes of NO_x. See also AMINES.

TDS000 CAS:6130-93-4 ***HR: 3***
2,2,6,6-TETRAMETHYLNITROSOPIPERIDINE
mf: $C_9H_{18}N_2O$ mw: 170.29

SYN: 1-NITROSO-2,2,6,6-TETRAMETHYLPIPERIDINE

TOXICITY DATA with REFERENCE
orl-rat TDLo:3375 mg/kg/50W-I:ETA IJCNAW 16,318,75
orl-mus LD50:300 mg/kg NATUAS 184,1707,59
ivn-mus LD50:25 mg/kg NATUAS 184,1707,59

SAFETY PROFILE: Poison by ingestion and intravenous routes. Questionable carcinogen with experimental tumorigenic data. When heated to decomposition it emits toxic fumes of NO_x. See also NITROSO COMPOUNDS.

TDS250 CAS:63907-04-0 ***HR: 3***
N,N,2,3-TETRAMETHYL-2-NORBORNANAMINE
mf: $C_{11}H_{21}N$ mw: 167.33

SYN: M & B 4620

TOXICITY DATA with REFERENCE
orl-mus LDLo:165 mg/kg 32ZCAI -,167,67
ivn-mus LD50:18 mg/kg 32ZCAI -,167,67

SAFETY PROFILE: Poison by ingestion and intravenous routes. When heated to decomposition it emits toxic fumes of NO_x.

TDS500 CAS:10031-58-0 ***HR: 3***
1,(5,5,7,7-TETRAMETHYL-2-OCTANYL)-2-METHYL-5-ETHYLPYRIDINIUM CHLORIDE
mf: $C_{20}H_{36}N \cdot Cl$ mw: 326.02

TOXICITY DATA with REFERENCE
skn-rbt 100 μg/24H open AIHAAP 23,95,62
orl-rat LD50:540 mg/kg AIHAAP 23,95,62
skn-rbt LD50:71 mg/kg AIHAAP 23,95,62

SAFETY PROFILE: Poison by skin contact. Moderately toxic by ingestion. A skin irritant. When heated to decomposition it emits very toxic fumes of NO_x and Cl^-.

TDS750 CAS:4466-77-7 ***HR: 3***
1:2:3:4-TETRAMETHYLPHENANTHRENE
mf: $C_{18}H_{18}$ mw: 234.36

TOXICITY DATA with REFERENCE
skn-mus TDLo:1340 mg/kg/56W-I:ETA PRLBA4 131,170,42

SAFETY PROFILE: Questionable carcinogen with experimental tumorigenic data. When heated to decomposition it emits acrid smoke and irritating fumes.

TDT000 CAS:704-01-8 ***HR: 3***
N,N,N',N'-TETRAMETHYL-o-PHENYLENEDIAMINE
mf: $C_{10}H_{16}N_2$ mw: 164.28

PROP: Crystals. Mp: 51-52°, bp: 260°. Sltly sol in cold water; more sol in hot water; very sol in alc, chloroform, ether, petr ether.

TOXICITY DATA with REFERENCE
ivn-mus LD50:320 mg/kg CSLNX* NX#03995

CONSENSUS REPORTS: Reported in EPA TSCA Inventory.

SAFETY PROFILE: Poison by intravenous route. When heated to decomposition it emits toxic fumes of NO_x.

TDT250 CAS:637-01-4 ***HR: 3***
N,N,N',N'-TETRAMETHYL-p-PHENYLENEDIAMINE DIHYDROCHLORIDE
mf: $C_{10}H_{16}N_2 \cdot 2ClH$ mw: 237.20

PROP: Crystals. Mp: 222-224° (decomp). Sol in water.

TOXICITY DATA with REFERENCE
scu-mus LDLo:20 mg/kg NDRC** -,13,44

CONSENSUS REPORTS: Reported in EPA TSCA Inventory.

SAFETY PROFILE: Poison by subcutaneous route. When heated to decomposition it emits very toxic fumes of NO_x and HCl.

TDT500 CAS:993-11-3 ***HR: 3***
TETRAMETHYLPHOSPHONIUM IODIDE
mf: $C_4H_{12}P \cdot I$ mw: 218.03

PROP: Crystals. Mp: 312-322°.

TOXICITY DATA with REFERENCE
scu-mus LDLo:100 mg/kg JPETAB 25,315,25
ivn-mus LDLo:12 mg/kg JPETAB 25,315,25

SAFETY PROFILE: Poison by subcutaneous and intravenous routes. When heated to decomposition it emits very toxic fumes of PO_x and I^-. See also IODIDES.

TDT750 CAS:768-66-1 ***HR: 3***
2,2,6,6-TETRAMETHYLPIPERIDINE
mf: $C_9H_{19}N$ mw: 141.29

PROP: Bp: 152°, refr index: 1.4440, d: 0.837, flash p: 76°F.

TOXICITY DATA with REFERENCE
orl-mus LD50:220 mg/kg NATUAS 184,1707,59
ivn-mus LD50:33 mg/kg NATUAS 184,1707,59

CONSENSUS REPORTS: Reported in EPA TSCA Inventory.

SAFETY PROFILE: Poison by ingestion and intravenous routes. Flammable liquid. When heated to decomposition it emits toxic fumes of NO_x.

TDU000 CAS:4168-79-0 ***HR: 3***
2,2,6,6-TETRAMETHYL-4-PIPERIDONE OXIME
mf: $C_9H_{18}N_2O$ mw: 170.29

TOXICITY DATA with REFERENCE
ivn-mus LD50:180 mg/kg CSLNX* NX#04513

CONSENSUS REPORTS: Reported in EPA TSCA Inventory.

SAFETY PROFILE: Poison by intravenous route. When heated to decomposition it emits toxic fumes of NO_x.

TDU250 CAS:22295-11-0 ***HR: 2***
TETRAMETHYLPLATINUM
mf: $C_4H_{12}Pt$ mw: 255.23

SAFETY PROFILE: Heat-sensitive weak explosive. When heated to decomposition it emits acrid smoke and irritating fumes. See also PLATINUM COMPOUNDS.

TDU500 CAS:110-95-2 ***HR: 3***
N,N,N',N'-TETRAMETHYL-1,3-PROPANEDIAMINE
mf: $C_7H_{18}N_2$ mw: 130.27

PROP: Liquid. Bp: 145-146°, refr index: 1.4234, flash p: 89° F

TOXICITY DATA with REFERENCE
orl-rat LD50:410 mg/kg AIHAAP 30,470,69
ihl-rat LCLo:375 ppm/2H WADTAA 55-250,10,55
skn-rbt LD50:300 mg/kg AIHAAP 30,470,69

CONSENSUS REPORTS: Reported in EPA TSCA Inventory.

SAFETY PROFILE: Poison by skin contact. Moderately toxic by inhalation and ingestion. Flammable liquid. When heated to decomposition it emits toxic fumes of NO_x.

TDU750 CAS:17471-59-9 ***HR: 3***
N,N,N',N'-TETRAMETHYL PROPENE-1,3-DIAMINE
mf: $C_7H_{16}N_2$ mw: 128.25
SYN: N,N,N',N'-TETRAMETHYL-1,3-PROPENYLDIAMINE

TOXICITY DATA with REFERENCE
ihl-rat LCLo:200 ppm/2H WADTAA 55-250,9,55
ivn-rbt LDLo:50 mg/kg WADTAA 55-250,3,55

SAFETY PROFILE: Poison by intravenous route. Moderately toxic by inhalation. When heated to decomposition it emits toxic fumes of NO_x.

TDV000 CAS:690-49-3 ***HR: 3***
TETRAMETHYLPYROPHOSPHATE
mf: $C_4H_{12}O_7P_2$ mw: 234.10
SYNS: DIPHOSPHORIC ACID, TETRAMETHYL ESTER ◇ METHYL PYROPHOSPHATE ◇ PYROPHOSPHORIC ACID, TETRAMETHYL ESTER

TOXICITY DATA with REFERENCE
orl-rat LDLo:100 mg/kg NCNSA6 5,38,53
ipr-mus LD50:589 μg/kg PHARAT 35,806,80

SAFETY PROFILE: Poison by ingestion and intraperitoneal routes. When heated to decomposition it emits toxic fumes of PO_x. See also PHOSPHATES.

TDV250 CAS:4567-22-0 ***HR: 3***
2,2,5,5-TETRAMETHYLPYRROLIDINE
mf: $C_8H_{17}N$ mw: 127.26

TOXICITY DATA with REFERENCE
orl-mus LD50:450 mg/kg MDCHAG 7,314,67
ivn-mus LD50:63 mg/kg MDCHAG 7,314,67

SAFETY PROFILE: Poison by intravenous route. Moderately toxic by ingestion. When heated to decomposition it emits toxic fumes of NO_x.

TDV500 CAS:75-76-3 ***HR: 3***
TETRAMETHYLSILANE
DOT: UN 2749
mf: $C_4H_{12}Si$ mw: 88.23

PROP: Mp: -99°, bp: 26-28°, refr index: 1.3580, d: 0.648, flash p: -17° F.

DOT Classification: Flammable Liquid; Label:Flammable Liquid

SAFETY PROFILE: A very dangerous fire hazard when exposed to heat or flame. Spontaneously flammable in air. Explosive reaction with chlorine + antimony trichloride above 100°C. When heated to decomposition it emits acrid smoke and irritating fumes. See also SILANE.

TDV725 CAS:1124-11-4 ***HR: 3***
TETRAMETHYLPYRAZINE
mf: $C_8H_{12}N_2$ mw: 136.22

PROP: White crystals or powder; fermented soybean odor. Mp: 85-90°, bp: 190°. Sol in alc, propylene glycol, and fixed oils; sltly sol in water.

SYNS: FEMA No. 3237 ◇ 2,3,5,6-TETRAMETHYL PYRAZINE (FCC)

TOXICITY DATA with REFERENCE
orl-rat LD50:1910 mg/kg DCTODJ 3,249,80
ipr-mus LD50:800 μg/kg IJCREE 23,119,85
ivn-mus LD50:239 mg/kg CMJODS 4,319,78

CONSENSUS REPORTS: Reported in EPA TSCA Inventory.

SAFETY PROFILE: Poison by intravenous and intraperitoneal routes. Moderately toxic by ingestion. When heated to decomposition it emits toxic fumes of NO_x.

TDV750 CAS:594-27-4 ***HR: 3***
TETRAMETHYLSTANNANE
mf: $C_4H_{12}Sn$ mw: 178.85

PROP: Colorless liquid. Bp: 78°, lel: 1.9%, d: 1.314 @ 0°/4°, vap d: 6.2, flash p: <69.8°F.

SYN: TETRAMETHYL TIN

TOXICITY DATA with REFERENCE
ihl-mus LCLo:2550 mg/m³ NDRC** NDCrc-132,Feb,42

CONSENSUS REPORTS: Reported in EPA TSCA Inventory.

OSHA PEL: TWA 0.1 mg(Sn)/m³ (skin)
ACGIH TLV: TWA 0.1 mg(Sn)/m³ (skin) (Proposed: TWA 0.1 mg(Sn)/m³; STEL 0.2 mg(Sn)/m³ (skin))
NIOSH REL: (Organotin Compounds) TWA 0.1 mg(Sn)/m³

SAFETY PROFILE: Moderately toxic by inhalation. A very dangerous fire and explosion hazard when exposed to heat, flame or oxidizers. Powerful explosive reaction with dinitrogen tetroxide. To fight fire, use water, foam, CO_2, dry chemical. When heated it emits acrid smoke and irritating fumes. See also TIN COMPOUNDS and ORGANOMETALS.

TDW000 CAS:2185-78-6 ***HR: 2***
TETRAMETHYLSTIBONIUM IODIDE
mf: $C_4H_{12}Sb \bullet I$ mw: 308.81

TOXICITY DATA with REFERENCE
scu-mus LDLo:1000 mg/kg JPETAB 25,315,25
ivn-mus LDLo:600 mg/kg JPETAB 25,315,25

CONSENSUS REPORTS: Antimony and its compounds are on the Community Right-To-Know List.

OSHA PEL: TWA 0.5 mg(Sb)/m³
ACGIH TLV: TWA 0.5 mg(Sb)/m³
NIOSH REL: (Antimony) TWA 0.5 mg(Sb)/m³

SAFETY PROFILE: Moderately toxic by subcutaneous and intravenous routes. Many antimony compounds are poisons by several routes. When heated to decomposition it emits toxic fumes of I^- and Sb. See also ANTIMONY COMPOUNDS and IODIDES.

TDW250 CAS:3333-52-6 ***HR: 3***
TETRAMETHYLSUCCINONITRILE
mf: $C_8H_{12}N_2$ mw: 136.22

PROP: Crystallizes in plates; almost no odor. Mp: 169° (sublimes).

SYN: TMSN

TOXICITY DATA with REFERENCE
ipr-ham TDLo:9944 μg/kg (8D preg):TER FAATDF 3,41,83
orl-rat LD50:38900 μg/kg FAATDF 7,41,86
ipr-mus LD50:17710 μg/kg JPETAB 223,635,82
ivn-mus LD50:18 mg/kg CSLNX* NX#00002

CONSENSUS REPORTS: Cyanide and its compounds are on the Community Right-To-Know List.

OSHA PEL: TWA 0.5 ppm (skin)
ACGIH TLV: TWA 0.5 ppm (skin)
DFG MAK: 0.5 ppm (3 mg/m³)
NIOSH REL: (Nitriles) CL 6 mg/m³/15M

SAFETY PROFILE: Poison by ingestion, intraperitoneal, and intravenous routes. An experimental teratogen. A human skin irritant and allergen. In the preparation of sponge rubber, an azo compound is used which decomposes to form tetramethylsuccinonitrile or TSN. Rats exposed to a concentration of 90 ppm exhibit their first convulsion after 1.5-2 hours or less. Rats exposed to concentration of 5.5 ppm exhibited their first convulsions in 27-31 hours and were dead in from 31-46 hours. Absorbed by skin. The fatal dose in humans is thought to be about 25 mg/kg of body weight. TSN is slowly detoxified by the body. This nitrile is different from other nitriles in that thiosulfate is a poor antidote for intoxication. When heated to decomposition it emits toxic fumes of CN^- and NO_x. See also NITRILES and CYANIDES.

TDW275 CAS:1073-91-2 ***HR: 3***
3,3,6,6-TETRAMETHYL-1,2,4,5-TETRAOXANE
mf: $C_6H_{12}O_4$ mw: 148.16

$$\overline{OOC(CH_3)_2OOC(CH_3)_2}$$

SAFETY PROFILE: A powerful explosive sensitive to impact, friction, or rapid heating. When heated to decomposition it emits acrid smoke and irritating fumes. See also PEROXIDES.

TDW300 CAS:6130-87-6 ***HR: 3***
TETRAMETHYL-2-TETRAZENE
mf: $C_4H_{12}N_4$ mw: 116.17

$(CH_3)_2NN=NN(CH_3)_2$

SAFETY PROFILE: Explodes when heated to its boiling point 130°C. Upon decomposition it emits toxic fumes of NO_x.

TDW500 CAS:108-62-3 ***HR: 3***
2,4,6,8-TETRAMETHYL-1,3,5,7-TETROXOCANE
DOT: UN 1332
mf: $C_8H_{16}O_4$ mw: 176.24

SYNS: ACETALDEHYDE, TETRAMER ◇ ANTIMILACE ◇ ARIOTOX ◇ CEKUMETA ◇ CORRY'S SLUG DEATH ◇ HALIZAN ◇ META ◇ METACETALDEHYDE ◇ METALDEHYD (GERMAN) ◇ METALDEHYDE (DOT) ◇ METALDEIDE (ITALIAN) ◇ METASON ◇ NAMEKIL ◇ SLUG-TOX

TOXICITY DATA with REFERENCE
mmo-smc 5 ppm RSTUDV 6,161,76
orl-rat TDLo:61 g/kg (MGN):REP TXCYAC 4,97,75
orl-chd LDLo:100 mg/kg:CNS 85GYAZ -,158,71
orl-rat LD50:227 mg/kg 85GMAT -,15,82
ihl-rat LC50:203 mg/m³/4H 85GMAT -,15,82
skn-rat LD50:2275 mg/kg 85GMAT -,15,82
orl-mus LD50:200 mg/kg 85GMAT -,15,82
ihl-mus LC50:348 mg/m³/2H 85GMAT -,15,82
orl-dog LD50:600 mg/kg GUCHAZ 6,331,73

DOT Classification: Flammable Solid; Label: Flammable Solid.

SAFETY PROFILE: Human poison by ingestion. Human systemic effects by ingestion: convulsions or effect on seizure threshold. Moderately toxic by inhalation and skin contact. Experimental reproductive effects. Mutation data reported. A flammable solid. When heated to decomposition it emits acrid smoke and irritating fumes. See also ALDEHYDES.

TDX000 CAS:2782-91-4 ***HR: 3***
TETRAMETHYLTHIOUREA
mf: $C_5H_{12}N_2S$ mw: 132.25

SYNS: NA-101 ◇ 1,1,3,3-TETRAMETHYLTHIOUREA ◇ TMTU

TOXICITY DATA with REFERENCE
cyt-hmn:lyms 100 µg/L CYGEDX 20(2),55,86
skn-rat TDLo:500 mg/kg (female 11-12D post):TER TXAPA9 41,35,77
orl-rat TDLo:1848 mg/kg/79W-C:CAR HunNJ# 10May77
orl-rat LD50:920 mg/kg HunNJ# 10May77
orl-mus LDLo:500 mg/kg TJADAB 23,335,81

CONSENSUS REPORTS: Reported in EPA TSCA Inventory.

SAFETY PROFILE: Moderately toxic by ingestion. Questionable carcinogen with experimental carcinogenic and teratogenic data. Human mutation data reported. When heated to decomposition it emits very toxic fumes of NO_x and SO_x. See also SULFIDES.

TDX250 CAS:632-22-4 ***HR: 2***
1,1,3,3-TETRAMETHYLUREA
mf: $C_5H_{12}N_2O$ mw: 116.19

PROP: Liquid, fat odor. Bp: 177°, mp: −1.2°, d: 0.969, flash p: 167°F. Very sol in alc and ether; misc with water.

SYNS: TEMUR ◇ TETRAMETHYLUREA ◇ TETRAMETHYLUREE (FRENCH) ◇ TMU

TOXICITY DATA with REFERENCE
dni-hmn:lym 10 mmol/L PNASA6 79,1171,82
orl-mus TDLo:1499 mg/kg (female 8D post):TER APFRAD 30,585,72
orl-rat TDLo:500 mg/kg (12D preg):REP TJADAB 23,335,81
ivn-rat LD50:1100 mg/kg ACIEAY 3,260,64
orl-mus LD50:2920 mg/kg AIPTAK 160,333,66
ivn-mus LD50:2230 mg/kg AIPTAK 160,333,66

CONSENSUS REPORTS: Reported in EPA TSCA Inventory.

SAFETY PROFILE: Moderately toxic by ingestion and intravenous routes. An experimental teratogen. Experimental reproductive effects. Human mutation data reported. Combustible when exposed to heat, flame, and oxidizers. To fight fire, use foam, mist, spray, dry chemicals. When heated to decomposition it emits toxic fumes of NO_x.

TDX500 CAS:80-12-6 ***HR: 3***
TETRAMINE
mf: $C_4H_8N_4O_4S_2$ mw: 240.28

SYNS: 2,6-DITHIA-1,3,5,7-TETRAZATRICYCLO(3.3.1.1^{3,7})DECANE-2,2,6,6-TETROXIDE ◇ TETRAMETHYLENEDISULFOTETRAMINE

TOXICITY DATA with REFERENCE
ipr-rat LD50:4850 mg/kg JAPMA8 46,239,57
orl-mus LDLo:200 µg/kg DMWOAX 75,183,50
ipr-mus LD50:210 mg/kg JPMSAE 50,858,61
scu-mus LDLo:100 µg/kg DMWOAX 75,183,50
orl-mam LD50:100 µg/kg PCOC** -,1122,66

SAFETY PROFILE: Poison by ingestion, subcutaneous, and intraperitoneal routes. When heated to decomposition it emits very toxic fumes of NO_x and SO_x. See also AMINES.

TDX750 CAS:5086-74-8 ***HR: 3***
TETRAMISOLE HYDROCHLORIDE
mf: $C_{11}H_{12}N_2S•ClH$ mw: 240.77

PROP: Crystals. Mp: 264-265°. Sol in water (21 g/100 ml at 20°), methanol, and propylene glycol. Sparingly sol in ethanol, chloroform, hexane, and acetone.

SYNS: ANTHELVET ◇ BAYER 9051 ◇ CITARIN ◇ MCN-JR-8299 ◇ NILVERM ◇ NILVEROM ◇ OROVERMOL ◇ dl-6-PHENYL-2,3,5,6-TETRAHYDROIMIDAZOLE(2,1-b)THIAZOLE-HYDROCHLORIDE◇ R 8299 ◇ RIPERCOL ◇ RIPEREOL ◇ SPARTAKON ◇ (±)-2,3,5,6-TETRAHYDRO-6-PHENYLIMIDAZO(2,1-b)THIAZOLEMONOHYDROCHLORIDE ◇ dl-TETRAMISOLE HYDROCHLORIDE ◇ TETRAMISOL HYDROCHLORIDE

TOXICITY DATA with REFERENCE
orl-hmn TDLo:2 mg/kg:SKN TXAPA9 20,602,71
orl-rat LD50:480 mg/kg NATUAS 209,1084,66
scu-rat LD50:130 mg/kg NATUAS 209,1084,66
ivn-rat LD50:24 mg/kg NATUAS 209,1084,66
orl-mus LD50:210 mg/kg NATUAS 209,1084,66
scu-mus LD50:84 mg/kg NATUAS 209,1084,66
ivn-mus LD50:22 mg/kg NATUAS 209,1084,66
orl-dck LD50:958 mg/kg VETNAL 54(5),64,78

SAFETY PROFILE: Poison by ingestion, subcutaneous, and intravenous routes. Human systemic effects by ingestion: skin dermatitis. When heated to decomposition it emits very toxic fumes of NO_x, SO_x, and HCl. Used as an anthelmintic and immunopotentiator.

TDX830 CAS:518-34-3 ***HR: 3***
TETRANDRINE
mf: $C_{38}H_{42}N_2O_6$ mw: 622.82

PROP: Needles. Mp: 217-218°. Practically insol in water, petr ether; sol in ether and some other organic solvents.

SYNS: (1-β)-6,6′,7,12-TETRAMETHOXY-2,2′-DIMETHYL-BERBAMAN (9CI) ◇ (+)-TETRANDRINE ◇ d-TETRANDRINE

TOXICITY DATA with REFERENCE
dni-mus:oth 100 mg/L BCPCA6 25,1887,76
oms-mus:oth 100 mg/L BCPCA6 25,1887,76
ipr-mus LD50:280 mg/kg CPBTAL 24,2413,76

SAFETY PROFILE: Poison by intraperitoneal route. Mutation data reported. When heated to decomposition it emits toxic fumes of NO_x.

TDX835 CAS:22445-73-4 ***HR: 3***
TETRANDRINE DIMETHIODIDE
mf: $C_{40}H_{48}N_2O_6•2I$ mw: 906.70

SYNS: N,N[1]-DIMETHYLPHAEANTHINE DIIODIDE ◇ 2,2′-DIMETHYLTETRANDRINIUM DIIODIDE ◇ SINOMENINE A BIS(METHYL IODIDE) ◇ SINOMENINE A DIMETHIODIDE ◇ (1-β)-6,6′,7,12-TETRAMETHOXY-2,2,2′,2′-TETRAMETHYL-BERBAMANIUM DIIODIDE (9CI)

TOXICITY DATA with REFERENCE
ipr-mus LD50:7 mg/kg CPBTAL 24,2413,76
scu-mus LD50:24600 μg/kg APSXAS 7,279,70
ivn-mus LD50:2630 μg/kg APSXAS 7,279,70

SAFETY PROFILE: Poison by subcutaneous, intravenous, and intraperitoneal routes. When heated to decomposition it emits toxic fumes of I^- and NO_x.

TDX840 CAS:7351-08-8 ***HR: 3***
TETRANGOMYCIN
mf: $C_{19}H_{14}O_5$ mw: 322.33

SYN: (+)-3,4-DIHYDRO-3,8-DIHYDROXY-3-METHYLBENZ(a)ANTHRACENE-1,7,12(2H)-TRIONE

TOXICITY DATA with REFERENCE
orl-mus LD50:221 mg/kg 85GDA2 3,181,80
ipr-mus LD50:41 mg/kg 85GDA2 3,181,80
scu-mus LD50:147 mg/kg 85GDA2 3,181,80

SAFETY PROFILE: Poison by ingestion, subcutaneous, and intraperitoneal routes. When heated to decomposition it emits acrid smoke and irritating fumes.

TDX860 CAS:12041-87-1 ***HR: 1***
TETRANICOTYLFRUCTOSE
mf: $C_{30}H_{24}N_4O_{10}$ mw: 600.58

SYNS: BRADILAN ◇ FRUCTOFURANOSE, TETRANICOTINATE ◇ TETRANICOTYLFRUCTOFURANOSE

TOXICITY DATA with REFERENCE
orl-mus TDLo:3 g/kg (female 7-12D post):TER TOIZAG 14,41,61
orl-mus LD50:23800 mg/kg TOIZAG 14,41,61

SAFETY PROFILE: Slightly toxic by ingestion. An experimental teratogen. When heated to decomposition it emits toxic fumes of NO_x.

TDY000 CAS:3698-54-2 ***HR: 3***
TETRANITROANILINE
mf: $C_6H_3N_5O_8$ mw: 273.14

PROP: Solid. Mp: 170°, bp: explodes @ 237°.

SYNS: TETRANITRANILINE (FRENCH) ◇ 2,3,4,6-TETRANITROANILINE ◇ TNA

TOXICITY DATA with REFERENCE
scu-dog LDLo:2500 mg/kg JIDHAN 2,247,20

SAFETY PROFILE: Moderately toxic by subcutaneous route. For fire hazard, see NITRATES. A severe explosion hazard when shocked or exposed to heat (at 237°C). Tetranitroaniline is a powerful and sensitive high explosive, similar to tetryl. It deteriorates in the presence of moisture. Incompatible with reducing materials. When heated to decomposition it emits toxic fumes of NO_x. See also EXPLOSIVES, HIGH; and NITRATES.

TDY050 ***HR: 3***
N,2,3,5-TETRANITROANILINE
mf: $C_6H_3N_5O_8$ mw: 273.12

$(O_2N)_3C_6H_2NHNO_2$

SAFETY PROFILE: The dry material explodes spontaneously. When heated to decomposition it emits toxic fumes of NO_x. See also NITRO COMPOUNDS of AROMATIC HYDROCARBONS.

TDY075 CAS:4591-46-2 ***HR: 3***
N,2,4,6-TETRANITROANILINE
mf: $C_6H_3N_5O_8$ mw: 273.12

$(O_2N)_3C_6H_2NHNO_2$

DOT Classification: Forbidden

SAFETY PROFILE: The impure material deflagrates (burns explosively) when heated to 50°C. When heated to decomposition it emits toxic fumes of NO_x. See also NITRO COMPOUNDS of AROMATIC HYDROCARBONS.

TDY250 CAS:509-14-8 ***HR: 3***
TETRANITROMETHANE
DOT: UN 1510
mf: CN_4O_8 mw: 196.05

PROP: Colorless or yellow liquid. Mp: 13°, bp: 125.7°, d: 1.650 @ 13°, vap press: 10 mm @ 22.7°. Insol in water; very sol in alc, ether.

SYNS: NCI-C55947 ◇ RCRA WASTE NUMBER P112 ◇ TNM

TOXICITY DATA with REFERENCE
orl-rat LD50:130 mg/kg NTIS** AD-A051-334
ihl-rat LC50:18 ppm/4H AMRL** TR-77-25,77
ivn-rat LD50:12600 μg/kg AMRL** TR-77-25,77
orl-mus LD50:375 mg/kg NTIS** AD-A051-334
ihl-mus LC50:54 ppm/4H AMRL** TR-77-25,77
ipr-mus LD50:53 mg/kg NTIS** AD-A051-334
ivn-mus LD50:63100 μg/kg AMRL** TR-77-25,77

CONSENSUS REPORTS: EPA Extremely Hazardous Substances List. Reported in EPA TSCA Inventory.

OSHA PEL: TWA 1 ppm
ACGIH TLV: TWA 1 ppm
DFG MAK: 1 ppm (8 mg/m^3)
DOT Classification: Oxidizer; Label: Oxidizer.

SAFETY PROFILE: Poison by ingestion, inhalation, intravenous, and intraperitoneal routes. Irritating to the skin, eyes, mucous membranes, and respiratory passages, and does serious damage to the liver. It occurs as an impurity in crude TNT, and is thought to be mainly responsible for the irritating properties of that material. It can cause pulmonary edema, mild methemoglobinemia, and fatty degeneration of the liver and kidneys.

A powerful oxidizer. A very dangerous fire hazard. A severe explosion hazard when shocked or exposed to heat. May explode during distillation. Potentially explosive reaction with ferrocene, pyridine, sodium ethoxide. Mixtures with amines (e.g., aniline) ignite spontaneously and may explode. Mixtures with cotton or toluene may explode when ignited. Forms sensitive and powerful explosive mixtures with nitrobenzene; 1-nitrotoluene; 4-nitrotoluene; 1,3-dinitrobenzene; 1-nitronaphtahlene; other oxygen-deficient explosives; hydrocarbons. Can react vigorously with oxidizing materials. Incompatible with aluminum. When heated to decomposition it emits highly toxic fumes of NO_x. Used as an oxidizer in rocket propellants and as an explosive. See also NITRATES, and EXPLOSIVES, HIGH.

TDY500 CAS:28995-89-3 ***HR: 3***
1,3,6,8-TETRANITRO NAPHTHALENE
mf: $C_{10}H_4N_4O_8$ mw: 308.2

$C_{10}H_4(NO_2)_4$

PROP: Crystals. Mp: 200° (approx), bp: explodes.

SAFETY PROFILE: A shock- and heat-sensitive explosive. A very dangerous fire hazard. Tetranitronaphthalene is a much used high explosive equal to but somewhat less sensitive to impact than TNT. Incompatible with reducing materials. When heated to decomposition it emits toxic fumes of NO_x. See also NITRO COMPOUNDS of AROMATIC HYDROCARBONS, and EXPLOSIVES, HIGH.

TDY600 CAS:641-16-7 ***HR: 3***
2,3,4,6-TETRANITROPHENOL
mf: $C_6H_4N_2O_9$ mw: 248.11

DOT Classification: Forbidden

SAFETY PROFILE: A powerful explosive. When heated to decomposition it emits toxic fumes of NO_x. See also NITRO COMPOUNDS of AROMATIC HYDROCARBONS.

TDY775 ***HR: 3***
trans-1,4,5,8-TETRANITRO-1,4,5,8-TETRAAZA-DECAHYDRONAPHTHALENE
mf: $C_6H_{10}N_8O_8$ mw: 322.19

SAFETY PROFILE: An explosive. When heated to decomposition it emits toxic fumes of NO_x. See also NITRO COMPOUNDS of AROMATIC HYDROCARBONS.

TEA250 CAS:866-97-7 ***HR: 1***
TETRAPENTYLAMMONIUM BROMIDE
mf: $C_{20}H_{44}N{\cdot}Br$ mw: 378.56

PROP: Mp: 100-101°.

SYNS: TETRA-N-AMYLAMMONIUM BROMIDE ◇ TETRA-N-PENTYLAMMONIUM BROMIDE

TOXICITY DATA with REFERENCE
dnd-mam:lym 50 mmol/L CBINA8 19,197,77

CONSENSUS REPORTS: Reported in EPA TSCA Inventory.

SAFETY PROFILE: Mutation data reported. An irritant. When heated to decomposition it emits very toxic fumes of NO_x, NH_3, and Br^-. Hygroscopic. See also BROMIDES.

TEA300 CAS:507-28-8 ***HR: 3***
TETRAPHENYLARSENIUM CHLORIDE
mf: $C_{24}H_{20}As{\cdot}Cl$ mw: 418.81

PROP: Crystals. Mp: 258-260°. Sol in water, alc, or methanol; sltly sol in acetone.

SYNS: ARSONIUM, TETRAPHENYL-, CHLORIDE ◇ TETRAPHENYLARSONIUM CHLORIDE

TOXICITY DATA with REFERENCE
ivn-mus LD50:32 mg/kg CSLNX* NX#00643

OSHA PEL: TWA 0.5 mg(As)/m^3
ACGIH TLV: TWA 0.2 mg(As)/m^3
SAFETY PROFILE: Poison by intravenous route. When heated to decomposition it emits toxic fumes of As and Cl^-.

TEA500 CAS:63469-15-8 ***HR: 3***
1,1,4,4-TETRAPHENYL-1,4-BUTANEDIOL
mf: $C_{28}H_{26}O_2$ mw: 394.54

TOXICITY DATA with REFERENCE
ivn-mus LD50:180 mg/kg CSLNX* NX#02184

CONSENSUS REPORTS: Reported in EPA TSCA Inventory.

SAFETY PROFILE: Poison by intravenous route. When heated to decomposition it emits acrid smoke and irritating fumes.

TEA750 CAS:630-76-2 ***HR: 3***
TETRAPHENYL METHANE
mf: $C_{25}H_{20}$ mw: 320.45

PROP: Rhombic crystals from cold benzene. Mp: 285°, bp: 431°. Insol in ligroin, alc, ether, acetic acid; sol in hot benzene.

SYN: 1,1',1'',1'''-METHANETHETRAYLTETRAKISBENZENE

TOXICITY DATA with REFERENCE
skn-mus TDLo:920 mg/kg/23W-I:ETA AJCAA7 26,754,36

SAFETY PROFILE: Questionable carcinogen with experimental tumorigenic data. When heated to decomposition it emits acrid smoke and irritating fumes.

TEB000 CAS:757-58-4 ***HR: 3***
TETRAPHOSPHORIC ACID, HEXAETHYL ESTER (dry mixture)
DOT: NA 2783
mf: $C_{12}H_{30}O_{13}P_4$ mw: 506.30

SYNS: HEXAETHYL TETRAPHOSPHATE MIXTURE, DRY (DOT) ◇ RCRA WASTE NUMBER P062

DOT Classification: Poison B; Label: Poison.

SAFETY PROFILE: A poison. When heated to decomposition it emits toxic fumes of PO_x. See also HEXAETHYLTETRAPHOSPHATE and ESTERS.

TEB250 CAS:757-58-4 ***HR: 3***
TETRAPHOSPHORIC ACID, HEXAETHYL ESTER (liquid mixture)
mf:$C_{12}H_{39}O_{13}P_4$ mw: 515.37
DOT: NA 2783

SYN: HEXAETHYL TETRAPHOSPHATE MIXTURE, liquid (DOT)

DOT Classification: Poison B; Label: Poison.

SAFETY PROFILE: A poisonous liquid. When heated to decomposition it emits toxic fumes of PO_x. See also ESTERS and HEXAETHYL TETRAPHOSPHOATE.

TEB500 CAS:757-58-4 ***HR: 3***
TETRAPHOSPHORIC ACID, HEXAETHYL ESTER, mixed with compressed gas
$C_{12}H_{39}O_{13}P_4$ mw: 515.37
DOT: UN 1612

SYN: HEXAETHYL TETRAPHOSPHATE and compressed gas mixture (DOT)

DOT Classification: Poison A; Label: Poison Gas.

SAFETY PROFILE: A poison gas. When heated to decomposition it emits toxic fumes of PO_x. See also HEXAETHYL TETRAPHOSPHATE.

TEB750 ***HR: 2***
TETRAPHOSPHORUS IODIDE
mf: IP_4 mw: 250.90

SAFETY PROFILE: Very irritating and toxic. Incompatible with nitric acid. When heated to decomposition it emits very toxic fumes of I^- and PO_x. See also PHOSPHIDES and IODIDES.

TEC000 CAS:1314-86-9 ***HR: 3***
TETRAPHOSPHORUS TRISELENIDE
mf: P_4Se_3 mw: 260.78

CONSENSUS REPORTS: Selenium and its compounds are on the Community Right-To-Know List.

OSHA PEL: TWA 0.2 mg(Se)/m^3
ACGIH TLV: TWA 0.2 mg(Se)/m^3
DFG MAK: 0.1 mg(Se)/m^3

SAFETY PROFILE: Many selenium compounds are poisons. Ignites when heated in air. When heated to decomposition it emits very toxic fumes of PO_x and Se. See also PHOSPHIDES and SELENIUM COMPOUNDS.

TEC250 CAS:14860-53-8 ***HR: 2***
TETRAPOTASSIUM ETIDRONATE
mf: $C_2H_4O_7P_2{\cdot}4K$ mw: 358.40

SYNS: ETHANE-1-HYDROXY-1,1-DIPHOSPHONIC ACID, TETRAPOTASSIUM SALT ◇ (1-HYDROXYETHYLIDENE)DIPHOSPHONIC ACID, TETRAPOTASSIUM SALT

TOXICITY DATA with REFERENCE
orl-rat LD50:520 mg/kg TXAPA9 22,661,72

CONSENSUS REPORTS: Reported in EPA TSCA Inventory.

SAFETY PROFILE: Moderately toxic by ingestion. When heated to decomposition it emits toxic fumes of PO_x and K_2O.

TEC500 CAS:13943-58-3 ***HR: 1***
TETRAPOTASSIUM HEXACYANOFERRATE
mf: $C_6FeN_6{\cdot}4K$ mw: 368.37

SYNS: FERRATE(4-), HEXACYANO-, TETRAPOTASSIUM ◇ FERRATE(4-), HEXAKIS(CYANO-C)-, TETRAPOTASSIUM, (OC-6-11)- ◇ POTASSIUM FERROCYANATE ◇ POTASSIUM FERROCYANIDE ◇ POTASSIUM HEXACYANOFERRATE ◇ POTASSIUM HEXACYANOFERRATE(II) ◇ TETRAPOTASSIUM FERROCYANIDE ◇ TETRAPOTASSIUM HEXACYANOFERRATE(II) ◇ TETRAPOTASSIUM HEXACYANOFERRATE(4-)

TOXICITY DATA with REFERENCE
ihl-rat TDLo:537 μg/m^3 (female 1-22D post):TER GISAAA 52(3),79,87
orl-rat LD50:6400 mg/kg GISAAA 51(4),23,86
orl-mus LD50:5 g/kg GISAAA 51(4),23,86

CONSENSUS REPORTS: Reported in EPA TSCA Inventory.

SAFETY PROFILE: Mildly toxic by ingestion. An experimental teratogen. When heated to decomposition it emits toxic fumes of NO_x.

TEC750 CAS:1941-30-6 ***HR: D***
TETRAPROPYLAMMONIUM BROMIDE
mf: $C_{12}H_{28}N{\cdot}Br$ mw: 266.32

PROP: Crystals. Mp: 280-285°.

TOXICITY DATA with REFERENCE
dnd-mam:lym 50 mmol/L CBINA8 19,197,77

CONSENSUS REPORTS: Reported in EPA TSCA Inventory.

SAFETY PROFILE: Mutation data reported. When heated to decomposition it emits very toxic fumes of NO_x, NH_3, and Br^-. See also BROMIDES.

TED000 CAS:4499-86-9 ***HR: 3***
TETRAPROPYLAMMONIUM HYDROXIDE
mf: $C_{12}H_{28}N{\cdot}HO$ mw: 203.42

PROP: Refr index: 1.3716, d: 1.012, flash p: none.

SYNS: TETRA-N-PROPYLAMMONIUM HYDROXIDE ◇ TETRAPROPYLAMMONIUM OXIDE ◇ N,N,N-TRIPROPYL-1-PROPANAMINIUM HYDROXIDE

TOXICITY DATA with REFERENCE
scu-mus LDLo:52 mg/kg JPETAB 28,367,26

CONSENSUS REPORTS: Reported in EPA TSCA Inventory.

SAFETY PROFILE: Poison by subcutaneous route. A corrosive. When heated to decomposition it emits toxic fumes of NO_x and NH_3.

TED250 CAS:631-40-3 ***HR: 3***
TETRAPROPYLAMMONIUM IODIDE
mf: $C_{12}H_{28}N{\cdot}I$ mw: 313.31

PROP: Mp: 283° (decomp). Hygroscopic.

SYN: TETRA-N-PROPYLAMMONIUMIODIDE

TOXICITY DATA with REFERENCE
ivn-mus LD50:5600 μg/kg CSLNX* NX#00822

CONSENSUS REPORTS: Reported in EPA TSCA Inventory.

SAFETY PROFILE: Poison by intravenous route. An irritant. When heated to decomposition it emits very toxic fumes of NO_x, NH_3, and I^-. See also IODIDES.

TED500 CAS:3244-90-4 ***HR: 3***
TETRA-n-PROPYL DITHIOPYROPHOSPHATE
mf: $C_{12}H_{28}O_5P_2S_2$ mw: 378.46

SYNS: A 42 ◇ ASP 51 ◇ ASPON ◇ BIS-O,O-DI-n-PROPYLPHOSPHOROTHIONIC ANHYDRIDE ◇ ENT 16,894 ◇ NPD ◇ PROPYL THIOPYROPHOSPHATE ◇ STAUFFER ASP-51 ◇ TETRA-n-PROPYL DITHIONOPYROPHOSPHATE ◇ O,O,O,O-TETRAPROPYL DITHIOPYROPHOSPHATE

TOXICITY DATA with REFERENCE
orl-rat LD50:450 mg/kg WRPCA2 9,119,70
skn-rat LD50:1800 mg/kg TXAPA9 2,88,60
ipr-rat LD50:1100 mg/kg AMIHBC 8,350,53

orl-ckn LD50:436 mg/kg TXAPA9 6,147,64
ipr-mus LD50:8 mg/kg CJCHAG 34,1819,56
ivn-mus LD50:3250 μg/kg CJCHAG 34,1819,56
ims-mus LD50:4410 μg/kg CJCHAG 34,1819,56
skn-rbt LD50:3830 mg/kg 31ZOAD 1,42,68

CONSENSUS REPORTS: EPA Genetic Toxicology Program.

SAFETY PROFILE: Poison by intramuscular, intravenous, and intraperitoneal routes. Moderately toxic by ingestion and skin contact. When heated to decomposition it emits very toxic fumes of PO_x and SO_x. Used as a pesticide to control chinch bugs and sod webworm in lawns and on turfs.

TED750 CAS:3440-75-3 ***HR: 3***
TETRAPROPYL LEAD
mf: $C_{12}H_{28}Pb$ mw: 379.59

PROP: Colorless liquid. D: 1.44, bp: 126° @ 13 mm. Sol in benzene.

TOXICITY DATA with REFERENCE
orl-rat LDLo:395 mg/kg BJIMAG 18,277,61
par-rat LD50:200 mg/kg AOHYA3 3,226,61

CONSENSUS REPORTS: Lead and its compounds are on the Community Right-To-Know List. Reported in EPA TSCA Inventory.

SAFETY PROFILE: Poison by ingestion and parenteral routes. When heated to decomposition it emits toxic fumes of Pb. See also LEAD COMPOUNDS.

TEE000 ***HR: 3***
TETRASELENIUM TETRANITRIDE
mf: N_4Se_4 mw: 371.87

CONSENSUS REPORTS: Selenium and its compounds are on the Community Right-To-Know List.

OSHA PEL: TWA 0.2 mg(Se)/m^3
ACGIH TLV: TWA 0.2 mg(Se)/m^3
DFG MAK: 0.1 mg(Se)/m^3

SAFETY PROFILE: Many selenium compounds are poisons. An explosive very sensitive to compression or heating to 130-230°C. Explodes on contact with bromine, chlorine, halogen derivatives, fuming hydrochloric acid. When heated to decomposition it emits very toxic fumes of NO_x and Se. See also SELENIUM COMPOUNDS.

TEE100 ***HR: 3***
TETRASILVER DIIMIDOTRIPHOSPHATE
mf: $Ag_4H_3N_2O_8P_3$ mw: 683.43

CONSENSUS REPORTS: Silver and its compounds are on The Community Right-To-Know List.

SAFETY PROFILE: An explosive sensitive to heat. Ignites on contact with sulfuric acid. When heated to decomposition it emits toxic fumes of PO_x and NO_x. See also SILVER COMPOUNDS.

TEE125 ***HR: 3***
TETRASILVER ORTHODIAMIDOPHOSPHATE
mf: $Ag_5H_2N_2O_3P$ mw: 648.34

CONSENSUS REPORTS: Silver and its compounds are on The Community Right-To-Know List.

SAFETY PROFILE: An explosive sensitive to heat, friction, or contact with sulfuric acid. When heated to decomposition it emits toxic fumes of PO_x and NO_x. See also SILVER COMPOUNDS.

TEE225 ***HR: 3***
TETRASODIUM BIS(CITRATE(3-)FERRATE(4-))
mf: $C_{12}H_{10}FeNa_4O_{14}$ mw: 526.03

SYNS: SCF ◇ TETRASODIUM BISCITRATO FERRATE

TOXICITY DATA with REFERENCE
orl-rbt TDLo:3900 mg/kg (female 6-18D post):REP KSRNAM 22,4681,88
orl-rat LD50:4 g/kg KSRNAM 19,5762,85
ipr-rat LD50:680 mg/kg KSRNAM 19,5762,85
scu-rat LD50:350 mg/kg KSRNAM 19,5762,85
orl-mus LD50:2400 mg/kg KSRNAM 19,5762,85
ipr-mus LD50:155 mg/kg KSRNAM 19,5762,85
scu-mus LD50:150 mg/kg KSRNAM 19,5762,85

SAFETY PROFILE: Poison by subcutaneous and intraperitoneal routes. Experimental reproductive effects. When heated to decomposition it emits acrid smoke and irritating fumes.

TEE250 CAS:3794-83-0 ***HR: 2***
TETRASODIUM ETIDRONATE
mf: $C_2H_4O_7P_2 \cdot 4Na$ mw: 293.96

SYNS: ETHANE-1-HYDROXY-1,1-DIPHOSPHONIC ACID, TRATRASODIUM SALT ◇ (1-HYDROXYETHYLIDENE)DIPHOSPHONIC ACID, TETRASODIUM SALT

TOXICITY DATA with REFERENCE
orl-rat LD50:990 mg/kg TXAPA9 22,661,72

CONSENSUS REPORTS: Reported in EPA TSCA Inventory.

SAFETY PROFILE: Moderately toxic by ingestion. When heated to decomposition it emits toxic fumes of PO_x and Na_2O.

TEE300 CAS:4719-75-9 ***HR: D***
TETRASODIUM FOSFESTROL
mf: $C_{18}H_{18}O_8P_2 \cdot 4Na$ mw: 516.26

SYNS: α,α'-DIETHYL-4,4'-STILBENEDIOL BIS(DIHYDROGEN PHOSPHATE) ◇ DIETHYLSTILBESTROL DIPHOSPHATE TETRASODIUM

◇ DIETHYLSTILBESTROL PHOSPHATE TETRASODIUM ◇ 4,4'-DIHYDROXY-α,α'-DIETHYL-STILBEN-DIPHOSPHATETETRASODIUM ◇ FOSFESTROL TETRASODIUM ◇ HONVAN TETRASODIUMTETRA-SODIUM SALT, (E)- ◇ TETRASODIUM DIETHYLSTILBESTROL PHOSPHATE

TOXICITY DATA with REFERENCE
dlt-mus-ipr 600 mg/kg ARTODN 42,171,79
spm-mus-ipr 300 mg/kg ARTODN 42,171,79
ipr-mus TDLo:150 mg/kg (1D male):TER ARTODN 42,171,79

CONSENSUS REPORTS: EPA Genetic Toxicology Program.

SAFETY PROFILE: An experimental teratogen. Mutation data reported. When heated to decomposition it emits toxic fumes of PO_x and Na_2O. See also DIETHYLSTILBESTEROL.

TEE500 CAS:7722-88-5 ***HR: 3***
TETRASODIUM PYROPHOSPHATE
mf: $O_7P_2 \cdot 4Na$ mw: 265.90

PROP: White crystalline powder. Mp: 988°, d: 2.534. Sol in water; insol in alc.

SYNS: NATRIUMPYROPHOSPHAT ◇ PHOSPHOTEX ◇ PYROPHOSPHATE ◇ SODIUM PYROPHOSPHATE (FCC) ◇ TETRANATRIUM-PYROPHOSPHAT (GERMAN) ◇ TETRASODIUM DIPHOSPHATE ◇ TETRASODIUM PYROPHOSPHATE, ANHYDROUS ◇ TSPP ◇ VICTOR TSPP

TOXICITY DATA with REFERENCE
orl-rat LD50:4000 mg/kg ARZNAD 7,172,57
ipr-rat LD50:59 mg/kg JPETAB 108,117,53
ivn-rat LD50:100 mg/kg ARZNAD 7,172,57
orl-mus LDLo:40 mg/kg AEPPAE 169,238,33
ipr-mus LD50:380 mg/kg REPMBN 10,391,62
scu-mus LD50:400 mg/kg ARZNAD 7,445,57
ivn-mus LD50:69 mg/kg ARZNAD 7,445,57
ivn-rbt LDLo:50 mg/kg AEPPAE 169,238,33

CONSENSUS REPORTS: Reported in EPA TSCA Inventory.

OSHA PEL: TWA 5 mg/m^3
ACGIH TLV: TWA 5 mg/m^3

SAFETY PROFILE: Poison by ingestion, intraperitoneal, intravenous, and subcutaneous routes. It is not a cholinesterase inhibitor. When heated to decomposition it emits toxic fumes of PO_x and Na_2O.

TEE750 CAS:32607-15-1 ***HR: 3***
TETRASULFUR DINITRIDE
mf: N_2S_4 mw: 156.27

S=NSSN=S

SAFETY PROFILE: Explodes when heated to 100°C. When heated to decomposition it emits very toxic fumes of SO_x and NO_x. See also NITRIDES.

TEF250 CAS:866-87-5 ***HR: 3***
TETRAVINYLLEAD
mf: $C_8H_{12}Pb$ mw: 315.33

$(H_2C{=}CH)_4Pb$

CONSENSUS REPORTS: Lead and its compounds are on the Community Right-To-Know List.

SAFETY PROFILE: Violent explosions occur in preparation. Potentially explosive reaction with diborane, phosphorus trichloride. When heated to decomposition it emits toxic fumes of Pb. See also LEAD COMPOUNDS.

TEF500 CAS:109-27-3 ***HR: 3***
TETRAZENE
DOT: UN 0114
mf: $C_2H_8N_{10}O$ mw: 188.20

PROP: Crystals.

SYNS: 4-AMIDINO-1-(NITROSAMINOAMIDINO)-1-TETRAZENE ◇ GUANYL NITROSAMINO GUANYL TETRAZENE, containing, by weight, at least 30% water (DOT) ◇ GUANYL NITROSAMINO GUANYL TETRAZENE (DOT) ◇ 1-GUANYL-4-NITROSAMINOGUANYLTETRAZENE ◇ INITIATING EXPLOSIVE-TETRAZENE (DOT) ◇ TETRACENE ◇ TETRACENE EXPLOSIVE

CONSENSUS REPORTS: Reported in EPA TSCA Inventory.

DOT Classification: Class A Explosive; Label: Explosive.

SAFETY PROFILE: Many nitrosamines are carcinogens. A very dangerous fire hazard. A shock- and heat-sensitive high explosive which evolves much flame. Highly dangerous. Upon decomposition it emits highly toxic fumes of NO_x. See also NITROSAMINES and EXPLOSIVES, HIGH.

TEF600 CAS:290-96-0 ***HR: 2***
1,2,4,5-TETRAZINE
mf: $C_2H_2N_4$ mw: 82.06

SAFETY PROFILE: The solid reacts violently on contact with sulfuric acid. When heated to decomposition it emits toxic fumes of NO_x.

TEF650 CAS:288-94-8 ***HR: 3***
TETRAZOL
mf: CH_2N_4 mw: 70.05

SAFETY PROFILE: Explodes when heated above 155°C. When heated to decomposition it emits toxic fumes of NO_x.

TEF675 CAS:27275-90-7 ***HR: 3***
TETRAZOLE-5-DIAZONIUM CHLORIDE
mf: $CHClN_6$ mw: 132.51

SAFETY PROFILE: The crystals are an extremely shock- and friction-sensitive explosive, which may explode when touched. Solutions in ether may explode. When heated to decomposition it emits toxic fumes of Cl^- and NO_x.

TEF700 CAS:63074-08-8 ***HR: 3***
TETRAZOSIN HYDROCHLORIDE DIHYDRATE
mf: $C_{19}H_{25}N_5O_4•ClH•2H_2O$ mw: 459.99

SYNS: A-45975 ◇ 1-(4-AMINO-6,7-DIMETHOXY-2-QUINAZOLINYL)-4-((TETRAHYDRO-2-FURANYL)CARBON YL)PIPERAZINE HCl $2H_2O$ ◇ HYTRIN ◇ PIPERAZINE, 1-(4-AMINO-6,7-DIMETHOXY-2-QUINAZOLINYL)-4-((TETRAHYDRO-2-FURANYL)CARBONYL)-, MONOHYDROCHLORIDE, DIHYDRATE

TOXICITY DATA with REFERENCE
orl-rat TDLo:650 mg/kg (female 17-22D post):REP IYKEDH 18,622,87
orl-rat TDLo:3300 mg/kg (female 7-17D post):TER IYKEDH 18,427,87
orl-rat LD50:5500 mg/kg NCDREP 4,1,86
scu-rat LD50:1050 mg/kg IYKEDH 18,505,87
ivn-rat LD50:255 mg/kg NCDREP 4,1,86
scu-mus LD50:956 mg/kg IYKEDH 18,505,87
ivn-mus LD50:237 mg/kg IYKEDH 18,505,87

SAFETY PROFILE: Poison by intravenous route. Moderately toxic by subcutaneous route. An experimental teratogen. Experimental reproductive effects. When heated to decomposition it emits toxic fumes of NO_x and HCl.

TEF725 CAS:73666-84-9 ***HR: 3***
TETROCARCIN A
mf: $C_{67}H_{96}N_2O_{24}$ mw: 1313.65

SYNS: ANTIBIOTIC DC 11 ◇ ANTLERMICIN A ◇ DC-11

TOXICITY DATA with REFERENCE
dni-bcs 40 mg/L ABCHA6 48,419,84
oms-bcs 10 mg/L JANTAJ 33,946,80
dni-mus:lym 30 μmol/L JANTAJ 35,1033,82
oms-mus:lym 30 μmol/L JANTAJ 35,1033,82
ipr-mus LD50:54200 μg/kg JANTAJ 35,1033,82
ivn-mus LD50:64 mg/kg JANTAJ 35,1033,82

SAFETY PROFILE: Poison by intravenous and intraperitoneal routes. Mutation data reported. When heated to decomposition it emits toxic fumes of NO_x.

TEF750 CAS:120-54-7 ***HR: 3***
TETRONE A
mf: $C_{12}H_{20}N_2S_6$ mw: 384.70

SYNS: BIS(PENTAMETHYLENETHIURAM)-TETRASULFIDE ◇ BIS(PIPERIDINOTHIOCARBONYL) TETRASULFIDE ◇ DIPENTAMETHYLENETHIURAM TETRASULFIDE ◇ USAF B-31

TOXICITY DATA with REFERENCE
ipr-mus LD50:200 mg/kg NTIS** AD277-689

CONSENSUS REPORTS: Reported in EPA TSCA Inventory.

SAFETY PROFILE: Poison by intraperitoneal route. When heated to decomposition it emits very toxic fumes of NO_x and SO_x. See also SULFIDES.

TEF775 CAS:83-93-2 ***HR: 3***
TETROPHINE
mf: $C_{18}H_{13}NO_2$ mw: 275.32

PROP: Pale yellow needles. Decomp at 252°. Insol in water; sol in glacial acetic acid; mod sol in tetralin; sparingly sol in alc; practically insol in benzene, ether. Mineral acids form incompletely sol, yellow salts which are hydrolyzed by water. Has strychnine-like activity.

SYNS: 5,6-DIHYDROBENZ(c)ACRIDINE-7-CARBOXYLICACID ◇ TETRAPHAN ◇ TETROFAN ◇ TETROPHAN

TOXICITY DATA with REFERENCE
ipr-mus LD50:760 mg/kg JPETAB 113,27,55
ivn-rbt LDLo:40 mg/kg JPETAB 113,371,55

SAFETY PROFILE: Poison by intravenous route. Moderately toxic by intraperitoneal route. When heated to decomposition it emits toxic fumes of NO_x.

TEG250 CAS:479-45-8 ***HR: 3***
TETRYL
DOT: UN 0208
mf: $C_7H_5N_5O_8$ mw: 287.17

PROP: Yellow, monoclinic crystals. Mp: 130°, bp: explodes @ 187°, d: 1.57 @ 19°.

SYNS: N-METHYL-N,2,4,6-TETRANITROANILINE ◇ PICRYLMETHYLNITRAMINE ◇ PICRYLNITROMETHYLAMINE ◇ TETRALITE ◇ N,2,4,5-TETRANITRO-N-METHYLANILINE ◇ 2,4,6-TETRYL ◇ TRINITROPHENYLMETHYLNITRAMINE ◇ 2,4,6-TRINITROPHENYLMETHYLNITRAMINE ◇ 2,4,6-TRINITROPHENYL-N-METHYLNITRAMINE

TOXICITY DATA with REFERENCE
mmo-sat 2500 ng/plate TOLED5 5,11,80
mma-sat 2500 ng/plate TOLED5 5,11,80
mmo-nsc 5 μg/plate TOLED5 5,11,80
mrc-smc 5 μg/plate TOLED5 5,11,80
scu-dog LDLo:5000 mg/kg JIDHAN 2,247,20

OSHA PEL: (Transitional: TWA 1.5 mg/m^3 (skin)) TWA 0.1 mg/m^3 (skin)
ACGIH TLV: TWA 1.5 mg/m^3

DOT Classification: Class A Explosive; Label: Explosive A.

SAFETY PROFILE: Mutation data reported. An irritant, sensitizer, and allergen. The chief effect from exposure is dermatitis. Conjunctivitis is followed by iridocyclitis, and keratitis can occur. Sensitization produced by exposure may play a part in these symptoms. Gastrointestinal effects and anemia have also been reported.

A powerful oxidant. A dangerous fire and explosion hazard. A high explosive sensitive to shock, friction, or heat. More sensitive to shock and friction than TNT. Explodes on contact with trioxygen difluoride. Ignites on contact with hydrazine. When heated to decomposition it emits toxic fumes of NO_x. See also NITRATES and EXPLOSIVES, HIGH.

TEG500 CAS:25265-77-4 ***HR: 2***
TEXANOL
mf: $C_{12}H_{24}O_3$ mw: 216.36

SYN: 2,2,4-TRIMETHYL-1,3-PENTANEDIOLMONOISOBUTYRATE

TOXICITY DATA with REFERENCE
orl-rat LDLo:3200 mg/kg KODAK* -,-,71
orl-mus LDLo:3200 mg/kg KODAK* -,-,71

CONSENSUS REPORTS: Reported in EPA TSCA Inventory.

SAFETY PROFILE: Moderately toxic by ingestion. When heated to decomposition it emits acrid smoke and irritating fumes.

TEG650 ***HR: 3***
TEXTILOTOXIN

SYNS: AUSTRALIAN BROWN SNAKE VENOM ◇ PSEUDONAJA TEXTILIS VENOM ◇ TEXTILON (OBS.) ◇ VENOM, AUSTRALIAN SNAKE, PSEUDONAJA TEXTILIS

TOXICITY DATA with REFERENCE
ipr-mus LD50:310 μg/kg TOXIA6 17(Suppl 1),121,79
scu-mus LD50:41 μg/kg TOXIA6 17,661,79
ivn-mus LD50:1 μg/kg TOXIA6 21,143,83
ims-frg LDLo:500 μg/kg TOXIA6 19,749,81

SAFETY PROFILE: A deadly poison by subcutaneous, intramuscular, intravenous, and intraperitoneal routes.

TEG750 CAS:102418-16-6 ***HR: 3***
THAIMYCIN B

PROP: One of a complex of antibiotics produced by *Streptomyces michiganensis var. amylolyticus* (85ERAY 2,1109,78).

TOXICITY DATA with REFERENCE
orl-mus LD50:250 mg/kg 85ERAY 2,1109,78
ipr-mus LD50:5 mg/kg 85ERAY 2,1109,78

SAFETY PROFILE: Poison by ingestion and intraperitoneal routes.

TEH000 CAS:102418-17-7 ***HR: 3***
THAIMYCIN C

PROP: One of a complex of antibiotics produced by *Streptomyces michiganensis var. amylolyticus* (85ERAY 2,1109,78).

TOXICITY DATA with REFERENCE
orl-mus LD50:250 mg/kg 85ERAY 2,1109,78
ipr-mus LD50:250 mg/kg 85ERAY 2,1109,78

SAFETY PROFILE: Poison by ingestion and intraperitoneal routes.

TEH250 CAS:5373-42-2 ***HR: 3***
THALICARPINE
mf: $C_{41}H_{48}N_2O_8$ mw: 696.91

PROP: Needles from ethyl acetate. Mp: 160-161°. Alkaloid from *Thalictrum dasycarpum* .

SYNS: NSC 68075 ◇ NY IV-34-1 ◇ TALIBLASTIN ◇ TALIBLASTINE ◇ THALIBLASTINE ◇ THALICARPIN ◇ (+)-THALICARPINE

TOXICITY DATA with REFERENCE
dni-rat:leu 20 mg/L MUREAV 147 303,85
dnd-mam:lym 125 μg/L JPMSAE 63,474,74
orl-rat LD50:1500 mg/kg DBANAD 28,709,75
ipr-rat LD50:140 mg/kg NTISP* JAN 86
ivn-rat LD50:78300 μg/kg DBANAD 28,709,75
orl-mus LD50:1542 mg/kg NCIAL* -,60,69
ipr-mus LD50:247 mg/kg TXAPA9 17,284,70
ivn-mus LD50:43700 μg/kg NTIS** PB82-166299
ivn-mky LDLo:42 mg/kg TXAPA9 22,304,72

SAFETY PROFILE: Poison by intraperitoneal and intravenous routes. Moderately toxic by ingestion. Mutation data reported. When heated to decomposition it emits toxic fumes of NO_x.

TEH500 CAS:50-35-1 ***HR: 3***
THALIDOMIDE
mf: $C_{13}H_{10}N_2O_4$ mw: 258.25

PROP: Needles. Mp: 269-271°. Sltly sol in water, methanol, ethanol, acetone, ethylacrylate; very sol in dioxane; sol in ether.

SYNS: ALGOSEDIV ◇ ASIDON 3 ◇ ASMADION ◇ ASMAVAL ◇ BONBRAIN ◇ CALMORE ◇ CALMOREX ◇ CONTERGAN ◇ CORRONAROBETIN ◇ 2,6-DIOXO-3-PHTHALIMIDOPIPERIDINE ◇ 2-(2-6-DIOXO-3-PIPERIDINYL)1H-ISOINDOLE-1,3(2H)-DIONE ◇ N-(2,6-DIOXO-3-PIPERIDYL)PHTHALIMIDE ◇ DISTAVAL ◇ DISTAXAL ◇ DISTOVAL ◇ ECTILURAN ◇ ENTEROSEDIV ◇ GASTRINIDE ◇ GLUPAN ◇ GLUTANON ◇ GRIPPEX ◇ HIPPUZON ◇ IMIDA-LAB ◇ IMIDAN (PEYTA) ◇ IMIDENE ◇ ISOMIN ◇ K 17 ◇ KEDAVON ◇ KEVADON ◇ LULAMIN ◇ NEAUFATIN ◇ NEO ◇ NEOSEDYN ◇ NEOSYDYN ◇ NEURODYN ◇ NEUROSEDIN ◇ NEVRODYN ◇ NIBROL ◇ NOCTOSEDIV ◇ NOXODYN ◇ NSC-66847 ◇ PANGUL ◇ PANTOSEDIV ◇ α-PHTHALIMIDOGLUTARIMIDE ◇ 2-PHTHALIMIDOGLUTARIMIDE ◇ α-(N-PHTHALIMIDO)GLUTARIMIDE ◇ 3-PHTHALIMIDOGLUTARIMIDE ◇ N-PHTHALOYLGLUT-

AMIMIDE ◇ N-PHTHALYLGLUTAMIC ACID IMIDE ◇ N-PHTHALYL-GLUTAMINSAEURE-IMID (GERMAN) ◇ α-N-PHTHALYLGLUTARAM-IDE ◇ POLY-GIRON ◇ POLYGRIPAN ◇ PREDNI-SEDIV ◇ PRO-BAN M ◇ PROFARMIL ◇ PSYCHOLIQUID ◇ PSYCHOTABLETS ◇ QUETIMID ◇ QUIETOPLEX ◇ SANDORMIN ◇ SEDALIS SEDI-LAB ◇ SEDIMIDE ◇ SEDIN ◇ SEDISPERIL ◇ SEDOVAL ◇ SHIN-NAITO S ◇ SHINNIBROL ◇ SLEEPAN ◇ SLIPRO ◇ SOFTENIL ◇ SOFTENON ◇ TALARGAN ◇ TALIMOL ◇ TELAGAN ◇ TELARGAN ◇ TELARGEAN ◇ TENSIVAL ◇ THALIN ◇ THALINETTE ◇ THE-OPHILCHOLINE ◇ ULCERFEN ◇ VALGIS ◇ VALGRAINE ◇ YODOMIN

TOXICITY DATA with REFERENCE
cyt-hmn:lym 1 mg/L AMSVAZ 177,783,65
dns-rat-ipr 80 mg/kg JPETAB 171,109,70
orl-wmn TDLo:17500 μg/kg (1-5W preg):TER BMJOAE 1,123,63
orl-rbt TDLo:400 mg/kg (female 7-16D post):REP ARZNAD 31,315,81
scu-mus TDLo:34 g/kg/57W-I:ETA NATUAS 200,1016,63
orl-rat LD50:113 mg/kg TXAPA9 14,515,69
skn-rat LD50:1550 mg/kg TXAPA9 14,515,69
orl-mus LD50:2000 mg/kg LIFSAK 3,721,64
ipr-mus LDLo:800 mg/kg TXAPA9 23,288,72

CONSENSUS REPORTS: EPA Genetic Toxicology Program.

SAFETY PROFILE: Poison by ingestion. Moderately toxic by skin contact and intraperitoneal routes. Human teratogenic effects by ingestion: developmental abnormalities of the musculoskeletal and cardiovascular systems. Experimental reproductive effects. Questionable carcinogen with experimental tumorigenic and teratogenic data. Human mutation data reported. It was commonly used as a prescription drug in Europe in the late 1950s and early 1960s. Its use was discontinued because it was discovered to cause serious congenital abnormalities in the fetus, notably amelia and phocomelia (absence or deformity of the limbs including hands and feet) when taken by a woman during early pregnancy. When heated to decomposition it emits toxic fumes of NO_x. Used as a sedative and hypnotic.

TEH510 CAS:2614-06-4 ***HR: 3***
(+)-THALIDOMIDE
mf: $C_{13}H_{10}N_2O_4$ mw: 258.25

SYNS: 1H-ISOINDOLE-1,3(2H)-DIONE,2-(2,6-DIOXO-3-PIPERIDINYL)-, (R)- ◇ R-(+)-THALIDOMIDE

TOXICITY DATA with REFERENCE
orl-rbt TDLo:900 mg/kg (7-12D preg):REP NATUAS 215,296,67
orl-rbt TDLo:900 mg/kg (7-12D preg):TER NATUAS 215,296,67
orl-mus LD50:400 mg/kg NATUAS 215,296,67

SAFETY PROFILE: Poison by ingestion. An experimental teratogen. Experimental reproductive effects. When heated to decomposition it emits toxic fumes of NO_x. See also THALIDOMIDE.

TEH520 CAS:841-67-8 ***HR: 2***
(−)-THALIDOMIDE
mf: $C_{13}H_{10}N_2O_4$ mw: 258.25

SYNS: (s)-2-(2,6-DIOXO-3-PIPERIDINYL)-1H-ISOINDOLE-1,3(2H)-DIONE ◇ (±)-THALIDOMIDE ◇ s-(−)-THALIDOMIDE

TOXICITY DATA with REFERENCE
orl-rbt TDLo:900 mg/kg (female 7-12D post):REP NATUAS 215,296,67
ipr-rat TDLo:100 mg/kg (female 10D post):TER ARZNAD 29,1640,79
orl-mus LD50:700 mg/kg NATUAS 215,266,67

SAFETY PROFILE: Moderately toxic by ingestion. An experimental teratogen. Experimental reproductive effects. When heated to decomposition it emits toxic fumes of NO_x. See also THALIDOMIDE.

TEI000 CAS:7440-28-0 ***HR: 3***
THALLIUM
af: Tl aw: 204.37

PROP: Bluish-white, soft, malleable metal. Mp: 303.5°, bp: 1457°, d: 11.85 @ 20°, vap press: 1 mm @ 825°.

SYN: RAMOR

TOXICITY DATA with REFERENCE
orl-man TDLo:5714 μg/kg:PNS,EYE,SKN ATXKA8 19,65,61
unr-man LDLo:4412 μg/kg 85DCAI 2,73,70

CONSENSUS REPORTS: Thallium and its compounds are on the Community Right-To-Know List. Reported in EPA TSCA Inventory.

OSHA PEL: TWA 0.1 mg(Tl)/m^3 (skin)
ACGIH TLV: TWA 0.1 mg(Tl)/m^3 (skin)
DFG MAK: 0.1 mg/m^3

SAFETY PROFILE: Human poison by unspecified route. Human systemic effects by ingestion: nerve or sheath structural changes, extra-ocular muscle changes, sweating, and other effects. Flammable in the form of dust when exposed to heat or flame. Violent reaction with F_2. When heated to decomposition it emits toxic fumes of Tl. Used as a rodenticide and fungicide, and in lenses and prisms, in high-density liquids. See also THALLIUM COMPOUNDS and POWDERED METALS.

TEI250 CAS:563-68-8 ***HR: 3***
THALLIUM ACETATE
mf: $C_2H_3O_2$•Tl mw: 263.42

PROP: Silk-white crystals. Mp: 110°, d: 3.68. Sol in water, alc.

SYNS: RCRA WASTE NUMBER U214 ◇ THALLIUM(1+) ACETATE ◇ THALLIUM(I) ACETATE ◇ THALLIUM MONOACETATE ◇ THALLOUS ACETATE

TOXICITY DATA with REFERENCE
otr-ham:emb 100 μmol/L CNREA8 39,193,79
orl-rat TDLo:30 mg/kg (6-15D preg):TER TJADAB 24(2),46A,81
orl-hmn LDLo:12 mg/kg YAKUD5 22,291,80
unr-hmn LDLo:26 mg/kg AJCPAI 13,422,43
unr-chd LDLo:8 mg/kg AJCPAI 13,422,43
unr-cld LDLo:1400 mg/kg LANCAO 2,1340,30
orl-rat LD50:41300 μg/kg 85ECAN 2,101,78
ipr-rat LD50:30 mg/kg AIHAAP 21,399,60
orl-mus LD50:35 mg/kg JFALAX 5,15,69
ipr-mus LD50:37 mg/kg TXAPA9 49,41,79
scu-mus LDLo:500 μg/kg EQSSDX 1,1,75
orl-dog LDLo:13 mg/kg HBAMAK 4,1406,35
scu-rbt LDLo:5 mg/kg EQSSDX 1,1,75
ivn-rbt LDLo:26 mg/kg AIHAAP 21,399,60

CONSENSUS REPORTS: Thallium and its compounds are on the Community Right-To-Know List. EPA Genetic Toxicology Program. Reported in EPA TSCA Inventory.

OSHA PEL: TWA 0.1 mg(Tl)/m^3 (skin)
ACGIH TLV: TWA 0.1 mg(Tl)/m^3 (skin)

SAFETY PROFILE: Human poison by ingestion. Experimental poison by ingestion, intravenous, intraperitoneal, and subcutaneous routes. An experimental teratogen. Mutation data reported. When heated to decomposition it emits toxic fumes of Tl. See also THALLIUM COMPOUNDS.

TEI500 CAS:13847-66-0 ***HR: 3***
THALLIUM(I) AZIDE
mf: N_3Tl mw: 246.39

CONSENSUS REPORTS: Thallium and its compounds are on the Community Right-To-Know List.

SAFETY PROFILE: Thallium compounds are poisons. Can explode on heavy impact or by heating to 350-400°C. A relatively stable azide. When heated to decomposition it emits toxic fumes of NO_x and Tl. See also THALLIUM COMPOUNDS and AZIDES.

TEI600 ***HR: 3***
THALLIUM(I) AZIDODITHIOCARBONATE
mf: CN_3S_2Tl mw: 322.53

SYN: THALLIUM(I) DITHIOCARBONAZIDATE

CONSENSUS REPORTS: Thallium and its compounds are on The Community Right-To-Know List.

SAFETY PROFILE: Many thallium compounds are poisons. A very unstable explosive salt. Explodes on contact with sulfuric acid. When heated to decomposition it emits toxic fumes of NO_x, Tl, and SO_x. See also THALLIUM COMPOUNDS and AZIDES.

TEI625 CAS:14550-84-6 ***HR: 3***
THALLIUM BROMATE
mf: BrO_3Tl mw: 332.28

CONSENSUS REPORTS: Thallium and its compounds are on The Community Right-To-Know List.

SAFETY PROFILE: Many thallium compounds are poisons. Explodes when heated to 140°C. When heated to decomposition it emits toxic fumes of Br^- and Tl. See also THALLIUM COMPOUNDS and BROMATES.

TEI750 CAS:7789-40-4 ***HR: 3***
THALLIUM BROMIDE
mf: BrTl mw: 284.28

PROP: Yellowish-white powder or crystals. Mp: 460-480° (approx), bp: 815°, d: 7.557, vap press: 10 mm @ 522°.

TOXICITY DATA with REFERENCE
orl-rat LDLo:35 mg/kg 85GMAT -,110,82
scu-mus LDLo:29 mg/kg TPKVAL 2,94,61

CONSENSUS REPORTS: Thallium and its compounds are on the Community Right-To-Know List. Reported in EPA TSCA Inventory.

OSHA PEL: TWA 0.1 mg(Tl)/m^3 (skin)
ACGIH TLV: TWA 0.1 mg(Tl)/m^3 (skin)

SAFETY PROFILE: Poison by ingestion and subcutaneous routes. Reacts violently with Na; K. When heated to decomposition it emits very toxic fumes of Br^- and Tl. See also BROMIDES and THALLIUM COMPOUNDS.

TEJ000 CAS:6533-73-9 ***HR: 3***
THALLIUM(I) CARBONATE (2:1)
mf: $CO_3•2Tl$ mw: 468.75

PROP: Monoclinic, colorless crystals. Mp: 273°, d: 7.11.

SYNS: CARBONIC ACID, DITHALLIUM(1+) SALT ◇ DITHALLIUM CARBONATE ◇ RCRA WASTE NUMBER U215 ◇ THALLOUS CARBONATE ◇ THIOCHROMAN-4-ONE, OXIME

TOXICITY DATA with REFERENCE
cyt-rat-orl 100 ng/kg GTPZAB 24(4),54,80
dnd-mus:emb 10 μmol/L MUREAV 124,163,83
orl-mus TDLo:366 μg/kg (male 26W pre):REP CHYCDW 21,141,87
orl-rat LDLo:23 mg/kg HYSAAV 29,26,64
skn-rat LD50:117 mg/kg GTPZAB 24(4),54,80
scu-rat LDLo:18 mg/kg TPKVAL 2,94,61

orl-mus LD50:21 mg/kg HYSAAV 29,26,64
scu-mus LD50:27 mg/kg TPKVAL 2,94,61

CONSENSUS REPORTS: Thallium and its compounds are on the Community Right-To-Know List. EPA Extremely Hazardous Substances List. Reported in EPA TSCA Inventory.

OSHA PEL: TWA 0.1 mg(Tl)/m^3 (skin)
ACGIH TLV: TWA 0.1 mg(Tl)/m^3 (skin)

SAFETY PROFILE: Poison by ingestion, skin contact, and subcutaneous routes. Experimental reproductive effects. Mutation data reported. When heated to decomposition it emits toxic fumes of Tl. See also THALLIUM COMPOUNDS.

TEJ100 CAS:13453-30-0 ***HR: 3***
THALLIUM CHLORATE
DOT: UN 2573
mf: ClO_3•Tl mw: 287.82

SYN: CHLORIC ACID, THALLIUM(1+) SALT

ACGIH TLV: TWA 0.1 mg(Tl)/m^3
DOT Classification: Oxidizer; Label: Oxidizer and Poison

SAFETY PROFILE: A poison and oxidizer. When heated to decomposition it emits toxic fumes of Tl and Cl^-.

TEJ250 CAS:7791-12-0 ***HR: 3***
THALLIUM CHLORIDE
mf: ClTl mw: 239.82

PROP: Colorless or white powder or crystals. Mp: 430°, bp: 720°, d: 7.00, vap press: 10 mm @ 517°. Sol in 260 parts cold water, 70 parts boiling water; insol in alc.

SYNS: RCRA WASTE NUMBER U216 ◇ THALLIUM(1+) CHLORIDE ◇ THALLIUM MONOCHLORIDE ◇ THALLOUS CHLORIDE

TOXICITY DATA with REFERENCE
otr-ham:emb 100 μmol/L CNREA8 39,193,79
orl-rat TDLo:30 mg/kg (6-15D preg):TER TJADAB 24(2),46A,81
orl-mus LD50:24 mg/kg HYSAAV 29,26,64
ipr-mus LD50:28 mg/kg COREAF 256,1043,63

CONSENSUS REPORTS: EPA Extremely Hazardous Substances List. Thallium and its compounds are on the Community Right-To-Know List. EPA Genetic Toxicology Program. Reported in EPA TSCA Inventory.

OSHA PEL: TWA 0.1 mg(Tl)/m^3 (skin)
ACGIH TLV: TWA 0.1 mg(Tl)/m^3 (skin)

SAFETY PROFILE: Poison by ingestion and intraperitoneal routes. An experimental teratogen. Mutation data reported. Incompatible with F_2. When heated to decomposition it emits very toxic fumes of Cl^- and Tl. See also THALLIUM COMPOUNDS and CHLORIDES.

TEJ500 ***HR: 3***
THALLIUM COMPOUNDS

CONSENSUS REPORTS: Thallium and its compounds are on the Community Right-To-Know List.

SAFETY PROFILE: Extremely toxic. The lethal dose for a man by ingestion is 0.5-1.0 gram. Effects are cumulative and with continuous exposure toxicity occurs at much lower levels. Major effects are on the nervous system, skin, and cardiovascular tract. The peripheral nervous system can be severely affected with dying-back of the longest sensory and motor fibers. Reproductive organs and the fetus are highly susceptible. Acute poisoning has followed the ingestion of toxic quantities of a thallium-bearing depilatory and accidental or suicidal ingestion of rat poison. Acute poisoning results in swelling of the feet and legs, arthralgia, vomiting, insomnia, hyperesthesia and paresthesia of the hands and feet, mental confusion, polyneuritis with severe pains in the legs and loins, partial paralysis of the legs with reaction of degeneration, angina-like pains, nephritis, wasting and weakness, and lymphocytosis and eosinophilia. About the 18th day, complete loss of the hair on the body and head may occur. Fatal poisoning has been known to occur. Recovery requires months and may be incomplete. Industrial poisoning is reported to have caused discoloration of the hair (which later falls out), joint pain, loss of appetite, fatigue, severe pain in the calves of the legs, albuminuria, eosinophilia, lymphocytosis, and optic neuritis followed by atrophy. Cases of industrial poisoning are rare, however. Thallium is an experimental teratogen. When heated to decomposition they emit highly toxic fumes of Tl. See also THALLIUM and specific compounds.

TEJ750 CAS:28625-02-7 ***HR: 3***
THALLIUM(I) FLUOBORATE
mf: BF_4•Tl mw: 291.18

SYN: TL 385

TOXICITY DATA with REFERENCE
orl-rat LDLo:50 mg/kg NCNSA6 5,28,53
ihl-mus LCLo:1870 mg/m^3/10M NDRC** NDCrc-132,Sept,42

CONSENSUS REPORTS: Thallium and its compounds are on the Community Right-To-Know List.

OSHA PEL: TWA 0.1 mg(Tl)/m^3 (skin); 2.5 mg(F)/m^3
ACGIH TLV: TWA 0.1 mg(Tl)/m^3 (skin)
NIOSH REL: (Inorganic Fluorides) TWA 2.5 mg(F)/m^3

SAFETY PROFILE: Poison by ingestion. Moderately toxic by inhalation. When heated to decomposition it emits very toxic fumes of Tl and F^-. See also BORON COMPOUNDS, THALLIUM COMPOUNDS, and FLUORIDES.

TEK000 CAS:7789-27-7 ***HR: 3***
THALLIUM(I) FLUORIDE
mf: FTl mw: 223.37

PROP: Colorless orthorhombic crystals which deliquesce when breathed upon and resolidify in dry air. Mp: 327°, d: 8.36. Very sol in water.

SYNS: THALLIUM MONOFLUORIDE ◇ THALLOUS FLUORIDE

TOXICITY DATA with REFERENCE
orl-rat LDLo:50 mg/kg NCNSA6 5,28,53

CONSENSUS REPORTS: Thallium and its compounds are on the Community Right-To-Know List. Reported in EPA TSCA Inventory.

OSHA PEL: TWA 0.1 mg(Tl)/m^3 (skin)
ACGIH TLV: TWA 0.1 mg(Tl)/m^3 (skin)

SAFETY PROFILE: Poison by ingestion. When heated to decomposition it emits very toxic fumes of F^- and Tl. See also THALLIUM COMPOUNDS and FLUORIDES.

TEK250 CAS:27685-40-1 ***HR: 3***
THALLIUM(I) FLUOSILICATE
mf: F_6Si•Tl mw: 346.46

SYN: HEXAFLUORO-SILICATE(2−), THALLIUM

TOXICITY DATA with REFERENCE
orl-rat LDLo:50 mg/kg NCNSA6 5,1,53

CONSENSUS REPORTS: Thallium and its compounds are on the Community Right-To-Know List.

OSHA PEL: 8H TWA 0.1 mg(Tl)/m^3 (skin)
ACGIH TLV: TWA 0.1 mg(Tl)/m^3
NIOSH REL: TWA 2.5 mg(F)/m^3

SAFETY PROFILE: Poison by ingestion. When heated to decomposition it emits toxic fumes of F^- and Tl. See also THALLIUM COMPOUNDS and FLUORIDES.

TEK300 CAS:20991-79-1 ***HR: 3***
THALLIUM FULMINATE
mf: CNOTl mw: 246.40

CONSENSUS REPORTS: Thallium and its compounds are on The Community Right-To-Know List.

SAFETY PROFILE: Many thallium compounds are poisons. An explosive more sensitive to shock and heat than mercury fulminate. When heated to decomposition it emits toxic fumes of NO_x and Tl. See also THALLIUM COMPOUNDS and FULMINATES.

TEK500 CAS:7790-30-9 ***HR: 3***
THALLIUM IODIDE
mf: ITl mw: 331.27

PROP: Triclinic yellow crystals. D: 7.1, mp: 440°, bp: 824°. Sol in KI soln; practically insol in water; insol in alc.

SYNS: THALLIUM(I) IODIDE ◇ THALLIUM(1+) IODIDE ◇ THALLIUM MONOIODIDE ◇ THALLOUS IODIDE

TOXICITY DATA with REFERENCE
orl-rat LDLo:55 mg/kg 85GMAT -,111,82
scu-mus LDLo:28 mg/kg TPKVAL 2,94,61

CONSENSUS REPORTS: Thallium and its compounds are on the Community Right-To-Know List. Reported in EPA TSCA Inventory.

OSHA PEL: TWA 0.1 mg(Tl)/m^3 (skin)
ACGIH TLV: TWA 0.1 mg(Tl)/m^3 (skin)

SAFETY PROFILE: Poison by ingestion and subcutaneous routes. When heated to decomposition it emits very toxic fumes of Tl and I^-. See also IODIDES and THALLIUM COMPOUNDS.

TEK525 ***HR: 3***
THALLIUM(I) IODOACETYLIDE
mf: C_2ITl mw: 355.31

TlC≡CI

CONSENSUS REPORTS: Thallium and its compounds are on The Community Right-To-Know List.

SAFETY PROFILE: Many thallium compounds are poisons. A shock- and friction-sensitive explosive. When heated to decomposition it emits toxic fumes of Tl and I^-. See also THALLIUM COMPOUNDS and ACETYLIDES.

TEK750 CAS:10102-45-1 ***HR: 3***
THALLIUM NITRATE
DOT: UN 2727
mf: NO_3•Tl mw: 266.38

PROP: Cubic crystals. Mp: 206°, bp: 430°, d: 5.55. Decomp @ 450°.

SYNS: NITRIC ACID, THALLIUM(1+) SALT ◇ RCRA WASTE NUMBER U217 ◇ THALLIUM MONONITRATE ◇ THALLOUS NITRATE

TOXICITY DATA with REFERENCE
mrc-bcs 1 mmol/L MUREAV 77,109,80
orl-man TDLo:73 mg/kg:GIT,MET SMEZA5 18(4),37,75
ipr-rat LD50:21 μg/kg AIPTAK 182,425,69
scu-rat LDLo:26 mg/kg EQSSDX 1,1,75
orl-mus LD50:33 mg/kg JFALAX 5,15,69
orl-dog LDLo:45 mg/kg HBAMAK 4,1406,35
scu-dog LDLo:15 mg/kg EQSSDX 1,1,75

ivn-dog LDLo:15 mg/kg EQSSDX 1,1,75
ivn-rbt LDLo:14 mg/kg EQSSDX 1,1,75

CONSENSUS REPORTS: Thallium and its compounds are on the Community Right-To-Know List. Reported in EPA TSCA Inventory.

OSHA PEL: TWA 0.1 mg(Tl)/m^3 (skin)
ACGIH TLV: TWA 0.1 mg(Tl)/m^3 (skin)
DOT Classification: Poison B; Label:Poison

SAFETY PROFILE: Poison by ingestion, intravenous, intraperitoneal, and subcutaneous routes. Human systemic effects by ingestion: hypermotility, diarrhea, nausea or vomiting, and dehydration. Mutation data reported. When heated to decomposition it emits very toxic fumes of Tl and NO_x. See also THALLIUM COMPOUNDS and NITRATES.

TEL000 ***HR: 3***
THALLIUM(I) NITRIDE
mf: NTl_3 mw: 627.12

CONSENSUS REPORTS: Thallium and its compounds are on the Community Right-To-Know List.

SAFETY PROFILE: Many thallium compounds are poisons. A powerful explosive sensitive to shock, heat, or contact with water or dilute acids. When heated to decomposition it emits toxic fumes of NO_x and Tl. See also THALLIUM COMPOUNDS and NITRIDES.

TEL050 CAS:1314-32-5 ***HR: 3***
THALLIUM(III) OXIDE
mf: O_3Tl_2 mw: 456.74

PROP: Hexagonal black crystals, amorph prisms, or powder. Mp: 717° ± 5°, bp: $-O_2$ @ 875°, d(amorph): 9.65 @ 21°, d(hexagonal): 10.19 @ 22°.

SYNS: DITHALLIUM TRIOXIDE ◇ RCRA WASTE NUMBER P113 ◇ THALLIC OXIDE ◇ THALLIUM OXIDE ◇ THALLIUM (3+) OXIDE ◇ THALLIUM PEROXIDE ◇ THALLIUM SESQUIOXIDE

TOXICITY DATA with REFERENCE
orl-rat LD50:44 mg/kg AIHAAP 21,399,60
ipr-rat LDLo:103 mg/kg AIHAAP 21,399,60
orl-dog LDLo:34 mg/kg AIHAAP 21,399,60
orl-rbt LDLo:34 mg/kg AIHAAP 21,399,60
ipr-rbt LDLo:67 mg/kg AIHAAP 21,399,60
ivn-rbt LDLo:44 mg/kg AIHAAP 21,399,60
orl-gpg LDLo:6 mg/kg AIHAAP 21,399,60
ipr-gpg LDLo:34 mg/kg AIHAAP 21,399,60

CONSENSUS REPORTS: Thallium and its compounds are on the Community Right-To-Know List. Reported in EPA TSCA Inventory.

OSHA PEL: TWA 0.1 mg(Tl)/m^3 (skin)
ACGIH TLV: TWA 0.1 mg(Tl)/m^3 (skin)

SAFETY PROFILE: Poison by ingestion, intraperitoneal, and intravenous routes. Combustible by chemical reaction. Evolves O_2 @ 875°. Mixtures with sulfur or antimony trisulfide explode when ground. Hydrogen sulfide ignites and may explode weakly on contact with the oxide. When heated to decomposition it emits toxic fumes of Tl. See also THALLIUM COMPOUNDS.

TEL100 ***HR: 3***
THALLIUM(I) PEROXODIBORATE
mf: $B_2O_7Tl_2{\cdot}H_2O$ mw: 560.40

CONSENSUS REPORTS: Thallium and its compounds are on The Community Right-To-Know List.

SAFETY PROFILE: Many thallium compounds are poisons. Explodes when warmed. Upon decomposition it emits toxic fumes of Tl. See also THALLIUM COMPOUNDS, BORON COMPOUNDS, and PEROXIDES.

TEL150 CAS:53847-48-6 ***HR: 3***
THALLIUM aci-PHENYLNITROMETHANIDE
mf: $C_7H_6NO_2Tl$ mw: 340.51

$C_6H_5CH{=}N(:O)OTl$

CONSENSUS REPORTS: Thallium and its compounds are on The Community Right-To-Know List.

SAFETY PROFILE: Many thallium compounds are poisons. A dangerously unstable explosive. When heated to decomposition it emits toxic fumes of Tl and NO_x. See also THALLIUM COMPOUNDS.

TEL250 CAS:7440-28-0 ***HR: 3***
THALLIUM SALT (solid)
DOT: NA 1707

CONSENSUS REPORTS: Thallium and its compounds are on the Community Right-To-Know List.

OSHA PEL: TWA 0.1 mg(Tl)/m^3 (skin)
ACGIH TLV: TWA 0.1 mg(Tl)/m^3 (skin)
DFG MAK: 0.1 mg/m^3
DOT Classification: Poison B; Label: Poison.

SAFETY PROFILE: A poison. When heated to decomposition it emits toxic fumes of Tl. See also THALLIUM COMPOUNDS.

TEL500 CAS:12039-52-0 ***HR: 3***
THALLIUM SELENIDE

PROP: Dark gray plates with a shiny appearance. Mp: 340°. Insol in water and acid.

SYNS: RCRA WASTE NUMBER P114 ◇ THALLIUM MONOSELENIDE

CONSENSUS REPORTS: Reported in EPA TSCA Inventory. Selenium and its compounds, as well as thallium and its compounds, are on the Community Right-To-Know List.

OSHA PEL: TWA 0.2 mg(Se)/m^3
ACGIH TLV: TWA 0.2 mg(Se)/m^3
DFG MAK: 0.1 mg(Se)/m^3

SAFETY PROFILE: Thallium compounds and selenium compounds are poisons. When heated to decomposition it emits very toxic fumes of Tl and Se. See also THALLIUM COMPOUNDS and SELENIUM COMPOUNDS.

TEL750 CAS:10031-59-1 ***HR: 3***
THALLIUM SULFATE
DOT: NA 1707
mf: $O_4S{\cdot}xTl$ mw: 1526.65

PROP: White or Colorless rhomboid crystals. Mp: 632°, bp: decomp, d: 6.77.

SYNS: RATOX ◇ SULFURIC ACID, THALLIUM SALT ◇ THALLIUM SULFATE, solid (DOT) ◇ ZELIO

TOXICITY DATA with REFERENCE
ipr-rat TDLo:7500 μg/kg (12-14D preg):TER TXAPA9 16,120,70
orl-wmn TDLo:2 mg/kg:CVS,GIT JTCTDW 19,259,82
orl-hmn LDLo:7 mg/kg PCOC** -,1126,66
orl-man TDLo:11 mg/kg:PNS,GIT,SKN JTCTDW 19,1015,82/83
orl-rat LD50:16 mg/kg JAMAAP 129,927,45
ivn-rat LD50:12 mg/kg TXAPA9 10,199,67
orl-mus LD50:15 mg/kg YKYUA6 28,329,77
scu-mus LDLo:40 mg/kg UDHU** -,-,36

CONSENSUS REPORTS: Thallium and its compounds are on the Community Right-To-Know List. EPA Extremely Hazardous Substances List. Reported in EPA TSCA Inventory.

OSHA PEL: TWA 0.1 mg(Tl)/m^3 (skin)
ACGIH TLV: TWA 0.1 mg(Tl)/m^3 (skin)
DOT Classification: Poison B; Label: Poison.

SAFETY PROFILE: Poison to humans by ingestion. Experimental poison by ingestion, intravenous, and subcutaneous routes. Human systemic effects: cardiomyopathy, diarrhea, effects on recordings from peripheral motor nerve, hair changes, hypermotility, nausea or vomiting. Its main hazard is due to its cumulation, especially in liver, brain, and skeletal muscle; readily absorbed by gastrointestinal tract and skin. A cellular toxicant like arsenic. Fatal human dose is about 500 mg of thallium. Intake of thallium causes depilation. Many reported fatalities. An experimental teratogen. When heated to decomposition it emits very toxic fumes of SO_x and Tl. Pesticide for control of rats, moles, and house mice. See also THALLIUM COMPOUNDS and SULFATES.

TEM000 CAS:7446-18-6 ***HR: 3***
THALLIUM(I) SULFATE (2:1)
mf: $O_4S{\cdot}2Tl$ mw: 504.80

PROP: Crystals. Mp: 632°, bp: (decomp), d: 6.77.

SYNS: C.F.S. ◇ CSF-GIFTWEIZEN ◇ DITHALLIUM SULFATE ◇ DITHALLIUM(1+) SULFATE ◇ ECCOTHAL ◇ M7-GIFTKOERNER ◇ RATTENGIFTKONSERVE ◇ RCRA WASTE NUMBER P115 ◇ SULFURIC ACID, DITHALLIUM(1+) SALT (8CI, 9CI) ◇ SULFURIC ACID, THALLIUM(1+) SALT (1:2) ◇ THALLOUS SULFATE

TOXICITY DATA with REFERENCE
orl-rat TDLo:57 mg/kg (male 60D pre):REP ENVRAL 40,531,86
orl-man LDLo:3 mg/kg:EYE,SYS 85ECAN 2,101,78
orl-man LDLo:7 μg/kg:CNS,GIT CTOXAO 17,133,80
orl-man LDLo:14 mg/kg:PNS,CNS,CVS JTSCDR 3,325,78
orl-hmn LDLo:2166 μg/kg YKYUA6 28,329,77
orl-rat LD50:16 mg/kg YAKUD5 22,291,80
scu-rat LDLo:13 mg/kg APTOA6 12,260,56
orl-mus LD50:29 mg/kg JFALAX 5,15,69
scu-mus LD50:26600 μg/kg MEPAAX 30,257,79

CONSENSUS REPORTS: Thallium and its compounds are on the Community Right-To-Know List. Reported in EPA TSCA Inventory.

OSHA PEL: TWA 0.1 mg(Tl)/m^3 (skin)
ACGIH TLV: TWA 0.1 mg(Tl)/m^3 (skin)

SAFETY PROFILE: Human poison by ingestion. Experimental poison by ingestion and subcutaneous routes. Human systemic effects by ingestion: ataxia, change in heart rate, excitement, eye changes, irritability, nausea or vomiting, nerve or sheath structural changes, somnolence, wakefulness. Experimental reproductive effects. When heated to decomposition it emits very toxic fumes of Tl and SO_x. Used as a rat poison, ant bait, and a reagent in analytical chemistry. See also THALLIUM COMPOUNDS and SULFATES.

TEM250 CAS:63906-56-9 ***HR: 3***
THALLIUM(II) SULFATE (1:1)
mf: $O_4S{\cdot}Tl$ mw: 300.43

SYN: SULFURIC ACID, THALLIUM(2+) SALT

TOXICITY DATA with REFERENCE
orl-rat LDLo:23 mg/kg HYSAAV 29,26,64
orl-mus LD50:24 mg/kg HYSAAV 29,26,64

CONSENSUS REPORTS: Thallium and its compounds are on the Community Right-To-Know List.

OSHA PEL: TWA 0.1 mg(Tl)m^3 (skin)
ACGIH TLV: TWA 0.1 mg(Tl)/m^3 (skin)

SAFETY PROFILE: Poison by ingestion. When heated to decomposition it emits very toxic fumes of Tl and SO_x. See also THALLIUM COMPOUNDS and SULFATES.

TEM399 CAS:2757-18-8 ***HR: 3***
THALLOUS MALONATE
mf: $C_3H_2O_4 \cdot 2Tl$ mw: 510.79

SYNS: FORMOMALENIC THALLIUM ◇ MALONIC ACID, THALLIUM SALT (1:2) ◇ PROPANEDIOIC ACID, DITHALLIUM SALT ◇ THALLIUM MALONATE

TOXICITY DATA with REFERENCE
orl-rat LD50:18800 μg/kg GTPZAB 20(8),35,76
skn-rat LD50:57700 μg/kg GTPZAB 20(8),35,76

CONSENSUS REPORTS: EPA Extremely Hazardous Substances List. Thallium compounds are on the Community Right-To-Know List.

SAFETY PROFILE: Poison by ingestion and skin contact. When heated to decomposition it emits acrid smoke and fumes. See also THALLIUM COMPOUNDS.

TEM500 CAS:77-86-1 ***HR: 2***
THAM
mf: $C_4H_{11}NO_3$ mw: 121.16

SYNS: ADDEX-THAM ◇ 2-AMINO-2-(HYDROXYMETHYL)PROPANE-1,3-DIOL ◇ 2-AMINO-2-(HYDROXYMETHYL)-1,3-PROPANEDIOL ◇ 2-AMINO-2-METHYLOL-1,3-PROPANEDIOL ◇ AMINOTRIMETHYLOMETHANE ◇ AMINOTRIS(HYDROXYMETHYL)METHANE ◇ PEHANORM ◇ TALATROL ◇ THAM-E ◇ THAM SET ◇ TRIMETHYLOLAMINOMETHANE ◇ TRIS ◇ TRISAMINE ◇ TRIS-AMINO ◇ TRISAMINOL ◇ TRIS BUFFER ◇ TRISHYDROXYMETHYL-AMINOMETHAN (GERMAN) ◇ TRISHYDROXYMETHYLAMINOMETHANE ◇ TRIS(HYDROXYMETHYL) METHYLAMINE ◇ 1,1,1-TRIS(HYDROXYMETHYL)METHYLAMINE ◇ TRISPUFFER ◇ TRIS-STERIL ◇ TROMETAMOL ◇ TROMETHAMINE ◇ TROMETHANE ◇ TROMETHANMIN ◇ TUTOFUSIN TRIS

TOXICITY DATA with REFERENCE
orl-rat LD50:5900 mg/kg BCFAAI 110,653,71
ivn-mus LD50:1210 mg/kg ABMGAJ 17,217,66
orl-rbt LDLo:1000 mg/kg JIHTAB 22,315,40

CONSENSUS REPORTS: Reported in EPA TSCA Inventory.

SAFETY PROFILE: Moderately toxic by ingestion and intravenous routes. When heated to decomposition it emits toxic fumes of NO_x.

TEN000 CAS:115-37-7 ***HR: 3***
THEBAINE
mf: $C_{19}H_{21}NO_3$ mw: 311.41

PROP: White to sltly yellowish, lustrous leaflets or prisms. Mp: 193°, d: 1.305.

SYNS: PARAMORPHINE ◇ 6,7,8,14-TETRADEHYDRO-4,5-α-EPOXY-3,6-DIMETHOXY-17-METHYLMORPHINAN

TOXICITY DATA with REFERENCE
ipr-mus LD50:42 mg/kg AIPTAK 132,255,61
scu-mus LD50:31 mg/kg 28ZNAE 138,31,38
scu-rbt LD50:13.9 mg/kg JPETAB 66,182,39
ivn-rbt LD50:3 mg/kg AIPTAK 132,255,61

SAFETY PROFILE: Poison by subcutaneous, intravenous, and intraperitoneal routes. Its action on humans resembles that of strychnine. When heated to decomposition it emits highly toxic fumes of NO_x.

TEN100 CAS:850-57-7 ***HR: 3***
THEBAINE HYDROCHLORIDE
mf: $C_{19}H_{21}NO_3 \cdot ClH$ mw: 347.87

SYNS: MORPHINAN,6,7,8,14-TETRADEHYDRO-4,5-α-EPOXY-3,6-DIMETHOXY-17-METHYL-, HYDROCHLORIDE ◇ 6,7,8,14-TETRADEHYDRO-4,5-α-EPOXY-3,6-DIMETHOXY-17-METHYLMORPHINAN HYDROCHLORIDE ◇ (-)-THEBAINE HYDROCHLORIDE ◇ THEBAIN HYDROCHLORID

TOXICITY DATA with REFERENCE
scu-ham TDLo:227 mg/kg (female 8D post):TER AJOGAH 123,705,75
ipr-mus LD50:25 mg/kg NYKZAU 61,396,65
scu-mus LD50:31 mg/kg NYKZAU 61,396,65
scu-rbt LDLo:9741 μg/kg JPETAB 66,182,39
ivn-rbt LD50:3351 μg/kg AIPTAK 132,255,61
ivn-gpg LDLo:55 mg/kg AEPPAE 131,171,28

SAFETY PROFILE: Poison by intravenous, subcutaneous, and intraperitoneal routes. An experimental teratogen. When heated to decomposition it emits toxic fumes of NO_x and HCl.

TEN725 CAS:65654-13-9 ***HR: D***
N-(2-THENOYL)GLYCINOHYDROXAMIC ACID
mf: $C_7H_8N_2O_3S$ mw: 200.23

SYN: N-(2-(HYDROXYAMINO)-2-OXOETHYL)-2-THIOPHENECARBOXAMIDE

TOXICITY DATA with REFERENCE
mmo-sat 1 μmol/plate JOPHDQ 3,557,80
dnr-bcs 10 μmol/disc JOPHDQ 3,557,80

SAFETY PROFILE: Mutation data reported. When heated to decomposition it emits toxic fumes of SO_x and NO_x.

TEN750 CAS:40828-46-4 ***HR: 3***
p-(2-THENOYL)HYDRATROPIC ACID
mf: $C_{14}H_{12}O_3S$ mw: 260.32

SYNS: MALDOCIL ◇ α-METHYL-4-(2-

THIENYLCARBONYL)BENZENEACETIC ACID ◇ R 25061 ◇ SUPROFEN ◇ SUTOPROFEN ◇ (±)-2-(p-(2-THENOYL)PHENYL)PROPIONIC ACID ◇ TN 762

TOXICITY DATA with REFERENCE
skn-rbt 150 mg/30D-I MLD YACHDS 15,3611,87
orl-rat TDLo:132 mg/kg (female 7-17D post):REP OYYAA2 26,449,83
orl-rat TDLo:2016 mg/kg (9W male/2W pre-7D preg):TER OYYAA2 26,441,83
orl-man TDLo:8571 μg/kg/4D-I:KID DICPBB 20,860,86
orl-man TDLo:2857 μg/kg:KID AIMEAS 105,799,86
orl-wmn TDLo:4 mg/kg:KID AIMEAS 106,235,87
orl-rat LD50:70600 μg/kg ARZNAD 25,1526,75
ipr-rat LD50:335 mg/kg YACHDS 12,1005,84
scu-rat LD50:500 mg/kg YACHDS 12,1005,84
orl-mus LD50:590 mg/kg DRFUD4 1,148,76
ipr-mus LD50:375 mg/kg YACHDS 12,1005,84
scu-mus LD50:710 mg/kg YACHDS 12,1005,84
ipr-dog LD50:300 mg/kg YACHDS 12,1005,84
scu-dog LD50:1250 mg/kg YACHDS 12,1005,84

SAFETY PROFILE: Poison by ingestion and intraperitoneal routes. Moderately toxic by subcutaneous route. Human systemic effects: changes in kidney tubules, hematuria. An experimental teratogen. Experimental reproductive effects. When heated to decomposition it emits toxic fumes of SO_x. Used as an analgesic and anti-inflammatory agent.

TEO000 CAS:958-93-0 ***HR: 3***
THENYLDIAMINE CHLORIDE
mf: $C_{14}H_{19}N_3S{\cdot}ClH$ mw: 297.88

SYNS: DETHYLANDIAMINE ◇ 2-((2-(DIMETHYLAMINO)ETHYL)-3-THENYL-AMINO)-PYRIDINE HYDROCHLORIDE ◇ N,N-DIMETHYL-N'-(3-THENYL)-N'-(2-PYRIDYL) ETHYLENEDIAMINE HYDROCHLORIDE ◇ THENFADIL HYDROCHLORIDE ◇ THENYLDIAMINE HYDROCHLORIDE ◇ WIN-2848

TOXICITY DATA with REFERENCE
orl-rat LD50:525 mg/kg 29ZVAB -,113,69
ivn-rat LD50:15 mg/kg JPETAB 97,371,49
orl-mus LD50:277 mg/kg JPETAB 97,371,49
ipr-mus LD50:55 mg/kg JPETAB 97,371,49
scu-mus LD50:36 mg/kg JPETAB 97,371,49
ivn-mus LD50:12200 μg/kg JPETAB 97,371,49
orl-dog LD50:60 mg/kg JPETAB 97,371,49
ivn-dog LD50:10 mg/kg JPETAB 97,371,49
ims-dog LD50:12 mg/kg JPETAB 97,371,49
ivn-rbt LD50:12 mg/kg JPETAB 97,371,49
ivn-ham LD50:9 mg/kg JPETAB 97,371,49

SAFETY PROFILE: Poison by ingestion, intraperitoneal, subcutaneous, intramuscular, and intravenous routes. When heated to decomposition it emits very toxic fumes of NO_x, SO_x, and HCl. Used as an antihistamine.

TEO250 CAS:91-80-5 ***HR: 3***
THENYLPYRAMINE
mf: $C_{14}H_{19}N_3S$ mw: 261.42

SYNS: A 3322 ◇ AH-42 ◇ 2-((2-(DIMETHYLAMINO)ETHYL)-2-THENYLAMINO)PYRIDINE ◇ N,N-DIMETHYL-N'-2-PYRIDINYL-N'-(2-THIENYLMETHYL)-1,2-ETHANEDIAMIDE ◇ N,N-DIMETHYL-N'-PYRID-2-YL-N'-2-THENYLETHYLENEDIAMINE ◇ DORMIN ◇ HISTADYL ◇ LULAMIN ◇ LULLAMIN ◇ METHAPYRILENE ◇ NCI-C55550 ◇ PARADORMALENE ◇ PYRATHYN ◇ N-(α-PYRIDYL)-N-(α-THENYL)-N',N'-DIMETHYLETHYLENEDIAMINE ◇ PYRINISTAB ◇ PYRINISTOL ◇ RCRA WASTE NUMBER U155 ◇ REST-ON ◇ RESTRYL ◇ SEMIKON ◇ SLEEPWELL ◇ TENALIN ◇ THENYLENE ◇ THIONYLAN

TOXICITY DATA with REFERENCE
dnd-esc 30 μmol/L MUREAV 89,95,81
msc-mus:lym 100 mg/L ENMUDM 5,420,83
scu-rat LD50:150 mg/kg CRSBAW 144,887,50
orl-mus LD50:182 mg/kg PSEBAA 80,458,52
ipr-mus LD50:77 mg/kg JPETAB 96,388,49
ivn-mus LD50:20 mg/kg PSEBAA 80,458,52
orl-gpg LD50:375 mg/kg PSEBAA 80,458,52

SAFETY PROFILE: Poison by ingestion, subcutaneous, intraperitoneal, and intravenous routes. Mutation data reported. When heated to decomposition it emits very toxic fumes of SO_x and NO_x. Used as an antihistamine.

TEO500 CAS:83-67-0 ***HR: 3***
THEOBROMINE
mf: $C_7H_8N_4O_2$ mw: 180.19

PROP: White powder or monoclinic needles, bitter tasting alkaloid. Mp: 357°, subl @ 290-295°. Moderately sol in ammonia; almost insol in benzene, ether, chloroform, carbon tetrachloride.

SYNS: 3,7-DIHYDRO-3,7-DIMETHYL-1H-PURINE-2,6-DIONE ◇ 3,7-DIMETHYLXANTHINE ◇ DIUROBROMINE ◇ SANTHEOSE ◇ SC 15090 ◇ TEOBROMIN ◇ THEOSALVOSE ◇ THEOSTENE ◇ THESAL ◇ THESODATE

TOXICITY DATA with REFERENCE
mmo-eug 600 mg/L MUREAV 32,169,75
sce-hmn:lym 100 mg/L MUREAV 169,105,86
orl-mus TDLo:6300 mg/kg (female 1-21D post):REP TOXID9 4,191,84
ipr-mus TDLo:500 mg/kg (female 13D post):TER OFAJAE 46,167,69
orl-hmn TDLo:26 mg/kg:CNS,GIT JPETAB 86,113,46
orl-rat LD50:1265 mg/kg GTPZAB 26(3),59,82
orl-mus LD50:837 mg/kg GTPZAB 26(3),59,82
scu-mus LD50:530 mg/kg ARZNAD 6,601,56
orl-dog LD50:300 mg/kg TXAPA9 53,481,80
scu-rbt LDLo:1 g/kg HBAMAK 4,1289,35

CONSENSUS REPORTS: Reported in EPA TSCA Inventory. EPA Genetic Toxicology Program.

SAFETY PROFILE: Poison by ingestion. Moderately toxic by subcutaneous route. An experimental teratogen. Human systemic effects by ingestion: central nervous system and gastrointestinal changes. Experimental reproductive effects. Human mutation data reported. When heated to decomposition it emits toxic fumes of NO_x. Used as a diuretic, smooth muscle relaxant, cardiac stimulant, and vasodilator.

TEO750 CAS:8048-31-5 ***HR: D***
THEOBROMINE SODIUM SALICYLATE
mf: $C_7H_8N_4O_2{\cdot}C_7H_6O_3{\cdot}2Na$ mw: 364.30

PROP: White odorless or almost odorless powder. Hygroscopic. Very sol in water; sltly sol in alc.

TOXICITY DATA with REFERENCE
ipr-mus TDLo:600 mg/kg (13D preg):TER OFAJAE 46,167,69

SAFETY PROFILE: An experimental teratogen. When heated to decomposition it emits toxic fumes of NO_x and Na_2O. See also THEOBROMINE.

TEO800 CAS:65098-93-3 ***HR: 2***
THEOESBERIVEN

TOXICITY DATA with REFERENCE
orl-rat LD50:864 mg/kg NIIRDN 6,471,82
ivn-rat LD50:3300 mg/kg OYYAA2 12,543,76
orl-mus LD50:2125 mg/kg NIIRDN 6,471,82
ivn-mus LD50:4250 mg/kg OYYAA2 12,543,76

SAFETY PROFILE: Moderately toxic by ingestion and intravenous routes.

TEP000 CAS:58-55-9 ***HR: 3***
THEOPHYLLINE
mf: $C_7H_8N_4O_2$ mw: 180.19

PROP: Monoclinic, odorless needles or thin monoclinic tablets; bitter taste. Mp: 270-274°. Sol in hot water, alkali hydroxides, ammonia, dil HCl, HNO_3; sltly sol in ether.

SYNS: ACET-THEOCIN ◇ AMINOPHYLLINE ◇ 3,7-DIHYDRO-1,3-DIMETHYL-1H-PURINE-2,6-DIONE ◇ 1,3-DIMETHYLXANTHINE ◇ ELIXICON ◇ ELIXOPHYLLIN ◇ ELIXOPHYLLINE ◇ LANOPHYLLIN ◇ LIQUOPHYLLINE ◇ NSC 2066 ◇ OPTIPHYLLIN ◇ PARKOPHYLLIN ◇ PSEUDOTHEOPHYLLINE ◇ SLO-PHYLLIN ◇ SOLOSIN ◇ TEFAMIN ◇ TEOFYLLAMIN ◇ THEAL TABL. ◇ THEOCIN ◇ THEOFOL ◇ THEOGRAD ◇ THEOLAIR ◇ THEOLIX ◇ THEOPHYL-225 ◇ THEOPHYLLIN ◇ THEOPHYLLINE, anhydrous

TOXICITY DATA with REFERENCE
mmo-eug 600 mg/L MUREAV 32,169,75
oms-hmn:lym 100 μmol/L MUREAV 15,197,72
orl-mus TDLo:4375 mg/kg (male 7D pre):REP FAATDF 10,525,88
orl-mus TDLo:3 g/kg (female 6-15D post):TER NTIS** PB86-103223
orl-man TDLo:129 mg/kg:SYS,GLN AIMEAS 104,284,86
ivn-man LDLo:3429 μg/kg JAMAAP 136,397,48
par-man LDLo:12 mg/kg DAKMAJ 80,510,1904
orl-wmn TDLo:5 mg/kg:CNS,CVS,GIT BMJOAE 288,1497,84
orl-wmn LDLo:130 mg/kg:SYS AIMEAS 101,457,84
orl-wmn TDLo:281 mg/kg/4W:CNS BMJOAE 281,1322,80
ivn-wmn TDLo:120 mg/kg/3D-C:CVS DICPBB 16,877,82
orl-cld TDLo:10 mg/kg SMJOAV 78,1000,85
ims-chd TDLo:50 mg/kg:CNS,CVS,PUL JOPDAB 90,827,77
orl-inf TDLo:348 mg/kg/4D-I PPHAD4 5,209,85
orl-hmn TDLo:5 mg/kg:CNS,GIT,MET 34ZIAG -,92,69
orl-hmn LDLo:20 mg/kg/2D-I LANCAO 2,610,83
scu-hmn TDLo:3500 μg/kg:CNS,GIT,MET 34ziag -,92,69
ivn-hmn TDLo:10 mg/kg/D:CNS,CVS,GIT JAMAAP 235,1983,76
rec-hmn TDLo:6 mg/kg:CNS,GIT,MET 34ZIAG -,92,69
orl-rat LD50:244 mg/kg GTPZAB 26(3),59,82
ivn-hmn LDLo:1430 μg/kg 32ZWAA 8,361,74
ivn-man LDLo:3429 μg/kg:CNS,PUL,GIT JAMAAP 136,397,48
orl-hmn LDLo:20 mg/kg/2D-I:CNS LANCAO 2,610,83
orl-rat LD50:244 mg/kg GTPZAB 26(3),59,82
ipr-rat LD50:188 mg/kg PEREBL 11,783,77
scu-rat LD20:75 mg/kg FAATDF 1,443,81
ivn-rat LDLo:240 mg/kg 27ZIAQ -,-,65
orl-mus LD50:252 mg/kg GTPZAB 26(3),59,82
ipr-mus LD50:70 mg/kg BJPCBM 73,887,81
scu-mus LD50:180 mg/kg THERAP 4,28,49
ivn-mus LD50:136 mg/kg PAHEAA 48,133,73
ims-mus LD50:271 mg/kg NIIRDN 6,34,82
rec-mus LD50:166 mg/kg PEREBL 11,783,77
orl-dog LDLo:290 mg/kg DAKMAJ 80,510,1904
orl-cat LDLo:800 mg/kg DAKMAJ 80,510,1904
scu-cat LDLo:600 mg/kg DAKMAJ 80,510,1904
scu-rbt LDLo:890 mg/kg DAKMAJ 80,510,1904
ivn-rbt LD50:150 mg/kg NIIRDN 6,34,82
scu-gpg LDLo:170 mg/kg HBTXAC 5,168,59

CONSENSUS REPORTS: Reported in EPA TSCA Inventory. EPA Genetic Toxicology Program.

SAFETY PROFILE: Human poison by ingestion, parenteral, intravenous, and rectal routes. Experimental poison by multiple routes. An experimental teratogen. Human systemic effects: coma, convulsions or effect on seizure threshold, cyanosis, EKG changes, fever and other metabolic effects, heart arrythmias, heart rate change, hyperglycemia, metabolic acidosis, nausea or vomiting, potassium level changes, respiratory stimulation, salivary gland changes, somnolence, tremor. Experimental reproductive effects. Human mutation data reported. Used as a diuretic, cardiac stimulant, smooth muscle relaxant, and to treat asthma. When heated to decomposition it emits toxic fumes of NO_x.

TEP500 CAS:317-34-0 ***HR: 3***
THEOPHYLLINE, compound with ETHYLENEDIAMINE (2:1)
mf: $C_7H_8N_4O_2 \cdot 1/2C_2H_8N_2$ mw: 210.25

SYNS: AMINOCARDOL ◇ AMINODUR ◇ AMINOFILINA (SPANISH) ◇ AMINOPHYLLINE ◇ AMMOPHYLLIN ◇ CARDIOFILINA ◇ CARDIOMIN ◇ CARDOPHYLIN ◇ CARDOPHYLLIN ◇ CARENA ◇ CARIOMIN ◇ DIAPHYLLINE ◇ 3,7-DIHYDRO-1,3-DIMETHYL-1H-PURINE-2,6-DIONE compounded with 1,2-ETHANEDIAMINE (2:1) ◇ DIOPHYLLIN ◇ DIUXANTHINE ◇ DOBO ◇ DURA-TAB S.M. AMINOPHYLLINE ◇ ETHOPHYLLINE ◇ ETHYLENEDIAMINE, compounded with THEOPHYLLINE (1:2) ◇ ETILEN-XANTISAN TABL. ◇ EUPHYLLIN ◇ EUPHYLLINE ◇ EUUFILIN ◇ EURPHYLLIN ◇ GENOPHYLLIN ◇ GRIFOMIN ◇ INOPHYLLINE ◇ LASODEX ◇ LINAMPHETA ◇ METAPHYLLIN ◇ METAPHYLLINE ◇ METHOPHYLLINE ◇ MINAPHIL ◇ MIOFILIN ◇ NEOPHYILINE ◇ PETERPHYLLIN ◇ PHYLCARDIN ◇ PHYLLINDON ◇ RECTALAD-AMINOPHYLLINE ◇ SOMOPHYLLIN ◇ SOMOPHYLLIN O ◇ STENOVASAN ◇ TAD ◇ TEFAMIN ◇ TH/100 ◇ THEODROX ◇ THEOLAMINE ◇ THEOMIN ◇ THEOPHYLDINE ◇ THEOPHYLLAMINE ◇ THEOPHYLLAMINIUM ◇ THEOPHYLLIN AETHYLENDIAMIN (GERMAN) ◇ THEOPHYLLINE ETHYLENEDIAMINE ◇ THEOPHYLLIN ETHYLENEDIAMINE ◇ THEPHYLDINE ◇ VASOFILINA

TOXICITY DATA with REFERENCE
cyt-mus:mmr 500 mg/L/24H-C JTSCDR 5,141,80
orl-man TDLo:19 mg/kg/3D:CNS,GIT BCPHBM 10,101,80
ivn-hmn TDLo:139 mg/kg/7D-C:CNS,CVS CLCHAU 24,1603,78
ivn-man LDLo:1420 μg/kg JAMAAP 123,1115,43
rec-chd TDLo:39 mg/kg/8H:CNS,GIT,MET JAMAAP 161,693,56
rec-chd LDLo:221 mg/kg/3D:CNS,GIT,MET JAMAAP 161,693,56
orl-rat LD50:243 mg/kg BCFAAI 110,653,71
ipr-rat LD50:200 mg/kg JAPMA8 36,248,47
scu-rat LD50:176 mg/kg AEEPAE 230,194,57
ivn-rat LD50:184 mg/kg ARZNAD 29,1013,79
ims-rat LD50:167 mg/kg ARZNAD 29,1013,79
orl-mus LD50:150 mg/kg JLCMAK 31,1337,46
ipr-mus LD50:217 mg/kg ARZNAD 29,1013,79
scu-mus LD50:186 mg/kg ARZNAD 4,649,54
ivn-mus LD50:146 mg/kg CTYAD8 11,555,80
orl-dog LDLo:200 mg/kg BTSRAF 29,69,42
rec-dog LDLo:200 mg/kg BTSRAF 29,69,42

CONSENSUS REPORTS: Reported in EPA TSCA Inventory.

SAFETY PROFILE: Human poison by intravenous and rectal routes. Experimental poison by ingestion, intraperitoneal, intravenous, intramuscular, subcutaneous, and rectal routes. Human systemic effects by ingestion, intravenous, and rectal routes: convulsions or effect on seizure threshold, withdrawals, muscle contraction or spasticity, coma, nausea or vomiting, arrythmias, dehydration and fever. Mutation data reported. When heated to decomposition it emits toxic fumes of NO_x. See also THEOPHYLLINE and ETHYLENE DIAMINE.

TEP750 CAS:6336-12-5 ***HR: 3***
8-THEOPHYLLINE MERCURIC ACETATE
mf: $C_9H_{10}HgN_4O_4$ mw: 438.82

SYNS: (ACETATO)(8-THEOPHYLLINYL)MERCURY ◇ 8-(ACETOXYMERCURIO)THEOPHYLLINE

TOXICITY DATA with REFERENCE
ivn-mus LD50:18 mg/kg CSLNX* NX#04387

CONSENSUS REPORTS: Mercury and its compounds are on the Community Right-To-Know List.

OSHA PEL: (Transitional: CL 1 mg/10m^3) CL 0.1 mg (Hg)/m^3 (skin)
ACGIH TLV: TWA 0.1 mg(Hg)/m^3 (skin)
NIOSH REL: (Inorganic Mercury) TWA 0.05 mg (Hg)/m^3

SAFETY PROFILE: Poison by intravenous route. When heated to decomposition it emits very toxic fumes of Hg and NO_x. See also MERCURY COMPOUNDS and THEOPHYLLINE.

TEQ000 CAS:8069-64-5 ***HR: 3***
THEOPHYLLINE METHOXYOXIMERCURIPROPYL SUCCINYLUREA
mf: $C_{16}H_{22}HgN_6O_7$ mw: 611.03

SYNS: MERALLURIDE ◇ MERCUHYDRIN

TOXICITY DATA with REFERENCE
scu-rat LD50:28 mg/kg TXAPA9 18,185,71
ims-rat LD50:37 mg/kg JPETAB 105,336,52
scu-mus LD50:256 mg/kg JPETAB 105,336,52

CONSENSUS REPORTS: Mercury and its compounds are on the Community Right-To-Know List.

OSHA PEL: (Transitional: CL 1 mg/10m^3) CL 0.1 mg (Hg)/m^3 (skin)
ACGIH TLV: TWA 0.1 mg(Hg)/m^3 (skin)
NIOSH REL: (Inorganic Mercury) TWA 0.05 mg (Hg)/m^3

SAFETY PROFILE: Poison by intramuscular and subcutaneous routes. When heated to decomposition it emits very toxic fumes of Hg and NO_x. Used as a diuretic. See also MERCURY COMPOUNDS and THEOPHYLLINE.

TEQ175 CAS:69975-86-6 ***HR: 3***
2-(7′-THEOPHYLLINEMETHYL)-1,3-DIOXOLANE
mf: $C_{11}H_{14}N_4O_4$ mw: 266.29

SYNS: ABC 12/3 ◇ 3,7-DIHYDRO-7-(1,3-DIOXOLAN-2-YLMETHYL)-1,3-DIMETHYL-1H-PURINE-2,6-DIONE ◇ 7-(1,3-DIOXOLAN-2-YLMETHYL)THEOPHYLLINE ◇ 2-(7′-TEOFILLINMETIL)-1,3-DIOSSOLANO (ITALIAN)

TOXICITY DATA with REFERENCE
orl-rat LD50:1022 mg/kg FRPSAX 36,201,81
ipr-rat LD50:445 mg/kg FRPSAX 36,201,81
orl-mus LD50:841 mg/kg FRPSAX 36,201,81
ivn-mus LD50:216 mg/kg FRPSAX 36,201,81

SAFETY PROFILE: Poison by intravenous route. Moderately toxic by ingestion and intraperitoneal routes. When heated to decomposition it emits toxic fumes of NO_x. See also THEOPHYLLINE.

TEQ250 CAS:8002-89-9 ***HR: 3***
THEOPHYLLINE SODIUM ACETATE
mf: $C_7H_8N_4O_2 \cdot C_2H_4O_2 \cdot Na$ mw: 263.24

PROP: White, odorless, crystalline powder; bitter taste. Sol in 25 parts water; insol in alc, chloroform, ether.

SYNS: THEACITIN ◇ THEOCIN SOLUBLE ◇ THEOPHYLLINE, compound with SODIUM ACETATE (1:1) ◇ THEOPHYLLINUM NATRIUM ACETICUM (GERMAN)

TOXICITY DATA with REFERENCE
orl-rat LD50:583 mg/kg AEPPAE 230,194,57
scu-rat LD50:188 mg/kg AEPPAE 230,194,57
ivn-rat LD50:311 mg/kg AEPPAE 230,194,57
ipr-mus LD50:184 mg/kg APJUA8 24,89,74
scu-mus LD50:286 mg/kg AEPPAE 197,193,41
ivn-mus LD50:257 mg/kg AEPPAE 230,194,57
orl-gpg LD50:205 mg/kg AEPPAE 230,194,57
scu-gpg LD50:180 mg/kg AEPPAE 230,194,57

SAFETY PROFILE: Poison by ingestion, intraperitoneal, intravenous, and subcutaneous routes. When heated to decomposition it emits toxic fumes of NO_x and Na_2O. See also THEOPHYLLINE.

TEQ500 CAS:54504-70-0 ***HR: 2***
1-(THEOPHYLLIN-7-YL)ETHYL-2-(2-(p-CHLOROPHENOXY)-2-METHYLPROPIONATE
mf: $C_{19}H_{21}ClN_4O_5$ mw: 420.89

SYNS: DUOLIP ◇ ETOFYLLINCLOFIBRAT (GERMAN) ◇ ETOFYLLINE CLOFIBRATE ◇ ML 1024 ◇ THEOFIBRATE ◇ 1-(THEOPHYLLIN-7-YL)AETHYL-2-(2-(p-CHLORPHENOXY)-2-METHYLPROPIONAT(GERMAN)

TOXICITY DATA with REFERENCE
orl-rbt TDLo:7800 mg/kg (female 6-18D post):TER ARZNAD 27,1173,77
orl-rat TDLo:4200 mg/kg (female 14D post):REP ARZNAD 30,2023,80
orl-rat LD50:17000 mg/kg ARZNAD 30,2023,80
ipr-rat LD50:2830 mg/kg ARZNAD 27,1173,77
orl-mus LD50:1730 mg/kg ARZNAD 25,1686,75

SAFETY PROFILE: Moderately toxic by ingestion and intraperitoneal routes. An experimental teratogen. Experimental reproductive effects. When heated to decomposition it emits very toxic fumes of Cl^- and NO_x. See also ESTERS.

TEQ700 CAS:5503-08-2 ***HR: 3***
THEPHORIN HYDROCHLORIDE
mf: $C_{19}H_{19}N \cdot ClH$ mw: 297.85

SYNS: 2-METHYL-9-PHENYL-2,3,4,9-TETRAHYDRO-1-PYRIDINDENE HYDROCHLORIDE ◇ NU 1604 HYDROCHLORIDE ◇ PHENINDAMINE HYDROCHLORIDE ◇ 2,3,4,9-TETRAHYDRO-2-METHYL-9-PHENYL-1H-INDENO(2,1-c)PYRIDINEHYDROCHLORIDE

TOXICITY DATA with REFERENCE
orl-rat LD50:280 mg/kg JPETAB 92,249,48
orl-mus LD50:255 mg/kg JPETAB 92,249,48
ipr-mus LD50:88 mg/kg JPETAB 92,249,48
scu-mus LD50:270 mg/kg JPETAB 92,249,48
ivn-mus LD50:22500 μg/kg JPETAB 92,249,48
ivn-dog LD50:33 mg/kg JPETAB 92,249,48
orl-rbt LD50:500 mg/kg JPETAB 92,249,48
ivn-rbt LD50:15 mg/kg JPETAB 92,249,48
ipr-gpg LD50:140 mg/kg JPETAB 92,249,48

SAFETY PROFILE: Poison by ingestion, subcutaneous, intravenous, and intraperitoneal routes. When heated to decomposition it emits toxic fumes of NO_x and HCl.

TER000 ***HR: 3***
"THERMIT"

PROP: Composition: Fe_2O_3 + Al.

SAFETY PROFILE: Dangerous when exposed to heat or flame. The violent reaction of Fe_2O_3 + Al is typical of a series of the oxide-metal "thermite" reactions. They are very difficult to stop, as they supply their own oxygen. They may attain a temperature of about 2500°. The presence of manganese dioxide may cause explosions. Keep away from combustible materials. See also ALUMINUM and IRON COMPOUNDS.

TER250 CAS:11018-93-2 ***HR: 3***
THEVETIN
mf: $C_{42}H_{66}O_{18}$ mw:859.08

PROP: Separated from the nuts of *Thevetin neriifolia* (JPETAB 49,561,33).

TOXICITY DATA with REFERENCE
ivn-cat LDLo:843 μg/kg ARZNAD 19,657,69

ivn-gpg LDLo:3560 μg/kg ARZNAD 17,1258,67
ivn-pgn LDLo:1400 μg/kg AIPTAK 126,412,60

SAFETY PROFILE: Poison by intravenous route. When heated to decomposition it emits acrid smoke and irritating fumes.

TER500 CAS:19525-20-3 **HR: 3**
THIABENDAZOLE HYDROCHLORIDE
mf: $C_{10}H_7N_3S \cdot ClH$ mw: 237.72

SYN: 2-(4-THIAZOLYL)-BENZIMIDAZOLE,HYDROCHLORIDE

TOXICITY DATA with REFERENCE
orl-rat LD50:3600 mg/kg TXAPA9 7,53,65
ipr-rat LD50:436 mg/kg TXAPA9 7,53,65
ivn-rat LD50:180 mg/kg TXAPA9 7,53,65
orl-mus LD50:2400 mg/kg TXAPA9 7,53,65
ipr-mus LD50:430 mg/kg TXAPA9 7,53,65
ivn-mus LD50:160 mg/kg TXAPA9 7,53,65

SAFETY PROFILE: Poison by intravenous route. Moderately toxic by ingestion and intraperitoneal routes. When heated to decomposition it emits very toxic fumes of HCl, SO_x, and NO_x.

TES000 **HR: 2**
1,3,4-THIADIAZOLE-2-ACETAMIDO
mf: $C_4H_5N_3OS$ mw: 143.18

PROP: Liquid. D: 1.200 @ 17/4°, bp: 116.8°. Sltly sol in water; sol in alc, ether.

SYNS: CL 1950675526 ◇ NSC 4729 ◇ X 134

TOXICITY DATA with REFERENCE
ipr-mus LD50:1022 mg/kg NCISP* JAN86

SAFETY PROFILE: Moderately toxic by intraperitoneal route. When heated to decomposition it emits very toxic fumes of NO_x and SO_x.

TES250 CAS:1072-71-5 **HR: 3**
1,3,4-THIADIAZOLE-2,5-DITHIOL
mf: $C_2H_2N_2S_3$ mw: 150.24

SYNS: 2,5-DIMERCAPTO-1,3,4-THIADIAZOLE ◇ USAF A-8354 ◇ USAF FA-4

TOXICITY DATA with REFERENCE
eye-rbt 100 mg SEV EPASR* 8EHQ-0181-0380
eye-rbt 100 mg/305 rns EPASR* 8EHQ-0181-0380
ipr-mus LD50:200 mg/kg NTIS** AD277-689

CONSENSUS REPORTS: Reported in EPA TSCA Inventory.

SAFETY PROFILE: Poison by intraperitoneal route. A severe eye irritant. When heated to decomposition it emits very toxic fumes of NO_x and SO_x. See also SULFIDES and MERCAPTANS.

TES500 CAS:467-36-7 **HR: 3**
THIALPENTON
mf: $C_{13}H_{16}N_2O_2S$ mw: 264.37

SYNS: 5-ALLYL-5-(2-CYCLOHEXEN-1-YL)-2-THIOBARBITURIC ACID ◇ 5-$\Delta^{2:3}$-CYCLOHEXENYL-5-ALLYL-2-THIOBARBITURIC ACID ◇ 5-(2-CYCLOHEXEN-1-YL)DIHYDRO-5-(2-PROPENYL)-2-THIOXO-4,6(1H,5H)-PYRIMIDINEDIONE (9CI) ◇ KEMITHAL ◇ THIALBARBITAL ◇ THIALBARBITONE

TOXICITY DATA with REFERENCE
ivn-rat LD50:175 mg/kg AIPTAK 106,437,56
orl-mus LD50:370 mg/kg BJPCAL 1,215,46
ipr-mus LD50:384 mg/kg BJPCAL 1,215,46
ivn-mus LD50:390 mg/kg BJPCAL 1,215,46

SAFETY PROFILE: Poison by ingestion, intraperitoneal and intravenous routes. When heated to decomposition it emits very toxic fumes of NO_x and SO_x. See also BARBITURATES.

TES750 CAS:59-43-8 **HR: 3**
THIAMINE CHLORIDE
mf: $C_{12}H_{17}N_4OS \cdot Cl$ mw: 300.84

SYNS: 3-((4-AMINO-2-METHYL-5-PYRIMIDINYL)METHYL)-5-(2-HYDROXYETHYL)-4-METHYLTHIAZOLIUM CHLORIDE ◇ ANEURINE ◇ APATATE DRAPE ◇ B-AMIN ◇ BEIVON ◇ BETABION ◇ BETALIN S ◇ BETAXIN ◇ BETHIAMIN ◇ BEWON ◇ ORYZANIN ◇ ORYZANINE ◇ THIAMIN ◇ THIAMINE MONOCHLORIDE ◇ VINOTHIAM ◇ VITAMIN B1 ◇ VITANEURON

TOXICITY DATA with REFERENCE
ivn-mus TDLo:20 mg/kg (8D preg):TER TOIZAG 8,175,61
scu-rat LD50:560 mg/kg ARZNAD 9,1,59
ivn-rat LD50:118 mg/kg ARZNAD 9,1,59
scu-mus LD50:301 mg/kg JPETAB 119,444,57
ivn-mus LD50:83 mg/kg ARZNAD 9,1,59

CONSENSUS REPORTS: Reported in EPA TSCA Inventory.

SAFETY PROFILE: Poison by subcutaneous and intravenous routes. An experimental teratogen. Experimental reproductive effects. When heated to decomposition it emits very toxic fumes of NO_x, SO_x, and Cl^-.

TET250 CAS:61227-05-2 **HR: 3**
THIAMINE, DISULFIDE, OROTATE
mf: $C_{24}H_{34}N_8O_4S_2 \cdot C_5H_4N_2O_4$ mw: 718.89

SYNS: THIAMINDISULFID-MONOOROTAT(GERMAN) ◇ THIOORATIN

TOXICITY DATA with REFERENCE
orl-mus LD50:7400 mg/kg ARZNAD 27,1998,77
scu-mus LD50:580 mg/kg ARZNAD 27,1998,77
ivn-mus LD50:250 mg/kg ARZNAD 27,1998,77

SAFETY PROFILE: Poison by intravenous route. Moderately toxic by subcutaneous route. Mildly toxic by ingestion. When heated to decomposition it emits very toxic fumes of NO_x and SO_x. See also SULFIDES.

TET300 CAS:67-03-8 ***HR: 3***
THIAMINE HYDROCHLORIDE
mf: $C_{12}H_{17}N_4OS{\bullet}ClH{\bullet}Cl$ mw: 337.30

PROP: Small white hygroscopic crystals or crystalline powder; nut-like odor. Mp: 248° (decomp). Sol in water, glycerin; sltly sol in alc; insol in ether, benzene.

SYNS: THIAMINE CHLORIDE HYDROCHLORIDE ◇ THIAMINE DICHLORIDE ◇ THIAMIN HYDROCHLORIDE ◇ THIAMINIUM CHLORIDE HYDROCHLORIDE ◇ USAF CB-20 ◇ VITAMIN B^1 ◇ VITAMIN B HYDROCHLORIDE

TOXICITY DATA with REFERENCE
orl-mus LD50:8224 mg/kg IZVIAK 37,82,67
ipr-mus LD50:200 mg/kg NTIS** AD277-689
ivn-mus LD50:89 mg/kg IZVIAK 37,82,67
ivn-rbt LD50:117 mg/kg PSEBAA 68,153,48

CONSENSUS REPORTS: Reported in EPA TSCA Inventory.

SAFETY PROFILE: Poison by intravenous and intraperitoneal routes. Mildly toxic by ingestion. The vitamin is destroyed by alkalies and alkaline drugs such as phenobarbital sodium and by oxidizing and reducing agents. When heated to decomposition it emits very toxic fumes of HCl, Cl^-, SO_x, and NO_x.

TET500 CAS:532-43-4 ***HR: 3***
THIAMINE MONONITRATE
mf: $C_{12}H_{17}N_4OS{\bullet}NO_3$ mw: 327.40

PROP: White crystals or crystalline powder; slt characteristic odor. Mp: 196-200° (decomp). Non-hygroscopic; sltly sol in water, alc, and chloroform.

SYNS: 3-(4-AMINO-2-METHYLPYRIMIDYL-5-METHYL)-4-METHYL-5,β-HYDROXYETHYLTHIAZOLIUM NITRATE ◇ THIAMINE NITRATE ◇ VITAMIN B1 MONONITRATE ◇ VITAMIN B1 NITRATE

TOXICITY DATA with REFERENCE
ivn-rbt LD50:113 mg/kg PSEBAA 68,153,48
ipr-mus LD50:387 mg/kg PSEBAA 68,153,48
ivn-mus LD50:84 mg/kg PSEBAA 68,153,48

CONSENSUS REPORTS: Reported in EPA TSCA Inventory.

SAFETY PROFILE: Poison by intravenous and intraperitoneal routes. A powerful oxidizer. When heated to decomposition it emits very toxic fumes of NO_x and SO_x. See also NITRATES.

TET750 CAS:154-87-0 ***HR: 3***
THIAMINE, TRIHYDROGEN PYROPHOSPHATE (ESTER)
mf: $C_{12}H_{19}N_4O_7P_2S{\bullet}Cl$ mw: 460.80

SYNS: THIAMINEPYROPHOSPHATE ◇ THIAMINEPYROPHOSPHATECHLORIDE ◇ THIAMINEPYROPHOSPHORICESTER ◇ THIAMINPYROPHOSPHATE ◇ TRIHYDROGENPYROPHOSPHATE(ESTER)THIAMINE

TOXICITY DATA with REFERENCE
scu-rat LD50:5000 mg/kg ARZNAD 9,1,59
ivn-rat LD50:465 mg/kg ARZNAD 9,1,59
scu-mus LD50:2500 mg/kg ARZNAD 9,1,59
ivn-mus LD50:360 mg/kg ARZNAD 9,1,59
ivn-cat LD50:510 mg/kg ARZNAD 9,1,59
ivn-rbt LD50:500 mg/kg ARZNAD 9,1,59
ivn-pig LD50:630 mg/kg ARZNAD 9,1,59
ivn-gpg LD50:335 mg/kg ARZNAD 9,1,59
scu-frg LD50:2600 mg/kg ARZNAD 9,1,59

CONSENSUS REPORTS: Reported in EPA TSCA Inventory.

SAFETY PROFILE: Poison by intravenous route. Moderately toxic by subcutaneous route. When heated to decomposition it emits very toxic fumes of NO_x, PO_x, SO_x and Cl^-. See also ESTERS.

TET780 CAS:51155-15-8 ***HR: 3***
THIAMPHENICOL AMINOACETATE HYDROCHLORIDE
mf: $C_{14}H_{18}Cl_2N_2O_6S{\bullet}ClH$ mw: 449.76

TOXICITY DATA with REFERENCE
ipr-rat LD50:1220 mg/kg NIIRDN 6,452,82
ivn-rat LD50:490 mg/kg NIIRDN 6,452,82
ipr-mus LD50:1298 mg/kg NIIRDN 6,452,82
ivn-mus LD50:290 mg/kg NIIRDN 6,452,82

SAFETY PROFILE: Poison by intravenous route. Moderately toxic by intraperitoneal route. When heated to decomposition it emits toxic fumes of SO_x, NO_x, and HCl. See also ESTERS.

TET800 CAS:55297-95-5 ***HR: 3***
THIAMUTILIN
mf: $C_{28}H_{47}NO_4S$ mw: 493.82

PROP: Crystals from acetone. Mp: 147-148° (after stirring in ethyl acetate and drying at 60° and 80° overnight).

SYNS: 14-DEOXY-14-((2-DIETHYLAMINOETHYL-THIO)-ACETOXY) MUTILINE ◇ 14-DESOSSI-14-((2-DIETILAMINOETIL)MERCAPTOACETOSSI)MUTILIN IDROGENO FUMARATO (ITALIAN) ◇ 14-DESOXY-14-((DIETHYLAMINOETHYL)-MERCAPTOACETOXYL)-MUTILIN HYDROGEN FUMARATE ◇ DYNALIN INJECTABLE ◇ DYNAMUTILIN ◇ 81723 HFU ◇ SQ 14055 ◇ SQ 22947 ◇ TIAMULIN ◇ TIAMULINA (ITALIAN)

TOXICITY DATA with REFERENCE
orl-rat LD50:2230 mg/kg RZOVBM 8,251,80
scu-rat LD50:4380 mg/kg RZOVBM 8,251,80
ivn-rat LD50:20 mg/kg RZOVBM 8,251,80

orl-mus LD50:710 mg/kg RZOVBM 8,251,80
scu-mus LD50:1020 mg/kg RZOVBM 8,251,80
ivn-mus LD50:49 mg/kg RZOVBM 8,251,80
orl-ckn LD50:1550 mg/kg RZOVBM 8,251,80
ivn-ckn LD50:25 mg/kg RZOVBM 8,251,80
ims-ckn LD50:250 mg/kg RZOVBM 8,251,80
orl-trk LD50:1345 mg/kg RZOVBM 8,251,80
ims-trk LD50:409 mg/kg RZOVBM 8,251,80
orl-mam LD50:812 mg/kg DRFUD4 2,274,77

SAFETY PROFILE: Poison by intramuscular and intravenous routes. Moderately toxic by ingestion and subcutaneous routes. When heated to decomposition it emits toxic fumes of SO_x and NO_x. See also MERCAPTANS.

TEU000 CAS:10215-25-5 ***HR: 3***
3,7-THIAXANTHENEDIAMINE-5,5-DIOXIDE
mf: $C_{13}H_{11}N_2O_2S$ mw: 259.32

TOXICITY DATA with REFERENCE
ivn-mus LD50:56 mg/kg CSLNX* NX#01363

CONSENSUS REPORTS: Reported in EPA TSCA Inventory.

SAFETY PROFILE: Poison by intravenous route. When heated to decomposition it emits very toxic fumes of NO_x and SO_x.

TEU250 CAS:3122-01-8 ***HR: 3***
THIAZESIUM HYDROCHLORIDE
mf: $C_{19}H_{22}N_2OS•ClH$ mw: 362.95

PROP: Crystals. Mp: 222-224°.

SYNS: 5-(2-DIMETHYLAMINOETHYL)-2,3-DIHYDRO-2-PHENYL-1,5-BENZOTHIAZEPIN-4-(5H)-ONE HYDROCHLORIDE ◇ THIAZESIM HYDROCHLORIDE

TOXICITY DATA with REFERENCE
orl-mus LD50:210 mg/kg 27ZQAG -,310,72
ivn-mus LD50:29 mg/kg 27ZQAG -,310,72

SAFETY PROFILE: Poison by ingestion and intravenous routes. When heated to decomposition it emits very toxic fumes of HCl, NO_x and SO_x.

TEV000 CAS:444-27-9 ***HR: 3***
4-THIAZOLIDINECARBOXYLIC ACID
mf: $C_4H_7NO_2S$ mw: 133.18

PROP: dl Form: Crystals. Mp: 195°.

SYNS: ATC ◇ 4-CARBOXYTHIAZOLIDINE ◇ HEPALIDINE ◇ HEPAREGENE ◇ NORGAMEM ◇ NSC 25855 ◇ T-4CA ◇ THIAZOLIDINECARBOXYLIC ACID ◇ THIAZOLIDINE-4-CARBOXYLIC ACID ◇ THIOPROLINE ◇ USAF A-3701 ◇ USAF C-1

TOXICITY DATA with REFERENCE
ipr-mus TDLo:30 mg/kg (female 1D pre):REP CYLPDN 8,453,87
orl-hmn TDLo:420 mg/kg LANCAO 1,778,81
orl-rat LD50:875 mg/kg DECRDP 13,399,87
ipr-rat LD50:494 mg/kg DECRDP 13,399,87
ipr-mus LD50:100 mg/kg NTIS** AD277-689

CONSENSUS REPORTS: Reported in EPA TSCA Inventory.

SAFETY PROFILE: Poison by intraperitoneal route. Moderately toxic by ingestion. Human systemic effects by ingestion: somnolence, tremors, and depressed renal function. When heated to decomposition it emits very toxic fumes of NO_x and SO_x.

TEV500 CAS:2295-31-0 ***HR: 3***
2,4-THIAZOLIDINEDIONE
mf: $C_3H_3NO_2S$ mw: 117.13

PROP: Crystals. Mp: 121-124°, bp: 178-179°/19 mm.

SYNS: 2,4-DIOXOTHIAZOLIDINE ◇ 2,4(3H,5H)-THIAZOLEDIONE ◇ THIAZOLIDINEDIONE-2,4 ◇ USAF EK-5496

TOXICITY DATA with REFERENCE
ipr-mus LD50:400 mg/kg NTIS** AD277-689
scu-mus LD50:335 mg/kg FRPSAX 9,274,54

CONSENSUS REPORTS: Reported in EPA TSCA Inventory.

SAFETY PROFILE: Poison by intraperitoneal and subcutaneous routes. When heated to decomposition it emits very toxic fumes of NO_x and SO_x.

TEW000 CAS:92760-57-1 ***HR: D***
1-(2-THIAZOLIDINYL)1,2,3,4,5-PENTANEPENTOL
mf: $C_8H_{17}NO_5S$ mw: 239.32

SYN: 2-(1,2,3,4,5-PENTAHYDROXY)-N-PENTYLTHIAZOLIDINE

TOXICITY DATA with REFERENCE
mmo-sat 10 mg/L JAFCAU 28,62,80
mma-sat 1 mg/L JAFCAU 28,62,80

SAFETY PROFILE: Mutation data reported. When heated to decomposition it emits very toxic fumes of NO_x and SO_x.

TEX000 CAS:148-79-8 ***HR: 3***
2-(THIAZOL-4-YL)BENZIMIDAZOLE
mf: $C_{10}H_7N_3S$ mw: 201.26

PROP: White-to-tan; odorless. Mp: 304°. Insol in water; sltly sol in alc, acetone; very sltly sol in ether, chloroform.

SYNS: APL-LUSTER ◇ ARBOTECT ◇ 4-(2-BENZIMIDAZOLYL)THIAZOLE ◇ BOVIZOLE ◇ EPROFIL ◇ EQUIZOLE ◇ LOMBRISTOP ◇ MERTEC ◇ METASOL TK-100 ◇ MINTEZOL ◇ MINZOLUM ◇ MK 360 ◇ MYCOZOL ◇ NEMAPAN ◇ OMNIZOLE ◇ POLIVAL ◇ TBDZ ◇ TECTO ◇ THIABEN ◇ THIABENDAZOLE (USDA) ◇ THIABENZOLE ◇ 2-(4-THIAZOLYL)BENZIMIDAZOLE ◇ 2-(4'-THIAZOLYL)

BENZIMIDAZOLE ◇ 2-(4-THIAZOLYL)-1H-BENZIMIDAZOLE ◇ THIBENZOLE ◇ TOP FORM WORMER

TOXICITY DATA with REFERENCE
oms-asn 80 μmol/L BBACAQ 543,82,78
sln-asn 20 mg/L EVHPAZ 31,81,79
orl-mus TDLo:11700 mg/kg (female 7-15D post):TER TRENAF 29,112,78
orl-rat TDLo:5336 mg/kg (female 8-15D post):REP TRENAF 37,421,86
orl-man TDLo:47619 μg/kg/1D-I JOPDAB 102,317,83
orl-rat LD50:3100 mg/kg TXAPA9 7,53,65
orl-mus LD50:1395 mg/kg FATOAO 34,483,71
orl-rbt LD50:3850 mg/kg 85DPAN -,-,71/76
orl-ckn LD50:4 g/kg VMDNAV 19(3),99,82
orl-dom LD50:400 mg/kg VMDNAV 19(3),99,82

CONSENSUS REPORTS: EPA Genetic Toxicology Program. Reported in EPA TSCA Inventory.

SAFETY PROFILE: Moderately toxic by ingestion. An experimental teratogen. Experimental reproductive effects. Mutation data reported. When heated to decomposition it emits toxic fumes of SO_x and NO_x. See also SULFIDES.

TEX200 CAS:27146-15-2 ***HR: D***
2-(4-THIAZOLYL)-5-BENZIMIDAZOLECARBAMIC ACID METHYL ESTER
mf: $C_{12}H_{10}N_4O_2S$ mw: 274.32

TOXICITY DATA with REFERENCE
oms-hmn:leu 1 mg/L THERAP 31,505,76
oms-hmn:oth 2 mg/L THERAP 31,505,76
orl-rat TDLo:110 mg/kg (8-15D preg):TER THERAP 31,505,76

SAFETY PROFILE: An experimental teratogen. Human mutation data reported. When heated to decomposition it emits toxic fumes of SO_x and NO_x. See also CARBAMATES and ESTERS.

TEX250 CAS:72-14-0 ***HR: 3***
N^1-2-THIAZOLYLSULFANILAMIDE
mf: $C_9H_9N_3O_2S_2$ mw: 255.33

PROP: Mp: 200-202°.

SYNS: 2-(p-AMINOBENZENESULFONAMIDO)THIAZOLE ◇ 2-(p-AMINOBENZENESULPHONAMIDO)THIAZOLE ◇ 4-AMINO-N-2-THIAZOLYLBENZENESULFONAMIDE ◇ AZOSEPTALE ◇ CERAZOL (suspension) ◇ CHEMOSEPT ◇ DUATOK ◇ ELEUDRON ◇ FORMOSULFATHIAZOLE ◇ M + B 760 ◇ NEOSTREPSAN ◇ NORSULFASOL ◇ NORSULFAZOLE ◇ PLANOMIDE ◇ POLISEPTIL ◇ RP 2990 ◇ STREPTOSILTHIAZOLE ◇ SULFAMUL ◇ 2-SULFANILAMIDOTHIAZOLE ◇ 2-(SULFANILYLAMINO)THIAZOLE ◇ SULFATHIAZOL ◇ SULFATHIAZOLE (USDA) ◇ 2-SULFONAMIDOTHIAZOLE ◇ SULPHATHIAZOLE ◇ SULZOL ◇ THIACOCCINE ◇ THIAZAMIDE ◇ THIOZAMIDE ◇ USAF SN-9

TOXICITY DATA with REFERENCE
dnd-esc 50 μmol/L MUREAV 89,95,81
pic-omi 5 mg/L JGMIAN 8,116,53
orl-rat TDLo:29400 mg/kg (male 6W pre):REP JRPFA4 81,259,87
par-rat TDLo:500 mg/kg:ETA ACRAAX 37,258,52
unr-man LDLo:250 mg/kg/23D-I:EYE,KID,SKN JPBAA7 59,501,47
ipr-rat LDLo:1250 mg/kg HBTXAC 5,164,59
par-rat LDLo:1000 mg/kg ACRAAX 37,258,52
orl-mus LD50:4500 mg/kg ARZNAD 21,571,71
ipr-mus LD50:400 mg/kg NTIS** AD277-689
scu-mus LD50:1450 mg/kg HBTXAC 5,164,59
ivn-mus LD50:990 mg/kg HBTXAC 5,164,59

CONSENSUS REPORTS: Reported in EPA TSCA Inventory. EPA Genetic Toxicology Program.

SAFETY PROFILE: Human poison by unspecified route. Experimental poison by intraperitoneal route. Moderately toxic by intravenous, subcutaneous, and parenteral routes. Mildly toxic by ingestion. Human systemic effects by unspecified route: conjuctiva irritation, tubule changes and allergic skin dermatitis. Experimental reproductive effects. Questionable carcinogen with experimental tumorigenic data. Mutation data reported. When heated to decomposition it emits very toxic fumes of NO_x and SO_x.

TEX500 CAS:144-74-1 ***HR: 2***
N^1-2-THIAZOLYSULFANILAMIDE SODIUM SALT
mf: $C_9H_9N_3O_2S_2$•Na mw: 278.32

SYNS: MONOSODIUM-2-SULFANILAMIDOTHIAZOLE ◇ SODIUM NORSULFAZOLE ◇ SODIUM-2-SULFANILAMIDOTHIAZOLE ◇ SODIUM SULFATHIAZOLE ◇ SOLUBLE SULFATHIAZOLE ◇ 2-SULFANILAMIDOTHIAZOLE SODIUM SALT ◇ 2-SULFATHIAZOLE SODIUM ◇ N^1-2-THIAZOLYLSULFANIDAMIDE SODIUM SALT ◇ (N^1-2-THIAZOLYLSULFANIDAMIDO)SODIUM

TOXICITY DATA with REFERENCE
ipr-rat LDLo:750 mg/kg PSEBAA 45,15,40
orl-mus LD50:3800 mg/kg AIPTAK 94,338,53
scu-mus LD50:1434 mg/kg JPETAB 71,138,41
ivn-mus LD50:950 mg/kg AIPTAK 94,338,53
par-mus LD50:1950 mg/kg PSEBAA 43,328,40

CONSENSUS REPORTS: Reported in EPA TSCA Inventory.

SAFETY PROFILE: Moderately toxic by ingestion, intraperitoneal, subcutaneous, intravenous, and parenteral routes. When heated to decomposition it emits very toxic fumes of NO_x, SO_x, and Na_2O.

TEX600 CAS:51707-55-2 ***HR: 2***
THIDIAZURON
mf: $C_9H_8N_4OS$ mw: 220.27

SYNS: DEFOLIT ◇ DROPP ◇ N-PHENYL-N'-1,2,3-THIADIAZOL-5-YL-UREA ◇ SN 49537 ◇ (N-1,2,3-THIADIAZOLYL-5)-N'-PHENYLUREA

TOXICITY DATA with REFERENCE
orl-rat TDLo:813 mg/kg (76D male):REP GISAAA 49(1),72,84
orl-rat LD50:5350 mg/kg GISAAA 49(1),72,84
ipr-rat LD50:4200 mg/kg FMCHA2 -,C91,83
orl-mus LD50:3740 mg/kg GISAAA 49(1),72,84
orl-rbt LD50:7100 mg/kg GISAAA 49(1),72,84
orl-gpg LD50:2813 mg/kg GISAAA 49(1),72,84

SAFETY PROFILE: Moderately toxic by ingestion. Experimental reproductive effects. When heated to decomposition it emits toxic fumes of SO_x and NO_x.

TEY000 CAS:50-59-9 ***HR: 3***
7-((2-THIENYL)ACETAMIDO)-3-(1-PYRIDYLMETHYL)CEPHALOSPORANIC ACID
mf: $C_{19}H_{17}N_3O_4S_2$ mw: 415.51

SYNS: ALIPORINA ◇ AMPLIGRAM ◇ BETAINE CEPHALORIDINE ◇ CEFALORIDIN ◇ CEFALORIDINE ◇ CEFALORIZIN ◇ CEFLORIN ◇ CEPALORIDIN ◇ CEPALORIN ◇ CEPH 87/4 ◇ CEPHALORIDIN ◇ CEPHALORIDINE ◇ CEPORAN ◇ CEPORINE ◇ FOR ◇ DEFLORIN ◇ FAREDINA ◇ FLORIDIN ◇ GLAXORIDIN ◇ INTRASPORIN ◇ KEFLODIN ◇ LILLY 40602 ◇ LLONCEFAL ◇ LORIDIN ◇ SCH 11527 ◇ SEFACIN ◇ 7-(α-(2-THIENYL)ACETAMIDO)-3-(1-PYRIDYLMETHYL)-3-CEPHEM-4-CARBOXYLIC ACID BETAINE ◇ 7-(THIOPHENE-2-ACETAMIDO)-3-(1-PYRIDYLMETHYL)-3-CEPHEM-4-CARBOXYLIC ACID BETAINE

TOXICITY DATA with REFERENCE
ims-rbt TDLo:3500 mg/kg (1-25D preg):REP TXAPA9 8,407,66
ivn-man LDLo:1137 mg/kg/4D:GIT,LIV JAMAAP 200,724,67
orl-rat LD50:2500 mg/kg TXAPA9 8,398,66
ipr-rat LD50:3170 mg/kg TXAPA9 18,185,71
scu-rat LD50:2500 mg/kg TXAPA9 8,398,66
ivn-rat LD50:1300 mg/kg AACHAX -,863,65
par-rat LD50:2500 μg/kg AACHAX -,863,65
scu-mus LD50:9600 mg/kg NIIRDN 6,409,82
ivn-mus LD50:2200 mg/kg TXAPA9 8,398,66
ims-mus LD50:7 g/kg BYYADW 3,220,78
ice-mus LD50:10 mg/kg CHTHBK 26,196,80
scu-gpg LD50:550 mg/kg TXAPA9 8,398,66

SAFETY PROFILE: Experimental poison by intracerebral route. Moderately toxic to humans by intravenous route. Moderately toxic experimentally by ingestion, subcutaneous, intraperitoneal, and intravenous routes. Human systemic effects by intravenous route: nausea or vomiting and fatty liver degeneration. Experimental reproductive effects. When heated to decomposition it emits very toxic fumes of NO_x and SO_x.

TEY250 CAS:139-86-6 ***HR: D***
2-THIENYLALANINE
mf: $C_7H_9NO_2S$ mw: 171.23

SYNS: α-AMINO-2-THIOPHENEPROPANOIC ACID ◇ β-2-THIENYLALANINE ◇ dl-β-2-THIENYLALANINE ◇ 2-THIOPHENEALANINE

TOXICITY DATA with REFERENCE
mmo-omi 100 mg/L MUREAV 12,349,71
dnd-hmn:hla 200 μmol/L ECREAL 107,191,77
ipr-mus TDLo:4800 mg/kg (10-12D preg):REP TJADAB 4,295,71

SAFETY PROFILE: Experimental reproductive effects, Human mutation data reported. When heated to decomposition it emits very toxic fumes of NO_x and SO_x.

TEY600 CAS:76175-45-6 ***HR: 3***
2-(2-THIENYL)MORPHOLINE HYDROCHLORIDE
mf: $C_8H_{11}NOS•ClH$ mw: 205.72

SYNS: M.G. 18590 ◇ 2-(2-TIENYL)MORFOLINA CLORIDRATE (ITALIAN)

TOXICITY DATA with REFERENCE
ipr-rat LD50:300 mg/kg FRPSAX 35,812,80
orl-mus LD50:1220 mg/kg FRPSAX 35,812,80
ipr-mus LD50:225 mg/kg FRPSAX 35,812,80

SAFETY PROFILE: Poison by intraperitoneal route. Moderately toxic by ingestion. When heated to decomposition it emits toxic fumes of SO_x, NO_x, and HCl.

TEY750 CAS:17689-16-6 ***HR: 2***
2-THIEPANONE
mf: $C_6H_{10}OS$ mw: 130.22

SYNS: THIEPAN-2-ONE ◇ epsilon-THIOCAPROLACTONE

TOXICITY DATA with REFERENCE
orl-rat LD50:600 mg/kg TXAPA9 28,313,74
skn-rbt LD50:500 mg/kg TXAPA9 28,313,74

SAFETY PROFILE: Moderately toxic by ingestion and skin contact. When heated to decomposition it emits toxic fumes of SO_x. See also SULFIDES.

TEZ000 CAS:1179-69-7 ***HR: 3***
THIETHYLPERAZINE MALEATE
mf: $C_{22}H_{29}N_3S_2•2C_4H_4O_4$ mw: 631.82

SYNS: 3-ETHYLMERCAPTO-10-(1'-METHYLPIPERAZINYL-4'-PROPYL)PHENOTHIAZINE DIMALEATE ◇ 2-ETHYLMERCAPTO-10-(3-(1-METHYL-4-PIPERAZINYL)PROPYL)PHENOTHIAZINEDIMALEATE ◇ 2-(ETHYLTHIO)-10-(3-(4-METHYL-1-PIPERAZINYL)PROPYL)-10-PHENOTHIAZINE-(Z)-2-BUTENEDIOATE ◇ 2-(ETHYLTHIO)-10-(3-(4-METHYL-1-PIPERAZINYL)PROPYL)PHENOTHIAZINEDIMALEATE ◇ GS-95 ◇ NSC-130044 ◇ THIETHYLPERAZINE DIMALEATE ◇ TORECAN ◇ TORECAN BIMALEATE ◇ TORECAN DIMALEATE ◇ TORECAN MALEATE ◇ TORESTEN ◇ TRESTEN

TOXICITY DATA with REFERENCE
orl-rat LD50:1260 mg/kg 27ZQAG -,51,72
ivn-rat LD50:90 mg/kg 27ZQAG -,51,72
orl-mus LD50:680 mg/kg 27ZQAG -,51,72
ipr-mus LD50:263 mg/kg 27ZQAG -,51,72
ivn-mus LD50:94 mg/kg 27ZQAG -,51,72
orl-rbt LD50:1050 mg/kg 27ZQAG -,51,72
ivn-rbt LD50:27 mg/kg 27ZQAG -,51,72

SAFETY PROFILE: Poison by intravenous and intraperitoneal routes. Moderately toxic by ingestion. When heated to decomposition it emits very toxic fumes of NO_x and SO_x. Used as an antiemetic. See also MERCAPTANS.

TFA000 CAS:62-55-5 ***HR: 3***
THIOACETAMIDE
mf: C_2H_5NS mw: 75.14

PROP: Colorless leaflets; mercaptan odor. Mp: 113°. Very sol in water; sltly sol in alc and ether.

SYNS: ACETOTHIOAMIDE ◇ ETHANETHIOAMIDE ◇ RCRA WASTE NUMBER U218 ◇ TAA ◇ THIACETAMIDE ◇ USAF CB-21 ◇ USAF EK-1719

TOXICITY DATA with REFERENCE
mmo-smc 19900 μmol/L MGGEAE 174,39,79
otr-rat:emb 30 mg/L JJIND8 67,1303,81
ipr-rat TDLo:150 mg/kg (9-11D preg):TER FHCYAI 8,11,70
ipr-rat TDLo:1 g/kg (7D preg):REP 85DJA5 -,95,71
orl-rat TDLo:7350 mg/kg/40W-C:CAR JJIND8 79,1047,87
orl-mus TDLo:10 g/kg/39W-C:NEO BJCAAI 24,498,70
orl-rat TD:6000 mg/kg/43W-C:ETA ONCOBS 38,249,81
orl-rat TD:5140 mg/kg/47W-C:ETA,TER JPBAA7 72,415,56
orl-rat LD50:301 mg/kg TXAPA9 27,380,74
ipr-mus LD50:300 mg/kg NTIS** AD277-689
scu-mus LDLo:2000 mg/kg AIPTAK 12,447,04

CONSENSUS REPORTS: NTP Fifth Annual Report on Carcinogens. IARC Cancer Review: Group 2B IMEMDT 7,56,87; Animal Sufficient Evidence IMEMDT 7,77,74. EPA Genetic Toxicology Program. Community Right-To-Know List. Reported in EPA TSCA Inventory.

SAFETY PROFILE: Confirmed carcinogen with experimental carcinogenic, neoplastigenic, tumorigenic, and teratogenic data. Poison by ingestion and intraperitoneal routes. Moderately toxic by subcutaneous route. Human mutation data reported. An experimental teratogen. Experimental reproductive effects. Exposure has caused liver damage. When heated to decomposition it emits very toxic fumes of NO_x and SO_x. See also SULFIDES and MERCAPTANS.

TFA250 CAS:637-53-6 ***HR: 3***
THIOACETANILIDE
mf: C_8H_9NS mw: 151.24

PROP: Needles from water. Mp: 75-76°, bp: decomp. Insol in water, acid; sol in alkali.

SYNS: N-PHENYLTHIOACETAMIDE ◇ USAF EK-1902

TOXICITY DATA with REFERENCE
orl-rat LD50:100 mg/kg JPETAB 90,260,47
ipr-mus LD50:300 mg/kg NTIS** AD277-689

CONSENSUS REPORTS: Reported in EPA TSCA Inventory.

SAFETY PROFILE: Poison by ingestion and intraperitoneal routes. When heated to decomposition it emits very toxic fumes of NO_x and SO_x.

TFA350 CAS:531-72-6 ***HR: 3***
THIOACETARAMIDE
mf: $C_{11}H_{12}AsNO_5S_2$ mw: 377.28

PROP: White, crystalline powder. Sltly sol in cold water, methanol, and eth. Insol in isopropyl ether.

SYNS: ACETIC ACID, 2,2'-(((4-(AMINOCARBONYL)PHENYL)ARSINIDENE)BIS(THIO))BIS-(9CI) ◇ (((4-(AMINOCARBONYL)PHENYL)ARSINIDINE)BIS(THIO))BIACETICACID ◇ ARSENAMIDE ◇ p-(BIS(CARBOXYMETHYLMERCAPTO)ARSINO)BENZAMIDE ◇ BIS(CARBOXYMETHYLMERCAPTO)(p-CARBAMYLPHENYL)-ARSINE ◇ CAPARISIDE ◇ CAPARSOLATE ◇ 4-CARBAMYLPHENYL BIS(CARBOXYMETHYL-THIO)ARSENITE ◇ DITHIOGLYCOLYL p-ARSENOBENZAMIDE ◇ THIACETARSAMIDE ◇ THIOARSENITE

TOXICITY DATA with REFERENCE
dnr-esc 16 μg/well ENMUDM 3,429,81

OSHA PEL: TWA 0.5 mg(As)/m^3
ACGIH TLV: TWA 0.2 mg(As)/m^3

SAFETY PROFILE: Possibly a poison. Mutation data reported. When heated to decomposition it emits toxic fumes of NO_x, SO_x, and As.

TFA500 CAS:507-09-5 ***HR: 3***
THIOACETIC ACID
DOT: UN 2436
mf: C_2H_4OS mw: 76.12

PROP: Colorless liquid or yellow fuming liquid, pungent, disagreeable odor. Mp: $< -17°$, bp: 93°, d: 1.074 @ 10/4°. flash p: $<73.4°$. Sol in water; misc in alc, ether.

SYNS: ACETYL MERCAPTAN ◇ ETHANETHIOIC ACID ◇ ETHANETHIOLIC ACID ◇ METHANECARBOTHIOLIC ACID ◇ THIACETIC ACID ◇ THIOLACETIC ACID ◇ THIONOACETIC ACID ◇ USAF EK-P-737

TOXICITY DATA with REFERENCE
ipr-mus LD50:75 mg/kg NTIS** AD691-490

CONSENSUS REPORTS: Reported in EPA TSCA Inventory.

DOT Classification: Flammable Liquid; Label: Flammable Liquid.

SAFETY PROFILE: Poison by intraperitoneal route. A very dangerous fire hazard when exposed to heat or flame. When heated to decomposition it emits toxic fumes of SO_x. See also SULFIDES and MERCAPTANS.

TFC250 CAS:100-68-5 ***HR: 3***
THIOANISOLE
mf: C_7H_8S mw: 124.21

PROP: Mp: -15°, bp: 188°, refr index: 1.5865, d: 1.058, flash p: 135°F.

SYN: (METHYLTHIO)BENZENE

TOXICITY DATA with REFERENCE
ivn-mus LD50:56 mg/kg CSLNX* NX#00184

CONSENSUS REPORTS: Reported in EPA TSCA Inventory.

SAFETY PROFILE: Poison by intravenous route. Combustible liquid. When heated to decomposition it emits toxic fumes of SO_x.

TFC500 CAS:1401-63-4 ***HR: 3***
THIOAURIN
mf: $C_{14}H_{12}N_4O_4S_4$ mw: 428.54

PROP: Isolated from *Streptomyces sp* (ANTCAO 3,382,53).

SYNS: ANTIBIOTIC HA-9 ◇ OROSOMYCIN

TOXICITY DATA with REFERENCE
scu-mus LD50:20 mg/kg ANTCAO 3,385,53
ivn-mus LD50:15 mg/kg ANTCAO 3,385,53
ivn-dog LDLo:7800 μg/kg ANTCAO 3,385,53

SAFETY PROFILE: Poison by subcutaneous and intravenous routes. When heated to decomposition it emits very toxic fumes of NO_x and SO_x.

TFC600 CAS:96-69-5 ***HR: 3***
4,4'-THIOBIS(6-tert-BUTYL-m-CRESOL)
mf: $C_{22}H_{30}O_2S$ mw: 358.58

PROP: Light gray to tan powder. Mp: 150°, d: 1.10.

SYNS: BIS(3-tert-BUTYL-4-HYDROXY-6-METHYLPHENYL)SULFIDE ◇ BIS(4-HYDROXY-5-tert-BUTYL-2-METHYLPHENYL) SULFIDE ◇ DISPERSE MB-61 ◇ SANTONOX ◇ SNATOWHITE CRYSTALS ◇ THIOALKOFEN BM 4 ◇ 4,4'-THIOBIS(2-tert-BUTYL-5-METHYLPHENOL) ◇ 4,4'-THIOBIS(6-tert-BUTYL-3-METHYLPHENOL) ◇ 4,4'-THIOBIS(3-METHYL-6-tert-BUTYLPHENOL) ◇ 1,1'-THIOBIS(2-METHYL-4-HYDROXY-5-tert-BUTYLBENZENE) ◇ USAF B-15 ◇ YOSHINOX S

TOXICITY DATA with REFERENCE
ipr-mus LD50:50 mg/kg NTIS** AD277-689

CONSENSUS REPORTS: Reported in EPA TSCA Inventory.

OSHA PEL: (Transitional: TWA Total Dust: 15 mg/m^3; Respirable Fraction: 5 mg/m^3) TWA Total Dust: 10 mg/m^3; Respirable Fraction: 5 mg/m^3
ACGIH TLV: TWA 10 mg/m^3

SAFETY PROFILE: Poison by intraperitoneal route and probably by ingestion and inhalation. Mutation data reported. See also SULFIDES. When heated to decomposition it emits highly toxic fumes of SO_x.

TFD000 CAS:96-66-2 ***HR: 1***
4,4'-THIOBIS(6-tert-BUTYL-o-CRESOL)
mf: $C_{22}H_{30}O_2S$ mw: 358.58

PROP: Light gray to tan powder. Mp: 150°, d: 1.10.

SYNS: ANTIOXIDANT 736 ◇ E 736 ◇ ETHANOX 736 ◇ ETHYL 736 ◇ ETHYL ANTIOXIDANT 736 ◇ TB 2 ◇ THIOALKOFEN BM ◇ 4,4'-THIOBIS(2-(1,1-DIMETHYLETHYL)-6-METHYLPHENOL ◇ TIOALKOFEN BM

TOXICITY DATA with REFERENCE
orl-mam LD50:6340 mg/kg RCTEA4 45,627,72

CONSENSUS REPORTS: Reported in EPA TSCA Inventory.

SAFETY PROFILE: Mildly toxic by ingestion. When heated to decomposition it emits toxic fumes of SO_x.

TFD250 CAS:97-18-7 ***HR: 3***
2,2'-THIOBIS(4,6-DICHLOROPHENOL)
mf: $C_{12}H_6Cl_4O_2S$ mw: 356.04

PROP: White crystalline powder; very faint phenolic odor. Mp: 187-188°, d: 1.61 @ 25°, vap press: 1.1×10^{-9} mm @ 37°.

SYNS: ACTAMER ◇ BIDIPHEN ◇ BIS(2-HYDROXY-3,5-DICHLOROPHENYL) SULFIDE ◇ BITHIONOL ◇ BITHIONOL SULFIDE ◇ BITIN ◇ CP 3438 ◇ 2,2'-DIHYDROXY-3,3',5,5'-TETRACHLORODIPHENYLSULFIDE ◇ 2-HYDROXY-3,5-DICHLOROPHENYL SULPHIDE ◇ LOROTHIDOL ◇ NCI-C60628 ◇ NEOPELLIS ◇ TBP ◇ USAF B-22 ◇ VANCIDE BL ◇ XL 7

TOXICITY DATA with REFERENCE
orl-mus TDLo:12 g/kg/78W-I:ETA NTIS** PB223-159
orl-rat LD50:7 mg/kg CLDND* 6,612,82
orl-mus LD50:760 mg/kg NIIRDN 6,612,82
ipr-mus LD50:100 mg/kg NTIS** AD277-689
ivn-mus LD50:18 mg/kg CSLNX* NX#01763

CONSENSUS REPORTS: EPA Extremely Hazardous

Substance List. Chlorophenol compounds are on the Community Right-To-Know List. Reported in EPA TSCA Inventory.

SAFETY PROFILE: Questionable carcinogen with experimental tumorigenic data. Poison by ingestion, intraperitoneal, and intravenous routes. A food additive permitted in feed and drinking water of animals and for the treatment of food-producing animals. Also a food additive permitted in food for human consumption. When heated to decomposition it emits very toxic fumes of Cl^- and SO_x. See also CHLOROPHENOLS.

TFD500 CAS:123-28-4 ***HR: 1***
THIOBIS(DODECYL PROPIONATE)
mf: $C_{30}H_{58}O_4S$ mw: 514.94

PROP: White crystalline flakes; characteristic sweetish odor. Sol in organic solvents; insol in water.

SYNS: BIS(DODECYLOXYCARBONYLETHYL)SULFIDE ◇ DIDODECYL-3,3'-THIODIPROPIONATE ◇ DILAURYLESTER KYSELINY β',β'-THIODIPROPIONOVE (CZECH) ◇ DILAURYL THIODIPROPIONATE ◇ DILAURYL-β-THIODIPROPIONATE ◇ DILAURYL-β',β'-THIODIPROPIONATE ◇ DILAURYL-3,3'-THIODIPROPIONATE

TOXICITY DATA with REFERENCE
eye-rbt 500 mg/24H MOD 28ZPAK -,174,72

CONSENSUS REPORTS: Reported in EPA TSCA Inventory.

SAFETY PROFILE: An eye irritant. When heated to decomposition it emits toxic fumes of SO_x.

TFD750 CAS:91-71-4 ***HR: 3***
THIOCARBAMIZINE
mf: $C_{21}H_{17}AsN_2O_5S_2$ mw: 516.44

PROP: White crystal powder. Sltly sol in water, alc, and alkali; insol in acids.

SYNS: 2,2'-(((4-((AMINOCARBONYL)AMINO)PHENYL) ARSINIDENEBIS(THIO))BIS) BENZOIC ACID ◇ p-(BIS(o-CARBOXYPHENYLMERCAPTO)-ARSINO)-PHENYLUREA ◇ p-CARBAMIDOPHENYL-BIS(2-CARBOXYPHENYLMERCAPTO)ARSINE ◇ 4-CARBAMIDOPHENYLBIS(o-CARBOXYPHENYLTHIO)ARSENITE ◇ p-CARBAMIDOPHENYL-DI(1'-CARBOXYPHENYL-2') THIOARSENITE ◇ S,S-DIESTER with DITHIO-p-UREIDOBENZENEARSONOUS ACID o-MERCAPTOBENZOIC ACID ◇ DIESTER with o-MERCAPTOBENZOIC ACID DITHIO-p-UREIDOBENZENEARSONOUS ACID ◇ o-MERCAPTOBENZOIC ACID, DIESTER with DITHIO-p-UREIDOBENZENEARSONOUS ACID ◇ THIOCARBAMISIN ◇ (p-UREIDOBENZENEARSYLENEDITHIO)DI-o-BENZOICACID ◇ (p-UREIDOPHENYLARSYLENEDITHIO)DI-o-BENZOIC ACID

TOXICITY DATA with REFERENCE
orl-rat LD50:1220 mg/kg JPETAB 91,112,47
ipr-rat LD50:76 mg/kg JPETAB 91,112,47
ivn-rat LD50:70 mg/kg JPETAB 91,112,47
ipr-mus LD50:265 mg/kg JPETAB 91,112,47
ivn-mus LD50:120 mg/kg JPETAB 91,112,47
ivn-rbt LDLo:100 mg/kg JPETAB 91,112,47

CONSENSUS REPORTS: Arsenic and its compounds are on the Community Right-To-Know List.

OSHA PEL: TWA 0.5 mg(As)/m^3
ACGIH TLV: TWA 0.2 mg(As)/m^3

SAFETY PROFILE: Poison by intraperitoneal and intravenous routes. Moderately toxic by ingestion. When heated to decomposition it emits very toxic fumes of As, SO_x, and NO_x. See also MERCAPTANS and ARSENIC COMPOUNDS.

TFE250 CAS:2231-57-4 ***HR: 3***
THIOCARBOHYDRAZIDE
mf: CH_6N_4S mw: 106.17

PROP: Crystals. Mp: 164°.

SYNS: CARBONOTHIOIC DIHYDRAZIDE ◇ HYDRAZINECARBOHYDRAZONOTHIOIC ACID ◇ TCH ◇ THIOCARBAZIDE ◇ THIOCARBONIC DIHYDRAZIDE ◇ THIOCARBONOHYDRAZIDE ◇ USAF EK-7372

TOXICITY DATA with REFERENCE
orl-rat LDLo:10 mg/kg NCNSA6 5,44,53
ipr-mus LDLo:5 mg/kg NTIS** AD277-689

CONSENSUS REPORTS: EPA Extremely Hazardous Substances List. Reported in EPA TSCA Inventory.

SAFETY PROFILE: Poison by ingestion and intraperitoneal routes. When heated to decomposition it emits very toxic fumes of NO_x and SO_x. See also SULFIDES.

TFE275 ***HR: 3***
THIOCARBONYL AZIDE THIOCYANATE
mf: $C_2N_4S_2$ mw: 144.17

SAFETY PROFILE: An unstable explosive. Explosive reaction with ammonia gas; concentrated hydrazine solutions. When heated to decomposition it emits toxic fumes of SO_x and NO_x. See also CARBONYLS, AZIDES, and THIOCYANATES.

TFE325 CAS:602-41-5 ***HR: 3***
10-THIOCOLCHICOSIDE
mf: $C_{27}H_{33}NO_{10}S$ mw: 563.67

SYN: THIOCOLCHICOSIDE

TOXICITY DATA with REFERENCE
ivn-rat LD50:10 mg/kg AIPTAK 109,386,57
ims-rat LD50:27500 μg/kg AIPTAK 109,386,57
ice-rat LDLo:1 μg/kg AIPTAK 109,386,57
par-rat LDLo:1 μg/kg FRPSAX 15,533,60
ipr-mus LD50:1 mg/kg FRPSAX 15,533,60

SAFETY PROFILE: Poison by intramuscular, intracerebral, parenteral, intravenous, and intraperitoneal routes. When heated to decomposition it emits toxic fumes of SO_x and NO_x.

TFE500 ***HR: D***
THIOCYANATES

SAFETY PROFILE: Variable toxicity. Thiocyanates are not normally dissociated into cyanide; they have a low acute toxicity. Prolonged absorption may produce various skin eruptions, running nose, and occasionally dizziness, cramps, nausea, vomiting, and mild or severe disturbances of the nervous system. Violent reactions have occurred when mixed with chlorates; nitrates; HNO_3; organic peroxides; peroxides; $KClO_3$; $NaClO_3$. Metal thiocyanates are oxidized explosively by chlorates, nitrates at 400° in intimate mixture, HNO_3, or spark or flame ignition. When heated to decomposition or on contact with acid or acid fumes they emit highly toxic fumes of CN^-.

TFF000 CAS:5349-27-9 ***HR: 2***
THIOCYANATOACETIC ACID CYCLOHEXYL ESTER
mf: $C_9H_{13}NO_2S$ mw: 199.29

SYNS: CYKLOHEXYLESTER KYSELINY THIOKYANOOCTOVE (CZECH) ◇ CYKLOHEXYLTHIOKYANOACETAT (CZECH)

TOXICITY DATA with REFERENCE
eye-rbt 500 mg/24H MLD 28ZPAK -,175,72
orl-rat LD50:1180 mg/kg 28ZPAK -,175,72

SAFETY PROFILE: Moderately toxic by ingestion. An eye irritant. When heated to decomposition it emits very toxic fumes of NO_x and SO_x. See also ESTERS and THIOCYANATES.

TFF100 CAS:5349-28-0 ***HR: 3***
THIOCYANATOACETIC ACID ETHYL ESTER
mf: $C_5H_7NO_2S$ mw: 145.19

SYNS: ETHYL THIOCYANATOACETATE ◇ ETHYL THIOCYANOACETATE ◇ REE ◇ THIOCYANO-ESSIGSAEURE-AETHYL-ESTER (GERMAN)

TOXICITY DATA with REFERENCE
ipr-rat LDLo:20 mg/kg ARZNAD 16,870,66
orl-mus LD50:52 mg/kg JJPAAZ 3,99,54
ipr-mus LD50:18300 μg/kg JJPAAZ 3,99,54
scu-mus LD50:39100 μg/kg JJPAAZ 3,99,54
ivn-mus LD50:6 mg/kg JJPAAZ 3,99,54
scu-rbt LD50:10 mg/kg JJPAAZ 3,99,54
ims-pgn LD50:20 mg/kg JJPAAZ 3,99,54

SAFETY PROFILE: Poison by ingestion, intramuscular, subcutaneous, intravenous, and intraperitoneal routes. When heated to decomposition it emits toxic fumes of SO_x, CN^-, and NO_x. See also ESTERS and THIOCYANATES.

TFF250 CAS:593-84-0 ***HR: 3***
THIOCYANIC ACID compounded with GUANIDINE (1:1)
mf: CH_5N_3•CHNS mw: 118.18

SYNS: GUANIDINE THIOCYANATE ◇ GUANIDINIUM THIOCYANATE ◇ USAF EK-705

TOXICITY DATA with REFERENCE
ipr-mus LD50:300 mg/kg NTIS** AD277-689

CONSENSUS REPORTS: Reported in EPA TSCA Inventory.

SAFETY PROFILE: Poison by intraperitoneal route. When heated to decomposition it emits very toxic fumes of NO_x, CN^-, and SO_x. See also THIOCYANATES.

TFF500 CAS:556-65-0 ***HR: 3***
THIOCYANIC ACID, LITHIUM SALT (1:1)
mf: CNS•Li mw: 58.08

PROP: White deliq crystals.

SYNS: LITHIUMRHODANID ◇ LITHIUM SULFOCYANATE ◇ LITHIUM THIOCYANATE

TOXICITY DATA with REFERENCE
orl-mus LDLo:210 mg/kg AEPPAE 169,429,33

CONSENSUS REPORTS: Reported in EPA TSCA Inventory.

SAFETY PROFILE: Poison by ingestion. Decomposes in air at 650°C to yield small amounts of cyanide. The presence of water or 1,3-dioxolane increase the amount of cyanide gas evolved. When heated to decomposition it emits very toxic fumes of CN^-, NO_x, SO_x, and Li_2O. See also LITHIUM COMPOUNDS and THIOCYANATES.

TFH500 CAS:7152-80-9 ***HR: 3***
4-THIOCYANO-N,N-DIMETHYLANILINE
mf: $C_9H_{10}N_2S$ mw: 178.27

SYNS: DEFOLIANT 2929 RP ◇ 4-(DIMETHYLAMINO)PHENYL ESTER THIOCYANIC ACID ◇ p-(DIMETHYLAMINO) PHENYLTHIOCYANATE ◇ DIMETHYLAMINORHODANBENZOL ◇ p-N,N-DIMETHYLAMINOTHIOCYANOBENZENE ◇ DIMETHYLAMINO-4-THIOCYANOBENZENE ◇ p-(DIMETHYLAMINO)THIOCYANOBENZENE ◇ 4-DIMETHYLAMINOTHIOCYANOBENZENE ◇ RP 2929 ◇ THIOCYANIC ACID-p-DIMETHYLAMINOPHENYL ESTER ◇ p-THIOCYANODIMETHYLANILINE

TOXICITY DATA with REFERENCE
orl-rat LD50:500 mg/kg TXAPA9 28,313,74
orl-mus LD50:150 mg/kg 28ZEAL 4,181,69
ipr-mus LD50:55 mg/kg JJPAAZ 3,99,54
scu-mus LD50:150 mg/kg JJPAAZ 3,99,54

ivn-mus LD50:56 mg/kg CSLNX* NX#02758
skn-rbt LD50:1260 mg/kg TXAPA9 28,313,74

SAFETY PROFILE: Poison by ingestion, intraperitoneal, subcutaneous, and intravenous routes. Moderately toxic by skin contact. When heated to decomposition it emits very toxic fumes of NO_x, CN^-, and SO_x. See also ESTERS.

TFH600 CAS:505-14-6 ***HR: 3***
THIOCYANOGEN
mf: $C_2N_2S_2$ mw: 116.16

SAFETY PROFILE: Polymerizes explosively above its melting point −2°C. When heated to decomposition it emits toxic fumes of NO_x, CN^- and SO_x.

TFH750 CAS:31895-22-4 ***HR: 3***
THIOCYCLAM HYDROGEN OXALATE
mf: $C_5H_{11}NS_3 \cdot C_2H_2O_4$ mw: 271.39

SYNS: 5-DIMETHYLAMINO-1,2,3-TRITHIANEHYDROGENOXALATE ◇ N,N-DIMETHYL-1,2,3-TRITHIAN-5-AMINE, ETHANEDIOATE (1:1) ◇ N,N-DIMETHYL-1,2,3-TRITHIAN-5-AMINE HYDROGENOXALATE ◇ N,N-DIMETHYL-1,2,3-TRITHIAN-5-YLAMMONIUM HYDROGEN OXALATE ◇ EVISECT ◇ EVISEKT ◇ SAN 155 ◇ SAN 1551 ◇ SAN 155 I ◇ THIOCYCLAM (ETHANEDIOATE 1:1)

TOXICITY DATA with REFERENCE
orl-rat LD50:195 mg/kg KSKZAN 16(2),59,78
skn-rat LD50:1000 mg/kg FMCHA2 -,C234,83
orl-mus LD50:273 mg/kg SPEADM 78-1,24,78
orl-dog LD50:1000 mg/kg SPEADM 78-1,24,78

SAFETY PROFILE: Poison by ingestion. Moderately toxic by skin contact. When heated to decomposition it emits very toxic fumes of NO_x, NH_3, and SO_x.

TFI000 CAS:139-65-1 ***HR: 3***
4,4'-THIODIANILINE
mf: $C_{12}H_{12}N_2S$ mw: 216.32

PROP: Needles. Mp: 108°.

SYNS: BIS(p-AMINOPHENYL)SULFIDE ◇ BIS(4-AMINOPHENYL) SULFIDE ◇ BIS(p-AMINOPHENYL)SULPHIDE ◇ BIS(4-AMINOPHENYL) SULPHIDE ◇ p,p'-DIAMINODIPHENYL SULFIDE ◇ 4,4'-DIAMINODIPHENYL SULFIDE ◇ p,p'-DIAMINODIPHENYL SULPHIDE ◇ DI(p-AMINOPHENYL) SULFIDE ◇ DI(p-AMINOPHENYL)SULPHIDE ◇ NCI-C01707 ◇ THIOANILINE ◇ 4,4'-THIOANILINE ◇ 4,4'-THIOBIS(ANILINE) ◇ 4,4'-THIOBISBENZENAMINE ◇ p,p-THIODIANILINE ◇ THIODI-p-PHENYLENEDIAMINE

TOXICITY DATA with REFERENCE
mmo-sat 10 μg/plate MUREAV 67,123,79
mma-sat 10 μg/plate MUREAV 67,123,79
orl-rat TDLo:250 mg/kg (1-5D preg):REP IJEBA6 4,120,66
orl-rat TDLo:3600 mg/kg/27D-I:CAR CNREA8 28,924,68
orl-mus TDLo:113 g/kg/54W-C:CAR NCITR* NCI-CG-TR-47,78
orl-rat TD:1500 mg/kg/30D-I:ETA IARC** 27,147,82
orl-rat LD50:1100 mg/kg MarJV# 29MAR77
orl-mus LD50:620 mg/kg 85GMAT -,43,82
ivn-mus LD50:180 mg/kg CSLNX* NX#01564

CONSENSUS REPORTS: IARC Cancer Review: Group 2B IMEMDT 7,56,87; Human Limited Evidence IMEMDT 27,147,82; Animal Sufficient Evidence IMEMDT 27,147,82; Animal Limited Evidence IMEMDT 16,343,78. NCI Carcinogenesis Bioassay (feed); Clear Evidence: mouse, rat NCITR* NCI-CG-TR-47,78. Reported in EPA TSCA Inventory. Community Right-To-Know List.

DFG MAK: Animal Carcinogen, Suspected Human Carcinogen.

SAFETY PROFILE: Confirmed carcinogen with experimental carcinogenic and tumorigenic data. Poison by intravenous route. Moderately toxic by ingestion. May be a human carcinogenic. Experimental reproductive effects. Mutation data reported. When heated to decomposition it emits very toxic fumes of NO_x and SO_x. See also SULFIDES.

TFI500 CAS:111-48-8 ***HR: 2***
β-THIODIGLYCOL
mf: $C_4H_{10}O_2S$ mw: 122.20

PROP: Syrupy, colorless liquid; characteristic odor. Mp: −11.2°, bp: 282°, flash p: 320°F (OC), d: 1.1847 @ 20/20°, vap d: 4.21. Misc in water; sol in chloroform.

SYNS: BIS(β-HYDROXYETHYL)SULFIDE ◇ BIS(2-HYDROXYETHYL)SULFIDE ◇ β,β'-DIHYDROXYDIETHYL SULFIDE ◇ β,β'-DIHYDROXYETHYL SULFIDE ◇ GLYECINE A ◇ β-HYDROXYETHYL SULFIDE ◇ KROMFAX SOLVENT ◇ 2,2'-THIODIETHANOL ◇ THIODIETHYLENE GLYCOL ◇ THIODIGLYCOL

TOXICITY DATA with REFERENCE
skn-rbt 500 mg open MLD UCDS** 11/3/71
eye-rbt 500 mg AJOPAA 29,1363,46
scu-rat LD50:4000 mg/kg JPETAB 93,1,48
scu-mus LD50:4000 mg/kg JPETAB 93,1,48
ivn-rbt LD50:3000 mg/kg JPETAB 93,1,48
orl-gpg LD50:3960 mg/kg JIHTAB 23,259,41

CONSENSUS REPORTS: Reported in EPA TSCA Inventory.

SAFETY PROFILE: Moderately toxic by ingestion, intravenous, and subcutaneous routes. A skin and eye irritant. Combustible when exposed to heat or flame. Reacts violently with acetone + sodium peroxide. Can react with oxidizing materials. To fight fire, use alcohol foam, CO_2, dry chemical. When heated to decomposition it emits highly toxic fumes of SO_x.

TFJ000 CAS:2664-63-3 ***HR: 2***
4,4'-THIODIPHENOL
mf: $C_{12}H_{10}O_2S$ mw: 218.28

PROP: Crystals. Mp:152°.

SYNS: BIS(4-HYDROXYPHENYL) SULFIDE ◇ 4,4'-DIHYDROXYDIPHENYL SULFIDE ◇ 4,4'-DIOXYDIPHENYLSULFIDE

TOXICITY DATA with REFERENCE
skn-rbt 500 mg MLD BIOFX* A408,71
eye-rbt 100 mg SEV BIOFX* A408,71
orl-rat LD50:3362 mg/kg BIOFX* A408,71
orl-mus LD50:5500 mg/kg TPKVAL 13,154,73

CONSENSUS REPORTS: Reported in EPA TSCA Inventory.

SAFETY PROFILE: Moderately toxic by ingestion. Corrosive. A skin and severe eye irritant. When heated to decomposition it emits toxic fumes of SO_x. See also SULFIDES.

TFJ100 CAS:68-11-1 ***HR: 3***
THIOGLYCOLIC ACID
DOT: UN 1940
mf: $C_2H_4O_2S$ mw: 92.12

PROP: Liquid, strong odor. Mp: −16.5°, bp: 108° @ 15 mm. Misc with water, alc, ether, chloroform, and benzene.

SYNS: ACIDE THIOGLYCOLIQUE (FRENCH) ◇ MERCAPTOACETATE ◇ MERCAPTOACETIC ACID ◇ 2-MERCAPTOACETIC ACID ◇ α-MERCAPTOACETIC ACID ◇ 2-THIOGLYCOLIC ACID ◇ THIOVANIC ACID ◇ USAF CB-35

TOXICITY DATA with REFERENCE
skn-hmn 3% INMEAF 15,669,46
orl-rat LD50:114 mg/kg ZHYGAM 20,575,74
ipr-rat LD50:70 mg/kg ZHYGAM 20,575,74
orl-mus LD50:242 mg/kg ZHYGAM 20,575,74
ipr-mus LD50:138 mg/kg ARZNAD 31,1713,81
scu-mus LDLo:1000 mg/kg AIPTAK 12,447,04
ivn-mus LD50:145 mg/kg NYKZAU 60,278,64
orl-rbt LD50:119 mg/kg ZHYGAM 20,575,74
skn-rbt LDLo:300 mg/kg GTPZAB 13,48,69
ivn-rbt LD50:100 mg/kg BIJOAK 41,325,47
orl-gpg LD50:126 mg/kg 34ZIAG -,586,69
ipr-gpg LD50:157 mg/kg ZHYGAM 20,575,74

CONSENSUS REPORTS: Reported in EPA TSCA Inventory.

OSHA PEL: TWA 1 ppm (skin)
ACGIH TLV: TWA 1 ppm (skin)

DOT Classification: Corrosive Material; Label: Corrosive.

SAFETY PROFILE: Poison by ingestion, skin contact, intraperitoneal, and intravenous routes. Moderately toxic by subcutaneous route. A corrosive irritant to skin, eyes, and mucous membranes. When heated to decomposition it emits toxic fumes of SO_x. See also MERCAPTANS and HYDROGEN SULFIDE.

TFJ250 CAS:64039-27-6 ***HR: 3***
β-THIOGUANINE DEOXYRIBOSIDE
mf: $C_{10}H_{13}N_5O_3S{\bullet}H_2O$ mw: 301.36

SYNS: 2-AMINO-9-(2-DEOXY-β-d-RIBOFURANOSYL)-9H-PURINE-6-THIOL HYDRATE ◇ β-DEOXYTHIOGUANOSINE ◇ β-2'-DEOXY-6-THIOGUANOSINE MONOHYDRATE ◇ NSC-71261 ◇ β-TGDR

TOXICITY DATA with REFERENCE
ipr-rat TDLo:1092 mg/kg/1Y-I:CAR NCITR* NCI-CG-TR-57,78
ipr-mus TDLo:175 mg/kg/8W-I:NEO CNREA8 33,3069,73
orl-mus LD50:467 mg/kg NCICP* -,522,69
ipr-mus LD50:354 mg/kg NCICP* -,522,69
ivn-dog LDLo:142 mg/kg CCYPBY 4,41,73

SAFETY PROFILE: Poison by intravenous and intraperitoneal routes. Moderately toxic by ingestion. Questionable carcinogen with experimental carcinogenic and neoplastigenic data. When heated to decomposition it emits very toxic fumes of NO_x and SO_x.

TFJ500 CAS:85-31-4 ***HR: 3***
6-THIOGUANOSINE
mf: $C_{10}H_{13}N_5O_4S$ mw: 299.34

PROP: Decomp: 224-227°.

SYNS: 2-AMINO-6-MERCAPTOPURINE RIBONUCLEOSIDE ◇ 2-AMINO-6-MERCAPTOPURINE RIBOSIDE ◇ 2-AMINO-6-MERCAPTO-9(β-d-RIBOFURANOSYL)PURINE ◇ 2-AMINO-9-(β-d-RIBOFURANOSYL)PURINE-6-THIOL ◇ 2-AMINO-9-β-d-RIBOFURANOSYL-9H-PURINE-6-THIOL ◇ 6-MERCAPTOGUANOSINE ◇ NSC-29422 ◇ RIBOSYLTHIOGUANINE ◇ SK 18615 ◇ SRI 759 ◇ TGR ◇ TGS ◇ 6-THIODEOXYGUANOSINE ◇ 6-THIOGUANINE RIBONUCLEOSIDE ◇ THIOGUANINE RIBOSIDE ◇ THIOGUANOSINE

TOXICITY DATA with REFERENCE
ipr-rat LD50:200 mg/kg ADTEAS 3,181,68
ipr-mus LD50:156 mg/kg NCISP* JAN86

SAFETY PROFILE: Poison by intraperitoneal route. When heated to decomposition it emits very toxic fumes of NO_x and SO_x. See also MERCAPTANS.

TFJ750 CAS:503-87-7 ***HR: 3***
2-THIOHYDANTOIN
mf: $C_3H_4N_2OS$ mw: 116.15

PROP: Mp: 229-231°, (decomp).

SYN: USAF BE-25

TOXICITY DATA with REFERENCE
ipr-mus LD50:100 mg/kg NTIS** AD277-689

CONSENSUS REPORTS: Reported in EPA TSCA Inventory.

SAFETY PROFILE: Poison by intraperitoneal route. When heated to decomposition it emits very toxic fumes of NO_x and SO_x.

TFJ825 CAS:574-25-4 ***HR: 2***
THIOINOSINE
mf: $C_{10}H_{12}N_4O_4S$ mw: 284.32

SYN: 9-β-d-RIBOFURANOSYL-9H-PURINE-6-THIOL

TOXICITY DATA with REFERENCE
mmo-sat 200 μg/plate TAKHAA 44,96,85
orl-rat TDLo:42 mg/kg (female 9-15D post):TER KSRNAM 6,2170,72
orl-rat TDLo:175 mg/kg (9-15D preg):REP KSRNAM 6,2170,72
orl-rat LD50:900 mg/kg IYKEDH 4,467,73
ipr-rat LD50:1240 mg/kg IYKEDH 4,467,73
scu-rat LD50:1310 mg/kg IYKEDH 4,467,73
ivn-rat LD50:1000 mg/kg IYKEDH 4,467,73
ipr-mus LD50:1270 mg/kg IYKEDH 4,467,73
scu-mus LD50:1840 mg/kg IYKEDH 4,467,73
ivn-mus LD50:1250 mg/kg IYKEDH 4,467,73

SAFETY PROFILE: Moderately toxic by ingestion, intravenous, intraperitoneal and subcutaneous routes. An experimental teratogen. Experimental reproductive effects. Mutation data reported. When heated to decomposition it emits toxic fumes of NO_x and SO_x.

TFK000 CAS:2196-13-6 ***HR: 3***
THIOISONICOTINAMIDE
mf: $C_6H_6N_2S$ mw: 138.20

TOXICITY DATA with REFERENCE
ipr-mus LD50:250 mg/kg NTIS** AD691-490

CONSENSUS REPORTS: Reported in EPA TSCA Inventory.

SAFETY PROFILE: Poison by intraperitoneal route. When heated to decomposition it emits very toxic fumes of NO_x and SO_x. See also SULFIDES.

TFK250 CAS:79-42-5 ***HR: 3***
2-THIOLACTIC ACID
DOT: UN 2936
mf: $C_3H_6O_2S$ mw: 106.15

PROP: Oil. Mp: 10° ±, bp: 98-99° @ 14 mm. Misc in water, alc, ether.

SYNS: α-MERCAPTOPROPANOIC ACID ◇ α-MERCAPTOPROPIONIC ACID ◇ 2-MERCAPTOPROPIONIC ACID

TOXICITY DATA with REFERENCE
orl-rat LDLo:50 mg/kg 14CYAT 2,1809,63
ihl-rat LCLo:700 ppm KODAK* -,-,71

CONSENSUS REPORTS: Reported in EPA TSCA Inventory.

DOT Classification: Poison B; Label:Poison

SAFETY PROFILE: Poison by ingestion. Mildly toxic by inhalation. When heated to decomposition it emits toxic fumes of SO_x. See also SULFIDES and MERCAPTANS.

TFK270 CAS:21259-76-7 ***HR: 3***
THIOMERIN SODIUM
mf: $C_{16}H_{27}HgNO_6S{\cdot}2Na$ mw: 608.07

SYNS: N-(Γ-CARBOXYMETHYLMERCAPTOMERCURI-β-METHOXY)PROPYLCAMPHORAMIC ACID DISODIUM SALT ◇ DISODIUM-N-(3-(CARBOXYMETHYLTHIOMERCURI)-2-METHOXYPROPYL)-α-CAMPHORAMATE ◇ DIUCARDYN SODIUM ◇ MERCAPTOMERIN SODIUM ◇ SODIUM MERCAPTOMERIN

TOXICITY DATA with REFERENCE
scu-rat LDLo:14 mg/kg CLDND* 8,316,49
ivn-rat LDLo:250 mg/kg JPETAB 99,149,50
ivn-rbt LDLo:125 mg/kg JPETAB 99,149,50

CONSENSUS REPORTS: Mercury and its compounds are on the Community Right-To-Know List.

OSHA PEL: (Transitional: CL 1 mg/10m^3) CL 0.1 mg(Hg)/m^3 (skin)
ACGIH TLV: TWA 0.1 mg(Hg)/m^3 (skin)
NIOSH REL: (Inorganic Mercury) TWA 0.05 mg (Hg)/m^3

SAFETY PROFILE: Poison by subcutaneous and intravenous routes. When heated to decomposition it emits very toxic fumes of Hg, NO_x, Na_2O, and SO_x. See also MERCURY COMPOUNDS and MERCAPTANS.

TFK300 CAS:2205-73-4 ***HR: 2***
THIOMESTERONE
mf: $C_{24}H_{34}O_4S_2$ mw: 450.70

SYNS: ANDROST-4-EN-3-ONE, 17-β-HYDROXY-1-α-7-α-DIMERCAPTO-17-METHYL-, 1,7-DIACETATE ◇ EMBADOL ◇ EMDABOL ◇ EMDABOLIN ◇ PROTABOL ◇ ST A 307 ◇ THIOMESTRONE ◇ TIOMESTERONE

TOXICITY DATA with REFERENCE
orl-rat TDLo:200 mg/kg (female 10D pre):REP ARZNAD 14,330,64
orl-mus LDLo:2150 mg/kg ARZNAD 14,330,64

SAFETY PROFILE: Moderately toxic by ingestion. Experimental reproductive effects. When heated to decomposition it emits toxic fumes of SO_x.

TFL000 CAS:7719-09-7 ***HR: 3***
THIONYL CHLORIDE
DOT: UN 1836
mf: Cl_2OS mw: 118.96

PROP: Colorless to yellow to red liquid; suffocating odor. Mp: −105°, bp: 78.8° @ 746 mm, d: 1.640 @ 15.5/15.5°, vap press: 100 mm @ 21.4°. Misc with benzene, chloroform, carbon tetrachloride.

SYNS: SULFINYL CHLORIDE ◇ SULFUR CHLORIDE OXIDE ◇ SULFUROUS DICHLORIDE ◇ SULFUROUS OXYCHLORIDE ◇ THIONYL DICHLORIDE

TOXICITY DATA with REFERENCE
ihl-rat LC50:500 ppm/1H NTIS** AD-A148-952

CONSENSUS REPORTS: Reported in EPA TSCA Inventory.

OSHA PEL: CL 1 ppm
ACGIH TLV: CL 1 ppm
DOT Classification: Corrosive Material; Label: Corrosive.

SAFETY PROFILE: Moderately toxic by inhalation. The material itself is more toxic than sulfur dioxide. Has a pungent odor similar to that of sulfur dioxide; it fumes upon exposure to air. Violent reaction with water releases hydrogen chloride and sulfur dioxide. Both these decomposition products constitute serious toxicity hazards. A corrosive irritant which causes burns to the skin and eyes. A powerful chlorinating agent. Potentially explosive reaction with ammonia; bis(dimethylamino)sulfoxide (above 80°C); chloryl perchlorate; 1,2,3-cyclo hexanetrione trioxime + sulfur dioxide; dimethyl sulfoxide; hexafluoroisopropylideneaminolithium. Violent reaction or ignition with 2,4-hexadiyn-1-6-diol; o-nitrobenzoyl acetic acid; o-nitrophenyl-acetic acid; sodium (ignites at 300°C). Incompatible with ammonia; dimethyl formamide + trace iron or zinc; linseed oil + quinoline; toluene + ethanol + water. When heated to decomposition it emits toxic fumes of SO_x and Cl^-. See also HYDROGEN CHLORIDE and SULFUR DIOXIDE.

TFL250 CAS:7783-42-8 ***HR: 2***
THIONYL FLUORIDE
mf: F_2OS mw: 86.06

PROP: Colorless gas; suffocating odor. Bp: −44°, mp: −130°. d (liq): 1.780 @ −100°, (solid): 2.095 @ −183°. Sol in ether, benzene.

SYNS: FLUORURE de THIONYLE (FRENCH) ◇ SULFUR DIFLUORIDE MONOXIDE ◇ SULFUR DIFLUORIDE OXIDE ◇ SULFUROUS OXYFLUORIDE ◇ THIONYL DIFLUORIDE

TOXICITY DATA with REFERENCE
ihl-mus LCLo:260 ppm/1H AMPMAR 34(10-11), 581,73
ihl-rbt LCLo:1000 ppm/1H AMPMAR 34(10-11),581,73

CONSENSUS REPORTS: Reported in EPA TSCA Inventory.

OSHA PEL: TWA 2.5 mg(F)/m^3
ACGIH TLV: TWA 2.5 mg(F)/m^3
NIOSH REL: (Inorganic Fluorides) TWA 2.5 mg(F)/m^3

SAFETY PROFILE: Moderately toxic by inhalation. A severe irritant to skin, eyes, and mucous membranes. When heated to decomposition or on contact with water or steam it emits highly toxic and corrosive fumes of SO_x and F^-. See also FLUORIDES.

TFL500 CAS:2487-40-3 ***HR: 2***
2-THIO-6-OXYPURINE
mf: $C_5H_4N_4OS$ mw: 168.19

SYN: 1,2,3,7-TETRAHYDRO-2-THIOXO-6H-PURIN-6-ONE

TOXICITY DATA with REFERENCE
ipr-mus LD50:625 mg/kg NTIS** AD691-490

CONSENSUS REPORTS: Reported in EPA TSCA Inventory.

SAFETY PROFILE: Moderately toxic by intraperitoneal route. When heated to decomposition it emits very toxic fumes of NO_x and SO_x.

TFM100 CAS:316-81-4 ***HR: 2***
THIOPERAZINE
mf: $C_{22}H_{30}N_4O_2S_2$ mw: 446.68

PROP: Crystals. Mp: 140°.

SYNS: MAJEPTYL ◇ MAZEPTYL ◇ MEGEPTIL ◇ 7843 R.P. ◇ RP 7843 ◇ SKF 5883 ◇ SULFENAZIN ◇ THIOPROPERAZIN ◇ THIOPROPERAZINE ◇ (THIOPROPERAZINE)-2-DIMETHYLSULFAMIDO-(10-(3-1-METHYLPIPERAZINYL-4)PROPYL)-PHENOTHIAZINE ◇ VONTIL

TOXICITY DATA with REFERENCE
cyt-hmn:lym 5 mg/L CYGEDX 9(5),14,75
dlt-mus-ipr 5 mg/kg CYGEDX 9(5),14,75
par-rat TDLo:200 mg/kg (1-8D preg):REP 15QWAW, 290,65
orl-mus LD50:830 mg/kg DNEUD5 7,45,80

SAFETY PROFILE: Moderately toxic by ingestion. Experimental reproductive effects. Human mutation data reported. When heated to decomposition it emits toxic fumes of SO_x and NO_x.

TFM250 CAS:110-02-1 ***HR: 3***
THIOPHENE
mf: C_4H_4S mw: 84.14

CH=CHSCH=CH (ring)

DOT: UN 2414

PROP: Clear, colorless liquid; slt aromatic odor similar to benzene. D: 1.0573 @ 25/4°, mp: −38.3°, bp: 84.4°, flash p: 21.2°F, vap press: 40 mm @ 12.5°, vap d: 2.9. Insol in water; misc with most organic solvents. May be heated to 850° without decomposition.

SYNS: CP 34 ◇ DIVINYLENE SULFIDE ◇ HUILE H50 ◇ THIACYCLOPENTADIENE ◇ THIAPHENE ◇ THIOFURAM ◇ THIOFURAN ◇ THIOFURFURAN ◇ THIOLE ◇ THIOPHEN ◇ THIOTETROLE ◇ USAF EK-1860

TOXICITY DATA with REFERENCE
ihl-mus LCLo:2900 ppm 34ZIAG -,587,69
ipr-mus LD50:100 mg/kg NTIS** AD277-689

CONSENSUS REPORTS: Reported in EPA TSCA Inventory.

DOT Classification: Flammable Liquid; Label: Flammable Liquid.

SAFETY PROFILE: Poison by intraperitoneal route. Mildly toxic by inhalation. A very dangerous fire hazard when exposed to heat or flame. Explosive reaction with N-nitrosoacetanilide. Violent or explosive reaction with nitric acid. Incompatible with oxidizing materials. To fight fire, use foam, CO_2, dry chemical. When heated to decomposition it emits highly toxic fumes of SO_x.

TFM500 CAS:98-03-3 ***HR: 3***
2-THIOPHENECARBOXALDEHYDE
mf: C_5H_4OS mw: 112.15

PROP: Liquid. Bp: 198°, d: 1.2, refr index: 1.5900, flash p. 172°.

SYNS: α-FORMYLTHIOPHENE ◇ 2-FORMYLTHIOPHENE ◇ 2-THIENYLALDEHYDE ◇ 2-THIENYLCARBOXALDEHYDE ◇ 2-THIOPHENEALDEHYDE ◇ α-THIOPHENECARBOXALDEHYDE

TOXICITY DATA with REFERENCE
ipr-mus LDLo:16 mg/kg CBCCT* 2,191,50

CONSENSUS REPORTS: Reported in EPA TSCA Inventory.

SAFETY PROFILE: Poison by intraperitoneal route. Combustible liquid. When heated to decomposition it emits toxic fumes of SO_x. See also ALDEHYDES.

TFN500 CAS:463-71-8 ***HR: 3***
THIOPHOSGENE
DOT: UN 2474
mf: CCl_2S mw: 114.97

PROP: Reddish liquid. Bp: 73.5°, d: 1.5085 @ 15°. Decomp in water, alc; sol in ether.

SYNS: CARBON CHLOROSULFIDE ◇ CARBONOTHIOIC DICHLORIDE ◇ DICHLOROTHIOCARBONYL ◇ THIOCARBONIC DICHLORIDE ◇ THIOCARBONYL CHLORIDE (DOT) ◇ THIOCARBONYL DICHLORIDE ◇ THIOFOSGEN (CZECH) ◇ THIOKARBONYLCHLORID (CZECH)

TOXICITY DATA with REFERENCE
skn-rbt 500 mg/24H MOD 28ZPAK -,13,72
eye-rbt 50 μg/24H SEV 28ZPAK -,13,72
orl-rat LD50:929 mg/kg 28ZPAK -,13,72
ivn-mus LD50:100 mg/kg CSLNX* NX#04557

CONSENSUS REPORTS: Reported in EPA TSCA Inventory.

DOT Classification: Poison B; Label: Poison.

SAFETY PROFILE: Poison by intravenous route. Moderately toxic by ingestion. A skin, mucous membrane, and severe eye irritant. When heated to decomposition it emits very toxic fumes of Cl^- and SO_x. See also PHOSGENE.

TFN750 CAS:3931-89-3 ***HR: 2***
THIOPHOSPHORYL BROMIDE
mf: Br_3PS mw: 302.76

PROP: Cubic yellow crystals. Mp: 38°, bp: decomp @ 175°, d: 2.85 @ 17°.

SYNS: PHOSPHORUS SULFOBROMIDE ◇ THIOPHOSPHOROUS TRIBROMIDE

TOXICITY DATA with REFERENCE
ihl-mus LC50:2600 mg/m³ NDRC** -,17,42

SAFETY PROFILE: Moderately toxic by inhalation. Will react with water or steam to produce toxic and corrosive fumes. When heated to decomposition it emits highly toxic fumes of Br^-, SO_x, and PO_x. See also BROMIDES and THIOPHOSPHORYL CHLORIDE.

TFO000 CAS:3982-91-0 ***HR: 2***
THIOPHOSPHORYL CHLORIDE
mf: Cl_3PS mw: 169.38
DOT: UN 1837

PROP: Colorless, mobile liquid; pungent odor. Bp: 125°, fp: −35°, flash p: none, d: 1.63 @ 25/4°, vap press: 22 mm @ 25°, vap d: 5.86.

SYNS: PHOSPHOROTHIOIC TRICHLORIDE ◇ PHOSPHOROTHIONIC TRICHLORIDE ◇ PHOSPHOROUS SULFOCHLORIDE ◇ PHOSPHOROUS THIOCHLORIDE ◇ PHOSPHOROUS TRICHLORIDE SULFIDE ◇ THIOPHOSPHORYL TRICHLORIDE ◇ TRICHLOROPHOSPHINE SULFIDE ◇ TL 262

TOXICITY DATA with REFERENCE
orl-rat LD50:750 mg/kg JJATDK 4,230,84
ihl-mus LCLo:3000 mg/m³/10M NDRC** NDCrc-132,June,42

CONSENSUS REPORTS: Reported in EPA TSCA Inventory.

DOT Classification: Corrosive Material; Label: Corrosive.

SAFETY PROFILE: Moderately toxic by ingestion and inhalation. A corrosive irritant to skin, eyes, and mucous membranes. Explosive reaction with methylmagnesium iodide. Explosive reaction with pentaerythritol + heat. Reacts with water or steam to produce toxic and corrosive fumes. When heated to decomposition it emits highly toxic fumes of PO_x, SO_x, and Cl^-.

TFO250 CAS:13706-10-0 ***HR: 2***
THIOPHOSPHORYL DIBROMIDEMONOFLUORIDE
mf: Br_2FPS mw: 241.85

SYN: TL 331

TOXICITY DATA with REFERENCE
ihl-mus LCLo:3100 mg/m^3/10M NDRC** NDCrc-132,Sept,42

OSHA PEL: TWA 2.5 mg(F)/m^3
ACGIH TLV: TWA 2.5 mg(F)/m^3
NIOSH REL: TWA 2.5 mg(F)/m^3

SAFETY PROFILE: Moderately toxic by inhalation. When heated to decomposition it emits very toxic fumes of Br^-, F^-, PO_x, and SO_x. See also BROMIDES, FLUORIDES, and THIOPHOSPHORYL CHLORIDE.

TFO500 CAS:13706-09-7 ***HR: 2***
THIOPHOSPHORYL DIFLUORIDEMONOBROMIDE
mf: BrF_2PS mw: 180.94

SYNS: MONOBROMODIFLUOROPHOSPHINESULFIDE ◇ THIOPHOSPHORYL MONOBROMODIFLUORIDE ◇ TL 263

TOXICITY DATA with REFERENCE
ihl-mus LCLo:3000 mg/m^3/10M NDRC** NDCrc-132,June,42

OSHA PEL: TWA 2.5 mg(F)/m^3
ACGIH TLV: TWA 2.5 mg(F)/m^3
NIOSH REL: TWA 2.5 mg(F)/m^3

SAFETY PROFILE: Moderately toxic by inhalation. When heated to decomposition it emits very toxic fumes of Br^-, F^-, PO_x, and SO_x. See also BROMIDES, FLUORIDES, and THIOPHOSPHORYL CHLORIDE.

TFO750 CAS:2404-52-6 ***HR: 3***
THIOPHOSPHORYL FLUORIDE
mf: PSF_3 mw: 120.05

PROP: Mp: 3.8° @ 7.6 atm, bp: decomp.

SAFETY PROFILE: Poison irritant to skin, eyes, and mucous membranes. Ignites or explodes on contact with air. Heated sodium ignites in the gas. When heated to decomposition or in reaction with water or steam it emits toxic and corrosive fumes of PO_x, SO_x, and F^-. See also FLUORIDES and THIOPHOSPHORYL CHLORIDE.

TFP000 CAS:75-18-3 ***HR: 3***
2-THIOPROPANE
mf: C_2H_6S mw: 62.14
DOT: UN 1164

PROP: Colorless liquid; disagreeable odor. Mp: −83.2°, lel: 2.2%, uel: 19.7%, flash p: <0°F, bp: 37.5-38°, d: 0.8458 @ 21/4°, vap d: 2.14, autoign temp: 403°F. Insol in water; sol in alc, ether.

SYNS: DIMETHYLSULFID (CZECH) ◇ DIMETHYL SULFIDE (DOT) ◇ DIMETHYL SULPHIDE ◇ DMS ◇ EXACT-S ◇ METHYL SULFIDE (DOT) ◇ METHYL SULPHIDE ◇ METHYLTHIOMETHANE ◇ SULFURE de METHYLE (FRENCH) ◇ 2-THIAPROPANE

TOXICITY DATA with REFERENCE
skn-rbt 500 mg/24H MLD FCTXAV 17,365,79
eye-rbt 250 μg/24H SEV 28ZPAK -,169,72
orl-rat LD50:535 mg/kg 28ZPAK -,169,72
ihl-rat LC50:40250 ppm LacHB# 09JUN78
orl-mus LD50:3700 mg/kg FCTXAV 17,365,79
ihl-mus LC50:31620 μg/m^3 FCTXAV 17,365,79
ipr-mus LD50:8000 mg/kg IJRBA3 3,41,61

CONSENSUS REPORTS: EPA Extremely Hazardous Substances List. Reported in EPA TSCA Inventory.

DOT Classification: Flammable Liquid; Label: Flammable Liquid.

SAFETY PROFILE: Poison by inhalation. Moderately toxic by ingestion. A skin and severe eye irritant. A very dangerous fire hazard when exposed to heat or flame. Explosive in the form of vapor when exposed to heat or flame. Can react vigorously with oxidizing materials. To fight fire, use CO_2, dry chemical. When heated to decomposition it emits highly toxic fumes of SO_x and may explode. See also SULFIDES.

TFP250 CAS:146-28-1 ***HR: 3***
THIOPROPAZATE DIHYDROCHLORIDE
mf: $C_{23}H_{28}ClN_3O_2S{\cdot}2ClH$ mw: 518.97

PROP: Crystals. Decomp @ 223-229°. Very sol in water; sol in alc, chloroform; nearly insol in ether.

SYNS: 1-(2-ACETOXYETHYL)-4-(3-(2-CHLORO-10-PHENOTHIAZINYL)PROPYL)PIPERAZIME DIHYDROCHLORIDE ◇ 2-CHLORO-10-(3-(4-(2-ACETOXYETHYL)PIPERAZINYL)PROPYL)PHENOTHIAZINE ◇ THIOPROPAZATE HYDROCHLORIDE

TOXICITY DATA with REFERENCE
orl-mus LD50:279 mg/kg 27ZQAG -,52,72
ipr-mus LD50:197 mg/kg 27ZQAG -,52,72
scu-mus LD50:1080 mg/kg 27ZQAG -,52,72

SAFETY PROFILE: Poison by ingestion and intraperitoneal routes. Moderately toxic by subcutaneous route. When heated to decomposition it emits very toxic fumes of SO_x, NO_x, and Cl^-.

TFQ000 CAS:79-19-6 ***HR: 3***
THIOSEMICARBAZIDE
mf: CH_5N_3S mw: 91.15

PROP: Needles from water. Mp: 182-184°. Sol in water, alc.

SYNS: N-AMINOTHIOUREA ◇ HYDRAZINECARBOTHIOAMIDE ◇ RCRA WASTE NUMBER P116 ◇ THIOCARBAMYLHYDRAZINE ◇ 3-THIOSEMICARBAZIDE ◇ TSC ◇ USAF EK-1275

TOXICITY DATA with REFERENCE
dnd-hmn:hla 20 μmol/L BCPCA6 25,821,76
orl-rat TDLo:1024 mg/kg/78W-C:ETA JJIND8 67,75,81
orl-rat LD50:9160 μg/kg MarJV# 29MAR77
orl-mus LDLo:94 mg/kg AECTCV 14,111,85
ipr-mus LD50:1 mg/kg NTIS** AD277-689
scu-mus LD50:16407 μg/kg ABMGAJ 21,635,68
ivn-mus LD50:13 mg/kg JPETAB 122,110,58
orl-dog LD50:10 mg/kg HBTXAC 5,155,59
orl-cat LD50:20 mg/kg PSEBAA 70,688,49
ipr-gpg LD50:24 mg/kg PSEBAA 70,688,49
orl-bwd LD50:9100 μg/kg TXAPA9 21,315,72

CONSENSUS REPORTS: EPA Extremely Hazardous Substances List. Reported in EPA TSCA Inventory.

SAFETY PROFILE: Poison by ingestion, intraperitoneal, and intravenous routes. Questionable carcinogen with experimental tumorigenic data. Human mutation data reported. When heated to decomposition it emits very toxic fumes of NO_x and SO_x.

TFQ250 CAS:1752-30-3 ***HR: 3***
THIOSEMICARBAZONE ACETONE
mf: $C_4H_9N_3S$ mw: 131.22

TOXICITY DATA with REFERENCE
orl-rat LDLo:10 mg/kg NCNSA6 5,43,53
ipr-mus LD50:23 mg/kg JPETAB 122,110,58

CONSENSUS REPORTS: EPA Extremely Hazardous Substances List. Reported in EPA TSCA Inventory.

SAFETY PROFILE: Poison by ingestion and intraperitoneal routes. When heated to decomposition it emits very toxic fumes of NO_x and SO_x.

TFQ275 CAS:1393-48-2 ***HR: 3***
THIOSTREPTON
mf: $C_{72}H_{85}N_{19}O_{18}S_5$ mw: 1665.06

PROP: Crystals from chloroform + methanol. Decomp 246-256°. Sol in chloroform, dioxane, pyridine, glacial acetic acid, DMF; practically insol in water, the lower alcohols, nonpolar organic solvents, dil aq acid or base. Dissolved by methanolic acid or base with decomp.

SYNS: A-8506 ◇ ANTIBIOTIC 6761-31 ◇ ANTIBIOTIC A 8506 ◇ ANTIBIOTIC X 146 ◇ BRYAMYCIN ◇ GARGON ◇ THIACTIN ◇ X 146

TOXICITY DATA with REFERENCE
ipr-mus LD50:2000 mg/kg 85GDA2 4(1),396,80
ivn-mus LD50:41 mg/kg 85GDA2 4(1),396,80
ims-mus LD50:1000 mg/kg 85GDA2 4(1),397,80

SAFETY PROFILE: Poison by intravenous route. Moderately toxic by intraperitoneal and intramuscular routes. When heated to decomposition it emits toxic fumes of SO_x and NO_x.

TFQ500 ***HR: 1***
THIOSULFATES

SAFETY PROFILE: Up to 12 grams of sodium thiosulfate can be taken daily by mouth with no ill effects except catharsis. Most of the thiosulfates are low in acute toxicity. When heated to decomposition they emit highly toxic fumes of SO_x.

TFQ600 CAS:58513-59-0 ***HR: 3***
THIOTHIXENE DIHYDROCHLORIDE
mf: $C_{23}H_{29}N_3O_2S_2 \cdot 2ClH$ mw: 516.59

TOXICITY DATA with REFERENCE
orl-rat LD50:410 mg/kg KSRNAM 3,709,69
ipr-rat LD50:145 mg/kg KSRNAM 3,709,69
scu-rat LD50:1537 mg/kg KSRNAM 3,709,69
orl-mus LD50:160 mg/kg KSRNAM 3,709,69
ipr-mus LD50:112 mg/kg KSRNAM 3,709,69
scu-mus LD50:550 mg/kg KSRNAM 3,709,69

SAFETY PROFILE: Poison by ingestion and intraperitoneal routes. Moderately toxic by subcutaneous route. When heated to decomposition it emits toxic fumes of SO_x, NO_x, and HCl.

TFQ750 CAS:52-24-4 ***HR: 3***
THIOTRIETHYLENEPHOSPHORAMIDE
mf: $C_6H_{12}N_3PS$ mw: 189.24

SYNS: CBC 806495 ◇ GIROSTAN ◇ NCI-C01649 ◇ NSC-6396 ◇ ONCOTEPA ◇ ONCOTIOTEPA ◇ 1,1',1''-PHOSPHINOTHIOYLIDYNETRISAZIRIDINE ◇ PHOSPHOROTHIOIC ACID TRIETHYLENETRIAMIDE ◇ SK 6882 ◇ TESPAMINE ◇ THIOFOZIL ◇ THIOPHOSPHAMIDE ◇ THIO-TEP ◇ TIOFOSFAMID ◇ TIOFOZIL ◇ TRIAZIRIDINYLPHOSPHINE SULFIDE ◇ N,N',N''-TRI-1,2-ETHANEDIYLPHOSPHOROTHIOIC TRIAMIDE ◇ N,N',N''-TRI-1,2-ETHANEDIYLTHIOPHOSPHORAMIDE ◇ TRI(ETHYLENEIMINO) THIOPHOSPHORAMIDE ◇ N,N',N''-TRIETHYLENEPHOSPHOROTHIOIC TRIAMIDE ◇ N,N',N''-TRIETHYLENETHIOPHOSPHAMIDE ◇ N,N',N''-TRIETHYLENETHIOPHOSPHORAMIDE ◇ TRIETHYLENETHIOPHOSPHOROTRIAMIDE ◇ TRIS(1-AZIRIDINYL)PHOSPHINE SULFIDE ◇ TRIS(ETHYLENIMINO)THIOPHOSPHATE ◇ TSPA

TOXICITY DATA with REFERENCE
mmo-sat 100 μg/plate TAKHAA 44,96,85
cyt-rbt-ivn 3 mg/kg BEXBAN 94,1118,82
ipr-rat TDLo:500 μg/kg (female 13D post):TER DANND6 (3),59,83

itt-ham TDLo:16 mg/kg (male 1D pre):REP CCPTAY 22,175,80
par-man TDLo:33 mg/kg/3Y-I:CAR,BLD CMAJAX 129,578,83
unr-wmn TDLo:17 mg/kg/56W-I:CAR,BLD LANCAO 2,775,70
ipr-rat TDLo:218 mg/kg/1Y-I:CAR NCITR* NCI-CG-TR-58,78
ivn-rat TDLo:52 mg/kg/1Y-I:CAR ARZNAD 20,1461,70
skn-mus TDLo:10 g/kg/17W-I:ETA GANNA2 57,295,66
ipr-mus TD:47 mg/kg/8W-I:NEO CNREA8 33,3069,73
par-man TDLo:631 μg/kg:PNS,BLD CANCAR 38,1471,76
ipr-rat LD50:8 mg/kg CPCHAQ 18,307,62
ivn-rat LD50:9400 μg/kg ARZNAD 8,1,58
orl-mus LD50:38 mg/kg 21ACAB -,129,68
ipr-mus LD50:18 mg/kg LIFSAK 36,1473,85
scu-mus LD50:19500 μg/kg NIIRDN 6,455,82
ivn-mus LD50:14500 μg/kg NIIRDN 6,455,82
ivn-dog LDLo:760 μg/kg CCSUBJ 2,203,65
ivn-mky LDLo:2 mg/kg CCSUBJ 2,203,65

CONSENSUS REPORTS: NTP Fifth Annual Report on Carcinogens. IARC Cancer Review: Group 2A IMEMDT 7,368,87; Human Limited Evidence IMEMDT 9,85,75; Animal Sufficient Evidence IMEMDT 9,85,75. NCI Carcinogenesis Bioassay (ipr); Clear Evidence: mouse, rat NCITR* NCI-CG-TR-58,78. EPA Genetic Toxicology Program.

SAFETY PROFILE: Confirmed human carcinogen producing leukemia. Poison by ingestion, intraperitoneal, intravenous, and subcutaneous routes. Experimental teratogenic data. Human systemic effects by parenteral route: paresthesia, bone marrow changes and leukemia. Experimental reproductive effects. Human mutation data reported. When heated to decomposition it emits very toxic fumes of PO_x, SO_x, and NO_x.

TFR000 CAS:79796-40-0 ***HR: 3***
THIOTRITHIAZYL NITRATE
mf: $N_4O_3S_4$ mw: 232.28

SNSNSNS•NO_3

SAFETY PROFILE: Explodes on friction or impact. When heated to decomposition it emits very toxic fumes of SO_x and NO_x. See also NITRATES.

TFR250 CAS:141-90-2 ***HR: 3***
2-THIOURACIL
mf: $C_4H_4N_2OS$ mw: 128.16

PROP: Small crystals; bitter taste. Mp: 300°. Practically insol in water, alc, ether, and acids; sol in alkalies.

SYNS: ANTAGOTHYROID ◇ ANTAGOTHYROIL ◇ DERACIL ◇ 2,3-DIHYDRO-2-THIOXO-4(1H)-PYRIMIDINONE ◇ 6-HYDROXY-2-MERCAPTOPYRIMIDINE ◇ 4-HYDROXY-2(1H)-PYRIMIDINETHIONE ◇ 2-MERCAPTO-4-HYDROXYPYRIMIDINE ◇ 2-MERCAPTO-4-PYRIMIDINOL ◇ 2-MERCAPTO-4-PYRIMIDONE ◇ 2-MERCAPTO-PYRIMID-4-ONE ◇ NOBILEN ◇ 2-THIO-6-OXYPYRIMIDINE ◇ 2-THIO-1,3-PYRIMIDIN-4-ONE ◇ THIOURACIL ◇ 6-THIOURACIL ◇ TIOURACYL (POLISH) ◇ TU ◇ 2-TU

TOXICITY DATA with REFERENCE
pic-esc 100 mmol/L MDMIAZ 31,11,79
unr-wmn TDLo:924 mg/kg (26-41W preg):REP JCENA4 7,767,47
unr-wmn TDLo:924 mg/kg (26-41W preg):TER JCENA4 7,767,47
orl-rat TDLo:10 g/kg/14W-C:ETA CANCAR 6,111,53
orl-mus TDLo:184 g/kg/73W-C:NEO PSEBAA 113,493,63
orl-mus LD50:3900 mg/kg JAPMA8 44,56,55
orl-rbt LDLo:3700 mg/kg MEIEDD 10,1342,83

CONSENSUS REPORTS: IARC Cancer Review: Group 3 IMEMDT 7,56,87; Animal Sufficient Evidence IMEMDT 7,85,74. Reported in EPA TSCA Inventory. EPA Genetic Toxicology Program.

SAFETY PROFILE: Moderately toxic by ingestion. Human teratogenic effects by unspecified routes: developmental abnormalities of the central nervous system, craniofacial area, and endocrine system. Human reproductive effects by unspecified route: effects on newborn including viability index changes. Experimental teratogenic effects. Other experimental reproductive effects. Questionable carcinogen with experimental neoplastigenic and tumorigenic data. Mutation data reported. When heated to decomposition it emits very toxic fumes of NO_x and SO_x. Used in the treatment of hyperthyroidism, angina pectoris, and congestive heart failure. See also MERCAPTANS and KETONES.

TFS250 CAS:96-53-7 ***HR: 2***
2-THIOZOLIDINETHIONE

SYNS: 2-MERCAPTOTHIAZOLINE ◇ 2-THIOTHIAZOLIDONE

TOXICITY DATA with REFERENCE
ipr-mus LDLo:600 mg/kg CURL** 21,24,56

CONSENSUS REPORTS: Reported in EPA TSCA Inventory.

SAFETY PROFILE: Moderately toxic by intraperitoneal route. When heated to decomposition it emits very toxic fumes of NO_x and SO_x. See also MERCAPTANS.

TFS350 CAS:137-26-8 ***HR: 3***
THIRAM
DOT: UN 2771
mf: $C_6H_{12}N_2S_4$ mw: 240.44

PROP: Crystals. Mp: 156°, d: 1.30, bp: 129° @ 20 mm. Insol in water; sol in alc, ether, acetone, and chloroform.

SYNS: AATACK ◇ ACCELERATOR THIURAM ◇ ACETO TETD ◇ ARASAN ◇ AULES ◇ BIS((DIMETHYLAMINO)CARBONOTHIOYL) DISULPHIDE ◇ BIS(DIMETHYL-THIOCARBAMOYL)-DISULFID (GERMAN) ◇ BIS(DIMETHYLTHIOCARBAMOYL) DISULFIDE ◇ CHIPCO THIRAM 75 ◇ CYURAM DS ◇ DISOLFURO DI TETRAMETILTIOURAME (ITALIAN) ◇ DISULFURE de TETRAMETHYLTHIOURAME (FRENCH) ◇ α,α'-DITHIOBIS(DIMETHYLTHIO)FORMAMIDE ◇ 1,1'-DITHIOBIS (N,N-DIMETHYLTHIO)FORMAMIDE ◇ N,N'-(DITHIODICARBONOTHIOYL)BIS(N-METHYLMETHANAMINE) ◇ EKAGOM TB ◇ FALITIRAM ◇ FERMIDE ◇ FERNACOL ◇ FERNASAN ◇ FERNIDE ◇ FLO PRO T SEED PROTECTANT ◇ HERMAL ◇ HERMAT TMT ◇ HERYL ◇ HEXATHIR ◇ KREGASAN ◇ MERCURAM ◇ METHYL THIRAM ◇ METHYL THIURAMDISULFIDE ◇ METHYL TUADS ◇ NOBECUTAN ◇ NOMERSAN ◇ NORMERSAN ◇ PANORAM 75 ◇ POLYRAM ULTRA ◇ POMARSOL ◇ POMASOL ◇ PURALIN ◇ RCRA WASTE NUMBER U244 ◇ REZIFILM ◇ ROYAL TMTD ◇ SADOPLON ◇ SPOTRETE ◇ SQ 1489 ◇ TERSAN ◇ TERAMETHYL THIURAM DISULFIDE ◇ TETRAMETHYLDIURANE SULPHITE ◇ TETRAMETHYLENETHIURAM DISULPHIDE ◇ TETRAMETHYLTHIOCARBAMOYLDISULPHIDE ◇ TETRAMETHYLTHIORAMDI- SULFIDE (DUTCH) ◇ TETRAMETHYL-THIRAM DISULFID (GERMAN) ◇ TETRAMETHYLTHIURAM BISULFIDE ◇ TETRAMETHYLTHIURAM DISULFIDE ◇ N,N-TETRAMETHYLTHIURAM DISULPHIDE ◇ N,N,N',N'-TETRAMETHYLTHIURAM DISULFIDE ◇ TETRAMETHYL THIURANE DISULFIDE ◇ TETRAPOM ◇ TETRASIPTON ◇ TETRATHIURAM DISULFIDE ◇ TETRATHIURAM DISULPHIDE ◇ THILLATE ◇ THIMER ◇ THIOSAN ◇ THIOTEX ◇ THIOTOX ◇ THIRAMAD ◇ THIRAME (FRENCH) ◇ THIRASAN ◇ THIULIX ◇ THIURAD ◇ THIURAM ◇ THIURAMIN ◇ THIURAMYL ◇ THYLATE ◇ TIRAMPA ◇ TIURAM (POLISH) ◇ TIURAMYL ◇ TMTD ◇ TMTDS ◇ TRAMETAN ◇ TRIDIPAM ◇ TRIPOMOL ◇ TTD ◇ TUADS ◇ TUEX ◇ TULISAN ◇ USAF B-30 ◇ USAF EK-2089 ◇ USAF P-5 ◇ VANCIDA TM-95 ◇ VANCIDE TM ◇ VUAGT-I-4 ◇ VULCAFOR TMTD ◇ VULKACIT MTIC ◇ VULKACIT THIURAM ◇ VULKACIT THIURAM/C

TOXICITY DATA with REFERENCE
pic-esc 1 ng/tube MUREAV 89,1,81
msc-ham:lng 5 mg/L FCTOD7 23,373,85
orl-rat TDLo:550 mg/kg (female 1-22D post):REP AEEDDS 2,215,76
orl-rat TDLo:300 mg/kg (female 15D post):TER GISAAA 43(6),37,78
orl-rat TDLo:108 mg/kg/1Y-C:ETA RPZHAW 31,67,80
ihl-hmn TCLo:30 $\mu g/m^3$/5Y-I:PUL VRDEA5 (10),136,71
orl-rat LD50:560 mg/kg TXAPA9 11,546,67
ihl-rat LC50:500 mg/m^3/4H 85JCAE-,1027,86
ipr-rat LD50:138 mg/kg JNPHAG 9,35,78
unr-rat LD50:740 mg/kg KOKABN 26,358,77
orl-mus LD50:1350 mg/kg HYSAAV 29(7),37,64
ipr-mus LD50:70 mg/kg DRUGAY 6,566,82
scu-mus LD50:1150 mg/kg FATOAO 32,356,69
unr-mus LD50:1150 mg/kg KOKABN 26,358,77
orl-cat LDLo:230 mg/kg JPETAB 17,349,21
orl-rbt LD50:210 mg/kg KOKABN 26,358,77
skn-rbt LDLo:1 g/kg 85GMAT-,110,82
orl-brd LD50:300 mg/kg AECTCV 12,355,83

CONSENSUS REPORTS: IARC Cancer Review: Group 3 IMEMDT 7,56,87; Human Inadequate Evidence IMEMDT 12,225,76; Animal Inadequate Evidence IMEMDT 12,225,76. EPA Genetic Toxicology Program. Reported in EPA TSCA Inventory.

OSHA PEL: TWA 5 mg/m^3
ACGIH TLV: TWA 1 mg/m^3
DFG MAK: 5 mg/m^3
DOT Classification: ORM-A; Label: None.

SAFETY PROFILE: Poison by ingestion and intraperitoneal routes. Questionable carcinogen with experimental tumorigenic and teratogenic data. Other experimental reproductive effects. Mutation data reported. Affects human pulmonary system. A mild allergen and irritant. Acute poisoning in experimental animals produced liver, kidney, and brain damage. Dangerous in a fire; see NITROGEN MONOXIDE and SULFUR DIOXIDE.

TFS500 CAS:504-90-5 ***HR: 3***
THIURAM DISULFIDE
mf: $C_2H_4N_2S_4$ mw: 184.32

SYN: BIS(THIOCARBAMOYL)DISULFIDE

TOXICITY DATA with REFERENCE
ihl-rat LD50:740 mg/kg EQSFAP 3,618,75
ipr-mus LD50:250 mg/kg APTOA6 8,329,52
ihl-rat LD50:740 mg/kg EQSFAP 3,618,75

SAFETY PROFILE: Poison by intraperitoneal route. Moderately toxic by inhalation. When heated to decomposition it emits very toxic fumes of NO_x and SO_x. See also BIS(DIMETHYLTHIOCARBAMYL) DISULFIDE.

TFS750 CAS:7440-29-1 ***HR: 3***
THORIUM
af: Th aw: 232.00
DOT: UN 2975

PROP: Silvery-white, air stable, soft, ductile metal. D: 11.72; mp: 1842 ± 30°. A radioactive material.

SYNS: THORIUM-232 ◇ THORIUM METAL, PYROPHORIC (DOT)

CONSENSUS REPORTS: Reported in EPA TSCA Inventory.

DOT Classification: Radioactive Material; Label: Radioactive and Flammable Solid.

SAFETY PROFILE: Suspected carcinogen. Taken internally as ThO_2, it has proven to be a carcinogenic due to its radioactivity. On an acute basis it has caused dermatitis. Flammable in the form of dust when exposed to heat or flame, or by chemical reaction with oxidizers. The powder may ignite spontaneously in air. Potentially hazardous reactions with chlorine, fluorine, bromine, oxygen, phosphorus, silver, sulfur, air, nitryl fluoride, peroxyformic acid.

TFT000 CAS:10026-08-1 ***HR: 3***
THORIUM CHLORIDE
mf: Cl_4Th mw: 373.80

PROP: White, odorless crystals. D: 4.59, mp: 770°, bp: 921°. Sol in water, alc.

SYNS: TETRACHLOROTHORIUM ◇ THORIUM TETRACHLORIDE

TOXICITY DATA with REFERENCE
ivn-rat LDLo:15 mg/kg JPETAB 43,61,31
ipr-rat LD50:1900 mg/kg EQSSDX 1,1,75
ivn-rat LDLo:28 mg/kg EQSSDX 1,1,75
ipr-mus LD50:534 mg/kg COREAF 256,1043,63
scu-mus LD50:4 g/kg EQSSDX 1,1,75
ivn-rbt LDLo:50 mg/kg EQSSDX 1,1,75

CONSENSUS REPORTS: Reported in EPA TSCA Inventory. EPA Genetic Toxicology Program.

SAFETY PROFILE: Poison by intravenous route. Moderately toxic by intraperitoneal and subcutaneous routes. When heated to decomposition it emits toxic fumes of Cl^-. See also THORIUM.

TFT100 CAS:12674-40-7 ***HR: 3***
THORIUM DICARBIDE
mf: C_2Th mw: 256.06

SAFETY PROFILE: Thorium is a radioactive material. Incandescent reaction with selenium, sulfur vapor, molten potassium chlorate, potassium nitrate, potassium hydroxide. See also THORIUM.

TFT250 CAS:15457-87-1 ***HR: 3***
THORIUM HYDRIDE
mf: H_4Th mw: 236.07

PROP: Black, metallic crystals. D: 8.24, mp: decomp explosively @ red heat. Reacts with water.

SAFETY PROFILE: Suspected carcinogen. Explodes on heating in air. The powder ignites spontaneously in air. See also THORIUM and HYDRIDES.

TFT500 CAS:13823-29-5 ***HR: 3***
THORIUM(IV) NITRATE
mf: $H_4N_4O_{12}$•Th mw: 484.08
DOT: UN 2976

PROP: White, crystalline mass. Sol in water, alc.

SYNS: NITRIC ACID, THORIUM(4+) SALT ◇ THORIUM (4+) NITRATE ◇ THORIUM TETRANITRATE

TOXICITY DATA with REFERENCE
scu-rat TDLo:44170 μg/kg (1D male):REP JRPFA4 7,21,64
ipr-rat LD50:60 mg/kg EQSSDX 1,1,75
ivn-rat LD50:47600 μg/kg EQSSDX 1,1,75
itr-rat LD50:85 mg/kg XEURAQ AECD-2283
orl-mus LD50:1760 mg/kg CNRMAW 26,303,48
ivn-mus LD50:45 mg/kg EQSSDX 1,1,75
ivn-dog LDLo:8400 μg/kg AJPHAP 18,426,07
ipr-rbt LDLo:500 mg/kg AJPHAP 18,426,07
ivn-rbt LDLo:57 mg/kg EQSSDX 1,1,75

CONSENSUS REPORTS: Reported in EPA TSCA Inventory.

DOT Classification: Radioactive Material; Label: Radioactive and Oxidizer.

SAFETY PROFILE: Poison by intraperitoneal, intravenous, and intratracheal routes. Moderately toxic by ingestion. Experimental reproductive effects. Radioactive. An oxidizing material; when in contact with readily combustible substances, will cause violent combustion or ignition. When heated to decomposition it emits toxic fumes of NO_x. See also THORIUM and NITRATES.

TFT750 CAS:1314-20-1 ***HR: 3***
THORIUM OXIDE
mf: O_2Th mw: 264.00

PROP: Heavy, white crystalline powder. D: 9.7, mp: 3390°. Insol in water, alkalies; slowly sol in acids.

SYNS: THORIA ◇ THORIUM DIOXIDE ◇ THOROTRAST ◇ THORTRAST ◇ UMBRATHOR

TOXICITY DATA with REFERENCE
par-wmn TDLo:1 g/kg:CAR,LIV ANYAA9 145,700,67
unr-hmn TDLo:2880 mg/kg:NEO,LIV AJRRAV 83,163,60
iat-hmn TDLo:490 mg/kg:CAR ANYAA9 145,776,67
ivn-rat TDLo:160 mg/kg:ETA ANYAA9 145,738,67
unr-ham TDLo:2 g/kg:CAR GANNA2 57,431,66
par-hmn TD:700 mg/kg:NEO,LIV,BLD ANYAA9 145,676,67
par-hmn TD:1260 mg/kg:NEO,LIV,BLD ANYAA9 145,676,67
par-wmn TD:2350 mg/kg:CAR,KID EUURAV 3,69,77
iat-man TD:1190 mg/kg:CAR,LIV,BLD APMIAL 53,147,61
iat-hmn TD:1302 mg/kg:CAR,LIV BJCAAI 41,446,80

CONSENSUS REPORTS: NTP Fifth Annual Report on Carcinogens. Community Right-To-Know List. Reported in EPA TSCA Inventory.

SAFETY PROFILE: Confirmed human carcinogen producing angiosarcoma, liver and kidney tumors, lymphoma and other tumors of the blood system, and tumors at the application site. See also THORIUM.

TFU000 CAS:12218-77-8 ***HR: 3***
THORIUM OXIDE SULFIDE
mf: OSTh mw: 280.10

SAFETY PROFILE: Ignites on contact with air. When

heated to decomposition it emits toxic fumes of SO_x. See also THORIUM and SULFIDES.

TFU500 CAS:299-75-2 ***HR: 3***
l-THREITOL-1,4-BISMETHANESULFONATE
mf: $C_6H_{14}O_6S_2$ mw: 246.32

SYNS: CB 2562 ◇ 1,4-DIMETHANESULFONATE THREITOL ◇ (2s,3s)-1,4-DIMETHANESULFONATE TREITOL ◇ NSC-39069 ◇ TREOSULFAN ◇ TRESULFAN

TOXICITY DATA with REFERENCE
sln-dmg-unr 160 mmol/L ANYAA9 160,228,69
sln-dmg-par 160 mmol/L GENTAE 46,447,61
sce-hmn-unr 28 mg/kg/D EJCODS 18,979,82
sce-hmn:lym 4500 μg/L EJCODS 18,979,82
ivn-dog LDLo:222 mg/kg CCSUBJ 2,203,65
ivn-mky LDLo:222 mg/kg CCSUBJ 2,203,65

CONSENSUS REPORTS: IARC Cancer Review: Group 1 IMEMDT 7,363,87; Human Sufficient Evidence IMEMDT 26,341,81. EPA Genetic Toxicology Program.

SAFETY PROFILE: Confirmed carcinogen. Poison by intravenous route. Human mutation data reported. When heated to decomposition it emits toxic fumes of SO_x.

TFU750 CAS:72-19-5 ***HR: 2***
l-THREONINE
mf: $C_4H_9NO_3$ mw: 119.14

PROP: An essential amino acid. Colorless crystals or white crystalline powder; slt sweet taste. Mp: 255-257° with decomp. Sol in water; very sol in hot water; insol in alc, chloroform, ether.

SYNS: l-2-AMINO-3-HYDROXYBUTYRIC ACID ◇ THREONINE

TOXICITY DATA with REFERENCE
ipr-rat LD50:3098 mg/kg ABBIA4 58,253,55

CONSENSUS REPORTS: Reported in EPA TSCA Inventory.

SAFETY PROFILE: Moderately toxic by intraperitoneal route. When heated to decomposition it emits toxic fumes of NO_x.

TFV750 CAS:672-76-4 ***HR: 3***
Γ-THUJAPLICIN
mf: $C_{10}H_{12}O_2$ mw: 164.22

SYN: 2-HYDROXY-5-(1-METHYLETHYL)-2,4,6-CYCLOHEPTATRIEN-1-ONE

TOXICITY DATA with REFERENCE
ipr-mus LD50:162 mg/kg JAPMA8 48,722,59
ivn-frg LDLo:100 mg/kg JAPMA8 48,722,59

CONSENSUS REPORTS: Reported in EPA TSCA Inventory.

SAFETY PROFILE: Poison by intraperitoneal and intravenous routes. When heated to decomposition it emits acrid smoke and irritating fumes.

TFW000 CAS:546-80-5 ***HR: 3***
THUJONE
mf: $C_{10}H_{16}O$ mw: 152.26

PROP: A flavor constituent. A major component of Wormwood Oil, (*Artemisia absinthium, L*) which is the principal ingredient of absinthe, a liquor. Occurs as α, l, (−) or β, d, (+) called isothujone.

SYNS: (1S-1-α,4-α,5-α)-4-METHYL-1-(1-METHYLETHYL)-BICYCLO(3.1.0)HEXAN-3-ONE ◇ (1S,4R,5R)-(−)-3-THUJANONE ◇ THUJON ◇ (−)-THUJONE ◇ α-THUJONE ◇ l-THUJONE

TOXICITY DATA with REFERENCE
orl-rat LD50:500 mg/kg FRXXBL #2448856
ipr-rat LDLo:120 mg/kg JPETAB 65,275,39
scu-mus LD50:2157 μg/kg JAPMA8 29,2,40
scu-rbt LD50:362 μg/kg JAPMA8 29,2,40
ivn-rbt LD50:31 μg/kg JAPMA8 29,2,40

CONSENSUS REPORTS: Reported in EPA TSCA Inventory.

SAFETY PROFILE: Poison by intravenous, intraperitoneal, and subcutaneous routes. Moderately toxic by ingestion. Serious physiological consequences from abuse of absinthe (mainly in France), led to its abolition in 1915. Wormwood is still used in concentrations of less than 10 ppm in flavored wines. Thujon at 30 mg/kg causes convulsions associated with lesions of the cerebral cortex. Little is known of Thujone metabolism. Both forms occur in Wormwood oil, Oak Moss. The α form is major constituent of Cedar Leaf Oil or Oil of Thuja, Sage. The β form occurs in Tansy, Yarrow. When heated to decomposition it emits acrid smoke and irritating fumes.

TFW250 ***HR: 3***
THULIUM
af: Tm aw: 168.934

PROP: A bright, silvery-gray, lustrous, soft, malleable, ductile metallic element. Mp: 1545°, bp: 1727°, d: 9.333.

SAFETY PROFILE: As a lanthanide it has probably at least a moderate degree of toxicity. Flammable in the form of dust when exposed to flame. Explosive in the form of dust when exposed to heat or flame. Violent reaction with air, halogens. See also POWDERED METALS and RARE EARTHS.

TFW500 CAS:13537-18-3 ***HR: 3***
THULIUM CHLORIDE
mf: Cl_3Tm mw: 275.28

TOXICITY DATA with REFERENCE
skn-rbt 500 mg/24H MOD TXAPA9 5,427,63
eye-rbt 50 mg TXAPA9 5,427,63
orl-mus LD50:4294 mg/kg EQSSDX 1,1,75
ipr-mus LD50:332 mg/kg EQSSDX 1,1,75
ipr-gpg LD50:144 mg/kg AEHLAU 5,437,62

CONSENSUS REPORTS: Reported in EPA TSCA Inventory.

SAFETY PROFILE: Poison by intraperitoneal route. Mildly toxic by ingestion. A skin and eye irritant. When heated to decomposition it emits toxic fumes of Cl^-. See also THULIUM and RARE EARTHS.

TFX000 CAS:63869-91-0 ***HR: 3***
THULIUM EDETATE
mf: $C_{10}H_{12}N_2O_8Tm•H$ mw: 458.18

TOXICITY DATA with REFERENCE
ipr-gpg LD50:89 mg/kg AEHLAU 5,437,62

SAFETY PROFILE: Poison by intraperitoneal route. When heated to decomposition it emits toxic fumes of NO_x. See also THULIUM and RARE EARTHS.

TFX250 CAS:35725-33-8 ***HR: 3***
THULIUM(III) NITRATE, HEXAHYDRATE (1:3:6)
mf: $N_3O_9•Tm•6H_2O$ mw: 463.08

SYN: NITRIC ACID, THULIUM(3+) SALT, HEXAHYDRATE

TOXICITY DATA with REFERENCE
ipr-rat LD50:285 mg/kg TXAPA9 5,750,63
ipr-mus LD50:255 mg/kg TXAPA9 5,750,63

SAFETY PROFILE: Poison by intraperitoneal route. When heated to decomposition it emits toxic fumes of NO_x. See also THULIUM, NITRATES, and RARE EARTHS.

TFX500 CAS:8007-46-3 ***HR: 2***
THYME OIL

PROP: From distillation of flowering plant *Thymus vulgaris* L. (Fam. *Labiatae*). Colorless to reddish-brown liquid; pleasant odor, sharp taste. D: 0.930 @ 25/25°, refr index: 1.495 @ 20°.

SYNS: OIL of THYME ◇ THYMIAN OEL (GERMAN) ◇ THYM OIL

TOXICITY DATA with REFERENCE
dnr-bcs 2 mg/disc TOFOD5 8,91,85
orl-rat LD50:2840 mg/kg PHARAT 14,435,59

CONSENSUS REPORTS: Reported in EPA TSCA Inventory.

SAFETY PROFILE: Moderately toxic by ingestion. Mutation data reported. An allergen and an irritant. Combustible when exposed to heat or flame. When heated to decomposition it emits acrid smoke and irritating fumes.

TFX750 CAS:8007-46-3 ***HR: 2***
THYME OIL RED

PROP: Main constituents are thymol, carvacrol. Found in plants *Thymus vulgaris L.* and *Thymus zygis L.* (FCTXAV 12,807,74).

SYN: SPANISH THYME OIL

TOXICITY DATA with REFERENCE
skn-mus 500 mg SEV FCTXAV 12,807,74
skn-rbt 500 mg/24H SEV FCTXAV 12,807,74
orl-rat LD50:4700 mg/kg FCTXAV 12,807,74

SAFETY PROFILE: Mildly toxic by ingestion. A severe skin irritant. When heated to decomposition it emits acrid smoke and irritating fumes.

TFX790 CAS:50-89-5 ***HR: 2***
THYMIDINE
mf: $C_{10}H_{14}N_2O_5$ mw: 242.26

PROP: Rosettes of needles. Mp: 185°. Sol in water, methanol, hot alc, hot acetone, hot ethyl acetate, pyridine, glacial acetic acid; sltly sol in hot chloroform.

SYNS: DEOXYTHYMIDINE ◇ 2′-DEOXYTHYMIDINE ◇ DT ◇ DTHYD ◇ 5-METHYLDEOXYURIDINE ◇ THYMIDIN ◇ THYMINEDEOXYRIBOSIDE ◇ THYMINE-2-DEOXYRIBOSIDE

TOXICITY DATA with REFERENCE
dni-hmn:oth 1 mmol/L CNREA8 35,2872,75
msc-ham:ovr 200 μmol/L BICMBE 64,809,82
par-mus TDLo:800 mg/kg (female 8D post):TER TJADAB 32,229,85
ipr-mus TDLo:20 g/kg (male 5D pre):REP MUREAV 200,249,88
ipr-mus LD50:2512 mg/kg JPETAB 207,504,78

CONSENSUS REPORTS: Reported in EPA TSCA Inventory. EPA Genetic Toxicology Program.

SAFETY PROFILE: Moderately toxic by intraperitoneal route. An experimental teratogen. Experimental reproductive effects. Human mutation data reported. When heated to decomposition it emits toxic fumes of NO_x.

TFX800 CAS:65-71-4 ***HR: 2***
THYMINE
mf: $C_5H_6N_2O_2$ mw: 126.13

PROP: Star-shaped plates or needles. Decomp @ 335°-337°. Sol in hot water; sltly sol in cold water, alc, ether; very sol in alkalies.

SYNS: 5-METHYLURACIL ◇ THYMIN (PURINE BASE)

TOXICITY DATA with REFERENCE
pic-esc 1 g/L ZAPOAK 12,583,72
orl-mus LD50:3500 mg/kg JAPMA8 44,56,55

CONSENSUS REPORTS: EPA Genetic Toxicology Program. Reported in EPA TSCA Inventory.

SAFETY PROFILE: Moderately toxic by intraperitoneal route. Mutation data reported. When heated to decomposition it emits toxic fumes of NO_x.

TFX810 CAS:89-83-8 ***HR: 3***
THYMOL
mf: $C_{10}H_{14}O$ mw: 150.24

PROP: Colorless, translucent crystals; pungent, caustic taste. Mp: 51°, bp: 233°, d: 0.972, vap press: 1 mm @ 64°. Sol in water, alkali; very sol in alc, ether, chloroform.

SYNS: p-CYMEN-3-OL ◇ 3-p-CYMENOL ◇ 3-HYDROXY-p-CYMENE ◇ 3-HYDROXY-1-METHYL-4-ISOPROPYLBENZENE ◇ ISOPROPYL CRESOL ◇ 6-ISOPROPYL-m-CRESOL ◇ 2-ISOPROPYL-5-METHYLPHENOL ◇ 1-METHYL-3-HYDROXY-4-ISOPROPYLBENZENE ◇ 5-METHYL-2-ISOPROPYL-1-PHENOL ◇ 5-METHYL-2-(1-METHYLETHYL)PHENOL ◇ THYME CAMPHOR ◇ THYMIC ACID ◇ m-THYMOL

TOXICITY DATA with REFERENCE
cyt-smc 500 μmol/tube HEREAY 33,457,47
scu-rat TDLo:20 mg/kg (4D pre):REP JSICAZ 19,264,60
orl-rat LD50:980 mg/kg FCTXAV 2,327,64
scu-rat LDLo:1600 mg/kg HBTXAC 5,172,59
orl-mus LD50:640 mg/kg OSDIAF 5,111,56
scu-mus LDLo:800 mg/kg HBAMAK 4,1289,35
ivn-mus LD50:100 mg/kg JMCMAR 23,1350,80
ivn-dog LDLo:150 mg/kg THERAP 3,109,48
orl-cat LDLo:250 mg/kg HBTXAC 5,172,59
orl-rbt LDLo:750 mg/kg HBTXAC 5,172,59
ivn-rbt LDLo:60 mg/kg HBTXAC 5,172,59

CONSENSUS REPORTS: Reported in EPA TSCA Inventory. EPA Genetic Toxicology Program.

SAFETY PROFILE: Poison by ingestion and intravenous routes. Moderately toxic by subcutaneous route. Experimental reproductive effects. Mutation data reported. An allergen. Incompatible with acetanilide. When heated to decomposition it emits acrid smoke and irritating fumes. An FDA over the counter drug used as an antibacterial and antifungal agent.

TFY000 CAS:964-52-3 ***HR: 3***
THYMOXAMINE HYDROCHLORIDE
mf: $C_{16}H_{25}NO_3 \cdot ClH$ mw: 315.88

SYNS: (2-(4-ACETOXY-2-ISOPROPYL-5-METHYLPHENOXY) ETHYL)DIMETHYLAMINE HYDROCHLORIDE ◇ CARLYTENE ◇ CHLORHYDRATE de ACETOXY-THYMOXY-ETHYL-DIMETHYLAMINE (FRENCH) ◇ 5-(2-(N,N-DIMETHYLAMINO) ETHOXY)CARVACROL ACETATE HYDROCHLORIDE ◇ M-101 ◇ MOXISYLYTE HYDROCHLORIDE ◇ OPILON HYDROCHLORIDE ◇ WV 365

TOXICITY DATA with REFERENCE
orl-rat TDLo:2040 mg/kg (17-22D preg/28D post):REP YACHDS 10,629,82
orl-rat LD50:740 mg/kg THERAP 26,775,71
ipr-rat LD50:96 mg/kg KSRNAM 16,1147,82
scu-rat LD50:190 mg/kg THERAP 26,775,71
ivn-rat LD50:32 mg/kg KSRNAM 16,1147,82
orl-mus LD50:255 mg/kg JPPMAB 32,209,80
ipr-mus LD50:81 mg/kg KSRNAM 16,1147,82
scu-mus LD50:152 mg/kg KSRNAM 16,1147,82
ivn-mus LD50:28 mg/kg KSRNAM 16,1147,82
orl-dog LDLo:700 mg/kg KSRNAM 16,1147,82
ivn-dog LD50:54500 μg/kg KSRNAM 16,1147,82
orl-mky LDLo:1500 mg/kg KSRNAM 16,1147,82
ivn-mky LDLo:40 mg/kg KSRNAM 16,1147,82
orl-cat LDLo:375 mg/kg KSRNAM 16,1147,82
ivn-cat LDLo:30 mg/kg KSRNAM 16,1147,82
ivn-rbt LD50:58600 μg/kg KSRNAM 16,1147,82

SAFETY PROFILE: Poison by ingestion, subcutaneous, intravenous and intraperitoneal routes. Experimental reproductive effects. Used as an adrenergic blocker. When heated to decomposition it emits toxic fumes of NO_x and HCl.

TFY100 ***HR: 2***
THYMUS KOTSCHYANUS, OIL EXTRACT

PROP: The main constituents of the ether oil are carvacrol (13.74%), thymol (10.07%), terpinen (7.6%), terpinden (6.72%), caryophilen (6.33%), terpineol (4.56%), α-pinen (3.36%), and β-pinen (3.2%) (DAZRA7 35(8),87,79).

TOXICITY DATA with REFERENCE
ipr-mus LD50:869 mg/kg DAZRA7 35(8),87,79
scu-mus LD50:991 mg/kg DAZRA7 35(8),87,79
ipr-gpg LD50:1767 mg/kg DAZRA7 35(8),87,79
ims-gpg LD50:1892 mg/kg DAZRA7 35(8),87,79

SAFETY PROFILE: Moderately toxic by intraperitoneal, subcutaneous, and intramuscular routes. See also individual components.

TFZ000 CAS:9007-12-9 ***HR: 3***
THYROCALCITONIN

SYN: TCT

TOXICITY DATA with REFERENCE
ipr-mus TDLo:120 mg/kg/33D-I:ETA ANREAK 175,462,73

SAFETY PROFILE: Questionable carcinogen with experimental tumorigenic data.

TFZ100 CAS:50809-32-0 ***HR: 1***
THYROID

SYN: THYRADIN ◇ TIROIDINA

TOXICITY DATA with REFERENCE
ipr-rat TDLo:2 g/kg (female 7-10D post):TER NAGZAC 41,332,66
orl-wmn TDLo:48 g/kg/26W-I:CVS,PUL,SKN AIMDAP 148,1450,88

SAFETY PROFILE: An experimental teratogen. Human systemic effects by ingestion: cardiomyopathy, dyspnea, and sweating. When heated to decomposition it emits acrid smoke and irritating fumes.

TFZ275 CAS:51-48-9 ***HR: 2***
l-THYROXIN
mf: $C_{15}H_{11}I_4NO_4$ mw: 776.87

PROP: dl-Thyroxine: Needle-like crystals. Decomp @ 231-233°. Insol in water, alc, and the other usual organic solvents, but in the presence of mineral acids or alkalies it dissolves in alc; sol in solns of the alkali hydroxides and in hot solns of the alkali carbonates. When alkali hydroxide solns of thyroxine are saturated with sodium chloride, the sodium salt of thyroxine separates. l-Thyroxine: Crystals, decomp @ 235-236°. d-Thyroxine, Debetrol: Crystals, decomp @ 237°.

SYNS: LEVOTHYROXINE ◇ l-T4 ◇ T4 (hormone) ◇ TETRAIODOTHYRONINE ◇ THX ◇ THYREOIDEUM ◇ THYROXIN ◇ THYROXINE ◇ (−)-THYROXINE ◇ l-THYROXINE

TOXICITY DATA with REFERENCE
dns-rat-scu 6250 μg/kg/2D-C BEXBAN 72,942,71
orl-rat TDLo:1400 μg/kg (female 16-22D post):REP ACEDAB 215,73,78
orl-rat TDLo:26250 μg/kg (1-21D preg):TER CRSBAW 145,525,51
orl-man TDLo:63 μg/kg JTCTDW 20,517,83
orl-wmn TDLo:400 μg/kg/2D-I ICMED9 13,33,87

SAFETY PROFILE: Human toxic effects at low oral doses. An experimental teratogen. Experimental reproductive effects. Mutation data reported. When heated to decomposition it emits toxic fumes of I^- and NO_x.

TFZ300 CAS:7488-70-2 ***HR: 2***
THYROXINE
mf: $C_{15}H_{11}I_4NO_4$ mw: 776.87

SYNS: 3-(4-(4-HYDROXY-3,5-DIIODOPHENOXY)-3,5-DIIODOPHENYL) ALANINE ◇ o-(4-HYDROXY-3,5-DIIODOPHENYL)-3,5-DIIODO-TYROSINE

TOXICITY DATA with REFERENCE
dns-rat-scu 3125 μg/kg BEXBAN 72,942,71
cyt-mus-ipr 250 μg/kg/7D RRENAR 10,149,73
orl-mus TDLo:5 mg/kg (female 10-13D post):TER TXAPA9 84,115,86
orl-rat TDLo:30 mg/kg (9-20D preg):REP AFPEAM 11,168,54
orl-man TDLo:30 μg/kg/3W-I:CVS BMJOAE 292,1185,86
orl-wmn TDLo:42 μg/kg/1W-I NEURAI 35,1792,85

SAFETY PROFILE: Human toxic effects at low oral doses. Human systemic effects by ingestion: EKG changes. An experimental teratogen. Experimental reproductive effects. Mutation data reported. When heated to decomposition it emits toxic fumes of I^- and NO_x.

TGA275 CAS:67418-30-8 ***HR: 3***
TIAPAMIL
mf: $C_{26}H_{37}NO_5S_2 \cdot ClH$ mw: 544.22

SYN: RO 11-1781

TOXICITY DATA with REFERENCE
orl-rat LD50:1803 mg/kg ARZNAD 31,1401,81
ipr-rat LD50:455 mg/kg ARZNAD 31,1401,81
scu-rat LD50:546 mg/kg ARZNAD 31,1401,81
ivn-rat LD50:84 mg/kg ARZNAD 31,1401,81
orl-mus LD50:581 mg/kg ARZNAD 31,1401,81
ipr-mus LD50:200 mg/kg ARZNAD 31,1401,81
scu-mus LD50:347 mg/kg ARZNAD 31,1401,81
ivn-mus LD50:40 mg/kg ARZNAD 31,1401,81

SAFETY PROFILE: Poison by subcutaneous, intravenous, and intraperitoneal routes. Moderately toxic by ingestion. When heated to decomposition it emits toxic fumes of SO_x, NO_x, and HCl.

TGA375 CAS:51012-33-0 ***HR: 3***
TIAPRIDE HYDROCHLORIDE
mf: $C_{15}H_{24}N_2O_4S \cdot ClH$ mw: 364.93

SYNS: N-(2-(DIETHYLAMINO)ETHYL)-2-METHOXY-5-(METHYLSULFONYL)BENZAMIDE HYDROCHLORIDE ◇ N-(2-(DIETHYLAMINO)ETHYL)-5-(METHYLSULFONYL)-o-ANISAMIDEHYDROCHLORIDE

TOXICITY DATA with REFERENCE
orl-rat TDLo:6125 mg/kg (7W male):TER KSRNAM 19,1961,85
orl-rat TDLo:2625 mg/kg (female 2W pre):REP KSRNAM 19,1961,85
orl-rat LD50:4840 mg/kg KSRNAM 19,1933,85
ipr-rat LD50:421 mg/kg KSRNAM 19,1933,85
scu-rat LD50:1130 mg/kg KSRNAM 19,1933,85
ivn-rat LD50:254 mg/kg KSRNAM 19,1933,85

orl-mus LD50:1340 mg/kg KSRNAM 19,1933,85
ipr-mus LD50:336 mg/kg KSRNAM 19,1933,85
scu-mus LD50:755 mg/kg KSRNAM 19,1933,85
ivn-mus LD50:189 mg/kg KSRNAM 19,1933,85
orl-dog LD50:240 mg/kg KSRNAM 19,1933,85

SAFETY PROFILE: Poison by ingestion, intravenous, and intraperitoneal routes. Moderately toxic by subcutaneous route. Experimental reproductive effects. When heated to decomposition it emits toxic fumes of SO_x, NO_x, and HCl.

TGA500 CAS:74-55-5 ***HR: 3***
TIBUTOL
mf: $C_{10}H_{24}N_2O_2$ mw: 204.36

SYNS: d,N,N'-BIS(1-HYDROXYMETHYLPROPYL)ETHYLENEDIAMINE ◇ DADIBUTOL ◇ EMB ◇ ETHAMBUTOL ◇ (R)-2,2'-(1,2-ETHANEDIYLDIIMINO)BIS-1-BUTANOL ◇ (+)-2,2'-(ETHYLENEDIIMINO)DI-1-BUTANOL

TOXICITY DATA with REFERENCE
orl-wmn TDLo:900 mg/kg/60D-I:LIV BMJOAE 292,866,86
orl-man TDLo:600 mg/kg:PNS,CNS,KID BMJOAE 2,1105,76
orl-man TDLo:1200 mg/kg:SKN,IMM BMJOAE 2,1105,76
orl-mus LD50:8700 mg/kg DPHFAK 23,463,71
ipr-mus LD50:1075 mg/kg DPHFAK 23,463,71
ivn-mus LD50:240 mg/kg DPHFAK 23,463,71

SAFETY PROFILE: Poison by intravenous route. Moderately toxic by intraperitoneal route. Mildly toxic by ingestion. Human systemic effects by ingestion: nerve or sheath structural changes, motor activity changes, cholestatic jaundice, changes in urine composition, skin dermatitis and allergic anaphylaxis. When heated to decomposition it emits toxic fumes of NO_x.

TGA520 CAS:4697-14-7 ***HR: 1***
TICARCILLIN SODIUM
mf: $C_{15}H_{14}N_2O_6S_2$•2Na mw: 428.41

TOXICITY DATA with REFERENCE
scu-rat TDLo:5500 mg/kg (7-17D preg):TER KSRNAM 12,556,78
orl-rat LD50:16 g/kg KSRNAM 12,556,78
ipr-rat LD50:7200 mg/kg KSRNAM 12,556,78
scu-rat LD50:10 g/kg KSRNAM 12,556,78
ivn-rat LD50:5350 mg/kg KSRNAM 12,556,78
ipr-mus LD50:7800 mg/kg KSRNAM 12,556,78
scu-mus LD50:10700 mg/kg KSRNAM 12,556,78
ivn-mus LD50:5200 mg/kg KSRNAM 12,556,78
scu-dog LD50:7 g/kg KSRNAM 12,556,78

SAFETY PROFILE: Mildly toxic by ingestion. An experimental teratogen. When heated to decomposition it emits toxic fumes of SO_x, NO_x, and Na_2O.

TGA525 CAS:53885-35-1 ***HR: 3***
TICLODONE
mf: $C_{14}H_{14}ClNS$•ClH mw: 300.26

SYNS: 4-C-32 ◇ 53-32C ◇ 5-(o-CHLOROBENZYL)-4,5,6,7-TETRAHYDROTHIENO(3,2-c)PYRIDINE HYDROCHLORIDE ◇ PANALDINE ◇ TICLID ◇ TICLODIX ◇ TICLOPIDINE HYDROCHLORIDE ◇ TIKLID

TOXICITY DATA with REFERENCE
orl-rat TDLo:220 mg/kg (female 7-17D post):REP IYKEDH 11,265,80
orl-rat TDLo:4400 mg/kg (female 7-17D post):TER IYKEDH 11,265,80
orl-rat LD50:1780 mg/kg MEPHDN 15,272,81
ivn-rat LD50:70 mg/kg IYKEDH 12,1204,81
orl-mus LD50:600 mg/kg MEPHDN 15,272,81
scu-mus LD50:2690 mg/kg MEPHDN 15,272,81
ivn-mus LD50:55 mg/kg USXXAM #4051141

SAFETY PROFILE: Poison by intravenous route. Moderately toxic by ingestion and subcutaneous routes. An experimental teratogen. Experimental reproductive effects. When heated to decomposition it emits toxic fumes of SO_x, NO_x, and HCl.

TGA600 CAS:40180-04-9 ***HR: 3***
TIENILIC ACID
mf: $C_{13}H_8Cl_2O_4S$ mw: 331.17

PROP: Crystals from 50% ethanol. Mp: 148-149°; also reported as mp: 157°.

SYNS: ANP 3624 ◇ CE 3624 ◇ (2,3-DICHLORO-4-(2-THENOYL) PHENOXY)ACETIC ACID ◇ (2,3-DICHLORO-4-(2-THIENYLCARBONYL)PHENOXY)ACETIC ACID ◇ (2,3-DICHLORO-4-(2-THIOPHENECARBONYL)PHENOXY)ACETIC ACID ◇ DIFLUREX ◇ FR 3068 ◇ SELACRYN ◇ SKF 62698 ◇ THIENYLIC ACID ◇ 4-(2-THIENYLKETO)-2,3-DICHLOROPHENOXYACETIC ACID ◇ TICREX ◇ TICRYNAFEN

TOXICITY DATA with REFERENCE
mma-ham:lng 10 μmol/L MUREAV 157,1,85
msc-ham:lng 100 μmol/L MUREAV 157,1,85
orl-man TDLo:6666 μg/kg:KID NEJMAG 301,1180,79
orl-wmn TDLo:5 mg/kg:KID,LIV NEJMAG 301,1179,79
orl-mus LD50:1275 mg/kg EJMCA5 9,625,74
ivn-mus LD50:225 mg/kg USXXAM #3758506

SAFETY PROFILE: Poison by intravenous route. Moderately toxic by ingestion. Human systemic effects by ingestion: nausea or vomiting, kidney damage and decreased urine volume or anuria. Mutation data reported. Used as a diuretic, uricosuric and antihypertensive. When heated to decomposition it emits toxic fumes of Cl^- and SO_x.

TGA700 CAS:80-59-1 ***HR: 2***
TIGLIC ACID
mf: $C_5H_8O_2$ mw: 100.13

PROP: Triclinic plates or rods; spicy odor. D: 0.972, mp: 63.5-64°, bp: 198.5°. Sltly sol in cold water; sol in hot water, alc, and eth.

SYNS: 2-BUTENOIC ACID, 2-METHYL-, (E)-(9CI) ◇ CEVADIC ACID ◇ CROTONIC ACID, 2-METHYL-, (E)- ◇ (E)-2,3-DIMETHYLACRYLIC ACID ◇ trans-α-β-DIMETHYLACRYLIC ACID ◇ trans-2,3-DIMETHYL-ACRYLIC ACID ◇ trans-2-METHYL-2-BUTENOIC ACID ◇ (E)-2-METHYLCROTONIC ACID ◇ trans-2-METHYLCROTONIC ACID ◇ TIGLINIC ACID

TOXICITY DATA with REFERENCE
skn-rbt 500 mg/24H SEV FCTOD7 20,837,82

CONSENSUS REPORTS: Reported in EPA TSCA Inventory.

SAFETY PROFILE: A severe skin irritant. When heated to decomposition it emits acrid smoke and irritating fumes.

TGB000 CAS:27591-69-1 ***HR: 3***
TILORONE HYDROCHLORIDE
mf: $C_{25}H_{34}N_2O_3 \cdot 2ClH$ mw: 483.53

SYNS: BIS-DEAE-FLUORENONE HYDROCHLORIDE ◇ 2,7-BIS(2-(DIETHYLAMINO)ETHOXY)-FLUOREN-9-ONEDIHYDROCHLORIDE ◇ 2,7-BIS(2-(DIETHYLAMINO)ETHOXY)-9H-FLUOREN-9-ONE DIHYDROCHLORIDE ◇ NSC-143969

TOXICITY DATA with REFERENCE
dnd-esc 20 μmol/L MUREAV 89,95,81
dni-mus:leu 14600 μg/L INNDDK 1,103,83
orl-hmn TDLo:4 mg/kg:CNS,GIT CCROBU 57,209,73
orl-rat LD50:852 mg/kg SCIEAS 169,1213,70
ipr-rat LD50:244 mg/kg SCIEAS 169,1213,70
orl-mus LD50:959 mg/kg SCIEAS 169,1213,70
ipr-mus LD50:145 mg/kg SCIEAS 169,1213,70

SAFETY PROFILE: Poison by intraperitoneal route. Moderately toxic by ingestion. Human systemic effects by ingestion: somnolence, anorexia, hypermotility and diarrhea. Mutation data reported. When heated to decomposition it emits very toxic fumes of HCl and NO_x.

TGB150 ***HR: 3***
TIMBER RATTLESNAKE VENOM

SYNS: C. HORRIDUS HORRIDUS VENOM ◇ CROTALUS HORRIDUS HORRIDUS VENOM ◇ VENOM, SNAKE, CROTALUS HORRIDUS HORRIDUS

TOXICITY DATA with REFERENCE
ipr-mus LD50:2844 μg/kg TOXIA6 9,131,71
scu-mus LD50:9150 μg/kg PAASAH 44,145,56
ivn-mus LD50:1644 μg/kg TOXIA6 9,131,71
ims-mus LD50:6 μg/kg JTASAG 55,96,80
ipr-mam LD50:2910 μg/kg CLPTAT 8,849,67
ivn-mam LD50:2630 μg/kg CLPTAT 8,849,67

SAFETY PROFILE: Poison by subcutaneous, intramuscular, intravenous, and intraperitoneal routes.

TGB160 CAS:35035-05-3 ***HR: 3***
TIMEPIDIUM BROMIDE
mf: $C_{17}H_{22}NOS_2 \cdot Br$ mw: 400.43

PROP: Colorless crystals from acetone/ether. Mp: 198-200°.

SYNS: 1,1-DIMETHYL-5-METHOXY-3-(DITHIEN-2-YLMETHYLENE) PIPERIDINIUM BROMIDE ◇ N-METHYL-5-METHOXY-3-PIPERIDYLIDENEDITHIENYLMETHANE METHOBROMIDE ◇ SA 50 Y ◇ SA 504 ◇ SESDEN

TOXICITY DATA with REFERENCE
orl-mus TDLo:960 mg/kg (female 7-12D post):REP OYYAA2 7,1293,73
ipr-rat TDLo:60 mg/kg (female 9-14D post):TER OYYAA2 7,1293,73
orl-rat LD50:1213 mg/kg JJPAAZ 22,685,72
scu-rat LD50:252 mg/kg JJPAAZ 22,685,72
ipr-rat LD50:55 mg/kg JJPAAZ 22,685,72
ivn-rat LD50:7 mg/kg OYYAA2 23,461,82
orl-mus LD50:713 mg/kg JMCMAR 15,914,72
ipr-mus LD50:97 mg/kg JMCMAR 15,914,72
scu-mus LD50:145 mg/kg JJPAAZ 22,685,72
ivn-mus LD50:12 mg/kg JJPAAZ 22,685,72
ims-mus LD50:199 mg/kg NIIRDN 6,351,82

SAFETY PROFILE: Poison by intramuscular, subcutaneous, intravenous, and intraperitoneal routes. Moderately toxic by ingestion. An experimental teratogen. Experimental reproductive effects. When heated to decomposition it emits toxic fumes of NO_x, SO_x, and Br^-.

TGB175 CAS:57648-21-2 ***HR: 2***
TIMIPERONE
mf: $C_{22}H_{24}FN_3OS$ mw: 397.55

PROP: Butyrophenone derivative with neuroleptic activity. Crystals from acetone. Mp: 201-203°. Sltly sol in water.

SYNS: 1-(1-(3-(p-FLUOROBENZOYL)PROPYL)-4-PIPERIDYL)-2-BENZIMIDAZOLINETHIONE ◇ 4'-FLUORO-4-(4-(2-THIOXOBENZIMIDAZOL-1-YL)PIPERIDINO)-BUTYROPHENONE ◇ TOLOPELON

TOXICITY DATA with REFERENCE
orl-rat TDLo:11 mg/kg (7-17D preg):TER IYKEDH 12,861,81
ivn-rat TDLo:1600 μg/kg (female 1D pre):REP TJADAB 32,34B,85
orl-mus LD50:500 mg/kg JMCMAR 21,1116,78

SAFETY PROFILE: Moderately toxic by ingestion. An experimental teratogen. Experimental reproductive effects. Used as a tranquilizer. When heated to decomposition it emits toxic fumes of F^-, SO_x, and NO_x.

TGB185 CAS:26921-17-5 ***HR: 3***
TIMOLOL MALEATE
mf: $C_{13}H_{24}N_4O_3S \cdot C_4H_4O_4$ mw: 432.55

SYNS: (−)-1-(tert-BUTYLAMINO)-3-((4-MORPHOLINO-1,2,5-THIADIAZOL-3-YL)OXY)-2-PROPANOL MALEATE ◇ MK 950 ◇ TIMACOR ◇ l-TIMOLOL MALEATE ◇ TIMOPTOL

TOXICITY DATA with REFERENCE
orl-rat LD50:1028 mg/kg NIIRDN 6,464,82
ipr-rat LD50:381 mg/kg NIIRDN 6,464,82
scu-rat LD50:881 mg/kg NIIRDN 6,464,82
orl-mus LD50:1137 mg/kg IYKEDH 12,1204,81
ipr-mus LD50:300 mg/kg IYKEDH 12,1204,81
scu-mus LD50:805 mg/kg IYKEDH 12,1204,81

SAFETY PROFILE: Poison by intraperitoneal route. Moderately toxic by ingestion and subcutaneous routes. When heated to decomposition it emits toxic fumes of NO_x and SO_x.

TGB250 CAS:7440-31-5 ***HR: 3***
TIN
af: Sn aw: 118.71

PROP: Cubic, gray, crystalline metallic element. Mp: 231.9°, stabilizes <18°, d: 7.31, vap press: 1 mm @ 1492°, bp: 2507°.

SYNS: SILVER MATT POWDER ◇ TIN (α) ◇ TIN FLAKE ◇ TIN POWDER ◇ ZINN (GERMAN)

TOXICITY DATA with REFERENCE
imp-rat TDLo:395 g/kg:ETA RCOCB8 18,201,77

CONSENSUS REPORTS: Reported in EPA TSCA Inventory.

OSHA PEL: Organic Compounds: TWA 0.1 mg(Sn)/m^3 (skin); Inorganic Compounds (except oxides): TWA 2 mg/m^3
ACGIH TLV: TWA metal, oxide and inorganic compounds (except SnH_4) as Sn 2 mg/m^3; organic compounds 0.1 mg/m^3 (skin) (Proposed: TWA 0.1 mg(Sn)/m^3; STEL 0.2 mg(Sn)/m^3 (skin))
DFG MAK: Inorganic 2 mg/m^3, organic 0.1 mg/m^3
NIOSH REL: (Organotin Compounds) TWA 0.1 mg(Sn)/m^3

SAFETY PROFILE: Questionable carcinogen with experimental tumorigenic data by implant route. Combustible in the form of dust when exposed to heat or by spontaneous chemical reaction with Br_2; BrF_3; Cl_2; ClF_3; $Cu(NO_3)$; K_2O_2; S. See also POWDERED METALS and TIN COMPOUNDS.

TGB475 CAS:2398-96-1 ***HR: 1***
TINACTIN
mf: $C_{19}H_{17}NOS$ mw: 307.43

PROP: Crystals from alc. Mp: 110.5-111.5°. Insol in water; sparingly sol in methanol, ethanol; sol in chloroform (1:1.5), acetone (1:7), CCl4 (1:9).

SYNS: AFTATE ◇ CHINOFUNGIN ◇ DERMOXIN ◇ m,N-DIMETHYLTHIOCARBANILIC ACID-o-2 NAPHTHYL ESTER ◇ FOCUSAN ◇ FUNGISTOP ◇ HI-ALAZIN ◇ METHYL (3-METHYLPHENYL)CARBAMOTHIOIC ACID, o-2-NAPHTHALENYL ESTER ◇ NAPHTHIOMATE T ◇ o-2-NAPHTHYL m,N-DIMETHYLTHIOCARBANILATE ◇ 2-NAPHTHYL N-METHYL-N-(3-TOLYL)THIONOCARBAMATE ◇ PITREX ◇ SORGOA ◇ SPORILINE ◇ TIMOPED ◇ TINADERM ◇ TOLNAFTATE ◇ TOLNAPHTHATE ◇ TOLSANIL ◇ TONOFTAL

TOXICITY DATA with REFERENCE
orl-rat TDLo:3500 mg/kg (8-14D preg):TER TXAPA9 8,386,66
orl-mus LD50:10 g/kg NIIRDN 6,530,82
scu-mus LD50:6000 mg/kg NIIRDN 6,530,82
ivn-mus LD50:4800 mg/kg NIIRDN 6,530,82

CONSENSUS REPORTS: EPA Genetic Toxicology Program. Reported in EPA TSCA Inventory.

SAFETY PROFILE: Mildly toxic by ingestion. An experimental teratogen. When heated to decomposition it emits toxic fumes of SO_x and NO_x. A fungicide used to control athletes foot. See also ESTERS and CARBAMATES.

TGB500 ***HR: 3***
TIN AZIDE TRICHLORIDE
mf: Cl_3N_3Sn mw: 267.07

SAFETY PROFILE: Explosive solid. When heated to decomposition it emits very toxic fumes of NO_x and Cl^-. See also AZIDES and TIN COMPOUNDS.

TGB750 CAS:7789-67-5 ***HR: 3***
TIN(IV) BROMIDE (1:4)
mf: Br_4Sn mw: 438.33

PROP: White crystalline mass. Mp: 31°, bp: 202°, d (liquid): 3.340 @ 35°, vap press: 10 mm @ 72.7°.

SYNS: STANNIC BROMIDE ◇ TIN PERBROMIDE ◇ TIN TETRABROMIDE

TOXICITY DATA with REFERENCE
ivn-mus LD50:18 mg/kg CSLNX* NX#02222

CONSENSUS REPORTS: Reported in EPA TSCA Inventory.

OSHA PEL: TWA 2 mg(Sn)/m^3
ACGIH TLV: TWA 2 mg(Sn)/m^3

SAFETY PROFILE: Poison by intravenous route. Violent reaction with NO_2Cl. When heated to decomposition it emits toxic fumes of Br^-. See also BROMIDES and TIN COMPOUNDS.

TGC000 CAS:7772-99-8 ***HR: 3***
TIN(II) CHLORIDE (1:2)
mf: Cl_2Sn mw: 189.59
DOT: NA 1759

PROP: Colorless crystals. D: 2.71, mp: 37-38°. Sol in less than its own weight of water; very sol in hydrochloric acid (dilute or conc); sol in alc, ethyl acetate, glacial acetic acid, sodium hydroxide solution.

SYNS: C.I. 77864 ◇ NCI-C02722 ◇ STANNOUS CHLORIDE (FCC) ◇ STANNOUS CHLORIDE, solid (DOT) ◇ TIN DICHLORIDE ◇ TIN PROTOCHLORIDE

TOXICITY DATA with REFERENCE
sln-smc 6 mmol/L MUTAEX 1,21,86
dnd-hmn:leu 10 μmol/L CBINA8 46,189,83
itt-rat TDLo:15167 μg/kg (1D male):REP JRPFA4 7,21,64
orl-rat LD50:700 mg/kg FOREAE 7,313,42
orl-mus LD50:1200 mg/kg FOREAE 7,313,42
ipr-mus LD50:105 mg/kg COREAF 256,1043,63
ivn-mus LD50:17800 μg/kg CSLNX* NX#02202
orl-dog LDLo:500 mg/kg HBAMAK 4,1420,35
scu-dog LDLo:159 mg/kg EQSSDX 1,1,75
ivn-dog LDLo:20 mg/kg HBAMAK 4,1420,35
orl-rbt LD50:10 g/kg FAONAU 48A,75,70
scu-gpg LDLo:400 mg/kg BMJOAE 2,217,13

CONSENSUS REPORTS: NTP Carcinogenesis Bioassay (feed); No Evidence: mouse, rat NTPTR* NTP-TR-231,82. Reported in EPA TSCA Inventory. EPA Genetic Toxicology Program.

OSHA PEL: TWA 2 mg(Sn)/m^3
ACGIH TLV: TWA 2 mg(Sn)/m^3

DOT Classification: ORM-B; Label: None.

SAFETY PROFILE: Poison by ingestion, intraperitoneal, intravenous, and subcutaneous routes. Experimental reproductive effects. Human mutation data reported. Potentially explosive reaction with metal nitrates. Violent reactions with hydrogen peroxide; ethylene oxide; hydrazine hydrate; nitrates; K; Na. Ignition on contact with bromine trifluoride. A vigorous reaction with calcium acetylide is initiated by flame. When heated to decomposition it emits toxic fumes of Cl^-. See also TIN COMPOUNDS.

TGC250 CAS:7646-78-8 ***HR: 3***
TIN(IV) CHLORIDE (1:4)
mf: Cl_4Sn mw: 260.49
DOT: UN 1827

PROP: Colorless, fuming caustic liquid or crystals. Mp: −33°, bp: 114.1°, d: 2.232, vap press: 10 mm @ 10°.

SYNS: ETAIN (TETRACHLORURE d') (FRENCH) ◇ LIBAVIUS FUMING SPIRIT ◇ STAGNO (TETRACLORURO di) (ITALIAN) ◇ STANNIC CHLORIDE, anhydrous (DOT) ◇ TIN CHLORIDE, fuming (DOT) ◇ TIN PERCHLORIDE (DOT) ◇ TIN TETRACHLORIDE, anhydrous (DOT) ◇ TINTETRACHLORIDE (DUTCH) ◇ ZINNTETRACHLORID (GERMAN)

TOXICITY DATA with REFERENCE
ihl-rat LC50:2300 mg/m^3/10M TOXID9 1,77,81
ipr-mus LD50:101 mg/kg COREAF 256,1043,63

CONSENSUS REPORTS: Reported in EPA TSCA Inventory. EPA Genetic Toxicology Program.

OSHA PEL: TWA 2 mg(Sn)/m^3
ACGIH TLV: TWA 2 mg(Sn)/m^3
DOT Classification: Corrosive Material; Label: Corrosive.

SAFETY PROFILE: Poison by intraperitoneal route. Moderately toxic by inhalation. A corrosive irritant to skin, eyes, and mucous membranes. Combustible by chemical reaction. Upon contact with moisture, considerable heat is generated. Violent reaction with K; Na; turpentine; ethylene oxide; alkyl nitrates. Dangerous; hydrochloric acid is liberated on contact with moisture or heat. When heated to decomposition it emits toxic fumes of Cl^-. See also HYDROCHLORIC ACID.

TGC275 CAS:10025-69-1 ***HR: 3***
TIN(II) CHLORIDE DIHYDRATE (1:2:2)
mf: $Cl_2Sn{\cdot}2H_2O$ mw: 225.63

SYNS: STANNOCHLOR ◇ STANNOUS CHLORIDE DIHYDRATE

TOXICITY DATA with REFERENCE
dnd-ham:ovr 50 μmol/L MUREAV 119,195,83
ivn-rat LD50:7830 μg/kg APYPAY 32,193,81
ivn-dog LDLo:20 mg/kg HBTXAC 1,282,55

OSHA PEL: TWA 2 mg(Sn)/m^3
ACGIH TLV: TWA 2 mg(Sn)/m^3

SAFETY PROFILE: Poison by intravenous route. Mutation data reported. When heated to decomposition it emits toxic fumes of Cl^-. See also TIN COMPOUNDS.

TGC280 CAS:13940-16-4 ***HR: 3***
TIN CHLORIDE IODIDE
mf: Cl_2I_2Sn mw: 443.39

SYN: STANNIC DICHLORIDE DIIODIDE

TOXICITY DATA with REFERENCE
ivn-mus LD50:56200 μg/kg CSLNX* NX#02301

OSHA PEL: TWA 2 mg(Sn)/m^3
ACGIH TLV: TWA 2 mg(Sn)/m^3

SAFETY PROFILE: Poison by intravenous route. When heated to decomposition it emits toxic fumes of Sn, I^-, and Cl^-.

TGC285 CAS:59178-29-9 ***HR: 2***
TIN(2+) CITRATE
mf: $C_6H_8O_7{\cdot}{}^3/_2Sn$

SYNS: 1,2,3-PROPANETRICARBOXYLIC ACID, 2-HYDROXY-,

TIN(2+) SALT (2:3) (9CI) ◇ STANNOUS CITRIC ACID ◇ TIN CITRATE (7CI)

TOXICITY DATA with REFERENCE
ivn-rbt LDLo:500 mg/kg JLCMAK 28,1344,43

OSHA PEL: TWA 2 mg(Sn)/m^3
ACGIH TLV: TWA 2 mg(Sn)/m^3

SAFETY PROFILE: Moderately toxic by intravenous route. When heated to decomposition it emits toxic fumes of Sn.

TGC300 ***HR: 3***
TIN, COMPLEX with PYROCATECHOL

PROP: Aromatic complex salt containing 9.5% Sn(II) (AEPPAE 114,39,26).

SYN: HEYDEN 768

TOXICITY DATA with REFERENCE
scu-mus LDLo:150 mg(Sn)/kg AEPPAE 114,39,26
scu-rbt LDLo:150 mg(Sn)/kg AEPPAE 114,39,26
scu-gpg LDLo:150 mg(Sn)/kg AEPPAE 114,39,26

OSHA PEL: TWA 2 mg(Sn)/m^3
ACGIH TLV: TWA 2 mg(Sn)/m^3

SAFETY PROFILE: Poison by subcutaneous route. When heated to decomposition it emits acrid smoke and irritating fumes. See also PYROCATECHOL and TIN COMPOUNDS.

TGC500 ***HR: D***
TIN COMPOUNDS

OSHA PEL: Organic Compounds: TWA 0.1 mg(Sn)/m^3 (skin); Inorganic Compounds (except oxides): TWA 2 mg/m^3
ACGIH TLV: TWA metal, oxide and inorganic compounds (except SnH_4) as Sn 2 mg/m^3; organic compounds 0.1 mg/m^3 (skin) (Proposed: TWA 0.1 mg(Sn)/m^3; STEL 0.2 mg(Sn)/m^3 (skin))
DFG MAK: Inorganic 2 mg/m^3, organic 0.1 mg/m^3
NIOSH REL: (Organotin Compounds) TWA 0.1 mg(Sn)/m^3

SAFETY PROFILE: Variable toxicity. Elemental tin and inorganic tin compounds have low toxicity and are poorly absorbed when ingested. Some inorganic tin salts are irritating or can liberate toxic fumes on decomposition. The latter is particularly true of tin halogens. Tin hydride is highly toxic with effects similar to arsenic hydride. Inhalation of tin dusts over a period of years may cause pneumoconiosis. Some of the organic tin compounds are strong poisons. Short chain alkyl tin compounds (e.g., ethyl and methyl compounds) are particularly toxic. Generally alkyl tin compounds are more toxic than aryl compounds and short chain compounds are more toxic than long chain compounds. The toxicity increases with the number of alkyl groups. Tetramethyl tin chloride and triethyl tin chloride are very toxic to the nervous system. They are lipid soluble and can be absorbed through the skin. Symptoms recede slowly. The concentration of tin in condensed milk from cans may reach 160 ppm and could become hazardous for babies. Some alkyl tin compounds have high ecotoxicity. They have been used in marine paints to prevent growth on boat hulls, but may have too much environmental effect for this purpose. See also TIN, ORGANO METALS, and specific compounds.

TGD000 CAS:58-14-0 ***HR: 3***
TINDURIN
mf: $C_{12}H_{13}ClN_4$ mw: 248.74

SYNS: CD ◇ CHLORIDIN ◇ CHLORIDINE ◇ 5-(4'-CHLOROPHENYL)-2,4-DIAMINO-6-ETHYLPYRIMIDINE ◇ 5-(4-CHLOROPHENYL)-6-ETHYL-2,4-PYRIMIDINEDIAMINE ◇ DARACLOR ◇ DARAPRAM ◇ DARAPRIM ◇ DARAPRIME ◇ 2,4-DIAMINO-5-p-CHLOROPHENYL-6-ETHYLPYRIMIDINE ◇ 2,4-DIAMINO-5-(4-CHLOROPHENYL)-6-ETHYLPYRIMIDINE ◇ DIAMINOPYRITAMIN ◇ ERBAPRELINA ◇ KHLORIDIN ◇ MALACID ◇ MALOCID ◇ MALOCIDE ◇ MALOPRIM ◇ NCI-C01683 ◇ NSC 3061 ◇ PIRIMECIDAN ◇ PIRIMETAMINA (SPANISH) ◇ 4753 R.P. ◇ WR 2978

TOXICITY DATA with REFERENCE
mmo-sat 10 μL/plate ANYAA9 76,475,58
cyt-mus-ipr 100 mg/kg GNKAA5 9,67,73
ipr-rat TDLo:1 mg/kg (female 9D post):REP BEXBAN 71,254,71
ipr-rat TDLo:75 mg/kg (female 12-14D post):TER TJADAB 35,58A,87
ipr-rat LD50:70 mg/kg 14XBAV -,367,64
orl-mus LD50:92 mg/kg BJPCAL 6,185,51
ipr-mus LD50:74 mg/kg EJMCA5 9,658,74
scu-mus LDLo:160 mg/kg JMCMAR 16,1399,73
orl-ham LDLo:250 mg/kg TJADAB 4,205,71

CONSENSUS REPORTS: IARC Cancer Review: Group 3 IMEMDT 7,56,87; Animal Limited Evidence IMEMDT 13,233,77. NCI Carcinogenesis Bioassay (feed); Inadequate Studies: mouse NCITR* NCI-CG-TR-77,78; (feed); No Evidence: rat NCITR* NCI-CG-TR-77,78. EPA Genetic Toxicology Program.

SAFETY PROFILE: Poison by ingestion, subcutaneous, and intraperitoneal routes. Experimental teratogenic and reproductive effects. Questionable carcinogen. Human mutation data reported. When heated to decomposition it emits very toxic fumes of Cl^- and NO_x. Used as an antimalarial drug for humans and to treat toxoplasmosis in hogs.

TGD100 CAS:7783-47-3 ***HR: 3***
TIN FLUORIDE
mf: F_2Sn mw: 156.69

PROP: Monoclinic, lamellar plates. Mp: 213°, d: (25) 4.57. Sol in water (about 30%). Forms an oxyfluoride, $SnOF_2$, on exposure to air.

SYNS: FLUORISTAN ◇ STANNOUS FLUORIDE ◇ TIN BIFLUORIDE ◇ TIN DIFLUORIDE

TOXICITY DATA with REFERENCE
mma-sat 4 μmol/plate MUREAV 90,91,81
sln-dmg-orl 15 pph/24H FLUOA4 6,113,73
orl-rat LD50:377 mg/kg IARC** 27,237,82
orl-mus LD50:210 mg/kg IARC** 27,237,82
ipr-mus LD50:16150 μg/kg DZZEA7 34,484,79

CONSENSUS REPORTS: IARC Cancer Review: Group 3 IMEMDT 7,208,87. EPA Genetic Toxicology Program. Reported in EPA TSCA Inventory.

OSHA PEL: TWA 2 mg(Sn)/m^3; 2.5 mg(F)/m^3
ACGIH TLV: TWA 2 mg(Sn)/m^3; 2.5 mg(F)/m^3
NIOSH REL: TWA 2.5 mg(F)/m^3

SAFETY PROFILE: Poison by ingestion and intraperitoneal routes. Questionable carcinogen. Mutation data reported. When heated to decomposition it emits toxic fumes of F^-. See also TIN COMPOUNDS.

TGD125 CAS:50802-21-6 ***HR: 3***
TINGENONE
mf: $C_{28}H_{36}O_3$ mw: 420.64

SYNS: MAITENIN ◇ MAYTENIN ◇ TINGENIN A ◇ TINGENON

TOXICITY DATA with REFERENCE
dni-omi 4700 nmol/L EXPEAM 41,646,85
oms-omi 19 μmol/L EXPEAM 41,646,85
ivn-mus LD50:15 mg/kg 85GDA2 8(1),91,82

SAFETY PROFILE: Poison by intravenous route. Mutation data reported. When heated to decomposition it emits acrid smoke and irritating fumes.

TGD250 CAS:19387-91-8 ***HR: 2***
TINIDAZOLE
mf: $C_8H_{13}N_3O_4S$ mw: 247.30

PROP: Colorless crystals from benzene. Mp: 127-128°.

SYNS: 1-(2-(ETHYLSULFONYL)-ETHYL)-2-METHYL-5-NITROIMIDAZOLE ◇ 1-(2-(ETHYLSULFONYL)-ETHYL)-2-METHYL-5-NITRO-1H-IMIDAZOLE

TOXICITY DATA with REFERENCE
mmo-sat 50 μg/plate MUREAV 77,301,80
mnt-mus-ipr 50 mg/kg AIVMBU 15,311,84
orl-rat TDLo:3 g/kg (9-14D preg):TER OYYAA2 8,421,74
orl-rat TDLo:5460 mg/kg (female 91D pre):REP OYYAA2 8,1089,74
orl-rat LD50:2710 mg/kg IYKEDH 11,811,80
ipr-rat LD50:2720 mg/kg IYKEDH 11,811,80
scu-rat LD50:3000 mg/kg IYKEDH 11,811,80
orl-mus LD50:3200 mg/kg JMCMAR 21,781,78
ipr-mus LD50:2730 mg/kg IYKEDH 11,811,80
scu-mus LD50:3940 mg/kg IYKEDH 11,811,80

SAFETY PROFILE: Moderately toxic by ingestion, intraperitoneal, and subcutaneous routes. An experimental teratogen. Experimental reproductive effects. Mutation data reported. When heated to decomposition it emits very toxic fumes of NO_x and SO_x.

TGD500 CAS:10294-70-9 ***HR: 3***
TIN(II) IODIDE
mf: I_2Sn mw: 372.49

PROP: Powder. Mp: 320°, bp: 720°, d: 5.21.

SYN: STANNOUS IODIDE

TOXICITY DATA with REFERENCE
ivn-mus LD50:100 mg/kg CSLNX* NX#02455

CONSENSUS REPORTS: Reported in EPA TSCA Inventory.

OSHA PEL: TWA 2 mg(Sn)/m^3
ACGIH TLV: TWA 2 mg(Sn)/m^3

SAFETY PROFILE: Poison by intravenous route. When heated to decomposition it emits toxic fumes of I^-. See also IODIDES and TIN COMPOUNDS.

TGD750 CAS:7790-47-8 ***HR: 3***
TIN(IV) IODIDE (1:4)
mf: I_4Sn mw: 626.29

PROP: Red cubic crystals. Mp: 144.5°, bp: 364°, d: 4.473 @ 0°.

SYNS: STANNIC IODIDE ◇ TIN TETRAIODIDE

TOXICITY DATA with REFERENCE
ivn-rat LDLo:200 mg/kg MEIEDD 11,1384,89

CONSENSUS REPORTS: Reported in EPA TSCA Inventory.

OSHA PEL: TWA 2 mg(Sn)/m^3
ACGIH TLV: TWA 2 mg(Sn)/m^3

SAFETY PROFILE: Poison by intravenous route. Strong reaction with NO_2Cl; (K + S); (Na + S). When heated to decomposition it emits toxic fumes of I^-. See also TIN COMPOUNDS and IODIDES.

TGE000 ***HR: 3***
TIN(II) NITRATE OXIDE
mf: $N_2O_7Sn_2$ mw: 377.39

SAFETY PROFILE: Self-explodes or organic dust. When heated to decomposition it emits very toxic fumes of NO_x. See also TIN COMPOUNDS.

TGE100 CAS:13863-31-5 ***HR: 1***
TINOPAL 5BM
mf: $C_{38}H_{36}N_{12}O_8S_2 \cdot 2Na$ mw: 898.96

SYN: 2,2'-STILBENEDISULFONIC ACID, 4,4'-BIS((4-ANILINO-6-((2-HYDROXYETHYL)METHYLAMINO)-s-TRIAZIN-2-YL)AMINO)-, DISODIUM SALT

TOXICITY DATA with REFERENCE
eye-rbt 100 mg MOD TXAPA9 27,494,74
eye-rbt 100 mg/2S RNS MLD TXAPA9 27,494,74
orl-rbt TDLo:110 mg/kg (female 6-16D post):REP TXAPA9 27,494,74

CONSENSUS REPORTS: Reported in EPA TSCA Inventory.

SAFETY PROFILE: Experimental reproductive effects. An eye irritant. When heated to decomposition it emits toxic fumes of NO_x and SO_x.

TGE150 CAS:27344-41-8 ***HR: 2***
TINOPAL CBS
mf: $C_{28}H_{20}O_6S_2 \cdot 2Na$ mw: 562.58

SYNS: 2,2'-((1,1'-BIPHENYL)-4,4'-DIYLDI-2,1-ETHENEDIYL)BIS-BENZENESULFONIC ACID DISODIUM SALT ◇ DISODIUM-4,4'-BIS(2-SULFOSTYRYL)BIPHENYL ◇ STILBENE 3 ◇ TINOPAL CBS-X

TOXICITY DATA with REFERENCE
skn-rbt 500 mg/24H MLD MVCRB3 2,193,73
eye-rbt 100 mg SEV TXAPA9 27,494,74
eye-rbt 100 mg/25 rns MLD TXAPA9 27,494,74
orl-rbt TDLo:110 mg/kg (6-16D preg):REP APFRAD 30,415,72
orl-rat LD50:5580 mg/kg MVCRB3 2,193,73

CONSENSUS REPORTS: EPA Genetic Toxicology Program. Reported in EPA TSCA Inventory.

SAFETY PROFILE: Mildly toxic by ingestion. A skin and severe eye irritant. Experimental reproductive effects. When heated to decomposition it emits toxic fumes of SO_x and Na_2O.

TGE155 CAS:6416-68-8 ***HR: 1***
TINOPAL RBS
mf: $C_{24}H_{16}N_3O_3S \cdot Na$ mw: 449.48

SYNS: C.I. 40645 ◇ C.I. FLUORESCENT BRIGHTENER 46 ◇ C.I. FLUORESCENT BRIGHTENING AGENT 46, SODIUM SALT ◇ FLUORESCENT BRIGHTENER 46 ◇ 4-(2H-NAPHTHO(1,2-d)TRIAZOL-2-YL)-2-STILBENESULFONIC ACID SODIUM SALT ◇ 2-(4-(2-PHENYL-ETHENYL)-3-SULFOPHENYL)-2H-NAPHTHO(1,2-d)TRIAZOLESODIUM SALT ◇ SODIUM-2-(4-STYRYL-3-SULFOPHENYL)-2H-NAPHTHO-(1,2-d)-TRIAZOLE ◇ TINOPAL RBS 200

TOXICITY DATA with REFERENCE
skn-rbt 500 mg/24H MLD MVCRB3 2,193,73
eye-rbt 100 mg MOD TXAPA9 27,494,74
eye-rbt 100 mg/2S rns MLD TXAPA9 27,494,74
eye-rbt 100 mg MLD MVCRB3 2,193,73
ihl-rat LCLo:4500 mg/m³/4H TXAPA9 27,494,74

CONSENSUS REPORTS: EPA Genetic Toxicology Program. Reported in EPA TSCA Inventory.

SAFETY PROFILE: Mildly toxic by inhalation. A skin and eye irritant. When heated to decomposition it emits toxic fumes of SO_x, NO_x, and Na_2O.

TGE165 CAS:25913-34-2 ***HR: 2***
TINORIDINE HYDROCHLORIDE
mf: $C_{17}H_{20}N_2O_2S \cdot ClH$ mw: 352.91

SYNS: 2-AMINO-3-ETHOXYCARBONYL-6-BENZYL-4,5,6,7-TETRAHYDROTHIENO(2,3-c)PYRIDINE HYDROCHLORIDE ◇ ETHYL 2-AMINO-6-BENZYL-4,5,6,7-TETRAHYDROTHIENO(2,3-c)PYRIDINE-3-CARBOXYLATE HYDROCHLORIDE ◇ NONFLAMIN ◇ Y-3642-HCl ◇ Y-3642 HYDROCHLORIDE

TOXICITY DATA with REFERENCE
orl-mus TDLo:600 mg/kg (7-12D preg):REP YKKZAJ 90,1447,70
orl-rat TDLo:4900 mg/kg (9-14D preg):TER YKKZAJ 90,1447,70
orl-rat LD50:1200 mg/kg YKKZAJ 99,240,79
ipr-rat LD50:1350 mg/kg JMGZAI 8(8),7,71
orl-mus LD50:1601 mg/kg YACHDS 2,1028,74
ipr-mus LD50:510 mg/kg JMGZAI 8(8),7,71

SAFETY PROFILE: Moderately toxic by ingestion and intraperitoneal routes. An experimental teratogen. Experimental reproductive effects. When heated to decomposition it emits toxic fumes of SO_x, NO_x, and HCl.

TGE250 CAS:814-94-8 ***HR: 2***
TIN(II) OXALATE
mf: $C_2H_2O_4 \cdot Sn$ mw: 208.73

PROP: Powder. Mp: 280° (decomp), d: 3.56.

SYNS: ETHANEDIOIC ACID, TIN(2+) SALT (1:1) (9CI) ◇ OXALIC ACID, TIN(2+) SALT (1:1) (8CI) ◇ STANNOUS OXALATE ◇ STAVELAN CINATY (CZECH) ◇ TIN(2+) OXALATE ◇ TIN OXALATE

TOXICITY DATA with REFERENCE
orl-rat LD50:3620 mg/kg MarJV# 29MAR77

CONSENSUS REPORTS: Reported in EPA TSCA Inventory.

OSHA PEL: TWA 2 mg(Sn)/m³
ACGIH TLV: TWA 2 mg(Sn)/m³

SAFETY PROFILE: Moderately toxic by ingestion. When heated to decomposition it emits acrid smoke and irritating fumes. See also TIN COMPOUNDS and OXALATES.

TGE300 CAS:1332-29-2 ***HR: 2***
TIN OXIDE
mf: O_2Sn mw: 150.69

SYNS: MESA ◇ STANNOXYL

CONSENSUS REPORTS: Reported in EPA TSCA Inventory.

OSHA PEL: TWA 2 mg/m^3
ACGIH TLV: TWA 2 mg(Sn)/m_3

SAFETY PROFILE: When heated to decomposition it emits acrid smoke and irritating fumes. See also TIN COMPOUNDS.

TGE500 CAS:25324-56-5 ***HR: 3***
TIN (IV) PHOSPHIDE
mw: PSn mw: 149.66
DOT: UN 1433

PROP: Silver-white crystals. D: 6.56.

SYN: STANNIC PHOSPHIDE (DOT)

CONSENSUS REPORTS: Reported in EPA TSCA Inventory.

OSHA PEL: TWA 2 mg(Sn)/m^3
ACGIH TLV: TWA 2 mg(Sn)/m^3

DOT Classification: Flammable Solid; Label: Flammable Solid and Dangerous When Wet.

SAFETY PROFILE: A flammable solid. Reacts with moisture or acid fumes to liberate highly toxic phosphine gas. When heated to decomposition it emits toxic fumes of PO_x. See also PHOSPHINE, PHOSPHIDES, and TIN COMPOUNDS.

TGE750 CAS:73926-79-1 ***HR: 3***
TIN POTASSIUM TARTRATE

SYN: STANNOUS POTASSIUM TARTRATE

TOXICITY DATA with REFERENCE
scu-mus LDLo:70 mg(Sn)/kg AEPPAE 114,39,26
scu-rbt LDLo:70 mg(Sn)/kg AEPPAE 114,39,26
scu-gpg LDLo:70 mg(Sn)/kg AEPPAE 114,39,26

OSHA PEL: TWA 2 mg(Sn)/m^3
ACGIH TLV: TWA 2 mg(Sn)/m^3

SAFETY PROFILE: Poison by subcutaneous route. When heated to decomposition it emits toxic fumes of K_2O. See also TIN COMPOUNDS.

TGF000 CAS:72378-89-3 ***HR: 3***
TIN SODIUM TARTRATE

SYN: SODIUM STANNOUS TARTRATE

TOXICITY DATA with REFERENCE
ivn-rat LDLo:52400 μg/kg EQSSDX 1,1,75
ivn-cat LDLo:105 mg/kg EQSSDX 1,1,75
scu-rbt LDLo:65500 μg/kg EQSSDX 1,1,75

OSHA PEL: TWA 2 mg(Sn)/m^3
ACGIH TLV: TWA 2 mg(Sn)/m^3

SAFETY PROFILE: Poison by intravenous and subcutaneous routes. When heated to decomposition it emits toxic fumes of Na_2O. See also TIN COMPOUNDS.

TGF025 CAS:102577-46-8 ***HR: 1***
TINUVIN 1130

TOXICITY DATA with REFERENCE
eye-rbt 100 mg MLD EPASR* 8EHQ-0986-0627

SAFETY PROFILE: An eye irritant. When heated to decomposition it emits acrid smoke and irritating fumes.

TGF050 CAS:65899-73-2 ***HR: 2***
TIOCONAZOLE
mf: $C_{16}H_{13}Cl_3N_2OS$ mw: 387.72

SYN: (±)-1-(2,4-DICHLORO-β-((2-CHLORO-3-ETHENYL)OXY)PHENETHYL)IMIDAZOLE

TOXICITY DATA with REFERENCE
scu-rat TDLo:330 mg/kg (female 7-17D post):REP YACHDS 10,3849,82
orl-rat LD50:770 mg/kg YKYUA6 35,1623,84
ipr-rat LD50:730 mg/kg YKYUA6 35,1623,84
orl-mus LD50:1870 mg/kg YKYUA6 35,1623,84
ipr-mus LD50:508 mg/kg YKYUA6 35,1623,84

SAFETY PROFILE: Moderately toxic by ingestion and intraperitoneal routes. Experimental reproductive effects. When heated to decomposition it emits toxic fumes of Cl^-, SO_x, and NO_x.

TGF075 CAS:71731-58-3 ***HR: 3***
TIQUIZIUM BROMIDE
mf: $C_{19}H_{24}NS_2{\bullet}Br$ mw: 410.47

PROP: Needles from methanol-acetone. Mp: 278-281° (decomp).

SYNS: HSR-902 ◇ 3-(DI(2-THIENYL)METHYLENE)-5-METHYLDECAHYDROQUINOLIZANIUM BROMIDE ◇ 3-(DI-2-THIENYLMETHYLENE)-5-METHYL-trans-QUINOLIZIDINIUMBROMIDE ◇ THIATON

TOXICITY DATA with REFERENCE
orl-rat TDLo:1350 mg/kg (female 17-22D post):REP KSRNAM 15,6215,81
orl-rbt TDLo:3900 mg/kg (female 6-18D post):TER AMBNAS 28,7,80
orl-rat LD50:1177 mg/kg OYYAA2 23,1,82
ipr-rat LD50:70500 μg/kg OYYAA2 23,1,82
scu-rat LD50:169 mg/kg OYYAA2 23,1,82

ivn-rat LD50:11400 μg/kg OYYAA2 23,1,82
orl-mus LD50:578 mg/kg OYYAA2 23,1,82
ipr-mus LD50:83700 μg/kg OYYAA2 23,1,82
scu-mus LD50:82100 μg/kg OYYAA2 23,1,82
ivn-mus LD50:10300 μg/kg OYYAA2 23,1,82
orl-dog LD50:662 mg/kg OYYAA2 23,1,82
ipr-dog LD50:56700 μg/kg OYYAA2 23,1,82
scu-dog LD50:129 mg/kg OYYAA2 23,1,82
ivn-dog LD50:14200 μg/kg OYYAA2 23,1,82

SAFETY PROFILE: Poison by subcutaneous, intravenous and intraperitoneal routes. Moderately toxic by ingestion. An experimental teratogen. Experimental reproductive effects. When heated to decomposition it emits toxic fumes of Br^-, NO_x, and SO_x.

TGF175 CAS:55837-29-1 ***HR: 3***
TIROPRAMIDE
mf: $C_{28}H_{41}N_3O_3$ mw: 467.72

PROP: Crystals from petr ether. Mp: 65-67°.

SYNS: dl-α-BENZAMIDO-p-(2-(DIETHYLAMINO)ETHOXY)-N,N-DIPROPYLHYDROCINNAMAMIDE ◇ (±)-α-(BENZOYLAMINO)-4-(2-(DIETHYLAMINO)ETHOXY)-N,N-DIPROPYLBENZENEPROPANAMIDE ◇ CR-605 ◇ o-(DIETHYLAMINOETHYL)-N-BENZOYL-dl-TYROSIL-DI-n-PROPYLAMIDE ◇ MAIORAD ◇ TIROPRAMIDA (SPANISH)

TOXICITY DATA with REFERENCE
orl-rat LD50:800 mg/kg DRFUD4 7,413,82
ipr-rat LD50:140 mg/kg DRFUD4 7,413,82
scu-rat LD50:430 mg/kg DRFUD4 7,413,82
ivn-rat LD50:33900 μg/kg USXXAM #4004008
orl-mus LD50:550 mg/kg DRFUD4 7,413,82
ipr-mus LD50:120 mg/kg DRFUD4 7,413,82
scu-mus LD50:245 mg/kg DRFUD4 7,413,82
ivn-mus LD50:40500 μg/kg DRFUD4 7,413,82

SAFETY PROFILE: Poison by subcutaneous, intravenous, and intraperitoneal routes. Moderately toxic by ingestion. When heated to decomposition it emits toxic fumes of NO_x.

TGF200 CAS:25168-15-4 ***HR: D***
2,4,5-T ISOOCTYL ESTER
mf: $C_{16}H_{21}Cl_3O_3$ mw: 367.72

SYNS: 2,4,5-TRICHLOROPHENOXY, ACETIC ACID, ISOOCTYL ESTER ◇ U 46T

TOXICITY DATA with REFERENCE
orl-mus TDLo:1471 mg/kg (12-15D preg):TER
AECTCV 6,33,77

SAFETY PROFILE: An experimental teratogen. When heated to decomposition it emits toxic fumes of Cl^-. See also 2,4,5-TRICHLOROPHENOXY ACETIC ACID.

TGF210 CAS:93-78-7 ***HR: 2***
2,4,5-T, ISOPROPYL ESTER
mf: $C_{11}H_{11}Cl_3O_3$ mw: 297.57

SYNS: (2,4,5,-TRICHLOROPHENOXY)ACETIC ACID, ISOPROPYL ESTER ◇ (2,4,5,-TRICHLOROPHENOXY)ACETIC ACID-1-METHYL ESTER (9CI)

TOXICITY DATA with REFERENCE
orl-rat LD50:495 mg/kg AJVRAH 15,622,54
orl-mus LD50:551 mg/kg AJVRAH 15,622,54
orl-gpg LD50:449 mg/kg AJVRAH 15,622,54

SAFETY PROFILE: Moderately toxic by ingestion. When heated to decomposition it emits toxic fumes of Cl^-. See also 2,4,5-TRICHLOROPHENOXYACETIC ACID and ESTERS.

TGF250 CAS:7440-32-6 ***HR: 3***
TITANIUM
DOT: UN 1352/UN 2546/UN 2878
af: Ti aw: 47.90

PROP: Dark gray amorphous powder or lustrous white metal. D: 4.5 @ 20°, autoign temp: 1200° for solid metal in air, 250° for powder, mp: 1677°, bp: 3277°.

SYNS: CONTIMET 30 ◇ C.P. TITANIUM ◇ IMI 115 ◇ NCI-C04251 ◇ OREMET ◇ TITANIUM ALLOY ◇ TITANIUM METAL POWDER, DRY (DOT) ◇ TITANIUM SPONGE GRANULES (DOT) ◇ TITANIUM SPONGE POWDERS (DOT)

TOXICITY DATA with REFERENCE
ims-rat TDLo:114 mg/kg/77W-I:ETA NCIUS* PH 43-64-886,JUL,68

CONSENSUS REPORTS: Reported in EPA TSCA Inventory.

DOT Classification: Flammable Solid; Label: Flammable Solid (UN1352, UN2546, UN2878); Flammable Solid; Label: Spontaneously Combustible (UN2546).

SAFETY PROFILE: Questionable carcinogen with experimental tumorigenic data. The dust may ignite spontaneously in air. Flammable when exposed to heat or flame or by chemical reaction. Titanium can burn in an atmosphere of carbon dioxide, nitrogen or air. Also reacts violently with BrF_3; CuO; PbO; (Ni + $KClO_3$); metaloxy salts; halocarbons; halogens; CO_2; metal carbonates; Al; water; AgF; O_2; nitryl fluoride; HNO_3; O_2; $KClO_3$; KNO_3; $KMnO_4$; steam @ 704°; trichloroethylene; trichlorotri-fluoroethane. Ordinary extinguishers are often ineffective against titanium fires. Such fires require special extinguishers designed for metal fires. In airtight enclosures, titanium fires can be controlled by the use of argon or helium. titanium, in the absence of moisture, burns slowly, but evolves much heat. The application of water to burning titanium can cause an explosion. Finely divided titanium dust and powders, like most metal powders, are potential explosion hazards when exposed to sparks, open flame or high heat sources. See also TITANIUM COMPOUNDS, POWDERED METALS, and MAGNESIUM.

TGF500 CAS:7440-32-6 ***HR: D***
TITANIUM (wet powder)

SYN: TITANIUM METAL POWDER, WET (DOT)

TOXICITY DATA with REFERENCE
orl-rat TDLo:158 mg/kg (MGN):TER AEHLAU 23,102,71

CONSENSUS REPORTS: Reported in EPA TSCA Inventory.

SAFETY PROFILE: An experimental teratogen. See also TITANIUM COMPOUNDS and TITANIUM.

TGF750 ***HR: 3***
TITANIUM AZIDE TRICHLORIDE
mf: Cl_3N_3Ti mw: 196.27

SAFETY PROFILE: Explosive solid. When heated to decomposition it emits very toxic fumes of Cl^- and NO_x. See also TITANIUM COMPOUNDS and AZIDES.

TGG000 CAS:12070-08-5 ***HR: 3***
TITANIUM CARBIDE
mf: CTi mw: 59.91
PROP: Solid. D: 4.930.

SAFETY PROFILE: The dust may explode if exposed to spark. Incompatible with air. See also TITANIUM COMPOUNDS.

TGG250 CAS:7705-07-9 ***HR: 3***
TITANIUM CHLORIDE
mf: Cl_3Ti mw: 154.25
DOT: UN 2441/UN 2869

PROP: Colorless to light yellow liquid; fumes in moist air. Mp: −30°, bp: 136.4°, d: 1.772 @ 25/25°, vap press: 10 mm @ 21.3°.

SYNS: TAC 121 ◇ TAC 131 ◇ TITANIUM (III) CHLORIDE ◇ TITANIUM TRICHLORIDE ◇ TITANIUM TRICHLORIDE, PYROPHORIC (DOT) ◇ TITANOUS CHLORIDE ◇ TRICHLORO TRITANIUM

TOXICITY DATA with REFERENCE
itt-rat TLDo:12341 μg/kg (1D male):REP JRPFA4 7,21,64

CONSENSUS REPORTS: Reported in EPA TSCA Inventory.

DOT Classification: Flammable Solid; Label: Spontaneously Combustible, Corrosive (UN2441); Corrosive Material; Label: Corrosive (UN2869).

SAFETY PROFILE: A corrosive irritant to skin, eyes, and mucous membranes. A severe corrosive because it liberates heat and hydrochloric acid upon contact with moisture. If spilled on skin, wipe off with dry cloth before applying water. May ignite spontaneously in air. Flammable when exposed to heat or flame. Reacts violently with K; HF. Experimental reproductive effects. When heated to decomposition it emits toxic fumes of Cl^-. See also TITANIUM COMPOUNDS.

TGG500 ***HR: D***
TITANIUM COMPOUNDS

SAFETY PROFILE: This material is generally considered to be physiologically inert. There are no reported cases in the literature where titanium as such has caused human intoxication. The dusts of titanium or most titanium compounds such as titanium oxide may be placed in the nuisance category. Titanium tetrachloride, however, is an irritant and corrosive material, because when exposed to moisture, it hydrolyzes to hydrogen chloride. See also TITANIUM and specific compounds.

TGG600 CAS:32006-07-8 ***HR: 3***
TITANIUM DIAZIDE DIBROMIDE
mf: $TiBr_2N_6$ mw: 291.73

SAFETY PROFILE: A highly explosive solid. When heated to decomposition it emits toxic fumes of Br^- and NO_x. See also TITANIUM COMPOUNDS, AZIDES, and BROMIDES.

TGG625 CAS:13783-04-5 ***HR: 3***
TITANIUM DIBROMIDE
mf: $TiBr_2$ mw: 207.69

SAFETY PROFILE: May ignite spontaneously in moist air. When heated to decomposition it emits toxic fumes of Br^-. See also TITANIUM COMPOUNDS and BROMIDES.

TGG750 CAS:10049-06-6 ***HR: 3***
TITANIUM DICHLORIDE
mf: Cl_2Ti mw: 118.81

PROP: Black crystals. Mp: 1035°, d: 3.13. Decomp by water; sol in alc; almost insol in chloroform, ether, carbon disulfide.

SAFETY PROFILE: Ignites spontaneously in air. Violent reaction with water liberates explosive hydrogen gas. When heated to decomposition it emits toxic fumes of Cl^-. See also TITANIUM COMPOUNDS.

TGG760 CAS:13463-67-7 ***HR: 1***
TITANIUM DIOXIDE
mf: O_2Ti mw: 79.90

PROP: White amorphous powder. Mp: 1860° (decomp),

d: 4.26. Insol in water, hydrochloric acid, dil sulfuric acid, and alc.

SYNS: 1700 WHITE ◇ A-FIL CREAM ◇ ATLAS WHITE TITANIUM DIOXIDE ◇ AUSTIOX ◇ BAYERITIAN ◇ BAYERTITAN ◇ BAYTITAN ◇ CALCOTONE WHITE T ◇ C.I. 77891 ◇ C.I. PIGMENT WHITE 6 ◇ COSMETIC WHITE C47-5175 ◇ C-WEISS 7 (GERMAN) ◇ FLAMENCO ◇ HOMBITAN ◇ HORSE HEAD A-410 ◇ KH 360 ◇ KRONOS TITANIUM DIOXIDE ◇ LEVANOX WHITE RKB ◇ NCI-C04240 ◇ RAYOX ◇ RUNA RH20 ◇ RUTILE ◇ TIOFINE ◇ TIOXIDE ◇ TITANDIOXID (SWEDEN) ◇ TITANIUM OXIDE ◇ TRIOXIDE(S) ◇ TRONOX ◇ UNITANE O-110 ◇ ZOPAQUE

TOXICITY DATA with REFERENCE
skn-hmn 300 μg/3D-I MLD 85DKA8 -,127,77
ihl-rat TCLo:250 mg/m^3/6H/2Y-I:CAR TXAPA9 79,179,85
ims-rat TDLo:360 mg/kg/2Y-I:NEO NCIUS* PH 43-64-886,JUL,68
ims-rat TD:260 mg/kg/84W-I:ETA NCIUS* PH 43-64-886,AUG,69

CONSENSUS REPORTS: NCI Carcinogenesis Bioassay (feed); No Evidence: mouse, rat NCITR* NCI-CG-TR-97,79. Reported in EPA TSCA Inventory. EPA Genetic Toxicology Program.

OSHA PEL: (Transitional: TWA Total Dust: 15 mg/m^3; Respirable Fraction: 5 mg/m^3) TWA Total Dust: 10 mg/m^3; Respirable Fraction: 5 mg/m^3
ACGIH TLV: TWA (nuisance particulate) 10 mg/m^3 of total dust (when toxic impurities are not present, e.g., quartz < 1%).
DFG MAK: 6 mg/m^3

SAFETY PROFILE: A human skin irritant. Questionable carcinogen with experimental carcinogenic, neoplastigenic, and tumorigenic data. A common air contaminant and nuisance dust. Violent or incandescent reaction with metals at high temperatures (e.g., aluminum; calcium; magnesium; potassium; sodium; zinc; lithium). See also TITANIUM COMPOUNDS.

TGG775 CAS:7245-18-3 ***HR: 3***
TITANIUM(III) METHOXIDE
mf: $C_3H_9O_3Ti$ mw: 140.98

$$Ti(OCH_3)_3$$

SAFETY PROFILE: Ignites spontaneously in air. When heated to decomposition it emits acrid smoke and irritating fumes. See also TITANIUM COMPOUNDS.

TGH250 CAS:13825-74-6 ***HR: 2***
TITANIUM SULFATE (solution)
DOT: NA 1760

CONSENSUS REPORTS: Reported in EPA TSCA Inventory.

DOT Classification: Corrosive Material; Label: Corrosive.

SAFETY PROFILE: A corrosive irritant to skin, eyes and mucous membranes. When heated to decomposition it emits toxic fumes of SO_x. See also TITANIUM COMPOUNDS and SULFATES.

TGH350 CAS:7550-45-0 ***HR: 3***
TITANIUM TETRACHLORIDE
DOT: UN 1838
mf: Cl_4Ti mw: 189.70

PROP: Colorless liquid, penetrating acid odor. D: 1.726, mp: -24.1°, bp: 136.4°. Sol in cold water and alc; dec by hot water.

SYNS: TETRACHLORURE de TITANE (FRENCH) ◇ TITAANTETRACHLORED (DUTCH) ◇ TITANE (TETRACHLORURE de) (FRENCH) ◇ TITANIO TETRACHLORURO di (ITALIAN) ◇ TITANIUM CHLORIDE ◇ TITANTETRACHLORID (GERMAN)

TOXICITY DATA with REFERENCE
ihl-rat LC50:460 mg/m^3/4H TOXID9 1,76,81
ihl-mus LC50:100 mg/m^3/2H 85GMAT -,111,82

CONSENSUS REPORTS: EPA Extremely Hazardous Substances List. Community Right-To-Know List. Reported in EPA TSCA Inventory.

DOT Classification: Corrosive Material; Label: Corrosive.

SAFETY PROFILE: Poison by inhalation. A corrosive irritant to skin, eyes, and mucous membranes. When heated to decomposition it emits toxic fumes of Cl^-. See also TITANIUM COMPOUNDS.

TGH500 CAS:1271-29-0 ***HR: 3***
TITANOCENE
mf: $C_{10}H_{10}Ti$ mw: 178.10

SYN: DI-pi-CYCLOPENTADIENYLTITANIUM

TOXICITY DATA with REFERENCE
ims-mus TDLo:75 mg/kg:ETA NCIUS* PH 43-64-886, SEPT,65
ims-rat LDLo:50 mg/kg NCIUS* PH-43-64-886
ims-ham LDLo:83 mg/kg NCIUS* PH-43-64-886

SAFETY PROFILE: Poison by intramuscular route. Questionable carcinogen with experimental tumorigenic data. When heated to decomposition it emits acrid smoke and irritating fumes. See also TITANIUM COMPOUNDS.

TGH600 CAS:64461-82-1 ***HR: 3***
TIZANIDINE HYDROCHLORIDE
mf: $C_9H_8ClN_5S{\cdot}ClH$ mw: 290.19

SYN: 5-CHLORO-4-(2-IMIDAZOLIN-2-YLAMINO)-2,1,3-BENZOTHIADIAZOLE HYDROCHLORIDE

TOXICITY DATA with REFERENCE
orl-rat TDLo:11 mg/kg (female 7-17D post):REP KSRNAM 19,5825,85
orl-rat TDLo:110 mg/kg (female 7-17D post):TER KSRNAM 19,5825,85
orl-rat LD50:414 mg/kg KSRNAM 19,5811,85
scu-rat LD50:282 g/kg KSRNAM 19,5811,85
ivn-rat LD50:35 mg/kg KSRNAM 19,5811,85
orl-mus LD50:235 mg/kg JKXXAF #81-02912

SAFETY PROFILE: Poison by ingestion, subcutaneous, and intravenous routes. An experimental teratogen. Experimental reproductive effects. When heated to decomposition it emits toxic fumes of SO_x, NO_x, and HCl.

TGH650 CAS:64049-85-0 ***HR: 3***
TL-1185
mf: $C_{12}H_{19}N_2O_2 \cdot CH_3O_4S$ mw: 334.43

SYNS: N-METHYLCARBAMIC ACID-2-METHYL-5-DIMETHYLAMINOPHENYL ESTER, METHOSULFATE ◇ METHYLCARBAMIC ACID-5-(TRIMETHYLAMMONIO)-o-TOLYL ESTER, METHYLSULFATE ◇ (3-(METHYLCARBAMOYLOXY)-p-TOLYL)TRIMETHYLAMMONIUM METHYLSULFATE

TOXICITY DATA with REFERENCE
scu-rat LDLo:200 μg/kg NTIS** PB158-508
scu-mus LD50:110 μg/kg NDRC** No.9-4-1-19,44
scu-dog LDLo:200 μg/kg NTIS** PB158-508
scu-cat LDLo:200 μg/kg NTIS** PB158-508
scu-rbt LDLo:200 μg/kg NTIS** PB158-508
scu-gpg LDLo:200 μg/kg NTIS** PB158-508
scu-dom LDLo:300 μg/kg NTIS** PB158-508

SAFETY PROFILE: Poison by subcutaneous route. When heated to decomposition it emits toxic fumes of SO_x, NH_3, and NO_x. See also CARBAMATES.

TGH655 CAS:64050-01-7 ***HR: 3***
TL-1186
mf: $C_{12}H_{19}N_2O_2 \cdot HO_4S$ mw: 320.40

SYNS: N-METHYLCARBAMIC ACID 2-METHYL-5-DIMETHYLAMINOPHENYL ESTER, METHOSULFURIC ACID ◇ METHYLCARBAMIC ACID 5-(TRIMETHYLAMMONIO)-o-TOLYL ESTER, BISULFATE ◇ (3-(METHYLCARBAMOYLOXY)-p-TOLYL)TRIMETHYLAMMONIUM BISULFATE

TOXICITY DATA with REFERENCE
scu-rat LDLo:200 μg/kg NTIS** PB158-508
scu-mus LD50:103 μg/kg NTIS** PB158-508
scu-dog LDLo:200 μg/kg NTIS** PB158-508
scu-rbt LDLo:200 μg/kg NTIS** PB158-508
scu-gpg LDLo:100 μg/kg NTIS** PB158-508

SAFETY PROFILE: Poison by subcutaneous route. When heated to decomposition it emits toxic fumes of SO_x, NH_3, and NO_x. See also CARBAMATES.

TGH660 CAS:64046-38-4 ***HR: 3***
TL-1188
mf: $C_{12}H_{19}N_2O_2 \cdot CH_3O_4S$ mw: 334.43

SYNS: N-METHYLCARBAMIC ACID-4-METHYL-3-DIMETHYLAMINOPHENYL ESTER, METHOSULFATE ◇ METHYLCARBAMIC ACID-3-(TRIMETHYLAMMONIO)-p-TOLYL ESTER, METHYLSULFATE ◇ ((2-METHYL-5-METHYLCARBAMOYLOXY)PHENYL)TRIMETHYLAMMONIUM METHYLSULFATE

TOXICITY DATA with REFERENCE
scu-rat LDLo:200 μg/kg NTIS** PB158-508
scu-mus LD50:200 μg/kg NTIS** PB158-508
scu-dog LDLo:200 μg/kg NTIS** PB158-508
scu-rbt LDLo:200 μg/kg NTIS** PB158-508
scu-gpg LDLo:200 μg/kg NTIS** PB158-508

SAFETY PROFILE: A deadly poison by subcutaneous route. When heated to decomposition it emits toxic fumes of SO_x, NH_3, and NO_x. See also CARBAMATES and ESTERS.

TGH665 CAS:64046-12-4 ***HR: 3***
TL-1299
mf: $C_{13}H_{21}N_2O_2 \cdot Cl$ mw: 272.81

SYNS: AR-16 ◇ N-METHYLCARBAMIC ACID-3-DIETHYLAMINOPHENYL ESTER, METHOCHLORIDE ◇ METHYLCARBAMIC ACID-m-(DIETHYLMETHYLAMMONIO)PHENYL ESTER, CHLORIDE ◇ (3-(N-METHYLCARBAMOYLOXY)PHENYL)DIETHYLMETHYLAMMONIUM CHLORIDE ◇ T-1123

TOXICITY DATA with REFERENCE
scu-mus LD50:90 μg/kg NTIS** PB158-508
ivn-mus LD80:100 μg/kg NTIS** PB158-508
scu-dog LDLo:100 μg/kg NTIS** PB158-508
imp-mky LDLo:100 μg/kg NTIS** PB158-508
imp-dom LDLo:300 μg/kg NTIS** PB158-508

SAFETY PROFILE: A deadly poison by subcutaneous, intravenous, and implant routes. When heated to decomposition it emits toxic fumes of NO_x, NH_3, and Cl^-. See also CARBAMATES and ESTERS.

TGH670 CAS:64046-11-3 ***HR: 3***
TL-1422
mf: $C_{14}H_{23}N_2O_2 \cdot Cl$ mw: 286.84

SYNS: N,N-DIMETHYLCARBAMIC ACID-3-DIETHYLAMINOPHENYL ESTER, METOCHLORIDE ◇ DIMETHYLCARBAMIC ACID-m-(DIETHYLMETHYLAMMONIO)PHENYL ESTER, CHLORIDE ◇ (3-(DIMETHYLCARBAMOYLOXY)PHENYL)DIETHYLMETHYLAMMONIUM CHLORIDE

TOXICITY DATA with REFERENCE
scu-mus LD50:58 μg/kg NTIS** PB158-508
scu-dog LDLo:100 μg/kg NTIS** PB158-508
scu-rbt LDLo:100 μg/kg NTIS** PB158-508

SAFETY PROFILE: A deadly poison by subcutaneous route. When heated to decomposition it emits toxic fumes of Cl^-, NH_3, and NO_x. See also CARBAMATES and ESTERS.

TGH675 CAS:64046-20-4 ***HR: 3***
TL-1434
mf: $C_{13}H_{21}N_2O_2 \cdot Br$ mw: 317.27

SYNS: N-METHYLCARBAMICACID-3-DIMETHYLAMINOPHENYL ESTER, PROPYL BROMIDE ◇ (3-(N-METHYLCARBAMOYLOXY)PHENYL)DIMETHYLPROPYLAMMONIUMBROMIDE

TOXICITY DATA with REFERENCE
scu-rat LDLo:200 μg/kg NTIS** PB158-508
scu-mus LD50:100 μg/kg NTIS** PB158-508
scu-dog LDLo:50 μg/kg NTIS** PB158-508
scu-cat LDLo:50 μg/kg NTIS** PB158-508
scu-rbt LDLo:100 μg/kg NTIS** PB158-508
scu-gpg LDLo:100 μg/kg NTIS** PB158-508

SAFETY PROFILE: A deadly poison by subcutaneous route. When heated to decomposition it emits toxic fumes of Br^-, NH_3, and NO_x. See also CARBAMATES and ESTERS.

TGH680 CAS:64046-07-7 ***HR: 3***
TL-1435
mf: $C_{13}H_{19}N_2O_2 \cdot Br$ mw: 315.25

SYNS: N-METHYLCARBAMICACID-3-DIMETHYLAMINOPHENYL ESTER, ALLYL BROMIDE ◇ (3-(N-METHYLCARBAMOYLOXY)PHENYL)ALLYLDIMETHYLAMMONIUMBROMIDE

TOXICITY DATA with REFERENCE
scu-rat LDLo:200 μg/kg NTIS** PB158-508
scu-mus LD50:102 μg/kg NTIS** PB158-508
scu-dog LDLo:200 μg/kg NTIS** PB158-508
scu-rbt LDLo:200 μg/kg NTIS** PB158-508
scu-gpg LDLo:100 μg/kg NTIS** PB158-508

SAFETY PROFILE: A deadly poison by subcutaneous route. When heated to decomposition it emits toxic fumes of Br^-, NH_3, and NO_x. See also CARBAMATES and ESTERS.

TGH685 CAS:64050-95-9 ***HR: 3***
TL-1448
mf: $C_{14}H_{21}N_2O_2 \cdot I$ mw: 276.27

SYNS: METHYLCARBAMICACID-(4-ALLYLDIMETHYLAMMONIO)-m-TOLYL ESTER, IODIDE ◇ N-METHYLCARBAMIC ACID-3-METHYL-4-DIMETHYLAMINOPHENYL ESTER, ALLYLIODIDE ◇ (4-METHYLCARBAMOYLOXY-o-TOLYL)ALLYLDIMETHYLAMMONIUMIODIDE

TOXICITY DATA with REFERENCE
scu-rat LDLo:500 μg/kg NTIS** PB158-508
scu-mus LD50:240 μg/kg NTIS** PB158-508
scu-rbt LDLo:500 μg/kg NTIS** PB158-508
scu-gpg LDLo:300 μg/kg NTIS** PB158-508

SAFETY PROFILE: A deadly poison by subcutaneous route. When heated to decomposition it emits toxic fumes of I^-, NH_3, and NO_x. See also CARBAMATES and ESTERS.

TGH690 CAS:52207-87-1 ***HR: 3***
TM 3
mf: $C_{17}H_{24}N_4O_2 \cdot ClH$ mw: 352.86

SYN: 2-PHENYL-4-DIMETHYLAMINOPROPYLAMINO-6-ETHOXY-3(2H)-PYRIDAZINONE HYDROCHLORIDE

TOXICITY DATA with REFERENCE
ipr-rat LD50:200 mg/kg NYKZAU 68,442,72
orl-mus LD50:1320 mg/kg NYKZAU 68,442,72
ipr-mus LD50:380 mg/kg NYKZAU 68,442,72
scu-mus LD50:520 mg/kg NYKZAU 68,442,72
ivn-mus LD50:100 mg/kg NYKZAU 68,442,72

SAFETY PROFILE: Poison by intravenous and intraperitoneal routes. Moderately toxic by ingestion and subcutaneous routes. When heated to decomposition it emits toxic fumes of NO_x and HCl.

TGH725 ***HR: D***
TOBACCO

SYNS: BURLEY TOBACCO ◇ NICOTIANA TABACUM ◇ TOBACCO LEAF, NICOTIANA TABACUM

TOXICITY DATA with REFERENCE
mnt-mus-ipr 20 mg/kg CRNGDP 5,501,84
msc-ham:lng 10 mg/L CRNGDP 5,501,84
orl-pig TDLo:1680 mg/kg (11-52D preg):TER AJVRAH 35,1071,74

SAFETY PROFILE: An experimental teratogen. Mutation data reported. See also other tobacco entries, SMOKELESS TOBACCO, and NICOTINE.

TGH750 CAS:8037-19-2 ***HR: 3***
TOBACCO LEAF ABSOLUTE

PROP: An extract of the cured leaves of *Nicotiana affinis* with petroleum or benzene and then with alc (FCTXAV 16,637,78).

SYN: BURLEY TOBACCO

TOXICITY DATA with REFERENCE
skn-gpg 500 mg/24H MLD FCTXAV 16,637,78
skn-mus TDLo:31 g/kg/40W-I:ETA CNREA8 28,2363,68

CONSENSUS REPORTS: Reported in EPA TSCA Inventory.

SAFETY PROFILE: Questionable carcinogen with experimental tumorigenic data. A skin irritant. When heated to decomposition it emits acrid smoke and irritating fumes. See also SMOKELESS TOBACCO, NICOTINE, and other tobacco entries.

TGI000 ***HR: D***
TOBACCO LEAF, NICOTIANA GLAUCA

SYN: NICOTIANA GLAUCA

TOXICITY DATA with REFERENCE
orl-pig TDLo:24 mg/kg (female 30-45D post):TER JTCTDW 20,47,83

SAFETY PROFILE: The smoke produced by burning tobacco contains the highly toxic alkaloid, nicotine, tars and phenols, carbon monoxide, cyanides, nitrates, nitrites, carcinogenic, co-carcinogenic and perhaps 100 other chemicals, alpha-emitters, etc. An experimental teratogen. A nicotine-containing dried leaf of the tobacco plant. Habitual inhalation of tobacco smoke is considered a leading cause of lung cancer and circulatory problems, cardiac problems, etc. Combustible when exposed to heat or flame. See also other tobacco entries, SMOKELESS TOBACCO, and NICOTINE.

TGI100 ***HR: 3***
TOBACCO PLANT

PROP: Large annual or perennial shrubs with leaves that are often broad, hairy, and sticky. The trumpet-shaped flowers are white, yellow, green-yellow or red. The seed capsule holds many small seeds. *N. tabacum* is the principal commercial tobacco in the western countries. *N. rustica*, native to South America, is found sporadically across the United States and is the most widely cultivated tobacco in the Orient. *N. longiflora* is commonly cultivated as a garden ornamental. *N. attenuata* grows in the region bounded by Idaho, Baja California, and Texas. *N. glauca* is native to South America and now grows in the southwestern United States, Hawaii, Mexico, and the West Indies.

SYNS: NICOTIANA ATTENUATA ◇ NICOTIANA GLAUCA ◇ NICOTIANA LONGIFLORA ◇ NICOTIANA RUSTICA ◇ NICOTIANA TABACUM ◇ PAKA (HAWAII) ◇ TABAC (FRENCH) ◇ TABACO (SPANISH)

SAFETY PROFILE: Confirmed human carcinogen by several routes. The whole plant contains poisonous nicotine and other chemically related alkaloids. The primary alkaloid in *N. tabacum* is nicotine. The primary alkaloid in *N. glauca* is anabasine. Ingestion of any part of the plant can cause salivation, nausea, vomiting, distorted perceptions, convulsions vasomotor collapse and respiratory failure. Most serious poisonings result from ingestion of the leaves in salad, use of infusions as enemas, or skin absorption of alkaloids during commercial harvesting. See also SMOKELESS TOBACCO, NICOTINE, and other tobacco entries.

TGI250 CAS:32986-56-4 ***HR: 3***
TOBRAMYCIN
mf: $C_{18}H_{37}N_5O_9$ mw: 467.60

PROP: Water-sol substance. Tobramycin was isolated from a fermentation of *Streptomyces tenebrarius* (TXAPA9 25,398,73).

SYNS: DISTOBRAM ◇ GERNEBCIN ◇ NEBRAMYCIN FACTOR 6 ◇ NF 6 ◇ OBRAMYCIN ◇ TOBRADISTIN ◇ TOBREX

TOXICITY DATA with REFERENCE
ims-mus TDLo:600 mg/kg (7-12D preg):TER NKRZAZ 23,1544,75
ipr-rat LD50:1030 mg/kg JIMRBV 2,100,74
scu-rat LD50:969 mg/kg TXAPA9 22,332,72
ivn-rat LD50:104 mg/kg JIMRBV 2,100,74
ims-rat LD50:913 mg/kg IYKEDH 8,107,77
ipr-mus LD50:445 mg/kg JIMRBV 2,100,74
scu-mus LD50:367 mg/kg TXAPA9 25,398,73
ivn-mus LD50:72500 μg/kg ANTBAL 29,361,84
ims-mus LD50:440 mg/kg YKYUA6 27,1467,76
scu-pig LD50:676 mg/kg TXAPA9 25,398,73

CONSENSUS REPORTS: EPA Genetic Toxicology Program.

SAFETY PROFILE: Poison by intravenous and subcutaneous routes. Moderately toxic by intramuscular and intraperitoneal routes. An experimental teratogen. When heated to decomposition it emits toxic fumes of NO_x. Used as an antibacterial and antibiotic drug. See also TOBRAMYCIN SULPHATE.

TGI500 CAS:49842-07-1 ***HR: 3***
TOBRAMYCIN SULPHATE
mf: $C_{18}H_{37}N_5O_9 \cdot xH_2O_4S$ mw: 1154.16

SYNS: NEBCIN ◇ OBRACINE ◇ TOBRA

TOXICITY DATA with REFERENCE
ivn-wmn LDLo:15 mg/kg/5D-I:CNS JAMAAP 247,1319,82
scu-rat LD50:1680 mg/kg ANTBAL 24,67,79
ivn-rat LD50:126 mg/kg ANTBAL 24,67,79
ipr-mus LD50:262 mg/kg ANTBAL 24,67,79
scu-mus LD50:560 mg/kg ANTBAL 24,67,79
ivn-mus LD50:77 mg/kg ANTBAL 24,67,79

SAFETY PROFILE: Poison by intravenous and intraperitoneal routes. Moderately toxic by subcutaneous route. Human systemic effects by intravenous route: wakefulness and hallucinations and distorted perceptions. When heated to decomposition it emits very toxic fumes of NO_x and SO_x. See also TOBRAMYCIN and SULFATES.

TGI699 CAS:41708-72-9 ***HR: 3***
TOCAINIDE
mf: $C_{11}H_{16}N_2O$ mw: 192.29

SYNS: ALANYL-2,6-XYLIDIDE ◇ 2-AMINO-2',6'-PROPIONOXYLIDIDE ◇ TONOCARD

TOXICITY DATA with REFERENCE
orl-hmn TDLo:5714 μg/kg:CNS,GIT DRFUD4 2,141,77
orl-man TDLo:2606 mg/kg/11W-I:BLD NEJMAG 314,583,86
orl-man LDLo:229 mg/kg:CNS,CVS BMJOAE 288,760,84
orl-mus LD50:529 mg/kg DRFUD4 2,141,77
orl-mus LD50:94 mg/kg JMCMAR 24,1059,81

SAFETY PROFILE: Human and experimental poison by ingestion. Human systemic effects by ingestion: mental disturbances, coma, cardiac damage, nausea or vomiting, and aplastic anemia. When heated to decomposition it emits toxic fumes of NO_x.

TGI725 ***HR: 3***
TOCHERGAMINE
mf: $C_{16}H_{24}N_2O$ mw: 260.42

SYNS: ACETAMIDE, N,N-DIETHYL-N'-(1,2,3,4-TETRAHYDRO-1-NAPHTHYL)- ◇ N,N-DIETHYL-N'-2-(TETRALYL)-GLYCINAMIDE ◇ N,N-DIETHYL-N'-(1,2,3,4-TETRAHYDRO-1-NAPHTHYL)ACETAMIDE ◇ N,N-DIETHYL-N'-2-(TETRAHYDRO-1,2,3,4-NAPHTHYL)-GLYCINAMIDE

TOXICITY DATA with REFERENCE
ims-rat TDLo:100 mg/kg (female 5D post):REP PSEBAA 100,555,59
ivn-rbt LD50:40 mg/kg AIPTAK 96,327,54

SAFETY PROFILE: Poison by intravenous route. Experimental reproductive effects. When heated to decomposition it emits toxic fumes of NO_x.

TGJ000 CAS:50465-39-9 ***HR: 1***
TOCOFIBRATE
mf: $C_{39}H_{59}ClO_4$ mw: 627.43

SYNS: 2-(4-CHLOROPHENOXY)-2-METHYLPROPANOIC ACID, 3,4-DIHYDRO-2,5,7,8-TETRAMETHYL-2-(4,8,12-TRIMETHYLTRIDECYL)-2H-1-BENZOPYRAN-6-YL-ESTER, (2r(4r,8r)) ◇ TOCOFEROL-2-(p-CHLOROPHENOXY)-2-METHYLPROPIONATE ◇ TOCOFIBRATO (SPANISH)

TOXICITY DATA with REFERENCE
orl-rbt TDLo:30 g/kg (30D male):REP OYYAA2 16,1,78
orl-rat LD50:15 g/kg DRFUD4 4,679,79
ipr-rat LD50:15 g/kg DRFUD4 4,679,79
orl-mus LD50:15 g/kg DRFUD4 4,679,79
ipr-mus LD50:15 g/kg DRFUD4 4,679,79

SAFETY PROFILE: Mildly toxic by ingestion. Experimental reproductive effects. When heated to decomposition it emits toxic fumes of Cl^-.

TGJ150 CAS:3778-76-5 ***HR: 3***
TODRALAZINE HYDROCHLORIDE HYDRATE
mf: $C_{11}H_{12}N_4O_2 \cdot ClH \cdot H_2O$ mw: 286.75

SYNS: APIRACHOL ◇ APIRACOHL ◇ BINAZIN ◇ BINAZINE ◇ 621-BT HYDROCHLORIDE HYDRATE ◇ ECARAZINE ◇ ECARAZINE HYDROCHLORIDE ◇ TODRALAZINE HYDROCHLORIDE

TOXICITY DATA with REFERENCE
mmo-sat 500 μg/plate RCOCB8 49,415,85
mma-sat 1 mg/plate RCOCB8 49,415,85
orl-mus TDLo:210 mg/kg (7-13D preg):TER YIKUAO 18,29,69
orl-rat LD50:598 mg/kg YIKUAO 18,29,69
ipr-rat LD50:200 mg/kg YIKUAO 18,29,69
scu-rat LD50:466 mg/kg YIKUAO 18,29,69
ivn-rat LD50:240 mg/kg YIKUAO 18,29,69
orl-mus LD50:516 mg/kg OYYAA2 3,97,69
ipr-mus LD50:501 mg/kg YIKUAO 18,29,69
scu-mus LD50:319 mg/kg OYYAA2 3,97,69
ivn-mus LD50:300 mg/kg OYYAA2 3,97,69
orl-rbt LD50:399 mg/kg YIKUAO 18,29,69
ipr-rbt LD50:450 mg/kg YIKUAO 18,29,69
ivn-rbt LD50:417 mg/kg YIKUAO 18,29,69

SAFETY PROFILE: Poison by ingestion, subcutaneous, intravenous, and intraperitoneal routes. An experimental teratogen. Mutation data reported. When heated to decomposition it emits toxic fumes of NO_x and HCl. See also ESTERS.

TGJ250 CAS:10488-36-5 ***HR: 3***
TOFENACINE HYDROCHLORIDE
mf: $C_{17}H_{21}NO \cdot ClH$ mw: 291.85

PROP: White crystals. Mp: 143-147°. Very sol in water, methanol, ethanol, chloroform; sltly sol in acetone; almost insol in ethyl ether.

SYNS: BS 7331 ◇ N-DEMETHYLORPHENADRINE HYDROCHLORIDE ◇ ELAMOL ◇ N-METHYLAMINOETHYL-2-METHYLBENZHYDRYL ETHER HYDROCHLORIDE ◇ N-METHYL-2-((o-METHYL-α-PHENYLBENZYL)OXY)ETHYLAMINE HYDROCHLORIDE ◇ N-METHYL-2-((2-METHYLPHENYL)PHENYLMETHOXY)ETHANAMINE HYDROCHLORIDE ◇ N-METHYL-2-(α-(2-TOLYLBENZYL)OXY) ETHYLAMINE HYDROCHLORIDE ◇ (2-(PHENYL-o-TOLYLMETHOXY)ETHYL)METHYLAMINE HYDROCHLORIDE ◇ PHENYL-(o-TOLYLMETHYL) METHYLAMINOETHYL ETHER HYDROCHLORIDE ◇ TOFENACIN HYDROCHLORIDE

TOXICITY DATA with REFERENCE
orl-rat LD50:400 mg/kg 27ZQAG -,375,72
ipr-rat LD50:72 mg/kg AIPTAK 177,28,69
ivn-rat LD50:15 mg/kg 27ZQAG -,375,72
orl-mus LD50:182 mg/kg AIPTAK 177,28,69
ipr-mus LD50:55 mg/kg AIPTAK 177,28,69
scu-mus LD50:82 mg/kg AIPTAK 177,28,69
ivn-mus LD50:32 mg/kg 27ZQAG -,375,72
orl-dog LD50:90 mg/kg 27ZQAG -,375,72
scu-gpg LD50:92 mg/kg AIPTAK 177,28,69

SAFETY PROFILE: Poison by ingestion, intraperitoneal, intravenous and subcutaneous routes. When heated to decomposition it emits very toxic fumes of HCl and NO_x. Used as an antidepressant.

TGJ350 CAS:8064-38-8 ***HR: 2***
TOGAL
mf: $C_{20}H_{24}N_2O_2 \cdot C_9H_8O_4 \cdot C_6H_5O_7 \cdot 2ClH \cdot 3Li$ mw: 786.47

SYN: CINCHONAN-9-OL, 6′-METHOXY-, DIHYDROCHLORIDE, (8-α-9R)-, mixt. with 2-(ACETYLOXY) BENZOIC ACID and 2-HYDROXY-1,2,3-PROPANETRICARBOXYLIC ACID, TRILITHIUM SALT

TOXICITY DATA with REFERENCE
orl-mus TDLo:10200 mg/kg (female 1-17D post):TER ARZNAD 24,1317,74
orl-mus LDLo:500 mg/kg ARZNAD 24,1305,74

SAFETY PROFILE: Moderately toxic by ingestion. An experimental teratogen. When heated to decomposition it emits toxic fumes of NO_x, Li, and HCl^-.

TGJ475 CAS:4024-34-4 ***HR: 3***
TOLADRYL
mf: $C_{18}H_{23}NO \cdot ClH$ mw: 305.88

SYNS: N,N-DIMETHYL-2-(α-(p-TOLYL)BENZYLOXY)ETHYLAMINE HYDROCHLORIDE ◇ NEO-BENODINE

TOXICITY DATA with REFERENCE
orl-mus LD50:350 mg/kg ARZNAD 14,964,64
ipr-mus LD50:150 mg/kg JPETAB 112,318,54
ivn-mus LD50:45 mg/kg ARZNAD 14,964,64

SAFETY PROFILE: Poison by ingestion, intravenous, and intraperitoneal routes. When heated to decomposition it emits toxic fumes of NO_x and HCl.

TGJ500 CAS:1156-19-0 ***HR: 2***
TOLAZAMIDE
mf: $C_{14}H_{21}N_3O_3S$ mw: 311.44

PROP: Crystals. Mp: 170-173°.

SYNS: N-((HEXAHYDRO-1H-AZEPIN-1-YL)-AMINO)CARBONYL)-4-METHYLBENZENESULFONAMIDE ◇ 1-(HEXAHYDRO-1-AZEPINYL)-3-p-TOLYLSULFONYLUREA ◇ 1-(HEXAHYDRO-1H-AZEPIN-1-YL)-3-(p-TOLYLSULFONYL)UREA ◇ NCI-C03327 ◇ NSC-70762 ◇ TOLINASE ◇ N-(p-TOLUENESULFONYL)-N′-HEXAMETHYLENIMINOUREA ◇ 4-(p-TOLYLSULFONYL)-1,1-HEXAMETHYLENESEMICARBAZIDE ◇ U-17835

TOXICITY DATA with REFERENCE
orl-mus LD50:1000 mg/kg JMCMAR 24,1521,81
ipr-mus LD50:1000 mg/kg JMCMAR 24,1521,81

CONSENSUS REPORTS: NCI Carcinogenesis Bioassay (feed); No Evidence: mouse, rat NCITR* NCI-CG-TR-51,78.

SAFETY PROFILE: Moderately toxic by ingestion and intraperitoneal routes. When heated to decomposition it emits very toxic fumes of NO_x and SO_x.

TGJ625 CAS:33697-73-3 ***HR: 3***
TOLCAINE HYCROCHLORIDE
mf: $C_{18}H_{29}NO_2 \cdot ClH$ mw: 327.94

SYNS: (−)-LOTUCAINE HYDROCHLORIDE ◇ MY 33-7 ◇ (−)-2,2,5,5-TETRAMETHYL-α-((o-TOLYLOXY)METHYL)-1-PYRROLIDINE-ETHANOL HYDROCHLORIDE ◇ (−)-1-(o-TOLOSSI-3-(2,2,5,5-TETRAMETIL-PIRROLIDIN-1-IL)-PROPAN-2-OLOCLORIDRATO (ITALIAN)

TOXICITY DATA with REFERENCE
orl-rat LD50:418 mg/kg BCFAAI 110,330,71
scu-rat LD50:795 mg/kg BCFAAI 110,330,71
ivn-rat LD50:32 mg/kg BCFAAI 110,330,71
orl-mus LD50:355 mg/kg BCFAAI 110,330,71
scu-mus LD50:729 mg/kg BCFAAI 110,330,71
ivn-mus LD50:22 mg/kg BCFAAI 110,330,71
ivn-cat LDLo:14 mg/kg ARZNAD 23,1596,73

SAFETY PROFILE: Poison by ingestion and intravenous routes. Moderately toxic by subcutaneous route. When heated to decomposition it emits toxic fumes of NO_x and HCl.

TGJ750 CAS:119-93-7 ***HR: 3***
o-TOLIDINE
mf: $C_{14}H_{16}N_2$ mw: 212.32

PROP: White to reddish crystals. Mp: 129-131°C. Very sltly sol in water; sol in alc, ether, acetic acid.

SYNS: BIANISIDINE ◇ 4,4′-BI-o-TOLUIDINE ◇ C.I. 37230 ◇ C.I. AZOIC DIAZO COMPONENT 113 ◇ (4,4′-DIAMINE-3,3′-DIMETHYL(1,1′-BIPHENYL) ◇ 4,4′-DIAMINO-3,3′-DIMETHYLBIPHENYL ◇ 4,4′-DIAMINO-3,3′-DIMETHYLDIPHENYL ◇ DIAMINODITOLYL ◇ 3,3′-DIMETHYLBENZIDIN ◇ 3,3′-DIMETHYLBENZIDINE ◇ 3,3′-DIMETHYL-4,4′-BIPHENYLDIAMINE ◇ 3,3′-DIMETHYLBIPHENYL-4,4′-DIAMINE ◇ 3,3′-DIMETHYL-(1,1′-BIPHENYL)-4,4′-DIAMINE ◇ 3,3′-DIMETHYL-4,4′-DIPHENYLDIAMINE ◇ 3,3′-DIMETHYLDIPHENYL-4,4′-DIAMINE ◇ 4,4′-DI-o-TOLUIDINE ◇ FAST DARK BLUE BASE R ◇ RCRA WASTE NUMBER U095 ◇ o-TOLIDIN ◇ 2-TOLIDIN (GERMAN) ◇ 2-TOLIDINA (ITALIAN) ◇ TOLIDINE ◇ 3,3′-TOLIDINE ◇ o,o′-TOLIDINE ◇ 2-TOLIDINE

TOXICITY DATA with REFERENCE
dns-hmn:hla 1 μmol/L CNREA8 38,2621,78
dns-rat:lvr 1 μmol/L MUREAV 136,255,84
orl-rat TDLo:4500 mg/kg/27D-I:CAR CNREA8 28,924,68
scu-rat TDLo:1650 mg/kg/33W-I:ETA VOONAW 20(2),53,74
orl-rat LD50:404 mg/kg 28ZPAK -,71,72
ipr-rat LDLo:125 mg/kg CBCCT* 6,64,54
ipr-mus LDLo:125 mg/kg CBCCT* 6,64,54
orl-dog LDLo:600 mg/kg AEXPBL 58,167,1907

CONSENSUS REPORTS: NTP Fifth Annual Report on Carcinogens. IARC Cancer Review: Group 2B IMEMDT 7,56,87; Animal Limited Evidence IMEMDT 1,87,72. EPA Genetic Toxicology Program. Community Right-To-Know List. Reported in EPA TSCA Inventory.

ACGIH TLV: Suspected Human Carcinogen
DFG MAK: Animal Carcinogen, Suspected Human Carcinogen.
NIOSH REL: (o-Toluidine) CL 0.02 mg/m^3/60M; avoid skin contact.

SAFETY PROFILE: Confirmed carcinogen with experimental carcinogenic and tumorigenic data. Poison by intraperitoneal route. Moderately toxic by ingestion. Human mutation data reported. When heated to decomposition it emits toxic fumes of NO_x.

TGJ850 CAS:26171-23-3 ***HR: 3***
TOLMETINE
mf: $C_{15}H_{15}NO_3$ mw: 257.31

PROP: Crystals from acetonitrile. Mp: 155-157° (decomp).

SYNS: ACIDO-1-METIL-5-(p-TOLNIL)-PIRROL-2-ACETICO(SPANISH) ◇ MCN 2559 ◇ 1-METHYL-5-(4-METHYLBENZOYL)-PYRROLE-2-ACETIC ACID ◇ 1-METHYL-5-p-TOLUOYL-PYRROLE-2-ACETIC ACID ◇ PYROGENS ◇ TOLMETIN

TOXICITY DATA with REFERENCE
orl-rat TDLo:20600 μg/kg (21D preg):TER TJADAB 30(1),25A,84
orl-rat LD50:293 mg/kg ARZNAD 25,1527,76
rec-rat LD50:744 mg/kg AFTOD7 5,257,79
orl-mus LD50:914 mg/kg FRPSAX 38,90,83
ipr-mus LD50:600 mg/kg FRPSAX 38,90,83

SAFETY PROFILE: Poison by ingestion. Moderately toxic by intraperitoneal and rectal routes. An experimental teratogen. When heated to decomposition it emits toxic fumes of NO_x. See also SODIUM TOLMETIN.

TGJ875 CAS:50454-68-7 ***HR: 2***
TOLNIDAMIDE
mf: $C_{16}H_{13}ClN_2O_2$ mw: 300.76

SYNS: 1-(4-CHLORO-2-METHYLBENZYL)-1H-INDAZOLE-3-CARBOXYLIC ACID ◇ 1-((2-METHYL-4-CHLOROPHENYL)METHYL)-1H-INDAZOLE-3-CARBOXYLIC ACID ◇ 1-((2-METHYL-4-CHLOROPHENYL)METHYL)-INDAZOLE-3-CARBOXYLICACID

TOXICITY DATA with REFERENCE
orl-rat TDLo:250 mg/kg (male 1D):REP JOAND3 6,171,85
orl-rat LD50:4 g/kg CHTHBK 27,91,81
ipr-rat LD50:1120 mg/kg CHTHBK 27,91,81
orl-mus LD50:1250 mg/kg CHTHBK 27,91,81
ipr-mus LD50:575 mg/kg CHTHBK 27,91,81

SAFETY PROFILE: Moderately toxic by ingestion and intraperitoneal routes. Experimental reproductive effects. When heated to decomposition it emits toxic fumes of Cl^- and NO_x.

TGJ885 CAS:57524-15-9 ***HR: 3***
TOLONIDINE NITRATE
mf: $C_{10}H_{12}ClN_3 \cdot HNO_3$ mw: 272.72

SYNS: CERM 10,137 ◇ 2-(2-CHLORO-p-TOLUIDINO)-2-IMIDAZOLINE NITRATE ◇ 4,5-DIHYDRO-N-(2-CHLORO-4-METHYLPHENYL)-1H-IMIDAZOL-2-AMINE MONONITRATE ◇ EUCTAN

TOXICITY DATA with REFERENCE
orl-rat LD50:420 mg/kg ARZNAD 25,1926,75
ivn-rat LD50:42 mg/kg ARZNAD 25,1926,75
orl-mus LD50:160 mg/kg ARZNAD 25,1926,75
ivn-mus LD50:18500 μg/kg ARZNAD 25,1926,75

SAFETY PROFILE: Poison by ingestion and intravenous routes. Used as an antihypertensive agent. When heated to decomposition it emits toxic fumes of Cl^- and NO_x. See also NITRATES.

TGK200 CAS:728-88-1 ***HR: 3***
TOLPERISONE
mf: $C_{16}H_{23}NO$ mw: 245.40

PROP: Crystals from methyl ethyl ketone. Mp:176-177°.

SYNS: 2,4'-DIMETHYL-3-PIPERIDINOPROPIOPHENONE◇ 2-METHYL-3-PIPERIDINO-1-p-TOLYLPROPAN-1-ONE◇ MYDETON ◇ MYDOCALM ◇ 1-PIPERIDINO-2-METHYL-3-(p-TOLYL)-3-PROPANONE ◇ PMP

TOXICITY DATA with REFERENCE
orl-rat LD50:1450 mg/kg OYYAA2 5,421,71
ipr-rat LD50:170 mg/kg OYYAA2 5,421,71
scu-rat LD50:645 mg/kg OYYAA2 5,421,71
orl-mus LD50:1350 mg/kg OYYAA2 5,421,71
scu-mus LD50:620 mg/kg ARZNAD 11,257,61
ivn-mus LD50:34 mg/kg OYYAA2 5,421,71

SAFETY PROFILE: Poison by intravenous and intraperitoneal routes. Moderately toxic by ingestion and subcutaneous routes. When heated to decomposition it emits toxic fumes of NO_x.

TGK225 CAS:6775-25-3 ***HR: 3***
TOLPRONINE HYDROCHLORIDE
mf: $C_{15}H_{21}NO_2 \cdot ClH$ mw: 283.83

PROP: Bitter taste, crystals. Decomp 136-137.5°.

SYNS: 3,6-DIHYDRO-α-((2-METHYLPHENOXY)METHYL)-1(2H)-PYRIDINEETHANOL HYDROCHLORIDE ◇ 1-Δ-^{3}-PIPERIDINO-3-o-TOLOXYPROPAN-2-OL HYDROCHLORIDE ◇ PROPONESIN HYDROCHLORIDE ◇ 1-(1,2,3,6-TETRAHYDROPYRIDINO)-3-o-TOLYLOXYPROPAN-2-OLHYDROCHLORIDE

TOXICITY DATA with REFERENCE
orl-rat LD50:340 mg/kg JPPMAB 10,60,58
scu-rat LD50:1030 mg/kg JPPMAB 10,60,58
orl-mus LD50:330 mg/kg JPPMAB 10,60,58

ipr-mus LD50:146 mg/kg MEIEDD 10,1363,83
scu-mus LD50:390 mg/kg JPPMAB 10,60,58 JPPMAB 10,60,58

SAFETY PROFILE: Poison by ingestion, subcutaneous, and intraperitoneal routes. When heated to decomposition it emits toxic fumes of NO_x and HCl.

TGK250 CAS:1334-78-7 ***HR: 2***
TOLUALDEHYDE
mf: C_8H_8O mw: 120.16

PROP: Mixture of o-, m-, and p-methylbenzaldehyde (FCTXSV 14,659,76).

SYNS: TOLUENECARBOXALDEHYDE ◇ TOLYL ALDEHYDE

TOXICITY DATA with REFERENCE
skn-rbt 500 mg/24H MOD FCTXAV 14,659,76
orl-rat LD50:2250 mg/kg FCTXAV 14,659,76

CONSENSUS REPORTS: Reported in EPA TSCA Inventory.

SAFETY PROFILE: Moderately toxic by ingestion. A skin irritant. When heated to decomposition it emits acrid smoke and irritating fumes. See also ALDEHYDES.

TGK500 CAS:73987-51-6 ***HR: 2***
TOLUALDEHYDE GLYCERYL ACETAL
mf: $C_{11}H_{12}O_2$ mw: 176.23

SYNS: 2-(METHYLPHENYL)-1,3-DIOXAN-5-OL (mixed isomers) ◇ TOLYLALDEHYDE GLYCERYL ACETAL

TOXICITY DATA with REFERENCE
skn-rbt 500 mg/24H MOD FCTXAV 14,887,76
orl-rat LD50:3400 mg/kg FCTXAV 14,887,76

SAFETY PROFILE: Moderately toxic by ingestion. A skin irritant. When heated to decomposition it emits acrid smoke and irritating fumes. See also ALDEHYDES.

TGK750 CAS:108-88-3 ***HR: 3***
TOLUENE
DOT: UN 1294
mf: C_7H_8 mw: 92.15

PROP: Colorless liquid; benzol-like odor. Mp: −95 to −94.5°, bp: 110.4°, flash p: 40°F (CC), ULC: 75-80, lel: 1.27%, uel: 7%, d: 0.866 @ 20°/4°, autoign temp: 996°F, vap press: 36.7 mm @ 30°, vap d: 3.14. Insol in water; sol in acetone; misc in abs alc, ether, chloroform.

SYNS: ANTISAL 1a ◇ BENZENE, METHYL- ◇ METHACIDE ◇ METHANE, PHENYL- ◇ METHYLBENZENE ◇ METHYLBENZOL ◇ NCI-C07272 ◇ PHENYLMETHANE ◇ RCRA WASTE NUMBER U220 ◇ TOLUEEN (DUTCH) ◇ TOLUEN (CZECH) ◇ TOLUOL (DOT) ◇ TOLUOLO (ITALIAN) ◇ TOLU-SOL

TOXICITY DATA with REFERENCE
eye-hmn 300 ppm JIHTAB 25,282,43
skn-rbt 435 mg MLD UCDS** 7/23/70
skn-rbt 500 MOD FCTOD7 20,563,82
eye-rbt 870 μg MLD UCDS** 7/23/70
eye-rbt 2 mg/24H SEV 28ZPAK -,23,72
eye-rbt 100 mg/30S rns MLD FCTOD7 20,573,82
oms-grh-ihl 562 mg/L MUREAV 113,467,83
cyt-rat-scu 12 g/kg/12D-I GTPZAB 17(3),24,73
ihl-mus TCLo:400 ppm/7H (female 7-16D post):REP FAATDF 6,145,86
orl-mus TDLo:9 g/kg (female 6-15D post):TER TJADAB 19,41A,79
orl-hmn LDLo:50 mg/kg YAKUD5 22,883,80
ihl-hmn TCLo:200 ppm:BRN,CNS,BLD JAMAAP 123,1106,43
ihl-man TCLo:100 ppm:CNS WEHRBJ 9,131,72
orl-rat LD50:5000 mg/kg AMIHAB 19,403,59
ihl-rat LCLo:4000 ppm/4H AIHAAP 30,470,69
ipr-rat LDLo:800 mg/kg TXAPA9 1,156,59
ivn-rat LD50:1960 mg/kg MELAAD 54,486,63
ihl-mus LC50:5320 ppm/8H JIHTAB 25,366,43
ipr-mus LD50:640 mg/kg ANYAA9 243,104,75
ihl-rbt LCLo:55000 ppm/40M JIDHAN 26,69,44
skn-rbt LD50:12124 mg/kg AIHAAP 30,470,69

CONSENSUS REPORTS: Community Right-To-Know List. Reported in EPA TSCA Inventory. EPA Genetic Toxicology Program.

OSHA PEL: (Transitional: TWA 200 ppm; CL 300 ppm; Pk 500 ppm/10M/8H) TWA 100 ppm; STEL 150 ppm
ACGIH TLV: TWA 100 ppm; STEL 150 ppm; (Proposed: TWA 50 ppm); BEI: 1 mg(toluene)/L in venous blood end of shift; 20 ppm toluene in end-exhaled air during shift.
DFG MAK: 100 ppm (380 mg/m^3); BAT: 340 μg/dl in blood at end of shift.
NIOSH REL: (Toluene) TWA 100 ppm; CL 200 ppm/10M
DOT Classification: Flammable Liquid; Label: Flammable Liquid.

SAFETY PROFILE: Poison by intraperitoneal route. Moderately toxic by intravenous and subcutaneous routes. Mildly toxic by inhalation. An experimental teratogen. Human systemic effects by inhalation: CNS recording changes, hallucinations or distorted perceptions, motor activity changes, antipsychotic, psychophysiological test changes and bone marrow changes. Experimental reproductive effects. Mutation data reported. A human eye irritant. An experimental skin and severe eye irritant.

Toluene is derived from coal tar, and commercial grades usually contain small amounts of benzene as an impurity. Inhalation of 200 ppm of toluene for 8 hours may cause impairment of coordination and reaction time; with higher concentrations (up to 800 ppm) these effects are increased and are observed in a shorter time. In the few cases of acute toluene poisoning reported, the effect has been that of a narcotic, the workman passing through a stage of intoxication into one of coma. Recovery following removal from exposure has been the rule. An occasional report of chronic poisoning describes an anemia and leucopenia, with biopsy showing a bone marrow hypoplasia. These effects, however, are less common in people working with toluene, and they are not as severe. At 200-500 ppm, headache, nausea, eye irritation, loss of appetite, a bad taste, lassitude, impairment of coordination and reaction time are reported, but are not usually accompanied by any laboratory or physical findings of significance. With higher concentrations, the above complaints are increased and in addition, anemia, leukopenia and enlarged liver may be found in rare cases. A common air contaminant, emitted from modern building materials. (CENEAR 69,22,91) Used in production of drugs of abuse.

Flammable liquid. A very dangerous fire hazard when exposed to heat, flame, or oxidizers. Explosive in the form of vapor when exposed to heat or flame. Explosive reaction with 1,3-dichloro-5,5-dimethyl-2,4-imidazolididione; dinitrogen tetraoxide; concentrated nitric acid; H_2SO_4 + HNO_3; N_2O_4; $AgClO_4$; BrF_3; UF_6, sulfur dichloride. Forms an explosive mixture with tetranitromethane. Can react vigorously with oxidizing materials. To fight fire, use foam, CO_2, dry chemical. When heated to decomposition it emits acrid smoke and irritating fumes.

TGL500 CAS:25376-45-8 ***HR: 3***
TOLUENEDIAMINE
mf: $C_7H_{10}N_2$ mw: 122.19
DOT: NA 1709

SYNS: DIAMINOTOLUENE ◇ METHYLPHENYLENEDIAMINE ◇ ar-METHYLBENZENEDIAMINE ◇ TOLYLENEDIAMINE

CONSENSUS REPORTS: Community Right-To-Know List. Reported in EPA TSCA Inventory.

DOT Classification: ORM-A; Label: None.

SAFETY PROFILE: Probably a poison. When heated to decomposition it emits toxic fumes of NO_x. See also other toluene diamine entries and AROMATIC AMINES.

TGL750 CAS:95-80-7 ***HR: 3***
TOLUENE-2,4-DIAMINE
mf: $C_7H_{10}N_2$ mw: 122.19
DOT: UN 1709

PROP: Prisms. Mp: 99°, bp: 280°, vap press: 1 mm @ 106.5°.

SYNS: 3-AMINO-p-TOLUIDINE ◇ 5-AMINO-o-TOLUIDINE ◇ AZOGEN DEVELOPER H ◇ BENZOFUR MT ◇ C.I. 76035 ◇ C.I. OXIDATION BASE ◇ DEVELOPER H ◇ 1,3-DIAMINO-4-METHYLBENZENE ◇ 2,4-DIAMINO-1-METHYLBENZENE ◇ 2,4-DIAMINOTOLUEN (CZECH) ◇ DIAMINOTOLUENE ◇ 2,4-DIAMINOTOLUENE ◇ 2,4-DIAMINO-1-TOLUENE ◇ 2,4-DIAMINOTOLUOL ◇ EUCANINE GB ◇ FOURAMINE ◇ FOURRINE M ◇ META TOLUYLENE DIAMINE ◇ 4-METHYL-1,3-BENZENEDIAMINE ◇ 4-METHYL-m-PHENYLENEDIAMINE ◇ MTD ◇ NAKO TMT ◇ NCI-C02302 ◇ PELAGOL GREY J ◇ PONTAMINE DEVELOPER TN ◇ RCRA WASTE NUMBER U221 ◇ RENAL MD ◇ TDA ◇ 2,4-TOLAMINE ◇ m-TOLUENEDIAMINE ◇ 2,4-TOLUENEDIAMINE ◇ m-TOLUYLENDIAMIN (CZECH) ◇ m-TOLUYLENEDIAMINE ◇ 2,4-TOLUYLENEDIAMINE (DOT) ◇ m-TOLYENEDIAMINE ◇ m-TOLYLENEDIAMINE ◇ TOLYLENE-2,4-DIAMINE ◇ 2,4-TOLYLENEDIAMINE ◇ 4-m-TOLYLENEDIAMINE ◇ ZOBA GKE ◇ ZOGEN DEVELOPHER H

TOXICITY DATA with REFERENCE
skn-rbt 500 mg/24H MLD 28ZPAK -,69,72
eye-rbt 100 mg/24H MOD 28ZPAK -,69,72
mma-sat 50 μg/plate ENMUDM 7,535,85
dnd-hmn:fbr 100 μmol/L MUREAV 127,107,84
orl-mus TDLo:1200 mg/kg (female 6-13D post):REP TCMUD8 7,29,87
orl-rat TDLo:2100 mg/kg/90W-C:CAR JJIND8 62,1107,79
orl-mus TDLo:8050 mg/kg/101W-C:CAR NCITR* NCI-CG-TR-162,79
ipr-rat LD50:325 mg/kg JEPTDQ 3(1-2),149,80
scu-rat LDLo:50 mg/kg ZEPTAT 17,59,15
ipr-mus LD50:480 mg/kg JEPTDQ 3(1-2),149,80
scu-dog LDLo:200 mg/kg XPHBAO 271,56,41
scu-rbt LDLo:400 mg/kg ZEPTAT 17,59,15

CONSENSUS REPORTS: NTP Fifth Annual Report on Carcinogens. IARC Cancer Review: Group 2B IMEMDT 7,56,87; Animal Sufficient Evidence IMEMDT 16,83,78. NCI Carcinogenesis Bioassay (feed); Clear Evidence: mouse, rat NCITR* NCI-CG-TR-162,79. Community Right-To-Know List. Reported in EPA TSCA Inventory. EPA Genetic Toxicology Program.

DFG MAK: Animal Carcinogen, Suspected Human Carcinogen.

DOT Classification: Poison B; Label: St. Andrews Cross.

SAFETY PROFILE: Confirmed carcinogen with experimental carcinogenic data. Poison by intraperitoneal and subcutaneous routes. Experimental reproductive effects. Human mutation data reported. A skin and eye irritant.

This material has a marked toxic action upon the liver and can cause fatty degeneration of that organ. When heated to decomposition it emits toxic fumes of NO_x. See also other toluene diamine entries and AROMATIC AMINES.

TGM000 CAS:95-70-5 ***HR: 3***
TOLUENE-2,5-DIAMINE
mf: $C_7H_{10}N_2$ mw: 122.19

PROP: Colorless, crystalline tablets. Mp: 64°, bp: 274°.

SYNS: 4-AMINO-2-METHYLANILINE ◇ C.I. 76042 ◇ 2,5-DIAMINOTOLUENE ◇ 2-METHYL-1,4-BENZENEDIAMINE ◇ 2-METHYL-p-PHENYLENEDIAMINE ◇ p-TOLUENEDIAMINE ◇ p-TOLUYLENDIAMINE ◇ TOLUYLENE-2,5-DIAMINE ◇ p,m-TOLYLENEDIAMINE

TOXICITY DATA with REFERENCE
skn-rbt 12500 μg/24H MLD FCTXAV 15,607,77
mma-sat 100 μg/plate PNASA6 72,2423,75
dns-rat:lvr 10 μmol/L MUREAV 136,255,84
dni-mus-ipr 40 mg/kg MRLEDH 91,75,81
dns-ham:lvr 10 μmol/L MUREAV 136,255,84
orl-rat LD50:102 mg/kg FCTXAV 15,607,77
scu-rat LDLo:50 mg/kg ZEPTAT 17,59,15
scu-rbt LDLo:100 mg/kg ZEPTAT 17,59,15
orl-mam LDLo:3600 mg/kg JIDHAN 13,87,31

CONSENSUS REPORTS: IARC Cancer Review: Group 3 IMEMDT 7,56,87; Animal Inadequate Evidence IMEMDT 16,97,78. Reported in EPA TSCA Inventory. EPA Genetic Toxicology Program.

SAFETY PROFILE: Poison by ingestion and subcutaneous routes. A skin irritant. Mutation data reported. Questionable carcinogen. Has a toxic action upon the liver and can cause fatty degeneration of that organ. Its total effect upon the body seems to take place three different ways. It is toxic to the central nervous system. It produces jaundice by action on the liver and spleen, and it produces anemia by destruction of the red blood cells. In this action it is quite similar to aniline, although by no means identical with it. Its high boiling point and the fact that the material is solid at room temperature makes it somewhat less hazardous than aniline, particularly at ordinary working temperatures. The literature contains a reference to a permanent injury to an eye due to the use of this material as an eyelash dye. It is considered to be an irritating dye material. When heated to decomposition it emits toxic fumes of NO_x. See also other toluene diamine entries and AROMATIC AMINES.

TGM100 CAS:823-40-5 ***HR: 3***
TOLUENE-2,6-DIAMINE
mf: $C_7H_{10}N_2$ mw: 122.19

SYNS: 2,6-DIAMINOTOLUENE ◇ 2-METHYL-1,2-BENZENEDIAMINE ◇ 2,6-TOLUYLENEDIAMINE ◇ 2,6-TOLYLENEDIAMINE

TOXICITY DATA with REFERENCE
mma-sat 1500 nmol/plate TXCYAC 18,219,80
dns-hmn:lvr 1 mmol/L PAACA3 24,69,83
dni-mus-ipr 30 mg/kg MRLEDH 91,75,81

CONSENSUS REPORTS: Reported in EPA TSCA Inventory.

SAFETY PROFILE: Probably a poison. Human mutation data reported. When heated to decomposition it emits toxic fumes of NO_x. See also other toluene diamine entries and AROMATIC AMINES.

TGM250 CAS:496-72-0 ***HR: 2***
TOLUENE-3,4-DIAMINE
mf: $C_7H_{10}N_2$ mw: 122.19

SYNS: 3,4-DIAMINOTOLUENE ◇ 4-METHYL-1,2-BENZENEDIAMINE ◇ 3,4-TOLUYLENEDIAMINE ◇ 3,4-TOLYLENEDIAMINE

TOXICITY DATA with REFERENCE
mnt-mus-ipr 488 mg/kg ARTODN 43,249,80
dni-mus-ipr 200 mg/kg MRLEDH 91,75,81
scu-rat LDLo:500 mg/kg ZEPTAT 17,59,15
scu-rbt LDLo:1800 mg/kg ZEPTAT 17,59,15

CONSENSUS REPORTS: Reported in EPA TSCA Inventory.

SAFETY PROFILE: Moderately toxic by subcutaneous route. Mutation data reported. When heated to decomposition it emits toxic fumes of NO_x. See also other toluene diamine entries and AROMATIC AMINES.

TGM400 CAS:6369-59-1 ***HR: 3***
p-TOLUENEDIAMINE SULFATE
mf: $C_7H_{10}N_2 \cdot 7H_2O_4S$ mw: 808.75

SYNS: C.I. 76043 ◇ C.I. OXIDATION BASE 4 ◇ FOURAMINE STD ◇ 2-METHYL-1,4-BENZENEDIAMINE ◇ NCI-C01832 ◇ 2,5-TDS ◇ 2,5-TOULENEDIAMINE SULFATE

TOXICITY DATA with REFERENCE
mma-sat 33300 ng/plate ENMUDM 7(Suppl 5),1,85
orl-mus TDLo:800 mg/kg (female 8-12D post):REP TCMUD8 6,361,86
orl-rat TDLo:800 mg/kg (female 6-15D post):TER TJADAB 33,31A,86
orl-mus TDLo:66 g/kg/78W-C:ETA NCITR* NCI-CG-TR-126,78

CONSENSUS REPORTS: IARC Cancer Review: Animal Inadequate Evidence IMEMDT 16,97,78. EPA Genetic Toxicology Program. Reported in EPA TSCA Inventory.

SAFETY PROFILE: An experimental teratogen. Experimental reproductive effects. Questionable carcinogen with experimental tumorigenic data. Mutation data reported. When heated to decomposition it emits toxic fumes of SO_x and NO_x. See also AROMATIC AMINES and SULFATES.

TGM425 CAS:54514-12-4 ***HR: 2***
2-TOLUENEDIAZONIUM BROMIDE
mf: $C_7H_7BrN_2$ mw: 199.05

$(CH_3)C_6H_4N_2^-Br^-$

SAFETY PROFILE: Reaction with copper results in violent nitrogen evolution. When heated to decomposition it emits toxic fumes of Br^- and NO_x. See also BROMIDES and other toluenediazonium compounds.

TGM525 ***HR: 3***
2-TOLUENEDIAZONIUM PERCHLORATE
mf: $C_7H_7ClN_2O_4$ mw: 218.60

$(CH_3)C_6H_4N_2^-ClO_4^-$

SYN: o-TOLUENEDIAZONIUM PERCHLORATE

SAFETY PROFILE: An explosive. It is sensitive even when wet. When heated to decomposition it emits toxic fumes of Cl^- and NO_x. See also PERCHLORATES and other toluenediazonium compounds.

TGM550 CAS:17333-74-3 ***HR: 2***
2-TOLUENEDIAZONIUM SALTS
mf: $C_7H_7N_2^-Z^-$

SAFETY PROFILE: Reaction with ammonium sulfide or hydrogen sulfide forms explosive products. When heated to decomposition they emit toxic fumes of NO_x. See also other toluenediazonium compounds.

TGM575 CAS:14604-29-6 ***HR: 2***
3-TOLUENEDIAZONIUM SALTS
mf: $C_7H_7N_2^-Z^-$

SAFETY PROFILE: Reaction with ammonium sulfide or hydrogen sulfide forms explosive products. Potentially explosive reaction with potassium o-ethyl dithiocarbonate. Vigorous reaction with potassium iodide. When heated to decomposition they emit toxic fumes of NO_x. See also other toluenediazonium compounds.

TGM580 CAS:14597-45-6 ***HR: 3***
4-TOLUENEDIAZONIUM SALTS
mf: $C_7H_7N_2^-Z^-$

SAFETY PROFILE: Reaction with ammonium sulfide or hydrogen sulfide forms explosive products. When heated to decomposition they emit toxic fumes of NO_x. See also other toluenediazonium compounds.

TGM590 CAS:68596-94-1 ***HR: 3***
4-TOLUENEDIAZONIUM TRIIODIDE
mf: $C_7H_7I_3N_2$ mw: 499.86

$CH_3C_6H_4N_2^-I_3^-$

SAFETY PROFILE: Unstable, probably explosive. When heated to decomposition it emits toxic fumes of NO_x and I^-. See also other toluenediazonium compounds and IODIDES.

TGM740 CAS:26471-62-5 ***HR: 3***
TOLUENE-1,3-DIISOCYANATE
DOT: UN 2078
mf: $C_9H_6N_2O_2$ mw: 174.17

SYN: BENZENE-, 1,3-DIISOCYANATOMETHYL- ◇ DESMODUR T100 ◇ DIISOCYANATOMETHYLBENZENE ◇ DIISOCYANATOTOLUENE ◇ HYLENE-T ◇ ISOCYANIC ACID, METHYLPHENYLENE ESTER ◇ METHYL-m-PHENYLENE DIISOCYANATE ◇ METHYLPHENYLENE ISOCYANATE ◇ MONDUR-TD ◇ MONDUR-TD-80 ◇ NACCONATE-100 ◇ NIAX ISOCYANATE TDI ◇ RCRA WASTE NUMBER U223 ◇ RUBINATE TDI ◇ RUBINATE TDI 80/20 ◇ T 100 ◇ TDI ◇ TDI-80 ◇ TDI 80-20 ◇ TOLUENE DIISOCYANATE ◇ TOLYLENE DIISOCYANATE ◇ TOLYLENE ISOCYANATE

TOXICITY DATA with REFERENCE
skn-rbt 500 mg open SEV UCDS** 7/11/67
mma-sat 100 μg/plate ENMUDM 9(Suppl 9),1,87
cyt-hmn:lyms 92 mg/L TOLED5 36,37,87
orl-rat TDLo:31800 mg/kg/2Y-I:CAR NTPTR* NTP-TR-251,86
orl-mus TDLo:63 g/kg/2Y-I:CAR NTPTR* NTP-TR-251,86
orl-rat TD:63600 mg/kg/2Y-I:NEO NTPTR* NTP-TR-251,86
orl-rat LD50:4130 mg/kg AIHAAP 43,89,82
ihl-rat LCLo:600 ppm/6H UCDS** 7/11/67
orl-mus LD50:1950 mg/kg TAKHAA 39,202,80
ihl-mus LC50:9700 ppb/4H AIHAAP 23,447,62
ihl-rbt LC50:11 ppm/4H AIHAAP 23,447,62
ihl-gpg LC50:12700 ppb/4H AIHAAP 23,447,62

CONSENSUS REPORTS: IARC Cancer Review: Group 2B IMEMDT 7,56,87, Animal Sufficient Evidence IMEMDT 39,287,86; Human Inadequate Evidence IMEMDT 39,287,86. NTP Carcinogenesis Studies (gavage): Clear Evidence: mouse, rat NTPTR* NTP-TR-251,86. Reported in EPA TSCA Inventory.

NIOSH REL: (TDI): 10H TWA 0.005 ppm;CL 0.02 ppm/10M
DOT Classification: Poison B; LABEL: Poison

SAFETY PROFILE: Suspected carcinogen with experimental carcinogenic and neoplastigenic data. Poison by inhalation. Moderately toxic by ingestion. Severe skin irritant. Human mutation data reported. Capable of producing severe dermatitis and bronchial spasm. A common air contaminant. Combustible when exposed to heat or flame. Explosive in the form of vapor when exposed to heat or flame. To fight fire, use dry chemical, CO_2. Potentially violent polymerization reaction with bases or acyl chlorides. Reaction with water releases carbon dioxide. Storage in polyethylene containers is hazardous due to absorption of water through the plastic. When heated to decomposition it emits highly toxic fumes of NO_x. See also ISOCYANATES.

TGM750 CAS:584-84-9 ***HR: 3***
TOLUENE-2,4-DIISOCYANATE
mf: $C_9H_6N_2O_2$ mw: 174.17

PROP: Clear, faintly yellow liquid; sharp, pungent odor. Mp: 19.5-21.5°, d (liquid): 1.2244 @ 20/4°, bp: 251°, flash p: 270°F (OC), vap d: 6.0, lel: 0.9%, uel: 9.5%. Misc with alc (decomp), ether, acetone, carbon tetrachloride, benzene, chlorobenzene, kerosene, olive oil.

SYNS: CRESORCINOL DIISOCYANATE ◇ DESMODUR T80 ◇ DI-ISO-CYANATE de TOLUYLENE ◇ DI-ISO-CYANATOLUENE ◇ 2,4-DIISOCYANATO-1-METHYLBENZENE (9CI) ◇ 2,4-DIISOCYANA-TOTOLUENE ◇ DIISOCYANAT-TOLUOL ◇ ISOCYANIC ACID, METHYLPHENYLENE ESTER ◇ ISOCYANIC ACID, 4-METHYL-m-PHENYLENE ESTER ◇ HYLENE T ◇ HYLENE TCPA ◇ HYLENE TLC ◇ HYLENE TM ◇ HYLENE TM-65 ◇ HYLENE TRF ◇ 4-METHYL-PHENYLENE DIISOCYANATE ◇ 4-METHYL-PHENYLENE ISOCYANATE ◇ MONDUR TD ◇ MONDUR TD-80 ◇ MONDUR TDS ◇ NACCONATE 100 ◇ NCI-C50533 ◇ NIAX TDI ◇ NIAX TDI-P ◇ RCRA WASTE NUMBER U223 ◇ RUBINATE TDI 80/20 ◇ TDI (OSHA) ◇ 2,4-TDI ◇ TDI-80 ◇ TOLUEEN-DIISOCYANAAT ◇ TOLUEN-DISOCIANATO ◇ TOLUENE DIISOCYANATE ◇ 2,4-TOLUENEDIISOCYANATE ◇ TOLUILENODWUIZOCYJANIAN ◇ TULUYLENDIISOCYANAT ◇ TOLUYLENE-2,4-DIISOCYANATE ◇ m-TOLYLENE DIISOCYANATE ◇ TOLYLENE-2,4-DIISOCYANATE ◇ 2,4-TOLYLENEDIISOCYANATE

TOXICITY DATA with REFERENCE
skn-rbt 500 mg open SEV UCDS** 3/8/73
skn-rbt 500 mg/24H MOD EJTXAZ 9,41,76
eye-rbt 100 mg SEV EJTXAZ 9,41,76
mma-sat 100 μg/plate ENMUDM 9(Suppl 9),1,87
ihl-wmn TCLO:300 ppt/8H/5D:PUL LANCAO 1,756,80
ihl-hmn TCLo:0.02 ppm/2Y:PUL PRSMA4 63,372,70
ihl-hmn TCLo:0.5 ppm:NOSE,PUL JOCMA7 1,448,59
ihl-hmn TCLo:80 ppb:NOSE,PUL,EYE ATXKA8 19,364,62
orl-rat LD50:5800 mg/kg AMIHAB 15,324,57
ihl-rat LC50:14 ppm/4H AIHAAP 23,447,62
orl-bwd LD50:100 mg/kg AECTCV 12,355,83
ihl-rat LC50:600 ppm/6H AMIHAB 15,324,57
ihl-mus LC50:10 ppm/4H AIHAAP 23,447,62
ivn-mus LD50:56 mg/kg CSLNX* NX#07807
ihl-rbt LC50:1500 ppm/3H AMIHAB 15,324,57
ihl-gpg LC50:13 ppm/4H AIHAAP 23,447,62

CONSENSUS REPORTS: NTP Fifth Annual Report on Carcinogens. IARC Cancer Review: Group 2B IMEMDT 7,56,87; Human Inadequate Evidence IMEMDT 39,287,86; Animal Sufficient Evidence IMEMDT 39,287,86. Community Right-To-Know List. EPA Extremely Hazardous Substances List. Reported in EPA TSCA Inventory.

OSHA PEL: (Transitional: CL 0.02 ppm) TWA 0.005 ppm; STEL 0.02 ppm
ACGIH TLV: TWA 0.005 ppm; STEL 0.02 ppm
DFG MAK: 0.01 ppm (0.07 mg/m^3
NIOSH REL: (Diisocyanates) TWA 0.005 ppm; CL 0.02 ppm/10M
DOT Classification: Poison B; Label: Poison.

SAFETY PROFILE: Confirmed carcinogen. Poison by ingestion, inhalation, and intravenous routes. Human systemic effects by inhalation: unspecified changes to the eyes and sense of smell, respiratory obstruction, cough, sputum, and other pulmonary and gastrointestinal changes. Mutation data reported. A severe skin and eye irritant. Capable of producing severe dermatitis and bronchial spasm. A common air contaminant. Combustible when exposed to heat or flame. Explosive in the form of vapor when exposed to heat or flame. To fight fire, use dry chemical, CO_2. Potentially violent polymerization reaction with bases or acyl chlorides. Reaction with water releases carbon dioxide. Storage in polyethylene containers is hazardous due to absorption of water through the plastic. When heated to decomposition it emits highly toxic fumes of NO_x. See also ISOCYANATES.

TGM800 CAS:91-08-7 ***HR: 3***
TOLUENE-2,6-DIISOCYANATE
mf: $C_9H_6N_2O_2$ mw: 174.17

SYNS: 2,6-DIISOCYANATO-1-METHYLBENZENE ◇ 2,6-DIISOCYANA- TOTOLUENE ◇ HYLENE TM ◇ 2-METHYL-m-PHENYLENE ESTER, ISOCYANIC ACID ◇ 2-METHYL-m-PHENYLENE ISOCYANATE ◇ NIAX TDI ◇ 2,6-TDI ◇ 2,6-TOLUENE DIISOCYANATE ◇ TOLYLENE-2,6-DIISOCYANATE ◇ m-TOLYLENE DIISOCYANATE

TOXICITY DATA with REFERENCE
ihl-hmn TCLo:50 ppb:NOSE,EYE,PUL ATXKA8 19,364,62
ihl-mus LC50:91 mg/m^3/4H TXAPA9 64,423,82
orl-bwd LD50:100 mg/kg AECTCV 12,355,83

CONSENSUS REPORTS: IARC Cancer Review: Group 2B IMEMDT 7,56,87; Human Inadequate Evidence IMEMDT 39,287,86; Animal Sufficient Evidence IMEMDT 39,287,86. Reported in EPA TSCA Inventory. Community Right-To-Know List. EPA Hazardous Substances List.

DFG MAK: 0.01 ppm (0.07 mg/m^3)
NIOSH REL: (Diisocyanates) TWA 0.005 ppm; CL 0.02 ppm/10M

SAFETY PROFILE: Suspected carcinogen. Poison by ingestion and inhalation. Human systemic effects by inhalation: olfactory, eye and pulmonary changes. When heated to decomposition it emits toxic fumes of NO_x. See also CYANATES.

TGN000 CAS:496-74-2 ***HR: 3***
TOLUENE-3,4-DITHIOL
mf: $C_7H_8S_2$ mw: 156.27

PROP: Crystals. Mp: 31°, bp: 185-187°.

SYNS: 1,2-DIMERCAPTO-4-METHYLBENZENE ◇ 4-METHYL-1,2-DIMERCAPTOBENZENE ◇ o-TOLUENESULFONYLAMIDE ◇ o-TOLUENSULFAMID (CZECH) ◇ USAF B-59

TOXICITY DATA with REFERENCE
ipr-mus LD50:50 mg/kg NTIS** AD277-689

CONSENSUS REPORTS: Reported in EPA TSCA Inventory.

SAFETY PROFILE: Poison by intraperitoneal route. When heated to decomposition it emits toxic fumes of SO_x. See also SULFIDES and MERCAPTANS.

TGN100 CAS:40560-76-7 ***HR: 3***
4-TOLUENESULFINYL AZIDE
mf: $C_7H_7N_3OS$ mw: 181.21

$CH_3C_6H_4S(O)N_3$

SAFETY PROFILE: Explodes at 8°C. It is thermally unstable, the pure material or concentrated solutions can explode when warmed. Upon decomposition it emits toxic fumes of SO_x and NO_x. See also AZIDES.

TGN250 CAS:88-19-7 ***HR: 3***
o-TOLUENESULFONAMIDE
mf: $C_7H_9NO_2S$ mw: 171.23

PROP: Tetragonal prisms. Mp: 156°. Sol in water, alc.

SYNS: o-METHYLBENZENESULFONAMIDE ◇ 2-METHYLBENZENE-SULFONAMIDE ◇ ONCO-CARBIDE ◇ ORTHO-TOLUOL-SULFONAMID (GERMAN) ◇ OTS ◇ OXYUREA ◇ TOLUENE-2-SULFONAMIDE

TOXICITY DATA with REFERENCE
eye-rbt 100 mg/24H MOD 28ZPAK -,199,72
mma-sat 70 μmol/plate 45OHAA -,170,80
sln-dmg-par 5 mmol/L MUREAV 56,163,77
sln-dmg-orl 2500 μmol/L 45OHAA -,170,80
orl-rat TDLo:4300 mg/kg (1-22D preg/21D post):REP TXAPA9 51,455,79
orl-rat TDLo:13 g/kg/96W-C:ETA ZKKOBW 91,19,78
orl-rat LD50:4870 mg/kg 28ZPAK -,199,72

CONSENSUS REPORTS: IARC Cancer Review: Group 2B IMEMDT 7,334,87; Animal Limited Evidence IMEMDT 22,111,80. Reported in EPA TSCA Inventory. EPA Genetic Toxicology Program.

SAFETY PROFILE: Suspected carcinogen with experimental tumorigenic data. Mildly toxic by ingestion. Experimental reproductive effects. Mutation data reported. An eye irritant. When heated to decomposition it emits very toxic fumes of NO_x and SO_x. Used as a chemical intermediate in the production of saccharin.

TGN500 CAS:70-55-3 ***HR: 3***
p-TOLUENESULFONAMIDE
mf: $C_7H_9NO_2S$ mw: 171.23

PROP: Monoclinic. Mp: 137°. Sol in water, alc.

SYNS: p-METHYLBENZENESULFONAMIDE ◇ 4-METHYLBENZENE-SULFONAMIDE ◇ p-TOLUENESULFANAMIDE ◇ 4-TOLUENESULFANAMIDE ◇ TOLUENE-4-SULFONAMIDE ◇ p-TOLUENESULFONYLAMIDE ◇ TOLUENE-p-SULPHONAMIDE ◇ TOLYLSULFONAMIDE ◇ p-TOLYLSULFONAMIDE ◇ TOSYLAMIDE ◇ p-TOSYLAMIDE

TOXICITY DATA with REFERENCE
mma-sat 40 μmol/plate 45OHAA -,170,80
sln-dmg-orl 2500 μmol/plate 45OHAA -,170,80
ipr-mus LD50:250 mg/kg NTIS** AD691-490
orl-bwd LD50:75 mg/kg TXAPA9 21,315,72

CONSENSUS REPORTS: Reported in EPA TSCA Inventory. EPA Genetic Toxicology Program.

SAFETY PROFILE: Poison by ingestion and intraperitoneal routes. Mutation data reported. Used as a fungicide. When heated to decomposition it emits very toxic fumes of NO_x and SO_x. See also AMIDES.

TGO000 CAS:104-15-4 ***HR: 3***
p-TOLUENESULFONIC ACID
DOT: UN 2583/UN 2584/UN 2585/UN 2586
mf: $C_7H_8O_3S$ mw: 172.21

PROP: Colorless leaflets. Mp: 107°, bp: 140° @ 20 mm. Sol in alc, ether, water.

SYNS: KYSELINA p-TOLUENESULFONOVA (CZECH) ◇ p-METHYLBENZENESULFONIC ACID ◇ 4-METHYLBENZENESULFONIC ACID ◇ p-METHYLPHENYLSULFONIC ACID ◇ TOLUENESULFONIC ACID ◇ 4-TOLUENESULFONIC ACID ◇ p-TOLUENESULPHONIC ACID ◇ p-TOLYLSULFONIC ACID ◇ TOSIC ACID

TOXICITY DATA with REFERENCE
orl-rat LD50:2480 mg/kg MarJV# 29MAR77
orl-mus LD50:400 mg/kg 14CYAT 2,1841,63

CONSENSUS REPORTS: Reported in EPA TSCA Inventory.

DOT Classification: Corrosive Material; Label: Corrosive

SAFETY PROFILE: Poison by ingestion. Highly irritating to skin and mucous membranes. Potentially explosive reaction with acetic anhydride + water. When heated to decomposition it emits toxic fumes of SO_x.

TGO100 CAS:25231-46-3 ***HR: 2***
TOLUENESULFONIC ACID (liquid)
DOT: UN 2583/UN 2584/UN 2585/UN 2586

DOT Classification: Corrosive Material; Label: Corrosive.

SAFETY PROFILE: A corrosive irritant to skin, eyes, and mucous membranes. When heated to decomposition it emits toxic fumes of SO_x.

TGO250 CAS:98-59-9 ***HR: 1***
p-TOLUENE SULFONYL CHLORIDE
mf: $C_7H_7ClO_2S$ mw: 190.64

PROP: Colorless to light yellow. Highly reactive, hygroscopic crystals; pungent odor. Slowly hydrolyzes to HCl + p-toluene sulfonic acid. Decomp @ >220°, mp: 69-71°, bp: 152° @ 20 mm, vap press, 1 mm Hg @ 88°, flash p: 230°F (COC), d: 1.33. Insol in H_2O; very sol in alc, benzene, ether.

SYNS: p-METHYL BENZENE SULFONYL CHLORIDE ◇ 4-METHYL BENZENE SYLFONYL CHLORIDE ◇ p-TOLUENE SULFONIC ACID CHLORIDE ◇ TOSYL CHLORIDE

SAFETY PROFILE: Mildly toxic by ingestion and skin contact. Most common hazards of handling p-toluene sulfonyl chloride are irritation of skin, eyes, and mucous membranes. Probably due to hydrolysis to HCl and toluene sulfonic acid. Considered a vesicant. Can do irreversible damage to eyes. Although its vapor pressure is low, handling the material may give rise to solid and liquid aerosols which can be inhaled or can expose skin and eyes. The 15 minute TWA is 5 mg/m^3 (AIHA-WEEL). Combustible when exposed to heat or flame. When heated to decomposition it emits very toxic fumes of SO_x and Cl^-.

TGO500 CAS:455-16-3 ***HR: 3***
p-TOLUENESULFONYL FLUORIDE
mf: $C_7H_7FO_2S$ mw: 174.20

PROP: Mp: 41-42°, bp: 112°/16 mm, flash p. 222° F.

SYN: TOSYL FLUORIDE

TOXICITY DATA with REFERENCE
orl-rat LDLo:500 mg/kg NCNSA6 5,1,53
ipr-rat LD50:200 mg/kg NATUAS 173,33,54

CONSENSUS REPORTS: Reported in EPA TSCA Inventory.

SAFETY PROFILE: Poison by intraperitoneal route. Moderately toxic by ingestion. A corrosive. Combustible. When heated to decomposition it emits very toxic fumes of F^- and SO_x. See also p-TOLUENE SULFONYL CHLORIDE and FLUORIDES.

TGO550 CAS:877-66-7 ***HR: 1***
4-TOLUENESULFONYL HYDRAZIDE
mf: $C_7H_{10}N_2O_2S$ mw: 186.23

$CH_3C_6H_4SO_2NHNH_2$

PROP: Crystals. Mp: 109-110°

SAFETY PROFILE: Decomposes exothermically when heated above 60°C. Used as a blowing agent. When heated to decomposition it emits toxic fumes of SO_x and NO_x.

TGO575 CAS:52913-14-1 ***HR: 3***
o-4-TOLUENE SULFONYL HYDROXYLAMINE
mf: $C_7H_9NO_3S$ mw: 187.21

$CH_3C_6H_4SO_2ONH_2$

SAFETY PROFILE: The dry material ignites spontaneously at room temperature. When heated to decomposition it emits toxic fumes of SO_x and NO_x.

TGO750 CAS:100-53-8 ***HR: 3***
α-TOLUENETHIOL
mf: C_7H_8S mw: 124.21

PROP: A water-white, mobile liquid; strong odor. Bp: 194.8°, flash p: 158°F (CC), d: 1.058 @ 20°, vap d: 4.28.

SYNS: BENZYL MERCAPTAN ◇ BENZYLTHIOL ◇ (MERCAPTOMETHYL)BENZENE ◇ α-MERCAPTOTOLUENE ◇ PHENYLMETHANETHIOL ◇ PHENYLMETHYL MERCAPTAN ◇ THIOBENZYL ALCOHOL ◇ α-TOLUOLTHIOL ◇ α-TOLYL MERCAPTAN ◇ USAF EK-1509

TOXICITY DATA with REFERENCE
eye-rbt 106 mg AIHAAP 19,171,58
skn-mus TDLo:16 g/kg/26W-I:ETA AJCAA7 15,2149,31
orl-rat LD50:493 mg/kg AIHAAP 19,171,58

ipr-rat LD50:373 mg/kg AIHAAP 19,171,58
ipr-mus LD50:100 mg/kg NTIS** AD277-689

CONSENSUS REPORTS: Reported in EPA TSCA Inventory.

SAFETY PROFILE: Poison by intraperitoneal route. Moderately toxic by ingestion. An eye irritant. Questionable carcinogen with experimental tumorigenic data. Flammable when exposed to heat or flame. Can react vigorously with oxidizing materials. To fight fire, use foam, CO_2, dry chemical, water spray, mist, fog. When heated to decomposition and on contact with acid or acid fumes it emits highly toxic fumes of SO_x. See also SULFIDES and MERCAPTANS.

TGP000 CAS:137-06-4 ***HR: 3***
o-TOLUENETHIOL
mf: C_7H_8S mw: 124.21

SYNS: o-MERCAPTOTOLUENE ◇ o-METHYLBENZENETHIOL ◇ 2-METHYLBENZENETHIOL ◇ o-METHYLTHIOPHENOL ◇ 2-METHYLTHIOPHENOL ◇ o-THIOCRESOL ◇ 2-TOLUENETHIOL ◇ o-TOLYL MERCAPTAN ◇ USAF EK-2676

TOXICITY DATA with REFERENCE
ipr-mus LD50:100 mg/kg NTIS** AD277-689

CONSENSUS REPORTS: Reported in EPA TSCA Inventory.

SAFETY PROFILE: Poison by intraperitoneal route. When heated to decomposition it emits toxic fumes of SO_x. See also α-TOLUENETHIOL and MERCAPTANS.

TGP250 CAS:106-45-6 ***HR: 3***
p-TOLUENETHIOL
mf: C_7H_8S mw: 124.21

SYNS: p-MERCAPTOTOLUENE ◇ p-METHYLBENZENETHIOL ◇ 4-METHYLBENZENETHIOL ◇ p-METHYLPHENYLMERCAPTAN ◇ 4-METHYLPHENYLMERCAPTAN ◇ p-METHYLTHIOPHENOL ◇ 4-METHYLTHIOPHENOL ◇ p-THIOCRESOL ◇ 4-THIOCRESOL ◇ 4-TOLUENETHIOL ◇ p-TOLYL MERCAPTAN ◇ p-TOLYLTHIOL ◇ USAF EK-510

TOXICITY DATA with REFERENCE
ipr-mus LD50:200 mg/kg NTIS** AD277-689

CONSENSUS REPORTS: Reported in EPA TSCA Inventory.

SAFETY PROFILE: Poison by intraperitoneal route. When heated to decomposition it emits toxic fumes of SO_x. See also α-TOLUENETHIOL and MERCAPTANS.

TGP750 CAS:99-04-7 ***HR: 2***
m-TOLUIC ACID
mf: $C_8H_8O_2$ mw: 136.16

PROP: Prisms in water. D: 1.054 @ 112/4°, mp: 110-111°, bp: 273°. Sol in water; very sol in alc, ether.

SYNS: m-METHYLBENZOIC ACID ◇ 3-METHYLBENZOIC ACID ◇ m-TOLUYLIC ACID

TOXICITY DATA with REFERENCE
orl-mus LD50:1630 mg/kg GTPZAB 18(7),57,74

CONSENSUS REPORTS: Reported in EPA TSCA Inventory.

SAFETY PROFILE: Moderately toxic by ingestion. When heated to decomposition it emits acrid smoke and irritating fumes.

TGQ000 CAS:118-90-1 ***HR: 3***
o-TOLUIC ACID
mf: $C_8H_8O_2$ mw: 136.16

PROP: Crystals from water. D: 1.062 @ 115/4°, mp: 104-105°, bp: 259° @ 751 mm. Sol in water, chloroform; very sol in alc.

SYNS: o-METHYLBENZOIC ACID ◇ 2-METHYLBENZOIC ACID ◇ ORTHOTOLUIC ACID ◇ o-TOLUYLIC ACID

TOXICITY DATA with REFERENCE
orl-rat LD50:400 mg/kg 14CYAT 2,1838,63

CONSENSUS REPORTS: Reported in EPA TSCA Inventory.

SAFETY PROFILE: Poison by ingestion. When heated to decomposition it emits acrid smoke and irritating fumes.

TGQ250 CAS:99-94-5 ***HR: 3***
p-TOLUIC ACID
mf: $C_8H_8O_2$ mw: 136.16

PROP: Crystals from water. Mp: 179-180°, bp: 274-275°. Sol in water; very sol in alc, ether.

SYN: p-TOLUYLIC ACID

TOXICITY DATA with REFERENCE
orl-rat LD50:400 mg/kg 14CYAT 2,1838,63
ipr-rat LD50:874 mg/kg KSRNAM 12,1893,78
orl-mus LD50:2340 mg/kg KSRNAM 12,1893,78
ipr-mus LD50:916 mg/kg KSRNAM 12,1893,78

CONSENSUS REPORTS: Reported in EPA TSCA Inventory.

SAFETY PROFILE: Poison by ingestion. Moderately toxic by intraperitoneal route. When heated to decomposition it emits acrid smoke and irritating fumes.

TGQ500 CAS:108-44-1 ***HR: 3***
m-TOLUIDINE
DOT: UN 1708
mf: C_7H_9N mw: 107.17

PROP: Colorless liquid. Mp: −50.5°, bp: 203.3°, d: 0.989 @ 20/4°, vap press: 1 mm @ 41°, vap d: 3.90. Sltly sol in water; sol in alc, ether.

SYNS: 3-AMINO-1-METHYLBENZENE ◇ 3-AMINOPHENYLMETHANE ◇ 3-AMINOTOLUEN (CZECH) ◇ m-AMINOTOLUENE ◇ 3-AMINOTOLUENE ◇ m-METHYLANILINE ◇ 3-METHYLANILINE ◇ m-METHYLBENZENAMINE ◇ 3-METHYLBENZENAMINE ◇ m-TOLUIDIN (CZECH) ◇ 3-TOLUIDINE ◇ m-TOLYLAMINE

TOXICITY DATA with REFERENCE
skn-rbt 500 mg/24H MLD 28ZPAK -,66,72
eye-rbt 20 mg/24H MOD 28ZPAK -,66,72
orl-rat LD50:450 mg/kg JPETAB 90,260,47
orl-mus LD50:740 mg/kg GTPZAB 25(8),50,81
ipr-mus LD50:116 mg/kg AISFAR 1,284,51
orl-bwd LD50:242 mg/kg AECTCV 12,355,83

CONSENSUS REPORTS: Reported in EPA TSCA Inventory.

OSHA PEL: TWA 2 ppm (skin)
ACGIH TLV: TWA 2 ppm (skin)
DOT Classification: Poison B; Label: Poison

SAFETY PROFILE: Poison by ingestion and intraperitoneal routes. A skin and eye irritant. Flammable when exposed to heat or flame. Can react vigorously on contact with oxidizing materials. To fight fire, use foam, CO_2, dry chemical. When heated to decomposition it emits highly toxic fumes of NO_x. See also ANILINE and o-TOLUIDINE.

TGQ750 CAS:95-53-4 ***HR: 3***
o-TOLUIDINE
mf: C_7H_9N mw: 107.17
DOT: UN 1708

PROP: Colorless liquid. Mp: −16.3°, bp: 200-202°, ULC: 20-25, flash p: 185° (CC), d: 1.004 @ 20/4°, autoign temp: 900°F, vap press: 1 mm @ 44°, vap d: 3.69. Sltly sol in water, dil acid; sol in alc and ether.

SYNS: 1-AMINO-2-METHYLBENZENE ◇ 2-AMINO-1-METHYLBENZENE ◇ o-AMINOTOLUENE ◇ 2-AMINOTOLUENE ◇ C.I. 37077 ◇ 1-METHYL-2-AMINOBENZENE ◇ 2-METHYL-1-AMINOBENZENE ◇ o-METHYLANILINE ◇ 2-METHYLANILINE ◇ o-METHYLBENZENAMINE ◇ 2-METHYLBENZENAMINE ◇ o-TOLUIDIN (CZECH) ◇ 2-TOLUIDINE ◇ o-TOLUIDYNA (POLISH) ◇ o-TOLYLAMINE

TOXICITY DATA with REFERENCE
skn-rbt 10 mg/24H open MLD AIHAAP 23,95,62
skn-rbt 500 mg/24H MLD 28ZPAK -,65,72
eye-rbt 750 μg/24H SEV 85JCAE -,460,86
slt-dmg-orl 1 mmol/L PMRSDJ 5,313,85
dns-hmn:hla 50 uL/L PMRSDJ 5,347,85
msc-mus:lym 10 mg/L PMRSDJ 5,587,85
orl-rat TDLo:109 g/kg/2Y-C:NEO FCTOD7 25,619,87
orl-rat TDLo:7250 mg/kg/23W-C:ETA APMIAL 26,447,49
scu-mus TDLo:320 mg/kg (15-21D preg):NEO,TER BEXBAN 78,1402,74
ihl-man TCLo:25 mg/m^3:KID,BLD GTPZAB 14(8),7,70
orl-rat LD50:670 mg/kg IARC** 27,155,82
orl-mus LD50:520 mg/kg IARC** 27,155,82
ipr-mus LD50:150 mg/kg NTIS** AD691-490
orl-cat LDLo:150 mg/kg XPHBAO 271,49,41
orl-rbt LD50:840 mg/kg IARC** 27,155,82
skn-rbt LD50:3250 mg/kg AIHAAP 23,95,62
orl-frg LDLo:151 mg/kg XPHBAO 271,49,41

CONSENSUS REPORTS: NTP Fifth Annual Report on Carcinogens. IARC Cancer Review: Group 2B IMEMDT 7,362,87; Human Inadequate Evidence IMEMDT 16,349,78; Human Limited Evidence IMEMDT 27,155,82; Animal Inadequate Evidence IMEMDT 16,349,78. EPA Genetic Toxicology Program. Community Right-To-Know List. Reported in EPA TSCA Inventory.

OSHA PEL: TWA 5 ppm (skin)
ACGIH TLV: TWA 2 ppm (skin); Suspected Human Carcinogen.
DFG MAK: Animal Carcinogen, Suspected Human Carcinogen.
DOT Classification: Poison B; Label: Poison.

SAFETY PROFILE: Confirmed carcinogen with experimental neoplastigenic and tumorigenic data. Poison by ingestion and intraperitoneal routes. Moderately toxic by skin contact. Human systemic effects by inhalation: urine volume increase, hematuria and blood methemoglobinemia-carboxhemoglobinemia. An experimental teratogen. Human mutation data reported. A skin and severe eye irritant. Human mucous membrane effects. Can produce severe systemic disturbances. The main portal of entry into the body is the respiratory tract, particularly in cases of industrial exposure. The symptoms produced are headache, weakness, difficulty in breathing, air hunger, psychic disturbances, and marked irritation of the kidneys and bladder. The literature does not yield any good data for comparing the toxicity of the o-, m- and p-isomers. Their behavior is generally comparable to that of aniline. It has been determined experimentally that a concentration of about 100 ppm is the maximum endurable for 1 hour without serious consequences and that from 6-23 ppm is endurable for several hours without serious disturbances.

Flammable when exposed to heat or flame. Hypergolic reaction with red fuming nitric acid. Can react with oxidizing materials. To fight fire, use foam, CO_2, dry chemical. When heated to decomposition it emits highly toxic fumes of NO_x. See also ANILINE.

TGR000 CAS:106-49-0 ***HR: 3***
p-TOLUIDINE
DOT: UN 1708
mf: C_7H_9N mw: 107.17

PROP: Colorless leaflets. Mp: 44.5°, bp: 200.4°, flash p: 188°F (CC), d: 1.046 @ 20/4°, autoign temp: 900°F, vap press: 1 mm @ 42°, vap d: 3.90. Sol in water, dil acid, CS_2; very sol in alc, ether.

SYNS: 4-AMINO-1-METHYLBENZENE ◇ 4-AMINOTOLUEN (CZECH) ◇ p-AMINOTOLUENE ◇ 4-AMINOTOLUENE ◇ C.I. 37107 ◇ C.I. AZOIC COUPLING COMPONENT 107 ◇ p-METHYLANILINE ◇ 4-METHYLANILINE ◇ p-METHYLBENZENAMINE ◇ 4-METHYLBENZENAMINE ◇ NAPHTOL AS-KG ◇ NAPHTOL AS-KGLL ◇ p-TOLUIDIN (CZECH) ◇ 4-TOLUIDINE ◇ TOLYLAMINE ◇ p-TOLYLAMINE

TOXICITY DATA with REFERENCE
skn-rbt 500 mg/24H SEV BIOFX* 31-4/73
skn-rbt 500 mg/24H MLD 28ZPAK -,66,72
eye-rbt 100 mg SEV BIOFX* 31-4/73
eye-rbt 20 mg/24H MOD 28ZPAK -,66,72
dns-rat:lvr 100 μmol/L ENMUDM 5,803,83
dnd-mus-ipr 35 mg/kg ATSUDG (5),355,82
dni-mus-orl 200 mg/kg MUREAV 46,305,77
orl-rat LD50:656 mg/kg BIOFX* 31-4/73
orl-mus LD50:330 mg/kg GTPZAB 25(8),50,81
ipr-mus LD50:50 mg/kg NTIS** AD691-490
orl-qal LD50:237 mg/kg AECTCV 12,355,83
orl-bwd LD50:42 mg/kg TXAPA9 21,315,72

CONSENSUS REPORTS: Reported in EPA TSCA Inventory. EPA Genetic Toxicology Program.

OSHA PEL: TWA 2 ppm (skin)
ACGIH TLV: TWA 2 ppm (skin); Suspected Human Carcinogen.
DOT Classification: Poison B; Label: Poison

SAFETY PROFILE: Poison by ingestion and intraperitoneal routes. Mutation data reported. A severe skin and eye irritant. Flammable when exposed to heat, flame, or oxidizers. Can react vigorously on contact with oxidizing materials. To fight fire, use foam, CO_2, dry chemical. When heated to decomposition it emits highly toxic fumes of NO_x. See also o-TOLUIDINE and ANILINE.

TGR250 CAS:64011-34-3 ***HR: 3***
m-TOLUIDINE ANTIMONYL TARTRATE

TOXICITY DATA with REFERENCE
ipr-mus LD50:29 mg(Sb)/kg AJTMAQ 25,263,45

CONSENSUS REPORTS: Antimony and its compounds are on the Community Right-To-Know List.

OSHA PEL: TWA 0.5 mg(Sb)/m^3
ACGIH TLV: TWA 0.5 mg(Sb)/m^3
NIOSH REL: TWA 0.5 mg/m^3

SAFETY PROFILE: Poison by intraperitoneal route. When heated to decomposition it emits very toxic fumes of Sb and NO_x. See also ANTIMONY COMPOUNDS.

TGR500 CAS:64011-35-4 ***HR: 3***
o-TOLUIDINE ANTIMONYL TARTRATE
SYN: 2-METHYLBENZENAMINE HYDROCHLORIDE

TOXICITY DATA with REFERENCE
ipr-mus LD50:20 mg(Sb)/kg AJTMAQ 25,263,45

CONSENSUS REPORTS: Antimony and its compounds are on the Community Right-To-Know List.

OSHA PEL: TWA 0.5 mg(Sb)/m^3
ACGIH TLV: TWA 0.5 mg(Sb)/m^3
NIOSH REL: TWA 0.5 mg/m^3

SAFETY PROFILE: Poison by intraperitoneal route. When heated to decomposition it emits very toxic fumes of NO_x and Sb. See also ANTIMONY COMPOUNDS.

TGR750 CAS:64011-36-5 ***HR: 3***
p-TOLUIDINE ANTIMONYL TARTRATE

TOXICITY DATA with REFERENCE
ipr-mus LD50:22 mg(Sb)/kg AJTMAQ 25,263,45

CONSENSUS REPORTS: Antimony and its compounds are on the Community Right-To-Know List.

OSHA PEL: TWA 0.5 mg(Sb)/m^3
ACGIH TLV: TWA 0.5 mg(Sb)/m^3
NIOSH REL: TWA 0.5 mg/m^3

SAFETY PROFILE: Poison by intraperitoneal route. When heated to decomposition it emits very toxic fumes of NO_x and Sb. See also ANTIMONY COMPOUNDS.

TGS000 CAS:3209-30-1 ***HR: 3***
TOLUIDINE BLUE
mf: $C_{28}H_{22}N_2O_{10}S_2 \cdot 2Na$ mw: 656.62

SYNS: C.I. 63340 ◇ 6,6'-((4,8-DIHYDROXY-1,5-ANTHRAQUINONYLENE)DIIMINO) DI-m-TOLUENE SULFONIC ACID DISODIUM SALT

TOXICITY DATA with REFERENCE
sln-dmg-unr 500 ppm 14ZYA8 -,115,65
ivn-rat LD50:29 mg/kg FEPRA7 10,337,51
ivn-mus LD50:28 mg/kg FEPRA7 10,337,51
ivn-rbt LD50:13 mg/kg FEPRA7 10,337,51

SAFETY PROFILE: Poison by intravenous route. Mutation data reported. When heated to decomposition it emits very toxic fumes of NO_x, Na_2O, and SO_x.

TGS250 CAS:638-03-9 ***HR: 3***
m-TOLUIDINE HYDROCHLORIDE
mf: $C_7H_9N{\cdot}ClH$ mw: 143.63

PROP: Leaves from water. Mp: 228°, bp: 250°. Sol in water and alc.

TOXICITY DATA with REFERENCE
orl-mus TDLo:475 g/kg/78W-C:ETA JEPTDQ 2,325,78
ipr-rat LD50:660 mg/kg JEPTDQ 2,325,78
ipr-mus LD50:650 mg/kg NCIBR* NIH-NCI-E-68-1311,10,73
scu-frg LDLo:574 mg/kg AEPPAE 130,250,28

SAFETY PROFILE: Moderately toxic by subcutaneous and intraperitoneal routes. Questionable carcinogen with experimental tumorigenic data. When heated to decomposition it emits very toxic fumes of NO_x and HCl. See also m-TOLUIDINE.

TGS500 CAS:636-21-5 ***HR: 3***
o-TOLUIDINE HYDROCHLORIDE
mf: $C_7H_9N{\cdot}ClH$ mw: 143.63

PROP: Monoclinic prisms. Mp: 218-220°, bp: 242°. Sol in water; sltly sol in alc.

SYNS: 1-AMINO-2-METHYLBENZENE HYDROCHLORIDE ◇ 2-AMINO-1-METHYLBENZENE HYDROCHLORIDE ◇ 2-AMINO-TOLUENE HYDROCHLORIDE ◇ o-AMINOTOLUENE HYDROCHLORIDE ◇ 1-METHYL-2-AMINOBENZENE HYDROCHLORIDE ◇ 2-METHYL-1-AMINOBENZENE HYDROCHLORIDE ◇ o-METHYL- ANILINE HYDROCHLORIDE ◇ 2-METHYLANILINE HYDROCHLORIDE ◇ o-METHYLBENZENAMINE HYDROCHLORIDE ◇ 2-METHYLBENZENAMINE HYDROCHLORIDE ◇ NCI-C02335 ◇ RCRA WASTE NUMBER U222 ◇ 2-TOLUIDINE HYDROCHLORIDE ◇ o-TOLYLAMINE HYDROCHLORIDE

TOXICITY DATA with REFERENCE
mma-sat 50 μg/plate PMRSDJ 1,351,81
dnr-esc 500 mg/L PMRSDJ 1,195,81
dnd-bcs 20 μL/disc PMRSDJ 1,175,81
sln-smc 50 mg/L PMRSDJ 1,468,81
orl-rat TDLo:117 g/kg/101W-C:CAR NCITR* NCI-CG-TR-153,79
orl-mus TDLo:146 g/kg/2Y-C:CAR NCITR* NCI-CG-TR-153,79
orl-rat LD50:2951 mg/kg JPETAB 167,223,69
ipr-rat LD50:150 mg/kg NCIBR* NCI-E-68-1311,73
orl-mus LD50:1100 mg/kg GTPZAB 25(8),50,81
ipr-mus LD50:113 mg/kg NCIBR* NCI-E-68-1311,73

CONSENSUS REPORTS: NTP Fifth Annual Report on Carcinogens. IARC Cancer Review: Animal Sufficient Evidence IMEMDT 27,155,82. NCI Carcinogenesis Bioassay (feed); Clear Evidence: mouse, rat NCITR* NCI-CG-TR-153,79. EPA Genetic Toxicology Program. Community Right-To-Know List. Reported in EPA TSCA Inventory.

SAFETY PROFILE: Confirmed carcinogen with experimental carcinogenic data. Poison by intraperitoneal route. Moderately toxic by ingestion. Mutation data reported. When heated to decomposition it emits very toxic fumes of HCl and NO_x. See also o-TOLUIDINE.

TGS750 CAS:540-23-8 ***HR: 3***
p-TOLUIDINE HYDROCHLORIDE
mf: $C_7H_9N{\cdot}ClH$ mw: 143.63

PROP: Needles from acetic ether. Mp: 243°, bp: 257.5°. Sol in water, alc; insol in ether, benzene.

SYNS: 4-AMINOTOLUENE HYDROCHLORIDE ◇ 4-METHYLANILINE HYDROCHLORIDE ◇ 4-METHYLBENZENAMINE HYDROCHLORIDE ◇ p-TOLUIDINIUM CHLORIDE

TOXICITY DATA with REFERENCE
orl-mus TDLo:86 g/kg/78W-C:CAR JEPTDQ 2(2),325,78
orl-mus TD:43 g/kg/78W-C:ETA JEPTDQ 2(2),325,78
orl-rat LD50:1285 mg/kg JPETAB 167,223,69
ipr-rat LD50:265 mg/kg NCIBR* NCI-E-68-1311,73
ipr-mus LD50:399 mg/kg NCIBR* NCI-E-68-1311,73

CONSENSUS REPORTS: Reported in EPA TSCA Inventory. EPA Genetic Toxicology Program.

SAFETY PROFILE: Poison by intraperitoneal route. Moderately toxic by ingestion. Questionable carcinogen with experimental carcinogenic and tumorigenic data. When heated to decomposition it emits very toxic fumes of NO_x and HCl. See also p-TOLUIDINE.

TGT000 CAS:136-80-1 ***HR: 2***
2-o-TOLUIDINOETHANOL
mf: $C_9H_{13}NO$ mw: 151.23

PROP: Liquid. Mp: −25°, bp: 297.1°, flash p: 290°F (OC), d: 1.0723 @ 20/20°.

SYNS: EMERY 5711 ◇ N-β-HYDROXYETHYL-o-TOLUIDINO ◇ o-TOLYL ETHANOLAMINE

TOXICITY DATA with REFERENCE
orl-rat LD50:2200 mg/kg JIHTAB 31,60,49
skn-rbt LD50:1000 mg/kg JIHTAB 31,60,49

CONSENSUS REPORTS: Reported in EPA TSCA Inventory.

SAFETY PROFILE: Moderately toxic by ingestion and skin contact. Combustible when exposed to heat or flame. Can react with oxidizing materials. To fight fire, use foam, CO_2, dry chemical. When heated to decomposition it emits toxic fumes of NO_x. See also AMINES and ALCOHOLS.

TGT250 CAS:620-22-4 ***HR: 2***
m-TOLUNITRILE
mf: C_8H_7N mw: 117.16

SYNS: NITRIL KYSELINY m-TOLUYLOVE (CZECH) ◇ m-TOLYNITRILE

TOXICITY DATA with REFERENCE
eye-rbt 500 mg/24H SEV 28ZPAK -,158,72
orl-rat LD50:4200 mg/kg 28ZPAK -,158,72

CONSENSUS REPORTS: Cyanide and its compounds are on the Community Right-To-Know List. Reported in EPA TSCA Inventory.

SAFETY PROFILE: Mildly toxic by ingestion. A severe eye irritant. When heated to decomposition it emits very toxic fumes of NO_x and CN^-. See also NITRILES.

TGT500 CAS:529-19-1 ***HR: 2***
o-TOLUNITRILE
mf: C_8H_7N mw: 117.16

PROP: Colorless liquid. D: 0.998 @ 15/15°, mp: −13°, bp: 205.2°. Insol in water; misc in alc, ether.

SYNS: 2-CYANOTOLUENE ◇ o-CYANOTOLUENE ◇ 2-METHYLBENZENECARBONITRILE ◇ o-METHYLBENZONITRILE ◇ 2-METHYLBENZONITRILE ◇ o-TOLUIC NITRILE ◇ o-TOLUONITRILE ◇ o-TOLYLNITRILE

TOXICITY DATA with REFERENCE
scu-rbt LDLo:600 mg/kg AIPTAK 5,161,1899
scu-frg LDLo:1000 mg/kg AIPTAK 5,161,1899

CONSENSUS REPORTS: Cyanide and its compounds are on the Community Right-To-Know List. Reported in EPA TSCA Inventory.

SAFETY PROFILE: Moderately toxic by subcutaneous route. When heated to decomposition it emits very toxic fumes of NO_x and CN^-. See also NITRILES.

TGT750 CAS:104-85-8 ***HR: 2***
p-TOLUNITRILE
mf: C_8H_7N mw: 117.16

PROP: Needles from ethyl alc. D: 0.981 @ 20/30°, mp: 29.5°, bp: 217.6°. Insol in water; very sol in alc, ether.

SYNS: p-CYANOTOLUENE ◇ 4-CYANOTOLUENE ◇ 4-METHYLBENZONITRILE ◇ p-METHYLBENZONITRILE ◇ 4-METHYLCYANOBENZENE ◇ NITRIL KYSELINY p-TOLUYLOVE (CZECH) ◇ p-TOLUENENITRILE ◇ p-TOLUIC NITRILE ◇ p-TOLUNITRIL (CZECH) ◇ 4-TOLUNITRILE ◇ p-TOLUONITRILE ◇ p-TOLYLNITRILE

TOXICITY DATA with REFERENCE
skn-rbt 500 mg/24H MOD 28ZPAK -,159,72
eye-rbt 500 mg/24H SEV 28ZPAK -,159,72
ipr-mus LDLo:512 mg/kg CBCCT* 3,129,51
orl-rat LD50:4060 mg/kg 28ZPAK -,159,72
scu-rbt LDLo:1080 mg/kg AIPTAK 5,161,1899

CONSENSUS REPORTS: Cyanide and its compounds are on the Community Right-To-Know List. Reported in EPA TSCA Inventory.

SAFETY PROFILE: Moderately toxic by ingestion, intraperitoneal, and subcutaneous routes. A skin and severe eye irritant. When heated to decomposition it emits very toxic fumes of NO_x and CN^-. See also NITRILES.

TGU000 CAS:874-60-2 ***HR: 3***
p-TOLUOYL CHLORIDE
mf: C_8H_7ClO mw: 154.60

TOXICITY DATA with REFERENCE
ivn-mus LD50:56 mg/kg CSLNX* NX#04246

CONSENSUS REPORTS: Reported in EPA TSCA Inventory.

SAFETY PROFILE: Poison by intravenous route. When heated to decomposition it emits toxic fumes of Cl^-.

TGU500 CAS:6424-34-6 ***HR: 2***
TOLUYLENE BLUE MONOHYDRATE
mf: $C_{15}H_{19}N_4{\bullet}Cl{\bullet}H_2O$ mw: 308.85

PROP: Monohydrate, prismatic, copper brown, shiny crystals. Gives blue solution with cold water, alc, or acetic acid.

SYNS: AMMONIUM,(4-((4,6-DIAMINO-m-TOLYL)IMINO)-2,5-CYCLOHEXADIEN-1-YLIDENE)DIMETHYL-, CHLORIDE, MONOHYDRATE ◇ (4-((4,6-DIAMINO-m-TOLYL)IMINO)-2,5-CYCLOHEXADIEN-1-YLIDENE)DIMETHYLAMMONIUM CHLORIDE H_2O

TOXICITY DATA with REFERENCE
scu-rat TDLo:940 mg/kg/46W-I:ETA GANNA2 45,447,54

SAFETY PROFILE: Questionable carcinogen with experimental neoplastigenic data. When heated to decomposition it emits toxic fumes of NO_x and Cl^-.

TGV000 CAS:16524-23-5 ***HR: 3***
N-(p-TOLYL)ANTHRANILIC ACID
mf: $C_{14}H_{13}NO_2$ mw: 227.28

SYNS: 4′-METHYL-2-DIPHENYLAMINECARBOXYLIC ACID ◇ 2-((4-METHYLPHENYL)AMINO)BENZOIC ACID ◇ N-(p-METHYLPHENYL) ANTHRANILIC ACID ◇ N-(4-METHYLPHENYL)ANTHRANILIC ACID

TOXICITY DATA with REFERENCE
ipr-mus LD50:250 mg/kg CHTPBA 6,346,71
ivn-mus LD50:94 mg/kg YKKZAJ 89,1392,69

SAFETY PROFILE: Poison by intraperitoneal and intravenous routes. When heated to decomposition it emits toxic fumes of NO_x.

TGV100 CAS:102395-95-9 ***HR: 3***
4-TOLYLARSENOUS ACID
mf: C_7H_7AsO mw: 182.06

SYNS: p-ARSENOSOTOLUENE ◇ TOLUENE, p-ARSENOSO-

TOXICITY DATA with REFERENCE
ivn-mus LDLo:1720 μg/kg PHBUA9 2,19,54

OSHA PEL: TWA 0.5 mg(As)/m^3

SAFETY PROFILE: Poison by intravenous route. When heated to decomposition it emits toxic fumes of As.

TGV250 CAS:829-65-2 **HR: 3**
N-(p-TOLYL)-1-AZIRIDINECARBOXAMIDE
mf: $C_{10}H_{12}N_2O$ mw: 176.24

SYN: p-TOLYL-N-CARBAMOYLAZIRIDINE

TOXICITY DATA with REFERENCE
ipr-mus TDLo:240 mg/kg/4W-I:NEO CNREA8 29,2184,69

SAFETY PROFILE: Questionable carcinogen with experimental neoplastigenic data. When heated to decomposition it emits toxic fumes of NO_x.

TGV500 CAS:64046-59-9 **HR: 3**
m-TOLYLAZOACETANILIDE
mf: $C_{15}H_{16}N_3O$ mw: 254.34

TOXICITY DATA with REFERENCE
orl-rat TDLo:4700 mg/kg/52W-C:ETA,REP JNCIAM 24,149,60

SAFETY PROFILE: Experimental reproductive effects. Questionable carcinogen with experimental tumorigenic data. When heated to decomposition it emits toxic fumes of NO_x.

TGV750 CAS:722-25-8 **HR: 3**
p-(p-TOLYLAZO)-ANILINE
mf: $C_{13}H_{13}N_3$ mw: 211.29

SYNS: 4'-METHYL-4-AMINOAZOBENZENE ◇ 4-((4-METHYLPHENYL)AZO)BENZENAMINE

TOXICITY DATA with REFERENCE
dni-mus-ipr 20 g/kg ARGEAR 51,605,81
orl-rat TDLo:8000 mg/kg:ETA CNREA8 5,235,45

SAFETY PROFILE: Questionable carcinogen with experimental tumorigenic data. Mutation data reported. When heated to decomposition it emits toxic fumes of NO_x.

TGW000 CAS:2646-17-5 **HR: 3**
1-(o-TOLYLAZO)-2-NAPHTHOL
mf: $C_{17}H_{14}N_2O$ mw: 262.33

SYNS: A.F.ORANGE No. 2 ◇ AIZEN FOOD ORANGE No. 2 ◇ ATUL OIL ORANGE T ◇ C.I. 12100 ◇ C.I. SOLVENT ORANGE 2 ◇ D&C ORANGE No. 2 ◇ DOLKWAL ORANGE SS ◇ EXTRACT D&C ORANGE No. 4 ◇ FAT ORANGE II ◇ HEXACOL OIL ORANGE SS ◇ LACQUER ORANGE V ◇ 1-((2-METHYLPHENYL)AZO)-2-NAPHTHALENOL ◇ OIL ORANGE O'PEL ◇ OIL ORANGE SS ◇ OLEAL ORANGE SS ◇ ORANGE 3R SOLUBLE IN GREASE ◇ ORGANOL ORANGE 2R ◇ TOLUENE-2-AZONAPHTHOL-2 ◇ o-TOLUENO-AZO-β-NAPHTHOL ◇ 1-(o-TOLYLAZO)-β-NAPHTHOL

TOXICITY DATA with REFERENCE
scu-mus TDLo:6 g/kg/52W:NEO BJCAAI 10,653,56
imp-mus TDLo:50 mg/kg:CAR BJCAAI 12,222,58
orl-rat LDLo:5000 mg/kg JAMAAP 109,493,37
ivn-dog LDLo:200 mg/kg JAMAAP 109,493,37
orl-rbt LDLo:5000 mg/kg JAMAAP 109,493,37
ivn-rbt LDLo:60 mg/kg JAMAAP 109,493,37

CONSENSUS REPORTS: IARC Cancer Review: Group 2B IMEMDT 7,56,87; Animal Sufficient Evidence IMEMDT 8,165,75. Reported in EPA TSCA Inventory. EPA Genetic Toxicology Program.

SAFETY PROFILE: Suspected carcinogen with experimental carcinogenic and neoplastigenic data. Poison by intravenous route. Mildly toxic by ingestion. When heated to decomposition it emits toxic fumes of NO_x. Used to color cosmetics, varnishes, oils, fats and waxes, petroleum products.

TGW500 CAS:63980-19-8 **HR: 3**
2-(o-TOLYLAZO)-p-TOLUIDINE
mf: $C_{14}H_{15}N_3$ mw: 225.32

SYN: 2'-AMINO-2:5'-AZOTOLUENE

TOXICITY DATA with REFERENCE
orl-rat TDLo:15 g/kg/57W-C:ETA BJCAAI 3,387,49
orl-mus TDLo:30 g/kg/57W-C:CAR BJCAAI 3,387,49

SAFETY PROFILE: Questionable carcinogen with experimental carcinogenic and tumorigenic data. When heated to decomposition it emits toxic fumes of NO_x.

TGW750 CAS:63980-18-7 **HR: 3**
4-(p-TOLYLAZO)-o-TOLUIDINE
mf: $C_{14}H_{15}N_3$ mw: 225.32

SYN: 4'-AMINO-4-3'-AZOTOLUENE

TOXICITY DATA with REFERENCE
orl-mus TDLo:30 g/kg/57W-C:ETA BJCAAI 3,387,49

SAFETY PROFILE: Questionable carcinogen with experimental tumorigenic data. When heated to decomposition it emits toxic fumes of NO_x.

TGX000 CAS:63980-27-8 **HR: 3**
1-((4-TOLYLAZO)TOLYLAZO)-2-NAPHTHOL
mf: $C_{24}H_{20}N_4O$ mw: 380.48

PROP: Mixture of m-, o-, and p- isomers (ZEKBAI 57,530,51).

SYN: D&C RED No. 14

TOXICITY DATA with REFERENCE
orl-rat TDLo:13 g/kg/39W-C:ETA ZEKBAI 57,530,51

SAFETY PROFILE: Questionable carcinogen with experimental tumorigenic data. When heated to decomposition it emits toxic fumes of NO_x.

TGX100 CAS:614-34-6 ***HR: 2***
p-TOLYL BENZOATE
mf: $C_{14}H_{12}O_2$ mw: 212.26

PROP: Solid. Mp: 70-71°.

SYNS: BENZOIC ACID, 4-METHYLPHENYL ESTER ◇ BENZOIC ACID, p-TOLYL ESTER ◇ p-CRESYL BENZOATE ◇ 4-METHYLPHENYL BENZOATE

TOXICITY DATA with REFERENCE
skn-rbt 500 mg/24H MLD FCTOD7 21,833,83
orl-rat LD50:2644 mg/kg FAVUAI 18,69,86

CONSENSUS REPORTS: Reported in EPA TSCA Inventory.

SAFETY PROFILE: Moderately toxic by ingestion. A skin irritant. When heated to decomposition it emits acrid smoke and irritating fumes.

TGX500 CAS:31642-65-6 ***HR: 3***
3-(α(o-TOLYL)BENZYLOXY)TROPANE HYDROBROMIDE
mf: $C_{22}H_{27}NO•BrH$ mw: 402.42

SYN: BS 6825

TOXICITY DATA with REFERENCE
orl-mus LD50:150 mg/kg ARZNAD 14,964,64
ivn-mus LD50:26 mg/kg ARZNAD 14,964,64

SAFETY PROFILE: Poison by ingestion and intravenous routes. When heated to decomposition it emits very toxic fumes of NO_x and HBr.

TGY075 CAS:106-43-4 ***HR: 2***
p-TOLYL CHLORIDE
DOT: UN 2238
mf: C_7H_7Cl mw: 126.59

PROP: Liquid. Bp: 162.4°, d: (20/4) 1.0697, mp: 7.5°. Sltly sol in water; sol in alc, benzene, chloroform, ether.

SYNS: 4-CHLORO-1-METHYLBENZENE ◇ 4-CHLOROTOLUENE ◇ p-CHLOROTOLUENE (DOT)

TOXICITY DATA with REFERENCE
orl-rat LD50:3600 mg/kg 85GMAT -,38,82
orl-mus LD50:1900 mg/kg 85GMAT -,38,82
ihl-mus LC50:34 g/m^3/2H 85GMAT -,38,82

CONSENSUS REPORTS: EPA Genetic Toxicology Program. Reported in EPA TSCA Inventory.

DOT Classification: Flammable or Combustible Liquid; Label: Flammable Liquid.

SAFETY PROFILE: Moderately toxic by ingestion. Mildly toxic by inhalation. Flammable when exposed to heat or flame. When heated to decomposition it emits toxic fumes of Cl^-. See also TOLYL CHLORIDE, and CHLORINATED HYDROCARBONS, AROMATIC.

TGY250 CAS:20854-03-9 ***HR: 3***
2-TOLYLCOPPER
mf: C_7H_7Cu mw: 154.67

CONSENSUS REPORTS: Copper and its compounds are on the Community Right-To-Know List.

SAFETY PROFILE: Explodes on contact with oxygen at 0°C. Explodes at 100°C in vacuum. The 2- and 3- isomers behave similarly. See also COPPER COMPOUNDS.

TGY750 CAS:26444-49-5 ***HR: 1***
TOLYL DIPHENYL PHOSPHATE
mf: $C_{19}H_{17}O_4P$ mw: 340.33

SYNS: CRESOL DIPHENYL PHOSPHATE ◇ CRESYL DIPHENYL PHOSPHATE ◇ DIPHENYL CRESOL PHOSPHATE ◇ DIPHENYL CRESYL PHOSPHATE ◇ DIPHENYL TOLYL PHOSPHATE ◇ METHYLPHENYL DIPHENYL PHOSPHATE ◇ MONOCRESYL DIPHENYL PHOSPHATE ◇ PHOSPHORIC ACID, METHYLPHENYL DIPHENYL ESTER (9CI)

TOXICITY DATA with REFERENCE
orl-rat LD50:6400 mg/kg NPIRI* 2,14,75

CONSENSUS REPORTS: Reported in EPA TSCA Inventory.

SAFETY PROFILE: Mildly toxic by ingestion. When heated to decomposition it emits toxic fumes of PO_x. See also ESTERS and PHOSPHATES.

TGZ000 CAS:536-50-5 ***HR: 2***
1-(p-TOLYL)ETHANOL
mf: $C_9H_{12}O$ mw: 136.21

PROP: Bp: 218-220°, d: 0.987, refr index: 1.522.

SYNS: BILAGEN ◇ α,4-DIMETHYLBENZENEMETHANOL ◇ p,α-DIMETHYLBENZYL ALCOHOL ◇ 4-(α-HYDROXYETHYL)TOLUENE ◇ 1-(p-METHYLPHENYL)ETHANOL ◇ 1-(4-(METHYLPHENYL)ETHANOL ◇ METHYL-p-TOLYLCARBINOL ◇ NORBILAN ◇ p-TOLYLMETHYLCARBINOL (GERMAN) ◇ TOMOBIL

TOXICITY DATA with REFERENCE
orl-rat LD50:2800 mg/kg ARZNAD 12,347,62
ims-rat LD50:1 g/kg AEPPAE 222,244,54

CONSENSUS REPORTS: Reported in EPA TSCA Inventory.

SAFETY PROFILE: Moderately toxic by ingestion and intramuscular routes. When heated to decomposition it emits acrid smoke and irritating fumes.

TGZ100 CAS:26447-14-3 ***HR: 3***
TOLYL GLYCIDYL ETHER
mf: $C_{10}H_{12}O_2$ mw: 164.20

SYNS: CRESOL FLYCIDYL ETHER ◇ CRESYLGLYCIDE ETHER ◇ CRESYL GLYCIDYL ETHER ◇ 1,2-EPOXY-3-(TOLYLOXY)PROPANE ◇ GLYCIDYL METHYLPHENYL ETHER ◇ ((METHYLPHENOXY) METHYL)OXIRANE ◇ OXIRANE ((METHYLPHENOXY)METHYL) (9CI)

TOXICITY DATA with REFERENCE
orl-rat LD50:5140 mg/kg GTPZAB 29(3),50,85
ihl-rat LC50:282 mg/m^3 GTPZAB 29(3),50,85
orl-mus LD50:1700 mg/kg GTPZAB 29(3),50,85
ihl-mus LC50:310 mg/m^3 GTPZAB 29(3),50,85
orl-gpg LD50:1650 mg/kg GTPZAB 29(3),50,85

CONSENSUS REPORTS: Reported in EPA TSCA Inventory.

SAFETY PROFILE: Poison by inhalation. Moderately toxic by ingestion. When heated to decomposition it emits acrid smoke and irritating fumes.

THA000 CAS:611-22-3 ***HR: D***
o-TOLYLHYDROXYLAMINE
mf: C_7H_9NO mw: 123.17

PROP: Colorless in benzene and ether. Mp: 44°. Sltly sol in ligroin; sol in alc, ether.

SYNS: N-(2-METHYLPHENYL)-HYDROXYLAMINE ◇ N-(o-TOLYL) HYDROXYLAMINE

TOXICITY DATA with REFERENCE
mma-sat 2 μmol/plate JMCMAR 22,981,79

SAFETY PROFILE: Mutation data reported. When heated to decomposition it emits toxic fumes of NO_x. See also AROMATIC AMINES.

THA250 CAS:103-93-5 ***HR: 2***
p-TOLYL ISOBUTYRATE
mf: $C_{11}H_{14}O_2$ mw: 178.25

PROP: Colorless liquid; characteristic odor. D: 0.990-0.996, refr index: 1.485, flash p: +212°F. Sol in alc; insol in water.

SYNS: p-CRESYL ISOBUTYRATE ◇ FEMA No. 3075 ◇ ISOBUTYRIC ACID, p-TOLYL ESTER ◇ PARACRESYL ISOBUTYRATE

TOXICITY DATA with REFERENCE
orl-rat LD50:4000 mg/kg FCTXAV 13,773,75
skn-rbt LD50:3970 mg/kg FCTXAV 13,773,75

CONSENSUS REPORTS: Reported in EPA TSCA Inventory.

SAFETY PROFILE: Moderately toxic by ingestion and skin contact. Combustible liquid. When heated to decomposition it emits acrid smoke and irritating fumes.

THB000 CAS:59558-23-5 ***HR: 2***
p-TOLYL OCTANOATE
mf: $C_{15}H_{22}O_2$ mw: 234.37

SYNS: p-CRESYL CAPRYLATE ◇ p-CRESYL OCTANOATE ◇ OCTANOIC ACID, p-TOLYL ESTER

TOXICITY DATA with REFERENCE
skn-rbt 500 mg/24H MOD FCTXAV 16,697,78
orl-rat LD50:1600 mg/kg FCTXAV 16,697,78

CONSENSUS REPORTS: Reported in EPA TSCA Inventory.

SAFETY PROFILE: Moderately toxic by ingestion. A skin irritant. When heated to decomposition it emits acrid smoke and irritating fumes. See also ESTERS.

THB250 CAS:21224-81-7 ***HR: 3***
S-2-((4-(p-TOLYLOXY)BUTYL)AMINO)ETHYL THIOSULFATE
mf: $C_{13}H_{21}NO_4S_2$ mw: 319.47

SYNS: THIOSULFURIC ACID, S-(2-((4-(p-TOLYLOXY)BUTYL) AMINO)ETHYL) ESTER ◇ 2-((4-(p-TOLYLOXY)BUTYL)AMINO)ETHANETHIOL HYDROGEN SULFATE (ESTER) ◇ WR 3121

TOXICITY DATA with REFERENCE
orl-mus LD50:1400 mg/kg JMCMAR 11,1190,68
ipr-mus LD50:35 mg/kg JMCMAR 11,1190,68

SAFETY PROFILE: Poison by intraperitoneal route. Moderately toxic by ingestion. When heated to decomposition it emits very toxic fumes of SO_x and NO_x.

THC250 CAS:69781-93-7 ***HR: 3***
(2-o-TOLYLOXYETHYL)HYDRAZINE HYDROCHLORIDE
mf: $C_9H_{14}N_2O•ClH$ mw: 202.71

TOXICITY DATA with REFERENCE
orl-mus LD50:250 mg/kg JMCMAR 6,63,63
ipr-mus LD50:200 mg/kg JMCMAR 6,63,63

SAFETY PROFILE: Poison by ingestion and intraperitoneal routes. When heated to decomposition it emits very toxic fumes of Cl^- and NO_x.

THC500 CAS:81866-63-9 ***HR: 3***
(2-p-TOLYLOXYETHYL)HYDRAZINE HYDROCHLORIDE
mf: $C_9H_{14}N_2O•ClH$ mw: 202.71

TOXICITY DATA with REFERENCE
orl-mus LD50:300 mg/kg JMCMAR 6,63,63
ipr-mus LD50:250 mg/kg JMCMAR 6,63,63

SAFETY PROFILE: Poison by ingestion and intraperitoneal routes. When heated to decomposition it emits very toxic fumes of Cl^- and NO_x.

THD275 CAS:61785-73-7 ***HR: D***
7-(4-(m-TOLYL)-1-PIPERAZINYL)-4-NITRO-BENZOFURAZAN-1-OXIDE
mf: $C_{17}H_{17}N_5O_4$ mw: 355.39

SYNS: B2740 ◇ 7-(4-(3-METHYLPHENYL)-1-PIPERAZINYL)-4-NITRO-BENZOFURAZAN-1-OXIDE

TOXICITY DATA with REFERENCE
mmo-sat 60 μg/plate MUREAV 48,145,77
mma-sat 50 μg/well CBINA8 19,77,77

CONSENSUS REPORTS: EPA Genetic Toxicology Program.

SAFETY PROFILE: Mutation data reported. When heated to decomposition it emits toxic fumes of NO_x.

THD300 CAS:23412-26-2 ***HR: 3***
TOLYLMYCIN Y
mf: $C_{43}H_{54}N_2O_{14}$ mw: 822.99

SYNS: B-2847-Y ◇ 2,3,4,6-TETRADEOXY-4-((5,6,11,12,13,14,15,16,16A,17,17A,18,21,22-TETRADECAHYDRO-2,12,14,16-TETRAHYDROXY-10-METHOXY-3,6,11,13,15,20-HEXAMETHYL-5,18,21,26-TETRAOXO-4,6-EPOXY-1,23-METHANOBANZO(d)CYCLOPROP(n)(1,9)OXAAZA-CYCLOTETRACOSIN-25(10H)-YLIDENE)AMINO)-l-erythro-HEXOPYRANOSE, 12-ACETATE

TOXICITY DATA with REFERENCE
orl-mus LD50:5700 mg/kg 85GDA2 2,468,80
scu-mus LD50:2200 mg/kg 85GDA2 2,468,80
ivn-mus LD50:330 mg/kg 85GDA2 2,468,80

SAFETY PROFILE: Poison by intravenous route. Moderately toxic by subcutaneous route. Mildly toxic by ingestion. When heated to decomposition it emits toxic fumes of NO_x.

THD750 CAS:99-72-9 ***HR: 2***
2-(p-TOLYL)PROPIONIC ALDEHYDE
mf: $C_{10}H_{12}O$ mw: 148.22

SYNS: p-METHYL HYDROTROPALDEHYDE ◇ 2-(p-METHYL-PHENYL)PROPIONALDEHYDE

TOXICITY DATA with REFERENCE
orl-rat LD50:3500 mg/kg FCTXAV 14(6),601,76

CONSENSUS REPORTS: Reported in EPA TSCA Inventory.

SAFETY PROFILE: Moderately toxic by ingestion. When heated to decomposition it emits acrid smoke and irritating fumes. See also ALDEHYDES.

THD850 CAS:607-88-5 ***HR: 2***
p-TOLYL SALICYLATE
mf: $C_{14}H_{12}O_3$ mw: 228.26

SYNS: BENZOIC ACID, 2-HYDROXY-, 4-METHYLPHENYL ESTER (9CI) ◇ p-CRESYL SALICYLATE ◇ 4-METHYLPHENYL 2-HYDROXYBENZOATE ◇ 4-METHYLPHENYL SALICYLATE ◇ SALICYLIC ACID, p-TOLYL ESTER

TOXICITY DATA with REFERENCE
skn-rbt 500 mg/24H MOD FCTOD7 21,835,83
orl-rat LD50:1300 mg/kg FCTOD7 21,835,83

CONSENSUS REPORTS: Reported in EPA TSCA Inventory.

SAFETY PROFILE: Moderately toxic by ingestion. A skin irritant. When heated to decomposition it emits acrid smoke and irritating fumes.

THE250 CAS:1576-35-8 ***HR: 2***
p-TOLYLSULFONYLHYDRAZINE
mf: $C_7H_{10}N_2O_2S$ mw: 186.25

SYNS: p-TOLUENESULFONIC ACID HYDRAZIDE ◇ N-TOLUOLSULPHONYL HYDRAZINE

TOXICITY DATA with REFERENCE
ipr-mus LD50:1200 mg/kg RPTOAN 36,27,73

CONSENSUS REPORTS: Reported in EPA TSCA Inventory.

SAFETY PROFILE: Moderately toxic by intraperitoneal route. When heated to decomposition it emits very toxic fumes of SO_x and NO_x.

THE500 CAS:80-11-5 ***HR: 3***
p-TOLYLSULFONYLMETHYLNITROSAMINE
mf: $C_8H_{10}N_2O_3S$ mw: 214.26

PROP: Yellow crystals from benzene and petr ether. Mp: 62°. Insol in water; sol in ether, petr ether, benzene, chloroform, and carbon tet.

SYNS: DIAZALE ◇ METHYLNITROSO-p-TOLUENESULFONAMIDE ◇ N-NITROSO-N-METHYL-4-TOLYLSULFONAMIDE ◇ TOLUENE-p-SULFONYLMETHYLNITROSAMIDE ◇ p-TOLYLSULFONYL-METHYL-NITROSAMID (GERMAN) ◇ p-TOLYLSULFONYLMETHYLNITROS-AMIDE

TOXICITY DATA with REFERENCE
mmo-sat 14 μmol/L ENMUDM 3,11,81
mmo-esc 14 μmol/L ENMUDM 3,11,81
slt-dmg-orl 2330 μmol/kg MUREAV 144,177,85
dns-rat:lvr 100 μmol/L ENMUDM 3,11,81
ipr-mus TDLo:15 mg/kg:ETA CNREA8 30,11,70
orl-rat LD50:2700 mg/kg NATWAY 48,165,61
ipr-mus LD50:19 mg/kg CNREA8 30,11,70

CONSENSUS REPORTS: Reported in EPA TSCA Inventory. EPA Genetic Toxicology Program.

SAFETY PROFILE: Poison by intraperitoneal route. Moderately toxic by ingestion. Questionable carcinogen with experimental tumorigenic data. Mutation data reported. Many nitrosamines are carcinogens. When heated to decomposition it emits very toxic fumes of NO_x and SO_x. See also NITROSAMINES.

THF250 CAS:614-78-8 ***HR: 3***
1-o-TOLYL-2-THIOUREA
mf: $C_8H_{10}N_2S$ mw: 166.26

PROP: Crystals from water. Mp: 151-152°. Very sol in hot water, alc; very sltly sol in ether.

SYN: o-TOLYL THIOUREA

TOXICITY DATA with REFERENCE
orl-rat LDLo:5 mg/kg NCNSA6 5,1,53
ipr-mus LD50:150 mg/kg NTIS** AD691-490

CONSENSUS REPORTS: EPA Extremely Hazardous Substances List. Reported in EPA TSCA Inventory.

SAFETY PROFILE: Poison by ingestion and intraperitoneal routes. When heated to decomposition it emits very toxic fumes of NO_x and SO_x.

THF750 CAS:63-99-0 ***HR: 2***
3-TOLYLUREA
mf: $C_8H_{10}N_2O$ mw: 150.20

PROP: Leaves from water. Mp: 142-143°.

SYNS: 3-METHYLPHENYLUREA ◇ m-TOLYLCARBAMIDE ◇ m-TOLYLUREA

TOXICITY DATA with REFERENCE
orl-rat LD50:1330 mg/kg IIFBA4 12,195,69
ipr-rat LD50:410 mg/kg IIFBA4 12,195,69
orl-mus LD50:665 mg/kg IIFBA4 12,195,69
ipr-mus LD50:538 mg/kg IIFBA4 12,195,69

SAFETY PROFILE: Moderately toxic by ingestion and intraperitoneal routes. When heated to decomposition it emits toxic fumes of NO_x.

THG000 CAS:622-51-5 ***HR: 3***
p-TOLYLUREA
mf: $C_8H_{10}N_2O$ mw: 150.20

PROP: Plates from alc. Mp: 188°. Very sltly sol in cold water; sol in hot alc.

SYNS: 4-METHYLPHENYLUREA ◇ NCI-C02153 ◇ p-TOLYCARBAMIDE ◇ p-TOLYUREA

TOXICITY DATA with REFERENCE
orl-mus TDLo:88 g/kg/1Y-C:CAR JEPTDQ 3(5-6),149,80
orl-rat LD50:1200 mg/kg NCIMR* NIH-71-E-2144,73

SAFETY PROFILE: Moderately toxic by ingestion. Questionable carcinogen with experimental carcinogenic data. When heated to decomposition it emits toxic fumes of NO_x.

THG250 CAS:17406-45-0 ***HR: 3***
TOMATINE
mf: $C_{50}H_{83}NO_{21}$ mw: 1034.34

PROP: Antifungal substance in wilt-resistant tomato plants (ARBIAE 18,467,48). Needles. Mp: 263-268°. Sol in ethanol, methanol, dioxane, propylene alc; almost insol in water, ether, petr ether.

SYNS: LYCOPERSICIN ◇ A″-TOMATIDINE ◇ TOMATIDINE GLYCOSIDE ◇ TOMATIN ◇ α-TOMATINE

TOXICITY DATA with REFERENCE
orl-rat LD50:900 mg/kg TXAPA9 3,39,61
orl-mus LD50:500 mg/kg FCTXAV 17,61,79
ipr-mus LD50:25 mg/kg SZAPAC 22,557,59
scu-mus LD50:1000 mg/kg 85GDA2 8(2),218,82
ivn-mus LD50:18 mg/kg TXAPA9 3,39,61

SAFETY PROFILE: Poison by intravenous and intraperitoneal routes. Moderately toxic by ingestion and subcutaneous routes. When heated to decomposition it emits toxic fumes of NO_x.

THG500 CAS:35050-55-6 ***HR: 3***
TOMAYMYCIN
mf: $C_{16}H_{20}N_2O_4$ mw: 304.38

PROP: Produced by *Steptomyces achromogenes var. tomaymyceticus* (85ERAY 2,1303,78).

SYN: 5H-PYRROLO(2,2-c)(1,4)BENZODIAZEPIN-5-ONE,2-ETHYLIDENE-1,2,3,10,11,11A-HEXAHYDRO-8-HYDROXY-7,11-DIMETHOXY-

TOXICITY DATA with REFERENCE
dnd-mam:lym 400 μmol/L BBACAQ 475,521,77
ipr-mus LD50:1 mg/kg TOLED5 18,337,83
scu-mus LD50:7400 μg/kg 85GDA2 5,186,81
ivn-mus LD50:3540 μg/kg 85GDA2 5,186,81

SAFETY PROFILE: Poison by intraperitoneal, subcutaneous, and intravenous routes. Mutation data reported. When heated to decomposition it emits toxic fumes of NO_x.

THG600 CAS:50602-44-3 ***HR: 3***
TOMIZINE
mf: $C_7H_8N_4OS{\cdot}ClH$ mw: 232.71

SYNS: 4-METHOXY-7H-PYRIMIDO(4,5-b)(1,4)THIAZIN-6-AMINE MONOHYDROCHLORIDE ◇ THOMIZINE

TOXICITY DATA with REFERENCE
sln-dmg-orl 400 mg/L PCJOAU 15,772,81
cyt-mus-ipr 3 mg/kg PCJOAU 15,772,81
spm-mus-ipr 90 mg/kg PCJOAU 15,772,81
ipr-rat LD50:170 mg/kg FATOAO 42,85,79
ipr-mus LD50:180 mg/kg PCJOAU 15,772,81

SAFETY PROFILE: Poison by intraperitoneal route.

Mutation data reported. When heated to decomposition it emits toxic fumes of NO_x, SO_x, and HCl.

THG700 CAS:76145-76-1 ***HR: 2***
TOMOXIPROLE
mf: $C_{21}H_{20}N_2O$ mw: 316.43

SYNS: 3-ISOPROPYL-2-(p-METHOXYPHENYL)-3H-NAPHTH(1,2-d)IMIDAZOLE ◇ MDL-035 ◇ 2-(4-METHOXYPHENYL)-3-(1-METHYLETHYL)-3H-NAPHTH(1,2-d)IMIDAZOLE ◇ TOMOXIPROL

TOXICITY DATA with REFERENCE
orl-rat LD50:15 g/kg EJDPD2 10,161,85
ipr-rat LD50:1500 mg/kg DRFUD4 10,133,85
ipr-mus LD50:1500 mg/kg DRFUD4 10,133,85
ipr-dog LD50:1500 mg/kg DRFUD4 10,144,85

SAFETY PROFILE: Moderately toxic by intraperitoneal route. Mildly toxic by ingestion. An anti-inflammatory and analgesic agent. When heated to decomposition it emits toxic fumes of NO_x.

THG750 ***HR: 2***
TONKA ABSOLUTE

PROP: Main constitutent is coumarin, found in seeds of fruit of tree *Dipteyx odorata* (FCTXAV 12,807,74).

TOXICITY DATA with REFERENCE
skn-mus 500 mg FCTXAV 12,807,74
skn-rbt 500 mg/24H FCTXAV 12,807,74
orl-rat LD50:1380 mg/kg FCTXAV 12,807,74

SAFETY PROFILE: Moderately toxic by ingestion. A skin irritant. See also COUMARIN.

THH000 CAS:1302-59-6 ***HR: 2***
TOPAZ

PROP: White powder; colored crystals.

SYNS: ALUMINUM HEXAFLUOROSILICATE ◇ FLUOSILICATE de ALUMINUM (FRENCH)

TOXICITY DATA with REFERENCE
orl-gpg LDLo:5000 mg/kg AMSSAQ 400,5,63
scu-gpg LDLo:4000 mg/kg AMSSAQ 400,5,63

CONSENSUS REPORTS: Reported in EPA TSCA Inventory.

SAFETY PROFILE: Moderately toxic by subcutaneous route. Mildly toxic by ingestion. When heated to decomposition it emits toxic fumes of F^-.

THH350 CAS:40507-23-1 ***HR: 2***
TORMOSYL
mf: $C_{18}H_{17}FN_2O$ mw: 296.37

SYNS: 4-(p-FLUOROPHENYL)-1-ISOPROPYL-7-METHYL-2(1H)-QUINAZOLINONE ◇ 4-(4-FLUOROPHENYL)-7-METHYL-1-(1-METHYLETHYL)-2(1H)-QUINAZOLINONE ◇ 4-FLUORPHENYL-1-ISOPROPYL-7-METHYL-2(1H)-CHINAZOLINON (GERMAN) ◇ FLUPROQUAZONE ◇ 1-ISOPROPYL-7-METHYL-4-(p-FLUOROPHENYL)-2(1H)-QUINAZOLINONE ◇ RF 46-790 ◇ SaH 46-790

TOXICITY DATA with REFERENCE
orl-rat TDLo:435 mg/kg (female 15-22D post):REP ARZNAD 31,882,81
orl-rat TDLo:500 mg/kg (female 6-15D post):TER ARZNAD 31,882,81
ipr-mus LD50:650 mg/kg ARZNAD 34,879,84
orl-rbt LD50:742 mg/kg ARZNAD 31,882,81

SAFETY PROFILE: Moderately toxic by ingestion and intraperitoneal routes. An experimental teratogen. Experimental reproductive effects. When heated to decomposition it emits toxic fumes of F^- and NO_x.

THH450 CAS:402-71-1 ***HR: D***
l-1-TOSYLAMIDO-2-PHENYLETHYL CHLOROMETHYL KETONE
mf: $C_{17}H_{18}ClNO_3S$ mw: 351.87

SYNS: BENZENESULFONAMIDE,N-(3-CHLORO-2-OXO-1-(PHENYLMETHYL)PROPYL)-4-METHYL-,(S)- ◇ l-N-(α-(CHLOROACETYL)PHENETHYL)-p-TOLUENESULFONAMIDE ◇ p-TOLUENESULFONAMIDE, N-(α-(CHLOROACETYL)PHENETHYL)-, (-)-

TOXICITY DATA with REFERENCE
iut-mus TDLo:160 mg/kg (female 2D post):REP FESTAS 25,954,74

CONSENSUS REPORTS: Reported in EPA TSCA Inventory.

SAFETY PROFILE: Experimental reproductive effects. When heated to decomposition it emits toxic fumes of NO_x, SO_x, and Cl^-.

THH500 CAS:2364-87-6 ***HR: 3***
N-α-TOSYL-l-LYSYL-CHLOROMETHYLKETONE
mf: $C_{14}H_{21}ClN_2O_3S$ mw: 332.88

SYNS: (s)-N-(5-AMINO-1-(CHLOROACETYL)PENTYL)-4-METHYLBENZENESULFONAMIDE ◇ TLCK ◇ TOSYLLYSINE CHLOROMETHYL KETONE ◇ TOSYL-l-LYSINE CHLOROMETHYL KETONE ◇ TOSYLLYSYL CHLOROMETHYL KETONE ◇ α-TOSYL-l-LYSYLCHLOROMETHYL KETONE

TOXICITY DATA with REFERENCE
ipr-mus TDLo:280 mg/kg (28D pre):REP JPMSAE 70,60,81
ipr-mus LD50:59 mg/kg DCTODJ 3,227,80
scu-mus LD50:64 mg/kg DCTODJ 3,227,80

SAFETY PROFILE: Poison by intraperitoneal and subcutaneous routes. Experimental reproductive effects. When heated to decomposition it emits very toxic fumes of Cl^-, NO_x, and SO_x. See also KETONES.

THH550 CAS:102516-65-4 ***HR: 3***
TOSYL-l-PHENYLALANYLCHLOROMETHYL KETONE
mf: $C_{18}H_{18}ClNO_5S$ mw: 395.88

SYNS: N-(CHLOROACETYL)-3-PHENYL-N-(p-TOLYLSULFONYL)ALANINE ◇ N-TOSYL-l-PHENYLALANINE CHLOROMETHYL KETONE

TOXICITY DATA with REFERENCE
ipr-mus TDLo:280 mg/kg (28D pre):REP JPMSAE 70,60,81
unr-mus LD50:83 mg/kg JPMSAE 67,1726,78
ivg-mus LD50:75 mg/kg JPMSAE 68,696,79

SAFETY PROFILE: Poison by intravaginal route. Experimental reproductive effects. When heated to decomposition it emits toxic fumes of Cl^-, SO_x, and NO_x.

THH560 CAS:102489-36-1 ***HR: 3***
TOXAPHENE TOXICANT A

TOXICITY DATA with REFERENCE
orl-rat TDLo:36 μg/kg (female 5-22D post):REP AECTCV 9,247,80
ipr-mus LD50:3100 μg/kg JAFCAU 22,653,74

SAFETY PROFILE: Poison by intraperitoneal route. Experimental reproductive effects. When heated to decomposition it emits toxic fumes of Cl^-.

THH575 CAS:51775-36-1 ***HR: 3***
TOXAPHENE TOXICANT B
mf: $C_{10}H_{11}Cl_7$ mw: 379.36

SYNS: BICYCLO(2.2.1)HEPTANE,2,2,5,6-TETRACHLORO-1,7,7-TRIS(CHLOROMETHYL)-, (5-endo,6-exo)- ◇ BORNANE, 2,2,5-endo,6-exo,8,9,10-HEPTACHLORO- ◇ 2,2,5-endo,6-exo,8,9,10-HEPTACHLOROBORNANE ◇ 5-endo,6-exo-2,2,5,6-TETRACHLORO-1,7,7-TRIS(CHLOROMETHYL)-BICYCLO(2.2.1)HEPTANE

TOXICITY DATA with REFERENCE
orl-rat TDLo:36 μg/kg (female 5-22D post):REP AECTCV 9,247,80
ipr-mus LD50:6600 μg/kg JAFCAU 22,653,74

SAFETY PROFILE: Poison by intraperitoneal route. Experimental reproductive effects. When heated to decomposition it emits toxic fumes of Cl^-.

THI000 CAS:6696-58-8 ***HR: 3***
TOXIFERINE DICHLORIDE
mf: $C_{40}H_{46}N_4O_2•2Cl$ mw: 685.80

PROP: Crystals. Sol in water.

SYN: C-TOXIFERINE 1

TOXICITY DATA with REFERENCE
ivn-mky LD50:8900 ng/kg TXAPA9 35,107,76
ims-mky LD50:18 μg/kg TXAPA9 35,107,76
ims-gpg LD50:14 μg/kg TXAPA9 35,107,76

SAFETY PROFILE: A deadly poison by intravenous and intramuscular routes. When heated to decomposition it emits very toxic fumes of NO_x and Cl^-.

THI250 CAS:26934-87-2 ***HR: 3***
TOXIN HT 2
mf: $C_{21}H_{32}O_8$ mw: 412.53

SYNS: 12,13-EPOXY-TRICHOTHEC-9-ENE-3-α,15-TETROL 15-ACETATE, 8-ISOVALERATE ◇ HT-2 TOXIN

TOXICITY DATA with REFERENCE
skn-gpg 330 ng MLD FAATDF 4(2,Pt 2),S124,84
orl-mus LD50:3800 μg/kg 85GDA2 6,191,81
ipr-mus LD50:5200 μg/kg 85GDA2 6,191,81
orl-ckn LD50:7220 μg/kg AEMIDF 35,636,78

SAFETY PROFILE: Poison by ingestion and intraperitoneal routes. A skin irritant. When heated to decomposition it emits acrid smoke and irritating fumes.

THI425 ***HR: D***
TOXOFACTOR

PROP: A toxin associated with *Toxoplasma gondii* infection that was obtained from the trophozoites and culture medium used to propagate parasites in cell cultures (INFIBR 42,1126,83).

TOXICITY DATA with REFERENCE
orl-mus TDLo:20 g/kg (1D pre):REP INFIBR 42,1126,83

SAFETY PROFILE: Experimental reproductive effects.

THJ100 CAS:3084-62-6 ***HR: 2***
2,4,5-T PROPYLENE GLYCOL BUTYL ETHER ESTER
mf: $C_{15}H_{19}Cl_3O_4$ mw: 369.69

SYNS: 2,4,5-T PGBEE ◇ 2,4,5-TRICHLOROPHENOXYACETIC ACID, PROPYLENE GLYCOL BUTYL ETHER ESTERS

TOXICITY DATA with REFERENCE
orl-mus TDLo:1479 mg/kg (12-15D preg):TER AECTCV 6,33,77
orl-rat LD50:500 mg/kg NTIS** PB85-143766

CONSENSUS REPORTS: Glycol ether compounds are on the Community Right-To-Know List.

SAFETY PROFILE: Moderately toxic by ingestion. Some glycol ethers have dangerous human reproductive effects. An experimental teratogen. When heated to decomposition it emits toxic fumes of Cl^-. See also 2,4,5-TRICHLOROPHENOXYACETIC ACID and GLYCOL ETHERS.

THJ250 CAS:9000-65-1 ***HR: 1***
TRAGACANTH GUM

PROP: from the shrub *Astragalus gummifier* Labillardiere. Powder is white, pieces are white to pale yellow, translucent, and horny; odorless with mucilaginous taste.

SYNS: GUM TRAGACANTH ◇ TRAGACANTH

TOXICITY DATA with REFERENCE
orl-rat LD50:16400 mg/kg 85AIAL -,45,73
orl-mus LD50:10000 mg/kg FDRLI* 124,-,76
orl-rbt LD50:7200 mg/kg FDRLI* 124,-,76
orl-ham LD50:8800 mg/kg FDRLI* 124,-,76

CONSENSUS REPORTS: Reported in EPA TSCA Inventory.

SAFETY PROFILE: Mildly toxic by ingestion. A mild allergen. Combustible when exposed to heat or flame. When heated to decomposition it emits acrid smoke and irritating fumes.

THJ500 CAS:27203-92-5 ***HR: 3***
TRAMADOL
mf: $C_{16}H_{25}NO_2$ mw: 263.42

SYNS: CG 315 ◇ (±)-trans-2-((DIMETHYLAMINO)METHYL-1-(m-METHOXYPHENYL)CYCLOHEXANOL ◇ TRAMAL

TOXICITY DATA with REFERENCE
orl-rat LDLo:228 mg/kg TXAPA9 18,185,71
scu-rat LDLo:286 mg/kg TXAPA9 18,185,71
orl-mus LD50:350 mg/kg ARZNAD 28,164,78
scu-mus LD50:200 mg/kg ARZNAD 28,164,78
ivn-mus LD50:68 mg/kg ARZNAD 28,164,78
orl-dog LD50:450 mg/kg ARZNAD 28,164,78
ivn-dog LD50:50 mg/kg ARZNAD 28,164,78
ims-dog LD50:100 mg/kg ARZNAD 28,164,78
orl-rbt LD50:500 mg/kg ARZNAD 28,164,78
ivn-rbt LD50:50 mg/kg ARZNAD 28,164,78
ims-rbt LD50:300 mg/kg ARZNAD 28,164,78
orl-gpg LD50:850 mg/kg ARZNAD 28,164,78
scu-gpg LD50:245 mg/kg ARZNAD 28,164,78

SAFETY PROFILE: Poison by ingestion, subcutaneous, intravenous, and intramuscular routes. When heated to decomposition it emits toxic fumes of NO_x. See also TRAMADOL HYDROCHLORIDE.

THJ750 CAS:73806-49-2 ***HR: 3***
TRAMADOL HYDROCHLORIDE
mf: $C_{16}H_{25}NO_2{\cdot}ClH$ mw: 299.88

PROP: White crystals. Mp: 180-181°. Sol in water.

SYNS: CRISPIN ◇ trans-2-(DIMETHYLAMINOMETHYL)-1-(m-METHOXYPHENYL)CYCLOHEXANOL HYDROCHLORIDE ◇ (E)-2-((DIMETHYLAMINO)METHYL)-1-(m-METHOXYPHENYL)-1-CYCLOHEXANOL HYDROCHLORIDE ◇ K-315 ◇ 1-(m-METHOXYPHENYL)-2-DIMETHYLAMINOMETHYL-CYCLOHEXAN-1-OL HYDROCHLORIDE ◇ trans-1-(m-METHOXYPHENYL)-2-DIMETHYLAMINOMETHYLCYCLOHEXAN-1-OL HYDROCHLORIDE ◇ TRAMAL

TOXICITY DATA with REFERENCE
scu-rat TDLo:2880 mg/kg (36 D male):REP KSRNAM 6,1427,72
orl-rat LD50:228 mg/kg ARZNAD 28,164,78
scu-rat LD50:286 mg/kg ARZNAD 28,164,78
ivn-rat LD50:57600 μg/kg NIIRDN 6,513,82
ims-rat LD50:244 mg/kg NIIRDN 6,513,82
orl-mus LD50:270 mg/kg CCCCAK 52,1340,87
scu-mus LD50:200 mg/kg ARZNAD 28,114,78
ivn-mus LD50:60450 μg/kg CCCCAK 52,1340,87
ims-mus LD50:258 mg/kg KSRNAM 6,1427,72

SAFETY PROFILE: Poison by ingestion, subcutaneous, intramuscular, and intravenous routes. Experimental reproductive effects. When heated to decomposition it emits very toxic fumes of NO_x and HCl. Used as an analgesic. See also TRAMADOL.

THJ755 CAS:46941-74-6 ***HR: 3***
TRAMADOL (2)
mf: $C_{16}H_{25}NO_2$ mw: 263.42

SYNS: trans-(±)-CYCLOHEXANOL-2-((DIMETHYLAMINO)METHYL)-1-(3-METHOXYPHENYL) ◇ (+)-trans-2-(DIMETHYLAMINOMETHYL)-1-(m-METHOXYPHENYL)CYCLOHEXANOL ◇ (+)-(E)-2-(DIMETHYLAMINOMETHYL)-1-(m-METHOXYPHENYL)CYCLOHEXANOL

TOXICITY DATA with REFERENCE
scu-mus LD50:198 mg/kg ARZNAD 28,114,78
ivn-mus LD50:47 mg/kg ARZNAD 28,114,78

SAFETY PROFILE: Poison by subcutaneous and intravenous routes. When heated to decomposition it emits toxic fumes of NO_x.

THJ825 CAS:3715-90-0 ***HR: 3***
TRAMAZOLINE HYDROCHLORIDE
mf: $C_{13}H_{17}N_3{\cdot}ClH$ mw: 251.79

PROP: Crystals from alc and ether or acetone and ether. Mp: 172-174°. Sol in water.

SYNS: 4,5-DIHYDRO-N-(4,6,7,8-TETRAHYDRO-1-NAPHTHALENYL)-1H-IMIDAZOL-2-AMINE MONOHYDROCHLORIDE ◇ KB 227 ◇ TETRAHYDRONAPHTHYLAMINOIMIDAZOLINE HYDROCHLORIDE ◇ 2'-(5,6,7,8-TETRAHYDRO-1-NAPHTHYLAMINO)IMIDAZOLINE HYDROCHLORIDE

TOXICITY DATA with REFERENCE
orl-mus LD50:195 mg/kg NIIRDN 6,513,82
ipr-mus LD50:57 mg/kg NIIRDN 6,513,82
scu-mus LD50:77 mg/kg NIIRDN 6,513,82

SAFETY PROFILE: Poison by ingestion, subcutaneous, and intraperitoneal routes. When heated to decomposition it emits toxic fumes of NO_x and HCl.

THK000 CAS:64-95-9 ***HR: 3***
TRANSENTINE
mf: $C_{20}H_{25}NO_2$ mw: 311.46

SYNS: ADIPHENIN ◇ 2-DIETHYLAMINOETHYL DIPHENYL ACETATE ◇ 2-DIETHYLAMINOETHYL ESTER KYSELINY DIFENYLOCTOVE (CZECH) ◇ 2-(DIETHYLAMINO)ETHYL ESTER-2-PHENYL BENZENE ACETIC ACID ◇ DIFACIL ◇ DIPHENYLACETIC ACID, 2-(DIETHYLAMINO)ETHYL ESTER ◇ DIPHENYLACETIC ACID DIETHYLAMINOETHYL ESTER ◇ DIPHENYLACETYLDIETHYLAMINOETHANOL ◇ ESTER DWETYLOAMINOETYLOWSKY KWASU DWUFENYLOOCTOWEGO (POLISH) ◇ PATROVINE ◇ SPASMOLYTON ◇ TRAZENTYNA (POLISH) ◇ VEGANTINE ◇ WEGANTYNA (POLISH)

TOXICITY DATA with REFERENCE
ipr-mus LD50:182 mg/kg PCJOAU 2,201,68
ivn-rat LD50:27 mg/kg AJDDAL 18,241,51
orl-mus LD50:600 mg/kg CLDND*
ivn-rbt LD50:30 mg/kg CLDND*
ivn-dog LD50:35 mg/kg JPETAB 89,131,47
ivn-mus LD50:21500 ug/kg JPETAB 104,269,52
orl-rat LDLo:1600 mg/kg APPNAH 1,4,50
scu-mus LD50:400 mg/kg BJPCAL 14,559,59
scu-rat LDLo:1600 mg/kg APPNAH 1,4,50

SAFETY PROFILE: Poison by intravenous, subcutaneous, and intraperitoneal routes. Moderately toxic by ingestion. When heated to decomposition it emits toxic fumes of NO_x. See also ESTERS.

THK600 CAS:298-51-1 ***HR: 3***
TRANSERGAN
mf: $C_{19}H_{22}N_2O_2S{\cdot}ClH$ mw: 378.95

SYNS: β-DIETHYLAMINOETHYLPHENOTHIAZINE-10-CARBOXYLATE HYDROCHLORIDE ◇ β-DIETHYLAMINOETHYL PHENOTHIAZINE-N-CARBOXYLATE HYDROCHLORIDE ◇ 10-PHENOTHIAZINECARBOXYLIC ACID β-DIETHYLAMINOETHYL ESTER HYDROCHLORIDE

TOXICITY DATA with REFERENCE
orl-mus LD50:440 mg/kg APTOA6 13,59,57
ipr-mus LD50:140 mg/kg APTOA6 13,59,57
scu-mus LD50:620 mg/kg APTOA6 13,59,57
ivn-mus LD50:26 mg/kg AIPTAK 90,241,52

SAFETY PROFILE: Poison by intraperitoneal and intravenous routes. Moderately toxic by ingestion and subcutaneous routes. When heated to decomposition it emits very toxic fumes of NO_x, SO_x, and HCl. See also ESTERS.

THK750 CAS:6452-73-9 ***HR: 3***
TRASICOR
mf: $C_{15}H_{23}NO_3{\cdot}ClH$ mw: 301.85

SYNS: 2-(o-ALLYLOXYPHENOXY)-2-HYDROXY-N-ISOPROPYL-1-PROPYLAMINE HYDROCHLORIDE ◇ 1-(o-ALLYLOXYPHENOXY)-3-ISOPROPYLAMINOPROPAN-2-OL HYDROCHLORIDE ◇ Ba-39089 ◇ C-39089-Ba ◇ CIBA 39089-Ba ◇ CORETAL ◇ OXPRENOLOL HYDROCHLORIDE

TOXICITY DATA with REFERENCE
ivn-rbt TDLo:10500 μg/kg (7D pre):REP CCPTAY 12,53,75
orl-wmn LDLo:90 mg/kg BMJOAE 1,552,77
orl-rat LD50:214 mg/kg ARZNAD 35,1236,85
ipr-rat LD50:147 mg/kg NIIRDN 6,158,82
scu-rat LD50:940 mg/kg ARZNAD 18,164,68
ivn-rat LD50:33 mg/kg ARZNAD 18,164,68
ipr-mus LD50:170 mg/kg PJPPAA 25,151,73
scu-mus LD50:245 mg/kg ARZNAD 18,164,68
ivn-mus LD50:20 mg/kg ARZNAD 27,1022,77
ivn-dog LD50:15 mg/kg ARZNAD 20,1890,70
ivn-rbt LDLo:20 mg/kg ARZNAD 20,1890,70

SAFETY PROFILE: Human poison by ingestion. Experimental poison by ingestion, intraperitoneal, and intravenous routes. Moderately toxic experimentally by subcutaneous route. Experimental reproductive effects. A beta-adrenergic blocker. When heated to decomposition it emits toxic fumes of NO_x and HCl. See also ALLYL COMPOUNDS and AMINES.

THK850 ***HR: 1***
TRAXANOX SODIUM PENTAHYDRATE
mf: $C_{13}H_5ClN_5O_2{\cdot}Na{\cdot}5H_2O$ mw: 411.77

SYNS: 9-CHLORO-5-OXO-7-(1H-TETRAZOL-5-YL)-5H-1-BENZOPYRANO(2,3-b)PYRIDINE SODIUM SALT PENTAHYDRATE ◇ 9-CHLORO-7-(1H-TETRAZOL-5-YL)-5H-1-BENZOPYRANO(2,3-b)PYRIDIN-5-ONE SODIUM PENTAHYDRATE

TOXICITY DATA with REFERENCE
orl-rat TDLo:13500 mg/kg (female 17-22D post):REP IYKEDH 14,693,83
orl-rat TDLo:5500 mg/kg (7-17D preg):TER IYKEDH 14,673,83
ipr-rat LD50:6685 mg/kg IYKEDH 14,709,83
ipr-mus LD50:9274 mg/kg IYKEDH 14,709,83

SAFETY PROFILE: An experimental teratogen. Experimental reproductive effects. When heated to decomposition it emits toxic fumes of Cl^-, NO_x, and Na_2O.

THK875 CAS:19794-93-5 ***HR: 3***
TRAZODONE
mf: $C_{19}H_{22}ClN_5O$ mw: 371.91

PROP: Crystals. Mp: 86-87°, pKa (50% ethanol): 6.14.

SYNS: 2-(3-(4-(3-CHLOROPHENYL)-1-PIPERAZINYL)PROPYL)-1,2,4-TRIAZOLO(4,3-a)PYRIDIN-3(2H)-ONE ◇ DESYREL ◇ TRAZODON

TOXICITY DATA with REFERENCE
orl-rat TDLo:11 g/kg (14D pre/1-22D preg):REP APTOD9 19,A90,80
orl-wmn TDLo:14 mg/kg:LIV AIMEAS 99,572,83
orl-rat LD50:690 mg/kg DRUGAY 21,401,81
ipr-rat LD50:178 mg/kg DRUGAY 21,401,81
ivn-rat LD50:91 mg/kg DRUGAY 21,401,81

orl-mus LD50:610 mg/kg DRUGAY 21,401,81
ipr-mus LD50:210 mg/kg DRUGAY 21,401,81
ivn-mus LD50:91 mg/kg DRUGAY 21,401,81

SAFETY PROFILE: Poison by intravenous and intraperitoneal routes. Moderately toxic by ingestion. Human systemic effects: cholestatic jaundice. Experimental reproductive effects. When heated to decomposition it emits toxic fumes of Cl^- and NO_x.

THK880 CAS:25332-39-2 ***HR: 3***
TRAZODONE HYDROCHLORIDE
mf: $C_{19}H_{22}ClN_5O{\cdot}ClH$ mw: 408.37

SYNS: AF 1161 ◇ 2-(3-(4-(3-CHLOROPHENYL)-1-PIPERAZINYL)PROPYL)-s-TRIAZOLO(4,3-a)PYRIDIN-3(2H)-ONE HCl ◇ DESYREL ◇ MOLIPAXIN ◇ PRAGMAZONE ◇ THOMBRAN ◇ TOMBRAN ◇ s-TRIAZOLO(4,3-a)PYRIDIN-3(2H)-ONE,2-(3-(4-(m-CHLOROPHENYL)-1-PIPERAZINYL)PROPYL)-, MONOHYDROCHLORIDE ◇ TRITTICO

TOXICITY DATA with REFERENCE
orl-man TDLo:159 mg/kg (male 7W pre):REP AJPSAO 140,1256,83
orl-wmn TDLo:750 μg/kg AJPSAO 141,434,84
orl-wmn TDLo:7500 μg/kg/5D-I AJPSAO 140,642,83
orl-man TDLo:46 mg/kg/8D-I JCPYDR 6,117,86
orl-man TDLo:667 μg/kg:CVS AJPSAO 141,1472,84
orl-rat LD50:690 mg/kg MPPPBK 9,76,74
ipr-rat LD50:178 mg/kg MPPPBK 9,76,74
ivn-rat LD50:91 mg/kg MPPPBK 9,76,74
orl-mus LD50:610 mg/kg MPPPBK 9,76,74
ivn-mus LD50:91 mg/kg MPPPBK 9,76,74
orl-dog LD50:500 mg/kg MPPPBK 9,76,74
ivn-mky LD50:25 mg/kg PBPSDY 3,94,81

SAFETY PROFILE: Poison by intravenous route. Moderately toxic by ingestion and intraperitoneal routes. Experimental reproductive effects. Human systemic effects by ingestion: cardiomyopathy. When heated to decomposition it emits toxic fumes of NO_x, HCl, and Cl^-.

THL500 CAS:1553-34-0 ***HR: 3***
TREMARIL HYDROCHLORIDE
mf: $C_{29}H_{23}NS{\cdot}ClH$ mw: 345.96

SYNS: METHIXENE HYDROCHLORIDE ◇ 9-(1-METHYL-3-PIPERIDYLMETHYL)THIAXANTHENE HYDROCHLORIDE ◇ 9-((N-METHYL-3-PIPERIDYL)METHYL)-THIOXANTHENHYDROCHLORID (GERMAN) ◇ 1-METHYL-3-(THIOXANTHEN-9-YLMETHYL)-1-PIPERIDINE HYDROCHLORIDE ◇ N 715 ◇ 3-(THIOXANTHEN-9-YL)METHYL-1-PIPECOLINE, HYDROCHLORIDE ◇ TREMARIL WONDER ◇ TREST

TOXICITY DATA with REFERENCE
orl-rat LD50:1460 mg/kg NIIRDN 6,824,82
ipr-rat LD50:295 mg/kg NIIRDN 6,824,82
scu-rat LD50:1519 mg/kg NIIRDN 6,824,82
ivn-rat LD50:24 mg/kg AIPTAK 141,331,63
orl-mus LD50:346 mg/kg NIIRDN 6,824,82
ipr-mus LD50:108 mg/kg NIIRDN 6,824,82
scu-mus LD50:238 mg/kg NIIRDN 6,824,82
ivn-mus LD50:18 mg/kg 27ZQAG -,80,72

SAFETY PROFILE: Poison by ingestion, intraperitoneal, subcutaneous and intravenous routes. When heated to decomposition it emits very toxic fumes of NO_x, SO_x, and HCl.

THL750 CAS:3736-86-5 ***HR: 3***
TRENTADIL HYDROCHLORIDE
mf: $C_{20}H_{27}N_5O_3{\cdot}ClH$ mw: 421.98

SYNS: BAMIFYLLINE HYDROCHLORIDE ◇ BAMIPHYLLINE HYDROCHLORIDE ◇ BAX 2793Z ◇ BENZETAMOPHYLLINE HYDROCHLORIDE ◇ 8-BENZYL-7-(2-(ETHYL(2-HYDROXYETHYL)AMINO)ETHYL)THEOPHYLLINE, HYDROCHLORIDE ◇ TRENTADIL

TOXICITY DATA with REFERENCE
orl-rat LD50:1139 mg/kg ARZNAD 18,460,68
ipr-rat LD50:131 mg/kg ARZNAD 18,460,68
ivn-rat LD50:64800 μg/kg ARZNAD 18,460,68
orl-mus LD50:246 mg/kg ARZNAD 18,460,68
ipr-mus LD50:89400 μg/kg ARZNAD 18,460,68
ivn-mus LD50:66800 μg/kg ARZNAD 18,460,68

SAFETY PROFILE: Poison by ingestion, intraperitoneal, and intravenous routes. When heated to decomposition it emits very toxic fumes of NO_x and HCl.

THM250 CAS:16800-47-8 ***HR: 3***
TRIACETONITRILE TUNGSTEN TRICARBONYL
mf: $C_9H_9N_3O_3W$ mw: 391.06

SYN: TRIS(ACETONITRILE)TRICARBONYLTUNGSTEN

TOXICITY DATA with REFERENCE
ivn-mus LD50:56200 μg/kg CSLNX* NX#02385

CONSENSUS REPORTS: Cyanide and its compounds are on the Community Right-To-Know List.

ACGIH TLV: TWA 5 mg(W)/m^3; STEL 10 mg(W)/m^3
NIOSH REL: TWA 1 mg(W)/m^3

SAFETY PROFILE: Poison by intravenous route. When heated to decomposition it emits very toxic fumes of NO_x and CN^-. See also TUNGSTEN COMPOUNDS, NITRILES, and CARBONYLS.

THM500 CAS:102-76-1 ***HR: 3***
TRIACETYL GLYCERIN
mf: $C_9H_{14}O_6$ mw: 218.23

PROP: Colorless oily liquid; slt fatty odor and taste. Mp: −78°, bp: 258°, flash p: 280°F (COC), d: 1.161, autoign temp: 812°F, vap d: 7.52. Sol in water; misc with alc, ether, chloroform.

SYNS: ENZACTIN ◇ FEMA No. 2007 ◇ FUNGACETIN ◇ GLYCERINE TRIACETATE ◇ GLYCEROL TRIACETATE ◇ GLYCERYL TRIACE-

TATE ◇ GLYPED ◇ KESSCOFLEX TRA ◇ KODAFLEX TRIACETIN ◇ 1,2,3-PROPANETRIOL TRIACETATE ◇ TRIACETIN (FCC) ◇ VANAY

TOXICITY DATA with REFERENCE
eye-rbt 116 mg JPETAB 82,377,44
orl-rat LD50:3000 mg/kg AMIHAB 21,28,60
ipr-rat LD50:2100 mg/kg FCTXAV 16,637,78
scu-rat LD50:2800 mg/kg PSEBAA 46,26,41
orl-mus LD50:1100 mg/kg FEPRA7 22,368,63
ipr-mus LD50:1400 mg/kg FEPRA7 22,368,63
scu-mus LD50:2300 mg/kg PSEBAA 46,26,41
ivn-mus LD50:1600 mg/kg APSCAX 40,338,57
ivn-dog LD50:1500 mg/kg FCTXAV 16,637,78
ivn-rbt LD50:750 mg/kg FCTXAV 16,637,78
orl-frg LDLo:150 mg/kg FCTXAV 16,637,78

CONSENSUS REPORTS: Reported in EPA TSCA Inventory.

SAFETY PROFILE: Poison by ingestion. Moderately toxic by intraperitoneal, subcutaneous, and intravenous routes. An eye irritant. Combustible when exposed to heat, flame, or powerful oxidizers. To fight fire, use alcohol foam, water, CO_2, dry chemical. When heated to decomposition it emits acrid smoke and irritating fumes.

THM750 CAS:2169-64-4 ***HR: 1***
2-(2′,3′,5′-TRIACETYL-β-d-RIBOFURANOSYL)-as-TRIAZINE-3,5-(2H,4H)-DIONE
mf: $C_{14}H_{17}N_3O_9$ mw: 371.34

SYNS: AZARIBINE ◇ CB 304 ◇ NSC-67239 ◇ 2-β-d-RIBOFURANOSYL-as-TRIAZINE-3,5(2H,4H)-DIONE 2′,3′,5′-TRIACETATE ◇ TA-AZUR ◇ TRIACETYL-6-AZAURIDINE ◇ 2′,3′,5′-TRIACETYL-6-AZAURIDINE ◇ 2′,3′,5′-TRI-o-ACETYL-6-AZAURIDINE ◇ TRIAZURE

TOXICITY DATA with REFERENCE
orl-hmn TDLo:5670 mg/kg/6W:CNS,BLD AJOGAH 108,272,70
orl-rat LD50:12 g/kg TXAPA9 17,511,70
orl-mus LD50:7800 mg/kg TXAPA9 17,511,70

SAFETY PROFILE: Mildly toxic by ingestion. Human systemic effects by ingestion: somnolence, convulsions or effect on seizure threshold and cell count changes. When heated to decomposition it emits toxic fumes of NO_x. Used to treat psoriasis.

THN000 CAS:102-70-5 ***HR: 3***
TRIALLYLAMINE
DOT: UN 2610
mf: $C_9H_{15}N$ mw: 137.25

PROP: Liquid. D: 0.800 @ 20°/4°, mp: < −70°, bp: 150-151°, flash p: 103°F (TOC).

SYN: N-N-DI-2-PROPENYL-2-PROPEN-1-AMINE

TOXICITY DATA with REFERENCE
skn-rbt 10 mg/24H open SEV AIHAAP 23,95,62
eye-rbt 50 mg/20S rns MLD AEHLAU 1,343,60
ihl-man TCLo:13 ppm/5M:PUL AEHLAU 1,343,60
orl-rat LD50:1030 mg/kg AIHAAP 23,95,62
ihl-rat LC50:2800 mg/4H 85GMAT -,112,82
orl-mus LD50:492 mg/kg AEHLAU 1,343,60
ipr-mus LD50:187 mg/kg AEHLAU 1,343,60
skn-rbt LD50:400 mg/kg AIHAAP 23,95,62
ihl-mam LC50:2800 mg/m^3 TPKVAL 14,80,75

CONSENSUS REPORTS: Reported in EPA TSCA Inventory.

DOT Classification: Flammable or Combustible Liquid; Label: Flammable Liquid.

SAFETY PROFILE: Poison by skin contact and intraperitoneal routes. Moderately toxic by ingestion and inhalation. An eye and severe skin irritant. Human systemic effects by inhalation: structural or functional changes in trachea or bronchi. Flammable when exposed to heat, flame or oxidizers. To fight fire, use foam, alcohol foam, fog. When heated to decomposition it emits toxic fumes of NO_x. See also AMINES and ALLYL COMPOUNDS.

THN250 CAS:1693-71-6 ***HR: 2***
TRIALLYL BORATE
DOT: UN 2609

TOXICITY DATA with REFERENCE
orl-mus LD50:1800 mg/kg USBCC* 32,-,58

CONSENSUS REPORTS: Reported in EPA TSCA Inventory.

SAFETY PROFILE: Moderately toxic by ingestion. When heated to decomposition it emits acrid smoke and irritating fumes of BO_x. See also BORON COMPOUNDS, ESTERS, and ALLYL COMPOUNDS.

THN500 CAS:101-37-1 ***HR: 3***
TRIALLYL CYANAURATE
mf: $C_{12}H_{15}N_3O_3$ mw: 243.24

PROP: Bp: 120° @ 5 mm, fp: 27.3°, flash p: >176°F (TOC), d: 1.1133 @ 30°, vap press: 1 mm @ 100°.

SYNS: TRIPROPARGYL CYANURATE ◇ 2,4,6-TRIPROP-2-YNYLOXY-s-TRIAZINE ◇ 2,4,6-TRIS(ALLYLOXY)TRIAZINE

TOXICITY DATA with REFERENCE
ivn-mus LD50:180 mg/kg CSLNX* NX#00905

CONSENSUS REPORTS: Reported in EPA TSCA Inventory.

SAFETY PROFILE: Poison by intravenous route. Flammable when exposed to heat, flame, or oxidizers. To fight fire, use spray, foam, dry chemical. When heated to decomposition or on contact with acid or acid fumes it emits highly toxic fumes of CN^- and NO_x. See also ESTERS and ALLYL COMPOUNDS.

THN750 CAS:1623-19-4 ***HR: 3***
TRIALLYL PHOSPHATE
mf: $C_9H_{15}O_4P$ mw: 218.21

SYNS: ALLYL PHOSPHATE ◇ PHOSPHORIC ACID, TRIALLYL ESTER ◇ PHOSPHORIC ACID, TRI-2-PROPENYL ESTER

TOXICITY DATA with REFERENCE
ivn-mus LD50:71 mg/kg CBCCT* 6,138,54

CONSENSUS REPORTS: Reported in EPA TSCA Inventory.

SAFETY PROFILE: Poison by intravenous route. Can explode on distillation. When heated to decomposition it emits toxic fumes of PO_x. See also PHOSPHATES, ESTERS, and ALLYL COMPOUNDS.

THN800 CAS:4000-16-2 ***HR: 2***
TRIAMINOGUANIDINE NITRATE
mf: $CH_8N_6 \cdot NO_3$ mw: 166.16

$$(H_2NNH)_2C{=}NN^-H_3NO_3^-$$

SYNS: CARBONOHYDRAZONIC DIHYDRAZIDE, MONONITRATE (9CI) ◇ TAGN

TOXICITY DATA with REFERENCE
mmo-sat 500 μg/plate NTIS** AD-A064-950
dns-hmn:emb 10 mg/L NTIS** AD-A064-950
ipr-rat TDLo:2 g/kg (6-15D preg):TER NTIS** AD-A095-424
ipr-rat TDLo:8 g/kg (6-15D preg):REP NTIS** AD-A095-424
ivn-mus LD50:3650 mg/kg NTIS** AD-A039-514

CONSENSUS REPORTS: Reported in EPA TSCA Inventory.

SAFETY PROFILE: Moderately toxic by intravenous route. An experimental teratogen. Experimental reproductive effects. Human mutation data reported. Decomposes violently at 230°C. When heated to decomposition it emits toxic fumes of NO_x. See also NITRATES and AMINES.

THO250 CAS:4104-85-2 ***HR: 2***
TRIAMINOGUANIDINIUM PERCHLORATE
mf: $CH_9ClN_6O_4$ mw: 204.58

SAFETY PROFILE: Violent decomposition at 217°C. When heated to decomposition it emits very toxic fumes of Cl^- and NO_x. See also PERCHLORATES and AMINES.

THO500 CAS:6334-30-1 ***HR: 3***
2,4,6-TRIAMINOPHENOL TRIHYDROCHLORIDE
mf: $C_6H_9N_3O \cdot 3ClH$ mw: 248.56

TOXICITY DATA with REFERENCE
ipr-mus LDLo:63 mg/kg CBCCT* 6,225,54

CONSENSUS REPORTS: Reported in EPA TSCA Inventory.

SAFETY PROFILE: Poison by intraperitoneal route. When heated to decomposition it emits very toxic fumes of NO_x and HCl. See also AROMATIC AMINES.

THO550 CAS:4232-84-2 ***HR: 3***
TRIAMINOPHENYL PHOSPHATE
mf: $C_{18}H_{18}N_3O_4P$ mw: 371.36

SYNS: p-AMINOPHENOL PHOSPHATE (3:1) (ester) ◇ 4-AMINOPHENOL PHOSPHATE (3:1) (ester) ◇ TRIS(4-AMINOPHENYL) PHOSPHATE

TOXICITY DATA with REFERENCE
orl-rat LD50:138 mg/kg GISAAA 49(10),82,84
orl-mus LD50:91 mg/kg GISAAA 49(10),82,84
orl-rbt LD50:400 mg/kg GISAAA 49(10),82,84
orl-gpg LD50:400 mg/kg GISAAA 49(10),82,84

SAFETY PROFILE: Poison by ingestion. When heated to decomposition it emits toxic fumes of PO_x and NO_x. See also AMINES.

THO750 CAS:37640-57-6 ***HR: 2***
2,4,6-TRIAMINO-s-TRIAZINE compounded with s-TRIAZINE-TRIOL
mf: $C_3H_6N_6 \cdot C_3H_3N_3O_3$ mw: 219.21

SYN: MELAMINKYANURAT (CZECH)

TOXICITY DATA with REFERENCE
eye-rbt 500 mg/24H MOD 28ZPAK -,154,72
orl-rat LDLo:2500 mg/kg 28ZPAK -,154,72
skn-rat LD50:5520 mg/kg GTPZAB 30(1),44,86
ipr-rat LD50:2020 mg/kg GTPZAB 30(1),44,86
orl-mus LD50:3460 mg/kg GTPZAB 30(1),44,86
ihl-mus LCLo:1240 mg/m³/2H GTPZAB 30(1),44,86
ipr-mus LD50:1130 mg/kg GTPZAB 30(1),44,86

SAFETY PROFILE: Moderately toxic by ingestion, inhalation, and intraperitoneal routes. Mildly toxic by skin contact. An eye irritant. When heated to decomposition it emits toxic fumes of NO_x. See also AMINES.

THO775 ***HR: 3***
1,3,5-TRIAMINOTRINITROBENZENE
mf: $C_6H_6N_6O_6$ mw: 258.15

$$(H_2N)_3C_6(NO_2)_3$$

SAFETY PROFILE: Mixtures with hydroxylaminium perchlorate are explosive. When heated to decomposition it emits toxic fumes of NO_x. See also NITRO COMPOUNDS of AROMATIC HYDROCARBONS.

THP000 CAS:548-61-8 ***HR: 3***
TRIAMINOTRIPHENYLMETHANE
mf: $C_{19}H_{19}N_3$ mw: 289.41

PROP: Leaves from water. Mp: 148°. Sltly sol in cold water; sol in abs alc and benzene.

SYNS: LEUCOPARAFUCHSIN ◇ LEUCOPARAFUCHSINE ◇ 4,4',4''-METHYLIDYNETRIANILINE ◇ 4,4',4''-METHYLIDYNETRIS-BENZENEAMINE ◇ p,p',p''-TRIAMINOTRIPHENYLMETHANE ◇ 4,4',4''-TRIAMINOTRIPHENYLMETHANE ◇ TRIS-4-AMINOFENYLMETHAN (CZECH)

TOXICITY DATA with REFERENCE
eye-rbt 100 mg/24H MOD 28ZPAK -,73,72
orl-rat TDLo:26 g/kg/47W-I:ETA VOONAW 22(9),66,76
orl-rat LD50:2640 mg/kg 28ZPAK -,73,72

SAFETY PROFILE: Moderately toxic by ingestion. An eye irritant. Questionable carcinogen with experimental tumorigenic data. When heated to decomposition it emits toxic fumes of NO_x. See also AMINES.

THP250 CAS:17168-85-3 ***HR: 3***
TRIAMMINEDIPEROXOCHROMIUM(IV)
mf: $CrH_9N_3O_4$ mw: 167.09

SAFETY PROFILE: Suspected carcinogen. Chromium compounds are generally poisons. May explode with heat or shock. May explode at 120°C. An oxidizer. When heated to decomposition it emits toxic fumes of NO_x. See also CHROMIUM COMPOUNDS, PEROXIDES, and AMINES.

THP500 ***HR: 3***
TRIAMMINE GOLDTRIHYDROXIDE
mf: $AuH_{12}N_3O_3$ mw: 209.07

SAFETY PROFILE: A potentially explosive compound. When heated to decomposition it emits toxic fumes of NO_x. See also GOLD COMPOUNDS.

THP750 CAS:17524-18-4 ***HR: 2***
TRIAMMINENITRATOPLANTINUM(II) NITRATE
mf: $H_9N_5O_6Pt$ mw: 370.20

SAFETY PROFILE: Decomposes violently on heating. When heated to decomposition it emits toxic fumes of NO_x. See also PLATINUM COMPOUNDS and NITRATES.

THQ000 CAS:58240-55-4 ***HR: 3***
cis-TRIAMMINETRICHLORORUTHENIUM(III)
mf: $Cl_3H_9N_3Ru$ mw: 258.54

TOXICITY DATA with REFERENCE
ipr-mus LD50:108 mg/kg TXAPA9 48,A112,79

SAFETY PROFILE: Poison by intraperitoneal route. When heated to decomposition it emits very toxic fumes of NO_x, Ru, and Cl^-. See also RUTHENIUM COMPOUNDS, AMINES, and CHLORIDES.

THQ250 CAS:13600-88-9 ***HR: 3***
TRIAMMINETRINITROCOBALT(III)
mf: $CoH_9N_6O_6$ mw: 248.05
$[(H_3N)_3Co(NO_2)_3]$

SAFETY PROFILE: Explodes at 305°C or on impact. When heated to decomposition it emits toxic fumes of NO_x. See also COBALT COMPOUNDS.

THQ500 CAS:7784-19-2 ***HR: 3***
TRIAMMONIUM HEXAFLUOROALUMINATE
mf: $AlF_6 \cdot 3H_4N$ mw: 195.13

SYNS: AMMONIUM ALUMINUM FLUORIDE ◇ AMMONIUM CRYOLITE ◇ AMMONIUM FLUOALUMINATE ◇ AMMONIUM HEXAFLUOROALUMINATE ◇ TRIAMMONIUM ALUMINUM HEXAFLUORIDE

TOXICITY DATA with REFERENCE
ivn-mus LD50:18 mg/kg CSLNX* NX#00136

CONSENSUS REPORTS: Reported in EPA TSCA Inventory.

OSHA PEL: TWA 2.5 mg(F)/m^3
ACGIH TLV: TWA 2 mg(Al)/m^3
NIOSH REL: TWA 2.5 mg(F)/m^3

SAFETY PROFILE: Poison by intravenous route. When heated to decomposition it emits very toxic fumes of F^-, NO_x, and NH_3. See also ALUMINUM COMPOUNDS and FLUORIDES.

THQ600 CAS:52851-26-0 ***HR: 3***
3,6,9-TRIAZATETRACYCLO[6.1.0.0^{2,4}.O^{5,7}] NONANE
mf: $C_6H_9N_3$ mw: 123.16

SYN: cis-BENZENE TRIIMINE

SAFETY PROFILE: Explodes when heated to 200°C. When heated to decomposition it emits toxic fumes of NO_x.

THQ750 CAS:15056-34-5 ***HR: D***
1-TRIAZENE
mf: H_3N_3 mw: 45.06

SYN: TRIAZENE

TOXICITY DATA with REFERENCE
sln-oin-dmg 500 umol/L/3D-I 35WYAM-,63,76

SAFETY PROFILE: Mutation data reported. Violent reaction with HNO_3. When heated to decomposition it emits very toxic fumes of NO_x and NH_3.

THQ900 CAS:5433-44-3 ***HR: 2***
p,p'-TRIAZENYLENEDIBENZENESULFONAMIDE
mf: $C_{12}H_{13}N_5O_4S_2$ mw: 355.42

SYNS: 1,3-DI(4-SULFAMOYLPHENYL)TRIAZENE ◇ DSPT

TOXICITY DATA with REFERENCE
cyt-dmg-orl 2800 μmol/L/3D-I CRNGDP 5,571,84
ipr-mus TDLo:66 mg/kg (9-11D post):TER CRNGDP 5,571,84

SAFETY PROFILE: An experimental teratogen. Questionable carcinogen with experimental tumorigenic data. Mutation data reported. When heated to decomposition it emits toxic fumes of NO_x and SO_x.

THR100 CAS:19708-47-5 ***HR: 3***
TRIAZIDOMETHYLIUM HEXACHLOROANTIMONATE
mf: CCl_6N_9Sb mw: 472.54

CONSENSUS REPORTS: Antimony and its compounds are on The Community Right-To-Know List.

SAFETY PROFILE: Most antimony compounds are poisons. A shock- and heat-sensitive explosive. When heated to decomposition it emits toxic fumes of Cl^-, NO_x, and Sb. See also ANTIMONY COMPOUNDS.

THR250 CAS:5637-83-2 ***HR: 3***
2,4,6-TRIAZIDO-1,3,5-TRIAZINE
mf: C_3N_{12} mw: 204.12

$N{=}C(N_3)N{=}C(N_3)N{=}CN_3$

SYNS: CYANURIC TRIAZIDE ◇ CYANURIC TRIAZIDE (DOT)

DOT Classification: Forbidden

SAFETY PROFILE: Explodes violently on impact, shock or rapid heating to 170-180°C. When heated to decomposition it emits toxic fumes of NO_x. See also AZIDES.

THR500 CAS:41191-04-2 ***HR: 3***
TRIAZINATE
mf: $C_{19}H_{21}ClN_6O_2 \cdot C_2H_6O_3S$ mw: 511.05

SYNS: α-(2-CHLORO-4-(4,6-DIAMINO-2,2-DIMETHYL-S-TRIAZINE-1(2H)-YL)PHENOXY)-N,N-DIMETHYL-m-TOLUAMIDEETHANESULFONATE ◇ BAF ◇ BAKER'S ANTIFOLANTE ◇ BAKER'S ANTIFOL SOLUBLE ◇ NSC 139105 ◇ TZT

TOXICITY DATA with REFERENCE
dni-hmn-leu 2 μg/L CNREA8 36,3659,76
ivn-hmn TDLo:8 mg/kg:CNS,BLD,SKN CNREA8 36,48,76
ivn-hmn TDLo:14 mg/kg:EYE,BLD,SKN CNREA8 36,48,76
ivn-hmn TDLo:24 mg/kg/5D:BLD CANCAR 38,690,76
ivn-hmn TDLo:20 mg/kg:GIT,SKN CANCAR 40,9,77
ipr-mus LD50:58600 μg/kg NCISP* JAN86

SAFETY PROFILE: Poison by intraperitoneal route. Human systemic effects by intravenous route: somnolence, changes in bone marrow, dermatitis, visual field changes, leukopenia, thrombocytopenia, hypermotility, diarrhea, nausea or vomiting. Human mutation data reported. When heated to decomposition it emits very toxic fumes of Cl^-, NO_x, and SO_x.

THR525 CAS:290-87-9 ***HR: 3***
1,3,5-TRIAZINE
mf: $C_3H_3N_3$ mw: 81.08

N=CHN=CHN=CH

PROP: Mp: 75-80°. Solubility of 1 g in 10 ml methanol.

SAFETY PROFILE: Explosive reaction with nitric acid + trifluoroacetic anhydride. When heated to decomposition it emits toxic fumes of NO_x.

THR750 CAS:461-89-2 ***HR: 3***
s-TRIAZINE-3,5(2H,4H)-DIONE
mf: $C_3H_3N_3O_2$ mw: 113.09

SYNS: 4(6)-AZAURACIL ◇ 6-AZAURACIL ◇ NSC 3425 ◇ USAF CB-30

TOXICITY DATA with REFERENCE
mma-sat 10 μg/plate GENEA3 25,327,84
mmo-omi 80 mg/L ZBPIA9 130,1,75
orl-rat TDLo:39 g/kg/1Y-I:ETA,REP JNCIAM 41,985,68
ipr-mus LD50:200 mg/kg NTIS** AD277-689
scu-mus LD50:2076 mg/kg BCPCA6 15,408,66

CONSENSUS REPORTS: Reported in EPA TSCA Inventory.

SAFETY PROFILE: Poison by intraperitoneal route. Experimental reproductive effects. Questionable carcinogen with experimental tumorigenic data. Mutation data reported. When heated to decomposition it emits toxic fumes of NO_x.

THS000 CAS:108-80-5 ***HR: 3***
s-TRIAZINE-2,4,6-TRIOL
mf: $C_3H_3N_3O_3$ mw: 129.09

N=C(OH)N=C(OH)NCOH

PROP: Off-white; odorless crystals. Mp: >360°, d: 2.500 @ 20°/4°.

SYNS: CYANURIC ACID ◇ ISOCYANURIC ACID ◇ KYSELINA KYANUROVA (CZECH) ◇ PSEUDOCYANURIC ACID ◇ sym-TRIAZINETRIOL ◇ s-2,4,6-TRIAZINETRIOL ◇ s-TRIAZINE-2,4,6(1H,3H,5H)-TRIONE ◇ TRICYANIC ACID ◇ TRIHYDROXYCYANIDINE ◇ 2,4,6-TRIHYDROXY-1,3,5-TRIAZINE

TOXICITY DATA with REFERENCE
eye-rbt 500 mg/24H MLD 28ZPAK -,152,72
eye-rbt 20 mg/24H rns MLD MONS** -,-,72
orl-rat TDLo:55 g/kg/82W-I:ETA VOONAW 16(1),82,70
orl-rat LD50:7700 mg/kg ZKMAAX 25,345,85
orl-mus LD50:3400 mg/kg ZKMAAX 25,345,85

CONSENSUS REPORTS: Reported in EPA TSCA Inventory.

SAFETY PROFILE: Moderately toxic by ingestion. Questionable carcinogen with experimental tumorigenic data. An eye irritant. Irritating to abraided skin. Violent reaction with ethanol. Reacts with chlorine to form a spontaneously explosive product. When heated to decomposition it emits very toxic fumes of NO_x and CN^-. Used to stabilize chlorine solutions used in swimming pools.

THS250 CAS:638-16-4 ***HR: 3***
s-TRIAZINE-2,4,6-TRITHIOL
mf: $C_3H_3N_3S_3$ mw: 177.27

SYNS: 1,3,5-TRIAZINE-2,4,6-TRIMERCAPTAN ◇ 2,4,6-TRIAZINE-TRITHIOL ◇ 1,3,5-TRIAZINE-2,4,6(1H,3H,5H)-TRITHIONE ◇ 1,3,5-TRIMERCAPTOTRIAZINE ◇ 2,4,6-TRIMERCAPTO-S-TRIAZINE ◇ TRITHIOCYANURIC ACID ◇ USAF TH-3

TOXICITY DATA with REFERENCE
orl-rat LD50:9500 mg/kg FCTOD7 21,495,83
ipr-mus LD50:200 mg/kg NTIS** AD277-689
ivn-mus LD50:180 mg/kg CSLNX* NX#03037

CONSENSUS REPORTS: Reported in EPA TSCA Inventory.

SAFETY PROFILE: Poison by intraperitoneal and intravenous routes. Mildly toxic by ingestion. When heated to decomposition it emits very toxic fumes of NO_x and SO_x. See also MERCAPTANS.

THS500 CAS:13046-06-5 ***HR: 1***
(s-TRIAZIN-2,4,6-TRIYLTRIAMINO)TRIS-METHANESULFONIC ACID TRISODIUM SALT
mf: $C_6H_9N_6O_9S_3 \cdot 3Na$ mw: 474.36

SYNS: MELAMIN-N,N',N''-TRIMETHYLSULFONSAURESNATRIUM (GERMAN) ◇ s-TRIAZIN-2,4,6-TRIYLTRIAMINOMETHANESULFONIC ACID TRISODIUM SALT

TOXICITY DATA with REFERENCE
ipr-rat LD50:7200 mg/kg ARPMAS 306,274,73
ipr-mus LD50:9660 mg/kg ARZNAD 16,734,66

SAFETY PROFILE: Mildly toxic by intraperitoneal route. When heated to decomposition it emits very toxic fumes of SO_x, Na_2O, and NO_x.

THS800 CAS:28911-01-5 ***HR: 2***
TRIAZOLAM
mf: $C_{17}H_{12}Cl_2N_4$ mw: 343.23

PROP: Tan crystals from 2-propanol. Mp: 233-235°.

SYNS: 8-CHLORO-6-(o-CHLOROPHENYL)-1-METHYL-4H-s-TRIAZOLO(4,3-a)(1,4)BENZODIAZEPINE ◇ 8-CHLORO-6-(2-CHLOROPHENYL)-1-METHYL-4H-(1,2,4)TRIAZOLO(4,3-a)(1,4)BENZODIAZEPINE ◇ HALCION ◇ NOVIDORM ◇ U-33,030

TOXICITY DATA with REFERENCE
orl-rat TDLo:810 mg/kg (female 17-22D post):REP IYKEDH 10,52,79
orl-rat TDLo:4400 mg/kg (female 7-17D post):TER OYYAA2 30,765,85
orl-man TDLo:100 μg/kg JCLPDE 47,50,86
orl-hmn TDLo:7 μg/kg:BRN,CNS JCPCBR 14,192,74
orl-man TDLo:107 μg/kg AIMDAP 145,663,85
ipr-mus LD50:1625 mg/kg IYKEDH 14,484,83

SAFETY PROFILE: Moderately toxic by intraperitoneal route. An experimental teratogen. Human systemic effects by ingestion: changes in brain EEG, distorted perceptions, and sleep. Experimental reproductive effects. Used as an hypnotic agent. Note: This is a controlled substance (depressant) listed in the U.S. Code of Federal Regulations, Title 21 Part 1308.14 (1985). When heated to decomposition it emits toxic fumes of Cl^- and NO_x.

THS850 CAS:27070-49-1 ***HR: 3***
1,2,3-TRIAZOLE
mf: $C_2H_3N_3$ mw: 69.07

PROP: Hygroscopic crstals or liquid. Bp: 203-210°, mp: 23°, d: 1.1861, refr index: 1.4975. Sol in water, ether, or acetone.

SAFETY PROFILE: The vapor explodes if heated above 200°C. When heated to decomposition it emits toxic fumes of NO_x.

THT000 CAS:3179-31-5 ***HR: 3***
1H-1,2,4-TRIAZOLE-3-THIOL
mf: $C_2H_3N_3S$ mw: 101.14

PROP: Crystalline. Mp: 220°-224°. Solubility of 1 g in 20 ml water.

SYN: 3-MERCAPTO-1H-1,2,4-TRIAZOLE

TOXICITY DATA with REFERENCE
ivn-mus LD50:180 mg/kg CSLNX* NX#03630

CONSENSUS REPORTS: Reported in EPA TSCA Inventory.

SAFETY PROFILE: Poison by intravenous route.

When heated to decomposition it emits very toxic fumes of NO_x and SO_x. See also MERCAPTANS.

THT250 ***HR: 3***
1,2,4-TRIAZOLO[4,3-a]PYRIDINE-SILVER NITRATE
mf: $C_6H_5N_3 \cdot AgNO_3$ mw: 289.00

CONSENSUS REPORTS: Silver and its compounds are on The Community Right-To-Know List.

SAFETY PROFILE: Explodes when heated to 228°C. When heated to decomposition it emits toxic fumes of NO_x. See also SILVER NITRATE.

THT350 CAS:1468-26-4 ***HR: 3***
1H-v-TRIAZOLO(4,5-d)PYRIMIDINE-5,7(4H,6H)-DIONE
mf: $C_4H_3N_5O_2$ mw: 153.12

SYNS: 8-AZAXANTHINE ◇ 2,6-DIOXY-8-AZAPURINE ◇ NSC 756 ◇ v-TRIAZOLO(4,5-d)PYRIMIDINE-5,7-DIOL ◇ USAF CB-26

TOXICITY DATA with REFERENCE
mmo-esc 4 g/L/3H CRSUBM 3,69,55
ipr-mus LD50:200 mg/kg NTIS** AD277-689

SAFETY PROFILE: Poison by intraperitoneal route. Mutation data reported. When heated to decomposition it emits toxic fumes of NO_x.

THT500 CAS:41083-11-8 ***HR: 3***
(1H-1,2,4-TRIAZOLYL-1-YL)TRICYCLOHEXYLSTANNANE
mf: $C_{20}H_{35}N_3Sn$ mw: 436.27

SYNS: AZOCYCLOTIN ◇ BAY BUE 1452 ◇ PEROPAL ◇ (1H-1,2,4-TRIAZOLYL)TRICYCLOHEXYLSTANNANE ◇ 1-(TRICYCLOHEXYLSTANNYL)-1H-1,2,4-TRIAZOLE

TOXICITY DATA with REFERENCE
orl-rat LD50:99 mg/kg FMCHA2 -,C182,83
skn-rat LD50:1000 mg/kg FMCHA2 -,C182,83
OSHA PEL: TWA 0.1 mg(Sn)/m^3 (skin)
ACGIH TLV: TWA 0.1 mg(Sn)/m^3 (skin) (Proposed: TWA 0.1 mg(Sn)/m^3; STEL 0.2 mg(Sn)/m^3 (skin))
NIOSH REL: (Organotin Compounds) TWA 0.1 mg (Sn)/m^3

SAFETY PROFILE: Poison by ingestion. Moderately toxic by skin contact. When heated to decomposition it emits toxic fumes of NO_x. See also TIN COMPOUNDS.

THT750 CAS:24017-47-8 ***HR: 3***
TRIAZOPHOS
mf: $C_{12}H_{16}N_3O_3PS$ mw: 313.34

SYNS: O,O-DIETHYLO-(1-PHENYL-1H-1,2,4-TRIAZOL-3-YL)PHOSPHOROTHIOATE ◇ HOE 2960 OJ ◇ HOSTATHION ◇ 1-PHENYL-3-(O,O-DIETHYL-THIONOPHOSPHORYL)-1,2,4-TRIAZOLE ◇ 1-PHENYL-1,2,4-TRIAZOLYL-3-(O,O-DIETHYLTHIONOPHOSPHATE) ◇ TRIAZOFOSZ (HUNGARIAN)

TOXICITY DATA with REFERENCE
orl-rat LD50:64 mg/kg FMCHA2 -,D164,80
ihl-rat LC50:280 mg/m^3/4H EGESAQ 24,173,80
skn-rat LD50:1100 mg/kg GUCHAZ 6,508,73
ipr-rat LD50:107 mg/kg GUCHAZ 6,508,73
orl-dog LD50:320 mg/kg 28ZEAL 5,227,76

SAFETY PROFILE: Poison by ingestion, inhalation and intraperitoneal routes. Moderately toxic by skin contact. When heated to decomposition it emits very toxic fumes of NO_x, PO_x, and SO_x. A pesticide.

THU000 CAS:5888-61-9 ***HR: 3***
TRIBENZYLARSINE
mf: $C_{21}H_{21}As$ mw: 348.39

CONSENSUS REPORTS: Arsenic and its compounds are on the Community Right-To-Know List.

SAFETY PROFILE: Arsenic compounds are poisons. Slow oxididation in air becomes violent through autocatalysis. When heated to decomposition it emits fumes of As. See also ARSENIC.

THU250 CAS:73926-83-7 ***HR: 3***
TRIBENZYLSULFONIUM IODIDE MERCURIC IODIDE

SYN: TRIBENZYLSULFONIUM IODIDE, compounded with MERCURY IDODIDE (1:1)

TOXICITY DATA with REFERENCE
ivn-mus LD50:32 mg/kg CSLNX* NX#01717

CONSENSUS REPORTS: Mercury and its compounds are on the Community Right-To-Know List.
NIOSH REL: TWA 0.05 mg(Hg)/m^3

SAFETY PROFILE: Poison by intravenous route. When heated to decomposition it emits very toxic fumes of Hg, I^-, and SO_x. See also MERCURY COMPOUNDS and IODIDES.

THU275 CAS:15538-67-7 ***HR: 3***
TRIBORON PENTAFLUORIDE
mf: B_3F_5 mw: 127.42

SAFETY PROFILE: Explosive reaction with air, water, tetrafluoroethylene. When heated to decomposition it emits toxic fumes of F^-. See also BORON COMPOUNDS and FLUORIDES.

THU500 CAS:507-42-6 ***HR: 3***
TRIBROMOALDEHYDE HYDRATE
mf: $C_2H_3Br_3O_2$ mw: 298.78

PROP: Crystals. D: 2.566 @ 40°/4°, mp: 53.5°. Sol in water, chloroform, alc, ether, glycerol cold.

SYN: BROMAL HYDRATE

TOXICITY DATA with REFERENCE
orl-rat LDLo:40 mg/kg JPETAB 63,453,38
ipr-rat LDLo:40 mg/kg JPETAB 63,453,38
ivn-rbt LDLo:30 mg/kg JPETAB 63,453,38

SAFETY PROFILE: Poison by ingestion, intraperitoneal and intravenous routes. When heated to decomposition it emits toxic fumes of Br^-. See also ALDEHYDES and BROMIDES.

THU750 CAS:147-82-0 ***HR: 2***
2,4,6-TRIBROMOANILINE
mf: $C_6H_4Br_3N$ mw: 329.84

PROP: Needles. Sltly sol in cold alc. D: 2.35, mp: 120-122°, bp: 300°.

SYNS: sym-TRIBROMOANILINE ◇ USAF DO-43

TOXICITY DATA with REFERENCE
ipr-mus LD50:500 mg/kg NTIS** AD277-689

CONSENSUS REPORTS: Reported in EPA TSCA Inventory.

SAFETY PROFILE: Moderately toxic by intraperitoneal route. When heated to decomposition it emits very toxic fumes of Br^- and NO_x. See also BROMIDES and ANILINE.

THV000 CAS:1329-86-8 ***HR: 3***
TRIBROMOETHANOL
mf: $C_2H_3Br_4O$ mw: 282.78

PROP: Crystals; ethereal odor, aromatic taste. Bp: 92° @ 10 mm, mp: 79-82°, decomp @ 70°. Sltly water sol; sol in alc, organic solvents.

SYNS: AVERTIN ◇ BROMETHOL ◇ ETHOBROM ◇ NARCOLAN ◇ NARKOLAN ◇ TRIBROMETHANOL ◇ TRIBROMOETHYL ALCOHOL

TOXICITY DATA with REFERENCE
ihl-mus TCLo:40 mg/kg/10M (1D pre):REP JRPFA4 49,167,77
orl-rat LDLo:300 mg/kg CRAAA7 17,258,38
ipr-rat LDLo:400 mg/kg JPETAB 63,453,38
orl-mus LDLo:500 mg/kg CRAAA7 17,258,38
ipr-mus LD50:546 mg/kg JPETAB 81,72,44
ivn-mus LD50:279 mg/kg JPETAB 81,72,44
orl-cat LDLo:150 mg/kg MEIEDD 10,1374,83
ivn-rbt LDLo:200 mg/kg JPETAB 63,453,38
rec-rbt LDLo:400 mg/kg CRAAA7 17,258,38
orl-qal LD50:422 mg/kg AECTCV 12,355,83
orl-bwd LD50:316 mg/kg AECTCV 12,355,83

SAFETY PROFILE: Poison by ingestion, intravenous, intraperitoneal, and rectal routes. Experimental reproductive effects. When heated to decomposition it emits toxic fumes of Br^-. An anesthetic drug.

THV250 CAS:724-31-2 ***HR: 3***
1,3,7-TRIBROMO-2-FLUORENAMINE
mf: $C_{13}H_8Br_3N$ mw: 417.95

SYN: 1,3,7-TRIBROMOFLUOREN-2-AMINE

TOXICITY DATA with REFERENCE
orl-rat TDLo:360 mg/kg/27D-I:ETA CNREA8 28,924,68

SAFETY PROFILE: Questionable carcinogen with experimental tumorigenic data. When heated to decomposition it emits very toxic fumes of Br^- and NO_x.

THV450 CAS:2034-22-2 ***HR: 3***
2,4,5-TRIBROMOIMIDAZOLE
mf: $C_3HBr_3N_2$ mw: 304.79

TOXICITY DATA with REFERENCE
orl-rat LD50:34 mg/kg ARTODN 56,109,84
ivn-rat LD50:15 mg/kg ARTODN 56,109,84
ivn-mus LD50:18 mg/kg CSLNX* NX#05092

SAFETY PROFILE: Poison by ingestion and intravenous routes. When heated to decomposition it emits toxic fumes of Br^- and NO_x.

THV500 CAS:73941-35-2 ***HR: 3***
2,4,5-TRIBROMOIMIDAZOLE CADMIUM SALT (2 : 1)
mf: $C_6Br_6N_4{\cdot}Cd$ mw: 719.96

SYN: CADMIUM salt of 2,4-5-TRIBROMOIMIDAZOLE

TOXICITY DATA with REFERENCE
ivn-mus LD50:56 mg/kg CSLNX* NX#06532

CONSENSUS REPORTS: Cadmium and its compounds are on the Community Right-To-Know List.

OSHA PEL: TWA 0.2 mg(Cd)/m^3; CL 0.6 mg(Cd)/m^3 (dust)
ACGIH TLV: TWA 0.05 mg(Cd)/m^3 (Proposed: TWA 0.01 mg(Cd)/m^3 (dust), Suspected Human Carcinogen; 0.002 mg(Cd)/m^3 (respirable dust), Suspected Human Carcinogen); BEI: 10 μg/g creatinine in urine; 10 μg/L in blood.
NIOSH REL: (Cadmium) Reduce to lowest feasible level.

SAFETY PROFILE: Confirmed human carcinogen. Poison by intravenous route. When heated to decomposition it emits very toxic fumes of Br^-, Cd, and NO_x. See also BROMIDES and CADMIUM COMPOUNDS.

THV750 CAS:118-79-6 ***HR: 2***
2,4,6-TRIBROMOPHENOL
mf: $C_6H_3Br_3O$ mw: 330.82

PROP: Long crystals. D: 2.55, mp: 94-96°, bp: 244°. Sol in 14,000 parts water @ 15°, alc, chloroform, ether, glycerols.

SYNS: BROMOL ◇ TRIBROMOPHENOL

TOXICITY DATA with REFERENCE
orl-rat LD50:2 g/kg 85JCAE-,524,86

CONSENSUS REPORTS: Reported in EPA TSCA Inventory.

SAFETY PROFILE: Slightly toxic by ingestion. A powerful irritant to skin, eyes, and mucous membranes. May be absorbed dermally. When heated to decomposition it emits toxic fumes of Br^-. See also BROMIDES and CHLOROPHENOLS.

THW750 CAS:87-10-5 ***HR: 2***
3,4',5-TRIBROMOSALICYLANILIDE
mf: $C_{13}H_8Br_3NO_2$ mw: 449.95

SYN: POLYBROMINATEDSALICYLANILIDE

TOXICITY DATA with REFERENCE
orl-rat LD50:410 mg/kg IMSUAI 39,56,70

CONSENSUS REPORTS: Reported in EPA TSCA Inventory.

SAFETY PROFILE: Moderately toxic by ingestion. When heated to decomposition it emits very toxic fumes of Br^- and NO_x. See also BROMIDES.

THX000 CAS:7789-57-3 ***HR: 3***
TRIBROMOSILANE
mf: Br_3HSi mw: 268.83

PROP: Mobile liquid. D: 2.7 @ 17°/4°, mp: −73.5°, bp: 112°, vap press: 8.8 mm @ 0°. Sol in chlorinated hydrocarbons.

SYN: SILICOBROMOFORM

SAFETY PROFILE: Readily hydrolyzes to liberate hydrogen bromide which is a powerful irritant. Spontaneously flammable in air. When heated to decomposition it emits toxic fumes of Br^-. See also HYDROBROMIC ACID and SILANE.

THX250 CAS:102-82-9 ***HR: 3***
TRIBUTYLAMINE
DOT: UN 2542
mf: $C_{12}H_{27}N$ mw: 185.40

PROP: A colorless liquid. Mp: −70°, bp: 213°, flash p: 187°F (OC) d: 0.78-0.79, vap d: 6.38. Insol in water; sol in alc, ether.

SYNS: TRI-n-BUTYLAMINE ◇ TRIS-N-BUTYLAMINE

TOXICITY DATA with REFERENCE
orl-rat LD50:540 mg/kg TXAPA9 28,313,74
ihl-rat LCLo:75 ppm/4H TXAPA9 28,313,74
scu-rat LDLo:380 mg/kg JPETAB 20,435,23
orl-mus LD50:114 mg/kg GISAAA 42(12),36,77
orl-rbt LD50:615 mg/kg GISAAA 42(12),36,77
skn-rbt LD50:250 mg/kg TXAPA9 28,313,74
orl-gpg LD50:350 mg/kg GISAAA 42(12),36,77

CONSENSUS REPORTS: Reported in EPA TSCA Inventory.

DOT Classification: Corrosive Material; Label: Corrosive.

SAFETY PROFILE: Poison by ingestion, inhalation, skin contact, and subcutaneous routes. A central nervous system stimulant, irritant, and sensitizer. A corrosive irritant to skin, eyes, and mucous membranes. Flammable when exposed to heat, flame, or oxidizers. Can react with oxidizing materials. To fight fire, use foam, CO_2, dry chemical. When heated to decomposition it emits toxic fumes of NO_x. See also AMINES.

THX500 CAS:122-56-5 ***HR: 3***
TRI-n-BUTYL BORANE
mf: $C_{12}H_{27}B$ mw: 182.20

PROP: Colorless pyroforic liquid. Mp: 34°, bp: 170° @ 222 mm, d: 0.747 @ 25°, vap press: 1 mm @ 20°, flash p: −32°F. Insol in water; sol in most organic solvents.

SYNS: BORIC ACID, TRIBUTYL ESTER ◇ TBB ◇ TRIBUTYLBORINE

TOXICITY DATA with REFERENCE
orl-rat LD50:1125 mg/kg NKOGAV 22,533,73
ivn-rat LD50:104 mg/kg NKOGAV 22,533,73

CONSENSUS REPORTS: Reported in EPA TSCA Inventory.

SAFETY PROFILE: Poison by intravenous route. Moderately toxic by ingestion. A very dangerous fire hazard when exposed to heat or flame; can ignite spontaneously. When heated to decomposition it emits acrid smoke and irritating fumes. See also BORANES.

THX750 CAS:688-74-4 ***HR: 2***
TRI-n-BUTYL BORATE
mf: $C_{12}H_{27}BO_3$ mw: 230.20

PROP: Colorless, mobile liquid; odor like n-butanol. Bp: 230°, fp: < −70°, flash p: 200°F (COC), d: 0.847 @ 28°, vap d: 7.95.

SYNS: BORESTER 2 ◇ BORIC ACID, TRI-sec-BUTYL ESTER ◇ BUTYL BORATE ◇ n-BUTYL BORATE ◇ TRIBUTOXYBORANE ◇ TRI-n-BUTOXYBORANE ◇ TRIBUTYL BORATE

TOXICITY DATA with REFERENCE
eye-rbt 100 mg MOD 14KTAK -,706,64
orl-mus LD50:1740 mg/kg USBCC* 14,-,58
ipr-mus LDLo:500 mg/kg CBCCT* 5,59,53

CONSENSUS REPORTS: Reported in EPA TSCA Inventory.

SAFETY PROFILE: Moderately toxic by ingestion and intraperitoneal routes. An eye irritant. Flammable when exposed to heat, flame, or oxidizers. To fight fire, use foam, CO_2, dry chemical. When heated to decomposition or on contact with acid or acid fumes it can emit toxic fumes; on contact with oxidizing materials it can react vigorously. See also BORANES and BORON COMPOUNDS.

THY000 CAS:22238-17-1 ***HR: 2***
TRI-sec-BUTYL BORATE
mf: $C_{12}H_{27}BO_3$ mw: 230.20

PROP: Colorless liquid; odor of sec-butanol. Bp: 184-192°, flash p: 165°F (COC), d: 0.829 @ 24°.

SYN: BORIC ACID, TRI-sec-BUTYL ESTER

TOXICITY DATA with REFERENCE
eye-rbt 100 mg MLD 14KTAK -,706,64
orl-mus LD50:2100 mg/kg 14KTAK -,706,64

SAFETY PROFILE: Moderately toxic by ingestion. An eye irritant. Combustible when exposed to heat or flame; can react with oxidizing materials. To fight fire, use foam, CO_2, dry chemical. When heated to decomposition it emits acrid smoke and irritating fumes. See also BORON COMPOUNDS and sec-BUTYL ALCOHOL.

THY500 CAS:115-78-6 ***HR: 3***
TRIBUTYL(2,4-DICHLOROBENZYL)PHOSPHONIUM CHLORIDE
mf: $C_{19}H_{32}Cl_2P{\cdot}Cl$ mw: 397.83

PROP: Crystals. Readily sol in water, acetone, ethanol.

SYNS: CBBP ◇ CHLORFONIUM ◇ CHLORPHONIUM CHLORIDE ◇ 2,4-DICHLOROBENZYLTRIBUTYLPHOSPHONIUM CHLORIDE ◇ FOSFON D ◇ PHOSFON D ◇ PHOSPHON D ◇ PHOSPHONE D

TOXICITY DATA with REFERENCE
orl-rat LD50:178 mg/kg PCOC** -,899,66
orl-mus LDLo:470 mg/kg AECTCV 14,111,85
ipr-mus LDLo:19 mg/kg TXAPA9 23,288,72

SAFETY PROFILE: Poison by ingestion and intraperitoneal routes. Used as a plant growth regulant. When heated to decomposition it emits very toxic fumes of Cl^- and PO_x.

THY750 CAS:681-99-2 ***HR: 3***
TRIBUTYLISOCYANATOSTANNANE
mf: $C_{13}H_{27}NOSn$ mw: 332.10

SYNS: TRIBUTYLSTANNYL ISOCYANATE ◇ TRIBUTYLTIN ISOCYANATE ◇ TRI-n-BUTYLTIN ISOCYANATE ◇ TRIBUTYLTIN ISOTHIOCYANATE

TOXICITY DATA with REFERENCE
ivn-mus LD50:6300 μg/kg CSLNX* NX#03473

OSHA PEL: TWA 0.1 mg(Sn)/m^3 (skin)
ACGIH TLV: TWA 0.1 mg(Sn)/m^3 (skin) (Proposed: TWA 0.1 mg(Sn)/m^3; STEL 0.2 mg(Sn)/m^3 (skin))
NIOSH REL: (Organotin Compounds) TWA 0.1 mg (Sn)/m^3

SAFETY PROFILE: Poison by intravenous route. When heated to decomposition it emits toxic fumes of NO_x. See also TIN COMPOUNDS and ISOCYANATES.

THY850 CAS:2587-82-8 ***HR: 3***
TRIBUTYLLEAD ACETATE
mf: $C_{14}H_{30}O_2Pb$ mw: 437.63

SYNS: ACETOXYTRIBUTYLPLUMBANE ◇ TRI-n-BUTYLPLUMBYL ACETATE

TOXICITY DATA with REFERENCE
orl-rat LD50:2 mg/kg JJATDK 1,247,81
ipr-rat LD50:13 mg/kg APFRAD 24,17,66
ipr-mus LD50:23 mg/kg APFRAD 24,17,66
ivn-mus LD50:8900 μg/kg CSLNX* NX#04713

CONSENSUS REPORTS: Lead and its compounds are on the Community Right-To-Know List.

SAFETY PROFILE: Poison by ingestion, intravenous, and intraperitoneal routes. When heated to decomposition it emits toxic fumes of Pb. See also LEAD COMPOUNDS.

THZ000 CAS:2155-70-6 ***HR: 3***
TRIBUTYL(METHACRYLOXY)STANNANE
mf: $C_{16}H_{32}O_2Sn$ mw: 375.17

SYNS: TRIBUTYL(METHACRYLOYLOXY)STANNANE◇ TRIBUTYL ((2-METHYL-1-OXO-2-PROPENYL)OXY)STANNANE◇ TRIBUTYLSTANNYL METHACRYLATE ◇ TRIBUTYLTIN METHACRYLATE

TOXICITY DATA with REFERENCE
ivn-mus LD50:18 mg/kg CSLNX* NX#02761
orl-rat LD50:160 mg/kg UBZHAZ 50,695,78
orl-mus LD50:160 mg/kg UBZHAZ 50,695,78

CONSENSUS REPORTS: Reported in EPA TSCA Inventory.

OSHA PEL: TWA 0.1 mg(Sn)/m^3 (skin)
ACGIH TLV: TWA 0.1 mg(Sn)/m^3 (skin) (Proposed: TWA 0.1 mg(Sn)/m^3; STEL 0.2 mg(Sn)/m^3 (skin))
NIOSH REL: (Organotin Compounds) TWA 0.1 mg (Sn)/m^3

SAFETY PROFILE: Poison by ingestion and intravenous routes. When heated to decomposition it emits acrid smoke and irritating fumes. See also TIN COMPOUNDS.

TIA000 CAS:3090-35-5 ***HR: 3***
TRIBUTYL(OLEOYLOXY)STANNANE
mf: $C_{30}H_{60}O_2Sn$ mw: 571.59

SYNS: ENT 27,261 ◇ 9-DECAOCTENOIC ACID, TRIBUTYLSTANNYL ESTER ◇ TRIBUTYL(OLEOYLOXY)TIN ◇ TRI-N-BUTYLTIN OLEATE ◇ TRI-N-BUTYL-ZINN OLEAT (GERMAN)

TOXICITY DATA with REFERENCE
orl-rat LD50:195 mg/kg ARZNAD 19,934,69
ivn-mus LD50:18 mg/kg CSLNX* NX#03810

OSHA PEL: TWA 0.1 mg(Sn)/m^3 (skin)
ACGIH TLV: TWA 0.1 mg(Sn)/m^3 (skin) (Proposed: TWA 0.1 mg(Sn)/m^3; STEL 0.2 mg(Sn)/m^3 (skin))
NIOSH REL: (Organotin Compounds) TWA 0.1 mg (Sn)/m^3

SAFETY PROFILE: Poison by ingestion and intravenous routes. When heated to decomposition it emits acrid smoke and irritating fumes. See also TIN COMPOUNDS and ESTERS.

TIA250 CAS:126-73-8 ***HR: 3***
TRIBUTYL PHOSPHATE
mf: $C_{12}H_{27}O_4P$ mw: 266.36

PROP: Colorless odorless liquid. Bp: 289° (decomp), mp: < −80°, flash p: 295°F (COC), d: 0.982 @ 20°, vap d: 9.20. Sol in water; misc in alc and ether.

SYNS: CELLUPHOS 4 ◇ TBP ◇ TRIBUTILFOSFATO (ITALIAN) ◇ TRIBUTYLE (PHOSPHATE de) (FRENCH) ◇ TRIBUTYLFOSFAAT (DUTCH) ◇ TRIBUTYLPHOSPHAT (GERMAN) ◇ TRI-n-BUTYL PHOSPHATE

TOXICITY DATA with REFERENCE
skn-rbt 10 mg/24H JIHTAB 26,269,44
eye-rbt 97 mg AJOPAA 29,1363,46
orl-rat TDLo:12600 mg/kg (63D male):REP TOLED5 13,29,82
orl-rat LD50:3000 mg/kg JIHTAB 26,269,44
ipr-rat LD50:251 mg/kg GTPZAB 15(8),30,71
ivn-rat LDLo:100 mg/kg NATUAS 179,154,57
orl-mus LD50:1189 mg/kg GTPZAB 15(8),30,71
ihl-mus LC50:1300 mg/m^3 GTPZAB 15(8),30,71
ipr-mus LD50:159 mg/kg GTPZAB 15(8),30,71
scu-mus LDLo:3 g/kg EDWU** -,-,37
ihl-cat LDLo:24510 mg/m^3/5H EDWU** -,-,37

CONSENSUS REPORTS: Reported in EPA TSCA Inventory.

OSHA PEL: (Transitional: TWA 5 mg/m^3) TWA 0.2 ppm
ACGIH TLV: TWA 0.2 ppm

SAFETY PROFILE: Poison by intraperitoneal and intravenous routes. Moderately toxic by ingestion, inhalation, and subcutaneous routes. Experimental reproductive effects. A skin, eye, and mucous membrane irritant. Combustible when exposed to heat or flame. To fight fire, use CO_2, dry chemical, fog, mist. When heated to decomposition it emits toxic fumes of PO_x.

TIA450 CAS:3084-50-2 ***HR: 2***
TRIBUTYLPHOSPHINE SULFIDE
mf: $C_{12}H_{27}PS$ mw: 234.42

PROP: Liquid.

SYNS: PHOSPHINE SULFIDE, TRIBUTYL- ◇ PHOSPHINE, TRIBUTYL-, SULFIDE ◇ TRIBUTYLFOSFINSULFID

TOXICITY DATA with REFERENCE
skn-rbt 10 mg/24H open MLD AIHAAP 23,95,62
skn-rbt 100 mg/24H MOD 85JCAE-,1119,86
eye-rbt 500 mg/24H MLD 85JCAE-,1119,86
orl-rat LD50:930 mg/kg AIHAAP 23,95,62
skn-rbt LD50:1000 mg/kg AIHAAP 23,95,62

SAFETY PROFILE: Moderately toxic by ingestion and skin contact. A skin and eye irritant. When heated to decomposition it emits toxic fumes of PO_x and SO_x

TIA750 CAS:102-85-2 ***HR: 2***
TRIBUTYL PHOSPHITE
mf: $C_{12}H_{27}O_3P$ mw: 250.36

PROP: Liquid. Decomp in water, flash p: 248°F (OC), d: 0.9, bp: 120° @ 7 mm.

TOXICITY DATA with REFERENCE
skn-rbt 10 mg/24H JIHTAB 26,269,44
eye-rbt 500 mg AJOPAA 29,1363,46
orl-rat LD50:3000 mg/kg JIHTAB 26,269,44

CONSENSUS REPORTS: Reported in EPA TSCA Inventory.

SAFETY PROFILE: Moderately toxic by ingestion. A skin and eye irritant. Has been known to damage eyes. Combustible when exposed to heat or flame. To fight fire, use CO_2, dry chemical. When heated to decomposition it emits toxic fumes of PO_x.

TIB000 CAS:5488-45-9 ***HR: 3***
TRIBUTYL(8-QUINOLINOLATO)TIN
mf: $C_{21}H_{33}NOSn$ mw: 434.24

SYN: (8-QUINOLINOLATO)TRIBUTYLSTANNANE

TOXICITY DATA with REFERENCE
ivn-mus LD50:8900 μg/kg CSLNX* NX#03564

OSHA PEL: TWA 0.1 mg(Sn)/m^3 (skin)
ACGIH TLV: TWA 0.1 mg(Sn)/m^3 (skin) (Proposed: TWA 0.1 mg(Sn)/m^3; STEL 0.2 mg(Sn)/m^3 (skin))
NIOSH REL: (Organotin Compounds) TWA 0.1 mg (Sn)/m^3

SAFETY PROFILE: Poison by intravenous route. When heated to decomposition it emits toxic fumes of NO_x. See also TIN COMPOUNDS.

TIB250 CAS:2179-92-2 ***HR: 3***
TRIBUTYLSTANNANECARBONITRILE
mf: $C_{13}H_{27}NSn$ mw: 316.10

SYN: TRI-n-BUTYLTIN CYANIDE

TOXICITY DATA with REFERENCE
ivn-mus LD50:18 mg/kg CSLNX* NX#05643

CONSENSUS REPORTS: Cyanide and its compounds are on the Community Right-To-Know List.

OSHA PEL: TWA 0.1 mg(Sn)/m^3 (skin)
ACGIH TLV: TWA 0.1 mg(Sn)/m^3 (skin) (Proposed: TWA 0.1 mg(Sn)/m^3; STEL 0.2 mg(Sn)/m^3 (skin))
NIOSH REL: (Organotin Compounds) TWA 0.1 mg (Sn)/m^3

SAFETY PROFILE: Poison by intravenous route. When heated to decomposition it emits very toxic fumes of NO_x and CN^-. See also TIN COMPOUNDS and NITRILES.

TIB500 CAS:688-73-3 ***HR: 2***
TRI-n-BUTYLSTANNANE HYDRIDE
mf: $C_{12}H_{28}Sn$ mw: 291.09

SYNS: TRIBUTYLSTANNIC HYDRIDE ◇ TRIBUTYLTIN HYDRIDE ◇ TRI-n-BUTYLTIN HYDRIDE

TOXICITY DATA with REFERENCE
ihl-mus LCLo:1460 mg/m^3 NDRC** NDCrc-132,Feb,42

CONSENSUS REPORTS: Reported in EPA TSCA Inventory.

OSHA PEL: TWA 0.1 mg(Sn)/m^3 (skin)
ACGIH TLV: TWA 0.1 mg(Sn)/m^3 (skin) (Proposed: TWA 0.1 mg(Sn)/m^3; STEL 0.2 mg(Sn)/m^3 (skin))
NIOSH REL: (Organotin Compounds) TWA 0.1 mg (Sn)/m^3

SAFETY PROFILE: Moderately toxic by inhalation. When heated to decomposition it emits acrid smoke and irritating fumes. See also TIN COMPOUNDS.

TIB750 CAS:2857-03-6 ***HR: 3***
TRIBUTYLTIN-p-ACETAMIDOBENZOATE
mf: $C_{21}H_{35}NO_3Sn$ mw: 468.26

SYN: ((p-ACETAMIDOBENZOYL)OXY)TRIBUTYLSTANNANE

TOXICITY DATA with REFERENCE
ivn-mus LD50:18 mg/kg CSLNX* NX#02820

OSHA PEL: TWA 0.1 mg(Sn)/m^3
ACGIH TLV: TWA 0.1 mg(Sn)/m^3 (skin) (Proposed: TWA 0.1 mg(Sn)/m^3; STEL 0.2 mg(Sn)/m^3 (skin))
NIOSH REL: (Organotin Compounds) TWA 0.1 mg (Sn)/m^3

SAFETY PROFILE: Poison by intravenous route. When heated to decomposition it emits toxic fumes of NO_x. See also TIN COMPOUNDS.

TIC000 CAS:56-36-0 ***HR: 3***
TRIBUTYLTIN ACETATE
mf: $C_{14}H_{30}O_2Sn$ mw: 349.13

PROP: Mp: 80-83°.

SYNS: ACETOXYTRIBUTYLSTANNANE ◇ TRI-n-BUTYL-ZINN-ACETAT (GERMAN)

TOXICITY DATA with REFERENCE
orl-rat LD50:99 mg/kg JPMSAE 56,240,67
ipr-rat LDLo:10 mg/kg BJPCAL 10,16,55
orl-mus LDLo:46 mg/kg ATXKA8 23,283,68
ivn-mus LD50:180 mg/kg CSLNX* NX#01672
orl-rbt LDLo:40 mg/kg BJIMAG 15,15,58
orl-gpg LDLo:20 mg/kg BJPCAL 10,16,55
unr-mam LD50:500 mg/kg 30ZDA9 -,301,71

OSHA PEL: TWA 0.1 mg(Sn)/m^3
ACGIH TLV: TWA 0.1 mg(Sn)/m^3 (skin) (Proposed: TWA 0.1 mg(Sn)/m^3; STEL 0.2 mg(Sn)/m^3 (skin))
NIOSH REL: (Organotin Compounds) TWA 0.1 mg (Sn)/m^3

SAFETY PROFILE: Poison by ingestion, intraperitoneal, and intravenous routes. Moderately toxic by an unspecified route. When heated to decomposition it emits acrid smoke and irritating fumes. See also TIN COMPOUNDS.

TIC250 CAS:1461-23-0 ***HR: 2***
TRI-n-BUTYLTIN BROMIDE
mf: $C_{12}H_{27}Sn \cdot Br$ mw: 369.99

PROP: Liquid. D: 1.3365, bp: 163°/12 mm, refr index: 1.5000.

SYN: BROMOTRIBUTYLSTANNANE

TOXICITY DATA with REFERENCE
ihl-mus LCLo:1030 mg/m^3 NDRC** NDCrc-132,Feb,42
orl-rbt LDLo:1336 mg/kg SAIGBL 15,3,73

skn-rbt LDLo:935 mg/kg SAIGBL 15,3,73

OSHA PEL: TWA 0.1 mg(Sn)/m^3 (skin)
ACGIH TLV: TWA 0.1 mg(Sn)/m^3 (skin) (Proposed: TWA 0.1 mg(Sn)/m^3; STEL 0.2 mg(Sn)/m^3 (skin))
NIOSH REL: (Organotin Compounds) TWA 0.1 mg (Sn)/m^3

SAFETY PROFILE: Moderately toxic by ingestion, skin contact, and inhalation. When heated to decomposition it emits toxic fumes of Br^-. See also BROMIDES and TIN COMPOUNDS.

TIC500 CAS:56573-85-4 ***HR: 2***
TRIBUTYLTIN CHLORIDE COMPLEX

SYN: TIN-SAN

TOXICITY DATA with REFERENCE
skn-mam LD50:10 g/kg FMCHA2 -,D309,80

OSHA PEL: TWA 0.1 mg(Sn)/m^3 (skin)
ACGIH TLV: TWA 0.1 mg(Sn)/m^3 (skin) (Proposed: TWA 0.1 mg(Sn)/m^3; STEL 0.2 mg(Sn)/m^3 (skin))
NIOSH REL: (Organotin Compounds) TWA 0.1 mg (Sn)/m^3

SAFETY PROFILE: Mildly toxic by skin contact. When heated to decomposition it emits toxic fumes of Cl^-.

TIC750 CAS:5847-52-9 ***HR: 3***
TRIBUTYLTIN CHLOROACETATE
mf: $C_{14}H_{29}ClO_2Sn$ mw: 383.57

SYN: (CHLOROACETOXY)TRIBUTYLSTANNANE

TOXICITY DATA with REFERENCE
ivn-mus LD50:8 mg/kg CSLNX* NX#02270

OSHA PEL: TWA 0.1 mg(Sn)/m^3 (skin)
ACGIH TLV: TWA 0.1 mg(Sn)/m^3 (skin) (Proposed: TWA 0.1 mg(Sn)/m^3; STEL 0.2 mg(Sn)/m^3 (skin))
NIOSH REL: (Organotin Compounds) TWA 0.1 mg (Sn)/m^3

SAFETY PROFILE: Poison by intravenous route. When heated to decomposition it emits toxic fumes of Cl^-. See also TIN COMPOUNDS.

TID000 CAS:33550-22-0 ***HR: 3***
TRIBUTYLTIN-Γ-CHLOROBUTYRATE
mf: $C_{16}H_{33}ClO_2Sn$ mw: 411.63

SYN: 4-CHLOROBUTYRIC ACID TRIBUTYLSTANNYL ESTER

TOXICITY DATA with REFERENCE
ivn-mus LD50:18 mg/kg CSLNX* NX#03746

OSHA PEL: TWA 0.1 mg(Sn)/m^3 (skin)
ACGIH TLV: TWA 0.1 mg(Sn)/m^3 (skin) (Proposed: TWA 0.1 mg(Sn)/m^3; STEL 0.2 mg(Sn)/m^3 (skin))
NIOSH REL: (Organotin Compounds) TWA 0.1 mg (Sn)/m^3

SAFETY PROFILE: Poison by intravenous route. When heated to decomposition it emits toxic fumes of Cl^-. See also TIN COMPOUNDS.

TID100 CAS:2669-35-4 ***HR: 3***
TRIBUTYLTIN CYCLOHEXANECARBOXYLATE
mf: $C_{19}H_{38}O_2Sn$ mw: 417.26

SYNS: CYCLOHEXANECARBOXYLIC ACID, TRIBUTYLSTANNYL ESTER ◇ ((CYCLOHEXYLCARBONYL)OXY)TRIBUTYLSTANNANE

TOXICITY DATA with REFERENCE
ivn-mus LD50:18 mg/kg CSLNX* NX#04268

OSHA PEL: TWA 0.1 mg(Sn)/m^3 (skin)
ACGIH TLV: TWA 0.1 mg(Sn)/m^3; STEL 0.2 mg/m^3 (skin)
NIOSH REL: (Organotin Compound):10H TWA 0.1 mg(Sn)/m^3

SAFETY PROFILE: Poison by intravenous route. When heated to decomposition it emits toxic fumes of Sn.

TID150 CAS:20369-63-5 ***HR: 3***
TRIBUTYLTIN DIMETHYLDITHIOCARBAMATE
mf: $C_{15}H_{33}NS_2Sn$ mw: 410.30

SYNS: N,N-DIMETHYLDITHIOCARBAMIC ACID S-TRIBUTYLSTANNYL ESTER ◇ STANNANE, ((DIMETHYLTHIOCARBAMOYL)THIO)TRIBUTYL-

TOXICITY DATA with REFERENCE
ivn-mus LD50:100 mg/kg CSLNX* NX#04269

OSHA PEL: TWA 0.1 mg(Sn)/m^3 (skin)
ACGIH TLV: TWA 0.1 mg(Sn)/m^3; STEL 0.2 mg/m^3 (skin)
NIOSH REL: (Organotin Compound):10H TWA 0.1 mg(Sn)/m^3

SAFETY PROFILE: Poison by intravenous route. When heated to decomposition it emits toxic fumes of NO_x, SO_x, and Sn.

TID250 CAS:5035-67-6 ***HR: 3***
TRIBUTYLTIN-2-ETHYLHEXANOATE
mf: $C_{20}H_{42}O_2Sn$ mw: 433.31

SYNS: ((2-ETHYLHEXANOYL)OXY)TRIBUTYLSTANNANE ◇ TRIBUTYL((2-ETHYLHEXANOYL)OXY)STANNANE ◇ TRIBUTYL(2-ETHYL-1-OXOHEXYL)OXY)STANNANE

TOXICITY DATA with REFERENCE
orl-mus LDLo:710 mg/kg AECTCV 14,111,85

ivn-mus LD50:180 mg/kg CSLNX* NX#02794

OSHA PEL: TWA 0.1 mg(Sn)/m^3 (skin)
ACGIH TLV: TWA 0.1 mg(Sn)/m^3 (skin) (Proposed: TWA 0.1 mg(Sn)/m^3; STEL 0.2 mg(Sn)/m^3 (skin))
NIOSH REL: (Organotin Compounds) TWA 0.1 mg (Sn)/m^3

SAFETY PROFILE: Poison by intravenous route. Moderately toxic by ingestion. When heated to decomposition it emits acrid smoke and irritating fumes. See also TIN COMPOUNDS.

TID500 CAS:1067-97-6 ***HR: 2***
TRIBUTYLTIN HYDROXIDE
mf: $C_{12}H_{28}OSn$ mw: 307.09

SYNS: (HYDROXY)TRIBUTYLSTANNANE ◇ TRIBUTYLHYDROXYSTANNANE

TOXICITY DATA with REFERENCE
unr-mam LD50:500 mg/kg 30ZDA9 -,301,71

OSHA PEL: TWA 0.1 mg(Sn)/m^3 (skin)
ACGIH TLV: TWA 0.1 mg(Sn)/m^3 (skin) (Proposed: TWA 0.1 mg(Sn)/m^3; STEL 0.2 mg(Sn)/m^3 (skin))
NIOSH REL: (Organotin Compounds) TWA 0.1 mg(Sn)/m^3

SAFETY PROFILE: Moderately toxic by unspecified routes. When heated to decomposition it emits acrid smoke and irritating fumes. See also TIN COMPOUNDS.

TID750 CAS:73927-91-0 ***HR: 3***
TRIBUTYLTIN IODOACETATE
mf: $C_{14}H_{29}IO_2Sn$ mw: 475.02

SYN: (IODOACETOXY)TRIBUTYLSTANNANE

TOXICITY DATA with REFERENCE
ivn-mus LD50:56 mg/kg CSLNX* NX#03469

OSHA PEL: TWA 0.1 mg(Sn)/m^3 (skin)
ACGIH TLV: TWA 0.1 mg(Sn)/m^3 (skin) (Proposed: TWA 0.1 mg(Sn)/m^3; STEL 0.2 mg(Sn)/m^3 (skin))
NIOSH REL: (Organotin Compounds) TWA 0.1 mg (Sn)/m^3

SAFETY PROFILE: Poison by intravenous route. When heated to decomposition it emits toxic fumes of I^-. See also IODIDES and TIN COMPOUNDS.

TIE000 CAS:73927-93-2 ***HR: 3***
TRIBUTYLTIN-o-IODOBENZOATE
mf: $C_{19}H_{31}IO_2Sn$ mw: 537.09

SYN: o-IODOBENZOIC ACID TRIBUTYLSTANNYL ESTER

TOXICITY DATA with REFERENCE
ivn-mus LD50:180 mg/kg CSLNX* NX#03747

OSHA PEL: TWA 0.1 mg(Sn)/m^3
ACGIH TLV: TWA 0.1 mg(Sn)/m^3 (skin) (Proposed: TWA 0.1 mg(Sn)/m^3; STEL 0.2 mg(Sn)/m^3 (skin))
NIOSH REL: (Organotin Compounds) TWA 0.1 mg (Sn)/m^3

SAFETY PROFILE: Poison by intravenous route. When heated to decomposition it emits toxic fumes of I^-. See also TIN COMPOUNDS, IODIDES, and ESTERS.

TIE250 CAS:73940-88-2 ***HR: 3***
TRIBUTYLTIN-p-IODOBENZOATE
mf: $C_{19}H_{31}IO_2Sn$ mw: 537.09

SYNS: p-IODOBENZOIC ACID TRIBUTYLSTANNYL ESTER ◇ (4-IODOBENZOYLOXY)TRIBUTYLSTANNANE

TOXICITY DATA with REFERENCE
ivn-mus LD50:56 mg/kg CSLNX* NX#03794

OSHA PEL: TWA 0.1 mg(Sn)/m^3 (skin)
ACGIH TLV: TWA 0.1 mg(Sn)/m^3 (skin) (Proposed: TWA 0.1 mg(Sn)/m^3; STEL 0.2 mg(Sn)/m^3 (skin))
NIOSH REL: (Organotin Compounds) TWA 0.1 mg(Sn)/m^3

SAFETY PROFILE: Poison by intravenous route. When heated to decomposition it emits toxic fumes of I^-. See also IODIDES and TIN COMPOUNDS.

TIE500 CAS:73927-95-4 ***HR: 3***
TRIBUTYLTIN-β-IODOPROPIONATE
mf: $C_{15}H_{31}IO_2Sn$ mw: 489.05

SYN: (IODOPROPIONYLOXY)TRIBUTYLSTANNANE

TOXICITY DATA with REFERENCE
ivn-mus LD50:180 mg/kg CSLNX* NX#03470

OSHA PEL: TWA 0.1 mg(Sn)/m^3 (skin)
ACGIH TLV: TWA 0.1 mg(Sn)/m^3 (skin) (Proposed: TWA 0.1 mg(Sn)/m^3; STEL 0.2 mg(Sn)/m^3 (skin))
NIOSH REL: (Organotin Compounds) TWA 0.1 mg (Sn)/m^3

SAFETY PROFILE: Poison by intravenous route. When heated to decomposition it emits toxic fumes of I^-. See also IODIDES and TIN COMPOUNDS.

TIE600 CAS:53404-82-3 ***HR: 3***
TRIBUTYLTIN ISOPROPYLSUCCINATE
mf: $C_{19}H_{38}O_4Sn$ mw: 449.26

SYNS: STANNANE, (ISOPROPYLSUCCINYLOXY)TRIBUTYL- ◇ SUCCINIC ACID, O-ISOPROPYL-O'-TRIBUTYLSTANNYL ESTER

TOXICITY DATA with REFERENCE
ivn-mus LD50:18 mg/kg CSLNX* NX#04264

OSHA PEL: TWA 0.1 mg(Sn)/m^3 (skin)
ACGIH TLV: TWA 0.1 mg(Sn)/m^3; STEL 0.2 mg/m^3 (skin)
NIOSH REL: (Organotin Compound):10H TWA 0.1 mg(Sn)/m^3

SAFETY PROFILE: Poison by intravenous route. When heated to decomposition it emits toxic fumes of Sn.

TIE750 CAS:3090-36-6 ***HR: 3***
TRIBUTYLTIN LAURATE
mf: $C_{24}H_{50}O_2Sn$ mw: 489.43

SYNS: (LAUROYLOXY)TRIBUTYLSTANNANE ◇ TRIBUTYL((1-OXODODECYL)OXY)STANNANE (9CI) ◇ TRIBUTYLTIN DODECANOATE ◇ TRIBUTYLTIN MONOLAURATE ◇ TRI-N-BUTYLZINN-LAURAT (GERMAN)

TOXICITY DATA with REFERENCE
orl-mus LD50:180 mg/kg ATXKA8 23,283,68
ivn-mus LD50:20 mg/kg CSLNX* NX#02760

OSHA PEL: TWA 0.1 mg(Sn)/m^3 (skin)
ACGIH TLV: TWA 0.1 mg(Sn)/m^3 (skin) (Proposed: TWA 0.1 mg(Sn)/m^3; STEL 0.2 mg(Sn)/m^3 (skin))
NIOSH REL: (Organotin Compounds) TWA 0.1 mg (Sn)/m^3

SAFETY PROFILE: Poison by ingestion and intravenous routes. When heated to decomposition it emits acrid smoke and irritating fumes. See also TIN COMPOUNDS.

TIF000 CAS:13302-06-2 ***HR: 3***
TRI-n-BUTYLTIN METHANESULFONATE
mf: $C_{13}H_{30}O_3SSn$ mw: 385.18

SYN: ((METHYLSULFONYL)OXY)TRIBUTYLSTANNANE

TOXICITY DATA with REFERENCE
ivn-mus LD50:5620 μg/kg CSLNX* NX#02291

OSHA PEL: TWA 0.1 mg(Sn)/m^3 (skin)
ACGIH TLV: TWA 0.1 mg(Sn)/m^3 (skin) (Proposed: TWA 0.1 mg(Sn)/m^3; STEL 0.2 mg(Sn)/m^3 (skin))
NIOSH REL: (Organotin Compounds) TWA 0.1 mg (Sn)/m^3

SAFETY PROFILE: Poison by intravenous route. When heated to decomposition it emits toxic fumes of SO_x. See also TIN COMPOUNDS and SULFONATES.

TIF250 CAS:28801-69-6 ***HR: 3***
TRIBUTYLTIN NEODECANOATE
mf: $C_{22}H_{46}O_2Sn$ mw: 461.37

SYNS: 4,4-DIMETHYLOCTANOIC ACID, TRIBUTYLSTANNYL ESTER ◇ (4,4-DIMETHYLOCTANOYLOXY)TRIBUTYLSTANNANE ◇ HYDROXYTRIBUTYLSTANNANE-4,4-DIMETHYLOCTANOATE ◇ TRIBUTYL(NEODECANOYLOXY)STANNANE

TOXICITY DATA with REFERENCE
orl-mus LDLo:1070 mg/kg AECTCV 14,111,85
ivn-mus LD50:180 mg/kg CSLNX* NX#04263

OSHA PEL: TWA 0.1 mg(Sn)/m^3
ACGIH TLV: TWA 0.1 mg(Sn)/m^3 (skin) (Proposed: TWA 0.1 mg(Sn)/m^3; STEL 0.2 mg(Sn)/m^3 (skin))
NIOSH REL: (Organotin Compounds) TWA 0.1 mg (Sn)/m^3

SAFETY PROFILE: Poison by intravenous route. Moderately toxic by ingestion. When heated to decomposition it emits acrid smoke and irritating fumes. See also TIN COMPOUNDS.

TIF500 CAS:4027-14-9 ***HR: 3***
TRIBUTYLTIN NONANOATE
mf: $C_{21}H_{44}O_2Sn$ mw: 447.34

SYNS: NONANOIC ACID, TRIBUTYLSTANNYL ESTER ◇ (NONANOYLOXY)TRIBUTYLSTANNANE

TOXICITY DATA with REFERENCE
ivn-mus LD50:32 mg/kg CSLNX* NX#04818

OSHA PEL: TWA 0.1 mg(Sn)/m^3 (skin)
ACGIH TLV: TWA 0.1 mg(Sn)/m^3 (skin) (Proposed: TWA 0.1 mg(Sn)/m^3; STEL 0.2 mg(Sn)/m^3 (skin))
NIOSH REL: (Organotin Compounds) TWA 0.1 mg (Sn)/m^3

SAFETY PROFILE: Poison by intravenous route. When heated to decomposition it emits acrid smoke and irritating fumes. See also TIN COMPOUNDS and ESTERS.

TIF600 CAS:26377-04-8 ***HR: 3***
TRIBUTYLTIN SULFATE
mf: $C_{24}H_{54}O_4SSn_2$ mw: 676.22

SYNS: (HEXABUTYL(mu-(SULFATO(2−)-O,O'':O',O''')))DI TIN ◇ HYDROXYTRIBUTYLSTANNANE, SULFATE (2:1)

TOXICITY DATA with REFERENCE
orl-rat LDLo:50 mg/kg BJPCAL 10,16,55
ipr-rat LDLo:10 mg/kg BJPCAL 10,16,55
orl-mus LDLo:710 mg/kg AECTCV 14,111,85
orl-rbt LDLo:60 mg/kg BJPCAL 10,16,55

OSHA PEL: TWA 0.1 mg(Sn)/m^3 (skin)
ACGIH TLV: TWA 0.1 mg(Sn)/m^3 (skin) (Proposed: TWA 0.1 mg(Sn)/m^3; STEL 0.2 mg(Sn)/m^3 (skin))
NIOSH REL: (Organotin Compounds) TWA 0.1 mg (Sn)/m^3

SAFETY PROFILE: Poison by ingestion and intraperitoneal routes. When heated to decomposition it emits toxic fumes of SO_x. See also TIN COMPOUNDS and SULFATES.

TIF750 CAS:73940-89-3 ***HR: 3***
TRIBUTYLTIN-α-(2,4,5-TRICHLOROPHENOXY)PROPIONATE
mf: $C_{21}H_{33}Cl_3O_3Sn$ mw: 558.58

SYN: 2-(2,4,5-TRICHLOROPHENOXY)PROPIONICACID TRIBUTYLSTANNYL ESTER

TOXICITY DATA with REFERENCE
ivn-mus LD50:18 mg/kg CSLNX* NX#03748

OSHA PEL: TWA 0.1 mg(Sn)/m^3 (skin)
ACGIH TLV: TWA 0.1 mg(Sn)/m^3 (skin) (Proposed: TWA 0.1 mg(Sn)/m^3; STEL 0.2 mg(Sn)/m^3 (skin))
NIOSH REL: (Organotin Compounds) TWA 0.1 mg(Sn)/m^3

SAFETY PROFILE: Poison by intravenous route. When heated to decomposition it emits toxic fumes of Cl^-. See also TIN COMPOUNDS.

TIG000 CAS:73927-98-7 ***HR: 3***
TRIBUTYL(2,4,5-TRICHLOROPHENOXY)TIN
mf: $C_{18}H_{29}Cl_3OSn$ mw: 486.51

SYN: TRIBUTYL(2,4,5-TRICHLOROPHENOXY)STANNANE

TOXICITY DATA with REFERENCE
ivn-mus LD50:18 mg/kg CSLNX* NX#03638

OSHA PEL: TWA 0.1 mg(Sn)/m^3 (skin)
ACGIH TLV: TWA 0.1 mg(Sn)/m^3 (skin) (Proposed: TWA 0.1 mg(Sn)/m^3; STEL 0.2 mg(Sn)/m^3 (skin))
NIOSH REL: (Organotin Compounds) TWA 0.1 mg (Sn)/m^3

SAFETY PROFILE: Poison by intravenous route. When heated to decomposition it emits toxic fumes of Cl^-. See also TIN COMPOUNDS.

TIG250 CAS:150-50-5 ***HR: 3***
S,S,S-TRIBUTYL TRITHIOPHOSPHITE
mf: $C_{12}H_{27}PS_3$ mw: 298.54

PROP: Colorless liquid; mild characteristic odor. Bp: 142-145° @ 4.5 mm, flash p: 295°F (COC), d: 0.987 @ 20°/4°.

SYNS: CHEMAGRO B-1776 ◇ DELEAF DEFOLIANT ◇ EASY OFF-D ◇ FOLEX ◇ MERPHOS ◇ PHOSPHOROTRITHIOUS ACID, S,S,S-TRIBUTYL ESTER ◇ TRIBUTYL PHOSPHOROTRITHIOITE ◇ S,S,S-TRIBUTYL PHOSPHOROTRITHIOITE

TOXICITY DATA with REFERENCE
orl-rat LD50:910 mg/kg TXAPA9 14,515,69
skn-rat LD50:615 mg/kg WRPCA2 9,119,70
ipr-rat LD50:150 mg/kg PSEBAA 114,509,63
ipr-mus LD50:1400 mg/kg BCPCA6 12,73,63

SAFETY PROFILE: Poison by intraperitoneal route. Moderately toxic by ingestion and skin contact. A cholinesterase inhibitor. Combustible when exposed to heat or flame. Can react vigorously with oxidizing materials. When heated to decomposition it emits highly toxic fumes of PO_x and SO_x. Used as a defoliant. See also PARATHION.

TIG500 CAS:69226-47-7 ***HR: 3***
TRIBUTYL(UNDECANOYLOXY)STANNANE
mf: $C_{23}H_{48}O_2Sn$ mw: 475.40

SYNS: TRI-n-BUTYLTIN UNDECYLATE ◇ TRIBUTYLTIN UNDECYLENATE ◇ TRI-n-BUTYL-ZINN UNDECYLAT (GERMAN) ◇ UNDECANOIC ACID, TRIBUTYLSTANNYL ESTER

TOXICITY DATA with REFERENCE
orl-rat LD50:205 mg/kg ARZNAD 19,934,69
ivn-mus LD50:56 mg/kg CSLNX* NX#04271

OSHA PEL: TWA 0.1 mg(Sn)/m^3 (skin)
ACGIH TLV: TWA 0.1 mg(Sn)/m^3 (skin) (Proposed: TWA 0.1 mg(Sn)/m^3; STEL 0.2 mg(Sn)/m^3 (skin))
NIOSH REL: (Organotin Compounds) TWA 0.1 mg(Sn)/m^3

SAFETY PROFILE: Poison by ingestion and intravenous routes. When heated to decomposition it emits acrid smoke and irritating fumes. See also TIN COMPOUNDS and ESTERS.

TIG750 CAS:60-01-5 ***HR: 3***
TRIBUTYRIN
mf: $C_{15}H_{26}O_6$ mw: 302.41

PROP: Colorless, oily liquid; bitter taste. Mp: −75°, d: 1.0356 @ 20/20°, bp: 305-310°, flash p: +212°F. Insol in water; very sol in alc, ether, chloroform.

SYNS: BUTANOIC ACID, 1,2,3-PROPANETRIYL ESTER ◇ BUTYRIC ACID TRIESTER with GLYCERIN ◇ BUTYRYL TRIGLYCERIDE ◇ FEMA No. 2223 ◇ GLYCEROL TRIBUTYRATE ◇ KODAFLEX ◇ TRIBUTYROIN

TOXICITY DATA with REFERENCE
orl-rat TDLo:177 g/kg/3W-C:ETA JNCIAM 10,361,49
orl-rat LDLo:3200 mg/kg KODAK* -,-,71
ivn-mus LD50:320 mg/kg APSCAX 40,338,57

CONSENSUS REPORTS: Reported in EPA TSCA Inventory.

SAFETY PROFILE: Poison by intravenous route. Moderately toxic by ingestion. Questionable carcinogen with experimental tumorigenic data. Combustible liquid. When heated to decomposition it emits acrid smoke and irritating fumes. See also ESTERS.

TIH000 CAS:12380-95-9 ***HR: 3***
TRICADMIUM DINITRIDE
mf: Cd_3N_2 mw: 365.21

SYN: CADMIUM NITRIDE

CONSENSUS REPORTS: Cadmium compounds are on the Community Right-To-Know List.

OSHA PEL: TWA 0.2 mg(Cd)/m^3; CL 0.6 mg(Cd)/m^3 (dust)
ACGIH TLV: TWA 0.05 mg(Cd)/m^3 (Proposed: TWA 0.01 mg(Cd)/m^3 (dust), Suspected Human Carcinogen; 0.002 mg(Cd)/m^3 (respirable dust), Suspected Human Carcinogen); BEI: 10 μg/g creatinine in urine; 10 μg/L in blood.
NIOSH REL: (Cadium) Reduce to lowest feasible level.

SAFETY PROFILE: Confirmed human carcinogen. Many cadmium compounds are poisons. Explodes violently on shock or heating. Explodes on contact with water, acids, or bases. When heated to decomposition it emits very toxic fumes of NO_x and Cd. See also NITRIDES and CADMIUM COMPOUNDS.

TIH250 CAS:12013-82-0 ***HR: 3***
TRICALCIUM DINITRIDE
mf: Ca_3N_2 mw: 148.25
SYN: CALCIUM NITRIDE

SAFETY PROFILE: Spontaneously flammable in air. Incandescent reaction with chlorine gas or bromine vapor. Incompatible with halogens. When heated to decomposition it emits toxic fumes of NO_x. See also NITRIDES and CALCIUM COMPOUNDS.

TIH600 ***HR: 1***
TRICALCIUM SILICATE

SAFETY PROFILE: A nuisance dust.

TIH750 CAS:12134-29-1 ***HR: 2***
TRICESIUM NITRIDE
mf: Cs_3N mw: 412.72

SYN: CESIUM NITRIDE

SAFETY PROFILE: Burns in air. Incompatible with chlorine, phosphorus or sulfur. When heated to decomposition it emits toxic fumes of NO_x. See also NITRIDES and CESIUM COMPOUNDS.

TIH800 CAS:363-20-2 ***HR: 3***
TRICETAMIDE
mf: $C_{16}H_{24}N_2O_5$ mw: 324.42

PROP: Crystals from water. Mp: 133-134°.

SYNS: N-((DIETHYLCARBAMOYL)METHYL)-3,4,5-TRIMETHOXYBENZAMIDE ◇ R-548 ◇ RIKER 548 ◇ TOE ◇ TRIMEGLAMIDE ◇ N-(3,4,5-TRIMETHOXYBENZOYL)GLYCINEDIETHYLAMIDE

TOXICITY DATA with REFERENCE
ipr-rat LD50:340 mg/kg AIPTAK 137,218,62
ipr-mus LD50:3200 mg/kg YKKZAJ 86,120,66

SAFETY PROFILE: Poison by intraperitoneal route. When heated to decomposition it emits toxic fumes of NO_x.

TIH825 CAS:1117-99-3 ***HR: 3***
TRICHLOROACETALDEHYDE OXIME
mf: $C_2H_2Cl_3NO$ mw: 162.40

SAFETY PROFILE: Explosive reaction with alkali forms carbon dioxide and the toxic gases hydrogen cyanide and hydrogen chloride. When heated to decomposition it emits toxic fumes of Cl^- and NO_x. See also ALDEHYDES.

TII000 CAS:594-65-0 ***HR: 3***
2,2,2-TRICHLOROACETAMIDE
mf: $C_2H_2Cl_3NO$ mw: 162.40

SYNS: TRICHLOROACETAMIDE ◇ α,α,α-TRICHLOROACETAMIDE

TOXICITY DATA with REFERENCE
orl-rat LDLo:500 mg/kg JPETAB 90,260,47
ipr-mus LDLo:1100 mg/kg JACSAT 63,1437,41
ivn-mus LD50:180 mg/kg CSLNX* NX#04129

CONSENSUS REPORTS: Reported in EPA TSCA Inventory.

SAFETY PROFILE: Poison by intravenous route. Moderately toxic by ingestion and intraperitoneal routes. When heated to decomposition it emits very toxic fumes of Cl^- and NO_x.

TII250 CAS:76-03-9 ***HR: 3***
TRICHLOROACETIC ACID
DOT: UN 1839/UN 2564
mf: $C_2HCl_3O_2$ mw: 163.38

PROP: Colorless, rhombic, deliq crystals. Bp: 197.5°, fp: 57.7°, flash p: none, d: 1.6298 @ 61°/4°, vap press: 1 mm @ 51.0°.

SYNS: ACETO-CAUSTIN ◇ ACIDE TRICHLORACETIQUE (FRENCH) ◇ ACIDO TRICLOROACETICO (ITALIAN) ◇ AMCHEM GRASS KILLER ◇ DOW SODIUM TCA INHIBITED ◇ KONESTA ◇ SODIUM TCA, solution ◇ TCA ◇ TRICHLOORAZIJNZUUR (DUTCH) ◇ TRICHLORESSIGSAEURE (GERMAN) ◇ TRICHLOROACETIC ACID, solid (DOT) ◇ TRICHLOROACETIC ACID, solution (DOT) ◇ TRICHLOROETHANOIC ACID ◇ VARITOX

TOXICITY DATA with REFERENCE
skn-rbt 210 μg MLD XEURAQ MDDC-1715
eye-rbt 3500 μg/5S SEV XEURAQ MDDC-1715
mmo-sat 250 μg/plate CNJGA8 22,35,80
ipr-mus TDLo:125 mg/kg (male 5D pre):REP MUREAV 188,215,87
orl-mus TDLo:427 g/kg/61W-C:CAR TXAPA9 90,183,87
orl-rat LD50:400 mg/kg PEMNDP 8,765,87
ipr-mus LDLo:500 mg/kg CBCCT* 6,214,54
scu-mus LD50:270 mg/kg NIIRDN 6,879,82

CONSENSUS REPORTS: Reported in EPA TSCA Inventory. EPA Genetic Toxicology Program.

OSHA PEL: TWA 1 ppm
ACGIH TLV: TWA 1 ppm
DOT Classification: Corrosive Material; Label: Corrosive, solid; Corrosive Material; Label: Corrosive, solution.

SAFETY PROFILE: Poison by ingestion and subcutaneous routes. Moderately toxic by intraperitoneal route. Questionable carcinogen with experimental carcinogenic data. Experimental reproductive effects. Mutation data reported. A corrosive irritant to skin, eyes, and mucous membranes. When heated to decomposition it emits toxic fumes of Cl^- and Na_2O. Used as an herbicide. See also TRICHLOROACETIC ACID SODIUM SALT.

TII500 CAS:650-51-1 ***HR: 2***
TRICHLOROACETIC ACID SODIUM SALT
mf: $C_2Cl_3O_2 \cdot Na$ mw: 185.36

PROP: Crystals. Water-sol.

SYNS: ACP GRASS KILLER ◇ ANTIPERZ ◇ DOW SODIUM TCA INHIBITED ◇ GREEN CROSS COUCH GRASS KILLER ◇ NATA ◇ NATRIUMTRICHLOORACETAAT (DUTCH) ◇ NATRIUMTRICHLORACETAT (GERMAN) ◇ SODIO(TRICLOROACETATO di) (ITALIAN) ◇ SODIUM TCA INHIBITED ◇ SODIUM (TRICHLORACETATE de) (FRENCH) ◇ SODIUM TRICHLOROACETATE ◇ STCA ◇ TCA ◇ TCA SODIUM ◇ TRICHLORESSIGSAURES NATRIUM (GERMAN) ◇ TRICHLOROCTAN SODNY (CZECH) ◇ VARITOX ◇ WEEDMASTER GRASS KILLER

TOXICITY DATA with REFERENCE
cyt-hmn:lym 100 mg/L TGANAK 18,318,84
orl-rat LD50:3320 mg/kg KSKZAN 18(3),46,80
orl-mus LD50:3600 mg/kg PCOC** -,1061,66
ivn-mus LD50:2370 mg/kg 28ZPAK -,91,72

CONSENSUS REPORTS: Reported in EPA TSCA Inventory.

SAFETY PROFILE: Moderately toxic by ingestion and intravenous routes. Human mutation data reported. Large doses cause central nervous system depression. Used as an herbicide. Bags of the salt can ignite spontaneously in storage. When heated to decomposition it emits very toxic fumes of Cl^- and Na_2O. Used as an herbicide.

TII550 CAS:921-03-9 ***HR: 3***
1,1,3-TRICHLOROACETONE
mf: $C_3H_3Cl_3O$ mw: 161.41

PROP: Mp: 13-15°, bp: 172°, refr index: 1.4892, d: 1.508, flash p: 175° F.

SYNS: α,α′,α′-TRICHLOROACETONE ◇ 1,1,3-TRICHLORO-2-PROPANONE

TOXICITY DATA with REFERENCE
mmo-sat 50 μL/plate ENMUDM 7,163,85
mrc-smc 5 μL/L MUREAV 155,53,85
ihl-rat LC50:390 mg/m³/2H 85GMAT -,113,82
ihl-mus LC50:360 mg/m³/2H 85GMAT -,113,82

SAFETY PROFILE: Poison by inhalation. Corrosive and lachrymator. Mutation data reported. Combustible liquid. When heated to decomposition it emits toxic fumes of Cl^-. See also KETONES.

TII750 CAS:545-06-2 ***HR: 3***
TRICHLOROACETONITRILE
mf: C_2Cl_3N mw: 144.38

PROP: Crystals; odor of chloral and hydrogen cyanide. Mp: 61°, bp: 220°.

SYNS: CYANOTRICHLOROMETHANE ◇ NITRILE TRICHLORACETIQUE (FRENCH) ◇ TRICHLOR-ACETONITRIL (GERMAN) ◇ TRICHLOROMETHYL CYANIDE ◇ TRICHLOROMETHYLNITRILE ◇ TRICHLOURACETONITRIL (DUTCH) ◇ TRITOX

TOXICITY DATA with REFERENCE
skn-rbt 100 μg/24H open AIHAAP 23,95,62
skn-rbt 5 mg/24H SEV 85JCAE -,907,86
eye-rbt 50 μg/24H SEV 85JCAE -,907,86
mma-sat 3333 μg/plate ENMUDM 8(Suppl 7),1,86
dnd-hmn:lym 50 μmol/L FAATDF 6,447,86
orl-rat TDLo:825 mg/kg (female 7-21D post):REP TXCYAC 46,83,87
orl-rat TDLo:97500 μg/kg (female 6-18D post):TER TJADAB 38,113,88
orl-rat LD50:250 mg/kg AIHAAP 23,95,62
ihl-rat LCLo:250 ppm/4H AIHAAP 23,95,62
ivn-mus LD50:56 mg/kg CSLNX* NX#06416
ihl-rbt LCLo:311 ppm/5H JIHTAB 31,235,49
skn-rbt LD50:900 mg/kg AIHAAP 23,95,62
ihl-gpg LCLo:311 ppm/5H JIHTAB 31,235,49

CONSENSUS REPORTS: Cyanide and its compounds are on the Community Right-To-Know List. Reported in EPA TSCA Inventory.

SAFETY PROFILE: Poison by ingestion and intravenous routes. Moderately toxic by inhalation and skin contact. Human mutation data reported. A skin and severe eye irritant. An experimental teratogen. Other experimental reproductive effects. When heated to decomposition or in reaction with water, steam, acid or acid fumes it produces toxic fumes of CN^-, Cl^-, and NO_x. Used as an insecticide. See also NITRILES and CYANIDES.

TIJ000 CAS:63041-25-8 ***HR: 3***
10-TRICHLOROACETYL-1,2-BENZANTHRACENE
mf: $C_{20}H_{11}Cl_3O$ mw: 373.66

SYNS: 1-(BENZ(a)ANTHRACEN-7-YL)-2,2,2-TRICHLOROETHANONE ◇ BENZ(a)ANTHRACEN-7-YL TRICHLOROMETHYL KETONE

TOXICITY DATA with REFERENCE
scu-rat TDLo:45 mg/kg:ETA XPHPAW 149,191,51

SAFETY PROFILE: Questionable carcinogen with experimental tumorigenic data. When heated to decomposition it emits toxic fumes of Cl^-. See also KETONES.

TIJ150 CAS:76-02-8 ***HR: 2***
TRICHLOROACETYL CHLORIDE
DOT: UN 2442
mf: C_2Cl_4O mw: 181.82

PROP: Mp: -146°, bp: 114-116°, refr index: 1.4700, d: 1.629, flash p: none.

SYNS: TRICHLOROACETIC ACID CHLORIDE ◇ TRICHLOROACETOCHLORIDE

TOXICITY DATA with REFERENCE
orl-rat LD50:600 mg/kg 85GMAT -,113,82
ihl-rat LC50:475 mg/m^3/4H 85GMAT -,113,82
ihl-mus LC50:445 mg/m^3 85GMAT -,113,82

CONSENSUS REPORTS: EPA Extremely Hazardous Substances List. Reported in EPA TSCA Inventory.

DOT Classification: Corrosive Material; Label: Corrosive.

SAFETY PROFILE: Moderately toxic by inhalation and ingestion. A corrosive irritant to skin, eyes, and mucous membranes. When heated to decomposition it emits toxic fumes of Cl^-.

TIJ175 CAS:354-13-2 ***HR: 2***
TRICHLOROACETYL FLUORIDE
mf: C_2Cl_3FO mw: 165.37

TOXICITY DATA with REFERENCE
ihl-mus LCLo:2000 mg/m^3 11FYAN 3,75,63

OSHA PEL: TWA 2.5 mg(F)/m^3
ACGIH TLV: TWA 2.5 mg(F)/m^3

SAFETY PROFILE: Moderately toxic by inhalation. When heated to decomposition it emits toxic fumes of F^- and Cl^-.

TIJ250 CAS:3787-28-8 ***HR: 3***
2,3,3-TRICHLOROACROLEIN
mf: C_3HCl_3O mw: 159.39

SYNS: 2,3,3-TRICHLOROPROPENAL ◇ 2,3,3-TRICHLORO-2-PROPENAL

TOXICITY DATA with REFERENCE
mmo-sat 1 nmol/plate MUREAV 78,113,80
mma-sat 120 ng/plate MUREAV 157,111,85
ipr-mus LD50:4 mg/kg JAFCAU 30,627,82

SAFETY PROFILE: Poison by intraperitoneal route. Mutation data reported. When heated to decomposition it emits toxic fumes of Cl^-.

TIJ500 CAS:815-58-7 ***HR: 3***
TRICHLOROACRYLOYL CHLORIDE
mf: C_3Cl_4O mw: 193.83

SYNS: TCAT ◇ TRICHLORACRYLYL CHLORIDE ◇ 2,3,3-TRICHLOROACRYLOYL CHLORIDE ◇ TRICHLOROACRYLYL CHLORIDE ◇ 2,3,4-TRICHLORO-2-PROPENOYL CHLORIDE

TOXICITY DATA with REFERENCE
skn-rbt 270 μg MLD XEURAQ MDDC-1715
eye-rbt 2560 μg/5S MLD XEURAQ MDDC-1715
mmo-sat 1 μL/plate MUREAV 117,21,83
mma-sat 1 μL/plate MUREAV 117,21,83
ihl-rat LC50:107 ppm/30M XEURAQ MDDC-1715
ipr-rat LD50:750 mg/kg XEURAQ MDDC-1715
scu-rat LDLo:1500 mg/kg XEURAQ MDDC-1715
ihl-mus LCLo:67 ppm/30M XEURAQ MDDC-1715
ihl-rbt LCLo:67 ppm/30M XEURAQ MDDC-1715

CONSENSUS REPORTS: Reported in EPA TSCA Inventory.

SAFETY PROFILE: Poison by inhalation. Moderately toxic by intraperitoneal and subcutaneous routes. Mutation data reported. A skin and eye irritant. When heated to decomposition it emits toxic fumes of Cl^-.

TIJ750 CAS:634-93-5 ***HR: 3***
2,4,6-TRICHLOROANILINE
mf: $C_6H_4Cl_3N$ mw: 196.46

PROP: Needles from liquid. Mp: 77.5-78.5°, bp: 262° @ 46 mm, Insol in H_3PO_4; sol in alc, ether.

SYNS: sym-TRICHLOROANILINE ◇ 2,4,6-TRICHLOROBENZENAMINE

TOXICITY DATA with REFERENCE
orl-mus TDLo:390 g/kg/78W-C:CAR JEPTDQ 2,325,78
orl-mus TD:780 g/kg/78W-C:CAR JEPTDQ 2,325,78
orl-rat LD50:3850 mg/kg GISAAA 50(3),83,85
orl-mus LD50:5800 mg/kg GISAAA 50(3),83,85

CONSENSUS REPORTS: Reported in EPA TSCA Inventory.

SAFETY PROFILE: Moderately toxic by ingestion. Irritant. Questionable carcinogen with experimental carcinogenic data. When heated to decomposition it emits very toxic fumes of Cl^- and NO_x. See also AROMATIC AMINES.

TIK000 CAS:2307-49-5 ***HR: 3***
3,5,6-TRICHLORO-o-ANISIC ACID
mf: $C_8H_5Cl_3O_3$ mw: 255.48

PROP: A powder. Mp: 138°. Freely sol in alc; sol in xylene. Its salts are very sol in water.

SYNS: BANVEL T ◇ 2-METHOXY-3,5,6-TRICHLOROBENZOIC ACID ◇ METRIBEN ◇ TRICAMBA ◇ 2,3,5-TRICHLORO-6-METHOXYBENZOIC ACID ◇ 3,5,6-TRICHLORO-2-METHOXYBENZOIC ACID ◇ VELSICOL ◇ VELSICOL COMPOUND C

TOXICITY DATA with REFERENCE
orl-rat LD50:300 mg/kg WRPCA2 4,36,65

SAFETY PROFILE: Poison by ingestion. When heated to decomposition it emits toxic fumes of Cl^-. Used as an herbicide.

TIK250 CAS:120-82-1 ***HR: 3***
1,2,4-TRICHLOROBENZENE
DOT: UN 2321
mf: $C_6H_3Cl_3$ mw: 181.44

PROP: Colorless liquid. Mp: 17°, bp: 213°, flash p: 230°F (CC), d: 1.454 @ 25°/25°, vap press: 1 mm @ 38.4°, vap d: 6.26. Sol in water.

SYNS: 1,2,4-TRICHLOROBENZENE, liquid (DOT) ◇ unsym-TRICHLOROBENZENE ◇ TROJCHLOROBENZEN (POLISH)

TOXICITY DATA with REFERENCE
skn-rbt 1950 mg/13W-I MOD AEHLAU 30,165,75
orl-rat TDLo:1800 mg/kg (9-13D preg):TER ENVRAL 31,362,83
ipr-rat TDLo:750 mg/kg (female 3D pre):REP JTEHD6 8,489,81
orl-rat LD50:756 mg/kg AOHYA3 12,209,69
orl-mus LD50:300 mg/kg NAIZAM 29,569,78
ipr-mus LD50:1223 mg/kg MUTAEX 2,111,87

CONSENSUS REPORTS: Community Right-To-Know List. Reported in EPA TSCA Inventory.

OSHA PEL: CL 5 ppm
ACGIH TLV: CL 5 ppm
DFG MAK: 5 ppm (40 mg/m^3)
DOT Classification: Poison B; Label: St. Andrews Cross.

SAFETY PROFILE: Poison by ingestion. Moderately toxic by intraperitoneal route. An experimental teratogen. Experimental reproductive effects. A skin irritant. Combustible when exposed to heat or flame. Can react vigorously with oxidizing materials. To fight fire, use water, foam, CO_2, dry chemical. When heated to decomposition it emits toxic fumes of Cl^-. See also CHLORINATED HYDROCARBONS, AROMATIC.

TIK500 CAS:50-31-7 ***HR: 3***
2,3,6-TRICHLOROBENZOIC ACID
mf: $C_7H_3Cl_3O_2$ mw: 225.45

SYNS: BENZABAR ◇ BENZAC ◇ FEN-ALL ◇ HC 1281 ◇ NCI-C60242 ◇ T-2 ◇ 2,3,6-TBA (herbicide) ◇ 2,3,6-TCB ◇ 2,3,6-TCBA ◇ TRIBAC ◇ 2,3,6-TRICHLORBENZOESAEURE (GERMAN) ◇ TRICHLOROBENZOIC ACID ◇ TRYBEN ◇ TRYSBEN 200 ◇ ZOBAR

TOXICITY DATA with REFERENCE
orl-rat LD50:650 mg/kg HYSAAV 35(7-9),14,70
ipr-rat LD50:1000 mg/kg GUCHAZ 6,515,73
orl-mus LD50:615 mg/kg HYSAAV 35(7-9),14,70
ipr-mus LD50:178 mg/kg JMCMAR 11,1020,68
scu-mus LD50:1500 mg/kg BCPCA6 13,1538,64
orl-rbt LD50:812 mg/kg HYSAAV 35(7-9),14,70
orl-gpg LD50:1218 mg/kg HYSAAV 35(7-9),14,70

SAFETY PROFILE: Poison by intraperitoneal route. Moderately toxic by ingestion and subcutaneous routes. When heated to decomposition it emits toxic fumes of Cl^-. Used as an herbicide.

TIL250 CAS:1344-32-7 ***HR: 2***
TRICHLOROBENZYL CHLORIDE
mf: $C_7H_4Cl_4$ mw: 229.91

SYN: TCBC

TOXICITY DATA with REFERENCE
orl-rat LD50:3075 mg/kg 28ZEAL 4,359,69

CONSENSUS REPORTS: Reported in EPA TSCA Inventory.

SAFETY PROFILE: Moderately toxic by ingestion. When heated to decomposition it emits toxic fumes of Cl^-. See also CHLORINATED HYDROCARBONS, AROMATIC.

TIL300 CAS:26445-82-9 ***HR: 2***
B-1,3,5-TRICHLOROBORAZINE
mf: $B_3Cl_3H_3N_3$ mw: 183.83

ClBN(H)B(Cl)N(H)B(Cl)NH (ring: first Cl-B bonded to final NH)

PROP: Mp: 84°, d: 1.58°.

SAFETY PROFILE: Reacts violently with water. When heated to decomposition it emits toxic fumes of Cl^- and NO_x. See also BORON COMPOUNDS and BORAZINE.

TIL350 CAS:2852-07-5 ***HR: 2***
1,1,2-TRICHLOROBUTADIENE
mf: $C_4H_3Cl_3$ mw: 157.42

SYN: 1,1,2-TRICHLORO-1,3-BUTADIENE

TOXICITY DATA with REFERENCE
eye-rbt 100 mg GISAAA 45(10),77,80
orl-rat LD50:680 mg/kg GISAAA 45(10),77,80
ihl-rat LC50:7000 mg/m^3 GISAAA 45(10),77,80
orl-mus LD50:1000 mg/kg GISAAA 45(10),77,80
ihl-mus LC50:3300 mg/m^3 GISAAA 45(10),77,80

SAFETY PROFILE: Moderately toxic by ingestion and

inhalation. An eye irritant. When heated to decomposition it emits toxic fumes of Cl^-. See also CHLORINATED HYDROCARBONS, ALIPHATIC.

TIL360 CAS:2431-50-7 ***HR: 3***
2,3,4-TRICHLOROBUTENE-1
mf: $C_4H_5C_{13}$ mw: 159.44

SYN: 1-BUTENE, 2,3,4-TRICHLORO-

TOXICITY DATA with REFERENCE
orl-rat LD50:341 mg/kg GISAAA 46(1),92,81

CONSENSUS REPORTS: Reported in EPA TSCA Inventory.
DFG MAK: Animal Carcinogen; Suspected Human Carcinogen.

SAFETY PROFILE: Confirmed carcinogen. Poison by ingestion. When heated to decomposition it emits toxic fumes of Cl^-.

TIL500 CAS:101-20-2 ***HR: 2***
3,4,4'-TRICHLOROCARBANILIDE
mf: $C_{13}H_9Cl_3N_2O$ mw: 315.59

SYN: N-(3,4-DICHLOROPHENYL)-N'-(4-CHLOROPHENYL)UREA

TOXICITY DATA with REFERENCE
ipr-mus LD50:2100 mg/kg LPPTAK 27,306,79

CONSENSUS REPORTS: Reported in EPA TSCA Inventory.

SAFETY PROFILE: Moderately toxic by intraperitoneal route. When heated to decomposition it emits very toxic fumes of Cl^- and NO_x.

TIL526 ***HR: D***
3,4,4'-TRICHLOROCARBANILIDE mixed with 3-TRIFLUOROMETHYL-4,4'-DICHLOROCARBANILIDE (2 : 1)
mf: $C_{26}H_{18}Cl_6N_4O_2 \cdot C_{14}H_9Cl_2F_3N_2O$ mw: 980.33

SYNS: TCC mixed with TFC (2:1) ◇ TFC mixed with TCC (1:2) ◇ 3-TRIFLUOROMETHYL-4,4'-DICHLOROCARBANILIDE mixed with 3,4,4'-TRICHLOROCARBANILIDE (1:2)

TOXICITY DATA with REFERENCE
orl-rat TDLo:7310 mg/kg (1-22D preg/21D post):REP TXAPA9 51,417,79

SAFETY PROFILE: Experimental reproductive effects. When heated to decomposition it emits toxic fumes of F^-, Cl^-, and NO_x.

TIL750 CAS:53555-01-4 ***HR: 3***
4,5,6-TRICHLORO-2-(2,4-DICHLOROPHENOXY) PHENOL
mf: $C_{12}H_5Cl_5O_2$ mw: 358.42

SYNS: C15-PREDIOXIN ◇ 2-(2,4-DICHLOROPHENOXY)-4,5,6-TRICHLOROPHENOL ◇ 6-(2,4-DICHLOROPHENOXY)-2,3,4-TRICHLOROPHENOL

TOXICITY DATA with REFERENCE
mrc-smc 50 mg/L EVSRBT 12,325,78
mmo-sat 50 mg/L EVSRBT 12,325,78
slt-mus-ipr 140 μmol/L EVSRBT 12,325,78
ipr-mus LD50:170 mg/kg JTEHD6 12,245,83

CONSENSUS REPORTS: Chlorophenol compounds are on the Community Right-To-Know List. EPA Genetic Toxicology Program.

SAFETY PROFILE: Poison by intraperitoneal route. Mutation data reported. When heated to decomposition it emits toxic fumes of Cl^-. See also CHLOROPHENOLS.

TIM000 CAS:354-21-2 ***HR: 1***
1,1,2-TRICHLORO-2,2-DIFLUOROETHANE
mf: $CHCl_3F_2$ mw: 157.37

SYNS: 1,1-DIFLUORO-1,2,2-TRICHLOROETHANE ◇ UCON FLUOROCARBON 122

TOXICITY DATA with REFERENCE
orl-rat LDLo:7500 mg/kg HXPHAU 20(Pt 1),459,66
ihl-rat LCLo:4000 ppm/4H UCMH** 15NOV62

SAFETY PROFILE: Mildly toxic by ingestion and inhalation. When heated to decomposition it emits very toxic fumes of F^- and Cl^-. See also CHLORINATED HYDROCARBONS, ALIPHATIC, and FLUORIDES.

TIM500 ***HR: 1***
TRICHLORO ESTERTIN

SYN: ESTERTRICHLOROSTANNANE

TOXICITY DATA with REFERENCE
unr-rat LD50:5500 mg/kg TIUSAD 107,1,76

OSHA PEL: TWA 0.1 mg(Sn)/m^3 (skin)
ACGIH TLV: TWA 0.1 mg(Sn)/m^3 (skin) (Proposed: TWA 0.1 mg(Sn)/m^3; STEL 0.2 mg(Sn)/m^3 (skin))
NIOSH REL: (Organotin Compounds) TWA 0.1 mg (Sn)/m^3

SAFETY PROFILE: When heated to decomposition it emits toxic fumes of Cl^-. See also TIN COMPOUNDS and ESTERS.

TIN000 CAS:79-00-5 ***HR: 3***
1,1,2-TRICHLOROETHANE
mf: $C_2H_3Cl_3$ mw: 133.40

PROP: Liquid; pleasant odor. Bp: 114°, fp: −35°, d: 1.4416 @ 20°/4°, vap press: 40 mm @ 35.2°.

SYNS: ETHANE TRICHLORIDE ◇ NCI-C04579 ◇ RCRA WASTE NUMBER U227 ◇ β-T ◇ 1,1,2-TRICHLORETHANE ◇ β-

TRICHLOROETHANE ◇ 1,2,2-TRICHLOROETHANE ◇ TROJCHLOROETAN(1,1,2) (POLISH) ◇ VINYL TRICHLORIDE

TOXICITY DATA with REFERENCE
skn-rbt 500 mg open MLD UCDS** 6/28/72
skn-rbt 810 mg/24H SEV JETOAS 9,171,76
eye-rbt 162 mg MLD JETOAS 9,171,76
skn-gpg 1440 mg/15M APTOA6 41,298,77
otr-mus:emb 25 mg/L CALEDQ 28,85,85
cyt-gpg-skn 2880 μg/kg APTOA6 41,298,77
orl-mus TDLo:532 mg/kg (14D male):REP DCTODJ 8,333,85
orl-mus TDLo:76 g/kg/78W-I:CAR NCITR* NCI-CG-TR-74,78
orl-mus TD:152 g/kg/78W-I:CAR NCITR* NCI-CG-TR-74,78
orl-rat LD50:836 mg/kg AIHAAP 30,470,69
ihl-rat LCLo:2000 ppm/4H JIDHAN 31,343,49
orl-mus LD50:378 mg/kg DCTODJ 8,333,85
ipr-mus LD50:494 mg/kg TXAPA9 9,139,66
scu-mus LD50:227 mg/kg JPETAB 123,224,58
orl-dog LDLo:500 mg/kg AJHYA2 16,325,32
ipr-dog LD50:450 mg/kg TXAPA9 10,119,67
ivn-dog LDLo:95 mg/kg QJPPAL 7,205,34
ihl-cat LCLo:13100 mg/m^3/4.5H AHBAAM 116,131,36
skn-rbt LD50:5377 mg/kg AIHAAP 30,470,69

CONSENSUS REPORTS: IARC Cancer Review: Group 3 IMEMDT 7,56,87; Animal Limited Evidence IMEMDT 20,533,79. NCI Carcinogenesis Bioassay (gavage); No Evidence: rat NCITR* NCI-CG-TR-74,78; (gavage); Clear Evidence: mouse NCITR* NCI-CG-TR-74,78. Community Right-To-Know List. Reported in EPA TSCA Inventory.

OSHA PEL: TWA 10 ppm (skin)
ACGIH TLV: TWA 10 ppm (skin)
DFG MAK: 10 ppm (55 mg/m^3); Suspected Carcinogen.

SAFETY PROFILE: Suspected carcinogen with experimental carcinogenic data. Poison by ingestion, intravenous, and subcutaneous routes. Moderately toxic by inhalation, skin contact, and intraperitoneal routes. Experimental reproductive effects. Mutation data reported. An eye and severe skin irritant. Has narcotic properties and acts as a local irritant to the eyes, nose, and lungs. It may also be injurious to the liver and kidneys. Incompatible with potassium. When heated to decomposition it emits toxic fumes of Cl^-. See also CHLORINATED HYDROCARBONS, ALIPHATIC; and other trichloroethane entries.

TIN500 CAS:115-20-8 ***HR: 3***
TRICHLOROETHANOL
mf: $C_2H_3Cl_3O$ mw: 149.40

PROP: Liquid. Mp: 17.8°, bp: 150° @ 765 mm, d: 1.54 @ 25/4°, vap press: 1 mm @ 20°, vap d: 5.16.

SYNS: TRICHLORETHANOL ◇ 2,2,2-TRICHLOROETHANOL ◇ TRICHLOROETHYL ALCOHOL ◇ 2,2,2-TRICHLOROETHYL ALCOHOL

TOXICITY DATA with REFERENCE
mmo-asn 5 μL/plate/2H CBINA8 30,9,80
sln-asn 10240 μmol/L MUREAV 155,105,85
sce-hmn:lym 178 g/L TOERD9 3,63,81
orl-rat LDLo:500 mg/kg CRAAA7 17,258,38
ipr-rat LDLo:300 mg/kg JPETAB 63,453,38
orl-mus LDLo:500 mg/kg CRAAA7 17,258,38
ivn-mus LD50:201 mg/kg 28ZPAK -,78,72
ivn-rbt LDLo:50 mg/kg JPETAB 63,453,38
rec-rbt LDLo:500 mg/kg CRAAA7 17,258,38

CONSENSUS REPORTS: EPA Genetic Toxicology Program. Reported in EPA TSCA Inventory.

SAFETY PROFILE: Poison by intravenous and intraperitoneal routes. Moderately toxic by ingestion and rectal routes. Human mutation data reported. Explosive reaction with concentrated sodium hydroxide solutions. When heated to decomposition it emits toxic fumes of Cl^-. Used as an hypnotic and anesthetic. See also CHLORINATED HYDROCARBONS, ALIPHATIC.

TIN750 CAS:75-94-5 ***HR: 3***
TRICHLOROETHENYLSILANE
DOT: UN 1305
mf: $C_2H_3Cl_3Si$ mw: 161.49

PROP: Fuming liquid. Bp: 90.6°, d: 1.265 @ 25/25°, flash p: 16°F.

SYNS: SILANE, VINYL TRICHLORO 1-150 ◇ TRICHLORO(VINYL) SILANE ◇ TRICHLOROVINYL SILICANE ◇ UNION CARBIDE A-150 ◇ VINYLSILICON TRICHLORIDE ◇ VINYL TRICHLOROSILANE (DOT) ◇ VINYL TRICHLOROSILANE, INHIBITED (DOT)

TOXICITY DATA with REFERENCE
skn-rbt 10 mg/24H open AMIHBC 10,61,54
skn-rbt 625 mg open SEV UCDS** 1/19/72
eye-rbt 50 μg open SEV AMIHBC 10,61,54
orl-rat LD50:1280 mg/kg AMIHBC 10,61,54
ihl-rat LCLo:500 ppm/4H UCDS** 1/19/72
skn-rbt LD50:680 mg/kg AMIHBC 10,61,54

CONSENSUS REPORTS: Reported in EPA TSCA Inventory.

DOT Classification: Flammable Liquid; Label: Flammable Liquid, Corrosive.

SAFETY PROFILE: Moderately toxic by ingestion, inhalation, and skin contact. A severe eye, and skin irritant. A corrosive irritant to skin, eyes, and mucous membranes. A very dangerous fire hazard when exposed to heat or flame. Reacts violently with water; moist air or steam to produce toxic and corrosive fumes. When heated to decomposition it emits toxic fumes of Cl^-. See also CHLOROSILANES.

TIO000 CAS:515-83-3 ***HR: 2***
2,2,2-TRICHLORO-1-ETHOXYETHANOL
mf: $C_4H_7Cl_3O_2$ mw: 193.46

PROP: Crystals. D: 1.143, mp: 47.5°, bp: 116°. Less sol in water than chloral hydrate; sol in organic solvents.

SYNS: CHLORAL ALCOHOLATE ◇ CHLORAL ETHYLALCOHOLATE ◇ CHLORAL, ETHYL HEMIACETAL ◇ TRICHLOROACETALDEHYDE MONOETHYLACETAL

TOXICITY DATA with REFERENCE
orl-rat LD50:880 mg/kg JPETAB 78,340,43
orl-dog LDLo:1200 mg/kg JPETAB 78,340,43
orl-cat LDLo:500 mg/kg JPETAB 78,340,43
orl-rbt LDLo:1100 mg/kg JPETAB 78,340,43

CONSENSUS REPORTS: Reported in EPA TSCA Inventory.

SAFETY PROFILE: Moderately toxic by ingestion. When heated to decomposition it emits toxic fumes of Cl^-. See also ALDEHYDES.

TIO500 CAS:107-69-7 ***HR: 3***
TRICHLOROETHYL CARBAMATE
mf: $C_3H_4Cl_3NO_2$ mw: 192.43

SYNS: CARBAMIC ACID, 2,2,2-TRICHLOROETHYL ESTER ◇ 2,2,2-TRICHLOROETHANOL CARBAMATE ◇ VOLUNTAL

TOXICITY DATA with REFERENCE
ipr-mus TDLo:3250 mg/kg/13W-I:NEO JNCIAM 8,99,47
orl-mus LDLo:750 mg/kg LDTU** -,-,31
ipr-mus LD50:500 mg/kg JNCIAM 8,99,47

SAFETY PROFILE: Moderately toxic by ingestion and intraperitoneal routes. Questionable carcinogen with experimental neoplastigenic data. When heated to decomposition it emits very toxic fumes of Cl^- and NO_x. See also ESTERS and CARBAMATES.

TIO750 CAS:79-01-6 ***HR: 3***
TRICHLOROETHYLENE
DOT: UN 1710
mf: C_2HCl_3 mw: 131.38

PROP: Clear, colorless, mobile liquid; characteristic sweet odor of chloroform. D: 1.4649 @ 20°/4°, bp: 86.7°, flash p: 89.6°F (but practically nonflammable), lel: 12.5%, uel: 90% @ > 30°, mp: −73°, fp: −86.8°, autoign temp: 788°F, vap press: 100 mm @ 32°, vap d: 4.53, refr index: 1.477 @ 20°. Immisc with water; misc with alc, ether, acetone, carbon tetrachloride.

SYNS: ACETYLENE TRICHLORIDE ◇ ALGYLEN ◇ ANAMENTH ◇ BENZINOL ◇ BLACOSOLV ◇ CECOLENE ◇ 1-CHLORO-2,2-DICHLOROETHYLENE ◇ CHLORYLEA ◇ CHORYLEN ◇ CIRCOSOLV ◇ CRAWHASPOL ◇ DENSINFLUAT ◇ 1,1-DICHLORO-2-CHLOROETHYLENE ◇ DOW-TRI ◇ DUKERON ◇ ETHINYL TRICHLORIDE ◇ ETHYLENE TRICHLORIDE ◇ FLECK-FLIP ◇ FLUATE ◇ GERMALGENE ◇ LANADIN ◇ LETHURIN ◇ NARCOGEN ◇ NARKOSOID ◇ NCI-C04546 ◇ NIALK ◇ PERM-A-CHLOR ◇ PETZINOL ◇ RCRA WASTE NUMBER U228 ◇ THRETHYLENE ◇ TRIAD ◇ TRIASOL ◇ TRICHLOORETHEEN (DUTCH) ◇ TRICHLOORETHYLEEN, TRI (DUTCH) ◇ TRICHLORAETHEN (GERMAN) ◇ TRICHLORAETHYLEN, TRI (GERMAN) ◇ TRICHLORAN ◇ TRICHLORETHENE (FRENCH) ◇ TRICHLORETHYLENE, TRI (FRENCH) ◇ TRICHLOROETHENE ◇ 1,2,2-TRICHLOROETHYLENE ◇ TRI-CLENE ◇ TRICLORETENE (ITALIAN) ◇ TRICLOROETILENE (ITALIAN) ◇ TRIELINA (ITALIAN) ◇ TRILENE ◇ TRIMAR ◇ TRIPLUS ◇ VESTROL ◇ VITRAN ◇ WESTROSOL

TOXICITY DATA with REFERENCE
skn-rbt 500 mg/24H SEV 28ZPAK -,28,72
eye-rbt 20 mg/24H MOD 28ZPAK -,28,72
mmo-asn 2500 ppm MUREAV 155,105,85
otr-ham:emb 5 mg/L CRNGDP 4,291,83
orl-rat TDLo:2688 mg/kg (1-22D preg/21D post):REP TOXID9 4,179,84
ihl-rat TCLo:100 ppm/4H (female 6-22D post):TER JPHYA7 276,24P,78
ihl-rat TCLo:150 ppm/7H/2Y-I:CAR INHEAO 21,243,83
orl-mus TDLo:455 g/kg/78W-I:CAR NCITR* NCI-CG-TR-2,76
ihl-ham TCLo:100 ppm/6H/77W-I:ETA ARTODN 43,237,80
orl-man TDLo:2143 mg/kg:GIT 34ZIAG -,602,69
ihl-hmn TCLo:6900 mg/m³/10M:CNS AHBAAM 116,131,36
ihl-hmn TCLo:160 ppm/83M:CNS AIHAAP 23,167,62
ihl-hmn TDLo:812 mg/kg:CNS,GIT,LIV BMJOAE 2,689,45
ihl-man TCLo:110 ppm/8H:EYE,CNS BJIMAG 28,293,71
ihl-man LCLo:2900 ppm NZMJAX 50,119,51
orl-hmn LDLo:7 g/kg ARTODN 35,295,76
ihl-rat LC50:25700 ppm/1H TXAPA9 42,417,77
orl-mus LD50:2402 mg/kg NTIS** AD-A080-636
ihl-mus LC50:8450 ppm/4H APTOA6 9,303,53
ivn-mus LD50:33900 μg/kg CBCCT* 6,141,54
ipr-dog LD50:1900 mg/kg TXAPA9 10,119,67
scu-dog LDLo:150 mg/kg HBTXAC 5,76,59
ivn-dog LDLo:150 mg/kg QJPPAL 7,205,34
orl-cat LDLo:5864 mg/kg HBTXAC 5,76,59
orl-rbt LDLo:7330 mg/kg HBTXAC 5,76,59
scu-rbt LDLo:1800 mg/kg QJPPAL 7,205,34
ihl-gpg LCLo:37200 ppm/40M HBTXAC 5,76,59

CONSENSUS REPORTS: IARC Cancer Review: Group 3 IMEMDT 7,364,87; Animal Limited Evidence IMEMDT 20,545,79; Human Inadequate Evidence IMEMDT 20,545,79; Animal Sufficient Evidence IMEMDT 11,263,76. NCI Carcinogenesis Bioassay (gavage); No Evidence: rat NCITR* NCI-CG-TR-2,76; (gavage); Clear Evidence: mouse NCITR* NCI-CG-TR-

2,76. Community Right-To-Know List. Reported in EPA TSCA Inventory. EPA Genetic Toxicology Program.

OSHA PEL: (Transitional: TWA 100 ppm; CL 200 ppm; Pk 300 ppm/5M/2H)TWA 50 ppm; STEL 200 ppm
ACGIH TLV: TWA 50 ppm; STEL 200 ppm; BEI: 320 mg(trichloroethanol)/g creatinine in urine at end of shift; 0.5 ppm trichloroethylene in end-exhaled air prior to shift and end of work week.
DFG MAK: Suspected Carcinogen; 50 ppm (270 mg/m^3); BAT: 500 μg/dL in blood at end of shift or work week.
NIOSH REL: (Trichloroethylene) TWA 250 ppm; (Waste Anesthetic Gases) CL 2 ppm/1H
DOT Classification: ORM-A; Label: None; Poison B; Label: St. Andrews Cross.

SAFETY PROFILE: Suspected carcinogen with experimental carcinogenic, tumorigenic, and teratogenic data. Experimental poison by intravenous and subcutaneous routes. Moderately toxic experimentally by ingestion and intraperitoneal routes. Mildly toxic to humans by ingestion and inhalation. Mildly toxic experimentally by inhalation. Human systemic effects by ingestion and inhalation: eye effects, somnolence, hallucinations or distorted perceptions, gastrointestinal changes and jaundice. Experimental reproductive effects. Human mutation data reported. An eye and severe skin irritant. Inhalation of high concentrations causes narcosis and anesthesia. A form of addiction has been observed in exposed workers. Prolonged inhalation of moderate concentrations causes headache and drowsiness. Fatalities following severe, acute exposure have been attributed to ventricular fibrillation resulting in cardiac failure. There is damage to liver and other organs from chronic exposure. A common air contaminant.

High concentrations of trichloroethylene vapor in high-temperature air can be made to burn mildly if plied with a strong flame. Though such a condition is difficult to produce, flames or arcs should not be used in closed equipment which contains any solvent residue or vapor. Reacts with alkali; epoxides [e.g., 1-chloro-2,3-epoxypropane; 1,4-butanediol mono-2,3-epoxypropylether; 1,4-butanediol di-2,3-epoxypropylether; 2,2-bis((4(2′,3′-epoxypropoxy)phenyl)propane] to form the spontaneously flammable gas dichloroacetylene. Can react violently with Al; Ba; N_2O_4; Li; Mg; liquid O_2; O_3; KOH; KNO_3; Na; NaOH; Ti. Reacts with water under heat and pressure to form HCl gas. When heated to decomposition it emits toxic fumes of Cl^-. See also CHLORINATED HYDROCARBONS, ALIPHATIC.

TIP000 ***HR: 2***
α-TRICHLOROETHYLIDENE GLYCEROL
mf: $C_5H_7Cl_3O_3$ mw: 221.47

SYN: α-2-(TRICHLOROMETHYL)-1,3-DIOXOLANE-4-METHANOL

TOXICITY DATA with REFERENCE
ipr-mus LD50:920 mg/kg JPETAB 81,72,44
ivn-mus LD50:520 mg/kg JPETAB 81,72,44

SAFETY PROFILE: Moderately toxic by intraperitoneal and intravenous routes. When heated to decomposition it emits toxic fumes of Cl^-.

TIP250 ***HR: 2***
β-TRICHLOROETHYLIDENE GLYCEROL
mf: $C_5H_7Cl_3O_3$ mw: 221.47

SYN: β-2-(TRICHLOROMETHYL)-1,3-DIOXOLANE-4-METHANOL

TOXICITY DATA with REFERENCE
ipr-mus LD50:959 mg/kg JPETAB 81,72,44
ivn-mus LD50:518 mg/kg JPETAB 81,72,44

SAFETY PROFILE: Moderately toxic by intraperitoneal and intravenous routes. When heated to decomposition it emits toxic fumes of Cl^-.

TIP500 CAS:75-69-4 ***HR: 2***
TRICHLOROFLUOROMETHANE
mf: CCl_3F mw: 137.36

PROP: Colorless liquid. Mp: −111°, bp: 24.1°, d: 1.484 @ 17.2°.

SYNS: ALGOFRENE TYPE 1 ◇ ARCTON 9 ◇ ELECTRO-CF 11 ◇ ESKIMON 11 ◇ FLUOROCARBON No. 11 ◇ FLUOROTRICHLOROMETHANE (OSHA) ◇ FLUOROTROJCHLOROMETAN (POLISH) ◇ FREON 11 ◇ FREON MF ◇ FRIGEN 11 ◇ GENETRON 11 ◇ HALOCARBON 11 ◇ ISCEON 131 ◇ ISOTRON 11 ◇ LEDON 11 ◇ MONOFLUOROTRICHLOROMETHANE ◇ NCI-C04637 ◇ RCRA WASTE NUMBER U121 ◇ TRICHLOROMONOFLUOROMETHANE ◇ UCON REFRIGERANT 11

TOXICITY DATA with REFERENCE
ihl-hmn TDLo:50000 ppm/30M:EYE,PUL,LIV EJTXAZ 9,385,76
ihl-rat LCLo:10 pph/20M AIHOAX 2,335,50
ihl-mus LC50:10 pph/30M EJTXAZ 9,385,76
ipr-mus LD50:1743 mg/kg TOIZAG 18,363,71
ihl-rbt LC50:25 pph/30M JETOAS 9,385,76
ihl-gpg LC50:25 pph/30M JETOAS 9,385,76

CONSENSUS REPORTS: NCI Carcinogenesis Bioassay (gavage); No Evidence: mouse NCITR* NCI-CG-TR-106,78; (gavage); Inadequate Studies: rat NCITR* NCI-CG-TR-106,78. Reported in EPA TSCA Inventory.

OSHA PEL: (Transitional: TWA 1000 ppm) CL 1000 ppm
ACGIH TLV: CL 1000 ppm
DFG MAK: 1000 ppm (5600 mg/m^3)

SAFETY PROFILE: High concentrations cause narcosis and anesthesia in humans. Human systemic effects by inhalation: conjunctiva irritation, fibrosing alveolitis and liver changes. Experimental poison by inhalation. Moderately toxic by intraperitoneal route. Reacts violently with aluminum, barium, or lithium. When heated to decomposition it emits highly toxic fumes of F^- and Cl^-. Used as an aerosol propellant, refrigerant, and blowing agent for polymeric foams. See also CHLORINATED HYDROCARBONS, ALIPHATIC; and FLUORIDES.

TIP750 CAS:4019-40-3 ***HR: 3***
4′,4″,5-TRICHLORO-2-HYDROXY-3-BIPHENYLCARBOXANILIDE
mf: $C_{19}H_{12}Cl_3NO_2$ mw: 392.67

SYNS: 3-(4-CHLOROPHENYL)-4′,5-DICHLOROSALICYLANILIDE ◇ CP 43858 ◇ 4′,5-DICHLORO-N-(4-CHLOROPHENYL)-2-HYDROXY-(1,1′-BIPHENYL)-3-CARBOXAMIDE ◇ ENT 27,139 ◇ MONSANTO CP-43858 ◇ OM-1463

TOXICITY DATA with REFERENCE
orl-rat LD50:2810 mg/kg ARSIM* 20,16,66
ipr-mus LD50:73 mg/kg BCPCA6 18,1389,69

SAFETY PROFILE: Poison by intraperitoneal route. Moderately toxic by ingestion. When heated to decomposition it emits very toxic fumes of Cl^- and NO_x.

TIQ000 CAS:3380-34-5 ***HR: 3***
2,4,4′-TRICHLORO-2′-HYDROXYDIPHENYL ETHER
mf: $C_{12}H_7Cl_3O_2$ mw: 289.54

SYNS: CH 3565 ◇ 5-CHLORO-2-(2,4-DICHLOROPHENOXYPHENOL) ◇ 2′-HYDROXY-2,4,4′-TRICHLORO-PHENYLETHER ◇ IRGASAN ◇ IRGASAN DP300 ◇ TCC ◇ TRICLOSAN

TOXICITY DATA with REFERENCE
skn-hmn 750 μg/3D-I MLD 85DKA8 -,127,77
dnr-bcs 5 mg/disc JOSCDQ 15,243,81
orl-rat TDLo:4400 mg/kg (female 7-17D post):TER ESKHA5 (105),28,87
orl-rat LD50:3700 mg/kg 26UZAB 6,245,68/70
skn-rat LD50:9300 mg/kg 26UZAB 6,245,68/70
scu-rat LD50:3900 mg/kg YKYUA6 28,985,77
ivn-rat LD50:19 mg/kg TXAPA9 42,1,77
orl-mus LD50:4530 mg/kg 26UZAB 6,245,68/70
ipr-mus LD50:84 mg/kg YKYUA6 28,985,77

CONSENSUS REPORTS: Chlorophenol compounds are on the Community Right-To-Know List. Reported in EPA TSCA Inventory. EPA Genetic Toxicology Program.

SAFETY PROFILE: Poison by intravenous and intraperitoneal routes. Moderately toxic by ingestion. Mildly toxic by skin contact. Mutation data reported. A human skin irritant. When heated to decomposition it emits toxic fumes of Cl^-. See also ETHERS and CHLOROPHENOLS.

TIQ250 CAS:52-68-6 ***HR: 3***
((2,2,2-TRICHLORO-1-HYDROXYETHYL) DIMETHYLPHOSPHONATE)
DOT: NA 2783
mf: $C_4H_8Cl_3O_4P$ mw: 257.44

SYNS: AEROL 1 (pesticide) ◇ AGROFOROTOX ◇ ANTHON ◇ BAY 15922 ◇ BAYER 15922 ◇ BAYER L 13/59 ◇ BILARCIL ◇ BOVINOX ◇ BRITON ◇ BRITTEN ◇ CEKUFON ◇ CHLORAK ◇ CHLORFOS ◇ CHLOROFOS ◇ CHLOROFTALM ◇ CHLOROPHOS ◇ CHLOROPHTHALM ◇ CHLOROXYPHOS ◇ CICLOSOM ◇ CLOROFOS (RUSSIAN) ◇ COMBOT EQUINE ◇ DANEX ◇ DEP (pesticide) ◇ DEPTHON ◇ DETF ◇ DIMETHOXY-2,2,2-TRICHLORO-1-HYDROXY-ETHYL-PHOSPHINE OXIDE ◇ O,O-DIMETHYL-(1-HYDROXY-2,2,2-TRICHLORAETHYL)PHOSPHONSAEUREESTER (GERMAN) ◇ O,O-DIMETHYL-(1-HYDROXY-2,2,2-TRICHLORATHYL)-PHOSPHAT (GERMAN) ◇ O,O-DIMETHYL-(1-HYDROXY-2,2,2-TRICHLORO)ETHYL PHOSPHATE ◇ DIMETHYL-1-HYDROXY-2,2,2-TRICHLOROETHYL PHOSPHONATE ◇ O,O-DIMETHYL-(1-HYDROXY-2,2,2-TRICHLOROETHYL)PHOSPHONATE◇ O,O-DIMETHYL-1-OXY-2,2,2-TRICHLOROETHYL PHOSPHONATE ◇ O,O-DIMETHYL-(2,2,2-TRICHLOOR-1-HYDROXY-ETHYL)-FOSFONAAT (DUTCH) ◇ O,O-DIMETHYL-(2,2,2-TRICHLOR-1-HYDROXY-AETHYL)PHOSPHONAT (GERMAN) ◇ DIMETHYLTRICHLORO-HYDROXYETHYL PHOSPHONATE ◇ DIMETHYL-2,2,2-TRICHLORO-1-HYDROXYETHYLPHOSPHONATE ô O,O-DIMETHYL-2,2,2-TRICHLORO-1-HYDRO- XTYETHYL PHOSPHONATE ◇ O,O-DIMETIL-(2,2,2-TRICLORO-1-IDROSSI-ETIL)-FOSFONATO(ITALIAN) ◇ DIMETOX ◇ DIPTERAX ◇ DIPTEREX ◇ DIPTEREX 50 ◇ DIPTEVUR ◇ DITRIFON ◇ DYLOX ◇ DYLOX-METASYSTOX-R ◇ DYREX ◇ DYVON ◇ ENT 19,763 ◇ EQUINO-ACID ◇ EQUINO-AID ◇ FLIBOL E ◇ FLIEGENTELLER ◇ FOROTOX ◇ FOSCHLOR ◇ FOSCHLOREM (POLISH) ◇ FOSCHLOR R-50 ◇ 1-HYDROXY-2,2,2-TRICHLOROETHYLPHOSPHONIC ACID DIMETHYL ESTER ◇ HYPODERMACID ◇ LEIVASOM ◇ LOISOL ◇ MASOTEN ◇ MAZOTEN ◇ METHYL CHLOROPHOS ◇ METIFONATE ◇ METRIFONATE ◇ METRIPHONATE ◇ NCI-C54831 ◇ NEGUVON ◇ NEGUVON A ◇ PHOSCHLOR R50 ◇ POLFOSCHLOR ◇ PROXOL ◇ RICIFON ◇ RITSIFON ◇ SATOX 20WSC ◇ SOLDEP ◇ SOTIPOX ◇ TRICHLOORFON (DUTCH) ◇ TRICHLORFON (USDA) ◇ 2,2,2-TRICHLORO-1-HYDROXYETHYL-PHOSPHONATE, DIMETHYL ESTER ◇ (2,2,2-TRICHLORO-1-HYDROXYETHYL)PHOSPHONIC ACID DIMETHYL ESTER ◇ TRICHLOROPHON ◇ TRICHLORPHENE ◇ TRICHLORPHON ◇ TRICHLORPHON FN ◇ TRINEX ◇ TUGON ◇ TUGON FLY BAIT ◇ TUGON STABLE SPRAY ◇ VERMICIDE BAYER 2349 ◇ VOLFARTOL ◇ VOTEXIT ◇ WEC 50 ◇ WOTEXIT

TOXICITY DATA with REFERENCE
eye-rbt 120 mg/6D-I MLD BUMMAB 9,7,55
mma-ssp 20 mmol/L MUREAV 117,139,83
dns-hmn:oth 4 mol/L PSSCBG 15,439,84
orl-rat TDLo:78400 μg/kg (16-22D preg):REP NUPOBT 17,135,79
orl-ham TDLo:400 mg/kg (female 8D post):TER EVHPAZ 30,105,79
orl-rat TDLo:186 mg/kg/6W-I:CAR ARGEAR 41,311,73
ims-rat TDLo:183 mg/kg/6W-I:CAR ARGEAR 41,311,73
skn-rat TDLo:1950 mg/kg/22W-I:ETA ERNFA7 16,515,72
ihl-hmn TCLo:1710 $\mu g/m^3$/90D-I:SYS BJIMAG 43,414,86
orl-rat LD50:250 mg/kg FMCHA2 -,C294,89
ihl-rat LC50:1300 $\mu g/m^3$ ARGEAR 48,112,78
skn-rat LD50:2000 mg/kg ARGEAR 48,112,78

ipr-rat LD50:160 mg/kg APCRAW 4,117,61
scu-rat LD50:400 mg/kg JEENAI 50(3),356,57
ims-rat LD50:395 mg/kg ARTODN 41,3,78
orl-mus LD50:300 mg/kg SPEADM 74-1,-,74
ipr-mus LD50:196 mg/kg TXAPA9 2,495,60
scu-mus LD50:267 mg/kg ARZNAD 31,555,81
ivn-mus LD50:290 mg/kg ARTODN 41,3,78

CONSENSUS REPORTS: IARC Cancer Review: Group 3 IMEMDT 7,56,87; Animal Inadequate Evidence IMEMDT 30,207,83. Community Right-To-Know List. EPA Genetic Toxicology Program.

DOT Classification: ORM-A; Label: None.

SAFETY PROFILE: Poison by ingestion, inhalation, intraperitoneal, subcutaneous, intravenous, and intramuscular routes. Moderately toxic by skin contact. Human systemic effects: true cholinesterase. Experimental teratogenic and reproductive effects. Questionable carcinogen with experimental carcinogenic and tumorigenic data. Human mutation data reported. An eye irritant. When heated to decomposition it emits very toxic fumes of Cl^- and PO_x.

TIQ750 CAS:87-90-1 ***HR: 2***
N,N',N"-TRICHLOROISOCYANURIC ACID
DOT: NA 2468
mf: $C_3Cl_3N_3O_3$ mw: 232.41

PROP: White crystals; chlorine odor. Mp: 225-230° (decomp). Moderately sol in water.

SYNS: ACL 85 ◇ CBD 90 ◇ FICHLOR 91 ◇ FI CLOR 91 ◇ ISOCYANURIC CHLORIDE ◇ KYSELINA TRICHLOISOKYANUROVA (CZECH) ◇ NSC-405124 ◇ SYMCLOSEN ◇ SYMCLOSENE ◇ TRICHLORINATED ISOCYANURIC ACID ◇ TRICHLOROCYANURIC ACID ◇ TRICHLOROISOCYANIC ACID ◇ TRICHLOROISOCYANURIC ACID ◇ 1,3,5-TRICHLOROISOCYANURIC ACID ◇ TRICHLORO-s-TRIAZINETRIONE ◇ 1,3,5-TRICHLORO-1,3,5-TRIAZINETRIONE ◇ TRICHLORO-s-TRIAZINE-2,4,6(1H,3H,5H)-TRIONE ◇ 1,3,5-TRICHLORO-2,4,6-TRIOXOHEXAHYDRO-s-TRIAZINE

TOXICITY DATA with REFERENCE
skn-rbt 500 mg/24H MOD 28ZPAK -,153,72
skn-rbt 500 mg SEV 34ZIAG -,167,69
eye-rbt 50 μg/24H SEV 28ZPAK -,153,72
eye-rbt 3125 mg MOD MONS** -,-,72
orl-hmn LDLo:3570 mg/kg:GIT 34ZIAG -,167,69
orl-rat LD50:406 mg/kg TXAPA9 42,417,77
skn-rbt LD50:20 g/kg TXAPA9 42,417,77

CONSENSUS REPORTS: Reported in EPA TSCA Inventory.

DOT Classification: Oxidizer; Label: Oxidizer.

SAFETY PROFILE: Moderately toxic to humans and experimentally by ingestion. Mildly toxic experimentally by skin contact. Human systemic effects by ingestion: ulceration or bleeding from stomach. A severe skin and eye irritant. Toxicity symptoms include emaciation, lethargy, weakness and delayed death. Autopsy shows inflammation of gastrointestinal tract, liver discoloration and kidney hyperemia.

A powerful oxidizer. Forms an explosive product with cyanuric acid + sodium hydroxide. Potentially violent reaction with combustible materials. When heated to decomposition it emits very toxic fumes of Cl^- and NO_x. Used to chlorinate swimming pools.

TIR250 CAS:2633-54-7 ***HR: 3***
TRICHLOROMETAPHOS-3
mf: $C_9H_{10}Cl_3O_3PS$ mw: 335.57

SYNS: O,O-DIMETHYL-2-ETHYLMERCAPTOETHYLTHIOPHOSPHATE ◇ O-METHYL-O-ETHYL-O-2,4,5-TRICHLOROPHENYL THIOPHOSPHATE ◇ TRICHLORMETAFOS-3 ◇ TRICHLORO-3-METHAPHOS

TOXICITY DATA with REFERENCE
orl-rat LD50:360 mg/kg HYSAAV 31,18,66
orl-rbt LDLo:400 mg/kg 85GMAT -,83,82

SAFETY PROFILE: Poison by ingestion. When heated to decomposition it emits very toxic fumes of Cl^-, PO_x, and SO_x. See also MERCAPTANS.

TIR750 CAS:64057-58-5 ***HR: 2***
TRICHLOROMETHYL ALLYL PERTHIOXANTHATE
mf: $C_5H_5Cl_3S_3$ mw: 267.63

SYNS: CARBONOTRITHIOICACID-2-PROPENYLTRICHLOROMETHYL ESTER ◇ PERTHIOXANTHATE, TRICHLOROMETHYL ALLYL

TOXICITY DATA with REFERENCE
orl-rat LD50:2180 mg/kg AIHAAP 30,470,69
skn-rbt LDLo:600 mg/kg AIHAAP 30,470,69

SAFETY PROFILE: Moderately toxic by ingestion and skin contact. When heated to decomposition it emits very toxic fumes of Cl^- and SO_x. See also ESTERS and ALLYL COMPOUNDS.

TIS500 CAS:25991-93-9 ***HR: 2***
TRICHLOROMETHYL METHYL PERTHIOXANTHATE
mf: $C_3H_3Cl_3S_3$ mw: 241.59

SYNS: CARBONOTRITHIOIC ACID, METHYL TRICHLOROMETHYL- ESTER ◇ PERTHIOXANTHATE, TRICHLOROMETHYL METHYL

TOXICITY DATA with REFERENCE
orl-rat LD50:1540 mg/kg AIHAAP 30,470,69
skn-rbt LDLo:1300 mg/kg AIHAAP 30,470,69

SAFETY PROFILE: Moderately toxic by ingestion and skin contact. When heated to decomposition it emits very toxic fumes of Cl^- and SO_x.

TIS750 ***HR: 3***
TRICHLOROMETHYL PERCHLORATE
mf: CCl_4O_4 mw: 217.82

Cl_3COClO_3

SAFETY PROFILE: Extremely explosive; self-reactive. When heated to decomposition it emits toxic fumes of Cl^-. See also PERCHLORATES.

TIT000 CAS:90-17-5 ***HR: 1***
TRICHLOROMETHYLPHENYLCARBINYL ACETATE
mf: $C_{10}H_9Cl_3O_2$ mw: 267.54

SYNS: ROSACETOL ◇ ROSE CRYSTALS ◇ α-(TRICHLOROMETHYL) BENZENEMETHANOL, ACETATE (9CI) ◇ α-(TRICHLOROMETHYL) BENZYL ALCOHOL, ACETATE (9CI)

TOXICITY DATA with REFERENCE
skn-rbt 500 mg/24H MLD FCTXAV 13,681,75
orl-rat LD50:6800 mg/kg FCTXAV 13,681,75

CONSENSUS REPORTS: Reported in EPA TSCA Inventory.

SAFETY PROFILE: Mildly toxic by ingestion. A skin irritant. When heated to decomposition it emits toxic fumes of Cl^-.

TIT250 CAS:133-07-3 ***HR: 3***
N-(TRICHLOROMETHYLTHIO)PHTHALIMIDE
mf: $C_9H_4Cl_3NO_2S$ mw: 296.55

SYNS: FOLPAN ◇ FOLPET ◇ FTALAN ◇ ORTHOPHALTAN ◇ PHALTAN ◇ PHTHALTAN ◇ THIOPHAL ◇ N-(TRICHLORMETHYLTHIO)-PHTHALAMID (GERMAN) ◇ N-(TRICHLOROMETHYLMERCPATO)PHTHALIMIDE ◇ 2-((TRICHLOROMETHYL) THIO)-1H-ISOINDOLE-1,3(2H)-DIONE ◇ TROYSAN ANTI-MILDEW O

TOXICITY DATA with REFERENCE
mmo-sat 16 nmol/plate CRNGDP 2,283,81
dlt-rat-orl 500 mg/kg/5D FCTXAV 10,363,72
ihl-mus TCLo:491 mg/m^3/4H (female 6-13D post):TER NTIS** PB84-128099
scu-mus TDLo:900 mg/kg (female 6-14D post):REP NTIS** PB223-160
scu-mus TDLo:1000 mg/kg:ETA NTIS** PB223-159
orl-rat LD50:7540 mg/kg GTPZAB 18(5),50,74
ipr-rat LD50:68400 μg/kg JTEHD6 9,867,82
orl-mus LD50:1546 mg/kg GTPZAB 18(5),50,74
orl-rbt LD50:1115 mg/kg GTPZAB 18(5),50,74

CONSENSUS REPORTS: Reported in EPA TSCA Inventory. EPA Genetic Toxicology Program.

SAFETY PROFILE: Poison by intraperitoneal route. Moderately toxic by ingestion. Questionable carcinogen with experimental tumorigenic and teratogenic data. Experimental reproductive effects. Human mutation data reported. When heated to decomposition it emits very toxic fumes of Cl^-, NO_x, and SO_x. Used as a fungicide.

TIT275 CAS:72385-44-5 ***HR: 3***
5-TRICHLOROMETHYL-1-TRIMETHYLSILYLTETRAZOLE
mf: $C_5H_9Cl_3N_4Si$ mw: 259.60

$(CH_3)_3SiNN{=}NN{=}CCCl_3$ (ring)

SAFETY PROFILE: Explodes when heated to 80-90°C. When heated to decomposition it emits toxic fumes of Cl^- and NO_x.

TIT500 CAS:1321-65-9 ***HR: 3***
TRICHLORONAPHTHALENE
mf: $C_{10}H_5Cl_3$ mw: 231.50

PROP: A white solid.

SYNS: HALOWAX ◇ NIBREN WAX ◇ SEEKAY WAX

CONSENSUS REPORTS: Reported in EPA TSCA Inventory.

OSHA PEL: TWA 5 mg/m^3 (skin)
ACGIH TLV: TWA 5 mg/m^3 (skin)
DFG MAK: 5 mg/m^3

SAFETY PROFILE: A poison. The chlorinated naphthalenes have toxic effects on the skin and liver. See also CHLORINATED HYDROCARBONS, ALIPHATIC; and POLYCHLORINATED BIPHENYLS.

TIT750 CAS:89-69-0 ***HR: 3***
2,4,5-TRICHLORONITROBENZENE
mf: $C_6H_2Cl_3NO_2$ mw: 226.44

SYN: 1,2,4-TRICHLORO-5-NITROBENZENE

TOXICITY DATA with REFERENCE
orl-mus LDLo:1070 mg/kg AECTCV 14,111,85
orl-bwd LD50:100 mg/kg TXAPA9 21,315,72

CONSENSUS REPORTS: Reported in EPA TSCA Inventory.

SAFETY PROFILE: Poison by ingestion. When heated to decomposition it emits very toxic fumes of Cl^- and NO_x. See also NITRO COMPOUNDS of AROMATIC HYDROCARBONS and CHLORINATED HYDROCARBONS, AROMATIC.

TIV275 CAS:7796-16-9 ***HR: 3***
TRICHLOROPEROXYACETIC ACID
mf: $C_2HCl_3O_3$ mw: 179.39

SAFETY PROFILE: Very unstable, it decomposes to yield the toxic gases phosgene; chlorine; carbon monoxide; and hydrogen chloride. See also PEROXIDES.

TIV500 CAS:933-75-5 **HR: 3**
2,3,6-TRICHLOROPHENOL
mf: $C_6H_3Cl_3O$ mw: 197.44

PROP: Colorless needles. Mp: 62°, bp: 253°.

TOXICITY DATA with REFERENCE
ipr-rat LD50:308 mg/kg BJPCAL 13,20,58

CONSENSUS REPORTS: Chlorophenol compounds are on the Community Right-To-Know List. Reported in EPA TSCA Inventory.

SAFETY PROFILE: Poison by intraperitoneal route. When heated to decomposition it emits toxic fumes of Cl^-. See also CHLOROPHENOLS.

TIV750 CAS:95-95-4 **HR: 3**
2,4,5-TRICHLOROPHENOL
mf: $C_6H_3Cl_3O$ mw: 197.44

PROP: Colorless needles or gray flakes; strong phenolic odor. Mp: 61-63°, bp: 252°, d: 1.678 @ 25/4°, vap press: 1 mm @ 72.0°. Insol in water; sol in CCl_4, alc, benzene, and ether.

SYNS: COLLUNOSOL ◇ DOWICIDE 2 ◇ DOWICIDE B ◇ NCI-C61187 ◇ NURELLE ◇ PREVENTOL I ◇ RCRA WASTE NUMBER U230

TOXICITY DATA with REFERENCE
mmo-sat 10 μg/plate TECSDY 14,143,87
mma-sat 10 μg/plate TECSDY 14,143,87
orl-mus TDLo:4 g/kg (8-12D preg):REP JTEHD6 10,541,82
skn-mus TDLo:6700 mg/kg/16W-I:NEO CNREA8 19,413,59
orl-rat LD50:820 mg/kg FEPRA7 2,76,43
ipr-rat LD50:355 mg/kg BJPCAL 13,20,58
scu-rat LD50:2260 mg/kg FEPRA7 2,76,43
orl-mus LD50:600 mg/kg PHARAT 30,147,75
ivn-mus LD50:56 mg/kg CSLNX* NX#03492

CONSENSUS REPORTS: IARC Cancer Review: Human Limited Evidence IMEMDT 41,319,86; Animal Inadequate Evidence IMEMDT 20,349,79. Chlorophenol compounds are on the Community Right-To-Know List. Reported in EPA TSCA Inventory.

SAFETY PROFILE: Suspected carcinogen with experimental neoplastigenic data. Poison by intraperitoneal and intravenous routes. Moderately toxic by ingestion and subcutaneous routes. Experimental reproductive effects. Mutation data reported. When heated to decomposition it emits toxic fumes of Cl^- and explodes. See also CHLOROPHENOLS.

TIW000 CAS:88-06-2 **HR: 3**
2,4,6-TRICHLOROPHENOL
mf: $C_6H_3Cl_3O$ mw: 197.44

PROP: Colorless needles or yellow solid; strong phenolic odor. Mp: 68°, bp: 244.5°, fp: 62°, d: 1.490 @ 75/4°, vap press: 1 mm @ 76.5°. Sol in water; very sol in alc and ether.

SYNS: DOWICIDE 2S ◇ NCI-C02904 ◇ OMAL ◇ PHENACHLOR ◇ RCRA WASTE NUMBER U231 ◇ 2,4,6-TRICHLORFENOL (CZECH)

TOXICITY DATA with REFERENCE
skn-rbt 500 mg/24H MOD 28ZPAK -,80,72
eye-rbt 50 μg/24H SEV 28ZPAK -,80,72
mma-sat 10 μg/plate TECSDY 14,143,87
msc-mus:lyms 80 mg/L EMMUEG 12,85,88
orl-rat TDLo:12500 mg/kg (female 2W pre):REP FAATDF 6,233,86
orl-rat TDLo:185 g/kg/2Y-C:CAR NCITR* NCI-CG-TR-155,79
orl-mus TDLo:441 g/kg/2Y-C:CAR NCITR* NCI-CG-TR-155,79
orl-rat LD50:820 mg/kg PCOC** -,1176,66
ipr-rat LD50:276 mg/kg BJPCAL 13,20,58
orl-mam LD50:454 mg/kg GISAAA 45(10),16,80
skn-mam LD50:700 mg/kg GISAAA 45(10),16,80

CONSENSUS REPORTS: NTP Fifth Annual Report on Carcinogens. IARC Cancer Review: Animal Inadequate Evidence IMEMDT 20,349,79; Human Limited Evidence IMEMDT 41,319,86. NCI Carcinogenesis Bioassay (feed); Clear Evidence: mouse, rat NCITR* NCI-CG-TR-155,79. Chlorophenol compounds are on the Community Right-To-Know List. Reported in EPA TSCA Inventory. EPA Genetic Toxicology Program.

SAFETY PROFILE: Confirmed carcinogen with experimental carcinogenic data. Poison by intraperitoneal route. Moderately toxic by ingestion and skin contact. A skin and severe eye irritant. Experimental reproductive effects. Mutation data reported. When heated to decomposition it emits toxic fumes of Cl^-. Used as a germicide and preservative. See also CHLOROPHENOLS.

TIW750 CAS:93-80-1 **HR: D**
4-(2,4,5-TRICHLOROPHENOXY)BUTYRIC ACID
mf: $C_{10}H_9Cl_3O_3$ mw: 283.54

TOXICITY DATA with REFERENCE
orl-mus TDLo:680 mg/kg (11-13D preg):TER AECTCV 6,33,77

SAFETY PROFILE: An experimental teratogen. When heated to decomposition it emits toxic fumes of Cl^-.

TIX000 CAS:2122-77-2 **HR: 3**
2-(2,4,5-TRICHLOROPHENOXY)ETHANOL
mf: $C_8H_7Cl_3O_2$ mw: 241.50

SYNS: KLORINOL ◇ TCPE

TOXICITY DATA with REFERENCE
sce-ham:ovr 100 mg/L CRNGDP 5,1725,84
orl-mus TDLo:3480 mg/kg/52W-I:ETA IARCCD 25,167,79
orl-rat LD50:1500 mg/kg NTOTDY 5,503,83
orl-mus LD50:1320 mg/kg PBPHAW 14,82,78

SAFETY PROFILE: Moderately toxic by ingestion. Questionable carcinogen with experimental tumorigenic data. Mutation data reported. When heated to decomposition it emits toxic fumes of Cl^-. Used as an agricultural chemical and pesticide.

TIX250 CAS:25056-70-6 ***HR: 2***
2,4,5-TRICHLOROPHENOXYETHYL-α,α,α-TRICHLOROACETATE
mf: $C_{10}H_6Cl_6O_3$ mw: 386.86

SYNS: HEXANATE ◇ TRICHLOROACETIC ACID-2-(2,4,5-TRICHLOROPHENOXY)ETHYL ESTER

TOXICITY DATA with REFERENCE
unr-rat LD50:2090 mg/kg HYSAAV 34,174,69
unr-mus LD50:1200 mg/kg HYSAAV 34,174,69
unr-rbt LD50:2200 mg/kg HYSAAV 34,174,69
unr-gpg LD50:2700 mg/kg HYSAAV 34,174,69

SAFETY PROFILE: Moderately toxic by unspecified routes. When heated to decomposition it emits toxic fumes of Cl^-.

TIX500 CAS:93-72-1 ***HR: 3***
α-(2,4,5-TRICHLOROPHENOXY)PROPIONIC ACID
mf: $C_9H_7Cl_3O_3$ mw: 269.51

PROP: Crystals. Mp: 182°. Sltly water-sol.

SYNS: ACIDE 2-(2,4,5-TRICHLORO-PHENOXY) PROPIONIQUE (FRENCH) ◇ ACIDO 2-(2,4,5-TRICLORO-FENOSSI)-PROPIONICO (ITALIAN) ◇ AMCHEM 2,4,5-TP ◇ AQUA-VEX ◇ COLOR-SET ◇ DED-WEED ◇ DOUBLE STRENGTH ◇ FENOPROP ◇ FENORMONE ◇ FRUITONE T ◇ HERBICIDES, SILVEX ◇ KURAN ◇ KURON ◇ KUROSAL ◇ MILLER NU SET ◇ PROPON ◇ RCRA WASTE NUMBER U233 ◇ SILVEX (USDA) ◇ SILVI-RHAP ◇ STA-FAST ◇ 2,4,5-TC ◇ 2,4,5-TCPPA ◇ 2,4,5-TP ◇ 2-(2,4,5-TRICHLOOR-FENOXY)-PROPIONZUUR (DUTCH) ◇ 2-(2,4,5-TRICHLOROPHENOXY)PROPIONIC ACID ◇ 2,4,5-TRICHLOROPHENOXY-α-PROPIONIC ACID ◇ 2-(2,4,5-TRICHLOR-PHENOXY)-PROPIONSAEURE (GERMAN) ◇ WEED-B-GON

TOXICITY DATA with REFERENCE
orl-mus TDLo:1617 mg/kg (12-15D preg):TER AECTCV 6,33,77
orl-rat LD50:650 mg/kg RREVAH 10,97,65
orl-mus LD50:276 mg/kg RPZHAW 31,373,80

CONSENSUS REPORTS: IARC Cancer Review: Human Limited Evidence IMEMDT 41,357,86.

SAFETY PROFILE: A suspected carcinogen. Poison by ingestion. An experimental teratogen. When heated to decomposition it emits toxic fumes of Cl^-.

TIX750 CAS:6047-17-2 ***HR: 2***
2-(2,4,5-TRICHLOROPHENOXY)PROPIONIC ACID PROPYLENE GLYCOL BUTYL ETHER ESTER
mf: $C_{16}H_{21}Cl_3O_4$ mw: 383.72

PROP: Liquid. Insol in water.

SYN: KURON

TOXICITY DATA with REFERENCE
orl-rat LD50:500 mg/kg PCOC** -,997,66
orl-mus LD50:2000 mg/kg PCOC** -,997,66
orl-rbt LD50:500 mg/kg PCOC** -,997,66
orl-gpg LD50:500 mg/kg PCOC** -,997,66
orl-ckn LD50:2000 mg/kg PCOC** -,997,66

CONSENSUS REPORTS: Glycol ether compounds are on the Community Right-To-Know List.

SAFETY PROFILE: Moderately toxic by ingestion. Has caused experimental liver and kidney damage. When heated to decomposition it emits toxic fumes of Cl^-. See also ESTERS and GLYCOL ETHERS.

TIY250 CAS:23399-90-8 ***HR: 2***
2,4,6-TRICHLOROPHENYL ACETATE
mf: $C_8H_5Cl_3O_2$ mw: 239.48

SYNS: 2,4,6-TRICHLORFENYLESTER KYSELINY OCTOVE (CZECH) ◇ 2,4,6-TRICHLOROPHENOL ACETATE

TOXICITY DATA with REFERENCE
skn-rbt 500 mg/24H MLD 28ZPAK -,93,72
eye-rbt 100 mg/24H MOD 28ZPAK -,93,72
orl-rat LD50:2650 mg/kg 28ZPAK -,93,72

CONSENSUS REPORTS: Chlorophenol compounds are on the Community Right-To-Know List.

SAFETY PROFILE: Moderately toxic by ingestion. A skin and eye irritant. When heated to decomposition it emits toxic fumes of Cl^-. See also CHLOROPHENOLS.

TIY500 CAS:85-34-7 ***HR: 2***
2,3,6-TRICHLOROPHENYLACETIC ACID
mf: $C_8H_5Cl_3O_2$ mw: 239.48

PROP: Colorless crystals. Mp: 160°.

SYNS: CHLORFENAC ◇ FENAB ◇ FENAC ◇ FENATROL ◇ KANEPAR ◇ TCPA ◇ 2,3,6-TRICHLOROBENZENEACETIC ACID ◇ 2,3,6-TRICHLORPHENYLESSIGSAEURE (GERMAN) ◇ TRI-FEN

TOXICITY DATA with REFERENCE
orl-rat LD50:1780 mg/kg FMCHA2 -,D136,80

CONSENSUS REPORTS: Reported in EPA TSCA Inventory.

SAFETY PROFILE: Moderately toxic by ingestion. When heated to decomposition it emits toxic fumes of Cl^-. An herbicide used for preemergence season-long control of weeds.

TIY750 CAS:5337-60-0 ***HR: 3***
TRI-o-CHLOROPHENYL BORATE
mf: $C_{18}H_{12}BCl_3O_3$ mw: 393.46

PROP: White solid; odor of o-chlorophenol. Mp: 47-49°, bp: 264-270° @ 14 mm.

SYN: BORIC ACID, TRI-o-CHLOROPHENYL ESTER

TOXICITY DATA with REFERENCE
eye-rbt 100 mg SEV 14KTAK -,706,64
orl-mus LDLo:230 mg/kg 14KTAK -,706,64

SAFETY PROFILE: Poison by ingestion. A severe eye irritant. When heated to decomposition it emits very toxic fumes of Cl^-. See also ESTERS and BORON COMPOUNDS.

TJA000 CAS:50355-74-3 ***HR: D***
2,4,6-TRICHLORO-PHENYL-DIMETHYLTRIAZENE
mf: $C_8H_8Cl_3N_3$ mw: 252.54

SYNS: 3,3-DIMETHYL-1-(2,4,6-TRICHLOROPHENYL)-TRIAZINE ◇ 1-(2,4,6-TRICHLOROPHENYL)-3,3-DIMETHYLTRIAZENE ◇ 2,4,6-TRICHLORO-PMDT ◇ 2,4,6-TRICl-PDMT

TOXICITY DATA with REFERENCE
sln-dmg-orl 500 μmol/L/3D-I ARTODN 43,201,80
mnt-mus-ipr 50 mg/kg/24H MUREAV 56,319,78

CONSENSUS REPORTS: EPA Genetic Toxicology Program.

SAFETY PROFILE: Mutation data reported. When heated to decomposition it emits very toxic fumes of Cl^- and NO_x.

TJA500 CAS:34320-82-6 ***HR: 2***
1-(2,4,6-TRICHLOROPHENYL)-3-p-NITRO-ANILINO-2-PYRAZOLIN-5-ONE
mf: $C_{15}H_{11}Cl_3N_4$ mw: 353.65

TOXICITY DATA with REFERENCE
ipr-rat LDLo:1600 mg/kg KODAK* -,-,71

CONSENSUS REPORTS: Reported in EPA TSCA Inventory.

SAFETY PROFILE: Moderately toxic by intraperitoneal route. When heated to decomposition it emits very toxic fumes of Cl^- and NO_x. See also NITRO COMPOUNDS of AROMATIC HYDROCARBONS.

TJA750 CAS:98-13-5 ***HR: 3***
TRICHLOROPHENYLSILANE
DOT: UN 1804
mf: $C_6H_5Cl_3Si$ mw: 211.55

PROP: Liquid. Bp: 201°, d: 1.321, refr index: 1.5247.

SYNS: PHENYLSILICON TRICHLORIDE ◇ PHENYL TRICHLOROSILANE (DOT) ◇ SILICON PHENYL TRICHLORIDE

TOXICITY DATA with REFERENCE
skn-rbt 10 mg/24H open AMIHBC 10,61,54
skn-rbt 500 mg/24H MOD 28ZPAK -,218,72
eye-rbt 250 μg SEV AMIHBC 10,61,54
eye-rbt 5 mg/24H SEV 28ZPAK -,218,72
orl-rat LD50:2390 mg/kg AMIHBC 10,61,54
ihl-mus LC50:330 mg/m³/2H TPKVAL (3),23,61
ivn-mus LD50:100 mg/kg CSLNX* NX#04051
skn-rbt LD50:890 mg/kg AMIHBC 10,61,54

CONSENSUS REPORTS: EPA Extremely Hazardous Substances List. Reported in EPA TSCA Inventory.

DOT Classification: Corrosive Material; Label: Corrosive.

SAFETY PROFILE: Poison by inhalation and intravenous routes. Moderately toxic by ingestion and skin contact. A corrosive irritant to skin, eyes, and mucous membranes. When heated to decomposition it emits toxic fumes of Cl^-. See also CHLOROSILANES.

TJB000 CAS:7789-89-1 ***HR: 2***
1,1,1-TRICHLOROPROPANE
mf: $C_3H_5Cl_3$ mw: 147.43

PROP: Oil. D: 1.372 @ 25°, mp: < −20°, bp: 140°. Insol in water; misc in alc, ether.

TOXICITY DATA with REFERENCE
eye-hmn 100 ppm/15M JIHTAB 28,262,46
skn-rbt 10 mg/24H MLD AMIHBC 10,61,54
eye-rbt 20 mg SEV AMIHBC 10,61,54
orl-rat LD50:7460 mg/kg AIHAAP 23,95,62
ihl-rat LCLo:8000 ppm/4H AIHAAP 23,95,62

SAFETY PROFILE: Mildly toxic by ingestion and inhalation routes. A human eye irritant. An experimental skin and severe eye irritant. When heated to decomposition it emits toxic fumes of Cl^-. See also CHLORINATED HYDROCARBONS, ALIPHATIC.

TJB250 CAS:598-77-6 ***HR: 2***
1,1,2-TRICHLOROPROPANE
mf: $C_3H_5Cl_3$ mw: 147.43

TOXICITY DATA with REFERENCE
skn-rbt 10 mg/24H open MLD AMIHBC 10,61,54
eye-rbt 20 mg open SEV AMIHBC 10,61,54
orl-rat LD50:1230 mg/kg AMIHBC 10,61,54

ihl-rat LC50:2000 ppm/4H AMIHBC 10,61,54
skn-rbt LD50:14100 mg/kg AMIHBC 10,61,54

CONSENSUS REPORTS: Reported in EPA TSCA Inventory.

SAFETY PROFILE: Moderately toxic by ingestion. Mildly toxic by inhalation and skin contact. A skin and severe eye irritant. When heated to decomposition it emits toxic fumes of Cl^-. See also CHLORINATED HYDROCARBONS, ALIPHATIC.

TJB500 CAS:3175-23-3 ***HR: 2***
1,2,2-TRICHLOROPROPANE
mf: $C_3H_5Cl_3$ mw: 147.43

TOXICITY DATA with REFERENCE
orl-rat LD50:1230 mg/kg UCDS**

CONSENSUS REPORTS: Reported in EPA TSCA Inventory.

SAFETY PROFILE: Moderately toxic by ingestion. A human eye irritant. When heated to decomposition it emits toxic fumes of Cl^-. See also other trichloropropane entries and CHLORINATED HYDROCARBONS, ALIPHATIC.

TJB600 CAS:96-18-4 ***HR: 3***
1,2,3-TRICHLOROPROPANE
mf: $C_3H_5Cl_3$ mw: 147.43

PROP: Bp: 142°, d: 1.414 @ 20°/20°, flash p: 180°F (OC).

SYNS: ALLYL TRICHLORIDE ◇ GLYCEROL TRICHLOROHYDRIN ◇ GLYCERYL TRICHLOROHYDRIN ◇ NCI-C60220 ◇ TRICHLOROHYDRIN

TOXICITY DATA with REFERENCE
skn-rbt 700 mg open MLD UCDS** 3/20/73
eye-rbt 140 mg SEV UCDS** 3/20/73
mmo-sat 100 nmol/plate ENMUDM 2,59,80
mma-sat 500 ng/plate ENMUDM 7(Suppl 3),15,85
orl-rat TDLo:400 mg/kg (5D male):REP MUREAV 101,321,82
orl-rat LD50:320 mg/kg UCDS** 3/20/73
ihl-rat LCLo:1000 ppm/4H AIHAAP 23,95,62
ihl-mus LC50:3400 mg/m^3/2H 85GMAT -,114,82
orl-dog LDLo:200 mg/kg AJHYA2 16,325,32
skn-rbt LD50:1770 mg/kg AIHAAP 23,95,62

CONSENSUS REPORTS: Reported in EPA TSCA Inventory.

OSHA PEL: (Transitional: TWA 50 ppm) TWA 10 ppm
ACGIH TLV: TWA 10 ppm (skin)
DFG MAK: 50 ppm (300 mg/m^3)

SAFETY PROFILE: Poison by ingestion. Moderately toxic by inhalation and skin contact. Experimental reproductive effects. A skin and severe eye irritant. Mutation data reported. Moderately flammable by heat, flames (sparks), or powerful oxidizers. See also ALLYL COMPOUNDS and CHLORINATED HYDROCARBONS, ALIPHATIC. When heated to decomposition it yields highly toxic Cl^-. To fight fire, use water (as a blanket), spray, mist, dry chemical.

TJB750 CAS:67664-94-2 ***HR: D***
1,2,3-TRICHLOROPROPANE-2,3-OXIDE
mf: $C_3H_3Cl_3O$ mw: 161.41

SYNS: TCPO ◇ TRICHLOROPROPYLENE

TOXICITY DATA with REFERENCE
mmo-sat 500 μmol/L TOLED5 4,103,79
mmo-klp 500 μmol/L MUREAV 89,269,81

CONSENSUS REPORTS: EPA Genetic Toxicology Program.

SAFETY PROFILE: Mutation data reported. When heated to decomposition it emits toxic fumes of Cl^-. See also CHLORINATED HYDROCARBONS, ALILPHATIC.

TJC000 CAS:96-19-5 ***HR: 2***
1,2,3-TRICHLOROPROPENE
mf: $C_3H_3Cl_3$ mw: 145.41

PROP: Bp: 142°, d: 1.414 @ 20/20°, flash p: 180°F (OC).

TOXICITY DATA with REFERENCE
skn-rbt 10 mg/24H open SEV AIHAAP 23,95,62
eye-rbt 50 mg MOD UCDS** 5/4/60
mmo-sat 100 nmol/plate ENMUDM 2,59,80
mma-sat 100 nmol/plate ENMUDM 2,59,80
orl-rat LD50:616 mg/kg UCDS** 5/4/60
ihl-rat LCLo:500 ppm/4H UCDS** 5/4/60
skn-rbt LD50:640 mg/kg AIHAAP 23,95,62

CONSENSUS REPORTS: Reported in EPA TSCA Inventory.

SAFETY PROFILE: Moderately toxic by ingestion, inhalation, and skin contact. Mutation data reported. An eye and severe skin irritant. Combustible when exposed to heat, flames (sparks) or powerful oxidizers. To fight fire, use water (as a blanket), spray, mist, dry chemical. When heated to decomposition it emits toxic fumes of Cl^-. See also CHLORINATED HYDROCARBONS, ALIPHATIC.

TJC250 CAS:3083-23-6 ***HR: 3***
1,1,1-TRICHLOROPROPENE-2,3-OXIDE
mf: $C_3H_3Cl_3O$ mw: 161.41

SYNS: 1,2-EPOXY-3,3,3-TRICHLOROPROPANE ◇ 1,1,1-TRICHLORO-2-3-EPOXYPROPANE ◇ (TRICHLOROMETHYL)OXIRANE ◇ TRICHLOROPROPANE OXIDE ◇ TRICHLOROPROPENE OXIDE

◇ 1,1,1-TRICHLOROPROPENE OXIDE ◇ 3,3,3-TRICHLOROPROPENE OXIDE ◇ 1,1,1-TRICHLOROPROPYLENE OXIDE ◇ 3,3,3-TRICHLOROPROPYLENE OXIDE

TOXICITY DATA with REFERENCE
mmo-sat 740 μg/plate APSXAS 17,189,80
sln-dmg-orl 500 ppm ENMUDM 7,677,85
sce-hmn:lym 50 μmol/L MUREAV 91,243,81
scu-mus TDLo:100 mg/kg (female 13D post):REP TCMUD8 7,159,87
ipr-rat LD50:142 mg/kg TXAPA9 52,422,80

CONSENSUS REPORTS: EPA Genetic Toxicology Program.

SAFETY PROFILE: Poison by intraperitoneal route. Human mutation data reported. Experimental reproductive effects. When heated to decomposition it emits toxic fumes of Cl^-. See also CHLORINATED HYDROCARBONS, ALIPHATIC.

TJC500 CAS:3266-39-5 ***HR: D***
2,3,3-TRICHLORO-2-PROPEN-1-OL
mf: $C_3H_3Cl_3O$ mw: 161.41

TOXICITY DATA with REFERENCE
mmo-sat 1 nmol/plate MUREAV 78,113,80
mma-sat 1 nmol/plate MUREAV 78,113,80

SAFETY PROFILE: Mutation data reported. When heated to decomposition it emits toxic fumes of Cl^-.

TJC750 CAS:7789-90-4 ***HR: 3***
2,2,3-TRICHLOROPROPIONALDEHYDE
mf: $C_3H_3Cl_3O$ mw: 161.41

TOXICITY DATA with REFERENCE
mmo-sat 1 nmol/plate MUREAV 78,113,80
mma-sat 1 nmol/plate MUREAV 78,113,80
orl-rat LD50:240 mg/kg AIHAAP 23,95,62
skn-rbt LDLo:710 mg/kg AIHAAP 23,95,62

SAFETY PROFILE: Poison by ingestion. Moderately toxic by skin contact. Mutation data reported. When heated to decomposition it emits toxic fumes of Cl^-. See also ALDEHYDES and CHLORINATED HYDROCARBONS, ALIPHATIC.

TJC800 CAS:12408-07-0 ***HR: 3***
TRICHLOROPROPIONITRILE
mf: $C_3H_2Cl_3N$ mw: 158.41

SYNS: PROPANENITRILE, TRICHLORO- ◇ PROPIONITRILE, TRICHLORO-

TOXICITY DATA with REFERENCE
orl-rat TDLo:103 mg/kg (multi) :REP HYSAAV 34(5),274,69
orl-rat TDLo:103 mg/kg (multi) :TER HYSAAV 34(5),274,69
orl-rat LD50:250 mg/kg HYSAAV 34(5),274,69

SAFETY PROFILE: Poison by ingestion. An experimental teratogen. Experimental reproductive effects. When heated to decomposition it emits toxic fumes of NO_x and Cl^-.

TJD500 CAS:10025-78-2 ***HR: 3***
TRICHLOROSILANE
DOT: UN 1295
mf: Cl_3HSi mw: 135.45

PROP: Colorless, very volatile liquid. Mp: −126.5°, bp: 31.8°, flash p: −18.4°F (OC), d: 1.35 @ 0°, vap press: 400 mm @ 14.5°, vap d: 4.7, autoign temp: 219°F. Sol in benzene, carbon disulfide, chloroform, carbon tetrachloride. Fumes in air. Decomp in water.

SYNS: SILICI-CHLOROFORME (FRENCH) ◇ SILICIUMCHLOROFORM (GERMAN) ◇ SILICOCHLOROFORM ◇ TRICHLOORSILAAN (DUTCH) ◇ TRICHLOROMONOSILANE ◇ TRICHLORSILAN (GERMAN) ◇ TRICLOROSILANO (ITALIAN)

TOXICITY DATA with REFERENCE
orl-rat LD50:1030 mg/kg JIHTAB 31,60,49
ihl-rat LCLo:1000 ppm/4H JIHTAB 31,343,49
ihl-mus LC50:1500 mg/m^3/2H 85GMAT -,114,82

CONSENSUS REPORTS: Reported in EPA TSCA Inventory.

DOT Classification: Flammable Liquid; Label: Flammable Liquid; Flammable Solid; Label: Dangerous When Wet, Flammable Liquid, Corrosive.

SAFETY PROFILE: Moderately toxic by ingestion and inhalation. A corrosive irritant to skin, eyes and mucous membranes. A very dangerous fire hazard when exposed to heat, flame, or by chemical reaction. May be ignited by spark or impact. Spontaneously flammable in air. Explosive reaction with acetonitrile + diphenyl sulfoxide. Will react with water or steam to produce heat and toxic and corrosive fumes. Can react vigorously with oxidizing materials. To fight fire, use CO_2, dry chemical. When heated to decomposition it emits toxic fumes of Cl^-. See also CHLOROSILANES.

TJD750 CAS:108-77-0 ***HR: 3***
2,4,6-TRICHLOROTRIAZINE
DOT: UN 2670
mf: $C_3Cl_3N_3$ mw: 184.41

PROP: Monoclinic, colorless crystals; pungent odor. Mp: 145.8°, bp: 190°, d: 1.32 @ 20/4°, vap press: 2 mm @ 70°, vap d: 6.36. Contains 96.9% cyanuric chloride, the remainder is cyanuric acid (VOONAW 12(4),78,66).

SYNS: CHLOROTRIAZINE ◇ CYANURCHLORIDE ◇ CYANURIC

ACID CHLORIDE ◇ CYANURIC CHLORIDE (DOT) ◇ CYANURIC TRICHLORIDE (DOT) ◇ CYANURYL CHLORIDE ◇ KYANURCHLORID (CZECH) ◇ s-TRIAZINE TRICHLORIDE ◇ TRICHLOROCYANIDINE ◇ TRICHLORO-s-TRIAZINE ◇ sym-TRICHLOROTRIAZINE ◇ 1,3,5-TRICHLOROTRIAZINE ◇ 2,4,6-TRICHLORO-s-TRIAZINE ◇ 2,4,6-TRICHLORO-1,3,5-TRIAZINE ◇ sym-TRICHLOTRIAZIN (CZECH) ◇ TRICYANOGEN CHLORIDE

TOXICITY DATA with REFERENCE
skn-rbt 500 mg/24H MOD 28ZPAK -,152,72
eye-rbt 50 μg/24H SEV 28ZPAK -,152,72
orl-rat TDLo:20 g/kg/73W-I:ETA,REP VOONAW 12(4),78,66
mul-rat TDLo:16 g/kg/73W-I:ETA VOONAW 12(4),78,66
orl-rat LD50:485 mg/kg 85GMAT -,114,82
orl-mus LD50:350 mg/kg 85GMAT -,114,82
ihl-mus LCLo:10 mg/m^3/2H 85GMAT -,114,82
ivn-mus LD50:18 mg/kg CSLNX* NX#07336

CONSENSUS REPORTS: Reported in EPA TSCA Inventory.

SAFETY PROFILE: Poison by ingestion, inhalation, and intravenous routes. Questionable carcinogen with experimental tumorigenic data. Experimental reproductive effects. A skin and severe eye irritant. An allergen. Has been reported as causing irritation of mucous membranes and heart rhythm disturbances in humans. Violent reaction with water (above 30°C); acetone + water; methanol; methanol + sodium hydrogencarbonate; 2-ethoxyethanol; dimethyl formamide; 3-butanone + sodium hydroxide + water; allyl alcohol + sodium hydroxide + water (at 28°C). When heated to decomposition it emits toxic fumes of Cl^- and NO_x. See also CHLORIDES.

TJE050 CAS:56943-26-1 ***HR: 3***
1,3,5-TRICHLORO-2,4,6-TRIFLUOROBORAZINE
mf: $B_3Cl_3F_3N_3$ mw: 237.80

NB(F)N(Cl)B(F)N(Cl)BF (ring)

SAFETY PROFILE: Explodes on contact with water. When heated to decomposition it emits toxic fumes of F^-, Cl^-, and NO_x. See also BORON COMPOUNDS and BORAZINE.

TJE750 CAS:6379-69-7 ***HR: 3***
TRICHOTHECIN
mf: $C_{19}H_{24}O_5$ mw: 332.43

PROP: Needles. Mp: 118°. Sltly sol in water; very sol in organic solvents.

SYNS: 12,13-EPOXY-4-HYDROXYTRICHOTHEC-9-EN-8-ONE CROTONATE ◇ 12,13-EPOXY-4-((1-OXO-2-BUTENYL)OXY) TRICHOTHEC-9-EN-8-ONE

TOXICITY DATA with REFERENCE
skn-gpg 332 ng MLD FAATDF 4(2, Pt 2),S124,84
ivn-mus LD50:300 mg/kg 85GDA2 6,182,81
scu-rbt LDLo:250 mg/kg JGMIAN 12,213,55

SAFETY PROFILE: Poison by intravenous and subcutaneous routes. A skin irritant. When heated to decomposition it emits acrid smoke and irritating fumes. See also FUSARENONE.

TJE870 CAS:35943-35-2 ***HR: 3***
TRICIRIBINE
mf: $C_{13}H_{16}N_6O_4$ mw: 320.35

SYNS: 3-AMINO-1,5-DIHYDRO-5-METHYL-1-β-d-RIBOFURANOSYL-1,4,5,6,8-PENTAAZAACENAPHTHYLENE ◇ 1,5-DIHYDRO-5-METHYL-1-β-d-RIBOFURANOSYL-1,4,5,6,8-PENTAAZAACENAPHTHYLEN-3-AMINE ◇ NSC-154020 ◇ PENTAAZACENTOPTHYLENE

TOXICITY DATA with REFERENCE
dni-mus:leu 90 nmol/L BCPCA6 27,233,78
oms-mus:leu 90 nmol/L BCPCA6 27,233,78
oms-mus:leu 100 nmol/L CNREA8 45,6355,85
ipr-mus LD50:150 mg/kg NCISP* JAN86
scu-mus LD50:199 mg/kg NCISP* JAN86

SAFETY PROFILE: Poison by subcutaneous and intraperitoneal routes. Mutation data reported. When heated to decomposition it emits toxic fumes of NO_x.

TJE875 ***HR: 3***
TRICIRIBINE PHOSPHATE HYDRATE
mf: $C_{13}H_{17}N_6O_7P\cdot H_2O$ mw: 418.35

SYNS: NSC-280594 HYDRATE ◇ PENTAAZAACENAPHTHYLENE-5'-PHOSPHATE ESTER MONOHYDRATE

TOXICITY DATA with REFERENCE
ivn-mus TDLo:234 mg/kg (1D male):REP NTIS** PB82-148404
ivn-mus LD50:257 mg/kg NTIS** PB82-148404
ivn-dog LDLo:33800 μg/kg NTIS** PB82-148404

SAFETY PROFILE: Poison by intravenous route. Experimental reproductive effects. When heated to decomposition it emits toxic fumes of PO_x and NO_x.

TJE880 CAS:79-90-3 ***HR: 3***
TRICLOBISONIUM CHLORIDE
mf: $C_{36}H_{74}N_2\cdot 2Cl$ mw: 606.02

PROP: White, crystalline powder. Mp: 243-253° (decomp). Sol in water, chloroform, alc.

SYNS: RO-5-0810/1 ◇ TRIBURON ◇ TRIBURON CHLORIDE

TOXICITY DATA with REFERENCE
orl-mus LD50:375 mg/kg ANTCAO 9,267,59
ipr-mus LD50:35 mg/kg ANTCAO 9,267,59
scu-mus LD50:154 mg/kg JMCMAR 6,780,63
ivn-mus LD50:12500 μg/kg ANTCAO 9,267,59
ivn-dog LDLo:8 mg/kg ANTCAO 9,267,59

SAFETY PROFILE: Poison by ingestion, subcutaneous, intravenous, and intraperitoneal routes. When heated to decomposition it emits toxic fumes of Cl^- and NO_x. See also CHLORIDES.

TJE890 CAS:55335-06-3 ***HR: 2***
TRICLOPYR
mf: $C_7H_4Cl_3NO_3$ mw: 256.47

PROP: Fluffy solid. Mp: 148-150°, vap. press: at 25°: 0.00000126 mm Hg. Subject to photolysis. Sol in water at 25°: 440 mg/L. Sol at 25° (g/kg): acetone 989, 1-octanol 307.

SYNS: DOWCO 233 ◇ GARLON ◇ 3,5,6-TRICHLORO-2-PYRIDYLOXY-ACETIC ACID

TOXICITY DATA with REFERENCE
orl-rat TDLo:1550 mg/kg (MGN):REP FAATDF 4,872,84
orl-rat TDLo:2 g/kg (6-15D preg):TER FAATDF 4,872,84
orl-rat LD50:630 mg/kg FMCHA2 -,C242,83
orl-rbt LD50:550 mg/kg PEMNDP 8,824,87
orl-gpg LD50:310 mg/kg PEMNDP 8,824,87

SAFETY PROFILE: Moderately toxic by ingestion. An experimental teratogen. Experimental reproductive effects. Used as an herbicide. When heated to decomposition it emits toxic fumes of Cl^- and NO_x.

TJF000 CAS:62973-76-6 ***HR: 2***
TRICLOSE
mf: $C_{10}H_{10}N_6O_2$ mw: 246.26

SYNS: 2-AMINO-4-((E)-2-(1-METHYL-5-NITRO-1H-IMIDAZOL-2-YL)ETHENYL)PYRIMIDINE ◇ (E)-2-AMINO-4-(2-(1-METHYL-5-NITROIMIDAZOL-2-YL)VINYL)PYRIMIDINE ◇ AZANIDAZOLE ◇ 4-((E)-2-(1-METHYL-5-NITRO-1H-IMIDAZOL-2-YL)-AETHENYL)-2-PYRIMIDINAMIN (GERMAN) ◇ 4-((E)-2-(1-METHYL-5-NITRO-1H-IMIDAZOL-2-YL)-ETHENYL)-2-PYRIMIDINAMINE ◇ NITROMIDINE

TOXICITY DATA with REFERENCE
mmo-sat 18 μmol/L TCMUD8 3,51,83
orl-rbt TDLo:1430 mg/kg (6-18D preg):REP ARZNAD 28,2251,78
orl-rat LD50:7600 mg/kg ARZNAD 28,2251,78
ipr-rat LD50:860 mg/kg ARZNAD 28,2251,78
orl-mus LD50:5100 mg/kg ARZNAD 28,2251,78
ipr-mus LD50:590 mg/kg ARZNAD 28,2251,78

SAFETY PROFILE: Moderately toxic by intraperitoneal route. Mildly toxic by ingestion. Experimental reproductive effects. Mutation data reported. When heated to decomposition it emits toxic fumes of NO_x.

TJF250 CAS:540-09-0 ***HR: 3***
12-TRICOSANONE
mf: $C_{23}H_{46}O$ mw: 338.69

PROP: Scales or plates. D: 0.809, mp: 69°. Insol in water; sol in alc.

SYN: DI-n-UNDECYL KETONE

TOXICITY DATA with REFERENCE
ivn-mus LD50:56 mg/kg CSLNX* NX#03438

CONSENSUS REPORTS: Reported in EPA TSCA Inventory.

SAFETY PROFILE: Poison by intravenous route. When heated to decomposition it emits acrid smoke and irritating fumes. See also KETONES.

TJF350 CAS:60318-52-7 ***HR: D***
TRICOSANTHIN

SYN: TRICHOSANTHIN

TOXICITY DATA with REFERENCE
ipr-mus TDLo:2 mg/kg (10D preg):TER CCPTAY 19,175,79
ims-mky TDLo:400 μg/kg (female 80D post):REP TWHPA3 22,156,76

SAFETY PROFILE: An experimental teratogen. Experimental reproductive effects.

TJF500 CAS:2665-12-5 ***HR: 3***
TRI-o-CRESYL BORATE
mf: $C_{21}H_{21}BO_6$ mw: 380.23

PROP: Straw-yellow liquid; odor of o-cresol. Bp: 189-195° @ 2 mm, flash p: 345°F (COC), d: 1.079 @ 22°, vap d: 11.4.

SYN: BORIC ACID, TRI-o-CRESYL ESTER

TOXICITY DATA with REFERENCE
eye-rbt 100 mg SEV 14KTAK -,706,64
orl-mus LD50:400 mg/kg 14KTAK -,706,64

SAFETY PROFILE: Poison by ingestion. A severe eye irritant. Combustible when exposed to heat or flame. When heated to decomposition it emits acrid smoke and irritating fumes. See also BORON COMPOUNDS.

TJF750 CAS:2622-08-4 ***HR: 3***
TRI-o-CRESYL PHOSPHITE
mf: $C_{21}H_{21}O_6P$ mw: 400.39

TOXICITY DATA with REFERENCE
scu-rat LDLo:10 mg/kg JPETAB 49,78,33
scu-cat LD50:100 mg/kg 14CYAT 2,1918,63
orl-ckn LDLo:1000 mg/kg JPETAB 49,78,33

SAFETY PROFILE: Poison by subcutaneous route. Moderately toxic by ingestion. When heated to decomposition it emits toxic fumes of PO_x. See also ESTERS.

TJG000 CAS:620-42-8 ***HR: 3***
TRI-p-CRESYL PHOSPHITE
mf: $C_{21}H_{21}O_6P$ mw: 400.39

TOXICITY DATA with REFERENCE
scu-rat LDLo:3000 mg/kg JPETAB 49,78,33
scu-cat LDLo:200 mg/kg JPETAB 49,78,33

CONSENSUS REPORTS: Reported in EPA TSCA Inventory.

SAFETY PROFILE: Poison by subcutaneous route. When heated to decomposition it emits toxic fumes of PO_x. See also ESTERS.

TJG225 ***HR: 3***
TRICYCLAMOL SULFATE
mf: $C_{20}H_{32}NO•CH_3SO_4$ mw: 413.63

SYNS: COMPOUND 14045 METHSULFATE ◇ 1-(3-CYCLOHEXYL-3-HYDROXY-3-PHENYLPROPYL)-1-METHYL-PYRROLIDINIUM METHYL SULFATE ◇ 1-CYCLOHEXYL-1-PHENYL-3-PYRROLIDINO-1-PROPANOL METHSULFATE ◇ ELORINE SULFATE

TOXICITY DATA with REFERENCE
orl-rat LD50:985 mg/kg JAPMA8 43,408,54
orl-mus LD50:554 mg/kg JAPMA8 43,408,54
ivn-mus LD50:15600 μg/kg JAPMA8 43,408,54

SAFETY PROFILE: Poison by intravenous route. Moderately toxic by ingestion. When heated to decomposition it emits toxic fumes of NO_x and SO_x.

TJG239 ***HR: 3***
TRICYCLIC ANTIDEPRESSANTS

SAFETY PROFILE: A group of antidepressant drugs that contain three fused rings in their chemical structure and that potentiate the action of catecholamines. They include imipramine, amitriptyline, nortriptyline, protriptyline, desipramine and doxepin. They have an atropine like action and can affect various systems. Central nervous system effects: dizziness, weakness, fatigue, headache, confusion, hallucinations, disturbed concentration, disorientation, delusions, excitement, anxiety, restlessness, insomnia, tremors; seizures. Cardiac effects: arrhythmias, sinus tachycardia, and prolongation of the conduction time. May precipitate myocardial infarction and stroke, hypotension, hypertension. Anticholinergic effects include: dry mouth, blurred vision.

TJG250 CAS:768-94-5 ***HR: 3***
TRICYCLO(3.3.1.1^{3,7})DECAN-1-AMINE
mf: $C_{10}H_{17}N$ mw: 151.28

SYNS: 1-ADAMANTAMINE ◇ 1-ADAMANTANAMINE ◇ AMANTADINE ◇ 1-AMINOADAMANTANE ◇ 1-AMINOADAMATANE ◇ 1-AMINOTRICYCLO(3.3.1.1^{3,7})DECANE ◇ EXP-105-1 ◇ PK-MERZ ◇ SYMMETREL

TOXICITY DATA with REFERENCE
dnd-esc 10 μmol/L MUREAV 89,95,81
orl-mus LD50:900 mg/kg DRFUD4 5,557,80
ipr-mus LD50:245 mg/kg DRFUD4 5,557,80

CONSENSUS REPORTS: Reported in EPA TSCA Inventory. EPA Genetic Toxicology Program.

SAFETY PROFILE: Poison by intraperitoneal route. Moderately toxic by ingestion. Mutation data reported. When heated to decomposition it emits toxic fumes of NO_x. Used as an antiviral agent. See also AMINES.

TJG500 CAS:4747-82-4 ***HR: 2***
TRICYCLODECANE(5.2.1.0^{2,6})-3,10-DIISOCYANATE
mf: $C_{12}H_8N_2O_2$ mw: 212.22

SYN: ISOCYANIC ACID, HEXAHYDRO-4,7-METHANOINDAN-1,8-YLENE ESTER

TOXICITY DATA with REFERENCE
orl-rat LD50:1090 mg/kg TXAPA9 28,313,74
skn-rbt LD50:450 mg/kg TXAPA9 28,313,74
NIOSH REL: (Diisocyanates) TWA 0.005 ppm; CL 0.02 ppm/10M

SAFETY PROFILE: Moderately toxic by ingestion and skin contact. When heated to decomposition it emits toxic fumes of NO_x. See also ISOCYANATES.

TJG750 CAS:5440-19-7 ***HR: 2***
TRI(2-CYCLOHEXYLCYCLOHEXYL)BORATE
mf: $C_{36}H_{63}BO_3$ mw: 554.80

PROP: White solid; odor of 2-cyclohexylcyclohexanol. Mp: 172-175°, bp: 230-250° @ 0.3 mm.

TOXICITY DATA with REFERENCE
eye-rbt 100 mg MOD 14KTAK -,706,64
orl-mus LD50:2050 mg/kg 14KTAK -,706,64

SAFETY PROFILE: Moderately toxic by ingestion. An eye irritant. Combustible when exposed to heat or flame; can react with oxidizing materials. When heated to decomposition it emits toxic fumes of BO_x. See also BORON COMPOUNDS and ESTERS.

TJH250 CAS:195-84-6 ***HR: 3***
TRICYCLOQUINAZOLINE
mf: $C_{21}H_{12}N_4$ mw: 320.37

TOXICITY DATA with REFERENCE
scu-rat TDLo:1600 mg/kg/17W-I:ETA BJCAAI 13,94,59
skn-mus TDLo:380 mg/kg/16W-I:CAR BJCAAI 13,94,59
skn-mus TD:960 mg/kg/40W-I:NEO BJCAAI 16,275,62

SAFETY PROFILE: Questionable carcinogen with experimental carcinogenic, neoplastigenic, and tumorigenic data. When heated to decomposition it emits toxic fumes of NO_x.

TJH500 CAS:629-50-5 ***HR: 2***
TRIDECANE
mf: $C_{13}H_{28}$ mw: 184.41

PROP: Colorless liquid. D: 0.757 @ 20/4°, mp: −6.2°, bp: 234°. Insol in water; very sol in alc, ether.

SYN: n-TRIDECANE

TOXICITY DATA with REFERENCE
ivn-mus LD50:1161 mg/kg JPMSAE 67,566,78

CONSENSUS REPORTS: Reported in EPA TSCA Inventory.

SAFETY PROFILE: Moderately toxic by intravenous route. When heated to decomposition it emits acrid smoke and irritating fumes.

TJH750 CAS:629-60-7 ***HR: 3***
TRIDECANENITRILE
mf: $C_{13}H_{25}N$ mw: 195.39

SYN: NC12

TOXICITY DATA with REFERENCE
ipr-mus LDLo:256 mg/kg CBCCT* 2,191,50

CONSENSUS REPORTS: Cyanide and its compounds are on the Community Right-To-Know List. Reported in EPA TSCA Inventory.

SAFETY PROFILE: Poison by intraperitoneal route. When heated to decomposition it emits toxic fumes of NO_x and CN^-. See also NITRILES.

TJI000 CAS:1070-01-5 ***HR: 2***
TRIDECANITRILE (mixed isomers)
mf: $C_{30}H_{63}N$ mw: 437.94

SYN: TRIS(DECYL)AMINE

TOXICITY DATA with REFERENCE
orl-rat LD50:3730 mg/kg AIHAAP 30,470,69
skn-rbt LD50:3180 mg/kg AIHAAP 30,470,69

CONSENSUS REPORTS: Cyanide and its compounds are on the Community Right-To-Know List. Reported in EPA TSCA Inventory.

SAFETY PROFILE: Moderately toxic by ingestion and skin contact. When heated to decomposition it emits toxic fumes of NO_x and CN^-. See also NITRILES.

TJI250 CAS:638-53-9 ***HR: 3***
TRIDECANOIC ACID
mf: $C_{13}H_{26}O_2$ mw: 214.39

PROP: Mp: 41-42°, bp: 236°/100 mm, flash p: >230°F.

SYNS: n-TRIDECOIC ACID ◇ TRIDECYLIC ACID

TOXICITY DATA with REFERENCE
ivn-mus LD50:130 mg/kg APTOA6 18,141,61

CONSENSUS REPORTS: Reported in EPA TSCA Inventory.

SAFETY PROFILE: Irritant. Poison by intravenous route. Combustible liquid. When heated to decomposition it emits acrid smoke and irritating fumes.

TJI500 CAS:63978-73-4 ***HR: 3***
TRIDECANOIC ACID-2,3-EPOXYPROPYL ESTER
mf: $C_{16}H_{30}O_3$ mw: 270.46

SYN: GLYCIDYL ESTER of DODECANOIC ACID

TOXICITY DATA with REFERENCE
scu-rat TDLo:4400 mg/kg/5W-I:ETA ANYAA9 68,750,58

SAFETY PROFILE: Questionable carcinogen with experimental tumorigenic data. When heated to decomposition it emits acrid smoke and irritating fumes. See also ESTERS.

TJI750 CAS:112-70-9 ***HR: 1***
1-TRIDECANOL
mf: $C_{13}H_{28}O$ mw: 200.41

PROP: General term for a commercial mixture of isomers of the formula $C_{12}H_{25}CH_2OH$. Water-white liquid; pleasant odor. Flash p: 250°F (COC), vap d: 6.9. D: 0.822 @ 21/4°, mp: 30.5°, bp: 155-156° @ 15 mm.

SYNS: TRIDECANOL ◇ n-TRIDECANOL ◇ TRIDECYL ALCOHOL ◇ n-TRIDECYL ALCOHOL

TOXICITY DATA with REFERENCE
skn-rbt 410 mg open MLD UCDS** 10/14/64
orl-rat LD50:17200 mg/kg NPIRI* 1,114,74
skn-rbt LD50:5600 mg/kg NPIRI* 1,114,74

CONSENSUS REPORTS: Reported in EPA TSCA Inventory.

SAFETY PROFILE: Mildly toxic by ingestion and skin contact. A skin irritant. Combustible when exposed to heat or flame. To fight fire, use mist, spray, dry chemical, foam. When heated to decomposition it emits acrid smoke and irritating fumes. See also ALCOHOLS.

TJJ250 CAS:24938-91-8 ***HR: 1***
TRIDECANOL condensed with 6 moles ETHYLENE OXIDE

SYNS: ALKYL(C-13) POLYETHOXYLATES(ETHOXY-6) ◇ POLYOXIETHYLENE (6) ALKYL (13) ETHER

TOXICITY DATA with REFERENCE
skn-rbt 2 g/4W MLD FCTXAV 15,319,77

CONSENSUS REPORTS: Reported in EPA TSCA Inventory.

SAFETY PROFILE: A skin irritant. When heated to decomposition it emits acrid smoke and irritating fumes. See also ETHERS.

TJJ500 CAS:81412-43-3 ***HR: 2***
TRIDEMORPH
mf: $C_{19}H_{39}NO$ mw: 297.59

SYNS: BAS 2203F ◇ CALIXIN ◇ COSMIC ◇ TRIDEMORF

TOXICITY DATA with REFERENCE
oth-mmo-omi 100 umol/L NNGADV 5,69,80
orl-rat TDLo:650 mg/kg (female 4-18D post):TER WZERDH 32(1-2),19,83
orl-rat LD50:650 mg/kg 85JCAE -,894,86
orl-rbt LD50:562 mg/kg 85JCAE -,894,86
skn-rbt LD50:1350 mg/kg 85JCAE -,894,86

SAFETY PROFILE: Moderately toxic by ingestion and skin contact. An experimental teratogen. Mutation data reported. When heated to decomposition it emits toxic fumes of NO_x.

TJJ750 CAS:73758-18-6 ***HR: 2***
TRI(DIISOBUTYLCARBINYL) BORATE
mf: $C_{27}H_{57}BO_3$ mw: 440.65

PROP: White crystals; odor of diisobutyl carbinol. Mp: 99-100°, bp: 198-209° @ 22 mm.

SYN: TRIS(3,5-DIMETHYL-4-HEPTYL)BORATE

TOXICITY DATA with REFERENCE
eye-rbt 10 mg MOD 14KTAK -,706,64
orl-mus LD50:3700 mg/kg 14KTAK -,706,64

SAFETY PROFILE: Moderately toxic by ingestion. An eye irritant. When heated to decomposition it emits acrid smoke and irritating fumes. See also BORON COMPOUNDS.

TJK000 CAS:467-63-0 ***HR: 3***
TRI(p-DIMETHYLAMINOPHENYL)METHANOL
mf: $C_{25}H_{31}N_3O$ mw: 389.59

SYNS: CARBINOLBASE DES KRISTALLVIOLETT (GERMAN) ◇ C.I. 42555B ◇ C.I. SOLVENT VIOLET 9 ◇ 4-(DIMETHYLAMINO)-α,α-BIS(4-(DIMETHYLAMINO)PHENYL)-BENZENEMETHANOL(9CI) ◇ METHYLROSANILINE

TOXICITY DATA with REFERENCE
orl-rat LD50:770 mg/kg ARZNAD 1,5,51
ipr-rat LD50:58 mg/kg ARZNAD 1,5,51
orl-mus LD50:1000 mg/kg ARZNAD 1,5,51
ipr-mus LD50:56 mg/kg ARZNAD 1,5,51
orl-rbt LD50:180 mg/kg ARZNAD 1,5,51
ipr-rbt LD50:120 mg/kg ARZNAD 1,5,51
idu-rbt LD50:870 mg/kg ARZNAD 1,5,51

CONSENSUS REPORTS: Reported in EPA TSCA Inventory.

SAFETY PROFILE: Poison by ingestion and intraperitoneal routes. Moderately toxic by intraduodenal route. When heated to decomposition it emits toxic fumes of NO_x.

TJK100 CAS:58138-08-2 ***HR: 2***
TRIDIPHANE
mf: $C_{10}H_7Cl_5O$ mw: 320.42

SYNS: (RS)-2-(3,5-DICHLOROPHENYL)-2-(2,2,2-TRICHLOROETHYL)OXIRANE ◇ DOWCO 356 ◇ NELPON ◇ OXIRANE, 2-(3,5-DICHLOROPHENYL)-2-(2,2,2-TRICHLOROETHYL)- ◇ TANDEM

TOXICITY DATA with REFERENCE
orl-mus TDLo:2 g/kg (female 8-15D post):REP FAATDF 8,179,87
orl-mus TDLo:2 g/kg (female 8-15D post):TER FAATDF 8,179,87
orl-rat LD50:1500 mg/kg FAATDF 8,179,87
orl-mus LD50:740 mg/kg FAATDF 8,179,87
skn-rbt LD50:3536 mg/kg PEMNDP 8,829,87

SAFETY PROFILE: Moderately toxic by ingestion and skin contact. An experimental teratogen. Experimental reproductive effects. When heated to decomposition it emits toxic fumes of Cl^-.

TJK250 CAS:2467-15-4 ***HR: 2***
TRI-n-DODECYL BORATE
mf: $C_{36}H_{75}BO_3$ mw: 566.92

PROP: Light straw-yellow, oily liquid. Bp: 479°, flash p: 465°F (COC), d: 0.845 @ 26.8°, vap d: 19.6.

SYN: BORIC ACID, TRIETHYL ESTER

TOXICITY DATA with REFERENCE
eye-rbt 100 mg MOD 14KTAK -,706,64
orl-mus LD50:1400 mg/kg USBCC* -,-,58

SAFETY PROFILE: Moderately toxic by ingestion. An eye irritant. Combustible when exposed to heat or flame. Can react with oxidizing materials. To fight fire, use foam, CO_2, dry chemical. When heated to decomposition it emits acrid smoke and irritating fumes. See also BORON COMPOUNDS and ESTERS.

TJK500 CAS:52338-90-6 ***HR: 3***
1,2,4,5,9,10-TRIEPOXYDECANE
mf: $C_{10}H_{16}O_3$ mw: 184.26

TOXICITY DATA with REFERENCE
skn-mus TDLo:864 mg/kg/72W-I:ETA JNCIAM 53,695,74
scu-mus TDLo:1040 mg/kg/26W-I:NEO JNCIAM 53,695,74

SAFETY PROFILE: Questionable carcinogen with experimental neoplastigenic and tumorigenic data. When heated to decomposition it emits acrid smoke and irritating fumes.

TJK750 CAS:283-56-7 ***HR: 1***
TRIETHANOLAMINE BORATE
mf: $C_6H_{18}BN_3O_3$ mw: 191.08

PROP: White, odorless solid. Mp: 235.5-238.5°.

SYN: BORIC ACID, TRIS(2-AMINOETHYL) ESTER

TOXICITY DATA with REFERENCE
eye-rbt 15 mg MLD 14KTAK -,706,64
orl-mus LD50:6200 mg/kg 14KTAK -,706,64

SAFETY PROFILE: Mildly toxic by ingestion. An eye irritant. When heated to decomposition it emits toxic fumes of NO_x. See also TRIHYDROXYTRIETHYLAMINE, ESTERS, and BORON COMPOUNDS.

TJL250 CAS:588-42-1 ***HR: 3***
TRIETHANOLAMINE TRINITRATE BIPHOSPHATE
mf: $C_6H_{12}N_4O_9 \bullet 2H_3O_4P$ mw: 480.22

SYNS: AMINOTRATE PHOSPHATE ◇ ANGITRIT ◇ BENTONYL ◇ DURONITRIN ◇ ETHANOL, 2,2',2''-NITRILOTRIS-, TRINITRATE (ester), PHOSPHATE (1:2)(SALT) (9CI) ◇ KARDIN ◇ METAMIN ◇ METAMINE ◇ NITRANOL ◇ NITRETAMIN ◇ NITRETAMIN PHOSPHATE ◇ 2,2',2''-NITRILOTRIETHANOL TRINITRATE PHOSPHATE ◇ NITROCARDIOL ◇ NITRODURAN ◇ ORTIN ◇ PRAENITRON ◇ PRAENITRONA ◇ PRENITRON ◇ THIBETINE ◇ TRIANATE ◇ TRICORYL ◇ TRIETHANOLAMINE TRINITRATE DIPHOSPHATE ◇ TRINITROTRIETHANOLAMINE ◇ TRISUSTAN ◇ TROLMINE ◇ TROLNITRATE PHOSPHATE ◇ VASOMED

TOXICITY DATA with REFERENCE
orl-rat LD50:130 mg/kg DRUGAY 6,535,82
ipr-rat LD50:23 mg/kg DRUGAY 6,535,82
scu-rat LD50:34 mg/kg DRUGAY 6,535,82
orl-mus LD50:330 mg/kg DRUGAY 6,535,82
ipr-mus LD50:82 mg/kg DRUGAY 6,535,82
scu-mus LD50:141 mg/kg DRUGAY 6,535,82
ivn-mus LD50:100 mg/kg YKYUA6 27,1401,76

SAFETY PROFILE: Poison by ingestion, subcutaneous, intravenous, and intraperitoneal routes. When heated to decomposition it emits very toxic fumes of PO_x and NO_x. See also NITRATES.

TJL500 CAS:1912-26-1 ***HR: 2***
TRIETHAZINE
mf: $C_9H_{16}ClN_5$ mw: 229.75

SYNS: 6-CHLORO-N,N,N'-TRIETHYL-1,3,5-TRIAZINE-2,4-DIAMINE ◇ TRIETAZINE

TOXICITY DATA with REFERENCE
orl-rat LD50:594 mg/kg WRPCA2 9,119,70

CONSENSUS REPORTS: Reported in EPA TSCA Inventory.

SAFETY PROFILE: Moderately toxic by ingestion. When heated to decomposition it emits very toxic fumes of Cl^- and NO_x.

TJL775 CAS:65232-69-1 ***HR: 3***
TRIETHOXYDIALUMINUM TRIBROMIDE
mf: $C_6H_{15}Al_2Br_3O_3$ mw: 428.86

$(CH_3CH_2O)_3Al \bullet AlBr_3$

SAFETY PROFILE: Ignites spontaneously in air. Explosive reaction with water, ethanol. When heated to decomposition it emits toxic fumes of Br^-. See also ALUMINUM COMPOUNDS and BROMIDES.

TJM000 CAS:101-33-7 ***HR: 1***
1,1,3-TRIETHOXYHEXANE
mf: $C_{12}H_{26}O_3$ mw: 218.38

PROP: Liquid. Insol in water. D: 0.8746 @ 20/20°, bp: 133° @ 50 mm, fp: −100°, flash p: 210°F (OC), vap d: 7.5.

SYN: 3-ETHOXYHEXANAL DIETHYL ACETAL

TOXICITY DATA with REFERENCE
skn-rbt 10 mg/24H open MLD AIHAAP 23,95,62
orl-rat LD50:17 g/kg AIHAAP 23,95,62

SAFETY PROFILE: Mildly toxic by ingestion. A skin irritant. Combustible when exposed to heat or flame. To fight fire, use foam, alcohol foam, fog. When heated to decomposition it emits acrid smoke and irritating fumes.

TJM250 CAS:7789-92-6 ***HR: 2***
1,3,3-TRIETHOXYPROPANE
mf: $C_9H_{20}O_3$ mw: 176.29

TOXICITY DATA with REFERENCE
skn-rbt 10 mg/24H open MLD AMIHBC 4,119,51
eye-rbt 750 mg open AMIHBC 4,119,51
orl-rat LD50:1600 mg/kg AMIHBC 4,119,51
skn-rbt LD50:8 g/kg AMIHBC 4,119,51

SAFETY PROFILE: Moderately toxic by ingestion. Mildly toxic by skin contact. A skin and eye irritant. When heated to decomposition it emits acrid smoke and irritating fumes.

TJM500 CAS:5444-80-4 ***HR: 3***
1,3,3-TRIETHOXY-1-PROPENE
mf: $C_9H_{18}O_3$ mw: 174.27

SYNS: 3-ETHOXY ACROLEIN DIETHYL ACETAL ◇ 1,3,3-TRIETHOXYPROPENE

TOXICITY DATA with REFERENCE
skn-rbt 10 mg/24H open SEV AMIHBC 4,119,51
eye-rbt 20 mg open SEV AMIHBC 4,119,51
orl-rat LD50:2460 mg/kg AMIHBC 4,119,51
ihl-rat LCLo:250 ppm/4H JIHTAB 31,343,49
skn-rbt LD50:370 mg/kg AMIHBC 4,119,51

SAFETY PROFILE: Poison by skin contact. Moderately toxic by ingestion and inhalation. A severe skin and eye irritant. When heated to decomposition it emits acrid smoke and irritating fumes.

TJM750 CAS:998-30-1 ***HR: 3***
TRIETHOXYSILANE
mf: $C_6H_{16}O_3Si$ mw: 164.31

PROP: D: 0.875, refr index: 1.3762, bp: 131.5, flash p: 80° F.

TOXICITY DATA with REFERENCE
ihl-mus LC50:500 mg/m³/2H 85GMAT -,115,82
ivn-mus LD50:180 mg/kg CSLNX* NX#00018

CONSENSUS REPORTS: EPA Extremely Hazardous Substances List. Reported in EPA TSCA Inventory.

SAFETY PROFILE: Poison by intravenous route. Moderately toxic by inhalation. Flammable Liquid. When heated to decomposition it emits acrid smoke and irritating fumes. See also SILANE.

TJN000 CAS:919-30-2 ***HR: 3***
3-(TRIETHOXYSILYL)-1-PROPANAMINE
mf: $C_9H_{23}NO_3Si$ mw: 221.42

SYNS: (Γ-AMINOPROPYL)TRIETHOXYSILANE ◇ (3-AMINOPROPYL)TRIETHOXYSILANE ◇ TRIETHOXY(3-AMINOPROPYL)SILANE ◇ 3-(TRIETHOXYSILYL)PROPYLAMINE

TOXICITY DATA with REFERENCE
skn-rbt 100 μg/24H open AIHAAP 23,95,62
eye-rbt 100 mg MLD UCDS** 1/19/72
orl-rat LD50:1780 mg/kg AIHAAP 23,95,62
ipr-mus LD50:260 mg/kg RCRVAB 38(12),975,69
skn-rbt LD50:4000 mg/kg UCDS** 1/19/72

CONSENSUS REPORTS: Reported in EPA TSCA Inventory.

SAFETY PROFILE: Poison by intraperitoneal route. Moderately toxic by ingestion and skin contact. A skin and eye irritant. When heated to decomposition it emits toxic fumes of NO_x. See also SILANE and AMINES.

TJN250 CAS:78-08-0 ***HR: 1***
TRIETHOXYVINYLSILANE
mf: $C_8H_{18}O_3Si$ mw: 190.35

SYNS: TRIETHOXYVINYLSILICANE ◇ VINYLTRIETHOXYSILANE

TOXICITY DATA with REFERENCE
skn-rbt 500 mg open MLD UCDS** 7/10/67
eye-rbt 500 mg AMIHBC 10,61,54
orl-rat LD50:23 g/kg AMIHBC 10,61,54
ihl-rat LCLo:4000 ppm/4H AMIHBC 10,61,54
skn-rbt LD50:10 g/kg AMIHBC 10,61,54

CONSENSUS REPORTS: Reported in EPA TSCA Inventory.

SAFETY PROFILE: Mildly toxic by ingestion, inhalation and skin contact. When heated to decomposition it emits acrid smoke and irritating fumes. See also SILANE.

TJN750 CAS:97-93-8 ***HR: 3***
TRIETHYLALUMINUM
DOT: UN 1102
mf: $C_6H_{15}Al$ mw: 114.19

PROP: Fp: −52.5°, d: 0.837 @ 20°, vap press: 4 mm @ 83°, flash p: < −63°F, bp: 194°.

SYN: TEA

CONSENSUS REPORTS: Reported in EPA TSCA Inventory.

ACGIH TLV: TWA 2 mg(Al)/m³
DOT Classification: Flammable Solid; Label: Spontaneously Combustible.

SAFETY PROFILE: Extremely destructive to living tissue. A very dangerous fire hazard when exposed to heat or flame. Ignites spontaneously in air. Explodes violently in water. To fight fire, use CO_2, dry sand, dry chemical. Do not use water, foam or halogenated fire-fighting agents. Explosive reaction with alcohols (e.g., methanol; ethanol; propynol); carbon tetrachloride; N,N-dimethylformamide + heat. Incompatible with halogenated hydrocarbons; triethyl borine. When heated to decomposition it emits acrid smoke and irritating fumes. See also ALUMINUM COMPOUNDS and ORGANOMETALS.

TJO000 CAS:121-44-8 ***HR: 3***
TRIETHYLAMINE
DOT: UN 1296
mf: $C_6H_{15}N$ mw: 101.22

PROP: Colorless liquid; ammonia odor. Mp: −114.8°, bp: 89.5°, flash p: 20°F (OC), d: 0.7255 @ 25/4°, vap d: 3.48, lel: 1.2%, uel: 8.0%. Misc in water, alc, ether.

SYNS: (DIETHYLAMINO)ETHANE ◇ N,N-DIETHYLETHANAMINE ◇ TEN ◇ TRIAETHYLAMIN (GERMAN) ◇ TRIETILAMINA (ITALIAN)

TOXICITY DATA with REFERENCE
skn-rbt 10 mg/24H open MLD AMIHBC 4,119,51

skn-rbt 365 mg open MLD UCDS** 3/23/70
eye-rbt 250 mg open SEV AMIHBC 4,119,51
eye-rbt 50 ppm/30D-I SEV AMIHBC 3,287,51
cyt-rat-ihl 1 mg/m^3 GISAAA 36(11),9,71
orl-rbt TDLo:6900 $\mu g/kg$ (1-3D preg):REP FESTAS 15,97,64
ihl-hmn TCLo:12 mg/m^3/11W-C:EYE IAEHDW 57,297,86
orl-rat LD50:460 mg/kg AMIHBC 4,119,51
ihl-rat LCLo:1000 ppm/4H AMIHBC 4,119,51
orl-mus LD50:546 mg/kg HYSAAV 30,351,65
skn-rbt LD50:570 mg/kg AMIHBC 4,119,51
ihl-gpg LCLo:1000 ppm/4H JIHTAB 30,2,48
ihl-mam LC50:6 g/m^3

CONSENSUS REPORTS: Reported in EPA TSCA Inventory.

OSHA PEL: (Transitional: TWA 25 ppm) TWA 10 ppm; STEL 15 ppm
ACGIH TLV: TWA 10 ppm; STEL 15 ppm
DFG MAK: 10 ppm (40 mg/m^3)
DOT Classification: Flammable Liquid; Label: Flammable Liquid.

SAFETY PROFILE: Moderately toxic by ingestion and skin contact. Mildly toxic by inhalation. Human systemic effects: visual field changes. Experimental reproductive effects. Mutation data reported. A skin and severe eye irritant. Can cause kidney and liver damage. A very dangerous fire hazard when exposed to heat, flame, or oxidizers. Explosive in the form of vapor when exposed to heat or flame. Complex with dinitrogen tetraoxide explodes below 0°C when undiluted with solvent. Exothermic reaction with maleic anhydride above 150°C. Can react with oxidizing materials. Incompatible with N_2O_4. To fight fire, use CO_2, dry chemical, alcohol foam. When heated to decomposition it emits toxic fumes of NO_x.

TJO100 CAS:27096-31-7 ***HR: 3***
TRIETHYL AMMONIUM NITRATE
mf: $C_6H_{16}N_2O_3$ mw: 164.20

$$(CH_3CH_2)_3N^-HNO_3^-$$

SAFETY PROFILE: Complex with dinitrogen tetraoxide + diethyl ether explodes when dried. Forms an unstable explosive complex with dinitrogen tetraoxide. When heated to decomposition it emits toxic fumes of NH_3 and NO_x. See also NITRATES.

TJO250 CAS:617-85-6 ***HR: 3***
TRIETHYLANTIMONY
mf: $C_6H_{15}Sb$ mw: 208.84

PROP: D: 1.324, refr index: 1.42, mp: -29°, bp: 159.5.

CONSENSUS REPORTS: Antimony and its compounds are on the Community Right-To-Know List.

SAFETY PROFILE: Most antimony compounds are poisons. Ignites spontaneously in air. When heated to decomposition it emits toxic fumes of Sb. See also ANTIMONY COMPOUNDS.

TJO500 ***HR: 3***
TRIETHYLARSINE
mf: $C_6H_{15}As$ mw: 162.01

PROP: D: 1.150, bp: 140°/736 mm. Insol in water; miscible in alc and ether.

CONSENSUS REPORTS: Arsenic and its compounds are on the Community Right-To-Know List.

SAFETY PROFILE: Arsenic compounds are poisons. Ignites spontaneously in air. When heated to decomposition it emits toxic fumes of As. See also ARSENIC COMPOUNDS.

TJO750 CAS:25340-18-5 ***HR: 1***
TRIETHYLBENZENE
mf: $C_{12}H_{18}$ mw: 162.30

PROP: Clear, colorless liquid. Mp: $< -70°$, bp: 218-219°, flash p: 181°F (CC), d: 870 @ 25/25°, vap d: 5.6.

SYN: TRIETHYL-BENZENE (mixed isomers)

TOXICITY DATA with REFERENCE
orl-rat LDLo:5000 mg/kg AMIHAB 19,403,59

CONSENSUS REPORTS: Reported in EPA TSCA Inventory.

SAFETY PROFILE: Mildly toxic by ingestion. Flammable when exposed to heat or flame. Can react with oxidizers. To fight fire, use foam, CO_2, dry chemical. When heated to decomposition it emits acrid smoke and irritating fumes.

TJP000 CAS:617-77-6 ***HR: 3***
TRIETHYLBISMUTH
mf: $C_6H_{15}Bi$ mw: 296.17

SAFETY PROFILE: Ignites spontaneously in air. Explodes at 150°C. When heated to decomposition it emits toxic fumes of Bi. See also BISMUTH COMPOUNDS.

TJP250 CAS:97-94-9 ***HR: 3***
TRIETHYLBORANE
mf: $C_6H_{15}B$ mw: 98.02

PROP: Colorless liquid. Mp: −93°, d: 0.6961 @ 23°.

SYN: TRIETHYLBORINE

TOXICITY DATA with REFERENCE
orl-rat LD50:235 mg/kg 14KTAK -,693,64
ihl-rat LC50:700 ppm/4H 14KTAK -,693,64
ipr-rat LD50:22700 μg/kg 14KTAK -,693,64

CONSENSUS REPORTS: Reported in EPA TSCA Inventory.

SAFETY PROFILE: Poison by ingestion and intraperitoneal routes. Mildly toxic by inhalation. Animal experiments show that the vapor is a poison which causes pulmonary irritation and convulsions. A very dangerous fire hazard by spontaneous chemical reaction with oxidizers. Spontaneously flammable in air. Explodes in oxygen atmospheres. Hypergolic reaction with triethylaluminum. Ignites on contact with chlorine, bromine, or other halogens. Will react with water or steam to produce toxic and flammable vapors. To fight fire, do NOT use halogenated extinguishing agents. When heated to decomposition or upon contact with air it emits toxic acrid smoke and irritating fumes. See also BORANES and BORON COMPOUNDS.

TJP500 CAS:150-46-9 ***HR: 2***
TRIETHYL BORATE
mf: $C_6H_{15}BO_3$ mw: 146.02

PROP: Liquid. D: 0.864 @ 20/40°, bp: 120°, flash p: 51.8°F. Decomp in water.

SYN: BORIC ACID, TRIETHYL ESTER

TOXICITY DATA with REFERENCE
eye-rbt 100 mg MLD 14KTAK -,706,64
orl-mus LD50:1800 mg/kg USBCC*

CONSENSUS REPORTS: Reported in EPA TSCA Inventory.

SAFETY PROFILE: Moderately toxic by ingestion. An eye irritant. A very dangerous fire hazard when exposed to heat or flame. When heated to decomposition it emits acrid smoke and irritating fumes. See also BORON COMPOUNDS and ESTERS.

TJP750 CAS:77-93-0 ***HR: 2***
TRIETHYL CITRATE
mf: $C_{12}H_{20}O_7$ mw: 276.32

PROP: Colorless oily liquid; odorless. Bp: 294°, flash p: 303°F (COC), d: 1.136 @ 25°, vap press: 1 mm @ 107.0°. Sltly sol in water; misc in alc, ether.

SYNS: CITROFLEX 2 ◇ ETHYL CITRATE ◇ 2-HYDROXY,1,2,3-PROPANETRICARBOXYLIC ACID, TRIETHYL ESTER ◇ TEC

TOXICITY DATA with REFERENCE
orl-rat LD50:5900 mg/kg IYKEDH 16,214,85
ihl-rat LC50:1300 ppm/6H FCTXAV 17,357,79
ipr-rat LD50:4 g/kg IYKEDH 16,214,85
scu-rat LD50:6600 mg/kg IYKEDH 16,214,85
ipr-mus LD50:1750 mg/kg JPMSAE 53,774,64
orl-cat LD50:35000 mg/kg FCTXAV 17,357,79

CONSENSUS REPORTS: Reported in EPA TSCA Inventory.

SAFETY PROFILE: Moderately toxic by intraperitoneal route. Mildly toxic by ingestion and inhalation. Combustible liquid when exposed to heat or flame. To fight fire, use dry chemical, CO_2. When heated to decomposition it emits acrid smoke and irritating fumes. See also ESTERS and CITRIC ACID.

TJP775 CAS:12075-68-2 ***HR: 3***
TRIETHYL DIALUMINUM TRICHLORIDE
DOT: UN 1925
mf: $C_6H_{15}Al_2Cl_3$ mw: 247.51

$(CH_3CH_2)_3Al{\bullet}AlCl_3$

DOT Classification: Flammable Solid; Label: Spontaneously Combustible

SAFETY PROFILE: Mixtures with carbon tetrachloride explode at room temperature. When heated to decomposition it emits toxic fumes of Cl^-. See also ALUMINUM COMPOUNDS.

TJP780 CAS:62133-36-2 ***HR: 3***
TRIETHYLDIBORANE
mf: $C_6H_{16}B_2$ mw: 109.81

SAFETY PROFILE: Ignites spontaneously in air. When heated to decomposition it emits acrid smoke and irritating fumes. See also BORANES and BORON COMPOUNDS.

TJQ000 CAS:112-27-6 ***HR: 3***
TRIETHYLENE GLYCOL
mf: $C_6H_{14}O_4$ mw: 150.20

PROP: Odorless, colorless liquid; hygroscopic. Fp: −7.3°, flash p: 350°F, d: 1.122 @ 25/25°, lel: 0.9%, uel: 9.2%, autoign temp: 700°F, vap press: 1 mm @ 114°, vap d: 5.17, bp: 285°. Misc in water, alc, benzene; insol in petr ether; very sltly sol in ether.

SYNS: DI-β-HYDROXYETHOXYETHANE ◇ 3,6-DIOXAOCTANE-1,8-DIOL ◇ 2,2′-(1,2-ETHANEDIYLBIS(OXY))BISETHANOL ◇ 2,2′-ETHYLENEDIOXYDIETHANOL ◇ 2,2′-ETHYLENEDIOXYETHANOL ◇ ETHYLENE GLYCOL-BIS-(2-HYDROXYETHYL ETHER) ◇ ETHYLENE GLYCOL DIHYDROXYDIETHYL ETHER ◇ GLYCOL BIS(HYDROXYETHYL) ETHER ◇ TEG ◇ TRIGEN ◇ TRIGLYCOL

TOXICITY DATA with REFERENCE
skn-rbt 500 mg/24H MOD FCTXAV 17,913,79
orl-mus TDLo:90160 mg/kg (7-14D preg):REP EVHPAZ 57,141,84
orl-hmn LDLo:5000 mg/kg FCTXAV 17,913,79

orl-rat LD50:17 g/kg JIHTAB 28,40,46
ivn-rat LD50:11700 μg/kg ARZNAD 18,1536,68
ims-rat LDLo:8400 mg/kg HBTXAC 5,172,59
orl-mus LDLo:18500 mg/kg PCOC** -,1186,66
ipr-mus LD50:8141 mg/kg FEPRA7 6,342,47
scu-mus LD50:8750 mg/kg JPETAB 65,89,39
ivn-mus LD50:6500 mg/kg JPETAB 65,89,39
orl-rbt LD50:8400 mg/kg JIHTAB 28,40,46
ivn-rbt LD50:1900 μg/kg ARZNAD 18,1536,68
orl-gpg LD50:7900 mg/kg JIHTAB 28,40,46
ivn-gpg LD50:10600 μg/kg ARZNAD 18,1536,68

CONSENSUS REPORTS: Glycol ether compounds are on the Community Right-To-Know List. Reported in EPA TSCA Inventory.

SAFETY PROFILE: Poison by intravenous route. Mildly toxic to humans by ingestion. Experimental reproductive effects. A skin irritant. Many glycol ether compounds have dangerous human reproductive effects. Combustible when exposed to heat or flame. Can react with oxidizing materials. Explosive in the form of vapor when exposed to heat, flame or spark. To fight fire, use alcohol foam, dry chemical. When heated to decomposition it emits acrid smoke and irritating fumes. See also ESTERS and GLYCOL ETHERS.

TJQ100 CAS:1680-21-3 ***HR: 2***
TRIETHYLENE GLYCOL DIACRYLATE
mf: $C_{12}H_{18}O_6$ mw: 258.30

SYNS: ACRYLIC ACID, DIESTER with TRIETHYLENE GLYCOL ◇ 2-PROPENOIC ACID, 1,2-ETHANEDIYLBIS(OXY-2,1-ETHANEDIYL) ESTER (9CI)

TOXICITY DATA with REFERENCE
skn-rbt 500 mg/2H4 SEV EPASR* 8EHQ-0981-0410
eye-rbt 100 mg SEV EPASR* 8EHQ-0981-0410
skn-mus TDLo:16 g/kg/80W-I:ETA JTEHD6 19,149,86
orl-rat LD50:500 mg/kg 85GMAT-,115,82
orl-mus LD50:700 mg/kg 85GMAT-,115,82
skn-rbt LD50:1900 mg/kg EPASR* 8EHQ-0981-0410

CONSENSUS REPORTS: Reported in EPA TSCA Inventory.

SAFETY PROFILE: Moderately toxic by ingestion and skin contact. Questionable carcinogen with experimental tumorigenic data. Severe skin and eye irritant. When heated to decomposition it emits acrid smoke and irritating fumes.

TJQ250 CAS:95-08-9 ***HR: 2***
TRIETHYLENE GLYCOL DI(2-ETHYL-BUTYRATE)
mf: $C_{18}H_{34}O_6$ mw: 346.52

PROP: Colorless liquid. Mp: −65°, bp: 197° @ 5 mm, flash p: 385°F (OC), d: 0.9945 @ 20/20°, vap press: 5.8 mm @ 200°, vap d: 11.95.

SYNS: 2,2'-(ETHYLENEDIOXY)DI(ETHYL2-ETHYLBUTYRATE) ◇ TRIETHYLENEGLYCOL DIETHYL BUTYRATE ◇ TRIGLYCOL DICAPROATE ◇ TRIGLYCOL DIHEXOATE

TOXICITY DATA with REFERENCE
orl-rat LD50:6000 mg/kg NPIRI* 2,102,75
orl-gpg LD50:3130 mg/kg JIHTAB 23,259,41

CONSENSUS REPORTS: Reported in EPA TSCA Inventory.

SAFETY PROFILE: Moderately toxic by ingestion. Combustible when exposed to heat or flame. Can react with oxidizing materials. To fight fire, use foam, CO_2, dry chemical. When heated to decomposition it emits acrid smoke and irritating fumes. See also ESTERS.

TJQ333 CAS:1954-28-5 ***HR: 3***
TRIETHYLENE GLYCOL DIGLYCIDYL ETHER
mf: $C_{12}H_{22}O_6$ mw: 262.34

SYNS: AYERST 62013 ◇ 2,2'-(2,5,8,11-TETRAOXA-1,12-DODECANEDIYL)BISOXIRANE ◇ 1,2-BIS(2,3-EPOXYPROPOXY)ETHOXY)ETHANE ◇ 1,2,15,16-DIEPOXY-4,7,10,13-TETRAOXAHEXADECANE ◇ DIGLYCIDYLTRIETHYLENE GLYCOL ◇ EPODYL ◇ ETHOGLUCID ◇ ETHOGLUCIDE ◇ ETOGLUCID ◇ ICI-32865 ◇ OXIRANE, 2,2'-(2,5,8,11-TETRAOXADODECANE-1,12-DIYL)BIS- (9CI) ◇ 2,2'-(2,5,8,11-TETRAOXA-1,2-DODECANEDIYL)BISOXIRANE ◇ TDE

TOXICITY DATA with REFERENCE
skn-rbt 10 mg/24H open MLD AIHAAP 23,95,62
mmo-sat 10 μg/plate MUREAV 111,99,83
ipr-mus TFLo:3800 mg/kg/4W:NEO JNCIAM 36,915,66
ipr-mus LDLo:700 mg/kg IARC** 11,209,76

CONSENSUS REPORTS: IARC Cancer Review: Group 3 IMEMDT 7,56,87; Animal Limited Evidence IMEMDT 11,209,76.

SAFETY PROFILE: Moderately toxic by intraperitoneal route. Questionable carcinogen with experimental neoplastigenic data. A skin irritant. When heated to decomposition it emits acrid smoke and irritating fumes.

TJQ500 CAS:111-22-8 ***HR: 2***
TRIETHYLENE GLYCOL, DINITRATE
mf: $C_6H_{12}O_4{\cdot}N_2O_4$ mw: 240.20

TOXICITY DATA with REFERENCE
ipr-gpg LD50:700 mg/kg:CAR AIHAAP 34,526,73
orl-rat LD50:1000 mg/kg AIHAAP 34,526,73
ipr-rat LD50:796 mg/kg AIHAAP 34,526,73
scu-rat LD50:2520 mg/kg AIHAAP 34,526,73
ipr-mus LD50:945 mg/kg AIHAAP 34,526,73

CONSENSUS REPORTS: Reported in EPA TSCA Inventory.

SAFETY PROFILE: Moderately toxic by ingestion, intraperitoneal, and subcutaneous routes. Questionable carcinogen with experimental carcinogenic data. When heated to decomposition it emits toxic fumes of NO_x. See also NITRATES.

TJQ750 CAS:112-35-6 ***HR: 1***
TRIETHYLENE GLYCOLMONOMETHYL ETHER
mf: $C_7H_{16}O_4$ mw: 164.23

PROP: Misc with water. D: 1.0494, bp: 249°, fp: −44°, flash p: 245°F (OC).

SYNS: DOWANOL TMAT ◇ 2-(2-(2-METHOXYETHOXY)ETHOXY)ETHANOL ◇ METHOXYTRIGLYCOL ◇ POLY-SOLV TM ◇ TRIGLYCOL MONOMETHYL ETHER

TOXICITY DATA with REFERENCE
skn-rbt 10 mg/24H open MLD AIHAAP 23,95,62
orl-rat LD50:11300 mg/kg AIHAAP 23,95,62
skn-rbt LD50:7100 mg/kg AIHAAP 23,95,62

CONSENSUS REPORTS: Glycol ether compounds are on the Community Right-To-Know List. Reported in EPA TSCA Inventory.

SAFETY PROFILE: Mildly toxic by ingestion and skin contact. A skin irritant. Many glycol ether compounds have dangerous human reproductive effects. Combustible when exposed to heat or flame. To fight fire, use alcohol foam, dry chemical. When heated to decomposition it emits acrid smoke and irritating fumes. See also GLYCOL ETHERS.

TJR000 CAS:112-24-3 ***HR: 3***
TRIETHYLENETETRAMINE
DOT: UN 2259
mf: $C_6H_{18}N_4$ mw: 146.28
$(H_2NC_2H_4NHCH_2-)_2$

PROP: Moderately viscous, yellowish liquid. Bp: 278°, mp: 12°; flash p: 275°F, d: 0.982, vap press: <0.01 mm @ 20°, autoign temp: 640°F. Very sol in water and ether.

SYNS: ARALDITE HARDENER HY 951 ◇ ARALDITE HY 951 ◇ N,N'-BIS(2-AMINOETHYL)-1,2-DIAMINOETHANE ◇ N,N'-BIS(2-AMINOETHYL)ETHYLENEDIAMINE ◇ N,N'-BIS(2-AMINOETHYL)-1,2-ETHYLENEDIAMINE ◇ DEH 24 ◇ 3,6-DIAZAOCTANE-1,8-DIAMINE ◇ HY 951 ◇ TECZA ◇ TETA ◇ 1,4,7,10-TETRAAZADECANE ◇ TRIEN ◇ TRIENTINE

TOXICITY DATA with REFERENCE
skn-rbt 490 mg open SEV UCDS** 12/12/66
eye-rbt 49 mg SEV UCDS** 12/12/66
mmo-sat 1 nmol/plate MUREAV 88,165,81
mma-sat 100 μg/plate ENMUDM 8(Suppl 7),1,86
skn-gpg TDLo:3667 mg/kg (female 10-56D post):TER AITEAT 22,123,74
orl-rat TDLo:9130 mg/kg (1-22D preg):REP LANCAO 1,1127,82
orl-rat LD50:2500 mg/kg 37ASAA 7,580,79
orl-mus LD50:1600 mg/kg KHZDAN 22,179,79
ivn-mus LD50:350 mg/kg EJMCA5 19,425,84
orl-rbt LD50:5500 mg/kg KHZDAN 22,179,79
skn-rbt LD50:805 mg/kg JIHTAB 31,60,49

CONSENSUS REPORTS: Reported in EPA TSCA Inventory.

DOT Classification: Corrosive Material; Label: Corrosive.

SAFETY PROFILE: Poison by intravenous route. Moderately toxic by ingestion and skin contact. An experimental teratogen. Experimental reproductive effects. Mutation data reported. A corrosive irritant to skin, eyes, and mucous membranes. Causes skin sensitization. Combustible when exposed to heat or flame. Ignites on contact with cellulose nitrate of high surface area. Can react with oxidizing materials. To fight fire, use CO_2, dry chemical, alcohol foam. When heated to decomposition it emits toxic fumes of NO_x.

TJR250 CAS:1115-99-7 ***HR: 3***
TRIETHYLGALLIUM
mf: $C_6H_{15}Ga$ mw: 156.91

PROP: D: 1.0576, mp: -82.3°, bp: 142.6.

SAFETY PROFILE: Ignites spontaneously in air. Explodes in water or nitric acid. See also GALLIUM COMPOUNDS.

TJR500 CAS:2467-13-2 ***HR: 2***
TRI(2-ETHYLHEXYL) BORATE
mf: $C_{24}H_{51}BO_3$ mw: 398.56

PROP: Colorless, mobile liquid; odor of 2-ethylhexanol. Bp: 350°-354°, flash p: 350°F (COC), d: 0.857 @ 23.6°, vap d: 13.8.

SYN: BORIC ACID, TRIS(2-ETHYLHEXYL) ESTER

TOXICITY DATA with REFERENCE
eye-rbt 100 mg MLD 14KTAK -,706,64
orl-rat LD50:2830 mg/kg USBCC* -,-,58
orl-mus LD50:3300 mg/kg 14KTAK -,693,64

SAFETY PROFILE: Moderately toxic by ingestion. An eye irritant. Combustible when exposed to heat or flame. Can react with oxidizing materials. To fight fire, use foam, CO_2, dry chemical. When heated to decomposition it emits acrid smoke and irritating fumes. See also ESTERS and BORON COMPOUNDS.

TJR750 CAS:923-34-2 ***HR: 3***
TRIETHYL INDIUM
mf: $C_6H_{15}In$ mw: 202

PROP: D: 1.260, refr index: 1.538, mp: -32°, bp: 144°.

SAFETY PROFILE: Spontaneously flammable in air. See also INDIUM.

TJS000 ***HR: 3***
TRIETHYL LEAD
mf: $Pb_2(C_2H_5)_6$ mw: 588.8

PROP: A liquid. D: 1.471, bp: decomp. Insol in water

SYN: HEXAETHYLDILEAD

CONSENSUS REPORTS: Lead and its compounds are on the Community Right-To-Know List.

SAFETY PROFILE: Poison by parenteral route. When heated to decomposition it emits toxic fumes of Pb. See also LEAD COMPOUNDS.

TJS250 CAS:1067-14-7 ***HR: 3***
TRIETHYL LEAD CHLORIDE
mf: $C_6H_{15}ClPb$ mw: 329.85

SYN: TRIETHYLCHLOROPLUMBANE

TOXICITY DATA with REFERENCE
sln-dmg-orl 4 mg/L AMBOCX 1,28,72
ipr-gpg TDLo:2500 μg/kg (female 62D post):TER TXCYAC 27,111,83
orl-mus TDLo:6600 μg/kg (3-5D preg):REP TXAPA9 39,359,77
ipr-rat LD50:11200 μg/kg BJIMAG 16,191,59
scu-rat LD50:11 mg/kg NTOTDY 4,671,82
par-rat LD50:11 mg/kg AOHYA3 3,226,61
ivn-rbt LDLo:7 mg/kg JPETAB 34,85,28

CONSENSUS REPORTS: Lead and its compounds are on the Community Right-To-Know List. Reported in EPA TSCA Inventory. EPA Genetic Toxicology Program.

SAFETY PROFILE: Poison by intraperitoneal, subcutaneous, parenteral, and intravenous routes. An experimental teratogen. Experimental reproductive effects. Mutation data reported. When heated to decomposition it emits very toxic fumes of Cl^- and Pb. See also LEAD COMPOUNDS and CHLORIDES.

TJS500 CAS:562-95-8 ***HR: 3***
TRIETHYL LEAD FLUOROACETATE
mf: $C_2H_2FO_2•C_6H_{15}Pb$ mw: 371.44

SYN: FLUOROACETIC ACID, TRIETHYLLEAD SALT

TOXICITY DATA with REFERENCE
ihl-hmn TCLo:1700 μg/m³/10M:PUL JCSOA9 -,1773,48
scu-mus LD50:15 mg/kg JCSOA9 -,1773,48

CONSENSUS REPORTS: Lead and its compounds are on the Community Right-To-Know List.

SAFETY PROFILE: Poison by subcutaneous route. Human systemic effects by inhalation: pulmonary system effects. When heated to decomposition it emits very toxic fumes of F^- and Pb. See also FLUORIDES and LEAD COMPOUNDS.

TJS750 CAS:73928-18-4 ***HR: 3***
TRIETHYL LEAD FUROATE
mf: $C_{11}H_{18}O_3Pb$ mw: 439.55

SYN: (FUROYLOXY)TRIETHYLPLUMBANE

TOXICITY DATA with REFERENCE
ivn-rat LDLo:15 mg/kg JPETAB 41,1,31
ims-rat LDLo:50 mg/kg JPETAB 41,1,31

CONSENSUS REPORTS: Lead and its compounds are on the Community Right-To-Know List.

SAFETY PROFILE: Poison by intravenous and intramuscular routes. When heated to decomposition it emits toxic fumes of Pb. See also LEAD COMPOUNDS.

TJT000 CAS:63916-98-3 ***HR: 3***
TRIETHYL LEAD OLEATE
mf: $C_{24}H_{48}O_2Pb$ mw: 575.91

TOXICITY DATA with REFERENCE
orl-rat LDLo:50 mg/kg NCNSA6 5,30,53
ipr-rat LDLo:25 mg/kg NCNSA6 5,30,53

CONSENSUS REPORTS: Lead and its compounds are on the Community Right-To-Know List.

SAFETY PROFILE: Poison by ingestion and intraperitoneal routes. When heated to decomposition it emits toxic fumes of Pb. See also LEAD COMPOUNDS.

TJT250 CAS:73928-21-9 ***HR: 3***
TRIETHYL LEAD PHENYL ACETATE
mf: $C_{13}H_{20}O_2Pb$ mw: 415.52

SYN: (PHENYLACETOXY)TRIETHYLPLUMBANE

TOXICITY DATA with REFERENCE
ivn-rat LDLo:15 mg/kg JPETAB 41,1,31
ims-rat LDLo:50 mg/kg JPETAB 41,1,31

CONSENSUS REPORTS: Lead and its compounds are on the Community Right-To-Know List.

SAFETY PROFILE: Poison by intravenous and intramuscular routes. When heated to decomposition it emits toxic fumes of Pb. See also LEAD COMPOUNDS.

TJT500 CAS:56267-87-9 ***HR: 3***
TRIETHYL LEAD PHOSPHATE

TOXICITY DATA with REFERENCE
ipr-rat LDLo:8500 μg(Pb)/kg JPETAB 38,161,30
scu-rat LDLo:8500 μg(Pb)/kg JPETAB 38,161,30

CONSENSUS REPORTS: Lead and its compounds are on the Community Right-To-Know List.

SAFETY PROFILE: Poison by intraperitoneal and subcutaneous routes. When heated to decomposition it emits very toxic fumes of Pb and PO_x. See also LEAD COMPOUNDS and PHOSPHATES.

TJT750 CAS:78-40-0 ***HR: 2***
TRIETHYL PHOSPHATE
mf: $C_6H_{15}O_4P$ mw: 182.18

PROP: Liquid. Mp: −56.5°, flash p: 240°F (OC), d: 1.067-1.072 @ 20/20°, vap press: 1 mm @ 39.6°, vap d: 6.28; bp: 215-216°. Sol in most organic solvents, water, alc, ether.

SYNS: ETHYL PHOSPHATE ◇ TEP

TOXICITY DATA with REFERENCE
mmo-sat 160 mmol/L MUREAV 21,175,73
mmo-klp 5000 ppm MUREAV 16,413,72
mmo-omi 160 mmol/L MUREAV 21,175,73
sln-dmg-orl 10 mmol/L MUREAV 21,175,73
orl-rat TDLo:57 g/kg (92D pre/1-22D preg):REP TXAPA9 12,360,68
orl-rat LDLo:1600 mg/kg 34ZIAG -,605,69
ipr-rat LDLo:800 mg/kg 34ZIAG -,605,69
ivn-rat LDLo:1000 mg/kg NATUAS 179,154,57
orl-mus LD50:1500 mg/kg 85JCAE -,1129,86
ipr-mus LD50:485 mg/kg THERAP 15,237,60
orl-gpg LDLo:1600 mg/kg 34ZIAG -,605,69
ipr-gpg LDLo:800 mg/kg 34ZIAG -,605,69

CONSENSUS REPORTS: Reported in EPA TSCA Inventory.

SAFETY PROFILE: Moderately toxic by ingestion, intraperitoneal, and intravenous routes. Experimental reproductive effects. Mutation data reported. Causes cholinesterase inhibition, but to a lesser extent than parathion. May be expected to cause nerve injury similar to that of other phosphate esters. Combustible when exposed to heat or flame. Can react vigorously with oxidizing materials. To fight fire, use CO_2, dry chemical, alcohol foam. When heated to decomposition it emits toxic fumes of PO_x. See also TRI-2-TOLYL PHOSPHATE and PARATHION.

TJT775 CAS:554-70-1 ***HR: 3***
TRIETHYL PHOSPHINE
mf: $C_6H_{15}P$ mw: 118.16

$P(CH_2CH_3)_3$

PROP: Pyrophoric; stench. Bp: 128-128°, refr index: 1.4560, d: 0.0800, flash p: 1° F.

SAFETY PROFILE: Highly flammable liquid. Reacts with oxygen at low temperatures to form an explosive product. When heated to decomposition it emits toxic fumes of PO_x. See also PHOSPHINE.

TJT780 ***HR: 3***
TRIETHYL PHOSPHINE GOLD NITRATE
mf: $C_6H_{15}AuNO_3P$ mw: 377.13

$(CH_3CH_2)_3P{\bullet}AuNO_3$

SAFETY PROFILE: The material explodes spontaneously when dry. When heated to decomposition it emits toxic fumes of PO_x and NO_x. See also PHOSPHINE, NITRATES, and GOLD.

TJT800 CAS:122-52-1 ***HR: 2***
TRIETHYL PHOSPHITE
DOT: UN 2323
mf: $C_6H_{15}O_3P$ mw: 166.18

PROP: Bp: 156°, refr index: 1.4130, d: 0.969, flash p: 130° F.

SYN: FOSFORYN TROJETYLOWY (CZECH)

TOXICITY DATA with REFERENCE
skn-mam 500 mg MLD MEPAAX 29,393,78
eye-mam 100 mg MLD MEPAAX 29,393,78
orl-rat LD50:3200 mg/kg ALBRW* #OPB-3,84

CONSENSUS REPORTS: Reported in EPA TSCA Inventory.

DOT Classification: Flammable or Combustible Liquid; Label: Flammable Liquid.

SAFETY PROFILE: Moderately toxic by ingestion. A skin and eye irritant. Combustible liquid, flammable when exposed to heat or flame. When heated to decomposition it emits toxic fumes of PO_x.

TJT900 CAS:2404-78-6 ***HR: 3***
O,S,S-TRIETHYL PHOSPHORODITHIOATE
mf: $C_6H_{15}O_2PS_2$ mw: 214.30

TOXICITY DATA with REFERENCE
orl-rat LD50:140 mg/kg DTESD7 8,631,80
ivn-rat LD50:95 mg/kg DTESD7 8,631,80
orl-mus LD50:126 mg/kg DTESD7 8,631,80
ivn-mus LD50:83 mg/kg DTESD7 8,631,80

SAFETY PROFILE: Poison by ingestion and intravenous routes. When heated to decomposition it emits toxic fumes of PO_x and SO_x.

TJU000 CAS:126-68-1 ***HR: 3***
TRIETHYL PHOSPHOROTHIOATE
mf: $C_6H_{15}O_3PS$ mw: 198.24

PROP: Colorless liquid; strong characteristic odor. Bp:

93.2-94° @ 10 mm, flash p: 225°F (COC), d: 1.074 @ 20/4°.

SYNS: O,O,O-TRIETHYLESTER KYSELINY THIOFOSFORECNE (CZECH) ◇ O,O,O-TRIETHYL PHOSPHOROTHIOATE ◇ TRIETHYLTHIOFOSFAT (CZECH)

TOXICITY DATA with REFERENCE
ihl-rat LCLo:41 ppm/4H 28ZPAK -,207,72
ivn-rat LDLo:250 mg/kg NATUAS 179,154,57

SAFETY PROFILE: Poison by inhalation and intravenous routes. A cholinesterase inhibitor. Combustible when exposed to heat or flame; can react with oxidizing materials. When heated to decomposition it emits very toxic fumes of PO_x and SO_x. See also PARATHION.

TJU150 CAS:2587-81-7 ***HR: 3***
TRIETHYLPLUMBYL ACETATE
mf: $C_8H_{18}O_2Pb$ mw: 353.45

SYN: ACETOXYTRIETHYLPLUMBANE

TOXICITY DATA with REFERENCE
orl-rat LD50:57 mg/kg APFRAD 24,17,66
ipr-rat LD50:30 mg/kg APFRAD 24,17,66
orl-mus LD50:25 mg/kg APFRAD 24,17,66
ipr-mus LD50:15 mg/kg APFRAD 24,17,66
ivn-mus LD50:14 mg/kg CSLNX* NX#04715

CONSENSUS REPORTS: Lead and its compounds are on the Community Right-To-Know List.

SAFETY PROFILE: Poison by ingestion, intravenous, and intraperitoneal routes. When heated to decomposition it emits toxic fumes of Pb. See also LEAD COMPOUNDS

TJU250 CAS:994-43-4 ***HR: 2***
TRIETHYLPROPYL GERMANE
mf: $C_9H_{22}Ge$ mw: 202.90

TOXICITY DATA with REFERENCE
orl-rat LDLo:4700 mg/kg CHDDAT 262,1302,66
ipr-rat LDLo:1430 mg/kg CHDDAT 262,1302,66

SAFETY PROFILE: Moderately toxic by intraperitoneal route. Mildly toxic by ingestion. When heated to decomposition it emits acrid smoke and irritating fumes. See also GERMANIUM COMPOUNDS.

TJU500 CAS:18244-91-2 ***HR: 3***
TRIETHYLSILYL PERCHLORATE
mf: $C_6H_{15}ClO_4Si$ mw: 214.72

$(CH_3CH_2)_3SiOClO_3$

SAFETY PROFILE: Explodes when heated. When heated to decomposition it emits toxic fumes of Cl^-. See also PERCHLORATES.

TJU600 CAS:12328-03-9 ***HR: 3***
TRIETHYLSULFONIUM IODIDE BIS(MERCURIC IODIDE) addition compound

SYN: TRIETHYL-SULFONIUM, IODIDE, compound with MERCURY IODIDE (1:2)

TOXICITY DATA with REFERENCE
ivn-mus LD50:18 mg/kg CSLNX* NX#01854

CONSENSUS REPORTS: Mercury and its compounds are on the Community Right-To-Know List.

NIOSH REL: TWA 0.05 mg(Hg)/m^3

SAFETY PROFILE: Poison by intravenous route. When heated to decomposition it emits very toxic fumes of SO_x, I^-, and Hg. See also IODIDES and MERCURY COMPOUNDS.

TJU750 CAS:19493-75-5 ***HR: 3***
TRIETHYLSULFONIUM IODIDE MERCURIC IODIDE addition compound

SYN: TRIETHYLSULFONIUM, IODIDE compounded with MERCURY IODIDE (1:1)

TOXICITY DATA with REFERENCE
ivn-mus LD50:18 mg/kg CSLNX* NX#01851

CONSENSUS REPORTS: Mercury and its compounds are on the Community Right-To-Know List.

NIOSH REL: TWA 0.05 mg(Hg)/m^3

SAFETY PROFILE: Poison by intravenous route. When heated to decomposition it emits very toxic fumes of SO_x, I^-, and Hg. See also MERCURY COMPOUNDS and IODIDES.

TJU800 CAS:1186-09-0 ***HR: 3***
O,O,S-TRIETHYL THIOPHOSPHATE
mf: $C_6H_{15}O_3PS$ mw: 198.24

SYNS: O,O-DIETHYL-S-ETHYL PHOSPHOROTHIOATE ◇ O,O,S-TRIETHYL PHOSPHOROTHIOATE

TOXICITY DATA with REFERENCE
orl-rat LD50:27 mg/kg DTESD7 8,631,80
ipr-rat LD50:27 mg/kg FAATDF 4(2,Pt 2),S215,84
ivn-rat LD50:27 mg/kg FAATDF 4(2,Pt 2),S215,84
orl-mus LD50:170 mg/kg DTESD7 8,631,80
ivn-mus LD50:90 mg/kg DTESD7 8,631,80

SAFETY PROFILE: Poison by ingestion, intravenous, and intraperitoneal routes. When heated to decomposition it emits toxic fumes of PO_x and SO_x.

TJU850 CAS:73926-90-6 ***HR: 3***
TRIETHYLTIN BROMIDE-2-PIPECOLINE
mf: $C_6H_{15}BrSn\cdot C_6H_{13}N$ mw: 385.01

SYNS: BROMOTRIETHYLSTANNANE compd. with 2-PIPECOLINE

(1:1) ◇ STANNANE, BROMOTRIETHYL-, compd. with 2-PIPECOLINE (1:1)

TOXICITY DATA with REFERENCE
ivn-mus LD50:56 mg/kg CSLNX* NX#05989

OSHA PEL: TWA 0.1 mg(Sn)/m^3 (skin)
ACGIH TLV: TWA 0.1 mg(Sn)/m^3; STEL 0.2 mg/m^3 (skin)
NIOSH REL: (Organotin Compound):10H TWA 0.1 mg(Sn)/m^3

SAFETY PROFILE: Poison by intravenous route. When heated to decomposition it emits toxic fumes of NO_x, Sn, and Br^-.

TJV000 CAS:994-31-0 ***HR: 3***
TRIETHYLTIN CHLORIDE
mf: $C_6H_{15}ClSn$ mw: 241.35

PROP: Colorless liquid. D: 1.428 @ 8°, mp: 10°, bp: 210°. Insol in water; sol in organic solvents.

SYNS: CHLOROTRIETHYLSTANNANE ◇ CHLOROTRIETHYLTIN ◇ TRIETHYLCHLOROSTANNANE ◇ TRIETHYLCHLOROTIN ◇ TRIETHYLSTANNYL CHLORIDE

TOXICITY DATA with REFERENCE
ipr-rat LD50:5160 μg/kg FCTXAV 7,47,69

OSHA PEL: TWA 0.1 mg(Sn)/m^3 (skin)
ACGIH TLV: TWA 0.1 mg(Sn)/m^3 (skin) (Proposed: TWA 0.1 mg(Sn)/m^3; STEL 0.2 mg(Sn)/m^3 (skin))
NIOSH REL: (Organotin Compounds) TWA 0.1 mg (Sn)/m^3

SAFETY PROFILE: Poison by intraperitoneal route. When heated to decomposition it emits toxic fumes of Cl^-. See also TIN COMPOUNDS and CHLORIDES.

TJV100 ***HR: 3***
TRIETHYLTIN HYDROPEROXIDE
mf: $C_6H_{16}O_2Sn$ mw: 238.90

$(CH_3CH_2)_3SnOOH$

SAFETY PROFILE: Forms a highly explosive addition compound with hydrogen peroxide. See also PEROXIDES and TIN COMPOUNDS.

TJV250 CAS:1529-30-2 ***HR: 3***
TRIETHYLTIN PHENOXIDE
mf: $C_{12}H_{20}OSn$ mw: 299.01

SYN: PHENOXYTRIETHYLSTANNANE

TOXICITY DATA with REFERENCE
ivn-mus LD50:7100 μg/kg CSLNX* NX#02823

OSHA PEL: TWA 0.1 mg(Sn)/m^3 (skin)
ACGIH TLV: TWA 0.1 mg(Sn)/m^3 (skin) (Proposed: TWA 0.1 mg(Sn)/m^3; STEL 0.2 mg(Sn)/m^3 (skin))
NIOSH REL: (Organotin Compounds) TWA 0.1 mg (Sn)/m^3

SAFETY PROFILE: Poison by intravenous route. When heated to decomposition it emits acrid smoke and irritating fumes. See also TIN COMPOUNDS.

TJV500 CAS:123-12-6 ***HR: 3***
3,6,9-TRIETHYL-3,6,9-TRIAZAUNDECANE
mf: $C_{14}H_{33}N_3$ mw: 243.50

TOXICITY DATA with REFERENCE
ivn-mus LD50:180 mg/kg CSLNX* NX#01684

CONSENSUS REPORTS: Reported in EPA TSCA inventory.

SAFETY PROFILE: Poison by intravenous route. When heated to decomposition it emits toxic fumes of NO_x.

TJV775 ***HR: 3***
TRIETHYNYL ALUMINUM
mf: C_6H_3Al mw: 92.07

$(HC{\equiv}C)_3Al$

SAFETY PROFILE: Forms explosive complexes with dioxane; trimethylamine; diethyl ether. See also ACETYLIDES and ALUMINUM COMPOUNDS.

TJV785 CAS:687-81-0 ***HR: 3***
TRIETHYNYL ANTIMONY
mf: C_6H_3Sb mw: 196.85

$(HC{\equiv}C)_3Sb$

CONSENSUS REPORTS: Antimony and its compounds are on The Community Right-To-Know List.

SAFETY PROFILE: Many antimony compounds are poisons. A friction-sensitive explosive. When heated to decomposition it emits toxic fumes of Sb. See also ANTIMONY COMPOUNDS and ACETYLENE COMPOUNDS.

TJW000 CAS:687-78-5 ***HR: 3***
TRIETHYNYLARSINE
mf: C_6H_3As mw: 150.01

$(HC{\equiv}C)_3As$

CONSENSUS REPORTS: Arsenic and its compounds are on the Community Right-To-Know List.

SAFETY PROFILE: Arsenic compounds are poisons. An unstable explosive sensitive to strong friction. When heated to decomposition it emits toxic fumes of As. See also ARSENIC COMPOUNDS and ACETYLENE COMPOUNDS.

TJW250 CAS:687-80-9 ***HR: 3***
TRIETHYNYLPHOSPHINE
mf: C_6H_3P mw: 106.06

$(HC{\equiv}C)_3P$

SAFETY PROFILE: A friction-sensitive explosive. A storage hazard, it decomposes at room temperature to form a spontaneously explosive product. When heated to decomposition it emits toxic fumes of PO_x. See also PHOSPHINE and ACETYLENE COMPOUNDS.

TJW500 CAS:69-23-8 ***HR: 3***
TRIFLUMETHAZINE
mf: $C_{22}H_{26}F_3N_3OS$ mw: 437.57

SYNS: ANATENSOL ◇ FLUPHENAZINE ◇ 10-(3-(2-HYDROXYETHYL)PIPERAZINOPROPYL)-2-(TRIFLUOROMETHYL)PHENOTHIAZINE ◇ 10-(3-(4-(2-HYDROXYETHYL)-1-PIPERAZINYL)PROPYL)-2-(TRIFLUOROMETHYL)PHENOTHIAZINE ◇ MODITEN ◇ OMCA ◇ PACINOL ◇ PERMITIL ◇ PROLIXINE ◇ SEVINOL ◇ 4-(3-(2-TRIFLUOROMETHYL-10-PHENOTHIAZYL)-PROPYL)-1-PIPERAZINEETHANOL ◇ YESPAZINE

TOXICITY DATA with REFERENCE
orl-rat TDLo:62 mg/kg (9D pre/1-22D preg):REP APEPA2 257,338,67
ipr-rat LD50:100 mg/kg TXAPA9 2,540,60
scu-rat LD50:640 mg/kg MDCHAG 4(2),199,67
orl-mus LD50:220 mg/kg TXAPA9 2,540,60
ipr-mus LD50:89 mg/kg TXAPA9 2,540,60
ivn-mus LD50:51 mg/kg TXAPA9 2,540,60

SAFETY PROFILE: Poison by ingestion, intraperitoneal, and intravenous routes. Moderately toxic by subcutaneous route. Experimental reproductive effects. When heated to decomposition it emits very toxic fumes of F^-, NO_x, and SO_x. See also FLUORIDES.

TJW600 CAS:605-75-4 ***HR: 3***
TRIFLUOPERAZINE DIMALEATE
mf: $C_{21}H_{24}F_3N_3S{\bullet}2C_4H_4O_4$ mw: 639.70

TOXICITY DATA with REFERENCE
orl-mus LD50:1150 mg/kg NIIRDN 6,524,82
ivn-mus LD50:30 mg/kg NIIRDN 6,524,82
ivn-dog LD50:50 mg/kg NIIRDN 6,524,82

SAFETY PROFILE: Poison by intravenous route. Moderately toxic by ingestion. When heated to decomposition it emits toxic fumes of F^-, SO_x, and NO_x.

TJX000 CAS:407-25-0 ***HR: 2***
TRIFLUOROACETIC ACID ANHYDRIDE
mf: $C_4F_6O_3$ mw: 210.04

$F_3CCO{\bullet}OCO{\bullet}CF_3$

SYNS: ANHYDRID KYSELINY TRIFLUOROCTOVE (CZECH) ◇ BIS(TRIFLUOROACETIC) ANHYDRIDE ◇ HEXAFLUOROACETIC ANHYDRIDE ◇ PERFLUOROACETIC ANHYDRIDE ◇ TRIFLUOROACETIC ANHYDRIDE ◇ TRIFLUOROACETYL ANHYDRIDE

TOXICITY DATA with REFERENCE
skn-rbt 500 mg/24H SEV 28ZPAK -,91,72
eye-rbt 5 mg/24H SEV 28ZPAK -,91,72

CONSENSUS REPORTS: Reported in EPA TSCA Inventory.

SAFETY PROFILE: A severe skin and eye irritant. Explosive reaction with dimethyl sulfoxide; nitric acid + 1,3,5-triazine (at 36°C); nitric acid + 1,3,5-triacetylhexahydro-1,3,5-triazine (at 30°C). Incompatible with lithium tetrahydroaluminate. When heated to decomposition it emits toxic fumes of F^-. See also FLUORIDES and ANHYDRIDES.

TJX250 CAS:429-30-1 ***HR: 3***
TRIFLUOROACETIC ACID TRIETHYLSTANNYL ESTER
mf: $C_8H_{15}F_3O_2Sn$ mw: 318.92

SYN: TRIETHYL(TRIFLUOROACETOXY)STANNANE

TOXICITY DATA with REFERENCE
ivn-mus LD50:11200 μg/kg CSLNX* NX#05827

OSHA PEL: TWA 0.1 mg(Sn)/m³ (skin)
ACGIH TLV: TWA 0.1 mg(Sn)/m³ (skin) (Proposed: TWA 0.1 mg(Sn)/m³; STEL 0.2 mg(Sn)/m³ (skin))
NIOSH REL: (Organotin Compounds) TWA 0.1 mg (Sn)/m³

SAFETY PROFILE: Poison by intravenous route. When heated to decomposition it emits toxic fumes of F^-. See also TIN COMPOUNDS, FLUORIDES, and ESTERS.

TJX350 CAS:56124-62-0 ***HR: 3***
TRIFLUOROACETYLADRIAMYCIN-14-VALERATE
mf: $C_{34}H_{36}F_3NO_{13}$ mw: 723.71

SYNS: AD 32 ◇ ANTIBIOTIC AD 32 ◇ NSC-246131 ◇ N-TRIFLUOROACETYLADRIAMYCIN-14-VALERATE

TOXICITY DATA with REFERENCE
dnd-hmn:lym 3 mg/L CJBIAE 58,720,80
oms-hmn:lym 1 mg/L CNREA8 41,2745,81
ipr-mus LD50:109 mg/kg NCISnP* JAN86

SAFETY PROFILE: Poison by intraperitoneal route. Human mutation data reported. When heated to decomposition it emits toxic fumes of F^- and NO_x.

TJX375 CAS:23292-52-6 ***HR: 3***
TRIFLUOROACETYL AZIDE
mf: $C_2F_3N_3O$ mw: 139.04

SAFETY PROFILE: An explosive sensitive to mechanical or thermal shock. When heated to decomposition it emits toxic fumes of F^- and NO_x. See also AZIDES and FLUORIDES.

TJX500 CAS:354-32-5 ***HR: 1***
TRIFLUOROACETYL CHLORIDE

SYN: PERFLUOROACETYL CHLORIDE

TOXICITY DATA with REFERENCE
ihl-rat LCLo:5000 mg/m^3 NDRC** -,7,43

CONSENSUS REPORTS: Reported in EPA TSCA Inventory.

SAFETY PROFILE: Mildly toxic by inhalation. When heated to decomposition it emits very toxic fumes of F^- and Cl^-. See also FLUORIDES and CHLORIDES.

TJX600 CAS:87050-94-0 ***HR: 3***
2-TRIFLUOROACETYL-1,3,4-DIOXAZALONE
mf: $C_3F_3NO_3$ mw: 155.03

SAFETY PROFILE: Explodes when heated. It is not sensitive to mechanical shock. When heated to decomposition it emits toxic fumes of F^- and NO_x. See also FLUORIDES.

TJX625 CAS:27961-70-2 ***HR: 3***
O-TRIFLUOROACETYL-S-FLUOROFORMYL THIOPEROXIDE
mf: $C_3F_4O_3S$ mw: 192.08

SAFETY PROFILE: May explode spontaneously. When heated to decomposition it emits toxic fumes of F^- and SO_x. See also PEROXIDES and FLUORIDES.

TJX650 CAS:65597-25-3 ***HR: 3***
TRIFLUOROACETYL HYPOCHLORITE
mf: $C_2ClF_3O_2$ mw: 148.47

SAFETY PROFILE: Thermally unstable. The gas is explosive. When heated to decomposition it emits toxic fumes of F^- and Cl^-. See also HYPOCHLORITES and FLUORIDES.

TJX750 CAS:359-46-6 ***HR: 3***
TRIFLUOROACETYL HYPOFLUORITE
mf: $C_2F_4O_2$ mw: 132.01

SAFETY PROFILE: An explosive sensitive to sparks or contact with aqueous potassium iodide. Incompatible with water. When heated to decomposition it emits toxic fumes of F^-. See also FLUORIDES and HYPOCHLORITES.

TJX775 ***HR: 3***
TRIFLUOROACETYLIMINOIODOBENZENE
mf: $C_8H_5F_3INO$ mw: 315.03

SAFETY PROFILE: Explodes when heated to 100°C. When heated to decomposition it emits toxic fumes of F^-, I^-, and NO_x. See also FLUORIDES and IODIDES.

TJX780 CAS:667-29-8 ***HR: 3***
TRIFLUOROACETYL NITRITE
mf: $C_2F_3NO_3$ mw: 143.02

SAFETY PROFILE: The vapor explodes at 160-200°C. When heated to decomposition it emits toxic fumes of F^- and NO_x. See also NITRITES and FLUORIDES.

TJX800 CAS:68602-57-3 ***HR: 3***
TRIFLUOROACETYL TRIFLUOROMETHANE SULFONATE
mf: $C_3F_6O_4S$ mw: 138.14

SAFETY PROFILE: Violent reaction with water. When heated to decomposition it emits toxic fumes of F^- and SO_x. See also FLUORIDES.

TJX825 CAS:667-49-2 ***HR: 3***
TRIFLUOROACRYLOYL FLUORIDE
mf: C_3F_4O mw: 128.03

SAFETY PROFILE: Reaction with sodium azide forms a highly explosive solid. When heated to decomposition it emits toxic fumes of F^-. See also FLUORIDES.

TJX900 CAS:3862-73-5 ***HR: 2***
2,3,4-TRIFLUOROANILINE
mf: $C_6H_4F_3N$ mw: 147.11

PROP: Bp: 92°/48 mm Hg, d: 1.395, flash p: 155° F.

SYNS: ANILINE, 2,3,4-TRIFLUORO- ◇ BENZENAMINE, 2,3,4-TRIFLUORO- ◇ TFA ◇ 2,3,4-TFA

TOXICITY DATA with REFERENCE
skn-rbt 500 mg MOD JACTDZ 1,61,90
eye-rbt 10 mg/30S RNS MOD JACTDZ 1,61,90
orl-rat LD50:699 mg/kg JACTDZ 1,61,90

SAFETY PROFILE: Moderately toxic by ingestion. A skin and eye irritant. Combustible liquid. When heated to decomposition it emits toxic fumes of NO_x and F^-.

TJY000 CAS:354-06-3 ***HR: 1***
1,1,2-TRIFLUORO-1-BROMO-2-CHLOROETHANE
mf: $C_2HBrClF_3$ mw: 197.39

TOXICITY DATA with REFERENCE
ihl-mus LCLo:35000 ppm/17M ANASAB 17,337,62

CONSENSUS REPORTS: Reported in EPA TSCA Inventory.

SAFETY PROFILE: Mildly toxic by inhalation. When heated to decomposition it emits very toxic fumes of F^-, Cl^-, and Br^-.

TJY100 CAS:75-63-8 ***HR: 1***
TRIFLUOROBROMOMETHANE
DOT: UN 1009
mf: $CBrF_3$ mw: 148.92

SYNS: BROMOFLUOROFORM ◇ BROMOTRIFLUOROMETHANE ◇ F-13B1 ◇ FREON 13B1 ◇ HALON 1301 ◇ TRIFLUOROMONOBROMOMETHANE

TOXICITY DATA with REFERENCE
ihl-rat LC50:416 g/m^3 GTPZAB 26(8),53,82
ihl-mus LC50:381 g/m^3 GTPZAB 26(8),53,82

CONSENSUS REPORTS: Reported in EPA TSCA Inventory.

OSHA PEL: TWA 1000 ppm
ACGIH TLV: TWA 1000 ppm
DFG MAK: 1000 ppm (6100 mg/m^3)
DOT Classification: Nonflammable Gas; Label: Nonflammable Gas.

SAFETY PROFILE: Mildly toxic by inhalation. Incompatible with aluminum. When heated to decomposition it emits toxic fumes of F^- and Br^-. See also BROMIDES and FLUORIDES.

TJY175 CAS:75-88-7 ***HR: 3***
2,2,2-TRIFLUOROCHLOROETHANE
mf: $C_2H_2ClF_3$ mw: 118.49

SYNS: CFC 133a ◇ 1-CHLORO-2,2,2-TRIFLUOROETHANE ◇ 2-CHLORO-1,1,1-TRIFLUOROETHANE ◇ FC 133a ◇ FREON 133a ◇ GENETRON 133a ◇ R 133a ◇ 1,1,1-TRIFLUORO-2-CHLOROETHANE ◇ 1,1,1-TRIFLUOROETHYL CHLORIDE

TOXICITY DATA with REFERENCE
orl-rat TDLo:78 g/kg/1Y-I:CAR,REP TXAPA9 72,15,84

CONSENSUS REPORTS: IARC Cancer Review: Group 3 IMEMDT 7,56,87; Animal Limited Evidence IMEMDT 41,253,86. Reported in EPA TSCA Inventory.

SAFETY PROFILE: Experimental reproductive effects. Questionable carcinogen with experimental carcinogenic data. When heated to decomposition it emits toxic fumes of F^-.

TJY200 ***HR: 2***
1,1,1-TRIFLUORO-3-CHLOROPROPANE
mf: $C_3H_4ClF_3$ mw: 132.52

SYNS: 3-CHLORO-1,1,1-TRIFLUOROPROPANE ◇ FREON 253

TOXICITY DATA with REFERENCE
ihl-rat LCLo:1800 mg/m^3/2H 85GMAT -,115,82
ihl-mus LC50:800 mg/m^3/2H 85GMAT -,115,82
ihl-rbt LCLo:2300 mg/m^3/2H 85GMAT -,115,82

SAFETY PROFILE: Moderately toxic by inhalation. When heated to decomposition it emits toxic fumes of F^- and Cl^-. See also CHLORINATED HYDROCARBONS, ALIPHATIC; and FLUORIDES.

TJY275 CAS:371-67-5 ***HR: 3***
2,2,2-TRIFLUORODIAZOETHANE
mf: $C_2HF_3N_2$ mw: 110.04

SAFETY PROFILE: An unstable explosive nearly as powerful as TNT. When heated to decomposition it emits toxic fumes of F^- and NO_x. See also FLUORIDES.

TJY500 CAS:306-83-2 ***HR: 3***
1,1,1-TRIFLUORO-2,2-DICHLOROETHANE
mf: $C_2HCl_2F_3$ mw: 152.93

TOXICITY DATA with REFERENCE
ihl-mus LCLo:14 pph/4M ANASAB 17,337,62

CONSENSUS REPORTS: Reported in EPA TSCA Inventory.

SAFETY PROFILE: Poison by inhalation. When heated to decomposition it emits very toxic fumes of F^- and Cl^-. See also CHLORINATED HYDROCARBONS, ALIPHATIC; and FLUORIDES.

TJY750 CAS:354-23-4 ***HR: 3***
1,1,2-TRIFLUORO-1,2-DICHLOROETHANE
mf: $C_2HCl_2F_3$ mw: 152.93

TOXICITY DATA with REFERENCE
ihl-mus LCLo:15 pph/2M ANASAB 16,3,61

CONSENSUS REPORTS: Reported in EPA TSCA Inventory.

SAFETY PROFILE: Poison by inhalation. When heated to decomposition it emits very toxic fumes of Cl^- and F^-. See also CHLORINATED HYDROCARBONS, ALIPHATIC; and FLUORIDES.

TJZ000 CAS:421-53-4 ***HR: 2***
2,2,2-TRIFLUORO-1,1-ETHANEDIOL
mf: $C_2H_3F_3O_2$ mw: 116.05

SYNS: FLUORAL HYDRATE ◇ TRIFLUOROACETALDEHYDE HYDRATE

TOXICITY DATA with REFERENCE
orl-mus LD50:600 mg/kg JMCMAR 13,1212,70

ipr-mus LD50:600 mg/kg JMCMAR 13,1212,70
ivn-mus LD50:660 mg/kg AMEBA7 46,242,68

CONSENSUS REPORTS: Reported in EPA TSCA Inventory. EPA Genetic Toxicology Program.

SAFETY PROFILE: Moderately toxic by ingestion, intravenous, and intraperitoneal routes. When heated to decomposition it emits very toxic fumes of F^-. See also ALDEHYDES and FLUORIDES.

TKA250 CAS:76-05-1 ***HR: 3***
TRIFLUOROETHANOIC ACID
DOT: UN 2699
mf: $C_2HF_3O_2$ mw: 114.03

PROP: Colorless liquid; strong pungent odor. Mp: −15.25°, bp: 71.1° @ 734 mm, d: 1.535 @ 0°.

SYNS: PERFLUOROACETIC ACID ◇ TRIFLUORACETIC ACID ◇ TRIFLUOROACETIC ACID (DOT)

TOXICITY DATA with REFERENCE
orl-rat LD50:200 mg/kg 14CYAT 2,1802,63
ihl-rat LC50:10 g/m^3 GTPZAB 10(3),13,66
ihl-mus LC50:13500 mg/m^3 GTPZAB 10(3),13,66
ipr-mus LDLo:150 mg/kg TXAPA9 15,83,69
ivn-mus LD50:1200 mg/kg AMEBA7 46,242,68

CONSENSUS REPORTS: Reported in EPA TSCA Inventory. EPA Genetic Toxicology Program.

DOT Classification: Corrosive Material; Label: Corrosive.

SAFETY PROFILE: Poison by ingestion and intraperitoneal routes. Moderately toxic by intravenous route. Mildly toxic by inhalation. A corrosive irritant to skin, eyes, and mucous membranes. When heated to decomposition it emits toxic fumes of F^-. Used as a strong organic acid catalyst.

TKA350 CAS:75-89-8 ***HR: 3***
2,2,2-TRIFLUOROETHANOL
mf: $C_2H_3F_3O$ mw: 100.05

PROP: Liquid. Bp: 77-80°, d: 1.288, refr index: < 1.3000.

SYN: TFE

TOXICITY DATA with REFERENCE
skn-rbt 500 mg/24H SEV 28ZPAK -,78,72
eye-rbt 20 mg/24H MOD 28ZPAK -,78,72
eye-rbt 100 mg/20S rns SEV NTIS** LMF-84
ihl-rat TCLo:100 ppm/6H (14D male):REP TOXID9 1,27,81
orl-rat LD50:240 mg/kg 34ZIAG -,607,69
skn-rat LD50:1680 mg/kg 34ZIAG -,607,69
ipr-rat LD50:210 mg/kg TXAPA9 71,84,83
orl-mus LD50:366 mg/kg TXAPA9 15,83,69
ihl-mus LC50:2900 mg/m^3 GTPZAB 13(10),29,69
ipr-mus LD50:158 mg/kg AMEBA7 46,242,68
ivn-mus LD50:250 mg/kg AMEBA7 46,242,68
ivn-dog LDLo:400 mg/kg TXAPA9 15,83,69

CONSENSUS REPORTS: Reported in EPA TSCA Inventory. EPA Genetic Toxicology Program.

SAFETY PROFILE: Poison by ingestion, intravenous, and intraperitoneal routes. Moderately toxic by inhalation and skin contact. Experimental reproductive effects. A severe skin and eye irritant. When heated to decomposition it emits toxic fumes of F^-.

TKA500 CAS:753-90-2 ***HR: 3***
2,2,2-TRIFLUOROETHYLAMINE
mf: $C_2H_4F_3N$ mw: 99.07

PROP: Liquid. Bp: 37-38°, refr index: 1.3010, d: 1.245, flash p: 2° F.

TOXICITY DATA with REFERENCE
ihl-mus LCLo:500 mg/m^3 NDRC** -,9,44

CONSENSUS REPORTS: Reported in EPA TSCA Inventory.

SAFETY PROFILE: Corrosive. Moderately toxic by inhalation. Highly flammable liquid. When heated to decomposition it emits very toxic fumes of F^- and NO_x. See also FLUORIDES and AMINES.

TKA750 CAS:373-88-6 ***HR: 2***
TRIFLUOROETHYLAMINE HYDROCHLORIDE
mf: $C_2H_4F_3N•ClH$ mw: 135.53

PROP: Hygroscopic. Mp: 220-222° (subl.)

TOXICITY DATA with REFERENCE
unr-mus LD50:476 mg/kg 11FYAN 3,81,63

CONSENSUS REPORTS: Reported in EPA TSCA Inventory.

SAFETY PROFILE: Moderately toxic by unspecified route. When heated to decomposition it emits very toxic fumes of F^-, NO_x, and HCl. See also FLUORIDES and AMINES.

TKB250 CAS:406-90-6 ***HR: 3***
2,2,2-TRIFLUOROETHYL VINYL ETHER
mf: $C_4H_5F_3O$ mw: 126.09

SYNS: FLOROXENE ◇ FLUOOXENE ◇ FLUOROMAR ◇ FLUOROXENE ◇ FLUORXENE ◇ FLUROXENE ◇ (2,2,2-TRIFLUOROETHOXY)ETHENE

TOXICITY DATA with REFERENCE
mmo-sat 3 pph/2H MUREAV 58,183,78
sln-dmg-ihl 2 pph/1H ANESAV 62,305,85

ihl-mky TCLo:8 pph/20M (22W preg):TER AANEAB 20,183,76
ihl-man TCLo:2 pph/90M:LIV ANESAV 37,462,72
ipr-rat LD50:5600 mg/kg TXAPA9 71,84,83
ihl-mus LCLo:4 pph/1H BJANAD 48,399,76

CONSENSUS REPORTS: Reported in EPA TSCA Inventory. EPA Genetic Toxicology Program.

NIOSH REL: (Waste Anesthetic Gases) CL 2 ppm/1H

SAFETY PROFILE: Poison by inhalation. Moderately toxic by intraperitoneal route. An experimental teratogen. Human systemic effects by inhalation: jaundice and liver function tests impaired. Experimental reproductive effects. Mutation data reported. When heated to decomposition it emits toxic fumes of F^-. Used as an anesthetic. See also FLUORIDES and ETHERS.

TKB275 CAS:31330-22-0 ***HR: 3***
N,N,N'-TRIFLUOROHEXANAMIDINE
mf: $C_6H_{11}F_3N_2$ mw: 168.16

$C_5H_{11}C(:NF)NF_2$

SAFETY PROFILE: A shock-sensitive explosive. When heated to decomposition it emits toxic fumes of F^- and NO_x. See also FLUORIDES.

TKB300 CAS:421-17-0 ***HR: 3***
TRIFLUOROMETHANESULFENYL CHLORIDE
mf: $CClF_3S$ mw: 136.52

SAFETY PROFILE: Explosive reaction with chlorine fluorides. When heated to decomposition it emits toxic fumes of F^-, Cl^-, and SO_x. See also FLUORIDES.

TKB310 CAS:1493-13-6 ***HR: 3***
TRIFLUOROMETHANE SULFONIC ACID
mf: CHF_3O_3S mw: 150.08

PROP: Hygroscopic. Bp: 162°, refr index: 1.3270, d: 1.696, flash p: none.

SYN: TRIFLIC ACID

SAFETY PROFILE: A corrosive irritant to the skin, eyes, and mucous membranes. A strong acid. Violent reaction with acyl chlorides or aromatic hydrocarbons evolves toxic hydrogen chloride gas. When heated to decomposition it emits toxic fumes of F^- and SO_x. See also FLUORIDES.

TKB325 CAS:59034-32-1 ***HR: 3***
1,3,3-TRIFLUORO-2-METHOXYCYCLOPROPENE
mf: $C_4H_3F_3O$ mw: 122.07

$FC{=}C(OCH_3)CF_2$ (ring)

SAFETY PROFILE: Explosive reaction with water or methanol. A preparative hazard. When heated to decomposition it emits toxic fumes of F^-. See also FLUORIDES.

TKB750 CAS:455-14-1 ***HR: 3***
p-TRIFLUOROMETHYLANILINE
mf: $C_7H_6F_3N$ mw: 161.14

SYNS: p-AMINOBENZOTRIFLUORIDE ◇ α,α,α-TRIFLUORO-p-TOLUIDINE

TOXICITY DATA with REFERENCE
mmo-sat 10 mg/L ENMUDM 5,803,83
mmo-esc 10 mg/L ENMUDM 5,803,83
ipr-mus LD50:101 mg/kg JMCMAR 17,900,74

CONSENSUS REPORTS: Reported in EPA TSCA Inventory.

SAFETY PROFILE: Poison by intraperitoneal route. Mutation data reported. When heated to decomposition it emits very toxic fumes of F^- and NO_x. See also FLUORIDES.

TKC000 CAS:52833-75-7 ***HR: 3***
4-TRIFLUOROMETHYL-6H-BENZO(e)(1)BENZOTHIOPYRANO(4,3-b)INDOLE
mf: $C_{20}H_{12}F_3NS$ mw: 355.39

TOXICITY DATA with REFERENCE
scu-mus TDLo:92 mg/kg/9W-I:NEO MUREAV 66,307,79

SAFETY PROFILE: Questionable carcinogen with experimental neoplastigenic data. When heated to decomposition it emits very toxic fumes of F^-, NO_x, and SO_x.

TKD000 CAS:57165-71-6 ***HR: 3***
6-TRIFLUOROMETHYLCYCLOPHOSPHAMIDE
mf: $C_8H_{14}Cl_2F_3N_2O_2P$ mw: 329.11

SYNS: 2-(BIS(2-CHLOROETHYL)AMINO)-6-TRIFLUOROMETHYLTETRAHYDRO-2H-1,3,2-OXAZAPHOSPHORINE 2-OXIDE ◇ TETRAHYDRO-2-(BIS(2-CHLOROETHYL)AMINO)-6-TRIFLUOROMETHYL-2H-1,3,2-OXAZAPHOSPHORINE

TOXICITY DATA with REFERENCE
ipr-rat LD50:224 mg/kg JMCMAR 18,1106,75
ipr-mus LD50:400 mg/kg JMCMAR 18,1106,75

SAFETY PROFILE: Poison by intraperitoneal route. When heated to decomposition it emits very toxic fumes of Cl^-, F^-, NO_x, and PO_x. See also FLUORIDES.

TKD325 ***HR: D***
3'-TRIFLUOROMETHYL-4-DIMETHYLAMINOAZOBENZENE
mf: $C_{15}H_{14}F_3N_3$ mw: 293.32

TOXICITY DATA with REFERENCE
ipr-mus TDLo:1 g/kg (8-9D preg):TER KAIZAN 37,179,62
ipr-mus TDLo:1 g/kg (8-9D preg):REP KAIZAN 37,179,62

SAFETY PROFILE: An experimental teratogen. Experimental reproductive effects. When heated to decomposition it emits toxic fumes of F^- and NO_x.

TKD350 CAS:32750-98-4 ***HR: 3***
TRIFLUOROMETHYL-3-FLUOROCARBONYL HEXAFLUORO-PEROXYBUTYRATE
mf: $C_6F_{10}O_4$ mw: 326.05

$FCO \cdot [CF_2]_3CO \cdot OOCF_3$

SAFETY PROFILE: Explodes when heated to 70°C. When heated to decomposition it emits toxic fumes of F^-. See also PEROXIDES and FLUORIDES.

TKD375 CAS:373-91-1 ***HR: 3***
TRIFLUOROMETHYL HYPOFLUORITE
mf: CF_4O mw: 104.00

SAFETY PROFILE: A powerful oxidant. Explodes on contact with acetylene, cyclopropane, ethylene, hydrogen containing solvents, polymers, and rubber. Solutions in benzene are spark- and UV light-sensitive explosives. Reacts with pyridine to form an explosive product. When heated to decomposition it emits toxic fumes of F^-. See also FLUORIDES and HYPOCHLORITES.

TKE500 CAS:117-89-5 ***HR: 3***
TRIFLUOROMETHYLPERAZINE
mf: $C_{21}H_{24}F_3N_3S$ mw: 407.54

SYNS: FLUOPERAZINE ◇ JATRONEURAL ◇ 10-(Γ-(N'-METHYL-PIPERAZINO)PROPYL)-2-TRIFLUOROMETHYL-PHENOTHIOZINE ◇ 10-(3-(4-METHYL-1-PIPERAZINYL)PROPYL)-2-(TRIFLUOROMETHYL) PHENOTHIAZINE ◇ STELAZINE ◇ STELLAZINE ◇ TERFLUZINE ◇ TRIFLUOPERAZINA (ITALIAN) ◇ TRIFLUOPERAZINE ◇ TRIFLUOROMETHYL-10-(3'-(1-METHYL-4-PIPERAZINYL)PROPYL)PHENOTHIAZINE ◇ TRIFLUPERAZINE ◇ TRIPHTHAZINE

TOXICITY DATA with REFERENCE
mmo-sat 2500 μg/L INJHA9 12,21,80
hma-mus/sat 144 mg/kg INJHA9 12,21,80
scu-rat TDLo:8160 μg/kg (female 51D pre):REP PSYPAG 16,5,69
orl-rbt TDLo:144 mg/kg (female 8-16D post):TER PSDTAP 10,235,69
ipr-mus LD50:175 mg/kg ARZNAD 21,1727,71

SAFETY PROFILE: Poison by intraperitoneal route. An experimental teratogen. Experimental reproductive effects. Mutation data reported. When heated to decomposition it emits very toxic fumes of F^-, NO_x, and SO_x. Used as an antipsychotic and sedative. See also FLUORIDES.

TKE525 CAS:50311-48-3 ***HR: 3***
TRIFLUOROMETHYL PEROXONITRATE
mf: CF_3NO_4 mw: 147.01

SAFETY PROFILE: A shock-sensitive explosive. When heated to decomposition it emits toxic fumes of F^- and NO_x. See also NITRATES, FLUORIDES, and PEROXIDES.

TKE550 CAS:33017-08-2 ***HR: 3***
TRIFLUOROMETHYL PEROXYACETATE
mf: $C_3H_3F_3O_3$ mw: 144.05

SAFETY PROFILE: Explodes violently at 22°C. When heated to decomposition it emits toxic fumes of F^-. See also PEROXIDES and FLUORIDES.

TKE750 CAS:98-17-9 ***HR: 3***
3-(TRIFLUOROMETHYL)PHENOL
mf: $C_7H_5F_3O$ mw: 162.12

SYN: α,α,α-TRIFLUORO-m-CRESOL

TOXICITY DATA with REFERENCE
ivn-rat LD50:57 mg/kg JAPMA8 38,570,49

CONSENSUS REPORTS: Reported in EPA TSCA Inventory.

SAFETY PROFILE: Poison by intravenous route. When heated to decomposition it emits toxic fumes of F^-. See also FLUORIDES.

TKF000 CAS:2338-76-3 ***HR: 3***
m-TRIFLUOROMETHYLPHENYLACETONITRILE
mf: $C_{19}H_6F_3N$ mw: 185.16

SYN: 3-(TRIFLUOROMETHYL)BENZENEACETONITRILE

TOXICITY DATA with REFERENCE
ipr-mus LD50:100 mg/kg NTIS** AD691-490

CONSENSUS REPORTS: Cyanide and its compounds are on the Community Right-To-Know List. Reported in EPA TSCA Inventory.

SAFETY PROFILE: Poison by intraperitoneal route. When heated to decomposition it emits very toxic fumes of F^-, NO_x, and CN^-. See also NITRILES.

TKF250 CAS:34929-08-3 ***HR: 3***
1-(m-TRIFLUOROMETHYLPHENYL)-3-(2'-HYDROXYETHYL)QUINAZOLINE-2,4-DIONE
mf: $C_{17}H_{13}F_3N_2O_3$ mw: 350.32

SYNS: H-88 ◇ 3-(2-HYDROXYETHYL)-1-(3-(TRIFLUOROMETHYL) PHENYL)-2,4(1H,3H)-QUINAZOLINEDIONE ◇ 1-(m-TRIFLUOROMETHYLPHENYL)-3-(2-HYDROXYETHYL)-QUINAZOLINE-2,4(1H,3H)-DIONE

TOXICITY DATA with REFERENCE
orl-rat LD50:810 mg/kg DRFUD4 4,201,79
ipr-rat LD50:275 mg/kg DRFUD4 4,201,79
orl-mus LD50:630 mg/kg OYYAA2 15,501,78

ipr-mus LD50:275 mg/kg OYYAA2 15,501,78
orl-gpg LD50:320 mg/kg OYYAA2 15,501,78

SAFETY PROFILE: Poison by ingestion and intraperitoneal routes. When heated to decomposition it emits very toxic fumes of F^- and NO_x. See also FLUORIDES.

TKF525 CAS:395-47-1 ***HR: 3***
2-TRIFLUOROMETHYLPHENYL MAGNESIUM BROMIDE
mf: $C_7H_4BrF_3Mg$ mw: 249.31

$F_3CC_6H_4MgBr$

SAFETY PROFILE: Solutions in organic solvents (e.g., ether, benzene) may explode above 40°C. When heated to decomposition it emits toxic fumes of F^- and Br^-. See also BROMIDES, MAGNESIUM COMPOUNDS, and FLUORIDES.

TKF530 CAS:402-26-6 ***HR: 3***
3-TRIFLUOROMETHYLPHENYL MAGNESIUM BROMIDE
mf: $C_7H_4BrF_3Mg$ mw: 249.31

$F_3CC_6H_4MgBr$

SAFETY PROFILE: Solutions in organic solvents (e.g., ether, benzene) may explode above 40°C. When heated to decomposition it emits toxic fumes of F^- and Br^-. See also BROMIDES, MAGNESIUM COMPOUNDS, and FLUORIDES.

TKF535 CAS:402-51-7 ***HR: 3***
4-TRIFLUOROMETHYLPHENYL MAGNESIUM BROMIDE
mf: $C_7H_4BrF_3Mg$ mw: 249.31

$F_3CC_6H_4MgBr$

SAFETY PROFILE: Solutions in organic solvents (e.g., ether; benzene) may explode above 40°C. When heated to decomposition it emits toxic fumes of F^- and Br^-. See also BROMIDES, MAGNESIUM COMPOUNDS, and FLUORIDES.

TKF699 CAS:23595-00-8 ***HR: 2***
1-(m-TRIFLUOROMETHYLPHENYL)-N-NITROSOANTHRANILIC ACID
mf: $C_{14}H_9F_3N_2O_3$ mw: 310.25

SYNS: ACIDO 1-(m-TRIFLUOROMETILFENIL)-N-NITROSO ANTRANILICO (ITALIAN) ◇ ITF 611 ◇ 2-(NITROSO(3-(TRIFLUOROMETHYL)PHENYL)AMINO)-BENZOIC ACID ◇ N-NITROSO-N-(α,α,α-TRIFLUORO-m-TOLYL)ANTHRANILIC ACID

TOXICITY DATA with REFERENCE
orl-rat LD50:900 mg/kg FRPSAX 26,525,71
ipr-rat LD50:500 mg/kg FRPSAX 26,525,71
orl-mus LD50:860 mg/kg FRPSAX 26,191,71
ipr-mus LD50:450 mg/kg FRPSAX 26,525,71

SAFETY PROFILE: Moderately toxic by ingestion and intraperitoneal routes. Many N-nitroso compounds are carcinogens. When heated to decomposition it emits toxic fumes of F^- and NO_x. See also N-NITROSO COMPOUNDS and AROMATIC AMINES.

TKF750 CAS:37924-13-3 ***HR: 2***
1,1,1-TRIFLUORO-N-(2-METHYL-4-(PHENYLSULFONYL)PHENYL)METHANESULFONAMIDE
mf: $C_{14}H_{12}F_3NO_4S_2$ mw: 379.39

SYNS: DESTUN ◇ MBR 8251 ◇ PERFLUIDONE ◇ N-(4-PHENYLSULFONYL-o-TOLYL)-1,1,1-TRIFLUOROMETHANESULFONAMIDE

TOXICITY DATA with REFERENCE
orl-rat LD50:633 mg/kg 85ARAE 2,215,77
orl-mus LD50:920 mg/kg FMCHA2 -,D235,80

SAFETY PROFILE: Moderately toxic by ingestion. When heated to decomposition it emits very toxic fumes of F^-, NO_x, and SO_x. See also FLUORIDES.

TKF775 CAS:420-52-0 ***HR: 3***
TRIFLUOROMETHYL PHOSPHINE
mf: CH_2F_3P mw: 102.00

SAFETY PROFILE: Ignites spontaneously in air. When heated to decomposition it emits toxic fumes of F^- and PO_x. See also PHOSPHINE and FLUORIDES.

TKG000 CAS:23779-99-9 ***HR: 3***
2-(8′-TRIFLUOROMETHYL-4′-QUINOLYLAMINO)BENZOIC ACID, 2,3-DIHYDROXY PROPYL ESTER
mf: $C_{20}H_{17}F_3N_2O_4$ mw: 406.39

SYNS: 4-(o-(2′,3′-DIHYDROXYPROPYLOXYCARBONYL)PHENYL)-AMINO-8-TRIFLUOROMETHYLQUINOLINE ◇ 2,3-DIHYDROXYPROPYL-N-(8-(TRIFLUOROMETHYL)-4-QUINOLYL)ANTHRANILATE ◇ DIRALGAN ◇ FLOCTAFENINE ◇ IDARAC ◇ NOVODOLAN ◇ R 4318 ◇ RU 15750 ◇ 8-TRIFLUOROMETHYL-7-DESCHLOROGLAFENINE

TOXICITY DATA with REFERENCE
orl-rbt TDLo:1300 mg/kg (female 6-18D post):REP YACHDS 9(Suppl 2),379,81
orl-mus TDLo:700 mg/kg (female 1-7D post):TER YACHDS 9(Suppl 2),325,81
orl-rat LD50:535 mg/kg YACHDS 9(Suppl 2),299,81
ipr-rat LD50:250 mg/kg YACHDS 9(Suppl 2),299,81
ivn-rat LD50:160 mg/kg TXAPA9 36,173,76
orl-mus LD50:1960 mg/kg YACHDS 9(Suppl 2),299,81
ipr-mus LD50:245 mg/kg TXAPA9 36,173,76
ivn-mus LD50:180 mg/kg TXAPA9 36,173,76
orl-rbt LD50:700 mg/kg TXAPA9 36,173,76

SAFETY PROFILE: Poison by intravenous and intraperitoneal routes. Moderately toxic by ingestion. An experimental teratogen. Experimental reproductive effects. When heated to decomposition it emits very toxic fumes of F^- and NO_x. Used as an analgesic.

TKG275 ***HR: 3***
N-(TRIFLUOROMETHYLSULFINYL) TRIFLUOROMETHYL IMIDOSULFINYL AZIDE
mf: $C_2F_6N_4O_2S_2$ mw: 290.16

SAFETY PROFILE: An explosive liquid. When heated to decomposition it emits toxic fumes of F^-, SO_x, and NO_x. See also AZIDES and FLUORIDES.

TKG525 CAS:3855-45-6 ***HR: 3***
TRIFLUOROMETHYLSULFONYL AZIDE
mf: $CF_3N_3O_2S$ mw: 175.09

SAFETY PROFILE: An explosive. When heated to decomposition it emits toxic fumes of F^-, SO_x, and NO_x. See also AZIDES and FLUORIDES.

TKG750 CAS:148-56-1 ***HR: 2***
TRIFLUOROMETHYLTHIAZIDE
mf: $C_8H_6F_3N_3O_4S_2$ mw: 329.29

SYNS: ADEMOL ◇ FLUMETHIAZIDE ◇ RONTYL ◇ ROUTRAX ◇ TRIFLUOMETHYLTHIAZIDE ◇ 6-(TRIFLUOROMETHYL)-1,2,4-BENZO-THIADIAZINE-7-SULFONAMIDE-1,1-DIOXIDE ◇ 6-(TRIFLUOROMETHYL)-1,4,2-BENZOTHIADIAZINE-7-SULFONAMIDO-1,1-DIOXIDE ◇ 6-TRIFLUOROMETHYL-7-SULFAMOYL-4H-1,4,2-BENZOTHIADIAZINE-1,1-DIOXIDE ◇ 6-TRIFLUOROMETHYL-7-SULFAMYL-1,2,4-BENZOTHIADIAZINE-1,1-DIOXIDE

TOXICITY DATA with REFERENCE
ipr-mus LD50:1760 mg/kg JPETAB 134,273,61
ivn-mus LD50:910 mg/kg JPETAB 128,405,60

SAFETY PROFILE: Moderately toxic by intraperitoneal and intravenous routes. When heated to decomposition it emits very toxic fumes of SO_x, NO_x, and F^-.

TKH000 CAS:318-22-9 ***HR: 3***
2,2,2-TRIFLUORO-N-(9-OXOFLUOREN-2-YL) ACETAMIDE
mf: $C_{15}H_8F_3NO_2$ mw: 291.24

SYN: 2-TRIFLUOROACETYLAMINOFLUOREN-9-ONE

TOXICITY DATA with REFERENCE
orl-rat TDLo:6400 mg/kg/35W-C:CAR CNREA8 22,1002,62

SAFETY PROFILE: Questionable carcinogen with experimental carcinogenic data. When heated to decomposition it emits very toxic fumes of F^- and NO_x. See also FLUORIDES.

TKH025 CAS:21372-60-1 ***HR: 3***
N,N,N'-TRIFLUOROPROPIONAMIDINE
mf: $C_3H_5F_3N_2$ mw: 126.08

$CH_3CH_2C(:NF)NF_2$

SAFETY PROFILE: A shock-sensitive explosive. When heated to decomposition it emits toxic fumes of F^- and NO_x. See also FLUORIDES.

TKH030 CAS:661-54-1 ***HR: 3***
3,3,3-TRIFLUOROPROPYNE
mf: C_3HF_3 mw: 94.04

SAFETY PROFILE: Explodes when heated. Upon decomposition it emits toxic fumes of F^-. See also ACETYLENE COMPOUNDS and FLUORIDES.

TKH050 CAS:2061-56-5 ***HR: D***
17-(3,3,3-TRIFLUORO-1-PROPYNYL)ESTRA-1,3,5(10)-TRIEN-3,17-β-DIOL
mf: $C_{21}H_{23}F_3O_2$ mw: 364.44

SYNS: BDH 6146 ◇ (17-β)-ESTRA-1,3,5(10)-TRIEN-3,17-DIOL, 3-METHOXY-17-(3,3,3-TRIFLUORO-1-PROPYNYL)

TOXICITY DATA with REFERENCE
orl-mus TDLo:270 μg/kg (female 1-5D post):REP ACENA7 53,443,66

SAFETY PROFILE: Experimental reproductive effects. When heated to decomposition it emits toxic fumes of F^-.

TKH250 CAS:59544-89-7 ***HR: 3***
TRIFLUORO SELENIUM HEXAFLUORO ARSENATE
mf: AsF_9Se mw: 324.9

CONSENSUS REPORTS: Selenium and its compounds, as well as arsenic and its compounds, are on the Community Right-To-Know List.

OSHA PEL: TWA 0.01 mg(As)/m^3; Cancer Hazard; TWA 0.2 mg(Se)/m^3
ACGIH TLV: TWA 0.2 mg(As)/m^3; TWA 0.2 mg(Se)/m^3
DFG TRK: 0.2 mg/m^3 calculated as arsenic in that portion of dust that can possibly be inhaled.
NIOSH REL: CL 2 μg(As)/m^3

SAFETY PROFILE: Arsenic compounds are poisons. Violent reaction with water. When heated to decomposition it emits very toxic fumes of As, F^- and Se. See also FLUORIDES, ARSENIC COMPOUNDS,and SELENIUM COMPOUNDS.

TKH300 CAS:101913-67-1 ***HR: 3***
TRIFLUOROSTANNITE HEXADECYLAMINE
mf: $C_{16}H_{35}N \cdot F_3HSn$ mw: 418.22

SYN: TRIFLUOROSTANNITE OF HEXADECYLAMINE

TOXICITY DATA with REFERENCE
ivn-mus LD50:100 mg/kg CSLNX* NX#04259

OSHA PEL: TWA 2 mg(Sn)/m^3
ACGIH TLV: TWA 2 mg(Sn)/m^3

SAFETY PROFILE: Poison by intravenous route. When heated to decomposition it emits toxic fumes of NO_x, Sn, and F^-.

TKH325 CAS:70-00-8 ***HR: 2***
TRIFLUOROTHYMIDINE
mf: $C_{10}H_{11}F_3N_2O_5$ mw: 296.23

PROP: Crystals from ethyl acetate. Mp: 186-189°.

SYNS: 1-(2-DEOXY-β-d-RIBOFURANOSYL)-5-(TRIFLUOROMETHYL)-2,4(1H,3H)-PYRIMIDINEDIONE ◇ 2'-DEOXY-5-(TRIFLUOROMETHYL) URIDINE ◇ F3DThd ◇ F3T ◇ F3TDR ◇ NSC 75520 ◇ TFDU ◇ TFT THILO ◇ 5-TRIFLUORO-2'-DEOXYTHYMIDINE ◇ TRIFLUOROMETHYLDEOXYURIDINE ◇ 5-(TRIFLUOROMETHYL)DEOXYURIDINE ◇ 5-TRIFLUOROMETHYL-2-DEOXYURIDINE ◇ 5-(TRIFLUOROMETHYL)-2'-DEOXYURIDINE ◇ α,α,α-TRIFLUOROTHYMIDINE ◇ TRIFLURIDINE ◇ VIROPHTA ◇ VIROPTIC

TOXICITY DATA with REFERENCE
mmo-sat 500 μg/L MUREAV 169,123,86
sce-hmn:lym 500 μg/L DRFUD4 7,520,82
ipr-mus LD50:1931 mg/kg NCISP* JAN86

CONSENSUS REPORTS: EPA Genetic Toxicology Program.

SAFETY PROFILE: Moderately toxic by intraperitoneal route. Human mutation data reported. When heated to decomposition it emits toxic fumes of F^- and NO_x.

TKH500 CAS:63980-13-2 ***HR: 3***
α,α,α-TRIFLUORO-m-TOLUIC ACID THALLIUM(I) SALT
mf: $C_8H_4F_3O_2{\cdot}Tl$ mw: 393.49

SYN: m-TRIFLUOROMETHYL BENZOIC ACID, THALLIUM SALT

TOXICITY DATA with REFERENCE
orl-rat LDLo:50 mg/kg NCNSA6 5,43,53

CONSENSUS REPORTS: Thallium and its compounds are on the Community Right-To-Know List.

OSHA PEL: TWA 0.1 mg(Tl)/m^3 (skin)
ACGIH TLV: TWA 0.1 mg(Tl)/m^3 (skin)

SAFETY PROFILE: Poison by ingestion. Thallium compounds are very toxic. When heated to decomposition it emits very toxic fumes of F^- and Tl. See also FLUORIDES and THALLIUM COMPOUNDS.

TKH750 CAS:530-78-9 ***HR: 3***
N-(α,α,α-TRIFLUORO-m-TOLYL)ANTHRANILIC ACID
mf: $C_{14}H_{10}F_3NO_2$ mw: 281.25

SYNS: ACHLESS ◇ ACIDO FLUFENAMICO (ITALIAN) ◇ ANSATIN ◇ ANT-1 ◇ ARLEF ◇ C.I. 440 ◇ CN-27,554 ◇ FLUFENAMIC ACID ◇ FLUFENAMINSAURE (GERMAN) ◇ FLUPHENAMIC ACID ◇ FULLSAFE ◇ INF 1837 ◇ MERALEN ◇ NSC-82699 ◇ PARAFLU ◇ PARLEF ◇ PARLIF ◇ PLOSTENE ◇ RISTOGEN ◇ SASTRIDEX ◇ SURIKA ◇ TECRAMINE ◇ 3-TRIFLUOROMETHYLDIPHENYLAMINE-2-CARBOXYLIC ACID ◇ N-(m-TRIFLUOROMETHYLPHENYL)-2-AMINOBENZOIC ACID ◇ N-(3-TRIFLUOROMETHYLPHENYL) ANTHRANILIC ACID

TOXICITY DATA with REFERENCE
dni-hmn-unr 504 mg/kg/8W STBIBN 50,172,75
orl-rbt TDLo:1 g/kg (female 1D pre):REP FESTAS 38,238,82
orl-wmn TDLo:2160 mg/kg/26W-I:GIT PGMJAO 62,773,86
orl-rat LD50:249 mg/kg AIPTAK 221,132,76
ipr-rat LD50:185 mg/kg OYYAA2 16,1011,78
ivn-rat LD50:98 mg/kg CMROCX 4,17,76
orl-mus LD50:490 mg/kg OYYAA2 16,1011,78
ipr-mus LD50:150 mg/kg NTIS** AD691-490
ivn-mus LD50:158 mg/kg YKKZAJ 89,1392,69

SAFETY PROFILE: Poison by ingestion, intravenous, and intraperitoneal routes. Experimental reproductive effects. Human systemic effects: hypermotility, diarrhea. Human mutation data reported. When heated to decomposition it emits very toxic fumes of F^- and NO_x. Used as an anti-inflammatory agent. See also FLUORIDES.

TKI000 CAS:3216-14-6 ***HR: 3***
4-(2-(α-(α,α,α-TRIFLUORO-m-TOLYL)BENZYLOXY)ETHYL)MORPHOLINE FUMARATE
mf: $C_{20}H_{22}F_3NO_2{\cdot}C_4H_4O_4$ mw: 481.51

TOXICITY DATA with REFERENCE
orl-mus LDLo:250 mg/kg ARZNAD 14,964,64
ivn-mus LDLo:15 mg/kg ARZNAD 14,964,64

SAFETY PROFILE: Poison by ingestion and intravenous routes. When heated to decomposition it emits very toxic fumes of F^- and NO_x. See also FLUORIDES.

TKI750 CAS:3560-78-9 ***HR: 3***
3-(α-(α,α,α-TRIFLUORO-p-TOLYL)BENZYLOXY)TROPANE FUMARATE
mf: $C_{22}H_{24}F_3NO{\cdot}C_4H_4O_4$ mw: 491.55

TOXICITY DATA with REFERENCE
orl-mus LDLo:500 mg/kg ARZNAD 14,964,64
ivn-mus LDLo:30 mg/kg ARZNAD 14,964,64

SAFETY PROFILE: Poison by intravenous route. Moderately toxic by ingestion. When heated to decomposition it emits very toxic fumes of F^- and NO_x. See also FLUORIDES.

TKJ250 CAS:329-01-1 ***HR: 2***
(α,α,α-TRIFLUORO-m-TOLYL) ISOCYANATE
mf: $C_8H_4F_3NO$ mw: 187.13

PROP: Bp: 54°/11 mm, refr index: 1.4700, d: 1.359, flash p: 138° F.

SYNS: ISOCYANIC ACID, (m-TRIFLUOROMETHYLPHENYL) ESTER ◇ TIC

TOXICITY DATA with REFERENCE
orl-rat LD50:975 mg/kg GTPZAB 20(3),53,76
ihl-rat LC50:3600 mg/m^3 GTPZAB 20(3),53,76
orl-mus LD50:975 mg/kg GTPZAB 20(3),53,76
ihl-mus LC50:3300 mg/m^3 GTPZAB 20(3),53,76
ipr-mus LD50:871 mg/kg CNREA8 39,2204,79
orl-gpg LD50:478 mg/kg GTPZAB 20(3),53,76

SAFETY PROFILE: Moderately toxic by ingestion, inhalation, and intraperitoneal routes. A lachrymator. When heated to decomposition it emits very toxic fumes of NO_x and F^-. See also ISOCYANATES and FLUORIDES.

TKJ500 CAS:30914-89-7 ***HR: 3***
2-(α,α,α-TRIFLUORO-m-TOLYL)MORPHOLINE
mf: $C_{11}H_{12}F_3NO$ mw: 231.24

SYNS: CERM-1841 ◇ TETRAHYDRO-2-(α,α,α-TRIFLUORO-m-TOLYL)-1,4-OXAZINE

TOXICITY DATA with REFERENCE
orl-rat LD50:271 mg/kg ARZNAD 28,642,78
orl-mus LD50:495 mg/kg ARZNAD 28,642,78

CONSENSUS REPORTS: Reported in EPA TSCA Inventory.

SAFETY PROFILE: Poison by ingestion. When heated to decomposition it emits very toxic fumes of F^- and NO_x. See also FLUORIDES.

TKJ750 CAS:3414-47-9 ***HR: 2***
5-(α,α,α-TRIFLUORO-m-TOLYOXYMETHYL)-2-OXAZOLIDINETHIONE
mf: $C_{10}H_{10}F_3NO_2S$ mw: 265.27

SYNS: 5-(((α,α,α-TRIFLUORO-m-TOLYL)OXY)METHYL)-2-OXAZOLIDINETHIONE ◇ U-11,634

TOXICITY DATA with REFERENCE
orl-rat TDLo:87500 μg/kg (1D pre/1-6D preg):REP TXAPA9 10,322,67
orl-rat LD50:524 mg/kg TXAPA9 10,322,67
scu-rat LD50:1197 mg/kg JRPFA4 11,85,66
ipr-mus LD50:451 mg/kg TXAPA9 10,322,67

SAFETY PROFILE: Moderately toxic by ingestion, intraperitoneal, and subcutaneous routes. Experimental reproductive effects. When heated to decomposition it emits very toxic fumes of F^-, NO_x, and SO_x. See also FLUORIDES.

TKK000 CAS:675-14-9 ***HR: 3***
2,4,6-TRIFLUORO-s-TRIAZINE
mf: $C_3F_3N_3$ mw: 135.06

SYN: CYANURIC FLUORIDE

TOXICITY DATA with REFERENCE
ihl-rat LC50:3.1 ppm/4H AIHAAP 33,382,72
skn-rbt LD50:160 mg/kg:CAR AIHAAP 33,382,72

CONSENSUS REPORTS: EPA Extremely Hazardous Substances List. Reported in EPA TSCA Inventory.

SAFETY PROFILE: Poison by skin contact and inhalation. Questionable carcinogen with experimental carcinogenic data. When heated to decomposition it emits very toxic fumes of F^- and NO_x. See also FLUORIDES.

TKK050 CAS:1423-11-6 ***HR: 3***
1,3,5-TRIFLUOROTRINITROBENZENE
mf: $C_6F_3N_3O_6$ mw: 267.08

$F_3C_6(NO_2)_3$

SAFETY PROFILE: Reacts with hydrazine to form an explosive product. When heated to decomposition it emits toxic fumes of F^- and NO_x. See also NITRO COMPOUNDS of AROMATIC HYDROCARBONS and FLUORIDES.

TKK250 CAS:440-17-5 ***HR: 3***
TRIFLUPERAZINE DIHYDROCHLORIDE
mf: $C_{21}H_{24}F_3N_3S\cdot 2ClH$ mw: 480.46

SYNS: ESKAZINE ◇ ESKAZINE DIHYDROCHLORIDE ◇ FLUOPERAZINE ◇ JATRONEURAL ◇ 10-(3-(4-METHYL-1-PIPERAZINYL)PROPYL)-2-TRIFLUOROMETHYLPHENOTHIAZINEDIHYDROCHLORIDE ◇ SKF 5019 ◇ STELAZINE ◇ STELAZINE DIHYDROCHLORIDE ◇ TERFLUZINE ◇ TERFLUZINE DIHYDROCHLORIDE ◇ TRIFLORPERAZINE DIHYDROCHLORIDE ◇ TRIFLUOPERAZINE HYDROCHLORIDE ◇ TRIFLUOROPERAZINE DIHYDROCHLORIDE ◇ TRIFLUOROPYRAZIN DIHYDROCHLORIDE ◇ TRIFTAZIN ◇ TRIPHTHAZINE ◇ TRIPHTHAZINE DIHYDROCHLORIDE ◇ TRYPTAZINE DIHYDROCHLORIDE

TOXICITY DATA with REFERENCE
cyt-hmn-unr 26 mg/kg/17W-I CYTOAN 35,552,70
mnt-mus-orl 80 μg/kg/24H FCTOD7 25,615,87
unr-rat TDLo:220 mg/kg (1-22D preg):REP FATOAO 36,165,73
orl-rat LD50:543 mg/kg ARZNAD 27,866,77
orl-mus LD50:424 mg/kg ARZNAD 27,866,77
ipr-mus LD50:185 mg/kg DCTODJ 8,495,85
ivn-mus LD50:82 mg/kg APTOA6 19,87,62
ivn-dog LD50:50 mg/kg 29ZSA2 -,157,59

CONSENSUS REPORTS: Reported in EPA TSCA Inventory. EPA Genetic Toxicology Program.

SAFETY PROFILE: by intravenous and intraperitoneal routes. Moderately toxic by ingestion. Experimental reproductive effects. Human mutation data reported. When heated to decomposition it emits very toxic fumes of F^-, NO_x, SO_x, and HCl. See also FLUORIDES.

TKK500 CAS:749-13-3 ***HR: 3***
TRIFLUPERIDOL
mf: $C_{22}H_{23}F_4NO_2$ mw: 409.46

SYNS: 4'-FLUORO-4-(4-HYDROXY-4-(α,α,α-TRIFLUORO-m-TOLYL) PIPERIDINO)BUTYROPHENONE ◇ 4-FLUORO-4,4-IDROSSI-4-(m-TRIFLUOROMETIL-FENIL)-PIPERIDINO-BUTIRROFENONE(ITALIAN) ◇ 1-(4-FLUOROPHENYL)-4-(4-HYDROXY-4-(3-(TRIFLUOROMETHYL)PHENYL)-1-PIPERIDINYL)-1-BUTANONE ◇ MCN-JR-2498 ◇ PSICOPERIDOL-R ◇ PSYCHOPERIDOL ◇ R-2498 ◇ TRIFLUPERIDOLO (ITALIAN) ◇ TRIPERIDOL

TOXICITY DATA with REFERENCE
ims-mus TDLo:64 mg/kg (female 10-13D post):REP SPERA2 118,245,68
orl-mus TDLo:75 mg/kg (10-12D preg):TER TOIZAG 28,621,81
orl-rat LD50:140 mg/kg TXAPA9 18,185,71
ipr-mus LDLo:150 mg/kg NTIS** AD691-490
scu-rat LD50:70 mg/kg MDCHAG 4(2),199,67
orl-mus LD50:110 mg/kg ARZNAD 24,1248,74
scu-mus LD50:50 mg/kg ARZNAD 24,1248,74

SAFETY PROFILE: Poison by ingestion, subcutaneous, and intraperitoneal routes. An experimental teratogen. Experimental reproductive effects. When heated to decomposition it emits very toxic fumes of F^- and NO_x. See also FLUORIDES and TRIFLUPERIDOL HYDROCHLORIDE.

TKK750 CAS:2062-77-3 ***HR: 3***
TRIFLUPERIDOL HYDROCHLORIDE
mf: $C_{22}H_{23}F_4NO_2 \cdot ClH$ mw: 445.92

PROP: Crystals from acetone. Mp: 200.5-201.3°. Sol in water.

SYNS: FLUMOPERONE HYDROCHLORIDE ◇ PSYCOPERIDOL HYDROCHLORIDE ◇ R 2498 ◇ TRIPERIDOL ◇ TRIPERIDOL HYDROCHLORIDE

TOXICITY DATA with REFERENCE
orl-rat LDLo:75 mg/kg THERAP 17,1053,62
scu-rat LD50:70 mg/kg ARZNAD 11,932,61
ivn-rat LD50:14 mg/kg 27ZQAG -,194,72
orl-mus LD50:99 mg/kg THERAP 17,1053,62
scu-mus LD50:80 mg/kg 27ZQAG -,194,72
ivn-mus LD50:17400 μg/kg THERAP 17,1053,62

SAFETY PROFILE: Poison by subcutaneous, intravenous, and ingestion routes. When heated to decomposition it emits very toxic fumes of F^-, NO_x, and HCl. See also FLUORIDES.

TKL000 CAS:146-54-3 ***HR: 3***
TRIFLUPROMAZINE
mf: $C_{18}H_{19}F_3N_2S$ mw: 352.45

PROP: Liquid. Bp: 176°, refr index: 1.5780.

SYNS: 10-(3-(DIMETHYLAMINO)PROPYL-2-(TRIFLUOROMETHYL) PHENOTHIAZINE ◇ N,N-DIMETHYL-2-(TRIFLUOROMETHYL)-10H-PHENOTHIAZINE-10-PROPANAMINE ◇ VESPRIN

TOXICITY DATA with REFERENCE
dlt-mus-ipr 25 mg/kg MUREAV 17,87,73
cyt-mam:kdy 10 mg/L/8H-C FCTXAV 8,617,70
ipr-mus TDLo:25 mg/kg (male 1D pre):REP MUREAV 17,87,73
orl-rat LD50:185 mg/kg ARZNAD 36,797,86
ipr-rat LD50:94 mg/kg TXAPA9 2,540,60
orl-mus LD50:245 mg/kg TXAPA9 2,540,60
ipr-mus LD50:100 mg/kg MUREAV 17,87,73
ivn-mus LD50:44 mg/kg TXAPA9 2,540,60

CONSENSUS REPORTS: EPA Genetic Toxicology Program.

SAFETY PROFILE: Poison by ingestion, intravenous, and intraperitoneal routes. Experimental reproductive effects. Mutation data reported. When heated to decomposition it emits very toxic fumes of F^-, NO_x, and SO_x. See also FLUORIDES.

TKL175 ***HR: 3***
TRIFORMYL-STROSPESIDE

SYN: STROSPESIDE TRIFORMATE

TOXICITY DATA with REFERENCE
orl-mus LD50:19030 μg/kg AIPTAK 153,436,65
scu-mus LD50:9200 μg/kg AIPTAK 153,436,65
orl-gpg LDLo:3480 μg/kg AIPTAK 153,436,65
ivn-gpg LDLo:348 μg/kg AIPTAK 153,436,65

SAFETY PROFILE: Poison by ingestion, subcutaneous, and intravenous routes.

TKL250 CAS:2589-01-7 ***HR: 2***
TRIGLYCIDYL CYANURATE
mf: $C_{12}H_{15}N_3O_6$ mw: 297.30

SYNS: CYANURIC ACID TRIGLYCIDYL ESTER ◇ s-TRIAZINE-2,4,6-TRIOL, TRI(2,3-EPOXYPROPYL) ESTER ◇ 1,3,5-TRIS(2,3-EPOXYPROPYL)TRIAZINE-2,4,6-TRIONE

TOXICITY DATA with REFERENCE
orl-rat LD50:1680 mg/kg SCCUR* -,9,61
ipr-rat LD50:595 mg/kg SCCUR* -,9,61
orl-mus LD50:1490 mg/kg SCCUR* -,9,61

SAFETY PROFILE: Moderately toxic by ingestion and

intraperitoneal routes. When heated to decomposition it emits very toxic fumes of NO_x. A preparative hazard, explosions may occur if chlorine, epoxy content, and pH are not correct. A component of epoxy resins. See also ESTERS.

TKL500 CAS:112-26-5 ***HR: 3***
TRIGLYCOL DICHLORIDE
mf: $C_6H_{12}Cl_2O_2$ mw: 187.08

PROP: Colorless liquid. Bp: 240°, fp: −31.5°, flash p: 250°F (OC), d: 1.197, vap press: 0.03 mm @ 20°.

SYNS: 1,2-BIS(2-CHLOROETHOXY)ETHANE ◇ 2-(2-CHLORETHOXY)ETHYL 2'-CHLORETHYL ETHER ◇ 2-(2-CHLOROETHOXY) ETHYL 2'-CHLOROETHYL ETHER ◇ TRIETHYLENE GLYCOL DICHLORIDE

TOXICITY DATA with REFERENCE
skn-rbt 500 mg open MLD UCDS** 8/28/57
eye-rbt 20 mg AJOPAA 29,1363,46
orl-rat LD50:250 mg/kg UCDS** 8/28/57
skn-rbt LD50:1410 mg/kg UCDS** 8/28/57
orl-gpg LD50:120 mg/kg JIHTAB 23,259,41

CONSENSUS REPORTS: Reported in EPA TSCA Inventory.

SAFETY PROFILE: Poison by ingestion. Moderately toxic by skin contact. A skin and eye irritant. Combustible when exposed to heat or flame; can react with oxidizing materials. To fight fire, use CO_2, dry chemical. When heated to decomposition it emits toxic fumes of Cl^-. See also ETHERS.

TKL750 CAS:143-22-6 ***HR: 2***
TRIGLYCOL MONOBUTYL ETHER
mf: $C_{10}H_{22}O_4$ mw: 206.32

PROP: Liquid. Completely sol in water. D: 1.0021 @ 20/20°, bp: decomp, fp: −47.4°, flash p: 290°F.

SYNS: 2-(2-(2-BUTOXYETHOXY)ETHOXY)ETHANOL ◇ BUTOXYTRIETHYLENE GLYCOL ◇ BUTOXYTRIGLYCOL ◇ POLY-SOLV TB ◇ TRIETHYLENE GLYCOL-n-BUTYL ETHER ◇ TRIETHYLENE GLYCOL MONOBUTYL ETHER

TOXICITY DATA with REFERENCE
skn-rbt 10 mg/24H open MLD AIHAAP 23,95,62
eye-rbt 50 mg SEV UCDS** 4/26/65
orl-rat LD50:6730 mg/kg UCDS** 4/26/65
skn-rbt LD50:3540 mg/kg AIHAAP 23,95,62

CONSENSUS REPORTS: Glycol ether compounds are on the Community Right-To-Know List. Reported in EPA TSCA Inventory.

SAFETY PROFILE: Moderately toxic by skin contact. Mildly toxic by ingestion. A skin and severe eye irritant. Many glycol ether compounds have dangerous human reproductive effects. Combustible when exposed to heat or flame. To fight fire, use water, foam, fog. When heated to decomposition it emits acrid smoke and irritating fumes. See also ETHERS and GLYCOL ETHERS.

TKL875 CAS:112-49-2 ***HR: 2***
TRIGLYME
mf: $C_8H_{18}O_4$ mw: 178.26

PROP: Liquid. D: (20/4) 0.990, flash p: 111°, mp: -45°, bp: (760) 216°, bp: (10) 103.5°, bp: (0.9) 20°, n (20/D) 1.4233. Misc with water, hydrocarbon solvents.

SYNS: GLYME-3 ◇ 2,5,8,11-TETRAOXADODECANE ◇ TRIETHYLENE GLYCOL DIMETHYL ETHER

TOXICITY DATA with REFERENCE
orl-rbt TDLo:3500 mg/kg (female 6-19D post):TER NTIS** PB87-181657
orl-mus TDLo:28 g/kg (female 6-13D post):REP TCMUD8 7,29,87

CONSENSUS REPORTS: Glycol ether compounds are on the Community Right-To-Know List.

SAFETY PROFILE: An experimental teratogen. Experimental reproductive effects. Many glycol ether compounds have dangerous human reproductive effects. Combustible when exposed to heat or flame. When heated to decomposition it emits acrid smoke and irritating fumes. See also GLYCOL ETHERS.

TKM000 CAS:5337-36-0 ***HR: 2***
TRI-n-HEXYL BORATE
mf: $C_{18}H_{39}BO_3$ mw: 314.38

PROP: Colorless liquid; odor of n-hexanol. Bp: 140-146° @ 2 mm, flash p: 300°F, d: 0.847 @ 28°, vap d: 10.8.

SYNS: BORIC ACID, TRIHEXYL ESTER ◇ TRIHEXYL BORATE

TOXICITY DATA with REFERENCE
eye-rbt 100 mg MOD 14KTAK -,706,64
orl-mus LD50:1800 mg/kg USBCC* -,-,58

CONSENSUS REPORTS: Reported in EPA TSCA Inventory.

SAFETY PROFILE: Moderately toxic by ingestion. An eye irritant. Combustible when exposed to heat or flame; can react with oxidizing materials. To fight fire, use foam, CO_2, dry chemical. When heated to decomposition it emits acrid smoke and irritating fumes. See also ESTERS and BORON COMPOUNDS.

TKM250 CAS:100-89-0 ***HR: 2***
TRIHEXYLENE GLYCOL BIBORATE
mf: $C_{18}H_{36}B_2O_6$ mw: 370.16

PROP: Colorless liquid; odor of hexylene glycol. Bp:

143-149° @ 2 mm, flash p: 345°F, d: 0.982 @ 21°, vap d: 12.8.

SYNS: 2-METHYL-2,4-PENTANEDIOL ESTER with BORIC ACID (H3BO3) (1:2) cyclic BIS(1,1,3-TRIMETHYLMETHYLENE) ESTER ◇ TRI(2-METHYL-2,4-PENTANEDIOL)BIBORATE ◇ 2,2'-((1,1,3-TRIMETHYLTRIMETHYLENE)DIOXY)BIS(4,4,6-TRIMETHYL)1,3,2-DIOXABORINANE ◇ USAF BO-1

TOXICITY DATA with REFERENCE
ipr-mus LD50:750 mg/kg NTIS** AD277-689

CONSENSUS REPORTS: Reported in EPA TSCA Inventory.

SAFETY PROFILE: Moderately toxic by intraperitoneal route. Combustible when exposed to heat or flame; can react with oxidizing materials. To fight fire, use foam, CO_2, dry chemical. When heated to decomposition it emits acrid smoke and irritating fumes. See also BORON COMPOUNDS.

TKM500 CAS:3084-48-8 ***HR: 3***
TRI-n-HEXYLPHOSPHINE OXIDE
mf: $C_{18}H_{39}OP$ mw: 302.54

TOXICITY DATA with REFERENCE
ivn-mus LD50:180 mg/kg CSLNX* NX#03141

CONSENSUS REPORTS: Reported in EPA TSCA Inventory.

SAFETY PROFILE: Poison by intravenous route. When heated to decomposition it emits toxic fumes of PO_x. See also PHOSPHINE OXIDE.

TKM750 ***HR: 3***
TRIHYDRAZINECOBALT(II) NITRATE
mf: $CoH_{12}N_8O_6$ mw: 279.09

CONSENSUS REPORTS: Cobalt and its compounds are on the Community Right-To-Know List.

SAFETY PROFILE: Explosive. When heated to decomposition it emits toxic fumes of NO_x. See also COBALT COMPOUNDS.

TKN000 ***HR: 3***
TRIHYDRAZINENICKEL(II) NITRATE
mf: $H_{12}N_8NiO_6$ mw: 278.86

CONSENSUS REPORTS: Nickel and its compounds are on the Community Right-To-Know List.

SAFETY PROFILE: Explodes violently (dry); spontaneously deflagrates (moist). When heated to decomposition it emits toxic fumes of NO_x. See also NITRATES and NICKEL COMPOUNDS.

TKN250 CAS:528-21-2 ***HR: 2***
2,3,4-TRIHYDROXYACETOPHENONE
mf: $C_8H_8O_4$ mw: 168.16

PROP: Colorless crystals in water. Mp: 173°. Sol in hot and cold water, alc; very sltly sol in benzene.

SYNS: C.I. 57000 ◇ GALLACETOPHENONE

TOXICITY DATA with REFERENCE
ipr-mus LD50:650 mg/kg JMCMAR 7,178,64

CONSENSUS REPORTS: Reported in EPA TSCA Inventory.

SAFETY PROFILE: Moderately toxic by intraperitoneal route. When heated to decomposition it emits acrid smoke and irritating fumes. See also KETONES.

TKN500 CAS:602-64-2 ***HR: D***
1,2,3-TRIHYDROXYANTHRAQUINONE
mf: $C_{14}H_8O_5$ mw: 256.22

PROP: Crystals from alc and acetic acid. Mp: decomp @ 310°, bp: subl 290°. Very sltly sol in water; sol in alc, H_2SO_4, ether.

SYNS: ANTHRACENE BROWN ◇ ANTHRAGALLIC ACID ◇ ANTHRAGALLOL ◇ 1,2,3-TRIHYDROXY-9,10,ANTHRACENEDIONE

TOXICITY DATA with REFERENCE
mmo-sat 100 μg/plate BCSTB5 5,1489,77
mma-sat 100 μg/plate BCSTB5 5,1489,77

SAFETY PROFILE: Mutation data reported. When heated to decomposition it emits acrid smoke and irritating fumes.

TKN750 CAS:81-54-9 ***HR: 1***
1,2,4-TRIHYDROXYANTHRAQUINONE
mf: $C_{14}H_8O_5$ mw: 256.22

PROP: Red needles from alc. Mp: 256-257°. Sltly sol in hot water; sol in water, ether.

SYNS: C.I. 58205 ◇ C.I. 75410 ◇ HYDROXYLIZARIC ACID ◇ PURPURIN ◇ PURPURINE ◇ SMOKE BROWN G ◇ 1,2,4-TRIHYDROXY-9,10-ANTHRACENEDIONE ◇ 1,2,4-TRIHYDROXYANTHRACHINON (CZECH) ◇ VERANTIN

TOXICITY DATA with REFERENCE
eye-rbt 500 mg/24H MLD 28ZPAK -,103,72
mmo-sat 10 μg/plate MUREAV 40,203,76
mma-sat 10 μg/plate MUREAV 40,203,76

SAFETY PROFILE: Mutation data reported. An eye irritant. When heated to decomposition it emits acrid smoke and irritating fumes.

TKO000 CAS:83-30-7 ***HR: 2***
2,4,6-TRIHYDROXYBENZOIC ACID
mf: $C_7H_6O_5$ mw: 170.13

PROP: Needles from water. Mp: 206° decomp, bp: subl in CO_2. Sol in water, alc; sltly sol in ether.

TOXICITY DATA with REFERENCE
ipr-mus LDLo:512 mg/kg CBCCT* 1,44,49

CONSENSUS REPORTS: Reported in EPA TSCA Inventory.

SAFETY PROFILE: Moderately toxic by intraperitoneal route. When heated to decomposition it emits acrid smoke and irritating fumes.

TKO250 CAS:1421-63-2 ***HR: 3***
2′,4′,5′-TRIHYDROXYBUTYROPHENONE
mf: $C_{10}H_{12}O_4$ mw: 196.22

PROP: Yellow-tan crystals. Mp: 149-153°, d: 6.0 lb/gal @ 20°. Very sltly sol in water; sol in alc, propylene glycol.

SYNS: THBP ◇ 2,4,5-TRIHYDROXYBUTYROPHENONE ◇ USAF EK

TOXICITY DATA with REFERENCE
mma-sat 167 μg/plate ENMUDM 8(Suppl 7),1,86
ipr-mus LD50:200 mg/kg NTIS** AD277-689

CONSENSUS REPORTS: Reported in EPA TSCA Inventory.

SAFETY PROFILE: Poison by intraperitoneal route. Mutation data reported. When heated to decomposition it emits acrid smoke and irritating fumes. See also KETONES.

TKO500 CAS:6807-96-1 ***HR: 3***
Z-(−)-4,6,8-TRIHYDROXY-3a,12a-DIHYDRO-ANTHRA (2,3-b)FURO(3,2-d)FURAN-5,10-DIONE
mf: $C_{18}H_{10}O_7$ mw: 338.28

SYN: VERSICOLORIN A

TOXICITY DATA with REFERENCE
mmo-sat 1 nmol/plate ENMUDM 4,19,82
dns-rat:lvr 500 nmol/L MUREAV 143,121,85
ivn-mus LD50:20 mg/kg 85GDA2 3,189,80

CONSENSUS REPORTS: EPA Genetic Toxicology Program.

SAFETY PROFILE: Poison by intravenous route. Mutation data reported. When heated to decomposition it emits acrid smoke and irritating fumes.

TKO750 CAS:68780-95-0 ***HR: D***
1,9,10-TRIHYDROXY-9,10-DIHYDRO-3-METHYLCHOLANTHRENE
mf: $C_{21}H_{18}O_3$ mw: 318.39

SYN: 9,10-DIHYDRO-3-METHYL-CHOLANTHRENE-1,9,10-TRIOL

TOXICITY DATA with REFERENCE
mma-sat 10 nmol/plate CNREA8 38,3398,78
mma-ham:lng 15 nmol/plate CNREA8 38,3398,78

SAFETY PROFILE: Mutation data reported. When heated to decomposition it emits acrid smoke and irritating fumes.

TKP000 CAS:533-87-9 ***HR: 3***
9,10,16-TRIHYDROXYHEXADECANOIC ACID
mf: $C_{16}H_{32}O_5$ mw: 304.48

SYNS: ALEURITIC ACID, tech ◇ dl-erthro-9,10,16-TRIHYDROXYHEXADECANOIC ACID ◇ 8,9,15-TRIHYDROXYPENTADECANE-1-CARBOXYLIC ACID

TOXICITY DATA with REFERENCE
ivn-mus LD50:178 mg/kg CSLNX* NX#00691

CONSENSUS REPORTS: Reported in EPA TSCA Inventory.

SAFETY PROFILE: Poison by intravenous route. When heated to decomposition it emits acrid smoke and irritating fumes.

TKP500 CAS:102-71-6 ***HR: 3***
TRIHYDROXYTRIETHYLAMINE
mf: $C_6H_{15}NO_3$ mw: 149.22

PROP: Pale yellow viscous liquid. Mp: 21.2°, bp: 360°, flash p: 355°F (CC), d: 1.1258 @ 20/20°, vap press: 10 mm @ 205°, vap d: 5.14.

SYNS: DALTOGEN ◇ NITRILO-2,2′,2″-TRIETHANOL ◇ 2,2′,2″-NITRILOTRIETHANOL ◇ STEROLAMIDE ◇ THIOFACO T-35 ◇ TRIAETHANOLAMIN-NG ◇ TRIETHANOLAMIN ◇ TRIETHANOLAMINE ◇ TRIETHYLOLAMINE ◇ TRI(HYDROXYETHYL)AMINE ◇ 2,2′,2″-TRIHYDROXYTRIETHYLAMINE ◇ TRIS(2-HYDROXYETHYL)AMINE ◇ TROLAMINE

TOXICITY DATA with REFERENCE
skn-hmn 15 mg/3D-I MLD 85DKA8 -,127,77
skn-rbt 560 mg/24H MLD TXAPA9 19,276,71
eye-rbt 10 mg MLD TXAPA9 55,501,80
orl-mus TDLo:16 g/kg/64W-C:CAR CNREA8 38,3918,78
orl-mus TD:154 g/kg/61W-C:CAR CNREA8 38,3918,78
orl-rat LD50:8 g/kg NTIS** PB158-507
orl-mus LD50:7400 mg/kg GTPZAB 26(8),53,82
ipr-mus LD50:1450 mg/kg RCRVAB 38,975,69
orl-gpg LD50:5300 mg/kg GISAAA 29(11),25,64

CONSENSUS REPORTS: Cyanide and its compounds are on the Community Right-To-Know List. Reported in EPA TSCA Inventory. EPA Genetic Toxicology Program.

ACGIH TLV: (Proposed: TWA 0.5 ppm)

SAFETY PROFILE: Moderately toxic by intraperitoneal route. Mildly toxic by ingestion. Liver and kidney damage has been demonstrated in animals from chronic exposure. A human and experimental skin irritant. An eye irritant. Questionable carcinogen with experimental carcinogenic data. Combustible liquid when exposed to heat or flame; can react vigorously with oxidizing materials. To fight fire, use alcohol foam, CO_2, dry chemical. When heated to decomposition it emits toxic fumes of NO_x and CN^-. See also NITRILES.

TKP750 ***HR: 2***
2,4,6-TRIHYDROXY-1,3,5,2,4,6-TRIOXATRIPHOSPHORINANE TRISODIUM SALT
mf: $Na_3O_6P_3$ mw: 257.88

SYNS: CYCLIC SODIUM TRIMETAPHOSPHATE ◇ CYCLISCHES TRINATRIUMMETAPHOSPHAT (GERMAN) ◇ SODIUM METAPHOSPHATE ◇ SODIUM PHOSPHATE ◇ SODIUM TRIMETAPHOSPHATE ◇ TRISODIUM TRIMETAPHOSPHATE

TOXICITY DATA with REFERENCE
orl-mus LD50:10 g/kg ARZNAD 7,445,57
scu-mus LD50:5940 mg/kg ARZNAD 7,445,57
ivn-mus LD50:1165 mg/kg ARZNAD 7,445,57

SAFETY PROFILE: Moderately toxic by intravenous route. Mildly toxic by ingestion. When heated to decomposition it emits toxic fumes of PO_x and Na_2O. See also PHOSPHATES.

TKP850 CAS:1160-36-7 ***HR: 2***
3,5,3'-TRIIODO-4'-ACETYLTHYROFORMIC ACID
mf: $C_{15}H_9I_3O_4$ mw: 633.94

SYNS: 3,5-DIIODO-4-(3'-IODO-4'-ACETOXYPHENOXY)BENZOIC ACID ◇ TBF-43

TOXICITY DATA with REFERENCE
orl-rat TDLo:750 mg/kg (female 30D pre):REP TAKHAA 29,207,70
orl-rat LD50:4600 mg/kg TAKHAA 29,207,70
ipr-rat LD50:480 mg/kg TAKHAA 29,207,70
scu-rat LD50:520 mg/kg TAKHAA 29,207,70
orl-mus LD50:3700 mg/kg TAKHAA 29,207,70
ipr-mus LD50:1 g/kg TAKHAA 29,207,70
scu-mus LD50:2300 mg/kg TAKHAA 29,207,70

SAFETY PROFILE: Moderately toxic by ingestion, intraperitoneal, and subcutaneous routes. Experimental reproductive effects. When heated to decomposition it emits toxic fumes of I^-.

TKQ000 CAS:68596-99-6 ***HR: 3***
3,4,5-TRIIODOBENZENEDIAZONIUM NITRATE
mf: $C_6H_2I_3N_3O_3$ mw: 544.80

SAFETY PROFILE: An unstable explosive sensitive to heat or contact with flame. When heated to decomposition it emits very toxic fumes of I^- and NO_x. See also IODINE COMPOUNDS and NITRATES.

TKQ250 CAS:88-82-4 ***HR: 2***
2,3,5-TRIIODOBENZOIC ACID
mf: $C_7H_3I_3O_2$ mw: 499.80

PROP: Prisms from alc. Mp: 223-224°. Insol in water, ether; sol in hot alc; very sltly sol in benzene.

SYNS: FLORALTONE ◇ JOHNKOLOR ◇ REGIM 8 ◇ REGIN 8 ◇ TIB ◇ TIBA ◇ 2,3,5-TIBA ◇ TRIIODOBENZOIC ACID

TOXICITY DATA with REFERENCE
orl-rat LD50:813 mg/kg GUCHAZ 6,504,73
orl-mus LD50:700 mg/kg QJPPAL 19,483,46
ipr-mus LD50:562 mg/kg JMCMAR 11,1020,68

CONSENSUS REPORTS: Reported in EPA TSCA Inventory. EPA Genetic Toxicology Program.

SAFETY PROFILE: Moderately toxic by ingestion and intraperitoneal routes. When heated to decomposition it emits toxic fumes of I^-.

TKR000 CAS:609-23-4 ***HR: 3***
2,4,6-TRIIODOPHENOL
mf: $C_6H_3I_3O$ mw: 471.79

PROP: Mp: 157-159°.

TOXICITY DATA with REFERENCE
orl-rat LD50:2000 mg/kg 14CYAT 2,1406,63
ivn-mus LD50:180 mg/kg CSLNX* NX#03493

CONSENSUS REPORTS: Reported in EPA TSCA Inventory.

SAFETY PROFILE: Poison by intravenous route. Moderately toxic by ingestion. When heated to decomposition it emits toxic fumes of I^-. See also IODIDES and PHENOL.

TKR500 CAS:100-99-2 ***HR: 3***
TRIISOBUTYL ALUMINUM
DOT: UN 1930
mf: $(C_4H_9)_3Al$ mw: 198.3

PROP: Clear, colorless liquid. D: 0.7859 @ 20°, vap press: 1 mm @ 47°, flash p: $<4°$, fp: 4.3, bp: decomp.

SYNS: TRIISOBUTYLALANE ◇ TRIS(2-METHYLPROPYL)ALUMINIUM

CONSENSUS REPORTS: Reported in EPA TSCA Inventory.

ACGIH TLV: TWA 2 mg(Al)/m^3
DOT Classification: Flammable Solid; Label: Spontaneously Combustible.

SAFETY PROFILE: A poison. Extremely destructive to living tissue. A very dangerous fire hazard; ignites on exposure to air. Incompatible with moisture; acids; air; alcohols; amines; halogens. To fight fire, use CO_2, dry sand, dry chemical. Do not use water, foam, or halogenated extinguishing agents. When heated to decomposition it emits acrid smoke and irritating fumes.

TKR750 CAS:13195-76-1 ***HR: 2***
TRIISOBUTYL BORATE
mf: $C_{12}H_{27}BO_3$ mw: 230.20

PROP: Colorless liquid; odor of isobutyl alc. Bp: 212°, flash p: 185°F (COC), d: 0.843 @ 23°.

SYN: BORIC ACID, TRIISOBUTYL ESTER

TOXICITY DATA with REFERENCE
eye-rbt 100 mg MLD 14KTAK -,706,64
orl-mus LD50:2020 mg/kg USBCC*

SAFETY PROFILE: Moderately toxic by ingestion. An eye irritant. Flammable when exposed to heat, flame or oxidizers. To fight fire, use water spray, foam, dry chemical. When heated to decomposition it emits acrid smoke and irritating fumes. See also ESTERS and BORON COMPOUNDS.

TKS000 CAS:68955-06-6 ***HR: 1***
TRIISOBUTYLENE OXIDE
mf: $C_{12}H_{24}O$ mw: 184.36

SYN: EP-1086

TOXICITY DATA with REFERENCE
eye-rbt 500 mg UCDS** 12/13/63
orl-rat LD50:6690 mg/kg UCDS** 12/13/63
skn-rbt LD50:14 g/kg UCDS** 12/13/63

CONSENSUS REPORTS: Reported in EPA TSCA Inventory.

SAFETY PROFILE: Mildly toxic by ingestion and skin contact. An eye irritant. When heated to decomposition it emits acrid smoke and irritating fumes.

TKS500 CAS:2757-28-0 ***HR: 2***
TRIISOOCTYLAMINE
mf: $C_{24}H_{51}N$ mw: 353.76

SYN: 6,6′,6″-TRIMETHYLTRIHEPTYLAMINE

TOXICITY DATA with REFERENCE
skn-rbt 100 μg/24H open AIHAAP 23,95,62
orl-rat LD50:1620 mg/kg AIHAAP 23,95,62
skn-rbt LD50:3180 mg/kg AIHAAP 23,95,62

CONSENSUS REPORTS: Reported in EPA TSCA Inventory.

SAFETY PROFILE: Moderately toxic by ingestion and skin contact. A skin irritant. When heated to decomposition it emits toxic fumes of NO_x. See also AMINES.

TKT000 CAS:25103-12-2 ***HR: 2***
TRIISOOCTYL PHOSPHITE
mf: $C_{24}H_{54}O_3P$ mw: 421.75

PROP: Liquid. Bp: 161-164°, d: 0.891.

TOXICITY DATA with REFERENCE
skn-rbt 10 mg/24H open MLD AIHAAP 23,95,62
orl-rat LD50:9200 mg/kg ALBRW* #OPB-3,84
skn-rbt LD50:3970 mg/kg AIHAAP 23,95,62

CONSENSUS REPORTS: Reported in EPA TSCA Inventory.

SAFETY PROFILE: Moderately toxic by skin contact. Mildly toxic by ingestion. A skin irritant. When heated to decomposition it emits toxic fumes of PO_x. See also PHOSPHATES.

TKT100 CAS:45173-31-7 ***HR: 2***
TRIISOPENTYLPHOSPHINE
mf: $C_{15}H_{33}P$ mw: 244.45

SYN: TRIS(3-METHYLBUTYL)PHOSPHINE

TOXICITY DATA with REFERENCE
orl-rat LD50:2236 mg/kg GISAAA 47(8),27,82
ihl-rat LC50:1109 g/m³ GISAAA 47(8),27,82
orl-mus LD50:2600 mg/kg GISAAA 47(8),27,82
ihl-mus LC50:988 g/m³ GISAAA 47(8),27,82

SAFETY PROFILE: Moderately toxic by ingestion. Mildly toxic by inhalation. When heated to decomposition it emits toxic fumes of PO_x. See also PHOSPHINE.

TKT200 CAS:101-00-8 ***HR: 1***
TRIISOPROPANOLAMINE BORATE
mf: $C_9H_{18}BNO_3$ mw: 199.09

SYN: BORIC ACID, TRIS(1-AMINO-2-PROPYL) ESTER

TOXICITY DATA with REFERENCE
eye-rbt 10 mg MLD 14KTAK-,693,64
orl-mus LD50:7200 mg/kg 14KTAK-,693,64

SAFETY PROFILE: Mildly toxic by ingestion. An eye irritant. When heated to decomposition it emits toxic fumes of NO_x and boron.

TKT500 CAS:116-17-6 ***HR: 3***
TRIISOPROPYL PHOSPHITE
mf: $C_9H_{21}O_3P$ mw: 208.27

PROP: Liquid. Bp: 63-64°/11 mm.

SYNS: PHOSPHOROUS ACID, TRIISOPROPYL ESTER ◇ PHOSPHOROUS ACID, TRIS(1-METHYLETHYL) ESTER

TOXICITY DATA with REFERENCE
mmo-sat 5 μL/plate MUREAV 28,405,75
sln-dmg-orl 50 mmol/L MUREAV 28,405,75

orl-rat LD50:167 mg/kg ALBRW* #OPB-3,84
ipr-mus LD50:500 mg/kg 14CYAT 2,1918,63

CONSENSUS REPORTS: Reported in EPA TSCA Inventory.

SAFETY PROFILE: Poison by ingestion. Moderately toxic by intraperitoneal route. Mutation data reported. When heated to decomposition it emits toxic fumes of PO_x. See also ESTERS and PHOSPHITES.

TKT750 CAS:19464-55-2 ***HR: 3***
TRIISOPROPYLTIN ACETATE
mf: $C_{11}H_{24}O_2Sn$ mw: 307.04

SYN: ACETOXYTRIISOPROPYLSTANNANE

TOXICITY DATA with REFERENCE
orl-rat LD50:44 mg/kg BJIMAG 15,15,58
ivn-rat LD50:12 mg/kg BJIMAG 15,15,58

OSHA PEL: TWA 0.1 mg(Sn)/m^3 (skin)
ACGIH TLV: TWA 0.1 mg(Sn)/m^3 (skin) (Proposed: TWA 0.1 mg(Sn)/m^3; STEL 0.2 mg(Sn)/m^3 (skin))
NIOSH REL: (Organotin Compounds) TWA 0.1 mg (Sn)/m^3

SAFETY PROFILE: Poison by ingestion and intravenous routes. When heated to decomposition it emits acrid smoke and irritating fumes. See also TIN COMPOUNDS and ESTERS.

TKT850 CAS:73928-00-4 ***HR: 3***
TRIISOPROPYLTIN UNDECYLENATE
mf: $C_{20}H_{42}O_2Sn$ mw: 433.31

SYNS: STANNANE, TRIISOPROPYL(UNDECANOYLOXY)- ◇ UNDECANOIC ACID, TRIISOPROPYLSTANNYL ESTER

TOXICITY DATA with REFERENCE
ivn-mus LD50:32 mg/kg CSLNX* NX#04267

OSHA PEL: TWA 0.1 mg(Sn)/m^3 (skin)
ACGIH TLV: TWA 0.1 mg(Sn)/m^3; STEL 0.2 mg/m^3 (skin)
NIOSH REL: (Organotin Compound):10H TWA 0.1 mg(Sn)/m^3

SAFETY PROFILE: Poison by intravenous route. When heated to decomposition it emits toxic fumes of Sn.

TKU000 CAS:7739-33-5 ***HR: 2***
N-TRI-ISOPROPYL-B-TRIETHYL BORAZOLE
mf: $C_{15}H_{36}B_3N_3$ mw: 290.97

SYN: 2,4,6-TRIETHYL-1,3,5-TRIISOPROPYLBORAZINE

TOXICITY DATA with REFERENCE
ipr-rat LD50:707 mg/kg SCCUR* -,8,61
orl-mus LD50:3530 mg/kg SCCUR* -,8,61
ipr-mus LD50:2460 mg/kg SCCUR* -,8,61

SAFETY PROFILE: Moderately toxic by ingestion and intraperitoneal routes. When heated to decomposition it emits toxic fumes of NO_x. See also ESTERS and BORON COMPOUNDS.

TKU250 ***HR: 3***
TRILEAD DINITRIDE
mf: N_2Pb_3 mw: 649.58

CONSENSUS REPORTS: Lead and its compounds are on the Community Right-To-Know List.

SAFETY PROFILE: Unstable; decomposes explosively. When heated to decomposition it emits toxic fumes of Pb and NO_x. See also LEAD COMPOUNDS.

TKU275 ***HR: 3***
N,N,4-TRILITHIOANILINE
mf: $C_6H_4Li_3N$ mw: 110.93

$Li_2NC_6H_4Li$

SAFETY PROFILE: Explodes on contact with air. When heated to decomposition it emits toxic fumes of NO_x. See also LITHIUM COMPOUNDS.

TKU500 CAS:919-16-4 ***HR: D***
TRILITHIUM CITRATE
mf: $C_6H_5O_7 \cdot 3Li$ mw: 209.93

SYN: CITRIC ACID, TRILITHIUM SALT

TOXICITY DATA with REFERENCE
mnt-mus-ipr 7800 nmol/kg MUREAV 66,33,79

CONSENSUS REPORTS: Reported in EPA TSCA Inventory.

SAFETY PROFILE: Mutation data reported. When heated to decomposition it emits acrid smoke and irritating fumes. See also LITHIUM COMPOUNDS.

TKU650 CAS:39133-31-8 ***HR: 3***
TRIMEBUTINE
mf: $C_{22}H_{29}NO_5$ mw: 387.48

PROP: Crystals from ethanol. Mp: 78-80°C. Sol in methylene chloride.

SYN: (±)-2-(DIMETHYLAMINO)-2-PHENYLBUTYL-3,4,5-TRIMETHOXYBENZOATE

TOXICITY DATA with REFERENCE
ipr-rat LD50:365 mg/kg IYKEDH 15,688,84
scu-rat LD50:3610 mg/kg IYKEDH 15,688,84
ivn-rat LD50:23400 μg/kg IYKEDH 15,688,84
orl-mus LD50:3230 mg/kg IYKEDH 15,688,84

ipr-mus LD50:260 mg/kg IYKEDH 15,688,84
ivn-mus LD50:47800 μg/kg IYKEDH 15,688,84

SAFETY PROFILE: Poison by intravenous and intraperitoneal routes. Moderately toxic by ingestion and subcutaneous routes. When heated to decomposition it emits toxic fumes of NO_x. See also ESTERS.

TKU675 CAS:34140-59-5 ***HR: D***
TRIMEBUTINE MALEATE
mf: $C_{22}H_{29}NO_5 \cdot C_4H_4O_4$ mw: 503.60

PROP: Crystals from water. Mp: 105-106°.

SYNS: DEBRIDAT ◇ DROMOSTAT ◇ TM 906

TOXICITY DATA with REFERENCE
orl-rat TDLo:625 mg/kg (female 17-21D post):REP KSRNAM 16,633,82
orl-rat TDLo:5500 mg/kg (7-17D preg):TER KSRNAM 16,633,82

SAFETY PROFILE: An experimental teratogen. Experimental reproductive effects. When heated to decomposition it emits toxic fumes of NO_x. See also ESTERS.

TKV000 CAS:552-30-7 ***HR: 3***
TRIMELLITIC ANHYDRIDE
mf: $C_9H_4O_5$ mw: 192.13

PROP: Crystals. Mp: 162°, bp: 240-245° @ 14 mm. Sol in acetone, ethyl acetate, dimethylformamide.

SYNS: ANHYDROTRIMELLIC ACID ◇ 1,2,4-BENZENETRICARBOXYLIC ACID ANHYDRIDE ◇ 1,2,4-BENZENETRICARBOXYLIC ACID, CYCLIC 1,2-ANHYDRIDE ◇ 1,2,4-BENZENETRICARBOXYLIC ANHYDRIDE ◇ 4-CARBOXYPHTHALIC ANHYDRIDE ◇ 1,3-DIHYDRO-1,3-DIOXO-5-ISOBENZOFURANCARBOXYLIC ACID ◇ 1,3-DIOXO-5-PHTHALANCARBOXYLIC ACID ◇ DIPHENYLMETHANE-4,4'-DIISOCYANATE-TRIMELLIC ANHYDRIDE-ETHOMID HT POLYMER ◇ NCI-C56633 ◇ TMA ◇ TMAN ◇ TRIMELLIC ACID ANHYDRIDE ◇ TRIMELLIC ACID-1,2-ANHYDRIDE ◇ TRIMELLITIC ACID CYCLIC-1,2-ANHYDRIDE

TOXICITY DATA with REFERENCE
orl-rat LD50:5600 mg/kg DTLVS* 4,415,80

CONSENSUS REPORTS: Reported in EPA TSCA Inventory.

OSHA PEL: TWA 0.005 ppm
ACGIH TLV: TWA 0.005 ppm; (Proposed: CL 0.01 mg/m^3)
DFG MAK: 0.005 ppm (0.04 mg/m^3)
NIOSH REL: (Trimellitic Anhydride): handle as extremely toxic.

SAFETY PROFILE: Mildly toxic by ingestion. Has caused pulmonary edema from inhalation. Irritant to lungs and air passages. May be a powerful allergen. Typical attack consists of breathlessness, wheezing, cough, running nose, immunological sensitization and asthma symptoms. When heated to decomposition it emits acrid smoke and irritating fumes. See also ANHYDRIDES.

TKW000 CAS:12136-15-1 ***HR: 3***
TRIMERCURY DINITRIDE
mf: Hg_3N_2 mw: 629.78

SYN: MERCURY NITRIDE

CONSENSUS REPORTS: Mercury and its compounds are on the Community Right-To-Know List.

DOT Classification: Forbidden

SAFETY PROFILE: Mercury compounds are poisons. An explosive sensitive to friction, impact, heating or contact with sulfuric acid. Incompatible with sulfuric acid. When heated to decomposition it emits very toxic fumes of Hg and NO_x. See also MERCURY COMPOUNDS and NITRIDES.

TKW100 ***HR: 3***
TRIMERESURUS FLAVOVIRIDIS VENOM

SYNS: T. FLAVOVIRIDIS VENOM ◇ VENOM, SNAKE, TRIMERESURUS FLAVOVIRIDIS

TOXICITY DATA with REFERENCE
ipr-mus LD50:800 μg/kg TOXIA6 18,384,80
scu-mus LD50:4300 μg/kg KDIZAA 8,974,57
ivn-mus LD50:3333 μg/kg TOXIA6 5,17,67
ims-mus LD50:7500 μg/kg JJEMAG 33,245,63
ims-rbt LD50:6 mg/kg TOXIA6 18,351,80

SAFETY PROFILE: Poison by subcutaneous, intramuscular, intravenous, and intraperitoneal routes.

TKW500 CAS:68-91-7 ***HR: 3***
TRIMETHIOPHANE
mf: $C_{22}H_{25}N_2OS \cdot C_{10}H_{15}O_4S$ mw: 596.86

SYNS: ARFONAD ◇ ARFONAD CAMPHORSULFONATE ◇ ARFONAD ROCHE ◇ ARPHONAD ◇ CAMFOSULFONATO del d-3-4-(1'DIBENZIL-2-CHETO-IMIDAZOLIDO)-1,2-TRIMETILTHIOPHANIUM (ITALIAN) ◇ 1,3-DIBENZYLDECAHYDRO-2-OXOIMIDAZO(4,5-c)THIENO(1,2-a)THIOLIUM 10-CAMPHORSULFONATE ◇ 1,3-DIBENZYLDECAHYDRO-2-OXO-IMIDAZO(4,5-c)THIENO(1,2-a)THIOLIUM-2-OXO-10-BORANESULFONATE ◇ d-3,4-(1',3'-DIBENZYL-2'-KETO-IMIDAZOLIDO)-1,2-TRIMETHYLENETHIOPHANIUM-d-CAMPHORSULFONATE ◇ METHIOPLEGIUM ◇ NU 2222 ◇ RO 2-2222 ◇ TRIMETAPHAN CAMPHORSULFONATE ◇ TRIMETHAPHAN CAMPHORSULFONATE ◇ TRIMETHAPHAN-10-CAMPHORSULFONATE ◇ TRIMETAPHAN CAMSILATE ◇ TRIMETHAPHAN CAMSYLATE

TOXICITY DATA with REFERENCE
ipr-rat TDLo:6240 mg/kg (13W male):REP YACHDS 3,1805,75
orl-rat LD50:2494 mg/kg NIIRDN 6,528,82
ipr-rat LD50:141 mg/kg NIIRDN 6,528,82
scu-rat LD50:2024 mg/kg NIIRDN 6,528,82
ivn-rat LD50:21 mg/kg CLDND*
orl-mus LD50:927 mg/kg NIIRDN 6,528,82

ipr-mus LD50:112 mg/kg FAATDF 6,35,86
scu-mus LD50:629 mg/kg NIIRDN 6,528,82
ivn-mus LD50:14400 μg/kg RPTOAN 37,7,74
ims-mus LD50:133 mg/kg YKYUA6 28,495,77
orl-dog LD50:400 mg/kg CLDND*
ivn-gpg LD50:13 mg/kg FRPSAX 10,1027,55

SAFETY PROFILE: Poison by ingestion, intraperitoneal, intravenous, and intramuscular routes. Moderately toxic by subcutaneous route. Experimental reproductive effects. When heated to decomposition it emits very toxic fumes of NO_x and SO_x.

TKW750 CAS:554-92-7 ***HR: 3***
TRIMETHOBENZAMIDE HYDROCHLORIDE
mf: $C_{21}H_{28}N_2O_5 \cdot ClH$ mw: 424.97

SYNS: N-(p-(2-(DIMETHYLAMINO)ETHOXY)BENZYL)-3,4,5-TRIMETHOXYBENZAMIDE HYDROCHLORIDE ◇ N-(p-(2-(DIMETHYLAMINO)ETHOXY)-BENZYL)-3,4,5-TRIMETHOXYBENZAMIDE MONOHYDROCHLORIDE ◇ 4-(2-DIMETHYLAMINOETHOXY)-N-(3,4,5-TRIMETHOXYBENZOYL)BENZYLAMINEHYDROCHLORIDE ◇ TIGAN ◇ TIGAN HYDROCHLORIDE

TOXICITY DATA with REFERENCE
orl-mus LD50:1600 mg/kg AIPTAK 174,350,68
ipr-mus LD50:350 mg/kg JPETAB 126,270,59
scu-mus LD50:564 mg/kg AIPTAK 174,350,68
ivn-mus LD50:122 mg/kg AIPTAK 174,350,68

SAFETY PROFILE: Poison by intraperitoneal and intravenous routes. Moderately toxic by ingestion and subcutaneous routes. When heated to decomposition it emits very toxic fumes of Cl^- and NO_x.

TKX000 CAS:8064-90-2 ***HR: 3***
TRIMETHOPRIM and SULPHAMETHOXAZOLE
mf: $C_{14}H_{18}N_4O_3 \cdot C_{10}H_{11}N_3O_3S$ mw:543.66

PROP: A mixture containing 16.7% 2,4-diamino-5-(3,4,5-trimethoxybenzyl)pyrimidine and 83.3% N′-(5-methyl-3-isoxazolyl)sulfanilamide (LANCAO 1,604,77).

SYNS: ABACIN ◇ ABACTRIM ◇ APOSULFATRIM ◇ BACTRAMIN ◇ BACTRIM ◇ BACTROMIN ◇ BAKTAR ◇ BISEPTOL ◇ CHEMITRIM ◇ CO-TRIMOXAZOLE ◇ DRYLIN ◇ ELTRIANYL ◇ EUSAPRIM ◇ FECTRIM ◇ GANTAPRIN ◇ GANTRIM ◇ KEPINOL ◇ LINARIS ◇ MICROTRIM ◇ MOMENTOL ◇ NOPIL ◇ OMSAT ◇ OXAPRIM ◇ PANTOPRIM ◇ SEPTRA ◇ SEPTRAN ◇ SEPTRIM ◇ SEPTRIN ◇ SIGAPRIN ◇ SULFAMETHOXAZOL-TRIMETHOPRIM ◇ SULFOTRIM ◇ SULFOTRIMIN ◇ SULPRIM ◇ SUMETROLIM ◇ SUPRIN ◇ TACUMIL ◇ TELEPRIN ◇ TMS 480 ◇ TRIGONYL ◇ TRIMESULF ◇ TRIMETHOPRIMSULFA ◇ TRIMFORTE ◇ TRIMOSULFA ◇ URO-SEPTRA

TOXICITY DATA with REFERENCE
orl-wmn TDLo:346 mg/kg:KID,SKN,MET LANCAO 1,604,77
ivn-man TDLo:80 mg/kg/4D-I:MET AIMEAS 103,161,85
orl-hmn LDLo:274 mg/kg/10D-I LANCAO 1,831,78
orl-rat LD50:5350 mg/kg KSRNAM 12,2716,78
ipr-rat LD50:1840 mg/kg NIIRDN 6,389,82
orl-mus LD50:3740 mg/kg KSRNAM 12,2716,78
ipr-mus LD50:2010 mg/kg NIIRDN 6,389,82

SAFETY PROFILE: Human poison by ingestion. Moderately toxic experimentally by ingestion and intraperitoneal routes. Human systemic effects by ingestion and intravenous routes: decreased urine volume or anuria, skin dermatitis, fever and other metabolic changes. When heated to decomposition it emits very toxic fumes of NO_x and SO_x.

TKX125 CAS:18559-63-2 ***HR: 3***
(±)-TRIMETHOQUINOL
mf: $C_{19}H_{23}NO_5 \cdot ClH$ mw: 381.89

SYNS: AQ 110 ◇ (±)-1,2,3,4-TETRAHYDRO-1-(3,4,5-TRIMETHOXYBENZYL)-6,7-ISOQUINOLINEDIOL HYDROCHLORIDE ◇ dl-1-(3,4,5-TRIMETHOXYBENZYL)-6,7-DIHYDROXY-1,2,3,4-TETRAHYDROISOQUINOLINE HYDROCHLORIDE ◇ (±)-TRIMETOQUINOL HYDROCHLORIDE ◇ dl-TRIMETOQUINOL HYDROCHLORIDE

TOXICITY DATA with REFERENCE
orl-rat TDLo:4 g/kg (7-14D preg):REP OYYAA2 2,196,68
orl-rat TDLo:4 g/kg (7-14D preg):TER OYYAA2 2,196,68
ipr-rat LD50:311 mg/kg EJPHAZ 5,303,68
ivn-rat LD50:174 mg/kg EJPHAZ 5,303,68
ipr-mus LD50:340 mg/kg EJPHAZ 5,303,68
ivn-mus LD50:130 mg/kg EJPHAZ 5,303,68
ipr-gpg LD50:660 mg/kg EJPHAZ 5,303,68

SAFETY PROFILE: Poison by intravenous and intraperitoneal routes. An experimental teratogen. Experimental reproductive effects. When heated to decomposition it emits toxic fumes of NO_x and HCl. An adrenergic beta-stimulant. See also INOLIN.

TKX250 CAS:635-41-6 ***HR: 2***
TRIMETHOXAZINE
mf: $C_{14}H_{19}NO_5$ mw: 281.34

SYNS: ABBOTT-22370 ◇ LG 50043 ◇ NSC-62939 ◇ PS 2383 ◇ TRIKSAZIN ◇ 4-(3,4,5-TRIMETHOXYBENZOYL)MORPHOLINE ◇ N-(3,4,5-TRIMETHOXYBENZOYL)MORPHOLINE ◇ 3,4,5-TRIMETHOXYBENZOYL-N-TETRAHYDROXAZINE ◇ 3,4,5-TRIMETHOXY-N-BENZOYLTETRAHYDROXAZINE ô N-(3,4,5-TRIMETHOXYBENZOYL) TETRAHYDRO-1,4-OXAZINE ◇ TRIMETOGINE ◇ TRIMETOZIN ◇ TRIMETOZINA ◇ TRIMETOZINE ◇ TRIOXAZIN ◇ V 7

TOXICITY DATA with REFERENCE
orl-mus TDLo:7 g/kg (female 7-13D post):REP YACHDS 4,2526,76
orl-rat TDLo:2100 mg/kg (female 9-15D post):TER YACHDS 4,2526,76
orl-rat LD50:1800 mg/kg ARZNAD 21,719,71
ipr-rat LD50:1080 mg/kg YACHDS 4,2487,76
scu-rat LD50:1400 mg/kg YACHDS 4,2487,76
orl-mus LD50:2400 mg/kg YACHDS 4,2487,76
ipr-mus LD50:1150 mg/kg YACHDS 4,2487,76
scu-mus LD50:1550 mg/kg THERAP 20,401,65

ivn-mus LD50:960 mg/kg 27ZQAG -,312,72
par-mus LD50:1200 mg/kg RPTOAN 33,70,70

SAFETY PROFILE: Moderately toxic by ingestion, intraperitoneal, subcutaneous, parenteral, and intravenous routes. An experimental teratogen. Experimental reproductive effects. When heated to decomposition it emits toxic fumes of NO_x.

TKX500 CAS:5688-80-2 ***HR: 3***
3,4,5-TRIMETHOXYAMPHETAMINE HYDROCHLORIDE
mf: $C_{12}H_{19}NO_3 \cdot ClH$ mw: 261.78

SYNS: α-METHYL-3,4,5-TRIMETHOXYPHENETHYLAMINEHYDROCHLORIDE ◇ 3,4,5-TRIMETHOXY-α-METHYL-β-PHENYLETHYLAMINE HYDROCHLORIDE ◇ 1-(3,4,5-TRIMETHOXYPHENYL)-2-AMINOPROPANE

TOXICITY DATA with REFERENCE
orl-man TDLo:880 μg/kg:CNS JMSCA9 101,317,55
ipr-rat LD50:149 mg/kg TXAPA9 25,299,73
ipr-mus LD50:250 mg/kg JMCMAR 13,26,70
ivn-dog LD50:23 mg/kg TXAPA9 25,299,73
ivn-mky LD50:31 mg/kg TXAPA9 25,299,73
ipr-gpg LD50:172 mg/kg TXAPA9 25,299,73

SAFETY PROFILE: Poison by intraperitoneal and intravenous routes. Human systemic effects by ingestion: central nervous system effects. When heated to decomposition it emits very toxic fumes of NO_x and HCl. See also BENZEDRINE.

TKX700 CAS:2169-44-0 ***HR: 3***
1,2,10-TRIMETHOXY-6a-α-APORPHIN-9-OL
mf: $C_{20}H_{23}NO_4$ mw: 341.44

SYNS: LAUROSCHOLTZINE ◇ N-METHYLLAUROTETANINE ◇ ROGERSINE ◇ (s)-5,6,6a,7-TETRAHYDRO-1,2,10-TRIMETHOXY-6-METHYL-4H-DIBENZO(de,g)QUINOLIN-9-OL

TOXICITY DATA with REFERENCE
orl-mus LD50:450 mg/kg APFRAD 38,537,80
ipr-mus LD50:170 mg/kg APFRAD 38,537,80
ivn-mus LD50:90 mg/kg APFRAD 38,537,80

SAFETY PROFILE: Poison by intravenous and intraperitoneal routes. Moderately toxic by ingestion. When heated to decomposition it emits toxic fumes of NO_x.

TKY000 CAS:3086-62-2 ***HR: 2***
3,4,5-TRIMETHOXYBENZAMIDE
mf: $C_{10}H_{13}NO_4$ mw: 211.24

TOXICITY DATA with REFERENCE
ipr-mus LD50:750 mg/kg BCPCA6 11,639,62

CONSENSUS REPORTS: Reported in EPA TSCA Inventory.

SAFETY PROFILE: Moderately toxic by intraperitoneal route. When heated to decomposition it emits toxic fumes of NO_x. See also AMIDES.

TKY250 CAS:621-23-8 ***HR: 2***
1,3,5-TRIMETHOXYBENZENE
mf: $C_9H_{12}O_3$ mw: 168.21

PROP: Mp: 51-53°, bp: 255°, flash p: 186° F.

SYN: PHLOROGLUCINOL TRIMETHYL ETHER

TOXICITY DATA with REFERENCE
orl-mus LD50:1480 mg/kg OYYAA2 3,187,69
ipr-mus LD50:580 mg/kg OYYAA2 3,187,69
scu-mus LD50:2800 mg/kg OYYAA2 3,187,69

CONSENSUS REPORTS: Reported in EPA TSCA Inventory.

SAFETY PROFILE: Moderately toxic by ingestion, intraperitoneal, and subcutaneous routes. When heated to decomposition it emits acrid smoke and irritating fumes. See also ETHERS.

TKZ000 CAS:738-70-5 ***HR: 3***
5-(3,4,5-TRIMETHOXYBENZYL)-2,4-DIAMINOPYRIMIDINE
mf: $C_{14}H_{18}N_4O_3$ mw: 290.36

SYNS: BW 56-72 ◇ 2,4-DIAMINO-5-(3,4,5-TRIMETHOXYBENZYL)PYRIMIDINE ◇ MONOPRIM ◇ NIH 204 ◇ NSC-106568 ◇ PROLOPRIM ◇ SYRAPRIM ◇ TIEMPE ◇ TRIMANYL ◇ TRIMETHOPRIM ◇ TRIMETHOPRIOM ◇ 5-((3,4,5-TRIMETHOXYPHENYL)-METHYL)-2,4-PYRIMIDINEDIAMINE ◇ TRIMOPAN ◇ TRIMPEX ◇ WELLCOPRIM

TOXICITY DATA with REFERENCE
dni-esc 5 mg/L CBINA8 17,113,77
sln-nsc 62 mg/L MUREAV 167,35,86
orl-rat TDLo:2250 mg/kg (11-13D preg):TER ANANAU 149,151,81
orl-rat TDLo:2250 mg/kg (11-13D preg):REP ANANAU 149,151,81
orl-rat LD50:200 mg/kg 14XBAV -,367,64
orl-mus LD50:3960 mg/kg KSRNAM 13,115,79
ipr-mus LD50:1870 mg/kg NKRZAZ 21,175,73
ivn-mus LD50:200 mg/kg BJPCAL 33,72,68

CONSENSUS REPORTS: EPA Genetic Toxicology Program.

SAFETY PROFILE: Poison by ingestion and intravenous routes. Moderately toxic by intraperitoneal route. An experimental teratogen. Experimental reproductive effects. Mutation data reported. When heated to decomposition it emits toxic fumes of NO_x.

TLA000 CAS:10138-89-3 ***HR: 2***
1,1,3-TRIMETHOXYBUTANE
mf: $C_7H_{16}O_3$ mw: 148.23

TOXICITY DATA with REFERENCE
skn-rbt 10 mg/24H open MLD AMIHBC 10,61,54
eye-rbt 500 mg open AMIHBC 10,61,54
orl-rat LD50:1480 mg/kg AMIHBC 10,61,54
ihl-rat LCLo:2000 ppm/4H AMIHBC 10,61,54

SAFETY PROFILE: Moderately toxic by ingestion. Mildly toxic by inhalation. A skin and eye irritant. When heated to decomposition it emits acrid smoke and irritating fumes.

TLA250 CAS:34346-90-2 ***HR: 3***
3,4,5-TRIMETHOXYCINNAMALDEHYDE
mf: $C_{12}H_{14}O_4$ mw: 222.26

SYNS: TMCA ◇ 3-(3,4,5-TRIMETHOXYPHENYL)-2-PROPENAL

TOXICITY DATA with REFERENCE
scu-rat TDLo:100 mg/kg:ETA JNCIAM 47,1037,71
mul-rat TDLo:250 mg/kg/7D-I:ETA,REP BJCAAI 26,504,72

SAFETY PROFILE: Experimental reproductive effects. Questionable carcinogen with experimental tumorigenic data. When heated to decomposition it emits acrid smoke and irritating fumes. See also ALDEHYDES.

TLA500 CAS:26219-22-7 ***HR: 3***
4-(3,4,5-TRIMETHOXYCINNAMOYL)-1-PIPERAZINEACETIC ACID ETHYL ESTER HYDROCHLORIDE
mf: $C_{20}H_{28}N_2O_6$•ClH mw: 428.96

SYN: 4-(3′,4′,5′-TRIMETHOXYCINNAMOYL)-1-(ETHOXYCARBONYLEMETHYL)PIPERAZINEHYDROCHLORIDE

TOXICITY DATA with REFERENCE
orl-mus LD50:1300 mg/kg CHTPBA 4,293,69
ivn-mus LD50:300 mg/kg USXXAM #3590034

SAFETY PROFILE: Poison by intravenous route. Moderately toxic by ingestion. When heated to decomposition it emits very toxic fumes of NO_x and HCl. Used to treat angina. See also ESTERS.

TLA525 CAS:24536-75-2 ***HR: 3***
4-(3,4,5-TRIMETHOXYCINNAMOYL)-1-PIPERAZINEACETIC ACID ETHYL ESTER MALEATE
mf: $C_{20}H_{28}N_2O_6$•$C_4H_4O_4$ mw: 508.58

SYN: ((TRIMETHOXY-3′,4′,5′ CINNAMOYL)-4 PIPERAZINYL)-2 ACETATE D'ETHYLE (MALEATE) (FRENCH)

TOXICITY DATA with REFERENCE
orl-rat LD50:4190 mg/kg THERAP 26,845,71
ivn-rat LD50:360 mg/kg THERAP 26,845,71
ims-rat LD50:1115 mg/kg THERAP 26,845,71
orl-mus LD59:1190 mg/kg THERAP 26,845,71
ivn-mus LD50:360 mg/kg THERAP 26,845,71
ims-mus LD50:845 mg/kg THERAP 26,845,71

SAFETY PROFILE: Poison by intravenous route. Moderately toxic by ingestion and intramuscular routes. When heated to decomposition it emits toxic fumes of NO_x.

TLA600 CAS:6163-73-1 ***HR: 1***
TRI-(2-METHOXYETHANOL)PHOSPHATE
mf: $C_9H_{21}O_7P$ mw: 272.27

SYNS: ETHANOL, 2-METHOXY-, PHOSPHATE (3:1) ◇ 2-METHOXYETHANOL PHOSPHATE (3:1) ◇ TRIS-(2-METHOXYETHYL)FOSFAT

TOXICITY DATA with REFERENCE
skn-rbt 500 mg/24H MLD 85JCAE-,1137,86
eye-rbt 500 mg/24H MLD 85JCAE-,1137,86
orl-rat LD50:17200 mg/kg TXAPA9 28,313,74

CONSENSUS REPORTS: Reported in EPA TSCA Inventory.

SAFETY PROFILE: Mildly toxic by ingestion. A skin and eye irritant. When heated to decomposition it emits toxic fumes of PO_x.

TLB750 CAS:2487-90-3 ***HR: 2***
TRIMETHOXY SILANE
mf: $C_3H_{10}O_3Si$ mw: 122.22

PROP: Liquid. Mp: -115°, bp: 81°, refr index: 1.3580, d: 0.960.

TOXICITY DATA with REFERENCE
orl-rat LD50:9330 mg/kg AIHAAP 30,470,69
ihl-rat LC50:125 ppm/4H EPASR* 8EHQ-0680-0347
skn-rbt LD50:6300 mg/kg EPASR* 8EHQ-0680-0347

CONSENSUS REPORTS: Reported in EPA TSCA Inventory.

SAFETY PROFILE: Moderately toxic by inhalation. Mildly toxic by ingestion and skin contact. When heated to decomposition it emits acrid smoke and irritating fumes. See also SILANE.

TLC000 CAS:4420-74-0 ***HR: 2***
TRIMETHOXYSILYLPROPANETHIOL
mf: $C_6H_{16}O_3SSi$ mw: 196.37

SYNS: Γ-MERCAPTOPROPYLTRIMETHOXYSILANE ◇ 3-MERCAPTOPROPYLTRIMETHOXYSILANE ◇ SILICONE A-189 ◇ UNION CARBIDE 1-189

TOXICITY DATA with REFERENCE
skn-rbt 500 mg open MLD UCDS** 4/2/71
orl-rat LD50:2940 mg/kg AIHAAP 30,470,69
ipr-mus LD50:633 mg/kg DANKAS 229(4),1011,76
skn-rbt LD50:5880 mg/kg AIHAAP 30,470,69

CONSENSUS REPORTS: Reported in EPA TSCA Inventory.

SAFETY PROFILE: Moderately toxic by ingestion and intraperitoneal routes. Mildly toxic by skin contact. A skin irritant. When heated to decomposition it emits toxic fumes of SO_x. See also MERCAPTANS and SILANE.

TLC250 CAS:2530-85-0 ***HR: 1***
3-(TRIMETHOXYSILYL)-1-PROPANOL METHACRYLATE
mf: $C_{10}H_{20}O_5Si$ mw: 248.39

SYNS: (3-HYDROXYPROPYL)TRIMETHOXYSILANEMETHACRYLATE ◇ Γ-METHACRYLOXYPROPYLTRIMETHOXYSILANE ◇ 2-METHYL-2-PROPENOICACID-3-(TRIMETHOXYSILYL)PROPYL ESTER ◇ SILICONE 1-174 ◇ TRIMETHOXYSILYL-3-PROPYLESTER KYSELINY METHAKRYLOVE (CZECH) ◇ 3-(TRIMETHOXYSILYL)PROPYL ESTER METHACRYLIC ACID ◇ UNION CARBIDE 1-174

TOXICITY DATA with REFERENCE
skn-rbt 500 mg/24H MOD 28ZPAK -,220,72
eye-rbt 500 mg/24H MLD 28ZPAK -,220,72

CONSENSUS REPORTS: Reported in EPA TSCA Inventory.

SAFETY PROFILE: A skin and eye irritant. When heated to decomposition it emits acrid smoke and irritating fumes. See also ESTERS and SILANE.

TLC500 CAS:1760-24-3 ***HR: 3***
N-(3-TRIMETHOXYSILYLPROPYL)-ETHYLENEDIAMINE
mf: $C_8H_{22}N_2O_3Si$ mw: 222.41

PROP: Bp: 146°/15 mm, refr index: 1.4450, d: 1.010, flash p: >230°F.

SYNS: (3-(2-AMINOETHYL)AMINOPROPYL)TRIMETHOXYSILANE ◇ SILICONE A-1120

TOXICITY DATA with REFERENCE
skn-rbt 500 mg open MLD UCDS** 4/2/71
eye-rbt 15 mg SEV UCDS** 4/2/71
orl-rat LD50:7460 mg/kg UCDS** 4/2/71
ivn-mus LD50:180 mg/kg CSLNX* NX#03517
skn-rbt LDLo:16 g/kg UCDS** 4/2/71

CONSENSUS REPORTS: Reported in EPA TSCA Inventory.

SAFETY PROFILE: Poison by intravenous route. Mildly toxic by ingestion and skin contact. A skin and severe eye irritant. When heated to decomposition it emits toxic fumes of NO_x. See also SILANE and AMINES.

TLC850 ***HR: 2***
3,4,5-TRIMETHOXY-α-VINYLBENZYL ALCOHOL ACETATE
mf: $C_{14}H_{18}O_5$ mw: 266.32

SYN: 1'-ACETOXYELEMICIN

TOXICITY DATA with REFERENCE
mmo-sat 2 μmol/plate CRNGDP 7,2089,86
ipr-mus TDLo:408 mg/kg/4D-I:CAR CNREA8 47,2275,87

SAFETY PROFILE: Questionable carcinogen with experimental carcinogenic data. Mutation data reported. When heated to decomposition it emits acrid smoke and irritating fumes.

TLD000 CAS:2768-02-7 ***HR: 2***
TRIMETHOXYVINYLSILANE
mf: $C_5H_{12}O_3Si$ mw: 148.26

PROP: Liquid. Bp: 123°, refr index: 1.3920, d: 1.130, flash p: 73° F.

SYN: VINYL TRIMETHOXY SILANE

TOXICITY DATA with REFERENCE
orl-rat LD50:11300 mg/kg AIHAAP 30,470,69
skn-rbt LD50:3540 mg/kg AIHAAP 30,470,69

CONSENSUS REPORTS: Reported in EPA TSCA Inventory.

SAFETY PROFILE: Moderately toxic by skin contact. Mildly toxic by ingestion. When heated to decomposition it emits acrid smoke and irritating fumes. See also SILANE.

TLD250 CAS:5096-21-9 ***HR: 3***
2,4,6-TRIMETHYLACETANILIDE
mf: $C_{11}H_{15}NO$ mw: 177.27

SYNS: ACETOMESIDIDE ◇ 2',4',6'-TRIMETHYLACETANILIDE ◇ N-(2,4,6-TRIMETHYLPHENYL)ACETAMIDE

TOXICITY DATA with REFERENCE
orl-rat TDLo:3500 mg/kg:ETA CNREA8 26,619,66
orl-rat LDLo:3500 mg/kg CNREA8 26,619,66
orl-mus LD50:750 mg/kg TXAPA9 19,20,71

SAFETY PROFILE: Moderately toxic by ingestion. Questionable carcinogen with experimental tumorigenic data. When heated to decomposition it emits toxic fumes of NO_x.

TLD272 CAS:75-24-1 ***HR: 3***
TRIMETHYLALUMINUM
DOT: UN 1103
mf: C_3H_9Al mw: 72.09

PROP: Mp: 15°, bp: 125-126°, d: 0.752.

SYNS: TRIMETHYLALANE ◇ TRIMETHYLALUMINIUM (DOT)

ACGIH TLV: TWA 2 mg(Al)/m^3
DOT Classification: Flammable Solid; Label: Spontaneously Combustible

SAFETY PROFILE: Extremely pyrophoric flammable solid. Mixtures with dichlorodi-μ-chlorobis(pentamethylcyclopentadienyl)dirhodium + air ignite and burn violently. See also ALUMINUM COMPOUNDS.

TLD500 CAS:75-50-3 ***HR: 3***
TRIMETHYLAMINE
DOT: UN 1083/UN 1297
mf: C_3H_9N mw: 59.13

PROP: Colorless gas. Pungent, fishy, ammoniacal odor; saline taste. Bp: 2.87°, lel: 2%, uel: 11.6%, fp: −117.1°, d: 0.662 @ −5°, autoign temp: 374°F, vap d: 2.0, flash p: 20°F (CC). Misc with alc; sol in ether, benzene, toluene, xylene, chloroform.

SYNS: N,N-DIMETHYLMETHANAMINE ◇ TMA ◇ TRIMETHYLAMINE, anhydrous (DOT) ◇ TRIMETHYLAMINE, aqueous solution (DOT) ◇ TRIMETHYLAMINE, aqueous solutions containing not more than 30% of trimethylamine (DOT)

TOXICITY DATA with REFERENCE
ihl-rat LCLo:3500 ppm/4H TOXID9 4,68,84
ivn-mus LD50:90 mg/kg MPHEAE 16,529,67
scu-mus LDLo:1000 mg/kg BBMS** -,-,48
scu-rbt LDLo:800 mg/kg CRSBAW 83,481,20
ivn-rbt LDLo:400 mg/kg BBMS** -,-,48
rec-rbt LDLo:800 mg/kg CRSBAW 83,481,20
scu-frg LDLo:2000 mg/kg SAPHAO 10,201,1900
ihl-mam LC50:19 g/m^3 TPKVAL 14,80,75

CONSENSUS REPORTS: Reported in EPA TSCA Inventory.

OSHA PEL: TWA 10 ppm; STEL 15 ppm
ACGIH TLV: TWA 10 ppm; STEL 15 ppm; (Proposed: TWA 5 ppm; STEL 15 ppm)
DOT Classification: Flammable Gas; Label: Flammable Gas, Anhydrous; Flammable Liquid; Label: Flammable Liquid.

SAFETY PROFILE: Poison by intravenous route. Moderately toxic by subcutaneous and rectal routes. Mildly toxic by inhalation. A very dangerous fire hazard when exposed to heat or flame. Self-reactive. Moderately explosive in the form of vapor when exposed to heat or flame. Can react with oxidizing materials. To fight fire, stop flow of gas. Potentially explosive reaction with bromine + heat; ethylene oxide; triethynylaluminum. When heated to decomposition it emits toxic fumes of NO_x. See also AMINES.

TLD750 CAS:75-50-3 ***HR: 3***
TRIMETHYLAMINE (anhydrous)

CONSENSUS REPORTS: Reported in EPA TSCA Inventory.

DOT Classification: Flammable Gas; Label: Flammable Gas.

SAFETY PROFILE: A poison. Flammable when exposed to heat or flame. See also TRIMETHYLAMINE.

TLE000 CAS:75-50-3 ***HR: 3***
TRIMETHYLAMINE, aqueous solution

PROP: A 25 wt % solution in water. D: 0.932, flash p: 38° F.

CONSENSUS REPORTS: Reported in EPA TSCA Inventory.

DOT Classification: Flammable Liquid; Label: Flammable Liquid.

SAFETY PROFILE: Corrosive. Probably a poison. Flammable liquid when exposed to heat or flame. See also TRIMETHYLAMINE.

TLE100 CAS:1184-78-7 ***HR: 3***
TRIMETHYLAMINE OXIDE
mf: C_3H_9NO mw: 75.11

PROP: Hygroscopic needles. Mp: 255-257°. Sol in water and alc.

SAFETY PROFILE: An unstable explosive. When heated to decomposition it emits toxic fumes of NO_x. See also TRIMETHYLAMINE OXIDE, DIHYDRATE and AMINES.

TLE250 CAS:62637-93-8 ***HR: 1***
TRIMETHYLAMINE OXIDE, DIHYDRATE
mf: $C_3H_9NO \cdot 2H_2O$ mw: 111.17

SYNS: N,N-DIMETHYLMETHANAMINE OXIDE, DIHYDRATE ◇ TRIMETHYLAMMONIUMOXID HYDRAT (CZECH)

TOXICITY DATA with REFERENCE
eye-rbt 500 mg/24H MOD 28ZPAK -,74,72
orl-rat LD50:8700 mg/kg 28ZPAK -,74,72
scu-rbt LDLo:3 g/kg HBAMAK 4,1289,35

SAFETY PROFILE: Mildly toxic by ingestion and subcutaneous routes. An eye irritant. Can explode during concentration. When heated to decomposition it emits toxic fumes of NO_x and NH_3. See also AMINES.

TLE500 CAS:22755-36-8 ***HR: 3***
TRIMETHYLAMINE-N-OXIDE PERCHLORATE
mf: $C_3H_{10}ClNO_5$ mw: 175.57

$(CH_3)_3N^+OHClO_4^-$

SYNS: TRIMETHYLAMINE OXIDE PERCHLORATE ◇ TRIMETHYLHYDROXYLAMMONIUM PERCHLORATE

SAFETY PROFILE: A shock- and heat-sensitive explosive. When heated to decomposition it emits toxic fumes of NO_x and Cl^-. See also PERCHLORATES and AMINES.

TLE750 CAS:54-88-6 ***HR: 3***
2,N,N-TRIMETHYL-4-AMINOAZOBENZENE
mf: $C_{15}H_{17}N_3$ mw: 239.35

SYNS: N,N-DIMETHYL-4-(PHENYLAZO)-m-TOLUIDINE ◇ 2-MeDAB ◇ 2-METHYL-DAB ◇ 2-METHYL-N,N-DIMETHYL-4-AMINOAZO- BENZENE ◇ 2-METHYL-4-DIMETHYLAMINOAZOBENZENE

TOXICITY DATA with REFERENCE
mma-sat 120 μg/plate PNASA6 72,5135,75
mrc-smc 5 pph JNCIAM 62,901,79
otr-rat:emb 1200 μg/L JJIND8 67,1303,81
dns-rat:lvr 1 μmol/L CNREA8 42,3010,82
dni-rat-orl 1080 mg/kg/30D-I CBINA8 48,221,84
oms-rat-orl 1080 mg/kg30D-I CBINA8 48,221,84
sce-rat:lvr 40 μmol/L MUREAV 93,409,82
scu-mus TDLo:40 mg/kg/5D-I:NEO JNCIAM 47,593,71
ipr-mus LD50:345 mg/kg JJIND8 62,911,79

CONSENSUS REPORTS: EPA Genetic Toxicology Program.

SAFETY PROFILE: Poison by intraperitoneal route. Questionable carcinogen with experimental neoplastigenic data. Mutation data reported. When heated to decomposition it emits toxic fumes of NO_x. See also AROMATIC AMINES.

TLF000 CAS:73728-79-7 ***HR: 3***
3,2',5'-TRIMETHYL-4-AMINODIPHENYL
mf: $C_{15}H_{17}N$ mw: 211.33

SYN: 3,2',5'-TRIMETHYLBIPHENYLAMINE

TOXICITY DATA with REFERENCE
scu-rat TDLo:1400 mg/kg/W-I:ETA BMBUAQ 14,141,58

SAFETY PROFILE: Questionable carcinogen with experimental tumorigenic data. When heated to decomposition it emits toxic fumes of NO_x. See also AROMATIC AMINES.

TLF250 CAS:7145-92-8 ***HR: 3***
TRIMETHYL-3-AMINOPHENYLARSONIUM CHLORIDE
mf: $C_9H_{15}AsN•Cl$ mw: 247.62

TOXICITY DATA with REFERENCE
ivn-rat LDLo:40 mg/kg JACSAT 63,1493,41

CONSENSUS REPORTS: Arsenic and its compounds are on the Community Right-To-Know List.

OSHA PEL: TWA 0.5 mg(As)/m^3

SAFETY PROFILE: Poison by intravenous route. When heated to decomposition it emits very toxic fumes of As, NO_x, and Cl^-. See also ARSENIC COMPOUNDS and CHLORIDES.

TLF500 CAS:513-10-0 ***HR: 3***
S-(2-(N,N,N-TRIMETHYLAMMONIO)ETHYL)-O,O-DIETHYLPHOSPHOROTHIOLATE IODIDE
mf: $C_9H_{23}NO_3PS•I$ mw: 383.26

SYNS: 2-DIAETHOXYPHOSPHINYL-THIOAETHYL-TRIMETHYL-AMMONIUM-JODID (GERMAN) ◇ N-(2-(DIETHOXYPHOSPHINYL-THIO)ETHYL)TRIMETHYLAMMONIUM IODIDE ◇ 2-DIETHOXY-PHOSPHINYLTHIOETHYL-TRIMETHYLAMMONIUM IODIDE ◇ DIETHOXYPHOSPHORYL-THIOCHOLINE IODIDE ◇ (2-(O,O-DIETHYLPHOSPHOROTHIO)ETHYL)TRIMETHYLAMMONIUM,IODIDE ◇ O,O-DIETHYL-S-2-TRIMETHYLAMMONIUM ETHYLPHOSPHONOTHIOLATE IODIDE ◇ S-(2-DIMETHYLAMINO- ETHYL)-O,O-DIETHYLPHOSPHORITHIOATE METHIODIDE ◇ ECHODIDE ◇ ECHOTHIOPHATE ◇ ECHOTHIOPHATE IODIDE ◇ ECOTHIOPATE IODIDE ◇ ECOTHIOPHATE IODIDE ◇ S-ESTER of (2-MERCAPTO-ETHYL)TRIMETHYLAMMONIUM IODIDE with O,O-DIETHYL PHOSPHOROTHIOATE ◇ (2-MERCAPTOETHYL) TRIMETHYLAMMONIUM IODIDE S-ESTER with O,O-DIETHYL PHOSPHOROTHIOATE ◇ 217 MI ◇ PHOSPHOLINE (pharmaceutical) ◇ PHOSPHOLINE IODIDE

TOXICITY DATA with REFERENCE
ipr-mus LD50:140 μg/kg TXAPA9 6,269,64
scu-mus LD50:130 μg/kg JMCMAR 19,810,76
ocu-rbt LD50:250 μg/kg AJOPAA 53,512,62

SAFETY PROFILE: Poison by intraperitoneal, subcutaneous, and ocular routes. When heated to decomposition it emits very toxic fumes of NO_x, PO_x, SO_x, NH_3, and I^-. See also IODIDES and MERCAPTANS.

TLF750 CAS:593-81-7 ***HR: 3***
TRIMETHYLAMMONIUM CHLORIDE
mf: $C_3H_9N•ClH$ mw: 95.59

PROP: Hygroscopic. Mp: 283-284° (dec.)

SYN: TRIMETHYLAMINE HYDROCHLORIDE

TOXICITY DATA with REFERENCE
ivn-mus LD50:325 mg/kg MPHEAE 16,529,67

CONSENSUS REPORTS: Reported in EPA TSCA Inventory.

SAFETY PROFILE: Poison by intravenous route. When heated to decomposition it emits very toxic fumes of NO_x, NH_3, and HCl. See also AMINES.

TLF800 ***HR: 3***
TRIMETHYLAMMONIUM PERCHLORATE
mf: $C_3H_{10}ClNO_4$ mw: 159.57

$(CH_3)_3N^+HClO_4^-$

SAFETY PROFILE: Burns violently. A more powerful rocket propellent than ammonium perchlorate. When heated to decomposition it emits toxic fumes of Cl^-, NH_3, and NO_x. See also PERCHLORATES.

TLG000 CAS:14149-43-0 ***HR: 3***
N-(3-TRIMETHYLAMMONIUMPROPYL)-N-METHYLCAMPHIDINIUM BIS(METHYLSULFATE)
mf: $C_{17}H_{36}N_2{\cdot}2CH_3O_4S$ mw: 490.75

SYNS: BARATOL ◇ CAMPHIDONIUM ◇ EUPREX ◇ HA 106 ◇ OSTENOL ◇ OSTENSIN ◇ TRIMETHIDINIUM BIMETHOSULFATE ◇ TRIMETHIDINIUM METHOSULFATE ◇ N-(Γ-TRIMETHYLAMMONIUMPROPYL)-N-METHYLCAMPHIDINIUM DIMETHYL SULFATE ◇ N-(Γ-TRIMETHYLAMMONIUMPROPYL)-N-METHYLCAMPHIDINIUM-DIMETHYLSULPHATE (GERMAN) ◇ N-(Γ-TRIMETHYLAMMONIUMPROPYL)-N-METHYLCAMPHIDINIUM METHYL SULFATE (GERMAN) ◇ WY-1395

TOXICITY DATA with REFERENCE
scu-rat LD50:540 mg/kg ARZNAD 7,123,57
ivn-rat LD50:66 mg/kg ARZNAD 7,123,57
orl-mus LD50:65 mg/kg ARZNAD 7,123,57
ipr-mus LD50:240 mg/kg ARZNAD 7,123,57
scu-mus LD50:104 mg/kg ARZNAD 7,123,57
ivn-mus LD50:25 mg/kg ARZNAD 7,123,57

SAFETY PROFILE: Poison by ingestion, intravenous, intraperitoneal, and subcutaneous routes. When heated to decomposition it emits very toxic fumes of NO_x, NH_3, and SO_x. See also SULFATES.

TLG250 CAS:137-17-7 ***HR: 3***
2,4,5-TRIMETHYLANILINE
mf: $C_9H_{13}N$ mw: 135.23

SYNS: 1-AMINO-2,4,5-TRIMETHYLBENZENE ◇ psi-CUMIDINE ◇ NCI-C02299 ◇ PSEUDOCUMIDINE ◇ 1,2,4-TRIMETHYL-5-AMINOBENZENE ◇ 2,4,5-TRIMETHYLANILIN (CZECH) ◇ 2,4,5-TRIMETHYLBENZENAMINE

TOXICITY DATA with REFERENCE
mma-sat 10 μg/plate ENMUDM 8(Suppl 7),1,86
dnd-ham:lng 10 mmol/L MUREAV 77,317,80
orl-rat TDLo:7000 mg/kg/101W-C:CAR NCITR* NCI-CG-TR-160,79
orl-mus TDLo:17 g/kg/2Y-C:CAR NCITR* NCI-CG-TR-160,79
orl-rat TD:141 g/kg/2Y-C:ETA IARC** 27,177,82
orl-rat LD50:1250 mg/kg MarJV# 29MAR77

CONSENSUS REPORTS: IARC Cancer Review: Group 3 IMEMDT 7,56,87; Animal Limited Evidence IMEMDT 27,177,82. NCI Carcinogenesis Bioassay (feed); Clear Evidence: mouse, rat NCITR* NCI-CG-TR-160,79.

DFG MAK: Animal Carcinogen, Suspected Human Carcinogen.

SAFETY PROFILE: Confirmed carcinogen with experimental carcinogenic and tumorigenic data. Moderately toxic by ingestion. Mutation data reported. When heated to decomposition it emits toxic fumes of NO_x. Used as a dye, pigment, and printing ink. See also ANILINE DYES.

TLG500 CAS:88-05-1 ***HR: 3***
2,4,6-TRIMETHYLANILINE
mf: $C_9H_{13}N$ mw: 135.23

PROP: Liquid. Mp: 233°, d: 0.96, refr index: 1.5510.

SYNS: AMINOMESITYLENE ◇ 2-AMINOMESITYLENE ◇ 1-AMINO-2,4,6-TRIMETHYLBENZEN (CZECH) ◇ 2-AMINO-1,3,5-TRIMETHYLBENZENE ◇ MESIDIN (CZECH) ◇ MESIDINE ◇ MESITYLAMINE ◇ MEZIDINE ◇ 2,4,6-TRIMETHYLBENZENAMINE

TOXICITY DATA with REFERENCE
skn-rbt 500 mg/24H MOD 28ZPAK -,66,72
eye-rbt 20 mg/24H SEV 28ZPAK -,66,72
dnd-ham:lng 3 mmol/L/2H MUREAV 77,317,80
orl-rat TDLo:4000 mg/kg/78W-C:CAR JNCIAM 58,377,77
orl-rat TD:4200 mg/kg/78W-C:CAR AICCA6 20,1364,64
orl-rat LD50:743 mg/kg MarJV# 29MAR77
orl-mus LD50:590 mg/kg 85GMAT -,20,82
ihl-mus LC50:290 mg/m^3/2H 85GMAT -,20,82

CONSENSUS REPORTS: IARC Cancer Review: Group 3 IMEMDT 7,56,87; Animal Inadequate Evidence IMEMDT 27,177,82. EPA Extremely Hazardous Substances List. Reported in EPA TSCA Inventory.

SAFETY PROFILE: Poison by inhalation. Moderately toxic by ingestion. A skin and severe eye irritant. Questionable carcinogen with experimental carcinogenic data. Mutation data reported. When heated to decomposition it emits toxic fumes of NO_x.

TLG750 CAS:21436-97-5 ***HR: 3***
2,4,5-TRIMETHYLANILINE HYDROCHLORIDE
mf: $C_9H_{13}N{\cdot}ClH$ mw: 171.69

SYNS: 1-AMINO-2,4,5-TRIMETHYLBENZENE HYDROCHLORIDE ◇ psi-CUMIDINE HYDROCHLORIDE ◇ PSEUDOCUMIDINE HYDROCHLORIDE ◇ 1,2,4-TRIMETHYL-5-AMINOBENZENE HYDROCHLORIDE ◇ 2,4,5-TRIMETHYLBENZENAMINE HYDROCHLORIDE

TOXICITY DATA with REFERENCE
orl-rat TDLo:32 g/kg/78W-C:NEO JEPTDQ 2,325,78
orl-mus TDLo:65 g/kg/78W-C:CAR JEPTDQ 2,325,78
orl-rat TD:65 g/kg/78W-C:ETA JEPTDQ 2,325,78
orl-mus TD:130 g/kg/78W-C:CAR JEPTDQ 2,325,78

orl-rat LD50:1585 mg/kg JPETAB 167,223,69
ipr-mus LD50:340 mg/kg NCIBR* NIH-NCI-E-68-1311,10,73

CONSENSUS REPORTS: IARC Cancer Review: Animal Inadequate Evidence IMEMDT 27,177,82.

SAFETY PROFILE: Poison by intraperitoneal route. Moderately toxic by ingestion. Questionable carcinogen with experimental carcinogenic, neoplastigenic, and tumorigenic data. When heated to decomposition it emits very toxic fumes of NO_x and HCl.

TLH000 CAS:6334-11-8 ***HR: 3***
2,4,6-TRIMETHYLANILINE HYDROCHLORIDE
mf: $C_9H_{13}N{\bullet}ClH$ mw: 171.69

SYNS: AMINOMESITYLENE HYDROCHLORIDE ◇ 2-AMINOMESITYLENE HYDROCHLORIDE ◇ 2-AMINO-1,3-5-TRIMETHYLBENZENE HYDROCHLORIDE ◇ MESIDINE HYDROCHLORIDE ◇ MESITYLAMINE HYDROCHLORIDE ◇ 2,4,6-TRIMETHYLBENZENAMINE HYDROCHLORIDE

TOXICITY DATA with REFERENCE
orl-rat TDLo:9500 mg/kg/78W-C:CAR JEPTDQ 2,325,78
orl-mus TDLo:22 g/kg/78W-C:CAR JEPTDQ 2,325,78
orl-rat TD:4700 mg/kg/78W-C:NEO JEPTDQ 2,325,78
orl-rat LD50:660 mg/kg JPETAB 167,223,69
ipr-rat LD50:338 mg/kg NCIBR* NIH-NCI-E-68-1311,10,73
ipr-mus LD50:260 mg/kg NCIBR* NIH-NCI-E-68-1311,10,73

CONSENSUS REPORTS: IARC Cancer Review: Animal Inadequate Evidence IMEMDT 27,177,82.

SAFETY PROFILE: Suspected carcinogen with experimental carcinogenic and neoplastigenic data. Poison by intraperitoneal route. Moderately toxic by ingestion. When heated to decomposition it emits very toxic fumes of NO_x and HCl.

TLH050 CAS:63018-94-0 ***HR: 3***
2,9,10-TRIMETHYLANTHRACENE
mf: $C_{17}H_{16}$ mw: 220.31

TOXICITY DATA with REFERENCE
mmo-sat 20 μg/plate CRNGDP 6,1483,85
mma-sat 5 μg/plate CRNGDP 6,1483,85
skn-mus TDLo:40 mg/kg/20D-I:CAR CRNGDP 6,1483,85

SAFETY PROFILE: Questionable carcinogen with experimental carcinogenic data. Mutation data reported. When heated to decomposition it emits acrid smoke and irritating fumes.

TLH100 CAS:594-10-5 ***HR: 3***
TRIMETHYL ANTIMONY
mf: C_3H_9Sb mw: 166.85

$(CH_3)_3Sb$

SYN: TRIMETHYLSTILBINE

CONSENSUS REPORTS: Antimony and its compounds are on The Community Right-To-Know List.

SAFETY PROFILE: Most antimony compounds are poisons. Ignites spontaneously in air. Mixtures with 2-iodoethanol explode at 150°C. Violent reaction with halogens. When heated to decomposition it emits toxic fumes of Sb. See also ANTIMONY COMPOUNDS.

TLH150 CAS:593-88-4 ***HR: 3***
TRIMETHYL ARSINE
mf: C_3H_9As mw: 120.03

$(CH_3)_3As$

CONSENSUS REPORTS: Arsenic and its compounds are on The Community Right-To-Know List.

SAFETY PROFILE: Arsenic compounds are poisons. Ignites spontaneously in air. Violent reaction with halogens. When heated to decomposition it emits toxic fumes of As. See also ARSINE and ARSENIC COMPOUNDS.

TLH250 CAS:41262-21-9 ***HR: 3***
TRIMETHYLARSINE SELENIDE
mf: C_3H_9AsSe mw: 199.00

SYN: ARSINE SELENIDE, TRIMETHYL-

TOXICITY DATA with REFERENCE
ivn-mus LDLo:8 mg/kg JPETAB 25,315,25

OSHA PEL: TWA 0.5 mg(As)/m^3; 0.2 mg(Se)/m^3
ACGIH TLV: TWA 0.2 mg(Se)/m^3; 0.2 mg(As)/m^3

SAFETY PROFILE: Poison by intravenous route. When heated to decomposition it emits toxic fumes of As and Se.

TLH350 CAS:63040-05-1 ***HR: 3***
3,5,9-TRIMETHYL-1:2-BENZACRIDINE
mf: $C_{20}H_{17}N$ mw: 271.38

SYNS: 1,6,10-TRIMETHYL,7:8 BENZACRIDINE (FRENCH) ◇ 5,7,11-TRIMETHYLBENZ(c)ACRIDINE

TOXICITY DATA with REFERENCE
skn-mus TDLo:250 mg/kg/21W-I:ETA AICCA6 7,184,50

SAFETY PROFILE: Questionable carcinogen with experimental tumorigenic data. When heated to decomposition it emits toxic fumes of NO_x.

TLH500 CAS:63040-01-7 ***HR: 3***
3,8,12-TRIMETHYLBENZ(a)ACRIDINE
mf: $C_{20}H_{17}N$ mw: 271.38

SYN: 1,10,3'-TRIMETHYL-5,6-BENZACRIDINE (FRENCH)

TOXICITY DATA with REFERENCE
scu-mus TDLo:40 mg/kg:ETA ACRSAJ 4,315,56

SAFETY PROFILE: Questionable carcinogen with experimental tumorigenic data. When heated to decomposition it emits toxic fumes of NO_x.

TLH750 CAS:63040-02-8 ***HR: 3***
5,7,8-TRIMETHYL-3:4-BENZACRIDINE
mf: $C_{20}H_{17}N$ mw: 271.38

SYN: 2,3,10 TRIMETHYL,5:6 BENZACRIDINE (FRENCH)

TOXICITY DATA with REFERENCE
skn-mus TDLo:270 mg/kg/22W-I:ETA AICCA6 7,184,50

SAFETY PROFILE: Questionable carcinogen with experimental tumorigenic data. When heated to decomposition it emits toxic fumes of NO_x.

TLI000 CAS:64038-40-0 ***HR: 3***
7,8,11-TRIMETHYLBENZ(c)ACRIDINE
mf: $C_{20}H_{17}N$ mw: 271.38

SYNS: 5,6,9-TRIMETHYL-1:2-BENZACRIDINE ◇ 1,4,10 TRIMETHYL, 7:8 BENZACRIDINE (FRENCH)

TOXICITY DATA with REFERENCE
skn-mus TDLo:270 mg/kg/25W-I:ETA ACRSAJ 4,315,56

SAFETY PROFILE: Questionable carcinogen with experimental tumorigenic data. When heated to decomposition it emits toxic fumes of NO_x.

TLI250 CAS:58430-01-6 ***HR: 3***
7,9,10-TRIMETHYLBENZ(c)ACRIDINE
mf: $C_{20}H_{17}N$ mw: 271.38

SYN: 2,3,10-TRIMETHYL-7:8-BENZACRIDINE (FRENCH)

TOXICITY DATA with REFERENCE
mma-sat 1 nmol/plate GANNA2 70,749,79
skn-mus TDLo:270 mg/kg/25W-I:ETA ACRSAJ 4,315,56

SAFETY PROFILE: Questionable carcinogen with experimental tumorigenic data. Mutation data reported. When heated to decomposition it emits toxic fumes of NO_x.

TLI500 CAS:51787-42-9 ***HR: 3***
7,9,11-TRIMETHYLBENZ(c)ACRIDINE
mf: $C_{20}H_{17}N$ mw: 271.38

SYN: 1,3,10-TRIMETHYL-7,8-BENZACRIDINE(FRENCH)

TOXICITY DATA with REFERENCE
mma-sat 1 nmol/plate GANNA2 70,749,79
skn-mus TDLo:336 mg/kg/28W-I:ETA ACRSAJ 4,315,56

SAFETY PROFILE: Questionable carcinogen with experimental tumorigenic data. Mutation data reported. When heated to decomposition it emits toxic fumes of NO_x.

TLI750 CAS:51787-43-0 ***HR: 3***
8,10,12-TRIMETHYLBENZ(a)ACRIDINE
mf: $C_{20}H_{17}N$ mw: 271.38

SYN: 1,3,10-TRIMETHYL-5,6-BENZACRIDINE(FRENCH)

TOXICITY DATA with REFERENCE
skn-mus TDLo:348 mg/kg/29W-I:ETA ACRSAJ 4,315,56

SAFETY PROFILE: Questionable carcinogen with experimental tumorigenic data. When heated to decomposition it emits toxic fumes of NO_x.

TLJ250 CAS:18429-71-5 ***HR: 3***
4,5,10-TRIMETHYLBENZ(a)ANTHRACENE
mf: $C_{21}H_{18}$ mw: 270.39

TOXICITY DATA with REFERENCE
scu-rat TDLo:18 mg/kg:ETA PSEBAA 128,720,68

SAFETY PROFILE: Questionable carcinogen with experimental tumorigenic data. When heated to decomposition it emits acrid smoke and irritating fumes.

TLJ500 CAS:35187-24-7 ***HR: 3***
4,7,12-TRIMETHYLBENZ(a)ANTHRACENE
mf: $C_{21}H_{18}$ mw: 270.39

TOXICITY DATA with REFERENCE
mma-sat 20 μg/plate CNREA8 36,4525,76
ims-rat TDLo:10 mg/kg:NEO NATUAS 273,566,78

SAFETY PROFILE: Questionable carcinogen with experimental neoplastigenic data. Mutation data reported. When heated to decomposition it emits acrid smoke and irritating fumes.

TLJ750 CAS:20627-33-2 ***HR: 3***
4,9,10-TRIMETHYL-1,2-BENZANTHRACENE
mf: $C_{21}H_{18}$ mw: 270.39

SYN: 6,7,12-TRIMETHYLBENZ(a)ANTHRACENE

TOXICITY DATA with REFERENCE
cyt-rat-ivn 50 mg/kg GANNA2 64,637,73
skn-mus TDLo:670 mg/kg/28W-I:ETA CRSBAW 148,812,54

SAFETY PROFILE: Questionable carcinogen with experimental tumorigenic data. Mutation data reported. When heated to decomposition it emits acrid smoke and irritating fumes.

TLK000 CAS:20627-32-1 ***HR: 3***
6,7,8-TRIMETHYLBENZ(a)ANTHRACENE
mf: $C_{21}H_{18}$ mw: 270.39

TOXICITY DATA with REFERENCE
ims-rat TDLo:50 mg/kg:NEO CNREA8 29,506,69

SAFETY PROFILE: Questionable carcinogen with ex-

perimental neoplastigenic data. When heated to decomposition it emits acrid smoke and irritating fumes.

TLK500 CAS:20627-34-3 ***HR: 3***
6,8,12-TRIMETHYLBENZ(a)ANTHRACENE
mf: $C_{21}H_{18}$ mw: 270.39

TOXICITY DATA with REFERENCE
cyt-rat-ivn 50 mg/kg GANNA2 64,637,73
cyt-rat-ipr 140 mg/kg/10D-I JEMEAV 131,331,70
ivn-rat TDLo:175 mg/kg/10W-I:ETA JEMEAV 131,321,70
ims-rat TDLo:50 mg/kg:NEO CNREA8 29,506,69

SAFETY PROFILE: Questionable carcinogen with experimental neoplastigenic and tumorigenic data. Mutation data reported. When heated to decomposition it emits acrid smoke and irritating fumes.

TLK600 CAS:24891-41-6 ***HR: 3***
6,9,12-TRIMETHYL-1,2-BENZANTHRACENE
mf: $C_{21}H_{18}$ mw: 270.39

SYN: 7,9,12-TRIMETHYLBENZ(a)ANTHRACENE

TOXICITY DATA with REFERENCE
ivn-rat TDLo:175 mg/kg/10W-I:ETA JEMEAV 131,321,70

SAFETY PROFILE: Questionable carcinogen with experimental tumorigenic data. When heated to decomposition it emits acrid smoke and irritating fumes.

TLK750 CAS:13345-64-7 ***HR: 3***
7,8,12-TRIMETHYLBENZ(a)ANTHRACENE
mf: $C_{21}H_{18}$ mw: 270.39

SYNS: 7,8,12-TMBA ◇ 5:9:10-TRIMETHYL-1:2-BENZANTHRACENE

TOXICITY DATA with REFERENCE
sln-dmg-par 5 mmol/L MUREAV 125,243,84
sce-rat:bmr 270 mg/L JJIND8 67,831,81
ivn-rat TDLo:175 mg/kg/10W-I:CAR JEMEAV 131,321,70
skn-mus TDLo:48 mg/kg/10W-I:ETA PRLBA4 129,439,40
ivn-mus TDLo:20 mg/kg:ETA,REP MOPMA3 4,427,68
ivn-rat LD50:125 mg/kg MOPMA3 4,427,68
ivn-mus LD50:50 mg/kg MOPMA3 4,427,68

SAFETY PROFILE: Poison by intravenous route. Experimental reproductive effects. Questionable carcinogen with experimental carcinogenic and tumorigenic data. Mutation data reported. When heated to decomposition it emits acrid smoke and irritating fumes.

TLL000 CAS:35187-27-0 ***HR: 3***
7,10,12-TRIMETHYLBENZ(a)ANTHRACENE
mf: $C_{21}H_{18}$ mw: 270.39

TOXICITY DATA with REFERENCE
ims-rat TDLo:50 mg/kg:ETA JMCMAR 14,940,71

SAFETY PROFILE: Questionable carcinogen with experimental tumorigenic data. When heated to decomposition it emits acrid smoke and irritating fumes.

TLL250 CAS:25551-13-7 ***HR: 1***
TRIMETHYL BENZENE
mf: C_9H_{12} mw: 120.21

SYN: TRIMETHYL BENZENE (mixed isomers)

TOXICITY DATA with REFERENCE
skn-rbt 500 mg/24H MOD 28ZPAK -,24,72
eye-rbt 500 mg/24H MLD 28ZPAK -,24,72
orl-rat LD50:8970 mg/kg 28ZPAK -,24,72

CONSENSUS REPORTS: Reported in EPA TSCA Inventory.

OSHA PEL: TWA 25 ppm
ACGIH TLV: TWA 25 ppm

SAFETY PROFILE: Mildly toxic by ingestion. A skin and eye irritant. Flammable when exposed to heat, flame, and oxidizers. When heated to decomposition it emits acrid smoke and irritating fumes. See also individual trimethyl benzene isomers.

TLL500 CAS:526-73-8 ***HR: 1***
1,2,3-TRIMETHYL BENZENE
mf: C_9H_{12} mw: 120.21

PROP: Mp: -25.4, bp: 176.1, d: 0.894, refr index: 1.5139, vap d: 4.15, flash p: 119° F, autoign temp: 878° F.

SYNS: HEMIMELLITENE ◇ TRIMETHYL BENZENE

TOXICITY DATA with REFERENCE
orl-rat LDLo:5000 mg/kg AMIHAB 19,403,59

CONSENSUS REPORTS: Reported in EPA TSCA Inventory.

SAFETY PROFILE: Mildly toxic by ingestion. Combustible liquid; flammable when exposed to heat, flame, or oxidizers. To fight fire, use water spray, mist, dry chemical, CO_2, foam. When heated to decomposition it emits acrid smoke and irritating fumes.

TLL750 CAS:95-63-6 ***HR: 2***
1,2,4-TRIMETHYL BENZENE
mf: C_9H_{12} mw: 120.21

PROP: Liquid. Mp: 120.19°, d: 0.888 @ 4/4°, fp: −61°, bp: 168.89°, flash p: 130°F, autoign temp: 959°F. Insol in water; sol in alc, benzene, and ether.

SYNS: ASYMMETRICAL TRIMETHYL BENZENE ◇ psi-CUMENE ◇ PSEUDOCUMENE ◇ PSEUDOCUMOL ◇ 1,2,5-TRIMETHYL BENZENE ◇ as-TRIMETHYL BENZENE

TOXICITY DATA with REFERENCE
ihl-rat LC50:18 g/m^3/4H GISAAA 44(5),15,79

ipr-rat LDLo:1752 mg/kg MEIEDD 10,1141,83
ipr-gpg LDLo:1788 mg/kg AMIHBC 9,227,54

CONSENSUS REPORTS: Reported in EPA TSCA Inventory. Community Right-To-Know List.

SAFETY PROFILE: Moderately toxic by intraperitoneal route. Mildly toxic by inhalation. Can cause central nervous system depression, anemia, bronchitis. Combustible liquid; flammable when exposed to heat, flame, or oxidizers. To fight fire, use foam, alcohol foam, mist. Emitted from modern building materials. (CENEAR 69,22,91) When heated to decomposition it emits acrid smoke and irritating fumes.

TLM000 ***HR: 2***
1,2,4-TRIMETHYL BENZENE mixed with MESITYLENE (5:3)

SYN: FLEET-X-DV-99

TOXICITY DATA with REFERENCE
ihl-man TCLo:10 ppm:CNS ZEPRAN 1,389,56
ipr-rat LD50:1500 mg/kg ZEPRAN 1,389,56

SAFETY PROFILE: Moderately toxic by intraperitoneal route. Human systemic effects by inhalation: central nervous system effects. When heated to decomposition it emits acrid smoke and irritating fumes. See also MESITYLENE and 1,2,4-TRIMETHYL BENZENE.

TLM050 CAS:108-67-8 ***HR: 3***
1,3,5-TRIMETHYL BENZENE
DOT: UN 2325
mf: C_9H_{12} mw: 120.21

PROP: A liquid; peculiar odor. Mp: −44.8°, d: 0.8637 @ 20°/4°, bp: 164.7°, autoign temp: 1022°F. Insol in water; misc in alc, benzene, and ether.

SYNS: FLEET-X ◇ MESITYLENE ◇ sym-TRIMETHYL BENZENE ◇ TRIMETHYL BENZENE ◇ TRIMETHYL BENZOL

TOXICITY DATA with REFERENCE
cyt-smc 350 μmol/tube HEREAY 33,457,47
ihl-hmn TCLo:10 ppm:CNS,PNS,PUL ZUBEAQ 49,265,56
ihl-rat LC50:24 mg/m³/4H GISAAA 44(5),15,79
ipr-gpg LDLo:1303 mg/kg AMIHBC 9,227,54

CONSENSUS REPORTS: Reported in EPA TSCA Inventory.

OSHA PEL: TWA 25 ppm
ACGIH TLV: TWA 25 ppm
DOT Classification: Flammable or Combustible Liquid; label: Flammable Liquid.

SAFETY PROFILE: Poison by inhalation. Moderately toxic by intraperitoneal route. Human systemic effects by inhalation: sensory changes involving peripheral nerves, somnolence (general depressed activity), and structural or functional change in trachea or bronchi. Reports of leukopenia and thrombocytopenia in experimental animals. Mutation data reported. Flammable when exposed to heat or flame; can react vigorously with oxidizing materials. Violent reaction with HNO_3. To fight fire, use water spray, fog, foam, CO_2. Emitted from modern building materials. (CENEAR 69,22,91) When heated to decomposition it emits acrid smoke and irritating fumes.

TLM250 CAS:56287-19-5 ***HR: D***
TRIMETHYL-N-(p-BENZENESULPHONAMIDO) PHOSPHORIMIDATE
mf: $C_9H_{15}N_2O_5PS$ mw: 294.23

TOXICITY DATA with REFERENCE
mmo-sat 5 μL/plate MUREAV 28,405,75
slr-dmg-orl 20 mmol/L MUREAV 28,405,75

SAFETY PROFILE: Mutation data reported. When heated to decomposition it emits very toxic fumes of NO_x, PO_x, and SO_x.

TLM500 CAS:16757-92-9 ***HR: 3***
1,3,6-TRIMETHYLBENZO(a)PYRENE
mf: $C_{23}H_{18}$ mw: 294.41

TOXICITY DATA with REFERENCE
scu-mus TDLo:72 mg/kg/13W-I:ETA IJCNAW 3,238,68

SAFETY PROFILE: Questionable carcinogen with experimental tumorigenic data. When heated to decomposition it emits acrid smoke and irritating fumes.

TLM750 CAS:593-91-9 ***HR: 3***
TRIMETHYL BISMUTH
mf: C_3H_9Bi mw: 254.06

PROP: Liquid. Bp: 110°, d: 2.300 @ 18°.

SYN: TRIMETHYLBISMUTHINE

TOXICITY DATA with REFERENCE
orl-dog LDLo:233 mg/kg JPETAB 67,17,39
ihl-dog LDLo:233 mg/kg JPETAB 67,17,39
skn-dog LDLo:233 mg/kg JPETAB 67,17,39
scu-dog LD50:182 mg/kg JPETAB 67,17,39
ivn-dog LD50:12 mg/kg JPETAB 67,17,39
orl-rbt LD50:484 mg/kg JPETAB 67,17,39
scu-rbt LD50:182 mg/kg JPETAB 67,17,39
ivn-rbt LDLo:11 mg/kg JPETAB 67,17,39
ivn-pgn LDLo:6 mg/kg JPETAB 67,17,39

SAFETY PROFILE: Poison by ingestion, inhalation, skin contact, intravenous, and subcutaneous routes. Can cause narcosis and central nervous system depression. Prolonged exposure can cause encephalopathy similar to that of organic lead compounds. Flammable when exposed to heat or flame; can react with oxidizing materials. Spontaneously flammable in air. Explodes at 110°C. Upon decomposition it emits toxic fumes of Bi. See also BISMUTH COMPOUNDS.

TLM775 CAS:593-90-8 ***HR: 3***
TRIMETHYLBORANE
mf: C_3H_9B mw: 55.91

$(CH_3)_3B$

SAFETY PROFILE: The gas ignites spontaneously in air or when mixed with chlorine. When heated to decomposition it emits acrid smoke and irritating fumes. See also BORANES and BORON COMPOUNDS.

TLN000 CAS:121-43-7 ***HR: 3***
TRIMETHYL BORATE
DOT: UN 2416
mf: $C_3H_9BO_3$ mw: 103.93

PROP: Colorless liquid. Decomp in water; misc in alc, ether. Mp: −29°, bp: 68°, flash p: <73°F, d: 0.92 @ 20°, vap d: 3.59.

SYNS: BORESTER O ◇ METHYL BORATE ◇ TRIMETHOXYBORINE

TOXICITY DATA with REFERENCE
eye-rbt 500 mg AJOPAA 29,1363,46
orl-rat LD50:6140 mg/kg 14KTAK -,693,64
ipr-rat LDLo:1600 mg/kg 14KTAK -,693,64
orl-mus LD50:1290 mg/kg 14KTAK -,693,64
ipr-mus LDLo:1000 mg/kg 14KTAK -,693,64
skn-rbt LD50:1980 mg/kg AIHAAP 23,95,62
ipr-rbt LDLo:1600 mg/kg 14KTAK -,693,64

CONSENSUS REPORTS: Reported in EPA TSCA Inventory.

DOT Classification: Flammable or Combustible Liquid; Label: Flammable Liquid.

SAFETY PROFILE: Moderately toxic by ingestion, skin contact, and intraperitoneal routes. An eye irritant. A very dangerous fire hazard when exposed to heat, flame, or oxidizers. Moderately explosive when exposed to flame. Will react with water or steam to produce toxic and flammable vapors. To fight fire, use dry chemical, CO_2, spray, foam. When heated to decomposition it emits acrid smoke and irritating fumes. See also ESTERS and BORON COMPOUNDS.

TLN100 CAS:5314-85-2 ***HR: 2***
2,4,6-TRIMETHYLBORAZINE
mf: $C_3H_{12}B_3N_3$ mw: 122.58

$HNB(CH_3)NHB(CH_3)HNBCH_3$ (ring)

SYN: B-TRIMETHYLBORAZINE

SAFETY PROFILE: Violent reaction with nitryl chloride. When heated to decomposition it emits toxic fumes of NO_x. See also BORON COMPOUNDS and BORAZINE.

TLN150 CAS:87-76-3 ***HR: 3***
TRIMETHYLCETYLAMMONIUM PENTACHLOROPHENATE
mf: $C_{19}H_{42}N•C_6HCl_5O$ mw: 550.94

SYN: HEXADECYLTRIMETHYLAMMONIUM PENTACHLOROPHENOL

TOXICITY DATA with REFERENCE
orl-rat LD50:790 mg/kg NIIRDN 6,529,82
ipr-rat LD50:58 mg/kg NIIRDN 6,529,82
scu-rat LD50:4130 mg/kg NIIRDN 6,529,82
orl-mus LD50:594 mg/kg NIIRDN 6,529,82
ipr-mus LD50:21 mg/kg NIIRDN 6,529,82
scu-mus LD50:800 mg/kg NIIRDN 6,529,82

CONSENSUS REPORTS: Chlorophenol compounds are on the Community Right-To-Know List.

SAFETY PROFILE: Poison by intraperitoneal route. Moderately toxic by ingestion and subcutaneous routes. When heated to decomposition it emits toxic fumes of Cl^-, NH_3, and NO_x. See also CHLOROPHENOLS.

TLN250 CAS:75-77-4 ***HR: 3***
TRIMETHYL CHLOROSILANE
DOT: UN 1298
mf: C_3H_9ClSi mw: 108.66

PROP: Colorless liquid. Bp: 57°, d: 0.854 @ 25/25°, flash p: −18°F. Sol in benzene, ether, perchloroethylene.

SYNS: CHLOROTRIMETHYLSILICANE ◇ TL 1163

TOXICITY DATA with REFERENCE
mmo-sat 1 mg/plate ENMUDM 8(Suppl 7),1,86
mma-sat 1666 μg/plate ENMUDM 8(Suppl 7),1,86
ipr-mus TDLo:1000 mg/kg/I:NEO JNCIAM 54,495,75
ihl-mus LCLo:500 mg/m^3/10M NDRC** No.9-4-1-19,44
ipr-mus LDLo:750 mg/kg StoGD# 27May75

CONSENSUS REPORTS: EPA Extremely Hazardous Substances List. Reported in EPA TSCA Inventory.

DOT Classification: Flammable Liquid; Label: Flammable Liquid and Flammable Liquid, Corrosive.

SAFETY PROFILE: Moderately toxic by inhalation and intraperitoneal routes. A corrosive irritant to skin, eyes, and mucous membranes. Questionable carcinogen with experimental neoplastigenic data. Mutation data reported. A very dangerous fire hazard when exposed to heat or flame. Violent reaction with water or hexafluoroisopropylideneamino lithium. A preparative hazard. To fight fire, use foam, alcohol foam, and fog. When heated to decomposition it emits toxic fumes of Cl^-. An intermediate in the production of silicones. See also CHLOROSILANES.

TLN500 CAS:24815-24-5 **HR: 3**
3,4,5-TRIMETHYLCINNAMOYL METHYL RESERPATE
mf: $C_{35}H_{42}N_2O_9$ mw: 634.79

SYNS: ANAPRAL ◇ ANAPREL ◇ CINAMINE ◇ CINATABS ◇ METHYL RESERPATE 3,4,5-TRIMETHOXYCINNAMIC ACID ESTER ◇ METHYL-18-o-(3,4,5-TRIMETHOXYCINNAMOYL RESERPATE ◇ MODERIL ◇ NORMORESCINA ◇ RAUPYROL ◇ RAURESCINE ◇ RECINNAMINE ◇ RECITENSINA ◇ RESCALOID ◇ RESCAMIN ◇ RESCIDAN ◇ RESCIN ◇ RESCINNAMINE ◇ RESCINPAL ◇ RESCINSAN ◇ RESCITEN ◇ RESERPINENE ◇ RESIPAL ◇ RESKINNAMIN ◇ SCINNAMINA ◇ TENAMINE ◇ TRIMETHOXYCINNAMOYL METHYL RESERPATE ◇ 3,4,5-TRIMETHYLCINNAMIC ACID, ESTER with METHYL RESERPATE ◇ TUAREG

TOXICITY DATA with REFERENCE
orl-dog TDLo:32 μg/kg (male 1D pre):REP NYKZAU 80,239,82
orl-hmn TDLo:4 μg/kg/D:NOSE,CNS 34ZIAG -,517,69
orl-rat LD50:1000 mg/kg NIIRDN 6,898,82
ipr-rat LD50:250 mg/kg NIIRDN 6,898,82
scu-rat LD50:540 mg/kg NIIRDN 6,898,82
orl-mus LD50:1420 mg/kg 27ZQAG -,109,72
ipr-mus LD50:505 mg/kg NIIRDN 6,898,82
scu-mus LD50:440 mg/kg NIIRDN 6,898,82
ivn-mus LD50:56 mg/kg CSLNX* NX#00965
ivn-gpg LDLo:35 mg/kg ARZNAD 23,600,73

SAFETY PROFILE: Poison by intravenous and intraperitoneal routes. Moderately toxic by ingestion and subcutaneous routes. Human systemic effects by ingestion: olfactory changes, somnolence and antipsychotic effects. Experimental reproductive effects. When heated to decomposition it emits toxic NO_x.

TLN750 CAS:3482-37-9 **HR: 3**
TRIMETHYLCOLCHICINIC ACID
mf: $C_{19}H_{21}NO_5$ mw: 343.41

SYNS: (s)-7-AMINO-6,7-DIHYDRO-10-HYDROXY-1,2,3-TRIMETHOXYBENZO(a)HEPTALEN-9(5H)-ONE ◇ DEACETYLCHOLCHICEINE ◇ N-DEACETYLCHOLCHICEINE ◇ DESACETYLCHOLCHICEINE ◇ TMCA

TOXICITY DATA with REFERENCE
oms-mus-par 200 mg/kg CANCAR 3,134,50
spm-mus-par 200 mg/kg CANCAR 2,134,50
ipr-mus LD50:200 mg/kg CANCAR 3,124,50

SAFETY PROFILE: Poison by intraperitoneal route. Mutation data reported. When heated to decomposition it emits toxic fumes of NO_x. See also COLCHICINE.

TLO000 CAS:3476-50-4 **HR: 3**
TRIMETHYLCOLCHICINIC ACID METHYL ETHER
mf: $C_{20}H_{23}NO_5$ mw: 357.44

SYNS: (s)-7-AMINO-6,7-DIHYDRO-1,2,3,10-TETRAMETHOXYBENZO(a)HEPTALEN-9(5H)-ONE ◇ COLCHINIC ACID TRIMETHYL ◇ DEACETYLCOLCHICINE ◇ N-DEACETYLCOLCHICINE ◇ DESACETYLCOLCHICINE ◇ N-DESACETYLCOLCHICINE ◇ TMCA METHYL ETHER ◇ TRIMETHYLCOLCHICINSAEUREMETHYL ESTER (GERMAN)

TOXICITY DATA with REFERENCE
oms-mus-par 46 mg/kg CANCAR 3,134,50
spm-mus-par 46 mg/kg CANCAR 3,134,50
ipr-mus LD50:46 mg/kg MDREP* No.204,49
ims-mus LD50:49 mg/kg JMCMAR 24,257,81

SAFETY PROFILE: Poison by intraperitoneal and intramuscular routes. Mutation data reported. When heated to decomposition it emits toxic fumes of NO_x. See also COLCHICINE and ETHERS.

TLO250 CAS:1845-38-1 **HR: 1**
3,3,5-TRIMETHYLCYCLOHEXANECARBOXALDEHYDE
mf: $C_{10}H_{18}O$ mw: 154.28

TOXICITY DATA with REFERENCE
skn-rbt 10 mg/24H open MLD AMIHBC 10,61,54
eye-rbt 500 mg AMIHBC 10,61,54
orl-rat LD50:4140 mg/kg AMIHBC 10,61,54
skn-rbt LD50:15800 mg/kg AMIHBC 10,61,54

SAFETY PROFILE: Mildly toxic by ingestion and skin contact. A skin and eye irritant. When heated to decomposition it emits acrid smoke and irritating fumes. See also ALDEHYDES.

TLO500 CAS:116-02-9 **HR: 2**
3,5,5-TRIMETHYLCYCLOHEXANOL
mf: $C_9H_{18}O$ mw: 142.27

PROP: Liquid. Fp: 37.0°, bp: 198°, flash p: 190°F (OC), d: 0.878 @ 40/20°, vap press: 0.1 mm @ 20°, vap d: 4.91.

SYNS: CYCLONOL ◇ DIHYDROISOPHOROL ◇ HOMOMENTHOL ◇ 3,3,5-TRIMETHYLCYCLOHEXANOL ◇ 3,3,5-TRIMETHYL-1-CYCLOHEXANOL

TOXICITY DATA with REFERENCE
skn-rbt 500 mg/24H MOD FCTXAV 12,807,74
eye-rbt 675 μg SEV AJOPAA 29,1363,46
orl-rat LD50:3250 mg/kg JIHTAB 31,60,49
skn-rbt LD50:2800 mg/kg JIHTAB 31,60,49

CONSENSUS REPORTS: Reported in EPA TSCA Inventory.

SAFETY PROFILE: Moderately toxic by ingestion and

skin contact. A skin and severe eye irritant. Combustible when exposed to heat, flame, or oxidizers. Can react with oxidizing materials. To fight fire, use alcohol, foam, CO_2, dry chemical. When heated to decomposition it emits acrid smoke and irritating fumes. See also ALCOHOLS.

TLO600 CAS:73987-16-3 ***HR: 2***
3,3,5-TRIMETHYLCYCLOHEXYL DIPROPYLENE GLYCOL
mf: $C_{15}H_{30}O_3$ mw: 258.45

SYNS: DIPROPYLENE GLYCOL, 3,3,5-TRIMETHYLCYCLOHEXYL ETHER ◇ 2-PROPANOL, 1-(2-((3,3,5-TRIMETHYLCYCLOHEXYL) OXY)PROPOXY)-(9CI)

TOXICITY DATA with REFERENCE
skn-rbt 500 mg/24H SEV JPETAB 82,377,44

SAFETY PROFILE: A severe skin irritant. When heated to decomposition it emits acrid smoke and irritating fumes.

TLP000 CAS:5831-11-8 ***HR: 3***
11,12-17-TRIMETHYL-15H-CYCLOPENTA(a) PHENANTHRENE
mf: $C_{20}H_{18}$ mw: 258.38

TOXICITY DATA with REFERENCE
mma-sat 20 μg/plate CNREA8 36,4525,76
skn-mus TDLo:108 mg/kg/1Y-I:CAR PEXTAR 11,69,69

CONSENSUS REPORTS: EPA Genetic Toxicology Program.

SAFETY PROFILE: Questionable carcinogen with experimental carcinogenic data. Mutation data reported. When heated to decomposition it emits acrid smoke and irritating fumes.

TLP250 CAS:63041-23-6 ***HR: 3***
3,8,13-TRIMETHYLCYCLOQUINAZOLINE
mf: $C_{24}H_{18}N_4$ mw: 362.46

TOXICITY DATA with REFERENCE
skn-mus TDLo:440 mg/kg/1Y-I:ETA BJCAAI 16,275,62

SAFETY PROFILE: Questionable carcinogen with experimental tumorigenic data. When heated to decomposition it emits toxic fumes of NO_x.

TLP275 CAS:21107-27-7 ***HR: 3***
TRIMETHYLDIBORANE
mf: $C_3H_{12}B_2$ mw: 69.75

$(CH_3)_2B{:}H_2{:}BHCH_3$

SAFETY PROFILE: Ignites spontaneously in air. When heated to decomposition it emits acrid smoke and irritating fumes. See also BORANES and BORON COMPOUNDS.

TLP500 CAS:147-47-7 ***HR: 2***
2,2,4-TRIMETHYL-1,2-DIHYDROQUINOLINE
mf: $C_{12}H_{15}N$ mw: 173.28

SYNS: ACETONE ANIL ◇ 1,2-DIHYDRO-2,2,4-TRIMETHYLQUINOLINE ◇ FLECTOL H ◇ NCI-C60902

TOXICITY DATA with REFERENCE
orl-rat LD50:2000 mg/kg HYSAAV 31,183,66
orl-mus LD50:1450 mg/kg HYSAAV 31,183,66

CONSENSUS REPORTS: Reported in EPA TSCA Inventory.

SAFETY PROFILE: Moderately toxic by ingestion. When heated to decomposition it emits toxic fumes of NO_x.

TLP750 CAS:127-48-0 ***HR: 3***
3,3,5-TRIMETHYL-2,4-DIKETOOXAZOLIDINE
mf: $C_6H_9NO_3$ mw: 143.16

SYNS: A 2297 ◇ ABSENTOL ◇ ABSETIL ◇ CONVENIXA ◇ EDION ◇ EPIDIONE ◇ EPIDONE ◇ ETYDION ◇ MINOALEUIATIN ◇ PETIDION ◇ PETIDON ◇ PETILEP ◇ PITMAL ◇ TIOXANONA ◇ TREDIONE ◇ TRICIONE ◇ TRIDILONA ◇ TRIDIONE ◇ TRIDONE ◇ TRILIDONA ◇ TRIMEDAL ◇ TRIMEDONE ◇ TRIMETADIONE ◇ TRIMETHADIONE ◇ TRIMETHIN ◇ 3,5,5-TRIMETHYL-2,4-OXAZOLIDINEDIONE ◇ TRIMETIN ◇ TRIOZANONA ◇ TROXIDONE ◇ TROMEDONE

TOXICITY DATA with REFERENCE
orl-wmn TDLo:6480 mg/kg (1-39W preg):REP AJDCAI 129,1229,75
orl-wmn TDLo:6480 mg/kg (1-39W preg):TER AJDCAI 129,1229,75
unr-man LDLo:88 mg/kg 85DCAI 2,73,70
orl-rat LD50:2140 mg/kg JPETAB 138,224,62
scu-rat LD50:2000 mg/kg 27ZQAG -,312,72
orl-mus LD50:2100 mg/kg ARZNAD 23,377,73
ipr-mus LD50:2 g/kg JPETAB 134,60,61
ivn-mus LD50:2000 mg/kg JLCMAK 31,1330,46
ipr-cat LDLo:2000 mg/kg JLCMAK 31,1330,46

SAFETY PROFILE: Human poison by unspecified routes. Moderately toxic experimentally by ingestion, subcutaneous, intraperitoneal, and intravenous routes. An experimental teratogen. Human reproductive effects by ingestion: effects on newborn including physical and other postnatal measures or effects. Human teratogenic effects by ingestion: developmental abnormalities of the craniofacial, musculoskeletal, cardiovascular, and urogenital systems. Experimental reproductive effects. When heated to decomposition it emits toxic fumes of NO_x. See also KETONES.

TLQ000 CAS:363-42-8 ***HR: 3***
TRIMETHYL(2-(2,6-DIMETHYLPHENOXY) PROPYL)AMMONIUM CHLORIDE MONO-HYDRATE
mf: $C_{14}H_{24}NO{\cdot}Cl{\cdot}H_2O$ mw: 275.86

SYNS: COMPOUND 6890 ◇ (2-(2,6-DIMETHYLPHENOXY)PROPYL)TRIMETHYLAMMONIUM CHLORIDE MONOHYDRATE ◇ 2-(2,6-DIMETHYLPHENOXY)-N,N,N-TRIMETHYL-1-PROPANAMINIUMHYDRATE ◇ β-TM10

TOXICITY DATA with REFERENCE
orl-mus LDLo:1400 mg/kg JPETAB 129,17,60
ivn-mus LD50:6620 μg/kg JPETAB 129,17,60

SAFETY PROFILE: Poison by intravenous route. Moderately toxic by ingestion. When heated to decomposition it emits very toxic fumes of NO_x, NH_3, and Cl^-.

TLQ250 CAS:5786-77-6 ***HR: 3***
N-DI-TRIMETHYL-3,3-DI-2-THIENYLALLYLAMINE HYDROCHLORIDE
mf: $C_{14}H_{17}NS_2{\cdot}ClH$ mw: 299.90

SYNS: N,N-DIMETHYL-4,4-DI-2-THIENYL-3-BUTEN-2-AMINEHYDROCHLORIDE ◇ DIMETHYLTHIAMBUTENE HYDROCHLORIDE ◇ 3,3-DI-2-THIENYL-N,N,1-TRIMETHYLALLYLAMINE HYDROCHLORIDE

TOXICITY DATA with REFERENCE
scu-rat LD50:170 mg/kg BJPCAL 8,2,53
orl-mus LD50:199 mg/kg JPETAB 107,385,53
scu-mus LD50:100 mg/kg BJPCAL 8,2,53
ivn-mus LD50:16 mg/kg BJPCAL 8,2,53
ivn-dog LD50:23 mg/kg CPBTAL 7,372,59

SAFETY PROFILE: Poison by ingestion, intravenous, and subcutaneous routes. When heated to decomposition it emits very toxic fumes of NO_x, SO_x, and HCl. See also ALLYL COMPOUNDS and AMINES.

TLQ500 CAS:56-97-3 ***HR: 3***
1,1'-TRIMETHYLENEBIS(4-FORMYLPYRIDINIUM BROMIDE)DIOXIME
mf: $C_{15}H_{18}N_4O_2{\cdot}2Br$ mw: 446.19

SYNS: 1,3-BIS(4-FORMYLPYRIDINIUM)-PROPANE BISOXIDE DIBROMIDE ◇ DIPYROXIME ◇ 1,3-PROPAN-BIS-(4-HYDROXYIMINOMETHYL-PYRIDINIUM-(1))-DIBROMIDS (GERMAN) ◇ 1,1'-(1,3-PROPANEDIYL)BIS(4-(HYDROXYIMINO)METHYLPYRIDINIUM,DIBROMIDE ◇ TMB-4 DIBROMIDE ◇ TMV-4 ◇ 1,3-TRIMETHYLEN-BIS-(4-HYDROXIMINOFORMYLPYRIDINIUM)-DIBROMID(GERMAN) ◇ N,N-TRIMETHYLEN-BIS-(PYRIDINIUM-4-ALDOXIM)-DIBROMID (GERMAN)

TOXICITY DATA with REFERENCE
ipr-rat LD50:192 mg/kg RPTOAN 38(4),168,75
ivn-rat LD50:89 mg/kg JPETAB 129,31,60
ims-rat LD50:123 mg/kg JPETAB 129,31,60
ipr-mus LD50:53500 μg/kg RPTOAN 38(4),168,75
scu-mus LD50:83 mg/kg RPTOAN 35(5),243,72
ivn-mus LD50:45 mg/kg ARZNAD 14,870,64
ims-mus LD50:102 mg/kg ARZNAD 14(1),5,64
ims-cat LD50:117 mg/kg RPTOAN 38(4),168,75
ivn-rbt LD50:44 mg/kg AEHLAU 5,21,62

SAFETY PROFILE: Poison by intraperitoneal, intramuscular, subcutaneous, and intravenous routes. When heated to decomposition it emits very toxic fumes of NO_x and Br^-. Used as an antidote for organophosphate poisoning. See also BROMIDES.

TLQ750 CAS:3613-82-9 ***HR: 3***
1,1'-TRIMETHYLENEBIS(4-FORMYLPYRIDINIUM CHLORIDE) DIOXIME
mf: $C_{15}H_{18}N_4O_2{\cdot}2Cl$ mw: 357.27

SYNS: 1,3-BIS(4-FORMYLPYRIDINIUM)-PROPANE BISOXIME DICHLORIDE ◇ 1,3-BIS(4-HYDROXYIMINOMETHYL-1-PYRIDINIO)PROPANE DICHLORIDE ◇ 1,1'-(1,3-PROPANEDIYL)BIS((4-HYDROXYIMINO)METHYL)-PYRIDINIUM DICHLORIDE ◇ TMB-4 DICHLORIDE ◇ TRIMEDOXIME DICHLORIDE ◇ 1,1'-TRIMETHYLENEBIS(4-FORMYLPYRIDINIUM) DIOXIME DICHLORIDE ◇ 1,1'-TRIMETHYLENEBIS(4-(HYDROXYIMINOMETHYL)PYRIDINIUMCHLORIDE) ◇ N,N-TRIMETHYLENE BIS(PYRIDINIUM-4-ALDOXIME)DICHLORIDE

TOXICITY DATA with REFERENCE
orl-hmn TDLo:45 mg/kg:CVS AEHLAU 15,599,67
ivn-hmn TDLo:30 mg/kg:CVS AEHLAU 15,599,67
ipr-mus LD50:88 mg/kg TXAPA9 16,194,70
ivn-mus LD50:56 mg/kg CSLNX* NX#00647
ivn-rbt LD50:44 mg/kg JPETAB 132,50,61

SAFETY PROFILE: Poison by intraperitoneal and intravenous routes. Human systemic effects by ingestion and intravenous routes: blood pressure decrease. When heated to decomposition it emits very toxic fumes of NO_x and HCl.

TLR000 CAS:109-64-8 ***HR: 2***
TRIMETHYLENE DIBROMIDE
mf: $C_3H_6Br_2$ mw: 201.91

PROP: Colorless liquid. Bp: 166.5°, fp: −33°, d: 1.977 @ 25/25°, vap d: 7.0. Mp: −36°. Sltly sol in water; sol in alc, ether.

SYNS: α,Γ-DIBROMOPROPANE ◇ ω,ω'-DIBROMOPROPANE ◇ 1,3-DIBROMOPROPANE ◇ TRIMETHYLENE BROMIDE

TOXICITY DATA with REFERENCE
mmo-sat 10 μmol/plate ENMUDM 2,59,80
mma-sat 10 μmol/plate ENMUDM 2,59,80
ipr-mus LD50:473 mg/kg JPCEAO 320,133,78
rec-rbt LDLo:1000 mg/kg JPETAB 34,223,28

CONSENSUS REPORTS: Reported in EPA TSCA Inventory. EPA Genetic Toxicology Program.

SAFETY PROFILE: Moderately toxic by intraperitoneal and rectal routes. Mutation data reported. Irritating and narcotic in high concentration. When heated to decomposition it emits toxic fumes of Br^-. Used as an herbicide. See also BROMIDES.

TLR250 CAS:15886-84-7 ***HR: 2***
TRIMETHYLENEDIMETHANESULFONATE
mf: $C_5H_{12}O_6S_2$ mw: 232.29

SYNS: 1:3-DIMETHANESULFONOXYPROPANE ◇ 1,3-DIMETHANESULPHONOXYPROPANE ◇ ENT 51,904 ◇ PROPANE-1,3-DIMETHANESULFONATE ◇ PROPYLENE DIMETHANESULFONATE ◇ TRIMETHYLENE DIMETHANESULPHONATE

TOXICITY DATA with REFERENCE
cyt-oin-par 20 μg AESAAI 63,422,70
dlt-oin-par 2000 ppm MUREAV 13,49,71
ipr-rat TDLo:50 mg/kg (15D preg):REP JRPFA4 17,325,68
ipr-rat TDLo:50 mg/kg (15D preg):TER JRPFA4 18,15,69
ipr-mus LD50:500 mg/kg PSEBAA 85,211,54

SAFETY PROFILE: Moderately toxic by intraperitoneal route. An experimental teratogen. Experimental reproductive effects. Mutation data reported. When heated to decomposition it emits toxic fumes of SO_x.

TLR500 CAS:544-13-8 ***HR: 3***
1,3-TRIMETHYLENEDINITRILE
mf: $C_5H_6N_2$ mw: 94.13

PROP: Colorless liquid. D: 0.989 @ 15/4°, mp: −29°, bp: 286.4°. Sol in water; insol in ether.

SYNS: 1,3-DICYANOPROPANE ◇ GLUTARIC ACID DINITRILE ◇ GLUTARODINITRILE ◇ GLUTARONITRILE ◇ PENTANEDINITRILE ◇ PYROTARTARIC ACID NITRILE

TOXICITY DATA with REFERENCE
orl-mus LD50:2670 mg/kg ARTODN 57,88,85
scu-dog LDLo:50 mg/kg AIPTAK 3,77,1897
scu-frg LDLo:3000 mg/kg AIPTAK 3,77,1897

CONSENSUS REPORTS: Cyanide and its compounds are on the Community Right-To-Know List. Reported in EPA TSCA Inventory.

SAFETY PROFILE: Poison by subcutaneous route. Moderately toxic by ingestion. When heated to decomposition it emits very toxic fumes of NO_x and CN^-. See also NITRILES.

TLR675 CAS:2825-82-3 ***HR: 3***
exo-TRIMETHYLENENORBORNANE
mf: $C_{10}H_{16}$ mw: 136.26

SYNS: exo-HEXAHYDRO-4,7-METHANOINDAN ◇ JP-10 ◇ exo-TETRAHYDROBICYCLOPENTADIENE ◇ exo-TETRAHYDRODI(CYCLOPENTADIENE) ◇ exo-TRICYCLO(5.2.1.$0^{2,6}$)DECANE ◇ exo-5,6-TRIMETHYLENENORBORNANE

TOXICITY DATA with REFERENCE
cyt-ham:ovr 1 mg/L NTIS** AD-A124-785
orl-rat TDLo:5 g/kg (6-15D preg):TER DCTODJ 6,181,83
orl-rat TDLo:10 g/kg (6-15D preg):REP DCTODJ 6,181,83
ihl-rat TCLo:100 ppm/6H/1Y-I:CAR NTIS** AD-A163-179
ihl-rat TC:556 mg/m³/6H/1Y-I:ETA AETODY 7,133,84
orl-mus LD50:3660 mg/kg NTIS** AD-A086-341
ihl-mus LCLo:900 ppm/4H DCTODJ 6,181,83

CONSENSUS REPORTS: Reported in EPA TSCA Inventory.

SAFETY PROFILE: Moderately toxic by ingestion. Mildly toxic by inhalation. An experimental teratogen. Experimental reproductive effects. Questionable carcinogen with experimental carcinogenic and tumorigenic data. Mutation data reported. Used as a major component of cruise missile fuel. When heated to decomposition it emits acrid smoke and irritating fumes.

TLR750 CAS:1073-05-8 ***HR: D***
TRIMETHYLENE SULFATE
mf: $C_3H_6O_4S$ mw: 138.15

SYNS: 1,3,2-DIOXATHIANE-2,2-DIOXIDE (9CI) ◇ 1,3-PROPYLENE SULFATE

TOXICITY DATA with REFERENCE
dnr-omi 690 μg/plate BIZNAT 95,463,76
mmo-sat 100 nmol/plate CBINA8 19,241,77

CONSENSUS REPORTS: Reported in EPA TSCA Inventory.

SAFETY PROFILE: Mutation data reported. When heated to decomposition it emits toxic fumes of SO_x. See also SULFATES.

TLS000 CAS:2055-46-1 ***HR: 2***
N,N'-TRIMETHYLENETHIOUREA
mf: $C_4H_8N_2S$ mw: 116.20

SYN: TETRAHYDRO-2(1H)-PYRIMIDINETHIONE

TOXICITY DATA with REFERENCE
ipr-mus LD50:560 mg/kg JMCMAR 11,214,68

CONSENSUS REPORTS: Reported in EPA TSCA Inventory.

SAFETY PROFILE: Moderately toxic by intraperitoneal route. When heated to decomposition it emits very toxic fumes of NO_x and SO_x.

TLS500 CAS:291-21-4 ***HR: 3***
TRIMETHYLENE TRISULFIDE
mf: $C_3H_6S_3$ mw: 138.27

SYNS: THIOFORM (CZECH) ◇ TRIMETHYLENTRISULFID (CZECH) ◇ 1,3,5-TRITHIACYCLOHEXANE ◇ sym-TRITHIAN (CZECH) ◇ 1,3,5-TRITHIANE ◇ TRITHIOFORMALDEHYDE

TOXICITY DATA with REFERENCE
skn-rbt 500 mg/24H MLD 28ZPAK -,204,72
eye-rbt 500 mg/24H MOD 28ZPAK -,204,72
ipr-mus LD50:250 mg/kg NTIS** AD691-490

CONSENSUS REPORTS: Reported in EPA TSCA Inventory.

SAFETY PROFILE: Poison by intraperitoneal route. An eye and skin irritant. When heated to decomposition it emits toxic fumes of SO_x. See also SULFIDES and ALDEHYDES.

TLT000 CAS:1445-79-0 ***HR: 3***
TRIMETHYLGALLIUM
mf: C_3H_9Ga mw: 114.80

SAFETY PROFILE: Ignites spontaneously in air. Reacts violently with water. When heated to decomposition it emits acrid smoke and irritating fumes. See also GALLIUM.

TLT100 CAS:20519-92-0 ***HR: 3***
TRIMETHYLGERMYL PHOSPHINE
mf: $C_3H_{11}GeP$ mw: 150.68

$(CH_3)_3GePH_2$

SAFETY PROFILE: Ignites on contact with oxygen. When heated to decomposition it emits toxic fumes of PO_x. See also GERMANIUM COMPOUNDS and PHOSPHINE.

TLT500 CAS:58430-94-7 ***HR: 1***
3,5,5-TRIMETHYLHEXYL ACETATE
mf: $C_{11}H_{22}O_2$ mw: 186.33

SYNS: ISONONYL ACETATE ◇ 3,5,5-TRIMETHYLHEXYL ACETIC ACID

TOXICITY DATA with REFERENCE
orl-rat LD50:4250 mg/kg FCTXAV 12,1009,74

CONSENSUS REPORTS: Reported in EPA TSCA Inventory.

SAFETY PROFILE: Mildly toxic by ingestion. When heated to decomposition it emits acrid smoke and irritating fumes. See also ESTERS.

TLT750 CAS:60597-20-8 ***HR: 3***
TRIMETHYLHYDRAZINE HYDROCHLORIDE
mf: $C_3H_{10}N_2 \cdot ClH$ mw: 110.61

TOXICITY DATA with REFERENCE
orl-mus TDLo:22 g/kg/37W-C:CAR JNCIAM 57,187,76
orl-ham TDLo:80 g/kg/27W-C:CAR JNCIAM 59,431,77

SAFETY PROFILE: Suspected carcinogen with experimental carcinogenic data. When heated to decomposition it emits very toxic fumes of HCl and NO_x.

TLT775 CAS:3385-78-2 ***HR: 3***
TRIMETHYL INDIUM
mf: C_3H_9In mw: 159.92

$(CH_3)_3In$

SAFETY PROFILE: Ignites spontaneously in air. When heated to decomposition it emits acrid smoke and irritating fumes. See also INDIUM.

TLU000 CAS:558-17-8 ***HR: D***
TRIMETHYLIODOMETHANE
mf: C_4H_9I mw: 172.02

SYNS: tert-BUTYL IODIDE ◇ 2-IODO-2-METHYLPROPANE

TOXICITY DATA with REFERENCE
dnr-esc 5 μL/16H CBINA8 15,219,76

CONSENSUS REPORTS: Reported in EPA TSCA Inventory.

SAFETY PROFILE: Mutation data reported. When heated to decomposition it emits toxic fumes of I^-. See also IODIDES.

TLU175 CAS:1520-78-1 ***HR: 3***
TRIMETHYL LEAD CHLORIDE
mf: C_3H_9ClPb mw: 287.76

PROP: Powder. Mp: 187° (subl).

SYNS: CHLOROTRIMETHYLPLUMBANE ◇ TriML

TOXICITY DATA with REFERENCE
orl-rat TDLo:30 mg/kg (female 12-14D post):TER TXAPA9 21,265,72
orl-rat TDLo:15 mg/kg (9-11D preg):REP TXAPA9 21,265,72
orl-rat LD50:80 mg/kg 85JCAE -,1258,86
ipr-rat LD50:26 mg/kg BJIMAG 18,277,61
orl-dck LD50:29900 μg/kg 51UDAB 2,714,83

CONSENSUS REPORTS: Lead and its compounds are on the Community Right-To-Know List.

SAFETY PROFILE: Poison by ingestion and intraperitoneal routes. An experimental teratogen. Experimental reproductive effects. When heated to decomposition it emits toxic fumes of Cl^- and Pb. See also LEAD COMPOUNDS.

TLU500 CAS:1498-88-0 ***HR: 3***
1,3,3-TRIMETHYL-6'-NITROINDOLINE-2-SPIRO-2'-BENZOPYRAN
mf: $C_{19}H_{18}N_2O_3$ mw: 322.39

TOXICITY DATA with REFERENCE
ivn-mus LD50:56 mg/kg CSLNX* NX#07770

CONSENSUS REPORTS: Reported in EPA TSCA Inventory.

SAFETY PROFILE: Poison by intravenous route. When heated to decomposition it emits toxic fumes of NO_x.

TLU750 CAS:3475-63-6 ***HR: 3***
1,1,3-TRIMETHYL-3-NITROSOUREA
mf: $C_4H_9N_3O_2$ mw: 131.16

SYNS: N-NITROSO-TRIMETHYLHARNSTOFF(GERMAN) ◇ NITROSOTRIMETHYLUREA ◇ N-NITROSOTRIMETHYLUREA ◇ TRIMETHYLNITROSOHARNSTOFF (GERMAN) ◇ N-TRIMETHYL-N-NITROSOUREA

TOXICITY DATA with REFERENCE
mmo-sat 7000 μmol/L/48H MUREAV 48,131,77
mma-sat 1 μg/plate MUREAV 51,319,78
ivn-rat TDLo:40 mg/kg (10D preg):TER APEPA2 257,296,67
orl-rat TDLo:1070 mg/kg/53W-C:ETA ZEKBAI 69,103,67
orl-rat LD50:240 mg/kg ZEKBAI 69,103,67
ivn-rat LD50:240 mg/kg ZEKBAI 69,103,67

CONSENSUS REPORTS: EPA Genetic Toxicology Program.

SAFETY PROFILE: Poison by ingestion and intravenous routes. Experimental teratogenic effects. Questionable carcinogen with experimental tumorigenic data. Mutation data reported. Many N-nitroso compounds are carcinogens. When heated to decomposition it emits toxic fumes of NO_x. See also N-NITROSO COMPOUNDS.

TLV000 CAS:123-17-1 ***HR: 1***
2,6,8-TRIMETHYLNONANOL-4
mf: $C_{12}H_{26}O$ mw: 186.38

PROP: Liquid. Bp: 225.2, fp: −60°, flash p: 200°F (OC), vap press: <0.01 mm @ 20°, d: 0.8193 @ 20/20°, vap d: 6.43.

SYN: 2,6,8-TRIMETHYL-4-NONANOL

TOXICITY DATA with REFERENCE
skn-rbt 500 mg open MLD UCDS** 6/28/72
eye-rbt 500 mg open AMIHBC 10,61,54
orl-rat LD50:17 g/kg AMIHBC 10,61,54
skn-rbt LD50:11 g/kg AMIHBC 10,61,54

CONSENSUS REPORTS: Reported in EPA TSCA Inventory.

SAFETY PROFILE: Mildly toxic by ingestion and skin contact. A skin and eye irritant. Flammable when exposed to heat or flame; can react with oxidizing materials. To fight fire, use foam, CO_2, dry chemical. When heated to decomposition it emits acrid smoke and irritating fumes.

TLV250 CAS:1331-50-6 ***HR: 1***
TRIMETHYL NONANONE
mf: $C_{12}H_{24}O$ mw: 184.36

PROP: Liquid. Mp: −75°, bp: 211-219°, flash p: 196°F (OC), d: 0.8165 @ 20/20°, vap d: 6.37.

TOXICITY DATA with REFERENCE
skn-rbt 10 mg/24H open MLD AMIHBC 4,119,51
skn-rbt 500 mg open MLD UCDS** 4/25/58
eye-rbt 500 mg open AMIHBC 4,119,51
orl-rat LD50:8470 mg/kg UCDS** 4/25/68
skn-rbt LD50:11 g/kg UCDS** 4/25/68

SAFETY PROFILE: Mildly toxic by ingestion and skin contact. A skin and eye irritant. Flammable when exposed to heat or flame; can react with oxidizing materials. To fight fire, use foam, CO_2, dry chemical. When heated to decomposition it emits acrid smoke and irritating fumes. See also KETONES.

TLW000 CAS:512-13-0 ***HR: 2***
(−)-endo-1,3,3-TRIMETHYL-2-NORBORNANOL
mf: $C_{10}H_{18}O$ mw: 154.28

PROP: D: 0.9641, mp: 48°, bp: 201°, flash p: 165° F.

SYNS: α-FENCHOL ◇ endo-FENCHOL ◇ α-FENCHYL ALCOHOL ◇ (1R-endo)-1,3,3-TRIMETHYLBICYCLO(2.2.1)HEPTAN-2-OL

TOXICITY DATA with REFERENCE
skn-rbt 500 mg/24H MOD FCTXAV 14,775,76

CONSENSUS REPORTS: Reported in EPA TSCA Inventory.

SAFETY PROFILE: A skin irritant. Combustible. When heated to decomposition it emits acrid smoke and irritating fumes.

TLW250 CAS:1195-79-5 ***HR: 2***
1,3,3-TRIMETHYL-2-NORBORNANONE
mf: $C_{10}H_{16}O$ mw: 152.26

SYNS: FENCHON (GERMAN) ◇ FENCHONE ◇ 1,3,3-TRIMETHYL-2-NORCAMPHANONE

TOXICITY DATA with REFERENCE
skn-rbt 500 mg/24H MLD FCTXAV 14,769,76
orl-rat LD50:4400 mg/kg FCTXAV 2,327,64
scu-mus LDLo:2100 mg/kg AEXPBL 50,199,1903
scu-frg LDLo:650 mg/kg AEXPBL 50,199,1903

CONSENSUS REPORTS: Reported in EPA TSCA Inventory.

SAFETY PROFILE: Moderately toxic by subcutaneous

route. Mildly toxic by ingestion. A skin irritant. When heated to decomposition it emits acrid smoke and irritating fumes. See also KETONES.

TLW500 CAS:112-03-8 ***HR: 3***
TRIMETHYLOCTADECYLAMMONIUM CHLORIDE
mf: $C_{21}H_{46}N{\bullet}Cl$ mw: 348.13

SYNS: OCTADECYLTRIMETHYLAMMONIUMCHLORIDE ◇ STEARYLTRIMETHYLAMMONIUM CHLORIDE ◇ TRIMETHYLSTEARYLAMMONIUM CHLORIDE

TOXICITY DATA with REFERENCE
unr-mus LDLo:50 mg/kg ATMPA2 32,177,38

CONSENSUS REPORTS: Reported in EPA TSCA Inventory.

SAFETY PROFILE: Poison by unspecified routes. When heated to decomposition it emits very toxic fumes of NO_x, NH_3, and Cl^-.

TLW750 CAS:1017-56-7 ***HR: 2***
TRIMETHYLOLMELAMINE
mf: $C_6H_{12}N_6O_3$ mw: 216.24

SYNS: N,N',N''-TRIHYDROXYMETHYLMELAMINE ◇ N,N',N''-TRIS(HYDROXYMETHYL)-1,3,5-TRIAZINE-2,4,6-TRIAMINE

TOXICITY DATA with REFERENCE
sln-dmg-unk 1 pph ZEVBA5 93,1,62
ipr-rat LD50:1750 mg/kg ARZNAD 16,1734,66
ipr-mus LDLo:1350 mg/kg BJPCAL 6,201,51

CONSENSUS REPORTS: Reported in EPA TSCA Inventory.

SAFETY PROFILE: Mutation data reported. Moderately toxic by intraperitoneal route. When heated to decomposition it emits toxic fumes of NO_x. See also AMINES.

TLX000 CAS:682-09-7 ***HR: 1***
TRIMETHYLOLPROPANE DIALLYL ETHER

TOXICITY DATA with REFERENCE
orl-rat LD50:6500 mg/kg IHFCAY 6,1,67

CONSENSUS REPORTS: Reported in EPA TSCA Inventory.

SAFETY PROFILE: Mildly toxic by ingestion. When heated to decomposition it emits acrid smoke and irritating fumes. See also ETHERS and ALLYL COMPOUNDS.

TLX100 CAS:682-11-1 ***HR: 1***
TRIMETHYLOPROPANE MONOALLYL ETHER
mf: $C_9H_{18}O_3$ mw: 174.27

SYN: 2,2-(DIHYDROXYMETHYL)-1-BUTANOL, MONOALLYL ETHER

TOXICITY DATA with REFERENCE
orl-rat LD50:4930 mg/kg IHFCAY 6,1,67

CONSENSUS REPORTS: Reported in EPA TSCA Inventory.

SAFETY PROFILE: Mildly toxic by ingestion. When heated to decomposition it emits acrid smoke and irritating fumes. See also ETHERS and ALLYL COMPOUNDS.

TLX175 CAS:15625-89-5 ***HR: 1***
TRIMETHYLOLPROPANE TRIACRYLATE
mf: $C_{15}H_{20}O_6$ mw: 296.35

SYNS: 2-ETHYL-2-(HYDROXYMETHYL)-1,3-PROPANEDIOL TRIACRYLATE ◇ MFA ◇ MFM ◇ NK ESTER A-TMPT ◇ SARTOMER SR 351 ◇ SR 351 ◇ TMPTA ◇ 1,1,1-(TRIHYDROXYMETHYL)PROPANE TRIESTER ACRYLIC ACID

TOXICITY DATA with REFERENCE
skn-hmn 1% AIHAAP 42,B-53,81
orl-rat LD50:5190 mg/kg TXAPA9 28,313,74
skn-rbt LD50:5170 mg/kg AIHAAP 42,B-53,81

CONSENSUS REPORTS: Reported in EPA TSCA Inventory.

SAFETY PROFILE: Mildly toxic by ingestion and skin contact. A human skin irritant. When heated to decomposition it emits acrid smoke and irritating fumes.

TLX250 CAS:3290-92-4 ***HR: 2***
TRIMETHYLOLPROPANE TRIMETHACRYLATE
mf: $C_{18}H_{26}O_6$ mw: 338.44

PROP: Liquid. Bp: >200°/1 mm, d: 0.97, refr index: 1.4700, flash p: 149° F

SYNS: 2-ETHYL-2-HYDROXYMETHYL-1,3-PROPANEDIOL TRIMETHACRYLATE ◇ TRIMETHYLOLPROPANE TRIMETHANCRYLATE

TOXICITY DATA with REFERENCE
ipr-mus LD50:2889 mg/kg JDREAF 51,526,72

CONSENSUS REPORTS: Reported in EPA TSCA Inventory.

SAFETY PROFILE: Moderately toxic by intraperitoneal route. Combustible liquid. When heated to decomposition it emits acrid smoke and irritating fumes.

TLX600 CAS:149-73-5 ***HR: 2***
TRIMETHYL ORTHOFORMATE
mf: $C_4H_{10}O_3$ mw: 106.14

PROP: Colorless liquid; pungent odor. Vap d: 3.67, flash p: 59°F.

SYNS: METHYLESTER KYSELINY ORTHOMRAVENCI (CZECH) ◇ METHYL ORTHOFORMATE ◇ ORTHOFORMIC ACID, TRIMETHYL

ESTER ◇ ORTHOMRAVENCAN METHYLNATY (CZECH) ◇ TRIMETHOXYMETHANE

TOXICITY DATA with REFERENCE
skn-rbt 500 mg/24H MLD 28ZPAK -,43,72
eye-rbt 100 mg/24H MOD 28ZPAK -,43,72
orl-rat LD50:3130 mg/kg 28ZPAK -,43,72
ihl-rat LC50:5000 ppm/4H 28ZPAK -,43,72

CONSENSUS REPORTS: Reported in EPA TSCA Inventory.

SAFETY PROFILE: Moderately toxic by ingestion. Mildly toxic by inhalation. A skin and eye irritant. A very dangerous fire hazard when exposed to heat or flame; can react with oxidizing materials. Hazardous to prepare. To fight fire, use CO_2, fog, haze. When heated to decomposition it emits acrid smoke and irritating fumes. See also ESTERS.

TLY000 CAS:64047-30-9 ***HR: 3***
TRIMETHYL-2-OXEPANONE (mixed isomers)
mf: $C_9H_{16}O_2$ mw: 156.25

SYN: TRIMETHYL-ε-LACTONE (mixed isomers)

TOXICITY DATA with REFERENCE
orl-rat LD50:7430 mg/kg AIHAAP 23,95,62
skn-rbt LD50:6300 mg/kg AIHAAP 23,95,62

CONSENSUS REPORTS: IARC Cancer Review: Animal Sufficient Evidence IMEMDT 19,303,79.

SAFETY PROFILE: Confirmed carcinogen. Mildly toxic by ingestion and skin contact. When heated to decomposition it emits acrid smoke and irritating fumes. See also KETONES.

TLY175 CAS:5076-19-7 ***HR: 2***
TRIMETHYLOXIRANE
mf: $C_5H_{10}O$ mw: 86.15

SYNS: β-ISOAMYLENE OXIDE ◇ TRIMETHYLETHYLENE OXIDE ◇ TRIMETHYLOXACYCLOPROPANE ◇ 2,2,3-TRIMETHYLOXIRANE

TOXICITY DATA with REFERENCE
orl-rat LD50:2635 mg/kg GTPZAB 24(10),49,80
ipr-rat LD50:1410 mg/kg GTPZAB 24(10),49,80
orl-mus LD50:2600 mg/kg GTPZAB 24(10),49,80
ipr-mus LD50:1513 mg/kg GTPZAB 24(10),49,80

SAFETY PROFILE: Moderately toxic by ingestion and intraperitoneal routes. When heated to decomposition it emits acrid smoke and irritating fumes.

TLY250 CAS:25351-18-2 ***HR: 3***
TRIMETHYL(4-OXYPENTHYL)AMMONIUM IODIDE
mf: $C_8H_{18}NO{\cdot}I$ mw: 271.17

SYNS: 4-KETOAMYLTRIMETHYLAMMONIUM IODIDE ◇ N-PENTAN-4-ONE-N,N,N-TRIMETHYLAMMONIUM IODIDE ◇ N,N,N-TRIMETHYL-4-OXO-1-PENTANAMINIUMIODIDE

TOXICITY DATA with REFERENCE
scu-mus LD50:13.5 mg/kg JPETAB 103,196,51
ivn-dog LDLo:1085 μg/kg JPETAB 103,196,51

SAFETY PROFILE: Poison by subcutaneous and intravenous routes. When heated to decomposition it emits very toxic fumes of NO_x, NH_3, and I^-. See also KETONES and IODIDES.

TLY500 CAS:540-84-1 ***HR: 3***
2,2,4-TRIMETHYLPENTANE
DOT: UN 1262
mf: C_8H_{18} mw: 114.26

PROP: Clear liquid; odor of gasoline. Bp: 99.2°, fp: −116°, flash p: 10°F, d: 0.692 @ 20/4°, autoign temp: 779°F, vap press: 40.6 mm @ 21°, vap d: 3.93, lel: 1.1%, uel: 6.0%.

SYNS: ISOBUTYLTRIMETHYLETHANE ◇ ISOOCTANE (DOT)

CONSENSUS REPORTS: Reported in EPA TSCA Inventory.

NIOSH REL: TWA (Alkanes) 350 mg/m^3
DOT Classification: Flammable Liquid; Label: Flammable Liquid.

SAFETY PROFILE: High concentrations can cause narcosis. A very dangerous fire hazard when exposed to heat, flame, oxidizers. Can react vigorously with reducing materials. Explosive in the form of vapor when exposed to heat or flame. To fight fire, use CO_2, dry chemical. When heated to decomposition it emits acrid smoke and irritating fumes. See also ALKANES.

TLY750 CAS:144-19-4 ***HR: 3***
2,2,4-TRIMETHYL-1,3-PENTANEDIOL
mf: $C_8H_{18}O_2$ mw: 146.26

PROP: White, crystalline solid. Mp: 49-51°, bp: 109-111° @ 4 mm, flash p: 235°.

SYN: TMPD

TOXICITY DATA with REFERENCE
skn-rbt 9370 μg/24H open MLD AIHAAP 23,95,62
orl-rat LDLo:2000 mg/kg KODAK* -,-,71
ipr-rat LDLo:800 mg/kg TXAPA9 29,87,74
ivn-rat LDLo:145 mg/kg TXAPA9 29,87,74
orl-mus LDLo:2200 mg/kg KODAK* -,-,71
ipr-mus LDLo:800 mg/kg TXAPA9 29,87,74
ivn-mus LDLo:145 mg/kg TXAPA9 29,87,74

CONSENSUS REPORTS: Reported in EPA TSCA Inventory.

SAFETY PROFILE: Poison by intravenous route.

Moderately toxic by ingestion and intraperitoneal routes. A skin irritant. An insect-repellent. Combustible when exposed to heat or flame; can react with oxidizing materials. To fight fire, use CO_2, fog, mist, dry chemical. When heated to decomposition it emits acrid smoke and irritating fumes.

TLZ000 CAS:6846-50-0 ***HR: 1***
2,2,4-TRIMETHYL-1,3-PENTANEDIOL DIISOBUTYRATE
mf: $C_{16}H_{30}O_4$ mw: 286.46

PROP: Bp: 536°F, d: 0.9, vap d: 9.9, flash p: 250°F (OC).

SYNS: ISOBUTYRIC ACID, 1-ISOPROPYL-2,2-DIMETHYLTRIMETHYLENE ESTER ◇ 2,2,4-TRIMETHYLPENTANEDIOL-1,3-DIISOBUTYRATE

TOXICITY DATA with REFERENCE
skn-gpg 5 mg/kg MLD TXAPA9 22,387,72

CONSENSUS REPORTS: Reported in EPA TSCA Inventory.

SAFETY PROFILE: A skin irritant. Combustible when exposed to heat or flame. To fight fire, use alcohol foam, spray, mist, dry chemical. When heated to decomposition it emits acrid smoke and irritating fumes.

TMA250 CAS:25167-70-8 ***HR: 2***
2,4,4-TRIMETHYL PENTENE
DOT: UN 2050
mf: C_8H_{16} mw: 112.24

PROP: A clear liquid. Bp: 104.5°, flash p: 35°F (TOC), fp: −106.4°, d: 0.724 @ 15.5/15.5°, vap press: 77.5 mm @ 38°, vap d: 3.9, autoign temp: 581°F.

SYNS: DIISOBUTENE ◇ DIISOBUTYLENE

CONSENSUS REPORTS: Reported in EPA TSCA Inventory.

SAFETY PROFILE: An irritant. Irritating and narcotic in high concentration. Has caused liver and kidney damage in experimental animals. A very dangerous fire hazard when exposed to heat or flame; can react vigorously with oxidizing materials such as oleum, chlorosulfonic acid, H_2SO_4. Keep away from heat and open flame. To fight fire, use foam, CO_2, dry chemical.

TMA500 CAS:19109-66-1 ***HR: 3***
TRIMETHYLPENTYLAMMONIUM IODIDE
mf: $C_8H_{20}N{\cdot}I$ mw: 257.19

SYNS: AMYLTRIMETHYLAMMONIUM IODIDE ◇ PENTYLTRIMETHYLAMMONIUM IODIDE ◇ N,N,N-TRIMETHYL-1-PENTANAMINIUM IODIDE

TOXICITY DATA with REFERENCE
ipr-mus LD50:18 mg/kg UCPHAQ 1,187,39
scu-mus LD50:25 mg/kg JPETAB 103,196,51
ivn-mus LD50:2400 μg/kg JPETAB 110,369,54
ivn-dog LDLo:1021 μg/kg JPETAB 103,196,51

SAFETY PROFILE: Poison by subcutaneous, intraperitoneal, and intravenous routes. When heated to decomposition it emits very toxic fumes of NO_x, NH_3, and I^-. See also IODIDES.

TMA750 CAS:73791-32-9 ***HR: 1***
1-(2,4,6-TRIMETHYLPHENYLAMINO)ANTHRAQUINONE
mf: $C_{23}H_{19}NO_2$ mw: 341.43

SYN: MODR MIDLONOVA STALA ER (CZECH)

TOXICITY DATA with REFERENCE
skn-rbt 500 mg/24H MLD 28ZPAK -,242,72
eye-rbt 100 mg/24H MOD 28ZPAK -,242,72

SAFETY PROFILE: A skin and eye irritant. When heated to decomposition it emits toxic fumes of NO_x.

TMB000 CAS:16056-11-4 ***HR: 3***
TRIMETHYLPHENYLAMMONIUM BROMIDE
mf: $C_9H_{14}N{\cdot}Br$ mw: 216.15

SYN: PHENYL TRIMETHYL AMMONIUM BROMIDE

TOXICITY DATA with REFERENCE
ivn-mus LD50:4 mg/kg JPETAB 99,16,50

CONSENSUS REPORTS: Reported in EPA TSCA Inventory.

SAFETY PROFILE: Poison by intravenous route. When heated to decomposition it emits very toxic fumes of NO_x, NH_3, and Br^-. See also BROMIDES.

TMB250 CAS:138-24-9 ***HR: 3***
TRIMETHYLPHENYL AMMONIUM CHLORIDE
mf: $C_9H_{14}N{\cdot}Cl$ mw: 171.69

SYNS: PHENYLTRIMETHYLAMMONIUM CHLORIDE ◇ TRIMETHYLANILINIUM CHLORIDE ◇ N,N,N-TRIMETHYLANILINIUM CHLORIDE

TOXICITY DATA with REFERENCE
orl-mus LDLo:200 mg/kg JPETAB 43,413,31
ivn-mus LDLo:15 mg/kg JPETAB 43,413,31

CONSENSUS REPORTS: Reported in EPA TSCA Inventory.

SAFETY PROFILE: Poison by ingestion and intravenous routes. When heated to decomposition it emits very toxic fumes of NO_x, NH_3, and Cl^-.

TMB500 CAS:1899-02-1 ***HR: 3***
TRIMETHYLPHENYLAMMONIUM HYDROXIDE
mf: $C_9H_{14}N•HO$ mw: 153.25

SYNS: PHENYL TRIMETHYL AMMONIUM HYDROXIDE ◇ TRIMETHYLANILINIUM HYDROXIDE

TOXICITY DATA with REFERENCE
scu-mus LDLo:49 mg/kg JPETAB 28,367,26

CONSENSUS REPORTS: Reported in EPA TSCA Inventory.

SAFETY PROFILE: Poison by subcutaneous route. When heated to decomposition it emits toxic fumes of NO_x and NH_3.

TMB750 CAS:98-04-4 ***HR: 3***
TRIMETHYLPHENYLAMMONIUM IODIDE
mf: $C_9H_{14}N•I$ mw: 263.14

PROP: Leaves from alc. Mp: 228-230° decomp, bp: subl. Sol in alc, water; insol in chloroform.

SYNS: N,N-DIMETHYLANILINE METHIODIDE ◇ PHENYLTRIMETHYLAMMONIUM IODIDE ◇ PHT ◇ TRIMETHYLANILINIUM IODIDE ◇ N,N,N-TRIMETHYLANILINIUM IODIDE ◇ N,N,N-TRIMETHYLBENZENAMINIUM IODIDE

TOXICITY DATA with REFERENCE
ipr-mus LD50:55 mg/kg UCPHAQ 2,161,44
scu-mus LD50:85 mg/kg JCSOA9 -,182,47
ivn-mus LD50:5620 μg/kg CSLNX* NX#02332

CONSENSUS REPORTS: Reported in EPA TSCA Inventory.

SAFETY PROFILE: Poison by intraperitoneal, subcutaneous, and intravenous routes. When heated to decomposition it emits very toxic fumes of NO_x, NH_3, and I^-. See also IODIDES.

TMC750 ***HR: 3***
TRIMETHYLPHENYL METHYLCARBAMATE
mf: $C_{11}H_{15}NO_2$ mw: 193.27

PROP: It is a mixture of the 3,4,5- and the 2,3,5-trimethyl phenyl methyl carbamate isomers, which are present in a ratio of 4:1 (SHELL*).

SYNS: LANDRIN ◇ METHYLCARBAMIC ACID, TRIMETHYLPHENYL ESTER

TOXICITY DATA with REFERENCE
orl-rat LD50:208 mg/kg SHELL*
ipr-rat LD50:94400 μg/kg BWHOA6 44(1-3),241,71
ivn-rat LD50:31800 μg/kg BWHOA6 44(1-3),241,71

SAFETY PROFILE: Poison by ingestion, intraperitoneal, and intravenous routes. When heated to decomposition it emits toxic fumes of NO_x. See also 3,4,5-TRIMETHYLPHENYL METHYLCARBAMATE and CARBAMATES.

TMD000 CAS:2686-99-9 ***HR: 3***
3,4,5-TRIMETHYLPHENYL METHYLCARBAMATE
mf: $C_{11}H_{15}NO_2$ mw: 193.27

SYNS: ENT 25,843 ◇ LANDRIN ◇ OMS-597 ◇ SD 8530 ◇ SHELL SD-8530

TOXICITY DATA with REFERENCE
orl-rat LD50:178 mg/kg TXAPA9 21,315,72
ipr-rat LDLo:136 mg/kg TXAPA9 25,569,73
ivn-rat LD50:32 mg/kg BJIMAG 22,317,65
ims-rat LD50:283 mg/kg BJIMAG 22,317,65
orl-mus LD50:101 mg/kg ARSIM* 20,20,66
orl-pgn LD50:168 mg/kg TXAPA9 20,57,71
orl-ckn LD50:50 mg/kg TXAPA9 11,49,67
orl-qal LD50:71 mg/kg TXAPA9 20,57,71
orl-dck LD50:22 mg/kg TXAPA9 20,57,71
orl-bwd LD50:10 mg/kg TXAPA9 21,315,72

SAFETY PROFILE: Poison by ingestion, intraperitoneal, intravenous, and intramuscular routes. When heated to decomposition it emits toxic fumes of NO_x. See also CARBAMATES.

TMD250 CAS:512-56-1 ***HR: 3***
TRIMETHYL PHOSPHATE
mf: $C_3H_9O_4P$ mw: 140.09

$(CH_3O)_3P(:O)$

PROP: Liquid. D: 1.97 @ 19.5/0°, bp: 197.2°. Sol in alc, water, ether.

SYNS: METHYL PHOSPHATE ◇ NCI-C03781 ◇ PHOSPHORIC ACID, TRIMETHYL ESTER ◇ TMP ◇ O,O,O-TRIMETHYL PHOSPHATE

TOXICITY DATA with REFERENCE
cyt-hmn:lym 100 mmol/L/5H MUREAV 65,121,79
trn-mus-ipr 1 g/kg MUREAV 157,205,85
ipr-mus TDLo:1 g/kg (male 1D pre):REP MUREAV 157,205,85
orl-mus TDLo:2500 mg/kg (male 5D pre):TER SCIEAS 168,584,70
orl-rat TDLo:31 g/kg/2Y-C:NEO NCITR* NCI-CG-TR-81,78
orl-mus TDLo:154 g/kg/2Y-C:CAR,TER NCITR* NCI-CG-TR-81,78
orl-rat TD:16 g/kg/2Y-I:ETA NCITR* NCI-CG-TR-81,78
orl-rat LD50:840 mg/kg NCILB* NIH-NCI-E-C-72-3252,73
ipr-rat LDLo:800 mg/kg JPPMAB 11,150,59
ivn-rat LDLo:2400 mg/kg NATUAS 179,154,57
orl-mus LD50:1470 mg/kg NCILB* NIH-NCI-E-C-72-3252,73
orl-rbt LD50:1050 mg/kg JPETAB 88,338,46
skn-rbt LD50:3388 mg/kg AIHAAP 30,470,69

CONSENSUS REPORTS: NCI Carcinogenesis Bioassay (gavage); Clear Evidence: mouse, rat NCITR* NCI-CG-TR-81,78. Reported in EPA TSCA Inventory. EPA Genetic Toxicology Program.

DFG MAK: Suspected Carcinogen.

SAFETY PROFILE: Suspected carcinogen with experimental carcinogenic, neoplastigenic, tumorigenic, and teratogenic data. Moderately toxic by ingestion, skin contact, intraperitoneal, and intravenous routes. Experimental reproductive effects. Human mutation data reported. Explodes when heat distilled. When heated to decomposition it emits toxic fumes of PO_x. See also ESTERS.

TMD275 CAS:594-09-2 ***HR: 3***
TRIMETHYLPHOSPHINE
mf: C_3H_9P mw: 76.08

$(CH_3)_3P$

PROP: Stench. Mp: -86°, bp: 38-39°, d: 0.735, flash p: -22° F.

SAFETY PROFILE: Flammable liquid. Extremely dangerous fire hazard. May ignite spontaneously in air. When heated to decomposition it emits toxic fumes of PO_x. See also PHOSPHINE.

TMD400 CAS:20819-54-9 ***HR: 3***
TRIMETHYLPHOSPHINE SELENIDE
mf: C_3H_9PSe mw: 155.05

SYN: PHOSPHINE SELENIDE, TRIMETHYL-

TOXICITY DATA with REFERENCE
ivn-mus LDLo:8 mg/kg JPETAB 25,315,25

OSHA PEL: TWA 0.2 mg(Se)/m^3
ACGIH TLV: TWA 0.2 mg(Se)/m^3

SAFETY PROFILE: Poison by intravenous route. When heated to decomposition it emits toxic fumes of PO_x and Se.

TMD500 CAS:121-45-9 ***HR: 2***
TRIMETHYL PHOSPHITE
DOT: UN 2329
mf: $C_3H_9O_3P$ mw: 124.09

$(CH_3O)_3P$

PROP: Colorless liquid. D: 1.046 @ 20/4°, vap d: 4.3, bp: 232-234°F, flash p: 130°F (OC). Insol in water; sol in hexane, benzene, acetone, alc, ether, carbon tetrachloride, kerosene.

SYNS: FOSFORYN TROJMETYLOWY (CZECH) ◇ METHYL PHOSPHITE ◇ PHOSPHOROUS ACID, TRIMETHYL ESTER ◇ TRIMETHOXYPHOSPHINE

TOXICITY DATA with REFERENCE
skn-rbt 500 mg SEV 34ZIAG -,609,69
eye-rbt 100 mg SEV 34ZIAG -,609,69
skn-mam 500 mg MLD MEPAAX 29,393,78
eye-mam 100 mg MLD MEPAAX 29,393,78
orl-rat TDLo:1640 mg/kg (6-15D preg):TER TXAPA9 72,119,84
orl-rat LD50:1600 mg/kg ALBRW* #OPB-3,84
ipr-mus LD50:4180 mg/kg ENVRAL 9,1,75
skn-rbt LDLo:2200 mg/kg 34ZIAG -,610,69

CONSENSUS REPORTS: Reported in EPA TSCA Inventory.

OSHA PEL: TWA 2 ppm
ACGIH TLV: TWA 2 ppm
DOT Classification: Flammable or Combustible Liquid; Label: Flammable Liquid.

SAFETY PROFILE: Moderately toxic by ingestion and skin contact. An experimental teratogen. A severe skin and eye irritant. Flammable when exposed to heat, flame, or oxidizers. To fight fire, use water, foam, fog, CO_2. Violent explosive reaction on contact with magnesium perchlorate or trimethyl platinum(IV) azide tetramer. When heated to decomposition it emits toxic fumes of PO_x. An intermediate in the production of pesticides, fire retardants, and organic phosphorous additives. See also ESTERS.

TMD625 CAS:22608-53-3 ***HR: 3***
O,S,S-TRIMETHYL PHOSPHORODITHIOATE
mf: $C_3H_9O_2PS_2$ mw: 172.21

TOXICITY DATA with REFERENCE
orl-rat LD50:26 mg/kg ARTODN 42,95,79
ipr-rat LD50:26 mg/kg FAATDF 4(2,Pt 2),S215,84
ivn-rat LD50:26 mg/kg FAATDF 4(2,Pt 2),S215,84
orl-mus LD50:120 mg/kg TXAPA9 75,219,84
ipr-mus LD50:38 mg/kg ARTODN 51,221,82
ivn-mus LD50:149 mg/kg DTESD7 8,631,80

SAFETY PROFILE: Poison by ingestion, intravenous, and intraperitoneal routes. When heated to decomposition it emits toxic fumes of PO_x and SO_x.

TMD699 CAS:681-71-0 ***HR: 3***
S,S,S-TRIMETHYL PHOSPHOROTRITHIOATE
mf: $C_3H_9OPS_3$ mw: 188.27

TOXICITY DATA with REFERENCE
orl-rat LD50:30 mg/kg FAATDF 4(2,Pt 2),S215,84
ipr-rat LD50:30 mg/kg FAATDF 4(2,Pt 2),S215,84
ivn-rat LD50:30 mg/kg FAATDF 4(2,Pt 2),S215,84
orl-mus LD50:30 mg/kg ARTODN 51,221,82

SAFETY PROFILE: Poison by ingestion, intravenous,

and intraperitoneal routes. When heated to decomposition it emits toxic fumes of PO_x and SO_x.

TME250 CAS:14477-33-9 ***HR: 3***
TRIMETHYLPLATINUM HYDROXIDE
mf: $C_3H_{10}OPt$ mw: 257.18

SAFETY PROFILE: Explodes on heating. See also PLATINUM COMPOUNDS and ORGANO METALS.

TME270 CAS:14667-55-1 ***HR: 2***
2,3,5-TRIMETHYLPYRAZINE
mf: $C_7H_{10}N_2$ mw: 122.19

PROP: Colorless to sltly yellow liquid; baked potato, peanut odor. D: 0.960-0.990 @ 20°, refr index: 1.503, flash p: +153°F. Sol in water and organic solvents.

SYNS: FEMA No. 3244 ◇ TRIMETHYLPYRAZINE

TOXICITY DATA with REFERENCE
orl-rat LD50:806 mg/kg DCTODJ 3,249,80

CONSENSUS REPORTS: Reported in EPA TSCA Inventory.

SAFETY PROFILE: Moderately toxic by ingestion. Combustible liquid. When heated to decomposition emits toxic fumes of NO_x.

TME275 CAS:940-93-2 ***HR: 3***
2,4,6-TRIMETHYLPYRILIUM PERCHLORATE
mf: $C_8H_{11}ClO_5$ mw: 212.55

SAFETY PROFILE: The dry crystalline solid is an impact- and friction-sensitive explosive. When heated to decomposition it emits toxic fumes of Cl^-. See also PERCHLORATES.

TME500 CAS:25930-79-4 ***HR: 3***
TRIMETHYLSELENONIUM
mf: C_3H_9Se mw: 124.08

SYNS: SELENONIUM, TRIMETHYL- ◇ TRIMETHYLSELENONIUM ION

TOXICITY DATA with REFERENCE
scu-rat TDLo:31 mg/kg ARTODN 45,207,80

OSHA PEL: TWA 0.2 mg(Se)/m^3
ACGIH TLV: TWA 0.2 mg(Se)/m^3

SAFETY PROFILE: Poison by subcutaneous route. When heated to decomposition it emits toxic fumes of Se.

TME600 CAS:18987-38-7 ***HR: 3***
TRIMETHYLSELENONIUM CHLORIDE
mf: $C_3H_9Se \cdot Cl$ mw: 159.53

SYN: SELENONIUM, TRIMETHYL-, CHLORIDE

TOXICITY DATA with REFERENCE
ipr-rat LD50:99 mg/kg CTOXAO 17,171,80

OSHA PEL: TWA 0.2 mg(Se)/m^3
ACGIH TLV: TWA 0.2 mg(Se)/m^3

SAFETY PROFILE: Poison by intraperitoneal route. When heated to decomposition it emits toxic fumes of Se and Cl^-.

TME750 CAS:13435-12-6 ***HR: 3***
N-TRIMETHYLSILYLACETAMIDE
mf: $C_5H_{13}NOSi$ mw: 131.28

PROP: Mp: 46-49°, bp: 84°/18 mm.

TOXICITY DATA with REFERENCE
ipr-mus LDLo:350 mg/kg StoGD# 27May75

CONSENSUS REPORTS: Reported in EPA TSCA Inventory.

SAFETY PROFILE: Poison by intraperitoneal route. When heated to decomposition it emits toxic fumes of NO_x. See also SILANE.

TMF000 CAS:10416-59-8 ***HR: 3***
N-(TRIMETHYLSILYL)ACETIMIDIC ACID, TRIMETHYLSILYL ESTER
mf: $C_8H_{21}NOSi_2$ mw: 203.48

SYNS: BIS(TRIMETHYLSILYL)ACETAMIDE ◇ N,o-BIS(TRIMETHYLSILYL)ACETAMIDE ◇ BSA ◇ N-(TRIMETHYLSILYL)ETHANIMIDIC ACID, TRIMETHYLSILYL ESTER

TOXICITY DATA with REFERENCE
ipr-mus TDLo:500 mg/kg/I:NEO JNCIAM 54,495,75
ipr-mus LDLo:750 mg/kg StoGD# 27May75

CONSENSUS REPORTS: Reported in EPA TSCA Inventory.

SAFETY PROFILE: Moderately toxic by intraperitoneal route. Questionable carcinogen with experimental neoplastigenic data. When heated to decomposition it emits toxic fumes of NO_x. See also ESTERS and SILANE.

TMF100 CAS:4648-54-8 ***HR: 3***
TRIMETHYLSILYL AZIDE
mf: $C_3H_9N_3Si$ mw: 115.21

$$(CH_3)_3SiN_3$$

PROP: D: 0.868, bp: 52-53°

SYN: AZIDOTRIMETHYLSILANE

SAFETY PROFILE: Explosive reaction with rhenium hexafluoride, selenium halides. Reacts with tungsten hexafluoride to form an explosive product. When heated to decomposition it emits toxic fumes of NO_x. See also AZIDES and SILANE.

TMF125 CAS:18230-75-6 ***HR: 2***
TRIMETHYL SILYL HYDROPEROXIDE
mf: $C_3H_{10}O_2Si$ mw: 106.20

$(CH_3)_3SiOOH$

SAFETY PROFILE: Potentially hazardous thermal decomposition when heated above 35°C. When heated to decomposition it emits acrid smoke and irritating fumes. See also PEROXIDES and SILANE.

TMF250 CAS:18156-74-6 ***HR: 3***
N-(TRIMETHYLSILYL)IMIDAZOLE
mf: $C_6H_{12}N_2Si$ mw: 140.29

PROP: Bp: 93-94°/14 mm, refr index: 1.470, d: 0.956, flash p: 42° F.

SYNS: (TRIMETHYLSILYL)IMIDAZOLE ◇ N-(TRIMETHYLSILYL)-IMIDAZOL ◇ 1-(TRIMETHYLSILYL)IMIDAZOLE ◇ 1-(TRIMETHYL-SILYL)-1H-IMIDAZOLE ◇ TSIM

TOXICITY DATA with REFERENCE
ipr-mus TDLo:1000 mg/kg/I:NEO JNCIAM 54,495,75
ipr-mus LDLo:750 mg/kg StoGD# 27May75

CONSENSUS REPORTS: Reported in EPA TSCA Inventory.

SAFETY PROFILE: Moderately toxic by intraperitoneal route. Questionable carcinogen with experimental neoplastigenic data. Flammable liquid. When heated to decomposition it emits toxic fumes of NO_x. See also SILANE.

TMF500 CAS:3219-63-4 ***HR: 2***
TRIMETHYLSILYLMETHANOL
mf: $C_4H_{12}OSi$ mw: 104.25

TOXICITY DATA with REFERENCE
ipr-mus LD50:1080 mg/kg DANKAS 229(4),1011,76

CONSENSUS REPORTS: Reported in EPA TSCA Inventory.

SAFETY PROFILE: Moderately toxic by intraperitoneal route. When heated to decomposition it emits acrid smoke and irritating fumes. See also SILANE and ALCOHOLS.

TMF600 CAS:39482-21-8 ***HR: D***
N-TRIMETHYLSILYLMETHYL-N-NITROSOUREA
mf: $C_5H_{13}N_3O_2Si$ mw: 175.30

SYN: N-NITROSO-N-((TRIMETHYLSILYL)METHYL)UREA

TOXICITY DATA with REFERENCE
mmo-sat 1 mmol/L MUREAV 157,87,85
msc-ham:lng 1200 µmol/L MUREAV 157,87,85

SAFETY PROFILE: Mutation data reported. Many N-nitroso compounds are carcinogens. When heated to decomposition it emits toxic fumes of NO_x. See also N-NITROSO COMPOUNDS and SILANE.

TMF625 CAS:18204-79-0 ***HR: 3***
TRIMETHYLSILYL PERCHLORATE
mf: $C_3H_9ClO_4Si$ mw: 172.64

$(CH_3)_3SiOClO_3$

SAFETY PROFILE: Explodes when heated. When heated to decomposition it emits toxic fumes of Cl^-. See also PERCHLORATES and SILANE.

TMF750 CAS:63019-09-0 ***HR: 3***
N,N,2'-TRIMETHYL-4-STILBENAMINE
mf: $C_{17}H_{19}N$ mw: 237.37

SYNS: 4-DIMETHYLAMINO-2'-METHYLSTILBENE ◇ N,N-DIMETHYL-2'-METHYLSTILBENAMINE ◇ 2'-METHYL-4-DIMETHYL-AMINOSTILBENE

TOXICITY DATA with REFERENCE
orl-rat TDLo:750 mg/kg/71W-C:ETA ABMGAJ 9,87,62
scu-rat TDLo:80 mg/kg/4W-I:NEO PTRMAD 241,147,48

SAFETY PROFILE: Questionable carcinogen with experimental tumorigenic and neoplastigenic data. When heated to decomposition it emits toxic fumes of NO_x. See also AMINES.

TMG000 CAS:63040-32-4 ***HR: 3***
N,N,3'-TRIMETHYL-4-STILBENAMINE
mf: $C_{17}H_{19}N$ mw: 237.37

SYN: 3'-METHYL-4-DIMETHYLAMINOSTILBENE

TOXICITY DATA with REFERENCE
orl-rat TDLo:560 mg/kg/53W-C:ETA ABMGAJ 9,87,62

SAFETY PROFILE: Questionable carcinogen with experimental tumorigenic data. When heated to decomposition it emits toxic fumes of NO_x. See also AMINES.

TMG250 CAS:7378-54-3 ***HR: 3***
N,N,4'-TRIMETHYL-4-STILBENAMINE
mf: $C_{17}H_{19}N$ mw: 237.37

SYN: 4'-METHYL-4-DIMETHYLAMINOSTILBENE

TOXICITY DATA with REFERENCE
orl-rat TDLo:305 mg/kg/29W-C:ETA ABMGAJ 9,87,62

SAFETY PROFILE: Questionable carcinogen with experimental tumorigenic data. When heated to decomposition it emits toxic fumes of NO_x. See also AMINES.

TMG500 CAS:2181-42-2 ***HR: 3***
TRIMETHYLSULFONIUM IODIDE
mf: C_3H_9IS mw: 204.08

PROP: Mp: 215-220°.

TOXICITY DATA with REFERENCE
ipr-rat LDLo:88 mg/kg TXAPA9 20,135,71
scu-mus LDLo:300 mg/kg JPETAB 25,315,25
ivn-mus LD50:18 mg/kg CSLNX* NX#02185

CONSENSUS REPORTS: Reported in EPA TSCA Inventory.

SAFETY PROFILE: Poison by intraperitoneal, subcutaneous, and intravenous routes. When heated to decomposition it emits very toxic fumes of I^- and SO_x. See also IODIDES.

TMG750 CAS:7575-48-6 ***HR: 3***
TRIMETHYLSULFONIUM IODIDE MERCURIC IODIDE addition compound

SYN: TRIMETHYLSULFONIUM, IODIDE, COMPOUND with MERCURY IODIDE (1:1)

TOXICITY DATA with REFERENCE
ivn-mus LD50:18 mg/kg CSLNX* NX#01850

CONSENSUS REPORTS: Mercury and its compounds are on the Community Right-To-Know List.

NIOSH REL: TWA 0.05 mg(Hg)/m^3

SAFETY PROFILE: Poison by intravenous route. When heated to decomposition it emits very toxic fumes of SO_x, I^-, and Hg. See also TRIMETHYLSULFONIUM IODIDE, MERCURY(II) IODIDE, IODIDES, and MERCURY COMPOUNDS.

TMG775 CAS:25596-24-1 ***HR: 2***
TRIMETHYLSULFOXONIUM BROMIDE
mf: C_3H_9BrOS mw: 173.07

$(CH_3)_3S^+OBr^-$

SAFETY PROFILE: Thermal decompositon at 180°C results in vigorous, potentially dangerous release of vapor. Solutions in dimethyl sulfoxide decompose similarly at 74-80°C. When heated to decomposition it emits toxic fumes of Br^- and SO_x. See also BROMIDES.

TMH250 CAS:3003-15-4 ***HR: 3***
TRIMETHYLTHALLIUM
mf: C_3H_9Tl mw: 265.48

CONSENSUS REPORTS: Thallium and its compounds are on the Community Right-To-Know List.

SAFETY PROFILE: Thallium compounds are cumulative poisons. Explodes above 90°C. Ignites spontaneously in air. Complex with diethyl ether explodes at 0°C. See also THALLIUM COMPOUNDS and ORGANO METALS.

TMH525 CAS:152-18-1 ***HR: 2***
TRIMETHYL THIOPHOSPHATE
mf: $C_3H_9O_3PS$ mw: 156.14

$(CH_3O)_3P(:S)$

SYN: TRIMETHYL PHOSPHOROTHIOATE

SAFETY PROFILE: Potentially explosive exothermic reaction with chlorine. When heated to decomposition it emits toxic fumes of PO_x and SO_x.

TMH750 CAS:2489-77-2 ***HR: 3***
1,1,3-TRIMETHYL-2-THIOUREA
mf: $C_4H_{10}N_2S$ mw: 118.22

PROP: Trimethylthiourea tested in NCITR* NCI-CG-TR-129 contained 15% 1,3-dimethyl-2-thiourea and 5% Zeolex 80 NCITR* NCI-CG-TR-129,79.

SYNS: NCI-C02186 ◇ TRIMETHYLTHIOUREA ◇ N,N,N'-TRIMETHYLTHIOUREA

TOXICITY DATA with REFERENCE
orl-rat TDLo:13 g/kg/77W-C:CAR NCITR* NCI-CG-TR-129,79
orl-rat LD50:316 mg/kg NCILB* NIH-NCI-E-C-72-3252
orl-mus LD50:215 mg/kg NCILB* NIH-NCI-E-C-72-3252

CONSENSUS REPORTS: NCI Carcinogenesis Bioassay (feed); No Evidence: mouse NCITR* NCI-CG-TR-129,79. Reported in EPA TSCA Inventory.

SAFETY PROFILE: Poison by ingestion. Questionable carcinogen with experimental carcinogenic data. When heated to decomposition it emits very toxic fumes of NO_x and SO_x. See also ISOTHIOUREA.

TMI000 CAS:1118-14-5 ***HR: 3***
TRIMETHYLTIN ACETATE
mf: $C_5H_{12}O_2Sn$ mw: 222.86

SYN: ACETOXYTRIMETHYLSTANNANE

TOXICITY DATA with REFERENCE
orl-rat LD50:9 mg/kg BJIMAG 15,15,58

OSHA PEL: TWA 0.1 mg(Sn)/m^3 (skin)
ACGIH TLV: TWA 0.1 mg(Sn)/m^3 (skin) (Proposed: TWA 0.1 mg(Sn)/m^3; STEL 0.2 mg(Sn)/m^3 (skin))
NIOSH REL: (Organotin Compounds) TWA 0.1 mg(Sn)/m^3

SAFETY PROFILE: Poison by ingestion. When heated to decomposition it emits acrid smoke and irritating fumes. See also TIN COMPOUNDS.

TMI100 CAS:73940-86-0 ***HR: 3***
TRIMETHYLTIN CYANATE
mf: C_4H_9NOSn mw: 205.83

SYNS: CYANIC ACID, TRIMETHYLSTANNYL ESTER ◇ STANNANE, CYANATOTRIMETHYL-

TOXICITY DATA with REFERENCE
ivn-mus LD50:5600 μg/kg CSLNX* NX#06276

OSHA PEL: TWA 0.1 mg(Sn)/m^3 (skin)
ACGIH TLV: TWA 0.1 mg(Sn)/m^3; STEL 0.2 mg/m^3 (skin)
NIOSH REL: (Organotin Compound):10H TWA 0.1 mg(Sn)/m^3

SAFETY PROFILE: Poison by intravenous route. When heated to decomposition it emits toxic fumes of NO_x and Sn.

TMI250 CAS:56-24-6 ***HR: 3***
TRIMETHYLTIN HYDROXIDE
mf: $C_3H_{10}OSn$ mw: 180.82

PROP: Colorless crystals. Mp: 118° (decomp). Sol in water and many organic solvents.

SYN: HYDROXYTRIMETHYLSTANNANE

TOXICITY DATA with REFERENCE
scu-mus LDLo:1800 μg/kg JPETAB 28,367,26

OSHA PEL: TWA 0.1 mg(Sn)/m^3 (skin)
ACGIH TLV: TWA 0.1 mg(Sn)/m^3 (skin) (Proposed: TWA 0.1 mg(Sn)/m^3; STEL 0.2 mg(Sn)/m^3 (skin))
NIOSH REL: (Organotin Compounds) TWA 0.1 mg(Sn)/m^3

SAFETY PROFILE: Poison by subcutaneous route. When heated to decomposition it emits acrid smoke and irritating fumes. See also TIN COMPOUNDS.

TMI500 CAS:63869-87-4 ***HR: 3***
TRIMETHYLTIN SULPHATE
mf: $C_3H_{10}O_4SSn$ mw: 260.88

SYN: TRIMETHYLSTANNANESULPHATE

TOXICITY DATA with REFERENCE
orl-rat LDLo:30 mg/kg BJPCAL 10,16,55
ipr-rat LDLo:16 mg/kg BJPCAL 10,16,55

OSHA PEL: TWA 0.1 mg(Sn)/m^3 (skin)
ACGIH TLV: TWA 0.1 mg(Sn)/m^3 (skin) (Proposed: TWA 0.1 mg(Sn)/m^3; STEL 0.2 mg(Sn)/m^3 (skin))
NIOSH REL: (Organotin Compounds) TWA 0.1 mg(Sn)/m^3

SAFETY PROFILE: Poison by ingestion and intraperitoneal routes. When heated to decomposition it emits toxic fumes of SO_x. See also SULFATES and TIN COMPOUNDS.

TMI750 CAS:4638-25-9 ***HR: 3***
TRIMETHYLTIN THIOCYANATE
mf: C_4H_9NSSn mw: 221.89

SYN: THIOCYANIC ACID, TRIMETHYLSTANNYL ESTER

TOXICITY DATA with REFERENCE
ivn-mus LD50:1800 μg/kg CSLNX* NX#03079

OSHA PEL: TWA 0.1 mg(Sn)/m^3 (skin)
ACGIH TLV: TWA 0.1 mg(Sn)/m^3 (skin) (Proposed: TWA 0.1 mg(Sn)/m^3; STEL 0.2 mg(Sn)/m^3 (skin))
NIOSH REL: (Organotin Compounds) TWA 0.1 mg(Sn)/m^3

SAFETY PROFILE: Poison by intravenous route. When heated to decomposition it emits very toxic fumes of NO_x, SO_x, and CN^-. See also TIN COMPOUNDS, THIOCYANATES, and ESTERS.

TMJ000 CAS:1709-70-2 ***HR: 2***
1,3,5-TRIMETHYL-2,4,6-TRIS(3,5-DI-tert-BUTYL-4-HYDROXYBENZYL) BENZENE
mf: $C_{54}H_{78}O_3$ mw: 775.32

SYNS: AHYDOL (RUSSIAN) ◇ ANTIOXIDANT 330 ◇ AO-40 ◇ ETHANOX 330 ◇ SANTOQUIN EMULSION ◇ SANTOQUIN MIXTURE 6

TOXICITY DATA with REFERENCE
orl-rat TDLo:1100 mg/kg (1-22D preg):TER GISAAA 43(9),31,78
orl-rat LD50:1500 mg/kg IPSTB3 3,93,76

CONSENSUS REPORTS: Reported in EPA TSCA Inventory.

SAFETY PROFILE: Moderately toxic by ingestion. An experimental teratogen. When heated to decomposition it emits acrid smoke and irritating fumes.

TMJ250 CAS:632-14-4 ***HR: 2***
TRIMETHYLUREA
mf: $C_4H_{10}N_2O$ mw: 102.16

PROP: Mp: 74-75°.

SYN: TRIMETHYLHARNSTOFF (GERMAN) ◇ 1,1,3-TRIMETHYLUREA

TOXICITY DATA with REFERENCE
orl-rat LD50:1250 mg/kg ARZNAD 19,1073,69
ipr-mus LDLo:3188 mg/kg JPETAB 54,188,35

SAFETY PROFILE: Moderately toxic by ingestion and intraperitoneal routes. When heated to decomposition it emits toxic fumes of NO_x.

TMJ750 CAS:86-21-5 ***HR: 3***
TRIMETON
mf: $C_{16}H_{20}N_2$ mw: 240.38

SYNS: p-AMINOSALICYLSAURES SALZ (GERMAN) ◇ 2-(α-(2-

DIMETHYLAMINOETHYL)BENZYL)PYRIDINE ◇ 2-(3-DIMETHYLAMINO-1-PHENYLPROPYL)PYRIDINE ◇ N,N-DIMETHYL-3-PHENYL-3-(2-PYRIDYL)PROPYLAMINE ◇ NCI-C60695 ◇ 1-PHENYL-1-(2-PYRIDYL)-3-DIMETHYLAMINOPROPANE ◇ 3-PHENYL-3-(2-PYRIDYL)-N,N-DIMETHYLPROPYLAMINE

TOXICITY DATA with REFERENCE
orl-wmn TDLo:14 mg/kg:CNS MJAUAJ 2(3),110,76
orl-mus LD50:343 mg/kg ARZNAD 7,237,57
ipr-mus LD50:179 mg/kg ARZNAD 7,237,57
ivn-mus LD50:48 mg/kg AEPPAE 211,328,50
ivn-rbt LDLo:30 mg/kg AEPPAE 211,328,50

SAFETY PROFILE: Poison by ingestion, intraperitoneal, and intravenous routes. Human systemic effects by ingestion: central nervous system effects. When heated to decomposition it emits toxic fumes of NO_x.

TMJ800 CAS:18559-60-9 ***HR: 3***
(+)-TRIMETOQUINOL HYDROCHLORIDE

SYNS: AQD ◇ (+)-1,2,3,4-TETRAHYDRO-1-(3,4,5-TRIMETHOXYBENZYL)-6,7-ISOQUINOLINEDIOL HYDROCHLORIDE ◇ R-(+)-TRIMETOQUINOL HYDROCHLORIDE

TOXICITY DATA with REFERENCE
orl-mus LD50:2450 mg/kg EJPHAZ 5,303,68
ipr-mus LD50:390 mg/kg EJPHAZ 5,303,68
scu-mus LD50:2250 mg/kg EJPHAZ 5,303,68
ivn-mus LD50:145 mg/kg EJPHAZ 5,303,68
ipr-gpg LD50:820 mg/kg EJPHAZ 5,303,68

SAFETY PROFILE: Poison by intravenous and intraperitoneal routes. Moderately toxic by ingestion and subcutaneous routes. When heated to decomposition it emits toxic fumes of Cl^-.

TMK000 CAS:132-20-7 ***HR: 3***
TRIMETOSE
mf: $C_{16}H_{20}N_2{\cdot}C_4H_4O_4$ mw: 356.46

SYNS: AVIL-RETARD ◇ DANERAL ◇ 2-(α-(2-(DIMETHYLAMINO)ETHYL)BENZYL)PYRIDINE, BIMALEATE ◇ 2-(α-(2-(DIMETHYLAMINO)ETHYL)BENZYL)PYRIDINE, MALEATE ◇ 1-(N,N-DIMETHYLAMINO)-3-(PHENYL-3-α-PYRIDYL)PROPANE MALEATE ◇ HO 11513 ◇ INHISTON ◇ PHENIRAMINE MALEATE ◇ PHENYL(2-PYRIDYL)(β-N,N-DIMETHYLAMINOMETHYL) METHANE MALEATE ◇ 1-PHENYL-1-(2-PYRIDYL)-3-DIMETHYLAMINOPROPANE MALEATE ◇ PROPHENPYRIDAMINE MALEATE ◇ TRIMETON MALEATE

TOXICITY DATA with REFERENCE
cyt-mus-orl 1120 μg/kg IJEBA6 19,516,81
orl-hmn LDLo:30 mg/kg ATXKA8 29,317,72
orl-rat LD50:520 mg/kg KIZAAL 43,168,80
scu-rat LDLo:200 mg/kg CRSBAW 144,887,50
orl-mus LD50:268 mg/kg KIZAAL 43,168,80
ivn-dog LDLo:111 mg/kg JPETAB 113,72,55
ivn-gpg LD50:72 mg/kg AIPTAK 113,313,58

CONSENSUS REPORTS: Reported in EPA TSCA Inventory.

SAFETY PROFILE: Human poison by ingestion. Experimental poison by ingestion, subcutaneous, and intravenous routes. Mutation data reported. Used as an antihistamine. When heated to decomposition it emits toxic fumes of NO_x.

TMK150 CAS:16378-22-6 ***HR: 3***
TRIMOL
mf: $C_{22}H_{25}N{\cdot}ClH$ mw: 339.94

SYNS: 3-(10,11-DIHYDRO-5H-DIBENZO(a,d)CYCLOHEPTEN-5-YLIDENE)-1-ETHYL-2-METHYL-PYRROLIDINEHYDROCHLORIDE ◇ PIROHEPTINE HYDROCHLORIDE

TOXICITY DATA with REFERENCE
orl-rat LD50:600 mg/kg IYKEDH 4,467,73
ipr-rat LD50:100 mg/kg IYKEDH 4,467,73
scu-rat LD50:330 mg/kg NIIRDN 6,647,82
ivn-rat LD50:16 mg/kg IYKEDH 4,467,73
orl-mus LD50:127 mg/kg NIIRDN 6,647,82
ipr-mus LD50:78 mg/kg NIIRDN 6,647,82
scu-mus LD50:91 mg/kg NIIRDN 6,647,82
ivn-mus LD50:19 mg/kg IYKEDH 4,467,73
orl-dog LD50:195 mg/kg KSRNAM 6,941,72
ivn-dog LD50:12500 μg/kg KSRNAM 6,941,72
orl-rbt LD50:383 mg/kg KSRNAM 6,941,72
ivn-rbt LD50:6200 μg/kg KSRNAM 6,941,72

SAFETY PROFILE: Poison by ingestion, subcutaneous, intravenous, and intraperitoneal routes. When heated to decomposition it emits toxic fumes of NO_x and HCl.

TMK250 CAS:630-72-8 ***HR: 3***
TRINITROACETONITRILE
mf: $C_2N_4O_6$ mw: 176.05

CONSENSUS REPORTS: Cyanide and its compounds are on the Community Right-To-Know List.

DOT Classification: Forbidden.

SAFETY PROFILE: An explosive sensitive to friction, impact, or rapid heating to 220°C. When heated to decomposition it emits toxic fumes of CN^- and NO_x. See also NITRILES and NITRO COMPOUNDS.

TMK500 CAS:99-35-4 ***HR: 3***
1,3,5-TRINITROBENZENE
mf: $C_6H_3N_3O_6$ mw: 213.12
DOT: UN 0214

PROP: Yellow crystals. Mp: 122°, bp: decomp, d: 1.760 @ 20/4°.

SYNS: RCRA WASTE NUMBER U234 ◇ TNB ◇ TRINITROBENZEEN (DUTCH) ◇ TRINITROBENZENE ◇ TRINITROBENZENE, dry (DOT) ◇ TRINITROBENZOL (GERMAN)

TOXICITY DATA with REFERENCE
mmo-sat 10 μg/plate ENMUDM 2,531,80
mma-sat 10 μg/plate ENMUDM 2,531,80
orl-rat LD50:450 mg/kg GISAAA 42(10),12,77
orl-mus LD50:572 mg/kg TNICS* 13,132,73
ivn-mus LD50:32 mg/kg CSLNX* NX#00192
orl-gpg LD50:730 mg/kg GISAAA 42(10),12,77

CONSENSUS REPORTS: Reported in EPA TSCA Inventory.

DOT Classification: Class A Explosive; Label: Explosive A.

SAFETY PROFILE: Poison by intravenous route. Moderately toxic by ingestion. Mutation data reported. A severe explosion hazard when shocked or exposed to heat. Trinitrobenzene is considered a powerful high explosive and has more shattering power than TNT. Although it is less sensitive to impact than TNT, it is not used much because it is difficult to produce. The complex with potassium trimethyl stannate explodes at room temperature. Forms heat-sensitive explosive complexes with alkyl or aryl metallates (e.g., lithium or potassium salts of trimethyl-; triethyl-; or triphenyl-germanate; -silanate; or -stannate). Can react vigorously with reducing materials. When heated to decomposition it emits highly toxic fumes of NO_x and explodes. See also NITRO COMPOUNDS of AROMATIC HYDROCARBONS.

TMK750 **HR: 3**
TRINITROBENZENE (wet)
mf: $C_6H_3N_3O_5$ mw:197.12
DOT: UN 0214/UN 1354

SYNS: RCRA WASTE NUMBER U234 ◇ TRINITROBENZENE, WET, containing at least 10% water (DOT) ◇ TRINITROBENZENE, WET, containing less than 30% water (DOT) ◇ TRINITROBENZENE, WET, at least 10% water, over 15 ounces in one outside packaging (DOT)

DOT Classification: Flammable Solid; Label: Flammable Solid; Class A Explosive; Label: Explosive A.

SAFETY PROFILE: Flammable and explosive. When heated to decomposition it emits toxic fumes of NO_x. See also 1,3,5-TRINITROBENZENE.

TMK775 **HR: 3**
2,3,5-TRINITROBENZENEDIAZONIUM-4-OXIDE
mf: $C_6HN_5O_7$ mw: 255.10

$$N_2^+(O_2N)_3C_6HO^-$$

SAFETY PROFILE: Extremely explosive. When heated to decomposition it emits toxic fumes of NO_x. See also NITRO COMPOUNDS of AROMATIC HYDROCARBONS.

TML000 CAS:129-66-8 **HR: 3**
TRINITROBENZOIC ACID (dry)
mf: $C_7H_3N_3O_8$ mw: 257.13
DOT: UN 0215/UN 1355

PROP: Orthorhombic crystals. Mp: 228.7°. Sol @ 25° (2.05% in water, 26.6% in alc, 14.7% in ether), sol in methanol; sltly sol in benzene.

SYN: 2,4,6-TRINITROBENZOIC ACID

DOT Classification: Flammable Solid; Label: Flammable Solid (UN1355); Class A Explosive; Label: Explosive A (UN0215)

SAFETY PROFILE: An explosive. A hazard in preparation. Reacts with heavy metals to form heat- or impact-sensitive explosive salts. When heated to decomposition it emits toxic fumes of NO_x. See also NITRO COMPOUNDS of AROMATIC HYDROCARBONS and EXPLOSIVES, HIGH.

TML250 CAS:129-66-8 **HR: 3**
TRINITROBENZOIC ACID (wet)
mf: $C_7H_3N_3O_8$ mw: 257.13
DOT: UN 0215/UN 1355

PROP: Containing at least 10% water (FEREAC 41,15972,76).

DOT Classification: Flammable Solid; Label: Flammable Solid; Class A Explosive; Label: Explosive A.

SAFETY PROFILE: Flammable and explosive. When heated to decomposition it emits toxic fumes of NO_x. See also TRINITROBENZOIC ACID (DRY).

TML325 CAS:28260-61-9 **HR: 3**
TRINITROCHLOROBENZENE
DOT: UN 0155
mf: $C_6H_2ClN_3O_6$ mw: 247.56

TOXICITY DATA with REFERENCE
dnd-mus-ipr 60 mg/kg BSIBAC 56,1680,80
dns-ham:lvr 5 μmol/L MUREAV 131,215,84

DOT Classification: Class A Explosive; Label: Explosive A.

SAFETY PROFILE: Mutation data reported. An explosive. When heated to decomposition it emits toxic fumes of Cl^- and NO_x. See also NITRO COMPOUNDS of AROMATIC HYDROCARBONS and EXPLOSIVES, HIGH.

TML500 CAS:602-99-3 **HR: 3**
2,4,6-TRINITRO-m-CRESOL
DOT: UN 0216
mf: $C_7H_5N_3O_7$ mw: 243.15

PROP: Yellow crystals. Mp: 106°, bp: explodes @ 150°.

SYNS: CRESYLITE ◇ 3-METHYL-2,4,6-TRINITROPHENOL ◇ TRINITRO-m-CRESOL ◇ TRINITRO-m-CRESOLIC ACID ◇ TRINITROMETACRESOL (DOT)

TOXICITY DATA with REFERENCE
ipr-mus LD50:168 mg/kg QJPPAL 7,205,34

DOT Classification: Class A Explosive; Label: Explosive A.

SAFETY PROFILE: Poison by intraperitoneal route. A severe explosion hazard when shocked or exposed to heat. Explodes when heated above 150°C. Trinitrocresol is not as powerful a high explosive as TNT or picric acid. Can react vigorously with oxidizing materials. When heated to decomposition it emits highly toxic fumes of NO_x and explodes. See also NITRO COMPOUNDS of AROMATIC HYDROCARBONS and EXPLOSIVES, HIGH.

TML750 CAS:81-15-2 ***HR: 1***
2,4,6-TRINITRO-1,3-DIMETHYL-5-tert-BUTYLBENZENE
mf: $C_{12}H_{15}N_3O_6$ mw: 297.30

SYNS: 1-tert-BUTYL-3,5-DIMETHYL-2,4,6-TRINITROBENZENE ◇ 5-tert-BUTYL-2,4,6-TRINITRO-m-XYLENE ◇ MUSK XYLDL ◇ MUSK XYLENE

TOXICITY DATA with REFERENCE
skn-hmn 5 mg/48H MLD FCTXAV 13,881,75

CONSENSUS REPORTS: Reported in EPA TSCA Inventory.

SAFETY PROFILE: A human skin irritant. When heated to decomposition it emits toxic fumes of NO_x. See also NITRO COMPOUNDS of AROMATIC HYDROCARBONS.

TMM000 CAS:918-54-7 ***HR: 3***
2,2,2-TRINITROETHANOL
mf: $C_2H_3N_3O_7$ mw: 181.08

SYN: TRINITROETHANOL (DOT)

TOXICITY DATA with REFERENCE
ipr-mus LD50:36 mg/kg KHFZAN 11(1),73,77

DOT Classification: Forbidden.

SAFETY PROFILE: Poison by intraperitoneal route. A shock-sensitive explosive. When heated to decomposition it emits toxic fumes of NO_x. See also NITRO COMPOUNDS.

TMM250 CAS:129-79-3 ***HR: 3***
2,4,7-TRINITROFLUOREN-9-ONE
mf: $C_{13}H_5N_3O_7$ mw: 315.21

PROP: Mp: 175-176°.

SYNS: 2,4,7-TRINITRO-9-FLUORENONE ◇ 2,4,7-TRINITROFLUORENONE (MAK)

TOXICITY DATA with REFERENCE
skn-rbt 500 mg/24H MLD EPASR* 8EHQ-0480-0339
eye-rbt 100 mg MLD EPASR* 8EHQ-0480-0339
mma-sat 10 μL/plate EPASR* 8EHQ-0280-0333
sce-hmn:lym 3 mg/L MUREAV 138,181,84
orl-rat TDLo:1000 mg/kg:ETA SCIEAS 137,257,62
orl-rat LD50:9910 mg/kg PESTC* 8,4,80

CONSENSUS REPORTS: Reported in EPA TSCA Inventory.

DFG MAK: Suspected Carcinogen.

SAFETY PROFILE: Suspected carcinogen with experimental tumorigenic data. Mildly toxic by ingestion. Human mutation data reported. A skin and eye irritant. When heated to decomposition it emits highly toxic fumes of NO_x. See also NITRO COMPOUNDS of AROMATIC HYDROCARBONS and KETONES.

TMM500 CAS:517-25-9 ***HR: 3***
TRINITROMETHANE
mf: CHN_3O_6 mw: 151.05

PROP: Mp: 15°, d: 1.469, bp: decomp > 25°. Sol in water.

SYN: NITROFORM

TOXICITY DATA with REFERENCE
orl-mus LDLo:300 mg/kg 85GMAT -,93,82
ihl-mus LC50:800 mg/m^3/2H 85GMAT -,93,82
ipr-mus LD50:115 mg/kg KHFZAN 10(6),53,76

CONSENSUS REPORTS: Reported in EPA TSCA Inventory.

DOT Classification: Forbidden.

SAFETY PROFILE: Poison by ingestion and intraperitoneal routes. Moderately toxic by inhalation. Irritating to skin, eyes, and mucous membranes. Inhalation can cause headache and nausea. Causes mild narcosis. A very dangerous explosion hazard; explodes when heated rapidly. Dissolution is exothermic and solutions of more than 50% can explode. Mixtures of 90% trinitromethane + 10% isopropyl alcohol in polyethylene bottles have exploded. Frozen mixtures with 2-propanol (10%) explode when thawed. Can explode during distillation. Mixtures with divinyl ketone can explode at 4°C. When heated to decomposition it emits toxic fumes of NO_x. See also NITRO COMPOUNDS.

TMM775 CAS:4328-17-0 ***HR: 3***
TRINITROPHLOROGLUCINOL
mf: $C_6H_3N_3O_9$ mw: 261.10

$(O_2N)_3C_6(OH)_3$

SYN: 2,4,6-TRINITROBENZENE-1,3,5-TRIOL

SAFETY PROFILE: Probably an eye, skin, and mucous membrane irritant. A powerful oxidant. Explodes when heated. May react with metals to form explosive salts. Upon decomposition it emits toxic fumes of NO_x. See also NITRO COMPOUNDS of AROMATIC HYDROCARBONS.

TMN000 CAS:75321-19-6 ***HR: 2***
1,3,6-TRINITROPYRENE
mf: $C_{16}H_7N_3O_6$ mw: 337.26

SYN: TRINITROPYRENE

TOXICITY DATA with REFERENCE
mmo-sat 1 nmol/plate CRNGDP 3,917,82
msc-ham:lng 2500 μg/L CRNGDP 3,917,82
msc-ham:ovr 200 μg/L MUREAV 119,387,83

DFG MAK: Suspected Carcinogen.

SAFETY PROFILE: Suspected carcinogen. Mutation data reported. When heated to decomposition it emits toxic fumes of NO_x. See also NITRO COMPOUNDS of AROMATIC HYDROCARBONS.

TMN400 CAS:610-25-3 ***HR: 3***
2,4,5-TRINITROTOLUENE
mf: $C_7H_5N_3O_6$ mw: 227.15

SYN: 1-METHYL-2,4,5-TRINITROBENZENE

TOXICITY DATA with REFERENCE
mmo-sat 10 μg/plate ENMUDM 4,163,82
mma-sat 10 μg/plate ENMUDM 4,163,82
scu-mus LD20:250 mg/kg 85GMAT -,117,82

SAFETY PROFILE: Poison by subcutaneous route. Mutation data reported. Reaction with sodium carbonate forms flammable and explosive products. When heated to decomposition it emits toxic fumes of NO_x. See also NITRO COMPOUNDS of AROMATIC HYDROCARBONS.

TMN490 CAS:118-96-7 ***HR: 3***
2,4,6-TRINITROTOLUENE
DOT: UN 0209/UN 1356
mf: $C_7H_5N_3O_6$ mw: 227.15

PROP: Colorless, monoclinic crystals. Mp: 80.7°, bp: 240° (explodes), flash p: explodes, d: 1.654. Sol in hot water, alc, ether.

SYNS: ENTSUFON ◇ NCI-C56155 ◇ TNT ◇ α-TNT ◇ TNT-TOLITE (FRENCH) ◇ TOLIT ◇ TOLITE ◇ 2,4,6-TRINITROTOLUEEN (DUTCH) ◇ TRINITROTOLUENE ◇ s-TRINITROTOLUENE ◇ TRINITROTOLUENE, dry (DOT) ◇ sym-TRINITROTOLUENE ◇ s-TRINITROTOLUOL ◇ sym-TRINITROTOLUOL ◇ 2,4,6-TRINITROTOLUOL (GERMAN) ◇ TRITOL ◇ TROJNITROTOLUEN (POLISH) ◇ TROTYL ◇ TROTYL OIL

TOXICITY DATA with REFERENCE
skn-rbt 500 mg/24H MLD NTIS** AD-B011-150
mmo-sat 10 μg/plate NTIS** AD-A080-146
orl-rat TDLo:5376 mg/kg (28D male):REP JTEHD6 9,565,82
orl-hmn LDLo:28 g/kg:CNS,PUL,GIT 34ZIAG -,610,69
orl-rat LD50:795 mg/kg JTEHD6 9,565,82
orl-mus LD50:660 mg/kg JTEHD6 9,565,82
orl-cat LDLo:1850 mg/kg MRCSAB 58,32,21
scu-cat LDLo:200 mg/kg MRCSAB 58,32,21
orl-rbt LDLo:500 mg/kg MRCSAB 58,32,21
scu-rbt LDLo:500 mg/kg MRCSAB 58,32,21

CONSENSUS REPORTS: Reported in EPA TSCA Inventory. EPA Genetic Toxicology Program.

OSHA PEL: (Transitional: TWA 1.5 mg/m^3 (skin)) TWA 0.5 mg/m^3 (skin)
ACGIH TLV: TWA 0.5 mg/m^3 (skin)
DFG MAK: 0.01 ppm (0.1 mg/m^3)
DOT Classification: Class A Explosive; Label: Explosive A (UN0209); Flammable Solid; Label: Flammable Solid (UN1356).

SAFETY PROFILE: Poison by subcutaneous route. Moderately toxic by ingestion. Human systemic effects by ingestion: hallucinations or distorted perceptions, cyanosis and gastrointestinal changes. Experimental reproductive effects. Mutation data reported. A skin irritant. Has been implicated in aplastic anemia. Can cause headache, weakness, anemia, liver injury. May be absorbed through skin.

Flammable or explosive when exposed to heat or flame. Moderate explosion hazard; will detonate under strong shock. It detonates at around 240°C but can be distilled safely under reduced pressure. It is a comparatively insensitive explosive. In small quantities it will burn quietly if not confined. However, sudden heating of any quantity will cause it to detonate; the accumulation of heat when large quantities are burning will cause detonation. In other respects it is one of the most stable of all high explosives and there are but a few restrictions for its handling. It is for this reason, from the military standpoint, that TNT is quantitatively the most used. It requires a fall of 130 cm for a 2 kg weight to detonate it. It is one of the most powerful high explosives. It can be detonated by the usual detonators and blasting caps (at least a No. 6). For full efficiency, the use of a high velocity initiator, such as tetryl, is required. TNT is one of those explosives containing an oxygen deficiency. In other words, the addition of products which are oxygen-rich can enhance its explosive power. Also mono- and dinitrotoluene may be added for reduction of the temperature of the explosion and to make the explosion flashless. Various materials are added to TNT to make what is known as permissible explosives. TNT may be regarded as the equivalent of 40% dynamite and can be used underwater. It is also used in the manufacture of a detonator fuse known as Cordeau Detonant. For the military, TNT finds use in all types of bursting charges, including armor-piercing types, although it is somewhat too sensitive to be ideal for this purpose, and has since been replaced to a great extent by ammonium picrate. It is a relatively expensive explosive and does not compete seriously with dynamite for general commercial use.

Highly dangerous; explodes with shock or heating to 297°C. Various materials can reduce the explosive temperature: red lead (to 192°C); sodium carbonate (to 218°C); potassium hydroxide (to 192°C). Mixtures with sodium dichromate + sulfuric acid may ignite spontaneously. Reacts with nitric acid + metals (e.g., lead or iron) to form explosive products more sensitive to shock, friction, or contact with nitric or sulfuric acids. Reacts with potassium hydroxide dissolved in methanol to form explosive aci-nitro salts. Bases (e.g., sodium hydroxide; potassium iodide; tetramethyl ammonium octahydrotriborate) induce deflagration in molten TNT. Can react vigorously with reducing materials. When heated to decomposition it emits highly toxic fumes of NO_x. See also NITRO COMPOUNDS of AROMATIC HYDROCARBONS and EXPLOSIVES, HIGH.

TMN500 CAS:118-96-7 ***HR: 3***
2,4,6-TRINITROTOLUENE (wet)
DOT: UN 0209/UN 1356
mf: $C_7H_5N_3O_6$ mw:227.15

SYNS: TNT ◇ α-TNT ◇ TOLIT ◇ TOLITE ◇ TRINITROTOLUENE ◇ sym-TRINITROTOLUENE ◇ TRINITROTOLUENE, WET containing at least 10% water (DOT) ◇ TRINITROTOLUENE, WET containing at least 10% water, over 16 ounces in one outside packaging ◇ TRINITROTOLUENE, wetted with not less than 30% water (DOT) ◇ sym-TRINITROTOLUOL ◇ TRITOL ◇ TROTYL OIL

CONSENSUS REPORTS: Reported in EPA TSCA Inventory. EPA Genetic Toxicology Program.

DOT Classification: Class A Explosive; Label: Explosive A; Flammable Solid; Label: Flammable Solid (UN 1356).

SAFETY PROFILE: An explosive. Flammable when exposed to heat or flame. When heated to decomposition it emits toxic fumes of NO_x. See also NITRO COMPOUNDS of AROMATIC HYDROCARBONS and other trinitrotoluene entries.

TMN750 CAS:5337-41-7 ***HR: 1***
TRIOCTADECYL BORATE
mf: $C_{54}H_{111}BO_3$ mw: 819.46

PROP: White solid; odor of stearyl alc. Mp: 49.8-54°, bp: 300-331° @ 0.3 mm.

SYNS: BORIC ACID, TRIOCTADECYL ESTER ◇ BORIC ACID, TRISTEARYL ESTER ◇ TRISTEARYL BORATE

TOXICITY DATA with REFERENCE
eye-rbt 100 mg MLD 14KTAK -,693,64
orl-mus LD50:6200 mg/kg 14KTAK -,693,64

SAFETY PROFILE: Mildly toxic by ingestion. An eye irritant. When heated to decomposition it emits acrid smoke and irritating fumes. See also BORON COMPOUNDS and 1-OCTADECANOL.

TMO000 CAS:538-23-8 ***HR: 3***
TRIOCTANOIN
mf: $C_{27}H_{50}O_6$ mw: 470.77

SYNS: CAPRYLIC ACID TRIGLYCERIDE ◇ GLYCEROL TRICAPRYLATE ◇ GLYCEROL TRIOCTANOATE ◇ GLYCERYL TRIOCTANOATE ◇ MCT ◇ OCTANOIC ACID, 1,2,3-PROPANETRIYL ESTER ◇ OCTANOIC ACID TRIGLYCERIDE ◇ RATO ◇ TRICAPRYLIC GLYCERIDE ◇ TRICAPRYLIN ◇ TRIOCTANOYLGLYCEROL

TOXICITY DATA with REFERENCE
orl-rat TDLo:250 g/kg (7D pre-21D post):REP IYKEDH 3,180,72
orl-rat LD50:33300 mg/kg OYYAA2 4,871,70
ipr-rat LD50:50 mg/kg NCIUS* PH 43-64-886,SEPT,65
ivn-rat LDLo:4 g/kg OYYAA2 4,871,70
orl-mus LD50:29600 mg/kg OYYAA2 4,871,70
ivn-mus LD50:3700 mg/kg APSCAX 40,338,57
ipr-rbt LDLo:3400 mg/kg JNCIAM 54,1439,75

CONSENSUS REPORTS: Reported in EPA TSCA Inventory. EPA Genetic Toxicology Program.

SAFETY PROFILE: Poison by intraperitoneal route. Moderately toxic by intravenous route. Mildly toxic by ingestion. Experimental reproductive effects. When heated to decomposition it emits acrid smoke and irritating fumes. See also ESTERS.

TMO250 CAS:2467-12-1 ***HR: 2***
TRI-n-OCTYL BORATE
mf: $C_{24}H_{51}BO_3$ mw: 398.56

PROP: Colorless liquid; odor of octyl alc. Bp: 192-194° @ 2 mm, flash p: 370°F (COC), d: 0.846 @ 23°, vap d: 13.7.

SYN: BORIC ACID, TRI-n-OCTYL ESTER

TOXICITY DATA with REFERENCE
eye-rbt 100 mg MOD 14KTAK -,693,64
orl-mus LD50:1290 mg/kg 14KTAK -,693,64

CONSENSUS REPORTS: Reported in EPA TSCA Inventory.

SAFETY PROFILE: Moderately toxic by ingestion. An

eye irritant. Combustible when exposed to heat or flame; can react with oxidizing materials. To fight fire, use foam, CO_2, dry chemical. When heated to decomposition it emits acrid smoke and irritating fumes. See also ESTERS and BORON COMPOUNDS.

TMO500 CAS:24848-81-5 ***HR: 2***
TRI(2-OCTYL)BORATE
mf: $C_{24}H_{51}BO_3$ mw: 398.56

PROP: Colorless liquid; odor of 2-octanol. Bp: 340-349°, flash p: 330°F (COC), d: 0.837 @ 24.5°, vap d: 13.8.

SYN: BORIC ACID, TRIS(1-METHYLHEPTYL) ESTER

TOXICITY DATA with REFERENCE
eye-rbt 100 mg MLD 14KTAK -,693,64
orl-mus LD50:3300 mg/kg 14KTAK -,693,64

SAFETY PROFILE: Moderately toxic by ingestion. An eye irritant. Combustible when exposed to heat or flame; can react with oxidizing materials. To fight fire, use foam, CO_2, dry chemical. When heated to decomposition it emits acrid smoke and irritating fumes. See BORON COMPOUNDS and ESTERS.

TMO600 CAS:78-30-8 ***HR: 3***
TRIORTHOCRESYL PHOSPHATE
mf: $C_{21}H_{21}O_4P$ mw: 368.39

PROP: Colorless liquid. Mp: −25 to −30°, bp: 410° (slt decomp), flash p: 437°F, d: 1.17, autoign temp: 725°F, vap d: 12.7. Insol in water; sol in alc and ether.

SYNS: o-CRESYL PHOSPHATE ◇ PHOSFLEX 179-C ◇ PHOSPHORIC ACID, TRI-o-CRESYL ESTER ◇ PHOSPHORIC ACID, TRIS(2-METHYLPHENYL) ESTER ◇ TOCP ◇ TOFK ◇ o-TOLYL PHOSPHATE ◇ TOTP ◇ TRICRESYL PHOSPHATE ◇ TRI-o-CRESYL PHOSPHATE ◇ o-TRIKESYLPHOSPHATE (GERMAN) ◇ TRI 2-METHYLPHENYL PHOSPHATE ◇ TRI-2-TOLYL PHOSPHATE ◇ TRIS(o-CRESYL)-PHOSPHATE ◇ TRIS(o-METHYLPHENYL)PHOSPHATE ◇ TRIS(o-TOLYL)-PHOSPHATE ◇ TRI-o-TOLYL PHOSPHATE ◇ TROJKREZYLU FOSFORAN (POLISH)

TOXICITY DATA with REFERENCE
orl-rat TDLo:630 mg/kg (male 63D pre):REP TXAPA9 89,49,87
orl-rat LD50:1160 mg/kg TOXID9 4,55,84
ipr-rat LD50:2500 mg/kg APCRAW 4,117,61
scu-mus LDLo:12500 mg/kg EDWU** -,-,37
scu-dog LDLo:100 mg/kg AEPPAE 168,473,32
scu-cat LDLo:185 mg/kg JHHBAI 52,39,33
ipr-rbt LDLo:100 mg/kg AEPPAE 168,473,32
scu-rbt LDLo:100 mg/kg AEPPAE 168,473,32
ivn-rbt LDLo:100 mg/kg AEPPAE 168,473,32
ims-rbt LDLo:135 mg/kg AEPPAE 171,439,33

CONSENSUS REPORTS: Reported in EPA TSCA Inventory.

OSHA PEL: (Transitional: TWA 0.1 mg/m^3) TWA 0.1 mg/m^3 (skin)
ACGIH TLV: TWA 0.1 mg/m^3 (skin)

SAFETY PROFILE: Poison by subcutaneous, intramuscular, intravenous, and intraperitoneal routes. Moderately toxic by ingestion. Most of the cases of tri-o-cresyl phosphate poisoning have followed its ingestion. In 1930, some 15,000 persons were affected in the United States, and of these, 10 died. The responsible material was found to be an alcoholic drink known as Jamaica ginger, or "jake." This beverage had been adulterated with about 2% of tri-o-cresyl phosphate. The affected persons developed a polyneuritis, which progressed, in many cases, with degeneration of the peripheral motor nerves, the anterior horn cells and the pyramidal tracts. Sensory changes were absent. Since 1930 there have been several other outbreaks of poisoning following ingestion of the material. Tri-o-cresyl phosphate is more toxic than the m-form, and much more so than tri-p-cresyl phosphate or triphenyl phosphate. Experimental reproductive effects.

Combustible when exposed to heat or flame. Can react with oxidizing materials. To fight fire, use CO_2, dry chemical. When heated to decomposition it emits highly toxic fumes of PO_x. See also PHOSPHATES.

TMO750 CAS:283-60-3 ***HR: 3***
2,8,9-TRIOXA-5-AZA-1-SILABICYCLO(3.3.3)UNDECANE
mf: $C_6H_{13}NO_3Si$ mw: 175.29

SYN: SILATRANE

TOXICITY DATA with REFERENCE
ipr-mus LD50:100 mg/kg RCRVAB 38(12),975,69

CONSENSUS REPORTS: Reported in EPA TSCA Inventory.

SAFETY PROFILE: Poison by intraperitoneal route. When heated to decomposition it emits toxic fumes of NO_x. See also SILICON COMPOUNDS.

TMO775 CAS:81781-28-4 ***HR: 3***
TRIOXACARCIN C
mf: $C_{42}H_{54}O_{20}$ mw: 878.96

SYNS: DC-45-B2 ◇ 7″-DEOXY-7″-HYDROXY-TRIOXACARCIN A

TOXICITY DATA with REFERENCE
dni-mus:lym 139 nmol/L JANTAJ 36,1216,83
oms-mus:lym 139 nmol/L JANTAJ 36,1216,83
ipr-mus LD50:1 mg/kg JANTAJ 36,1216,83

SAFETY PROFILE: Poison by intraperitoneal route. Mutation data reported. When heated to decomposition it emits acrid smoke and irritating fumes.

TMP000 CAS:110-88-3 ***HR: 3***
s-TRIOXANE
mf: $C_3H_6O_3$ mw: 90.09

$OCH_2OCH_2OCH_2$ (cyclic)

PROP: Stable, cyclic trimer of formaldehyde, having characteristic ethanol and chloroform-like odors. Crystalline solid. Mp: 64°, bp: 114.5°, subl readily, lel: 3.6%, uel: 28.7%, flash p: 113°F (OC), d: 1.17, @ 65°, autoign temp: 777°F, vap press: 13 mm @ 25°, vap d: 3.1. Very sol in water, alc, ketones, ether, acetone, chlorinated and aromatic hydrocarbons, organic solvents; sltly sol in pentane, petr ether.

SYNS: POLYOXYMETHYLENE ◇ TRIOSSIMETHLENE (ITALIAN) ◇ TRIOXANE ◇ 1,3,5-TRIOXANE ◇ TRIOXYMETHYLEEN (DUTCH) ◇ TRIOXYMETHYLEN (GERMAN) ◇ TRIOXYMETHYLENE

TOXICITY DATA with REFERENCE
skn-rbt 500 mg/24H SEV BIOFX*
eye-rbt 100 mg SEV BIOFX*
eye-rbt 100 mg IHFCAY 6,1,67
orl-rat LD50:800 mg/kg 28ZEAL 4,308,69
skn-rbt LDLo:10,000 mg/kg BIOFX*

CONSENSUS REPORTS: Reported in EPA TSCA Inventory.

SAFETY PROFILE: Moderately toxic by ingestion. Mildly toxic by skin contact. A severe eye and skin irritant. Can evolve toxic formaldehyde fumes when heated strongly or in contact with strong acids or acid fumes. Flammable when exposed to heat, flame, or oxidizers. May explode when heated. Explosive in the form of vapor when exposed to heat or flame. Explodes on impact, possibly due to peroxide contamination. Mixtures with hydrogen peroxide are explosives sensitive to heat, shock, or contact with lead. Mixtures with liquid oxygen are highly explosive. Incompatible with oxidizing materials. To fight fire, use foam, CO_2, or dry chemical. When heated to decomposition it emits acrid smoke and irritating fumes. See also FORMALDEHYDE.

TMP175 CAS:68307-81-3 ***HR: D***
TRIOXIFENE MESYLATE
mf: $C_{30}H_{31}NO_3 \cdot CH_4O_3S$ mw: 549.73

SYNS: LILLY COMPOUND LY133314 ◇ LY133314

TOXICITY DATA with REFERENCE
dni-hmn:mmr 100 mmol/L CNREA8 45,1611,85
scu-rat TDLo:275 μg/kg (1-11D preg):REP JMCMAR 22,962,79

SAFETY PROFILE: Experimental reproductive effects. Human mutation data reported. When heated to decomposition it emits toxic fumes of SO_x and NO_x. See also KETONES.

TMP250 CAS:752-58-9 ***HR: 3***
(2,4,6-TRIOXO-s-TRIAZINETRIYLTRIS(TRIBUTYLSTANNANE)
mf: $C_{39}H_{81}N_3O_3Sn_3$ mw: 996.30

SYN: 1,3,5-TRIS(TRIBUTYLTIN)-s-TRIAZINE-2,4,6-TRIONE

TOXICITY DATA with REFERENCE
ivn-mus LD50:5 mg/kg CSLNX* NX#03567

OSHA PEL: TWA 0.1 mg(Sn)/m^3 (skin)
ACGIH TLV: TWA 0.1 mg(Sn)/m^3 (skin) (Proposed: TWA 0.1 mg(Sn)/m^3; STEL 0.2 mg(Sn)/m^3 (skin))
NIOSH REL: (Organotin Compounds) TWA 0.1 mg(Sn)/m^3

SAFETY PROFILE: Poison by intravenous route. When heated to decomposition it emits toxic fumes of NO_x. See also TIN COMPOUNDS.

TMP500 CAS:78-41-1 ***HR: 2***
TRIPARANOL
mf: $C_{27}H_{32}ClNO_2$ mw: 438.05

PROP: Crystals. Mp: 102-104°. Sol in alc; sltly sol in olive oil; practically insol in water.

SYNS: α-(p-CHLOROBENZYL)-4-DIETHYLAMINOETHOXY-4-METHYLBENZHYDROL ◇ 2-(p-CHLOROPHENYL)-1-(p-(β-DIETHYLAMINOETHOXY)PHENYL)-1-(p-TOLYL)ETHANOL◇ 2-p-CHLOROPHENYL-1-(p-(2-DIETHYLAMINOETHOXY)PHENYL)-1-p-TOLYLETHANOL ◇ 1-(p-(β-DIETHYLAMINOETHOXY)PHENYL)-1-(p-TOLYL)-2-(p-CHLOROPHENYL)ETHANOL◇ 1-(4-(2-(DIETHYLAMINO)ETHOXY)PHENYL)-1-(p-TOLYL)-2-(p-CHLOROPHENYL)ETHANOL ◇ MER 29 ◇ METASQUALENE

TOXICITY DATA with REFERENCE
scu-rat TDLo:70 mg/kg (female 14D pre):REP ENDOAO 74,64,64
orl-rat TDLo:350 mg/kg (female 2-15D post):TER CRSBAW 167,1523,73
orl-rat LD50:2000 mg/kg CRSBAW 155,2255,61

SAFETY PROFILE: Moderately toxic by ingestion. An experimental teratogen. Experimental reproductive effects. When heated to decomposition it emits very toxic fumes of NO_x and Cl^-.

TMP750 CAS:91-81-6 ***HR: 3***
TRIPELENNAMINE
mf: $C_{16}H_{21}N_3$ mw: 255.40

PROP: Oily liquid; amine odor. Bp: 167-172° @ 0.1 mm. Freely sol in water and alc; sltly sol in ether; practically insol in benzene and chloroform.

SYNS: BENZOXALE ◇ 2-(BENZYL(2-DIMETHYL AMINOETHYL) AMINO)PYRIDINE ◇ N-BENZYL-N',N'-DIMETHYL-N-2-PYRIDYL-ETHYLENE DIAMINE ◇ BENZYL-(α-PYRIDYL)-DIMETHYLAETHYL-ENDIAMIN (GERMAN) ◇ CIZARON ◇ DEHISTIN ◇ β-DIMETHYL-AMINO ETHYL-2-PYRIDYLAMINOTOLUENE ◇ β-DIMETHYLAMINO-ETHYL-2-PYRIDYLBENZYLAMINE ◇ N,N-DIMETHYL-N'-BENZYL-N'-

(α-PYRIDYL)ETHYLENEDIAMINE ◇ NCI-C60662 ◇ PBZ ◇ PIRIBENZIL ◇ PYRIBENZAMINE ◇ PYRINAMINE BASE ◇ RESISTAMINE ◇ TONARIL ◇ TRIPELENAMINE ◇ TRIPELENNAMINA (ITALIAN)

TOXICITY DATA with REFERENCE
dnd-hmn:lvr 33 μmol/L MUREAV 173,229,86
dns-hmn:lvr 10 μmol/L MUREAV 173,229,86
ipr-rat LDLo:37 mg/kg TXAPA9 1,156,59
orl-mus LD50:152 mg/kg ARZNAD 7,237,57
ipr-mus LD50:43 mg/kg ARZNAD 7,237,57
ipr-gpg LD50:64 mg/kg THERAP 28,767,73

SAFETY PROFILE: Poison by ingestion and intraperitoneal routes. Human mutation data reported. Has been implicated in aplastic anemia. Used as an antihistamine. Addicts have added it to paregoric to make "blue velvet," which can cause a euphoria by injection. When heated to decomposition it emits toxic fumes of NO_x.

TMQ000 CAS:621-78-3 ***HR: 2***
TRI-n-PENTYL BORATE
mf: $C_{15}H_{33}BO_3$ mw: 272.29

SYNS: BORIC ACID, TRI-n-AMYL ESTER ◇ BORIC ACID, TRI-n-PENTYL ESTER ◇ TRI-n-AMYL BORATE

TOXICITY DATA with REFERENCE
eye-rbt 100 mg MOD 14KTAK -,693,64
orl-mus LD50:1060 mg/kg USBCC* -,-,58

CONSENSUS REPORTS: Reported in EPA TSCA Inventory.

SAFETY PROFILE: Moderately toxic by ingestion. An eye irritant. When heated to decomposition it emits acrid smoke and irritating fumes. See also BORON COMPOUNDS and ESTERS.

TMQ250 CAS:6304-33-2 ***HR: 3***
2,3,3-TRIPHENYLACRYLONITRILE
mf: $C_{21}H_{15}N$ mw: 281.37

SYNS: α,β-DIPHENYLCINNAMONITRILE ◇ α-(DIPHENYLMETHYLENE)BENZENEACETIC ACID ◇ TRIPHENYLACRYLONITRILE ◇ α,β,β-TRIPHENYLACRYLONITRILE ◇ TRIPHENYLCYANOETHYLENE

TOXICITY DATA with REFERENCE
scu-mus TDLo:94 mg/kg/26W-I:CAR MMJJAI 11,95,61
orl-rat LD50:284 mg/kg TXAPA9 14,340,69
ivn-mus LD50:180 mg/kg CSLNX* NX#04868

CONSENSUS REPORTS: Cyanide and its compounds are on the Community Right-To-Know List.

SAFETY PROFILE: Poison by ingestion and intravenous routes. Questionable carcinogen with experimental carcinogenic data. When heated to decomposition it emits toxic fumes of NO_x and CN^-. See also NITRILES.

TMQ500 CAS:603-34-9 ***HR: 2***
TRIPHENYLAMINE
mf: $C_{18}H_{15}N$ mw: 245.34

PROP: Monoclinic crystals. D: 0.774 @ 0/0°, mp: 127°, bp: 365°.

SYN: N,N-DIPHENYLANILINE

TOXICITY DATA with REFERENCE
orl-rat LD50:3200 mg/kg 85INA8 5,612,86
orl-mus LD50:1600 mg/kg 85INA8 5,612,86

CONSENSUS REPORTS: Reported in EPA TSCA Inventory.

OSHA PEL: TWA 5 mg/m^3
ACGIH TLV: TWA 5 mg/m^3

SAFETY PROFILE: Moderately toxic by ingestion. When heated to decomposition it emits toxic fumes of NO_x. See also AROMATIC AMINES.

TMQ550 CAS:4756-75-6 ***HR: 3***
TRIPHENYLANTIMONY OXIDE
mf: $C_{18}H_{15}OSb$ mw: 369.08

PROP: Crystals.

SYN: STIBINE OXIDE, TRIPHENYL-

TOXICITY DATA with REFERENCE
ivn-mus LD50:180 mg/kg CSLNX* NX#01710

OSHA PEL: TWA 0.5 $mg(Sb)/m^3$
ACGIH TLV: TWA 0.5 $mg(Sb)/m^3$
NIOSH REL: (Antimony):10H TWA 0.5 $mg(Sb)/m^3$

SAFETY PROFILE: Poison by intravenous route. When heated to decomposition it emits toxic fumes of Sb.

TMQ600 CAS:3958-19-8 ***HR: 3***
TRIPHENYL ANTIMONY SULFIDE
mf: $C_{18}H_{15}SSb$ mw: 385.14

PROP: Crystals. Mp: 120°.

SYN: STIBINE SULFIDE, TRIPHENYL-

TOXICITY DATA with REFERENCE
ivn-mus LD50:320 mg/kg CSLNX* NX#02058

OSHA PEL: TWA 0.5 $mg(Sb)/m^3$
ACGIH TLV: TWA 0.5 $mg(Sb)/m^3$
NIOSH REL: (Antimony):10H TWA 0.5 $mg(Sb)/m^3$

SAFETY PROFILE: Poison by intravenous route. When heated to decomposition it emits toxic fumes of SO_x and Sb.

TMR000 CAS:612-71-5 ***HR: 3***
1,3,5-TRIPHENYLBENZENE
mf: $C_{24}H_{18}$ mw: 306.42

PROP: Rhombic crystals. D: 1.205, mp: 170-171°C. Very sol in benzene; sol in abs alc, ether.

SYNS: 5'-PHENYL-m-TERPHENYL ◇ TRIPHENYLBENZENE ◇ sym-TRIPHENYLBENZENE

TOXICITY DATA with REFERENCE
scu-mus TDLo:1400 mg/kg/35W-I:ETA AJCAA7 26,754,36

SAFETY PROFILE: Questionable carcinogen with experimental tumorigenic data. When heated to decomposition it emits acrid smoke and irritating fumes.

TMR250 CAS:603-33-8 ***HR: 3***
TRIPHENYLBISMUTHINE
mf: $C_{18}H_{15}Bi$ mw: 440.31

PROP: Monoclinic crystals. Mp: 78°, bp: 242° @ 14 mm, d: 1.585.

SYN: TRIPHENYLBISMUTH

TOXICITY DATA with REFERENCE
ipr-mus LDLo:250 mg/kg CBCCT* 4,317,52
orl-mus LDLo:320 mg/kg AECTCV 14,111,85
ivn-mus LD50:180 mg/kg CSLNX* NX#01712

CONSENSUS REPORTS: Reported in EPA TSCA Inventory.

SAFETY PROFILE: Poison by ingestion, intraperitoneal, and intravenous routes. When heated to decomposition it emits toxic fumes of Bi. See also BISMUTH COMPOUNDS.

TMR500 CAS:1095-03-0 ***HR: 3***
TRIPHENYL BORATE
mf: $C_{18}H_{15}BO_3$ mw: 290.14

PROP: White to pink solid, odor of phenol, decomp by water. Mp: 35°, bp: >360°.

SYN: PHENYL BORATE

TOXICITY DATA with REFERENCE
eye-rbt 5 mg SEV 14KTAK -,706,64
orl-mus LD50:200 mg/kg 14KTAK -,706,64

SAFETY PROFILE: Poison by ingestion. A severe eye irritant. Flammable when exposed to heat or flame; can react vigorously with oxidizing materials. Reacts with water or steam to form toxic fumes of phenol. To fight fire, use foam, CO_2, dry chemical. When heated to decomposition it emits acrid smoke and fumes. See also PHENOL and BORON COMPOUNDS.

TMR750 CAS:63732-31-0 ***HR: 2***
TRIPHENYLCYCLOHEXYL BORATE
mf: $C_{36}H_{45}BO_3$ mw: 536.62

SYN: BORIC ACID, TRIS(PHENYLCYCLOHEXYL) ESTER

TOXICITY DATA with REFERENCE
eye-rbt 5 mg SEV 14KTAK -,693,64
orl-mus LD50:1240 mg/kg 14KTAK -,693,64

SAFETY PROFILE: Moderately toxic by ingestion. A severe eye irritant. When heated to decomposition it emits acrid smoke and irritating fumes. See also ESTERS and BORON COMPOUNDS.

TMS000 CAS:217-59-4 ***HR: D***
TRIPHENYLENE
mf: $C_{18}H_{12}$ mw: 228.30

PROP: Long needles from alc or chloroform; solns have blue fluorescence. Subl; d: 1.302, Mp: 199°; bp: 425°.

SYNS: 9,10-BENZOPHENANTHRENE ◇ BENZO(l)PHENANTHRENE ◇ 9,10-BENZPHENANTHRENE ◇ 1,2,3,4-DIBENZNAPHTHALENE ◇ ISOCHRYSENE

TOXICITY DATA with REFERENCE
mmo-sat 1 nmol/L CNREA8 40,1985,80
mma-sat 100 mg/L/72H FCTXAV 17,141,79

CONSENSUS REPORTS: IARC Cancer Review: Group 3 IMEMDT 7,56,87; Animal Inadequate Evidence IMEMDT 32,447,83.

SAFETY PROFILE: Questionable carcinogen. Mutation data reported. When heated to decomposition it emits acrid smoke and irritating fumes.

TMS250 CAS:58-72-0 ***HR: 3***
TRIPHENYLETHYLENE
mf: $C_{20}H_{16}$ mw: 256.36

PROP: Mp: 69-71°.

SYN: 1,1,2-TRIPHENYLETHYLENE

TOXICITY DATA with REFERENCE
scu-rat TDLo:3 mg/kg (3D pre):REP AIPTAK 151,475,64
scu-mus TD:7200 mg/kg/36W-I:ETA,REP CNREA8 3,92,43
scu-mus TD:4800 mg/kg/40W-I:ETA,REP JPBAA7 56,15,44
scu-mus TD:4920 mg/kg/41W-I:ETA,REP JPBAA7 54,149,42

CONSENSUS REPORTS: Reported in EPA TSCA Inventory.

SAFETY PROFILE: Experimental reproductive effects. Questionable carcinogen with experimental tumorigenic data. Human mutation data reported. When heated to decomposition it emits acrid smoke and irritating fumes.

TMS500 CAS:101-01-9 ***HR: 3***
TRIPHENYLGUANIDINE
mf: $C_{19}H_{17}N_3$ mw: 287.39

PROP: Plates. Mp: 131°C. Sltly sol in benzene; sol in alc, ether.

SYN: N,N′,N″-TRIPHENYLGUANIDINE

TOXICITY DATA with REFERENCE
orl-rat LDLo:250 mg/kg NCNSA6 5,15,53
ivn-mus LD50:56 mg/kg CSLNX* NX#01711
orl-mam LDLo:350 mg/kg JIDHAN 13,87,31

CONSENSUS REPORTS: Reported in EPA TSCA Inventory.

SAFETY PROFILE: Poison by ingestion and intravenous routes. When heated to decomposition it emits toxic fumes of NO_x.

TMS750 CAS:484-47-9 ***HR: 3***
2,4,5-TRIPHENYLIMIDAZOLE

SYN: LOPHINE

TOXICITY DATA with REFERENCE
ivn-mus LD50:100 mg/kg CSLNX* NX#04415

CONSENSUS REPORTS: Reported in EPA TSCA Inventory.

SAFETY PROFILE: Poison by intravenous route. When heated to decomposition it emits toxic fumes of NO_x.

TMT000 CAS:1162-06-7 ***HR: 3***
TRIPHENYLLEAD ACETATE
mf: $C_{20}H_{18}O_2Pb$ mw: 497.57

SYNS: ACETOXYTRIPHENYLLEAD ◇ (ACETYLOXY)TRIPHENYLPLUMBANE

TOXICITY DATA with REFERENCE
orl-rat LD50:200 mg/kg BIJOAK 127,24P,72
ipr-rat LD50:2800 μg/kg JJATDK 1,247,81
ivn-rat LD50:5 mg/kg JJATDK 1,247,81

CONSENSUS REPORTS: Lead and its compounds are on the Community Right-To-Know List.

SAFETY PROFILE: Poison by ingestion, intravenous, and intraperitoneal routes. When heated to decomposition it emits toxic fumes of Pb. See also LEAD COMPOUNDS.

TMT250 CAS:27679-98-7 ***HR: 3***
TRIPHENYL LEAD(1+) HEXAFLUOROSILICATE

TOXICITY DATA with REFERENCE
orl-rat LDLo:100 mg/kg NCNSA6 5,30,53

CONSENSUS REPORTS: Lead and its compounds are on the Community Right-To-Know List.

SAFETY PROFILE: Poison by ingestion. When heated to decomposition it emits very toxic fumes of Pb and F^-. See also LEAD COMPOUNDS.

TMT500 CAS:3695-77-0 ***HR: 3***
TRIPHENYLMETHANETHIOL
mf: $C_{19}H_{16}S$ mw: 276.41

PROP: Mp: 104-106°.

SYN: TRITYLTHIOL

TOXICITY DATA with REFERENCE
ivn-mus LD50:180 mg/kg CSLNX* NX#04016

CONSENSUS REPORTS: Reported in EPA TSCA Inventory.

SAFETY PROFILE: Poison by intravenous route. When heated to decomposition it emits toxic fumes of SO_x. See also SULFIDES.

TMT750 CAS:115-86-6 ***HR: 3***
TRIPHENYL PHOSPHATE
mf: $C_{18}H_{15}O_4P$ mw: 326.30

PROP: Colorless, odorless, crystalline solid. Mp: 49-50°, bp: 245° @ 11 mm, flash p: 428°F (CC), d: 1.268 @ 60°, vap press: 1 mm @ 193.5°. Insol in water; sol in alc, benzene, ether, chloroform and acetone.

SYNS: CELLUFLEX TPP ◇ PHOSPHORIC ACID, TRIPHENYL ESTER ◇ TPP

TOXICITY DATA with REFERENCE
orl-rat LD50:3800 mg/kg DTLVS* 4,420,80
orl-mus LD50:1320 mg/kg DTLVS* 4,420,80
scu-mky LDLo:500 mg/kg DTLVS* 4,420,80
scu-cat LD50:100 mg/kg 14CYAT 2,1916,63

CONSENSUS REPORTS: Reported in EPA TSCA Inventory.

OSHA PEL: TWA 3 mg/m^3
ACGIH TLV: TWA 3 mg/m^3

SAFETY PROFILE: Poison by subcutaneous route. Moderately toxic by ingestion. Absorbed slowly, particularly by skin contact. Not a potent cholinesterase inhibitor. Combustible when exposed to heat or flame. To fight fire, use CO_2, dry chemical. When heated to decomposition it emits toxic fumes of PO_x. See also TRITOLYL PHOSPHATE.

TMU000 CAS:603-35-0 ***HR: 2***
TRIPHENYLPHOSPHINE
mf: $C_{18}H_{15}P$ mw: 262.30

PROP: Odorless crystals. Mp: 79°, bp: >360°, d: 1.194,

flash p: 356°F (OC), vap d: 9.0. Insol in water; sol in HCl, benzene; sltly sol in alc; very sol in ether.

TOXICITY DATA with REFERENCE
orl-rat LD50:800 mg/kg 14CYAT 2,1918,63
ihl-rat LC50:1135 ppm/4H AIHAM* -,-,69

CONSENSUS REPORTS: Reported in EPA TSCA Inventory.

SAFETY PROFILE: Moderately toxic by ingestion. Mildly toxic by inhalation. Combustible when exposed to heat or flame. Slight explosion hazard in the form of vapor when exposed to flame. Can react vigorously with oxidizing materials. To fight fire, use dry chemical, fog, CO_2. When heated to decomposition it emits highly toxic fumes of phosphine and PO_x. See also PHOSPHINE and PHENOL.

TMU250 CAS:101-02-0 ***HR: 3***
TRIPHENYL PHOSPHITE
mf: $C_{18}H_{15}O_3P$ mw: 310.30

PROP: Water white to pale yellow solid or oily liquid; clean and pleasant odor. D: 1.184 @ 25/25°, mp: 22-25°, bp: 155-160° @ 0.1 mm, flash p: 425°F (OC). Insol in water.

SYNS: EFED ◇ PHOSPHOROUS ACID, TRIPHENYL ESTER ◇ TRIFENOXYFOSFIN (CZECH) ◇ TRIFENYLFOSFIT (CZECH)

TOXICITY DATA with REFERENCE
skn-hmn 125 mg/48H SEV AMIHBC 5,311,52
skn-rbt 500 mg SEV AMIHBC 5,311,52
skn-rbt 500 mg/24H MOD 28ZPAK -,205,72
eye-rbt 500 mg/24H MLD 28ZPAK -,205,72
orl-rat LD50:1600 mg/kg 14CYAT 2,1918,63
scu-rat LDLo:2000 mg/kg JPETAB 49,78,33
orl-mus LD50:1333 mg/kg GTPZAB 17(10),38,73
ipr-mus LD50:1167 mg/kg GTPZAB 17(10),38,73
scu-cat LDLo:300 mg/kg JPETAB 49,78,33
orl-ckn LDLo:1000 mg/kg JPETAB 49,78,33
ipr-mam LD50:250 mg/kg AMIHBC 5,311,52

CONSENSUS REPORTS: Reported in EPA TSCA Inventory.

SAFETY PROFILE: Poison by intraperitoneal and subcutaneous routes. Moderately toxic by ingestion. An experimental eye and severe human skin irritant. Combustible when exposed to heat or flame. To fight fire, use CO_2, mist, dry chemical. When heated to decomposition it emits toxic fumes of PO_x. See also PHENOL.

TMU750 CAS:78218-49-2 ***HR: 3***
1,3,4-TRIPHENYLPYRAZOLE-5-ACETIC ACID SODIUM SALT
mf: $C_{23}H_{17}N_2O_2{\bullet}Na$ mw: 376.41

TOXICITY DATA with REFERENCE
orl-rat LD50:13 mg/kg AIPTAK 238,305,79
orl-mus LD50:215 mg/kg AIPTAK 238,305,79

SAFETY PROFILE: Poison by ingestion. When heated to decomposition it emits toxic fumes of NO_x and Na_2O.

TMV000 CAS:910-06-5 ***HR: 3***
TRIPHENYLSTANNYL BENZOATE
mf: $C_{25}H_{20}O_2Sn$ mw: 471.15

SYNS: BENZOYLOXYTRIPHENYLSTANNANE ◇ TRIPHENYLTIN BENZOATE

TOXICITY DATA with REFERENCE
ivn-mus LD50:56 mg/kg CSLNX* NX#02984

OSHA PEL: TWA 0.1 mg(Sn)/m^3 (skin)
ACGIH TLV: TWA 0.1 mg(Sn)/m^3 (skin) (Proposed: TWA 0.1 mg(Sn)/m^3; STEL 0.2 mg(Sn)/m^3 (skin))
NIOSH REL: (Organotin Compounds) TWA 0.1 mg(Sn)/m^3

SAFETY PROFILE: Poison by intravenous route. When heated to decomposition it emits acrid smoke and irritating fumes. See also TIN COMPOUNDS and ESTERS.

TMV250 CAS:603-36-1 ***HR: 3***
TRIPHENYL STIBINE
mf: $C_{18}H_{15}Sb$ mw: 353.08

PROP: Crystals. Mp: 50°, d: 1.4343 @ 25°, bp: > 360°. Water-insol; sol in organic solvents.

SYN: TRIPHENYLANTIMONY

TOXICITY DATA with REFERENCE
orl-rat LD50:183 mg/kg MarJV# 29MAR77
ipr-rat LD50:168 mg/kg AMRL** TR-74-78,74
orl-mus LD50:650 mg/kg AMRL** TR-74-78,74
ipr-mus LDLo:500 mg/kg CBCCT* 6,229,54

CONSENSUS REPORTS: Antimony and its compounds are on the Community Right-To-Know List. Reported in EPA TSCA Inventory.

OSHA PEL: TWA 0.5 mg(Sb)/m^3
ACGIH TLV: TWA 0.5 mg(Sb)/m^3
NIOSH REL: (Antimony) TWA 0.5 mg(Sb)/m^3

SAFETY PROFILE: Poison by ingestion and intraperitoneal routes. Flammable when exposed to heat or flame. Can react vigorously with oxidizing materials. To fight fire, use water, foam, mist. When heated to decomposition it emits toxic fumes of Sb. See also ANTIMONY COMPOUNDS.

TMV500 CAS:298-96-4 ***HR: 3***
2,3,5-TRIPHENYL-2H-TETRAZOLIUM CHLORIDE
mf: $C_{19}H_{15}N_4 \cdot Cl$ mw: 334.83

PROP: Colorless needles. Decomp @ 243°. Sol in water, alc, acetone; insol in ether.

TOXICITY DATA with REFERENCE
ivn-mus LD50:5600 μg/kg CSLNX* NX#00925

CONSENSUS REPORTS: Reported in EPA TSCA Inventory.

SAFETY PROFILE: Poison by intravenous route. When heated to decomposition it emits very toxic fumes of NO_x and Cl^-.

TMV750 CAS:7224-23-9 ***HR: 3***
TRIPHENYLTHIOCYANATOSTANNANE
mf: $C_{19}H_{15}NSSn$ mw: 408.10

SYNS: THIOCYANIC ACID, TRIPHENYLSTANNYL ESTER ◇ TRIPHENYL TIN THIOCYANATE

TOXICITY DATA with REFERENCE
ipr-rat LDLo:100 mg/kg NCNSA6 5,46,53
ivn-mus LD50:56 mg/kg CSLNX* NX#04262

OSHA PEL: TWA 0.1 mg(Sn)/m^3 (skin)
ACGIH TLV: TWA 0.1 mg(Sn)/m^3 (skin) (Proposed: TWA 0.1 mg(Sn)/m^3; STEL 0.2 mg(Sn)/m^3 (skin))
NIOSH REL: (Organotin Compounds) TWA 0.1 mg(Sn)/m^3

SAFETY PROFILE: Poison by intraperitoneal and intravenous routes. When heated to decomposition it emits very toxic fumes of NO_x, SO_x, and CN^-. See also THIOCYANATES, TIN COMPOUNDS, and ESTERS.

TMV775 CAS:892-20-6 ***HR: 3***
TRIPHENYLTIN
mf: $C_{18}H_{16}Sn$ mw: 351.03

SYNS: TRIPHENYLSTANNANE ◇ TRIPHENYLSTANNYL HYDRIDE ◇ TRIPHENYLTIN HYDRIDE

TOXICITY DATA with REFERENCE
orl-rat LD50:491 mg/kg 34ZIAG -,591,69
ipr-rat LD50:8500 μg/kg 34ZIAG -,591,69
orl-mus LD50:81 mg/kg 34ZIAG -,591,69
ipr-mus LD50:7900 μg/kg 34ZIAG -,591,69

SAFETY PROFILE: Poison by ingestion and intraperitoneal routes. When heated to decomposition it emits acrid smoke and irritating fumes. See also TIN COMPOUNDS.

TMV800 CAS:2847-65-6 ***HR: 3***
TRIPHENYLTIN p-ACETAMIDOBENZOATE
mf: $C_{27}H_{23}NO_3Sn$ mw: 528.20

SYN: STANNANE,((p-ACETAMIDOBENZOYL)OXY)TRIPHENYL-

TOXICITY DATA with REFERENCE
ivn-mus LD50:18 mg/kg CSLNX* NX#02971

OSHA PEL: TWA 0.1 mg(Sn)/m^3 (skin)
ACGIH TLV: TWA 0.1 mg(Sn)/m^3; STEL 0.2 mg/m^3 (skin)
NIOSH REL: (Organotin Compound):10H TWA 0.1 mg(Sn)/m^3

SAFETY PROFILE: Poison by intravenous route. When heated to decomposition it emits toxic fumes of NO_x and Sn.

TMV825 CAS:73927-89-6 ***HR: 3***
TRIPHENYLTIN CYANOACETATE
mf: $C_{21}H_{17}NO_2Sn$ mw: 434.08

SYNS: ACETIC ACID, CYANO-, TRIPHENYLSTANNYL ESTER ◇ STANNANE, (CYANOACETOXY)TRIPHENYL-

TOXICITY DATA with REFERENCE
ivn-mus LD50:56200 μg/kg CSLNX* NX#05971

OSHA PEL: TWA 0.1 mg(Sn)/m^3 (skin)
ACGIH TLV: TWA 0.1 mg(Sn)/m^3; STEL 0.2 mg/m^3 (skin)
NIOSH REL: (Organotin Compound):10H TWA 0.1 mg(Sn)/m^3

SAFETY PROFILE: Poison by intravenous route. When heated to decomposition it emits toxic fumes of NO_x and Sn.

TMW000 CAS:4150-34-9 ***HR: 3***
TRIPHENYLTIN HYDROPEROXIDE
mf: $C_{18}H_{16}O_2Sn$ mw: 383.02

SAFETY PROFILE: A powerful oxidizer. It explodes at 75°C. Upon decomposition it emits acrid smoke and fumes. See also TIN COMPOUNDS and PEROXIDES.

TMW250 CAS:23292-85-5 ***HR: 3***
TRIPHENYLTIN LEVULINATE
mf: $C_{23}H_{22}O_3Sn$ mw: 465.14

SYNS: LEVULINIC ACID, TRIPHENYLSTANNYL ESTER ◇ (4-OXOVALERYLOXY)TRIPHENYLSTANNANE

TOXICITY DATA with REFERENCE
ivn-mus LD50:18 mg/kg CSLNX* NX#04820

OSHA PEL: TWA 0.1 mg(Sn)/m^3 (skin)
ACGIH TLV: TWA 0.1 mg(Sn)/m^3 (skin) (Proposed: TWA 0.1 mg(Sn)/m^3; STEL 0.2 mg(Sn)/m^3 (skin))
NIOSH REL: (Organotin Compounds) TWA 0.1 mg(Sn)/m^3

SAFETY PROFILE: Poison by intravenous route.

When heated to decomposition it emits acrid smoke and irritating fumes. See also TIN COMPOUNDS.

TMW500 CAS:13302-08-4 ***HR: 3***
TRIPHENYLTIN METHANESULFONATE
mf: $C_{19}H_{18}O_3SSn$ mw: 445.12

SYN: ((METHYLSULFONYL)OXY)TRIPHENYLSTANNANE

TOXICITY DATA with REFERENCE
ivn-mus LD50:56200 μg/kg CSLNX* NX#02305

OSHA PEL: TWA 0.1 mg(Sn)/m^3 (skin)
ACGIH TLV: TWA 0.1 mg(Sn)/m^3 (skin) (Proposed: TWA 0.1 mg(Sn)/m^3; STEL 0.2 mg(Sn)/m^3 (skin))
NIOSH REL: (Organotin Compounds) TWA 0.1 mg(Sn)/m^3

SAFETY PROFILE: Poison by intravenous route. When heated to decomposition it emits toxic fumes of SO_x. See also TIN COMPOUNDS.

TMW600 CAS:67410-20-2 ***HR: 3***
TRIPHENYLTIN PROPIOLATE
mf: $C_{21}H_{16}O_2Sn$ mw: 419.06

SYNS: (ACETYLENECARBONYLOXY)TRIPHENYLTIN ◇ PROPIOLIC ACID, TRIPHENYLSTANNYL ESTER ◇ STANNANE, (ACETYLENECARBONYLOXY)TRIPHENYL-

TOXICITY DATA with REFERENCE
ivn-mus LD50:18 mg/kg CSLNX* NX#05130

OSHA PEL: TWA 0.1 mg(Sn)/m^3 (skin)
ACGIH TLV: TWA 0.1 mg(Sn)/m^3; STEL 0.2 mg/m^3 (skin)
NIOSH REL: (Organotin Compound):10H TWA 0.1 mg(Sn)/m^3

SAFETY PROFILE: Poison by intravenous route. When heated to decomposition it emits toxic fumes of Sn.

TMX000 CAS:974-29-8 ***HR: 3***
TRIPHENYL-1H-1,2,4-TRIAZOL-1-YL TIN
mf: $C_{20}H_{17}N_3Sn$ mw: 418.09

SYN: 1H-1,2,4-TRIAZOL-1-YL)TRIPHENYLSTANNANE

TOXICITY DATA with REFERENCE
ivn-mus LD50:18 mg/kg CSLNX* NX#05668

OSHA PEL: TWA 0.1 mg(Sn)/m^3 (skin)
ACGIH TLV: TWA 0.1 mg(Sn)/m^3 (skin) (Proposed: TWA 0.1 mg(Sn)/m^3; STEL 0.2 mg(Sn)/m^3 (skin))
NIOSH REL: (Organotin Compounds) TWA 0.1 mg(Sn)/m^3

SAFETY PROFILE: Poison by intravenous route. When heated to decomposition it emits toxic fumes of NO_x. See also TIN COMPOUNDS.

TMX250 CAS:4441-17-2 ***HR: 3***
TRIPIPERIDINOPHOSPHINE OXIDE
mf: $C_{15}H_{30}N_3OP$ mw: 299.45

PROP: Mp: 40-42°, bp: 273°.

TOXICITY DATA with REFERENCE
ivn-mus LD50:56 mg/kg CSLNX* NX#05845

CONSENSUS REPORTS: Reported in EPA TSCA Inventory.

SAFETY PROFILE: Poison by intravenous route. When heated to decomposition it emits very toxic fumes of NO_x and PO_x. See also PHOSPHINE.

TMX350 CAS:68541-88-8 ***HR: 3***
TRIPIPERIDINOPHOSPHINE SELENIDE
mf: $C_{15}H_{30}N_3PSe$ mw: 362.41

SYN: PHOSPHINE SELENIDE, TRIPIPERIDINO-

TOXICITY DATA with REFERENCE
ivn-mus LD50:56 mg/kg CSLNX* NX#05677

OSHA PEL: TWA 0.2 mg(Se)/m^3
ACGIH TLV: TWA 0.2 mg(Se)/m^3

SAFETY PROFILE: Poison by intravenous route. When heated to decomposition it emits toxic fumes of NO_x, PO_x, and Se.

TMX500 CAS:1317-95-9 ***HR: 3***
TRIPOLI

PROP: Finely granulated white or gray siliceous rock. A form of crystalline silica.

OSHA PEL: (Transitional: TWA Respirable: 10 mg/m^3/2(%SiO_2+2); Total Dust: TWA 30 mg/m^3/2 (%SiO_2+2)) TWA 0.1 mg/m^3
ACGIH TLV: TWA 0.1 mg/m^3 (of contained respirable quartz dust)
NIOSH REL: (Silica, Crystalline): 10H TWA 0.05 mg/m^3

SAFETY PROFILE: The prolonged inhalation of dusts containing free silica may result in the development of a disabling pulmonary fibrosis known as silicosis. See also SILICA.

TMX600 CAS:14023-90-6 ***HR: 3***
TRIPOTASSIUM HEXACYANOMANGANATE(3-)
mf: $C_6MnN_6 \cdot 3K$ mw: 328.36

SYNS: MANGANATE(3-), HEXACYANO-, TRIPOTASSIUM ◇ MANGANATE(3-), HEXAKIS(CYANO-C)-, TRIPOTASSIUM, (OC-6-11)- ◇ POTASSIUM MANGANOCYANIDE

TOXICITY DATA with REFERENCE
orl-mus LD50:275 mg/kg JPMSAE 52,59,63

OSHA PEL: CL 5 mg(Mn)/m^3
ACGIH TLV: TWA 5 mg(Mn)/m^3

SAFETY PROFILE: Poison by ingestion. When heated to decomposition it emits toxic fumes of NO_x and Mn.

TMX750 CAS:2399-85-1 ***HR: 2***
TRIPOTASSIUM NITRILOTRIACETATE
mf: $C_6H_6NO_6{\cdot}3K$ mw: 305.43

TOXICITY DATA with REFERENCE
orl-rat LD50:1220 mg/kg TXAPA9 18,398,71

CONSENSUS REPORTS: Reported in EPA TSCA Inventory.

SAFETY PROFILE: Moderately toxic by ingestion. When heated to decomposition it emits very toxic fumes of NO_x and K_2O.

TMX775 CAS:550-70-9 ***HR: 3***
TRIPROLIDINE HYDROCHLORIDE
mf: $C_{19}H_{22}N_2{\cdot}ClH$ mw: 314.89

SYNS: ACTIDILAT ◇ ACTIDOL ◇ 295 C 51 ◇ ENTRA ◇ trans-1-(4'-METHYLPHENYL)-1-(2'-PYRIDYL)-3-PYRROLIDINOPROP-1-N E HYDROCHLORIDE ◇ (E)-2-(1-(4-METHYLPHENYL)-3-(1-PYRROLIDINYL)-1-PROPENYL)-PYRIDINE MONOHYDROCHLORIDE ◇ PRO-ACTIDIL ◇ trans-2-(3-(1-PYRROLIDINYL)-1-p-TOLYLPROPENYL)PYRIDINE MONOHYDROCHLORIDE

TOXICITY DATA with REFERENCE
orl-mus LD50:495 mg/kg NIIRDN 6,525,82
scu-mus LD50:247 mg/kg NIIRDN 6,525,82
ivn-mus LD50:21 mg/kg BJPCAL 8,171,53

SAFETY PROFILE: Poison by subcutaneous and intravenous routes. Moderately toxic by ingestion. When heated to decomposition it emits toxic fumes of NO_x and HCl.

TMY000 CAS:139-45-7 ***HR: 2***
TRIPROPIONIN
mf: $C_{12}H_{20}O_6$ mw: 260.32

SYNS: GLYCERINE TRIPROPIONATE ◇ GLYCERYL TRIPROPIONATE ◇ TRIPROPIONINE

TOXICITY DATA with REFERENCE
ivn-mus LD50:840 mg/kg APSCAX 40,338,57

CONSENSUS REPORTS: Reported in EPA TSCA Inventory.

SAFETY PROFILE: Moderately toxic by intravenous route. When heated to decomposition it emits acrid smoke and irritating fumes.

TMY100 CAS:102-67-0 ***HR: 2***
TRIPROPYLALUMINUM
DOT: UN 2718
mf: $C_9H_{21}Al$ mw: 156.28

PROP: Liquid. Mp: -107°, bp: 82-84°/2 mm Hg, d: 0.823.

SYNS: ALUMINUM, TRIPROPYL- ◇ TRIPROPYLALUMINUM (DOT)

CONSENSUS REPORTS: Reported in EPA TSCA Inventory.

ACGIH TLV: TWA 2 mg(Al)/m^3
DOT Classification: Flammable Solid; Label: Spontaneously Combustible

SAFETY PROFILE: Pyrophoric, moisture sensitive, flammable solid. Danger from spontaneous combustion. When heated to decomposition it emits toxic fumes of Al.

TMY250 CAS:102-69-2 ***HR: 3***
TRI-N-PROPYLAMINE
mf: $C_9H_{21}N$ mw: 143.31
DOT: UN 2260

PROP: Liquid. Mp: −93°, bp: 156°, flash p: 105°F (OC), d: 0.75, vap d: 4.9. Very sltly sol in water.

SYNS: N,N-DIPROPYL-1-PROPANAMINE ◇ TRIPROPYLAMINE (DOT)

TOXICITY DATA with REFERENCE
orl-rat LD50:72 mg/kg AIHAAP 30,470,69
ihl-rat LCLo:250 ppm/4H AIHAAP 30,470,69
ihl-mus LC50:3800 mg/m^3/2H 85GMAT -,118,82
skn-rbt LD50:429 mg/kg AIHAAP 30,470,69
ihl-mam LC50:5100 mg/m^3 TPKVAL 14,80,75

CONSENSUS REPORTS: Reported in EPA TSCA Inventory.

DOT Classification: Flammable Liquid; Label: Flammable Liquid, Corrosive; Flammable or Combustible Liquid; Label: Flammable Liquid, Corrosive.

SAFETY PROFILE: Poison by ingestion. Moderately toxic by skin contact and inhalation. A corrosive irritant to skin, eyes, and mucous membranes. Flammable when exposed to heat, flame, or oxidizers. Can react with oxidizing materials. To fight fire, use foam, CO_2, dry chemical. When heated to decomposition it emits toxic fumes of NO_x. See also AMINES.

TMY750 CAS:688-71-1 ***HR: 2***
TRI-n-PROPYL BORATE
mf: $C_9H_{21}BO_3$ mw: 188.11

PROP: Colorless liquid; odor of n-propanol. Bp: 176°, flash p: 155°F (COC), d: 0.856 @ 24°.

TOXICITY DATA with REFERENCE
eye-rbt 100 mg MLD 14KTAK -,706,64
orl-mus LD50:2080 mg/kg 14KTAK -,706,64

CONSENSUS REPORTS: Reported in EPA TSCA Inventory.

SAFETY PROFILE: Moderately toxic by ingestion. An eye irritant. Combustible when exposed to heat or flame; can react with oxidizing materials. When heated to decomposition it emits acrid smoke and irritating fumes. See also ESTERS and BORON COMPOUNDS.

TMY850 CAS:67445-50-5 ***HR: 3***
TRIPROPYL(BUTYLTHIO)STANNANE
mf: $C_{13}H_{30}SSn$ mw: 337.18

SYNS: (BUTYLTHIO)TRIPROPYLSTANNANE ◇ STANNANE, (BUTYLTHIO)TRIPROPYL-

TOXICITY DATA with REFERENCE
ipr-mus LD50:27 mg/kg RPTOAN 42,73,79

OSHA PEL: TWA 0.1 mg(Sn)/m^3 (skin)
ACGIH TLV: TWA 0.1 mg(Sn)/m^3; STEL 0.2 mg/m^3 (skin)
NIOSH REL: (Organotin Compound):10H TWA 0.1 mg(Sn)/m^3

SAFETY PROFILE: Poison by intraperitoneal route. When heated to decomposition it emits toxic fumes of SO_x and Sn.

TMZ000 CAS:24800-44-0 ***HR: 2***
TRIPROPYLENE GLYCOL
mf: $C_9H_{20}O_4$ mw: 192.29

PROP: Colorless liquid. Mp: does not crystallize, bp: 267°, flash p: 285°F, d: 1.023 @ 25/25°, vap press: 1 mm @ 96.0°, vap d: 6.63.

SYN: 2-(2-(2-HYDROXYPROPOXY)PROPOXY-1-PROPANOL

TOXICITY DATA with REFERENCE
orl-rat LD50:3000 mg/kg 14CYAT 2,1522,63

CONSENSUS REPORTS: Reported in EPA TSCA Inventory.

SAFETY PROFILE: Moderately toxic by ingestion. Combustible when exposed to heat or flame; can react with oxidizing materials. To fight fire, use water, foam, CO_2, dry chemical. When heated to decomposition it emits acrid smoke and irritating fumes.

TNA000 CAS:10213-77-1 ***HR: 2***
TRIPROPYLENE GLYCOL, METHYL ETHER
mf: $C_{10}H_{22}O_4$ mw: 206.32

PROP: Bp: 243°, flash p: 250°F, d: 0.967 @ 25/25°, vap d: 7.1

SYNS: DOWANOL 62B ◇ 2-(2-(2-METHOXYPROPOXY)PROPOXY PROPANOL

TOXICITY DATA with REFERENCE
eye-rbt 243 mg MLD AMIHBC 9,509,54
orl-rat LD50:3300 mg/kg AMIHBC 9,509,54
orl-dog LDLo:5000 mg/kg JPETAB 102,79,51

CONSENSUS REPORTS: Glycol ether compounds are on the Community Right-To-Know List. Reported in EPA TSCA Inventory.

SAFETY PROFILE: Moderately toxic by ingestion. An eye irritant. Many glycol ether compounds have dangerous human reproductive effects. Combustible when exposed to heat or flame; can react with oxidizing materials. To fight fire, use foam, CO_2, dry chemical. When heated to decomposition it emits acrid smoke and irritating fumes. See also GLYCOL ETHERS.

TNA250 CAS:3015-98-3 ***HR: 3***
TRIPROPYL INDIUM
mf: $C_9H_{21}In$ mw: 244

SAFETY PROFILE: Spontaneously flammable in air. When heated to decomposition it emits acrid smoke and irritating fumes. See also INDIUM.

TNA500 CAS:6618-03-7 ***HR: 3***
TRIPROPYL LEAD
mf: $C_9H_{22}Pb$ mw: 337.50

SYNS: LEAD TRIPROPYL ◇ TRIPROPYL PLUMBANE

TOXICITY DATA with REFERENCE
par-rat LDLo:20 mg/kg AOHYA3 3,226,61

CONSENSUS REPORTS: Lead and its compounds are on the Community Right-To-Know List.

SAFETY PROFILE: Poison by parenteral route. Highly flammable. when heated to decomposition it emits toxic fumes of Pb. See also LEAD COMPOUNDS.

TNA750 CAS:1520-71-4 ***HR: 3***
TRI-n-PROPYL LEAD CHLORIDE
mf: $C_9H_{21}ClPb$ mw: 371.94

SYNS: CHLOROTRIPROPYLPLUMBANE ◇ TRIPROPYL LEAD CHLORIDE

TOXICITY DATA with REFERENCE
orl-rat LD50:27 mg/kg BJIMAG 18,277,61
ipr-rat LDLo:5380 μg/kg JPETAB 38,161,30
scu-rat LDLo:11 mg/kg JPETAB 38,161,30
ivn-mus LD50:22 mg/kg CSLNX* NX#03647

CONSENSUS REPORTS: Lead and its compounds are on the Community Right-To-Know List.

SAFETY PROFILE: Poison by ingestion, intraperitoneal, subcutaneous, and intravenous routes. When heated to decomposition it emits very toxic fumes of Cl^-

and Pb. See also LEAD COMPOUNDS and CHLORIDES.

TNB000 CAS:3267-78-5 **HR: 3**
TRIPROPYLTIN ACETATE
mf: $C_{11}H_{24}O_2Sn$ mw: 307.04

SYN: ACETOXYTRIPROPYLSTANNANE

TOXICITY DATA with REFERENCE
orl-rat LD50:118 mg/kg BJIMAG 15,15,58
ivn-rat LDLo:24 mg/kg BJIMAG 15,15,58
orl-mus LDLo:210 mg/kg AECTCV 14,111,85

OSHA PEL: TWA 0.1 mg(Sn)/m^3 (skin)
ACGIH TLV: TWA 0.1 mg(Sn)/m^3 (skin) (Proposed: TWA 0.1 mg(Sn)/m^3; STEL 0.2 mg(Sn)/m^3 (skin))
NIOSH REL: (Organotin Compounds) TWA 0.1 mg(Sn)/m^3

SAFETY PROFILE: Poison by ingestion and intravenous routes. When heated to decomposition it emits acrid smoke and irritating fumes. See also TIN COMPOUNDS.

TNB250 CAS:7342-45-2 **HR: 3**
TRIPROPYLTIN IODIDE
mf: $C_9H_{21}ISn$ mw: 374.89

PROP: Colorless liquid. D: 1.692 @ 16°, mp: −53°, bp: 262°. Sol in organic solvents.

SYN: IODOTRIPROPYLSTANNANE

TOXICITY DATA with REFERENCE
ivn-mus LD50:4470 μg/kg CSLNX* NX#02335

OSHA PEL: TWA 0.1 mg(Sn)/m^3 (skin)
ACGIH TLV: TWA 0.1 mg(Sn)/m^3 (skin) (Proposed: TWA 0.1 mg(Sn)/m^3; STEL 0.2 mg(Sn)/m^3 (skin))
NIOSH REL: (Organotin Compounds) TWA 0.1 mg(Sn)/m^3

SAFETY PROFILE: Poison by intravenous route. When heated to decomposition it emits toxic fumes of I^-. See also TIN COMPOUNDS and IODIDES.

TNB500 CAS:73927-92-1 **HR: 3**
TRIPROPYLTIN IODOACETATE
mf: $C_{11}H_{23}IO_2Sn$ mw: 432.93

SYN: (IODOACETOXY)TRIPROPYLSTANNANE

TOXICITY DATA with REFERENCE
ivn-mus LD50:18 mg/kg CSLNX* NX#03452

OSHA PEL: TWA 0.1 mg(Sn)/m^3 (skin)
ACGIH TLV: TWA 0.1 mg(Sn)/m^3 (skin) (Proposed: TWA 0.1 mg(Sn)/m^3; STEL 0.2 mg(Sn)/m^3 (skin))
NIOSH REL: (Organotin Compounds) TWA 0.1 mg(Sn)/m^3

SAFETY PROFILE: Poison by intravenous route. When heated to decomposition it emits toxic fumes of I^-. See also TIN COMPOUNDS.

TNB750 CAS:31709-32-7 **HR: 3**
TRIPROPYLTIN ISOTHIOCYANATE
mf: $C_{10}H_{21}NSSn$ mw: 306.07

SYN: (ISOTHIOCYANATO)TRIPROPYLSTANNANE

TOXICITY DATA with REFERENCE
ivn-mus LD50:5 mg/kg CSLNX* NX#03420

OSHA PEL: TWA 0.1 mg(Sn)/m^3 (skin)
ACGIH TLV: TWA 0.1 mg(Sn)/m^3 (skin) (Proposed: TWA 0.1 mg(Sn)/m^3; STEL 0.2 mg(Sn)/m^3 (skin))
NIOSH REL: (Organotin Compounds) TWA 0.1 mg(Sn)/m^3

SAFETY PROFILE: Poison by intravenous route. When heated to decomposition it emits very toxic fumes of NO_x and SO_x. See also TIN COMPOUNDS and THIOCYANATES.

TNC000 CAS:73927-99-8 **HR: 3**
TRIPROPYLTIN TRICHLOROACETATE
mf: $C_{11}H_{21}Cl_3O_2Sn$ mw: 410.36

SYNS: TRICHLOROACETIC ACID TRIPROPYLSTANNYL ESTER ◇ (TRICHLOROACETOXY)TRIPROPYLSTANNANE

TOXICITY DATA with REFERENCE
ivn-mus LD50:5600 μg/kg CSLNX* NX#06281

OSHA PEL: TWA 0.1 mg(Sn)/m^3 (skin)
ACGIH TLV: TWA 0.1 mg(Sn)/m^3 (skin) (Proposed: TWA 0.1 mg(Sn)/m^3; STEL 0.2 mg(Sn)/m^3 (skin))
NIOSH REL: (Organotin Compounds) TWA 0.1 mg(Sn)/m^3

SAFETY PROFILE: Poison by intravenous route. When heated to decomposition it emits toxic fumes of Cl^-. See also TIN COMPOUNDS and CHLORIDES.

TNC175 CAS:38748-32-2 **HR: 3**
TRIPTOLIDE
mf: $C_{20}H_{24}O_6$ mw: 360.44

SYN: TRIPTOLID

TOXICITY DATA with REFERENCE
ipr-mus LD50:900 μg/kg CYLPDN 2,70,81
ivn-mus LD50:800 μg/kg CYLPDN 2,70,81
ivn-dog LDLo:160 μg/kg CYLPDN 2,70,81

SAFETY PROFILE: Poison by intravenous and intraperitoneal routes. When heated to decomposition it emits acrid smoke and irritating fumes.

TNC250 CAS:64011-26-3 ***HR: 2***
(TRI-2-PYRIDYL)STIBINE
mf: $C_{15}H_{12}N_3Sb$ mw: 356.05

SYN: TL 365

TOXICITY DATA with REFERENCE
orl-rat LDLo:500 mg/kg NCNSA6 5,23,53
ihl-mus LCLo:1100 mg/m^3/10M NDRC** NDCrc-132,Dec,42

CONSENSUS REPORTS: Antimony and its compounds are on the Community Right-To-Know List.

OSHA PEL: TWA 0.5 mg(Sb)/m^3
ACGIH TLV: TWA 0.5 mg(Sb)/m^3
NIOSH REL: (Antimony) TWA 0.5 mg(Sb)/m^3

SAFETY PROFILE: Moderately toxic by ingestion and inhalation. When heated to decomposition it emits very toxic fumes of NO_x and Sb. See also ANTIMONY COMPOUNDS.

TNC500 CAS:126-72-7 ***HR: 3***
TRIS
mf: $C_9H_{15}Br_6O_4P$ mw: 697.67

PROP: Crystals. D: 2.24, flash p: > 112°.

SYNS: ANFRAM 3PB ◇ APEX 462-5 ◇ BROMKAL P 67-6HP ◇ 2,3-DIBROMO-1-PROPANOL, PHOSPHATE (3:1) ◇ 2,3-DIBROMO-1-PROPANOL PHOSPHATE ◇ (2,3-DIBROMOPROPYL) PHOSPHATE ◇ FIREMASTER T23P-LV ◇ FLACAVON R ◇ FLAMMEX AP ◇ FYROL HB32 ◇ NCI-C03270 ◇ PHOSPHORIC ACID, TRIS(2,3-DIBROMOPROPYL) ESTER ◇ RCRA WASTE NUMBER U235 ◇ TDBP (CZECH) ◇ TRIS (flame retardant) ◇ TRIS(DIBROMOPROPYL)PHOSPHATE ◇ TRIS(2,3-DIBROMOPROPYL) PHOSPHATE ◇ TRIS(2,3-DIBROMOPROPYL) PHOSPHORIC ACID ESTER ◇ TRIS-2,3-DIBROMPROPYL ESTER KYSELINY FOSFORECNE (CZECH) ◇ USAF DO-41 ◇ ZETIFEX ZN

TOXICITY DATA with REFERENCE
skn-rbt 500 mg/24H SEV 28ZPAK -,206,72
eye-rbt 500 mg/24H MLD 28ZPAK -,206,72
mma-esc 100 μg/plate ENMUDM 7(Suppl 5),1,85
dnd-hmn:oth 2 mg/L MUREAV 56,89,77
orl-rat TDLo:450 mg/kg (female 7-15D post):REP JTSCDR 4,296,79
orl-rat TDLo:250 mg/kg (female 6-15D post):TER FCTXAV 19,67,81
orl-rat TDLo:1330 mg/kg/76W-C:CAR NCITR* NCI-CG-TR-76,78
orl-mus TDLo:39 g/kg/92W-C:CAR NCITR* NCI-CG-TR-76,78
orl-rat TD:3600 mg/kg/103W-C:NEO NCITR* NCI-CG-TR-76,78
orl-rat TD:1820 mg/kg/2Y-C:ETA FCTXAV 18,743,80
orl-rat LD50:1010 mg/kg 28ZPAK -,206,72
ipr-mus LD50:300 mg/kg NTIS** AD277-689

CONSENSUS REPORTS: NTP Fifth Annual Report on Carcinogens. IARC Cancer Review: Group 2A IMEMDT 7,341,87; Animal Sufficient Evidence IMEMDT 20,575,79; Human Limited Evidence IMEMDT 20,575,79. NCI Carcinogenesis Bioassay (feed); Clear Evidence: mouse, rat NCITR* NCI-CG-TR-76,78. Community Right-To-Know List. Reported in EPA TSCA Inventory. EPA Genetic Toxicology Program.

SAFETY PROFILE: Confirmed carcinogen with experimental carcinogenic, neoplastigenic, tumorigenic, and teratogenic data. Poison by intraperitoneal route. Moderately toxic by ingestion. Experimental reproductive effects. Human mutation data reported. An eye and severe skin irritant. Can cause testicular atrophy and sterility. Once used as a flame retardant additive to synthetic textiles and plastics, particularly in children's sleepwear. Use discontinued because it can be absorbed by human skin, or chewed or sucked off of sleepwear by infants. May be flammable when exposed to heat or flame. When heated to decomposition it emits very toxic fumes of Br^- and PO_x.

TNC725 CAS:2706-47-0 ***HR: 2***
TRIS(p-AMINOPHENYL)CARBONIUM PAMOATE
mf: $C_{23}H_{14}O_6 \cdot 2C_{19}H_{18}N_3$ mw: 963.17

SYNS: 4,4'-METHYLENEBIS(3-HYDROXY-2-NAPHTHOIC ACID) with TRIS-(p-AMINOPHENYL)CARBONIUM SALT (1:2) ◇ 4,4'-METHYLENEBIS(3-HYDROXY-2-NAPHTHOIC ACID) with TRIS-(p-AMINOPHENYL)METHYLIUM SALT (1:2) ◇ METHYLIUM, TRIS(4-AMINOPHENYL)-, 4,4'-METHYLENEBIS(3-HYDROXY-NAPHTHOATE)(2:1) ◇ TACP ◇ TRIS(p-AMINOPHENYL)CARBONIUM SALT with 4,4-METHYLENEBIS(3-HYDROXY-2-NAPHTHOIC ACID) (2:1) ◇ TRIS(p-AMINOPHENYL)METHYLIUM SALT with 4,4'-METHYLENEBIS(3-HYDROXY-2-NAPHTHOIC ACID) (2:1)

TOXICITY DATA with REFERENCE
orl-rat TDLo:100 mg/kg/78W-C:NEO CNREA8 25,1919,65

SAFETY PROFILE: Questionable carcinogen with experimental neoplastigenic data. When heated to decomposition it emits toxic fumes of NO_x.

TNC750 CAS:2706-47-0 ***HR: 3***
TRIS(p-AMINOPHENYL)METHYLIUM SALT with 4,4'-METHYLENEBIS(3-HYDROXY-2-NAPHTHOIC ACID) (2:1)
mf: $C_{19}H_8N_3 \cdot 1/2C_{24}H_{14}O_6$ mw: 471.48

SYNS: 4,4'-METHYLENEBIS(3-HYDROXY-2-NAPHTHOIC ACID) with TRIS-(p-AMINOPHENYL)CARBONIUM SALT (1:2) ◇ 4,4'-METHYLENEBIS (3-HYDROXY-2-NAPHTHOIC ACID) with TRIS-(p-AMINOPHENYL) METHYLIUM SALT (1:2) ◇ TRIS(p-AMINOPHENYL)CARBONIUM PAMOATE ◇ TRIS-(p-AMINOPHENYL)CARBONIUM SALT with 4,4-METHYLENEBIS(3-HYDRO XY-2-NAPHTHOIC ACID) (2:1)

TOXICITY DATA with REFERENCE
orl-rat TDLo:100 mg/kg/78W-C:NEO CNREA8 25,1919,65

SAFETY PROFILE: Questionable carcinogen with experimental neoplastigenic data. When heated to decomposition it emits toxic fumes of NO_x.

TNC800 CAS:84928-99-4 ***HR: 3***
TRIS(2-AZIDOETHYL)AMINE
mf: $C_6H_{12}N_{10}$ mw: 224.23

$(N_3C_2H_4)_3N$

SAFETY PROFILE: An impact-sensitive explosive. When heated to decomposition it emits toxic fumes of NO_x. See also AZIDES.

TNC825 CAS:31044-86-7 ***HR: 2***
1,1,1-TRIS(AZIDOMETHYL)ETHANE
mf: $C_5H_9N_9$ mw: 195.19

$(N_3CH_2)_3CCH_3$

SAFETY PROFILE: Potentially explosive reaction with hydrogen + palladium catalyst. When heated to decomposition it emits toxic fumes of NO_x. See also AZIDES.

TND000 CAS:68-76-8 ***HR: 3***
TRIS(1-AZIRIDINYL)-p-BENZOQUINONE
mf: $C_{12}H_{13}N_3O_2$ mw: 231.28

SYNS: BAYER 3231 ◇ 1,1′,1″-(3,6-DIOXO-1,4-CYCLOHEXADIENE-1,2,4-TRIYL)TRISAZIRIDINE ◇ NSC-29215 ◇ ONCOVEDEX ◇ PRENIMON ◇ RIKER 601 ◇ 10257 R.P. ◇ TEIB ◇ TRENIMON ◇ TRIAZICHON (GERMAN) ◇ TRIAZIQUINONE ◇ TRIAZIQUONE ◇ 2,3,5-TRI-(1-AZIRIDINYL)-p-BENZOQUINONE ◇ 2,3,5-TRIETHYLENEIMINO-1,4-BENZOQUINONE ◇ TRIETHYLENIMINOBENZOQUINONE ◇ TRISAETHYLENIMINOBENZOCHINON (GERMAN) ◇ 2,3,5-TRIS(AZIRIDINO)-1,4-BENZOQUINONE ◇ 2,3,5-TRIS(1-AZIRIDINO)-p-BENZOQUINONE ◇ TRIS(AZIRIDINYL)-p-BENZOQUINONE ◇ 2,3,5-TRIS(1-AZIRIDINYL)-p-BENZOQUINONE ◇ 2,3,5-TRIS(AZIRIDINYL)-1,4-BENZOQUINONE ◇ 2,3,5-TRIS(1-AZIRIDINYL)-2,5-CYLOHEXADIENE-1,4-DIONE ◇ 2,3,5-TRISETHYLENEIMINOBENZOQUINONE ◇ TRISETHYLENEIMINOQUINONE ◇ 2,3,5-TRIS(ETHYLENIMINO)BENZOQUINONE ◇ 2,3,5-TRIS(ETHYLENIMINO)-p-BENZOQUINONE ◇ 2,3,5-TRIS(ETHYLENIMINO)-1,4-BENZOQUINONE

TOXICITY DATA with REFERENCE
sln-dmg-orl 50000 ppm MUREAV 2,29,65
cyt-hmn:lym 50 nmol/L MUREAV 149,83,85
ipr-mus TDLo:125 μg/kg (male 1D pre):TER MUREAV 54,175,78
ipr-mus TDLo:125 μg/kg (male 1D pre):REP HUMAA7 7,43,69
ivn-rat TDLo:1560 μg/kg/1Y-I:CAR ARZNAD 20,1461,70
ipr-rat LD50:500 μg/kg ARZNAD 16,1533,66
ivn-rat LD50:500 μg/kg ARZNAD 20,1467,70
par-rat LD50:500 μg/kg RRCRBU 52,76,75

CONSENSUS REPORTS: IARC Cancer Review: Group 3 IMEMDT 7,367,87; Animal Sufficient Evidence IMEMDT 9,67,75; Human Inadequate Evidence IMEMDT 9,67,75. Community Right-To-Know List. EPA Genetic Toxicology Program.

SAFETY PROFILE: Poison by intraperitoneal, intravenous, and parenteral routes. Experimental teratogenic and reproductive effects. Questionable carcinogen with experimental experimental carcinogenic data. Human mutation data reported. When heated to decomposition it emits toxic fumes of NO_x. Used as a drug for the treatment of neoplastic diseases.

TND250 CAS:545-55-1 ***HR: 3***
TRIS-(1-AZIRIDINYL)PHOSPHINE OXIDE
DOT: UN 2501
mf: $C_6H_{12}N_3OP$ mw: 173.18

PROP: Colorless crystals. Mp: 41°, bp: 90° @ 23 mm. Sol in water, alc, ether.

SYNS: APHOXIDE ◇ APO ◇ 1-AZIRIDINYL PHOSPHINE OXIDE (TRIS) (DOT) ◇ CBC 906288 ◇ ENT 24,915 ◇ IMPERON FIXER T ◇ NSC 9717 ◇ 1,1′,1″-PHOSPHINYLIDYNETRISAZIRIDINE ◇ PHOSPHORIC ACID TRIETHYLENE IMIDE ◇ PHOSPHORIC ACID TRIETHYLENEIMINE (DOT) ◇ SK-3818 ◇ TEF ◇ TEPA ◇ TRIAETHYLENPHOSPHORSAEUREAMID (GERMAN) ◇ TRIAZIRIDINOPHOSPHINE OXIDE ◇ TRI(AZIRIDINYL)PHOSPHINE OXIDE ◇ TRI-1-AZIRIDINYL)PHOSPHINE OXIDE ◇ N,N′,N″-TRI-1,2-ETHANEDIYL PHOSPHORIC TRIMIDE ◇ TRIETHYLENEPHOSPHOROTRIAMIDE ◇ TRIS(1-AZIRIDINE)PHOSPHINE OXIDE ◇ TRIS(N-ETHYLENE)PHOSPHOROTRIAMIDATE

TOXICITY DATA with REFERENCE
sln-dmg-par 500 μg/kg EVSRBT 25,917,81
cyt-hmn:leu 100 μmol/L CHROAU 24,314,68
ipr-mus TDLo:2500 μg/kg (male 1D pre):REP FOBLAN 16,367,70
ipr-mus TDLo:2500 μg/kg (male 1D pre):TER FOBLAN 16,367,70
orl-rat TDLo:1300 μg/kg/1Y-I:CAR JNCIAM 41,985,68
orl-rat TD:3120 μg/kg/1Y-I:NEO JNCIAM 41,985,68
orl-rat LD50:37 mg/kg BWHOA6 31,737,64
skn-rat LD50:87 mg/kg BWHOA6 31,737,64
orl-mus LD50:420 mg/kg KOBUA3 24,788,75
ipr-mus LD50:25500 μg/kg BJPCAL 25,223,65
ivn-mus LD50:178 mg/kg CSLNX* NX#02129
ivn-dog LDLo:430 μg/kg CCSUBJ 2,202,65
ivn-mky LDLo:870 μg/kg CCSUBJ 2,202,65
orl-ckn LD50:151 mg/kg TXAPA9 16,100,70
orl-qal LD50:237 mg/kg JRPFA4 48,371,76

CONSENSUS REPORTS: IARC Cancer Review: Group 3 IMEMDT 7,56,87; Animal Inadequate Evidence IMEMDT 9,75,75. EPA Genetic Toxicology Program.

DOT Classification: Label: Corrosive; Poison B; Label: Poison.

SAFETY PROFILE: Poison by ingestion, skin contact, intravenous, and intraperitoneal routes. Experimental teratogenic and reproductive effects. Questionable carcinogen with experimental carcinogenic and neoplastigenic data. Human mutation data reported. A corrosive irritant to the skin, eyes, and mucous membranes. When heated to decomposition it emits very toxic fumes of PO_x and NO_x. Used as an acaricide and in the permanent press treatment of cotton.

TND500 CAS:51-18-3 ***HR: 3***
TRISAZIRIDINYLTRIAZINE
mf: $C_9H_{12}N_6$ mw: 204.27

PROP: Small crystals. Water-sol. Decomp @ 139°.

SYNS: DRP 859025 ◇ ENT 25,296 ◇ M-9500 ◇ NSC 9706 ◇ PERSISTOL ◇ R-246 ◇ SEM (cytostatic) ◇ SK1133 ◇ TRETAMINE ◇ TRIAETHYLENMELAMIN (GERMAN) ◇ TRIAMELIN ◇ 1,1',1''-s-TRIAZINE-2,4,6-TRIYLTRISAZIRIDINE ◇ TRIAZIRIDINYL TRIAZINE ◇ TRIETHANOMELAMINE ◇ 2,4,6-TRI(ETHYLENEIMINO)-1,3,5-TRIAZINE ◇ 2,4,6-TRIETHYLENEIMINO-s-TRIAZINE ◇ TRIETHYLENEMELAMINE ◇ 2,4,6-TRIETHYLENIMINO-s-TRIAZINE ◇ 2,4,6-TRIETHYLENIMINO-1,3,5-TRIAZINE ◇ 2,4,6-TRIS(1-AZIRIDINYL)-s-TRIAZINE ◇ 2,4,6-TRIS(1'-AZIRIDINYL)-1,3,5-TRIAZINE ◇ TRIS(ETHYLENEIMINO)TRIAZINE ◇ 2,4,6-TRIS(ETHYLENEIMINO)-s-TRIAZINE ◇ TRISETHYLENEIMINO-1,3,5-TRIAZINE ◇ 2,4,6-TRIS(ETHYLENIMINO)-s-TRIAZINE

TOXICITY DATA with REFERENCE
mmo-sat 25 μg/plate TAKHAA 44,96,85
trn-mus-ipr 750 μg/kg/5W-I CIHPDR 6,425,84
orl-mus TDLo:500 μg/kg (male 5D pre):REP TXAPA9 23,277,72
ipr-mus TDLo:200 μg/kg (male 1D pre):TER TAKHAA 40,37,81
scu-rat TDLo:10 mg/kg/I:ETA ANYAA9 68,750,58
skn-mus TDLo:10 mg/kg:NEO BJCAAI 9,177,55
orl-rat LD50:1 mg/kg BWHOA6 31,721,64
ipr-rat LD50:1 mg/kg JPETAB 100,398,50
ivn-rat LD50:1110 mg/kg ARZNAD 6,539,56
ims-rat LD50:1500 μg/kg CLDND*
orl-mus LD50:15 mg/kg JPETAB 100,398,50
ipr-mus LD50:2800 μg/kg JPETAB 100,398,50
scu-mus LD50:1871 μg/kg NCISP* JAN86
ivn-dog LDLo:400 μg/kg JPETAB 100,398,50
ivn-mky LDLo:100 μg/kg CCSUBJ 2,202,65

CONSENSUS REPORTS: IARC Cancer Review: Group 3 IMEMDT 7,56,87; Animal Sufficient Evidence IMEMDT 9,95,75. EPA Genetic Toxicology Program.

SAFETY PROFILE: Poison by ingestion, intraperitoneal, intramuscular, intravenous, and subcutaneous routes. Experimental teratogenic and reproductive effects. Questionable carcinogen with experimental neoplastigenic and tumorigenic data. Human mutation data reported. Can cause gastrointestinal tract disturbances and bone marrow depression. When heated to decomposition it emits highly toxic fumes of NO_x. Used as an antineoplastic agent and as an insect sterilant.

TND750 CAS:14751-89-4 ***HR: 3***
TRIS-2,2'-BIPYRIDINE CHROMIUM
mf: $C_{30}H_{24}CrN_6$ mw: 520.41

CONSENSUS REPORTS: Chromium and its compounds are on the Community Right-To-Know List.

SAFETY PROFILE: Ignites spontaneously in air. When heated to decomposition it emits toxic fumes of NO_x. See also CHROMIUM COMPOUNDS.

TNE000 CAS:15388-46-2 ***HR: 3***
TRIS-2,2'-BIPYRIDINE CHROMIUM(II) PERCHLORATE
mf: $C_{30}H_{24}Cl_2CrN_6O_8$ mw: 699.31

CONSENSUS REPORTS: Chromium and its compounds are on the Community Right-To-Know List.

SAFETY PROFILE: Many chromium compounds are poisons. Explodes violently when heated to 250°C or when exposed to sparks. When heated to decomposition it emits very toxic fumes of Cl^- and NO_x. See also PERCHLORATES and CHROMIUM COMPOUNDS.

TNE250 ***HR: 3***
TRIS-2,2'-BIPYRIDINESILVER(II) PERCHLORATE
mf: $C_{30}H_{24}AgCl_2N_6O_8$ mw: 775.19

CONSENSUS REPORTS: Silver and its compounds are on the Community Right-To-Know List.

SAFETY PROFILE: Explodes on heating. When heated to decomposition it emits very toxic fumes of NO_x and Cl^-. See also SILVER COMPOUNDS and PERCHLORATES.

TNE275 CAS:22755-34-6 ***HR: 3***
2,4,6-TRIS(BROMOAMINO)-1,3,5-TRIAZINE
mf: $C_3H_3Br_3N_6$ mw: 362.81

SAFETY PROFILE: A powerful oxidizer. Its use as a brominating agent may cause violent or explosive reactions. Violent reaction with allyl alcohol. When heated to decomposition it emits toxic fumes of Br^- and NO_x. See also BROMIDES.

TNE500 CAS:7328-28-1 ***HR: D***
TRIS(1-BROMO-3-CHLOROISOPROPYL)PHOSPHATE
mf: $C_9H_{15}Br_3Cl_3O_4P$ mw: 564.29

SYN: TRIS(2-BROMO-1-CHLORO-2-PROPYL)PHOSPHATE

TOXICITY DATA with REFERENCE
mmo-sat 5 μL/plate MUREAV 28,405,75
mmo-esc 5 μL/plate MUREAV 28,405,75

SAFETY PROFILE: Mutation data reported. When heated to decomposition it emits very toxic fumes of PO_x, Cl^-, and Br^-. See also PHOSPHATES, BROMIDES, and CHLORIDES.

TNE600 CAS:27568-90-7 ***HR: 3***
TRIS(2-BROMOETHYL)PHOSPHATE
mf: $C_6H_{12}Br_3O_4P$ mw: 418.88

SYN: 2-BROMO-ETHANEPHOSPHORIC ACID BIS(2-BROMOETHYL) ESTER

TOXICITY DATA with REFERENCE
mmo-sat 100 nmol/plate MUREAV 66,373,79
mma-sat 100 nmol/plate MUREAV 66,373,79
ihl-rat LCLo:260 mg/m^3 NDRC** -,9,43

SAFETY PROFILE: Poison by inhalation. Mutation data reported. When heated to decomposition it emits toxic fumes of Br^- and PO_x. See also PHOSPHATES and ESTERS.

TNE750 CAS:814-29-9 ***HR: 3***
TRISBUTYLPHOSPHINE OXIDE
mf: $C_{12}H_{27}OP$ mw: 218.36

TOXICITY DATA with REFERENCE
eye-rbt 250 μg AJOPAA 29,1363,46
ivn-mus LD50:320 mg/kg CSLNX* NX#03085

CONSENSUS REPORTS: Reported in EPA TSCA Inventory.

SAFETY PROFILE: Poison by intravenous route. An eye irritant. When heated to decomposition it emits toxic fumes of PO_x. See also PHOSPHINE.

TNE775 CAS:7673-09-8 ***HR: 3***
2,4,6-TRIS(CHLOROAMINO)-1,3,5-TRIAZINE
mf: $C_3H_3Cl_3N_6$ mw: 229.46

SYN: TRICHLOROMELAMINE

SAFETY PROFILE: Ignites on contact with acetone or bases (e.g., ammonia; aniline; diphenylamine). When heated to decomposition it emits toxic fumes of Cl^- and NO_x.

TNF000 CAS:10138-79-1 ***HR: 3***
TRIS(2-CHLOROETHOXY)SILANE
mf: $C_6H_{13}Cl_3O_3Si$ mw: 267.63

TOXICITY DATA with REFERENCE
skn-rbt 10 mg/24H MLD AMIHBC 10,61,54
eye-rbt 20 mg SEV AMIHBC 10,61,54
orl-rat LD50:190 mg/kg AMIHBC 10,61,54
skn-rbt LD50:89 mg/kg AMIHBC 10,61,54

CONSENSUS REPORTS: Reported in EPA TSCA Inventory.

SAFETY PROFILE: Poison by ingestion and skin contact. A skin and severe eye irritant. When heated to decomposition it emits toxic fumes of Cl^-. See also CHLOROSILANES.

TNF250 CAS:555-77-1 ***HR: 3***
TRIS(2-CHLOROETHYL)AMINE
mf: $C_6H_{12}Cl_3N$ mw: 204.54

SYNS: TL 145 ◇ TRICHLORMETHINE ◇ TRI-(2-CHLOROETHYL) AMINE ◇ 2,2′,2″-TRICHLOROTRIETHYLAMINE ◇ TRIS(β-CHLOROETHYL)AMINE ◇ TS 160

TOXICITY DATA with REFERENCE
eye-rbt 400 μg AJOPAA 29,1553,46
sln-dmg-ihl 100 pph/5M PREBA3 62B,284,46/47
dni-ham:lng 500 μg/L MUREAV 116,431,83
scu-rat TDLo:18 mg/kg/26W-C:CAR NEOLA4 28,565,81
scu-rat TD:45 mg/kg/26W-C:CAR NEOLA4 28,565,81
ihl-hmn LC50:1000 mg/m^3 SCJUAD 4,33,67
orl-rat LD50:5 mg/kg NTIS** PB158-507
ihl-rat LC50:200 mg/m^3/10M NTIS** PB158-508
skn-rat LD50:2 mg/kg NTIS** PB158-507
scu-rat LD50:2 mg/kg NTIS** PB158-507
ihl-mus LC50:120 mg/m^3/10M NTIS** PB158-508
skn-mus LD50:7 mg/kg JPETAB 91,224,47
scu-mus LD50:6900 μg/kg NTIS** PB158-507
ihl-dog LC50:100 mg/m^3/10M NTIS** PB158-508
skn-dog LD50:10 mg/kg NTIS** PB158-507
ivn-dog LDLo:1 mg/kg NTIS** PB158-507
skn-rbt LD50:5 mg/kg NTIS** PB158-507

CONSENSUS REPORTS: EPA Extremely Hazardous Substances List. EPA Genetic Toxicology Program. Reported in EPA TSCA Inventory.

SAFETY PROFILE: Experimental poison by ingestion, inhalation, skin contact, subcutaneous, and intravenous routes. Moderately toxic to humans by inhalation. Questionable carcinogen with experimental carcinogenic data. An eye irritant. Human mutation data reported. When heated to decomposition it emits very toxic fumes of Cl^- and NO_x. See also AMINES.

TNF500 CAS:817-09-4 ***HR: 3***
TRIS(2-CHLOROETHYL)AMMONIUM CHLORIDE
mf: $C_6H_{12}Cl_3N•ClH$ mw: 241.00

SYNS: LEKAMIN ◇ NSC-30211 ◇ R-47 ◇ SINALOST ◇ SK-100 ◇ TRICHLORMETHINE ◇ TRICHLORMETHINIUM CHLORIDE ◇ TRI(β-CHLOROETHYL)AMINE HYDROCHLORIDE ◇ TRI-(2-CHLOROETHYL)AMINE HYDROCHLORIDE ◇ 2,2′,2″-

TRICHLOROTRIETHYLAMINE HYDROCHLORIDE ◇ TRICHLORTRIAETHYLAMIN-HYDROCHLORID (GERMAN) ◇ TRILLEKAMIN ◇ TRIMITAN ◇ TRIMUSTINE ◇ TRIMUSTINE HYDROCHLORIDE ◇ TRIS(β-CHLOROETHYL)AMINE HYDROCHLORIDE ◇ TRIS(2-CHLOROETHYL)AMINE HYDROCHLORIDE ◇ TRIS(2-CHLOROETHYL)AMINE MONOHYDROCHLORIDE ◇ TRIS-N-LOST ◇ TS-160

TOXICITY DATA with REFERENCE
pic-esc 200 mg/L ARMKA7 51,9,65
dnd-mus-ipr 20 mg/kg FOBLAN 25,380,79
dlt-mus-ipr 5 mg/kg NEOLA4 25,523,78
ivn-rat TDLo:2400 μg/kg (1D male):REP JEZOAO 121,225,52
scu-mus TDLo:10 mg/kg/9W-I:CAR BJCAAI 3,118,49
orl-hmn TDLo:214 μg/kg:BLD NTIS** PB158-507
orl-hmn TDLo:30 μg/kg:CNS,GIT NTIS** PB158-507
ivn-hmn TDLo:100 μg/kg:CNS,CVS,GIT NTIS** PB158-507
orl-rat LDLo:5 mg/kg NCNSA6 5,19,53
ipr-rat LD50:750 μg/kg CPBTAL 8,807,60
scu-rat LD50:2 mg/kg NTIS** PB158-507
ivn-rat LD50:700 μg/kg NTIS** PB158-507
ipr-mus LD50:1600 μg/kg AEPPAE 230,559,57
scu-mus LD50:2 mg/kg JPETAB 91,224,47
ivn-rbt LD50:2500 μg/kg JPETAB 91,224,47

CONSENSUS REPORTS: IARC Cancer Review: Group 3 IMEMDT 7,56,87; Animal Inadequate Evidence IMEMDT 9,229,75. EPA Genetic Toxicology Program.

SAFETY PROFILE: Poison by ingestion, subcutaneous, intravenous, and intraperitoneal routes. Human systemic effects by ingestion and intravenous routes: somnolence, anorexia, headache, thrombosis distant from injection site, nausea or vomiting, and leukopenia. Experimental reproductive effects. Mutation data reported. Questionable carcinogen with experimental carcinogenic data. When heated to decomposition it emits very toxic fumes of Cl^-, NH_3, and NO_x. Used as an antineoplastic agent.

TNG000 CAS:13674-84-5 ***HR: 3***
TRIS(2-CHLOROISOPROPYL)PHOSPHATE
mf: $C_9H_{18}Cl_3O_4P$ mw: 327.59

SYN: PHOSPHORIC ACID, TRIS(2-CHLORO-1-METHYLETHYL) ESTER

TOXICITY DATA with REFERENCE
ivn-mus LD50:56 mg/kg CSLNX* NX#05768

CONSENSUS REPORTS: Reported in EPA TSCA Inventory.

SAFETY PROFILE: Poison by intravenous route. When heated to decomposition it emits very toxic fumes of Cl^- and PO_x.

TNG050 CAS:427-45-2 ***HR: 3***
TRIS(p-CHLOROPHENYL)TIN FLUORIDE
mf: $C_{18}H_{12}Cl_3FSn$ mw: 472.34

SYN: STANNANE, FLUOROTRIS(p-CHLOROPHENYL)-

TOXICITY DATA with REFERENCE
ivn-mus LD50:100 mg/kg CSLNX* NX#05959

OSHA PEL: TWA 0.1 mg(Sn)/m^3 (skin)
ACGIH TLV: TWA 0.1 mg(Sn)/m^3; STEL 0.2 mg/m^3 (skin)
NIOSH REL: (Organotin Compound):10H TWA 0.1 mg(Sn)/m^3

SAFETY PROFILE: Poison by intravenous route. When heated to decomposition it emits toxic fumes of Sn, Cl^-, and F^-.

TNG100 CAS:15246-55-6 ***HR: 3***
TRIS(1,2-DIAMINOETHANE)CHROMIUM(III) PERCHLORATE
mf: $C_6H_{24}Cl_3CrN_6O_{12}$ mw: 530.64

$[(C_2H_8N_2)_3Cr][ClO_4]_3$

CONSENSUS REPORTS: Chromium and its compounds are on The Community Right-To-Know List.

SAFETY PROFILE: An extremely sensitive high explosive. When heated to decomposition it emits toxic fumes of Cl^- and NO_x. See also CHROMIUM COMPOUNDS and PERCHLORATES.

TNG150 CAS:6865-68-5 ***HR: 3***
TRIS(1,2-DIAMINOETHANE) COBALT(III) NITRATE
mf: $C_6H_{24}CoN_9O_9$ mw: 425.24

$[(C_2H_8N_2)_3Co][NO_3]_3$

CONSENSUS REPORTS: Cobalt and its compounds are on The Community Right-To-Know List.

SAFETY PROFILE: An impact-sensitive explosive. When heated to decomposition it emits toxic fumes of NO_x. See also COBALT COMPOUNDS and NITRATES.

TNG250 CAS:34333-07-8 ***HR: 3***
TRIS(DIBUTYLBIS(HYDROXYETHYLTHIO)TIN) BIS(BORIC ACID ESTER)
mf: $C_{36}H_{78}B_2O_6S_6Sn_3$ mw: 1177.19

SYN: THIOBORIC ACID, ESTER with 2,2'-((DIBUTYLSTANNYLENE)DIOXY)DIETHANETHIOL (2:3)

TOXICITY DATA with REFERENCE
ivn-mus LD50:18 mg/kg CSLNX* NX#02225

OSHA PEL: TWA 0.1 mg(Sn)/m^3 (skin)
ACGIH TLV: TWA 0.1 mg(Sn)/m^3 (skin) (Proposed: TWA 0.1 mg(Sn)/m^3; STEL 0.2 mg(Sn)/m^3 (skin))
NIOSH REL: (Organotin Compounds) TWA 0.1 mg(Sn)/m^3

SAFETY PROFILE: Poison by intravenous route. When heated to decomposition it emits toxic fumes of SO_x. See also TIN COMPOUNDS, BORON COMPOUNDS, and ESTERS.

TNG275 CAS:2428-04-8 ***HR: 2***
2,4,6-TRIS(DICHLOROAMINO)-1,3,5-TRIAZINE
mf: $C_3Cl_6N_6$ mw: 332.79

$N{=}C(NCl_2)N{=}C(NCl_2)N{=}C(NCl_2)$ (ring)

SYN: HEXACHLOROMELAMINE

SAFETY PROFILE: Probably a severe eye, skin and mucous membrane irritant. A powerful oxidizer. Energetic reactions (possibly leading to ignition) with acetone; ammonia; aniline; and diphenylamine could become explosive if large amounts were confined. When heated to decomposition it emits toxic fumes of Cl^- and NO_x.

TNG750 CAS:78-43-3 ***HR: 2***
TRIS-DICHLOROPROPYLPHOSPHATE
mf: $C_9H_{15}Cl_6O_4P$ mw: 430.91

SYNS: CELLUFLEX FR-2 ◇ 2,3-DICHLOROPROPANOL PHOSPHATE (3:1)

TOXICITY DATA with REFERENCE
mma-sat 100 μg/plate JEPTDQ 3,207,79
cyt-mus:lym 50 μL/L JEPTDQ 3,207,79
sce-mus:lym 10 μL/L JEPTDQ 3,207,79
cyt-ham:lng 250 mg/L/27H MUREAV 66,277,79
orl-rat LD50:2830 mg/kg NPIRI* 2,118,75

CONSENSUS REPORTS: Reported in EPA TSCA Inventory.

SAFETY PROFILE: Moderately toxic by ingestion. Mutation data reported. When heated to decomposition it emits very toxic fumes of Cl^- and PO_x. See also PHOSPHATES and CHLORIDES.

TNG775 CAS:14362-68-6 ***HR: 3***
TRIS(DIFLUOROAMINE)FLUOROMETHANE
mf: CF_7N_3 mw: 187.02

PROP: Bp: 5.6°C

SAFETY PROFILE: A shock-sensitive explosive, especially as the liquid. Mixtures with pentaborane(9) are powerful explosives and can explode on contact with air. When heated to decomposition it emits toxic fumes of F^- and NO_x. See also FLUORIDES and AMINES.

TNG780 CAS:7289-92-1 ***HR: 3***
TRIS(DIMETHYLAMINO)ANTIMONY
mf: $C_6H_{18}N_3Sb$ mw: 253.98

$[(CH_3)_2N]_3Sb$

CONSENSUS REPORTS: Antimony and its compounds are on The Community Right-To-Know List.

SAFETY PROFILE: Most antimony compounds are poisons. Explosive reaction with ethyl diazoacetate at room temperature. When heated to decomposition it emits toxic fumes of Sb and NO_x. See also ANTIMONY COMPOUNDS.

TNH000 CAS:90-72-2 ***HR: 2***
2,4,6-TRIS(DIMETHYLAMINOMETHYL)PHENOL
mf: $C_{15}H_{27}N_3O$ mw: 265.45

SYNS: 2,4,6-TRIS-N,N-DIMETHYLAMINOMETHYLFENOL(CZECH) ◇ 2,4,6-TRI(DIMETHYLAMINOMETHYL)PHENOL

TOXICITY DATA with REFERENCE
skn-rbt 500 mg/24H SEV 28ZPAK -,112,72
eye-rbt 50 μg/24H SEV 28ZPAK -,112,72
orl-rat LD50:1200 mg/kg RPTOAN 37,130,74
skn-rat LD50:1280 mg/kg ROHM**

CONSENSUS REPORTS: Reported in EPA TSCA Inventory.

SAFETY PROFILE: Moderately toxic by ingestion and skin contact. A severe skin and eye irritant. When heated to decomposition it emits toxic fumes of NO_x.

TNH250 CAS:15875-13-5 ***HR: 2***
N,N',N''-TRIS(DIMETHYLAMINOPROPYL)-s-HEXAHYDROTRIAZINE
mf: $C_{18}H_{42}N_6$ mw: 342.66

SYN: HEXAHYDRO-1,3,5-TRIS(DIMETHYLAMINOPROPYL)-s-TRIAZINE

TOXICITY DATA with REFERENCE
orl-rat LD50:3250 mg/kg TXAPA9 28,313,74
skn-rbt LD50:2020 mg/kg TXAPA9 28,313,74

CONSENSUS REPORTS: Reported in EPA TSCA Inventory.

SAFETY PROFILE: Moderately toxic by ingestion and skin contact. When heated to decomposition it emits toxic fumes of NO_x.

TNH500 CAS:20248-45-7 ***HR: 2***
1,2-(TRISDIMETHYLAMINOSILYL)ETHANE
mf: $C_{14}H_{40}N_6Si_2$ mw: 348.78

TOXICITY DATA with REFERENCE
orl-rat LD50:1230 mg/kg TXAPA9 28,313,74
skn-rbt LD50:790 mg/kg TXAPA9 28,313,74

SAFETY PROFILE: Moderately toxic by ingestion and skin contact. When heated to decomposition it emits toxic fumes of NO_x. See also SILANE.

TNH750 CAS:7276-58-6 ***HR: 3***
TRIS(1-DODECYL-3-METHYL-2-PHENYLBENZI-MIDAZOLIUM)FERRICYANIDE
mf: $C_{84}H_{111}FeN_{12}$ mw: 1344.92

SYNS: BAY 32394 ◇ BAYER 32394 ◇ 1H-BENZIMIDAZOLIUM HEXAKIS-1-DODECYL-3-METHYL-2-PHENYL-(CYANO-C)FERRATE(1−) ◇ B 169-FERRICYANIDE ◇ 3-DODECYL-1-METHYL-2-PHENYLBENZIMIDAZOLIUM FERRICYANIDE ◇ 1-DODECYL-3-METHYL-2-PHENYL-1H-BENZIMIDAZOLIUM,HEXACYANOFERRATE(III) ◇ ENT 25,678 ◇ FERRICYANURE de TRI(1-DODECYL-2-PHENYL-3-METHYL-1,3-BENZIMIDAZOLIUM (FRENCH) ◇ FUNGILON ◇ HEXACYANOTRIS(3-DODECYL-1-METHYL-2-PHENYLBENZIMIIMIDAZOLINIUM FERRATE (3−) ◇ 1-METHYL-2-PHENYL-3-DODECYLBENZIMIDAZOLINIUM FERROCYANIDE ◇ 1-METHYL-2-PHENYL-3-m-DODECYLBENZIMIDAZOLIUM HEXACYANOFERRATE ◇ TRI-(3-DODECYL-1-METHYL-2-PHENYLBENZIMIDAZOLIUM) FERRICYANIDE ◇ TRIS(1-DODECIL-3-METIL-2-FENIL-1,3-BENZIMIDAZOLIO)-FERRICIANURO(ITALIAN) ◇ TRIS(1-DODECYL-3-METHYL-2-FENYL-1,3-BENZIMIDAZOLIUM)-HEXACYANOFERRAAT(III) (DUTCH) ◇ TRIS(1-DODECYL-3-METHYL-2-PHENYL-1,3-BENZIMIDAZOLIUM)-HEXACYANOFERRAT(III)(GERMAN)

TOXICITY DATA with REFERENCE
orl-rat LD50:500 mg/kg 28ZEAL 4,384,69
ipr-rat LD50:10 mg/kg 30ZDA9 -,417,71

CONSENSUS REPORTS: Cyanide and its compounds are on the Community Right-To-Know List.

SAFETY PROFILE: Poison by intraperitoneal route. Moderately toxic by ingestion. When heated to decomposition it emits very toxic fumes of NO_x and CN^-. See also CYANIDE.

TNH850 CAS:6939-83-9 ***HR: 3***
TRIS(DODECYLTHIO)ANTIMONY
mf: $C_{36}H_{75}S_3Sb$ mw: 726.04

SYN: STIBINE, TRIS(DODECYLTHIO)-

TOXICITY DATA with REFERENCE
ivn-mus LD50:180 mg/kg CSLNX* NX#02626

ACGIH TLV: TWA 0.5 mg(Sb)/m^3
OSHA PEL: TWA 0.5 mg(Sb)/m^3
NIOSH REL: (Antimony):10H TWA 0.5 mg(Sb)/m^3

SAFETY PROFILE: Poison by intravenous route. When heated to decomposition it emits toxic fumes of SO_x and Sb.

TNI000 CAS:18924-91-9 ***HR: 3***
2,4,6-TRIS-(1-(2-ETHYLAZIRIDINYL)-1,3,5-TRIAZINE
mf: $C_{15}H_{24}N_6$ mw: 288.45

SYN: NRL-18-B

TOXICITY DATA with REFERENCE
ipr-mus LD50:24 mg/kg NTIS** AD441-640

CONSENSUS REPORTS: Reported in EPA TSCA Inventory.

SAFETY PROFILE: Poison by intraperitoneal route. When heated to decomposition it emits toxic fumes of NO_x.

TNI250 CAS:78-42-2 ***HR: 3***
TRIS(2-ETHYLHEXYL)PHOSPHATE
mf: $C_{24}H_{51}O_4P$ mw: 434.72

PROP: Liquid. Mp: −74°, bp: 216° @ 5 mm, flash p: 405°F (OC), d: 0.9262 @ 20/20°, vap d: 14.95.

SYNS: DISFLAMOLL TOF ◇ 2-ETHYL-1-HEXANOL PHOSPHATE ◇ FLEXOL TOF ◇ KRONITEX TOF ◇ NCI-C54751 ◇ PHOSPHORIC ACID, TRIS(2-ETHYLHEXYL) ESTER ◇ TOF ◇ TRIETHYLHEXYL PHOSPHATE ◇ TRI(2-ETHYLHEXYL)PHOSPHATE ◇ TRIOCTYL PHOSPHATE

TOXICITY DATA with REFERENCE
skn-rbt 250 mg MOD 34ZIAG -,606,69
orl-mus TDLo:520 g/kg/2Y-I:CAR EVHPAZ 65,271,86
orl-rat LD50:37 g/kg JIDHAN 30,63,48
skn-rbt LD50:20 g/kg 85JCAE-,1133,86
ihl-gpg LC50:450 mg/m^3/30M 85JCAE-,1133,86

CONSENSUS REPORTS: NTP Carcinogenesis Studies (gavage); Some Evidence: mouse NTPTR* NTP-TR-274,84; Equivocal Evidence: rat NTPTR* NTP-TR-274,84. Reported in EPA TSCA Inventory.

SAFETY PROFILE: Questionable carcinogen with experimental carcinogenic data. Mildly toxic by ingestion and skin contact. A skin and eye irritant. Combustible when exposed to heat or flame. Can react with oxidizing materials. To fight fire, use foam, CO_2, dry chemical. When heated to decomposition it emits toxic fumes of PO_x. See also PHOSPHATES.

TNI500 CAS:23319-66-6 ***HR: 3***
TRIS(2-HYDROXYETHYL)PHENYLMERCURI-AMMONIUM LACTATE
mf: $C_{12}H_{20}HgNO_3 \cdot C_3H_5O_3$ mw: 516.00

PROP: White, crystalline solid. Sol in water.

SYNS: LACTIC ACID, TRIS(2-HYDROXYETHYL)(PHENYLMERCURI)AMMONIUM derivative ◇ LACTIC ACID, ion(1−), TRIS(2-HYDROXYETHYL)PHENYLMERCURIO)AMMONIUM ◇ (2,2′,2″-NITRILOTRIETHANOL)PHENYLMERCURY(1+) LACTATE (salt) ◇ PHENYLMERCURIC TRIETHANOLAMMONIUM LACTATE

◇ PHENYLMERCURITRIETHANOLAMMONIUM LACTATE ◇ PHENYLMERCURY TRIETHANOLAMINE LACTATE ◇ PTAB ◇ PURATIZED ◇ PURATIZED AGRICULTURAL SPRAY ◇ PURATIZEDAT AGRICULTURAL SPRAY ◇ PURATIZED N5E ◇ PURATURF

TOXICITY DATA with REFERENCE
orl-rat LD50:30 mg/kg GUCHAZ 5,378,68

CONSENSUS REPORTS: Mercury and its compounds are on the Community Right-To-Know List.

OSHA PEL: (Transitional: CL 1 mg/10m^3) CL 0.1 mg (Hg)/m^3 (skin)
ACGIH TLV: TWA 0.1 mg(Hg)/m^3 (skin)
NIOSH REL: (Inorganic Mercury) TWA 0.05 mg(Hg)/m^3

SAFETY PROFILE: Poison by ingestion. When heated to decomposition it emits very toxic fumes of Hg, NH_3, and NO_x. See also MERCURY COMPOUNDS.

TNI750 CAS:824-11-3 ***HR: 3***
1,1,1-TRISHYDROXYMETHYLPROPANE BICYCLIC PHOSPHITE
mf: $C_6H_{11}O_3P$ mw: 162.14

SYNS: 4-AETHYL-1-PHOSPHA-2,6,7-TRIOXABICYCLO(2.2.2)OCTAN (GERMAN) ◇ 4-ETHYL-1-PHOSPHA-2,6,7-TRIOXABICYCLO(2.2.2)OCTANE ◇ 4-ETHYL-2,6,7-TRIOXA-1-PHOSPHABICYCLO(2.2.2)OCTANE

TOXICITY DATA with REFERENCE
orl-rat LD50:8390 μg/kg ARTODN 35,149,76
ihl-rat LCLo:10 ppm/4H BJIMAG 27,1,70
skn-rat LD50:929 mg/kg ARTODN 35,149,76
ipr-rat LD50:1020 μg/kg ARTODN 35,149,76
orl-mus LD50:7 mg/kg SCIEAS 182,1135,73
skn-mus LD50:4 mg/kg SCIEAS 182,1135,73
ipr-mus LD50:1100 μg/kg CENEAR 52(1),56,74
ivn-mus LDLo:1 mg/kg EJMCA5 13,207,78
orl-dog LD50:5 mg/kg ARTODN 35,149,76

CONSENSUS REPORTS: EPA Extremely Hazardous Substances List. Reported in EPA TSCA Inventory.

SAFETY PROFILE: Poison by ingestion, inhalation, skin contact, intraperitoneal, and intravenous routes. When heated to decomposition it emits toxic fumes of PO_x.

TNJ250 CAS:13862-16-3 ***HR: 3***
TRISILYLAMINE
mf: H_9NSi_3 mw: 107.34

PROP: Mp: −105.6°, bp: 52°, d: 0.895 @ −106°.

SYN: NITRILOTRISILANE

SAFETY PROFILE: Liquid ignites spontaneously in air. Vigorous reaction with ammonia, hydrogen, water or steam produces toxic and flammable vapors. When heated to decomposition it emits toxic fumes of NO_x. See also AMINES and SILANE.

TNJ275 ***HR: 3***
TRISILYLARSINE
mf: AsH_9Si_3 mw: 168.25

CONSENSUS REPORTS: Arsenic and its compounds are on The Community Right-To-Know List.

SAFETY PROFILE: Arsenic compounds are poisons. Ignites spontaneously in air. When heated to decomposition it emits toxic fumes of As. See also ARSINE, ARSENIC COMPOUNDS, and SILANE.

TNJ500 CAS:1067-53-4 ***HR: 2***
TRIS(METHOXYETHOXY)VINYLSILANE
mf: $C_{11}H_{24}O_6Si$ mw: 280.44

SYNS: 6-ETHENYL-6-(METHOXYETHOXY)-2,5,7,10-TETRAOXA-6-SILAUNDECANE ◇ (TRIS(β-METHOXYETHOXY))VINYLSILANE ◇ TRIS(2-METHOXYETHOXY)VINYLSILANE ◇ VINYLTRIS (METHOXYETHOXY)SILANE ◇ VINYLTRIS(β-METHOXYETHOXY) SILANE ◇ VINYLTRIS(2-METHOXYETHOXY)SILANE

TOXICITY DATA with REFERENCE
skn-rbt 500 mg open MLD UCDS** 1/30/64
orl-rat LD50:2960 mg/kg UCDS** 1/30/64
skn-rbt LD50:1500 mg/kg UCDS** 1/30/64

CONSENSUS REPORTS: Reported in EPA TSCA Inventory.

SAFETY PROFILE: Moderately toxic by ingestion and skin contact. A skin irritant. When heated to decomposition it emits acrid smoke and irritating fumes. See also SILANE.

TNJ750 CAS:568-69-4 ***HR: 3***
2,3,3-TRIS(p-METHOXYPHENYL)-N,N-DIMETHYALLYLAMINE HYDROCHLORIDE
mf: $C_{26}H_{29}NO_3$•ClH mw: 440.02

SYNS: AMINOOXYTRIPHENE HYDROCHLORIDE ◇ AMOTRIPHENE HYDROCHLORIDE ◇ 3-DIMETHYLAMINO-1,1,2-TRIS(4-METHOXYPHENYL)-1-PROPENE HYDROCHLORIDE ◇ 4-METHOXY-α-(BIS(4-METHOXYPHENYL)METHYLENE)-N,N-DIMETHYLBENZENEETHANAMINE HCl ◇ MYORDIL

TOXICITY DATA with REFERENCE
orl-mus LD50:385 mg/kg MEIEDD 10,83,83
ivn-mus LD50:30 mg/kg MEIEDD 10,83,83

SAFETY PROFILE: Poison by ingestion and intravenous routes. When heated to decomposition it emits very toxic fumes of NO_x and HCl. Used as a coronary vasodilator.

TNJ825 CAS:2827-46-5 ***HR: 3***
2,4,6-TRIS(METHYLAMINO)-s-TRIAZINE
mf: $C_6H_{12}N_6$ mw: 168.24

SYNS: N^2,N^4,N^6-TRIMETHYLMELAMINE ◇ 2,4,6-TRIS-METHYLAMINO-1,3,5-TRIAZINE

TOXICITY DATA with REFERENCE
cyt-rat-ipr 200 mg/kg BJPCAL 6,357,51
ipr-mus LD50:56 mg/kg BCPCA6 26,2385,77
ivn-mus LD50:320 mg/kg CSLNX* NX#03985

SAFETY PROFILE: Poison by intravenous and intraperitoneal routes. Mutation data reported. When heated to decomposition it emits toxic fumes of NO_x.

TNK000 CAS:13009-91-1 ***HR: 3***
2,4,6-TRIS((1-(2-METHYLAZIRIDINYL))-1,3,5-TRIAZINE)
mf: $C_{12}H_{18}N_6$ mw: 246.36

SYNS: METHYLTRETAMINE ◇ TMAT ◇ 2,4,6-TRIS(2-METHYL-1-AZIRIDINYL)-s-TRIAZINE

TOXICITY DATA with REFERENCE
dlt-oin-par 150 μmol/L MUREAV 4,225,67
ipr-mus LD50:10 mg/kg NTIS** AD441-640

CONSENSUS REPORTS: Reported in EPA TSCA Inventory.

SAFETY PROFILE: Poison by intraperitoneal route. Mutation data reported. When heated to decomposition it emits toxic fumes of NO_x.

TNK250 CAS:57-39-6 ***HR: 3***
TRIS(1-METHYLETHYLENE)PHOSPHORIC TRIAMIDE
mf: $C_9H_{18}N_3OP$ mw: 215.27

PROP: Amber-colored liquid; amine odor. Bp: 118-125° @ 1 mm, d: 1.079 @ 25/25°. Misc with water and all organic solvents.

SYNS: C 3172 ◇ ENT 50,003 ◇ MAPO ◇ METEPA ◇ METHAPHOXIDE ◇ METHYL APHOXIDE ◇ 1,1',1''-PHOSPHINYLIDYNETRIS(2-METHYL)AZRIDINE ◇ TRIS(2-METHYL-1-AZIRIDINYL)PHOSPHINE OXIDE ◇ TRIS(2-METHYLAZIRIDIN-1-YL)PHOSPHINE OXIDE ◇ N,N',N''-TRIS(1-METHYLETHYLENE)PHOSPHORAMIDE

TOXICITY DATA with REFERENCE
dni-esc 10 mmol/L IJEBA6 22,453,84
cyt-hmn:lym 20 mg/L SOGEBZ 8,783,72
ipr-mus TDLo:25 mg/kg (female 5D pre):TER MUREAV 70,109,80
ipr-rat TDLo:30 mg/kg (female 12D post):REP BWHOA6 34,317,66
orl-rat TDLo:1000 mg/kg/60W-I:CAR BWHOA6 34,317,66
orl-rat LD50:136 mg/kg TXAPA9 14,515,69
skn-rat LD50:183 mg/kg BWHOA6 31,737,64
orl-mus LD50:292 mg/kg INHEAO 8,54,70
skn-mus LD50:375 mg/kg INHEAO 8,54,70
ipr-mus LDLo:3125 μg/kg TXAPA9 23,288,72
scu-mus LD50:140 mg/kg INHEAO 8,54,70
orl-ckn LD50:329 mg/kg TXAPA9 16,100,70
orl-dom LDLo:50 mg/kg ANYAA9 111,715,64

CONSENSUS REPORTS: IARC Cancer Review: Group 3 IMEMDT 7,56,87; Animal Inadequate Evidence IMEMDT 9,107,75. Reported in EPA TSCA Inventory. EPA Genetic Toxicology Program.

SAFETY PROFILE: Poison by ingestion, skin contact, intraperitoneal, and subcutaneous routes. Experimental teratogenic and reproductive effects. Experimental reproductive effects. Questionable carcinogen with experimental carcinogenic data. Animal experiments suggest cholinesterase inhibition, possibly due to metabolic products of this material in the body. When heated to decomposition it emits very toxic fumes of NO_x and PO_x.

TNK400 CAS:38668-83-6 ***HR: 3***
TRIS(OCTAMETHYLPYROPHOSPHORAMIDE)MANGANESE(2+), DIPERCHLORATE
mf: $C_{24}H_{72}MnN_{12}O_9P_6 \cdot 2ClO_4$ mw: 1112.74

SYNS: MANGANESE(2+),TRIS(OCTAMETHYLDIPHOSPHORAMIDE-Op,Op')-, (OC-6-11)-,DIPERCHLORATE ◇ PERCHLORIC ACID, MANGANESE(2+) SALT, compounded with 3 mols. of OCTAMETHYLPYROPHOSPHORAMIDE

TOXICITY DATA with REFERENCE
ipr-mus LD50:14 mg/kg JAFCAU 14,512,66

OSHA PEL: CL 5 mg(Mn)/m^3
ACGIH TLV: TWA 5 mg(Mn)/m^3

SAFETY PROFILE: Poison by intraperitoneal route. When heated to decomposition it emits toxic fumes of NO_x, Mn, PO_x, and ClO^-.

TNL000 CAS:68-04-2 ***HR: 3***
TRISODIUM CITRATE
mf: $C_6H_5O_7 \cdot 3Na$ mw: 258.08

SYNS: CITROSODINE ◇ SODIUM CITRATE, anhydrous

TOXICITY DATA with REFERENCE
ipr-rat LD50:1548 mg/kg JPETAB 94,65,48
ipr-mus LD50:1364 mg/kg JPETAB 94,65,48
ivn-mus LD50:170 mg/kg JPETAB 94,65,48
ivn-rbt LD50:449 mg/kg JPETAB 94,65,48

CONSENSUS REPORTS: Reported in EPA TSCA Inventory.

SAFETY PROFILE: Poison by intravenous route. Moderately toxic by intraperitoneal route. When heated to decomposition it emits toxic fumes of Na_2O. See also MONOSODIUM CITRATE.

TNL250 CAS:150-38-9 ***HR: 3***
TRISODIUM EDETATE
mf: $C_{10}H_{13}N_2O_8 \cdot 3Na$ mw: 358.22

SYNS: EDETATE TRISODIUM ◇ EDTA TRISODIUM SALT ◇ N,N'-1,2-ETHANEDIYLBIS(N-CARBOXYMETHYL)GLYCINE,TRISODIUM SALT ◇ ETHYLENEDIAMINEACETIC ACID TRISODIUM SALT ◇ ETHYLENEDIAMINETETRAACETICACID, TRISODIUM SALT ◇ NCI-C03974 ◇ NEVANAID-B POWDER ◇ PERMA KLEER 50, TRISODIUM SALT ◇ SEQUESTRENE Na3 ◇ SEQUESTRENE TRISODIUM ◇ SEQUESTRENE TRISODIUM SALT ◇ TRILON AO ◇ TRISODIUM EDTA ◇ TRISODIUM ETHYLENEDIAMINETETRAACETATE ◇ TRISODIUM HYDROGEN ETHYLENEDIAMINETETRAACETATE ◇ TRISODIUM HYDROGEN (ETHYLENEDINITRILO)TETRAACETATE ◇ TRISODIUM VERSENATE ◇ VERSENE 9

TOXICITY DATA with REFERENCE
orl-rat LD50:2150 mg/kg NCILB* NIH-NCI-E-C-72-3252
orl-mus LD50:2150 mg/kg NCILB* NIH-NCI-E-C-72-3252
ipr-mus LD50:300 mg/kg REPMBN 10,391,62

CONSENSUS REPORTS: Reported in EPA TSCA Inventory.

SAFETY PROFILE: Poison by intraperitoneal route. Moderately toxic by ingestion. When heated to decomposition it emits toxic fumes of NO_x and Na_2O.

TNL500 ***HR: D***
TRISODIUM ETHYLENEDIAMINETETRAACETATE TRIHYDRATE
mf: $C_{10}H_{13}N_2O_8 \cdot 3Na \cdot 3H_2O$ mw: 412.28

SYNS: EDTA TRISODIUM SALT (TRIHYDRATE) ◇ (ETHYLENEDINITRILO)-TETRAACETIC ACID ◇ NCI-C03974

CONSENSUS REPORTS: NCI Carcinogenesis Bioassay (feed); No Evidence: mouse, rat NCITR* NCI- CG-TR-11,77.

SAFETY PROFILE: When heated to decomposition it emits very toxic fumes of Na_2O and NO_x. See also TRISODIUM EDETATE.

TNL750 CAS:2666-14-0 ***HR: 2***
TRISODIUM ETIDRONATE
mf: $C_2H_5O_7P_2 \cdot 3Na$ mw: 271.98

SYNS: ETHANE-1-HYDROXY-1,1-DIPHOSPHONIC ACID, TRISODIUM SALT ◇ (1-HYDROXYETHYLIDENE)DIPHOSPHONIC ACID, TRISODIUM SALT

TOXICITY DATA with REFERENCE
orl-rat LD50:1280 mg/kg TXAPA9 22,661,72

CONSENSUS REPORTS: Reported in EPA TSCA Inventory.

SAFETY PROFILE: Moderately toxic by ingestion. When heated to decomposition it emits toxic fumes of PO_x and Na_2O.

TNM000 CAS:6358-69-6 ***HR: 2***
TRISODIUM-1-HYDROXY-3,6,8-PYRENETRISULFONATE
mf: $C_{16}H_7O_{10}S_3 \cdot 3Na$ mw: 524.38

SYNS: C.I. 59040 ◇ C.I. SOLVENT GREEN 7 ◇ D&C GREEN No. 8 ◇ 8-HYDROXYPYRENE-1,3,6-TRISULFONIC ACID SODIUM SALT ◇ 8-HYDROXY-1,3,6-PYRENETRISULFONIC ACID TRISODIUM SALT

TOXICITY DATA with REFERENCE
ivn-mus LD50:1050 mg/kg TXAPA9 44,225,78

CONSENSUS REPORTS: Reported in EPA TSCA Inventory.

SAFETY PROFILE: Moderately toxic by intravenous route. When heated to decomposition it emits toxic fumes of SO_x and Na_2O.

TNM750 CAS:10101-88-9 ***HR: 3***
TRISODIUM THIOPHOSPHATE
mf: $O_3PS \cdot 3Na$ mw: 180.00

SYNS: SODIUM PHOSPHOROTHIOATE ◇ SODIUM THIOPHOSPHATE ◇ THIOPHOSPHORIC ACID, TRISODIUM SALT ◇ TRISODIUM MONOTHIOPHOSPHATE ◇ TRISODIUM PHOSPHOROTHIOATE

TOXICITY DATA with REFERENCE
ivn-mus LD50:100 mg/kg CSLNX* NX#02869

CONSENSUS REPORTS: Reported in EPA TSCA Inventory.

SAFETY PROFILE: Poison by intravenous route. When heated to decomposition it emits very toxic fumes of PO_x, Na_2O, and SO_x. See also PHOSPHATES.

TNM850 CAS:11082-38-5 ***HR: D***
TRISODIUM ZINC DTPA
mf: $C_{14}H_{18}N_3O_{10}Zn \cdot 3Na$ mw: 522.69

SYNS: ZINC TRISODIUM DIETHYLENETRIAMINEPENTAACETATE ◇ ZINC TRISODIUM DTPA

TOXICITY DATA with REFERENCE
dni-rat:lvr 30 mmol/L BCPCA6 23,901,74
oms-rat:lvr 40 mmol/L BCPCA5 23,901,74
scu-mus TDLo:126 g/kg (female 1-21D post):REP HLTPAO 36,524,79

CONSENSUS REPORTS: Zinc and its compounds are on the Community Right-To-Know List.

SAFETY PROFILE: Experimental reproductive effects. Mutation data reported. When heated to decomposition it emits toxic fumes of NO_x, ZnO, and Na_2O. See also ZINC COMPOUNDS.

TNN000 CAS:13963-57-0 ***HR: 3***
TRIS(2,4-PENTANEDIONATO)ALUMINUM
mf: $C_{15}H_{21}AlO_6$ mw: 324.34

PROP: Amorph white powder. Mp: decomp.

SYNS: ALUMINUM ACETYLACETONATE ◇ ALUMINUM(III) ACETYLACETONATE ◇ ALUMINUM TRIACETYLACETONATE ◇ ALUMINUM TRIS(ACETYLACETONATE) ◇ TRIS (ACETYLACETONATO)ALUMINUM ◇ TRIS(ACETYLACETONE)ALUMINUM ◇ TRIS(ACETYLACETONYL)ALUMINUM ◇ TRIS(2,4-PENTANEDIONE)ALUMINUM

TOXICITY DATA with REFERENCE
ivn-mus LD50:178 mg/kg CSLNX* NX#02376

CONSENSUS REPORTS: Reported in EPA TSCA Inventory.

SAFETY PROFILE: Poison by intravenous route. A weak sensitizer and mild irritant. Local contact may cause contact dermatitis. When heated to decomposition it emits acrid smoke and irritating fumes. See also ALUMINUM COMPOUNDS.

TNN250 CAS:21679-31-2 ***HR: 2***
TRIS(2,4-PENTANEDIONATO)CHROMIUM
mf: $C_{15}H_{21}CrO_6$ mw: 349.36

SYNS: CHROMIC ACETYLACETONATE ◇ CHROMIUM ACETYLACETONATE ◇ CHROMIUM(3+) ACETYLACETONATE ◇ CHROMIUM(III) ACETYLACETONATE ◇ CHROMIUM TRIACETYLACETONATE ◇ CHROMIUM TRIS(ACETYLACETONATE) ◇ CHROMIUM TRIS(2,4-PENTANEDIONATE) ◇ TRIS(ACETYLACETONATO)CHROMIUM ◇ TRIS(ACETYLACETONATO)CHROMIUM(III) ◇ TRIS(2,4-PENTANEDIONATO)CHROMIUM(3+)

TOXICITY DATA with REFERENCE
orl-rat LD50:3360 mg/kg TXAPA9 28,313,74
skn-rbt LD50:6350 mg/kg TXAPA9 28,313,74

CONSENSUS REPORTS: Chromium and its compounds are on the Community Right-To-Know List. Reported in EPA TSCA Inventory.

OSHA PEL: TWA 0.5 mg(Cr)/m^3
ACGIH TLV: TWA 0.5 mg(Cr)/m^3

SAFETY PROFILE: Moderately toxic by ingestion. Mildly toxic by skin contact. When heated to decomposition it emits acrid smoke and irritating fumes. See also CHROMIUM COMPOUNDS.

TNN500 CAS:16432-36-3 ***HR: 3***
TRIS(1-PHENYL-1,3-BUTANEDIONO)CHROMIUM(III)
mf: $C_{30}H_{27}CrO_6$ mw: 535.57

SYNS: CHROMIUM TRIS(BENZOYLACETONATE) ◇ TRIS(BENZOYLACETONATO)CHROMIUM ◇ TRIS(1-PHENYL-1,3-BUTANEDIONATO)CHROMIUM ◇ TRIS(1-PHENYL-1,3-BUTANEDIONATO)CHROMIUM(3+) ◇ TRIS(1-PHENYL-1,3-BUTANEDIONATO-O,O') CHROMIUM

TOXICITY DATA with REFERENCE
ivn-mus LD50:180 mg/kg CSLNX* NX#01651

CONSENSUS REPORTS: Chromium and its compounds are on the Community Right-To-Know List. Reported in EPA TSCA Inventory.

OSHA PEL: TWA 0.5 mg(Cr)/m^3
ACGIH TLV: TWA 0.5 mg(Cr)/m^3

SAFETY PROFILE: Poison by intravenous route. When heated to decomposition it emits acrid smoke and irritating fumes. See also CHROMIUM COMPOUNDS.

TNN750 CAS:1038-95-5 ***HR: 3***
TRIS-(p-TOLYL)PHOSPHINE
mf: $C_{21}H_{21}P$ mw: 304.39

TOXICITY DATA with REFERENCE
ivn-mus LD50:180 mg/kg CSLNX* NX#02043

CONSENSUS REPORTS: Reported in EPA TSCA Inventory.

SAFETY PROFILE: Poison by intravenous route. When heated to decomposition it emits toxic fumes of PO_x. See also PHOSPHINE.

TNN775 CAS:432-04-2 ***HR: 3***
TRIS(TRIFLUOROMETHYL)PHOSPHINE
mf: C_3F_9P mw: 237.99

SAFETY PROFILE: Ignites on contact with oxygen. When heated to decomposition it emits toxic fumes of F^- and PO_x. See also FLUORIDES and PHOSPHINE.

TNO000 CAS:64048-98-2 ***HR: 3***
TRIS(3,3,5-TRIMETHYLCYCLOHEXYL)ARSINE
mf: $C_{27}H_{51}As$ mw: 450.70

TOXICITY DATA with REFERENCE
skn-rbt 500 mg MLD SCCUR* -,9,61
orl-rat LD50:650 mg/kg SCCUR* -,9,61
ipr-rat LD50:250 mg/kg SCCUR* -,9,61
orl-mus LD50:107 mg/kg SCCUR* -,9,61

CONSENSUS REPORTS: Arsenic and its compounds are on the Community Right-To-Know List.

OSHA PEL: TWA 0.5 mg(As)/m^3

SAFETY PROFILE: Poison by ingestion and intraperitoneal routes. A skin irritant. When heated to decomposition it emits toxic fumes of As. See also ARSENIC COMPOUNDS.

TNO250 CAS:12164-01-1 ***HR: 3***
TRITELLURIUM TETRANITRIDE
mf: N_4Te_3 mw: 438.83

PROP: Probably polymeric, with inner-ring bonding.

SYN: TETRATELLURIUM TETRANITRIDE

SAFETY PROFILE: The black form explodes on impact. The yellow form explodes at 200°C. When heated to decomposition it emits very toxic fumes of Te and NO_x. See also TELLURIUM COMPOUNDS and NITRIDES.

TNO275 CAS:8053-39-2 ***HR: 3***
TRITERPENE SAPONINS mixture from AESCULUS HIPPOCASTONUM

PROP: The mixture of triterpene saponins obtained from the seeds of *Aesculus hippocastonum L.* (IYKEDH 13,349,82).

TOXICITY DATA with REFERENCE
orl-rat LD50:862 mg/kg IYKEDH 13,349,82
ipr-rat LD50:10100 μg/kg IYKEDH 13,349,82
scu-rat LD50:150 mg/kg IYKEDH 13,349,82
ivn-rat LD50:2800 μg/kg IYKEDH 13,349,82
orl-mus LD50:164 mg/kg IYKEDH 13,349,82
ipr-mus LD50:6700 μg/kg IYKEDH 13,349,82
scu-mus LD50:38500 μg/kg IYKEDH 13,349,82
ivn-mus LD50:4800 μg/kg IYKEDH 13,349,82

SAFETY PROFILE: Poison by ingestion, subcutaneous, intravenous, and intraperitoneal routes. See also SAPONIN.

TNO300 CAS:23657-27-4 ***HR: 3***
3,4,5-TRITHIATRICYCLO(5.2.1.0^{2,6})DECANE
mf: $C_7H_8S_3$ mw: 188.33

SYN: 4,7-METHANOBENZOTRITHIOLE, HEXAHYDRO-

TOXICITY DATA with REFERENCE
skn-rbt 20 mg/24H MOD 85JCAE-,1091,86
eye-rbt 500 mg/24H MLD 85JCAE-,1091,86
orl-rat LD50:300 mg/kg TXAPA9 28,313,74

SAFETY PROFILE: Poison by ingestion. A skin and eye irritant. When heated to decomposition it emits toxic fumes of SO_x.

TNP250 CAS:786-19-6 ***HR: 3***
TRITHION
mf: $C_{11}H_{16}ClO_2PS_3$ mw: 342.87

PROP: Amber liquid. Bp: 82° @ 0.1 mm, d: 1.29 @ 20°. Essentially insol in water; misc in common solvents.

SYNS: ACARITHION ◇ AKARITHION ◇ CARBOFENOTHION (DUTCH) ◇ S-((p-CHLOROPHENYLTHIO)METHYL)-O,O-DIETHYL PHOSPHORODITHIOATE ◇ S-(4-CHLOROPHENYLTHIOMETHYL)DIETHYL PHOSPHOROTHIOLOTHIONATE ◇ DAGADIP ◇ O,O-DIAETHYL-S-((4-CHLOR-PHENYL-THIO)-METHYL)DITHIOPHOSPHAT (GERMAN) ◇ O,O-DIETHYL-S-(4-CHLOOR-FENYL-THIO)-METHYL)-DITHIOFOSFAAT (DUTCH) ◇ O,O-DIETHYL-S-p-CHLORFENYLTHIOMETHYLESTER KYSELINY DITHIOFOSFORECNE (CZECH) ◇ O,O-DIETHYL-S-p-CHLORLPHENYLTHIOMETHYL DITHIOPHOSPHATE ◇ O,O-DIETHYL-P-CHLOROPHENYLMERCAPTOMETHYL DITHIOPHOSPHATE ◇ O,O-DIETHYL-S-(4-CHLOROPHENYLTHIOMETHYL) DITHIOPHOSPHATE ◇ O,O-DIETHYL-S-(p-CHLOROPHENYLTHIOMETHYL)PHOSPHORODITHIOATE ◇ O,O-DIETHYL- DITHIOPHOSPHORIC ACID, p-CHLOROPHENYLTHIOMETHYL ESTER ◇ O,O-DIETIL-S-((4-CLORO-FENILTIO)-METILE)-DITIOFOSFATO (ITALIAN) ◇ DITHIOPHOSPHATE de O,O-DIETHYLE et de (4-CHLORO-PHENYL) THIOMETHYLE (FRENCH) ◇ ENDYL ◇ ENT 23,708 ◇ GARRATHION ◇ LETHOX ◇ NEPHOCARP ◇ OLEOAKARITHION ◇ R-1303 ◇ STAUFFER R-1,303 ◇ TRITHION MITICIDE

TOXICITY DATA with REFERENCE
orl-rat LD50:6800 μg/kg FMCHA2 -,C246,83
skn-rat LD50:27 mg/kg TXAPA9 2,88,60
ipr-rat LD50:40 mg/kg PSEBAA 114,509,63
orl-mus LD50:218 mg/kg ARSIM* 20,6,66
ipr-mus LD50:27 mg/kg PSEBAA 129,699,68
orl-rbt LD50:1250 mg/kg SPEADM 78-1,40,78
skn-rbt LD50:1270 mg/kg PCOC** -,200,66
orl-ckn LD50:57 mg/kg TXAPA9 11,49,67
scu-ckn LD50:640 mg/kg BCPCA6 12,1377,63
orl-bwd LD50:5600 μg/kg TXAPA9 21,315,72

CONSENSUS REPORTS: EPA Farm Worker Field Reentry FEREAC 39,16888,74. EPA Extremely Hazardous Substances List.

SAFETY PROFILE: Poison by ingestion, skin contact, and intraperitoneal routes. Moderately toxic by subcutaneous route. A cholinesterase inhibitor. When heated to decomposition it emits very toxic fumes of SO_x, PO_x, and Cl^-. See also PARATHION, ESTERS, and MERCAPTANS.

TNP275 CAS:35619-65-9 ***HR: 2***
TRITIOZINE
mf: $C_{14}H_{19}NO_4S$ mw: 297.40

PROP: Pale yellow solid from ethanol. Mp: 141-143°.

SYNS: ISF 2001 ◇ 4-(THIOXO(3,4,5-TRIMETHOXYPHENYL) METHYL)MORPHOLINE ◇ TRESANIL ◇ 4-(3,4,5-TRIMETHOXYTHIOBENZOYL)MORPHOLINE ◇ 4-(3,4,5-TRIMETHOXYTHIOBENZOYL)TETRAHYDRO-1,4-OXAZINE ◇ TRITHIOZINE ◇ TRITIOZINA (ITALIAN)

TOXICITY DATA with REFERENCE
orl-rat TDLo:13500 mg/kg (15-22D preg/20D post):REP ATSUDG 4,284,80
orl-rat LD50:670 mg/kg FRPSAX 38,811,83
ipr-mus LD50:2000 mg/kg CHTPBA 8,462,73

SAFETY PROFILE: Moderately toxic by ingestion and intraperitoneal routes. Experimental reproductive effects. When heated to decomposition it emits toxic fumes of SO_x and NO_x.

TNP500 CAS:1330-78-5 ***HR: 3***
TRITOLYL PHOSPHATE
DOT: UN 2574
mf: $C_{21}H_{21}O_4P$ mw: 368.39

PROP: Oily, flame-resistant liquid. D: 1.16, bp: 265°, pour point: 28°, flash p: 410°F. Insol in water; misc with all common organic solvents and thinners, linseed oil, china wood oil, and caster oil.

SYNS: CELLUFLEX 179C ◇ CRESYL PHOSPHATE ◇ DISFLAMOLL TKP ◇ DURAD ◇ FLEXOL PLASTICIZER TCP ◇ FYRQUEL 150 ◇ IMOL S 140 ◇ KRONITEX ◇ LINDOL ◇ NCI-C61041 ◇ PHOSPHATE de TRICRESYLE (FRENCH) ◇ PHOSPHORIC ACID, TRITOLYL ESTER ◇ TRICRESILFOSFATI (ITALIAN) ◇ TRICRESYLFOSFATEN (DUTCH) ◇ TRICRESYL PHOSPHATE ◇ TRICRESYLPHOSPHATE, with more than 3% ortho isomer (DOT) ◇ TRIKRESYLPHOSPHATE (GERMAN) ◇ TRIS(TOLYLOXY)PHOSPHINE OXIDE

TOXICITY DATA with REFERENCE
skn-rbt 500 mg open MLD UCDS** 12/29/64
eye-rbt 500 mg/24H MLD 28ZPAK -,207,72
orl-mus TDLo:2250 mg/kg (male 7D pre):TER FAATDF 10,344,88
orl-mus TDLo:4464 mg/kg (male 7D pre):REP FAATDF 10,344,88
orl-wmn TDLo:70 mg/kg/14D:PNS,CNS LANCAO 1,88,81
orl-rat LD50:5190 mg/kg 28ZPAK -,207,72
orl-mus LD50:3900 mg/kg 85GMAT -,114,82
orl-dog LDLo:500 mg/kg 29ZWAE -,339,68
skn-cat LD50:1500 mg/kg TOLED5 1000(Sp.1),141,80
orl-rbt LDLo:100 mg/kg 29ZWAE -,339,68

CONSENSUS REPORTS: Reported in EPA TSCA Inventory.

DOT Classification: Poison B; Label: Poison.

SAFETY PROFILE: Poison by ingestion. Moderately toxic by skin contact. Human systemic effects by ingestion: flaccid paralysis without anesthesia, motor activity changes and muscle weakness. An experimental teratogen. Experimental reproductive effects. An eye and skin irritant. Combustible. When heated to decomposition it emits toxic fumes of PO_x.

TNR475 CAS:2799-07-7 ***HR: 3***
3-TRITYLTHIO-l-ALANINE
mf: $C_{22}H_{21}NO_2S$ mw: 363.50

SYNS: NSC 83265 ◇ S-TRIPHENYLMETHYL-l-CYSTEINE ◇ S-TRITYLCYSTEINE ◇ S-TRITYL-l-CYSTEINE ◇ 3-(TRITYLTHIO)-l-ALANINE

TOXICITY DATA with REFERENCE
orl-mus LD50:936 mg/kg NCISP* JAN86
ipr-mus LD50:396 mg/kg NCISP* JAN86
orl-dog LDLo:750 mg/kg NTIS** PB81-109993
ivn-dog LDLo:150 mg/kg NTIS** PB81-109993

SAFETY PROFILE: Poison by intravenous and intraperitoneal routes. Moderately toxic by ingestion. When heated to decomposition it emits toxic fumes of NO_x and SO_x.

TNR485 CAS:3605-01-4 ***HR: 3***
TRIVASTAN
mf: $C_{16}H_{18}N_4O_2$ mw: 298.38

PROP: Crystals from anhydrous ethanol. Mp: 98°.

SYNS: 2-(4-(1,3-BENZODIOXOL-5-YLMETHYL)-1-PIPERAZINYL)PYRIMIDINE ◇ ET 495 ◇ EU 4200 ◇ 2-(4-(3,4-METHYLENEDIOXYBENZYL)PIPERAZINO)PYRIMIDINE ◇ 1-(3,4-METHYLENEDIOXYBENZYL)-4-(2-PYRIMIDYL)PIPERAZINE ◇ 2-(4-PIPERONYL-1-PIPERAZINYL)PYRIMIDINE ◇ PIRIBEDIL ◇ 1-(2-PYRIMIDYL)-4-PIPERONYLPIPERAZINE ◇ 1-(2''-PYRIMIDYL)-4-PIPERONYL PIPERAZINE ◇ TRIVASTAL

TOXICITY DATA with REFERENCE
orl-mus LD50:1460 mg/kg EJPHAZ 6,75,67
ipr-mus LD50:690 mg/kg JMCMAR 11,1151,68
ivn-mus LD50:88 mg/kg EJPHAZ 6,75,67

SAFETY PROFILE: Poison by intravenous route. Moderately toxic by ingestion and intraperitoneal routes. When heated to decomposition it emits toxic fumes of NO_x.

TNR490 CAS:5613-68-3 ***HR: 3***
TRIVINYLANTIMONY
mf: C_6H_9Sb mw: 202.89

$(H_2C{=}CH)_3Sb$

CONSENSUS REPORTS: Antimony and its compounds are on The Community Right-To-Know List.

SAFETY PROFILE: Most antimony compounds are poisons. Ignites spontaneously in air. When heated to decomposition it emits toxic fumes of Sb. See also ANTIMONY COMPOUNDS.

TNR500 CAS:65313-35-1 ***HR: 3***
TRIVINYLBISMUTH
mf: C_6H_9Bi mw: 290.09

$(H_2C{=}CH)_3Bi$

SAFETY PROFILE: Ignites spontaneously in air. When heated to decomposition it emits toxic fumes of Bi. See also BISMUTH COMPOUNDS.

TNR625 CAS:56305-04-5 ***HR: 2***
TROLOX C
mf: $C_{14}H_{18}O_4$ mw: 250.32

SYNS: 6-HYDROXY-2,5,7,8-TETRAMETHYLCHROMAN-2-CARBOXYLIC ACID ◇ TROLOX

TOXICITY DATA with REFERENCE
orl-rat LD50:4300 mg/kg JAOCA7 52,174,75
ipr-rat LD50:1800 mg/kg JAOCA7 52,174,75

orl-mus LD50:1630 mg/kg JAOCA7 52,174,75
ipr-mus LD50:1700 mg/kg JAOCA7 52,174,75
scu-mus LD50:1930 mg/kg JAOCA7 52,174,75

SAFETY PROFILE: Moderately toxic by ingestion, intraperitoneal, and subcutaneous routes. When heated to decomposition it emits acrid smoke and irritating fumes.

TNS200 CAS:637-23-0 ***HR: 3***
TROPACAINE HYDROCHLORIDE
mf: $C_{15}H_{19}NO_2{\bullet}ClH$ mw: 281.81

PROP: Strongly refractive prisms from alc. Decomposes at 283°. Sol in water; sltly sol in abs alc; practically insol in ether.

SYNS: BENZOYLPSEUDOTROPINE HYDROCHLORIDE ◇ O-BENZOYLTROPINE HYDROCHLORIDE ◇ 8-METHYL-8-AZABICYCLO(3.2.1)OCTAN-3-OL BENZOATE (ester), HYDROCHLORIDE, exo- ◇ PSEUDOTROPINE BENZOATE HYDROCHLORIDE ◇ TROPACOCAINE HYDROCHLORIDE ◇ TROPAKOKAIN HYDROCHLORID (GERMAN) ◇ TROPOCOCAIN HYDROCHLORIDE

TOXICITY DATA with REFERENCE
scu-mus TDLo:120 mg/kg (7D preg):TER RCSADO 5,279,84
ipr-mus LDLo:350 mg/kg QJPPAL 7,227,34
scu-mus LD50:465 mg/kg RCSADO 5,279,84
ivn-mus LDLo:42 mg/kg WDMU** -,-,36

SAFETY PROFILE: Poison by intraperitoneal and intravenous routes. Moderately toxic by subcutaneous route. An experimental teratogen. When heated to decomposition it emits toxic fumes of NO_x and HCl.

TNT500 CAS:22089-22-1 ***HR: 3***
TROPHOSPHAMIDE
mf: $C_9H_{18}Cl_3N_2O_2P$ mw: 323.61

SYNS: A-4828 ◇ ASTA Z 4828 ◇ 2-(BIS(2-CHLOROETHYL)AMINO)-3-(2-CHLOROETHYL)TETRAHYDRO-2H-1,3,2-OXAPHOSPHORINE-2-OXIDE ◇ 3-(2-CHLOROETHYL)-2-(BIS(2-CHLOROETHYL)AMINO)PERHYDRO-2H-1,3,2-OXAZAPHOSPHORINE-2-OXIDE◇ CYCLOPHOSPHAMIDE-N-MONOCHLOROETHYL derivative ◇ IXOTEN ◇ NSC 109723 ◇ TFF ◇ N,N,N'-TRIS(2-CHLORAETHYL)-N',O-PROPYLEN-PHOSPHORSAUREESTER-DIAMID (GERMAN) ◇ N,N,N'-TRIS(2-CHLOROETHYL)-N',O-PROPYLENE PHOSPHORIC ACID ESTER DIAMIDE ◇ N,N,3-TRIS(2-CHLOROETHYL)TETRAHYDRO-2H-1,3,2-OXAPHOSPHORIN-2-AMINE-2-OXIDE◇ TRIFOSFAMIDE ◇ TRILOFOSFAMIDA ◇ TRILOPHOSPHAMIDE ◇ TRISFOSFAMIDE ◇ TRISPHOSPHAMIDE ◇ TROFOSFAMID ◇ TROPHOSPHAMID ◇ Z 4828

TOXICITY DATA with REFERENCE
mma-esc 10 mmol/L ARTODN 33,225,75
sln-dmg-unr 1 mmol/L/24H MUREAV 33,221,75
cyt-hmn:leu 100 mg/L HUMAA7 5,321,68
cyt-rat-ipr 25 mg/kg HUMAA7 5,321,68
mnt-mus-ipr 80 mg/kg/24H MUREAV 56,319,78
cyt-ham-ipr 3300 μg/kg ARTODN 38,35,77
unr-hmn TDLo:50 mg/kg:BLD,KID SAMJAF 53,886,78
orl-rat LD50:202 mg/kg MDACAP 13,115,77
scu-rat LD50:210 mg/kg ARZNAD 23,922,73
ivn-rat LD50:90 mg/kg ARZNAD 23,922,73
orl-mus LD50:464 mg/kg MDACAP 13,115,77
ipr-mus LD50:212 mg/kg ARZNAD 24,1149,74

CONSENSUS REPORTS: EPA Genetic Toxicology Program.

SAFETY PROFILE: Poison by intraperitoneal, subcutaneous, and intravenous routes. Moderately toxic by ingestion. Human mutation data reported. Human systemic effects by unspecified routes: hematuria, luekopenia and thrombocytopenia. When heated to decomposition it emits very toxic fumes of Cl^-, NO_x, and PO_x.

TNU000 CAS:132-17-2 ***HR: 3***
TROPINE BENZOHYDRYL ETHER METHANESULFONATE
mf: $C_{21}H_{25}NO{\bullet}CH_4O_3S$ mw: 403.58

SYNS: BENZATROPINE METHANESULFONATE ◇ BENZOTROPINE MESYLATE ◇ BENZOTROPINE METHANESULFONATE ◇ BENZTROPINE MESYLATE ◇ BENZTROPINE METHANESULFONATE ◇ 3-DIPHENYLMETHOXYTROPANE MESYLATE ◇ 3-DIPHENYL-METHOXYTROPANE METHANESULFONATE

TOXICITY DATA with REFERENCE
dnd-esc 10 umol/L MUREAV 89,95,81
orl-hmn TDLo:100 μg/kg:PSY PSDTAP 8,59,67
ipr-mus LD50:65 mg/kg AIPTAK 144,555,63
scu-rat LD50:353 mg/kg 27ZQAG -,208,72
orl-mus LD50:91 mg/kg DRUGAY 6,775,82
scu-mus LD50:103 mg/kg 27ZQAG -,208,72
ivn-mus LD50:24 mg/kg 27ZQAG -,208,72
ivn-cat LDLo:33 mg/kg CLDND*

SAFETY PROFILE: Poison by ingestion, intravenous, subcutaneous, and intraperitoneal routes. Human systemic effects by ingestion: psychotropic effects. Mutation data reported. When heated to decomposition it emits very toxic fumes of NO_x and SO_x. See also ETHERS.

TNV550 CAS:533-75-5 ***HR: 3***
TROPOLONE
mf: $C_7H_6O_2$ mw: 122.13

PROP: Mp: 51-54°, flash p: >230°. Hygroscopic.

SYN: PURPUROCATECHOL

TOXICITY DATA with REFERENCE
ipr-rat LD50:190 mg/kg TXCYAC 14,217,79
ipr-mus LD50:212 mg/kg TXCYAC 14,217,79
scu-mus LD50:233 mg/kg YKKZAJ 91,550,71
ivn-mus LD50:106 mg/kg YKKZAJ 92,19,72

SAFETY PROFILE: Poison by subcutaneous, intravenous, and intraperitoneal routes. Combustible. When

heated to decomposition it emits acrid smoke and irritating fumes.

TNV575 CAS:25230-72-2 ***HR: 3***
TROPYLIUM PERCHLORATE
mf: $C_7H_7ClO_4$ mw: 190.58

SAFETY PROFILE: An explosive sensitive to friction and pressure. When heated to decomposition it emits toxic fumes of Cl^-. See also PERCHLORATES.

TNV625 CAS:391-70-8 ***HR: 3***
TROXONIUM TOSYLATE
mf: $C_{18}H_{30}NO_5 \cdot C_7H_7O_3S$ mw: 511.69

SYNS: FWH 399 ◇ FWH 429 ◇ TRIETHYL(2-HYDROXYETHYL)AMMONIUM-p-TOLUENESULFONATE3,4,5-TRIMETHOXYBENZOATE ◇ TRIETHYL-2-(3,4,5-TRIMETHOXYBENZOYLOXY)ETHYLAMMONIUM-p-TOLUENESULFATE ◇ TRIETHYL-2-(3,4,5-TRIMETHOXYBENZOYLOXY)ETHYLAMMONIUM TOSYLATE ◇ TROXONIUM TOSILATE

TOXICITY DATA with REFERENCE
orl-mus LD50:850 mg/kg 27ZQAG -,381,72
ipr-mus LD50:25 mg/kg APSXAS 11,401,74
ivn-mus LD50:8 mg/kg APSXAS 11,401,74

SAFETY PROFILE: Poison by intravenous and intraperitoneal routes. Moderately toxic by ingestion. When heated to decomposition it emits toxic fumes of SO_x, NO_x, and NH_3.

TNW000 CAS:9002-07-7 ***HR: 3***
TRYPSIN

PROP: Yellow to grayish-yellow powder or crystals. Sol in water: practically insol in alc or glycerol. Readily sol in Sorenson's sodium phosphate buffer soln.

SYNS: PARENZYME ◇ PARENZYMOL ◇ TRYPTAR ◇ TRYPURE ◇ U-4858

TOXICITY DATA with REFERENCE
dnd-ham:fbr 100 ppm SFCRAO 23,346,70
ivn-rat LD50:36 mg/kg FATOAO 45(6),78,82
ipr-mus LD50:100 mg/kg FATOAO 45(6),78,82
ivn-mus LD50:11100 μg/kg NIIRDN 6,523,82
ivn-rbt LD50:2200 μg/kg NIIRDN 6,523,82
ivn-gpg LDlo:30000 units/kg AIPTAK 106,164,56

CONSENSUS REPORTS: Reported in EPA TSCA Inventory.

SAFETY PROFILE: Poison by intraperitoneal and intravenous routes. Mutation data reported. When heated to decomposition it emits acrid smoke and irritating fumes.

TNW250 CAS:153-94-6 ***HR: 1***
d-TRYPTOPHAN
mf: $C_{11}H_{12}N_2O_2$ mw: 204.25

SYN: d-TRYPTOPHANE

TOXICITY DATA with REFERENCE
ipr-rat LD50:4289 mg/kg ABBIA4 64,319,56

CONSENSUS REPORTS: Reported in EPA TSCA Inventory.

SAFETY PROFILE: Mildly toxic by intraperitoneal route. When heated to decomposition it emits toxic fumes of NO_x.

TNW500 CAS:54-12-6 ***HR: 3***
dl-TRYPTOPHAN
mf: $C_{11}H_{12}N_2O_2$ mw: 204.25

PROP: White crystals or crystalline powder; odorless. Sol in water, dil acids, alkalies; sltly sol in alc. Optically inactive.

TOXICITY DATA with REFERENCE
orl-rat TDLo:844 g/kg/92W-C:CAR,REP CNREA8 39,1207,79

CONSENSUS REPORTS: Reported in EPA TSCA Inventory. EPA Genetic Toxicology Program.

SAFETY PROFILE: Experimental reproductive effects. Questionable carcinogen with experimental carcinogenic data. When heated to decomposition it emits toxic fumes of NO_x.

TNW950 ***HR: 2***
l-TRYPTOPHAN, pyrolyzate

PROP: Smoke condensate obtained by pyrolysis of l-TRYPTOPHAN (CALEDQ 2,335,77).

TOXICITY DATA with REFERENCE
otr-ham:emb 50 mg/L MUREAV 49,145,78
cyt-ham:emb 30 mg/L/24H MUREAV 49,145,78
orl-rat TDLo:77600 mg/kg/2Y-C:NEO CALEDQ 13,181,81
imp-mus TDLo:80 mg/kg:CAR CALEDQ 17,101,82

SAFETY PROFILE: Questionable carcinogen with experimental carcinogenic and neoplastigenic data. Mutation data reported. When heated to decomposition it emits acrid smoke and irritating fumes.

TNX000 CAS:73-22-3 ***HR: 3***
l-TRYPTOPHANE
mf: $C_{11}H_{12}N_2O_2$ mw: 204.25

PROP: Leaflets or plates from dil alc. An essential amino acid occurring in isomeric forms. Mp: decomp 289°. The l and dl forms: white crystals or crystalline powder; slt bitter taste. dl form: sltly sol in water. l form:

sol in water, hot alc, and alkali hydroxides; insol in chloroform.

SYNS: l-α-AMINO-3-INDOLEPROPRIONIC ACID ◇ α′-AMINO-3-INDOLEPROPRIONIC ACID ◇ α-AMINO-INDOLE-3-PROPRIONIC ACID ◇ 2-AMINO-3-INDOL-3-YL-PROPRIONIC ACID ◇ EH 121 ◇ INDOLE-3-ALANINE ◇ 1-β-3-INDOLYLALANINE ◇ NCI-C01729 ◇ (−)-TRYPTOPHAN ◇ l-TRYPTOPHAN (FCC) ◇ TRYPTOPHANE

TOXICITY DATA with REFERENCE
dni-hmn:lym 1 mmol/L PNASA6 79,1171,82
dni-rat:lvr 100 μmol/L CNREA8 45,337,85
ipr-rat TDLo:15 g/kg (30D pre):REP IYKEDH 11,646,80
scu-rat TDLo:9500 mg/kg/2Y-C:ETA,TER VOONAW 20(8),75,74
imp-mus TDLo:80 mg/kg:ETA CALEDQ 17,101,82
orl-man TDLo:300 mg/kg BMJOAE 2,701,76
ipr-rat LD50:1634 mg/kg ABBIA4 58,253,55
ipr-mus LD50:4800 mg/kg IYKEDH 11,635,80

CONSENSUS REPORTS: NCI Carcinogenesis Bioassay (feed); No Evidence: mouse, rat NCITR* NCI-CG-TR-71,78. Reported in EPA TSCA Inventory.

SAFETY PROFILE: Moderately toxic by intraperitoneal route. Experimental teratogenic and reproductive effects. Human mutation data reported. Questionable carcinogen with experimental tumorigenic data. When heated to decomposition it emits toxic fumes of NO_x.

TNX275 CAS:62450-06-0 ***HR: 3***
TRYPTOPHAN P1
mf: $C_{13}H_{13}N_3$ mw: 211.29

SYNS: 3-AMINO-1,4-DIMETHYL-Γ-CARBOLINE ◇ 3-AMINO-1,4-DIMETHYL-5H-PYRIDO(4,3-b)INDOLE ◇ 1,4-DIMETHYL-5H-PYRIDO(4,3-b)INDOL-3-AMINE ◇ TRP-P-1 ◇ dl-TRYPTOPHAN, pyrolyzate 1

TOXICITY DATA with REFERENCE
mma-sat 250 ng/plate JJCREP 76,835,85
dns-hmn:fbr 20 mg/L JRARAX 24,356,83
orl-rat TDLo:2 g/kg/49W-C:CAR EVHPAZ 67,129,86
scu-rat TDLo:150 mg/kg/20W-I:CAR PPTCBY 9,159,79
scu-mus TD:50 mg/kg:NEO CRNGDP 8,1721,87
orl-rat LD50:100 mg/kg PPTCBY 9,159,79
orl-mus LD50:200 mg/kg PPTCBY 9,159,79
orl-ham LD50:380 mg/kg PPTCBY 9,159,79

CONSENSUS REPORTS: IARC Cancer Review: Group 2B IMEMDT 7,56,87; Animal Sufficient Evidence IMEMDT 31,247,83. EPA Genetic Toxicology Program.

SAFETY PROFILE: Suspected carcinogen with experimental carcinogenic and neoplastigenic data. Poison by ingestion. Human mutation data reported. When heated to decomposition it emits toxic fumes of NO_x.

TNX375 CAS:8064-18-4 ***HR: 2***
TSELATOX
mf: $C_{13}H_{17}ClO_3{\cdot}C_{13}H_{15}Cl_3O_3$ mw: 582.38

SYNS: CELATOX ◇ TSLT

TOXICITY DATA with REFERENCE
orl-rat LDLo:800 mg/kg GNAMAP 9,72,70
orl-mus LD50:1300 mg/kg GNAMAP 9,72,70
orl-gpg LD50:1600 mg/kg GNAMAP 9,72,70

SAFETY PROFILE: Moderately toxic by ingestion. When heated to decomposition it emits toxic fumes of Cl^-.

TNX400 CAS:24305-27-9 ***HR: 2***
TSH-RELEASING HORMONE
mf: $C_{16}H_{22}N_6O_4$ mw: 362.44

PROP: Partially sol in chloroform; highly sol in abs methanol; insol in pyridine.

SYNS: FDA 1725 ◇ LOPREMONE ◇ PROTIRELIN ◇ l-PYROGLUTAMYL-l-HISTIDYL-l-PROLINEAMIDE ◇ RIFATHYROIN ◇ RO 8-6270/9 ◇ SYNTHETIC TRF ◇ SYNTHETIC TRH ◇ SYNTHETIC TSH-RELEASING FACTOR ◇ SYNTHETIC TSH-RELEASING HORMONE ◇ THYROID RELEASING HORMONE ◇ THYROLIBERIN ◇ THYROTROPIC-RELEASING FACTOR ◇ THYROTROPIC RELEASING HORMONE ◇ THYROTROPIN-RELEASING FACTOR ◇ THYROTROPIN-RELEASING HORMONE ◇ TRF ◇ TSH-RELEASING FACTOR ◇ TSH-RF

TOXICITY DATA with REFERENCE
ipr-rat TDLo:200 mg/kg (female 7-16D post):REP OYYAA2 8,807,74
ipr-mus TDLo:150 mg/kg (female 6-15D post):TER OYYAA2 8,807,74
ivn-man TDLo:5714 ng/kg/1M-C MJAUAJ 143,264,85
ivn-man TDLo:5333 ng/kg/2M-C BMJOAE 287,532,83
ivn-wmn TDLo:8 μg/kg/2M-C BMJOAE 287,532,83
ivn-rat LD50:514 mg/kg NIIRDN 6,727,82
ivn-mus LD50:921 mg/kg NIIRDN 6,727,82

SAFETY PROFILE: Moderately toxic by intravenous route. An experimental teratogen. Experimental reproductive effects. When heated to decomposition it emits toxic fumes of NO_x.

TNX650 CAS:11054-63-0 ***HR: 3***
TSUSHIMYCIN
mf: $C_{59}H_{93}N_{13}O_{20}$ mw: 1304.65

TOXICITY DATA with REFERENCE
ipr-mus LD50:50 mg/kg 85GDA2 4(1),318,80
scu-mus LD50:50 mg/kg 85GDA2 4(1),318,80
ivn-mus LD50:50 mg/kg 85GDA2 4(1),318,80

SAFETY PROFILE: Poison by subcutaneous, intravenous, and intraperitoneal routes. When heated to decomposition it emits toxic fumes of NO_x.

TNX750 CAS:123-82-0 ***HR: 3***
TUAMINE
mf: $C_7H_{17}N$ mw: 115.25

SYNS: dl-2-AMINOHEPTANE ◇ ARMEEN L-7 ◇ HEPTAMINE ◇ 2-HEPTANAMINE ◇ HEPTEDRINE ◇ 2-HEPTYLAMINE ◇ 1-METHYLHEXYLAMINE ◇ RINEPTIL ◇ TUAMINOHEPTANE

TOXICITY DATA with REFERENCE
scu-rat LD50:130 mg/kg JPETAB 85,119,45
ipr-mus LDLo:60 mg/kg JAPMA8 30,623,41
scu-mus LD50:115 mg/kg FEPRA7 4,139,45

CONSENSUS REPORTS: Reported in EPA TSCA Inventory.

SAFETY PROFILE: Poison by subcutaneous and intraperitoneal routes. When heated to decomposition it emits toxic fumes of NO_x.

TNY250 ***HR: 3***
TUBERACTINOMYCIN-N SULFATE
mf: $C_{25}H_{43}N_{13}O_{10}$•3/2 H_2O_4S mw: 933.59

SYNS: ENVIOMYCIN SULFATE ◇ TUBERACTIN SULFATE ◇ VIOMYCIN SULFATE SALT (2:3)

TOXICITY DATA with REFERENCE
ims-rat TDLo:15 g/kg (female 30D pre):REP OYYAA2 12,585,76
ivn-rat LD50:640 mg/kg OYYAA2 8,817,74
ivn-mus LD50:420 mg/kg YAKUD5 17,1217,75

SAFETY PROFILE: Poison by intravenous route. Experimental reproductive effects. When heated to decomposition it emits very toxic fumes of NO_x and Cl^-.

TNY500 CAS:69-33-0 ***HR: 3***
TUBERCIDIN
mf: $C_{11}H_{14}N_4O_4$ mw: 266.29

PROP: Needles. Decomp @ 247-248°. Sol in acidic and alkaline soln; almost insol in acetone, ethyl acetate, chloroform, benzene, petr ether.

SYNS: 4-AMINO-7-(β-d-RIBOFURANOSYL)-PYRROLO(2,3-D)PYRIMIDINE ◇ 4-AMINO-7-β-d-RIBOFURANOSYL-7H-PYRROLO(2,3-D)PYRIMIDINE ◇ 7-DEAZAADENOSINE ◇ NSC 56408 ◇ 7-β-d-RIBOFURANOSYL-7H-PYRROLO(2,3-D)PYRIMIDINE-4-AMINE

TOXICITY DATA with REFERENCE
dlt-mus-ipr 500 μg/kg MUREAV 54,226,78
orl-rat LD50:16 mg/kg CNREA8 29,116,69
ipr-rat LD50:1 mg/kg CNREA8 29,116,69
orl-mus LD50:28320 μg/kg NCISP* JAN86
ipr-mus LD50:6 mg/kg UPJOH* 2(6),-,71
ivn-mus LD50:45 mg/kg JAJAAA 10,201,57
orl-dog LDLo:48 mg/kg CNREA8 29,116,69
ivn-dog LDLo:48 mg/kg CNREA8 29,116,69

SAFETY PROFILE: Poison by ingestion, intraperitoneal and intravenous routes. Mutation data reported. When heated to decomposition it emits toxic fumes of NO_x.

TNY750 CAS:57-95-4 ***HR: 3***
d-TUBOCURARINE
mf: $C_{38}H_{44}N_2O_6$ mw: 624.84

SYNS: TUBOCURARIN ◇ (+)-TUBOCURARINE ◇ TUBOCURARINE

TOXICITY DATA with REFERENCE
ipr-rat LD50:210 μg/kg TXAPA9 14,67,69
ims-rat LD50:500 μg/kg JPPMAB 10,638,58
ipr-mus LD50:410 μg/kg AIPTAK 153,308,65
scu-mus LD50:560 μg/kg AIPTAK 152,277,64
ivn-mus LD50:130 μg/kg ARZNAD 15,130,65
ivn-dog LD50:500 μg/kg AIPTAK 80,172,49
ims-dog LD50:250 μg/kg JPPMAB 10,638,58
ivn-rbt LD50:20 μg/kg RISSAF 13,339,50
ivn-ckn LD50:700 μg/kg AIPTAK 122,152,59

SAFETY PROFILE: A deadly poison by subcutaneous, intraperitoneal, intramuscular, and intravenous routes. Human toxicity: Large doses and overdoses may cause respiratory paralysis and hypotension. When heated to decomposition it emits toxic fumes of NO_x.

TNZ000 CAS:6989-98-6 ***HR: 3***
d-TUBOCURARINE CHLORIDE PENTAHYDRATE
mf: $C_{38}H_{44}N_2O_6$•2Cl•5H_2O mw: 785.84

PROP: l Form: Needles. Mp: 268° (effervescence).

SYN: (+)-TUBOCURARINE DICHLORIDE PENTAHYDRATE

TOXICITY DATA with REFERENCE
orl-mus LD50:150 mg/kg JPETAB 118,395,56
scu-mus LD50:600 μg/kg JPETAB 100,333,50
ivn-mus LD50:130 μg/kg JPETAB 118,388,56
orl-cat LDLo:18 mg/kg JPETAB 118,395,56
ivn-cat LD50:400 μg/kg JPETAB 118,395,56
ivn-rbt LD50:146 μg/kg JPETAB 118,395,56

SAFETY PROFILE: Poison by ingestion, subcutaneous and intravenous routes. Human Toxicity: Large doses and overdoses may cause respiratory paralysis and hypotension. When heated to decomposition it emits very toxic fumes of NO_x and Cl^-.

TOA000 CAS:57-94-3 ***HR: 3***
TUBOCURARINE HYDROCHLORIDE
mf: $C_{38}H_{44}N_2O_6$•2Cl mw: 694.74

SYNS: AMERIZOL ◇ CURARIN-HAF ◇ DELACURARINE ◇ DEXTROTUBOCURARINE CHLORIDE ◇ d-7',12'-DIHYDROXY-6,6'-DIMETHOXY-2,2',2'-TRIMETHYLTUBOCURARANIUM CHLORIDE ◇ INTOCOSTRIN ◇ d-PARACURARINE CHLORIDE ◇ TUBADIL ◇ TUBARINE ◇ TUBOCURARINE CHLORIDE ◇ (+)-TUBOCURARINE CHLORIDE ◇ d-TUBOCURARINE CHLORIDE ◇ TUBOCURARINE, CHLORIDE, HYDROCHLORIDE, (+)- (8CI) ◇ d-TUBOCURARINE DI-

CHLORIDE ◇ d-TUBOCURARINE HYDROCHLORIDE ◇ (+)-TUBOCURARINE HYDROCHLORIDE

TOXICITY DATA with REFERENCE
orl-rat LD50:28 mg/kg PSEBAA 120,511,65
ipr-rat LD50:340 μg/kg OYYAA2 9,117,75
scu-rat LD50:310 μg/kg OYYAA2 9,117,75
ivn-rat LD50:66 μg/kg OYYAA2 9,117,75
orl-mus LD50:33 mg/kg PSEBAA 120,511,65
ipr-mus LD50:420 μg/kg RPTOAN 32,74,69
scu-mus LD50:640 μg/kg OYYAA2 9,117,75
ivn-mus LD50:97 μg/kg AIPTAK 109,191,57
ivn-rbt LD50:190 μg/kg OYYAA2 9,117,75
ivn-gpg LD50:66 μg/kg JLCMAK 34,516,49

CONSENSUS REPORTS: EPA Genetic Toxicology Program.

SAFETY PROFILE: Poison by ingestion, intravenous, intraperitoneal, and subcutaneous routes. Human toxicity: Large doses and overdoses may cause respiratory paralysis and hypotension. When heated to decomposition it emits very toxic fumes of NO_x and Cl^-. Used as a muscle relaxant.

TOA275 ***HR: 2***
TUNG NUT

PROP: A medium-sized flowering tree with large lobed leaves. At maturity the fruit is a brown capsule 2 to 3 inches long with 3 to 7 hard seeds. The seed resembles an unshelled hickory nut, and the meat inside looks like a chestnut.

SYNS: ACEITE CHINO (CUBA) ◇ A. CORDATA ◇ A. FORDII ◇ ALEURITES (VARIOUS SPECIES) ◇ A. MONTANA ◇ A. TRISPERMA ◇ AVELLANO (DOMINICAN REPUBLIC) ◇ BANUCALAD (HAWAII) ◇ CANDLEBERRY ◇ CANDLENUT ◇ CHINAWOOD OIL TREE ◇ COUNTRY WALNUT ◇ INDIAN WALNUT ◇ JAMAICAN WALNUT ◇ JAPAN OIL TREE ◇ KUKUI (HAWAII, GUAM) ◇ LUMBANG (GUAM) ◇ MU OIL TREE ◇ NOGAL de LA INDIA (CUBA) ◇ NOIX de BANCOUL (GUADELOUPE) ◇ NOIX des MOLLUQUES (GUADELOUPE) ◇ NOISETTE des GRANDS-FONDS (GUADELOUPE) ◇ NUEZ de la INDIA (PUERTO RICO) ◇ NUEZ NOGAL (PUERTO RICO) ◇ OTAHEITE WALNUT (VIRGIN ISLANDS) ◇ PALO de NUEZ (PUERTO RICO) ◇ RAGUAR (GUAM) ◇ TUNG OIL TREE

SAFETY PROFILE: The whole plant contains an irritating phorbol derivative toxin. Ingestion of the seed is the most common cause of poisoning, resulting in nausea, vomiting, severe abdominal pain and diarrhea. See also PHORBOL.

TOA500 ***HR: 2***
TUNG NUT MEALS

SAFETY PROFILE: Toxic by ingestion. Contact causes dermatitis. Ingestion causes nausea, vomiting, cramps, diarrhea and tenesmus, thirst, dizziness, lethargy and disorientation. Large doses can cause fever, tachycardia and respiratory effects. Combustible in the form of dust when exposed to heat or flame. Processed material must be cooled thoroughly before storage so as not to over dry; can react with oxidizing materials. See also SAPONIN and TUNG NUT.

TOA510 ***HR: 2***
TUNG NUT OIL

PROP: Pale yellow liquid: characteristic disagreeable odor. Sol in chloroform, ether, carbon disulfide, and oils. Polymerized product is practically insol in organic solvents.

SYN: CHINAWOOD OIL

SAFETY PROFILE: Toxic by ingestion. Contact causes dermatitis. Ingestion causes nausea, vomiting, cramps, diarrhea and tenesmus, thirst, dizziness, lethargy and disorientation. Large doses can cause fever, tachycardia and respiratory effects. Combustible when exposed to heat or flame. Can react with oxidizing materials.

TOA750 CAS:7440-33-7 ***HR: 3***
TUNGSTEN
af: W aw: 183.85

PROP: A steely-gray to white, cuttable, forgeable and spinnable metallic element. Mp: 3410°, d: 19.3 @ 20°, bp: 5900°.

SYN: WOLFRAM

TOXICITY DATA with REFERENCE
skn-rbt 500 mg/24H MLD 28ZPAK -,19,72
eye-rbt 500 mg/24H MLD 28ZPAK -,19,72
orl-rat TDLo:1210 μg/kg (35W pre):REP GISAAA 42(8),30,77
orl-rat TDLo:1210 μg/kg (35W pre):TER GISAAA 42(8),30,77
unr-rat LD50:2 g/kg GISAAA 48(7),71,83

CONSENSUS REPORTS: Reported in EPA TSCA Inventory.

OSHA PEL: TWA (insoluble compounds) 5 mg(W)/m^3; STEL 10 mg(W)/m^3; (soluble compounds) 1 mg(W)/m^3; STEL 3 mg(W)/m^3
ACGIH TLV: TWA (insoluble compounds) 5 mg (W)/m^3; STEL 10 mg(W)/m^3; (soluble compounds) 1 mg(W)/m^3; STEL 3 mg(W)/m^3
NIOSH REL: (Tungsten, Insoluble) TWA 5 mg(W)/m^3

SAFETY PROFILE: Mildly toxic by an unspecified route. An experimental teratogen. Experimental reproductive effects. A skin and eye irritant. Flammable in the form of dust when exposed to flame. The powdered metal may ignite on contact with air or oxidants (e.g., bromine pentafluoride; bromine; chlorine trifluoride; potassium perchlorate; potassium dichromate; nitryl fluoride; fluorine; oxygen difluoride; iodine pentafluoride;

hydrogen sulfide; sodium peroxide; lead(IV) oxide. See also TUNGSTEN COMPOUNDS and POWDERED METALS.

TOB000 ***HR: 3***
TUNGSTEN AZIDE PENTABROMIDE
mf: Br_5N_3W mw: 625.42

OSHA PEL: TWA 1 mg(W)/m^3; STEL 3 mg(W)/m^3
ACGIH TLV: TWA 1 mg(W)/m^3; STEL 3 mg(W)/m^3

SAFETY PROFILE: Extremely explosive. When heated to decomposition it emits very toxic fumes of NO_x and Br^-. See also TUNGSTEN COMPOUNDS, AZIDES, and BROMIDES.

TOB250 ***HR: 3***
TUNGSTEN AZIDE PENTACHLORIDE
mf: Cl_5N_3W mw: 403.21

OSHA PEL: TWA 1 mg(W)/m^3; STEL 3 mg(W)/m^3
ACGIH TLV: TWA 1 mg(W)/m^3; STEL 3 mg(W)/m^3

SAFETY PROFILE: Explosive. When heated to decomposition it emits very toxic fumes of NO_x and Cl^-. See also TUNGSTEN COMPOUNDS and AZIDES.

TOB500 CAS:12070-12-1 ***HR: 3***
TUNGSTEN CARBIDE
mf: WC mw: 195.9.

CONSENSUS REPORTS: Reported in EPA TSCA Inventory.

OSHA PEL: TWA 5 mg(W)/m^3; STEL 10 mg(W)/m^3
ACGIH TLV: TWA 5 mg(W)/m^3; STEL 10 mg(W)/m^3
NIOSH REL: (Tungsten, insoluble) TWA 5 mg(W)/m^3

SAFETY PROFILE: Chronic inhalation causes lung damage in humans. Ignites at 600°C in nitrogen oxide atmospheres. Violent reaction with F_2; ClF_3; NO_x; IF_5; PbO_2; NO_2; N_2O. See also TUNGSTEN COMPOUNDS.

TOB750 CAS:11107-01-0 ***HR: 3***
TUNGSTEN CARBIDE, mixed with COBALT (85% : 15%)
SYN: BK 15

TOXICITY DATA with REFERENCE
itr-rat LDLo:50 mg/kg NTIS** AEC-TR-6710

OSHA PEL: TWA 5 mg(W)/m^3; STEL 10 mg(W)/m^3
ACGIH TLV: TWA 5 mg(W)/m^3; STEL 10 mg(W)/m^3
NIOSH REL: (Tungsten, insoluble) TWA 5 mg(W)/m^3

CONSENSUS REPORTS: Cobalt and its compounds are on the Community Right-To-Know List.

SAFETY PROFILE: Poison by intratracheal route. See also TUNGSTEN CARBIDE, TUNGSTEN COMPOUNDS, and COBALT COMPOUNDS.

TOC000 CAS:12718-69-3 ***HR: 3***
TUNGSTEN CARBIDE, mixed with COBALT (92% : 8%)

SYN: BK8

TOXICITY DATA with REFERENCE
itr-rat LDLo:75 mg/kg NTIS** AEC-TR-6710

CONSENSUS REPORTS: Cobalt and its compounds are on the Community Right-To-Know List.

OSHA PEL: TWA 5 mg(W)/m^3; STEL 10 mg(W)/m^3
ACGIH TLV: TWA 5 mg(W)/m^3; STEL 10 mg(W)/m^3
NIOSH REL: (Tungsten, insoluble) TWA 5 mg(W)/m^3

SAFETY PROFILE: Poison by intratracheal route. See also TUNGSTEN COMPOUNDS, TUNGSTEN CARBIDE, and COBALT COMPOUNDS.

TOC250 CAS:37329-49-0 ***HR: 3***
TUNGSTEN CARBIDE, mixed with COBALT and TITANIUM (78% : 14% : 8%)

SYN: TIL4K8

TOXICITY DATA with REFERENCE
itr-rat LDLo:75 mg/kg NTIS** AEC-TR-6710

CONSENSUS REPORTS: Cobalt and its compounds are on the Community Right-To-Know List.

OSHA PEL: TWA 5 mg(W)/m^3; STEL 10 mg(W)/m^3
ACGIH TLV: TWA 5 mg(W)/m^3; STEL 10 mg(W)/m^3
NIOSH REL: (Tungsten, insoluble) TWA 5 mg(W)/m^3

SAFETY PROFILE: Poison by intratracheal route. See also TUNGSTEN CARBIDE, COBALT, TITANIUM, and TUNGSTEN COMPOUNDS.

TOC500 ***HR: 2***
TUNGSTEN COMPOUNDS

OSHA PEL: TWA (insoluble compounds) 5 mg(W)/m^3; STEL 10 mg(W)/m^3; (soluble compounds) 1 mg(W)/m^3; STEL 3 mg(W)/m^3
ACGIH TLV: TWA (insoluble compounds) 5 mg(W)/m^3; STEL 10 mg(W)/m^3; (soluble compounds) 1 mg(W)/m^3; STEL 3 mg(W)/m^3

SAFETY PROFILE: Tungsten compounds are considered somewhat more toxic than those of molybdenum. However, industrially, this element does not constitute an important health hazard. Exposure is related chiefly to the dust arising from the crushing and milling of the two chief ores of tungsten, namely, scheelite and wolframite. The feeding of 2, 5, and 10% of diet as tungsten metal over a period of 70 days has shown no marked effect upon the growth of rats, as measured in terms of gain in weight. Sodium tungstate (Na_2WO_4), the most soluble salt, is moderately toxic by ingestion. Large overdoses cause central nervous system disturbances, diarrhea, respiratory failure, and death in experimental animals. Ammonium-p-tungstate has been found to be much less toxic to rats upon ingestion than either tungstic oxide or sodium tungstate. Tungsten carbide (WC) is chronically toxic to humans by inhalation although the effect may be due to cobalt content. Heavy exposure to the dust or the ingestion of large amounts of the soluble compounds produces changes in body weight, behavior, blood cells, choline esterase activity and sperm in experimental animals. See also specific compounds.

TOC550 CAS:7783-82-6 ***HR: 3***
TUNGSTEN HEXAFLUORIDE
DOT: UN 2196
mf: F_6W mw: 297.85

PROP: Colorless gas or pale yellow liquid (orthorhombic crystals when solid). Mp: 2.3°, bp: 17.5°.

SYN: TUNGSTEN FLUORIDE

CONSENSUS REPORTS: Reported in EPA TSCA Inventory.

OSHA PEL: TWA 2.5 mg(F)/m^3
ACGIH TLV: TWA 2.5 mg(F)/m^3; ACGIH TLV: TWA 5 mg(W)/m^3; STEL 10 mg(W)/m^3
DOT Classification: Corrosive Material; Label: Corrosive; Poison A; Label: Poison Gas

SAFETY PROFILE: A poison and corrosive liquid or gas.

TOC600 ***HR: 3***
TUNGSTEN-NICKEL CATALYST DUST

PROP: Composed of 50% tungstic oxide, 24% nickel oxide, and 24% sulfur (NTIMBF 115,30,72).

TOXICITY DATA with REFERENCE
orl-rat LD50:4000 mg/kg NTIMBF 115,30,72
itr-rat LDLo:120 μg/kg NTIMBF 115,30,72
orl-mus LD50:2150 mg/kg NTIMBF 115,30,72

CONSENSUS REPORTS: Nickel and its compounds are on the Community Right-To-Know List.

OSHA PEL: TWA 5 mg(W)/m^3; STEL 10 mg(W)/m^3
ACGIH TLV: TWA 5 mg(W)/m^3; STEL 10 mg(W)/m^3

SAFETY PROFILE: Poison by intratracheal route. Moderately toxic by ingestion. See TUNGSTEN OXIDE, TUNGSTEN COMPOUNDS, NICKEL MONOXIDE, NICKEL COMPOUNDS, and SULFUR.

TOC750 CAS:1314-35-8 ***HR: 2***
TUNGSTEN OXIDE
mf: O_3W mw: 231.85

PROP: Heavy, yellow powder. Insol in water; sol in caustic alkalies; very sltly sol in acids.

SYNS: C.I. 77901 ◇ TUNGSTEN BLUE ◇ TUNGSTEN TRIOXIDE ◇ TUNGSTIC ANHYDRIDE ◇ TUNGSTIC OXIDE ◇ WOLFRAMITE

TOXICITY DATA with REFERENCE
orl-rat LDLo:840 mg/kg HYSAAV 31,197,66

CONSENSUS REPORTS: Reported in EPA TSCA Inventory.

OSHA PEL: TWA 5 mg(W)/m^3
ACGIH TLV: TWA 5 mg(W)/m^3
NIOSH REL: TWA 5 mg(W)/m^3

SAFETY PROFILE: Moderately toxic by ingestion. Can react violently with ClF_3, Li, Cl_2. See also TUNGSTEN COMPOUNDS.

TOD000 CAS:11105-11-6 ***HR: 2***
TUNGSTIC ACID
mf: H_2O_4W mw: 249.87

PROP: Yellow-green powder. Insol in water and acid except hydrofluoric acid; slowly sol in solns of caustic alkalies.

SYN: KYSELINA WOLFRAMOVA (CZECH)

TOXICITY DATA with REFERENCE
eye-rbt 500 mg/24H SEV 28ZPAK -,19,72

OSHA PEL: TWA 1 mg(W)/m^3; STEL 3 mg(W)/m^3
ACGIH TLV: TWA 1 mg(W)/m^3; STEL 3 mg(W)/m^3
NIOSH REL: TWA 1 mg(W)/m^3

SAFETY PROFILE: A severe eye irritant. See also TUNGSTEN COMPOUNDS.

TOD500 CAS:8002-33-3 ***HR: 1***
TURKEY-RED OIL

PROP: A reddish, viscous liquid; characteristic odor. Flash p: 476°F (CC), d: 0.95, autoign temp: 833°F.

SYNS: RED OIL ◇ SULFATED CASTOR OIL ◇ SULFONATED CASTOR OIL

CONSENSUS REPORTS: Reported in EPA TSCA Inventory.

SAFETY PROFILE: An irritant. Combustible when exposed to heat or flame. To fight fire, use alcohol foam, CO_2, dry chemical. When heated to decomposition it emits toxic fumes of SO_x. See also CASTOR OIL.

TOD625 ***HR: D***
TURMERIC

PROP: From solvent extraction of dried ground rhizome of *Curcuma ionga L.* Bright yellow powder or yellow-orange to brown liquid; mustard taste. Misc in water.

SYN: OLEORESIN TUMERIC

TOXICITY DATA with REFERENCE
cyt-hmn:lym 6 mg/L FCTXAV 14,9,76
dni-ham:fbr 10 mg/L FCTXAV 14,9,76

SAFETY PROFILE: Human mutation data reported. When heated to decomposition it emits acrid smoke and irritating fumes.

TOD750 CAS:8006-64-2 ***HR: 3***
TURPENTINE
DOT: UN 1299

PROP: Colorless liquid, characteristic odor. Bp: 154-170°, lel: 0.8%, flash p: 95°F (CC), d: 0.854-0.868 @ 25/25°, autoign temp: 488°F, vap d: 4.84, ULC: 40-50.

SYNS: OIL of TURPENTINE ◇ OIL of TURPENTINE, rectified ◇ SPIRIT of TURPENTINE ◇ SPIRITS of TURPENTINE ◇ TEREBENTHINE (FRENCH) ◇ TERPENTIN OEL (GERMAN) ◇ TURPENTINE OIL, rectifier ◇ TURPENTINE, steam distilled

TOXICITY DATA with REFERENCE
eye-hmn 175 ppm JIHTAB 25,282,43
skn-mus TDLo:240 g/kg/20W-I:ETA CNREA8 19,413,59
orl-inf TDLo:874 mg/kg:CNS ADCHAK 28,475,53
orl-wmn TDLo:560 mg/kg:KID ADCHAK 28,475,53
ihl-hmn TCLo:175 ppm:NOSE,EYE,PUL JIHTAB 25,282,43
ihl-hmn TCLo:6 g/m^3/3H:EAR,CNS AHYGAJ 83,239,14
orl-inf LDLo:1748 mg/kg ADCHAK 28,475,53
unr-man LDLo:441 mg/kg 85DCAI 2,73,70
orl-rat LD50:5760 mg/kg PHARAT 14,435,59
ihl-mus LC50:29 g/m^3/2H TXAPA9 6,360,64
ivn-mus LD50:1180 μg/kg TXAPA9 6,360,64
ihl-gpg LCLo:16 g/m^3/1H 85GMAT -,119,82

CONSENSUS REPORTS: Reported in EPA TSCA Inventory.

OSHA PEL: TWA 100 ppm
ACGIH TLV: TWA 100 ppm
DFG MAK: 100 ppm (560 mg/m^3)

DOT Classification: Flammable Liquid; Label: Flammable Liquid; Combustible Liquid; Label: None; Flammable or Combustible Liquid; Label: Flammable Liquid.

SAFETY PROFILE: An experimental poison by intravenous route. Moderately toxic to humans by ingestion. Mildly toxic experimentally by ingestion and inhalation. Human systemic effects by ingestion and inhalation: conjunctiva irritation, other olfactory and eye effects, hallucinations or distorted perceptions, antipsychotic, headache, pulmonary and kidney changes. A human eye irritant. Irritating to skin and mucous membranes. Can cause serious irritation of kidneys. Questionable carcinogen with experimental tumorigenic data. A common air contaminant. A very dangerous fire hazard when exposed to heat or flame; can react vigorously with oxidizing materials. Avoid impregnation of combustibles with turpentine. Keep cool and ventilated. Spontaneous heating is possible. Moderate explosion hazard in the form of vapor when exposed to flame; can react violently with $Ca(OCl)_2$; Cl_2; CrO_3; $Cr(OCl)_2$; $SnCl_4$; hexachloromelamine; trichloromelamine. To fight fire, use foam, CO_2, dry chemical. When heated to decomposition it emits acrid smoke and irritating fumes.

TOE150 CAS:891-33-8 ***HR: 3***
TUTOCAINE
mf: $C_{14}H_{22}N_2O_2$ mw: 250.38

SYN: 4-(DIMETHYLAMINO)-3-METHYL-2-BUTANOLp-AMINOBENZOATE (ester)

TOXICITY DATA with REFERENCE
ivn-rat LDLo:180 mg/kg HDKU** -,-,33
scu-mus LDLo:350 mg/kg PHREA7 12,190,32
ivn-mus LDLo:50 mg/kg PHREA7 12,190,32
ipr-dog LDLo:80 mg/kg BDHU** -,-,36
ivn-dog LDLo:15 mg/kg PHREA7 12,190,32
scu-rbt LDLo:200 mg/kg PHREA7 12,190,32
ivn-rbt LDLo:15 mg/kg PHREA7 12,190,32
ipr-gpg LDLo:250 mg/kg PHREA7 12,190,32
scu-gpg LDLo:193 mg/kg PHREA7 12,190,32
ivn-gpg LDLo:30 mg/kg PHREA7 12,190,32

SAFETY PROFILE: Poison by intravenous, subcutaneous, and intraperitoneal routes. When heated to decomposition it emits toxic fumes of NO_x. See also ESTERS and AMINES.

TOE175 CAS:2571-22-4 ***HR: 3***
TUTU
mf: $C_{15}H_{18}O_6$ mw: 294.33

SYNS: 1a-β,1b,5a6,6a7a-β-HEXAHYDRO-1b-α,6-β-DIHYDROXY-8a-ISOPROPENYL-6a-α-METHYL-SPIRO(2,5-METHANO-7H-OXIRENO(3,4)CYCLOPENT(1,2-d)OXEPIN-7,2-OXIRAN)-3(2aH)-ONE ◇ TOOT POISON ◇ TUTIN ◇ TUTINE

TOXICITY DATA with REFERENCE
ipr-mus LD50:3 mg/kg JMCMAR 11,729,68
scu-mus LD50:3613 μg/kg JAPMA8 29,2,40
scu-rbt LD50:1521 μg/kg JAPMA8 29,2,40
ivn-rbt LD50:1244 μg/kg JAPMA8 29,2,40

SAFETY PROFILE: Poison by subcutaneous, intravenous, and intraperitoneal routes. When heated to decomposition it emits acrid smoke and irritating fumes.

TOE250 CAS:9005-70-3 ***HR: 1***
TWEEN 85

PROP: D: 1.055, viscosity 250-450 mPa.

SYNS: GLYCOSPERSE TO-20 ◇ POLYOXYETHYLENE (20) SORBITAN TRIOLEATE ◇ SORBIMACROGOL TRIOLEATE 300

TOXICITY DATA with REFERENCE
skn-hmn 15 mg/3D-I MLD 85DKA8 -,127,77

CONSENSUS REPORTS: Reported in EPA TSCA Inventory.

SAFETY PROFILE: A human skin irritant. A surfactant. See also SURFACTANTS.

TOE600 CAS:1401-69-0 ***HR: 3***
TYLOSIN
mf: $C_{45}H_{77}NO_{17}$ mw: 904.23

PROP: Crystals from water. Mp: 128-132°. Sol in water at 25°: 5 mg/mL. Sol in lower alc, esters, and ketones, in chlorinated hydrocarbons, benzene, ether.

SYNS: TYLAN ◇ TYLON

TOXICITY DATA with REFERENCE
orl-mus LD50:10 g/kg 85GDA2 2,135,80
ipr-mus LD50:594 mg/kg FCTXAV 4,1,66
ivn-mus LD50:400 mg/kg 85FZAT -,669,67
orl-ckn LDLo:2122 mg/kg AAGAAW -,595,60

SAFETY PROFILE: Poison by intravenous route. Moderately toxic by ingestion and intraperitoneal routes. When heated to decomposition it emits toxic fumes of NO_x. See also TYLOSIN HYDROCHLORIDE.

TOE750 CAS:11032-12-5 ***HR: 3***
TYLOSIN HYDROCHLORIDE
mf: $C_{45}H_{77}NO_{17}$•ClH mw: 940.69

PROP: Crystals. Mp: 141-145°.

TOXICITY DATA with REFERENCE
ivn-mus LD50:582 mg/kg FCTXAV 4,1,66
ivn-gpg LD50:4 mg/kg FCTXAV 4,1,66

SAFETY PROFILE: Poison by intravenous route. When heated to decomposition it emits very toxic fumes of NO_x and HCl.

TOF750 CAS:60-19-5 ***HR: 3***
TYRAMINE MONOCHLORIDE
mf: $C_8H_{11}NO$•Cl mw: 172.65

SYNS: 4-(2-AMINOETHYL)PHENOL HYDROCHLORIDE ◇ p-(2-AMINOETHYL)PHENOL MONOCHLORIDE

TOXICITY DATA with REFERENCE
ipr-mus LD50:710 mg/kg TXAPA9 28,227,74
ivn-mus LD50:208 mg/kg EJPHAZ 9,289,70
scu-gpg LDLo:1044 mg/kg JPETAB 47,339,33

CONSENSUS REPORTS: Reported in EPA TSCA Inventory.

SAFETY PROFILE: Poison by intravenous route. Moderately toxic by intraperitoneal and subcutaneous routes. When heated to decomposition it emits very toxic fumes of NO_x and Cl^-.

TOF825 CAS:865-28-1 ***HR: 3***
TYROCIDIN B
mf: $C_{68}H_{88}N_{14}O_{13}$•ClH mw: 1346.16

SYNS: 9-l-TRYPTOPHAN-TRYCODINE A HYDROCHLORIDE (9CI) ◇ TYROCIDINE B, HYDROCHLORIDE

TOXICITY DATA with REFERENCE
orl-mus LD50:1000 mg/kg 85GDA2 4(1),269,80
ipr-mus LD50:40 mg/kg 85GDA2 4(1),269,80
ivn-mus LD50:15 mg/kg 85GDA2 4(1),269,80

SAFETY PROFILE: Poison by intravenous and intraperitoneal routes. Moderately toxic by ingestion. When heated to decomposition it emits toxic fumes of NO_x and HCl.

TOG250 CAS:51-67-2 ***HR: 3***
TYROSAMINE
mf: $C_8H_{11}NO$ mw: 137.20

PROP: Leaves from benzene. Mp: 161°, bp: 175-181° @ 8 mm. Sol in hot water, alc, benzene; sltly sol in hot xylene.

SYNS: p-(2-AMINOETHYL)PHENOL ◇ p-β-AMINOETHYLPHENOL ◇ p-HYDROXYPHENETHYLAMINE ◇ p-HYDROXY-β-PHENETHYLAMINE ◇ 4-HYDROXYPHENETHYLAMINE ◇ α-(4-HYDROXYPHENYL)-β-AMINOETHANE ◇ β-HYDROXYPHENYLETHYLAMINE ◇ 2-(p-HYDROXYPHENYL)ETHYLAMINE ◇ 4-HYDROXYPHENYLETHYLAMINE ◇ SYSTOGENE ◇ TENOSIN-WIRKSTOFF ◇ TOCOSINE ◇ TYRAMINE ◇ p-TYRAMINE ◇ UTERAMINE

TOXICITY DATA with REFERENCE
ipr-mus LDLo:800 mg/kg JPHYA7 76,224,32
scu-mus LDLo:225 mg/kg 27ZIAQ -,-,65
ivn-mus LD50:229 mg/kg APTOA6 38,474,76
scu-cat LDLo:30 mg/kg HBAMAK 4,1412,35
scu-cat LDLo:30 mg/kg HBAMAK 4,1412,35

SAFETY PROFILE: Poison by subcutaneous and intravenous routes. Moderately toxic by intraperitoneal route. When heated to decomposition it emits toxic fumes of NO_x. See also AMINES.

TOG275 CAS:775-06-4 ***HR: 3***
dl-m-TYROSINE
mf: $C_9H_{11}NO_3$ mw: 181.21

SYNS: d,l-METATYROSINE ◇ m-TYROSINE, dl-

TOXICITY DATA with REFERENCE
orl-rat TDLo:1400 mg/kg (female 1-7D post):TER TXAPA9 38,251,76
ivn-mus LD50:320 mg/kg CSLNX* NX#02549

SAFETY PROFILE: Poison by intraperitoneal route. An experimental teratogen. When heated to decomposition it emits toxic fumes of NO_x.

TOG300 CAS:60-18-4 ***HR: D***
l-TYROSINE
mf: $C_9H_{11}NO_3$ mw: 181.21

PROP: Colorless, silky needles or white crystalline powder. Sol in water, dil mineral acids, alkaline solutions; sltly sol in alc.

SYNS: l-β-(p-HYDROXYPHENYL)ALANINE ◇ TYROSINE ◇ l-p-TYROSINE ◇ p-TYROSINE

TOXICITY DATA with REFERENCE
orl-rbt LDLo:3 g/kg (29-31D preg):REP BINEAA 12,282,68
orl-rat LDLo:20750 mg/kg (125-19D preg):TER TOIZAG 30,518,83

CONSENSUS REPORTS: Reported in EPA TSCA Inventory.

SAFETY PROFILE: An experimental teratogen. Experimental reproductive effects. When heated to decomposition it emits acrid smoke and irritating fumes.

TOG500 CAS:1404-88-2 ***HR: 3***
TYROTHRICIN

PROP: Gray to brown powder. Decomp @ 215-220°. Almost insol in water; sol in alc, methanol.

SYNS: BACTRATYCIN ◇ COLTIROT ◇ DUBOS CRUDE CRYSTALS ◇ HYDROTRICINE ◇ INTRADERM TYROTHRICIN

TOXICITY DATA with REFERENCE
ipr-mus LD50:100 mg/kg ARZNAD 7,98,57
ivn-mus LDLo:1200 μg/kg JPETAB 74,75,42

SAFETY PROFILE: Poison by intraperitoneal and intravenous routes. Incompatible with alkalies, strong acids.

U

UAG000 CAS:3737-72-2 ***HR: 2***
U-0290
mf: $C_{23}H_{31}NO_4 \cdot ClH$ mw: 422.01

SYN: p-(2-((DIETHYLAMINO)METHYL)BUTOXY)BENZOICACID, p-METHOXYPHENYL ESTER HYDROCHLORIDE

TOXICITY DATA with REFERENCE
skn-rbt 2500 ppm MLD AIPTAK 137,410,62
eye-rbt 2500 ppm MLD AIPTAK 137,410,62
ipr-mus LD50:479 mg/kg AIPTAK 137,410,62

SAFETY PROFILE: Moderately toxic by intraperitoneal route. A skin and eye irritant. When heated to decomposition it emits toxic fumes of NO_x and HCl. See also ESTERS.

UAG025 CAS:95004-22-1 ***HR: 3***
U-1804
mf: $C_{21}H_{29}NO_3 \cdot ClH$ mw: 379.97

SYN: p-(2-CYCLOHEXEN-1-YLOXY)BENZOIC ACID, 3-(2-METHYL-1-PYRROLIDINYL)PROPYL ESTER

TOXICITY DATA with REFERENCE
skn-rbt 2500 ppm MLD AIPTAK 137,410,62
eye-rbt 2500 ppm MLD AIPTAK 137,410,62
ipr-mus LD50:182 mg/kg AIPTAK 137,410,62

SAFETY PROFILE: Poison by intraperitoneal route. A skin and eye irritant. When heated to decomposition it emits toxic fumes of NO_x and HCl. See also ESTERS.

UAG050 ***HR: 1***
U-2363
mf: $C_{20}H_{31}NO_3 \cdot ClH$ mw: 369.98

SYN: 2,6-DIMETHYL-4-PROPOXY-BENZOIC ACID 2-METHYL-2-(1-PYRROLIDINYL)PROPYL ESTER

TOXICITY DATA with REFERENCE
skn-rbt 5 pph MLD AIPTAK 137,410,62
eye-rbt 5000 ppm MLD AIPTAK 137,410,62
ipr-mus LD50:55700 mg/kg AIPTAK 137,410,62

SAFETY PROFILE: A skin and eye irritant. When heated to decomposition it emits toxic fumes of NO_x and HCl. See also ESTERS.

UAG075 CAS:3737-66-4 ***HR: 3***
U-2397
mf: $C_{19}H_{29}NO_3 \cdot ClH$ mw: 355.95

SYN: 2,6-DIMETHYL-4-PROPOXY-BENZOIC ACID 2-(1-PYRROLIDINYL)PROPYL ESTER HYDROCHLORIDE

TOXICITY DATA with REFERENCE
skn-rbt 1 pph MLD AIPTAK 137,410,62
eye-rbt 2500 ppm MLD AIPTAK 137,410,62
ipr-mus LD50:152 mg/kg AIPTAK 137,410,62

SAFETY PROFILE: Poison by intraperitoneal route. A skin and eye irritant. When heated to decomposition it emits toxic fumes of NO_x and HCl. See also ESTERS.

UAH000 CAS:303-98-0 ***HR: 3***
UBIQUINONE 10
mf: $C_{59}H_{90}O_4$ mw: 863.49

SYNS: COENZYME Q_{10} ◇ CoQ_{10} ◇ UBIDECARENONE ◇ UBIQUINONE 50

TOXICITY DATA with REFERENCE
orl-mus TDLo:42 mg/kg (female 7-13D post):REP IYKEDH 3,306,72
orl-mus TDLo:4200 mg/kg (female 7-13D post):TER IYKEDH 3,306,72
orl-rat LDLo:4000 mg/kg NIIRDN 6,862,82
scu-rat LDLo:500 mg/kg NIIRDN 6,862,82
ivn-rat LDLo:250 mg/kg NIIRDN 6,862,82
ims-rat LDLo:500 mg/kg NIIRDN 6,862,82
orl-mus LDLo:4000 mg/kg NIIRDN 6,862,82
scu-mus LDLo:500 mg/kg NIIRDN 6,862,82
ivn-mus LDLo:500 mg/kg NIIRDN 6,862,82
ims-mus LDLo:500 mg/kg NIIRDN 6,862,82

CONSENSUS REPORTS: Reported in EPA TSCA Inventory.

SAFETY PROFILE: Poison by intravenous route. Moderately toxic by ingestion, subcutaneous, and intramuscular routes. An experimental teratogen. Experimental reproductive effects. When heated to decomposition it emits acrid smoke and irritating fumes.

UAK000 CAS:68442-69-3 ***HR: 2***
UCANE ALKYLATE 12

TOXICITY DATA with REFERENCE
skn-rbt 500 mg/1D SEV UCDS** 11/8/71
eye-rbt 5 mg SEV UCDS** 11/8/71
orl-rat LD50:2460 mg/kg UCDS** 11/8/71

CONSENSUS REPORTS: Reported in EPA TSCA Inventory.

SAFETY PROFILE: Moderately toxic by ingestion. A severe skin and eye irritant.

UBA000 CAS:107-98-2 ***HR: 1***
UCAR TRIOL HG-170

TOXICITY DATA with REFERENCE
skn-rbt 500 mg open MLD UCDS** 8/7/62
orl-rat LD50:36 g/kg UCDS** 8/7/62

CONSENSUS REPORTS: Reported in EPA TSCA Inventory.

SAFETY PROFILE: Mildly toxic by ingestion. A skin irritant.

UBJ000 CAS:102646-51-5 ***HR: 1***
UCON FLUID AP-1

TOXICITY DATA with REFERENCE
skn-rbt 500 mg open MLD UCDS** 11/17/69
orl-rat LD50:4850 mg/kg UCDS** 11/17/69
skn-rbt LD50:20 g/kg UCDS** 11/17/69

SAFETY PROFILE: Mildly toxic by ingestion and skin contact. A skin irritant.

UBS000 CAS:9038-95-3 ***HR: 2***
UCON 50-HB-55

TOXICITY DATA with REFERENCE
skn-rbt 500 mg open MLD UCDS** 3/10/70
eye-rbt 50 mg SEV UCDS** 3/10/70
orl-rat LD50:8530 mg/kg UCDS** 3/10/70

CONSENSUS REPORTS: Reported in EPA TSCA Inventory.

SAFETY PROFILE: Mildly toxic by ingestion. A skin and severe eye irritant.

UCA000 CAS:9038-95-3 ***HR: 1***
UCON 50-HB-100

TOXICITY DATA with REFERENCE
skn-rbt 500 mg open MLD UCDS** 12/12/68
orl-rat LD50:9170 mg/kg UCDS** 12/12/68
skn-rbt LD50:14 g/kg UCDS** 12/12/68

CONSENSUS REPORTS: Reported in EPA TSCA Inventory.

SAFETY PROFILE: Mildly toxic by ingestion and skin contact. A skin irritant.

UCJ000 CAS:9038-95-3 ***HR: 2***
UCON 50-HB-260

TOXICITY DATA with REFERENCE
skn-rbt 500 mg open MLD UCDS** 3/20/73
orl-rat LD50:4000 mg/kg UCDS** 3/20/73
orl-rbt LD50:1770 mg/kg TXAPA9 16,675,70

CONSENSUS REPORTS: Reported in EPA TSCA Inventory.

SAFETY PROFILE: Moderately toxic by ingestion. A skin irritant.

UDA000 CAS:9038-95-3 ***HR: 1***
UCON 50-HB-400

TOXICITY DATA with REFERENCE
skn-rbt 500 mg open MLD UCDS** 5/17/68
orl-rat LD50:5370 mg/kg UCDS** 5/17/68

CONSENSUS REPORTS: Reported in EPA TSCA Inventory.

SAFETY PROFILE: Mildly toxic by ingestion. A skin irritant.

UDJ000 CAS:9038-95-3 ***HR: 1***
UCON 50-HB-660

TOXICITY DATA with REFERENCE
skn-rbt 500 mg open MLD UCDS** 4/12/65
orl-rat LD50:18 g/kg UCDS** 4/12/65

CONSENSUS REPORTS: Reported in EPA TSCA Inventory.

SAFETY PROFILE: Mildly toxic by ingestion. A skin irritant.

UDS000 CAS:9038-95-3 ***HR: 1***
UCON 50-HB-2000

TOXICITY DATA with REFERENCE
skn-rbt 500 mg open MLD UCDS**
orl-rat LD50:21 g/kg UCDS**

CONSENSUS REPORTS: Reported in EPA TSCA Inventory.

SAFETY PROFILE: Mildly toxic by ingestion. A skin irritant.

UEA000 CAS:9038-95-3 ***HR: 1***
UCON 50-HB-3520

PROP: D: 1.04.

TOXICITY DATA with REFERENCE
skn-rbt 500 mg open MLD UCDS** 3/30/65
orl-rat LD50:38 g/kg UCDS** 3/30/65

CONSENSUS REPORTS: Reported in EPA TSCA Inventory.

SAFETY PROFILE: Mildly toxic by ingestion. A skin irritant.

UEJ000 CAS:9038-95-3 ***HR: 1***
UCON 50-HB-5100

TOXICITY DATA with REFERENCE
skn-rbt 500 mg open MLD TXAPA9 16,657,70
orl-rat LD50:49 g/kg TXAPA9 16,657,70

CONSENSUS REPORTS: Reported in EPA TSCA Inventory.

SAFETY PROFILE: Mildly toxic by ingestion. A skin irritant.

UFA000 CAS:9038-95-3 ***HR: 1***
UCON 50-HB-280-X

TOXICITY DATA with REFERENCE
skn-rbt 500 mg open MLD UCDS** 10/29/68
orl-rat LD50:6130 mg/kg UCDS** 10/29/68

CONSENSUS REPORTS: Reported in EPA TSCA Inventory.

SAFETY PROFILE: Mildly toxic by ingestion. A skin irritant.

UHA000 CAS:52581-71-2 ***HR: 1***
UCON LO-500
mf: $(C_3H_6O)_n \cdot C_{18}H_{36}O$

SYN: α-9-OCTADECENYL-ω-HYDROXYPOLY(OXY(METHYL-1,2-ETHANEDIYL))

TOXICITY DATA with REFERENCE
skn-rbt 500 mg open MLD UCDS** 7/8/69
orl-rat LD50:45 g/kg UCDS** 7/8/69

CONSENSUS REPORTS: Reported in EPA TSCA Inventory.

SAFETY PROFILE: Mildly toxic by ingestion. A skin irritant. When heated to decomposition it emits acrid smoke and irritating fumes.

UIA000 CAS:25736-79-2 ***HR: 1***
UCON LUBRICANT DLB-62-E

TOXICITY DATA with REFERENCE
skn-rbt 500 mg open MLD UCDS** 10/14/64
orl-rat LD50:20 g/kg UCDS** 10/14/64

CONSENSUS REPORTS: Reported in EPA TSCA Inventory.

SAFETY PROFILE: Mildly toxic by ingestion. A skin irritant.

UIJ000 CAS:25736-79-2 ***HR: 1***
UCON LUBRICANT DLB-140-E

TOXICITY DATA with REFERENCE
skn-rbt 500 mg open MLD UCDS** 10/14/64
orl-rat LD50:57 g/kg UCDS** 10/14/64

CONSENSUS REPORTS: Reported in EPA TSCA Inventory.

SAFETY PROFILE: Mildly toxic by ingestion. A skin irritant.

UIS000 CAS:25736-79-2 ***HR: 1***
UCON LUBRICANT DLB-200-E

TOXICITY DATA with REFERENCE
skn-rbt 500 mg open MLD UCDS** 10/14/64
skn-rbt LD50:20 g/kg UCDS** 10/14/64

CONSENSUS REPORTS: Reported in EPA TSCA Inventory.

SAFETY PROFILE: Mildly toxic by skin contact. A skin irritant.

UIS300 ***HR: 1***
ULATKAMBAL ROOT EXTRACT

PROP: Indian plant belonging to the family *Sterculiaceae* IJMRAQ 63,378,75

SYN: ABROMA AUGUSTA Linn., root extract

TOXICITY DATA with REFERENCE
orl-mus TDLo:300 mg/kg (female 1-6D post):REP IJMRAQ 63,378,75
ipr-mus LD50:1 g/kg IJEBA6 18,594,80

SAFETY PROFILE: Slightly toxic by intraperitoneal route. Experimental reproductive effects. When heated to decomposition it emits acrid smoke and irritating fumes.

UJA200 CAS:57455-37-5 ***HR: 1***
ULTRAMARINE BLUE
mf: $Na_7Al_6Si_6O_{24}S_3$ mw: 971.50

PROP: Calcined mixture of kaolin, sulfur, sodium carbonate, and carbon above 700°.

SAFETY PROFILE: A nuisance dust.

UJA800 ***HR: 1***
Γ-UNDECALACTONE
mf: $C_{11}H_{20}O_2$ mw: 184.28

PROP: Colorless to slightly yellow liquid; peach odor. D: 0.825, refr index: 1.430, flash p: 279°F. Sol in fixed oils, propylene glycol; insol in glycerine, water @ 223°.

SYNS: ALDEHYDE C-14 PURE ◇ FEMA No. 3091 ◇ PEACH ALDEHYDE

SAFETY PROFILE: Combustible liquid. When heated to decomposition it emits acrid smoke and irritating fumes.

UJJ000 CAS:112-44-7 ***HR: 1***
1-UNDECANAL
mf: $C_{11}H_{22}O$ mw: 170.33

PROP: Colorless to sltly yellow liquid; sweet, fatty, floral odor. Mp: −4°, bp: 117° @ 18 mm, flash p: 235°F (COC), d: 0.830 @ 20/4°, refr index: 1.430, vap press: 0.04 mm @ 20°, vap d: 5.94. Sol in fixed oils, propylene glycol; glycerin, water @ 223°. Reported in lemon and mandarin oils (FCTXAV 11,477,73).

SYNS: ALDEHYDE-14 ◇ 1-DECYL ALDEHYDE ◇ FEMA No. 3092 ◇ HENDECANAL ◇ HENDECANALDEHYDE ◇ UNDECANAL ◇ n-UNDECANAL ◇ UNDECANALDEHYDE ◇ UNDECYL ALDEHYDE ◇ N-UNDECYL ALDEHYDE ◇ UNDECYLIC ALDEHYDE

TOXICITY DATA with REFERENCE
skn-rbt 500 mg MLD FCTXAV 11,1079,73

CONSENSUS REPORTS: Reported in EPA TSCA Inventory.

SAFETY PROFILE: A skin irritant. Combustible liquid when exposed to heat or flame. To fight fire, use CO_2, dry chemical. When heated to decomposition it emits acrid smoke and irritating fumes. See also ALDEHYDES.

UJS000 CAS:1120-21-4 ***HR: 2***
UNDECANE
DOT: UN 2330
mf: $C_{11}H_{24}$ mw: 156.35

PROP: Colorless liquid. D: 0.7402 @ 20/4°, fp: −25.75°, bp: 195.6°, flash p: 149°F (OC), vap d: 5.4. Insol in water.

SYNS: HENDECANE ◇ n-UNDECANE

TOXICITY DATA with REFERENCE
ivn-mus LD50:517 mg/kg JPMSAE 67,566,78

CONSENSUS REPORTS: Reported in EPA TSCA Inventory.

DOT Classification: Flammable or Combustible Liquid; Label: Flammable Liquid.

SAFETY PROFILE: Moderately toxic by intravenous route. Combustible when exposed to heat, flame or oxidizers. To fight fire, use foam, mist, dry chemical. Emitted from modern building materials. (CENEAR 69,22,91) When heated to decomposition it emits acrid smoke and irritating fumes. See also ALKANES

UKA000 CAS:112-37-8 ***HR: 3***
UNDECANOIC ACID
mf: $C_{11}H_{22}O_2$ mw: 186.33

SYNS: 1-DECANECARBOXYLIC ACID ◇ HENDECANOIC ACID ◇ n-UNDECOIC ACID ◇ UNDECYLIC ACID

TOXICITY DATA with REFERENCE
skn-rbt 150 mg/24H MLD TXAPA9 21,369,72
ivn-mus LD50:140 mg/kg APTOA6 18,141,61

CONSENSUS REPORTS: Reported in EPA TSCA Inventory.

SAFETY PROFILE: Poison by intravenous route. A skin irritant. When heated to decomposition it emits acrid smoke and irritating fumes.

UKJ000 CAS:710-04-3 ***HR: 1***
UNDECANOLIDE-1,5
mf: $C_{11}H_{20}O_2$ mw: 184.31

SYNS: 5-HYDROXYUNDECANOIC ACID LACTONE ◇ Δ-UNDECALACTONE

TOXICITY DATA with REFERENCE
skn-rbt 500 mg/24H MOD FCTXAV 14,659,76

CONSENSUS REPORTS: Reported in EPA TSCA Inventory.

SAFETY PROFILE: A skin irritant. When heated to decomposition it emits acrid smoke and irritating fumes.

UKS000 CAS:112-12-9 ***HR: 2***
2-UNDECANONE
mf: $C_{11}H_{22}O$ mw: 170.33

PROP: Colorless liquid. Mp: 12°, bp: 223°, flash p: 192°F (CC), d: 0.829 @ 30°, vap d: 5.9. Insol in water.

SYNS: 2-HENDECANONE ◇ METHYL NONYL KETONE ◇ METHYL-n-NONYL KETONE ◇ MGK DOG AND CAT REPELLENT ◇ NONYL METHYL KETONE

TOXICITY DATA with REFERENCE
orl-rat LD50:5000 mg/kg FMCHA2 -,D200,80
orl-mus LD50:3880 mg/kg APJUA8 12,79,62

CONSENSUS REPORTS: Reported in EPA TSCA Inventory.

SAFETY PROFILE: Moderately toxic by ingestion. Combustible when exposed to heat or flame; can react with oxidizing materials. To fight fire, use CO_2, dry chemical. When heated to decomposition it emits acrid smoke and irritating fumes. See also KETONES.

ULA000 CAS:927-49-1 ***HR: 3***
6-UNDECANONE
mf: $C_{11}H_{22}O$ mw: 170.33

SYNS: DIAMYL KETONE ◇ DIPENTYL KETONE ◇ 6-OXOUNDECANE

TOXICITY DATA with REFERENCE
orl-rat LDLo:1000 mg/kg CTOXAO 17,271,80
ivn-mus LD50:117 mg/kg JPMSAE 67,566,78

CONSENSUS REPORTS: Reported in EPA TSCA Inventory.

SAFETY PROFILE: Poison by intravenous route. Moderately toxic by ingestion. When heated to decomposition it emits acrid smoke and irritating fumes. See also KETONES.

ULJ000 CAS:112-45-8 ***HR: 1***
10-UNDECENAL
mf: $C_{11}H_{20}O$ mw: 168.31

PROP: Colorless to light yellow liquid; rose odor. D: 0.840-0.850, refr index: 1.441-1.447, flash p: 212°F. Sol in fixed oils, propylene glycol; insol in water @ 235°, glycerin.

SYNS: ALDEHYDE C-11, UNDECYLENIC ◇ FEMA No. 3095 ◇ HENDECENAL ◇ 1-UNDECEN-10-AL ◇ UNDECYLENALDEHYDE ◇ 10-UNDECYLENEALDEHYDE ◇ UNDECYLENIC ALDEHYDE

TOXICITY DATA with REFERENCE
skn-rbt 500 mg MLD FCTXAV 11,1079,73

CONSENSUS REPORTS: Reported in EPA TSCA Inventory.

SAFETY PROFILE: A skin irritant. Combustible liquid. When heated to decomposition it emits acrid smoke and irritating fumes. See also ALDEHYDES.

ULS000 CAS:112-38-9 ***HR: 2***
10-UNDECENOIC ACID
mf: $C_{11}H_{20}O_2$ mw: 184.31

PROP: Bright, clear, mobile liquid or crystals. Mp: 24.5°, bp: 160° @ 10 mm, flash p: 295°F (COC), d: 0.910 @ 25/25°.

SYNS: DESENEX ◇ 10-HENEDECENOIC ACID ◇ UNDECYLENIC ACID ◇ UNDECYL-10-ENIC ACID ◇ 9-UNDECYLENIC ACID ◇ 10-UNDECYLENIC ACID

TOXICITY DATA with REFERENCE
skn-rbt 500 mg/24H SEV FCTXAV 16,637,78
orl-rat LD50:2500 mg/kg 28ZEAL 4,386,69
orl-mus LD50:8500 mg/kg JIDEAE 13,145,49
ipr-mus LD50:960 mg/kg JIDEAE 13,145,49

CONSENSUS REPORTS: Reported in EPA TSCA Inventory.

SAFETY PROFILE: Moderately toxic by ingestion and intraperitoneal routes. A severe skin irritant. Ingestion can cause nausea, vomiting, and urticaria (hives). Combustible when exposed to heat or flame. To fight fire, use foam, CO_2, dry chemical. When heated to decomposition it emits acrid smoke and irritating fumes.

ULS400 CAS:5760-50-9 ***HR: 1***
9-UNDECENOIC ACID, METHYL ESTER
mf: $C_{12}H_{22}O_2$ mw: 198.34

SYN: METHYL 9-UNDECENOATE ◇ METHYL 10-UNDECENOATE ◇ METHYL UNDECYLENATE

TOXICITY DATA with REFERENCE
skn-rbt 500 mg/24H MOD FCTOD7 20,767,82
orl-rat LD50:3 g/kg FCTOD7 20,767,82

CONSENSUS REPORTS: Reported in EPA TSCA Inventory.

SAFETY PROFILE: A skin irritant. When heated to decomposition it emits acrid smoke and irritating fumes.

UMA000 CAS:112-43-6 ***HR: 1***
10-UNDECENOL

PROP: Found in leaves of *Litsea odorifera val.* (FCTXAV 11,95,73).
mf: $C_{11}H_{22}O$ mw: 170.33

SYNS: ALCOHOL C-11 ◇ 1-UNDECEN-11-OL ◇ ω-UNDECENYL ALCOHOL ◇ UNDECYLENIC ALCOHOL

TOXICITY DATA with REFERENCE
skn-rbt 500 mg MLD FCTXAV 11,1079,73

CONSENSUS REPORTS: Reported in EPA TSCA Inventory.

SAFETY PROFILE: A skin irritant. When heated to decomposition it emits acrid smoke and irritating fumes. See also ALCOHOLS.

UMJ000 CAS:38460-95-6 ***HR: 3***
10-UNDECENOYL CHLORIDE
mf: $C_{11}H_{19}ClO$ mw: 202.75

SYNS: ω-UNDECYLENIC ACID CHLORIDE ◇ 10-UNDECYLENOYL CHLORIDE

TOXICITY DATA with REFERENCE
ivn-mus LD50:56 mg/kg CSLNX* NX#08967

CONSENSUS REPORTS: Reported in EPA TSCA Inventory.

SAFETY PROFILE: Poison by intravenous route. When heated to decomposition it emits toxic fumes of Cl^-. See also CHLORINATED HYDROCARBONS, ALIPHATIC.

UMS000 CAS:112-19-6 ***HR: 1***
UNDECENYL ACETATE
mf: $C_{13}H_{24}O_2$ mw: 212.37

SYNS: ACETATE C-11 ◇ 10-HENDECEN-1-YL ACETATE ◇ 10-UNDECENYL ACETATE

TOXICITY DATA with REFERENCE
skn-rbt 500 mg/24H MLD FCTXAV 14,659,76

CONSENSUS REPORTS: Reported in EPA TSCA Inventory.

SAFETY PROFILE: A skin irritant. When heated to decomposition it emits acrid smoke and irritating fumes.

UNA000 CAS:112-42-5 ***HR: 2***
UNDECYL ALCOHOL
mf: $C_{11}H_{24}O$ mw: 172.35

PROP: Colorless liquid; fatty-floral odor. D: 0.820-0.840, refr index: 1.437-1.443, mp: 19°, bp: 131° @ 15 mm, flash p: 234°F. Sol in fixed oils; insol in water.

SYNS: ALCOHOL C-11 ◇ FEMA No. 3097 ◇ HENDECANOIC ALCOHOL ◇ 1-HENDECANOL ◇ HENDECYL ALCOHOL ◇ n-HENDECYLENIC ALCOHOL ◇ n-UNDECANOL

TOXICITY DATA with REFERENCE
skn-rbt 10 mg/24H JIHTAB 26,269,44
skn-rbt 500 mg/24H MOD FCTXAV 16,637,78
orl-rat LD50:3000 mg/kg JIHTAB 26,269,44

CONSENSUS REPORTS: Reported in EPA TSCA Inventory.

SAFETY PROFILE: Moderately toxic by ingestion. A skin irritant. Combustible liquid. When heated to decomposition it emits acrid smoke and irritating fumes. See also ALCOHOLS.

UNA100 CAS:67785-74-4 ***HR: 1***
UNDECYLENIC ALDEHYDE DIGERANYL ACETAL
mf: $C_{31}H_{54}O_2$ mw: 458.85

SYNS: 11,11-BIS-((3,7-DIMETHYL-2,6-OCTADIENYL)OXY)-1-UNDECENE ◇ 11,11-DIGERANYLOXY-1-UNDECENE ◇ 10-UNDECENAL DIGERANYL ACETAL ◇ 1-UNDECENE, 11,11-BIS((3,7-DIMETHYL-2,6-OCTADIENYL)OXY)-

TOXICITY DATA with REFERENCE
skn-rbt 500 mg/24H MOD FCTOD7 20,845,82

CONSENSUS REPORTS: Reported in EPA TSCA Inventory.

SAFETY PROFILE: A skin irritant. When heated to decomposition it emits acrid smoke and irritating fumes.

UNJ800 CAS:66-22-8 ***HR: 3***
URACIL
mf: $C_4H_4N_2O_2$ mw: 112.10

PROP: Needles from water. Mp: 335° with effervescence. Freely sol in hot water; sparingly sol in cold water (100 parts of water at 25° dissolves 0.358 part of uracil); almost insol in alc, ether; sol in ammonia water and in other alkalies.

SYNS: 2,4-DIHYDROXYPYRIMIDINE ◇ 2,4-DIOXOPYRIMIDINE ◇ HYBAR X ◇ PIROD ◇ 2,4-PYRIMIDINEDIOL ◇ 2,4-PYRIMIDINEDIONE ◇ 2,4(1H,3H)-PYRIMIDINEDIONE (9CI) ◇ PYROD

TOXICITY DATA with REFERENCE
pic-esc 1 g/L ZAPOAK 12,583,72
cyt-mus-ipr 15 mg/kg NULSAK 19,40,76
orl-rat TDLo:18 g/kg (17-22D preg/21D post):REP OYYAA2 22,109,81
orl-rat TDLo:15400 mg/kg (7-17D post):TER OYYAA2 22,85,81
orl-rat TDLo:235 g/kg/20W-C:ETA CALEDQ 34,249,87
ipr-mus LD50:1513 mg/kg JPETAB 207,504,78

CONSENSUS REPORTS: EPA Genetic Toxicology Program. Reported in EPA TSCA Inventory.

SAFETY PROFILE: Moderately toxic by intraperitoneal route. An experimental teratogen. Experimental reproductive effects. Questionable carcinogen with experimental tumorigenic data. Mutation data reported. When heated to decomposition it emits toxic fumes of NO_x.

UNJ810 CAS:74578-38-4 ***HR: 3***
URACIL mixture with TEGAFUR (4:1)
mf: $C_8H_9FN_2O_3•4C_4H_4N_2O_2$ mw: 648.59

SYNS: FT mixture with URACIL (1:4) ◇ TEGAFUR mixture with URACIL (1:4) ◇ 1-(2-TETRAHYDROFURYL)-5-FLUOROURACIL mixture with URACIL (1:4) ◇ UFT ◇ URACIL mixture with FT (4:1) ◇ URACIL mixture with 1-(2-TETRAHYDROFURYL)-5-FLUOROURACIL (4:1)

TOXICITY DATA with REFERENCE
orl-rat TDLo:891 mg/kg (7-17D preg):TER OYYAA2 22,85,81
orl-rat TDLo:891 mg/kg (7-17D preg):REP OYYAA2 22,85,81
orl-rat LD50:1580 mg/kg OYYAA2 20,1009,80
orl-mus LD50:1275 mg/kg OYYAA2 20,1009,80
orl-dog LD50:150 mg/kg OYYAA2 20,1009,80
orl-rbt LD50:242 mg/kg OYYAA2 20,1009,80

SAFETY PROFILE: Poison by ingestion. An experimental teratogen. Experimental reproductive effects. When heated to decomposition it emits toxic fumes of F^- and NO_x. See also URACIL.

UNS000 CAS:7440-61-1 ***HR: 3***
URANIUM
DOT: UN 2979
af: U aw: 238.00

PROP: A heavy, silvery-white, malleable, ductile, softer-than-steel, metallic element. Mp: 1132°, bp: 3818°, d: 18.95 (ca). Radioactive material.

SYN: URANIUM METAL, PYROPHORIC (DOT)

CONSENSUS REPORTS: Reported in EPA TSCA Inventory.

OSHA PEL: (Transitional: TWA Soluble Compounds: 0.05 mg(U)/m^3; Insoluble Compounds 0.25 mg(U)/m^3) TWA Soluble Compounds: 0.05 mg(U)/m^3; Insoluble Compounds 0.2 mg(U)/m^3; STEL 0.6 mg(U)/m^3
ACGIH TLV: TWA 0.2 mg(U)/m^3; STEL 0.6 mg (U)/m^3
DFG MAK: 0.25 mg/m^3
DOT Classification: Radioactive Material; Label: Radioactive and Flammable.

SAFETY PROFILE: A highly toxic element on an acute basis. The permissible levels for soluble compounds are based on chemical toxicity, while the permissible body level for insoluble compounds is based on radiotoxicity. The high chemical toxicity of uranium and its salts is largely shown in kidney damage which may not be reversible. Acute arterial lesions may occur after acute exposures. The most soluble uranium compounds are UF_6, $UO_2(NO_3)_2$, UO_2Cl_2, UO_2F_2, and uranyl acetates, sulfates, and carbonates. Some moderately soluble compounds are UF_4, UO_2, UO_4, $(NH_4)_2U_2O_7$, UO_3, and uranyl nitrates. The rapid passage of soluble uranium compounds through the body tends to allow relatively large amounts to be absorbed. Soluble uranium compounds may be absorbed through the skin. The least soluble compounds are high-fired UO_2, U_3O_8, and uranium hydrides and carbides. The high toxicity effect of insoluble compounds is largely due to lung irradiation by inhaled particles. This material is transferred from the lungs of animals quite slowly.

A very dangerous fire hazard in the form of a solid or dust when exposed to heat or flame. It can react violently with air; Cl_2; F_2; HNO_3; NO; Se; S; water; NH_3; BrF_3; trichloroethylene; nitryl fluoride. During storage it may form a pyrophoric surface due to effects of air and moisture. Depleted uranium (the ^{238}U-by-product of the uranium enrichment process, with relatively low radioactivity) is used in armor-piercing shells, ship or aircraft ballast, and counterbalances. Uranium is also used in making colored ceramic glazes.

UOA000 CAS:55042-15-4 ***HR: 3***
URANIUM AZIDE PENTACHLORIDE

OSHA PEL: (Transitional: TWA 0.05 mg(U)/m^3) TWA 0.05 mg(U)/m^3
ACGIH TLV: TWA 0.2 mg(U)/m^3; STEL 0.6 mg (U)/m^3
mf: Cl_5N_3U mw: 457.32

SAFETY PROFILE: A radioactive material. An explosive. When heated to decomposition it emits very toxic fumes of Cl^- and NO_x. See also URANIUM and AZIDES.

UOB100 CAS:12070-09-6 ***HR: 3***
URANIUM CARBIDE
mf: UC mw: 250.04

OSHA PEL: (Transitional: TWA 0.25 mg(U)/m^3) TWA 0.2 mg(U)/m^3; STEL 0.6 mg(U)/m^3
ACGIH TLV: TWA 0.2 mg(U)/m^3; STEL 0.6 mg(U)/m^3

SAFETY PROFILE: A radioactive material. The powdered carbide ignites spontaneously in air. See also URANIUM.

UOC200 CAS:12071-33-9 ***HR: 3***
URANIUM DICARBIDE
mf: C_2U mw: 262.05

OSHA PEL: (Transitional: TWA 0.25 mg(U)/m^3) TWA 0.2 mg(U)/m^3; STEL 0.6 mg(U)/m^3
ACGIH TLV: TWA 0.2 mg(U)/m^3; STEL 0.6 mg (U)/m^3

SAFETY PROFILE: A radioactive material. Ignites when ground or when heated in air to 400°C. Reacts violently with warm water. Incandescent reaction in fluorine; chlorine (at 300°C); bromine (at 390°C). See also URANIUM.

UOJ000 CAS:7783-81-5 ***HR: 3***
URANIUM FLUORIDE (fissile)
DOT: UN 2977/UN 2978
mf: F_6U mw: 352.00

PROP: Containing more than 1% U-235 (DOT).

SYN: URANIUM HEXAFLUORIDE, FISSILE (containing more than 1% U-235) (DOT)

OSHA PEL: (Transitional: TWA 0.05 mg(U)/m^3) TWA Soluble Compounds: 0.05 mg(U)/m^3
ACGIH TLV: TWA 0.2 mg(U)/m^3; STEL 0.6 mg (U)/m^3; 2.5 mg(F)/m^3
DOT Classification: Radioactive Material; Label: Radioactive and Corrosive.

SAFETY PROFILE: Radioactive poison. A corrosive irritant to skin, eyes and mucous membranes. Violent reaction with hydroxy compounds (e.g., ethanol, water). Vigorous reaction with aromatic hydrocarbons (e.g., benzene, toluene, xylene). When heated to decomposition it emits toxic fumes of F^-. See also FLUORIDES and URANIUM.

UOS000 CAS:7783-81-5 ***HR: 3***
URANIUM FLUORIDE (low specific activity)
DOT: UN 2978
mf: F_6U mw: 352.00

PROP: Containing 0.7% or less U-235 (FEREAC 41,15972,76).

SYN: URANIUM HEXAFLUORIDE, LOW SPECIFIC ACTIVITY (containing 0.7% or less U-235) (DOT)

OSHA PEL: (Transitional: TWA 0.05 mg(U)/m^3) TWA Soluble Compounds: 0.05 mg(U)/m^3
ACGIH TLV: TWA 0.2 mg(U)/m^3; STEL 0.6 mg (U)/m^3; 2.5 mg(F)/m^3
DOT Classification: Radioactive Material; Label: Radioactive and Corrosive.

SAFETY PROFILE: Radioactive toxicity. A corrosive irritant to skin, eyes, and mucous membranes. When heated to decomposition it emits toxic fumes of F^-. See also URANIUM and FLUORIDES.

UPA000 CAS:13598-56-6 ***HR: 3***
URANIUM(III) HYDRIDE
mf: H_3U mw: 241.06

SAFETY PROFILE: A radioactive material. The powder ignites spontaneously in air or on contact with water. Potentially explosive reaction with halocarbons. See also HYDRIDES and URANIUM.

UPJ000 CAS:1344-57-6 ***HR: 3***
URANIUM(IV) OXIDE
mf: O_2U mw: 270.03

SAFETY PROFILE: A radioactive material. It ignites spontaneously in heated air and burns brilliantly. See also URANIUM.

UPS000 CAS:541-09-3 ***HR: 3***
URANIUM OXYACETATE
DOT: NA 9180
mf: $C_4H_6O_6U{\cdot}2H_2O$ mw: 424.19

PROP: Mp: loses $2H_2O$ @ 110°, bp: 275° (decomp), d: 2.893 @ 15°.

SYNS: URANIUM ACETATE ◇ URANYL ACETATE

TOXICITY DATA with REFERENCE
ipr-mus LD50:400 mg/kg REPMBN 10,391,62

CONSENSUS REPORTS: Reported in EPA TSCA Inventory.

OSHA PEL: TWA 0.05 mg(U)/m^3
ACGIH TLV: TWA 0.2 mg(U)/m^3
DOT Classification: Radioactive Material; Label: Radioactive.

SAFETY PROFILE: Poison by intraperitoneal route. A radioactive material. See also URANIUM.

UQA000 CAS:13536-84-0 ***HR: 3***
URANIUM OXYFLUORIDE
mf: F_2O_2U mw: 308.00

SYNS: URANIUM FLUORIDE OXIDE ◇ URANYL FLUORIDE

TOXICITY DATA with REFERENCE
ivn-rat LD50:40 mg/kg 14CYAT 2,1167,63

CONSENSUS REPORTS: Reported in EPA TSCA Inventory.

OSHA PEL: TWA 0.05 mg(U)/m^3; 2.5 mg(F)/m^3
ACGIH TLV: TWA 0.2 mg(U); 2.5 mg(F)/m^3
NIOSH REL: TWA 2.5 mg(F)/m^3

SAFETY PROFILE: Poison by intravenous route. When heated to decomposition it emits toxic fumes of F^-. See also FLUORIDES and URANIUM.

UQJ000 CAS:10026-10-5 ***HR: 3***
URANIUM TETRACHLORIDE
mf: Cl_4U mw: 379.80

PROP: Cubic, dark green-gray deliquescent crystals. Mp: 590°, bp: 791°, d: 4.725 @ 25/4°. Freely sol in water (decomp); insol in hydrocarbons, ethyl, ether. Should be stored in sealed ampules.

SYN: URANIUM(IV) CHLORIDE

TOXICITY DATA with REFERENCE
ipr-rat LD50:335 mg/kg 28ZFAO -,-,49

CONSENSUS REPORTS: Reported in EPA TSCA Inventory.

OSHA PEL: TWA 0.05 mg(U)/m^3
ACGIH TLV: TWA 0.2 mg(U)/m^3; STEL 0.6 mg (U)/m^3

SAFETY PROFILE: Poison by intraperitoneal route. When heated to decomposition it emits toxic fumes of Cl^-. See also URANIUM.

UQS000 ***HR: 3***
URANIUM(III) TETRAHYDROBORATE
mf: $B_3H_{12}U$ mw: 282.55

SAFETY PROFILE: Very toxic. Ignites spontaneously in air. Explodes on heating. A radioactive material. See URANIUM and BORON COMPOUNDS.

UQT300 ***HR: 3***
URANIUM(IV) TETRAHYDROBORATE
mf: $B_4H_{16}U$ mw: 297.40

SAFETY PROFILE: A radioactive material. Adduct with dimethyl ether ignites on contact with water. The diethyl ether and bis-tetrahydrofuran adducts explode on contact with water. See also BORON COMPOUNDS and URANIUM.

UQT700 CAS:6159-44-0 ***HR: 3***
URANYL ACETATE DIHYDRATE
mf: $C_4H_6O_6U \cdot 2H_2O$ mw: 424.14

SYNS: BIS(ACETO)DIOXOURANIUM DIHYDRATE ◇ BIS(ACETO-O) DIOXOURANIUM DIHYDRATE ◇ URANIUM, BIS(ACETATO)DIOXO-, DIHYDRATE ◇ URANIUM, BIS(ACETO-O)DIOXO-, DIHYDRATE (9CI)

TOXICITY DATA with REFERENCE
orl-rat TDLo:50 mg/kg (female 6-15D post):TER TXCYAC 55,143,89
orl-rat LD50:204 mg/kg BECTA6 39,168,87
scu-rat LD50:8300 μg/kg BECTA6 39,168,87
orl-mus LD50:242 mg/kg BECTA6 39,168,87
scu-mus LD50:20400 μg/kg BECTA6 39,168,87

SAFETY PROFILE: Poison by ingestion and subcutaneous routes. An experimental teratogen. When heated to decomposition it emits toxic fumes of uranium.

URA000 CAS:7791-26-6 ***HR: 3***
URANYL CHLORIDE
mf: Cl_2O_2U mw: 340.90

PROP: Yellow, deliq crystals. Mp: less than red heat. Very hygroscopic. Volatile above 775°. Very sol in water; sol in alc, acetone; insol in benzene. Unstable in aq solutions.

TOXICITY DATA with REFERENCE
ipr-mus LD50:10 mg/kg COREAF 256,1043,63

CONSENSUS REPORTS: Reported in EPA TSCA Inventory.

OSHA PEL: TWA 0.05 mg(U)/m^3
ACGIH TLV: TWA 0.2 mg(U)/m^3; STEL 0.6 mg (U)/m^3

SAFETY PROFILE: Poison by intraperitoneal route. A radioactive material. When heated to decomposition it emits toxic fumes of Cl^-. See also URANIUM.

URA100 CAS:36478-76-9 ***HR: 3***
URANYL NITRATE
mf: N_2O_8U mw: 394.02

SYN: URANIUM, BIS(NITRATO-O,O')DIOXO-, (OC-6-11)-

TOXICITY DATA with REFERENCE
scu-mus TDLo:31522 μg/kg (male 30D pre):REP JRPFA4 7,21,64
ivn-rat LD50:1655 μg/kg 38MKAJ 2A,2000,81
ivn-mus LD50:200 μg/kg 38MKAJ 2A,2000,81
ivn-rbt LD50:166 μg/kg 38MKAJ 2A,2000,81

ACGIH TLV: TWA 0.2 mg(U)/m^3; STEL 0.6 mg(U)/m^3

SAFETY PROFILE: Poison by intravenous route. Experimental reproductive effects. When heated to decomposition it emits toxic fumes of NO_x and U.

URA200 CAS:10102-06-4 ***HR: 3***
URANYL NITRATE (solid)
DOT: UN 2981
mf: N_2O_8U mw: 394.02

SYN: BIS(NITRATO-O,O')DIOXO URANIUM (solid)

TOXICITY DATA with REFERENCE
cyt-hmn:leu 100 μg/L DBLRAC 17(4),375,73

CONSENSUS REPORTS: Reported in EPA TSCA Inventory.

OSHA PEL: (Transitional: TWA0.25 mg(U)/m^3) TWA 0.2 mg(U)/m^3; STEL 0.6 mg(U)/m^3
ACGIH TLV: TWA 0.2 mg(U)/m^3; STEL 0.6 mg (U)/m^3
DOT Classification: Label: Radioactive and Oxidizer.

SAFETY PROFILE: Poison by ingestion and inhalation. Human mutation data reported. A corrosive irritant to skin, eyes, and mucous membranes. A radioactive material. A powerful explosive and oxidizer. Incompatible with cellulose. Ether solutions in sunlight may explode. When heated to decomposition it emits toxic fumes of NO_x. See also URANYL NITRATE HEXAHYDRATE and URANIUM.

URS000 CAS:13520-83-7 ***HR: 3***
URANYL NITRATE HEXAHYDRATE
DOT: UN 2980
mf: $N_2O_8U \cdot 6H_2O$ mw: 502.14

PROP: Rhombic, deliquescent, yellow crystals. Mp: 60.2°, bp: 118°, decomp @ 100°, d: 2.807 @ 13°.

SYNS: BIS(NITRATO)DIOXOURANIUM HEXAHYDRATE ◇ DINITRATODIOXOURANIUM, HEXAHYDRATE ◇ URANYL NITRATE HEXAHYDRATE, solution (DOT)

TOXICITY DATA with REFERENCE
cyt:ham:lng 180 mg/L MUREAV 85,288,81
ipr-rat LD50:135 mg/kg EQSSDX 1,1,75
ivn-rat LDLo:2 mg/kg 28ZFAO -,307,49
ipr-mus LD50:42 mg/kg 28ZFAO -,305,49
ivn-mus LDLo:21 mg/kg 28ZFAO -,307,49
orl-dog LDLo:12 mg/kg HBTXAC 1,310,56
scu-dog LD50:4 mg/kg 28ZFAO -,282,49
ivn-dog LDLo:6750 μg/kg EQSSDX 1,1,75
orl-cat LDLo:238 mg/kg HBTXAC 1,310,56
scu-rbt LD50:1470 μg/kg 28ZFAO -,282,49
ivn-rbt LDLo:210 μg/kg 28ZFAO -,307,49
ivn-gpg LDLo:630 μg/kg EQSSDX 1,1,75

OSHA PEL: TWA 0.05 mg(U)/m^3
ACGIH TLV: TWA 0.2 mg(U)/m^3; STEL 0.6 mg (U)/m^3
DOT Classification: Radioactive Material; Label: Radioactive and Corrosive.

SAFETY PROFILE: Poison by ingestion, subcutaneous, intravenous, and intraperitoneal routes. Mutation data reported. A corrosive irritant to skin, eyes, and mucous membranes. A radioactive material. When heated to decomposition it emits toxic fumes of NO_x. See also URANIUM.

USJ000 CAS:34661-75-1 ***HR: 3***
URAPIDIL
mf: $C_{20}H_{29}N_5O_3$ mw: 387.54

PROP: Crystals from water. Mp: 156-158°.

SYNS: B-66256 ◇ EBRANTIL ◇ 6-(3-(4-(o-METHOXYPHENYL)-1-PIPERAZINYL)PROPYLAMINO)-1,3-DIMETHYLURACIL

TOXICITY DATA with REFERENCE
orl-rat TDLo:2160 mg/kg (female 17-22D post):REP OYYAA2 33,535,87
orl-mus TDLo:224 mg/kg (female 14D pre):TER ARZNAD 27,1919,77
orl-rat LD50:520 mg/kg ARZNAD 27,1919,77
ivn-rat LD50:140 mg/kg ARZNAD 27,1919,77
orl-mus LD50:508 mg/kg OYYAA2 33,453,87
ivn-mus LD50:203 mg/kg OYYAA2 33,453,87

SAFETY PROFILE: Poison by intravenous route. Moderately toxic by ingestion. An experimental teratogen. Experimental reproductive effects. When heated to decomposition it emits toxic fumes of NO_x. Used as an antihypertensive agent.

USJ075 CAS:2445-07-0 ***HR: 3***
URBACIDE
mf: $C_7H_{15}AsN_2S_4$ mw: 330.40

SYNS: BIS(DIMETHYLTHIOCARBAMOYLTHIO)METHYL-ARSINE ◇ DITHIO-METHANEARSONOUS ACID BIS(ANHYDROSULFIDE) with DIMETHYLDITHIOCARBAMIC ACID ◇ METHYLARSENIC DIMETHYL DITHIOCARBAMATE ◇ METHYL ARSINE-BIS(DIMETHYLDITHIOCARBAMATE) ◇ METHYLBIS(DIMETHYLDITHIOCARBAMOYLTHIO)ARSINE ◇ MONZET ◇ TUZET ◇ URBACID ◇ URBAZID

TOXICITY DATA with REFERENCE
mnt-mus-orl 60 mg/kg CHYCDW 19,150,85
orl-rat TDLo:300 mg/kg (7-16D preg):TER CHYCDW 19,150,85
orl-rat LD50:100 mg/kg AAREAV 23,299,66
orl-mus LD50:221 mg/kg CHYCDW 19,150,85
ihl-mam LCLo:500 mg/m^3 AAREAV 23,299,66

CONSENSUS REPORTS: Arsenic and its compounds are on the Community Right-To-Know List.

OSHA PEL: TWA 0.5 mg(As)m^3

SAFETY PROFILE: Poison by ingestion. Moderately toxic by inhalation. An experimental teratogen. Mutation data reported. When heated to decomposition it emits toxic fumes of SO_x, NO_x, and As. See also ARSENIC COMPOUNDS and CARBAMATES.

USJ100 CAS:2375-03-3 ***HR: 2***
URBASON SOLUBLE
mf: $C_{26}H_{34}O_8$•Na mw: 497.59

SYNS: METHYLPREDNISOLONE SODIUM SUCCINATE ◇ 6-α-METHYLPREDNISOLONE SODIUM SUCCINATE ◇ MPS ◇ SOLU-MEDROL ◇ SOLU-MEDRONE ◇ 11-β,17,21-TRIHYDROXY-6-α-METHYLPREGNA-1,4-DIENE-3,20-DIONE, 21-(HYDROGEN SUCCINATE), MONOSODIUM SALT ◇ 11-β,17,21-TRIHYDROXY-6-α-METHYL-1,4-PREGNADIENE-3,20-DIONE 21-(SODIUM SUCCINATE) ◇ U 9088

TOXICITY DATA with REFERENCE
ipr-rat TDLo:80 mg/kg (female 15-22D post):REP OYYAA2 16,203,78
ipr-rat TDLo:690 mg/kg (female 61D pre):TER OYYAA2 16,193,78
ivn-man TDLo:43 mg/kg/3D-I:PUL AIMEAS 104,58,86
ivn-wmn TDLo:60 μg/kg/3D-I:CVS SJRHAT 15,302,86
ipr-rat LD50:640 mg/kg IYKEDH 11,181,80
scu-rat LD50:750 mg/kg IYKEDH 11,181,80
ivn-rat LD50:640 mg/kg IYKEDH 11,181,80
ipr-mus LD50:880 mg/kg NIIRDN 6,833,82
scu-mus LD50:860 mg/kg OYYAA2 8,633,74
ivn-mus LD50:750 mg/kg IYKEDH 11,181,80

SAFETY PROFILE: Moderately toxic by intraperitoneal, subcutaneous, and intravenous routes. Human systemic effects by intravenous route: respiratory system effects, pulse rate decrease, fall in blood pressure. An experimental teratogen. Experimental reproductive effects. When heated to decomposition it emits toxic fumes of Na_2O.

USS000 CAS:57-13-6 ***HR: 3***
UREA
mf: CH_4N_2O mw: 60.07

PROP: White crystals. Mp: 132.7°, bp: decomp, d: (solid) 1.335. Sol in water and alc; sltly sol in ether.

SYNS: CARBAMIDE ◇ CARBAMIDE RESIN ◇ CARBAMIMIDIC ACID ◇ CARBONYL DIAMIDE ◇ CARBONYLDIAMINE ◇ ISOUREA ◇ NCI-C02119 ◇ PRESPERSION, 75 UREA ◇ PSEUDOUREA ◇ SUPERCEL 3000 ◇ UREAPHIL ◇ UREOPHIL ◇ UREVERT ◇ VARIOFORM II

TOXICITY DATA with REFERENCE
skn-hmn 22 mg/3D-I MLD 85DKA8 -,127,77
cyt-hmn:leu 50 mmol/L CNREA8 25,980,65
ipc-wmn TDLo:1400 mg/kg (16W preg):REP OBGNAS 43,765,74
orl-rat TDLo:821 g/kg/1Y-C:NEO JEPTDQ 3(5-6),149,80
orl-mus TDLo:394 g/kg/1Y-C:CAR JEPTDQ 3(5-6),149,80

orl-rat LD50:8471 mg/kg GISAAA 51(6),8,86
scu-rat LD50:8200 mg/kg OYYAA2 13,749,77
ivn-rat LD50:5300 mg/kg OYYAA2 13,749,77
scu-mus LD50:9200 mg/kg OYYAA2 13,749,77
ivn-mus LD50:4600 mg/kg OYYAA2 13,749,77
scu-dog LDLo:3000 mg/kg HBAMAK 4,1353,35
ivn-dog LDLo:3000 mg/kg HBAMAK 4,1353,35

CONSENSUS REPORTS: Reported in EPA TSCA Inventory. EPA Genetic Toxicology Program.

SAFETY PROFILE: Moderately toxic by intravenous and subcutaneous routes. Human reproductive effects by intraplacental route: fertility effects. Experimental reproductive effects. Human mutation data reported. A human skin irritant. Questionable carcinogen with experimental carcinogenic and neoplastigenic data. Reacts with sodium hypochlorite or calcium hypochlorite to form the explosive nitrogen trichloride. Incompatible with $NaNO_2$; P_2Cl_5; nitrosyl perchlorate. Preparation of the ^{15}N-labeled urea is hazardous. When heated to decomposition it emits toxic fumes of NO_x.

UTA000 CAS:64024-08-4 ***HR: 3***
UREA ANTIMONYL TARTRATE

TOXICITY DATA with REFERENCE
ipr-mus LD50:14 mg(Sb)/kg AJTMAQ 25,263,45

CONSENSUS REPORTS: Antimony and its compounds are on the Community Right-To-Know List.

OSHA PEL: TWA 0.5 mg(Sb)/m^3
ACGIH TLV: TWA 0.5 mg(Sb)/m^3
NIOSH REL: (Antimony) TWA 0.5 mg/m^3

SAFETY PROFILE: Poison by intraperitoneal route. When heated to decomposition it emits very toxic fumes of NO_x and Sb. See also ANTIMONY COMPOUNDS.

UTJ000 CAS:124-47-0 ***HR: 3***
UREA NITRATE (wet)
DOT: UN 0220/UN 1357
mf: $CH_5N_3O_4$ mw: 123.09

PROP: Colorless minerals or prisms. Mp: 152° decomp. Very sltly sol in hot water; sol in alc; insol in HNO_3.

SYNS: UREA MONONITRATE (8CI, 9CI) ◇ UREA NITRATE ◇ URONIUM NITRATE

DOT Classification: Flammable Solid; Label: Flammable Solid (UN1357); Class A Explosive; Label: Explosive A (UN0220).

SAFETY PROFILE: A mild irritant. Flammable when exposed to heat or flame. The dry nitrate may explode when heated. The presence of heavy metals (e.g., lead, iron) catalyses the thermal decompositon of urea nitrate. When heated to decomposition it emits toxic fumes of NO_x.

UTU400 CAS:18727-07-6 ***HR: 3***
UREA PERCHLORATE
mf: $CH_5ClN_2O_5$ mw: 160.51

SYN: URONIUM PERCHLORATE

SAFETY PROFILE: Aqueous solutions dissolve some explosives (e.g., picric acid; nitromethane) and form extremely powerful, high velocity explosives. When heated to decomposition it emits toxic fumes of Cl^- and NO_x. See also PERCHLORATES and EXPLOSIVES, HIGH.

UVA000 CAS:51-79-6 ***HR: 3***
URETHANE
mf: $C_3H_7NO_2$ mw: 89.11

$CH_3CH_2OCO{\cdot}NH_2$

PROP: Colorless, odorless crystals. Mp: 49°, bp: 184°, d: 0.9862, vap press: 10 mm @ 77.8°, vap d: 3.07. Very sol in water, alc, ether.

SYNS: A 11032 ◇ AETHYLCARBAMAT (GERMAN) ◇ AETHYLURETHAN (GERMAN) ◇ CARBAMIC ACID, ETHYL ESTER ◇ CARBAMIDSAEURE-AETHYLESTER (GERMAN) ◇ ESTANE 5703 ◇ ETHYL CARBAMATE ◇ ETHYLURETHAN ◇ ETHYL URETHANE ◇ o-ETHYLURETHANE ◇ LEUCETHANE ◇ LEUCOTHANE ◇ NSC 746 ◇ PRACARBAMIN ◇ PRACARBAMINE ◇ RCRA WASTE NUMBER U238 ◇ U-COMPOUND ◇ URETAN ETYLOWY (POLISH) ◇ URETHAN

TOXICITY DATA with REFERENCE
dnd-hmn:fbr 3 mmol/L ENMUDM 7,267,85
sce-hmn:lym 10 μmol/L MUREAV 89,75,81
otr-mus:emb 1100 μmol/L MUREAV 152,113,85
ipr-mus TDLo:1 g/kg (female 18D post):REP TXCYAC 24,251,82
ivn-dog TDLo:1 g/kg (female 20D post):TER JZKEDZ 6,37,80
orl-rat TDLo:30 g/kg/52W-C:ETA CNREA8 7,107,47
ipr-rat TDLo:500 mg/kg (19D preg):ETA,TER CNREA8 30,2552,70
ipr-rat TDLo:500 mg/kg:NEO RRCRBU 52,29,75
orl-mus TDLo:12 g/kg/15D-C:CAR TUMOAB 53,81,67
skn-mus TDLo:90 g/kg/56W-I:CAR CRNGDP 5,911,84
ipr-mus TDLo:2500 mg/kg (7-11D preg):NEO,TER IARCCD 4,14,73
ipr-mus TDLo:500 mg/kg (19D preg):NEO,TER CNREA8 38,137,78
scu-mus TDLo:1 g/kg (11D preg):CAR,TER CALEDQ 18,131,83
scu-mus TDLo:1000 mg/kg (15D preg):CAR,TER CNREA8 34,2217,74
scu-mus TDLo:200 mg/kg:NEO,TER CNREA8 34,2217,74
ivn-mus TDLo:1000 mg/kg (18D preg):NEO,TER JNCIAM 8,63,47

ivn-mus TDLo:1 g/kg CNREA8 22,299,62
unr-mus TDLo:1 g/kg (17D preg):ETA,TER BCSTB5 2,710,74
ipr-mus TD:1000 mg/kg (18D preg):NEO,TER JNCIAM 8,63,47
orl-rat LDLo:2 g/kg JPETAB 121,136,57
ipr-rat LD50:1500 mg/kg CNREA8 26,1448,66
scu-rat LDLo:1800 mg/kg AEPPAE 182,348,36
ims-rat LD50:1400 mg/kg ZKKOBW 84,227,75
orl-mus LD50:2500 mg/kg ARZNAD 9,595,59
ipr-mus LD50:1539 mg/kg PMRSDJ 1,682,81
scu-mus LD50:1750 mg/kg GANNA2 63,731,72
par-mus LDLo:1000 mg/kg NCISA* PH-43-62-483
ivn-rbt LDLo:2000 mg/kg 27ZIAQ -,272,73

CONSENSUS REPORTS: NTP Fifth Annual Report on Carcinogens. IARC Cancer Review: Group 2B IMEMDT 7,56,87; Animal Sufficient Evidence IMEMDT 7,111,74. Community Right-To-Know List. Reported in EPA TSCA Inventory. EPA Genetic Toxicology Program.

DFG MAK: Animal Carcinogen, Suspected Human Carcinogen.

SAFETY PROFILE: Confirmed carcinogen with experimental carcinogenic, neoplastigenic, and tumorigenic data. A transplacental carcinogen. Moderately toxic by ingestion, intraperitoneal, subcutaneous, intramuscular, parenteral, and intravenous routes. An experimental teratogen. Experimental reproductive effects. Human mutation data reported. Causes depression of bone marrow and occasionally focal degeneration in the brain. Can also produce central nervous system depression, nausea and vomiting. Has been found in over 1000 beverages sold in the United States. The most heavily contaminated liquors are bourbons, sherries, and fruit brandies (some had 1,000 to 12,000 ppb urethane). Many whiskeys, table and dessert wines, brandies and liqueurs contain potentially hazardous amounts of urethane. The allowable limit for urethane in alcoholic beverages is 125 ppb. It is formed as a side product during processing.

Hot aqueous acids or alkalies decompose urethane to ethanol, carbon dioxide, and ammonia. Reacts with phosphorus pentachloride to form an explosive product. When heated it emits toxic fumes of NO_x. Used as an intermediate in the manufacture of pharmaceuticals, pesticides, and fungicides. See also CARBAMATES.

UVA150 CAS:2611-61-2 ***HR: D***
URFAMICIN HYDROCHLORIDE
mf: $C_{14}H_{18}Cl_2N_2O_6S{\bullet}ClH$ mw: 449.76

SYNS: NEOMYSON G HYDROCHLORIDE ◇ THIAMPHENICOL GLYCINATE HYDROCHLORIDE ◇ THIOPHENICOL GLYCINATE HYDROCHLORIDE ◇ TPG HYDROCHLORIDE

TOXICITY DATA with REFERENCE
ipr-rat TDLo:300 mg/kg (female 9-14D post):REP OYYAA2 7,859,73
ipr-mus TDLo:4200 mg/kg (female 7-12D post):TER OYYAA2 7,859,73

SAFETY PROFILE: An experimental teratogen. Experimental reproductive effects. When heated to decomposition it emits toxic fumes of SO_x, NO_x, and HCl. See also ESTERS.

UVA400 CAS:69-93-2 ***HR: D***
URIC ACID
mf: $C_5H_4N_4O_3$ mw: 168.13

SYNS: LITHIC ACID ◇ 1H-PURINE-2,6,8(3H)-TRIONE, 7,9-DIHYDRO-(9CI) ◇ 2,6,8-TRIHYDROXYPURINE ◇ 2,6,8-TRIOXOPURINE ◇ 2,6,8-TRIOXYPURINE

TOXICITY DATA with REFERENCE
oth-hmn:lyms 10 mmol/L CYTBAI 4,87,71
orl-rat TDLo:5040 mg/kg (male 4W pre):REP TOIZAG 32,36,85

CONSENSUS REPORTS: Reported in EPA TSCA Inventory.

SAFETY PROFILE: Experimental reproductive effects. Mutation data reported. When heated to decomposition it emits toxic fumes of NO_x.

UVJ000 CAS:58-96-8 ***HR: 1***
β-URIDINE
mf: $C_9H_{12}N_2O_6$ mw: 244.23

PROP: Needles. Mp: 165°. Sol in water.

SYNS: 1-β-d-RIBOFURANOSYLURACIL ◇ URACIL RIBOSIDE ◇ URIDINE

TOXICITY DATA with REFERENCE
dnd-mam:lym 100 mmol/L PNASA6 48,686,62
ipr-mus LD50:5100 mg/kg RPTOAN 40,66,77

CONSENSUS REPORTS: Reported in EPA TSCA Inventory.

SAFETY PROFILE: Mildly toxic by intraperitoneal route. Mutation data reported. When heated to decomposition it emits toxic fumes of NO_x.

UVJ400 CAS:1470-35-5 ***HR: 3***
URIDION
mf: $C_{15}H_9BrO_2$ mw: 301.15

SYNS: 5-BROMO-2-PHENYLINDAN-1,3-DIONE ◇ 5-BROMO-2-PHENYL-1,3-INDANDIONE ◇ 5-BROMO-2-PHENYL-1H-INDENE-1,3(2H)-DIONE ◇ 2-FENIL-5-BROMO-INDANDIONE (ITALIAN)

TOXICITY DATA with REFERENCE
orl-rbt TDLo:440 mg/kg (7-17D preg):TER ARZNAD 24,1609,74

orl-rbt TDLo:440 mg/kg (7-17D preg):REP ARZNAD 24,1609,74
orl-rat LD50:150 mg/kg ARZNAD 24,1609,74
orl-mus LD50:200 mg/kg BCFAAI 112,401,73
ipr-mus LD50:120 mg/kg ARZNAD 25,873,75

SAFETY PROFILE: Poison by ingestion and intraperitoneal routes. An experimental teratogen. Experimental reproductive effects. Promotes the excretion of uric acid in the urine. When heated to decomposition it emits toxic fumes of Br^-.

UVJ425 ***HR: 3***
UROCALUM

PROP: Extracted from *C. salicina. oerst* (NIIRDN 6,96,82).

TOXICITY DATA with REFERENCE
ipr-rat LD50:370 mg/kg NIIRDN 6,96,82
scu-rat LD50:18500 mg/kg NIIRDN 6,96,82
orl-mus LD50:24200 mg/kg NIIRDN 6,96,82
ipr-mus LD50:535 mg/kg NIIRDN 6,96,82
scu-mus LD50:1320 mg/kg NIIRDN 6,96,82

SAFETY PROFILE: Poison by intraperitoneal route. Moderately toxic by subcutaneous route. Mildly toxic by ingestion.

UVJ450 CAS:396-01-0 ***HR: 3***
UROCAUDAL
mf: $C_{12}H_{11}N_7$ mw: 253.30

PROP: Yellow plates from butanol. Mp: 316°. Also reported as crystals from DMF. Mp: 327°.

SYNS: ADEMINE ◇ DIREN ◇ DITAK ◇ DYREN ◇ DYRENIUM ◇ DYTAC ◇ JATROPUR ◇ NCI C56042 ◇ NORIDIL ◇ NORIDYL ◇ 6-PHENYL-2,4,7-PTERIDINETRIAMINE ◇ 6-PHENYL-2,4,7-TRIAMINOPTERIDINE ◇ PTEROFEN ◇ PTEROPHENE ◇ SKF 8542 ◇ TATURIL ◇ TERIAM ◇ TERIDIN ◇ 2,4,7-TRIAMINO-6-FENILPTERIDINA (ITALIAN) ◇ 2,4,7-TRIAMINO-6-PHENYLPTERIDINE ◇ TRIAMPUR ◇ TRIAMTEREN ◇ TRIAMTERENE ◇ TRIAMTERIL ◇ TRIAMTERIL COMPLEX ◇ TRI-SPAN ◇ TRITEREN

TOXICITY DATA with REFERENCE
otr-ham:emb 25 mg/L ENMUDM 8(Suppl 6),4,86
cyt-ham:lng 2800 μg/L GMCRDC 27,95,81
orl-mus LD50:285 mg/kg FRXXBL #2314719
ipr-mus LD50:249 mg/kg NIIRDN 6,519,82
scu-mus LD50:620 mg/kg NIIRDN 6,519,82

SAFETY PROFILE: Poison by ingestion and intraperitoneal routes. Moderately toxic by subcutaneous route. Mutation data reported. When heated to decomposition it emits toxic fumes of NO_x.

UVJ475 CAS:9010-53-1 ***HR: 2***
UROGASTRONE

PROP: Mouse EGF-URO: Heat stable and non-dialyzable. Biological activity stable in boiling water but destroyed by heating in dil acid or alkali. Human EGF-URO, anthelone, anthelone U, uroanthelone, uroenterone. Very sol in water; sol in methanol, ethylene gycol.

SYNS: ANTHELONE U ◇ EGF-UROGASTRONE ◇ KUTROL ◇ UROANTHELONE ◇ UROENTERONE ◇ UROGASTRON

TOXICITY DATA with REFERENCE
ivn-rat LD50:1990 mg/kg NIIRDN 6,96,82
orl-mus LD50:12600 mg/kg NIIRDN 6,96,82
ipr-mus LD50:5080 mg/kg NIIRDN 6,96,82
ivn-mus LD50:2850 mg/kg NIIRDN 6,96,82
ims-mus LD50:7900 mg/kg NIIRDN 6,96,82

SAFETY PROFILE: Moderately toxic by intravenous route. Mildly toxic by ingestion.

UVS500 CAS:9039-53-6 ***HR: D***
UROKINASE

SYNS: UROKINASE (ENZYME-ACTIVATING) ◇ WIN 22005 ◇ WIN-KINASE

TOXICITY DATA with REFERENCE
ipr-rat TDLo:4242 μg/kg (7-13D preg):REP OYYAA2 8,981,74
ipr-mus TDLo:14141 μg/kg (female 7-13D post):TER OYYAA2 8,981,74

CONSENSUS REPORTS: Reported in EPA TSCA Inventory.

SAFETY PROFILE: An experimental teratogen. Experimental reproductive effects. Used in the treatment of diseases caused by blood clots.

UWJ000 CAS:125-46-2 ***HR: 3***
USNEIN
mf: $C_{18}H_{16}O_7$ mw: 344.34

SYNS: 2,6-DIACETYL-7,9-DIHYDROXY-8,9b-DIMETHYL-1,3(2H,9bH)-DIBENZOFURANDIONE ◇ USNIACIN ◇ USNIC ACID ◇ USNINIC ACID ◇ USNINSAEURE (GERMAN)

TOXICITY DATA with REFERENCE
dnd-esc 20 μmol/L MUREAV 89,95,81
cyt-mam:oth 3750 μg/L PHMCAA 12,280,70
ivn-mus LD50:25 mg/kg MEIEDD 10,1414,83
scu-mus LD50:75 mg/kg 85GDA2 9,89,82
ivn-mus LD50:25 mg/kg MEIEDD 10,1414,83
orl-rbt LD50:500 mg/kg ARZNAD 5,507,55

CONSENSUS REPORTS: Reported in EPA TSCA Inventory.

SAFETY PROFILE: Poison by subcutaneous and intravenous routes. Moderately toxic by ingestion. Mutation data reported. When heated to decomposition it emits acrid smoke and fumes. Used as an antibacterial and antimitotic agent.

UWJ100 CAS:7562-61-0 ***HR: 2***
USNIC ACID, (R)-(8CI)
mf: $C_{18}H_{16}O_7$ mw: 344.34

SYNS: 1,3(2H,9bH)-DIBENZOFURANDIONE,2,6-DIACETYL-7,9-DIHYDROXY-8,9b-DIMETHYL-, (9bR)- ◇ 3(9bH)-DIBENZOFURANONE, 2,6-DIACETYL-8,9b-DIMETHYL-1,7,9-TRIHYDROXY-, D- ◇ 3(9bH)-DIBENZOFURANONE,2,6-DIACETYL-1,7,9-TRIHYDROXY-8,9b-DIMETHYL-, D- ◇ (+)-USNIC ACID ◇ d-USNIC ACID ◇ (+)-USNINIC ACID ◇ d-USNINIC ACID

TOXICITY DATA with REFERENCE
orl-mus TDLo:1500 μg/kg (female 1D post):REP NATUAS 176,249,55
orl-mus LD50:838 mg/kg TAKHAA 31,247,72

SAFETY PROFILE: Moderately toxic by ingestion. Experimental reproductive effects. When heated to decomposition it emits acrid smoke and irritating fumes.

V

VAD000 CAS:54965-21-8 ***HR: 2***
VALBAZEN
mf: $C_{12}H_{15}N_3O_2S$ mw: 265.36

PROP: Colorless crystals. Mp: 208-210°.

SYNS: ALBENDAZOLE (USDA) ◇ METHYL 5-(PROPYLTHIO)-2-BENZIMIDAZOLECARBAMATE ◇ ((PROPYLTHIO)-5-1H-BENZIMIDAZOLYL-2) CARBAMATE de METHYLE (FRENCH) ◇ (5-(PROPYLTHIO)-1H-BENZIMIDAZOL-2-YL)CARBAMIC ACID METHYL ESTER ◇ 5-(PROPYLTHIO)-2-CARBOMETHOXYAMINOBENZIMIDAZOLE ◇ SKF 62979 ◇ ZENTAL

TOXICITY DATA with REFERENCE
orl-dom TDLo:20 mg/kg (17D preg):REP AMSHAR 28,226,80
orl-rat TDLo:85 mg/kg (8-15D preg):TER ARCVBP 12,159,81
orl-rat LD50:2400 mg/kg APFRAD 40,55,82

SAFETY PROFILE: Moderately toxic by ingestion. An experimental teratogen. Experimental reproductive effects. When heated to decomposition it emits toxic fumes of SO_x and NO_x. See also CARBAMATES and ESTERS.

VAG000 CAS:110-62-3 ***HR: 3***
n-VALERALDEHYDE
DOT: UN 2058
mf: $C_5H_{10}O$ mw: 86.15

PROP: Liquid. Flash p: 53.6°F, bp: 102-103°, d: 0.8095 @ 20/4°. Very sltly sol in water; misc with organic solvents.

SYNS: AMYL ALDEHYDE ◇ BUTYL FORMAL ◇ PENTANAL ◇ n-PENTANAL ◇ VALERAL ◇ VALERIANIC ALDEHYDE ◇ VALERIC ACID ALDEHYDE ◇ VALERIC ALDEHYDE ◇ VALERYLALDEHYDE

TOXICITY DATA with REFERENCE
skn-rbt 500 mg/24H MOD FCTXAV 17(Suppl.),695,79
eye-rbt 100 mg/24H SEV FCTXAV 17,919,79
skn-gpg 100% SEV FCTXAV 17,919,79
orl-rat LD50:3200 mg/kg 14CYAT 2,1968,63
ihl-rat LCLo:4000 ppm/4H AIHAAP 30,470,69
orl-mus LD50:6400 mg/kg FCTXAV 17,919,79
skn-rbt LD50:4857 mg/kg AIHAAP 30,47,69
skn-gpg LD50:20 g/kg FCTXAV 17,919,79

CONSENSUS REPORTS: Reported in EPA TSCA Inventory.

OSHA PEL: TWA 50 ppm
ACGIH TLV: TWA 50 ppm
DOT Classification: Flammable Liquid; Label: Flammable Liquid.

SAFETY PROFILE: Moderately toxic by ingestion. Mildly toxic by inhalation and skin contact. A severe eye and skin irritant. A very dangerous fire hazard when exposed to heat or flame. When heated to decomposition it emits acrid smoke and irritating fumes. See also ALDEHYDES.

VAQ000 CAS:109-52-4 ***HR: 2***
VALERIC ACID
DOT: NA 1760
mf: $C_5H_{10}O_2$ mw: 102.15

PROP: Colorless, mobile liquid; penetrating, rancid odor. D: 0.940 @ 20/4°, refr index: 1.405-1.14 @ 25°, mp: −34.5°, bp: 186.4°, flash p: 203°F. Sol in water; misc in alc, ether.

SYNS: BUTANECARBOXYLIC ACID ◇ 1-BUTANECARBOXYLIC ACID ◇ FEMA No. 3101 ◇ PENTANOIC ACID ◇ n-PENTANOIC ACID ◇ PROPYLACETIC ACID ◇ VALERIANIC ACID ◇ n-VALERIC ACID

TOXICITY DATA with REFERENCE
orl-mus LD50:600 mg/kg 85GMAT -,119,82
ihl-mus LC50:4100 mg/m^3/2H 85GMAT -,119,82
ivn-mus LD50:1290 mg/kg APTOA6 18,141,61
scu-mus LD50:3590 mg/kg JPPMAB 21,85,69

CONSENSUS REPORTS: Reported in EPA TSCA Inventory.

DOT Classification: Corrosive Material; Label: Corrosive.

SAFETY PROFILE: Moderately toxic by ingestion, intravenous, and subcutaneous routes. Mildly toxic by inhalation. A corrosive irritant to skin, eyes, and mucous membranes. Combustible liquid. When heated to decomposition it emits acrid smoke and irritating fumes. Used in perfumes.

VAV000 CAS:108-29-2 ***HR: 2***
4-VALEROLACTONE
mf: $C_5H_8O_2$ mw: 100.13

PROP: Colorless, mobile liquid; sweet, herbaceous odor. Mp: −31°, bp: 205-206.5°, flash p: 205°F (COC), d: 1.047-1.054, refr index: 1.43, vap d: 3.45. Misc in alc, fixed oils, water.

SYNS: FEMA No. 3103 ◇ 4-HYDROXYPENTANOIC ACID LACTONE ◇ 4-HYDROXYVALERIC ACID LACTONE ◇ Γ-METHYL-Γ-BUTYROLACTONE ◇ 4-METHYL-Γ-BUTYROLACTONE ◇ Γ-PENTALACTONE ◇ 4-PENTANOLIDE ◇ Γ-VALEROLACTONE (FCC)

TOXICITY DATA with REFERENCE
skn-rbt 500 mg/24H MLD FCTOD7 20(Suppl),847,82
orl-rat LD50:8800 mg/kg JIHTAB 27,263,45
orl-rbt LD50:2480 mg/kg JIHTAB 27,263,45

CONSENSUS REPORTS: Reported in EPA TSCA Inventory.

SAFETY PROFILE: Moderately toxic by ingestion. A skin irritant. Combustible liquid when exposed to heat or flame; can react with oxidizing materials. To fight fire, use water, foam, CO_2, dry chemical. When heated to decomposition it emits acrid smoke and irritating fumes.

VBA000 CAS:638-29-9 ***HR: 2***
VALERYL CHLORIDE
DOT: UN 2502
mf: C_5H_9ClO mw: 120.59

PROP: Fp: 32°, bp: 125-127°, d: 1.016

CONSENSUS REPORTS: Reported in EPA TSCA Inventory.

DOT Classification: Corrosive Material; Label: Corrosive.

SAFETY PROFILE: A corrosive irritant to skin, eyes, and mucous membranes. When heated to decomposition it emits toxic fumes of Cl^-.

VBK000 CAS:90-22-2 ***HR: 3***
VALETHAMATE BROMIDE
mf: $C_{19}H_{32}NO_2 \cdot Br$ mw: 386.43

PROP: Crystals from ethanol and ether or acetone. Mp. 100-101°. Freely sol in water and alc; practically insol in ether.

SYNS: 2-DIETHYLAMINOETHYL-3-METHYL-2-PHENYLVALERATE METHYLBROMIDE ◇ 2-DIETHYLAMINOETHYL-2-PHENYL-3-METHYLVALERATE METHYL BROMIDE ◇ DIETHYL(2-HYDROXYETHYL)METHYLAMMONIUM-3-METHYL-2-PHENYLVALERATEBROMIDE ◇ EDIPOSIN ◇ EPIDOSIN ◇ EPIDOZIN ◇ 3-METHYL-2-PHENYLVALERIC ACID-2-DIETHYLAMINOETHYL ESTER METHYL BROMIDE ◇ 3-METHYL-2-PHENYLVALERIC ACID DIETHYL(2-HYDROXYETHYL)METHYLAMMONIUM BROMIDE ESTER ◇ 2-(((3-METHYL-2-PHENYLVALERYL)OXY)-N,N-DIETHYL-N-METHYLETHANAMINIUM BROMIDE ◇ MUREL ◇ PHENYLMETHYLVALERIANSAEURE-β-DIAETHYLAMINOAETHYLESTER-BROMMETHYLAT(GERMAN) ◇ RESITAN ◇ VALETHAMATE

TOXICITY DATA with REFERENCE
orl-rat LD50:1260 mg/kg NIIRDN 6,352,82
scu-rat LD50:575 mg/kg NIIRDN 6,352,82
ivn-rat LD50:4200 μg/kg NIIRDN 6,352,82
orl-mus LD50:330 mg/kg ARZNAD 5,599,55
scu-mus LD50:105 mg/kg OYYAA2 8,245,74
ivn-mus LD50:4200 μg/kg NIIRDN 6,352,82
ivn-rbt LD50:9500 μg/kg 29ZVAB -,123,69

SAFETY PROFILE: Poison by ingestion, subcutaneous, and intravenous routes. See also ESTERS and BROMIDES. When heated to decomposition it emits very toxic fumes of NO_x, NH_3, and Br^-.

VBP000 CAS:72-18-4 ***HR: 1***
VALINE
mf: $C_5H_{11}NO_2$ mw: 117.17

PROP: White, crystalline solid; characteristic taste. Mp (dl): 298° (decomp), mp (l): 315°, d (l): 1.230. Sol in water; very sltly sol in alc; insol in ether. An essential amino acid.

SYNS: l-(+)-α-AMINOISOVALERIC ACID ◇ l-VALINE (FCC)

TOXICITY DATA with REFERENCE
oms-omi 10 mmol/L CBINA8 16,201,77
ipr-rat LD50:5390 mg/kg ABBIA4 58,253,55

CONSENSUS REPORTS: Reported in EPA TSCA Inventory.

SAFETY PROFILE: Mutation data reported. When heated to decomposition it emits toxic fumes of NO_x.

VBU000 CAS:640-68-6 ***HR: 1***
d-VALINE
mf: $C_5H_{11}NO_2$ mw: 117.17

PROP: Hexagonal leaf or prisms. Sol in water.

TOXICITY DATA with REFERENCE
ipr-rat LD50:6093 mg/kg ABBIA4 64,319,56

CONSENSUS REPORTS: Reported in EPA TSCA Inventory.

SAFETY PROFILE: When heated to decomposition it emits toxic fumes of NO_x.

VBZ000 CAS:2001-95-8 ***HR: 3***
VALINOMYCIN
mf: $C_{54}H_{90}N_6O_{18}$ mw: 1111.50

PROP: Shiny, rectangular platelets. Mp: 190°. Almost insol in water; very sol in petr ether, ether, benzene, chloroform, glacial acetic acid, butyl acetate, acetone.

SYNS: ANTIBIOTIC N-329 B ◇ NSC 122023

TOXICITY DATA with REFERENCE
orl-rat LD50:4 mg/kg DCTODJ 8,451,85
ipr-rat LD50:800 μg/kg DCTODJ 8,451,85

orl-mus LD50:2500 μg/kg 85ERAY 1,325,78
ipr-mus LD50:390 μg/kg NCISP* JAN86
scu-mus LD50:4140 μg/kg 85ERAY 1,325,78
skn-rbt LD50:5 mg/kg DCTODJ 8,451,85

CONSENSUS REPORTS: EPA Extremely Hazardous Substances List. Reported in EPA TSCA Inventory.

SAFETY PROFILE: Poison by ingestion, skin contact, intraperitoneal, and subcutaneous routes. When heated to decomposition it emits toxic fumes of NO_x.

VCA000 CAS:2152-44-5 ***HR: 3***
VALISONE
mf: $C_{27}H_{37}FO_6$ mw: 476.64

SYNS: BETAMETHASONE VALERATE ◇ BETAMETHASONE 17-VALERATE ◇ BETNOVATE ◇ BETNOVATEAT ◇ CELESTODERM ◇ 9-FLUORO-11-β,17,21-TRIHYDROXY-16-β-METHYLPREGNA-1,4-DIENE-3,20,DIONE-17-VALERATE ◇ β-METHASONE-17-VALERATE

TOXICITY DATA with REFERENCE
skn-rat TDLo:19800 μg/kg (7-17D preg):REP YACHDS 9,3045,81
skn-rat TDLo:19800 μg/kg (7-17D preg):TER YACHDS 9,3045,81
scu-rat LDLo:2000 mg/kg ARZNAD 27,2102,77
orl-mus LD50:4067 mg/kg SKIZAB 29,153,73
ipr-mus LD50:632 SKIZAB 29,153,73
scu-mus LD50:496 mg/kg SKIZAB 29,153,73
scu-rbt LD50:61200 μg/kg OYYAA2 28,687,84

SAFETY PROFILE: Poison by subcutaneous route. Moderately toxic by intraperitoneal route. Mildly toxic by ingestion. An experimental teratogen. Experimental reproductive effects. When heated to decomposition it emits toxic fumes of F^-.

VCK000 ***HR: 3***
VALONEA TANNIN

SYNS: QUERCUS AEGILOPS L. TANNIN ◇ TANNIN from VALONEA

TOXICITY DATA with REFERENCE
scu-rat TDLo:750 mg/kg/2W-I:CAR BJCAAI 14,147,60
ipr-mus LD50:110 mg/kg JPPMAB 9,98,57
scu-mus LD50:170 mg/kg JPPMAB 9,98,57
ivn-mus LD50:50 mg/kg JPPMAB 9,98,57
ims-mus LD50:280 mg/kg JPPMAB 9,98,57

SAFETY PROFILE: Poison by intraperitoneal, intravenous, subcutaneous, and intramuscular routes. Questionable carcinogen with experimental carcinogenic data. See also TANNIC ACID.

VCK100 CAS:4093-35-0 ***HR: 3***
VALOPRIDE
mf: $C_{14}H_{22}BrN_3O_2$ mw: 344.30

SYNS: 4-AMINO-5-BROMO-N-(2-(DIETHYLAMINO)ETHYL)-o-ANISAMIDE ◇ ARTOMEY ◇ BENZAMIDE, 4-AMINO-5-BROMO-N-(2-(DIETHYLAMINO)ETHYL)-2-METHOXY-(9CI) ◇ BROMOPRIDA ◇ BROMOPRIDE ◇ N-(DIETHYLAMINOETHYL)-2-METHOXY-4-AMINO-5-BROMOBENZAMIDE ◇ VAL 13081

TOXICITY DATA with REFERENCE
scu-rat TDLo:2520 mg/kg (lactating female 21D post):REP PHBHA4 45,1081,89
orl-rat LD50:545 mg/kg YAKUD5 21,516,79
ipr-rat LD50:125 mg/kg YAKUD5 21,516,79
orl-mus LD50:310 mg/kg YAKUD5 21,516,79
ipr-mus LD50:105 mg/kg YAKUD5 21,516,79

SAFETY PROFILE: Poison by ingestion and intraperitoneal routes. Experimental reproductive effects. When heated to decomposition it emits toxic fumes of NO_x and Br^-.

VCP000 CAS:7440-62-2 ***HR: 3***
VANADIUM
af: V aw: 50.94

PROP: A bright, white, soft, ductile metal; sltly radioactive. Bp: 3000°, d: 6.11 @ 18.7°, mp: 1917°. Insol in water.

TOXICITY DATA with REFERENCE
ims-rat TDLo:340 mg/kg/43W-I:ETA NCIUS* PH 43-64-886,SEPT,71
scu-rbt LD50:59 mg/kg FATOAO 28,83,65

CONSENSUS REPORTS: Reported in EPA TSCA Inventory.

OSHA PEL: (Transitional: Respirable Dust: Cl 0.5 $mg(V_2O_5)/m^3$; Fume: Cl 0.1 $mg(V_2O_5)/m^3$) Respirable Dust and Fume: TWA 0.05 $mg(V_2O_5)/m^3$
NIOSH REL: TWA 1.0 $mg(V)/m^3$

SAFETY PROFILE: Poison by subcutaneous route. Questionable carcinogen with experimental tumorigenic data. Flammable in dust form from heat, flame, or sparks. Violent reaction with BrF_3; Cl_2; lithium; nitryl fluoride; oxidants. When heated to decomposition it emits toxic fumes of VO_x. See also VANADIUM COMPOUNDS.

VCU000 ***HR: 3***
VANADIUM AZIDE TETRACHLORIDE
mf: Cl_4N_3V mw: 234.76

CONSENSUS REPORTS: Community Right-To-Know List.

SAFETY PROFILE: Explosive. When heated to decomposition it emits very toxic fumes of VO_x, Cl^-, and NO_x. See also VANADIUM COMPOUNDS, AZIDES, and CHLORIDES.

VCZ000 ***HR: D***
VANADIUM COMPOUNDS
NIOSH REL: (Vanadium Compounds) CL 0.05 mg (V)/m^3/15M

SAFETY PROFILE: Variable toxicity. Vanadium compounds act chiefly as an irritant to the conjunctivae and respiratory tract. Acute and chronic exposure can give rise to conjunctivitis, rhinitis, reversible irritation of the respiratory tract, and to bronchitis, bronchospasms, and asthma-like diseases in more severe cases. There is still some controversy as to the effects of industrial exposure on other systems of the body. Responses are mostly acute, seldom chronic. The first report of human vanadium poisoning described rather widespread systemic effects, consisting of polycythemia, followed by red blood cell destruction and anemia, loss of appetite, pallor and emaciation, albuminuria and hematuria, gastrointestinal disorders, nervous complaints and cough, sometimes severe enough to cause hemoptysis. More recent reports describe symptoms which, for the most part, are restricted to the conjunctivae and respiratory system, no evidence being found of disturbances of the gastrointestinal tract, kidneys, blood or central nervous system. Vanadate (VO_3^-) is a potent inhibitor of the sodium pump, an enzyme universally present in eukaryotic organisms. The absorption of V_2O_5 by inhalation is nearly 100%. Though certain workers believe that it is only the pentoxide which is harmful, other investigators have found that patronite dust (chiefly vanadium sulfide) is quite toxic to animals, causing acute pulmonary edema. Acute poisoning in animals by ingestion of vanadium compounds causes nervous disturbances, paralysis of legs, respiratory failure, convulsions, bloody diarrhea, and death. Poisoning by inhalation causes bleeding of the nose and acute bronchitis. Some compounds have reported mutation effects. VF_5 and the oxyhalogenides of pentavalent vanadium (VOF_3, $VOCl_3$, $VOBr_3$) are volatile. Vanadium compounds are common air contaminants. The fumes are highly toxic. The major use of vanadium and its alloys is in the steel industry. When heated to decomposition it emits toxic fumes of VO_x. See also specific compounds.

VDA000 CAS:10580-52-6 ***HR: 3***
VANADIUM DICHLORIDE
mf: Cl_2V mw: 121.84

PROP: Hexagonal, green, deliquescent plates. D: 3.23 @ 18°. Sol in abs alc, glacial acetic acid.

TOXICITY DATA with REFERENCE
orl-rat LD50:540 mg/kg AIHAAP 30,470,69

ACGIH TLV: TWA 0.05 mg(V_2O_5)/m^3
NIOSH REL: (Vanadium Compounds) CL 0.05 mg (V)/m^3/15M

SAFETY PROFILE: Moderately toxic by ingestion. Will react with water or steam to produce toxic and corrosive fumes and explosive hydrogen gas. Platinum accelerates the reaction to violence. When heated to decomposition it emits toxic fumes of VO_x. See also HYDROCHLORIC ACID, VANADIUM COMPOUNDS, and CHLORIDES.

VDF000 ***HR: 2***
VANADIUM ORE

TOXICITY DATA with REFERENCE
ihl-hmn TDLo:4 μg/kg:PUL AMIHAB 12,635,55

OSHA PEL: (Transitional: Respirable Dust: Cl 0.5 mg (V_2O_5)/m^3; Fume: Cl 0.1 mg(V_2O_5)/m^3) Respirable Dust and Fume: TWA 0.05 mg(V_2O_5)/m^3
NIOSH REL: TWA 1.0 mg(V)/m^3
ACGIH TLV: TWA 0.05 mg(V_2O_5)/m^3
NIOSH REL: (Vanadium Compounds) CL 0.05 mg (V)/m^3/15M

SAFETY PROFILE: Human systemic effects by inhalation: pulmonary system effects. When heated to decomposition it emits toxic fumes of VO_x. See also VANADIUM COMPOUNDS.

VDK000 CAS:19120-62-8 ***HR: 3***
VANADIUM OXIDE TRIISOBUTOXIDE
mf: $C_{12}H_{27}O_4V$ mw: 287.14

SYNS: ISOBUTYL ORTHOVANADATE ◇ TRIISOBUTOXYOXOVANADIUM ◇ TRIISOBUTYL ORTHOVANADATE ◇ TRIISOBUTYL VANADATE ◇ TRIISOPROPOXYVANADIUM OXIDE

TOXICITY DATA with REFERENCE
skn-rbt 500 mg/24H SEV PESTC* 9,5,80
eye-rbt 100 mg SEV PESTC* 9,5,80
orl-rat LD50:293 mg/kg PESTC* 9,5,80
skn-rbt LD50:1930 mg/kg PESTC* 9,5,80

SAFETY PROFILE: Poison by ingestion. Moderately toxic by skin contact. A severe skin and eye irritant. When heated to decomposition it emits acrid smoke and irritating fumes of VO_x. See also VANADIUM COMPOUNDS.

VDP000 CAS:7727-18-6 ***HR: 3***
VANADIUM OXYTRICHLORIDE
DOT: UN 2443
mf: Cl_3OV mw: 173.29

PROP: Yellow, deliquescent liquid. Mp: −77° ± 2°, bp: 126.7°, d: 1.811 @ 32°.

SYNS: TRICHLOROOXOVANADIUM ◇ VANADIUM TRICHLORIDE OXIDE ◇ VANADYL TRICHLORIDE

TOXICITY DATA with REFERENCE
orl-rat LD50:140 mg/kg AIHAAP 30,470,69

CONSENSUS REPORTS: Reported in EPA TSCA Inventory.

ACGIH TLV: TWA 0.05 mg(V_2O_5)/m^3
NIOSH REL: (Vanadium Compounds) CL 0.05 mg (V)/m^3/15M
DOT Classification: Corrosive Material; Label: Corrosive.

SAFETY PROFILE: Poison by ingestion. A corrosive irritant to skin, eyes, and mucous membranes. Explosive reaction with sodium. Violently hygroscopic. Violent reaction with rubidium (at 60°C); potassium. When heated to decomposition it emits toxic fumes of VO_x and Cl^-. See also VANADIUM COMPOUNDS and HYDROCHLORIC ACID.

VDU000 CAS:1314-62-1 ***HR: 3***
VANADIUM PENTOXIDE (dust)
DOT: UN 2862
mf: O_5V_2 mw: 181.88

PROP: Yellow to red, crystalline powder. Mp: 690°, bp: decomp @ 1750°, d: 3.357 @ 18°.

SYNS: ANHYDRIDE VANADIQUE (FRENCH) ◇ C.I. 77938 ◇ RCRA WASTE NUMBER P120 ◇ VANADIC ANHYDRIDE ◇ VANADIO, PENTOSSIDO di (ITALIAN) ◇ VANADIUM DUST and FUME (ACGIH) ◇ VANADIUM(V) OXIDE ◇ VANADIUM PENTAOXIDE ◇ VANADIUM-PENTOXID (GERMAN) ◇ VANADIUM PENTOXIDE, non-fused form (DOT) ◇ VANADIUM, PENTOXYDE de (FRENCH) ◇ VANADIUMPENTOXYDE (DUTCH) ◇ WANADU PIECIOTLENEK (POLISH)

TOXICITY DATA with REFERENCE
mrc-bcs 500 mmol/L MUREAV 77,109,80
ivn-mus TDLo:10900 mg/kg (8D preg):TER ENVRAL 33,47,84
ihl-hmn TCLo:346 mg/m^3:PUL AMIHAB 19,497,59
ihl-hmn TCLo:1 mg/m^3/8H:PUL,EYE AEHLAU 14,709,67
orl-rat LD50:10 mg/kg ATXKA8 16,182,56
ihl-rat LCLo:70 mg/m^3/2H NTIS** AEC-TR-6710
ipr-rat LD50:12 mg/kg ATXKA8 16,182,56
scu-rat LD50:14 mg/kg ATXKA8 16,182,56
itr-rat LDLo:25 mg/kg NTIS** AEC-TR-6710
orl-mus LD50:23 mg/kg 85GMAT -,119,82
scu-mus LD50:10 mg/kg ZVKOA6 19,186,74
ihl-cat LCLo:500 mg/m^3/23M 30ZIAO -,140,64
scu-rbt LDLo:20 mg/kg 27ZWAY 3.3,1541,-
ivn-rbt LDLo:10 mg/kg 27ZWAY 3.3,1541,-
scu-gpg LDLo:20 mg/kg 30ZIAO -,140,64

CONSENSUS REPORTS: Reported in EPA TSCA Inventory. EPA Genetic Toxicology Program.

OSHA PEL: (Transitional: Respirable Dust: Cl 0.5 mg (V_2O_5)/m^3; Fume: Cl 0.1 mg(V_2O_5)/m^3) Respirable Dust and Fume: TWA 0.05 mg(V_2O_5)/m^3
ACGIH TLV: TWA 0.05 mg(V_2O_5)/m^3
DFG MAK: (fine dust) 0.05 mg/m^3
NIOSH REL: (Vanadium Compounds) CL 0.05 mg (V)/m^3/15M
DOT Classification: ORM-E; Label: None; Poison B; Label: Poison.

SAFETY PROFILE: Poison by ingestion, inhalation, intraperitoneal, subcutaneous, intratracheal, and intravenous routes. An experimental teratogen. Human systemic effects by inhalation: bronchiolar constriction, including asthma, cough, dyspnea, sputum, and conjunctiva irritation. Experimental reproductive effects. Mutation data reported. A respiratory irritant; causes skin pallor, greenish-black tongue, chest pain, cough, dyspnea, palpitation, lung changes. When ingested it causes gastrointestinal tract disturbances. May also cause a papular skin rash. Mixtures with calcium + sulfur + water may ignite spontaneously. The absorption of V_2O_5 by inhalation is nearly 100%. Incompatible with ClF_3; Li; peroxyformic acid. When heated to decomposition it emits acrid smoke and irritating fumes of VO_x. See also VANADIUM COMPOUNDS.

VDZ000 CAS:1314-62-1 ***HR: 3***
VANADIUM PENTOXIDE (fume)
mf: O_5V_2 mw: 181.88

SYN: VANADIUM DUST and FUME (ACGIH)

CONSENSUS REPORTS: EPA Extremely Hazardous Substances List. Reported in EPA TSCA Inventory. EPA Genetic Toxicology Program.

OSHA PEL: (Transitional: Fume: Cl 0.1 mg(V_2O_5)/m^3) Respirable Dust and Fume: TWA 0.05 mg(V_2O_5)/m^3
ACGIH TLV: TWA 0.05 mg(V_2O_5)/m^3
NIOSH REL: (Vanadium Compound) CL 0.05 mg (V)/m^3/15M

SAFETY PROFILE: A poison by several routes. Can react violently with (Ca + S + H_2O); ClF_3; Li. When heated to decomposition it emits toxic fumes of VO_x. See also VANADIUM PENTOXIDE (DUST).

VEA000 CAS:1314-34-7 ***HR: 3***
VANADIUM SESQUIOXIDE
DOT: UN 2860
mf: O_3V_2 mw: 149.88

PROP: Black crystals. Mp: 1970°, d: 4.87 @ 18°.

SYNS: VANADIC OXIDE ◇ VANADIUM OXIDE ◇ VANADIUM TRIOXIDE

TOXICITY DATA with REFERENCE
itr-rat LDLo:125 mg/kg NTIS** AEC-TR-6710

orl-mus LD50:382 mg/kg GTPZAB 8,25,64
scu-mus LD50:130 mg/kg ZVKOA6 19,186,74

CONSENSUS REPORTS: Reported in EPA TSCA Inventory.

ACGIH TLV: TWA 0.05 mg(V_2O_5)/m^3
NIOSH REL: (Vanadium Compound) CL 0.05 mg (V)/m^3/15M

SAFETY PROFILE: Poison by ingestion, subcutaneous, and intratracheal routes. Ignites when heated in air. When heated to decomposition it emits toxic fumes of VO_x. See also VANADIUM COMPOUNDS.

VEA100 CAS:16785-81-2 ***HR: 2***
VANADIUM SULFATE
mf: $H_2O_4S{\bullet}xV$ mw: 454.66

SYNS: SULFURIC ACID, VANADIUM SALT ◇ VANADIUM SULPHATE

TOXICITY DATA with REFERENCE
itt-rat TDLo:11921 μg/kg (male 1D pre):REP JRPFA4 7,21,64

ACGIH TLV: TWA 0.05 mg(V_2O_5)/m^3

SAFETY PROFILE: Experimental reproductive effects. When heated to decomposition it emits toxic fumes of SO_x and V_2O_5.

VEF000 CAS:7632-51-1 ***HR: 3***
VANADIUM TETRACHLORIDE
DOT: UN 2444
mf: Cl_4V mw: 192.74

PROP: Reddish-brown liquid. Mp: −28 ± 2°, bp: 148.5°, d: 1.816 @ 30°.

SYN: VANADIUM CHLORIDE

TOXICITY DATA with REFERENCE
orl-rat LD50:160 mg/kg AIHAAP 30,470,69

CONSENSUS REPORTS: Reported in EPA TSCA Inventory.

ACGIH TLV: TWA 0.05 mg(V_2O_5)/m^3
NIOSH REL: (Vanadium Compounds) CL 0.05 mg(V)/m^3/15M
DOT Classification: Corrosive Material; Label: Corrosive.

SAFETY PROFILE: Poison by ingestion. A corrosive irritant to skin, eyes, and mucous membranes. When heated to decomposition it emits toxic fumes of VO_x and Cl^-. See also VANADIUM COMPOUNDS and HYDROCHLORIC ACID.

VEK000 CAS:13470-26-3 ***HR: 3***
VANADIUM TRIBROMIDE
mf: Br_3V mw: 290.67

PROP: Green-black, deliq crystals. Mp: decomp.

SYN: VANADIUM BROMIDE

TOXICITY DATA with REFERENCE
scu-rbt LDLo:20 mg/kg 27ZWAY 3,3,1541,-

CONSENSUS REPORTS: Reported in EPA TSCA Inventory.

ACGIH TLV: TWA 0.05 mg(V_2O_5)/m^3
NIOSH REL: CL 0.05 mg(V)/m^3/15M

SAFETY PROFILE: Poison by subcutaneous route. When heated to decomposition it emits toxic fumes of VO_x and Br^-. See also VANADIUM COMPOUNDS and BROMIDES.

VEK100 CAS:13520-90-6 ***HR: 3***
VANADIUM TRIBROMIDE OXIDE
mf: Br_3OV mw: 306.65

SAFETY PROFILE: Reacts violently with water. When heated to decomposition it emits toxic fumes of Br^- and VO_x. See also VANADIUM OXYTRICHLORIDE, VANADIUM COMPOUNDS, and BROMIDES.

VEP000 CAS:7718-98-1 ***HR: 3***
VANADIUM TRICHLORIDE
DOT: UN 2475
mf: Cl_3V mw: 157.29

PROP: Pink crystals. Mp: decomp, d: 3.00 @ 18°.

SYN: VANADIUM(III)CHLORIDE

TOXICITY DATA with REFERENCE
orl-rat LD50:350 mg/kg AIHAAP 30,470,69
scu-rbt LDLo:20 mg/kg EQSSDX 1,1,75

CONSENSUS REPORTS: Reported in EPA TSCA Inventory.

ACGIH TLV: TWA 0.05 mg(V_2O_5)/m^3
NIOSH REL: (Vanadium Compounds) CL 0.05 mg (V)/m^3/15M
DOT Classification: Corrosive Material; Label: Corrosive.

SAFETY PROFILE: Poison by ingestion and subcutaneous routes. A corrosive irritant to skin, eyes, and mucous membranes. Extremely violent reaction with methyl magnesium iodide and other Grignard reagents. When heated to decomposition it emits toxic fumes of VO_x and Cl^-. See also VANADIUM COMPOUNDS and HYDROCHLORIC ACID.

VEU000 ***HR: 3***
VANADYL AZIDE DICHLORIDE
mf: Cl_2N_3OV mw: 179.87

SAFETY PROFILE: Explosive. When heated to decomposition it emits very toxic fumes of VO_x, Cl^-, and NO_x. See also AZIDES, VANADIUM COMPOUNDS, and CHLORIDES.

VEZ000 CAS:27774-13-6 ***HR: 3***
VANADYL SULFATE
DOT: UN 2931/NA 9152
mf: O_5SV mw: 163.00

PROP: Blue crystals.

SYNS: C.I. 77940 ◇ OXYSULFATOVANADIUM

TOXICITY DATA with REFERENCE
mrc-smc 6 mmol/L MUTAEX 1,21,86
sln-smc 4 mmol/L MUTAEX 1,21,86
scu-rat LDLo:140 mg/kg AJSNAO 1,347,17
ipr-mus LD50:144 mg/kg COREAF 256,1043,63
scu-mus LD50:560 mg/kg RPTOAN 34(3),135,71
ivn-rbt LDLo:16 mg/kg AJSNAO 1,347,17
scu-gpg LDLo:31 mg/kg AJSNAO 1,347,17

CONSENSUS REPORTS: Reported in EPA TSCA Inventory.

ACGIH TLV: TWA 0.05 mg(V_2O_5)/m^3
NIOSH REL: (Vanadium Compounds) CL 0.05 mg (V)/m^3/15M
DOT Classification: Poison B; Label: Poison

SAFETY PROFILE: Poison by intravenous, intraperitoneal, and subcutaneous routes. Mutation data reported. When heated to decomposition it emits toxic fumes of VO_x and SO_x. See also SULFATES and VANADIUM COMPOUNDS.

VEZ100 CAS:12439-96-2 ***HR: 3***
VANADYL SULFATE PENTAHYDRATE
mf: $O_5SV \cdot 5H_2O$ mw: 253.07

SYN: OXOSULFATOVANADIUM PENTAHYDRATE

TOXICITY DATA with REFERENCE
orl-rat LD50:448 mg/kg TOLED5 23,227,84
ipr-rat LD50:74 mg/kg TOLED5 23,227,84
orl-mus LD50:467 mg/kg TOLED5 23,227,84
ipr-mus LD50:113 mg/kg TOLED5 23,227,84

SAFETY PROFILE: Poison by intraperitoneal route. Moderately toxic by ingestion. When heated to decomposition it emits toxic fumes of VO_x and SO_x. See VANADYL SULFATE and VANADIUM COMPOUNDS.

VEZ925 CAS:149-17-7 ***HR: 2***
VANCIDE
mf: $C_{14}H_{13}N_3O_3$ mw: 271.30

SYNS: FTIVAZID ◇ FTIVAZIDE ◇ ISONICOTINIC ACID, VANILLYLIDENEHYDRAZIDE ◇ PHTIVAZID ◇ PHTIVAZIDE ◇ 4-PYRIDINECARBOXYLIC ACID, ((4-HYDROXY-3-METHOXYPHENYL) METHYLENE)HYDRAZIDE ◇ VANICID ◇ VANILLABERON ◇ VANIZIDE

TOXICITY DATA with REFERENCE
orl-mus TDLo:24960 mg/kg/1Y-I:ETA VOONAW 18(6),50,72

SAFETY PROFILE: Questionable carcinogen with experimental tumorigenic data. When heated to decomposition it emits toxic fumes of NO_x.

VFA000 CAS:1404-90-6 ***HR: 2***
VANCOCIN
mf: $C_{66}H_{75}C_{12}N_9O_{24}$ mw: 1449.40

SYN: VANCOMYCIN

TOXICITY DATA with REFERENCE
ivn-wmn TDLo:15 mg/kg/90M-C:SKN NEJMAG 313,756,85
ipr-mus LD50:1734 mg/kg 85FZAT -,675,67
scu-mus LD50:5000 mg/kg 85GDA2 1,315,80

SAFETY PROFILE: Moderately toxic by intraperitoneal route. Human systemic effects by intravenous route: allergic skin dermatitis. When heated to decomposition it emits toxic fumes of NO_x.

VFA050 CAS:1404-93-9 ***HR: 3***
VANCOCINE HYDROCHLORIDE
mf: $C_{66}H_{75}Cl_2N_9O_{24} \cdot ClH$ mw: 1485.86

SYN: VANCOCIN HYDROCHLORIDE

TOXICITY DATA with REFERENCE
ivn-wmn TDLo:20 mg/kg:CVS AIMEAS 101,880,84
ivn-man TDLo:3571 μg/kg/15M-C:CVS,SKN AIMEAS 104,285,86
ipr-rat LD50:2218 mg/kg IYKEDH 12,933,81
ivn-rat LD50:319 mg/kg IYKEDH 12,933,81
ipr-mus LD50:1734 mg/kg IYKEDH 12,933,81
ivn-mus LD50:489 mg/kg IYKEDH 12,933,81

SAFETY PROFILE: Poison by intravenous route. Moderately toxic by intraperitoneal routes. Human systemic effects by intravenous route: cardiac arrythmias, blood pressure lowering, and allergic dermatitis. When heated to decomposition it emits toxic fumes of NO_x and HCl.

VFA200 CAS:8047-24-3 ***HR: 1***
VANILLA TINCTURE

TOXICITY DATA with REFERENCE
skn-rbt 500 mg/24H MOD FCTOD7 20,849,82

SAFETY PROFILE: A skin irritant. When heated to decomposition it emits acrid smoke and irritating fumes.

VFF000 CAS:121-34-6 ***HR: 1***
VANILLIC ACID
mf: $C_8H_8O_4$ mw: 168.16

PROP: Odorless needles from water. Bp: subl, mp: 210°. Sol in water, ether; very sol in alc.

SYNS: ACIDE VANILLIQUE ◇ 4-HYDROXY-m-ANISIC ACID ◇ 4-HYDROXY-3-METHOXYBENZOIC ACID ◇ 3-METHOXY-4-HYDROXYBENZOIC ACID ◇ PROTOCATECHUIC ACID, 3-METHYL ESTER ◇ VA ◇ p-VANILLIC ACID

TOXICITY DATA with REFERENCE
cyt-ham:ovr 50 g/L CALEDQ 14,251,81
ipr-rat LD50:5020 mg/kg COREAF 243,609,56

CONSENSUS REPORTS: Reported in EPA TSCA Inventory.

SAFETY PROFILE: Mutation data reported. When heated to decomposition it emits acrid smoke and irritating fumes.

VFK000 CAS:121-33-5 ***HR: 2***
VANILLIN
mf: $C_8H_8O_3$ mw: 152.16

$HO(CH_3O)C_6H_3CO{\bullet}H$

PROP: White, crystalline needles; vanilla odor. D: 1.056, bp: 285°, mp: 80-81°. Sol in 125 parts water, 20 parts glycerin, 2 parts 95% alc, chloroform, ether.

SYNS: FEMA No. 3107 ◇ 4-HYDROXY-m-ANISALDEHYDE ◇ 4-HYDROXY-3-METHOXYBENZALDEHYDE ◇ LIOXIN ◇ 3-METHOXY-4-HYDROXYBENZALDEHYDE ◇ METHYLPROTOCATECHUALDEHYDE ◇ VANILLA ◇ VANILLALDEHYDE ◇ VANILLIC ALDEHYDE ◇ p-VANILLIN ◇ ZIMCO

TOXICITY DATA with REFERENCE
sce-hmn:lym 750 μmol/L MUREAV 169,129,86
scu-rat TDLo:20 mg/kg (4D pre):REP JSICAZ 19,264,60
orl-rat LD50:1580 mg/kg FCTXAV 2,327,64
ipr-rat LD50:1160 mg/kg COREAF 243,609,56
scu-rat LD50:1500 mg/kg RMSRA6 16,449,1896
ipr-mus LD50:475 mg/kg FAONAU 44A,79,67
ivn-dog LDLo:1320 mg/kg COREAF 236,2549,53
orl-rbt LDLo:3000 mg/kg JAPMA8 29,425,40
orl-gpg LD50:1400 mg/kg FCTXAV 2,327,64
ipr-gpg LD50:1190 mg/kg COREAF 236,2549,53

CONSENSUS REPORTS: Reported in EPA TSCA Inventory.

SAFETY PROFILE: Moderately toxic by ingestion, intraperitoneal, subcutaneous, and intravenous routes. Experimental reproductive effects. Human mutation data reported. Can react violently with Br_2; $HClO_4$; potassium-tert-butoxide; tert-chlorobenzene + NaOH; formic acid + thallium nitrate. When heated to decomposition it emits acrid smoke and irritating fumes. See also ALDEHYDES.

VFP000 CAS:148-53-8 ***HR: 3***
o-VANILLIN
mf: $C_8H_8O_3$ mw: 152.16

PROP: D: 1.056, mp: 82-83.5°, bp: 285° (in CO_2). Sol in water; very sol in alc, benzene, ether, and CS_2.

SYNS: 6-FORMYLGUAIACOL ◇ 2-HYDROXY-m-ANISALDEHYDE ◇ 2-HYDROXY-3-METHOXYBENZALDEHYDE ◇ 3-METHOXYSALICYLALDEHYDE ◇ ORTHOVANILLINE ◇ OXY-2-METHOXY-3-BENZALDEHYDE (FRENCH)

TOXICITY DATA with REFERENCE
skn-rbt 500 mg FCTOD7 20,563,82
eye-rbt 100 mg FCTOD7 20,573,82
eye-rbt 100 mg/4S rns MLD FCTOD7 20,573,82
ipr-rat LD50:347 mg/kg COREAF 243,609,56

CONSENSUS REPORTS: Reported in EPA TSCA Inventory.

SAFETY PROFILE: Poison by intraperitoneal route. An eye and skin irritant. When heated to decomposition it emits acrid smoke and irritating fumes. See also VANILLIN and ALDEHYDES.

VFP100 CAS:122-48-5 ***HR: 2***
VANILLYL ACETONE
mf: $C_{11}H_{14}O_3$ mw: 194.25

PROP: Crystals. Mp: 40-41°, Bp: 187-188°. Sltly sol in water, petr ether; sol in ether.

SYNS: 2-BUTANONE,4-(4-HYDROXY-3-METHOXYPHENYL)- ◇ GINGERONE ◇ 4-(4-HYDROXY-3-METHOXYPHENYL)-2-BUTANONE ◇ (4-HYDROXY-3-METHOXYPHENYL)ETHYL METHYL KETONE ◇ 3-METHOXY-4-HYDROXY-BENZYLACETONE ◇ (0)-PARADOL ◇ ZINGERONE ◇ ZINGIBERONE

TOXICITY DATA with REFERENCE
skn-rbt 500 mg/24H MLD FCTOD7 20,851,82
orl-rat LD50:2580 mg/kg FCTOD7 20,851,82

CONSENSUS REPORTS: Reported in EPA TSCA Inventory.

SAFETY PROFILE: Moderately toxic by ingestion. A skin irritant. When heated to decomposition it emits acrid smoke and irritating fumes.

VFP200 CAS:78100-57-9 ***HR: 3***
VANILOL
mf: $C_{15}H_{23}NO_5$ mw: 297.39

SYNS: 4-(3-ISOPROPYLAMINO)-2-HYDROXYPROPOXY)-3-METHOXY-BENZOIC ACID METHYL ESTER ◇ 1-(2-METHOXY-4-METHOXYCARBONYL-1-PHENOXY)-3-ISOPROPYLAMINO-2-PROPANOL

TOXICITY DATA with REFERENCE
orl-mus LD50:401 mg/kg CYLPDN 2(2),97,81

ipr-mus LD50:207 mg/kg CYLPDN 2(2),97,81
ivn-mus LD50:50 mg/kg CYLPDN 2(2),97,81

SAFETY PROFILE: Poison by intravenous and intraperitoneal routes. Moderately toxic by ingestion. When heated to decomposition it emits toxic fumes of NO_x. See also ESTERS.

VFU000 CAS:137-42-8 ***HR: 3***
VAPAM
mf: $C_2H_4NS_2 \cdot Na$ mw: 129.18

SYNS: BASAMID-FLUID ◇ CARBAM ◇ CARBATHIONE ◇ KARBATION ◇ MAPOSOL ◇ METAM-SODIUM (DUTCH, FRENCH, GERMAN, ITALIAN) ◇ METHAM SODIUM ◇ N-METHYLDITHIOCARBAMATE de SODIUM (FRENCH) ◇ METHYLDITHIOCARBAMIC ACID, SODIUM SALT ◇ N-METHYLDITHIOCARBAMIC ACID, SODIUM SALT ◇ N-METIL-DITIOCARBAMMATO di SODIO (ITALIAN) ◇ NATRIUM-N-METHYL-DITHIOCARBAMAAT (DUTCH) ◇ NATRIUM-N-METHYL-DITHIOCARBAMAT (GERMAN) ◇ SISTAN ◇ SMDC ◇ SODIUM METHYLDITHIOCARBAMATE ◇ SODIUM N-METHYLDITHIOCARBAMATE ◇ TRAPEX ◇ TRIMATON ◇ VDM ◇ VPM

TOXICITY DATA with REFERENCE
orl-rat LD50:1700 mg/kg FMCHA2 -,C151,83
skn-rat LD50:636 mg/kg 85GMAT -,106,82
orl-mus LD50:50 mg/kg RREVAH 10,97,65
orl-rbt LD50:320 mg/kg HYSAAV 32,169,67
skn-rbt LD50:800 mg/kg WRPCA2 9,119,70
orl-gpg LD50:815 mg/kg HYSAAV 32,169,67

SAFETY PROFILE: Poison by ingestion. Moderately toxic by skin contact. Irritating to skin and mucous membranes. Accompanied by alcohol intake it causes violent vomiting and shock. A general-purpose soil fumigant. When heated to decomposition it emits very toxic fumes of NO_x, SO_x, and Na_2O. See also CARBAMATES.

VFW009 CAS:6734-80-1 ***HR: 3***
VAPOROOTER
mf: $C_2H_4NS_2 \cdot Na \cdot 2H_20$ mw: 165.22

SYNS: HERBATIM ◇ KARBATION ◇ MAPOSOL ◇ METAM ◇ METHAM SODIUM ◇ MONAM ◇ NATRIUMMETHYLDITHIOCARBAMAT (GERMAN) ◇ SMDC ◇ SODIUM N-METHYLDITHIOCARBAMATE DIHYDRATE ◇ TRIMATON ◇ VAPAM ◇ VPM

TOXICITY DATA with REFERENCE
orl-rat LD50:108 mg/kg PCOC** -,1047,66
orl-mus LD50:285 mg/kg 85GYAZ -,121,71
unr-mam LD50:450 mg/kg GISAAA 45(5),29,80

SAFETY PROFILE: Poison by ingestion. Moderately toxic by an unspecified route. When heated to decomposition it emits toxic fumes of SO_x and NO_x. See also CARBAMATES.

VGA000 CAS:6159-55-3 ***HR: 3***
VASICINE
mf: $C_{11}H_{10}N_2O$ mw: 186.23

PROP: dl-Form: Needles. Mp: 210°. Sol in acetone, alc, chloroform; sltly sol in water, ether, benzene. l-Form: Needles. Mp: 212°.

SYNS: 1,2,3,9-TETRAHYDROPYRROLO(2,1-B)QUINAZOLIN-3-OL ◇ VASICIN (GERMAN)

TOXICITY DATA with REFERENCE
ipr-gpg TDLo:150 mg/kg (56-60D preg):REP IJEBA6 16,1075,78
orl-rat LD50:640 mg/kg ARZNAD 13,474,63
ipr-rat LD50:115 mg/kg ARZNAD 13,474,63
scu-rat LD50:335 mg/kg ARZNAD 13,474,63
orl-mus LD50:290 mg/kg ARZNAD 13,474,63
ipr-mus LD50:79 mg/kg IJMRAQ 66,680,77
scu-mus LD50:200 mg/kg ARZNAD 13,474,63

SAFETY PROFILE: Poison by ingestion, intraperitoneal, and subcutaneous routes. Experimental reproductive effects. When heated to decomposition it emits toxic fumes of NO_x.

VGA025 CAS:80039-73-2 ***HR: 2***
VASICINONE HYDROCHLORIDE
mf: $C_{11}H_{10}N_2O_2 \cdot ClH$ mw: 238.69

SYN: 2,3-DIHYDRO-3-HYDROXYPYRROLO(2,1-b)QUINAZOLIN-9(1H)-ONE HYDROCHLORIDE

TOXICITY DATA with REFERENCE
orl-mus LD50:1100 mg/kg DDIPD8 8,833,82
ipr-mus LD50:520 mg/kg DDIPD8 8,833,82
ivn-mus LD50:440 mg/kg DDIPD8 8,833,82

SAFETY PROFILE: Moderately toxic by ingestion, intraperitoneal and intravenous routes. When heated to decomposition it emits toxic fumes of NO_x and HCl. See also KETONES.

VGA100 CAS:37244-86-3 ***HR: 3***
VASOBRIX 32
mf: $C_{12}H_{11}I_3N_2O_5 \cdot C_{12}H_{11}I_3N_2O_5 \cdot C_7H_{17}NO_5 \cdot C_2H_7NO$ mw: 1544.25

SYN: AG. 5895

TOXICITY DATA with REFERENCE
ivn-rat LD50:11500 mg/kg THERAP 26,595,71
ipr-mus LD50:18900 mg/kg THERAP 26,595,71
ivn-mus LD50:15100 mg/kg THERAP 26,595,71
ice-mus LD50:160 mg/kg THERAP 26,595,71

SAFETY PROFILE: Poison by intracerebral route. When heated to decomposition it emits toxic fumes of I^- and NO_x.

VGA300 CAS:579-56-6 ***HR: 3***
VASODILAN
mf: $C_{18}H_{23}NO_3 \cdot ClH$ mw: 337.88

SYNS: DILAVASE ◇ DUVADILAN ◇ 4-HYDROXY-α-(1-((1-METHYL-2-PHENOXYETHYL)AMINO)ETHYL)-BENZENEMETHANOLHYDRO-

CHLORIDE ◇ 1-(p-HYDROXYPHENYL)-2-(1'-METHYL-2'-PHENOXY) ETHYLAMINOPROPANOL-1 HYDROCHLORIDE ◇ ISOPLAIT ◇ ISOXSUPRINE HYDROCHLORIDE ◇ ISOXSUPRIN HYDROCHLORIDE ◇ SUPRILENT ◇ VADOSILAN ◇ VASOPLEX ◇ VASOTRAN

TOXICITY DATA with REFERENCE
ivn-rat TDLo:4400 μg/kg/3H (female 21D post):TER JPEMAO 9,293,81
orl-rat LD50:1750 mg/kg TXAPA9 18,185,71
ipr-rat LD50:164 mg/kg TXAPA9 18,185,71
orl-mus LD50:1100 mg/kg NIIRDN 6,71,82
ipr-mus LD50:185 mg/kg NIIRDN 6,71,82
ivn-mus LD50:61 mg/kg NIIRDN 6,71,82
ivn-dog LD50:57 mg/kg NIIRDN 6,71,82

SAFETY PROFILE: Poison by intravenous and intraperitoneal routes. Moderately toxic by ingestion. An experimental teratogen. Used as a vasodilator. When heated to decomposition it emits toxic fumes of NO_x and HCl.

VGF000 CAS:395-28-8 ***HR: 3***
VASODILIAN
mf: $C_{18}H_{23}NO_3$ mw: 301.42

SYNS: DILAVASE ◇ DUVADILAN ◇ p-HYDROXY-N-(1-METHYL-2-PHENOXYETHYL)NOREPHEDRINE ◇ 1-(4-HYDROXYPHENYL)-2-(1-METHYL-2-PHENOXYETHYLAMINO)PROPANOL ◇ 1-(p-HYDROXYPHENYL)-2-(1'-METHYL-2'-PHENOXYETHYLAMINO)PROPANOL-2-HYDROCHLORIDE ◇ ISOXSUPRINE ◇ 2-(PHENOXY-2-PROPYLAMINO)-1-(p-HYDROXYPHENYL)-1-PROPANOL HYDROCHLORIDE ◇ VASODILAN

TOXICITY DATA with REFERENCE
ivn-wmn TDLo:350 μg/kg (female 26W post):REP JOPDAB 94,444,79
ivn-wmn TDLo:350 μg/kg (female 32W post):TER JOPDAB 94,444,79
orl-mus LD50:200 mg/kg ARZNAD 21,1992,71
ipr-mus LD50:118 mg/kg FRPSAX 38,571,83
ivn-mus LD50:48 mg/kg EJMCA5 10,291,75

SAFETY PROFILE: Poison by ingestion, intravenous and intraperitoneal routes. A human teratogen. Human reproductive effects by ingestion: uterus, cervix and vagina effects. An experimental teratogen. Experimental reproductive effects. When heated to decomposition it emits toxic fumes of NO_x. Used to delay premature parturition (labor) and to accelerate fetal lung development.

VGK000 CAS:26328-04-1 ***HR: 2***
VASODISTAL
mf: $C_{22}H_{31}N_3O_5{\cdot}C_4H_4O_4$ mw: 533.64

SYNS: BRENDIL ◇ CINEPAZIDE MALEATE ◇ MALEATE de CINEPAZIDE (FRENCH) ◇ MD 67350 ◇ 1-((1-PYRROLIDINYLCARBONYL)METHYL)-4-(3,4,5-TRIMETHOXYCINNAMOYL)PIPERAZINEMALEATE ◇ 1-(4-((3',4',5'-TRIMETHOXYCINNAMOYL)-1-PIPERAZINYL)ACETYL)PYRROLIDINE MALEATE ◇ 4-(3,4,5-TRIMETHOXYCINNAMOYL)-1-(1-PYRROLIDINYL)CARBONYLEMETHYLPIPERAZINEMALEATE

TOXICITY DATA with REFERENCE
orl-mus TDLo:672 mg/kg (15-21D preg/21D post):REP IYKEDH 10,559,79
orl-rbt TDLo:312 mg/kg (female 6-18D post):TER IYKEDH 10,572,79
orl-rat LD50:1310 mg/kg THERAP 29,29,74
scu-rat LD50:710 mg/kg IYKEDH 10,407,79
ivn-rat LD50:414 mg/kg THERAP 29,29,74
orl-mus LD50:1000 mg/kg THERAP 29,29,74
scu-mus LD50:946 mg/kg IYKEDH 10,407,79
ivn-mus LD50:617 mg/kg THERAP 29,29,74

SAFETY PROFILE: Moderately toxic by ingestion, subcutaneous, and intravenous routes. An experimental teratogen. Experimental reproductive effects. When heated to decomposition it emits toxic fumes of NO_x. Used as a vasodilator.

VGP000 CAS:51-43-4 ***HR: 3***
VASOTONIN
mf: $C_9H_{13}NO_3$ mw: 183.23

SYNS: ADNEPHRINE ◇ ADRENAL ◇ 1-ADRENALIN ◇ ADRENALIN-MEDIHALER ◇ ADRENAMINE ◇ ADRENAN ◇ ADRENAPAX ◇ ADRENASOL ◇ ADRENATRATE ◇ ADRENODIS ◇ ADRENOHORMA ◇ ADRENUTOL ◇ ADRINE ◇ ASMATANE MIST ◇ ASTHMA METER MIST ◇ ASTMAHALIN ◇ BALMADREN ◇ BERNARENIN ◇ BIORENINE ◇ BOSMIN ◇ BREVIRENIN ◇ BRONKAID MIST ◇ CHELAFRIN ◇ CORISOL ◇ 3,4-DIHYDROXY-α-((METHYLAMINO)METHYL)BENZYL ALCOHOL ◇ 1-1-(3,4-DIHYDROXYPHENYL)-2-METHYLAMINOETHANOL ◇ DRENAMIST ◇ DYLEPHRIN ◇ DYSPNE-INHAL ◇ EPIFRIN ◇ EPINEPHRAN ◇ EPINEPHRINE ◇ (−)-EPINEPHRINE ◇ (R)-EPINEPHRINE ◇ 1-EPINEPHRINE ◇ 1-EPINEPHRINE (synthetic) ◇ EPIRENAMINE ◇ EPIRENAN ◇ EPITRATE ◇ ESPHYGMOGENINA ◇ EXADRIN ◇ GLYCIRENAN ◇ HAEMOSTASIN ◇ HEKTALIN ◇ HEMISINE ◇ HEMOSTASIN ◇ (R)-4-(1-HYDROXY-2-(METHYLAMINO)ETHYL)-1,2-BENZENEDIOL (9CI) ◇ HYPERNEPHRIN ◇ HYPORENIN ◇ INTRANEFRIN ◇ KIDOLINE ◇ LEVORENIN ◇ LYOPHRIN ◇ MEDIHALER-EPI ◇ METANEPHRIN ◇ METHYLARTERENOL ◇ MUCIDRINA ◇ MYOSTHENINE ◇ MYTRATE ◇ NEPHRIDINE ◇ NIERALINE ◇ PARANEPHRIN ◇ PRIMATENE MIST ◇ RCRA WASTE NUMBER P042 ◇ RENAGLADIN ◇ RENALEPTINE ◇ RENALINA ◇ RENOFORM ◇ RENOSTYPRICIN ◇ RENOSTYPTIN ◇ SCURENALINE ◇ SINDRENINA ◇ SOLADREN ◇ SPHYGMOGENIN ◇ STRYPTIRENAL ◇ SUPRACAPSULIN ◇ SUPRADIN ◇ SUPRANEPHRANE ◇ SUPRANEPHRINE ◇ SUPRANOL ◇ SUPRARENIN ◇ SUPREL ◇ SURENINE ◇ SUSPHRINE ◇ SYMPATHIN I ◇ TAKAMINA ◇ TOKAMINA ◇ TONOGEN ◇ VAPONEFRIN ◇ VASOCONSTRICTINE ◇ VASOCONSTRICTOR ◇ VASODRINE ◇ VASOTON

TOXICITY DATA with REFERENCE
mmo-sat 500 μg/plate ABCHA6 45,327,81
oms-mus:oth 2500 ng/L ECREAL 35,629,64
ipr-rat TDLo:6 mg/kg (female 7-14D post):REP JCPPAV 58,309,64
ipr-rat TDLo:1 mg/kg (female 13D post):TER DANND6 (3),59,83
scu-man LDLo:735 μg/kg 85DCAI 2,73,70
scu-man TDLo:8571 ng/kg/80M-I:CVS AHJOA2 111,1193,86
ivn-wmn TDLo:6 μg/kg:CVS BMJOAE 286,519,83

orl-rat LDLo:30 mg/kg 85IXA4 -,22,48
skn-rat LD50:62 mg/kg GTPZAB 8(4),30,64
ipr-rat LDLo:10 mg/kg JPETAB 88,268,46
scu-rat LD50:5 mg/kg SMWOAS 71,554,41
ivn-rat LD50:150 μg/kg AIPTAK 41,365,31
ims-rat LD50:3500 mg/kg 27ZIAQ -,105,73
orl-mus LDLo:50 mg/kg 85IXA4 -,22,48
ipr-mus LD50:4 mg/kg JPETAB 90,110,47
scu-mus LD50:1470 μg/kg AEPPAE 202,658,43
ivn-mus LD50:217 μg/kg APTOA6 38,474,76
scu-dog LD50:5 mg/kg NIIRDN 6,120,82
ivn-dog LD50:100 μg/kg NIIRDN 6,120,82
scu-cat LDLo:20 mg/kg BBMS** -,-,48
ivn-cat LDLo:500 μg/kg 85IXA4 -,22,48

CONSENSUS REPORTS: Reported in EPA TSCA Inventory. EPA Genetic Toxicology Program.

SAFETY PROFILE: Human poison by subcutaneous route. Experimental poison by ingestion, skin contact, subcutaneous, intraperitoneal, intravenous, and intramuscular routes. Human systemic effects: cardiomyopathy including infarction, arrhythmias. An experimental teratogen. Experimental reproductive effects. Mutation data reported. When heated to decomposition it emits toxic fumes of NO_x. Used as an adrenergic, sympathomimetic, vasoconstrictor, bronchodilator, and cardiac stimulant.

VGU000 CAS:6198-57-8 ***HR: 3***
VEATCHINE HYDROCHLORIDE
mf: $C_{22}H_{33}NO_2 \cdot ClH$ mw: 380.02

PROP: Crystals. Decomp @ 267-271°. Sol in water. Alkaloid isolated from *Garrya veatchii* (JAPMA8 45,733,56).

TOXICITY DATA with REFERENCE
ivn-mus LD50:13 mg/kg JAPMA8 45,733,56
ivn-cat LDLo:2 mg/kg JAPMA8 45,733,56

SAFETY PROFILE: Poison by intravenous route. When heated to decomposition it emits very toxic fumes of NO_x and HCl.

VGU075 CAS:50700-72-6 ***HR: 3***
VECURONIUM BROMIDE
mf: $C_{34}H_{57}N_2O_4 \cdot Br$ mw: 637.84

PROP: Crystals. Mp: 227-229°.

SYNS: NORCURON ◇ ORG NC 45 ◇ 16β-PIPECOLINIO-2β-PIPERIDINO-5α-ANDROSTAN-3α,17β-DIOL BROMIDE DIACETATE

TOXICITY DATA with REFERENCE
orl-rat LD50:455 mg/kg KSRNAM 20,807,86
ipr-rat LD50:2630 μg/kg KSRNAM 20,807,86
scu-rat LD50:1730 μg/kg KSRNAM 20,807,86
ivn-rat LD50:200 μg/kg KSRNAM 20,807,86
orl-mus LD50:41 mg/kg KSRNAM 20,807,86
ipr-mus LD50:144 μg/kg KSRNAM 20,807,86
scu-mus LD50:148 μg/kg KSRNAM 20,807,86
ivn-mus LD50:51 μg/kg KSRNAM 20,807,86

SAFETY PROFILE: Poison by ingestion, subcutaneous, intravenous and intraperitoneal routes. When heated to decomposition it emits toxic fumes of NO_x and Br^-.

VGU200 CAS:68956-68-3 ***HR: 1***
VEGETABLE OIL

SYN: VEGETABLE OIL MIST (OSHA)

OSHA PEL: TWA 15 mg/m^3, total dust; TWA 5 mg/m^3, respirable fraction

CONSENSUS REPORTS: Reported in EPA TSCA Inventory.

SAFETY PROFILE: A nuisance mist. When heated to decomposition it emits acrid smoke and irritating fumes.

VGU700 CAS:9046-56-4 ***HR: D***
VENACIL

PROP: Colorless substance when pure, having a light powdery texture when in the freeze-dried state. Sol in physiological saline. Absorbable on weakly basic anion exchange materials.

SYNS: ABBOTT 38414 ◇ A 38414 (ENZYME) ◇ ANCROD ◇ ARVIN ◇ ARWIN ◇ IRC-50 ARVIN

TOXICITY DATA with REFERENCE
ims-rbt TDLo:30 iu/kg (18-20D preg):REP TXAPA9 20,460,71
ims-rbt TDLo:30 iu/kg (7-9D preg):TER TXAPA9 20,460,71
ivn-rat LDLo:120 units/kg LANCAO 1,486,68
ivn-rbt LDLo:2.5 units/kg LANCAO 1,486,68

SAFETY PROFILE: An experimental teratogen. Experimental reproductive effects.

VGU750 CAS:59917-39-4 ***HR: 3***
VENDESINE SULFATE
mf: $C_{43}H_{55}N_5O_7 \cdot H_2O_4S$ mw: 852.11

SYNS: 23-AMINO-O[4]-DEACETYL-23-DEMETHOXYVINCALEUKOBLASTINE SULFATE ◇ DAVA ◇ DESACETYLVINBLASTINE AMIDE SULFATE ◇ ELDESINE ◇ LILLY 99094 ◇ VINDESINA SULFATO (SPANISH)

TOXICITY DATA with REFERENCE
cyt-hmn:lym 50 mg/L CYTOAN 50,311,85
oms-ham:ovr 2300 pmol/L CNREA8 38,2886,78
ivn-rat TDLo:2500 μg/kg (10D pre):REP KSRNAM 17,1859,83
ipr-rat LD50:1050 μg/kg KSRNAM 17,1549,83
scu-rat LD50:1790 μg/kg KSRNAM 17,1549,83

ivn-rat LD50:1920 μg/kg KSRNAM 17,1549,83
ipr-mus LD50:3500 μg/kg KSRNAM 17,1549,83
scu-mus LD50:6800 μg/kg KSRNAM 17,1549,83
ivn-mus LD50:6300 μg/kg JTEHD6 1,843,76

SAFETY PROFILE: Poison by subcutaneous, intravenous, and intraperitoneal routes. Experimental reproductive effects. Human mutation data reported. When heated to decomposition it emits toxic fumes of SO_x and NO_x.

VGZ000 CAS:606-58-6 ***HR: 3***
VENGICIDE
mf: $C_{12}H_{13}N_5O_4$ mw: 291.30

SYNS: AHYGROSCOPIN-B ◇ 4-AMINO-5-CYANO-7-(d-RIBOFURANOSYL)-7H-PYRROLO(2,3-d)PYRIMIDINE ◇ 4-AMINO-7-β-d-RIBOFURANOSYL-7H-PYRROLO(2,3-d)PYRIMIDINE-5-CARBONITRILE ◇ ANTIBIOTIC 1037 ◇ ANTIBIOTIC A-399-Y4 ◇ ANTIBIOTIC E212 ◇ CYANOTUBERICIDIN ◇ E-212 ◇ NARITHERACIN ◇ NSC-63701 ◇ SIROMYCIN ◇ TOYOCAMYCIN NUCLEOSIDE ◇ UNAMYCIN-B ◇ URAMYCIN B

TOXICITY DATA with REFERENCE
dnd-esc 50 μmol/L MUREAV 89,95,81
oms-ckn:emb 100 μg/L CNREA8 29,1707,69
orl-mus LD50:8 mg/kg 85FZAT -,805,67
ipr-mus LD50:20 mg/kg 85GDA2 5,315,81
scu-mus LD50:10 mg/kg 85ERAY 2,1087,78
ivn-mus LD50:1500 μg/kg 85GDA2 5,318,81

CONSENSUS REPORTS: Cyanide and its compounds are on the Community Right-To-Know List.

SAFETY PROFILE: Poison by ingestion, intraperitoneal, subcutaneous, and intravenous routes. Mutation data reported. When heated to decomposition it emits toxic fumes of NO_x and CN^-. Used as an antibiotic. See also NITRILES.

VHA275 ***HR: 2***
VENOPIRIN

TOXICITY DATA with REFERENCE
orl-rat LD50:4350 mg/kg YACHDS 6,1275,78
scu-rat LD50:1860 mg/kg YACHDS 6,1275,78
ivn-rat LD50:1525 mg/kg YACHDS 6,1275,78
orl-mus LD50:3270 mg/kg YACHDS 6,1275,78
scu-mus LD50:1840 mg/kg YACHDS 6,1275,78
ivn-mus LD50:950 mg/kg YACHDS 6,1275,78

SAFETY PROFILE: Moderately toxic by ingestion, subcutaneous, and intravenous routes.

VHA350 CAS:21898-19-1 ***HR: 3***
VENTIPULMIN
mf: $C_{12}H_{18}Cl_2N_2O{\cdot}ClH$ mw: 313.68

SYNS: 4-AMINO-α-((tert-BUTYLAMINO)METHYL)-3,5-DICHLOROBENZYL ALCOHOL HYDROCHLORIDE ◇ 4-AMINO-α-((tert-BUTYLAMINO)METHYL)-3,5-DICHLOROBENZYLALKOHOL-HYDROCHLORID (GERMAN) ◇ CLENBUTEROL HYDROCHLORIDE ◇ NAB 365 ◇ NAB 365Cl ◇ SPIROPENT

TOXICITY DATA with REFERENCE
orl-rat TDLo:480 μg/kg (female 17-22D post):REP IYKEDH 15,597,84
orl-rbt TDLo:26 mg/kg (female 6-18D post):TER IYKEDH 15,590,84
orl-rat LD50:159 mg/kg IYKEDH 15,741,84
ipr-rat LD50:67 mg/kg OYYAA2 9,675,75
scu-rat LD50:148 mg/kg IYKEDH 15,741,84
ivn-rat LD50:35300 μg/kg ARZNAD 26,1420,76
orl-mus LD50:147 mg/kg IYKEDH 15,741,84
ipr-mus LD50:46 mg/kg OYYAA2 9,675,75
scu-mus LD50:63 mg/kg IYKEDH 15,741,75
ivn-mus LD50:27600 μg/kg ARZNAD 26,1420,76
orl-dog LDLo:500 mg/kg ARZNAD 26,1420,76
ivn-dog LDLo:45 mg/kg ARZNAD 26,1420,76
scu-rbt LD50:67100 μg/kg ARZNAD 26,1404,76
ivn-rbt LD50:12600 μg/kg ARZNAD 26,1404,76
orl-gpg LD50:67100 μg/kg ARZNAD 26,1420,76
scu-gpg LD50:74 mg/kg ARZNAD 26,1404,76
ivn-gpg LD50:12600 μg/kg ARZNAD 26,1420,76

SAFETY PROFILE: Poison by ingestion, subcutaneous, intravenous, and intraperitoneal routes. An experimental teratogen. Experimental reproductive effects. When heated to decomposition it emits toxic fumes of NO_x and HCl.

VHA450 CAS:152-11-4 ***HR: 3***
VERAPAMIL HYDROCHLORIDE
mf: $C_{27}H_{38}N_2O_4{\cdot}ClH$ mw: 491.13

PROP: Viscous, pale yellow oil. Bp. 243-246°. Practically insol in water; sparingly sol in hexane; sol in benzene, ether, the lower alcs, acetone, ethyl acetate, and chloroform.

SYNS: CALAN ◇ CORDILOX ◇ IPROVERATRIL HYDROCHLORIDE ◇ ISOPTIN ◇ IZOPTIN ◇ IZOPTIN HYDROCHLORIDE ◇ VASOLAN

TOXICITY DATA with REFERENCE
orl-man TDLo:128 mg/kg (28D male):REP AIMDAP 143,1248,83
orl-man TDLo:26 mg/kg:CVS,PUL HUTODJ 4,327,85
orl-man TDLo:143 μg/kg/4H-C:CVS ICMED9 8,55,82
ivn-man LDLo:256 μg/kg/1H-I:CVS AEMED3 14,159,85
ivn-man TDLo:71 μg/kg/5M-C SMJOAV 75,1429,82
orl-wmn TDLo:135 mg/kg:EYE,PUL AJCDAG 58,1142,86
ivn-wmn TDLo:200 μg/kg/15S-C:CVS,PUL NYSJAM 83,1181,83
orl-inf TDLo:80 mg/kg PEDIAU 73,543,84
ivn-inf TDLo:1026 μg/kg:CVS ADCHAK 58,465,83
orl-rat LD50:108 mg/kg NIIRDN 6,766,82
ipr-rat LD50:60 mg/kg PCJOAU 15,813,81
scu-rat LD50:107 mg/kg NIIRDN 6,766,82

ivn-rat LD50:16 mg/kg NIIRDN 6,766,82
ims-rat LD50:118 mg/kg NIIRDN 6,766,82
orl-mus LD50:163 mg/kg NIIRDN 6,766,82
ipr-mus LD50:46 mg/kg PCJOAU 15,813,81
scu-mus LD50:68 mg/kg NIIRDN 6,766,82
ivn-mus LD50:6 mg/kg PCJOAU 15,813,81
ims-dog LD50:25 mg/kg NIIRDN 6,766,82

SAFETY PROFILE: Poison by ingestion, subcutaneous, intramuscular, intravenous, and intraperitoneal routes. Human systemic effects by arrhythmias, cardiomyopathy, coma, cyanosis, dyspnea, fall in blood pressure, mydriasis, pulse rate decrease. Human reproductive effects: impotence. When heated to decomposition it emits toxic fumes of NO_x and HCl.

VHF000 CAS:63951-45-1 ***HR: 3***
VERATENSINE
mf: $C_{37}H_{59}NO_{11}$ mw: 693.97

PROP: An alkaloid ester separated from *Veratrum album*. (JPETAB 82,162,44).

SYNS: CEVANE-3-β,4-β,7-α,14,15-α,16-β,20-HEPTOL,4,9-EPOXY-, 15-((+)-2-HYDROXY-2-METHYLBUTYRATE)3-((−)-2-METHYLBUTYRATE) ◇ GERMERIN (GERMAN)

TOXICITY DATA with REFERENCE
orl-rat LD50:30 mg/kg AEPPAE 189,397,38
scu-rat LD50:3700 μg/kg AEPPAE 189,397,38
scu-cat LDLo:500 μg/kg AEPPAE 189,397,38
scu-rbt LDLo:2 mg/kg AEPPAE 189,397,38
ivn-rbt LDLo:300 μg/kg AEPPAE 189,397,38
scu-frg LD50:9 mg/kg AEPPAE 189,397,38

SAFETY PROFILE: Poison by ingestion, subcutaneous, and intravenous routes. An alkaloid poison. When heated to decomposition it emits toxic fumes of NO_x.

VHK000 CAS:120-14-9 ***HR: 2***
VERATRALDEHYDE
mf: $C_9H_{10}O_3$ mw: 166.19

PROP: Needles; odor of vanilla beans. Mp: 42-43°, bp: 281°. Sltly sol in hot water; freely sol in alc, ether.

SYNS: 3,4-DIMETHOXYBENZALDEHYDE ◇ 3,4-DIMETHOXYBENZENECARBONAL ◇ METHYLVANILLIN ◇ 4-o-METHYLVANILLIN ◇ PROTOCATECHUALDEHYDE DIMETHYL ETHER ◇ PROTOCATECHUIC ALDEHYDE DIMETHYL ETHER ◇ VANILLIN METHYL ETHER ◇ VERATRIC ALDEHYDE ◇ VERATRYL ALDEHYDE

TOXICITY DATA with REFERENCE
skn-rbt 500 mg/24H MLD FCTXAV 13,681,75
orl-rat LD50:2000 mg/kg FCTXAV 13,681,75

CONSENSUS REPORTS: Reported in EPA TSCA Inventory.

SAFETY PROFILE: Moderately toxic by ingestion. A skin irritant. When heated to decomposition it emits acrid smoke and irritating fumes. See also ALDEHYDES and ETHERS.

VHP500 CAS:60-70-8 ***HR: 3***
VERATRAMINE
mf: $C_{27}H_{39}NO_2$ mw: 409.67

PROP: Crystals. Mp: 206-107°. Sltly sol in water; sol in methanol, alc. Precipitated by digitonin.

TOXICITY DATA with REFERENCE
orl-rbt TDLo:16700 μg/kg (7D preg):TER PSEBAA 136,1174,71
scu-mus LD50:4500 μg/kg JPETAB 113,89,55
ivn-mus LD50:3100 μg/kg CKFRAY 2,418,53
orl-rbt LDLo:33 mg/kg TJADAB 3,175,70

SAFETY PROFILE: Poison by ingestion, intravenous, and subcutaneous routes. An experimental teratogen. When heated to decomposition it emits toxic fumes of NO_x.

VHU000 CAS:71-62-5 ***HR: 3***
VERATRIDINE
mf: $C_{36}H_{51}NO_{11}$ mw: 673.88

PROP: Yellow-white powder. Mp: 180°. Sol in water; sltly sol in ether.

SYNS: 4,9-EPOXYCEVANE-3,4,12,14,16,17,20-HEPTOL3-(3,4-DIMETHOXYBENZOATE) ◇ VERATRINE (AMORPHOUS) ◇ 3-VERATROYLVERACEVINE

TOXICITY DATA with REFERENCE
ipr-rat LD50:3500 μg/kg JPETAB 78,238,43
ipr-mus LD50:1350 μg/kg PSEBAA 76,847,51
scu-mus LD50:6300 μg/kg JPETAB 113,89,55
ivn-mus LD50:420 μg/kg JPETAB 82,167,44

SAFETY PROFILE: Poison by intraperitoneal, subcutaneous, and intravenous routes. Combustible when exposed to heat or flame. When heated to decomposition it emits toxic fumes of NO_x. See also VERATRINE.

VHZ000 CAS:8051-02-3 ***HR: 3***
VERATRINE

PROP: A powder from the plant *Schoenocaulon officinale*. A botanical insecticide. The active ingredients are a group of alkaloids known as veratrin, i.e., cevadine and veratridine.

SYNS: ASAGRAEA OFFICINALIS ◇ CAUSTIC BARLEY ◇ CEVADILLA ◇ CEVADINE ◇ ENT 123 ◇ SABACIDE ◇ SABADILLA ◇ SABANE DUST ◇ VERATRIDINE ◇ VERATRIN (GERMAN)

TOXICITY DATA with REFERENCE
orl-rbt TDLo:4450 μg/kg (7D preg):TER PSEBAA 136,1174,71
orl-hmn LDLo:143 mg/kg 34ZIAG -,522,69
orl-rat LD50:4000 mg/kg WRPCA2 9,119,70

ipr-mus LD50:7500 μg/kg PSEBAA 76,847,51
scu-mus LDLo:10 mg/kg HDTU** -,-,33
orl-dog LDLo:2 mg/kg HBAMAK 4,1289,35
orl-cat LDLo:2500 μg/kg HBAMAK 4,1289,35
orl-bwd LD50:17800 μg/kg AECTCV 12,355,83

SAFETY PROFILE: Human poison by ingestion. Experimental poison by ingestion, intraperitoneal, and subcutaneous routes. An experimental teratogen. Ingestion causes severe gastrointestinal tract disturbances, burning in the mouth, vomiting, diarrhea, and cramps. Also produces headache, dizziness, slow pulse and weakness. Large doses cause death by circulatory and respiratory failure. It is a powerful irritant to skin and mucous membranes. Less toxic than rotenone. Inhalation causes violent sneezing. When heated to decomposition it emits toxic fumes of NO_x. Used to kill lice.

VIA875 CAS:97805-00-0 ***HR: D***
N-(o-VERATROYL)GLYCINOHYDROXAMIC ACID
mf: $C_{11}H_{14}N_2O_5$ mw: 254.27

TOXICITY DATA with REFERENCE
mmo-sat 1 μmol/plate JOPHDQ 3,557,80
dnr-bcs 10 μmol/disc JOPHDQ 3,557,80

SAFETY PROFILE: Mutation data reported. When heated to decomposition it emits toxic fumes of NO_x.

VIK050 CAS:5763-61-1 ***HR: 3***
VERATRYLAMINE
mf: $C_9H_{13}NO_2$ mw: 167.23

SYNS: BENZENEMETHANAMINE, 3,4-DIMETHOXY- ◇ BENZYLAMINE, 3,4-DIMETHOXY- ◇ 3,4-DIMETHOXYBENZYLAMINE

TOXICITY DATA with REFERENCE
orl-rat TDLo:30 mg/kg (female 1D post):TER BEXBAN 77,646,74
ivn-mus LD50:178 mg/kg CSLNX* NX#00651

SAFETY PROFILE: Poison by intravenous route. An experimental teratogen. When heated to decomposition it emits toxic fumes of NO_x.

VIK100 CAS:93-17-4 ***HR: 3***
VERATRYL CYANIDE
mf: $C_{10}H_{11}NO_2$ mw: 177.22

PROP: Mp: 60-62.

SYNS: 3,4-DIMETHOXY-BENZENEACETONITRILE (9CI) ◇ 3,4-DIMETHOXYBENZYL CYANIDE ◇ 3,4-DIMETHOXYPHENYLACETONITRILE ◇ HOMOVERATRONITRILE

TOXICITY DATA with REFERENCE
orl-mus LD50:1029 mg/kg GTPZAB 26(2),55,82
ivn-mus LD50:178 mg/kg CSLNX* NX#00302

CONSENSUS REPORTS: Cyanide and its compounds are on the Community Right-To-Know List.

SAFETY PROFILE: Poison by intravenous route. Moderately toxic by ingestion. When heated to decomposition it emits toxic fumes of NO_x and CN^-. See also NITRILES.

VIK150 CAS:135-85-3 ***HR: 3***
VERATRYLHYDRAZINE
mf: $C_9H_{14}N_2O_2$ mw: 182.25

SYNS: 3,4-DIMETHOXYBENZYLHYDRAZINE ◇ ((3,4-DIMETHOXYPHENYL)METHYL)HYDRAZINE ◇ TAC-28 ◇ VETRAZIN ◇ VETRAZINE

TOXICITY DATA with REFERENCE
scu-mus LD50:127 mg/kg FATOAO 26,75,63
ivn-mus LD50:146 mg/kg FATOAO 26,75,63

SAFETY PROFILE: Poison by subcutaneous and intravenous routes. Used as a uterine stimulant. When heated to decomposition it emits toxic fumes of NO_x.

VIK200 CAS:93088-18-7 ***HR: 3***
1-VERATRYLPIPERAZINE DIHYDROCHLORIDE
mf: $C_{13}H_{20}N_2O_2$ mw: 236.35

SYN: DICHLORHYDRATE de DIMETHOXY-3,4 BENZYL PIPERAZINE (FRENCH)

TOXICITY DATA with REFERENCE
orl-mus LD50:740 mg/kg AIPTAK 128,17,60
ipr-mus LD50:310 mg/kg AIPTAK 128,17,60
ivn-mus LD50:130 mg/kg AIPTAK 128,17,60

SAFETY PROFILE: Poison by intravenous and intraperitoneal routes. Moderately toxic by ingestion. When heated to decomposition it emits toxic fumes of NO_x.

VIK400 ***HR: 3***
VERBENA HYBRIDA Cornol. & Rpl., extract

PROP: Indian plant belonging to the family *Verbenaceae* IJEBA6 22,312,84

TOXICITY DATA with REFERENCE
orl-rat TDLo:150 mg/kg (female 12-14D post):REP IJEBA6 22,312,84
ipr-mus LD50:316 mg/kg IJEBA6 22,312,84

SAFETY PROFILE: Poison by intraperitoneal route. Experimental reproductive effects. When heated to decomposition it emits acrid smoke and irritating fumes.

VIP000 CAS:18309-32-5 ***HR: 3***
d-VERBENONE
mf: $C_{10}H_{14}O$ mw: 150.24

PROP: Oil; characteristic odor. Mp: 6.5, d: 0.9780. Practically insol in water.

SYN: (1R,5R)-(+)-2-PINEN-4-ONE

TOXICITY DATA with REFERENCE
ipr-mus LDLo:250 mg/kg CBCCT* 7,794,55

CONSENSUS REPORTS: Reported in EPA TSCA Inventory.

SAFETY PROFILE: Poison by intraperitoneal route. When heated to decomposition it emits acrid smoke and irritating fumes.

VIZ000 CAS:65072-04-0 ***HR: 3***
VERILOID

SYNS: ALKALOIDS, VERATRUM ◇ ALKAVERVIR ◇ AMERICAN HELLEBORE ◇ AMERICAN VERATRUM ◇ GREEN HELLEBORE ◇ INDIAN POKE ◇ VERATRUM VIRIDE ◇ VERATRUM VIRIDE ALKALOIDS EXTRACT ◇ VERTAVIS

TOXICITY DATA with REFERENCE
orl-rat LD50:12 mg/kg FEPRA7 9,257,50
ipr-rat LD50:1690 μg/kg PSEBAA 76,847,51
scu-rat LD50:1590 μg/kg JAPMA8 39,610,50
ivn-rat LD50:440 μg/kg PSEBAA 76,847,51
orl-mus LD50:4500 μg/kg PSEBAA 76,847,51
ipr-mus LD50:3 mg/kg PSEBAA 76,847,51
scu-mus LD50:1120 μg/kg JAPMA8 39,610,50
ivn-mus LD50:430 μg/kg PSEBAA 85,400,54
orl-dog LDLo:3 mg/kg CLDND* 71,725,49
orl-rbt LD50:18 mg/kg PSEBAA 76,847,51
ivn-rbt LD50:270 μg/kg PSEBAA 76,847,51

SAFETY PROFILE: Poison by ingestion, intravenous, subcutaneous, and intraperitoneal routes.

VIZ100 CAS:37244-00-1 ***HR: 2***
VERMICULIN
mf: $C_{20}H_{24}O_8$ mw: 392.44

SYN: VERMICULINE

TOXICITY DATA with REFERENCE
dni-omi 20 mg/L FOMIAZ 23,389,78
oms-omi 20 mg/L FOMIAZ 23,389,78
ipr-mus LD50:420 mg/kg 85GDA2 2,390,80

SAFETY PROFILE: Moderately toxic by intraperitoneal route. Mutation data reported. When heated to decomposition it emits acrid smoke and irritating fumes.

VIZ200 CAS:6875-10-1 ***HR: 3***
VERODOXIN
mf: $C_{31}H_{46}O_{10}$ mw: 578.77

SYNS: GITALOXIGENIN + DIGITALOSE (GERMAN) ◇ STROSPESIDE-16-FORMATE

TOXICITY DATA with REFERENCE
orl-mus LD50:51870 μg/kg AIPTAK 153,436,65
scu-mus LD50:2890 μg/kg AIPTAK 153,436,65
ivn-cat LDLo:231 μg/kg JMPCAS 5,988,62
orl-gpg LDLo:2370 μg/kg AIPTAK 153,436,65
ivn-gpg LDLo:237 μg/kg AIPTAK 153,436,65

SAFETY PROFILE: Poison by ingestion, subcutaneous, and intravenous routes.

VIZ250 CAS:69598-87-4 ***HR: 1***
VEROS 030

TOXICITY DATA with REFERENCE
skn-rbt 500 mg/24H MLD 28ZPAK -,309,72
eye-rbt 500 mg/24H MLD 28ZPAK -,309,72
orl-rat LD50:11800 mg/kg 28ZPAK -,309,72

SAFETY PROFILE: Mildly toxic by ingestion. A skin and eye irritant.

VIZ400 CAS:60-40-2 ***HR: 3***
VERSAMINE
mf: $C_{11}H_{21}N$ mw: 167.33

PROP: dl-Form: Oily liquid. Bp: 72°, n (25/D) 1.4881. Sltly sol in water.

SYNS: INVERSINE ◇ MECAMILAMINA (ITALIAN) ◇ MECAMINE ◇ MECAMYLAMINE ◇ MEKAMINE ◇ 2-METHYLAMINOISOCAMPHANE ◇ 3-METHYLAMINOISOCAMPHANE ◇ 3-β-METHYLAMINO-2,2,3-TRIMETHYLBICYCLO(2.2.1)HEPTANE ◇ 2-METHYLAMINO-2,3,3-TRIMETHYLNORBORANE ◇ 2-METHYLAMINO-2,3,3-TRIMETHYLNORBORNANE ◇ N-METHYL-2-ISOCAMPHANAMINE ◇ MEVASINE ◇ PLEGANGIN ◇ REVERTINA ◇ N,2,3,3-TETRAMETHYL-2-NORBORNAMINE ◇ N,2,3,3-TETRAMETHYL-2-NORCAMPHANAMINE

TOXICITY DATA with REFERENCE
orl-mus LD50:90 mg/kg BCFAAI 103,490,64
ipr-mus LD50:40 mg/kg AITEAT 10,905,62
scu-mus LD50:56 mg/kg FRPSAX 20,482,65

SAFETY PROFILE: Poison by ingestion, subcutaneous, and intraperitoneal routes. When heated to decomposition it emits toxic fumes of NO_x.

VIZ500 CAS:71700-95-3 ***HR: 1***
VERSATIC 9-11 ACID

SYNS: VERSATIC 9-11 ◇ VERSATIC ACID 911

TOXICITY DATA with REFERENCE
skn-rbt 3000 mg/3D MOD BJIMAG 23,137,66
skn-rbt 11500 mg open MOD BJIMAG 23,137,66

SAFETY PROFILE: A skin and eye irritant. When heated to decomposition it emits acrid smoke and irritating fumes.

VJP800 CAS:4331-22-0 ***HR: D***
VERSICOLORIN B
mf: $C_{18}H_{12}O_7$ mw: 340.30

TOXICITY DATA with REFERENCE
mmo-sat 500 nmol/plate MUREAV 143,121,85
dns-rat:lvr 500 nmol/L MUREAV 143,121,85

SAFETY PROFILE: Mutation data reported. When heated to decomposition it emits acrid smoke and irritating fumes.

VJP900 ***HR: 3***
VESPA ORIENTALIS VENOM

SYNS: VENOM, ORIENTAL HORNET, VESPA ORIENTALIS ◇ V. ORIENTALIS VENOM

TOXICITY DATA with REFERENCE
orl-mus LD50:85 mg/kg TOXIA6 15,307,77
ipr-mus LD50:2500 μg/kg TOXIA6 18,469,80
ivn-mus LD50:1900 μg/kg TOXIA6 21,166,83

SAFETY PROFILE: Poison by ingestion, intravenous, and intraperitoneal routes.

VJU000 CAS:8016-96-4 ***HR: 1***
VETIVERT OIL

PROP: From steam distillation of roots of *Vetiveria zizanoides stapf* (FCTXAV 12,807,74).

SYN: OIL of VETIVER

TOXICITY DATA with REFERENCE
skn-rbt 500 mg/24H MOD FCTXAV 12,807,74

CONSENSUS REPORTS: Reported in EPA TSCA Inventory.

SAFETY PROFILE: A skin irritant. When heated to decomposition it emits acrid smoke and irritating fumes.

VJZ000 CAS:53-10-1 ***HR: 3***
VIADRIL
mf: $C_{25}H_{36}O_6 \cdot Na$ mw: 455.60

PROP: Lyophilized, fluffy white powder. Decomp 193-203°. Sol in water, in mildly alkaline buffer solns, acetone, chloroform.

SYNS: 21-(3-CARBOXY-1-OXOPROPOXY)-5-β-PREGNANE-3,20-DIONE SODIUM SALT ◇ HYDROXYDIONE ◇ HYDROXYDIONE SODIUM ◇ HYDROXYDIONE SUCCINATE ◇ 21-HYDROXYPREGNANE-3,20-DIONE SODIUM HEMISUCCINATE ◇ 21-HYDROXY-5-β-PREGNANE-3,20-DIONE SODIUM HEMISUCCINATE ◇ 21-HYDROXY-5-β-PREGNANE-3,20-DIONE, SODIUM SALT, HEMISUCCINATE ◇ P 55 ◇ PRESUREN ◇ SUCCINATE SODIQUE de 21-HYDROXYPREGNANDIONE (FRENCH)

TOXICITY DATA with REFERENCE
orl-rat LD50:700 mg/kg JPETAB 115,432,55
ipr-rat LD50:190 mg/kg THERAP 32,375,77
ivn-rat LD50:190 mg/kg JPETAB 115,432,55
orl-mus LD50:1200 mg/kg JPETAB 115,432,55
ipr-mus TDLo:640 mg/kg JMCMAR 11,117,68
scu-mus LD50:310 mg/kg AIPTAK 107,159,56
ivn-mus LD50:250 mg/kg JPETAB 115,432,55
ivn-rbt LD50:95 mg/kg JPETAB 115,432,55

SAFETY PROFILE: Poison by subcutaneous, intravenous, and intraperitoneal routes. Moderately toxic by ingestion. A steroid. When heated to decomposition it emits toxic fumes of Na_2O.

VKA600 CAS:2185-86-6 ***HR: 3***
VICTORIA LAKE BLUE R
mf: $C_{29}H_{32}ClN_3$ mw: 458.09

SYNS: AIZEN VICTORIA BLUE BOH ◇ BASIC BLUEK ◇ C.I. 44040 ◇ HIDACO VICTORIA BLUE R ◇ N,N'-(N,N'-TETRAMETHYL)-1-DIAMINODIPHENYLNAPHTHYLAMINOMETHANEHYDROCHLORIDE ◇ VICTORIA BLUE R ◇ VICTORIA BLUE RS

TOXICITY DATA with REFERENCE
orl-rat TDLo:4944 mg/kg/73W-I:CAR GISAAA 47(4),30,82
scu-rat TDLo:2361 mg/kg/66W-I:CAR GISAAA 47(4),30,82
orl-rat LD50:960 mg/kg GISAAA 47(4),30,82
scu-rat LD50:1408 mg/kg GISAAA 47(4),30,82

SAFETY PROFILE: Moderately toxic by ingestion and subcutaneous routes. Questionable carcinogen with experimental carcinogenic data. When heated to decomposition it emits toxic fumes of Cl^- and NO_x.

VKA650 CAS:93165-23-2 ***HR: D***
VIDANGA DRIED BERRY EXTRACT

TOXICITY DATA with REFERENCE
orl-rat TDLo:500 mg/kg (1-5D preg):REP JRIMAO 6,107,71

SAFETY PROFILE: Experimental reproductive effects.

VKA675 ***HR: D***
VIDR-2GD

PROP: Extracted from the powdered kernel of the seed of *Ensete superbum, cheesm, musaceae (bannakadali)* (FESTAS 21,247,70).

TOXICITY DATA with REFERENCE
orl-mus TDLo:3200 μg/kg (female 1-4D post):REP FESTAS 21,247,70

SAFETY PROFILE: Experimental reproductive effects.

VKA875 CAS:46817-91-8 ***HR: 3***
VILOXAZINE
mf: $C_{13}H_{19}NO_3$ mw: 237.33

SYNS: 2-((2-ETHOXYPHENOXY)METHYL)MORPHOLINE◇ 2-(2-ETHOXYPHENOXYMETHYL)TETRAHYDRO-1,4-OXAZINE◇ ICI-58834 ◇ VILOXAZIN

TOXICITY DATA with REFERENCE
scu-rat TDLo:130 mg/kg (8-20D preg):REP ATSUDG 7,504,84
orl-rat LD50:2000 mg/kg HEPHD2 55,527,80
ivn-rat LD50:60 mg/kg HEPHD2 55,527,80
orl-mus LD50:1000 mg/kg HEPHD2 55,527,80
ivn-mus LD50:60 mg/kg HEPHD2 55,527,80

SAFETY PROFILE: Poison by intravenous route. Moderately toxic by ingestion. Experimental reproductive effects. When heated to decomposition it emits toxic fumes of NO_x. See also VILOXAZINE HYDROCHLORIDE.

VKF000 CAS:35604-67-2 ***HR: 3***
VILOXAZINE HYDROCHLORIDE
mf: $C_{13}H_{19}NO_3 \cdot ClH$ mw: 273.79

PROP: Mp: 185-186°

SYNS: 2-((o-ETHOXYPHENOXY)METHYL)MORPHOLINEHYDROCHLORIDE ◇ 2-((2-ETHOXYPHENOXY)METHYL)MORPHOLINE HYDROCHLORIDE ◇ 2-(2-ETHOXYPHENOXYMETHYL)TETRAHYDRO-1,4-OXAZINE HYDROCHLORIDE ◇ ICI 58,834 ◇ VICILAN ◇ VIVALAN

TOXICITY DATA with REFERENCE
orl-hmn TDLo:25 mg/kg/6D:CNS BMJOAE 2,96,77
ipr-rat LD50:162 mg/kg FRPSAX 35,812,80
orl-mus LD50:1 g/kg MEIEDD 10,1427,83
ipr-mus LD50:162 mg/kg FRPSAX 35,812,80
ivn-mus LD50:60 mg/kg MEIEDD 10,1427,83

SAFETY PROFILE: Poison by intravenous and intraperitoneal routes. Moderately toxic by ingestion. Human systemic effects by ingestion: hallucinations, distorted perceptions and convulsions or effect on seizure threshold. When heated to decomposition it emits very toxic fumes of NO_x and HCl. Used as an antidepressant. See also VILOXAZINE.

VKP000 CAS:125-44-0 ***HR: 3***
VINBARBITAL SODIUM
mf: $C_{11}H_{15}N_2O_3 \cdot Na$ mw: 246.27

PROP: Hygroscopic crystals; bitter taste. Sltly sol in ether and chloroform.

SYNS: DELVINAL SODIUM ◇ 5-ETHYL-5-(1-METHYL-1-BUTENYL) BARBITURIC ACID SODIUM SALT ◇ 5-ETHYL-5-(1-METHYL-1-BUTENYL)-2,4,6(1H,3H,5H)-PYRIMIDINETRIONE SODIUM SALT ◇ SODIUM DELVINAL ◇ SODIUM-5-ETHYL-5-(1-METHYL-1-BUTENYL) BARBITURATE ◇ SODIUM VINBARBITAL

TOXICITY DATA with REFERENCE
orl-rat LD50:130 mg/kg MEIEDD 10,1427,83
ipr-rat LD50:80 mg/kg JAPMA8 32,180,43
orl-dog LD50:66 mg/kg JPETAB 68,22,40

SAFETY PROFILE: Poison by ingestion and intraperitoneal routes. Used as a sedative. When heated to decomposition it emits toxic fumes of Na_2O and NO_x. See also BARBITURATES.

VKZ000 CAS:865-21-4 ***HR: 3***
VINCALEUKOBLASTINE
mf: $C_{46}H_{58}N_4O_9$ mw: 811.08

SYNS: NCI-C04842 ◇ NDC 002-1452-01 ◇ NINCALUICOLFLASTINE ◇ NSC 47842 ◇ VINBLASTIN ◇ VINBLASTINE ◇ VINCALEUCOBLASTIN ◇ VINCOBLASTINE ◇ VLB

TOXICITY DATA with REFERENCE
dni-hmn:oth 200 μg/L 26QZAP 2,377,72
cyt-mus-ipr 900 μg/kg ENMUDM 8,273,86
ivn-ham TDLo:100 μg/kg (female 8D post):TER SCIEAS 141,426,63
ivn-rbt TDLo:500 μg/kg (female 1D post):REP FESTAS 18,7,67
ipr-rat TDLo:2 mg/kg/7W-I:ETA CANCAR 40(Suppl 4),1935,77
ivn-man LDLo:2319 μg/kg/38W-I:CVS LANCAO 2,692,80
ocu-hmn TDLo:14 μg/kg:EYE BJOPAL 62,97,78
unr-man TDLo:80 μg/kg:BLD CCROBU 50,219,66
ipr-rat LD50:1 mg/kg VINIT* #3713-83

CONSENSUS REPORTS: NCI Carcinogenesis Studies (ipr); No Evidence: mouse CANCAR 40,1935,77; (ipr); Clear Evidence: rat CANCAR 40,1935,77. EPA Genetic Toxicology Program.

SAFETY PROFILE: Human poison by intravenous route. Experimental poison by intraperitoneal route. Human systemic effects by intravenous and ocular routes: visual field changes, conjunctiva irritation and other eye effects, cardiomyopathy including infarction, and changes in bone marrow. Experimental teratogenic and reproductive effects. Questionable carcinogen with experimental tumorigenic data. Human mutation data reported. When heated to decomposition it emits toxic fumes of NO_x. Used as an antineoplastic agent. See also VINCALEUKOBLASTINE SULFATE (1 : 1) (salt).

VLA000 CAS:143-67-9 ***HR: 3***
VINCALEUKOBLASTINE SULFATE (1:1) (SALT)
mf: $C_{46}H_{58}N_4O_9 \cdot H_2O_4S$ mw: 909.16

SYNS: EXAL ◇ 29060 LE ◇ NSC 49842 ◇ VELBAN ◇ VELBE ◇ VINBLASTINE SULFATE ◇ VINCALEUKOBLASTINE SULFATE ◇ VLB MONOSULFATE

TOXICITY DATA with REFERENCE
pic-esc 500 mg/L APMBAY 12,234,64
cyt-hmn:lym 3750 μg/L CUSCAM 54,807,85
ipr-mus TDLo:350 μg/kg (female 9D post):TER CNJGA8 15,491,73
ivn-rbt TDLo:500 μg/kg (female 1D pre):REP FESTAS 38,238,82
ivn-hmn TDLo:557 μg/kg:BLD,SKN CNCRA6 29,111,63

orl-rat LD50:305 mg/kg OYYAA2 3,68,69
ipr-rat LD50:2200 μg/kg NIIRDN 6,650,82
ipr-mus LD50:5600 μg/kg NIIRDN 6,650,82
ivn-mus LD50:15 mg/kg NIIRDN 6,650,82
ipr-ham LD50:4300 μg/kg CALEDQ 2,267,77

CONSENSUS REPORTS: IARC Cancer Review: Group 3 IMEMDT 7,371,87; Animal Inadequate Evidence IMEMDT 26,349,81; Human Inadequate Evidence IMEMDT 26,349,81. EPA Genetic Toxicology Program.

SAFETY PROFILE: Poison by ingestion, intraperitoneal, and intravenous routes. An experimental teratogen. Human systemic effects by intravenous route: blood luekopenia and hair changes. Experimental reproductive effects. Questionable carcinogen. Human mutation data reported. When heated to decomposition it emits very toxic fumes of NO_x and SO_x. See also VINCALEUKOBLASTINE and SULFATES.

VLF000 CAS:1617-90-9 ***HR: 3***
VINCAMINE
mf: $C_{21}H_{26}N_2O_3$ mw: 354.49

PROP: Yellow crystals from acetone or methanol. Mp: 232-233°.

SYNS: ANASCLEROL ◇ ANGIOPAC ◇ ARTERIOVINCA ◇ DECINCAN ◇ DEVINCAN ◇ 14,15,DIHYDRO-14-HYDROXYEBURNAMENINE-14-CARBOXYLIC ACID METHYL ESTER ◇ EQUIPUR ◇ NOVICET ◇ OCU-VINC ◇ OXYGERON ◇ PERVAL ◇ PERVINCAMINE ◇ PERVONE ◇ TRIPERVAN ◇ VINCADAR ◇ VINCAFOLINA ◇ VINCAFOR ◇ VINCAGIL ◇ VINCAMIDOL ◇ (+)-VINCAMINE ◇ VINCAPAN ◇ VINCAPRONT ◇ VINCASAUNIER ◇ VINCIMAX ◇ VINODREL RETARD

TOXICITY DATA with REFERENCE
ipr-rat LD50:253 mg/kg ARZNAD 32,601,82
orl-mus LD50:1 g/kg ARZNAD 10,811,60
ipr-mus LD50:215 mg/kg EJMCA5 16,191,81
ivn-mus LD50:47740 μg/kg AGSOA6 19,211,78

SAFETY PROFILE: Poison by intravenous and intraperitoneal routes. Moderately toxic by ingestion. When heated to decomposition it emits toxic fumes of NO_x. Used as a vasodilator.

VLF300 ***HR: 3***
VINCA MINOR L., TOTAL ALKALOIDS

PROP: Total alkaloid extract of the leaves of *Vinca minor L.* (APFRAD 12,799,54).

TOXICITY DATA with REFERENCE
orl-mus LD50:500 mg/kg APFRAD 12,799,54
ipr-mus LD50:76 mg/kg APFRAD 12,799,54
ivn-mus LD50:24 mg/kg APFRAD 12,799,54
ivn-gpg LDLo:16 mg/kg APFRAD 12,799,54

SAFETY PROFILE: Poison by intravenous and intraperitoneal routes. Moderately toxic by ingestion.

VLU200 CAS:83768-87-0 ***HR: 3***
VINTHIONINE
mf: $C_6H_{11}NO_2S$ mw: 161.24

SYNS: S-ETHENYL-dl-HOMOCYSTEINE ◇ S-VINYL-dl-HOMOCYSTEINE

TOXICITY DATA with REFERENCE
mmo-sat 50 nmol/plate BBRCA9 88,395,79
mma-sat 50 nmol/plate BBRCA9 88,395,79
orl-rat TDLo:18 g/kg/86W-C:CAR CNREA8 42,4364,82
orl-rat TD:36 g/kg/86W-C:CAR CNREA8 42,4364,82
ipr-rat TDLo:5625 mg/kg/12W-I:CAR,REP CNREA8 42,4364,82

SAFETY PROFILE: Experimental reproductive effects. Suspected carcinogen with experimental carcinogenic data. Mutation data reported. When heated to decomposition it emits toxic fumes of SO_x and NO_x.

VLU210 CAS:70858-14-9 ***HR: D***
l-VINTHIONINE
mf: $C_6H_{11}NO_2S$ mw: 161.24

SYNS: S-ETHENYL-l-HOMOCYSTEINE ◇ S-VINYL-l-HOMOCYSTEINE

TOXICITY DATA with REFERENCE
mmo-sat 5 nmol/plate CNREA8 42,4364,82
dnd-rat-ipr 31 mg/kg CNREA8 42,4364,82

SAFETY PROFILE: Mutation data reported. When heated to decomposition it emits toxic fumes of SO_x and NO_x. See also VINTHIONINE.

VLU250 CAS:108-05-4 ***HR: 3***
VINYL ACETATE
DOT: UN 1301
mf: $C_4H_6O_2$ mw: 86.10

$$H_2C{=}CHOCO{\bullet}CH_3$$

PROP: Colorless, mobile liquid; polymerizes to solid on exposure to light. Mp: −92.8°, bp: 73°, flash p: 18°F, d: 0.9335 @ 20°, autoign temp: 800°F, vap press: 100 mm @ 21.5°, lel: 2.6%, uel: 13.4%, vap d: 3.0. Misc in alc, ether. Somewhat sol in water.

SYNS: ACETIC ACID ETHENYL ESTER ◇ ACETIC ACID VINYL ESTER ◇ 1-ACETOXYETHYLENE ◇ ETHENYL ACETATE ◇ OCTAN WINYLU (POLISH) ◇ VAC ◇ VINILE (ACETATO di) (ITALIAN) ◇ VINYL A MONOMER ◇ VINYLACETAT (GERMAN) ◇ VINYLACETAAT (DUTCH) ◇ VINYLE (ACETATE de) (FRENCH) ◇ VYAC ◇ ZESET T

TOXICITY DATA with REFERENCE
eye-hmn 22 ppm AIHAAP 30,449,69
skn-rbt 10 mg/24H open JIHTAB 30,63,48
eye-rbt 500 mg open JIHTAB 30,63,48

eye-rbt 500 mg/24H MLD 85JCAE -,354,86
cyt-hmn:lym 250 μmol/L MUREAV 159,109,86
sce-ham:ovr 125 μmol/L CNREA8 45,4816,85
orl-rat TDLo:500 mg/kg/D (multi) :REP EPASR* 8EHQ-0185-0543
orl-rat TDLo:100 g/kg/2Y-C:CAR TXAPA9 68,43,83
ihl-rat TCLo:600 ppm/6H/5D/2Y-I:ETA EPASR* 8EHQ-0187-0650
orl-rat LD50:2920 mg/kg UCDS** 4/25/58
ihl-rat LC50:4000 ppm/2H DUPON* ES-3574,75
orl-mus LD50:1613 mg/kg GISAAA 31(8),19,66
ihl-mus LC50:1550 ppm/4H DUPON* ES-3574,75
ihl-rbt LC50:2500 ppm/4H 85INA8 5,621,86
skn-rbt LD50:2335 mg/kg DUPON* ES-3574,75

CONSENSUS REPORTS: IARC Cancer Review: Group 3 IMEMDT 7,56,87; Animal Inadequate Evidence IMEMDT 19,341,79; IMEMDT 39,113,86; Human Inadequate Evidence IMEMDT 39,113,86. Reported in EPA TSCA Inventory. Community Right-To-Know List. EPA Extremely Hazardous Substances List.

OSHA PEL: TWA 10 ppm; STEL 20 ppm
ACGIH TLV: TWA 10 ppm; STEL 20 ppm; (Proposed: 10 ppm; Suspected Human Carcinogen)
DFG MAK: 10 ppm (35 mg/m^3)
NIOSH REL: (Vinyl Acetate) CL 15 mg/m^3/15M
DOT Classification: Label: Flammable Liquid.

SAFETY PROFILE: Moderately toxic by ingestion, inhalation, and intraperitoneal routes. A skin and eye irritant. Questionable carcinogen with experimental carcinogenic and tumorigenic data. Experimental reproductive effects. Human mutation data reported. Highly dangerous fire hazard when exposed to heat, flame, or oxidizers. A storage hazard, it may undergo spontaneous exothermic polymerization. Reaction with air or water to form peroxides which catalyze an exothermic polymerization reaction has caused several large industrial explosions. Reaction with hydrogen peroxide forms the explosive peracetic acid. Reacts with oxygen above 50°C to form an unstable explosive peroxide. Reacts with ozone to form the explosive vinyl acetate ozonide. Solution polymerization of the acetate dissolved in toluene has resulted in large industrial explosions. Polymerization reaction with dibenzoyl peroxide + ethyl acetate may release ignitable and explosive vapors. The vapor may react vigorously with dessicants (e.g., silica gel or alumina). Incompatible (explosive) with 2-amino ethanol; chlorosulfonic acid; ethylenediamine; ethyleneimine; HCl; HF; HNO_3; oleum; peroxides; H_2SO_4. See also ESTERS.

VLU310 ***HR: 3***
VINYL ACETATE OZONIDE
mf: $C_4H_6O_5$ mw: 134.09

$CH_3CO{\bullet}OCHOOCH_2O$

SYN: 3-ACETOXY-1,2,4-TRIOXOLANE

SAFETY PROFILE: The dry material is explosive. When heated to decomposition it emits acrid smoke and irritating fumes.

VLU400 CAS:593-67-9 ***HR: 2***
VINYLAMINE
mf: C_2H_5N mw: 43.08

PROP: Liquid. D: 0.832, bp: 56°. Sol in water and alc.

SYNS: AMINOETHYLENE ◇ ETHENEAMINE ◇ ETHYLENAMINE ◇ ETHYLENEAMINE

TOXICITY DATA with REFERENCE
skn-rbt 500 mg SEV 34ZIAG -,692,69
eye-rbt 100 mg SEV 34ZIAG -,692,69
orl-rat LD50:1850 mg/kg 34ZIAG -,692,69
skn-rbt LD50:560 mg/kg 34ZIAG -,692,69

SAFETY PROFILE: Moderately toxic by ingestion and skin contact. A severe eye and skin irritant. Reacts explosively with isoprene. When heated to decomposition it emits toxic fumes of NO_x. See also AMINES.

VLY300 CAS:7570-25-4 ***HR: 3***
VINYL AZIDE
mf: $C_2H_3N_3$ mw: 69.07

SAFETY PROFILE: A shock-sensitive explosive. When heated to decomposition it emits toxic fumes of NO_x. See also AZIDES.

VLZ000 CAS:5628-99-9 ***HR: 3***
1-VINYL AZIRIDINE
mf: C_4H_7N mw: 69.12

SYNS: 1-ETHENYLAZIRIDINE ◇ N-VINYLETHYLENEIMINE

TOXICITY DATA with REFERENCE
orl-rat LD50:88 mg/kg TXAPA9 28,313,74
ihl-rat LCLo:100 ppm/4H TXAPA9 28,313,74
skn-rbt LD50:20 mg/kg TXAPA9 28,313,74

SAFETY PROFILE: Poison by ingestion, inhalation, and skin contact. When heated to decomposition it emits toxic fumes of NO_x.

VMA000 CAS:3691-16-5 ***HR: 3***
α-VINYL-1-AZIRIDINEETHANOL
mf: $C_6H_{11}NO$ mw: 113.18

SYNS: AETHOXEN ◇ AETHYLENIMINO-2-OXYBUTEN (GERMAN) ◇ 4-AZIRIDINYL-3-HYDROXY-1,2-BUTENE ◇ 2-(1-AZIRIDINYL)-1-VINYLETHANOL ◇ α-ETHENYL-1-AZIRIDINEETHANOL (9CI) ◇ ETHOXENE ◇ 1-ETHYLENEIMINO-2-HYDROXY-3-BUTENE ◇ 1-ETHYLENIMINO-2-HYDROXYBUTENE ◇ 1-(2-HYDROXYBUT-1-ENYL)AZIRIDINE ◇ NSC-26806 ◇ TETRAMIN ◇ α-VINYL AE

TOXICITY DATA with REFERENCE
ipr-mus LD50:33 mg/kg NCISA* PH-43-63-1132
ivn-dog LDLo:3 mg/kg CCSUBJ 2,202,65
ivn-mky LDLo:11 mg/kg CCSUBJ 2,202,65

SAFETY PROFILE: Poison by intraperitoneal and intravenous routes. When heated to decomposition it emits toxic fumes of NO_x.

VMF000 CAS:61695-70-3 ***HR: 3***
7-VINYLBENZ(a)ANTHRACENE
mf: $C_{20}H_{14}$ mw: 254.34

TOXICITY DATA with REFERENCE
scu-rat TDLo:10 mg/kg/40D-I:ETA CALEDQ 1,339,76

SAFETY PROFILE: Questionable carcinogen with experimental tumorigenic data. When heated to decomposition it emits acrid smoke and irritating fumes.

VMK000 CAS:769-78-8 ***HR: 2***
VINYL BENZOATE
mf: $C_9H_8O_2$ mw: 148.17

PROP: Bp: 72-76°.

SYN: BENZOIC ACID, VINYL ESTER

TOXICITY DATA with REFERENCE
skn-rbt 10 mg/24H open AMIHBC 10,61,54
eye-rbt 500 mg open AMIHBC 10,61,54
orl-rat LD50:3250 mg/kg AMIHBC 10,61,54

SAFETY PROFILE: Moderately toxic by ingestion. A skin and eye irritant. When heated to decomposition it emits acrid smoke and irritating fumes. See also ESTERS.

VMP000 CAS:593-60-2 ***HR: 3***
VINYL BROMIDE
DOT: UN 1085
mf: C_2H_3Br mw: 106.96

PROP: A gas. Mp: −138°, bp: 15.6°, d: 1.51. Insol in water; misc in alc, ether.

SYNS: BROMOETHENE ◇ BROMOETHYLENE ◇ BROMURE de VINYLE (FRENCH) ◇ VINILE (BROMURO di) (ITALIAN) ◇ VINYLBROMID (GERMAN) ◇ VINYL BROMIDE, inhibited (DOT) ◇ VINYLE (BROMURE de) (FRENCH)

TOXICITY DATA with REFERENCE
mma-sat 2 pph/16H ARTODN 41,249,79
mmo-sat 2 pph/16H ARTODN 41,249,79
otr-rat-ihl 2000 ppm/14W-I ARTODN 47,71,81
ihl-rat TCLo:10 ppm/6H/2Y-I:CAR TXAPA9,64,367,82
ihl-rat TC:10 ppm:ETA CHWKA9 123(24),25,78
ihl-rat TCLo:250 ppm/1Y:NEO CHWKA9 121(20),40,77
ihl-rat TC:52 ppm/6H/2Y-I:CAR TXAPA9 64,367,82
orl-rat LD50:500 mg/kg DOWCC* -,12,66

CONSENSUS REPORTS: IARC Cancer Review: Group 2A IMEMDT 7,56,87; Animal Sufficient Evidence IMEMDT 39,133,86; Animal Inadequate Evidence IMEMDT 19,367,79. Community Right-To-Know List. Reported in EPA TSCA Inventory. EPA Genetic Toxicology Program.

OSHA PEL: TWA 5 ppm
ACGIH TLV: TWA 5 ppm; Suspected Human Carcinogen.
DFG MAK: Human Carcinogen.
NIOSH REL: (Vinyl Bromide) Lowest Detectable Level
DOT Classification: Flammable Gas; Label: Flammable Gas.

SAFETY PROFILE: Confirmed carcinogen with experimental carcinogenic, neoplastigenic, and tumorigenic data. Moderately toxic by ingestion. Mutation data reported. A very dangerous fire hazard when exposed to heat or flame. Can react violently with oxidizing materials. May polymerize in sunlight. To fight fire, use CO_2, dry chemical or water spray. When heated to decomposition it emits toxic fumes of Br^-. See also BROMIDES and VINYL CHLORIDE.

VMU000 CAS:4223-11-4 ***HR: 2***
VINYL-2-(BUTOXYETHYL) ETHER
mf: $C_8H_{16}O_2$ mw: 144.24

SYNS: 2-BUTOXYETHYLVINYL ETHER ◇ 1-BUTOXY-2-(VINYLOXY)ETHANE ◇ 1-(2-(ETHENYLOXY)ETHOXY)BUTANE

TOXICITY DATA with REFERENCE
skn-rbt 10 mg/24H open MLD AMIHBC 10,61,54
eye-rbt 20 mg open SEV AMIHBC 10,61,54
orl-rat LD50:3100 mg/kg AMIHBC 10,61,54
ihl-rat LCLo:2000 ppm/8H AMIHBC 10,61,54
skn-rbt LD50:3000 mg/kg AMIHBC 10,61,54

SAFETY PROFILE: Moderately toxic by ingestion and skin contact. Mildly toxic by inhalation. A skin and severe eye irritant. When heated to decomposition it emits acrid smoke and irritating fumes. See also ETHERS.

VMZ000 CAS:111-34-2 ***HR: 3***
VINYL BUTYL ETHER
DOT: UN 2352
mf: $C_6H_{12}O$ mw: 100.18

PROP: Liquid. Mp: −112.7°, bp: 94.2°, flash p: −9°, d: 0.7803 @ 20°/20°, vap d: 3.45.

SYNS: BUTOXYETHENE ◇ BUTYL VINYL ETHER ◇ BUTYL VINYL ETHER (inhibited) ◇ 1-(ETHENYLOXY) BUTANE ◇ VINYL-n-BUTYL ETHER

TOXICITY DATA with REFERENCE
skn-rbt 500 mg open MLD UCDS** 6/28/73
eye-rbt 500 mg open AMIHBC 10,61,54

orl-rat LD50:10 g/kg UCDS** 6/28/72
ihl-mus LC50:62 g/m^3/2H 85GMAT -,119,82
skn-rbt LD50:4240 mg/kg AMIHBC 10,61,54

CONSENSUS REPORTS: Reported in EPA TSCA Inventory.

DOT Classification: Flammable Liquid; Label: Flammable Liquid.

SAFETY PROFILE: Mildly toxic by ingestion, skin contact, and inhalation. A skin and eye irritant. A very dangerous fire hazard when exposed to heat or flame. To fight fire, use foam, CO_2, dry chemical, alcohol foam. Moderately explosive by spontaneous chemical reaction. Can react with oxidizing materials. When heated to decomposition it emits acrid smoke and irritating fumes. See also ETHERS.

VNA000 CAS:6607-49-4 ***HR: 3***
VINYL 2-(BUTYLMERCAPTOETHYL) ETHER
mf: $C_8H_{16}OS$ mw: 160.30

SYNS: 2-(BUTYLMERCAPTO)ETHYL VINYL ETHER ◇ 2-(BUTYLTHIO)ETHYL VINYL ETHER

TOXICITY DATA with REFERENCE
skn-rbt 10 mg/24H open SEV AIHAAP 23,95,62
orl-rat LDLo:2830 mg/kg AIHAAP 23,95,62
skn-rbt LD50:10 mg/kg AIHAAP 23,95,62

SAFETY PROFILE: Poison by skin contact. Moderately toxic by ingestion. A severe skin irritant. Flammable when exposed to heat or flame. When heated to decomposition it emits toxic fumes of SO_x. See also MERCAPTANS and ETHERS.

VNF000 CAS:123-20-6 ***HR: 3***
VINYL BUTYRATE
DOT: UN 2838
mf: $C_6H_{10}O_2$ mw: 114.16

PROP: D: 0.9, vap d: 4.0, bp: 116°, flash p: 68°F (OC), lel: 1.4%, uel: 8.8%.

SYNS: BUTYRIC ACID, VINYL ESTER ◇ VINYL BUTYRATE, INHIBITED (DOT)

TOXICITY DATA with REFERENCE
skn-rbt 500 mg open MLD UCDS** 3/24/70
eye-rbt 500 mg open AMIHBC 4,119,51
orl-rat LD50:8530 mg/kg UCDS** 3/24/70
ihl-rat LCLo:4000 ppm/4H AMIHBC 4,119,51

DOT Classification: Flammable Liquid; Label: Flammable Liquid.

SAFETY PROFILE: Mildly toxic by inhalation and ingestion. A skin and eye irritant. A very dangerous fire hazard when exposed to heat, flame, or oxidizers. Explosive in the form of vapor when exposed to heat or flame. To fight fire, use alcohol foam, fog, mist, CO_2. When heated to decomposition it emits acrid smoke and irritating fumes. See also ESTERS.

VNK000 CAS:15805-73-9 ***HR: 3***
VINYL CARBAMATE
mf: $C_3H_5NO_2$ mw: 87.09

SYN: CARBAMIC ACID, VINYL ESTER

TOXICITY DATA with REFERENCE
sce-hmn:lym 10 mmol/L MUREAV 89,75,81
sce-rat-ipr 25 mg/kg MUREAV 126,159,84
skn-mus TDLo:200 mg/kg/1W-I:NEO CNREA8 38,3793,78
ipr-mus TD:3 mg/kg:ETA CNREA8 46,4911,86
ipr-mus LDLo:125 mg/kg CNREA8 38,3793,78

SAFETY PROFILE: Poison by intraperitoneal route. Questionable carcinogen with experimental neoplastigenic data. Human mutation data reported. When heated to decomposition it emits toxic fumes of NO_x. See also ESTERS and CARBAMATES.

VNP000 CAS:75-01-4 ***HR: 3***
VINYL CHLORIDE
DOT: UN 1086
mf: C_2H_3Cl mw: 62.50

PROP: Colorless liquid or gas (when inhibited); faintly sweet odor. Mp: −160°; bp: −13.9°, lel: 4%, uel: 22%; flash p: 17.6°F (COC), fp: −159.7°, d (liquid): 0.9195 @ 15/4°, vap press: 2600 mm @ 25°, vap d: 2.15, autoign temp: 882°F. Sltly sol in water; sol in alc; very sol in ether.

SYNS: CHLORETHENE ◇ CHLORETHYLENE ◇ CHLOROETHENE ◇ CHLOROETHYLENE ◇ CHLORURE de VINYLE (FRENCH) ◇ CLORURO di VINILE (ITALIAN) ◇ ETHYLENE MONOCHLORIDE ◇ MONOCHLOROETHENE ◇ MONOCHLOROETHYLENE (DOT) ◇ RCRA WASTE NUMBER U043 ◇ TROVIDUR ◇ VC ◇ VCM ◇ VINILE (CLORURO di) (ITALIAN) ◇ VINYLCHLORID (GERMAN) ◇ VINYL CHLORIDE MONOMER ◇ VINYL C MONOMER ◇ VINYLE(CHLORURE de) (FRENCH) ◇ WINYLU CHLOREK (POLISH)

TOXICITY DATA with REFERENCE
mma-sat 1 pph CBTOE2 1,159,85
cyt-hmn:hla 10 mmol/L TXCYAC 9,21,78
ihl-man TCLo:30 mg/m^3 (5Y male):REP GTPZAB 24(5),28,80
ihl-rat TCLo:500 ppm/7H (female 6-15D post):TER TXAPA9 33,134,75
ihl-man TCLo:200 ppm/14Y-I:CAR,LIV VAPHDQ 372,195,76
orl-rat TDLo:3463 mg/kg/52W-I:CAR EVHPAZ 41,3,81
ihl-rat TCLo:1 ppm/4H/52W-I:CAR EVHPAZ 41,3,81
ihl-rat TCLo:10000 ppm/4H (12-18D preg):CAR,TER CSHCAL 4,119,77
ipr-rat TDLo:21 mg/kg/65W-I:ETA APDCDT 3,216,76

ihl-hmn TC:300 mg/m³/W-C:CAR,BLD GTPZAB 26(1),28,82
orl-rat LD50:500 mg/kg DOWCC*

CONSENSUS REPORTS: NTP Fifth Annual Report on Carcinogens. IARC Cancer Review: Group 1 IMEMDT 7,373,87; Animal Sufficient Evidence IMEMDT 19,377,79; IMEMDT 7,291,74; Human Limited Evidence IMEMDT 7,291,74; Human Sufficient Evidence IMEMDT 19,377,79. Community Right-To-Know List. Reported in EPA TSCA Inventory. EPA Genetic Toxicology Program.

OSHA PEL: Cancer Suspect Agent
ACGIH TLV: TWA 5 ppm; Human Carcinogen.
DFG TRK: Existing installations: 3 ppm, Human Carcinogen; Others: 2 ppm.
NIOSH REL: (Vinyl Chloride) Lowest Detectable Level
DOT Classification: Flammable Gas; Label: Flammable Gas.

SAFETY PROFILE: Confirmed human carcinogen producing liver and blood tumors. Moderately toxic by ingestion. Experimental teratogenic data. Experimental reproductive effects. Human reproductive effects by inhalation: changes in spermatogenesis. Human mutation data reported. A severe irritant to skin, eyes, and mucous membranes. Causes skin burns by rapid evaporation and consequent freezing. In high concentration it acts as an anesthetic. Chronic exposure has shown liver injury. Circulatory and bone changes in the fingertips have been reported in workers handling unpolymerized materials.

A very dangerous fire hazard when exposed to heat, flame, or oxidizers. Large fires of this material are practically inextinguishable. A severe explosion hazard in the form of vapor when exposed to heat or flame. Long-term exposure to air may result in formation of peroxides which can initiate explosive polymerization of the chloride. Can react vigorously with oxidizing materials. Can explode on contact with oxides of nitrogen. Obtain instructions for its use from the supplier storing or handling this material. To fight fire, stop flow of gas. When heated to decomposition it emits highly toxic fumes of Cl^-. See also CHLORINATED HYDROCARBONS, ALIPHATIC.

VNU000 CAS:14861-06-4 ***HR: 1***
VINYL CROTONATE
mf: $C_6H_8O_2$ mw: 112.14

PROP: Sltly sol in water. D: 0.9, vap d: 4.0, bp: 134°, flash p: 78°F (OC).

SYNS: 2-BUTENOIC ACID, ETHENYL ESTER ◇ CROTONIC ACID, VINYL ESTER ◇ VINYL 2-BUTENOATE

TOXICITY DATA with REFERENCE
skn-rbt 500 mg open MLD UCDS** 11/15/71
eye-rbt 500 mg open AMIHBC 10,61,54
orl-rat LD50:6500 mg/kg UCDS** 11/15/71
ihl-rat LCLo:4000 ppm/4H AMIHBC 10,61,54

SAFETY PROFILE: Mildly toxic by ingestion and inhalation. A skin and eye irritant. A very dangerous fire hazard when exposed to heat, flame or oxidizers. To fight fire, use alcohol foam. When heated to decomposition it emits acrid smoke and irritating fumes. Used as a crosslinking agent for acrylic and polyolefin plastics. See also ESTERS.

VNZ000 CAS:106-86-5 ***HR: 2***
VINYLCYCLOHEXANE MONOXIDE
mf: $C_8H_{12}O$ mw: 124.20

PROP: Liquid. D: 0.9598 @ 20°/20°, bp: 169°, flash p: 136°F, fp: −100°. Very sltly sol in water.

SYNS: 1,2-EPOXY-4-VINYLCYCLOHEXANE ◇ 4-VINYLCYCLOHEXENE-1,2-EPOXIDE ◇ VINYLCYCLOHEXENE MONOXIDE ◇ 4-VINYLCYCLOHEXENE MONOXIDE ◇ 1-VINYL-3,4-EPOXYCYCLOHEXANE ◇ 3-VINYL-7-OXABICYCLO(4.1.0)HEPTANE

TOXICITY DATA with REFERENCE
skn-rbt 10 mg/24H open MLD AIHAAP 23,95,62
orl-rat LD50:2000 mg/kg UCDS** 6/11/63
skn-rbt LD50:2830 mg/kg UCDS** 6/11/63

CONSENSUS REPORTS: Reported in EPA TSCA Inventory.

SAFETY PROFILE: Moderately toxic by ingestion and skin contact. A skin irritant. Combustible when exposed to heat or flame. To fight fire, use foam, alcohol foam, mist. When heated to decomposition it emits acrid smoke and irritating fumes.

VOA000 CAS:106-87-6 ***HR: 3***
VINYL CYCLOHEXENE DIOXIDE
mf: $C_8H_{12}O_2$ mw: 140.20

PROP: Colorless liquid. D: 1.098 @ 20/20°, bp: 227°, flash p: 230°F.

SYNS: CHISSONOX 206 ◇ EP-206 ◇ 1,2-EPOXY-4-(EPOXYETHYL) CYCLOHEXANE ◇ 1-EPOXYETHYL-3,4-EPOXYCYCLOHEXANE ◇ 3-(EPOXYETHYL)-7-OXABICYCLO(4.1.0)HEPTANE ◇ 3-(1,2-EPOXYETHYL)-7-OXABICYCLO(4.1.0)HEPTANE ◇ 4-(1,2-EPOXYETHYL)-7-OXABICYCLO(4.1.0)HEPTANE ◇ 4-(EPOXYETHYL)-7-OXABICYCLO(4.1.0)HEPTANE ◇ ERLA-2270 ◇ ERLA-2271 ◇ 1-ETHYLENEOXY-3,4-EPOXYCYCLOHEXANE ◇ NCI-C60139 ◇ 3-OXIRANYL-7-OXABICYCLO(4.1.0)HEPTENE ◇ UCET TEXTILE FINISH 11-74 (OBS.) ◇ UNOX EPOXIDE 206 ◇ VINYL CYCLOHEXENE DIEPOXIDE ◇ 4-VINYLCYCLOHEXENE DIEPOXIDE ◇ 4-VINYL-1-CYCLOHEXENE DIEPOXIDE ◇ 4-VINYL-1,2-CYCLOHEXENE DIEPOXIDE ◇ 1-VINYL-3-CYCLOHEXENE DIOXIDE ◇ 4-VINLYCYCLOHEXENE DIOXIDE ◇ 4-VINYL-1-CYCLOHEXENE DIOXIDE (MAK)

TOXICITY DATA with REFERENCE
skn-rbt 545 mg open MLD UCDS** 9/19/72
skn-rbt 500 mg SEV SCCUR* -9,61
mmo-klp 1 mmol/L MUREAV 89,269,81
mmo-smc 25 mmol/L BSIBAC 56,1803,80
mrc-smc 25 mmol/L BSIBAC 56,1803,80
ipr-rat TDLo:5000 mg/kg/10W-I:ETA BJPCAL 6,235,51
skn-mus TDLo:56 g/kg/47W-I:CAR JNCIAM 31,41,63
orl-rat LD50:2130 mg/kg SCCUR* -,9,61
ihl-rat LC50:800 ppm/4H SCCUR* -,9,61
skn-rbt LD50:620 mg/kg UCDS** 8/29/75
unr-rbt LD50:680 μg/kg BSIBAC 56,1803,80

CONSENSUS REPORTS: IARC Cancer Review: Group 3 IMEMDT 7,56,87; Animal Sufficient Evidence IMEMDT 11,141,76. Reported in EPA TSCA Inventory.

OSHA PEL: TWA 10 ppm (skin)
ACGIH TLV: TWA 10 ppm (skin); Suspected Human Carcinogen.
DFG MAK: Animal Carcinogen, Suspected Human Carcinogen.

SAFETY PROFILE: Confirmed carcinogen with experimental carcinogenic and tumorigenic data. Poison by unspecified route. Moderately toxic by ingestion and skin contact. Mildly toxic by inhalation. Mutation data reported. A severe skin irritant. Combustible when exposed to heat or flame. To fight fire, use water, foam, dry chemical. When heated to decomposition it emits acrid smoke and irritating fumes.

VOA550 CAS:55520-67-7 ***HR: D***
5-VINYL-DEOXYURIDINE
mf: $C_{11}H_{14}N_2O_5$ mw: 254.27

SYN: 2′-DEOXY-5-VINYLURIDINE

TOXICITY DATA with REFERENCE
sce-hmn:lym 5 mg/L BMJOAE 283,817,81
sce-hmn:lng 50 μg/L MUREAV 117,317,83

SAFETY PROFILE: Human mutation data reported. When heated to decomposition it emits toxic fumes of NO_x.

VOF000 CAS:3622-76-2 ***HR: 3***
VINYL-2-(N,N-DIMETHYLAMINO)ETHYL ETHER
mf: $C_6H_{13}NO$ mw: 115.20

SYN: 2-(N,N-DIMETHYLAMINO)ETHYL VINYL ETHER

TOXICITY DATA with REFERENCE
orl-rat LD50:180 mg/kg TXAPA9 28,313,74
ihl-rat LCLo:500 ppm/4H TXAPA9 28,313,74
skn-rbt LD50:1120 mg/kg TXAPA9 28,313,74

SAFETY PROFILE: Poison by ingestion. Moderately toxic by inhalation and skin contact. When heated to decomposition it emits toxic fumes of NO_x. See also ETHERS.

VOK000 CAS:872-36-6 ***HR: 3***
VINYLENE CARBONATE
mf: $C_3H_2O_3$ mw: 86.05

SYNS: CARBONIC ACID, cyclic VINYLENE ESTER ◇ 1,3-DIOXOL-4-EN-2-ONE ◇ 1,3-DIOXOL-2-ONE

TOXICITY DATA with REFERENCE
scu-rat TDLo:1760 mg/kg/44W-I:NEO BJCAAI 19,392,65

SAFETY PROFILE: Questionable carcinogen with experimental neoplastigenic data. When heated to decomposition it emits acrid smoke and irritating fumes.

VOP000 CAS:109-93-3 ***HR: 3***
VINYL ETHER
DOT: UN 1167
mf: C_4H_6O mw: 70.10

$(H_2C{=}CH)_2O$

PROP: Colorless liquid; very volatile. Bp: 29°, ULC: 100, lel: 1.7%, uel: 27%, flash p: < −22°F (CC), d: 0.774 @ 20/20°, autoign temp: 680°F, vap d: 2.41. Very sltly sol in water; misc in alc, ether.

SYNS: DIVINYL ETHER (DOT) ◇ DIVINYL ETHER, inhibited (DOT) ◇ DIVYNYL OXIDE ◇ ETHENYLOXYETHENE ◇ 1,1′-OXYBISETHENE ◇ VINESTHENE ◇ VINESTHESIN ◇ VINETHEN ◇ VINETHENE ◇ VINETHER ◇ VINIDYL ◇ VINYDAN

TOXICITY DATA with REFERENCE
mmo-sat 1 pph BJANAD 51,417,79
mma-sat 1 pph BJANAD 51,417,79
sce-ham:ovr 19900 ppm ANESAV 50,426,79
ihl-mus LCLo:51233 ppm MEIEDD 10,1430,83

DOT Classification: Flammable Liquid; Label: Flammable Liquid.

SAFETY PROFILE: Mildly toxic by inhalation. Mutation data reported. Prolonged exposure causes liver injury. A very dangerous fire hazard when exposed to heat or flame; can react vigorously with oxidizing materials. A severe explosion hazard in the form of vapor when exposed to heat or flame. Forms peroxides when exposed to air or oxygen. Hypergolic reaction with concentrated nitric acid. To fight fire, use CO_2, dry chemical. When heated to decomposition it emits acrid smoke and irritating fumes. Used as an inhalation anesthestic. See also ETHERS.

VOU000 CAS:94-04-2 ***HR: 2***
VINYL-2-ETHYLHEXOATE
mf: $C_{10}H_{18}O_2$ mw: 170.28

PROP: Liquid. Flash p: 165°F (OC), d: 0.8751, bp: 185.2°, fp: −90°, vap d: 6.0. Insol in water.

SYNS: 2-ETHYLHEXANOIC ACID, VINYL ESTER ◇ 2-ETHYLHEXOIC ACID, VINYL ESTER

TOXICITY DATA with REFERENCE
skn-rbt 500 mg open SEV UCDS** 4/25/58
eye-rbt 500 mg open AMIHBC 10,61,54
orl-rat LD50:4290 mg/kg AMIHBC 10,61,54

SAFETY PROFILE: Mildly toxic by ingestion. An eye and severe skin irritant. Combustible when exposed to heat, flame, or oxidizers. To fight fire, use foam, alcohol foam, mist. When heated to decomposition it emits acrid smoke and irritating fumes. See also ESTERS.

VPA000 CAS:75-02-5 ***HR: 3***
VINYL FLUORIDE
DOT: UN 1860
mf: CH_2:CHF mw: 46

PROP: Colorless gas. Mp: −160.5°, bp: −72°, lel: 2.6%, uel: 21.7%. Insol in water; sol in alc, ether.

SYNS: FLUOROETHENE ◇ FLUOROETHYLENE ◇ MONOFLUOROETHYLENE

TOXICITY DATA with REFERENCE
otr-rat-ihl 2000 ppm/14W-I ARTODN 47,71,81

CONSENSUS REPORTS: Reported in EPA TSCA Inventory.
NIOSH REL: (Vinyl Chloride) TWA 1 ppm; CL 5 ppm/15M
DOT Classification: Label: Flammable Gas.

SAFETY PROFILE: A poison. Mutation data reported. A very dangerous fire hazard. To fight fire, stop flow of gas. When heated to decomposition it emits toxic fumes of F^-. See also FLUORIDES.

VPF000 CAS:692-45-5 ***HR: 2***
VINYL FORMATE
mf: $C_3H_4O_2$ mw: 72.07

PROP: Flash p: <32°F.

TOXICITY DATA with REFERENCE
skn-rbt 500 mg open MLD UCDS** 11/15/71
eye-rbt 1 mg SEV UCDS** 11/15/71
orl-rat LD50:2820 mg/kg UCDS** 11/15/71
skn-rbt LD50:3170 mg/kg AIHAAP 23,95,62

SAFETY PROFILE: Moderately toxic by ingestion and skin contact. A severe eye and mild skin irritant. A very dangerous fire hazard when exposed to heat or flame. When heated to decomposition it emits acrid smoke and irritating fumes. See also ACRYLIC ACID.

VPK000 CAS:75-35-4 ***HR: 3***
VINYLIDENE CHLORIDE
DOT: UN 1303
mf: $C_2H_2Cl_2$ mw: 96.94

PROP: Colorless, volatile liquid. Bp: 31.6°, lel: 7.3%, uel: 16.0%, fp: −122°, flash p: 0°F (OC), d: 1.213 @ 20°/4°, autoign temp: 1058°F.

SYNS: CHLORURE de VINYLIDENE (FRENCH) ◇ 1-1-DCE ◇ 1,1-DICHLOROETHENE ◇ 1,1-DICHLOROETHYLENE ◇ NCI-C54262 ◇ RCRA WASTE NUMBER U078 ◇ SCONATEX ◇ VDC ◇ VINYLIDENE CHLORIDE (II) ◇ VINYLIDENE DICHLORIDE ◇ VINYLIDINE CHLORIDE

TOXICITY DATA with REFERENCE
mmo-sat 5 pph MUREAV 57,141,78
dns-mus-ihl 50 ppm TXAPA9 53,357,80
orl-rat TDLo:200 mg/kg (6-15D preg):TER TXAPA9 49,189,79
ihl-rat TCLo:55 ppm/6H (55D pre):REP JTEHD6 3,965,77
ihl-rat TCLo:55 ppm/6H/52W-I:ETA JTEHD6 4,15,78
ihl-mus TCLo:25 ppm/4H/52W-I:CAR MELAAD 68,241,77
skn-mus TDLo:4840 mg/kg:NEO JJIND8 63,1433,79
ihl-hmn TCLo:25 ppm:CNS,LIV,KID CHINAG (11),463,76
orl-rat LD50:200 mg/kg DCTODJ 1,63,77
ihl-rat LC50:6350 ppm/4H TXAPA9 18,168,71
orl-mus LD50:194 mg/kg BJCAAI 37,411,78
orl-dog LDLo:5750 mg/kg QJPPAL 7,205,34
ivn-dog LDLo:225 mg/kg QJPPAL 7,205,34
scu-rbt LDLo:3700 mg/kg QJPPAL 7,205,34

CONSENSUS REPORTS: IARC Cancer Review: Group 3 IMEMDT 7,376,87; Human Inadequate Evidence IMEMDT 39,195,86, IMEMDT 19,439,79; Animal Limited Evidence IMEMDT 39,195,86; Animal Sufficient Evidence IMEMDT 19,439,79. EPA Genetic Toxicology Program. Reported in EPA TSCA Inventory. Community Right-To-Know List.

OSHA PEL: TWA 1 ppm
ACGIH TLV: TWA 5 ppm; STEL 20 ppm
DFG MAK: Suspected Carcinogen.
NIOSH REL: (Vinyl Halides): TWA reduce to lowest detectable level
DOT Classification: Flammable Liquid; Label: Flammable Liquid

SAFETY PROFILE: Suspected carcinogen with experimental carcinogenic, neoplastigenic, tumorigenic, and teratogenic data. Poison by inhalation, ingestion, and intravenous routes. Moderately toxic by subcutaneous route. Human systemic effects by inhalation: general anesthesia, liver and kidney changes. Experimental reproductive effects. Mutation data reported. See also VINYL CHLORIDE. A very dangerous fire hazard when exposed to heat or flame. Moderately explosive in the form of gas when exposed to heat or flame. It forms explosive peroxides upon exposure to air. Potentially explosive reaction with chlorotrifluoroethylene at 180°C. Reaction with ozone forms dangerous products. Explosive reaction with perchloryl fluoride when heated above 100°C. Also can explode spontaneously. Reacts violently with chlorosulfonic acid; HNO_3; oleum. Can react vigorously with oxidizing materials. To fight fire, use alcohol foam, CO_2, dry chemical. When heated to decomposition it emits toxic fumes of Cl^-. See also CHLORINATED HYDROCARBONS, ALIPHATIC.

VPP000 CAS:75-38-7 ***HR: 3***
VINYLIDENE FLUORIDE
DOT: UN 1959
mf: $C_2H_2F_2$ mw: 64.04

PROP: Colorless gas. Bp: < −70°, lel: 5.5%, uel: 21.3%.

SYNS: 1,1-DIFLUOROETHYLENE (DOT, MAK) ◇ HALOCARBON 1132A ◇ NCI-C60208 ◇ VDF

TOXICITY DATA with REFERENCE
mma-sat 50 pph/24H ARTODN 41,249,79
orl-rat TDLo:1930 mg/kg/52W-I:NEO MELAAD 70,363,79
ihl-rat LCLo:128000 ppm/4H JOCMA7 4,262,62

CONSENSUS REPORTS: IARC Cancer Review: Group 3 IMEMDT 7,56,87; Animal Inadequate Evidence IMEMDT 39,227,86. Reported in EPA TSCA Inventory.

DFG MAK: Suspected Carcinogen.
DOT Classification: Flammable Gas; Label: Flammable Gas.
NIOSH REL: (Vinyl Halides): TWA reduce to lowest detectable level

SAFETY PROFILE: Suspected carcinogen with experimental neoplastigenic data. Mildly toxic by inhalation. Mutation data reported. A very dangerous fire hazard when exposed to heat, flames, or oxidizers. Explosive in the form of vapor when exposed to heat or flame. Violent reaction with hydrogen chloride when heated under pressure. To fight fire, stop flow of gas. When heated to decomposition it emits toxic fumes of F^-. See also FLUORIDES.

VPU000 CAS:917-57-7 ***HR: 3***
VINYLLITHIUM
mf: C_2H_3Li mw: 33.99

SAFETY PROFILE: Ignites spontaneously and burns violently in air. See also LITHIUM COMPOUNDS and ORGANOMETALS.

VPZ000 CAS:1663-35-0 ***HR: 2***
VINYL-2-METHOXYETHYL ETHER
mf: $C_5H_{10}O_2$ mw: 102.15

PROP: Liquid. Mp: −82.8°, bp: 108.8°, flash p: 65°F (OC), d: 0.8967, vap d: 3.53.

SYNS: 2-METHOXYETHYL VINYL ETHER ◇ 1-METHOXY-2-(VINYLOXY)ETHANE

TOXICITY DATA with REFERENCE
eye-rbt 500 mg open AMIHBC 10,61,54
orl-rat LD50:3900 mg/kg AMIHBC 10,61,54
ihl-rat LCLo:8000 ppm/4H AMIHBC 10,61,54
skn-rbt LD50:7130 mg/kg AMIHBC 10,61,54

SAFETY PROFILE: Moderately toxic by ingestion. Mildly toxic by inhalation and skin contact. An eye irritant. May form dangerous peroxides in storage. A very dangerous fire hazard when exposed to heat or flame; can react with oxidizing materials. To fight fire, use foam, CO_2, dry chemical. When heated to decomposition it emits acrid smoke and irritating fumes. See also ETHERS.

VQA000 CAS:3048-64-4 ***HR: 1***
5-VINYL-2-NORBORNENE
mf: C_9H_{12} mw: 120.21

SYNS: 5-ETHENYLBICYCLO(2.2.1)HEPT-2-ENE ◇ VINYLNORBORNENE ◇ 2-VINYLNORBORNENE ◇ 5-VINYLNORBORNENE

TOXICITY DATA with REFERENCE
orl-rat LD50:4365 mg/kg AIHAAP 30,470,6
ihl-rat LCLo:4000 ppm/4H AIHAAP 30,470,69
ihl-mus LC50:18 g/m³ GTPZAB 18(10),52,74
skn-rbt LD50:13372 mg/kg AIHAAP 30,470,69

CONSENSUS REPORTS: Reported in EPA TSCA Inventory.

SAFETY PROFILE: Mildly toxic by ingestion, inhalation, and skin contact. When heated to decomposition it emits acrid smoke and irritating fumes.

VQA100 CAS:1072-93-1 ***HR: 2***
R-5-VINYL-2-OXAZOLIDINETHIONE
mf: C_5H_7NOS mw: 129.19

SYNS: BA 51-090278 ◇ D-GOITRIN ◇ (R)-GOITRIN ◇ 2-OXAZOLIDINETHIONE, 5-ETHENYL-, (R)-(9CI) ◇ 2-OXAZOLIDINETHIONE, 5-VINYL-, (R)-

TOXICITY DATA with REFERENCE
scu-rat TDLo:200 mg/kg (female 8-9D post):TER FCTXAV 18,159,80
uns-mus LD50:1260 mg/kg JAFCAU 17,483,69

SAFETY PROFILE: Moderately toxic by unspecified route. An experimental teratogen. When heated to decomposition it emits toxic fumes of NO_x and SO_x.

VQA200 CAS:2628-17-3 ***HR: 3***
4-VINYLPHENOL
mf: C_8H_8O mw: 120.16

SYNS: p-HYDROXYSTYRENE ◇ 4-HYDROXYSTYRENE ◇ PHENOL, 4-ETHENYL- (9CI) ◇ PHENOL, p-VINYL- ◇ p-VINYLPHENOL

TOXICITY DATA with REFERENCE
eye-rbt 100 mg SEV EPASR* 8EHQ-1285-0579S
orl-uns TDLo:1400 mg/kg (female 7D pre):REP SCIEAS 195,575,77
skn-rbt LDLo:200 mg/kg EPASR* 8EHQ-1285-0579S
ocu-rbt LDLo:50 mg/kg EPASR* 8EHQ-1285-0579S

SAFETY PROFILE: Poison by skin and eye contact. Severe eye irritant. Experimental reproductive effects. When heated to decomposition it emits acrid smoke and irritating fumes.

VQA400 CAS:36885-49-1 ***HR: 2***
VINYL PHOSPHATE
mf: $C_2H_5O_4P$ mw: 124.04

SYNS: MONOVINYL PHOSPHATE ◇ VINYL DIHYDROGEN PHOSPHATE

TOXICITY DATA with REFERENCE
orl-rat LD50:2550 mg/kg GNAMAP 17,33,78
orl-ckn LD50:1500 mg/kg TIVSAI 58,88,77
orl-brd LD50:823 mg/kg TIVSAI 58,88,77

SAFETY PROFILE: Moderately toxic by ingestion. When heated to decomposition it emits toxic fumes of PO_x.

VQF000 CAS:115-98-0 ***HR: 2***
VINYLPHOSPHONIC ACID BIS(2-CHLOROETHYL) ESTER
mf: $C_6H_{11}Cl_2O_3P$ mw: 233.04

SYN: BIS-(2-CHLORETHYL)VINYLFOSFONAT(CZECH)

TOXICITY DATA with REFERENCE
orl-rat LD50:990 mg/kg MarJV# 29MAR77

CONSENSUS REPORTS: Reported in EPA TSCA Inventory.

SAFETY PROFILE: Moderately toxic by ingestion. When heated to decomposition it emits very toxic fumes of PO_x and Cl^-. See also ESTERS.

VQK000 CAS:105-38-4 ***HR: 2***
VINYL PROPIONATE
mf: $C_5H_8O_2$ mw: 100.13

PROP: Liquid. D: 0.9173 @ 20/20°, bp: 95°, fp: −81.1°, flash p: 34°F (OC), vap d: 3.3. Almost insol in water.

SYN: PROPANOIC ACID, ETHENYL ESTER

TOXICITY DATA with REFERENCE
skn-rat 500 mg MLD AMIHBC 2,582,50
skn-rbt 10 mg/24H open MLD AIHAAP 23,95,62
eye-rbt 10 mg MLD AMIHBC 2,582,50
orl-rat LD50:4760 mg/kg AIHAAP 23,95,62
ihl-rat LCLo:4000 ppm/4H AIHAAP 23,95,62

CONSENSUS REPORTS: Reported in EPA TSCA Inventory.

SAFETY PROFILE: Mildly toxic by ingestion and inhalation. A skin and eye irritant. A very dangerous fire hazard when exposed to heat or flame. To fight fire, use alcohol foam, mist, fog. When heated to decomposition it emits acrid smoke and irritating fumes. See also ESTERS.

VQK550 CAS:1337-81-1 ***HR: 3***
VINYL PYRIDINE
mf: C_7H_7N mw: 105.14

SAFETY PROFILE: Spontaneous polymerization may be explosive. When heated to decomposition it emits toxic fumes of NO_x.

VQK650 CAS:25013-15-4 ***HR: 3***
VINYL TOLUENE
mf: C_9H_{10} mw: 118.19

SYNS: METHYL STYRENE ◇ NCI-C56406 ◇ VINYLTOLUENE ◇ VINYL TOLUENES (mixed isomers), inhibited (DOT)

TOXICITY DATA with REFERENCE
skn-rbt 100% MOD AMIHAB 14,387,56
eye-rbt 90 mg MLD AMIHAB 14,387,56
ipr-rat TDLo:3750 mg/kg (1-15D preg):TER SWEHDO 7(Suppl 4),66,81
ipr-rat TDLo:3750 mg/kg (1-15D preg):REP SWEHDO 7(Suppl 4),66,81
ihl-hmn TCLo:400 ppm:NOSE,EYE AMIHAB 14,387,56
orl-rat LD50:4 g/kg AMIHAB 14,387,56
orl-mus LD50:3160 mg/kg HYSAAV 34(7-9),334,69
ihl-mus LC50:3020 mg/m^3 HYSAAV 34(7-9),334,69

CONSENSUS REPORTS: Reported in EPA TSCA Inventory.

OSHA PEL: TWA 100 ppm
ACGIH TLV: TWA 50 ppm; STEL 100 ppm
DOT Classification: Flammable or Combustible Liquid; Label: Flammable Liquid.

SAFETY PROFILE: Moderately toxic by ingestion and inhalation. An experimental teratogen. Human systemic effects by inhalation: eye and olfactory effects. Experimental reproductive effects. Mutation data reported. A skin and eye irritant. Flammable when exposed to heat or flame; can react vigorously with oxidizing materials.

When heated to decomposition it emits acrid smoke and irritating fumes.

VQR300 CAS:463-88-7 ***HR: 3***
VINYLTRIMETHYLAMMONIUM HYDROXIDE
mf: $C_5H_{12}N{\bullet}HO$ mw: 103.19

PROP: Syrupy liquid; fishy odor. Forms a crystalline trihydrate. Readily absorbs CO_2 from the air. Sol in water, alc. Decomp readily forming trimethylamine. Forms an HCl salt.

SYNS: NEIRINE ◇ NEURIN ◇ NEURINE ◇ N,N,N-TRIMETHYL-ETHENAMINIUM HYDROXIDE ◇ TRIMETHYL VINYL AMMONIUM HYDROXIDE ◇ VITALOID

TOXICITY DATA with REFERENCE
ipr-mus LDLo:100 mg/kg HBAMAK 4,1289,35
scu-mus LDLo:46 mg/kg JPETAB 28,367,26
orl-rbt LDLo:90 mg/kg HBAMAK 4,1289,35
ipr-gpg LDLo:30 mg/kg HBAMAK 4,1289,35

SAFETY PROFILE: Poison by ingestion, subcutaneous, and intraperitoneal routes. When heated to decomposition it emits toxic fumes of NO_x and NH_3.

VQU000 CAS:10141-19-2 ***HR: 2***
VINYL-2,6,8-TRIMETHYLNONYL ETHER
mf: $C_{14}H_{28}O$ mw: 212.42

PROP: Mp: −90°, bp: 223.5°, flash p: 200°F (OC), vap d: 7.33, d: 0.8075 @ 20/20°.

SYN: 2,6,8-TRIMETHYLNONYL VINYL ETHER

TOXICITY DATA with REFERENCE
skn-rbt 10 mg/24H open SEV AMIHBC 10,61,54
eye-rbt 500 mg open AMIHBC 10,61,54
orl-rat LD50:1220 mg/kg AMIHBC 10,61,54
skn-rbt LD50:5 g/kg AMIHBC 10,61,54

SAFETY PROFILE: Moderately toxic by ingestion. Mildly toxic by skin contact. An eye and severe skin irritant. Combustible when exposed to heat, flame, or oxidizing agents. To fight fire, use foam, CO_2, dry chemical. When heated to decomposition it emits acrid smoke and irritating fumes. See also ETHERS.

VQU450 CAS:1404-96-2 ***HR: 3***
VIOLACETIN
mf: $C_{18}H_{28}N_{10}O_{16}{\bullet}2ClH$ mw: 713.48

TOXICITY DATA with REFERENCE
orl-mus LD50:375 mg/kg 85GDA2 4(1),197,80
ipr-mus LD50:45 mg/kg 85GDA2 4(1),197,80
scu-mus LD50:75 mg/kg 85GDA2 4(1),197,80
ivn-mus LD50:37 mg/kg 85GDA2 4(1),197,80

SAFETY PROFILE: Poison by ingestion, subcutaneous, intravenous, and intraperitoneal routes. When heated to decomposition it emits toxic fumes of NO_x and HCl.

VQU500 CAS:80539-34-0 ***HR: D***
VIOLET BNP
mf: $C_{37}H_{36}N_3O_6S_2{\bullet}Na$ mw: 705.87

SYNS: C.I. 42581 ◇ C.I. FOOD VIOLET 3

TOXICITY DATA with REFERENCE
mma-sat 100 μg/plate FCTXAV 19,419,81
mrc-smc 76700 nmol/L FCTXAV 19,419,81

SAFETY PROFILE: Mutation data reported. When heated to decomposition it emits toxic fumes of SO_x, NO_x, and Na_2O.

VQZ000 CAS:32988-50-4 ***HR: 3***
VIOMYCIN
mf: $C_{25}H_{43}N_{13}O_{10}$ mw: 685.81

PROP: Purple crystals strong base. Sol in water.

SYNS: CELIOMYCIN ◇ FLORIMYCIN ◇ FLOROMYCIN ◇ TUBERACTINOMYCIN B ◇ VINACETIN A ◇ VIOACTANE ◇ VIOCIN ◇ VIOMICINAE (ITALIAN)

TOXICITY DATA with REFERENCE
scu-rat TDLo:7 g/kg (6-10D preg):TER OSDIAF 14,107,65
scu-rat TDLo:7 g/kg (6-10D preg):REP OSDIAF 14,107,65
ipr-rat LD50:1075 mg/kg 85GMAT -,69,82
scu-rat LD50:1750 mg/kg 85GMAT -,69,82
ivn-rat LD50:340 mg/kg OYYAA2 8,817,74
ims-rat LD50:1300 mg/kg OYYAA2 8,817,74
orl-mus LD50:1637 mg/kg 85GMAT -,69,82
ipr-mus LD50:973 mg/kg ANTBAL 8,910,63
scu-mus LD50:1184 mg/kg ANTBAL 8,910,63
ivn-mus LD50:150 mg/kg ANTBAL 8,910,63
ims-mus LD50:840 mg/kg OYYAA2 8,817,74

SAFETY PROFILE: Poison by intravenous route. Moderately toxic by ingestion, intraperitoneal, intramuscular and subcutaneous routes. An experimental teratogen. Other experimental reproductive effects. Used as an antibiotic. When heated to decomposition it emits toxic fumes of NO_x.

VQZ100 ***HR: 2***
VIOMYCIN SULFATE SALT (2:3)
mf: $C_{25}H_{43}N_{13}O_{10}{\bullet}3/2H_2O_4S$ mw: 933.59

SYNS: ENVIOMYCIN SULFATE ◇ TUBERACTINOMYCIN-N SULFATE ◇ TUBERACTIN SULFATE ◇ VIOMYCIN, 1-(threo-4-HYDROXY-L-3,6-DIAMINOHEXANOICACID)-6-(L-2-(2-AMINO-1,4,5,6-TETRAHYDRO-4-PYRIMIDINYL)GLYCINE)-, (R)-, SESQUISULFATE

TOXICITY DATA with REFERENCE
ims-rat TDLo:15 g/kg (female 30D pre):REP OYYAA2 12,585,76

ipr-rat LD50:640 mg/kg YAKUD5 17,1217,75
ipr-mus LD50:420 mg/kg YAKUD5 17,1217,75

SAFETY PROFILE: Moderately toxic by intraperitoneal route. Experimental reproductive effects. When heated to decomposition it emits toxic fumes of NO_x and SO_x.

VQZ425 ***HR: 3***
VIPERA AMMODYTES VENOM

SYNS: V. AMMODYTES VENOM ◇ VENOM, SNAKE, VIPERA AMMODYTES

TOXICITY DATA with REFERENCE
par-rat LD50:850 ng/kg TOXIA6 20,191,82
ipr-mus LD50:1400 μg/kg TOXIA6 11,47,73
scu-mus LD50:2 mg/kg JOIMA3 67,299,51
ivn-mus LD50:10 μg/kg 29QKAZ 3,863,73

SAFETY PROFILE: Poison by subcutaneous, parenteral, intravenous, and intraperitoneal routes.

VQZ475 ***HR: 3***
VIPERA BERUS VENOM

SYN: VENOM, SNAKE, VIPERA ASPIS

TOXICITY DATA with REFERENCE
scu-mus TDLo:8 mg/kg (8D preg):REP ACATA5 88,11,74
scu-mus TDLo:8 mg/kg (8D preg):TER ACATA5 88,11,74
scu-mus LD50:1 mg/kg JOIMA3 67,299,51
ivn-mus LD50:850 μg/kg ACPMAP 24,179,72
par-mus LD50:200 μg/kg ACATA5 108,226,80
ivn-rbt LDLo:175 μg/kg SCIEAS 117,47,53

SAFETY PROFILE: Poison by subcutaneous, intravenous, and parenteral routes. An experimental teratogen. Experimental reproductive effects.

VQZ500 ***HR: 3***
VIPERA BORNMULLERI VENOM

SYN: VENOM, SNAKE, VIPERA BORNMULLERI

TOXICITY DATA with REFERENCE
ipr-mus LD50:1920 μg/kg TOXIA6 22,265,84
scu-mus LD50:6250 μg/kg TOXIA6 22,625,84
ivn-mus LD50:605 μg/kg TOXIA6 22,625,84

SAFETY PROFILE: Poison by subcutaneous, intravenous, and intraperitoneal routes.

VQZ525 ***HR: 3***
VIPERA LATIFII VENOM

SYN: VENOM, SNAKE, VIPERA LATIFII

TOXICITY DATA with REFERENCE
ipr-mus LD50:2070 μg/kg TOXIA6 22,625,84
scu-mus LD50:4610 μg/kg TOXIA6 22,625,84
ivn-mus LD50:224 μg/kg TOXIA6 22,373,84

SAFETY PROFILE: Poison by subcutaneous, intravenous, and intraperitoneal routes.

VQZ550 ***HR: 3***
VIPERA LEBETINA VENOM

SYN: VENOM, SNAKE, VIPERA LEBETINA

TOXICITY DATA with REFERENCE
scu-mus LD50:2720 μg/kg AIPSAH 34,100,56
ivn-mus LD50:568 μg/kg TOXIA6 22,373,84
ivn-rbt LDLo:2500 μg/kg AIPSAH 34,100,56

SAFETY PROFILE: Poison by subcutaneous and intravenous routes.

VQZ575 ***HR: 3***
VIPERA PALESTINAE VENOM

SYN: VENOM, SNAKE, VIPERA PALESTINAE

TOXICITY DATA with REFERENCE
ims-rat LDLo:2500 μg/kg NATUAS 189,320,61
ipr-mus LD50:1900 μg/kg TOXIA6 6,11,68
ivn-mus LD50:500 μg/kg TOXIA6 4,205,66
ims-rbt LDLo:2 mg/kg NATUAS 189,320,61

SAFETY PROFILE: Poison by intramuscular, intravenous, and intraperitoneal routes.

VQZ625 ***HR: 3***
VIPERA RUSSELLII FORMOSENSIS VENOM

SYN: VENOM, SNAKE, VIPERA RUSSELLII FORMOSENSIS

TOXICITY DATA with REFERENCE
ipr-mus LD50:489 μg/kg TOXIA6 9,131,71
scu-mus LD50:1400 μg/kg TIHHAH 61,239,62
ivn-mus LD50:178 μg/kg TOXIA6 9,131,71

SAFETY PROFILE: Poison by subcutaneous, intravenous, and intraperitoneal routes.

VQZ635 ***HR: 3***
VIPERA RUSSELLII VENOM

SYNS: RUSSELL'S VIPER VENOM ◇ VENOM, SNAKE, VIPERA RUSSELLII

TOXICITY DATA with REFERENCE
ipr-mus LD50:400 μg/kg TOXIA6 9,131,71
scu-mus LD50:8250 μg/kg JOIMA3 95,867,65
ivn-mus LD50:35 μg/kg JPPMAB 16,79,64
ivn-dog LDLo:100 μg/kg 19DDA6 1,269,67
ivn-rbt LDLo:25 μg/kg SCIEAS 117,47,53
ivn-mam LD50:80 μg/kg CLPTAT 8,849,67

SAFETY PROFILE: Poison by subcutaneous, intravenous, and intraperitoneal routes.

VQZ650 ***HR: 3***
VIPERA XANTHINA PALAESTINAE VENOM

SYN: VENOM, SNAKE, VIPERA XANTHINA PALAESTINAE

TOXICITY DATA with REFERENCE
scu-mus LD50:2333 μg/kg HAREA6 53,309,57
ipr-mus LD50:2500 μg/kg AJTHAB 6,180,57
ivn-mus LD50:200 μg/kg TOXIA6 14,146,76
ivn-rbt LDLo:750 μg/kg TOXIA6 2,5,64

SAFETY PROFILE: Poison by subcutaneous, intravenous, and intraperitoneal routes.

VQZ675 ***HR: 1***
VIPER'S BUGLOSS

PROP: Bristly biennials which grow to 2 feet and are speckled with red. The prickly leaves are oblong, about 6 inches long and alternate on the stem. They produce bright blue flowers on spikes and small nuts. They are native to Eurasia but are common in the eastern United States and most of Canada and Hawaii.

SYNS: BLUE DEVIL WEED ◇ ECHIUM PLANTAGINEUM ◇ ECHIUM VULGARE ◇ SNAKE FLOWER ◇ VIPERINE (CANADA)

SAFETY PROFILE: The whole plant contains poisonous pyrrolizidine alkaloids. The plant is used in herbal teas. Ingestion of plant parts or tea can cause nausea, vomiting, and diarrhea. Chronic ingestion can result in liver damage.

VRA000 CAS:53762-93-9 ***HR: 3***
VIRACTIN

PROP: An antibiotic produced by a strain of *Streptomyces griseus* (85ERAY 2,1247,78). Liquid. Bp: 105-135° @ 200 mm.

TOXICITY DATA with REFERENCE
scu-mus LD50:300 mg/kg MEIEDD 10,1432,83
ivn-mus LD50:300 mg/kg 85ERAY 2,1247,78

SAFETY PROFILE: Poison by subcutaneous and intravenous routes.

VRA700 CAS:3131-03-1 ***HR: 3***
VIRGIMYCIN
mf: $C_{45}H_{54}N_8O_{10}$ mw: 867.07

SYNS: ANTIBIOTIC 899 ◇ ANTIBIOTIC PA 11481 ◇ ANTIBIOTIC PA 114 B1 ◇ ESKALIN V ◇ MIKAMYCIN B ◇ MIKAMYCIN IA ◇ OSTREOGRYCIN B ◇ PA 114B ◇ PRISTINAMYCIN IA ◇ SKF 7988 ◇ STAFAC ◇ STAPHYLOMYCIN ◇ STAPHYLOMYCINE ◇ STREPTOGRAMIN B ◇ VERNAMYCIN BA

TOXICITY DATA with REFERENCE
orl-mus LD50:200 mg/kg 85FZAT -,750,67
ipr-mus LD50:350 μg/kg 85ERAY 1,367,78
scu-mus LD50:200 mg/kg 85FZAT -,750,67

SAFETY PROFILE: Poison by ingestion, subcutaneous, and intraperitoneal routes. When heated to decomposition it emits toxic fumes of NO_x.

VRF000 CAS:11006-76-1 ***HR: 2***
VIRGINIAMYCIN

PROP: White powder. Decomp @ 138-140°. Sltly sol in water and dil acid; sol in methanol, ethanol, acetone, benzene; almost insol in ligroin.

SYNS: ANTIBIOTIC No. 899 ◇ ESKALIN V ◇ MIKAMYCIN ◇ OSTREOGRYCIN ◇ PATRICIN ◇ PRISTINAMYCIN ◇ PYOSTACINE ◇ RP7293 ◇ SKF 7988 ◇ STAFAC ◇ STAPHYLOMYCIN ◇ STAPYOCINE ◇ STREPTOGRAMIN ◇ VERNAMYCIN ◇ VIRGIMYCIN

TOXICITY DATA with REFERENCE
orl-mus LD50:2100 mg/kg 85ERAY 1,383,78
ipr-mus LD50:450 mg/kg MEIEDD 10,1432,83
scu-mus LD50:2500 mg/kg 85ERAY 1,383,78

SAFETY PROFILE: Moderately toxic by ingestion, intraperitoneal, and subcutaneous routes. Used as an antibiotic.

VRP000 CAS:39277-41-3 ***HR: 3***
VIRIDICATUMTOXIN
mf: $C_{30}H_{31}NO_{10}$ mw: 565.62

TOXICITY DATA with REFERENCE
mma-sat 25 μg/plate MUREAV 58,193,78
orl-rat LD50:122 mg/kg TXAPA9 24,507,73
ipr-rat LD50:90 mg/kg TOLED5 22,287,84
ipr-mus LD50:80 mg/kg TOLED5 22,287,84

CONSENSUS REPORTS: EPA Genetic Toxicology Program.

SAFETY PROFILE: Poison by ingestion and intraperitoneal routes. Mutation data reported. When heated to decomposition it emits toxic fumes of NO_x.

VRP200 CAS:35483-50-2 ***HR: 3***
VIRIDITOXIN
mf: $C_{30}H_{26}O_6$ mw: 482.56

SYNS: (8,8'-BI-1H-NAPHTHO(2,3-c)PYRAN)-3,3'-DIACETIC ACID, 3,3',4,4'-TETRAHYDRO-9,9',10,10'-TETRAHYDRO-7,7'-DIMETHOXY-1,1'-DIOXO-, DIMETHYL ESTER ◇ CROTALUS VIRIDIS VIRIDIS TOXIN ◇ C. VIRIDIS VIRIDIS TOXIN ◇ SC 28762

TOXICITY DATA with REFERENCE
ipr-mus TDLo:3500 μg/kg (female 10D post):TER FCTXAV 14,175,76
ipr-mus LD50:2800 μg/kg 85GDA2 5,408,81
ims-mus LD50:50 μg/kg TOXIA6 25,1329,87

SAFETY PROFILE: Poison by intramuscular and intraperitoneal routes. An experimental teratogen. When heated to decomposition it emits acrid smoke and irritating fumes.

VRP775 CAS:84777-85-5 **HR: 3**
VIRUSTOMYCIN A
mf: $C_{48}H_{71}NO_{14}$ mw: 886.20

SYNS: AM-2604 A ◇ ANTIBIOTIC AM-2604 A

TOXICITY DATA with REFERENCE
dni-omi 300 μg/L JANTAJ 36,1755,83
oms-omi 30 μg/L JANTAJ 36,1755,83
ipr-mus LDLo:10 mg/kg JANTAJ 35,1632,82

SAFETY PROFILE: Poison by intraperitoneal route. Mutation data reported. When heated to decomposition it emits toxic fumes of NO_x.

VRU000 CAS:76822-96-3 **HR: 3**
VISCOTOXIN
mf: $C_{36}H_{63}N_{10}O_{21}S$ mw: 1004.15

TOXICITY DATA with REFERENCE
ivn-rat LD50:260 μg/kg AEPPAE 209,165,50
ivn-rbt LDLo:500 μg/kg AEPPAE 209,165,50

SAFETY PROFILE: Poison by intravenous route. When heated to decomposition it emits very toxic fumes of NO_x and SO_x.

VRZ000 CAS:522-48-5 **HR: 3**
VISINE
mf: $C_{13}H_{16}N_2{\cdot}ClH$ mw: 236.77

SYNS: 4,5-DIHYDRO-2-(1,2,3,4-TETRAHYDRO-1-NAPHTHALENYL)-1H-IMIDAZOLE MONOHYDROCHLORIDE ◇ 2-(1,2,3,4-TETRAHYDRO-1-NAPHTHYL-2-IMIDAZOLINE HYDROCHLORIDE ◇ 2-(1,2,3,4-TETRA-HYDRO-1-NAPHTHYL)-2-IMIDAZOLINEMONOHYDROCHLORIDE ◇ TETRAHYDROZOLINE HYDROCHLORIDE ◇ dl-TETRAHYDROZO-LINE HYDROCHLORIDE ◇ TYZANOL HYDROCHLORIDE ◇ TYZINE ◇ TYZINE HYDROCHLORIDE ◇ VISINE HYDROCHLORIDE

TOXICITY DATA with REFERENCE
unr-inf TDLo:80 mg/kg:CNS,PUL,MET SCMBE9 29,17,55
orl-rat LD50:785 mg/kg NIIRDN 6,495,82
ipr-rat LD50:122 mg/kg NIIRDN 6,495,82
scu-rat LD50:500 mg/kg NIIRDN 6,495,82
ivn-rat LD50:35 mg/kg NIIRDN 6,495,82
orl-mus LD50:345 mg/kg NIIRDN 6,495,82
ipr-mus LD50:110 mg/kg CLDND* 6,495,82
ivn-mus LD50:39 mg/kg 29ZVAB -,113,69
ims-brd LD50:150 mg/kg CLDND* 6,495,82

CONSENSUS REPORTS: Reported in EPA TSCA Inventory.

SAFETY PROFILE: Poison by ingestion, intraperitoneal, intravenous, subcutaneous, and intramuscular routes. Human systemic effects by an unspecified route: sleep changes, dyspnea and body temperature decrease. When heated to decomposition it emits very toxic fumes of HCl and NO_x.

VSA000 CAS:13523-86-9 **HR: 3**
VISKEN
mf: $C_{14}H_{20}N_2O_2$ mw: 248.36

SYNS: CALVISKEN ◇ CARDILATE ◇ 4-(2-HYDROXY-3-ISOPROPYL-AMINOPROPOXY)-INDOLE ◇ 1-(4-INDOLYLOXY)-3-ISOPROPYL-AMINO)-2-PROPANOL ◇ 1-(1H-INDOL-4-YLOXY)-3-((1-METHYLETH-YL)AMINO)-2-PROPANOL ◇ LB-46 ◇ PINDOLOL ◇ PRINODOLOL

TOXICITY DATA with REFERENCE
orl-rat TDLo:2200 mg/kg (7-17D preg):TER OYYAA2 29,747,85
orl-rat TDLo:6300 mg/kg (30D male):REP YACHDS 9,3573,81
orl-rat LD50:263 mg/kg NIGZAY 84,438,70
ipr-rat LD50:110 mg/kg IYKEDH 4,90,73
scu-rat LD50:251 mg/kg NIGZAY 84,438,70
ivn-rat LD50:51 mg/kg NIGZAY 84,438,70
orl-mus LD50:235 mg/kg JTSCDR 6,301,81
ipr-mus LD50:80 mg/kg YACHDS 9,3573,81
scu-mus LD50:336 mg/kg NIGZAY 84,438,70
ivn-mus LD50:22600 μg/kg ARZNAD 28,794,78
ivn-rbt LD50:10 mg/kg ARZNAD 27,1022,77

SAFETY PROFILE: Poison by ingestion, subcutaneous, intraperitoneal, and intravenous routes. An experimental teratogen. Experimental reproductive effects. When heated to decomposition it emits toxic fumes of NO_x.

VSF000 CAS:1244-76-4 **HR: 3**
VISTARIL HYDROCHLORIDE
mf: $C_{21}H_{27}ClN_2O_2{\cdot}ClH$ mw: 411.41

SYNS: ATARAX HYDROCHLORIDE ◇ 1-(p-CHLOROBENZHYDRYL)-4-(2-(2-HYDROXYETHOXY)ETHYL)DIETHYLENEDIAMINEHYDRO-CHLORIDE ◇ HYDROXYZINE HYDROCHLORIDE

TOXICITY DATA with REFERENCE
orl-rat TDLo:240 mg/kg (female 5-16D post):TER RCOCB8 7,701,74
orl-rat TDLo:800 mg/kg (8-15D preg):REP PSDTAP 9,-,68
orl-rat LD50:690 mg/kg NIIRDN 6,621,82
ipr-rat LD50:126 mg/kg TXAPA9 18,185,71
orl-mus LD50:515 mg/kg JPETAB 127,318,59
ipr-mus LD50:137 mg/kg NIIRDN 6,621,82
ivn-mus LD50:56 mg/kg NIIRDN 6,621,82

SAFETY PROFILE: Poison by intraperitoneal and intravenous routes. Moderately toxic by ingestion. An experimental teratogen. Experimental reproductive effects. When heated to decomposition it emits very toxic fumes of Cl^- and NO_x. Used as a tranquilizer.

VSF400 CAS:20231-45-2 **HR: 3**
VITACAMPHER
mf: $C_{10}H_{14}O_2$ mw: 166.24

SYN: trans-pi-OXOCAMPHOR

TOXICITY DATA with REFERENCE
orl-rat TDLo:22750 mg/kg (26W pre):REP OYYAA2 5,789,71
orl-rat LD50:3100 mg/kg NIIRDN 6,514,82
ipr-rat LD50:890 mg/kg NIIRDN 6,514,82
scu-rat LD50:1650 mg/kg NIIRDN 6,514,82
orl-mus LD50:400 mg/kg NIIRDN 6,514,82
ipr-mus LD50:260 mg/kg NIIRDN 6,514,82
scu-mus LD50:550 mg/kg NIIRDN 6,514,82

SAFETY PROFILE: Poison by ingestion and intraperitoneal routes. Moderately toxic by subcutaneous route. Experimental reproductive effects. When heated to decomposition it emits acrid smoke and irritating fumes. See also ALDEHYDES.

VSK000 CAS:12629-02-6 ***HR: 3***
VITALLIUM

PROP: Alloy of chromium, cobalt, and molybdenum (CNREA8 16,439,56).

SYNS: CHROMIUM-COBALT-MOLYBDENUM ALLOY ◇ COBALT-CHROMIUM-MOLYBDENUM ALLOY ◇ MOLYBDENUM-COBALT-CHROMIUM ALLOY ◇ STELLITE

TOXICITY DATA with REFERENCE
ims-rat TDLo:140 mg/kg:ETA JBJSB4 55-B,759,73

CONSENSUS REPORTS: Cobalt and its compounds, as well as chromium and its compounds, are on the Community Right-To-Know List.

OSHA PEL: (Transitional: TWA Total Dust: 15 mg/m^3; Respirable Fraction: 5 mg/m^3) TWA Total Dust: 10 mg/m^3; Respirable Fraction: 5 mg/m^3
ACGIH TLV: TWA 10 mg(Mo)/m^3
NIOSH REL: COBALT-air: Insufficient evidence for recommending limit.

SAFETY PROFILE: Questionable carcinogen with experimental tumorigenic data.

VSK600 CAS:68-26-8 ***HR: 3***
VITAMIN A
mf: $C_{20}H_{30}O$ mw: 286.50

PROP: Light yellow to red oil; mild fishy odor. Very sol in chloroform, ether; sol in abs alc, vegetable oil; insol in glycerin, water.

SYNS: ACON ◇ AFAXIN ◇ AGIOLAN ◇ ALPHALIN ◇ ALPHASTEROL ◇ ANATOLA ◇ ANTI-INFECTIVE VITAMIN ◇ ANTI-XEROPHTHALMIC VITAMIN ◇ AORAL ◇ APEXOL ◇ AQUASYNTH ◇ AVIBON ◇ AVITA ◇ AVITOL ◇ BIOSTEROL ◇ CHOCOLA A ◇ 3,7-DIMETHYL-9-(2,6,6-TRIMETHYL-1-CYCLOHEXEN-1-YL)-2,4,6,8-NONATETRAEN-1-OL ◇ DISATABS TABS ◇ DOFSOL ◇ EPITELIOL ◇ HI-A-VITA ◇ LARD FACTOR ◇ MYVPACK ◇ OLEOVITAMIN A ◇ OPHTHALAMIN ◇ PREPALIN ◇ RETINOL ◇ all-trans RETINOL ◇ RETRO-VITAMIN A ◇ TESTAVOL ◇ VAFLOL ◇ VI-ALPHA ◇ VITAMIN A1 ◇ VITAMIN A1 ALCOHOL ◇ all-trans-VITAMIN A ALCOHOL ◇ VITAVEL-A ◇ VITPEX ◇ VOGAN ◇ VOGAN-NEU

TOXICITY DATA with REFERENCE
oms-hmn:lym 4 mg/L EJCODS 21,1089,85
sce-hmn:lym 4 mg/L EJCODS 21,1089,85
dni-rat:mmr 3 μmol/L JJIND8 70,949,83
orl-wmn TDLo:200 mg/kg (8W preg):TER JGHUAY 23,135,75
orl-rat TDLo:76560 μg/kg (8-10D preg):REP NTOTDY 3,1,81
orl-rat LD50:2000 mg/kg AVSUAR 74,29,75
orl-mus LD50:1510 mg/kg 51ZKAW 2,287,84

CONSENSUS REPORTS: Reported in EPA TSCA Inventory. EPA Genetic Toxicology Program.

SAFETY PROFILE: Moderately toxic by ingestion. Human teratogenic effects by ingestion: developmental abnormalities of the craniofacial area and urogenital system. An experimental teratogen. Experimental reproductive effects. Human mutation data reported. When heated to decomposition it emits acrid smoke and irritating fumes.

VSK900 CAS:127-47-9 ***HR: 3***
VITAMIN A ACETATE
mf: $C_{22}H_{32}O_2$ mw: 328.54

SYNS: CRYSTALETS ◇ MYVAK ◇ MYVAX ◇ RETINOL ACETATE ◇ RETINYL ACETATE ◇ all-trans-RETINYL ACETATE ◇ trans-VITAMIN A ACETATE ◇ VITAMIN A ALCOHOL ACETATE

TOXICITY DATA with REFERENCE
dni-rat:mmr 3 μmol/L JJIND8 70,949,83
orl-rat TDLo:480 mg/kg (6-19D preg):REP TOXID9 4,84,84
orl-rat TDLo:826 mg/kg (female 12-15D post):TER TJADAB 18,277,78
orl-rat TDLo:51800 mg/kg/2Y-C:NEO JJIND8 74,715,85
orl-mus LDLo:1000 mg/kg APMIAL 70,398,67

CONSENSUS REPORTS: Reported in EPA TSCA Inventory.

SAFETY PROFILE: Moderately toxic by ingestion. Experimental teratogenic and reproductive effects. Questionable carcinogen with experimental neoplastigenic data. Mutation data reported. When heated to decomposition it emits acrid smoke and irritating fumes. See also all-trans RETINOL.

VSK950 CAS:302-79-4 ***HR: 3***
VITAMIN A ACID
mf: $C_{20}H_{28}O_2$ mw: 300.48

PROP: Mp: 180-182°.

SYNS: ABEREL ◇ 3,7-DIMETHYL-9-(2,6,6-TRIMETHYL-1-CYCLOHEXEN-1-YL-2,4,6,8-NONATETRAENOIC ACID ◇ NSC-122758 ◇ β-RA

◇ RETIN-A ◇ RETINOIC ACID ◇ β-RETINOIC ACID ◇ all-trans-RETINOIC ACID ◇ TRETINOIN

TOXICITY DATA with REFERENCE
skn-hmn 525 mg/21D-I MLD AVSUAR 74,128,75
dni-hmn:leu 1 μmol/L CNREA8 46,1388,86
oms-hmn-skn 1000 ppm 26UYA8 -,335,71
unr-rat TDLo:12 mg/kg (female 14-16D post):REP TJADAB 25(2),64A,82
orl-ham TDLo:60 mg/kg (female 8D post):TER TJADAB 28,341,83
skn-mus TDLo:8400 mg/kg/30W-I:NEO CALEDQ 7,85,79
orl-rat LD50:1960 mg/kg KSRNAM 7,3194,73
ipr-rat LD50:96 mg/kg KSRNAM 7,3194,73
scu-rat LD50:53 mg/kg KSRNAM 7,3194,73
ivn-rat LD50:78 mg/kg KSRNAM 7,3194,73
orl-mus LD50:216 mg/kg VOONAW 25(12),84,79
ipr-mus LD50:394 mg/kg KSRNAM 7,3194,73
scu-mus LD50:253 mg/kg KSRNAM 7,3194,73
ivn-mus LD50:191 mg/kg KSRNAM 7,3194,73

CONSENSUS REPORTS: Reported in EPA TSCA Inventory. EPA Genetic Toxicology Program.

SAFETY PROFILE: Poison by ingestion, intraperitoneal, subcutaneous, and intravenous routes. Experimental reproductive effects. Questionable carcinogen with experimental neoplastigenic and teratogenic data. Human mutation data reported. A human skin irritant. When heated to decomposition it emits acrid smoke and irritating fumes. Used to treat acne and other skin problems.

VSK955 CAS:4759-48-2 ***HR: 3***
13-cis-VITAMIN A ACID
mf: $C_{20}H_{28}O_2$ mw: 300.48

SYNS: ISOTRETINOIN ◇ NEOVITAMIN A ACID ◇ 13-RA ◇ 13-cis-RETINOIC ACID ◇ RO-4-3780

TOXICITY DATA with REFERENCE
sce-hmn:lym 50 μmol/L BLFSBY 29A,333,84
sce-mus:emb 7100 nmol/L ANYAA9 359,237,81
orl-wmn TDLo:24 mg/kg (female 1-4W post):REP CMAJAX 133,208,85
orl-wmn TDLo:157 mg/kg (female 2W pre):TER AJDCAI 141,263,87
orl-man TDLo:37 mg/kg/5W-I:SKN,ALR ARDEAC 122,815,86
orl-man TDLo:24 mg/kg/4W-I:GIT GASTAB 93,606,87
orl-wmn TDLo:56 mg/kg/8W-I:SKN CUTIBC 37,115,86
unr-man TDLo:21 mg/kg/3W-I:SKN BMJOAE 290,820,85
orl-cld TDLo:360 mg/kg/26W-I:SKN CUTIBC 38,275,86
orl-mus LD50:3389 mg/kg 51ZKAW 2,287,84
ipr-mus LD50:138 mg/kg 51ZKAW 2,287,84
orl-rbt LD50:1960 mg/kg 51ZKAW 2,287,84

CONSENSUS REPORTS: Reported in EPA TSCA Inventory.

SAFETY PROFILE: Poison by intraperitoneal route. Moderately toxic by ingestion. A human teratogen by ingestion with fetal developmental abnormalities of the skin and appendages and other postnatal effects. Human reproductive effects. Human systemic effects: decreased immune response, diarrhea, hypermotility, irritative dermatitis, sweating. Human mutation data reported. An experimental teratogen. Other experimental reproductive effects. When heated to decomposition it emits acrid smoke and irritating fumes.

VSK975 CAS:514-85-2 ***HR: D***
9-cis-VITAMIN A ALDEHYDE
mf: $C_{20}H_{28}O$ mw: 284.48

SYNS: 9-cis-3,7-DIMETHYL-9-(2,6,6-TRIMETHYL-1-CYCLOHEXEN-1-YL)-2,4,6,8-NONATETRA ENAL ◇ ISORETINENE a ◇ 9-cis-RETINAL ◇ 9-cis-RETINALDEHYDE

TOXICITY DATA with REFERENCE
orl-ham TDLo:25 mg/kg (female 8D post):REP TXAPA9 83,563,86

CONSENSUS REPORTS: Reported in EPA TSCA Inventory.

SAFETY PROFILE: Experimental reproductive effects. When heated to decomposition it emits acrid smoke and irritating fumes.

VSP000 CAS:79-81-2 ***HR: 1***
VITAMIN A PALMITATE
mf: $C_{36}H_{60}O_2$ mw: 524.96

SYNS: AQUASOL ◇ AROVIT ◇ RETINOL PALMITATE ◇ RETINYL PALMITATE

TOXICITY DATA with REFERENCE
sce-hmn:fbr 27500 μg/L MUREAV 58,317,78
orl-rat TDLo:16500 μg/kg (female 8-10D post):REP TJADAB 10,269,74
ipr-rat TDLo:413 mg/kg (female 10D post):TER ANANAU 152,329,82
orl-mus LD50:4760 mg/kg VOONAW 25(12),84,79

CONSENSUS REPORTS: Reported in EPA TSCA Inventory. EPA Genetic Toxicology Program.

SAFETY PROFILE: Mildly toxic by ingestion. An experimental teratogen. Experimental reproductive effects. Human mutation data reported. When heated to decomposition it emits acrid smoke and irritating fumes.

VSU000 CAS:65-22-5 ***HR: 3***
VITAMIN B_6 HYDROCHLORIDE
mf: $C_8H_9NO_3{\cdot}ClH$ mw: 203.64

SYNS: 3-HYDROXY-5-(HYDROXYMETHYL)-22-

METHYLISONICOTINALDEHYDE, HYDROCHLORIDE ◇ 2-METHYL-3-HYDROXY-4-FORMYL-5-HYDROXYMETHYLPYRIDINEHYDROCHLORIDE ◇ PYRIDOXAL HYDROCHLORIDE

TOXICITY DATA with REFERENCE
orl-rat LDLo:6000 mg/kg HBTXAC 5,177,59
scu-rat LD50:530 mg/kg ARZNAD 11,922,61
ivn-rat LD50:320 mg/kg ARZNAD 11,922,61
orl-mus LD50:1800 mg/kg ARZNAD 11,922,61
ipr-mus LD50:400 mg/kg NTIS** AD691-490
scu-mus LD50:530 mg/kg ARZNAD 11,922,61
ivn-mus LD50:390 mg/kg ARZNAD 11,922,61
ivn-cat LD50:160 mg/kg ARZNAD 11,922,61
ims-cat LD50:152 mg/kg ARZNAD 11,922,61
ivn-rbt LD50:465 mg/kg ARZNAD 11,922,61
ivn-pgn LD50:262 mg/kg ARZNAD 11,922,61

CONSENSUS REPORTS: Reported in EPA TSCA Inventory.

SAFETY PROFILE: Poison by intramuscular, intravenous, and intraperitoneal routes. Moderately toxic by ingestion and subcutaneous routes. When heated to decomposition it emits very toxic fumes of NO_x and HCl. See also ALDEHYDES.

VSU100 CAS:58-85-5 ***HR: D***
VITAMIN B_7
mf: $C_{10}H_{16}N_2O_3S$ mw: 244.34

SYNS: BIOEPIDERM ◇ BIOS II ◇ BIOTIN ◇ (+)-BIOTIN ◇ d-BIOTIN ◇ d-(+)-BIOTIN ◇ COENZYME R ◇ FACTOR S ◇ FACTOR S (vitamin) ◇ 1H-THIENO(3,4-d)IMIDAZOLE-4-PENTANOIC ACID, HEXAHYDRO-2-OXO-, (3aS-(3a-α-4-β, 6a-α))- ◇ VITAMIN H

TOXICITY DATA with REFERENCE
scu-rat TDLo:200 mg/kg (female 14-15D post):TER JNSVA5 22,181,76
scu-rat TDLo:100 mg/kg (male 1D pre):REP CUSCAM 42,613,73

CONSENSUS REPORTS: Reported in EPA TSCA Inventory.

SAFETY PROFILE: An experimental teratogen. Experimental reproductive effects. When heated to decomposition it emits toxic fumes of NO_x and SO_x.

VSZ000 CAS:68-19-9 ***HR: 3***
VITAMIN B_{12} COMPLEX
mf: $C_{63}H_{88}CoN_{14}O_{14}P$ mw: 1355.55

PROP: The anti-pernicious anemia vitamin. All vitamin B_{12} compounds contain the cobalt atom in its trivalent state. There are at least three active forms: cyanocobalamin, hydroxycobalamin, and nitrocobalamin. Dark red crystals or crystalline powder. Very hygroscopic; sltly sol in water; sol in alc; insol in acetone, chloroform, ether.

SYNS: ANACOBIN ◇ B-12 ◇ BERUBIGEN ◇ BETALIN 12 CRYSTALLINE ◇ BEVATINE-12 ◇ BEVIDOX ◇ BYLADOCE ◇ CABADON M ◇ COBADOCE FORTE ◇ COBALIN ◇ COBAMIN ◇ COBIONE ◇ COTEL ◇ COVIT ◇ CRYSTAMIN ◇ CRYSTWEL ◇ CYANO-B12 ◇ CYANOCOBALAMIN ◇ CYCOLAMIN ◇ CYKOBEMINET ◇ CYREDIN ◇ CYTACON ◇ CYTAMEN ◇ CYTOBION ◇ DEPINAR ◇ DIMETHYLBENZIMIDAZOLYCOBAMIDE ◇ 5,6-DIMETHYLBENZIMIDAZOLYCOBAMIDE CYANIDE ◇ DISTIVIT (B12 PEPTIDE) ◇ DOBETIN ◇ DOCEMINE ◇ DOCIBIN ◇ DOCIGRAM ◇ DODECABEE ◇ DODECAVITE ◇ DODEX ◇ DUCOBEE ◇ DUODECIBIN ◇ EMBIOL ◇ EMOCICLINA ◇ ERITRONE ◇ ERYCYTOL ◇ ERYTHROTIN ◇ EUHAEMON ◇ EXTRINSIC FACTOR ◇ FACTOR II (VITAMIN) ◇ FRESMIN ◇ HEMO-B-DOZE ◇ HEMOMIN ◇ HEPAGON ◇ HEPAVIS ◇ HEPCOVITE ◇ LACTOBACILLUS LACTIS DORNER FACTOR ◇ LLD FACTOR ◇ MACRABIN ◇ MEGABION ◇ MEGALOVEL ◇ MILBEDOCE ◇ NAGRAVON ◇ NORMOCYTIN ◇ PERNAEMON ◇ PERNAEVIT ◇ PERNIPURON ◇ PLECYAMIN ◇ POYAMIN ◇ REBRAMIN ◇ REDAMINA ◇ REDISOL ◇ RHODACRYST ◇ RUBESOL ◇ RUBRAMIN ◇ RUBRIPCA ◇ RUBROCITOL ◇ SYTOBEX ◇ VIBALT ◇ VIBISONE ◇ VIRUBRA ◇ VITAMIN B12 (FCC) ◇ VITARUBIN ◇ VITA-RUBRA ◇ VITRAL ◇ VI-TWEL

TOXICITY DATA with REFERENCE
orl-rat TDLo:115 mg/kg (1-22D preg):REP NATUAS 242,263,73
ims-mus TDLo:84 mg/kg (11D preg):TER JPMSAE 67,377,78
ipr-mus LDLo:1364 mg/kg ARPAAQ 49,278,50
scu-mus LDLo:3 mg/kg ARPAAQ 49,278,50

CONSENSUS REPORTS: Cobalt and its compounds are on the Community Right-To-Know List. Reported in EPA TSCA Inventory. EPA Genetic Toxicology Program.

NIOSH REL: COBALT-air: Insufficient evidence for recommending limit.

SAFETY PROFILE: Poison by subcutaneous route. Moderately toxic by intraperitoneal route. An experimental teratogen. Experimental reproductive effects. When heated to decomposition it emits very toxic fumes of PO_x and NO_x. See also COBALT COMPOUNDS.

VSZ050 CAS:13422-55-4 ***HR: 2***
VITAMIN B_{12}, METHYL
mf: $C_{63}H_{91}N_{13}O_{14}P{\cdot}Co$ mw: 1344.57

SYNS: COBALT-METHYLCOBALAMIN ◇ COBINAMIDE, COBALT-METHYL derivative, HYDROXIDE, DIHYDROGEN PHOSPHATE (ester), inner salt, 3′-ESTER with 5,6-DIMETHYL-1-α-D-RIBOFURANOSYLBENZIMIDAZOLE ◇ MECOBALAMIN ◇ METHYCOBAL ◇ METHYL-B_{12} ◇ METHYLCOBALAMIN ◇ METHYL COBALAMINE

TOXICITY DATA with REFERENCE
ivn-rat TDLo:5500 μg/kg (female 7-17D post):REP KSRNAM 22,3899,88
ivn-rat TDLo:550 mg/kg (female 7-17D post):TER KSRNAM 22,3899,88
ims-mus TDLo:10 mg/kg (14-20D post):NEO BEBMAE 101,471,86

SAFETY PROFILE: Questionable carcinogen with experimental neoplastigenic data. An experimental teratogen. Other experimental reproductive effects. When heated to decomposition it emits toxic fumes of NO_x and PO_x.

VSZ100 CAS:50-14-6 ***HR: 3***
VITAMIN D2
mf: $C_{28}H_{44}O$ mw: 396.72

PROP: White crystals; odorless. Mp: 115-118°. Insol in water; sol in alc, chloroform, ether, and fatty oils.

SYNS: d-ARTHIN ◇ CALCIFEROL ◇ CALCIFERON 2 ◇ CONDACAPS ◇ CONDOCAPS ◇ CONDOL ◇ CRTRON ◇ CRYSTALLINA ◇ DARAL ◇ DAVITAMON D ◇ DAVITIN ◇ DECAPS ◇ DEE-OSTEROL ◇ DEE-RON ◇ DEE-RONAL ◇ DEE-ROUAL ◇ DELTALIN ◇ DERATOL ◇ DETALUP ◇ DIACTOL ◇ DIVIT URTO ◇ DORAL ◇ DRISDOL ◇ ERGOCALCIFEROL ◇ ERGORONE ◇ ERGOSTEROL, activated ◇ ERGOSTEROL, irradiated ◇ ERTRON ◇ FORTODYL ◇ GELTABS ◇ HIDERATOL ◇ INFRON ◇ IRRADIATED ERGOSTA-5,7,22-TRIEN-3-β-OL ◇ METADEE ◇ MULSIFEROL ◇ MYKOSTIN ◇ OLEOVITAMIN D ◇ OSTELIN ◇ RADIOSTOL ◇ RADSTERIN ◇ 9,10,SECOERGOSTA-5,7,10(19),22-TETRAEN-3-β-OL ◇ SHOCK-FEROL ◇ STEROGYL ◇ VIGANTOL ◇ VIOSTEROL ◇ VITAVEL-D

TOXICITY DATA with REFERENCE
ims-rbt TDLo:9375 μg/kg (female 2-31D post):REP PEDIAU 43,12,69
orl-rat TDLo:22500 μg/kg (female 13-21D post):TER ARPAAQ 73,371,62
orl-wmn TDLo 12600 mg/kg/72W:CNS,GIT,MET LANCAO 1,1164,80
orl-rat LD50:56 mg/kg 85JFAN A685,85
orl-mus LD50:23700 μg/kg PEMNDP 8,117,87
orl-dog LDLo:4 mg/kg ZGEMAZ 116,138,50
ipr-dog LDLo:10 mg/kg ZGEMAZ 116,138,50
ivn-dog LDLo:5 mg/kg ZGEMAZ 116,138,50
ims-dog LDLo:5 mg/kg ZGEMAZ 116,138,50
orl-cat LDLo:5 mg/kg NIIRDN 6,128,82
orl-gpg LDLo:40 mg/kg NIIRDN 6,128,82

CONSENSUS REPORTS: EPA Extremely Hazardous Substances List.

SAFETY PROFILE: Poison by ingestion, intraperitoneal, intravenous, and intramuscular routes. An experimental teratogen. Human systemic effects by ingestion: anorexia, nausea or vomiting, and weight loss. Experimental reproductive effects. When heated to decomposition it emits acrid smoke and irritating fumes.

VSZ450 CAS:59-02-9 ***HR: D***
VITAMIN E
mf: $C_{29}H_{50}O_2$ mw: 430.79

PROP: dl-Form: Sltly viscous, pale yellow oil; d-form: red liquid; odorless. Natural α-tocopherol has been crystallized. Mp: 2.5-3.5°, d: (25°/4°) 0.950, bp: (0.1) 200-220°. Practically insol in water; freely sol in oils, fats, acetone, alc, chloroform, ether, other fat solvents. Gradually darkens on exposure to light.

SYNS: ALMEFROL ◇ ANTISTERILITY VITAMIN ◇ COVI-OX ◇ DENAMONE ◇ EMIPHEROL ◇ ENDO E ◇ EPHYNAL ◇ EPROLIN ◇ EPSILAN ◇ ESORB ◇ ETAMICAN ◇ ETAVIT ◇ EVION ◇ EVITAMINUM ◇ ILITIA ◇ PHYTOGERMINE ◇ PROFECUNDIN ◇ SPAVIT ◇ SYNTOPHEROL ◇ d-α-TOCOPHEROL (FCC) ◇ dl-α-TOCOPHEROL (FCC) ◇ (R,R,R)-α-TOCOPHEROL ◇ α-TOCOPHEROL ◇ (2R,4'R,8'R)-α-TOCOPHEROL ◇ TOKOPHARM ◇ 5,7,8-TRIMETHYLTOCOL ◇ VASCUALS ◇ VERROL ◇ VITAPLEX E ◇ VITAYONON ◇ VITEOLIN

TOXICITY DATA with REFERENCE
dnd-rat-ivn 27 nmol/kg EXPEAM 31,1023,75
dni-rat:lvr 100 μmol/L CNREA8 45,337,85
orl-rat TDLo:7500 mg/kg (1-20D preg):REP NYKZAU 69,293,73

CONSENSUS REPORTS: Reported in EPA TSCA Inventory.

SAFETY PROFILE: Experimental reproductive effects. Mutation data reported. When heated to decomposition it emits acrid smoke and irritating fumes.

VSZ500 CAS:12001-79-5 ***HR: 2***
VITAMIN K

TOXICITY DATA with REFERENCE
ims-mus TDLo:13500 μg/kg (female 6-11D post):TER CAJPBD 8,46,68
scu-mus LD50:700 mg/kg ARZNAD 8,25,58

CONSENSUS REPORTS: Reported in EPA TSCA Inventory.

SAFETY PROFILE: Moderately toxic by subcutaneous route. An experimental teratogen. When heated to decomposition it emits acrid smoke and irritating fumes.

VTA000 CAS:84-80-0 ***HR: 2***
VITAMIN K1
mf: $C_{31}H_{46}O_2$ mw: 450.77

SYNS: ANTIHEMORRHAGIC VITAMIN ◇ AQUA MEPHYTON ◇ COMBINAL K1 ◇ KATIV N ◇ KEPHTON ◇ KINADION ◇ KONAKION ◇ MEPHYTON ◇ 2-METHYL-3-PHYTHYL-1,4-NAPHTHOCHINON (GERMAN) ◇ 2-METHYL-3-(3,7,11,15-TETRAMETHYL-2-HEXADECENYL)-1,4-NAPHTHALENEDIONE ◇ MONODION ◇ MONO-KAY ◇ PHYLLOCHINON (GERMAN) ◇ PHYLLOQUINONE ◇ α-PHYLLOQUINONE ◇ trans-PHYLLOQUINONE ◇ PHYTOMENADIONE ◇ PHYTONADIONE

TOXICITY DATA with REFERENCE
orl-mus LD50:25 g/kg ARZNAD 17,1339,67
scu-mus LD50:1000 mg/kg ARZNAD 8,25,58

CONSENSUS REPORTS: Reported in EPA TSCA Inventory.

SAFETY PROFILE: Moderately toxic by subcutaneous route. Mildly toxic by ingestion. An FDA proprietary drug. Used as a vitamin. When heated to decomposition it emits acrid smoke and irritating fumes.

VTA650 CAS:863-61-6 ***HR: D***
VITAMIN MK 4
mf: $C_{31}H_{40}O_2$ mw: 444.71

SYNS: K2$_{20}$ ◇ KAYTWO ◇ MENAQUINONE-4 ◇ MENAQUINONE K_4 ◇ MENATETRENONE ◇ 2-METHYL-3-(3,7,11,15-TETRAMETHYL-2,6,10,14-HEXADECATETRAENYL)-1,4-NAPHTHOQUINONE ◇ 2-METHYL-3-trans-TETRAMETHYL-1,4-NAPHTHQUINONE ◇ MK_4 ◇ VITAMIN K2$_{20}$

TOXICITY DATA with REFERENCE
orl-rat TDLo:270 mg/kg (multi) :REP KSRNAM 15,1143,81
orl-rat TDLo:6 g/kg (9-14D preg):TER OYYAA2 5,469,71

SAFETY PROFILE: An experimental teratogen. Experimental reproductive effects. When heated to decomposition it emits acrid smoke and irritating fumes.

VTA750 CAS:610-88-8 ***HR: 3***
VIVOTOXIN
mf: $C_{10}H_{15}NO_3$ mw: 197.26

PROP: Pale brown, viscous, gummy substance. Bp: (0.035) 117°. Readily sol in organic solvents including petr ether; sparingly sol in water. On long standing, changes into the crystalline iso-form.

SYNS: 3-ACETYL-5-sec-BUTYL-4-HYDROXY-3-PYRROLIN-2-ONE ◇ 3-ACETYL-1,5-DIHYDRO-4-HYDROXY-5-(1-METHYLPROPYL)-2H-PYRROL-2-ONE ◇ TENUAZONIC ACID ◇ l-TENUAZONIC ACID

TOXICITY DATA with REFERENCE
orl-mus LD50:225 mg/kg 85GDA2 5,65,81
ipr-mus LD50:81 mg/kg EVHPAZ 4,87,73
scu-mus LD50:145 mg/kg 85GDA2 5,65,81
ivn-mus LD50:125 mg/kg 85GDA2 5,65,81

SAFETY PROFILE: Poison by ingestion, subcutaneous, intravenous, and intraperitoneal routes. When heated to decomposition it emits toxic fumes of NO_x.

VTF000 CAS:595-33-5 ***HR: 3***
VOLIDAN
mf: $C_{24}H_{32}O_4$ mw: 384.56

SYNS: 17-α-ACETOXY-6-DEHYDRO-6-METHYLPROGESTERONE ◇ 17-ACETOXY-6-METHYLPREGNA-4,6-DIENE-3,20-DIONE ◇ 17-α-ACETOXY-6-METHYLPREGNA-4,6-DIENE-3,20-DIONE◇ 17-α-ACETOXY-6-METHYL-4,6-PREGNADIENE-3,20-DIONE◇ BDH 1298 ◇ 6-DEHYDRO-6-METHYL-17-α-ACETOXYPROGESTERONE ◇ DMAP ◇ 17-HYDROXY-6-METHYLPREGNA-4,6-DIENE-3,20-DIONE ACETATE ◇ MEGACE ◇ MEGESTROL ACETATE ◇ MEGESTRYL ACETATE ◇ 6-METHYL-17-α-ACETOXYPREGNA-4,6-DIENE-3,20-DIONE ◇ 6-METHYL-6-DEHYDRO-17-α-ACETOXYPROGESTERONE ◇ 6-METHYL-6-DEHYDRO-17-α-ACETYLPROGESTERONE ◇ 6-METHYL-17-α- HYDROXY-Δ^6-PROGESTERONE ACETATE ◇ 6-METHYL-$\Delta^{4,6}$-PREGNADIEN-17-α-OL-3,20-DIONE ACETATE ◇ NSC-71423 ◇ OVABAN ◇ SC10363

TOXICITY DATA with REFERENCE
dns-mus-scu 200 mg/kg JOENAK 60,167,74
dni-mus-scu 200 mg/kg JOENAK 60,167,74
orl-wmn TDLo:1825 μg/kg (52W pre):REP BMJOAE 2,730,69
orl-rat TDLo:9 mg/kg (female 15-20D post):TER JRPFA4 5,331,63
orl-dog TDLo:256 mg/kg/7Y-C:CAR JTEHD6 3,167,77
orl-dog TD:182 mg/kg/2Y-C:ETA JAMAAP 219,1601,72
ivn-mus LD50:56 mg/kg CSLNX* NX#00931

CONSENSUS REPORTS: IARC Cancer Review: Animal Limited Evidence IMEMDT 21,431,79.

SAFETY PROFILE: Suspected carcinogen with experimental carcinogenic and teratogenic data. Poison by intravenous route. Human reproductive effects by ingestion and implant routes: effects on ovaries and fallopian tubes, menstrual cycle changes and female fertility index changes. Mutation data reported. Experimental reproductive effects. When heated to decomposition it emits acrid smoke and irritating fumes. An FDA proprietary drug used to treat endometriosis and breast cancer. A steroid.

VTF500 CAS:51481-10-8 ***HR: 3***
VOMITOXIN
mf: $C_{15}H_{20}O_6$ mw: 296.35

PROP: Fine needles from ethyl acetate + petr ether. Mp: 151-153°.

SYNS: DEHYDRONIVALENOL ◇ DEOXYNIVALENOL ◇ 4-DEOXYNIVALENOL ◇ DESOXYNIVALENOL

TOXICITY DATA with REFERENCE
skn-gpg 148 μg MLD FAATDF 4(2, Pt 2),S124,84
orl-mus TDLo:20 mg/kg (female 8-11D post):TER BECTA6 29,487,82
orl-mus TDLo:20 mg/kg (female 8-11D post):REP BECTA6 29,487,82
orl-mus LD50:46 mg/kg FAATDF 4(2, Pt 2),S124,84
ipr-mus LD50:43 mg/kg TOXIA6 24,985,86
scu-mus LD50:45 mg/kg TOXIA6 24,985,86
scu-dog LD50:27 mg/kg VHTODE 25,335,83

SAFETY PROFILE: Poison by ingestion, subcutaneous, and intraperitoneal routes. An experimental teratogen. Experimental reproductive effects. A skin irritant. When heated to decomposition it emits acrid smoke and fumes.

W

WAJ000 CAS:31232-27-6 ***HR: 3***
WA 335 HYDROCHLORIDE
mf: $C_{20}H_{21}NO \cdot ClH$ mw: 327.88

SYNS: 9,10-DIHYDRO-10-(1-METHYL-4-PIPERIDINYLIDENE)-9-ANTHRACENOL HYDROCHLORIDE ◇ (1-METHYL-4-PIPERIDYLIDENE)-9-ANTHROL-9,10-DIHYDRO-10-HYDROCHLORIDE

TOXICITY DATA with REFERENCE
orl-rat LD50:135 mg/kg ARZNAD 25,1723,75
ipr-rat LD50:27500 μg/kg ARZNAD 25,1723,75
ivn-rat LD50:2700 μg/kg ARZNAD 25,1723,75
orl-mus LD50:164 mg/kg ARZNAD 25,1723,75
ipr-mus LD50:56300 μg/kg ARZNAD 25,1723,75
ivn-mus LD50:8900 μg/kg ARZNAD 35,1723,75
ivn-rbt LD50:16100 μg/kg ARZNAD 25,1723,75
ipr-gpg LD50:126 mg/kg ARZNAD 25,1723,75

SAFETY PROFILE: Poison by ingestion, intravenous, and intraperitoneal routes. When heated to decomposition it emits toxic fumes of NO_x and HCl.

WAK000 CAS:91-84-9 ***HR: 3***
WAIT'S GREEN MOUNTAIN ANTIHISTAMINE
mf: $C_{17}H_{23}N_3O$ mw: 285.43

SYNS: AFKO-HIST ◇ ANHISTABS ◇ ANHISTOL ◇ ANTALERGAN ◇ ANTAMINE ◇ ANTHISAN ◇ COPSAMINE ◇ CORADON ◇ N-DIMETHYLAMINO-AETHYL-N-p-METHOXY-BENZYL-α-AMINO-PYRIDIN-MALEAT (GERMAN) ◇ 2-((2-(DIMETHYLAMINO)ETHYL)-(p-METHOXYBENZYL)AMINO)PYRIDINE ◇ DIPANE ◇ DORANTAMIN ◇ ENRUMAY ◇ HARVAMINE ◇ HISTACAP ◇ HISTALON ◇ HISTAN ◇ HISTAPYRAN ◇ HISTASAN ◇ ISAMIN ◇ KRIPTIN ◇ MARANHIST ◇ MEPYRAMIN (GERMAN) ◇ MEPYREN ◇ MINIHIST ◇ N-p-METHOXYBENZYL-N',N'-DIMETHYL-N-α-PYRIDYLETHYLENEDIAMINE ◇ N-(p-METHOXYBENZYL)-N',N'-DIMETHYL-N-2-PYRIDYLETHYLENEDIAMINE ◇ NCI-C60651 ◇ NEOANTERGAN ◇ NEOBRIDAL ◇ NYSCAPS ◇ PARAMINYL ◇ PARMAL ◇ PYMAFED ◇ PYRA ◇ PYRAMAL ◇ PYRANISAMINE ◇ PYRILAMINE ◇ R.D. 2786 ◇ RP 2786 ◇ STAMINE ◇ STANGEN ◇ STATOMIN ◇ THYLOGEN

TOXICITY DATA with REFERENCE
eye-rbt 1% OPHTAD 143,154,62
eye-rbt 100 mg MLD FCTOD7 20,573,82
ims-rat TDLo:50 mg/kg (5D preg):REP PSEBAA 100,555,59
ipr-rat TDLo:350 mg/kg (10-16D preg):TER JPHYA7 164,138,62
unr-hmn TDLo:714 μg/kg:CNS JOALAS 19,313,48
orl-mus LD50:220 mg/kg THERAP 26,1203,71
ipr-mus LD50:107 mg/kg ARZNAD 25,1723,75
scu-mus LD50:100 mg/kg ARZNAD 14(8),940,64
ivn-mus LD50:23 mg/kg THERAP 26,1203,71

SAFETY PROFILE: Human poison by an unspecified route. An experimental poison by ingestion, intraperitoneal, subcutaneous, and intravenous routes. An experimental teratogen. Human systemic effects by unspecified route: sleep effects, somnolence and muscle contraction or spasticity. Experimental reproductive effects. An eye irritant. When heated to decomposition it emits toxic fumes of NO_x. Used as an antihistamine.

WAT000 CAS:481-39-0 ***HR: 3***
WALNUT EXTRACT
mf: $C_{10}H_6O_3$ mw: 174.16

SYNS: C.I. 75500 ◇ C.I. NATURAL BROWN 7 ◇ 5-HYDROXY-1,4-NAPHTHALENEDIONE ◇ 5-HYDROXY-1,4-NAPHTHOQUINONE

TOXICITY DATA with REFERENCE
hma-mus/ast 10 mg/kg PSEBAA 126,583,67
skn-mus TDLo:394 mg/kg/53W-I:NEO JMCMAR 21,26,78
orl-mus LD50:2500 μg/kg SCIEAS 134,1617,61

SAFETY PROFILE: Poison by ingestion. Questionable carcinogen with experimental neoplastigenic data. Mutation data reported. When heated to decomposition it emits acrid smoke and irritating fumes.

WAT100 ***HR: 3***
WALTERINNESIA AEGYPTIA VENOM

SYNS: VENOM, SNAKE, WALTERINNESIA AEGYPTIA ◇ W. AEGYPTIA VENOM

TOXICITY DATA with REFERENCE
ipr-mus LD50:140 μg/kg TOXIA6 14,275,76
scu-mus LD50:400 μg/kg TOXIA6 5,47,67
ims-dog LDLo:30 μg/kg TOXIA6 1,77,63
ims-rbt LDLo:400 μg/kg TOXIA6 1,77,63

SAFETY PROFILE: Poison by subcutaneous, intramuscular, and intraperitoneal routes.

WAT200 CAS:81-81-2 ***HR: 3***
WARFARIN
mf: $C_{19}H_{16}O_4$ mw: 308.35

PROP: Colorless, odorless, tasteless crystals. Mp: 161°. Sol in acetone, dioxane; sltly sol in methanol, ethanol; very sol in alkaline aqueous sol; insol in water and benzene.

SYNS: 3-(α-ACETONYLBENZYL)-4-HYDROXYCOUMARIN ◇ ARAB RAT DETH ◇ ATHROMBINE-K ◇ BRUMIN ◇ COMPOUND 42 ◇ d-CON ◇ CO-RAX ◇ COUMADIN ◇ COUMAFENE ◇ DETHMORE ◇ EASTERN

STATES DUOCIDE ◇ 4-HYDROXY-3-(3-OXO-1-FENYL-BUTYL) CUMARINE (DUTCH) ◇ 4-HYDROXY-3-(3-OXO-1-PHENYL-BUTYL)-CUMARIN (GERMAN) ◇ 4-IDROSSI-3-(3-OXO-)-FENIL-BUTIL)-CUMARINE (ITALIAN) ◇ KUMADER ◇ LIQUA-TOX ◇ MOUSE PAK ◇ 3-(α-PHENYL-β-ACETYLETHYL)-4-HYDROXYCOUMARIN ◇ 3-(1'-PHENYL-2'-ACETYLETHYL)-4-HYDROXYCOUMARIN ◇ (PHENYL-1 ACETYL-2 ETHYL)-3-HYDROXY-4 COUMARINE (FRENCH) ◇ PROTHROMADIN ◇ RAT-A-WAY ◇ RAT-B-GON ◇ RAT-GARD ◇ RAT & MICE BAIT ◇ RATS-NO-MORE ◇ RCRA WASTE NUMBER P001 ◇ RO-DETH ◇ ROUGH & READY MOUSE MIX ◇ SOLFARIN ◇ SPRAY-TROL BRANCH RODEN-TROL ◇ TWIN LIGHT RAT AWAY ◇ WARFARINE (FRENCH) ◇ ZOOCOUMARIN (RUSSIAN)

TOXICITY DATA with REFERENCE
orl-wmn TDLo:33600 μg/kg (1-32W preg):REP AJOGAH 127,191,77
ims-wmn TDLo:12 mg/kg (27-31W preg):TER JMSHAO 26,562,59
orl-man TDLo:10.2 mg/kg:BLD CMEP** -,1,56
orl-wmn TDLo:15 mg/kg/21W-I:GIT SMJOAV 75,242,82
orl-hmn LDLo:6667 μg/kg YKYUA6 28,329,77
orl-rat LD50:1600 μg/kg TXAPA9 11,327,67
ihl-rat LD50:320 mg/m^3 GTPZAB 22(7),49,78
skn-rat LD50:1400 mg/kg GTPZAB 22(7),49,78
ipr-rat LDLo:420 mg/kg TXAPA9 1,156,59
orl-mus LD50:60 mg/kg YKYUA6 28,329,77
scu-mus LDLo:800 mg/kg TIVSAI 58,122,77
ivn-mus LD50:165 mg/kg 27ZIAQ -,274,73

CONSENSUS REPORTS: Reported in EPA TSCA Inventory. EPA Extremely Hazardous Substances List.

OSHA PEL: TWA 0.1 mg/m^3
ACGIH TLV: TWA 0.1 mg/m^3
DFG MAK: 0.5 mg/m^3

SAFETY PROFILE: A human poison by ingestion. Poison by ingestion, inhalation, and intravenous routes. Moderately toxic by skin contact, subcutaneous, and intraperitoneal routes. Human systemic effects by ingestion: hemorrhage, ulceration or bleeding from small intestine, blood clotting factor change. Human reproductive effects by ingestion, and intramuscular routes: fetal death and physical abnormalities at birth. Human teratogenic effects include developmental abnormalities of the craniofacial area, musculoskeletal system, and respiratory system. An experimental teratogen. Other experimental reproductive effects. Used as an oral anticoagulant and as a rodenticide. When heated to decomposition it emits acrid smoke and fumes. See also COUMADIN SODIUM.

WAT209 CAS:2610-86-8 ***HR: 3***
WARFARIN POTASSIUM
mf: $C_{19}H_{15}O_4$•K mw: 346.44

SYNS: ANTROMBIN K ◇ 4-HYDROXY-3-(3-OXO-1-PHENYLBUTYL)-2H-1-BENZOPYRAN-2-ONE POTASSIUM SALT ◇ POTASSIUM WARFARIN ◇ WARFARIN K

TOXICITY DATA with REFERENCE
orl-rat LD50:58 mg/kg NIIRDN 6,918,82
ivn-rat LD50:186 mg/kg NIIRDN 6,918,82
orl-mus LD50:760 mg/kg NIIRDN 6,918,82
ivn-mus LD50:232 mg/kg NIIRDN 6,918,82
orl-dog LD50:200 mg/kg NIIRDN 6,918,82
ivn-dog LD50:200 mg/kg NIIRDN 6,918,82
orl-rbt LD50:800 mg/kg NIIRDN 6,918,82
ivn-rbt LD50:100 mg/kg NIIRDN 6,918,82

CONSENSUS REPORTS: Reported in EPA TSCA Inventory.

SAFETY PROFILE: Poison by ingestion and intravenous routes. When heated to decomposition it emits toxic fumes of K_2O. See also COUMADIN.

WAT220 CAS:129-06-6 ***HR: 3***
WARFARIN SODIUM
mf: $C_{19}H_{15}O_4$•Na mw: 330.33

SYNS: 3-(α-ACETONYLBENZYL)-4-HYDROXY-COUMARINSODIUM SALT ◇ ATHROMBIN ◇ COUMADIN SODIUM ◇ 4-HYDROXY-3-(3-OXO-1-PHENYLBUTYL)-2H-1-BENZOPYRAN-2-ONESODIUMSALT (9CI) ◇ MAREVAN (SODIUM SALT) ◇ PANWARFIN ◇ PROTHROMBIN ◇ RATSUL SOLUBLE ◇ SODIUM COUMADIN ◇ SODIUM WARFARIN ◇ TINTORANE ◇ VARFINE ◇ WARAN ◇ WARCOUMIN ◇ WARFILONE

TOXICITY DATA with REFERENCE
dni-mus :leu 100 nmol/L ONCOBS 28,232,73
oms-mus:leu 100 nmol/L ONCOBS 28,232,73
unr-wmn TDLo:12 mg/kg (female 1-35W post):REP AJDCAI 129,360,75
orl-wmn TDLo:37 mg/kg (1-35W preg):TER AJDCAI 129,356,75
orl-wmn TDLo:300 μg/kg/2D:SKN ARDEAC 116,444,80
orl-rat LD50:8700 μg/kg PLRCAT 10,445,78
ivn-rat LD50:25 mg/kg 29ZVAB -,123,69
orl-mus LD50:374 mg/kg JAPMA8 42,379,53
ivn-mus LD50:160 mg/kg 29ZVAB -,123,69

CONSENSUS REPORTS: Reported in EPA TSCA Inventory. EPA Extremely Hazardous Substances List.

SAFETY PROFILE: Poison to humans by ingestion. Experimental poison by ingestion and intravenous routes. Human systemic effects by ingestion: dermatitis. Human reproductive effects by ingestion: fetotoxicity, abnormal condition of newborn at birth, other newborn physical effects, and teratogenic effects including developmental abnormalities of the eye and ear, craniofacial area, skin and appendages, musculoskeletal system, cardiovascular system, and gastrointestinal system of the fetus. An experimental teratogen. Other experimental reproductive effects. Mutation data reported. An anticoagulant drug. When heated to decomposition it emits toxic fumes of Na_2O. See also COUMARIN.

WAT259 CAS:7732-18-5 ***HR: 1***
WATER
mf: H_2O mw: 18.02

PROP: Odorless, colorless, tasteless liquid. Allotropic forms are ice (solid) and steam (vapor). D: (at atmospheric pressure) 1.00 (4°C), fp: 0°C (32°F) with 10% expansion, viscosity: 0.01002 poise (20°C), sp heat: 1 calorie/g, vap press: 760 mm (100°C), surface tension: 73 dynes/cm @ 20°C, latent heat of fusion (ice): 80 cal/g, latent heat of condensation (steam): 540 cal/g, bulk d: 8.337 lbs/gal (62.3/lb/cu ft), refr index: 1.333. Water is a polar liquid with high dielectric constant (81 @ 17°C) which largely accounts for its solvent power.

SYN: DIHYDROGEN OXIDE

TOXICITY DATA with REFERENCE
orl-man TDLo:42.86 g/kg:CNS JPETAB 29,135,26
orl-inf TDLo:333 g/kg:CNS,GIT,MET ADCHAK 54,551,79
rec-wmn LDLo:180 g/kg/28H JAMAAP 104,1569,35
ipr-mus LD50:190 g/kg NTIS** QD628-313
ivn-mus LD50:25 g/kg MIVRA6 8,320,74
orl-dog LDLo:629 g/kg JPETAB 29,135,26
orl-cat LDLo:320 g/kg JPETAB 29,135,26
orl-rbt LDLo:368 g/kg JPETAB 29,135,26
ivn-rbt LDLo:13 g/kg JPETAB 29,135,26
rec-rbt LDLo:450 g/kg JAMAAP 104,1569,35
orl-gpg LDLo:429 g/kg JPETAB 29,135,26
ipr-mus LD50:25 g/kg MIVRA6 8,320,74

CONSENSUS REPORTS: EPA Genetic Toxicology Program. Reported in EPA TSCA Inventory.

SAFETY PROFILE: Human systemic effects by ingestion of very large amounts: convulsions, tremors, muscle contractions or spasticity; hypermotility, diarrhea; fever. Human and experimental death reported by various routes at sufficiently large doses.

WAT300 ***HR: 2***
WATER ARUM

PROP: A small plant with 4- to 6-inch heart-shaped leaves on 10-inch stems. It forms thick clusters of red berries. It is found in marshy areas from Alaska to Florida.

SYNS: CALLA PALUSTRIS ◇ FEMALE WATER DRAGON ◇ WATER DRAGON ◇ WILD CALLA

SAFETY PROFILE: The whole plant contains toxic calcium oxalate raphides. Chewing any part of the plant results in burning pain in the lips, mouth, and throat, possibly followed by inflammation and blistering. Systemic effects are usually not seen because of the insolubility of calcium oxalate, however, ingestion may cause inflammation of the stomach and intestines. See also OXALATES.

WAT315 ***HR: 3***
WATER DROPWART

PROP: A perennial which grows to 5 feet. The roots are finger-shaped and contain a white sap which turns orange on exposure to air. The hollow stems bear compound leaves and ball-shaped clusters of white flowers. It is native to Europe and is now found in marshy areas around Washington D.C.

SYNS: DEAD MEN'S FINGERS ◇ HEMLOCK WATER DROPWORT ◇ OENANTHE CROCATA

SAFETY PROFILE: The whole plant and especially the roots contain the poison oenanthotoxin, an unsaturated aliphatic compound. It is chemically related to cicutoxin, found in the water hemlock. Ingestion of any part of the plant may cause salivation, vomiting, and convulsions within minutes.

WAT325 ***HR: 3***
WATER HEMLOCK

PROP: The various species of Cicuta may grow only to 6 feet with compound leaves, with small, whitish, strongly scented flowers, tuberous roots, and an oily yellow sap that smells of parsnip. They are found throughout North America but only in wet, marshy soil.

SYNS: BEAVER POISON ◇ CHILDREN'S BANE ◇ CICUTA BULBIFERA L. ◇ CICUTA DOUGLASII ◇ CICUTA MACULATA ◇ CICUTAIRE (CANADA) ◇ DEATH-OF-MAN ◇ MUSQUASH POISON ◇ MUSQUASH ROOT ◇ SPOTTED COWBANE

SAFETY PROFILE: The whole plant contains the poison cicutoxin. Human systemic effects by ingestion occur within 1 hour and include: nausea, salivation, vomiting, muscle spasms in the jaw, convulsions, and death. Survivors may experience prolonged mental deficits and abnormal electroencephalograms. See also CICUTOXIN.

WAT350 ***HR: D***
WATER-PEPPER HERB

SYN: POLYGONUM HYDROPIPER L., dry powdered whole plant

TOXICITY DATA with REFERENCE
orl-mus TDLo:4200 g/kg (female 15W pre):REP JOENAK 12,252,55

SAFETY PROFILE: Experimental reproductive effects.

WBA000 CAS:84929-34-0 ***HR: 3***
WAX MYRTLE

PROP: Tannin containing fraction of bark used (JNCIAM 57,207,76).

SYNS: MYRICA CERIFERA ◇ SOUTHERN BAYBERRY ◇ SWEET MYRTLE ◇ TANNIN from WAX MYRTLE

TOXICITY DATA with REFERENCE
scu-rat TDLo:560 mg/kg/69W-I:NEO JNCIAM 57,207,76

SAFETY PROFILE: Questionable carcinogen with experimental neoplastigenic data. See also TANNIC ACID.

WBA600 CAS:49561-54-8 ***HR: 2***
WD 67/2
mf: $C_{19}H_{21}N_3O\cdot ClH$ mw: 343.89

SYN: 4-(2-(α-METHYLPHENETHYLAMINO)ETHYL)-3-PHENYL-1,2,4-OXADIAZOLE MONOHYDROCHLORIDE

TOXICITY DATA with REFERENCE
orl-rat LD50:649 mg/kg BCFAAI 112,273,73
orl-mus LD50:1455 mg/kg BCFAAI 112,273,73
scu-mus LD50:3315 mg/kg BCFAAI 112,273,73

SAFETY PROFILE: Moderately toxic by ingestion and subcutaneous routes. When heated to decomposition it emits toxic fumes of NO_x and HCl.

WBJ000 ***HR: 3***
WELDING FUMES
ACGIH TLV: TWA 5 mg/m^3

SAFETY PROFILE: When welding is done on a surface coated with cadmium, toxic fumes of cadmium are evolved. When zinc-coated surfaces are welded, toxic quantities of zinc oxide may be liberated. When painted surfaces are welded, lead or other pigment fumes may be liberated. And when fluoride fluxes are used in welding, very toxic fluoride fumes are evolved. When oily surfaces are welded, offensive and toxic fumes can be liberated, and when the welding torch is improperly ignited, carbon monoxide, which is very toxic, may be evolved. Also, NO_x is formed. It is therefore considered hazardous to inhale excessive amounts of welding fumes. It is also possible to inhale sufficient quantities of iron oxide from welding to cause siderosis. Metal fume fever is a common reaction. It is characterized by chills, fever, sweating, and leukocytosis coming on several hours after exposure. Recovery is usually complete in 24-48 hours and there are no significant after effects. Safety goggles are required to protect against spatter. Light-filtering goggles are required to shield the eyes against the intense UV light from the arc. See also specific metals or their compounds (e.g., CADMIUM and CADMIUM COMPOUNDS).

WBJ500 CAS:31677-93-7 ***HR: 3***
WELLBATRIN
mf: $C_{13}H_{18}ClNO\cdot ClH$ mw: 276.23

SYNS: BUPROPION HYDROCHLORIDE ◇ (±)-α-tert-BUTYLAMINO-3-CHLOROPROPIOPHENONE HYDROCHLORIDE ◇ (+)-1-(3-CHLOROPHENYL)-2-((1,1-DIMETHYLETHYL)AMINO)1-PROPANONEHYDROCHLORIDE (9CI)

TOXICITY DATA with REFERENCE
orl-rat LD50:600 mg/kg JPPMAB 29,767,77
ipr-rat LD50:210 mg/kg JPPMAB 29,767,77
orl-mus LD50:575 mg/kg JPPMAB 29,767,77
ipr-mus LD50:230 mg/kg JPPMAB 29,767,77

SAFETY PROFILE: Poison by intraperitoneal route. Moderately toxic by ingestion. When heated to decomposition it emits toxic fumes of NO_x and Cl^-.

WBJ600 ***HR: 3***
WESTERN DIAMONDBACK RATTLESNAKE VENOM

SYNS: C. ATROX VENOM ◇ CROTALUS ATROX VENOM ◇ VENOM, SNAKE, CROTALUS ATROX

TOXICITY DATA with REFERENCE
ipr-rat LD50:172 mg/kg TOXIA6 17,601,79
ipr-mus LD50:3710 μg/kg 14FHAR -,409,63
scu-mus LD50:7800 μg/kg TOXIA6 24,71,86
ivn-mus LD50:2666 μg/kg TOXIA6 9,131,71
ims-mus LD50:19040 μg/kg AJMSA9 239,1,60
ivn-dog LDLo:500 μg/kg 19DDA6 1,269,67
par-dog LDLo:12 mg/kg 14FHAR -,399,63
ivn-mky LDLo:2500 μg/kg TOXIA6 8,33,70
ims-mky LDLo:6 mg/kg TOXIA6 8,33,70

SAFETY PROFILE: Poison by subcutaneous, intramuscular, parenteral, intravenous, and intraperitoneal routes.

WBJ700 CAS:68917-73-7 ***HR: 1***
WHEAT GERM OIL

PROP: Bland yellow oil. Misc with chloroform, ether, petr ether, and benzene, sltly sol in alc.

SYNS: BRAN ABSOLUTE ◇ CAV-ECOL ◇ MERIT ◇ MYOPONE ◇ OILS, WHEAT GERM ◇ UNIDERM WGO ◇ WHEAT HUSK OIL

TOXICITY DATA with REFERENCE
skn-rbt 500 mg MLD JEPTDQ 4(4),33,80
eye-rbt 100 mg MLD JEPTDQ 4(4),33,80

CONSENSUS REPORTS: Reported in EPA TSCA Inventory.

SAFETY PROFILE: A skin and eye irritant. When heated to decomposition it emits acrid smoke and irritating fumes.

WBS000 ***HR: 3***
WHISKEY

PROP: Light yellow-amber liquid. Pleasant to fruity odor. D: 0.923-0.935 @ 15.56°; 47%-53% of ethanol, by volume, flash p: 80.0°F (CC). Made by distillation of fermented malted grains, i.e., corn, rye, or barley. After distillation, whiskey is aged in wooden containers for up to several years. The aging extracts such components as acids and esters from the wood and promotes oxidation of components of raw whiskey and some reactions between organic components to form new flavors.

SAFETY PROFILE: The carcinogen urethane is sometimes found in whiskey. The whiskey equivalent of 1 ounce of pure ethanol per capita per day has been cited as healthful to adults to relieve stress and promote relaxation. However, it is often abused which can lead to habituation with consequent liver damage, malnutrition, and a wide variety of other physical and mental problems, including the development of cancer. A fire hazard when exposed to heat or flame. To fight fire, use water, water spray, alcohol foam, CO_2, dry chemical. See also ETHANOL and URETHANE.

WBS675 CAS:63394-00-3 ***HR: 1***
WHITE SPIRIT

SYN: SKDN

TOXICITY DATA with REFERENCE
ihl-rat TCLo:950 ppm/6H (female 3-20D post):TER TJADAB 34,415,86
ihl-hmn TCLo:600 mg/m^3/8H:EYE,PUL,GIT TPKVAL 10,116,68
ihl-mus LCLo:50000 mg/m^3 TPKVAL 10,116,68

SAFETY PROFILE: Slightly toxic by inhalation. An experimental teratogen. Human systemic effects by inhalation: conjunctive eye irritation, cough, and gastrointestinal changes. When heated to decomposition it emits acrid smoke and irritating fumes.

WBS700 ***HR: 1***
WHITE SPIRIT, DILUTINE 5

TOXICITY DATA with REFERENCE
skn-rbt 500 mg MLD FCTOD7 20,563,82
eye-rbt 100 mg MLD FCTOD7 20,573,82
eye-rbt 100 mg/30S rns MLD FCTOD7 20,573,82

SAFETY PROFILE: A skin and eye irritant. When heated to decomposition it emits acrid smoke and irritating fumes.

WBS850 ***HR: 1***
WILD ONION

PROP: A common field weed found throughout the United States and much of Canada. It has an onion or garlic odor when bruised.

SYNS: AIL du CANADA (CANADA) ◇ AJO ◇ 'AKA'AKAI (HAWAII) ◇ 'AKA'AKAI-PILAU (HAWAII) ◇ ALLIUM (VARIOUS SPECIES) ◇ CEBOLLA ◇ CLOWN TREACLE ◇ COW GARLIC ◇ FIELD GARLIC ◇ GARLIC ◇ LAI (HAITI) ◇ MEADOW GARLIC ◇ MEADOW ROSE LEEK ◇ ONION ◇ ONION TREE ◇ POOR MAN'S TREACLE ◇ WILD GARLIC ◇ ZONGNON (HAITI)

SAFETY PROFILE: The bulbs, flowers, and stems contain N-propyl sulfide, methyl disulfide, and allyl disulfide. Ingestion causes gastroenteritis and is common in children. Chronic ingestion reduces iodine uptake.

WBS855 CAS:102612-94-2 ***HR: 3***
WIN 2661
mf: $C_{17}H_{24}N_2O_3S•ClH$ mw: 372.95

SYN: 3-(3-(ISOBUTYLAMINO)PROPYL)-2-(3,4-METHYLENEDIOXYPHENYL)-4-THIAZOLIDINONEHYDROCHLORIDE

TOXICITY DATA with REFERENCE
eye-rbt 2% MLD JAPMA8 40,132,51
scu-mus LD50:320 mg/kg JAPMA8 40,132,51
ivn-mus LD50:40 mg/kg JPETAB 123,269,58
ivn-gpg LD50:20 mg/kg JAPMA8 40,132,51

SAFETY PROFILE: Poison by subcutaneous and intravenous routes. An eye irritant. When heated to decomposition it emits toxic fumes of SO_x, NO_x, and HCl.

WBS860 CAS:102612-93-1 ***HR: 3***
WIN 2663
mf: $C_{18}H_{24}N_2O_3S•ClH$ mw: 384.96

SYN: 3-(2-(CYCLOHEXYLAMINO)ETHYL)-2-(3,4-METHYLENEDIOXYPHENYL)-4-THIAZOLIDINONEHYDROCHLORIDE

TOXICITY DATA with REFERENCE
eye-rbt 2% MLD JAPMA8 40,132,51
eye-rbt 4% MOD JAPMA8 40,132,51
scu-mus LD50:190 mg/kg JAPMA8 40,132,51
ivn-mus LD50:30 mg/kg JPETAB 123,269,58
ivn-gpg LD50:18 mg/kg JAPMA8 40,132,51

SAFETY PROFILE: Poison by subcutaneous and intravenous routes. An eye irritant. When heated to decomposition it emits toxic fumes of SO_x, NO_x, and HCl.

WCA000 ***HR: 2***
WINE

PROP: An alcoholic beverage made from the fermented juice of grapes, other fruits or plants. Contains from 7-20% ethanol by volume. Concentrations of alcohol higher than those produced naturally are obtained by fortifying with pure ethanol. The distinctive colors, tastes, bouquets of wines are usually produced by adding coloring matter, sugar, acetic acid, salts, and higher fatty acids.

SAFETY PROFILE: Some wines contain the carcinogen urethane. The wine equivalent of 1 ounce of pure ethanol per capita per day has been cited as healthful to adults to relieve stress and promote relaxation. However, it is often abused which can lead to habituation with consequent liver damage, malnutrition, and a wide variety of other physical and mental problems, including the development of cancer. Some of the additives to wines have been known to cause allergic reactions in humans. See also ETHANOL and URETHANE.

WCA450 ***HR: 2***
WISTERIA

PROP: Woody vines which produce large clusters of blue, pink, or white flowers. The seed pods stay on the plant throughout the winter. They grow most commonly from the southeastern United States to Texas, but may be found in the northern United States.

SYNS: KIDNEY BEAN TREE ◇ WISTARIA ◇ WISTERIA FLORIBUNDA ◇ WISTERIA SINENSIS

SAFETY PROFILE: All parts of the plant contain the glycoside wistarine and an unidentified lectin. Ingestion of any of these plant parts and particularly the bark, can cause nausea, abdominal pain, and persistent vomiting. Ingestion of large amounts may cause shock due to fluid loss.

WCB000 CAS:68916-39-2 ***HR: 3***
WITCH HAZEL

SYNS: HAMAMELIS ◇ NCI-C50544 ◇ SNAPPING HAZEL ◇ SPOTTED ALDER ◇ STRIPED ALDER ◇ TOBACCO WOOD ◇ WINTER BLOOM

TOXICITY DATA with REFERENCE
mmo-sat 5 mg/plate ENMUDM 8(Suppl 7),1,86
mma-sat 100 μg/plate ENMUDM 8(Suppl 7),1,86
scu-rat TDLo:2920 mg/kg/73W-I:ETA JNCIAM 60,683,78

CONSENSUS REPORTS: Reported in EPA TSCA Inventory.

SAFETY PROFILE: Questionable carcinogen with experimental tumorigenic data. Mutation data reported. A mild irritant. Combustible when exposed to heat or flame; can react with oxidizing materials. Used as an ingredient in cosmetics. When heated to decomposition it emits acrid smoke and fumes.

WCB100 CAS:56749-17-8 ***HR: 2***
WITEPSOL E-75

TOXICITY DATA with REFERENCE
skn-rat 100 mg/24H MLD CTOIDG 94(8),41,79
skn-rbt 100 mg/24H SEV CTOIDG 94(8),41,79
skn-gpg 100 mg/24H MOD CTOIDG 94(8),41,79

SAFETY PROFILE: A severe skin irritant.

WCJ000 CAS:13983-17-0 ***HR: 3***
WOLLASTONITE
mf: $O_3Si{\cdot}Ca$ mw: 116.17

PROP: A calcium silicate mineral.

SYNS: CAB-O-LITE 100 ◇ CAB-O-LITE 130 ◇ CAB-O-LITE 160 ◇ CAB-O-LITE F 1 ◇ CAB-O-LITE P 4 ◇ CASIFLUX VP 413-004 ◇ DAB-O-LITE P 4 ◇ F 1 ◇ FW 50 ◇ FW 200 (mineral) ◇ NCI-C55470 ◇ NYAD 10 ◇ NYAD 325 ◇ NYA G ◇ NYCOR 200 ◇ NYCOR 300 ◇ VANSIL W 10 ◇ VANSIL W 20 ◇ VANSIL W 30 ◇ WOLLASTOKUP

TOXICITY DATA with REFERENCE
imp-rat TDLo:200 mg/kg:ETA JJIND8 67,965,81

CONSENSUS REPORTS: IARC Cancer Review: Group 3 IMEMDT 7,377,87; Animal Limited Evidence IMEMDT 42,145,87; Human Inadequate Evidence IMEMDT 42,145,87.

SAFETY PROFILE: Questionable carcinogen with experimental tumorigenic data. When heated to decomposition it emits acrid smoke and irritating fumes.

WCJ750 CAS:21062-28-2 ***HR: 3***
WR 81844
mf: $C_{19}H_{25}Cl_2N_7$ mw: 422.41

SYN: N-(3,4-DICHLOROPHENYL)-N'-(4-((1-ETHYL-3-PIPERIDYL)AMINO)-6-METHYL-2-PYRIMIDINYL)GUANIDINE

TOXICITY DATA with REFERENCE
orl-rat LD50:1041 mg/kg JMCMAR 17,75,74
ipr-rat LD50:65 mg/kg JMCMAR 17,75,74
orl-mus LD50:1128 mg/kg JMCMAR 17,75,74
ipr-mus LD50:53 mg/kg JMCMAR 17,75,74
scu-mus LDLo:640 mg/kg JMCMAR 17,75,74
orl-gpg LD50:261 mg/kg JMCMAR 17,75,74
ipr-gpg LD50:28 mg/kg JMCMAR 17,75,74

SAFETY PROFILE: Poison by ingestion and intraperitoneal routes. Moderately toxic by subcutaneous route. When heated to decomposition it emits toxic fumes of Cl^- and NO_x.

X

XAH000 **HR: 3**
XAMOTEROLFUMARATE
mf: $C_{16}H_{25}N_3O_5 \cdot 1/2C_4H_4O_4$ mw: 397.36

SYNS: CORWIN ◇ (±)-N-(2-(HYDROXY-3-(4-HYDROXYPHENOXY) PROPYLAMINO)ETHYL)MORPHOLINE-4-CARBOXAMIDE FUMARATE ◇ ICI 118587 ◇ 4-MORPHOLINECARBOXAMIDE, N-(2-((2-HYDROXY-3-(4-HYDROXYPHENOXY)PROPYL)AMINO)ETHYL)-,(E)-2-BUTENEDIOATE (2:1) (salt)

TOXICITY DATA with REFERENCE
orl-rat TDLo:7 g/kg (male 10W pre):REP YACHDS 16,1157,88
orl-rbt TDLo:650 mg/kg (female 6-18D post):TER YACHDS 16,1157,88
ivn-rat LD50:172 mg/kg YACHDS 16,1157,88
ivn-mus LD50:240 mg/kg YACHDS 16,1157,88

SAFETY PROFILE: Poison by intravenous route. An experimental teratogen. Experimental reproductive effects. When heated to decomposition it emits toxic fumes of NO_x.

XAJ000 CAS:28981-97-7 **HR: 3**
XANAX
mf: $C_{17}H_{13}ClN_4$ mw: 308.79

PROP: Crystals from ethyl acetate. Mp: 228-228.5°. Sol in alc; insol in water.

SYNS: ALPRAZOLAM ◇ 8-CHLORO-1-METHYL-6-PHENYL-4H-s-TRIAZOLO(4,3-a)(1,4)BENZODIAZEPINE ◇ D 65MT ◇ TAFIL ◇ TUS-1 ◇ U 31889

TOXICITY DATA with REFERENCE
orl-rbt TDLo:39 mg/kg (6-18D preg):REP JZKEDZ 7,79,81
orl-rat TDLo:550 mg/kg (7-17D preg):TER JZKEDZ 7,65,81
orl-man TDLo:160 μg/kg/4D-I AJPSAO 142,859,85
orl-wmn TDLo:20 μg/kg/1D-I AJPSAO 141,1127,84
orl-rat LD50:1220 mg/kg YACHDS 8,4695,80
ipr-rat LD50:355 mg/kg YACHDS 8,4695,80
orl-mus LD50:812 mg/kg YHTPAD 21,538,86
ipr-mus LD50:380 mg/kg YACHDS 8,4687,80

SAFETY PROFILE: Poison by intraperitoneal route. Moderately toxic by ingestion. An experimental teratogen. Experimental reproductive effects. *Caution:* Abuse leads to habituation or addiction. When heated to decomposition it emits toxic fumes of Cl^- and NO_x. See also DIAZEPAM.

XAK000 CAS:86-40-8 **HR: 3**
XANTHACRIDINE
mf: $C_{14}H_{14}N_3 \cdot Cl$ mw: 259.76

SYNS: ACRIFLAVINE ◇ ACRIFLAVINE NEUTRAL ◇ ACRIFLAVON ◇ AF ◇ AVLON ◇ BURNOL ◇ CHROMOFLAVINE ◇ C.I. 46000 ◇ 2,8-DIAMINO-10-METHYLACRIDINIUM CHLORIDE ◇ 3,6-DIAMINO-10-METHYLACRIDINIUM CHLORIDE ◇ EUFLAVINE ◇ FLAVIN ◇ FLAVINE ◇ FLAVOSAN ◇ GONACRINE ◇ GONOCRIN ◇ NEUTRAL ACRIFLAVINE ◇ NEUTROFLAVINE ◇ PANFLAVIN ◇ TRYPAFLAVIN

TOXICITY DATA with REFERENCE
mma-esc 10 mg/L MUREAV 140,13,84
sln-dmg-orl 10 pph MUREAV 121,199,83
dni-hmn:hla 10 μmol/L RAREAE 37,334,69
ivn-hmn LDLo:1500 μg/kg JPETAB 38,145,30
ipr-mus LDLo:11 mg/kg TXAPA9 23,388,72
scu-mus LD50:14 mg/kg BJEPA5 28,1,47
ivn-mus LD50:40 mg/kg BJEPA5 28,1,47
ivn-dog LDLo:25 mg/kg JPETAB 38,145,30
ivn-cat LDLo:7353 μg/kg LANCAO 196,838,19
ivn-rbt LDLo:25 mg/kg JPETAB 38,145,30
ipr-gpg LDLo:250 mg/kg HBAMAK 4,1292,35
scu-gpg LDLo:250 mg/kg HBAMAK 4,1292,35
ivn-gpg LDLo:40 mg/kg HBAMAK 4,1292,35
scu-frg LDLo:800 mg/kg ZGEMAZ 12,195,21
ipr-mam LDLo:65 mg/kg JPETAB 80,217,44

CONSENSUS REPORTS: Reported in EPA TSCA Inventory.

SAFETY PROFILE: A human poison by intravenous route. Poison experimentally by intraperitoneal, intravenous, and subcutaneous routes. Human mutation data reported. When heated to decomposition it emits very toxic fumes of NO_x and Cl^-.

XAT000 CAS:92-83-1 **HR: 2**
XANTHENE
mf: $C_{13}H_{10}O$ mw: 182.23

PROP: Mp: 100.5°; bp: 315°. Sltly sol in water; sol in alc.

SYNS: 10H-9-OXAANTHRACENE ◇ 9H-XANTHENE

TOXICITY DATA with REFERENCE
scu-mus LD50:690 mg/kg ARZNAD 8,107,58

CONSENSUS REPORTS: Reported in EPA TSCA Inventory.

SAFETY PROFILE: Moderately toxic by subcutaneous

route. When heated to decomposition it emits acrid smoke and irritating fumes.

XBA000 CAS:37971-99-6 ***HR: 3***
1,1'-(9H-XANTHENE-2,7-DIYL)BIS(2-(DIMETHYLAMINO)ETHANONE - DIHYDROCHLORIDE SESQUIHYDRATE
mf: $C_{21}H_{24}N_2O_3{\cdot}2ClH{\cdot}3/2H_2O$ mw: 452.39

SYN: RMI 11513 DA

TOXICITY DATA with REFERENCE
orl-mus LD50:1410 mg/kg ALACBI 12,77,79
scu-mus LD50:304 mg/kg ALACBI 12,77,79

SAFETY PROFILE: Poison by subcutaneous route. Moderately toxic by ingestion. When heated to decomposition it emits very toxic fumes of NO_x and HCl.

XBJ000 CAS:90-46-0 ***HR: 2***
XANTHEN-9-OL
mf: $C_{13}H_{10}O_2$ mw: 198.23

PROP: Colorless liquid. D: 0.880 @ 20/4°, mp: −25.2°, bp: 144.4°. Insol in water; misc in abs alc, ether.

SYNS: 9-HYDROXYXANTHENE ◇ XANTHANOL ◇ XANTHYDROL

TOXICITY DATA with REFERENCE
mmo-sat 100 μg/plate MUREAV 150,141,85
mma-sat 100 μg/plate MUREAV 150,141,85
orl-rat LDLo:500 mg/kg NCNSA6 5,26,53

CONSENSUS REPORTS: Reported in EPA TSCA Inventory.

SAFETY PROFILE: Moderately toxic by ingestion. Mutation data reported. When heated to decomposition it emits acrid smoke and irritating fumes. See also ALCOHOLS.

XBS000 CAS:90-47-1 ***HR: 3***
9-XANTHENONE
mf: $C_{13}H_8O_2$ mw: 196.21

PROP: Needles in alc. Mp: 173-174°, bp: 349-350° @ 730 mm. Sltly sol in hot water, ether, ligroin; sol in chloroform, alc, benzene.

TOXICITY DATA with REFERENCE
ivn-mus LD50:180 mg/kg CSLNX* NX#01611

CONSENSUS REPORTS: Reported in EPA TSCA Inventory.

SAFETY PROFILE: Poison by intravenous route. When heated to decomposition it emits acrid smoke and irritating fumes. See also KETONES.

XCA000 CAS:69-89-6 ***HR: 3***
XANTHINE
mf: $C_5H_4N_4O_2$ mw: 152.13

PROP: Yellow or white powder, scales or plates. Decomp on heating without melting, partial subl. Sol in water and mineral acids; less sol in alc; very sol in NH_4OH and NaOH solns.

SYNS: 3,7-DIHYDRO-1H-PURINE-2,6-DIONE ◇ 2,6-DIOXOPURINE ◇ ISOXANTHINE ◇ PSEUDOXANTHINE ◇ PURINE-2,6-DIOL ◇ 9H-PURINE-2,6-DIOL ◇ 2,6(1,3)-PURINEDION ◇ PURINE-2,6-(1H,3H)-DIONE ◇ USAF CB-17 ◇ XAN ◇ XANTHIC OXIDE

TOXICITY DATA with REFERENCE
scu-rat TDLo:3600 mg/kg/18W-I:NEO JNCIAM 24,109,60
ipr-mus LD50:500 mg/kg NTIS** AD277-689

CONSENSUS REPORTS: Reported in EPA TSCA Inventory. EPA Genetic Toxicology Program.

SAFETY PROFILE: Moderately toxic by intraperitoneal route. Questionable carcinogen with experimental neoplastigenic data. When heated to decomposition it emits toxic fumes of NO_x.

XCJ000 CAS:53-46-3 ***HR: 3***
XANTHINE BROMIDE
mf: $C_{21}H_{26}NO_3{\cdot}Br$ mw: 420.39

SYNS: ASABAINE ◇ AVAGAL ◇ BANTHIN ◇ BANTHINE ◇ BANTHINE BROMIDE ◇ β-DIETHYLAMINOETHYL XANTHENE-9-CARBOXYLATE METHOBROMIDE ◇ β-DIETHYLAMINOETHYL 9-XANTHENECARBOXYLATE METHOBROMIDE ◇ DIETHYL(2-HYDROXYETHYL) METHYLAMMONIUMBROMIDE XANTHENE-9-CARBOXYLATE ◇ DOLADENE ◇ ETHANAMINIUM, N,N-DIETHYL-N-METHYL-2-((9H-XANTHEN-9-YLCARBONYL)OXY)-, BROMIDE (9CI) ◇ FRENOGASTRICO ◇ GASTRON ◇ GASTROSEDAN ◇ MANTHELINE ◇ METANTYL ◇ METAXAN ◇ METHANIDE ◇ METHANTHELINE BROMIDE ◇ METHANTHELINIUM BROMIDE ◇ METHANTHINE BROMIDE ◇ METHELINA ◇ MTB 51 ◇ RESOBANTIN ◇ SC 2910 ◇ ULCINE ◇ ULCUDEXTER ◇ VAGAMIN ◇ VAGANTIN ◇ XANTELINE ◇ XANTHENE-9- CARBOXYLIC ACID, ESTER with DIETHYL(2-HYDROXYETHYL) METHYLAMMONIUM BROMIDE

TOXICITY DATA with REFERENCE
orl-rat LD50:1660 mg/kg NIIRDN 6,357,82
orl-mus LD50:460 mg/kg NIIRDN 6,357,82
ipr-mus LD50:46 mg/kg JPETAB 106,141,52
scu-mus LDLo:600 mg/kg ARZNAD 8,107,58
ivn-mus LD50:4300 μg/kg AIPTAK 105,221,56
ivn-dog LDLo:23 mg/kg PSEBAA 78,576,51

SAFETY PROFILE: Poison by intraperitoneal and intravenous routes. Moderately toxic by ingestion and subcutaneous routes. Unspecified human reproductive effects. When heated to decomposition it emits very toxic fumes of NO_x, NH_3, and Br^-.

XCS000 CAS:437-74-1 ***HR: 2***
XANTHINOL NICOTINATE
mf: $C_{13}H_{21}N_5O_4{\cdot}C_6H_5NO_2$ mw: 434.51

PROP: Crystals. Mp: 180°. Freely sol in water.

SYNS: ANGIOMIN ◇ COMPLAMEX ◇ COMPLAMIN ◇ 7-(3-(N-(2-HYDROXYETHYL)AMINO)-2-HYDROXYPROPYL)THIOPHYLLINE NICOTINATE ◇ 7-(2-HYDROXY-3-((2-HYDROXYETHYL)METHYL-AMINO)PROPYL)THEOPHYLLINE, compound with NICOTINIC ACID ◇ SADAMIN ◇ SK 331 A ◇ XANTHINOL NIACINATE ◇ XAVIN ◇ XN

TOXICITY DATA with REFERENCE
orl-rat TDLo:1920 mg/kg (7-14D preg):REP OYYAA2 8,1145,74
orl-rat TDLo:1920 mg/kg (7-14D preg):TER OYYAA2 8,1145,74
orl-rat LD50:14130 mg/kg OYYAA2 9,601,75
ipr-rat LD50:3028 mg/kg OYYAA2 9,601,75
scu-rat LD50:5255 mg/kg OYYAA2 9,601,75
ivn-rat LD50:690 mg/kg OYYAA2 9,601,75
orl-mus LD50:17350 mg/kg OYYAA2 9,601,75
scu-mus LD50:4260 mg/kg OYYAA2 9,601,75
ivn-rbt LD50:500 mg/kg OYYAA2 9,601,75

SAFETY PROFILE: Moderately toxic by intraperitoneal and intravenous routes. Mildly toxic by ingestion. An experimental teratogen. Experimental reproductive effects. When heated to decomposition it emits toxic fumes of NO_x. Used as a peripheral vasodilator. See also NICOTINIC ACID.

XCS400 ***HR: 3***
XANTHIUM CANADENSE, KERNAL EXTRACT

TOXICITY DATA with REFERENCE
ipr-mus LDLo:250 mg/kg JAPMA8 39,202,50
ipr-pig LDLo:175 mg/kg JAPMA8 39,202,50
ipr-gpg LDLo:150 mg/kg JAPMA8 39,202,50
ipr-ckn LDLo:375 mg/kg JAPMA8 39,202,50

SAFETY PROFILE: Poison by intraperitoneal route.

XCS680 CAS:38965-69-4 ***HR: 3***
XANTHOCILLIN Y 1
mf: $C_{18}H_{12}N_2O_3$ mw: 304.32

PROP: Clusters of yellow needles from alc; yellow rhombs from ethyl acetate. Chars at about 710°. Sol in alc, ether acetate, and dioxane; freely sol in alkaline aq solns; practically insol in water, petr ether, benzene, and chloroform.

SYN: 4-(4-(4-HYDROXYPHENYL)-2,3-DIISOCYANO-1,3-BUTADIENYL)-1,2-BENZENEDIOL

TOXICITY DATA with REFERENCE
orl-mus LD50:100 mg/kg 85GDA2 6,277,81
ipr-mus LD50:15 mg/kg 85GDA2 6,277,81
ivn-mus LD50:14 mg/kg 85GDA2 6,277,81

SAFETY PROFILE: Poison by ingestion, intravenous, and intraperitoneal routes. When heated to decomposition it emits toxic fumes of NO_x.

XCS700 CAS:38965-70-7 ***HR: 3***
XANTHOCILLIN Y 2
mf: $C_{18}H_{12}N_2O_4$ mw: 320.32

SYN: 4,4′-(2,3-DIISOCYANO-1,3-BUTADIENE-1,4-DIYL)BIS-1,2-BENZENEDIOL

TOXICITY DATA with REFERENCE
orl-mus LD50:100 mg/kg 85GDA2 6,277,81
ipr-mus LD50:20500 μg/kg 85GDA2 6,277,81
ivn-mus LD50:18 mg/kg 85GDA2 6,277,81

SAFETY PROFILE: Poison by ingestion, intravenous, and intraperitoneal routes. When heated to decomposition it emits toxic fumes of NO_x.

XCS800 ***HR: 2***
XANTHOSOMA (various species)

PROP: Variegated ornamentals with elongated leaves and thick tubers. They are cultivated in the southern United States, Hawaii, Guam, and the West Indies.

SYNS: BLUE 'APE (HAWAII) ◇ BLUE TARO (HAWAII) ◇ CARAIBE (HAITI) ◇ MALANGA (CUBA) ◇ YAUTIA (PUERTO RICO)

SAFETY PROFILE: The leaves contain poisonous crystals of calcium oxalate. Chewing these plant parts results in burning pain in the lips, mouth, and throat, possibly followed by inflammation and blistering. Systemic effects are usually not seen because of the insolubility of calcium oxalate. See also OXALATES.

XDJ000 CAS:298-81-7 ***HR: 3***
XANTHOTOXIN
mf: $C_{12}H_8O_4$ mw: 216.20

SYNS: AMMOIDIN ◇ 6-HYDROXY-7-METHOXY-5-BENZOFURAN-ACRYLIC ACID Δ-LACTONE ◇ MELADININ ◇ MELADININE ◇ MELOXINE ◇ METHOXA-DOME ◇ METHOXSALEN ◇ 8-METHOXY-(FURANO-3′.2′:6.7-COUMARIN) ◇ 9-METHOXY-7H-FURO(3,2-g)BENZO-PYRAN-7-ONE ◇ 8-METHOXY-2′,3′,6,7-FUROCOUMARIN ◇ 8-ME-THOXY-4′,5′,6,7-FUROCOUMARIN ◇ 8-METHOXYPSORALEN ◇ 9-METHOXYPSORALEN ◇ 8-MOP ◇ 8-MP ◇ NCI-C55903 ◇ OX-SORALEN ◇ OXYPSORALEN ◇ PRORALONE-MOP

TOXICITY DATA with REFERENCE
dnd-esc 20 μmol/L CBINA8 21,103,78
cyt-hmn:lym 100 μmol/L PLMEAA 42,333,81
orl-rat LD50:791 mg/kg DCTODJ 2,309,79
ipr-rat LD50:470 mg/kg JPETAB 131,394,61
orl-mus LD50:423 mg/kg DCTODJ 2,309,79
ipr-mus LDLo:60 mg/kg JMCMAR 28,1001,85
scu-mus LD50:860 mg/kg NIIRDN 6,837,82

CONSENSUS REPORTS: NTP Fifth Annual Report on Carcinogens. IARC Cancer Review: Group 1 IMEMDT 7,243,87; Human Inadequate Evidence IMEMDT 24, 101,80; Animal Inadequate Evidence IMEMDT 24,101, 80. Reported in EPA TSCA Inventory. EPA Genetic Toxicology Program.

SAFETY PROFILE: Confirmed carcinogen. Poison by intraperitoneal route. Moderately toxic by ingestion and subcutaneous routes. Human mutation data reported. When heated to decomposition it emits acrid smoke and irritating fumes. A drug used to treat skin diseases.

XDS000 CAS:7440-63-3 ***HR: 1***
XENON
DOT: UN 2036/UN 2591
af: Xe aw: 131.29

PROP: Colorless, gaseous element. D (gas): 5.8878 g/L, d (liq): 3.52 @ −109°, mp: −112°, bp: −107°.

SYN: XENON, refrigerated liquid (DOT)

CONSENSUS REPORTS: Reported in EPA TSCA Inventory.

DOT Classification: Nonflammable Gas; Label: Nonflammable Gas.

SAFETY PROFILE: An inert gas which acts as a simple asphyxiant. For a discussion of toxicity effects, see ARGON. A common air contaminant.

XEA000 ***HR: 3***
XENON(II) FLUORIDE METHANESULFONATE
mf: CH_3FO_3SXe mw: 245.36

SAFETY PROFILE: Explodes spontaneously at 0°C. Upon decomposition it emits very toxic fumes of SO_x and F^-. See also FLUORIDES.

XEJ000 CAS:25710-89-8 ***HR: 3***
XENON(II) FLUORIDE TRIFLUOROACETATE
mf: CF_4O_2Xe mw: 251.40

SAFETY PROFILE: Explodes when exposed to thermal or mechanical shock. When heated to decomposition it emits toxic fumes of F^-. See also FLUORIDES.

XEJ100 CAS:39274-39-0 ***HR: 3***
XENON(II) FLUORIDE TRIFLUOROMETHANE-SULFONATE
mf: CF_4O_3SXe mw: 299.35

SAFETY PROFILE: Explodes violently at room temperature. Upon decomposition it emits toxic fumes of SO_x and F^-. See also FLUORIDES.

XEJ300 CAS:13693-09-9 ***HR: 3***
XENON HEXAFLUORIDE
mf: F_6Xe mw: 245.28

PROP: Colorless solid, greenish yellow vapor. Mp: 49.48°; bp: 75.57°.

SAFETY PROFILE: Violent reaction with water forms the explosive xenon trioxide. Violent reaction with hydrogen; water + fluorides (e.g., sodium fluoride; potassium fluoride; rubidium fluoride; cesium fluoride; nitrosyl fluoride). When heated to decomposition it emits toxic fumes of F^-. See also FLUORIDES and XENON TRIOXIDE.

XES000 CAS:25523-79-9 ***HR: 3***
XENON(II) PERCHLORATE
mf: Cl_2O_8Xe mw: 330.206

SAFETY PROFILE: Can explode violently. When heated to decomposition it emits toxic fumes of Cl^-. See also PERCHLORATES.

XFA000 CAS:13709-61-0 ***HR: 3***
XENON TETRAFLUORIDE
mf: F_4Xe mw: 207.3

PROP: Colorless crystals. D: 4.55. Triple point 117.10°.

SAFETY PROFILE: Reacts with water to form a very shock-sensitive explosive. May explode on contact with acetone; aluminum; pentacarbonyliron; styrene; polyethylene; lubricants; paper; sawdust; wool or other combustible materials. Vigorous reaction with ethanol, potassium iodate, potassium permanganate. When heated to decomposition it emits toxic fumes of F^-. See also FLUORIDES and XENON TRIOXIDE.

XFJ000 ***HR: 3***
XENON(II) TRIFLUOROACETATE
mf: $C_2F_6O_4Xe$ mw: 333.31

SAFETY PROFILE: Shock-sensitive explosive. When heated to decomposition it emits toxic fumes of F^-. See also FLUORIDES.

XFS000 CAS:13776-58-4 ***HR: 3***
XENON TRIOXIDE
mf: O_3Xe mw: 179.30

PROP: Colorless, hygroscopic solid. D: 4.55.

SAFETY PROFILE: Powerful explosive. Formed by the reaction of xenon tetrafluoride or xenon hexaflluoride with water. Used as an epoxidation reagent.

XFS600 CAS:1394-04-3 ***HR: 3***
XEROSIN

SYNS: ANTIBIOTIC APM ◇ APM

TOXICITY DATA with REFERENCE
ipr-mus LD50:400 mg/kg 85GDA2 4(2),290,80
scu-mus LD50:800 mg/kg 85GDA2 4(2),290,80
ivn-mus LD50:200 mg/kg 85GDA2 4(2),290,80

SAFETY PROFILE: Poison by intravenous and in-

traperitoneal routes. Moderately toxic by subcutaneous route.

XGA000 ***HR: 3***
XF-408

SYN: DOW CORNING SILICONE FLUID and FLUOROHYDROCARBON

TOXICITY DATA with REFERENCE
orl-rat LD50:2840 mg/kg MRLR** No.256,54
ihl-rat LCLo:2100 mg/m^3 XAWPA2 CWL 2-10,58
ivn-rat LD50:230 mg/kg MRLR** No.256,54
ihl-dog LCLo:2100 mg/m^3 XAWPA2 CWL 2-10,58
ivn-rbt LD50:130 mg/kg MRLR** No.256,54

SAFETY PROFILE: Poison by intravenous route. Moderately toxic by ingestion and inhalation. See also SILICONES.

XGA500 ***HR: 3***
XIBENOL HYDROCHLORIDE
mf: $C_{15}H_{25}NO_2{\cdot}ClH$ mw: 287.83

SYN: (±)-1-((1,1-DIMETHYLETHYL)AMINO)-3-(2,3-DIMETHYLPHENOXY)-2-PROPANOL HYDROCHLORIDE

TOXICITY DATA with REFERENCE
orl-rat LD50:575 mg/kg YACHDS 12(Suppl 6),969,84
scu-rat LD50:219 mg/kg YACHDS 12(Suppl 6),969,84
ivn-rat LD50:24 mg/kg YACHDS 12(Suppl 6),969,84
orl-mus LD50:325 mg/kg YACHDS 12(Suppl 6),969,84
scu-mus LD50:284 mg/kg YACHDS 12(Suppl 6),969,84
ivn-mus LD50:28 mg/kg YACHDS 12(Suppl 6),969,84
orl-dog LD50:405 mg/kg YACHDS 12(Suppl 6),969,84
ivn-dog LD50:9200 μg/kg YACHDS 12(Suppl 6),969,84
orl-rbt LD50:425 mg/kg YACHDS 12(Suppl 6),969,84

SAFETY PROFILE: Poison by ingestion, subcutaneous, and intravenous routes. When heated to decomposition it emits toxic fumes of NO_x and HCl.

XGA725 CAS:50528-97-7 ***HR: 3***
XILOBAM
mf: $C_{14}H_{19}N_3O$ mw: 245.36

SYNS: N-(2,6-DIMETHYLPHENYL)-N'-(1-METHYL-2-PYRROLIDINYLIDENE)UREA ◇ MCN-3113 ◇ 1-(1-METHYL-2-PYRROLIDINYLIDENE)-3-(2,6-XYLYL)UREA

TOXICITY DATA with REFERENCE
orl-rat LD50:830 mg/kg AIPTAK 233,326,78
ipr-rat LD50:128 mg/kg AIPTAK 233,326,78
orl-mus LD50:320 mg/kg JMCMAR 21,1044,78
ipr-mus LD50:110 mg/kg AIPTAK 233,326,78

SAFETY PROFILE: Poison by ingestion and intraperitoneal routes. When heated to decomposition it emits toxic fumes of NO_x.

XGS000 CAS:1330-20-7 ***HR: 3***
XYLENE
DOT: UN 1307
mf: C_8H_{10} mw: 106.18

PROP: A clear liquid. Bp: 138.5°, flash p: 100°F (TOC), d: 0.864 @ 20°/4°, vap press: 6.72 mm @ 21°. Composition: as nonaromatics 0.07%, toluene 14%, ethyl benzene 19.27%, p-xylene 7.84%, m-xylene 65.01%, o-xylene 7.63%, C9 and aromatics 0.04% (TXAPA9 33,543,75).

SYNS: DIMETHYLBENZENE ◇ KSYLEN (POLISH) ◇ METHYL TOLUENE ◇ NCI-C55232 ◇ RCRA WASTE NUMBER U239 ◇ VIOLET 3 ◇ XILOLI (ITALIAN) ◇ XYLENEN (DUTCH) ◇ XYLOL (DOT) ◇ XYLOLE (GERMAN)

TOXICITY DATA with REFERENCE
eye-hmn 200 ppm JIHTAB 25,282,43
skn-rbt 100% MOD AMIHAB 14,387,56
skn-rbt 500 mg/24H MOD 28ZPAK -,24,72
eye-rbt 87 mg MLD AMIHAB 14,387,56
eye-rbt 5 mg/24H SEV 28ZPAK -,24,72
cyt-smc 1 mmol/tube HEREAY 33,457,47
ihl-rat TCLo:50 mg/m^3/6H (female 1-21D post):REP JOHYAY 27,337,83
ihl-rat TCLo:50 mg/m^3/6H (female 1-21D post):TER JOHYAY 27,337,83
orl-hmn LDLo:50 mg/kg YAKUD5 22,883,80
ihl-man LCLo:10000 ppm/6H BMJOAE 3,442,70
ihl-hmn TCLo:200 ppm:NOSE,EYE,PUL JIHTAB 25,282,43
orl-rat LD50:4300 mg/kg AMIHAB 14,387,56
ihl-rat LC50:5000 ppm/4H NPIRI* 1,123,74
ipr-rat LD50:2459 mg/kg ENVRAL 40,411,86

CONSENSUS REPORTS: Reported in EPA TSCA Inventory. EPA Genetic Toxicology Program. Community Right-To-Know List.

OSHA PEL: (Transitional: TWA 100 ppm) TWA 100 ppm; STEL 150 ppm
ACGIH TLV: TWA 100 ppm; STEL 150 ppm; BEI: 1.5 g(methyl hippuric acids)/g creatinine in urine end of shift.
DFG MAK: (all isomers) 100 ppm (440 mg/m^3); BAT: 150 μg/dL in blood at end of shift.
NIOSH REL: (Xylene) TWA 100 ppm; CL 200 ppm/10M

DOT Classification: Flammable Liquid; Label: Flammable Liquid; Flammable or Combustible Liquid; Label: Flammable Liquid.

SAFETY PROFILE: Moderately toxic by intraperitoneal and subcutaneous routes. Mildly toxic by ingestion and inhalation. An experimental teratogen. Human systemic effects by inhalation: olfactory changes, conjunctiva irritation and pulmonary changes. Experimental reproductive effects. Mutation data reported. A human eye irritant. An experimental skin and severe eye irritant. Some temporary corneal effects are noted, as well as some conjunctival irritation by instillation (adding drops to the eyes one at a time). Irritation can start @ 200 ppm. A very dangerous fire hazard when exposed to heat or flame; can react with oxidizing materials. To fight fire, use foam, CO_2, dry chemical. When heated to decomposition it emits acrid smoke and irritating fumes. See also other xylene entries.

XHA000 CAS:108-38-3 ***HR: 3***
m-XYLENE
DOT: UN 1307
mf: C_8H_{10} mw: 106.18

PROP: Colorless liquid. Mp: −47.9°, bp: 139°, lel: 1.1%, uel: 7.0%, flash p: 77°F, d: 0.864 @ 20/4°, vap press: 10 mm @ 28.3°, vap d: 3.66, autoign temp: 986°F. Insol in water; misc with alc, ether, and some organic solvents.

SYNS: m-DIMETHYLBENZENE ◇ 1,3-DIMETHYLBENZENE ◇ 1,3-XYLENE ◇ m-XYLOL (DOT)

TOXICITY DATA with REFERENCE
skn-rbt 10 μg/24H open SEV AIHAAP 23,95,62
ihl-rbt TCLo:500 mg/m³/24H (female 7-20D post):TER ARTODN 8,425,85
orl-mus TDLo:30 mg/kg (6-15D preg):REP APTOD9 19,A22,80
ihl-man TCLo:424 mg/m³/6H/6D:CNS TOLED5 1000(Sp. Iss. I),74,8
ihl-man TCLo:870 mg/m³/4H-I:CNS ATSUDG 7,412,84
orl-rat LD50:5 g/kg YAKUD5 22,883,80
ihl-rat LCLo:8000 ppm/4H AIHAAP 23,95,62
ihl-mus LCLo:2010 ppm/24H JPBAA7 46,95,38
ipr-mus LD50:1739 mg/kg ARTODN 58,106,85
skn-rbt LD50:14100 mg/kg AIHAAP 23,95,62

CONSENSUS REPORTS: Community Right-To-Know List. Reported in EPA TSCA Inventory.

OSHA PEL: (Transitional: TWA 100 ppm) TWA 100 ppm; STEL 150 ppm
ACGIH TLV: TWA 100 ppm; STEL 150 ppm; BEI: methyl hippuric acids in urine end of shift 1.5 g/g creatinine
NIOSH REL: (Xylene) TWA 100 ppm; CL 200 ppm/10M
DOT Classification: Flammable or Combustible Liquid; Label: Flammable Liquid; Flammable Liquid; Label: Flammable Liquid.

SAFETY PROFILE: Moderately toxic by intraperitoneal route. Mildly toxic by ingestion, skin contact, and inhalation. An experimental teratogen. Human systemic effects by inhalation: motor activity changes, ataxia and irritability. Experimental reproductive effects. A severe skin irritant. A common air contaminant. A very dangerous fire hazard when exposed to heat or flame; can react with oxidizing materials. Explosive in the form of vapor when exposed to heat or flame. To fight fire, use foam, CO_2, dry chemical. Emitted from modern building materials. (CENEAR 69,22,91) When heated to decomposition it emits acrid smoke and irritating fumes. See also other xylene entries.

XHJ000 CAS:95-47-6 ***HR: 3***
o-XYLENE
DOT: UN 1307
mf: C_8H_{10} mw: 106.18

PROP: Colorless liquid. D: 0.880 @ 20/4°, mp: −25.2°, bp: 144.4°, flash p: 62.6°F. lel: 1.0%, uel: 6.0%. Insol in water; misc in abs alc, ether.

SYNS: o-DIMETHYLBENZENE ◇ 1,2-DIMETHYLBENZENE ◇ o-METHYLTOLUENE ◇ 1,2-XYLENE ◇ o-XYLOL (DOT)

TOXICITY DATA with REFERENCE
ihl-rat TCLo:150 mg/m³/24H (7-14D preg):TER TXCYAC 18,61,80
ipr-rat TDLo:500 mg/kg (male 2D pre):REP ARANDR 11,233,83
ihl-hmn LCLo:6125 ppm/12H YAKUD5 22,883,80
orl-rat LDLo:5 g/kg YAKUD5 22,883,80
ihl-rat LCLo:6125 ppm/12H JPBAA7 46,95,38
ihl-mus LCLo:30 g/m³ AEPPAE 143,223,29
ipr-mus LD50:1364 mg/kg ARTODN 58,106,85

CONSENSUS REPORTS: Community Right-To-Know List. Reported in EPA TSCA Inventory.

OSHA PEL: (Transitional: TWA 100 ppm) TWA 100 ppm; STEL 150 ppm
ACGIH TLV: TWA 100 ppm; STEL 150 ppm; BEI: methyl hippuric acids in urine end of shift 1.5 g/g creatinine
NIOSH REL: (Xylene) TWA 100 ppm; CL 200 ppm/10M

DOT Classification: Flammable or Combustible Liquid; Label: Flammable Liquid; Flammable Liquid; Label: Flammable Liquid.

SAFETY PROFILE: Moderately toxic by intraperitoneal route. Mildly toxic by ingestion and inhalation. An experimental teratogen. A common air contaminant. A very dangerous fire hazard when exposed to heat or flame. Explosive in the form of vapor when exposed to heat or flame. To fight fire, use foam, CO_2, dry chemical. Incompatible with oxidizing materials. When heated to decomposition it emits acrid smoke and irritating fumes. Emitted from modern building materials. (CENEAR 69,22,91) See also other xylene entries.

XHS000 CAS:106-42-3 ***HR: 3***
p-XYLENE
DOT: UN 1307
mf: C_8H_{10} mw: 106.18

PROP: Clear plates. Bp: 138.3°, lel: 1.1%, uel: 7.0%, flash p: 77°F (CC), d: 0.8611 @ 20/4°, vap press: 10 mm @ 27.3°, vap d: 3.66, autoign temp: 986°F, mp: 13-14°. Insol in water; sol in alc, ether, organic solvents.

SYNS: CHROMAR ◇ p-DIMETHYLBENZENE ◇ 1,4-DIMETHYLBENZENE ◇ p-METHYLTOLUENE ◇ SCINTILLAR ◇ 1,4-XYLENE ◇ p-XYLOL (DOT)

TOXICITY DATA with REFERENCE
ihl-rbt TCLo:1 g/m^3/24H (female 7-20D post):TER ARTODN 8,425,85
ihl-rbt TCLo:1 g/m^3/24H (female 7-20D post):REP ARTODN 8,425,85
orl-rat LD50:5 g/kg YAKUD5 22,883,80
ihl-rat LC50:4550 ppm/4H 36YFAG -,302,77
ipr-rat LD50:3810 mg/kg 36YFAG -,302,77
ihl-mus LCLo:15 g/m^3 AEPPAE 143,223,29
ipr-mus LD50:2110 mg/kg ARTODN 58,106,85

CONSENSUS REPORTS: Community Right-To-Know List. Reported in EPA TSCA Inventory.

OSHA PEL: (Transitional: TWA 100 ppm) TWA 100 ppm; STEL 150 ppm
ACGIH TLV: TWA 100 ppm; STEL 150 ppm; BEI: methyl hippuric acids in urine end of shift 1.5 g/g creatinine
NIOSH REL: (Xylene) TWA 100 ppm; CL 200 ppm/10M
DOT Classification: Flammable Liquid; Label: Flammable Liquid; Flammable or Combustible Liquid; Label: Flammable Liquid.

SAFETY PROFILE: Moderately toxic by intraperitoneal route. Mildly toxic by ingestion and inhalation. An experimental teratogen. Experimental reproductive effects. May be narcotic in high concentrations. Chronic toxicity not established, but is less toxic than benzene. A very dangerous fire hazard when exposed to heat or flame; can react with oxidizing materials. Explosive in the form of vapor when exposed to heat or flame. To fight fire, use foam, CO_2, dry chemical. Potentially explosive reaction with acetic acid + air; 1,3-dichloro-5,5-dimethyl-2,4-imidazolidindione; nitric acid + pressure. When heated to decomposition it emits acrid smoke and irritating fumes. See also other xylene entries.

XHS800 CAS:1477-55-0 ***HR: 2***
m-XYLENE-α,α'-DIAMINE
mf: $C_8H_{12}N_2$ mw: 136.22

SYNS: 1,3-BIS-AMINOMETHYLBENZEN (CZECH) ◇ MXDA ◇ m-PHENYLENEBIS(METHYLAMINE) ◇ m-XYLYLENDIAMIN (CZECH)

TOXICITY DATA with REFERENCE
skn-rbt 200 mg/24H SEV 28ZPAK -,64,72
eye-rbt 50 μg/24H SEV 28ZPAK -,64,72
orl-rat LD50:1600 mg/kg 28ZPAK -,64,72
ihl-rat LC50:700 ppm/1H DTLVS* 4,440,80
skn-rbt LD50:2000 mg/kg DTLVS* 4,440,80

CONSENSUS REPORTS: Reported in EPA TSCA Inventory.

OSHA PEL: TWA CL 0.1 mg/m^3 (skin)
ACGIH TLV: TWA CL 0.1 mg/m^3 (skin)

SAFETY PROFILE: Moderately toxic by skin contact and ingestion. Mildly toxic by inhalation. A severe skin and eye irritant. When heated to decomposition it emits toxic fumes of NO_x. Used to make polyamide fibers and resins and as a curing agent.

XIJ000 CAS:3634-83-1 ***HR: 2***
m-XYLENE-α,α'-DIISOCYANATE
mf: $C_{10}H_8N_2O_2$ mw: 188.20

SYN: XYLYLENDIISOKYANAT (CZECH)

TOXICITY DATA with REFERENCE
skn-rbt 500 mg/24H SEV 28ZPAK -,166,72
eye-rbt 5 mg/24H SEV 28ZPAK -,166,72
orl-rat LD50:5350 mg/kg 28ZPAK -,166,72
NIOSH REL: (Diisocyanates) TWA 0.005 ppm; CL 0.02 ppm/10M

SAFETY PROFILE: Mildly toxic by ingestion. A severe skin and eye irritant. When heated to decomposition it emits very toxic fumes of NO_x. See also ISOCYANATES.

XJA000 CAS:25321-41-9 ***HR: 2***
XYLENESULFONIC ACID
mf: $C_8H_{10}O_3S$ mw: 186.24

TOXICITY DATA with REFERENCE
ipr-mus LD50:500 mg/kg 14CYAT 2,1841,63

CONSENSUS REPORTS: Reported in EPA TSCA Inventory.

SAFETY PROFILE: Moderately toxic by intraperitoneal route. When heated to decomposition it emits toxic fumes of SO_x.

XJJ000 CAS:88-61-9 ***HR: 2***
2,4-XYLENESULFONIC ACID
mf: $C_8H_{10}O_3S$ mw: 186.24

SYNS: 2,4-DIMETHYLBENZENESULFONIC ACID ◇ m-XYLENESULFONIC ACID ◇ m-XYLENE-4-SULFONIC ACID

TOXICITY DATA with REFERENCE
ipr-mus LDLo:500 mg/kg CBCCT* 6,375,54

CONSENSUS REPORTS: Reported in EPA TSCA Inventory.

SAFETY PROFILE: Moderately toxic by intraperitoneal route. When heated to decomposition it emits toxic fumes of SO_x.

XKA000 CAS:1300-71-6 ***HR: 3***
XYLENOL
DOT: UN 2261
mf: $C_8H_{10}O$ mw: 122.18

PROP: The six isomers of xylenol are sltly sol in water; very sol in alc, chloroform, ether, benzene; sol in NaOH soln.

SYNS: DIMETHYLPHENOL ◇ XILENOLI (ITALIAN) ◇ XYLENOLEN (DUTCH)

CONSENSUS REPORTS: Reported in EPA TSCA Inventory.

DOT Classification: Poison B; Label: Poison.

SAFETY PROFILE: A poison. When heated to decomposition it emits acrid smoke and irritating fumes. See also other xylenol entries.

XKJ000 CAS:526-75-0 ***HR: 3***
2,3-XYLENOL
DOT: UN 2261
mf: $C_8H_{10}O$ mw: 122.18

PROP: Needles in water. Mp: 75°, bp: 218°. Sol in water, alc.

SYNS: 2,3-DIMETHYLPHENOL ◇ o-XYLENOL (DOT)

TOXICITY DATA with REFERENCE
ivn-mus LD50:56 mg/kg CSLNX* NX#00158

CONSENSUS REPORTS: Reported in EPA TSCA Inventory.

DOT Classification: Poison B; Label: Poison.

SAFETY PROFILE: Poison by intravenous route. When heated to decomposition it emits acrid smoke and irritating fumes. See also other xylenol entries.

XKJ500 CAS:105-67-9 ***HR: 3***
2,4-XYLENOL
DOT: UN 2261
mf: $C_8H_{10}O$ mw: 122.18

PROP: Mp: 26°, bp: 211.5°. Sol in water and alc.

SYNS: 2,4-DIMETHYLPHENOL ◇ 4,6-DIMETHYLPHENOL ◇ 1-HYDROXY-2,4-DIMETHYLBENZENE ◇ RCRA WASTE NUMBER U101 ◇ m-XYLENOL ◇ m-XYLENOL (DOT)

TOXICITY DATA with REFERENCE
skn-mus TDLo:16 g/kg/39W-I:CAR CNREA8 19,413,59
orl-rat LD50:3200 mg/kg GTPZAB 18(2),58,74
skn-rat LD50:1040 mg/kg GTPZAB 18(2),58,74
orl-mus LD50:809 mg/kg GTPZAB 18(2),58,74
skn-mus LD50:1040 mg/kg 85GMAT -,119,82
ipr-mus LD50:183 mg/kg JMCMAR 868,75
ivn-mus LD50:100 mg/kg JMCMAR 23,1350,80

CONSENSUS REPORTS: Reported in EPA TSCA Inventory.

DOT Classification: DOT-IMO: Poison B; Label: Poison.

SAFETY PROFILE: Poison by intravenous and intraperitoneal routes. Moderately toxic by ingestion and skin contact. Questionable carcinogen with experimental carcinogenic data. When heated to decomposition it emits acrid smoke and irritating fumes. See also other xylenol entries.

XKS000 CAS:95-87-4 ***HR: 3***
2,5-XYLENOL
DOT: UN 2261
mf: $C_8H_{10}O$ mw: 122.18

PROP: Crystals. Mp: 74.5°, bp: 211.5-213.5°.

SYNS: 2,5-DIMETHYLPHENOL ◇ 3,6-DIMETHYLPHENOL ◇ 2,5-DMP ◇ 6-METHYL-m-CRESOL ◇ p-XYLENOL (DOT) ◇ 1,2,5-XYLENOL

TOXICITY DATA with REFERENCE
skn-mus TDLo:4000 mg/kg/20W-I:ETA CNREA8 19,413,59
orl-rat LD50:444 mg/kg HYSAAV 33(9),329,68
unr-rat LD50:730 mg/kg JPETAB 53,227,35
orl-mus LD50:383 mg/kg HYSAAV 33(9),329,68
orl-rbt LD50:938 mg/kg HYSAAV 33(9),329,68

CONSENSUS REPORTS: Reported in EPA TSCA Inventory.

DOT Classification: Poison B; Label: Poison.

SAFETY PROFILE: Poison by ingestion. Moderately toxic by an unspecified route. When heated to decomposition it emits acrid smoke and irritating fumes. Questionable carcinogen with experimental tumorigenic data. Used in disinfectants, solvents, pharmaceuticals, plasticizers, and wetting agents. See also other xylenol entries.

XLA000 CAS:576-26-1 ***HR: 3***
2,6-XYLENOL
mf: $C_8H_{10}O$ mw: 122.18

PROP: Colorless leaflets or needles. Mp: 48-49°, bp: 203°. Sol in hot water and alc.

SYNS: 2,6-DIMETHYLPHENOL ◇ 2,6-DMP

TOXICITY DATA with REFERENCE
eye-rbt 100 mg IHFCAY 6,1,67

skn-mus TDLo:4000 mg/kg/20W-I:ETA CNREA8 19,413,59
orl-rat LD50:296 mg/kg HYSAAV 33(9),329,68
orl-mus LD50:980 mg/kg GTPZAB 18,58,74
skn-mus LD50:920 mg/kg GTPZAB 18,58,74
ipr-mus LD50:150 mg/kg NTIS** AD691-490
ivn-mus LD50:80 mg/kg JMCMAR 23,1350,80
orl-rbt LD50:700 mg/kg HYSAAV 33(9),329,68
skn-rbt LD50:1000 mg/kg IHFCAY 6,1,67

CONSENSUS REPORTS: Reported in EPA TSCA Inventory.

SAFETY PROFILE: Questionable carcinogen with experimental tumorigenic data. Poison by ingestion, intravenous, and intraperitoneal routes. Moderately toxic by skin contact. An eye irritant. When heated to decomposition it emits acrid smoke and irritating fumes. Used in disinfectants, solvents, pharmaceuticals, and as an antioxidant in gas, oils, and elastomers. See also other xylenol entries.

XLJ000 CAS:95-65-8 ***HR: 3***
3,4-XYLENOL
mf: $C_8H_{10}O$ mw: 122.18

PROP: Prisms from ligroin. D: 1.076 @ 17.5°, mp: 62.5°, bp: 225°C. Very sltly sol in water; sol in alc, ether.

SYNS: 3,4-DIMETHYLPHENOL ◇ 4,5-DIMETHYLPHENOL ◇ 3,4-DMP ◇ 1,3,4-XYLENOL

TOXICITY DATA with REFERENCE
skn-mus TDLo:4000 mg/kg/20W-I:ETA CNREA8 19,413,59
unr-rat LD50:727 mg/kg GTPPAF 8,145,72
orl-mus LD50:400 mg/kg HYSAAV 33(9),329,68
orl-rbt LD50:800 mg/kg HYSAAV 33(9),329,68

CONSENSUS REPORTS: Reported in EPA TSCA Inventory.

SAFETY PROFILE: Questionable carcinogen with experimental tumorigenic data. Poison by ingestion. Moderately toxic by unspecified route. When heated to decomposition it emits acrid smoke and irritating fumes. Used in production of sulfur dyes, disinfectants, pharmaceuticals, solvents, and as an antioxidant. See also other xylenol entries.

XLS000 CAS:108-68-9 ***HR: 3***
3,5-XYLENOL
mf: $C_8H_{10}O$ mw: 122.18

PROP: White crystals. Mp: 64°, bp: 219.5°, d: 1.0362, vap press: 1 mm @ 62°. Sltly sol in water; sol in alc.

SYNS: 3,5-DIMETHYLPHENOL ◇ 3,5-DMP ◇ 1,3,5-XYLENOL

TOXICITY DATA with REFERENCE
eye-rbt 726 μg SEV AJOPAA 29,1363,46
skn-mus TDLo:4000 mg/kg/20W-I:ETA CNREA8 19,413,59
orl-rat LD50:608 mg/kg GTPPAF 8,145,72
orl-mus LD50:477 mg/kg HYSAAV 33(9),329,68
ipr-mus LD50:156 mg/kg JMCMAR 18,868,75
orl-rbt LD50:1313 mg/kg HYSAAV 33(9),329,68

CONSENSUS REPORTS: Reported in EPA TSCA Inventory. EPA Genetic Toxicology Program.

SAFETY PROFILE: Poison by intraperitoneal route. Moderately toxic by ingestion. A severe eye irritant. Questionable carcinogen with experimental tumorigenic data. When heated to decomposition it emits acrid smoke and irritating fumes. See also other xylenol entries.

XLS300 CAS:57021-61-1 ***HR: 1***
2,6-XYLIDIDE of 2-PYRIDONE-3-CARBOXYLIC ACID
mf: $C_{14}H_{14}NO_2$ mw: 228.29

SYNS: 1,2-DIHYDROXY-2-OXO-N-(2,6-XYLYL)-3-PYRIDINECARBOXAMIDE ◇ ISONIXIN ◇ ISONIXINE ◇ NIXYN ◇ NIXYN HERMES ◇ 3-PYRIDINECARBOXAMIDE, 1,2-DIHYDRO-2-OXO-N-(2,6-XYLYL)- ◇ 3-PYRIDINECARBOXAMIDE,N-(2,6-DIMETHYLPHENYL)-1,2-DI-HYDRO-2-OXO-(9CI) ◇ 2,6-XYLIDID der 2-PYRIDON-3-CARBOXYLSAURE

TOXICITY DATA with REFERENCE
orl-rat TDLo:250 g/kg (female 26W pre):REP ARZNAD 27,1460,77
orl-mus LD50:7000 mg/kg ARZNAD 27,1457,77

SAFETY PROFILE: Mildly toxic by ingestion. Experimental reproductive effects. When heated to decomposition it emits toxic fumes of NO_x.

XMA000 CAS:1300-73-8 ***HR: 3***
XYLIDINE
DOT: UN 1711
mf: $C_8H_{11}N$ mw: 121.20

PROP: Usually liquid (except for o-4-xylidine). Bp: 213-226°, flash p: 206° (CC), d: 0.97-0.99, vap d: 4.17. Sltly sol in water; sol in alc.

SYNS: ACID LEATHER BROWN 2G ◇ ACID ORANGE 24 ◇ AMINODIMETHYLBENZENE ◇ 11460 BROWN ◇ DIMETHYLANILINE ◇ DIMETHYLPHENYLAMINE ◇ RESORCINE BROWN J ◇ RESORCINE BROWN R ◇ XILIDINE (ITALIAN) ◇ XYLIDINEN (DUTCH)

TOXICITY DATA with REFERENCE
orl-rat LDLo:610 mg/kg JIHTAB 31,1,49
ivn-cat LDLo:120 mg/kg JIHTAB 31,1,49
orl-rbt LD50:600 mg/kg 34ZIAG -,637,69
ivn-rbt LDLo:240 mg/kg JIHTAB 31,1,49

CONSENSUS REPORTS: Reported in EPA TSCA Inventory.

OSHA PEL: (Transitional: TWA 5 ppm (skin)) TWA 0.2 ppm (skin)
ACGIH TLV: TWA 0.5 ppm (skin); Suspected Human Carcinogen
DFG MAK: (all isomers except 2,4-xylidene) 5 ppm (25 mg/m^3)
DOT Classification: Poison B; Label: Poison.

SAFETY PROFILE: Suspected human carcinogen. Poison by intravenous route. Moderately toxic by ingestion. This material, which so closely resembles aniline in its toxic effects, is actually twice as toxic as aniline. It can cause injury to the blood and the liver. It does not necessarily give any alarm or warning, such as cyanosis, headache, and dizziness which characterizes aniline poisoning. Thus, it may be considered a more insidious poison than aniline, and severe and possibly fatal intoxication may come about through skin absorption. Combustible when exposed to heat or flame. Can react vigorously with oxidizing materials. To fight fire, use foam, CO_2, dry chemical. When heated to decomposition it emits toxic fumes of NO_x. See also ANILINE and other xylidine entries.

XMJ000 CAS:87-59-2 ***HR: 3***
2,3-XYLIDINE
DOT: UN 1711
mf: $C_8H_{11}N$ mw: 121.20

PROP: Liquid. D: 0.991 @ 15°, mp: < −15°, bp: 220°. Very sltly sol in water; sol in alc, ether.

SYNS: 2,3-DIMETHYLANILINE ◇ 2,3-DIMETHYLBENZENAMINE ◇ 2,3-DIMETHYLPHENYLAMINE ◇ o-XYLIDINE (DOT) ◇ 2,3-XYLYLAMINE

TOXICITY DATA with REFERENCE
mma-sat 5 μmol/plate MUREAV 77,317,80
orl-rat LD50:933 mg/kg NTIS** PB214-270
orl-mus LD50:1072 mg/kg NTIS** PB214-270

CONSENSUS REPORTS: Reported in EPA TSCA Inventory.
DFG MAK: (all isomers except 2,4-xylidene) 5 ppm (25 mg/m^3)
DOT Classification: Poison B; Label: Poison.

SAFETY PROFILE: A poison. Moderately toxic by ingestion. Mutation data reported. When heated to decomposition it emits toxic fumes of NO_x. See also other xylidine entries.

XMS000 CAS:95-68-1 ***HR: 3***
2,4-XYLIDINE
DOT: UN 1711
mf: $C_8H_{11}N$ mw: 121.20

PROP: Liquid. Bp: 214°, mp: 16°, d: 0.978 @ 19.6/4°. Very sltly sol in water.

SYNS: 1-AMINO-2,4-DIMETHYLBENZENE ◇ 4-AMINO-1,3-DIMETHYLBENZENE ◇ 4-AMINO-3-METHYLTOLUENE ◇ 4-AMINO-1,3-XYLENE ◇ 2,4-DIMETHYLANILINE ◇ 2,4-DIMETHYLBENZENAMINE ◇ 2,4-DIMETHYLPHENYLAMINE ◇ 2-METHYL-p-TOLUIDINE ◇ 4-METHYL-o-TOLUIDINE ◇ 2,4-XYLIDENE (MAK) ◇ m-XYLIDINE (DOT) ◇ m-4-XYLIDINE

TOXICITY DATA with REFERENCE
mma-sat 5 μmol/plate MUREAV 77,317,80
dni-mus-orl 200 mg/kg MUREAV 46,305,77
orl-rat LD50:467 mg/kg NTIS** PB214-270
orl-mus LD50:250 mg/kg NTIS** PB214-270

CONSENSUS REPORTS: IARC Cancer Review: Group 3 IMEMDT 7,56,87; Animal Inadequate Evidence IMEMDT 16,367,78. Reported in EPA TSCA Inventory.
DFG MAK: 5 ppm (25 mg/m^3); Suspected Carcinogen.
DOT Classification: Poison B; Label: Poison.

SAFETY PROFILE: Suspected carcinogen. Poison by ingestion. Mutation data reported. When heated to decomposition it emits toxic fumes of NO_x. See also other xylidine entries.

XNA000 CAS:95-78-3 ***HR: 3***
2,5-XYLIDINE
DOT: UN 1711
mf: $C_8H_{11}N$ mw: 121.20

PROP: Colorless oil. Bp: 214°, d: 0.979 @ 21/4°, mp: 155°. Very sltly sol in water.

SYNS: 1-AMINO-2,5-DIMETHYLBENZENE ◇ 3-AMINO-1,4-DIMETHYLBENZENE ◇ 2-AMINO-1,4-XYLENE ◇ 2,5-DIMETHYLANILINE ◇ 2,5-DIMETHYLBENZENAMINE ◇ 2,5-DIMETHYLPHENYLAMINE ◇ 5-METHYL-o-TOLUIDINE ◇ 6-METHYL-m-TOLUIDINE ◇ p-XYLIDINE (DOT)

TOXICITY DATA with REFERENCE
mma-sat 1 μmol/plate MUREAV 77,317,80
dni-mus-orl 200 mg/kg MUREAV 46,305,77
orl-rat LD50:1297 mg/kg NTIS** PB214-270
orl-mus LD50:841 mg/kg NTIS** PB214-270

CONSENSUS REPORTS: IARC Cancer Review: Group 3 IMEMDT 7,56,87; Animal Inadequate Evidence IMEMDT 16,377,78. Reported in EPA TSCA Inventory.
DFG MAK: (all isomers except 2,4-xylidene) 5 ppm (25 mg/m^3)
DOT Classification: Poison B; Label: Poison.

SAFETY PROFILE: A poison. Moderately toxic by ingestion. Questionable carcinogen. Mutation data reported. When heated to decomposition it emits toxic fumes of NO_x. See also other xylidine entries.

XNJ000 CAS:87-62-7 ***HR: 2***
2,6-XYLIDINE
mf: $C_8H_{11}N$ mw: 121.20

PROP: Liquid. D: 0.980 @ 15°, mp: 10-12°, bp: 216-217°.

SYNS: 2,6-DIMETHYLANILINE ◇ 2,6-DIMETHYLBENZENAMINE ◇ NCI-C56188 ◇ o-XYLIDINE ◇ 2,6-XYLYLAMINE

TOXICITY DATA with REFERENCE
orl-rat LD50:840 mg/kg TXAPA9 22,153,72
orl-mus LD50:707 mg/kg NTIS** PB214-270

CONSENSUS REPORTS: Community Right-To-Know List. Reported in EPA TSCA Inventory.
DFG MAK: (all isomers except 2,4-xylidene) 5 ppm (25 mg/m^3)

SAFETY PROFILE: Moderately toxic by ingestion. When heated to decomposition it emits toxic fumes of NO_x. See also other xylidine entries.

XNS000 CAS:95-64-7 ***HR: 3***
3,4-XYLIDINE
mf: $C_8H_{11}N$ mw: 121.20

PROP: Mp: 49°, bp: 226°. Insol in water; sol in petroleum ether.

SYNS: 3,4-DIMETHYLAMINOBENZENE ◇ 3,4-DIMETHYLANILINE ◇ 3,4-DIMETHYLPHENYLAMINE ◇ 3,4-XYLYLAMINE

TOXICITY DATA with REFERENCE
mma-sat 500 nmol/plate MUREAV 77,317,80
orl-rat LD50:812 mg/kg AMRL** TR-72-62,72
orl-mus LD50:707 mg/kg NTIS** PB214-270
orl-bwd LD50:5 mg/kg TXAPA9 21,315,72

CONSENSUS REPORTS: Reported in EPA TSCA Inventory.
DFG MAK: (all isomers except 2,4-xylidene) 5 ppm (25 mg/m^3)

SAFETY PROFILE: Poison by ingestion. Mutation data reported. When heated to decomposition it emits toxic fumes of NO_x. See also other xylidine entries.

XOA000 CAS:108-69-0 ***HR: 2***
3,5-XYLIDINE
mf: $C_8H_{11}N$ mw: 121.20

PROP: An oil. D: 0.972 @ 20/4°, bp: 221-222°.

SYNS: 3,5-DIMETHYLANILINE ◇ 3,5-DIMETHYLBENZENAMINE ◇ 3,5-DIMETHYLPHENYLAMINE ◇ 3,5-XYLYLAMINE

TOXICITY DATA with REFERENCE
orl-rat LD50:707 mg/kg NTIS** PB214-270
orl-mus LD50:421 mg/kg NTIS** PB214-270

CONSENSUS REPORTS: Reported in EPA TSCA Inventory.
DFG MAK: (all isomers except 2,4-xylidene) 5 ppm (25 mg/m^3)

SAFETY PROFILE: Moderately toxic by ingestion. When heated to decomposition it emits toxic fumes of NO_x. See also other xylidine entries.

XOJ000 CAS:21436-96-4 ***HR: 3***
2,4-XYLIDINE HYDROCHLORIDE
mf: $C_8H_{11}N{\cdot}ClH$ mw: 157.66

SYNS: 1-AMINO-2,4-DIMETHYLBENZENE HYDROCHLORIDE ◇ 4-AMINO-1,3-DIMETHYLBENZENE HYDROCHLORIDE ◇ 4-AMINO-3-METHYLTOLUENE HYDROCHLORIDE ◇ 4-AMINO-1,3-XYLENE HYDROCHLORIDE ◇ 2,4-DIMETHYLANILINE HYDROCHLORIDE ◇ 2,4-DIMETHYLBENZENAMINE HYDROCHLORIDE ◇ 4-METHYL-o-TOLUIDINE HYDROCHLORIDE ◇ 2-METHYL-p-TOLUIDINE HYDROCHLORIDE ◇ m-XYLIDINE HYDROCHLORIDE

TOXICITY DATA with REFERENCE
orl-mus TDLo:12 g/kg/78W-C:NEO JEPTDQ 2,325,78
orl-rat LD50:1259 mg/kg JPETAB 167,223,69
ipr-rat LD50:545 mg/kg NCIBR* NIH-NCI-E-68-1311,10,73
ipr-mus LD50:420 mg/kg NCIBR* NIH-NCI-E-68-1311,10,73

SAFETY PROFILE: Moderately toxic by ingestion and intraperitoneal routes. Questionable carcinogen with experimental neoplastigenic data. When heated to decomposition it emits very toxic fumes of NO_x and HCl. See also other xylidine entries.

XOS000 CAS:51786-53-9 ***HR: 3***
2,5-XYLIDINE HYDROCHLORIDE
mf: $C_8H_{11}N{\cdot}ClH$ mw: 157.66

SYNS: 1-AMINO-2,5-DIMETHYLBENZENE HYDROCHLORIDE ◇ 3-AMINO-1,4-DIMETHYLBENZENE HYDROCHLORIDE ◇ 5-AMINO-1,4-DIMETHYLBENZENE HYDROCHLORIDE ◇ 2-AMINO-4-METHYLTOLUENE HYDROCHLORIDE ◇ 2-AMINO-1,4-XYLENE HYDROCHLORIDE ◇ 2,5-DIMETHYLANILINE HYDROCHLORIDE ◇ 2,5-DIMETHYLBENZENAMINE HYDROCHLORIDE ◇ 5-METHYL-o-TOLUIDINE HYDROCHLORIDE ◇ 6-METHYL-m-TOLUIDINE HYDROCHLORIDE ◇ p-XYLIDINE HYDROCHLORIDE

TOXICITY DATA with REFERENCE
orl-rat TDLo:124 g/kg/78W-C:ETA JEPTDQ 2,325,78
orl-mus TDLo:390 g/kg/78W-C:CAR JEPTDQ 2,325,78
orl-mus TD:780 g/kg/78W-C:CAR JEPTDQ 2,325,78
ipr-rat LD50:770 mg/kg NCIBR* NIH-NCI-E-68-1311,10,73
ipr-mus LD50:800 mg/kg NCIBR* NIH-NCI-E-68-1311,10,73

SAFETY PROFILE: Moderately toxic by intraperitoneal route. Questionable carcinogen with experimental carcinogenic and tumorigenic data. When heated to decomposition it emits very toxic fumes of NO_x and HCl. See also other xylidine entries.

XPJ000 CAS:87-99-0 ***HR: 2***
XYLITOL
mf: $C_5H_{12}O_5$ mw: 152.17

PROP: White crystals or crystalline powder; sweet taste with cooling sensation. Mp: 92-96°. Sol in water; sltly sol in alc.

SYNS: KLINIT ◇ XYLITE (SUGAR)

TOXICITY DATA with REFERENCE
orl-mus LD50:22 g/kg TOLED5 18(Suppl 1),37,83
ipr-mus LD50:22100 mg/kg RPTOAN 34,124,71
ivn-mus LD50:3770 mg/kg RPTOAN 34,124,71
ivn-rbt LD50:4000 mg/kg FEPRA7 31,726,72

CONSENSUS REPORTS: Reported in EPA TSCA Inventory.

SAFETY PROFILE: Moderately toxic by intravenous route. Mildly toxic by ingestion. When heated to decomposition it emits acrid smoke and irritating fumes. A sugar.

XQJ000 CAS:137-18-8 ***HR: 3***
p-XYLOQUINONE
mf: $C_8H_8O_2$ mw: 136.16

PROP: Yellow triclinic crystals from alc. Mp: 125°. Sltly sol in hot water; sol in alc.

SYNS: 2,5-DIMETHYL-p-BENZOQUINONE ◇ 2,5-DIMETILBENZOCHINONE (1:4) (ITALIAN) ◇ FLORONE (ITALIAN) ◇ PHLORONE

TOXICITY DATA with REFERENCE
orl-rat LDLo:500 mg/kg NCNSA6 5,39,53
orl-mus LD50:290 mg/kg BCFAAI 97,533,58
ipr-mus LD50:11 mg/kg BCFAAI 97,533,58
scu-mus LD50:43 mg/kg BCFAAI 97,533,58

CONSENSUS REPORTS: Reported in EPA TSCA Inventory.

SAFETY PROFILE: Poison by ingestion, intraperitoneal, and subcutaneous routes. When heated to decomposition it emits acrid smoke and irritating fumes. See also KETONES.

XQJ650 CAS:25546-65-0 ***HR: 3***
XYLOSTATIN
mf: $C_{17}H_{34}N_4O_{10}$ mw: 454.55

PROP: Colorless needles from methanol. Mp: 192-195°. Sol in water; sltly sol in methanol. Practically insol in acetone, n-butanol, ethyl acetate, benzene, hexane, and ether.

SYNS: ANTIBIOTIC SF 733 ◇ o-2,6-DIAMINO-2,6-DIDEOXY-α-d-GLUCOPYRANOSYL-(1-4)-O-(β-d-RIBOFURANOSYL-(1-5))-2-DEOXY-d-STREPTAMINE ◇ RIBOSTAMYCIN ◇ SF 733 ◇ VISTAMYCIN

TOXICITY DATA with REFERENCE
ims-rat TDLo:9 g/kg (10-15D preg):REP KSRNAM 4,2502,70
ims-rat TDLo:9 g/kg (10-15D preg):TER KSRNAM 4,2502,70
ipr-rat LD50:4400 mg/kg KSRNAM 4,2464,70
ivn-rat LD50:535 mg/kg KSRNAM 4,2464,70
ims-rat LD50:1850 mg/kg NKRZAZ 32,949,84
orl-mus LD50:7000 mg/kg 85GDA2 1,141,80
ipr-mus LD50:2830 mg/kg KSRNAM 4,2464,70
scu-mus LD50:3350 mg/kg KSRNAM 4,2464,70
ivn-mus LD50:300 mg/kg KSRNAM 4,2464,70
ims-mus LD50:1600 mg/kg 85GDA2 1,141,80

CONSENSUS REPORTS: EPA Genetic Toxicology Program.

SAFETY PROFILE: Poison by intravenous route. Moderately toxic by intramuscular and intraperitoneal routes. Mildly toxic by ingestion. An experimental teratogen. Experimental reproductive effects. When heated to decomposition it emits toxic fumes of NO_x.

XQS000 CAS:61-68-7 ***HR: 3***
N-(2,3-XYLYL)ANTHRANILIC ACID
mf: $C_{15}H_{15}NO_2$ mw: 241.31

SYNS: BAFHAMERITIN-M ◇ BONABOL ◇ CI-473 ◇ CN-35355 ◇ COSLAN ◇ 2',3'-DIMETHYL-2-DIPHENYLAMINECARBOXYLIC ACID ◇ 2-((2,3-DIMETHYLPHENYL)AMINO)BENZOIC ACID ◇ N-(2,3-DIMETHYLPHENYL)ANTHRANILIC ACID ◇ INF 3355 ◇ LYSALGO ◇ MEFENAMIC ACID ◇ MEPHENAMINIC ACID ◇ METHENAMIC ACID ◇ NAMPHEN ◇ PARKEMED ◇ PONALAR ◇ PONSTAN ◇ PONSTEL ◇ PONSTIL ◇ PONSTYL ◇ PONTAL ◇ TANSTON ◇ VIALIDON ◇ N-(2,3-XYLYL)-2-AMINOBENZOIC ACID

TOXICITY DATA with REFERENCE
orl-wmn TDLo:30 mg/kg (2D pre):REP JRPMAP 28,592,83
orl-rbt TDLo:10 mg/kg (21D preg):TER OYYAA2 27,117,84
orl-man TDLo:840 mg/kg/6W-I:GIT BMJOAE 287,1626,83
orl-man TDLo:257 mg/kg/12D-I:KID BMJOAE 291,661,85
orl-wmn TDLo:450 mg/kg:EYE,CNS SAMJAF 67,823,85
orl-wmn TDLo:20 mg/kg/4D-I:GIT CMAJAX 126,894,82
unr-wmn TDLo:120 mg/kg/4D LANCAO 2,745,80
orl-rat LD50:740 mg/kg TOIZAG 28,99,81
ipr-rat LD50:327 mg/kg TXAPA9 18,185,71
ivn-rat LD50:112 mg/kg CMROCX 4,17,76
orl-mus LD50:525 mg/kg JNPHAG 2,259,71
ipr-mus LD50:120 mg/kg ARZNAD 19,36,69
ivn-mus LD50:96 mg/kg YKKZAJ 89(10),1392,69
ivn-cat LD50:100 mg/kg CMROCX 4,17,76

SAFETY PROFILE: Poison by intraperitoneal and intravenous routes. Moderately toxic by ingestion. Human systemic effects by: changes in kidney tubules, changes in structure or function of exocrine pancreas, convulsions or effect on seizure threshold, diarrhea, hypermotility, mydriasis, toxic psychosis, ulceration or bleeding from large intestine. Human reproductive effects by ingestion: menstrual cycle changes or disorders. An experimental teratogen. When heated to decomposition it emits toxic fumes of NO_x. Used as an anti-inflammatory agent.

XRA000 CAS:3118-97-6 ***HR: 3***
1-(2,4-XYLYLAZO)-2-NAPHTHOL
mf: $C_{18}H_{16}N_2O$ mw: 276.36

SYNS: A.F. RED No. 5 ◇ AIZEN FOOD RED No. 5 ◇ BRASILAZINA OIL SCARLET 6G ◇ BRILLIANT OIL SCARLET B ◇ CALCO OIL SCARLET BL ◇ CERES ORANGES RR ◇ CERISOL SCARLET G ◇ CEROTINSCHARLACH G ◇ C.I. 12140 ◇ C.I. SOLVENT ORANGE 7 ◇ 1-((2,4-DIMETHYLPHENYL)AZO)-2-NAPHTHALENOL ◇ EXTRACT D&C RED No. 14 ◇ FAST OIL ORANGE II ◇ FAT RED (YELLOWISH) ◇ FAT SCARLET 2G ◇ FETTORANGE B ◇ GRASAN ORANGE 3R ◇ LACQUER ORANGE VR ◇ MOTIROT G ◇ OIL ORANGE KB ◇ OIL ORANGE N EXTRA ◇ OIL ORANGE R ◇ OIL ORANGE 2R ◇ OIL ORANGE X ◇ OIL ORANGE XO ◇ OIL RED GRO ◇ OIL RED O ◇ OIL RED RO ◇ OIL RED XO ◇ OIL SCARLET ◇ OIL SCARLET 371 ◇ OIL SCARLET APYO ◇ OIL SCARLET BL ◇ OIL SCARLET 6G ◇ OIL SCARLET L ◇ OIL SCARLET YS ◇ ORANGE INSOLUBLE OLG ◇ ORANGE INSOLUBLE RR ◇ ORANGE OIL KB ◇ PONCEAU INSOLUBLE OLG ◇ PYRONALROT R ◇ RED B ◇ RED No. 5 ◇ RESIN SCARLET 2R ◇ RESOFORM ORANGE R ◇ ROT B ◇ ROT GG FETTLOESLICH ◇ SOMALIA ORANGE A2R ◇ SOMALIA ORANGE 2R ◇ SOUDAN II ◇ SUDAN AX ◇ SUDAN ORANGE ◇ SUDAN ORANGE RPA ◇ SUDAN ORANGE RRA ◇ SUDAN RED ◇ SUDAN SCARLET 6G ◇ SUDAN X ◇ WAXAKOL VERMILION L ◇ 1-XYLYLAZO-2-NAPHTHOL ◇ 1-(o-XYLYLAZO)-2-NAPHTHOL

TOXICITY DATA with REFERENCE
mma-sat 50 μg/plate MUREAV 44,9,77
imp-mus TDLo:80 mg/kg:CAR BJCAAI 22,825,68

CONSENSUS REPORTS: IARC Cancer Review: Group 3 IMEMDT 7,56,87; Animal Sufficient Evidence IMEMDT 8,233,75. Reported in EPA TSCA Inventory. Community Right-To-Know List. EPA Genetic Toxicology Program.

SAFETY PROFILE: Questionable carcinogen with experimental carcinogenic data. Mutation data reported. When heated to decomposition it emits toxic fumes of NO_x.

XRS000 CAS:28258-59-5 ***HR: 3***
XYLYL BROMIDE
DOT: UN 1701
mf: C_8H_9Br mw: 185.08

PROP: Colorless liquid. Bp: 212-215° (slt decomp), d: 1.371 @ 23°. Almost insol in water; sol in alc, ether.

SYN: BROMURE de XYLYLE (FRENCH)

TOXICITY DATA with REFERENCE
ihl-hmn LCLo:75 ppm/10M NTIS** PB214-270

DOT Classification: Irritating Material; Label: Irritant; Poison B; Label: Poison.

SAFETY PROFILE: A human poison by inhalation. A powerful irritant. When heated to decomposition it emits toxic fumes of Br^-.

XSJ000 CAS:102584-86-1 ***HR: D***
1,1'-(p-XYLYLENE)BIS(3-(1-AZIRIDINYL)UREA)
mf: $C_{14}H_{20}N_6O_2$ mw: 304.40

SYN: N,N'-BIS-AZIRIDINYLFORMYL-1,4-XYLENEDIAMINE

TOXICITY DATA with REFERENCE
mmo-sat 3290 mg/L MUREAV 31,115,75
mmo-esc 1880 μg/plate MUREAV 31,115,75

SAFETY PROFILE: Mutation data reported. When heated to decomposition it emits toxic fumes of NO_x. See also AMINES.

XSS000 CAS:1519-47-7 ***HR: 3***
p-XYLYLENEBIS(TRIPHENYLPHOSPHONIUM CHLORIDE)
mf: $C_{44}H_{38}P_2{\cdot}2Cl$ mw: 699.66

TOXICITY DATA with REFERENCE
ivn-mus LD50:891 μg/kg CSLNX* NX#00609

CONSENSUS REPORTS: Reported in EPA TSCA Inventory.

SAFETY PROFILE: Poison by intravenous route. When heated to decomposition it emits very toxic fumes of PO_x and Cl^-.

XSS250 CAS:28347-13-9 ***HR: 3***
XYLYLENE CHLORIDE
mf: $C_8H_8Cl_2$ mw: 175.06

SYNS: BIS(CHLOROMETHYL)BENZENE ◇ DICHLOROXYLENE ◇ α,α'-DICHLOROXYLENE ◇ XYLYLENE DICHLORIDE

TOXICITY DATA with REFERENCE
orl-rat LD50:1 g/kg 85GMAT -,26,82
ihl-rat LC50:200 mg/m³/4H 85GMAT -,26,82
orl-mus LD50:470 mg/kg 85GMAT -,26,82
ihl-mus LC75:75 mg/m³/2H 85GMAT -,26,82

CONSENSUS REPORTS: EPA Extremely Hazardous Substances List.

SAFETY PROFILE: Poison by inhalation. Moderately toxic by ingestion. When heated to decomposition it emits toxic fumes of Cl^-. See CHLORINATED HYDRDOCARBONS, AROMATIC.

XSS375 CAS:2014-25-7 ***HR: 3***
s,6-XYLYL ESTER of 1-PIPERIDINEACETIC ACID HYDROCHLORIDE
mf: $C_{15}H_{21}NO_2{\cdot}ClH$ mw: 283.83

SYN: FC 403

TOXICITY DATA with REFERENCE
skn-rbt 200 mg MOD BCFAAI 107,310,68
eye-rbt 1 g MOD BCFAAI 107,310,68
scu-mus LD50:1400 mg/kg BCFAAI 107,310,68
ivn-mus LD50:32 mg/kg BCFAAI 107,310,68

SAFETY PROFILE: Poison by intravenous route. Moderately toxic by subcutaneous route. An eye and skin irritant. When heated to decomposition it emits toxic fumes of NO_x and HCl.

XSS900 CAS:669-49-8 ***HR: 3***
2,6-XYLYL ETHER BROMIDE
mf: $C_{13}H_{22}NO \cdot Br$ mw: 288.27

SYNS: CHOLINE-2,6-XYLYL ETHER BROMIDE ◇ 2-(2,6-DIMETHYLPHENOXY)-N,N,N-TRIMETHYL-ETHANAMINIUM BROMIDE (9CI) ◇ 1 DM 10 ◇ TM 10 ◇ TRIMETHYL(2-(2,6-XYLYLOXY)ETHYL)AMMONIUM BROMIDE ◇ XYLOCHOLINE ◇ XYLOCHOLINE BROMIDE

TOXICITY DATA with REFERENCE
ipr-mus LD50:95 mg/kg BJPCAL 9,471,54
scu-mus LD50:240 mg/kg BJPCAL 12,312,57
ivn-mus LD50:8500 μg/kg BJPCAL 12,312,57

SAFETY PROFILE: Poison by subcutaneous, intravenous, and intraperitoneal routes. When heated to decomposition it emits toxic fumes of Br^-, NO_x, and NH_3. See also BROMIDES.

XTJ000 CAS:2425-10-7 ***HR: 3***
3,4-XYLYL METHYLCARBAMATE
mf: $C_{10}H_{13}NO_2$ mw: 179.24

SYNS: 3,4-DIMETHYLPHENOL METHYLCARBAMATE ◇ 3,4-DIMETHYLPHENYL-N-METHYLCARBAMATE ◇ MEOBAL ◇ N-METHYLCARBAMIC ACID, (3,4-DIMETHYLPHENYL)ESTER ◇ METHYLCARBAMIC ACID-3,4-XYLYL ESTER ◇ MPCM ◇ S-1046 ◇ V 17004

TOXICITY DATA with REFERENCE
cyt-ham:fbr 125 mg/L/48H MUREAV 48,337,77
cyt-ham:lng 340 mg/L GMCRDC 27,95,81
orl-rat LD50:290 mg/kg GUCHAZ 6,218,73
orl-mus LD50:60 mg/kg GUCHAZ 6,218,73

CONSENSUS REPORTS: Reported in EPA TSCA Inventory. EPA Genetic Toxicology Program.
SAFETY PROFILE: Poison by ingestion. Mutation data reported. When heated to decomposition it emits toxic fumes of NO_x. An insecticide used to control hoppers on rice and scales on fruit. See also CARBAMATES.

XTS000 ***HR: D***
3,4-XYLYL-N-METHYLCARBAMATE, nitrosated
mf: $C_{10}H_{12}N_2O_3$ mw: 208.24

SYNS: MEOBAL, NITROSATED (JAPANESE) ◇ METHYLNITROSOCARBAMIC ACID 3,4-XYLYL ESTER

TOXICITY DATA with REFERENCE
mmo-esc 10 μg/plate BECTA6 14,389,75
cyt-ham:lng 5 mg/L GMCRDC 27,95,81

SAFETY PROFILE: Mutation data reported. Many nitroso compounds are carcinogens. When heated to decomposition it emits toxic fumes of NO_x. See also NITROSO COMPOUNDS.

XVA000 CAS:69781-89-1 ***HR: 3***
(2-(3,5-XYLYLOXY)ETHYL)HYDRAZINE HYDROCHLORIDE
mf: $C_{10}H_{16}N_2O \cdot ClH$ mw: 216.74

TOXICITY DATA with REFERENCE
orl-mus LD50:250 mg/kg JMCMAR 6,63,63
ipr-mus LD50:250 mg/kg JMCMAR 6,63,63

SAFETY PROFILE: Poison by ingestion and intraperitoneal routes. When heated to decomposition it emits very toxic fumes of Cl^- and NO_x.

XVJ000 CAS:97196-24-2 ***HR: 3***
(2-(3,4-XYLYLOXY)ETHYL)HYDRAZINE MALEATE
mf: $C_{10}H_{16}N_2O \cdot C_4H_4O_4$ mw: 296.36

TOXICITY DATA with REFERENCE
orl-mus LD50:750 mg/kg JMCMAR 6,63,63
ipr-mus LD50:375 mg/kg JMCMAR 6,63,63

SAFETY PROFILE: Poison by intraperitoneal route. Moderately toxic by ingestion. When heated to decomposition it emits toxic fumes of NO_x.

XVS000 CAS:1665-48-1 ***HR: 2***
5-((3,5-XYLYLOXY)METHYL)-2-OXAZOLIDINONE
mf: $C_{12}H_{15}NO_3$ mw: 221.28

SYNS: 5-((3,5-DIMETHYLPHENOXY)METHYL)-2-OXAZOLIDINONE ◇ METAXALONE ◇ METAZALONE ◇ METAZOLONE ◇ METHOXOLONE ◇ SKELAXIN ◇ 5-(3,5-XYLOLOXYMETHYL)OXAZOLIDIN-2-ONE ◇ ZORANE

TOXICITY DATA with REFERENCE
orl-rat LD50:775 mg/kg 27ZQAG -,261,72
ipr-rat LD50:515 mg/kg ARZNAD 17,242,67
orl-mus LD50:1690 mg/kg AIPTAK 233,326,78
ipr-mus LD50:490 mg/kg 27ZQAG -,261,72

SAFETY PROFILE: Moderately toxic by ingestion and intraperitoneal routes. When heated to decomposition it emits toxic fumes of NO_x. See also KETONES.

XWS000 CAS:78371-75-2 ***HR: 3***
2-(2,6-XYLYLOXY)TRIETHYLAMINE HYDROCHLORIDE
mf: $C_{14}H_{23}NO \cdot ClH$ mw: 257.84

SYN: C 3219

TOXICITY DATA with REFERENCE
eye-rbt 2% MLD ARZNAD 8,708,58
ipr-rat LD50:67 mg/kg ARZNAD 8,708,58
scu-mus LD50:155 mg/kg ARZNAD 8,708,58

SAFETY PROFILE: Poison by intraperitoneal and subcutaneous routes. An eye irritant. When heated to decomposition it emits very toxic fumes of NO_x and HCl. See also AMINES.

Y

YAG000 ***HR: 2***
YAM BEAN

PROP: A vine with large tuberous roots and violet flowers which grow in long stands. The seed pods are about 5 inches long and contain flat seeds which are yellow, brown, or red. They grow wild in Florida, Hawaii, Guam, and the West Indies. They are cultivated in the Gulf coast states.

SYNS: CHOPSUI POTATO ◇ HABILLA (PUERTO RICO) ◇ JICAMA de AQUA (CUBA) ◇ JICAMO (MEXICO) ◇ PACHYRHIZUS EROSUS ◇ POIS COCHON (HAITI) ◇ POIS MANIOC (HAITI) ◇ SARGOTT ◇ WILD YAM BEAN

SAFETY PROFILE: The seeds and mature pods contain the poison saponin. They also contain the insecticidal rotenone and pachyrrhizin. The roots and immature pods are edible. Ingestion of half of a seed produces strong diarrhea which may lead to dehydration and electrolyte loss, especially in children. See also SAPONIN and ROTENONE.

YAK000 ***HR: 1***
YEAST (active)

PROP: In liquid or pressed form (FEREAC 41,15972,76).

DOT Classification: ORM-C; Label: None.

SAFETY PROFILE: A nuisance dust. When heated to decomposition it emits acrid smoke and irritating fumes.

YAK100 ***HR: 2***
YELLOW JESSAMINE

PROP: A climbing evergreen with 4-inch long, lance-shaped paired leaves. The funnel-shaped flowers are bright yellow and fragrant. The seeds have wings. It grows wild in wooded areas in the region bounded by Virginia, Florida, Texas, and Arkansas, and is cultivated in the same areas and in southern California.

SYNS: CAROLINA JASMINE ◇ CAROLINA YELLOW JASMINE ◇ CAROLINA WILD WOODBINE ◇ EVENING TRUMPET FLOWER ◇ GELSEMIUM SEMPERVIRENS ◇ MADRESELVA (MEXICO) ◇ WOOD VINE ◇ YELLOW FALSE JESSAMINE

SAFETY PROFILE: The whole plant contains the poisons gelsemine, gelsemicine, and other alkaloids. Ingestion of any part of the plant and particularly the flowers may cause headache, dizziness, visual disturbances, and muscular weakness. Some effects are similar to mild strychnine poisoning. See also GELSEMINE.

YAK300 ***HR: 3***
YELLOW NIGHTSHADE

PROP: A shrub-like vine which produces clusters of yellow flowers. The flowers have 5 petals which may have red marks on the inside. The winged seeds are contained in 8-inch long, narrow seed pods. It grows wild in Florida, the Bahamas, and the Lesser Antilles.

SYNS: BABEIRO AMARILLO (PUERTO RICO) ◇ BEJUCO AHOJA VACA (DOMINICAN REPUBLIC) ◇ CATESBY'S VINE (BAHAMAS) ◇ CORNE CABRITE (HAITI) ◇ CURAMAGUEY (CUBA) ◇ NIGHTSAGE (JAMAICA) ◇ URECHITES LUTEA ◇ WILD ALLAMANDA (FLORIDA) ◇ WILD NIGHTSHADE ◇ WILD UNCTION (BAHAMAS)

SAFETY PROFILE: The leaf contains the poisonous urechitoxin, a cardiac glycoside. Human systemic effects by ingestion include: mouth pain, nausea, vomiting, abdominal pain and cramps, diarrhea. Cardiac glycosides may cause death by their effect on heart function. See also DIGITALIS.

YAK350 ***HR: 3***
YELLOW OLEANDER

PROP: A small tree which grows to 20 feet. It produces a pink-tinged, yellow flower and a 1-inch, clam-shaped seed pod which holds up to 4 flat seeds. The tree grows wild in southern Florida, the southwestern United States, Hawaii, Guam, and the West Indies.

SYNS: AHOUAI des ANTILLES ◇ BE-STILL TREE ◇ CABLONGA ◇ FLOR del PERU ◇ LUCKY NUT ◇ NOHO-MALIE (HAWAII) ◇ RETAMA ◇ SERPENT ◇ THEVETIA PERUVIANA

SAFETY PROFILE: All parts of the plant and especially the seeds contain poisonous digitalis-like glycosides. Human systemic effects by ingestion include: mouth pain, nausea, vomiting, abdominal pain and cramps, diarrhea. Cardiac glycosides may cause death by their effect on heart function. See also DIGITALIS.

YAK500 ***HR: 3***
YEW

PROP: Evergreen trees and shrubs with a thin, red-brown, scaley bark and needle-like, 1-inch leaves. The hard seeds range from green to black in color and are contained in a red cup. The various species grow throughout most of North America.

SYNS: BUIS de SAPIA (CANADA) ◇ GROUND HEMLOCK ◇ TAXUS (VARIOUS SPECIES)

SAFETY PROFILE: Most of the plant, including the seeds, contains poisonous taxine alkaloids. Ingestion of these plant parts may cause dizziness, dilated pupils, abdominal cramps, vomiting, slowed heartbeat and cardiac arrhythmias, low blood pressure, labored breathing, coma and death by cardiac or respiratory failure. See also TAXINE.

YAT000 CAS:8006-81-3 ***HR: 1***
YLANG YLANG OIL

PROP: Light yellow, very fragrant liquid. D: 0.930-0.950 @ 20/20°. From steam distillation of the flowers of *Cananga odorata hook F. et al* (FCTXAV 12,807,74).

TOXICITY DATA with REFERENCE
skn-rbt 500 mg/24H MLD FCTXAV 12,807,74

CONSENSUS REPORTS: Reported in EPA TSCA Inventory.

SAFETY PROFILE: A skin irritant.

YBJ000 CAS:146-48-5 ***HR: 3***
YOHIMBINE
mf: $C_{21}H_{26}N_2O_3$ mw: 354.49

PROP: Colorless needles from water and alc. Mp: 234°. Sltly sol in water, ether; sol in alc, chloroform, hot benzene.

SYNS: APHRODINE ◇ APHROSOL ◇ CORYNINE ◇ 17-HYDROXY-YOHIMBAN-16-CARBOXYLIC ACID METHYL ESTER ◇ QUEBRACHIN ◇ QUEBRACHINE ◇ YOHIMBIC ACID METHYL ESTER

TOXICITY DATA with REFERENCE
orl-mus LD50:51 mg/kg PSSCBG 11,555,80
scu-mus LDLo:200 mg/kg HBAMAK 4,1418,35
ivn-mus LDLo:16 mg/kg AIPTAK 50,241,35
scu-dog LDLo:20 mg/kg LDBU** -,3,32
orl-cat LD50:43 mg/kg ARZNAD 5,432,55
ipr-cat LD50:16 mg/kg ARZNAD 5,432,55
scu-cat LD50:37 mg/kg ARZNAD 5,432,55
scu-rbt LDLo:50 mg/kg HBAMAK 4,1418,35
ivn-rbt LDLo:11 mg/kg HBAMAK 4,1418,35

SAFETY PROFILE: Poison by ingestion, subcutaneous, intravenous, and intraperitoneal routes. Cases of poisoning have occurred from its use as an aphrodisiac. Upon local application it produces anesthesia. However, absorption of it can give rise to toxic symptoms, such as salivation, increased respiration, and repeated defecation. With reference to the circulatory system, there may be a fall in blood pressure and sometimes myocardial damage, involving particularly the conduction system of the heart with a resultant decrease in the efficiency of the heart. An adrenergic blocker used to treat arteriosclerosis and angina pectoris. Formerly used as a local anesthetic and mydriatic (pupillary dilator). When heated to decomposition it emits toxic fumes of NO_x.

YBS000 CAS:65-19-0 ***HR: 3***
YOHIMBINE HYDROCHLORIDE
mf: $C_{21}H_{26}N_2O_3•ClH$ mw: 390.95

PROP: Orthorhombic needles. Decomp @ 302°.

SYNS: APHRODINE HYDROCHLORIDE ◇ YOHIMBINE MONOHYDROCHLORIDE

TOXICITY DATA with REFERENCE
ipr-rat LD50:55 mg/kg AIPTAK 110,20,57
orl-mus LDLo:50 mg/kg LDBU** -,3,32
scu-dog LDLo:20 mg/kg LDBU** -,3,32
scu-frg LD50:34 mg/kg CRSBAW 137,305,43

CONSENSUS REPORTS: Reported in EPA TSCA Inventory.

SAFETY PROFILE: Poison by ingestion, intraperitoneal, and subcutaneous routes. When heated to decomposition it emits toxic fumes of NO_x and HCl. See also YOHIMBINE.

YCA000 CAS:6211-32-1 ***HR: 3***
α-YOHIMBIN HYDROCHLORIDE
mf: $C_{21}H_{26}N_2O_3•ClH$ mw: 390.95

PROP: Crystals. Mp: 288°.

SYNS: 17-α-HYDROXY-20-α-YOHIMBAN-16-β-CARBOXYLIC ACID, METHYL ESTER, HYDROCHLORIDE ◇ RAUWOLSCINE HYDROCHLORIDE

TOXICITY DATA with REFERENCE
orl-mus LDLo:125 mg/kg LDBU** -,3,32
scu-dog LDLo:20 mg/kg LDBU** -,3,32

SAFETY PROFILE: Poison by ingestion and subcutaneous routes. When heated to decomposition it emits very toxic fumes of NO_x and HCl. See also YOHIMBINE.

YCJ000 CAS:3458-22-8 ***HR: 3***
YOSHI 864
mf: $C_8H_{19}NO_6S_2•ClH$ mw: 325.86

SYNS: N,N-BIS(METHYLSULFONEPROPOXY)AMINEHYDROCHLORIDE ◇ COMPOUND 864 ◇ 3,3′-IMIDODI-1-PROPANOL, DIMETHANESULFONATE (ester), HYDROCHLORIDE ◇ IPD ◇ NCI-C01547 ◇ NSC 102627 ◇ SAKURAI No. 864

TOXICITY DATA with REFERENCE
oms-hmn:lym 100 mg/L EJCAAH 14,741,78
ipr-rat TDLo:3744 mg/kg/52W-I:NEO NCITR* NCI-CG-TR-18,78

ivn-hmn TDLo:5400 μg/kg/2D-I:CNS,GIT CANCAR 35,1145,75
ivn-rat LD50:75 mg/kg ARZNAD 24,1139,74
ipr-mus LD10:170 mg/kg JMCMAR 20,515,77

CONSENSUS REPORTS: NCI Carcinogenesis Bioassay (ipr); Clear Evidence: mouse, rat NCITR* NCI-CG-TR-18,78

SAFETY PROFILE: Poison by intraperitoneal and intravenous routes. Human systemic effects by intravenous route: somnolence, hypermotility, diarrhea, nausea or vomiting. Questionable carcinogen with experimental neoplastigenic data. Human mutation data reported. When heated to decomposition it emits very toxic fumes of NO_x, SO_x, and HCl.

YCJ200 CAS:13171-25-0 ***HR: 3***
YOSIMILON
mf: $C_{14}H_{22}N_2O_3 \cdot 2ClH$ mw: 339.30

SYNS: KYURINETT ◇ S 4004 ◇ 1-(2,3,4-TRIMETHOXYBENZYL)PIPERAZINE DIHYDROCHLORIDE ◇ 1-((2,3,4-TRIMETHOXYPHENYL)METHYL)PIPERAZINE DIHYDROCHLORIDE (9CI) ◇ TRIMETAZIDINE DIHYDROCHLORIDE ◇ TRIMETAZIDINE HYDROCHLORIDE ◇ VASTAREL

TOXICITY DATA with REFERENCE
orl-rat LD50:1700 mg/kg NIIRDN 6,527,82
ipr-rat LD50:345 mg/kg NIIRDN 6,527,82
scu-rat LD50:1500 mg/kg NIIRDN 6,527,82
orl-mus LD50:1550 mg/kg NIIRDN 6,527,82
ipr-mus LD50:310 mg/kg MEIEDD 10,1386,83
scu-mus LD50:410 mg/kg NIIRDN 6,527,82
ivn-mus LD50:150 mg/kg MEIEDD 10,1386,83

SAFETY PROFILE: Poison by intravenous and intraperitoneal routes. Moderately toxic by ingestion and subcutaneous routes. When heated to decomposition it emits toxic fumes of NO_x and HCl.

YDA000 CAS:7440-64-4 ***HR: 3***
YTTERBIUM
af: Yb aw: 173.04

PROP: A bright, silvery, lustrous soft, malleable, ductile, and fairly stable element. Mp: 824°, bp: 1193°, d: 6.977. A rare earth.

TOXICITY DATA with REFERENCE
imp-mus TDLo:25 g/kg:ETA PSEBAA 135,426,70

CONSENSUS REPORTS: Reported in EPA TSCA Inventory.

SAFETY PROFILE: As a lanthanon it may have an anticoagulant action on blood. Questionable carcinogen with experimental tumorigenic data. Flammable in the form of dust when reacted with air, halogens. See also LANTHANUM and RARE EARTHS.

YDJ000 CAS:10361-91-8 ***HR: 3***
YTTERBIUM CHLORIDE
mf: Cl_3Yb mw: 279.39

PROP: Hexahydrate, deliq needles or crystals. D: 2.575; mp: 150-155°.

SYN: YTTERBIUM TRICHLORIDE

TOXICITY DATA with REFERENCE
skn-rbt 500 mg/24H MOD TXAPA9 5,427,63
eye-rbt 50 mg TXAPA9 5,427,63
ivn-ham TDLo:50 mg/kg (8D preg):REP TJADAB 11,289,75
ivn-ham TDLo:100 mg/kg (female 8D post):TER TJADAB 11,289,75
orl-mus LD50:4836 mg/kg EQSSDX 1,1,75
ipr-mus LD50:300 mg/kg EQSSDX 1,1,75
ipr-gpg LD50:132 mg/kg AEHLAU 5,437,62

CONSENSUS REPORTS: Reported in EPA TSCA Inventory.

SAFETY PROFILE: Poison by intraperitoneal route. Mildly toxic by ingestion. An experimental teratogen. A skin and eye irritant. When heated to decomposition it emits toxic fumes of Cl^-. See also YTTERBIUM, RARE EARTHS, and CHLORIDES.

YDS800 CAS:13768-67-7 ***HR: 3***
YTTERBIUM NITRATE
mf: $NO_3 \cdot Yb$ mw: 235.05
PROP: Crystals.

SYN: NITRIC ACID, YTTERBIUM(3+) SALT

TOXICITY DATA with REFERENCE
itt-rat TDLo:18884 μg/kg (1D male):REP JRPFA4 7,21,64
orl-rat LD50:1623 mg/kg EQSSDX 1,1,75
ipr-rat LD50:128 mg/kg EQSSDX 1,1,75
orl-mus LD50:126 mg/kg EQSSDX 1,1,75

CONSENSUS REPORTS: Reported in EPA TSCA Inventory.

SAFETY PROFILE: Poison by ingestion and intraperitoneal routes. Experimental reproductive effects. When heated to decomposition it emits toxic fumes of NO_x. See also YTTERBIUM, RARE EARTHS, and NITRATES.

YEA000 CAS:13839-85-5 ***HR: 3***
YTTERBIUM(III) NITRATE, HEXAHYDRATE (1:3:6)
mf: $N_3O_9 \cdot Yb \cdot 6H_2O$ mw: 467.19

SYN: NITRIC ACID, YTTERBIUM(3+) SALT, HEXAHYDRATE

TOXICITY DATA with REFERENCE
orl-rat LD50:3100 mg/kg TXAPA9 5,750,63

ipr-rat LD50:255 mg/kg TXAPA9 5,750,63
ipr-mus LD50:250 mg/kg TXAPA9 5,750,63

SAFETY PROFILE: Poison by intraperitoneal route. Moderately toxic by ingestion. When heated to decomposition it emits toxic fumes of NO_x. See also YTTERBIUM NITRATE.

YEJ000 CAS:7440-65-5 ***HR: 3***
YTTRIUM
af: Y aw: 88.9059

PROP: Hexagonal, gray-black, metallic, rare earth element. Mp: 1509°, bp: 3200°, d: 4.472.

SYN: YTTRIUM-89

CONSENSUS REPORTS: Reported in EPA TSCA Inventory.

OSHA PEL: TWA 1 $mg(Y)/m^3$
ACGIH TLV: TWA 1 $mg(Y)/m^3$
DFG MAK: 5 $mg(Y)/m^3$

SAFETY PROFILE: As a lanthanon, it may have an anticoagulant effect on the blood. Flammable in the form of dust when reacted with air; halogens. See also LANTHANUM and RARE EARTHS.

YES000 CAS:10361-92-9 ***HR: 3***
YTTRIUM CHLORIDE
mf: Cl_3Y mw: 195.26

PROP: Hexahydrate, colorless, deliq crystals. Sol in water, alc.

SYN: YTTRIUM TRICHLORIDE

TOXICITY DATA with REFERENCE
ipr-rat LD50:45 mg/kg AMIHAB 16,475,57
ipr-mus LD50:88 mg/kg AMIHAB 16,475,57
ipr-gpg LD50:85 mg/kg AEHLAU 5,437,62

CONSENSUS REPORTS: Reported in EPA TSCA Inventory.
ACGIH TLV: TWA 1 $mg(Y)/m^3$

SAFETY PROFILE: Poison by intraperitoneal route. When heated to decomposition it emits toxic fumes of Cl^-. See also YTTRIUM and RARE EARTHS.

YFA000 CAS:63938-20-5 ***HR: 3***
YTTRIUM CITRATE
mf: $C_6H_8O_7 \cdot 1/3Y$ mw: 221.75

SYN: CITRIC ACID, YTTRIUM SALT (3:1)

TOXICITY DATA with REFERENCE
ipr-mus LD50:79 mg/kg:ETA AEHLAU 5,437,62
ipr-gpg LD50:44 mg/kg AEHLAU 5,437,62
ACGIH TLV: TWA 1 $mg(Y)/m^3$

SAFETY PROFILE: Poison by intraperitoneal route. Questionable carcinogen with experimental tumorigenic data. When heated to decomposition it emits acrid smoke and irritating fumes. See also YTTRIUM and RARE EARTHS.

YFA100 CAS:12558-71-3 ***HR: 3***
YTTRIUM EDETATE complex
mf: $C_{10}H_{13}N_2O_8 \cdot Y$ mw: 378.16

SYN: ACETIC ACID, (ETHYLENEDINITRILO)TETRA-, YTTRIUM complex

TOXICITY DATA with REFERENCE
ipr-gpg LD50:107 mg/kg AEHLAU 5,437,62
ACGIH TLV: TWA 1 $mg(Y)/m^3$

SAFETY PROFILE: Poison by intraperitoneal route. When heated to decomposition it emits toxic fumes of NO_x and Y.

YFJ000 CAS:10361-93-0 ***HR: 3***
YTTRIUM(III) NITRATE (1:3)
mf: $N_3O_9 \cdot Y$ mw: 274.94

PROP: Hexahydrate, deliq crystals. Sol in water.

SYN: NITRIC ACID, YTTRIUM(3+) SALT

TOXICITY DATA with REFERENCE
spm-dom-itr 5 mg/kg IJEBA6 11,143,73
itt-dom TDLo:5 mg/kg (1D male):REP IJEBA6 11,143,73
orl-mus TDLo:1300 mg/kg/60W-C:ETA JONUAI 101,1431,71
ipr-rat LD50:350 mg/kg AIHOAX 1,637,50
ipr-mus LD50:1710 mg/kg EQSFAP 1,1,75
ivn-rbt LD50:515 mg/kg EQSSDX 1,1,75

CONSENSUS REPORTS: Reported in EPA TSCA Inventory.
ACGIH TLV: TWA 1 $mg(Y)/m^3$

SAFETY PROFILE: Poison by intraperitoneal route. Moderately toxic by intravenous route. Experimental reproductive effects. Questionable carcinogen with experimental tumorigenic data. Mutation data reported. When heated to decomposition it emits toxic fumes of NO_x. See also NITRATES, YTTRIUM, and RARE EARTHS.

YFS000 CAS:13494-98-9 ***HR: 3***
YTTRIUM(III) NITRATE HEXAHYDRATE (1:3:6)
mf: $N_3O_9 \cdot Y \cdot 6H_2O$ mw: 383.06

PROP: Colorless to redish triclinic crystals. Deliq. D: 2.68, mp: $-3H_2O$. Sol in water and alc.

SYNS: NITRIC ACID, YTTRIUM(3+)SALT, HEXAHYDRATE ◇ YTTRIUMNITRAT (GERMAN)

TOXICITY DATA with REFERENCE
scu-mus LDLo:1660 mg/kg AEPPAE 141,273,29
scu-frg LDLo:350 mg/kg AEPPAE 141,273,29
ACGIH TLV: TWA 1 mg(Y)/m^3

SAFETY PROFILE: Poison by subcutaneous route. When heated to decomposition it emits toxic fumes of NO_x. See also YTTRIUM NITRATE.

YGA000 CAS:1314-36-9 ***HR: 2***
YTTRIUM OXIDE
mf: O_3Y_2 mw: 225.82

PROP: White powder. D: 4.84, mp: 2410. Ins in water.

SYN: YTTRIA

TOXICITY DATA with REFERENCE
ipr-rat LD50:500 mg/kg AIHOAX 1,637,50

CONSENSUS REPORTS: Reported in EPA TSCA Inventory.

ACGIH TLV: TWA 1 mg(Y)/m^3

SAFETY PROFILE: Moderately toxic by intraperitoneal route. See also YTTRIUM and RARE EARTHS.

YGA700 CAS:89194-77-4 ***HR: 3***
YUTAC
mf: $C_{17}H_{23}ClN_2O_2 \cdot ClH$ mw: 359.29

SYNS: 4-CHLOROBENZOICACID-3-ETHYL-7-METHYL-3,7-DIAZABICYCLO(3.3.1)NON-9-YL ESTER HYDROCHLORIDE ◇ 3-ETHYL-7-METHYL-9-α-(4'-CHLOROBENZOYLOXY)-3,7-DIAZABICYCLO(3.3.1) NONAME HYDROCHLORIDE ◇ RGH 2957

TOXICITY DATA with REFERENCE
orl-rat LD50:383 mg/kg DRFUD4 10,837,85
ivn-rat LD50:17600 μg/kg DRFUD4 10,837,85
orl-mus LD50:378 mg/kg DRFUD4 10,837,85
ivn-mus LD50:25900 μg/kg DRFUD4 10,837,85

SAFETY PROFILE: Poison by ingestion and intravenous routes. When heated to decomposition it emits toxic fumes of NO_x and HCl. See also ESTERS.

Z

ZAK000 ***HR: 3***
ZAMIA DEBILIS

PROP: Dried, ground-up zamia tubers were used (85CVA2 5,197,70).

TOXICITY DATA with REFERENCE
orl-rat TDLo:650 g/kg/77W-C:ETA 85CVA2 5,197,70

SAFETY PROFILE: Questionable carcinogen with experimental tumorigenic data.

ZAK300 CAS:17560-51-9 ***HR: 3***
ZAROXOLYN
mf: $C_{16}H_{16}ClN_3O_3S$ mw: 365.86

PROP: Crystals from ethanol. Mp: 253-259°.

SYNS: 7-CHLORO-1,2,3,4-TETRAHYDRO-2-METHYL-3-(2-METHYLPHENYL)-4-OXO-6-QUINAZOLINESULFONAMIDE ◇ DIULO ◇ METENIX ◇ 2-METHYL-3-o-TOLYL-6-SULFAMYL-7-CHLORO-1,2,3,4-TETRAHYDRO-4-QUINAZOLINONE ◇ METOLAZONE ◇ OLDREN ◇ SR 720-22

TOXICITY DATA with REFERENCE
orl-rat TDLo:22 mg/kg (7-17D preg):REP KSRNAM 12,3394,78
orl-rat TDLo:2750 mg/kg (7-17D preg):TER KSRNAM 12,3394,78
orl-wmn TDLo:150 μg/kg:CNS BMJOAE 1,1381,76

SAFETY PROFILE: Human systemic effects by ingestion: general anesthesia, convulsions, and muscle contractions. An experimental teratogen. Experimental reproductive effects. When heated to decomposition it emits Cl^-, SO_x, and NO_x.

ZAT000 CAS:17924-92-4 ***HR: 3***
ZEARALENONE
mf: $C_{18}H_{22}O_5$ mw: 318.40

PROP: l-Form: Crystals. Mp: 164-165°. Sol in aq alkali, ether, benzene, alc; almost insol in water. dl-Form: Crystals. Mp: 187-189°.

SYNS: COMPOUND F-2 ◇ 14,16-DIHYDROXY-3-METHYL-7-OXO-trans-BENZOXACYCLOTETRADEC-11-EN-1-ONE ◇ FES ◇ F-2 TOXIN ◇ FUSARIUM TOXIN ◇ trans-6-(10-HYDROXY-6-OXO-1-UNDECENYL)-mu-LACTONE, RESORCYLIC ACID ◇ 6-(10-HYDROXY-6-OXO-trans-1-UNDECENYL)-β-RESORCYCLIC ACID-N-LACTONE ◇ MYCOTOXIN F2 ◇ NCI-C50226 ◇ (−)-ZEARALENONE ◇ (s)-ZEARALENONE ◇ (10s)-ZEARALENONE ◇ trans-ZEARALENONE

TOXICITY DATA with REFERENCE
skn-gpg 50 mg/24H SEV JANCA2 57,1121,74
dnr-bcs 2500 mg/L IRLCDZ 7,204,79
mrc-bcs 100 μg/disc CNREA8 36,445,76
orl-pig TDLo:123 mg/kg (female 6D pre):REP AJVRAH 40,1260,79
orl-mus TDLo:20 mg/kg (female 9D post):TER ACVTA8 22,524,81
orl-mus TDLo:8652 mg/kg/2Y-C:NEO NTPTR* NTP-TR-235,82
orl-mus TD:4326 mg/kg/2Y-C:ETA NTPTR* NTP-TR-235,82

CONSENSUS REPORTS: IARC Cancer Review: Group 3 IMEMDT 7,56,87; Human Inadequate Evidence IMEMDT 31,279,83; Animal Limited Evidence IMEMDT 31,279,83. NTP Carcinogenesis Bioassay (feed); Clear Evidence: mouse NTPTR* NTP-TR-235,82; (feed); No Evidence: rat NTPTR* NTP-TR-235,82. Reported in EPA TSCA Inventory. EPA Genetic Toxicology Program.

SAFETY PROFILE: Experimental reproductive effects. Questionable carcinogen with experimental neoplastigenic, tumorigenic, and teratogenic data. Mutation data reported. A severe skin irritant. When heated to decomposition it emits acrid smoke and irritating fumes.

ZBA000 ***HR: D***
ZETAR EMULSION

PROP: A shampoo containing coal tar derivatives (TOLED5 3,325,79).

SYN: ZET

TOXICITY DATA with REFERENCE
mmo-sat 5 mg/plate PHMGBN 20,1,80
mma-sat 10 μg/plate TOLED5 3,325,79

SAFETY PROFILE: Mutation data reported.

ZBA500 CAS:56775-88-3 ***HR: 3***
ZIMELIDINE
mf: $C_{16}H_{17}BrN_2$ mw: 317.26

PROP: Crystals. Mp: 193°

SYNS: 3-(p-BROMOPHENYL)-N,N-DIMETHYL-3-(3-PYRIDYL)ALLYLAMINE ◇ 3-(4-BROMOPHENYL)-N,N-DIMETHYL-3-(3-PYRIDINYL)-2-PROPEN-1-AMINE ◇ (Z)-3-(4′-BROMOPHENYL)-3-(3″-PYRIDYL) DIMETHYLALLYLAMINE ◇ cis-H 102.09 ◇ cis-ZIMELIDINE ◇ (Z)-ZIMELIDINE

TOXICITY DATA with REFERENCE
scu-rat TDLo:280 mg/kg (female 14D pre):REP NEAGDO 7,9,86
orl-wmn TDLo:56 mg/kg/14D:GIT,CNS BMJOAE 285,1009,82
orl-rat LD50:900 mg/kg DRUGAY 24,169,82
ivn-rat LD50:50 mg/kg DRUGAY 24,169,82
orl-mus LD50:800 mg/kg DRUGAY 24,169,82
ivn-mus LD50:60 mg/kg DRUGAY 24,169,82

SAFETY PROFILE: Poison by intravenous route. Experimental reproductive effects. Moderately toxic by ingestion. Human systemic effects by ingestion: muscle weakness, headache and nausea. When heated to decomposition it emits toxic fumes of Br^- and NO_x.

ZBA525 CAS:60525-15-7 ***HR: 3***
ZIMELIDINE DIHYDROCHLORIDE
mf: $C_{16}H_{17}BrN_2$•2ClH mw: 390.18

SYNS: (Z)-3-(4-BROMOPHENYL)-N,N-DIMETHYL-3-(3-PYRIDINYL)-2-PROPEN-1-AMINE DIHYDROCHLORIDE ◇ H102/09 HYDROCHLORIDE ◇ ZIMELIDINE HYDROCHLORIDE

TOXICITY DATA with REFERENCE
orl-rat TDLo:12 mg/kg (6-17D preg):REP KSRNAM 17,1833,83
orl-rat TDLo:120 mg/kg (6-17D preg):TER KSRNAM 17,1833,83
orl-man TDLo:36 mg/kg BMJOAE 287,1672,83
orl-wmn TDLo:12 mg/kg/6D-I:PNS HUTODJ 3,141,84
orl-wmn TDLo:36 mg/kg/9D-I:GIT,LIV BMJOAE 287,1181,83
orl-rat LD50:844 mg/kg APSXAS 20,295,83
ipr-rat LD50:99800 μg/kg APSXAS 20,295,83
scu-rat LD50:227 mg/kg APSXAS 20,295,83
ivn-rat LD50:45800 μg/kg KSRNAM 17,1833,83
orl-mus LD50:341 mg/kg KSRNAM 17,1833,83
ipr-mus LD50:84400 μg/kg APSXAS 20,295,83
scu-mus LD50:154 mg/kg KSRNAM 17,1833,83
ivn-mus LD50:27700 μg/kg KSRNAM 17,1833,83
orl-dog LD50:271 mg/kg KSRNAM 17,1833,83
ivn-dog LD50:57 mg/kg KSRNAM 17,1833,83
orl-cat LDLo:80 mg/kg APSXAS 20,295,83
orl-rbt LD50:300 mg/kg KSRNAM 17,1833,83
ivn-rbt LD50:50800 μg/kg KSRNAM 17,1833,83

SAFETY PROFILE: Poison by ingestion, subcutaneous, intravenous, and intraperitoneal routes. Human systemic effects: ataxia, diarrhea, fasciculations, hypermotility, jaundice. An experimental teratogen. Experimental reproductive effects. When heated to decomposition it emits toxic fumes of Br^-, NO_x, and HCl. See also ZIMELIDINE.

ZBJ000 CAS:7440-66-6 ***HR: 3***
ZINC
DOT: UN 1383/UN 1436
af: Zn aw: 65.37

PROP: Bluish-white, lustrous, metallic element. Mp: 419.8°, bp: 908°, d: 7.14 @ 25°, vap press: 1 mm @ 487°. Stable in dry air.

SYNS: BLUE POWDER ◇ C.I. 77945 ◇ C.I. PIGMENT BLACK 16 ◇ C.I. PIGMENT METAL 6 ◇ EMANAY ZINC DUST ◇ GRANULAR ZINC ◇ JASAD ◇ MERRILLITE ◇ PASCO ◇ ZINC DUST ◇ ZINC POWDER ◇ ZINC, POWDER or DUST, non-pyrophoric (DOT) ◇ ZINC, POWDER or DUST, pyrophoric (DOT)

TOXICITY DATA with REFERENCE
skn-hmn 300 μg/3D-I:MLD 85DKA8 -,127,77
ihl-hmn TCLo:124 mg/m^3/50M:PUL,SKN AHYGAJ 72,358,10

CONSENSUS REPORTS: Zinc and its compounds are on the Community Right-To-Know List. Reported in EPA TSCA Inventory. EPA Genetic Toxicology Program.

DOT Classification: Flammable Solid; Label: Dangerous When Wet, non-pyrophoric; Flammable Solid; Label: Spontaneously Combustible, pyrophoric.

SAFETY PROFILE: Human systemic effects by ingestion: cough, dyspnea, and sweating. A human skin irritant. Pure zinc powder, dust, fume is relatively nontoxic to humans by inhalation. The difficulty arises from oxidation of zinc fumes immediately prior to inhalation or presence of impurities such as Cd, Sb, As, Pb. Inhalation may cause sweet taste, throat dryness, cough, weakness, generalized aches, chills, fever, nausea, vomiting.

Flammable in the form of dust when exposed to heat or flame. May ignite spontaneously in air when dry. Explosive in the form of dust when reacted with acids. Incompatible with NH_4NO_3; BaO_2; $Ba(NO_3)_2$; Cd; CS_2; chlorates; Cl_2; ClF_3; CrO_3; (ethyl acetoacetate + tribromoneopentyl alcohol); F_2; hydrazine mononitrate; hydroxylamine; $Pb(N_3)_2$; (Mg + $Ba(NO_3)_2$ + BaO_2); $MnCl_2$; HNO_3; performic acid; $KClO_3$; KNO_3; K_2O_2; Se; $NaClO_3$; Na_2O_2; S; Te; H_2O; $(NH_4)_2S$; As_2O_3; CS_2; $CaCl_2$; NaOH; chlorinated rubber; catalytic metals; halocarbons; o-nitroanisole; nitrobenzene; non-metals; oxidants; paint primer base; pentacarbonyliron; transition metal halides; seleninyl bromide. To fight fire, use special mixtures of dry chemical. When heated to decomposition it emits toxic fumes of ZnO. See also ZINC COMPOUNDS.

ZBS000 CAS:557-34-6 ***HR: 3***
ZINC ACETATE
mf: $C_4H_6O_4$•Zn mw: 183.47

PROP: Crystals; astringent taste. D: 1.735, mp: 237°. Very sol in water; somewhat sol in alc.

SYNS: ACETIC ACID, ZINC SALT ◇ DICARBOMETHOXYZINC ◇ ZINC DIACETATE

TOXICITY DATA with REFERENCE
ivg-rbt TDLo:10525 μg/kg (1D pre):REP CCPTAY 22,659,80
orl-rat LD50:2510 mg/kg MarJV# 29MAR77
ipr-mus LD50:57 mg/kg TXAPA9 49,41,79
orl-rbt LDLo:976 mg/kg 27ZWAY 3.3,1925,-
ivn-rbt LDLo:5 mg/kg AIMDAP 37,641,26

CONSENSUS REPORTS: Zinc and its compounds are on the Community Right-To-Know List. Reported in EPA TSCA Inventory.

SAFETY PROFILE: Poison by intraperitoneal and intravenous routes. Moderately toxic by ingestion. Experimental reproductive effects. Incompatible with zinc salts; alkalies and their carbonates; oxalates; phosphates; sulfides. When heated to decomposition it emits toxic fumes of ZnO. See also ZINC COMPOUNDS.

ZCA000 CAS:5970-45-6 ***HR: 2***
ZINC ACETATE, DIHYDRATE
mf: $C_4H_6O_4 \cdot Zn \cdot 2H_2O$ mw: 219.51

PROP: Crystals; slt acetone odor; sltly efflorescent, astringent taste. D: 1.735, mp: 237°, loses $2H_2O$ @ >100°. Very sol in water.

SYNS: ACETIC ACID, ZINC SALT, DIHYDRATE ◇ OCTAN ZINECNATY (CZECH) ◇ ZINC DIACETATE, DIHYDRATE

TOXICITY DATA with REFERENCE
skn-rbt 500 mg/24H MLD 28ZPAK -,10,72
eye-rbt 20 mg/24H MOD 28ZPAK -,10,72
cyt-hmn:lym 7 mg/L CYGEDX 12(3),46,78
orl-rat LD50:2170 mg/kg 28ZPAK -,10,72

CONSENSUS REPORTS: Zinc and its compounds are on the Community Right-To-Know List.

SAFETY PROFILE: Moderately toxic by ingestion. Human mutation data reported. A skin and eye irritant. Keep in well closed containers. Incompatible with zinc salts in general. When heated to decomposition it emits toxic fumes of ZnO. See also ZINC ACETATE and ZINC COMPOUNDS.

ZCJ000 CAS:14024-63-6 ***HR: 3***
ZINC ACETOACETONATE
mf: $C_{10}H_{14}O_4Zn$ mw: 263.61

PROP: Powder.

SYNS: BIS(2,4-PENTANEDIONATO-O,O')ZINC ◇ ZINC 2,4-PENTANEDIONATE

TOXICITY DATA with REFERENCE
ipr-rat LD50:50 mg/kg NCIUS* PH 43-64-886,SEPT,70

CONSENSUS REPORTS: Zinc and its compounds are on the Community Right-To-Know List. Reported in EPA TSCA Inventory.

SAFETY PROFILE: Poison by intraperitoneal route. When heated to decomposition it emits toxic fumes of ZnO. See also ZINC COMPOUNDS.

ZCS000 CAS:63904-83-6 ***HR: 3***
ZINC ALLYL DITHIO CARBAMATE
mf: $C_8H_{12}N_2S_4Zn$ mw: 329.83

SYN: BIS(ALLYLDITHIOCARBAMATO)ZINC

TOXICITY DATA with REFERENCE
orl-rat LD50:375 mg/kg SCCUR* -,9,61
orl-mus LD50:440 mg/kg SCCUR* -,9,61
orl-rbt LDLo:420 mg/kg SCCUR* -,9,61

CONSENSUS REPORTS: Zinc and its compounds are on the Community Right-To-Know List.

SAFETY PROFILE: Poison by ingestion. When heated to decomposition it emits very toxic fumes of NO_x, SO_x, and ZnO. See also ZINC COMPOUNDS and CARBAMATES.

ZDA000 CAS:63885-01-8 ***HR: 3***
ZINC AMMONIUM NITRITE
DOT: UN 1512

PROP: Solid.

CONSENSUS REPORTS: Zinc and its compounds are on the Community Right-To-Know List.

DOT Classification: Oxidizer; Label: Oxidizer.

SAFETY PROFILE: Flammable by spontaneous chemical reaction. A powerful oxidizing agent. When heated to decomposition it emits toxic fumes of NO_x, NH_3, and ZnO. See also NITRITES and ZINC COMPOUNDS.

ZDJ000 CAS:1303-39-5 ***HR: 3***
ZINC ARSENATE
DOT: UN 1712
mf: $As_4O_{15} \cdot 5Zn$ mw: 866.53

PROP: White, odorless powder.

SYNS: ARSENIC ACID, ZINC SALT ◇ ZINC ARSENATE, BASIC ◇ ZINC ARSENATE, solid (DOT)

CONSENSUS REPORTS: Arsenic and its compounds, as well as zinc and its compounds, are on the Community Right-To-Know List.

OSHA PEL: TWA 0.01 mg(As)/m^3; Cancer Hazard
NIOSH REL: CL 0.002 mg(As)/m^3/15M

DOT Classification: Poison B; Label: Poison.

SAFETY PROFILE: Confirmed human carcinogen. A poison. When heated to decomposition it emits toxic fumes of As and ZnO. See also ARSENIC COMPOUNDS and ZINC COMPOUNDS.

ZDS000 CAS:10326-24-6 ***HR: 3***
ZINC-m-ARSENITE
DOT: UN 1712
mf: $AsHO_2$•1/2Zn mw: 140.61

PROP: A white powder.

SYNS: ARSENIOUS ACID, ZINC SALT ◇ ZINC ARSENITE, solid (DOT) ◇ ZINC METAARSENITE ◇ ZINC METHARSENITE ◇ ZMA

CONSENSUS REPORTS: Arsenic and its compounds, as well as zinc and its compounds, are on the Community Right-To-Know List.

OSHA PEL: TWA 0.01 mg(As)/m^3: Cancer Hazard
ACGIH TLV: TWA 0.2 mg(As)/m^3
NIOSH REL: (Inorganic Arsenic) CL 0.002 mg(As)/m^3/15M

DOT Classification: Poison B; Label: Poison.

SAFETY PROFILE: Confirmed human carcinogen. A poison. When heated to decomposition it emits toxic fumes of As and ZnO. See also ARSENIC COMPOUNDS and ZINC COMPOUNDS.

ZEA000 CAS:16509-79-8 ***HR: 3***
ZINC BIS(DIMETHYLDITHIOCARBAMATE) CYCLOHEXYLAMINE COMPLEX
mf: $C_{12}H_{25}N_3S_4Zn$ mw: 405.01

SYNS: ZINC, DIMETHYLDITHIOCARBAMATE CYCLOHEXYLAMINE COMPLEX ◇ ZIRAM CYCLOHEXYLAMINE COMPLEX

TOXICITY DATA with REFERENCE
orl-rat LD50:1400 mg/kg TXAPA9 21,315,72
orl-bwd LD50:32 mg/kg TXAPA9 21,315,72

CONSENSUS REPORTS: Zinc and its compounds are on the Community Right-To-Know List.

SAFETY PROFILE: Poison by ingestion. When heated to decomposition it emits very toxic fumes of NO_x, SO_x, and ZnO. See also ZINC COMPOUNDS and CARBAMATES.

ZEJ000 CAS:557-09-5 ***HR: 3***
ZINC CAPRYLATE
mf: $C_{16}H_{30}O_4$•Zn mw: 351.83

PROP: Lustrous scales. Mp: 136°. Sltly sol in boiling water; mod sol in boiling alc.

SYN: OCTANOIC ACID, ZINC SALT (2:1)

TOXICITY DATA with REFERENCE
itr-rat LDLo:10 mg/kg JHEMA2 18,144,74
orl-mus LD50:2370 mg/kg JHEMA2 18,144,74

CONSENSUS REPORTS: Zinc and its compounds are on the Community Right-To-Know List. Reported in EPA TSCA Inventory.

SAFETY PROFILE: Poison by intratracheal route. Moderately toxic by ingestion. Used as a fungicide. When heated to decomposition it emits toxic fumes of ZnO. See also ZINC COMPOUNDS.

ZEJ050 CAS:3486-35-9 ***HR: D***
ZINC CARBONATE (1:1)
mf: CO_3•Zn mw: 125.38

SYN: CARBONIC ACID, ZINC SALT (1:1)

TOXICITY DATA with REFERENCE
orl-mus TDLo:2800 mg/kg (lactating female 14D post):TER 32XPAD -,83,75

CONSENSUS REPORTS: Reported in EPA TSCA Inventory.

SAFETY PROFILE: An experimental teratogen. When heated to decomposition it emits toxic fumes of Co and Zn.

ZES000 CAS:10361-95-2 ***HR: 3***
ZINC CHLORATE
DOT: UN 1513
mf: Cl_2O_6•Zn mw: 232.27

PROP: Colorless, very deliq crystals.

CONSENSUS REPORTS: Zinc and its compounds are on the Community Right-To-Know List. Reported in EPA TSCA Inventory.

DOT Classification: Oxidizer; Label: Oxidizer.

SAFETY PROFILE: A powerful oxidizer. Probably a skin, eye, and mucous membrane irritant. The tetrahydrated salt explodes at 60°C. Explosive reaction with copper(II) sulfide. Can react violently with Al, Sb_2S_3, As, C, charcoal, Cu, MnO_2, metal sulfides, dibasic organic acids, organic matter, P, S, H_2SO_4. Incandescent reaction with antimony(III) sulfide, arsenic (III) sulfide, tin(II) sulfide, tin(IV) sulfide. When heated to decomposition it emits toxic fumes of Cl^- and ZnO. See also CHLORATES and ZINC COMPOUNDS.

ZFA000 CAS:7646-85-7 ***HR: 3***
ZINC CHLORIDE
DOT: UN 1840/UN 2331
mf: Cl_2Zn mw: 136.27

PROP: Odorless, cubic, white, highly deliq crystals.

Mp: 290°, bp: 732°, d: 2.91 @ 25°, vap press: 1 mm @ 428°.

SYNS: BUTTER of ZINC ◇ CHLORURE de ZINC (FRENCH) ◇ TINNING GLUX (DOT) ◇ ZINC CHLORIDE, anhydrous (DOT) ◇ ZINC CHLORIDE, solid (DOT) ◇ ZINC CHLORIDE, solution (DOT) ◇ ZINC (CHLORURE de) (FRENCH) ◇ ZINC DICHLORIDE ◇ ZINC MURIATE, solution (DOT) ◇ ZINCO (CLORURO di) (ITALIAN) ◇ ZINKCHLORID (GERMAN) ◇ ZINKCHLORIDE (DUTCH)

TOXICITY DATA with REFERENCE
mma-sat 90 mmol/L SOGEBZ 13,1010,77
dni-hmn:lym 360 μmol/L IAAAAM 77,461,85
ipr-rat TDLo:30 g/kg (female 7-8D post):TER TJADAB 29(3),23A,84
ivg-rbt TDLo:29184 μg/kg (female 1D pre):REP CCPTAY 22,659,80
par-ham TDLo:17 mg/kg:ETA CNREA8 34,2612,74
par-ckn TDLo:15 mg/kg:ETA,REP CANCAR 6,464,53
ihl-man TCLo:4800 mg/m^3/30M:PUL SinJF# 10JAN74
ihl-hmn TCLo:4800 mg/m^3/3H YAKUD5 22,291,80
orl-rat LD50:350 mg/kg FOREAE 7,313,42
ihl-rat LCLo:1960 mg/m^3/10M ARTODN 59,160,86
ivn-rat LDLo:30 mg/kg FEPRA7 9,260,50
orl-mus LD50:350 mg/kg FOREAE 7,313,42
ipr-mus LD50:24 mg/kg TXAPA9 63,461,82

CONSENSUS REPORTS: Zinc and its compounds are on the Community Right-To-Know List. Reported in EPA TSCA Inventory. EPA Genetic Toxicology Program.

OSHA PEL: Fume: (Transitional: TWA 1 mg/m^3) TWA 1 mg/m^3; STEL 2 mg/m^3
ACGIH TLV: TWA 1 mg/m^3; STEL 2 mg/m^3 (fume)

DOT Classification: Corrosive Material; Label: Corrosive and Corrosive, solution; ORM-E; Label: None, solid.

SAFETY PROFILE: Poison by ingestion, intravenous, and intraperitoneal routes. Human systemic effects by inhalation: pulmonary changes. An experimental teratogen. Experimental reproductive effects. Questionable carcinogen with experimental tumorigenic data. Human mutation data reported. A corrosive irritant to skin, eyes, and mucous membranes. Exposure to $ZnCl_2$ fumes or dusts can cause dermatitis, boils, conjunctivitis, gastrointestinal tract upsets. The fumes are highly toxic. Incompatible with potassium. Mixtures of the powdered chloride and powdered zinc are flammable. When heated to decomposition it emits toxic fumes of Cl^- and ZnO. See also ZINC COMPOUNDS and CHLORIDES.

ZFJ000 CAS:7646-85-7 ***HR: 2***
ZINC CHLORIDE (solution)
mf: Cl_2Zn mw: 136.27

SYNS: ZINC CHLORIDE, solution (DOT) ◇ ZINC MURIATE, solution (DOT)

CONSENSUS REPORTS: Zinc and its compounds are on the Community Right-To-Know List. Reported in EPA TSCA Inventory.

DOT Classification: Label: Corrosive.

SAFETY PROFILE: A corrosive irritant to skin, eyes, and mucous membranes. When heated to decomposition it emits toxic fumes of Cl^- and ZnO. See also ZINC CHLORIDE.

ZFJ100 CAS:13530-65-9 ***HR: 3***
ZINC CHROMATE
mf: $CrH_2O_4{\cdot}Zn$ mw: 183.39

PROP: Lemon-yellow prisms.

SYNS: BASIC ZINC CHROMATE ◇ BUTTERCUP YELLOW ◇ CHROMIC ACID, ZINC SALT ◇ CHROMIUM ZINC OXIDE ◇ C.I. 77955 ◇ C.I. PIGMENT YELLOW 36 ◇ CITRON YELLOW ◇ C.P. ZINC YELLOW X-883 ◇ PRIMROSE YELLOW ◇ PURE ZINC CHROME ◇ ZINC CHROMATE(VI) HYDROXIDE ◇ ZINC CHROME YELLOW ◇ ZINC CHROMIUM OXIDE ◇ ZINC HYDROXYCHROMATE ◇ ZINC TETRAOXYCHROMATE 76A ◇ ZINC YELLOW

TOXICITY DATA with REFERENCE
mmo-sat 800 ng/plate MUREAV 156,219,85
oms-hmn:oth 500 mg/L BJCAAI 44,219,81
ihl-man TCLo:5 mg/m^3/8H/7Y-I:CAR,PUL BJIMAG 32,62,75
scu-rat TDLo:135 mg/kg:ETA PBPHAW 14,47,78
imp-rat TDLo:12928 μg/kg:CAR BJIMAG 43,243,86
ivn-mus LDLo:30 mg/kg AQMOAC #70-15,70

CONSENSUS REPORTS: NTP Fifth Annual Report on Carcinogens. IARC Cancer Review: Group 1 IMEMDT 7,165,87; Human Sufficient Evidence IMEMDT 23, 205,80; Animal Sufficient Evidence IMEMDT 23,205, 80. EPA Genetic Toxicology Program. Reported in EPA TSCA Inventory. Zinc and chromium and their compounds, are on the Community Right-To-Know List.

OSHA PEL: (Transitional: 1 $mg(CrO_3)/10m^3$) CL 0.1 mg $(CrO_3)/m^3$
ACGIH TLV: TWA 0.01 $mg(Cr)/M^3$; Confirmed Human Carcinogen
DFG TRK: 0.1 mg/m^3; Human Carcinogen.
NIOSH REL: (Chromium (VI)) TWA 0.001 mg $(Cr(VI))/m^3$

SAFETY PROFILE: Confirmed human carcinogen producing lung tumors. A poison via intravenous route. Human mutation data reported. See also CHROMIUM COMPOUNDS and ZINC COMPOUNDS.

ZFJ120 CAS:37300-23-5 ***HR: 3***
ZINC CHROMATE with ZINC HYDROXIDE and CHROMIUM OXIDE (9:1)
mf: $CrO_4 \cdot Zn \cdot H_4O_2Zn \cdot CrO_3$ mw: 183.39

SYN: ZINC YELLOW

TOXICITY DATA with REFERENCE
mmo-sat 90 nmol/plate CRNGDP 2,283,81

CONSENSUS REPORTS: Reported in EPA TSCA Inventory.

OSHA PEL: (Transitional: 1 mg(CrO_3)/10m^3) CL 0.1 mg (CrO_3)/m^3
ACGIH TLV: TWA 0.01 mg(Cr)/M^3; Confirmed Human Carcinogen
DFG TRK: 0.1 mg/m^3; Human Carcinogen.
NIOSH REL: (Chromium (VI)) TWA 0.001 mg (Cr(VI))/m^3

SAFETY PROFILE: Confirmed human carcinogen producing lung tumors. Mutation data reported. See also CHROMIUM COMPOUNDS and ZINC COMPOUNDS.

ZFJ150 ***HR: 3***
ZINC CHROMATE, POTASSIUM DICHROMATE and ZINC HYDROXIDE (3:1:1)
mf: $CrK_2O_4 \cdot 3CrO_4Zn \cdot H_2O_2Zn$ mw: 837.70

SYN: POTASSIUM DICHROMATE, ZINC CHROMATE and ZINC HYDROXIDE (1:3:1)

TOXICITY DATA with REFERENCE
imp-rat TDLo:8 mg/kg:CAR CRNGDP 7,831,86

OSHA PEL: (Transitional: 1 mg(CrO_3)/10m^3) CL 0.1 mg (CrO_3)/m^3
ACGIH TLV: TWA 0.01 mg(Cr)/M^3; Confirmed Human Carcinogen
DFG TRK: 0.1 mg/m^3; Human Carcinogen.
NIOSH REL: (Chromium (VI)) TWA 0.001 mg (Cr (VI))/m^3

SAFETY PROFILE: Confirmed human carcinogen with experimental carcinogenic data.

ZFJ250 CAS:546-46-3 ***HR: D***
ZINC CITRATE
mf: $C_6H_5O_7 \cdot 3/_2Zn$ mw: 290.18

SYN: CITRIC ACID, ZINC SALT (2:3) ◇ 1,2,3-PROPANETRICARBOXYLIC ACID, 2-HYDROXY-, ZINC SALT (2:3) (9CI)

TOXICITY DATA with REFERENCE
ivg-rbt TDLo:64366 μg/kg (female 1D pre):REP CCPTAY 22,659,80

CONSENSUS REPORTS: Reported in EPA TSCA Inventory.

SAFETY PROFILE: Experimental reproductive effects. When heated to decomposition it emits acrid smoke and irritating fumes.

ZFS000 ***HR: D***
ZINC COMPOUNDS

CONSENSUS REPORTS: Zinc and its compounds are on the Community Right-To-Know List.

SAFETY PROFILE: Variable toxicity, but generally of low toxicity. However, zinc salts, such as chromates and arsenates, are experimental carcinogens. Zinc is not inherently a toxic element. However, when heated, it evolves a fume of zinc oxide which, when inhaled fresh, can cause a disease known as "brass founders" "ague," or "brass chills," sweet taste, throat dryness, cough, weakness, generalized aching, fever, nausea, and vomiting. It is possible for people to become immune to it, but this immunity can be broken by cessation of exposure of only a few days. Zinc oxide dust which is not freshly formed is virtually innocuous. There is no cumulative effect from the inhalation of zinc fumes. Exposure to zinc chloride fumes can cause damage to the mucous membranes of the nasopharynx and respiratory tract, and give rise to a pale gray cyanosis; fatalities have resulted. Soluble salts of zinc have a harsh metallic taste; small doses can cause nausea and vomiting, while larger doses cause violent vomiting and purging. Some cases of intoxication have been reported due to drinking liquids stored in galvanized containers and in dialysis patients using a dialyzate prepared with water that had been stored in a galvanized tank. In general, the continued administration of zinc salts in small doses has no effect in humans except those of disordered digestion and constipation. Workers in zinc refining have been reported to suffer from a variety of non-specific intestinal, respiratory, and nervous symptoms. Ulceration of the nasal septum and eczematous dermatosis are also reported. It has been stated that zinc oxide or zinc stearate dust can block the ducts of the sebaceous glands and give rise to a papular, pustular eczema in workers engaged in packing these compounds into barrels. Sensitivity to zinc oxide in humans is extremely rare. Zinc chloride and zinc sulfate, because of caustic action, can cause ulceration of the fingers, hands, and forearms of those who use them as a flux in soldering or other industrial use. This condition has even been observed in men who handle railway ties which have been impregnated with this material. Common air contaminants. When heated to decomposition it emits toxic fumes of ZnO.

ZGA000 CAS:557-21-1 ***HR: 3***
ZINC CYANIDE
DOT: UN 1713
mf: C_2N_2Zn mw: 117.41

PROP: Rhombic, colorless crystals. Mp: decomp @ 800°. Insol in water; sol in solns of alkali cyanides; decomp by dil mineral acid.

SYNS: CYANURE de ZINC (FRENCH) ◇ RCRA WASTE NUMBER P121 ◇ ZINC DICYANIDE

TOXICITY DATA with REFERENCE
ipr-rat LDLo:100 mg/kg NCNSA6 5,28,53

CONSENSUS REPORTS: Zinc and its compounds, as well as cyanide and its compounds, are on the Community Right-To-Know List. Reported in EPA TSCA Inventory.

DOT Classification: Poison B; Label: Poison.

SAFETY PROFILE: Poison by intraperitoneal route. Can react violently with Mg. When heated to decomposition it emits toxic fumes of CN^-, ZnO, and NO_x. Used in electroplating operations. See also CYANIDE and ZINC COMPOUNDS.

ZGJ000 ***HR: 3***
ZINC DIHYDRAZIDE
mf: H_6N_4Zn mw: 127.45

CONSENSUS REPORTS: Zinc and its compounds are on the Community Right-To-Know List.

SAFETY PROFILE: Explodes at 70°C. When heated to decomposition it emits toxic fumes of NO_x and ZnO. See also ZINC COMPOUNDS and AZIDES.

ZGS000 CAS:19210-06-1 ***HR: 2***
ZINC DITHIOPHOSPHATE
mf: $H_3O_2PS_2 \cdot xZn$ mw: 587.71

SYN: PHOSPHORODITHIOIC ACID, ZINC SALT

TOXICITY DATA with REFERENCE
orl-rbt LDLo:2130 mg/kg AEHLAU 6,324,63

CONSENSUS REPORTS: Zinc and its compounds are on the Community Right-To-Know List. Reported in EPA TSCA Inventory.

SAFETY PROFILE: Moderately toxic by ingestion. When heated to decomposition it emits very toxic fumes of SO_x, ZnO, and PO_x. See also ZINC COMPOUNDS.

ZGS100 CAS:15954-98-0 ***HR: 3***
ZINC(II) EDTA COMPLEX
mf: $C_{10}H_{16}N_2O_8Zn$ mw: 357.65

SYN: ACETIC ACID, (ETHYLENEDINITRILO)TETRA-, ZINC(II) COMPLEX

TOXICITY DATA with REFERENCE
scu-rat TDLo:2910 mg/kg (female 11-15D post):TER TXAPA9 82,426,86
ipr-mus LD50:85 mg(Zn)/kg PABIAQ 11,853,63

SAFETY PROFILE: Poison by intraperitoneal route. An experimental teratogen. When heated to decomposition it emits toxic fumes of NO_x and Zn.

ZGW100 CAS:3851-22-7 ***HR: 3***
ZINC ETHOXIDE
mf: $C_4H_{10}O_2Zn$ mw: 155.51

$Zn(OCH_2CH_3)_2$

CONSENSUS REPORTS: Zinc and its compounds are on The Community Right-To-Know List.

SAFETY PROFILE: Mixtures with nitric acid may explode at room temperature. When heated to decomposition it emits toxic fumes of ZnO. See also ZINC COMPOUNDS.

ZHA000 CAS:14634-93-6 ***HR: D***
ZINC ETHYLPHENYLTHIOCARBAMATE
mf: $C_{18}H_{20}N_2S_4Zn$ mw: 458.01

SYNS: ACCELERATOR EFK ◇ BIS(N-ETHYLDITHIOCARBANILATO)ZINC ◇ BIS(ETHYLPHENYLCARBAMODITHIOATO-S,S')-(T-4)-ZINC ◇ HERMAT FEDK ◇ VULKACIT P EXTRA N ◇ ZINC ETHYLPHENYLDITHIOCARBAMATE

TOXICITY DATA with REFERENCE
mmo-sat 100 μg/plate MUREAV 68,313,79
mma-sat 100 μg/plate PCBRD2 141,407,84

CONSENSUS REPORTS: Zinc and its compounds are on the Community Right-To-Know List. Reported in EPA TSCA Inventory.

SAFETY PROFILE: Mutation data reported. When heated to decomposition it emits very toxic fumes of NO_x, ZnO, and SO_x. See also ZINC COMPOUNDS and CARBAMATES.

ZHJ000 CAS:14881-92-6 ***HR: 3***
ZINC-N-FLUOREN-2-YLACETOHYDROXAMATE
mf: $C_{30}H_{24}N_2O_4 \cdot Zn$ mw: 541.93

SYNS: N-FLUOREN-2-YLACETOHYDROXAMIC ACID, ZINC COMPLEX ◇ N-HYDROXY-2-ACETYLAMINOFLUORENE, ZINC CHELATE

TOXICITY DATA with REFERENCE
scu-rat TDLo:160 mg/kg/4W-I:NEO CNREA8 25,527,65

CONSENSUS REPORTS: Zinc and its compounds are on the Community Right-To-Know List.

SAFETY PROFILE: Questionable carcinogen with experimental neoplastigenic data. When heated to decomposition it emits toxic fumes of NO_x and ZnO. See also ZINC COMPOUNDS.

ZHS000 CAS:7783-49-5 ***HR: 3***
ZINC FLUORIDE
mf: F_2Zn mw: 103.37

PROP: Tetragonal needles or white crystalline mass. D: 5.00 @ 25°, mp: 872°, bp: 1500°, vap press: 1 mm @ 970°. Sltly sol in aq HF; sol in HCl, HNO_3, and NH_4OH.

SYN: ZINC FLUORURE (FRENCH)

TOXICITY DATA with REFERENCE
scu-frg LDLo:280 mg/kg CRSBAW 124,133,37

CONSENSUS REPORTS: Zinc and its compounds are on the Community Right-To-Know List. Reported in EPA TSCA Inventory.

OSHA PEL: TWA 2.5 mg(F)/m^3
ACGIH TLV: TWA 2.5 mg(F)/m^3
NIOSH REL: (Fluorides, Inorganic) TWA 2.5 mg (F)/m^3

SAFETY PROFILE: Poison by subcutaneous route. Can react violently with potassium. A fluorination agent. When heated to decomposition it emits toxic fumes of F^- and ZnO. See also FLUORIDES and ZINC COMPOUNDS.

ZIA000 CAS:16871-71-9 ***HR: 3***
ZINC FLUOSILICATE
DOT: UN 2855
mf: $F_6Si{\bullet}Zn$ mw: 207.46

SYNS: FLUOSILICATE de ZINC ◇ ZINC HEXAFLUOROSILICATE

TOXICITY DATA with REFERENCE
orl-rat LDLo:100 mg/kg NCNSA6 5,28,53
scu-frg LDLo:280 mg/kg CRSBAW 124,133,37

CONSENSUS REPORTS: Zinc and its compounds are on the Community Right-To-Know List. Reported in EPA TSCA Inventory.
NIOSH REL: TWA 2.5 mg(F)/m^3

DOT Classification: Poison B; Label: St. Andrews Cross

SAFETY PROFILE: Poison by ingestion and subcutaneous routes. When heated to decomposition it emits toxic fumes of F^- and ZnO. See also ZINC COMPOUNDS and HEXAFLUOROSILICATE(2−) DIHYDROGEN.

ZIA750 CAS:4468-02-4 ***HR: 1***
ZINC GLUCONATE
mf: $C_{12}H_{22}O_4Zn$ mw: 295.71

PROP: White granular or crystalline powder. Sol in water; very sltly sol in alc.

SYN: ZINC, BIS(d-GLUCONATO-O^1),O^2))- (9CI)

TOXICITY DATA with REFERENCE
ivg-rbt TDLo:90616 μg/kg (female 1D pre):REP CCPTAY 22,659,80

CONSENSUS REPORTS: Reported in EPA TSCA Inventory.

SAFETY PROFILE: Experimental reproductive effects. When heated to decomposition it emits toxic fumes of ZnO.

ZIJ000 CAS:14018-82-7 ***HR: 3***
ZINC HYDRIDE
mf: H_2Zn mw: 67.39

CONSENSUS REPORTS: Zinc and its compounds are on the Community Right-To-Know List.

SAFETY PROFILE: May ignite spontaneously in air. Violent reaction with aqueous acids. Slow reaction with water. When heated to decomposition it emits toxic fumes of ZnO. See also ZINC COMPOUNDS and HYDRIDES.

ZIS000 CAS:3030-80-6 ***HR: 2***
ZINC MERCAPTOBENZIMIDAZOLE
mf: $C_{14}H_{12}N_4S_2{\bullet}Zn$ mw: 365.79

SYNS: ALTERUNGSSCHUTZMITTEL ZMB ◇ ANTIOXIDANT ZMB ◇ BIS(MERCAPTOBENZIMIDAZOLATO)ZINC ◇ MERCAPTOBENZIMIDAZOLE ZINC SALT ◇ 2-MERCAPTOBENZIMIDAZOLE ZINC SALT (2:1) ◇ ZINC BENZIMIDAZOLE-2-THIOLATE ◇ ZINC BIS(1H-BENZIMIDAZOLE-2-THIOLATE) ◇ ZINC MERCAPTOBENZIMIDAZOLATE

TOXICITY DATA with REFERENCE
orl-rat LD50:540 mg/kg HYSAAV 31,183,66
orl-mus LD50:860 mg/kg HYSAAV 31,183,66

CONSENSUS REPORTS: Zinc and its compounds are on the Community Right-To-Know List. Reported in EPA TSCA Inventory.

SAFETY PROFILE: Moderately toxic by ingestion. When heated to decomposition it emits very toxic fumes of NO_x, SO_x, and ZnO. See also ZINC COMPOUNDS and MERCAPTANS.

ZJA000 CAS:22323-45-1 ***HR: 3***
ZINC MERCURY CHROMATE COMPLEX
mf: $7ZnO{\bullet}2HgO{\bullet}2CrO_3{\bullet}7H_2O$ mw: 1328.91

SYNS: CHROMIC ACID, MERCURY ZINC COMPLEX ◇ EXPERIMENTAL FUNGICIDE 224 (UNION CARBIDE) ◇ MERCURY ZINC CHROMATE COMPLEX

TOXICITY DATA with REFERENCE
orl-rat LD50:630 mg/kg 28ZEAL 4,393,69

CONSENSUS REPORTS: Zinc, mercury, chromium, and their compounds are on the Community Right-To-Know List.

OSHA PEL: (Transitional: 1 mg/10m^3) CL 0.1 mg $(CrO_3)/m^3$
ACGIH TLV: TWA 0.05 mg(Cr)/m^3, Confirmed Human Carcinogen.
DFG MAK: Animal Carcinogen, Suspected Human Carcinogen.
NIOSH REL: TWA 0.025 mg(Cr(VI))/m^3; CL 0.05/15M; 0.05 mg(Hg)/m^3

SAFETY PROFILE: Confirmed carcinogen. Moderately toxic by ingestion. When heated to decomposition it emits very toxic fumes of Hg and ZnO. See also MERCURY COMPOUNDS, ZINC COMPOUNDS, and CHROMIUM COMPOUNDS.

ZJJ000 CAS:7779-88-6 ***HR: 3***
ZINC NITRATE
DOT: UN 1514
mf: $N_2O_6 \cdot Zn$ mw: 189.39

PROP: A: needles; B: tetragonal, colorless crystals; A: trihydrate; B: hexahydrate; d: (B) 2.065 @ 14°; mp: (A) 42.5°; mp: (B): 36.4°; bp: (B): loses $6H_2O$ @ 105-131°. Very sol in alc; sol in water.

SYNS: NITRATE de ZINC (FRENCH) ◇ NITRIC ACID, ZINC SALT

CONSENSUS REPORTS: Zinc and its compounds are on the Community Right-To-Know List. Reported in EPA TSCA Inventory.

DOT Classification: Oxidizer; Label: Oxidizer.
SAFETY PROFILE: A powerful oxidizer. Can react violently with C; Cu; metal sulfides; organic matter; P; S. When heated to decomposition it emits toxic fumes of NO_x and ZnO. See also NITRATES and ZINC COMPOUNDS.

ZJJ200 ***HR: 1***
ZINC NITRILOTRIMETHYLPHOSPHONIC ACID TRISODIUM TETRAHYDRATE
mf: $C_3H_6NO_9P_3Zn \cdot 3Na \cdot 4H_2O$ mw: 499.43

SYN: NITRILOTRIMETHYLPHOSPHONIC ACID ZINC complex TRISODIUM TETRAHYDRATE

TOXICITY DATA with REFERENCE
orl-rat LD50:21 g/kg GISAAA 49(11),73,84
orl-mus LD50:16500 mg/kg GISAAA 49(11),73,84
orl-gpg LD50:27 g/kg GISAAA 49(11),73,84

CONSENSUS REPORTS: Zinc and its compounds are on the Community Right-To-Know List.

SAFETY PROFILE: Mildly toxic by ingestion. When heated to decomposition it emits toxic fumes of NO_x, PO_x, Na_2O, and ZnO. See also ZINC COMPOUNDS.

ZJJ400 CAS:14709-62-7 ***HR: 3***
ZINC NITROSYLPENTACYANOFERRATE
mf: $C_5FeN_6O \cdot Zn$ mw: 281.33

SYNS: PENTACYANONITROSYLFERRATE ZINC ◇ ZINC PENTACYANONITROSYLFERRATE(2−)

TOXICITY DATA with REFERENCE
orl-rat LD50:125 mg/kg ARZNAD 24,308,74
ipr-rat LD50:11 mg/kg ARZNAD 24,308,74
orl-mus LD50:56 mg/kg ARZNAD 24,308,74
ipr-mus LD50:10800 μg/kg ARZNAD 24,308,74
ivn-dog LDLo:24760 μg/kg ARZNAD 24,308,74
ivn-rbt LDLo:5500 μg/kg ARZNAD 24,308,74

CONSENSUS REPORTS: Zinc and its compounds, as well as cyanide and its compounds, are on the Community Right-To-Know List.

SAFETY PROFILE: Poison by ingestion, intravenous, and intraperitoneal routes. When heated to decomposition it emits toxic fumes of NO_x, CN^-, and ZnO. See also ZINC COMPOUNDS and CYANIDE.

ZJS000 CAS:557-07-3 ***HR: 3***
ZINC OLEATE (1:2)
mf: $C_{36}H_{68}O_4 \cdot Zn$ mw: 630.41

PROP: White, dry, greasy powder. Insol in water; sol in alc, ether, carbon disulfide, benzene, petr ether.

SYN: OLEIC ACID, ZINC SALT

TOXICITY DATA with REFERENCE
orl-mus TDLo:1080 g/kg/1Y-C:ETA FCTXAV 3,271,65

CONSENSUS REPORTS: Zinc and its compounds are on the Community Right-To-Know List. Reported in EPA TSCA Inventory.

SAFETY PROFILE: Questionable carcinogen with experimental tumorigenic data. When heated to decomposition it emits toxic fumes of ZnO. See also ZINC COMPOUNDS.

ZJS300 CAS:8066-21-5 ***HR: 1***
ZINCOP

SYNS: BI-CURVAL ◇ COPPER LONACOL ◇ COPPER OXYCHLORIDE-ZINEB mixture ◇ COPRAMAT ◇ CUPROCIN ◇ CUPROZAN ◇ CYNKOMIEDZIAN ◇ KHOMECIN ◇ KHOMEZIN ◇ KUPROTSIN ◇ LUNACOL ◇ MILTOX ◇ N 2038 ◇ NEW BLITANE ◇ POLYCHOM ◇ POLYCHOME ◇ THIOZIN ◇ TIOXIN ◇ ZIKHOM ◇ ZINEB-COPPER OXYCHLORIDE mixture

TOXICITY DATA with REFERENCE
orl-rat LD50:4430 mg/kg GISAAA 49(2),32,84
orl-mus LD50:5250 mg/kg GISAAA 49(2),32,84
orl-rbt LD50:8700 mg/kg GISAAA 49(2),32,84

CONSENSUS REPORTS: Zinc, copper, and their compounds are on the Community Right-To-Know List.

SAFETY PROFILE: Mildly toxic by ingestion. When heated to decomposition it emits toxic fumes of ZnO and SO_x. See also ETHYLENEBIS(DITHIOCARBAMATO)ZINC (zineb), ZINC COMPOUNDS, CARBAMATES, and COPPER COMPOUNDS.

ZKA000 CAS:1314-13-2 ***HR: 3***
ZINC OXIDE
mf: OZn mw: 81.37

PROP: Odorless, white or yellowish powder. Mp: >1800°, d: 5.47. Insol in water and alc; sol in dil acetic or mineral acids, ammonia.

SYNS: AKRO-ZINC BAR 85 ◇ AMALOX ◇ AZO-33 ◇ AZODOX-55 ◇ CALAMINE (spray) ◇ CHINESE WHITE ◇ C.I. 77947 ◇ C.I. PIGMENT WHITE 4 ◇ CYNKU TLENEK (POLISH) ◇ EMANAY ZINC OXIDE ◇ EMAR ◇ FELLING ZINC OXIDE ◇ FLOWERS of ZINC ◇ GREEN SEAL-8 ◇ HUBBUCK'S WHITE ◇ KADOX-25 ◇ K-ZINC ◇ OZIDE ◇ OZLO ◇ PASCO ◇ PERMANENT WHITE ◇ PHILOSOPHER'S WOOL ◇ PROTOX TYPE 166 ◇ RED-SEAL-9 ◇ SNOW WHITE ◇ WHITE SEAL-7 ◇ ZINCITE ◇ ZINCOID ◇ ZINC OXIDE FUME (MAK) ◇ ZINC WHITE

TOXICITY DATA with REFERENCE
skn-rbt 500 mg/24H MLD 28ZPAK -,10,72
eye-rbt 500 mg/24H MLD 28ZPAK -,10,72
dnd-esc 3000 ppm MUREAV 89,95,81
cyt-rat-ihl 100 $\mu g/m^3$ CYGEDX 12(3),46,78
orl-rat TDLo:6846 mg/kg (1-22D preg):REP JONUAI 98,303,69
orl-rat TDLo:6846 mg/kg (1-22D preg):TER JONUAI 98,303,69
orl-hmn LDLo:500 mg/kg YAKUD5 22,291,80
ihl-hmn TCLo:600 mg/m^3:PUL JIDHAN 9,88,27
ipr-rat LD50:240 mg/kg ZDKAA8 38(9),18,78
orl-mus LD50:7950 mg/kg GISAAA 51(4),89,86
ihl-mus LC50:2500 mg/m^3 IPSTB3 3,93,76

CONSENSUS REPORTS: Zinc and its compounds are on the Community Right-To-Know List. Reported in EPA TSCA Inventory.

OSHA PEL: Fume: (Transitional: TWA 5 mg/m^3) TWA 5 mg/m^3; STEL 10 mg/m^3; Dust: (Transitional: Total Dust: 15 mg/m^3; Respirable Fraction: 5 mg/m^3;) TWA Total Dust: 10 mg/m^3; Respirable Fraction: 5 mg/m^3
ACGIH TLV: Fume: TWA 5 mg/m^3; STEL 10 mg/m^3; Dust: 10 mg/m^3 of total dust (when toxic impurities are not present, e.g., quartz < 1%).
DFG MAK: 5 mg/m^3
NIOSH REL: TWA (Zinc Oxide) 5 mg/m^3; CL 15 mg/m^3/15M

SAFETY PROFILE: Moderately toxic to humans by ingestion. Poison experimentally by intraperitoneal route. An experimental teratogen. Other experimental reproductive effects. Human systemic effects by inhalation of freshly formed fumes: metal fume fever with chills, fever, tightness of chest, cough, dyspnea, and other pulmonary changes. Mutation data reported. A skin and eye irritant. Has exploded when mixed with chlorinated rubber. Violent reaction with Mg, linseed oil. When heated to decomposition it emits toxic fumes of ZnO. See also ZINC COMPOUNDS.

ZKJ000 CAS:8051-03-4 ***HR: 1***
ZINC OXIDE (ointment)

TOXICITY DATA with REFERENCE
skn-hmn 300 μg/3D-I MLD 85DKA8 -,127,77

CONSENSUS REPORTS: Zinc and its compounds are on the Community Right-To-Know List.

SAFETY PROFILE: A skin irritant. When heated to decomposition it emits toxic fumes of ZnO. See also ZINC OXIDE and ZINC COMPOUNDS.

ZKS000 CAS:65979-81-9 ***HR: 3***
ZINC PANTOTHENATE
mf: $C_{18}H_{32}N_2O_{10}$•Zn mw: 501.89

SYNS: (T-4-(R)(R)-BIS(N-(2-DIHYDROXY-3,3-DIMETHYL-1-OXOBUTYL-β-ALANIN-ATO)ZINC ◇ (R)-N-(2,4-DIHYDROXY-3,3-DIMETHYLBUTYRYL)-β-ALANINE ZINC SALT (2 : 1) ◇ PANTOTHENATE de ZINC (FRENCH) ◇ PANTOTHENIC ACID, ZINC SALT

TOXICITY DATA with REFERENCE
orl-rat LD50:3763 mg/kg DICPBB 13,611,79
ipr-rat LD50:489 mg/kg DICPBB 13,611,79
orl-mus LD50:2161 mg/kg DICPBB 13,611,79
ipr-mus LD50:342 mg/kg DICPBB 13,611,79

CONSENSUS REPORTS: Zinc and its compounds are on the Community Right-To-Know List.

SAFETY PROFILE: Poison by intraperitoneal route. Moderately toxic by ingestion. When heated to decomposition it emits toxic fumes of NO_x and ZnO. See also ZINC COMPOUNDS.

ZLA000 CAS:23414-72-4 ***HR: 1***
ZINC PERMANGANATE
DOT: UN 1515
mf: Mn_2O_8•Zn mw: 303.25

PROP: Violet-brown or black, hygroscopic crystals.

CONSENSUS REPORTS: Zinc, manganese, and their compounds are on the Community Right-To-Know List.
ACGIH TLV: TWA 5 mg(Mn)/m^3

DOT Classification: Oxidizer; Label: Oxidizer.

SAFETY PROFILE: Probably a skin, eye, and mucous membrane irritant. Flammable by chemical reaction with reducing agents. A powerful oxidizing agent. When heated to decomposition it emits toxic fumes of ZnO. Used as an antiseptic. See also MANGANESE COMPOUNDS and ZINC COMPOUNDS.

ZLJ000 CAS:1314-22-3 ***HR: 3***
ZINC PEROXIDE
DOT: UN 1516
mf: O_2Zn mw: 97.37

PROP: Odorless, yellow-white powder. D: 1.571 (theoretical). Decomp > 150°. Sol in dil acids.

SYN: ZINC SUPEROXIDE

CONSENSUS REPORTS: Zinc and its compounds are on the Community Right-To-Know List. Reported in EPA TSCA Inventory.

DOT Classification: Oxidizer; Label: Oxidizer.

SAFETY PROFILE: Systemic toxicity is similar to zinc oxide. Flammable when exposed to heat or by chemical reaction with reducing materials. Finely divided powder is slightly soluble in water, decomposes rapidly at 150°. A powerful oxidizer and dangerous when mixed with highly combustible materials. A very dangerous explosion hazard when exposed to heat. Explodes at 212°. Can react violently with Al, and Zn. Very dangerous, will react with water or steam to produce heat. Vigorous reaction with reducing materials. When heated to decomposition it emits toxic fumes of ZnO. See also PEROXIDES and ZINC COMPOUNDS.

ZLS000 CAS:1314-84-7 ***HR: 3***
ZINC PHOSPHIDE
DOT: UN 1714
mf: P_2Zn_3 mw: 258.05

PROP: Cubic, dark gray crystals or powder. Mp: 420°, bp: 1100°, d: 4.55 @ 13°. Insol in water, alc; sol in benzene, carbon disulfide.

SYNS: BLUE-OX ◇ KILRAT ◇ MOUS-CON ◇ PHOSPHURE de ZINC (FRENCH) ◇ PHOSVIN ◇ RCRA WASTE NUMBER P122 ◇ RUMETAN ◇ ZINCO(FOSFURO di) (ITALIAN) ◇ ZINC(PHOSPHURE de) (FRENCH) ◇ ZINC-TOX ◇ ZINKFOSFIDE (DUTCH) ◇ ZINKPHOSPHID (GERMAN) ◇ ZP

TOXICITY DATA with REFERENCE
orl-wmn LDLo:80 mg/kg:GIT ZACCAL 23,144,48
orl-rat LD50:12 mg/kg MAGJAL 52(2),166,79
orl-mus LD50:40 mg/kg YKYUA6 31,1247,80
orl-cat LDLo:250 mg/kg JAPMA8 42,468,52
orl-rbt LDLo:40 mg/kg JAPMA8 42,468,52
orl-bwd LD50:23700 μg/kg AECTCV 12,355,83

CONSENSUS REPORTS: Zinc and its compounds are on the Community Right-To-Know List. Reported in EPA TSCA Inventory.

DOT Classification: Flammable Solid; Label: Flammable Solid and Dangerous When Wet; Poison B; Label: Poison.

SAFETY PROFILE: Human poison by ingestion causing nausea, vomiting, death. Flammable when exposed to heat or flame. This material is stable while kept dry. In moist air, it decomposes slowly. Reacts violently with acids or acid fumes to emit the highly toxic and flammable phosphine. Violent reaction with concentrated sulfuric acid, nitric acid, and oxidizing materials. Incompatible with HCl, H_2SO_4. When heated to decomposition it emits toxic fumes of PO_x and ZnO. Used as an acute rodenticide. See also PHOSPHIDES and ZINC COMPOUNDS.

ZLS200 CAS:14332-59-3 ***HR: 2***
ZINC PHOSPHITE
mf: $H_3O_3P{\bullet}Zn$ mw: 147.37

SYNS: NERA ◇ NERA EMULZE (CZECH) ◇ SECONDARY ZINC PHOSPHITE

TOXICITY DATA with REFERENCE
skn-rbt 500 mg/24H SEV 28ZPAK -,285,72
eye-rbt 20 mg/24H MOD 28ZPAK -,285,72
orl-rat LD50:506 mg/kg 28ZPAK -,285,72

CONSENSUS REPORTS: Zinc and its compounds are on the Community Right-To-Know List.

SAFETY PROFILE: Moderately toxic by ingestion. An eye and severe skin irritant. When heated to decomposition it emits toxic fumes of PO_x and ZnO. See also ZINC COMPOUNDS.

ZMA000 CAS:12071-83-9 ***HR: 2***
ZINC (N,N'-PROPYLENE-1,2-BIS(DITHIOCARBAMATE))
mf: $(C_5H_8N_2S_4Zn)_x$

SYNS: AIRONE ◇ ANTRACOL ◇ BAY 46131 ◇ BAYER 46131 ◇ LH 3012 ◇ LH 30/Z ◇ (((1-METHYL-1,2-ETHANEDIYL)BIS(CARBAMODITHIOATO))(2−))ZINC HOMOPOLYMER ◇ METHYL ZINEB ◇ MEZINEB ◇ PROPINEB ◇ PROPINEBE ◇ PROPYLENEBIS(DITHIOCARBAMATO)ZINC ◇ TAIFEN ◇ TSIPROMAT (RUSSIAN) ◇ ZINK-(N,N'-PROPYLEN-1,2-BIS(DITHIOCARBAMAT)) (GERMAN) ◇ ZIPROMAT

TOXICITY DATA with REFERENCE
mrc-smc 2500 ppm MUREAV 10,533,70
cyt-rat-par 2600 mg/kg/5D-I PRKHDK 4,151,79
orl-rat TDLo:2 g/kg (female 6-15D post):TER BSIBAC 61,271,85
orl-rat TDLo:2300 mg/kg (11D preg):REP TJADAB 14,171,76
orl-rat LD50:8500 mg/kg FMCHA2 -,C198,83
orl-rbt LD50:2500 mg/kg 85GYAZ -,123,71

CONSENSUS REPORTS: Zinc and its compounds are on the Community Right-To-Know List. EPA Genetic Toxicology Program.

SAFETY PROFILE: Moderately toxic by ingestion. An experimental teratogen. Other experimental reproductive effects. Mutation data reported. When heated to decomposition it emits very toxic fumes of NO_x, SO_x and ZnO. Used as a fungicide. See also ETHYLENEBIS(DITHIOCARBAMATO)ZINC (zineb), ZINC COMPOUNDS, and CARBAMATES.

ZMJ000 CAS:13463-41-7 ***HR: 3***
ZINC PYRIDINE-2-THIOL-1-OXIDE
mf: $C_{10}H_8N_2O_2S_2 \cdot Zn$ mw: 317.69

SYNS: BIS(1-HYDROXY-2(1H)-PYRIDINETHIONATO)ZINC ◇ BIS(2-PYRIDYLTHIO)ZINC, 1,1'-DIOXIDE ◇ OM-1563 ◇ OMADINE ZINC ◇ 2-PYRIDINETHIOL-1-OXIDE, ZINC SALT ◇ PYRITHIONE ZINC ◇ VANCIDE P ◇ ZINC OMADINE ◇ ZINCPOLYANEMINE ◇ ZINC PT ◇ ZINC PYRIDINETHIONE ◇ ZINC PYRION ◇ ZINC PYRITHIONE

TOXICITY DATA with REFERENCE
eye-rbt 1 mg/48H JANCA2 56,905,73
skn-rat TDLo:1065 mg/kg (female 8W pre):REP FCTXAV 17,639,79
orl-rbt TDLo:65 mg/kg (female 6-18D post):TER FCTXAV 17,639,79
orl-rat LD50:177 mg/kg TOANDB 3,1,79
orl-mus LD50:160 mg/kg CTOXAO 13,1,78
ipr-mus LD50:26800 μg/kg OYYAA2 8,1067,74
scu-mus LD50:730 mg/kg CTOXAO 13,1,78
orl-dog LD50:600 mg/kg CTOXAO 13,1,78
ivn-dog LDLo:25 mg/kg TXAPA9 9,269,66
ivn-mky LDLo:25 mg/kg TXAPA9 9,269,66
skn-rbt LD50:100 mg/kg YKYUA6 32,965,81
ivn-rbt LDLo:10 mg/kg TXAPA9 9,269,66

CONSENSUS REPORTS: Zinc and its compounds are on the Community Right-To-Know List. Reported in EPA TSCA Inventory.

SAFETY PROFILE: Poison by ingestion, skin contact, intraperitoneal and intravenous routes. Moderately toxic by subcutaneous route. An experimental teratogen. Experimental reproductive effects. An eye irritant. When heated to decomposition it emits very toxic fumes of NO_x, SO_x, and ZnO. Used as an anti-dandruff agent in shampoos. See also ZINC COMPOUNDS and SULFIDES.

ZMS000 CAS:557-05-1 ***HR: 3***
ZINC STEARATE
mf: $Zn(C_{18}H_{35}O_2)_2$ mw: 632.30

PROP: White powder. Mp: 130°, flash p: 530°F (OC), autoign temp: 790°F. Insol in water, alc, ether; sol in benzene. Decomp by dil acids.

SYNS: DIBASIC ZINC STEARATE ◇ OCTADECANOIC ACID, ZINC SALT ◇ STEARIC ACID, ZINC SALT ◇ ZINC DISTERATE ◇ ZINC OCTADECANOATE

TOXICITY DATA with REFERENCE
itr-rat LDLo:250 mg/kg BJIMAG 15,130,58

CONSENSUS REPORTS: Zinc and its compounds are on the Community Right-To-Know List. Reported in EPA TSCA Inventory.

OSHA PEL: (Transitional: TWA Total Dust: 15 mg/m^3; Respirable Fraction: 5 mg/m^3) TWA Total Dust: 10 mg/m^3; Respirable Fraction: 5 mg/m^3
ACGIH TLV: TWA 10 mg/m^3 of total dust when toxic impurities are not present, e.g., quartz < 1%

SAFETY PROFILE: Poison by intratracheal route. Inhalation of zinc stearate has been reported as causing pulmonary fibrosis. A nuisance dust. Combustible when exposed to heat or flame. To fight fire, use water, foam, CO_2, dry chemical. When heated to decomposition it emits toxic fumes of ZnO. See also ZINC COMPOUNDS.

ZNA000 CAS:7733-02-0 ***HR: 3***
ZINC SULFATE
DOT: NA 9161
mf: $O_4S \cdot Zn$ mw: 161.43

PROP: Rhombic, colorless crystals or crystalline powder. Mp: decomp @ 740°, d: 3.74 @ 15°. Sol in water; almost insol in alc.

SYNS: BONAZEN ◇ BUFOPTO ZINC SULFATE ◇ OP-THAL-ZIN ◇ SULFATE de ZINC (FRENCH) ◇ SULFURIC ACID, ZINC SALT (1:1) ◇ VERAZINC ◇ WHITE COPPERAS ◇ WHITE VITRIOL ◇ ZINC SULPHATE ◇ ZINC VITRIOL ◇ ZINKOSITE

TOXICITY DATA with REFERENCE
eye-rbt 420 μg MOD JAPMA8 45,474,56
mmo-smc 100 mmol/L MUREAV 117,149,83
otr-ham:emb 200 μmol/L CNREA8 39,193,79
orl-dom TDLo:193 g/kg (female 6-20W post):REP ENVRAL 20,1,79
scu-ham TDLo:15 mg/kg (female 8D post):TER ENVRAL 35,405,84
scu-rbt TDLo:3625 μg/kg/5D-C:ETA COREAF 236,1387,53
orl-hmn TDLo:45 mg/kg/7D-C:CVS,GIT,BLD BMJOAE 1,754,78
orl-hmn TDLo:106 mg/kg:CVS,PUL,GIT BMJOAE 1,1390,77
orl-man TDLo:180 mg/kg/6W-I:BLD JAMAAP 252,1443,84
orl-wmn TDLo:3120 mg/kg/43W-I:BLD,SYS GASTAB 94,508,88
orl-rat LD50:2949 mg/kg TOERD9 1,371,78
scu-rat LDLo:330 mg/kg EQSSDX 1,1,75
ivn-rat LDLo:50 mg/kg EQSSDX 1,1,75
orl-mus LD50:57 mg/kg IPSTB3 3,93,76
ipr-mus LD50:71750 μg/kg COREAF 256,1043,63
scu-mus LDLo:1500 μg/kg TJIZAF 48,313,78
scu-dog LDLo:78 mg/kg EQSSDX 1,1,75
ivn-dog LDLo:66 mg/kg EQSSDX 1,1,75

CONSENSUS REPORTS: Zinc and its compounds are on the Community Right-To-Know List. Reported in EPA TSCA Inventory. EPA Genetic Toxicology Program.

SAFETY PROFILE: Poison by ingestion, intraperitoneal, subcutaneous, and intravenous routes. Human systemic effects by ingestion: acute pulmonary edema, agranulocytosis, blood pressure decrease, diarrhea and other gastrointestinal changes, hypermotility, increased pulse rate without blood pressure decrease, level changes for metals other than Na/K/Fe/Ca/P/Cl, microcytosis with or without anemia, normocytic anemia. Experimental teratogenic and reproductive effects. Questionable carcinogen with experimental tumorigenic data. Human mutation data reported. An eye irritant. When heated to decomposition it emits toxic fumes of SO_x and ZnO. See also SULFATES and ZINC COMPOUNDS.

ZNJ000 CAS:7446-20-0 ***HR: 3***
ZINC SULFATE HEPTAHYDRATE (1:1:7)
mf: $O_4SZn\cdot 7H_2O$ mw: 287.57

PROP: Colorless crystals or crystalline powder; odorless. D: 1.97; mp: 100°. Decomp >500°. Insol in alc; glycerin.

SYNS: SULFURIC ACID, ZINC SALT (1:1), HEPTAHYDRATE ◇ WHITE VITRIOL ◇ ZINC SULFATE ◇ ZINC SULFATE (1:1) HEPTAHYDRATE ◇ ZINC VITRIOL

TOXICITY DATA with REFERENCE
sln-dmg-orl 5 mmol/L MUREAV 90,91,81
dni-mus-ipr 20 g/kg ARGEAR 51,605,81
orl-mus TDLo:328 g/kg (female 13W pre):REP NNGADV 6,327,81
unr-man LDLo:221 mg/kg 85DCAI 2,73,70
orl-rat LD50:2150 mg/kg ARTODN 54,275,83
scu-rat LDLo:330 mg/kg HBAMAK 4,1419,35
ivn-rat LDLo:49 mg/kg HBAMAK 4,1419,35
orl-mus LD50:2200 mg/kg BSPBAD 116,47,77
ipr-mus LD50:260 mg/kg BSPBAD 116,47,77
scu-dog LDLo:78 mg/kg HBAMAK 4,1419,35
ivn-dog LDLo:66 mg/kg HBAMAK 4,1419,35

CONSENSUS REPORTS: Zinc and its compounds are on the Community Right-To-Know List.

SAFETY PROFILE: Human poison by an unspecified route. Poison experimentally by subcutaneous, intravenous, and intraperitoneal routes. Moderately toxic by ingestion. Experimental reproductive effects. When heated to decomposition it emits toxic fumes of SO_x and ZnO. See also ZINC SULFATE.

ZNS000 CAS:102916-22-3 ***HR: 3***
ZINC TRIFLUOROSTANNITE, HEPTAHYDRATE
mf: $F_6Sn_2\cdot Zn\cdot 7H_2O$ mw: 542.89

TOXICITY DATA with REFERENCE
ivn-mus LD50:32 mg/kg CSLNX* NX#00132

CONSENSUS REPORTS: Zinc and its compounds are on the Community Right-To-Know List.

OSHA PEL: TWA 2.5 mg(F)/m^3; 2 mg(Sn)/m^3
ACGIH TLV: TWA 2 mg(Sn)/m^3
NIOSH REL: (Fluorides, Inorganic) TWA 2.5 mg (F)/m^3

SAFETY PROFILE: Poison by intravenous route. When heated to decomposition it emits toxic fumes of F^- and ZnO. See also FLUORIDES, TIN COMPOUNDS, and ZINC COMPOUNDS.

ZNS200 ***HR: 3***
ZINGIBER ROSEUM (Roxb.) Rosc., extract

PROP: Indian plant belonging to the family *Zingiberaceae* IJEBA6 22,312,84

TOXICITY DATA with REFERENCE
orl-rat TDLo:150 mg/kg (female 12-14D post):REP IJEBA6 22,312,84
ipr-mus LD50:388 mg/kg IJEBA6 22,312,84

SAFETY PROFILE: Poison by intraperitoneal route. Experimental reproductive effects. When heated to decomposition it emits acrid smoke and irritating fumes.

ZOA000 CAS:7440-67-7 ***HR: 3***
ZIRCONIUM
DOT: UN 1308/UN 1358/UN 2008/UN 2009/UN 2858
af: Zr aw: 91.224

PROP: A grayish-white, lustrous, metallic element; very sltly radioactive. Mp: 1852°, bp: 3577°, d: 6.506 @ 20°.

SYNS: ZIRCAT ◇ ZIRCONIUM METAL ◇ ZIRCONIUM METAL, dry (DOT) ◇ ZIRCONIUM SHAVINGS ◇ ZIRCONIUM SHEETS (DOT) ◇ ZIRCONIUM TURNINGS

CONSENSUS REPORTS: Reported in EPA TSCA Inventory.

OSHA PEL: (Transitional: TWA 5 mg(Zr)/m^3) TWA 5 mg(Zr)/m^3; STEL 10 mg(Zr)/m^3
ACGIH TLV: TWA 5 mg(Zr)/m^3; STEL 10 mg(Zr)/m^3
DFG MAK: 5 mg(Zr)/m^3

DOT Classification: Flammable Solid; Label: Flammable Solid; Flammable Solid; Label: Spontaneously Combustible; Flammable Liquid; Label: Flammable Liquid.

SAFETY PROFILE: A very dangerous fire hazard in the form of dust when exposed to heat or flame or by chemical reaction with oxidizers. May ignite spontaneously. A dangerous explosion hazard in the form of dust by chemical reaction with air, alkali hydroxides, alkali metal chromates, dichromates, molybdates, sulfates, tungstates, borax, CCl_4, CuO, Pb, PbO, P, $KClO_3$, KNO_3, nitrylfluoride. Explosive range: 0.16 g/L in air. To fight fire, use special mixtures, dry chemical, salt, or dry sand. See also ZIRCONIUM COMPOUNDS.

ZOS000 CAS:7440-67-7 ***HR: 3***
ZIRCONIUM (wet)
af: Zr aw: 91.224

PROP: Chemically produced, finer than 20 mesh particle size or mechanically produced, finer than 270 mesh particle size (FEREAC 41,15972,76).

SYN: ZIRCONIUM METAL, WET (DOT)

CONSENSUS REPORTS: Reported in EPA TSCA Inventory.

OSHA PEL: (Transitional: TWA 5 mg(Zr)/m^3) TWA 5 mg(Zr)/m^3; STEL 10 mg(Zr)/m^3
ACGIH TLV: TWA 5 mg(Zr)/m^3; STEL 10 mg(Zr)/m^3
DFG MAK: 5 mg(Zr)/m^3
DOT Classification: Label: Flammable Solid.

SAFETY PROFILE: Flammable when exposed to heat or flame. See also ZIRCONIUM and ZIRCONIUM COMPOUNDS.

ZPA000 CAS:10026-11-6 ***HR: 3***
ZIRCONIUM CHLORIDE
DOT: UN 2503
mf: Cl_4Zr mw: 233.02

PROP: White, lustrous crystals. Mp: subls @ 300°, bp: 331°, d: 2.80, vap press: 1 mm @ 190°.

SYNS: ZIRCONIUM(IV) CHLORIDE (1:4) ◇ ZIRCONIUM TETRACHLORIDE (DOT) ◇ ZIRCONIUM TETRACHLORIDE, solid (DOT)

TOXICITY DATA with REFERENCE
orl-rat LD50:1688 mg/kg HYSAAV 31,328,66
orl-mus LD50:489 mg/kg JNPHAG 14,437,83

CONSENSUS REPORTS: Reported in EPA TSCA Inventory.

OSHA PEL: (Transitional: TWA 5 mg(Zr)/m^3) TWA 5 mg(Zr)/m^3; STEL 10 mg(Zr)/m^3
ACGIH TLV: TWA 5 mg(Zr)/m^3; STEL 10 mg(Zr)/m^3
DFG MAK: 5 mg(Zr)/m^3
DOT Classification: Corrosive Material; Label: Corrosive.

SAFETY PROFILE: Moderately toxic by ingestion. A corrosive irritant to skin, eyes, and mucous membranes. Ignites spontaneously in air. When heated to decomposition it emits toxic fumes of Cl^-. See also ZIRCONIUM COMPOUNDS and HYDROCHLORIC ACID.

ZPJ000 CAS:10119-31-0 ***HR: 1***
ZIRCONIUM CHLORIDE HYDROXIDE
mf: ClHOZr mw: 143.68

SYNS: ZIRCONIUM CHLOROHYDRATE ◇ ZIRCONIUM HYDROXYCHLORIDE ◇ ZIRCONYL HYDROXYCHLORIDE

TOXICITY DATA with REFERENCE
skn-hmn 45 mg/3D-I MOD 85DKA8 -,127,77

CONSENSUS REPORTS: Reported in EPA TSCA Inventory.

OSHA PEL: (Transitional: TWA 5 mg(Zr)/m^3) TWA 5 mg(Zr)/m^3; STEL 10 mg(Zr)/m^3
ACGIH TLV: TWA 5 mg(Zr)/m^3; STEL 10 mg(Zr)/m^3
DFG MAK: 5 mg(Zr)/m^3

SAFETY PROFILE: A human skin irritant. When heated to decomposition it emits toxic fumes of Cl^-. Used as an antiperspirant. See also ZIRCONIUM COMPOUNDS and CHLORIDES.

ZPS000 CAS:13520-92-8 ***HR: 3***
ZIRCONIUM CHLORIDE OXIDE OCTAHYDRATE
mf: $Cl_2OZr•8H_2O$ mw: 322.28

SYN: ZIRCONYL CHLORIDE OCTAHYDRATE

TOXICITY DATA with REFERENCE
idr-mus TDLo:800 μg/kg:ETA CNREA8 33,287,73

OSHA PEL: (Transitional: TWA 5 mg(Zr)/m^3) TWA 5 mg(Zr)/m^3; STEL 10 mg(Zr)/m^3
ACGIH TLV: TWA 5 mg(Zr)/m^3; STEL 10 mg(Zr)/m^3
DFG MAK: 5 mg(Zr)/m^3

SAFETY PROFILE: Questionable carcinogen with experimental tumorigenic data. When heated to decomposition it emits toxic fumes of Cl^-. See also ZIRCONIUM COMPOUNDS and CHLORIDES.

ZQA000 ***HR: 2***
ZIRCONIUM COMPOUNDS

OSHA PEL: (Transitional: TWA 5 mg(Zr)/m^3) TWA 5 mg(Zr)/m^3; STEL 10 mg(Zr)/m^3
ACGIH TLV: TWA 5 mg(Zr)/m^3; STEL 10 mg(Zr)/m^3
DFG MAK: 5 mg(Zr)/m^3

SAFETY PROFILE: Zirconium is not an important industrial poison, however, poisoning may occur due to excessive exposure to zirconium salts. Deaths in rabbits have been caused by intravenous injection of 150 mg/kg of body weight. Inhalation of $ZrCl_4$ (6 mg Zr/m^3) for 60 days produces slight decreases in hemoglobin and red blood cell count in dogs and increases mortality in rats and guinea pigs. Most zirconium compounds in common use are insoluble and considered inert. Pulmonary granuloma in zirconium workers has been reported and sodium zirconium lactate has been held responsible for skin granulomas. Avoid inhalation of Zr-containing aerosols, which can cause lung granulomas. Zirconium-containing drugs or cosmetic products are being controlled by the FDA.

ZQB100 CAS:24621-17-8 ***HR: 3***
ZIRCONIUM DIBROMIDE
mf: Br_2Zr mw: 251.03

PROP: Black powder.

SAFETY PROFILE: Ignites spontaneously in air. Violent reaction with water or steam. When heated to decomposition it emits toxic fumes of Br^-. See also ZIRCONIUM COMPOUNDS and BROMIDES.

ZQC200 CAS:12070-14-3 ***HR: 2***
ZIRCONIUM DICARBIDE
mf: C_2Zr mw: 115.25

SAFETY PROFILE: Ignites in cold fluoride; chlorine (at 250°C); bromine (at 300°C); iodine (at 400°C). See also ZIRCONIUM COMPOUNDS.

ZQJ000 CAS:13762-26-0 ***HR: 3***
ZIRCONIUM DICHLORIDE
mf: Cl_2Zr mw: 162.13

SYN: ZIRCONIUM(II) CHLORIDE

OSHA PEL: (Transitional: TWA 5 mg(Zr)/m^3) TWA 5 mg(Zr)/m^3; STEL 10 mg(Zr)/m^3
ACGIH TLV: TWA 5 mg(Zr)/m^3; STEL 10 mg(Zr)/m^3
DFG MAK: 5 mg(Zr)/m^3

SAFETY PROFILE: If warm, it ignites in air. When heated to decomposition it emits toxic fumes of Cl^-. See also ZIRCONIUM COMPOUNDS and CHLORIDES.

ZQS000 CAS:7783-64-4 ***HR: 3***
ZIRCONIUM FLUORIDE
mf: F_4Zr mw: 167.22

PROP: Refractive crystals. Water-sol. D: 4.6 @ 16°, subls @ 600°. Very sol in HF.

SYN: ZIRCONIUM TETRAFLUORIDE

TOXICITY DATA with REFERENCE
ivn-mus LD50:98 mg/kg 19UQAS -,30,65

CONSENSUS REPORTS: Reported in EPA TSCA Inventory.

OSHA PEL: (Transitional: TWA 5 mg(Zr)/m^3) TWA 5 mg(Zr)/m^3; STEL 10 mg(Zr)/m^3
ACGIH TLV: TWA 5 mg(Zr)/m^3; STEL 10 mg(Zr)/m^3
DFG MAK: 5 mg(Zr)/m^3
NIOSH REL: (Fluorides, Inorganic): 10H TWA 2.5 mg (F)/m^3

SAFETY PROFILE: Poison by intravenous route. When heated to decomposition it emits toxic fumes of F^-. See also ZIRCONIUM COMPOUNDS and FLUORIDES.

ZQS100 CAS:70983-41-4 ***HR: 3***
ZIRCONIUM GLUCONATE

TOXICITY DATA with REFERENCE
ipr-rat LD50:247 mg/kg NTIS** AEC-TR-6710

OSHA PEL: TWA 5 mg(Zr)/m^3; STEL 10 mg(Zr)/m^3
ACGIH TLV: TWA 5 mg(Zr)/m^3; STEL 10 mg(Zr)/m^3

CONSENSUS REPORTS: Reported in EPA TSCA Inventory.

SAFETY PROFILE: Poison by intraperitoneal route. When heated to decomposition it emits toxic fumes of Zr.

ZRA000 CAS:7704-99-6 ***HR: 3***
ZIRCONIUM HYDRIDE
DOT: UN 1437
mf: H_2Zr mw: 93.24

PROP: Metallic dark gray to black powder. D: 5.6, autoign temp: 270° (in air).

CONSENSUS REPORTS: Reported in EPA TSCA Inventory.

OSHA PEL: (Transitional: TWA 5 mg(Zr)/m^3) TWA 5 mg(Zr)/m^3; STEL 10 mg(Zr)/m^3
ACGIH TLV: TWA 5 mg(Zr)/m^3; STEL 10 mg(Zr)/m^3
DFG MAK: 5 mg(Zr)/m^3
DOT Classification: Flammable Solid; Label: Flammable Solid and Dangerous When Wet; Flammable Solid; Label: Flammable Solid.

SAFETY PROFILE: A powerful reducing agent. Flammable when dry or wet. Very dangerous to handle; can explode. Incandesces when heated in air. See also HYDRIDES and ZIRCONIUM COMPOUNDS.

ZRJ000 CAS:63919-14-2 ***HR: 2***
ZIRCONIUM(III) LACTATE (1:3)
mf: $C_9H_9O_9 \cdot H_4OZr$ mw: 372.44

SYN: LACTIC ACID, ZIRCONIUM SALT (3:1)

TOXICITY DATA with REFERENCE
skn-mus TDLo:20 μg/kg JAMAAP 190,940,64
ipr-rat LD50:670 mg/kg AIHAAP 24,131,63

OSHA PEL: (Transitional: TWA 5 mg(Zr)/m^3) TWA 5 mg(Zr)/m^3; STEL 10 mg(Zr)/m^3
ACGIH TLV: TWA 5 mg(Zr)/m^3; STEL 10 mg(Zr)/m^3
DFG MAK: 5 mg(Zr)/m^3

SAFETY PROFILE: Moderately toxic by intraperitoneal route. When heated to decomposition it emits acrid smoke and irritating fumes. See also ZIRCONIUM COMPOUNDS.

ZRS000 CAS:60676-90-6 ***HR: 2***
ZIRCONIUM(IV) LACTATE
mf: $Zr \cdot C_3H_5O_3$ mw: 180.30

PROP: White, sltly moist pulp. Very sltly sol in water and common organic solvents; sol in aq alkali solns with the formation of salts.

SYN: LACTIC ACID, ZIRCONIUM SALT (4:1)

TOXICITY DATA with REFERENCE
idr-man TDLo:170 μg/kg/I:SKN JIDEAE 38,223,62

OSHA PEL: (Transitional: TWA 5 mg(Zr)/m^3) TWA 5 mg(Zr)/m^3; STEL 10 mg(Zr)/m^3
ACGIH TLV: TWA 5 mg(Zr)/m^3; STEL 10 mg(Zr)/m^3
DFG MAK: 5 mg(Zr)/m^3

SAFETY PROFILE: Human systemic effects by intradermal route: after topical application causes primary irritation, allergic dermatitis and is corrosive. Prolonged inhalation of dust has caused granulomas, interstitial pneumonia. A powerful skin allergen. See also ZIRCONIUM COMPOUNDS.

ZSA000 CAS:13746-89-9 ***HR: 2***
ZIRCONIUM NITRATE
DOT: UN 2728
mf: $N_4O_{12} \cdot Zr$ mw: 339.26

PROP: White hygroscopic crystals. Very sol in water; sol in alc.

SYN: DUSICNAN ZIRKONICITY (CZECH)

TOXICITY DATA with REFERENCE
orl-rat LD50:2290 mg/kg MarJV# 29MAR77
ihl-rat LCLo:500 mg/m^3/30M NTIS** AEC-TR-6710

CONSENSUS REPORTS: Reported in EPA TSCA Inventory.

OSHA PEL: (Transitional: TWA 5 mg(Zr)/m^3) TWA 5 mg(Zr)/m^3; STEL 10 mg(Zr)/m^3
ACGIH TLV: TWA 5 mg(Zr)/m^3; STEL 10 mg(Zr)/m^3
DFG MAK: 5 mg(Zr)/m^3

DOT Classification: Oxidizer; Label: Oxidizer.

SAFETY PROFILE: Moderately toxic by inhalation and ingestion. A powerful oxidizer. When heated to decomposition it emits toxic fumes of NO_x. See also NITRATES and ZIRCONIUM COMPOUNDS.

ZSJ000 CAS:7699-43-6 ***HR: 3***
ZIRCONIUM OXYCHLORIDE
mf: Cl_2OZr mw: 178.12

PROP: Crystals. D: 1.91. Very sol in water, alc.

SYNS: BASIC ZIRCONIUM CHLORIDE ◇ CHLOROZIRCONYL ◇ DICHLOROOXOZIRCONIUM ◇ NCI-C60811 ◇ ZIRCONYL CHLORIDE

TOXICITY DATA with REFERENCE
idr-mus TDLo:800 μg/kg:NEO JIDEAE 57,411,71
orl-rat LD50:3500 mg/kg AIHOAX 1,637,50
ipr-rat LD50:400 mg/kg AIHOAX 1,637,50
scu-rat LDLo:500 mg/kg NTIS** AEC-TR-6710
scu-rat LD50:1227 mg/kg JNPHAG 14,437,83
ipr-mus LD50:335 mg/kg COREAF 256,1043,63

CONSENSUS REPORTS: Reported in EPA TSCA Inventory.

OSHA PEL: (Transitional: TWA 5 mg(Zr)/m^3) TWA 5 mg(Zr)/m^3; STEL 10 mg(Zr)/m^3
ACGIH TLV: TWA 5 mg(Zr)/m^3; STEL 10 mg(Zr)/m^3
DFG MAK: 5 mg(Zr)/m^3

SAFETY PROFILE: Poison by intraperitoneal route. Moderately toxic by ingestion and subcutaneous routes. Questionable carcinogen with experimental neoplastigenic data. When heated to decomposition it emits toxic fumes of Cl^-. Used as an antiperspirant. See also ZIRCONIUM COMPOUNDS and CHLORIDES.

ZSS000 CAS:14940-68-2 ***HR: 2***
ZIRCONIUM(IV) SILICATE (1:1)
mf: O_4SiZr mw: 183.31

PROP: Red or varying colored tetragonal bipyramidal crystals. D: 4.56, mp: 2550.

SYNS: HYACINTH ◇ SILICIC ACID, ZIRCONIUM(4+) SALT (1:1) ◇ ZIRCON

CONSENSUS REPORTS: Reported in EPA TSCA Inventory.

OSHA PEL: (Transitional: TWA 5 mg(Zr)/m^3) TWA 5 mg(Zr)/m^3; STEL 10 mg(Zr)/m^3
ACGIH TLV: TWA 5 mg(Zr)/m^3; STEL 10 mg(Zr)/m^3
DFG MAK: 5 mg(Zr)/m^3

SAFETY PROFILE: See SILICATES and ZIRCONIUM COMPOUNDS.

ZTA000 CAS:63904-82-5 ***HR: 3***
ZIRCONIUM SODIUM LACTATE
mf: $C_9H_{15}NaO_{10}Zr$ mw: 397.45

PROP: Straw colored liquid. D: 1.28.

SYN: SODIUM HYDROGEN TRILACTATOZIRCONYLATE

TOXICITY DATA with REFERENCE
idr-mus TDLo:200 mg/kg:NEO JIDEAE 57,411,71

OSHA PEL: (Transitional: TWA 5 mg(Zr)/m^3) TWA 5 mg(Zr)/m^3; STEL 10 mg(Zr)/m^3
ACGIH TLV: TWA 5 mg(Zr)/m^3; STEL 10 mg(Zr)/m^3
DFG MAK: 5 mg(Zr)/m^3

SAFETY PROFILE: Inhalation produced bronchiolar abcesses, lobar pneumonia, and peribronchial granulomas experimentally. Questionable carcinogen with experimental neoplastigenic data. When heated to decomposition it emits acrid smoke and irritating fumes of Na_2O. See also ZIRCONIUM COMPOUNDS.

ZTJ000 CAS:14644-61-2 ***HR: 3***
ZIRCONIUM(IV) SULFATE (1:2)
DOT: NA 9163
mf: $O_8S_2 \cdot Zr$ mw: 283.34

PROP: Tetrahydrate, crystalline solid.

SYNS: DISULFATOZIRCONIC ACID ◇ SULFURIC ACID, ZIRCONIUM(4+) SALT (2:1) ◇ ZIRCONYL SULFATE

TOXICITY DATA with REFERENCE
dns-mus:lyms 20 umol/L TOLED5 30,89,86
itt-rat TDLo:22667 μg/kg (male 1D pre):REP JRPFA4 7,21,64
orl-rat LD50:3500 mg/kg AIHOAX 1,637,50
ipr-rat LD50:175 mg/kg AIHOAX 1,637,50
scu-rat LDLo:500 mg/kg NTIS** AEC-TR-6710

CONSENSUS REPORTS: Reported in EPA TSCA Inventory.

OSHA PEL: (Transitional: TWA 5 mg(Zr)/m^3) TWA 5 mg(Zr)/m^3; STEL 10 mg(Zr)/m^3
ACGIH TLV: TWA 5 mg(Zr)/m^3; STEL 10 mg(Zr)/m^3
DFG MAK: 5 mg(Zr)/m^3
DOT Classification: ORM-B; Label:None

SAFETY PROFILE: Poison by intraperitoneal route. Moderately toxic by ingestion and subcutaneous routes. Experimental reproductive effects. Mutation data reported. When heated to decomposition it emits toxic fumes of SO_x. See also SULFATES and ZIRCONIUM COMPOUNDS.

ZTK300 CAS:23840-95-1 ***HR: 3***
ZIRCONIUM(IV) TETRAHYDROBORATE
mf: $B_4H_{16}Zr$ mw: 150.59

SAFETY PROFILE: Violent reaction with air. See also BORON COMPOUNDS and ZIRCONIUM COMPOUNDS.

ZTK400 CAS:1291-32-3 ***HR: 3***
ZIRCONOCENE, DICHLORIDE
mf: $C_{10}H_{10}Cl_2Zr$ mw: 292.32

SYN: ZIRCONIUM, DICHLORO-DI-pi-CYCLOPENTADIENYL-

TOXICITY DATA with REFERENCE
mmo-sat 333 μg/plate EMMUEG 11(Suppl 12),1,88
mma-sat 10 mg/plate EMMUEG 11(Suppl 12),1,88
ipr-rat LD50:30 mg/kg NCIBR* PH43-64-886,JUL68

OSHA PEL: TWA 5 mg(Zr)/m^3; STEL 10 mg(Zr)/m^3
ACGIH TLV: TWA 5 mg(Zr)/m^3; STEL 10 mg(Zr)/m^3

CONSENSUS REPORTS: Reported in EPA TSCA Inventory.

SAFETY PROFILE: Poison by intraperitoneal route. Mutation data reported. When heated to decomposition it emits toxic fumes of Zr and Cl^-.

ZTS000 CAS:20645-04-9 ***HR: 3***
ZIRCONYL ACETATE
mf: $C_4H_6O_5 \cdot Zr$ mw: 225.32

PROP: D: 1.46.

SYNS: BIS(ACETATO-O,O')OXOZIRCONIUM ◇ DIACETATOZIRCONIC ACID

TOXICITY DATA with REFERENCE
orl-rat LD50:4100 mg/kg AIHOAX 1,637,50
ipr-rat LD50:300 mg/kg AIHOAX 1,637,50

OSHA PEL: (Transitional: TWA 5 mg(Zr)/m^3) TWA 5 mg(Zr)/m^3; STEL 10 mg(Zr)/m^3
ACGIH TLV: TWA 5 mg(Zr)/m^3; STEL 10 mg(Zr)/m^3
DFG MAK: 5 mg(Zr)/m^3

SAFETY PROFILE: Poison by intraperitoneal route. Moderately toxic by ingestion. When heated to decomposition it emits acrid smoke and irritating fumes. See also ZIRCONIUM COMPOUNDS.

ZTS100 CAS:24735-35-1 ***HR: 1***
ZIRCONYL SODIUM SULPHATE
mf: $Na_2O_{18}S_4Zr_2$ mw: 644.66

SYNS: OXOBIS(SULFATO(2-)-O)-ZIRCONATE(2-), DISODIUM (9CI) ◇ SODIUM ZIRCONIUM OXIDE SULFATE ◇ SODIUM ZIRCONYL SULPHATE ◇ ZIRCONATE(2-), OXODISULFATO-, DISODIUM (8CI)

TOXICITY DATA with REFERENCE
orl-rat LD50:10 g/kg EQSFAP 1,1,75
ipr-rat LD50:4100 mg/kg AIHOAX 1,637,50

OSHA PEL: TWA 5 mg(Zr)/m^3; STEL 10 mg(Zr)/m^3
ACGIH TLV: TWA 5 mg(Zr)/m^3; STEL 10 mg(Zr)/m^3

SAFETY PROFILE: Mildly toxic by intraperitoneal route. When heated to decomposition it emits toxic fumes of SO_x and Zr.

ZTS600 ***HR: D***
ZOAPATLE, crude leaf extract

SYN: MONTANOA TOMENTOSA, leaf extract, crude

TOXICITY DATA with REFERENCE
orl-gpg TDLo:250 mg/kg (female 22D post):TER CCPTAY 23,133,81
orl-mus TDLo:3 g/kg (female 1-6D post):REP CCPTAY 23,133,81

SAFETY PROFILE: An experimental teratogen. Experimental reproductive effects.

ZTS625 ***HR: D***
ZOAPATLE, semi-purified leaf extract

SYN: MONTANOA TOMENTOSA, leaf extract, semi-purified

TOXICITY DATA with REFERENCE
orl-gpg TDLo:53 mg/kg (female 22D post):TER CCPTAY 23,133,81
orl-mus TDLo:1200 mg/kg (1-6D preg):REP CCPTAY 23,133,81

SAFETY PROFILE: An experimental teratogen. Experimental reproductive effects.

ZUA000 CAS:1155-03-9 ***HR: 3***
ZOLAMINE HYDROCHLORIDE
mf: $C_{15}H_{21}N_3OS{\cdot}ClH$ mw: 327.91

PROP: Odorless crystals; sltly bitter taste. Mp: 167.5-167.8°. Sol in water.

SYNS: 2-((2-(DIMETHYLAMINO)ETHYL)(p-METHOXYBENZYL) AMINO) THIAZOLE HYDROCHLORIDE ◇ N,N-DIMETHYL-N'-(p-METHOXYBENZYL)-N'-(2-THIAZOLYL)-ETHYLENEDIAMINE MONOHYDROCHLORIDE

TOXICITY DATA with REFERENCE
orl-rat LDLo:570 mg/kg CLDND*
scu-rat LDLo:275 mg/kg CLDND*
scu-mus LDLo:140 mg/kg CLDND*
ivn-mus LD50:40 mg/kg CLDND*

SAFETY PROFILE: Poison by subcutaneous and intravenous routes. Moderately toxic by ingestion. When heated to decomposition it emits very toxic fumes of NO_x, SO_x, and HCl.

ZUA200 CAS:1222-57-7 ***HR: 2***
ZOLIRIDINE
mf: $C_{14}H_{12}N_2O_2S$ mw: 272.34

PROP: Crystals. Mp: 242-244°.

SYNS: 2-(4-(METHYLSULFONYL)PHENYL)-IMIDAZO(1,2-a)PYRIDINE ◇ SOLIMIDIN ◇ ZOLIMIDIN ◇ ZOLIMIDINE

TOXICITY DATA with REFERENCE
orl-rat LD50:3710 mg/kg USXXAM #3318880
ipr-rat LD50:950 mg/kg ARZNAD 33,1655,83
ipr-mus LD50:1468 mg/kg ARZNAD 33,1655,83

SAFETY PROFILE: Moderately toxic by ingestion and intraperitoneal routes. An anti-ulcer agent. When heated to decomposition it emits toxic fumes of SO_x and NO_x.

ZUA300 CAS:64092-48-4 ***HR: 3***
ZOMEPIRAC SODIUM
mf: $C_{15}H_{13}ClNO_3{\cdot}Na{\cdot}2H_2O$ mw: 349.77

SYNS: 5-(4-CHLOROBENZOYL)-1,4-DIMETHYL-1H-PYRROLE-2-ACETIC ACID SODIUM SALT DIHYDRATE ◇ MCN 2783-21-98 ◇ SODIUM-5-(4-CHLOROBENZOYL)-1,4-DIMETHYL-1H-PYRROLE-2-ACETATEDIHYDRATE ◇ SODIUM ZOMEPIRAC ◇ ZOMAX ◇ ZOMEPIRAC ◇ ZOMEPIRAC SODIUM SALT

TOXICITY DATA with REFERENCE
orl-man TDLo:1429 µg/kg:PUL ANAEA3 48,233,82
orl-rat LD50:27 mg/kg DRUGAY 23,250,82
orl-mus LD50:63 mg/kg DRUGAY 23,250,82
orl-ham LD50:743 mg/kg DRUGAY 23,250,82

SAFETY PROFILE: Poison by ingestion. Human systemic effects by ingestion: cyanosis. When heated to decomposition it emits toxic fumes of Cl^-, Na_2O, and NO_x.

ZUA450 CAS:43200-80-2 ***HR: 3***
ZOPICLONE
mf: $C_{17}H_{17}ClH_6O_3$ mw: 388.85

PROP: Crystals from acetonitrile/diisopropyl ether (1:1). Mp: 178°.

SYNS: IMOVANCE ◇ IMOVANE ◇ 1-PIPERAZINECARBOXYLIC ACID,4-METHYL-, 6-(5-CHLORO-2-PYRIDINYL)-6,7-DIHYDRO-7-OXO-5H-PYRROLO(3,4-b)PYRAZIN-5-YL ESTER ◇ 27267 R.P.

TOXICITY DATA with REFERENCE
orl-rat TDLo:1300 mg/kg (17-22D preg/20D post):REP JZKEDZ 9,145,83
orl-rat LD50:827 mg/kg OYYAA2 26,935,83
ipr-rat LD50:771 mg/kg OYYAA2 26,935,83
scu-rat LD50:540 mg/kg OYYAA2 26,935,83
ivn-rat LD50:280 mg/kg OYYAA2 26,935,83
ims-rat LD50:295 mg/kg OYYAA2 26,935,83
orl-mus LD50:2174 mg/kg OYYAA2 26,935,83
ipr-mus LD50:1325 mg/kg OYYAA2 26,935,83
scu-mus LD50:888 mg/kg OYYAA2 26,935,83
ivn-mus LD50:321 mg/kg OYYAA2 26,935,83
ims-mus LD50:541 mg/kg OYYAA2 26,935,83

SAFETY PROFILE: Poison by intramuscular and intravenous routes. Moderately toxic by ingestion, intraperitoneal and subcutaneous routes. Experimental reproductive effects. When heated to decomposition it emits toxic fumes of Cl^-.

ZUJ000 CAS:26615-21-4 ***HR: 3***
ZOTEPINE
mf: $C_{18}H_{18}ClNOS$ mw: 331.88

PROP: Mp: 90-01°.

SYNS: 2-CHLOR-11-(2-DIMETHYAMINOAETHOXY)-DIBENZO(b,f)-THIEPIN (GERMAN) ◇ 2-((8-CHLORODIBENZO(b,f)THIEPIN-10-YL) OXY-N,N-DIMETHYLETHANAMINE ◇ 2-CHLORO-11-(2-(DIMETHYLAMINO)ETHOXY)DIBENZO(b,f)THIEPIN ◇ 2-CHLORO-11-(2-DIMETHYLAMINOETHOXY)DIBENZO(b,f)THIEPINE ◇ LODOPIN

TOXICITY DATA with REFERENCE
orl-rat TDLo:44 mg/kg (7-17D preg):REP ARZNAD 29,1600,79
orl-rat TDLo:176 mg/kg (7-17D preg):TER ARZNAD 29,1600,79
orl-rat LD50:306 mg/kg ARZNAD 29,1600,79
ipr-rat LD50:97 mg/kg ARZNAD 29,1600,79
scu-rat LD50:1290 mg/kg ARZNAD 29,1600,79
ivn-rat LD50:36800 μg/kg NIIRDN 6,APP-12,82
orl-mus LD50:108 mg/kg ARZNAD 29,1600,79
ipr-mus LD50:36200 μg/kg ARZNAD 29,1600,79
scu-mus LD50:84900 μg/kg ARZNAD 29,1600,79
ivn-mus LD50:43300 μg/kg ARZNAD 29,1600,79
ivn-dog LD50:26600 μg/kg ARZNAD 29,1600,79
orl-rbt LD50:250 mg/kg ARZNAD 29,1600,79
ivn-rbt LD50:23800 μg/kg ARZNAD 29,1600,79

SAFETY PROFILE: Poison by ingestion, intravenous, subcutaneous and intraperitoneal routes. An experimental teratogen. Experimental reproductive effects. When heated to decomposition it emits very toxic fumes of SO_x, Cl^-, and NO_x. Used as a tranquilizer.

ZUS000 CAS:22144-77-0 ***HR: 3***
ZYGOSPORIN A
mf: $C_{30}H_{37}NO_6$ mw: 507.68

SYNS: 3-BENZYL-3,3-α,4,5,6,6-α,9,10,12,15-DECAHYDRO-6,12,15-TRI-HYDROXY-4,10,12-TRIMETHYL-5-METHYLENE-1H-CYCLOUNDEC(d) ISOINDOLE-1,11(2H)-DIONE, 15-ACETATE ◇ CYTOCHALASIN D

TOXICITY DATA with REFERENCE
cyt-hmn:hla 1 mg/L ECREAL 91,47,75
cyt-hmn:oth 1 mg/L ECREAL 91,47,75
orl-mus TDLo:1500 μg/kg (female 8D post):TER TJADAB 35,87,87
orl-mus TDLo:1500 μg/kg (female 8D post):REP TJADAB 35,87,87
ipr-rat LD50:900 μg/kg TJADAB 15,27A,77
orl-mus LD50:36 mg/kg JJEMAG 48,105,78
ipr-mus LD50:2 mg/kg TJADAB 15,27A,77
scu-mus LD50:1850 μg/kg JJEMAG 48,105,78

SAFETY PROFILE: Poison by ingestion, subcutaneous, and intraperitoneal routes. An experimental teratogen. Experimental reproductive effects. Human mutation data reported. When heated to decomposition it emits toxic fumes of NO_x.

ZVA000 CAS:3563-92-6 ***HR: 3***
ZYLOFURAMINE
mf: $C_{14}N_{21}NO$ mw: 219.36

SYN: d-THREO-α-BENZYL-N-ETHYLTETRAHYDROFURFURYLAMINE

TOXICITY DATA with REFERENCE
orl-mus LD50:475 mg/kg AIPTAK 146,392,63
scu-mus LD50:155 mg/kg AIPTAK 146,392,63
ivn-mus LD50:32 mg/kg AIPTAK 146,392,63

SAFETY PROFILE: Poison by subcutaneous and intravenous routes. Moderately toxic by ingestion. When heated to decomposition it emits toxic fumes of NO_x.

ZVJ000 CAS:315-30-0 ***HR: 3***
ZYLOPRIM
mf: $C_5H_4N_4O$ mw: 136.13

SYNS: ADENOCK ◇ AL-100 ◇ ALLOPURINOL ◇ ALLOZYM ◇ ALLURAL ◇ ALOSITOL ◇ ALULINE ◇ ANOPROLIN ◇ ANZIEF ◇ APURIN ◇ APUROL ◇ BLEMINOL ◇ BLOXANTH ◇ BW 56-158 ◇ CAPLENAL ◇ CELLIDRIN ◇ DABROSIN ◇ 1,5-DIHYDRO-4H-PYRAZOLO(3,4-d)PYRIMIDIN-4-ONE ◇ EMBARIN ◇ EPIDROPAL ◇ FOLIGAN ◇ GICHTEX ◇ HPP ◇ 4'-HYDROXYPYRAZOLOL(3,4-d)PYRIMIDINE ◇ 4-HYDROXY-1H-PYRAZOLO(3,4-d)PYRIMIDINE ◇ 4-HYDROXY-3,4-PYRAZOLOPYRIMIDINE ◇ 4-HYDROXYPYRAZOLO (3,4-d)PYRIMIDINE ◇ 4-HYDROXYPYRAZOLYL(3,4-d)PYRIMIDINE ◇ KETANRIFT ◇ KETOBUN-A ◇ LOPURIN ◇ LYSURON ◇ MINIPLANOR ◇ MONARCH ◇ NEKTROHAN ◇ NSC-1390 ◇ 4H-PYRAZOLO(3,4-d)PYRIMIDIN-4-ONE ◇ REMID ◇ RIBALL ◇ SUSPENDOL ◇ TAKANARUMIN ◇ URBOL ◇ URICEMIL ◇ URITAS ◇ UROBENYL ◇ UROSIN ◇ XANTURAT ◇ ZYLORIC

TOXICITY DATA with REFERENCE
ipr-mus TDLo:100 mg/kg (10D preg):TER JJPAAZ 22,201,72
orl-wmn TDLo:42 mg/kg/7D-I:KID,SKN AJMEAZ 76,47,84
orl-wmn LDLo:88 mg/kg/22D-I:BLD ARDIAO 40,245,81
orl-man TDLo:21429 μg/kg/5D-I:CNS,LIV,BLD ARDIAO 40,245,81
ipr-rat LD50:900 mg/kg ADTEAS 3,181,68
ipr-mus LD50:214 mg/kg NYKZAU 64,108,68
scu-mus LD50:298 mg/kg YAKUD5 23,715,81

CONSENSUS REPORTS: Reported in EPA TSCA Inventory.

SAFETY PROFILE: Human poison by ingestion. Poison experimentally by intraperitoneal and subcutaneous routes. An experimental teratogen. Human systemic effects by ingestion: blood leukopenia, dermatitis, jaundice, muscle weakness, thrombocytopenia. When heated to decomposition it emits toxic fumes of NO_x. An FDA proprietary drug used as a xanthine oxidase inhibitor.

ZVS000 CAS:102583-71-1 ***HR: 2***
ZZL-0810

TOXICITY DATA with REFERENCE
skn-rbt 500 mg open MLD UCDS** 2/24/69
orl-rat LD50:2050 mg/kg UCDS** 2/24/69
skn-rbt LD50:2520 mg/kg UCDS** 2/24/69

SAFETY PROFILE: Moderately toxic by ingestion and skin contact. A skin irritant.

A Guide to Using This Book

Entry Number – Entries are indexed in order by this alphanumeric code.
See Introduction: paragraph 1, *p.* xiii.

Entry Name – A complete entry name and synonym cross-index is located in Section 2.
See Introduction: paragraph 2, *p.* xiii.

DOT: – The four digit hazard code assigned by the U.S. DOT.
See Introduction: paragraph 5, *p.* xiii.

mf: – The molecular formula
mw: – The molecular weight
See Introduction: paragraphs 6, *and* 7, *p.* xiv.

PROP: – Physical properties including solubility and flammability data. May contain a definition of the entry.
See Introduction: paragraph 9, *p.* xiv.

SYNS: – Synonyms for the entry. A complete synonym cross-index is located in Section 2.
See Introduction: paragraph 10, *p.* xiv.

Toxicity Data: – Data for skin and eye irritation, mutation, teratogenic, reproductive, carcinogenic, human, and acute lethal effects.
See Introduction: paragraphs 11, 12, 13, 14, and 15, *p.* xiv-xxiv.

Consensus Reports: – Supply additional information to enable the reader to make knowledgeable evaluations of potential chemical hazards.
See Introduction: paragraph 17, *p.* xxiv.

Standards and Recommendations: Here are listed the OSHA PEL, ACGIH TLV, DFG MAK, and NIOSH REL workplace air levels. U.S. DOT classification and labels are also listed.
See Introduction: paragraph 18, *p.* xxv.